"I like the writing style. **I didn't feel like I was reading a textbook but rather having a conversation with a professor.** It felt more personal…It kept my attention."
—*Sara DiTursi, SUNY Geneseo*

"The chapter's **focus on problem solving strategies largely addressed my main goals**. Many of my students reported putting pen to paper (i.e., to work through mechanisms, etc) while reading this chapter more so than when they were reading our normal text."
—*Professor Kevin Caran, James Madison University*

"The text is the closest I have seen to my style and approach to organic chemistry. The description of each mechanistic step correlates to the points I try to stress. The text also offers multiple connections to previous material and the concepts covered in the text to **limit the amount of memorization, and maximize the amount of understanding.**"
—*Professor Mackay Steffensen, Southern Utah University*

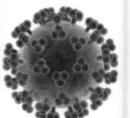

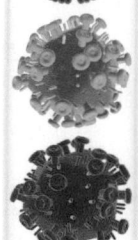

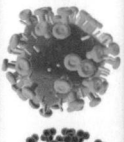

"I thought the sample chapter was very **well organized** and **clearly explained complicated topics.** It was much easier to follow than my current text and gave **much more relevant examples.**"
—*Jacob Faucheaux, Louisiana Tech University*

Reviewed by over 315 Professors, Class tested at 75 schools, with 4,000 students.

Organic Chemistry

David Klein
Johns Hopkins University

WILEY
John Wiley & Sons, Inc.

ASSOCIATE PUBLISHER Petra Recter
DIRECTOR OF DEVELOPMENT Barbara Heaney
SENIOR DEVELOPMENT EDITOR Leslie Kraham
SPONSORING EDITOR Joan Kalkut
MARKETING MANAGER Kristine Ruff
CREATIVE DIRECTOR Harold Nolan
INTERIOR DESIGN Carole Anson/DeMarinis Design LLC
COVER DESIGN DeMarinis Design LLC
SENIOR ILLUSTRATION EDITOR Sandra Rigby
SENIOR PHOTO EDITOR Lisa Gee
EXECUTIVE MEDIA EDITOR Thomas Kulesa
MEDIA EDITOR Marc Wezdecki
EDITORIAL ASSISTANT Lauren Stauber
MEDIA PRODUCTION EDITOR Evelyn Levich
CONTENT MANAGER Lucille Buonocore

This book was set in 10/12 Garamond by Prepare and printed and bound by Quad/Graphics Dubuque. The cover was printed by Lehigh-Phoenix.

Cover photo credits: *X-Ray of eight glass test tubes in a test tube holder,* Untitled X-Ray/Nick Veasey/Getty Images, Inc.; *cherry tomatoes,* Natalie Erhova (summerky)/Shutterstock; *tomato leaves,* Jennifer Stone/iStockphoto; *brass skeleton key hanging from string,* Gary S. Chapman/Photographer's Choice/Getty Images, Inc.; *medicine illustrations-capsule, tablet, gel cap,* Norph/Shutterstock; *popcorn evolution-corn kernel, partial popped corn, fullypopped corn,* Bill Grove/iStockphoto; *hiv virus (purple-maroon),* Sebastian Kaulitzki/Shutterstock; *hiv virus (blue),* Sebastian Kaulitzki/Shutterstock; *hiv virus (green),* Sebastian Kaulitzki/Shutterstock; *chemistry molecule,* David Klein; *curl of smoke,* stav klem/Shutterstock

This book is printed on acid free paper.

Founded in 1807, John Wiley & Sons, Inc. has been a valued source of knowledge and understanding for more than 200 years, helping people around the world meet their needs and fulfill their aspirations. Our company is built on a foundation of principles that include responsibility to the communities we serve and where we live and work. In 2008, we launched a Corporate Citizenship Initiative, a global effort to address the environmental, social, economic, and ethical challenges we face in our business. Among the issues we are addressing are carbon impact, paper specifications and procurement, ethical conduct within our business and among our vendors, and community and charitable support. For more information, please visit our website: www.wiley.com/go/citizenship.

Library of Congress Cataloging-in-Publication Data
Klein, David R.
Organic chemistry / David R. Klein.
p. cm.
Includes index.
ISBN 978-0-471-75614-9 (hardback)
Binder-ready version 978-0-470-91780-0
1. Organic chemistry—Textbooks. I. Title.
QD253.2.K55 2012
547--dc22 2010034742

Printed in the United States of America
10 9 8 7 6 5 4 3 2 1

Brief Contents

Contents

4

Alkanes and Cycloalkanes 136

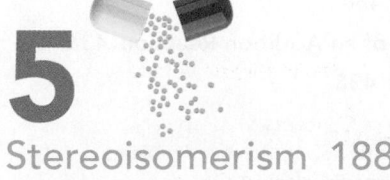

5

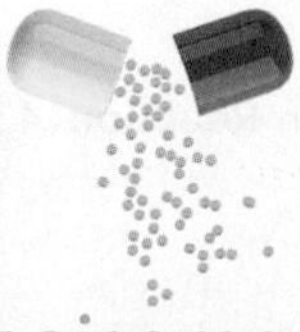

Stereoisomerism 188

6

Chemical Reactivity and Mechanisms 233

7 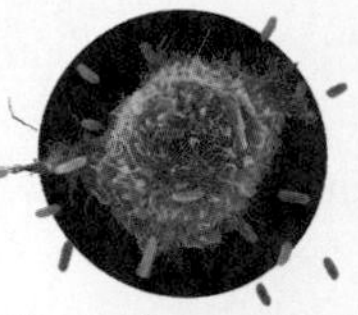

Substitution Reactions 281

8 

Alkenes: Structure and Preparation via Elimination Reactions 333

9 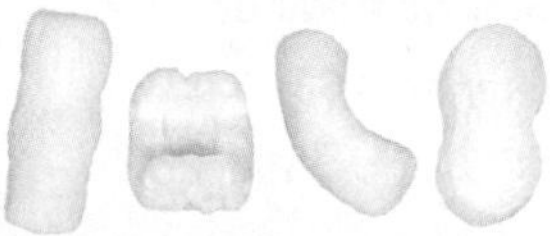

Addition Reactions of Alkenes 394

14 Ethers and Epoxides; Thiols and Sulfides 622

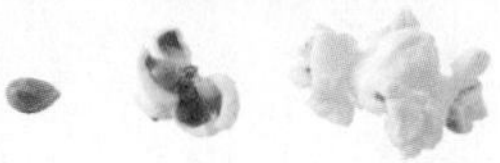

15 Infrared Spectroscopy and Mass Spectrometry 671

16 Nuclear Magnetic Resonance Spectroscopy 718

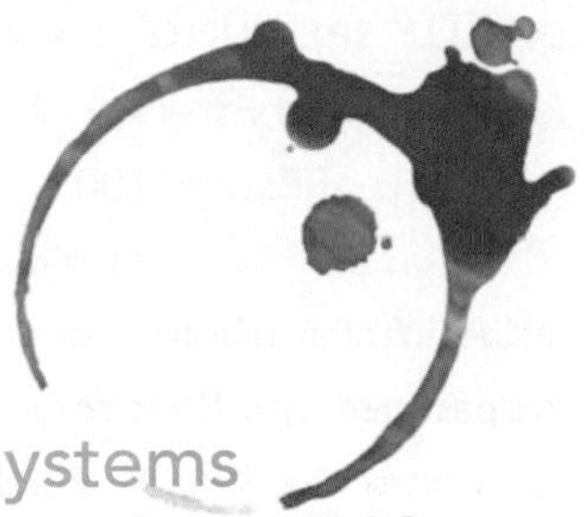

17 Conjugated Pi Systems and Pericyclic Reactions 768

18 Aromatic Compounds 817

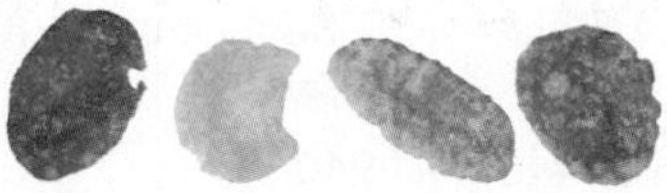

19 Aromatic Substitution Reactions 858

20

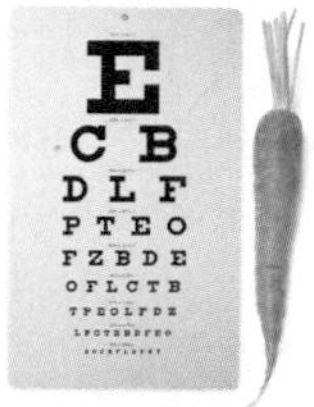

Aldehydes and Ketones 915

21

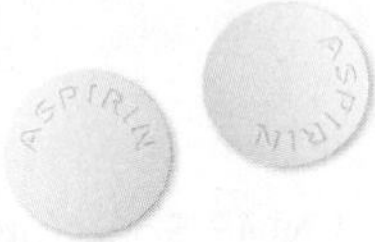

Carboxylic Acids and Their Derivatives 970

22

Alpha Carbon Chemistry: Enols and Enolates 1030

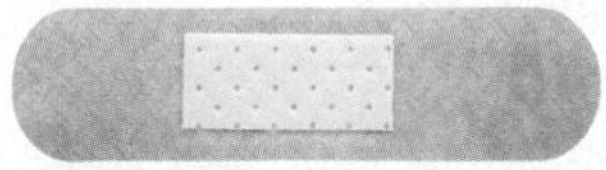

Preface

Why Did I Write This Book?

Students often say, "I studied 40 hours for this exam and I still didn't do well. Where did I go wrong?" Most instructors hear this complaint every year. In many cases, it is true that the student invested countless hours, only to produce abysmal results. Often, inefficient study habits are to blame. The important question is: why do so many students have difficulty preparing themselves for organic chemistry exams? There are certainly several factors at play here, but perhaps the most dominant factor is a fundamental disconnect between what students learn and the tasks expected of them. To illustrate this disconnect, consider the following analogy.

Imagine that a prestigious university offers a one-week course entitled "Bike-Riding 101." On the first day of the course, a physics professor delivers a fascinating lecture on the physics of bicycle riding. On the second day, an engineering professor delivers a captivating lecture on how bicycles have been engineered to minimize air-resistance. The week continues with exciting lectures, and the students invest significant time reviewing all of the information delivered throughout the week. On the last day of class, the final exam is unveiled: each student must ride a bike for a distance of 100 feet. Perhaps one or two students in the class have innate talents and are able to accomplish the task without falling even once. A large number of students fall several times, but are able to get back on the bike and slowly make it to the finish line, bruised and hurt. Other students, however, simply cannot ride for even one second without falling, even though they invested countless hours studying the material. There is a disconnect between what the students learned and what they were expected to do for their exam. A similar disconnect exists in organic chemistry textbook instruction.

The Current State of Organic Chemistry Instruction

Current organic chemistry textbooks as well as lectures provide the students with extensive coverage of the principles, but exams focus on very specific problem-solving tasks. It is often expected that students will independently develop the necessary skills for solving problems. This expectation is not much different than the expectation that students will be able to ride a bike without falling after attending a week of fascinating lectures. Organic chemistry is much like bicycle riding. It requires constant practice of very specific skills. There are key skills necessary to predict products, propose mechanisms, propose syntheses, etc. Although a few students have innate talents and are able to develop the necessary skills independently, most students require guidance. This guidance is not consistently integrated within current textbooks.

I firmly believe that the scientific discipline of organic chemistry is NOT merely a compilation of principles, but rather, it is a disciplined method of thought and analysis. Students must certainly understand the concepts and principles, but more importantly, students must learn to think like organic chemists . . . that is, they must learn to become proficient at approaching new situations methodically, based on a repertoire of skills. That is the true essence of organic chemistry.

A Skills-Based Approach to Organic Chemistry Instruction

To address the disconnect in organic chemistry instruction, I have developed a textbook that utilizes a skills-based approach to instruction. The textbook includes all of the concepts typically covered in an organic chemistry textbook, complete with *conceptual checkpoints* that promote mastery of the concepts, but special emphasis is placed on skills development to support these concepts. This emphasis upon skills development will provide students with a greater opportunity to develop proficiency in the key skills necessary to succeed in organic chemistry. Certainly, not all necessary skills can be covered in a textbook. However, there are certain skills that are fundamental to all other skills.

As an example, resonance structures are used repeatedly throughout the course, and students must become masters of resonance structures early in the course. Therefore, a significant portion of Chapter 2 is devoted to pattern-recognition for drawing resonance structures. Rather than just providing a list of rules and then a few follow-up problems, the skills-based approach provides students with a series of skills, each of which must be mastered in sequence. Each skill is reinforced with numerous practice problems. The sequence of skills is designed to foster and develop proficiency in drawing resonance structures.

As another example of the skills-based approach, Chapter 7, Substitution Reactions, places special emphasis on the skills necessary for drawing all of the mechanistic steps for S_N2 and S_N1 processes. Students are often confused when they see an S_N1 process whose mechanism is comprised of four or five mechanistic steps (proton transfers, carbocation rearrangements, etc.). This chapter contains a novel approach that trains students to identify the number of mechanistic steps required in a substitution process. Students are provided with numerous examples and are given ample opportunity to practice drawing mechanisms.

The skills-based approach to organic chemistry instruction is a unique approach. Certainly, other textbooks contain tips for problem solving, but no other textbook consistently presents skill development as one of the primary vehicles for instruction.

Pedagogical Tools that Support a Skills-Based Approach

The textbook utilizes several pedagogical tools designed to integrate the skills-based approach consistently throughout all chapters. Each chapter begins with a thought-provoking question followed by **Do You Remember?**, which is a list of relevant skills from previous chapters that should have been mastered before proceeding with the current chapter.

DO YOU REMEMBER?

Before you go on, be sure you understand the following topics.
If necessary, review the suggested sections to prepare for this chapter.

- Identifying Lone Pairs (Section 2.5)
- Curved Arrows in Drawing Resonance Structures (Section 2.8)
- Drawing Resonance Structures Via Pattern Recognition (Section 2.10)
- The Flow of Electron Density: Curved Arrow Notation (Section 3.2)

The main body of each chapter contains numerous **SkillBuilders**, each of which is designed to foster a specific skill. For example, SkillBuilder 6.6 focuses on drawing carbocation rearrangements. Each SkillBuilder contains three parts.

SKILLBUILDER

6.6 PREDICTING CARBOCATION REARRANGEMENTS

LEARN THE SKILL
A solved problem that demonstrates a particular skill.

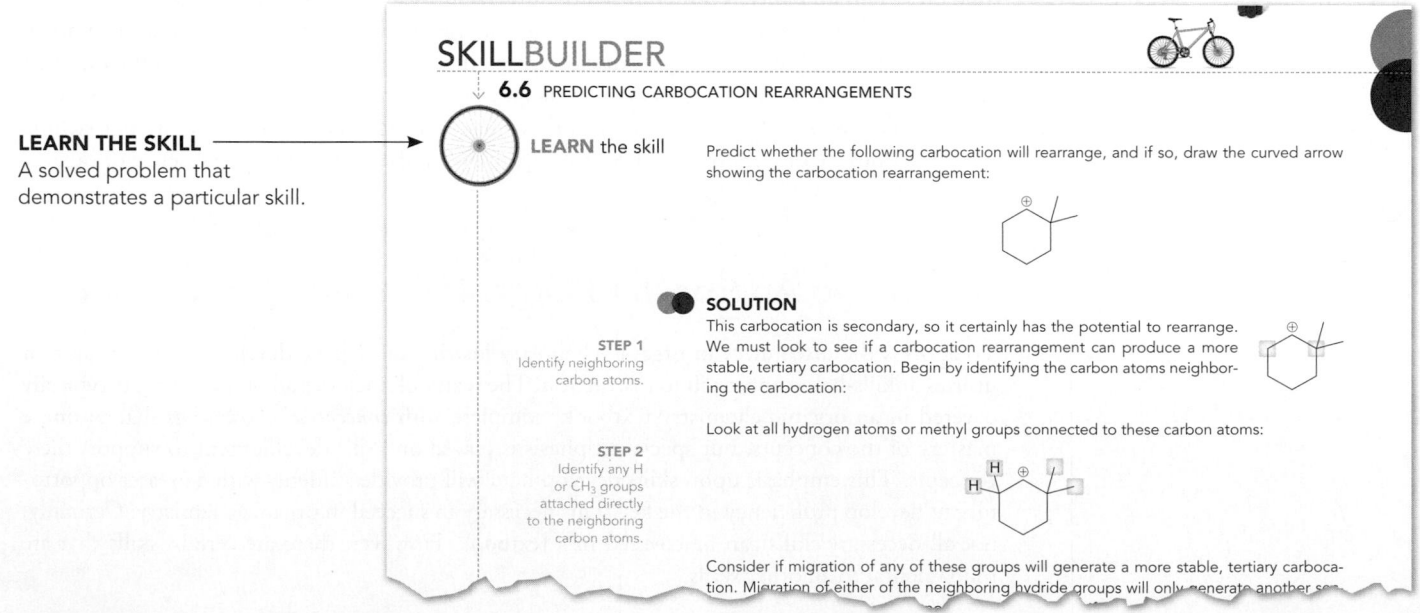

LEARN the skill

Predict whether the following carbocation will rearrange, and if so, draw the curved arrow showing the carbocation rearrangement:

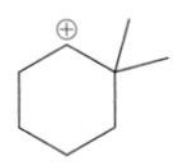

SOLUTION

STEP 1
Identify neighboring carbon atoms.

This carbocation is secondary, so it certainly has the potential to rearrange. We must look to see if a carbocation rearrangement can produce a more stable, tertiary carbocation. Begin by identifying the carbon atoms neighboring the carbocation:

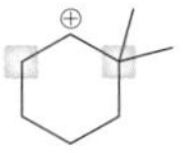

STEP 2
Identify any H or CH_3 groups attached directly to the neighboring carbon atoms.

Look at all hydrogen atoms or methyl groups connected to these carbon atoms:

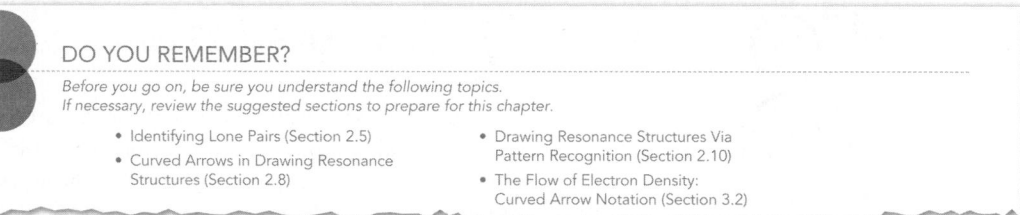

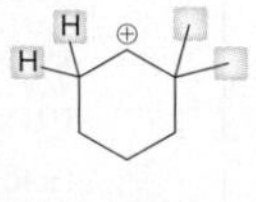

Consider if migration of any of these groups will generate a more stable, tertiary carbocation. Migration of either of the neighboring hydride groups will only generate another sec-

PRACTICE THE SKILL
Numerous problems (similar to the Solved Problem) that give students valuable opportunity to practice and master the skill.

PRACTICE the skill **6.18** For each of the following carbocations determine if it will rearrange, and if so, draw the carbocation rearrangement with a curved arrow:

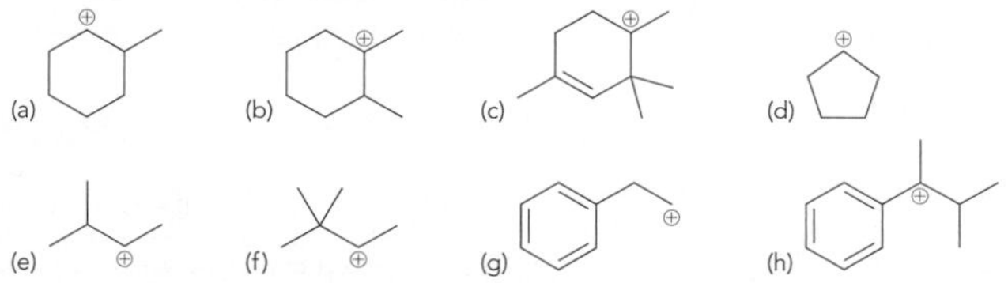

APPLY THE SKILL
One or two more challenging problems in which the student must apply the skill in a slightly different environment. These problems include conceptual, cumulative and applied problems that encourage students to think outside of the box. Sometimes problems that foreshadow concepts introduced in later chapters are also included.

APPLY the skill **6.19** Occasionally, carbocation rearrangements can be accomplished via the migration of a carbon atom other than a methyl group. Such an example follows. Identify the group that is migrating, and draw the curved arrow that shows the migration:

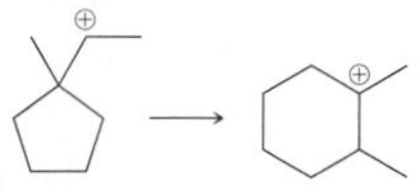

need more **PRACTICE?** **Try Problem 6.48**

NEED MORE PRACTICE?
Suggests end of chapter exercise(s) that students can work to practice the skill.

All SkillBuilders are visually summarized at the end of each chapter within the **SkillBuilder Review.**

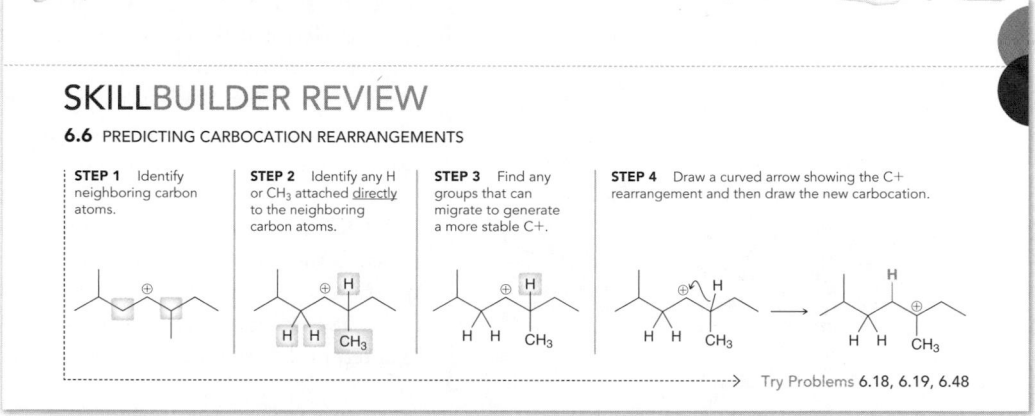

SKILLBUILDER REVIEW

6.6 PREDICTING CARBOCATION REARRANGEMENTS

STEP 1	STEP 2	STEP 3	STEP 4
STEP 1 Identify neighboring carbon atoms.	**STEP 2** Identify any H or CH_3 attached directly to the neighboring carbon atoms.	**STEP 3** Find any groups that can migrate to generate a more stable C+.	**STEP 4** Draw a curved arrow showing the C+ rearrangement and then draw the new carbocation.

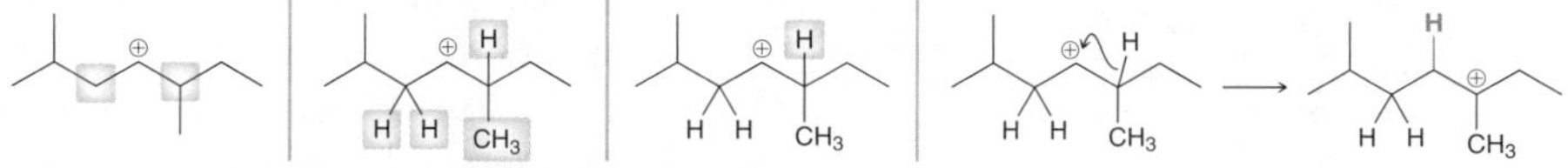

Try Problems **6.18, 6.19, 6.48**

In addition to the Skillbuilders, additional practice is provided within the chapter by way of the **Conceptual Checkpoints**.

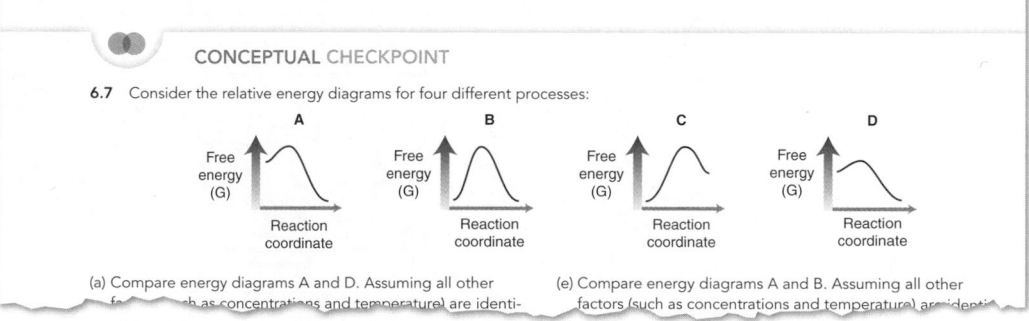

CONCEPTUAL CHECKPOINT

6.7 Consider the relative energy diagrams for four different processes:

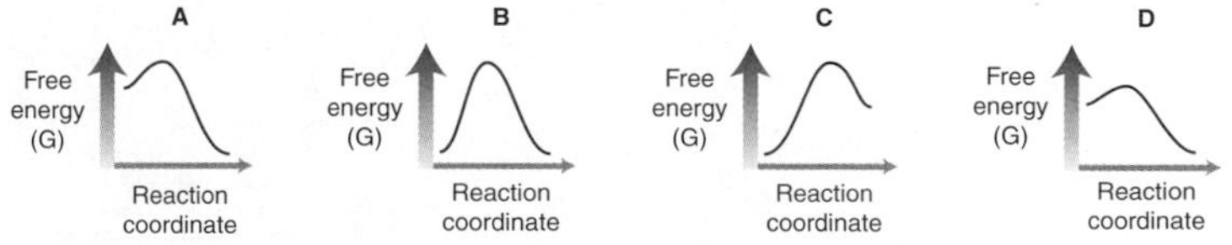

(a) Compare energy diagrams A and D. Assuming all other fa... ...ch as concentrations and temperature) are identi-

(e) Compare energy diagrams A and B. Assuming all other factors (such as concentrations and temperature) are identi...

End-of-chapter **Practice Problems** provide valuable additional skill-based exercises for continuing skills development. The **Practice Problems** are followed by **Integrated Problems**, which give the students opportunities to combine skills from the current chapter with skills from previous chapters. The end-of-chapter problems conclude with a set of **Challenge Problems**, which are designed to provide students with opportunities to apply their skills in more challenging situations. These **Challenge Problems** require students to think outside of the box and demonstrate skills mastery.

The SkillBuilder approach within the textbook is reinforced by the meaningful practice available within ***WileyPLUS***, an innovative online environment for effective teaching and learning.

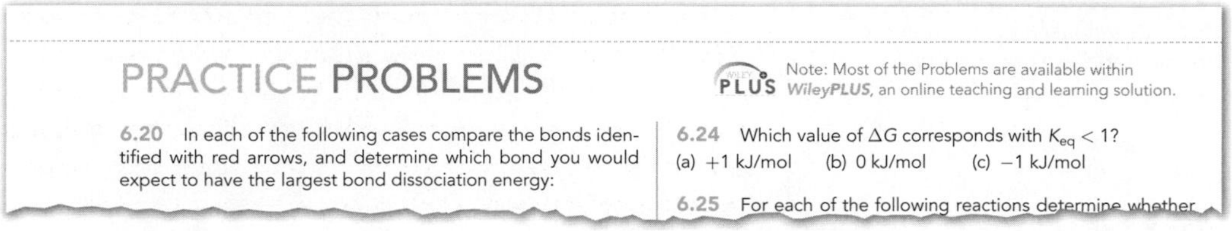
PRACTICE PROBLEMS

Note: Most of the Problems are available within *WileyPLUS*, an online teaching and learning solution.

6.20 In each of the following cases compare the bonds identified with red arrows, and determine which bond you would expect to have the largest bond dissociation energy:

6.24 Which value of ΔG corresponds with $K_{eq} < 1$?
(a) +1 kJ/mol (b) 0 kJ/mol (c) −1 kJ/mol

6.25 For each of the following reactions determine whether

There is breadth and depth of assessment within ***WileyPLUS***, consisting of **Practice the Skill** and **Learn the Skill** (from the SkillBuilders), many of the end-of-chapter problems, and a rich testbank. In addition, **Reaction Explorer** is a vast database of mechanism and synthesis questions. Professors can create online homework assignments or quizzes from these different assessment types and student responses are automatically graded. There are many forms of assistance available to students as they work questions within ***WileyPLUS***, including link(s) to the relevant section of the text and guided problem-solving tutorials.

Applications to Illustrate Relevancy

Alongside the numerous exercises built in to facilitate student understanding of the skills and concepts of organic chemistry, many real-world applications are featured throughout the chapters to demonstrate the relevance of organic chemistry in our everyday lives. These applications appear in a variety of ways, including ***Medically Speaking*** boxes which feature medicinal and pharmaceutical applications, as well as ***Practically Speaking*** boxes which feature commercial applications of organic chemistry. These real-world applications are highlighted in the detailed Table of Contents.

Each chapter begins by posing a question ("Did You Ever Wonder") which is then revisited within that chapter. For instance, in Chapter 10, a question is posed about Parkinson's disease.

The chapter opener question is always revisited and explored within that chapter, in this case within a "Medically Speaking" application box.

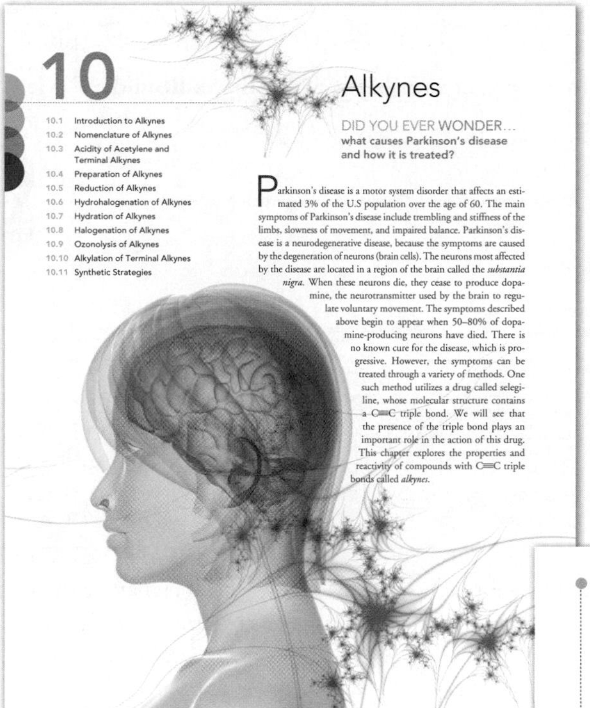
10

Alkynes

10.1 Introduction to Alkynes
10.2 Nomenclature of Alkynes
10.3 Acidity of Acetylene and Terminal Alkynes
10.4 Preparation of Alkynes
10.5 Reduction of Alkynes
10.6 Hydrohalogenation of Alkynes
10.7 Hydration of Alkynes
10.8 Halogenation of Alkynes
10.9 Ozonolysis of Alkynes
10.10 Alkylation of Terminal Alkynes
10.11 Synthetic Strategies

DID YOU EVER WONDER…
what causes Parkinson's disease and how it is treated?

Parkinson's disease is a motor system disorder that affects an estimated 3% of the U.S population over the age of 60. The main symptoms of Parkinson's disease include trembling and stiffness of the limbs, slowness of movement, and impaired balance. Parkinson's disease is a neurodegenerative disease, because the symptoms are caused by the degeneration of neurons (brain cells). The neurons most affected by the disease are located in a region of the brain called the *substantia nigra.* When these neurons die, they cease to produce dopamine, the neurotransmitter used by the brain to regulate voluntary movement. The symptoms described above begin to appear when 50–80% of dopamine-producing neurons have died. There is no known cure for the disease, which is progressive. However, the symptoms can be treated through a variety of methods. One such method utilizes a drug called selegiline, whose molecular structure contains a C≡C triple bond. We will see that the presence of the triple bond plays an important role in the action of this drug. This chapter explores the properties and reactivity of compounds with C≡C triple bonds called *alkynes.*

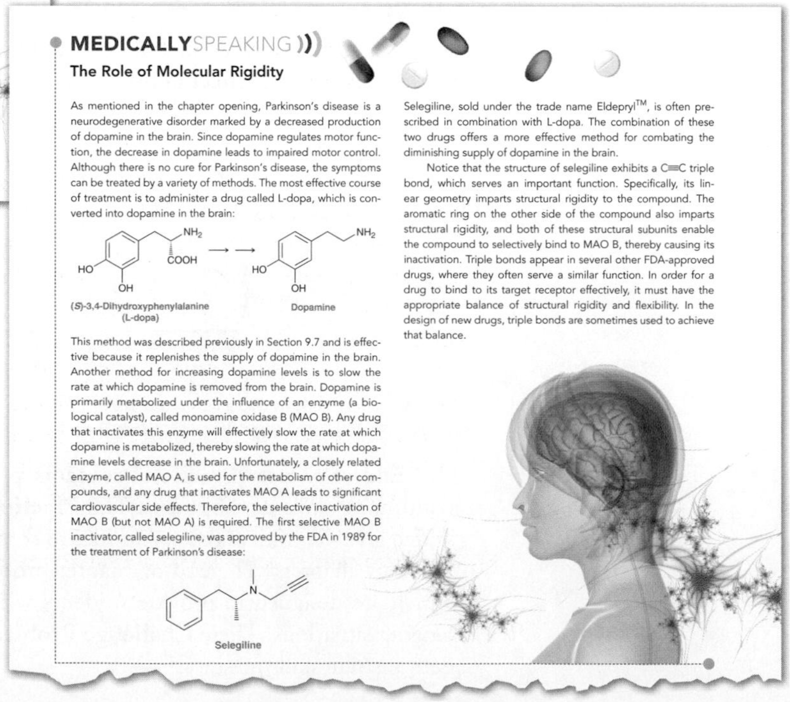
MEDICALLYSPEAKING

The Role of Molecular Rigidity

As mentioned in the chapter opening, Parkinson's disease is a neurodegenerative disorder marked by a decreased production of dopamine in the brain. Since dopamine regulates motor function, the decrease in dopamine leads to impaired motor control. Although there is no cure for Parkinson's disease, the symptoms can be treated by a variety of methods. The most effective course of treatment is to administer a drug called L-dopa, which is converted into dopamine in the brain:

(*S*)-3,4-Dihydroxyphenylalanine (L-dopa) → → Dopamine

This method was described previously in Section 9.7 and is effective because it replenishes the supply of dopamine in the brain. Another method for increasing dopamine levels is to slow the rate at which dopamine is removed from the brain. Dopamine is primarily metabolized under the influence of an enzyme (a biological catalyst), called monoamine oxidase B (MAO B). Any drug that inactivates this enzyme will effectively slow the rate at which dopamine is metabolized, thereby slowing the rate at which dopamine levels decrease in the brain. Unfortunately, a closely related enzyme, called MAO A, is used for the metabolism of other compounds, and any drug that inactivates MAO A leads to significant cardiovascular side effects. Therefore, the selective inactivation of MAO B (but not MAO A) is required. The first selective MAO B inactivator, called selegiline, was approved by the FDA in 1989 for the treatment of Parkinson's disease:

Selegiline

Selegiline, sold under the trade name Eldepryl™, is often prescribed in combination with L-dopa. The combination of these two drugs offers a more effective method for combating the diminishing supply of dopamine in the brain.

Notice that the structure of selegiline exhibits a C≡C triple bond, which serves an important function. Specifically, its linear geometry imparts structural rigidity to the compound. The aromatic ring on the other side of the compound also imparts structural rigidity, and both of these structural subunits enable the compound to selectively bind to MAO B, thereby causing its inactivation. Triple bonds appear in several other FDA-approved drugs, where they often serve a similar function. In order for a drug to bind to its target receptor effectively, it must have the appropriate balance of structural rigidity and flexibility. In the design of new drugs, triple bonds are sometimes used to achieve that balance.

Text Organization

The sequence of chapters and topics in *Organic Chemistry* do not differ markedly from that of other organic chemistry textbooks. Indeed, the topics are presented in the traditional order, based on functional groups (alkenes, alkynes, alcohols, ethers, aldehydes and ketones, carboxylic acid derivatives, etc.). Despite this traditional order, a strong emphasis is placed on mechanisms, with a focus on pattern recognition to illustrate the similarities between reactions that would otherwise appear unrelated (for example, ketal formation and enamine formation, which are mechanistically quite similar). No shortcuts were taken in any of the mechanisms, and all steps are clearly illustrated, including all proton transfer steps.

Two chapters (6 and 12) are devoted almost entirely to skill development and are generally not found in other textbooks. Chapter 6, Chemical Reactivity and Mechanisms, emphasizes skills that are necessary for drawing mechanisms, while Chapter 12, Synthesis, prepares the students for proposing syntheses. These two chapters are strategically positioned within the traditional order described above and can be assigned to the students for independent study. That is, these two chapters do not need to be covered during precious lecture hours, but can be, if so desired.

The traditional order allows instructors to adopt the skills-based approach without having to change their lecture notes or methods. For this reason, the spectroscopy chapters (chapters 15 and 16) were written to be stand-alone and portable, so that instructors can cover these chapters in any order desired. In fact, two of the chapters (Chapters 13 and 14) that precede the spectroscopy chapters include end-of-chapter spectroscopy problems, for those students who covered spectroscopy earlier. Spectroscopy coverage appears in subsequent functional group chapters, specifically Chapter 18 (Aromatic Compounds), Chapter 20 (Aldehydes and Ketones), Chapter 21 (Carboxylic Acids and Their Derivatives), Chapter 23 (Amines), Chapter 24 (Carbohydrates), and Chapter 25 (Amino Acids, Peptides and Proteins).

ACKNOWLEDGEMENTS

Certainly, books are not written in a vacuum, and this textbook is no exception. Over the last four years, I have benefitted greatly from the many, many reviewers and accuracy checkers who engaged with the materials as they were being developed and polished. I also benefited enormously from the extensive class test program in which 75 professors and over 4,000 students participated and offered valuable user feedback. Each individual played a crucial role in allowing us to develop a very high quality product.

This book could not have been created without the incredible efforts of the following people at John Wiley and Sons, Inc. Illustration Editor Sandra Rigby helped create a compelling art program. Photo Editor Lisa Gee helped identify exciting photos. Freelance designers Carole Anson and Anne DeMarinis conceived of a visually refreshing and compelling interior design and cover. Production Editor Elizabeth Swain, kept this book on schedule and was vital to ensuring such a high-quality product. The development team, consisting of Leslie Kraham, Barbara Heaney and Joan Kalkut, were invaluable in the creation of this first edition. The tireless efforts of Joan and Leslie, together with their day-to-day guidance and insight, made this project possible. Freelance editor Deena Cloud provided thoughtful and constructive guidance about how best to frame and present my ideas. Media Editors Tom Kulesa, Marc Wezdecki and Geraldine Osnato conceived of and built a compelling ***WileyPLUS*** course. Marketing Manager Kristine Ruff enthusiastically created an exciting message for this book. Editorial assistants Cathy Donovan and Lauren Stauber helped to manage many facets of the review and supplements process. Associate Editor Jennifer Yee helped with the PowerPoint slides. Associate Publisher Petra Recter provided strong vision and guidance in bringing this book to market.

Despite my best efforts, as well as the best efforts of the reviewers, accuracy checkers and class testers, errors may still exist. I take full responsibility for any such errors and would encourage those using my textbook to contact me with any errors that you may find.

David R. Klein, Ph.D.
Johns Hopkins University
DavidKlein@wiley.com

The Development Story

REVIEWED AND TESTED BY OVER 315 PROFESSORS AND 4,000 STUDENTS!

The creation, development, and execution of this first edition is the result of incredibly extensive professor and student involvement, every step of the way. Our solution (the textbook and the accompanying ***WileyPLUS*** course) has been carefully designed to meet the dual needs of faculty and students. Over the past four years, we have exhaustively tested and reviewed the manuscript, the art program, the design, the problems, and the value of the SkillBuilder approach to ensure accuracy and effectiveness in the classroom.

Adopters of the Preliminary Edition To meet customer demand, we created a preliminary edition that seven schools adopted for Fall 2010.

Class Testers Seventy-five instructors and over 4,000 students class tested materials in 2009 and Fall 2010 prior to publication. Their feedback was overwhelmingly supportive and enthusiastic, with over 98% of all instructors stating that the Klein materials met their course goals. They offered valuable suggestions that can only come from use in the classroom, and their comments factored into each decision that was made to produce the final textbook.

Accuracy Checkers Ten professors participated in accuracy checking at each stage of the manuscript development and proof process to ensure that the end product is as error-free as possible.

Reviewers More than 240 professors across the United States and Canada reviewed the manuscript to ensure the content was clear and precise and facilitated student engagement and understanding.

5 Professor Focus Groups Thirty-two professors participated to provide invaluable feedback on the art program and interior design, our *WileyPLUS* solution, and the specific needs of different schools.

Multiple Student Focus Groups Over 50 students participated in a variety of focus groups to provide feedback on the design and insights into their preferred learning style. These students confirmed that our design and pedagogy were appropriate and added value.

Developmental Review Over the past four years, a team of development editors, including line editors and art editors, have worked with the author to hone his distinctive voice, test explanation of concepts in the classroom, and confirm that the pedagogy was consistent and added value to the learning process.

Adopters of the Preliminary Edition (Fall 2010/Spring 2011)

Allegany College of Maryland
Augustana College
Florida Southern College
Long Beach City College
Nicholls State University
University of Tampa
Virginia Wesleyan College

Class Testers

Philip Albiniak, Ball State University
Laura Anna, Millersville University
Satinder Bains, Paradise Valley Community College
C. Eric Ballard, University of Tampa
Edie Banner, Florida Southern College
James Beil, Lorain County Community College
Peter Bell, Tarleton State University
Lea Blau, Yeshiva University
Cindy Browder, Northern Arizona University
Kathleen Brunke, Christopher Newport University
Kevin Cannon, Penn State University, Abington
Kevin Caran, James Madison University
David Cartrette, South Dakota State University
Ashley Causton, University of Calgary
Emma Chow, Palm Beach Community College
Marilyn Cox, Louisiana Tech University
Mapi Cuevas, Sante Fe Community College
Joyce Easter, Virginia Wesleyan College
Raymond Fong, City College of San Francisco
Andrew Frazer, University of Central Florida
Nathanial Grove, University of North Carolina, Wilmington
Sapna Gupta, Palm Beach State College
Kevin Gwaltney, Kennesaw State University
Scott Handy, Middle Tennessee State University
Eric Helms, SUNY Geneseo
Maged Henary, Georgia State University, Langate
Patricia Hill, Millersville University
Roxanne Hulet, Skagit Valley College
Christopher Hyland, California State University, Fullerton
Danielle Jacobs, Rider University
Dell Jensen, Augustana College
Marlon Jones, Long Beach City College
Reni Joseph, St. Louis Community College, Meramec Campus
Adam Keller, Columbus State Community College
Cindy Lamberty, Cloud County Community College, Geary County Campus
William Lavell, Camden County College
Iyun Lazik, San Jose City College
Rena Lou, Cerritos College
Vivian Mativo, Georgia Perimeter College, Clarkston
Dominic McGrath, University of Arizona
Kevin Minbiole, James Madison University
Richard Narske, Augustana College
Nasri Nesnas, Florida Institute of Technology
Edmond O'Connell, Fairfield University
Hasan Palandoken, California Polytechnic State University, San Luis Obispo
Sapan Parikh, Manhattanville College
Keith O. Pascoe, Georgia State University

Libbie Pelter, Purdue University, Calumet
Christine Pruis, Arizona State University
Cathrine Reck, Indiana University
Robert D. Rossi, Gloucester County College
Gillian Rudd, Northwestern State University
Steve Samuel, SUNY Old Westbury
Gita Sathianathan, California State University, Fullerton
Amber Schaefer, Texas A&M University
Jason Serin, Glendale Community College
Gary Shankweiler, California State University, Long Beach
Richard Shreve, Palm Beach State College
Douglas Smith, California State University, San Bernardino
Vadim Soloshonok, University of Oklahoma
Mackay Steffensen, Southern Utah University
Corey Stephenson, Boston University
Nhu Y. Stessman, California State University Stanislaus
James Stickler, Allegany College of Maryland
Cynthia Tidwell, University of Montevallo
Melissa Van Alstine, Adelphi University
Shirley Wachowich-Sgarbi, Langara College (Canada)
Edmir Wade, University of Southern Indiana
Vidyullata Waghulde, St. Louis Community College
Linda Waldman, Cerritos College
Kenneth Walsh, University of Southern Indiana
Reuben Walter, Tarleton State University
Leyte Winfield, Spelman College
Penny Workman, University of Wisconsin, Marathon County
Stephen Wuerz, Highland Community College

Accuracy checkers

C. Eric Ballard, University of Tampa
Kevin Caran, James Madison University
Jeffrey Carney, Christopher Newport University
James T. Fletcher, Creighton University
Eric Kantorowski, California Polytechnic State University, San Luis Obispo
Michael Leonard, Washington & Jefferson College
Kevin Minbiole, James Madison University
John Pollard, University of Arizona
Harold Rogers, California State University, Fullerton
Regina Zibuck, Wayne State University

Reviewers

ALABAMA
Silas Blackstock, University of Alabama
Edward Parish, Auburn University
Kevin Shaughnessy, The University of Alabama
Timothy Snowden, University of Alabama
Stephen Woski, University of Alabama

ALASKA
Tom Green, University of Alaksa, Fairbanks

ARIZONA
Cindy Browder, Northern Arizona University
Silvia Kölchens, Pima County Community College
Douglas Loy, University Of Arizona
Dominic McGrath, University of Arizona
Anne B. Padias, University of Arizona
John Pollard, University of Arizona
Jason Serin, Glendale Community College
Edward Skibo, Arizona State University

CALIFORNIA
Dianne Bennet, Sacramento City College
Daniel Bernier, Riverside Community College
Megan Bolitho, University of San Francisco
David Brook, San Jose State University
Philip J. Costanzo, California Polytechnic State University, San Luis Obispo
Peter de Lijser, California State University, Fullerton
John Flygare, Stanford University
Annaliese Franz, University of California, Davis
Ling Huang, Sacramento City College
Eric Kantorowski, California Polytechnic State University, San Luis Obispo
Barbara Mayer, California State University, Fresco
Thomas Minehan, California State University, Northridge
James Miranda, California State University, Sacramento
Barbara Murray, University of Redlands
William Nguyen, Santa Ana College
James Nowick, University of California, Irvine
Hasan Palandoken, California Polytechnic State University, San Luis Obispo
Harold Rogers, California State University, Fullerton
Douglas Smith, California State University, San Bernadino
Nhu Y. Stessman, California State University, Stanislaus
Ana Tontcheva, El Camino College
Christopher Vanderwal, University of California, Irvine
Andrew Wells, Chabot College
Jinsong Zhang, California State University, Chico

COLORADO
David Anderson, University of Colorado, Colorado Springs

CONNECTICUT
Adiel Coca, Southern Connecticut State University
Andrew Karatjas, Southern Connecticut State University
Edmond J. O'Connell, Fairfield University
Ronald Wikholm, University of Connecticut, Storrs

DISTRICT OF COLUMBIA
Jennifer Swift, Georgetown University

FLORIDA
C. Eric Ballard, University of Tampa
Edie Banner, Florida Southern College
Andrew Frazer, University of Central Florida
Cynthia Judd, Palm Beach State College
Kathleen Laurenzo, Florida State College
Stephen Milczanowski, Florida State College
Edith Onyeozili, Florida Agricultural & Mechanical University
Tchao Podona, Miami Dade College
Mary Roslonowski, Brevard Community College
Thomas Russo, Florida State College, Kent Campus
Kirk Schanze, University of Florida

GEORGIA
Scott Davis, Mercer University, Macon
Kevin Gwaltney, Kennesaw State University
Jason Locklin, University of Georgia
Keith Pascoe, Georgia State University
Michelle Smith, Georgia Southwestern State University
Matthew Weinschenk, Emory University
Leyte Winfield, Spelman College

ILLINOIS
David Crich, University of Illinois at Chicago
Theodore Dolter, Southwestern Illinois College
Steve Gentemann, Southwestern Illinois College
Dell Jensen, Augustana College
Richard Narske, Augustana College
Owen Priest, Northwestern University, Evanston
Preet-Pal S. Saluja, Triton College
Carole Szpunar, Loyola University Chicago

INDIANA
Philip Albiniak, Ball State University
Kyungsoo Oh, Indiana University, Purdue University Indianapolis
Michael Pelter, Purdue University, Calumet
Harold Pinnick, Purdue University, Calumet
Cathrine Reck, Indiana University Bloomington
Sergey Savinov, Purdue University, West Lafayette
Edmir Wade, University of Southern Indiana
Ken Walsh, University of Southern Indiana
Anne Wilson, Butler University

IOWA
Jeffrey Elbert, University of Northern Iowa
Kirk Manfredi, University of Northern Iowa
Chris Pigge, University of Iowa
Olga Rinco, Luther College
Brad Chamberlain, Luther College

KANSAS
Cindy Lamberty, Cloud County Community College, Geary County Campus
Sam Leung, Washburn University

KENTUCKY
Arthur Cammers, University of Kentucky, Lexington
Frederick A. Luzzio, University of Louisville
Lili Ma, Northern Kentucky University
Chad Snyder, Western Kentucky University

LOUISIANA
Marilyn Cox, Louisiana Tech University
Scott Grayson, Tulane University
Rober Hammer, Louisiana State University

MAINE
Jennifer Koviach-Côté, Bates College

MARYLAND
Timothy Brunker, Towson University
Bonnie Dixon, University of Maryland, College Park
Susan Ensel, Hood College
Lee Friedman, University of Maryland, College Park
Tiffany Gierasch, University Of Maryland, Baltimore County
Ray A. Gross, Jr., Prince George's Community College
Jesse More, Loyola College
H. Mark Perks, University of Maryland, Baltimore County
Lev Ryzhkov, Towson University
James Stickler, Allegany College of Maryland
Angela Winstead, Morgan State University

MASSACHUSETTS
Frank Day, North Shore Community College
Sivappa Rasapalli, University of Massachusetts, Dartmouth
Eriks Rozners, Northeastern University
Corey Stephenson, Boston University

MICHIGAN
Cory Emal, Eastern Michigan University
Kimberly Greve, Kalamazoo Valley Community College
Matthew Hart, Grand Valley State University
James Kiddle, Western Michigan University
Dalila Kovacs, Grand Valley State University
Harriet Lindsay, Eastern Michigan University
Regina Zibuck, Wayne State University

MINNESOTA
Thomas Nalli, Winona State University

MISSOURI
Sarah Chavez, Washington University
Michael Lewis, Saint Louis University
Sarah Mounter, Columbia College of Missouri

NEBRASKA
James T. Fletcher, Creighton University

NEVADA
Christopher S. Jeffrey, University of Nevada, Reno

NEW HAMPSHIRE
Ivan Aprahamian, Dartmouth College
Gordon Gribble, Dartmouth College
Richard Johnson, University of New Hampshire
Catherine Owens Welder, Dartmouth College

NEW JERSEY
Thomas Berke, Brookdale Community College
Danielle Jacobs, Rider University
Mushtaq Khan, Union County College
Massimiliano Lamberto, Monmouth University
William Lavell, Camden County College
Shanthi Rajaraman, Richard Stockton College Of New Jersey
Robert Rossi, Gloucester County College
Paul Schueler, Raritan Valley Community College
John Sowa, Seton Hall University
John Taylor, Rutgers University, New Brunswick
Peter Wepplo, Monmouth University`

NEW MEXICO
Robert Long, Eastern New Mexico University
Lisa Whalen, University of New Mexico

NEW YORK
Michael Aldersley, Rensselaer Polytechnic Institute
Lea Blau, Yeshiva University
Phillip Chung, Montefiore Medical Center
Jeremy Cody, Rochester Institute of Technology
Brahmadeo Dewprashad, Borough of Manhattan Community College
Preeti Dhar, SUNY New Paltz
Frantz Folmer-Andersen, SUNY New Paltz
Yi Guo, Montefiore Medical Center
Eric Helms, SUNY Geneseo
Kevin Kittredge, Siena College
Dina Merrer, Barnard College
Sapan Parikh, Manhattanville College
Martin Pulver, Bronx Community College
Scott Snyder, Columbia University
Bruce Toder, University of Rochester
Hanying Xu, Kingsborough Community College of CUNY

NORTH CAROLINA
Banita Brown, University Of North Carolina, Charlotte
Nathaniel Grove, University of North Carolina, Wilmington
Paul J. Kropp, University of North Carolina, Chapel Hill
Brian Love, East Carolina University
Andrew Morehead, East Carolina University
James Parise Jr., Duke University
Pamela Seaton, University of North Carolina, Wilmington
Erland Stevens, Davidson College

NORTH DAKOTA
Irina Smoliakova, University of North Dakota

OHIO
James Beil, Lorain County Community College
Jared Butcher, Ohio University
Steven Chung, Bowling Green State University

Wyatt Cotton, Cincinnati State College
Adam Keller, Columbus State Community College
Deborah Lieberman, University of Cincinnati
David Modarelli, University of Akron
Mike Rennekamp, Columbus State Community College
Richard Taylor, Miami University

OKLAHOMA
Steven Meier, University of Central Oklahoma
Donna Nelson, University of Oklahoma
Vadim Soloshonok, University of Oklahoma

OREGON
Paul Chamberlain, George Fox University
Ron Swisher, Oregon Institute of Technology

PENNSYLVANIA
Laura Anna, Millersville University
Kevin Cannon, Pennsylvania State University, Abington
Mark Forman, Saint Joseph's University
Michael Leonard, Washington & Jefferson College
Robert Stockland, Bucknell University
Eric Tillman, Bucknell University
Michael Wilson, Temple University

SOUTH CAROLINA
Gautam Bhattacharyya, Clemson University
J. Derek Elgin, Coastal Carolina University
John Shugart, Coastal Carolina University
Rhett Smith, Clemson University

SOUTH DAKOTA
David Cartrette, South Dakota University

TENNESSEE
Phillip Cook, East Tennessee State University
Norma Dunlap, Middle Tennessee State University
Scott Handy, Middle Tennessee State University
Yu Lin Jiang, East Tennessee State University
Aleskey Vasiliev, East Tennessee State University

TEXAS
Thomas Albright, University of Houston
Peter Bell, Tarleton State University
Narayan Bhat, University of Texas Pan American
Sergio Cortes, University of Texas at Dallas
Frank Foss, University of Texas at Arlington
Donovan Haines, Sam Houston State University
Christopher Hansen, Midwestern State University
Kenn Harding, Texas A&M University
Javier Macossay-Torres, University of Texas Pan American
Nancy Mills, Trinity University
Shizue Mito, University Of Texas at El Paso
Paul Primrose, Baylor University
Ron Reese, Victoria College
Patricio Santander, Texas A&M University
Emery Shier, Amarillo College
Claudia Taenzler, University of Texas at Dallas
Adam Urbach, Trinity University

UTAH
Merritt Andrus, Brigham Young University
Steven Castle, Brigham Young University
Tom Chang, Utah State University
Mackay Steffensen, Southern Utah University
Richard Steiner, University Of Utah
Heidi Vollmer-Snarr, Brigham Young University

VERMONT
Matthias Brewer, The University of Vermont

VIRGINIA
Kathleen Brunke, Christopher Newport University
Kevin Caran, James Madison University
Jeffrey Carney, Christopher Newport University
Jason Chruma, University of Virginia
Christine Hermann, Radford University
Scott Lewis, James Madison University
Kevin Minbiole, James Madison University

WASHINGTON
Jeff Corkill, Eastern Washington University

WEST VIRGINIA
John Hubbard, Marshall University

WISCONSIN
Asif Habib, University of Wisconsin, Waukesha
Alan Schwabacher, University of Wisconsin, Milwaukee
Linfeng Xie, University Of Wisconsin, Oshkosh

CANADA
Ashley Causton, University of Calgary
Michael Chong, University of Waterloo
Andrew Dicks, University of Toronto
Torsten Hegmann, University of Manitoba
Ian Hunt, University of Calgary
Norman Hunter, University of Manitoba
Michael Pollard, York University
Stanislaw Skonieczny, University of Toronto
Jackie Stewart, University of British Columbia
Shirley Wacowich- Sgarbi, Langara College

Focus Group Participants

Yiyan Bai, Houston Community College
Satinder Bains, Paradise Valley Community College
Peter Bell, Tarleton State University
Ashley Causton, University of Calgary
Brahmadeo Dewprashad, Borough of Manhattan Community College
Andrew Dicks, University of Toronto
Jeffrey Elbert, University of Northern Iowa
Derek Elgin, Coastal Carolina University
David Flanigan, Hillsborough Community College
Frank Foss, University of Texas, Arlington
Thomas Green, University of Alaska, Fairbanks
Nathaniel Grove, University of North Carolina, Wilmington
Donovan Haines, Sam Houston State University
Jack Hayes, State Fair Community College
Silvia Kölchens, Pima Community College
Dalila Kovacs, Grand Valley State University
Jens-Uwe Kuhn, Santa Barbara City College
Douglas Loy, University of Arizona
Dominic McGrath, University of Arizona
James Miranda, California State University, Sacramento
Javier Macossay-Torres, University of Texas Pan American
Sarah Mounter, Columbia College of Missouri
Anne Padias, University of Arizona
Chandrakant Panse, Massachusetts Bay Community College
H. Mark Perks, University of Maryland, Baltimore County
John Pollard, University of Arizona
Paul Primrose, Baylor University
Melinda Ripper, Butler County Community College
Robert Rossi, Gloucester County College
Jason Serin, Glendale Community College
Eric Tillman, Bucknell University
Bruce Toder, University of Rochester

Teaching and Learning Resources

WileyPLUS

An innovative, research-based, online environment for effective teaching and learning. *WileyPLUS* builds students' confidence because it takes the guesswork out of studying by providing students with a clear roadmap to success (what to do, how to do it, if they did it right). *WileyPLUS* does this through an innovative **design** that fosters **engagement**, which leads to improved learning **outcomes**. Students take more initiative so that instructors can have greater impact.

What Do Instructors Receive With *WileyPLUS*?

Customizable Course Plan: *WileyPLUS* comes with a pre-created Course Plan designed by a subject matter expert exclusively for this course. Simple drag-and-drop tools make it easy to assign the course plan as-is or modify it to reflect your course syllabus.

Gradebook: WileyPLUS provides instant access to reports on trends in class performance, student use of course materials, and progress towards learning objectives, thereby helping inform instructors' decisions and drive classroom discussions.

Breadth and Depth of Assessment: Four unique silos of assessment are available to Instructors for creating online homework and quizzes.

WILEYPLUS ASSESSMENT

- End of Chapter Content — SkillBuilders and end of chapter questions are coded for online assessment.
- Concept Mastery Assignments — Prebuilt Concept Mastery Assignments (from a database of over 25,000 questions)
- Reaction Explorer — Meaningful practice of mechanisms and synthesis problems (a database of over 100,000 questions)
- Test Bank — Rich Testbank, consisting of over 2,500 questions

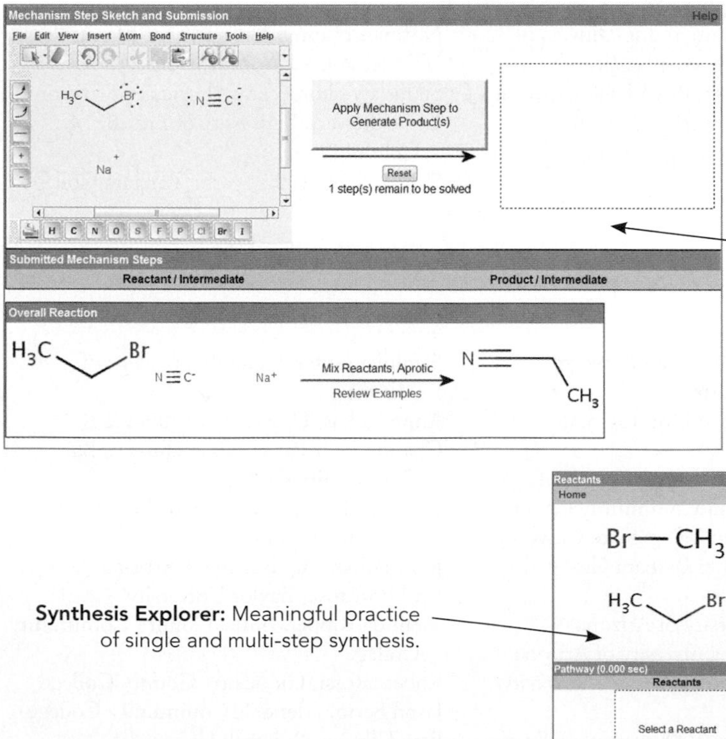

Reaction Explorer is an interactive system for **learning and practicing reactions, syntheses,** and **mechanisms** in organic chemistry, with advanced support for the automatic generation of random problems and curved-arrow mechanism diagrams.

Mechanism Explorer: valuable practice of reactions and mechanisms.

Synthesis Explorer: Meaningful practice of single and multi-step synthesis.

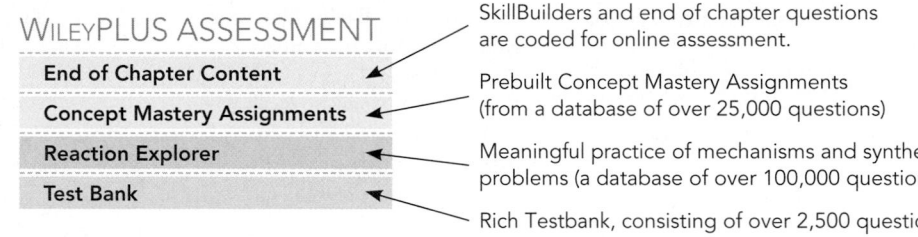

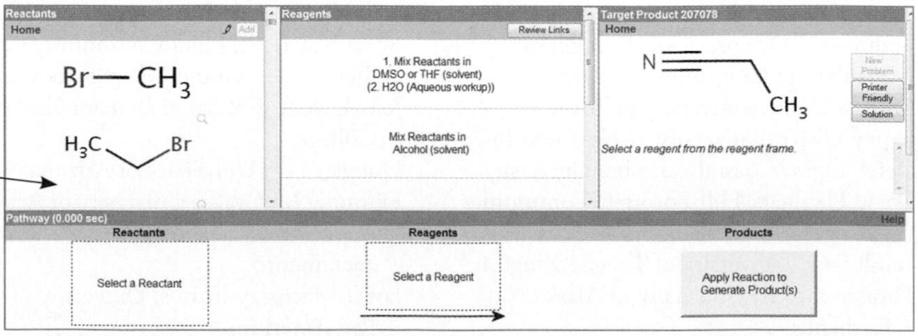

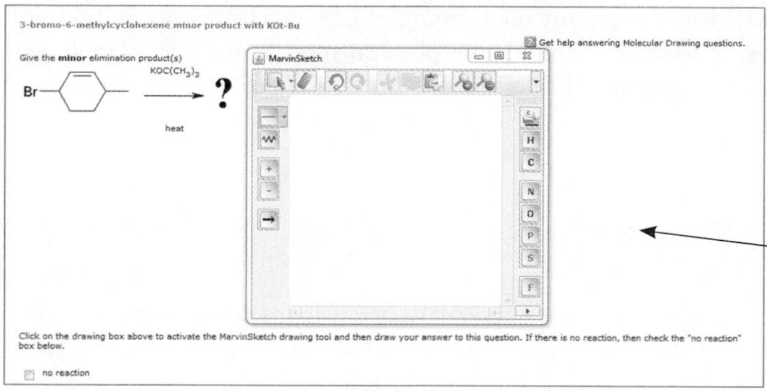

Prebuilt concept mastery assignments.

Prebuilt concept mastery assignments are included for each chapter, organized by topic, featuring feedback for incorrect answers. These assignments pull from a unique database of over 25,000 questions, over half of which require students to draw a structure using MarvinSketch.

What Do Students Receive With *WileyPLUS*?

A Research-based Design. *WileyPLUS* provides an online environment that integrates relevant resources, including the entire digital textbook, in an easy-to-navigate framework that helps students study more effectively.

One-on-One Engagement. With *WileyPLUS* students receive 24/7 access to resources that promote positive learning outcomes. Students engage with related examples (in various media) and practice items.

Measurable Outcomes. Throughout each study session, students can assess their progress and gain immediate feedback. *WileyPLUS* provides precise reporting of strengths and weaknesses, as well as individualized quizzes, so that students are confident they are spending their time on the right things. With *WileyPLUS*, students always know the exact outcome of their efforts.

How Do I Access *WileyPLUS*?

To access *WileyPLUS*, students need a *WileyPLUS* registration code. This can be purchased stand alone, or the code can be bundled with a textbook. For more information and/ or to request a *WileyPLUS* demonstration, contact your local Wiley sales representative or visit **www.wileyplus.com**

Additional Instructor Resources

All of these resources can be accessed within *WileyPLUS* or by contacting your local Wiley sales representative

- **PowerPoint Lecture Slides**
- **PowerPoint Art Slides**
- **Testbank** – printed and computerized version

Additional Student Resources

Study Guide and Solutions Manual (ISBN: 978-0-471-75739-9) Authored by David Klein, the Study Guide and Solutions Manual contains:

- Concept Review Exercises
- SkillBuilder Review exercises
- Reaction Review exercises
- Detailed solutions to all the problems in the text.

Molecular Visions™ Model Kit To support the learning of organic chemistry concepts and allow students the tactile experience of manipulating physical models, we offer a molecular modeling kit from the Darling Company. The model kit can be bundled with the textbook or purchased stand alone.

Customization and Flexible Options to Meet Your Needs

Wiley Custom Select allows you to create a textbook with precisely the content you want, in a simple, three-step online process that brings your students a cost-efficient alternative to a traditional textbook. Select from an extensive collection of content at **http://customselect.wiley.com,** upload your own materials as well, and select from multiple delivery formats—full color or black and white print with a variety of binding options, or eBook. Preview the full text online, get an instant price quote, and submit your order; we'll take it from there.

WileyFlex offers content in flexible and cost-saving options to students. Our goal is to deliver our learning materials to our customers in the formats that work best for them, whether it's traditional text, eTextbook, *WileyPLUS*, loose-leaf binder editions, or customized content through Wiley Custom Select.

About the Author

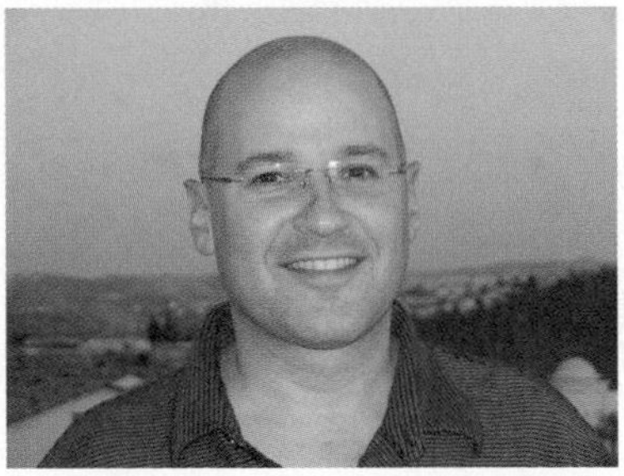

David Klein is a Senior Lecturer in the Department of Chemistry at The Johns Hopkins University where he has taught organic chemistry since 1999. Having worked with thousands of students, he has intense first-hand knowledge of how they learn and the difficulties they encounter. He received his bachelor's degree in chemistry from The Johns Hopkins University and his PhD from the University of California at Los Angeles under the supervision of Professor Orville Chapman. Motivated by his experiences teaching organic chemistry as a graduate student at UCLA, David wrote *Organic Chemistry as a Second Language* (John Wiley & Sons, 2004, updated 2nd edition published in 2008), which has become a highly valued student study resource. David has received numerous teaching awards at both UCLA and Johns Hopkins for his unique, skill-building approach to organic chemistry instruction. David is married, with five children, and enjoys skiing, scuba-diving, and Tae Kwon Do.

A Review of General Chemistry

1

ELECTRONS, BONDS, AND MOLECULAR PROPERTIES

DID YOU EVER WONDER…
what causes lightning?

Believe it or not, the answer to this question is still the subject of debate (that's right… scientists have not yet figured out everything, contrary to popular belief). There are various theories that attempt to explain what causes the buildup of electric charge in clouds. One thing is clear, though—lightning involves a flow of electrons. By studying the nature of electrons and how electrons flow, it is possible to control where lightning will strike. A tall building can be protected by installing a lightning rod (a tall metal column at the top of the building) that attracts any nearby lightning bolt, thereby preventing a direct strike on the building itself. The lightning rod on the top of the Empire State Building is struck over a hundred times each year.

Just as scientists have discovered how to direct electrons in a bolt of lightning, chemists have also discovered how to direct electrons in chemical reactions. We will soon see that although organic chemistry is literally defined as the study of compounds containing carbon atoms, its true essence is actually the study of electrons, not atoms. Rather than thinking of reactions in terms of the motion of atoms, we

continued >

must recognize that *reactions occur as a result of the motion of electrons*. For example, in the following reaction the curved arrows represent the motion, or flow, of electrons. This flow of electrons causes the chemical change shown:

$$:\ddot{\underset{..}{Cl}}:^{\ominus} \quad \underset{H}{\overset{H}{\underset{H}{|}}}C\!-\!\ddot{\underset{..}{Br}}: \quad \longrightarrow \quad :\ddot{\underset{..}{Cl}}\!-\!\underset{H}{\overset{H}{\underset{H}{C}}} \quad + \quad :\ddot{\underset{..}{Br}}:^{\ominus}$$

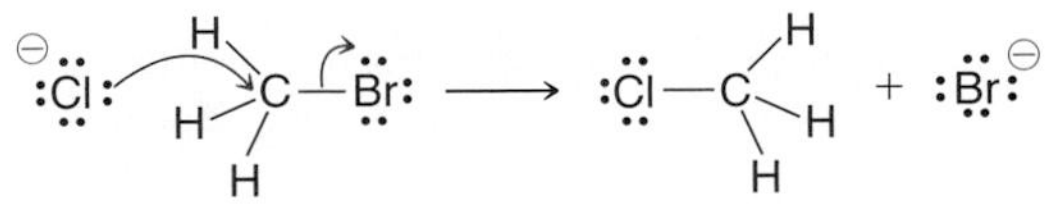

Throughout this course, we will learn how, when, and why electrons flow during reactions. We will learn about the barriers that prevent electrons from flowing, and we will learn how to overcome those barriers. In short, we will study the behavioral patterns of electrons, enabling us to predict, and even control, the outcomes of chemical reactions.

This chapter reviews some relevant concepts from your general chemistry course that should be familiar to you. Specifically, we will focus on the central role of electrons in forming bonds and influencing molecular properties.

1.1 Introduction to Organic Chemistry

In the early nineteenth century, scientists classified all known compounds into two categories: *organic compounds* were derived from living organisms (plants and animals), while *inorganic compounds* were derived from nonliving sources (minerals and gases). This distinction was fueled by the observation that organic compounds seemed to possess different properties than inorganic compounds. Organic compounds were often difficult to isolate and purify, and upon heating, they decomposed more readily than inorganic compounds. To explain these curious observations, many scientists subscribed to a belief that compounds obtained from living sources possessed a special "vital force" that inorganic compounds lacked. This notion, called vitalism, stipulated that it should be impossible to convert inorganic compounds into organic compounds without the introduction of an outside vital force. Vitalism was dealt a serious blow in 1828 when German chemist Friedrich Wöhler demonstrated the conversion of ammonium cyanate (a known inorganic salt) into urea, a known organic compound found in urine:

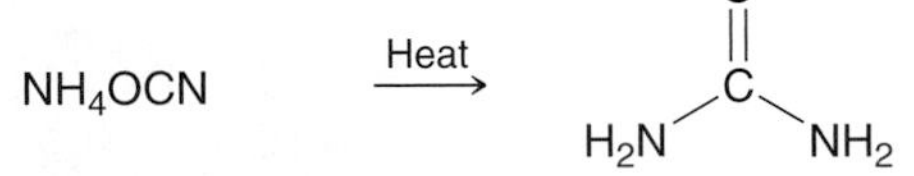

$$NH_4OCN \quad \xrightarrow{\text{Heat}} \quad \underset{H_2N}{\overset{\overset{\displaystyle O}{\|}}{\underset{}{C}}}\!\!\!\underset{NH_2}{}$$

Ammonium cyanate **Urea**
(Inorganic) (Organic)

Over the decades that followed, other examples were found, and the concept of vitalism was gradually rejected. The downfall of vitalism shattered the original distinction between organic and inorganic compounds, and a new definition emerged. Specifically, organic compounds became defined as those compounds containing carbon atoms, while inorganic compounds generally were defined as those compounds lacking carbon atoms.

Organic chemistry occupies a central role in the world around us, as we are surrounded by organic compounds. The food that we eat and the clothes that we wear are comprised of organic compounds. Our ability to smell odors or see colors results from the behavior of organic compounds. Pharmaceuticals, pesticides, paints, adhesives, and plastics are all made from organic compounds. In fact, our bodies are constructed mostly from organic compounds (DNA, RNA, proteins, etc.) whose behavior and function are determined by the guiding principles of organic chemistry. The responses of our bodies to pharmaceuticals are the results of reactions guided by the principles of organic chemistry. A deep understanding of those principles enables the design of new drugs that fight disease and improve the overall quality of life and longevity. Accordingly, it is not surprising that organic chemistry is required knowledge for anyone entering the health professions.

1.2 The Structural Theory of Matter

In the mid-nineteenth century three individuals, working independently, laid the conceptual foundations for the structural theory of matter. August Kekulé, Archibald Scott Couper, and Alexander M. Butlerov each suggested that substances are defined by a specific arrangement of atoms. As an example, consider the structures of ammonium cyanate and urea from Wöhler's experiment:

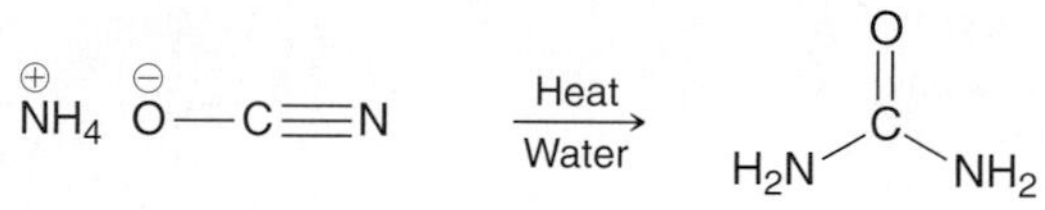

Ammonium cyanate **Urea**

These compounds have the same molecular formula (CH_4N_2O), yet they differ from each other in the way the atoms are connected—that is, they differ in their **constitution**. As a result, they are called **constitutional isomers**. Constitutional isomers have different physical properties and different names. Consider the following two compounds:

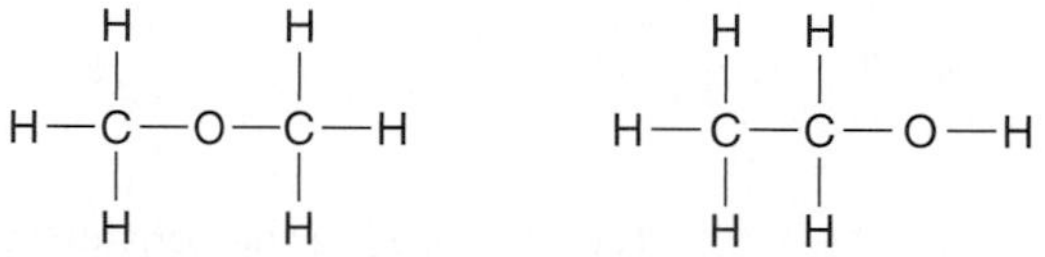

Dimethyl ether
Boiling point = −23°C

Ethanol
Boiling point = 78.4°C

These compounds have the same molecular formula (C_2H_6O) but different constitution, so they are constitutional isomers. The first compound is a colorless gas used as an aerosol spray propellant, while the second compound is a clear liquid, commonly referred to as "alcohol," found in alcoholic beverages.

According to the structural theory of matter, each element will generally form a predictable number of bonds. The term **valence** describes the number of bonds usually formed by each element. For example, carbon generally forms four bonds and is therefore said to be **tetravalent**. Nitrogen generally forms three bonds and is therefore **trivalent**. Oxygen forms two bonds and is **divalent**, while hydrogen and the halogens form one bond and are **monovalent** (Figure 1.1).

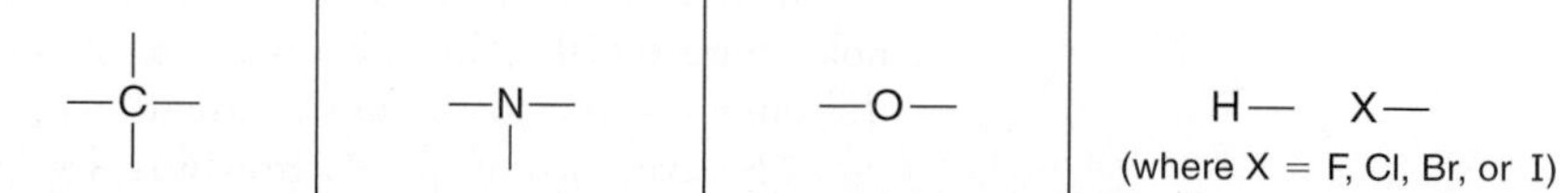

Tetravalent	Trivalent	Divalent	Monovalent
Carbon generally forms ***four*** bonds.	Nitrogen generally forms ***three*** bonds.	Oxygen generally forms ***two*** bonds.	H— X— (where X = F, Cl, Br, or I) Hydrogen and halogens generally form ***one*** bond.

FIGURE 1.1
Valencies of some common elements encountered in organic chemistry.

SKILLBUILDER

1.1 DETERMINING THE CONSTITUTION OF SMALL MOLECULES

LEARN the skill

There is only one compound that has molecular formula C_2H_5Cl. Determine the constitution of this compound.

SOLUTION

The molecular formula indicates which atoms are present in the compound. In this example, the compound contains two carbon atoms, five hydrogen atoms, and one chlorine atom. Begin by determining the valency of each atom that is present in the compound. Each carbon atom is expected to be tetravalent, while the chlorine and hydrogen atoms are all expected to be monovalent:

STEP 1
Determine the valency of each atom in the compound.

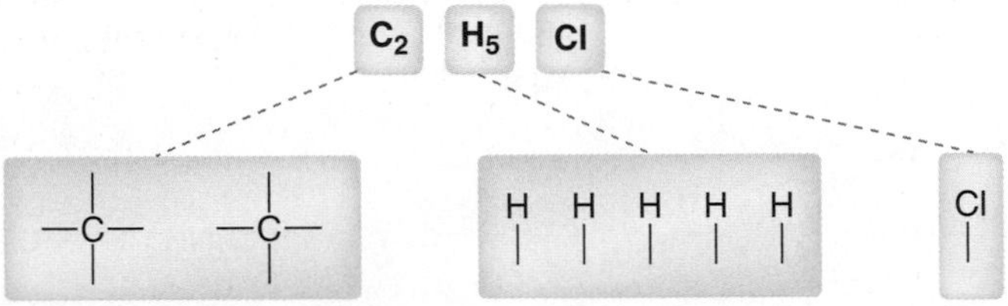

Now we must determine how these atoms are connected. The atoms with the most bonds (the carbon atoms) are likely to be in the center of the compound. In contrast, the chlorine atom and hydrogen atoms can each form only one bond, so those atoms must be placed at the periphery. In this example, it does not matter where the chlorine atom is placed. All six possible positions are equivalent.

STEP 2
Determine how the atoms are connected—atoms with the highest valency should be placed at the center and monovalent atoms should be placed at the periphery.

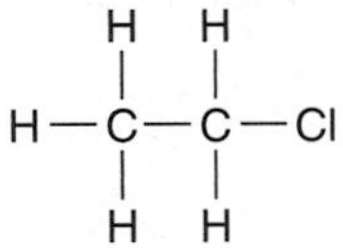

PRACTICE the skill

1.1 Determine the constitution of the compounds with the following molecular formulas:

(a) CH_4O (b) CH_3Cl (c) C_2H_6 (d) CH_5N

(e) C_2F_6 (f) C_2H_5Br (g) C_3H_8

APPLY the skill

1.2 Draw two constitutional isomers that have molecular formula C_3H_7Cl.

1.3 Draw three constitutional isomers that have molecular formula C_3H_8O.

1.4 Draw all constitutional isomers that have molecular formula $C_4H_{10}O$.

need more **PRACTICE?** **Try Problems 1.34, 1.46, 1.47, 1.54**

1.3 Electrons, Bonds, and Lewis Structures

What Are Bonds?

As mentioned, atoms are connected to each other by bonds. That is, bonds are the "glue" that hold atoms together. But what is this mysterious glue and how does it work? In order to answer this question, we must focus our attention on electrons.

The existence of the electron was first proposed in 1874 by George Johnstone Stoney (National University of Ireland), who attempted to explain electrochemistry by suggesting the existence of a particle bearing a unit of charge. Stoney coined the term *electron* to describe this particle. In 1897, J. J. Thomson (Cambridge University) demonstrated evidence supporting the existence of Stoney's mysterious electron and is credited with discovering the electron. In 1916,

Gilbert Lewis (University of California, Berkeley) defined a **covalent bond** as the result of *two atoms sharing a pair of electrons.* As a simple example, consider the formation of a bond between two hydrogen atoms:

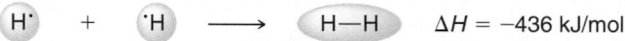

$$H^{\cdot} + {}^{\cdot}H \longrightarrow H{-}H \qquad \Delta H = -436 \text{ kJ/mol}$$

Each hydrogen atom has one electron. When these electrons are shared to form a bond, there is a decrease in energy, indicated by the negative value of ΔH. The energy diagram in Figure 1.2 plots the total energy of the two hydrogen atoms as a function of the distance between them. Focus on the right side of the diagram, which represents the hydrogen atoms separated by a large distance. Moving toward the left on the diagram, the hydrogen atoms approach each other, and there are several forces that must be taken into account: (1) the force of repulsion between the two negatively charged electrons, (2) the force of repulsion between the two positively charged nuclei, and (3) the forces of attraction between the positively charged nuclei and the negatively charged electrons. As the hydrogen atoms get closer to each other, all of these forces get stronger. Under these circumstances, the electrons are capable of moving in such a way so as to minimize the repulsive forces between them while maximizing their attractive forces with the nuclei. This provides for a net force of attraction, which lowers the energy of the system. As the hydrogen atoms move still closer together, the energy continues to be lowered until the nuclei achieve a separation (internuclear distance) of 0.74 angstroms (Å). At that point, the force of repulsion between the nuclei begins to overwhelm the forces of attraction, causing the energy of the system to increase. The lowest point on the curve represents the lowest energy (most stable) state. This state determines both the bond length (0.74 Å) and the bond strength (436 kJ/mol).

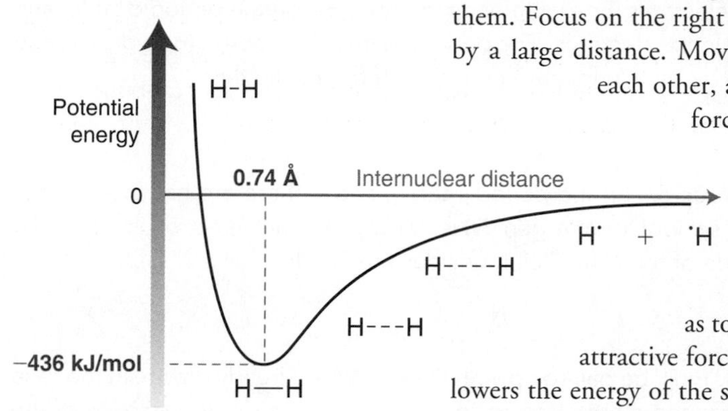

FIGURE 1.2
An energy diagram showing the total energy as a function of the internuclear distance between two hydrogen atoms.

Drawing the Lewis Structure of an Atom

Armed with the idea that a bond represents a pair of shared electrons, Lewis then devised a method for drawing structures. In his drawings, called **Lewis structures**, the electrons take center stage. We will begin by drawing individual atoms, and then we will draw Lewis structures for small molecules. First, we must review a few simple features of atomic structure:

- The nucleus of an atom is comprised of protons and neutrons. Each proton has a charge of +1, and each neutron is electrically neutral.
- For a neutral atom, the number of protons is balanced by an equal number of electrons, which have a charge of −1 and exist in shells. The first shell, which is closest to the nucleus, can contain two electrons, and the second shell can contain up to eight electrons.
- The electrons in the outermost shell of an atom are called the **valence electrons**. The number of valence electrons in an atom is identified by its group number in the periodic table (Figure 1.3).

The Lewis dot structure of an individual atom indicates the number of valence electrons, which are placed as dots around the periodic symbol of the atom (C for carbon, O for oxygen, etc.). The placement of these dots is illustrated in the following SkillBuilder.

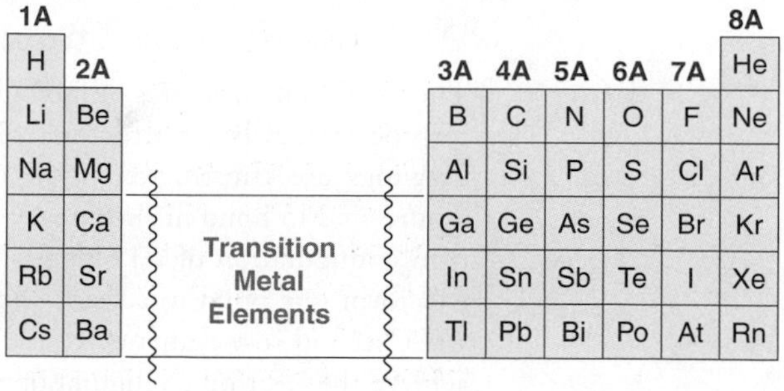

FIGURE 1.3
A periodic table showing group numbers.

SKILLBUILDER

1.2 DRAWING THE LEWIS DOT STRUCTURE OF AN ATOM

LEARN the skill

Draw the Lewis dot structure of (a) a boron atom and (b) a nitrogen atom.

SOLUTION

STEP 1
Determine the number of valence electrons.

(a) In a Lewis dot structure, only valence electrons are drawn, so we must first determine the number of valence electrons. Boron belongs to group 3A on the periodic table, and it therefore has three valence electrons. The periodic symbol for boron (B) is drawn, and each electron is placed by itself (unpaired) on a side of the B, like this:

STEP 2
Place one valence electron by itself on each side of the atom.

(b) Nitrogen belongs to group 5A on the periodic table, and it therefore has five valence electrons. The periodic symbol for nitrogen (N) is drawn, and each electron is placed by itself (unpaired) on a side of the N until all four sides are filled:

STEP 3
If the atom has more than four valence electrons, the remaining electrons are paired with the electrons already drawn.

Any remaining electrons must be paired up with the electrons already drawn. In the case of nitrogen, there is only one more electron to place, so we pair it up with one of the four unpaired electrons (it doesn't matter which one we choose):

PRACTICE the skill

1.5 Draw a Lewis dot structure for each of the following atoms:

(a) Carbon (b) Oxygen (c) Fluorine (d) Hydrogen
(e) Bromine (f) Sulfur (g) Chlorine (h) Iodine

APPLY the skill

1.6 Compare the Lewis dot structure of nitrogen and phosphorus and explain why you might expect these two atoms to exhibit similar bonding properties.

1.7 Name one element that you would expect to exhibit bonding properties similar to boron. Explain.

1.8 Draw a Lewis structure of a carbon atom that is missing one valence electron (and therefore bears a positive charge). Which second-row element does this carbon atom resemble in terms of the number of valence electrons?

1.9 Draw a Lewis structure of a carbon atom that has one extra valence electron (and therefore bears a negative charge). Which second-row element does this carbon atom resemble in terms of the number of valence electrons?

Drawing the Lewis Structure of a Small Molecule

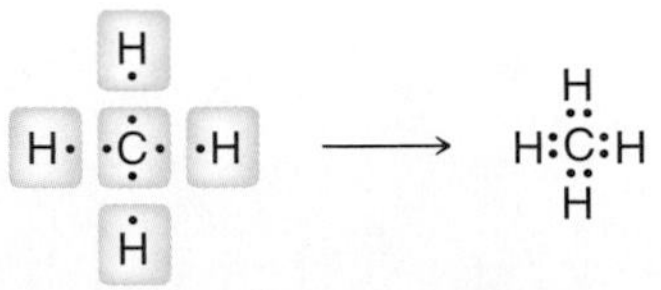

The Lewis dot structures of individual atoms are combined to produce Lewis dot structures of small molecules. These drawings are constructed based on the observation that atoms tend to bond in such a way so as to achieve the electron configuration of a noble gas. For example, hydrogen will form one bond to achieve the electron configuration of helium (two valence electrons), while second-row elements (C, N, O, and F) will form the necessary number of bonds so as to achieve the electron configuration of Neon (eight valence electrons).

This observation, called the **octet rule**, explains why carbon is tetravalent. As just shown, it can achieve an octet of electrons by using each of its four valence electrons to form a bond. The octet rule also explains why nitrogen is trivalent. Specifically, it has five valence electrons and requires three bonds in order to achieve an octet of electrons. Notice that the nitrogen contains one pair of unshared, or nonbonding electrons, called a **lone pair**.

In the next chapter, we will discuss the octet rule in more detail; in particular, we will explore when it can be violated and when it cannot be violated. For now, let's practice drawing Lewis structures.

SKILLBUILDER

1.3 DRAWING THE LEWIS STRUCTURE OF A SMALL MOLECULE

LEARN the skill

Draw the Lewis structure of CH_2O.

SOLUTION

There are four discrete steps when drawing a Lewis structure: First determine the number of valence electrons for each atom.

STEP 1
Draw all individual atoms.

Then, connect any atoms that form more than one bond. Hydrogen atoms only form one bond each, so we will save those for last. In this case, we connect the C and the O.

STEP 2
Connect atoms that form more than one bond.

Next, connect all hydrogen atoms. We place the hydrogen atoms next to carbon, because carbon has more unpaired electrons than oxygen.

STEP 3
Connect the hydrogen atoms.

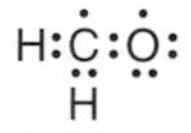

Finally, check to see if each atom (except hydrogen) has an octet. In fact, neither the carbon nor the oxygen has an octet, so in a situation like this, the unpaired electrons are shared as a double bond between carbon and oxygen.

STEP 4
Pair any unpaired electrons so that each atom achieves an octet.

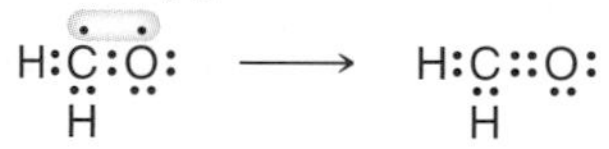

Now all atoms have achieved an octet. When drawing Lewis structures, remember that you cannot simply add more electrons to the drawing. For each atom to achieve an octet the existing electrons must be shared. The total number of valence electrons should be correct when you are finished. In this example, there was one carbon atom, two hydrogen atoms, and one oxygen atom, giving a total of 12 valence electrons ($4 + 2 + 6$). The drawing above MUST have 12 valence electrons, no more and no less.

PRACTICE the skill

1.10 Draw a Lewis structure for each of the following compounds:

(a) C_2H_6 (b) C_2H_4 (c) C_2H_2

(d) C_3H_8 (e) C_3H_6 (f) CH_3OH

APPLY the skill

1.11 Borane (BH_3) is very unstable and quite reactive. Draw a Lewis structure of borane and explain the source of the instability.

1.12 There are four constitutional isomers with molecular formula C_3H_9N. Draw a Lewis structure for each isomer and determine the number of lone pairs on the nitrogen atom in each case.

need more **PRACTICE?** **Try Problems 1.35, 1.38, 1.42**

1.4 Identifying Formal Charges

A **formal charge** is associated with any atom that does not exhibit the appropriate number of valence electrons. When such an atom is present in a Lewis structure, the formal charge must be drawn. Identifying a formal charge requires two discrete tasks:

1. Determine the appropriate number of valence electrons for an atom.
2. Determine whether the atom exhibits the appropriate number of electrons.

The first task can be accomplished by inspecting the periodic table. As mentioned earlier, the group number indicates the appropriate number of valence electrons for each atom. For example, carbon is in group 4A and therefore has four valence electrons. Oxygen is in group 6A and has six valence electrons.

After identifying the appropriate number of electrons for each atom in a Lewis structure, the next task is to determine if any of the atoms in the Lewis structure exhibit an unexpected number of electrons. For example, consider this structure:

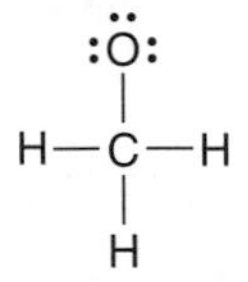

Remember that each bond represents two shared electrons. We split each bond apart equally, and then count the number of electrons on each atom:

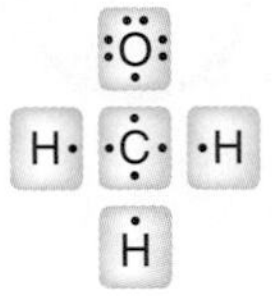

Each hydrogen atom exhibits one valence electron, as expected. The carbon atom also exhibits the appropriate number of valence electrons (four), but the oxygen atom does not. The oxygen atom in this structure exhibits seven valence electrons, but it should only have six. In this case, the oxygen atom has one extra electron, and it must therefore bear a negative formal charge, which is indicated like this:

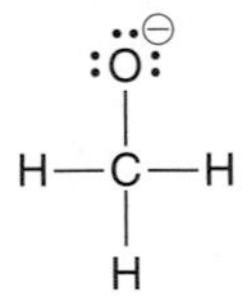

SKILLBUILDER

1.4 CALCULATING FORMAL CHARGE

LEARN the skill

Consider the nitrogen atom in the compound below and determine if it has a formal charge:

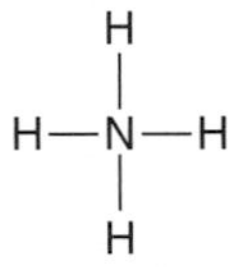

SOLUTION

STEP 1 Determine the appropriate number of valence electrons.

We begin by determining the appropriate number of valence electrons for a nitrogen atom. Nitrogen is in group 5A of the periodic table, and it should therefore have five valence electrons.

STEP 2 Determine the actual number of valence electrons in this case.

Next, we count how many valence electrons are exhibited by the nitrogen atom in this particular example:

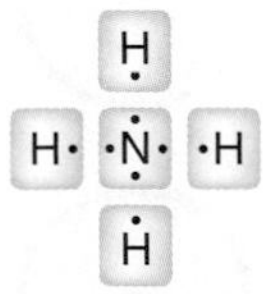

In this case, the nitrogen atom exhibits only four valence electrons. It is missing one electron, so it must bear a positive charge, which is shown like this:

STEP 3 Assign a formal charge.

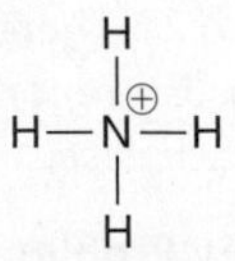

PRACTICE the skill **1.13** Identify any formal charges in the compounds below:

(a) 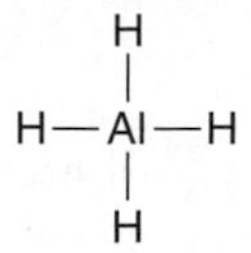
(b) 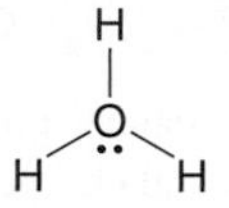
(c) 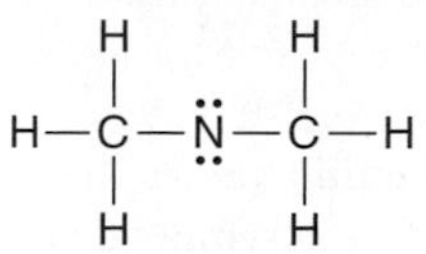
(d) 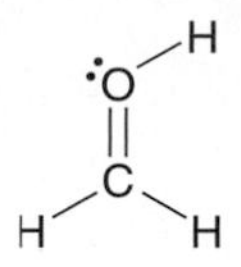
(e) 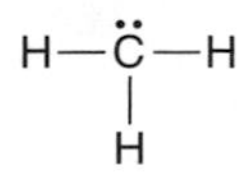

(f) 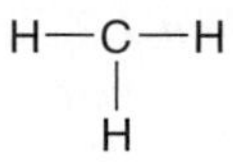
(g) 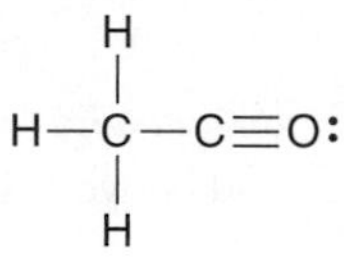
(h) 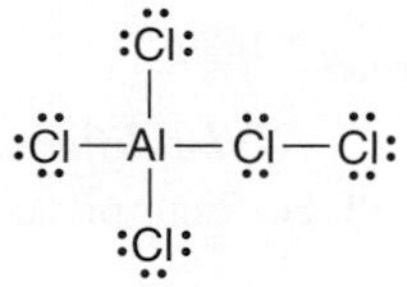
(i) 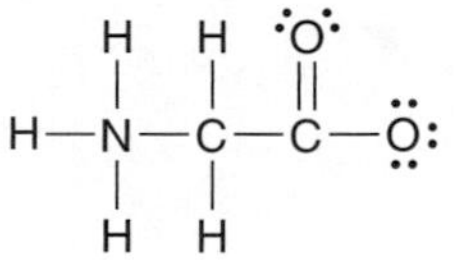

APPLY the skill **1.14** Draw a Lewis structure for each of the following ions; in each case, indicate which atom possesses the formal charge:

(a) $\mathbf{BH_4^-}$ (b) $\mathbf{NH_2^-}$ (c) $\mathbf{C_2H_5^+}$

need more **PRACTICE?** **Try Problem 1.41**

1.5 Induction and Polar Covalent Bonds

Chemists classify bonds into three categories: (1) covalent, (2) polar covalent, and (3) ionic. These categories emerge from the electronegativity values of the atoms sharing a bond. **Electronegativity** is a measure of the ability of an atom to attract electrons. Table 1.1 gives the electronegativity values for elements commonly encountered in organic chemistry.

TABLE 1.1 ELECTRONEGATIVITY VALUES OF SOME COMMON ELEMENTS

Increasing electronegativity →

			H 2.1			
Li 1.0	**Be** 1.5	**B** 2.0	**C** 2.5	**N** 3.0	**O** 3.5	**F** 4.0
Na 0.9	**Mg** 1.2	**Al** 1.5	**Si** 1.8	**P** 2.1	**S** 2.5	**Cl** 3.0
K 0.8						**Br** 2.8

↑ Increasing electronegativity

When two atoms form a bond, there is one critical question that allows us to classify the bond: What is the difference in the electronegativity values of the two atoms? Below are some rough guidelines:

If the difference in electronegativity is less than 0.5, the electrons are considered to be equally shared between the two atoms, resulting in a covalent bond. Examples include C—C and C—H:

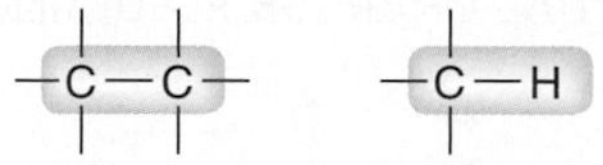

The C—C bond is clearly covalent, because there is no difference in electronegativity between the two atoms forming the bond. Even a C—H bond is considered to be covalent, because the difference in electronegativity between C and H is less than 0.5.

If the difference in electronegativity is between 0.5 and 1.7, the electrons are not shared equally between the atoms, resulting in a **polar covalent bond**. For example, consider a bond between carbon and oxygen (C—O). Oxygen is significantly more electronegative (3.5) than carbon (2.5), and therefore oxygen will more strongly attract the electrons of the bond.

The withdrawal of electrons toward oxygen is called **induction**, which is often indicated with an arrow like this:

$$\overset{\longmapsto}{\text{C}-\text{O}}$$

Induction causes the formation of partial positive and partial negative charges, symbolized by the Greek symbol delta (δ). The partial charges that result from induction will be very important in upcoming chapters.

$$\overset{\delta+ \quad \delta-}{\text{C}-\text{O}}$$

If the difference in electronegativity is greater than 1.7, the electrons are not shared at all. For example, consider the bond between sodium and oxygen in sodium hydroxide (NaOH):

$$\overset{\oplus}{\text{Na}} \quad \overset{\ominus\,\cdot\cdot}{\text{:}\ddot{\text{O}}\text{H}}$$

The difference in electronegativity between O and Na is so great that both electrons of the bond are possessed solely by the oxygen atom, rendering the oxygen negatively charged and the sodium positively charged. The bond between the oxygen and sodium, called an **ionic bond**, is the result of the force of attraction between the two oppositely charged ions.

The cutoff numbers (0.5 and 1.7) should be thought of as rough guidelines. Rather than viewing them as absolute, we must view the various types of bonds as belonging to a spectrum without clear cutoffs (Figure 1.4).

Covalent		Polar covalent			Ionic		
C—C	C—H	N—H	C—O	Li—C	Li—N	Na—Cl	Na—O
Small difference in electronegativity					**Large difference in electronegativity**		

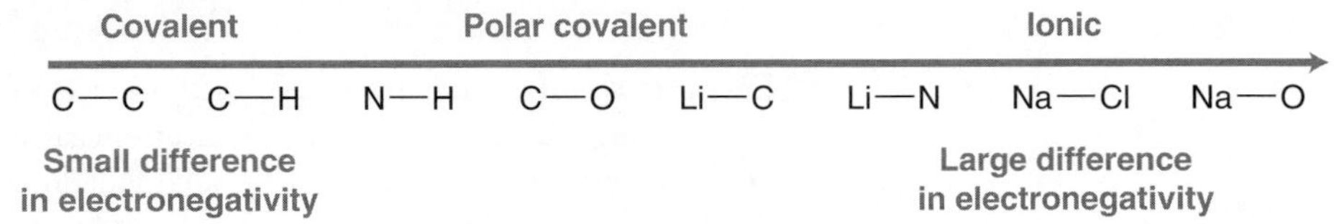

FIGURE 1.4
The nature of various bonds commonly encountered in organic chemistry.

This spectrum has two extremes: covalent bonds on the left and ionic bonds on the right. Between these two extremes are the polar covalent bonds. Some bonds fit clearly into one category, such as C—C bonds (covalent), C—O bonds (polar covalent), or Na—O bonds (ionic).

However, there are many cases that are not so clear-cut. For example, a C—Li bond has a difference in electronegativity of 1.5, and this bond is often drawn either as polar covalent or as ionic. Both drawings are acceptable.

$$-\overset{|}{\underset{|}{\text{C}}}-\text{Li} \quad \textit{or} \quad -\overset{|}{\underset{|}{\text{C}}}\text{:}^{\ominus} \; {}^{\oplus}\text{Li}$$

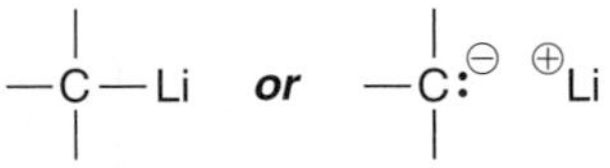

Another reason to avoid absolute cutoff numbers when comparing electronegativity values is that the electronegativity values shown above are obtained via a method developed by Linus Pauling. However, there are at least seven other methods for calculating electronegativity values, each of which provides slightly different values. Strict adherence to the Pauling scale would suggest that C—Br and C—I bonds are covalent, but these bonds will be treated as polar covalent throughout this course.

SKILLBUILDER

1.5 LOCATING PARTIAL CHARGES RESULTING FROM INDUCTION

LEARN the skill

Consider the structure of methanol. Identify all polar covalent bonds and show any partial charges that result from inductive effects.

$$\text{H}-\overset{\overset{\textstyle \text{H}}{|}}{\underset{\underset{\textstyle \text{H}}{|}}{\text{C}}}-\ddot{\text{O}}-\text{H}$$

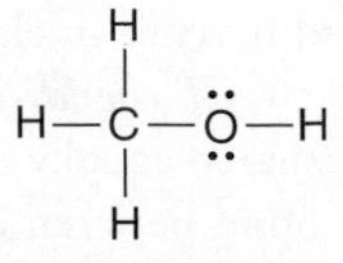

Methanol

SOLUTION

First identify all polar covalent bonds. The C—H bonds are considered to be covalent because the electronegativity values for C and H are fairly close. It is true that carbon is more electronegative than hydrogen, and therefore, there is a small inductive effect for each C—H bond. However, we will generally consider this effect to be negligible for C—H bonds.

The C—O bond and the O—H bond are both polar covalent bonds:

STEP 1
Identify all polar covalent bonds.

$$H-\overset{\overset{\displaystyle H}{|}}{\underset{\underset{\displaystyle H}{|}}{C}}-\overset{\cdot\cdot}{\underset{\cdot\cdot}{O}}-H$$

Polar covalent

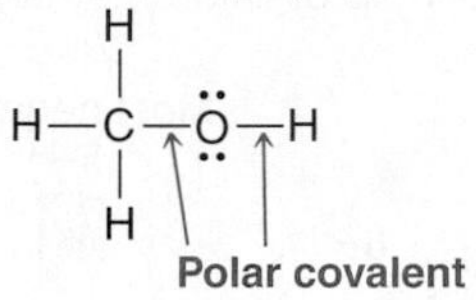

Now determine the direction of the inductive effects. Oxygen is more electronegative than C or H, so the inductive effects are shown like this:

STEP 2
Determine the direction of each dipole.

$$H-\overset{\overset{\displaystyle H}{|}}{\underset{\underset{\displaystyle H}{|}}{C}}-\overset{\longrightarrow\,\cdot\cdot\,\longleftarrow}{\underset{\cdot\cdot}{O}}-H$$

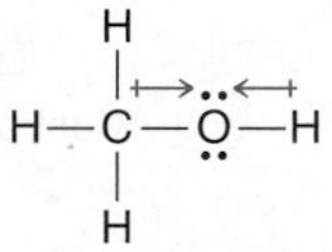

These inductive effects dictate the locations of the partial charges:

STEP 3
Indicate the location of partial charges.

$$H-\overset{\overset{\displaystyle H}{|}}{\underset{\underset{\displaystyle H}{|}}{C}}\overset{\delta+}{-}\overset{\delta-}{\underset{\cdot\cdot}{O}}\overset{\delta+}{-}H$$

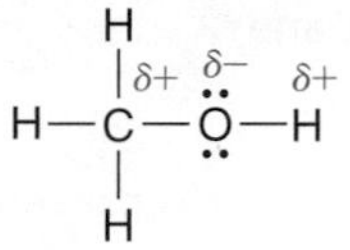

PRACTICE the skill **1.15** For each of the following compounds, identify any polar covalent bonds by drawing δ+ and δ− symbols in the appropriate locations.

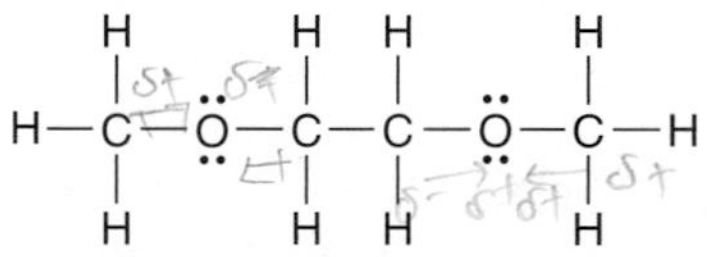

(a) $H-\overset{H}{\underset{H}{C}}-\overset{\cdot\cdot}{\underset{\cdot\cdot}{O}}-\overset{H}{\underset{H}{C}}-\overset{H}{\underset{H}{C}}-\overset{\cdot\cdot}{\underset{\cdot\cdot}{O}}-\overset{H}{\underset{H}{C}}-H$

(b) $H-\overset{\overset{\cdot\cdot}{\underset{\cdot\cdot}{F}}}{\underset{H}{C}}-\overset{\cdot\cdot}{\underset{\cdot\cdot}{Cl}}:$

(c) $H-\overset{H}{\underset{H}{C}}-Mg-\overset{\cdot\cdot}{\underset{\cdot\cdot}{Br}}:$

(d) $H-\overset{\cdot\cdot}{\underset{\cdot\cdot}{O}}-\overset{H}{\underset{H}{C}}-\overset{\overset{H}{|}\,\overset{\cdot\cdot}{\underset{\cdot\cdot}{O}}\,\overset{H}{|}}{C}-\overset{H}{\underset{H}{C}}-\overset{\cdot\cdot}{\underset{\cdot\cdot}{O}}-H$

(e) $H-\overset{H}{\underset{H}{C}}-\overset{\overset{\cdot\cdot}{O}}{C}-\overset{H}{\underset{H}{C}}-H$

(f) $:\overset{\cdot\cdot}{\underset{\cdot\cdot}{Cl}}-\overset{\overset{\cdot\cdot}{\underset{\cdot\cdot}{Cl}}:}{\underset{\underset{\cdot\cdot}{\underset{\cdot\cdot}{Cl}}:}{C}}-\overset{\cdot\cdot}{\underset{\cdot\cdot}{Cl}}:$

APPLY the skill **1.16** The regions of δ+ in a compound are the regions most likely to be attacked by an anion, such as hydroxide (HO⁻). In the compound below, identify the two carbon atoms that are most likely to be attacked by a hydroxide ion:

$$H-\overset{\overset{\displaystyle H}{|}}{\underset{\underset{\displaystyle H}{|}}{C}}-\overset{\overset{\displaystyle \overset{\cdot\cdot}{O}}{||}}{C}-\overset{\overset{\displaystyle H}{|}}{\underset{\underset{\displaystyle H}{|}}{C}}-\overset{\overset{\displaystyle H}{|}}{\underset{\underset{\displaystyle H}{|}}{C}}-\overset{\overset{\displaystyle H}{|}}{\underset{\underset{\displaystyle H}{|}}{C}}-\overset{\cdot\cdot}{\underset{\cdot\cdot}{Cl}}:$$

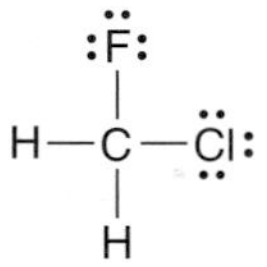

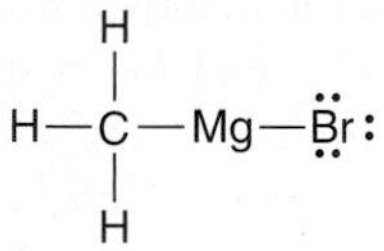

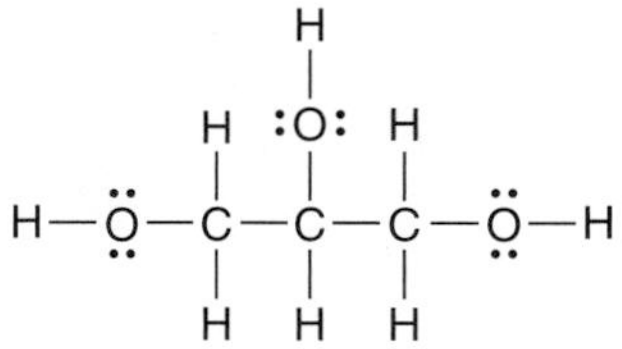

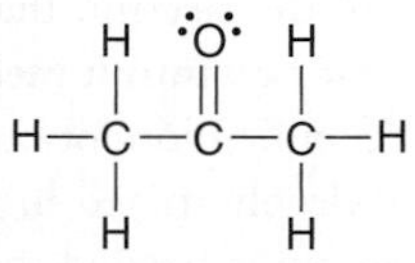

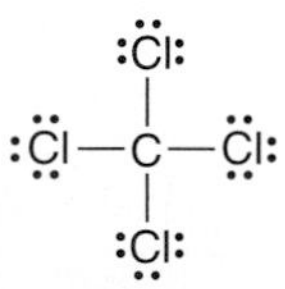

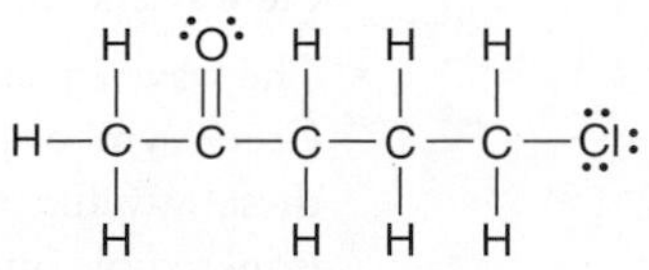

need more PRACTICE? **Try Problems 1.36, 1.37, 1.48, 1.57**

PRACTICALLYSPEAKING)))

Electrostatic Potential Maps

Partial charges can be visualized with three-dimensional, rainbow-like images called **electrostatic potential maps**. As an example, consider the electrostatic potential map of chloromethane.

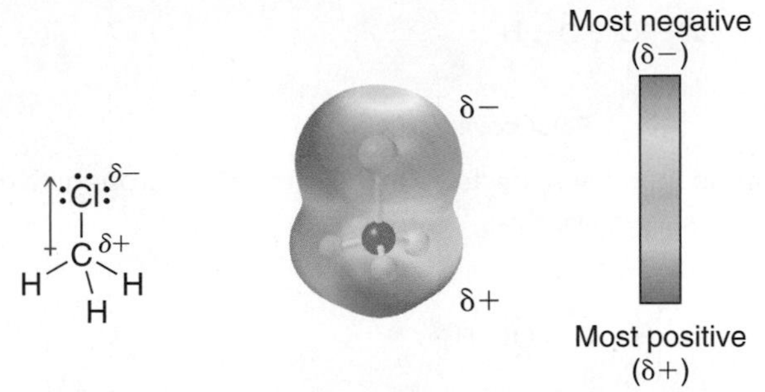

Chloromethane **Electrostatic potential map of chloromethane** **Color scale**

In the image, a color scale is used to represent areas of δ− and δ+. As indicated, red represents a region that is δ−, while blue represents a region that is δ+. In reality, electrostatic potential maps are rarely used by practicing organic chemists when they communicate with each other; however, these illustrations can often be helpful to students who are learning organic chemistry. Electrostatic potential maps are generated by performing a series of calculations. Specifically, an imaginary point positive charge is positioned at various locations, and for each location, we calculate the potential energy associated with the attraction between the point positive charge and the surrounding electrons. A large attraction indicates a position of δ−, while a small attraction indicates a position of δ+. The results are then illustrated using colors, as shown.

A comparison of any two electrostatic potential maps is only valid if both maps were prepared using the same color scale. Throughout this book, care has been taken to use the same color scale whenever two maps are directly compared to each other. However, it will not be useful to compare two maps from different pages of this book (or any other book), as the exact color scales are likely to be different.

1.6 Atomic Orbitals

Quantum Mechanics

By the 1920s, vitalism had been discarded. Chemists were aware of constitutional isomerism and had developed the structural theory of matter. The electron had been discovered and identified as the source of bonding, and Lewis structures were used to keep track of shared and unshared electrons. But the understanding of electrons was about to change dramatically.

In 1924, French physicist Louis de Broglie suggested that electrons, heretofore considered as particles, also exhibited wavelike properties. Based on this assertion, a new theory of matter was born. In 1926, Erwin Schrödinger, Werner Heisenberg, and Paul Dirac independently proposed a mathematical description of the electron that incorporated its wavelike properties. This new theory, called *wave mechanics,* or **quantum mechanics**, radically changed the way we viewed the nature of matter and laid the foundation for our current understanding of electrons and bonds.

Quantum mechanics is deeply rooted in mathematics and represents an entire subject by itself. The mathematics involved is beyond the scope of our course, and we will not discuss it here. However, in order to understand the nature of electrons, it is critical to understand a few simple highlights from quantum mechanics:

- An equation is constructed to describe the total energy of a hydrogen atom (i.e., one proton plus one electron). This equation, called the **wave equation**, takes into account the wavelike behavior of an electron that is in the electric field of a proton.
- The wave equation is then solved to give a series of solutions called **wavefunctions**. The Greek symbol psi (ψ) is used to denote each wavefunction (ψ_1, ψ_2, ψ_3, etc.). Each of these wavefunctions corresponds to an allowed energy level for the electron. This result is incredibly important because it suggests that an electron, when contained in an atom, can only exist at discrete energy levels (ψ_1, ψ_2, ψ_3, etc.). In other words, the energy of the electron is *quantized*.

- Each wavefunction is a function of spatial location. It provides information that allows us to assign a numerical value for each location in three-dimensional space relative to the nucleus. The square of that value (ψ^2 for any particular location) has a special meaning. It indicates the probability of finding the electron in that location. Therefore, a three-dimensional plot of ψ^2 will generate an image of an atomic orbital (Figure 1.5).

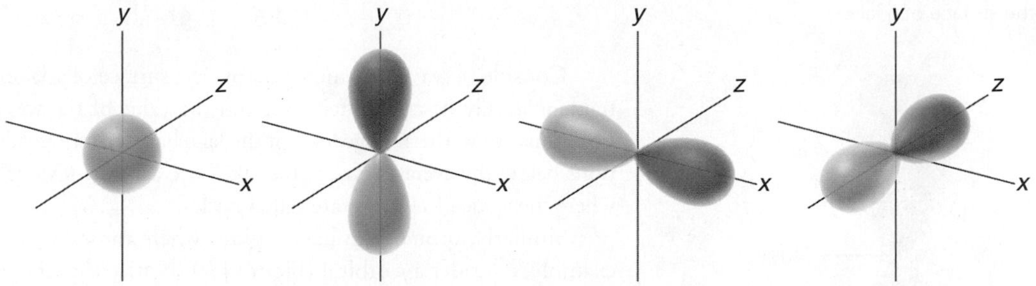

FIGURE 1.5
Illustrations of an *s* orbital and three *p* orbitals.

Electron Density and Atomic Orbitals

An *orbital* is a region of space that can be occupied by an electron. But care must be taken when trying to visualize this. There is a statement from the previous section that must be clarified because it is potentially misleading: "*ψ^2 represents the probability of finding an electron in a particular location.*" This statement seems to treat an electron as if it were a particle flying around within a specific region of space. But remember that an electron is not purely a particle—it has wavelike properties as well. Therefore, we must construct a mental image that captures both of these properties. That is not easy to do, but the following analogy might help. We will treat an occupied orbital as if it is a cloud—similar to a cloud in the sky. No analogy is perfect, and there are certainly features of clouds that are very different from orbitals. However, focusing on some of these differences between electron clouds (occupied orbitals) and real clouds makes it possible to construct a better mental model of an electron in an orbital:

- Clouds in the sky can come in any shape or size. However, electron clouds only come in a small number of shapes and sizes (as defined by the orbitals).
- A cloud in the sky is comprised of billions of individual water molecules. An electron cloud is not comprised of billions of particles. We must think of an electron cloud as a single entity, even though it can be thicker in some places and thinner in other places. This concept is critical and will be used extensively throughout the course in explaining reactions.
- A cloud in the sky has edges, and it is possible to define a region of space that contains 100% of the cloud. In contrast, an electron cloud does not have defined edges. We frequently use the term **electron density**, which is associated with the probability of finding an electron in a particular region of space. The "shape" of an orbital refers to a region of space that contains 90 – 95% of the electron density. Beyond this region, the remaining 5 – 10 % of the electron density tapers off but never ends. In fact, if we want to consider the region of space that contains 100% of the electron density, we must consider the entire universe.

In summary, we must think of an orbital as a region of space that can be occupied by electron density. An occupied orbital must be treated as a *cloud of electron density*. This region of space is called an **atomic orbital** (AO), because it is a region of space defined with respect to the nucleus of a single atom. Examples of atomic orbitals are the *s*, *p*, *d*, and *f* orbitals that were discussed in your general chemistry textbook.

Phases of Atomic Orbitals

Our discussion of electrons and orbitals has been based on the premise that electrons have wavelike properties. As a result, it will be necessary to explore some of the characteristics of simple waves in order to understand some of the characteristics of orbitals.

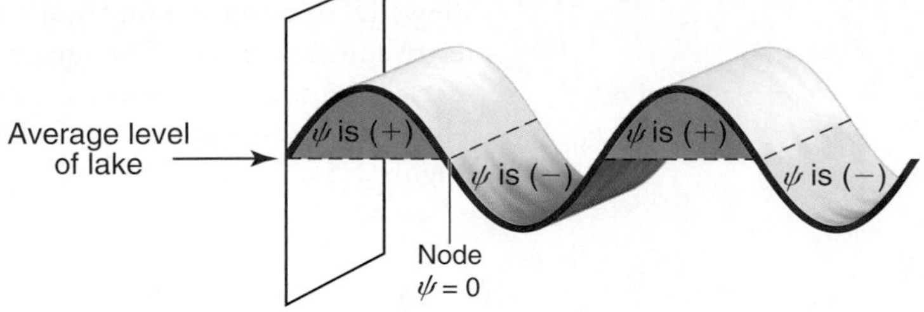

FIGURE 1.6
Phases of a wave moving across the surface of a lake.

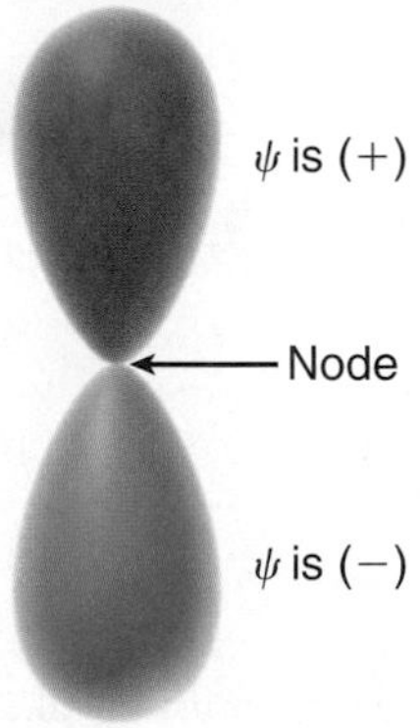

FIGURE 1.7
The phases of a *p* orbital.

Consider a wave that moves across the surface of a lake (Figure 1.6). The wavefunction (ψ) mathematically describes the wave, and the value of the wavefunction is dependent on location. Locations above the average level of the lake have a positive value for ψ (indicated in red), and locations below the average level of the lake have a negative value for ψ (indicated in blue). Locations where the value of ψ is zero are called **nodes**.

Similarly, orbitals can have regions where the value of ψ is positive, negative, or zero. For example, consider a *p* orbital (Figure 1.7). Notice that the *p* orbital has two lobes: the top lobe is a region of space where the values of ψ are positive, while the bottom lobe is a region where the values of ψ are negative. Between the two lobes is a location where $\psi = 0$. This location represents a node.

Be careful not to confuse the sign of ψ (+ or −) with electrical charge. A positive value for ψ does not imply a positive charge. The value of ψ (+ or −) is a mathematical convention that refers to the *phase* of the wave (just like in the lake). Although ψ can have positive or negative values, nevertheless ψ^2 (which describes the electron density as a function of location) will always be a positive number. At a node, where $\psi = 0$, the electron density (ψ^2) will also be zero. This means that there is no electron density located at a node.

From this point forward, we will draw the lobes of an orbital with colors (red and blue) to indicate the phase of ψ for each region of space.

Filling Atomic Orbitals with Electrons

The energy of an electron depends on the type of orbital that it occupies. Most of the organic compounds that we will encounter will be composed of first- and second-row elements (H, C, N, and O). These elements utilize the 1*s* orbital, the 2*s* orbital, and the three 2*p* orbitals. Our discussions will therefore focus primarily on these orbitals (Figure 1.8). Electrons are lowest in energy when they occupy a 1*s* orbital, because the 1*s* orbital is closest to the nucleus and it has no nodes (the more nodes that an orbital has, the greater its energy). The 2*s* orbital has one node and is farther away form the nucleus; it is therefore higher in energy than the 1*s* orbital. After the 2*s* orbital, there are three 2*p* orbitals that are all equivalent in energy to one another. Orbitals with the same energy level are called **degenerate orbitals**.

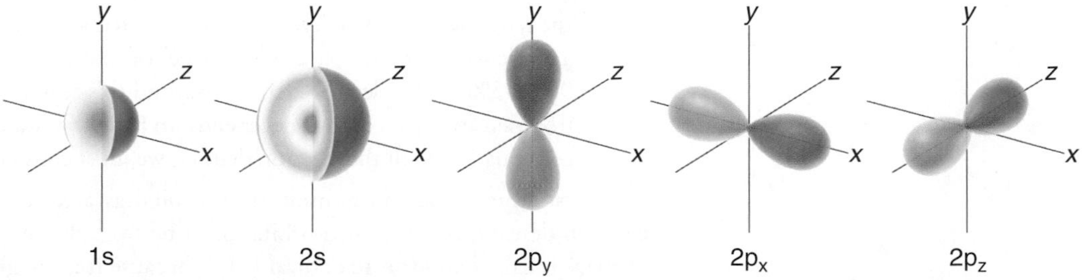

FIGURE 1.8
Illustrations of *s* orbitals and three *p* orbitals.

As we move across the periodic table, starting with hydrogen, each element has one more electron than the element before it (Figure 1.9). The order in which the orbitals are filled by electrons is determined by just three simple principles:

1. The **Aufbau principle.** The lowest-energy orbital is filled first.
2. The **Pauli exclusion principle.** Each orbital can accommodate a maximum of two electrons that have opposite spin. To understand what "spin" means, we can imagine an

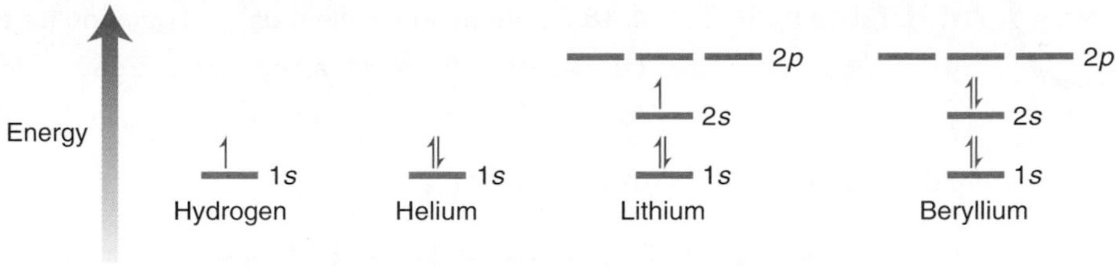

FIGURE 1.9
Energy diagrams showing the electron configurations for H, He, Li, and Be.

electron spinning in space (although this is an oversimplified explanation of the term "spin"). For reasons that are beyond the scope of this course, electrons only have two possible spin states (designated by ↓ or ↑). In order for the orbital to accommodate two electrons, the electrons must have opposite spin states.

3. **Hund's Rule.** When dealing with degenerate orbitals, such as *p* orbitals, one electron is placed in each degenerate orbital first, before electrons are paired up.

The application of the first two principles can be seen in the electron configurations shown in Figure 1.9 (H, He, Li, and Be). The application of the third principle can be seen in the electron configurations for the remaining second-row elements (Figure 1.10).

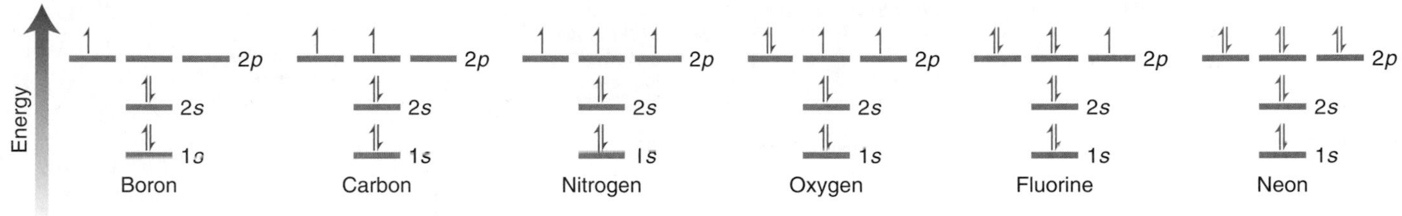

FIGURE 1.10
Energy diagrams showing the electron configurations for B, C, N, O, F, and Ne.

SKILLBUILDER

1.6 IDENTIFYING ELECTRON CONFIGURATIONS

LEARN the skill Identify the electron configuration of a nitrogen atom.

SOLUTION

STEP 1 Place the valence electrons in atomic orbitals using the Aufbau principle, the Pauli exclusion principle, and Hund's rule.

The electron configuration indicates which atomic orbitals are occupied by electrons. Nitrogen has a total of seven electrons. These electrons occupy atomic orbitals of increasing energy, with two electrons being placed in each orbital:

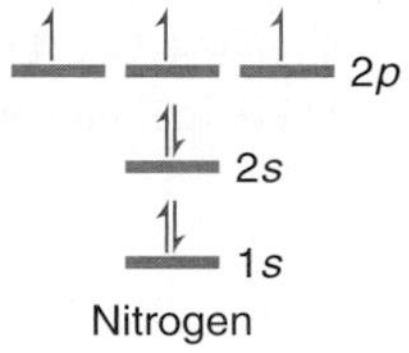

STEP 2 Identify the number of valence electrons in each atomic orbital.

Two electrons occupy the 1*s* orbital, two electrons occupy the 2*s* orbital, and three electrons occupy the 2*p* orbitals. This is summarized using the following notation:

$$1s^2 2s^2 2p^3$$

PRACTICE the skill **1.17** Determine the electron configuration for each of the following atoms:

(a) Carbon (b) Oxygen (c) Boron (d) Fluorine (e) Sodium (f) Aluminum

APPLY the skill

1.18 Determine the electron configuration for each of the following ions:

(a) A carbon atom with a negative charge

(b) A carbon atom with a positive charge

(c) A nitrogen atom with a positive charge

(d) An oxygen atom with a negative charge

need more **PRACTICE?** **Try Problem 1.44**

1.7 Valence Bond Theory

With the understanding that electrons occupy regions of space called orbitals, we can now turn our attention to a deeper understanding of covalent bonds. Specifically, a covalent bond is formed from the overlap of atomic orbitals. There are two commonly used theories for describing the nature of atomic orbital overlap: valence bond theory and molecular orbital (MO) theory. The valence bond approach is more simplistic in its treatment of bonds, and therefore we will begin our discussion with valence bond theory.

If we are going to treat electrons as waves, then we must quickly review what happens when two waves interact with each other. Two waves that approach each other can interfere in one of two possible ways—constructively or destructively. Similarly, when atomic orbitals overlap, they can interfere either constructively (Figure 1.11) or destructively (Figure 1.12).

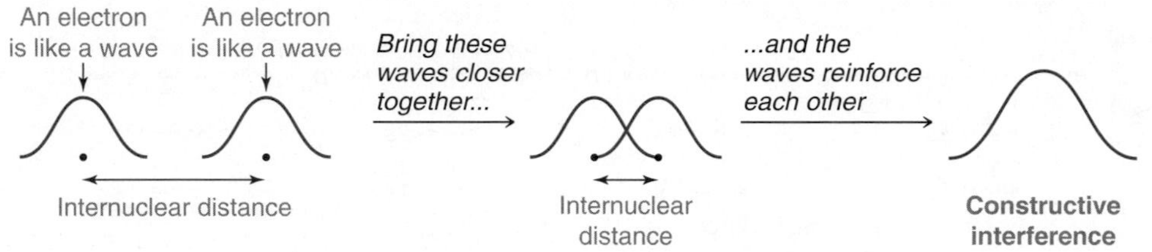

FIGURE 1.11 Constructive interference resulting from the interaction of two electrons.

Constructive interference produces a wave with larger amplitude. In contrast, **destructive interference** results in waves canceling each other, which produces a node (Figure 1.12).

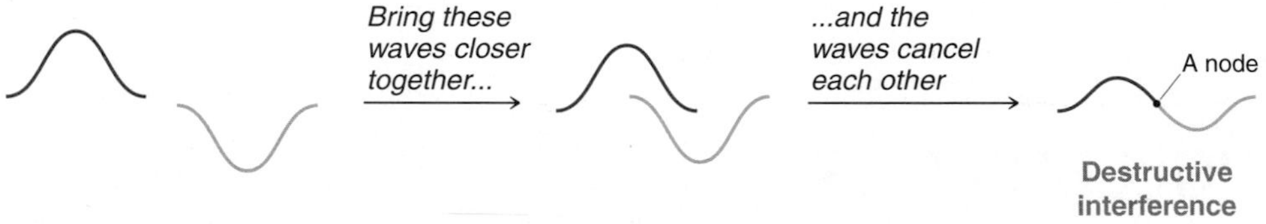

FIGURE 1.12 Destructive interference resulting from the interaction of two electrons.

According to **valence bond theory**, a bond is simply the sharing of electron density between two atoms as a result of the constructive interference of their atomic orbitals. Consider, for example, the bond that is formed between the two hydrogen atoms in molecular hydrogen (H_2). This bond is formed from the overlap of the 1*s* orbitals of each hydrogen atom (Figure 1.13). The electron density of this bond is primarily located on the bond axis (the line that can be drawn between the two hydrogen atoms). This type of bond is called a **sigma (σ) bond** and is characterized by circular symmetry with respect to the bond axis. To visualize what this means, imagine a plane that is drawn perpendicular to the bond axis. This plane will carve out a circle (Figure 1.14). This is the defining feature of σ bonds and will be true of all purely single bonds. Therefore, *all single bonds are σ bonds.*

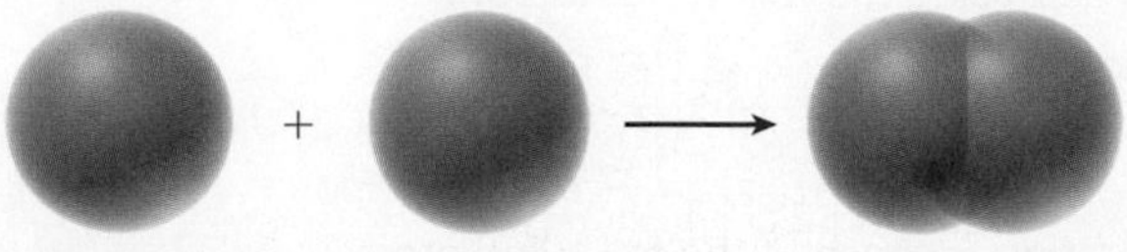

FIGURE 1.13 The overlap of the 1*s* atomic orbitals of two hydrogen atoms, forming molecular hydrogen (H_2).

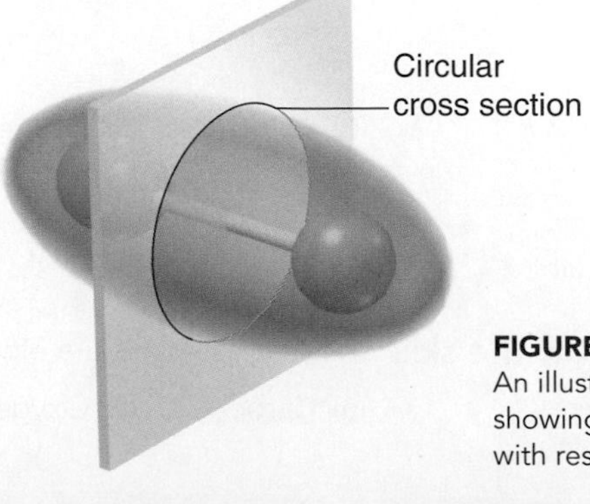

FIGURE 1.14 An illustration of a sigma bond, showing the circular symmetry with respect to the bond axis.

1.8 Molecular Orbital Theory

In most situations, valence bond theory will be sufficient for our purposes. However, there will be cases in the upcoming chapters where valence bond theory will be inadequate to describe the observations. In such cases, we will utilize molecular orbital theory, a more sophisticated approach to viewing the nature of bonds.

Much like valence bond theory, **molecular orbital (MO) theory** also describes a bond in terms of the constructive interference between two overlapping atomic orbitals. However, MO theory goes one step further and uses mathematics as a tool to explore the consequences of atomic orbital overlap. The mathematical method is called the **linear combination of atomic orbitals (LCAO)**. According to this theory, atomic orbitals are mathematically combined to produce new orbitals, called **molecular orbitals.**

It is important to understand the distinction between atomic orbitals and molecular orbitals. Both types of orbitals are used to accommodate electrons, but an atomic orbital is a region of space associated with an individual atom, while a molecular orbital is associated with an entire molecule. That is, the molecule is considered to be a single entity held together by many electron clouds, some of which can actually span the entire length of the molecule. These molecular orbitals are filled with electrons in a particular order in much the same way that atomic orbitals are filled. Specifically, electrons first occupy the lowest energy orbitals, with a maximum of two electrons per orbital. In order to visualize what it means for an orbital to be associated with an entire molecule, we will explore two molecules: molecular hydrogen (H_2) and bromomethane (CH_3Br).

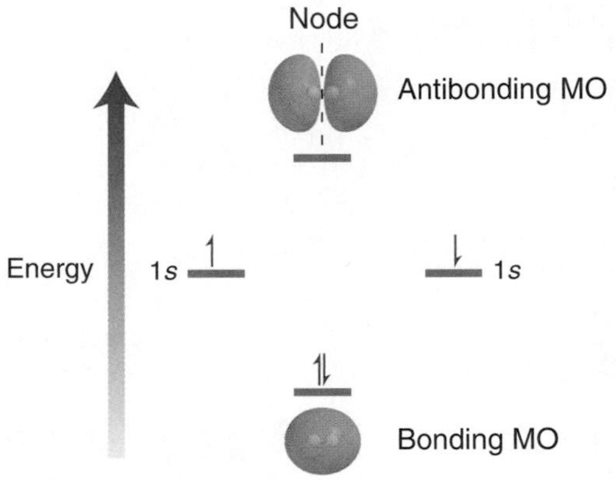

FIGURE 1.15
An energy diagram showing the relative energy levels of bonding and antibonding molecular orbitals.

Consider the bond formed between the two hydrogen atoms in molecular hydrogen. This bond is the result of the overlap of two atomic orbitals (*s* orbitals), each of which is occupied by one electron. According to MO theory, when two atomic orbitals overlap, they cease to exist. Instead, they are replaced by two molecular orbitals, each of which is associated with the entire molecule (Figure 1.15).

In the energy diagram shown in Figure 1.15, the individual atomic orbitals are represented on the right and left, with each atomic orbital having one electron. These atomic orbitals are combined mathematically (using the LCAO method) to produce two molecular orbitals. The lower energy molecular orbital, or **bonding MO**, is the result of constructive interference of the original two atomic orbitals. The higher energy molecular orbital, or **antibonding MO**, is the result of destructive interference. Notice that the antibonding MO has one node, which explains why it is higher in energy. Both electrons occupy the bonding MO in order to achieve a lower energy state. This lowering in energy is the essence of the bond. For an H—H bond, the lowering in energy is equivalent to 436 kJ/mol. This energy corresponds with the bond strength of an H—H bond (as shown in Figure 1.2).

Now let's consider a molecule such as CH_3Br, which contains more than just one bond. Valence bond theory continues to view each bond separately, with each bond being formed from two overlapping atomic orbitals. In contrast, MO theory treats the bonding electrons as being associated with the entire molecule. The molecule has many molecular orbitals, each of which can be occupied by two electrons. Figure 1.16 illustrates two of the many molecular orbitals of CH_3Br. Each of the two images in Figure 1.16 represents a molecular orbital capable of accommodating up to two electrons. In each molecular orbital, red and blue regions indicate the different phases, as described in Section 1.6. As we saw with molecular hydrogen,

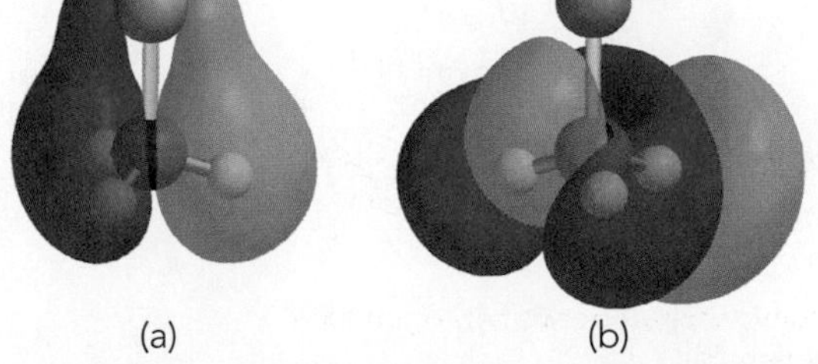

FIGURE 1.16
(a) A low-energy molecular orbital of CH_3Br.
(b) A high-energy molecular orbital of CH_3Br.

not all molecular orbitals will be occupied. The bonding electrons will occupy the lower energy molecular orbitals (such as the one shown in Figure 1.16a), while the higher energy molecular orbitals (as in Figure 1.16b) remain unoccupied. For every molecule, two of its molecular orbitals will be of particular interest: (1) the highest energy orbital from among the occupied orbitals is called the **highest occupied molecular orbital**, or **HOMO**, and (2) the lowest energy orbital from among the unoccupied orbitals is called the **lowest unoccupied molecular orbital**, or **LUMO**. For example, in Chapter 7, we will explore a reaction in which CH_3Br is attacked by a hydroxide ion (HO^-). In order for this process to occur, the hydroxide ion must transfer its electron density into the lowest energy, empty molecular orbital, or LUMO, of CH_3Br (Figure 1.17). The nature of the LUMO (i.e., number of nodes, location of nodes, etc.) will be useful in explaining the preferred direction from which the hydroxide ion will attack.

FIGURE 1.17
The LUMO of CH_3Br.

We will use MO theory several times in the chapters that follow. Most notably, in Chapter 17, we will investigate the structure of compounds containing several double bonds. For those compounds, valence bond theory will be inadequate, and MO theory will provide a more meaningful understanding of the bonding structure. Throughout this textbook, we will continue to develop both valence bond theory and MO theory.

1.9 Hybridized Atomic Orbitals

Methane and sp^3 Hybridization

Let us now apply valence bond theory to the bonds in methane:

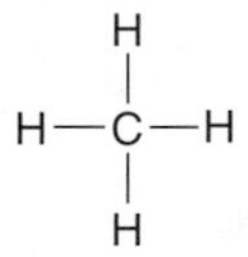

Methane

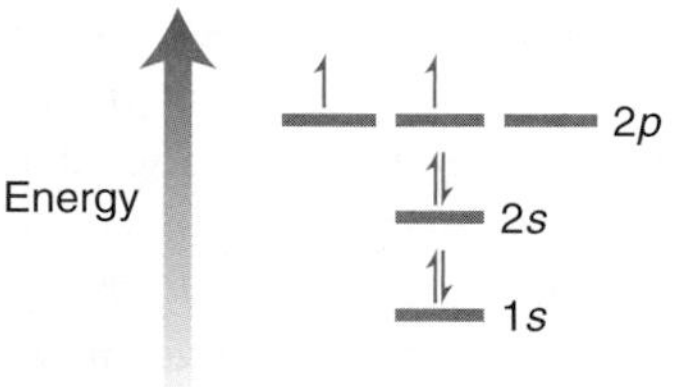

FIGURE 1.18
An energy diagram showing the electron configuration of carbon.

Recall the electron configuration of carbon (Figure 1.18). This electron configuration cannot satisfactorily describe the bonding structure of methane (CH_4), in which the carbon atom has four separate C—H bonds, because the electron configuration shows only two atomic orbitals capable of forming bonds (each of these orbitals has one unpaired electron). This would imply that the carbon atom will form only two bonds, but we know that it forms four bonds. We can solve this problem by imagining an excited state of carbon (Figure 1.19): a state in which a 2*s* electron has been promoted to a higher energy 2*p* orbital. Now the carbon atom has four atomic orbitals capable of forming bonds, but there is yet another problem here. The geometry of the 2*s* and three 2*p* orbitals does not satisfactorily explain the observed three-dimensional geometry of methane (Figure 1.20). All bond angles are 109.5°, and the four bonds point away from each other in a perfect tetrahedron. This geometry cannot be explained by an excited state of carbon because the *s* orbital and the three *p* orbitals do not occupy a tetrahedral geometry. The *p* orbitals are separated from each other by only 90° (as seen in Figure 1.5) rather than 109.5°.

This problem was solved in 1931 by Linus Pauling, who suggested that the electronic configuration of the carbon atom in methane does not necessarily have to be the same as the electronic configuration of a free carbon atom. Specifically, Pauling mathematically averaged, or *hybridized*, the 2*s* orbital and the three 2*p* orbitals, giving four degenerate **hybridized atomic orbitals** (Figure 1.21). The hybridization process in Figure 1.21 does not represent a real physical process that the orbitals undergo. Rather, it is a mathematical procedure that is used to arrive

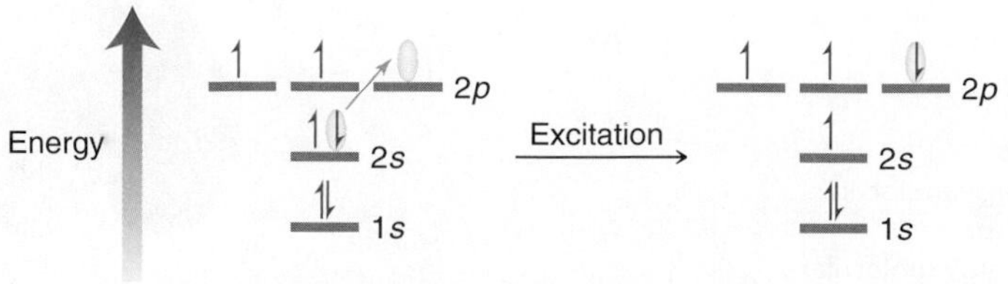

FIGURE 1.19
An energy diagram showing the electronic excitation of an electron in a carbon atom.

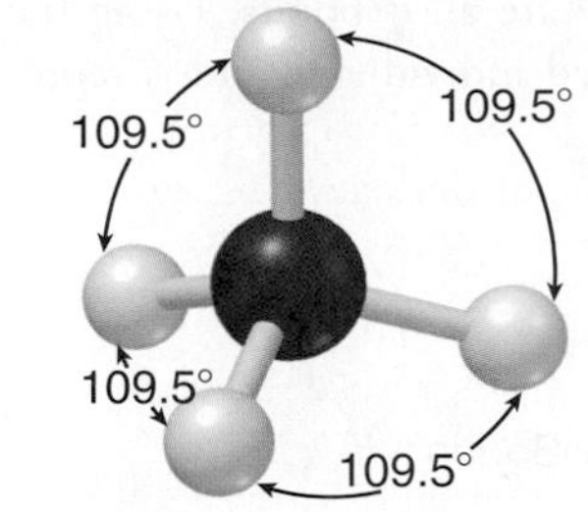

FIGURE 1.20
The tetrahedral geometry of methane. All bond angles are 109.5°.

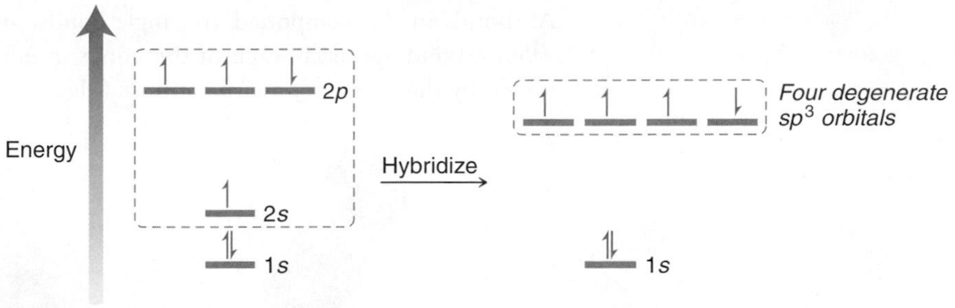

FIGURE 1.21
An energy diagram showing four degenerate hybridized atomic orbitals.

at a satisfactory description of the observed bonding. This procedure gives us four orbitals that were produced by averaging one *s* orbital and three *p* orbitals, and therefore we refer to these atomic orbitals as ***sp^3*-hybridized orbitals**. Figure 1.22 shows an sp^3-hybridized orbital. If we use these hybridized atomic orbitals to describe the bonding of methane, we can successfully explain the observed geometry of the bonds. The four sp^3-hybridized orbitals are equivalent in energy (degenerate) and will therefore position themselves as far apart from each other as possible, achieving a tetrahedral geometry. Also notice that hybridized atomic orbitals are unsymmetrical. That is, hybridized atomic orbitals have a larger front lobe (shown in red in Figure 1.22) and a smaller back lobe (shown in blue). The larger front lobe enables hybridized atomic orbitals to be more efficient than *p* orbitals in their ability to form bonds.

FIGURE 1.22
An illustration of an sp^3-hybridized atomic orbital.

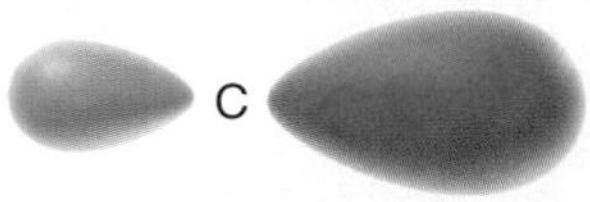

Using valence bond theory, each of the four bonds in methane is represented by the overlap between an sp^3-hybridized atomic orbital from the carbon atom and an *s* orbital from a hydrogen atom (Figure 1.23). For purposes of clarity the back lobes (blue) have been omitted from the images in Figure 1.23.

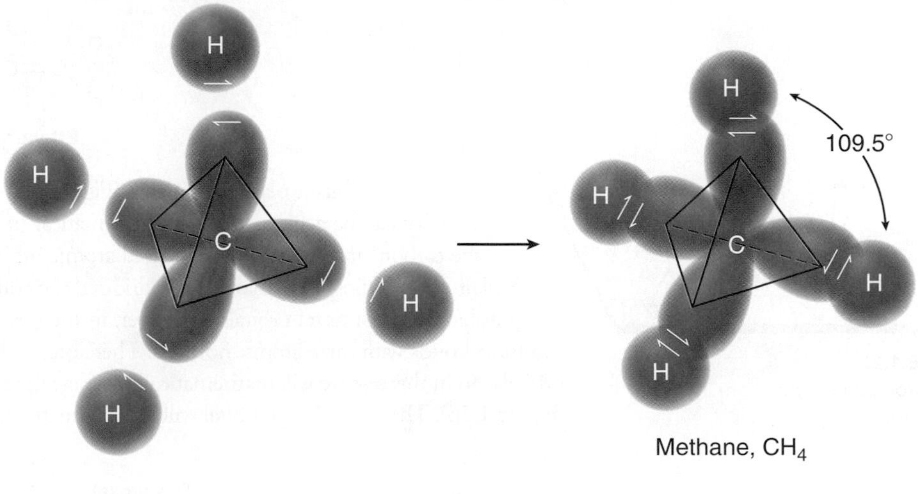

FIGURE 1.23
A tetrahedral carbon atom using each of its four sp^3-hybridized orbitals to form a bond.

The bonding in ethane is treated in much the same way:

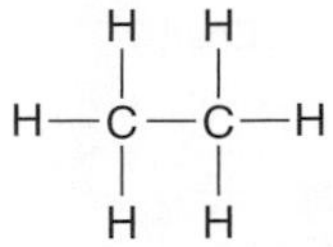

Ethane

All bonds in this compound are single bonds, and therefore they are all σ bonds. Using the valence bond approach, each of the bonds in ethane can be treated individually and is represented by the overlap of atomic orbitals (Figure 1.24).

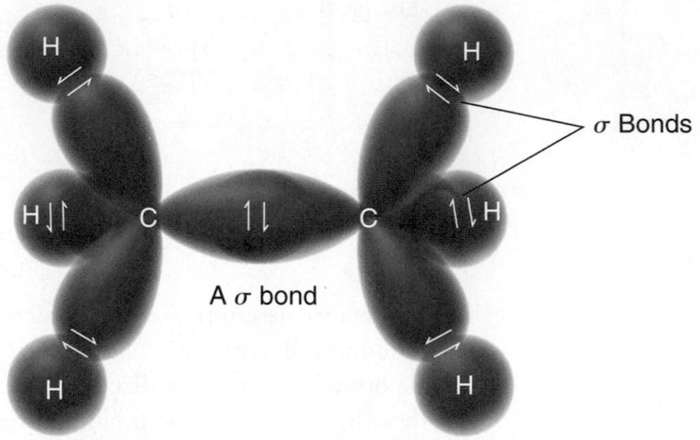

FIGURE 1.24
A valence bond picture of the bonding in ethane.

CONCEPTUAL CHECKPOINT

1.19 Cyclopropane is a compound in which the carbon atoms form a three-membered ring:

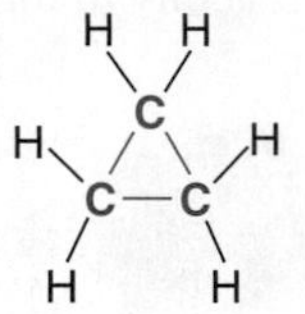

Cyclopropane

Each of the carbon atoms in cyclopropane is sp^3 hybridized. Cyclopropane is more reactive than other cyclic compounds (four-membered rings, five-membered rings, etc.). Analyze the bond angles in cyclopropane and explain why cyclopropane is so reactive.

Double Bonds and sp^2 Hybridization

Now let's consider the structure of a compound bearing a double bond. The simplest example is ethylene:

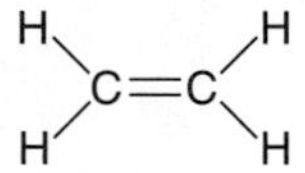

Ethylene

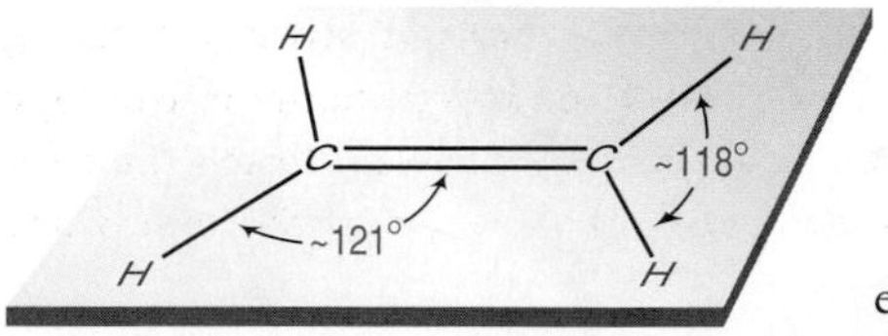

FIGURE 1.25
All six atoms of ethylene are in one plane.

Ethylene exhibits a planar geometry (Figure 1.25). A satisfactory model for explaining this geometry can be achieved by the mathematical maneuver of hybridizing the *s* and *p* orbitals of the carbon atom to obtain hybridized atomic orbitals. When we did this procedure before to explain the bonding in methane, we hybridized the *s* orbital and all three *p* orbitals to produce four equivalent sp^3-hybridized orbitals. However, in the case of ethylene, each carbon atom only needs to form bonds with three atoms, not four. Therefore, each carbon atom only needs three hybridized orbitals. So in this case we will mathematically average the *s* orbital with only two of the three *p* orbitals (Figure 1.26). The remaining *p* orbital will remain unaffected by our mathematical procedure.

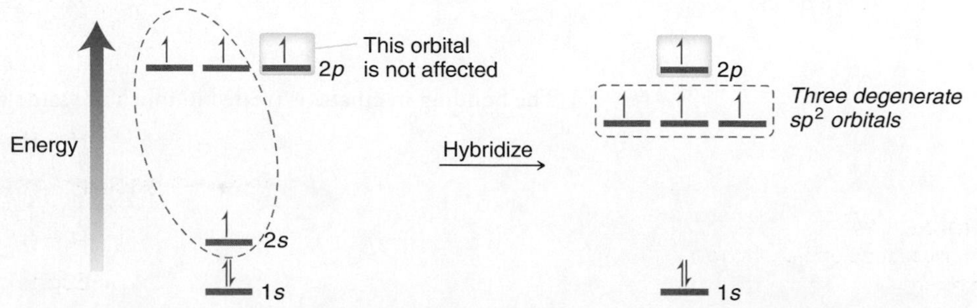

FIGURE 1.26
An energy diagram showing three degenerate sp^2-hybridized atomic orbitals.

The result of this mathematical operation is a carbon atom with one *p* orbital and three ***sp*²-hybridized orbitals** (Figure 1.27). In Figure 1.27 the *p* orbital is shown in red and blue, and the hybridized orbitals are shown in gray (for clarity, only the front lobe of each hybridized orbital is shown). They are called sp^2-hybridized orbitals to indicate that they were obtained by averaging one *s* orbital and two *p* orbitals. As shown in Figure 1.27, each of the carbon atoms in ethylene is sp^2 hybridized, and we can use this hybridization state to explain the bonding structure of ethylene.

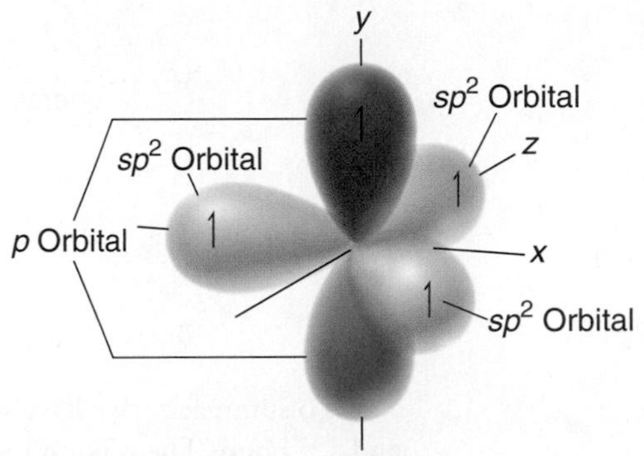

FIGURE 1.27
An illustration of an sp^2-hybridized carbon atom.

Each carbon atom in ethylene has three sp^2-hybridized orbitals available to form σ bonds (Figure 1.28). One σ bond forms between the two carbon atoms, and then each carbon atom also forms a σ bond with each of its neighboring hydrogen atoms.

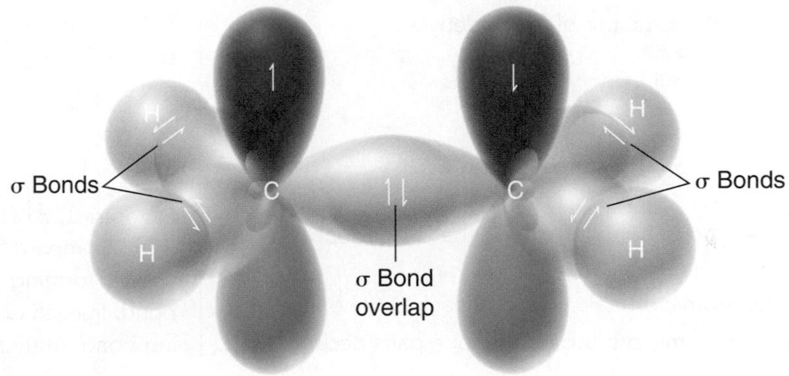

FIGURE 1.28
An illustration of the σ bonds in ethylene.

In addition, each carbon atom has one *p* orbital (shown in Figure 1.28 with blue and red lobes). These *p* orbitals actually overlap with each other as well, which is a separate bonding interaction called a **pi (π) bond** (Figure 1.29). Do not be confused by the nature of this type of bond. It is true that the π overlap occurs in two places—above the plane of the molecule (in red) and below the plane (in blue). Nevertheless, these two regions of overlap represent only one interaction called a π bond.

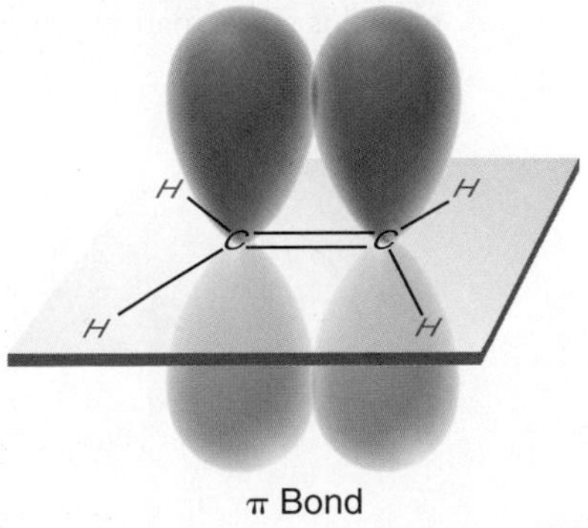

FIGURE 1.29
An illustration of the π bond in ethylene.

The picture of the π bond in Figure 1.29 is based on the valence bond approach (the *p* orbitals are simply drawn overlapping each other). Molecular orbital theory provides a fairly similar image of a π bond. Compare Figure 1.29 with the bonding MO in Figure 1.30.

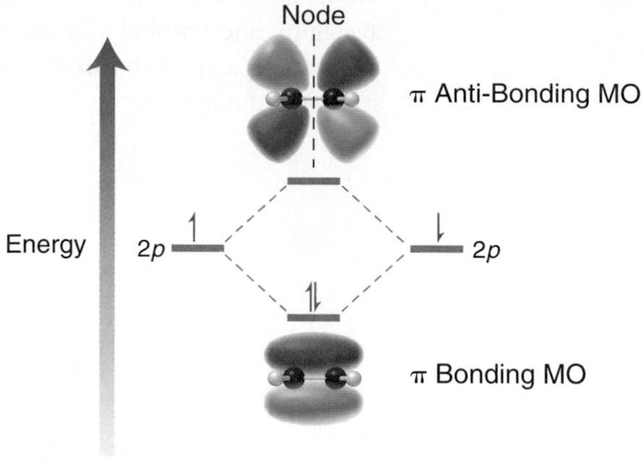

FIGURE 1.30 An energy diagram showing images of bonding and antibonding MOs in ethylene.

To summarize, we have seen that the carbon atoms of ethylene are connected via a σ bond and a π bond. The σ bond results from the overlap of sp^2-hybridized atomic orbitals, while the π bond results from the overlap of *p* orbitals. These two separate bonding interactions (σ and π) comprise the double bond of ethylene.

CONCEPTUAL CHECKPOINT

1.20 Consider the structure of formaldehyde:

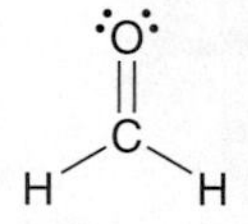

Formaldehyde

(a) Identify the type of bonds that form the C=O double bond.

(b) Identify the atomic orbitals that form each C—H bond.

(c) What type of atomic orbitals do the lone pairs occupy?

1.21 Sigma bonds experience free rotation at room temperature:

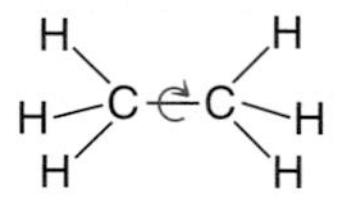

In contrast, π bonds do not experience free rotation. Explain. (*Hint*: Compare Figures 1.24 and 1.29, focusing on the orbitals used in forming a σ bond and the orbitals used in forming a π bond. In each case, what happens to the orbital overlap during bond rotation?)

Triple Bonds and *sp* Hybridization

Now let's consider the bonding structure of a compound bearing a triple bond, such as acetylene:

H—C≡C—H

Acetylene

A triple bond is formed by *sp*-hybridized carbon atoms. To achieve ***sp* hybridization**, one *s* orbital is mathematically averaged with only one *p* orbital (Figure 1.31). This leaves two *p* orbit-

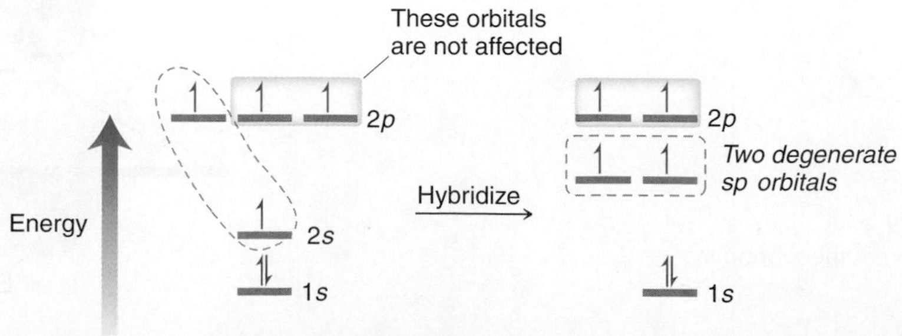

FIGURE 1.31 An energy diagram showing two degenerate *sp*-hybridized atomic orbitals.

als unaffected by the mathematical operation. As a result, an *sp*-hybridized carbon atom has two *sp* orbitals and two *p* orbitals (Figure 1.32).

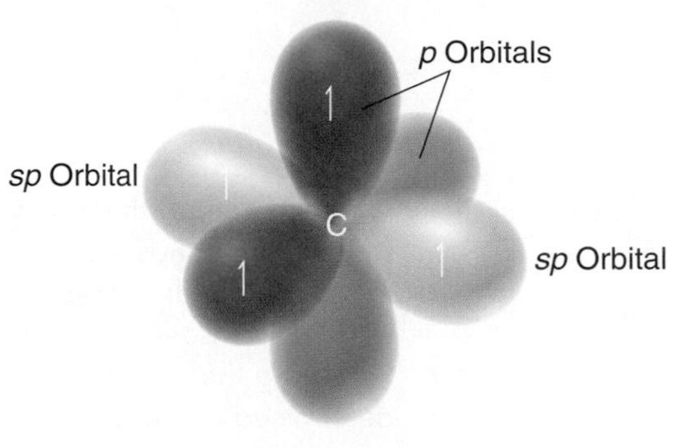

FIGURE 1.32
An illustration of an *sp*-hybridized carbon atom. The *sp*-hybridized orbitals are shown in gray.

The two *sp*-hybridized orbitals are available to form σ bonds (one on either side), and the two *p* orbitals are available to form π bonds, giving the bonding structure for acetylene shown in Figure 1.33. A triple bond between two carbon atoms is therefore the result of three separate bonding interactions: one σ bond and two π bonds. The σ bond results from the overlap of *sp* orbitals, while each of the two π bonds result from overlapping *p* orbitals. As shown in Figure 1.33, the geometry of the triple bond is linear.

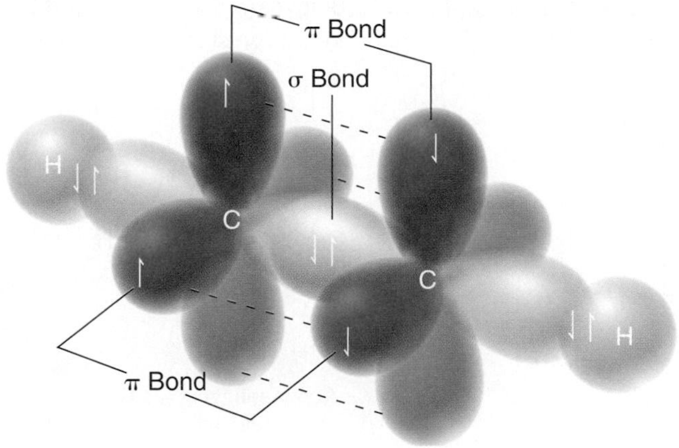

FIGURE 1.33
An illustration of the σ bonds and π bonds in acetylene.

SKILLBUILDER

1.7 IDENTIFYING HYBRIDIZATION STATES

LEARN the skill

Identify the hybridization state of each carbon atom in the following compound:

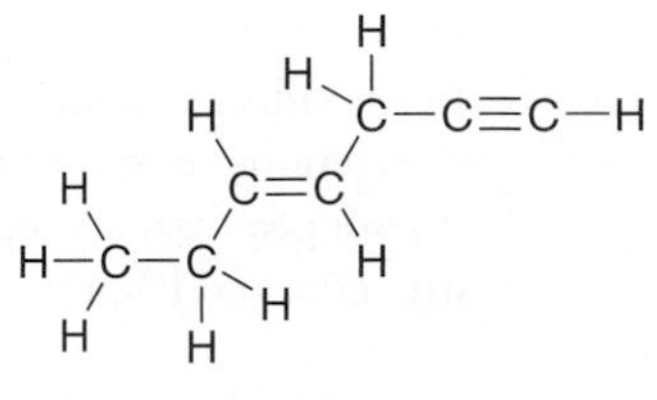

SOLUTION

To determine the hybridization state of an uncharged carbon atom, simply count the number of σ bonds and π bonds:

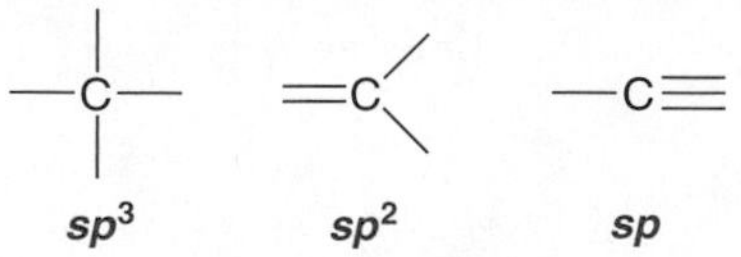

A carbon atom with four single bonds (four σ bonds) will be sp^3 hybridized. A carbon atom with three σ bonds and one π bond will be sp^2 hybridized. A carbon atom with two σ bonds and two π bonds will be *sp* hybridized. Carbon atoms bearing a positive or negative charge will be discussed in more detail in the upcoming chapter.

Using the simple scheme above, the hybridization state of most carbon atoms can be determined instantly:

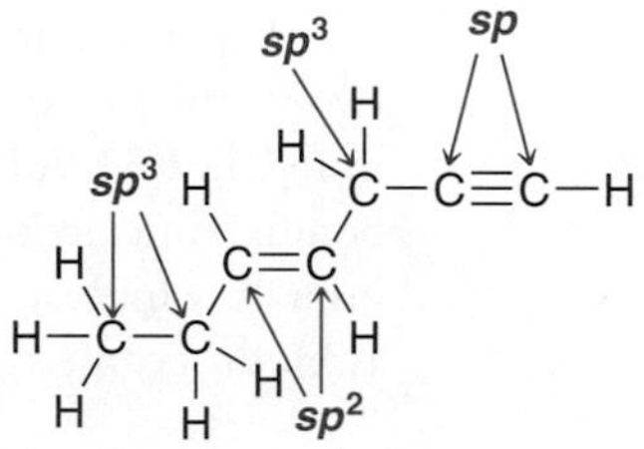

PRACTICE the skill **1.22** Below are the structures of two common over-the-counter pain relievers. Determine the hybridization state of each carbon atom in these compounds:

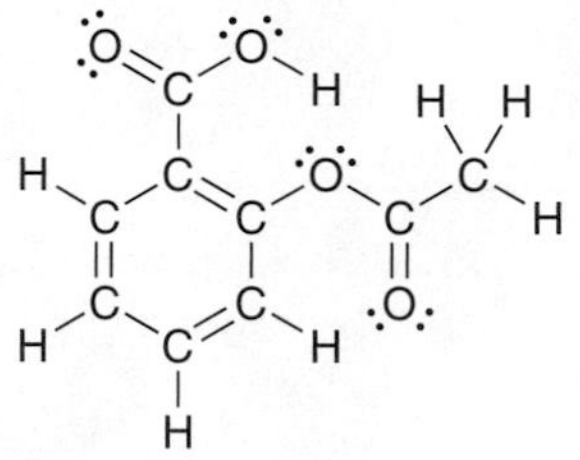

(a) **Acetylsalicyclic acid** (aspirin)

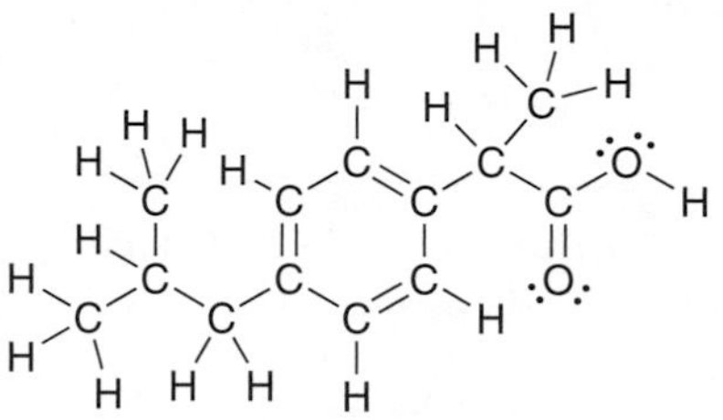

(b) **Ibuprofen** (Advil or Motrin)

APPLY the skill **1.23** Determine the hybridization state of each carbon atom in the following compounds:

(a) 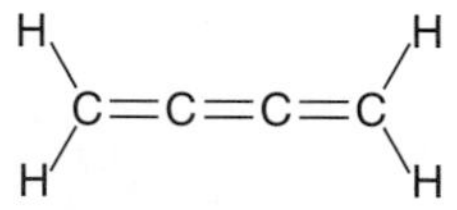

(b) 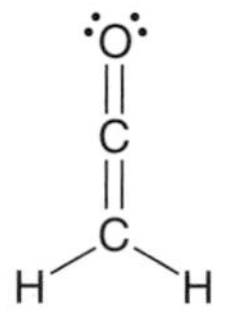

need more **PRACTICE?** **Try Problems 1.55, 1.56, 1.58**

Bond Strength and Bond Length

The information we have seen in this section allows us to compare single bonds, double bonds, and triple bonds. A single bond has only one bonding interaction (a σ bond), a double bond has two bonding interactions (one σ bond and one π bond), and a triple bond has three bonding interactions (one σ bond and two π bonds). Therefore, it is not surprising that a triple bond is stronger than a double bond, which in turn is stronger than a single bond. Compare the strengths and lengths of the C—C bonds in ethane, ethylene, and acetylene (Table 1.2).

TABLE 1.2 COMPARISON OF BOND LENGTHS AND BOND ENERGIES FOR ETHANE, ETHYLENE, AND ACETYLENE

	ETHANE	ETHYLENE	ACETYLENE
Structure	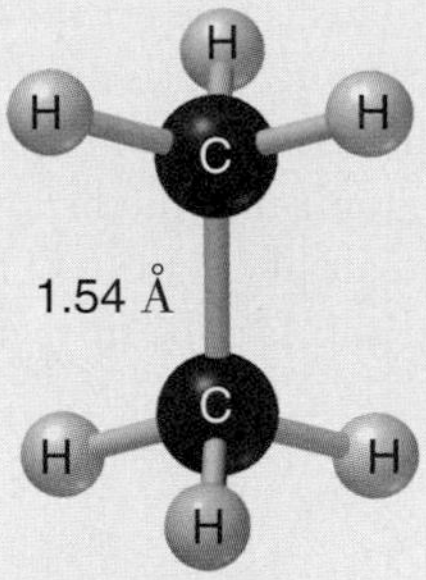	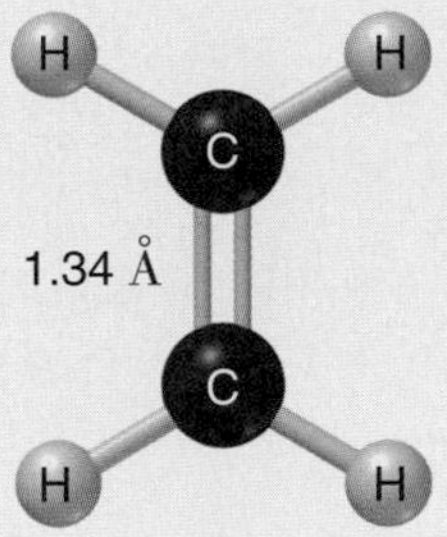	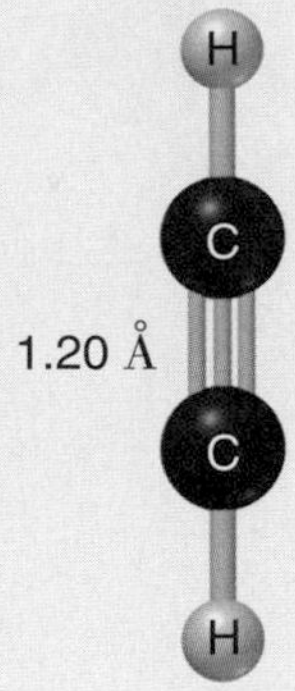
C—C bond length	1.54 Å	1.34 Å	1.20 Å
Bond energy	368 kJ/mol	632 kJ/mol	820 kJ/mol

CONCEPTUAL CHECKPOINT

1.24 Rank the indicated bonds in terms of increasing bond length:

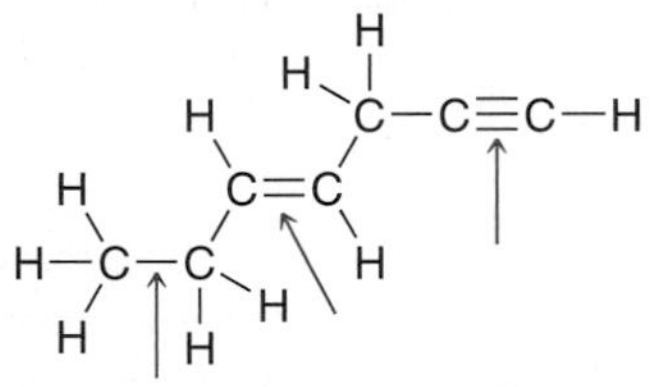

1.10 VSEPR Theory: Predicting Geometry

In order to predict the geometry of a small compound, we focus on the central atom and count the number of σ bonds and lone pairs. The total (σ bonds plus lone pairs) is called the **steric number**. Figure 1.34 gives several examples in which the steric number is 4 in each case.

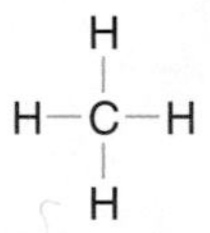

# of σ bonds = 4
# of lone pairs = 0
Steric number = 4

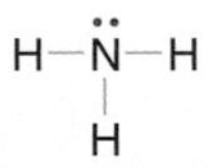

# of σ bonds = 3
# of lone pairs = 1
Steric number = 4

# of σ bonds = 2
# of lone pairs = 2
Steric number = 4

FIGURE 1.34
Calculation of the steric number for methane, ammonia, and water.

The steric number indicates the number of electron pairs (bonding and nonbonding) that are repelling each other. The repulsion causes the electron pairs to arrange themselves in three-dimensional space so as to achieve maximal distance from each other. As a result, the geometry of the central atom will be determined by the steric number. This principle is called the ***v**alence **s**hell **e**lectron **p**air **r**epulsion* (**VSEPR**) theory. Let's take a closer look at the geometry of each of the compounds above.

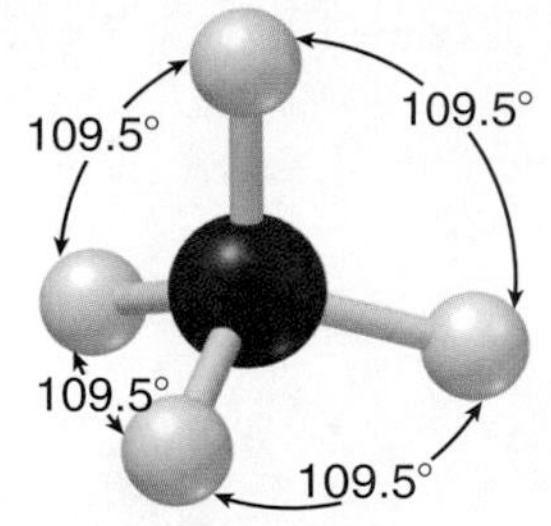

FIGURE 1.35
The tetrahedral geometry of methane.

Geometries Resulting from sp^3 Hybridization

In all of the previous examples, there are four pairs of electrons (steric number 4). In order for an atom to accommodate four electron pairs, it must use four orbitals and is therefore sp^3 hybridized. Recall that the geometry of methane is **tetrahedral** (Figure 1.35). In fact, any sp^3-hybridized atom will have four sp^3-hybridized orbitals arranged in a shape approximating a tetrahedron. This is true for the nitrogen atom in ammonia as well (Figure 1.36). The nitrogen atom is using four orbitals and is therefore sp^3 hybridized. As a result, its orbitals are arranged in a tetrahedron (shown on the left in Figure 1.36). However, there is one important difference between ammonia and methane. In the case of ammonia, one of the four orbitals is housing a nonbonding pair of electrons (a lone pair). This lone pair repels the other bonds more strongly, causing bond angles to be smaller than 109.5°. Bond angles for ammonia have been determined to be 107°.

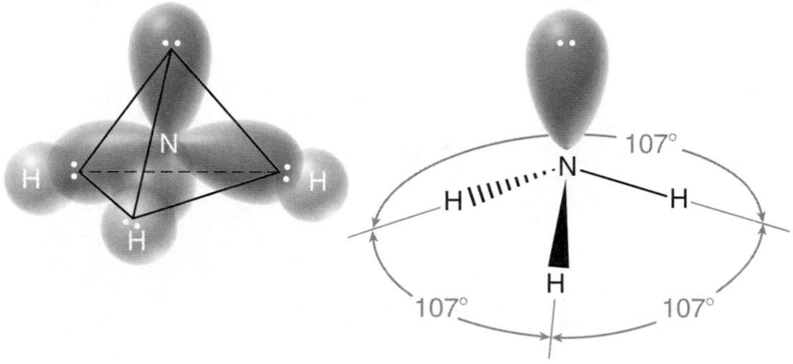

FIGURE 1.36
The orbitals of ammonia are arranged in a tetrahedral geometry.

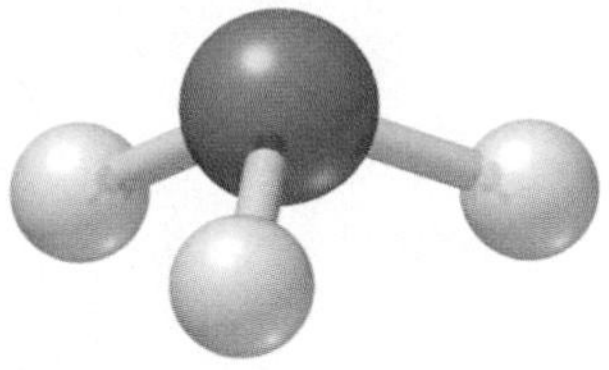

FIGURE 1.37
The geometry of ammonia is trigonal pyramidal.

The term "geometry" does not refer to the arrangement of electron pairs. Rather, it refers to the arrangement of atoms. When we show only the positions of the atoms (ignoring the lone pairs), ammonia appears as in Figure 1.37. This geometry is called **trigonal pyramidal** (Figure 1.37). "Trigonal" indicates the nitrogen atom is connected to three other atoms, and "pyramidal" indicates the compound is shaped like a pyramid, with the nitrogen atom sitting at the top of the pyramid.

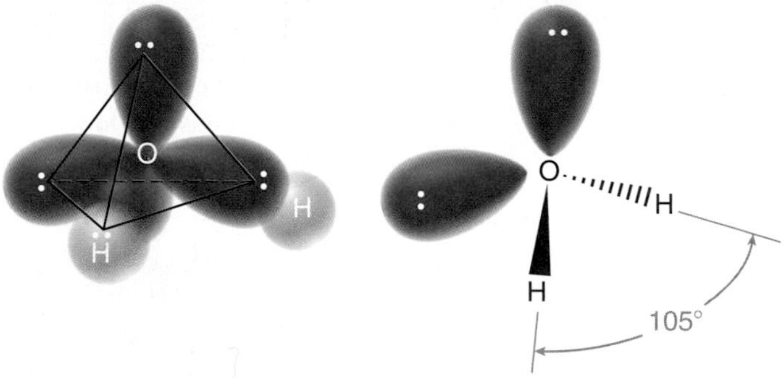

FIGURE 1.38
The orbitals of H_2O are arranged in a tetrahedral geometry.

Another example of sp^3 hybridization is water (H_2O). The oxygen atom has a steric number of 4, and it therefore requires the use of four orbitals. As a result, it must be sp^3 hybridized, with its four orbitals in a tetrahedral arrangement (Figure 1.38). Once again, lone pairs repel each other more strongly than bonds, causing the bond angle between the two O—H bonds to be even smaller than the bond angles in ammonia. The bond angle of water has been determined to be 105°. In order to describe the geometry, we ignore the lone pairs and focus only on the arrangement of atoms, which gives a **bent** geometry in this case (Figure 1.39). In summary,

FIGURE 1.39
The geometry of water is bent.

there are only three different types of geometry arising from sp^3 hybridization: tetrahedral, trigonal pyramidal, and bent. In all cases, the electrons were arranged in a tetrahedron, but the lone pairs were ignored when describing geometry. Table 1.3 summarizes this information.

TABLE 1.3 GEOMETRIES RESULTING FROM sp^3 HYBRIDIZATION

EXAMPLE	STERIC NUMBER	HYBRIDIZATION	ARRANGEMENT OF ELECTRON PAIRS	ARRANGEMENT OF ATOMS (GEOMETRY)
CH_4	4	sp^3	Tetrahedral	Tetrahedral
NH_3	4	sp^3	Tetrahedral	Trigonal pyramidal
H_2O	4	sp^3	Tetrahedral	Bent

Geometries Resulting from sp^2 Hybridization

When the central atom of a small compound has a steric number of 3, it will be sp^2 hybridized. As an example, consider the structure of BF_3. Boron has three valence electrons, each of which is used to form a bond. The result is three bonds and no lone pairs, giving a steric number of 3. The central boron atom therefore requires three orbitals, rather than four, and must be sp^2 hybridized. Recall that sp^2-hybridized orbitals achieve maximal separation in a **trigonal planar** arrangement (Figure 1.40): "trigonal" because the boron is connected to three other atoms and "planar" because all atoms are found in the same plane (as opposed to trigonal pyramidal).

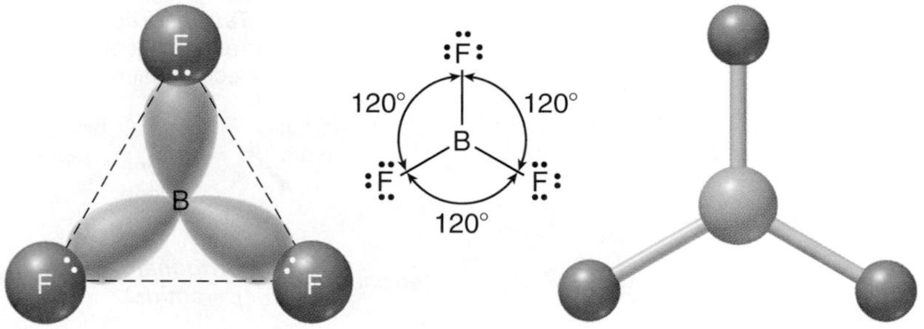

FIGURE 1.40
The geometry of BF_3 is trigonal planar.

As another example, consider the nitrogen atom of an imine:

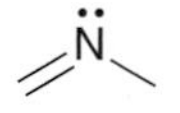

An imine

LOOKING AHEAD
We will explore imines in more detail in Chapter 20.

To determine the geometry of the nitrogen atom, we first consider the steric number, which is not affected by the presence of the π bond. Why not? Recall that a π bond results from the overlap of *p* orbitals. The steric number of an atom is meant to indicate how many hybridized orbitals are necessary (*p* orbitals are not included in this count). The steric number in this case is 3 (Figure 1.41). As a result, the nitrogen atom must be sp^2 hybridized. The sp^2 hybridization state is always characterized by a trigonal planar arrangement of electron pairs, but when describing geometry, we focus only on the atoms (ignoring any lone pairs). The geometry of this nitrogen atom is therefore bent.

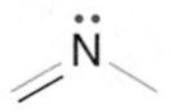

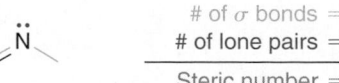

\# of σ bonds = 2
\# of lone pairs = 1
Steric number = 3

FIGURE 1.41
The steric number of the nitrogen atom of an imine.

Geometry Resulting from *sp* Hybridization

When the central atom of a small compound has a steric number of 2, the central atom will be *sp* hybridized. As an example, consider the structure of BeH_2. Beryllium has two valence electrons, each of which is used to form a bond. The result is two bonds and no lone pairs, giving a steric number of 2. The central beryllium atom therefore requires only two orbitals and must be *sp* hybridized. Recall that *sp*-hybridized orbitals achieve maximal separation when they are **linear** (Figure 1.42).

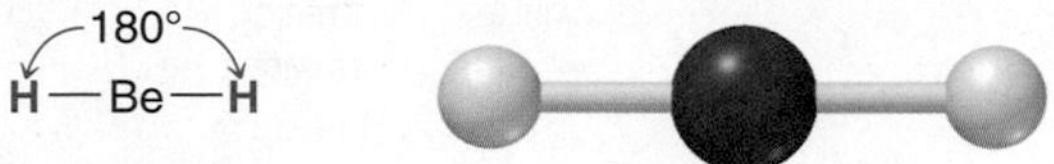

Linear geometry of BeH_2

FIGURE 1.42
The geometry of BeH_2 is linear.

As another example of *sp* hybridization, consider the structure of CO_2:

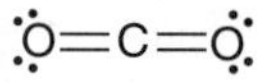

Once again, the π bonds do *not* impact the calculation of the steric number, so the steric number is 2. The carbon atom must be *sp* hybridized and is therefore linear.

As summarized in Figure 1.43, the three hybridization states give rise to five common geometries.

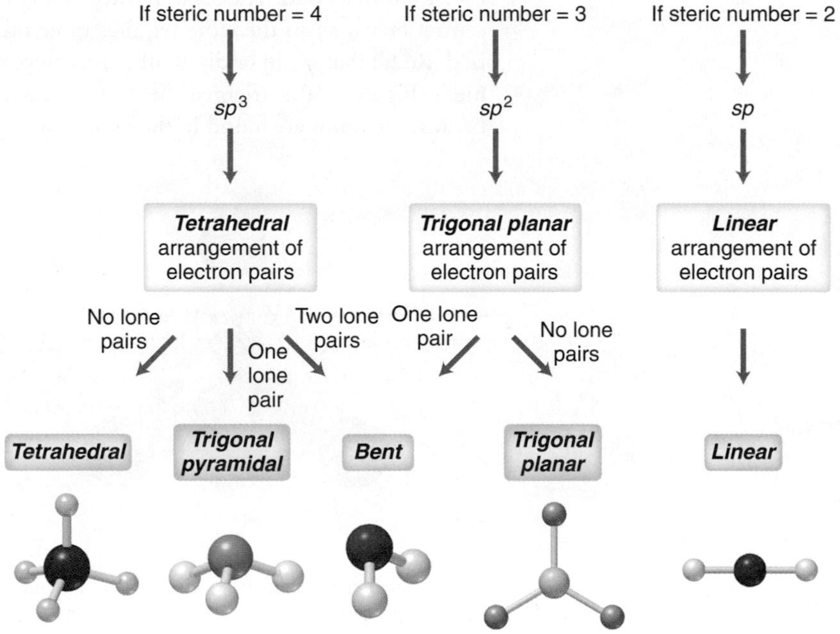

FIGURE 1.43
A decision tree for determining geometry.

SKILLBUILDER

1.8 PREDICTING GEOMETRY

LEARN the skill

Predict the geometry for all atoms (except hydrogens) in the compound below:

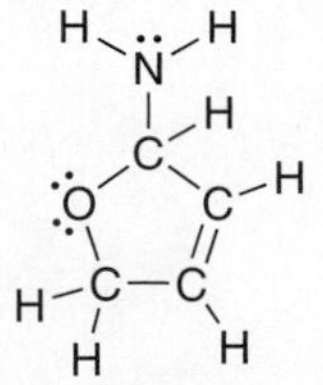

SOLUTION

For each atom, the following three steps are followed:

STEP 1 Determine the steric number.

1. Determine the steric number by counting the number of lone pairs and σ bonds.

STEP 2 Determine the hybridization state and electronic arrangement.

2. Use the steric number to determine the hybridization state and electronic arrangement:
   - If the steric number is 4, then the atom will be sp^3 hybridized, and the electronic arrangement will be tetrahedral.
   - If the steric number is 3, then the atom will be sp^2 hybridized, and the electronic arrangement will be trigonal planar.
   - If the steric number is 2, then the atom will be sp hybridized, and the electronic arrangement will be linear.

STEP 3 Ignore lone pairs and describe the resulting geometry.

3. Ignore any lone pairs and describe the geometry only in terms of the arrangement of atoms:

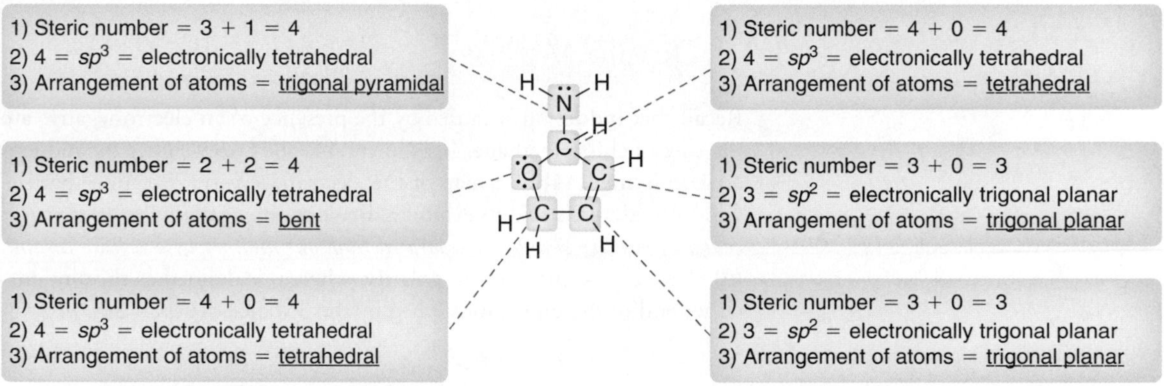
1) Steric number = 3 + 1 = 4
2) 4 = sp³ = electronically tetrahedral
3) Arrangement of atoms = trigonal pyramidal
1) Steric number = 2 + 2 = 4
2) 4 = sp³ = electronically tetrahedral
3) Arrangement of atoms = bent
1) Steric number = 4 + 0 = 4
2) 4 = sp³ = electronically tetrahedral
3) Arrangement of atoms = tetrahedral
1) Steric number = 4 + 0 = 4
2) 4 = sp³ = electronically tetrahedral
3) Arrangement of atoms = tetrahedral
1) Steric number = 3 + 0 = 3
2) 3 = sp² = electronically trigonal planar
3) Arrangement of atoms = trigonal planar
1) Steric number = 3 + 0 = 3
2) 3 = sp² = electronically trigonal planar
3) Arrangement of atoms = trigonal planar

It is not necessary to describe the geometry of hydrogen atoms. Each hydrogen atom is monovalent, so the geometry is irrelevant. Geometry is only relevant when an atom is connected to at least two other atoms. For our purposes, we can also disregard the geometry of the oxygen atom in a C═O double bond because it is connected to only one atom:

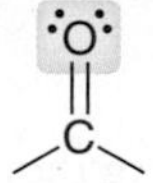

can disregard the geometry of this oxygen

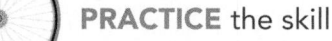

PRACTICE the skill **1.25** Predict the geometry for the central atom in each of the compounds below:

(a) NH_3 (b) H_3O^+ (c) BH_4^- (d) BCl_3 (e) BCl_4^- (f) CCl_4 (g) $CHCl_3$ (h) CH_2Cl_2

1.26 Predict the geometry for all atoms except hydrogen in the compounds below:

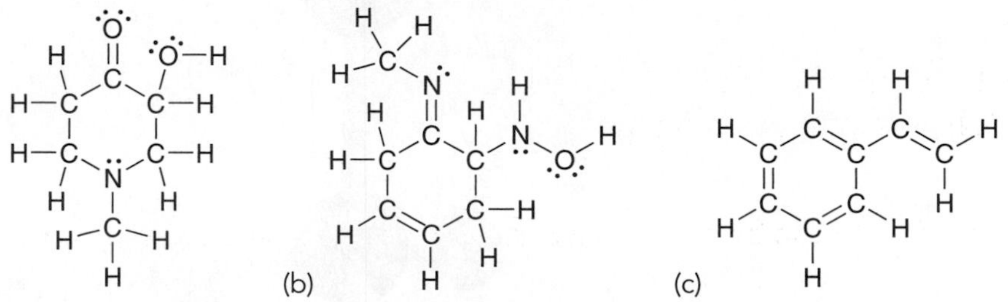

(a) (b) (c)

APPLY the skill **1.27** Compare the structures of a carbocation and a carbanion:

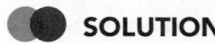

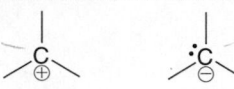

Carbocation Carbanion

In one of these ions, the central carbon atom is trigonal planar; in the other it is trigonal pyramidal. Assign the correct geometry to each ion.

1.28 Identify the hybridization state and geometry of each carbon atom in benzene. Use that information to determine the geometry of the entire molecule:

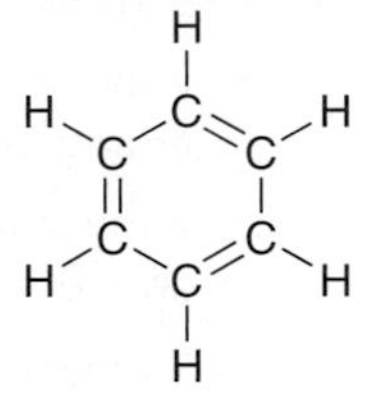

Benzene

need more **PRACTICE?** **Try Problems 1.39–1.41, 1.50, 1.55, 1.56, 1.58**

1.11 Dipole Moments and Molecular Polarity

Recall that induction is caused by the presence of an electronegative atom, as we saw earlier in the case of chloromethane. In Figure 1.44a the arrow shows the inductive effect of the chlorine atom. Figure 1.44b is a map of the electron density, revealing that the molecule is polarized. Chloromethane is said to exhibit a dipole moment, because *the center of negative charge and the center of positive charge are separated from one another by a certain distance*. The **dipole moment** (μ) is used as an indicator of polarity, where μ is defined as the amount of partial charge (δ) on either end of the dipole multiplied by the distance of separation (d):

$$\mu = \delta \times d$$

Partial charges ($\delta+$ and $\delta-$) are generally on the order of 10^{-10} esu (electrostatic units) and the distances are generally on the order of 10^{-8} cm. Therefore, for a polar compound, the dipole moment (μ) will generally have an order of magnitude of around 10^{-18} esu · cm. The dipole moment of chloromethane, for example, is 1.87×10^{-18} esu • cm. Since most compounds will have a dipole moment on this order of magnitude (10^{-18}), it is more convenient to report dipole moments with a new unit, called a **debye (D)**, where

$$1 \text{ debye} = 10^{-18} \text{ esu} \cdot \text{cm}.$$

Using these units, the dipole moment of chloromethane is reported as 1.87 D. The debye unit is named after Dutch scientist Peter Debye, whose contributions to the fields of chemistry and physics earned him a Nobel Prize in 1936.

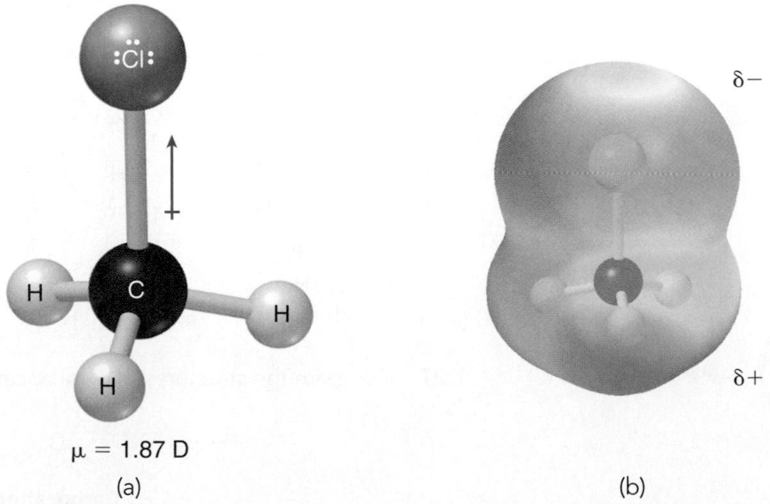

FIGURE 1.44
(a) Ball-and-stick model of chloromethane showing the dipole moment.
(b) An electrostatic potential map of chloromethane.

Measuring the dipole moment of a particular bond allows us to calculate the percent **ionic character** of that bond. As an example, let's analyze a C—Cl bond. This bond has a bond length of 1.772×10^{-8} cm, and an electron has a charge of 4.80×10^{-10} esu. If the bond were 100% ionic, then the dipole moment would be

$$\begin{aligned} \mu &= e \times d \\ &= (4.80 \times 10^{-10}\ \text{esu}) \times (1.772 \times 10^{-8}\ \text{cm}) \\ &= 8.51 \times 10^{-18}\ \text{esu} \cdot \text{cm} \end{aligned}$$

or 8.51 D. In reality, the bond is not 100% ionic. The experimentally observed dipole moment is measured at 1.87 D, and we can use this value to calculate the percent ionic character of a C—Cl bond:

$$\frac{1.87\ \text{D}}{8.51\ \text{D}} \times 100\% = \boxed{22\%}$$

Table 1.4 shows the percent ionic character for a few of the bonds that we will frequently encounter in this text. Take special notice of the C═O bond. It has considerable ionic character, rendering it extremely reactive. Chapters 20-22 are devoted exclusively to the reactivity of compounds containing C═O bonds.

TABLE 1.4 PERCENT IONIC CHARACTER FOR SEVERAL BONDS

BOND	BOND LENGTH ($\times 10^{-8}$ CM)	OBSERVED μ (D)	PERCENT IONIC CHARACTER
C—O	1.41	0.7 D	$\frac{(0.7 \times 10^{-18}\ \text{esu} \cdot \text{cm})}{(4.80 \times 10^{-10}\ \text{esu})(1.41 \times 10^{-8}\ \text{cm})} \times 100\% = \boxed{10\%}$
O—H	0.96	1.5 D	$\frac{(1.5 \times 10^{-18}\ \text{esu} \cdot \text{cm})}{(4.80 \times 10^{-10}\ \text{esu})(0.96 \times 10^{-8}\ \text{cm})} \times 100\% = \boxed{33\%}$
C═O	1.227	2.4 D	$\frac{(2.4 \times 10^{-18}\ \text{esu} \cdot \text{cm})}{(4.80 \times 10^{-10}\ \text{esu})(1.23 \times 10^{-8}\ \text{cm})} \times 100\% = \boxed{41\%}$

Chloromethane was a simple example, because it has only one polar bond. When dealing with a compound that has more than one polar bond, it is necessary to take the vector sum of the individual dipole moments. The vector sum is called the **molecular dipole moment**, and it takes into account both the magnitude and the direction of each individual dipole moment. For example, consider the structure of dichloromethane (Figure 1.45). The individual dipole moments partially cancel, but not completely. The vector sum produces a dipole moment of 1.14 D, which is significantly smaller than the dipole moment of chloromethane because the two dipole moments here partially cancel each other.

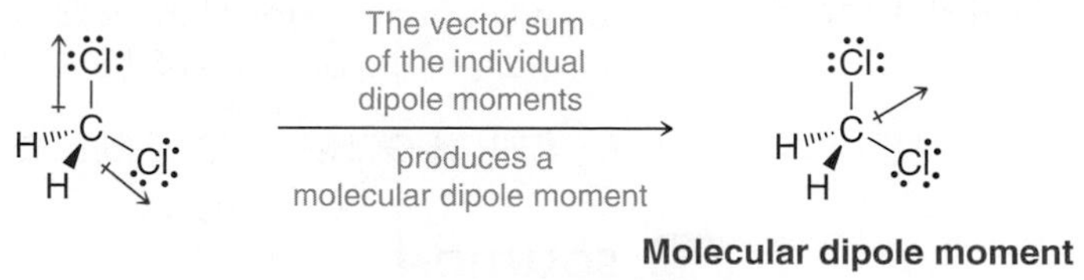
The vector sum
of the individual
dipole moments
produces a
molecular dipole moment
Molecular dipole moment

FIGURE 1.45
The molecular dipole moment of dichloromethane is the net sum of all dipole moments in the compound.

The presence of a lone pair has a significant effect on the molecular dipole moment. The two electrons of a lone pair are balanced by two positive charges in the nucleus, but the lone pair is separated from the nucleus by some distance. There is, therefore, a dipole moment associated with every lone pair. Common examples are ammonia and water (Figure 1.46).

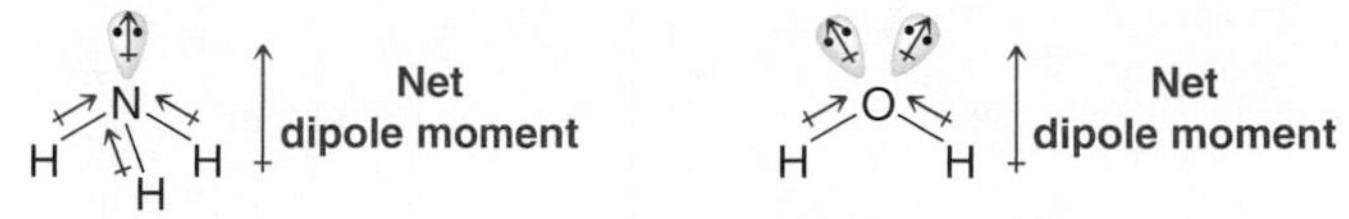
Net
dipole moment
Net
dipole moment

FIGURE 1.46
The net dipole moments of ammonia and water.

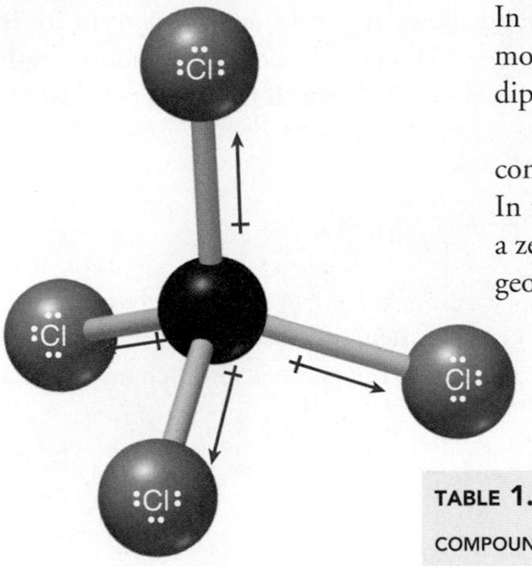

FIGURE 1.47
A ball-and-stick model of carbon tetrachloride. The individual dipole moments cancel to give a zero net dipole moment.

In this way, the lone pairs contribute significantly to the magnitude of the molecular dipole moment, although they do not contribute to its direction. That is, the direction of the molecular dipole moment would be the same with or without the contribution of the lone pairs.

Table 1.5 shows experimentally observed molecular dipole moments (at 20°C) for several common solvents. Notice that carbon tetrachloride (CCl_4), has no molecular dipole moment. In this case, the individual dipole moments cancel each other completely to give the molecule a zero net dipole moment ($\mu = 0$). This example (Figure 1.47) demonstrates that we must take geometry into account when assessing molecular dipole moments.

TABLE 1.5 DIPOLE MOMENTS FOR SOME COMMON SOLVENTS (AT 20°C)

COMPOUND	STRUCTURE	DIPOLE MOMENT	COMPOUND	STRUCTURE	DIPOLE MOMENT
Methanol	CH_3OH	2.87 D	Ammonia	$:NH_3$	1.47 D
Acetone	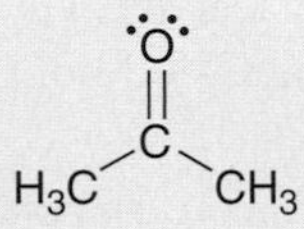	2.69 D	Diethyl ether	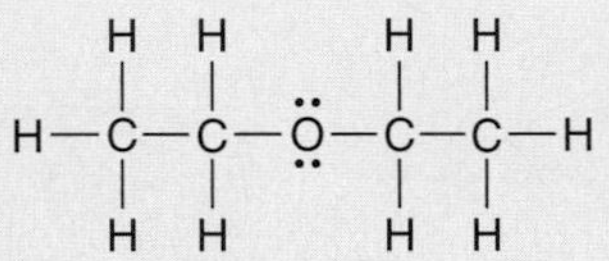	1.15 D
Chloromethane	CH_3Cl	1.87 D	Methylene chloride	CH_2Cl_2	1.14 D
Water	H_2O	1.85 D	Pentane	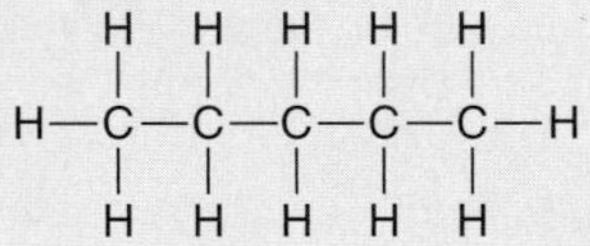	0 D
Ethanol	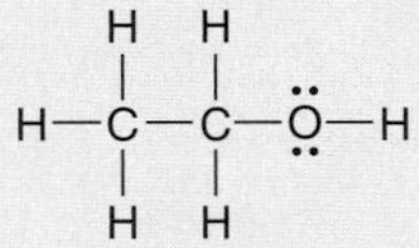	1.66 D	Carbon tetrachloride	CCl_4	0 D

SKILLBUILDER

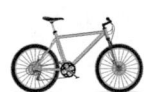

1.9 IDENTIFYING THE PRESENCE OF MOLECULAR DIPOLE MOMENTS

LEARN the skill

Identify whether each of the following compounds exhibits a molecular dipole moment. If so, indicate the direction of the net molecular dipole moment:

(a) $CH_3CH_2OCH_2CH_3$ (b) CO_2

SOLUTION

(a) In order to determine whether the individual dipole moments cancel each other completely, we must first predict the molecular geometry. Specifically, we need to know if the geometry around the oxygen atom is linear or bent:

STEP 1
Predict the molecular geometry.

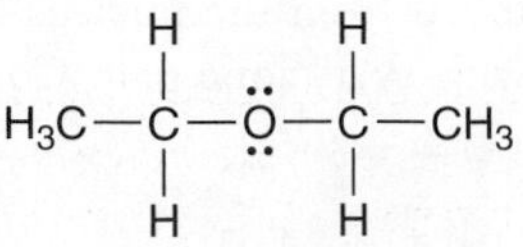

Linear

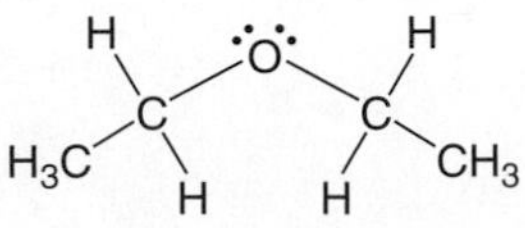

Bent

To make this determination, we use the three-step method from the previous section:

1. The steric number is 4.
2. Therefore, the hybridization state must be sp^3, and the arrangement of electron pairs must be tetrahedral.
3. Ignore the lone pairs, and the oxygen has a bent geometry.

After determining the molecular geometry, now draw all dipole moments and determine whether they cancel each other. In this case, they do not fully cancel each other:

STEP 2
Identify the direction of all dipole moments.

STEP 3
Draw the net dipole moment.

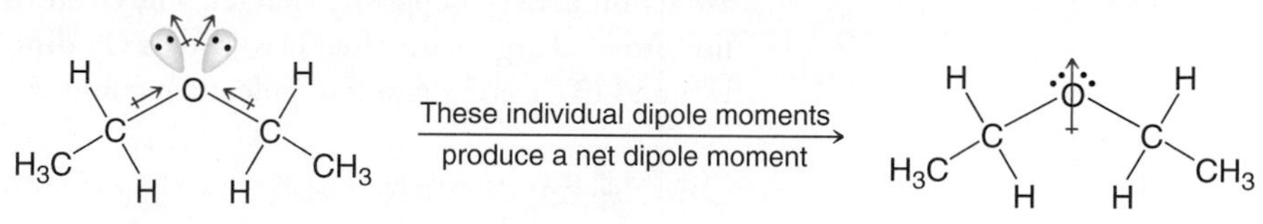

This compound does in fact have a net molecular dipole moment, and the direction of the moment is shown above.

(b) Carbon dioxide (CO_2) has two C═O bonds, each of which exhibits a dipole moment. In order to determine whether the individual dipole moments cancel each other completely, we must first predict the molecular geometry. We apply our three-step method: the steric number is 2, the hybridization state is *sp*, and the compound has a linear geometry. As a result, we expect the dipole moments to fully cancel each other:

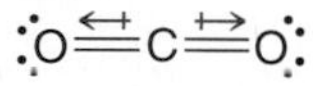

In a similar way, the dipole moments associated with the lone pairs also cancel each other, and therefore CO_2 does not have a net molecular dipole moment.

PRACTICE the skill **1.29** Identify whether each of the following compounds exhibits a molecular dipole moment. For compounds that do, indicate the direction of the net molecular dipole moment:

(a) $CHCl_3$ (b) CH_3OCH_3 (c) NH_3 (d) CCl_2Br_2

(e) 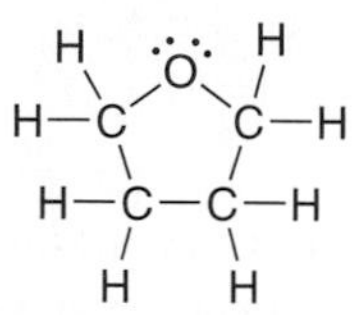 (f) 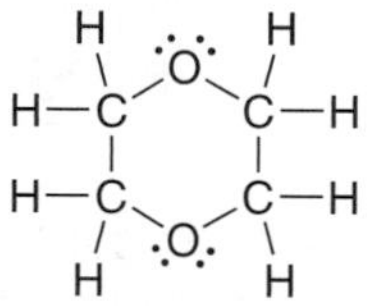 (g) 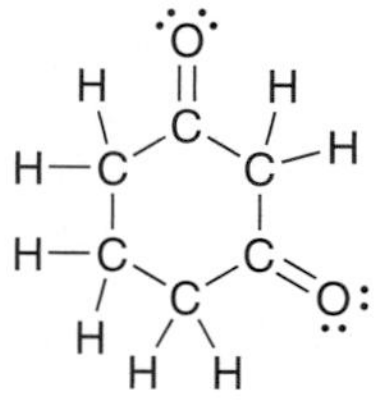 (h) 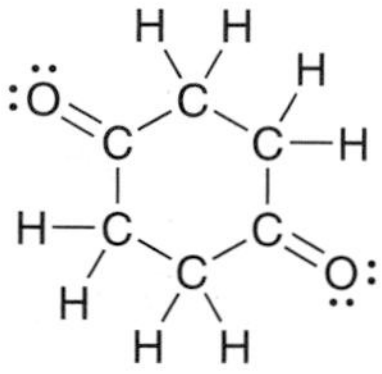

(i) 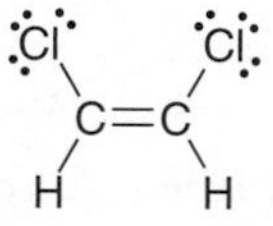 (j) 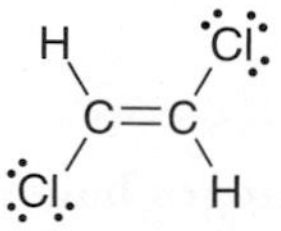 (k) 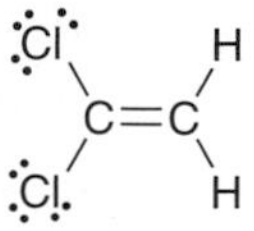 (l) 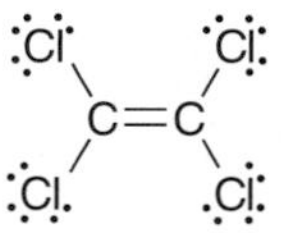

APPLY the skill **1.30** Which of the following compounds has the larger dipole moment? Explain your choice:

$CHCl_3$ or $CBrCl_3$

1.31 Bonds between carbon and oxygen (C—O) are more polar than bonds between sulfur and oxygen (S—O). Nevertheless, sulfur dioxide (SO_2) exhibits a dipole moment while carbon dioxide (CO_2) does not. Explain this apparent anomaly.

need more **PRACTICE?** **Try Problems 1.37, 1.40, 1.43, 1.61, 1.62**

1.12 Intermolecular Forces and Physical Properties

The physical properties of a compound are determined by the attractive forces between the individual molecules, called **intermolecular forces**. It is often difficult to use the molecular structure alone to predict a precise melting point or boiling point for a compound. However, a few simple trends will allow us to compare compounds to each other in a relative way, for example, to predict which compound will boil at a higher temperature.

All intermolecular forces are *electrostatic*—that is, these forces occur as a result of the attraction between opposite charges. The electrostatic interactions for neutral molecules (with no formal charges) are often classified as (1) **dipole-dipole interactions**, (2) hydrogen bonding, and (3) fleeting dipole-dipole interactions.

Dipole-Dipole Interactions

Compounds with net dipole moments can either attract each other or repel each other, depending on how they approach each other in space. In the solid phase, the molecules align so as to attract each other (Figure 1.48).

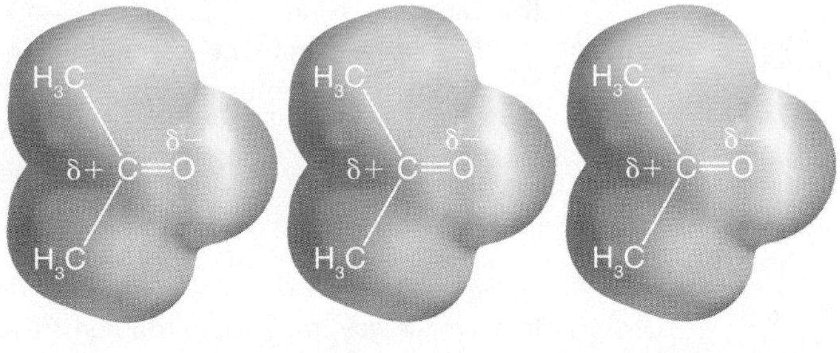

FIGURE 1.48
In solids, molecules align themselves so that their dipole moments experience attractive forces.

In the liquid phase, the molecules are free to tumble in space, but they do tend to move in such a way so as to attract each other more often then they repel each other. The resulting net attraction between the molecules results in an elevated melting point and boiling point. To illustrate this, compare the physical properties of isobutylene and acetone:

Isobutylene
Melting point = −140.3°C
Boiling point = −6.9°C

Acetone
Melting point = −94.9°C
Boiling point = 56.3°C

Isobutylene lacks a significant dipole moment, but acetone does have a net dipole moment. Therefore, acetone molecules will experience greater attractive interactions than isobutylene molecules. As a result, acetone has a higher melting point and higher boiling point than isobutylene.

Hydrogen Bonding

The term **hydrogen bonding** is misleading. A hydrogen bond is not actually a "bond" but is just a specific type of dipole-dipole interaction. When a hydrogen atom is connected to an electronegative atom, the hydrogen atom will bear a partial positive charge (δ+) as a result of induction. This δ+ can then interact with a lone pair from an electronegative atom of another molecule. This can be illustrated with water or ammonia (Figure 1.49). This attractive interac-

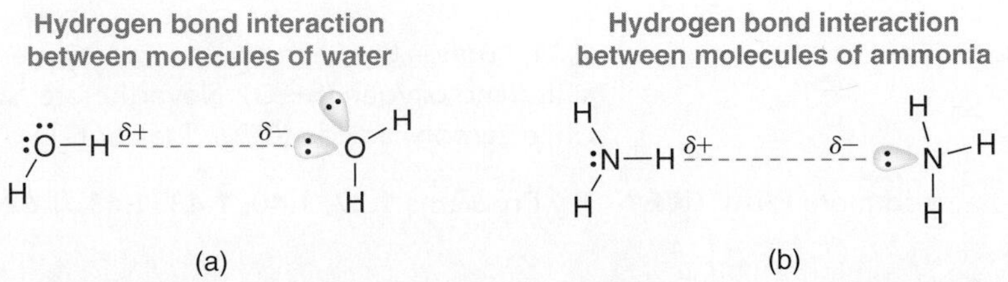

FIGURE 1.49
(a) Hydrogen bonding between molecules of water. (b) Hydrogen bonding between molecules of ammonia.

tion can occur with any **protic** compound, that is, any compound that has a proton connected to an electronegative atom. Ethanol, for example, exhibits the same kind of attractive interaction (Figure 1.50).

FIGURE 1.50
Hydrogen bonding between molecules of ethanol.

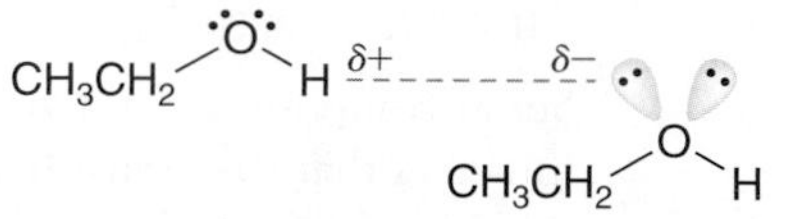

This type of interaction is quite strong because hydrogen is a relatively small atom, and as a result, the partial charges can get very close to each other. In fact, the effect of hydrogen bonding on physical properties is quite dramatic. At the beginning of this chapter, we briefly mentioned the difference in properties between the following two constitutional isomers:

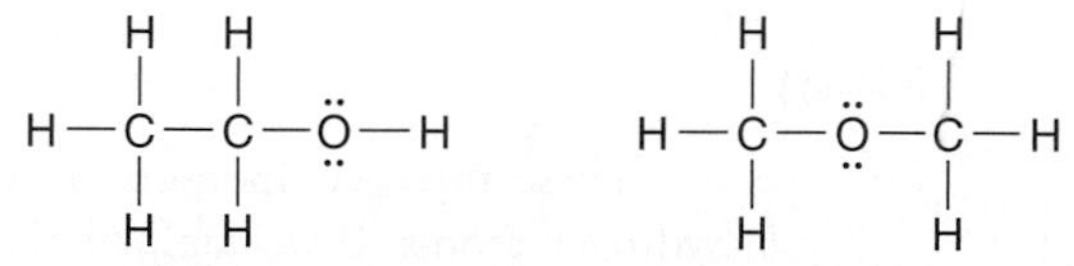

Ethanol
Boiling point = 78.4°C

Methoxymethane
Boiling point = −23°C

These compounds have the same molecular formula, but they have very different boiling points. Ethanol experiences intermolecular hydrogen bonding, giving rise to a very high boiling point. Methoxymethane does not experience intermolecular hydrogen bonding, giving rise to a relatively lower boiling point. A similar trend can be seen in a comparison of the following amines:

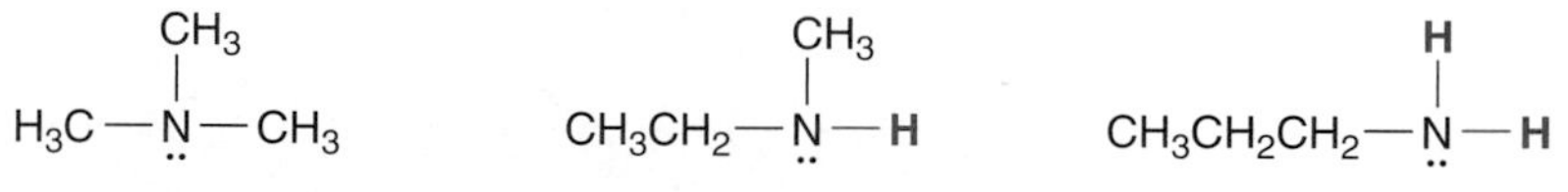

Trimethylamine
Boiling point = 3.5°C

Ethylmethylamine
Boiling point = 37°C

Propylamine
Boiling point = 49°C

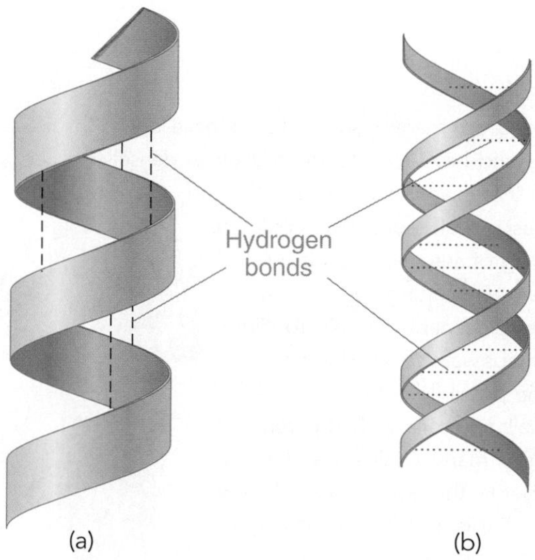

FIGURE 1.51
(a) An alpha helix of a protein.
(b) The double helix in DNA.

LOOKING AHEAD
The structure of DNA is explored in more detail in Section 24.9.

Once again, all three compounds have the same molecular formula (C_3H_9N), but they have very different properties as a result of the extent of hydrogen bonding. Trimethylamine does not exhibit any hydrogen bonding and has a relatively low boiling point. Ethylmethylamine does exhibit hydrogen bonding and therefore has a higher boiling point. Finally, propylamine, which has the highest boiling point of the three compounds, has two N—H bonds and therefore exhibits even more hydrogen-bonding interactions.

Hydrogen bonding is incredibly important in determining the shapes and interactions of biologically important compounds. Chapter 25 will focus on proteins, which are long molecules that coil up into specific shapes under the influence of hydrogen bonding (Figure 1.51a). These shapes ultimately determine their biological function. Similarly, hydrogen bonds hold together individual strands of DNA to form the familiar double-helix structure.

As mentioned earlier, hydrogen "bonds" are not really bonds. To illustrate this, compare the energy of a real bond with the energy of a hydrogen-bonding interaction. A typical single bond (C—H, N—H, O—H) has a bond strength of approximately 400 kJ/mol. In contrast, a hydrogen-bonding interaction has an average strength of approximately 20 kJ/mol. This leaves us with the obvious question: why do we call them hydrogen *bonds* instead of just hydrogen *interactions*? To answer this question, consider the double-helix structure of DNA (Figure 1.51b). The two strands are joined by hydrogen bonding interactions that function like rungs of a very long, twisted ladder. The net sum of these interactions is a significant factor that contributes to the structure of the double helix, in which the hydrogen-bonding interactions appear *as if* they were actually bonds. Nevertheless, it is relatively easy to "unzip" the double helix and retrieve the individual strands.

Fleeting Dipole-Dipole Interactions

Some compounds have no permanent dipole moments, and yet analysis of boiling points indicates that they must have fairly strong intermolecular attractions. To illustrate this point, consider the following compounds:

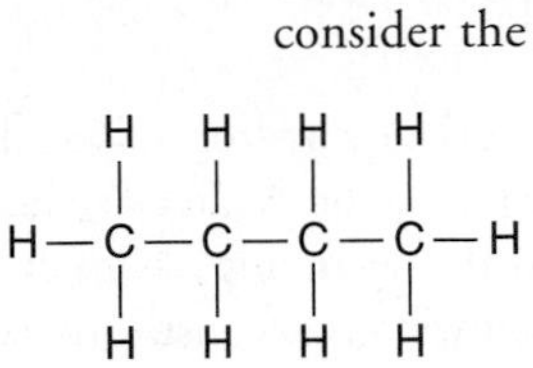

Butane
(C_4H_{10})
Boiling point = 0°C

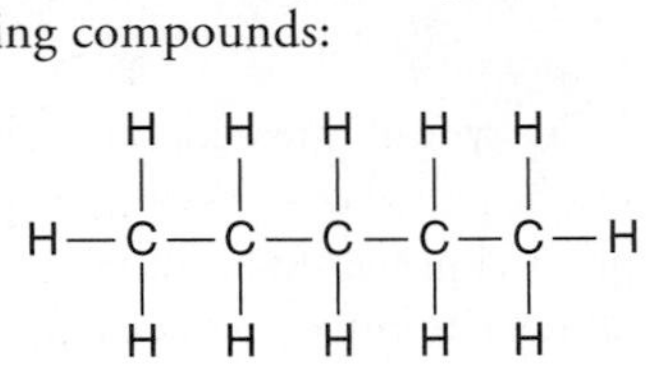

Pentane
(C_5H_{12})
Boiling point = 36°C

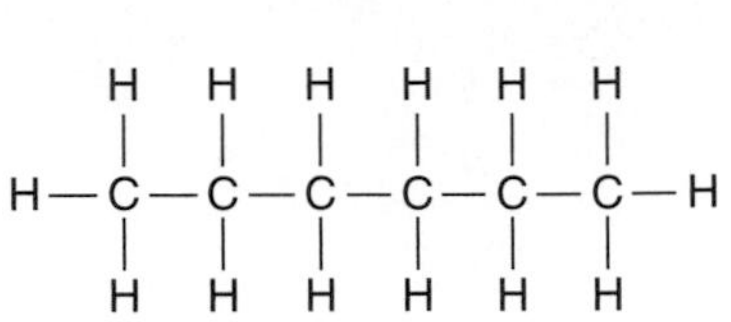

Hexane
(C_6H_{14})
Boiling point = 69°C

LOOKING AHEAD
Hydrocarbons will be discussed in more detail in Chapters 4, 17, and 18.

These three compounds are **hydrocarbons**, compounds that contain only carbon and hydrogen atoms. If we compare the properties of the hydrocarbons above, an important trend becomes apparent. Specifically, the boiling point appears to increase with increasing molecular weight. This trend can be justified by considering the fleeting, or transient, dipole moments that are more prevalent in larger hydrocarbons. To understand the source of these temporary dipole moments, we consider the electrons to be in constant motion, and therefore, the center of negative charge is also constantly moving around within the molecule. On average, the center of negative charge coincides with the center of positive charge, resulting in a zero dipole moment. However, at any given instant, the center of negative charge and the center of positive charge might not coincide. The resulting transient dipole moment can then induce a separate transient dipole moment in a neighboring molecule, initiating a fleeting attraction between the two mol-

PRACTICALLYSPEAKING)))

Biomimicry and Gecko Feet

The term biomimicry describes the notion that scientists often draw creative inspiration from studying nature. By investigating some of nature's processes, it is possible to mimic those processes and to develop new technology. One such example is based on the way that geckos can scurry up walls and along ceilings. Until recently, scientists were baffled by the curious ability of geckos to walk upside down, even on very smooth surfaces such as polished glass.

As it turns out, geckos do not use any chemical adhesives, nor do they use suction. Instead, their abilities arise from the intermolecular forces of attraction between the molecules in their feet and the molecules in the surface on which they are walking. When you place your hand on a surface, there are certainly intermolecular forces of attraction between the molecules of your hand and the surface, but the microscopic topography of your hand is quite bumpy. As a result, your hand only makes contact with the surface at perhaps a few thousand points. In contrast, the foot of a gecko has approximately half a million microscopic flexible hairs, called *setae*, each of which has even smaller hairs.

When a gecko places its foot on a surface, the flexible hairs allow the gecko to make extraordinary contact with the surface, and the resulting London dispersion forces are collectively strong enough to support the gecko.

In the last decade, many research teams have drawn inspiration from geckos and have created materials with densely packed microscopic hairs. For example, some scientists are developing adhesive bandages that could be used in the healing of surgical wounds, while other scientists are developing special gloves and boots that would enable people to climb up walls (and perhaps walk upside down on ceilings). Imagine the possibility of one day being able to walk on walls and ceilings like Spiderman.

There are still many challenges that we must overcome before these materials will show their true potential. It is a technical challenge to design microscopic hairs that are strong enough to prevent the hairs from becoming tangled but flexible enough to allow the hairs to stick to any surface. Many researchers believe that these challenges can be overcome, and if they are right, we might have the opportunity to see the world turned literally upside down within the next decade.

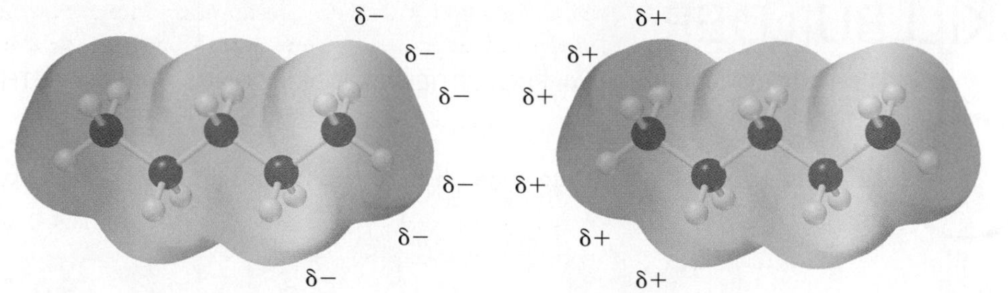

FIGURE 1.52 The fleeting attractive forces between two molecules of pentane.

ecules (Figure 1.52). These attractive forces are called **London dispersion forces,** named after German-American physicist Fritz London. Large hydrocarbons have more surface area than smaller hydrocarbons and therefore experience these attractive forces to a larger extent.

London dispersion forces are stronger for higher molecular weight hydrocarbons because these compounds have larger surface areas that can accommodate more interactions. As a result, compounds of higher molecular weight will generally boil at higher temperatures. Table 1.6 illustrates this trend.

A branched hydrocarbon generally has a smaller surface area than its corresponding straight-chain isomer, and therefore, branching causes a decrease in boiling point. This trend can be seen by comparing the following constitutional isomers of C_5H_{12}:

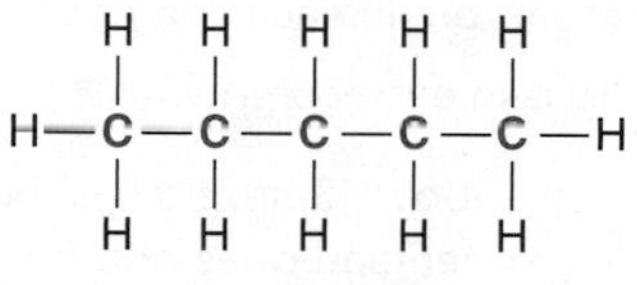

Pentane
Boiling point = 36°C

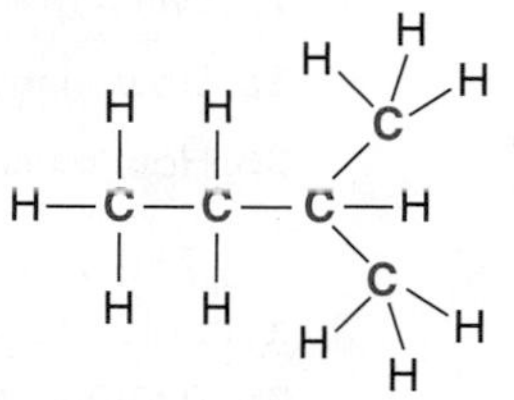

2-Methylbutane
Boiling point = 28°C

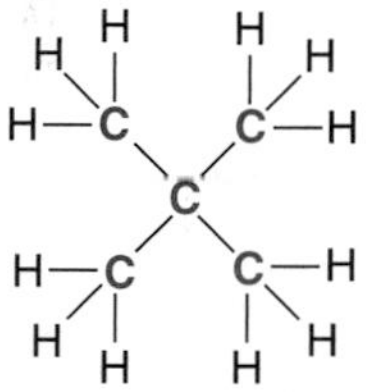

2,2-Dimethylpropane
Boiling point = 10°C

TABLE 1.6 BOILING POINTS FOR HYDROCARBONS OF INCREASING MOLECULAR WEIGHT

STRUCTURE	BOILING POINT (°C)	STRUCTURE	BOILING POINT (°C)
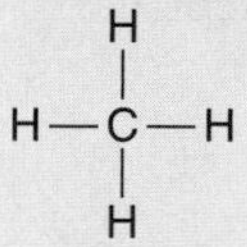	−164	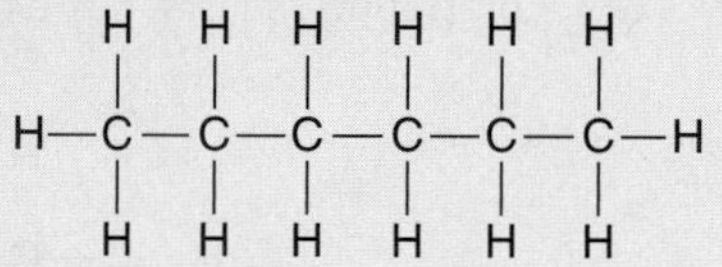	69
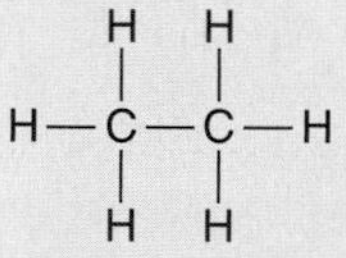	−89	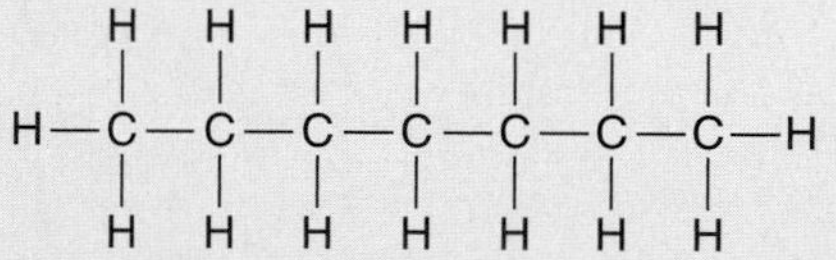	98
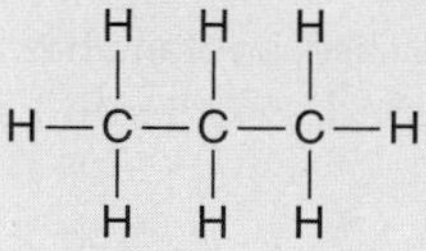	−42	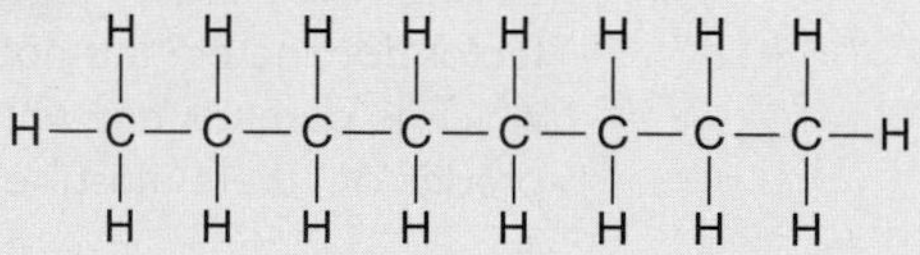	126
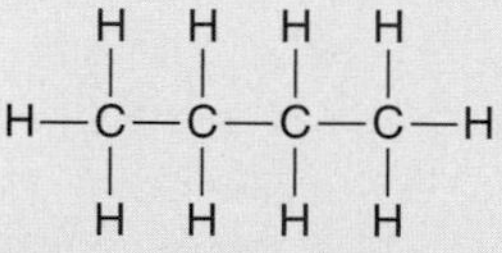	0	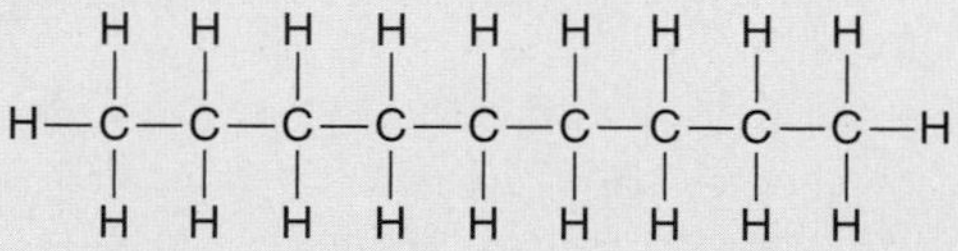	151
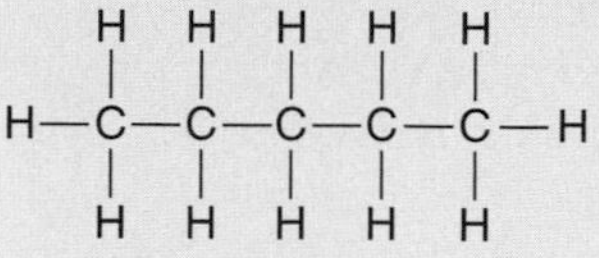	36	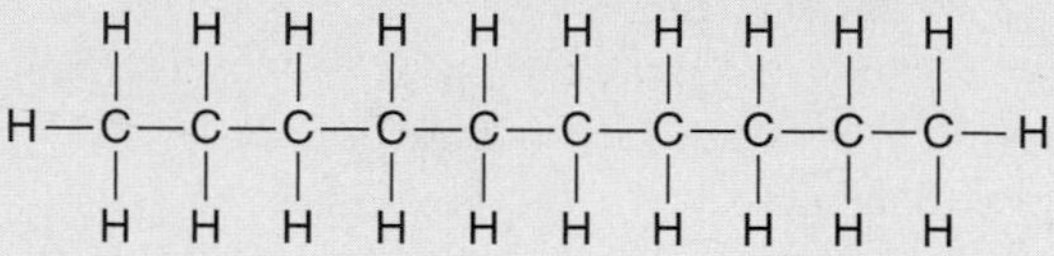	174

SKILLBUILDER

1.10 PREDICTING PHYSICAL PROPERTIES OF COMPOUNDS BASED ON THEIR MOLECULAR STRUCTURE

LEARN the skill

Determine which compound has the higher boiling point, neopentane or 3-hexanol:

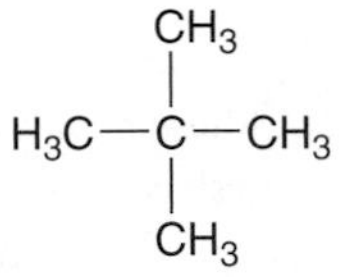

Neopentane

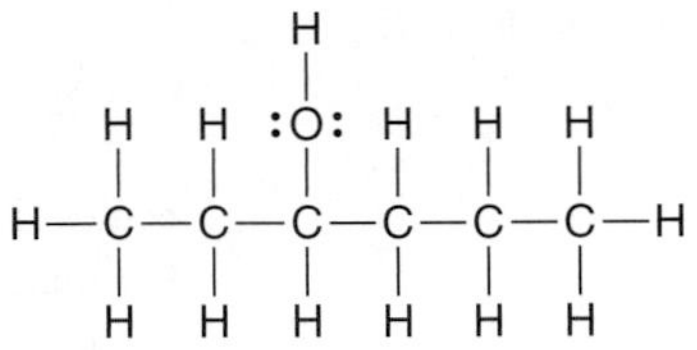

3-Hexanol

SOLUTION

When comparing boiling points of compounds, we look for the following factors:

STEP 1 Identify all dipole-dipole interactions in both compounds.

STEP 2 Identify all H-bonding interactions in both compounds.

STEP 3 Identify the number of carbon atoms and extent of branching in both compounds.

1. Are there any dipole-dipole interactions in either compound?

2. Will either compound form hydrogen bonds?

3a. How many carbon atoms are in each compound?

3b. How much branching is in each compound?

The second compound above (3-hexanol) is the winner in all of these categories. It has a dipole moment, while neopentane does not. It will experience hydrogen bonding, while neopentane will not. It has six carbon atoms, while neopentane only has five. And finally, it has a straight chain, while neopentane is highly branched. Each of these factors alone would suggest that 3-hexanol should have a higher boiling point. When we consider all of these factors together, we expect that the boiling point of 3-hexanol will be significantly higher than neopentane.

When comparing two compounds, it is important to consider all four factors. However, it is not always possible to make a clear prediction because in some cases there may be competing factors. For example, compare ethanol and heptane:

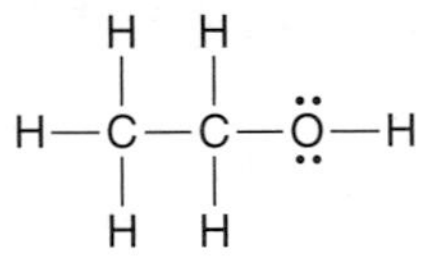

Ethanol

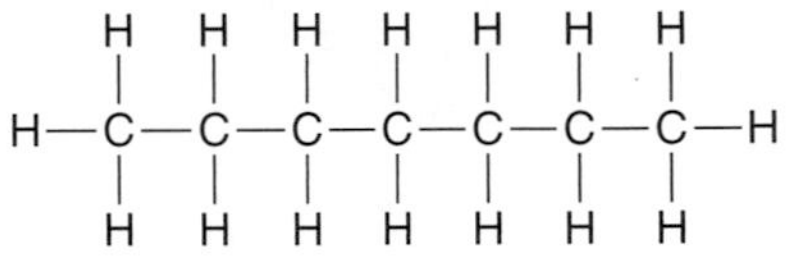

Heptane

Ethanol will exhibit hydrogen bonding, but heptane has many more carbon atoms. Which factor dominates? It is not easy to predict. In this case, heptane has the higher boiling point, which is perhaps not what we would have guessed. In order to use the trends to make a prediction, there must be a clear winner.

PRACTICE the skill **1.32** For each of the following pairs of compounds, identify the higher boiling compound and justify your choice:

(a)

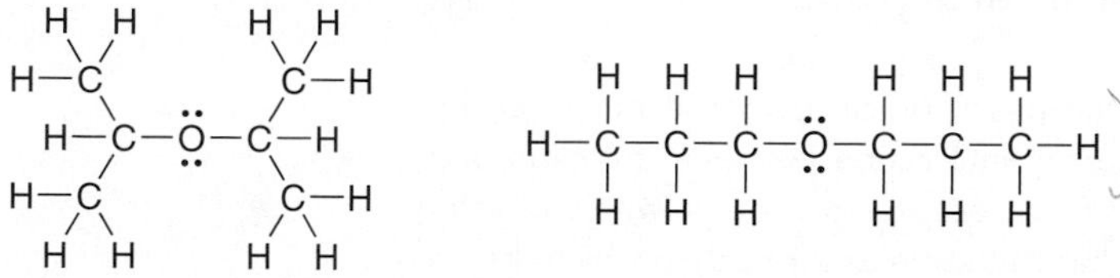

less branched

(b)

more c atoms

(c)

OH bond present

(d)

less branched

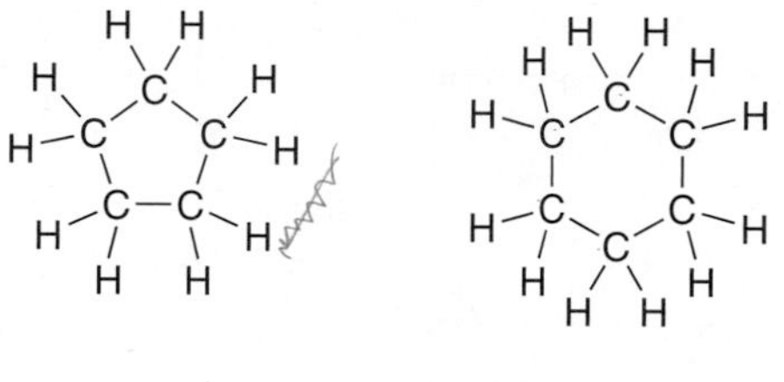

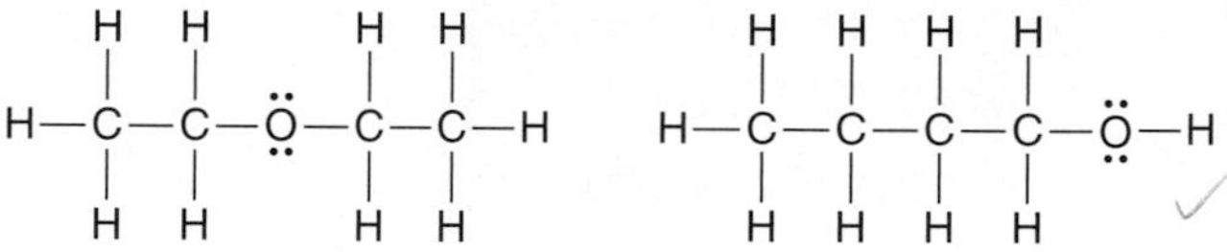

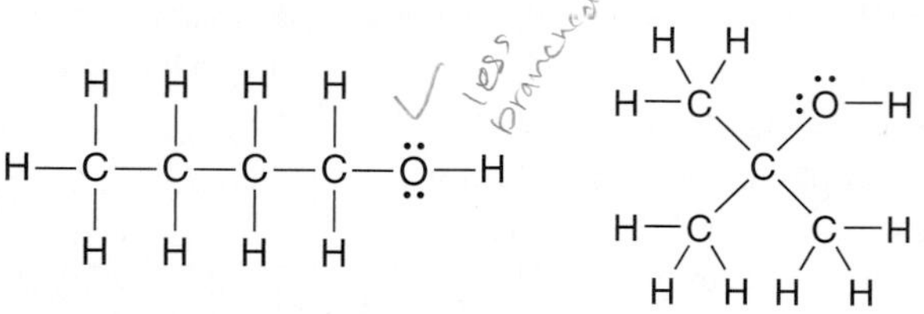

APPLY the skill **1.33** Arrange the following compounds in order of increasing boiling point:

- 2
- 3
- 5
- 1
- 4

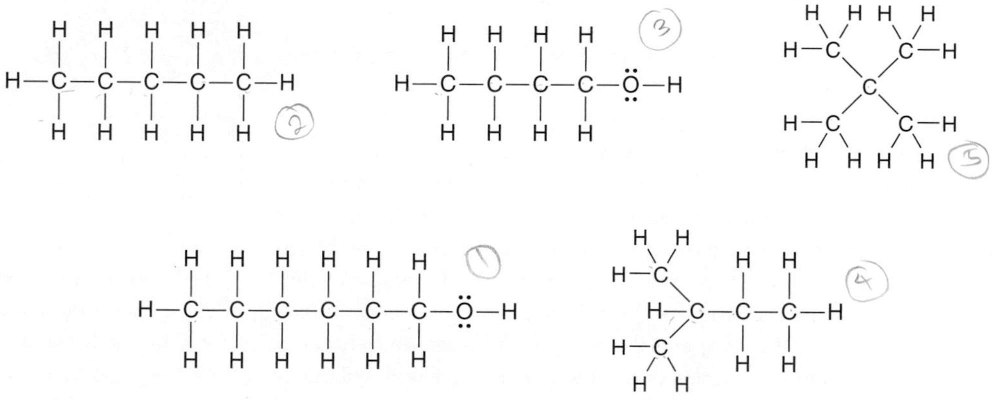

need more **PRACTICE?** Try Problems 1.52, 1.53, 1.60

MEDICALLYSPEAKING)))

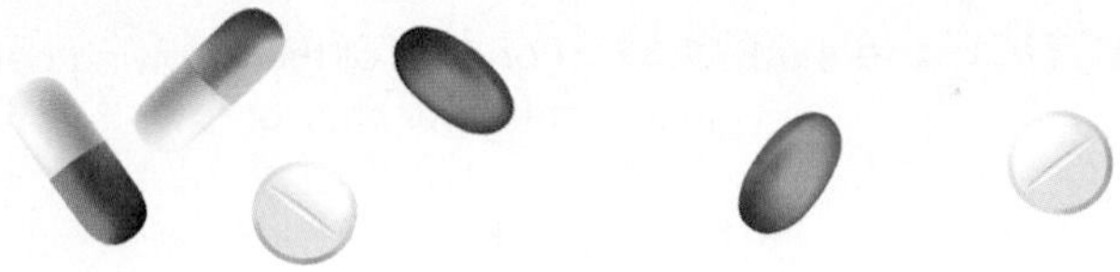

Drug-Receptor Interactions

In most situations, the physiological response produced by a drug is attributed to the interaction between the drug and a biological receptor site. A *receptor* is a region within a biological macromolecule that can serve as a pouch in which the drug molecule can fit:

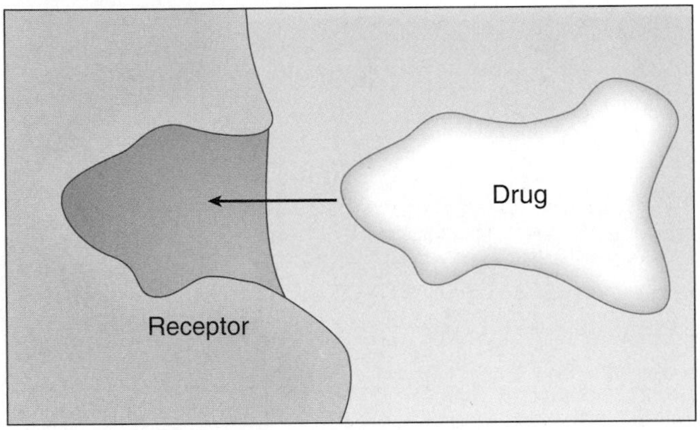

Initially, this mechanism was considered to work much like a lock and key. That is, a drug molecule would function as a key, either fitting or not fitting into a particular receptor. Extensive research on drug-receptor interactions has forced us to modify this simple lock-and-key model. It is now understood that both the drug and the receptor are flexible, constantly changing their shapes. As such, drugs can bind to receptors with various levels of efficiency, with some drugs binding more strongly and other drugs binding more weakly.

How does a drug bind to a receptor? In some cases, the drug molecule forms covalent bonds with the receptor. In such cases, the binding is indeed very strong (approximately 400 kJ/mol for each covalent bond) and therefore irreversible. We will see an example of irreversible binding when we explore a class of anticancer agents called nitrogen mustards (Chapter 7). For most drugs, however, the desired physiological response is meant to be temporary, which can only be accomplished if a drug can bind *reversibly* with its target receptor. This requires a weaker interaction between the drug and the receptor (at least weaker than a covalent bond). Examples of weak interactions include hydrogen-bonding interactions (20 kJ/mol) and London dispersion forces (approximately 4 kJ/mol for each carbon atom participating in the interaction). As an example, consider the structure of a benzene ring, which is incorporated as a structural subunit in many drugs.

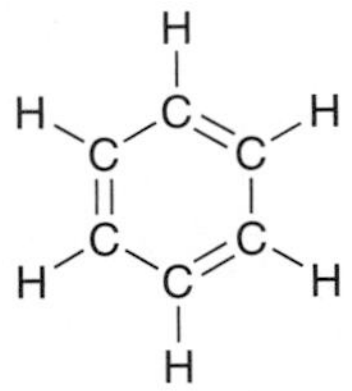

Benzene

In the benzene ring, each carbon is sp^2 hybridized and therefore trigonal planar. As a result, a benzene ring represents a flat surface:

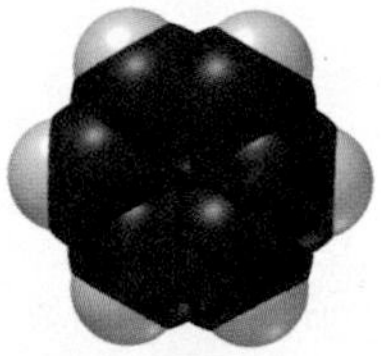

If the receptor also has a flat surface, the resulting London dispersion forces can contribute to the reversible binding of the drug to the receptor site:

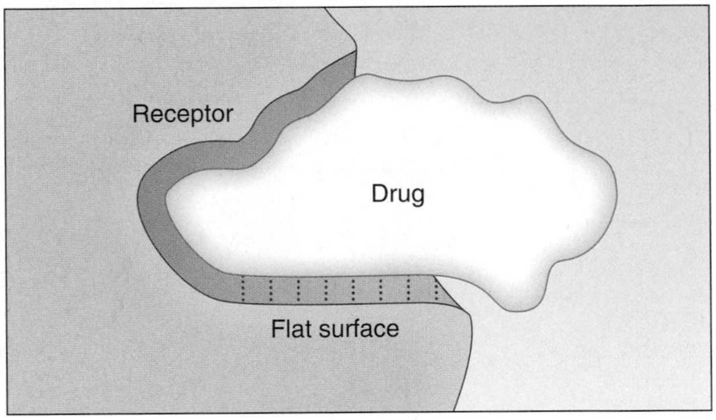

This interaction is roughly equivalent to the strength of a single hydrogen-bonding interaction. The binding of a drug to a receptor is the result of the sum of the intermolecular forces of attraction between a portion of the drug molecule and the receptor site. We will have more to say about drugs and receptors in the upcoming chapters. In particular, we will see how drugs make their journey to the receptor, and we will explore how drugs flex and bend when interacting with a receptor site.

1.13 Solubility

Solubility is based on the principle that "like dissolves like." In other words, polar compounds are soluble in polar solvents, while nonpolar compounds are soluble in nonpolar solvents. Why is this so? A polar compound experiences dipole-dipole interactions with the molecules of a polar solvent, allowing the compound to dissolve in the solvent. Similarly, a nonpolar compound experiences London dispersion forces with the molecules of a nonpolar solvent. Therefore, if an article of clothing is stained with a polar compound, the stain can generally be washed away with water (like dissolves like). However, water will be insufficient for cleaning clothing stained with nonpolar compounds, such as oil or grease. In a situation like this, the clothes can be cleaned with soap or by dry cleaning.

Soap

Soaps are compounds that have a polar group on one end of the molecule and a nonpolar group on the other end (Figure 1.53).

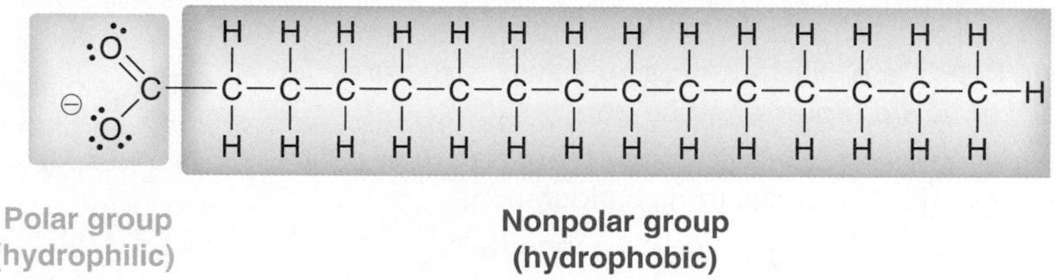

FIGURE 1.53
The hydrophilic and hydrophobic ends of a soap molecule.

The polar group represents the **hydrophilic** region of the molecule (literally, "loves water"), while the nonpolar group represents the **hydrophobic** region of the molecule (literally, "afraid of water"). Oil molecules are surrounded by the hydrophobic tails of the soap molecules, forming a **micelle** (Figure 1.54).

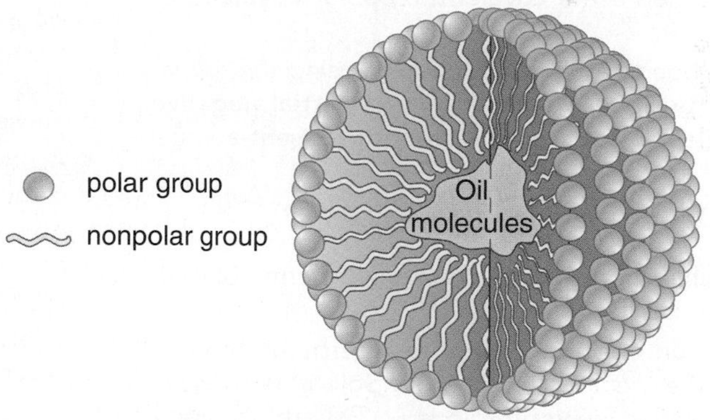

FIGURE 1.54
A micelle is formed when the hydrophobic tails of soap molecules surround the nonpolar oil molecules.

The surface of the micelle is comprised of all of the polar groups, rendering the micelle water soluble. This is a clever way to dissolve the oil in water, but this technique only works for clothing that can be subjected to water and soap. Some clothes will be damaged in soapy water, and in those situations, dry cleaning is the preferred method.

Dry Cleaning

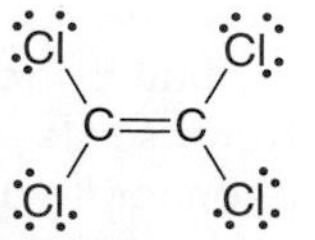

Tetrachloroethylene

Rather than surrounding the nonpolar compound with a micelle so that it will be water soluble, it is actually conceptually simpler to use a nonpolar solvent. This is just another application of the principle of "like dissolves like." Dry cleaning utilizes a nonpolar solvent, such as tetrachloroethylene, to dissolve the nonpolar compounds. This compound is nonflammable, making it an ideal choice as a solvent. Dry cleaning allows clothes to be cleaned without coming into contact with water or soap.

REVIEW OF CONCEPTS AND VOCABULARY

SECTION 1.1

- Organic compounds contain carbon atoms.

SECTION 1.2

- **Constitutional isomers** share the same molecular formula but have different connectivity of atoms and different physical properties.
- The predictable number of bonds usually formed by an atom of an element is its **valence**. Carbon is generally **tetravalent**, nitrogen **trivalent**, oxygen **divalent**, and hydrogen and the halogens **monovalent**.

SECTION 1.3

- A **covalent bond** results when two atoms share a pair of electrons.
- Covalent bonds are illustrated using **Lewis structures**, in which electrons are represented by dots.
- Second-row elements generally obey the **octet rule**, bonding to achieve noble gas electron configuration.
- A pair of unshared electrons is called a **lone pair**.

SECTION 1.4

- A **formal charge** occurs when atoms do not exhibit the appropriate number of valence electrons; formal charges must be drawn in Lewis structures.

SECTION 1.5

- Bonds are classified as (1) **covalent**, (2) **polar covalent**, or (3) **ionic.**
- Polar covalent bonds exhibit **induction**, causing the formation of **partial positive charges** ($\delta+$) and **partial negative charges** (δ-). **Electrostatic potential maps** present a visual illustration of partial charges.

SECTION 1.6

- **Quantum mechanics** describes electrons in terms of their wavelike properties.
- A **wave equation** describes the total energy of an electron when in the vicinity of a proton. Solutions to wave equations are called **wavefunctions** (ψ), where ψ^2 represents the probability of finding an electron in a particular location.
- **Atomic orbitals** are represented visually by generating three-dimensional plots of ψ^2; nodes indicate that the value of ψ is zero.
- An occupied orbital can be thought of as a cloud of **electron density**.
- Electrons fill orbitals following three principles: (1) the **Aufbau principle**, (2) the **Pauli exclusion principle**, and (3) **Hund's rule**. Orbitals with the same energy level are called degenerate orbitals.

SECTION 1.7

- **Valence bond theory** treats every bond as the sharing of electron density between two atoms as a result of the constructive interference of their atomic orbitals. **Sigma** (σ) **bonds** are formed when the electron density is located primarily on the bond axis.

SECTION 1.8

- **Molecular orbital theory** uses a mathematical method called the **linear combination of atomic orbitals** (LCAO) to form molecular orbitals. Each molecular orbital is associated with the entire molecule, rather than just two atoms.
- The bonding MO of molecular hydrogen results from constructive interference between its two atomic orbitals. The antibonding MO results from destructive interference.
- An **atomic orbital** is a region of space associated with an individual atom, while a molecular orbital is associated with an entire molecule.
- Two molecular orbitals are the most important to consider: (1) the **highest occupied molecular orbita**l, or HOMO, and (2) the **lowest unoccupied molecular orbital**, or LUMO.

SECTION 1.9

- Methane's tetrahedral geometry can be explained using four degenerate ***sp*^{3}-hybridized orbitals** to achieve its four single bonds.
- Ethylene's planar geometry can be explained using three degenerate ***sp*^{2}-hybridized orbitals**. The remaining *p* orbitals overlap to form a separate bonding interaction, called a pi (π) bond. The carbon atoms of ethylene are connected via a σ bond, resulting from the overlap of sp^2 hybridized atomic orbitals, and via a π bond, resulting from the overlap of *p* orbitals, both of which comprise the double bond of ethylene.
- Acetylene's linear geometry is achieved via ***sp*-hybridized** carbon atoms in which a triple bond is created from the bonding interactions of one σ bond, resulting from overlapping *sp* orbitals, and two π bonds, resulting from overlapping *p* orbitals.
- Triple bonds are stronger and shorter than double bonds, which are stronger and shorter than single bonds.

SECTION 1.10

- The geometry of small compounds can be predicted using valence shell electron pair repulsion (**VSEPR**) theory, which focuses on the number of σ bonds and lone pairs exhibited by each atom. The total, called the steric number, indicates the number of electron pairs that repel each other.
- A tetrahedral arrangement of orbitals indicates ***sp*^{3} hybridization** (steric number 4). A compound's geometry depends on the number of lone pairs and can be tetrahedral, trigonal pyramidal, or bent.
- A trigonal planar arrangement of orbitals indicates ***sp*^{2} hybridization** (steric number 3); however, the geometry may be bent, depending on the number of lone pairs.
- Linear geometry indicates ***sp* hybridization** (steric number 2).

SECTION 1.11

- **Dipole moments** (μ) occur when the center of negative charge and the center of positive charge are separated from one another by a certain distance; the dipole moment is used as an indicator of polarity (measured in debyes).
- The percent ionic character of a bond is determined by measuring its dipole moment. The vector sum of individual dipole moments in a compound determines the **molecular dipole moment**.

SECTION 1.12

- The physical properties of compounds are determined by intermolecular forces, the attractive forces between molecules.
- **Dipole-dipole interactions** occur between two molecules that possess permanent dipole moments. Hydrogen bonding, a special type of dipole-dipole interaction, occurs when the lone pairs of an electronegative atom interact with an electron-poor hydrogen atom. Compounds that exhibit hydrogen bonding have higher boiling points than similar compounds that lack hydrogen bonding.
- **London dispersion forces** result from the interaction between transient dipole moments and are stronger for larger alkanes due to their larger surface area and ability to accommodate more interactions.

SECTION 1.13

- Polar compounds are soluble in polar solvents; nonpolar compounds are soluble in nonpolar solvents.
- Soaps are compounds that contain both **hydrophilic** and **hydrophobic** regions. The hydrophobic tails surround nonpolar compounds, forming a water-soluble micelle.

KEY TERMINOLOGY

antibonding MO 17
atomic orbitals 13
Aufbau principle 14
bent 26
bonding MO 17
constitution 3
constitutional isomers 3
constructive interference 16
covalent bond 5
debye 30
degenerate orbitals 14
destructive interference 16
dipole-dipole interactions 34
dipole moment 30
divalent 3
electron density 13
electronegativity 9
electrostatic potential maps 12
formal charge 8
HOMO 18
Hund's rule 15
hybridized atomic orbitals 18
hydrocarbons 36
hydrogen bonding 34
hydrophilic 41
hydrophobic 41
induction 10
intermolecular forces 34
ionic bond 10
ionic character 31
Lewis structures 5
linear 28
linear combination of atomic orbitals 17
London dispersion forces 37
lone pair 7
LUMO 18
micelle 41
molecular dipole moment 31
molecular orbital theory 17
molecular orbitals 17
monovalent 3
nodes 14
octet rule 7
Pauli exclusion principle 14
pi (π) bond 21
polar covalent bond 9
protic 35
quantum mechanics 12
sigma (σ) bond 16
***sp*-hybridized orbitals** 22
***sp*^{2}-hybridized orbitals** 21
***sp*^{3}-hybridized orbitals** 19
steric number 25
tetrahedral 26
tetravalent 3
trigonal planar 27
trigonal pyramidal 26
trivalent 3
valence 3
valence bond theory 16
valence electrons 5
VSEPR theory 25
wave equation 12
wavefunction 12

SKILLBUILDER REVIEW

1.1 DETERMINING THE CONSTITUTION OF SMALL MOLECULES

STEP 1 Determine the valency of each atom in the compound.

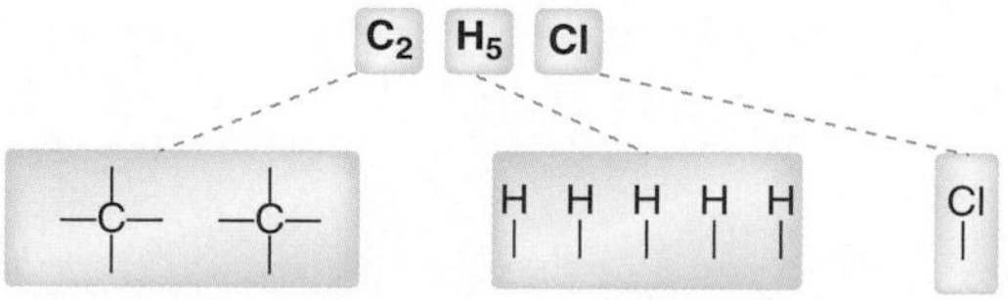

STEP 2 Determine how the atoms are connected. Atoms with highest valency should be placed at the center, and monovalent atoms should be placed at the periphery.

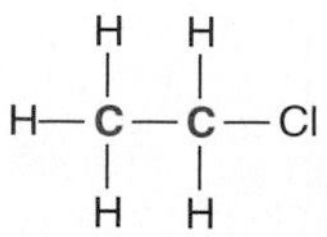

Try Problems 1.1–1.4, 1.34, 1.46, 1.47, 1.54.

1.2 DRAWING THE LEWIS DOT STRUCTURE OF AN ATOM

STEP 1 Determine the number of valence electrons.

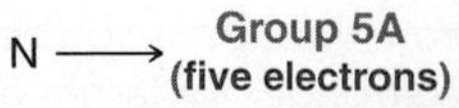

STEP 2 Place one electron by itself on each side of the atom.

STEP 3 If the atom has more than four valence electrons, the remaining electrons are paired with the electrons already drawn.

Try Problems **1.5–1.9.**

1.3 DRAWING THE LEWIS STRUCTURE OF A SMALL MOLECULE

STEP 1 Draw all individual atoms.

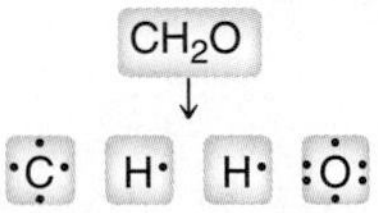

STEP 2 Connect atoms that form more than one bond.

STEP 3 Connect hydrogen atoms.

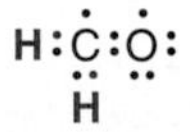

STEP 4 Pair any unpaired electrons, so that each atom achieves an octet.

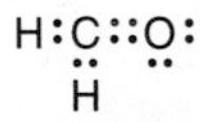

Try Problems **1.10–1.12, 1.35, 1.38, 1.42.**

1.4 CALCULATING FORMAL CHARGE

STEP 1 Determine appropriate number of valence electrons.

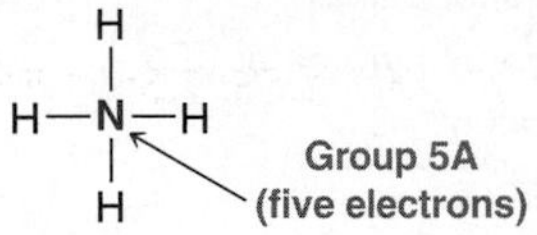

STEP 2 Determine the number of valence electrons in this case.

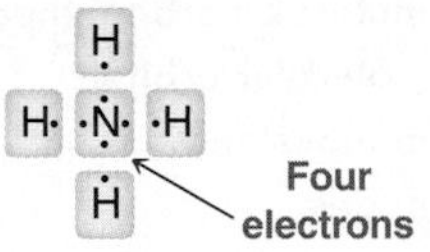

STEP 3 Assign a formal charge.

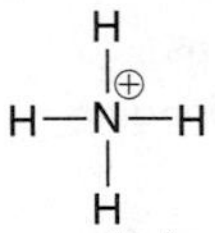

... missing an electron.

Try Problems **1.13, 1.14, 1.41.**

1.5 LOCATING PARTIAL CHARGES

STEP 1 Identify all polar covalent bonds.

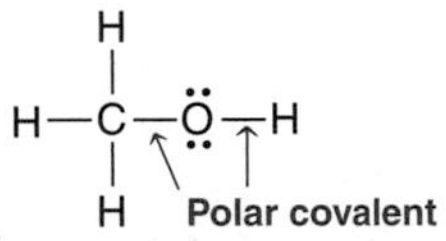

STEP 2 Determine the direction of each dipole.

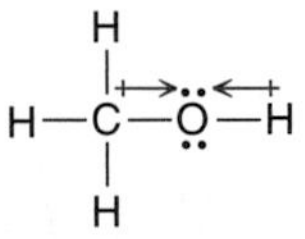

STEP 3 Indicate location of partial charges.

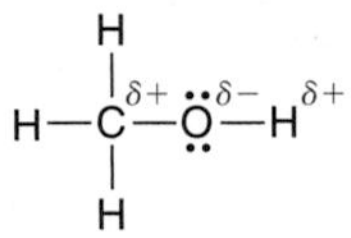

Try Problems **1.15, 1.16, 1.36, 1.37, 1.48, 1.57.**

1.6 IDENTIFYING ELECTRON CONFIGURATIONS

STEP 1 Fill orbitals using Aufbau principle, Pauli exclusion principle, and Hund's rule.

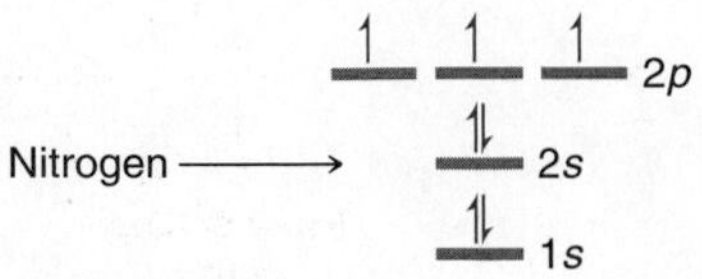

STEP 2 Summarize using the following notation:

$1s^22s^22p^3$

Try Problems **1.17, 1.18, 1.44.**

1.7 IDENTIFYING HYBRIDIZATION STATES

Four single bonds	A double bond	A triple bond
	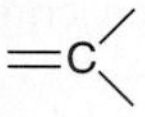	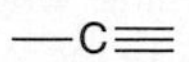
***sp*^{3}**	***sp*^{2}**	***sp***

Try Problems **1.22, 1.23, 1.55, 1.56, 1.58.**

1.8 PREDICTING GEOMETRY

STEP 1 Determine the steric number by adding the number of σ bonds and lone pairs.

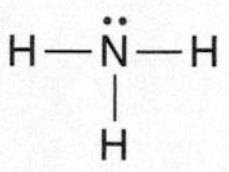

\# of σ bonds = 3

# of lone pairs = 1

***Steric number* = 4**

STEP 2 Use steric number to determine hybridization state and electronic geometry.

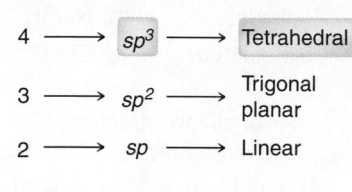

4 ⟶ *sp*^{3} ⟶ Tetrahedral

3 ⟶ *sp*^{2} ⟶ Trigonal planar

2 ⟶ *sp* ⟶ Linear

STEP 3 Ignore lone pairs and describe geometry.

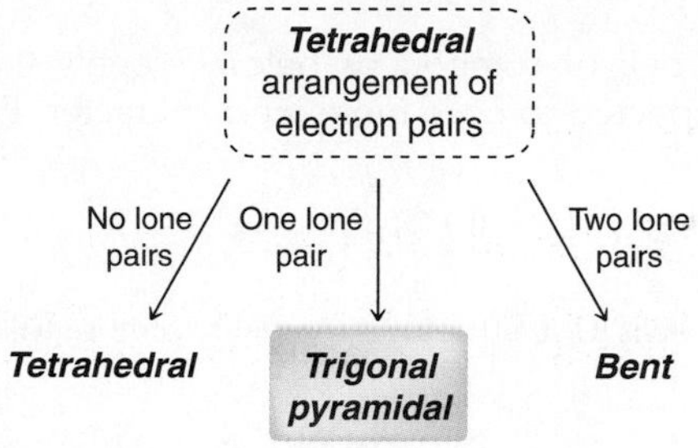

Try Problems **1.25–1.28, 1.39–1.41, 1.50, 1.55, 1.56, 1.58.**

1.9 IDENTIFYING MOLECULAR DIPOLE MOMENTS

STEP 1 Predict geometry

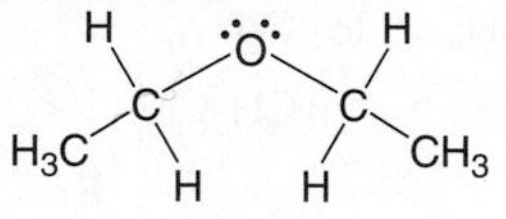

Bent

STEP 2 Identify direction of all dipole moments.

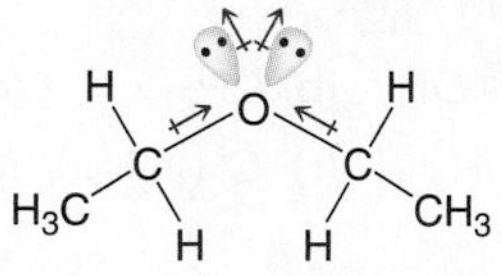

STEP 3 Draw net dipole moment.

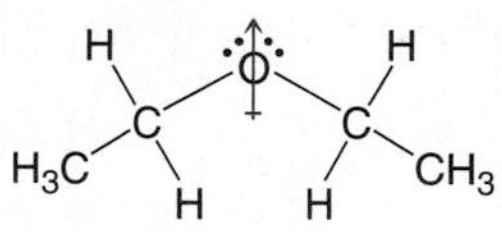

Try Problems **1.29–1.31, 1.37, 1.40, 1.43, 1.61, 1.62.**

1.10 PREDICTING PHYSICAL PROPERTIES

STEP 1 Identify dipole-dipole interactions.

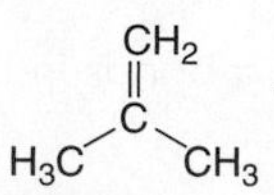 vs. 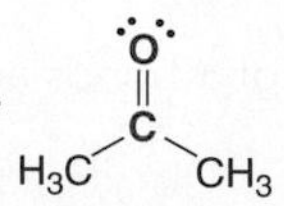

Higher boiling point

STEP 2 Identify H-bonding interactions.

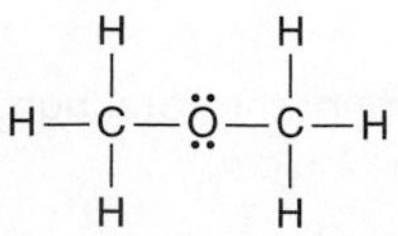 vs. 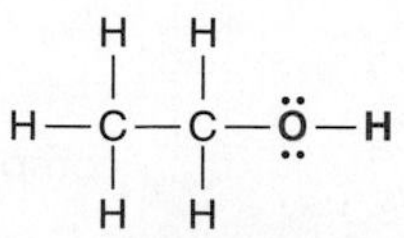

Higher boiling point

STEP 3 Identify number of carbon atoms and extent of branching.

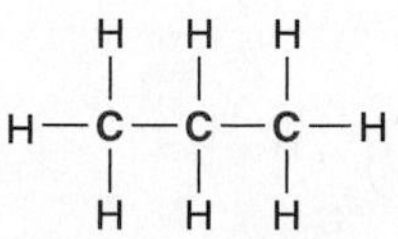 vs. 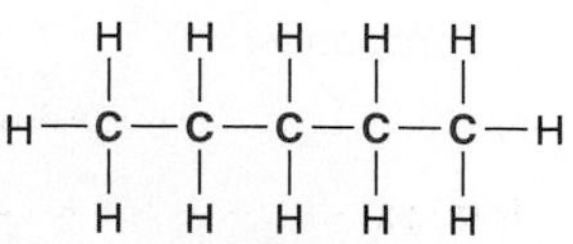

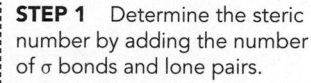

Higher boiling point

Try Problems **1.32, 1.33, 1.52, 1.53, 1.60.**

PRACTICE PROBLEMS

Note: Most of the Problems are available within WileyPLUS, an online teaching and learning solution.

1.34 Draw structures for all constitutional isomers with the following molecular formulas:

(a) C_4H_{10} (b) C_5H_{12} (c) C_6H_{14}
(d) C_2H_5Cl (e) $C_2H_4Cl_2$ (f) $C_2H_3Cl_3$

1.35 Draw structures for all constitutional isomers with the molecular formula C_4H_8 that have:

(a) Only single bonds (b) One double bond

1.36 For each compound below, identify any polar covalent bonds, and indicate the direction of the dipole moment using the symbols $\delta+$ and $\delta-$:

(a) HBr (b) HCl (c) H_2O (d) CH_4O

1.37 For each pair of compounds below, identify the one that would be expected to have more ionic character. Explain your choice.

(a) NaBr or HBr (b) BrCl or FCl

1.38 Draw a Lewis dot structure for each of the following compounds:

(a) CH_3CH_2OH (b) CH_3CN

1.39 Predict the geometry of each atom except hydrogen in the compounds below:

(a) 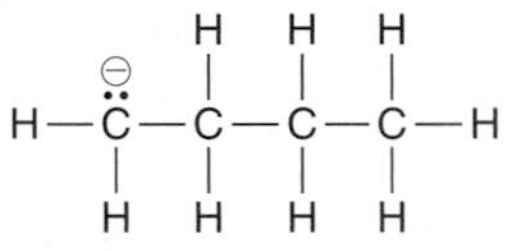

(b) 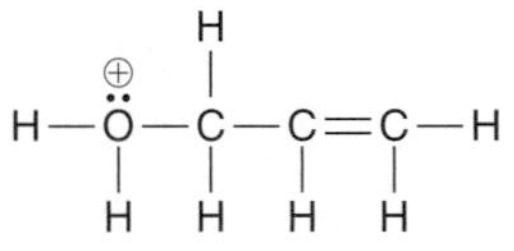

(c) 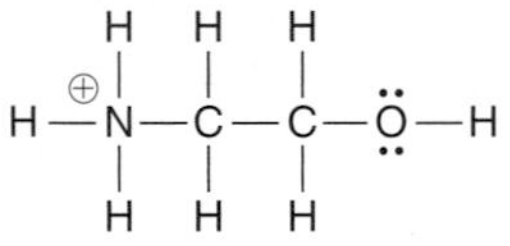

(d) 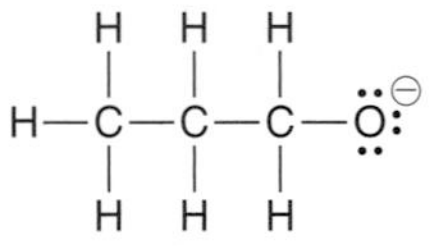

1.40 Draw a Lewis structure for a compound with molecular formula $C_4H_{11}N$ in which three of the carbon atoms are bonded to the nitrogen atom. What is the geometry of the nitrogen atom in this compound? Does this compound exhibit a molecular dipole moment? If so, indicate the direction of the dipole moment.

1.41 Draw a Lewis structure of the anion $AlBr_4^-$ and determine its geometry.

1.42 Draw the structure for the only constitutional isomer of cyclopropane:

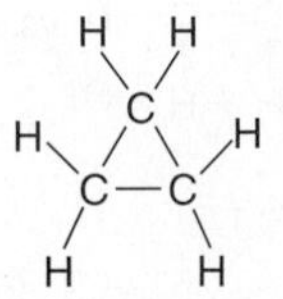

Cyclopropane

1.43 Determine whether each compound below exhibits a molecular dipole moment:

(a) CH_4 (b) NH_3 (c) H_2O
(d) CO_2 (e) CCl_4 (f) CH_2Br_2

1.44 Identify the neutral element that corresponds with each of the following electronic configurations:

(a) $1s^22s^22p^4$ (b) $1s^22s^22p^5$ (c) $1s^22s^22p^2$
(d) $1s^22s^22p^3$ (e) $1s^22s^22p^63s^23p^5$

1.45 In the compounds below, classify each bond as covalent, polar covalent, or ionic:

(a) NaBr (b) NaOH (c) $NaOCH_3$
(d) CH_3OH (e) CH_2O

1.46 Draw structures for all constitutional isomers with the following molecular formulas:

(a) C_2H_6O (b) $C_2H_6O_2$ (b) $C_2H_4Br_2$

1.47 Draw structures for any five constitutional isomers with molecular formula $C_2H_6O_3$.

1.48 For each type of bond below, determine the direction of the expected dipole moment.

(a) C—O (b) C—Mg (c) C—N (d) C—Li
(e) C—Cl (f) C—H (g) O—H (h) N—H

1.49 Predict the bond angles for all bonds in the following compounds:

(a) CH_3CH_2OH (b) CH_2O (c) C_2H_4 (d) C_2H_2
(e) CH_3OCH_3 (f) CH_3NH_2 (g) C_3H_8 (h) CH_3CN

1.50 Identify the expected hybridization state and geometry for the central atom in each of the following compounds:

(a) 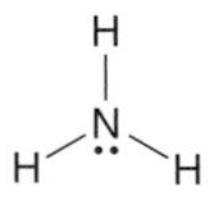

(b) 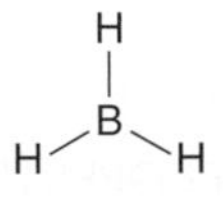

(c) 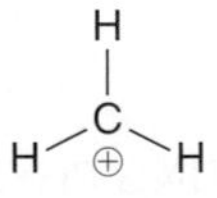

(d) 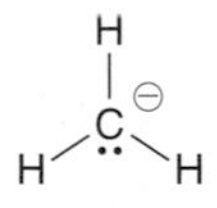

(e) 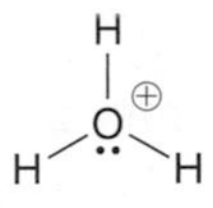

1.51 Count the total number of σ bonds and π bonds in the compound below:

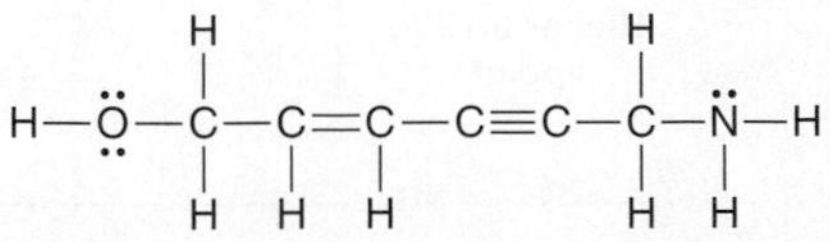

1.52 For each pair of compounds below, predict which compound will have the higher boiling point and explain your choice:

(a) $CH_3CH_2CH_2OCH_3$ or $CH_3CH_2CH_2CH_2OH$

(b) $CH_3CH_2CH_2CH_3$ or $CH_3CH_2CH_2CH_2CH_3$

(c) 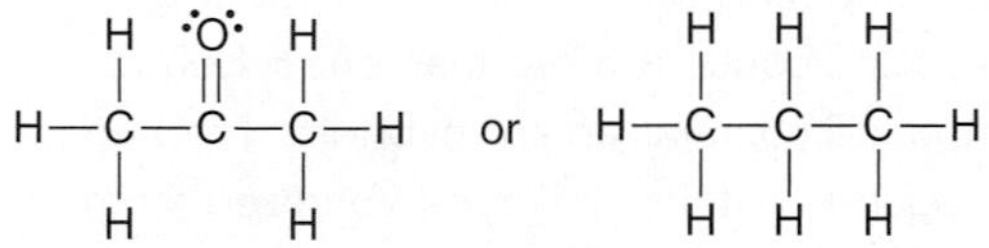

1.53 Which of the following pure compounds will exhibit hydrogen bonding?

(a) CH_3CH_2OH (b) CH_2O (c) C_2H_4 (d) C_2H_2
(e) CH_3OCH_3 (f) CH_3NH_2 (g) C_3H_8 (h) NH_3

1.54 For each case below, identify the most likely value for *x*:

(a) BH_x (b) CH_x (c) NH_x (d) CH_2Cl_x

1.55 Identify the hybridization state and geometry of each carbon atom in the following compounds:

(a) 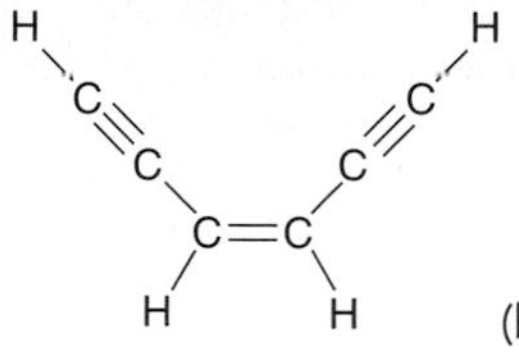

(b) 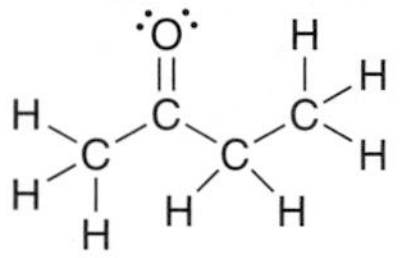

(c) 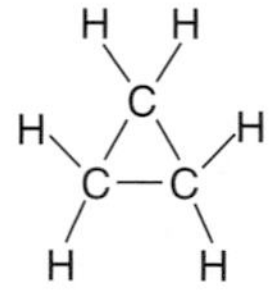

1.56 Ambien™ is a sedative used in the treatment of insomnia. It was discovered in 1982 and brought to market in 1992 (it takes a long time for new drugs to undergo the extensive testing required to receive approval from the Food and Drug Administration). Identify the hybridization state and geometry of each carbon atom in the structure of this compound:

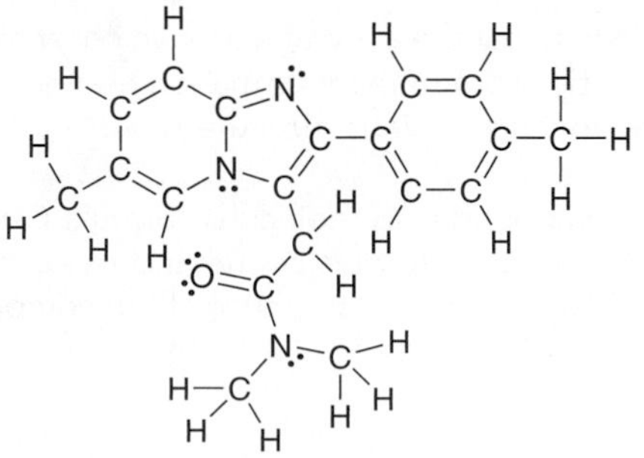

Zolpidem
(Ambien™)

1.57 Identify the most electronegative element in each of the following compounds:

(a) $CH_3OCH_2CH_2NH_2$ (b) CH_2ClCH_2F (c) CH_3Li

1.58 Nicotine is an addictive substance found in tobacco. Identify the hybridization state and geometry of each of the nitrogen atoms in nicotine:

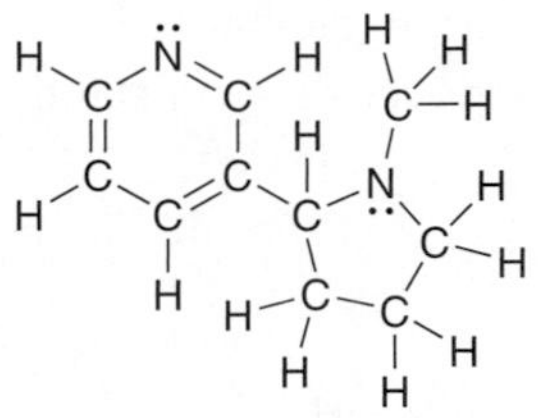

Nicotine

1.59 Below is the structure of caffeine, but its lone pairs are not shown. Identify the location of all lone pairs in this compound:

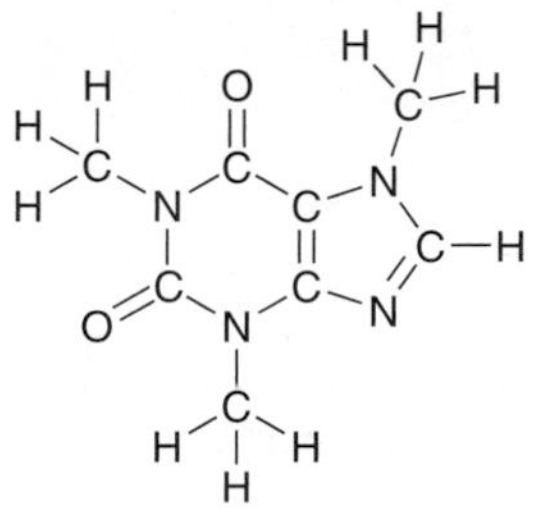

Caffeine

1.60 There are two different compounds with molecular formula C_2H_6O. One of these isomers has a much higher boiling point than the other. Explain why.

1.61 Identify which compounds below possess a molecular dipole moment and indicate the direction of that dipole moment:

(a) 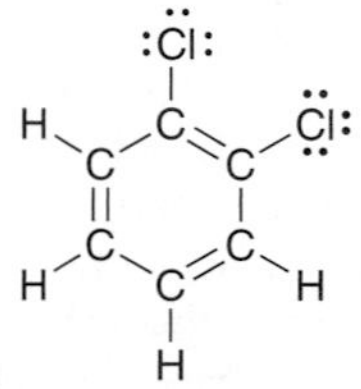

(b) 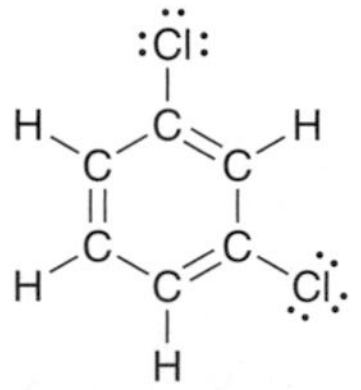

(c) 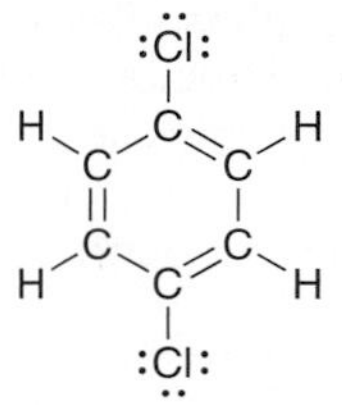

(d) 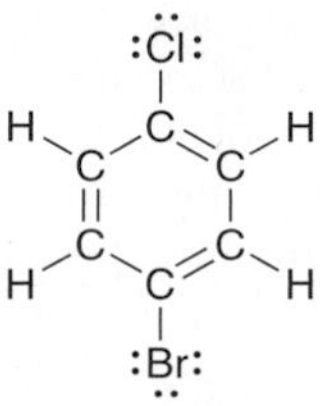

1.62 Methylene chloride (CH_2Cl_2) has fewer chlorine atoms than chloroform ($CHCl_3$). Nevertheless, methylene chloride has a larger molecular dipole moment than chloroform. Explain.

INTEGRATED PROBLEMS

1.63 Consider the three compounds shown below and then answer the questions that follow:

(a) Which two compounds are constitutional isomers?

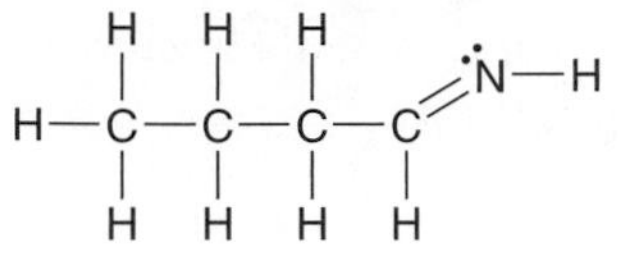

Compound A

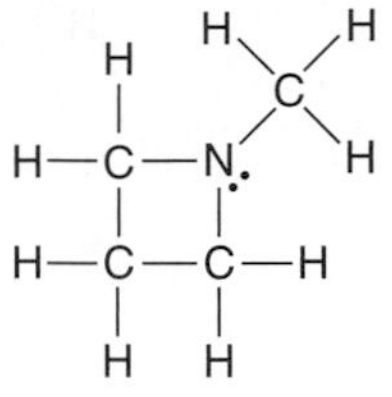

Compound B

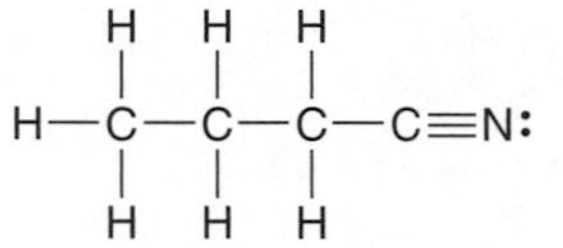

Compound C

(b) Which compound contains a nitrogen atom with trigonal pyramidal geometry?

(c) Identify the compound with the greatest number of σ bonds.

(d) Identify the compound with the fewest number of σ bonds.

(e) Which compound contains more than one π bond?

(f) Which compound contains an sp^2-hybridized carbon atom?

(g) Which compound contains only sp^3-hybridized atoms (in addition to hydrogen atoms)?

(h) Which compound do you predict will have the highest boiling point? Explain.

1.64 Propose at least two different structures for a compound with six carbon atoms that exhibits the following features:

(a) All six carbon atoms are sp^2 hybridized.

(b) Only one carbon atom is *sp* hybridized, and the remaining five carbon atoms are all sp^3 hybridized (remember that your compound can have elements other than carbon and hydrogen).

(c) There is a ring, and all of the carbon atoms are sp^3 hybridized.

(d) All six carbon atoms are *sp* hybridized, and the compound contains no hydrogen atoms (remember that a triple bond is linear and therefore cannot be incorporated into a ring of six carbon atoms).

CHALLENGE PROBLEMS

1.65 Draw all constitutional isomers with molecular formula C_5H_{10} that possess one π bond.

1.66 With current spectroscopic techniques (discussed in Chapters 15–17), chemists are generally able to determine the structure of an unknown organic compound in just one day. These techniques have only been available for the last several decades. In the first half of the twentieth century, structure determination was a very slow and painful process in which the compound under investigation would be subjected to a variety of chemical reactions. The results of those reactions would provide chemists with clues about the structure of the compound. With enough clues, it was sometimes (but not always) possible to determine the structure. As an example, try to determine the structure of an unknown compound, using the following clues:

- The molecular formula is $C_4H_{10}N_2$.
- There are no π bonds in the structure.
- The compound has no net dipole moment.
- The compound exhibits very strong hydrogen bonding.

You should find that there are at least two constitutional isomers that are consistent with the information above. (*Hint*: Consider incorporating a ring in your structure.)

1.67 A compound with molecular formula $C_5H_{11}N$ has no π bonds. Every carbon atom is connected to exactly two hydrogen atoms. Determine the structure of the compound.

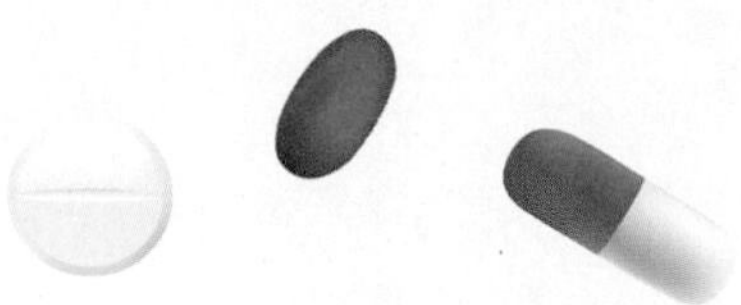

2

Molecular Representations

DID YOU EVER WONDER...
how new drugs are designed?

Scientists employ many techniques in the design of new drugs. One such technique, called lead modification, enables scientists to identify the portion of a compound responsible for its medicinal properties and then to design similar compounds with better properties. We will see an example of this technique, specifically, where the discovery of morphine led to the development of a whole family of potent analgesics (codeine, heroin, methadone, and many others).

In order to compare the structures of the compounds being discussed, we will need a more efficient way to draw the structures of organic compounds. Lewis structures are only efficient for small molecules, such as those we considered in the previous chapter. The goal of this chapter is to master the skills necessary to use and interpret the drawing method most often utilized by organic chemists and biochemists. These drawings, called bond-line structures, are fast to draw and easy to read, and they focus our attention on the reactive centers in a compound. In the second half of this chapter, we will see that bond-line structures are inadequate in some circumstances, and we will explore the technique that chemists employ to deal with the inadequacy of bond-line structures.

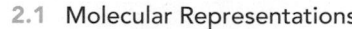

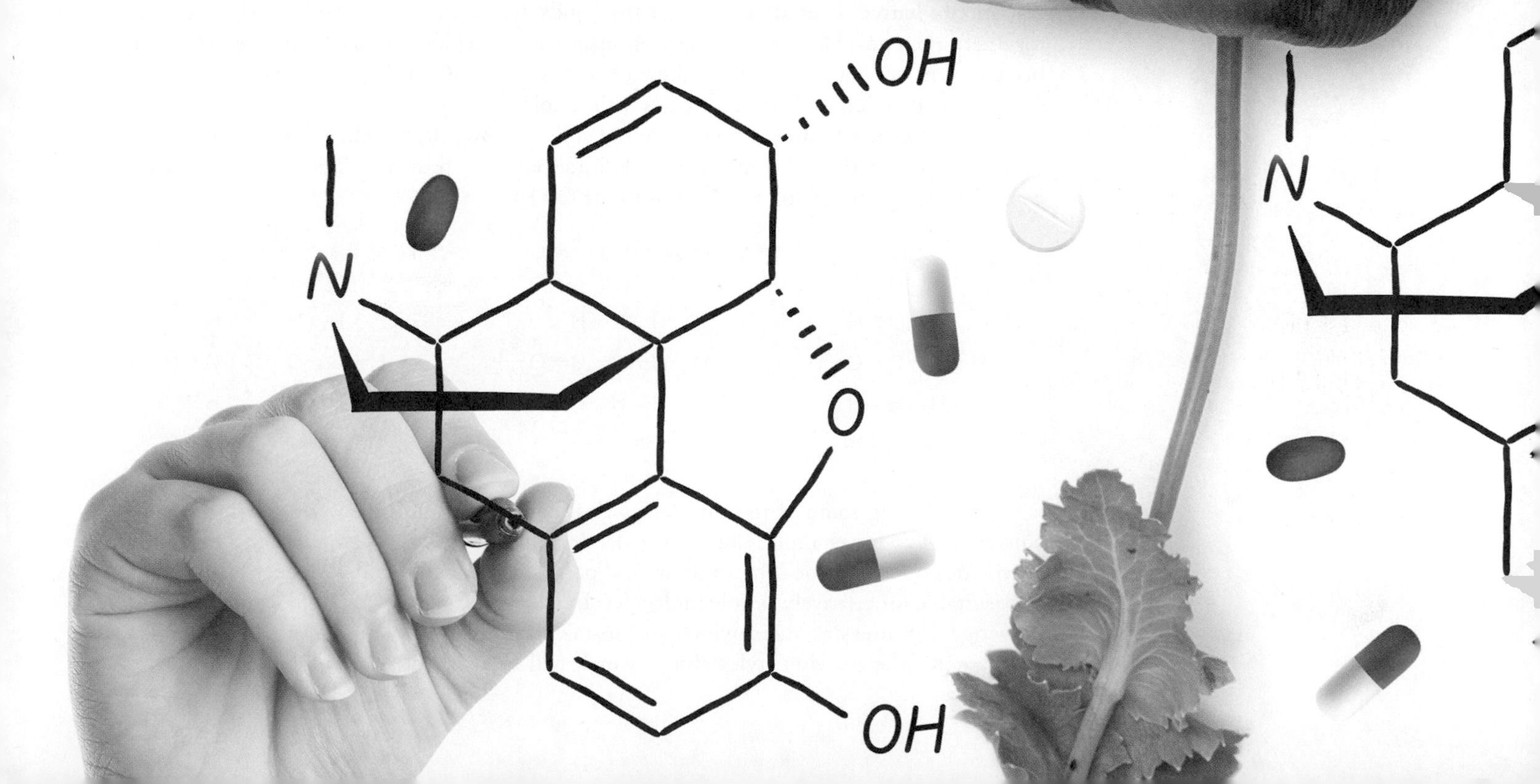

DO YOU REMEMBER?

Before you go on, be sure you understand the following topics.
If necessary, review the suggested sections to prepare for this chapter:

- Electrons, Bonds, and Lewis Structures (Section 1.3)
- Identifying Formal Charges (Section 1.4)
- Molecular Orbital Theory (Section 1.8)

WILEY PLUS Visit www.wileyplus.com to check your understanding and for valuable practice.

2.1 Molecular Representations

Chemists use many different styles to draw molecules. Let's consider the structure of isopropanol, also called isopropyl rubbing alcohol, which is used as a disinfectant in sterilizing pads. The structure of this compound is shown below in a variety of drawing styles:

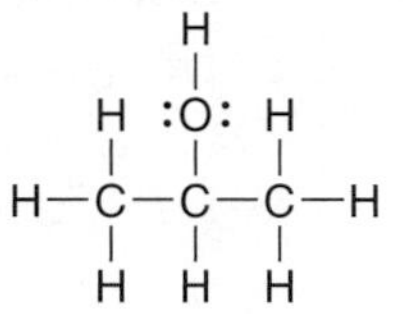

Lewis structure

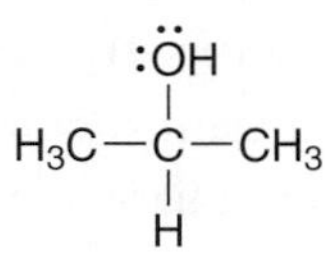

Partially condensed structure

$(CH_3)_2CHOH$

Condensed structure

C_3H_8O

Molecular formula

Lewis structures were discussed in the previous chapter. The advantage of Lewis structures is that all atoms and bonds are explicitly drawn. However, Lewis structures are only practical for very small molecules. For larger molecules, it becomes extremely burdensome to draw out every bond and every atom.

In **partially condensed structures**, the C—H bonds are not all drawn explicitly. For example, in the drawing above, CH_3 refers to a carbon atom with bonds to three hydrogen atoms. Once again, this drawing style is only practical for small molecules.

In **condensed structures**, none of the bonds are drawn. Groups of atoms are clustered together, when possible. For example, isopropanol has two CH_3 groups, both of which are connected to the central carbon atom, shown like this: $(CH_3)_2CHOH$. Once again, this drawing style is only practical for small molecules with simple structures.

The molecular formula of a compound simply shows the number of each type of atom in the compound (C_3H_8O). No structural information is provided. There are actually three constitutional isomers with molecular formula C_3H_8O:

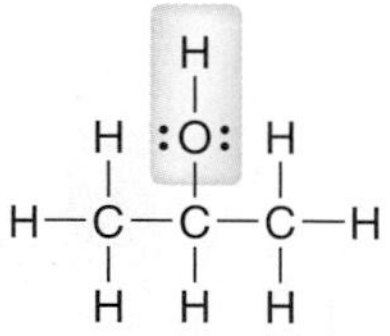

Isopropanol

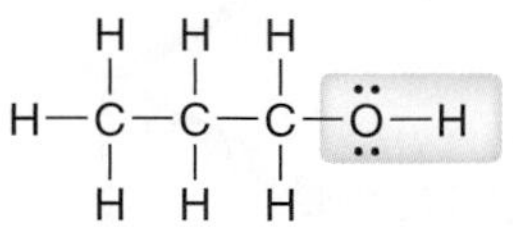

Propanol

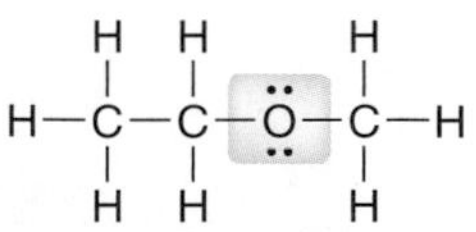

Ethyl methyl ether

In reviewing some of the different styles for drawing molecules, we see that none are convenient for larger molecules. Molecular formulas do not provide enough information, Lewis structures take too long to draw, and partially condensed and condensed drawings are only suitable for relatively simple molecules. In upcoming sections, we will learn the rules for drawing bond-line structures, which are most commonly used by organic chemists. For now, let's practice the drawing styles above, which will be used for small molecules throughout the course.

SKILLBUILDER

2.1 CONVERTING BETWEEN DIFFERENT DRAWING STYLES

LEARN the skill

Draw a Lewis structure for the following compound:

$$(CH_3)_2CH\ddot{\underset{..}{O}}CH_2CH_3$$

SOLUTION

This compound is shown in condensed format. In order to draw a Lewis structure, begin by drawing out each group separately, showing a partially condensed structure:

STEP 1
Draw each group separately.

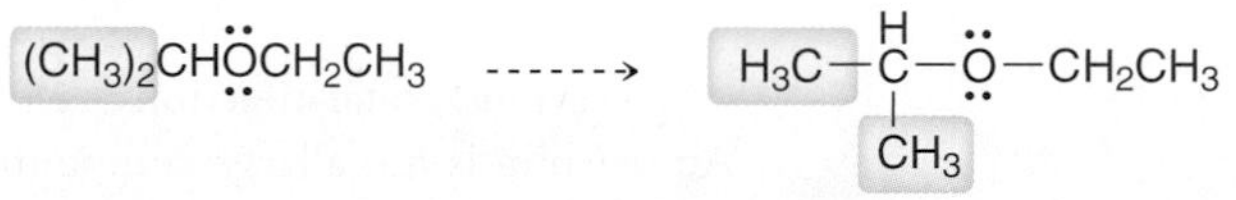

Condensed structure **Partially condensed structure**

Then draw all C—H bonds:

STEP 2
Draw all C—H bonds.

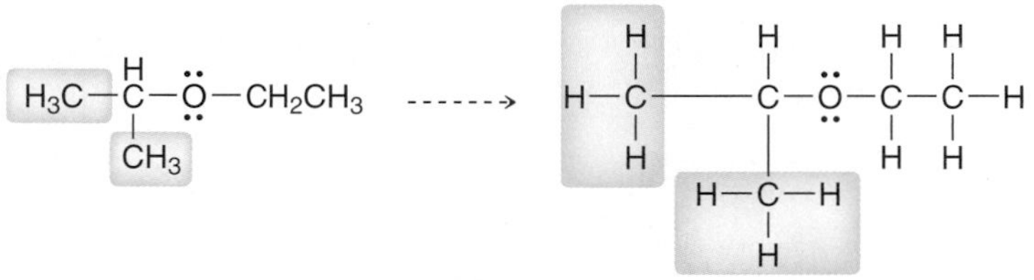

Partially condensed structure **Lewis structure**

PRACTICE the skill

2.1 Draw a Lewis structure for each of the compounds below:

(a) $CH_2{=}CHOCH_2CH(CH_3)_2$
(b) $(CH_3CH_2)_2CHCH_2CH_2OH$
(c) $(CH_3CH_2)_3COH$
(d) $(CH_3)_2C{=}CHCH_2CH_3$
(e) $CH_2{=}CHCH_2OCH_2CH(CH_3)_2$
(f) $(CH_3CH_2)_2C{=}CH_2$
(g) $(CH_3)_3CCH_2CH_2OH$
(h) $CH_3CH_2CH_2CH_2CH_2CH_3$
(i) $CH_3CH_2CH_2OCH_3$
(j) $(CH_3CH_2CH_2)_2CHOH$
(k) $(CH_3CH_2)_2CHCH_2OCH_3$
(l) $(CH_3)_2CHCH_2OH$

APPLY the skill

2.2 Identify which two compounds below are constitutional isomers:

$$(CH_3)_3C\ddot{\underset{..}{O}}CH_3 \qquad (CH_3)_2CH\ddot{\underset{..}{O}}CH_3 \qquad (CH_3)_2CH\ddot{\underset{..}{O}}CH_2CH_3$$

2.3 Identify the number of sp^3-hybridized carbon atoms in the following compound:

$$(CH_3)_2C{=}CHC(CH_3)_3$$

2.4 Cyclopropane (shown below) has only one constitutional isomer. Draw a condensed structure of that isomer.

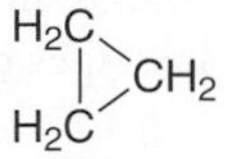

need more **PRACTICE?** **Try Problems 2.49, 2.50**

2.2 Bond-Line Structures

It is not practical to draw Lewis structures for all compounds, especially large ones. As an example, consider the structure of amoxicillin, one of the most commonly used antibiotics in the penicillin family:

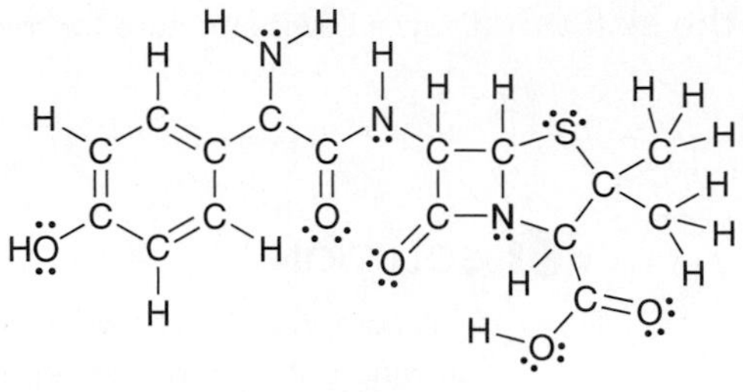

Amoxicillin

Previously fatal infections have been rendered harmless by antibiotics such as the one above. Amoxicillin is not a large compound, yet drawing this compound is time consuming. To deal with this problem, organic chemists have developed an efficient drawing style that can be used to draw molecules very quickly. **Bond-line structures** not only simplify the drawing process but also are easier to read. The bond-line structure for amoxicillin is

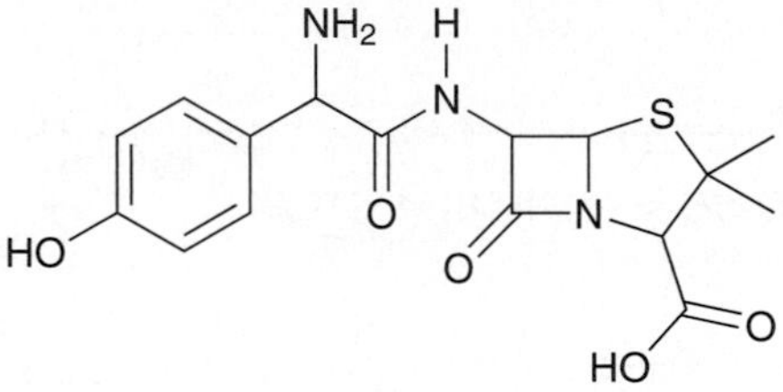

Most of the atoms are not drawn, but with practice, these drawings will become very user-friendly. Throughout the rest of this textbook, most compounds will be drawn in bond-line format, and therefore, it is absolutely critical to master this drawing technique. The following sections are designed to develop this mastery.

BY THE WAY
You may find it worthwhile to purchase or borrow a molecular model set. There are several different kinds of molecular model sets on the market, and most of them are comprised of plastic pieces that can be connected to generate models of small molecules. Any one of these model sets will help you to visualize the relationship between molecular structures and the drawings used to represent them.

How to Read Bond-Line Structures

Bond-line structures are drawn in a zigzag format (), where each corner or endpoint represents a carbon atom. For example, each of the following compounds has six carbon atoms (count them!):

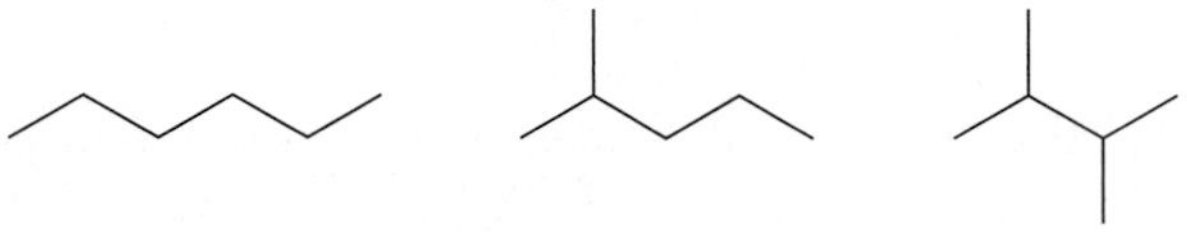

Double bonds are shown with two lines, and triple bonds are shown with three lines:

Notice that triple bonds are drawn in a linear fashion rather than in a zigzag format, because triple bonds involve *sp*-hybridized carbon atoms, which have linear geometry (Section 1.9). The two carbon atoms of a triple bond and the two carbon atoms connected to them are drawn in a straight line. All other bonds are drawn in a zigzag format; for example, the following compound has eight carbon atoms.

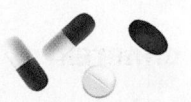

Hydrogen atoms bonded to carbon are also not shown in bond-line structures, because it is assumed that each carbon atom will possess enough hydrogen atoms so as to achieve a total of four bonds. For example, the highlighted carbon atom below appears to have only two bonds:

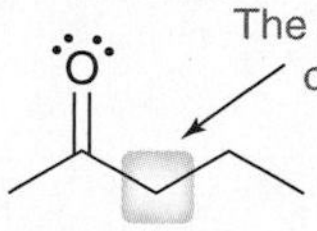
The drawing indicates only two bonds connected to this carbon atom

Therefore, we can infer that there must be two more bonds to hydrogen atoms that have not been drawn (to give a total of four bonds). In this way, all hydrogen atoms are inferred by the drawing:

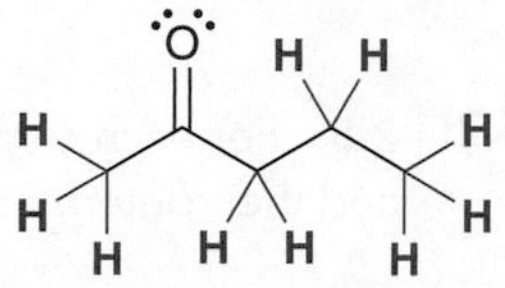

With a bit of practice, it will no longer be necessary to count bonds. Familiarity with bond-line structures will allow you to "see" all of the hydrogen atoms even though they are not drawn. This level of familiarity is absolutely essential, so let's get some practice.

SKILLBUILDER

2.2 READING BOND-LINE STRUCTURES

LEARN the skill

Consider the structure of diazepam, first marketed by the Hoffmann-La Roche Company under the trade name Valium:

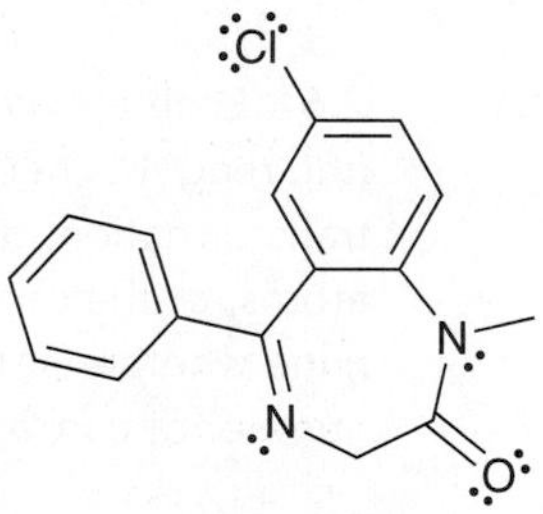

Diazepam
(Valium)

Valium is a sedative and muscle relaxant used in the treatment of anxiety, insomnia, and seizures. Identify the number of carbon atoms in diazepam, then fill in all the missing hydrogen atoms that are inferred by the drawing.

SOLUTION

Remember each corner and each endpoint represents a carbon atom. This compound therefore has 16 carbon atoms, highlighted below:

STEP 1
Count the carbon atoms, which are represented by corners or endpoints.

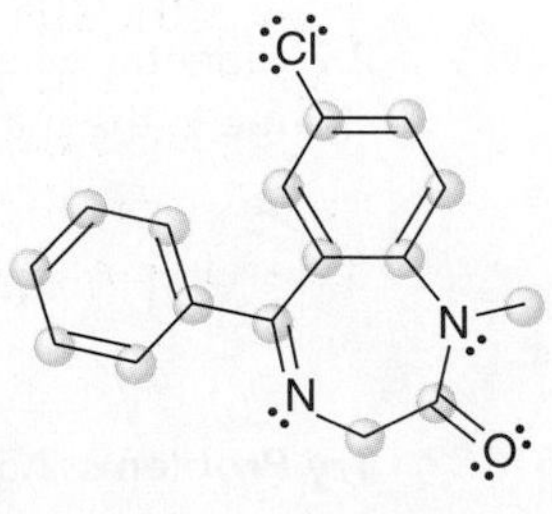

Each carbon atom should have four bonds. We therefore draw enough hydrogen atoms in order to give each carbon atom a total of four bonds. Any carbon atoms that already have four bonds will not have any hydrogen atoms:

STEP 2
Count the hydrogen atoms. Each carbon atom will have enough hydrogen atoms to have exactly four bonds.

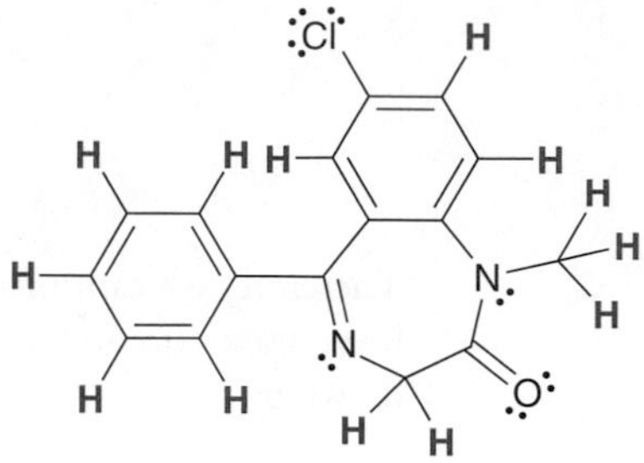

PRACTICE the skill **2.5** For each of the following molecules, determine the number of carbon atoms present, and then determine the number of hydrogen atoms connected to each carbon atom:

(a) 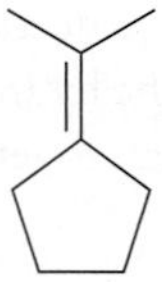 (b) 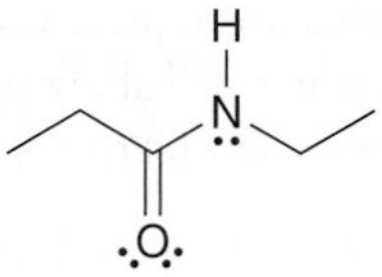 (c)  (d) 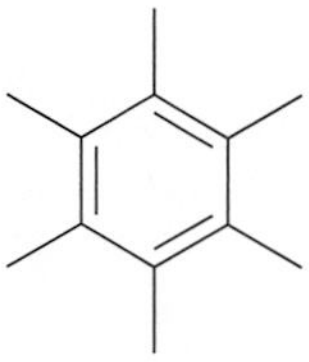

(e) 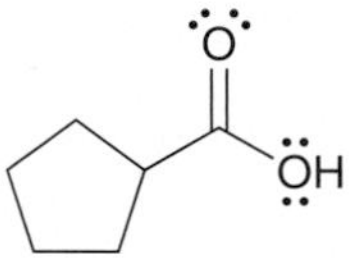 (f) 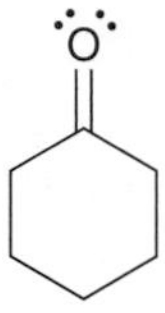 (g) 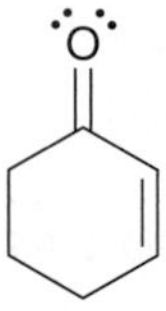 (h) 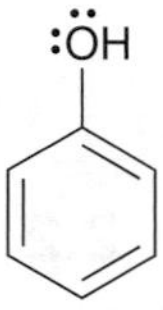

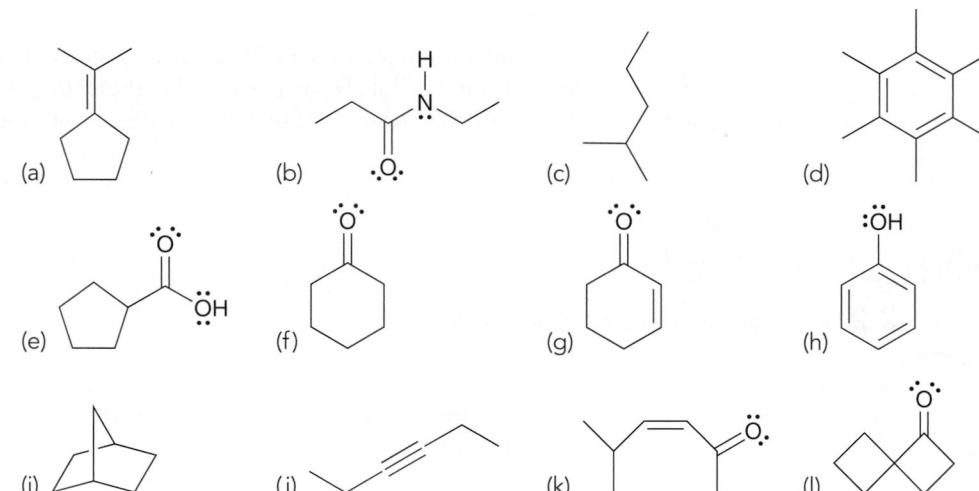

(i)  (j) 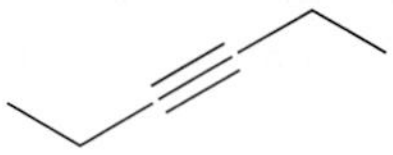 (k) 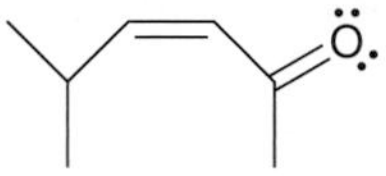 (l) 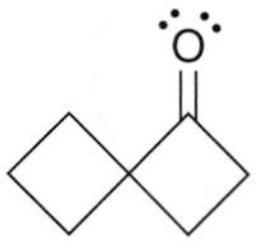

APPLY the skill **2.6** Each transformation below shows a starting material being converted into a product (the reagents necessary to achieve the transformation have not been shown). For each transformation, determine whether the product has more carbon atoms, fewer carbon atoms, or the same number of carbon atoms as the starting material. In other words, determine whether each transformation involves an increase, decrease, or no change in the number of carbon atoms.

(a) 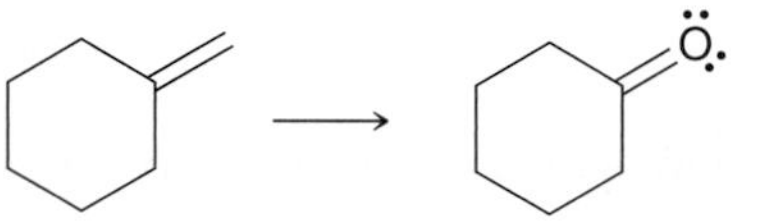 (b) 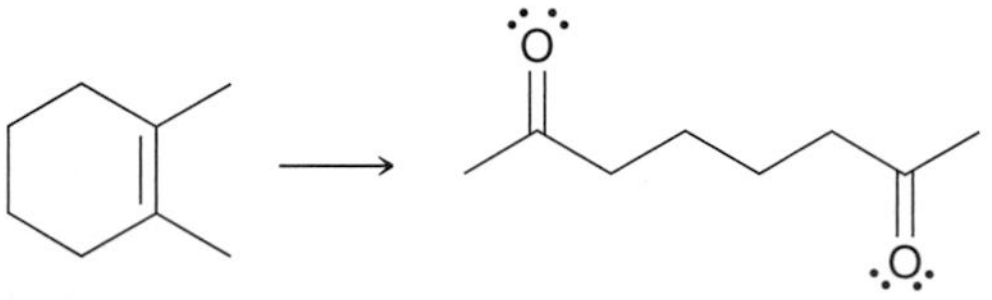

(c) 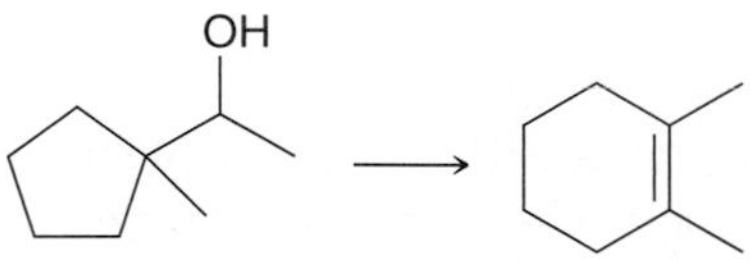 (d) 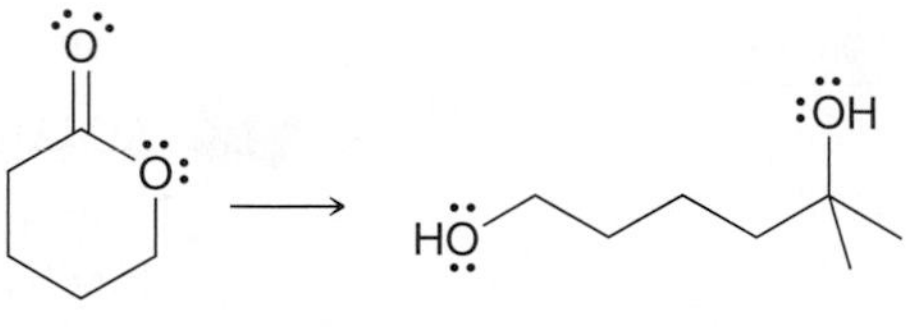

2.7 Identify whether each transformation below involves an increase, a decrease, or no change in the number of hydrogen atoms:

(a) 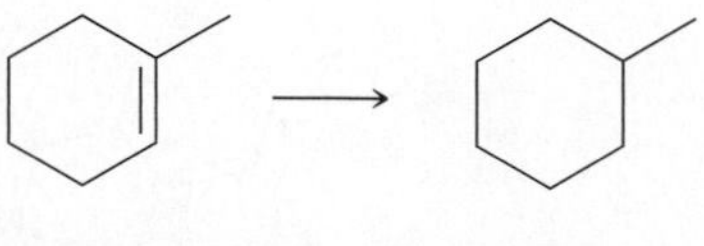 (b) 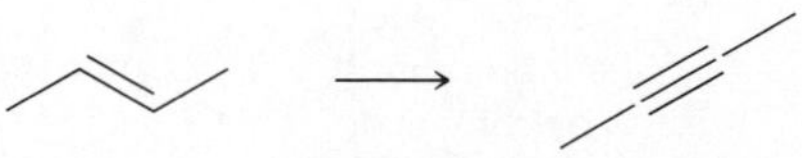

need more **PRACTICE?** **Try Problems 2.39, 2.49, 2.52**

How to Draw Bond-Line Structures

It is certainly important to be able to read bond-line structures fluently, but it is equally important to be able to draw them proficiently. When drawing bond-line structures, the following rules should be observed:

1. Carbon atoms in a straight chain should be drawn in a zigzag format:

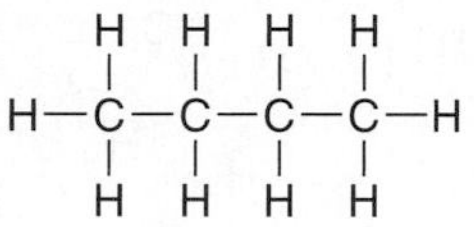

is drawn like this:

2. When drawing double bonds, draw all bonds as far apart as possible:

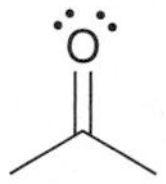

is much better than

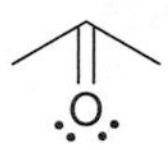

Bad

3. When drawing single bonds, the direction in which the bonds are drawn is irrelevant:

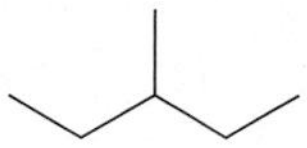

is the same as

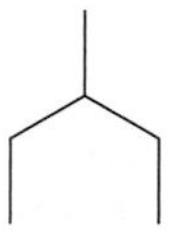

These two drawings do not represent constitutional isomers—they are just two drawings of the same compound. Both are perfectly acceptable.

4. All *heteroatoms* (atoms other than carbon and hydrogen) must be drawn, and any hydrogen atoms attached to a heteroatom must also be drawn. For example:

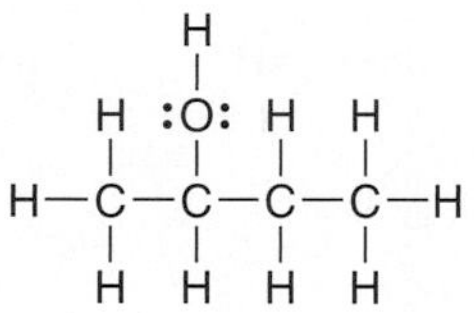

is drawn like this:

This H must be drawn:

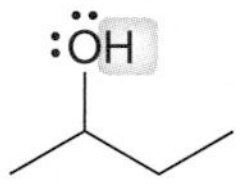

5. Never draw a carbon atom with more than four bonds. Carbon only has four orbitals in its valence shell, and therefore carbon atoms can only form four bonds.

SKILLBUILDER

2.3 DRAWING BOND-LINE STRUCTURES

LEARN the skill

Draw a bond-line structure for the following compound:

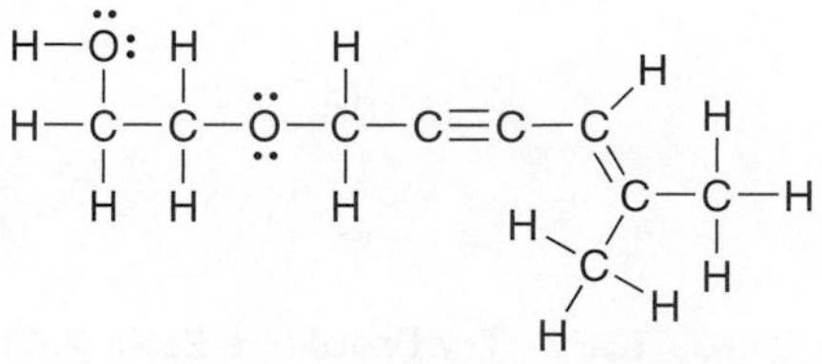

SOLUTION

Drawing a bond-line structure requires just a few conceptual steps. First, delete all hydrogen atoms except for those connected to heteroatoms:

STEP 1
Delete hydrogen atoms, except for those connected to heteroatoms.

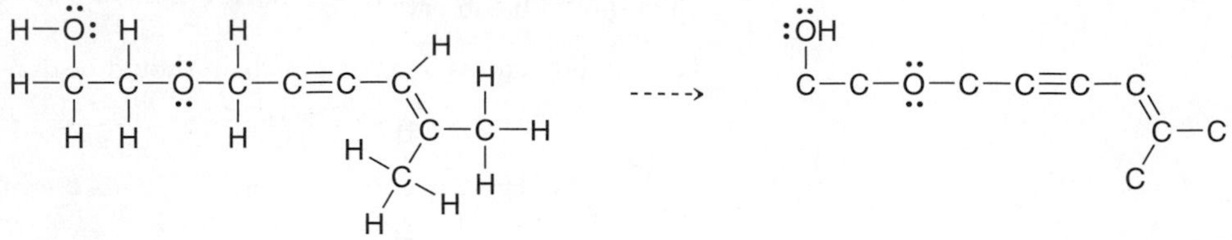

Then, place the carbon skeleton in a zigzag arrangement, making sure that any triple bonds are drawn as linear:

STEP 2
Draw in zigzag format, keeping triple bonds linear.

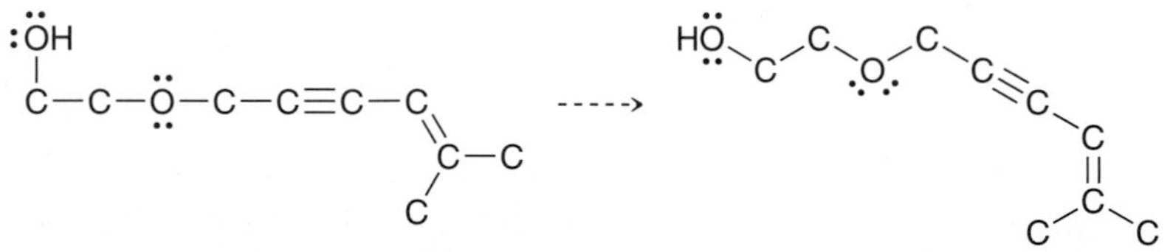

Finally, delete all carbon atoms:

STEP 3
Delete carbon atoms.

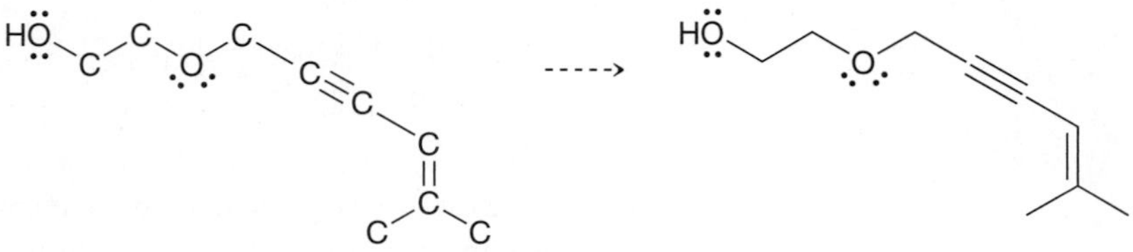

PRACTICE the skill **2.8** Draw a bond-line structure for each of the following compounds:

(a) 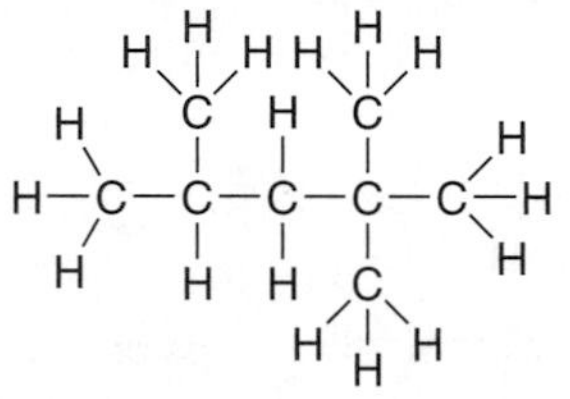

(b) 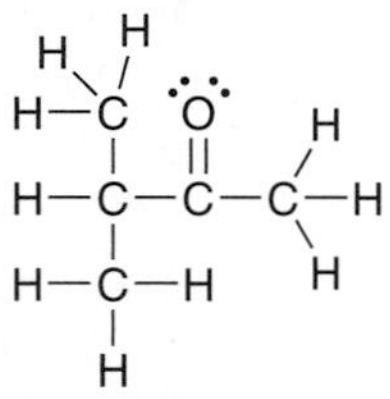

(c) 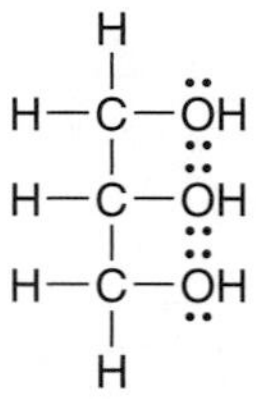

(d) $(CH_3)_3C{-}C(CH_3)_3$
(e) $CH_3CH_2CH(CH_3)_2$
(f) $(CH_3CH_2)_3COH$
(g) $(CH_3)_2CHCH_2OH$
(h) $CH_3CH_2CH_2OCH_3$
(i) $(CH_3CH_2)_2C{=}CH_2$
(j) $CH_2{=}CHOCH_2CH(CH_3)_2$
(k) $(CH_3CH_2)_2CHCH_2CH_2NH_2$
(l) $CH_2{=}CHCH_2OCH_2CH(CH_3)_2$
(m) $CH_3CH_2CH_2CH_2CH_2CH_3$
(n) $(CH_3CH_2CH_2)_2CHCl$
(o) $(CH_3)_2C{=}CHCH_2CH_3$
(p) $(CH_3CH_2)_2CHCH_2OCH_3$
(q) $(CH_3)_3CCH_2CH_2OH$
(r) $(CH_3CH_2CH_2)_3COCH_2CH_2CH{=}CHCH_2CH_2OC(CH_2CH_3)_3$

APPLY the skill **2.9** Draw bond-line structures for all constitutional isomers of the following compound:

$CH_3CH_2CH(CH_3)_2$

2.10 In each of the following compounds, identify all carbon atoms that you expect will be deficient in electron density (δ+). If you need help, refer to Section 1.5.

(a) 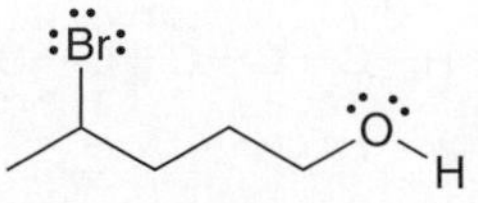

(b) 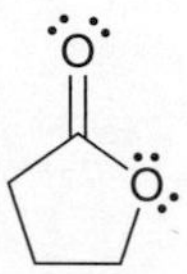

(c) 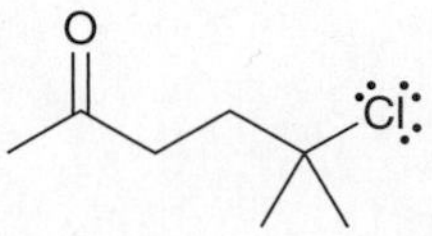

need more **PRACTICE?** **Try Problems 2.40, 2.41, 2.54, 2.58**

2.3 Identifying Functional Groups

Bond-line drawings are the preferred drawing style used by practicing organic chemists. In addition to being more efficient, bond-line drawings are also easier to read. As an example, consider the following reaction:

$$(CH_3)_2CHCHC(CH_3)_2 \xrightarrow[Pt]{H_2} (CH_3)_2CHCH_2CH(CH_3)_2$$

When the reaction is presented in this way, it is somewhat difficult to see what is happening. It takes time to digest the information being presented. However, when we redraw the same reaction using bond-line structures, it becomes very easy to identify the transformation taking place:

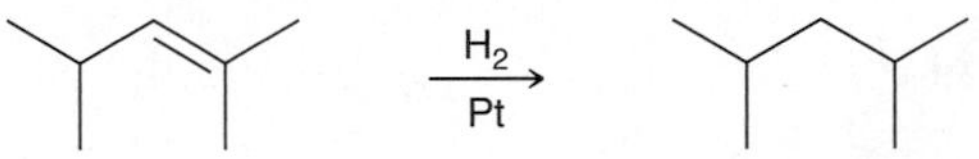

It is immediately apparent that a double bond is being converted into a single bond. With bond-line drawings, it is easier to identify the functional group. A **functional group** is a characteristic group of atoms/bonds that possess a predictable chemical behavior. In the reactions below, the starting material has a carbon-carbon double bond, which is a functional group. Compounds with carbon-carbon double bonds typically react with molecular hydrogen (H_2) in the presence of a catalyst (such as Pt). Both of the starting materials below have a carbon-carbon double bond, and consequently, they exhibit similar chemical behavior.

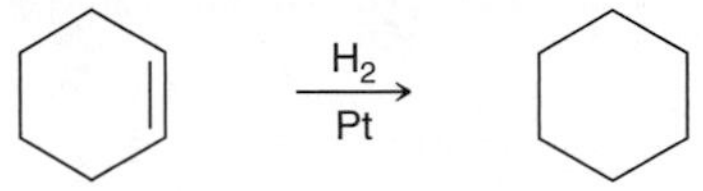

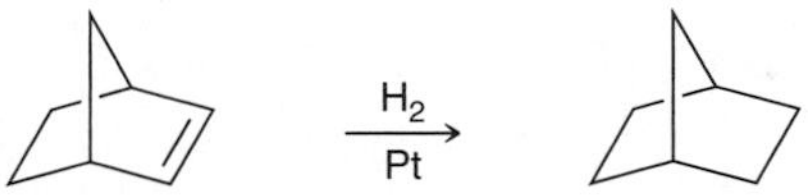

The chemistry of every organic compound is determined by the functional groups present in the compound. Therefore, the classification of organic compounds is based on their functional groups. For example, compounds with carbon-carbon double bonds are classified as *alkenes*, while compounds possessing an OH group are classified as *alcohols*. Many of the chapters in this book are organized by functional group. Table 2.1 provides a list of common functional groups and the corresponding chapters in which they appear.

CONCEPTUAL CHECKPOINT

2.11 Atenolol and enalapril are drugs used in the treatment of heart disease. Both of these drugs lower blood pressure (albeit in different ways) and reduce the risk of heart attack. Using Table 2.1, identify and label all functional groups in these two compounds:

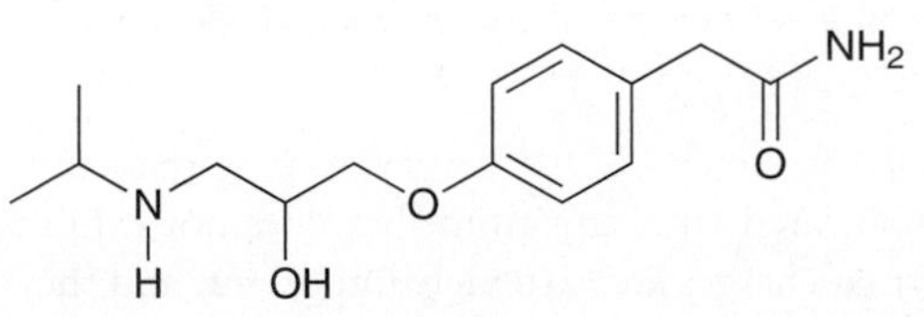

Atenolol

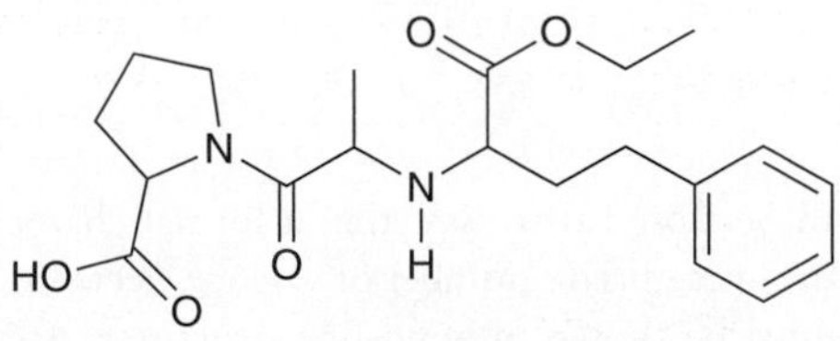

Enalapril

TABLE 2.1 EXAMPLES OF COMMON FUNCTIONAL GROUPS

FUNCTIONAL GROUP*	CLASSIFICATION	EXAMPLE	CHAPTER	FUNCTIONAL GROUP*	CLASSIFICATION	EXAMPLE	CHAPTER
R—X: (X = Cl, Br or I)	Alkyl halide	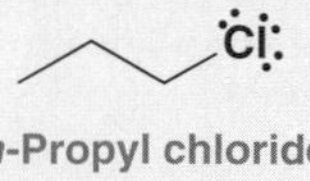 **n-Propyl chloride**	7	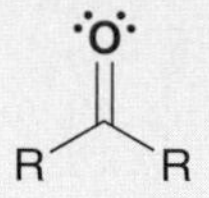	Ketone	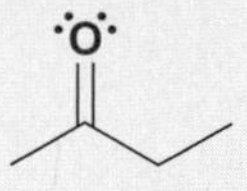 **2-Butanone**	20
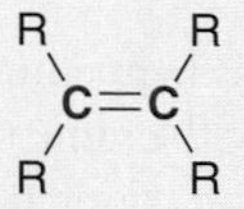	Alkene	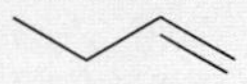 **1-Butene**	8, 9	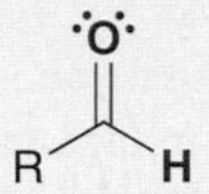	Aldehyde	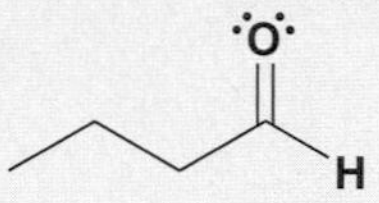 **Butanal**	20
R—C≡C—R	Alkyne	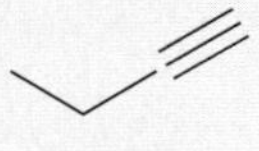 **1-Butyne**	10	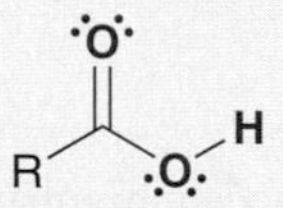	Carboxylic acid	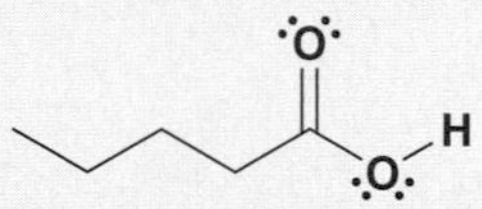 **Pentanoic acid**	21
R—OH	Alcohol	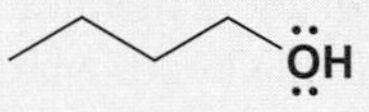 **1-Butanol**	13	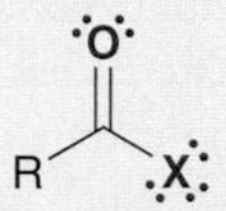	Acyl halide	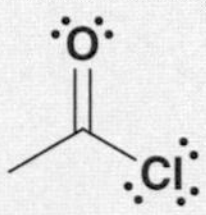 **Acetyl chloride**	21
R—O—R	Ether	 **Diethyl ether**	14	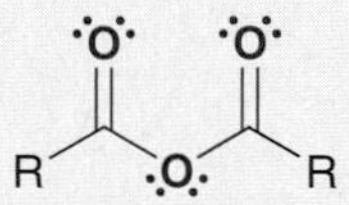	Anhydride	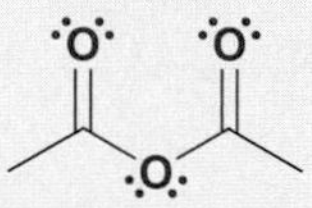 **Acetic anhydride**	21
R—SH	Thiol	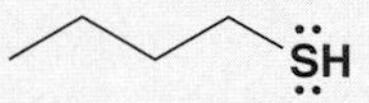 **1-Butanethiol**	14	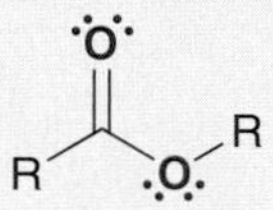	Ester	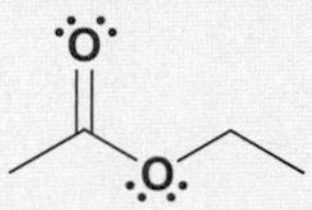 **Ethyl acetate**	21
R—S—R	Sulfide	 **Diethyl sulfide**	14	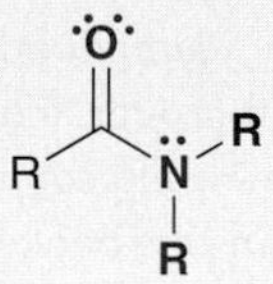	Amide	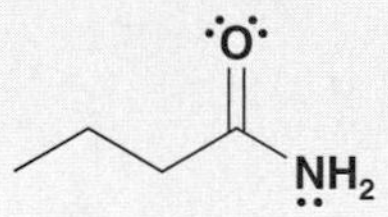 **Butanamide**	21
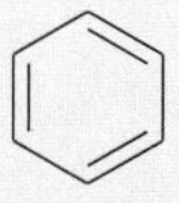	Aromatic (or arene)	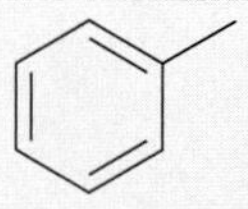 **Methylbenzene**	18, 19	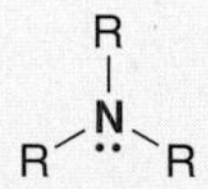	Amine	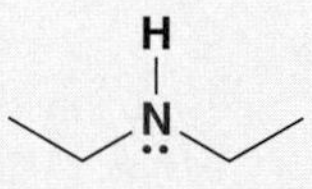 **Diethylamine**	23

* The "R" refers to the remainder of the compound, usually carbon and hydrogen atoms.

2.4 Carbon Atoms with Formal Charges

In Section 1.4 we saw that a formal charge is associated with any atom that does not exhibit the appropriate number of valence electrons. Formal charges are extremely important, and they must be shown in bond-line structures. A missing formal charge renders a bond-line structure incorrect and therefore useless. Accordingly, let's quickly practice identifying formal charges in bond-line structures.

CONCEPTUAL CHECKPOINT

2.12 For each of the compounds below determine whether any of the nitrogen atoms bear a formal charge:

(a) 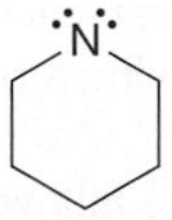

(b) 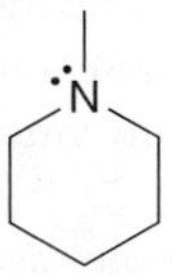

(c) 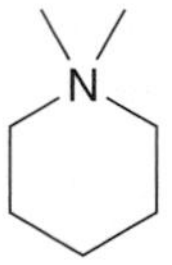

(d) 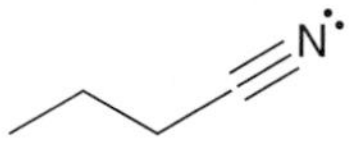

2.13 For each of the compounds below determine whether any of the oxygen atoms bear a formal charge:

(a) 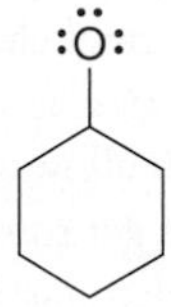

(b) 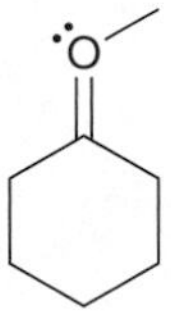

(c) 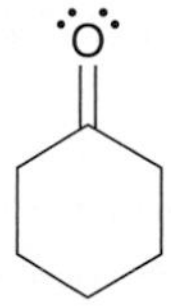

(d) 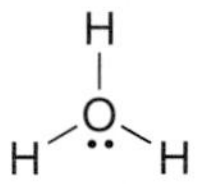

Now let's consider formal charges on carbon atoms. We have seen that carbon generally has four bonds, which allows us to "see" all of the hydrogen atoms even though they are not explicitly shown in bond-line structures. Now we must modify that rule: *A carbon atom will generally have four bonds only when it does not have a formal charge.* When a carbon atom bears a formal charge, either positive or negative, it will have three bonds rather than four. To understand why, let's first consider C^+, and then we will consider C^-.

Recall that the appropriate number of valence electrons for a carbon atom is four. In order to have a positive formal charge, a carbon atom must be missing an electron. In other words, it must have only three valence electrons. Such a carbon atom can only form three bonds. This must be taken into account when counting hydrogen atoms:

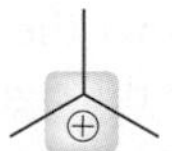

No hydrogen atoms on this C^+

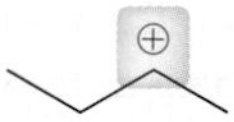

One hydrogen atom on this C^+

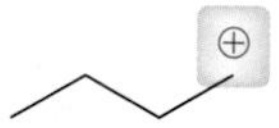

Two hydrogen atoms on this C^+

Now let's focus on negatively charged carbon atoms. In order to have a negative formal charge, a carbon atom must have one extra electron. In other words, it must have five valence electrons. Two of those electrons will form a lone pair, and the other three electrons will be used to form bonds:

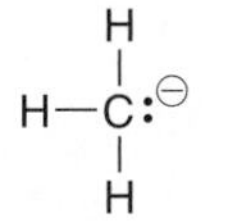

In summary, both C^+ and C^- will have only three bonds. The difference between them is the nature of the fourth orbital. In the case of C^+, the fourth orbital is empty. In the case of C^-, the fourth orbital holds a lone pair of electrons.

2.5 Identifying Lone Pairs

In order to determine the formal charge on an atom, we must know how many lone pairs it has. On the flip side, we must know the formal charge in order to determine the number of lone pairs on an atom. To understand this, let's examine a case where neither the lone pairs nor the formal charges are drawn:

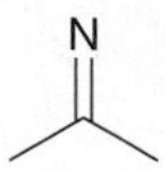 **could either be** 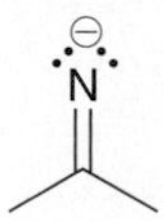 **or** 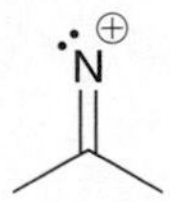

If the lone pairs were shown, then we could determine the charge (two lone pairs would mean a negative charge, and one lone pair would mean a positive charge). Alternatively, if the formal charge were shown, then we could determine the number of lone pairs (a negative charge would mean two lone pairs, and a positive charge would mean one lone pair).

Therefore, a bond-line structure will only be clear if it contains either all of the lone pairs or all of the formal charges. Since there are typically many more lone pairs than formal charges in any one particular structure, chemists have adopted the convention of always drawing formal charges, which allows us to leave out the lone pairs.

WATCH OUT
Formal charges must always be drawn and can never be omitted, unlike lone pairs, which may be omitted from a bond-line structure.

Now let's get some practice identifying lone pairs when they are not drawn. The following example will demonstrate the thought process:

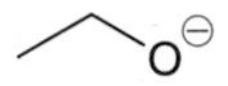

In order to determine the number of lone pairs on the oxygen atom, we simply use the same two-step process described in Section 1.4 for calculating formal charges:

1. *Determine the appropriate number of valence electrons for the atom.* Oxygen is in column 6A of the periodic table, and therefore, it should have six valence electrons.
2. *Determine if the atom actually exhibits the appropriate number of electrons.* This oxygen atom has a negative formal charge, which means it must have one extra electron. Therefore, this oxygen atom must have $6 + 1 = 7$ valence electrons. One of those electrons is being used to form the C—O bond, which leaves six electrons to be housed as lone pairs. This oxygen atom must therefore have three lone pairs:

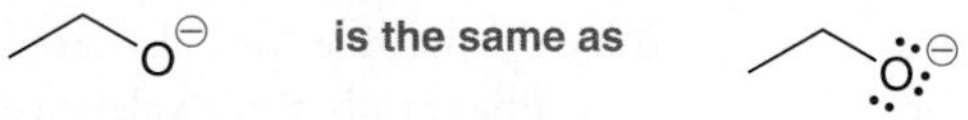

The process above represents an important skill; however, it is even more important to become familiar enough with atoms that the process becomes unnecessary. There are just a handful of patterns to recognize. Let's go through them methodically, starting with oxygen. Table 2.2 summarizes the important patterns that you will encounter for oxygen atoms.

TABLE 2.2 FORMAL CHARGE ON AN OXYGEN ATOM ASSOCIATED WITH A PARTICULAR NUMBER OF BONDS AND LONE PAIRS

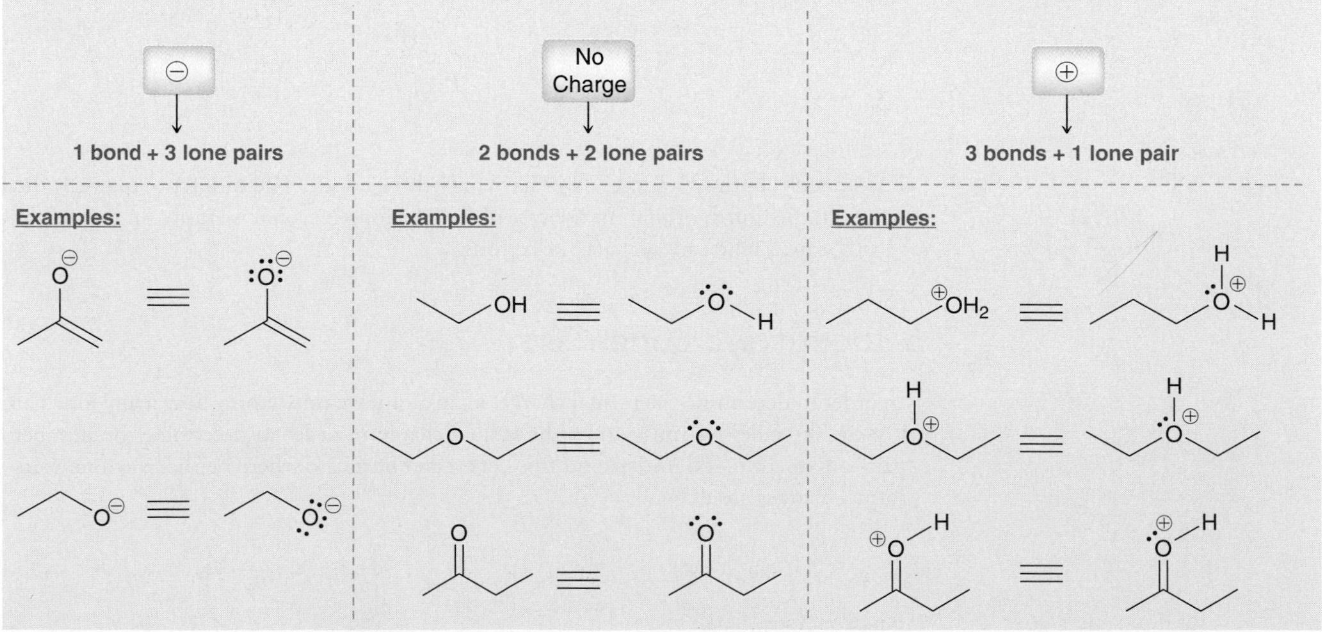

- A negative charge corresponds with one bond and three lone pairs.
- The absence of charge corresponds with two bonds and two lone pairs.
- A positive charge corresponds with three bonds and one lone pair.

SKILLBUILDER

2.4 IDENTIFYING LONE PAIRS ON OXYGEN ATOMS

LEARN the skill

Draw all of the lone pairs in the following structure:

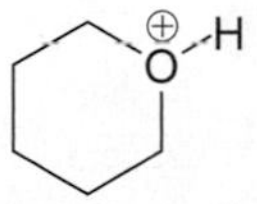

SOLUTION

STEP 1 Determine the appropriate number of valence electrons.

STEP 2 Analyze the formal charge, and determine the actual number of valence electrons.

STEP 3 Count the number of bonds, and determine how many of the actual valence electrons must be lone pairs.

The oxygen atom above has a positive formal charge and three bonds. It is preferable to recognize the pattern—that a positive charge and three bonds must mean that the oxygen has just one lone pair:

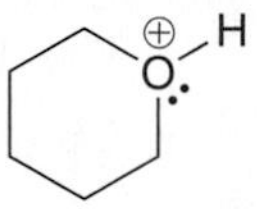

Alternatively, and less preferably, it is possible to calculate the number of lone pairs using the following two steps. First, determine the appropriate number of valence electrons for the atom. Oxygen is in column 6A of the periodic table, and therefore, it should have six valence electrons. Then, determine if the atom actually exhibits the appropriate number of electrons. This oxygen atom has a positive charge, which means it is missing an electron: $6 - 1 = 5$ valence electrons. Three of these five electrons are being used to form bonds, which leaves just two electrons for a lone pair. This oxygen atom has only one lone pair.

PRACTICE the skill

2.14 Draw all lone pairs on each of the oxygen atoms in the compounds below. Before doing this, review Table 2.2, and then come back to these problems. Try to identify all lone pairs without having to count. Then, count to see if you were right.

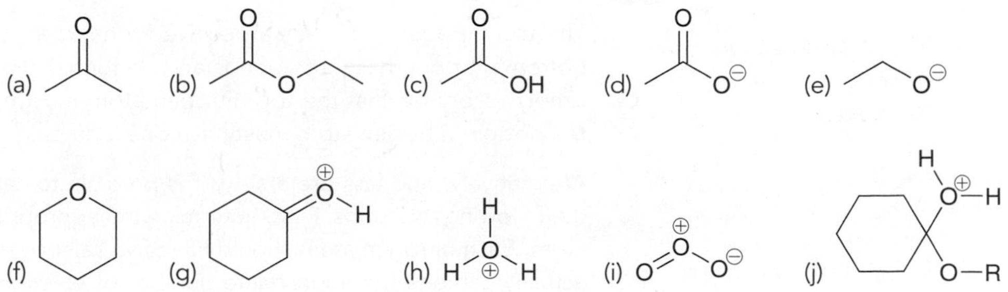

APPLY the skill

2.15 A carbene is a highly reactive intermediate in which a carbon atom bears a lone pair and no formal charge:

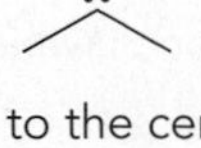

How many hydrogen atoms are attached to the central carbon atom above?

need more **PRACTICE?** **Try Problem 2.43**

TABLE 2.3 FORMAL CHARGE ON A NITROGEN ATOM ASSOCIATED WITH A PARTICULAR NUMBER OF BONDS AND LONE PAIRS

⊖	No Charge	⊕
2 bonds + 2 lone pairs	3 bonds + 1 lone pair	4 bonds + 0 lone pairs
Examples:	Examples:	Examples:
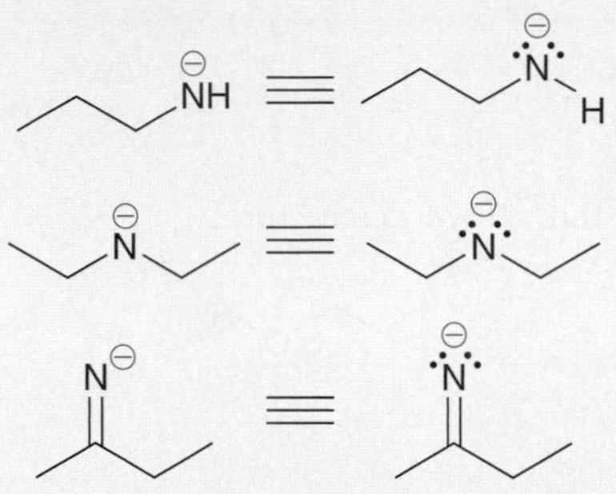	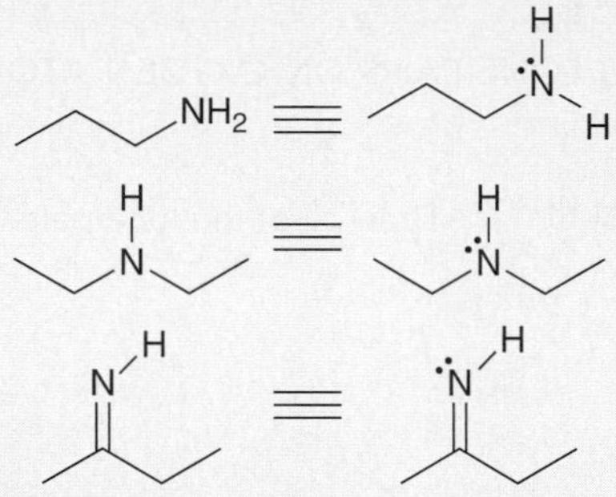	 has no lone pairs
		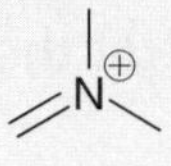 has no lone pairs
		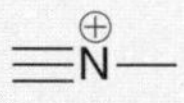 has no lone pairs

Now let's look at the common patterns for nitrogen atoms. Table 2.3 shows the important patterns that you will encounter with nitrogen atoms. In summary:

- A negative charge corresponds with two bonds and two lone pairs.
- The absence of charge corresponds with three bonds and one lone pair.
- A positive charge corresponds with four bonds and no lone pairs.

SKILLBUILDER

2.5 IDENTIFYING LONE PAIRS ON NITROGEN ATOMS

LEARN the skill

Draw any lone pairs associated with the nitrogen atoms in the following structure:

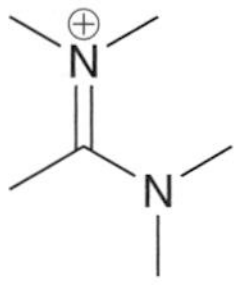

SOLUTION

STEP 1 Determine the appropriate number of valence electrons.

STEP 2 Analyze the formal charge and determine the actual number of valence electrons.

STEP 3 Count the number of bonds, and determine how many of the actual valence electrons must be lone pairs.

The top nitrogen atom has a positive formal charge and four bonds. The bottom nitrogen has three bonds and no formal charge. It is preferable to simply recognize that the top nitrogen atom must have no lone pairs and the bottom nitrogen atom must have one lone pair:

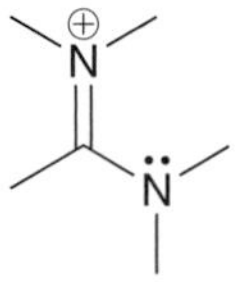

Alternatively, and less preferably, it is possible to calculate the number of lone pairs using the following two steps. First, determine the appropriate number of valence electrons for the atom. Each nitrogen atom should have five valence electrons. Next, determine if each atom actually exhibits the appropriate number of electrons. The top nitrogen atom has a positive charge, which means it is missing an electron. This nitrogen atom actually has only four valence electrons. Since the nitrogen atom has four bonds, it is using each of its four electrons to form a bond. This nitrogen atom does not possess a lone pair. The bottom nitrogen atom has no formal charge, so this nitrogen atom must be using five valence electrons. It has three bonds, which means that there are two electrons left over, forming one lone pair.

PRACTICE the skill **2.16** Draw all lone pairs on each of the nitrogen atoms in the compounds below. First, review Table 2.3, and then come back to these problems. Try to identify all lone pairs *without* having to count. Then, count to see if you were right.

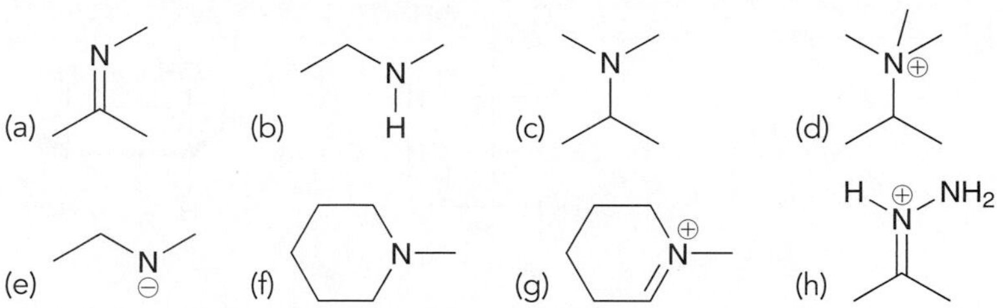

APPLY the skill **2.17** Each of the following compounds contains both oxygen and nitrogen atoms. Identify all lone pairs in each of the following compounds:

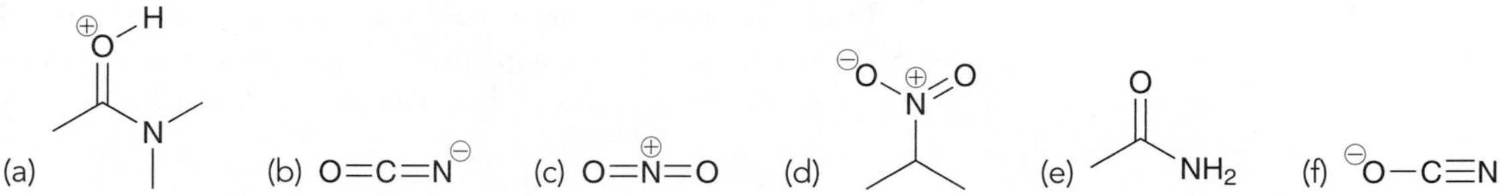

2.18 Identify the number of lone pairs in each of the following compounds:

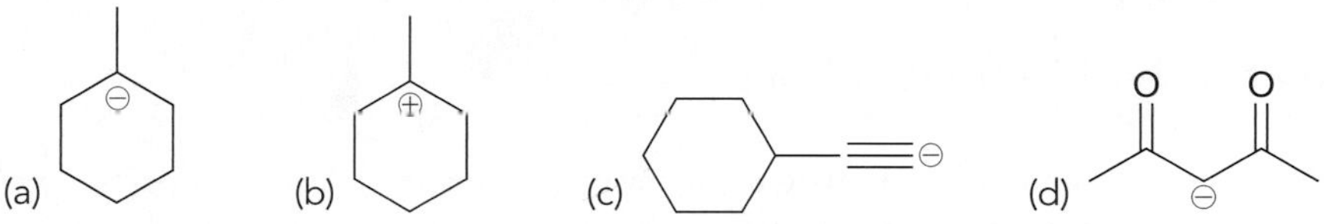

2.19 Amino acids are biological compounds with the following structure, where the R group can vary. The structure and biological function of amino acids will be discussed in Chapter 25. Identify the total number of lone pairs present in an amino acid, assuming that the R group does not contain any atoms with lone pairs.

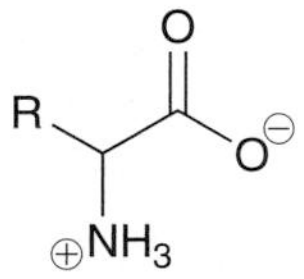

An amino acid

need more **PRACTICE?** **Try Problem 2.39**

2.6 Three-Dimensional Bond-Line Structures

Throughout this book, we will use many different kinds of drawings to represent the three-dimensional geometry of atoms. The most common method is a bond-line structure that includes **wedges** and **dashes** to indicate three dimensionality. These structures are used for all types of compounds, including acyclic, cyclic, and bicyclic compounds (Figure 2.1). In the drawings in Figure 2.1, a wedge represents a group coming out of the page, and a dash represents a group going behind the page. We will use wedges and dashes extensively in Chapter 5 and thereafter.

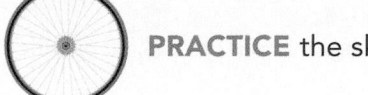

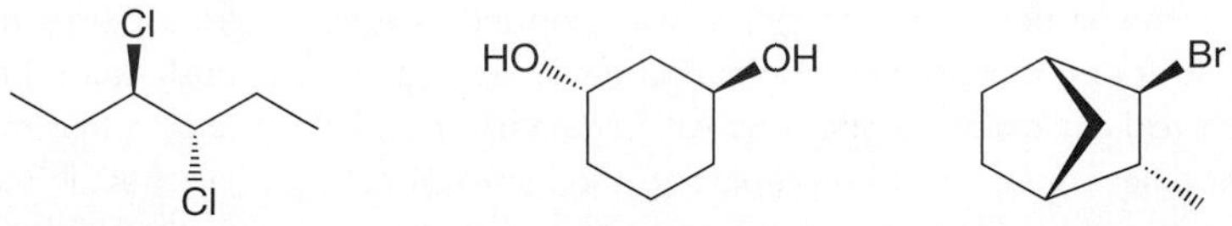

Acyclic (No ring) **Cyclic** (One ring) **Bicyclic** (Two rings)

FIGURE 2.1 Bond-line structures with wedges and dashes to indicate three dimensionality.

In certain circumstances, there are other types of drawings that can be used, all of which also indicate three-dimensional geometry (Figure 2.2).

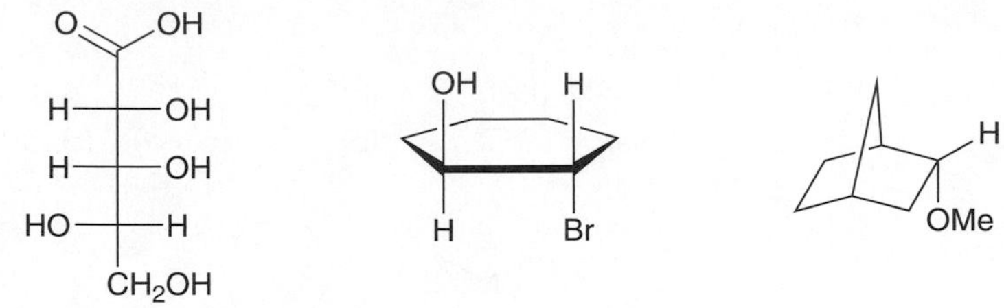

Fischer projection (Used only for acyclic compounds)

Haworth projection (Used only for cyclic compounds)

(Used only for bicyclic compounds)

FIGURE 2.2 Common drawing styles that show three dimensionality for acylic, cyclic, and bicyclic compounds.

Fischer projections are used for acyclic compounds while **Haworth projections** are used exclusively for cyclic compounds. Each of these drawing styles will be used several times throughout this book, particularly in Chapters 5, 9, and 24.

MEDICALLYSPEAKING)))

Identifying the Pharmacophore

As mentioned in the chapter opener, there are many techniques that scientists employ in the design of new drugs. One such technique is called *lead modification*, which involves modifying the structure of a compound known to exhibit desirable medicinal properties. The known compound "leads" the way to the development of other similar compounds and is therefore called the *lead compound*. The story of morphine provides a good example of this process.

Morphine is a very potent analgesic (pain reliever) that is known to act on the central nervous system as a depressant (causing sedation and slower respiratory function) and as a stimulant (relieving symptoms of anxiety and causing an overall state of euphoria). Because morphine is addictive, it is primarily used for the short-term treatment of acute pain and for terminally ill patients suffering from extreme pain. The analgesic properties of morphine have been exploited for over a millennium. It is the major component of opium, obtained from the unripe seed pods of the poppy plant, *Papaver somniferum*. Morphine was first isolated from opium in 1803, and by the mid-1800s, it was used heavily to control pain during and after surgical procedures. By the end of the 1800s, the addictive properties of morphine became apparent, which fueled the search for nonaddictive analgesics.

In 1925, the structure of morphine was correctly determined. This structure functioned as a lead compound and was modified to produce other compounds with analgesic properties. Early modifications focused on replacing the hydroxyl (OH) groups with other functional groups. Examples include heroin and codeine. Heroin exhibits stronger activity than morphine and is extremely addictive. Codeine shows less activity than morphine and is less addictive. Codeine is currently used as an analgesic and cough suppressant.

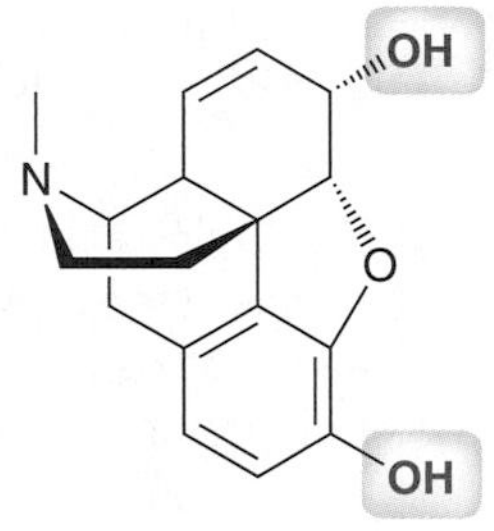

Morphine

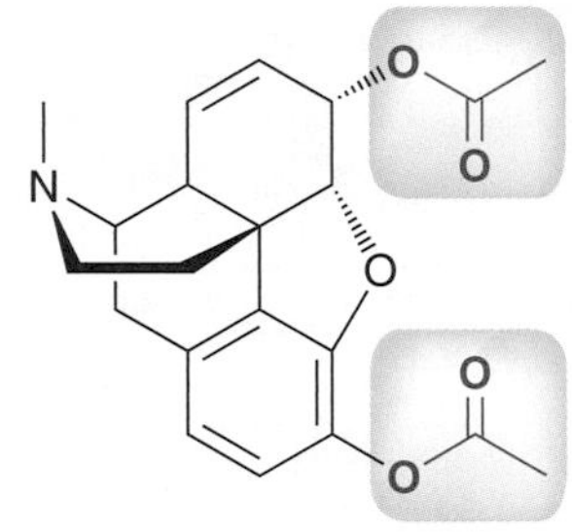

Heroin

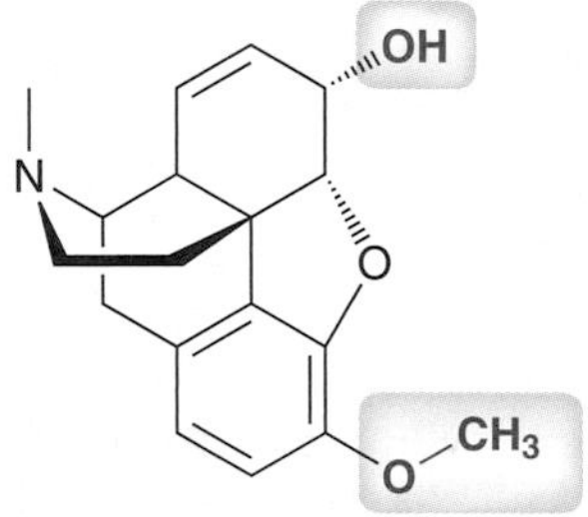

Codeine

In 1938, the analgesic properties of meperidine, also known as Demerol, were fortuitously discovered. As the story goes, meperidine was originally prepared to function as an antispasmodic agent (to suppress muscle spasms). When administered to mice, it curiously caused the tails of the mice to become erect. It was already known that morphine and related compounds produced a similar effect in mice, so meperidine was further tested and found to exhibit analgesic properties. This discovery generated much interest by providing new insights in the search for other analgesics. By comparing the structures of morphine, meperidine, and their derivatives, scientists were able to determine which structural features are essential for analgesic activity, shown in red on the next page:

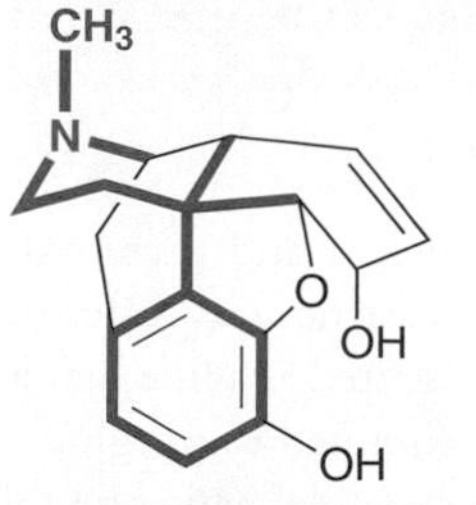

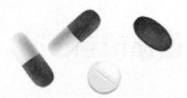

Morphine

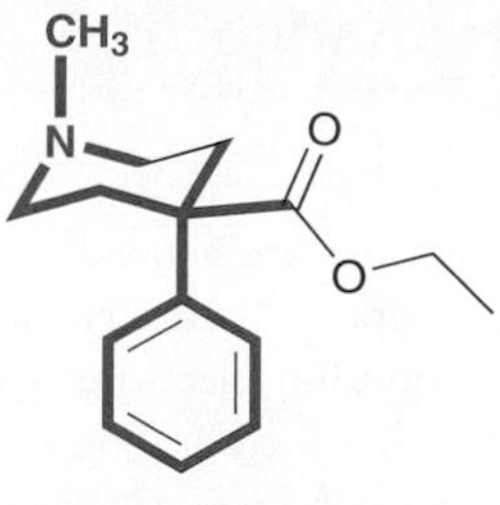

Meperidine

When morphine is drawn in this way, its structural similarity to meperidine becomes more apparent. Specifically, the bonds indicated in red represent the portion of each compound responsible for the analgesic activity. This part of the compound is called the *pharmacophore*. If any part of the pharmacophore is removed or changed, the resulting compound will not be capable of binding to the appropriate biological receptor, and the compound will not exhibit analgesic properties. The term *auxophore* refers to the rest of the compound (the bonds shown in black above). Removing any of these bonds may or may not affect the strength with which the pharmacophore binds to the receptor, thereby affecting the compound's analgesic potency. When modifying a lead compound, the auxophoric regions are the portions targeted for modification. For example, the auxophoric regions of morphine were modified to develop methadone and etorphine.

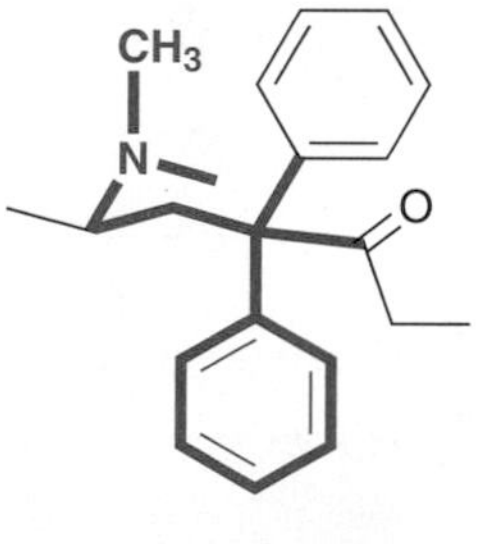

Methadone

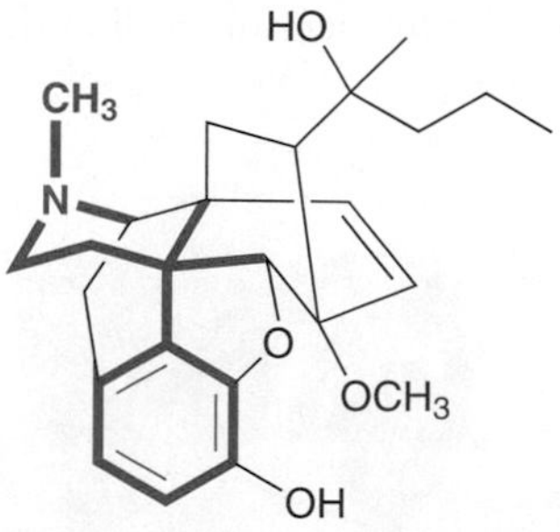

Etorphine

Methadone, developed in Germany during World War II, is used to treat heroin addicts suffering from withdrawal symptoms. Methadone binds to the same receptor as heroin, but it has a longer retention time in the body, thereby enabling the body to cope with the decreasing levels of drug that normally cause withdrawal symptoms. Etorphine is over 3000 times more potent than morphine and is used exclusively in veterinary medicine to immobilize elephants and other large mammals.

Scientists are constantly searching for new lead compounds. In 1992, researchers at NIH (National Institutes of Health) in Bethesda, Maryland, isolated epibatidine from the skin of the Ecuadorian frog, *Epipedobates tricolor*. Epibatidine was found to be an analgesic that is 200 times more potent than morphine. Further studies indicated that epibatidine and morphine bind to different receptors. This discovery is very exciting, because it means that epibatidine can serve as a new lead compound. Although this compound is too toxic for clinical use, a significant number of researchers are currently working to identify the pharmacophore of epibatidine and to develop nontoxic derivatives. This area of research indeed looks promising.

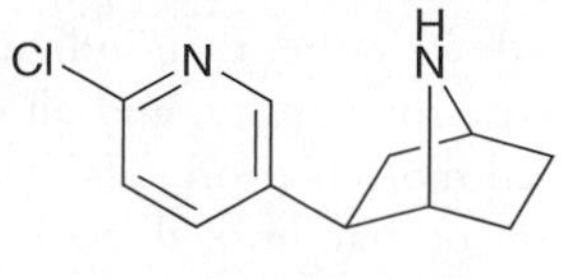

Epibatidine

CONCEPTUAL CHECKPOINT

2.20 Troglitazone, rosiglitazone, and pioglitazone, all antidiabetic drugs introduced to the market in the late 1990s, are believed to act on the same receptor:

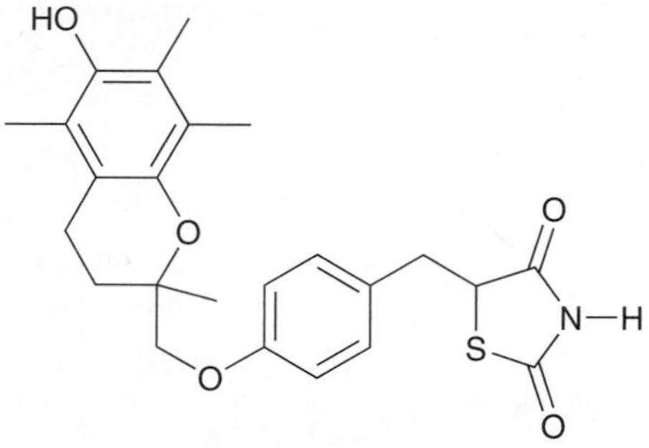

Troglitazone

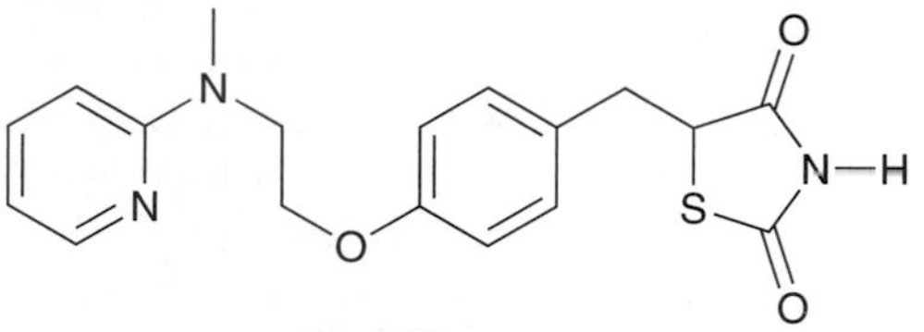

Rosiglitazone

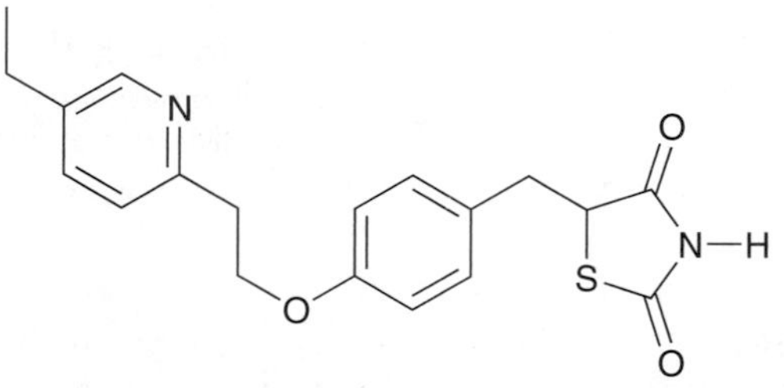

Pioglitazone

(a) Based on these structures, try to identify the likely pharmacophore that is responsible for the antidiabetic activity of these drugs.

(b) Consider the structure of rivoglitazone (below). This compound is currently being studied for potential antidiabetic activity. Based on your analysis of the likely pharmacophore, do you believe that rivoglitazone will exhibit antidiabetic properties?

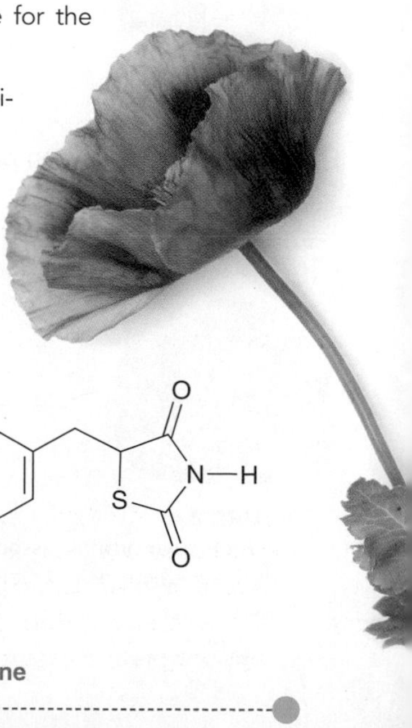

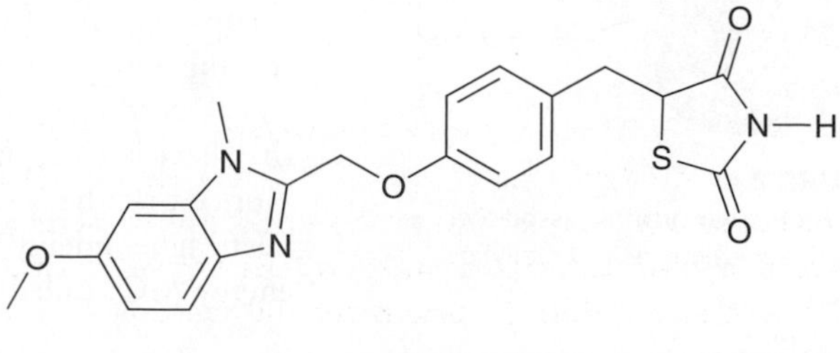

Rivoglitazone

2.7 Introduction to Resonance

The Inadequacy of Bond-Line Structures

We have seen that bond-line structures are generally the most efficient and preferred way to draw the structure of an organic compound. Nevertheless, bond-line structures suffer from one major defect. Specifically, a pair of bonding electrons is always represented as a line that is drawn between two atoms, which implies that the bonding electrons are confined to a region of space directly in between two atoms. In some cases, this assertion is acceptable, as in the following structure:

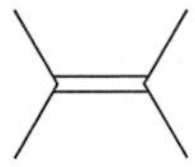

In this case, the π electrons are in fact located where they are drawn, in between the two central carbon atoms. But in other cases, the electron density is spread out over a larger region of the molecule. For example, consider the following ion, called an *allyl carbocation*:

It might seem from the drawing above that there are two π electrons on the left side and a positive charge on the right side. But this is not the entire picture, and the drawing above is inadequate. Let's take a closer look and first analyze the hybridization states. Each of the three carbon atoms above is sp^2 hybridized. Why? The two carbon atoms on the left side are each sp^2 hybridized because each of those carbon atoms is utilizing a p orbital to form the π bond (Section 1.9). The third carbon atom, bearing the positive charge, is also sp^2 hybridized because it has an empty p orbital. Now let's draw all of the p orbitals (Figure 2.3).

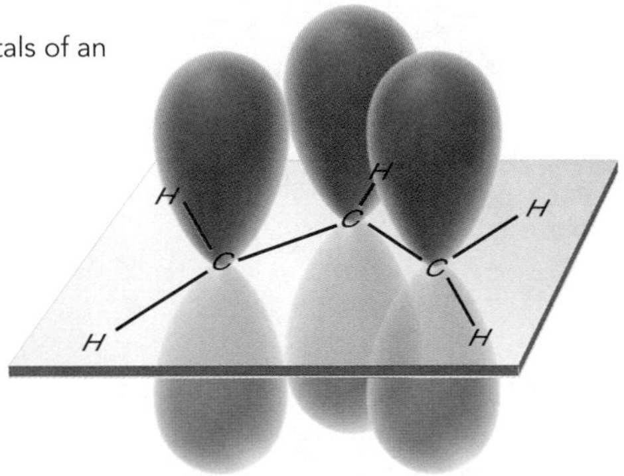

FIGURE 2.3
The overlapping *p* orbitals of an allylic carbocation.

This image focuses our attention on the continuous system of p orbitals, which functions as a "conduit," allowing the two π electrons to be associated with all three carbon atoms. Valence bond theory is inadequate for analysis of this system because it treats the electrons as if they were confined between only two atoms. A more appropriate analysis of the allyl cation requires the use of molecular orbital (MO) theory (Section 1.8), in which electrons are associated with the molecule as a whole, rather than individual atoms. Specifically, in MO theory, the entire molecule is treated as one entity, and all of the electrons in the entire molecule occupy regions of space called molecular orbitals. Two electrons are placed in each orbital, starting with the lowest energy orbital, until all electrons occupy orbitals.

According to MO theory, the three p orbitals shown in Figure 2.3 no longer exist. Instead, they have been replaced by three MOs, illustrated in Figure 2.4 in order of increasing energy. Notice that the lowest energy MO, called the *bonding molecular orbital*, has no nodes. The next higher energy MO, called the *nonbonding molecular orbital*, has one node. The highest energy MO, called the *antibonding molecular orbital*, has two nodes. The π electrons of the

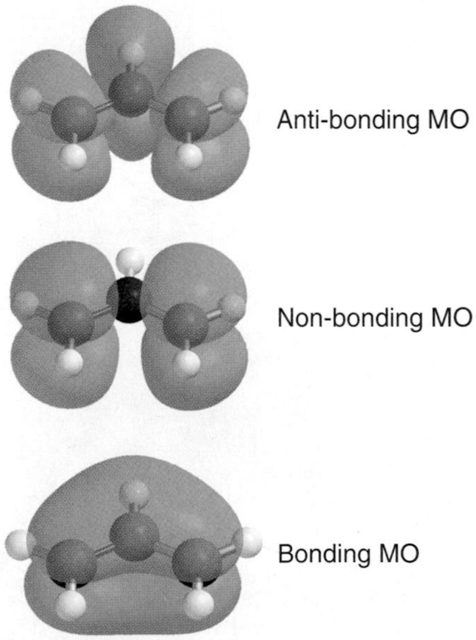

FIGURE 2.4
The molecular orbitals associated with the π electrons of an allylic system.

allyl system will fill these MOs, starting with the lowest energy MO. How many π electrons will occupy these MOs? The allyl carbocation has only two π electrons, rather than three, because one of the carbon atoms bears a positive formal charge indicating that one electron is missing. The two π electrons of the allyl system will occupy the lowest energy MO (the bonding MO). If the missing electron were to return, it would occupy the next higher energy MO, which is the nonbonding MO. Focus your attention on the nonbonding MO.

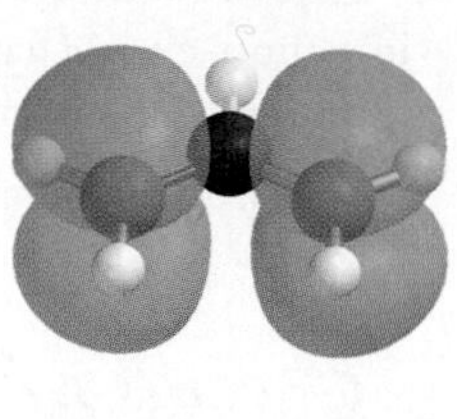

The nonbonding molecular orbital (from Figure 2.4) ssociated with the π electrons of an allylic system.

There should be an electron occupying this nonbonding MO, but the electron is missing. Therefore, the colored lobes are empty and represent regions of space that are electron deficient. In conclusion, MO theory suggests that the positive charge of the allyl carbocation is associated with the two ends of the system, rather than just one end.

In a situation like this, any single bond-line structure that we draw will be inadequate. How can we draw a positive charge that is spread out over two locations, and how can we draw two π electrons that are associated with three carbon atoms?

Resonance

The approach that chemists use to deal with the inadequacy of bond-line structures is called **resonance**. According to this approach, we draw more than one bond-line structure and then mentally meld them together:

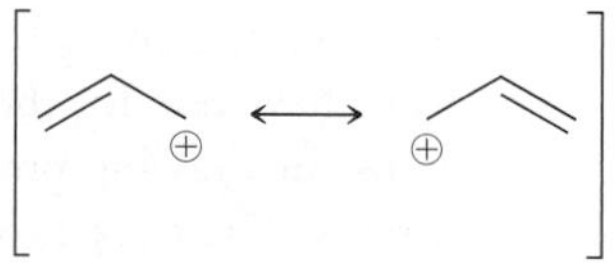

These drawings are called **resonance structures**, and they show that the positive charge is spread over two locations. Notice that we separate resonance structures with a straight, two-headed arrow, and we place brackets around the structures. The arrow and brackets indicate that the drawings are resonance structures *of one entity*. This one entity, called a **resonance hybrid**, is *not* flipping back and forth between the different resonance structures. To better understand this, consider the following analogy: A person who has never before seen a nectarine asks a farmer to describe a nectarine. The farmer answers:

> Picture a *peach* in your mind, and now picture a *plum* in your mind. Well, a *nectarine* has features of both fruits: the inside tastes like a peach, the outside is smooth like a plum, and the color is somewhere in between the color of a peach and the color of a plum. So take your image of a peach together with your image of a plum and *meld them together* in your mind into one image. That's a nectarine.

Here is the important feature of the analogy: the nectarine does not vibrate back and forth every second between being a peach and being a plum. A nectarine is a nectarine all of the time. The image of a peach by itself is not adequate to describe a nectarine. Neither is the image of a plum. But by combining certain characteristics of a peach with certain characteristics of a plum, it is possible to imagine the features of the hybrid fruit. Similarly, with resonance structures, no single drawing adequately describes the nature of the electron density spread out over the molecule. To deal with this problem, we draw several drawings and then meld them together in our minds to obtain one image, or hybrid, just like the nectarine.

Don't be confused by this important point: The term "resonance" does not describe something that is happening. Rather, it is a term that describes the way we deal with the inadequacy of our bond-line drawings.

Resonance Stabilization

We developed the concept of resonance using the allyl cation as an example, and we saw that the positive charge of an allyl cation is spread out over two locations. This spreading of charge, called **delocalization**, is a stabilizing factor. That is, *the delocalization of either a positive charge or a negative charge stabilizes a molecule.* This stabilization is often referred to as **resonance stabilization**, and the allyl cation is said to be *resonance stabilized.* Resonance stabilization plays a major role in the outcome of many reactions, and we will invoke the concept of resonance in almost every chapter of this textbook. The study of organic chemistry therefore requires a thorough mastery of drawing resonance structures, and the following sections are designed to foster the necessary skills.

2.8 Curved Arrows

In this section, we will focus on **curved arrows**, which are the tools necessary to draw resonance structures properly. Every curved arrow has a *tail* and *head:*

Curved arrows used for drawing resonance structures do not represent the motion of electrons—they are simply tools that allow us to draw resonance structures with ease. These tools treat the electrons *as if* they were moving, even though the electrons are actually not moving at all. In Chapter 3, we will encounter curved arrows that actually do represent the flow of electrons. For now, keep in mind that all curved arrows in this chapter are just tools and do not represent a flow of electrons.

It is essential that the tail and head of every arrow be drawn in precisely the proper location. The tail shows where the electrons are coming from, and the head shows where the electrons are going (remember, the electrons aren't really going anywhere, but we treat them as if they were for the purpose of drawing the resonance structures). We will soon learn patterns for drawing proper curved arrows. But, first, we must learn where not to draw curved arrows. There are two rules that must be followed when drawing curved arrows for resonance structures:

1. *Avoid breaking a single bond.*
2. *Never exceed an octet for second-row elements.*

Let's explore each of these rules:

1. *Avoid breaking a single bond when drawing resonance structures.* By definition, resonance structures must have all the same atoms connected in the same order. Breaking a single bond would change this—hence the first rule:

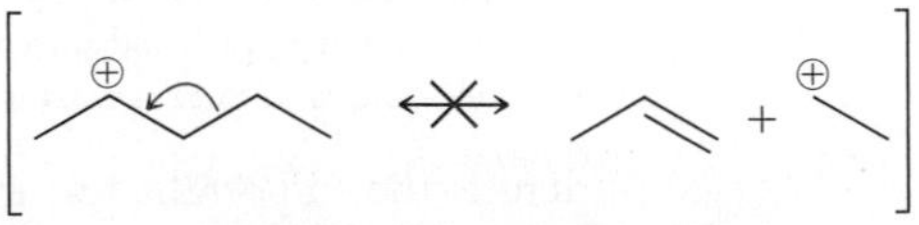

Don't break a single bond

There are very few exceptions to this rule, and we will only violate it two times in this textbook (both in Chapter 9). Each time, we will explain why it is permissible in that case. In all other cases, the tail of an arrow should never be placed on a single bond.

2. *Never exceed an octet for second-row elements.* Elements in the second row (C, N, O, F) have only four orbitals in their valence shell. Each orbital can either form a bond or hold a lone pair. Therefore, for second-row elements the total of the number of bonds plus the number of lone pairs can never be more than four. They can never have five or six bonds; the most is four. Similarly, they can never have four bonds and a lone pair, because this would also require five orbitals. For the same reason, they can never have three bonds and two lone pairs. Let's see some

examples of curved arrows that violate this second rule. In each of these drawings, the central atom cannot form another bond because it does not have a fifth orbital that can be used.

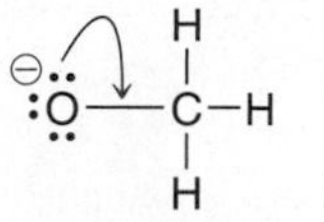

Bad arrow

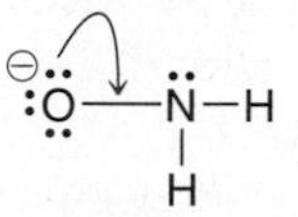

Bad arrow

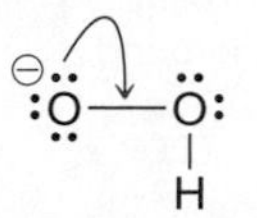

Bad arrow

The violation in each example above is clear, but with bond-line structures, it can be more difficult to see the violation because the hydrogen atoms are not drawn (and, very often, neither are the lone pairs). Care must be taken to "see" the hydrogen atoms even when they are not drawn:

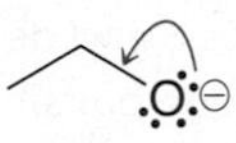

Bad arrow

is the same as

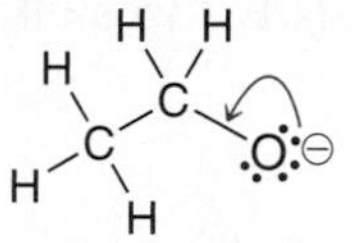

Bad arrow

At first it is difficult to see that the curved arrow on the left structure is violating the second rule. But when we count the hydrogen atoms, it becomes clear that the curved arrow above would create a carbon atom with five bonds.

From now on, we will refer to the second rule as the *octet rule*. But be careful—for purposes of drawing resonance structures, it is only considered a violation if a second-row element has *more* than an octet of electrons. However, it is not a violation if a second-row element has *less* than an octet of electrons. For example:

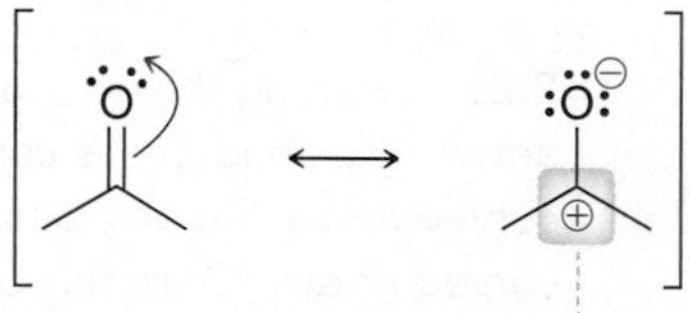

This carbon atom does not have an octet

This second drawing above is perfectly acceptable, even though the central carbon atom has only six electrons surrounding it. For our purposes, we will only consider the octet rule to be violated if we exceed an octet.

Our two rules (avoid breaking a single bond and never exceed an octet for a second-row element) reflect the two features of a curved arrow: the tail and the head. A poorly placed arrow tail violates the first rule, and a poorly directed arrow head violates the second rule.

SKILLBUILDER

2.6 IDENTIFYING VALID RESONANCE ARROWS

LEARN the skill

For the compound below, look at the arrow drawn on the structure and determine whether it violates either of the two rules for drawing curved arrows:

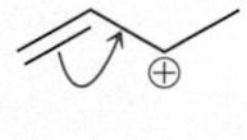

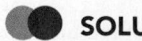

SOLUTION

In order to determine if either rule has been broken, we must look carefully at the tail and the head of the curved arrow. The tail is placed on a double bond, and therefore, this curved arrow does not break a single bond. So the first rule is not violated.

STEP 1
Make sure that the tail of the curved arrow is not located on a single bond.

STEP 2
Make sure that the head of the curved arrow does not violate the octet rule.

Next, we look at the head of the arrow: Has the octet rule been violated? Is there a fifth bond being formed here? Remember that a carbocation (C^+) only has three bonds, not four. Two of the bonds are shown, which means that the C^+ has only one bond to a hydrogen atom:

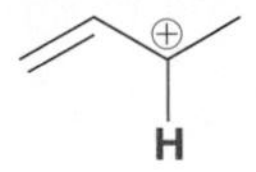

Therefore, the curved arrow will give the carbon atom a fourth bond, which does not violate the octet rule.

The curved arrow is valid, because the two rules were not violated. Both the tail and head of the arrow are acceptable.

PRACTICE the skill **2.21** For each of the problems below, determine whether each curved arrow violates either of the two rules, and describe the violation, if any. (Don't forget to count all hydrogen atoms and all lone pairs.)

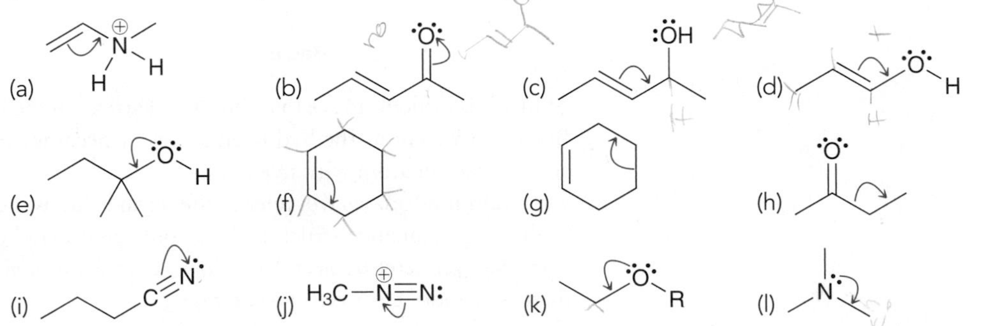

APPLY the skill **2.22** Drawing the resonance structure of the following compound requires one curved arrow. The head of this curved arrow is placed on the oxygen atom, and the tail of the curved arrow can only be placed in one location without violating the rules for drawing curved arrows. Draw this curved arrow.

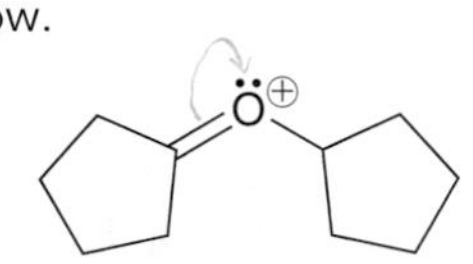

need more **PRACTICE?** **Try Problem 2.51**

Whenever more than one curved arrow is used, all curved arrows must be taken into account in order to determine if any of the rules have been violated. For example, the following arrow violates the octet rule:

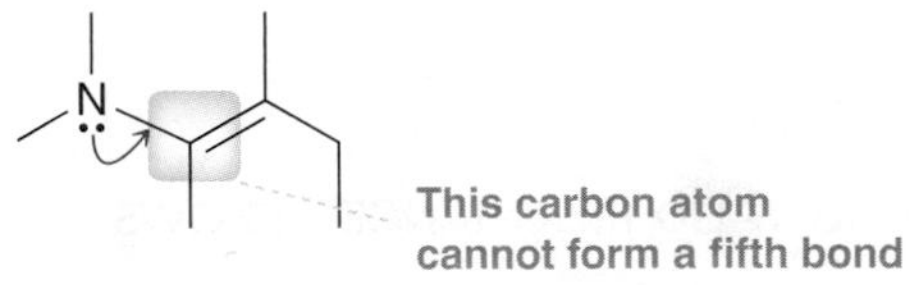
This carbon atom
cannot form a fifth bond

However, by adding another curved arrow, we remove the violation:

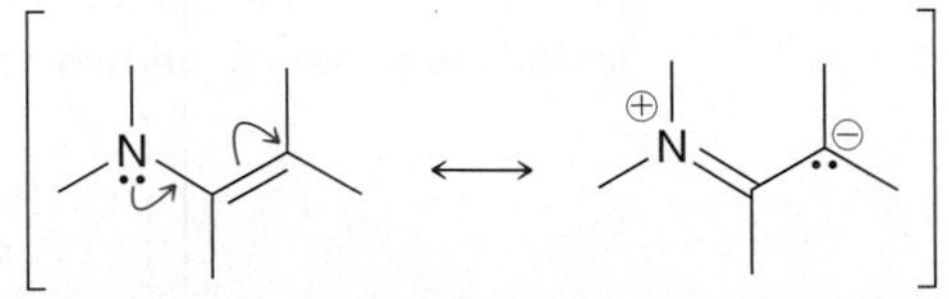

The second curved arrow removes the violation of the first curved arrow. In this example, both arrows are acceptable, because taken together, they do not violate our rules.

Arrow pushing is much like bike riding. The skill of bike riding cannot be learned by watching someone else ride. Learning to ride a bike requires practice. Falling occasionally is a necessary part of the learning process. The same is true with arrow pushing. The only way to learn is with practice. This chapter is designed to provide ample opportunity for practicing and mastering resonance structures.

2.9 Formal Charges in Resonance Structures

In Section 1.4, we learned how to calculate formal charges. Resonance structures very often contain formal charges, and it is absolutely critical to draw them properly. Consider the following example:

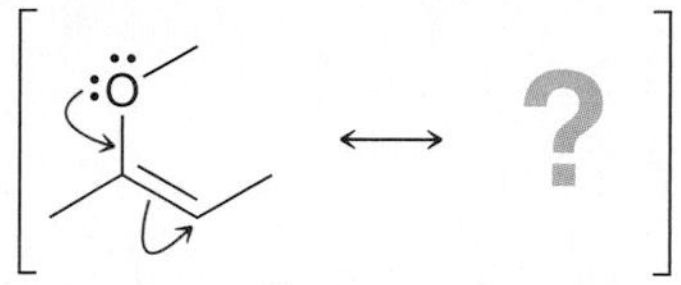

WATCH OUT
The electrons are not really moving. We are just treating them as if they were.

In this example, there are two curved arrows. The first arrow pushes one of oxygen's lone pairs to form a bond, and the second arrow pushes the π bond to form a lone pair on a carbon atom. When both arrows are pushed at the same time, neither of the rules is violated. So, let's focus on how to draw the resonance structure by following the instructions provided by the curved arrows. We delete one lone pair from the oxygen and place a π bond between the carbon and oxygen. Then we must delete the C—C π bond and place a lone pair on the carbon:

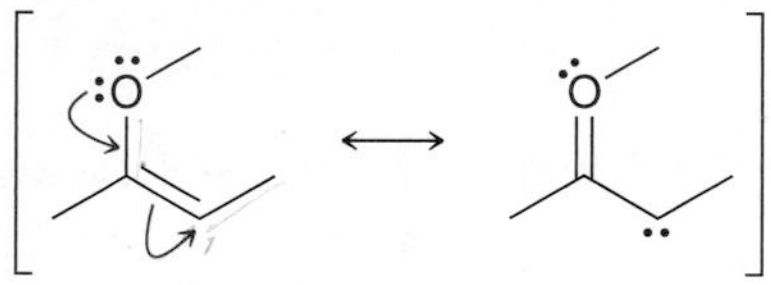

The arrows are really a language, and they tell us what to do. However, the structure is not complete without drawing formal charges. If we apply the rules of assigning formal charges, the oxygen acquires a positive charge and the carbon acquires a negative charge:

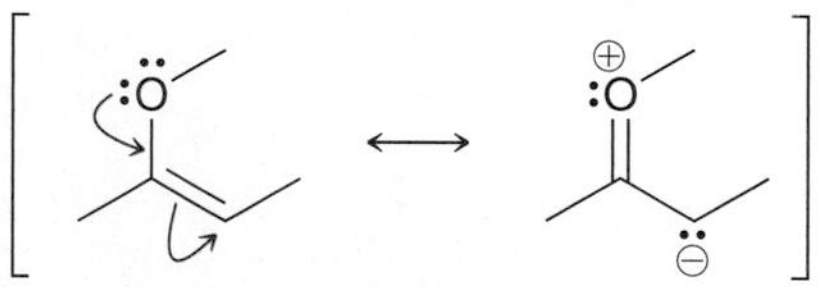

Another way to assign formal charges is to think about what the arrows are indicating. In this case, the curved arrows indicate that the oxygen atom is losing a lone pair and gaining a bond. In other words, it is losing two electrons and only gaining one back. The net result is the loss of one electron, indicating that oxygen must incur a positive charge in the resonance structure. A similar analysis for the carbon atom on the bottom right shows that it must incur a negative charge. Let's practice assigning formal charges in resonance structures.

SKILLBUILDER

2.7 ASSIGNING FORMAL CHARGES IN RESONANCE STRUCTURES

LEARN the skill

Draw the resonance structure below. Be sure to include formal charges.

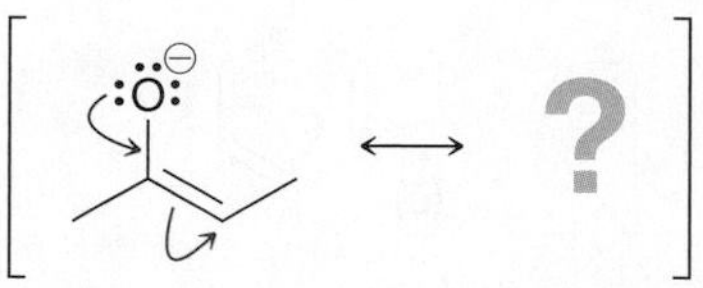

SOLUTION

The arrows indicate that one of the lone pairs on the oxygen is coming down to form a bond, and the C═C double bond is being pushed to form a lone pair on a carbon atom. This is very similar to the previous example. The arrows indicate that we must delete one lone pair on the oxygen, place a double bond between the carbon and oxygen, delete the carbon-carbon double bond, and place a lone pair on the carbon:

STEP 1
Carefully read what the curved arrows indicate.

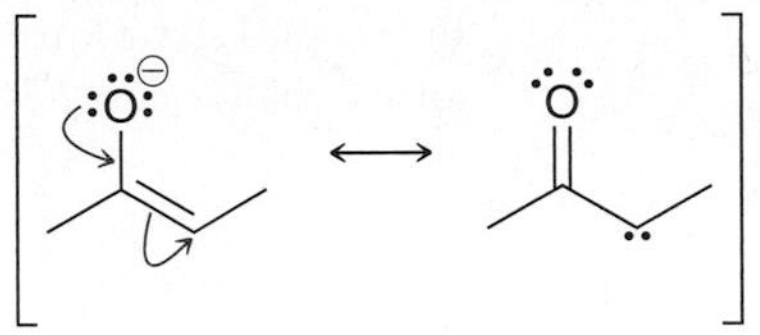

Finally, we must assign formal charges. In this case, the oxygen started with a negative charge, and this charge has now been pushed down (as the arrows indicate) onto the carbon. Therefore, the carbon must now bear the negative charge:

STEP 2
Assign formal charges.

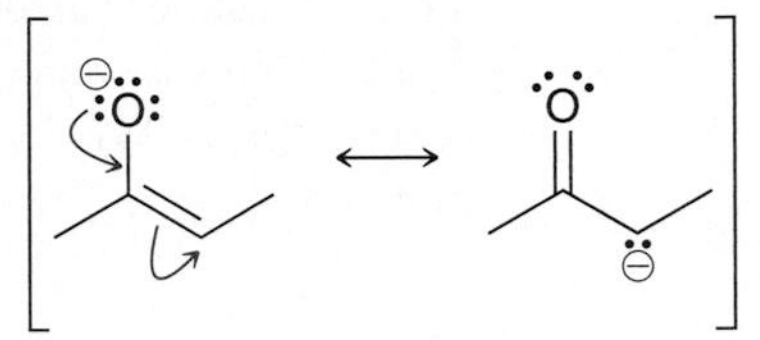

Earlier in this chapter, we said that it is not necessary to draw lone pairs, because they are implied by bond-line structures. In the example above, the lone pairs are shown for clarity. This raises an obvious question. Look at the first curved arrow above: the tail is drawn on a lone pair. If the lone pairs had not been drawn, how would the curved arrow be drawn? In situations like this, organic chemists will sometimes draw the curved arrow coming from the negative charge:

Nevertheless, you should avoid this practice, because it can easily lead to mistakes in certain situations. It is highly preferable to draw the lone pairs and then place the tail of the curved arrow on a lone pair, rather than placing it on a negative charge.

After drawing a resonance structure and assigning formal charges, it is always a good idea to count the total charge on the resonance structure. This total charge MUST be the same as on the original structure (conservation of charge). If the first structure had a negative charge, then the resonance structure must also have a net negative charge. If it doesn't, then the resonance structure cannot possibly be correct. The total charge on a compound must be the same for all resonance structures, and there are no exceptions to this rule.

PRACTICE the skill **2.23** For each of the structures below, draw the resonance structure that is indicated by the curved arrows. Be sure to include formal charges.

(a) 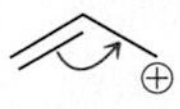 (b) 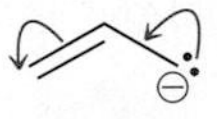 (c) 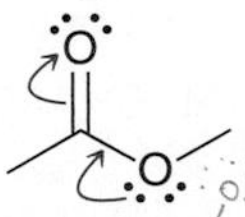 (d) 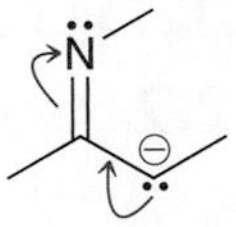

(e) 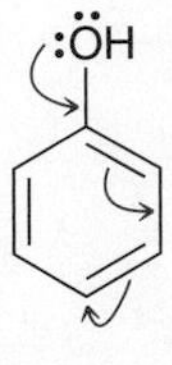 (f) 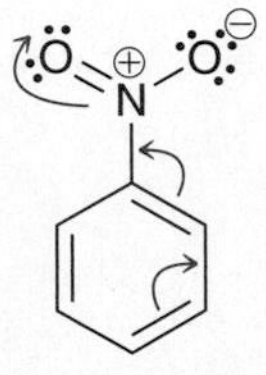 (g) 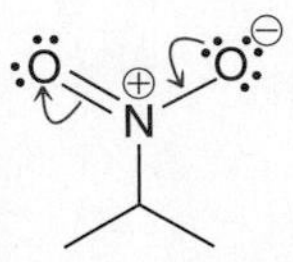 (h) 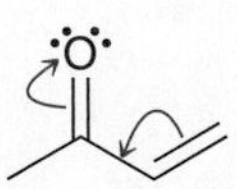

APPLY the skill

2.24 In each case below, draw the curved arrow(s) required in order to convert the first resonance structure into the second resonance structure. In each case, begin by drawing all lone pairs, and then use the formal charges to guide you.

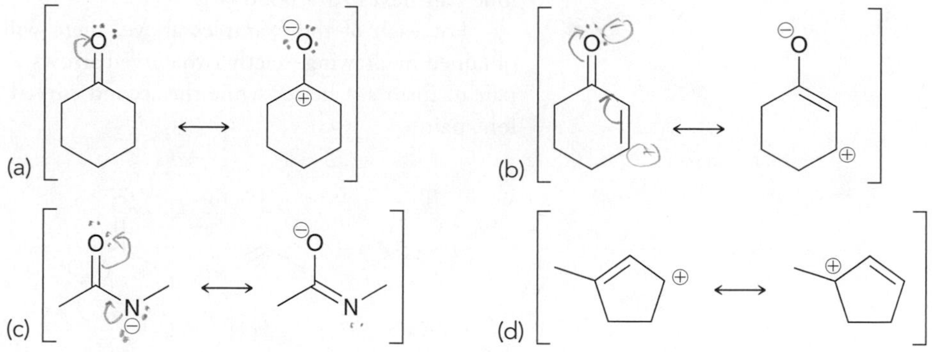

need more **PRACTICE?** **Try Problems 2.44, 2.53**

2.10 Drawing Resonance Structures via Pattern Recognition

In order to become truly proficient at drawing resonance structures, we must learn to recognize the following five patterns: (1) an allylic lone pair, (2) an allylic positive charge, (3) a lone pair adjacent to a positive charge, (4) a π bond between two atoms of differing electronegativity, and (5) conjugated π bonds in a ring.

We will now explore each of these five patterns, with examples and practice problems.

1. *An allylic lone pair.* Let's begin with some important terminology that we will use frequently throughout the remainder of the text. When a compound contains a carbon-carbon double bond, the two carbon atoms bearing the double bonds are called **vinylic** positions, while the atoms connected directly to the vinylic positions are called **allylic** positions:

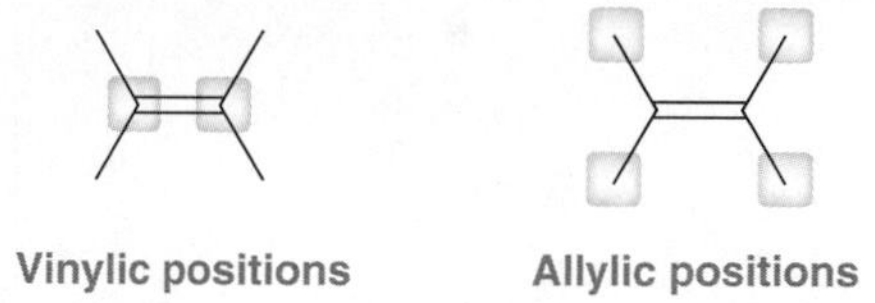

We are specifically looking for lone pairs in an allylic position. As an example, consider the following compound, which has two lone pairs:

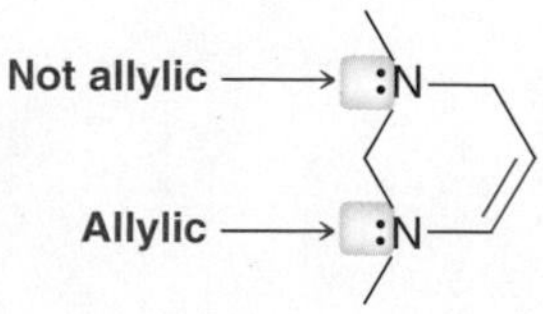

We must learn to identify lone pairs in allylic positions. Here are several examples:

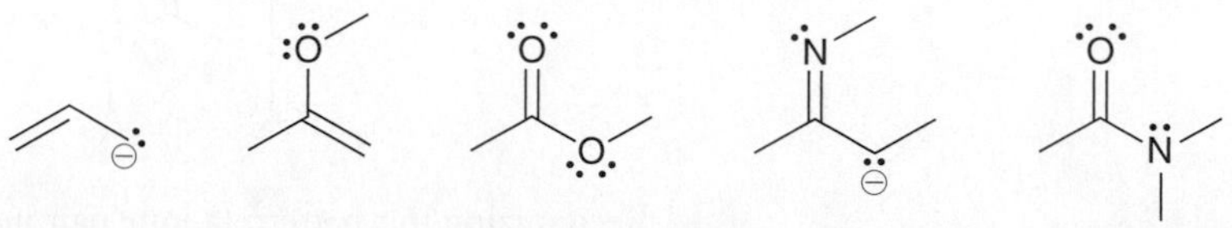

In the last three cases above, the lone pairs are not next to a *carbon-carbon* double bond and are technically not allylic lone pairs (an allylic position is the position next to a carbon-carbon double bond and not any other type of double bond). Nevertheless, for purposes of drawing resonance structures, we will treat these lone pairs in the same way that we treat allylic lone pairs. Specifically, all of the examples above exhibit at least one lone pair next to a π bond.

For each of the examples above, there will be a resonance structure that can be obtained by drawing exactly two curved arrows. The first curved arrow goes from the lone pair to form a π bond, while the second curved arrow goes from the π bond to form a lone pair:

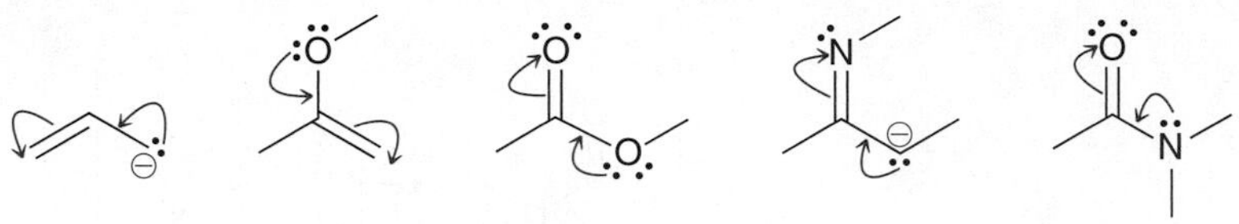

Let's carefully consider the formal charges produced in each of the cases above. When the atom with the lone pair has a negative charge, then it transfers its negative charge to the atom that ultimately receives a lone pair:

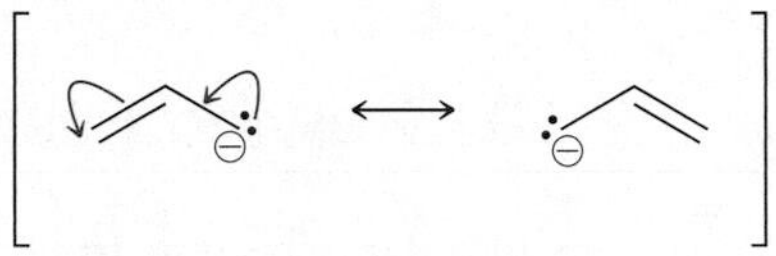

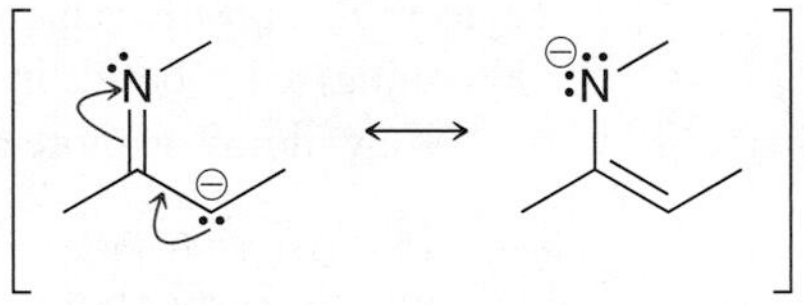

When the atom with the lone pair does not have a negative charge, then it will incur a positive charge, while the atom receiving the lone pair will incur a negative charge:

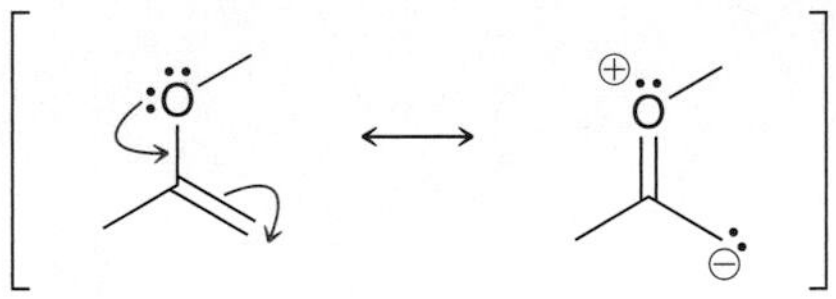

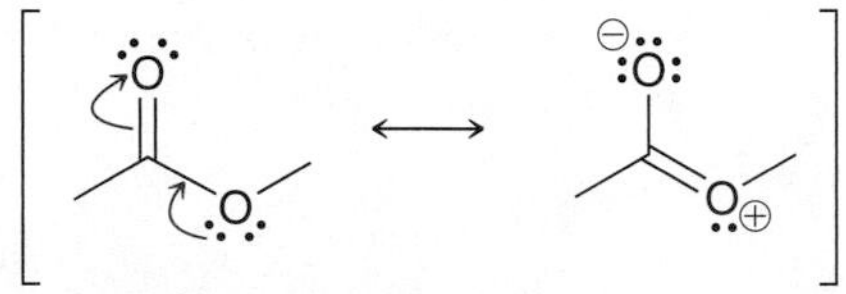

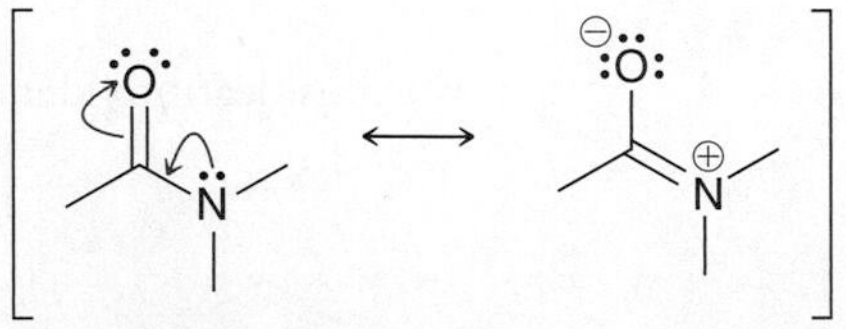

Recognizing this pattern (a lone pair next to a π bond) will save time in calculating formal charges and determining if the octet rule is being violated.

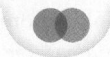

CONCEPTUAL CHECKPOINT

2.25 For each of the compounds below, locate the pattern we just learned (lone pair next to a π bond) and draw the appropriate resonance structure:

(a) 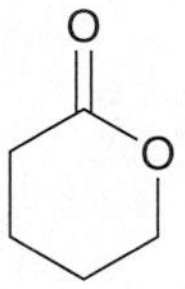

(b) 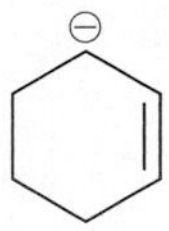

(c) 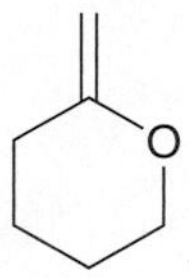

(d) 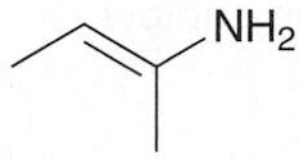

(e) 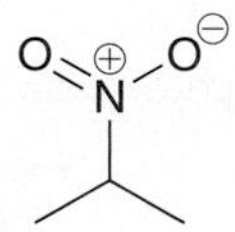

(f) 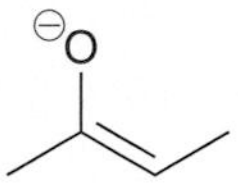

(g) 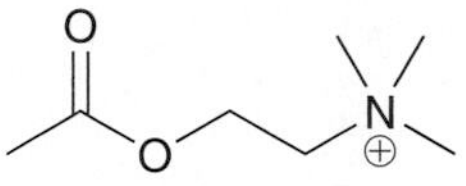

Acetylcholine (a neurotransmitter)

(h) 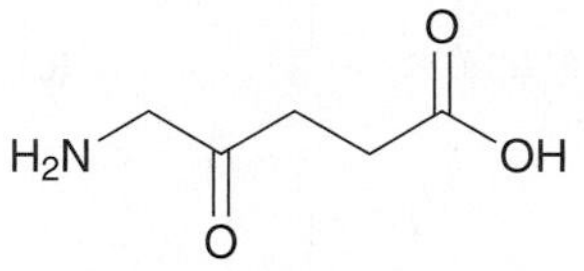

5-Amino-4-oxopentanoic acid (used in therapy and diagnosis of hepatic tumors)

2. *An allylic positive charge.* Again we are focusing on allylic positions, but this time, we are looking for a positive charge located in an allylic position:

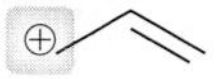

Allylic positive charge

When there is an allylic positive charge, only one curved arrow will be required; this arrow goes from the π bond to form a new π bond:

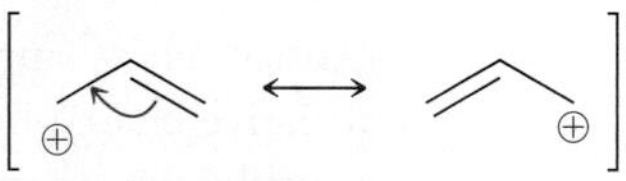

Notice what happens to the formal charge in the process. The positive charge is moved to the other end of the system.

In Chapter 17, we will explore **conjugated π bonds**, which are systems comprised of alternating double and single bonds. In some cases, we will encounter a positive charge next to a conjugated system.

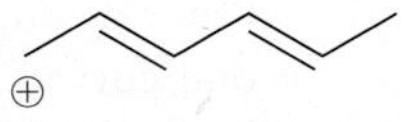

When this happens, we push each of the double bonds over one at a time:

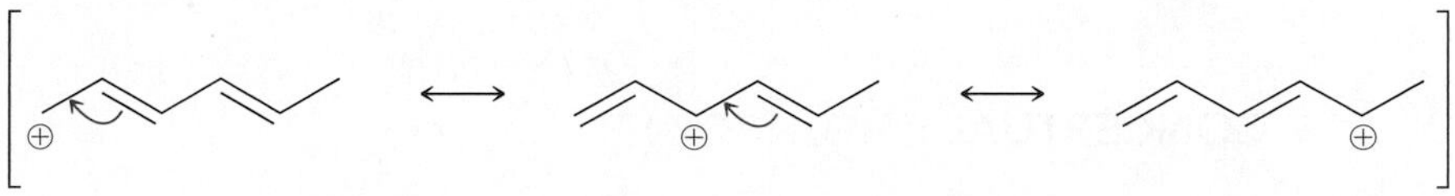

It is not necessary to waste time recalculating formal charges for each resonance structure, because the arrows indicate what is happening. Think of a positive charge as a hole of electron density—a place that is missing an electron. When we push π electrons to plug up the hole, a new hole is created nearby. In this way, the hole is simply moved from one location to another. Notice that in the above structures the tails of the curved arrows are placed on the π bonds, not on the positive charge. *Never place the tail of a curved arrow on a positive charge* (that is a common mistake).

CONCEPTUAL CHECKPOINT

2.26 Draw the resonance structure(s) for each of the compounds below:

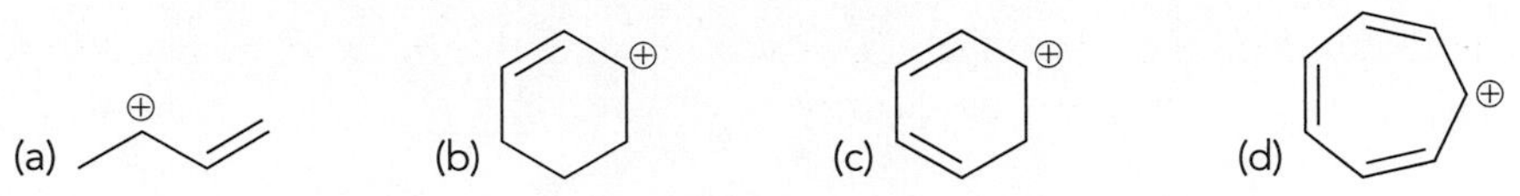

3. *A lone pair adjacent to a positive charge.* Consider the following example:

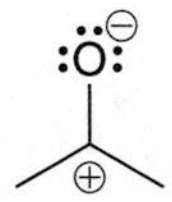

The oxygen atom exhibits three lone pairs, all of which are adjacent to the positive charge. This pattern requires only one curved arrow. The tail of the curved arrow is placed on a lone pair, and the head of the arrow is placed to form a π bond between the lone pair and the positive charge:

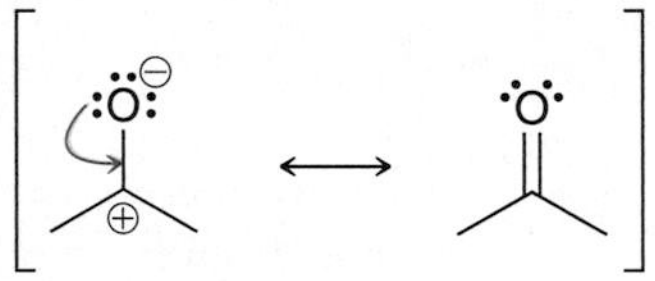

Notice what happens with the formal charges above. The atom with the lone pair has a negative charge in this case, and therefore the charges end up canceling each other. Let's consider what happens with formal charges when the atom with the lone pair does not bear a negative charge. For example, consider the following:

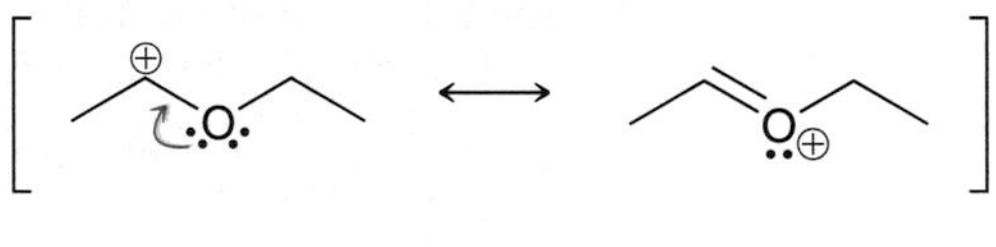

Once again, there is a lone pair adjacent to a positive charge. Therefore, we draw only one curved arrow: the tail goes on the lone pair, and the head is placed to form a π bond. In this case, the oxygen atom did not start out with a negative charge. Therefore, it will incur a positive charge in the resonance structure (remember conservation of charge).

CONCEPTUAL CHECKPOINT

2.27 For each of the compounds below, locate the lone pair adjacent to a positive charge and draw the resonance structure:

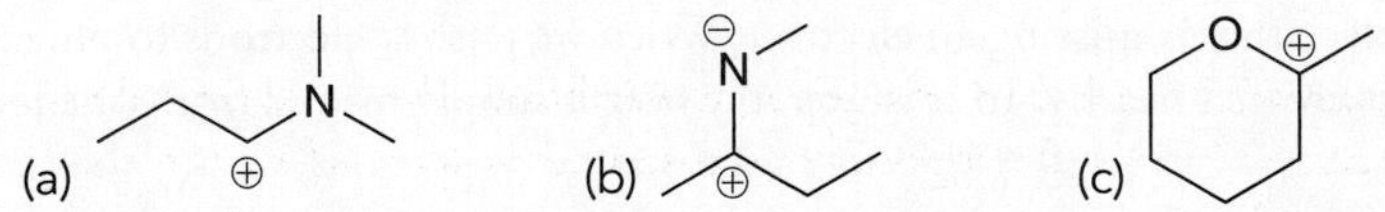

In one of the problems above, a negative charge and a positive charge are seen canceling each other to become a double bond. However, there is one situation where it is not possible to combine charges to form a double bond—this occurs with the nitro group. The structure of the nitro group looks like this:

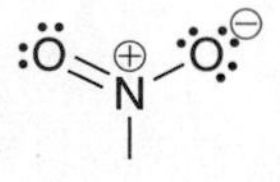

In this case, there is a lone pair adjacent to a positive charge, yet we cannot draw a single curved arrow to cancel out the charges:

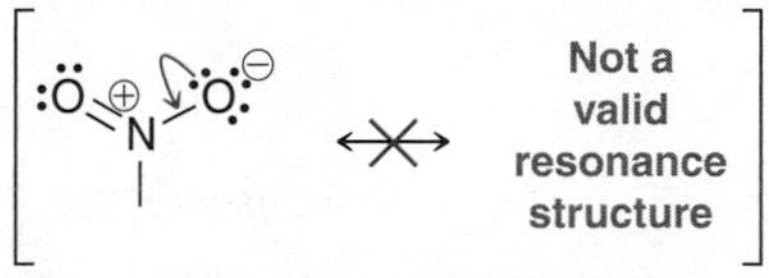

Why not? The curved arrow shown above violates the octet rule, because it would give the nitrogen atom five bonds. Remember that second-row elements can never have more than four bonds. There is only one way to draw the curved arrow above without violating the octet rule—we must draw a second curved arrow, like this:

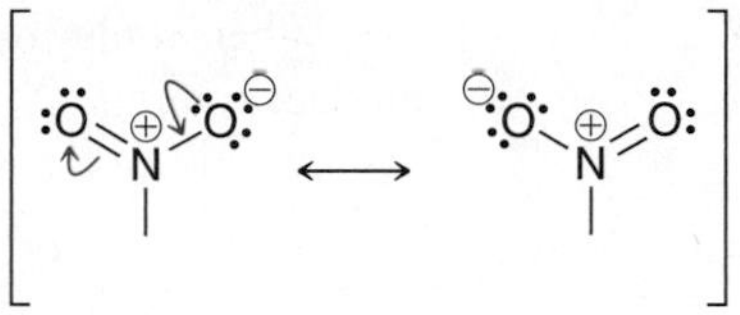

Look carefully. These two curved arrows are simply our first pattern (a lone pair next to a π bond). Notice that the charges have not been canceled. Rather, the location of the negative charge has moved from one oxygen atom to the other. The two resonance structures above are the only two valid resonance structures for a nitro group. In other words, the nitro group must be drawn with charge separation, even though the nitro group is overall neutral. The structure of the nitro group cannot be drawn without the charges.

4. *A π bond between two atoms of differing electronegativity.* Recall that electronegativity measures the ability of an atom to attract electrons. A chart of electronegativity values can be found in Section 1.11. For purposes of recognizing this pattern, we will focus on C=O and C=N double bonds.

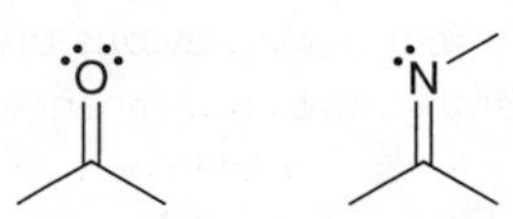

In these situations, we move the π bond up onto the electronegative atom to become a lone pair:

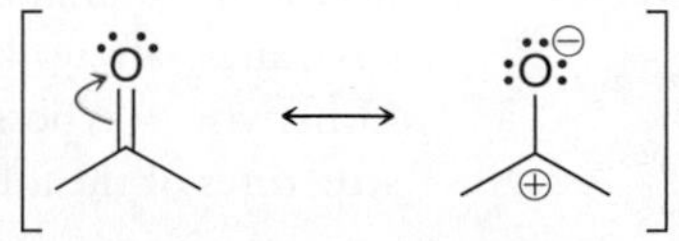

Notice what happens with the formal charges. A double bond is being separated into a positive and negative charge (this is the opposite of our second pattern, where the charges came together to form a double bond).

CONCEPTUAL CHECKPOINT

2.28 Draw a resonance structure for each of the compounds below.

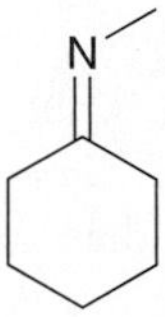
(a)

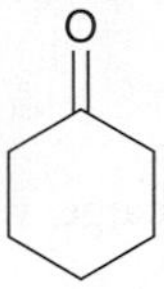
(b)

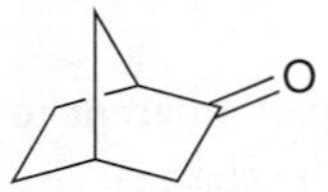
(c)

2.29 Draw a resonance structure of the compound below, which was isolated from the fruits of *Ocotea corymbosa*, a native plant of the Brazilian Cerrado.

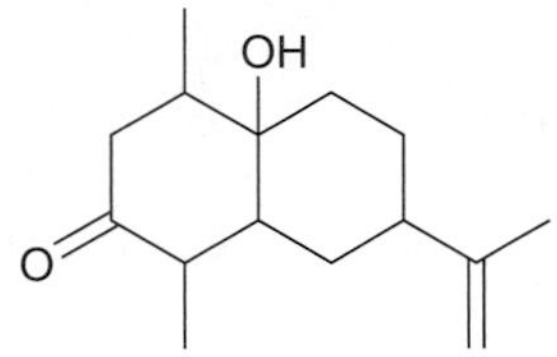

2.30 Draw a resonance structure of the compound shown below, called 2-heptanone, which is found in some kinds of cheese.

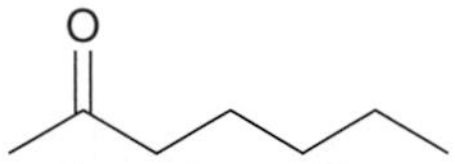

5. *Conjugated π bonds enclosed in a ring.* In one of the previous patterns, we introduced the term *conjugation* to refer to a system of alternating double and single bonds.

LOOKING AHEAD
In this molecule, called benzene, the electrons are delocalized. As a result, benzene exhibits significant resonance stabilization. We will explore the pronounced stability of benzene in Chapter 18.

When conjugated π bonds are enclosed in a ring, we push all of the π bonds over by one position:

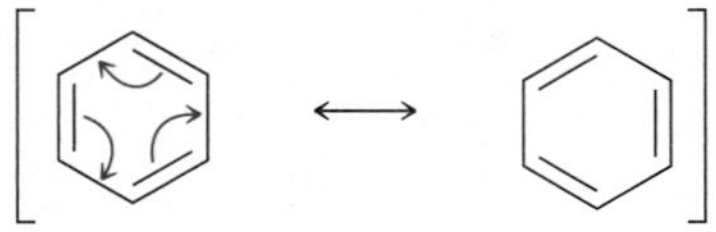

When drawing the resonance structure above, all of the π bonds can be pushed clockwise or they can all be pushed counterclockwise. Either way achieves the same result.

CONCEPTUAL CHECKPOINT

2.31 Fingolimod is a novel drug that has recently been developed for the treatment of multiple sclerosis. In April of 2008, researchers reported the results of phase III clinical trials of fingolimod, in which 70% of patients who took the drug daily for three years were relapse free. This is a tremendous improvement over previous drugs that only prevented relapse in 30% of patients. Draw a resonance structure of fingolimod:

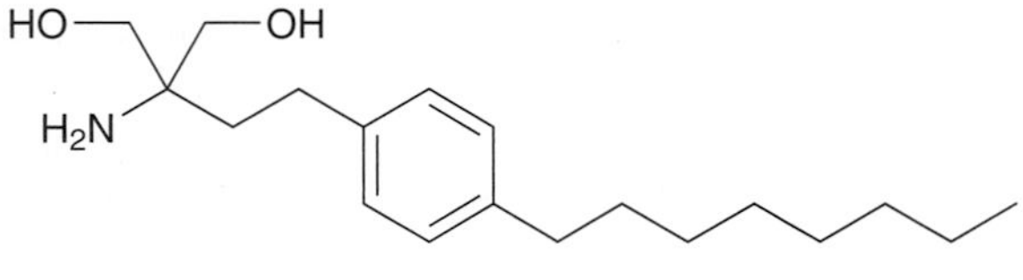
Fingolimod

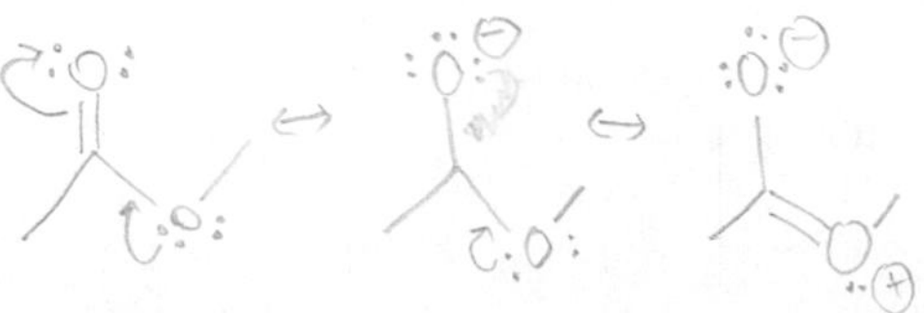

Figure 2.5 summarizes the five patterns for drawing resonance structures. Take special notice of the number of curved arrows used for each pattern. When drawing resonance structures, always begin by looking for the patterns that utilize only one curved arrow. Otherwise, it is possible to miss a resonance structure. For example, consider the resonance structures of the following compound:

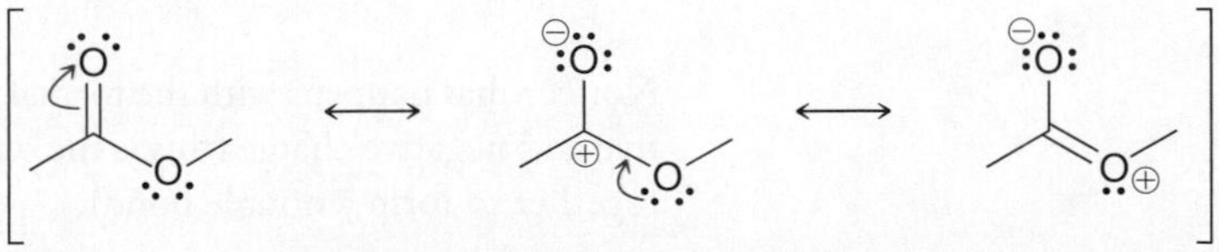

Notice that each pattern used in this example involves only one curved arrow. If we had started by recognizing a lone pair next to a π bond (which utilizes two curved arrows), then we might have missed the middle resonance structure above:

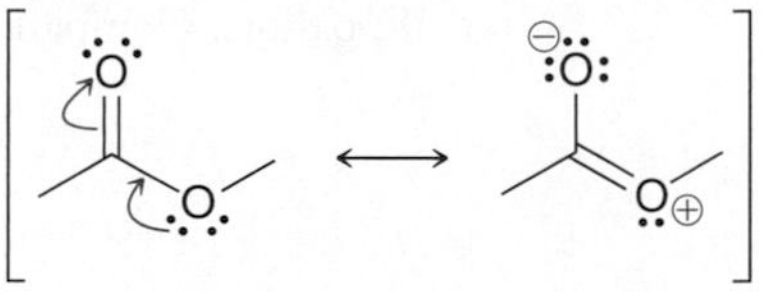

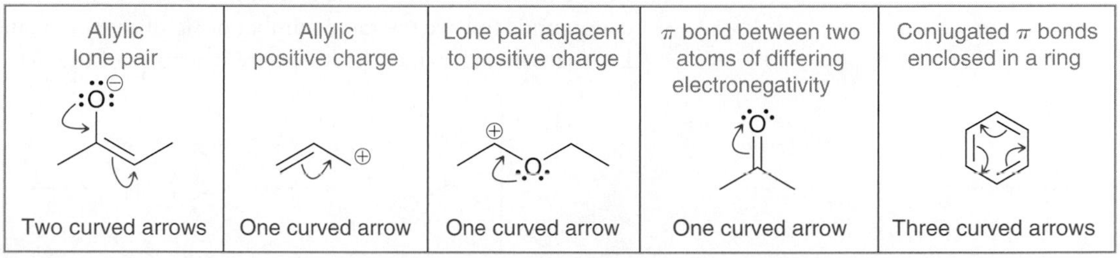
Allylic lone pair
Allylic positive charge
Lone pair adjacent to positive charge
π bond between two atoms of differing electronegativity
Conjugated π bonds enclosed in a ring
Two curved arrows
One curved arrow
One curved arrow
One curved arrow
Three curved arrows

FIGURE 2.5 A summary of the five patterns for drawing resonance structures.

CONCEPTUAL CHECKPOINT

2.32 Draw resonance structures for each of the following compounds:

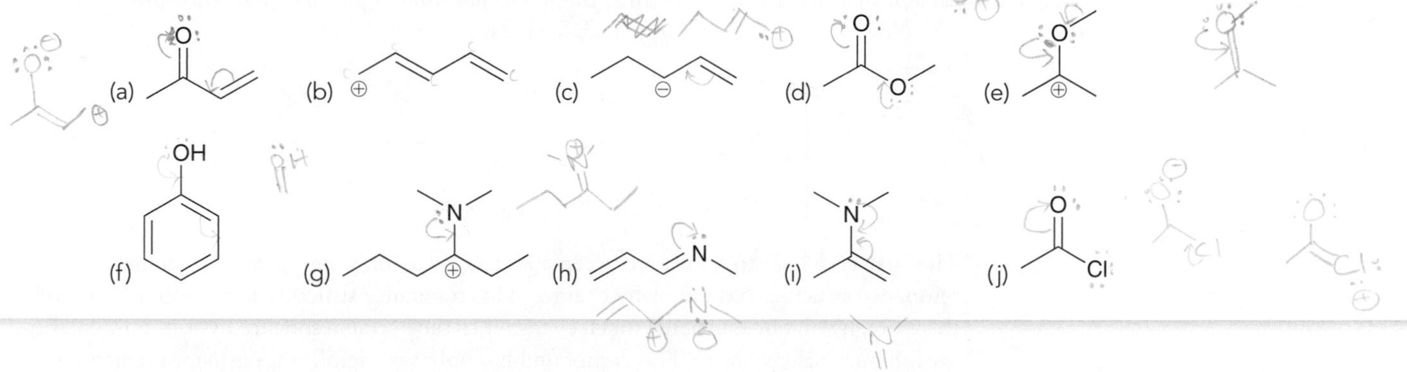

2.11 Assessing Relative Importance of Resonance Structures

Not all resonance structures are equally significant. A compound might have many valid resonance structures (structures that do not violate the two rules), but it is possible that one or more of the structures is insignificant. To understand what we mean by "insignificant," let's revisit the analogy used at the beginning of the chapter.

Recall the nectarine analogy (being a hybrid between a peach and plum) to explain the concept of resonance. Now, imagine that we create a new type of fruit that is a hybrid between *three* fruits: a peach, a plum, and a kiwi. Suppose that the hybrid fruit has the following character: 65% peach character, 34% plum character, and 1% kiwi character. This hybrid fruit will look almost exactly like a nectarine, because the amount of kiwi character is too small to affect the nature of the resulting hybrid. Even though the new fruit is actually a hybrid of all three fruits, it will look like a hybrid of only two fruits—because the kiwi character is *insignificant*.

A similar concept exists when comparing resonance structures. For example, a compound could have three resonance structures, but the three structures might not contribute equally to the overall resonance hybrid. One resonance structure might be the major contributor (like the peach), while another might be insignificant (like the kiwi). In order to understand the true nature of the compound, we must be able to compare resonance structures and determine which structures are major contributors and which structures are not significant.

Three rules will guide us in determining the significance of resonance structures:

1. *Minimize charges.* The best kind of structure is one without any charges. It is acceptable to have one or two charges, but structures with more than two charges should be avoided, if possible. Compare the following two cases:

Both compounds have a lone pair next to a C=O double bond. So we might expect these compounds to have the same number of significant resonance structures. But they do not. Let's see why. Consider the resonance structures of the first compound:

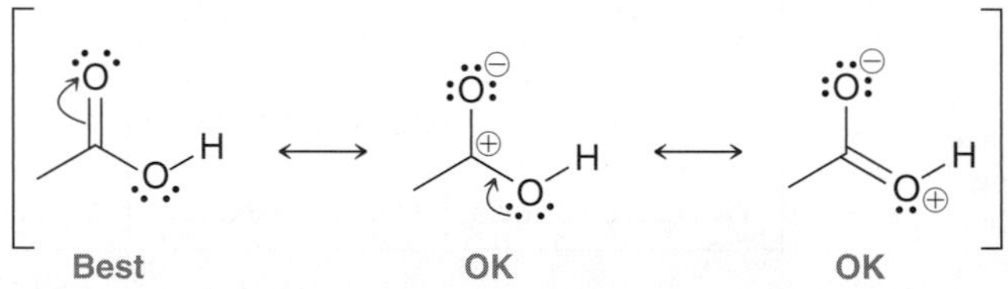

The first resonance structure is the major contributor to the overall resonance hybrid, because it has no charge separation. The other two drawings have charge separation, but there are only two charges in each drawing, so they are both significant resonance structures. They might not contribute as much character as the first resonance structure does, but they are still significant. Therefore, this compound has three significant resonance structures.

Now, let's try the same approach for the other compound:

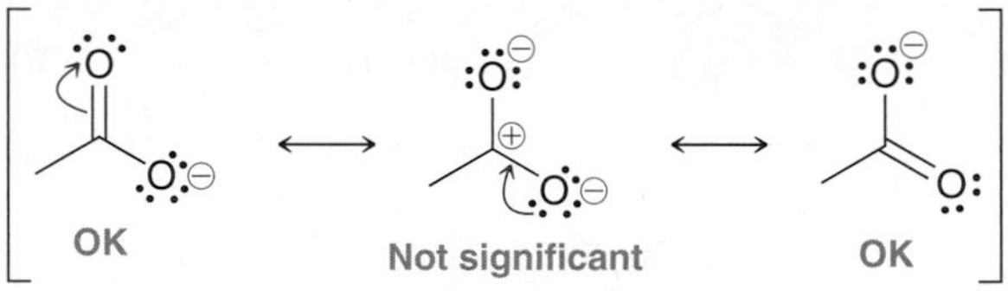

The first and last structures are acceptable (each has only one charge), but the middle resonance structure has too many charges. This resonance structure is not significant, and therefore, it will not contribute much character to the overall resonance hybrid. It is like the kiwi in our analogy above. This compound has only two significant resonance structures.

One notable exception to this rule involves compounds containing the nitro group ($-NO_2$), which have resonance structures with more than two charges. Why? We saw earlier that the structure of the nitro group must be drawn with charge separation in order to avoid violating the octet rule:

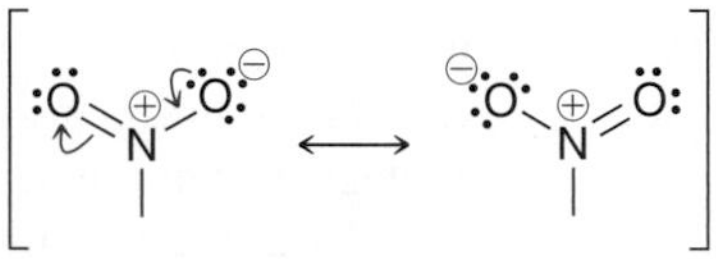

Therefore, the two charges of a nitro group don't really count when we are counting charges. Consider the following case as an example:

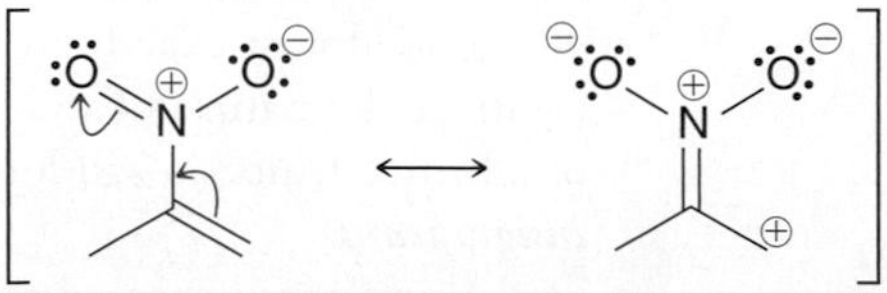

If we apply the rule about limiting charge separation to no more than two charges, then we might say that the second resonance structure above appears to have too many charges to be significant. But it actually is significant, because the two charges associated with the nitro group are not included in the count. We would consider the resonance structure above as if it only had two charges, and therefore it is significant.

2. *Electronegative atoms, such as N, O, and Cl, can bear a positive charge, but only if they possess an octet of electrons.* Consider the following as an example:

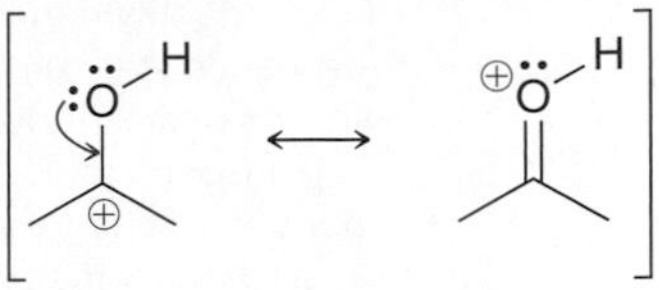

The second resonance structure is significant, even though it has a positive charge on oxygen. Why? Because the positively charged oxygen has an octet of electrons (three bonds plus one lone pair = 6 + 2 = 8 electrons). In fact, the second resonance structure above is even more significant than the first resonance structure. We might have thought otherwise, because the first resonance structure has a positive charge on carbon, which is generally much better than having a positive charge on an electronegative atom.

Nevertheless, the second resonance structure is more significant because all of its atoms achieve an octet. In the first structure, the oxygen has its octet, but the carbon only has six electrons. In the second resonance structure, both oxygen and carbon have an octet, which makes that structure more significant, even though the positive charge is on oxygen.

Here is another example, this time with the positive charge on nitrogen:

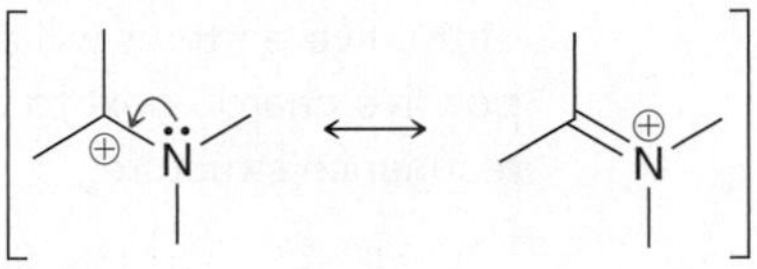

Once again, the second structure is significant, in fact, even more significant than the first. *In summary, the most significant resonance structures are generally those in which all atoms have an octet.*

3. *Avoid drawing a resonance structure in which two carbon atoms bear opposite charges.* Such resonance structures are generally insignificant, for example:

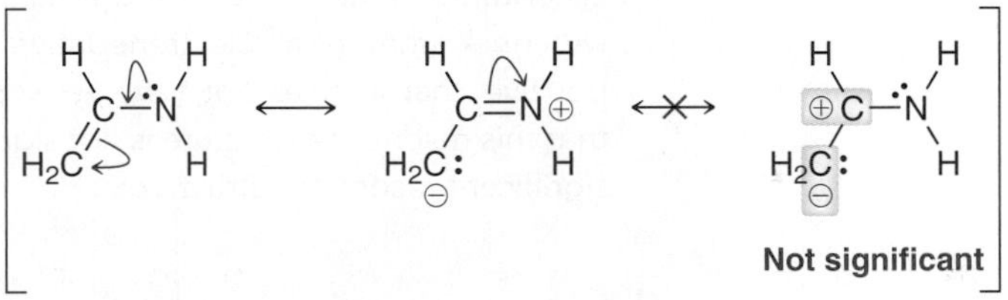

In this case, the third resonance structure is insignificant because it has both a C+ and a C−. The presence of carbon atoms with opposite charges, whether close to each other (as in the example above) or far apart, renders the structure insignificant. Throughout this text, we will see only one exception to this rule (in problem 18.54).

SKILLBUILDER

2.8 DRAWING SIGNIFICANT RESONANCE STRUCTURES

LEARN the skill Draw all significant resonance structures of the following compound:

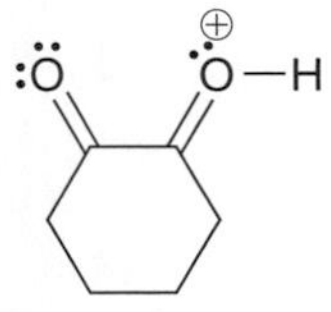

SOLUTION

STEP 1 Using the five patterns, identify a resonance structure.

We begin by looking for any of the five patterns. This compound contains a C=O bond (a π bond between two atoms of differing electronegativity), and we can therefore draw the following resonance structure:

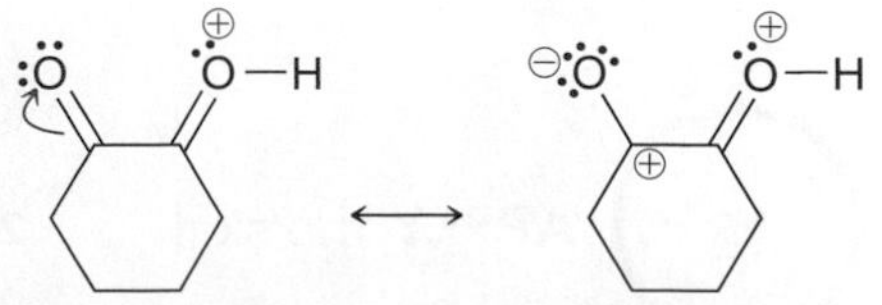

This resonance structure is valid, because it was generated using one of the five patterns. However, it has too many charges, and it is therefore not significant. In general, try to avoid drawing resonance structures with three or more charges.

STEP 2
Identify if the resonance structure is significant by inspecting the number of charges and the number of electrons on heteroatoms.

Next, we look at the other C═O bond, and we try the same pattern:

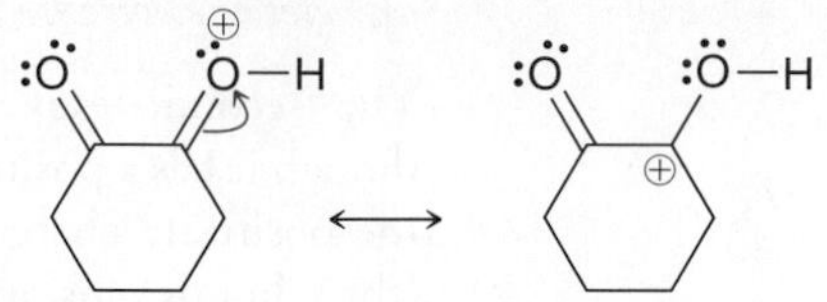

To determine if this resonance structure is significant, we ask three questions:

1. *Does this structure have an acceptable number of charges?* Yes, it has only one charge (on the carbon atom), which is perfectly acceptable.

2. *Do all electronegative atoms have an octet?* Yes, both oxygen atoms have an octet of electrons.

3. *Does the structure avoid having carbon atoms with opposite charges?* Yes.

This resonance structure passes the test, and therefore it is a significant resonance structure.

Now that we have found a significant resonance structure, we analyze it to see if any of the five patterns will allow us to draw another resonance structure. In this case, there is a positive charge next to a π bond. So we draw one curved arrow, generating the following resonance structure:

STEP 3
Repeat steps 1 and 2 to identify any other significant resonance structures.

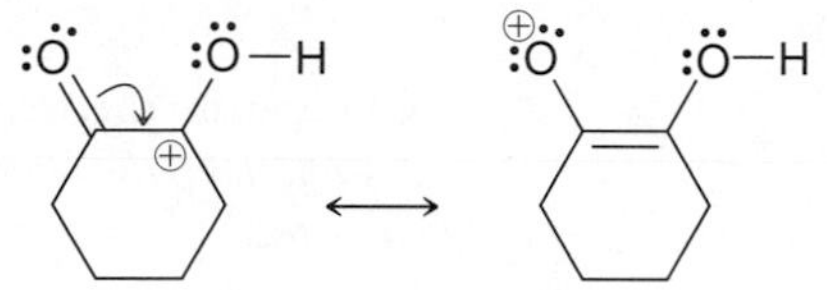

To determine whether the structure is significant, we first check to see whether it has an acceptable number of charges. It has only one charge, which is perfectly acceptable. Next we check whether all electronegative atoms have an octet. The oxygen atom bearing the positive charge does not have an octet of electrons, which is not acceptable and means that this resonance structure is not significant. In summary, this compound has the following significant resonance structure:

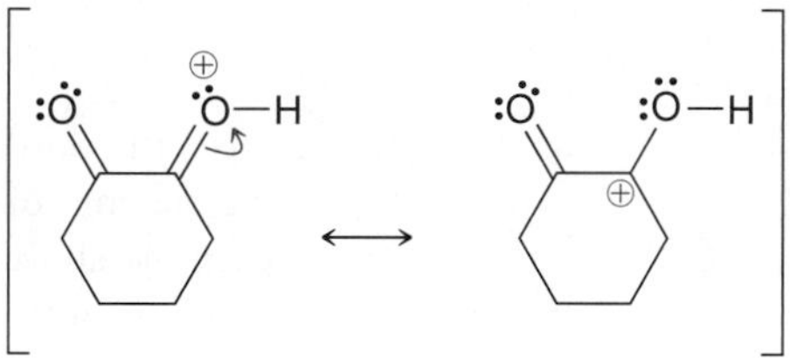

PRACTICE the skill **2.33** Draw all significant resonance structures for each of the following compounds:

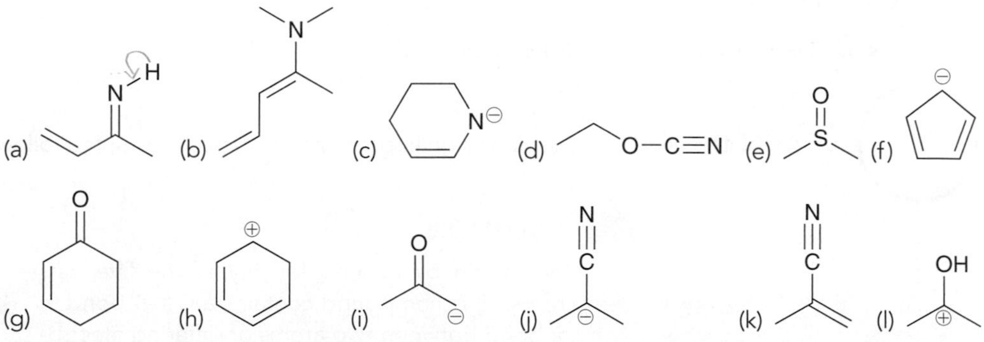

APPLY the skill **2.34** Use resonance structures to help you identify all sites of low electron density (δ+) in the following compound:

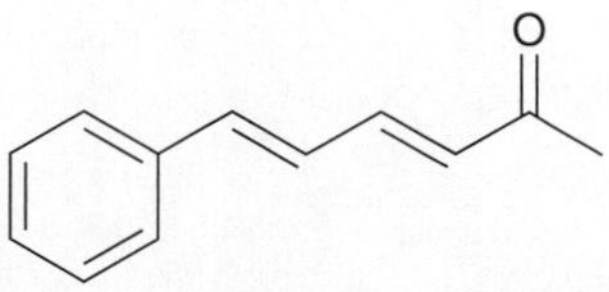

2.35 Use resonance structures to help you identify all sites of high electron density (δ−) in the following compound:

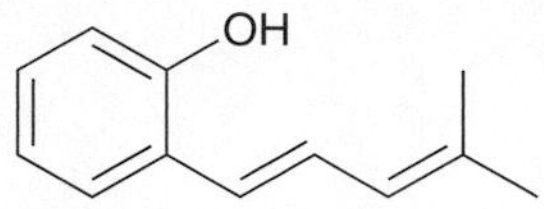

need more **PRACTICE?** **Try Problems 2.45, 2.48, 2.59, 2.60, 2.62, 2.65, 2.66**

2.12 Delocalized and Localized Lone Pairs

In this section, we will explore some important differences between lone pairs that participate in resonance and lone pairs that do not participate in resonance.

Delocalized Lone Pairs

Recall that one of our five patterns was a lone pair that is allylic to a π bond. Such a lone pair will participate in resonance and is said to be **delocalized.** When an atom possesses a delocalized lone pair, the geometry of that atom is affected by the presence of the lone pair. As an example, consider the structure of an amide:

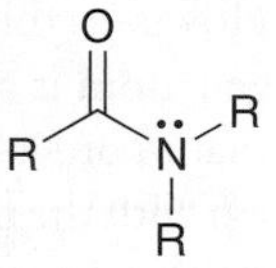

An amide

The rules we learned in Section 1.10 would suggest that the nitrogen atom should be sp^3 hybridized and trigonal pyramidal, but this is not correct. Instead, the nitrogen atom is actually sp^2 hybridized and trigonal planar. Why? The lone pair is participating in resonance and is therefore delocalized:

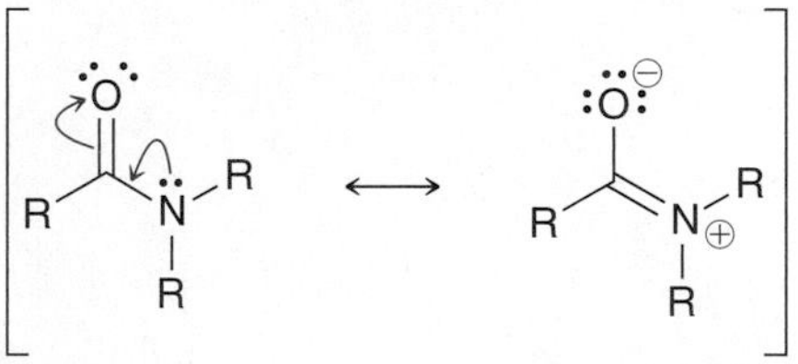

In the second resonance structure above, the nitrogen atom does not bear a lone pair. Rather, the nitrogen atom bears a *p* orbital being used to form a π bond. In that resonance structure, the nitrogen atom is clearly sp^2 hybridized. This creates a conflict: How can the nitrogen atom be sp^3 hybridized in one resonance structure and sp^2 hybridized in the other structure? That would imply that the geometry of the nitrogen atom is flipping back and forth between trigonal pyramidal and trigonal planar. This cannot be the case, because resonance is not a physical process. The nitrogen atom is actually sp^2 hybridized and trigonal planar in both resonance structures. How? The nitrogen atom has a delocalized lone pair, and it therefore occupies a *p* orbital (rather than a hybridized orbital), so that it can overlap with the *p* orbitals of the π bond (Figure 2.6).

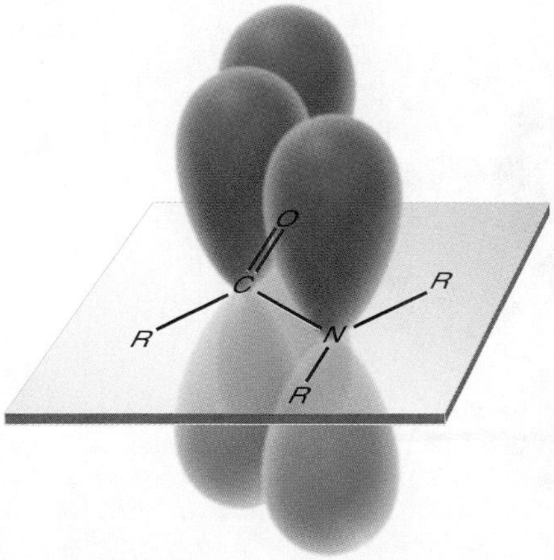

FIGURE 2.6
An illustration of the overlapping atomic *p* orbitals of an amide.

Whenever a lone pair participates in resonance, it will occupy a *p* orbital rather than a hybridized orbital, and this must be taken into account when predicting geometry. This will be extremely important in Chapter 25 when we discuss the three-dimensional shape of proteins.

Localized Lone Pairs

A **localized** lone pair, by definition, is a lone pair that does not participate in resonance. In other words, the lone pair is not allylic to a π bond:

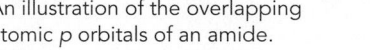

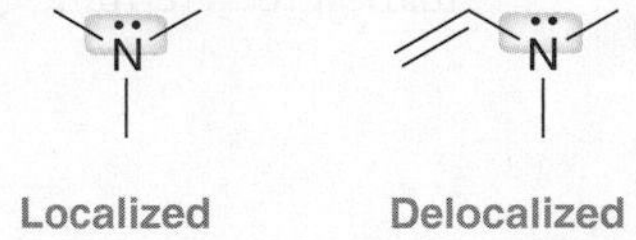

In some cases, a lone pair might appear to be delocalized even though it is actually localized. For example, consider the structure of pyridine:

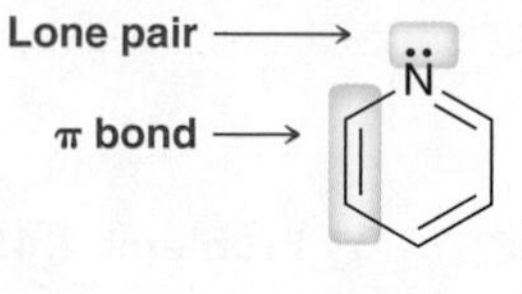

The lone pair in pyridine appears to be allylic to a π bond, and it is tempting to use our pattern to draw the following resonance structure:

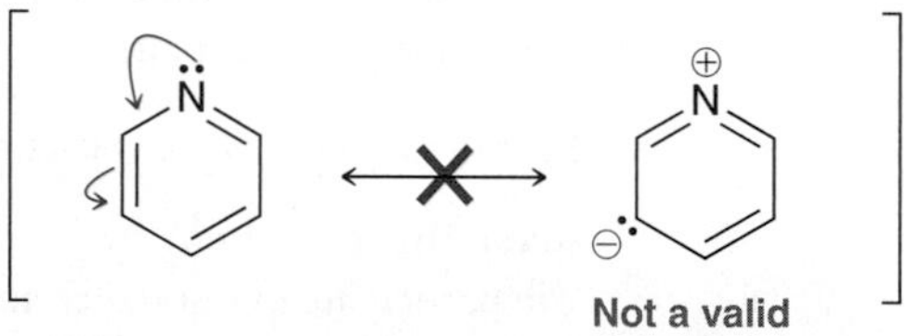

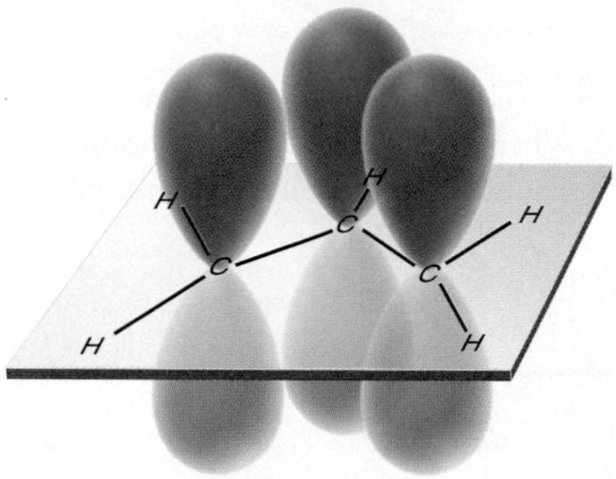

FIGURE 2.7
Resonance applies to systems that involve overlapping *p* orbitals that form a "conduit."

However, this resonance structure is not valid. Why not? In this case, the lone pair on the nitrogen atom is actually not participating in resonance, even though it is next to a π bond. Recall, that in order for a lone pair to participate in resonance, it must occupy a *p* orbital that can overlap with the neighboring *p* orbitals, forming a "conduit" (Figure 2.7). In the case of pyridine, the nitrogen atom is already using a *p* orbital for the π bond (Figure 2.8). The nitrogen atom can only use one *p* orbital to join in the conduit shown in Figure 2.8, and that *p* orbital is already being utilized by the π bond. As a result, the lone pair cannot join in the conduit, and therefore it cannot participate in resonance. In this case, the lone pair occupies an sp^2-hybridized orbital, which is in the plane of the ring.

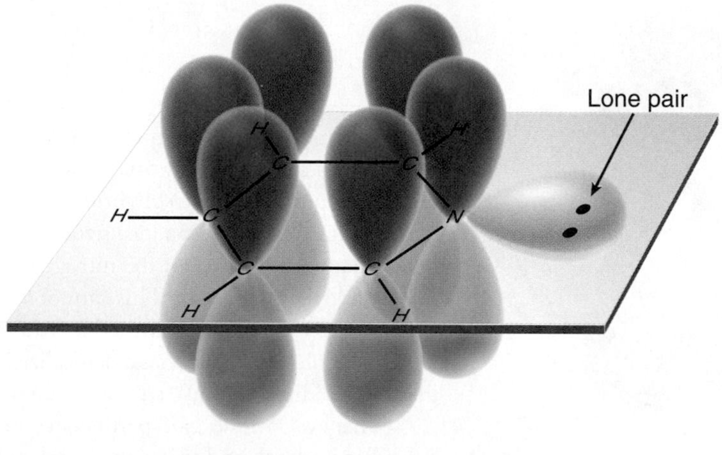

FIGURE 2.8
The overlapping *p* orbitals of pyridine.

Here is the bottom line: *Whenever an atom possesses both a π bond and a lone pair, they will not both participate in resonance.* In general, only the π bond will participate in resonance, and the lone pair will not.

Let's get some practice identifying localized and delocalized lone pairs and using that information to determine geometry.

SKILLBUILDER

2.9 IDENTIFYING LOCALIZED AND DELOCALIZED LONE PAIRS

LEARN the skill

Histamine is a compound that plays a key role in many biological functions. Most notably, it is involved in immune responses, where it triggers the symptoms of allergic reactions:

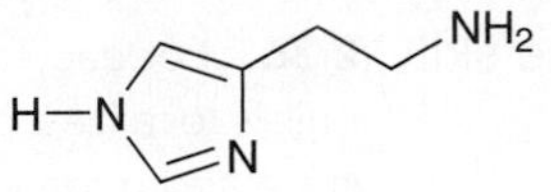

Histamine

Each nitrogen atom exhibits a lone pair. In each case, identify whether the lone pair is localized or delocalized, and then use that information to determine the hybridization state and geometry for each nitrogen atom in histamine.

SOLUTION

Let's begin with the nitrogen on the right side of the compound. This lone pair is localized, and therefore we can use the method outlined in Section 1.10 to determine the hybridization state and geometry:

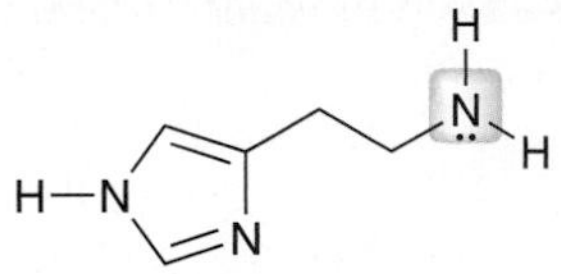

There are 3 bonds and 1 lone pair, and therefore:

1) Steric number = 3 + 1 = 4
2) 4 = sp^3 = electronically tetrahedral
3) Arrangement of atoms = <u>trigonal pyramidal</u>

This lone pair is not participating in resonance, so our method accurately predicts the geometry to be trigonal pyramidal.

Now, let's consider the nitrogen atom on the left side of the compound. The lone pair on that nitrogen atom is delocalized by resonance:

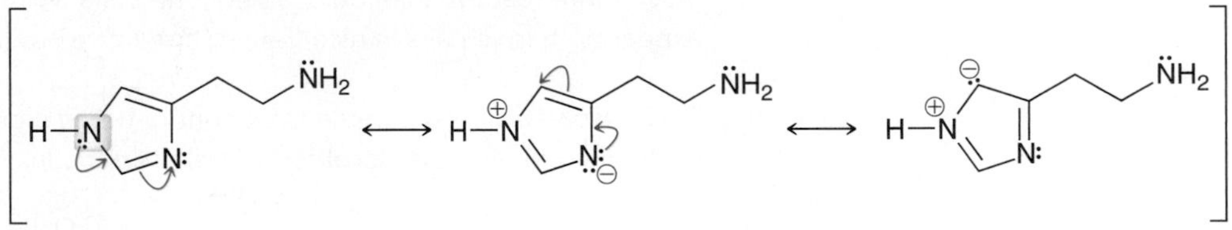

Therefore, this lone pair is actually occupying a *p* orbital, rendering the nitrogen atom sp^2 hybridized, rather than sp^3 hybridized. As a result, the geometry is trigonal planar.

Now let's consider the remaining nitrogen atom:

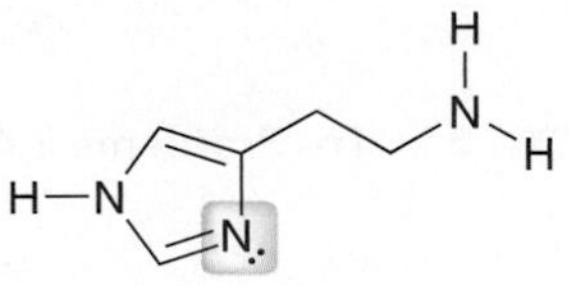

This nitrogen atom already has a π bond participating in resonance. Therefore, the lone pair cannot also participate in resonance. In this case, the lone pair must be localized. The nitrogen atom is in fact sp^2 hybridized and exhibits bent geometry.

To summarize, each of the nitrogen atoms in histamine has a different geometry:

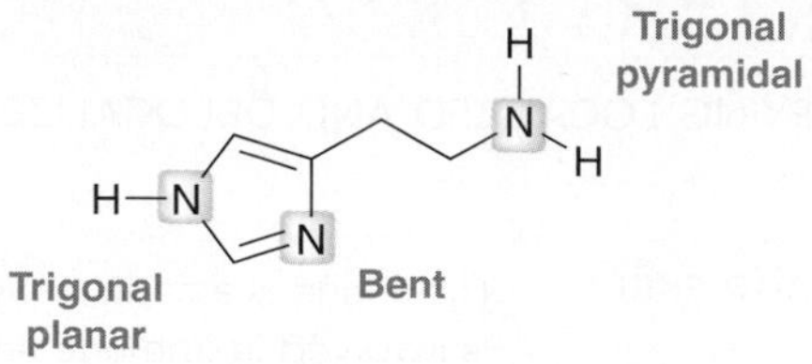

PRACTICE the skill **2.36** For each compound below, identify all lone pairs and indicate whether each lone pair is localized or delocalized. Then, use that information to determine the hybridization state and geometry for each atom that exhibits a lone pair.

(a) 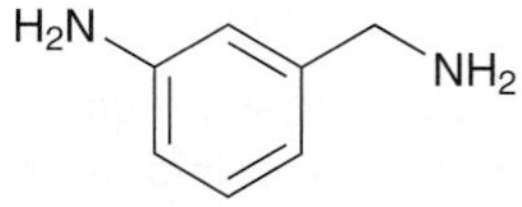

(b) 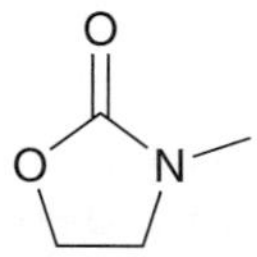

(c) 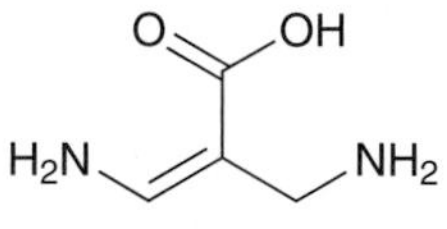

(d) 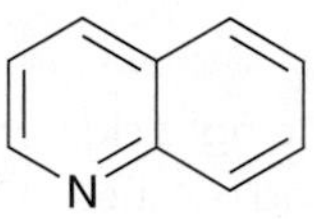

(e) 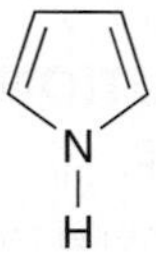

(f) 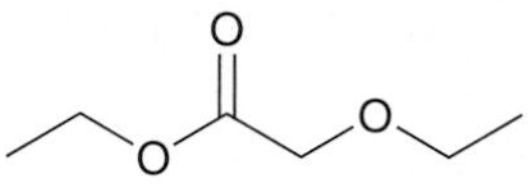

APPLY the skill **2.37** Nicotine is a toxic substance present in tobacco leaves.

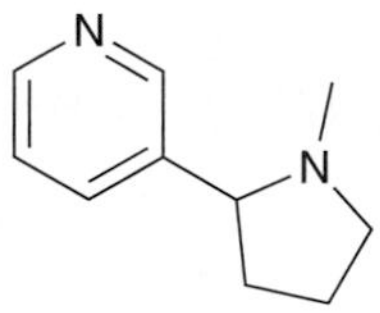

Nicotine

There are two lone pairs in the structure of nicotine. In general, localized lone pairs are much more reactive than delocalized lone pairs. With this information in mind, do you expect both lone pairs in nicotine to be reactive? Justify your answer.

2.38 Isoniazid is used in the treatment of tuberculosis and multiple sclerosis. Identify each lone pair as either localized or delocalized. Justify your answer in each case.

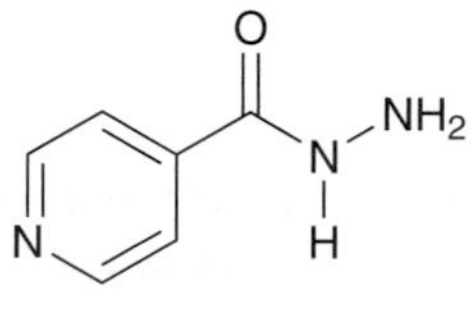

Isoniazid

need more **PRACTICE?** **Try Problems 2.47, 2.61**

REVIEW OF CONCEPTS AND VOCABULARY

SECTION 2.1

- Chemists use many different drawing styles to communicate structural information, including Lewis structures, **partially condensed structures**, and **condensed structures**.
- The molecular formula does not provide structural information.

SECTION 2.2

- In **bond-line structures,** carbon atoms and most hydrogen atoms are not drawn.
- Bond-line structures are faster to draw and easier to interpret than other drawing styles.

SECTION 2.3

- A **functional group** is a characteristic group of atoms/bonds that show a predictable chemical behavior.
- The chemistry of every organic compound is determined by the functional groups present in the compound.

SECTION 2.4

- A formal charge is associated with any atom that does not exhibit the appropriate number of valence electrons.
- When a carbon atom bears either a positive or negative charge, it will have only three, rather than four, bonds.

SECTION 2.5

- Lone pairs are often not drawn in bond-line structures. It is important to recognize that these lone pairs are present.

SECTION 2.6

- In bond-line structures, a **wedge** represents a group coming out of the page, and a **dash** represents a group behind the page.
- Other drawings used to show three dimensionality include **Fischer projections** and **Haworth projections**.

SECTION 2.7

- Bond-line structures are inadequate in some situations, and an approach called **resonance** is required.
- **Resonance structures** are separated by double-headed arrows and surrounded by brackets:

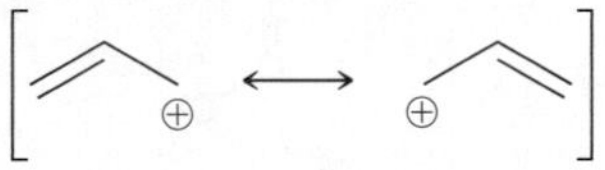

- **Resonance stabilization** refers to the **delocalization** of either a positive charge or a negative charge via resonance.

SECTION 2.8

- **Curved arrows** are tools for drawing resonance structures.
- When drawing curved arrows for resonance structures, avoid breaking a single bond and never exceed an octet for second-row elements.

SECTION 2.9

- All formal charges must be shown when drawing resonance structures.

SECTION 2.10

- Resonance structures are most easily drawn by looking for the following five patterns:
  1. An **allylic** lone pair
  2. An allylic positive charge
  3. A lone pair adjacent to a positive charge
  4. A π bond between two atoms of differing electronegativity
  5. **Conjugated π bonds** enclosed in a ring.
- When drawing resonance structures, always begin by looking for the patterns that utilize only one curved arrow.

SECTION 2.11

- There are three rules for identifying significant **resonance structures**:
  1. Minimize charge.
  2. Electronegative atoms (N, O, Cl, etc.) can bear a positive charge, but only if they possess an octet of electrons.
  3. Avoid drawing a resonance structure in which two carbon atoms bear opposite charges.

SECTION 2.12

- A **delocalized lone pair** participates in resonance and occupies a *p* orbital.
- A **localized lone pair** does not participate in resonance.

KEY TERMINOLOGY

SKILLBUILDER REVIEW

2.1 CONVERTING BETWEEN DIFFERENT DRAWING STYLES

Draw the Lewis structure of this compound.

$(CH_3)_3C\ddot{\underset{\cdot\cdot}{O}}CH_3$

STEP 1 Draw each group separately.

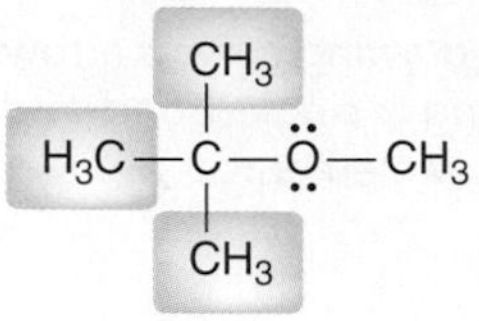

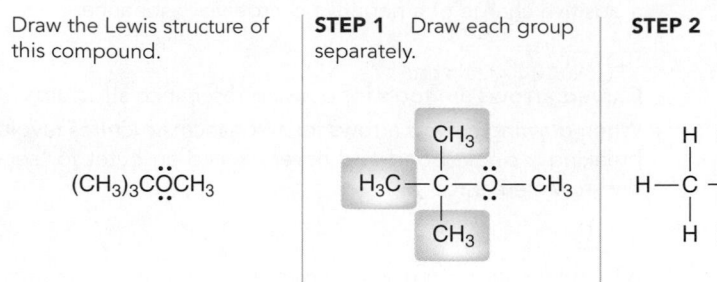

STEP 2 Draw all C—H bonds.

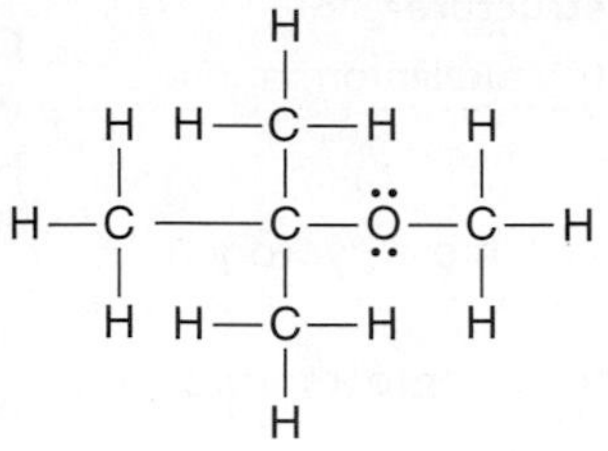

Try Problems 2.1–2.4, 2.49, 2.50

2.2 READING BOND-LINE STRUCTURES

Identify all carbon atoms and hydrogen atoms.

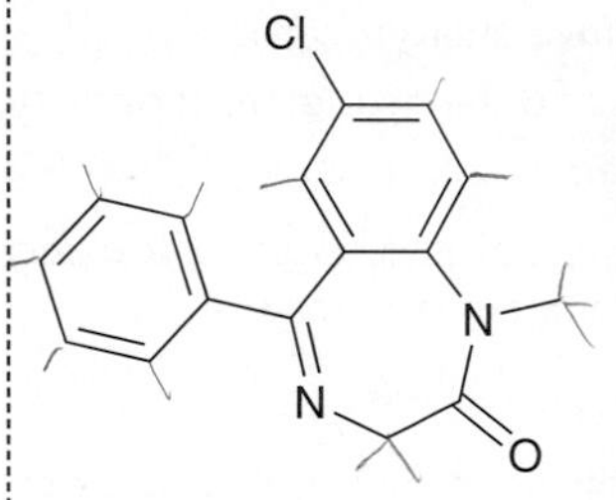

STEP 1 The end of every line represents a carbon atom.

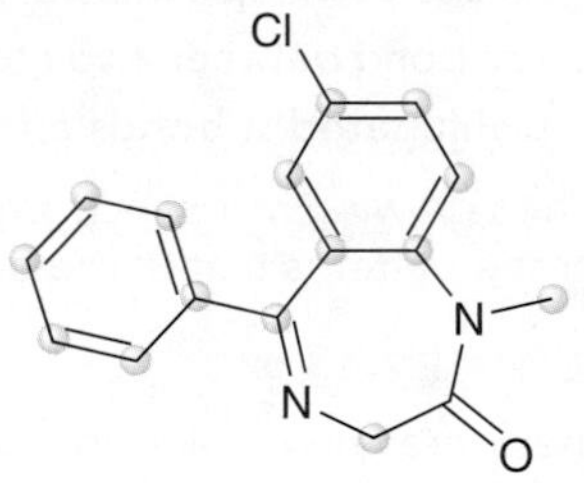

STEP 2 Each carbon atom will possess enough hydrogen atoms in order to achieve four bonds.

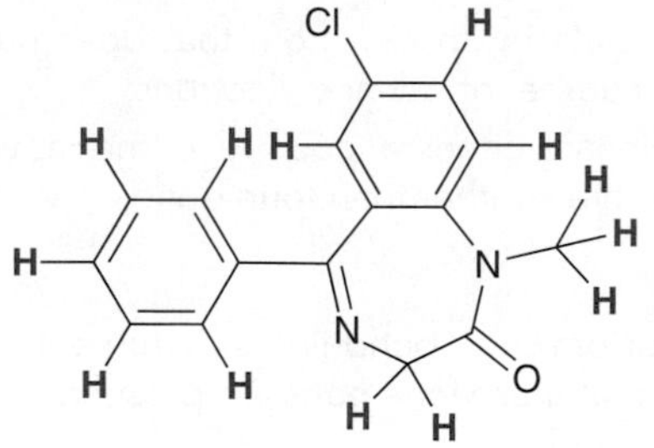

Try Problems 2.5–2.7, 2.39, 2.49, 2.52

2.3 DRAWING BOND-LINE STRUCTURES

Draw a bond-line drawing of this compound.

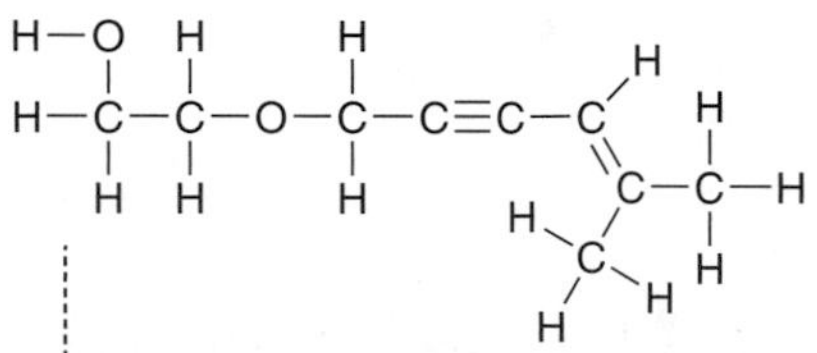

STEP 1 Delete all hydrogen atoms except for those connected to heteroatoms.

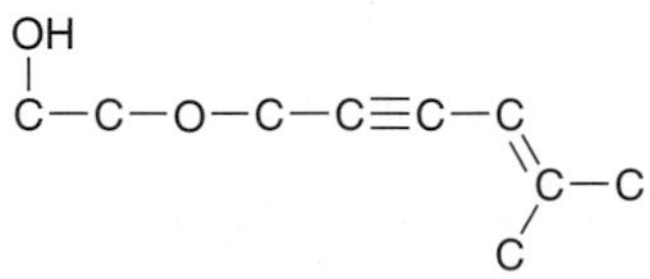

STEP 2 Draw in zig-zag format, keeping triple bonds linear.

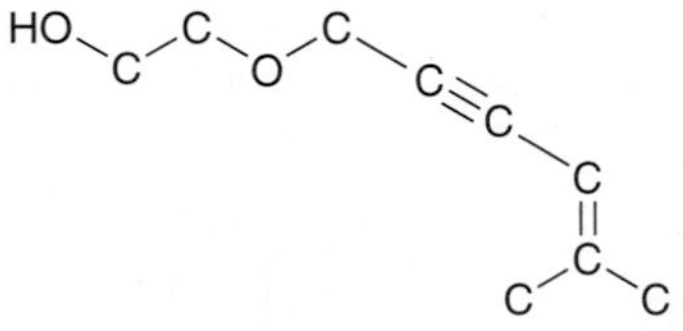

STEP 3 Delete all carbon atoms.

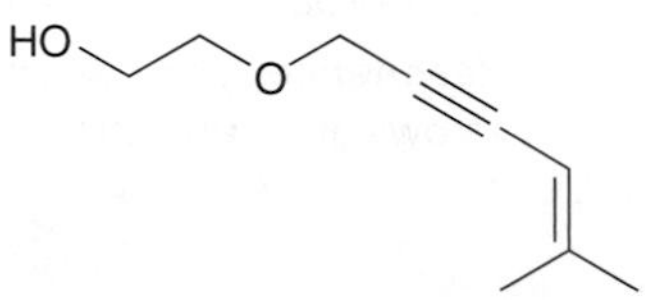

Try Problems 2.8–2.10, 2.40, 2.41, 2.54, 2.58

2.4 IDENTIFYING LONE PAIRS ON OXYGEN ATOMS

Oxygen with a negative charge...

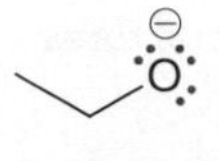

...has three lone pairs.

A neutral oxygen atom...

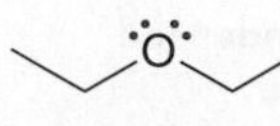

...has two lone pairs.

Oxygen with a positive charge...

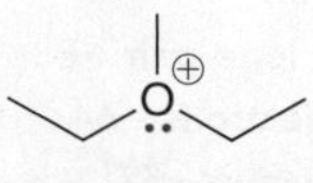

...has one lone pair.

Try Problems 2.14, 2.15, 2.43

2.5 IDENTIFYING LONE PAIRS ON NITROGEN ATOMS

Nitrogen with a negative charge...

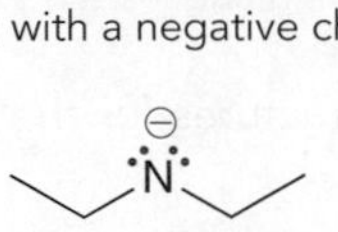

...has two lone pairs.

A neutral nitrogen atom...

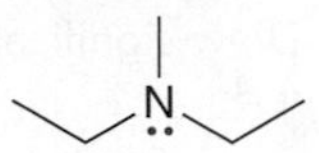

...has one lone pair.

Nitrogen with a positive charge...

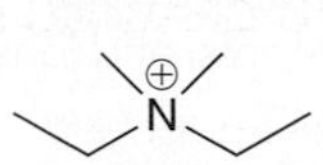

...has no lone pairs.

Try Problems 2.16–2.19, 2.39

2.6 IDENTIFYING VALID RESONANCE ARROWS

RULE 1: The tail of a curved arrow cannot be placed on a single bond.

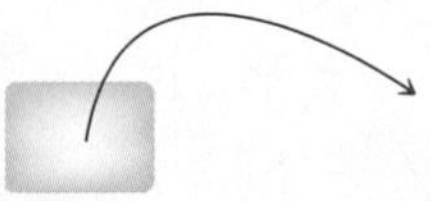

Tail

RULE 2: The head of a curved arrow cannot result in a bond causing a second-row element to exceed an octet.

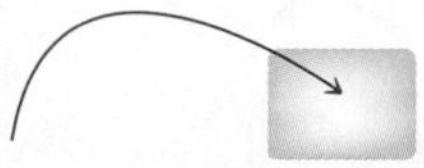

Head

Try Problems 2.21, 2.22, 2.51

2.7 ASSIGNING FORMAL CHARGES IN RESONANCE STRUCTURES

Read the curved arrows.

This negative charge...

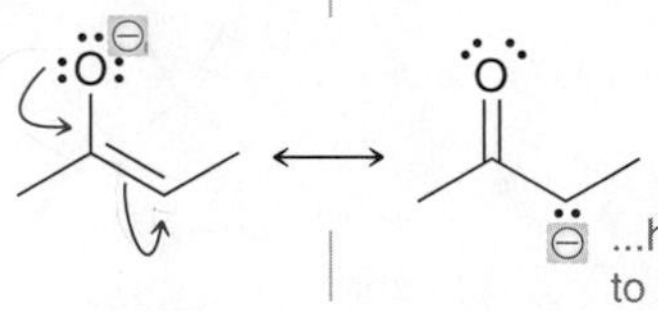

...has been pushed to here.

Try Problems 2.23, 2.24, 2.44, 2.53

2.8 DRAWING SIGNIFICANT RESONANCE STRUCTURES

Too many charges | Oxygen does not have an octet.

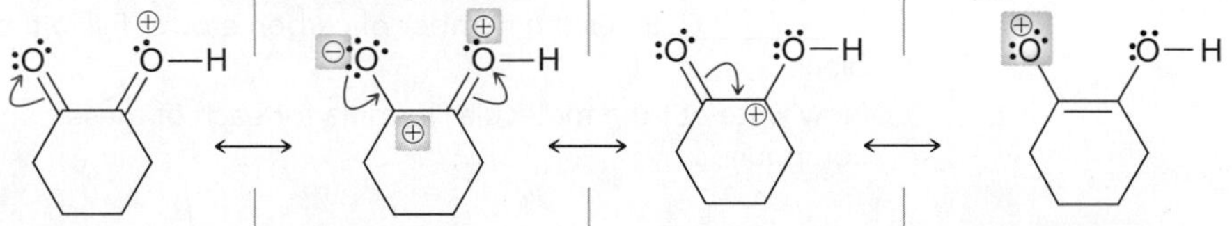

Significant | Not significant | Significant | Not significant

Try Problems 2.33–2.35, 2.45, 2.48, 2.59, 2.60, 2.62, 2.65, 2.66

2.9 IDENTIFYING LOCALIZED AND DELOCALIZED LONE PAIRS

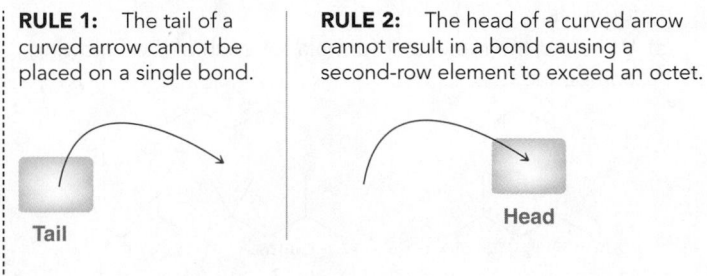

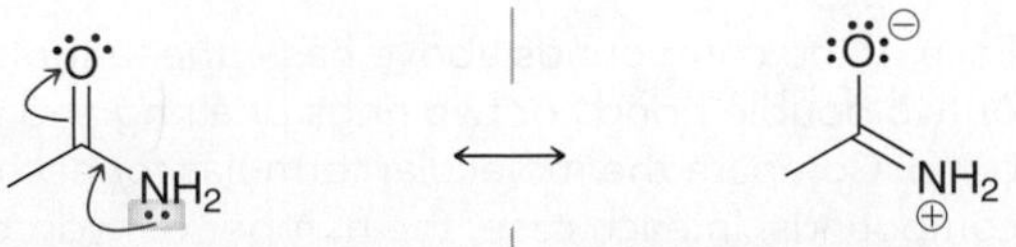

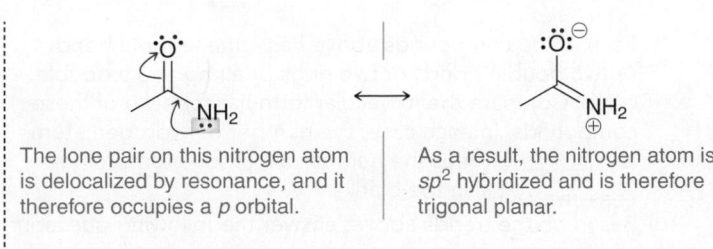

The lone pair on this nitrogen atom is delocalized by resonance, and it therefore occupies a *p* orbital.

As a result, the nitrogen atom is sp^2 hybridized and is therefore trigonal planar.

Try Problems 2.36–2.38, 2.47, 2.61

PRACTICE PROBLEMS

Note: Most of the Problems are available within *WileyPLUS*, an online teaching and learning solution.

2.39 Draw all carbon atoms, hydrogen atoms, and lone pairs for the following compounds:

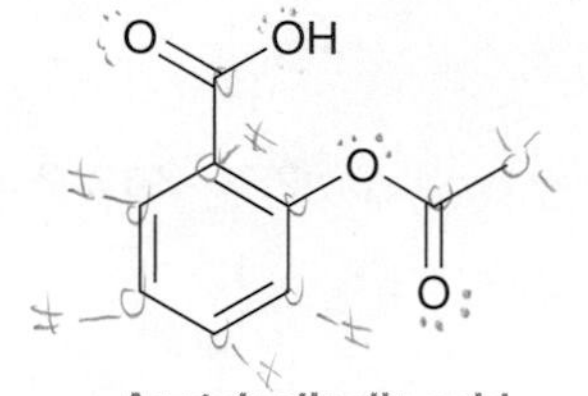

Acetylsalicylic acid (aspirin)

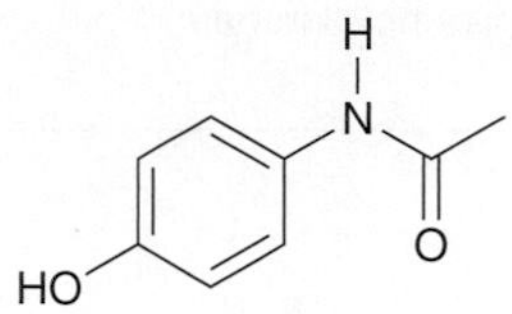

Acetaminophen (Tylenol)

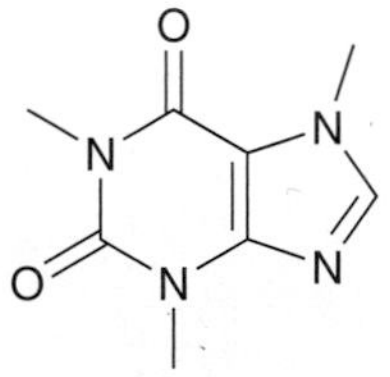

Caffeine

2.40 Draw bond-line structures for all constitutional isomers of C_4H_{10}.

2.41 Draw bond-line structures for all constitutional isomers of C_5H_{12}.

2.42 Draw bond-line structures for vitamin A and vitamin C:

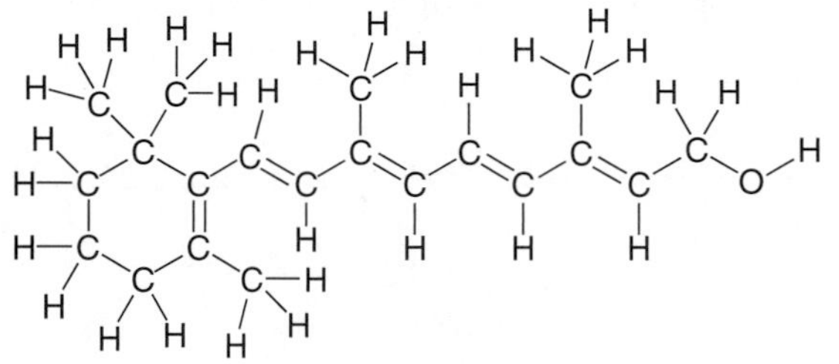

Vitamin A

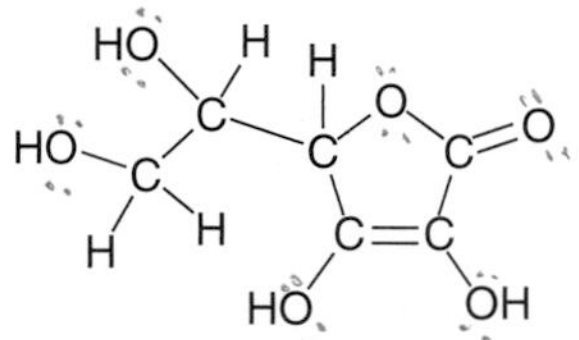

Vitamin C

2.43 How many lone pairs are found in the structure of vitamin C?

2.44 Identify the formal charge in each case below:

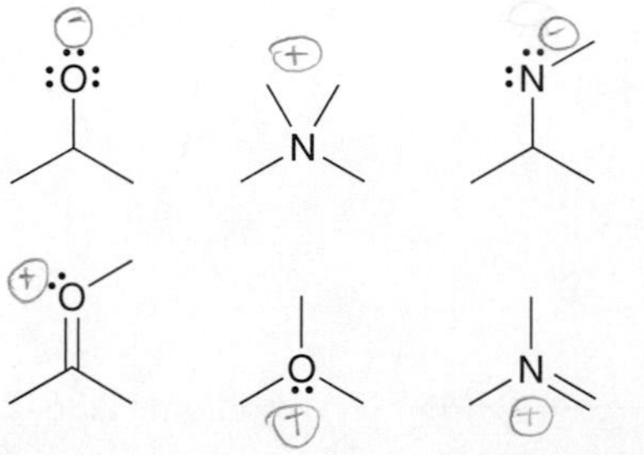

2.45 Draw significant resonance structures for the following compound:

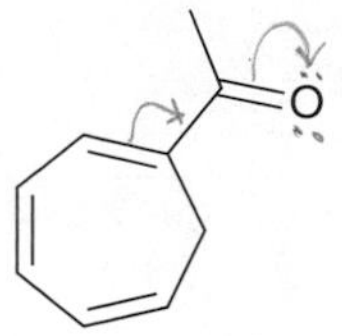

2.46 *Learning to extract structural information from molecular formulas:*

(a) Write out the molecular formula for each of the following compounds:

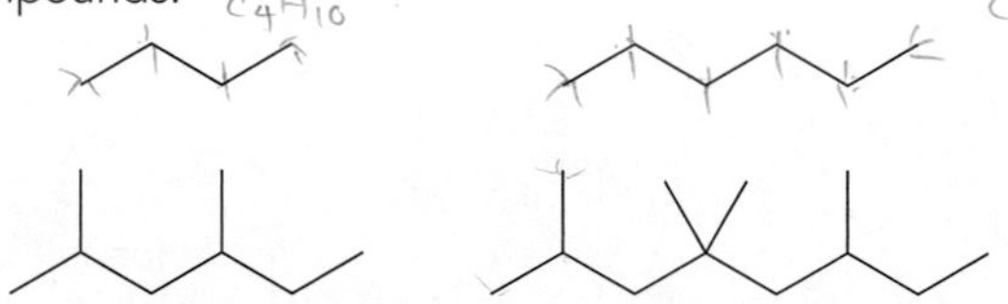

Compare the molecular formulas for the above compounds and fill in the blanks in the following sentence: The number of hydrogen atoms is equal to ________ times the number of carbon atoms, plus ________.

(b) Now write out the molecular formula for each of these compounds:

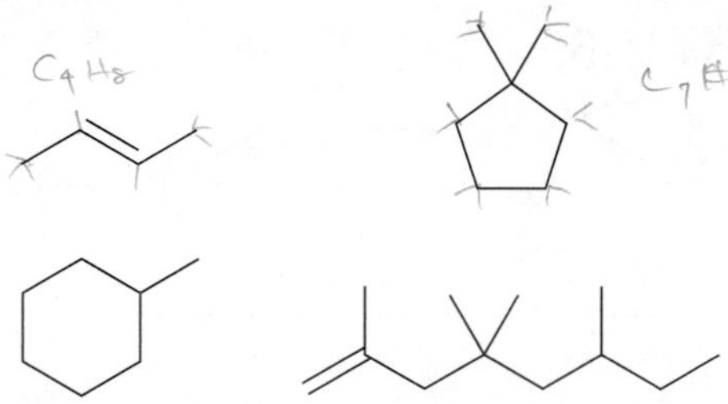

Each of the compounds above has *either* a double bond *or* a ring. Compare the molecular formulas for each of these compounds. In each case, the number of hydrogen atoms is ________ times the number of carbon atoms. Fill in the blank.

(c) Now write out the molecular formula for each of these compounds:

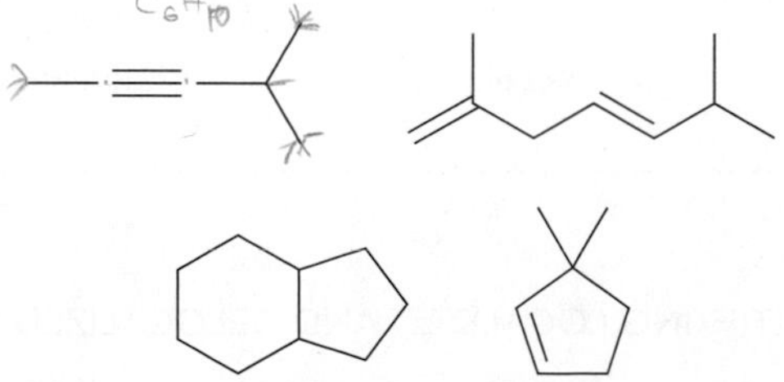

Each of the compounds above has *either* a triple bond *or* two double bonds *or* two rings *or* a ring and a double bond. Compare the molecular formulas for each of these compounds. In each case, the number of hydrogen atoms is ________ times the number of carbon atoms minus ________. Fill in the blanks.

(d) Based on the trends above, answer the following questions about the structure of a compound with molecular formula

$C_{24}H_{48}$. Is it possible for this compound to have a triple bond? Is it possible for this compound to have a double bond? no

(e) Draw all constitutional isomers that have the molecular formula C_4H_8.

2.47 Each compound below exhibits one lone pair. In each case, identify the type of atomic orbital in which the lone pair is contained.

(a) 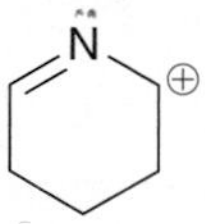 (b) 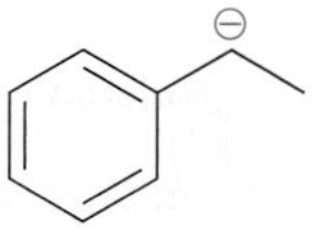 (c) 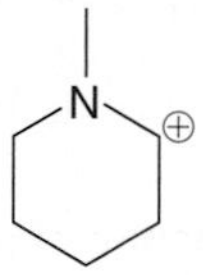

2.48 Draw all significant resonance structures for each of the following compounds:

(a) 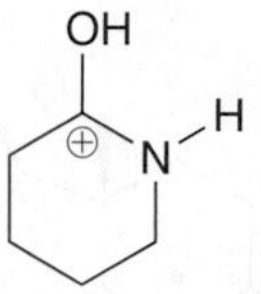 (b) 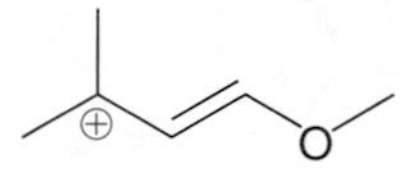 (c) 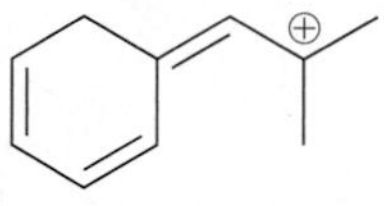

2.49 Write a condensed structural formula for each of the following compounds:

(a) 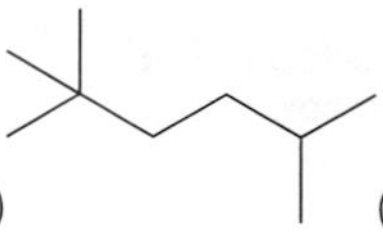 (b) 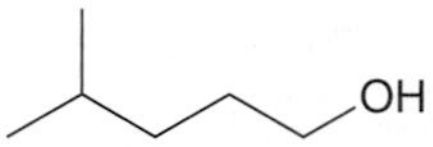 (c) 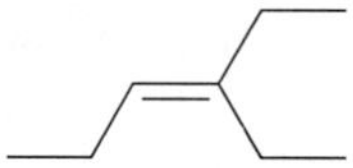

2.50 What is the molecular formula for each compound in the previous problem?

2.51 Which of the following drawings is not a resonance structure for 1-nitrocyclohexene? Explain why it cannot be a valid resonance structure.

(a) 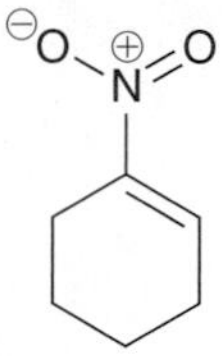 (b) 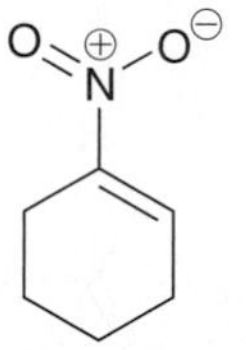 (c) 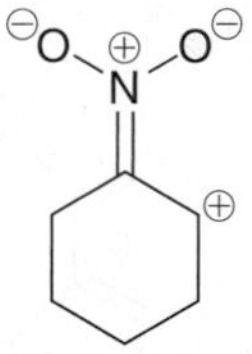 (d) 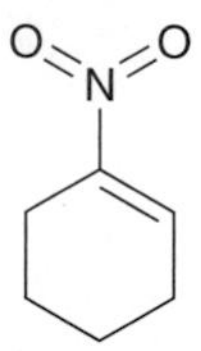

2.52 Identify the number of carbon atoms and hydrogen atoms in the compound below:

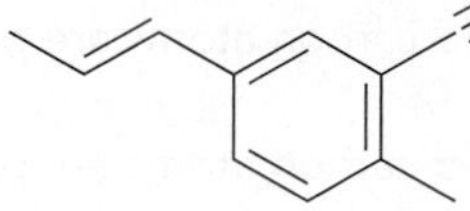

2.53 Identify any formal charges in the following structures:

(a) 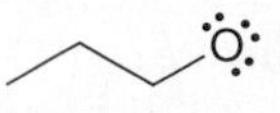 (b) 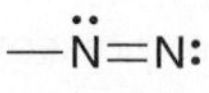

(c) 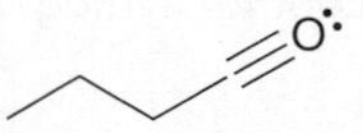 (d) 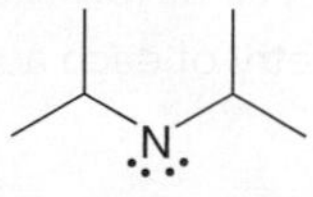

2.54 Draw bond-line structures for all constitutional isomers with molecular formula C_4H_9Cl.

2.55 Draw resonance structures for each of the following compounds:

(a) 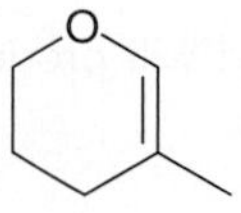 (b) 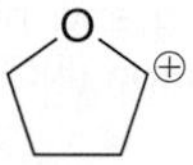 (c) 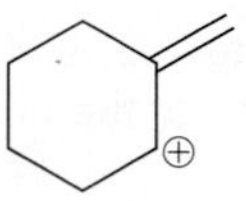

(d) 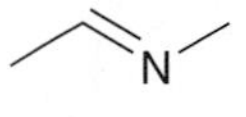 (e) 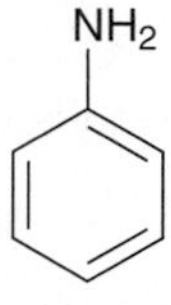 (f) 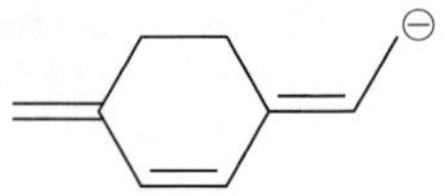

(g) 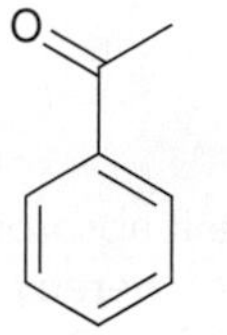 (h) 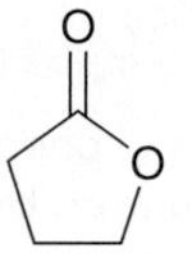 (i) 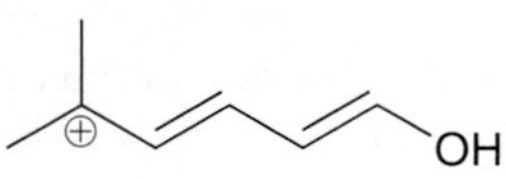

(j) 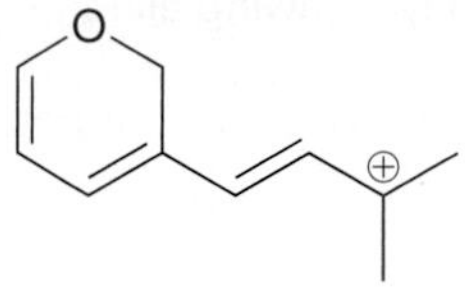

2.56 Determine the relationship between the two structures below. Are they resonance structures or are they constitutional isomers?

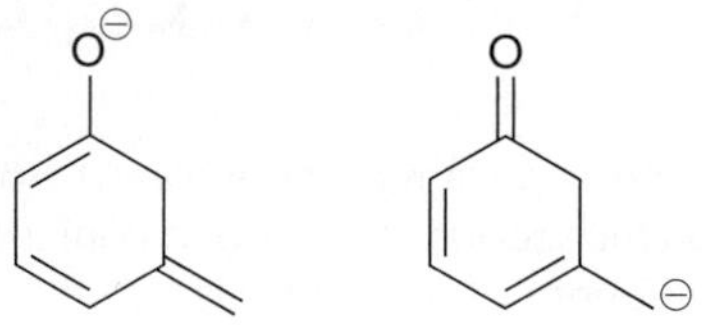

2.57 Consider each pair of compounds below, and determine whether the pair represent the same compound, constitutional isomers, or different compounds that are not isomeric at all:

(a) 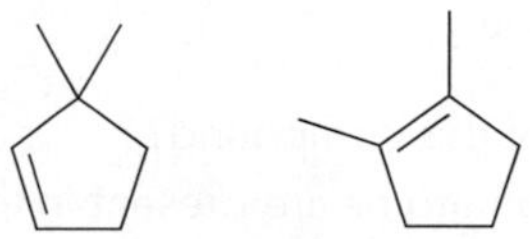

(b) 

(c) 

(d) 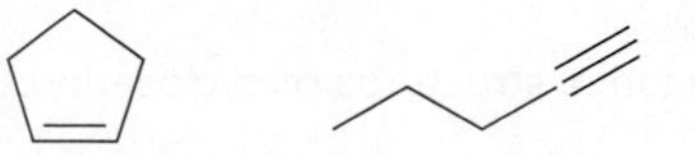

2.58 Draw a bond-line structure for each of the following compounds:

(a) $CH_2{=}CHCH_2C(CH_3)_3$ (b) $(CH_3CH_2)_2CHCH_2CH_2OH$
(c) $CH{\equiv}COCH_2CH(CH_3)_2$ (d) $CH_3CH_2OCH_2CH_2OCH_2CH_3$
(e) $(CH_3CH_2)_3CBr$ (f) $(CH_3)_2C{=}CHCH_3$

2.59 A mixture of sulfuric acid and nitric acid will produce small quantities of the nitronium ion (NO_2^+):

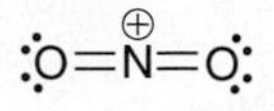

Does the nitronium ion have any significant resonance structures? Why or why not?

2.60 Consider the structure of ozone:

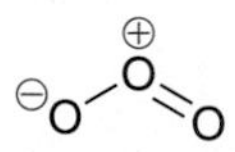

Ozone is formed in the upper atmosphere, where it absorbs short-wavelength UV radiation emitted by the sun, thereby protecting us from harmful radiation. Draw all significant resonance structures for ozone (*Hint*: Begin by drawing all lone pairs).

2.61 Melatonin is an animal hormone believed to have a role in regulating the sleep cycle:

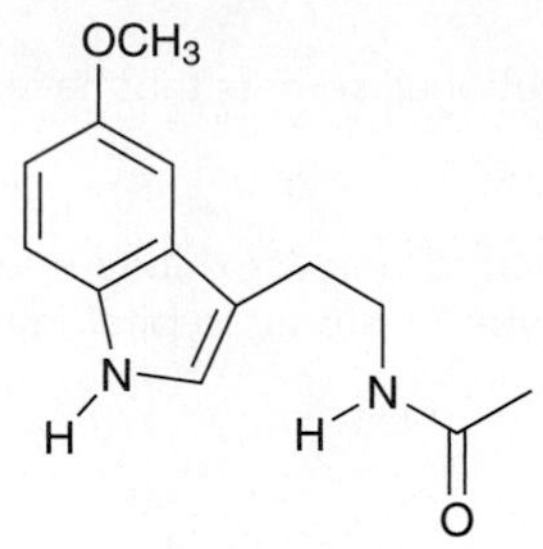

Melatonin

The structure of melatonin incorporates two nitrogen atoms. What are the hybridization state and geometry of each nitrogen atom? Explain your answer.

2.62 Draw all significant resonance structures for each of the following compounds:

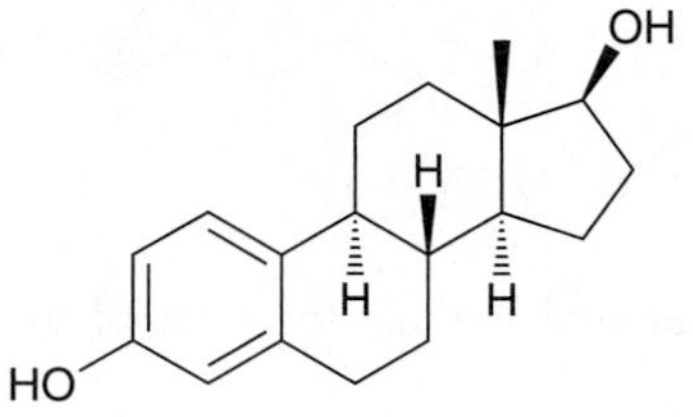

Estradiol
(Female sex hormone)

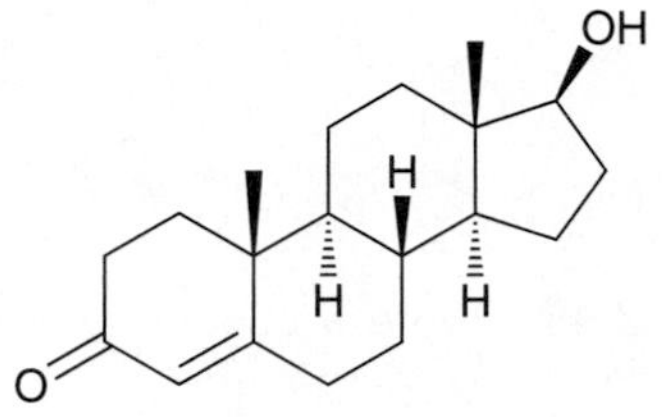

Testosterone
(Male sex hormone)

INTEGRATED PROBLEMS

2.63 Cycloserine is an antibiotic isolated from the microbe *Streptomyces orchidaceous*. It is used in conjunction with other drugs for the treatment of tuberculosis.

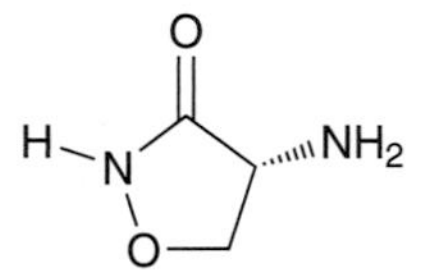

Cycloserine

(a) What is the molecular formula of this compound?
(b) How many sp^3-hybridized carbon atoms are present in this structure?
(c) How many sp^2-hybridized carbon atoms are present in this structure?
(d) How many *sp*-hybridized carbon atoms are present in this structure?
(e) How many lone pairs are present in this structure?
(f) Identify each lone pair as localized or delocalized.
(g) Identify the geometry of each atom (except for hydrogen atoms).
(h) Draw all significant resonance structures of cycloserine.

2.64 Ramelteon is a hypnotic agent used in the treatment of insomnia:

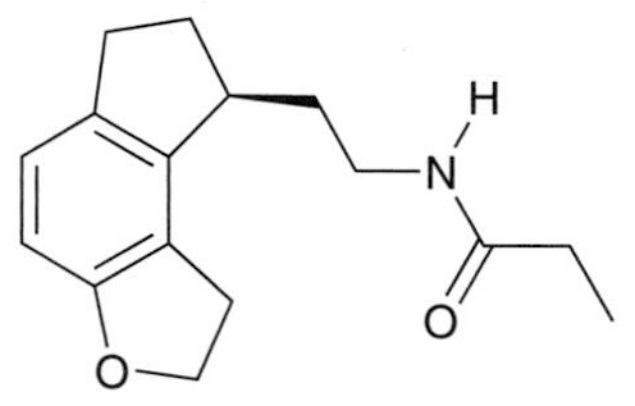

Ramelteon

(a) What is the molecular formula of this compound?
(b) How many sp^3-hybridized carbon atoms are present in this structure?
(c) How many sp^2-hybridized carbon atoms are present in this structure?
(d) How many *sp*-hybridized carbon atoms are present in this structure?
(e) How many lone pairs are present in this structure?
(f) Identify each lone pair as localized or delocalized.
(g) Identify the geometry of each atom (except for hydrogen atoms).

CHALLENGE PROBLEMS

2.65 In the compound below, identify all carbon atoms that are electron deficient (δ+) and all carbon atoms that are electron rich (δ−). Justify your answer with resonance structures.

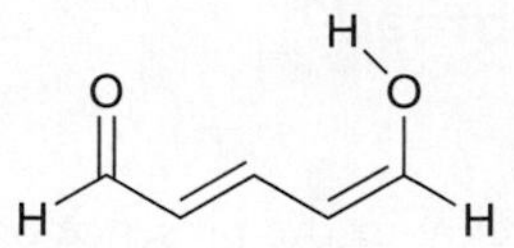

2.66 Consider the following two compounds:

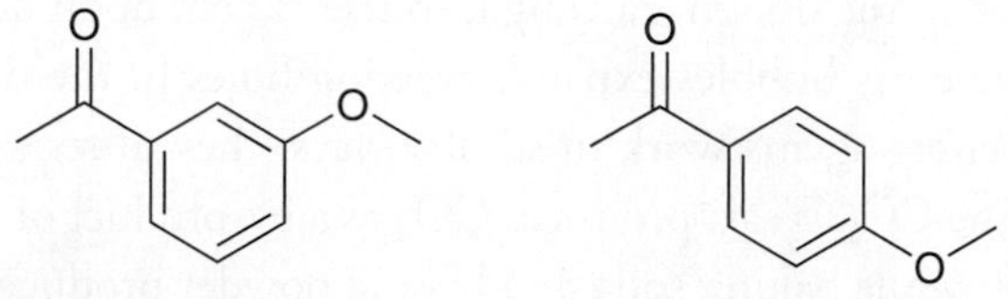

Compound A   Compound B

(a) Identify which of these two compounds has greater resonance stabilization.

(b) Would you expect compound C (below) to have a resonance stabilization that is more similar to compound A or to compound B?

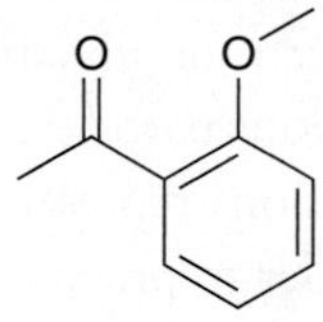

Compound C

2.67 Single bonds generally experience free rotation at room temperature (as will be discussed in more detail in Chapter 4):

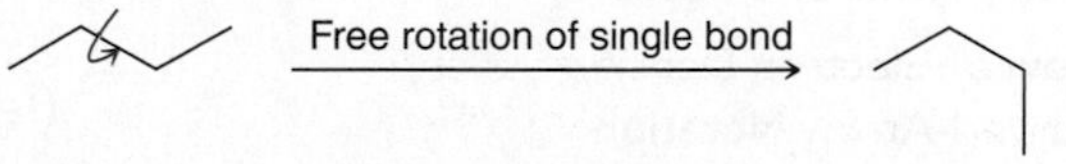

Nevertheless, the "single bond" shown below exhibits a large barrier to rotation. In other words, the energy of the system is greatly increased if that bond is rotated. Explain the source of this energy barrier. (*Hint:* Think about the atomic orbitals being used to form the "conduit.")

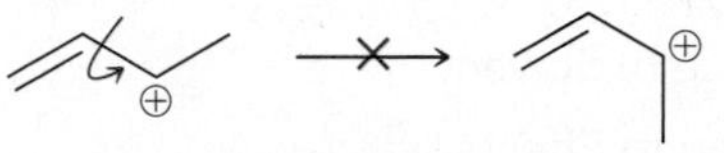

3

Acids and Bases

DID YOU EVER WONDER...
how dough rises to produce fluffy (leavened) rolls and bread?

Dough rises fairly quickly in the presence of a leavening agent, such as yeast, baking powder, or baking soda. All of these leavening agents work by producing bubbles of carbon dioxide gas that get trapped in the dough, causing it to rise. Then, upon heating in an oven, these gas bubbles expand, creating holes in the dough. Although leavening agents work in similar ways, they differ in how they produce the CO_2. Yeast produces CO_2 as a by-product of metabolic processes, while baking soda and baking powder produce CO_2 as a by-product of acid-base reactions. Later in this chapter, we will take a closer look at the acid-base reactions involved, and we will discuss the difference between baking soda and baking powder. An understanding of the relevant reactions will lead to a greater appreciation of food chemistry.

In this chapter, our study of acids and bases will serve as an introduction to the role of electrons in ionic reactions. An **ionic reaction** is a reaction in which ions participate as reactants, intermediates, or products. These reactions represent 95% of the reactions covered in this textbook. In order to prepare ourselves for the study of ionic reactions, it is critical to be able to identify acids and bases. We will learn how to draw acid-base reactions and to compare the acidity or basicity of compounds. These tools will enable us to predict when acid-base reactions are likely to occur and to choose the appropriate reagent to carry out any specific acid-base reaction.

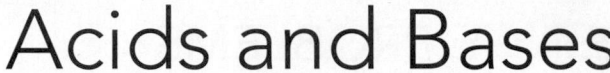

DO YOU REMEMBER?

Before you go on, be sure you understand the following topics.
If necessary, review the suggested sections to prepare for this chapter.

- Drawing and Interpreting Bond-Line Structures (Section 2.2)
- Identifying Formal Charges (Sections 1.4 and 2.4)
- Identifying Lone Pairs (Section 2.5)
- Drawing Resonance Structures (Section 2.10)

WILEY PLUS Visit www.wileyplus.com to check your understanding and for valuable practice.

3.1 Introduction to Brønsted-Lowry Acids and Bases

This chapter will focus primarily on Brønsted-Lowry acids and bases. There is also a brief section dealing with Lewis acids and bases, a topic that is revisited in Chapter 6 and subsequent chapters.

The definition of **Brønsted-Lowry acids and bases** is based on the transfer of a proton (H^+). An **acid** is defined as a *proton donor*, while a **base** is defined as a *proton acceptor*. As an example, consider the following acid-base reaction:

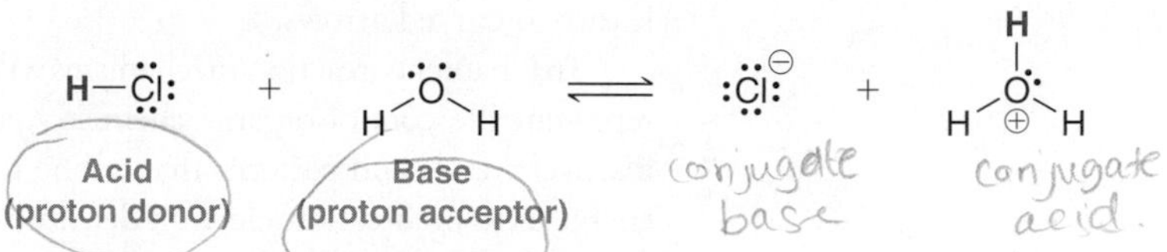

In the reaction above, HCl functions as an acid because it donates a proton to H_2O, while H_2O functions as a base because it accepts the proton from HCl. The products of a proton transfer reaction are called the **conjugate acid** and the **conjugate base**:

HCl	+	H_2O	⇌	Cl^-	+	H_3O^+
Acid		Base		Conjugate base		Conjugate acid

In this reaction, Cl^- is the conjugate base of HCl. In other words, the conjugate base is what remains of the acid after it has been deprotonated. Similarly, in the reaction above, H_3O^+ is the conjugate acid of H_2O. We will use this terminology throughout the rest of this chapter, so it is important to know these terms well.

In the example above, H_2O served as a base by accepting a proton, but in other situations, it can serve as an acid by donating a proton. For example:

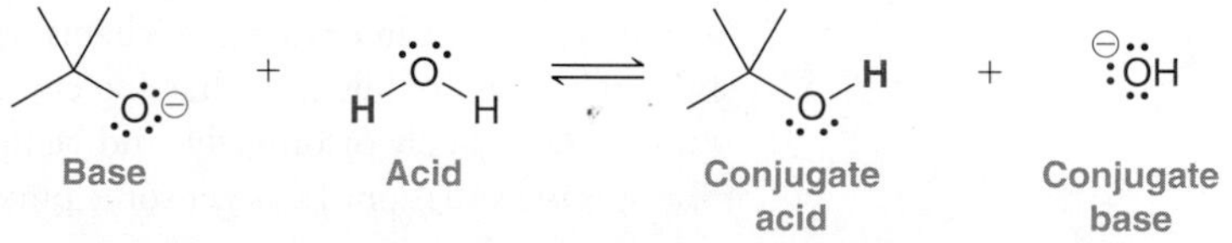

In this case, water functions as an acid rather than a base. Throughout this course, we will see countless examples of water functioning either as a base or as an acid, so it is important to understand that both are possible and very common. When water functions as an acid, as in the reaction above, the conjugate base is HO^-.

3.2 Flow of Electron Density: Curved-Arrow Notation

All reactions are accomplished via a flow of electron density (the motion of electrons). Acid-base reactions are no exception. The flow of electron density is illustrated with curved arrows:

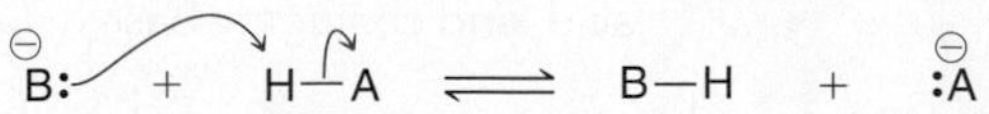

Although these curved arrows look exactly like the curved arrows used for drawing resonance structures, there is an important difference. When drawing resonance structures, curved arrows are used simply as tools and do not represent any real physical process. But in the reaction above, the curved arrows do represent an actual physical process. There is a flow of electron density that causes a proton to be transferred from one reagent to another; the curved arrows illustrate this flow. The arrows show the **reaction mechanism**, that is, they show how the reaction occurs in terms of the motion of electrons.

Notice that the mechanism of a proton transfer reaction involves electrons from a base deprotonating an acid. This is an important point, because acids do not lose protons without the help of a base. It is necessary for a base to abstract the proton. Here is a specific example:

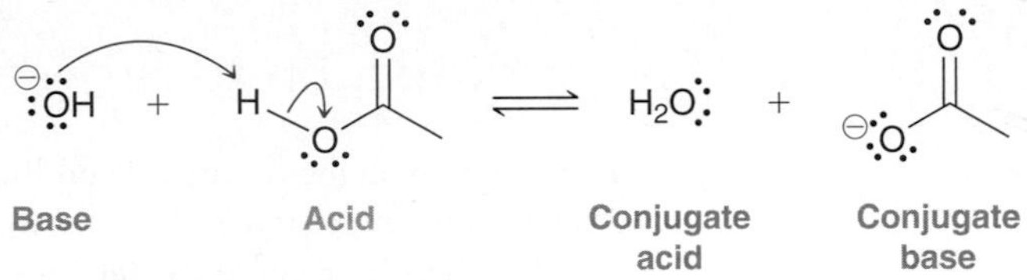

In this example, hydroxide (HO^-) functions as a base to abstract a proton from the acid. Notice that there are exactly two curved arrows. The mechanism of a proton transfer always involves at least two curved arrows.

In Chapter 6, reaction mechanisms will be introduced and explored in more detail. Mechanisms represent the core of organic chemistry, and by proposing and comparing mechanisms, we will discover trends and patterns that define the behavior of electrons. These trends and patterns will enable us to predict how electron density flows and to explain new reactions. For almost every reaction throughout this book, we will propose a mechanism and then analyze it in detail.

Most mechanisms involve one or more proton transfer steps. For example, one of the first reactions to be covered (Chapter 8) is called an elimination reaction, and it is believed to occur via the following mechanism:

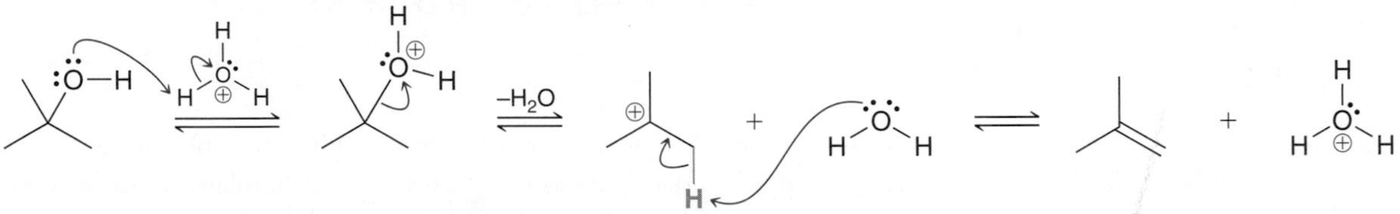

This mechanism has many steps, each of which is shown with curved arrows. Notice that the first and last steps are simply protons transfers. In the first step, H_3O^+ functions as an acid, and in the last step, water is acting a base.

Clearly, proton transfers play an integral role in reaction mechanisms. Therefore, in order to become proficient in drawing mechanisms, it is essential to master proton transfers. Important skills to be mastered include drawing curved arrows properly, being able to predict when a proton transfer is likely or unlikely, and being able to determine which acid or base is appropriate for a specific situation. Let's get some practice drawing the mechanism of a proton transfer.

SKILLBUILDER

3.1 DRAWING THE MECHANISM OF A PROTON TRANSFER

LEARN the skill

Show the mechanism for the following acid-base reaction. Label the acid, base, conjugate acid, and conjugate base:

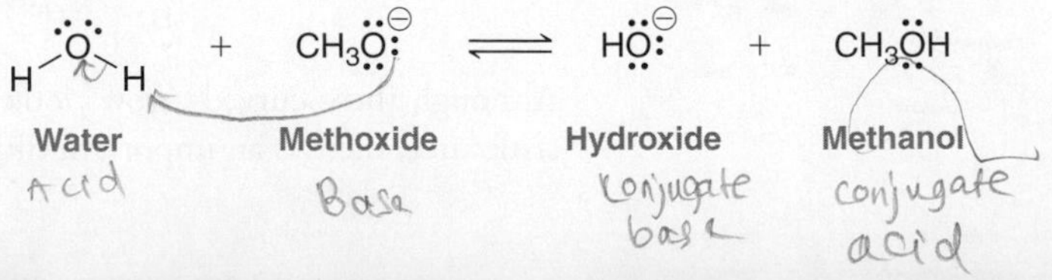

SOLUTION

STEP 1
Identify the acid and the base.

We begin by identifying the acid and the base. Water is losing a proton to form hydroxide. Therefore, water is functioning as a proton donor, rendering it an acid. Methoxide (CH_3O^-) is accepting the proton to form methanol (CH_3OH). Therefore, methoxide is the base.

To draw the mechanism properly, remember that there must be two curved arrows. The tail of the first curved arrow is placed on a lone pair of the base and the head is placed on the proton of the acid. This first curved arrow shows the base abstracting the proton. The next curved arrow always comes from the X—H bond being broken and goes to the atom connected to the proton:

STEP 2
Draw the first curved arrow.

STEP 3
Draw the second curved arrow.

$$H-\overset{..}{\underset{..}{O}}-H \quad + \quad CH_3\overset{..}{\underset{..}{O}}{:}^{\ominus} \quad \rightleftharpoons \quad H\overset{..}{\underset{..}{O}}{:}^{\ominus} \quad + \quad CH_3\overset{..}{O}H$$

 Acid **Base**

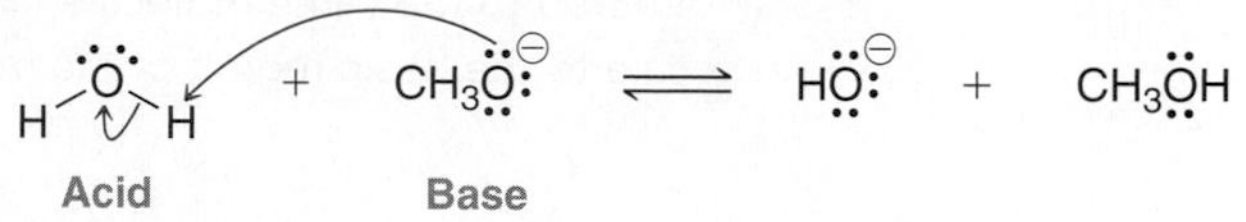

Make sure that the head and tail of each arrow are positioned in exactly the right place or the mechanism will be incorrect.

When water loses a proton, hydroxide is generated. Therefore, hydroxide is the conjugate base of water. When methoxide receives the proton, methanol is generated. Methanol is therefore the conjugate acid of methoxide:

$$H-\overset{O}{\diagdown}-H \quad + \quad CH_3O^{\ominus} \quad \rightleftharpoons \quad HO^{\ominus} \quad + \quad CH_3OH$$

 Acid **Base** **Conjugate** **Conjugate**
 base **acid**

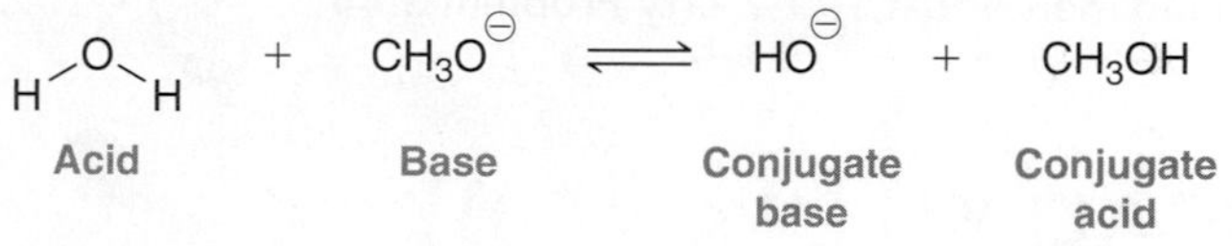

PRACTICE the skill **3.1** All of the following acid-base reactions are reactions that we will study in greater detail in the chapters to follow. For each one, draw the mechanism and then clearly label the acid, base, conjugate acid, and conjugate base:

(a), (b), (c), (d) [reaction mechanisms with handwritten annotations labeling acid, base, conjugate base, and conjugate acid]

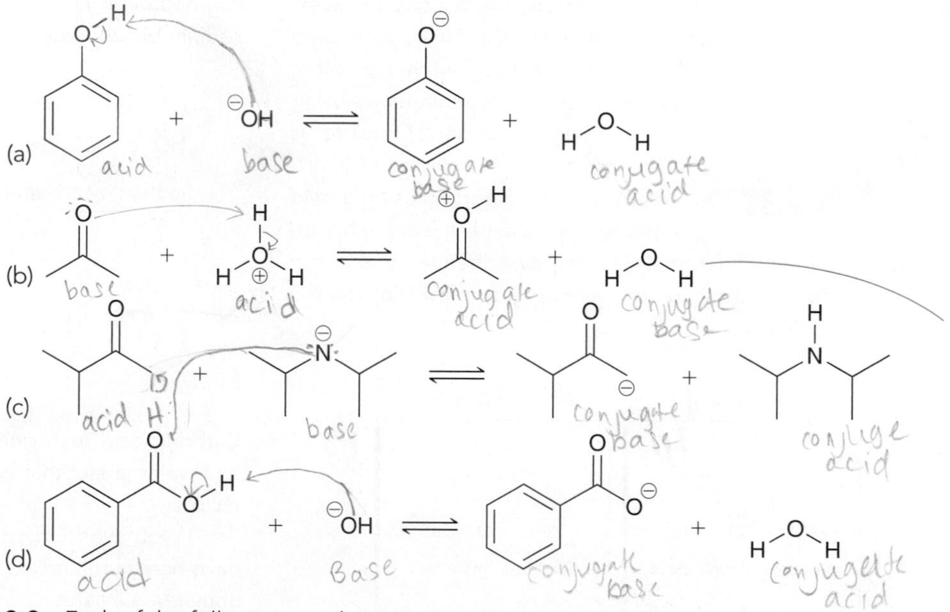

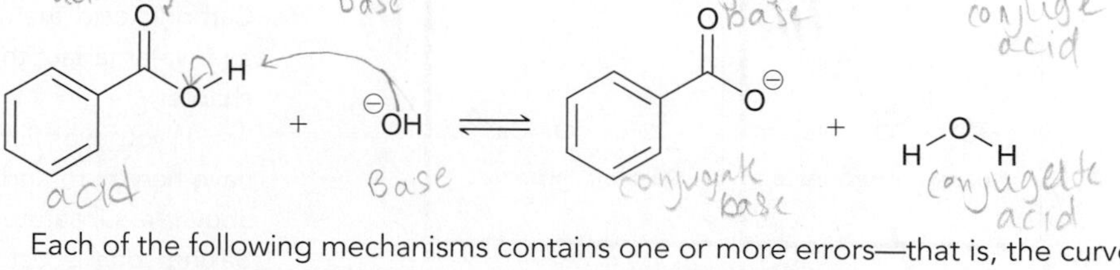

APPLY the skill **3.2** Each of the following mechanisms contains one or more errors—that is, the curved arrows may or may not be correct. In each case, identify the errors and then describe what modification would be necessary in order to make the curved arrows correct. Explain your suggested modification in each case (the following examples represent common student errors, so it is in your best interests to identify these errors, recognize them, and then avoid these mistakes):

(a) [reaction mechanism]

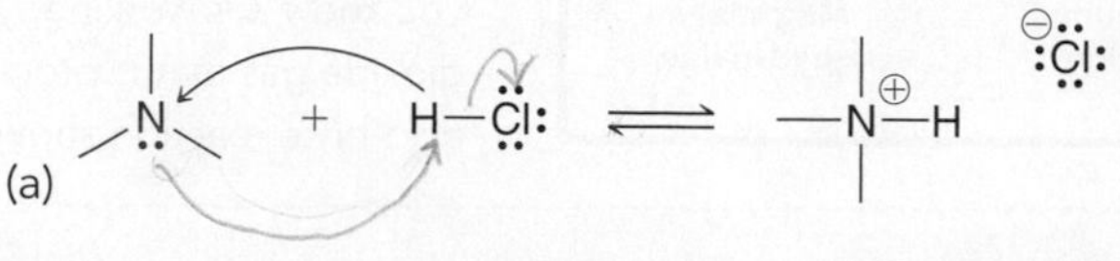

(b) 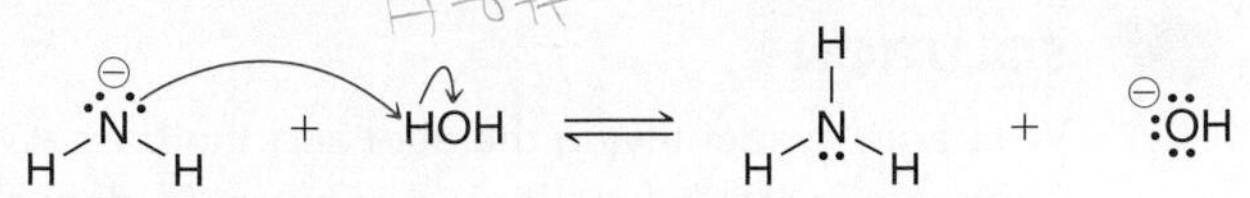

(c) 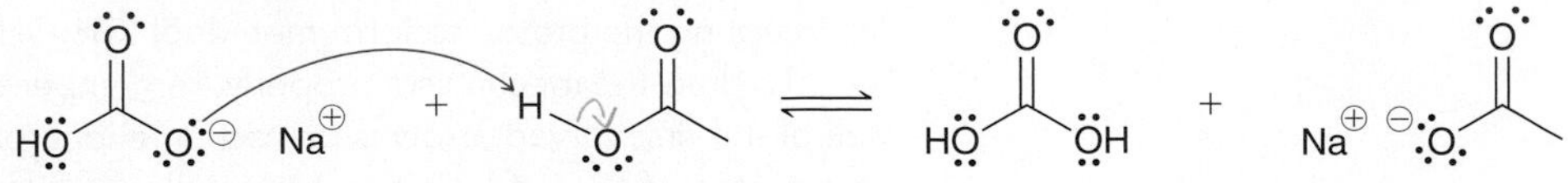

3.3 In an *intramolecular proton transfer reaction*, the acidic site and the basic site are tethered to the same molecule, and a proton is passed from the acidic region of the molecule to the basic region of the molecule, as shown below:

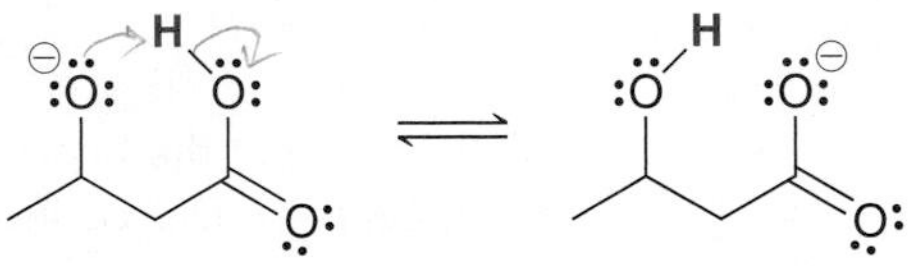

Draw a mechanism for this process.

need more **PRACTICE?** **Try Problem 3.44**

MEDICALLYSPEAKING

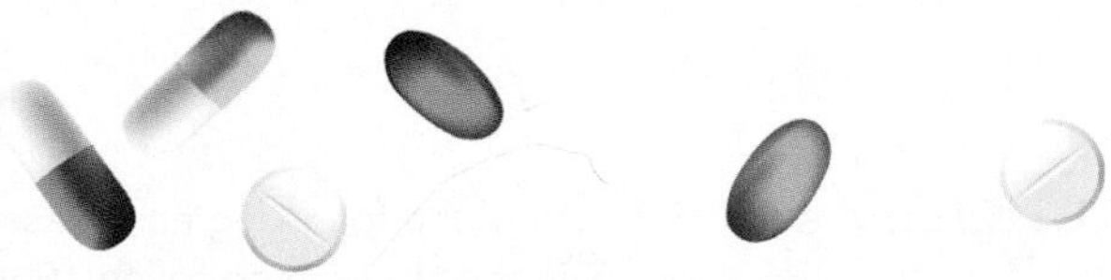

Antacids and Heartburn

Most of us have experienced occasional heartburn, especially after eating pizza. Heartburn is caused by the buildup of excessive amounts of stomach acid (primarily HCl). This acid is used to digest the food we eat, but it can often back up into the esophagus, causing the burning sensation referred to as heartburn. The symptoms of heartburn can be treated by using a mild base to neutralize the excess hydrochloric acid. Many different antacids are on the market and can be purchased over the counter. Here are just a few that you will probably recognize:

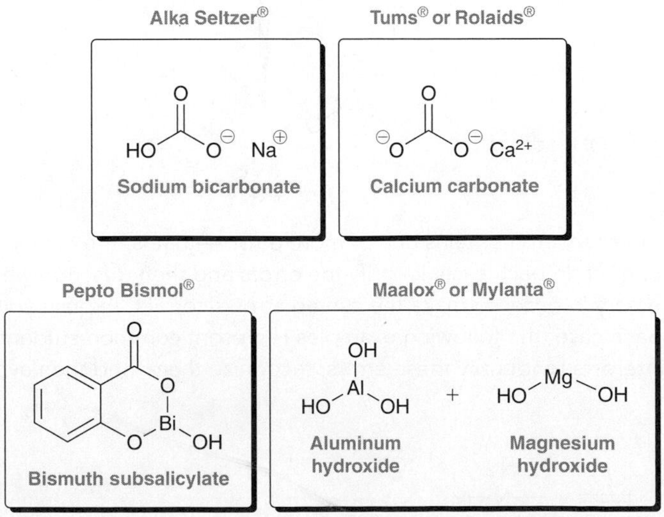

All of these work in similar ways. They are all mild bases that can neutralize HCl in a proton transfer reaction. For example, sodium bicarbonate deprotonates HCl to form carbonic acid:

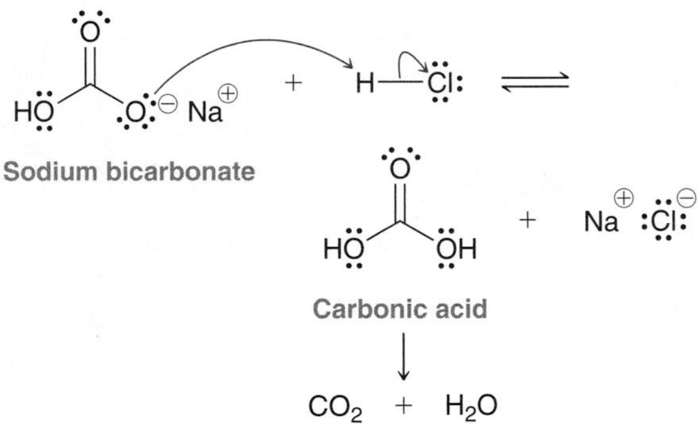

Carbonic acid then quickly degrades into carbon dioxide and water (a fact that we will discuss again later in this chapter).

If you ever find yourself in a situation where you have heartburn and no access to any of the antacids above, a substitute can be found in your kitchen. Baking soda is just sodium bicarbonate (the same compound found in Alka Seltzer®). Take a teaspoon of baking soda, dissolve it in a glass of water by stirring with a spoon, and then drink it down. The solution will taste salty, but it will alleviate the burning sensation of heartburn. Once you start burping, you know it is working; you are releasing the carbon dioxide gas that is produced as a by-product of the acid-base reaction shown above.

3.3 Brønsted-Lowry Acidity: Quantitative Perspective

There are two ways to predict when a proton transfer reaction will occur: (1) via a quantitative approach (comparing pK_a values) or (2) via a qualitative approach (analyzing the structures of the acids). It is essential to master both methods. In this section, we focus on the first method, and in the upcoming sections we will focus on the second method.

Using pK_a Values to Compare Acidity

The terms K_a and pK_a were defined in your general chemistry textbook, but it is worthwhile to quickly review their definitions. Consider the following general acid-base reaction between HA (an acid) and H_2O (functioning as a base in this case):

$$HA + H_2O \rightleftharpoons A^- + H_3O^+$$

The reaction is said to have reached **equilibrium** when there is no longer an observable change in the concentrations of reactants and products. At equilibrium, the rate of the forward reaction is exactly equivalent to the rate of the reverse reaction, which is indicated with two arrows pointing in opposite directions, as shown above. The position of equilibrium is described by the term $\boldsymbol{K_{eq}}$, which is defined in the following way:

$$K_{eq} = \frac{[H_3O^+][A^-]}{[HA][H_2O]}$$

It is the product of the equilibrium concentrations of the products divided by the product of the equilibrium concentrations of the reactants. When an acid-base reaction is carried out in dilute aqueous solution, the concentration of water is fairly constant (55.5M) and can therefore be removed from the expression. This gives us a new term, $\boldsymbol{K_a}$:

$$K_a = K_{eq}[H_2O] = \frac{[H_3O^+][A^-]}{[HA]}$$

The value of K_a measures the strength of the acid. Very strong acids can have a K_a on the order of 10^{10} (or 10,000,000,000), while very weak acids can have a K_a on the order of 10^{-50} (or 0.00000000000000000000000000000000000000000000000001). The values of K_a are often very small or very large numbers. To deal with this, chemists often express pK_a values, rather than K_a values, where pK_a is defined as:

$$pK_a = -\log K_a$$

When pK_a is used as the measure of acidity, the values will generally range from -10 to 50. We will deal with pK_a values extensively throughout this chapter, and there are two things to keep in mind: (1) A strong acid will have a low pK_a value, while a weak acid will have a high pK_a value. For example, an acid with a pK_a of 10 is more acidic than an acid with a pK_a of 16. (2) Each unit represents an order of magnitude. An acid with a pK_a of 10 is six orders of magnitude (one million times) more acidic than an acid with a pK_a of 16. Table 3.1 provides pK_a values for many of the compounds commonly encountered in this course.

TABLE 3.1 pK_a VALUES OF COMMON COMPOUNDS AND THEIR CONJUGATE BASES

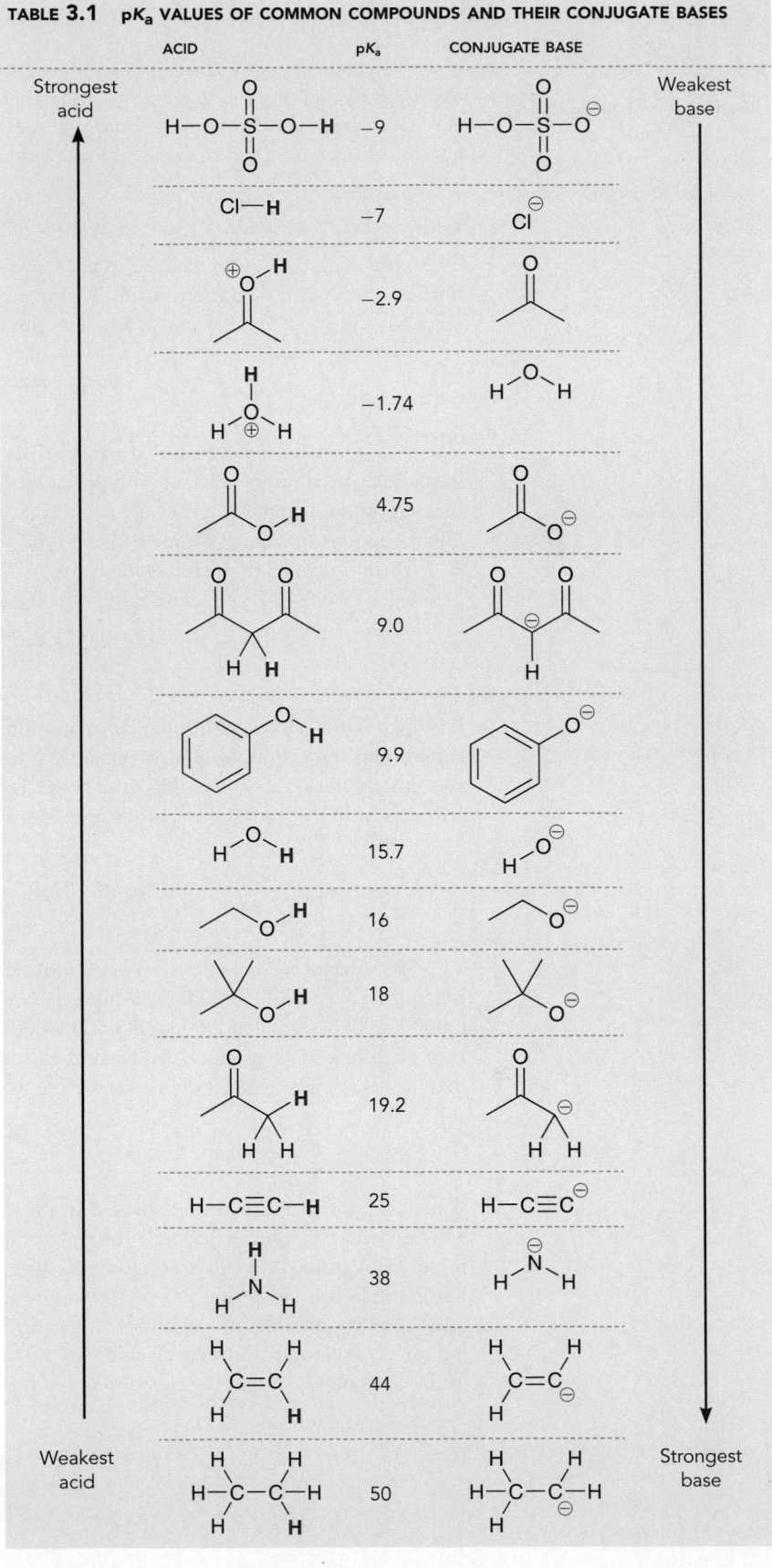

Strongest acid ↑ (top) … Weakest acid (bottom); Weakest base (top) … Strongest base ↓ (bottom)

ACID	pK_a	CONJUGATE BASE
	−9	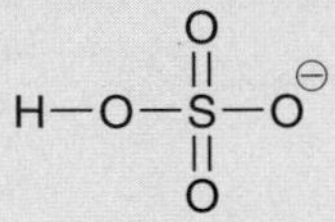
$Cl-H$	−7	Cl^-
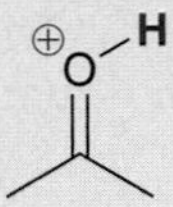	−2.9	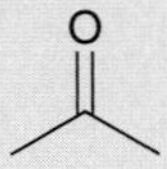
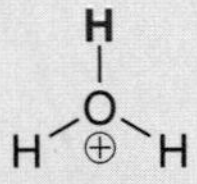	−1.74	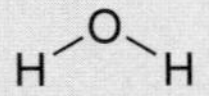
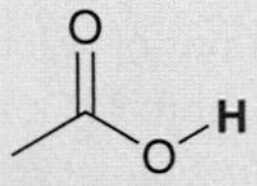	4.75	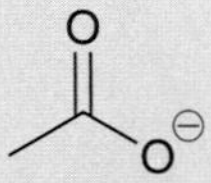
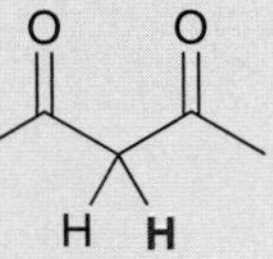	9.0	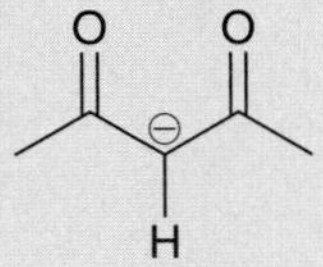
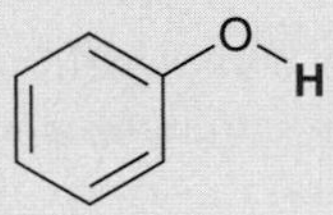	9.9	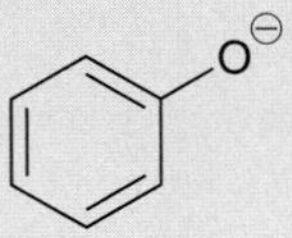
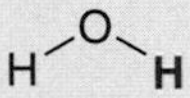	15.7	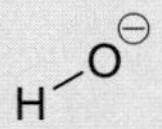
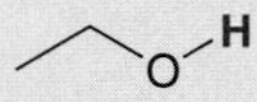	16	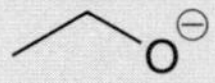
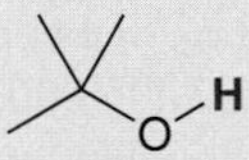	18	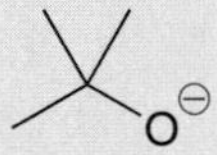
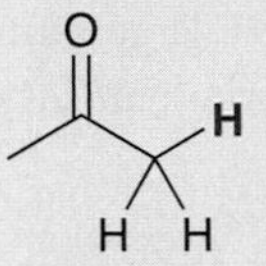	19.2	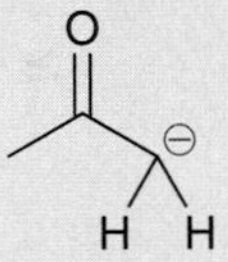
$H-C\equiv C-H$	25	$H-C\equiv C^-$
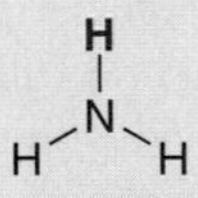	38	
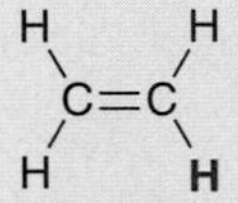	44	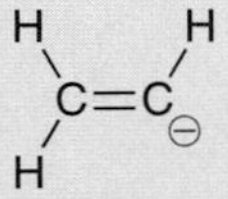
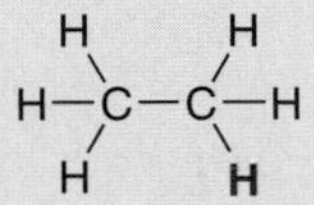	50	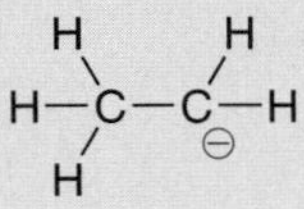

SKILLBUILDER

3.2 USING pK_a VALUES TO COMPARE ACIDS

LEARN the skill

Acetic acid is the main constituent in vinegar solutions and acetone is a solvent often used in nail polish remover:

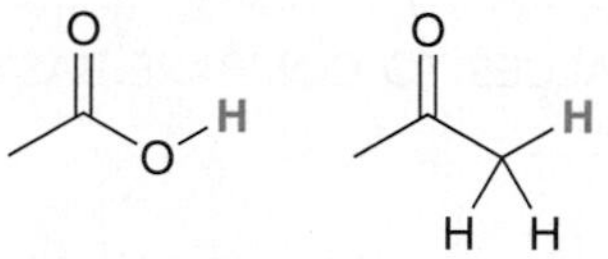

Acetic acid **Acetone**

Using the pK_a values in Table 3.1, identify which of these two compounds is more acidic.

 SOLUTION

Acetic acid has a pK_a of 4.75, while acetone has a pK_a of 19.2. The compound with the lower pK_a is more acidic, and therefore, acetic acid is more acidic. In fact, when we compare the pK_a values, we see that acetic acid is approximately 14 orders of magnitude (10^{14}) more acidic than acetone (or approximately 100,000,000,000,000 times more acidic). We will discuss the reason for this in the upcoming sections of this chapter.

PRACTICE the skill **3.4** For each pair of compounds below, identify the more acidic compound:

(a) 9.9 15.7

(b) 18 15.7

(c) 38 25

(d) −1.74 −7

(e) 50 25

(f) −2.9 −9

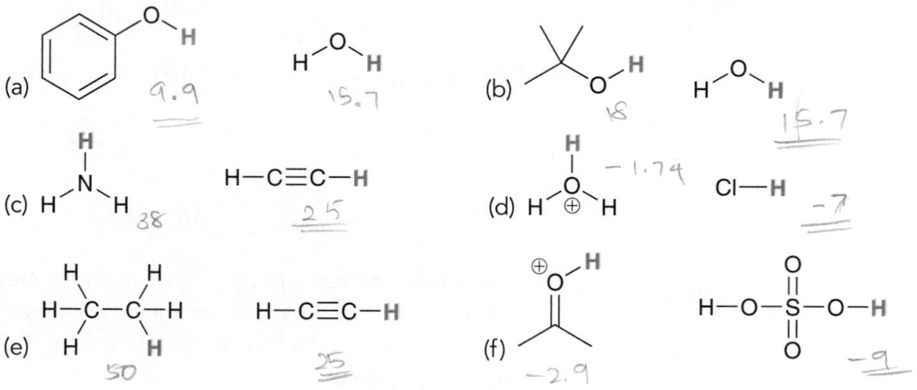

APPLY the skill

3.5 Propanolol is an antihypertensive agent (used to treat high blood pressure). Using Table 3.1, identify the two most acidic protons in the compound, and indicate the approximate expected pK_a for each proton:

pKa = 38

more acidic

pKa = 16

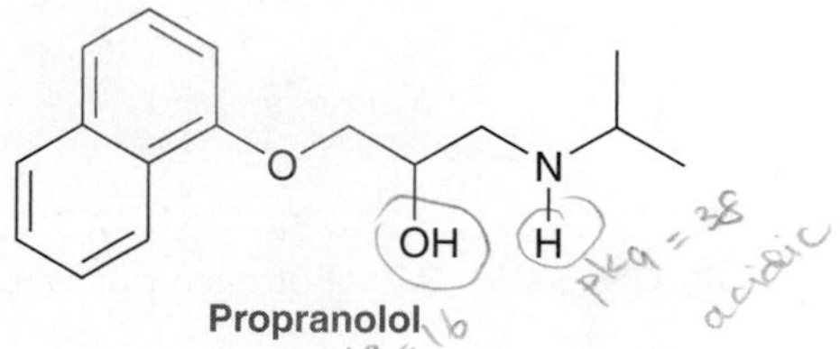

Propranolol

3.6 L-dopa is used in the treatment of Parkinson's disease. Using Table 3.1, identify the four most acidic protons in the compound, and then arrange them in order of increasing acidity (two of the protons will be very similar in acidity and difficult to distinguish at this point in time):

4.75

36

9.9

9.9

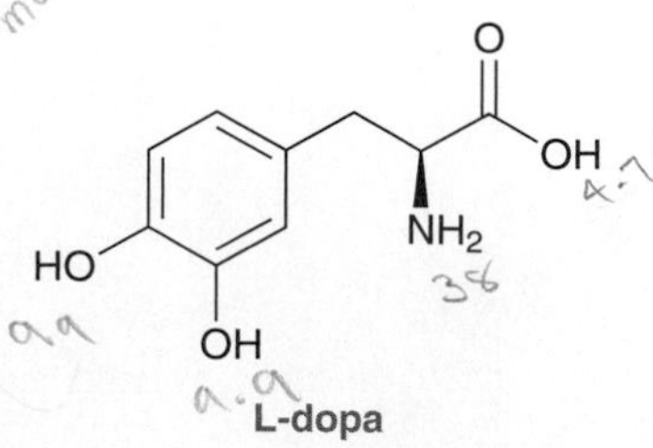

L-dopa

------> need more **PRACTICE?** Try Problem 3.38

Using pK_a Values to Compare Basicity

We have seen how to use pK_a values to compare acids, but it is also possible to use pK_a values to compare bases to one another. It is not necessary to use a separate chart of pK_b values. The following example will demonstrate how to use pK_a values to compare basicity.

SKILLBUILDER

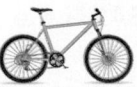

3.3 USING pK_a VALUES TO COMPARE BASICITY

LEARN the skill

Using the pK_a values in Table 3.1, identify which compound is a stronger base:

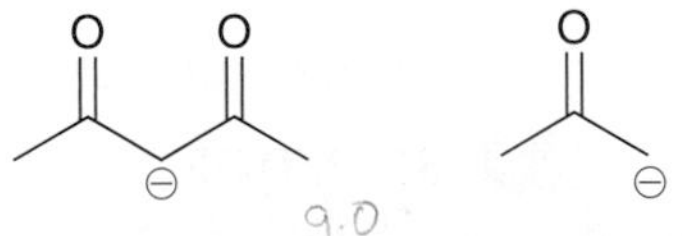

SOLUTION

Each of these bases can be thought of as the conjugate base of some acid. We just need to compare the pK_a values of those acids. To do this, we imagine protonating each of the bases above, producing the following conjugate acids:

STEP 1 Draw the conjugate acid of each base.

We then look up the pK_a values of these acids and compare them:

STEP 2 Compare pK_a values.

pK_a = 9.0 pK_a = 19.2

STEP 3 Identify the stronger base.

The first compound has a lower pK_a value and is therefore a stronger acid than the second compound. Remember that the stronger acid always generates the weaker base. As a result, the conjugate base of the first compound will be a weaker base than the conjugate base of the second compound:

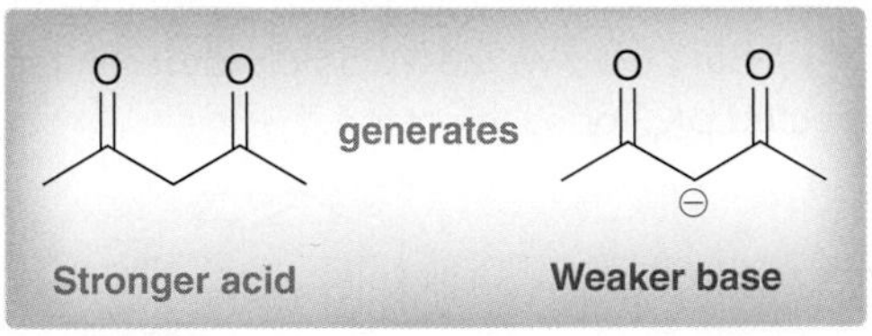

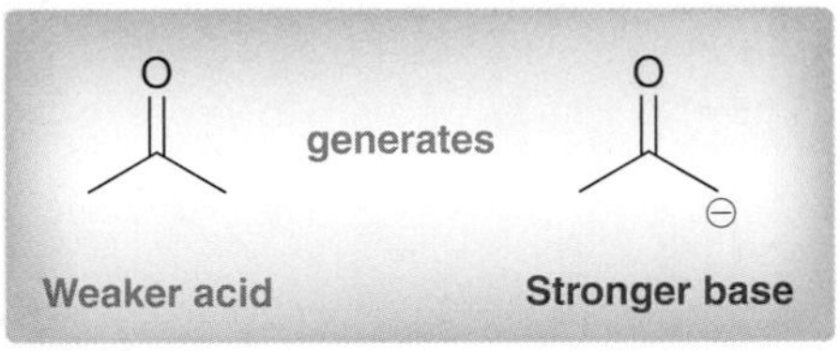

PRACTICE the skill

3.7 For each pair of compounds below, identify the stronger base:

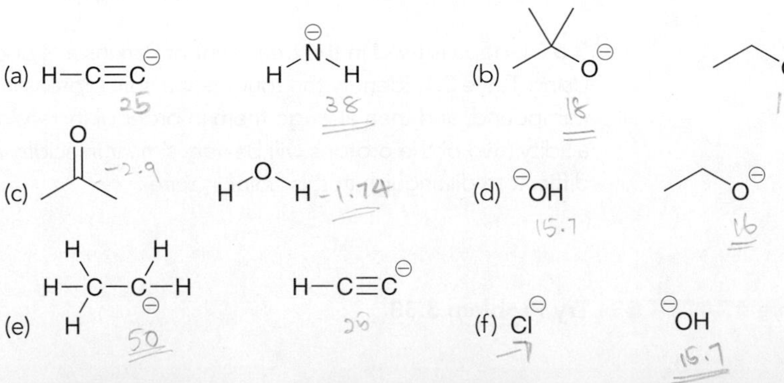

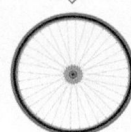

APPLY the skill

3.8 The following compound has three nitrogen atoms:

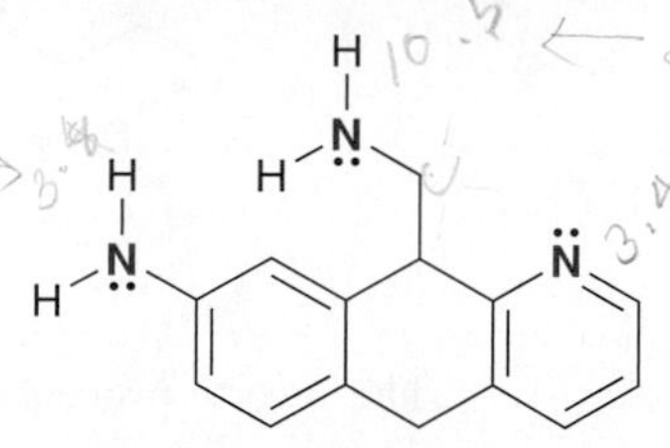

Each of the nitrogen atoms exhibits a lone pair that can function as a base (to abstract a proton from an acid). Rank these three nitrogen atoms in terms of increasing base strength using the following information:

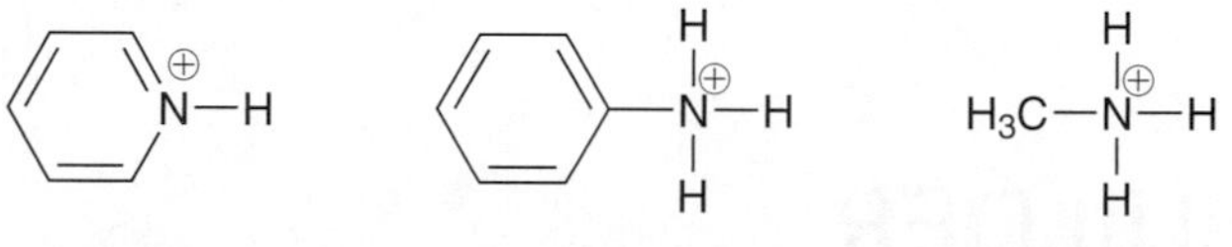

pK_a = 3.4 **pK_a = 3.8** **pK_a = 10.5**

3.9 Consider the following pK_a values, and then answer the following questions:

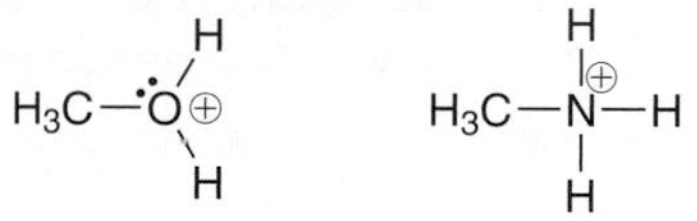

pK_a = −2.5 **pK_a = 10.5**

(a) For the following compound, will the lone pair on the nitrogen atom be more or less basic than the lone pair on the oxygen atom?

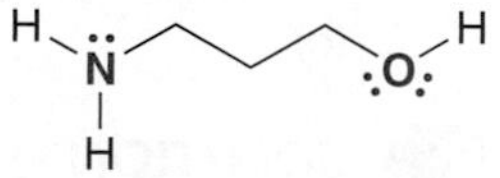

(b) Fill in the blanks: the lone pair on the ________ atom is ________ orders of magnitude more basic than the lone pair on the ________ atom.

need more **PRACTICE?** **Try Problem 3.50**

Using pK_a Values to Predict the Position of Equilibrium

Using the chart of pK_a values, we can also predict the position of equilibrium for any acid-base reaction. The equilibrium will always favor formation of the weaker acid (higher pK_a value). For example, consider the following acid-base reaction:

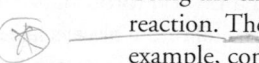

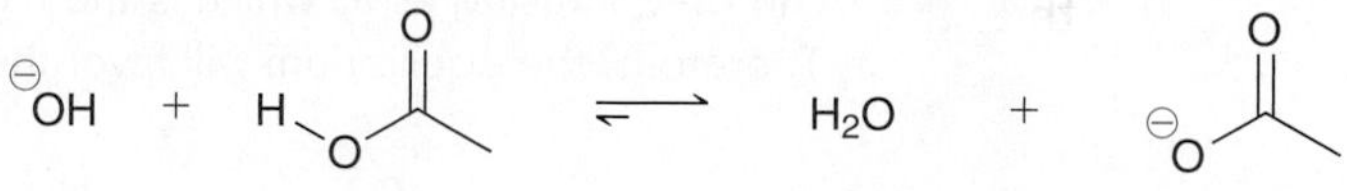

Base **Acid** **Conjugate acid** **Conjugate base**

pK_a = 4.75 **pK_a = 15.7**

The equilibrium for this reaction will lean to the right side, favoring formation of the weaker acid.

For some reactions, the pK_a values are so vastly different that for practical purposes the reaction is treated not as an equilibrium process but rather as one that goes to completion. For example, consider the following reaction:

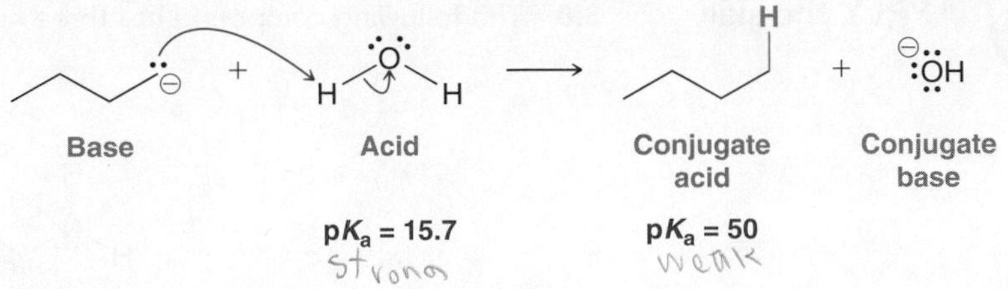

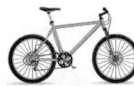

The reverse process is negligible, and for such reactions, organic chemists often draw a one-directional arrow separating reactants from products, rather than the traditional equilibrium arrows. Technically, it is true that all proton transfers are equilibrium processes, but in the case above, the pK_a values are so vastly different (34 orders of magnitude) that we can essentially ignore the reverse reaction.

SKILLBUILDER

3.4 USING pK_a VALUES TO PREDICT THE POSITION OF EQUILIBRIUM

LEARN the skill

Using pK_a values from Table 3.1, determine the position of equilibrium for each of the following two proton transfer reactions:

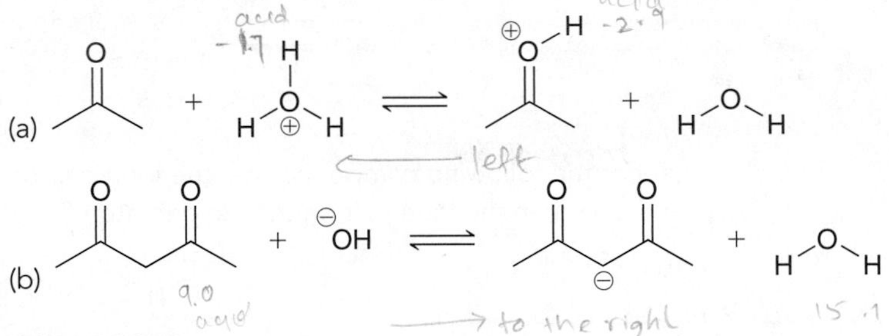

SOLUTION

(a) We begin by identifying the acid on either side of the reaction, and then we compare their pK_a values:

STEP 1 Identify the acid on each side of the equilibrium.

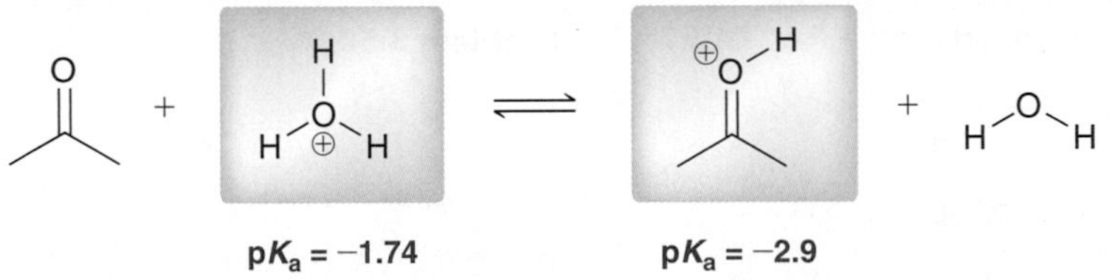

STEP 2 Compare pK_a values.

In this reaction, the C═O bond receives a proton (protonation of C═O bonds will be discussed in more detail in Chapter 20). The equilibrium always favors the weaker acid (the acid with the higher pK_a value). The pK_a values shown above are both negative numbers, so it can be confusing as to which is the higher pK_a value: −1.74 is a larger number than −2.9. Therefore, the equilibrium will favor the left side of the reaction, which is drawn like this:

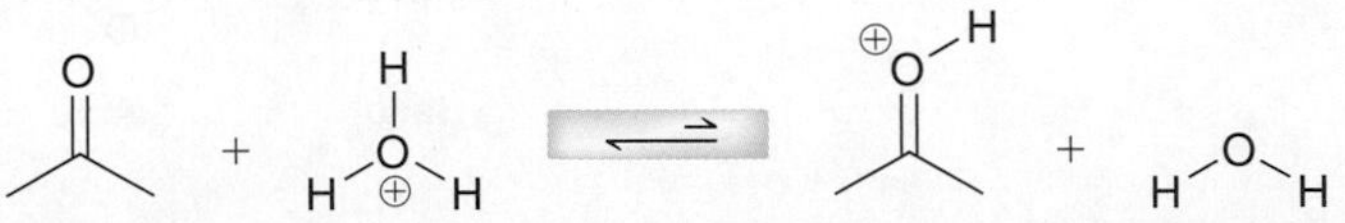

The difference in pK_a values is only one order of magnitude. This means that at any given moment in time approximately one out of every ten C═O bonds will bear a proton. When we study this reaction later on, we will see that protonation of C═O bonds serves as a way to catalyze a number of reactions. For catalytic purposes, it is sufficient to have only a small percentage of the C═O bonds protonated.

(b) We must first identify the acid on either side of the reaction, and then compare their pK_a values:

STEP 1
Identify the acid on each side of the equilibrium.

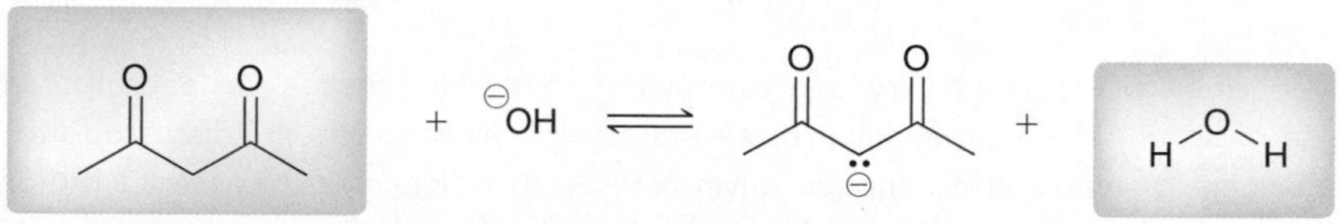

$pK_a = 9.0$ $pK_a = 15.7$

STEP 2
Compare pK_a values.

This reaction shows the deprotonation of a β-diketone (a compound with two C=O bonds separated from each other by one carbon atom). Once again, this is a proton transfer that we will study in more depth later in the course. The equilibrium will favor the side of the weaker acid (the side with the higher pK_a value). Therefore, the equilibrium will favor the right side of the reaction:

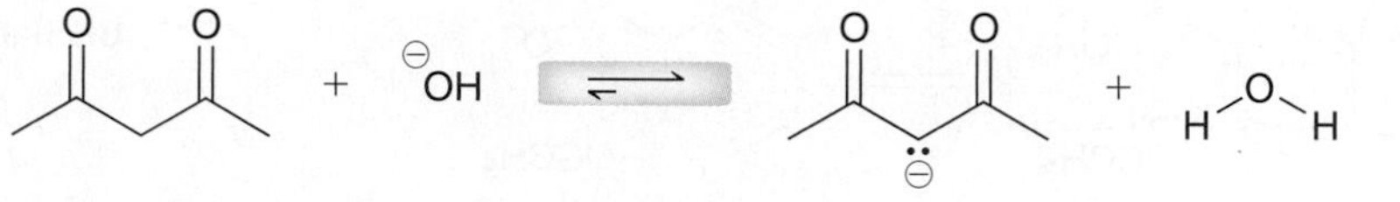

The difference in pK_a values is six orders of magnitude. In other words, when hydroxide is used to deprotonate a β-diketone, the vast majority of the diketone molecules are deprotonated (only one out of every million is not deprotonated). We can conclude from this analysis that hydroxide is a suitable base to accomplish this deprotonation.

PRACTICE the skill **3.10** Determine the position of equilibrium for each acid-base reaction below:

(a)

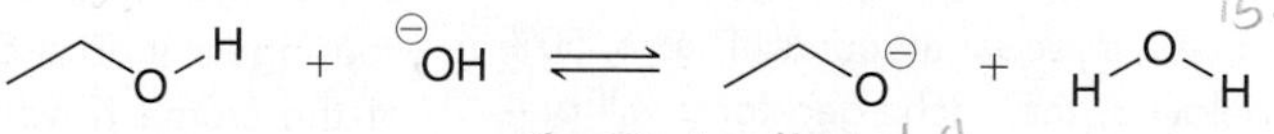

(b)

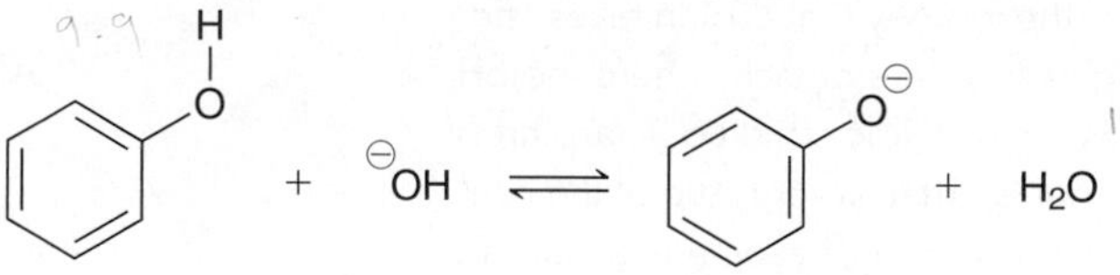

(c)

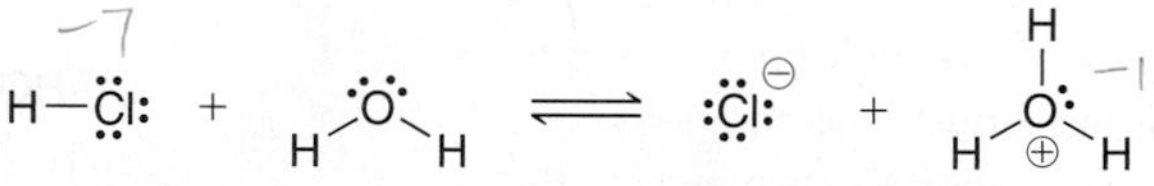

(d) H—C≡C—H + $^-NH_2$ ⇌ H—C≡C:$^-$ + NH_3

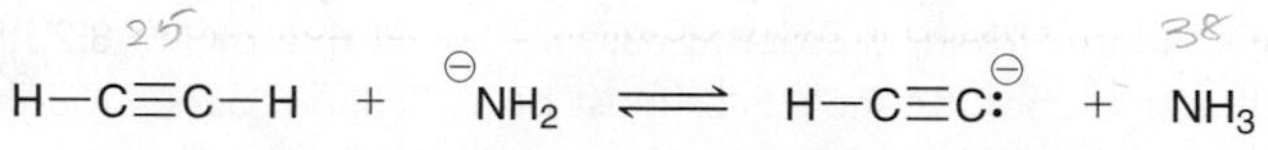

APPLY the skill **3.11** Hydroxide is not a suitable base for deprotonating acetylene:

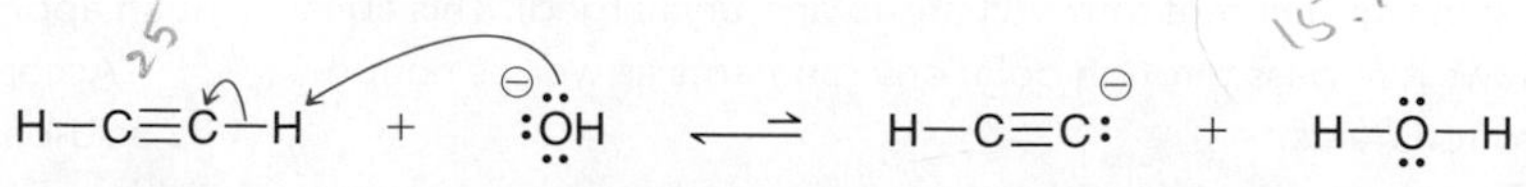

Acetylene **Hydroxide**

Explain why not. Can you propose a base that would be suitable?

------> need more **PRACTICE?** **Try Problem 3.48**

MEDICALLYSPEAKING

Drug Distribution and pK_a

Most drugs will endure a very long journey before reaching their site of action. This journey involves several transitions between polar environments and nonpolar environments. In order for a drug to reach its intended target, it must be capable of being distributed in both types of environments along the journey. A drug's ability to transition between environments is, in most cases, a direct result of the drug's acid-base properties. In fact, most drugs today are acids or bases and, as such, are in equilibrium between charged and uncharged forms. As an example, consider the structure of aspirin and its conjugate base:

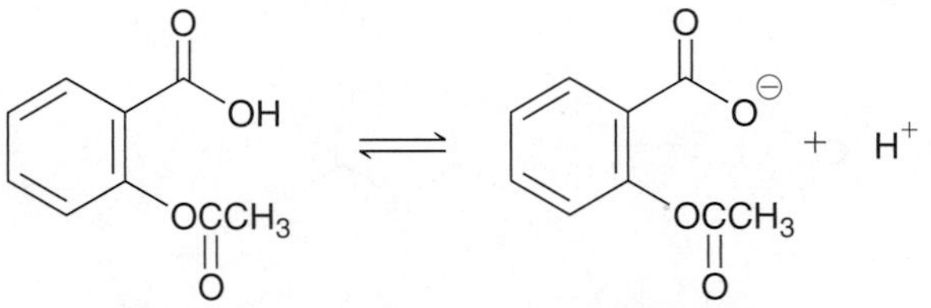

Aspirin
Uncharged form

Conjugate base
Charged form

In the equilibrium above, the left side represents the uncharged form of aspirin, while the right side represents the charged form (the conjugate base). The position of this equilibrium, or *percent ionization*, will depend on the pH of the solution. The pK_a of aspirin is approximately 3.0. At a pH of 3.0 (when pH = pK_a), aspirin and its conjugate base will be present in equal amounts. That is, 50% ionization occurs. At a pH below 3, the uncharged form will predominate. At a pH above 3, the charged form will predominate.

With this in mind, consider the journey that aspirin takes after you ingest it. This journey begins in your stomach, where the pH can be as low as 2. Under these very acidic conditions, aspirin is mostly in its uncharged form. That is, there is very little of the conjugate base present. The uncharged form of aspirin is absorbed by the nonpolar environment of the gastric mucosa in your stomach and the intestinal mucosa in the intestinal tract. After passing through these nonpolar environments, the molecules of aspirin enter the blood, which is a polar (aqueous) environment with a pH of approximately 7.4. At that pH, aspirin exists mainly in the charged form (the conjugate base), and it is distributed throughout the circulatory system in this form. Then, in order to pass the blood-brain barrier, or a cell membrane, the molecules must be converted once again into the uncharged form so that they can pass through the necessary nonpolar environments. The drug is capable of successfully reaching the target because of its ability to exist in two different forms (charged and uncharged). This ability allows it to pass through polar environments as well as nonpolar environments.

The case above (aspirin) was an example of an acid that achieves biodistribution as a result of its ability to lose a proton. In contrast, some drugs are bases, and they achieve biodistribution as a result of their ability to gain a proton. For example, codeine (discussed in the previous chapter) can function as a base and accept a proton:

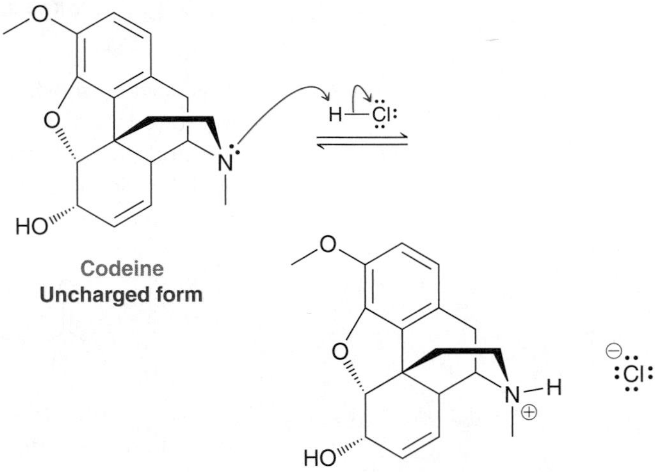

Codeine
Uncharged form

Codeine
Charged form

Once again, the drug exists in two forms: charged and uncharged. But in this case a low pH favors the charged form rather than the uncharged form. Consider the journey that codeine takes after you ingest it. The drug first encounters the acidic environment of the stomach, where it is protonated and exists mostly in its charged form:

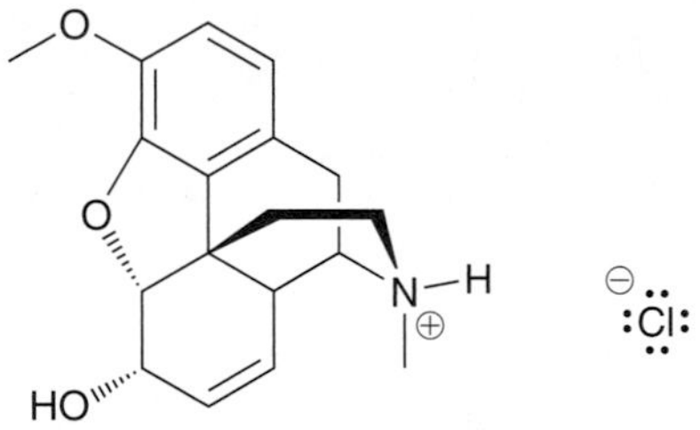

pK_a = 8.2

With a pK_a of 8.2, the charged form predominates at low pH. It cannot pass through nonpolar environments and is therefore not absorbed by the gastric mucosa in the stomach. When the drug reaches the basic conditions of the intestines, it is deprotonated, and the uncharged form predominates. Only then can it transition at an appreciable rate into a nonpolar environment.

Accordingly, the efficacy of any drug is highly dependent on its acid-base properties. This must be taken into account in the design of new drugs. It is certainly important that a drug can bind with its designated receptor, but it is equally important that its acid-base properties allow it to reach the receptor efficiently.

CONCEPTUAL CHECKPOINT

3.12 Amino acids, such as glycine, are the key building blocks of proteins and will be discussed in greater detail in Chapter 25. At the pH of the stomach, glycine exists predominantly in a protonated form in which there are two acidic protons of interest. The pK_a values for these protons are shown. Using this information, draw the form of glycine that will predominate at physiological pH of 7.4.

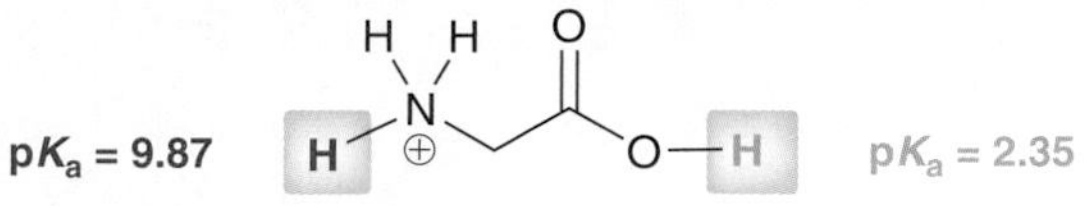

Glycine

3.4 Brønsted-Lowry Acidity: Qualitative Perspective

In the previous section, we learned how to compare acids or bases by comparing pK_a values. In this section, we will now learn how to make such comparisons by analyzing and comparing their structures and without the use of pK_a values.

Conjugate Base Stability

In order to compare acids without the use of pK_a values, we must look at the conjugate base of each acid:

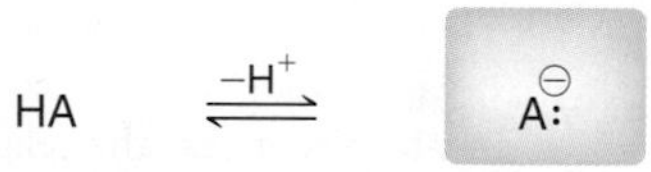

Conjugate base of HA

If A^- is very stable (weak base), then HA must be a strong acid. If, on the other hand, A^- is very unstable (strong base), then HA must be a weak acid. As an illustration of this point let's consider the deprotonation of HCl:

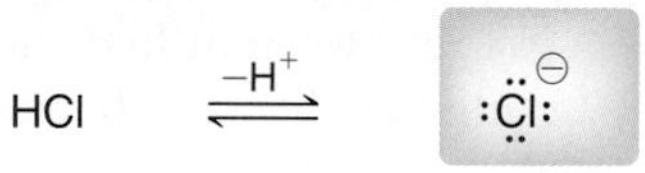

Conjugate base

Chlorine is an electronegative atom, and it can therefore stabilize a negative charge. The chloride ion (Cl^-) is in fact very stable, and therefore, HCl is a strong acid. HCl can serve as a proton donor because the conjugate base left behind is stabilized.

Let's look at one more example. Consider the structure of butane:

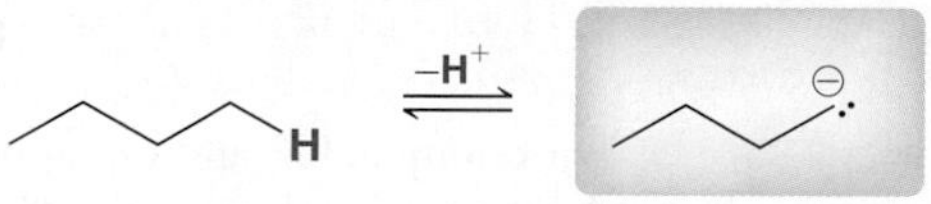

Conjugate base

When butane is deprotonated, a negative charge is generated on a carbon atom. Carbon is not a very electronegative element and is generally not capable of stabilizing a negative charge. Since this C^- is very unstable, we can conclude that butane is not very acidic.

LOOKING AHEAD
In Section 22.2, we will see an exceptional case of a stabilized negative charge on a carbon atom. Until then, most instances of C^- are considered to be very unstable.

This approach can be used to compare the acidity of two compounds, HA and HB. We simply look at their conjugate bases, A^- and B^-, and compare them to each other:

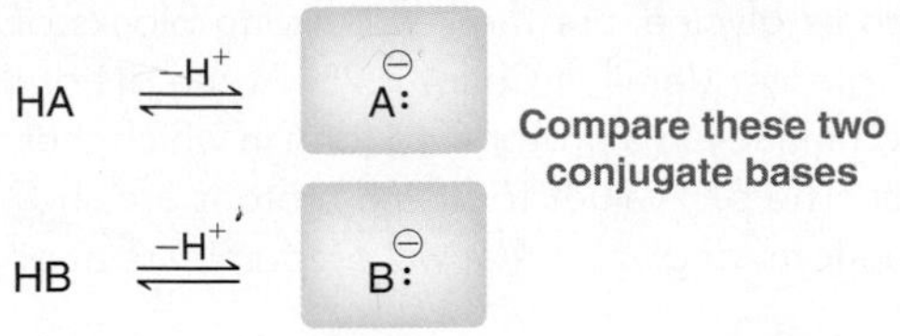
HA
−H⁺
A:
HB
−H⁺
B:
Compare these two conjugate bases

By determining the more stable conjugate base, we can identify the stronger acid. For example, if we determine that A^- is more stable than B^-, then HA must be a stronger acid than HB. This approach does not allow us to predict exact pK_a values, but it does allow us to compare the relative acidity of two compounds quickly, without the need for a chart of pK_a values.

Factors Affecting the Stability of Negative Charges

A qualitative comparison of acidity requires a comparison of the stability of negative charges. The following discussion will develop a methodical approach for comparing negative charge stability. Specifically, there are four factors to consider: (1) which atom is the charge on, (2) resonance, (3) induction, and (4) orbitals.

1. ***Which atom is the charge on?*** The first factor involves comparing the atoms bearing the negative charge in each conjugate base. For example, let's compare butane and propanol:

H
Butane
O
H
Propanol

In order to assess the relative acidity of these two compounds, we must first deprotonate each of these compounds and draw the conjugate bases:

Now we compare these conjugate bases by looking at where the negative charge is located in each case. In the first conjugate base, the negative charge is on a carbon atom. In the second conjugate base, the negative charge is on an oxygen atom. To determine which of these is more stable, we must find out whether these elements are in the same row or in the same column of the periodic table (Figure 3.1).

FIGURE 3.1
Examples of elements in the same row or in the same column of the periodic table.

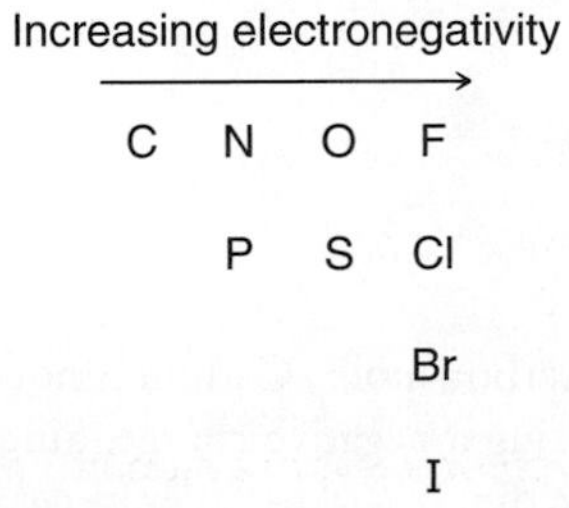

FIGURE 3.2
Electronegativity trends in the periodic table.

For example, C^- and O^- appear in the same row of the periodic table. When two atoms are in the same row, electronegativity is the dominant effect. Recall that electronegativity measures an atom's affinity for electrons (how willing the atom is to accept a new electron), and electronegativity increases across a row (Figure 3.2).Oxygen is more electronegative than carbon, so oxygen is more capable of stabilizing the negative charge. Therefore, a proton on oxygen is more acidic than a proton on carbon:

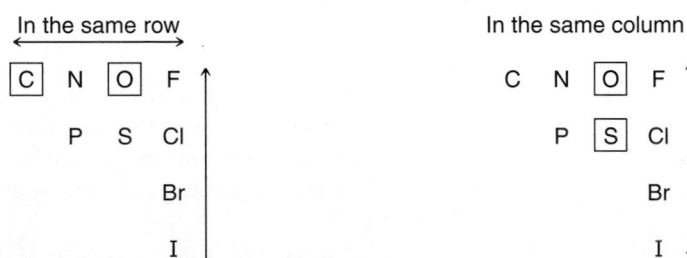

H
O
H
More acidic

The story is different when comparing two atoms in the same column of the periodic table. For example, let's compare the acidity of water and hydrogen sulfide:

In order to assess the relative acidity of these two compounds, we deprotonate each of them and compare their conjugate bases:

$$\ominus\overset{..}{\underset{..}{O}}H \qquad \ominus\overset{..}{\underset{..}{S}}H$$

more stable because bigger

In this example, we are comparing O^- and S^-, which appear in the same column of the periodic table. In such a case, electronegativity is not the dominant effect. Instead, the dominant effect is size (Figure 3.3). Sulfur is larger than oxygen and can therefore better stabilize a negative charge by spreading the charge over a larger volume of space. The HS^- is more stable than HO^-, and therefore, H_2S is a stronger acid than H_2O. We can verify this prediction by looking at pK_a values (the pK_a of H_2S is 7.0, while the pK_a of H_2O is 15.7).

electronegativity increases.

| C | N | O | F |
| | P | S | Cl | → Increasing electronegativity
| | | | Br |
| | | | I |

| C | N | O | F |
| | P | S | Cl | → Increasing size
| | | | Br |
| | | | I |

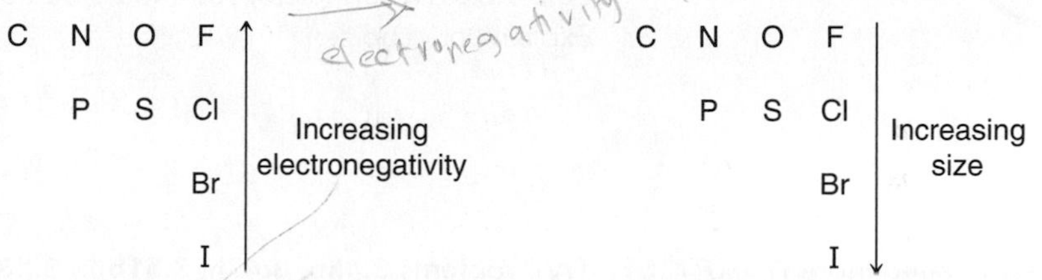

FIGURE 3.3
Competing trends in the periodic table: size vs. electronegativity.

To summarize, there are two important trends: electronegativity (for comparing atoms in the same row) and size (for comparing atoms in the same column).

SKILLBUILDER

3.5 ASSESSING RELATIVE STABILITY: FACTOR 1—ATOM

LEARN the skill

Compare the two protons that are shown in the following compound. Which one is more acidic?

more acidic because more electronegative than nitrogen

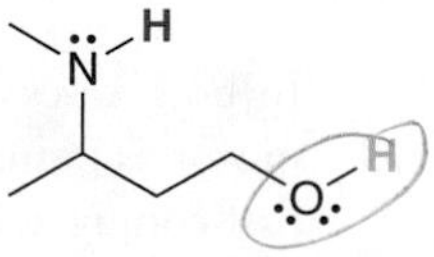

SOLUTION

The first step is to deprotonate each location and draw the two possible conjugate bases:

STEP 1
Draw the conjugate bases.

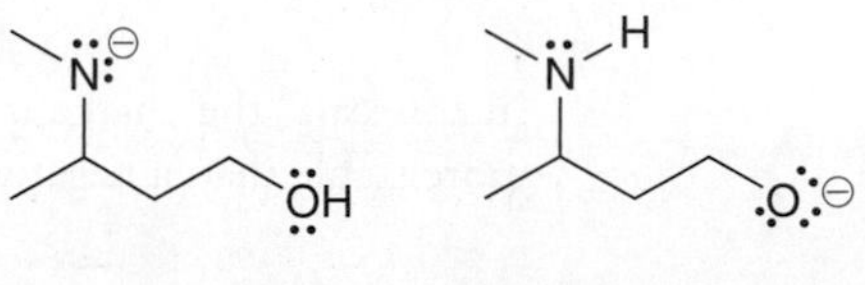

STEP 2
Compare the location of the charge in each case.

We now compare these two conjugate bases and determine which one is more stable. The first conjugate base has a negative charge on nitrogen, while the second conjugate base has a negative charge on oxygen. Nitrogen and oxygen are in the same row of the periodic table, so electronegativity is the determining factor. Oxygen is more electronegative than

STEP 3
The more stable conjugate base corresponds with the more acidic proton.

nitrogen and can better stabilize the negative charge. Therefore, the proton on the oxygen can be removed with greater ease than the proton on the nitrogen:

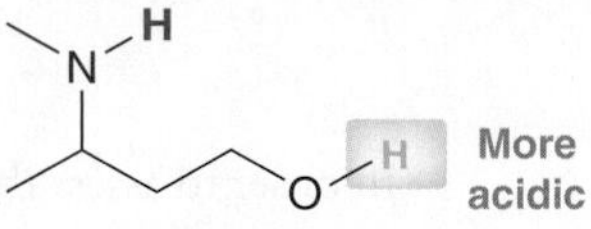

PRACTICE the skill **3.13** In each compound below, two protons are clearly identified. Determine which of the two protons is more acidic.

(a) 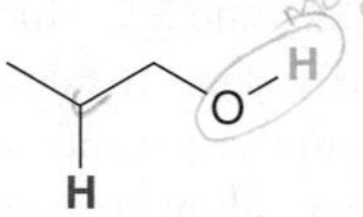 (b) 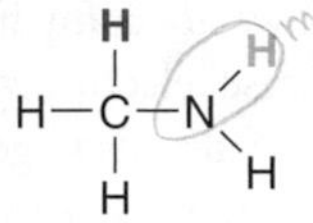 (c) 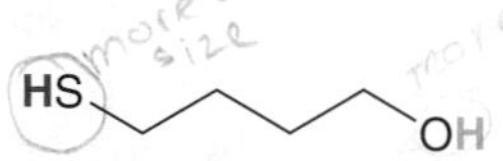 (d) 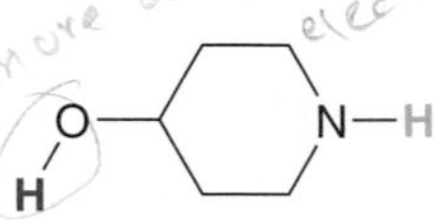

APPLY the skill **3.14** Nitrogen and sulfur are neither in the same row nor in the same column of the periodic table. Nevertheless, you should be able to identify which proton below is more acidic. Explain your choice:

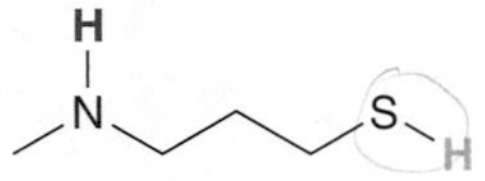

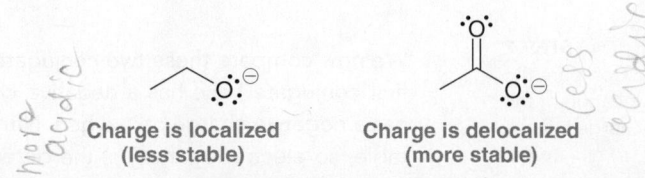

need more **PRACTICE?** **Try Problems 3.45b, 3.47h, 3.51b,h, 3.53**

2. ***Resonance.*** The second factor for comparing conjugate base stability is resonance. To illustrate the role of resonance in charge stability, let's consider the structures of ethanol and acetic acid:

Ethanol **Acetic acid**

In order to compare the acidity of these two compounds, we must deprotonate each of them and draw the conjugate bases:

In both cases, the negative charge is on oxygen. Therefore, factor 1 does not indicate which proton is more acidic. But there is a critical difference between these two negative charges. The first conjugate base has no resonance structures, while the second conjugate base does:

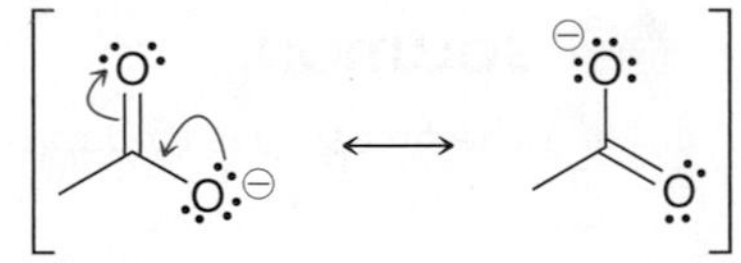

In this case, the charge is delocalized over both oxygen atoms. Such a negative charge will be more stable than a negative charge localized on one oxygen atom:

Charge is localized (less stable) **Charge is delocalized (more stable)**

For this reason, compounds containing a C═O bond directly next to an OH are generally mildly acidic, because their conjugate bases are resonance stabilized:

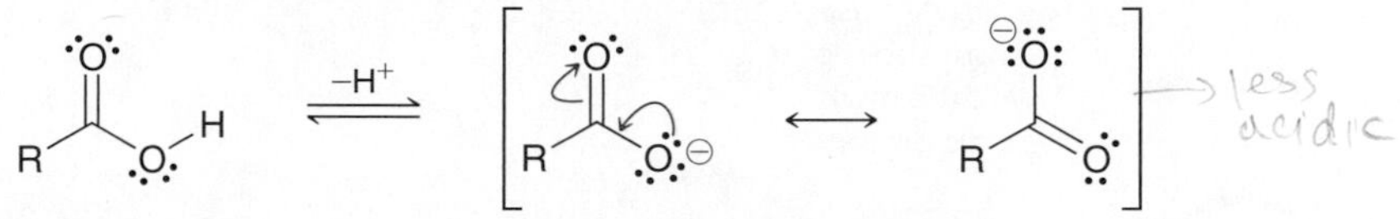

A carboxylic acid **Resonance-stabilized conjugate base**

These compounds are called *carboxylic acids*. The R group above simply refers to the rest of the molecule that has not been drawn. Carboxylic acids are actually not very acidic at all compared with inorganic acids such as H_2SO_4 or HCl. Carboxylic acids are only considered to be acidic when compared with other organic compounds. The "acidity" of carboxylic acids highlights the fact that *acidity is relative*.

SKILLBUILDER

3.6 ASSESSING RELATIVE STABILITY: FACTOR 2—RESONANCE

LEARN the skill

Compare the two protons shown in the following compound. Which one is more acidic?

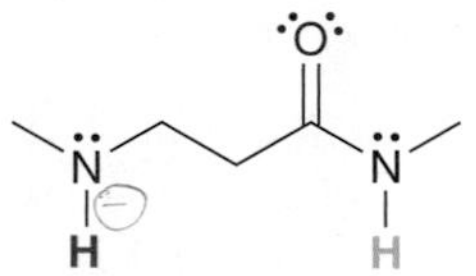

SOLUTION

Begin by drawing the respective conjugate bases:

STEP 1
Draw the conjugate bases.

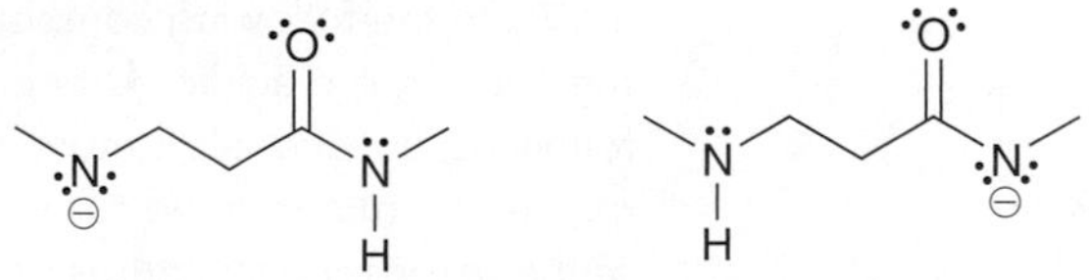

In the first conjugate base, the charge is localized on nitrogen:

STEP 2
Look for resonance stabilization.

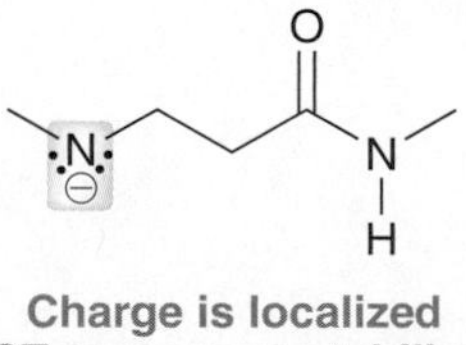

Charge is localized
NOT resonance stabilized

In the second conjugate base, the charge is delocalized by resonance:

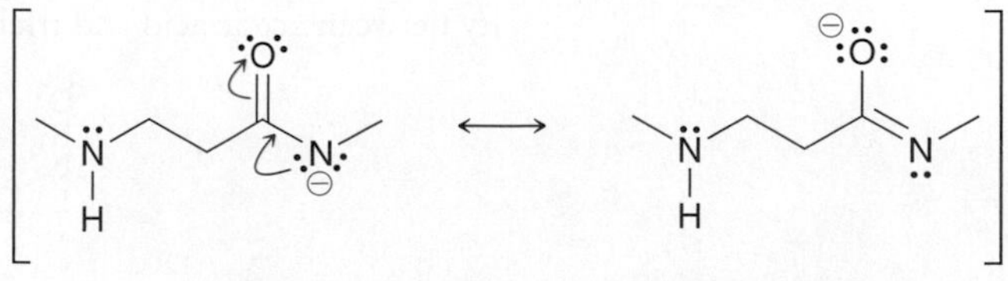

Charge is resonance stabilized

STEP 3 The more stable conjugate base corresponds with the more acidic proton.

The charge is distributed over two atoms, N and O. The delocalization of the charge makes it more stable, and therefore, this proton is more acidic:

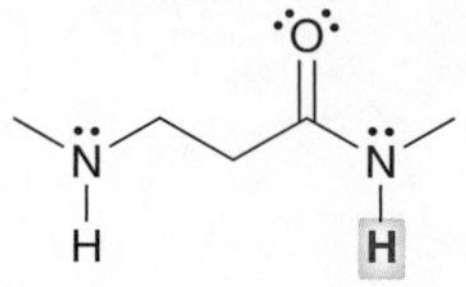

PRACTICE the skill **3.15** In each compound below, two protons are clearly identified. Determine which of the two protons is more acidic.

(a) 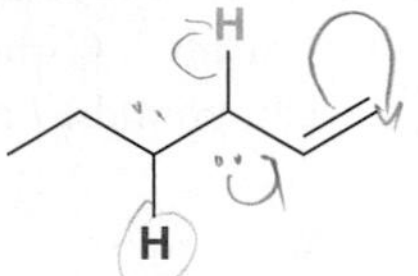 (b) 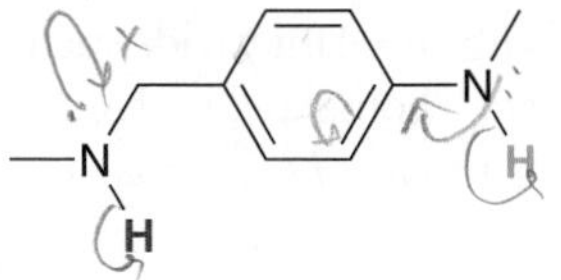 (c) 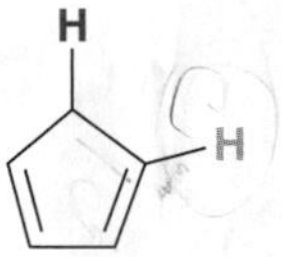

(d) 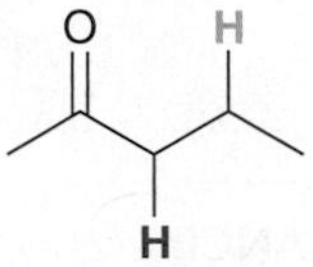 (e) 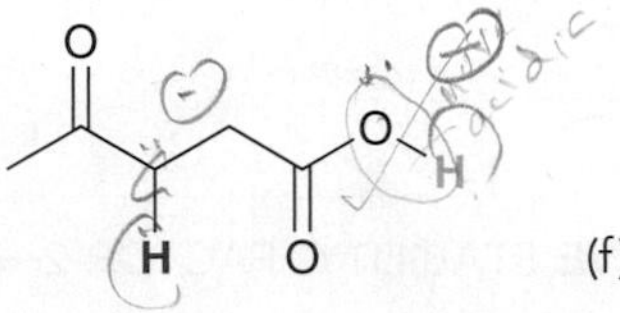 (f) 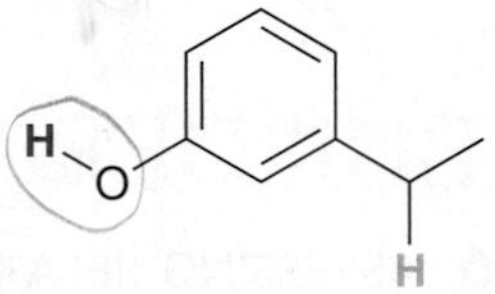

APPLY the skill

3.16 Ascorbic acid (vitamin C) does not contain a traditional carboxylic acid group, but it is, nevertheless, still fairly acidic ($pK_a = 4.2$). Identify the acidic proton, and explain your choice using resonance structures, if necessary.

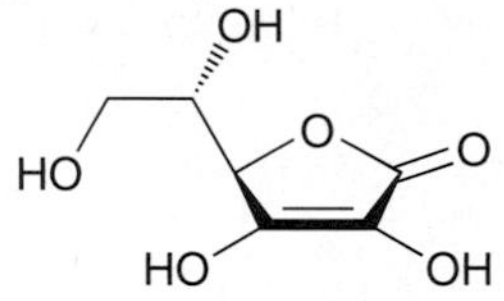

Ascorbic acid
(Vitamin C)

3.17 In the following compound two protons are clearly identified. Determine which of the two is more acidic. After comparing the conjugate bases, you should get stuck on the following question: Is it more stabilizing for a negative charge to be spread out over one oxygen atom and three carbon atoms or to be spread out over two oxygen atoms? Draw all of the resonance structures of each conjugate base, and then take a look at the pK_a values listed in Table 3.1.

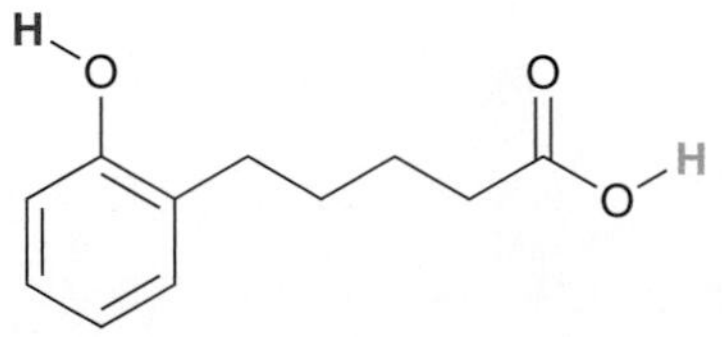

need more **PRACTICE?** **Try Problems 3.45a, 3.46a, 3.47b,e-g, 3.51c-f**

3. ***Induction.*** The two factors we have examined so far do not explain the difference in acidity between acetic acid and trichloroacetic acid:

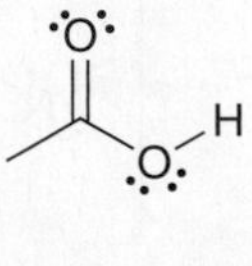

Acetic acid

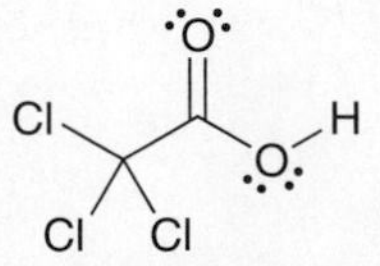

Trichloroacetic acid

Which compound is more acidic? In order to answer this question without help from a chart of pK_a values, we must draw the conjugate bases of the two compounds and then compare them:

Delocalized
- e⁻ can be moved

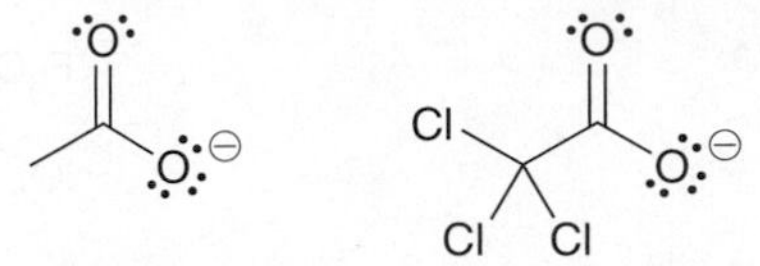

Localized
- e⁻ can't be moved

Factor 1 does not answer the question because the negative charge is on oxygen in both cases. Factor 2 also does not answer the question because there are resonance structures that delocalize the charge over two oxygen atoms in both cases. The difference between these compounds is clearly the chlorine atoms. Recall that each chlorine atom withdraws electron density via induction:

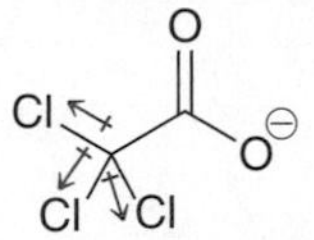

The net effect of the chlorine atoms is to withdraw electron density away from the negatively charged region of the compound, thereby stabilizing the negative charge. Therefore the conjugate base of trichloroacetic acid is more stable than the conjugate base of acetic acid:

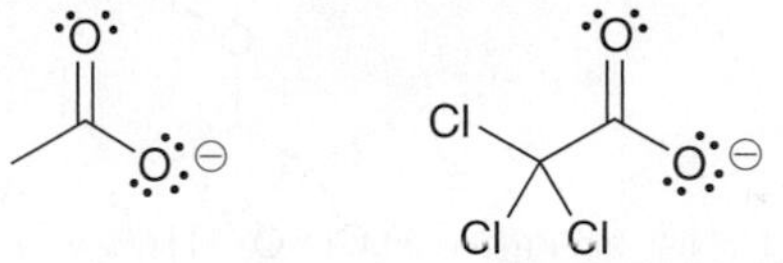

More stable

From this, we can conclude that trichloroacetic acid is more acidic:

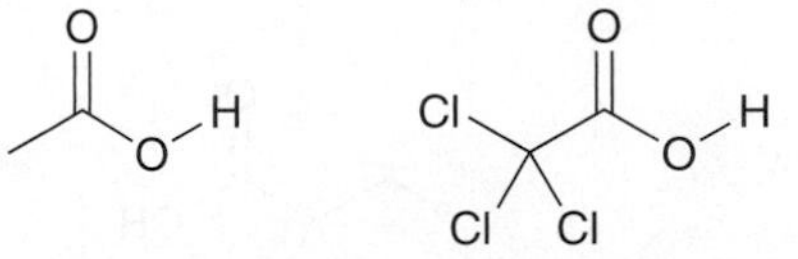

More acidic

We can verify this prediction by looking up pK_a values. In fact, we can use pK_a values to verify the individual effect of each Cl:

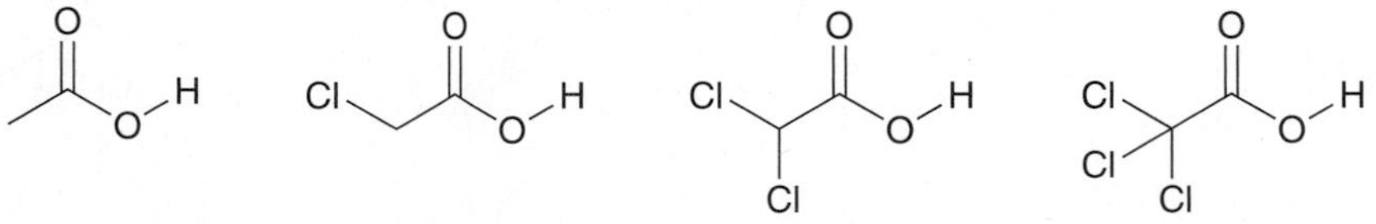

$pK_a = 4.75$ $pK_a = 2.87$ $pK_a = 1.25$ $pK_a = 0.70$

Notice the trend. With each additional Cl, the compound becomes more acidic.

SKILLBUILDER

3.7 ASSESSING RELATIVE STABILITY: FACTOR 3—INDUCTION

LEARN the skill

Identify which of the protons shown below is more acidic:

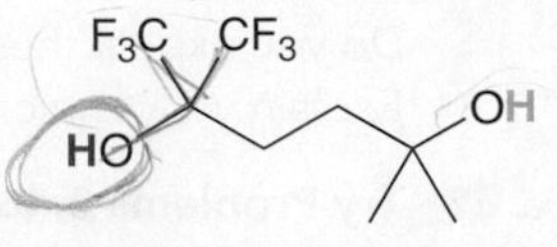

SOLUTION

Begin by drawing the respective conjugate bases:

STEP 1
Draw the conjugate bases.

STEP 2
Look for inductive effects.

In the conjugate base on the left, the charge is stabilized by the inductive effects of the nearby fluorine atoms. In contrast, the conjugate base on the right lacks this stabilization. Therefore, we predict that the conjugate base on the left is more stable. And as a result, we conclude that the proton near the fluorine atoms will be more acidic:

STEP 3
The more stable conjugate base corresponds with the more acidic proton.

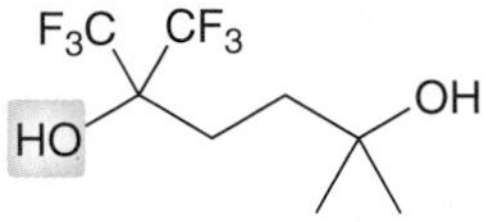

PRACTICE the skill

3.18 Identify the most acidic proton in each of the following compounds and explain your choice:

(a) 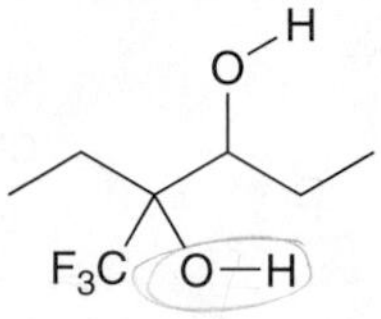

(b) 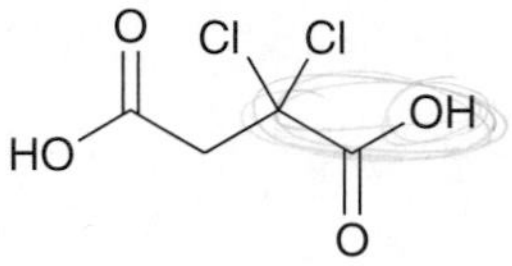

3.19 For each pair of compounds below, identify which compound is more acidic and explain your choice:

(a) 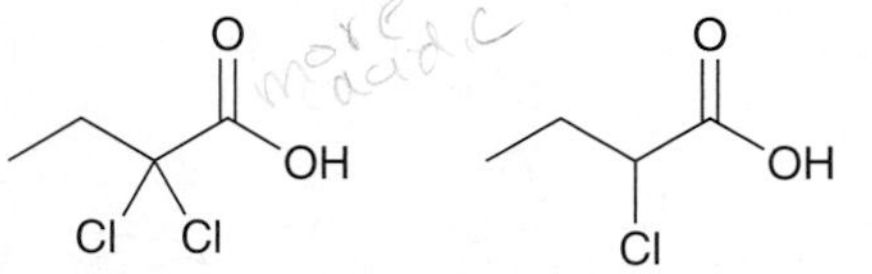

(b) 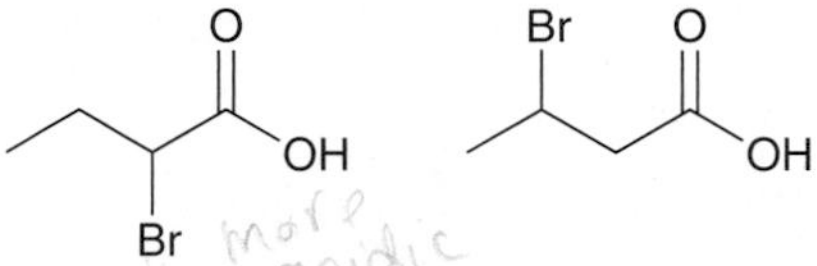

APPLY the skill

3.20 Consider the structure of 2,3-dichloropropanoic acid:

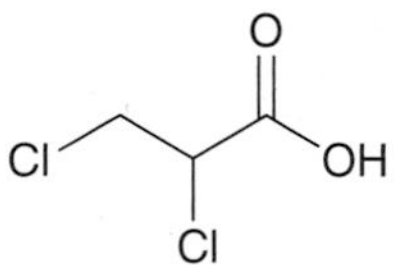

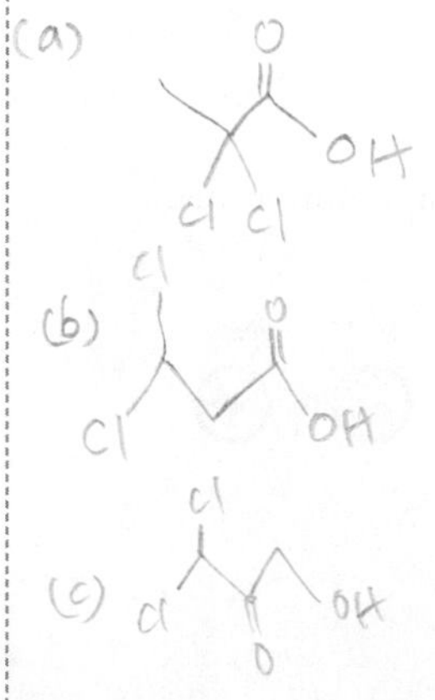

This compound has many constitutional isomers.

(a) Draw a constitutional isomer that is slightly more acidic and explain your choice.

(b) Draw a constitutional isomer that is slightly less acidic and explain your choice.

(c) Draw a constitutional isomer that is significantly (at least 10 orders of magnitude) less acidic and explain your choice.

3.21 Consider the two protons highlighted in the following compound:

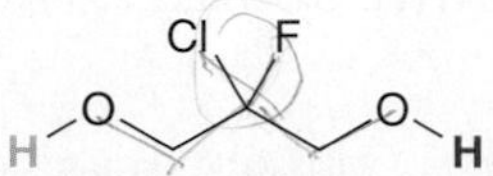

Do you expect these protons to be equivalent, or is one proton more acidic than the other? Explain your choice. (*Hint:* Think carefully about the geometry at the central carbon atom.)

need more **PRACTICE?** **Try Problems 3.46b, 3.47c, 3.51g**

4. ***Orbitals***. The three factors we have examined so far will not explain the difference in acidity between the two identified protons in the following compound:

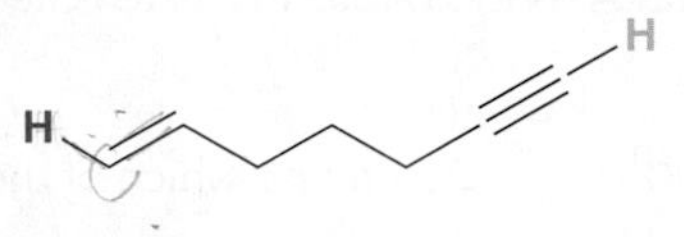

Draw the conjugate bases to compare them:

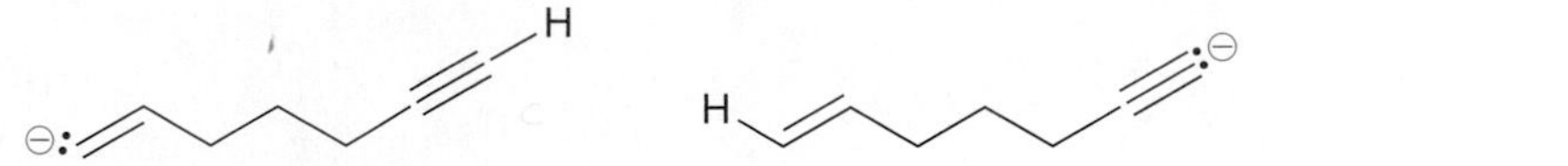

In both cases, the negative charge is on a carbon atom, so factor 1 does not help. In both cases, the charge is not stabilized by resonance, so factor 2 does not help. In both cases, there are no inductive effects to consider, so factor 3 does not help. The answer here comes from looking at the hybridization states of the orbitals that accommodate the charges. Recall from Chapter 1 that a carbon with a triple bond is *sp* hybridized, a carbon with a double bond is sp^2 hybridized, and a carbon with all single bonds is sp^3-hybridized. The first conjugate base (above left) has a negative charge on an sp^2-hybridized carbon atom, while the second conjugate base (above right) has a negative charge on an *sp*-hybridized carbon atom. What difference does this make? Let's quickly review the shapes of hybridized orbitals (Figure 3.4).

FIGURE 3.4 Relative shapes of hybridized orbitals.

A pair of electrons in an *sp*-hybridized orbital is held closer to the nucleus than a pair of electrons in an sp^2- or sp^3-hybridized orbital. As a result, electrons residing in an *sp* orbital are stabilized by being close to the nucleus. Therefore, a negative charge on an *sp*-hybridized carbon is more stable than a negative charge on an sp^2-hybridized carbon:

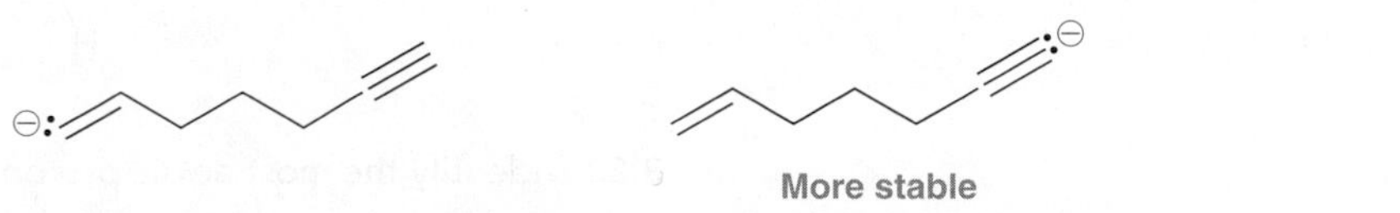

We conclude that a proton on a triple bond will be more acidic than a proton on a double bond, which in turn will be more acidic than a proton on a carbon with all single bonds. We can verify this trend by looking at the pK_a values in Figure 3.5. These pK_a values suggest that this effect is very significant; acetylene is 19 orders of magnitude more acidic than ethylene.

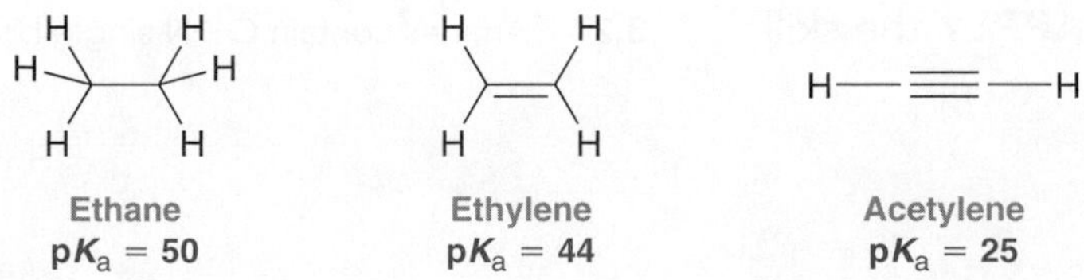

FIGURE 3.5 pK_a values for ethane, ethylene, and acetylene.

SKILLBUILDER

3.8 ASSESSING RELATIVE STABILITY: FACTOR 4—ORBITALS

LEARN the skill

Determine which of the protons identified below is more acidic:

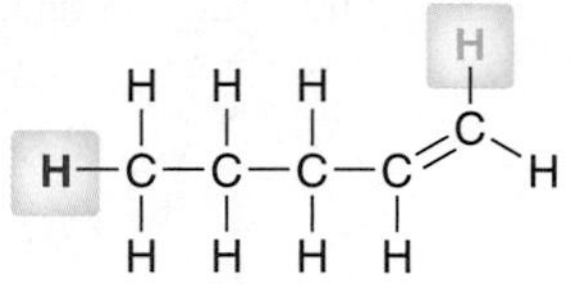

1-pentene

SOLUTION

Begin by drawing the respective conjugate bases:

STEP 1
Draw the conjugate bases.

STEP 2
Analyze the orbital that accommodates the charge in each case.

In both cases, the negative charge is on a carbon atom, so factor 1 does not help. In both cases, the charge is not stabilized by resonance, so factor 2 does not help. In both cases, there are no inductive effects to consider, so factor 3 does not help. The answer here comes from looking at the hybridization states of the orbitals that accommodate the charges. The first conjugate base has the negative charge in an sp^3-hybridized orbital, while the second conjugate base has the negative charge in an sp^2-hybridized orbital. An sp^2-hybridized orbital is closer to the nucleus than an sp^3-hybridized orbital and therefore better stabilizes a negative charge. So we conclude that the vinylic proton is more acidic:

STEP 3
The more stable conjugate base corresponds with the more acidic proton.

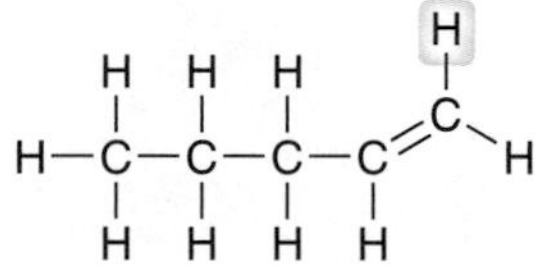

PRACTICE the skill

3.22 Identify which of the following compounds is more acidic. Explain your choice.

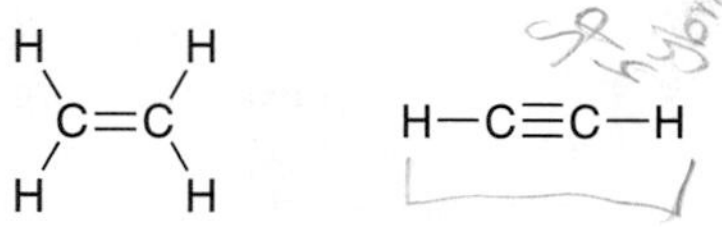

3.23 Identify the most acidic proton in each of the following compounds:

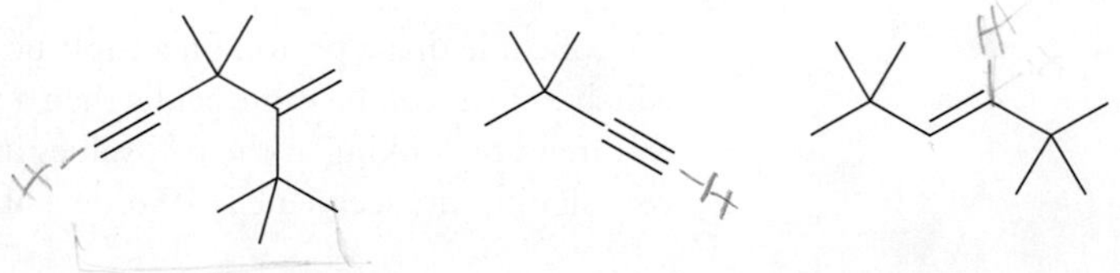

APPLY the skill

3.24 Amines contain C—N single bonds, while imines contain C—N double bonds:

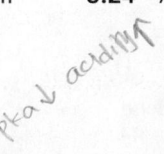

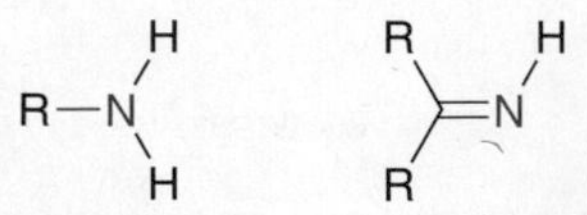

Amine **Imine**

Most simple amines have a pK_a somewhere in the range between 35 and 45. Based on this information, predict which statement is most likely to be true, and explain the reasoning behind your selection:

(a) Most imines will have a pK_a below 35.

(b) Most imines will have a pK_a above 45.

(c) Most imines will have a pK_a in the range between 35 and 45.

------> need more **PRACTICE?** **Try Problem 3.45c**

Ranking the Factors That Affect the Stability of Negative Charges

We have thus far examined each of the four factors that affect the stability of negative charges. We must now consider their order of priority—in other words, which factor takes precedence when two or more factors are present?

Generally speaking, the order of priority is the order in which the factors were presented:

1. *Atom.* Which atom is the charge on? (How do the atoms compare in terms of electronegativity and size? Remember the difference between comparing atoms in the same row vs. atoms in the same column.)
2. *Resonance.* Are there any resonance effects that make one conjugate base more stable than the other?
3. *Induction.* Are there any inductive effects that stabilize one of the conjugate bases?
4. *Orbital.* In what orbital do we find the negative charge for each conjugate base?

A helpful way to remember the order of these four factors is to take the first letter of each factor, giving the following mnemonic device: *ARIO*.

As an example, let's compare the protons shown in the following two compounds:

Ethanol **Propylene**

We compare these compounds by drawing their conjugate bases:

Factor 1 suggests that the first conjugate base is more stable (O^- better than C^-). However, factor 2 suggests that the second conjugate base is more stable (resonance that delocalizes the charge). This leaves us with an important question: Is a negative charge more stable when it is localized on one oxygen atom or is a negative charge more stable when it is delocalized over two carbon atoms? The answer is: In general, factor 1 beats factor 2. A negative charge is more stable on one oxygen than on two carbon atoms. We can verify this assertion by comparing pK_a values (Figure 3.6).

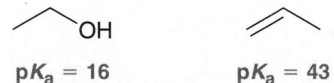

pK_a = 16 **pK_a = 43**

FIGURE 3.6
pK_a values for ethanol and ethylene.

In fact, the pK_a values indicate that a negative charge on one oxygen atom is 27 orders of magnitude (a billion billion billion times) more stable than a negative charge on two carbon atoms.

This prioritization scheme (ARIO) will often be helpful, but strict adherence to it can sometimes produce the wrong prediction. In other words, there are many exceptions. As an example, compare the structures of acetylene and ammonia:

$$H{-}C{\equiv}C{-}H \qquad :NH_3$$

Acetylene **Ammonia**

To determine which compound is more acidic, we draw the conjugate bases:

$$H{-}C{\equiv}C:^{\ominus} \qquad ^{\ominus}\!\!:\!\ddot{N}H_2$$

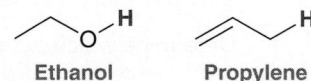

When comparing these two negative charges, there are two competing factors. Factor 1 suggests that the second conjugate base is more stable (N^- is more stable than C^-), but factor 4 suggests that the first conjugate base is more stable (an sp-hybridized orbital can stabilize a negative charge better than an sp^3-hybridized orbital). In general, factor 1 wins over the others. But this case is an exception, and factor 4 (orbitals) actually predominates here. In this case, the negative charge is more stable on the carbon atom, even though nitrogen is more electronegative than carbon.

$$H-C\equiv C\text{:}^{\ominus} \qquad\qquad {}^{\ominus}\ddot{\text{:}}NH_2$$
More stable

In fact, for this reason, H_2N^- is often used as a base to deprotonate a triple bond:

$$H-C\equiv C-H \;+\; \overset{\ominus}{\ddot{\text{:}}}NH_2 \;\rightleftharpoons\; H-C\equiv C\text{:}^{\ominus} \;+\; \text{:}NH_3$$
$$\mathbf{p}K_a = 25 \qquad\qquad\qquad\qquad\qquad \mathbf{p}K_a = 38$$

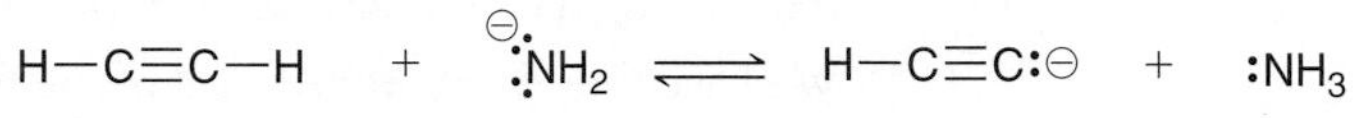

We see from the pK_a values that acetylene is 13 orders of magnitude more acidic than ammonia. This explains why H_2N^- is a suitable base for deprotonating acetylene.

There are, of course, other exceptions to the ARIO prioritization scheme, but the exception shown above is the most common. In the vast majority of cases, it would be a safe bet to apply the four factors in the order ARIO to provide a qualitative assessment of acidity. However, to be certain, it is always best to look up pK_a values and verify your prediction.

SKILLBUILDER

3.9 ASSESSING RELATIVE STABILITY: ALL FOUR FACTORS

LEARN the skill

Determine which of the two protons identified below is more acidic and explain why:

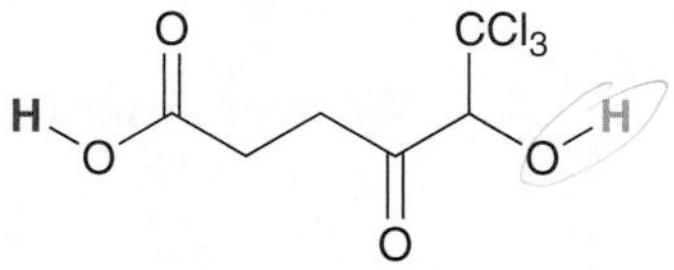

SOLUTION

We always begin by drawing the respective conjugate bases:

STEP 1
Draw the
conjugate
bases.

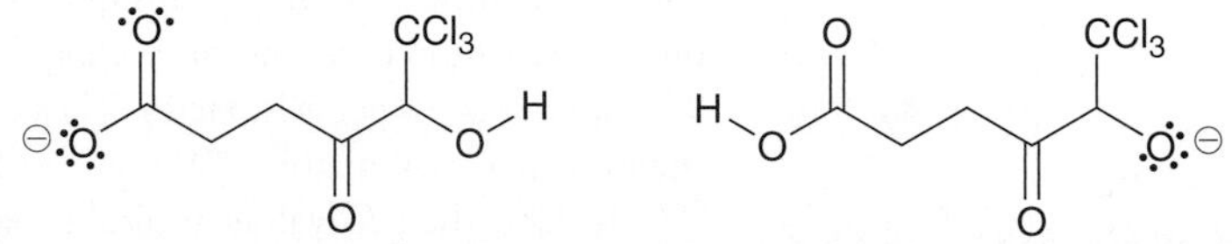

STEP 2
Analyze all four
factors.

Now we consider all four factors (ARIO) in comparing the stability of these negative charges:

1. *Atom.* In both cases, the charge is on an oxygen atom, so this factor doesn't help.

2. *Resonance.* The conjugate base on the left is resonance stabilized while the conjugate base on the right is not. Based on this factor alone, we would say the conjugate base on the left is more stable.

3. *Induction.* The conjugate base on the right has an inductive effect that stabilizes the charge while the conjugate base on the left does not. Based on this factor alone, we would say the conjugate base on the right is more stable.

4. *Orbital.* This factor does not help.

Our analysis reveals a competition between two factors. In general, resonance will beat induction. Based on this, we predict that the conjugate base on the left is more stable. Therefore, we conclude that the following proton is more acidic:

STEP 3
The more stable conjugate base corresponds with the more acidic proton.

ARSO has more resonance structures

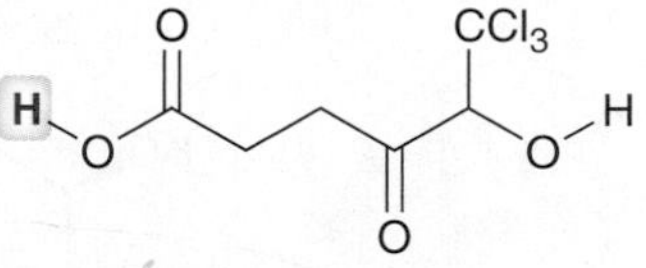

PRACTICE the skill **3.25** In each compound below, two protons are clearly identified. Determine which of the two protons is more acidic.

(a)

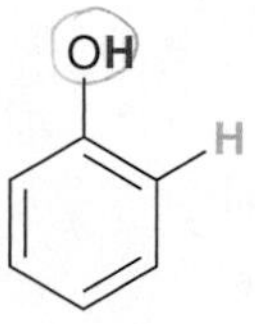

(b)

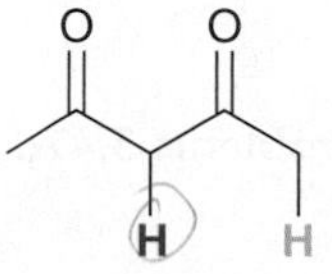

(c)

AR10

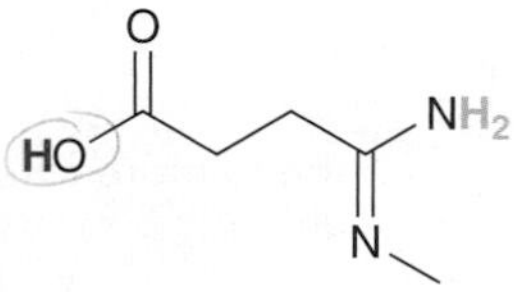

(d)

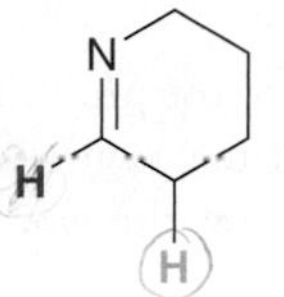

(e)

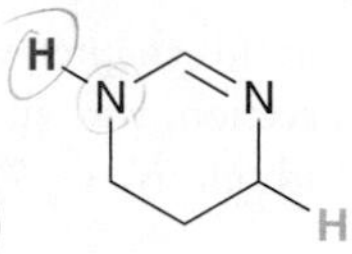

(f)

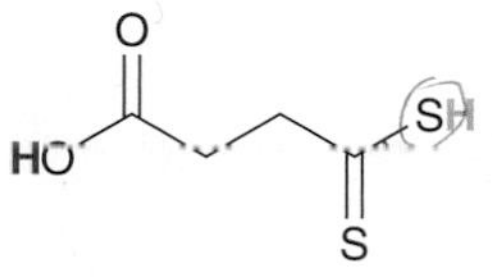

(g)

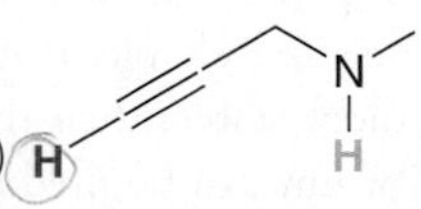

(h)

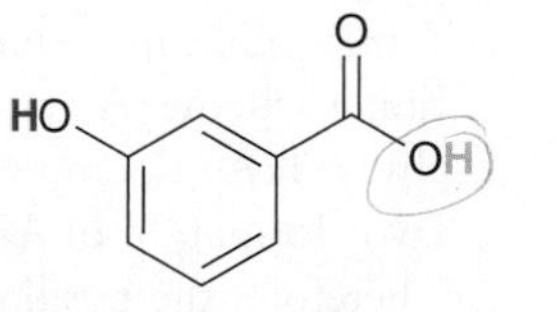

(i)

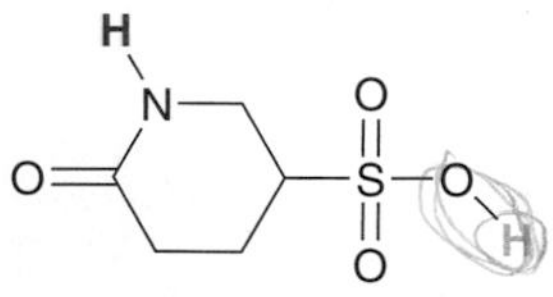

3.26 For each pair of compounds below, predict which will be more acidic:

(a) HCl (HBr) (b) H₂O (H₂S) (c) (NH₃) CH₄

(d) H—≡—H H H (e) Cl₃C—OH—CCl₃ OH

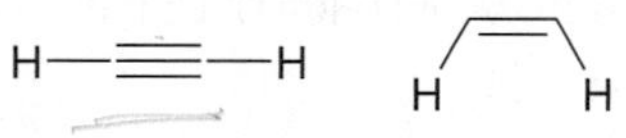

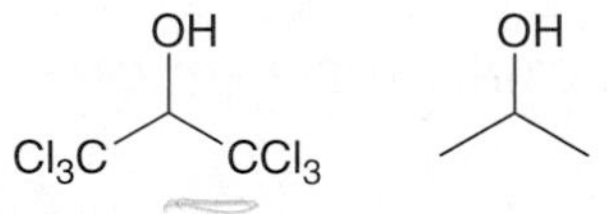

APPLY the skill **3.27** The following compound is one of the strongest known acids:

replace with S—H ···=S=···

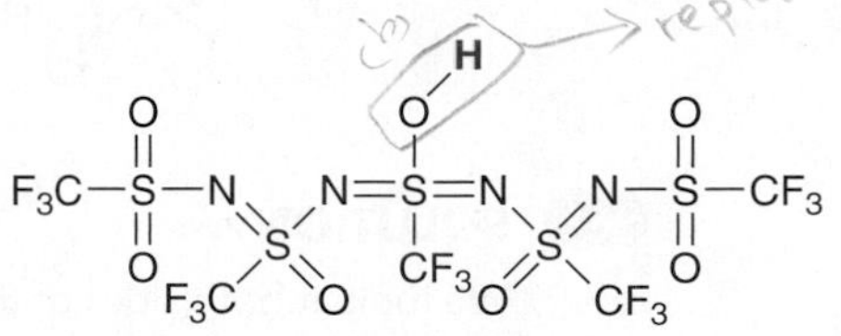

(a) Explain why it is such a strong acid. *3 CF₃ groups, negative charge spread over 4 nitrogen atoms when proton is removed*

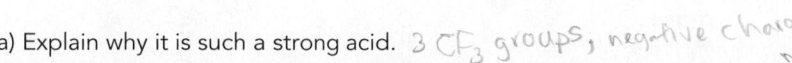

(b) Suggest a modification to the structure that would render the compound even more acidic.

3.28 Amphotericin B is a powerful antifungal agent used for intravenous treatment of severe fungal infections. Identify the most acidic proton in this compound:

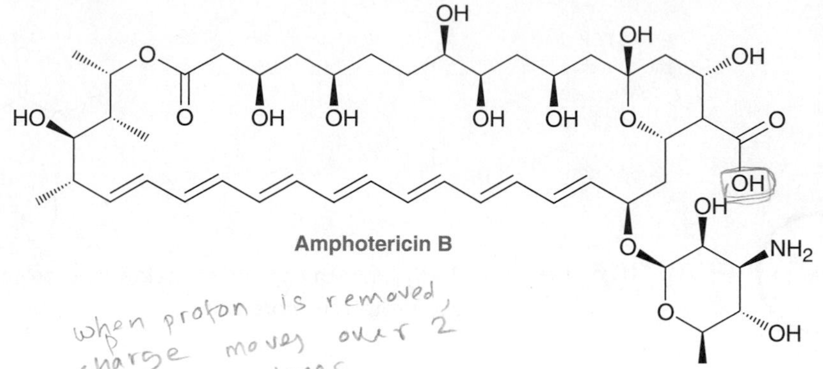

Amphotericin B

when proton is removed, charge moves over 2 oxygen atoms

need more **PRACTICE?** Try Problems 3.47d, 3.51a, 3.57–3.61

3.5 Position of Equilibrium and Choice of Reagents

Earlier in this chapter, we learned how to use pK_a values to determine the position of equilibrium. In this section, we will learn to predict the position of equilibrium just by comparing conjugate bases without using pK_a values. To see how this works, let's examine a generic acid-base reaction:

$$H{-}A + B^{\ominus} \rightleftharpoons A^{\ominus} + HB$$

This equilibrium represents the competition between two bases (A^- and B^-) for H^+. The question is whether A^- or B^- is more capable of stabilizing the negative charge. The equilibrium will always favor the more stabilized negative charge. If A^- is more stable, then the equilibrium will favor formation of A^-. If B^- is more stable, then the equilibrium will favor formation of B^-. Therefore, the position of equilibrium can be predicted by comparing the stability of A^- and B^-. Let's see an example of this.

SKILLBUILDER

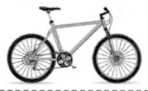

3.10 PREDICTING THE POSITION OF EQUILIBRIUM WITHOUT USING pK_a VALUES

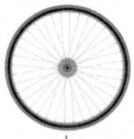

LEARN the skill

Predict the position of equilibrium for the following reaction:

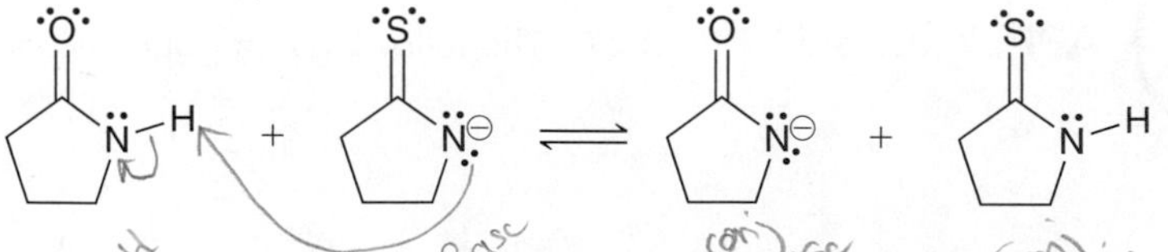

SOLUTION

We look at both sides of the equilibrium and compare the stability of the base on each side:

STEP 1
Identify the base on either side of the equilibrium.

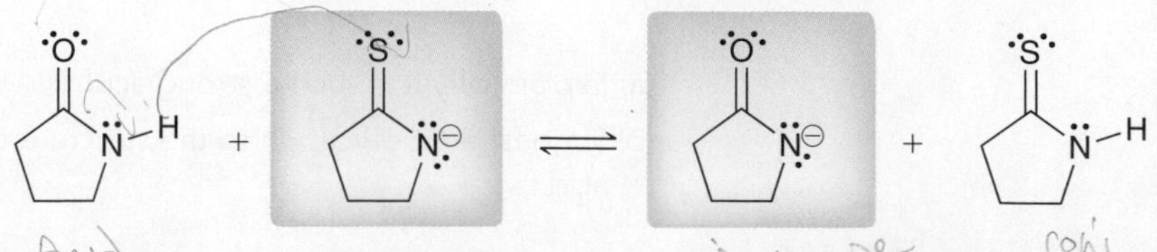

To determine which of these bases is more stable, we use the four factors (ARIO):

1. *Atom.* In both cases, the negative charge appears to be on a nitrogen atom, so this factor is not relevant.
2. *Resonance.* Both of these bases are resonance stabilized, but we do expect one of them to be more stable:

STEP 2
Compare the stability of these two bases using all four factors.

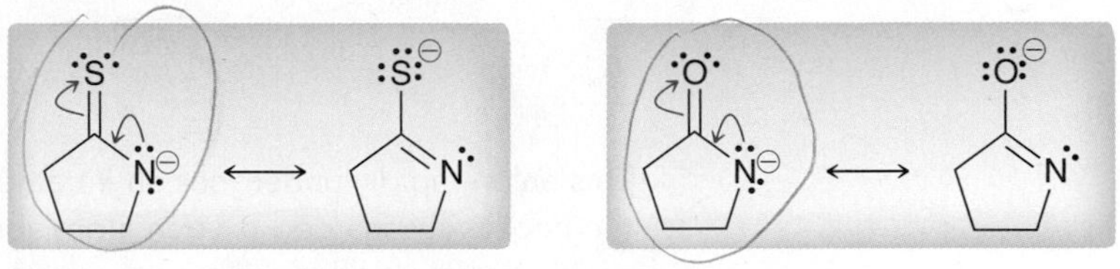

The first base has the negative charge delocalized over N and S. The second base has the negative charge delocalized over N and O. Because of its size, sulfur is more efficient than oxygen at stabilizing a negative charge, so we expect the first base to be more stable.

3. *Induction.* Neither of these bases is stabilized by inductive effects.
4. *Orbital.* Not a relevant factor in this case.

Based on factor 2, we conclude that the base on the left side is more stable, and therefore the equilibrium favors the left side of the reaction. Our prediction can be verified if we look up pK_a values for the acid on either side of the reaction:

STEP 3
The equilibrium will favor the more stable base.

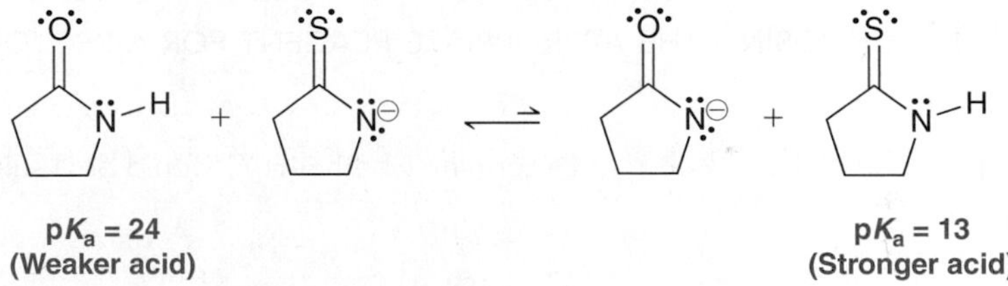

The equilibrium favors production of the weaker acid (higher pK_a), so the left side is favored, just as we predicted.

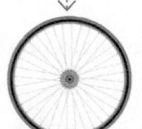

PRACTICE the skill **3.29** Predict the position of equilibrium for each of the following reactions:

(a)

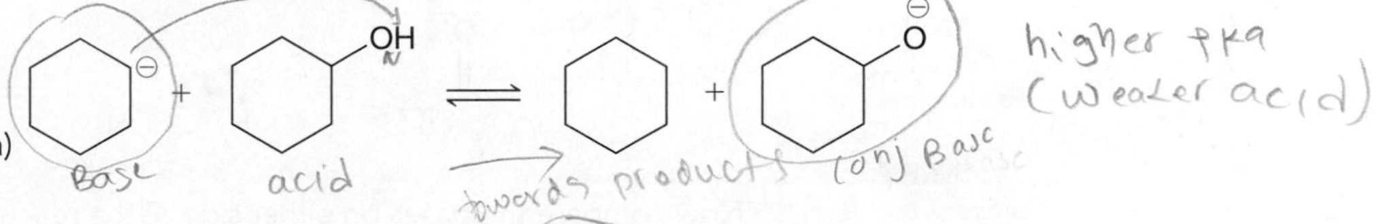

(b)

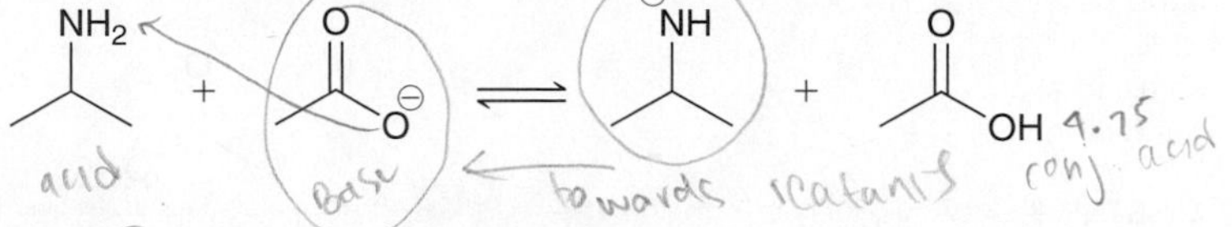

(c)

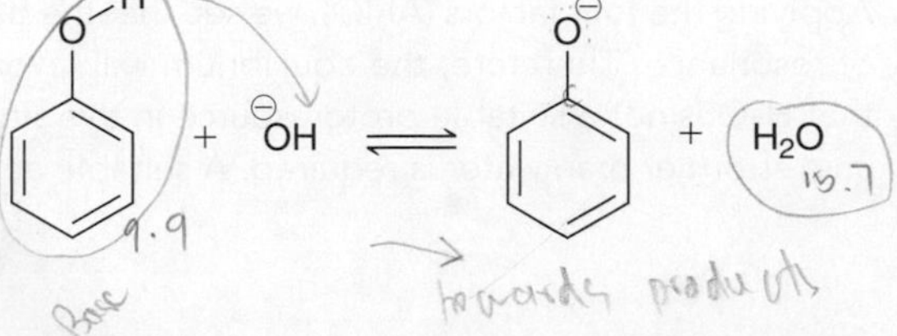

APPLY the skill

3.30 As we will learn in Chapter 21, treating a lactone (a cyclic ester) with sodium hydroxide will initially produce an anion:

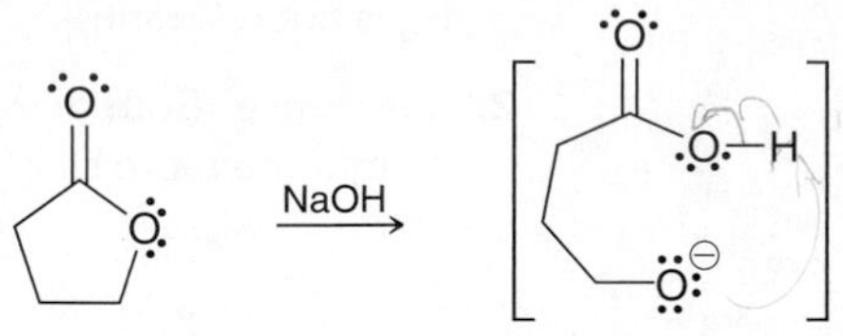

Initially formed

This anion rapidly undergoes an intramolecular proton transfer (see Problem 3.3), in which the negatively charged oxygen atom abstracts the nearby acidic proton. Draw the product of this intramolecular acid-base process, and then identify which side of the equilibrium is favored. Explain your answer.

need more **PRACTICE?** **Try Problems 3.49, 3.52**

The above process can also be used to determine whether a specific reagent is suitable for accomplishing a particular proton transfer, as shown in SkillBuilder 3.11.

SKILLBUILDER

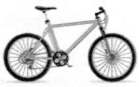

3.11 CHOOSING THE APPROPRIATE REAGENT FOR A PROTON TRANSFER REACTION

LEARN the skill

Determine whether H_2O would be a suitable reagent for protonating the acetate ion:

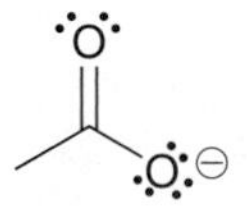

SOLUTION

We begin by drawing the acid-base reaction that occurs when the base is protonated by water:

STEP 1 Draw the equilibrium.

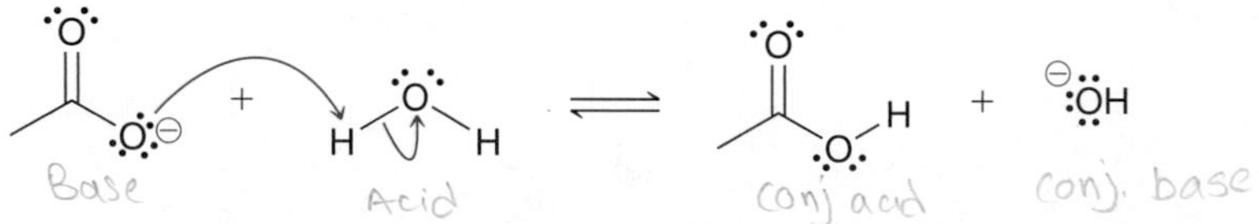

STEP 2 Compare the stability of the bases on either side of the equilibrium using all four factors.

Now compare the bases on either side of the reaction and ask which is more stable:

STEP 3 The reagent is only suitable if the equilibrium favors the desired products.

Applying the four factors (ARIO), we see that the base on the left side is more stable because of resonance. Therefore, the equilibrium will favor the left side of the reaction. This means that H_2O is not a suitable proton source in this situation. In order to protonate this base, an acid stronger than water is required. A suitable acid would be H_3O^+.

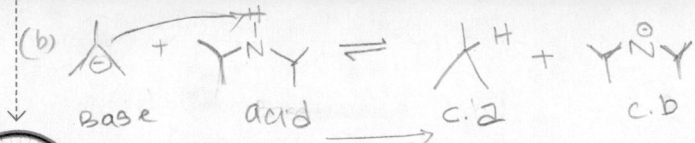

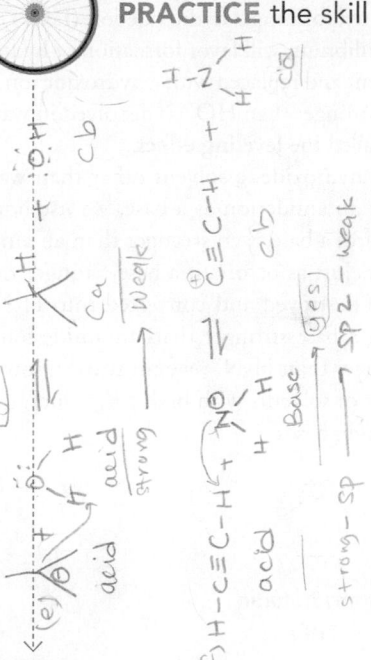

PRACTICE the skill **3.31** In each of the following cases, identify whether the reagent shown is suitable to accomplish the task described. Explain why or why not.

(a) **To protonate** 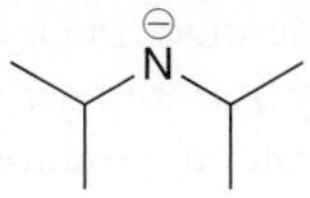 **using H_2O**

(b) **To protonate** 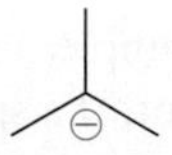 **using** 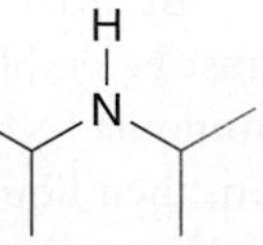

(c) **To deprotonate** 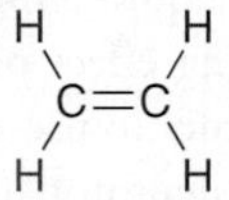 **using** $^{\ominus}NH_2$

(d) **To protonate** 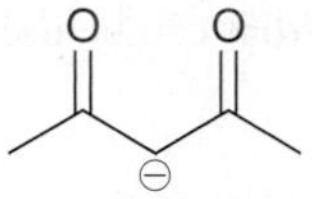 **using** H_2O

(e) **To protonate** 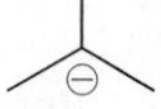 **using** H_2O

(f) **To deprotonate** H—C≡C—H **using** $^{\ominus}NH_2$

APPLY the skill

3.32 We will learn all of the following reactions in upcoming chapters. For each of these reactions, notice that the product is an anion (ignore the positively charged ion in each case). In order to obtain a neutral product, this anion must be treated with a proton source in a process called "working up the reaction." For each of the following reactions, identify whether water will be a suitable proton source for working up the reaction:

(a) 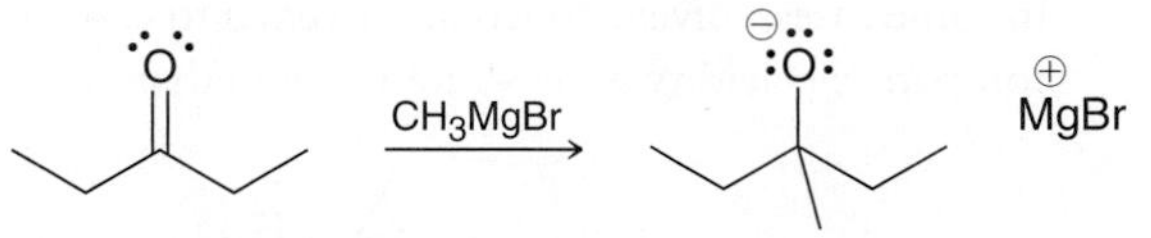

Chapter 20

(b) 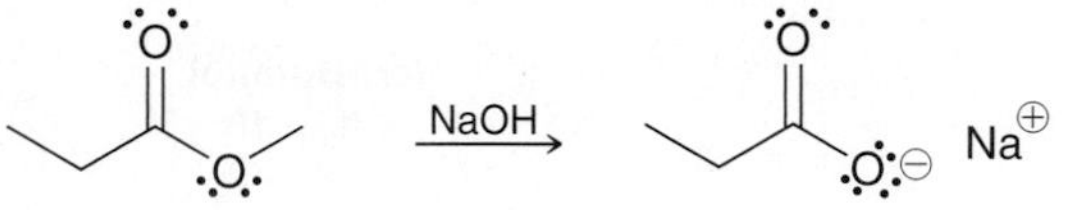

Chapter 21

(c) 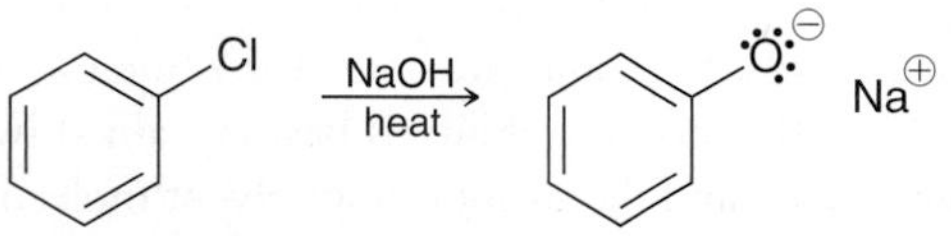

Chapter 19

need more **PRACTICE? Try Problems 3.40, 3.42**

3.6 Choice of Solvent

Bases stronger than hydroxide cannot be used when the solvent is water. To illustrate why, consider what happens if we mix the amide ion (H_2N^-) and water:

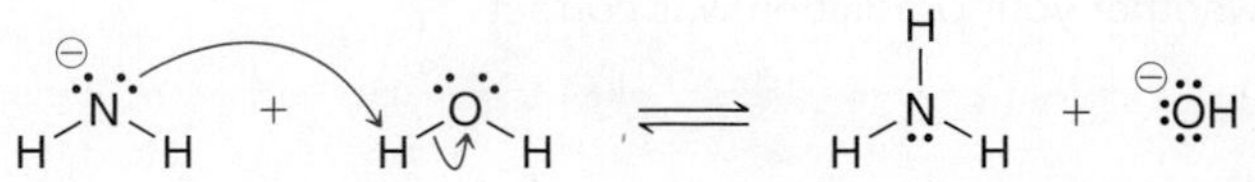

The amide ion is a strong enough base to deprotonate water, forming a hydroxide ion (HO^-). A hydroxide ion is more stable than an amide ion, so the equilibrium will favor formation of hydroxide. In other words, the amide ion is destroyed by the solvent and replaced with a hydroxide ion. In fact, this is true of any base stronger than HO^-. If a base stronger than HO^- is dissolved in water, the base reacts with water to produce hydroxide. This is called the **leveling effect**.

In order to work with bases that are stronger than hydroxide, a solvent other than water must be employed. For example, in order to work with an amide ion as a base, we use liquid ammonia (NH_3) as a solvent. If a specific situation requires a base even stronger than an amide ion, then liquid ammonia cannot be used as the solvent. Just as before, if a base stronger than H_2N^- is dissolved in liquid ammonia, the base will be destroyed and converted into H_2N^-. Once again, the leveling effect prevents us from having a base stronger than an amide ion in liquid ammonia. In order to use a base that is even stronger than H_2N^-, we must use a solvent that cannot be readily deprotonated. There are a number of solvents with high pK_a values, such as hexane and THF, that can be used to dissolve very strong bases:

Hexane

**Tetrahydrofuran
(THF)**

Throughout the course, we will see other examples of solvents suitable for working with very strong bases.

3.7 Solvating Effects

In some cases, solvent effects are invoked to explain small difference in pK_a values. For example, compare the acidity of *tert*-butanol and ethanol:

tert-**Butanol**
$pK_a = 18$
less acidic

Ethanol
$pK_a = 16$ *more acidic*

The pK_a values indicate that *tert*-butanol is less acidic than ethanol by two orders of magnitude. In other words, the conjugate base of *tert*-butanol is less stable than the conjugate base of ethanol. This difference in stability is best explained by considering the interactions between each conjugate base and the surrounding solvent molecules (Figure 3.7). Compare the way in which each conjugate base interacts with solvent molecules. The *tert*-butoxide ion is very bulky, or **sterically hindered**, and is less capable of interacting with the solvent. The ethoxide ion is not as sterically hindered so it can accommodate more solvent interactions. As a result, ethoxide is better solvated and is therefore more stable than *tert*-butoxide (Figure 3.7).

CONCEPTUAL CHECKPOINT

3.33 Predict which of the following compounds is more acidic.

After making your prediction, use the pK_a values from Table 3.1 to determine whether your prediction was correct.

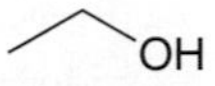

Ethanol **Water**

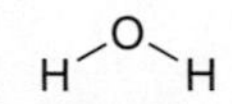

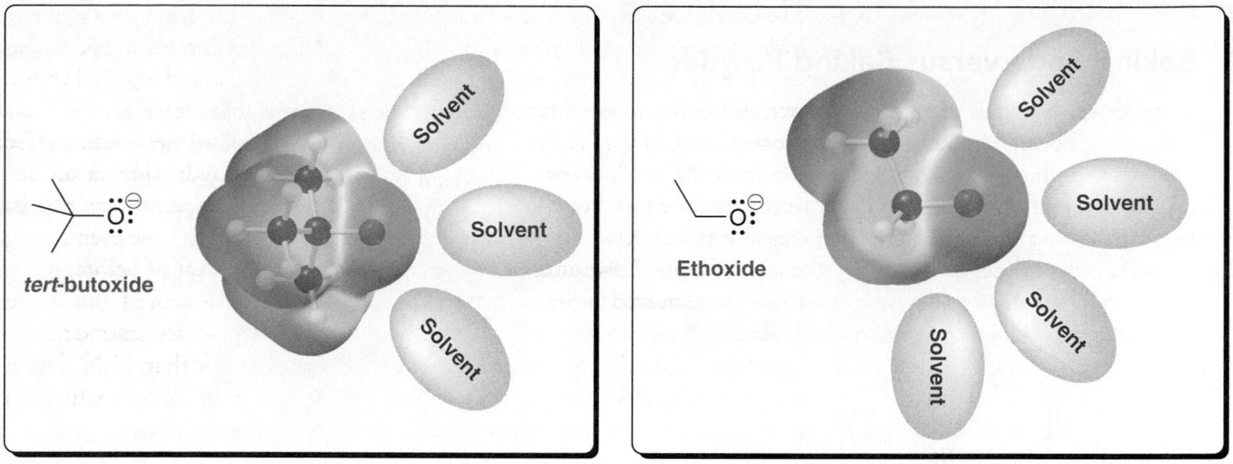

FIGURE 3.7
Electrostatic potential maps of *tert*-butoxide and ethoxide.

3.8 Counterions

LOOKING BACK
Your general chemistry textbook likely used the term *spectator ion* to refer to a counterion.

Negatively charged bases are always accompanied by positively charged species, called **cations** (pronounced CAT-EYE-ONZ). For example, HO^- must be accompanied by a counterion, such as Li^+, Na^+, or K^+. We will often see the following reagents: LiOH, or NaOH, or KOH. Don't be alarmed. All of these reagents are simply HO^- with the counterion indicated. Sometimes it is shown; sometimes not. Even when the counterion is not shown, it is still there. It is just not indicated because it is largely irrelevant. Up to this point in this chapter, counterions have not been shown, but from here on they will be. For example, consider the following equilibrium:

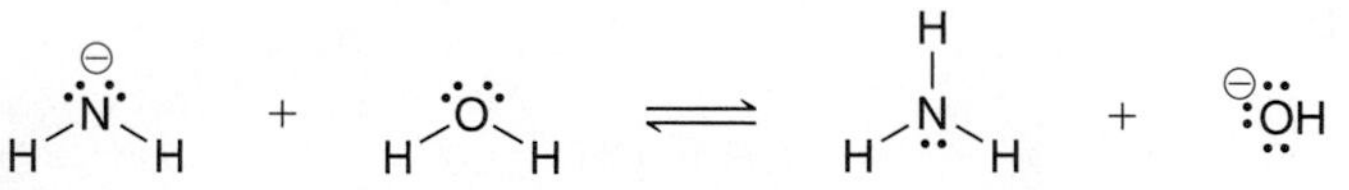

This reaction might be shown like this:

$$NaNH_2 + H_2O \rightleftharpoons NH_3 + NaOH$$

It is important to become accustomed to ignoring the cations when they are indicated and to focus on the real players—the bases. Although counterions generally do not play a significant role in reactions, they can, under some circumstances, influence the course of a reaction. We will only see one or two such examples throughout this course. The overwhelming majority of reactions that we encounter are not significantly affected by the choice of counterion.

PRACTICALLYSPEAKING

Baking Soda versus Baking Powder

In the opening of this chapter, we mentioned that baking soda and baking powder are both leavening agents. That is, they both produce CO_2 that will make a dough or batter fluffy. We will now explore how each of these compounds accomplishes its task, beginning with baking soda. As mentioned earlier in this chapter, baking soda is the household name for sodium bicarbonate. Because sodium bicarbonate is mildly basic, it will react with an acid to produce carbonic acid, which in turn degrades into CO_2 and water:

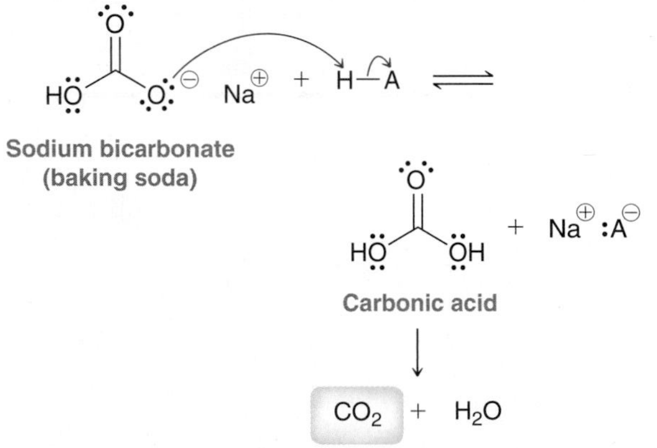

The mechanism for the conversion of carbonic acid into CO_2 and water will be discussed in Chapter 21. Looking at the reaction above, it is clear that an acid must be present in order for baking soda to do its job. Many breads and pastries include ingredients that naturally contain acids. For example, buttermilk, honey, and citrus fruits (such as lemons) all contain naturally occurring organic acids:

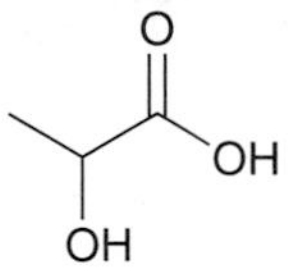
Lactic acid
(Found in buttermilk)

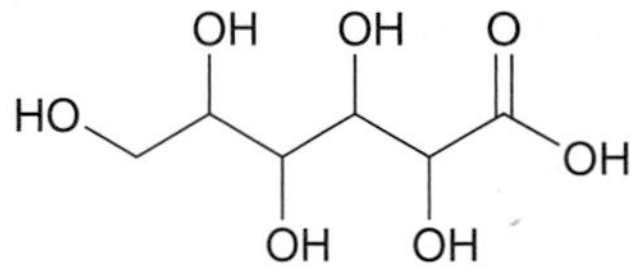
Gluconic acid
(Found in honey)

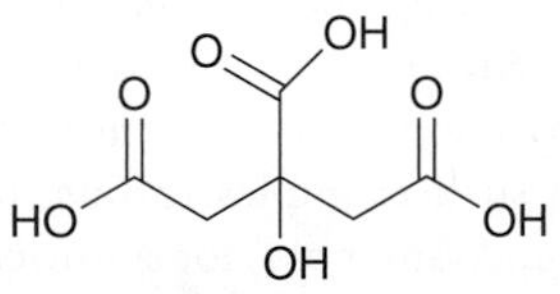
Citric acid
(Found in citrus fruits)

When an acidic compound is present in the dough or batter, baking soda can be protonated, causing liberation of CO_2. However, when acidic ingredients are absent, the baking soda cannot be protonated, and CO_2 is not produced. In such a situation, we must add both the base (baking soda) and some acid. Baking powder does exactly that. It is a powder mixture that contains both sodium bicarbonate and an acid salt, such as potassium bitartrate:

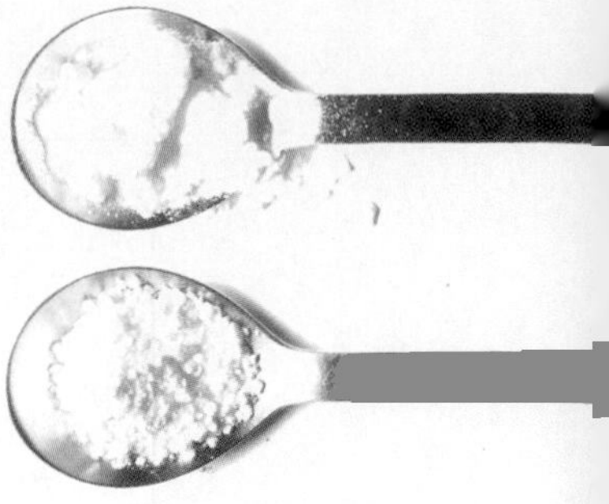

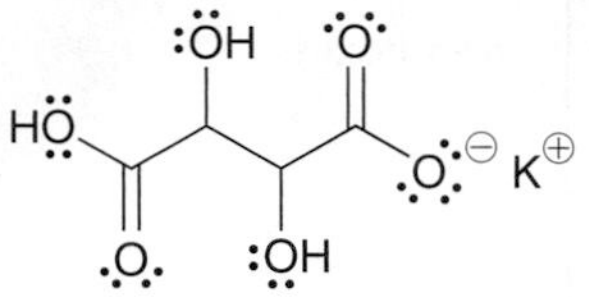
Potassium bitartrate

Baking powder also contains some starch to keep the mixture dry, which prevents the acid and base from reacting with each other. When mixed with water, the acid and the base can react with each other, ultimately producing CO_2:

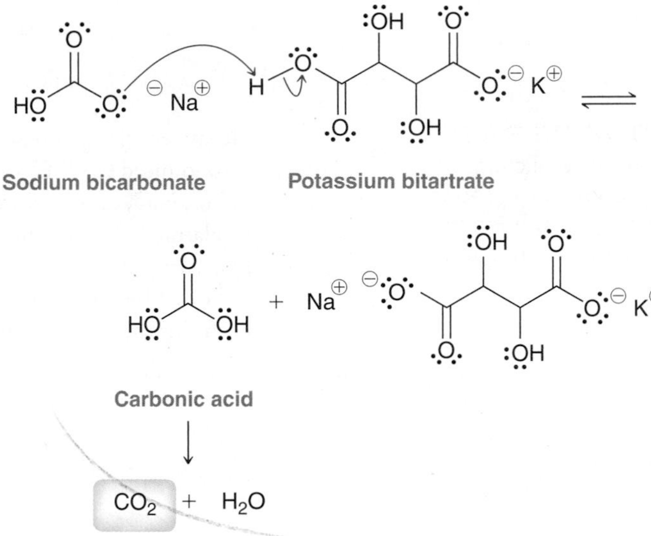

Baking powder is often used when making pancakes, muffins, and waffles. It is an essential ingredient in the recipe if you want your pancakes to be fluffy. In any recipe, the exact ratio of acid and base is important. Excess base (sodium bicarbonate) will impart a bitter taste, while excess acid will impart a sour taste. In order to get the ratio just right, a recipe will often call for some specific amount of baking soda and some specific amount of baking powder. The recipe is taking into account the amount of acidic compounds present in the other ingredients, so that the final product will not be unnecessarily bitter or sour. Baking is truly a science!

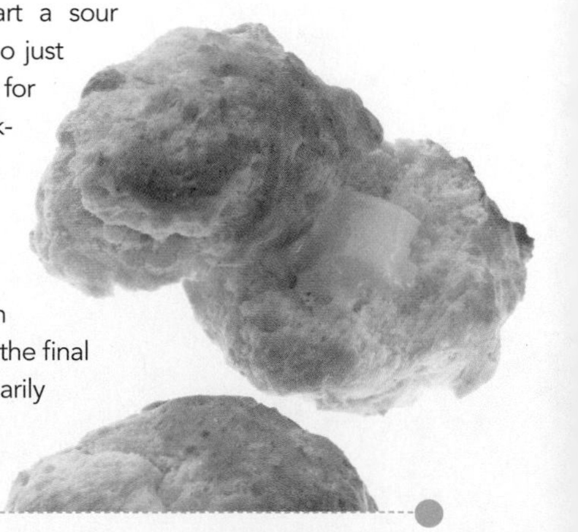

3.9 Lewis Acids and Bases

The Lewis definition of acids and bases is broader than the Brønsted-Lowry definition. According to the Lewis definition, acidity and basicity are described in terms of electrons, rather than protons. A **Lewis acid** is defined as an *electron acceptor,* while a **Lewis base** is defined as an *electron donor.* As an illustration, consider the following Brønsted-Lowry acid-base reaction:

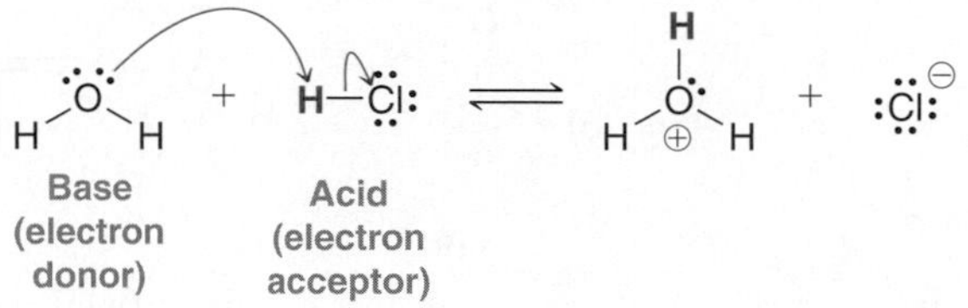

HCl is an acid according to either definition. It is a Lewis acid because it serves as an electron acceptor, and it is a Brønsted-Lowry acid because it serves as a proton donor. But the Lewis definition is an expanded definition of acids and bases, because it includes reagents that would otherwise not be classified as acids or bases. For example, consider the following reaction:

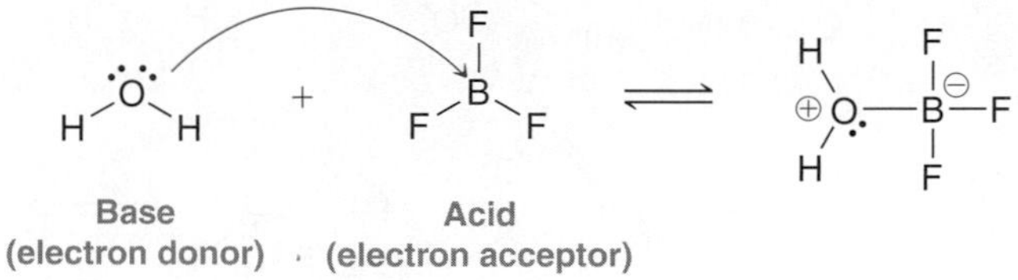

According to the Brønsted-Lowry definition, BF_3 is not considered an acid because it is has no protons and cannot serve as a proton donor. However, according to the Lewis definition, BF_3 can serve as an electron acceptor, and it is therefore a Lewis acid. In the reaction above, H_2O is a Lewis base because it serves as an electron donor.

Take special notice of the curved-arrow notation. There is only one curved arrow in the reaction above, not two.

Chapter 6 will introduce the skills necessary to analyze reactions, and in Section 6.7 we will revisit the topic of Lewis acids and bases. In fact, we will see that most of the reactions in this textbook occur as the result of the reaction between a Lewis acid and a Lewis base. For now, let's get some practice identifying Lewis acids and Lewis bases.

SKILLBUILDER

3.12 IDENTIFYING LEWIS ACIDS AND LEWIS BASES

LEARN the skill

Identify the Lewis acid and the Lewis base in the reaction between BH_3 and THF:

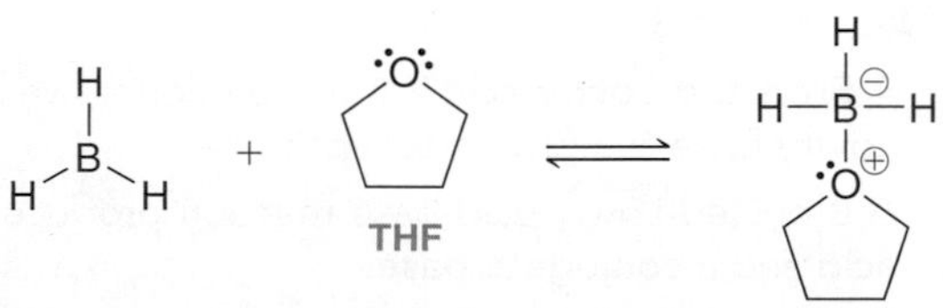

SOLUTION

We must first decide the direction of electron flow. Which reagent is serving as the electron donor and which reagent is serving as the electron acceptor? To answer this question, we analyze each reagent and look for a lone pair of electrons. Boron is in the third column of the periodic table and only has three valence electrons. It is using all three valence electrons to form bonds, which means that it does not have a lone pair of electrons. Rather, it has an empty *p* orbital (for a review of the structure of BH_3, see Section 1.10). Oxygen does have a lone pair. So, we conclude that oxygen attacks boron:

STEP 1 Identify the direction of the flow of electrons.

STEP 2 Identify the electron acceptor as the Lewis acid and the electron donor as the Lewis base.

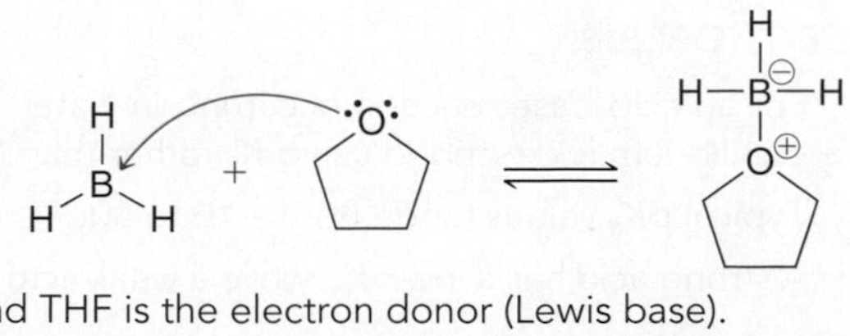

BH_3 is the electron acceptor (Lewis acid), and THF is the electron donor (Lewis base).

PRACTICE the skill **3.34** In each case below, identify the Lewis acid and the Lewis base:

(a)

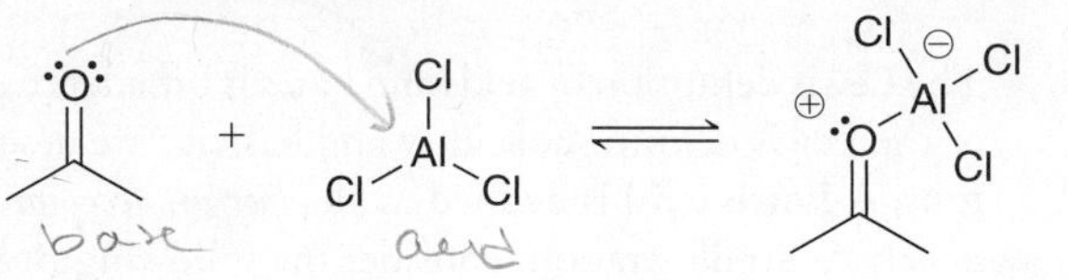

(b)

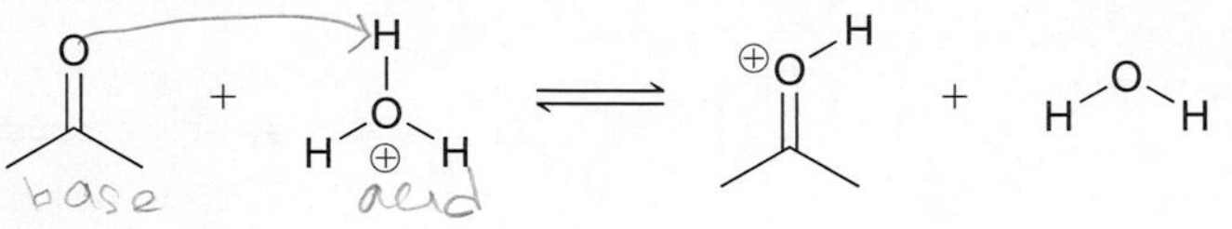

(c)

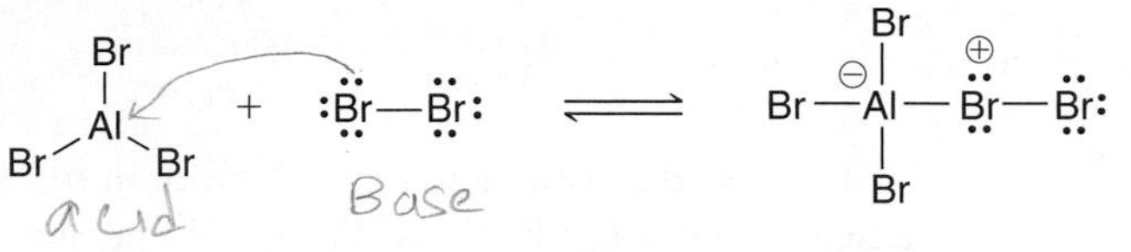

(d)

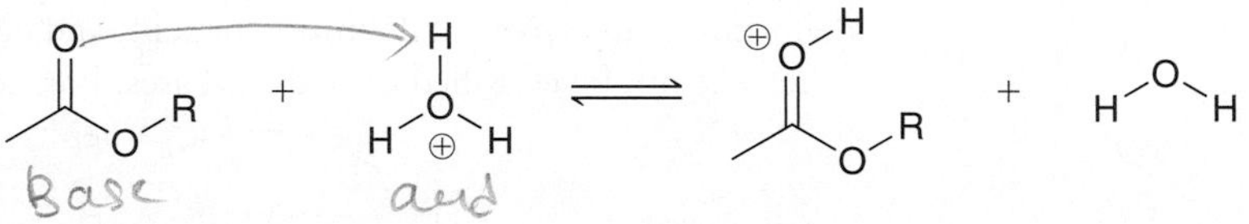

(e)

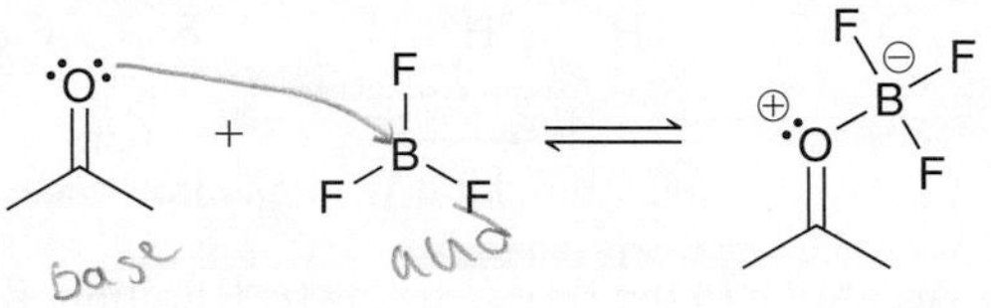

APPLY the skill **3.35** Identify the compounds below that can function as Lewis bases:

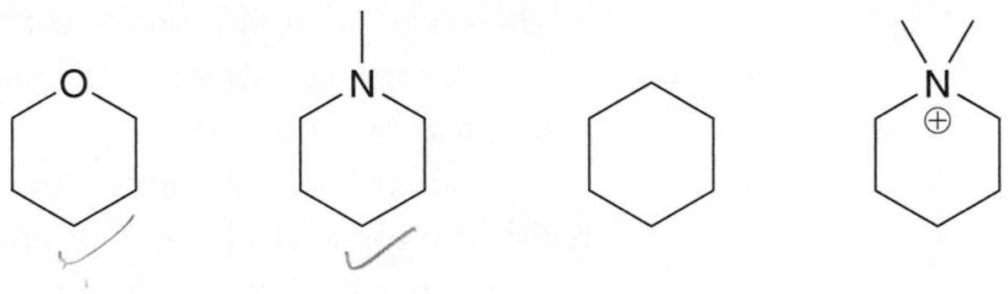

need more **PRACTICE?** **Try Problem 3.39**

REVIEW OF CONCEPTS AND VOCABULARY

SECTION 3.1

- A **Brønsted-Lowry acid** is a proton donor, while a **Brønsted-Lowry base** is a proton acceptor.
- A Brønsted-Lowry acid-base reaction produces a **conjugate acid** and a **conjugate base.**

SECTION 3.2

- Curved arrows show the **reaction mechanism**.
- The mechanism of **proton transfer** always involves at least two curved arrows.

SECTION 3.3

- For an acid-base reaction occurring in water, the position of equilibrium is described using K_a rather than K_{eq}.
- Typical pK_a values range from −10 to 50.
- A strong acid has a low pK_a, while a weak acid has a high pK_a.
- **Equilibrium** always favors formation of the weaker acid (higher pK_a).

SECTION 3.4

- Relative acidity can be predicted (qualitatively) by analyzing the structure of the conjugate base. If A^- is very stable, then HA must be a strong acid. If A^- is very unstable, then HA must be a weak acid.
- To compare the acidity of two compounds, HA and HB, simply compare the stability of their conjugate bases.
- There are four factors to consider when comparing the stability of conjugate bases:
  1. Which atom is the charge on? For elements in the same row of the periodic table, electronegativity is the dominant effect. For elements in the same column, size is the dominant effect.

2. Resonance—a negative charge is stabilized by resonance.
3. Induction—electron-withdrawing groups, such as halogens, stabilize a nearby negative charge via induction.
4. Orbital—a negative charge in an *sp*-hybridized orbital will be closer to the nucleus and more stable than a negative charge in an sp^3-hybridized orbital.

- When multiple factors compete, ARIO (atom, resonance, induction, orbital) is generally the order of priority, but there are exceptions.

SECTION 3.5

- The equilibrium of an acid-base reaction always favors the more stabilized negative charge.

SECTION 3.6

- A base stronger than hydroxide cannot be used when the solvent is water because of the leveling effect. If a stronger base is desired, a solvent other than water must be employed.

SECTION 3.7

- In some cases, solvent effects explain small difference in pK_a's. For example, bases that are bulky, or **sterically hindered**, are generally less efficient at forming stabilizing solvent interactions.

SECTION 3.8

- Negatively charged bases are always accompanied by positively charged species called **cations.**
- The choice of the counterion does not affect most reactions encountered in this book.

SECTION 3.9

- A **Lewis acid** is an electron acceptor, while a **Lewis base** is an electron donor.

KEY TERMINOLOGY

SKILLBUILDER REVIEW

3.1 DRAWING THE MECHANISM OF A PROTON TRANSFER

STEP 1 Identity the acid and the base.

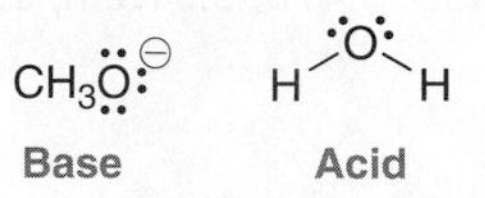

STEP 2 Draw the first curved arrow....
(a) Place tail on lone pair (of base).
(b) Place head on proton (from acid).

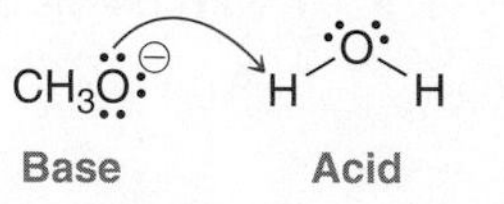

STEP 3 Draw the second curved arrow....
(a) Place tail on O—H bond.
(b) Place head on O.

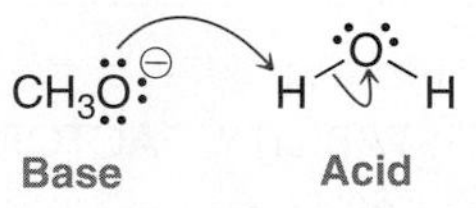

Try Problems 3.1–3.3, 3.44

3.2 USING pK_a VALUES TO COMPARE ACIDS

The compound with the lower pK_a is more acidic.

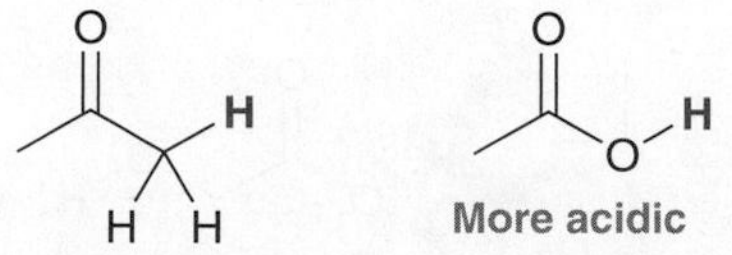

More acidic

$pK_a = 19.2$ $pK_a = 4.75$

Try Problems 3.4–3.6, 3.38

3.3 USING pK_a VALUES TO COMPARE BASICITY

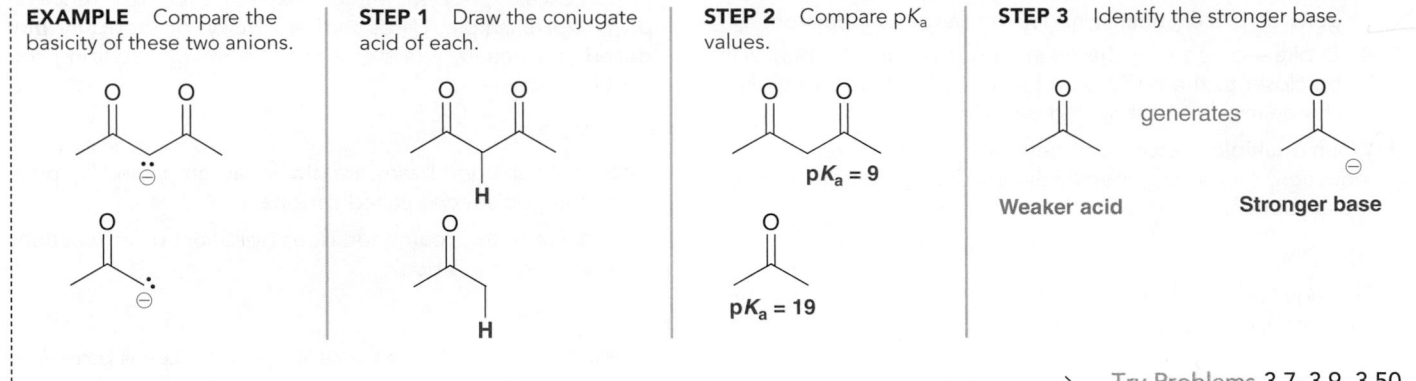

EXAMPLE Compare the basicity of these two anions.

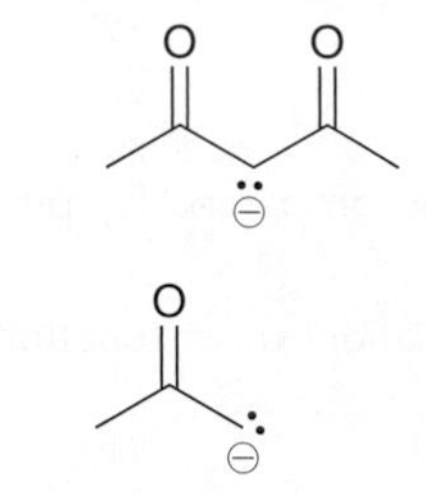

STEP 1 Draw the conjugate acid of each.

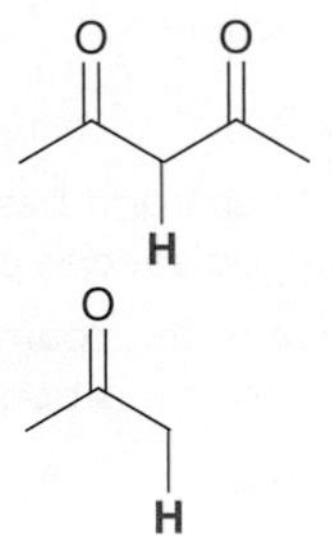

STEP 2 Compare pK_a values.

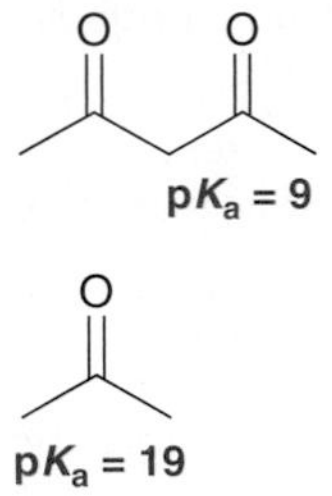

STEP 3 Identify the stronger base.

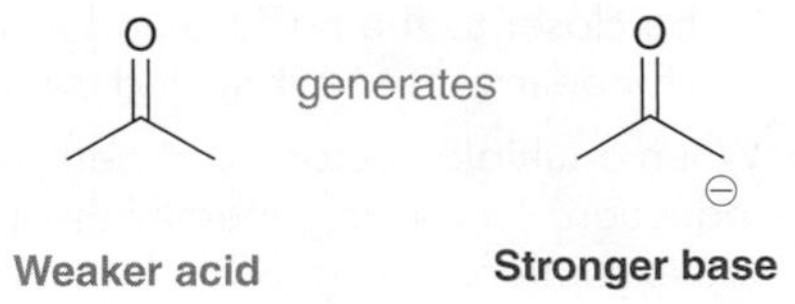

Try Problems 3.7–3.9, 3.50

3.4 USING pK_a VALUES TO PREDICT THE POSITION OF EQUILIBRIUM

STEP 1 Identify the acid on each side of the equilibrium.

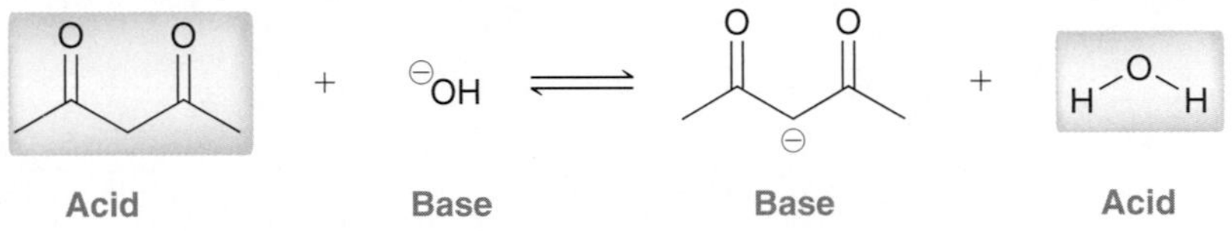

STEP 2 Compare pK_a values.

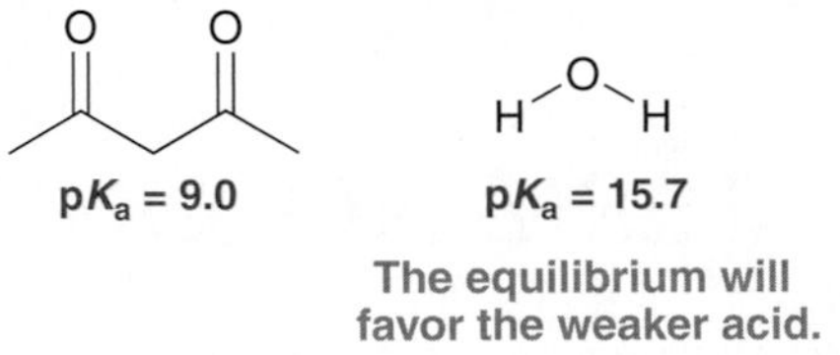

Try Problems 3.10–3.11, 3.48

3.5 ASSESSING RELATIVE STABILITY: FACTOR 1—ATOM

STEP 1 Draw the conjugate bases...

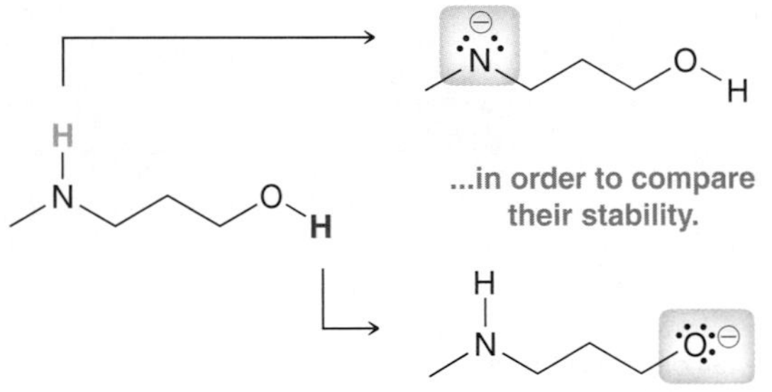

STEP 2 Compare location of charge taking into account two trends.

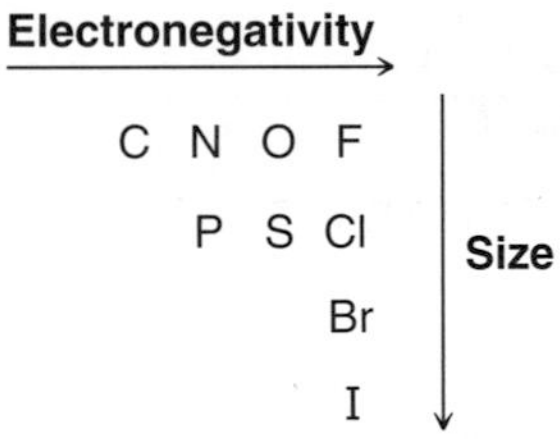

STEP 3 The more stable conjugate base.

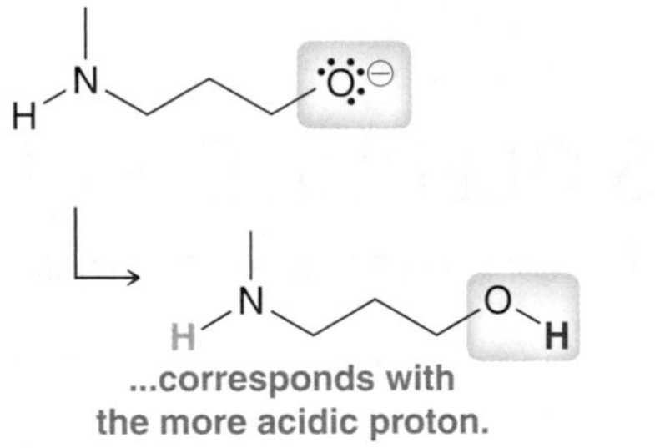

Try Problems 3.13, 3.14, 3.45b, 3.47h, 3.51b, h, 3.53

3.6 ASSESSING RELATIVE STABILITY: FACTOR 2—RESONANCE

STEP 1 Draw the conjugate bases...

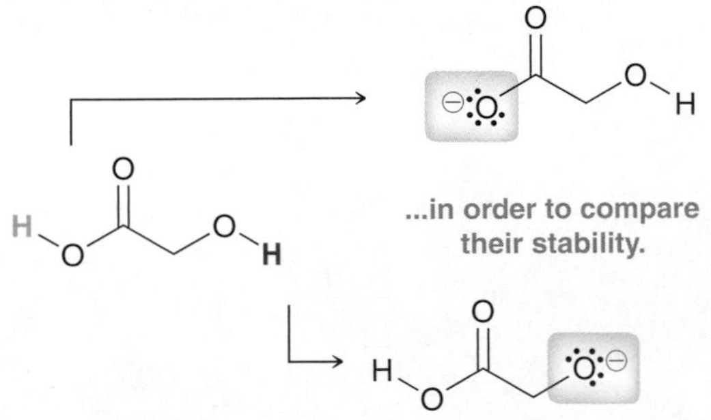

STEP 2 Look for resonance stabilization.

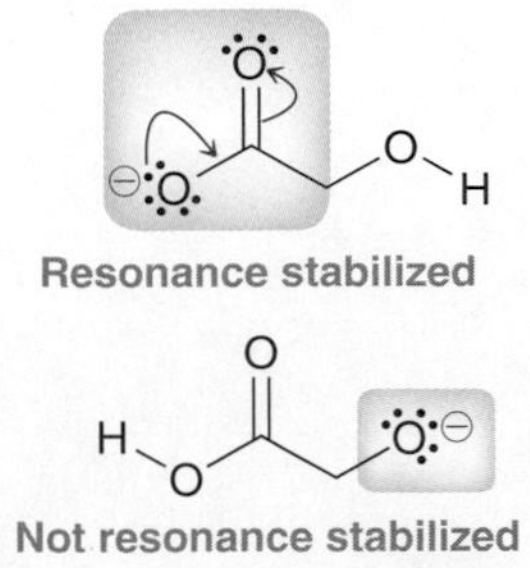

STEP 3 The more stable conjugate base.

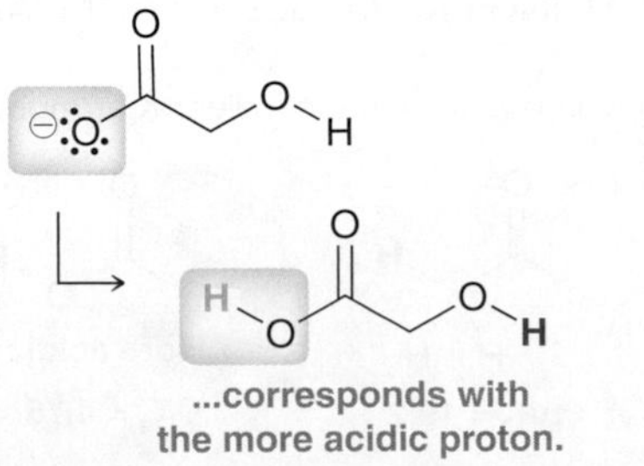

Try Problems 3.15–3.17, 3.45a, 3.46a, 3.47b,e–g, 3.51c–f

3.7 ASSESSING RELATIVE STABILITY: FACTOR 3—INDUCTION

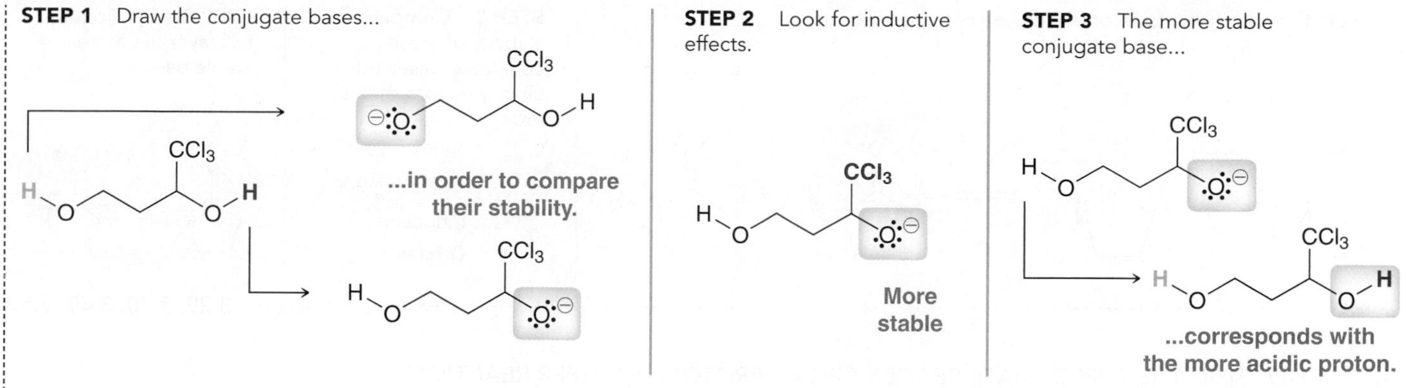

Try Problems 3.18–3.21, 3.46b, 3.47c, 3.51g

3.8 ASSESSING RELATIVE STABILITY: FACTOR 4—ORBITAL

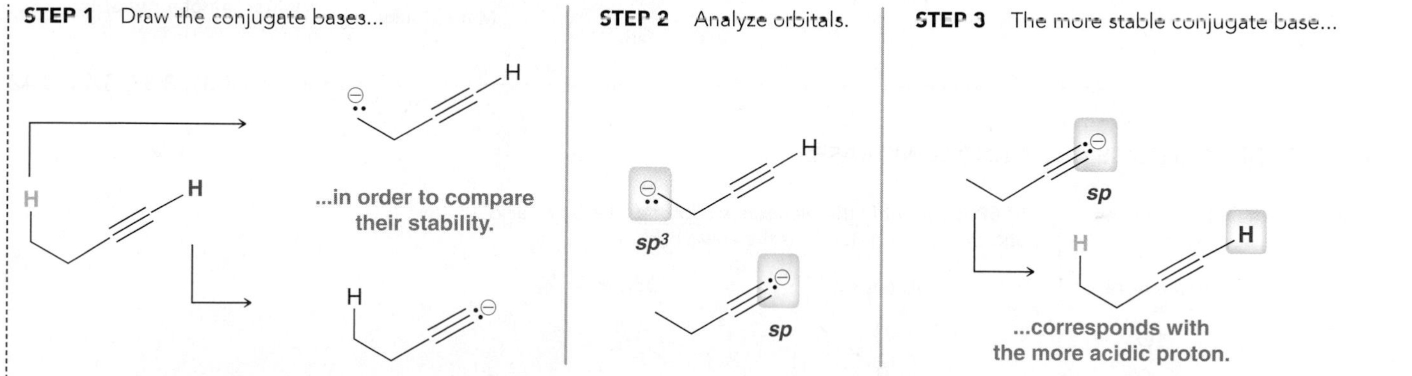

Try Problems 3.22–3.24, 3.45c

3.9 ASSESSING RELATIVE STABILITY: USING ALL FOUR FACTORS

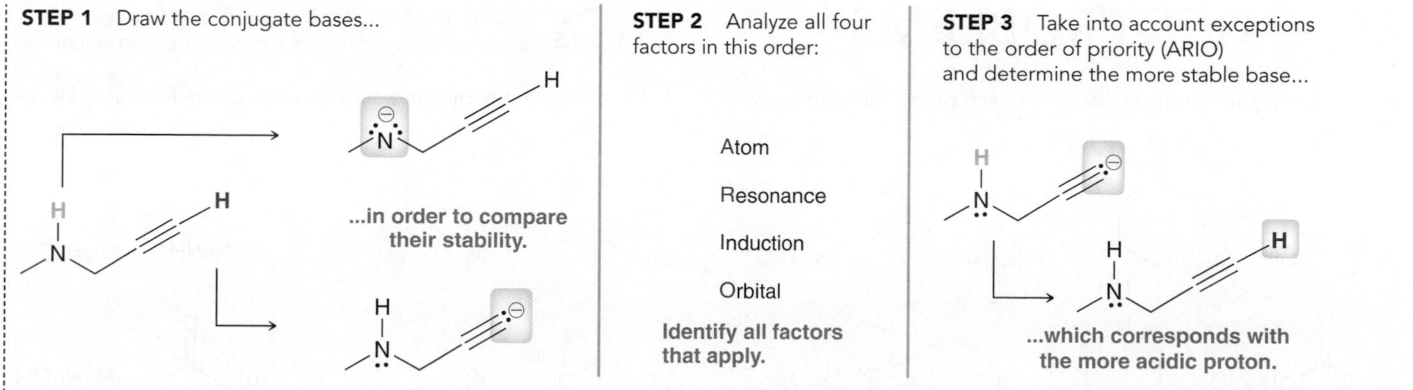

STEP 1 Draw the conjugate bases...
...in order to compare their stability.
STEP 2 Analyze all four factors in this order:
Atom
Resonance
Induction
Orbital
Identify all factors that apply.
STEP 3 Take into account exceptions to the order of priority (ARIO) and determine the more stable base...
...which corresponds with the more acidic proton.

Try Problems 3.25–3.28, 3.47d, 3.51a, 3.57–3.61

3.10 PREDICTING THE POSITION OF EQUILIBRIUM WITHOUT THE USE OF pK_a VALUES

STEP 1 Identify the base on either side of the equilibrium.

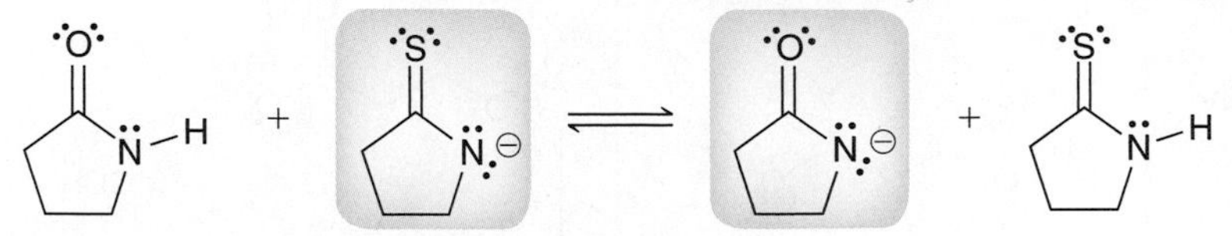

STEP 2 Compare the stability of these conjugate bases using all four factors, in this order:

Atom
Resonance
Induction
Orbital

STEP 3 Equilibrium will favor the more stable base.

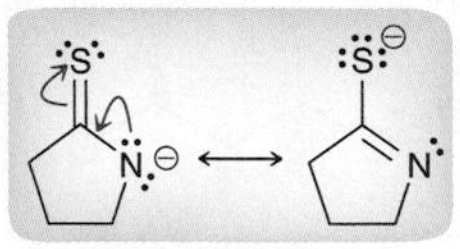

Try Problems **3.29, 3.30, 3.49, 3.52**

3.11 CHOOSING THE APPROPRIATE REAGENT FOR A PROTON TRANSFER REACTION

STEP 1 Draw the equilibrium and identify the base on either side

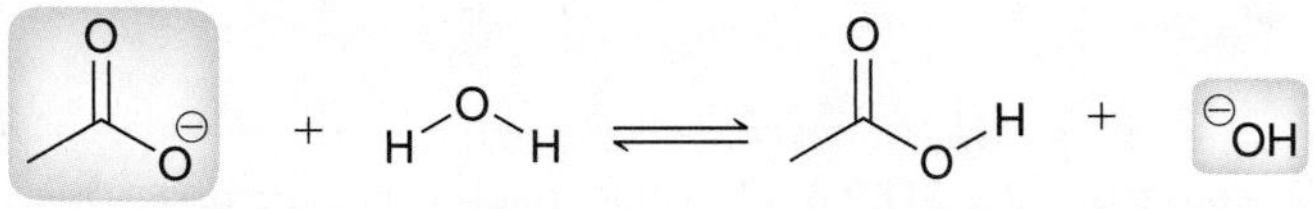

STEP 2 Compare the stability of these conjugate bases using all four factors, in this order:

Atom
Resonance →
Induction
Orbital

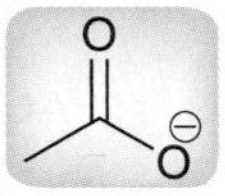

More stable

STEP 3 Equilibrium will favor the more stable base. If products are favored, the reaction is useful. If starting materials are favored, the reaction is not useful.

In this case, water is not a suitable proton source.

Try Problems **3.31, 3.32, 3.40, 3.42**

3.12 IDENTIFYING LEWIS ACIDS AND LEWIS BASES

STEP 1 Identify the direction of the flow of electrons.

To here **From here**

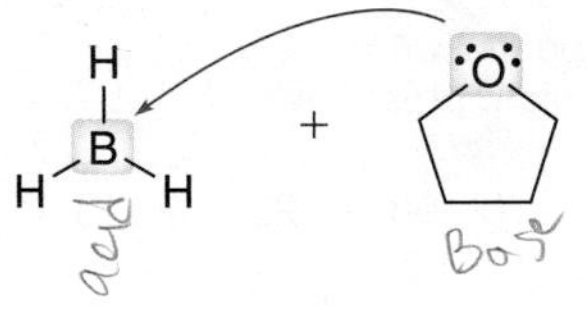

STEP 2 Identify the electron acceptor as the Lewis acid and the electron donor as the Lewis base.

Acceptor **Donor**

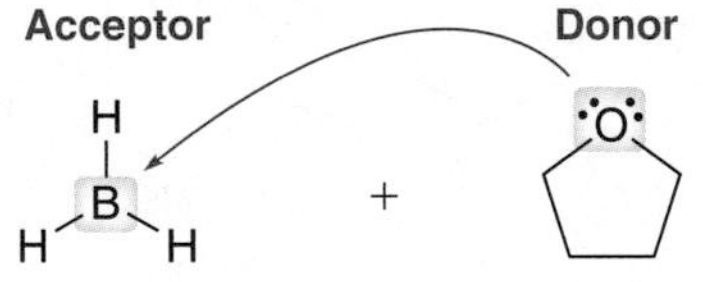

Lewis acid **Lewis base**

Try Problems **3.34, 3.35, 3.39**

PRACTICE PROBLEMS

Note: Most of the Problems are available within *WileyPLUS*, an online teaching and learning solution.

3.36 Draw the conjugate base for each of the following acids:

(a) 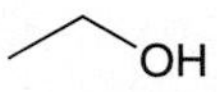 (b) 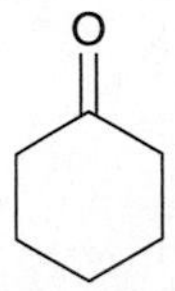 (c) NH_3 (d) H_3O^+

(e) 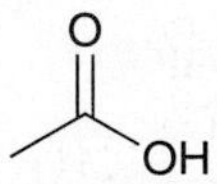 (f) 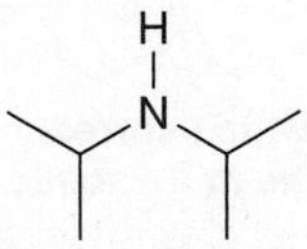 (g) 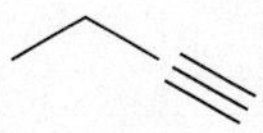 (h) NH_4^+

3.37 Draw the conjugate acid for each of the following bases:

(a) 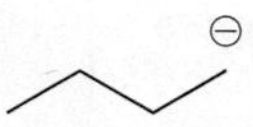 (b) 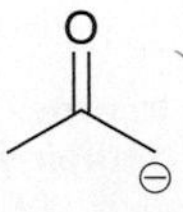 (c) $NaNH_2$ (d) H_2O

(e) 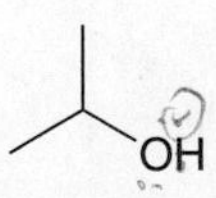 (f)  (g) 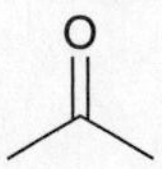 (h) NaOH

3.38 Compound A has a pK_a of 7 and compound B has a pK_a of 10. Compound A is how many times more acidic than compound B?

(a) 3 (b) 3000 (c) 1000

3.39 In each reaction, identify the Lewis acid and the Lewis base:

(a) 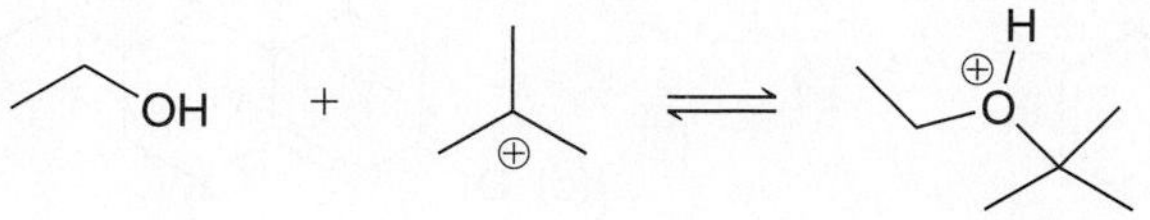

(b) 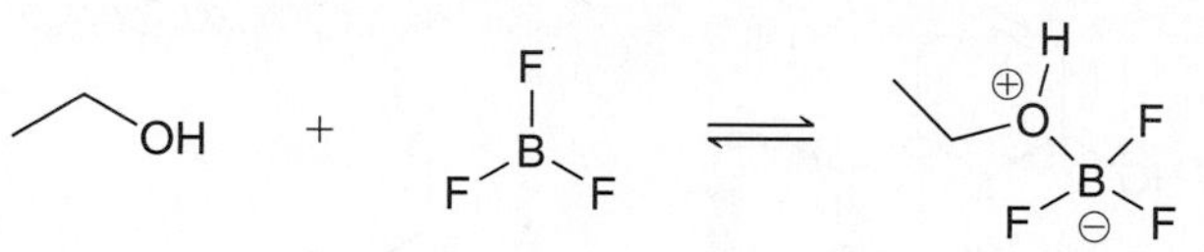

(c) 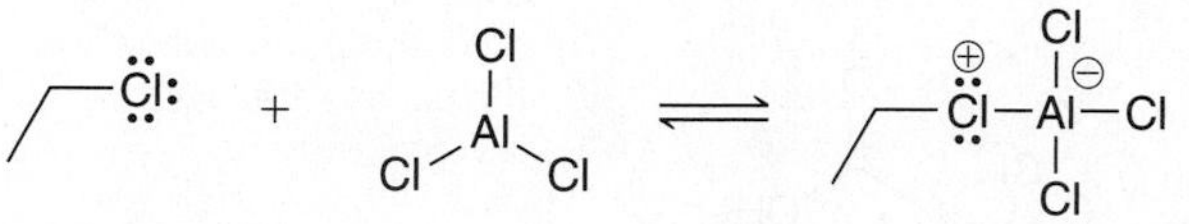

3.40 What reaction will take place if H_2O is added to a mixture of $NaNH_2/NH_3$?

3.41 Would ethanol (CH_3CH_2OH) be a suitable solvent in which to perform the following proton transfer? Explain your answer:

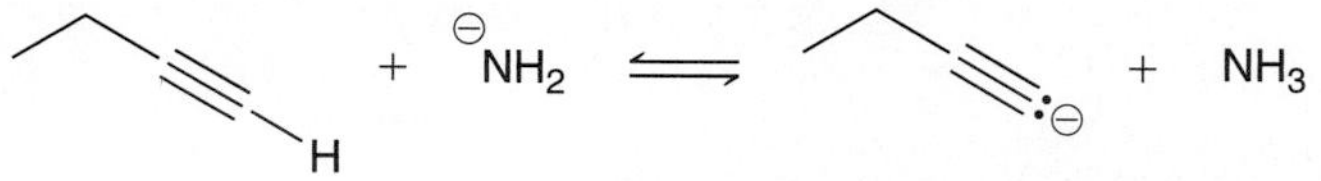

3.42 Would water be a suitable proton source to protonate the following compound?

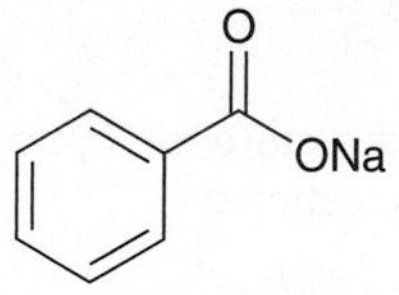

3.43 Write an equation for the proton transfer reaction that occurs when each of the following acids reacts with water. In each case, draw curved arrows that show the mechanism of the proton transfer:

(a) HBr (b) 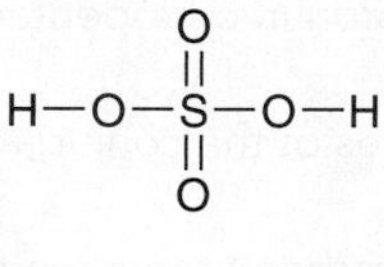 (c) 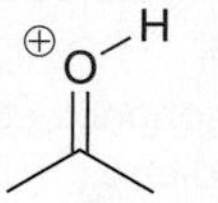 (d) 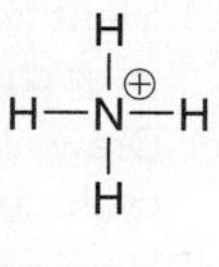

3.44 Write an equation for the proton transfer reaction that occurs when each of the following bases reacts with water. In each case, draw curved arrows that show the mechanism of the proton transfer:

(a) 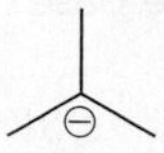 (b) 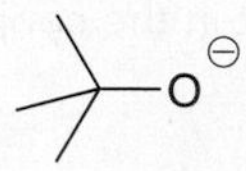 (c) 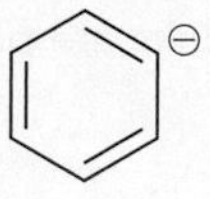 (d) 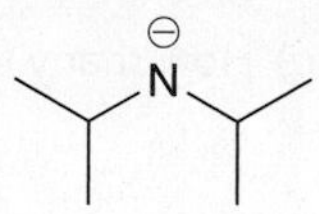

3.45 In each case, identify the more stable anion. Explain why it is more stable.

(a) 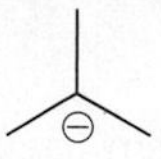 vs. 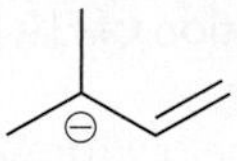

(b) 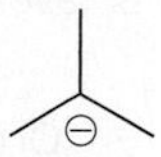 vs. 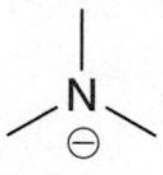

(c) 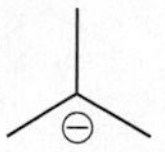 vs. 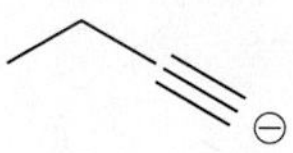

3.46 In each group of compounds below, select the most acidic compound:

(a) 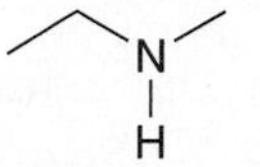 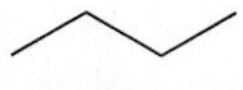 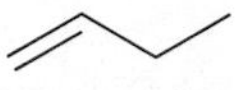 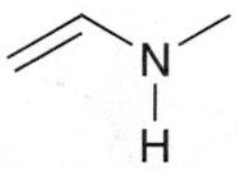

(b) 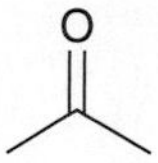 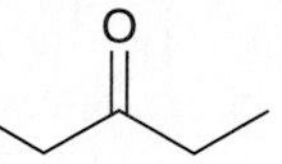 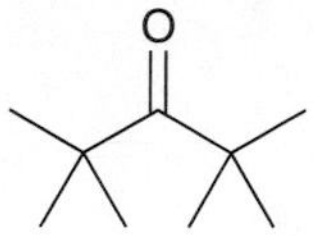 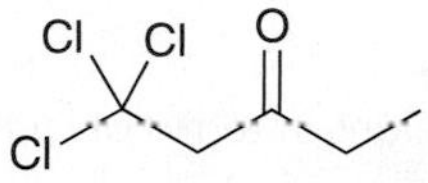

3.47 For each pair of compounds below, identify the more acidic compound:

(a) 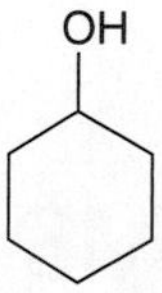 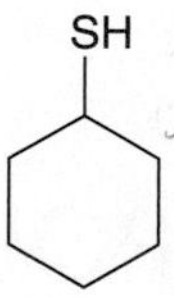 (b) 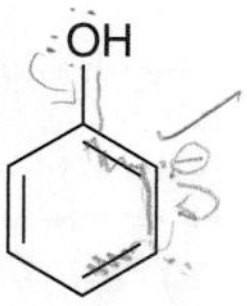 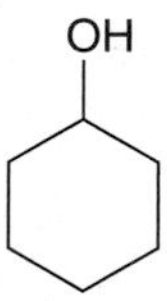

(c) 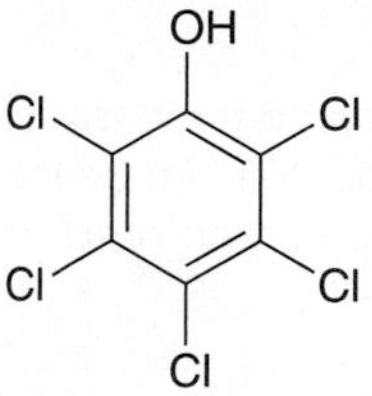 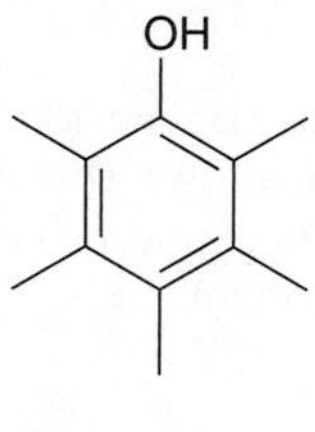

(d) 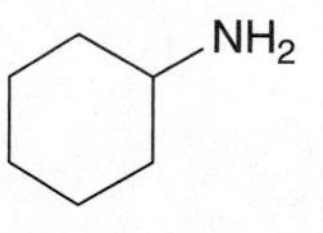 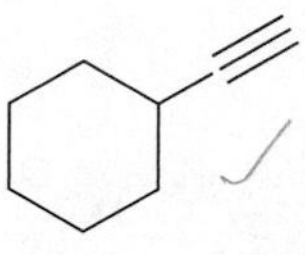

(e) 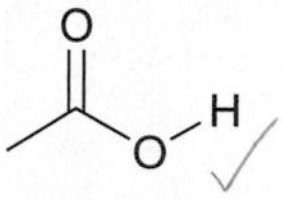 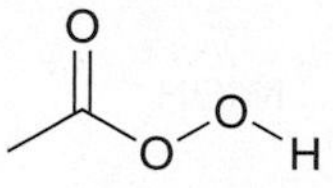

(f) 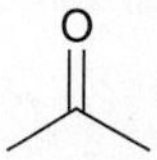 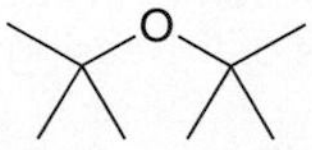

(g)  (h) 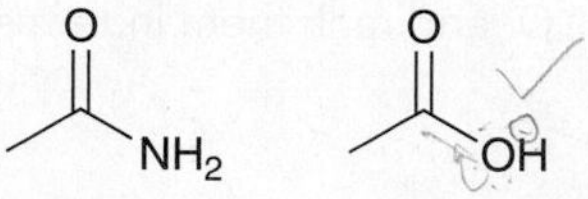

3.48 HA has a pK_a of 15, while HB has a pK_a of 5. Draw the equilibrium that would result upon mixing HB with NaA. Does the equilibrium favor formation of HA or of HB?

3.49 For each reaction below, draw the mechanism (curved arrows) and then predict which side of the reaction is favored under equilibrium conditions.

(a) 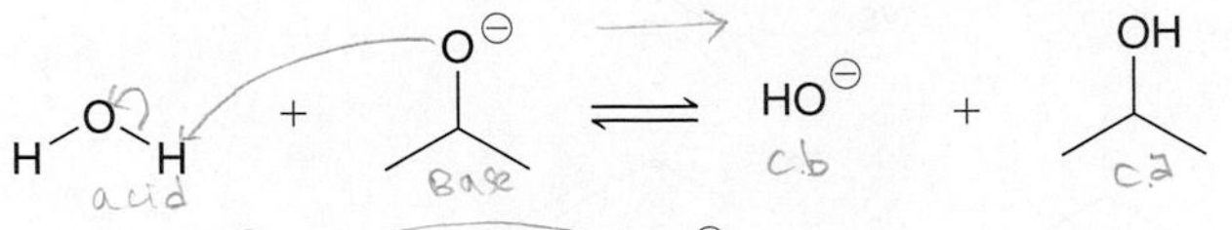

(b) 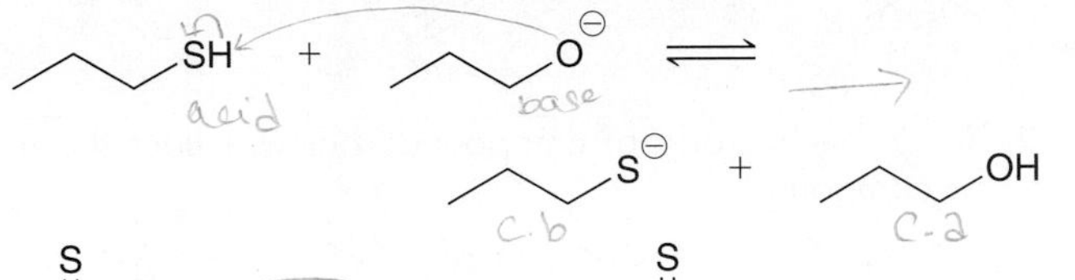

(c) 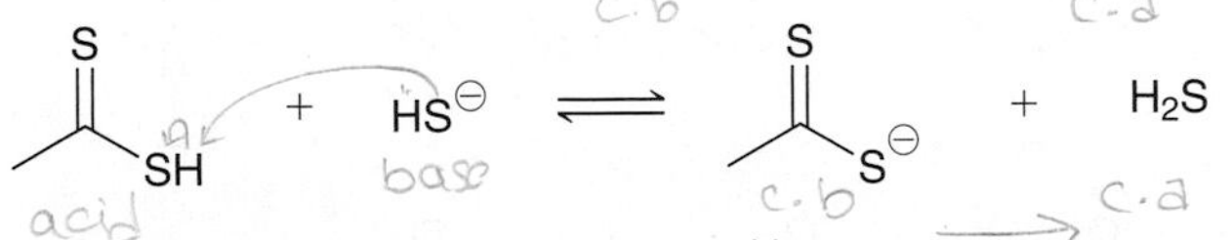

(d) 

3.50 Rank the following anions in terms of increasing basicity:

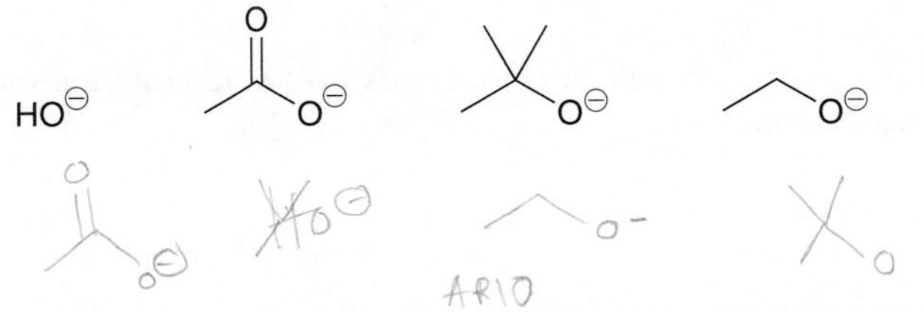

3.51 For each compound below, identify the most acidic proton in the compound:

(a) 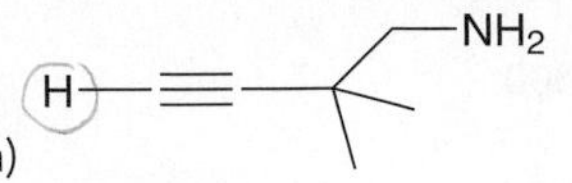

(b) 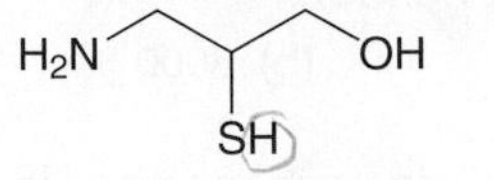

(c) 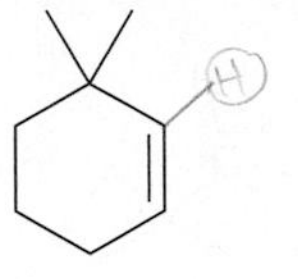

(d) 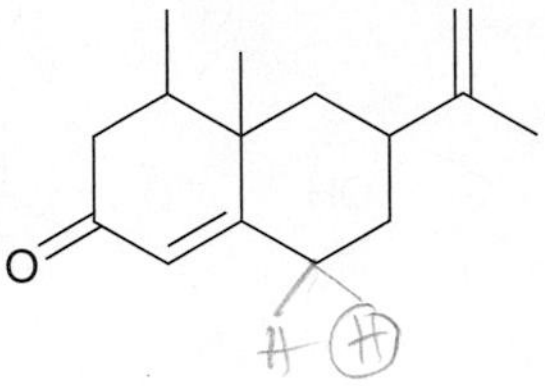

(e) 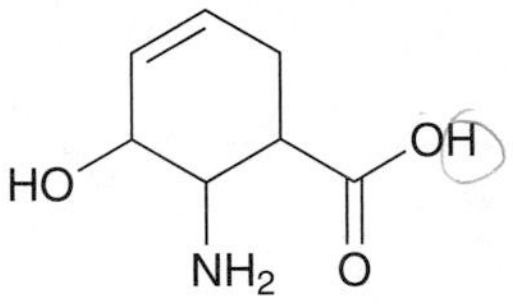

(f) 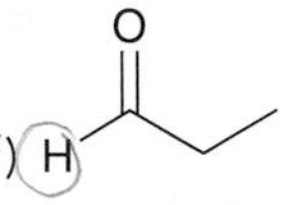

(g) 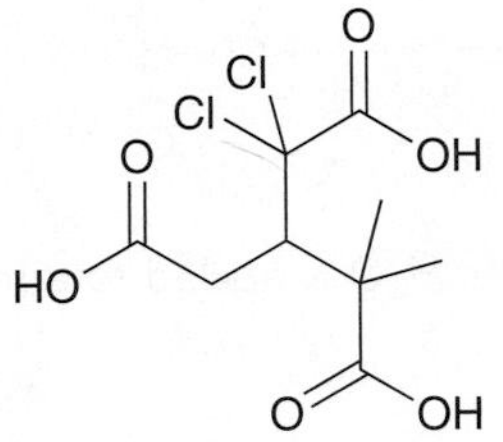

(h) 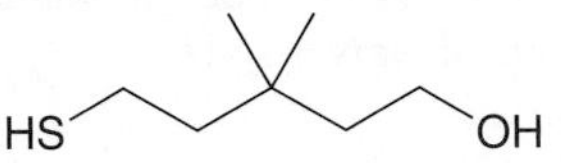

INTEGRATED PROBLEMS

3.52 In each case below, identify the acid and the base. Then draw the curved arrows showing a proton transfer reaction. Draw the products of that proton transfer, and then predict the position of equilibrium.

(a) 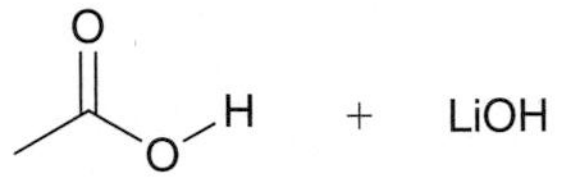

(b) 

(c) 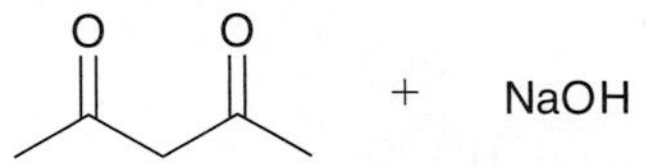

3.53 Draw all constitutional isomers with molecular formula C_2H_6S, and rank them in terms of increasing acidity.

3.54 Draw all constitutional isomers with molecular formula C_3H_8O, and rank them in terms of increasing acidity.

3.55 Consider the structure of cyclopentadiene and then answer the following questions:

Cyclopentadiene

(a) How many sp^3-hybridized carbon atoms are present in the structure of cyclopentadiene?
(b) Identify the most acidic proton in cyclopentadiene. Justify your choice.
(c) Draw all resonance structures of the conjugate base of cyclopentadiene.
(d) How many sp^3-hybridized carbon atoms are present in the conjugate base?
(e) What is the geometry of the conjugate base?
(f) How many hydrogen atoms are present in the conjugate base?
(g) How many lone pairs are present in the conjugate base?

CHALLENGE PROBLEMS

3.56 Consider the pK_a values of the following constitutional isomers:

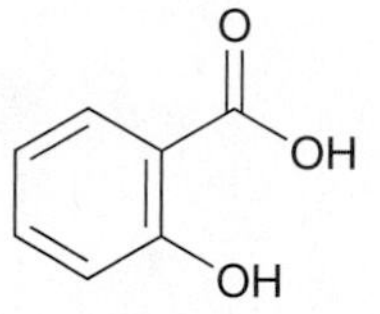

Salicylic acid
pK_a = 3.0

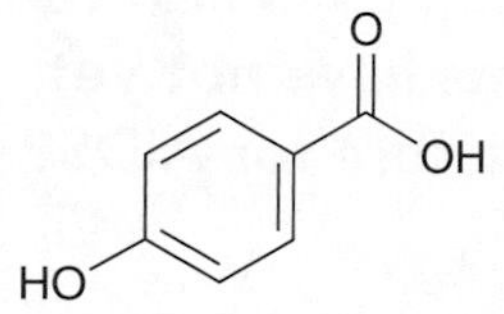

***para*-hydroxybenzoic acid**
pK_a = 4.6

Using the rules that we developed in this chapter (ARIO), we might have expected these two compounds to have the same pK_a. Nevertheless, they are different. Salicylic acid is apparently more acidic than its constitutional isomer. Can you offer an explanation for this observation?

3.57 Consider the following compound with molecular formula C_4H_8O:

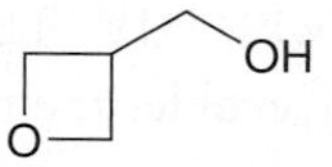

C_4H_8O

(a) Draw a constitutional isomer that you expect will be approximately one trillion (10^{12}) times more acidic than the compound above.
(b) Draw a constitutional isomer that you expect will be less acidic than the compound above.
(c) Draw a constitutional isomer that you expect will have approximately the same pK_a as the compound above.

3.58 There are only four constitutional isomers with molecular formula $C_4H_9NO_2$ that contain a nitro group (—NO_2). Three of these isomers have similar pK_a values, while the fourth isomer has a much higher pK_a value. Draw all four isomers, and identify which one has the higher pK_a. Explain your choice.

3.59 Predict which of the following compounds is more acidic, and explain your choice.

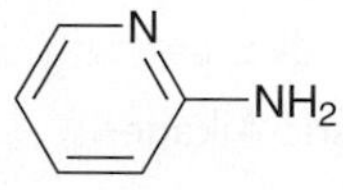

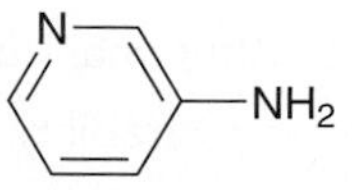

3.60 Below is the structure of rilpivirine, a promising new anti-HIV drug that combats resistant strains of HIV. Its ability to side-step resistance will be discussed in the upcoming chapter.

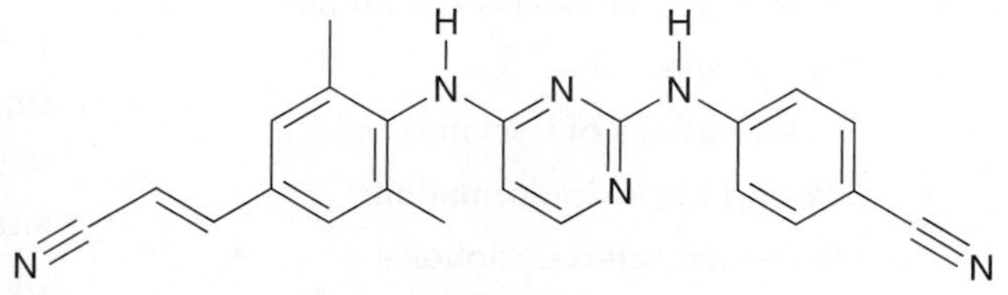

Rilpivirine

(a) Identify the two most acidic protons in rilpivirine.
(b) Identify which of these two protons is more acidic. Explain your choice.

3.61 Most common amines (RNH_2) exhibit pK_a values between 35 and 45. R represents the rest of the compound (generally carbon and hydrogen atoms). However, when R is a cyano group, the pK_a is found to be drastically lower:

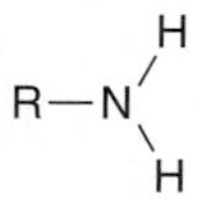

pK_a = 35-45

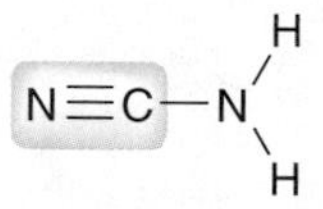

pK_a = 17

(a) Explain why the presence of the cyano group so drastically impacts the pK_a.
(b) Can you suggest a different replacement for R that would lead to an even stronger acid (pK_a lower than 17)?

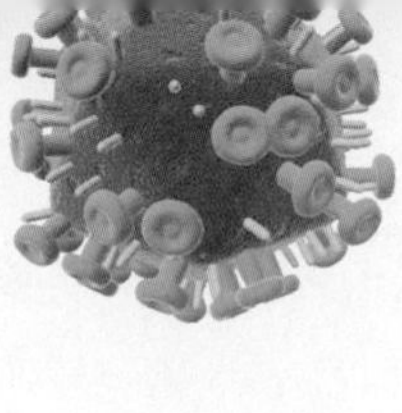

4 Alkanes and Cycloalkanes

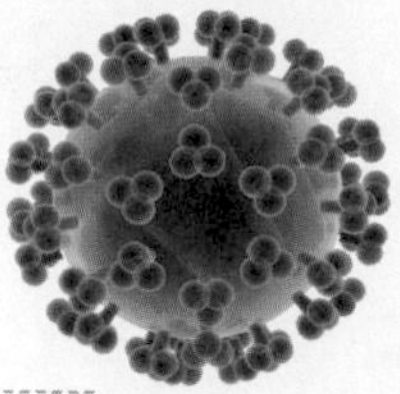

DID YOU EVER WONDER...

why scientists have not yet developed a cure for AIDS?

As you probably know, acquired immunodeficiency syndrome (AIDS) is caused by the human immunodeficiency virus (HIV). Although scientists have not yet developed a way to destroy HIV in infected people, they have developed drugs that significantly slow the progress of the virus and the disease. These drugs interfere with the various processes by which the virus replicates itself. Yet, anti-HIV drugs are not 100% effective, primarily because HIV has the ability to mutate into forms that are drug resistant. Recently, scientists have developed a class of drugs that show great promise in the treatment of patients infected with HIV. These drugs are designed to be *flexible*, which apparently enables them to evade the problem of drug resistance.

A flexible molecule is one that can adopt many different shapes, or conformations. The study of the three-dimensional shapes of molecules is called conformational analysis. This chapter will introduce only the most basic principles of conformational analysis, which we will use to analyze the flexibility of molecules. To simplify our discussion, we will explore compounds that lack a functional group, called alkanes and cycloalkanes. Analysis of these compounds will enable us to understand how molecules achieve flexibility. Specifically, we will explore how alkanes and cycloalkanes change their three-dimensional shape as a result of the rotation of C—C single bonds.

Our discussion of conformational analysis will involve the comparison of many different compounds and will be more efficient if we can refer to compounds by name. A system of rules for naming alkanes and cycloalkanes will be developed prior to our discussion of molecular flexibility.

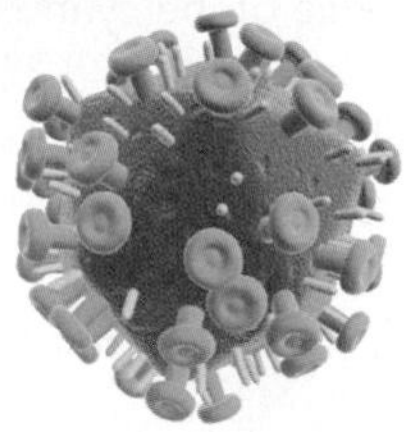

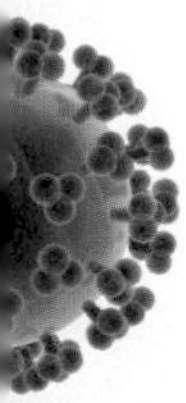

DO YOU REMEMBER?

Before you go on, be sure you understand the following topics.
If necessary, review the suggested sections to prepare for this chapter.

- Molecular Orbital Theory (Section 1.8)
- Predicting Geometry (Section 1.10)
- Bond-Line Structures (Section 2.2)
- Three-Dimensional Bond-Line Structures (Section 2.6)

WILEY PLUS Visit www.wileyplus.com to check your understanding and for valuable practice.

4.1 Introduction to Alkanes

Recall that hydrocarbons are compounds comprised of just C and H; for example:

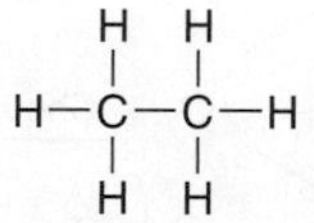

Ethane
C_2H_6

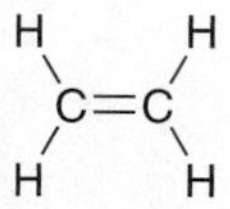

Ethylene
C_2H_4

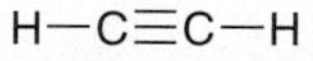

Acetylene
C_2H_2

Benzene
C_6H_6

Ethane, is unlike the other examples in that it has no π bonds. Hydrocarbons that lack π bonds are called **saturated hydrocarbons**, or **alkanes.** The names of these compounds usually end with the suffix "-ane," as seen in the following examples:

Propane

Butane

Pentane

This chapter will focus on alkanes, beginning with a procedure for naming them. The system of naming chemical compounds, or **nomenclature**, will be developed and refined throughout the remaining chapters of this book.

4.2 Nomenclature of Alkanes

An Introduction to IUPAC Nomenclature

In the early nineteenth century, organic compounds were often named at the whim of their discoverers. Here are just a few examples:

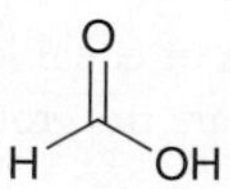

Formic acid
Isolated from ants and named after the Latin word for ant, *formica*

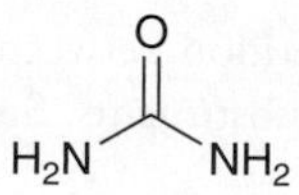

Urea
Isolated from urine

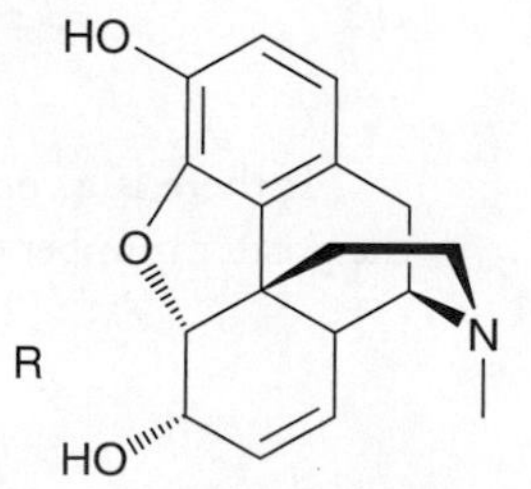

Morphine
A painkiller named after the Greek God of dreams, Morpheus

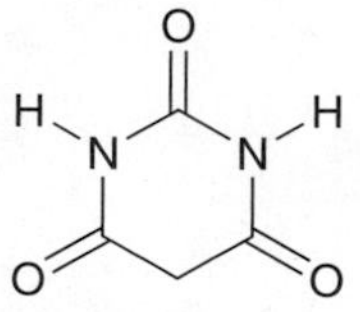

Barbituric acid
Adolf von Baeyer named this compound in honor of a woman named Barbara

A large number of compounds were given names that became part of the common language shared by chemists. Many of these common names are still in use today.

As the number of known compounds grew, a pressing need arose for a systematic method for naming compounds. In 1892, a group of 34 European chemists met in Switzerland and developed a system of organic nomenclature called the Geneva rules. The group ultimately became known as the International Union of Pure and Applied Chemistry, or **IUPAC** (pronounced "I–YOU–PACK"). The original Geneva rules have been regularly revised and updated and are now called IUPAC nomenclature. The most recent IUPAC recommendations were released in 2004.

Names produced by IUPAC rules are called **systematic names.** There are many rules, and we cannot possibly study all of them. The upcoming sections are meant to serve as an introduction to IUPAC nomenclature.

Selecting the Parent Chain

The first step in naming an alkane is to identify the longest chain, called the parent chain:

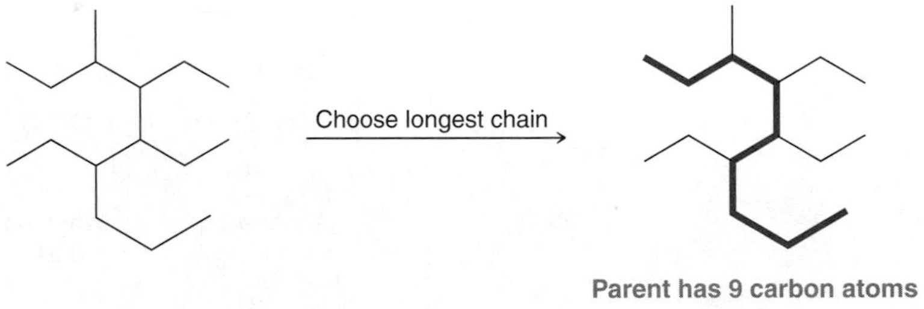

In this example, the parent chain has nine carbon atoms. When naming the parent chain of a compound, the names in Table 4.1 are used. These names will be used very often in this course. Parent chains of more than 10 carbon atoms will be less common, so it is essential to commit to memory at least the first 10 parents on the list in Table 4.1.

TABLE 4.1 PARENT NAMES FOR ALKANES

NUMBER OF CARBON ATOMS	PARENT	NAME OF ALKANE	NUMBER OF CARBON ATOMS	PARENT	NAME OF ALKANE
1	*meth*	methane	11	*undec*	undecane
2	*eth*	ethane	12	*dodec*	dodecane
3	*prop*	propane	13	*tridec*	tridecane
4	*but*	butane	14	*tetradec*	tetradecane
5	*pent*	pentane	15	*pentadec*	pentadecane
6	*hex*	hexane	20	*eicos*	eicosane
7	*hept*	heptane	30	*triacont*	triacontane
8	*oct*	octane	40	*tetracont*	tetracontane
9	*non*	nonane	50	*pentacont*	pentacontane
10	*dec*	decane	100	*hect*	hectane

If there is a competition between two chains of equal length, then choose the chain with the greater number of substituents. **Substituents** are the groups connected to the parent chain:

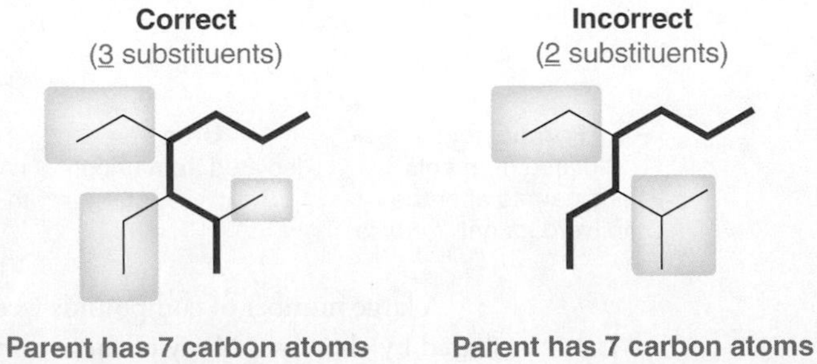

The term "cyclo" is used to indicate the presence of a ring in the structure of an alkane. For example, these compounds are called **cycloalkanes**:

△ □ ⬠

Cyclopropane **Cyclobutane** **Cyclopentane**

SKILLBUILDER

4.1 IDENTIFYING THE PARENT

LEARN the skill

Identify and provide a name for the parent in the following compound:

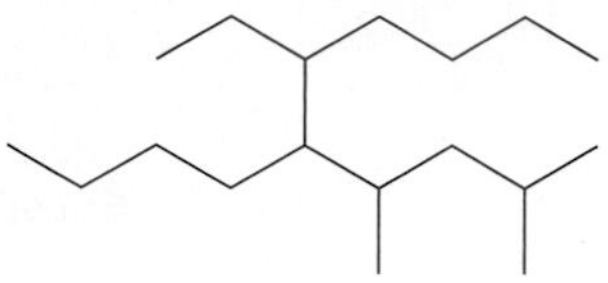

SOLUTION

Locate the longest chain. In this case, there are two choices that have 10 carbon atoms:

STEP 1
Choose the longest chain.

Either way, the parent will be *decane*, but we must choose the correct parent chain. The correct parent is the one with the greatest number of substituents:

Correct Incorrect

STEP 2
When two chains compete, choose the chain with more substituents.

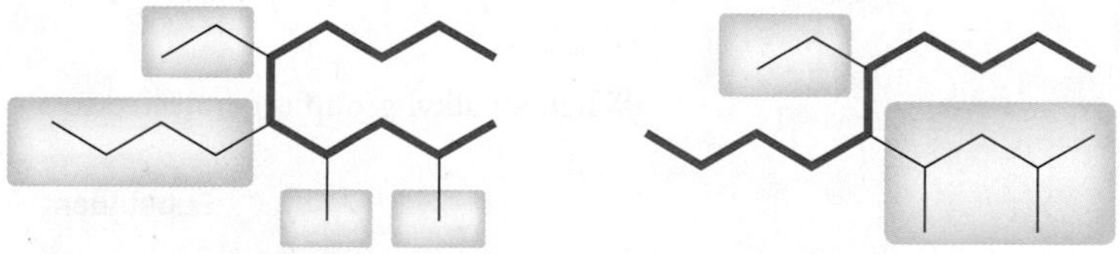

4 Substituents **2 Substituents**

PRACTICE the skill **4.1** Identify and name the parent in each of the following compounds:

3-methyl hexane 4-ethyl heptane

(a) (b) (c) (d)

3,4-dimethyl heptane

5-propyl nonane

(e) (f) 3,5-dimethyl-4-propyl heptane (g) (h) (i)

4,5-dimethyl octane 1-methyl cyclo-pentane 1-dimethyl

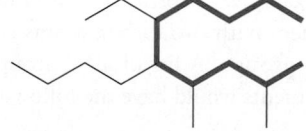

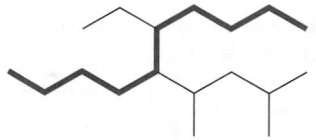

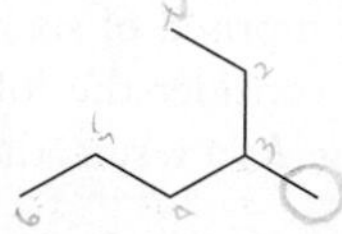

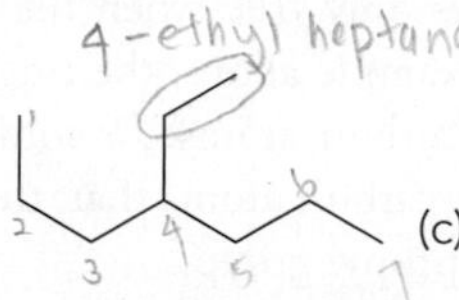

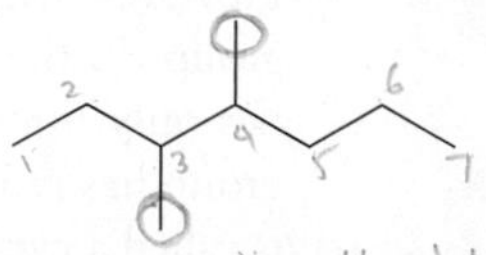

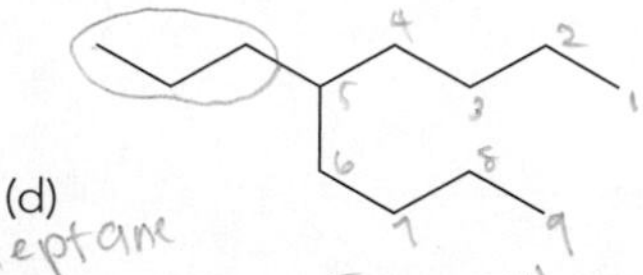

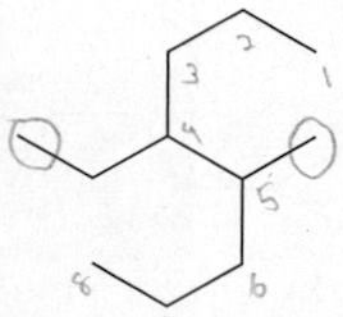

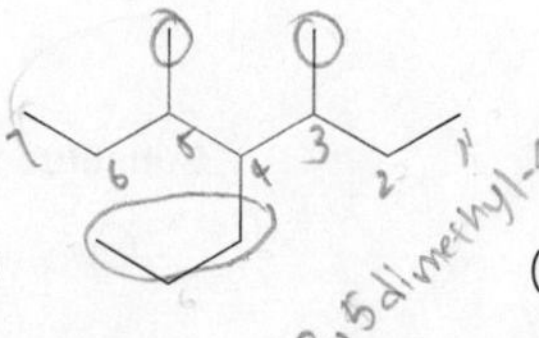

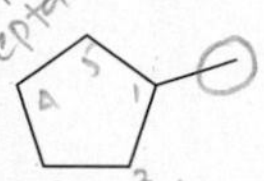

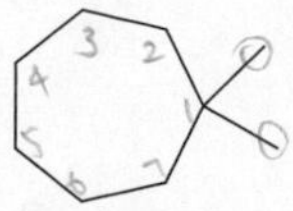

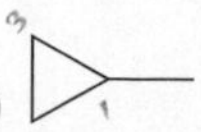

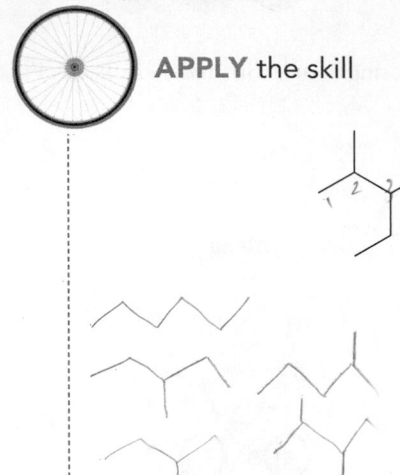

APPLY the skill

4.2 Identify the two compounds below that have the same parent:

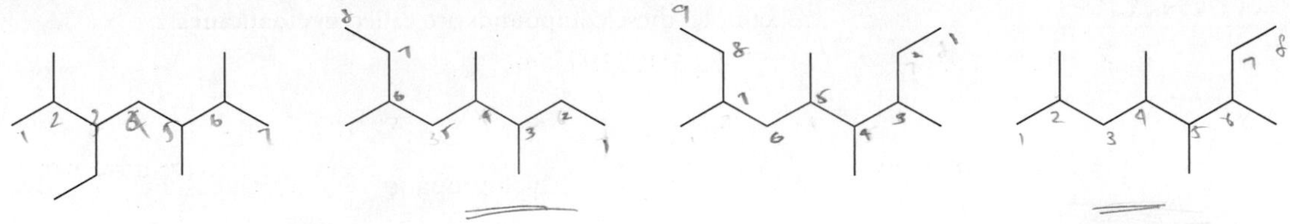

4.3 There are five constitutional isomers with molecular formula C_6H_{14}. Draw a bond-line structure for each isomer and identify the parent in each case.

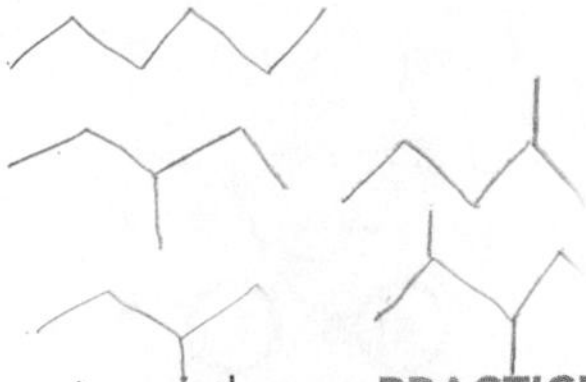

4.4 There are 18 constitutional isomers with molecular formula C_8H_{18}. *Without drawing all 18 isomers*, determine how many of the isomers will have a parent name of heptane.

need more **PRACTICE?** **Try Problem 4.39**

Naming Substituents

Once the parent has been identified, the next step is to list all of the substituents:

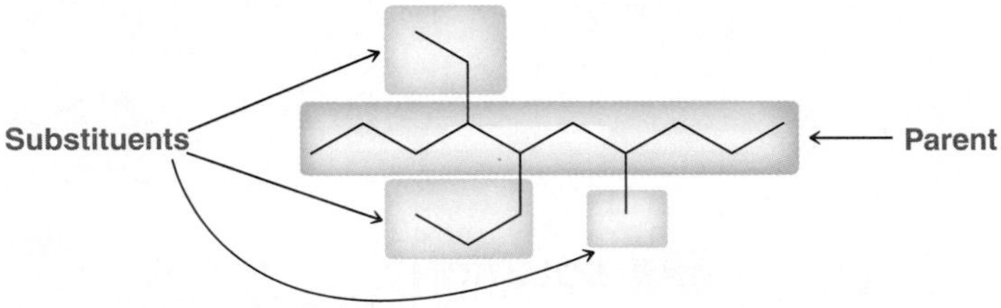

TABLE 4.2 NAMES OF ALKYL GROUPS

NUMBER OF CARBON ATOMS IN SUBSTITUENT	TERMINOLOGY
1	*Methyl*
2	*Ethyl*
3	*Propyl*
4	*Butyl*
5	*Pentyl*
6	*Hexyl*
7	*Heptyl*
8	*Octyl*
9	*Nonyl*
10	*Decyl*

Substituents are named with the same terminology used for naming parents; only we add the letters "yl." For example, a substituent with one carbon atom (a CH_3 group) is called a methyl group. A substituent with two carbon atoms is called an ethyl group. These groups are generically called **alkyl groups**. A list of alkyl groups is given in Table 4.2. In the example shown earlier, the substituents would have the following names:

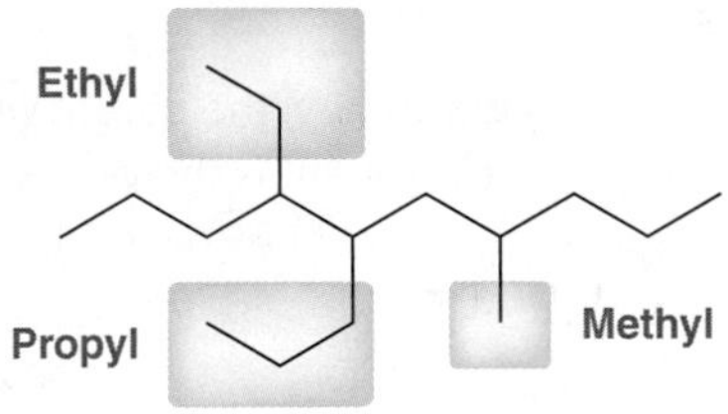

When an alkyl group is connected to a ring, the ring is generally treated as the parent:

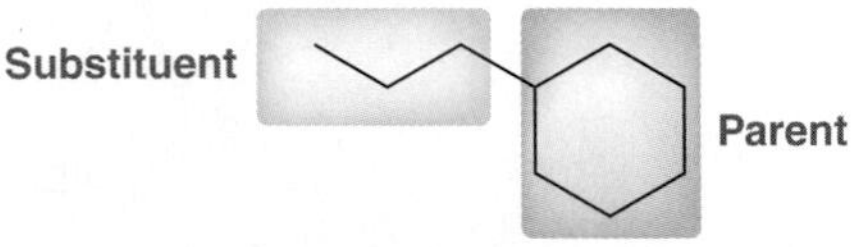

Propyl cyclohexane

However, this is only true when the ring is comprised of more carbon atoms than the alkyl group. In the example above, the ring is comprised of six carbon atoms, while the alkyl group has only three carbon atoms. In contrast, consider the following example in which the alkyl group has more carbon atoms than the ring. As a result, the ring is named as a substituent and is called a cyclopropyl group.

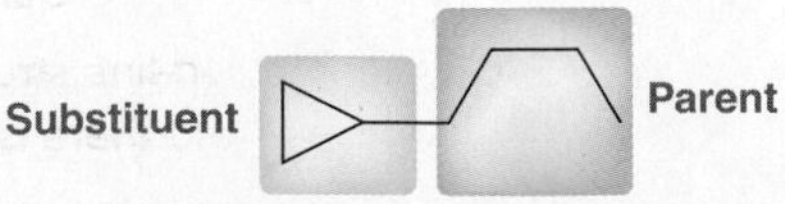

1-Cyclopropyl butane

SKILLBUILDER

4.2 IDENTIFYING AND NAMING SUBSTITUENTS

LEARN the skill

Identify and name all substituents in the following compound:

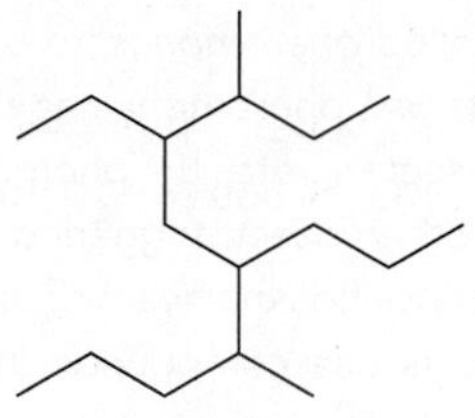

SOLUTION

First identify the parent by looking for the longest chain:

STEP 1
Identify the parent.

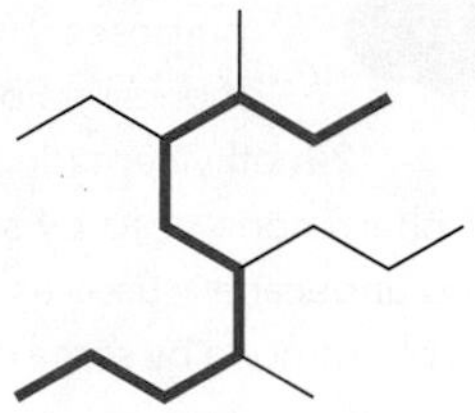

In this case, the parent has 10 carbon atoms (decane). Everything connected to the chain is a substituent, and we use the names from Table 4.2 to name each substituent:

STEP 2
Identify all alkyl substituents connected to the parent.

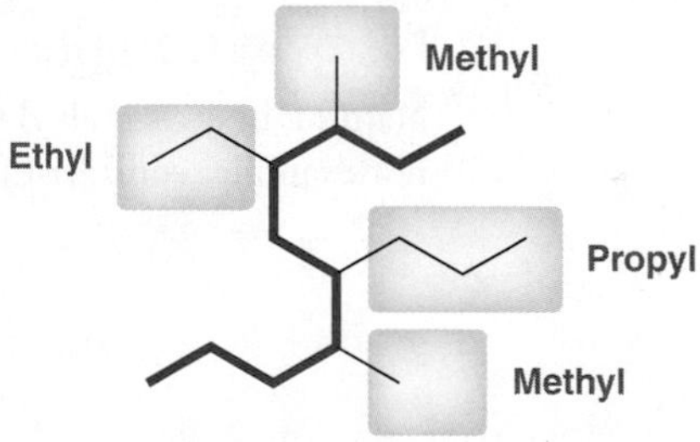

PRACTICE the skill

4.5 For each of the following compounds, identify all groups that would be considered substituents, and then indicate how you would name each substituent.

(a) 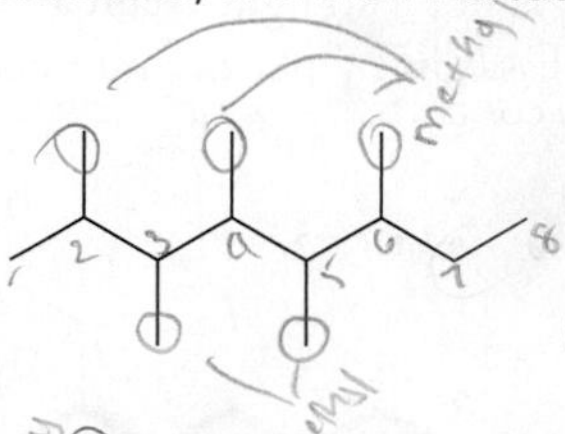
(b) 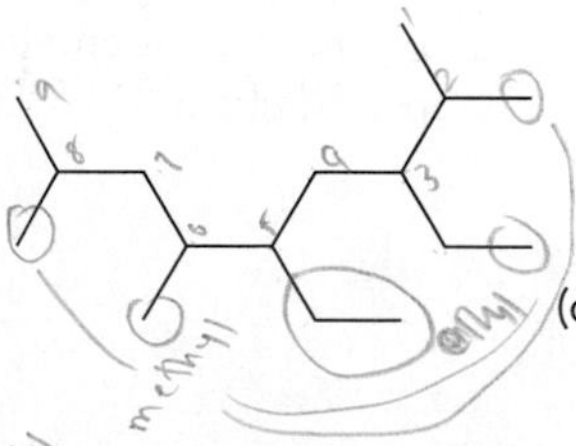
(c) 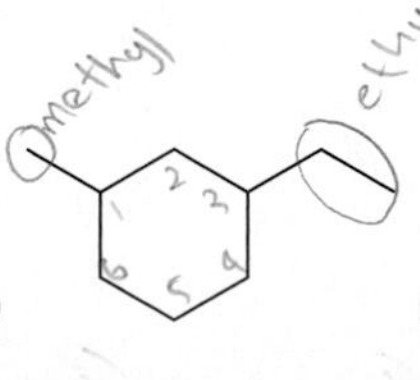
(d) 

(e) 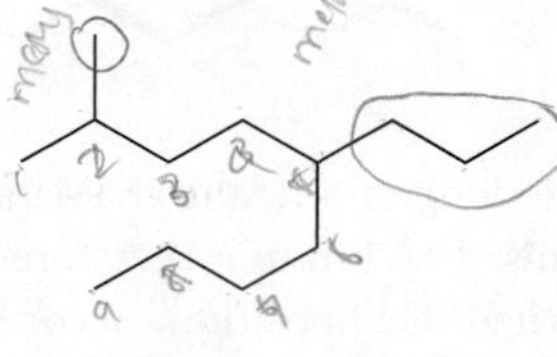
(f) 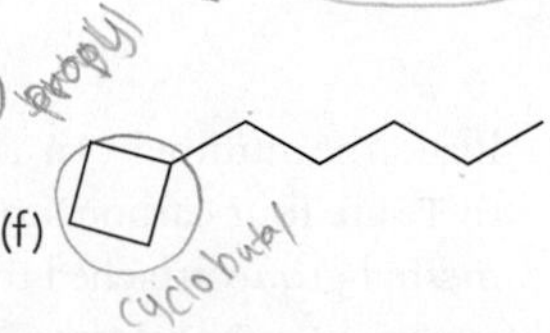
(g) 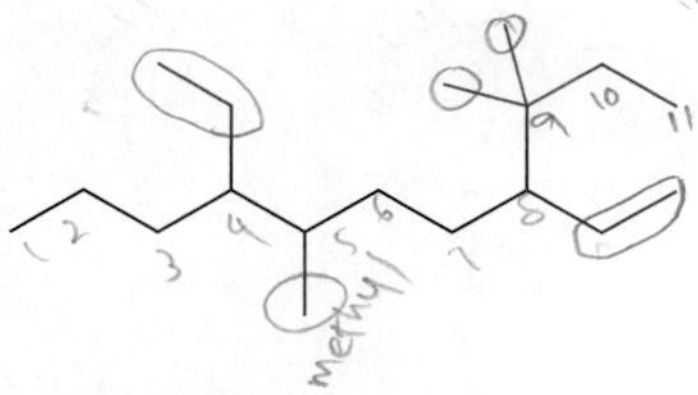

APPLY the skill

4.6 There are nine constitutional isomers with molecular formula C_7H_{16}.

(a) Draw a bond-line structure for all isomers that fit the following criteria: the parent = pentane, and there are two methyl groups connected to the parent.

(b) Draw a bond-line structure for the isomer that fits the following criteria: the parent = pentane, and there is one ethyl group connected to the parent.

need more **PRACTICE?** **Try Problems 4.40a, 4.40c**

PRACTICALLYSPEAKING)))

Pheromones: Chemical Messengers

Many animals use chemicals, called *pheromones*, to communicate with each other. In fact, insects use pheromones as their primary means of communication. An insect secretes the pheromone, which then binds to a receptor in another insect, triggering a biological response. Some insects use compounds that warn of danger (alarm pheromones), while other insects use compounds that promote aggregation among members of the same species (aggregation pheromones). Pheromones are also used by many insects to attract members of the opposite sex for mating purposes (sex pheromones). For example, 2-methylheptadecane is a sex pheromone used by some moths, and undecane is used as an aggregation pheromone by some cockroaches.

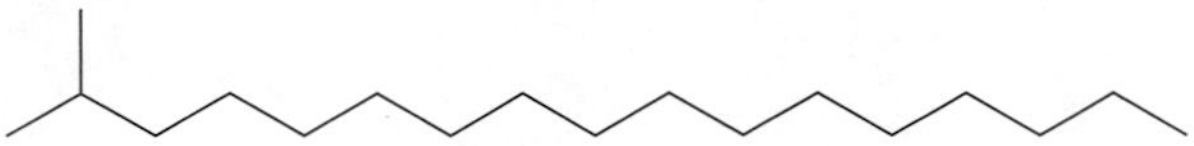

2-Methylheptadecane
Sex pheromone of the female tiger moth

Undecane
Aggregation pheromone of the *Blaberus cranifer* cockroach

Both of these examples above are alkanes, but many pheromones exhibit one or more functional groups. Examples of such pheromones will appear throughout this book.

Naming Complex Substituents

Naming branched alkyl substituents is more complex than naming straight-chain substituents. For example, consider the following substituent:

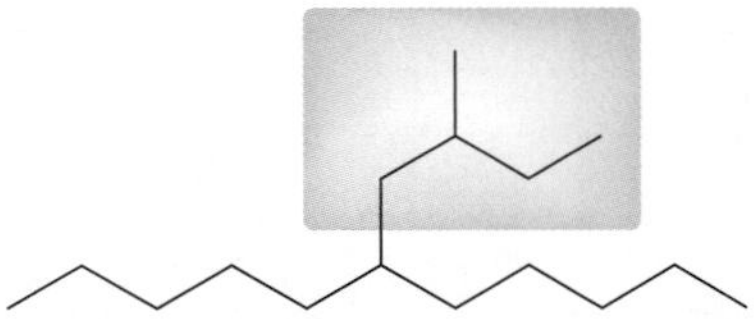

How do we name this substituent? It has five carbon atoms, but it cannot simply be called a pentyl group, because it is not a straight-chain alkyl group. In situations like this, the following method is employed: Begin by placing numbers on the substituent, going *away* from the parent chain:

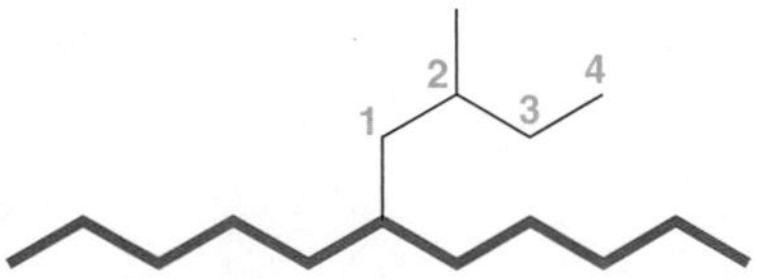

Place the numbers on the longest straight chain present in the substituent, and in this case, there are four carbon atoms. This group is therefore considered to be a butyl group that has one methyl group attached to it at the 2 position. Accordingly, this group is called a (2-methylbutyl) group. In essence, treat the complex substituent as if it is a miniparent with its own substituents. When naming a complex substituent, place parentheses around the name of the substituent. This will avoid confusion, as we will soon place numbers on the parent chain and we don't want to confuse those numbers with the numbers on the substituent chain.

Some complex substituents have common names. These common names are so well entrenched that IUPAC allows them. It would be wise to commit the following common names to memory, as they will be used frequently throughout the course.

An alkyl group bearing three carbon atoms can only be branched in one way, and it is called an *isopropyl group*:

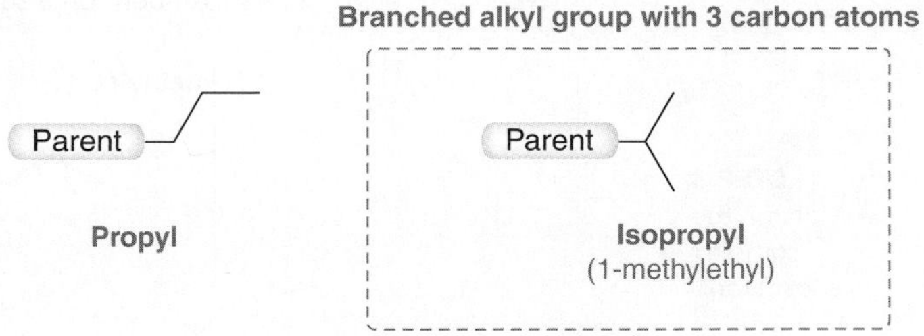

Alkyl groups bearing four carbon atoms can be branched in three different ways:

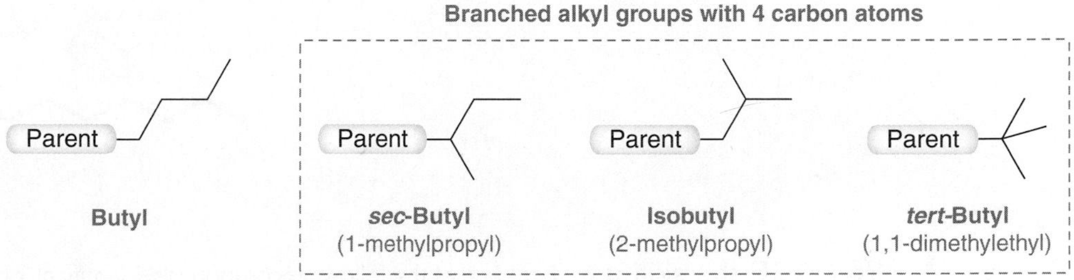

Alkyl groups bearing five carbon atoms can be branched in many more ways. Here are two common ways:

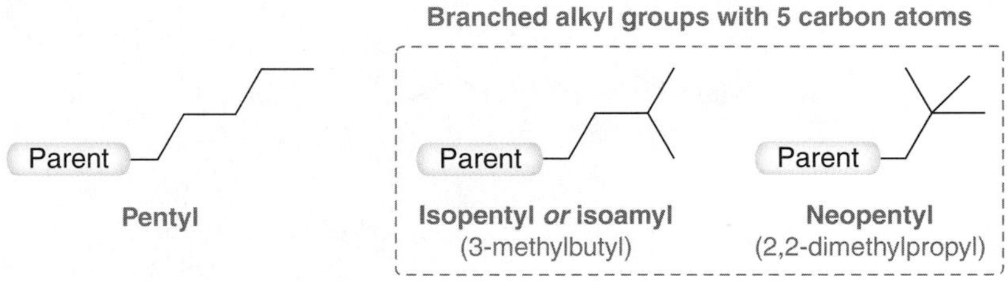

SKILLBUILDER

4.3 IDENTIFYING AND NAMING COMPLEX SUBSTITUENTS

LEARN the skill

In the following compound, identify all groups that would be considered substituents, and then indicate the systematic name as well as the common name for each substituent:

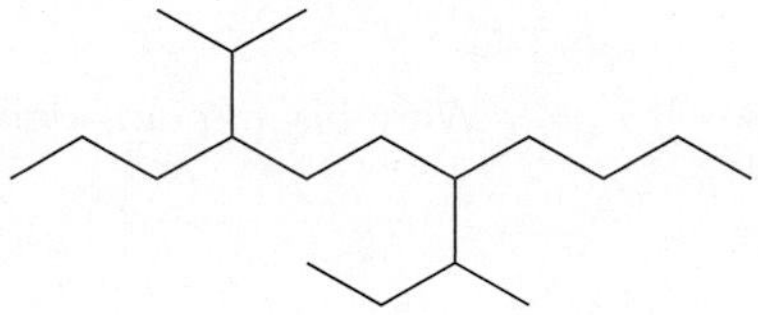

SOLUTION

First identify the parent:

STEP 1
Identify the parent.

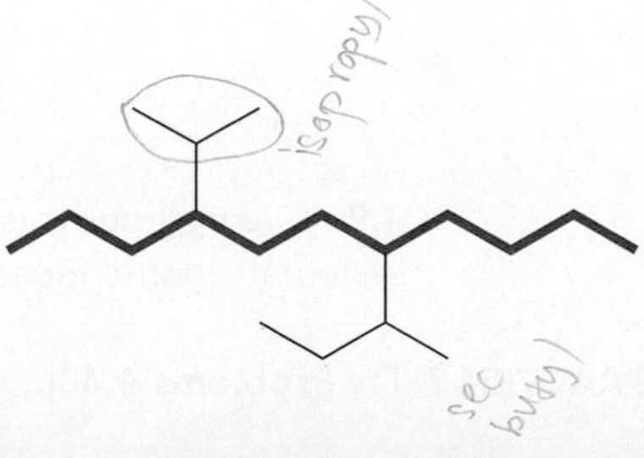

STEP 2
Place numbers on the complex substituent.

In this example, there are two complex substituents. To name them, we place numbers on each substituent, going away from the parent in each case. Then, we consider each complex substituent to be composed of "a substituent on a substituent," and we use the numbers for that purpose:

STEP 3
Name the complex substituent.

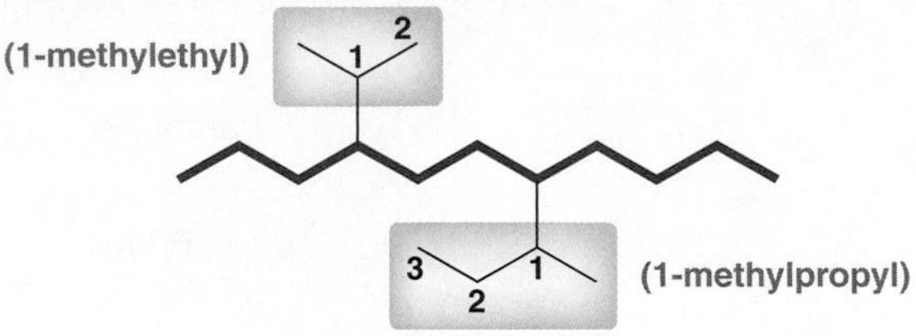

Alternatively, IUPAC rules allow us to use the following common names:

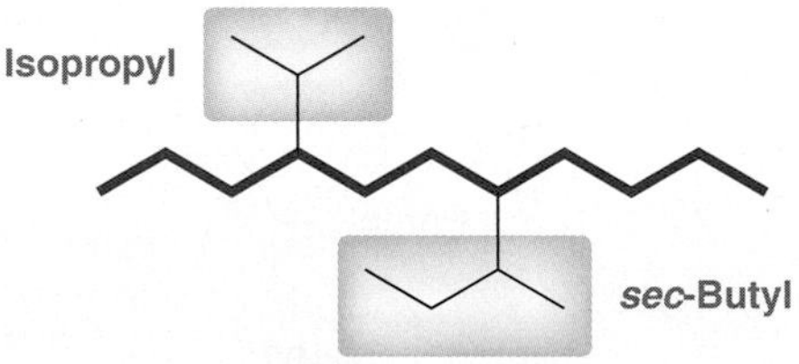

PRACTICE the skill **4.7** For each of the following compounds, identify all groups that would be considered substituents, and then indicate the systematic name as well as the common name for each substituent:

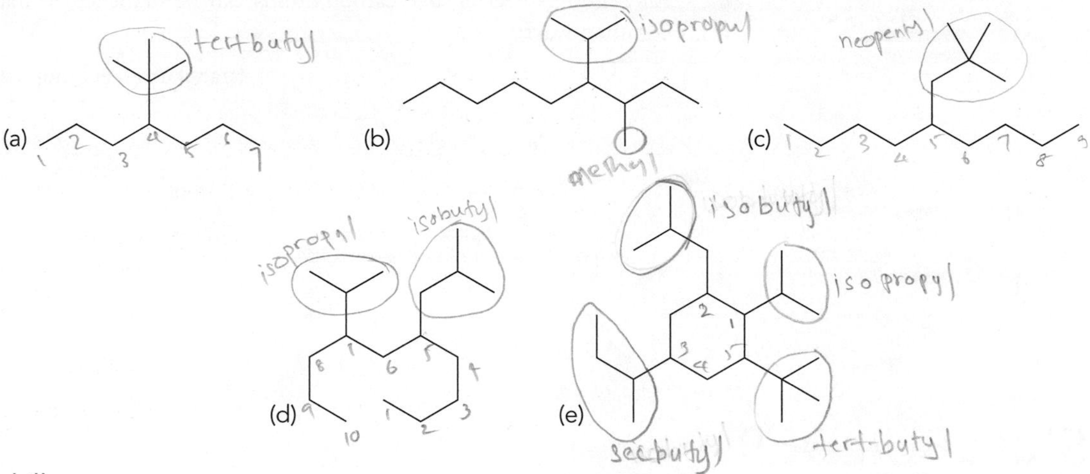

APPLY the skill **4.8** The following substituent is called a phenyl group:

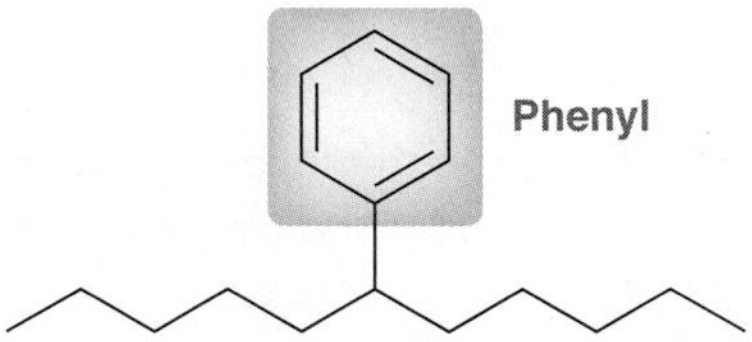

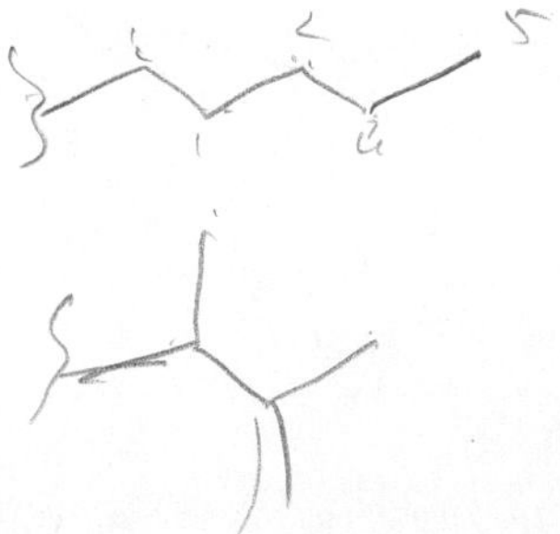

With this in mind, identify the systematic name for each substituent below:

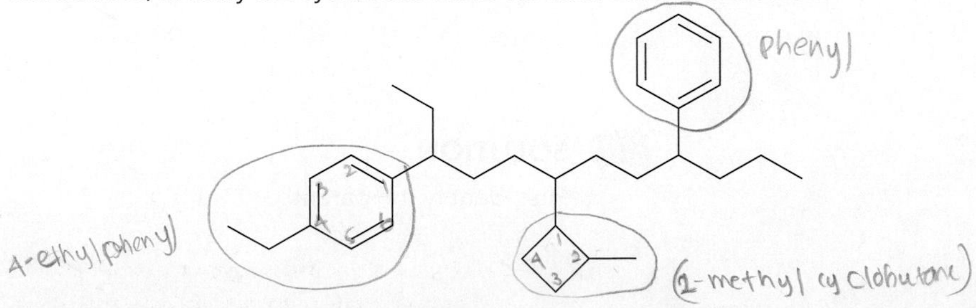

4.9 Draw all possible alkyl substituents that possess exactly five carbon atoms. Provide a systematic name for each of these substituents.

need more **PRACTICE?** **Try Problems 4.40b, 4.40d**

Assembling the Systematic Name of an Alkane

In order to assemble the systematic name of an alkane, the carbon atoms of the parent chain are numbered, and these numbers are used to identify the location of each substituent. As an example, consider the following two compounds:

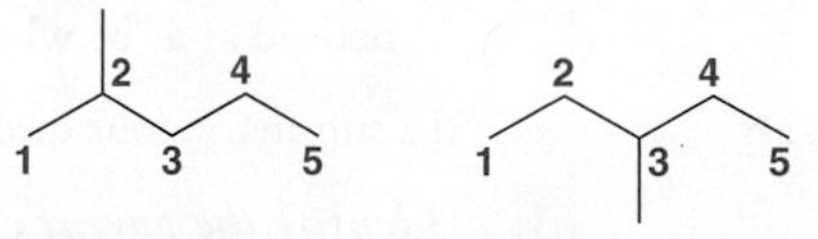

2-Methylpentane **3-Methylpentane**

In each case, the location of the methyl group is clearly identified with a number, called a **locant**. In order to assign the correct locant, we must number the parent chain properly, which can be done by following just a few rules:

- If one substituent is present, it should be assigned the lower possible number. In this example, we place the numbers so that the methyl group is at C2 rather than C6.

Correct **Incorrect**

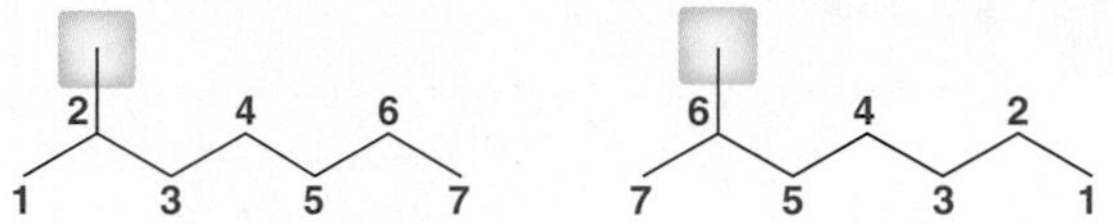

- When multiple substituents are present, assign numbers so that the first substituent receives the lower number. In the following case, we number the parent chain so that the substituents are 2,5,5 rather than 3,3,6 because we want the first locant to be as low as possible.

Correct **Incorrect**

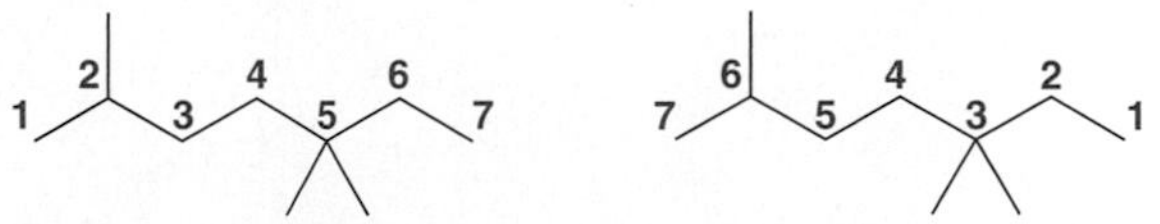

2, 5, 5 beats 3, 3, 6

- If there is a tie, then the second locant should be as low as possible:

Correct **Incorrect**

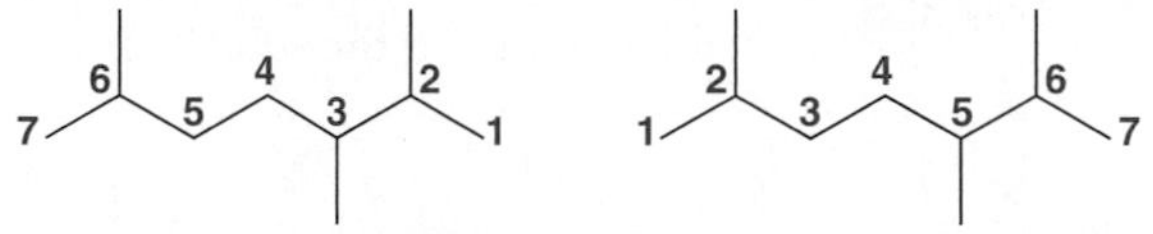

2, 3, 6 beats 2, 5, 6

- When dealing with cycloalkanes, all of the same rules apply, for example:

Correct **Incorrect**

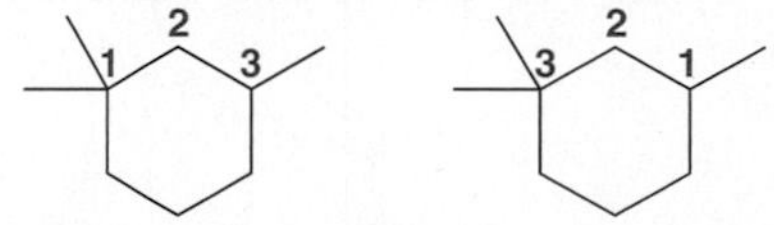

1, 1, 3 beats 1, 3, 3

- When a substituent appears more than once in a compound, a prefix is used to identify how many times that substituent appears in the compound (di = 2, tri = 3, tetra = 4, penta = 5, and hexa = 6). For example, the compound above would be called ***1,1,3-trimethyl***cyclohexane. Notice that a hyphen is used to separate numbers from letters, while commas are used to separate two numbers from each other.

- Once all substituents have been identified and assigned the proper locants, they are placed in alphabetical order. Prefixes (di, tri, tetra, penta, and hexa) are not included as part of the alphabetization scheme. In other words, "dimethyl" is alphabetized as if it started with the letter "m" rather than "d." Similarly, sec and tert are also ignored for alphabetization purposes, however, iso is not ignored. In other words, *sec*-butyl is alphabetized as a "b" while isobutyl is alphabetized as an "i."

In summary, four discrete steps are required when assigning the name of an alkane:

1. ***Identify the parent chain***: Choose the longest chain. For two chains of equal length, the parent chain should be the chain with the greater number of substituents.
2. ***Identify and name the substituents.***
3. ***Number the parent chain and assign a locant to each substituent***: Give the first substituent the lower possible number. If there is a tie, choose the chain in which the second substituent has the lower number.
4. ***Arrange the substituents alphabetically.*** Place locants in front of each substituent. For identical substituents, use di, tri, or tetra, which are ignored in alphabetizing.

SKILLBUILDER

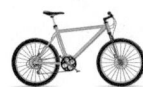

4.4 ASSEMBLING THE SYSTEMATIC NAME OF AN ALKANE

LEARN the skill

Provide a systematic name for the following compound:

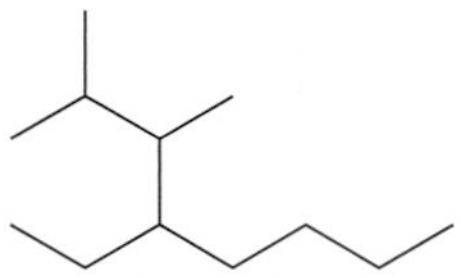

SOLUTION

Assembling a systematic name requires four discrete steps. The first two steps are to identify the parent and the substituents:

STEP 1
Identify the parent.

STEP 2
Identify and name substituents.

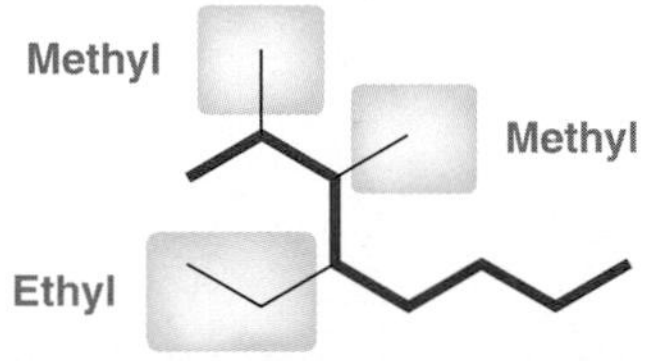

STEP 3
Assign locants.

Then, assign a locant to each substituent, and arrange the substituents alphabetically:

STEP 4
Arrange the substituents alphabetically.

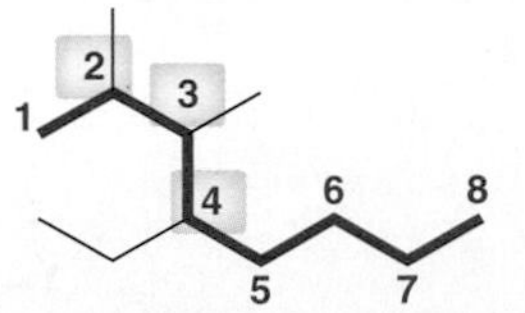

4-Ethyl-2,3-dimethyloctane

Notice that ethyl is placed before dimethyl in the name. Also, make sure that hyphens separate letters from numbers, while commas separate numbers from each other.

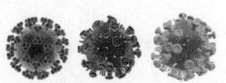

PRACTICE the skill **4.10** Provide a systematic name for each of the following compounds below:

(a) 3,4,6 – trimethyl octane

(b) secbutyl cyclohexane

(c) 3-ethyl-2-methyl heptane

(d) 3-isopropyl-2,4 methyl pentane

(e) 3-ethyl-2,2-dimethyl hexane

(f) 2-cyclohexane-4-ethyl-5,6 dimethyl octane

(g) 3-ethyl-2,5 dimethyl-4 propyl heptane

(h) 5-secbutyl-4-ethyl-2 methyl decane

(i) 2,2,6,6,7,7-hexamethyl nonane

(j) 4,5 dimethyl nonane

(k) 2-4,4-6-tetramethyl heptane

(l) 2,2-4 trimethylpentane

(m) 4-tertbutyl hexane

 3,5-diethyl-2 methyl octane

(n) 3-ethyl-6-isopropyl-2,4 dimethyl decane

(o)

(p) 1,3-diisopropyl cyclopentane

(q) 3-ethyl-2,5-dimethyl heptane

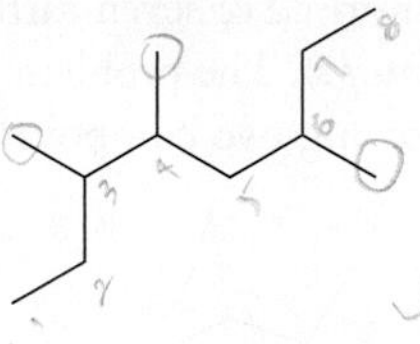

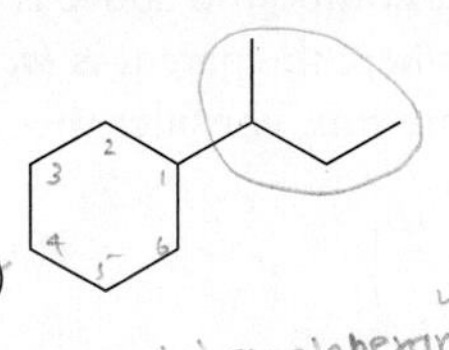

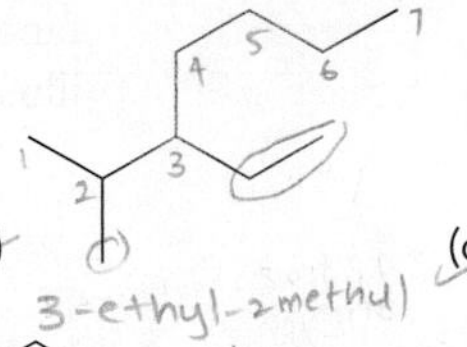

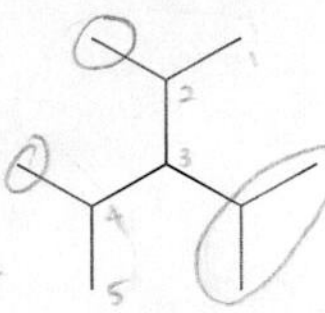

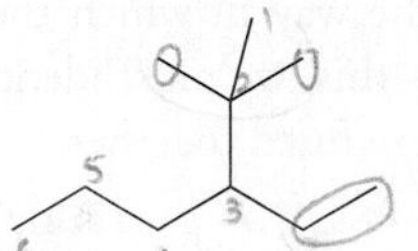

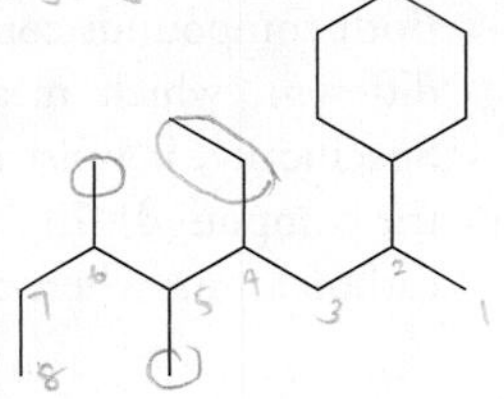

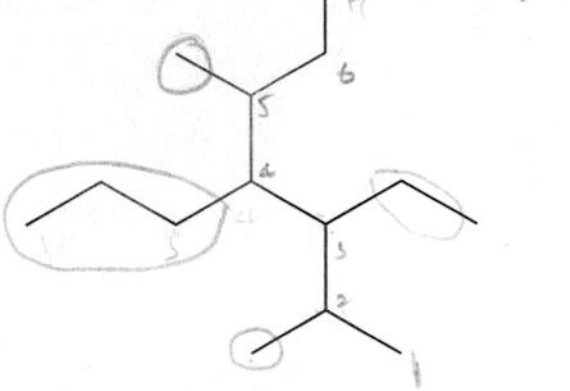

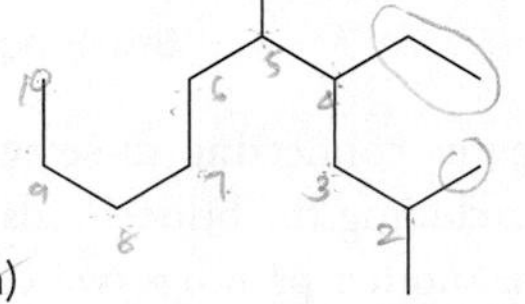

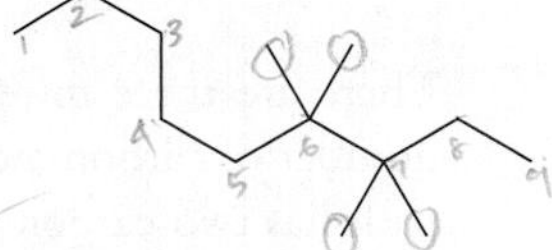

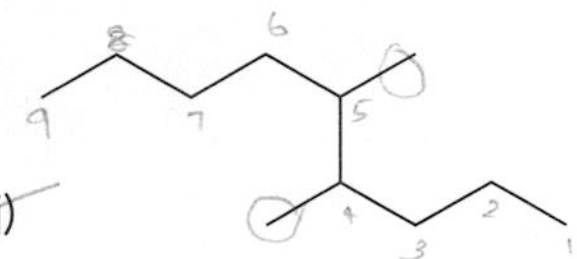

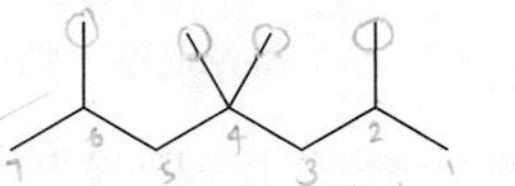

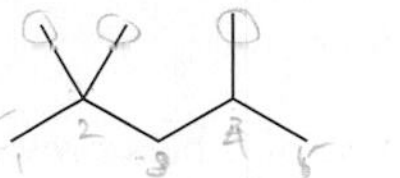

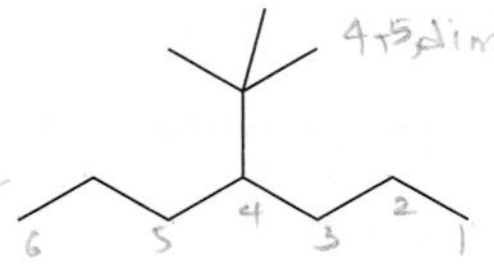

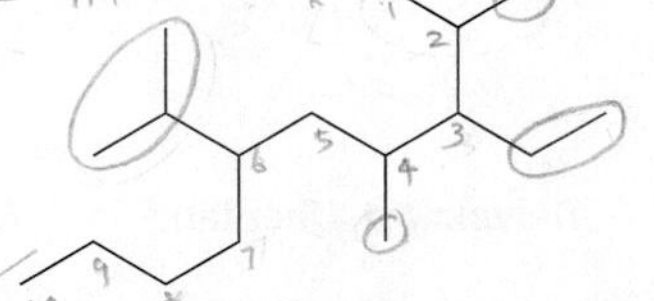

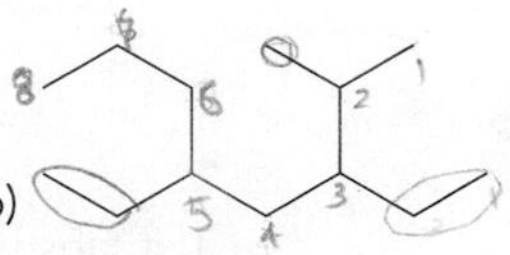

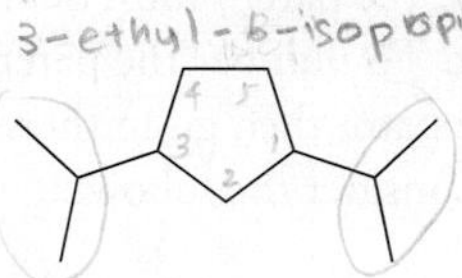

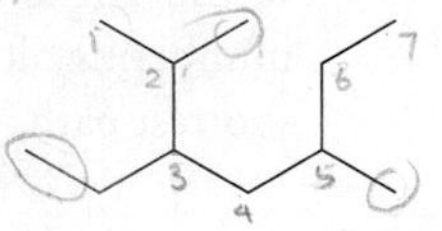

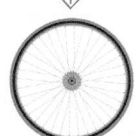

APPLY the skill **4.11** Draw a bond-line drawing for each of the following compounds:

(a) 3-Isopropyl-2,4-dimethylpentane (b) 4-Ethyl-2-methylhexane

(c) 1,1,2,2-Tetramethylcyclopropane

need more **PRACTICE?** **Try Problems 4.41a, 4.41b, 4.45a, 4.45b**

Naming Bicyclic Compounds

Compounds that contain two fused rings are called **bicyclic** compounds, and they can be drawn in different ways:

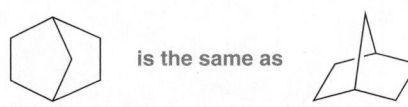

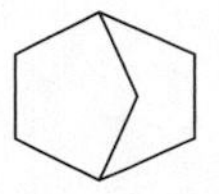

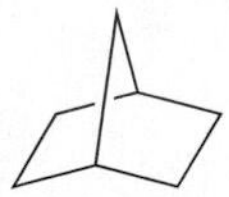

The second drawing style implies the three dimensionality of the molecule, a topic that will be covered in more detail in the upcoming chapter. For now, we will focus on naming bicyclic systems, which is very similar to naming alkanes and cycloalkanes. We follow the same four-step procedure outlined in the previous section, but there are differences in naming and numbering the parent. Let's start with naming the parent.

4.11
(a)

(b)

(c)

For bicyclic systems, the term "bicyclo" is introduced in the name of the parent; for example, the compound above is comprised of seven carbon atoms and is therefore called bicycloheptane (where the parent is *bicyclohept*). The problem is that this parent is not specific enough. To illustrate this, consider the following two compounds, both of which are called bicycloheptane:

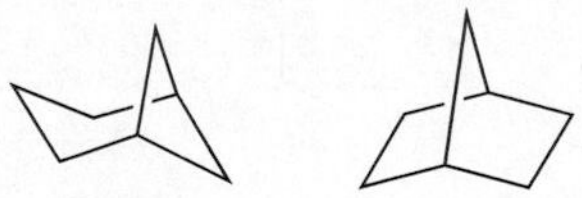

Both compounds consist of two rings and seven carbon atoms. Yet, the compounds are clearly different, which means that the name of the parent needs to contain more information. Specifically, it must indicate the way in which the rings are constructed (the constitution of the compound). In order to do this, we must identify the two **bridgeheads**. These are the two carbon atoms where the rings are fused together:

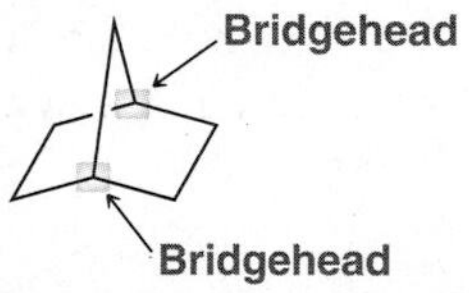

There are three different paths connecting these two bridgeheads. For each path, count the number of carbon atoms, excluding the bridgeheads themselves. In the compound above, one path has two carbon atoms, another path has two carbon atoms, and the third (shortest path) has only one carbon atom. These three numbers, ordered from largest to smallest, [2.2.1], are then placed in the middle of the parent surrounded by brackets:

Bicyclo[2.2.1]heptane

These numbers provide the necessary specificity to differentiate the compounds shown earlier:

Bicyclo[3.1.1]heptane **Bicyclo[2.2.1]heptane**

If a substituent is present, the parent must be numbered properly in order to assign the correct locant to the substituent. To number the parent, start at one of the bridgeheads and begin numbering along the longest path, then go to the second longest path, and finally go along the shortest path. For example, consider the following bicyclic system:

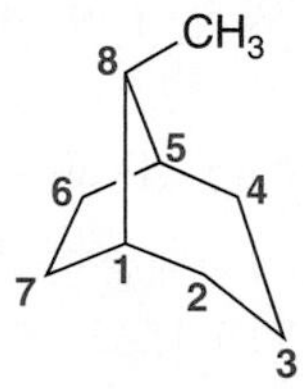

In this example, the methyl substituent did not get a low number. In fact, it got the highest number possible because of its location. Specifically, it is on the shortest path connecting the bridgeheads. Regardless of the position of substituents, the parent must be numbered beginning with the longest path first. The only choice is which bridgehead will be counted as C1; for example:

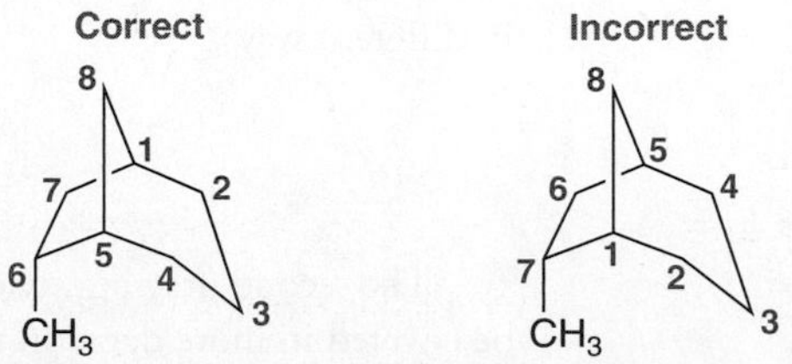

Either way, the numbers begin along the longest path. However, we must start numbering at the bridgehead that gives the substituent the lowest possible number. In the example above, the correct path places the substituent at C6 rather than at C7. The name of this compound is 6-methylbicyclo[3.2.1]octane.

SKILLBUILDER

4.5 ASSEMBLING THE NAME OF A BICYCLIC COMPOUND

LEARN the skill

Assign a complete name for the following compound:

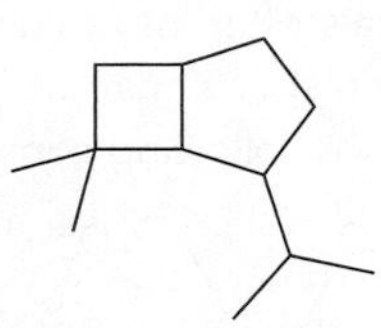

SOLUTION

STEP 1 Identify the parent.

Once again, we use our four-step procedure. First identify the parent. In this case, we are dealing with a bicyclic system, so we count the total number of carbon atoms comprising both ring systems. There are seven carbon atoms in both rings combined, so the parent must be "bicyclo-hept". The two bridgeheads are highlighted. Now count the number of carbon atoms along each of the three possible paths that connect the bridgeheads. The longest path (on the right side) has three carbon atoms in between the bridgeheads. The second longest path (on the left side) has two carbon atoms in between the bridgeheads. The shortest path has no carbon atoms in between the bridgeheads; that is, the bridgeheads are connected directly to each other. Therefore the parent is "bicyclo[3.2.0]hept." Next, identify and name the substituents:

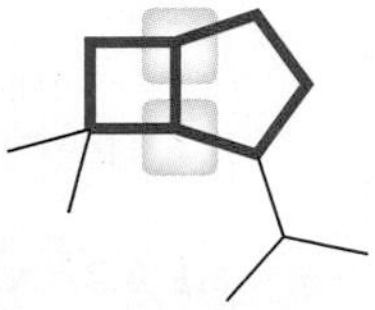

STEP 2 Identify and name substituents.

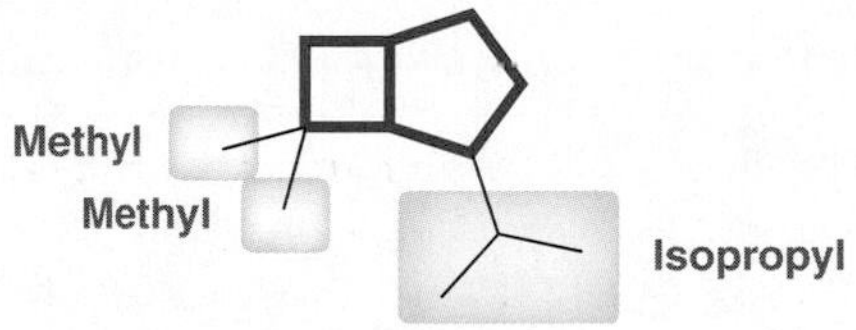

STEP 3 Assign locants.

Then number the parent, and assign a locant to each substituent. Start at one of the bridgeheads, and continue numbering along the longest path that connects the bridgeheads. In this case, we start at the lower bridgehead, so as to give the isopropyl group the lower number:

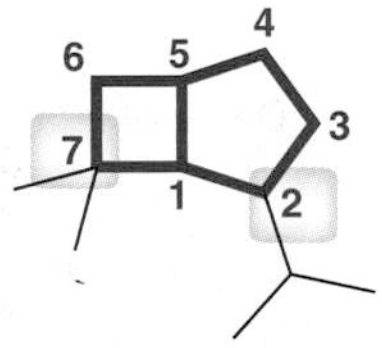

STEP 4 Arrange the substituents alphabetically.

Finally, arrange the substituents alphabetically:

2-Isopropyl-7,7-dimethylbicyclo[3.2.0]heptane

PRACTICE the skill

4.12 Name each of the following compounds:

(a) 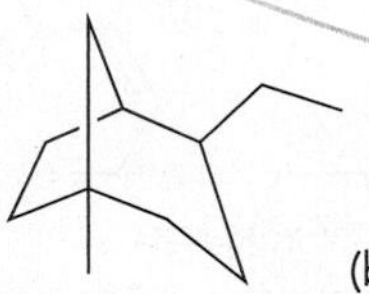
(b) 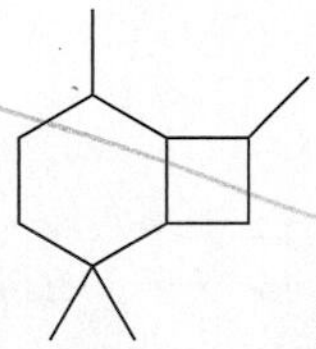
(c) 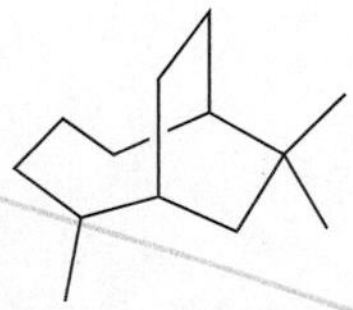
(d) 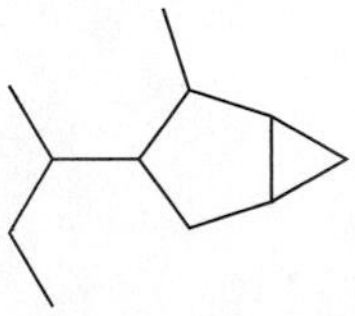
(e) 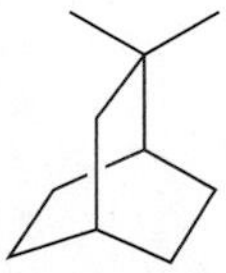

(f) 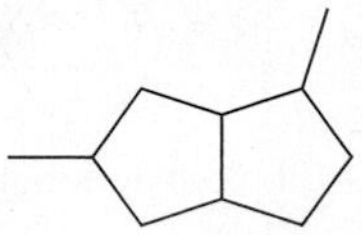
(g) 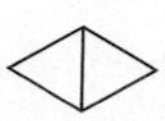
(h) 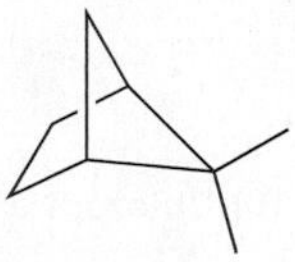
(i) 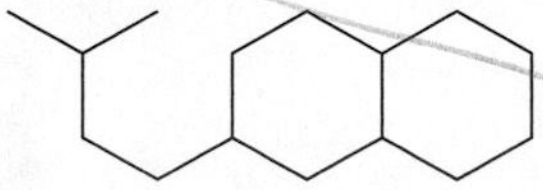

APPLY the skill

4.13 Draw a bond-line structure for each of the following compounds:

(a) 2,2,3,3-Tetramethylbicyclo[2.2.1]heptane

(b) 8,8-Diethylbicyclo[3.2.1]octane

(c) 3-Isopropylbicyclo[3.2.0]heptane

need more **PRACTICE?** **Try Problems 4.41c, 4.41d, 4.45c**

MEDICALLYSPEAKING)))

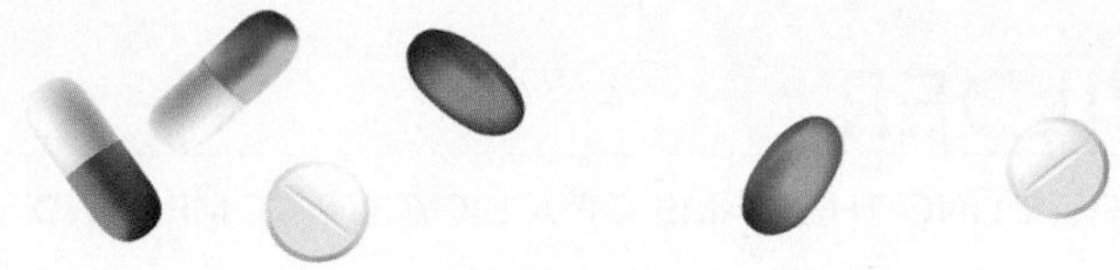

Naming Drugs

Pharmaceuticals often have cumbersome IUPAC names and are therefore given shorter names, called *generic names*. For example, consider the following compound:

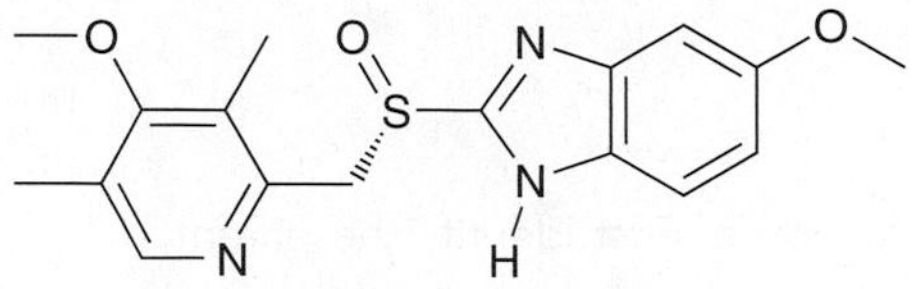

(*S*)-5-Methoxy-2-[(4-methoxy-3,5-dimethyl pyridin-2-yl)methylsulfinyl]-3*H*-benzoimidazole

The IUPAC name for this compound is quite a mouthful, so a generic name, *esomeprazole*, has been assigned and accepted by the international community. For marketing purposes, drug companies will also select a catchy name, called a *trade name*. The trade name of esomeprazole is Nexium®. This compound is a proton-pump inhibitor used in the treatment of reflux disease.

In summary, pharmaceuticals have three important names: (1) trade names, (2) generic names, and (3) systematic IUPAC names. Table 4.3 lists several common drugs whose trade names are likely to sound familiar.

TABLE 4.3 NAMES OF COMMON PHARMACEUTICALS

TRADE NAME	GENERIC NAME	STRUCTURE AND IUPAC NAME	USES
Aspirin®	Acetylsalicylic acid	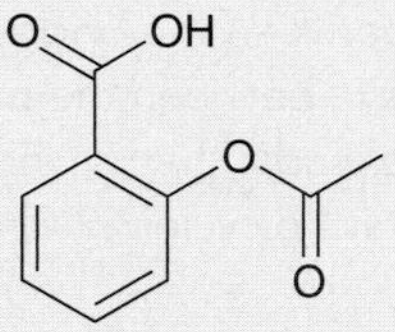 **2-Acetoxybenzoic acid**	Analgesic, antipyretic (reduces fever), anti-inflammatory
Advil® or Motrin®	Ibuprofen	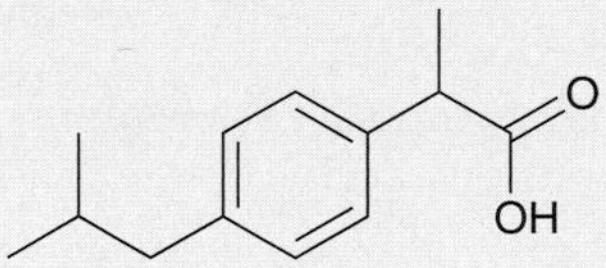 **2-[4-(2-Methylpropyl)phenyl]propanoic acid**	Analgesic, antipyretic, anti-inflammatory
Demerol®	Pethidine	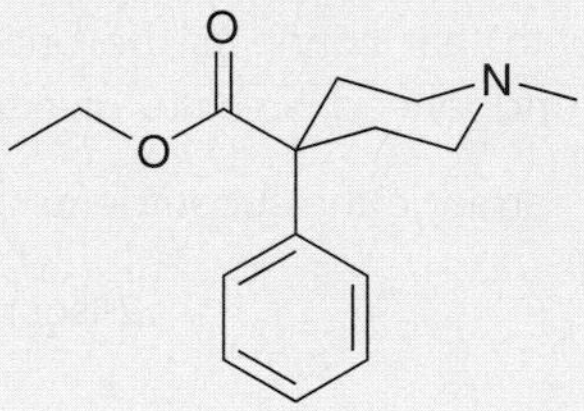 **Ethyl 1-methyl-4-phenylpiperidine-4-carboxylate**	Analgesic
Dramamine®	Meclizine	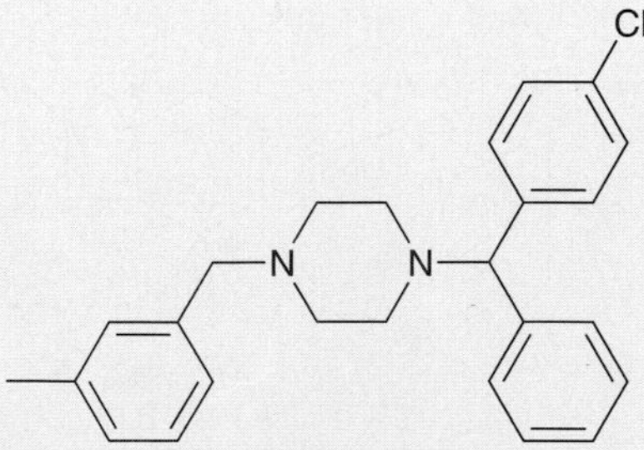 **1-[(4-Chlorophenyl)-phenyl-methyl]-4-[(3-methylphenyl)methyl]piperazine**	Antiemetic (inhibits nausea and vomiting)
Tylenol®	Acetaminophen	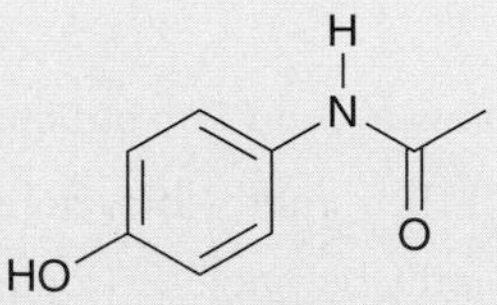 ***N*-(4-hydroxyphenyl)ethanamide**	Analgesic, antipyretic

4.3 Constitutional Isomers of Alkanes

For an alkane, the number of possible constitutional isomers increases with increasing molecular size. Table 4.4 illustrates this trend.

TABLE 4.4 NUMBER OF CONSTITUTIONAL ISOMERS FOR VARIOUS ALKANES

MOLECULAR FORMULA	NUMBER OF CONSTITUTIONAL ISOMERS
C_3H_8	1
C_4H_{10}	2
C_5H_{12}	3
C_6H_{14}	5
C_7H_{16}	9
C_8H_{18}	18
C_9H_{20}	35
$C_{10}H_{22}$	75
$C_{15}H_{32}$	4,347
$C_{20}H_{42}$	366,319
$C_{30}H_{62}$	4,111,846,763
$C_{40}H_{82}$	62,481,801,147,341

When drawing the constitutional isomers of an alkane, make sure to avoid drawing the same isomer twice. As an example, consider C_6H_{14}, for which there are five constitutional isomers. When drawing these isomers, it might be tempting to draw more than five structures. For example:

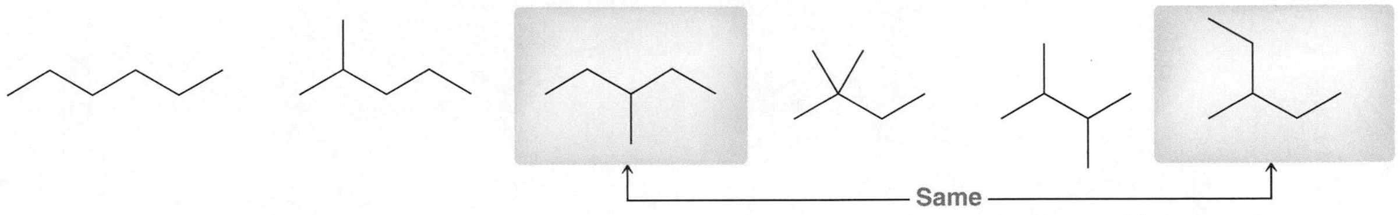

At first glance, the two highlighted compounds seem to be different. But upon further inspection, it becomes apparent that they are actually the same compound. To avoid drawing the same compound twice, it is helpful to use IUPAC rules to name each compound. If there are duplicates, it will become apparent:

3-Methylpentane

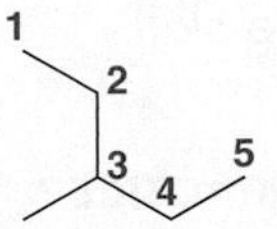

3-Methylpentane

These two drawings generate the same name, and therefore, they must be the same compound. Even without formally naming these compounds, it is helpful to simply look at molecules from an IUPAC point of view. In other words, each of these compounds should be viewed as a parent chain of five carbon atoms, with a methyl group at C3. Viewing molecules in this way (from an IUPAC point of view) will prove to be helpful in some cases.

SKILLBUILDER

4.6 IDENTIFYING CONSTITUTIONAL ISOMERS

LEARN the skill

Identify whether the following two compounds are constitutional isomers or whether they are simply different drawings of the same compound:

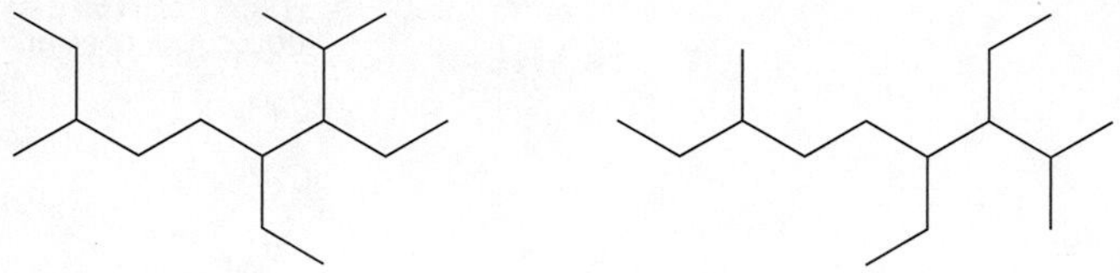

SOLUTION

Use the rules of nomenclature to name each compound. In each case, identify the parent, locate the substituents, number the parent, and assemble a name:

STEP 1 Name each compound.

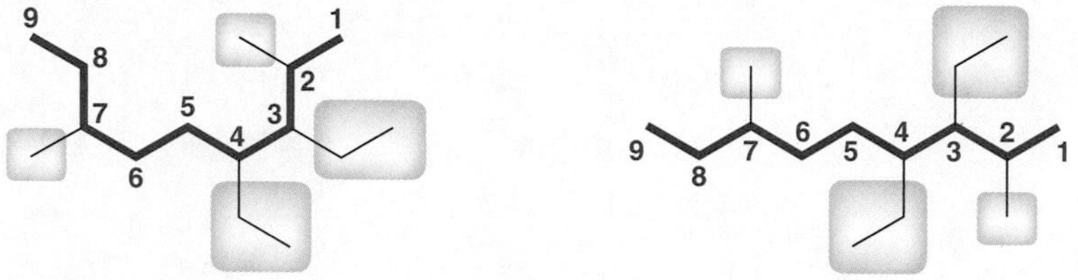

3,4-Diethyl-2,7-dimethylnonane **3,4-Diethyl-2,7-dimethylnonane**

STEP 2 Compare the names.

These compounds have the same name, and therefore, they are not constitutional isomers. They are actually two representations of the same compound.

PRACTICE the skill

4.14 For each pair of compounds, identify whether they are constitutional isomers or two representations of the same compound:

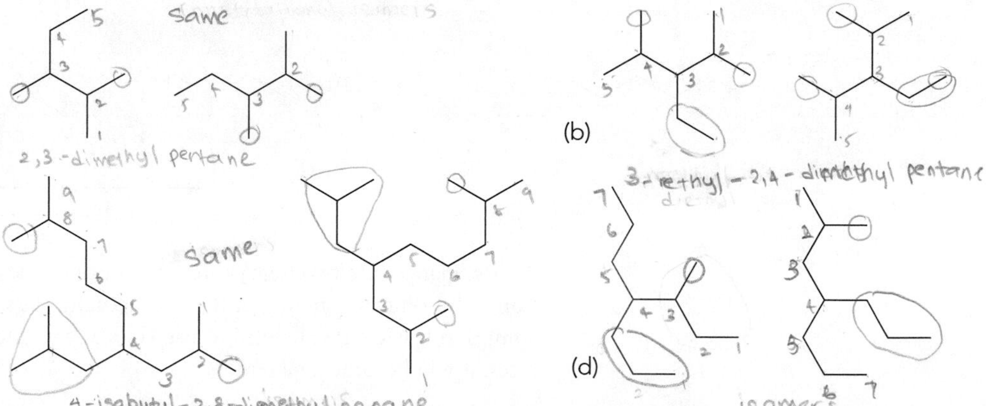

APPLY the skill

4.15 Table 4.4 indicates the number of constitutional isomers with molecular formula C_7H_{16}. Draw each of the isomers, making sure not to draw the same compound twice.

need more **PRACTICE?** **Try Problems 4.42, 4.66b,d,k,l**

4.4 Relative Stability of Isomeric Alkanes

In order to compare the stability of constitutional isomers, we look at the heat liberated when they each undergo combustion. For an alkane, combustion describes a reaction in which the alkane reacts with oxygen to produce CO_2 and water. Consider the following example:

 $+ 8\ O_2 \longrightarrow 5\ CO_2 + 6\ H_2O \qquad \Delta H° = -3509$ kJ/mol

In this reaction, an alkane (pentane) is ignited in the presence of oxygen, and the resulting reaction is called combustion. The value shown above, $\Delta H°$ for this reaction, is the **change in enthalpy** associated with the complete combustion of 1 mol of pentane in the presence of oxygen. We will revisit the concept of enthalpy in more detail in Chapter 6, but for now, we will simply think of it as the heat given off during the reaction. For a combustion process, $-\Delta H°$ is called the **heat of combustion**.

Combustion can be conducted under experimental conditions using a device called a calorimeter, which can measure heats of combustion accurately. Careful measurements reveal that the heats of combustion for two isomeric alkanes are different, even though the products of the reactions are identical:

$$+\ 12\tfrac{1}{2}\ O_2 \longrightarrow 8\ CO_2 + 9\ H_2O \qquad -\Delta H° = \mathbf{5470\ kJ/mol}$$

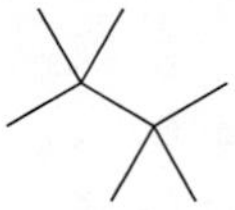

$$+\ 12\tfrac{1}{2}\ O_2 \longrightarrow 8\ CO_2 + 9\ H_2O \qquad -\Delta H° = \mathbf{5452\ kJ/mol}$$

Notice that although the two reactions shown above yield the same number of moles of CO_2 and water, the heats of combustion for the two reactions are different. We can use this difference to compare the stability of isomeric alkanes (Figure 4.1). By comparing the amount of heat

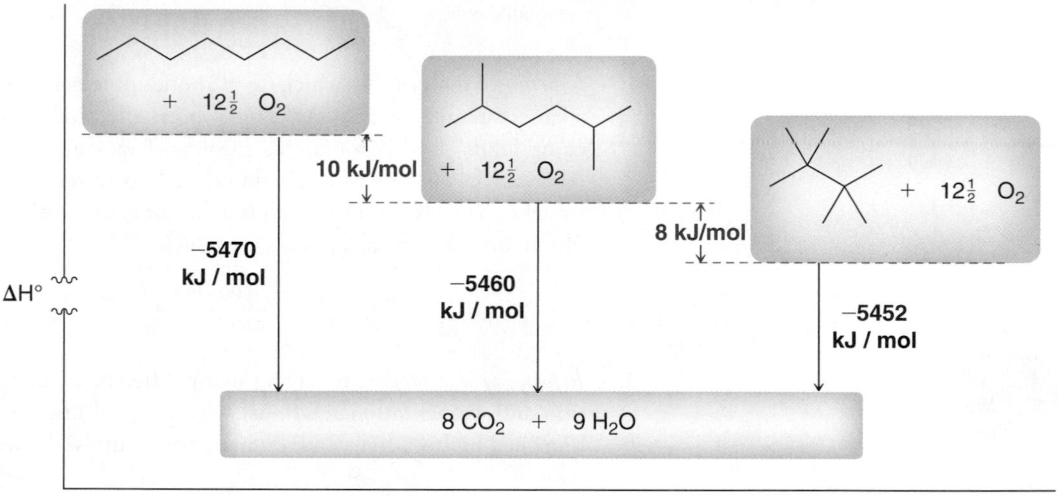

FIGURE 4.1
An energy diagram comparing the heats of combustion for three constitutional isomers of octane.

given off by each combustion process, we can compare the potential energy that each isomer had before combustion. This analysis leads to the conclusion that branched alkanes are lower in energy (more stable) than straight-chain alkanes.

Heats of combustion are an important way to determine the relative stability of compounds. We will use this technique several times throughout this book to compare the stability of compounds.

4.5 Sources and Uses of Alkanes

The main source of alkanes is petroleum, which comes from the Latin words *petro* ("rock") and *oleum* ("oil"). Petroleum is a complex mixture of hundreds of hydrocarbons, most of which are alkanes (ranging in size and constitution). It is believed that Earth's petroleum deposits were formed slowly, over millions of years, by the decay of prehistoric bioorganic material (such as plants and forests).

The first oil well was drilled in Pennsylvania in 1859. The petroleum obtained was separated into its various components by distillation, the process by which the components of a mixture are separated from each other based on differences in their boiling points. At the time, kerosene, one of the high-boiling fractions, was considered to be the most important product obtained from petroleum. Automobiles based on the internal combustion engine were still waiting to be invented (some 50 years later), so gasoline was not yet a coveted petroleum product. There was a large market opportunity for kerosene, as kerosene lamps produced better night light than standard candles. Over time, other uses were found for the other fractions of petroleum. Today, every precious drop of petroleum is put to use in one way or another (Table 4.5). The process of separating crude oil (petroleum) into commercially available products is called *refining.* A typical refinery can process 100,000 barrels of crude oil a day (1 barrel = 42 gallons). The most important product is currently the gasoline fraction ($C_5 - C_{12}$), yet this fraction only represents approximately 19% of the crude oil. This amount does not satisfy the current demand for gasoline, and therefore, two processes are employed that increase the yield of gasoline from every barrel of crude oil.

TABLE 4.5 INDUSTRIAL USES OF PETROLEUM FRACTIONS

BOILING RANGE OF FRACTION (° C)	NUMBER OF CARBON ATOMS IN MOLECULES	USE
Below 20	$C_1 - C_4$	Natural gas, petrochemicals, plastics
20 – 100	$C_5 - C_7$	Solvents
20 – 200	$C_5 - C_{12}$	Gasoline
200 – 300	$C_{12} - C_{18}$	Kerosene, jet fuel
200 – 400	C_{12} and higher	Heating oil, diesel
Nonvolatile liquids	C_{20} and higher	Lubricating oil, grease
Nonvolatile solids	C_{20} and higher	Wax, asphalt, tar

1. *Cracking* is a process by which C—C bonds of larger alkanes are broken, producing smaller alkanes suitable for gasoline. This process effectively converts more of the crude oil into compounds suitable for use as gasoline. Cracking can be achieved at high temperature (*thermal cracking*) or with the aid of catalysts (*catalytic cracking*). Cracking generally yields straight-chain alkanes. Although suitable for gasoline, these alkanes tend to give rise to preignition, or *knocking,* in automobile engines.

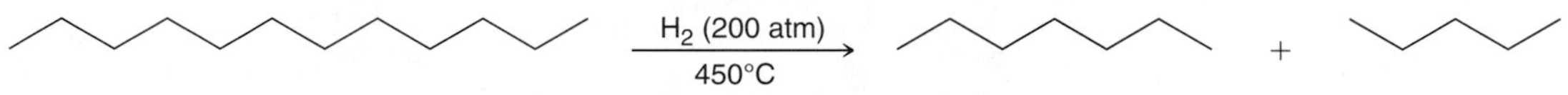

2. *Reforming* is a process involving many different types of reactions (such as dehydrogenation and isomerization reactions) with the goal of converting straight-chain alkanes into branched hydrocarbons and aromatic compounds (discussed in Chapter 18):

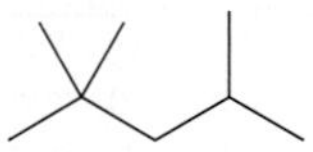

2,2,4-Trimethylpentane
A branched alkane

Benzene
An aromatic compound

Branched hydrocarbons and aromatic hydrocarbons show less of a tendency for knocking. It is therefore desirable to convert some petroleum into branched alkanes and aromatic compounds and then blend them with straight-chain alkanes. The combination of cracking and reforming effectively increases the gasoline yield from 19 to 47% for every barrel of crude oil. Gasoline is therefore a sophisticated blend of straight-chain alkanes, branched alkanes, and aromatic hydrocarbons. The precise blend is dependent on a number of conditions. In colder climates, for example, the blend must be appropriate for temperatures below zero. Therefore gasoline used in Chicago is not the same as the gasoline used in Houston.

Petroleum is not a renewable energy source. At our current rate of consumption it is estimated that the Earth's supply of petroleum will be exhausted by 2060. We may find more petroleum deposits, but that will just delay the inevitable. Petroleum is also the primary source of a wide variety of organic compounds used for making plastics, pharmaceuticals, and numerous other products. It is vital that Earth's supply of petroleum not be completely exhausted, although such a dire picture is unlikely. As the supply of petroleum dwindles and the demand increases, the price of crude oil will rise (and so will the price of gasoline at the pump). Eventually, the price of petroleum will surpass the price of alternative energy sources, at which point a major shift will occur. What do you think will replace petroleum as the next global energy source?

PRACTICALLYSPEAKING)))

An Introduction to Polymers

Table 4.5 reveals that low-molecular-weight alkanes (such as methane or ethane) are gases at room temperature, alkanes of slightly higher molecular weight (such as hexane and octane) are liquids at room temperature, and alkanes of very high molecular weight (such as hectane, with 100 carbon atoms) are solids at room temperature. This trend is explained by the increased London dispersion forces experienced by higher molecular weight alkanes (as described in Section 1.12). With this in mind, consider an alkane composed of approximately 100,000 carbon atoms. It should not be surprising that such an alkane should be a hard solid at room temperature. This material, called polyethylene, is used for a variety of purposes, including garbage containers, plastic bottles, packaging material, bulletproof vests, and toys. As its name implies, polyethylene is produced from polymerization of ethylene:

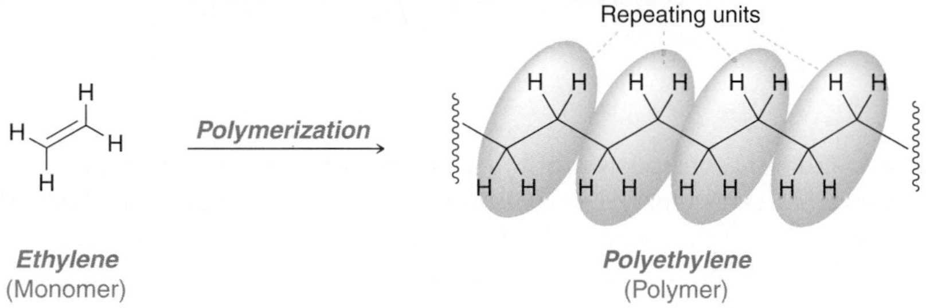

Polyethylene is an example of a *polymer*, because it is created by the joining of small molecules called *monomers*. Over 100 billion pounds of polyethylene are produced worldwide each year.

In our everyday lives, we are surrounded by a variety of polymers. From carpet fibers to plumbing pipes, our society has clearly become dependent on polymers. Polymers are discussed in more detail in later chapters.

4.6 Drawing Newman Projections

We will now turn our attention to the way in which molecules change their shape with time. Rotation about C—C single bonds allows a compound to adopt a variety of possible three-dimensional shapes, called **conformations**. Some conformations are higher in energy, while others are lower in energy. In order to draw and compare conformations, we will need to use a new kind of drawing—one specially designed for showing the conformation of a molecule. This type of drawing is called a **Newman projection** (Figure 4.2). To understand what a Newman projection represents, consider the wedge and dash drawing of ethane in Figure 4.2. Begin rotating it about the vertical axis drawn in gray so that all of the red H's come out in front of the page

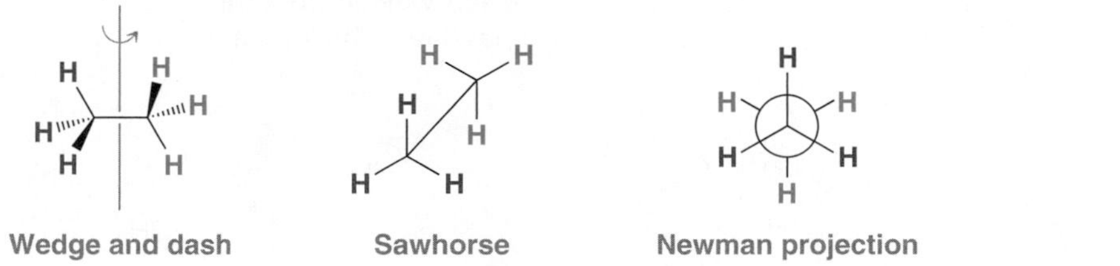

FIGURE 4.2
Three drawings of ethane: (a) wedge and dash, (b) sawhorse, and (c) a Newman projection.

and all of the blue H's go back behind the page. The second drawing (the sawhorse) represents a snapshot after 45° of rotation, while the Newman projection represents a snapshot after 90° of rotation. One carbon is directly in front of the other, and each carbon atom has three H's attached to it (Figure 4.3). The point at the center of the drawing in Figure 4.3 represents the front carbon atom, while the circle represents the back carbon. We will use Newman projections extensively throughout the rest of this chapter, so it is important to master both drawing and reading them.

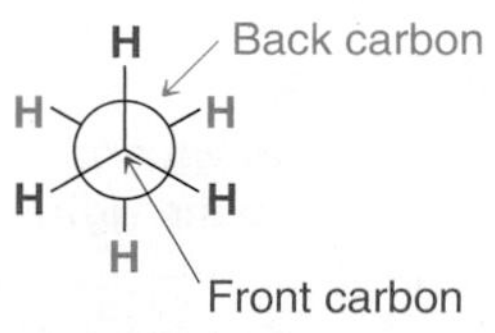

FIGURE 4.3
A Newman projection of ethane, showing the front carbon and the back carbon.

SKILLBUILDER

4.7 DRAWING NEWMAN PROJECTIONS

LEARN the skill

Draw a Newman projection of the following compound, as viewed from the angle indicated:

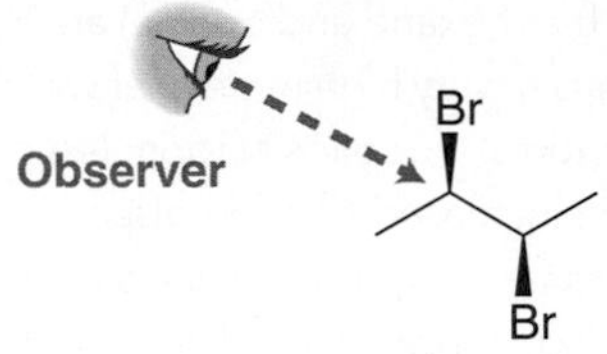

SOLUTION

Identify the front and back carbon atoms. From the angle of the observer, the front and back carbon atoms are:

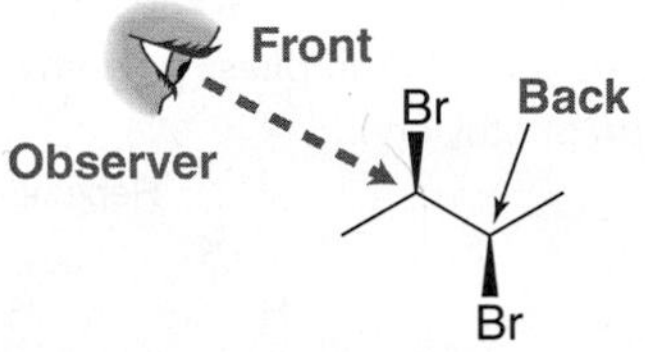

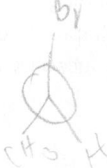

Now we must ask: From the perspective of the observer, what is connected to the front carbon? The observer will see a methyl group, a bromine, and a hydrogen atom. Remember that a wedge is coming out of the page, and a dash is going back behind the page. So, from the perspective of the observer, the front carbon atom looks like this:

STEP 1 Identify the three groups connected to the front carbon atom.

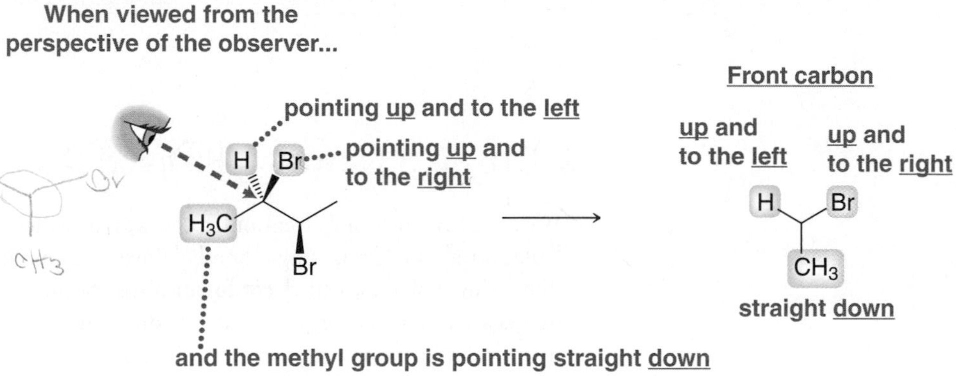

Now let's focus our attention on the back carbon atom. The back carbon has one CH_3 group, one Br, and one H. From the perspective of the observer, it looks like this:

STEP 2 Identify the three groups connected to the back carbon atom.

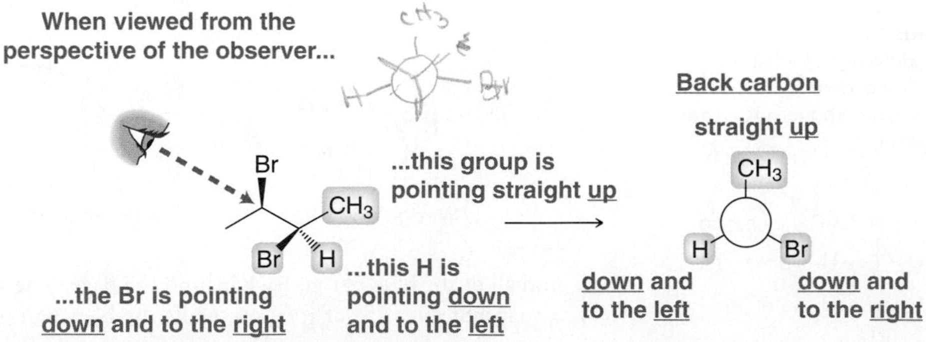

STEP 3 Draw the Newman projection.

Now we put both pieces of our drawing together:

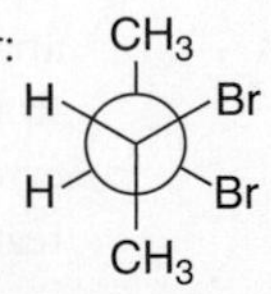

PRACTICE the skill **4.16** In each case below, draw a Newman projection as viewed from the angle indicated:

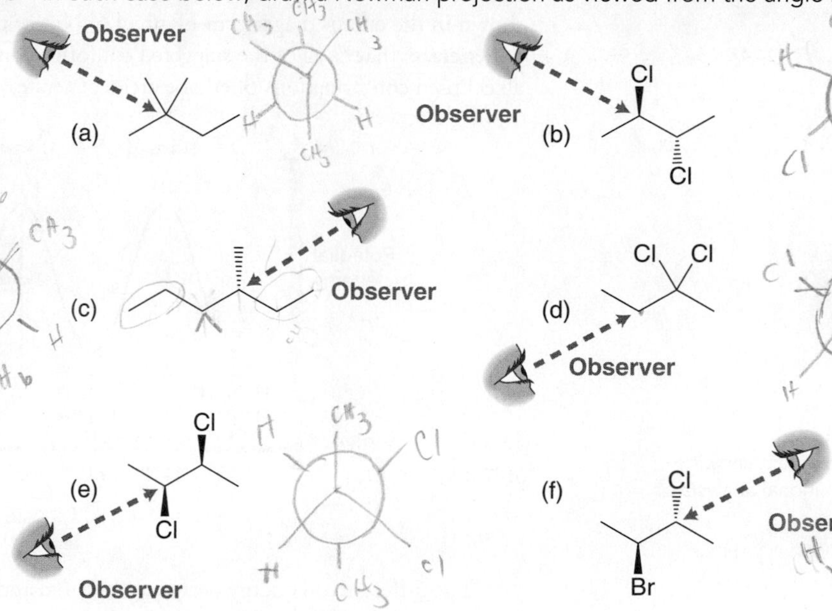

APPLY the skill **4.17** Draw a bond-line structure for each of the following compounds:

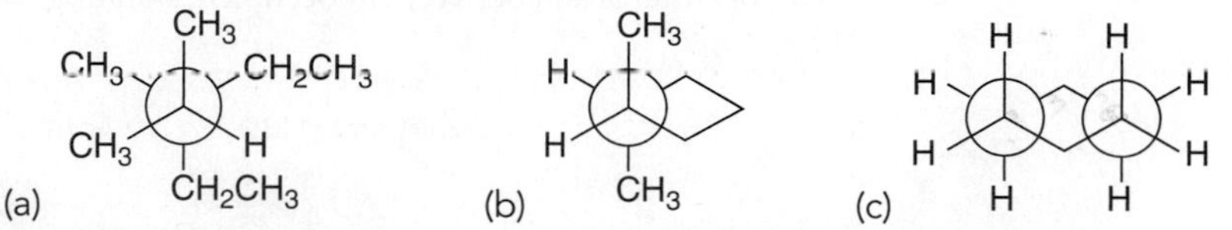

4.18 Determine whether the following compounds are constitutional isomers:

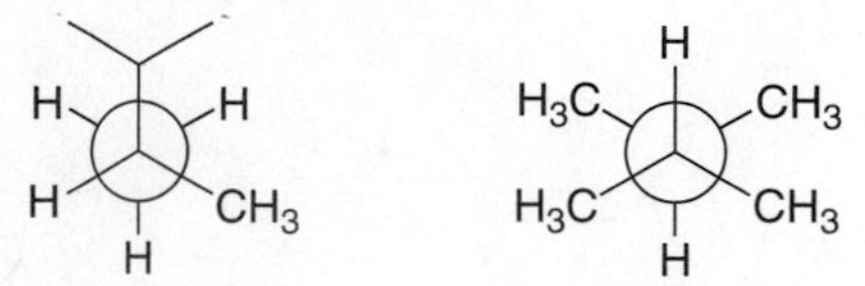

----> need more **PRACTICE?** **Try Problem 4.56**

4.7 Conformational Analysis of Ethane and Propane

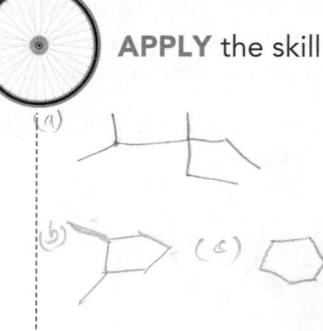

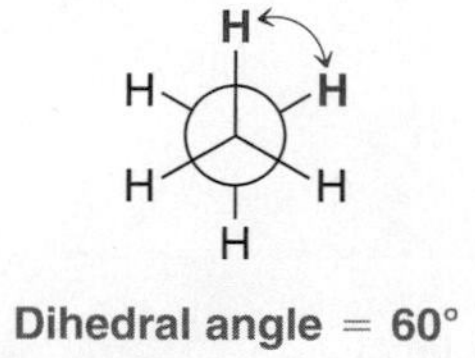

FIGURE 4.4 The dihedral angle between two hydrogen atoms in a Newman projection of ethane.

Consider the two hydrogen atoms shown in red in the Newman projection of ethane (Figure 4.4). These two hydrogen atoms appear to be separated by an angle of 60°. This angle is called the **dihedral angle** or **torsional angle**. This dihedral angle changes as the C—C bond rotates—for example, if the front carbon rotates clockwise while the back carbon is held stationary. The value for the dihedral angle between two groups can be any value between 0° and 180°. Therefore, there are an infinite number of possible conformations. Nevertheless, there are two conformations that require our special attention: the lowest energy conformation and the highest energy conformation (Figure 4.5). The **staggered conformation** is the lowest in energy, while the **eclipsed conformation** is the highest in energy.

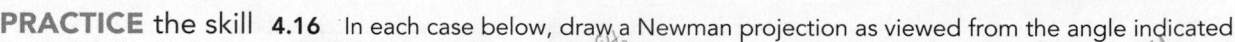

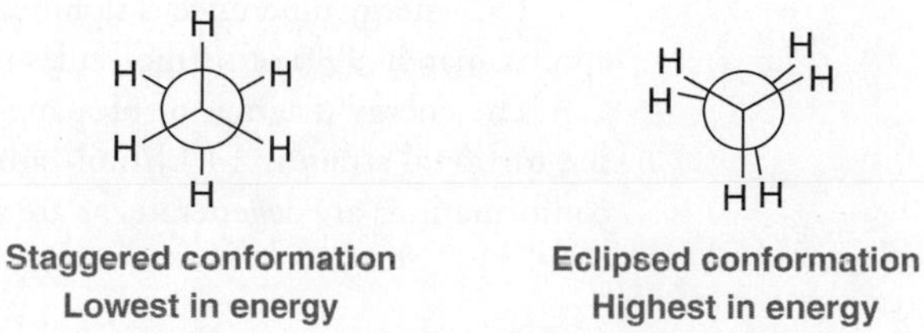

FIGURE 4.5 Staggered and eclipsed conformations of ethane.

The difference in energy between staggered and eclipsed conformations of ethane is 12 kJ/mol, as shown in the energy diagram in Figure 4.6. Notice that all staggered conformations of ethane are **degenerate**; that is, all of the staggered conformations have the same amount of energy. Similarly, all eclipsed conformations of ethane are degenerate.

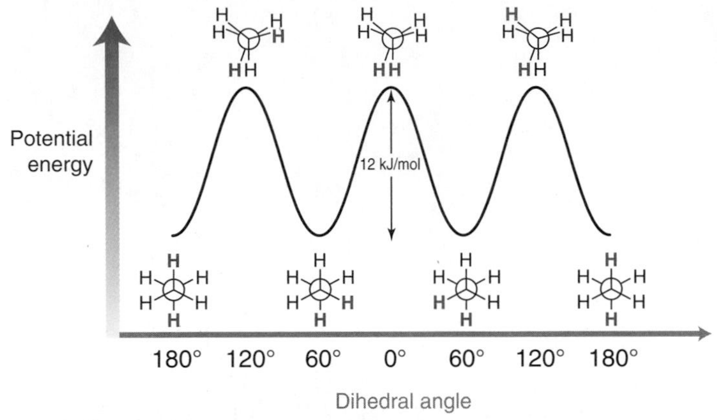

FIGURE 4.6
An energy diagram showing the conformational analysis of ethane.

LOOKING BACK
For a review of bonding and antibonding molecular orbitals, see Section 1.8.

The difference in energy between staggered and eclipsed conformations of ethane is referred to as **torsional strain**, and its cause has been somewhat debated over the years. Based on recent quantum mechanical calculations, it is now believed that the staggered conformation possesses a favorable interaction between an occupied, bonding MO and an unoccupied, antibonding MO (Figure 4.7).

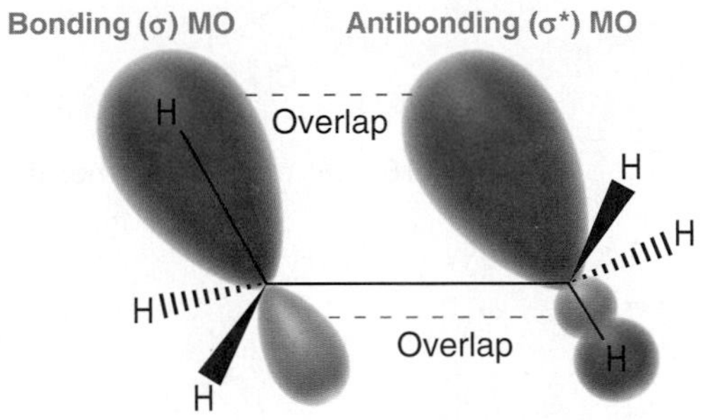

FIGURE 4.7
In the staggered conformation, favorable overlap occurs between a bonding MO and an antibonding MO.

This interaction lowers the energy of the staggered conformation. This favorable interaction is only present in the staggered conformation. When the C—C bond is rotated (going from a staggered to an eclipsed conformation), the favorable overlap above is temporarily disrupted, causing an increase in energy. In ethane, this increase amounts to 12 kJ/mol. Since there are three separate eclipsing interactions, it is reasonable to assign 4 kJ/mol to each pair of eclipsing H's (Figure 4.8).

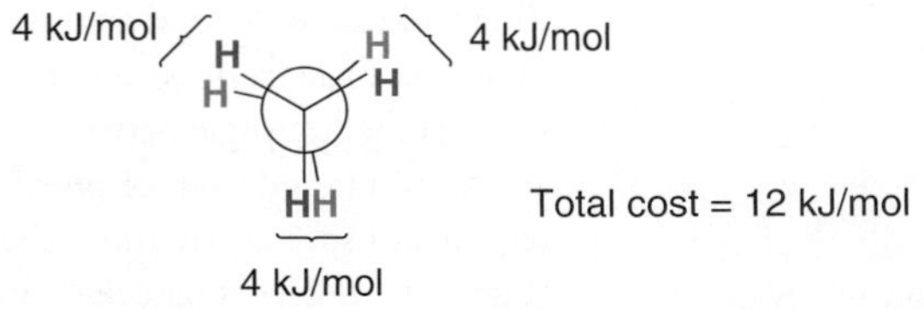

FIGURE 4.8
The total energy cost associated with the eclipsed conformation of ethane (relative to the staggered conformation) amounts to 12 kJ/mol.

This energy difference is significant. At room temperature, a sample of ethane gas will have approximately 99% of its molecules in staggered conformations at any given instant.

The energy diagram of propane (Figure 4.9) is very similar to that of ethane, except that the torsional strain is 14 kJ/mol rather than 12 kJ/mol. Once again, notice that all staggered conformations are degenerate, as are all eclipsed conformations.

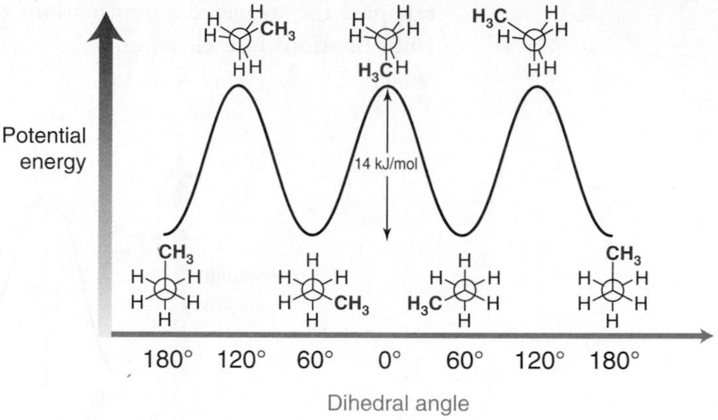
Potential energy
14 kJ/mol
180° 120° 60° 0° 60° 120° 180°
Dihedral angle

FIGURE 4.9
An energy diagram showing the conformational analysis of propane.

We already assigned 4 kJ/mol to each pair of eclipsing H's. If we know that the torsional strain of propane is 14 kJ/mol, then it is reasonable to assign 6 kJ/mol to the eclipsing of an H and a methyl group. This calculation is illustrated in Figure 4.10.

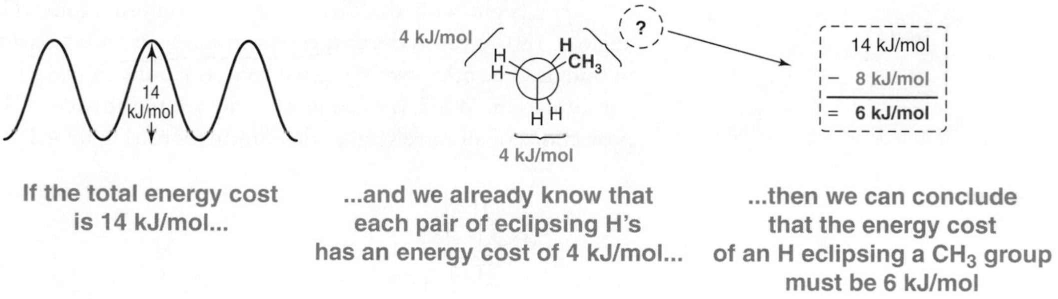
14 kJ/mol
4 kJ/mol
?
4 kJ/mol
14 kJ/mol
− 8 kJ/mol
= 6 kJ/mol
If the total energy cost is 14 kJ/mol...
...and we already know that each pair of eclipsing H's has an energy cost of 4 kJ/mol...
...then we can conclude that the energy cost of an H eclipsing a CH3 group must be 6 kJ/mol

FIGURE 4.10
The energy cost associated with a methyl group eclipsing a hydrogen atom amounts to 6 kJ/mol.

CONCEPTUAL CHECKPOINT

4.19 For each of the following compounds, predict the energy barrier to rotation (looking down any one of the C—C bonds). Draw a Newman projection and then compare the staggered and eclipsed conformations. Remember that we assigned 4 kJ/mol to each pair of eclipsing H's and 6 kJ/mol to an H eclipsing a methyl group:

(a) 2,2-Dimethylpropane (b) 2-Methylpropane

4.8 Conformational Analysis of Butane

Conformational analysis of butane is a bit more complex than the conformational analysis of either ethane or propane. Look carefully at the shape of the energy diagram for butane (Figure 4.11), and then we will analyze it step by step.

The three highest energy conformations are the eclipsed conformations, while the three lowest energy conformations are the staggered conformations. In this way, the energy diagram above is similar to the energy diagrams of ethane and propane. But in the case of butane, notice that one eclipsed conformation (where dihedral angle = 0) is higher in energy than the other two eclipsed conformations. In other words, the three eclipsed conformations are not degenerate. Similarly, one staggered conformation (where dihedral angle = 180) is lower in energy than the other two staggered conformations. Clearly, we need to compare the staggered conformations to each other, and we need to compare the eclipsed conformations to each other.

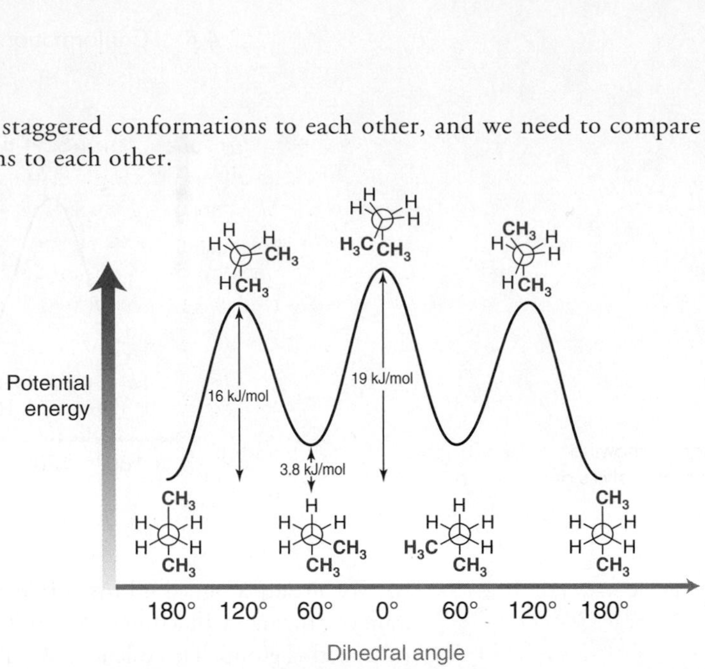

FIGURE 4.11 An energy diagram showing the conformational analysis of butane.

Let's begin with the three staggered conformations. The conformation with a dihedral angle of 180° is called the **anti conformation**, and it represents the lowest energy conformation of butane. The other two staggered conformations are 3.8 kJ/mol higher in energy than the anti conformation. Why? We can more easily see the answer to this question by drawing Newman projections of all three staggered conformations (Figure 4.12).

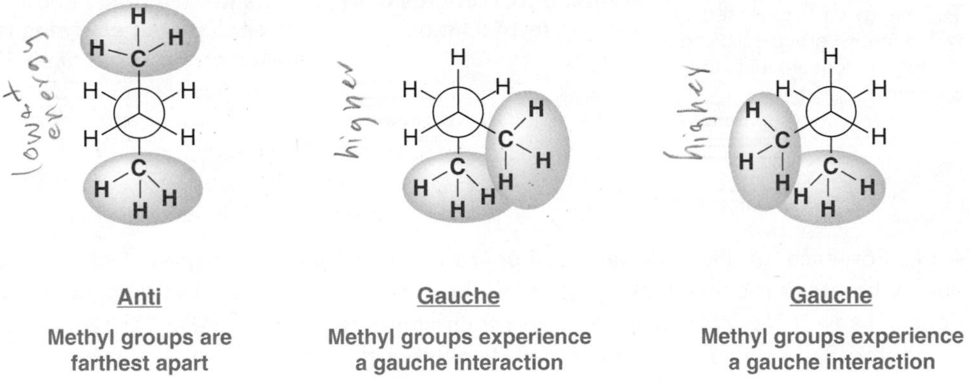

FIGURE 4.12 Two of the three staggered conformations of butane exhibit gauche interactions.

In the anti conformation, the methyl groups achieve maximum separation from each other. In the other two conformations, the methyl groups are closer to each other. Their electron clouds are repelling each other (trying to occupy the same region of space), causing an increase in energy of 3.8 kJ/mol. This unfavorable interaction, called a **gauche interaction**, is a form of steric hindrance, and it is different from the concept of torsional strain. The two conformations above that exhibit this interaction are called **gauche conformations**, and they are degenerate (Figure 4.13).

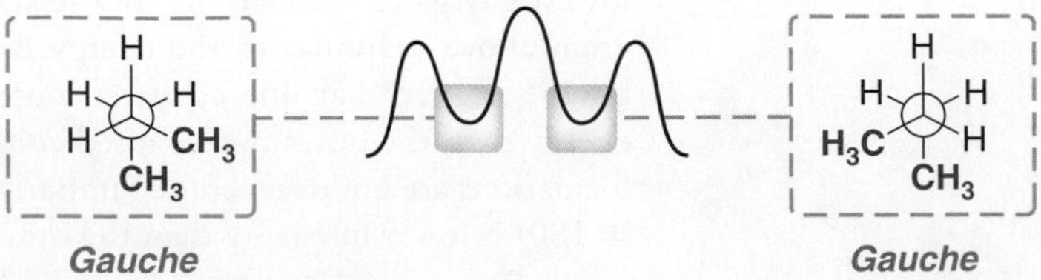

FIGURE 4.13 The two staggered conformations that exhibit gauche interactions are degenerate.

Now let's turn our attention to the three eclipsed conformations. One eclipsed conformation is higher in energy than the other two. Why? In the highest energy conformation, the methyl groups are eclipsing each other. Experiments suggest that this conformation has a total energy cost of 19 kJ/mol. Since we already assigned 4 kJ/mol to each H—H eclipsing interaction, it is reasonable to assign 11 kJ/mol to the eclipsing interaction of two methyl groups. This calculation is illustrated in Figure 4.14. The

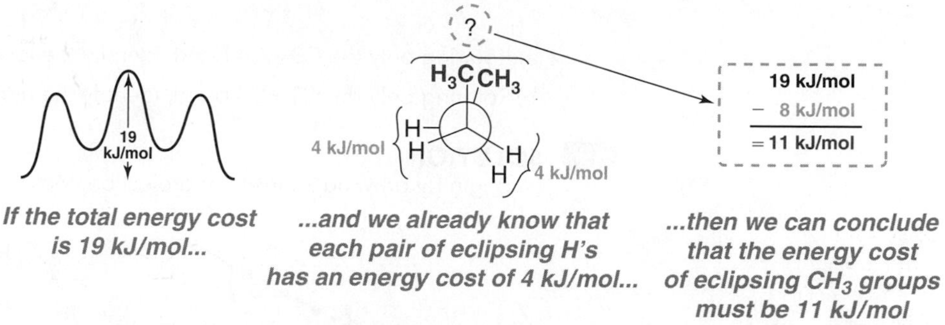

FIGURE 4.14
The energy cost associated with two methyl groups eclipsing each other amounts to 11 kJ/mol.

conformation with the two methyl groups eclipsing each other is the highest energy conformation. The other two eclipsed conformations are degenerate (Figure 4.15).

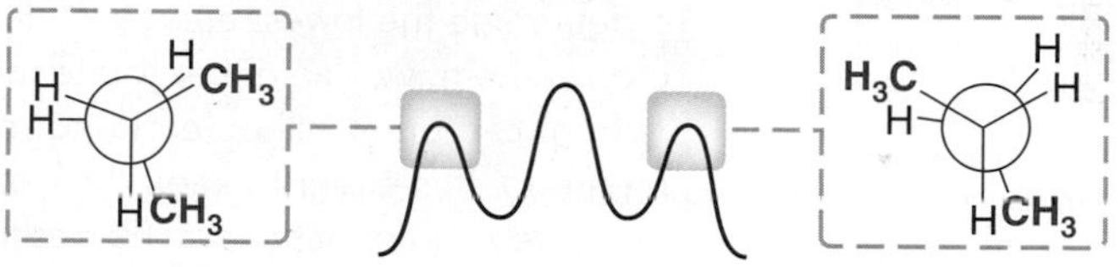

FIGURE 4.15
Two of the eclipsed conformations of butane are degenerate.

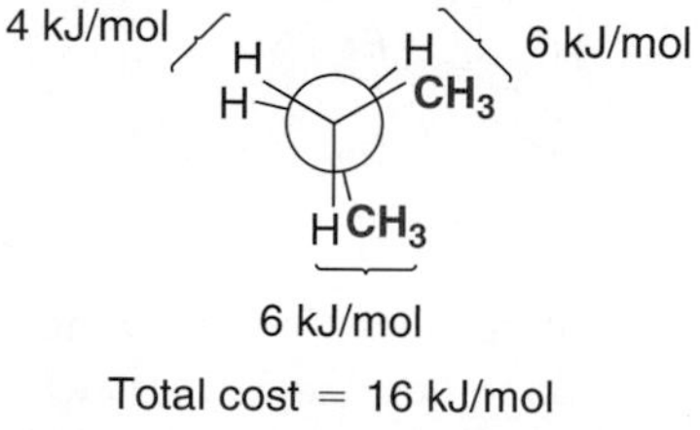

FIGURE 4.16
The total energy cost associated with the degenerate eclipsed conformations of butane amounts to 16 kJ/mol.

In each case, there is one pair of eclipsing H's and two pairs of eclipsing H/CH_3. We have all the information necessary to calculate the energy of these conformations. We know that eclipsing H's are 4 kJ/mol, and each set of eclipsing H/CH_3 is 6 kJ/mol. Therefore, we calculate a total energy cost of 16 kJ/mol (Figure 4.16).

To summarize, we have seen just a few numbers that can be helpful in analyzing energy costs. With these numbers, it is possible to analyze an eclipsed conformation or a staggered conformation and determine the energy cost associated with each conformation. Table 4.6 summarizes these numbers.

TABLE 4.6 ENERGY COSTS FOR COMPARING THE RELATIVE ENERGY OF CONFORMATIONS

INTERACTION	TYPE OF STRAIN	ENERGY COST (KJ/MOL)
 H/H Eclipsed	Torsional strain	4
 CH_3/H Eclipsed	Torsional strain	6
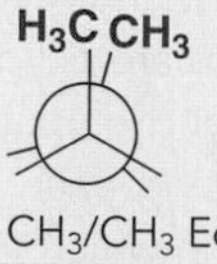 CH_3/CH_3 Eclipsed	Torsional strain + steric hindrance	11
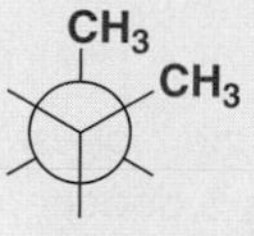 CH_3/CH_3 Gauche	Steric hindrance	3.8

SKILLBUILDER

4.8 IDENTIFYING RELATIVE ENERGY OF CONFORMATIONS

LEARN the skill

Consider the following compound:

(a) Rotating only the C3—C4 bond, identify the lowest energy conformation.

(b) Rotating only the C3—C4 bond, identify the highest energy conformation.

SOLUTION

STEP 1
Draw a Newman projection.

(a) Begin by drawing a Newman projection, looking along the C3—C4 bond:

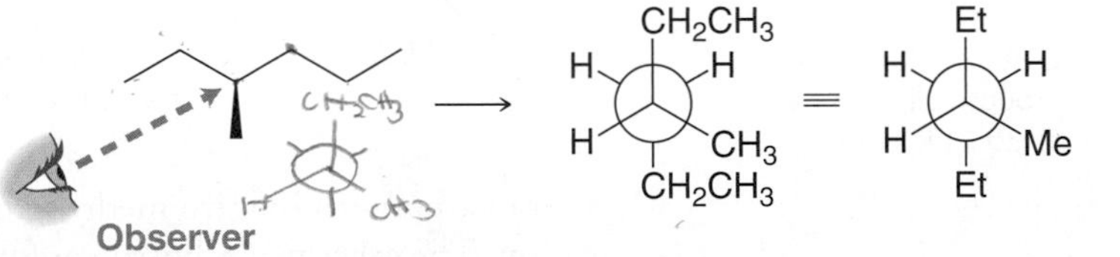

BY THE WAY
The symbol "Et" is commonly used for an ethyl group and the symbol "Me" is used for a methyl group.

To determine the lowest energy conformation, compare all three staggered conformations. To draw them, we can either rotate the groups on the back carbon or rotate the groups on the front carbon. It will be easier to rotate the back carbon since the back carbon has only one group. It is easier to keep track of only one group. Notice that the difference between these three conformations is the position of the ethyl group on the back carbon:

STEP 2
Compare all three staggered conformations. Look for the fewest or least severe gauche interactions.

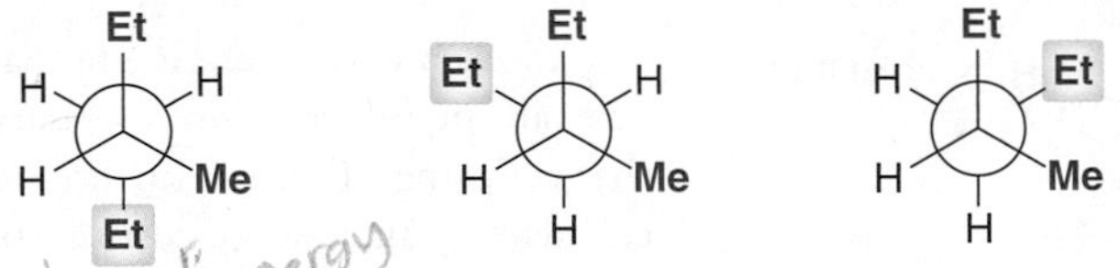

Now compare these three conformations by looking for gauche interactions. The first conformation has one Et/Me gauche interaction. The second conformation has one Et/Et gauche interaction. The third conformation has two gauche interactions: one Et/Et interaction and one Et/Me interaction. Choose the one with the fewest and least severe gauche interactions. The first conformation (with one Et/Me interaction) will be the lowest in energy.

(b) To determine the highest energy conformation, we will need to compare all three *eclipsed* conformations. To draw them, simply take the three staggered conformations and turn each of them into an eclipsed conformation by rotating the back carbon 60°. Notice once again that the difference between these three conformations is the position of the Et group on the back carbon:

STEP 3
Compare all three eclipsed conformations. Look for the highest energy interactions.

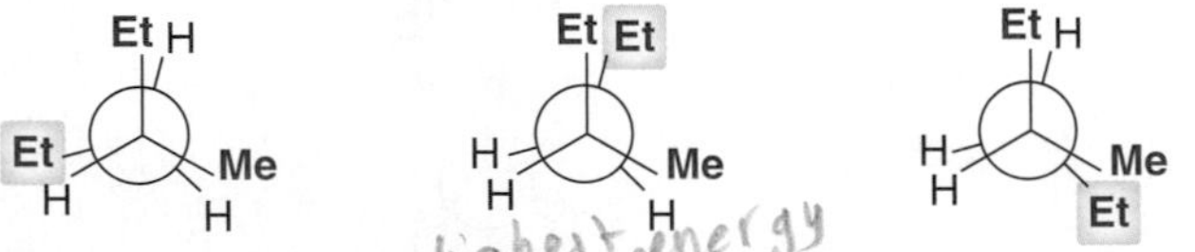

Now compare these three conformations by looking for eclipsing interactions. In the first conformation, none of the alkyl groups are eclipsing each other (they are all eclipsed by H's). In the second conformation, the ethyl groups are eclipsing each other. In the third conformation, a methyl and ethyl are eclipsing each other. Of the three possibilities, the highest energy conformation will be the one in which the two ethyl groups are eclipsing each other (the second conformation).

PRACTICE the skill

4.20 In each case below, identify the highest and lowest energy conformations. In cases where two or three conformations are degenerate, draw only one as your answer.

(a) 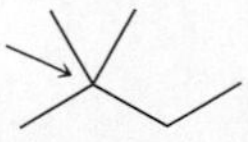 (b) 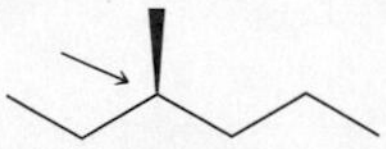 (c) 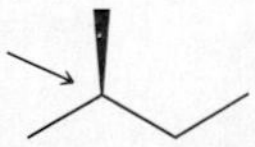 (d) 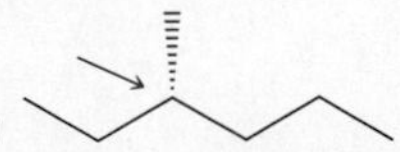

APPLY the skill

4.21 Compare the three staggered conformations of ethylene glycol. The anti conformation of ethylene glycol is not the lowest energy conformation. The other two, staggered conformations are actually lower in energy than the *anti* conformation. Suggest an explanation.

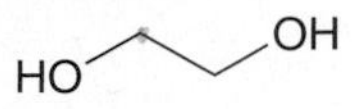

Ethylene glycol

need more **PRACTICE?** **Try Problems 4.47, 4.59**

MEDICALLYSPEAKING

Drugs and Their Conformations

Recall from Chapter 2 that a drug will bind with a biological receptor if the drug possesses a specific three-dimensional arrangement of functional groups, called a pharmacophore. For example, the pharmacophore of morphine is shown in red:

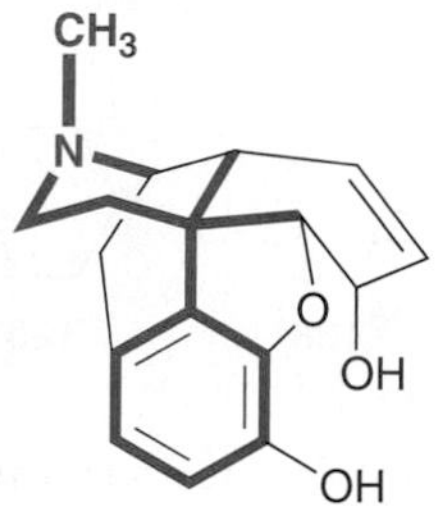

Morphine

Morphine is a very rigid molecule, because it has very few bonds that undergo free rotation. As a result, the pharmacophore is locked in place. In contrast, flexible molecules are capable of adopting a variety of conformations, and only some of those conformations can bind to the receptor. For example, methadone has many single bonds, each of which undergoes free rotation:

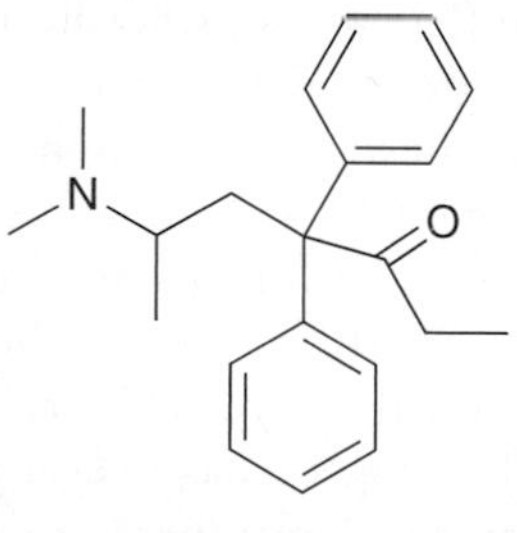

Methadone

Methadone is used to treat heroin addicts suffering from withdrawal symptoms. Methadone binds to the same receptor as heroin, and it is widely believed that the active conformation is the one in which the position of the functional groups matches the pharmacophore of heroin (and morphine):

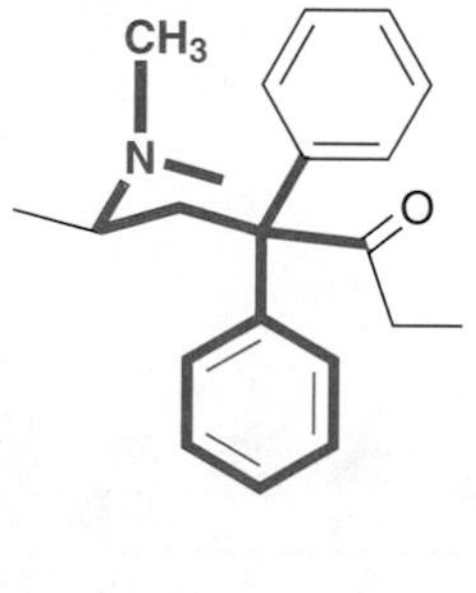

Methadone

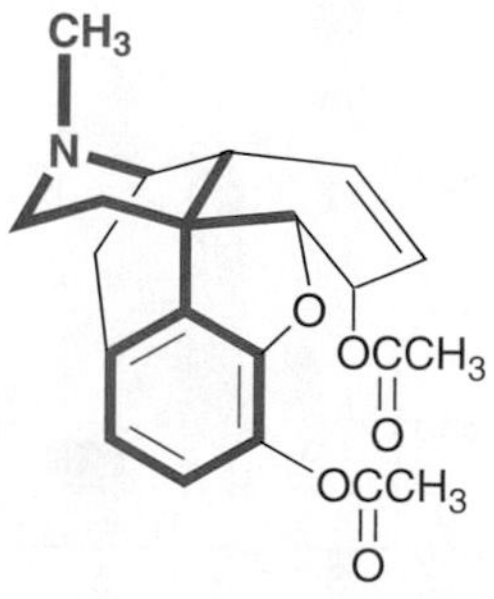

Heroin

Other, more open conformations of methadone are probably incapable of binding to the receptor.

This explains how it is possible for one drug to produce several physiological effects. In many cases, one conformation binds to one receptor, while another conformation binds to an entirely different receptor. Conformational flexibility is therefore an important consideration in the study of how drugs behave in our bodies.

As mentioned in the chapter opener, conformational flexibility has recently received much attention in the design of novel compounds to treat viral infections. To treat the symptoms of a virus, we must study the structure and behavior of that particular virus and then design drugs that interfere with the key steps in the replication process for that virus.

The vast majority of antiviral research has focused on designing drugs to treat those viral infections that are life threatening, such as HIV. Many anti-HIV drugs have been developed over the last few decades. Nevertheless, these drugs are not 100% effective, because HIV can undergo genetic mutations that effectively change the geometry of the cavity where the drugs are supposed to bind. The new strain of the virus is then drug resistant, because the drugs cannot bind with their intended receptors.

A new class of compounds, exhibiting conformational flexibility, appears to evade the problem of drug resistance. One such example, called rilpivirine, exhibits five single bonds whose rotation would lead to a conformational change:

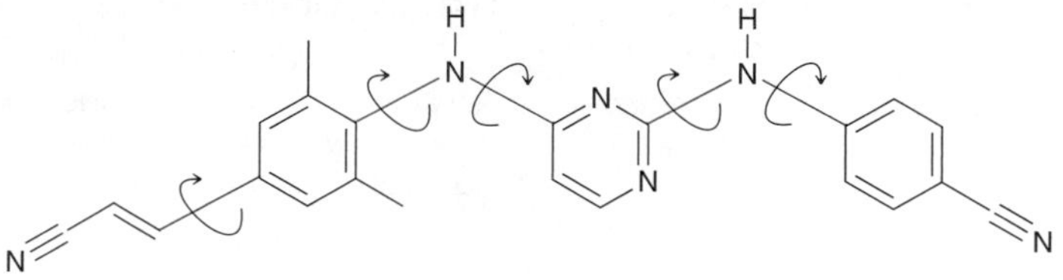

Rilpivirine

Bonds shown in red can undergo rotation without a significant energy cost, rendering the compound very flexible. Rilpivirine was recently tested as an anti-HIV drug in phase III clinical trials, which represent the last major hurdle in obtaining approval from the FDA (Food and Drug Administration). The results were extremely positive. The flexibility of rilpivirine enables it to bind to the desired receptor and tolerate changes to the geometry of the cavity resulting from virus mutation. In this way, rilpivirine makes it more difficult for the virus to develop resistance to it.

Compounds like rilpivirine have changed the way scientists approach drug design of antiviral agents. It is now clear that conformational flexibility plays an important role in the design of effective drugs.

4.9 Cycloalkanes

In the nineteenth century, chemists were aware of many compounds containing five-membered rings and six-membered rings, but no compounds with smaller rings were known. Many unfruitful attempts to synthesize smaller or larger rings fueled speculation regarding the feasibility of ever creating such compounds. Toward the end of the nineteenth century Adolph von Baeyer proposed a theory describing cycloalkanes in terms of **angle strain**, the increase in energy associated with a bond angle that has deviated from the preferred angle of 109.5°. Baeyer's theory was based on the angles found in geometric shapes (Figure 4.17). Baeyer reasoned that five-membered rings should contain almost no angle strain, while other rings would be strained (both smaller rings and larger rings). He also reasoned that very large cycloalkanes cannot exist, because the angle strain associated with such large bond angles would be prohibitive.

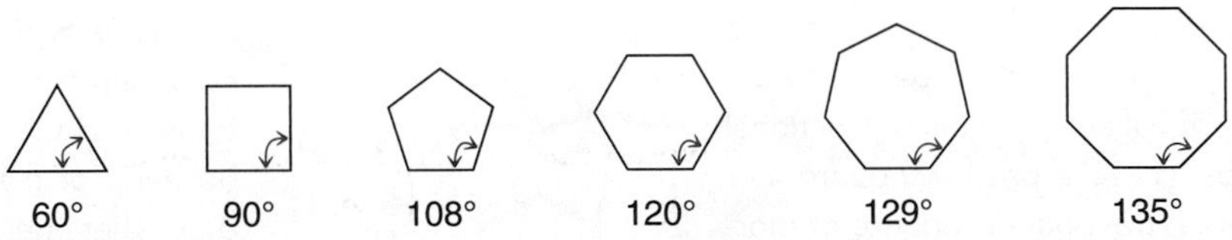

FIGURE 4.17
Bond angles found in geometric shapes.

Evidence refuting Baeyer's conclusions came from thermodynamic experiments. Recall from earlier in this chapter that heats of combustion can be used to compare isomeric compounds in terms of their total energy. It is not fair to compare the heats of combustion for rings of different sizes since heats of combustion are expected to increase with each additional CH_2 group. We can more accurately compare rings of different sizes by dividing the heat of combustion by the number of CH_2 groups in the compound, giving a heat of combustion per CH_2 group. Table 4.7 shows heats of combustion per CH_2 group for various ring sizes. The conclusions from these data are more easily seen when plotted (Figure 4.18). Notice that a six-membered ring is lower in energy than a five-membered ring, in contrast with Baeyer's theory. In addition, the relative energy level does not increase with increasing ring size, as Baeyer predicted. A 12-membered ring is in fact much lower in energy than an 11-membered ring.

TABLE 4.7 HEATS OF COMBUSTION PER CH_2 GROUP FOR CYCLOALKANES

CYCLOALKANE	NUMBER OF CH_2 GROUPS	HEAT OF COMBUSTION (KJ / MOL)	HEAT OF COMBUSTION PER CH_2 GROUP (KJ / MOL)
Cyclopropane	3	2091	697
Cyclobutane	4	2721	680
Cyclopentane	5	3291	658
Cyclohexane	6	3920	653
Cycloheptane	7	4599	657
Cyclooctane	8	5267	658
Cyclononane	9	5933	659
Cyclodecane	10	6587	659
Cycloundecane	11	7273	661
Cyclododecane	12	7845	654

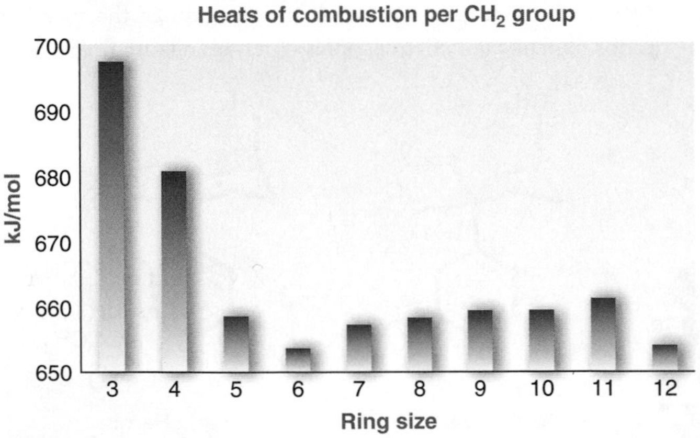

FIGURE 4.18
Heats of combustion per CH_2 group for cycloalkanes.

Baeyer's conclusions did not hold because they were based on the incorrect assumption that cycloalkanes are planar, like the geometric shapes shown earlier. In reality, the bonds of a larger cycloalkane can position themselves three dimensionally so as to achieve a conformation that minimizes the total energy of the compound. We will soon see that angle strain is only one factor that contributes to the energy of a cycloalkane. We will now explore the main factors contributing to the energy of various ring sizes, starting with cyclopropane.

Cyclopropane

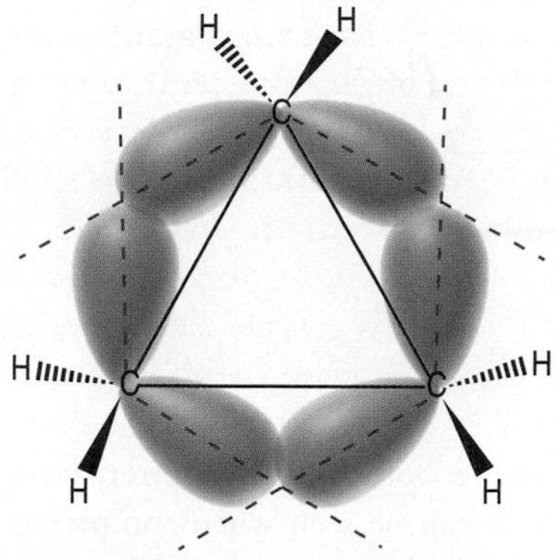

FIGURE 4.19
The C—C bonds of cyclopropane bend outward (on the dotted red lines) to alleviate some of the angle strain.

The angle strain in cyclopropane is severe. Some of this strain can be alleviated if the orbitals making up the bonds bend outward, as in Figure 4.19. Not all of the angle strain is removed, however, because there is an increase in energy associated with inefficient overlap of the orbitals. Although some of the angle strain is reduced, cyclopropane still has significant angle strain.

In addition, cyclopropane also exhibits significant torsional strain, which can best be seen in a Newman projection:

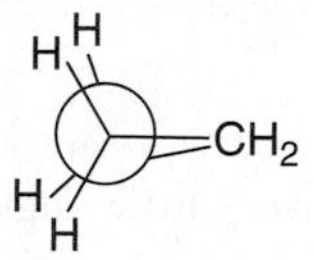

Notice that the ring is locked in an eclipsed conformation, with no possible way of achieving a staggered conformation.

In summary, cyclopropane has two main factors contributing to its high energy: angle strain (from small bond angles) and torsional strain (from eclipsing H's). This large amount of strain makes three-membered rings highly reactive and very susceptible to ring-opening reactions. In Chapter 14, we will explore many ring-opening reactions of a special class of three-membered rings called epoxides:

An epoxide

Cyclobutane

Cyclobutane has less angle strain than cyclopropane. However, it has more torsional strain, because there are four sets of eclipsing H's rather than just three. To alleviate some of this additional torsional strain, cyclobutane can adopt a slightly puckered conformation without gaining too much angle strain:

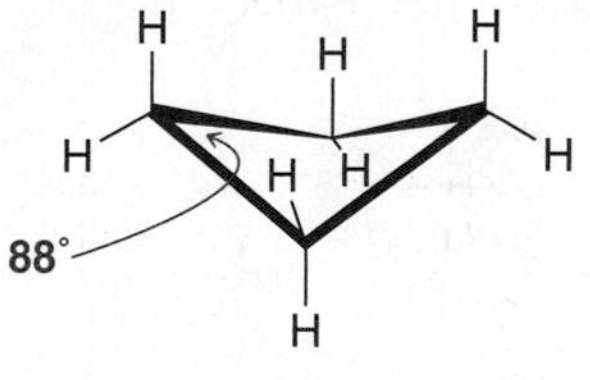

Cyclopentane

Cyclopentane has much less angle strain than cyclobutane or cyclopropane. It can also reduce much of its torsional strain by adopting the following conformation:

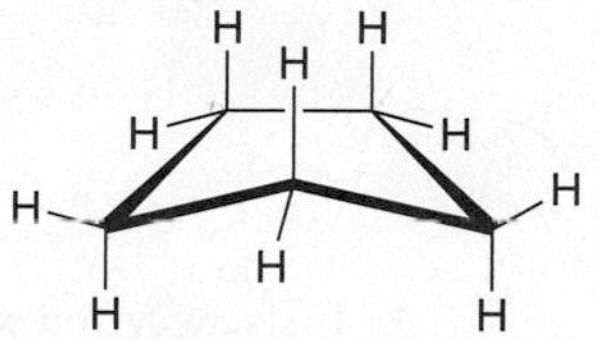

In total, cyclopentane has much less total strain than cyclopropane or cyclobutane. Nevertheless, cyclopentane does exhibit some strain. This is in contrast with cyclohexane, which can adopt a conformation that is nearly strain free. We will spend the remainder of the chapter discussing conformations of cyclohexane.

4.10 Conformations of Cyclohexane

Cyclohexane can adopt many conformations, as we will soon see. For now, we will explore two conformations: the **chair conformation** and the **boat conformation** (Figure 4.20). In both con-

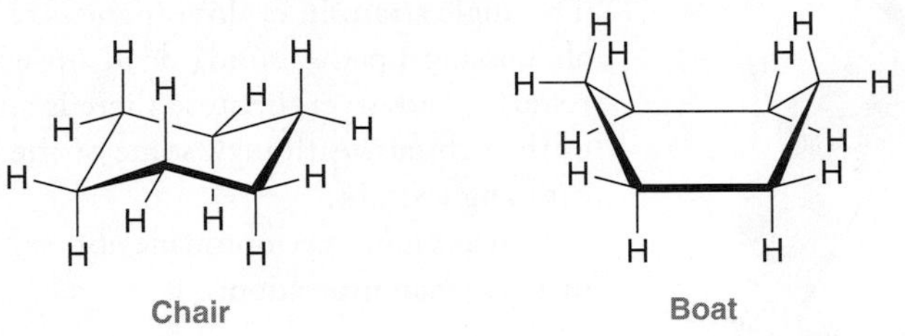

FIGURE 4.20
The chair and boat conformations of cyclohexane.

formations, the bond angles are fairly close to 109.5°, and therefore, both conformations possess very little angle strain. The significant difference between them can be seen when comparing torsional strain. The chair conformation has no torsional strain. This can best be seen with a Newman projection (Figure 4.21). Notice that all H's are staggered. None are eclipsed. This

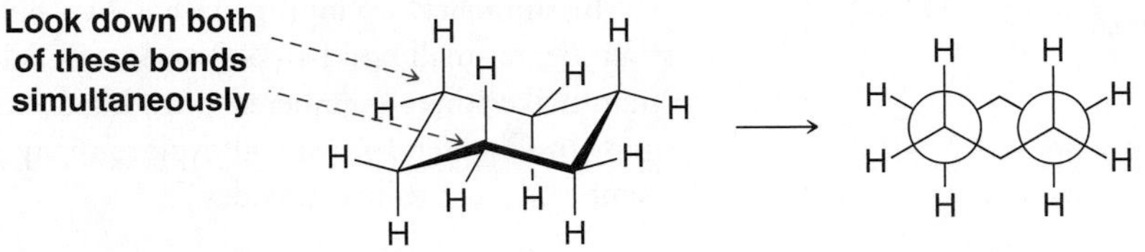

FIGURE 4.21
A Newman projection of cyclohexane in a chair conformation.

is not the case in a boat conformation, which has two sources of torsional strain (Figure 4.22). Many of the H's are eclipsed (Figure 4.22a), and the H's on either side of the ring experience steric interactions called **flagpole interactions,** as shown in Figure 4.22 b. The boat can allevi-

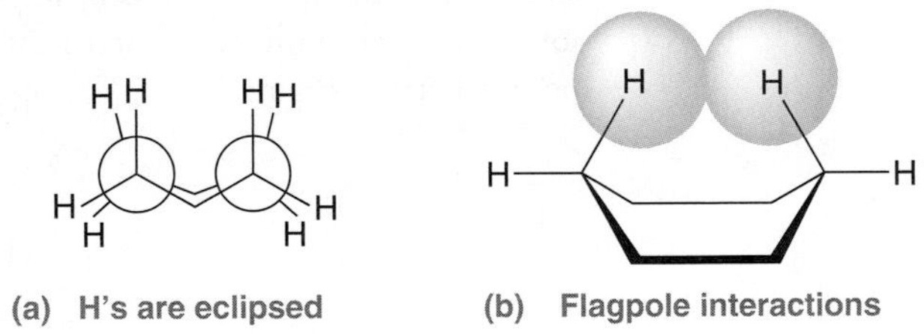

FIGURE 4.22
(a) A Newman projection of cyclohexane in a boat conformation. (b) Flagpole interactions in the boat conformation.

ate some of this torsional strain by twisting (very much the way cyclobutane puckers to alleviate some of its torsional strain), giving a conformation called a **twist boat** (Figure 4.23).

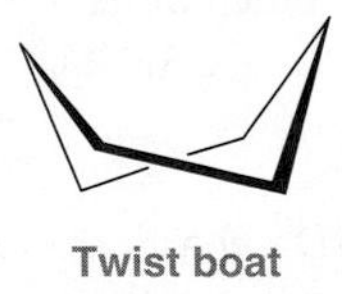

FIGURE 4.23
The twist boat conformation of cyclohexane.

In fact, cyclohexane can adopt many different conformations, but the most important is the chair conformation. There are actually two different chair conformations that rapidly interchange via a pathway that passes through many different conformations, including a high-energy half-chair conformation, as well as twist boat and boat conformations. This is illustrated in Figure 4.24, which is an energy diagram summarizing the relative energy levels of the various conformations of cyclohexane.

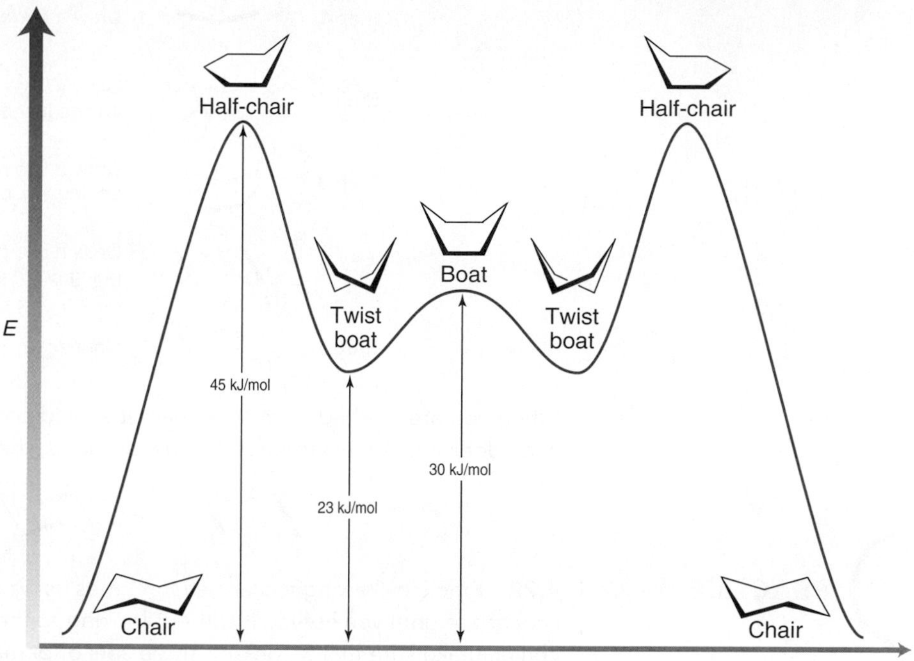

FIGURE 4.24
An energy diagram showing the conformational analysis of cyclohexane.

The lowest energy conformations are the two chair conformations, and therefore, cyclohexane will spend the majority of its time in a chair conformation. Accordingly, the remainder of our treatment of cyclohexane will focus on chair conformations. Our first step is to master drawing them.

4.11 Drawing Chair Conformations

When drawing a chair conformation, it is important to draw it precisely. Make sure that you avoid drawing sloppy chairs, because it will be difficult to draw the substituents correctly if the skeleton is not precise.

Drawing the Skeleton of a Chair Conformation

Let's get some practice drawing chairs.

SKILLBUILDER

4.9 DRAWING A CHAIR CONFORMATION

LEARN the skill

Draw a chair conformation of cyclohexane:

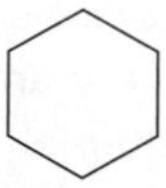

SOLUTION

The following procedure outlines a step-by-step method for drawing the skeleton of a chair conformation precisely:

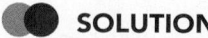

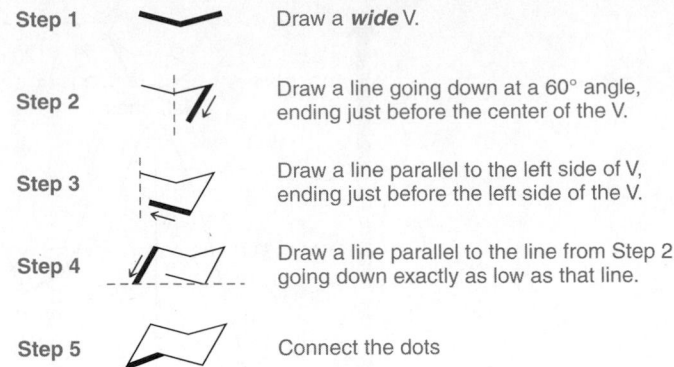

Step 1  Draw a ***wide*** V.

Step 2  Draw a line going down at a 60° angle, ending just before the center of the V.

Step 3 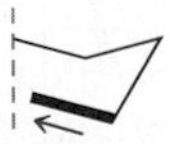 Draw a line parallel to the left side of V, ending just before the left side of the V.

Step 4  Draw a line parallel to the line from Step 2, going down exactly as low as that line.

Step 5  Connect the dots

When you are finished drawing a chair, it should contain three sets of parallel lines. If your chair does not contain three sets of parallel lines, then it has been drawn incorrectly.

PRACTICE the skill **4.22** Practice drawing a chair several times using a blank piece of paper. Repeat the procedure until you can do it without looking at the instructions above. For each of your chairs, make sure that it contains three sets of parallel lines.

APPLY the skill **4.23** Draw a chair conformation for each of the following compounds:

(a) 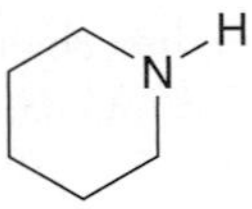

(b) 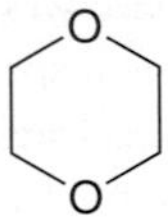

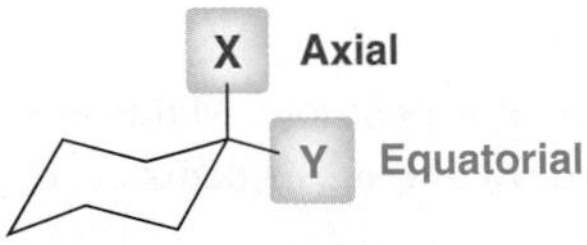

FIGURE 4.25 Axial and equatorial positions in a chair conformation.

Drawing Axial and Equatorial Substituents

Each carbon atom in a cyclohexane ring can bear two substituents (Figure 4.25). One group is said to occupy an **axial position**, which is parallel to a vertical axis passing through the center of the ring. The other group is said to occupy an **equatorial position**, which is positioned approximately along the equator of the ring (Figure 4.25). In order to draw a substituted cyclohexane, we must first practice drawing all axial and equatorial positions properly.

SKILLBUILDER

4.10 DRAWING AXIAL AND EQUATORIAL POSITIONS

LEARN the skill Draw all axial and all equatorial positions on a chair conformation of cyclohexane.

SOLUTION

Let's begin with the axial positions, as they are easier to draw. Begin at the right side of the V and draw a vertical line pointing up. Then, go around the ring, drawing vertical lines, alternating in direction (up, down, up, etc.):

STEP 1 Draw all axial positions as vertical lines alternating in direction.

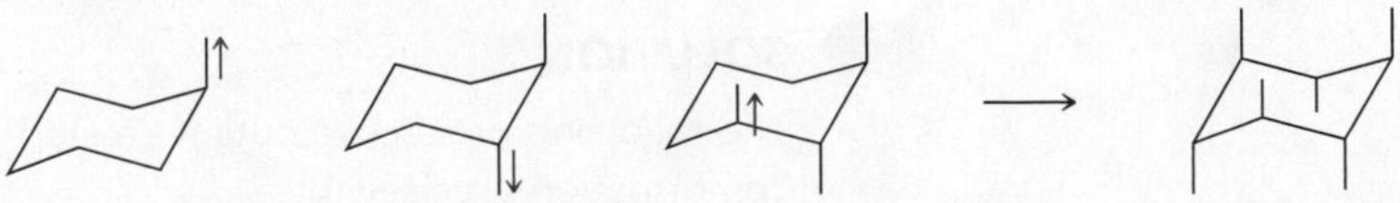

These are the six axial positions. All six lines are vertical.

Now let's draw the six equatorial positions. The equatorial positions are more difficult to draw properly, but mistakes can be avoided in the following way. We saw earlier that a properly drawn chair skeleton is composed of three pairs of parallel lines:

STEP 2
Draw all equatorial positions as pairs of parallel lines.

Now we will use these pairs of parallel lines to draw the equatorial positions. In between each pair of red lines above, draw two equatorial groups that are parallel to (but not directly touching) the red lines:

Notice that all equatorial positions are drawn going to the outside of, or away from, the ring; not going into the ring. Now let's summarize by drawing all six axial positions and all six equatorial positions:

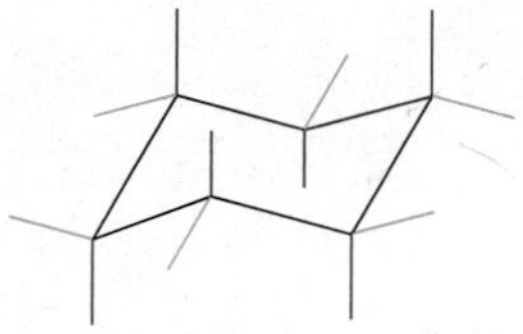

PRACTICE the skill

4.24 Practice drawing a chair conformation with all six axial positions. Repeat until you can draw all six positions without looking at the instructions above.

4.25 Practice drawing a chair conformation with all six equatorial positions. Repeat until you can draw all six positions without looking at the instructions above.

4.26 Practice drawing a chair conformation with all 12 positions (6 axial and 6 equatorial). Practice several times on a blank piece of paper. Repeat until you can draw all 12 positions without looking at the instructions above.

APPLY the skill

4.27 In the following compound, identify the number of hydrogen atoms that occupy axial positions as well as the number of hydrogen atoms that occupy equatorial positions:

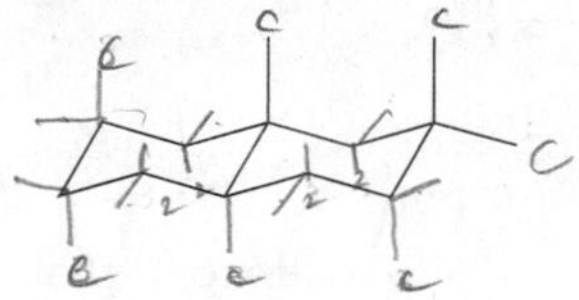

need more **PRACTICE?** **Try Problem 4.62**

4.12 Monosubstituted Cyclohexane

Drawing Both Chair Conformations

Consider a ring containing only one substituent. Two possible chair conformations can be drawn: The substituent can be in an axial position or in an equatorial position. These two possibilities represent two different conformations that are in equilibrium with each other:

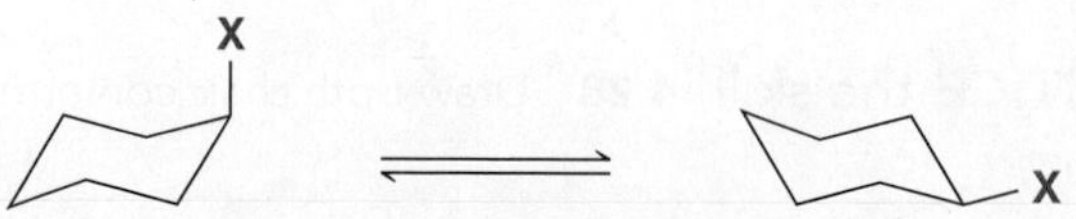

The term "ring flip" is used to describe the conversion of one chair conformation into the other. This process is not accomplished by simply flipping the molecule like a pancake.

Rather, a **ring flip** is a conformational change that is accomplished only through a rotation of all C—C single bonds. This can be seen with a Newman projection (Figure 4.26). Let's get some practice drawing ring flips.

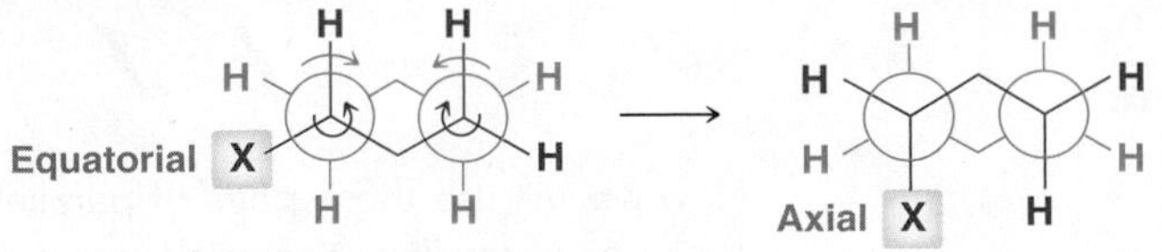

FIGURE 4.26
A ring flip drawn with Newman projections.

SKILLBUILDER

4.11 DRAWING BOTH CHAIR CONFORMATIONS OF A MONOSUBSTITUTED CYCLOHEXANE

LEARN the skill

Draw both chair conformations of bromocyclohexane:

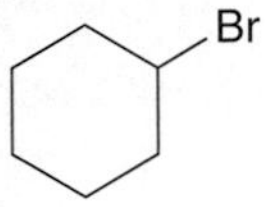

SOLUTION

STEP 1
Draw a chair conformation.

STEP 2
Place the substituent.

Begin by drawing the first chair conformation. Then, place the bromine in any position:

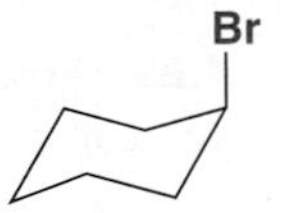

In this chair conformation, the bromine occupies an axial position. In order to draw the other chair conformation (the result of a ring flip), redraw the skeleton of the chair. Only this time, we will draw the skeleton differently. The first chair was drawn by following these steps:

The second chair must now be drawn by ***mirroring*** these steps:

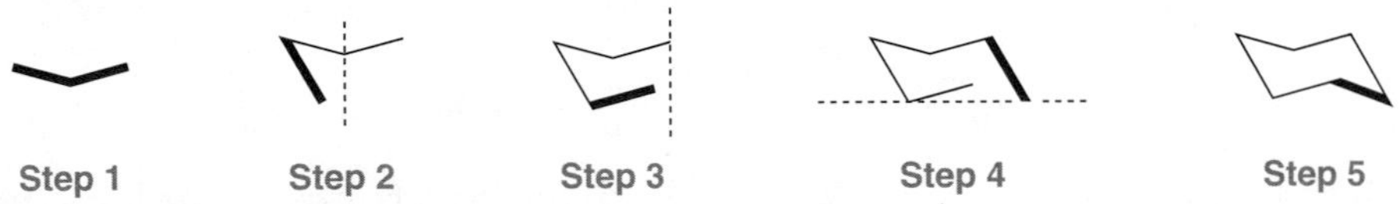

STEP 3
Draw a ring flip, and the axial group should become equatorial.

In the second chair, the bromine will occupy an equatorial position:

PRACTICE the skill **4.28** Draw both chair conformations for each of the following compounds:

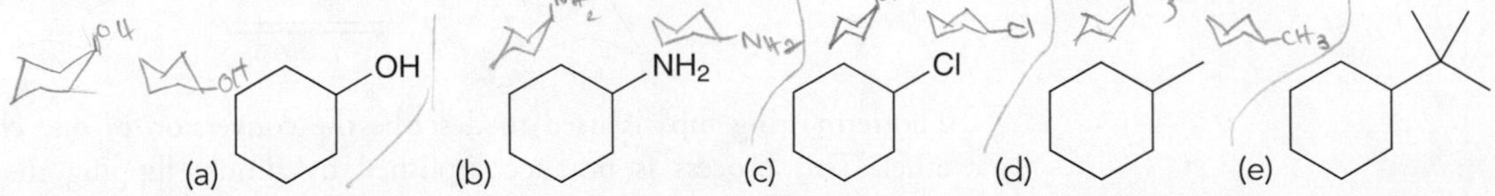

APPLY the skill

4.29 Consider the following chair conformation of bromocyclohexane:

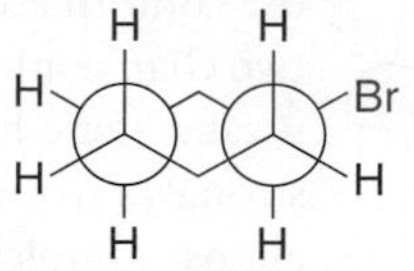

(a) Identify whether the bromine atom occupies an axial position or an equatorial position in the conformation above.

(b) Draw a bond-line drawing of this chair conformation (without Newman projections).

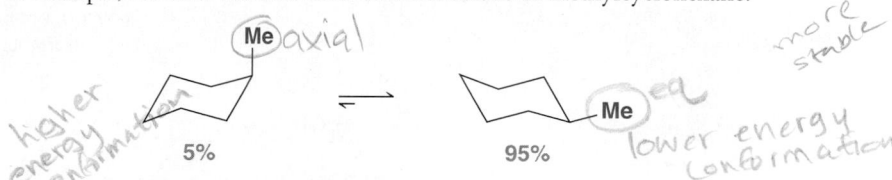

(c) Draw a bond-line drawing of the other chair conformation (after a ring flip).

need more **PRACTICE?** **Try Problem 4.54a**

Comparing the Stability of Both Chair Conformations

When two chair conformations are in equilibrium, the lower energy conformation will be favored. For example, consider the two chair conformations of methylcyclohexane.

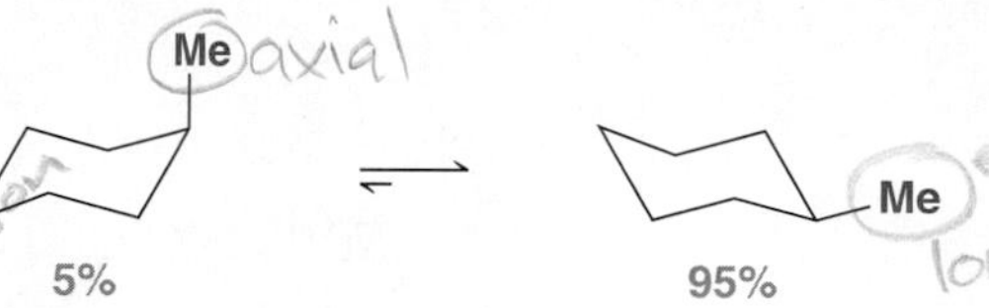

At room temperature, 95% of the molecules will be in the chair conformation that has the methyl group in an equatorial position. This must therefore be the lower energy conformation, but why? When the substituent is in an axial position, there are steric interactions with the other axial H's on the same side of the ring (Figure 4.27).

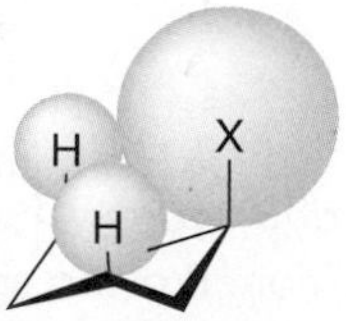

FIGURE 4.27 Steric interactions that occur when a substituent occupies an axial position.

The substituent's electron cloud is trying to occupy the same region of space as the H's that are highlighted, causing steric hindrance. These interactions are called **1,3-diaxial interactions**, where the numbers "1,3" describe the distance between the substituent and each of the H's. When the chair conformation is drawn in a Newman projection, it becomes clear that 1,3-diaxial interactions are nothing more than gauche interactions. Compare the gauche interaction in butane with one of the 1,3-diaxial interactions in methylcyclohexane (Figure 4.28).

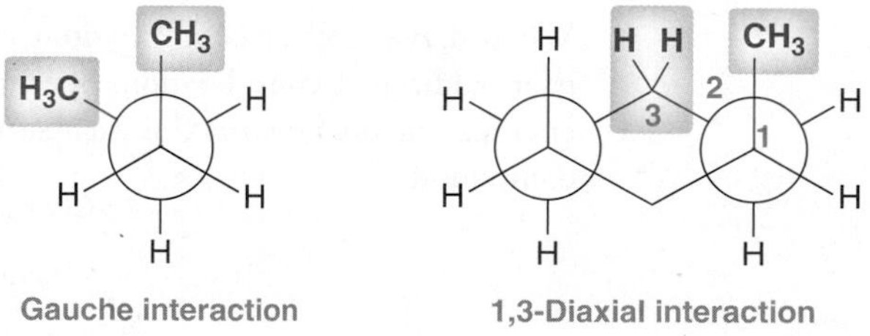

FIGURE 4.28 An illustration showing that 1,3-diaxial interactions are really just gauche interactions.

The presence of 1,3-diaxial interactions causes the chair conformation to be higher in energy when the substituent is in an axial position. In contrast, when the substituent is in an equatorial position, these 1,3-diaxial (gauche) interactions are absent (Figure 4.29).

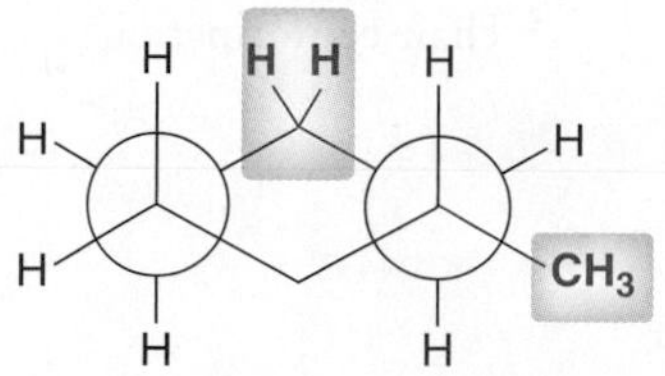

FIGURE 4.29 When a substituent is in an equatorial position, it experiences no gauche interactions.

For this reason, the equilibrium between the two chair conformations will generally favor the conformation with the equatorial substituent. The exact equilibrium concentrations of the two chair conformations will depend on the size of the substituent. Larger groups will experience greater steric hindrance resulting from 1,3-diaxial interactions, and the equilibrium will more strongly favor the equatorial substituent. For example, the equilibrium of *tert*-butylcyclohexane almost completely favors the chair conformation with an equatorial *tert*-butyl group:

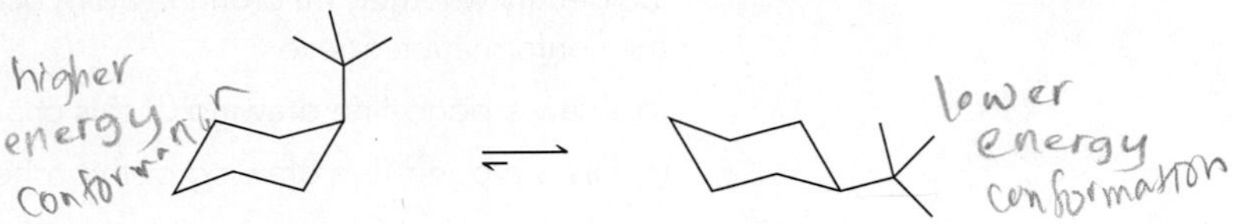

0.01% 99.99%

Table 4.8 shows the steric hindrance associated with various groups as well as the equilibrium concentrations that are achieved.

TABLE 4.8 1.3-DIAXIAL INTERACTIONS FOR SEVERAL COMMON SUBSTITUENTS

SUBSTITUENT	STERIC HINDRANCE FROM 1,3-DIAXIAL INTERACTIONS (KJ/MOL)	AXIAL-EQUATORIAL RATIO (AT EQUILIBRIUM)
—Cl	2.0	70 : 30
—OH	4.2	83 : 17
$—CH_3$	7.6	95 : 5
$—CH_2CH_3$	8.0	96 : 4
$—CH(CH_3)_2$	9.2	97 : 3
$—C(CH_3)_3$	22.8	9999 : 1

CONCEPTUAL CHECKPOINT

4.30 The most stable conformation of 5-hydroxy-1,3-dioxane has the OH group in an axial position, rather than an equatorial position. Provide an explanation for this observation.

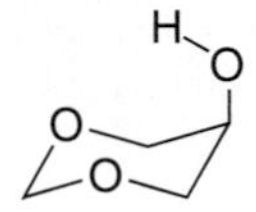

4.13 Disubstituted Cyclohexane

Drawing Both Chair Conformations

When drawing chair conformations of a compound that has two or more substituents, there is an additional consideration. Specifically, we must also consider the three-dimensional orientation, or *configuration*, of each substituent. To illustrate this point, consider the following compound:

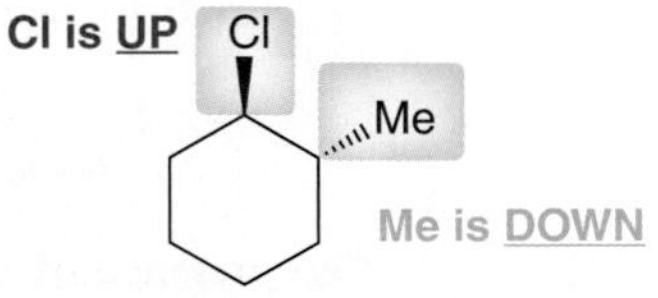

Notice that the chlorine atom is on a wedge, which means that it is coming out of the page: it is UP. The methyl group is on a dash, which means that it is below the ring, or DOWN. The two chair conformations for this compound are as follows:

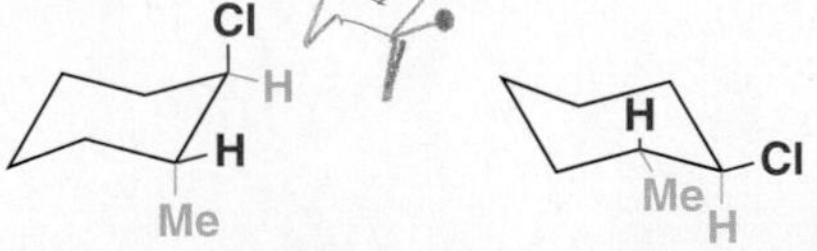

Notice that the chlorine atom is above the ring (UP) in both chair conformations, and the methyl group is below the ring (DOWN) in both chair conformations. The configuration (i.e., UP or DOWN) does not change during a ring flip. It is true that the chlorine is axial in one conformation and equatorial in the other conformation, but a ring flip does not change configuration. The chlorine atom must be UP in both chair conformations. Similarly, the methyl group must be DOWN in both chair conformations. Let's get some practice using these new descriptors (UP and DOWN) when drawing chair conformations.

SKILLBUILDER

4.12 DRAWING BOTH CHAIR CONFORMATIONS OF DISUBSTITUTED CYCLOHEXANES

LEARN the skill

Draw both chair conformations of the following compound:

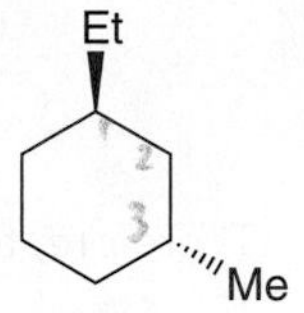

SOLUTION

Begin by numbering the ring and identifying the location and three-dimensional orientation of each substituent:

STEP 1
Determine the location and configuration of each substituent.

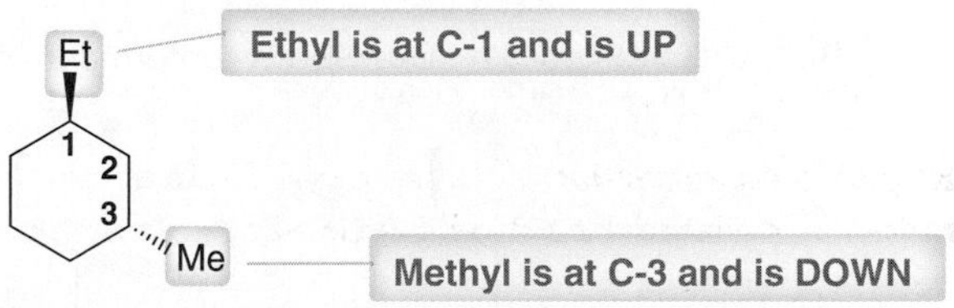

This numbering system does not need to be in accordance with IUPAC rules. It does not matter where the numbers are placed; these numbers are just tools used to compare positions in the original drawing and in the chair conformation to ensure that all substituents are placed correctly. The numbers can be placed either clockwise or counterclockwise, but they must be consistent. If the numbers are placed clockwise in the compound, then they must be placed clockwise as well when drawing the chair.

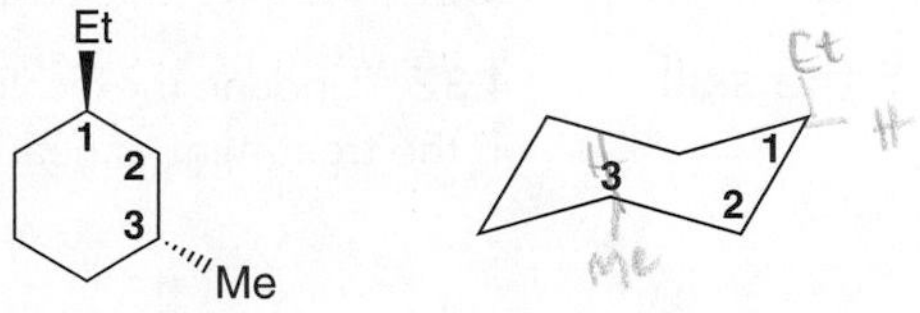

Once the numbers have been assigned, place the substituents in the correct locations and with the correct configuration. Ethyl is at C-1 and must be UP, while the methyl group is at C-3 and must be DOWN:

STEP 2
Place the substituents on the first chair using the information from step 1.

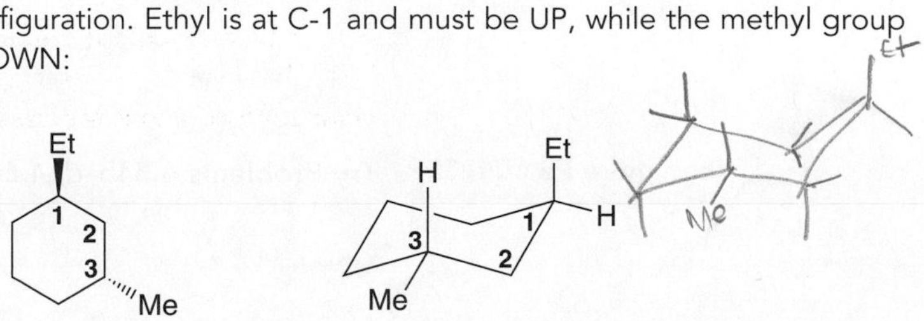

Notice that it is not possible to draw substituents properly without being able to draw all 12 positions on the cyclohexane ring. If you do not feel comfortable drawing all 12 positions, it would be a worthwhile investment of time to go back to that section of the chapter and practice. The drawing above represents the first chair conformation. In order to draw the second chair conformation, begin by drawing the other skeleton and numbering it:

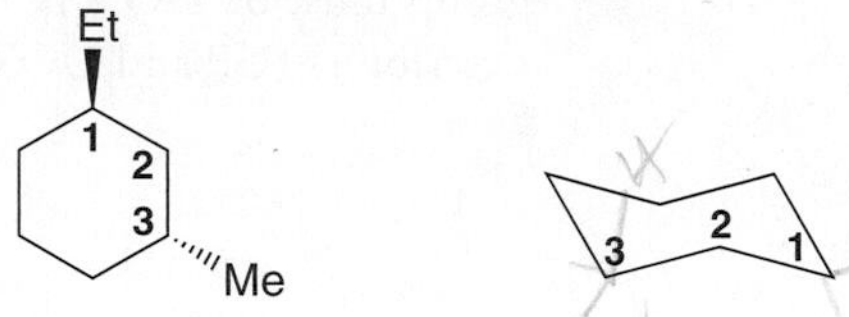

Then, once again, place the substituents so that the ethyl is at C-1 and is UP, while the methyl is at C-3 and is DOWN:

STEP 3
Place the substituents on the second chair using the information from step 1.

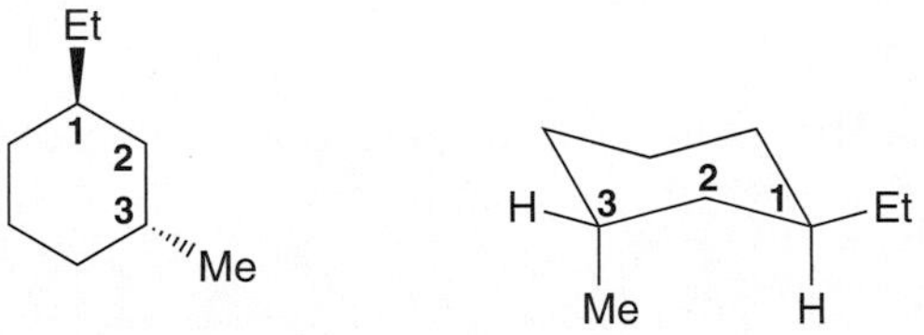

Therefore, the two chair conformations of this compound are:

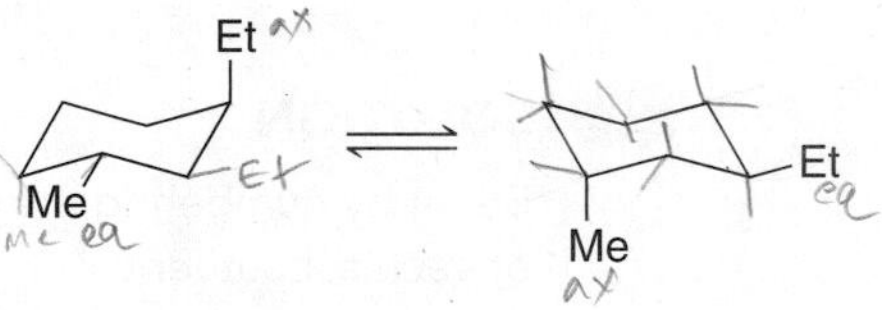

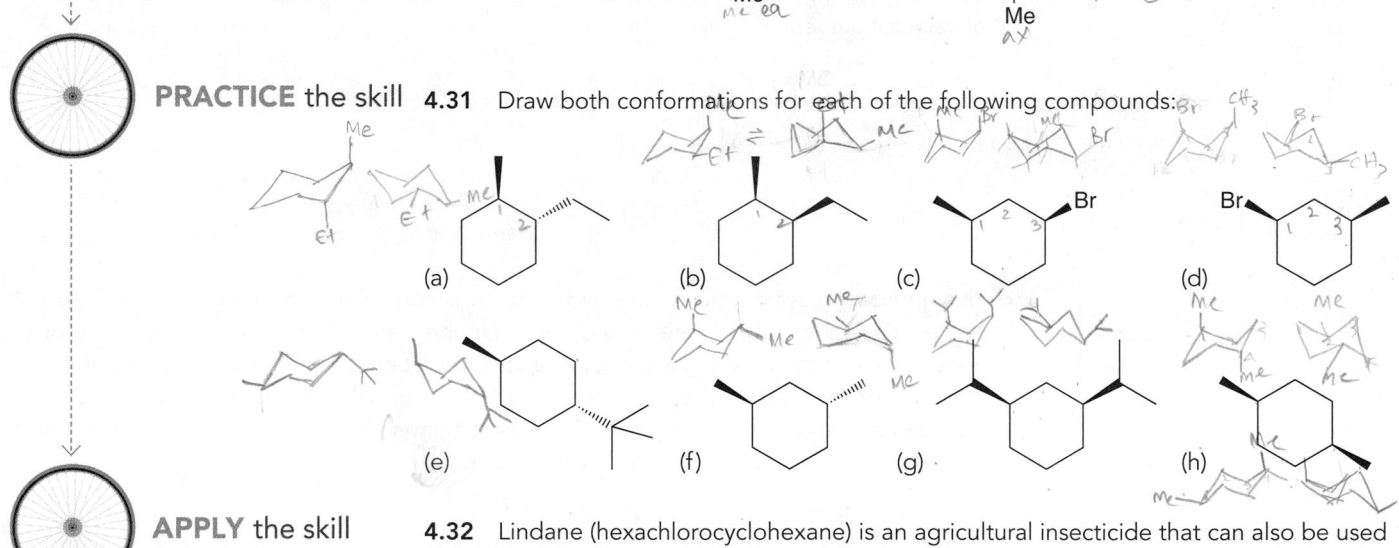

PRACTICE the skill **4.31** Draw both conformations for each of the following compounds:

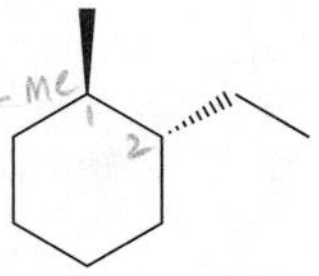
(a)

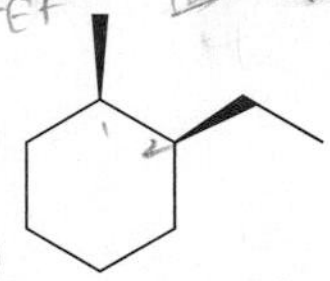
(b)

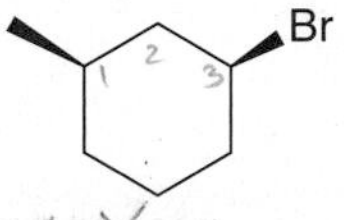

(c)

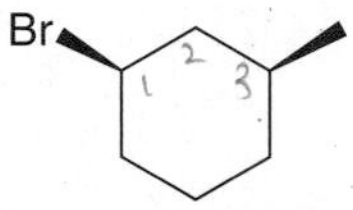

(d)

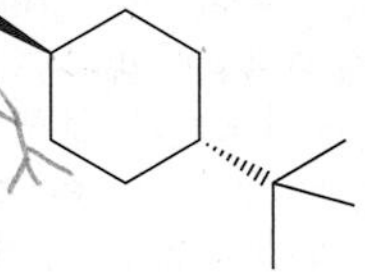
(e)

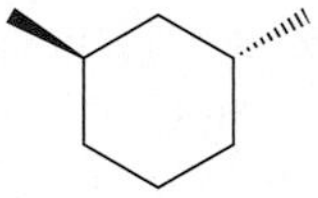
(f)

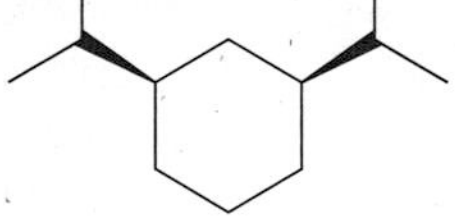
(g)

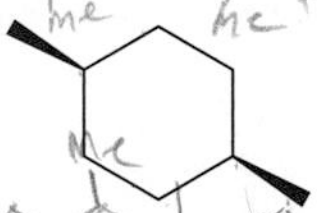
(h)

APPLY the skill **4.32** Lindane (hexachlorocyclohexane) is an agricultural insecticide that can also be used in the treatment of head lice. Draw both chair conformations of lindane.

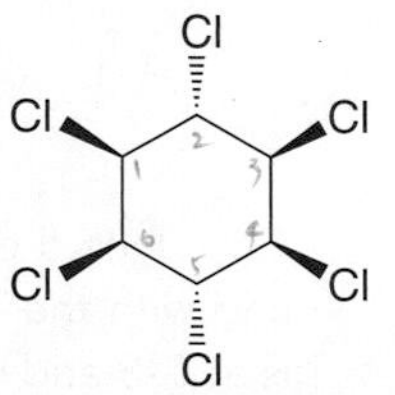

Lindane

need more **PRACTICE?** **Try Problems 4.54b–d, 4.66g**

Comparing the Stability of Chair Conformations

Let's compare the stability of chair conformations once again, this time for compounds that bear more than one substituent. Consider the following example:

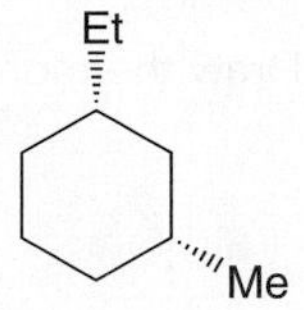

The two chair conformations of this compound are:

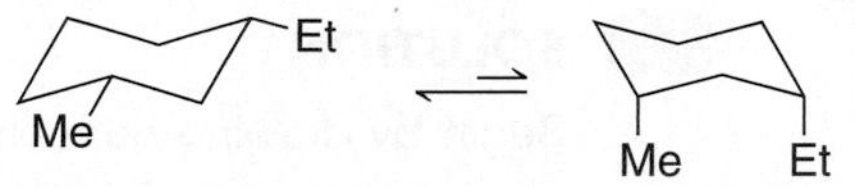

In the first conformation, both groups are equatorial. In the second conformation, both groups are axial. In the previous section, we saw that chair conformations will be lower in energy when substituents are in equatorial positions (avoiding 1,3-diaxial interactions). Therefore, the first chair will certainly be more stable.

In some cases, two groups might be in competition with each other. For example, consider the following compound:

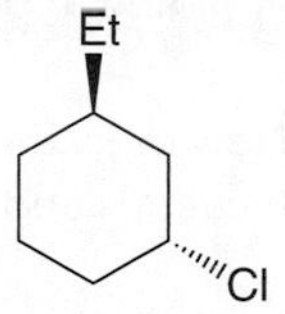

The two chair conformations of this compound are:

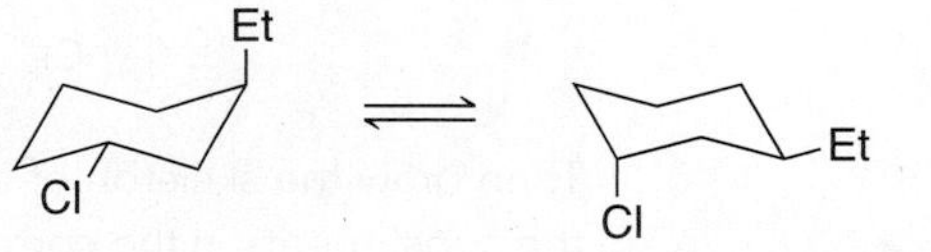

In this example, neither conformation has two equatorial substituents. In the first conformation, the chlorine is equatorial, but the ethyl group is axial. In the second conformation, the ethyl group is equatorial, but the chlorine is axial. In a situation like this, we must decide which group exhibits a greater preference for being equatorial: the chlorine atom or the ethyl group. To do this, we use the numbers from Table 4.8:

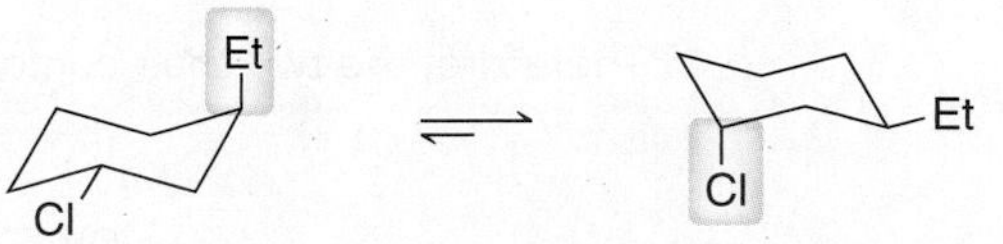

Axial ethyl = 8 kJ/mol **Axial chlorine = 2 kJ/mol**

Both conformations will exhibit 1,3-diaxial interactions, but these interactions are less pronounced in the second conformation. The energy cost of having a chlorine atom in an axial position is lower than the energy cost of having an ethyl group in an axial position. Therefore, the second conformation is lower in energy. Let's get some practice with this.

SKILLBUILDER

4.13 DRAWING THE MORE STABLE CHAIR CONFORMATION OF POLYSUBSTITUTED CYCLOHEXANES

LEARN the skill Draw the more stable chair conformation of the following compound:

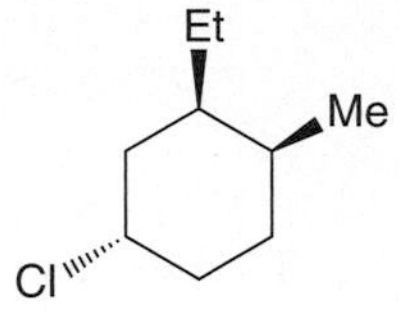

SOLUTION

Begin by drawing both chair conformations using the method from the previous section. Place numbers on the ring, and for each substituent, identify its location and configuration:

STEP 1 Determine the location and configuration of each substituent.

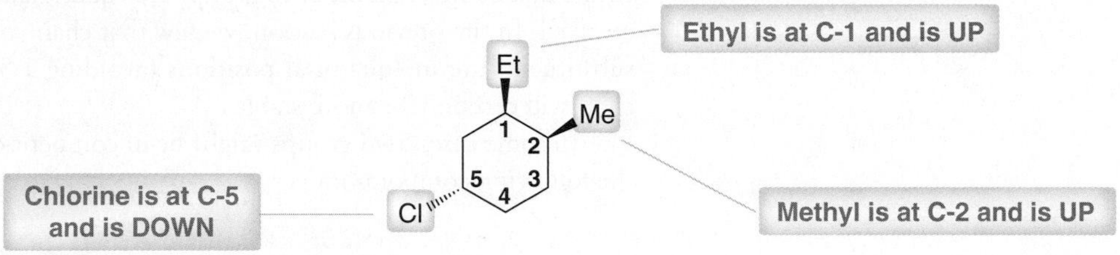

Now draw the skeleton of the first chair conformation, placing the substituents in the correct locations and with the correct configuration:

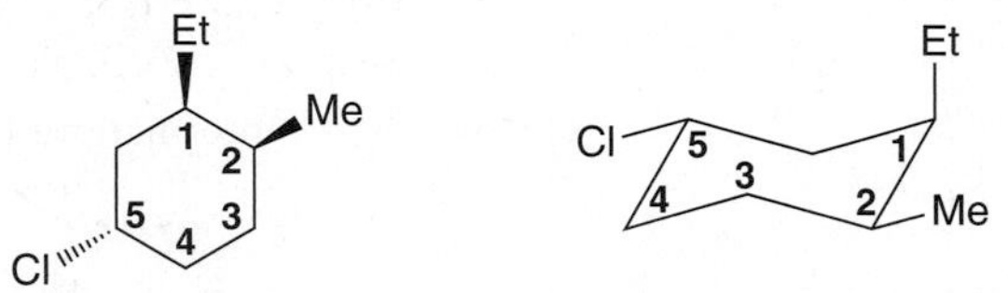

Then draw the skeleton of the second chair conformation, number it, and once again, place the substituents in the correct locations and with the correct configuration:

STEP 2 Draw both chair conformations.

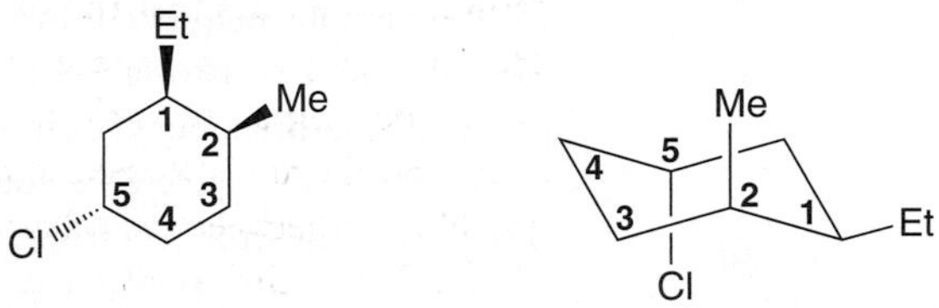

Therefore, the two chair conformations of this compound are:

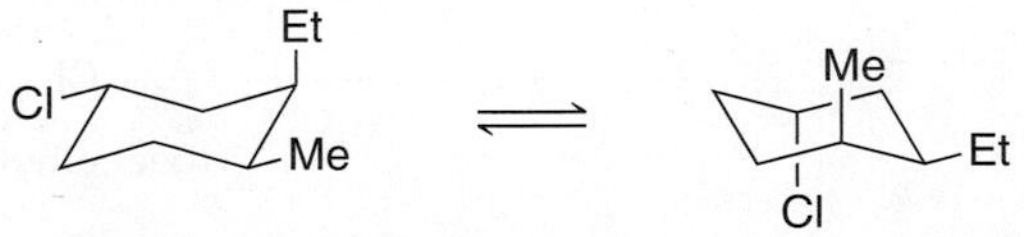

STEP 3 Assess the energy cost of each axial group.

Now we can compare the relative energy of these two chair conformations. In the first conformation, there is one ethyl group in an axial position. According to Table 4.8, the energy cost associated with an axial ethyl group is 8.0 kJ/mol. In the second conformation, two groups are in axial positions: a methyl group and a chlorine. According to Table 4.8, the total energy cost is 7.6 kJ/mol + 2.0 kJ/mol = 9.6 kJ/mol. According to this calculation, the energy cost is lower for the first conformation (with an axial ethyl group). The first conformation is therefore lower in energy (more stable).

PRACTICE the skill **4.33** Draw the lowest energy conformation for each of the following compounds:

(a) 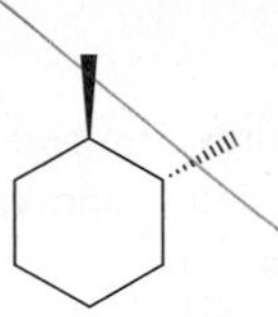 (b) 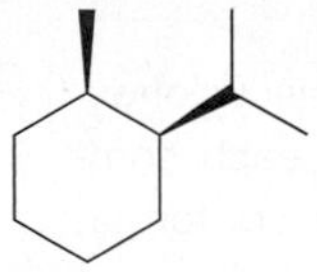 (c) 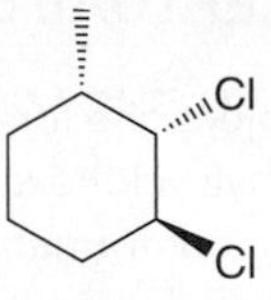

(d) 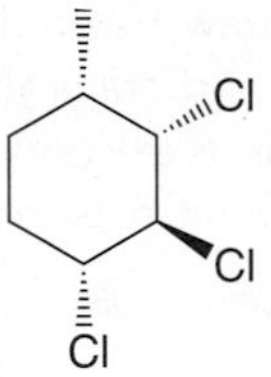 (e) 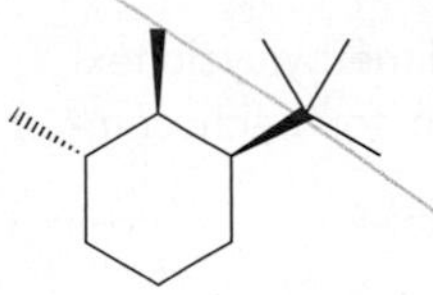 (f) 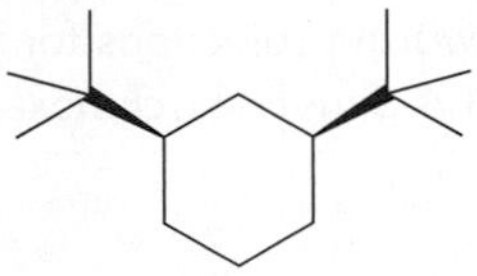

APPLY the skill **4.34** In Problem 4.32, you drew the two chair conformations of lindane. Carefully inspect them, and predict the difference in energy between them, if any.

4.35 Compound A exists predominantly in a chair conformation, while compound B exists predominantly in a twist boat conformation. Explain.

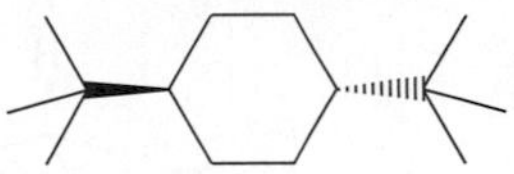
Compound A

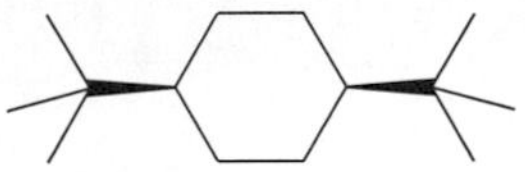
Compound B

need more **PRACTICE?** **Try Problems 4.53, 4.55, 4.57, 4.61, 4.69**

4.14 *cis-trans* Stereoisomerism

When dealing with cycloalkanes, the terms *cis* and *trans* are used to signify the relative spatial relationship of similar substituents:

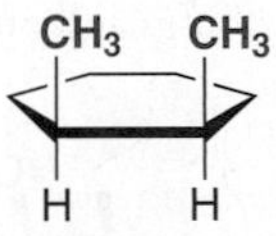

***cis*-1,2-Dimethylcyclohexane**

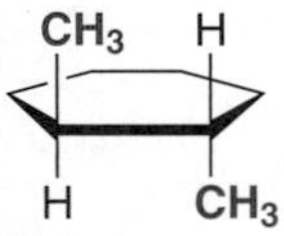

***trans*-1,2-Dimethylcyclohexane**

The term *cis* is used to signify that the two groups are on the same side of the ring, while the term *trans* signifies that the two groups are opposite sides of the ring. The drawings above are Haworth projections (as seen in section 2.6) and are used to clearly identify which groups are above the ring and which groups are below the ring. These drawings are planar representations and do not represent conformations. Each compound above is better represented as an equilibrium between two chair conformations (Figure 4.30). *cis*-1,2-Dimethylcyclohexane and *trans*-1,2-dimethylcyclohexane are **stereoisomers** (as we will see in the next chapter). They are different compounds with different physical properties, and they cannot be interconverted via a conformational change. *trans*-1,2-Dimethylcyclohexane is more stable, because it can adopt a chair conformation in which both methyl groups are in equatorial positions.

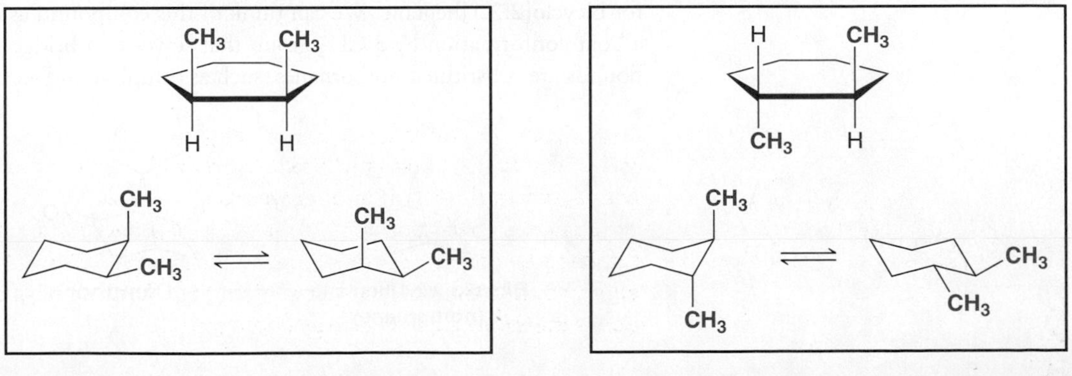

FIGURE 4.30 Each stereoisomer of 1,2-dimethylcyclohexane has two chair conformations.

CONCEPTUAL CHECKPOINT

4.36 Draw Haworth projections for *cis*-1,3-dimethylcyclohexane and *trans*-1,3-dimethylcyclohexane. Then, for each compound, draw the two chair conformations. Use these conformations to determine whether the *cis* isomer or the *trans* isomer is more stable.

4.37 Draw Haworth projections for *cis*-1,4-dimethylcyclohexane and *trans*-1,4-dimethylcyclohexane. Then for each compound, draw the two chair conformations. Use these conformations to determine whether the *cis* isomer or the *trans* isomer is more stable.

4.38 Draw Haworth projections for *cis*-1,3-di-*tert*-butylcyclohexane and *trans*-1,3-di-*tert*-butylcyclohexane. One of these compounds exists in a chair conformation, while the other exists primarily in a twist boat conformation. Offer an explanation.

4.15 Polycyclic Systems

Decalin is a bicyclic system composed of two fused six-membered rings. The structures of *cis*-decalin and *trans*-decalin are as follows:

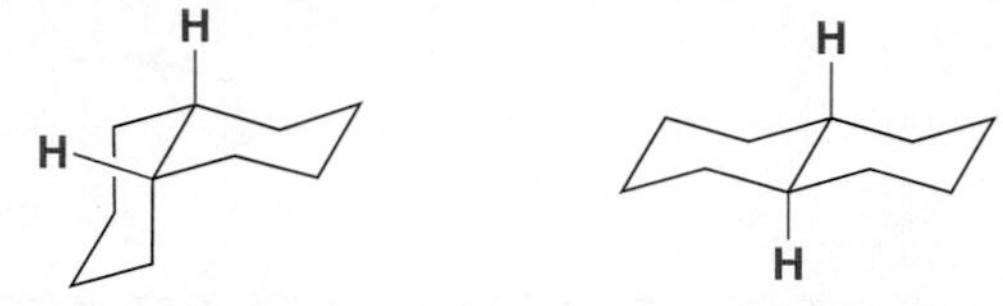

cis-Decalin *trans*-Decalin

The relationship between these compounds is stereoisomeric (as in the previous section). These two compounds are not interconvertible by ring flipping. They are two different compounds with different physical properties. Many naturally occurring compounds, such as steroids, incorporate decalin systems into their structures. Steroids are a class of compounds comprised of four fused rings (three six-membered rings and one five-membered ring). Below are two examples of steroids:

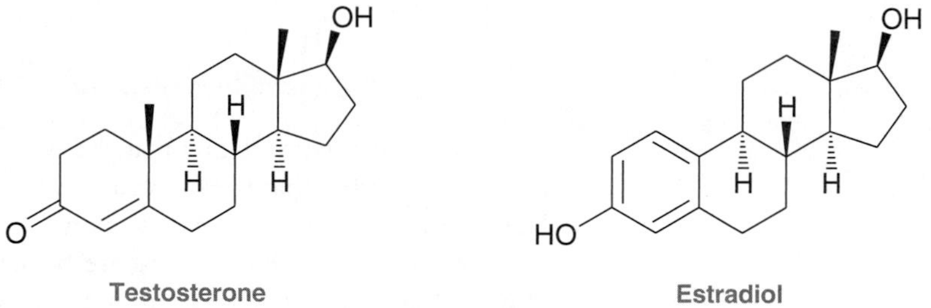

Testosterone Estradiol

Testosterone is an androgenic hormone (male sex hormone) produced in the testes, and estradiol is an estrogenic hormone (female sex hormone) produced from testosterone in the ovaries. Both compounds play a number of biological roles, ranging from the development of secondary sex characteristics to the promotion of tissue and muscle growth.

Another common polycyclic system is **norbornane**. Norbornane is the common name for bicyclo[2.2.1]heptane. We can think of this compound as a six-membered ring locked into a boat conformation by a CH_2 group that serves as a bridge. Many naturally occurring compounds are substituted norbornanes, such as camphor and camphene:

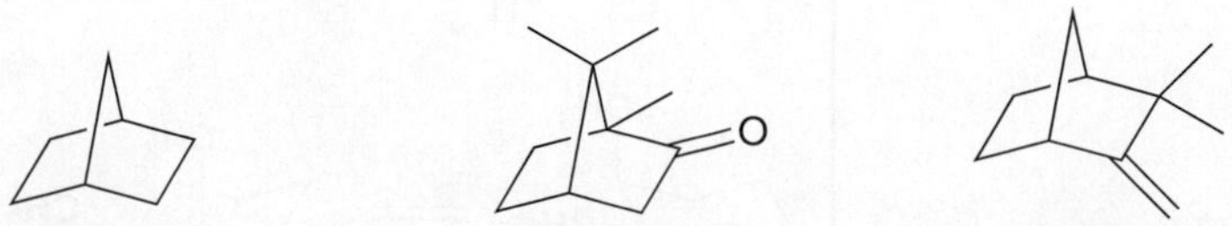

Bicyclo[2.2.1]heptane (norbornane) Camphor Camphene

Camphor is a strongly scented solid that is isolated from evergreen trees in Asia. It is used as a spice as well as for medicinal purposes. Camphene is a minor constituent in many natural oils, such as pine oil and ginger oil. It is used in the preparation of fragrances.

Polycyclic systems based on six-membered rings are also found in nonbiological materials. Most notably, the structure of diamond is based on fused six-membered rings locked in chair conformations. The drawing in Figure 4.31 represents a portion of the diamond structure. Every carbon atom is bonded to four other carbon atoms forming a three-dimensional lattice of chair conformations. In this way, a diamond is one large molecule. Diamonds are one of the hardest known substances, because cutting a diamond requires the breaking of billions of C—C single bonds.

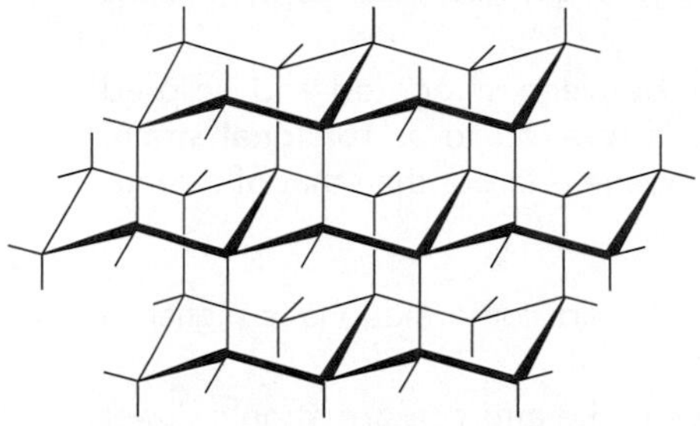

FIGURE 4.31
The structure of diamonds.

REVIEW OF CONCEPTS AND VOCABULARY

SECTION 4.1

- Hydrocarbons that lack π bonds are called **saturated hydrocarbons** or **alkanes**.
- The system of rules for naming compounds is called **nomenclature**.

SECTION 4.2

- Although **IUPAC** rules provide a systematic way for naming compounds, many common names are still in use. Assigning a **systematic name** involves four discrete steps:
  1. Identify the parent compound. Alkanes containing a ring are called **cycloalkanes**.
  2. Name the **substituents**, which can be either simple **alkyl groups** or branched alkyl groups, called complex substituents. Many common names for complex substituents are allowed according to IUPAC rules.
  3. Number the carbon atoms of the parent and assign a **locant** to each substituent.
  4. Assemble the substituents alphabetically, placing locants in front of each substituent. For identical substituents, use di, tri, tetra, penta, or hexa, which are ignored when alphabetizing.
- **Bicyclic** compounds are named just like alkanes and cycloalkanes, with just two subtle differences:
  1. The term "bicyclo" is used, and bracketed numbers indicate how the **bridgeheads** are connected.
  2. To number the parent, travel first along the longest path connecting the bridgeheads.

SECTION 4.3

- The number of possible constitutional isomers for an alkane increases with increasing molecular size.
- When drawing constitutional isomers, use IUPAC rules to avoid drawing the same compound twice.

SECTION 4.4

- For an alkane, the **heat of combustion** is the negative of the **change in enthalpy** ($-\Delta H°$) associated with the complete combustion of 1 mol of alkane in the presence of oxygen.
- Heats of combustion can be measured experimentally and used to compare the stability of isomeric alkanes.

SECTION 4.5

- Petroleum is a complex mixture of hydrocarbons, most of which are alkanes (ranging in size and constitution).
- These compounds are separated into fractions via distillation (separation based on differences in boiling points). The refining of crude oil separates it into many commercial products.
- The yield of useful gasoline can be improved in two ways:
  1. Cracking is a process by which C—C bonds of larger alkanes are broken, producing smaller alkanes suitable for gasoline.
  2. Reforming is a process in which straight-chain alkanes are converted into branched hydrocarbons and aromatic compounds, which exhibit less *knocking* during combustion.

SECTION 4.6

- Rotation about C—C single bonds allows a compound to adopt a variety of **conformations**.

- **Newman projections** are often used to draw the various conformations of a compound.

SECTION 4.7

- In a Newman projection, the **dihedral angle**, or **torsional angle,** describes the relative positions of one group on the back carbon and one group on the front carbon.
- **Staggered conformations** are lower in energy, while **eclipsed conformations** are higher in energy.
- In the case of ethane, all staggered conformations are **degenerate** (equivalent in energy), and all eclipsed conformations are degenerate.
- The difference in energy between staggered and eclipsed conformations of ethane is referred to as **torsional strain.** The torsional strain for propane is larger than that of ethane.

SECTION 4.8

- For butane, one of the eclipsed conformations is higher in energy than the other two.
- One staggered conformation (the **anti conformation**) is lower in energy than the other two staggered conformations, because they possess **gauche interactions**, a form of steric hindrance.

SECTION 4.9

- **Angle strain** occurs in cycloalkanes when bond angles deviate from the preferred 109.5°.
- Angle strain and torsional strain are components of the total energy of a cycloalkane, which can be assessed by measuring heats of combustion per CH_2 group.

SECTION 4.10

- The **chair conformation** of cyclohexane has no torsional strain and very little angle strain.
- The **boat conformation** of cyclohexane has significant torsional strain (from eclipsing H's as well as **flagpole interactions**). The boat can alleviate some of its torsional strain by twisting, giving a conformation called a **twist boat**.
- Cyclohexane is found in a chair conformation most of the time.

SECTION 4.11

- Each carbon atom in a cyclohexane ring can bear two substituents.
- One substituent is said to occupy an **axial position**, while the other substituent is said to occupy an **equatorial position**.

SECTION 4.12

- When a ring has one substituent, the substituent can occupy either an axial or an equatorial position. These two possibilities represent two different conformations that are in equilibrium with each other.
- The term **ring flip** is used to describe the conversion of one chair conformation into the other.
- The equilibrium will favor the chair conformation with the substituent in the equatorial position, because an axial substituent generates **1,3-diaxial interactions,** a form of steric hindrance.

SECTION 4.13

- To draw the two chair conformations of a disubstituted cyclohexane, each substituent must be identified as being either UP or DOWN. The three-dimensional orientation of the substituents (UP or DOWN) does not change during a ring flip.
- After drawing both chair conformations, the relative energy levels can be determined by comparing the energy cost associated with all axial groups.

SECTION 4.14

- The terms *cis* and *trans* signify the relative spatial relationship of similar substituents, as can be seen most clearly in Haworth projections.
- *cis*-1,2-Dimethylcyclohexane and *trans*-1,2-dimethylcyclohexane are **stereoisomers**. They are different compounds with different physical properties, and they cannot be interconverted via a conformational change.

SECTION 4.15

- The relationship between *cis*-decalin and *trans*-decalin is stereoisomeric. These two compounds are not interconvertable by ring flipping.
- **Norbornane** is the common name for bicyclo[2.2.1]heptane and is a commonly encountered bicyclic system.
- Polycyclic systems based on six-membered rings are also found in the structure of diamonds.

KEY TERMINOLOGY

SKILLBUILDER REVIEW

4.1 IDENTIFYING THE PARENT

STEP 1 Choose the longest chain.

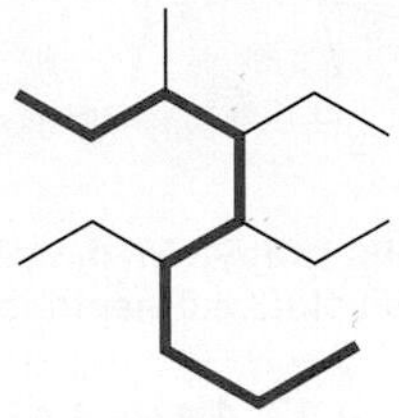

STEP 2 In a case where two chains compete, choose the chain with more substituents.

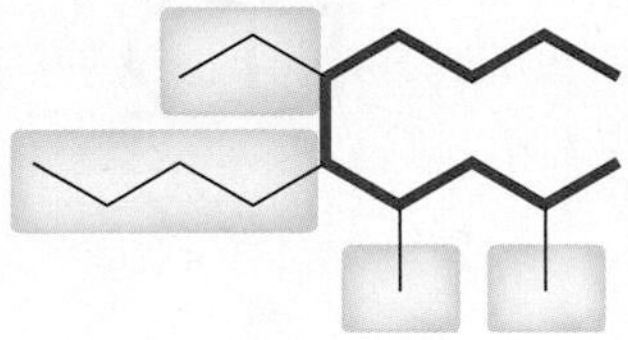

Correct

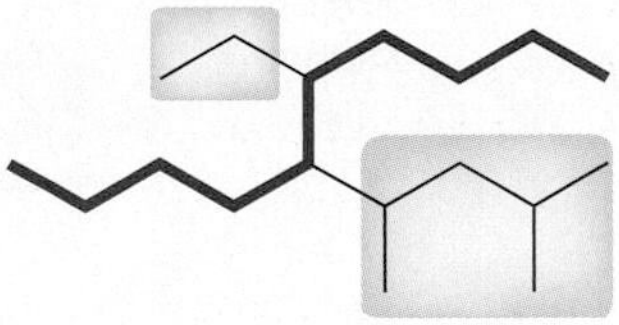

Incorrect

Try Problems **4.1–4.4, 4.39**

4.2 IDENTIFYING AND NAMING SUBSTITUENTS

STEP 1 Identify the parent.

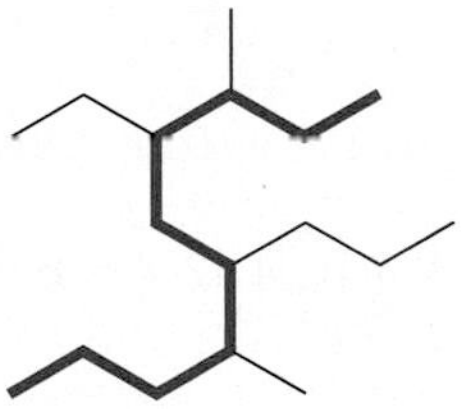

STEP 2 Identify all alkyl substituents connected to the parent.

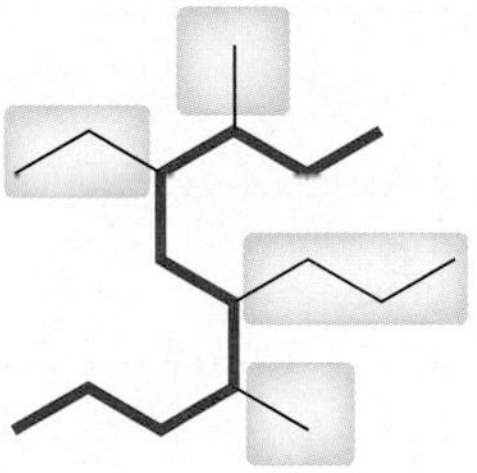

STEP 3 Name each substituent using the names in Table 4.2.

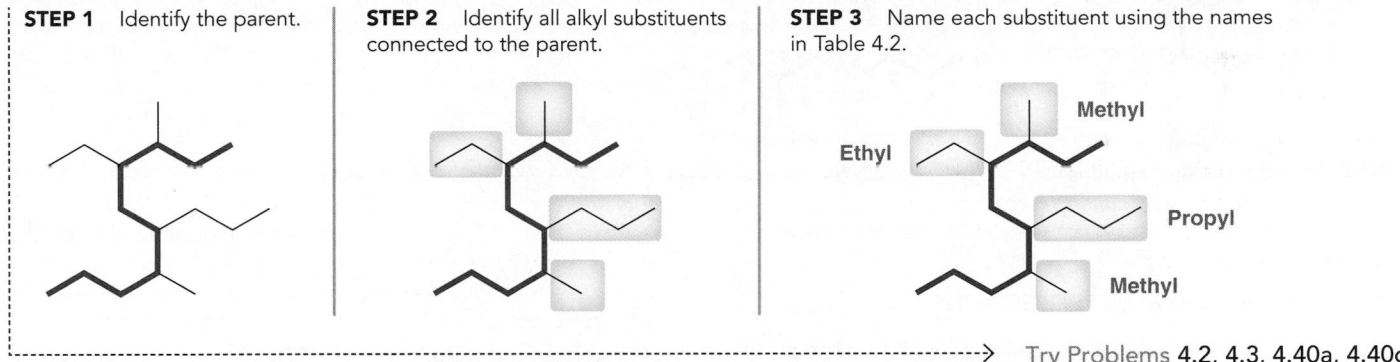

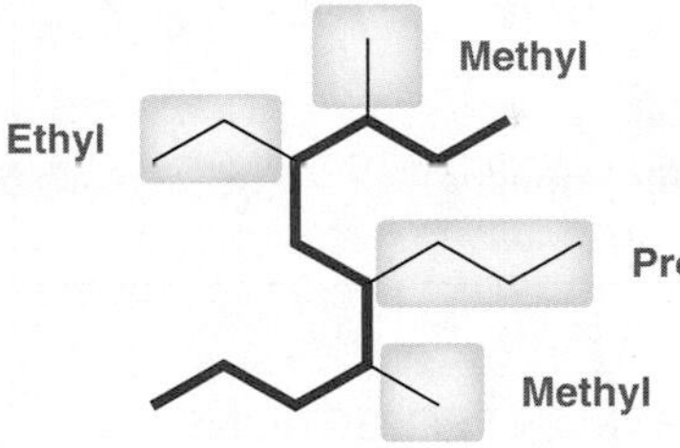

Try Problems **4.2, 4.3, 4.40a, 4.40c**

4.3 IDENTIFYING AND NAMING COMPLEX SUBSTITUENTS

STEP 1 Identify the parent.

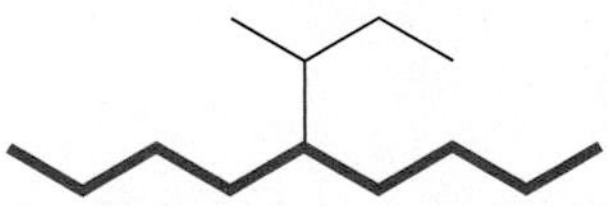

STEP 2 Place numbers on the complex substituent, going away from the parent chain.

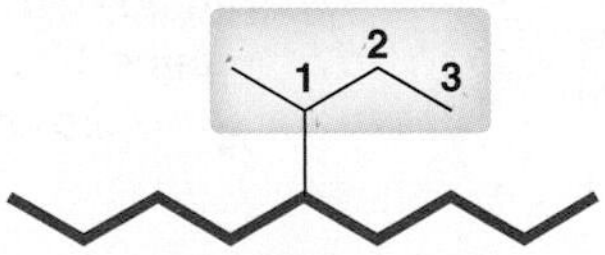

STEP 3 Treat the entire group as a substituent on a substituent.

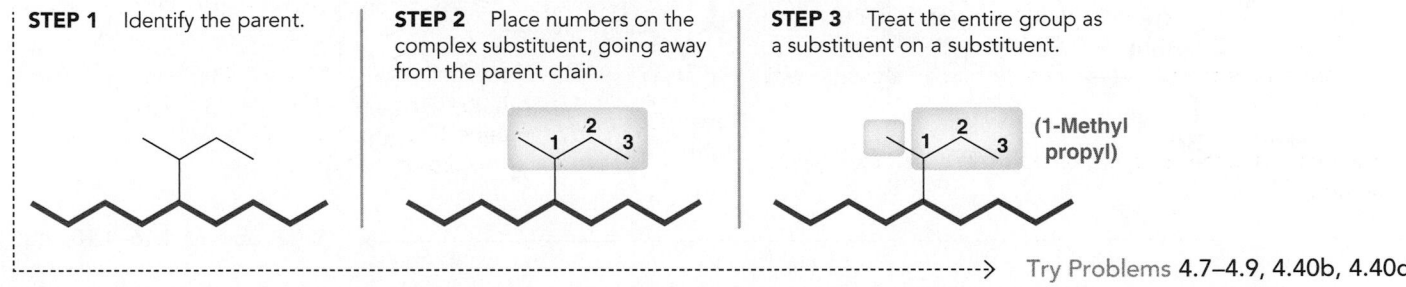

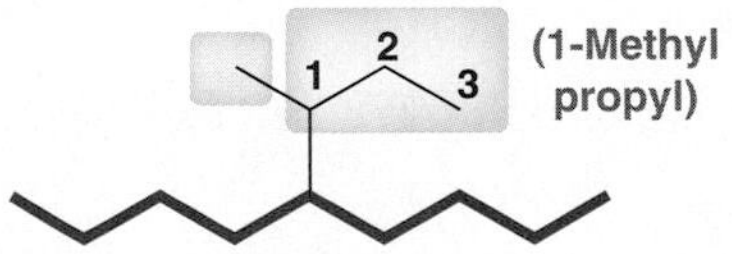

Try Problems **4.7–4.9, 4.40b, 4.40d**

4.4 ASSEMBLING THE SYSTEMATIC NAME OF AN ALKANE

STEP 1 Identify the parent.

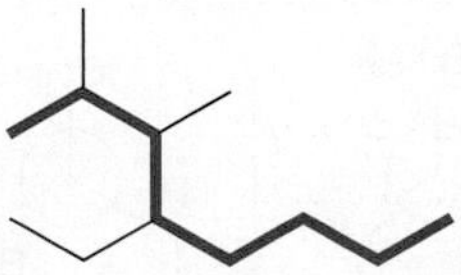

STEP 2 Identify and name substituents.

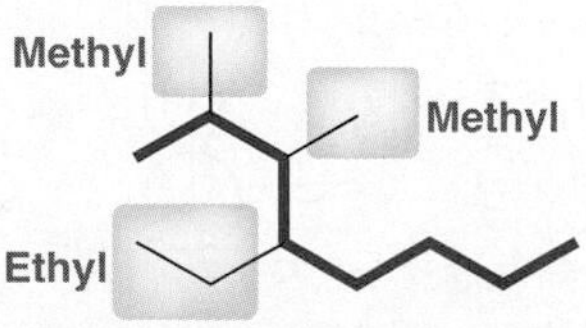

STEP 3 Number the parent chain and assign a locant to each substituent.

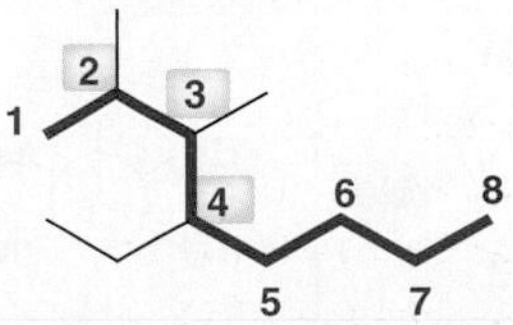

STEP 4 Arrange the substituents alphabetically.

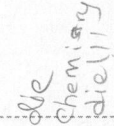

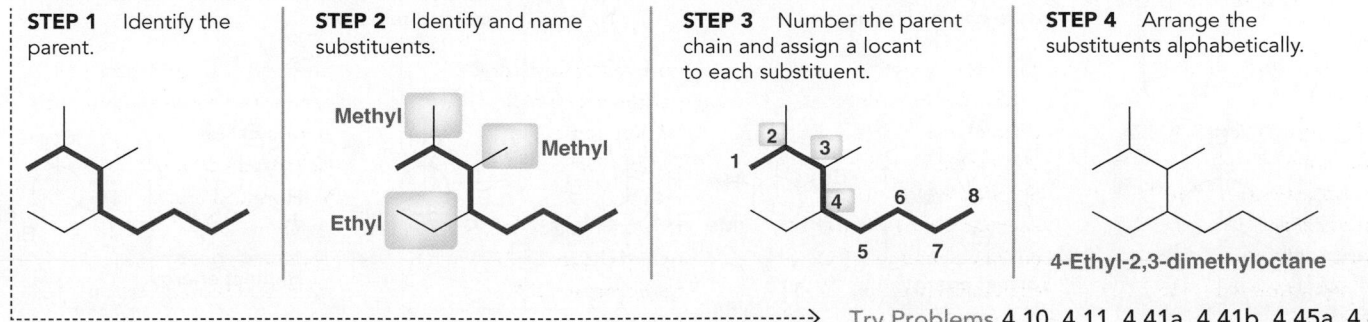

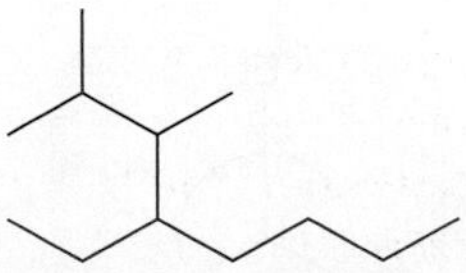

4-Ethyl-2,3-dimethyloctane

Try Problems **4.10, 4.11, 4.41a, 4.41b, 4.45a, 4.45b**

4.5 ASSEMBLING THE NAME OF A BICYCLIC COMPOUND

STEP 1 Identify thebicyclo parent, and then indicate how the bridgeheads are connected.

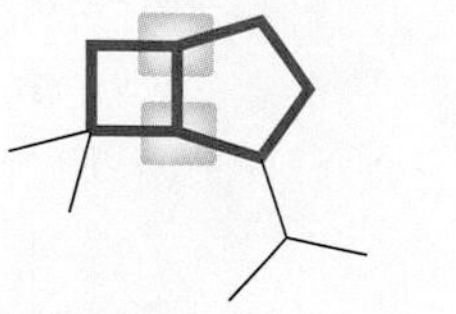

Bicyclo[3.2.0]heptane

STEP 2 Identify and name substituents.

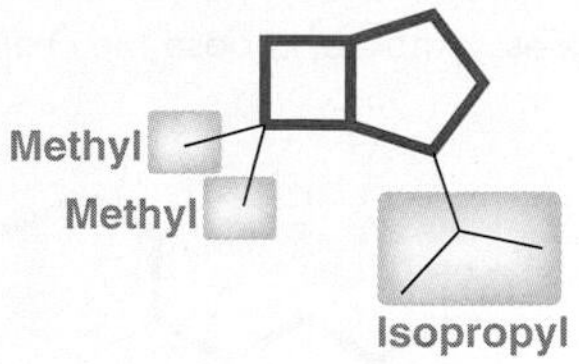

STEP 3 Number the parent chain (start with longest bridge) and assign a locant to each substituent.

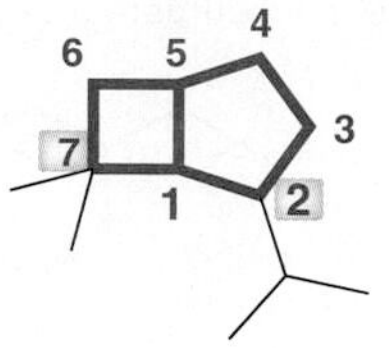

STEP 4 Assemble the substituents alphabetically.

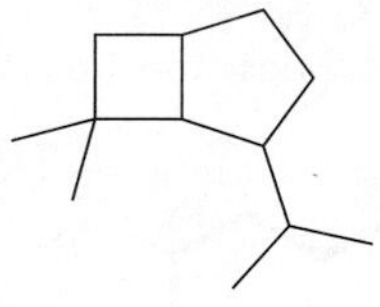

2-Isopropyl-7,7-dimethyl bicyclo[3.2.0]heptane

Try Problems 4.12, 4.13, 4.41c, 4.41d, 4.45c

4.6 IDENTIFYING CONSTITUTIONAL ISOMERS

Constitutional isomers have different names. If two compounds have the same name, then they are the same compound.

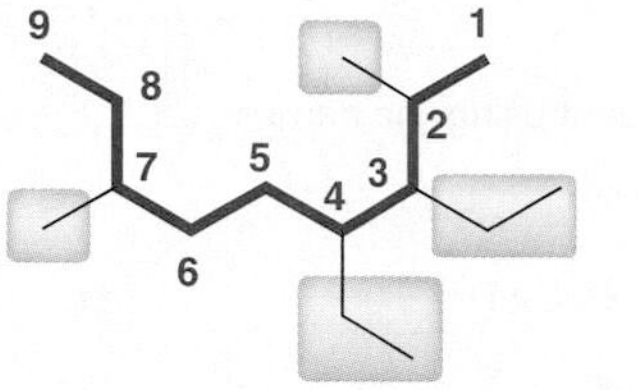

3,4-Diethyl-2,7-dimethylnonane

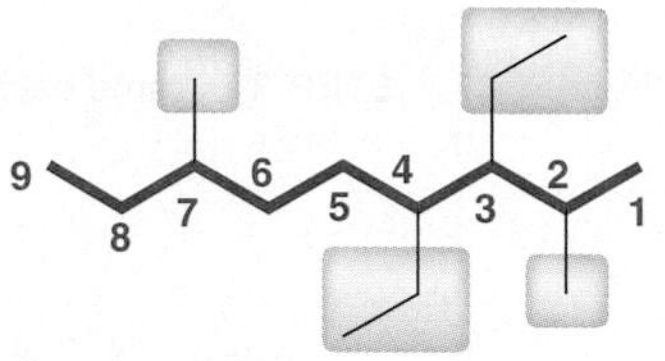

3,4-Diethyl-2,7-dimethylnonane

Try Problems 4.14, 4.15, 4.42, 4.66b,d,k,l

4.7 DRAWING NEWMAN PROJECTIONS

STEP 1 Identify the three groups connected to the front carbon atom.

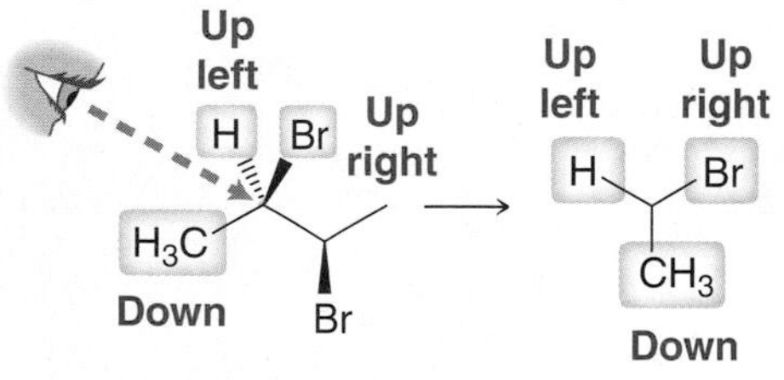

STEP 2 Identify the three groups connected to the back carbon atom.

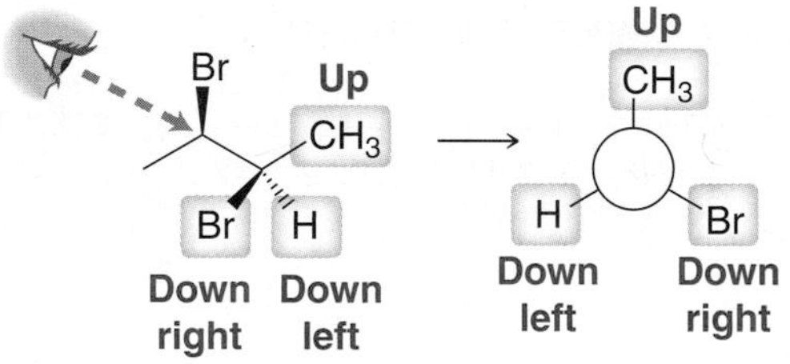

STEP 3 Assemble the Newman projection from the two pieces obtained in the previous steps.

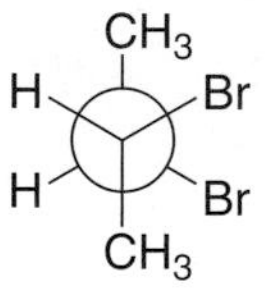

Try Problems 4.16–4.18, 4.56

4.8 IDENTIFYING RELATIVE ENERGY OF CONFORMATIONS

STEP 1 Draw a Newman projection.

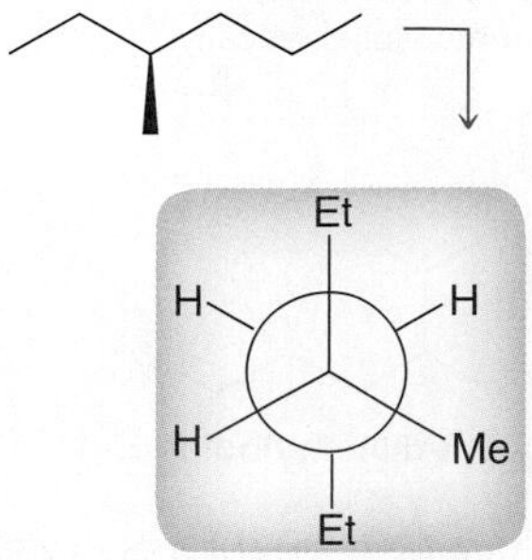

STEP 2 Draw all three staggered conformations and determine which one has the fewest or least severe gauche interactions.

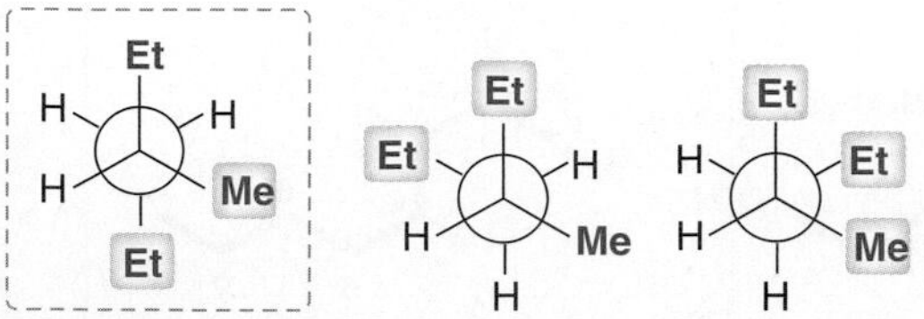

Lowest energy

STEP 3 Draw all three eclipsed conformations and determine which one has the highest energy interactions.

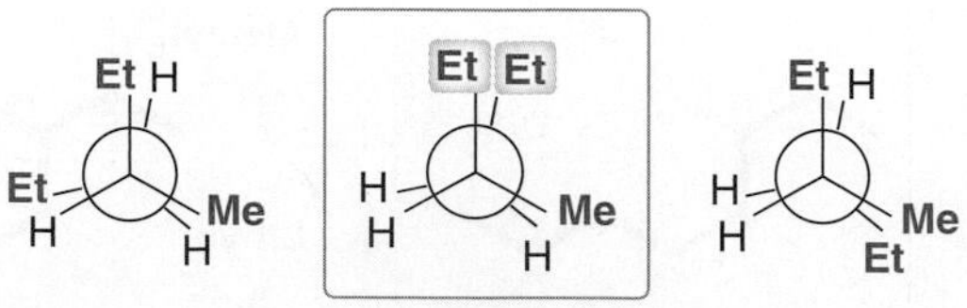

Highest energy

Try Problems 4.20, 4.21, 4.47, 4.59

4.9 DRAWING A CHAIR CONFORMATION

STEP 1 Draw a wide V.

STEP 2 Draw a line going down at a 60° angle, ending just before the center of the V.

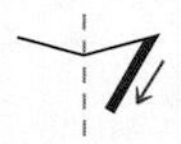

STEP 3 Draw a line parallel to the left side of the V ending just before the left side of the V.

STEP 4 Draw a line parallel to the line from step 2, going down exactly as low as that line.

STEP 5 Connect the dots.

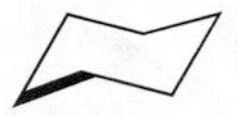

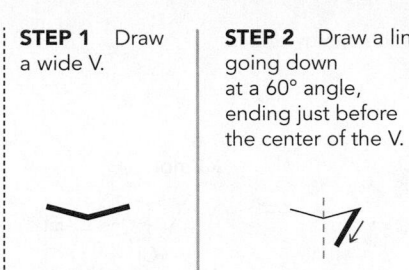

Try Problems **4.22, 4.23**

4.10 DRAWING AXIAL AND EQUATORIAL POSITIONS

STEP 1 Draw all axial positions as parallel lines, alternating in directions.

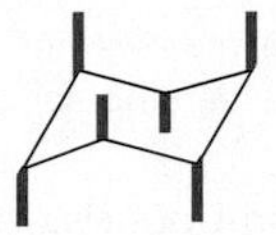

STEP 2 Draw all equatorial positions as pairs of parallel lines.

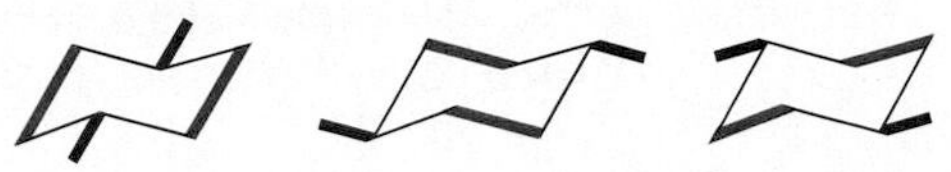

SUMMARY All substituents are drawn like this:

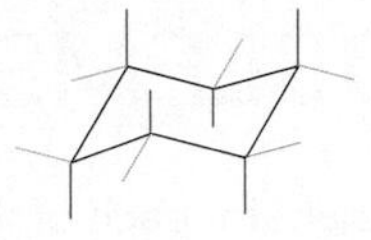

Try Problems **4.24–4.27, 4.62**

4.11 DRAWING BOTH CHAIR CONFORMATIONS OF A MONOSUBSTITUTED CYCLOHEXANE

STEP 1 Draw a chair conformation.

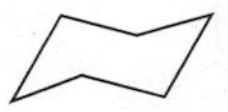

STEP 2 Place the substituent in an axial position.

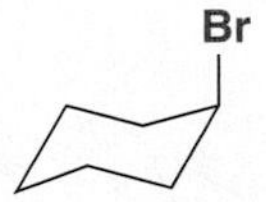

Axial

STEP 3 Draw the ring flip and the axial group becomes equatorial.

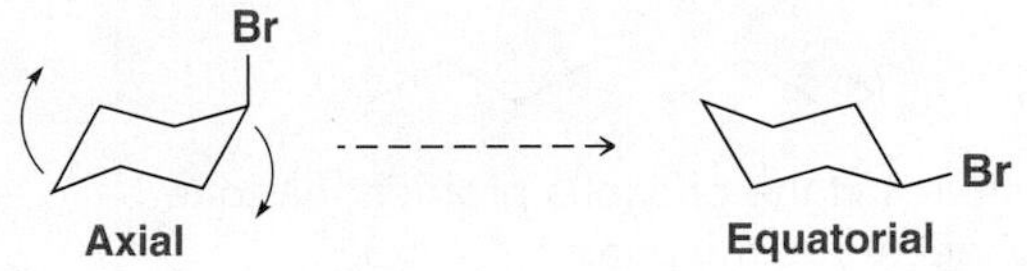

Axial **Equatorial**

Try Problems **4.28, 4.29, 4.54a**

4.12 DRAWING BOTH CHAIR CONFORMATIONS OF DISUBSTITUTED CYCLOHEXANES

STEP 1 Using a numbering system, determine the location and configuration of each substituent.

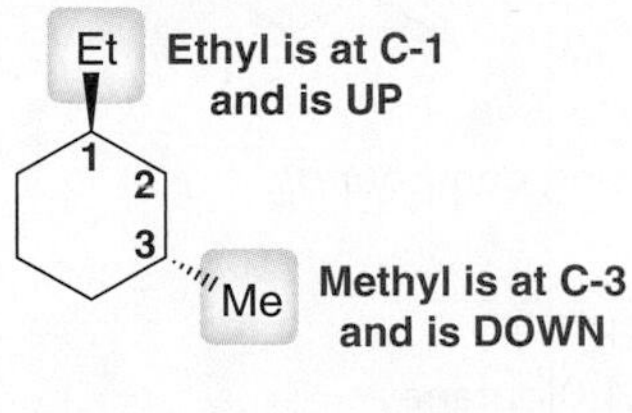

STEP 2 Place the substituents on the first chair using the information from step 1.

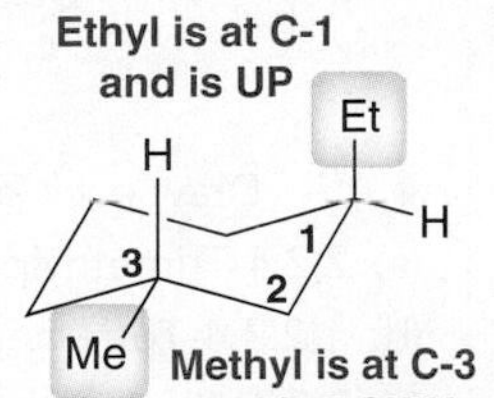

STEP 3 Draw the second chair skeleton, and place substituents using the information from step 1.

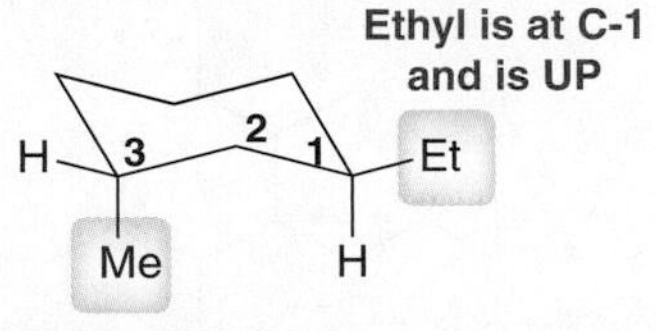

Try Problems **4.31, 4.32, 4.54b-d, 4.66g**

4.13 DRAWING THE MORE STABLE CHAIR CONFORMATION OF POLYSUBSTITUTED CYCLOHEXANES

STEP 1 Using a numbering system, determine the location and configuration of each substituent.

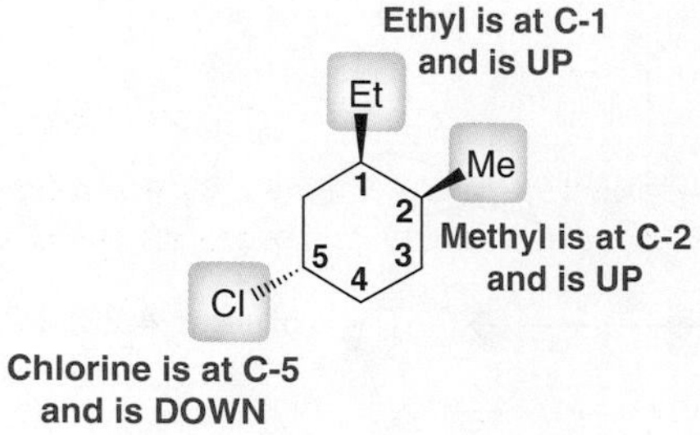

STEP 2 Using the information from step 1, draw both chair conformations.

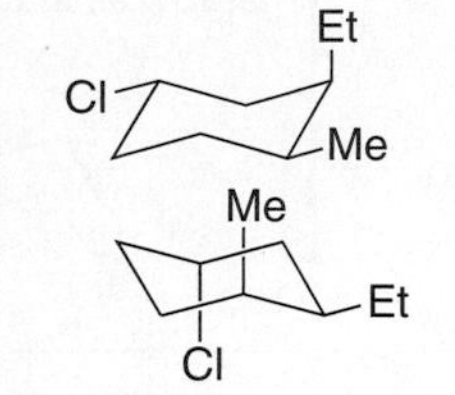

STEP 3 Assess the energy cost of each axial group.

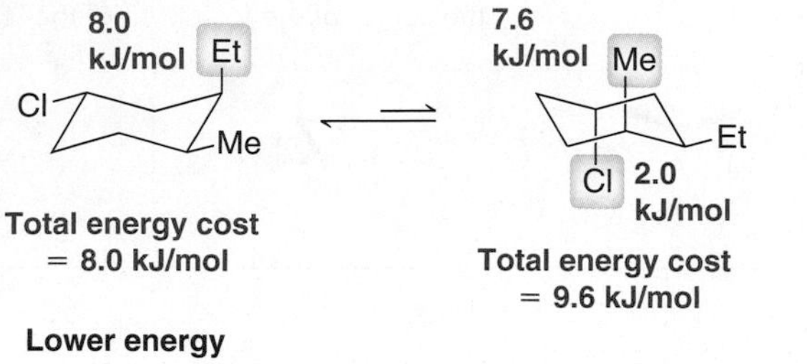

Try Problems 4.33–4.35, 4.53, 4.55, 4.57, 4.61, 4.69

PRACTICE PROBLEMS

Note: Most of the Problems are available within *WileyPLUS*, an online teaching and learning solution.

4.39 Identify the name of the parent for each of the following compounds:

(a) 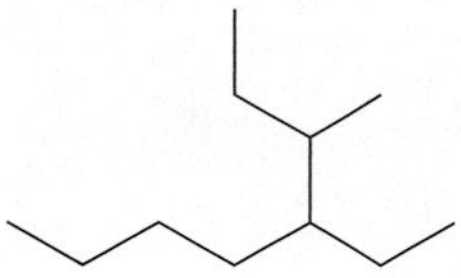 (b) 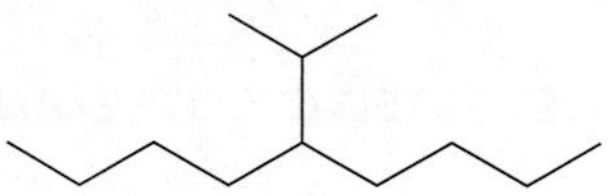

(c) 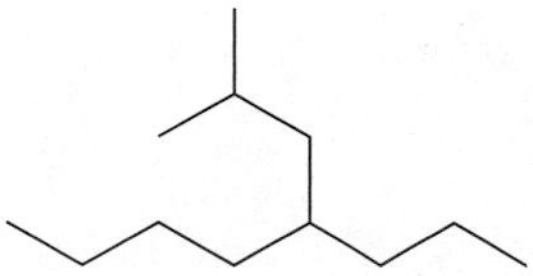 (d) 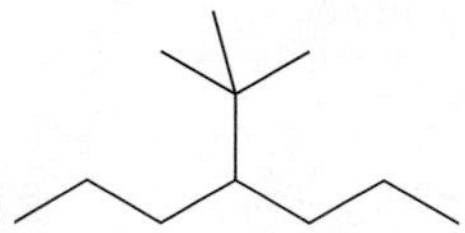

4.40 Each of the structures in the previous problem has one or more substituents connected to the parent.

(a) Identify the name of each substituent in 4.39a.

(b) Identify the common name and the IUPAC name of the complex substituent in 4.39b.

(c) Identify the name of each substituent in 4.39c.

(d) Identify the common name and the IUPAC name of the complex substituent in 4.39d.

4.41 What is the systematic name for each of the following compounds:

(a) 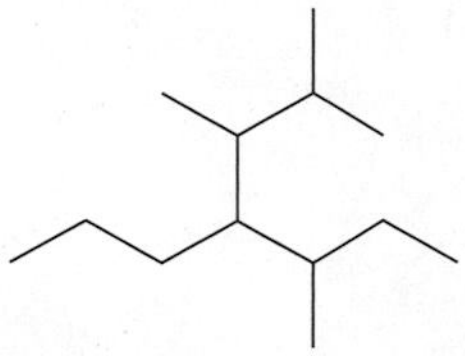 (b) 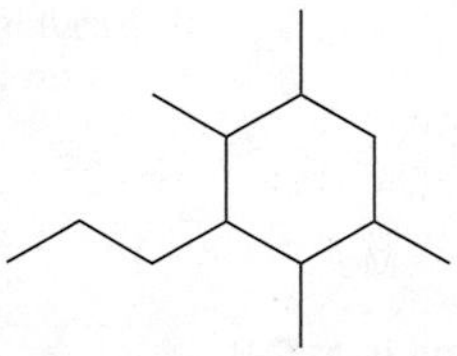

(c) 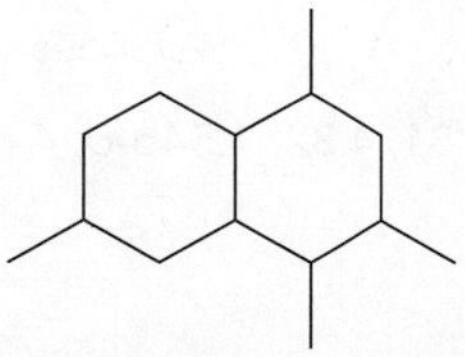 (d) 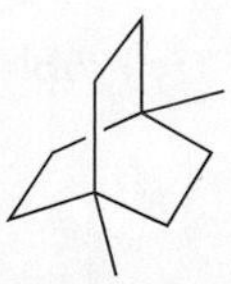

4.42 For each of the following pairs of compounds, identify whether the compounds are constitutional isomers or different representations of the same compound:

(a) 

(b) 

(c) 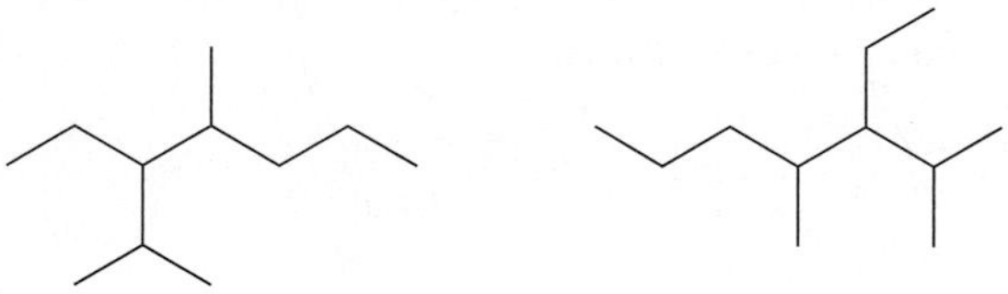

4.43 Use a Newman projection to draw the most stable conformation of 3-methylpentane, looking down the C2—C3 bond.

4.44 Identify which of the following compounds is expected to have the larger heat of combustion:

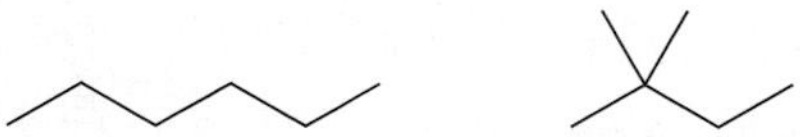

4.45 Draw each of the following compounds:

(a) 2,2,4-Trimethylpentane

(b) 1,2,3,4-Tetramethylcycloheptane

(c) 2,2,4,4-Tetraethylbicyclo[1.1.0]butane

4.46 Sketch an energy diagram that shows a conformational analysis of 2,2-dimethylpropane. Does the shape of this energy diagram more closely resemble the shape of the energy diagram for ethane or for butane?

4.47 What are the relative energy levels of the three staggered conformations of 2,3-dimethylbutane when looking down the C2—C3 bond?

4.48 Draw the ring flip for each of the following compounds:

(a) 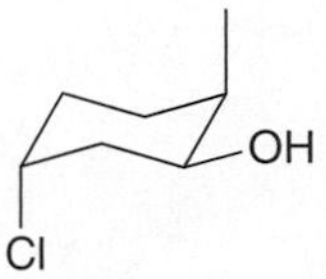

(b) 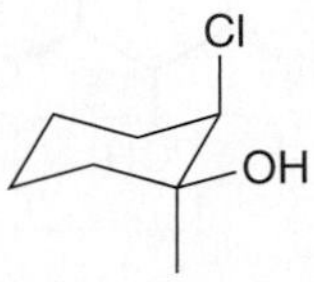

(c) 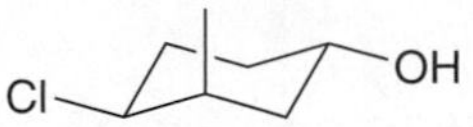

4.49 For each of the following pairs of compounds, identify the compound that would have the higher heat of combustion:

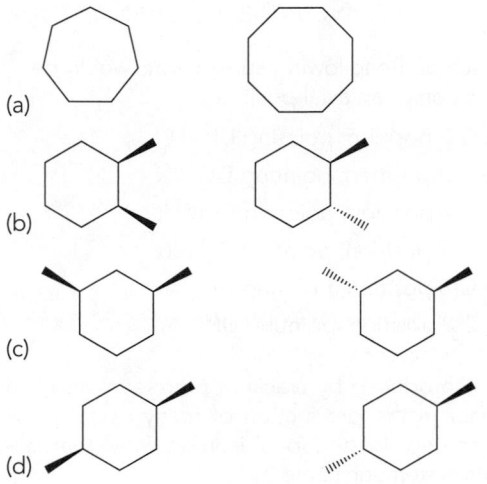

(a) 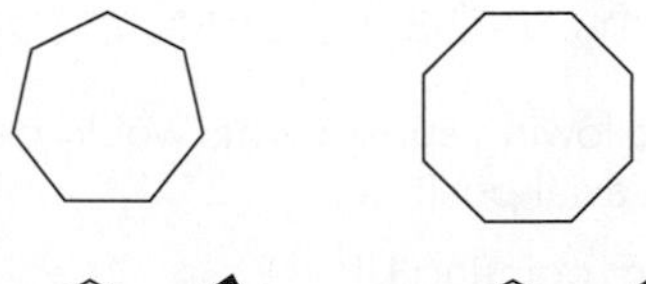

(b) 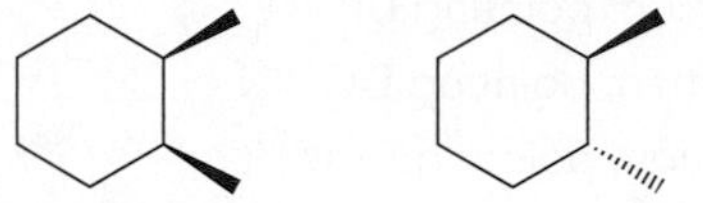

(c) 

(d) 

4.50 Draw a relative energy diagram showing the conformational analysis of 1,2-dichloroethane. Clearly label all staggered conformations and all eclipsed conformations with the corresponding Newman projections.

4.51 Assign IUPAC names for each of the following compounds:

(a) 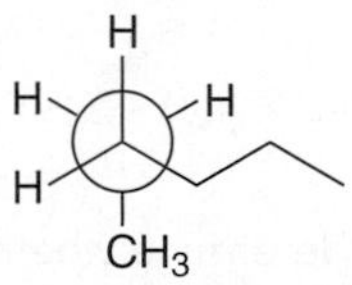

(b) 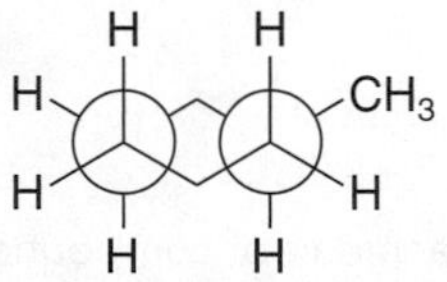

(c) 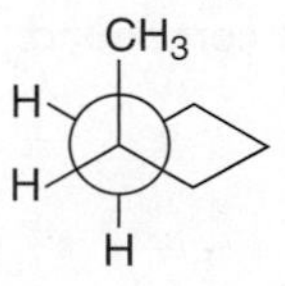

(d) 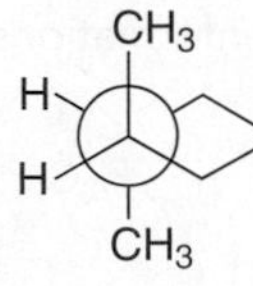

4.52 The barrier to rotation of bromoethane is 15 kJ/mol. Based on this information, determine the energy cost associated with the eclipsing interaction between a bromine atom and a hydrogen atom.

4.53 Menthol, isolated from various mint oils, is used in the treatment of minor throat irritation. Draw both chair conformations of menthol, and indicate which conformation is lower in energy.

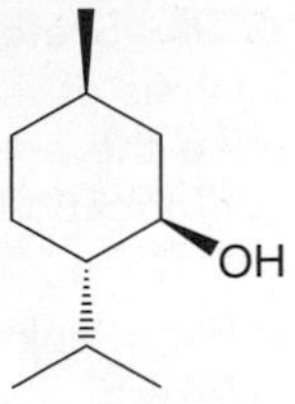

Menthol

4.54 Draw both chair conformations for each of the following compounds. In each case, identify the more stable chair conformation:

(a) Methylcyclohexane
(b) *trans*-1,2-Diisopropylcyclohexane
(c) *cis*-1,3-Diisopropylcyclohexane
(d) *trans*-1,4-Diisopropylcyclohexane

4.55 For each of the following pairs of compounds, determine which compound is more stable (you may find it helpful to draw out the chair conformations):

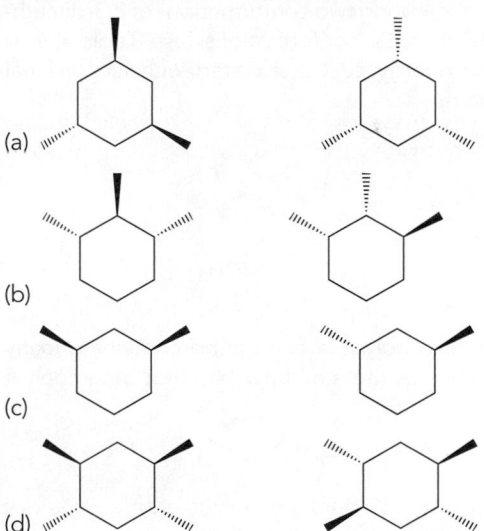

(a) 

(b) 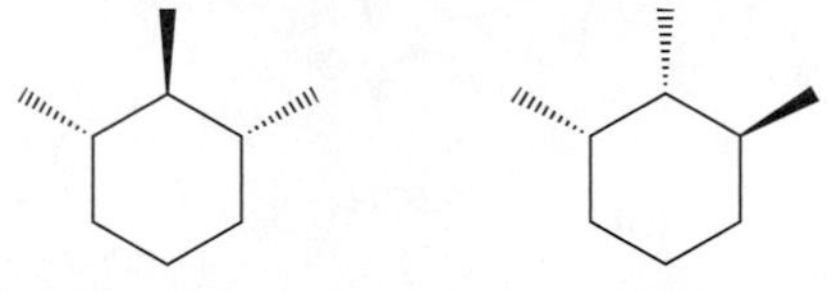

(c) 

(d) 

4.56 Draw a Newman projection of the following compound, as viewed from the angle indicated.

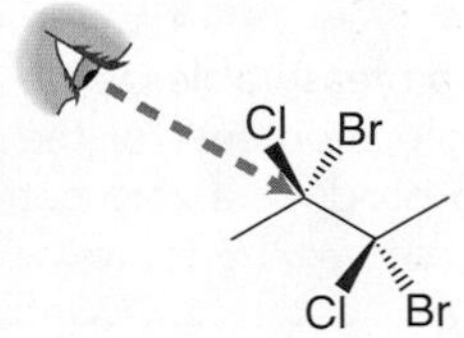

4.57 Glucose (a sugar) is produced by photosynthesis and is used by cells to store energy. Draw the most stable conformation of glucose:

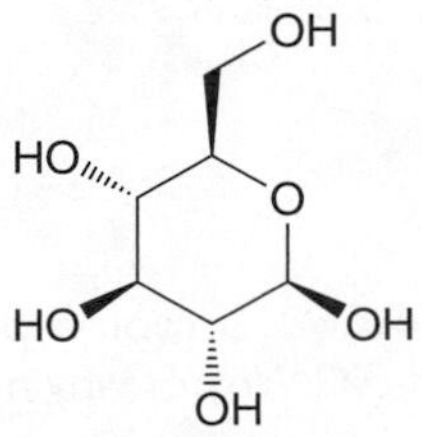

Glucose

4.58 Sketch an energy diagram showing the conformational analysis of 2,2,3,3-tetramethylbutane. Use Table 4.6 to determine the energy difference between staggered and eclipsed conformations of this compound.

4.59 Rank the following conformations in order of increasing energy:

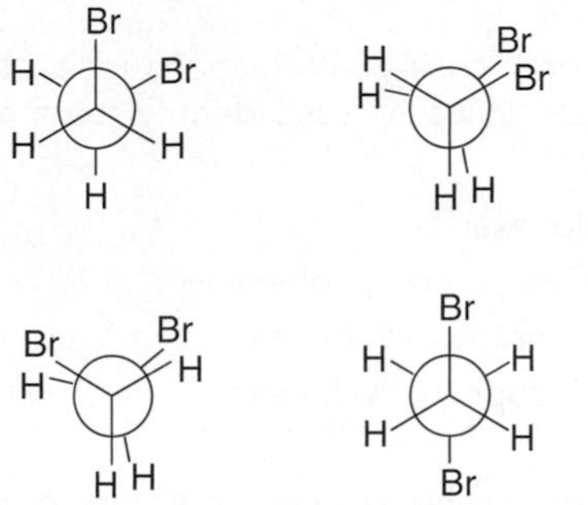

4.60 Consider the following two conformations of 2,3-dimethylbutane. For each of these conformations, use Table 4.6 to determine the total energy cost associated with all torsional strain and steric strain.

(a)

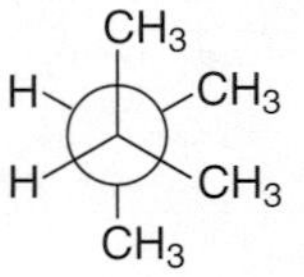

(b)

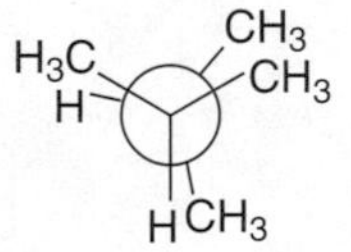

4.61 *myo*-Inositol is a polyol (a compound containing many OH groups) that serves as the structural basis for a number of secondary messengers in eukaryotic cells. Draw the more stable chair conformation of *myo*-inositol.

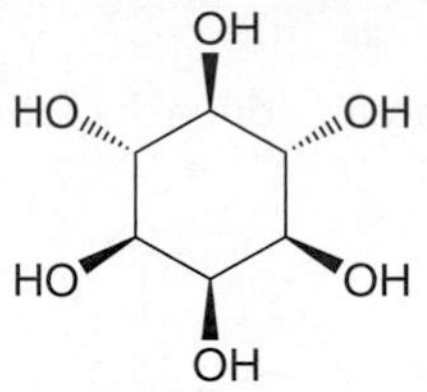

4.62 Below is the numbered skeleton of *trans*-decalin:

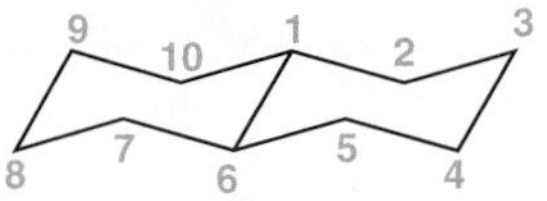

Identify whether each of the following substituents would be in an equatorial position or an axial position:

(a) A group at the C-2 position, pointing UP
(b) A group at the C-3 position, pointing DOWN
(c) A group at the C-4 position, pointing DOWN
(d) A group at the C-7 position, pointing DOWN
(e) A group at the C-8 position, pointing UP
(f) A group at the C-9 position, pointing UP

4.63 Propylene is produced by cracking petroleum and is a very useful precursor in the production of many useful polymers. Propylene has one constitutional isomer. Draw that isomer, and identify its systematic name.

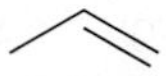

Propylene

INTEGRATED PROBLEMS

4.64 *trans*-1,3-Dichlorocyclobutane has a measurable dipole moment. Explain why the individual dipole moments of the C—Cl bonds do not cancel each other to produce a zero net dipole moment.

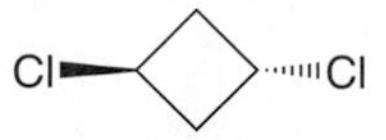

***trans*-1,3-Dichlorocyclobutane**

4.65 Do you expect cyclohexene to adopt a chair conformation? Why or why not? Explain.

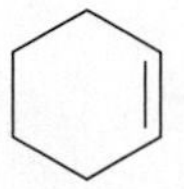

Cyclohexene

4.66 For each pair of compounds below, determine whether they are identical compounds, constitutional isomers, stereoisomers, or different conformations of the same compound:

(a)

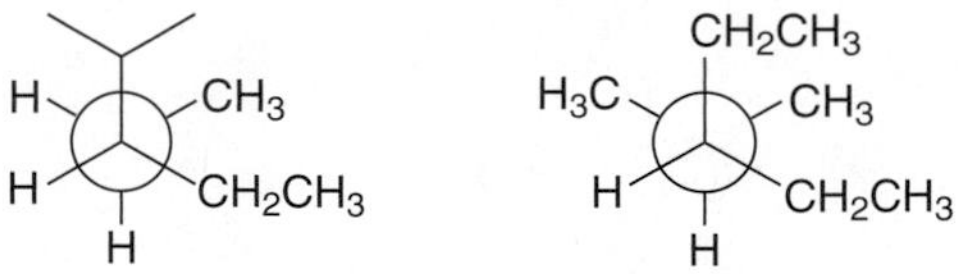

(b)

(c)

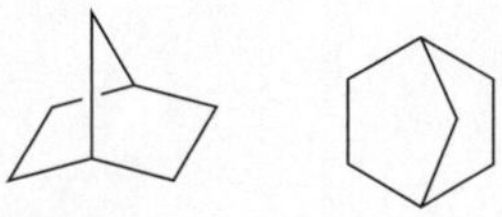

(d) 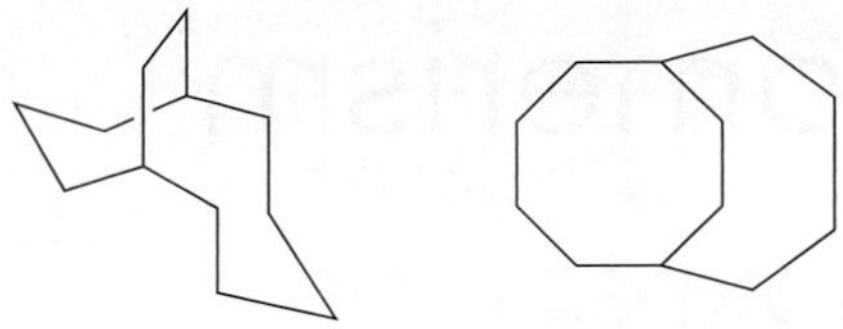

(e) 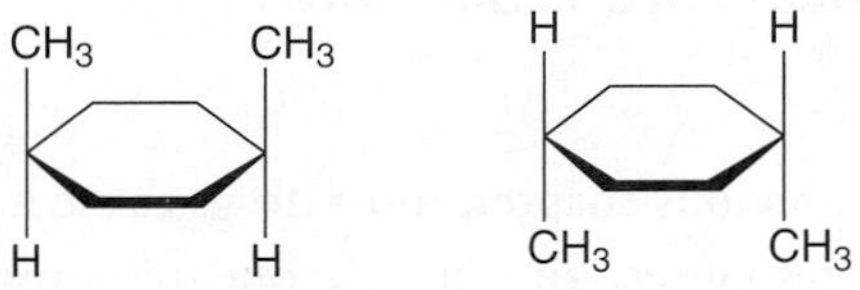

(f) 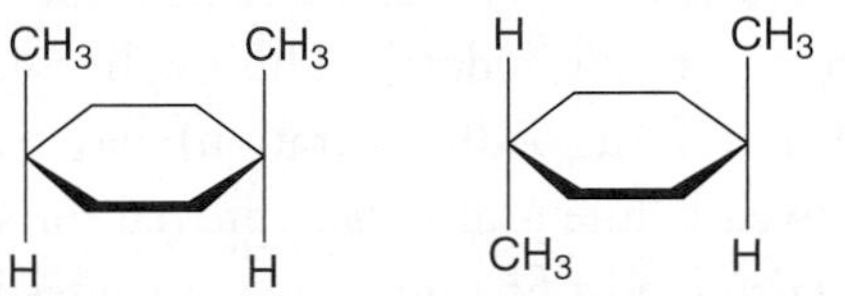

(g) 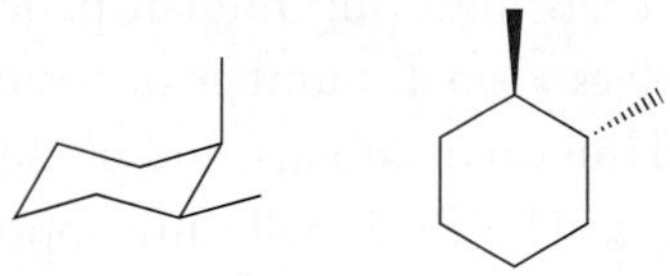

(h) 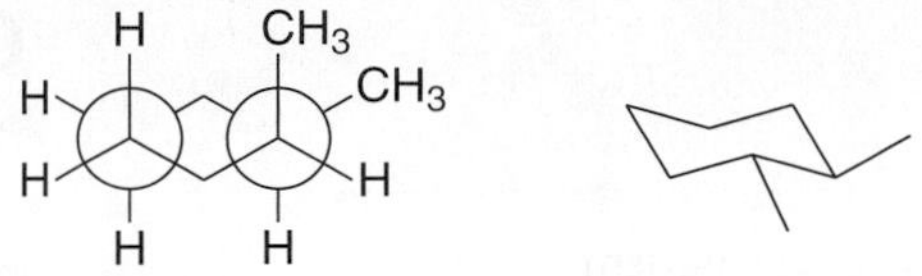

(i) 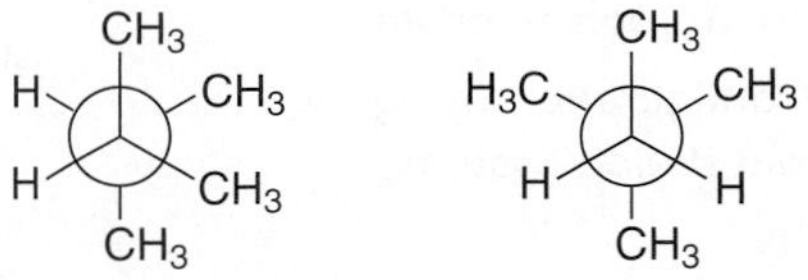

(j) 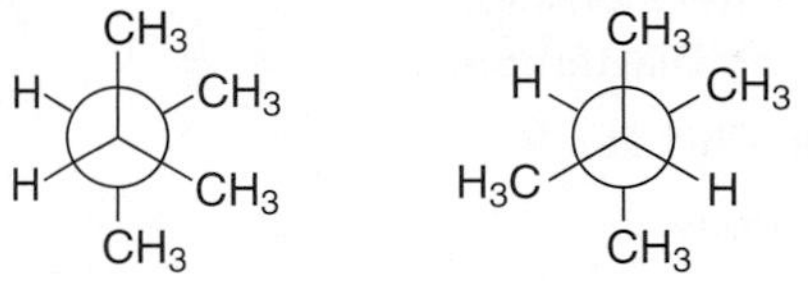

(k) 

(l) 

CHALLENGE PROBLEMS

4.67 Consider the structures of *cis*-1,2-dimethylcyclopropane and *trans*-1,2-dimethylcyclopropane:

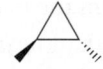

(a) Which compound would you expect to be more stable? Explain your choice.

(b) Predict the difference in energy between these two compounds.

4.68 If we compare the sizes of the halogens, we find that they increase in size from fluorine to iodine. Nevertheless, fluoroethane, chloroethane, bromoethane, and iodoethane all have very similar barriers to rotation about the C—C bond. Can you offer a possible explanation?

COMPOUND	BARRIER TO ROTATION (KJ / MOL)
Fluoroethane	13.8
Chloroethane	15.5
Bromoethane	15.5
Iodoethane	13.4

4.69 Consider the following tetra-substituted cyclohexane:

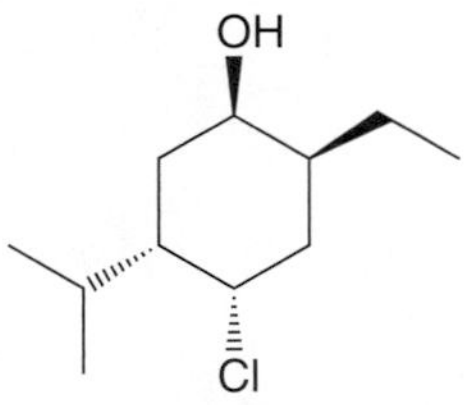

(a) Draw both chair conformations of this compound.

(b) Determine which conformation is more stable.

(c) At equilibrium, would you expect the compound to spend more than 95% of its time in the more stable chair conformation?

4.70 Consider the structures of *cis*-decalin and *trans*-decalin:

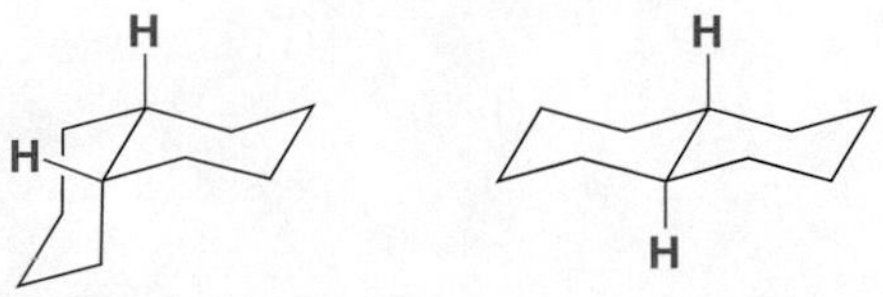

***cis*-Decalin** ***trans*-Decalin**

(a) Which of these compounds would you expect to be more stable?

(b) One of these two compounds is incapable of ring flipping. Identify it and explain your choice.

5

Stereoisomerism

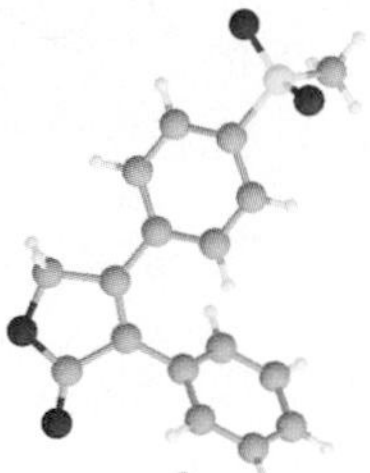

DID YOU EVER WONDER...
whether pharmaceuticals are really safe?

As mentioned in the previous chapter, most drugs cause multiple physiological responses. In general, one response is desirable, while the rest are undesirable. The undesirable responses can range in severity and can even result in death. This might sound scary, but the FDA (Food and Drug Administration) has strict guidelines that must be followed before a drug is approved for sale to the public. Every potential drug must first undergo animal testing followed by three separate clinical trials involving human patients (phases I, II, and III). Phase I involves a small number of patients (20–80), phase II can involve several hundred patients, and phase III can involve several thousand patients. The FDA will only approve drugs that have passed all three phases of clinical trials. Even after receiving approval to be sold, the effects of a drug are still monitored for potential long-term adverse effects. In some cases, a drug must be pulled off the market after adverse side effects are observed in a small percentage of the population. One such example is Vioxx, an anti-inflammatory drug that was heavily used in the treatment of osteoarthritis and acute pain. Vioxx was approved by the FDA in 1999 and instantly became very popular, providing its manufacturer (Merck & Co.) with over two billion dollars in annual sales. In 2004, Merck had to remove Vioxx from the market because of concerns that long-term use increased the risk of heart attacks and strokes. This example highlights the fact that pharmaceutical safety cannot be absolutely guaranteed. Nevertheless, the development of pharmaceuticals has vastly improved our quality of life and longevity, and the positive effects significantly outweigh the rare examples of negative effects.

continued >

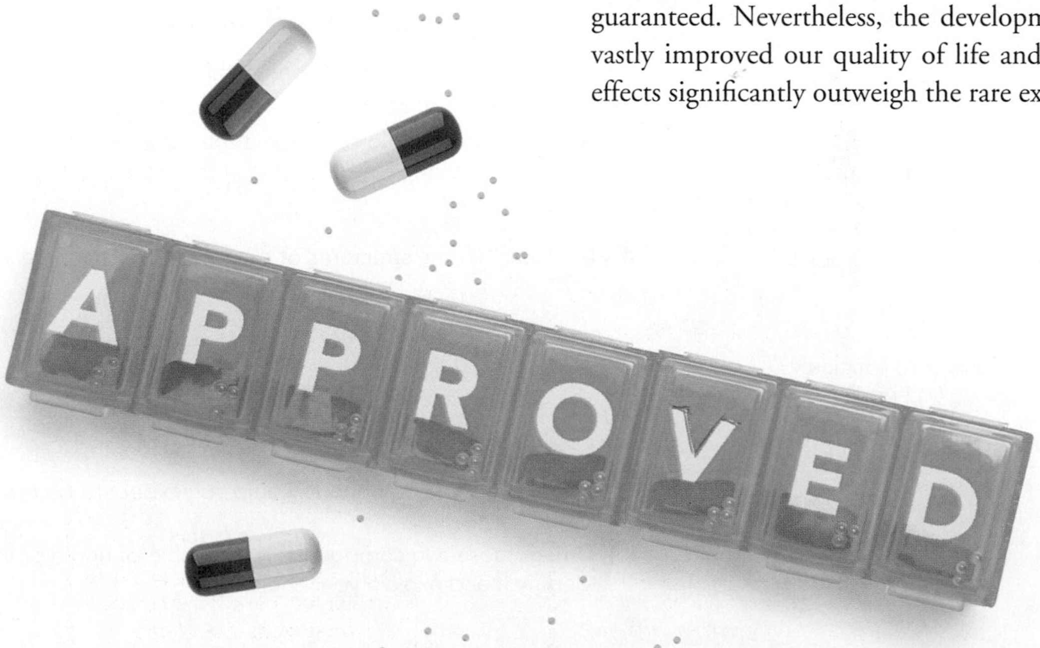

The effects of any particular drug are determined by a host of factors. We have already seen many of these factors, including the role of the pharmacophore, conformational flexibility, and acid-base properties. In this chapter, we will take a closer look at the three-dimensional structure of compounds, and we will see that this characteristic is arguably one of the most important factors to be considered when designing drugs and assessing their safety. In particular, we will explore compounds that differ from each other only in the three-dimensional, spatial arrangement of their atoms, but not in the connectivity of their atoms. Such compounds are called **stereoisomers**, and we will explore the connection between stereoisomerism and drug action.

This chapter will focus on the different kinds of stereoisomers. We will learn to identify stereoisomers, and we will learn several drawing styles that will allow us to compare stereoisomers. The upcoming chapters will focus on reactions that produce stereoisomers.

DO YOU REMEMBER?

Before you go on, be sure you understand the following topics.
If necessary, review the suggested sections to prepare for this chapter.

- Constitutional Isomerism (Section 1.2)
- Tetrahedral Geometry (Section 1.10)
- Three-Dimensional Representations (Section 2.1)
- Drawing and Interpreting Bond-Line Structures (Section 2.2)

PLUS Visit www.wileyplus.com to check your understanding and for valuable practice.

5.1 Overview of Isomerism

The term *isomers* comes from the Greek words *isos* and *meros*, meaning "made of the same parts". That is, isomers are compounds that are constructed from the same atoms (same molecular formula) but that still differ from each other. We have already seen two kinds of isomers: constitutional isomers (Section 4.3) and stereoisomers (Section 4.14), as illustrated in Figure 5.1. Constitutional isomers differ in the connectivity of their atoms; for example:

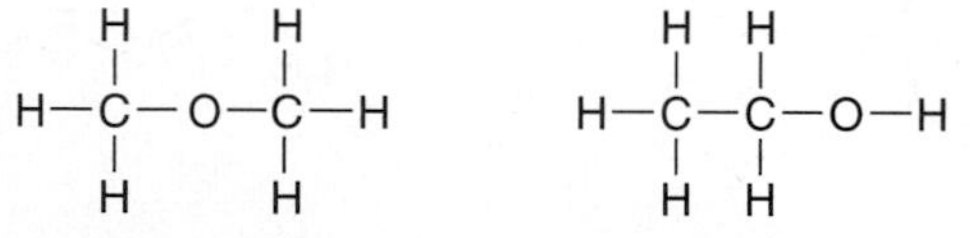

$$H-\overset{\displaystyle H}{\underset{\displaystyle H}{C}}-O-\overset{\displaystyle H}{\underset{\displaystyle H}{C}}-H \qquad H-\overset{\displaystyle H}{\underset{\displaystyle H}{C}}-\overset{\displaystyle H}{\underset{\displaystyle H}{C}}-O-H$$

Methoxymethane **Ethanol**
Boiling point = −23°C Boiling point = 78.4°C

The two compounds above have the same molecular formula, but they differ in their constitution. As a result, they are different compounds with different physical properties.

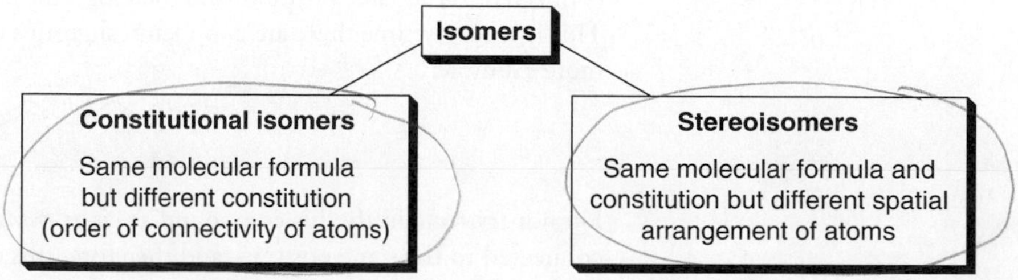

FIGURE 5.1
The main categories of isomers.

Stereoisomers are compounds that have the same constitution but differ in the spatial arrangement of their atoms. In the previous chapters, we discussed one example of *cis-trans* stereoisomerism among substituted cycloalkanes:

***cis*-** **1,2-Dimethylcyclohexane** ***trans*-** **1,2-Dimethylcyclohexane**

The *cis* stereoisomer exhibits groups on the same side of the ring, while the *trans* stereoisomer exhibits groups on opposite sides of the ring. In addition to the examples above, the terms *cis* and *trans* are also used to describe stereoisomerism among double bonds:

***cis*-2-Butene** Boiling point = 4°C ***trans*-2-Butene** Boiling point = 1°C

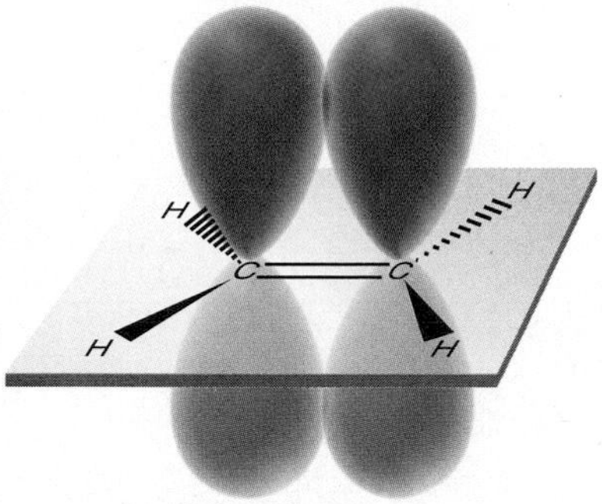

π Bond

FIGURE 5.2 An illustration of *p* orbitals overlapping to form a π bond.

The *cis* stereoisomer exhibits groups on the same side of the double bond, while the *trans* stereoisomer exhibits groups on opposite sides of the double bond. The two drawings above represent different compounds with different physical properties, because the double bond does not experience free rotation as single bonds do. Why not? Recall that a π bond is formed from the overlap of two *p* orbitals (Figure 5.2). Rotation about the C—C double bond would effectively destroy the overlap between the *p* orbitals. Therefore, the C—C double bond does not experience free rotation at room temperature.

In order to use the *cis-trans* terminology to differentiate stereoisomers, there must be two identical groups to compare:

Two fluorine atoms are *cis* **Two ethyl groups are *trans***

Hydrogen atoms can also be used to assign *cis-trans* terminology; for example:

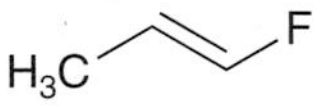

is *trans* because of the H's:

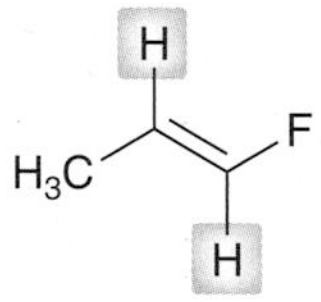

When two identical groups are connected to the same position, there cannot be *cis-trans* isomerism. For example, consider the following compound:

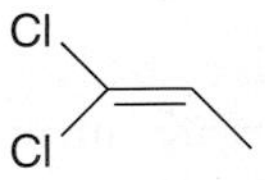

is the same as

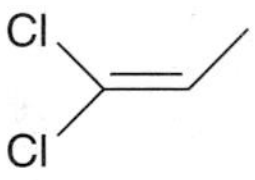

These two drawings represent the same compound. This compound has two chlorine atoms connected to the same position, and therefore, the compound does not exhibit stereoisomerism. This is true any time there are two identical groups connected to the same position. Here is one more example:

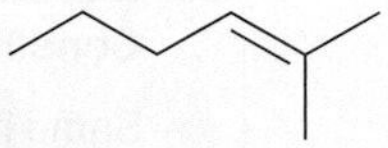

Do not try to identify this compound as *cis* or *trans*. Two identical groups (methyl groups) are connected to the same position, and therefore, this compound is neither *cis* nor *trans*.

SKILLBUILDER

5.1 IDENTIFYING *CIS-TRANS* STEREOISOMERISM

LEARN the skill

Determine whether the stereoisomer shown below exhibits a *cis* configuration or a *trans* configuration:

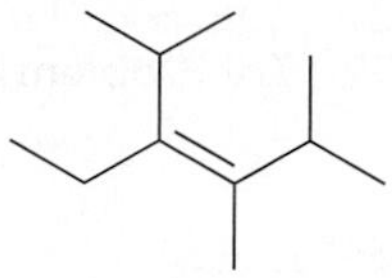

SOLUTION

Begin by circling the four groups attached to the double bond, and try to name them:

STEP 1
Identify and name all four groups attached to the double bond.

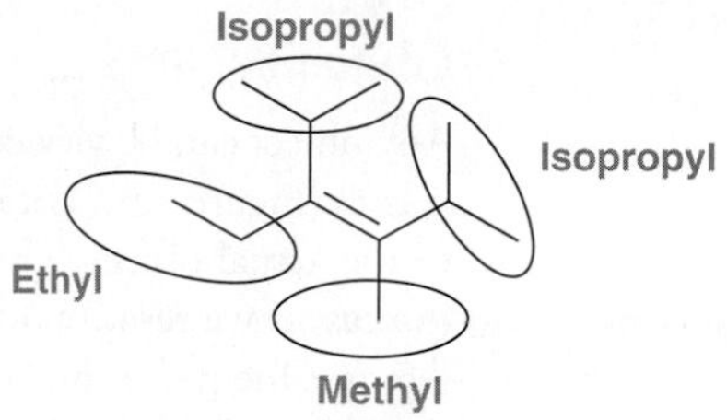

STEP 2
Look for two identical groups on different vinylic positions, and assign the configuration as *cis* or *trans*.

Naming the four groups helps to identify identical groups. There are always four groups on any double bond (even if some of those groups are just hydrogen atoms). In this case, naming the groups makes it evident that there are two isopropyl groups that are *cis* to each other.

PRACTICE the skill

5.1 For each of the following compounds determine whether it exhibits a *cis* configuration or a *trans* configuration or whether it is simply not stereoisomeric.

(a) 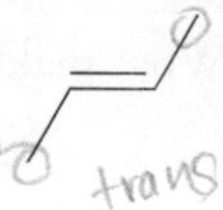

(b) 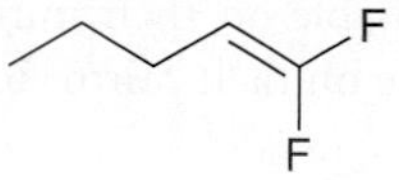

(c) 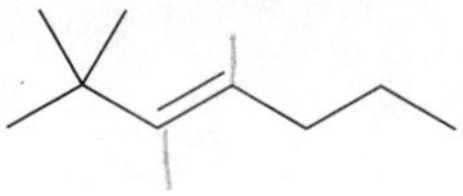

(d) 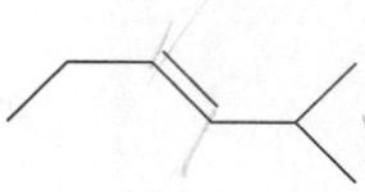

(e) 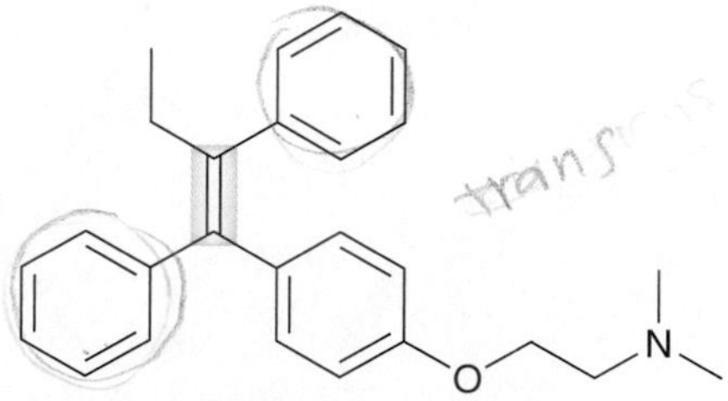

Tamoxifen
Used in treatment of breast cancer

(f) 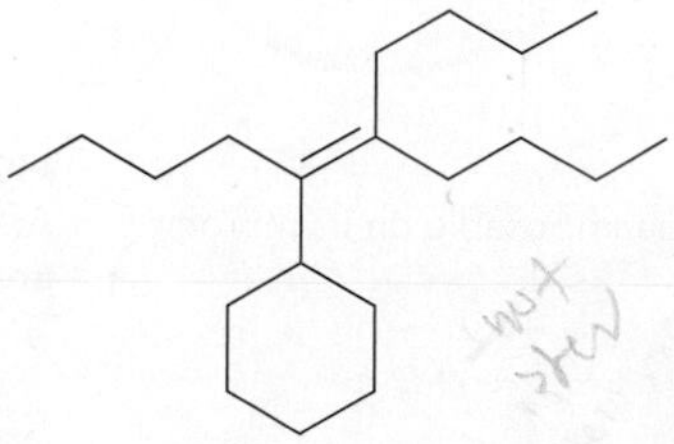

(g) 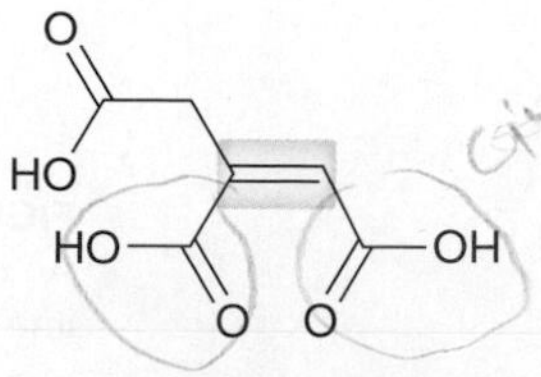

Aconitic acid
Involved in metabolism

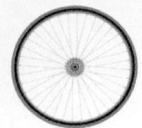

APPLY the skill

5.2 Identify the number of stereoisomers that are possible for a compound with the following constitution: $H_2CCHCH_2CH_2CH_2CHCH_2$.

5.3 Compound X and compound Y are constitutional isomers with molecular formula C_5H_{10}. Compound X possesses a carbon-carbon double bond in the *trans* configuration, while compound Y possesses a carbon-carbon double bond that is not stereoisomeric:

(a) Identify the structure of compound X.

(b) Identify four possible structures for compound Y.

need more **PRACTICE?** **Try Problem 5.35**

5.2 Introduction to Stereoisomerism

In the previous section, we reviewed *cis-trans* stereoisomerism, but there are many other kinds of stereoisomers. We begin our exploration of the various kinds of stereoisomers by investigating the relationship between an object and its mirror image.

Chirality

Any object can be viewed in a mirror, revealing its mirror image. Take, for example, a pair of sunglasses (Figure 5.3). For many objects, like the sunglasses in Figure 5.3 the mirror image is identical to the actual object. The object and its mirror image are said to be **superimposable**. This is not the case if we remove one of the lenses (Figure 5.4). The object and its mirror image are now different. One pair is missing the right lens, while the other pair is missing the left lens. In this case, the object and its mirror image are *nonsuperimposable*. Many familiar objects, such as hands, are nonsuperimposable on their mirror images. A right hand and a left hand are mirror images of one another, but they are not identical; they are not superimposable on one another. A left hand will not fit into a right-handed glove, and a right hand will not fit into a left-handed glove.

Objects, like hands, that are not superimposable on their mirror images are called **chiral** objects, from the Greek word *cheir* (meaning "hand"). All three-dimensional objects can be classified as either chiral or achiral. Molecules are three-dimensional objects and can therefore also be classified into one of these two categories. Chiral molecules are like hands: they are nonsuperimposable on their mirror images. Achiral molecules are not like hands: they are superimposable on their mirror images. What makes a molecule chiral?

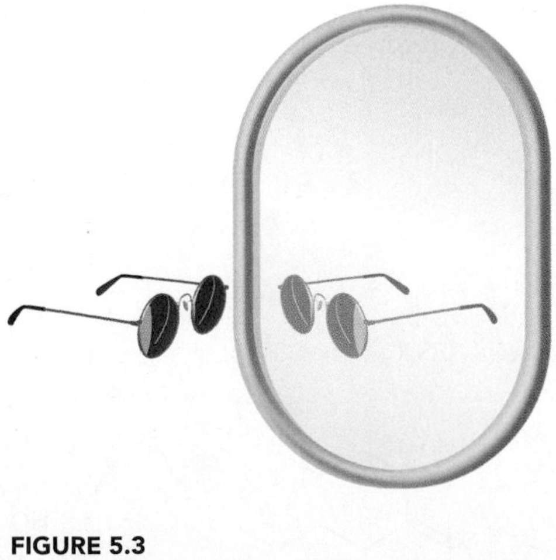

FIGURE 5.3
An object that is superimposable on its mirror image.

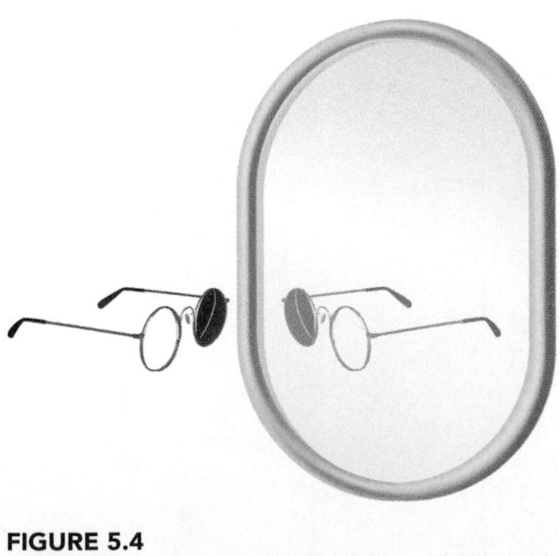

FIGURE 5.4
An object that is not superimposable on its mirror image.

chiral images

Chirality Centers

The most common source of molecular chirality is the presence of a carbon atom bearing four different groups. There are two different ways to arrange four groups around a central carbon atom (Figure 5.5). These two arrangements are nonsuperimposable mirror images.

LOOKING AHEAD
For other sources of molecular chirality, see the Challenge Problems at the end of this chapter.

FIGURE 5.5
There are two ways to arrange four different groups around a carbon atom.

Consider, for example, the structure of 2-butanol, which can be arranged in two ways in three-dimensional space:

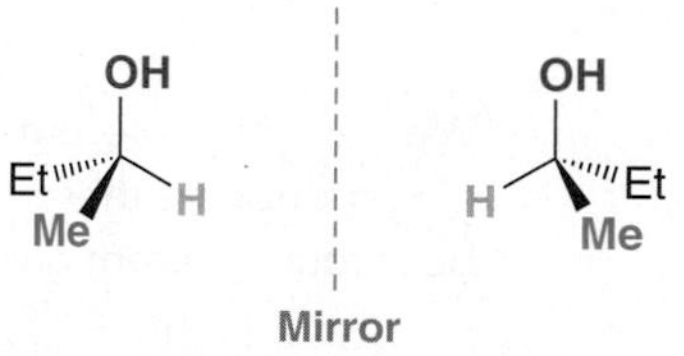

BY THE WAY
To see that these two compounds are nonsuperimposable, build a molecular model using any one of the commercially available molecular model kits.

These two compounds are nonsuperimposable mirror images, and they represent two different compounds. These compounds differ from each other only in the spatial arrangement of their atoms, and therefore, they are stereoisomers.

In 1996, the IUPAC recommended that a tetrahedral carbon bearing four different groups be called a **chirality center.** Many other names in common use include *chiral center, stereocenter, stereogenic center*, and *asymmetric center*. For the remainder of our discussion, we will use the IUPAC-recommended term. Below are several examples of chirality centers:

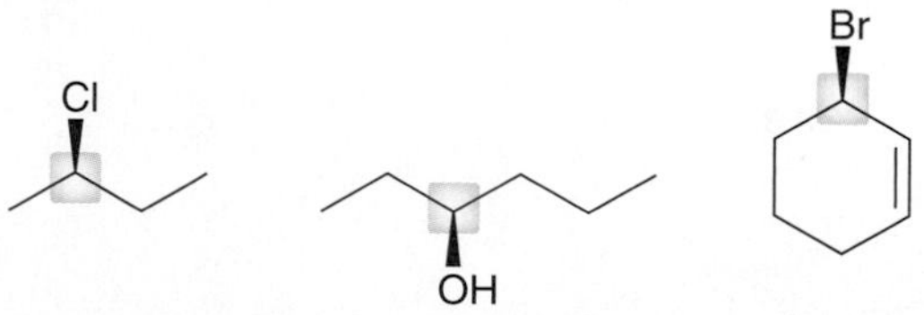

Each of the highlighted carbon atoms bears four different groups. In the last compound, the carbon atom is a chirality center because one path around the ring is different than the other path (one path encounters the double bond sooner). In contrast, the following compound does not have a chirality center. In this case, the clockwise path and the counterclockwise path are identical.

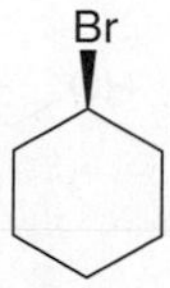

BY THE WAY
Despite the IUPAC recommendation to use the term "chirality center," the terms stereocenter and stereogenic center are more commonly used. However, these terms have a broader definition: a stereocenter, or stereogenic center, is defined as a location at which the interchange of two substituents will generate a stereoisomer. This definition includes chirality centers, but it also includes *cis* and *trans* double bonds, which are not chirality centers.

SKILLBUILDER

5.2 LOCATING CHIRALITY CENTERS

LEARN the skill

Propoxyphene, sold under the trade name Darvon™, is an analgesic (painkiller) and antitussive (cough suppressant). Identify all chirality centers in propoxyphene:

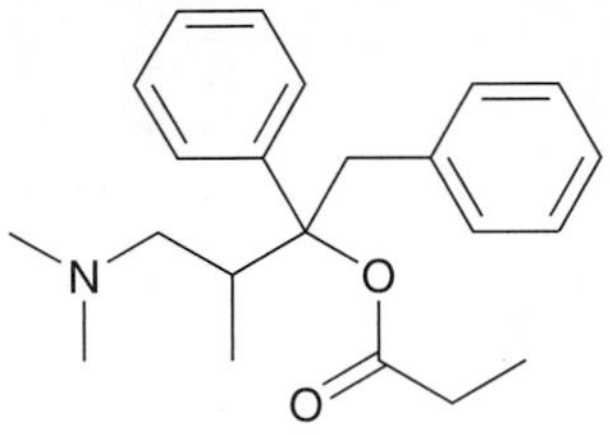

SOLUTION

STEP 1
Ignore sp^2- and sp-hybridized carbon atoms.

We are looking for a tetrahedral carbon atom that bears four different groups. Each of the sp^2-hybridized carbon atoms are connected to only three groups, rather than four. So, none of these carbon atoms can be chirality centers:

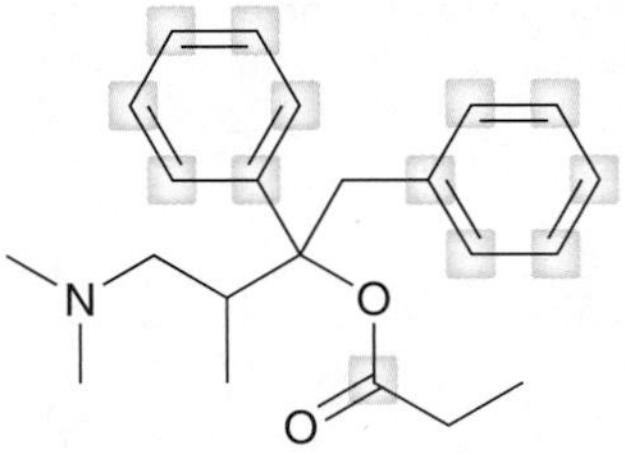

These carbon atoms cannot be chirality centers

STEP 2
Ignore CH_2 and CH_3 groups.

We can also rule out any CH_2 groups or CH_3 groups, as these carbon atoms do not bear four *different* groups:

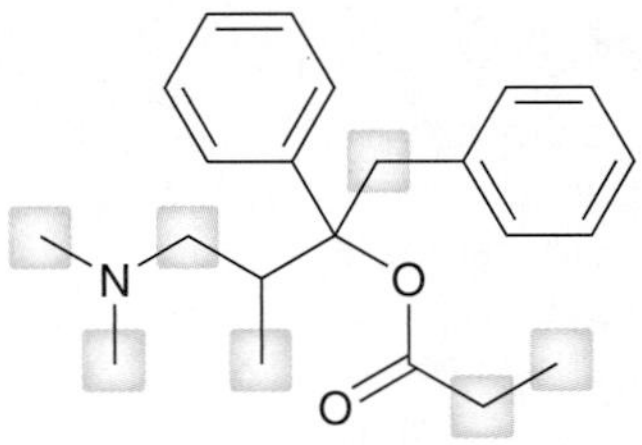

These carbon atoms cannot be chirality centers

It is helpful to become proficient at identifying carbon atoms that cannot be chirality centers. This will make it easier to spot the chirality centers more quickly. In this example, there are only two carbon atoms worth considering:

STEP 3
Identify any carbon atoms bearing four different groups.

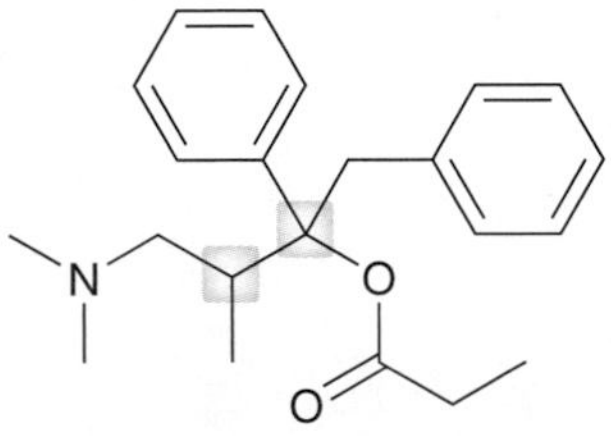

Each of these positions has four different groups, so these two positions are chirality centers.

PRACTICE the skill

5.4 Identify all chirality centers in each of the following compounds:

(a)

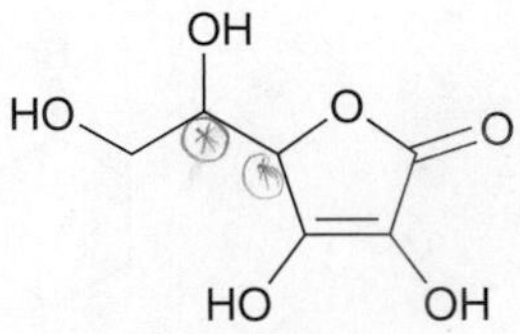

Ascorbic acid (Vitamin C)

(b)

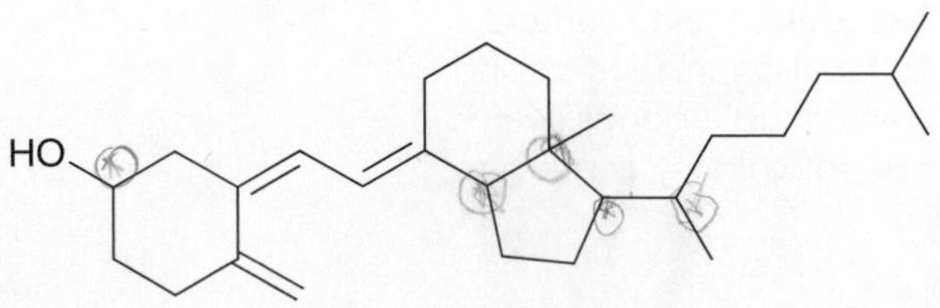

Vitamin D_3

(c)

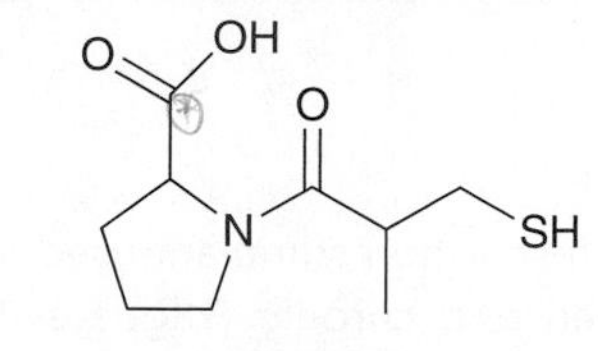

Captopril
Used to treat high blood pressure

(d)

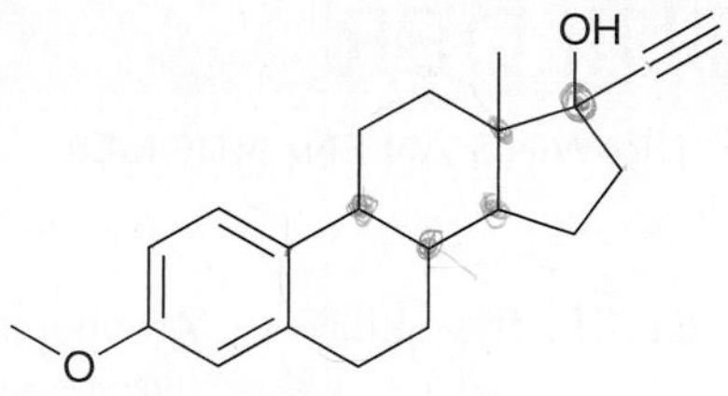

Mestranol
An oral contraceptive

(e)

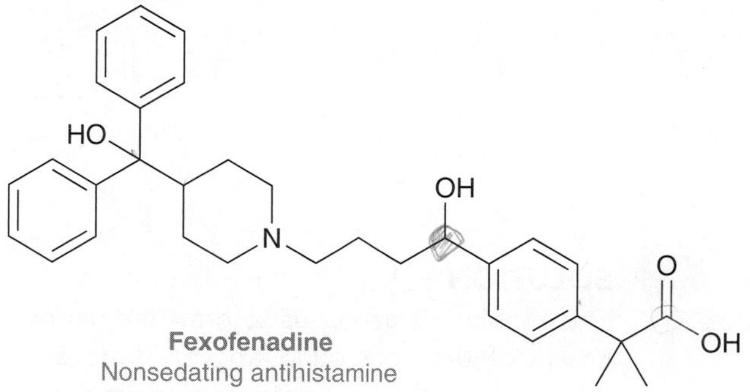

Fexofenadine
Nonsedating antihistamine

(f)

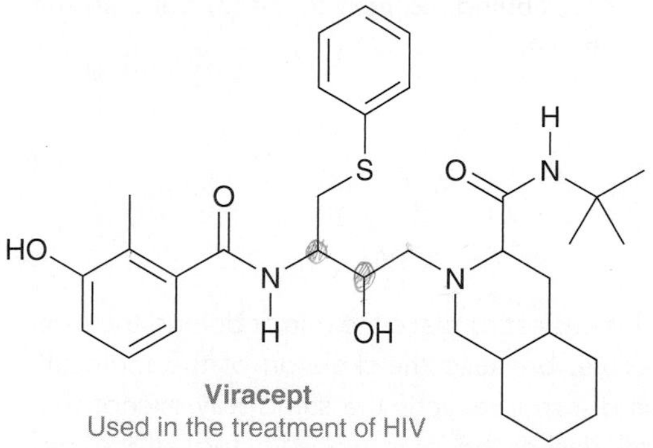

Viracept
Used in the treatment of HIV

APPLY the skill

5.5 Draw all constitutional isomers of C_4H_9Br, and identify the isomer(s) that possess chirality centers.

5.6 Do you expect the following compound to be chiral? Explain your answer (consider whether this compound is superimposable on its mirror image).

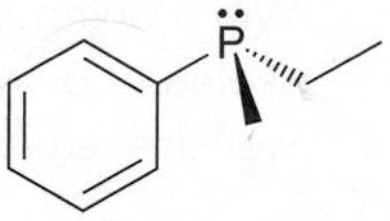

need more **PRACTICE?** **Try Problems 5.34b, 5.48**

Enantiomers

When a compound is chiral, it will have one nonsuperimposable mirror image, called its **enantiomer** (from the Greek word meaning "opposite"). The compound and its mirror image are said to be a pair of enantiomers. The word "enantiomer" is used in speech in the same way that the word "twin" is used in speech. When two children are *a pair of twins*, each one is said to be *the twin of the other*. Similarly, when two compounds are *a pair of enantiomers*, each compound is said to be *the enantiomer of the other*. A chiral compound will have exactly one enantiomer; never more and never less. Let's practice drawing the enantiomer of chiral compounds.

SKILLBUILDER

5.3 DRAWING AN ENANTIOMER

LEARN the skill

Amphetamine is a prescription stimulant used in the treatment of ADHD (attention-deficit hyperactivity disorder) and chronic fatigue syndrome. During World War II it was used heavily by soldiers to reduce fatigue and increase alertness. Draw the enantiomer of amphetamine.

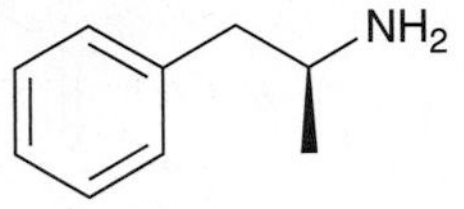

Amphetamine

SOLUTION

This problem is asking us to draw the mirror image of the compound above. There are three ways to do this, because there are three places where we can imagine placing a mirror: (1) behind, (2) next to, or (3) beneath the molecule:

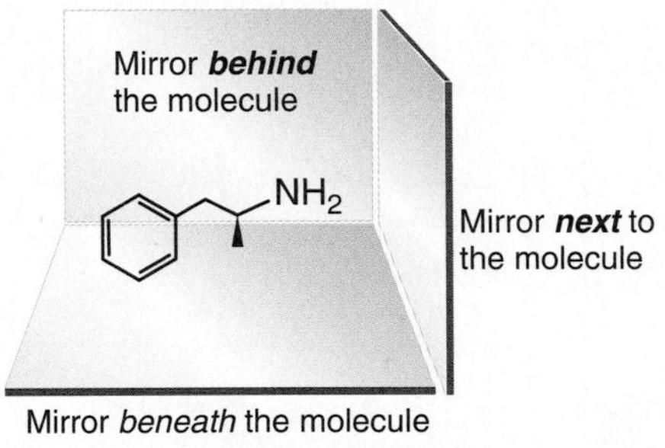

BY THE WAY
In most cases, it is easiest to draw a mirror image by placing the mirror behind the molecule.

It is easiest to place the mirror behind the molecule, because the skeleton of the molecule is drawn in exactly the same way except that *all dashes become wedges and all wedges become dashes.* Every other aspect of the molecule is drawn in exactly the same way.

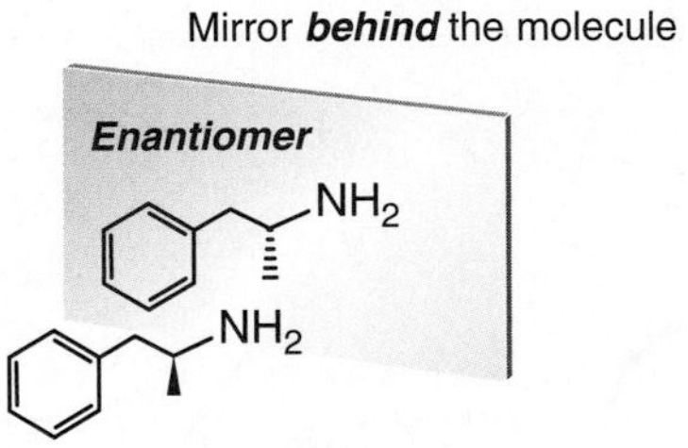

The second way to draw an enantiomer is to place the mirror on the side of the molecule. When doing so, draw the mirror image of the skeleton, but all dashes remain dashes and all wedges remain wedges:

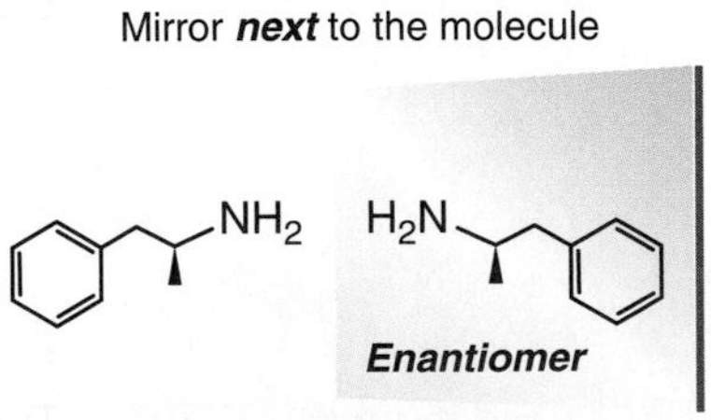

Finally, we can place the mirror under the molecule. When doing so, draw the mirror image of the skeleton; once again, all dashes remain dashes and all wedges remain wedges:

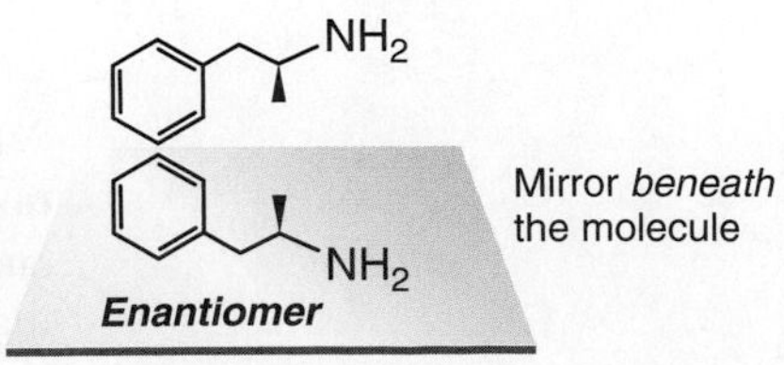

Of the three ways to draw an enantiomer, only the first method involves interchanging wedges and dashes. Even so, make sure to understand that all three of these methods produce the same answer:

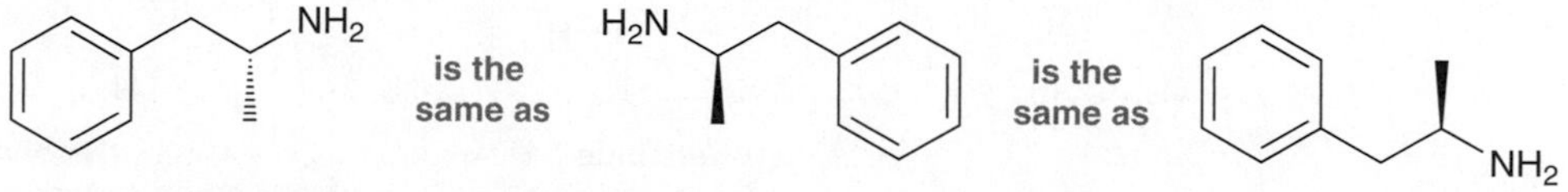

All three drawings represent the same compound. They might look different, but rotation of any one drawing will generate the other drawings. Remember that a molecule can only have one enantiomer, so all three methods must produce the same answer.

In general, it is easiest to use the first method. Simply redraw the compound, replacing all dashes with wedges and all wedges with dashes. However, this method will not work in all situations. In some molecular drawings, wedges and dashes are not drawn, because the three-dimensional geometry is implied by the drawing. This is the case for bicyclic compounds. When dealing with bicyclic compounds, it will be easier to use one of the other two methods (placing the mirror either on the side of the molecule or below the molecule). For example:

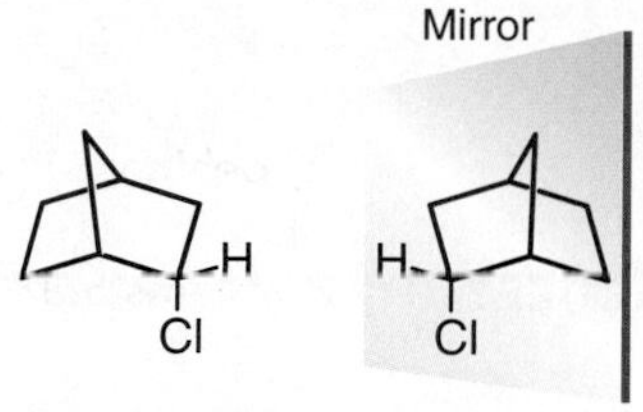

Compound A ***Compound B***

In this case, it is easiest to place the mirror on the side of compound A, which allows us to draw compound B (the enantiomer of compound A).

PRACTICE the skill **5.7** Draw the enantiomer of each of the following compounds:

(a)
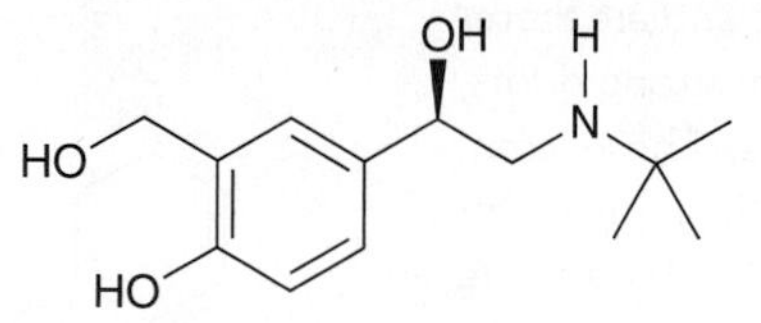

Albuterol
(sold under the trade name Ventolin™)
A bronchodilator used in the treatment of asthma

(b)
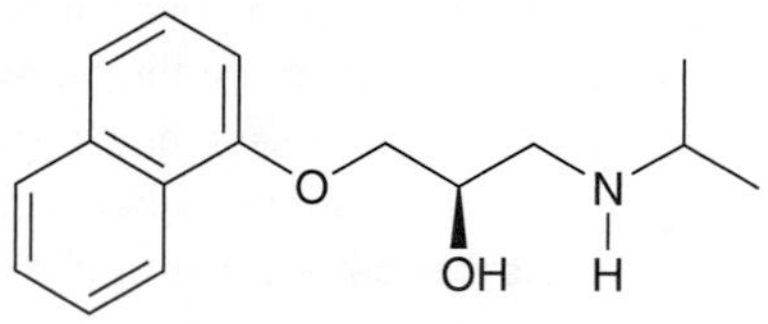

Propranolol
A beta blocker used in the treatment of hypertension

(c)
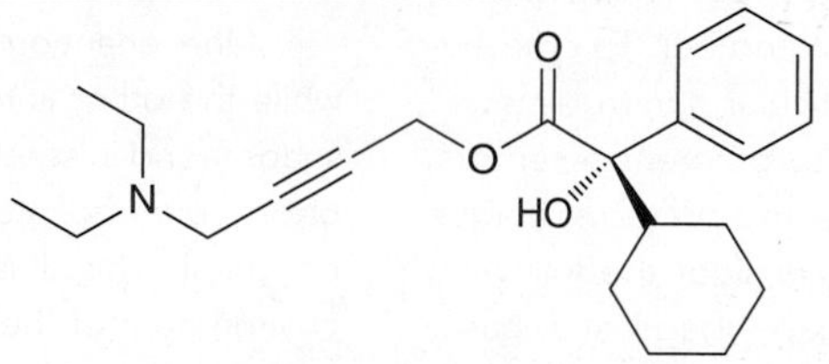

Oxybutynin
Used to treat urinary and bladder disorders

(d)
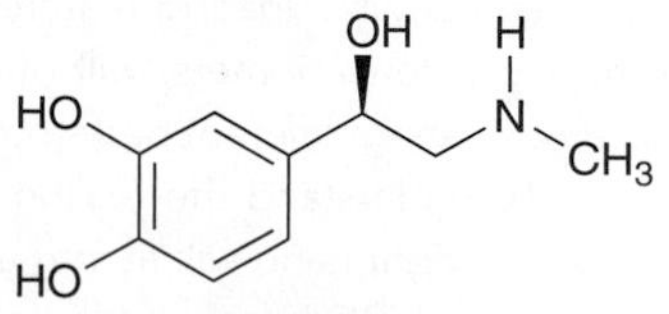

Adrenaline
A hormone that acts as a bronchodilator

(e) 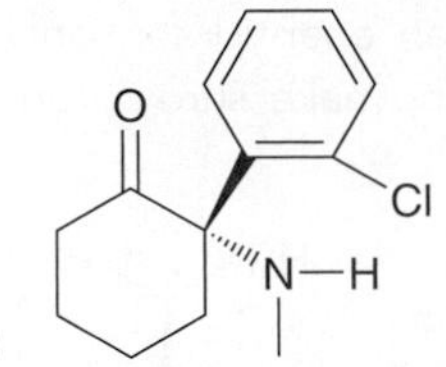

Ketamine
An anesthetic

(f) 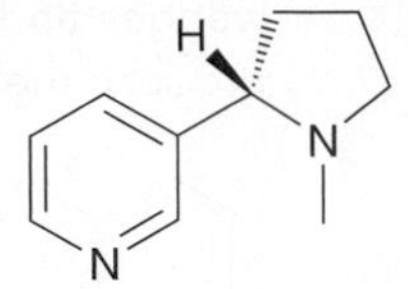

Nicotine
An addictive substance present in tobacco

(g) 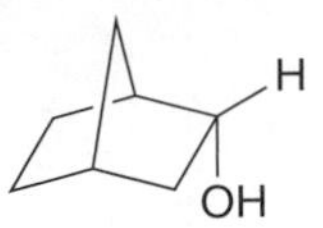

APPLY the skill

5.8 Ixabepilone is a cytotoxic compound approved by the FDA in 2007 for the treatment of advanced breast cancer. Bristol-Myers Squibb is marketing this new drug under the trade name Ixempra. Draw the enantiomer of this compound:

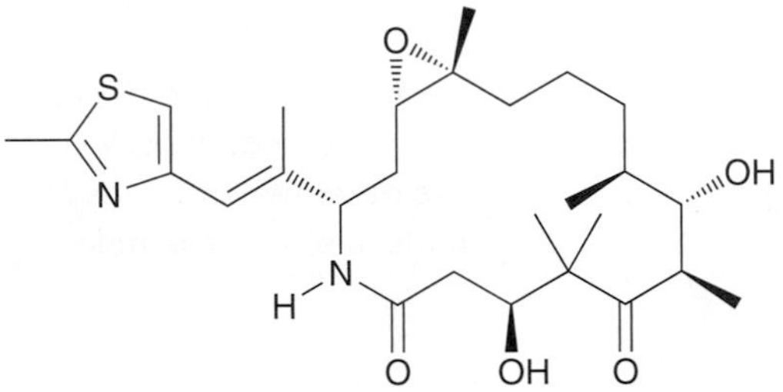

Ixabepilone

need more **PRACTICE?** **Try Problems 5.33, 5.34a, 5.38a-f, i-l**

PRACTICALLYSPEAKING)))

The Sense of Smell

In order for an object to have an odor, it must release organic compounds into the air. Most plastic and metal items do not release molecules into the air at room temperature and are therefore odorless. Spices, on the other hand, have very strong odors, because they release many organic compounds. These compounds enter your nose when you inhale, where they encounter receptors that detect their presence. The compounds bind to the receptors, causing the transmission of nerve signals that the brain interprets as an odor.

A specific compound can bind to a number of different receptors, creating a pattern that the brain identifies as a particular odor. Different compounds generate different patterns, enabling us to distinguish over 10,000 odors from one another. This mechanism has many fascinating features. In particular, "mirror-image" compounds (enantiomers) will often bind to different receptors, thereby creating different patterns which are interpreted as unique odors. To understand the reason for this, consider the following analogy. A right hand will fit into a brown paper bag just as easily as a left hand will, because the bag is not chiral. However, a right hand will not fit into a left-handed glove just as easily as a left hand will, because the glove itself is chiral. Similarly, if a receptor is chiral, as is usually the case, then we might expect that only one enantiomer of a pair will effectively bind to it.

As an example, carvone has one chirality center and therefore has two enantiomeric forms:

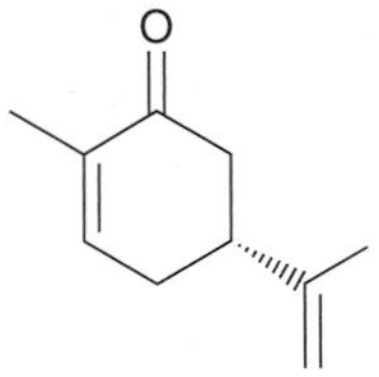

(*R*)-Carvone
(Odor of spearmint)

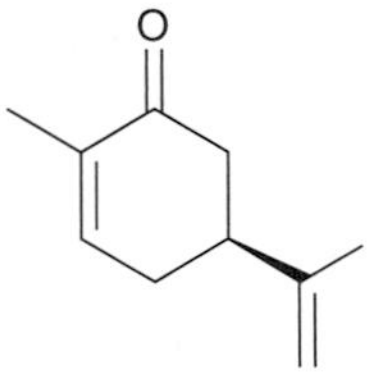

(*S*)-Carvone
(Odor of caraway seeds)

One enantiomer is responsible for the odor of spearmint, while the other is responsible for the odor of caraway seeds (the seeds found in seeded rye bread). In order for us to detect different odors, our receptors must themselves be chiral. This illustrates the chiral environment of the human body (a sea of chiral molecules interacting with each other). We will soon explore how drug action is also determined by the chiral nature of most biological receptors.

5.3 Designating Configuration Using the Cahn-Ingold-Prelog System

In the previous section, we used bond-line structures to illustrate the difference between a pair of enantiomers:

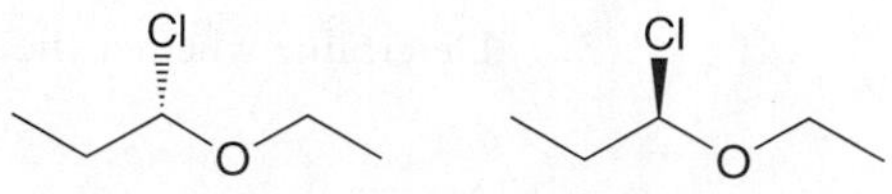

A pair of enantiomers

LOOKING AHEAD
A summary of the Cahn-Ingold-Prelog system appears immediately prior to SkillBuilder 5.4.

In order to communicate more effectively, we also need a system of nomenclature for identifying each enantiomer individually. This system is named after the chemists who devised it: Cahn, Ingold, and Prelog. The first step of this system involves assigning priorities to each of the four groups attached to the chirality center, based on atomic number. The atom with the highest atomic number is assigned the highest priority (1), while the atom with the lowest atomic number is assigned the lowest priority (4). As an example, consider one of the enantiomers from above:

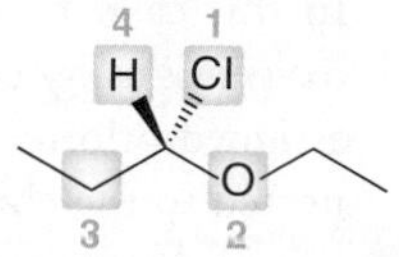

Of the four atoms attached to the chirality center, chlorine has the highest atomic number so it is assigned priority 1. Oxygen has the second highest atomic number, so it is assigned priority 2. Carbon is assigned priority 3, and finally, hydrogen is priority 4 because it has the lowest atomic number.

After assigning priorities to all four groups, we then rotate the molecule so that the fourth priority is directed behind the page (on a dash):

BY THE WAY
To visualize this rotation, you may find it helpful to build a molecular model using any one of the commercially available molecular model kits.

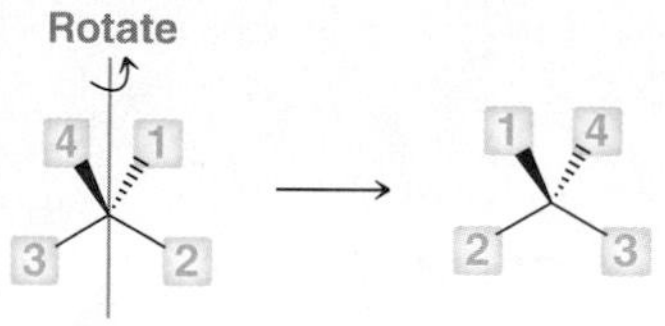

Then, we look to see if the sequence 1-2-3 is clockwise or counterclockwise. In this enantiomer, the sequence is counterclockwise:

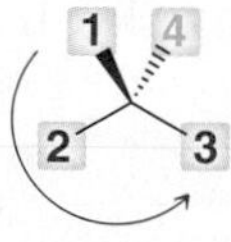

Counterclockwise = ***S***

A counterclockwise sequence is designated as ***S*** (from the Latin word *sinister*, meaning "left"). The enantiomer of this compound will have a clockwise sequence and is designated as ***R*** (from the Latin word *rectus*, meaning "right"):

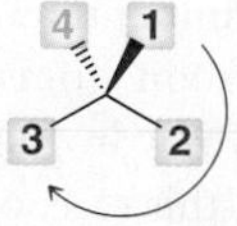

Clockwise = ***R***

The descriptors *R* and *S* are used to describe the **configuration** of a chirality center. Assigning the configuration of a chirality center therefore requires three distinct steps. The following sections will deal with the subtle nuances of these steps.

1. Prioritize all four groups connected to the chirality center.
2. If necessary, rotate the molecule so that the fourth priority group is on a dash.
3. Determine whether the sequence 1-2-3 is clockwise or counterclockwise.

Assigning Priorities to All Four Groups

The previous example was fairly straightforward because all four atoms attached to the chirality center were different (Cl, O, C, and H). It is more common to encounter two or more atoms with the same atomic number. For example:

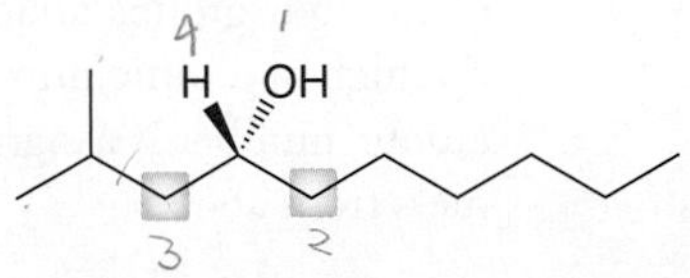

In this case, two carbon atoms are directly connected to the chirality center. Certainly, the oxygen is assigned priority 1 and the hydrogen is assigned priority 4. But which carbon atom is assigned priority 2? In order to make this determination, notice that each carbon atom is connected to the chirality center *and to three other atoms*. Make a list of the three atoms on either side in decreasing order of atomic number and compare:

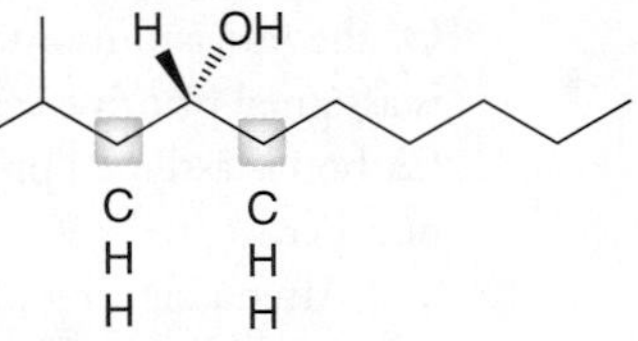

These lists are identical, so we move farther away from the chirality center and repeat the process:

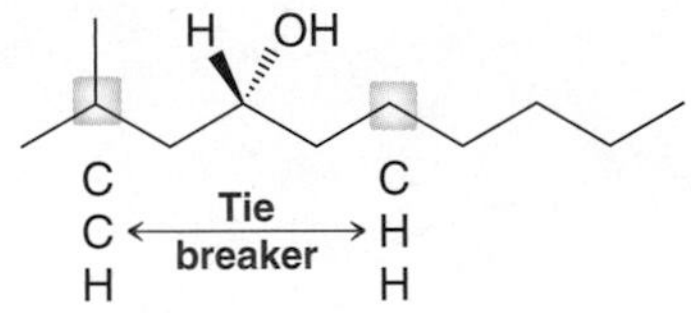

We are specifically looking for the first point of difference. In this case, the left side will be assigned the higher priority, because C has a higher atomic number than H.

Here is another example that illustrates an important point:

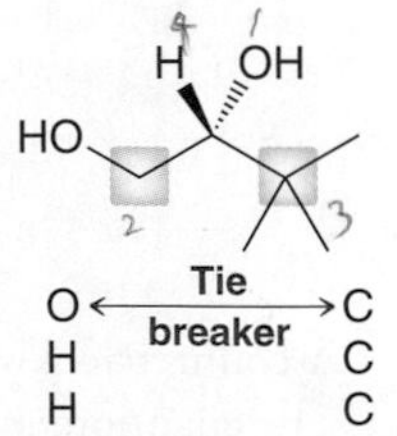

In this case, the left side will be assigned the higher priority, because oxygen has a higher atomic number than carbon. We do not compare the sum total of each list. It might be true that three carbon atoms have a greater sum total of atomic numbers (6+6+6) than one oxygen atom and two hydrogen atoms (8+1+1). However, the deciding factor is the first point of difference, and in this case, oxygen beats carbon.

When assigning priorities, a double bond is considered as two separate single bonds. The following carbon atom (highlighted) is treated as if it is connected to two oxygen atoms. The same rule is applied for any type of double bond.

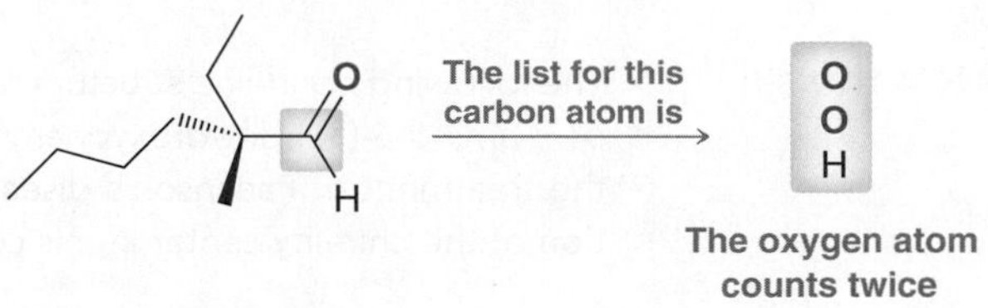

Rotating the molecule

Some students have trouble rotating a molecule and redrawing it in the proper perspective (with the fourth priority on a dash). If you are having trouble, there is a helpful technique based on the following principle: Switching any two groups on a chirality center will invert the configuration:

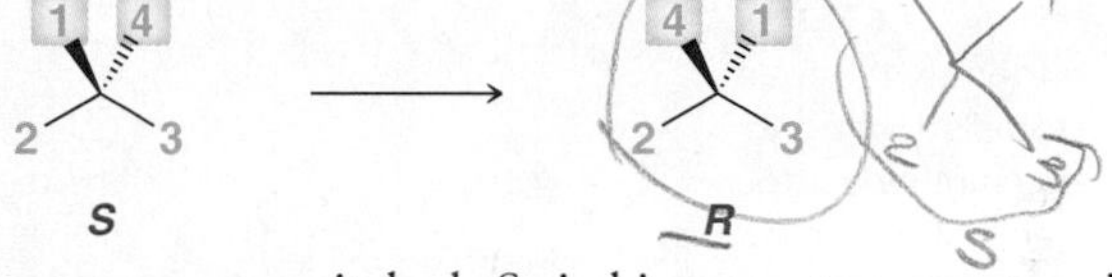

It doesn't matter which two groups are switched. Switching any two groups will change the configuration. Then, switching any two groups a second time will switch the configuration back. Based on this idea, we can devise a simple procedure that will allow us to rotate a molecule without actually having to visualize the rotation. Consider the following example:

To assign the configuration of this chirality center, first find the fourth position and switch it with the group that is on the dash. Then, switch the other two groups:

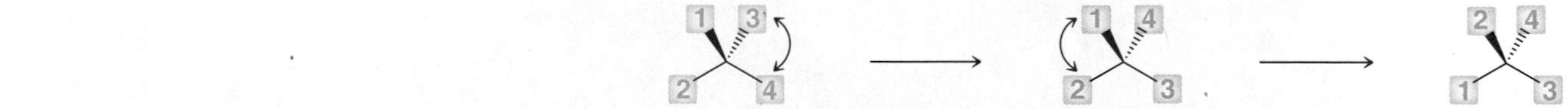

Two switches have been performed, so the final drawing must have the same configuration as the original drawing. But now, the molecule has been redrawn in a perspective that exhibits the fourth priority on a dash. This technique provides a method for redrawing the molecule in the appropriate perspective (as if the molecule had been rotated). In this example, the sequence 1-2-3 is clockwise, so the configuration is *R*. A summary of the procedure for assigning configuration follows.

A REVIEW OF CAHN-INGOLD-PRELOG RULES: ASSIGNING THE CONFIGURATION OF A CHIRALITY CENTER

STEP 1	STEP 2	STEP 3	STEP 4	STEP 5
Identify the four atoms directly attached to the chirality center.	Assign a priority to each atom based on its atomic number. The highest atomic number receives priority 1, and the lowest atomic number (often a hydrogen atom) receives priority 4.	If two atoms have the same atomic number, move away from the chirality center looking for the first point of difference. When constructing lists to compare, remember that a double bond is treated as two separate single bonds.	Rotate the molecule so that the fourth priority is on a dash (going behind the plane of the page).	Determine whether the sequence 1-2-3 follows a clockwise order (*R*) or a counterclockwise order (*S*).

SKILLBUILDER

5.4 ASSIGNING THE CONFIGURATION OF A CHIRALITY CENTER

LEARN the skill

The following bond-line structure represents one enantiomer of 2-amino-3-(3,4-dihydroxyphenyl)propanoic acid, used in the treatment of Parkinson's disease. Assign the configuration of the chirality center in this compound.

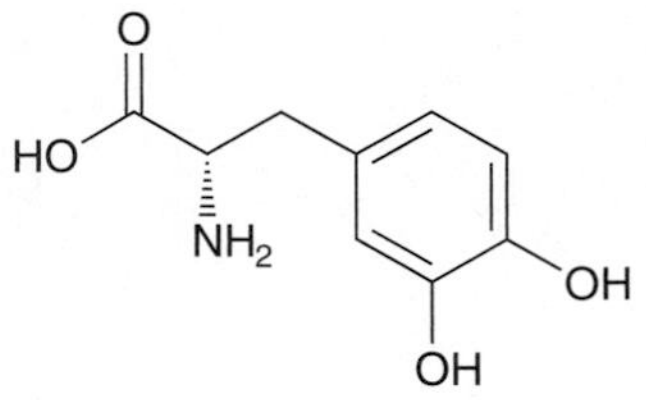

SOLUTION

Begin by identifying the four atoms directly attached to the chirality center:

STEP 1
Identify the four atoms attached to the chirality center and prioritize by atomic number.

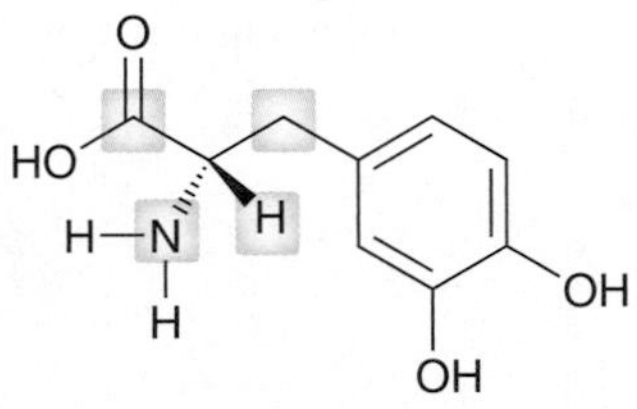

STEP 2
If two or more atoms are carbon, look for the first point of difference.

The four atoms are N, C, C, and H. Nitrogen has the highest atomic number, so it is assigned priority 1. Hydrogen is assigned priority 4. We must now decide which carbon atom is assigned priority 2. Make a list in each case and look for the first point of difference:

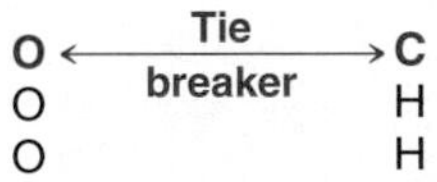

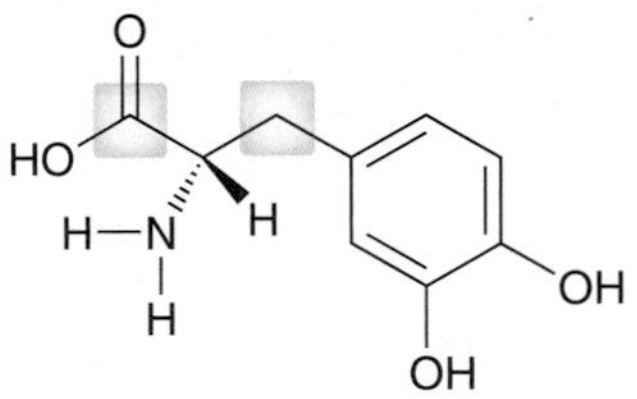

STEP 3
Redraw the chirality center showing only priorities.

Oxygen has a higher atomic number than carbon, so the left side is assigned the higher priority. The priorities are therefore arranged in the following way:

STEP 4
Rotate the molecule so that the fourth priority is on a dash.

In this case, the fourth priority is not on a dash, so the molecule must be rotated into the appropriate perspective:

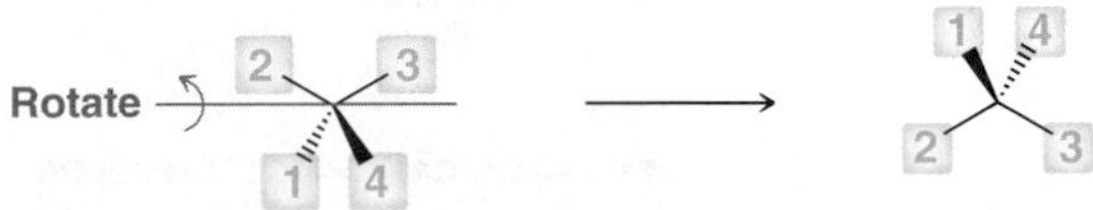

STEP 5
Assign the configuration based on the order of the sequence 1-2-3.

The sequence of 1-2-3 is counterclockwise, so the configuration is *S*. If the rotation step is difficult to see, the technique described earlier will help. Switch the 4 and the 1, so that the 4 is on a dash. Then switch the 2 and the 3. After two successive switches, the configuration remains the same:

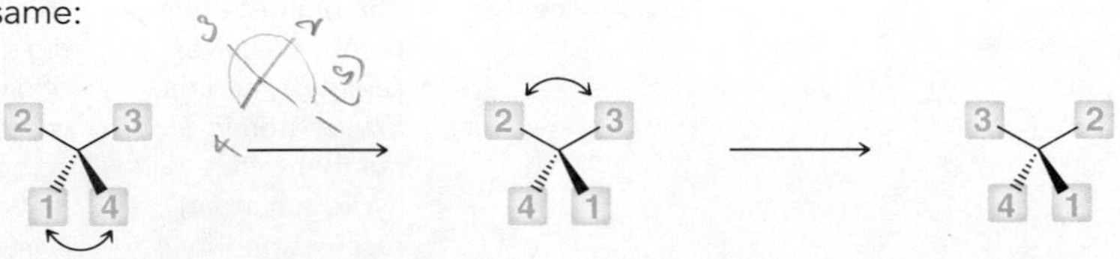

Now the 4 is on a dash, where it needs to be. The sequence of 1-2-3 is counterclockwise, so the configuration is *S*.

PRACTICE the skill **5.9** Each of the following compounds possesses carbon atoms that are chirality centers. Locate each of these chirality centers, and identify the configuration of each one:

(a)

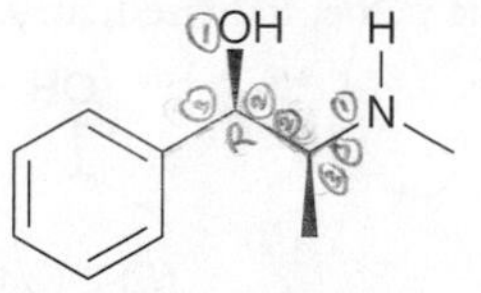

Ephedrine
A bronchodilator and decongestant obtained from the Chinese plant *Ephedra sinica*

(b)

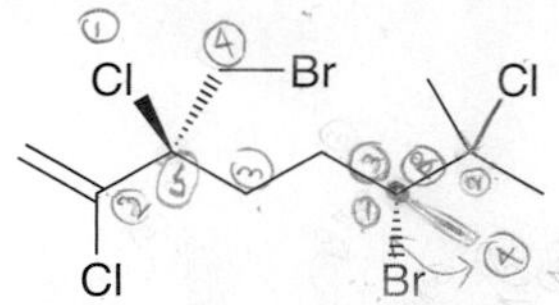

Halomon
An antitumor agent isolated from marine organisms

(c)

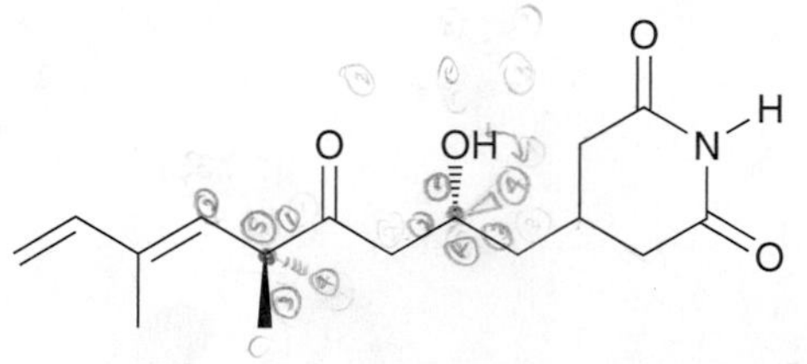

Streptimidone
An antibiotic

(d)

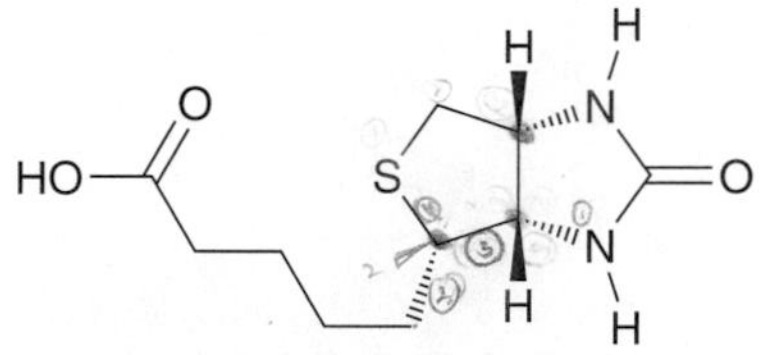

Biotin
(vitamin B_7)

(e)

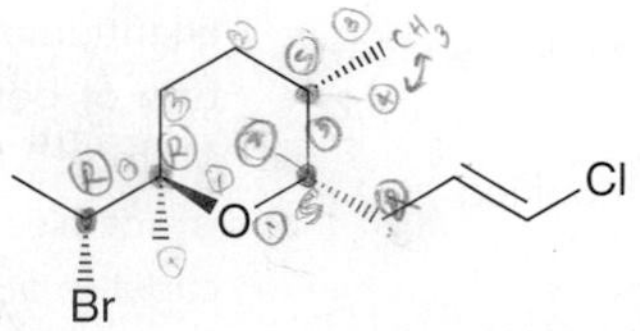

Kumepaloxane
A signal agent produced by *Haminoea cymbalum*, a snail indigenous to Guam

(f)

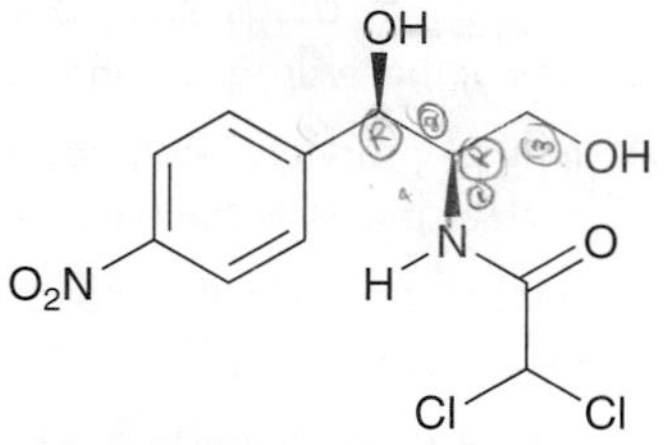

Chloramphenicol
An antibiotic agent isolated from the *Streptomyces venzuelae* bacterium

APPLY the skill **5.10** Assign the configuration of the chirality center in the following compound:

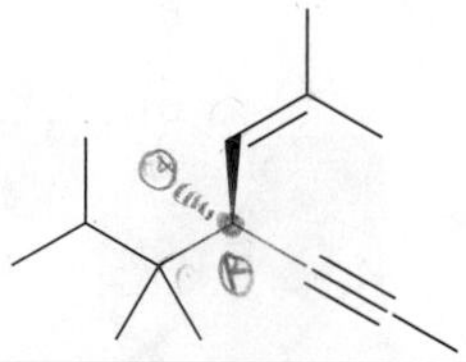

5.11 Carbon is not the only element that can function as a chirality center. In Problem 5.6 we saw an example in which a phosphorus atom is a chirality center. In such a case, the lone pair is always assigned the fourth priority. Using this information, assign the configuration of the chirality center in this compound.

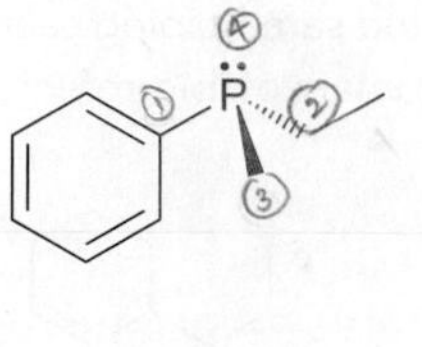

need more **PRACTICE?** **Try Problems 5.32, 5.39a-g, i, 5.45**

Designating Configuration in IUPAC Nomenclature

When naming a chiral compound, the configuration of the chirality center is indicated at the beginning of the name, italicized, and surrounded by parentheses:

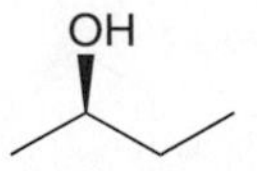

(*R*)-2-Butanol

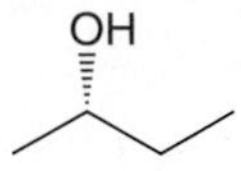

(*S*)-2-Butanol

When multiple chirality centers are present, each configuration must be preceded by a locant (a number) to indicate its location on the parent chain:

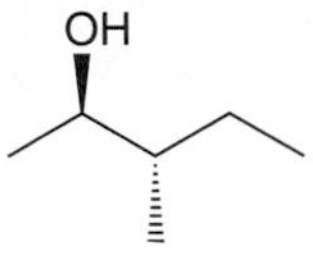

(2*R*,3*S*)-3-Methyl-2-pentanol

MEDICALLYSPEAKING

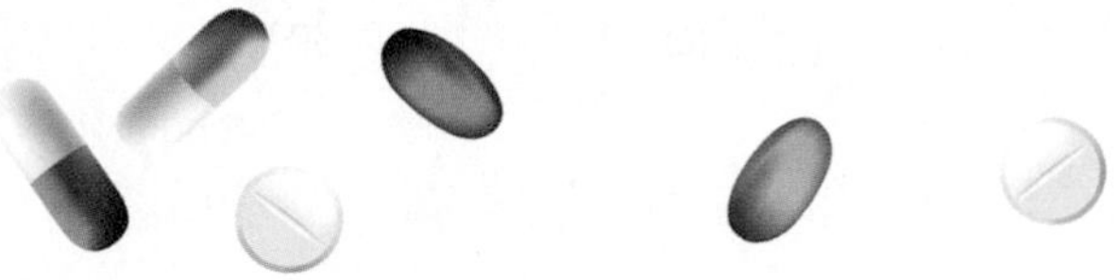

Chiral Drugs

Thousands of drugs are marketed throughout the world. The origins of these drugs fall into three categories:

1. natural products—compounds isolated from natural sources, such as plants or bacteria,
2. natural products that have been chemically modified in the laboratory, or
3. synthetic compounds (made entirely in the laboratory).

Most drugs obtained from natural sources consist of a single enantiomer. It is important to realize that a pair of enantiomers will rarely exhibit the same potency. We have seen in previous chapters that drug action is usually the result of drug-receptor binding. If the drug binds to the receptor in at least three places (called three-point binding), then one enantiomer of the drug may be more capable of binding with the receptor:

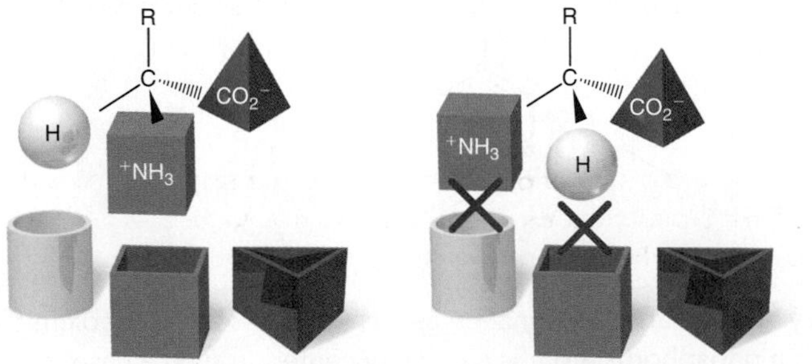

The first compound (left) can bind with the receptor, while its enantiomer (right) cannot bind with the receptor. For this reason, enantiomers will rarely produce the same biological response. As an example, consider the enantiomers of ibuprofen:

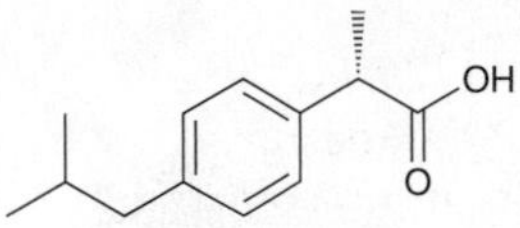

(*S*)-Ibuprofen

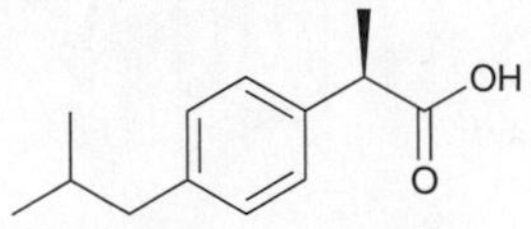

(*R*)-Ibuprofen

Ibuprofen is an analgesic (painkiller) with anti-inflammatory properties. The *S* enantiomer is the active agent, while the *R* enantiomer is inactive. Nevertheless, ibuprofen is sold as a mixture of both enantiomers (under the trade names Advil™ and Motrin™), because the benefit of separating the enantiomers is not clear. In fact, there is evidence that the human body is capable of slowly converting the *R* enantiomer into the desired *S* enantiomer. Many synthetic drugs are sold as a mixture of enantiomers, because of the high cost and difficulty associated with separating enantiomers.

In many cases, enantiomers can trigger entirely different physiological responses. For example, consider the enantiomers of Timolol:

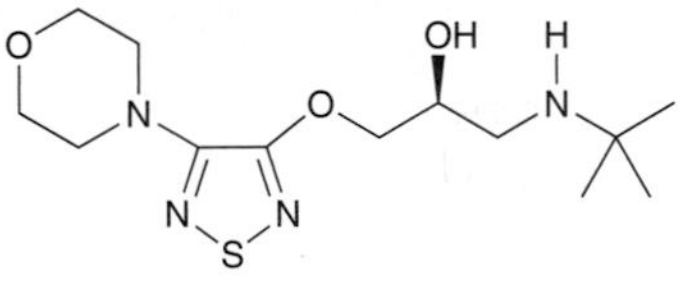

(*S*)-Timolol

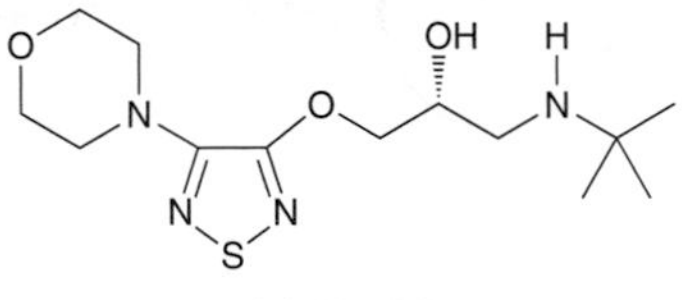

(*R*)-Timolol

The *S* enantiomer treats angina and high blood pressure, while the *R* enantiomer is useful in treating glaucoma. In this example, both enantiomers produce desirable, albeit different, results. In other cases, one enantiomer can produce an undesirable response. For example, consider the enantiomers of penicillamine:

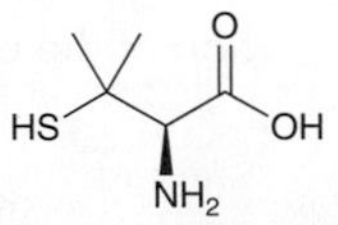

(*R*)-Penicillamine

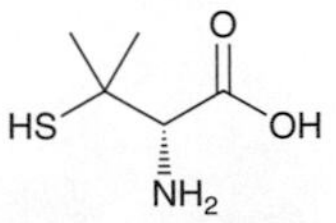

(*S*)-Penicillamine

The *S* enantiomer is used to treat chronic arthritis, while the *R* enantiomer is highly toxic. In such a case, the drug cannot be sold as a mixture of enantiomers. Another example is naproxen:

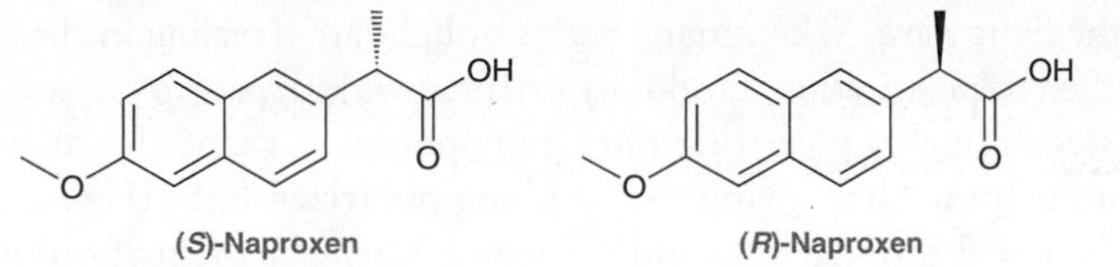

The *S* enantiomer is an anti-inflammatory agent, while the *R* enantiomer is a liver toxin.

In these cases where one enantiomer of the drug is known to be a toxin, the drug is sold as a single enantiomer. However, the vast majority of drugs on the market are sold as mixtures of enantiomers. The FDA has recently encouraged development of single enantiomers by allowing drug companies to apply for new patents that cover single enantiomers of drugs previously sold as mixtures (provided that they demonstrate the advantages of the single enantiomer over the mixture of enantiomers). Recent advances in enantioselective synthesis (discussed in Section 8.8) have opened new doorways to the production of single-enantiomer drugs. This is reflected by the fact that most new drugs entering the market are sold as a single enantiomer.

For more examples of enantiomers that produce different biological responses, see *J. Chem. Ed.*, 1996, (73), 481.

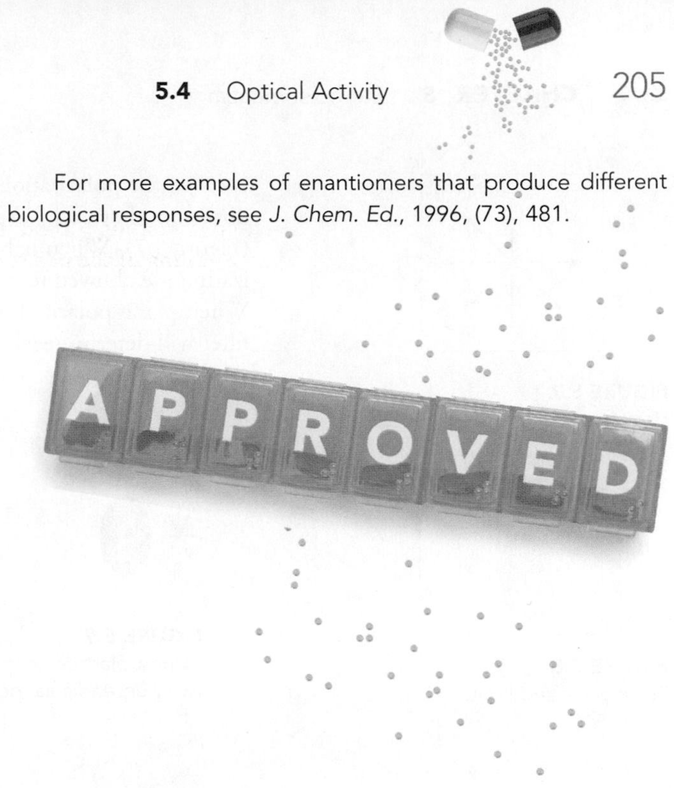

5.4 Optical Activity

Enantiomers exhibit identical physical properties. For example compare the melting and boiling points for the enantiomers of carvone:

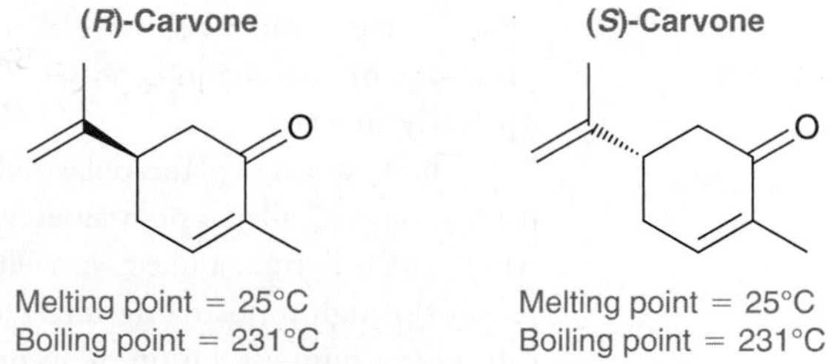

This should make sense, because physical properties are determined by intermolecular interactions, and the intermolecular interactions of one enantiomer are just the mirror image of the intermolecular interactions of the other enantiomer. Nevertheless, enantiomers do exhibit different behavior when exposed to plane-polarized light. To explore this difference, let's first quickly review the nature of light.

Plane-Polarized Light

Electromagnetic radiation (light) is comprised of oscillating electric and magnetic fields propagating through space (Figure 5.6). Notice that each oscillating field is located in a plane, and these planes are perpendicular to each other. The orientation of the electric field (shown in red)

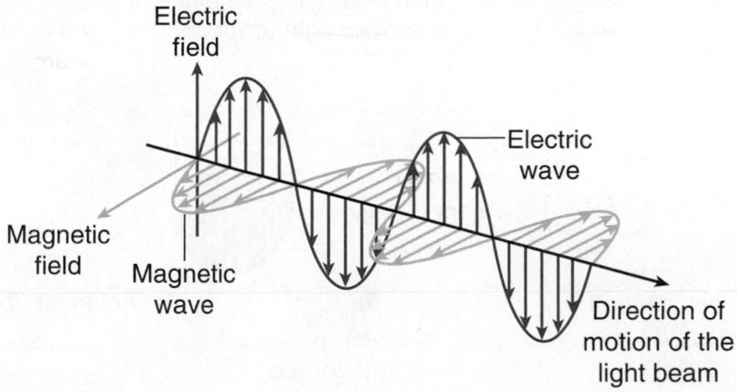

FIGURE 5.6
Light waves consist of perpendicular, oscillating electric and magnetic fields.

is called the polarization of the light wave. When many waves of light are traveling in the same direction, they each have a different polarization, randomly oriented with respect to one another (Figure 5.7). When light passes through a polarizing filter, only photons of a particular polarization are allowed to pass through the filter, giving rise to **plane-polarized light** (Figure 5.8). When plane-polarized light is passed through a second polarizing filter, the orientation of the filter will determine whether light passes through or is blocked (Figure 5.9).

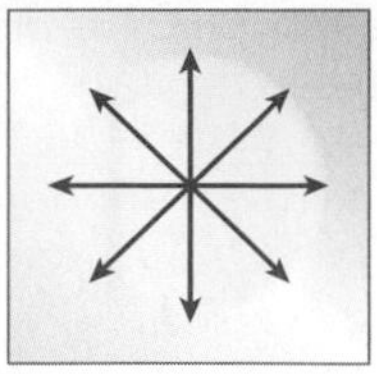

FIGURE 5.7
Unpolarized light.

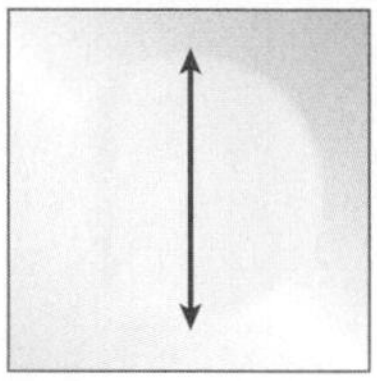

FIGURE 5.8
Plane-polarized light.

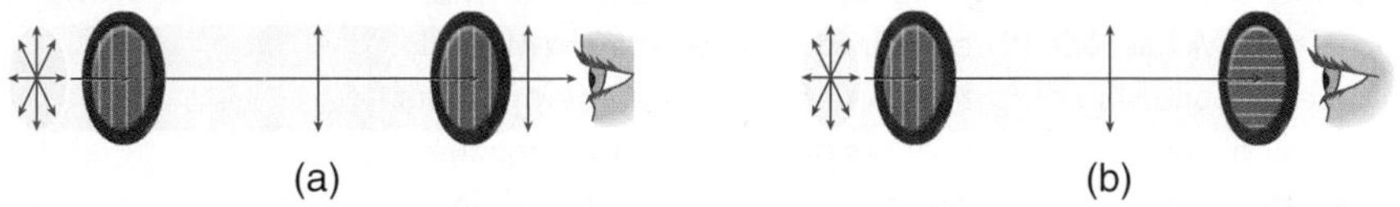

FIGURE 5.9
Plane-polarized light passing through (a) two parallel polarizing filters or (b) two perpendicular polarizing filters.

Polarimetry

In 1815, French scientist Jean Baptiste Biot was exploring the nature of light by passing plane-polarized light through various solutions of organic compounds. In so doing, he discovered that solutions of certain organic compounds (such as sugar) rotate the plane of plane-polarized light. These compounds were therefore said to be **optically active**. He also noted that only some organic compounds possess this quality. Compounds lacking this ability were said to be **optically inactive**.

The rotation of plane-polarized light caused by optically active compounds can be measured using a device called a **polarimeter** (Figure 5.10). The light source is generally a sodium lamp, which emits light at a fixed wavelength of 589 nm, called the D line of sodium. This light then passes through a polarizing filter, and the resulting plane-polarized light continues through a tube containing a solution of an optically active compound, which causes the plane to rotate. The polarization of the emerging light can then be determined by rotating the second filter and observing the orientation that allows the light to pass.

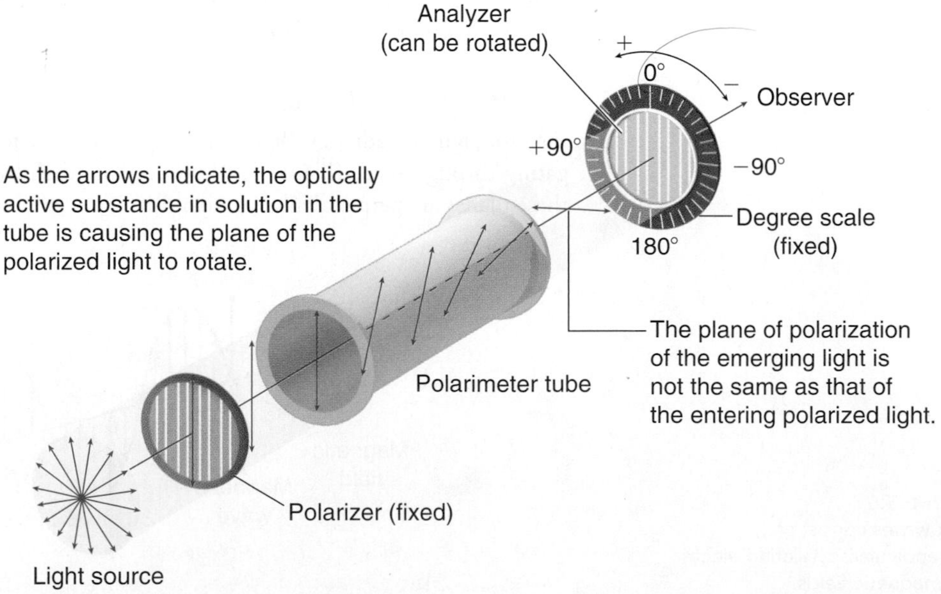

FIGURE 5.10
The components of a polarimeter.

Source of Optical Activity

In 1847, an explanation for the source of optical activity was proposed by French scientist Louis Pasteur. Pasteur's investigation of tartrate salts (discussed later in this chapter) led him to the conclusion that optical activity is a direct consequence of chirality. That is, chiral compounds are optically active, while achiral compounds are not. Moreover, Pasteur noted that enantiomers (nonsuperimposable mirror images) will rotate the plane of plane-polarized light in equal amounts but in opposite directions. We will now explore this idea in more detail.

Specific Rotation

When a solution of a chiral compound is placed in a polarimeter, the **observed rotation** (symbolized by the Greek letter alpha, α) will be dependent on the number of molecules that the light encounters as it travels through the solution. If the concentration of the solution is doubled, the observed rotation will be doubled. The same is true for the distance that the light travels through the solution (the pathlength). If the pathlength is doubled, the observed rotation will be doubled. In order to compare the rotations for various compounds, scientists had to choose a set of standard conditions. By using a standard concentration (1 g/mL) and a standard pathlength (1 dm) for measuring rotations, it is possible to make meaningful comparisons between compounds. The **specific rotation** for a compound is defined as the observed rotation under these standard conditions.

It is not always practical to use a concentration of 1 g/mL when measuring the optical activity of a compound, because chemists often work with very small quantities of a compound (milligrams, rather than grams). Very often, optical activity must be measured using a very dilute solution. To account for the use of nonstandard conditions, the specific rotation can be calculated in the following way:

$$\text{Specific rotation} = [\alpha] = \frac{\alpha}{c \times l}$$

where [α] is the specific rotation, α is the observed rotation, *c* is the concentration (measured in grams per milliliter), and *l* is the pathlength (measured in decimeters, where 1 dm = 10 cm). In this way, the specific rotation for a compound can be calculated for nonstandard concentrations or pathlengths. The specific rotation of a compound is a physical constant, much like its melting point or boiling point.

The specific rotation of a compound is also sensitive to temperature and wavelength, but these factors cannot be incorporated into the equation, because the relationship between these factors and the specific rotation is not a linear relationship. In other words, doubling the temperature does not necessarily double the observed rotation; in fact, it can sometimes cause a decrease in the observed rotation. The wavelength chosen can also affect the observed rotation in a nonlinear fashion. Therefore, these two factors are simply reported in the following format:

$$[\alpha]_{\lambda}^{T}$$

where *T* is the temperature (in degrees Celsius) and λ is the wavelength of light used. Some examples are as follows:

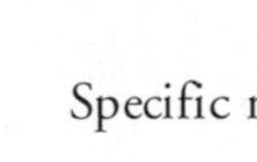

(*R*)-2-Bromobutane

$[\alpha]_{D}^{20} = -23.1$

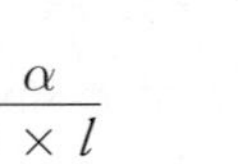

(*S*)-2-Bromobutane

$[\alpha]_{D}^{20} = +23.1$

where D stands for the D line of sodium (589 nm). Notice that the specific rotations for enantiomers are equal in magnitude but opposite in direction. A compound exhibiting a positive rotation (+) is called **dextrorotatory,** or ***d***, while a compound exhibiting a negative rotation (−) is called **levorotatory,** or ***l***. In the example above, the first enantiomer is levorotatory and is therefore called (−)-2-bromobutane, while the second enantiomer is dextrorotatory and is therefore called (+)-2-bromobutane. There is no direct relationship between the *R/S* system of nomenclature and the sign of the specific rotation (+ or −). The descriptors (*R*) and (*S*) refer to the configuration (the three-dimensional arrangement) of a chirality center, which is independent of conditions. In contrast, the descriptors (+) and (−) refer to the direction in which plane-polarized light is rotated, which is dependent on conditions. As an example, consider the structure of (*S*)-aspartic acid:

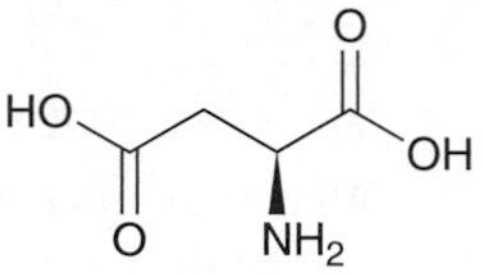

The configuration of the chirality center above is *S*, regardless of temperature. But the optical activity of this compound is very sensitive to temperature: it is levorotatory at 20°C, but it is dextrorotatory at 100°C. This example illustrates that we cannot use the configuration (*R* or *S*) to predict with certainty whether a compound will be (+) or (−). The magnitude and direction of optical activity can only be determined experimentally.

SKILLBUILDER

5.5 CALCULATING SPECIFIC ROTATION

LEARN the skill

When 0.300 g of sucrose is dissolved in 10.0 mL of water and placed in a sample cell 10.0 cm in length, the observed rotation is +1.99° (using the D line of sodium at 20°C). Calculate the specific rotation of sucrose.

SOLUTION

The following equation can be used to calculate the specific rotation:

STEP 1 Use the following equation.

$$\text{Specific rotation} = [\alpha] = \frac{\alpha}{c \times l}$$

and the problem provides all of the necessary values to plug into this equation. We just need to make sure that the units are correct: *Concentration* (c) must be in units of grams per milliliter. The problem states that 0.300 g is dissolved in 10.0 mL. Therefore, the concentration is 0.300 g/10.0 mL = 0.03 g/mL. *Pathlength* (*l*) must be reported in decimeters (where 1 dm = 10 cm). The problem states that the pathlength is 10.0 cm, which is equivalent to 1.00 dm. Now, simply plug these values into the equation:

STEP 2 Plug in the given values.

$$\text{Specific rotation} = [\alpha] = \frac{\alpha}{c \times l} = \frac{+1.99°}{0.03 \text{ g/mL} \times 1.00 \text{ dm}} = +66.3$$

The temperature (20°C) and wavelength (D line of sodium) are reported as a superscript and subscript, respectively. Notice that the specific rotation is generally reported as if it were a unitless number.

$$[\alpha]_D^{20} = +66.3$$

PRACTICE the skill

5.12 When 0.575 g of monosodium glutamate (MSG) is dissolved in 10.0 mL of water and placed in a sample cell 10.0 cm in length, the observed rotation at 20°C (using the D line of sodium) is +1.47°. Calculate the specific rotation of MSG.

5.13 When 0.095 g of cholesterol is dissolved in 1.00 mL of ether and placed in a sample cell 10.0 cm in length, the observed rotation at 20°C (using the D line of sodium) is −2.99°. Calculate the specific rotation of cholesterol.

5.14 When 1.30 g of menthol is dissolved in 5.00 mL of ether and placed in a sample cell 10.0 cm in length, the observed rotation at 20°C (using the D line of sodium) is +0.57°. Calculate the specific rotation of menthol.

APPLY the skill

5.15 Predict the value for the specific rotation of the following compound. Explain your answer.

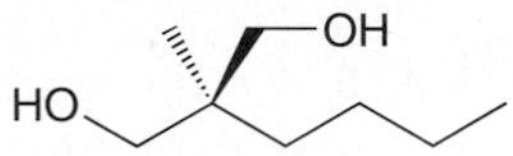

5.16 The specific rotation of (*S*)-2-butanol is +13.5. If 1.00 g of its enantiomer is dissolved in 10.0 mL of ethanol and placed in a sample cell with a length of 1.00 dm, what observed rotation do you expect?

need more **PRACTICE?** **Try Problems 5.44, 5.50c, 5.52**

Enantiomeric Excess

A solution containing a single enantiomer is said to be **optically pure**, or **enantiomerically pure**. That is, the other enantiomer is entirely absent.

A solution containing equal amounts of both enantiomers is called a **racemic mixture** and will be optically inactive. When plane-polarized light passes through such a mixture, the light encounters one molecule at a time, rotating slightly with each interaction. Since there are equal amounts of both enantiomers, the net rotation will be zero degrees. Although the individual compounds are optically active, the mixture is not optically active.

A solution containing both enantiomers in unequal amounts will be optically active. For example, imagine a solution of 2-butanol, containing 70% (*R*)-2-butanol and 30% (*S*)-2-butanol. In such a case, there is a 40% excess of the *R* stereoisomer (70% − 30%). The remainder of the solution is a racemic mixture of both enantiomers. In this case, the **enantiomeric excess** (*ee*) is 40%

The % *ee* of a compound is experimentally measured in the following way:

$$\% \, ee = \frac{|\text{observed } [\alpha]|}{|[\alpha] \text{ of pure enantiomer}|} \times 100\%$$

Let's get some practice using this expression to calculate % *ee*.

SKILLBUILDER

5.6 CALCULATING % *ee*

LEARN the skill

The specific rotation of optically pure adrenaline in water (at 25°C) is −53. A chemist devised a synthetic route to prepare optically pure adrenaline, but it was suspected that the product was contaminated with a small amount of the undesirable enantiomer. The observed specific rotation was found to be −45°. Calculate the % *ee* of the product.

SOLUTION

STEP 1
Use the following equation.

We use the following equation to calculate the % *ee*:

$$\% \, ee = \frac{|\text{observed } [\alpha]|}{|[\alpha] \text{ of pure enantiomer}|} \times 100\%$$

and then plug in the given values:

STEP 2
Plug in the given values.

$$\% \, ee = \frac{45}{53} \times 100\%$$
$$= 85\%$$

An 85% *ee* indicates that 85% of the product is adrenaline, while the remaining 15% is a racemic mixture of adrenaline and its enantiomer (7.5% adrenaline and 7.5% of the enantiomer). An 85% *ee* therefore indicates that the product is comprised of 92.5% adrenaline and 7.5% of the undesired enantiomer.

PRACTICE the skill

5.17 The specific rotation of L-dopa in water (at 15°C) is −39.5. A chemist prepared a mixture of L-dopa and its enantiomer, and this mixture had a specific rotation of −37. Calculate the % *ee* of this mixture.

5.18 The specific rotation of ephedrine in ethanol (at 20°C) is −6.3. A chemist prepared a mixture of ephedrine and its enantiomer, and this mixture had a specific rotation of −6.0. Calculate the % *ee* of this mixture.

5.19 The specific rotation of vitamin B_7 in water (at 22°C) is +92. A chemist prepared a mixture of vitamin B_7 and its enantiomer, and this mixture had a specific rotation of +85. Calculate the % *ee* of this mixture.

APPLY the skill

5.20 The specific rotation of L-alanine in water (at 25°C) is +2.8. A chemist prepared a mixture of L-alanine and its enantiomer, and 3.50 g of the mixture was dissolved in 10.0 mL of water. This solution was then placed in a sample cell with a pathlength of 10.0 cm and the observed rotation was +0.78. Calculate the % *ee* of the mixture.

need more **PRACTICE?** **Try Problems 5.40, 5.42, 5.50a, b**

5.5 Stereoisomeric Relationships: Enantiomers and Diastereomers

In Section 5.1, we saw that stereoisomers have the same constitution (connectivity of atoms) but are nonsuperimposable (they differ in their spatial arrangement of atoms). Stereoisomers can be subdivided into two categories, as shown in Figure 5.11.

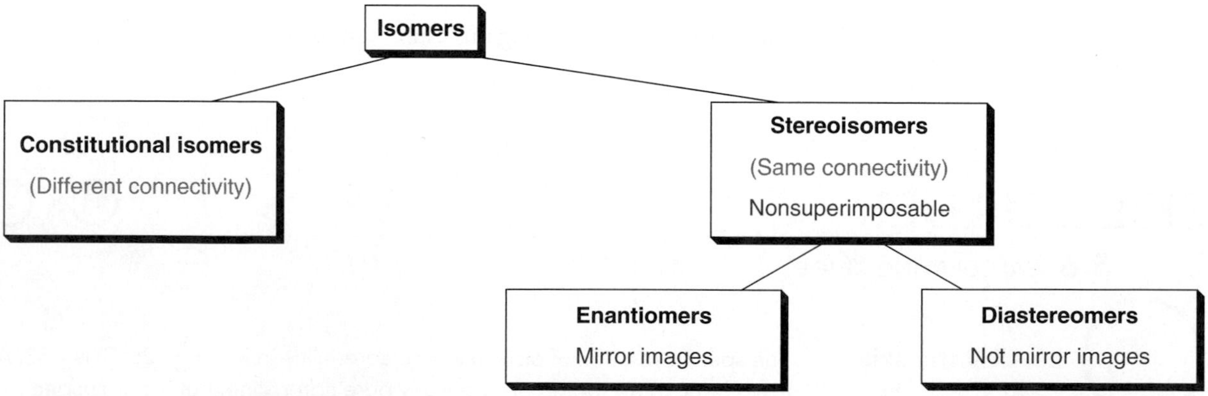

FIGURE 5.11
The main categories of stereoisomers.

Enantiomers are stereoisomers that are mirror images of one another, while **diastereomers** are stereoisomers that are not mirror images of one another. According to these definitions, we can understand why *cis-trans* isomers (discussed at the beginning of this chapter) are said to be diastereomers, rather than enantiomers.

Consider, once again, the structures of *cis*-2-butene and *trans*-2-butene:

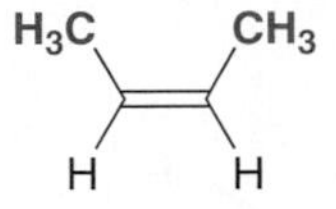

cis-2-Butene

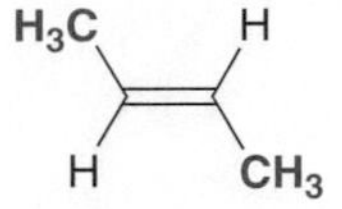

trans-2-Butene

They are stereoisomers, but they are not mirror images of one another and are therefore diastereomers. An important difference between enantiomers and diastereomers is that enantiomers have the same physical properties (as seen in Section 5.4), while diastereomers have different physical properties (as seen in Section 5.1).

The difference between enantiomers and diastereomers becomes especially relevant when we consider compounds containing more than one chirality center. As an example, consider the following structure:

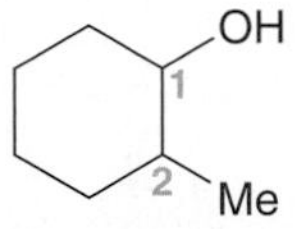

This compound has two chirality centers. Each one can have either the *R* configuration or the *S* configuration, giving rise to four possible stereoisomers (two pairs of enantiomers):

A pair of enantiomers

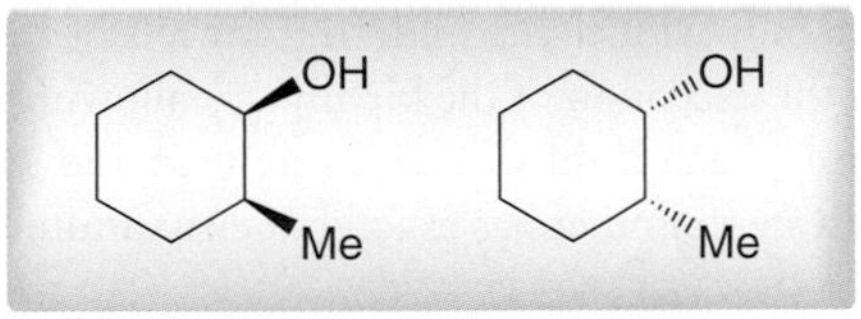

1*R*, 2*S* 1*S*, 2*R*

A pair of enantiomers

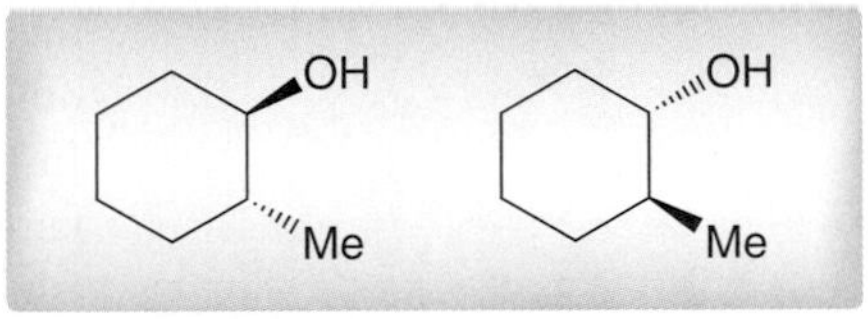

1*R*, 2*R* 1*S*, 2*S*

To describe the relationship between these four stereoisomers, we will look at one of them and describe its relationship to the other three stereoisomers. The first stereoisomer listed above has the configuration (*1R, 2S*). This stereoisomer has only one mirror image, or enantiomer, which has the configuration (*1S, 2R*). The third stereoisomer is not a mirror image of the first stereoisomer and is therefore its diastereomer. In a similar way, there is a diastereomeric relationship between the first stereoisomer and the fourth stereoisomer above because they are not mirror images of one another.

To help better visualize the relationship between the four compounds above we will use an analogy. Imagine a family with four children (two sets of twins). The first pair of twins are identical to each other in almost every way, except for the placement of one birthmark. One child has the birthmark on the right cheek, while the other child has the birthmark on the left cheek. These twins can be distinguished from each other based on the position of the birthmark. They are nonsuperimposable mirror images of each other. The second pair of twins look very different from the first pair. But the second pair of twins are once again identical to each other in every way, except the position of the birthmark on the cheek. They are nonsuperimposable mirror images of each other.

In this family of four children, each child has one twin and two other siblings. The same relationship exists for the four stereoisomers shown above. In this molecular family, each stereoisomer has exactly one enantiomer (mirror-image twin) and two diastereomers (siblings).

Now consider a case with three chirality centers:

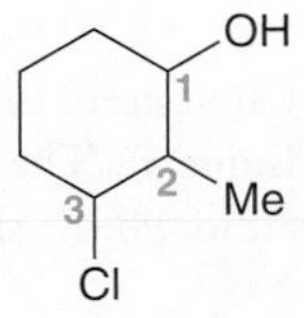

Once again, each chirality center can have either the *R* configuration or the *S* configuration, giving rise to a family of eight possible stereoisomers:

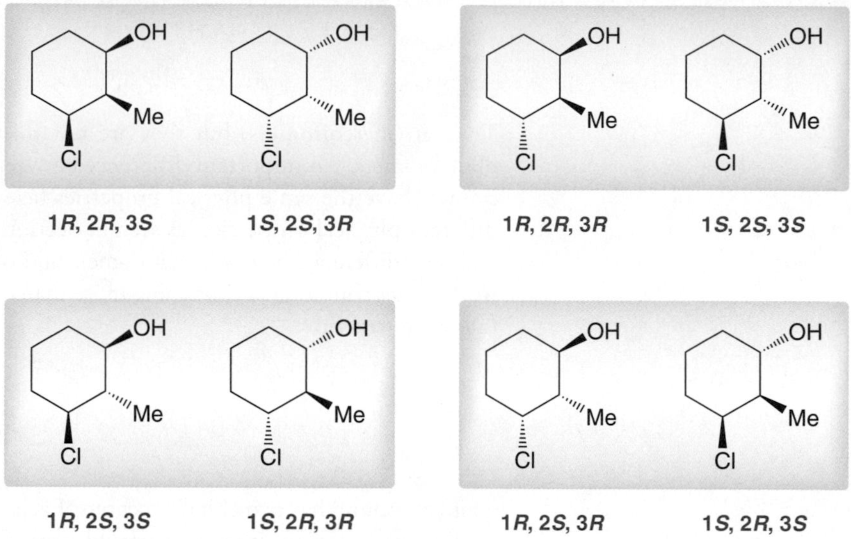

These eight stereoisomers are arranged above as four pairs of enantiomers. To help visualize this, imagine a family with eight children (four sets of twins). Each pair of twins are identical to each other with the exception of the birthmark, allowing them to be distinguished from one another. In this family, each child will have one twin and six other siblings. Similarly, in the molecular family, each stereoisomer has exactly one enantiomer (mirror-image twin) and six diastereomers (siblings).

Note that the presence of three chirality centers produces a family of four pairs of enantiomers. A compound with four chirality centers will generate a family of eight pairs of enantiomers. This begs the obvious question: What is the relationship between the number of chirality centers and the number of stereoisomers in the family? This relationship can be summarized as follows:

$$\text{Maximum number of stereoisomers} = 2^n$$

where n refers to the number of chirality centers. A compound with four chirality centers can have a maximum of 2^4 stereoisomers = 16 stereoisomers, or eight pairs of enantiomers. As another example, consider the structure of cholesterol:

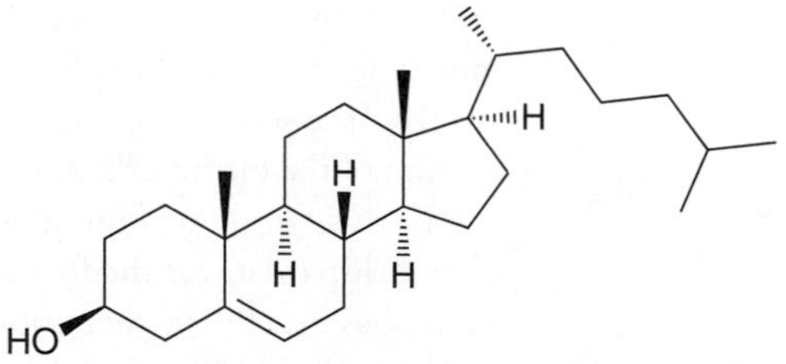

Cholesterol

Cholesterol has eight chirality centers, giving rise to a family of 2^8 stereoisomers (256 stereoisomers). The specific stereoisomer shown has only one enantiomer and 254 diastereomers, although the structure shown is the only stereoisomer produced by nature.

SKILLBUILDER

5.7 DETERMINING THE STEREOISOMERIC RELATIONSHIP BETWEEN TWO COMPOUNDS

LEARN the skill

Identify whether each pair of compounds are enantiomers or diastereomers:

(a)

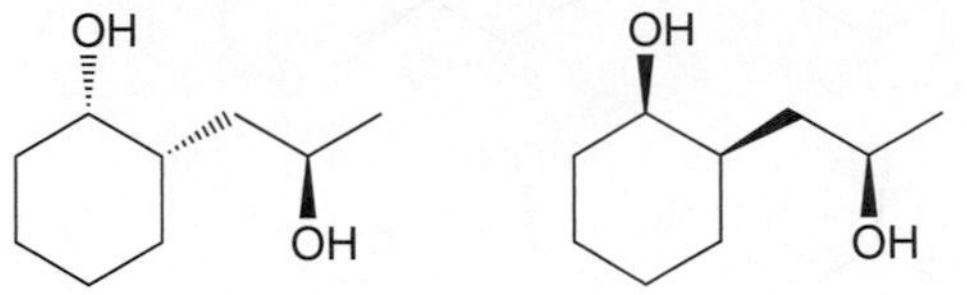

(b)

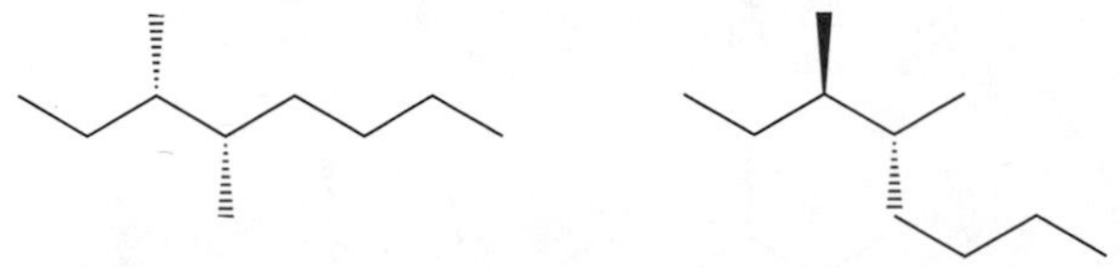

SOLUTION

(a) In SkillBuilder 5.2, we saw how to draw the enantiomer of a compound. In particular, we saw that the easiest method is to redraw the compound and invert all chirality centers. That is, all dashes are redrawn as wedges, and all wedges are redrawn as dashes. Therefore, two compounds can only be enantiomers if all of their chirality centers have opposite configuration. Let's carefully analyze all three chirality centers in each compound. Two of the chirality centers do have opposite configurations:

STEP 1 Compare the configuration of each chirality center.

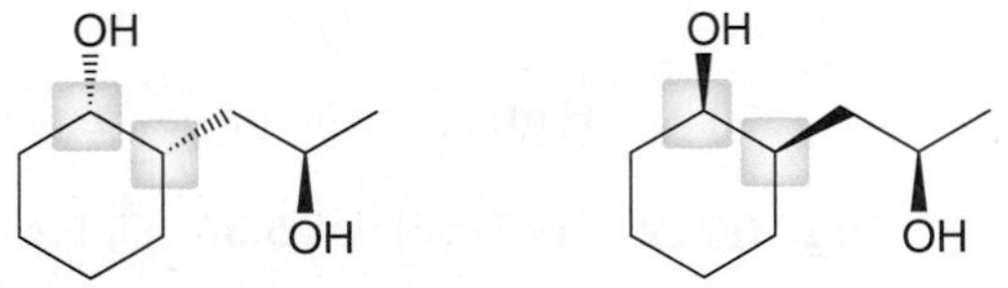

But there remains one chirality center that has the same configuration in both stereoisomers:

STEP 2 If only some of the chirality centers have opposite configuration, then the compounds are diastereomers.

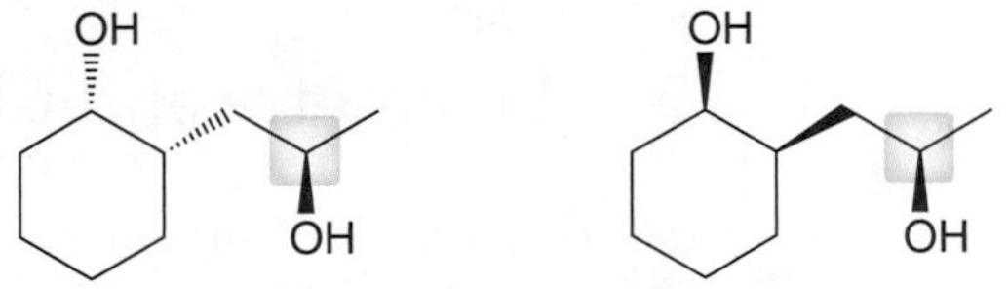

Therefore, these stereoisomers are not mirror images of each other; they must be diastereomers.

(b) Once again, compare the configurations of the chirality centers in both compounds. The first chirality center is easily identified as having a different configuration in each compound:

STEP 1 Compare the configuration of each chirality center.

The second chirality center is harder to compare with a brief inspection, because the skeleton is drawn in a different conformation (remember that single bonds are always free to rotate). When in doubt, you can always just assign a configuration (*R* or *S*) to each chirality center and compare them that way:

STEP 2 If all chirality centers have opposite configuration, then the compounds are enantiomers.

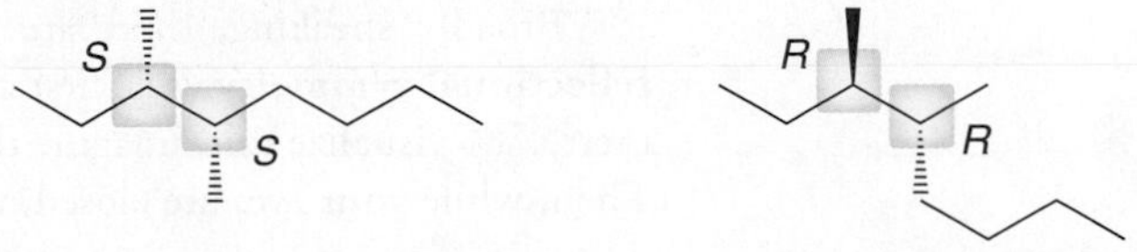

When comparing these stereoisomers, both chirality centers have opposite configurations, so these compounds are enantiomers.

PRACTICE the skill **5.21** Identify whether each of following pairs of compounds are enantiomers or diastereomers:

(a) 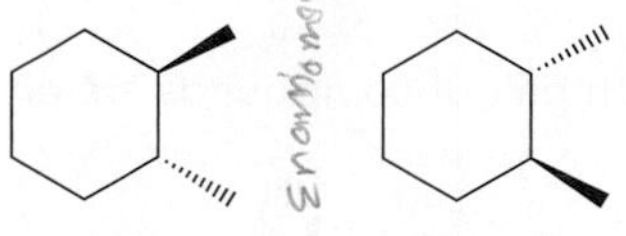

(b) 

(c) 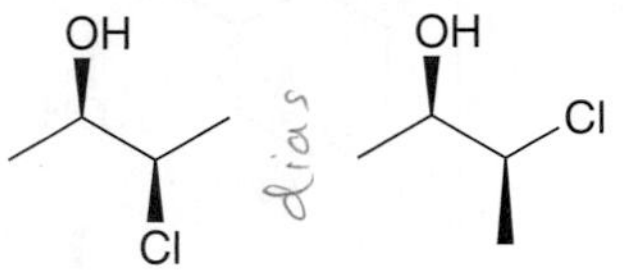

(d) 

(e) 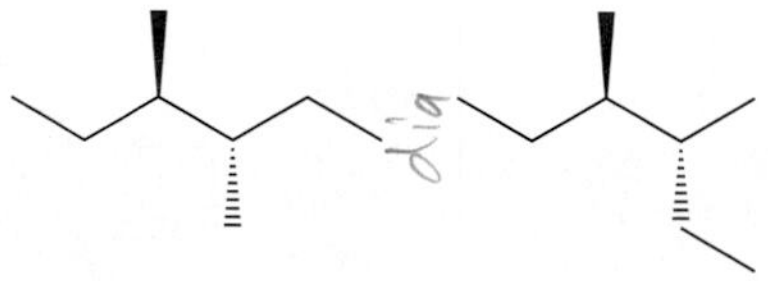

(f) 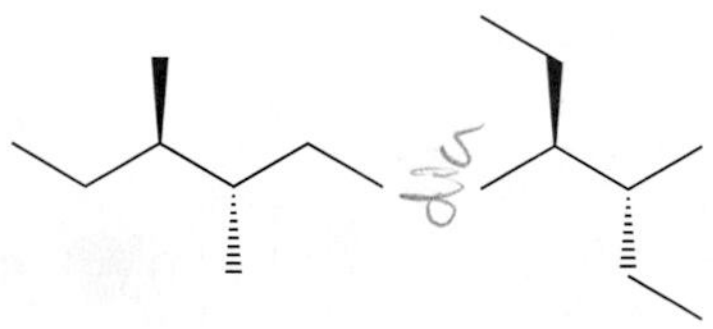

APPLY the skill **5.22** Identify the relationship between the following two compounds.

Hint: Take careful notice of the number of chirality centers.

need more **PRACTICE?** **Try Problems 5.36a-j,l, 5.41c,e-k, 5.49a,c,e-j, 5.57**

5.6 Symmetry and Chirality

Rotational Symmetry Versus Reflectional Symmetry

In this section we will learn how to determine whether a compound is chiral or achiral. It is true that any compound with a single chirality center must be chiral. But the same statement is not necessarily true for compounds with two chirality centers. Consider the *cis* and *trans* isomers of 1,2-dimethylcyclohexane:

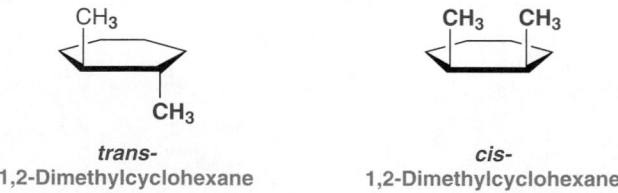

***trans*-**
1,2-Dimethylcyclohexane

***cis*-**
1,2-Dimethylcyclohexane

Each of these compounds has two chirality centers, but the *trans* isomer is chiral, while the *cis* isomer is not chiral. To understand why, we must explore the relationship between symmetry and chirality. Let's quickly review the different types of symmetry.

Broadly speaking, there are only two types of symmetry: **rotational symmetry** and **reflectional symmetry**. The first compound above (the *trans* isomer) exhibits rotational symmetry. To visualize this, imagine that the cyclohexane ring is pierced with an imaginary stick. Then, while your eyes are closed, the stick is rotated:

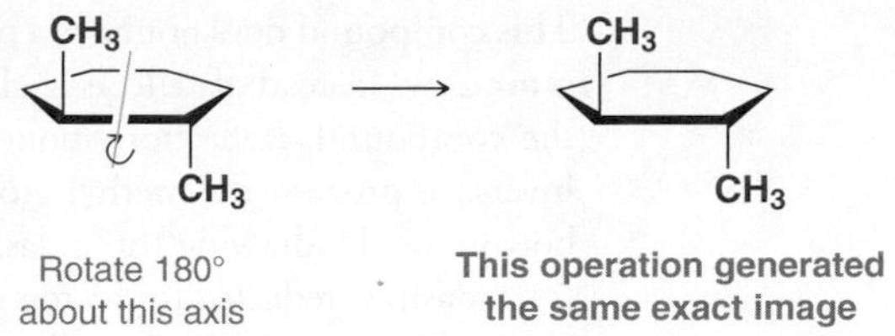

If, after opening your eyes, it is impossible to determine whether or not the rotation occurred, then the molecule possesses rotational symmetry. The imaginary stick is called an **axis of symmetry**.

Now let's consider the structure of the *cis* isomer:

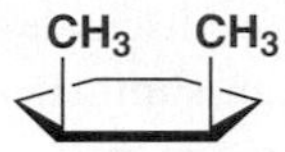

This compound does not have the same axis of symmetry that the *trans* isomer exhibits. If we pierce the molecule with an imaginary stick, we would have to rotate 360° in order to regenerate the exact same image. The same is true if we pierce the molecule from any angle. Therefore, this compound does not exhibit rotational symmetry. However, the compound does exhibit reflectional symmetry. Imagine closing your eyes while the molecule is reflected about the plane shown in Figure 5.12. That is, everything on the right side is reflected to the left side, and everything on the left side is reflected to the right side. When you open your eyes, you will not be able to determine whether or not the reflection took place. This molecule possesses a **plane of symmetry**.

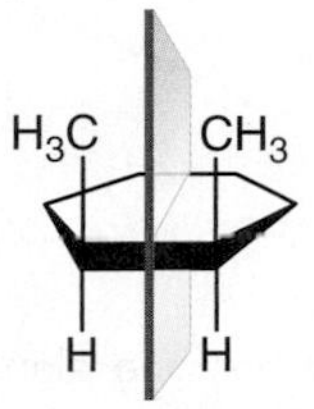

FIGURE 5.12
An illustration showing the plane of symmetry that is present in *cis*-1,2-dimethylcyclohexane.

To summarize, a molecule with an axis of symmetry possesses rotational symmetry, and a molecule with a plane of symmetry possesses reflectional symmetry. With an understanding of these two types of symmetry, we can now explore the relationship between symmetry and chirality. In particular, *chirality is not dependent in any way on rotational symmetry*. That is, the presence or absence of an axis of symmetry is completely irrelevant when determining whether a compound is chiral or achiral. We saw that *trans*-1,2-dimethylcyclohexane possesses rotational symmetry; nevertheless, the compound is still chiral and exists as a pair of enantiomers:

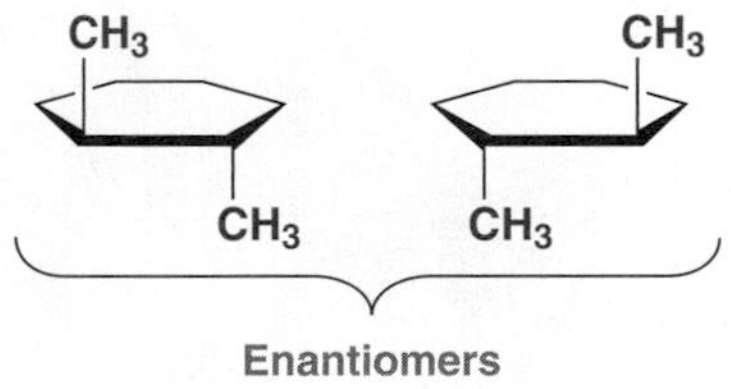

Even the presence of several different axes of symmetry does not indicate whether the compound is chiral or achiral. Chirality is only dependent on the presence or absence of reflectional symmetry. *Any compound that possesses a plane of symmetry in any conformation will be achiral.* We saw that the *cis* isomer of 1,2-dimethylcyclohexane exhibits a plane of symmetry, and therefore, the compound is achiral. It is identical to its mirror image. It does not have an enantiomer.

Although the presence of a plane of symmetry renders a compound achiral, the converse is not always true—that is, the absence of a plane of symmetry does not necessarily mean that the compound is chiral. Why? Because a plane of symmetry is only one type of reflectional symmetry. There are other types of reflectional symmetry, and the presence of any kind of reflectional symmetry will render a compound achiral. As an example, consider the following compound:

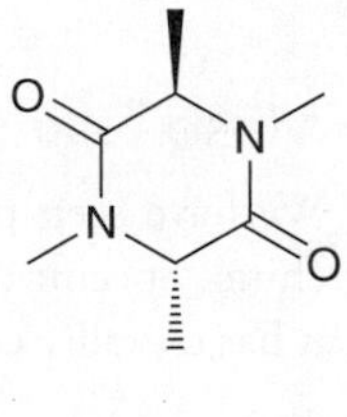

This compound does not have a plane of symmetry, but it does have another kind of reflectional symmetry. Instead of reflecting about a plane, imagine reflecting about a point at the center of the compound. Reflection about a point (rather than a plane) is called *inversion*. During the inversion process, the methyl group at the top of the drawing (on a wedge) is reflected to the bottom of the drawing (on a dash). Similarly, the methyl group at the bottom of the drawing (on a dash) is reflected to the top of the drawing (on a wedge). All other groups are also reflected about the center of the compound. If your eyes are closed when the inversion takes place, you would not be able to determine whether anything had happened. The compound is said to exhibit a center of inversion, which endows it with reflectional symmetry. As a result, the compound is achiral even though it lacks a plane of symmetry.

There are also other types of reflectional symmetry, but they are beyond the scope of our discussion. For our purposes, it will be sufficient to look for a plane of symmetry. We can summarize the relationship between symmetry and chirality with the following three statements:

- The presence or absence of rotational symmetry is irrelevant to chirality.
- A compound that has a plane of symmetry will be achiral.
- A compound that lacks a plane of symmetry will most likely be chiral (although there are rare exceptions, which can mostly be ignored for our purposes).

We will now practice identifying planes of symmetry.

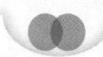

CONCEPTUAL CHECKPOINT

5.23 For each of the following objects determine whether or not it possesses a plane of symmetry:

(a) 

(b) 

(c) 

(d) 

(e) 

(f) 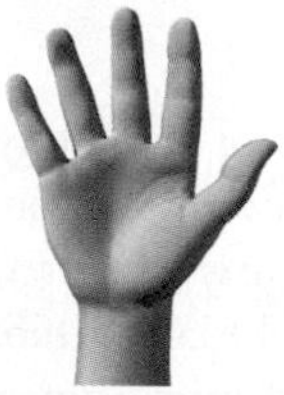

5.24 One object above has three planes of symmetry. Identify that object.

5.25 Each of the following molecules has one plane of symmetry. Find the plane of symmetry in each case: (Hint: A plane of symmetry can slice atoms in half.)

(a) 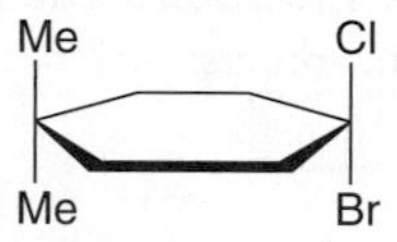

(b) 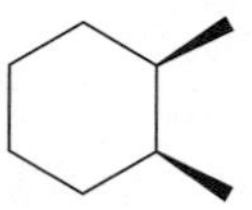

(c) 

(d) 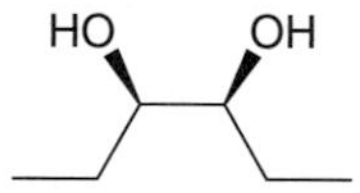

(e) 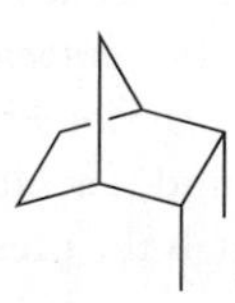

(f) 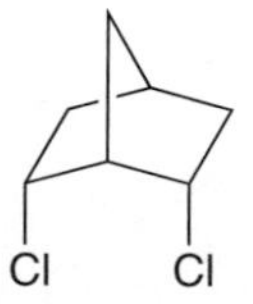

Meso Compounds

We have seen that the presence of chirality centers does not necessarily render a compound chiral. Specifically, a compound that exhibits reflectional symmetry will be achiral even though it has chirality centers. Such compounds are called ***meso* compounds**. A family of stereoisomers containing a *meso* compound will have fewer than 2^n stereoisomers. As an example, consider the following compound:

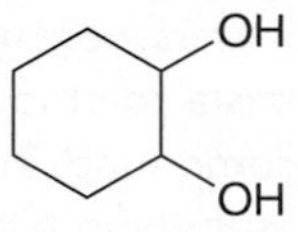

This compound has two chirality centers and therefore we would expect 2^2 (=4) stereoisomers. In other words, we expect two pairs of enantiomers:

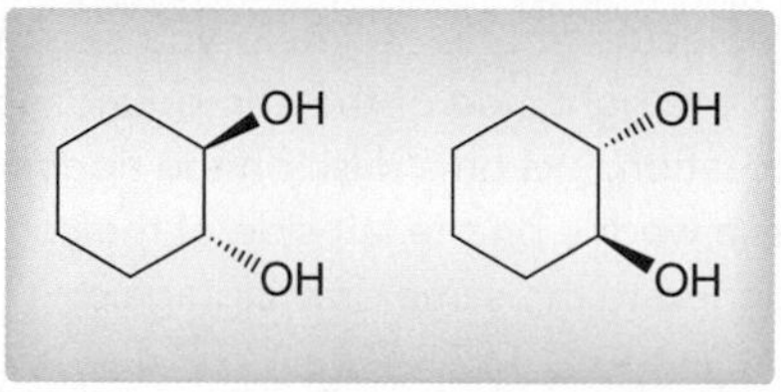

A pair of enantiomers

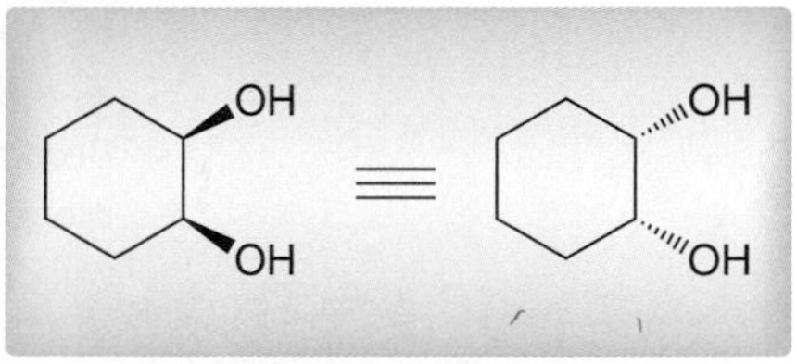

But these two drawings represent the same compound

The first pair of enantiomers meets our expectations. But the second pair is actually just one compound. It has reflectional symmetry and is therefore a *meso* compound:

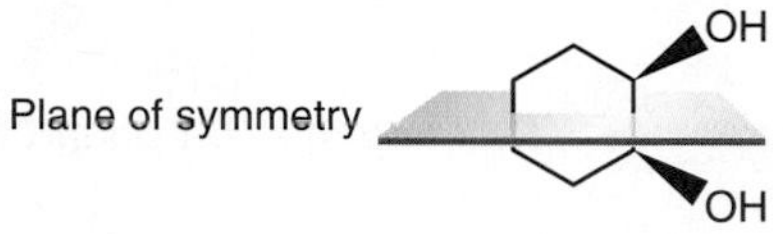

The compound is not chiral, and it does not have an enantiomer. To help visualize this, imagine a family with three children: a pair of twins and a single child. The pair of twins are identical to each other in almost every way, except for the placement of one birthmark. One child has the birthmark on the right cheek, while the other child has the birthmark on the left cheek. They are nonsuperimposable mirror images of each other. In contrast, the single child has two birthmarks: one on each cheek. This child is superimposable on his own mirror image; he does not have a twin. This family of three children is similar to the molecular family of three stereoisomers shown above.

SKILLBUILDER

5.8 IDENTIFYING *MESO* COMPOUNDS

LEARN the skill

Draw all possible stereoisomers of 1,3-dimethylcyclopentane:

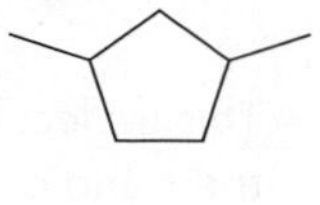

SOLUTION

This compound has two chirality centers, so we expect four possible stereoisomers (two pairs of enantiomers):

Before concluding that there are four stereoisomers here, we must look to see if either pair of enantiomers is actually just a meso compound. The first pair does in fact represent a pair of enantiomers.

Students often have trouble seeing that these two compounds are nonsuperimposable. It is a common mistake to believe that rotating the first compound will generate the second compound. This is not the case! Recall that a methyl group on a dash is pointing away from us (behind the page) while a methyl group on a wedge is pointing toward us (in front of the page). When the first compound above is rotated about a vertical axis, the methyl group on the left side ends up on the right side of the compound, but it does not remain as a wedge. Why? Before the rotation, it was pointing toward us, but after the rotation, it is pointing away from us (a molecular model will help you see this). As a result, that methyl group ends up as a dash on the right side of the compound. The fate of the other methyl group is similar. Before the rotation, it is on a dash on the right side of the compound, but after the rotation, it ends up as a wedge on the left side of the compound. As a result, rotating this compound will only regenerate the same drawing again:

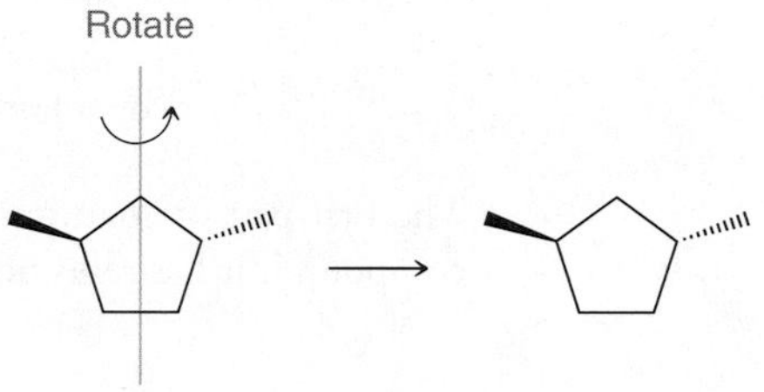

Rotating this compound will never give its enantiomer:

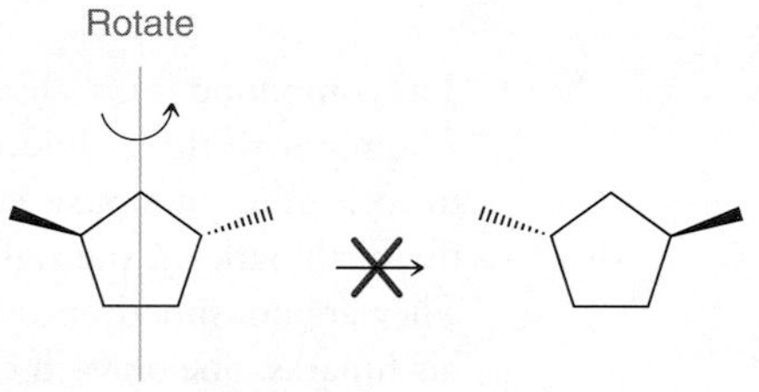

These compounds are enantiomers because they lack reflectional symmetry. They only possess rotational symmetry, which is completely irrelevant.

In contrast, the second pair of compounds do not represent a pair of enantiomers; instead, they represent two drawings of the same compound (a *meso* compound):

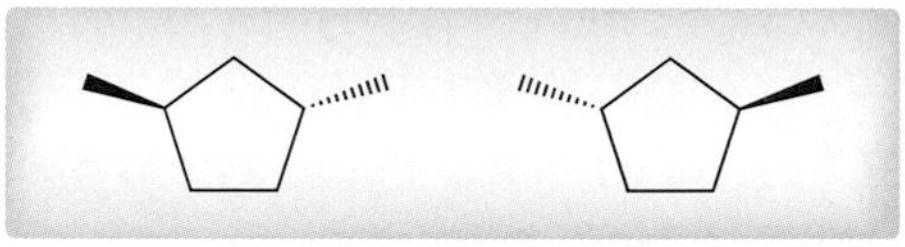

A pair of enantiomers

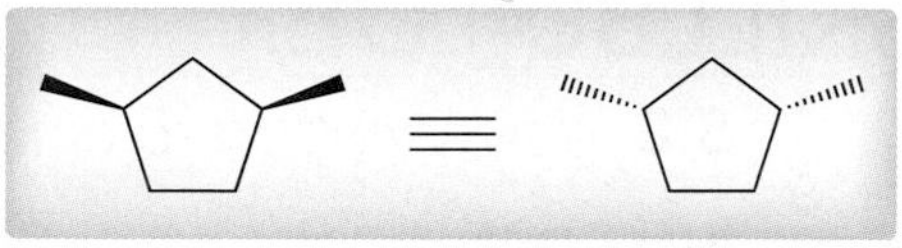

These two drawings represent the same compound

This molecular family is therefore comprised of only three stereoisomers—a pair of enantiomers and one *meso* compound:

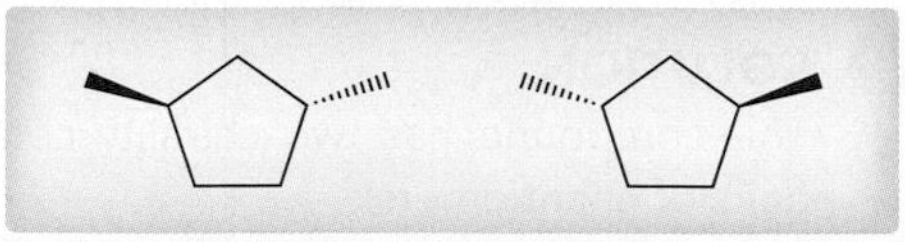

A pair of enantiomers

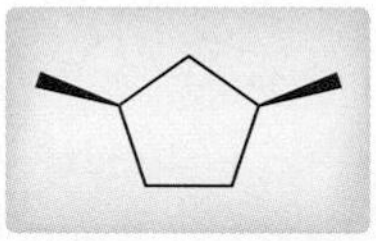

A *meso* compound

PRACTICE the skill **5.26** Draw all possible stereoisomers for each of the following compounds. Each possible stereoisomer should be drawn only once:

(a) 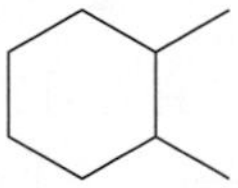

(b) 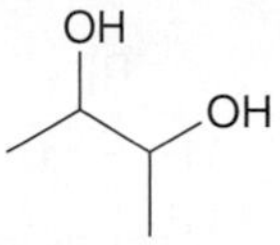

(c) 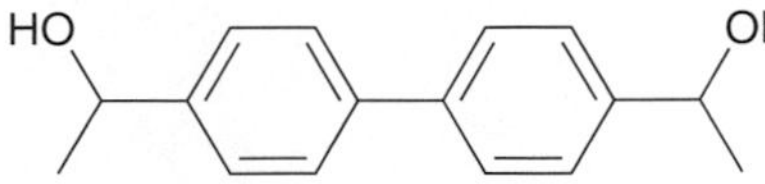

(d) 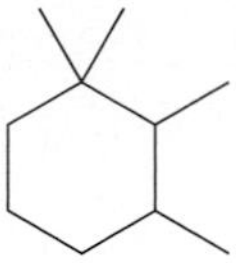

(e) 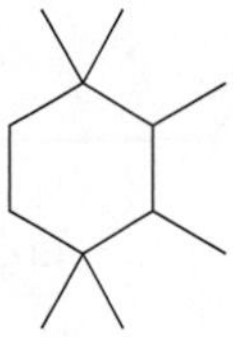

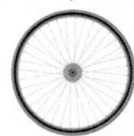

APPLY the skill **5.27** There are only two stereoisomers of 1,4-dimethylcyclohexane. Draw them, and explain why only two stereoisomers are observed.

5.28 How many stereoisomers do you expect for the following compound? Draw all of the stereoisomers.

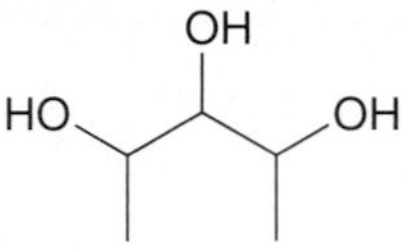

need more **PRACTICE?** **Try Problems 5.37, 5.46, 5.47, 5.51, 5.53a,b,d,f-h**

5.7 Fischer Projections

There is another drawing style that is often used when dealing with compounds bearing multiple chirality centers. These drawings, called **Fischer projections**, were devised by the German chemist Emil Fischer in 1891. Fischer was investigating sugars, which have multiple chirality centers. In order to quickly draw these compounds, he developed a faster method for drawing chirality centers:

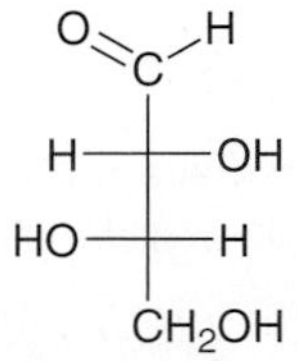

Two chirality centers

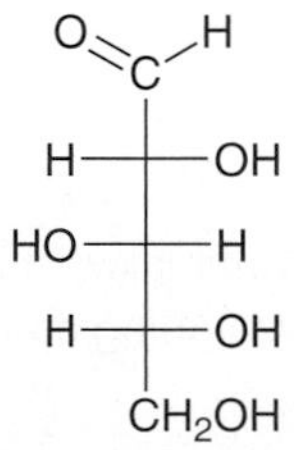

Three chirality centers

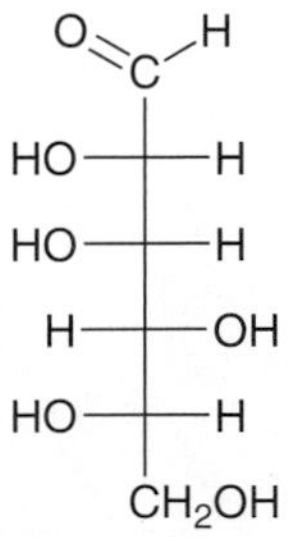

Four chirality centers

For each chirality center in a Fischer projection, the horizontal lines are considered to be coming out of the page, and the vertical lines are considered to be going behind the page:

Fischer projections are primarily used for analyzing sugars (Chapter 24). In addition, Fischer projections are also helpful for quickly comparing the relationship between stereoisomers:

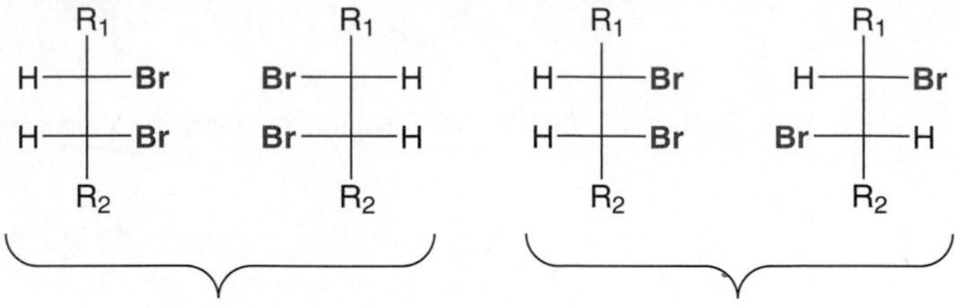

Enantiomers   Diastereomers

The first pair of compounds are enantiomers because all chirality centers have opposite configuration. The second pair of compounds are diastereomers. Recall that two compounds will be enantiomers only if all of their chirality centers have opposite configurations. Students often have trouble assigning configurations to chirality centers in a Fischer projection, so let's get some practice with that skill.

SKILLBUILDER

5.9 ASSIGNING CONFIGURATION FROM A FISCHER PROJECTION

LEARN the skill

Assign the configuration of the following chirality center:

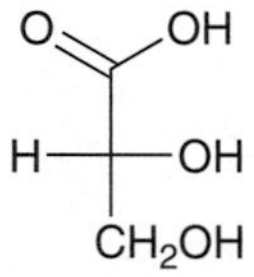

SOLUTION

Remember what a Fischer projection represents. All horizontal lines are wedges and all vertical lines are dashes:

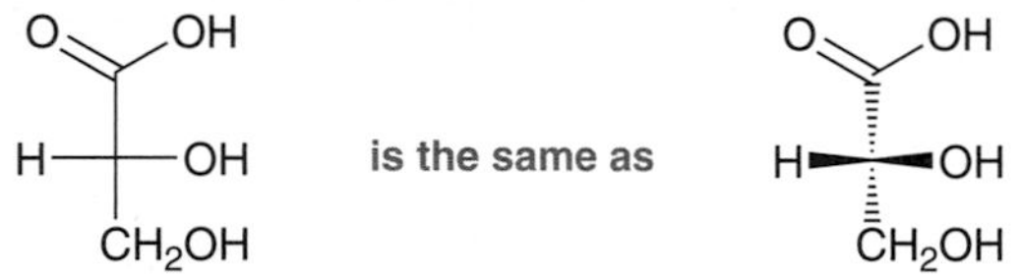

Until now, chirality centers have not been shown in this way. We are more accustomed to assigning configurations when a chirality center is shown with only one dash and one wedge:

STEP 1 Draw one horizontal line as a wedge, and draw one vertical line as a dash.

The chirality center of a Fischer projection can easily be converted into this more familiar type of drawing. Simply choose one horizontal line in the Fischer projection and redraw it as a wedge; then choose one vertical line and draw it as a dash. The answer will be the same, regardless of which two lines are chosen:

STEP 2 Assign the configuration using the steps in SkillBuilder 5.4.

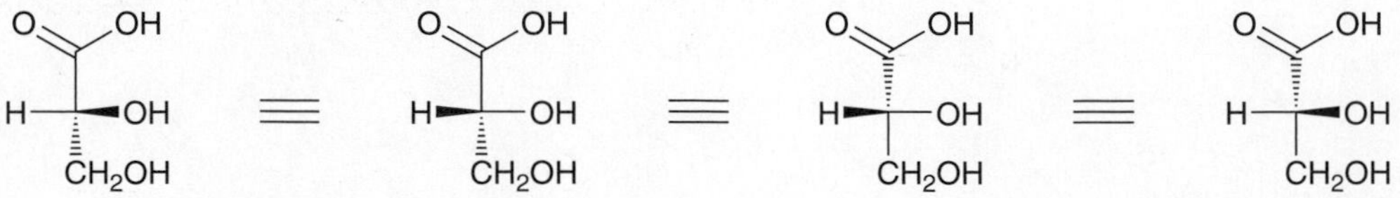

The configuration of this chirality center will be *R* in all of the drawings. To see why this works, build a molecular model and rotate it around to convince yourself.

When a Fischer projection has multiple chirality centers, this process is simply repeated for each chirality center.

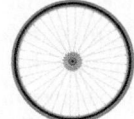

PRACTICE the skill **5.29** Identify the configuration of the chirality centers shown below:

(a) 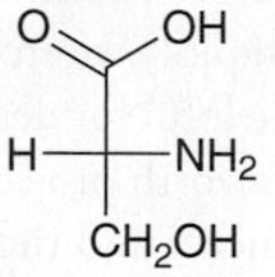

(b) 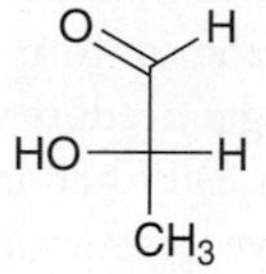

(c) 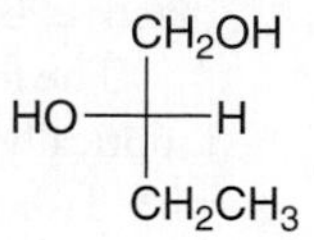

(d) 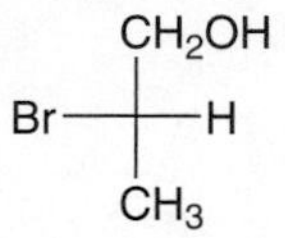

APPLY the skill **5.30** Determine the configuration for every chirality center in each of the following compounds.

(a) 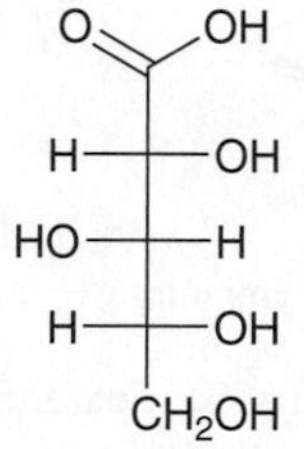

(b) 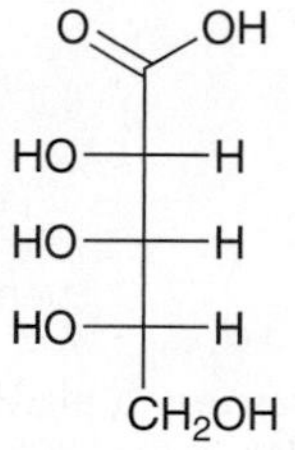

(c) 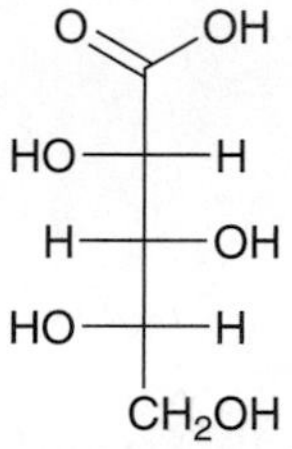

5.31 Draw the enantiomer of each compound in the previous problem.

need more **PRACTICE?** **Try Problems 5.39h, 5.41a,b, 5.43c, 5.53c,e, 5.54–5.56**

5.8 Conformationally Mobile Systems

In the previous chapter, we learned how to draw Newman projections, and we used them to compare the various conformations of butane. Recall that butane can adopt two staggered conformations that exhibit gauche interactions:

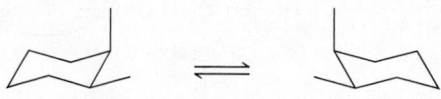

These two conformations are nonsuperimposable mirror images of each other, and their relationship is therefore enantiomeric. Nevertheless, butane is not a chiral compound. It is optically inactive, because these two conformations are constantly interconverting via single-bond rotation (which occurs with a very low energy barrier). The temperature would have to be extremely low to prevent interconversion between these two conformations. In contrast, a chirality center cannot invert its configuration via single-bond rotations. (*R*)-2-Butanol cannot be converted into (*S*)-2-butanol via a conformational change.

In the previous chapter, we also saw that substituted cyclohexanes adopt various conformations. Consider, for example, (*cis*)-1,2-dimethylcyclohexane. Below are both chair conformations:

The second chair has been rotated in space to illustrate the relationship between these two chair conformations. These conformations are mirror images, yet they are not superimposable on one another. To see this more clearly, compare them in the following way: Traveling *clockwise* around the ring, the first chair has an axial methyl group first, followed by an equatorial methyl group. The second chair, going *clockwise* again, has an equatorial methyl group first, followed by an axial methyl. The order has been reversed. These two conformations are distinguishable from one another; that is, they are not superimposable. The relationship between these two chair conformations is therefore enantiomeric; however, this compound is optically inactive, because the conformations rapidly interconvert at room temperature.

The following procedure will be helpful for determining whether or not a cyclic compound is optically active: (1) Either draw a Haworth projection of the compound or simply draw a ring with dashes and wedges for all substituents and then (2) look for a plane of symmetry in either one of these drawings. For example, (*cis*)-1,2-dimethylcyclohexane can be drawn in either of the following ways:

Internal plane of symmetry Internal plane of symmetry

In each drawing, a plane of symmetry is apparent, and the presence of that symmetry plane indicates that the compound is not optically active at room temperature.

5.9 Resolution of Enantiomers

As mentioned earlier, enantiomers have the same physical properties (boiling point, melting point, solubility, etc.). Since traditional separation techniques generally rely on differences in physical properties, they cannot be used to separate enantiomers from each other. The **resolution** (separation) of enantiomers can be achieved in a variety of other ways.

Resolution via Crystallization

The first resolution of enantiomers occurred in 1847, when Pasteur successfully separated enantiomeric tartrate salts from each other. Tartaric acid is a naturally occurring, optically active compound found in grapes and easily obtained during the wine-making process:

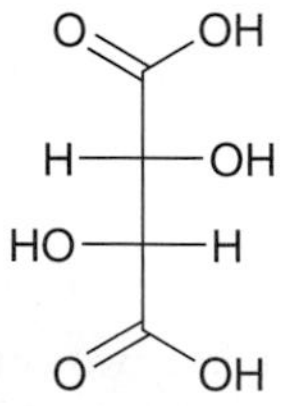

Only this stereoisomer is found in nature, yet Pasteur was able to obtain a racemic mixture of tartrate salts from the owner of a chemical plant:

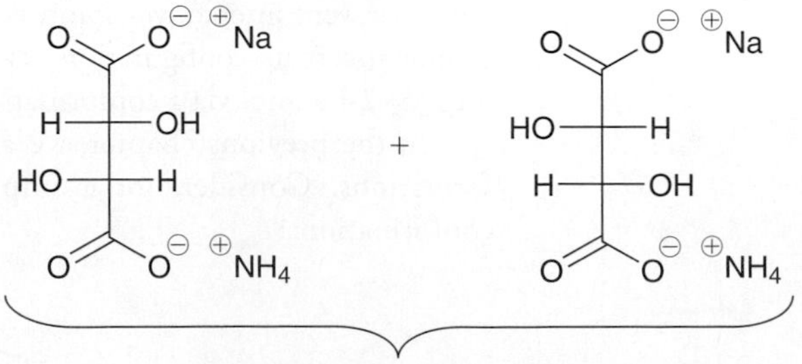

Racemic mixture of tartrate salts

The tartrate salts were then allowed to crystallize, and Pasteur noticed that the crystals had two distinct shapes that were nonsuperimposable mirror images of each other. Using only a pair of tweezers, he then physically separated the crystals into two piles. He dissolved each pile in water and placed each solution in a polarimeter to discover that their specific rotations were equal in amount but opposite in sign. Pasteur correctly concluded that the molecules themselves must be nonsuperimposable mirror images of each other. He was the first to describe molecules as having this property and is therefore credited with discovering the relationship between enantiomers.

Most racemic mixtures are not easily resolved into mirror-image crystals when allowed to crystallize, so other methods of resolution are required. Two common ways will now be discussed.

Chiral Resolving Agents

When a racemic mixture is treated with a single enantiomer of another compound, the resulting reaction produces a pair of diastereomers (rather than enantiomers).

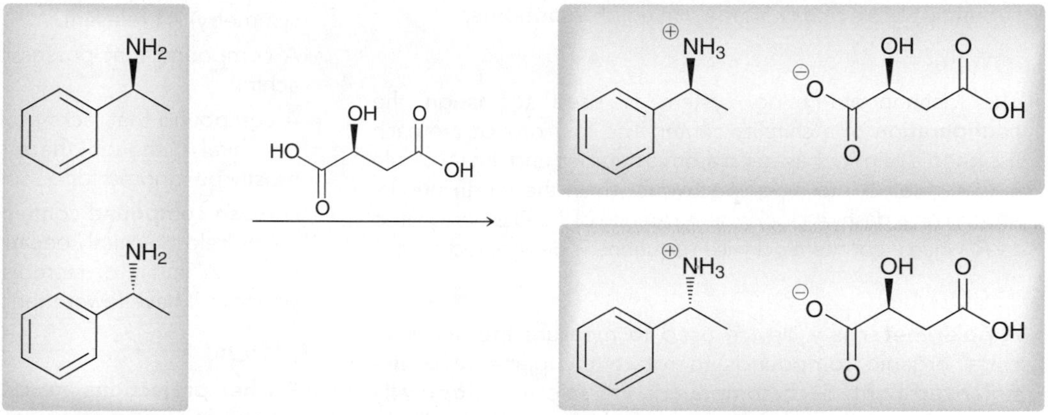

A pair of enantiomers **Diastereomeric salts**

On the left are the enantiomers of 1-phenylethylamine. When a racemic mixture of these enantiomers is treated with (*S*)-malic acid, a proton transfer reaction produces diastereomeric salts. Diastereomers have different physical properties and can therefore be separated by conventional means (such as crystallization). Once separated, the diastereomeric salts can then be converted back into the original enantiomers by treatment with a base. Thus (*S*)-malic acid is said to be a **resolving agent** in that it makes it possible to resolve the enantiomers of 1-phenylethylamine.

Chiral Column Chromatography

Resolution of enantiomers can also be accomplished with **column chromatography**. In column chromatography, compounds are separated from each other based on a difference in the way they interact with the medium (the adsorbent) through which they are passed. Some compounds interact strongly with the adsorbent and move very slowly through the column; other compounds interact weakly and travel at a faster rate through the column. When enantiomers are passed through a traditional column, they travel at the same rate because their properties are identical. However, if a chiral adsorbent is used, the enantiomers interact with the adsorbent differently, causing them to travel through the column at different rates. Enantiomers are often separated in this way.

REVIEW OF CONCEPTS AND VOCABULARY

SECTION 5.1

- Constitutional isomers have the same molecular formula but differ in their connectivity of atoms.
- **Stereoisomers** have the same connectivity of atoms but differ in their spatial arrangement. The terms *cis* and *trans* are used to differentiate stereoisomeric alkenes as well as disubstituted cycloalkanes.

SECTION 5.2

- **Chiral** objects are not **superimposable** on their mirror images.
- The most common source of molecular chirality is the presence of a **chirality center,** a carbon atom bearing four different groups.
- A compound with one chirality center will have one nonsuperimposable mirror image, called its **enantiomer**.

SECTION 5.3

- The Cahn-Ingold-Prelog system is used to assign the **configuration** of a chirality center. The four groups are each assigned a priority, based on atomic number, and the molecule is then rotated into a perspective in which the fourth priority atom is on a dash. A clockwise sequence of 1-2-3 is designated as ***R***, while a counterclockwise sequence is designated as ***S***.

SECTION 5.4

- A **polarimeter** is a device used to measure the ability of chiral organic compounds to rotate the plane of **plane-polarized light**. Such compounds are said to be **optically active**. Compounds that do not rotate plane-polarized light are said to be **optically inactive**.
- Enantiomers rotate plane-polarized light in equal amounts but in opposite directions. The **specific rotation** of a substance is a physical property. It is determined experimentally by measuring the **observed rotation** and dividing by the concentration of the solution and the pathlength.
- Compounds exhibiting a positive rotation (+) are said to be **dextrorotatory**, while compounds exhibiting a negative rotation (−) are said to be **levorotatory.**
- A solution containing a single enantiomer is **optically pure**, while a solution containing equal amounts of both enantiomers is called a **racemic mixture.**
- A solution containing a pair of enantiomers in unequal amounts is described in terms of **enantiomeric excess.**

SECTION 5.5

- For a compound with multiple chirality centers, a family of stereoisomers exists. Each stereoisomer will have at most one enantiomer, with the remaining members of the family being **diastereomers.**
- The number of stereoisomers of a compound can be no larger than 2^n, where $n =$ the number of chirality centers.
- Enantiomers are mirror images. Diastereomers are not mirror images.

SECTION 5.6

- There are two kinds of symmetry: **rotational symmetry** and **reflectional symmetry**.
- The presence or absence of an **axis of symmetry** (rotational symmetry) is irrelevant.
- A compound that possesses a **plane of symmetry** will be achiral.
- A compound that lacks a plane of symmetry will most likely be chiral (although there are rare exceptions, which can mostly be ignored for our purposes).
- A ***meso* compound** contains multiple chirality centers but is nevertheless achiral, because it possesses reflectional symmetry. A family of stereoisomers containing a *meso* compound will have fewer than 2^n stereoisomers.

SECTION 5.7

- **Fischer projections** are drawings that convey the configurations of chirality centers, without the use of wedges and dashes. All horizontal lines are understood to be wedges (coming out of the page) and all vertical lines are understood to be on dashes (going behind the page).

SECTION 5.8

- Some compounds, such as butane and (*cis*)-1,2-dimethylcyclohexane, can adopt enantiomeric conformations. These compounds are optically inactive because the enantiomeric conformations are in equilibrium.

SECTION 5.9

- **Resolution** (separation) of enantiomers can be accomplished in a number of ways, including the use of chiral **resolving agents** and chiral **column chromatography**.

KEY TERMINOLOGY

axis of symmetry 215
chiral 192
chirality center 193
column chromatography 223
configuration 200
dextrorotatory 207
diastereomers 210
enantiomer 195
enantiomerically pure 209
enantiomeric excess 209
Fischer projections 219
levorotatory 207
***meso* compound** 216
observed rotation 207
optically active 206
optically inactive 206
optically pure 209
plane of symmetry 215
plane-polarized light 206
polarimeter 206
R 199
racemic mixture 209
reflectional symmetry 214
resolution 222
resolving agents 223
rotational symmetry 214
S 199
specific rotation 207
stereoisomers 189
superimposable 192

SKILLBUILDER REVIEW

5.1 IDENTIFYING *CIS-TRANS* STEREOISOMERISM

STEP 1 Identify and name all four groups attached to the π bond.

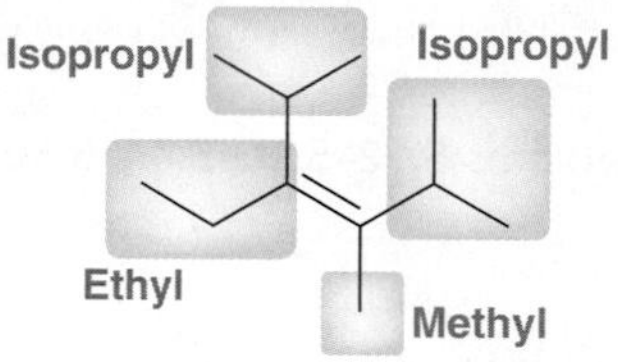

STEP 2 Look for two identical groups on different vinylic positions, and assign the configuration as *cis* or *trans*.

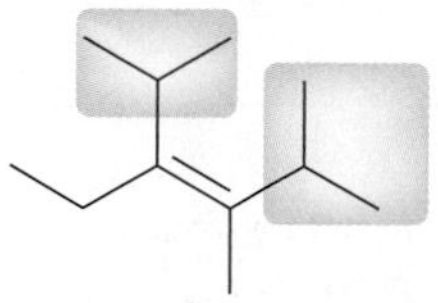

cis

Try Problems 5.1–5.3, 5.35

5.2 LOCATING CHIRALITY CENTERS

STEP 1 Ignore sp^2 and *sp* hybridized centers

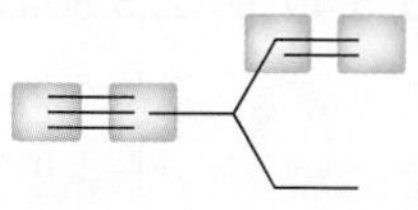

STEP 2 Ignore CH_2 and CH_3 groups

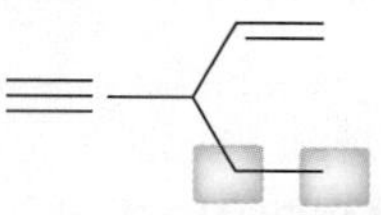

STEP 3 Identify any carbon atoms bearing four different groups

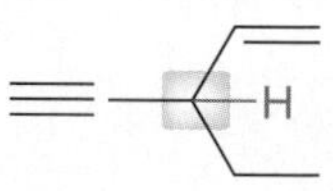

Try Problems 5.4–5.6, 5.34b, 5.48

5.3 DRAWING AN ENANTIOMER

Either place the mirror behind the compound...

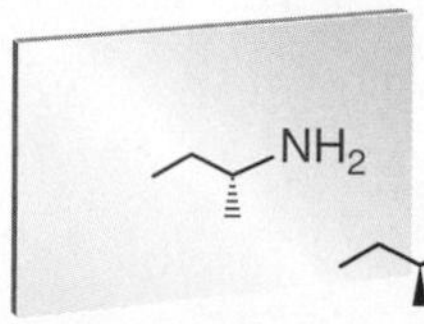

or place the mirror on the side of the compound...

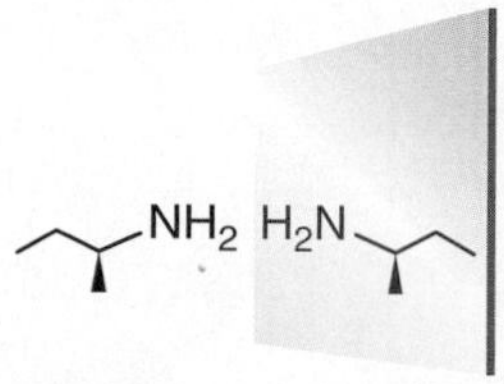

or place the mirror below the compound.

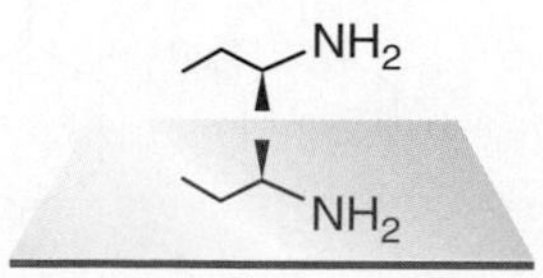

Try Problems 5.7, 5.8, 5.33, 5.34a, 5.38a-f,i-l

5.4 ASSIGNING CONFIGURATION

STEP 1 Identify the four atoms attached to the chirality center and prioritize by atomic number.

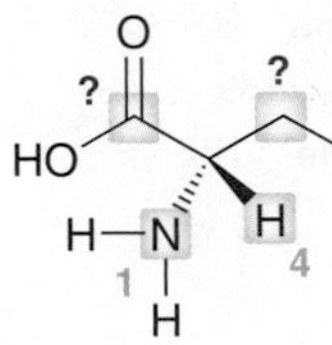

STEP 2 If two (or more) atoms are identical, make a list of substituents, and look for the first point of difference.

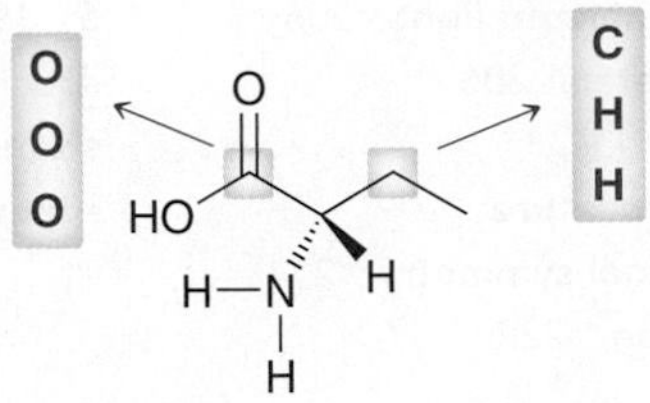

STEP 3 Redraw the chirality center, showing only the priorities.

STEP 4 Rotate molecule so that fourth priority is on a dash.

STEP 5 Identify direction of 1-2-3 sequence: clockwise is *R*, and counterclockwise is *S*.

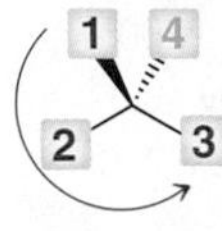

Counterclockwise = *S*

Try Problems 5.8–5.10, 5.32, 5.39a-g,i, 5.45

5.5 CALCULATING SPECIFIC ROTATION

EXAMPLE Calculate the specific rotation given the following information:

- 0.300 g sucrose dissolved in 10.0 mL of water
- Sample cell = 10.0 cm
- Observed rotation = +1.99°

STEP 1 Use the following equation:

$$\text{Specific rotation} = [\alpha] = \frac{\alpha}{c \times l}$$

STEP 2 Plug in the given values:

$$= \frac{+1.99°}{0.03\ \text{g/mL} \times 1.00\ \text{dm}}$$

$$= +66.3$$

Try Problems 5.12–5.16, 5.44, 5.50c, 5.52

5.6 CALCULATING % *ee*

EXAMPLE Given the following information, calculate the enantiomeric excess:

- The specific rotation of optically pure adrenaline is −53. A mixture of (*R*)- and (*S*)-adrenaline was found to have a specific rotation of −45. Calculate the % ee of the mixture.

STEP 1 Use the following equation:

$$\%\ \text{ee} = \frac{\text{observed } [\alpha]}{[\alpha] \text{ of pure enantiomer}} \times 100\%$$

STEP 2 Plug in the given values:

$$= \frac{-45}{-53} \times 100\% = 85\%$$

Try Problems 5.17–5.20, 5.40, 5.42, 5.50a,b

5.7 DETERMINING THE STEREOISOMERIC RELATIONSHIP BETWEEN TWO COMPOUNDS

STEP 1 Compare the configuration of each chirality center

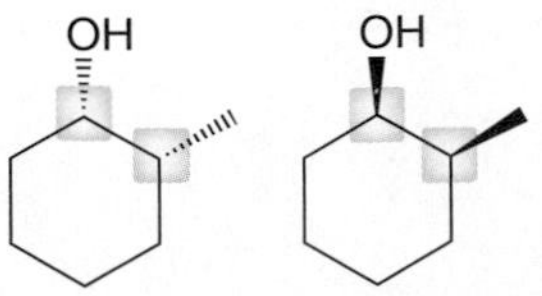

STEP 2 If all chirality centers have opposite configuration, the compounds are enantiomers. If only some of the chirality centers have opposite configuration, then the compounds are diastereomers.

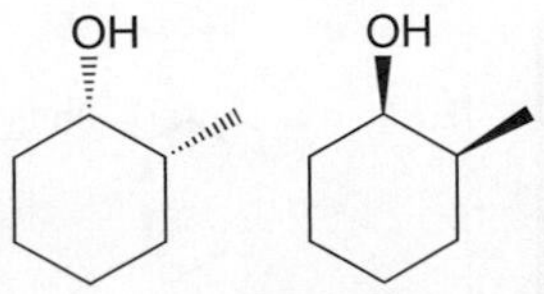

Enantiomers

Try Problems 5.21, 5.22, 5.36a-j,l, 5.41c,e-k, 5.49a,c,e-j

5.8 IDENTIFYING *MESO* COMPOUNDS

Example

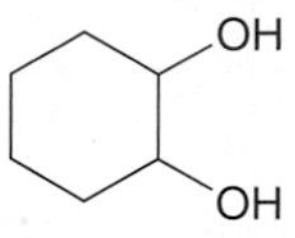

Draw all possible stereoisomers, and then look for a plane of symmetry in any of the drawings. The presence of a plane of symmetry indicates a meso compound.

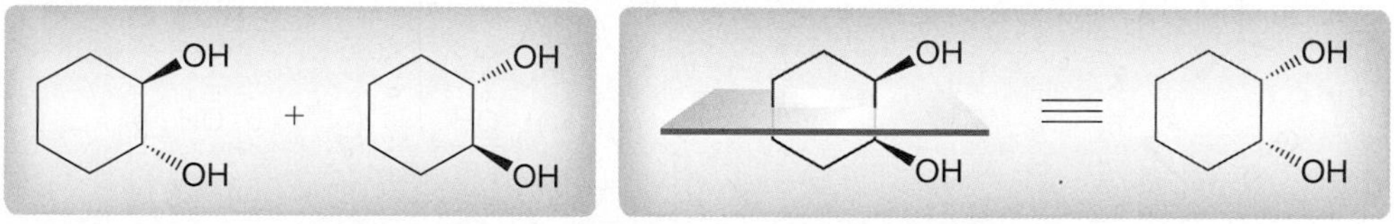

Enantiomers **Meso**

Try Problems 5.26–5.28, 5.37, 5.46, 5.47, 5.51, 5.53a,b,d,f-h

5.9 ASSIGNING CONFIGURATION FROM A FISCHER PROJECTION

EXAMPLE Assign the configuration of this chirality center.

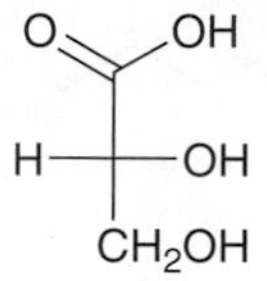

STEP 1 Choose one horizontal line and draw it as a wedge. Choose one vertical line and draw it as a dash.

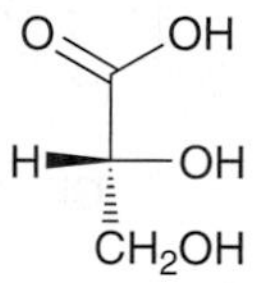

STEP 2 Prioritize.

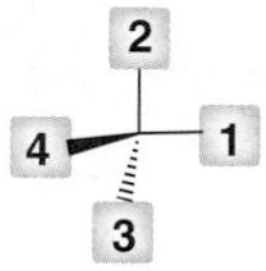

STEP 3 Rotate so that fourth priority is on a dash.

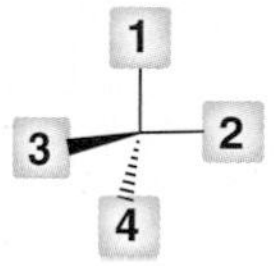

STEP 4 Assign configuration.

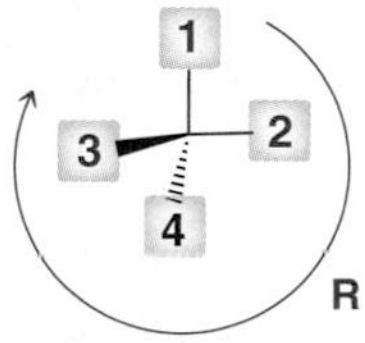

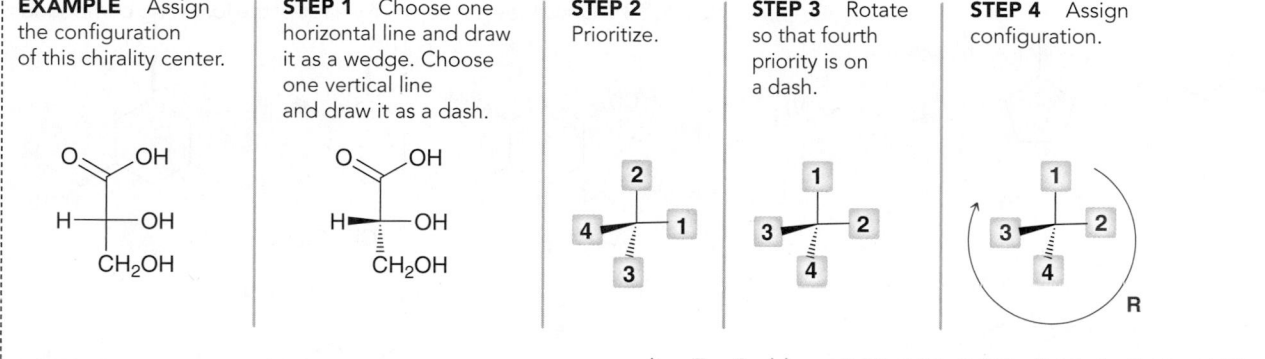

Try Problems 5.29–5.31, 5.39h, 5.41a,b, 5.43c, 5.53c,e, 5.54-5.56

PRACTICE PROBLEMS

Note: Most of the Problems are available within *WileyPLUS*, an online teaching and learning solution.

5.32 Atorvastatin is sold under the trade name Lipitor and is used for lowering cholesterol. Annual global sales of this compound exceed $13 billion. Assign a configuration to each chirality center in atorvastatin:

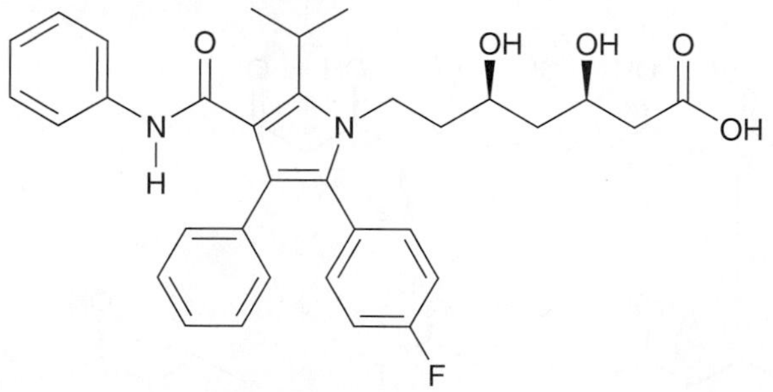

5.33 Atropine, extracted from the plant *Atropa belladonna*, has been used in the treatment of bradycardia (low heart rate) and cardiac arrest. Draw the enantiomer of atropine:

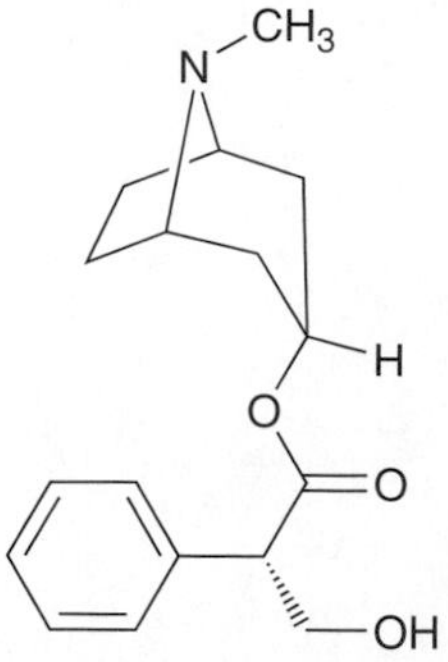

5.34 Paclitaxel (marketed under the trade name Taxol™) is found in the bark of the Pacific yew tree, *Taxus berevifolia*, and is used in the treatment of cancer:

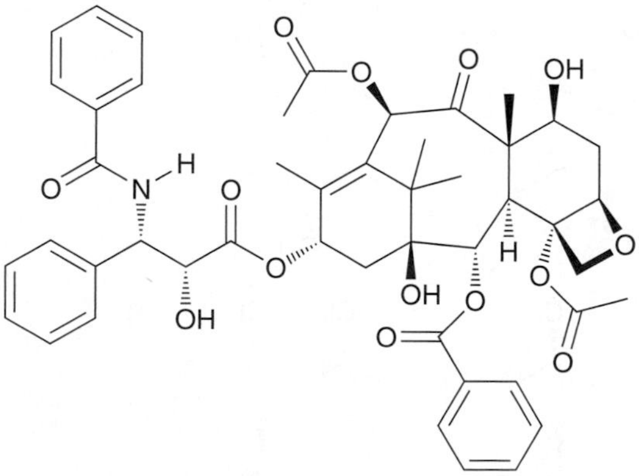

(a) Draw the enantiomer of paclitaxel. (b) How many chirality centers does this compound possess?

5.35 Classify each of the following compounds below as *cis*, *trans*, or not stereoisomeric:

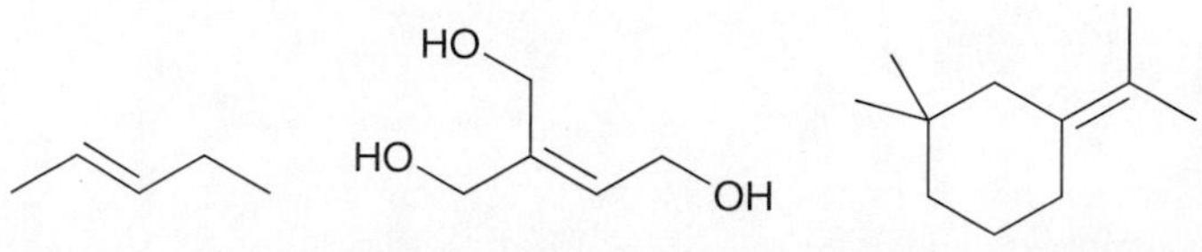

5.36 For each of the following pairs of compounds, determine the relationship between the two compounds:

(a) 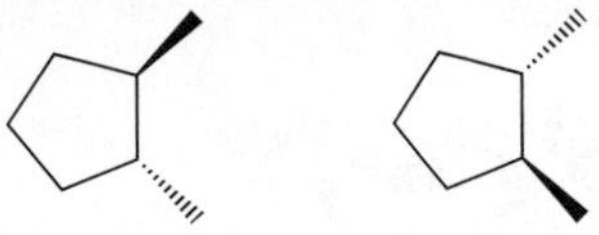

(b) 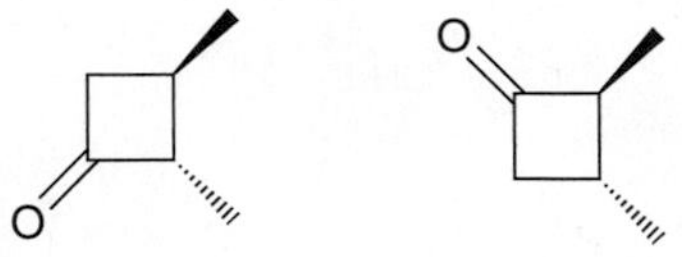

(c) 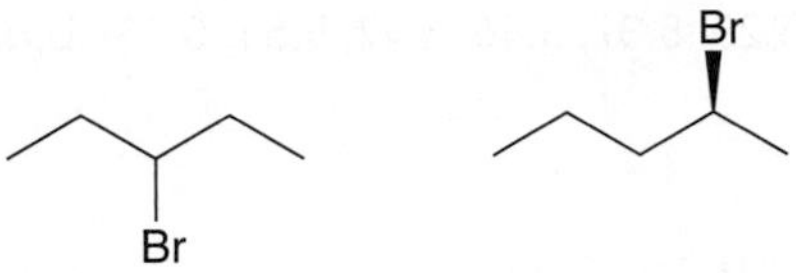

(d) 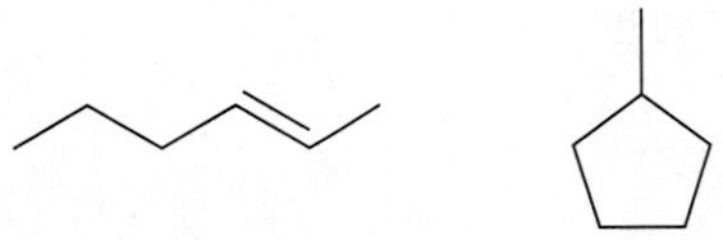

(e) 

(f) 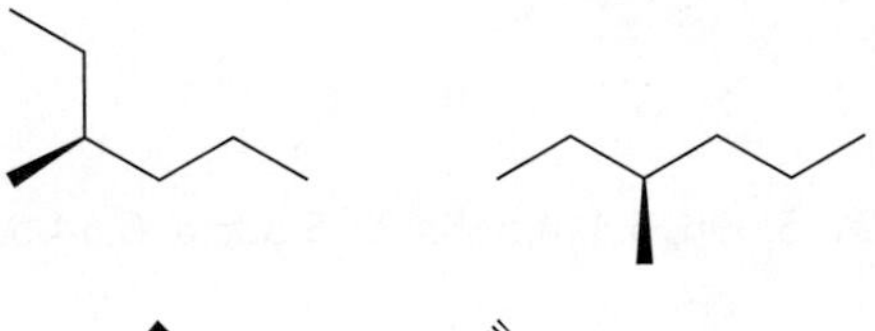

(g) 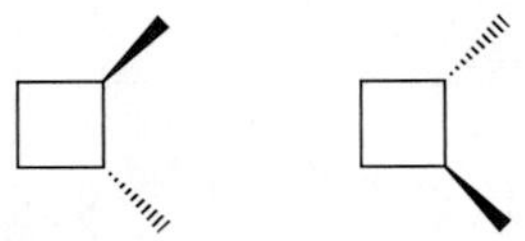

(h) 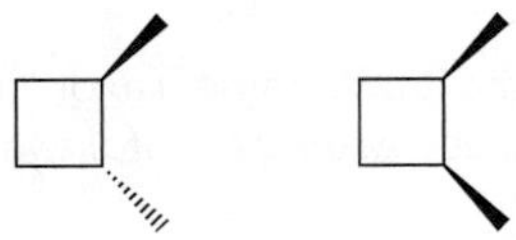

(i) 

(j) 

(k) 

(l) 

5.37 Identify the number of stereoisomers expected for each of the following:

(a) 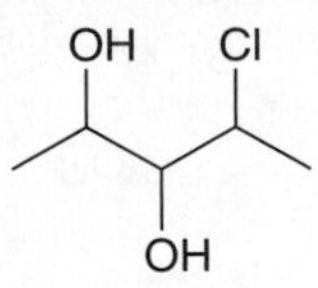

(b) 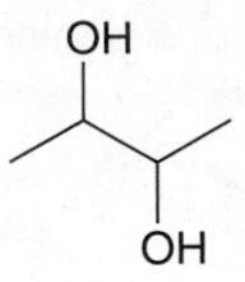

(c) 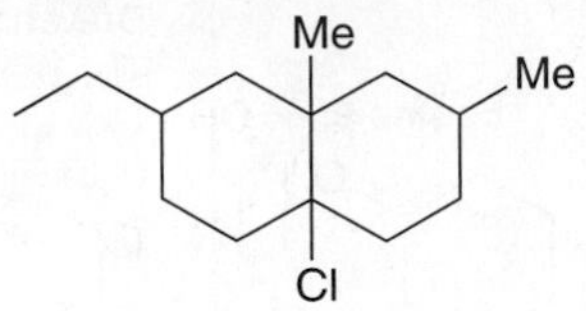

(d) 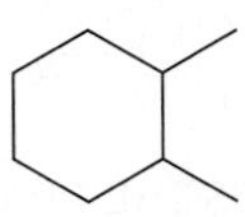

(e) 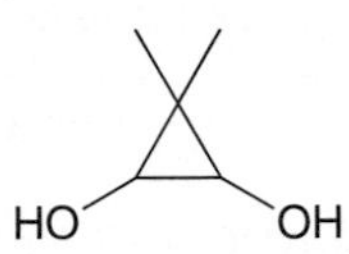

(f) 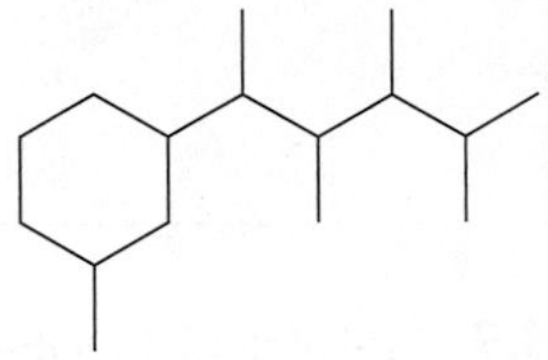

5.38 Draw the enantiomer for each of the following compounds:

(a) 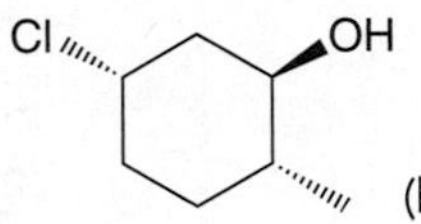

(b) 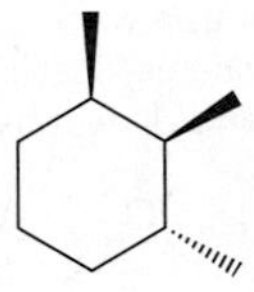

(c) 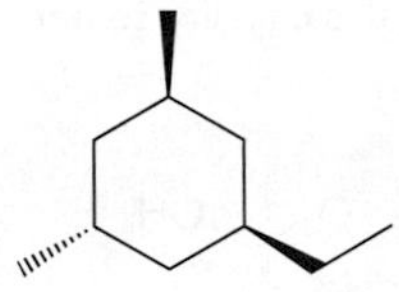

(d) 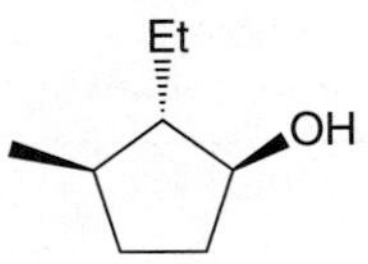

(e) 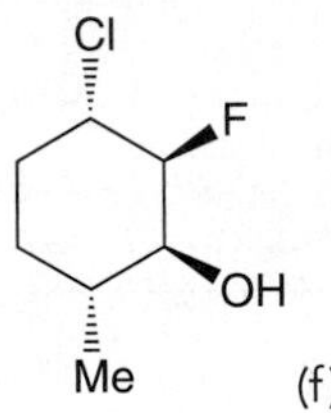

(f) 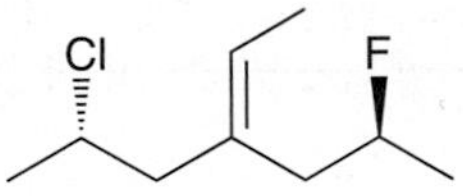

(g) 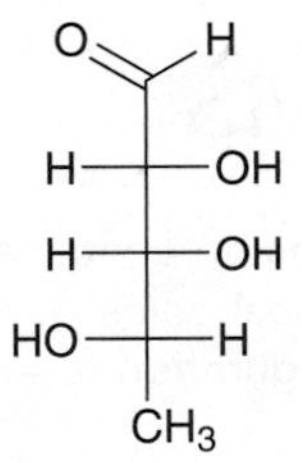

(h) 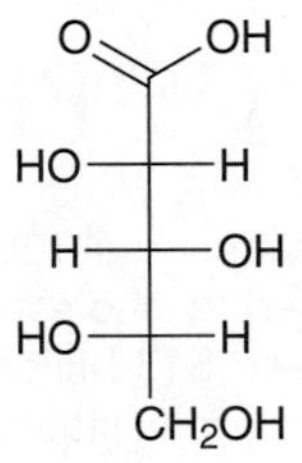

(i) 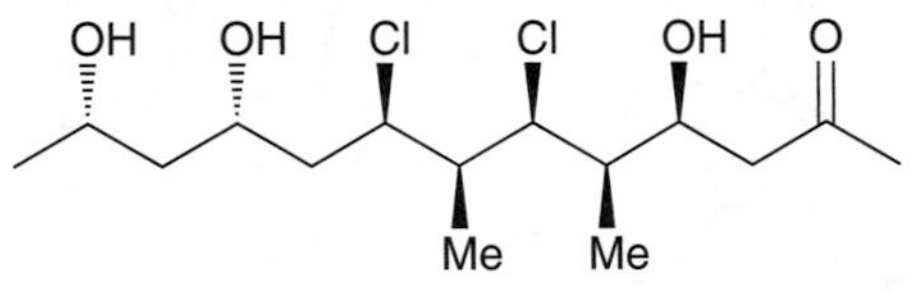

(j) 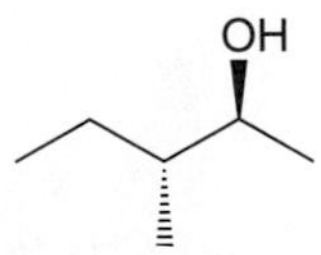

(k) 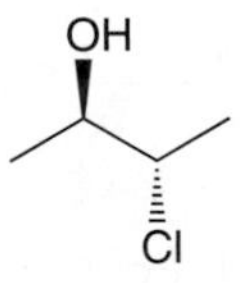

(l) 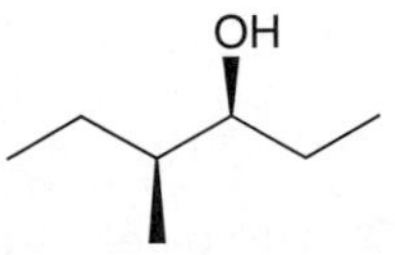

5.39 Identify the configuration of each chirality center in the following compounds:

(a) 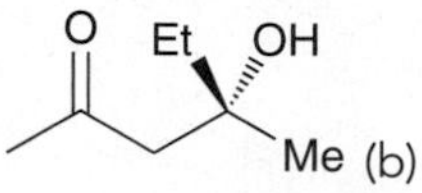
(b) 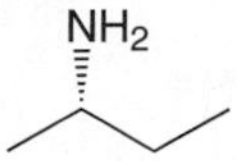
(c) 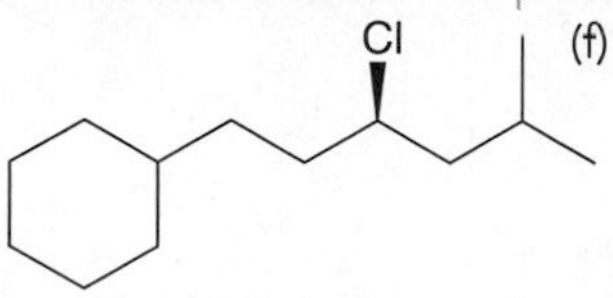
(d) 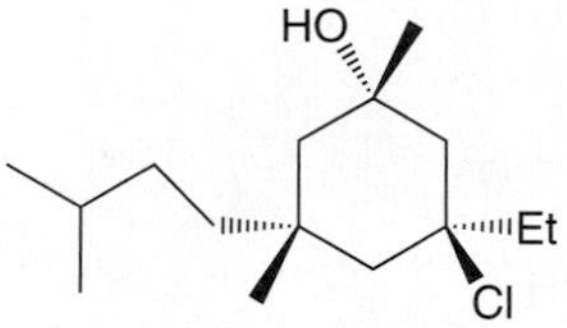
(e) 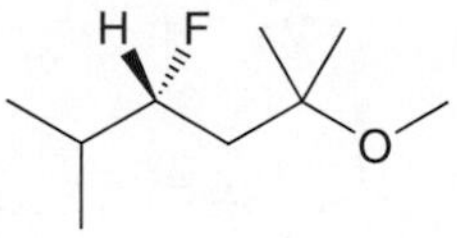
(f) 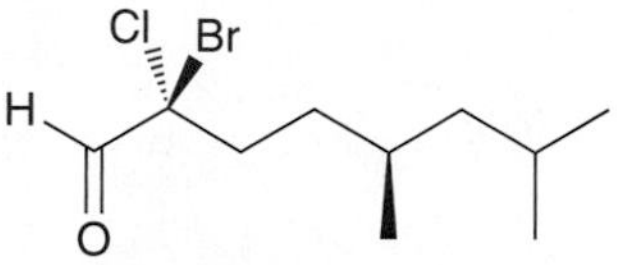
(g) 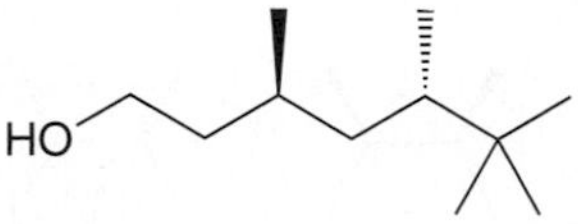
(h) 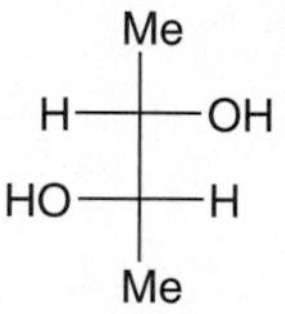
(i) 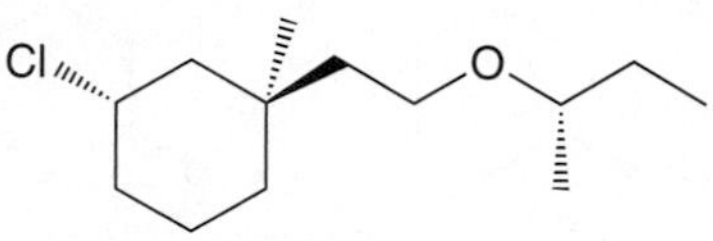

5.40 You are given a solution containing a pair of enantiomers (A and B). Careful measurements show that the solution contains 98% A and 2% B. What is the *ee* of this solution?

5.41 For each of the following pairs of compounds, determine the relationship between the two compounds:

(a) 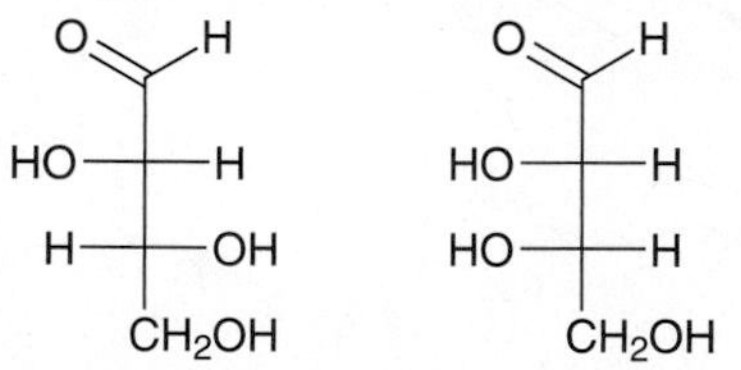
(b) 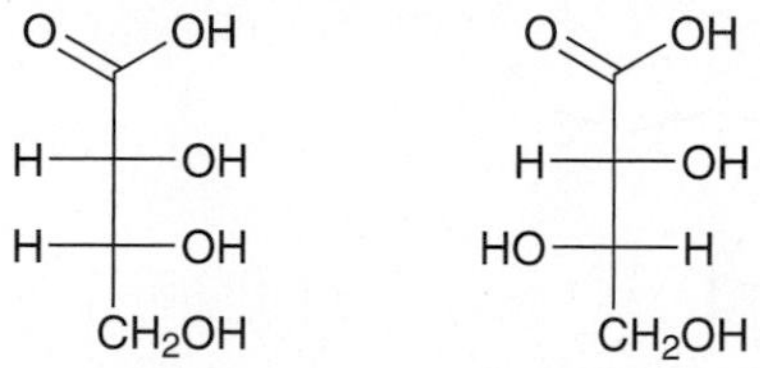
(c) 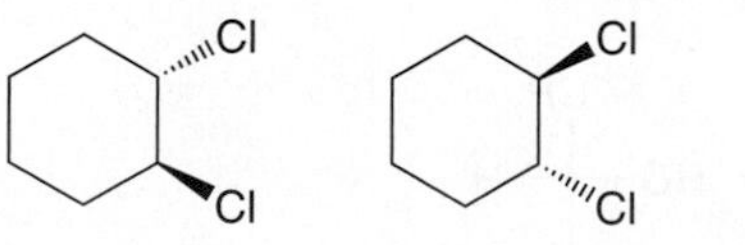
(d) 
(e) 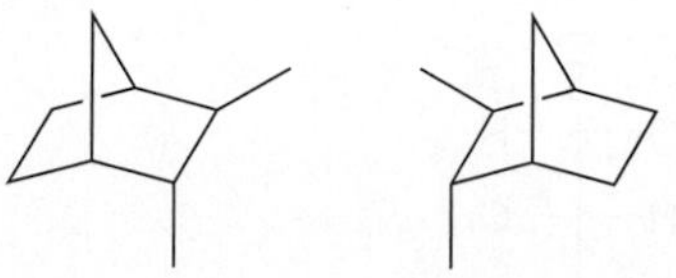
(f) 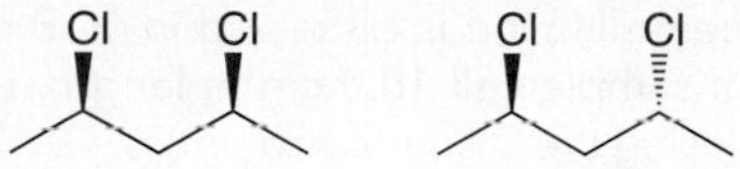
(g) 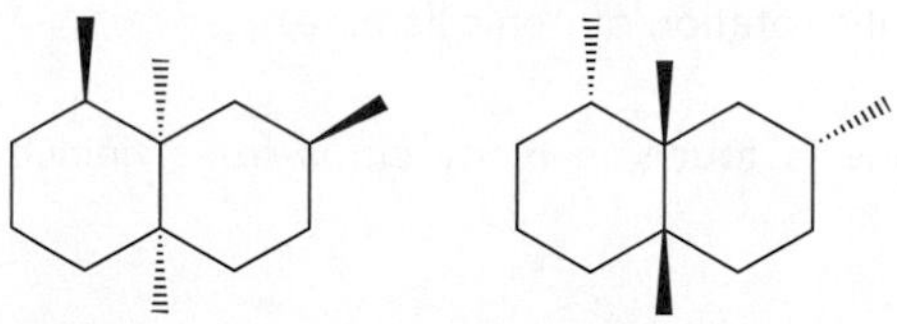
(h) 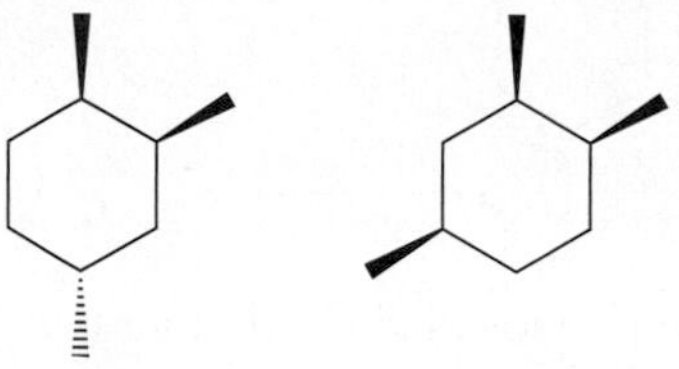
(i) 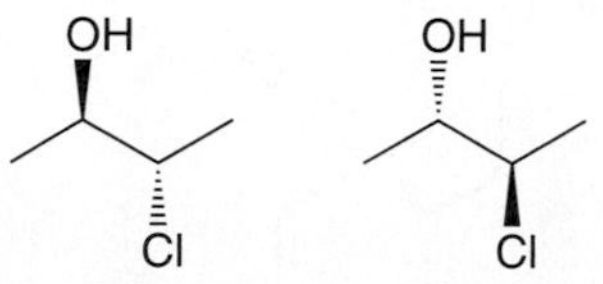
(j) 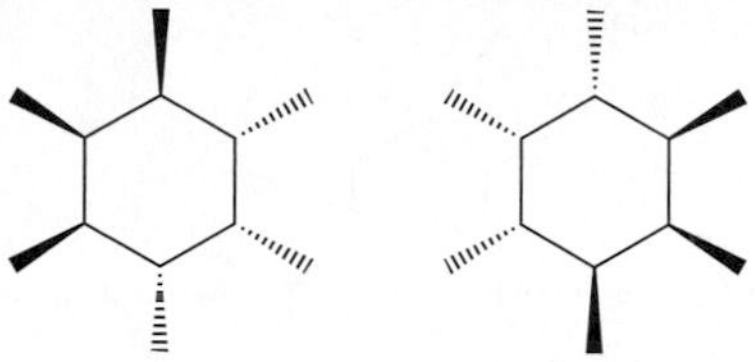
(k) 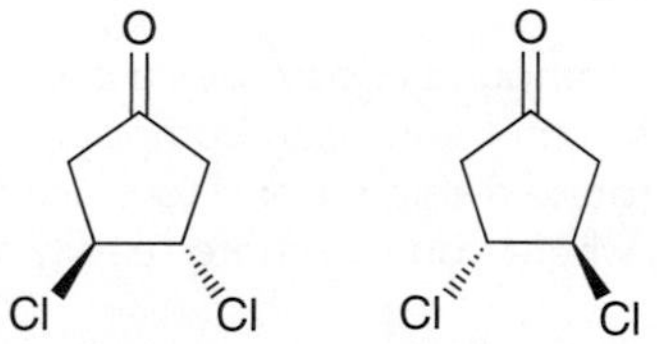
(l) 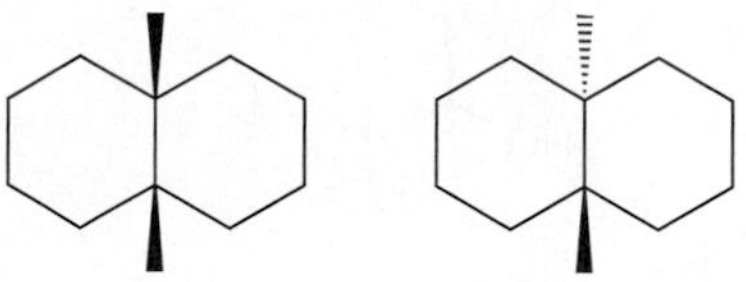

5.42 The specific rotation of (*R*)-carvone (at 20°C) is −61. A chemist prepared a mixture of (*R*)-carvone and its enantiomer, and this mixture had a specific rotation of −55. Calculate the % *ee* of this mixture.

5.43 Determine whether each statements is true or false:

(a) A racemic mixture of enantiomers is optically inactive.

(b) A *meso* compound will have exactly one nonsuperimposable mirror image.

(c) Rotating the Fischer projection of a molecule with a single chirality center by 90° will generate the enantiomer of the original Fischer projection:

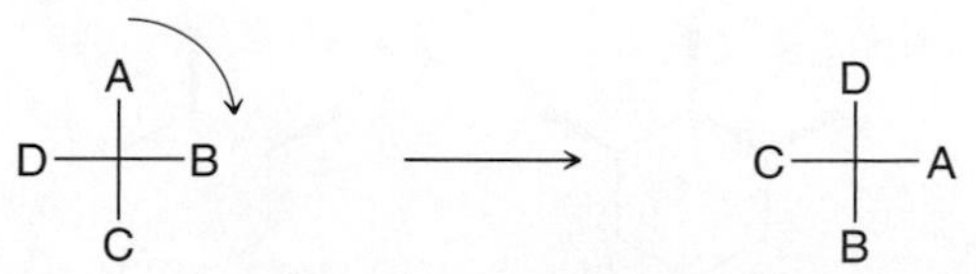

5.44 When 0.075 g of penicillamine is dissolved in 10.0 mL of pyridine and placed in a sample cell 10.0 cm in length, the observed rotation at 20°C (using the D line of sodium) is −0.47°. Calculate the specific rotation of penicillamine.

5.45 (*R*)-Limonene is found in many citrus fruits, including oranges and lemons:

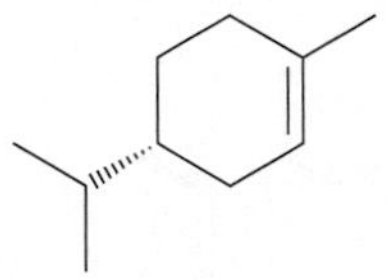

For each of the following compounds identify whether it is (*R*)-limonene or its enantiomer, (*S*)-limonene:

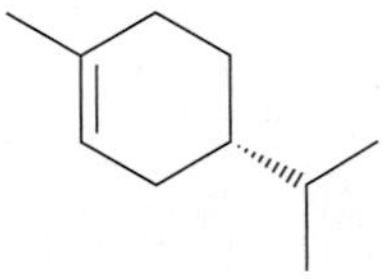
(a)

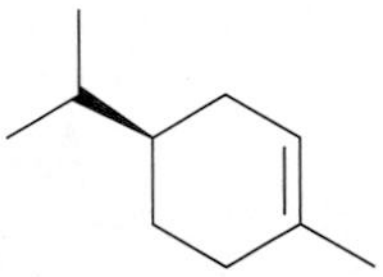
(b)

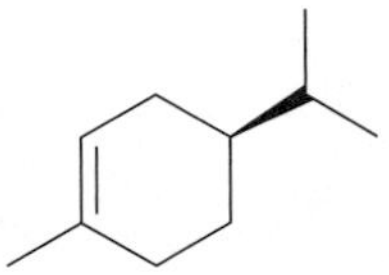
(c)

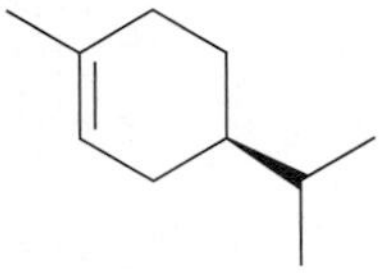
(d)

5.46 Each of the following compounds possesses a plane of symmetry. Find the plane of symmetry in each compound. In some cases, you will need to rotate a single bond to place the molecule into a conformation where you can more readily see the plane of symmetry.

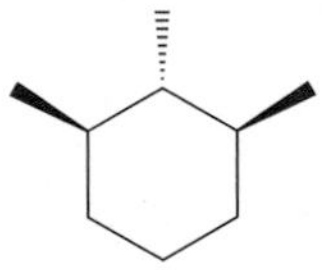
(a)

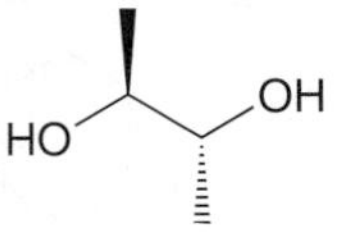

(b)

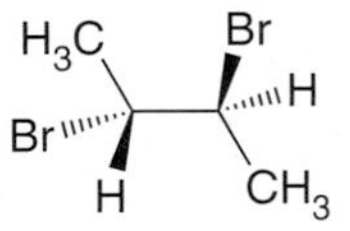

(c)

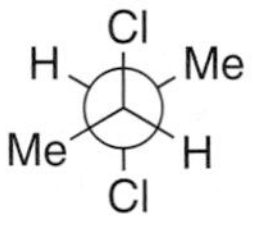

(d)

5.47 *cis*-1,3-Dimethylcyclobutane has two planes of symmetry. Draw the compound and identify both planes of symmetry.

5.48 Consider the following two compounds. These compounds are stereoisomers of 1,2,3-trimethylcyclohexane. One of these compounds has three chirality centers, while the other compound has only two chirality centers. Identify which compound has only two chirality centers and explain why.

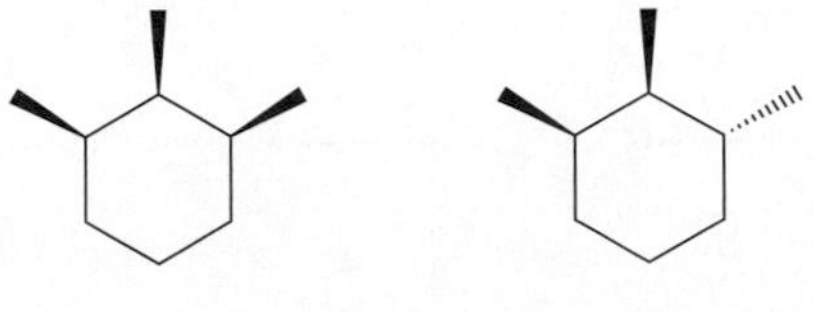

5.49 For each of the following pairs of compounds, determine the relationship between the two compounds.

(a)
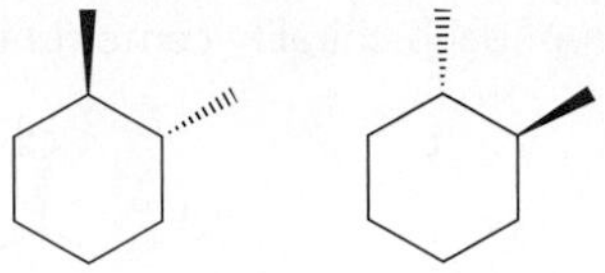

(b)
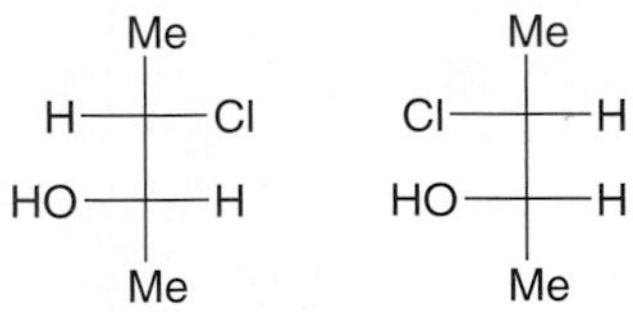

(c)
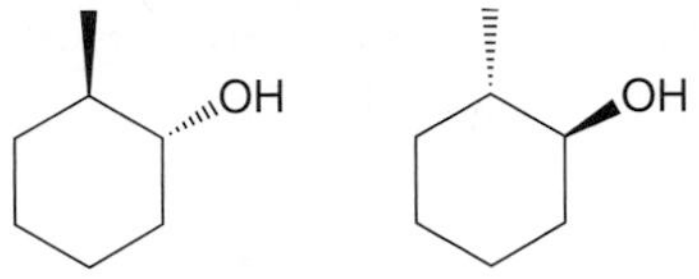

(d)
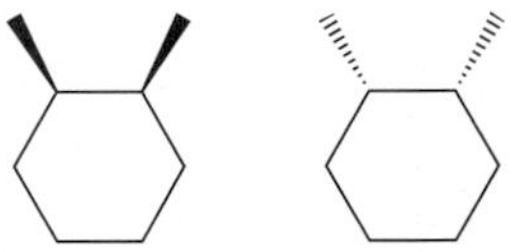

(e)

(f)
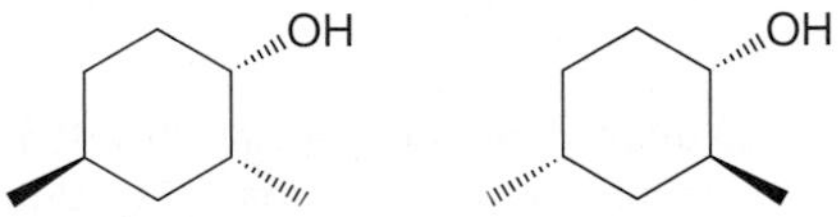

(g)
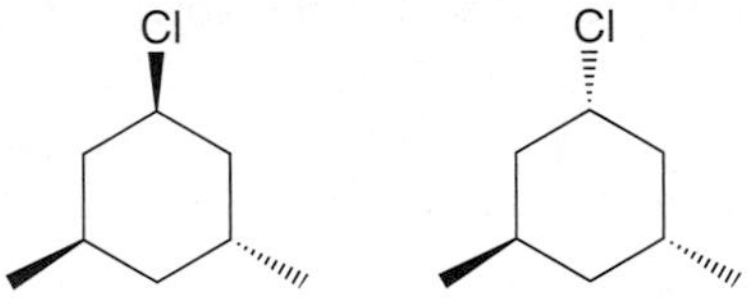

(h)
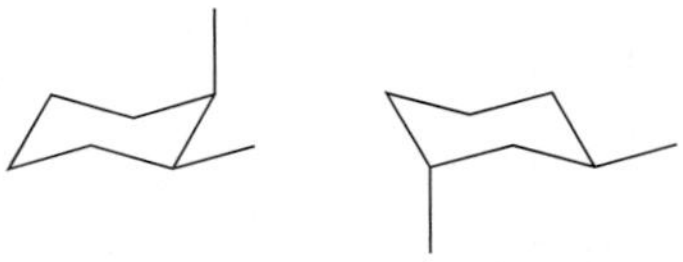

(i)
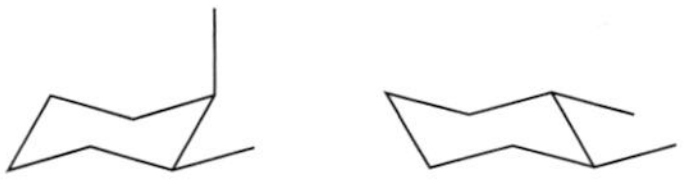

(j)
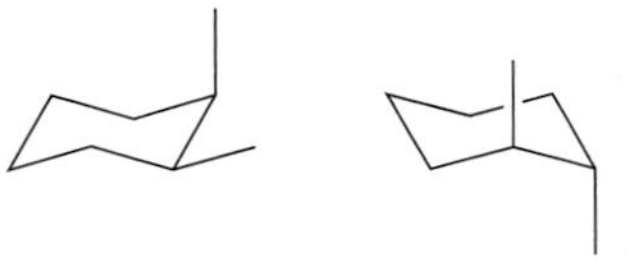

(k)
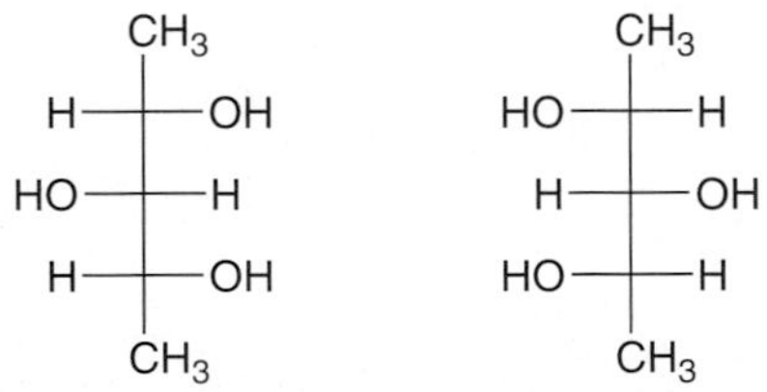

(l)
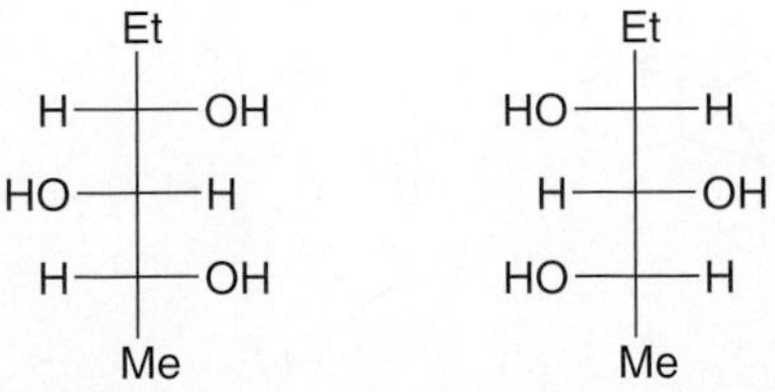

5.50 The specific rotation of (*S*)-carvone (at 20°C) is +61. A chemist prepared a mixture of (*R*)-carvone and its enantiomer, and this mixture had an observed rotation of −55°.

(a) What is the specific rotation of (*R*)-carvone at 20°C?

(b) Calculate the % *ee* of this mixture.

(c) What percentage of the mixture is (*S*)-carvone?

5.51 Identify whether each of the following compounds is chiral or achiral:

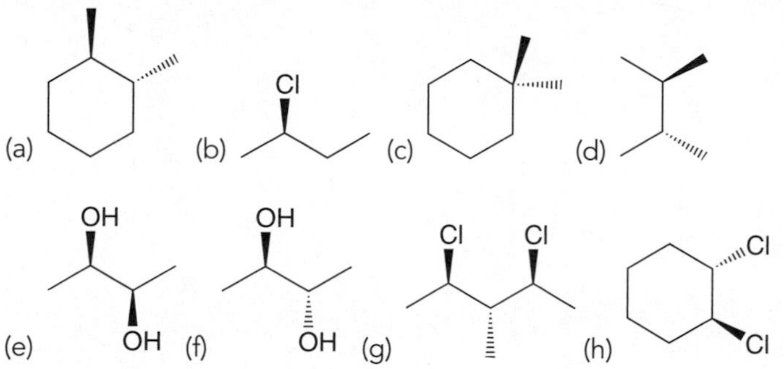

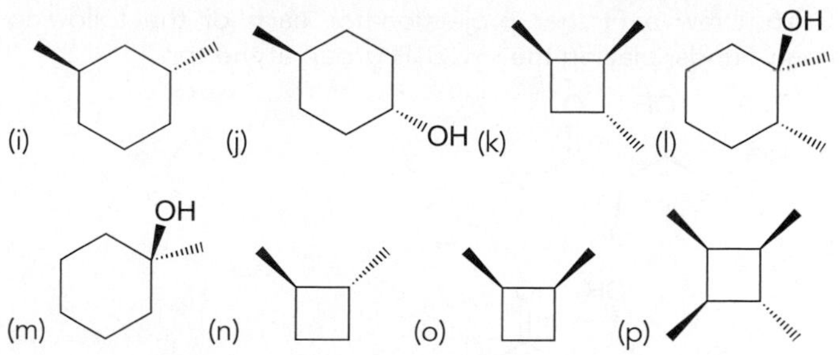

5.52 The specific rotation of vitamin C (using the D line of sodium, at 20°C) is +24. Predict what the observed rotation would be for a solution containing 0.100 g of vitamin C dissolved in 10.0 mL of ethanol and placed in a sample cell with a length of 1.00 dm.

INTEGRATED PROBLEMS

5.53 Determine whether each of the following compounds is optically active or optically inactive:

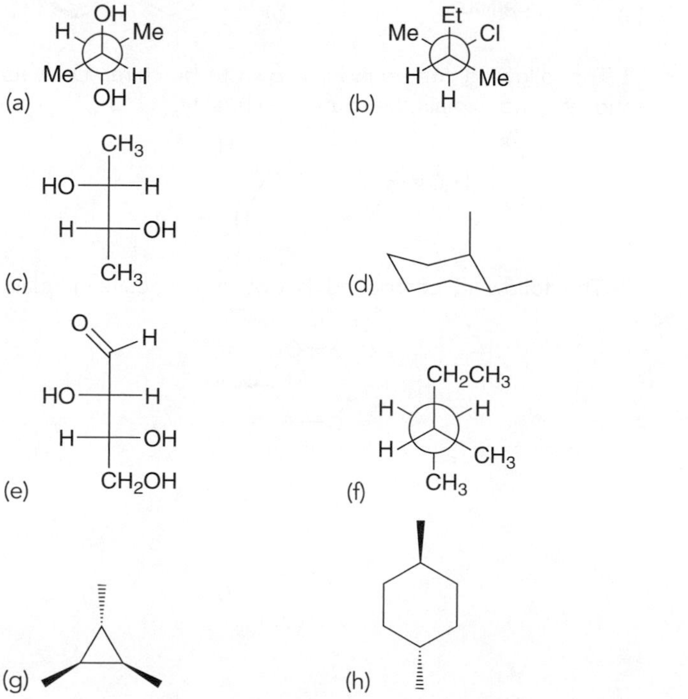

5.54 Draw bond-line structures using wedges and dashes for the following compounds:

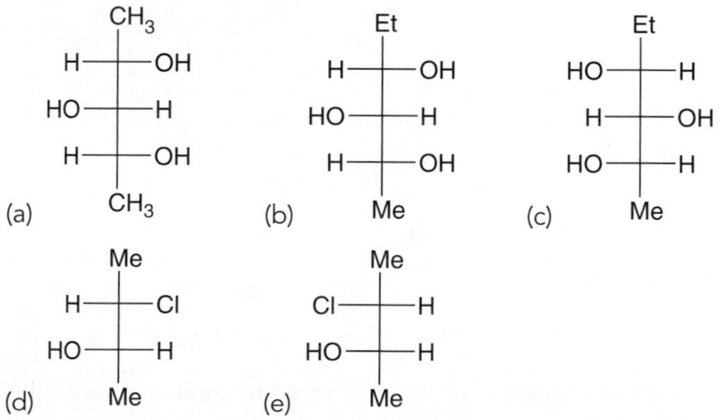

5.55 The following questions apply to the five compounds in Problem 5.54.

(a) Which compound is *meso*?

(b) Would an equal mixture of compounds b and c be optically active?

(c) Would an equal mixture of compounds d and e be optically active?

5.56 Draw a Fischer projection for each of the following compounds, placing the $-CO_2H$ group at the top.

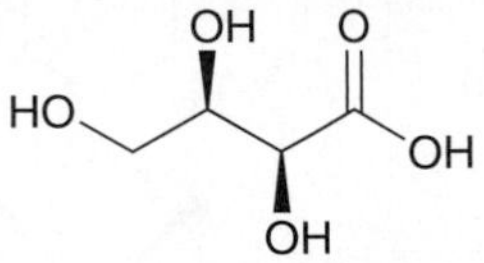

(a)

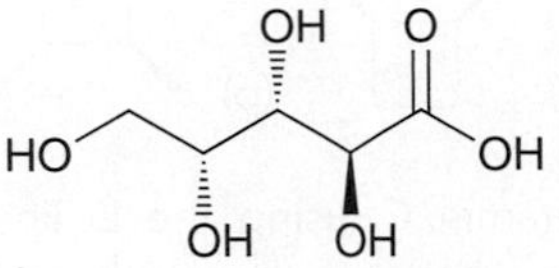

(b)

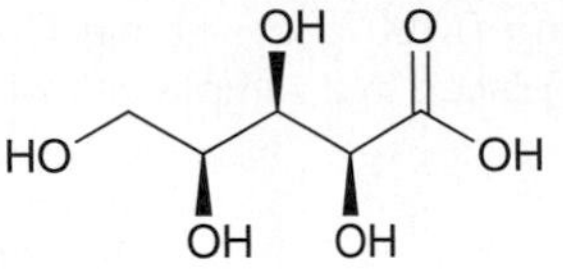

(c)

5.57 For each of the following pairs of compounds, determine the relationship between the two compounds.

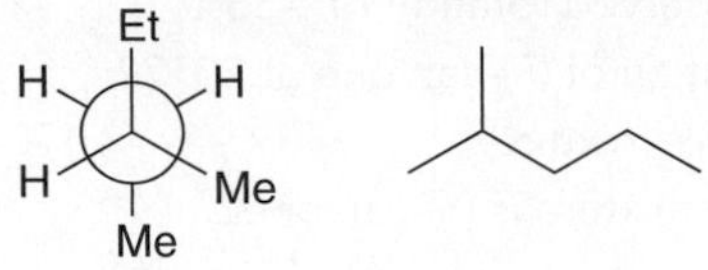

(a)

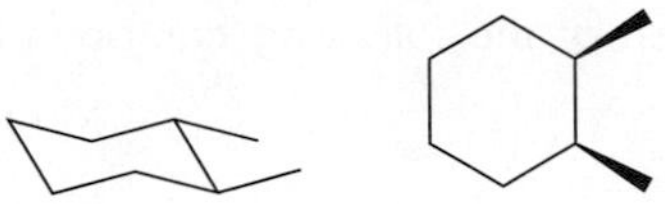

(b)

CHALLENGE PROBLEMS

5.58 It is possible for a compound to be chiral even though it lacks a carbon atom with four different groups. For example, consider the structure of the following compound which belongs to a class of compounds called allenes. This allene is chiral. Draw its enantiomer, and explain why this compound is chiral.

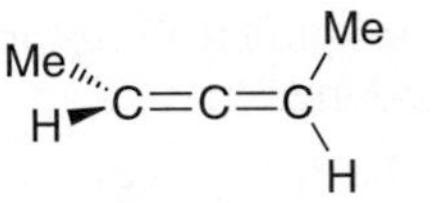

5.59 Based on your analysis in the previous problem, determine whether the following allene is expected to be chiral:

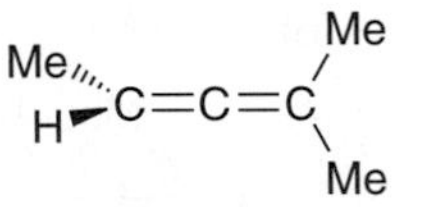

5.60 As mentioned in problem 5.58, some molecules are chiral even though they lack a chirality center. For example, consider the following two compounds shown, and explain the source of chirality in each case:

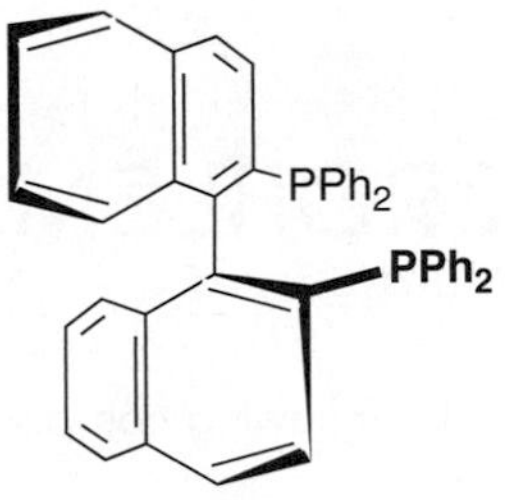

(a) (*S*)-BINAP

(b)

5.61 The following compound is known to be chiral. Draw its enantiomer, and explain the source of chirality.

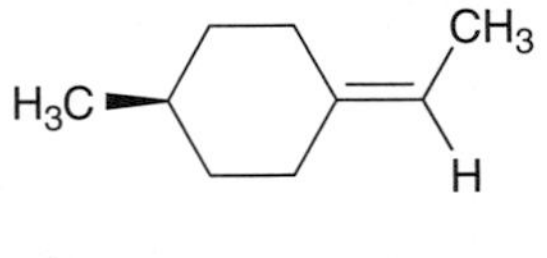

5.62 The following compound is optically inactive. Explain why.

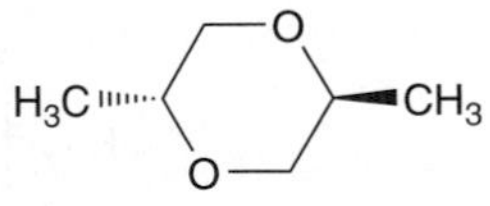

Chemical Reactivity and Mechanisms

6

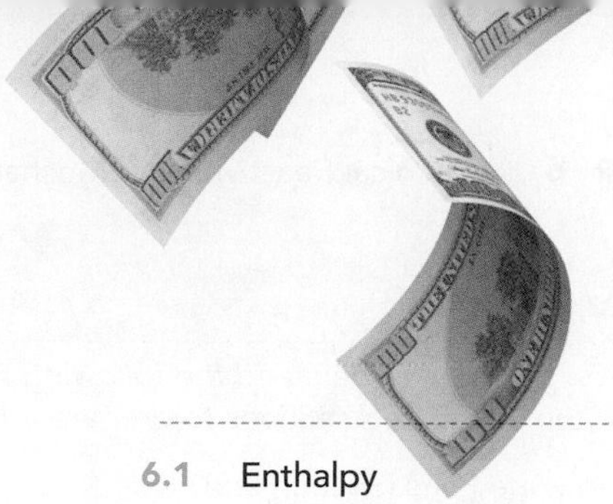

DID YOU EVER WONDER...

how Alfred Nobel made the fortune that he used to fund all of the Nobel Prizes (each of which includes a cash prize of over one million dollars)?

As we will see later in this chapter, Alfred Nobel earned his vast fortune by developing and marketing dynamite. Many of the principles that inform the design of explosives are the same principles that govern all chemical reactions. In this chapter, we will explore some of the key features of chemical reactions (including explosions). We will explore the factors that cause reactions to occur as well as the factors that speed up reactions. We will practice drawing and analyzing energy diagrams, which will be used heavily throughout this book to compare reactions. Most importantly, we will focus on the skills necessary to draw and interpret the individual steps of a reaction. This chapter will develop the core concepts and critical skills necessary for understanding chemical reactivity.

DO YOU REMEMBER?

Before you go on, be sure you understand the following topics.
If necessary, review the suggested sections to prepare for this chapter.

- Identifying Lone Pairs (Section 2.5)
- Curved Arrows in Drawing Resonance Structures (Section 2.8)
- Drawing Resonance Structures Via Pattern Recognition (Section 2.10)
- The Flow of Electron Density: Curved Arrow Notation (Section 3.2)

WileyPLUS Visit www.wileyplus.com to check your understanding and for valuable practice.

6.1 Enthalpy

In Chapter 1, we discussed the nature of electrons and their ability to form bonds. In particular, we saw that electrons achieve a lower energy state when they occupy a bonding molecular orbital (Figure 6.1). It stands to reason, then, that breaking a bond requires an input of energy.

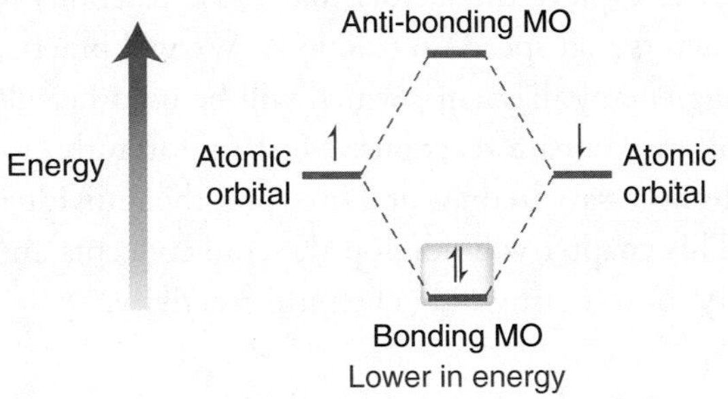

FIGURE 6.1 An energy diagram showing that bonding electrons occupy a bonding MO.

For a bond to break, the electrons in the bonding MO must receive energy from their surroundings. Specifically, the surrounding molecules must transfer some of their kinetic energy to the system (the bond being broken). The term **enthalpy** is used to measure this exchange of energy:

$$\Delta H = q \quad \text{(at constant pressure)}$$

The change in enthalpy (ΔH) for any process is defined as the exchange of kinetic energy, also called heat (q), between a system and its surroundings under conditions of constant pressure. For a bond-breaking reaction, ΔH is primarily determined by the amount of energy necessary to break the bond *homolytically*. **Homolytic bond cleavage** generates two uncharged species, called **radicals**, each of which bears an unpaired electron (Figure 6.2). Notice the use of single-headed curved arrows, often called fish-hooks. Radicals and fish-hook arrows will be discussed in more detail in Chapter 11. In contrast, **heterolytic bond cleavage** is illustrated with a two-headed curved arrow, generating charged species, called ions (Figure 6.3).

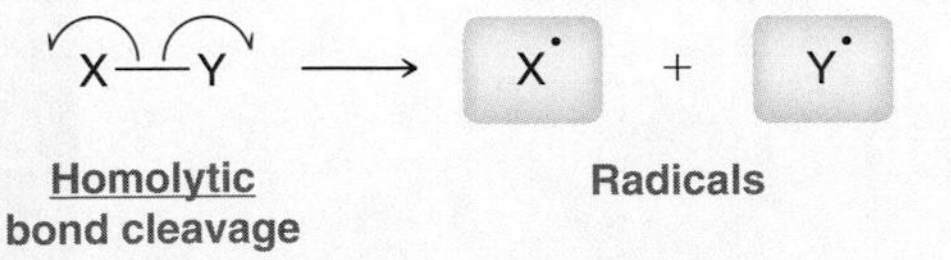

FIGURE 6.2 Homolytic bond cleavage produces two radicals.

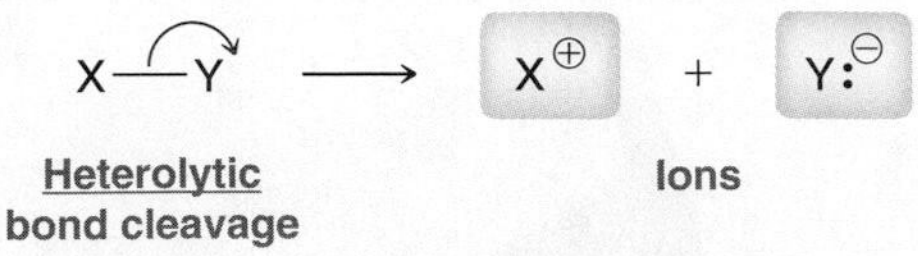

FIGURE 6.3 Heterolytic bond cleavage produces ions.

The energy required to achieve homolytic bond cleavage (generating radicals) is the primary factor that determines the value of ΔH for a bond-breaking reaction. Every bond has an associated ΔH, often referred to as the **bond dissociation energy**. The term $\Delta H°$ (with the "naught" symbol, or small circle, next to the H) refers to the bond dissociation energy when measured under standard conditions (i.e., where the pressure is 1 atm and the compound is in its standard state: a gas, a pure liquid, or a solid). Table 6.1 gives $\Delta H°$ values for a variety of bonds commonly encountered in this text.

TABLE 6.1 BOND DISSOCIATION ENERGIES ($\Delta H°$) OF COMMON BONDS

	KJ/MOL	KCAL/MOL
Bonds to H		
H—H	435	104
H—CH_3	435	104
H—CH_2CH_3	410	98
H—$CH(CH_3)_2$	397	95
H—$C(CH_3)_3$	381	91
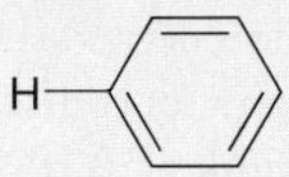	473	113
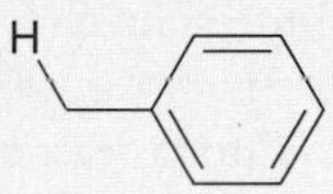	356	85
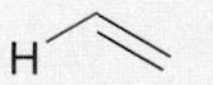	464	111
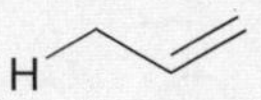	364	87
H—F	569	136
H—Cl	431	103
H—Br	368	88
H—I	297	71
H—OH	498	119
H—OCH_2CH_3	435	104
C—C bonds		
CH_3—CH_3	368	88
CH_3CH_2—CH_3	356	85
$(CH_3)_2CH$—CH_3	351	84

	KJ/MOL	KCAL/MOL
$H_2C{=}CH$—CH_3	385	92
$HC{\equiv}C$—CH_3	489	117
Bonds to methyl		
CH_3—H	435	104
CH_3—F	456	109
CH_3—Cl	351	84
CH_3—Br	293	70
CH_3—I	234	56
CH_3—OH	381	91
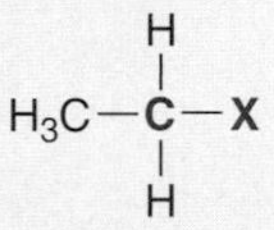		
CH_3CH_2—H	410	98
CH_3CH_2—F	448	107
CH_3CH_2—Cl	339	81
CH_3CH_2—Br	285	68
CH_3CH_2—I	222	53
CH_3CH_2—OH	381	91
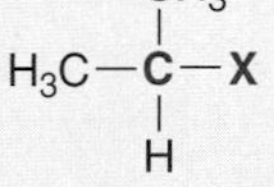		
$(CH_3)_2CH$—H	397	95

	KJ/MOL	KCAL/MOL
$(CH_3)_2CH$—F	444	106
$(CH_3)_2CH$—Cl	335	80
$(CH_3)_2CH$—Br	285	68
$(CH_3)_2CH$—I	222	53
$(CH_3)_2CH$—OH	381	91
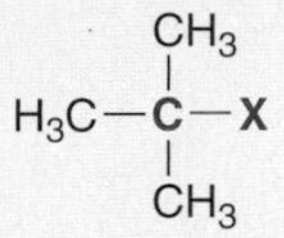		
$(CH_3)_3C$—H	381	91
$(CH_3)_3C$—F	444	106
$(CH_3)_3C$—Cl	331	79
$(CH_3)_3C$—Br	272	65
$(CH_3)_3C$—I	209	50
$(CH_3)_3C$—OH	381	91
X—X bonds		
F—F	159	38
Cl—Cl	242	58
Br—Br	192	46
I—I	151	36
HO—OH	213	51

Most reactions involve the breaking and forming of several bonds. In such cases, we must take into account each bond being broken or formed. The total change in enthalpy ($\Delta H°$) for the reaction is referred to as the **heat of reaction**. The sign of $\Delta H°$ for a reaction (whether it is positive or negative) indicates the direction in which the energy is exchanged and is determined from the perspective of the system. A positive $\Delta H°$ indicates that the system increased in energy (it received energy from the surroundings), while a negative $\Delta H°$ indicates that the system decreased in energy (it gave energy to the surroundings). The direction of energy exchange is described by the terms *endothermic* and *exothermic*. In an **exothermic** process, the system gives energy to the surroundings ($\Delta H°$ is negative). In an **endothermic** process, the system receives energy from the surroundings ($\Delta H°$ is positive). This is best illustrated with energy diagrams

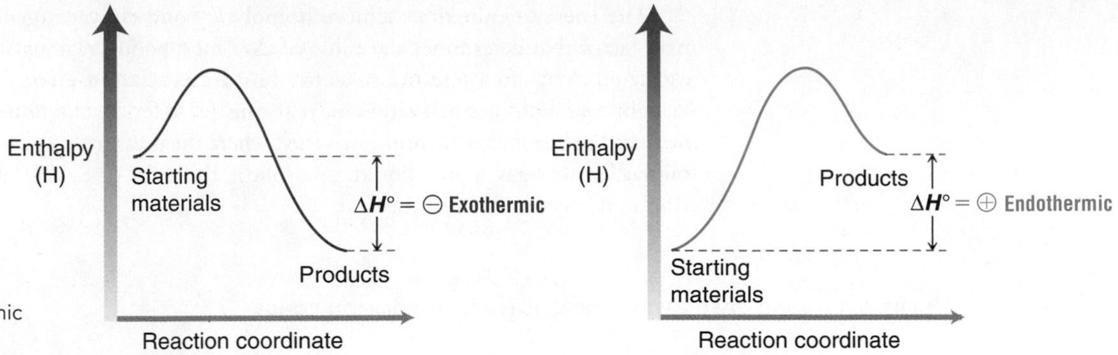

FIGURE 6.4
Energy diagrams of exothermic and endothermic processes.

(Figure 6.4) in which the curve represents the change in energy of the system as the reaction proceeds. In these diagrams, the progress of the reaction (the x axis of the diagram) is referred to as the *reaction coordinate*. Students are often confused by the sign of $\Delta H°$, and there is a valid reason for this confusion. In physics, the signs are the reverse of those shown here. Physicists think of $\Delta H°$ in terms of the surroundings rather than the system. They care about how devices have an impact on the environment—how much work the device can perform. Chemists, on the other hand, think in terms of the system. When chemists run a reaction, they care about the reactants and the products; they don't care about how the reaction ever so slightly changes the temperature in the laboratory. If you are currently enrolled in a physics course and find yourself confused about the sign of $\Delta H°$, just remember that chemists think of $\Delta H°$ from the perspective of the reaction (the system). For a chemist, an exothermic process involves the system losing energy to the surroundings, so $\Delta H°$ is negative.

Let's get some practice predicting the sign and value of $\Delta H°$ for a reaction.

SKILLBUILDER

6.1 PREDICTING $\Delta H°$ OF A REACTION

LEARN the skill

Predict the sign and magnitude of $\Delta H°$ for the following reaction. Give your answer in units of kilojoules per mole, and identify whether the reaction is expected to be endothermic or exothermic.

SOLUTION

Identify all bonds that are either broken or formed:

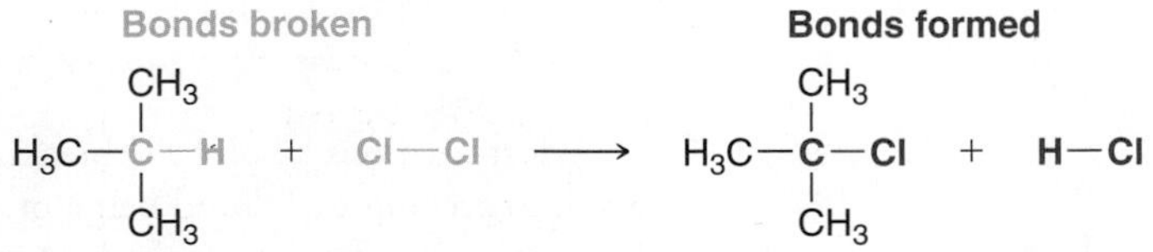

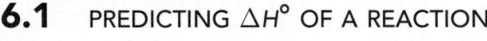

STEP 1
Identify the bond dissociation energy of each bond that is broken or formed.

Then use Table 6.1 to find the bond dissociation energies for each of these bonds:

Bond	kJ/mol
$H—C(CH_3)_3$	381
Cl—Cl	242
$(CH_3)_3C—Cl$	331
H—Cl	431

STEP 2
Determine the appropriate sign for each value from Step 1.

Now we must decide what sign (+ or −) to place in front of each value. Remember that $\Delta H°$ is defined with respect to the system. For each bond broken, the system must receive energy in order for the bond to break, so $\Delta H°$ must be positive. For each bond formed, the electrons are going to a lower energy state, and the system releases energy to the surroundings, so $\Delta H°$ must be negative. Therefore, $\Delta H°$ for this reaction will be the sum total of the following numbers:

STEP 3
Add the values for all bonds broken and formed.

Bonds broken	kJ/mol	Bonds formed	kJ/mol
H—C(CH_3)$_3$	+ 381	(CH_3)$_3$C—Cl	− 331
Cl—Cl	+ 242	H—Cl	− 431

That is, $\Delta H° = -139$ kJ/mol. For this reaction $\Delta H°$ is negative, which means that the system is losing energy. It is giving off energy to the environment, so the reaction is exothermic.

PRACTICE the skill

6.1 Using the data in Table 6.1, predict the sign and magnitude of $\Delta H°$ for each of the following reactions. In each case, identify whether the reaction is expected to be endothermic or exothermic:

(a)
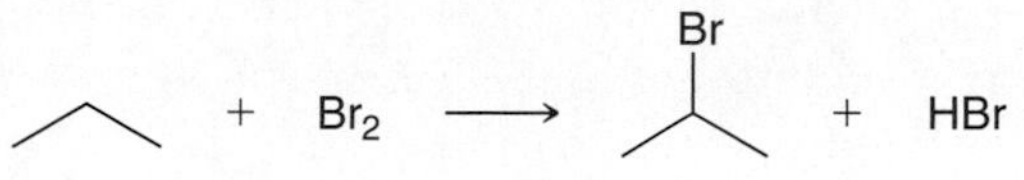

(b)
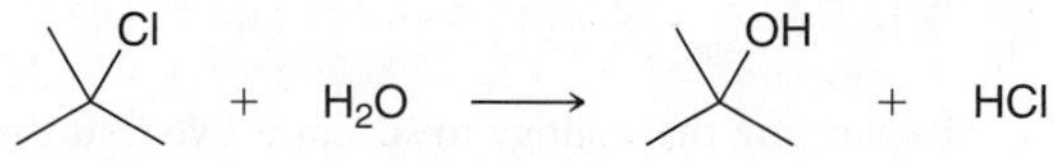

(c)
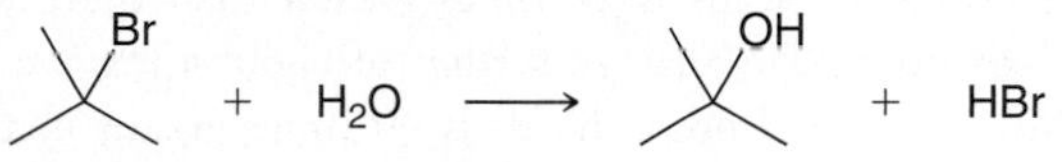

(d)
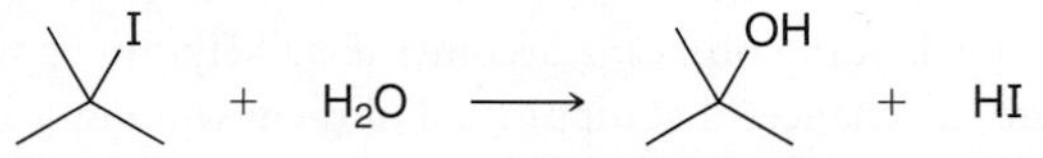

APPLY the skill

6.2 Recall that a C=C bond is comprised of a σ bond and a π bond. These two bonds together have a combined BDE (bond dissociation energy) of 632 kJ/mol. Use this information to predict whether the following reaction is exothermic or endothermic:

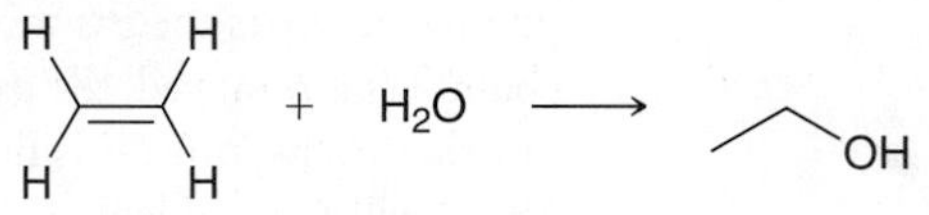

need more **PRACTICE?** **Try Problem 6.21a**

6.2 Entropy

The sign of $\Delta H°$ is not the ultimate measure of whether or not a reaction can or will occur. Although exothermic reactions are common, there are still plenty of examples of endothermic reactions that readily occur. This begs the question: What is the ultimate measure for determining whether or not a reaction can occur? The answer to this question is *entropy*, which is the underlying principle guiding all physical, chemical, and biological processes. **Entropy** is informally defined as the measure of disorder associated with a system, although this definition is overly simplistic. Entropy is more accurately described in terms of probabilities. To understand this, consider the following analogy. Imagine four coins, lined up in a row. If we toss all four coins, the chances that exactly half of the coins will land on heads are much greater than the chances that all four coins will land on heads. Why? Compare the number of possible ways to achieve each result (Figure 6.5). There is only one state in which all four coins land on heads, yet there are six different states in which exactly two coins land on heads. The probability of exactly two coins landing on heads is six times greater than the probability of all four coins landing on heads.

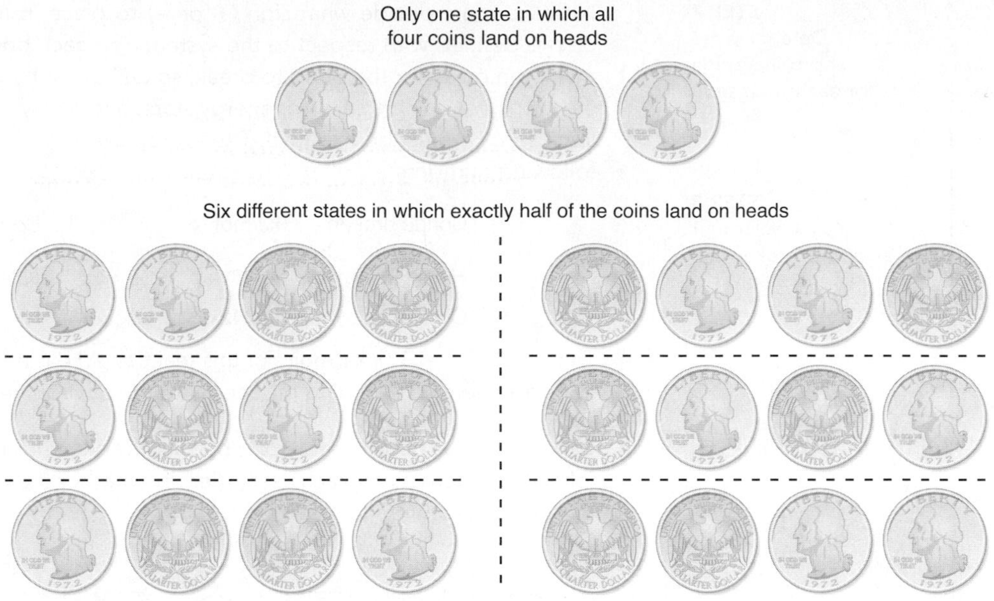
Only one state in which all four coins land on heads
Six different states in which exactly half of the coins land on heads

FIGURE 6.5
Comparing the number of ways to achieve various outcomes in a coin-tossing experiment.

Expanding the analogy to six coins, we find that the probability of exactly half of the coins landing on heads is 20 times greater than the probability of all six coins landing on heads. Expanding the analogy further to eight coins, we find that the probability of exactly half of the coins landing on heads is 70 times greater than the probability of all eight coins landing on heads. A trend is apparent: As the number of coins being tossed increases, the chances of all coins landing on heads becomes less likely. Imagine a floor covered with a billion coins. What are the chances that flipping all of them will result in all heads? The chances are miniscule (you have better chances of winning the lottery a hundred times in a row). It is much more likely that approximately half of the coins will land on heads because there are so many more ways of accomplishing that.

Now apply the same principle to describe the behavior of gas molecules in a two-chamber system, as shown in Figure 6.6. In the initial condition, one of the chambers is empty, and a divider prevents the gas molecules from entering that chamber. When the divider between the chambers is removed, the gas molecules undergo a free expansion. Free expansion occurs readily, but the reverse process is never observed. Once spread out into the two chambers, the gas molecules will not suddenly collect back into the first chamber, leaving the second chamber empty. This scenario is very similar to the coin-tossing analogy. At any moment in time, each molecule can be in either chamber 1 or chamber 2 (just like each coin can be either heads or tails). As we increase the number of molecules, the chances become less likely that all of the molecules will be found in one chamber. When dealing with a mole (6×10^{23} molecules), the chances are practically negligible, and we do not observe the molecules suddenly collecting into one chamber (at least not in our lifetimes).

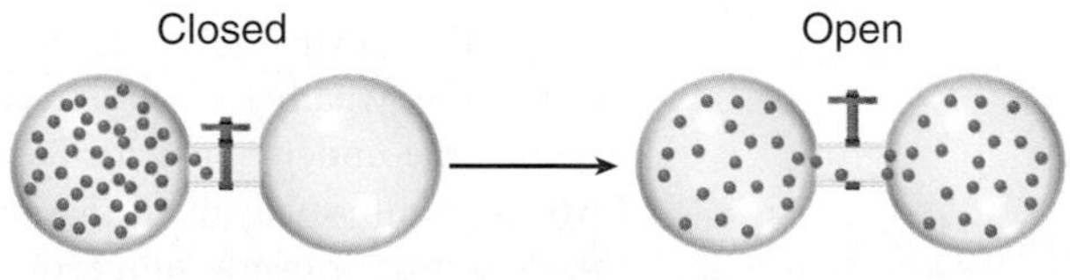
Closed
Open

FIGURE 6.6
Free expansion of a gas.

Free expansion is a classic example of entropy. When molecules occupy both chambers, the system is said to be at a higher state of entropy, because the number of states in which the molecules are spread between both chambers is so much greater than the number of states in which the molecules are all found in one chamber. Entropy is really nothing more than an issue of likelihood and probability.

A process that involves an increase in entropy is said to be **spontaneous**. That is, the process can and will occur, given enough time. Chemical reactions are no exception, although the considerations are slightly more complex than in a simple free expansion. In the case of free expansion, we only had to consider the change in entropy of the system (of the gas particles). The surroundings were unaffected by the free expansion. However, in a chemical reaction, the surroundings are affected. We must take into account not only the change in entropy of the system but also the change in entropy of the surroundings:

$$\Delta S_{tot} = \Delta S_{sys} + \Delta S_{surr}$$

where ΔS_{tot} is the total change in entropy associated with the reaction. In order for a process to be spontaneous, the total entropy must increase. The entropy of the system (the reaction) can actually decrease, as long as the entropy of the surroundings increases by an amount that offsets the decreased entropy of the system. As long as ΔS_{tot} is positive, the reaction will be spontaneous. Therefore, if we wish to assess whether a particular reaction will be spontaneous, we must assess the values of ΔS_{sys} and ΔS_{surr}. For now, we will focus our attention on ΔS_{sys}; we will discuss ΔS_{surr} in the next section.

The value of ΔS_{sys} is affected by a number of factors. The two most dominant factors are shown in Figure 6.7. In the first example in Figure 6.7, one mole of reactant produces two moles of product. This represents an increase in entropy, because the number of possible ways to arrange the molecules increases when there are more molecules (as we saw when we expanded our coin-tossing analogy by increasing the number of coins). In the second example, a cyclic compound is being converted into an acyclic compound. Such a process also represents an increase in entropy, because acyclic compounds have more freedom of motion than cyclic compounds. An acyclic compound can adopt a larger number of conformations than a cyclic compound, and once again, the larger number of possible states corresponds with a larger entropy.

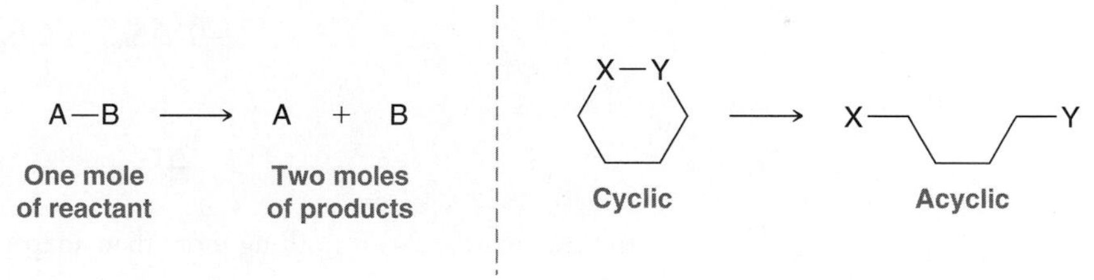

FIGURE 6.7
Two ways in which the entropy of a chemical system can increase.

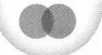

CONCEPTUAL CHECKPOINT

6.3 For each of the following processes predict the sign of ΔS for the reaction. In other words, will ΔS_{sys} be positive (an increase in entropy) or negative (a decrease in entropy)?

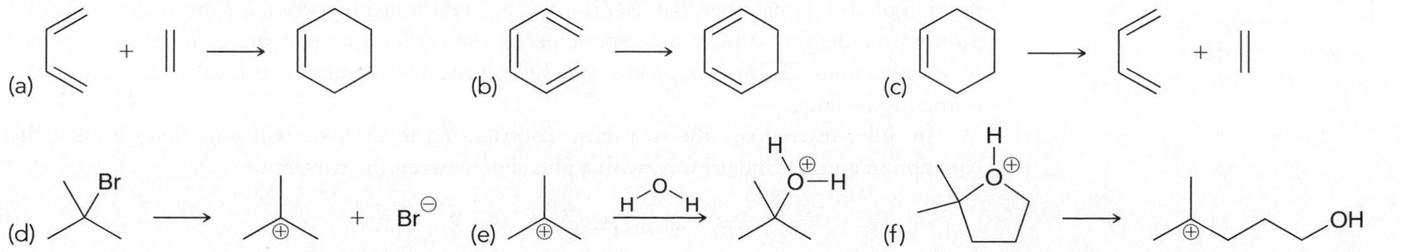

6.3 Gibbs Free Energy

In the previous section, we saw that entropy is the one and only criterion that determines whether or not a chemical reaction will be spontaneous. But it is not enough to consider ΔS of the system alone; we must also take into account ΔS of the surroundings:

$$\Delta S_{tot} = \Delta S_{sys} + \Delta S_{surr}$$

The total change in entropy (system plus surroundings) must be positive in order for the process to be spontaneous. It is fairly straightforward to assess ΔS_{sys} using tables of standard entropy values (a skill that you likely learned in your general chemistry course). However, the assessment of ΔS_{surr} presents more of a challenge. It is certainly not possible to observe the entire universe, so how can we possibly measure ΔS_{surr}? Fortunately, there is a clever solution to this problem.

Under conditions of constant pressure and temperature, it can be shown that:

BY THE WAY
Organic chemists generally carry out reactions under constant pressure (atmospheric pressure) and at a constant temperature. For this reason, the conditions most relevant to practicing organic chemists are conditions of constant pressure and temperature.

$$\Delta S_{surr} = -\frac{\Delta H_{sys}}{T}$$

Notice that ΔS_{surr} is now defined *in terms of the system*. Both ΔH_{sys} and T (temperature in Kelvin) are easily measured, which means that ΔS_{surr} can in fact be measured. Plugging this expression for ΔS_{surr} into the equation for ΔS_{tot}, we arrive at a new equation for ΔS_{tot} for which all terms are measurable:

$$\Delta S_{tot} = \left(-\frac{\Delta H_{sys}}{T}\right) + \Delta S_{sys}$$

This expression is still the ultimate criterion for spontaneity (ΔS_{tot} must be positive). As a final step, we multiply the entire equation by $-T$, which gives us a new term, called **Gibbs free energy**:

$$-T\Delta S_{tot} = \Delta H_{sys} - T\Delta S_{sys}$$

$$-T\Delta S_{tot} \rightarrow \Delta G$$

In other words, ΔG is nothing more than a repackaged way of expressing total entropy:

$$\Delta G = \Delta H - T\Delta S$$

Associated with the change in entropy of the surroundings (ΔH)   **Associated with the change in entropy of the system** ($T\Delta S$)

In this equation, the first term (ΔH) is associated with the change in entropy of the surroundings (a transfer of energy to the surroundings increases the entropy of the surroundings). The second term ($T\Delta S$) is associated with the change in entropy of the system. The first term (ΔH) is often much larger than the second term ($T\Delta S$), so for most processes, ΔH will determine the sign of ΔG. Remember that ΔG is just ΔS_{tot} multiplied by *negative T*. So if ΔS_{tot} must be positive in order for a process to be spontaneous, then ΔG must be negative. *In order for a process to be spontaneous, ΔG for that process must be negative*. This requirement is called the second law of thermodynamics.

BY THE WAY
You may recall from your general chemistry course that the first law of thermodynamics is the requirement for conservation of energy during any process.

In order to compare the two terms contributing to ΔG, we will sometimes present the equation in a non-standard way, with a plus sign between the two terms.

$$\Delta G = \Delta H + (-T\Delta S)$$

PRACTICALLYSPEAKING)))

Explosives

Explosives are compounds that can be detonated, generating sudden large pressures and liberating excessive heat. Explosives have several defining characteristics:

1. In the presence of oxygen, an explosive will produce a reaction for which both expressions contribute to a very favorable ΔG:

$$\Delta G = \Delta H - T\Delta S$$

That is, the changes in entropy of both the surroundings and the system are extremely favorable. As a result, both terms contribute to a very large and negative value of ΔG. This makes the reaction extremely favorable.

2. Explosives must generate large quantities of gas very quickly in order to produce the sudden increase in pressure observed during an explosion. Compounds containing multiple nitro groups are capable of liberating gaseous NO_2. Several examples are shown:

3. When detonated, the reaction must occur very rapidly. This aspect of explosives will be discussed later in this chapter.

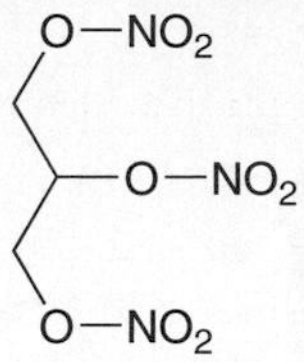

Nitroglycerin

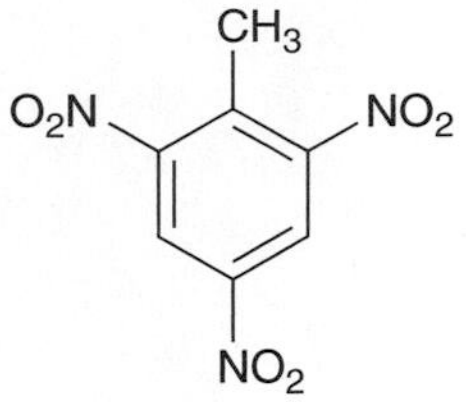

Trinitrotoluene (TNT)

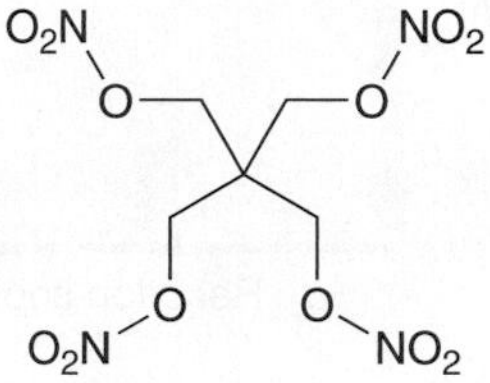

Pentaerythritol tetranitrate (PETN)

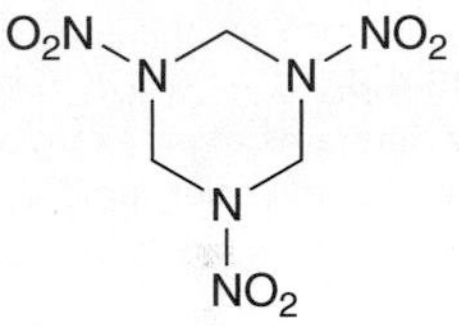

Cyclotrimethylenetrinitramine (RDX)

In some cases, this non-standard presentation will allow for a more efficient analysis of the competition between the two terms. As an example, consider the following reaction:

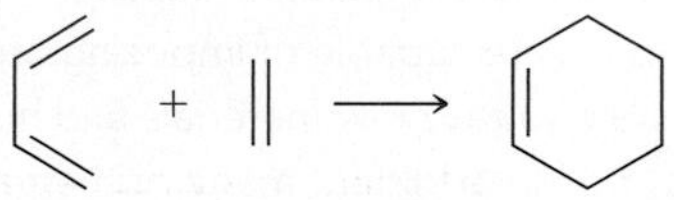

In this reaction, two molecules are converted into one molecule, so ΔS_{sys} is not favorable. However, ΔS_{surr} is favorable. Why? ΔS_{surr} is determined by ΔH of the reaction. In the first section of this chapter, we learned how to estimate ΔH for a reaction by looking at the bonds being formed and broken. In the reaction above, three π bonds are broken while one π bond and two σ bonds are formed. In other words, two π bonds have been transformed into two σ bonds. Sigma bonds are stronger (lower in energy) than π bonds, and therefore, ΔH for this reaction must be negative (exothermic). Energy is transferred to the surroundings, and this increases the entropy of the surroundings. To recap, ΔS_{sys} is not favorable, but ΔS_{surr} is favorable. Therefore, this process exhibits a competition between the change in entropy of the system and the change in entropy of the surroundings:

if (−) then rxn is spontaneous.

$$\Delta G = \underset{\ominus}{\Delta H} + \underset{\oplus}{(-T\Delta S)}$$

Gibbs free energy

Remember that ΔG must be negative in order for a process to be spontaneous. The first term is negative, which is favorable, but the second term is positive, which is unfavorable. The term that is larger will determine the sign of ΔG. If the first term is larger, then ΔG will be negative, and the reaction will be spontaneous. If the second term is larger, then ΔG will be positive, and the reaction will not be spontaneous. Which term dominates? The value of the second term is dependent on T, and therefore, the outcome of this reaction will be very sensitive to temperature. Below a certain T, the process will be spontaneous. Above a certain T, the reverse process will be favored.

Any process with a negative ΔG will be spontaneous. Such processes are called **exergonic**. Any process with a positive ΔG will not be spontaneous. Such processes are called **endergonic** (Figure 6.8).

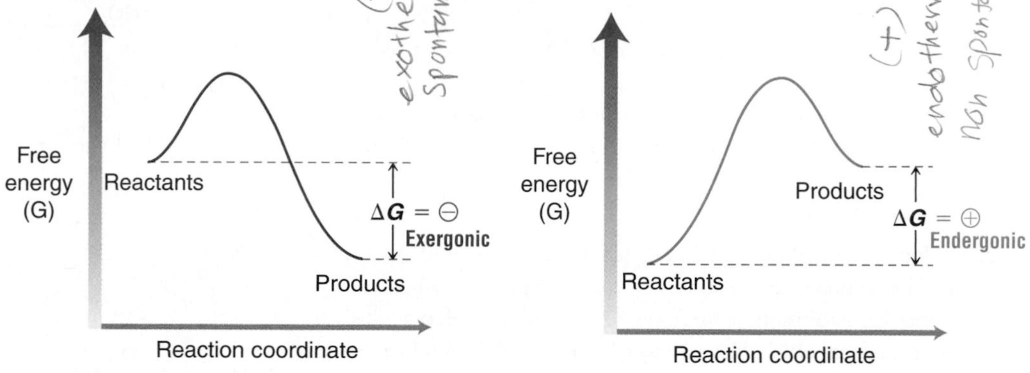

FIGURE 6.8 Energy diagrams of an exergonic process and an endergonic process.

These energy diagrams are incredibly useful when analyzing reactions, and we will use these diagrams frequently throughout the course. The next few sections of this chapter deal with some of the more specific details of these diagrams.

PRACTICALLYSPEAKING)))

Do Living Organisms Violate the Second Law of Thermodynamics?

Students often wonder whether life is a violation of the second law of thermodynamics. It is certainly a legitimate question. Our bodies are highly ordered, and we are capable of imposing order on our surroundings. We are able to take raw materials and build tall skyscrapers. How can that be? In addition, many of the reactions necessary to synthesize the macromolecules necessary for life, such as DNA and proteins, are endergonic processes (ΔG is positive). How can those reactions occur?

Earlier in this chapter, we saw that it is possible for a reaction to exhibit a decrease in entropy and still be spontaneous if and only if the entropy of the surroundings increases such that ΔS_{tot} for the process is positive. In other words, we have to take the whole picture into account when determining if a process will be spontaneous. Yes, it is true that many of the reactions employed by life are not spontaneous *by themselves*. But when they are coupled with other highly favorable reactions, such as the metabolism of food, the total entropy (system plus surroundings) does actually increase. As an example, consider the metabolism of glucose:

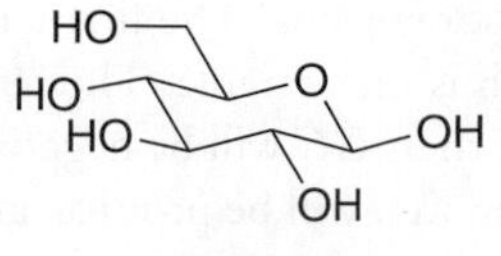

Glucose

$$+ \ 6\,O_2 \longrightarrow 6\,CO_2 + 6\,H_2O \qquad \Delta G° = -2880 \text{ kJ/mol}$$

This transformation has a large and negative ΔG. Processes like this one are ultimately responsible for driving otherwise nonspontaneous reactions that are necessary for life.

Living organisms do not violate the second law of thermodynamics—quite the opposite, in fact. Living organisms are prime examples of entropy at its finest. Yes, we do impose order on our environment, but that is only permitted because we are *entropy machines*. We give off heat to our surroundings and we consume highly ordered molecules (food) and break them down into smaller, more stable compounds with much lower free energy. These features allow us to impose order on our surroundings, because our net effect is to increase the entropy of the universe.

CONCEPTUAL CHECKPOINT

6.4 For each of the following reactions predict the sign of ΔG. If a prediction is not possible because the sign of ΔG will be temperature dependent, describe how ΔG will be affected by raising the temperature.

(a) An endothermic reaction for which the system exhibits an increase in entropy

(b) An exothermic reaction for which the system exhibits an increase in entropy

(c) An endothermic reaction for which the system exhibits a decrease in entropy

(d) An exothermic reaction for which the system exhibits a decrease in entropy

6.5 At room temperature, molecules spend most of their time in lower energy conformations. In fact, there is a general tendency for any system to move toward lower energy. As another example, electrons form bonds because they "prefer" to achieve a lower energy state. We can now appreciate that the reason for this preference is based on entropy. When a compound assumes a lower energy conformation or when electrons assume a lower energy state, ΔS_{tot} increases. Explain.

6.4 Equilibria

Consider the energy diagram in Figure 6.9, showing a reaction in which the reactants, A and B, are converted into products, C and D. The reaction exhibits a negative ΔG and therefore will be spontaneous. Accordingly, we might expect a mixture of A and B to be converted completely into C and D. But this is not the case. Rather, an equilibrium is established in which all four compounds are present. Why should this be the case? If C and D are truly lower in free energy than A and B, then why is there any amount of A and B present when the reaction is complete?

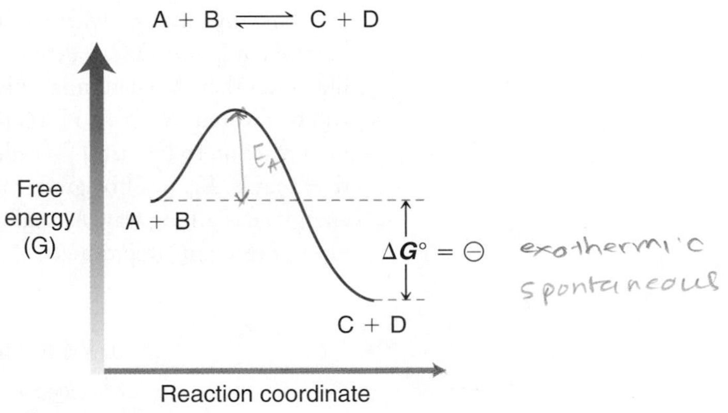

FIGURE 6.9
The energy diagram of an exergonic reaction in which reactants (A and B) are converted into products (C and D).

To answer this question, we must consider the effect of having a large number of molecules. The energy diagram in Figure 6.9 describes the reactions between one molecule of A and one molecule of B. However, when dealing with moles of A and B, the changing concentrations have an effect on the value of ΔG. When the reaction begins, only A and B are present. As the reaction proceeds, the concentrations of A and B decrease, and the concentrations of C and D increase. This has an effect on ΔG, which can be illustrated with the diagram in Figure 6.10. As the reaction proceeds, the free energy decreases until it reaches a minimum value at very particular concentrations of reactants and products. If the reaction were to proceed further in either direction, the result would be an increase in free energy, which is not spontaneous. At this point, no further change is observed, and the system is said to have reached **equilibrium**. The exact position of equilibrium for any reaction is described by the equilibrium constant, K_{eq},

$$K_{eq} = \frac{[\text{products}]}{[\text{reactants}]} = \frac{[\text{C}]\,[\text{D}]}{[\text{A}]\,[\text{B}]}$$

where K_{eq} is defined as the equilibrium concentrations of products divided by the equilibrium concentrations of reactants.

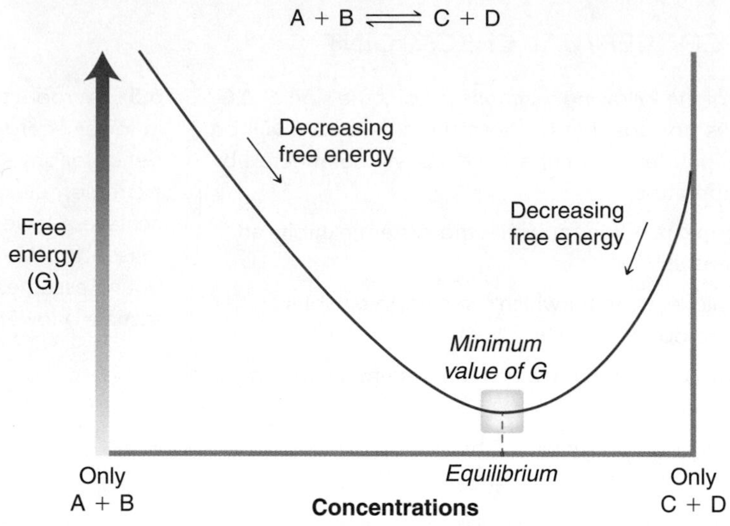

FIGURE 6.10
An energy diagram illustrating that ΔG is dependent on concentration. A minimum in free energy is achieved at particular concentrations (equilibrium).

If the concentration of products is greater than the concentration of reactants, then K_{eq} will be greater than 1. On the other hand, if the concentration of products is less than the concentration of reactants, then K_{eq} will be less than 1. The term K_{eq} indicates the exact position of the equilibrium, and it is related to ΔG in the following way, where R is the gas constant (8.314 J/mol · K) and T is the temperature measured in Kelvin.

$$\Delta G = -RT \ln K_{eq}$$

For any process, whether it is a reaction or a conformational change, the relationship between K_{eq} and ΔG is defined by the equation above. Table 6.2 gives a few examples and illustrates that ΔG determines the maximum yield of products. If ΔG is negative, the products will be favored ($K_{eq} > 1$). If ΔG is positive, then the reactants will be favored ($K_{eq} < 1$). In order for a reaction to be useful (in order for products to dominate over reactants), ΔG must be negative; that is, K_{eq} must be greater than 1. The values in Table 6.2 indicate that a small amount of energy goes a long way. A small difference in free energy can have a significant impact on the ratio of reactants to products.

TABLE 6.2 SAMPLE VALUES OF ΔG AND CORRESPONDING K_{eq}

$\Delta G°$ (KJ/MOL)	K_{eq}	% PRODUCTS AT EQUILIBRIUM
−17	10^3	99.9%
−11	10^2	99%
−6	10^1	90%
0	1	50%
+6	10^{-1}	10%
+11	10^{-2}	1%
+17	10^{-3}	0.1%

In this section, we have investigated the relationship between ΔG and equilibrium—a topic that falls within the realm of **thermodynamics**. Thermodynamics is the study of how energy is distributed under the influence of entropy. For chemists, the thermodynamics of a reaction specifically refers to the study of the relative energy levels of reactants and products (Figure 6.11). This difference in free energy (ΔG) ultimately determines the yield of products that can be expected for any reaction.

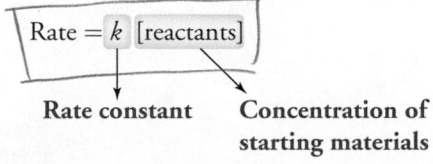

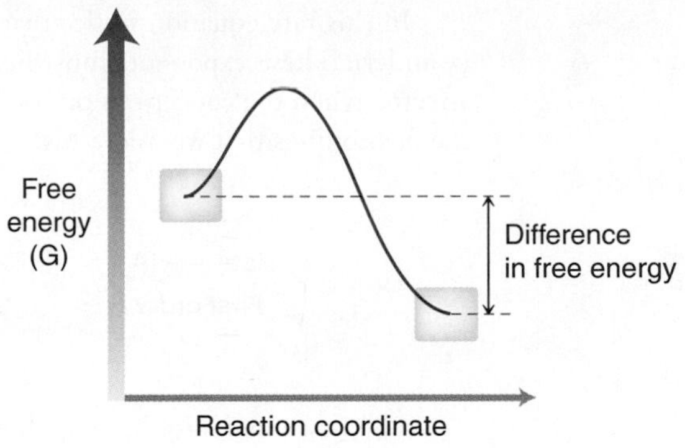

FIGURE 6.11 Thermodynamics of a reaction is based on the difference in energy between starting materials and products.

CONCEPTUAL CHECKPOINT

6.6 In each of the following cases, use the data given to determine whether the reaction favors reactants or products:

(a) A reaction for which $\Delta G = +1.52$ kJ/mol

(b) A reaction for which $K_{eq} = 0.5$

(c) A reaction carried out at 298 K, for which $\Delta H = +33$ kJ/mol and $\Delta S = +150$ J/mol · K

(d) An exothermic reaction with a positive value for ΔS_{sys}

(e) An endothermic reaction with a negative value for ΔS_{sys}

6.5 Kinetics

In the previous sections, we saw that a reaction will be spontaneous if ΔG for the reaction is negative. The term *spontaneous* does not mean that the reaction will occur suddenly. Rather, it means that the reaction is thermodynamically favorable; that is, the reaction favors formation of products. Spontaneity has nothing to do with the speed of the reaction. For example, the conversion of diamonds into graphite is a spontaneous process at standard pressure and temperature. In other words, all diamonds are turning into graphite at this very moment. But even though this process is spontaneous, it is nevertheless very, very slow. It will take millions of years, but ultimately, all diamonds will eventually turn into graphite!

Why is it that some spontaneous processes are fast, like explosions, while others are slow, like diamonds turning into graphite? The study of reaction rates is called **kinetics**. In this section, we will explore issues related to reaction rates.

Rate Equations

The rate of any reaction is described by a **rate equation,** which has the following general form:

$$\text{Rate} = k\,[\text{reactants}]$$

k: **Rate constant**; [reactants]: **Concentration of starting materials**

The general equation above indicates that the rate of a reaction is dependent on the rate constant (k) and on the concentration of the reactants. The rate constant (k) is a value that is specific to each reaction and is dependent on a number of factors. We will explore those factors in the next section. For now, we will focus on the effect of concentrations on the rate.

A reaction is the result of a collision between reactants, and therefore, it makes sense that increasing the concentrations of reactants should increase the frequency of collisions that lead to a reaction, thereby increasing the rate of the reaction. However, the precise effect of concentration on rate must be determined experimentally:

$$\text{Rate} = k\,[\text{A}]^x\,[\text{B}]^y$$

In this rate equation notice that the concentrations of A and B are shown with exponents (x and y). These exponents must be experimentally determined, by exploring how the rate is affected when the concentrations of A and B are each separately doubled. Here are just a few of the possibilities that we might discover for any particular reaction (Figure 6.12).

Rate $= k[A]$	Rate $= k[A]\,[B]$	Rate $= k[A]^2\,[B]$
First order	**Second order**	**Third order**

FIGURE 6.12
Rate equations for first-order, second-order, and third-order reactions

LOOKING AHEAD
If you are having trouble understanding how the rate would not be affected by the concentration of a reactant, don't worry. This will be explained in Chapter 7.

The first possibility has [B] absent from the rate equation. This is a situation where doubling the concentration of A has the effect of doubling the rate of the reaction, but doubling the concentration of B has no effect at all. In such a case, the sum of the exponents is 1, and the reaction is said to be **first order.**

The second possibility above represents a situation where doubling [A] has the effect of doubling the rate, while doubling [B] also has the effect of doubling the rate. In such a case, both [A] and [B] are present in the rate equation, and the exponent of each is 1. The sum of the exponents in this case is 2, and the reaction is said to be **second order**. The third possibility above represents a situation where doubling [A] has the effect of *quadrupling* the rate, while doubling [B] has the effect of doubling the rate. In such a case, the exponent of [A] is 2 and the exponent of [B] is 1. The sum of the exponents in this case is 3, and the reaction is said to be **third order**.

Factors Affecting the Rate Constant

As seen in the previous section, the rate of a reaction is dependent on the rate constant (k):

$$\text{Rate} = k\,[A]^x\,[B]^y$$

A relatively fast reaction is associated with a large rate constant, while a relatively slow reaction is associated with a small rate constant. The value of the rate constant is dependent on three factors: the energy of activation, temperature, and steric considerations.

1. **Energy of Activation.** The energy barrier (the hump) between the reactants and the products is called the **energy of activation**, or E_a (Figure 6.13). This energy barrier represents the minimum amount of energy required for a reaction to occur between two reactants that collide. If a collision between the reactants does not involve this much energy, they will not react with each other to form products. The number of successful collisions is therefore dependent on the number of molecules that have a certain threshold kinetic energy.

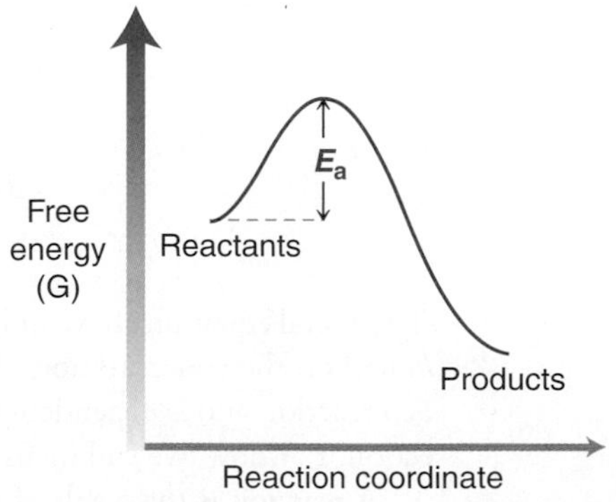

FIGURE 6.13
An energy diagram showing the energy of activation (E_a) of a reaction.

At any specific temperature, the reactants will have a specific average kinetic energy, but not all molecules will possess this average energy. In fact, most molecules have either less than the average or more than the average, giving rise to a distribution as shown in Figure 6.14. Notice that only a certain number of the molecules will have the minimum

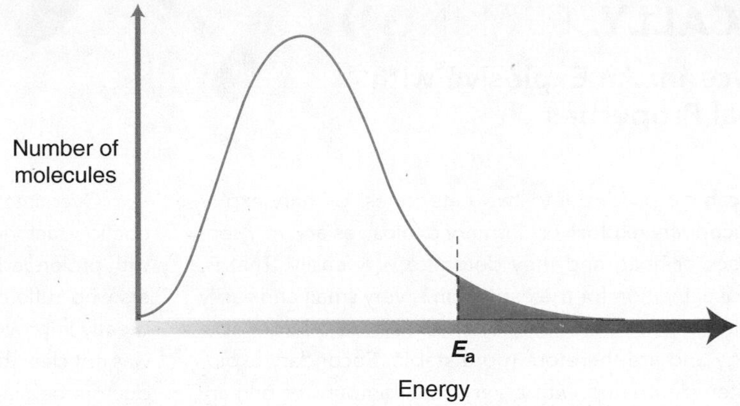

FIGURE 6.14
Distribution of kinetic energy. Shown in blue is the fraction of molecules that possess enough energy to produce a reaction.

energy necessary to produce a reaction. The number of molecules with this energy will be dependent on the value of E_a. If E_a is small, then a large percentage of the molecules will have the threshold energy necessary to produce a reaction. Therefore, a low E_a will lead to a faster reaction (Figure 6.15).

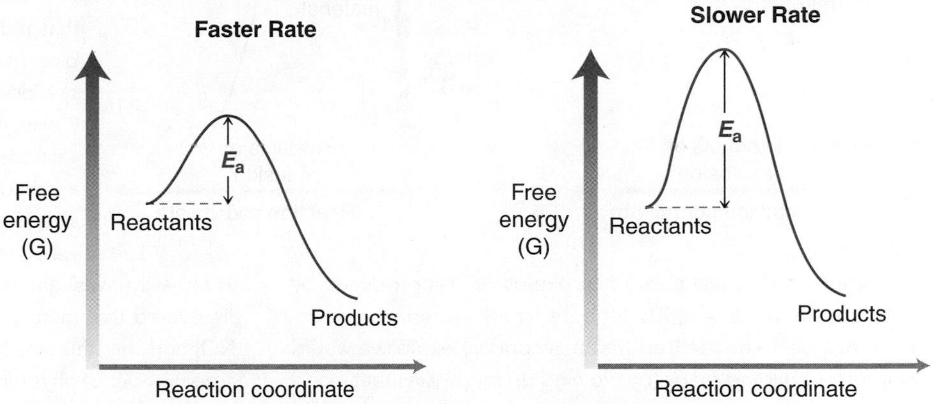

FIGURE 6.15
The rate of a reaction is dependent on the size of E_a.

2. **Temperature.** The rate of a reaction is also very sensitive to temperature (Figure 6.16). Raising the temperature of a reaction will cause the rate to increase, because the molecules will have more kinetic energy at a higher temperature. At higher temperature, a larger number of molecules will have the kinetic energy sufficient to produce a reaction. As a rule of thumb, raising the temperature by 10 °C causes the rate to double.

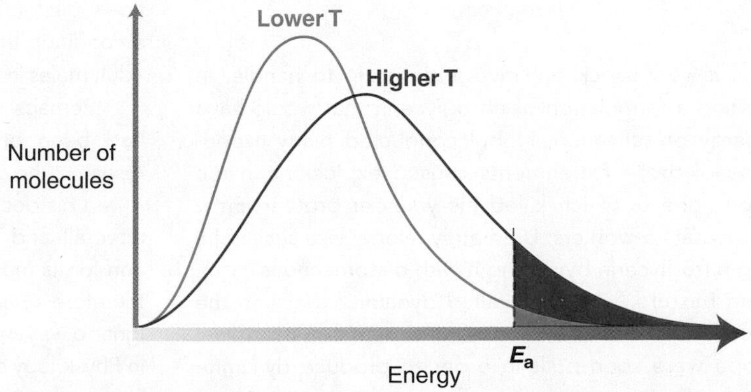

FIGURE 6.16
Increasing the temperature increases the number of molecules with sufficient energy to produce a reaction.

3. **Steric Considerations.** The geometry of the reactants and the orientation of their collision can also have an impact on the frequency of collisions that lead to a reaction. This factor will be explored in greater detail in Section 7.4.

MEDICALLYSPEAKING

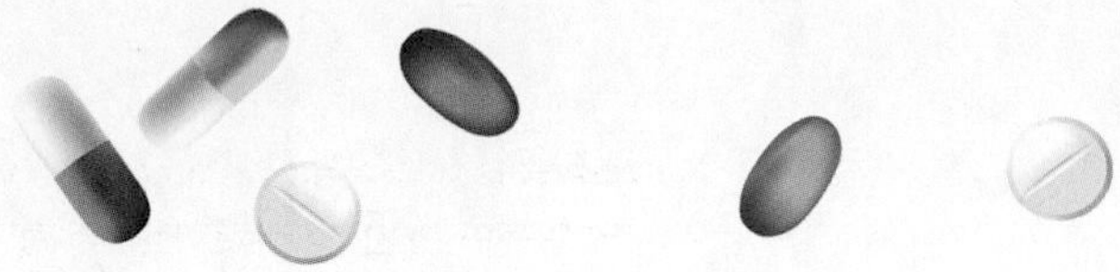

Nitroglycerin: An Explosive with Medicinal Properties

Explosives can be divided into two categories: primary explosives and secondary explosives. Primary explosives are very sensitive to shock or heat, and they detonate very easily. That is, the energy of activation for the explosion is very small and easily overcome. In contrast, secondary explosives have a larger activation energy and are therefore more stable. Secondary explosives are often detonated with a very small amount of primary explosive.

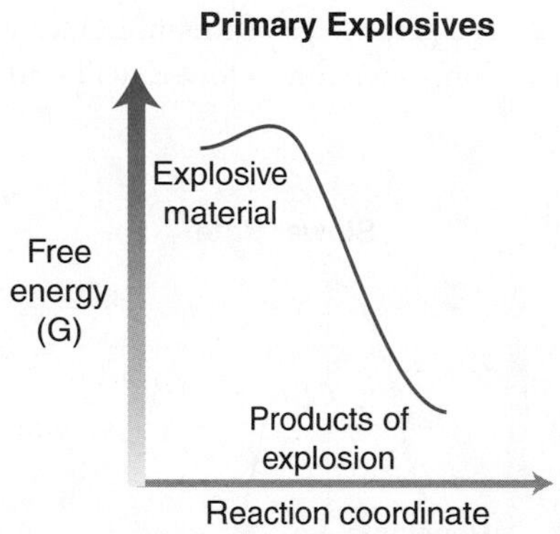

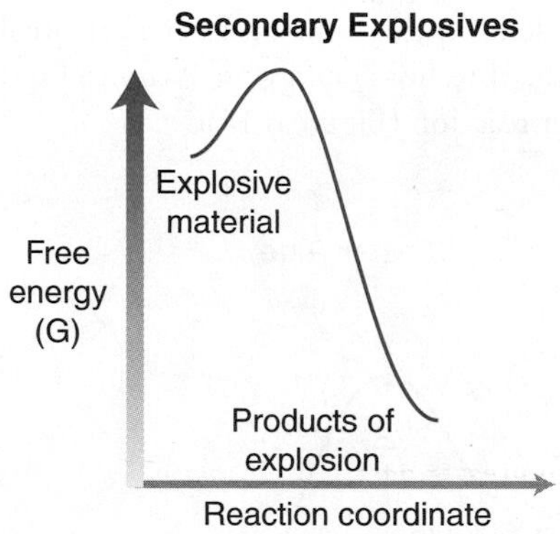

The first commercial secondary explosive was produced by Alfred Nobel in the mid-1800s. Nobel's family owned a construction company, and he realized that a secondary explosive would be helpful for building roads (by blowing up mountains that stood in the way). However, at the time, there were no strong explosives safe enough to handle. Nobel focused his efforts on finding a way to stabilize nitroglycerin:

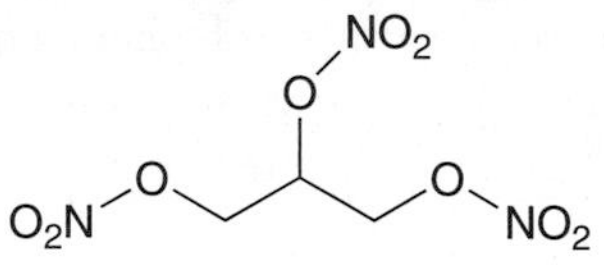

Nitroglycerin

Nitroglycerin is very shock sensitive and unsafe to handle. In an effort to find a formulation of nitroglycerin that would have a larger energy of activation, Nobel conducted many experiments. Many of those experiments caused explosions in the Nobel factory, one of which killed his younger brother, Emil, along with several co-workers. Ultimately, Nobel was successful in stabilizing nitroglycerin by mixing it with diatomaceous earth. The resulting mixture, which he called dynamite, became the first commercial secondary explosive, and numerous factories across Europe were soon built in order to produce dynamite in large quantities. As mentioned in the chapter opener, Alfred Nobel became fabulously wealthy from his invention and used some of the proceeds to fund the Nobel prizes, with which we are all so familiar.

Over time, it was discovered that workers in the dynamite production factories had several physiological responses associated with prolonged exposure to nitroglycerin. Most importantly, workers who suffered from heart problems found that their conditions greatly improved. For many decades, the reason for this response was not clear, but a clear pattern had been established. As a result, doctors began to treat patients experiencing heart problems by giving them small quantities of nitroglycerin to ingest. Ultimately, Nobel himself suffered from heart problems, and his doctor suggested that he eat nitroglycerin. Nobel refused to ingest what he considered to be an explosive, and he ultimately died of heart complications.

With the passing decades, it became clear that nitroglycerin serves as a vasodilator (dilates blood vessels) and therefore reduces the chances of blockage that leads to a heart attack. However, it was not known *how* nitroglycerin functions as a vasodilator. The study of drug action belongs to a broader field of study called pharmacology. A scientist at UCLA by the name of Louis Ignarro was interested in this pharmacological question and heavily investigated the action of nitroglycerin in the body. He discovered that metabolism of nitroglycerin produces nitric oxide (NO) and that this small compound is ultimately responsible for a large number of physiological processes. At first, his discovery was met with skepticism by the scientific community since nitric oxide was known to be an atmospheric contaminant (present in smog), and it was difficult to believe that the very same compound could be responsible for the medicinal value of nitroglycerin. Ultimately, his ideas were verified, and he was credited with discovering the mechanism for the physiological effects of nitroglycerin. Ignarro's discovery ultimately led to the development of many new commercial drugs, including Viagra. Viagra is a vasodilator that treats impotency, a condition that afflicts 9% of all adult males in the United States.

Perhaps if Alfred Nobel had been privy to Ignarro's research, he might have followed his doctor's instructions after all and eaten nitroglycerin for its medicinal value. It is therefore quite interesting that Ignarro was awarded the 1998 Nobel Prize in Physiology or Medicine, a prize that was funded by Nobel with the fortune that he accrued for his discovery of stabilized nitroglycerin. It appears that history does in fact have a sense of irony.

Catalysts and Enzymes

A **catalyst** is a compound that can speed up the rate of a reaction without itself being consumed by the reaction. A catalyst works by providing an alternate pathway with a smaller activation energy (Figure 6.17). Notice that a catalyst does not change the free energy of reactants or products, and therefore, the position of equilibrium is not affected by the presence of a catalyst. Only the rate of reaction is affected by the catalyst. We will see many examples of catalysts in the coming chapters.

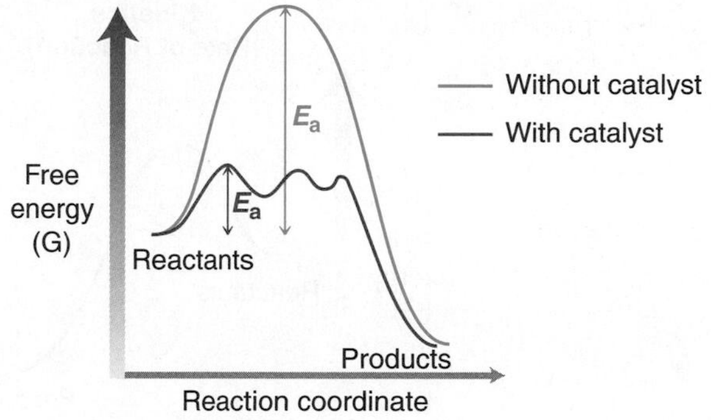

FIGURE 6.17 An energy diagram showing an uncatalyzed pathway and a catalyzed pathway.

Nature employs catalysis in many biological functions as well. Enzymes are naturally occurring compounds that catalyze very specific biologically important reactions. Enzymes will be discussed in greater detail in Chapter 25.

PRACTICALLYSPEAKING)))

Beer Making

Beer making relies on the fermentation of sugars (produced from grains such as wheat or barley). The fermentation process produces ethanol and is thermodynamically favorable (ΔG is negative). The direct conversion of sugar into ethanol has a very large activation energy and therefore does not occur by itself at an appreciable rate:

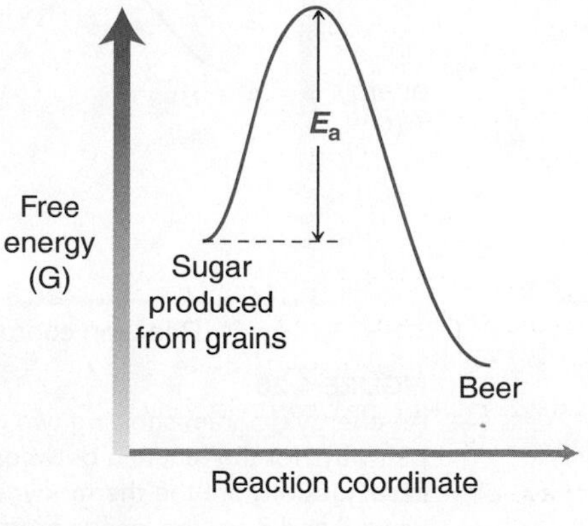

When yeast is added to the mixture, the energy of activation is lowered, and the process takes place at an observable rate. Yeast is a microorganism that utilizes many enzymes that catalyze many different reactions. Some of these enzymes enable the yeast to metabolize sugars, producing ethanol and CO_2 as waste products. The ethanol generated in the process is toxic to the yeast, so as the concentration of ethanol rises, the fermentation slows. As a result, it is difficult to achieve an alcohol concentration greater than 12% when using standard brewing yeast.

Without the catalytic action of yeast, beer might never have been discovered.

6.6 Reading Energy Diagrams

As we begin to study reactions in the next chapter, we will rely heavily on the use of energy diagrams. Let's quickly review some of the features of energy diagrams.

Kinetics Versus Thermodynamics

Don't confuse kinetics and thermodynamics— they are two entirely separate concepts (Figure 6.18).

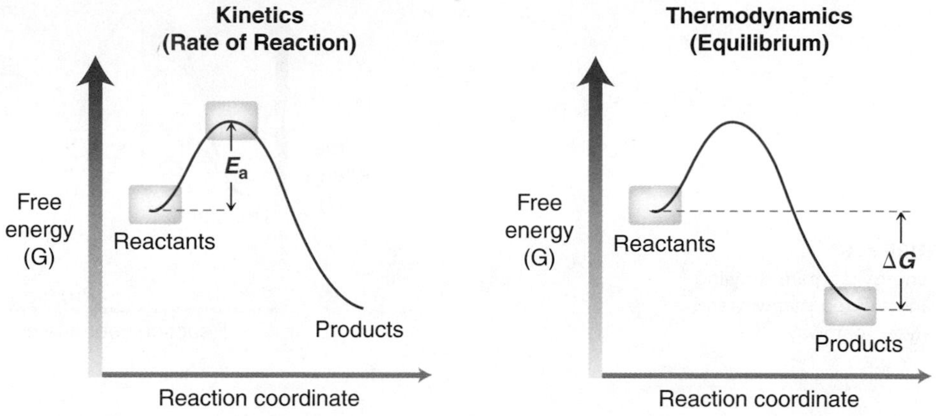

FIGURE 6.18
Energy diagrams showing the difference between kinetics and thermodynamics.

Kinetics refers to the rate of a reaction, while thermodynamics refers to the equilibrium concentrations of reactants and products. Suppose that two compounds, A and B, can react with each other in one of two possible pathways (Figure 6.19). The pathway determines the products. Notice that products C and D are thermodynamically favored over products E and F, because C and D are lower in energy. In addition, C and D are also kinetically favored over E and F because formation of C and D involves a smaller activation energy. In summary, C and D are favored by thermodynamics as well as kinetics. When reactants can react with each other in two possible ways, it is often the case that one reaction pathway is both thermodynamically and kinetically favored, although there are many cases in which thermodynamics and kinetics oppose each other. Consider the energy diagram in Figure 6.20 showing two possible reactions between A and B. In this case, products C and D are favored by thermodynamics because they are lower in energy. However, products E and F are favored by kinetics because formation of E and F involves a lower energy of activation. In such a case, temperature will play a pivotal role. At low temperature, the reaction that forms E and F will be more rapid, even though this reac-

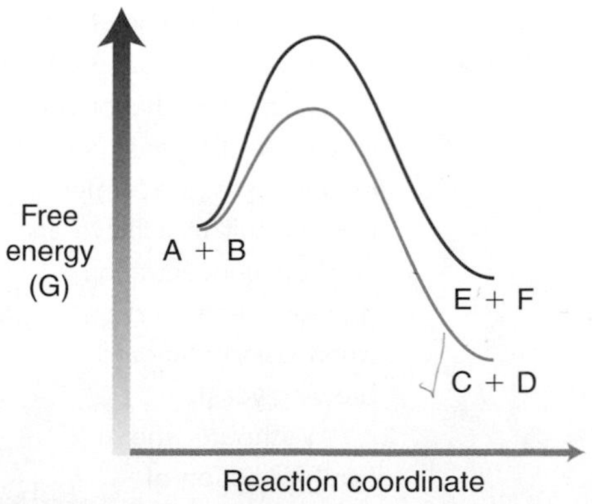

FIGURE 6.19
An energy diagram showing two possible pathways for the reaction between A and B.

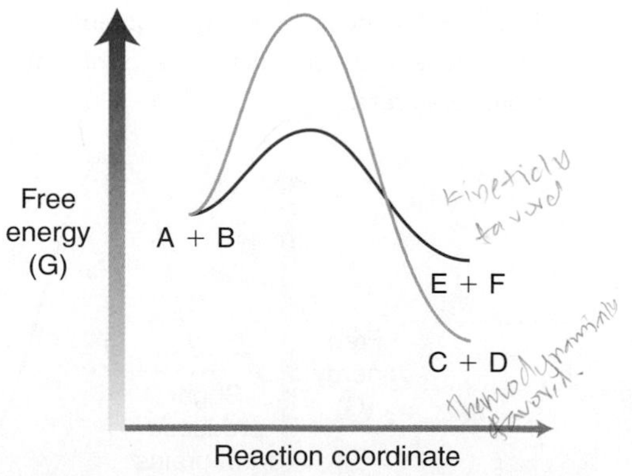

FIGURE 6.20
An energy diagram showing two possible pathways for the reaction between A and B. In this case, C and D are the thermodynamic products, while E and F are the kinetic products.

tion does not produce the most stable products. At high temperature, equilibrium concentrations will be quickly achieved, favoring formation of C and D. We will see several examples of kinetics versus thermodynamics throughout this course.

Transition States Versus Intermediates

Reactions often involve multiple steps. In the energy diagram of a multistep process, all local minima (valleys) represent *intermediates*, while all local maxima (peaks) represent *transition states* (Figure 6.21). It is important to understand the difference between transition states and intermediates.

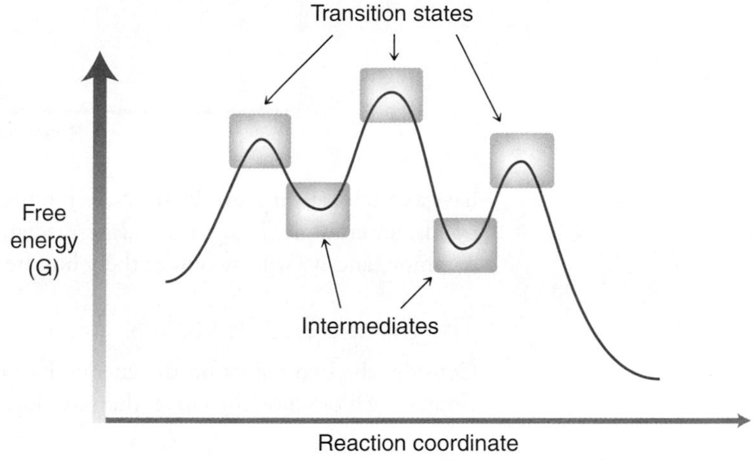

FIGURE 6.21
In an energy diagram, all peaks are transition states, and all valleys are intermediates.

A **transition state**, as the name implies, is a state through which the reaction passes. Transition states cannot be isolated. In this high-energy state, bonds are in the process of being broken and/or formed simultaneously, as shown in Figure 6.22.

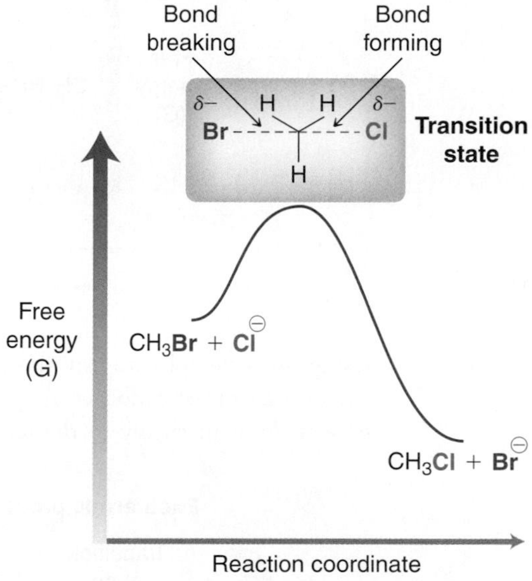

FIGURE 6.22
In a transition state, bonds are being broken and/or formed.

As a crude analogy for the difference between a transition state and an intermediate, consider jumping in the air as high as you can while your friend takes a photograph of the highest point of your trajectory. The picture shows the height that you achieved, although it would not be fair to say that you were able to spend any considerable amount of time at that height (to hover in midair). It was a state through which you passed. In contrast, imagine jumping onto a desk and then jumping back down. A photograph of you standing on the desk will be very different from the previous picture. It is actually possible to stand on a desk for a period of time, but it is not possible to remain hovering in the air for any reasonable period of time. The picture of you standing on the desk is similar to the picture of a reaction intermediate**. Intermediates**

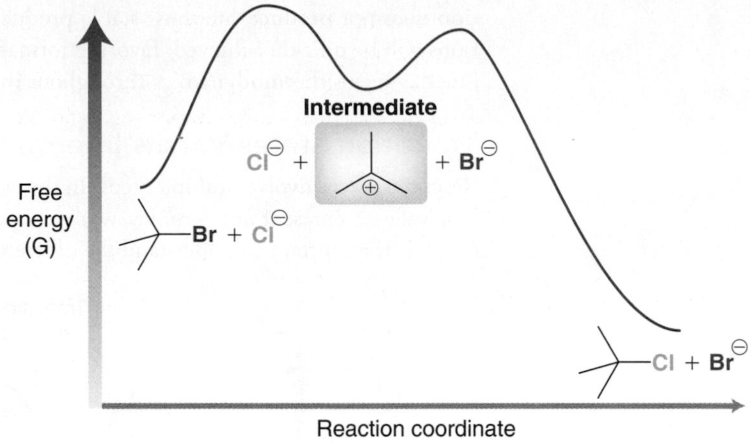

FIGURE 6.23 An energy diagram of a reaction that exhibits an intermediate characterized by a valley in the energy diagram.

have a certain, albeit short, lifetime. An intermediate is not in the process of forming or breaking bonds, for example (Figure 6.23). Intermediates, such as the one shown in Figure 6.23, are very common, and we will encounter them hundreds of times throughout this text.

The Hammond Postulate

Consider the two points on the energy diagram in Figure 6.24. Because these two points are close to each other on the curve, they are close in energy and are therefore structurally similar.

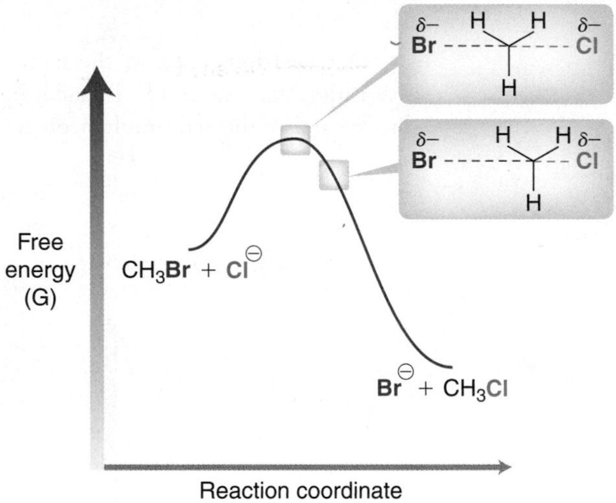

FIGURE 6.24 On an energy diagram, two points that are close together will represent structurally similar states.

Using this principle, we can make a generalization about the structure of a transition state in any exothermic or endothermic process (Figure 6.25). In an exothermic process the transition state is closer in energy to the reactants than to the products, and therefore, the structure of

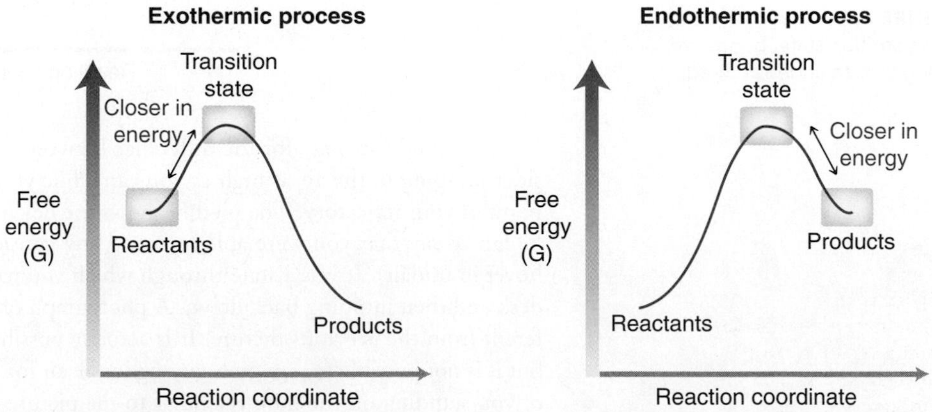

FIGURE 6.25 A transition state will be closer in energy to the starting materials in an exothermic process, but it will be closer in energy to the products in an endothermic process.

the transition state more closely resembles the reactants. In contrast, the transition state in an endothermic process is closer in energy to the products, and therefore, the transition state more closely resembles the products. This principle is called the **Hammond postulate**. We will use this principle many times in our discussions of transition states in the upcoming chapters.

CONCEPTUAL CHECKPOINT

6.7 Consider the relative energy diagrams for four different processes:

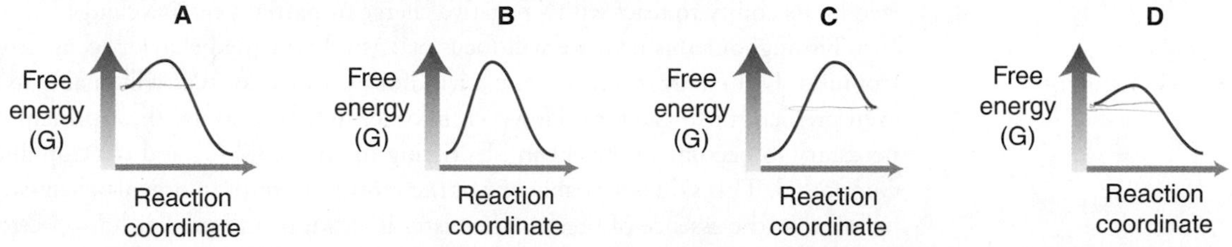

(a) Compare energy diagrams A and D. Assuming all other factors (such as concentrations and temperature) are identical for the two processes, identify which process will occur more rapidly. Explain. D . less Ea

(b) Compare energy diagrams A and B. Which process will more greatly favor products at equilibrium? Explain. A . exothermic

(c) Do any of the processes exhibit an intermediate? Do any of the processes exhibit a transition state? Explain. No. no local minima. yes local maxima

(d) Compare energy diagrams A and C. In which case will the transition state resemble reactants more than products? Explain. A. transition state closer to products.

(e) Compare energy diagrams A and B. Assuming all other factors (such as concentrations and temperature) are identical for the two processes, identify which process will occur more rapidly. Explain. A . less Ea

(f) Compare energy diagrams B and D. Which process will more greatly favor products at equilibrium? Explain. D exothermic exergonic

(g) Compare energy diagrams C and D. In which case will the transition state resemble products more than reactants? Explain. C . transition state more closer to products in C than in D.

6.7 Nucleophiles and Electrophiles

Ionic reactions, also called **polar reactions**, involve the participation of ions as reactants, intermediates, or products. In most cases, the ions are present as intermediates. These reactions represent most (approximately 95%) of the reactions that we will encounter in this text. The other two major categories, radical reactions and pericyclic reactions, occupy a much smaller focus in the typical undergraduate organic chemistry course but will be discussed in upcoming chapters. The remainder of this chapter will focus on ionic reactions.

Ionic reactions occur when one reactant has a site of high electron density and the other reactant has a site of low electron density. For example, consider the electrostatic potential maps of methyl chloride and methyllithium (Figure 6.26). Each compound exhibits an inductive effect, but in opposite directions.

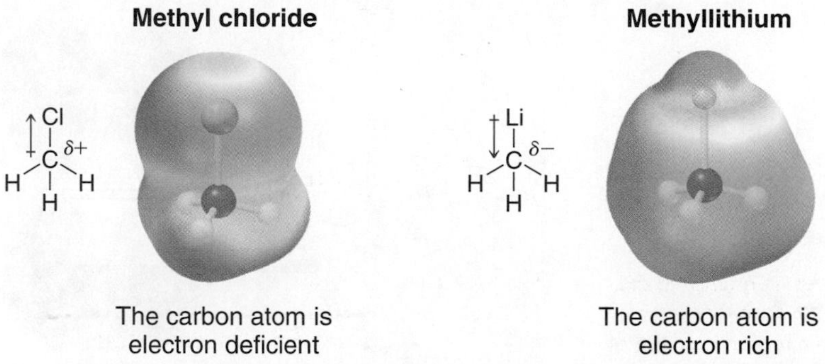

FIGURE 6.26
Electrostatic potential maps of methyl chloride and methyllithium, clearly indicating the inductive effects.

LOOKING BACK
For a review of inductive effects, see Section 1.11.

The carbon atom in methyl chloride represents a site of low electron density, while the carbon atom in methyllithium represents a site of high electron density. Since opposite charges attract, these two compounds will react with each other. The type of reaction that occurs between methyl chloride and methyllithium will be explored in the upcoming chapter. For now, we will focus on the nature of each reactant.

An electron-rich center, such as the carbon atom in methyllithium, is called a **nucleophile**, which comes from Greek meaning "nucleus lover." That is, a nucleophilic center is characterized by its ability to react with a positive charge or partial positive charge. In contrast, an electron-deficient center, such as the carbon atom in methyl chloride, is called an **electrophile**, which comes from Greek meaning "electron lover." That is, an electrophilic center is characterized by its ability to react with a negative charge or partial negative charge.

Throughout this text, we will focus extensively on the behavior of nucleophiles and electrophiles. Ultimately, there are just a handful of principles that will enable us to explain, and even predict, most reactions. However, in order to learn and use these principles, it will first be necessary to become proficient in identifying the nucleophilic and electrophilic centers in any compound. This skill is arguably one of the most important in organic chemistry. We have said before that the essence of organic chemistry is to study and predict how electron density flows during a reaction. It will not be possible to make any intelligent predictions without knowing where the electron density can be found and where it is likely to go. The following sections will explore the nature of nucleophiles and electrophiles in more depth.

Nucleophiles

LOOKING BACK
For a review of Lewis bases, see Section 3.9.

A nucleophilic center is an electron-rich atom that is capable of donating a pair of electrons. Notice that this definition is very similar to the definition of a Lewis base. In fact, the terms "nucleophile" and "Lewis base" are synonymous.

Two examples of nucleophiles are as follows:

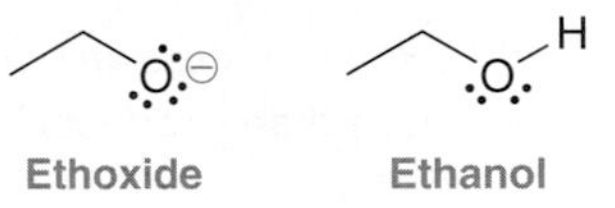

Each of these examples has lone pairs on an oxygen atom. Ethoxide bears a negative charge and is therefore more nucleophilic than ethanol. Nevertheless, ethanol can still function as a nucleophile (albeit weak), because the lone pairs in ethanol represent regions of high electron density. Any atom that possesses a localized lone pair can be nucleophilic.

We will see in Chapter 9 that π bonds can also function as nucleophiles, because a π bond is a region in space of high electron density (Figure 6.27).

The strength of a nucleophile is affected by many factors, including polarizability. **Polarizability**, loosely defined, describes the ability of an atom to distribute its electron density unevenly in response to external influences. Polarizability is directly related to the size of the atom (and more specifically, the number of electrons that are distant from the nucleus). For example, sulfur is very large and has many electrons that are distant from the nucleus, and its electron density can be unevenly distributed when it comes near an electrophile. Iodine shares the same feature. As a result, I^- and HS^- are particularly strong nucleophiles. This fact will be revisited in the upcoming chapter.

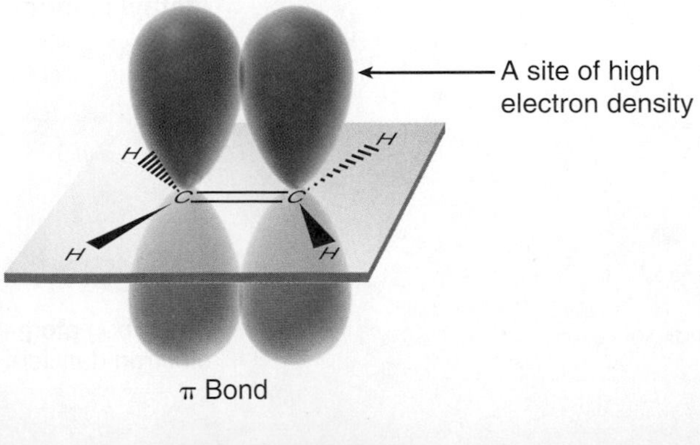

FIGURE 6.27
A π bond is a region in space of high electron density.

Electrophiles

An electrophilic center is an electron-deficient atom that is capable of accepting a pair of electrons. Notice that this definition is very similar to the definition for a Lewis acid. In fact, the terms "electrophile" and "Lewis acid" are synonymous.

Two examples of electrophiles are as follows:

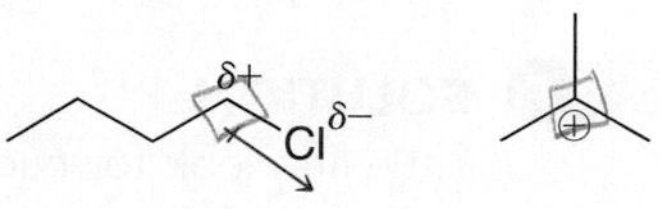

LOOKING BACK
For a review of Lewis acids, see Section 3.9.

The first compound exhibits an electrophilic carbon atom as a result of the inductive effects of the chlorine atom. The second example exhibits a positively charged carbon atom and is called a **carbocation**. A carbocation has an empty *p* orbital (Figure 6.28). The empty *p* orbital functions as a site that can accept a pair of electrons, rendering the compound electrophilic. We will discuss carbocations in more depth later in this chapter.

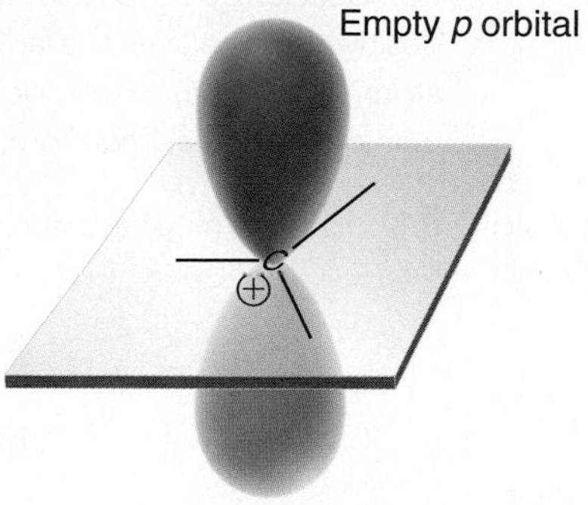

FIGURE 6.28
The empty *p* orbital of a carbocation.

Table 6.3 provides a summary of the common features that render a compound nucleophilic or electrophilic.

TABLE 6.3 A SUMMARY OF COMMON NUCLEOPHILIC CENTERS AND ELECTROPHILIC CENTERS

NUCLEOPHILES		ELECTROPHILES	
FEATURE	EXAMPLE	FEATURE	EXAMPLE
Inductive effects	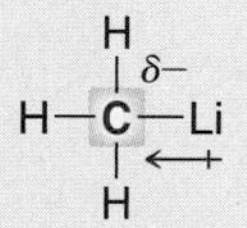	Inductive effects	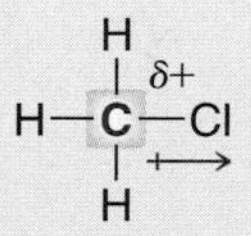
Lone pair		Empty *p* orbital	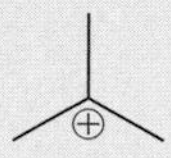
π bond	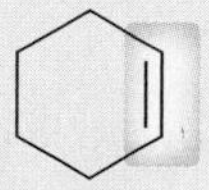		

SKILLBUILDER

6.2 IDENTIFYING NUCLEOPHILIC AND ELECTROPHILIC CENTERS

LEARN the skill

Identify all nucleophilic centers and all electrophilic centers in the following compound.

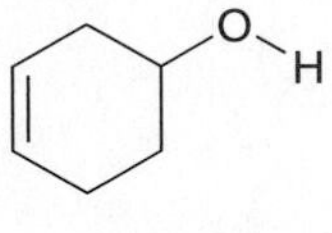

SOLUTION

STEP 1 Identify nucleophilic centers by looking for inductive effects, lone pairs, or π bonds.

Let's first look for nucleophilic centers. Specifically, we are looking for inductive effects, lone pairs, or π bonds. In this compound, there are two nucleophilic centers:

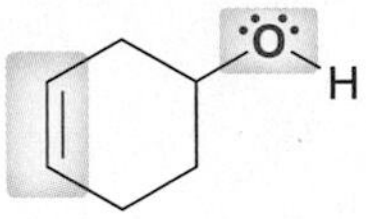

STEP 2 Identify electrophilic centers by looking for inductive effects or empty *p* orbitals.

Next, we look for electrophilic centers. Specifically, we are looking for inductive effects or an empty *p* orbital. None of the atoms in this compound exhibit an empty *p* orbital, but there are inductive effects:

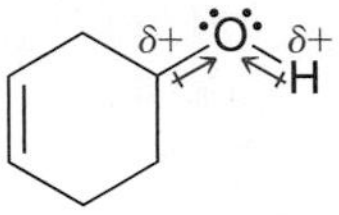

The oxygen atom is electronegative, causing the adjacent atoms to be electron deficient. In this case, there are two adjacent atoms that incur a partial positive charge (δ+): the adjacent carbon atom and the adjacent hydrogen atom. In general, a hydrogen atom with a partial positive charge is described as acidic, rather than electrophilic. Therefore, this compound is considered to have only one electrophilic center.

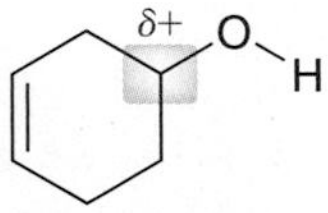

PRACTICE the skill

6.8 Identify all of the nucleophilic centers in each of the following compounds:

(a) 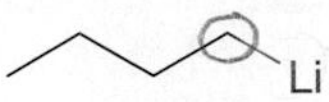

(b) 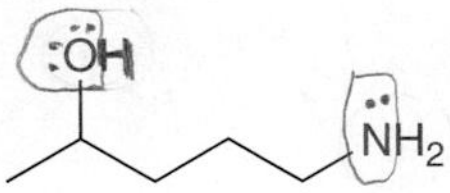

(c) 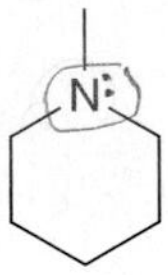

(d) 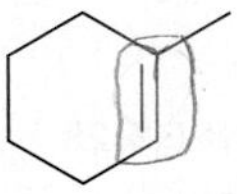

6.9 Identify all of the electrophilic centers in each of the following compounds:

(a) 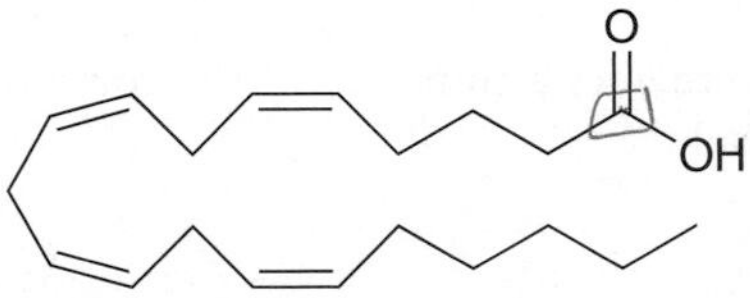

Arachidonic acid
A precursor in the biosynthesis of many hormones

(b) 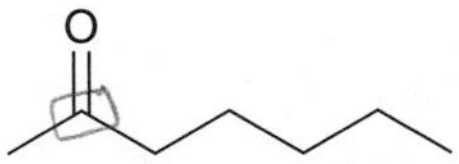

2-Heptanone
Used to control the population of Varroa mites in honey bee colonies

(c) 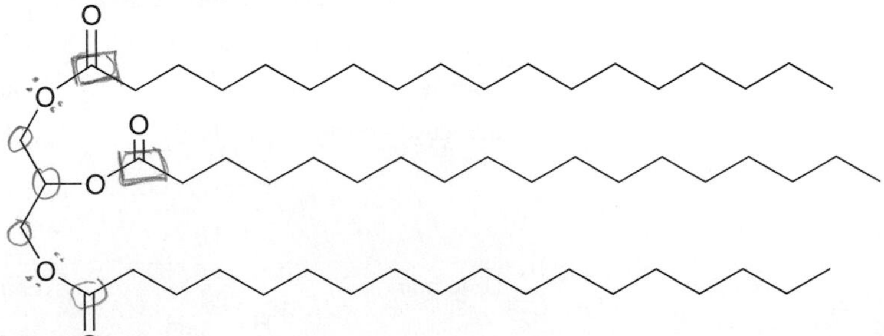

Hydrogenated animal fat

APPLY the skill

6.10 The following hypothetical compound cannot be prepared or isolated, because it has a very reactive nucleophilic center and a very reactive electrophilic center, and the two sites would react with each other rapidly. Identify the nucleophilic center and electrophilic center in this hypothetical compound:

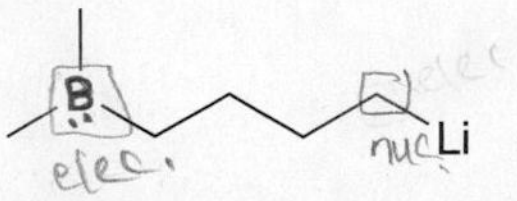

6.11 Each of the following compounds exhibits two electrophilic centers. Identify both centers in each compound. (Hint: You will need to draw resonance structures in each case.)

(a)
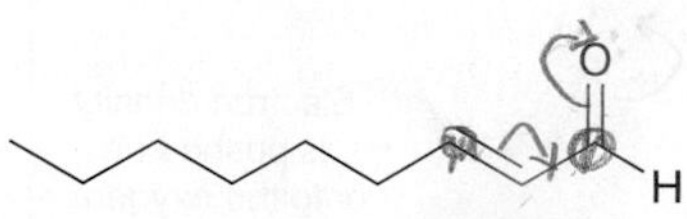

A cockroach repellant found in cucumbers

(b)
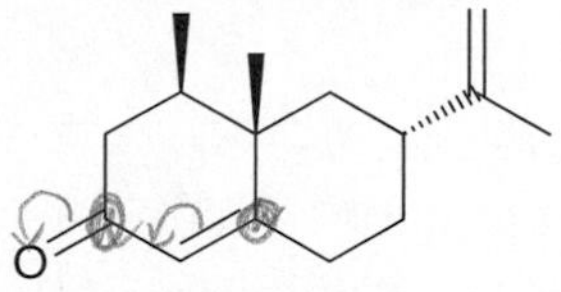

Nootkatone
Found in grapefruits

(c)
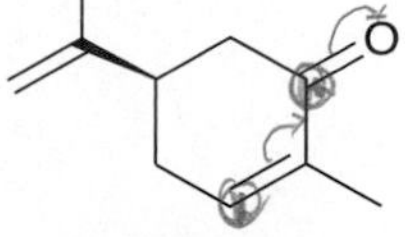

(*R*)-Carvone
Responsible for the odor of spearmint

need more **PRACTICE?** **Try Problem 6.53**

6.8 Mechanisms and Arrow Pushing

Recall from Chapter 3 that a mechanism shows how a reaction takes place using curved arrows to illustrate the flow of electrons. The tail of every curved arrow shows where the electrons are coming from, and the head of every curved arrow shows where the electrons are going:

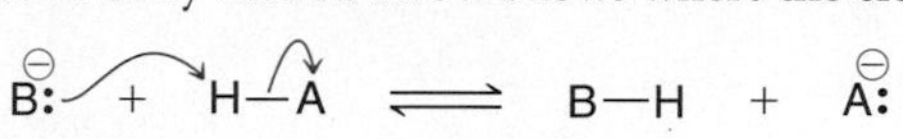

LOOKING BACK
Recall that we also used a handful of patterns for arrow pushing to master resonance structures in Chapter 2.

In order to master ionic mechanisms, it will be helpful to become familiar with characteristic patterns for arrow pushing. We will now learn patterns of electron flow, and these patterns will empower us to understand reaction mechanisms and even propose new mechanisms. There are only four characteristic patterns, and all ionic mechanisms are simply combinations of these four steps. Let's go through them one by one.

Nucleophilic Attack

The first pattern is characterized by a nucleophile attacking an electrophile; for example:

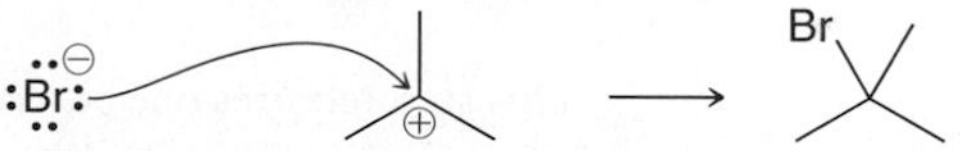

Nucleophile **Electrophile**

Bromide is a nucleophile because it possesses a lone pair, and the carbocation is an electrophile because of its empty *p* orbital. In this example, the attack of the nucleophile on the electrophile requires just one curved arrow. The tail of this curved arrow is placed on the nucleophilic center and the head is placed on the electrophilic center. It is also common to see a nucleophilic attack that utilizes more than one curved arrow; for example:

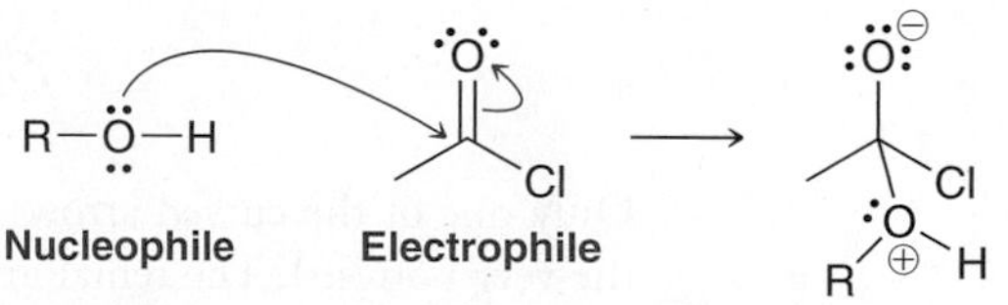

Nucleophile **Electrophile**

In this case, there are two curved arrows. The first shows the nucleophile attacking the electrophile, but what is the function of the second curved arrow? There are a couple of ways to view this second curved arrow. We can simply think of it as a resonance arrow: We can imagine first drawing the resonance structure of the electrophile and then having the nucleophile attack:

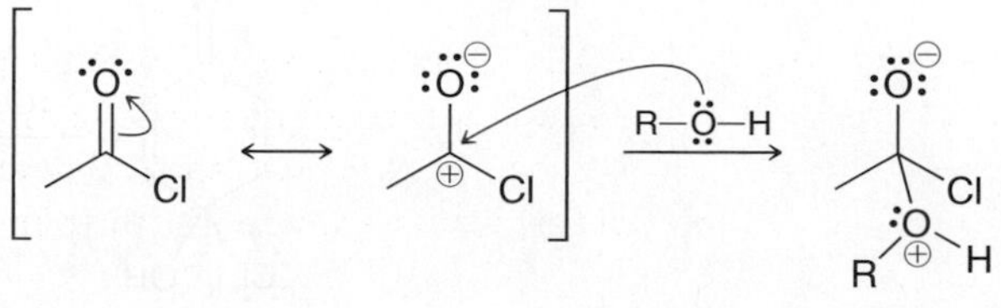

From this perspective, it seems that only one of the curved arrows is actually showing the nucleophilic attack. The other curved arrow can be thought of as a resonance curved arrow.

Alternatively, we can think of the second curved arrow as an actual flow of electron density that goes up onto the oxygen when the nucleophile attacks:

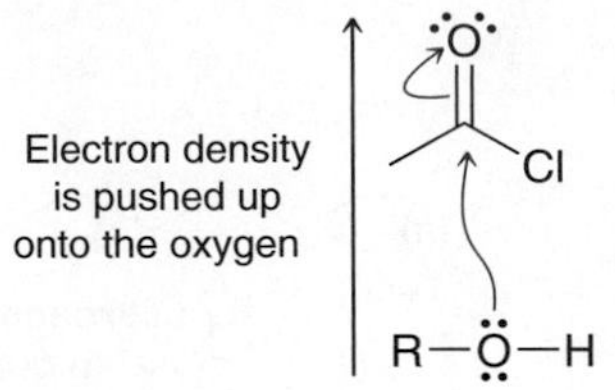
Electron density
is pushed up
onto the oxygen
Cl
R—O—H

This perspective is perhaps more accurate, but we must keep in mind that both curved arrows are showing just one arrow-pushing pattern: *nucleophilic attack.*

π bonds can also serve as nucleophiles; for example:

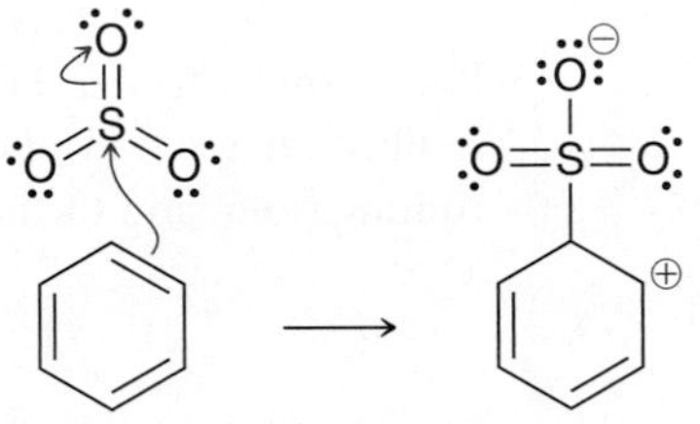

Loss of a Leaving Group

The second pattern for arrow pushing is characterized by the loss of a leaving group; for example:

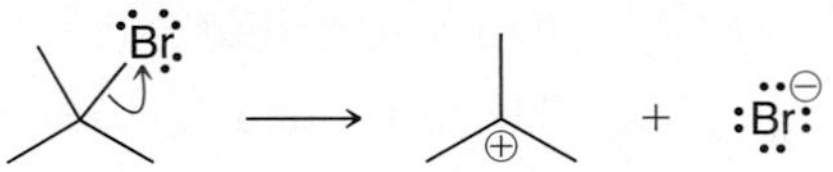

This step requires one curved arrow, although it is common to see more than one curved arrow being used to show the loss of a leaving group. For example:

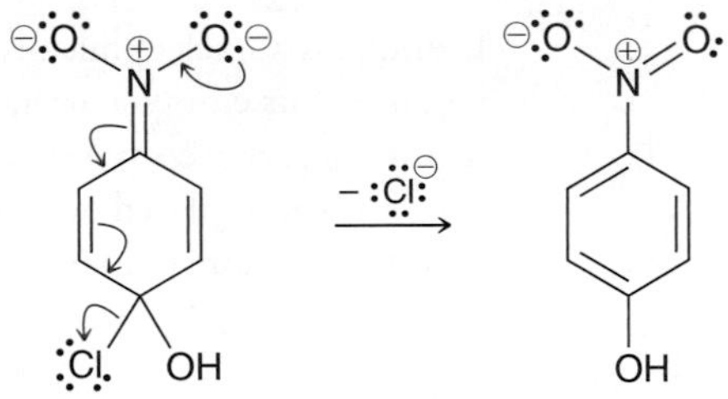

Only one of the curved arrows above actually shows the chloride leaving (the curved arrow at the very bottom). The remaining curved arrows can be viewed in two different ways (just as in the previous section). We can view the other curved arrows as resonance arrows, drawn after the leaving group leaves:

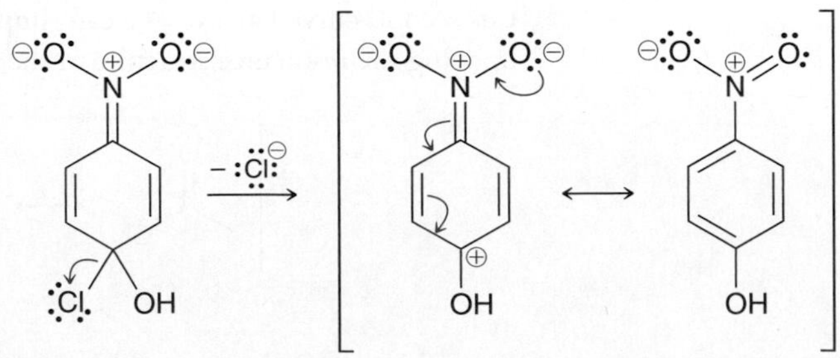

or we can view the process as a flow of electron density that pushes out the leaving group:

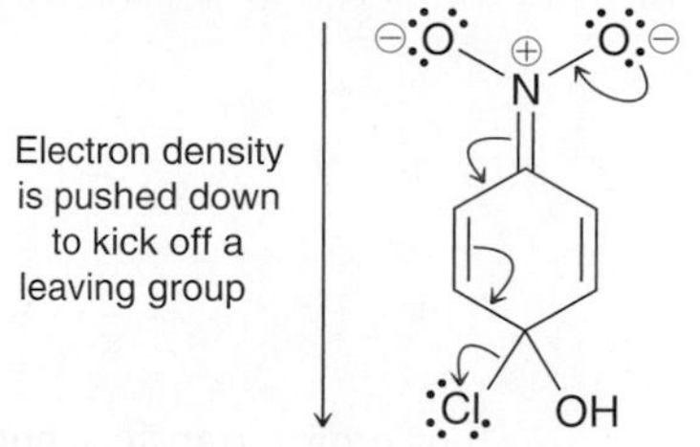
Electron density
is pushed down
to kick off a
leaving group

Regardless of how we view the process, it is important to recognize that all of these curved arrows together show only one arrow-pushing pattern: *loss of a leaving group*.

Proton Transfers

The third pattern for arrow pushing has already been discussed in detail in Chapter 3. Recall that a proton transfer is characterized by two curved arrows:

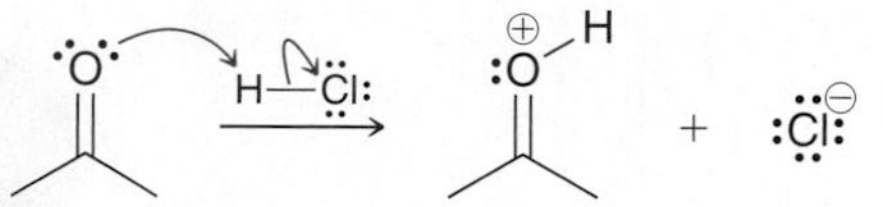

In this example, a ketone is being protonated, which is shown with two curved arrows. The first is drawn from the ketone to the proton. The second curved arrow shows what happens to the electrons that were previously holding the proton.

A proton transfer step is illustrated with two curved arrows, whether the compound is protonated (as above) or deprotonated, as shown below:

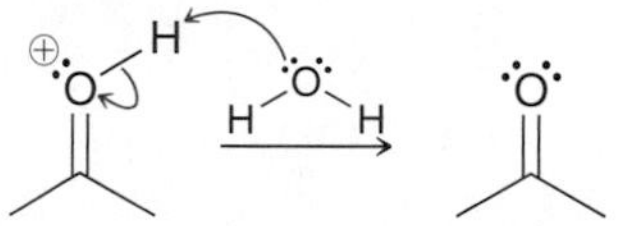

Sometimes, proton transfers are shown with only one curved arrow:

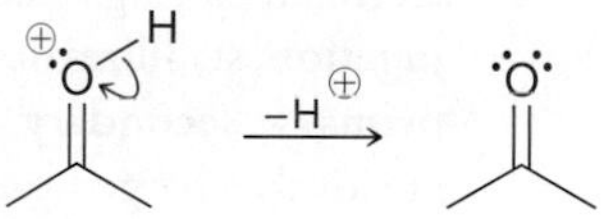

In this case, the base that grabs the proton has not been indicated. Chemists will sometimes use this approach for clarity of presentation, even though a proton doesn't fall off into space by itself. In order for a compound to lose a proton, a base must be involved in order to deprotonate the compound. In general, it is preferable to show the base involved and to use *two* curved arrows.

At times, *more* than two curved arrows are used for a proton transfer step. For example, the following case has three curved arrows:

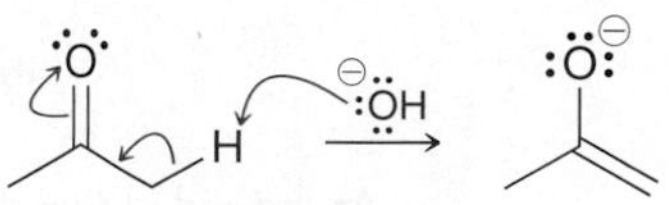

Once again, this step can be viewed in one of two ways. We can view this as a simple proton transfer followed by drawing a resonance structure:

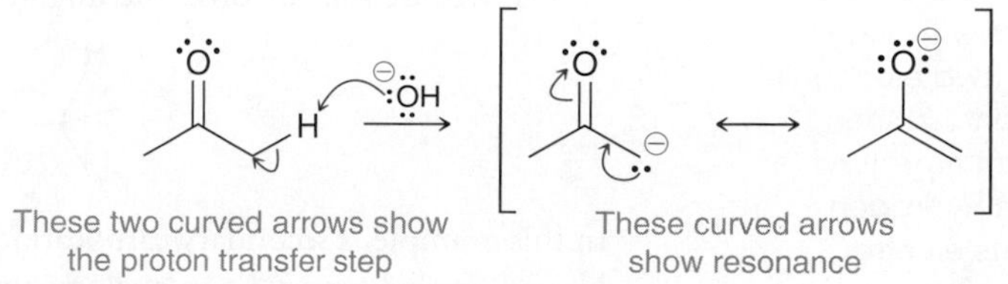
These two curved arrows show
the proton transfer step
These curved arrows
show resonance

or we can argue that all of the curved arrows together show the flow of electron density that takes place during the proton transfer:

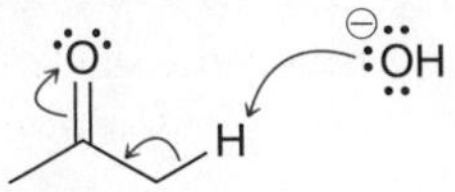

Electron density flows up onto the oxygen of the ketone

For proton transfers, both perspectives are equally valid.

Rearrangements

The fourth, and final, pattern is characterized by a **rearrangement**. There are several kinds of rearrangements, but at this point, we will focus exclusively on **carbocation rearrangements**. In order to discuss carbocation rearrangements, we must first explore one feature of carbocations. Specifically, carbocations are stabilized by neighboring alkyl groups (Figure 6.29).

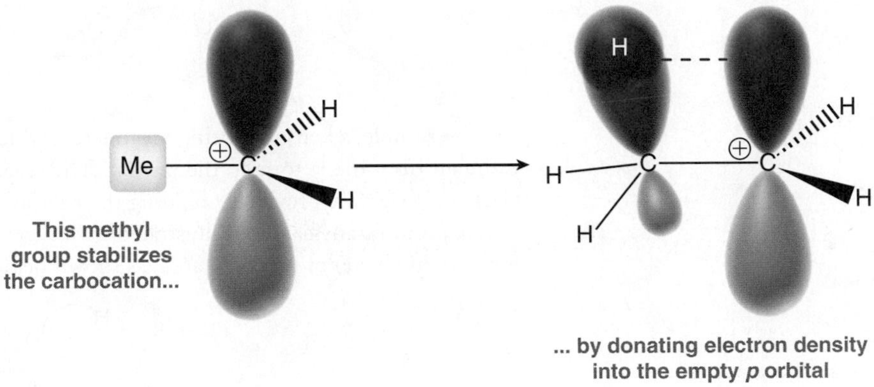

FIGURE 6.29
Neighboring alkyl groups will stabilize a carbocation through hyperconjugation.

The bonding MO associated with a neighboring C—H bond slightly overlaps with the empty *p* orbital, placing some of its electron density in the empty *p* orbital. This effect, called **hyperconjugation**, stabilizes the empty *p* orbital. This explains the observed trend in Figure 6.30. The terms **primary**, **secondary**, and **tertiary** refer to the number of alkyl groups attached directly to the positively charged carbon atom. Tertiary carbocations are more stable (lower in energy) than secondary carbocations, which are more stable than primary carbocations.

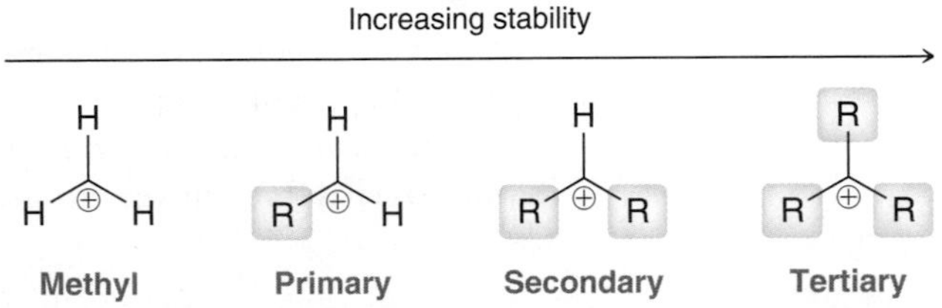

FIGURE 6.30
Carbocations with more alkyl groups will be more stable than carbocations with fewer alkyl groups.

At the end of this chapter, we will learn to predict when rearrangements are likely to occur. For now, we will just focus on recognizing carbocation rearrangements. There are two common ways in which carbocation rearrangements are accomplished: via either a **hydride shift** or a **methyl shift**.

A hydride shift involves the migration of H^-:

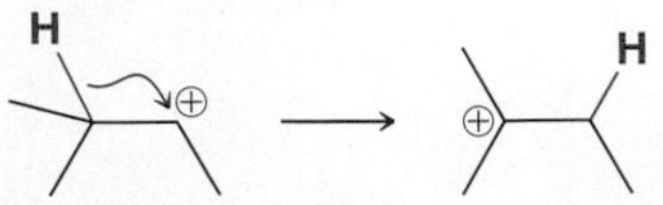

In this example, a secondary carbocation is transformed into a more stable, tertiary carbocation. How does this occur? As an analogy, imagine that there is a hole in the ground. Now imagine digging another hole nearby and using the dirt to fill up the first hole. The first hole can be filled,

WATCH OUT

Hydride is H:⁻. It is a hydrogen atom with an extra electron (two electrons total). Don't confuse the hydride ion with a proton (H^+), which is a hydrogen atom missing its electron.

but in its place, a new hole has been created nearby. A hydride shift is a similar concept. The carbocation is a hole (a place where electron density is missing). The neighboring hydrogen atom takes both of its electrons and migrates over to fill the empty *p* orbital. This process generates a new, more stable, empty *p* orbital nearby.

A carbocation rearrangement can also be accomplished via a methyl shift:

Once again, a secondary carbocation is being converted into a tertiary carbocation. But this time, it is a methyl group (rather than a hydride) that migrates with its two electrons to plug up the hole. In order for a methyl shift to occur, the methyl group must be attached to the carbon atom that is adjacent to the carbocation.

The two examples above (hydride shift and methyl shift) are the most common types of carbocation rearrangements.

In summary, we have seen only four characteristic patterns for arrow pushing in ionic reactions: (1) nucleophilic attack, (2) loss of a leaving group, (3) proton transfer, and (4) rearrangement. Let's get some practice identifying them.

SKILLBUILDER

6.3 IDENTIFYING AN ARROW-PUSHING PATTERN

LEARN the skill

Consider the following step:

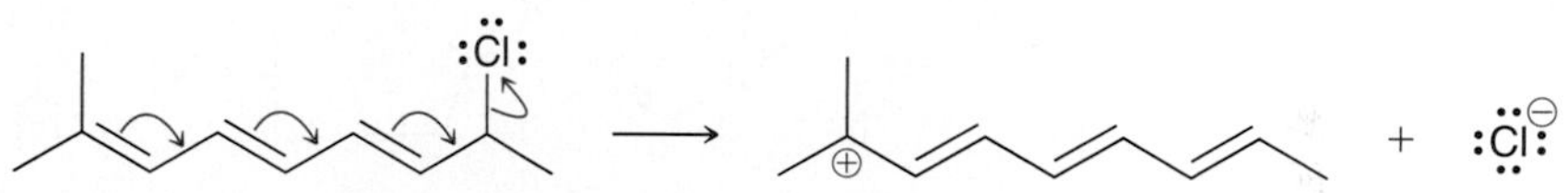

Identify which arrow-pushing pattern is utilized in this case.

SOLUTION

Read the curved arrows. The first three curved arrows show double bonds moving over, and the last curved arrow shows chloride being ejected from the compound:

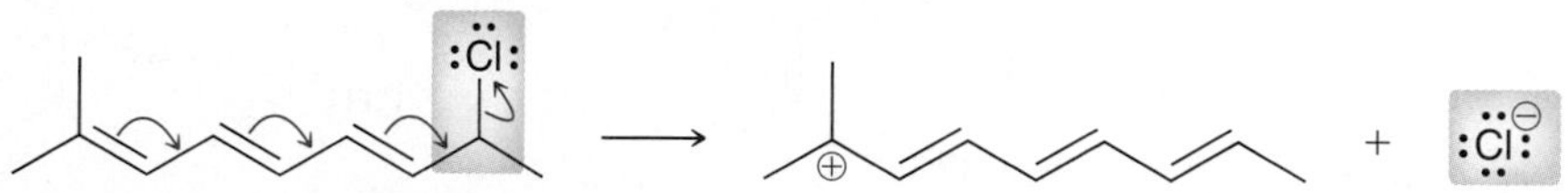

In other words, chloride is functioning as a leaving group. The other curved arrows can be viewed as resonance arrows:

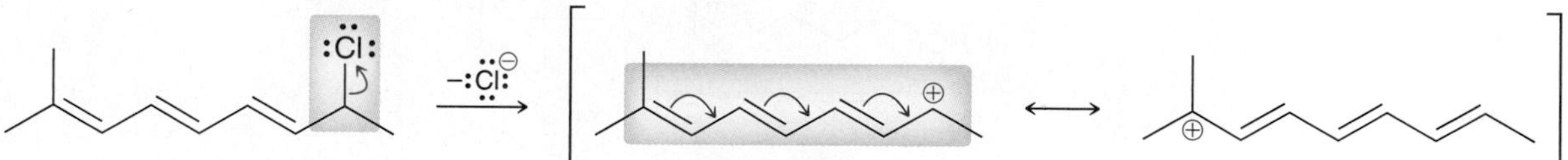

Loss of a leaving group **Resonance**

Alternatively, we can view the π bonds as "pushing out" the chloride:

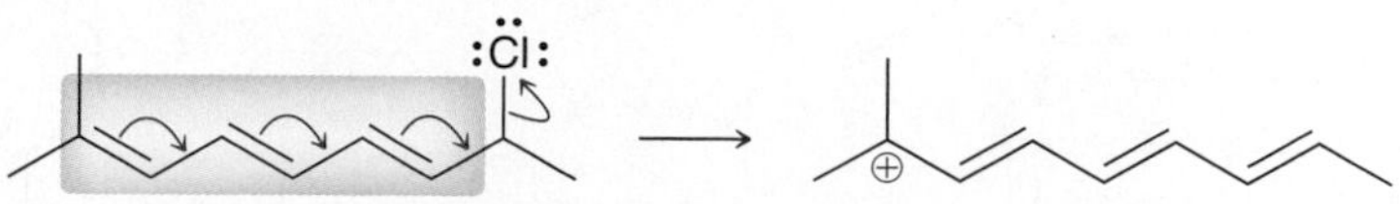

π bonds pushing out the chloride

Either way, there is only one arrow-pushing pattern being utilized here: *loss of a leaving group.*

PRACTICE the skill **6.12** For each of the following cases, read the curved arrows and identify which arrow-pushing pattern is utilized:

(a)

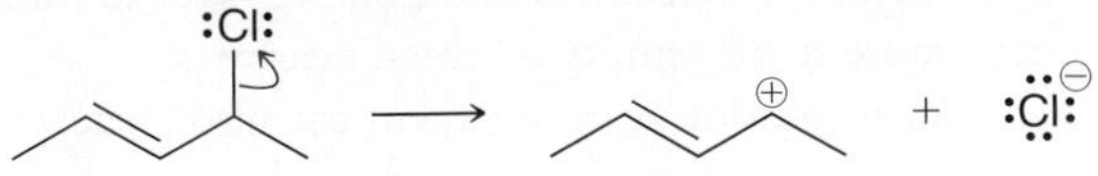

(b)

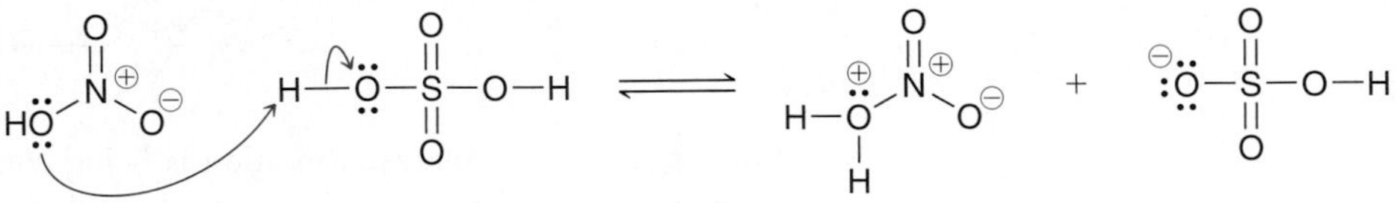

(c)

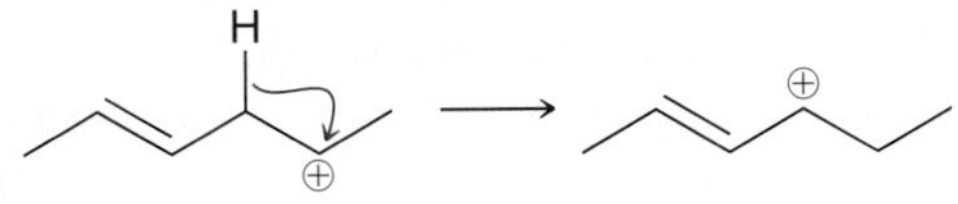

(d)

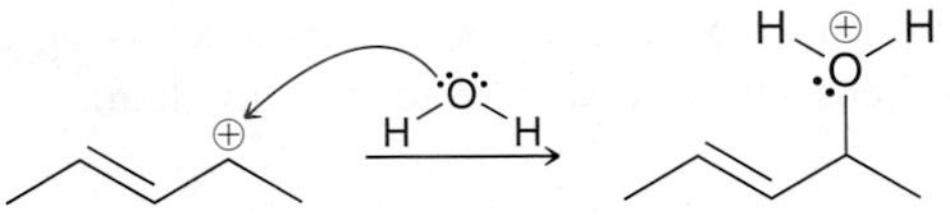

(e)

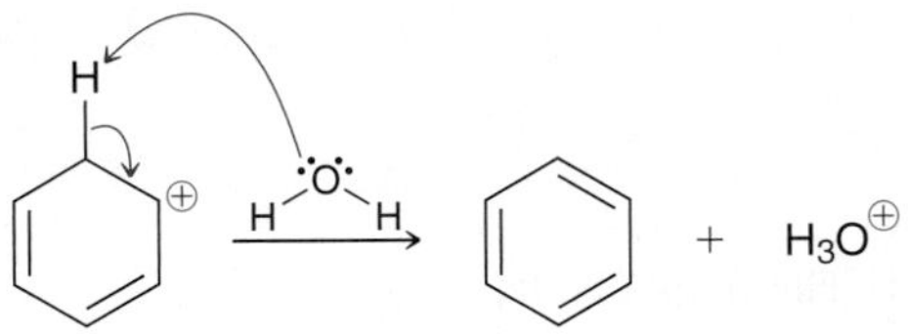

(f)

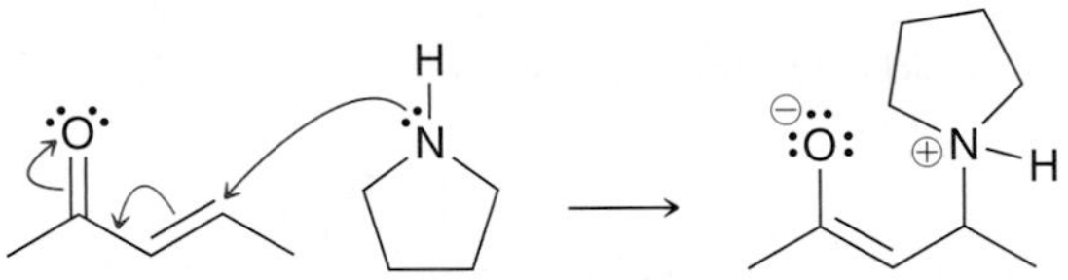

(g)

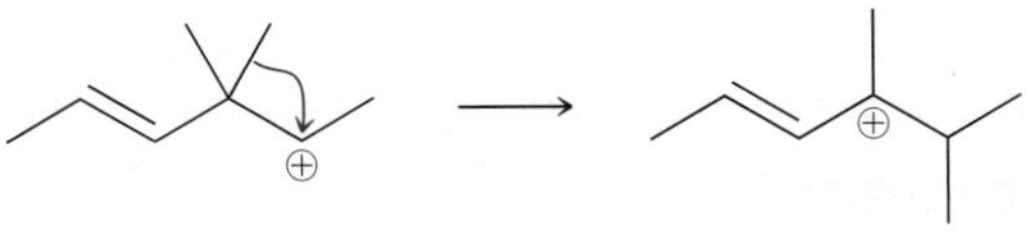

(h)

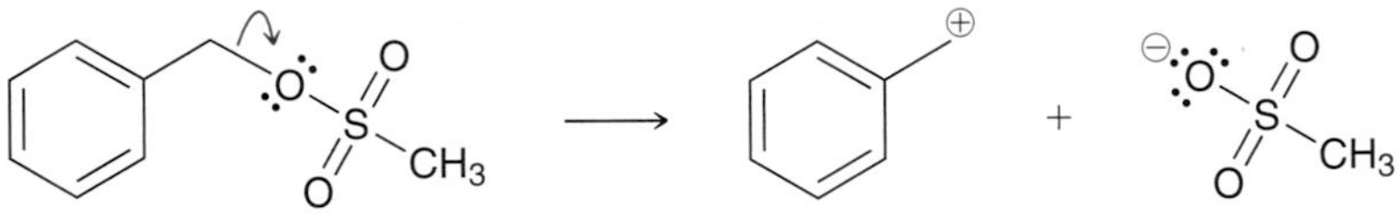

(i)

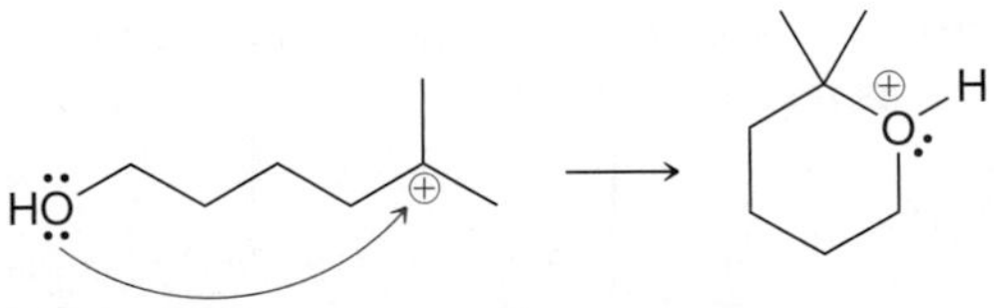

APPLY the skill **6.13** Identify which arrow-pushing pattern is utilized in the following step:

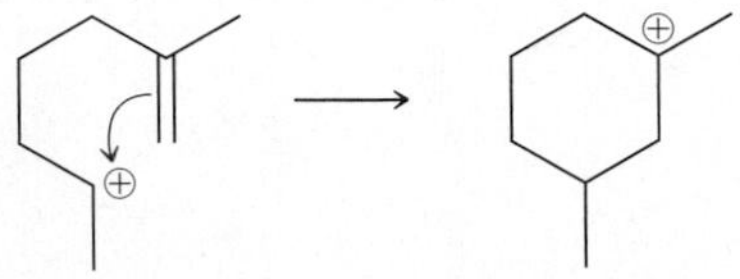

need more **PRACTICE?** **Try Problem 6.30**

6.9 Combining the Patterns of Arrow Pushing

All ionic mechanisms, regardless of how complex, are just different combinations of the four characteristic patterns seen in the previous section. As an example, consider the reaction in Figure 6.31, which we will explore in greater detail in Chapter 7.

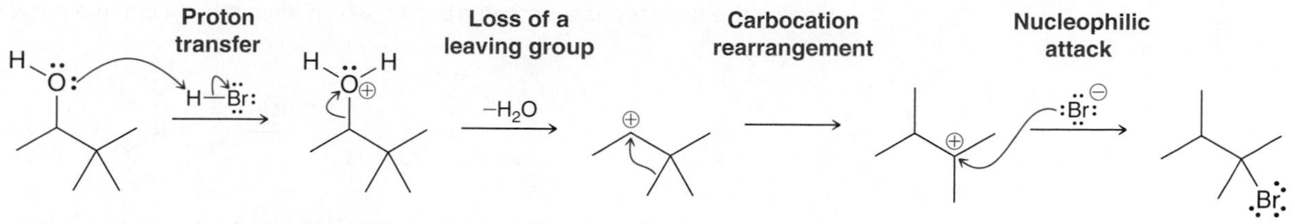

FIGURE 6.31
A reaction containing all four characteristic patterns for arrow pushing.

The mechanism in Figure 6.31 has four steps, each of which is one of the characteristic patterns from the previous section. Sometimes, a single mechanistic step will involve two simultaneous patterns; for example:

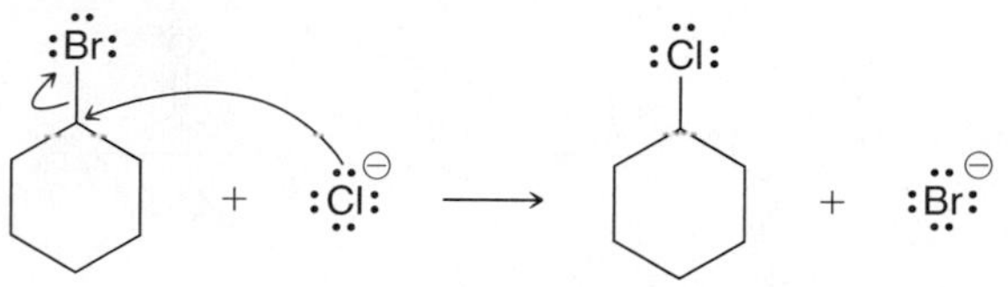

There are two curved arrows shown here. One curved arrow, coming from the chloride, shows a nucleophilic attack. The second curved arrow shows loss of a leaving group. In this case, we are utilizing two arrow-pushing patterns simultaneously. This is called a concerted process. In the next chapter, we will explore some of the important differences between concerted and stepwise mechanisms. For now, we will simply focus on recognizing the various patterns that can arise as a result of combining the various arrow-pushing patterns. By doing so, similarities will emerge between apparently different mechanisms.

SKILLBUILDER

6.4 IDENTIFYING A SEQUENCE OF ARROW-PUSHING PATTERNS

LEARN the skill

Identify the sequence of arrow-pushing patterns in the following reaction:

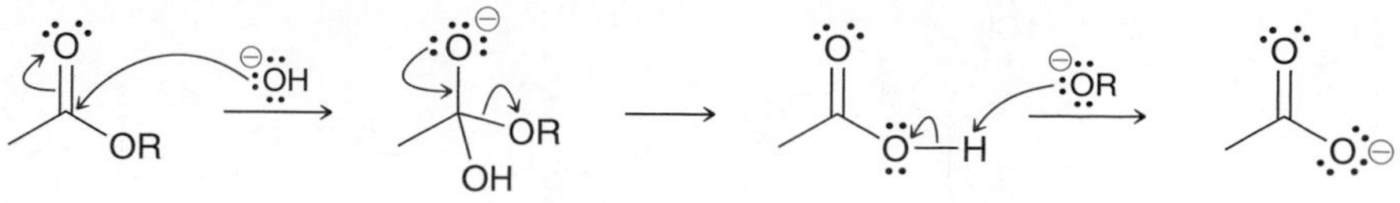

SOLUTION

In the first step, hydroxide is attacking the C=O in a nucleophilic attack:

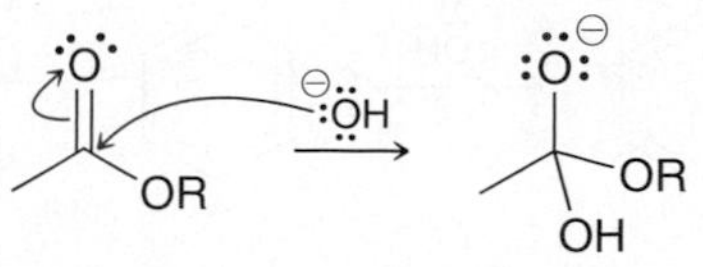

In the next step, an alkoxide (RO^-) is being ejected as a leaving group:

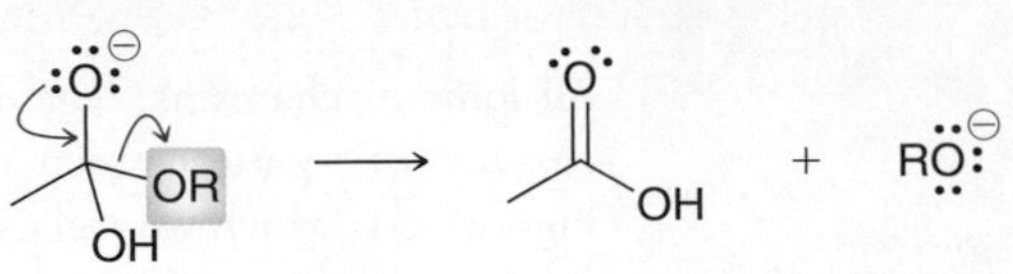

Finally, the last step is a proton transfer, which always involves two curved arrows:

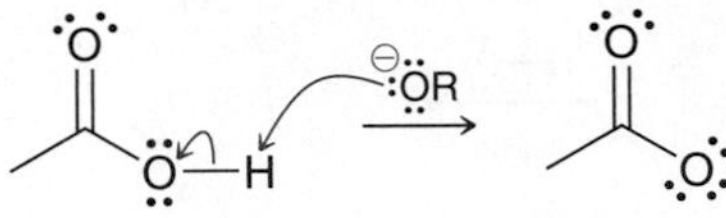

The sequence of this reaction is therefore (1) nucleophilic attack, (2) loss of a leaving group, and (3) proton transfer. Throughout this book, and especially in Chapter 21, we will see dozens of reactions that follow this three-step sequence. By viewing all of these reactions in this format (as a sequence of characteristic patterns), it will be easier to see the similarities between apparently different reactions.

PRACTICE the skill **6.14** For each of the following multistep reactions, read the curved arrows and identify the sequence of arrow–pushing patterns:

(a)

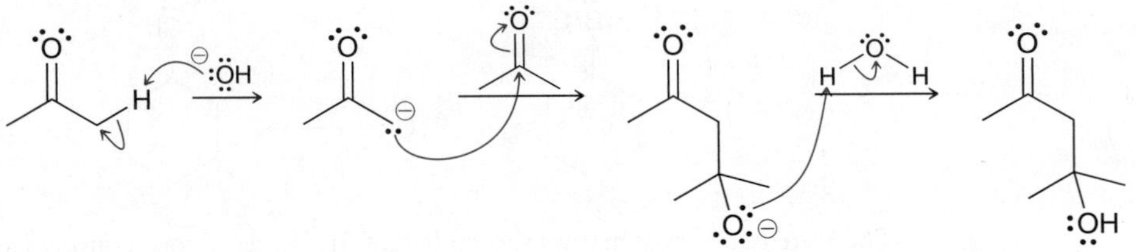

(b)

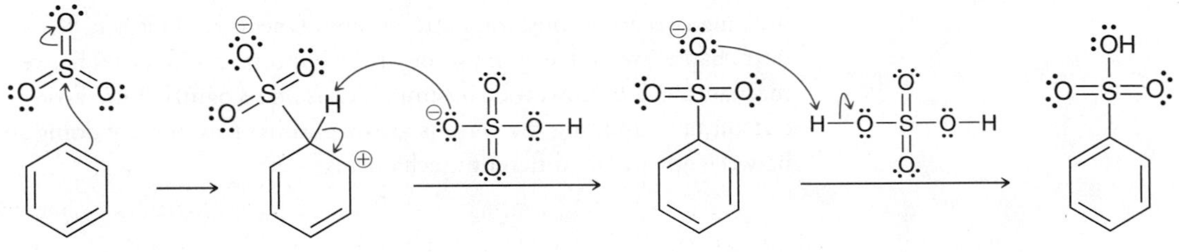

(c)

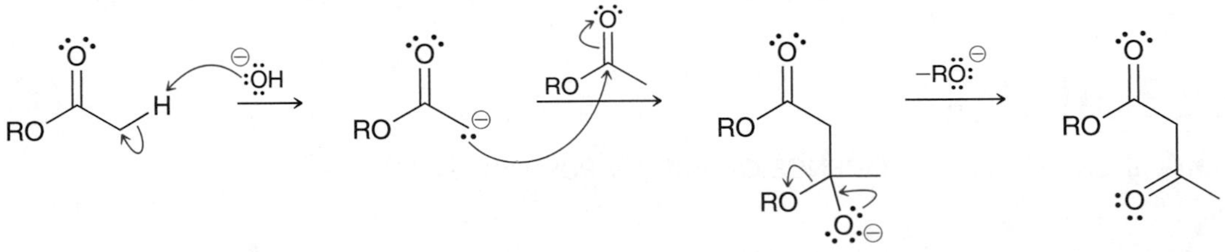

(d)

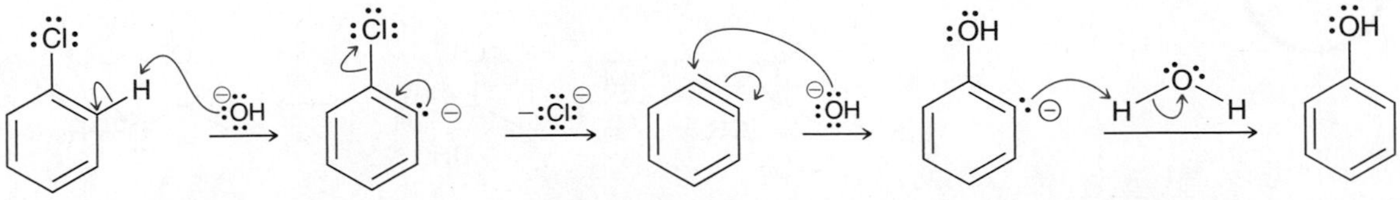

(e)

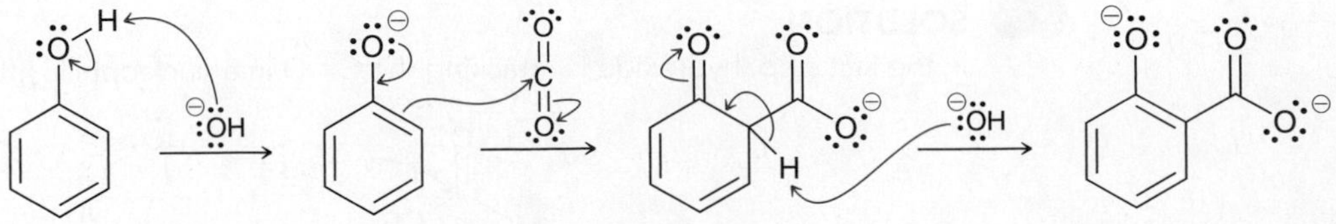

APPLY the skill

6.15 The following two reactions will be explored in different chapters. Yet, they are very similar. Identify and compare the sequences of arrow-pushing patterns for the two reactions.

Reaction 1 (Chapter 20)

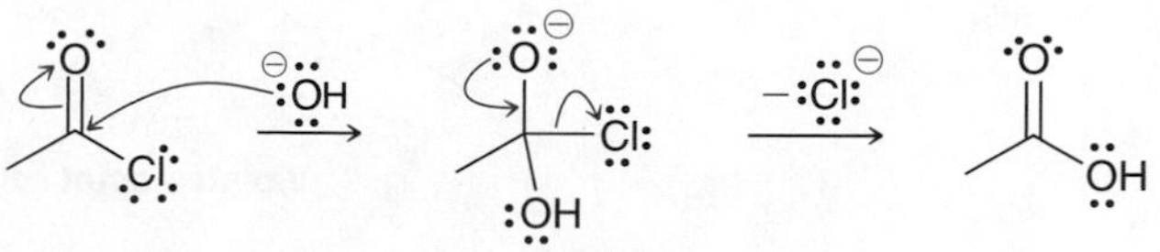

Reaction 2 (Chapter 18)

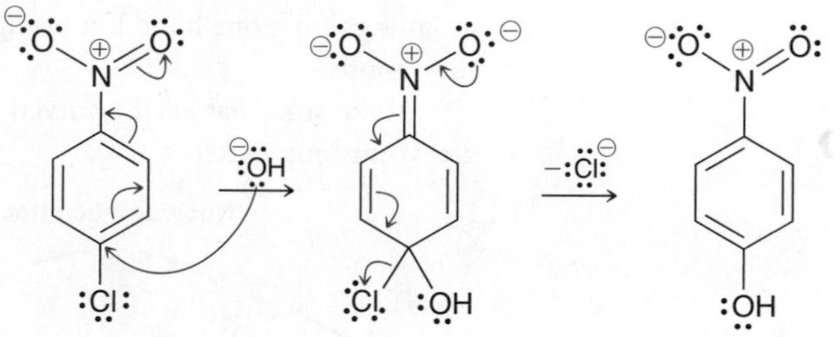

need more **PRACTICE?** **Try Problems 6.32 – 6.41**

There may be more than 100 mechanisms in a full organic chemistry course, but there are fewer than a dozen different sequences of arrow-pushing patterns in these mechanisms. As we proceed through the chapters, we will learn rules for determining when each pattern can and cannot be utilized. These rules will empower us to propose mechanisms for new reactions.

6.10 Drawing Curved Arrows

Curved arrows have a very precise meaning, and they must be drawn precisely. Avoid sloppy arrows. Make sure to be deliberate when drawing the tail and the head of every arrow. The tail must be placed on either a bond or a lone pair. For example, the following reaction employs two curved arrows. One of the curved arrows has its tail placed on a lone pair, and the other has its tail placed on a bond:

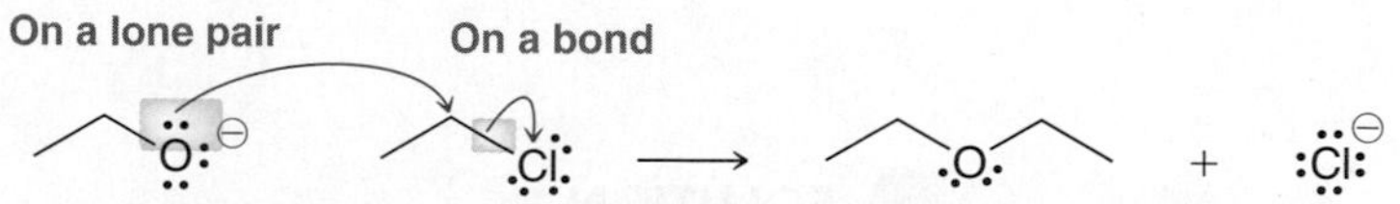

These are the only two possible locations where the tail of a curved arrow can be placed. The tail shows where electrons are coming from, and electrons can only be found in lone pairs or bonds. *Never place the tail of a curved arrow on a positive charge.*

The head of a curved arrow must be placed so that it shows either the formation of a bond or the formation of a lone pair:

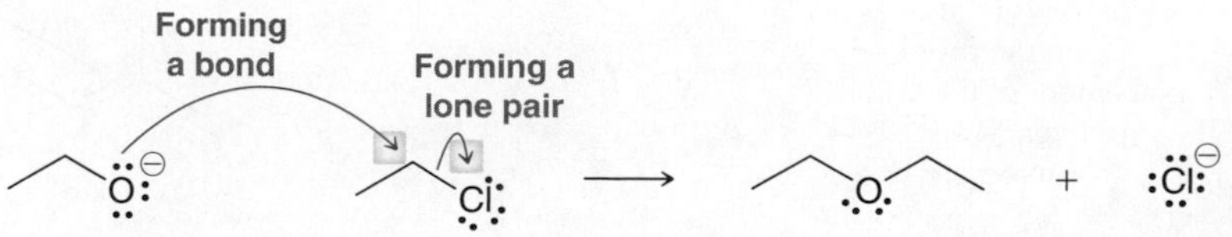

WATCH OUT
C, N, and O are the second-row elements that require special attention. Never give one of these elements more than four orbitals.

When drawing the head of a curved arrow, make sure to avoid drawing an arrow that violates the octet rule. Specifically, never draw an arrow that gives more than four orbitals to a second-row element:

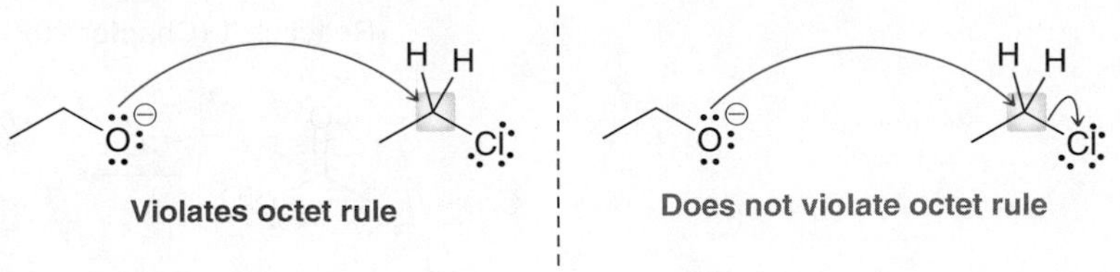

In the first example the head of the curved arrow is giving a fifth bond to the carbon atom. This violates the octet rule. The second example does not violate the octet rule, because the carbon atom is gaining one bond but losing another. In the end, the carbon atom never has more than four bonds.

Make sure that all the curved arrows you draw accomplish one of the four characteristic arrow-pushing patterns.

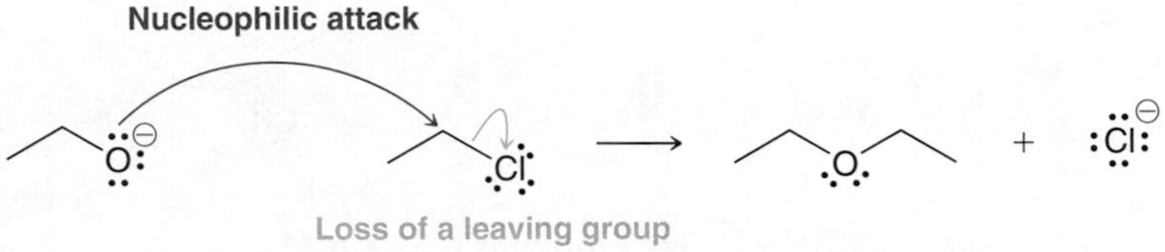

Avoid drawing arrows like the following. This arrow violates the octet rule, and it does not accomplish any one of the four characteristic arrow-pushing patterns.

SKILLBUILDER

6.5 DRAWING CURVED ARROWS

LEARN the skill

Draw the curved arrows that accomplish the following transformation:

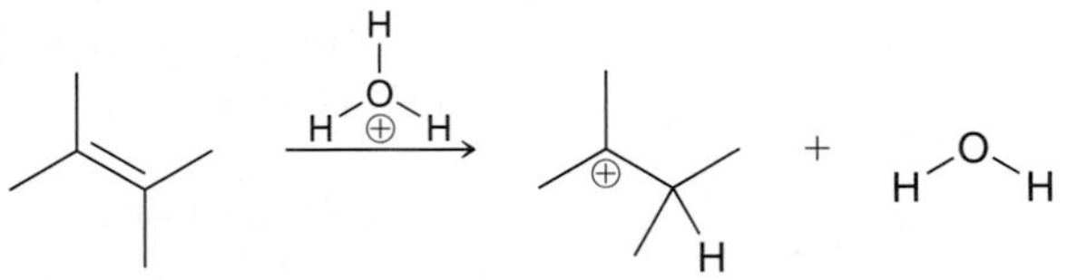

SOLUTION

STEP 1
Identify which of the four arrow-pushing patterns to use.

STEP 2
Draw the curved arrows focusing on the proper placement of the tail and head of each curved arrow.

We begin by identifying which characteristic arrow-pushing pattern to use in this case. Look carefully, and notice that H_3O^+ is losing a proton. That proton has been transferred to a carbon atom. Therefore, this is a proton transfer step, where the π bond is functioning as the base to deprotonate H_3O^+. A proton transfer step requires two curved arrows. Make sure to properly place the head and the tail of each curved arrow. The first curved arrow must originate on the π bond and end on a proton:

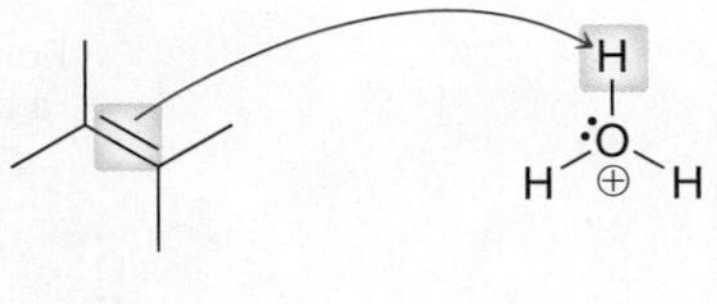

Don't forget to draw the second curved arrow (students will often leave out the second arrow, producing an incomplete mechanism). The second curved arrow shows what happens to the electrons that were previously holding the proton. Specifically, the tail is placed on the O—H bond, and the head is placed on the oxygen:

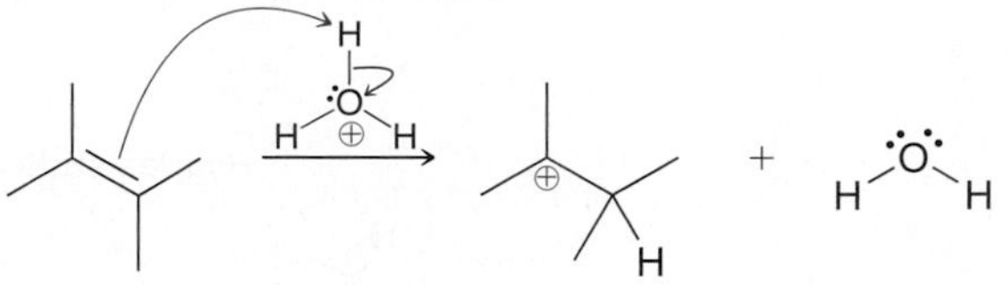

PRACTICE the skill **6.16** Draw the curved arrows that accomplish each of the following transformations:

(a) 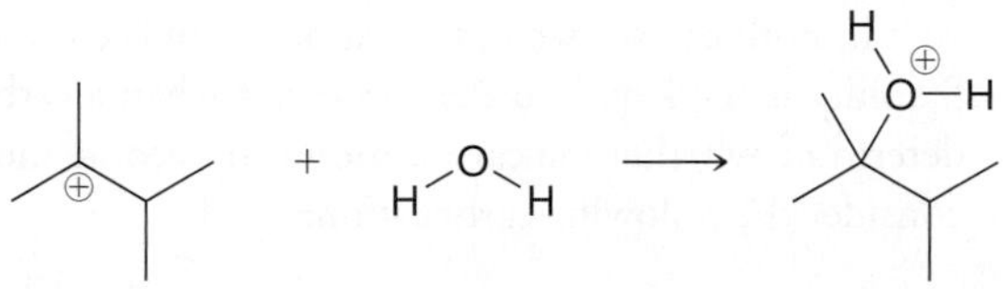

(b) 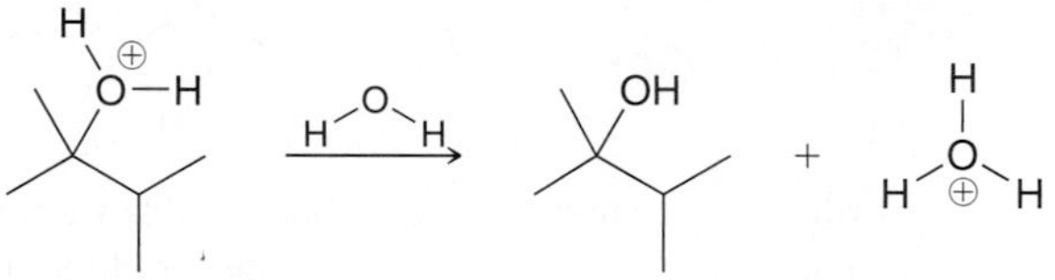

(c) 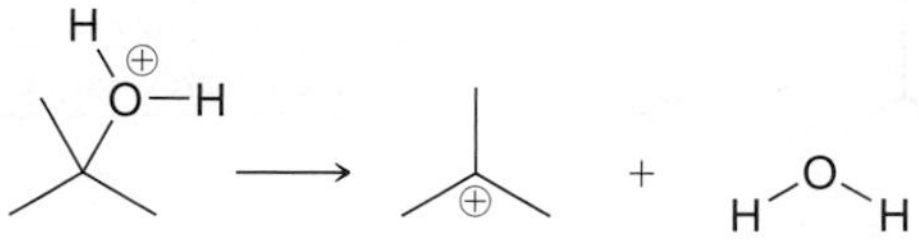

APPLY the skill **6.17** The following four reactions will be the focus of the upcoming chapters (substitution and elimination reactions). Draw the curved arrows that accomplish each of the transformations shown:

(a) 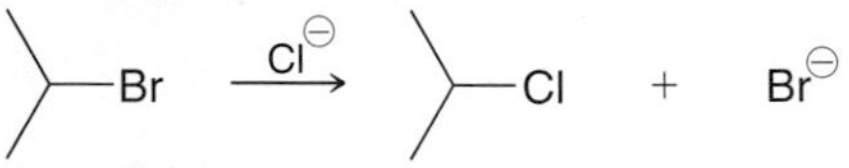

(b) 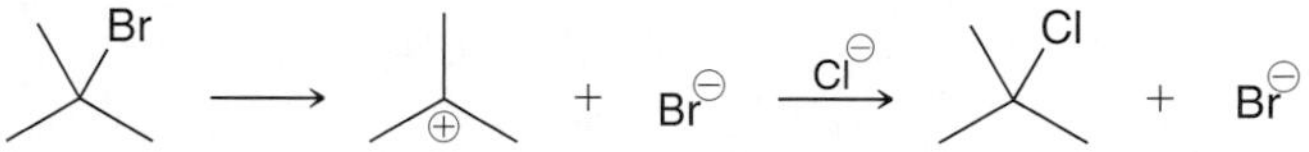

(c) 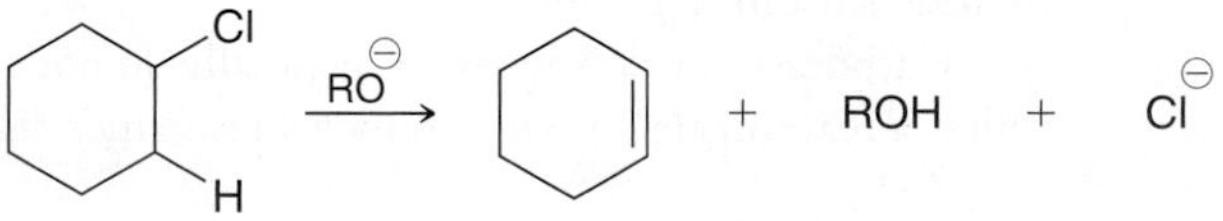

(d) 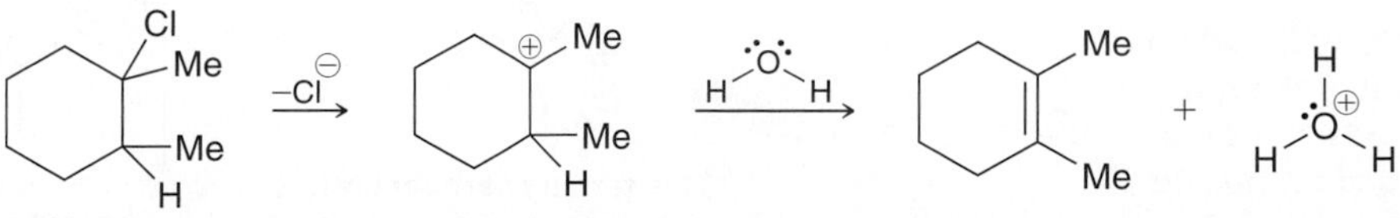

need more **PRACTICE?** **Try Problems 6.42–6.47, 6.54**

6.11 Carbocation Rearrangements

Throughout this course, we will encounter many examples of carbocation rearrangements, so we must be able to predict when these rearrangements will occur. Recall that the two common types of carbocation rearrangement are hydride shifts and methyl shifts (Figure 6.32).

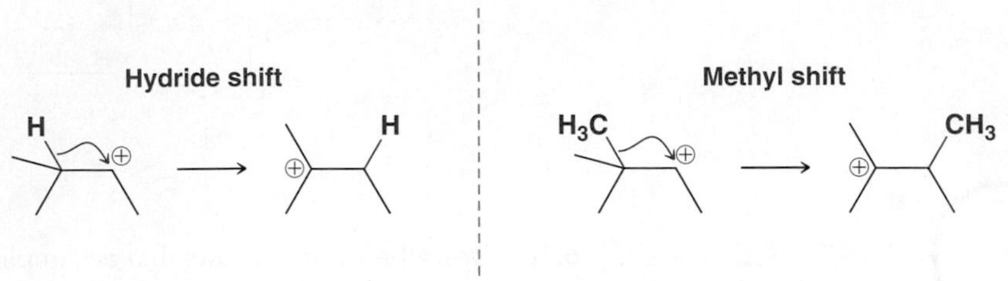

FIGURE 6.32 Hydride shifts and methyl shifts are the two most common types of carbocation rearrangements.

In both cases, a secondary carbocation is converted into a more stable, tertiary carbocation. Stability is the key. In order to predict when a carbocation rearrangement will occur, we must determine whether the carbocation can become more stable via a rearrangement. For example, consider the following carbocation:

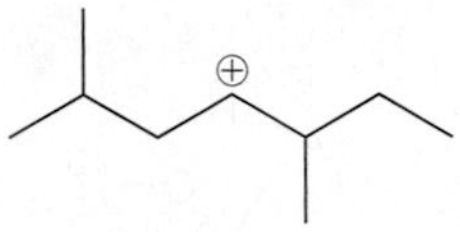

In order to determine if this carbocation can undergo a rearrangement, we must identify any hydrogen atoms or methyl groups attached directly to the neighboring carbon atoms:

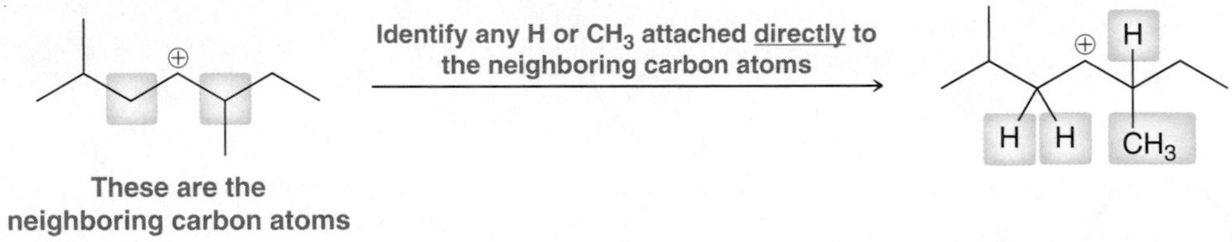

There are four candidates. Now imagine each one of these groups shifting over to plug the carbocation, generating a new carbocation. Would the new carbocation be more stable? In this example only one group can migrate to produce a more stable carbocation:

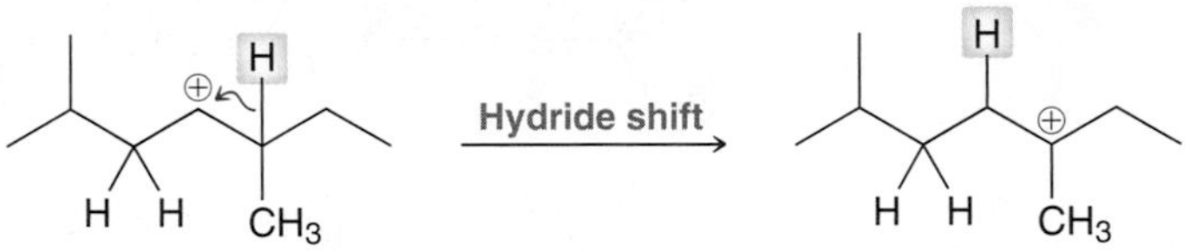

If this hydride migrates, it will generate a new, tertiary carbocation. Therefore, we do expect a hydride shift in this case.

Carbocation rearrangements generally do not occur when the carbocation is already tertiary unless a rearrangement will produce a resonance-stabilized carbocation; for example:

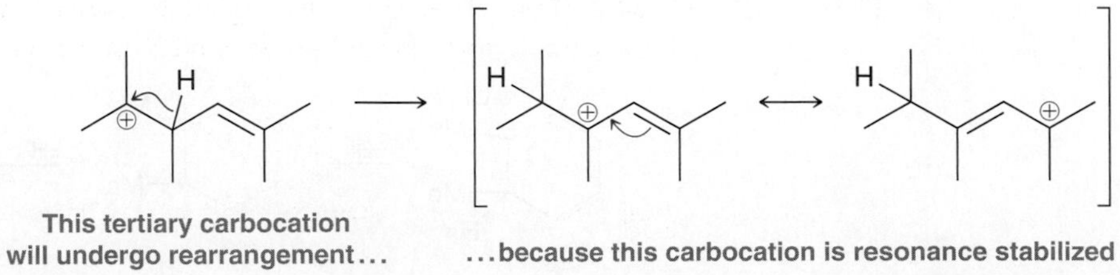

In this case, the original carbocation is tertiary. However, the newly formed carbocation is tertiary and it is stabilized by resonance. Such a carbocation is called an **allylic carbocation** because the positive charge is located at an allylic position.

LOOKING BACK

Recall that the term allylic describes the positions connected directly to the π bond:

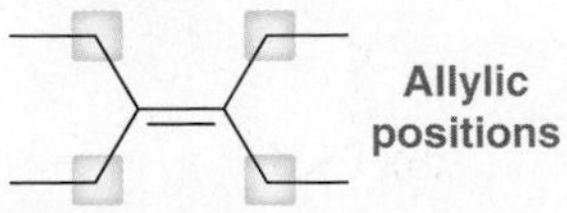

SKILLBUILDER

6.6 PREDICTING CARBOCATION REARRANGEMENTS

LEARN the skill

Predict whether the following carbocation will rearrange, and if so, draw the curved arrow showing the carbocation rearrangement:

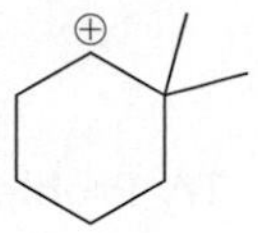

SOLUTION

STEP 1
Identify neighboring carbon atoms.

This carbocation is secondary, so it certainly has the potential to rearrange. We must look to see if a carbocation rearrangement can produce a more stable, tertiary carbocation. Begin by identifying the carbon atoms neighboring the carbocation:

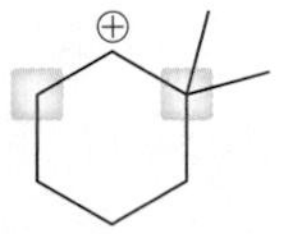

STEP 2
Identify any H or CH_3 groups attached directly to the neighboring carbon atoms.

Look at all hydrogen atoms or methyl groups connected to these carbon atoms:

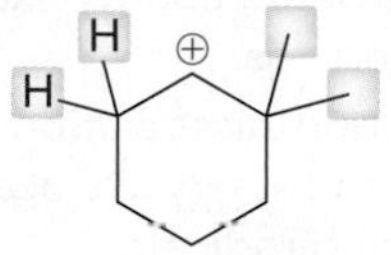

STEP 3
Determine which of these groups can migrate to generate a more stable carbocation.

Consider if migration of any of these groups will generate a more stable, tertiary carbocation. Migration of either of the neighboring hydride groups will only generate another secondary carbocation. So, we do not expect a hydride shift to occur.

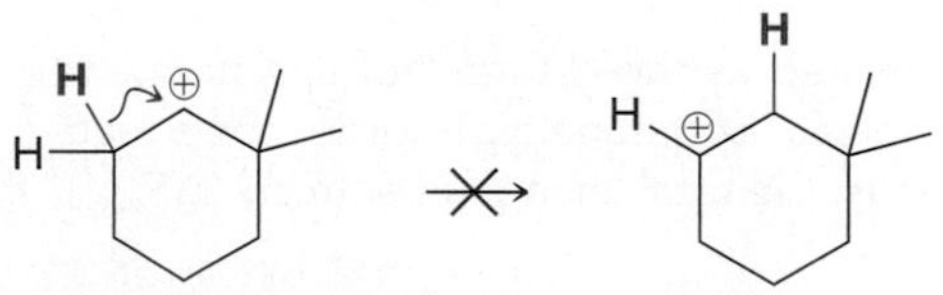

Secondary C+ Secondary C+

STEP 4
Draw a curved arrow showing the carbocation rearrangement and draw the new carbocation.

However, if either of the methyl groups migrates, a tertiary carbocation is generated. Therefore, we expect a methyl shift to take place, generating a more stable tertiary carbocation.

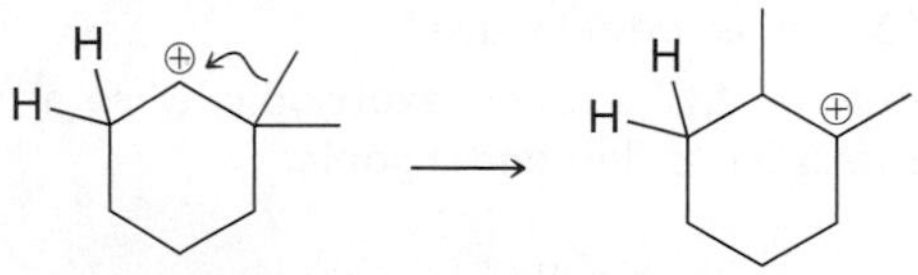

Secondary C+ Tertiary C+

PRACTICE the skill

6.18 For each of the following carbocations determine if it will rearrange, and if so, draw the carbocation rearrangement with a curved arrow:

(a) 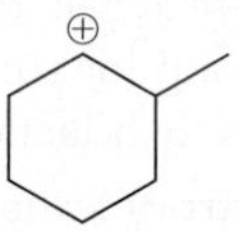 (b) 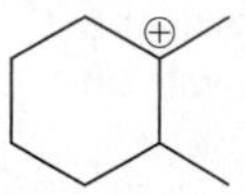 (c) 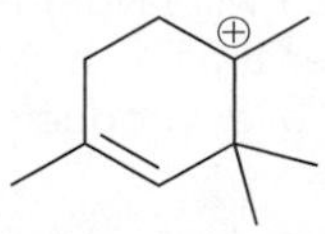 (d) 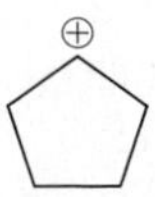

(e) 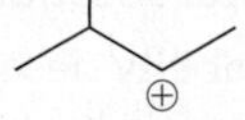 (f) 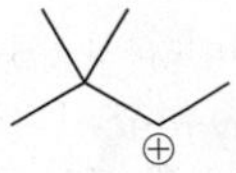 (g) 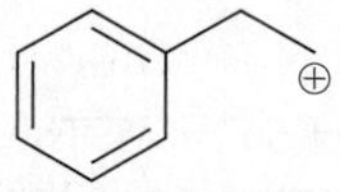 (h) 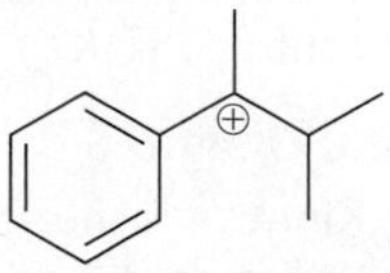

APPLY the skill

6.19 Occasionally, carbocation rearrangements can be accomplished via the migration of a carbon atom other than a methyl group. Such an example follows. Identify the group that is migrating, and draw the curved arrow that shows the migration:

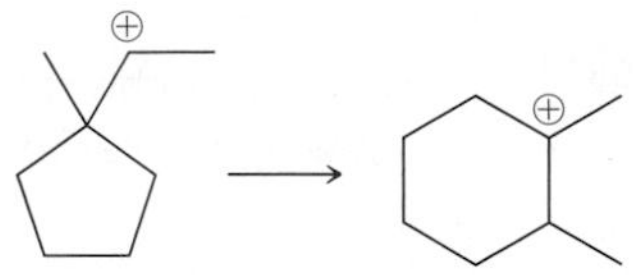

need more PRACTICE? Try Problem 6.48

REVIEW OF CONCEPTS AND VOCABULARY

SECTION 6.1

- The total change in **enthalpy** for any reaction (ΔH), also called the **heat of reaction**, is a measure of the energy exchanged between the system and its surroundings.
- Each type of bond has a unique **bond dissociation energy**, which is the amount of energy necessary to accomplish **homolytic bond cleavage**, producing **radicals**.
- **Exothermic** reactions involve a transfer of energy from the system to the surroundings, while **endothermic** reactions involve a transfer of energy from the surroundings to the system.

SECTION 6.2

- **Entropy** is loosely defined as the disorder of a system and is the ultimate criterion for spontaneity. In order for a reaction to be **spontaneous**, the total change in entropy ($\Delta S_{sys} + \Delta S_{surr}$) must be positive.
- Reactions with a positive ΔS_{sys} involve an increase in the number of molecules, or an increase in the amount of conformational freedom.

SECTION 6.3

- In order for a process to be spontaneous, the change in **Gibbs free energy** (ΔG) must be negative.
- A reaction with a negative ΔG is called **exergonic**, while a reaction with a positive ΔG is called **endergonic**.

SECTION 6.4

- Equilibrium concentrations of a reaction represent the point of lowest free energy available to the system.
- The exact position of equilibrium is described by the equilibrium constant, K_{eq}, and is a function of ΔG.
- If ΔG is negative, the reaction will favor products over reactants, and K_{eq} will be greater than 1. If ΔG is positive, the reaction will favor reactants over products, and K_{eq} will be less than 1.
- The study of relative energy levels (ΔG) and equilibrium concentrations (K_{eq}) is called **thermodynamics**.

SECTION 6.5

- **Kinetics** is the study of reaction rates. The rate of any reaction is described by a **rate equation.** A reaction can be **first order, second order**, or **third order**, depending on whether the sum of the exponents in the rate equation is 1, 2, or 3, respectively.
- A low **energy of activation, E_a**, corresponds with a fast rate
- Increasing the temperature will increase the number of molecules that possess the minimum necessary kinetic energy for a reaction, thereby increasing the rate of reaction.
- **Catalysts** speed up the rate of a reaction by providing an alternate pathway with a lower E_a.

SECTION 6.6

- Kinetics refers to the rate of a reaction and is dependent on the relative energy levels of the reactants and the transition state.
- Thermodynamics refers to the equilibrium concentrations and is dependent on the relative energy levels of the reactants and products.
- On an energy diagram, each peak represents a **transition state** while each valley represents an **intermediate**. Transition states cannot be isolated; intermediates have a finite lifetime.
- The **Hammond postulate** states that a transition state will resemble the reactants for an exergonic process, but will resemble the products for an endergonic process.

SECTION 6.7

- **Ionic reactions**, also called **polar reactions**, involve the participation of ions as reactants, intermediates, or products—in most cases, as intermediates.
- A **nucleophile** has an electron-rich atom that is capable of donating a pair of electrons. Nucleophilic centers include atoms with lone pairs, π bonds, or atoms that are electron rich due to inductive effects.
- An **electrophile** has an electron-deficient atom that is capable of accepting a pair of electrons. Electrophilic centers include atoms that are electron deficient due to inductive effects, as well as **carbocations**, which have an empty *p* orbital.
- **Polarizability** describes the ability of an atom to distribute its electron density unevenly as a result of external influences.

SECTION 6.8

- In drawing a mechanism, the tail of a curved arrow shows where the electrons are coming from, and the head of the arrow shows where the electrons are going.
- There are four characteristic arrow-pushing patterns: (1) **nucleophilic attack**, (2) **loss of a leaving group**, (3) **proton transfer**, and (4) **rearrangement**.
- The most common type of rearrangement is a carbocation rearrangement, in which a carbocation undergoes either a **hydride shift** or a **methyl shift** to produce a more stable carbocation. As a result of **hyperconjugation**, **tertiary** carbocations are more stable than **secondary** carbocations, which are more stable than **primary** carbocations.

SECTION 6.9

- All ionic mechanisms, regardless of complexity, are different combinations of the four characteristic arrow-pushing patterns.

- When a single mechanistic step involves two simultaneous arrow-pushing patterns, it is called a concerted process.

SECTION 6.10

- The tail of a curved arrow must be placed either on a bond or a lone pair, while the head of a curved arrow must be placed so that it shows either formation of a bond or of a lone pair.
- Never draw an arrow that gives a fifth orbital to a second-row element (C, N, O, F).

SECTION 6.11

- A carbocation rearrangement will occur if it leads to a more stable carbocation.
- Tertiary carbocations generally do not rearrange, unless a rearrangement will produce a resonance-stabilized carbocation, such as an **allylic carbocation.**

KEY TERMINOLOGY

allylic carbocation 268	enthalpy 234	homolytic bond cleavage 234	primary carbocation 260
bond dissociation energy 235	entropy 237	hydride shift 260	radicals 234
carbocation 255	equilibrium 243	hyperconjugation 260	rate equation 245
carbocation rearrangement 260	exergonic 242	intermediate 251	secondary carbocation 260
catalyst 249	exothermic 235	ionic reactions 253	second order 246
electrophile 254	first order 246	kinetics 245	spontaneous 239
endergonic 242	Gibbs free energy 240	methyl shift 260	tertiary carbocation 260
endothermic 235	Hammond postulate 253	nucleophile 254	thermodynamics 244
energy of activation 246	heat of reaction 235	polarizability 254	third order 246
	heterolytic bond cleavage 234	polar reactions 253	transition state 251

SKILLBUILDER REVIEW

6.1 PREDICTING ΔH° OF A REACTION

EXAMPLE Calculate ΔH° for this reaction.

STEPS 1 AND 2 Identify the bond dissociation energy of each bond broken and formed, and then determine the appropriate sign for each value.

Broken

+381 kJ/mol +242 kJ/mol

+623 kJ/mol

Formed

−331 kJ/mol −431 kJ/mol

−762 kJ/mol

STEP 3 Take the sum of Steps 1 and 2.

$$\left(\begin{array}{c}+623\\ \text{kJ/mol}\end{array}\right) + \left(\begin{array}{c}-762\\ \text{kJ/mol}\end{array}\right)$$

−139 kJ/mol
Exothermic

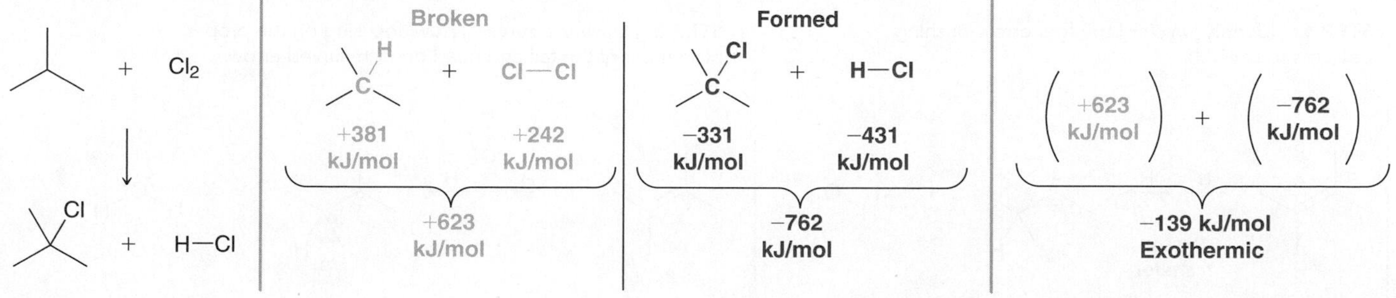

Try Problems 6.1, 6.2, 6.21a

6.2 IDENTIFYING NUCLEOPHILIC AND ELECTROPHILIC CENTERS

STEP 1 Identify nucleophilic centers by looking for inductive effects, lone pairs, or π bonds:

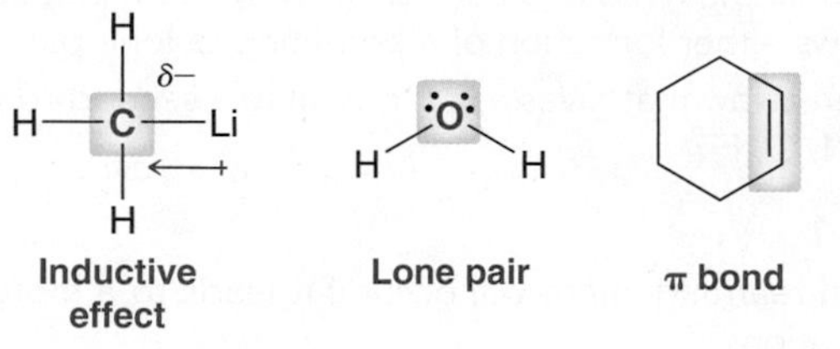

STEP 2 Identify electrophilic centers by looking for inductive effects or empty *p* orbitals:

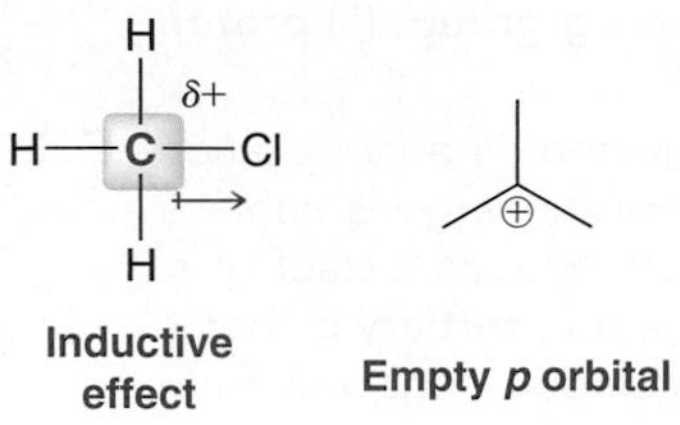

Try Problems **6.8–6.11, 6.53**

6.3 IDENTIFYING AN ARROW-PUSHING PATTERN

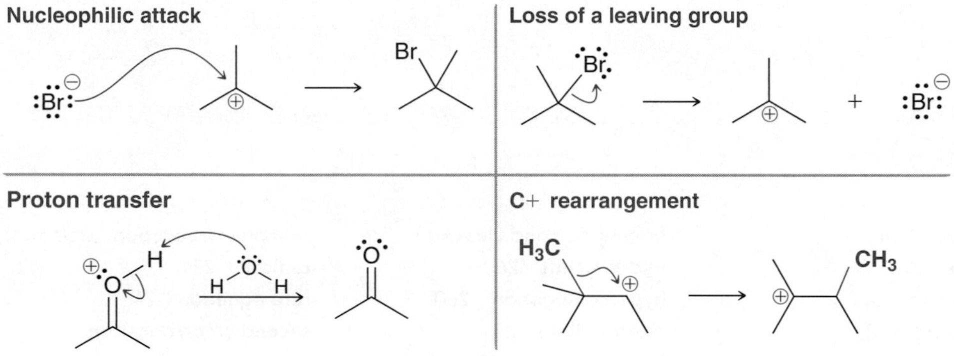

Try Problems **6.12, 6.13, 6.30**

6.4 IDENTIFYING A SEQUENCE OF ARROW-PUSHING PATTERNS

Identify each characteristic pattern of arrow pushing.

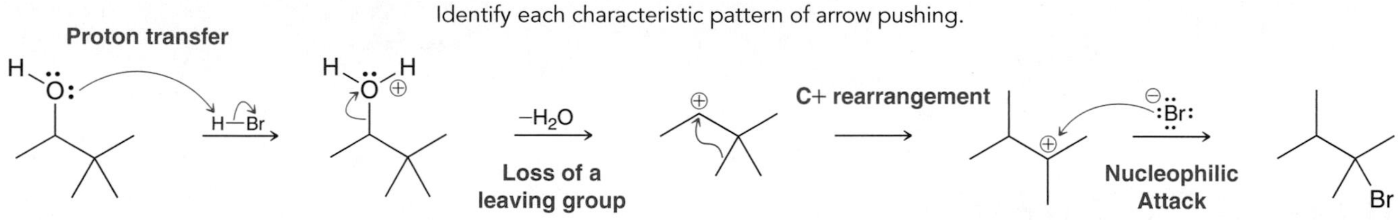

Try Problems **6.14, 6.15, 6.32–6.41**

6.5 DRAWING CURVED ARROWS

STEP 1 Identify which of the four arrow-pushing patterns to use.

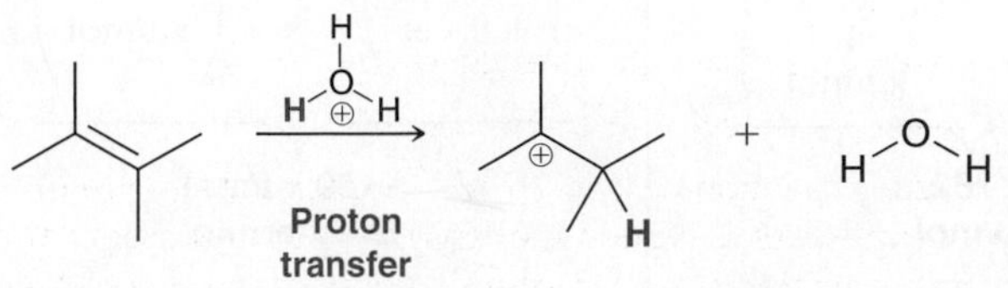

STEP 2 Draw the curved arrow, focusing on the proper placement of the tail and head of each curved arrow.

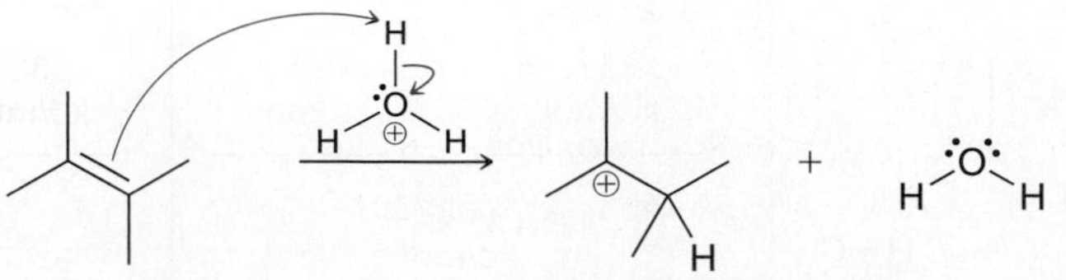

Try Problems **6.16, 6.17, 6.42–6.47, 6.54**

6.6 PREDICTING CARBOCATION REARRANGEMENTS

STEP 1 Identify neighboring carbon atoms.

STEP 2 Identify any H or CH_3 attached directly to the neighboring carbon atoms.

STEP 3 Find any groups that can migrate to generate a more stable C+.

STEP 4 Draw a curved arrow showing the C+ rearrangement and then draw the new carbocation.

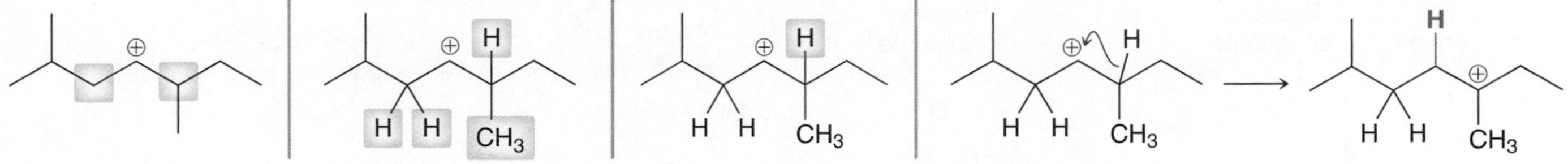

Try Problems 6.18, 6.19, 6.48

PRACTICE PROBLEMS

6.20 In each of the following cases compare the bonds identified with red arrows, and determine which bond you would expect to have the largest bond dissociation energy:

(a) 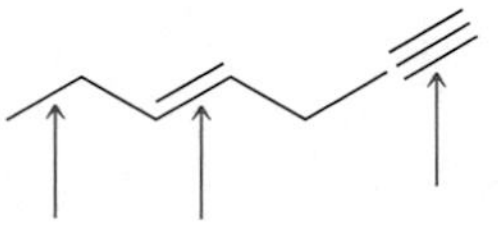

(b) 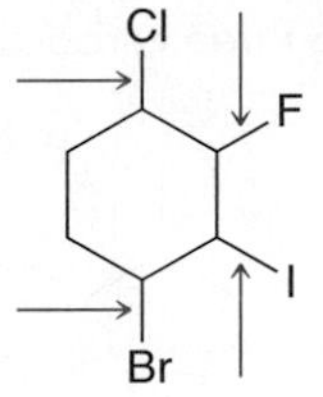

6.21 Consider the following reaction:

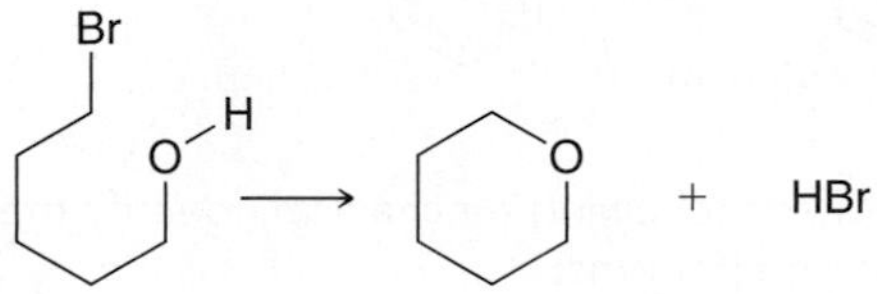

(a) Use Table 6.1 to estimate ΔH for this reaction.
(b) ΔS of this reaction is positive. Explain.
(c) Determine the sign of ΔG.
(d) Is the sign of ΔG dependent on temperature?
(e) Is the magnitude of ΔG dependent on temperature?

6.22 For each of the following cases use the information given to determine whether or not the equilibrium will favor products over reactants:
(a) A reaction with $K_{eq} = 1.2$
(b) A reaction with $K_{eq} = 0.2$
(c) A reaction with a positive ΔG
(d) An exothermic reaction with a positive ΔS
(e) An endothermic reaction with a negative ΔS

6.23 Which value of ΔG corresponds with $K_{eq} = 1$?
(a) +1 kJ/mol (b) 0 kJ/mol (c) −1 kJ/mol

6.24 Which value of ΔG corresponds with $K_{eq} < 1$?
(a) +1 kJ/mol (b) 0 kJ/mol (c) −1 kJ/mol

6.25 For each of the following reactions determine whether ΔS for the reaction (ΔS_{sys}) will be positive, negative, or approximately zero:

(a) 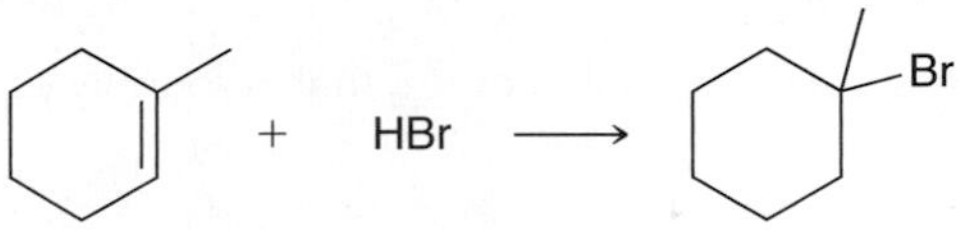

(b) 

(c) 

(d) 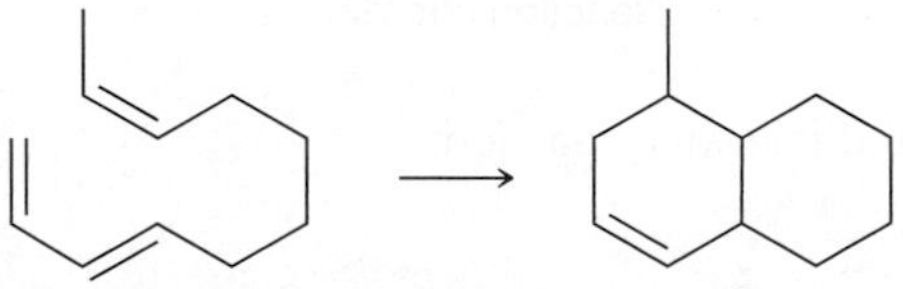

(e) 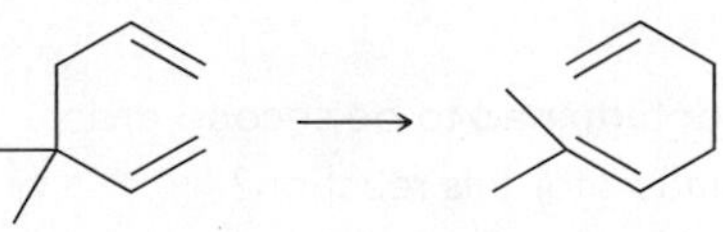

6.26 Draw an energy diagram of a reaction with the following characteristics:
(a) A one-step reaction with a negative ΔG
(b) A one-step reaction with a positive ΔG
(c) A two-step reaction with an overall negative ΔG, where the intermediate is higher in energy than the reactants and the first transition state is higher in energy than the second transition state.

6.27 Consider the following four energy diagrams:

A

Free
energy
(G)

Reaction
coordinate

B

Free
energy
(G)

Reaction
coordinate

C

Free
energy
(G)

Reaction
coordinate

D

Free
energy
(G)

Reaction
coordinate

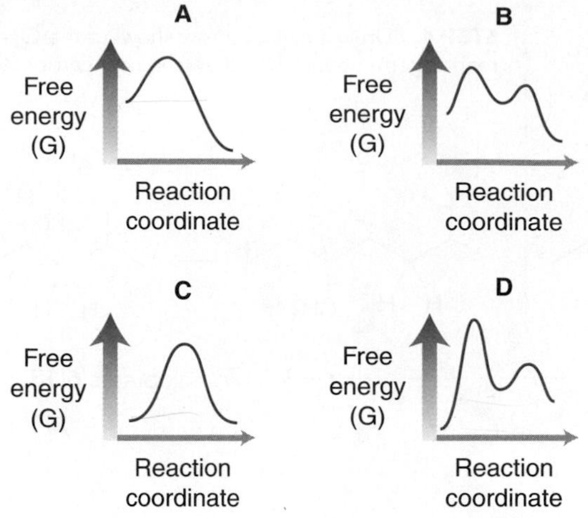

(a) Which diagrams correspond with a two-step mechanism? B, D
(b) Which diagrams correspond with a one-step mechanism? A, C
(c) Compare energy diagrams A and C. Which has a relatively larger E_a? C
(d) Compare diagrams A and C. Which has a negative ΔG? A
(e) Compare diagrams A and D. Which has a positive ΔG? D
(f) Compare all four energy diagrams. Which one exhibits the largest E_a? D
(g) Which processes will have a value of K_{eq} that is greater than 1?
(h) Which process will have a value of K_{eq} that is roughly equal to 1?

6.28 Identify all transition states and intermediates on the following energy diagram:

Free
energy
(G)

Reaction coordinate

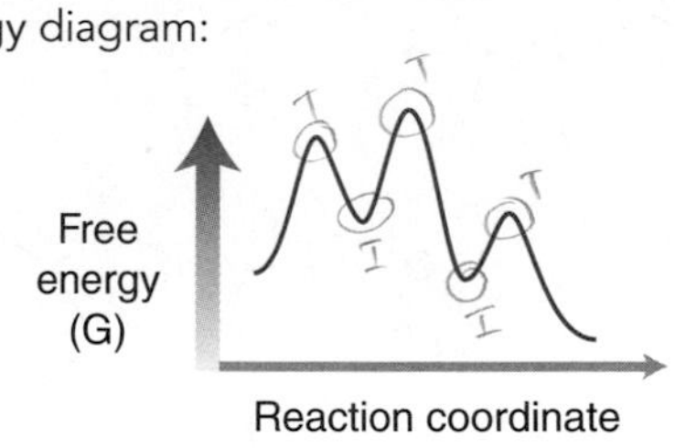

6.29 Consider the following reaction:

reactants

:ÖH ̄ —Cl ⟶ ⌃⌄OH + Cl ̄

reactants

This reaction has been determined to be second order.
(a) What is the rate equation for this reaction? Rate = $k[OH^{\ominus}][CH_3CH_2Cl]$
(b) How will the rate be affected if the concentration of hydroxide is tripled? 3
(c) How will the rate be affected if the concentration of chloroethane is tripled? 3
(d) How will the rate be affected if the temperature is raised by 40°C? increase

$2 \times 2 \times 2 \times 2 = 16$ times

6.30 For each of the following reactions identify the arrow-pushing pattern that is being utilized:

(a)

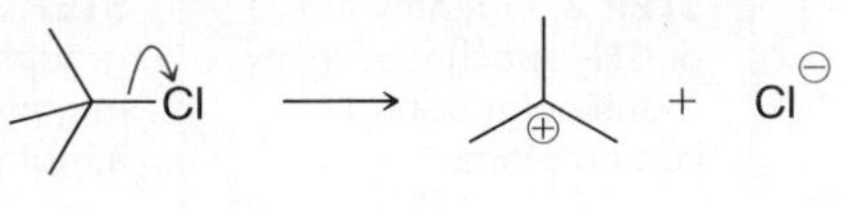

(b)

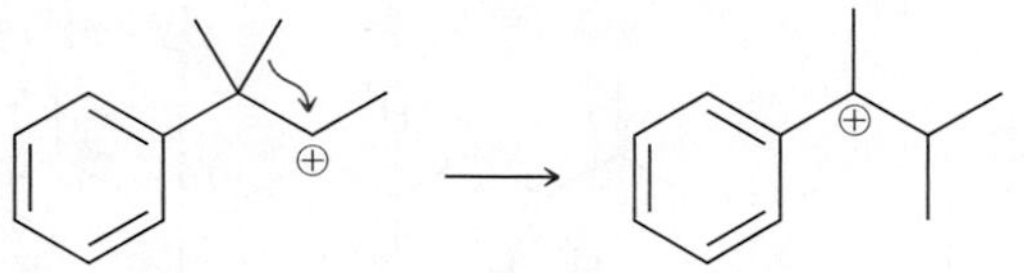

(c)

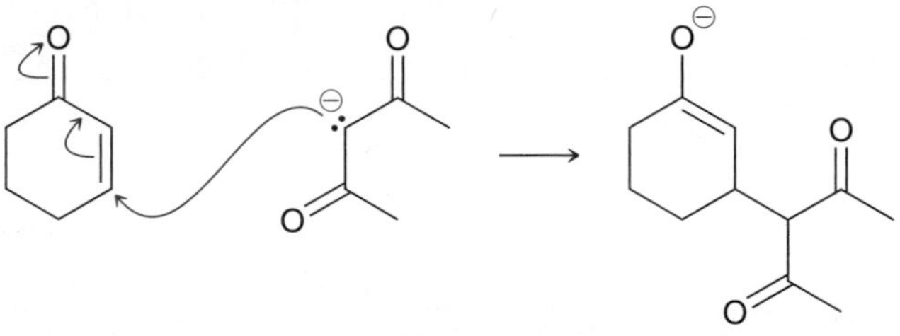

(d)

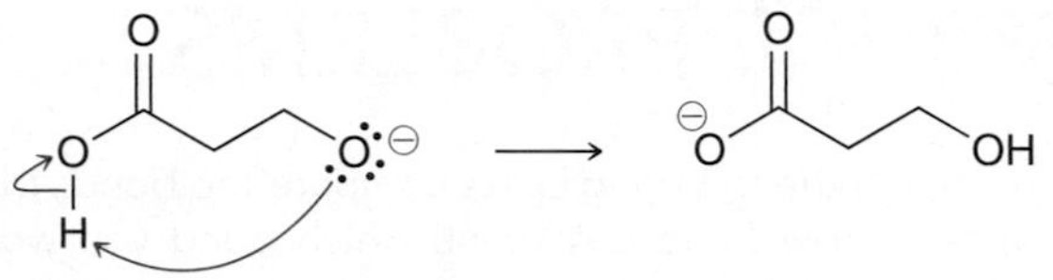

6.31 Rank the three carbocations shown in terms of increasing stability:

(a)

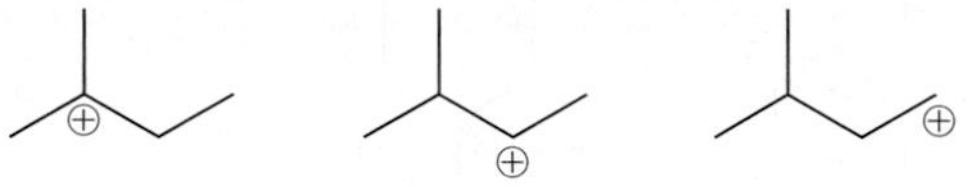

(b)

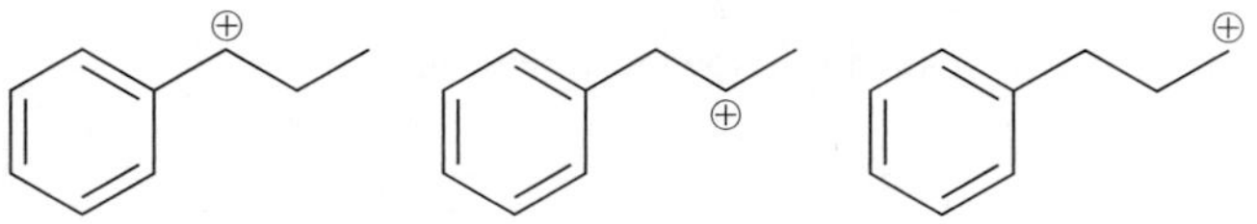

6.32 For the following mechanism, identify the sequence of arrow-pushing patterns:

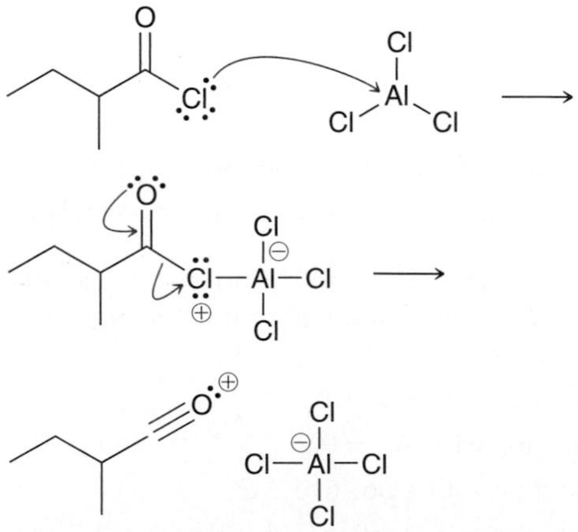

6.33 For the following mechanism, identify the sequence of arrow-pushing patterns:

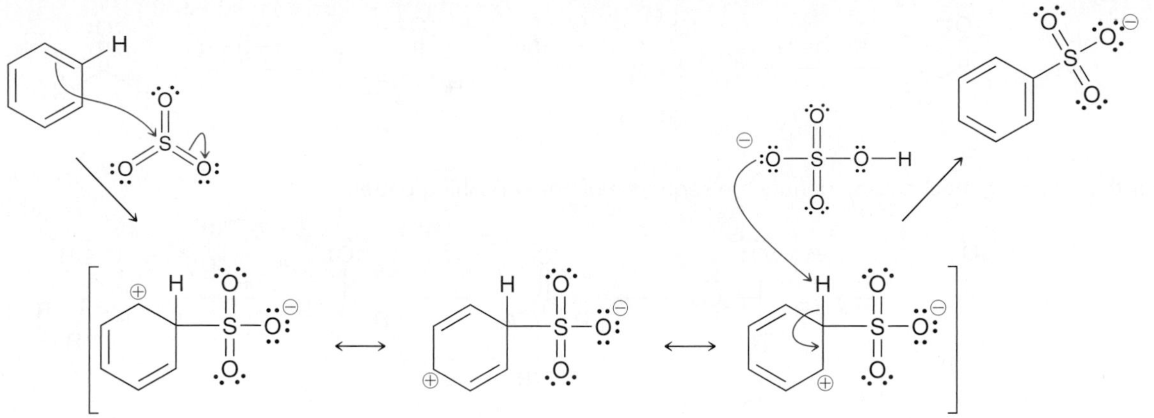

6.34 For the following mechanism, identify the sequence of arrow-pushing patterns:

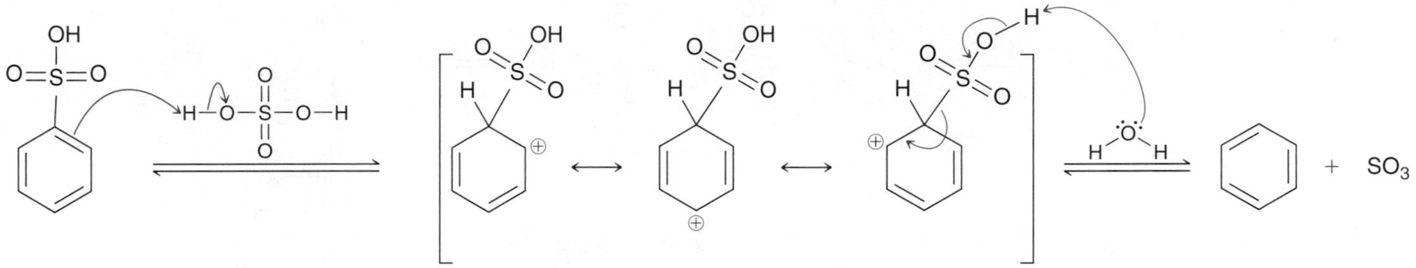

6.35 For the following mechanism, identify the sequence of arrow-pushing patterns:

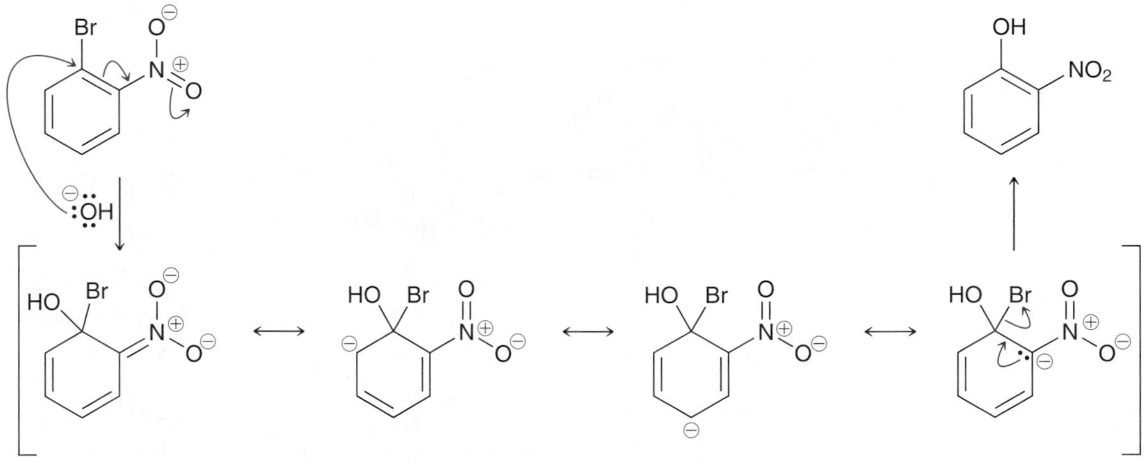

6.36 For the following mechanism, identify the sequence of arrow-pushing patterns:

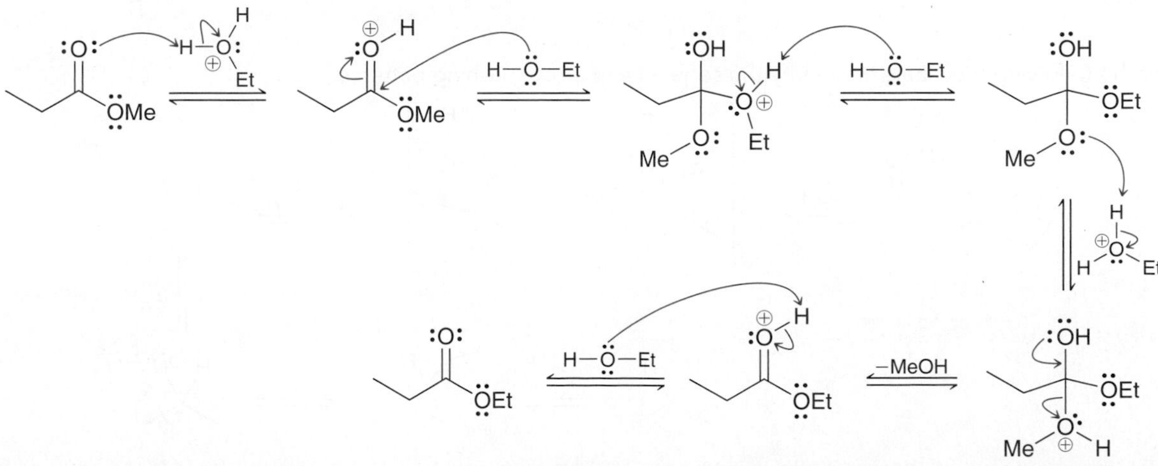

6.37 For the following mechanism, identify the sequence of arrow-pushing patterns:

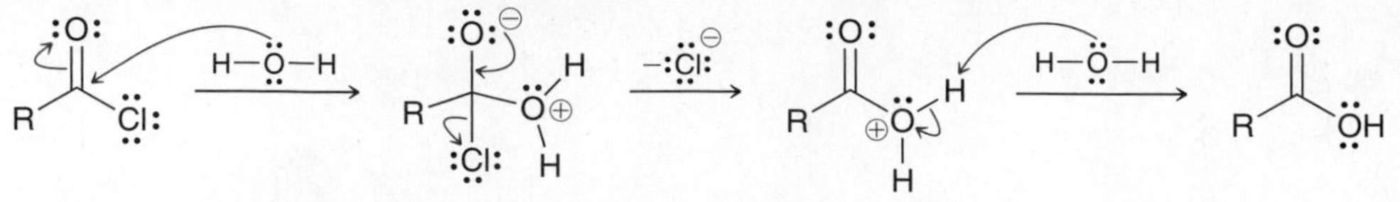

6.38 For the following mechanism, identify the sequence of arrow-pushing patterns:

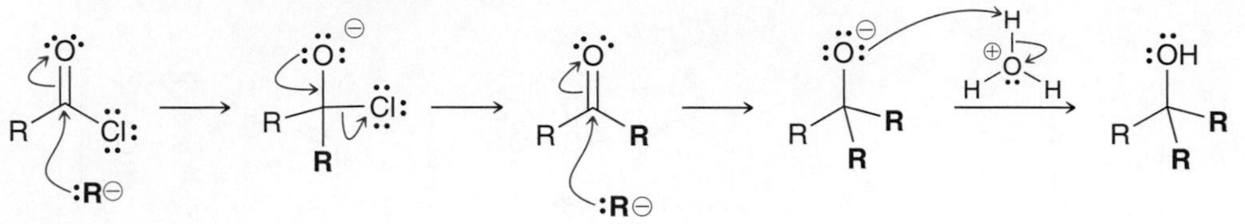

6.39 For the following mechanism, identify the sequence of arrow-pushing patterns:

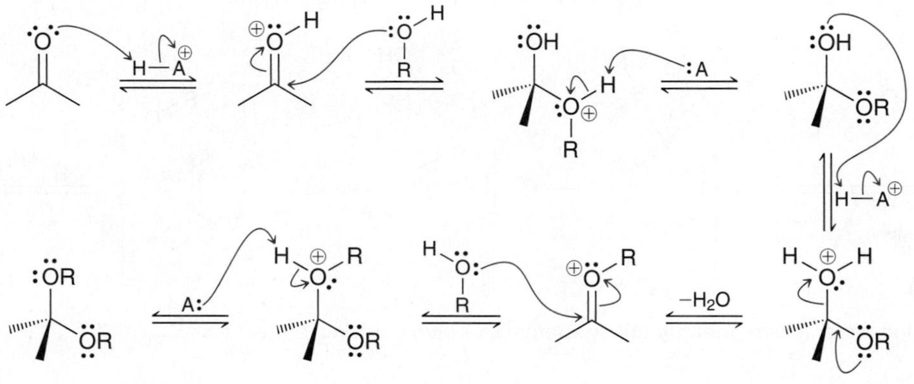

6.40 For the following mechanism, identify the sequence of arrow-pushing patterns:

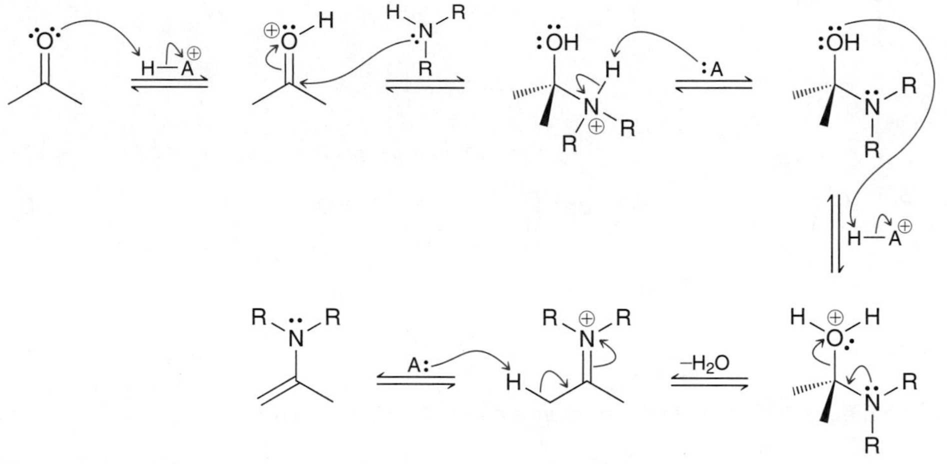

6.41 For the following mechanism, identify the sequence of arrow-pushing patterns:

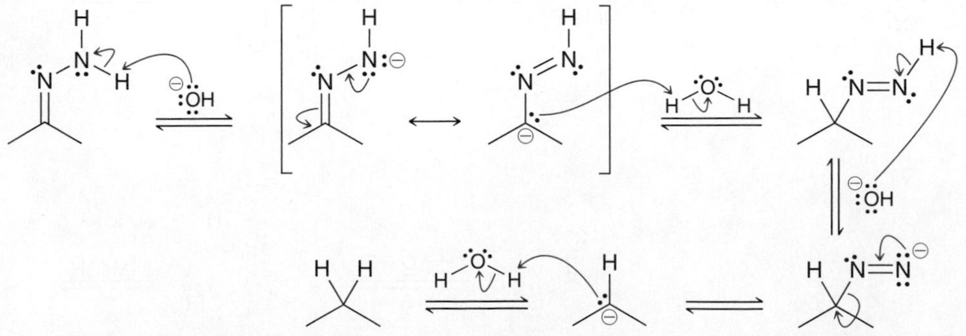

6.42 Draw curved arrows for each step of the following mechanism:

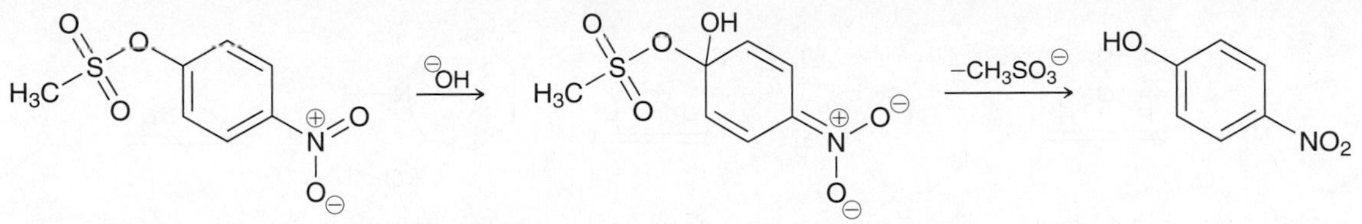

6.43 Draw curved arrows for each step of the following mechanism:

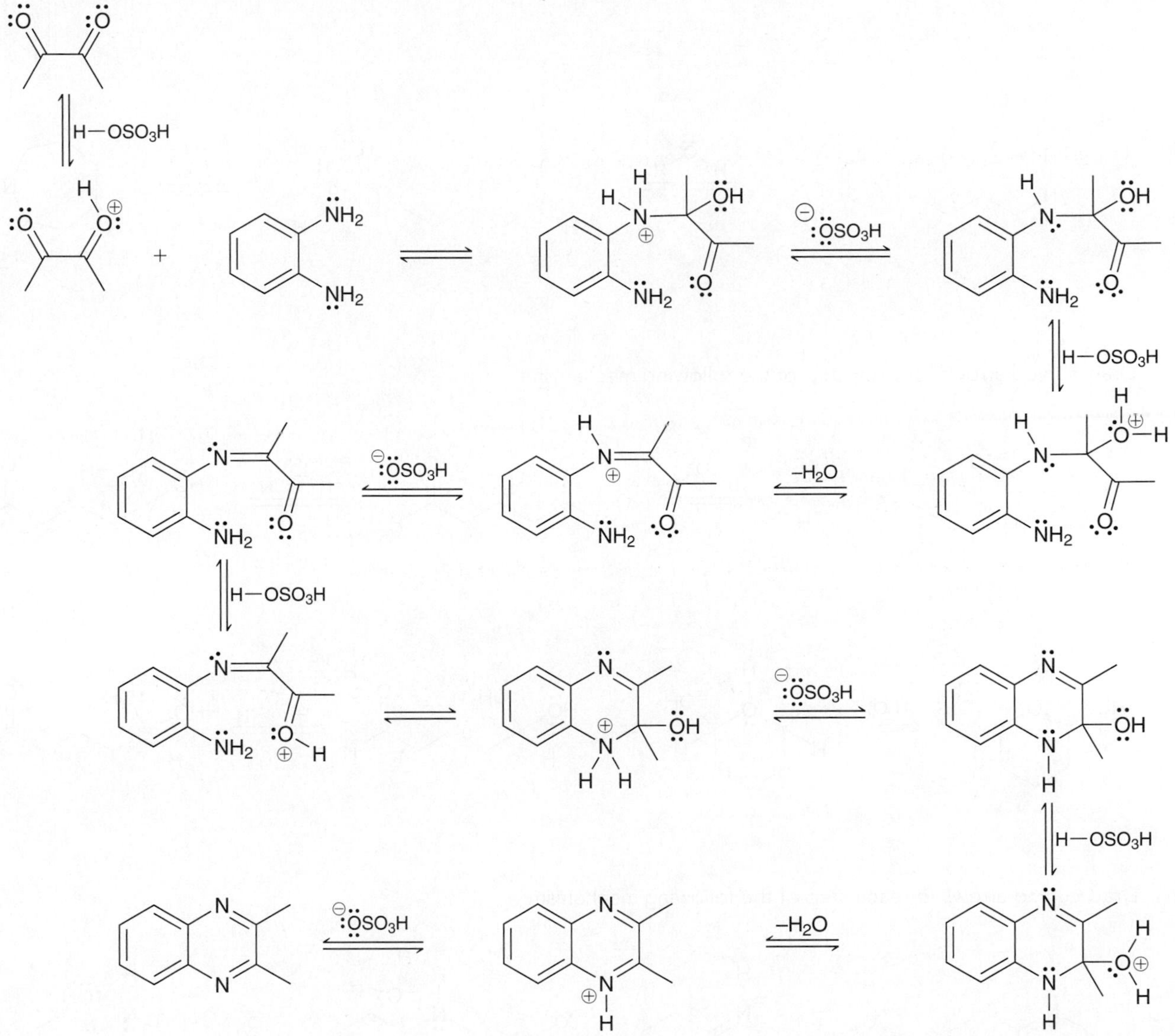

6.44 Draw curved arrows for each step of the following mechanism:

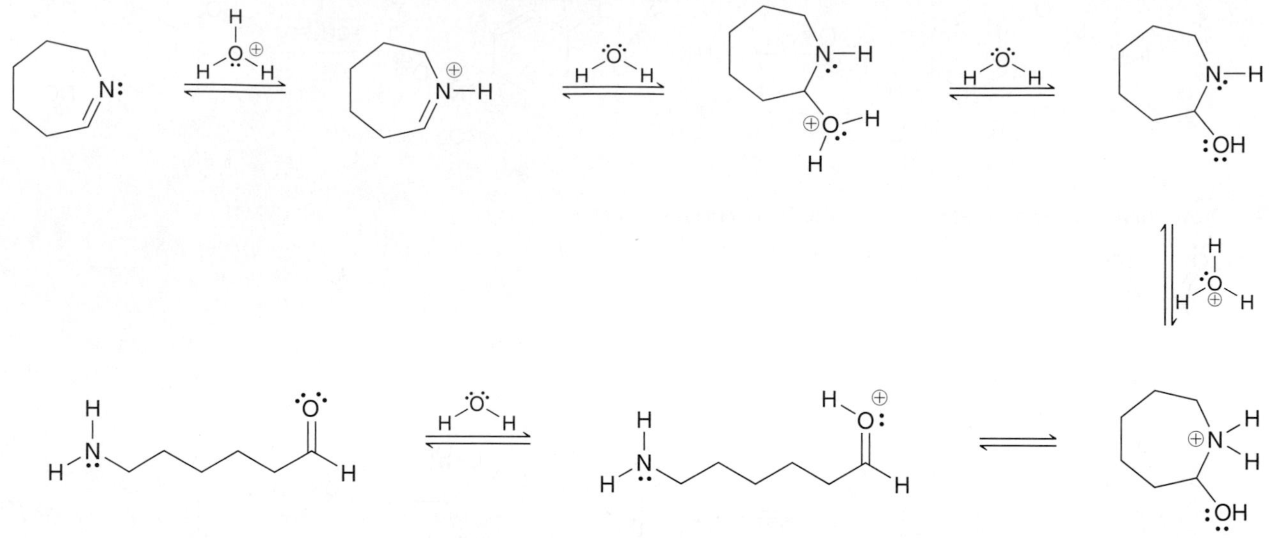

6.45 Draw curved arrows for each step of the following mechanism:

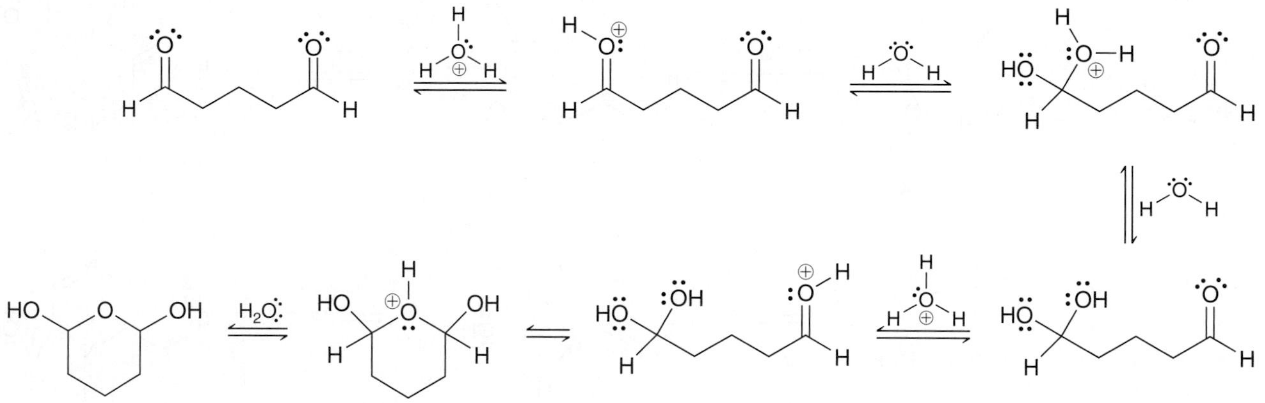

6.46 Draw curved arrows for each step of the following mechanism:

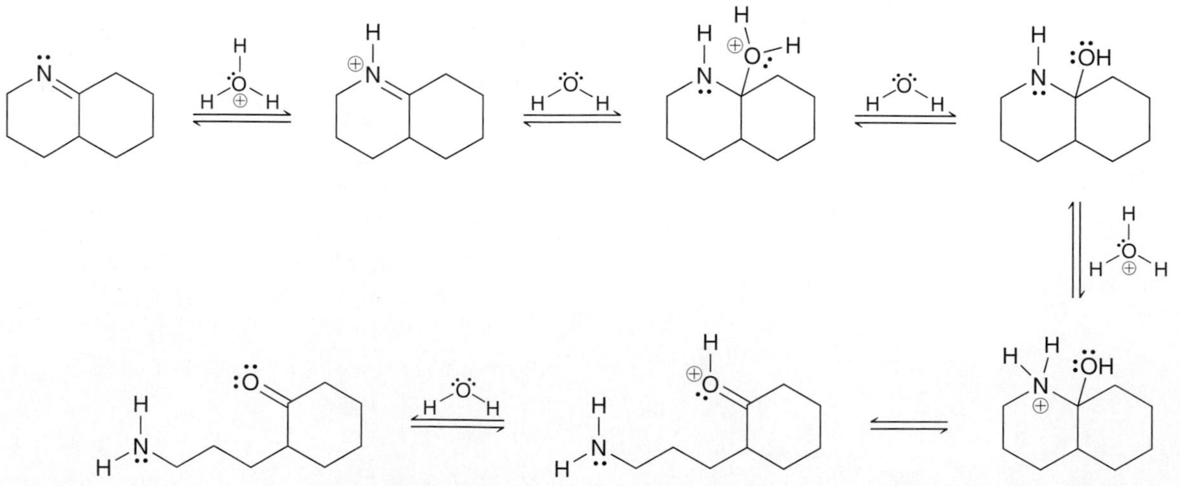

6.47 Draw curved arrows for each step of the following mechanism:

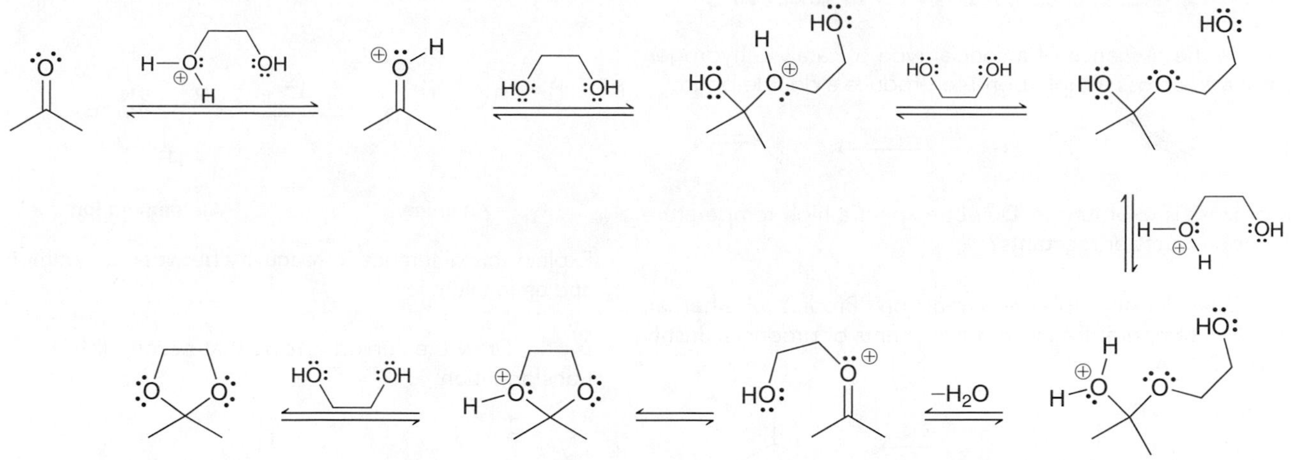

6.48 Predict whether each of the following carbocations will rearrange. If so, draw the expected rearrangement using curved arrows.

(a) (b) (c) (d) (e) (f) (g)

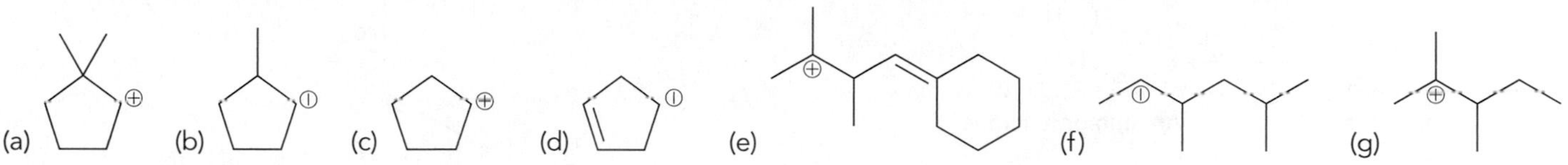

INTEGRATED PROBLEMS

6.49 Consider the following reaction:

$$\text{H—O}^{\ominus} + \text{H}_3\text{C—C(H)—Br} \longrightarrow \text{H—O—C(H)(CH}_3\text{)—H} + \text{:Br}^{\ominus}$$

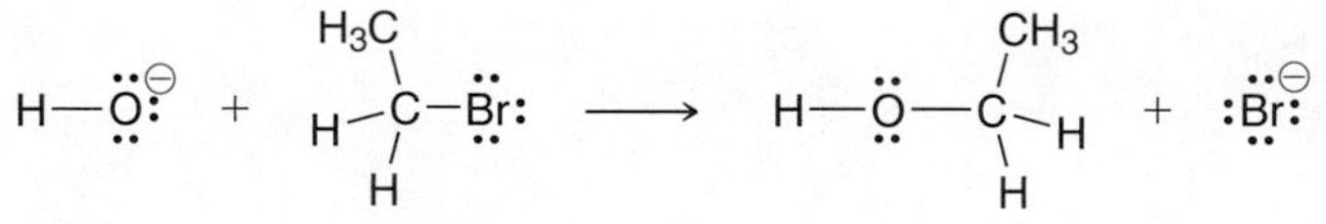

The following rate equation has been experimentally established for this process:

$$\text{Rate} = k[\text{HO}^-][\text{CH}_3\text{CH}_2\text{Br}]$$

The energy diagram for this process is shown below:

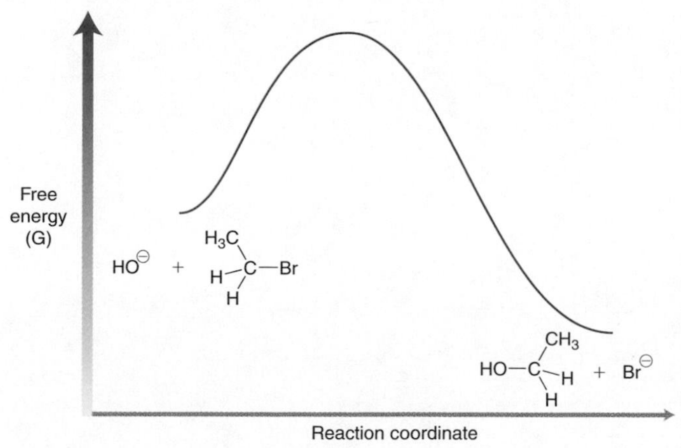

(a) Draw the curved arrows showing a mechanism for this process.

(b) Identify the two characteristic arrow-pushing patterns that are required for this mechanism.

(c) Would you expect this process to be exothermic or endothermic? Explain.

(d) Would you expect ΔS_{sys} for this process to be positive, negative, or approximately zero?

(e) Is ΔG for this process positive or negative?

(f) Would you expect an increase in temperature to have a significant impact on the position of equilibrium (equilibrium concentrations)? Explain.

(g) Draw the transition state of this process, and identify its location on the energy diagram.

(h) Is the transition state closer in structure to the reactants or products? Explain.

(i) Is the reaction first order or second order?

(j) How will the rate be affected if the concentration of hydroxide is doubled?

(k) Will the rate be affected by an increase in temperature?

6.50 Identify whether each of the following factors will affect the rate of a reaction:

(a) K_{eq} (b) ΔG (c) Temperature
(d) ΔH (e) E_a (f) ΔS

CHALLENGE PROBLEMS

6.51 In the presence of a special type of catalyst, hydrogen gas will add across a triple bond to produce a double bond:

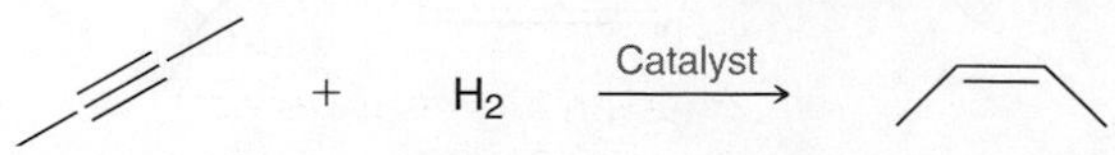

The process is exothermic. Do you expect a high temperature to favor products or reactants?

6.52 Consider the following reaction. Predict whether an increase in temperature will favor reactants or products. Justify your prediction.

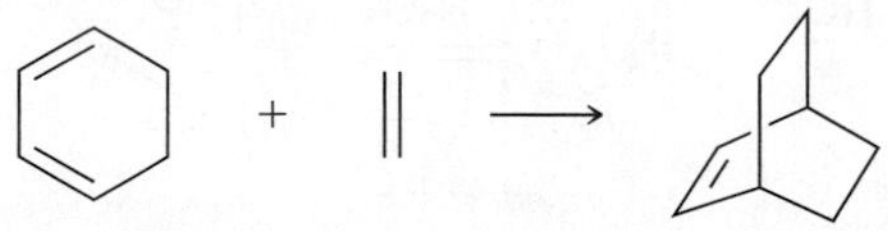

6.53 When an amine is protonated, the resulting ammonium ion is not electrophilic:

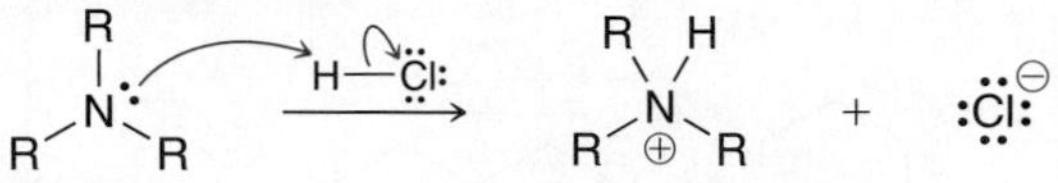

An amine — An ammonium ion

However, when an imine is protonated, the resulting iminium ion is highly electrophilic:

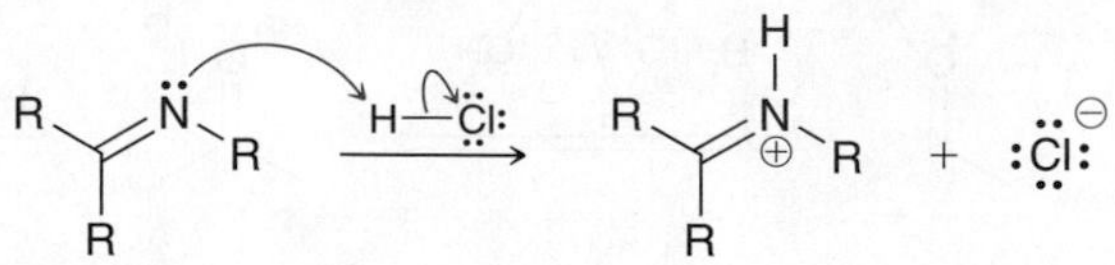

An imine — An iminium ion

Explain this difference in reactivity between an ammonium ion and an iminium ion.

6.54 Draw the curved arrows that accomplish the following transformation:

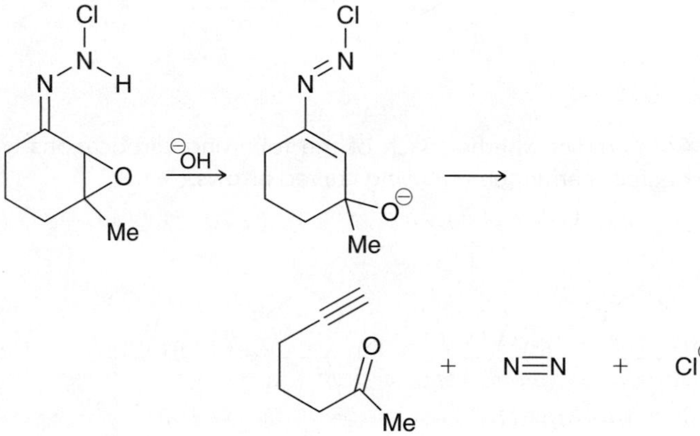

7 Substitution Reactions

DID YOU EVER WONDER...

what chemotherapy is?

As its name implies, chemotherapy is the use of chemical agents in the treatment of cancer. The picture below shows a cancer cell (red) surrounded by many smaller cells (green) that scientists have filled with a cocktail of chemotherapeutic agents. These smaller cells have been engineered to function as transport vehicles that will deliver their cargo to the cancer cell. Dozens of chemotherapy drugs are currently in clinical use, and researchers around the world are currently working on the design and development of new drugs for treating cancer. The primary goal of most chemotherapeutic agents is to cause irreparable damage to cancer cells while causing only minimal damage to normal, healthy cells. Since cancer cells grow much faster than most other cells, many anticancer drugs have been designed to interrupt the growth cycle of fast-growing cells. Unfortunately, some healthy cells are also fast growing, such as hair follicles and skin cells. For this reason, chemotherapy patients often experience a host of side effects, including hair loss and rashes.

The field of chemotherapy began in the mid-1930s, when scientists realized that a chemical warfare agent (sulfur mustard) could be modified and used to attack tumors. The action of sulfur mustard (and its derivatives) was thoroughly investigated and was found to involve a series of reactions called substitution reactions. Throughout this chapter, we will explore many important features of substitution reactions. At the end of the chapter, we will revisit the topic of chemotherapy by exploring the rational design of the first chemotherapeutic agents.

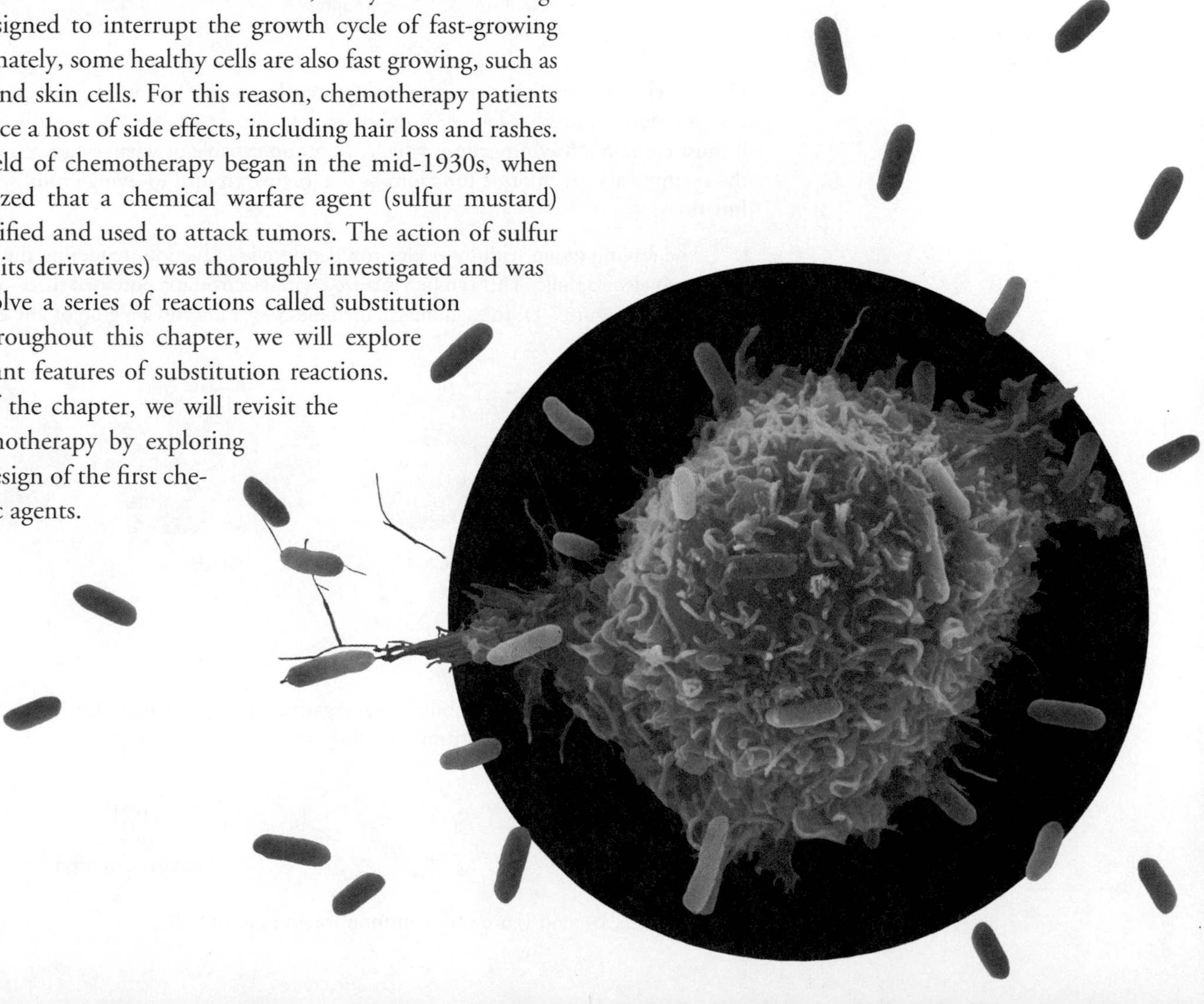

DO YOU REMEMBER?

Before you go on, be sure you understand the following topics.
If necessary, review the suggested sections to prepare for this chapter.

- The Cahn-Ingold-Prelog System (Section 5.3)
- Kinetics and Energy Diagrams (Sections 6.5, 6.6)
- Nucleophiles and Electrophiles (Section 6.7)
- Arrow Pushing and Carbocation Rearrangements (Sections 6.8–6.11)

7.1 Introduction to Substitution Reactions

Substitution reactions involve the exchange of one functional group for another:

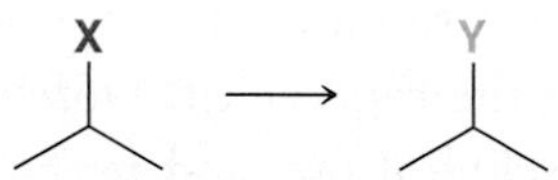

In every substitution reaction, there is an electrophile and a nucleophile:

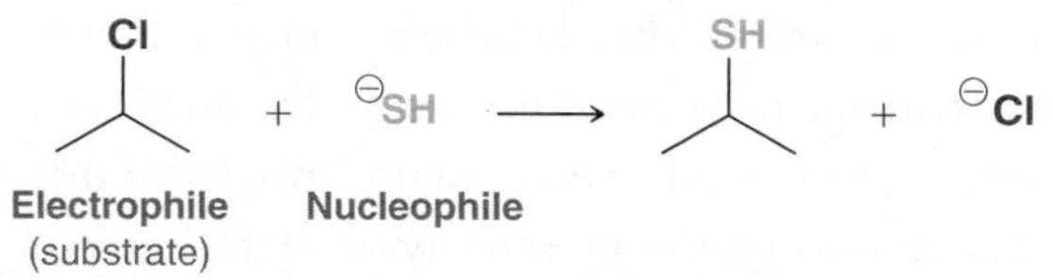

Organic chemists often use the term **substrate** when referring to the electrophile in a substitution reaction. In order for an electrophile to function as a substrate in a substitution reaction, it must contain a **leaving group**, which is a group capable of separating from the substrate. In the example above, chloride functions as the leaving group. A leaving group serves two critical functions:

1. The leaving group withdraws electron density via induction, rendering the adjacent carbon atom electrophilic. This can be visualized with electrostatic potential maps of various methyl halides (Figure 7.1). In each image, the blue color indicates a region of low electron density.

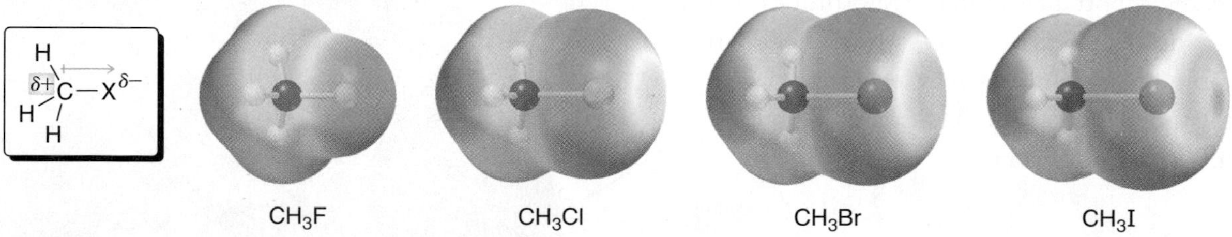

FIGURE 7.1
Electrostatic potential maps of methyl halides.

2. The leaving group can stabilize any negative charge that may develop as the result of the leaving group separating from the substrate:

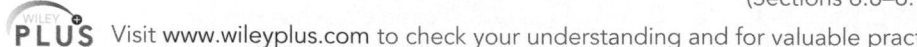

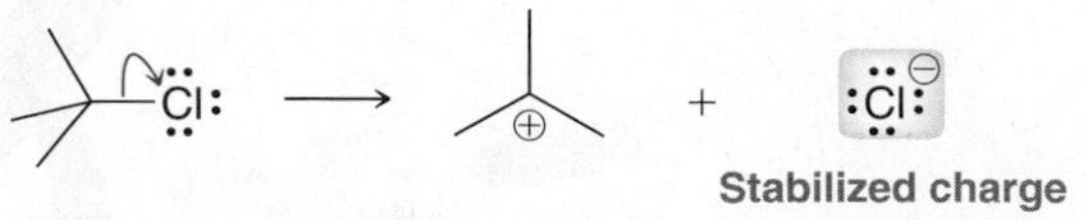

Halogens (Cl, Br, and I) are very common leaving groups.

7.2 Alkyl Halides

Halogenated organic compounds are commonly used as electrophiles in substitution reactions. Although other compounds can also serve as electrophiles, we will focus our attention for now on compounds containing halogens.

Naming Halogenated Organic Compounds

Recall from Section 4.2 that systematic (IUPAC) names of alkanes are assigned using four discrete steps:

1. Identify and name the parent.

2. Identify and name the substituents.

3. Number the parent chain and assign a locant to each substituent.

4. Assemble the substituents alphabetically.

The same exact four-step procedure is used to name compounds that contain halogens, and all of the rules discussed in Chapter 4 apply here as well. Halogens are simply treated as substituents and receive the following names: fluoro-, chloro-, bromo-, and iodo-. Below are two examples:

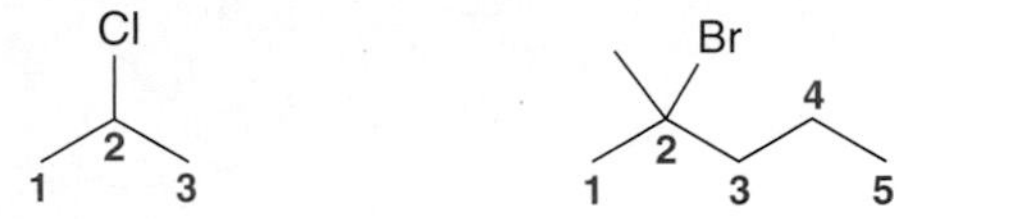

2-chloropropane **2-bromo-2-methylpentane**

As we saw in Chapter 4, the parent is the longest chain, and it should be numbered so that the first substituent receives the lower number:

Correct **Incorrect**

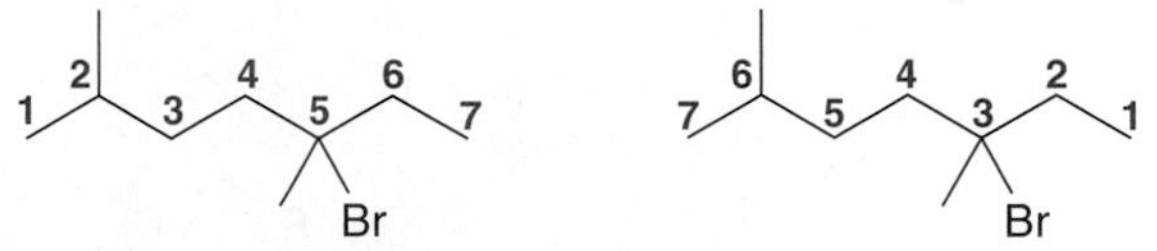

2, 5, 5 beats 3, 3, 6

CONCEPTUAL CHECKPOINT

7.1 Assign a systematic name for each of the following compounds:

(a) _4-chloro-4-ethylheptane_

(b) _1-bromo-1 methylcyclo - hexano_

(c) _4,4-dibromo-1 chloropentane_

(d) _(d)-5-floro-2,2 dimethyl hexane_

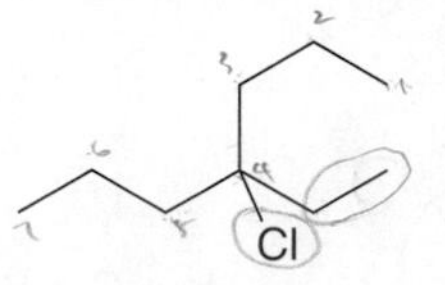

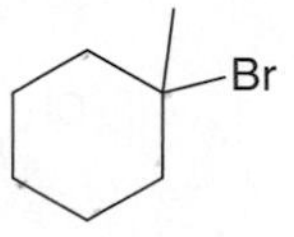

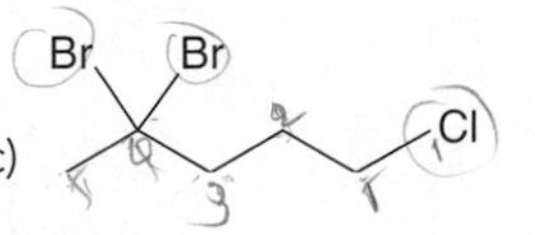

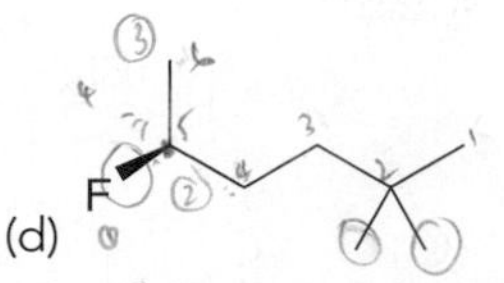

When a chirality center is present in the compound, the configuration must be indicated at the beginning of the name:

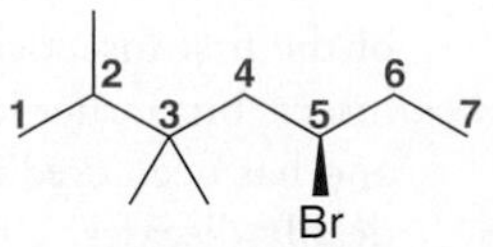

(R)-5-bromo-2,3,3-trimethylheptane

In addition to systematic names, IUPAC nomenclature also recognizes common names for many halogenated organic compounds.

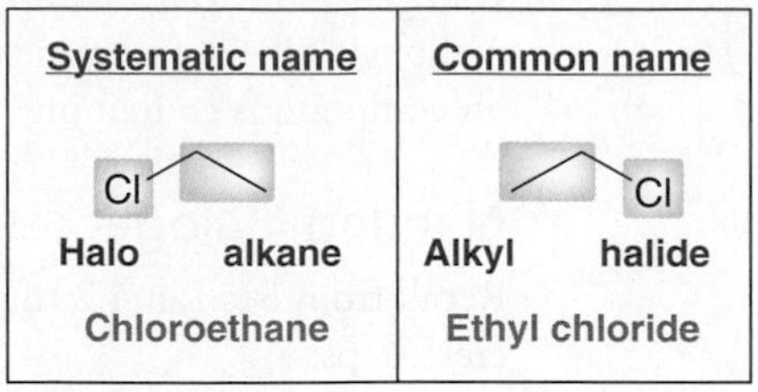

The systematic name treats a halogen as a substituent, calling the compound a **haloalkane.** The common name treats the compound as an alkyl substituent connected to a halide, and the compound is called an **alkyl halide** or an **organohalide**.

Structure of Alkyl Halides

Each carbon atom is described in terms of its proximity to the halogen using letters of the Greek alphabet. The **alpha** (α) **position** is the carbon atom connected directly to the halogen, while the **beta** (β) **positions** are the carbon atoms connected to the alpha position:

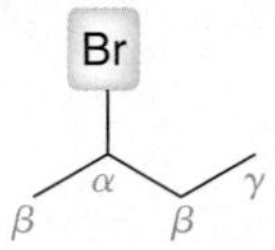

An alkyl halide will have only one α position, but there can be as many as three β positions. This chapter focuses on reactions that occur at the α position, and the next chapter will focus on reactions involving the β position.

Alkyl halides are classified as **primary** (1°), **secondary** (2°), or **tertiary** (3°) based on the number of alkyl groups connected to the α position.

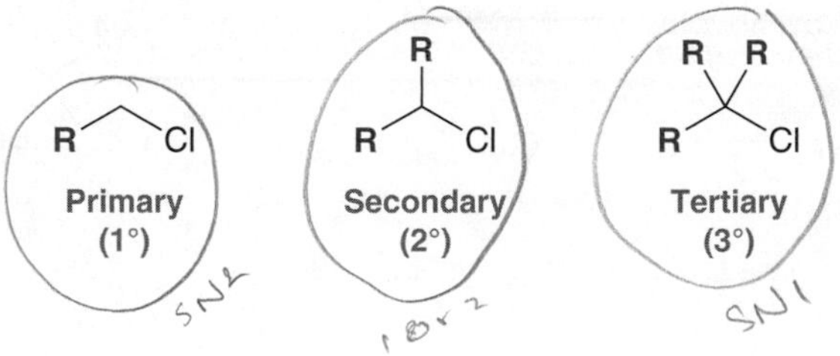

FIGURE 7.2
Classification of alkyl halides as primary, secondary, or tertiary.

Uses of Organohalides

Many organohalides are toxic and have been used as insecticides:

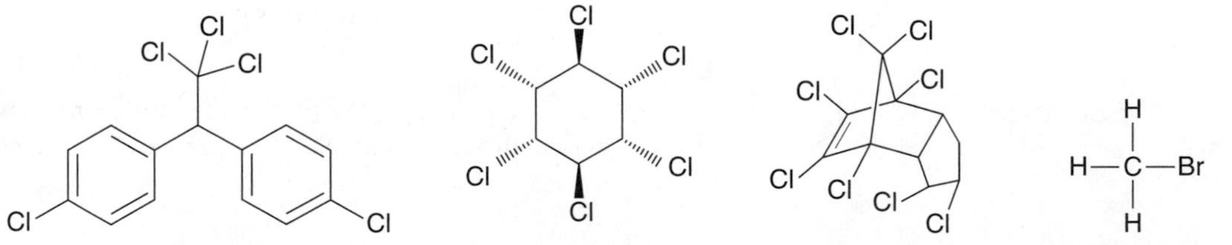

DDT **Lindane** **Chlordane** **Methyl bromide**

DDT (dichlorodiphenyltrichloroethane) was developed in the late 1930s and became one of the first insecticides to be used around the globe. It was found to exhibit strong toxicity for insects but rather low toxicity for mammals. DDT was used as an insecticide for many decades and has been credited with saving more than half a billion lives by killing mosquitos that carry deadly diseases. Unfortunately, it was found that DDT does not degrade quickly and persists in the environment. Rising concentrations of DDT in wildlife began to threaten the survival of many species. In response, the Environmental Protection Agency (EPA) banned the use of DDT in 1972, and it was replaced with other, environmentally safer, insecticides.

Lindane is used in shampoos designed to treat head lice, while chlordane and methyl bromide have been used to prevent and treat termite infestations. The use of methyl bromide has recently been regulated due to its role in the destruction of the ozone layer (for more on the hole in the ozone layer, see Section 11.8).

Organohalides are particularly stable compounds, and many of them, like DDT, persist and accumulate in the environment. PCBs (polychlorinated biphenyls) represent another well-known example. Biphenyl is a compound that can have up to 10 substituents:

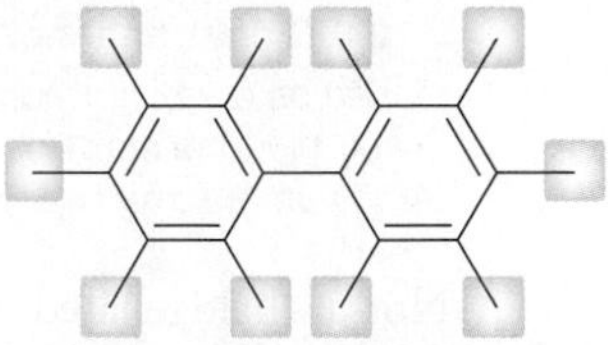

Biphenyl

PCBs are compounds in which many of these positions contain chlorine atoms. PCBs were originally produced as coolants and insulating fluids for industrial transformers and capacitors. They were also used as hydraulic fluids and as flame retardants. But their accumulation in the environment began to threaten wildlife, and their use was banned.

The above examples have contributed to the bad reputation of organohalides. As a result, organohalides are often viewed as man-made poisons. However, research over the last 20 years has indicated that organohalides are actually more common in nature than had previously been thought. For example, methyl chloride is the most abundant organohalide in the atmosphere. It is produced in large quantities by evergreen trees and marine organisms, and it is consumed by many bacteria, such as *Hyphominocrobium* and *Methylobacterium*, that convert methyl chloride into CO_2 and Cl^-.

Many organohalides are also produced by marine organisms. Over 4000 such compounds have already been identified, and several hundred new compounds are discovered each year. Here are two examples:

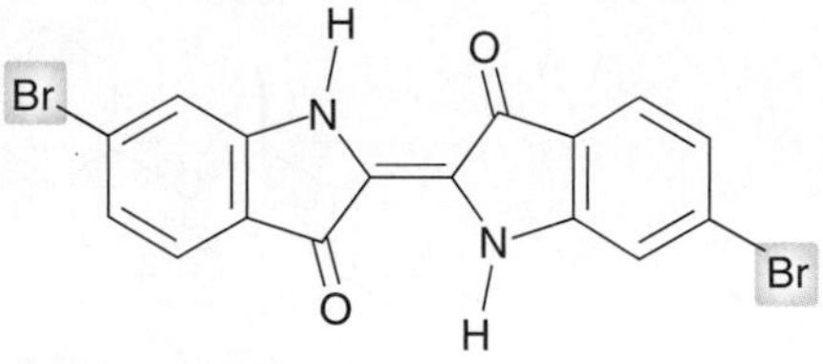

Tyrian purple

Isolated from the sea snail *Hexaplex trunculus*, this compound is one of the oldest known dyes and was used to make royal clothing thousands of years ago

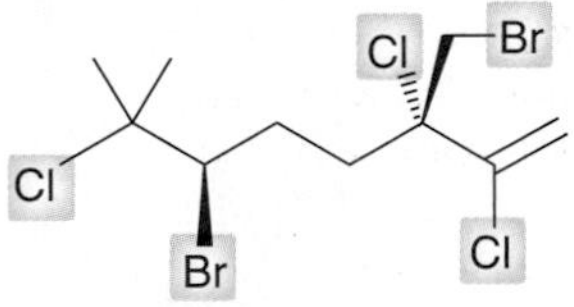

Halomon

Isolated from the red algae *Portieria hornemann* this compound is currently in clinical trials for use as an antitumor agent

Organohalides serve a variety of functions in living organisms. In sponges, corals, snails, and seaweeds organohalides are used as a defense mechanism against predators (a form of chemical warfare). Here are two such examples:

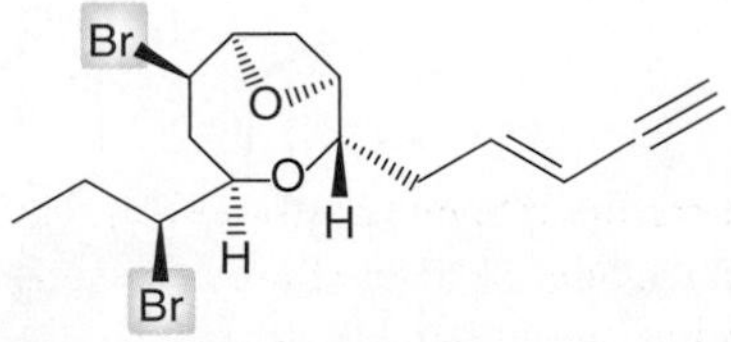

(3*E*)-Laureatin

Used by the red algae *Laurencia nipponina*

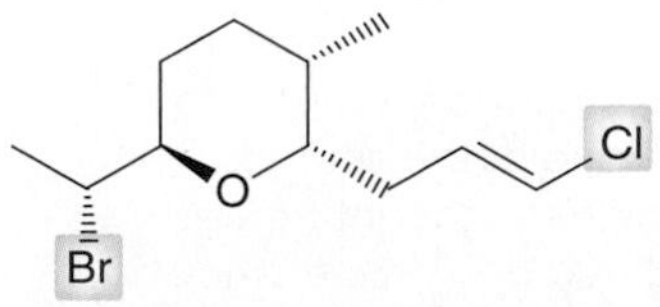

Kumepaloxane

Used by the snail *Haminoea cymbalum*

Both of these compounds are used to ward off predators. In many kinds of organisms organohalides act as hormones (chemical messengers that act only on specific target cells). Examples include the following:

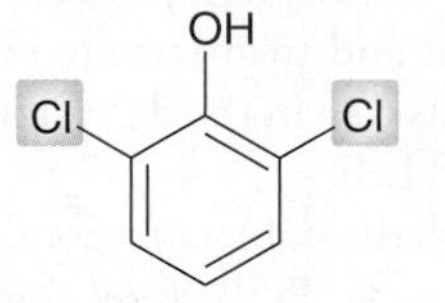

2,6-Dichlorophenol
Used as a sex hormone by the lone star tick *Amblyomma americanum*

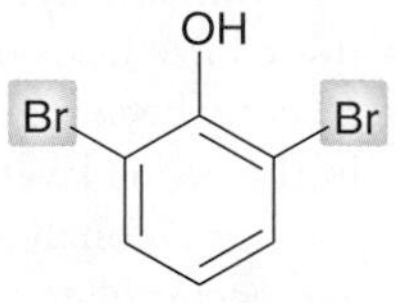

2,6-Dibromophenol
Isolated from the acorn worm *Balanoglossus biminiensis,* likely used as a hormone

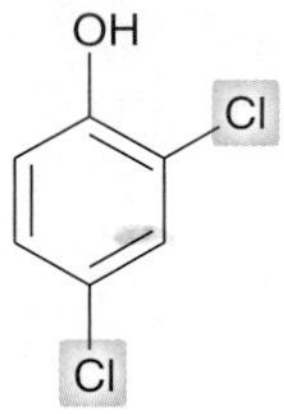

2,4-Dichlorophenol
Used as a growth hormone by *Penicillium* molds

Not all halogenated compounds are toxic. In fact, many organohalides have clinical applications. For example, the following compounds are widely used and have contributed much to the improvement of physical and psychological health:

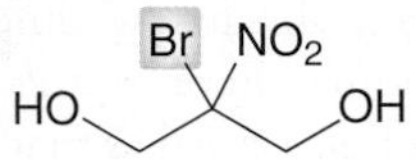

Bronopol
(2-Bromo-2-nitropropane-1,3-diol)

A powerful antimicrobial compound safe enough to use in baby-wipes

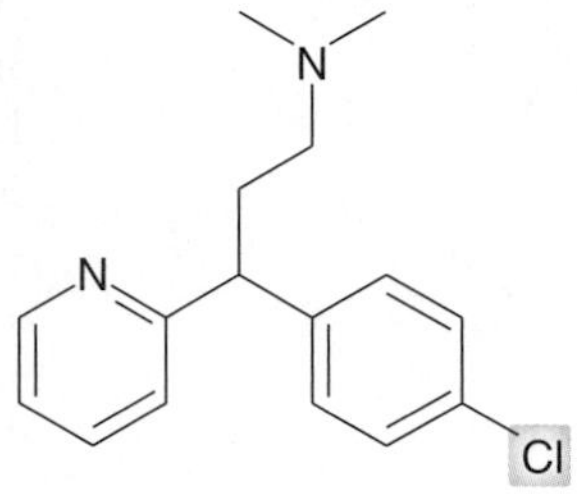

Chlorpheniramine

An antihistamine, sold under the trade name Chlor-Trimeton®

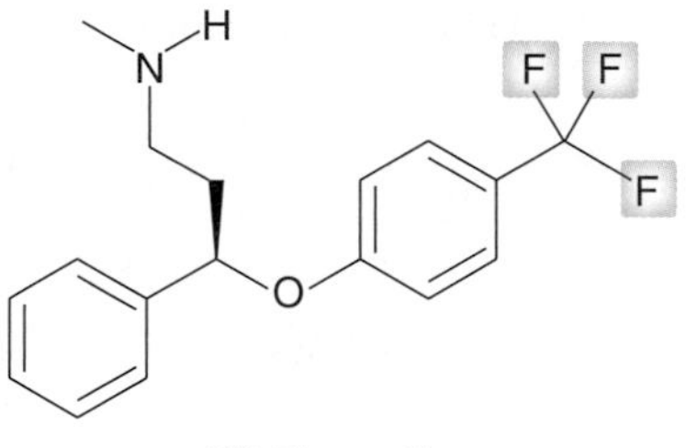

(*R*)-Fluoxetine

An antidepressant, sold under the trade name Prozac®

BY THE WAY
Sucralose was discovered by accident in 1976 when a British company (Tate and Lyle) was conducting research on potential uses of chlorinated sugars. A foreign graduate student participating in the research misunderstood a request to "test" one of the compounds and instead thought he was being asked to "taste" the compound. The graduate student reported an intensely sweet taste, and it was later found to be safe to consume.

Some organohalides have even been used in the food industry. Consider, for example, the structure of sucralose, shown here. Sucralose contains three chlorine atoms, but it is known not to be toxic. It is several hundred times sweeter than sugar and is sold as an artificial, low-calorie sweetener under the trade name Splenda®.

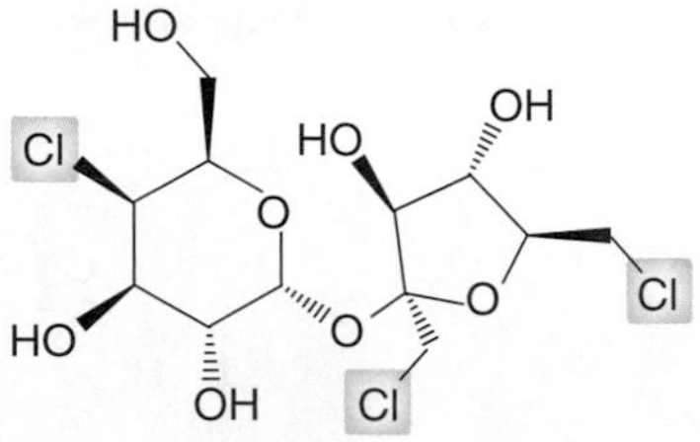

Sucralose
An artificial sweetener, sold under the trade name Splenda®

7.3 Possible Mechanisms for Substitution Reactions

Recall from Chapter 6 that ionic mechanisms are comprised of only four types of arrow-pushing patterns (Figure 7.3). All four of these steps will be used in this chapter, so it might be wise to review Sections 6.7–6.10.

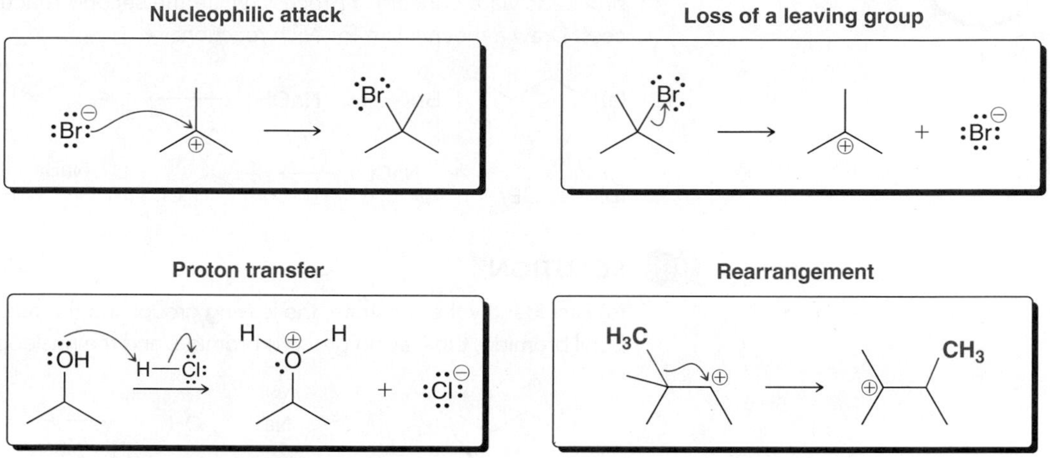

FIGURE 7.3 The four arrow-pushing patterns for ionic processes.

Every substitution reaction exhibits at least two of the four patterns—nucleophilic attack and loss of a leaving group:

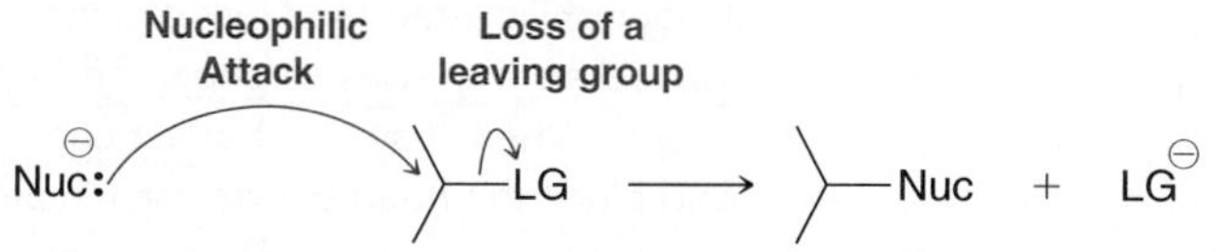

But consider the order of these events. Do they occur simultaneously (in a concerted fashion), as shown above, or do they occur in a stepwise fashion, as shown below?

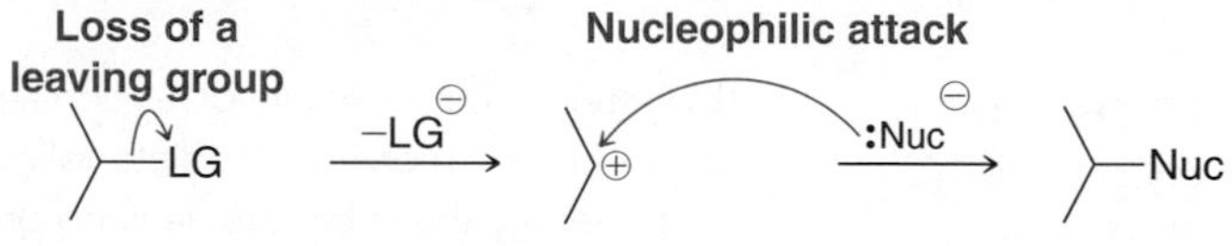

In the stepwise mechanism, the leaving group leaves, generating an intermediate carbocation, which is then attacked by the nucleophile. The nucleophile cannot attack before the leaving group leaves, because that would violate the octet rule:

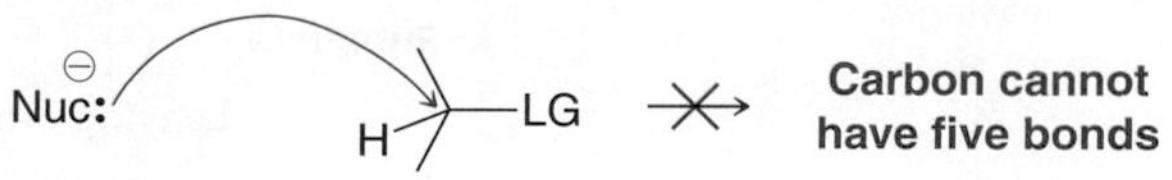

Therefore, there are only two possible mechanisms for a substitution reaction:

- In a *concerted process*, nucleophilic attack and loss of the leaving group occur simultaneously.
- In a *stepwise process*, loss of the leaving group occurs first followed by nucleophilic attack.

We will see that both of these mechanisms do occur, but under different conditions. We will explore each mechanism in the next section, but first let's practice drawing the curved arrows for the two mechanisms.

SKILLBUILDER

7.1 DRAWING THE CURVED ARROWS OF A SUBSTITUTION REACTION

LEARN the skill

Below are two substitution reactions. Experimental evidence suggests that the first reaction proceeds via a concerted process, while the second reaction proceeds via a stepwise process. Draw a mechanism for each reaction:

(a) 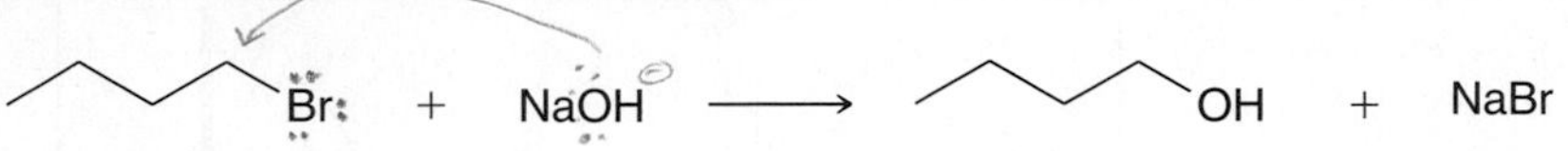
Br
+ NaOH
OH
+ NaBr

(b) 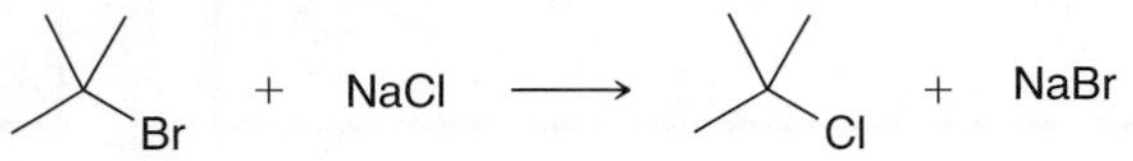
Br
+ NaCl
Cl
+ NaBr

SOLUTION

(a) First identify the substrate, the leaving group, and the nucleophile. Here, the substrate is butyl bromide, the leaving group is bromide, and the nucleophile is a hydroxide ion:

Step 1
Identify the substrate, leaving group, and nucleophile.

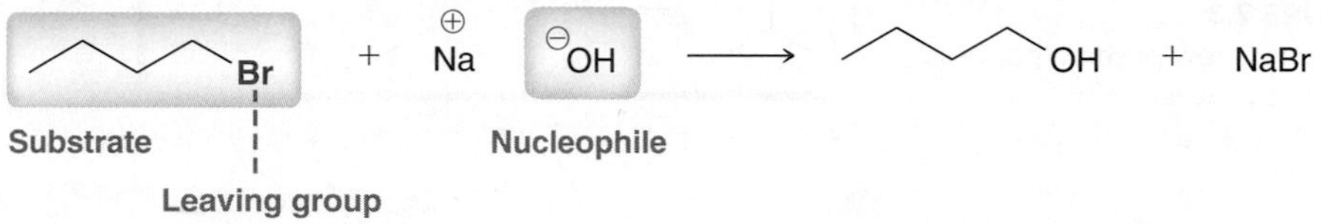
Br
Na
OH
OH
+ NaBr
Substrate
Nucleophile
Leaving group

When you see NaOH, remember that the reagent is a hydroxide ion ($HO^{\ominus}$). Na^+ is the counter-ion, and its role in the reaction does not concern us in most cases. In a concerted process, nucleophilic attack and loss of a leaving group occur simultaneously. This process requires two curved arrows—one to show the nucleophilic attack and one to show the loss of the leaving group. When drawing the first curved arrow, place the tail on a lone pair of the nucleophile, and place the head on the carbon atom bearing the leaving group:

Step 2
Draw two curved arrows, showing nucleophilic attack and loss of the leaving group.

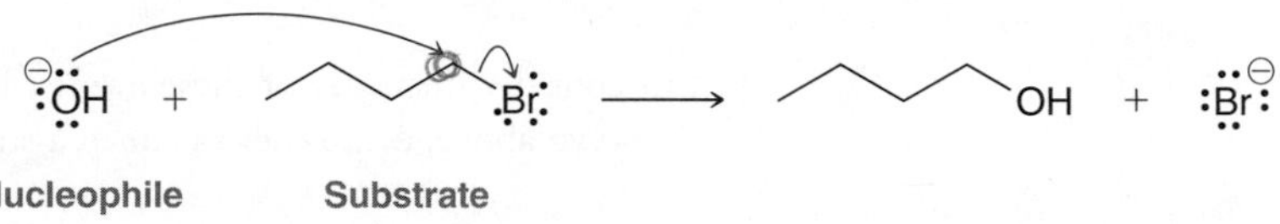
OH
+
Br
OH
+
Br
Nucleophile
Substrate

WATCH OUT
Be very precise in placing the head and tail of every curved arrow.

(b) A stepwise process involves two separate mechanistic steps: (1) loss of a leaving group to form a carbocation intermediate followed by (2) nucleophilic attack. To draw these steps, we must identify the substrate, leaving group, and nucleophile. Here, the substrate is *tert*-butyl bromide, the leaving group is a bromide ion, and the nucleophile is a chloride ion:

Step 1
Identify the substrate, leaving group, and nucleophile.

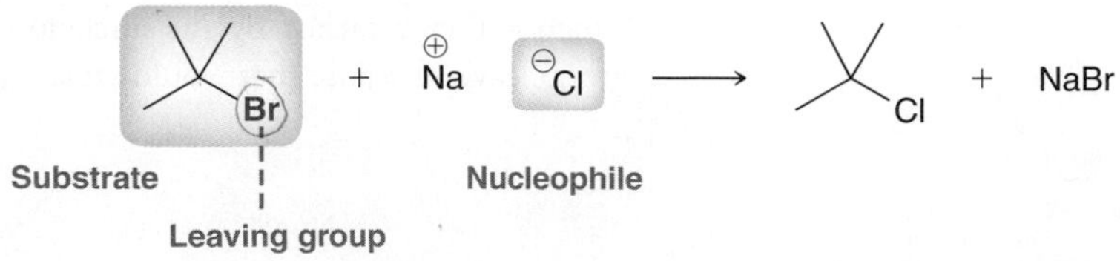
Br
+ Na
Cl
Cl
+ NaBr
Substrate
Nucleophile
Leaving group

Step 2
Draw a curved arrow showing loss of the leaving group, and then draw the resulting carbocation.

The first step of the mechanism requires one curved arrow showing the loss of the leaving group. The tail of this curved arrow is placed on the bond that is broken (the C—Br bond); the head of the arrow is placed on the bromine atom.

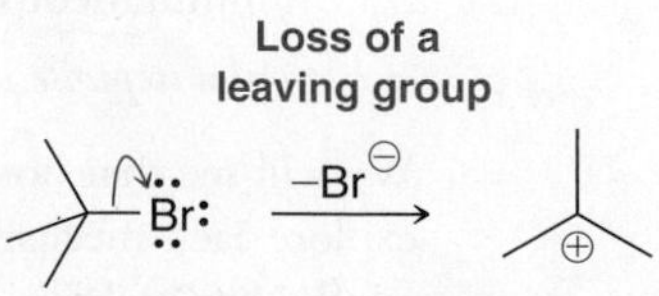
Loss of a
leaving group
Br
−Br

Step 3
Draw a curved arrow showing a nucleophilic attack.

The second step of the mechanism requires one curved arrow showing the nucleophilic attack in which the carbocation intermediate is captured by the nucleophile (chloride):

Nucleophilic attack

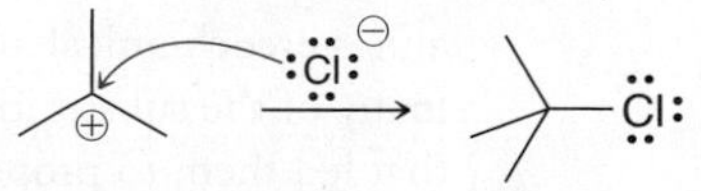

BY THE WAY
Draw all three groups of a tertiary carbocation as far apart from each other as possible:

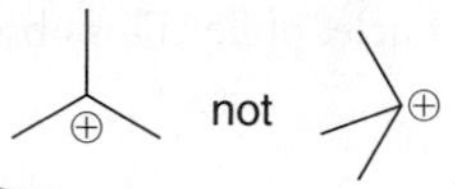

The complete mechanism can therefore be drawn like this:

Loss of a leaving group **Nucleophilic attack**

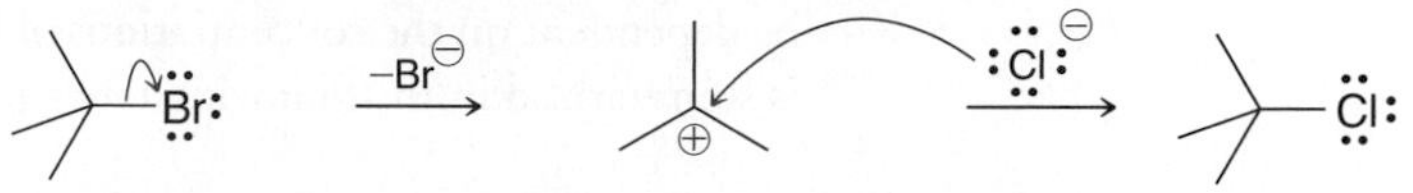

PRACTICE the skill **7.2** For each of the following reactions, assume a concerted process is taking place and draw the mechanism:

(a) (b)

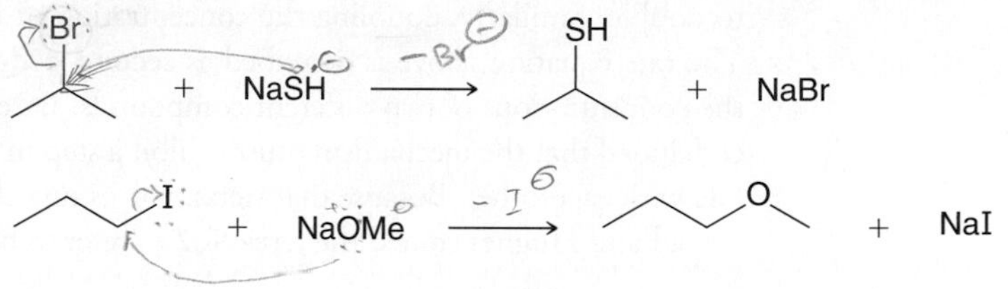

7.3 For each of the following reactions assume a stepwise process is taking place and draw the mechanism:

(a) (b)

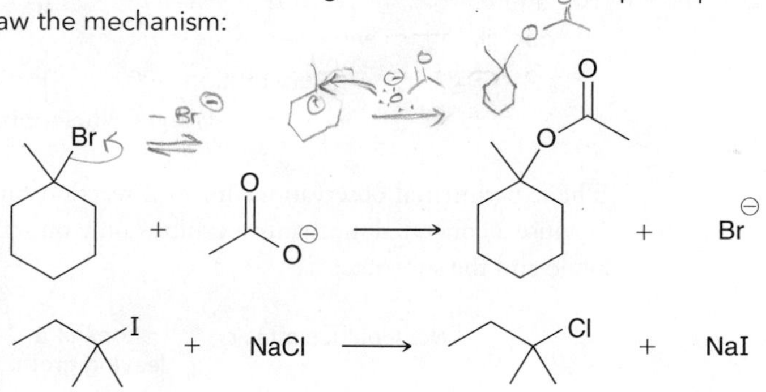

APPLY the skill **7.4** When a nucleophile and electrophile are tethered to each other (that is, both present in the same compound), an *intramolecular substitution reaction* can occur, as shown. Assume that this reaction occurs via a concerted process and draw the mechanism.

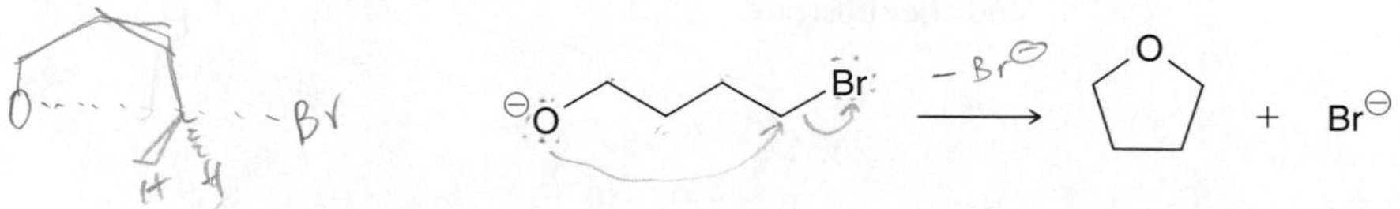

7.5 For the substitution reaction shown below, assume a stepwise process is taking place and draw the mechanism. (HINT: Review the rules for drawing resonance structures, Section 2.10)

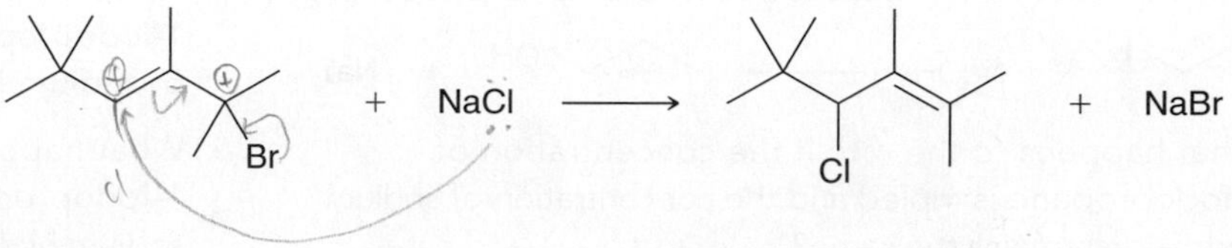

need more **PRACTICE?** **Try Problem 7.64a**

7.4 The S_N2 Mechanism

During the 1930s, Sir Christopher Ingold and Edward D. Hughes (University College, London) investigated substitution reactions in an effort to elucidate their mechanism. Based on kinetic and stereochemical observations, Ingold and Hughes proposed a concerted mechanism for many of the substitution reactions that they investigated. We will now explore the observations that led them to propose a concerted mechanism.

Kinetics

For most of the reactions that they investigated, Ingold and Hughes found the rate of reaction to be dependent on the concentrations of both the substrate and the nucleophile. This observation is summarized in the following rate equation:

$$\text{Rate} = k\,[\text{substrate}]\,[\text{nucleophile}]$$

Specifically, they found that doubling the concentration of the nucleophile caused the reaction rate to double. Similarly, doubling the concentration of the substrate also caused the rate to double. The rate equation above is described as **second order**, because the rate is linearly dependent on the concentrations of two different compounds. Based on their observations, Ingold and Hughes concluded that the mechanism must exhibit a step in which the substrate and the nucleophile collide with each other. Because that step involves two chemical entities, it is said to be **bimolecular**. Ingold and Hughes coined the term $\mathbf{S_N2}$ to refer to bimolecular substitution reactions:

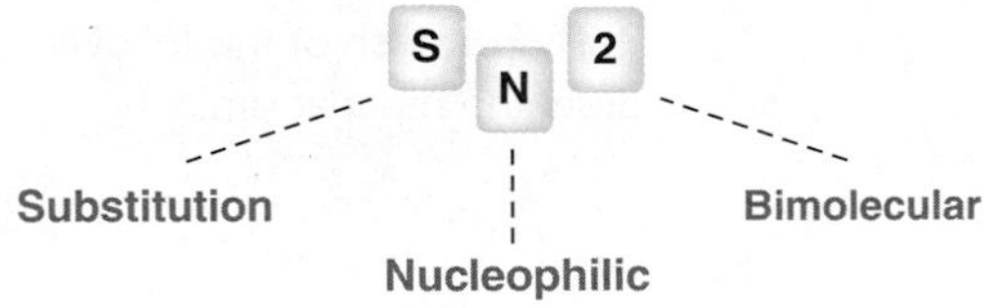

The experimental observations for S_N2 reactions are consistent with a concerted mechanism, because a concerted mechanism exhibits only one mechanistic step, involving both the nucleophile and the substrate:

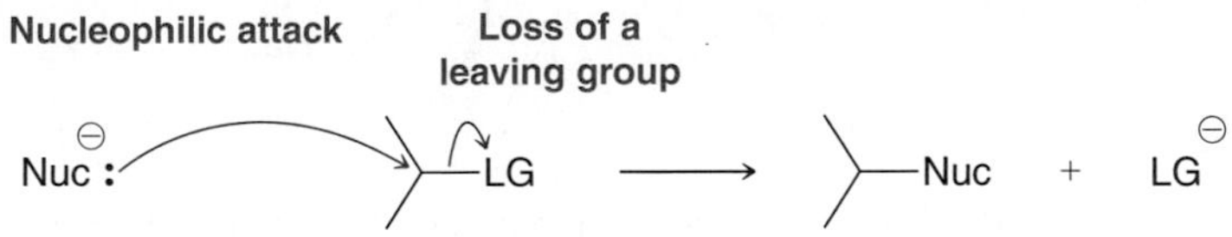

It makes sense that the rate should be dependent on the concentrations of both the nucleophile and the substrate.

CONCEPTUAL CHECKPOINT

7.6 The reaction below exhibits a second-order rate equation:

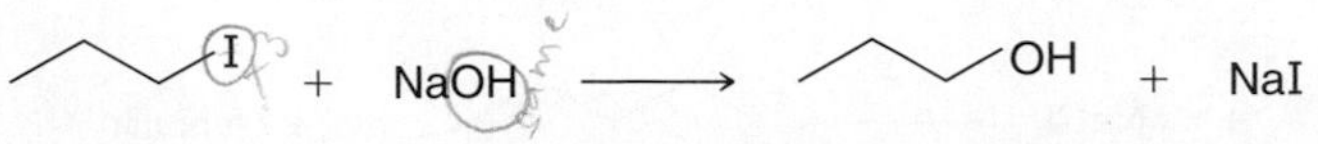

(a) What happens to the rate if the concentration of 1-iodopropane is tripled and the concentration of sodium hydroxide remains the same? substrate is tripled so the whole rxn is tripled.

(b) What happens to the rate if the concentration of 1-iodopropane remains the same and the concentration of sodium hydroxide is doubled? doubled

(c) What happens to the rate if the concentration of 1-iodopropane is doubled and the concentration of sodium hydroxide is tripled? 6 times faster because the rate depens on both the nucleophile & the substrate.

Stereospecificity of S_N2 Reactions

There is another crucial piece of evidence that led Ingold and Hughes to propose the concerted mechanism. When the α position is a chirality center, a change in configuration is generally observed, as illustrated in the following example:

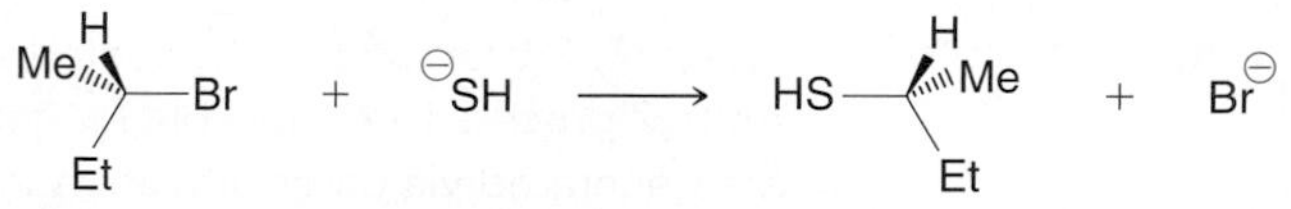

(*S*) (*R*)

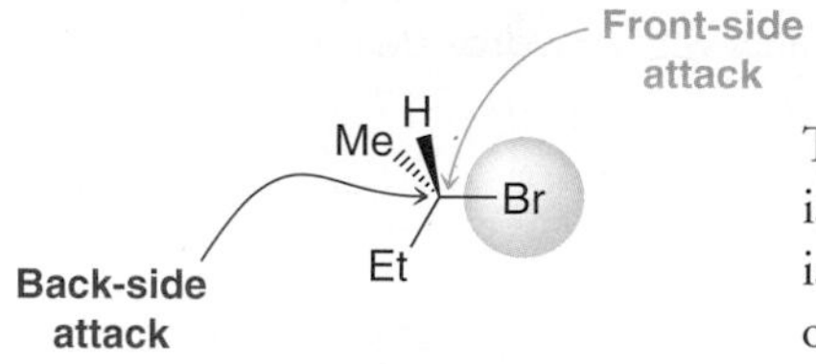

FIGURE 7.4
The front and back sides of a substrate.

The reactant exhibits the *S* configuration, while the product exhibits the *R* configuration. That is, this reaction is said to proceed with **inversion of configuration**. This stereochemical outcome is often called a Walden inversion, named after Paul Walden, the German chemist who first observed it.

The requirement for inversion of configuration means that the nucleophile can only attack from the back side (the side opposite the leaving group), and never from the front side (Figure 7.4). There are two ways to explain why the reaction proceeds through **back-side attack**:

1. The lone pairs of the leaving group create regions of high electron density that effectively block the front side of the substrate, so the nucleophile can only approach from the back side.
2. Molecular orbital (MO) theory provides a more sophisticated answer. Recall that molecular orbitals are associated with the entire molecule (as opposed to *atomic* orbitals, which are associated with individual atoms). According to MO theory, the electron density flows from the HOMO of the nucleophile into the LUMO of the electrophile. As an example let's focus our attention on the LUMO of methyl bromide (Figure 7.5). If a nucleophile attacks methyl bromide from the front side, the nucleophile will encounter a node, and as a result, no net bonding will result from the overlap between the HOMO of the nucleophile and the LUMO of the electrophile. In contrast, nucleophilic attack from the back side allows for efficient overlap between the HOMO of the nucleophile and the LUMO of the electrophile.

FIGURE 7.5
The lowest unoccupied molecular orbital (LUMO) of methyl bromide.

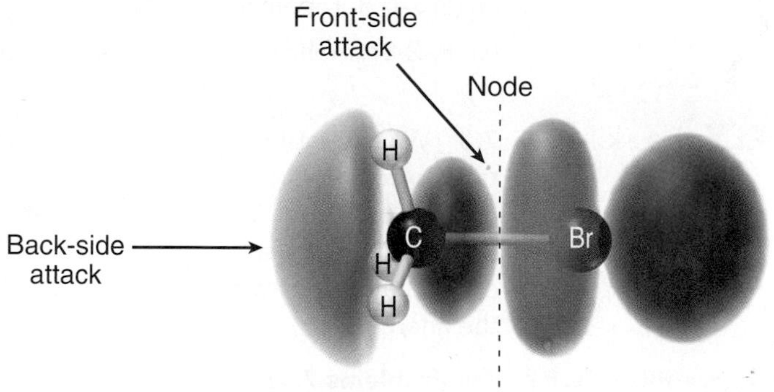

The observed stereochemical outcome for an S_N2 process (inversion of configuration) is consistent with a concerted mechanism. The nucleophile attacks with simultaneous loss of the leaving group. This causes the chirality center to behave like an umbrella flipping in the wind:

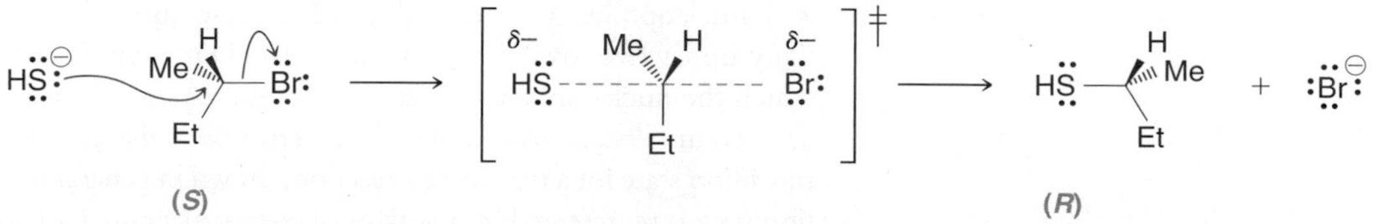

(*S*) (*R*)

The transition state (drawn in brackets) will be discussed in more detail in the coming section. This reaction is said to be **stereospecific,** because the configuration of the product is dependent on the configuration of the starting material.

SKILLBUILDER

7.2 DRAWING THE PRODUCT OF AN S_N2 PROCESS

LEARN the skill

When (*R*)-2-bromobutane is treated with a hydroxide ion, a mixture of products is obtained. An S_N2 process is responsible for generating one of the products, while the other products are generated via other processes that will be discussed in the next chapter. Draw the S_N2 product that is obtained when (*R*)-2-bromobutane reacts with a hydroxide ion.

SOLUTION

First draw the reagents described in the problem:

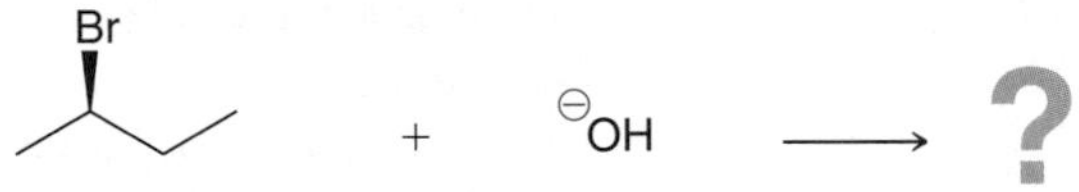

(*R*)-2-bromobutane **Hydroxide**

Now identify the nucleophile and the substrate. Bromobutane is the substrate and hydroxide is the nucleophile. When hydroxide attacks, it will eject the bromide ion as a leaving group. The net result is that the Br will be replaced with an OH group:

In this case, the α position is a chirality center, so we expect inversion:

WATCH OUT

In an S_N2 reaction, if the α position is a chirality center, make sure to draw an inversion of configuration in the product.

PRACTICE the skill

7.7 Draw the product for each of the following S_N2 reactions:

(a) (*S*)-2-Chloropentane and NaSH (b) (*R*)-3-Iodohexane and NaCl

(c) (*R*)-2-Bromohexane and sodium hydroxide

APPLY the skill

7.8 When (*S*)-1-bromo-1-fluoroethane reacts with sodium methoxide, an S_N2 reaction takes place in which the bromine atom is replaced by a methoxy group (OMe). The product of this reaction is (*S*)-1-fluoro-1-methoxyethane. How can it be that the starting material and the product both have the *S* configuration? Shouldn't S_N2 involve a change in the configuration? Draw the starting material and the product of inversion, and then explain the anomaly.

need more **PRACTICE?** **Try Problems 7.45, 7.56, 7.61**

Structure of the Substrate

For S_N2 reactions, Ingold and Hughes also found the rate to be sensitive to the nature of the starting alkyl halide. In particular, methyl halides and primary alkyl halides react most quickly with nucleophiles. Secondary alkyl halides react more slowly, and tertiary alkyl halides are essentially unreactive toward S_N2 (Figure 7.6). This trend is consistent with a concerted process in which the nucleophile is expected to encounter steric hindrance as it approaches the substrate.

To understand the nature of the steric effects that govern S_N2 reactions, we must explore the transition state for a typical S_N2 reaction, shown in general form in Figure 7.7. Recall that a transition state is represented by a peak in an energy diagram. Consider, for example, an energy diagram showing the reaction between a cyanide ion and methyl bromide (Figure 7.8). The highest point on the curve represents the transition state. The superscript symbol that looks like a telephone pole outside the brackets indicates that the drawing shows a transition state rather than an intermediate. The relative energy of this transition state determines the rate of the reaction. If the transition

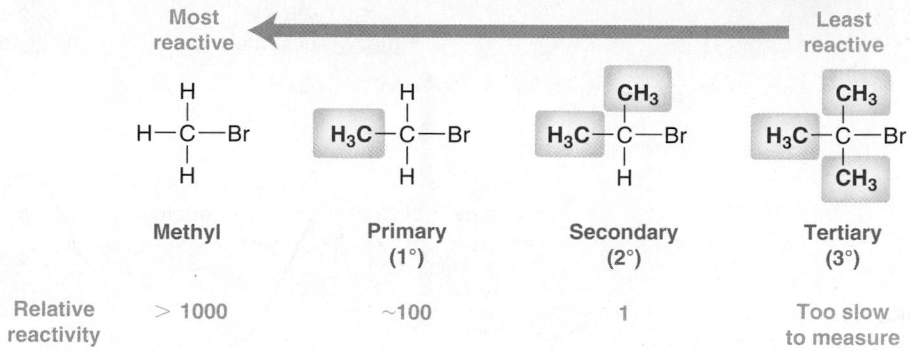

FIGURE 7.6
The relative reactivity of various substrates toward S_N2.

state is high in energy, then E_a will be large, and the rate will be slow. If the transition state is low in energy, then E_a will be small, and the rate will be fast. With this in mind, we can now explore the effects of steric hindrance in slowing down the reaction rate and explain why tertiary substrates are unreactive.

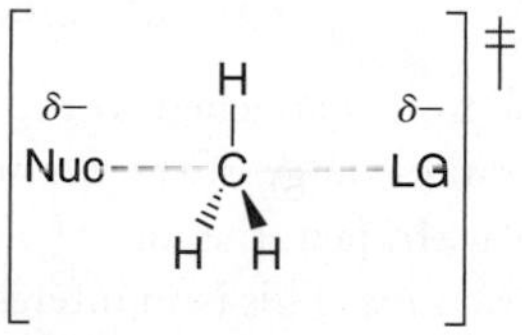

FIGURE 7.7
The generic form of a transition state in an S_N2 process.

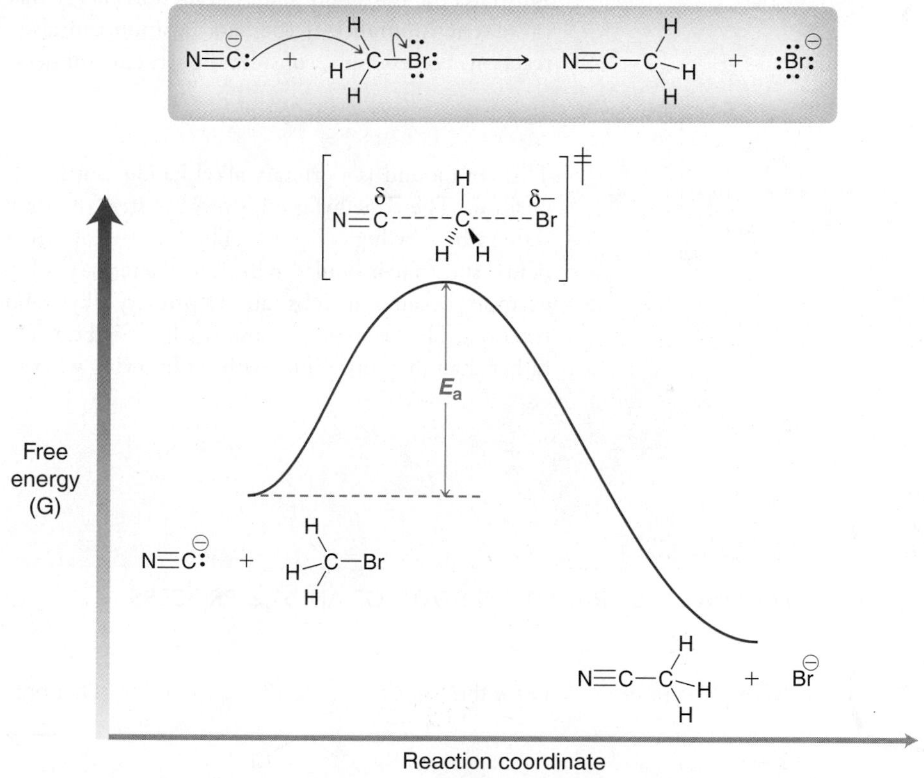

FIGURE 7.8
An energy diagram of the S_N2 reaction that occurs between methyl bromide and a cyanide ion.

Take a close look at the transition state. The nucleophile is in the process of forming a bond with the substrate, and the leaving group is in the process of breaking its bond with the substrate. Notice that there is a partial negative charge on either side of the transition state. This can be seen more clearly in an electrostatic potential map of the transition state (Figure. 7.9).

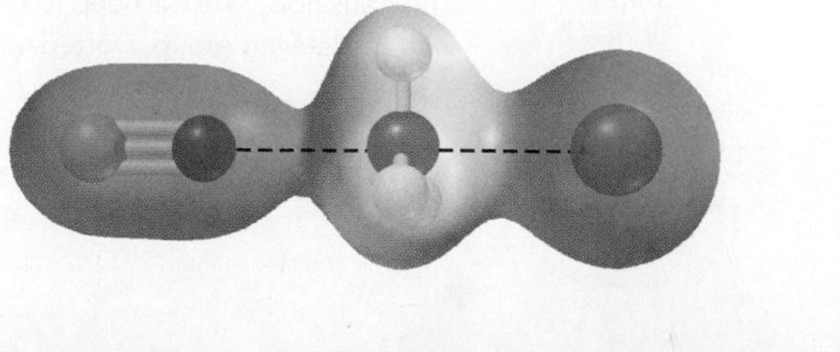

FIGURE 7.9
An electrostatic potential map of the transition state from Figure 7.7.

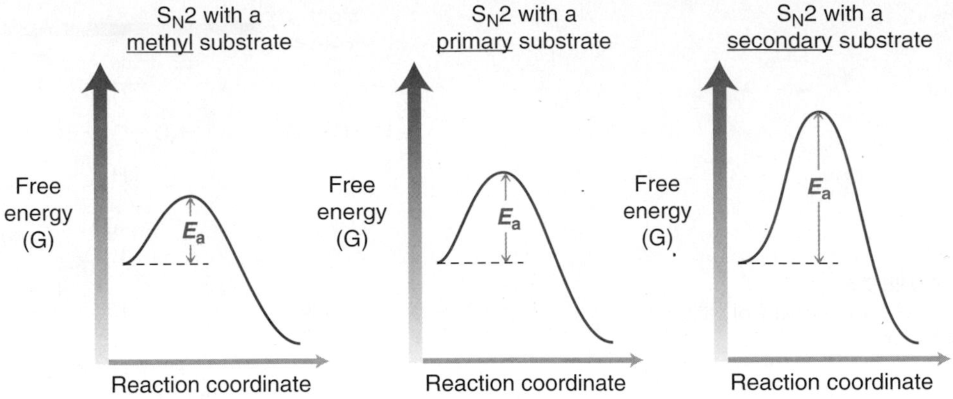

FIGURE 7.10
Energy diagrams comparing S_N2 processes for methyl, primary, and secondary substrates.

If the hydrogen atoms in Figure 7.9 are replaced with alkyl groups, steric interactions cause the transition state to be higher in energy, raising E_a for the reaction. Compare the energy diagrams for reactions involving methyl, primary, and secondary substrates (Figure 7.10). With a tertiary substrate, the transition state is so high in energy that the reaction occurs too slowly to observe.

Steric hindrance at the beta position can also decrease the rate of reaction. For example, consider the structure of neopentyl bromide:

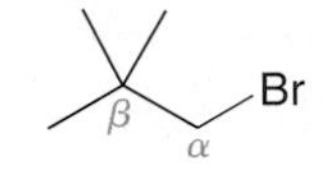

Neopentyl bromide

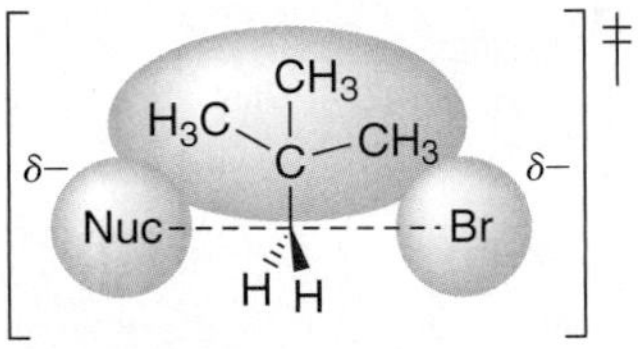

FIGURE 7.11
The transition state for an S_N2 process involving a neopentyl substrate.

This compound is a primary alkyl halide, but it has three methyl groups attached to the beta position. These methyl groups provide steric hindrance that causes the energy of the transition state to be very high (Figure 7.11). Once again, the rate is very slow. In fact, the rate of a neopentyl substrate is similar to the rate of a tertiary substrate in S_N2 reactions. This is an interesting example, because the substrate is a primary alkyl halide that essentially does not undergo an S_N2 reaction. This example illustrates why it is best to understand concepts in organic chemistry rather than memorize rules without knowing what they mean.

SKILLBUILDER

7.3 DRAWING THE TRANSITION STATE OF AN S_N2 PROCESS

LEARN the skill

Draw the transition state of the following reaction:

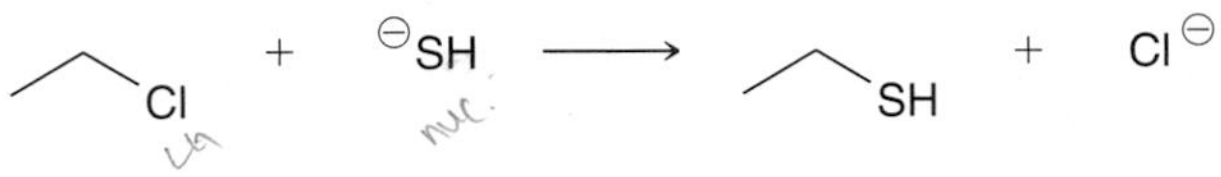

SOLUTION

Step 1 Identify the nucleophile and the leaving group.

First identify the nucleophile and the leaving group. These are the two groups that will be on either side of the transition state:

Step 2 Draw a carbon atom connected by dotted lines to the nucleophile and the leaving group.

The transition state will need to show a bond forming with the nucleophile and a bond breaking with the leaving group. Dotted lines are used to show the bonds that are breaking and forming:

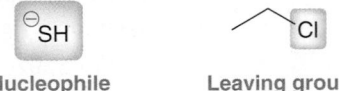

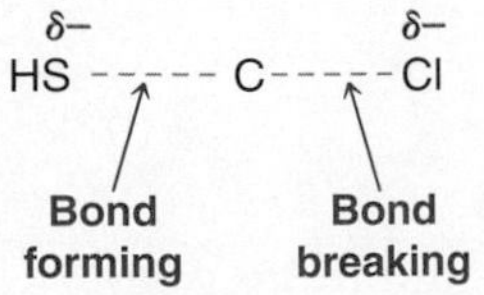

The δ− symbol is placed on both the incoming nucleophile and the outgoing leaving group to indicate that the negative charge is spread out over both locations.

Step 3 Draw the three groups attached to the carbon atom.

Now we must draw all of the alkyl groups connected to the α position. In our example, the α position has one CH_3 group and two H's:

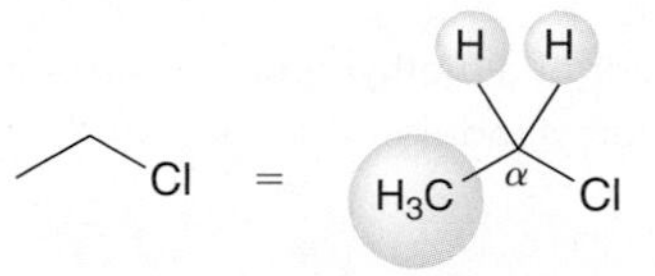

So we draw these groups in the transition state connected to the α position. One group is placed on a straight line, and the other two groups are placed on a wedge and on a dash:

Step 4 Place brackets as well as the symbol that indicates a transition state.

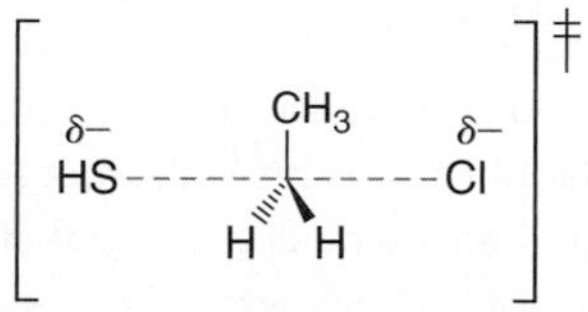

It does not matter whether the CH_3 group is placed on the line, wedge, or dash. But don't forget to indicate that the drawing is a transition state by surrounding it with brackets and using the symbol that indicates a transition state (the telephone pole).

PRACTICE the skill **7.9** Draw the transition state for each of the following S_N2 reactions:

(a)
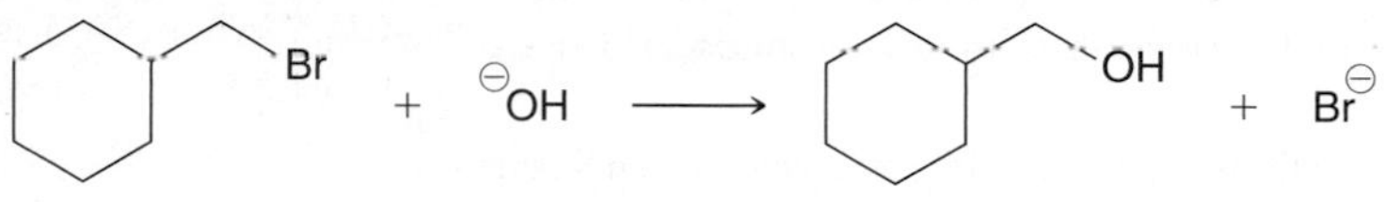

(b)

(c)
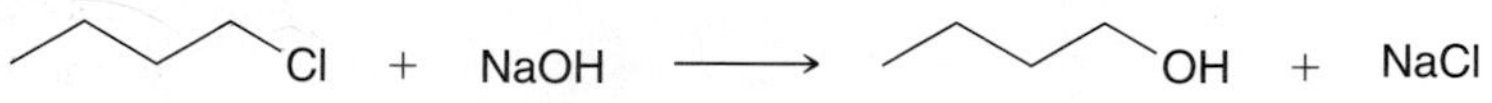

(d)
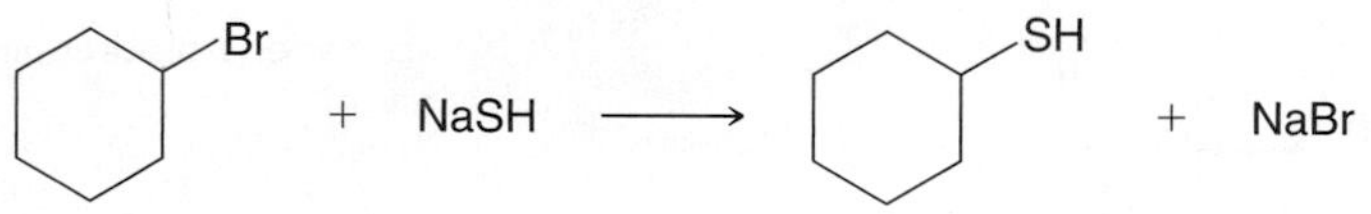

APPLY the skill

7.10 In Problem 7.4, we saw that an *intramolecular* substitution reaction can occur when the nucleophilic center and electrophilic center are present in the same compound. Draw the transition state of the reaction in Problem 7.4.

7.11 Treatment of 5-hexen-1-ol with bromine affords a cyclic product:

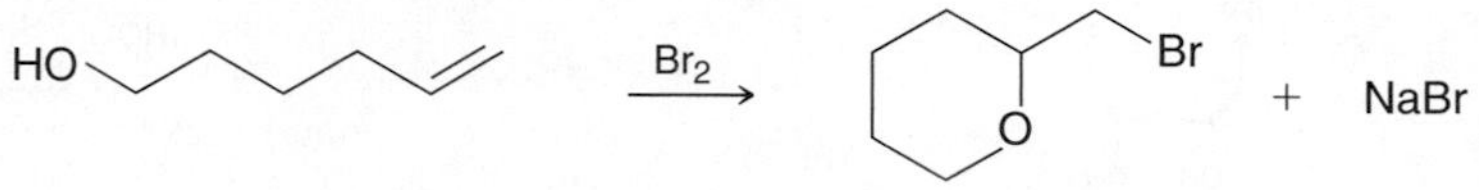

The mechanism of this reaction involves several steps, one of which is an intramolecular S_N2 process:

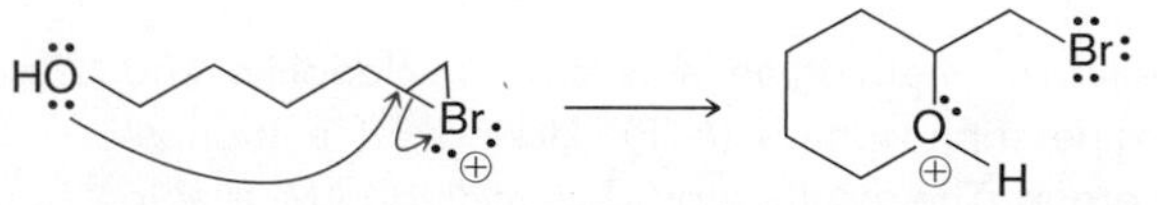

In this step, a bond is in the process of breaking, while another bond is in the process of forming. Draw the transition state of this S_N2 process, and identify which bond is being broken and which bond is being formed. Can you offer an explanation as to why this step is favorable?

need more **PRACTICE?** **Try Problems 7.46, 7.64e**

PRACTICALLYSPEAKING)))

S_N2 Reactions in Biological Systems—Methylation

In the laboratory, the transfer of a methyl group is accomplished via an S_N2 process using methyl iodide:

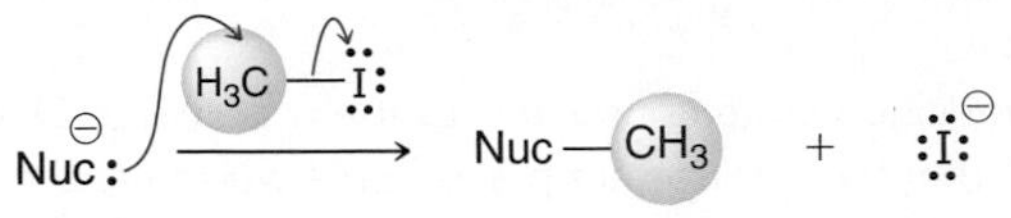

This process is called *alkylation*, because an alkyl group has been transferred to the nucleophile. It is an S_N2 process, which means there are limitations on the type of alkyl group that can be used. Tertiary alkyl groups cannot be transferred. Secondary alkyl groups can be transferred, but slowly. Primary alkyl groups and methyl groups are transferred most readily. The alkylation process shown above is the transfer of a methyl group and is therefore called *methylation*. Methyl iodide is ideally suited for this task, because iodide is an excellent leaving group and because methyl iodide is a liquid at room temperature. This makes it easier to work with than methyl chloride or methyl bromide, which are gases at room temperature.

Methylation reactions also occur in biological systems, but instead of CH_3I, the methylating agent is a compound called SAM (*S*-adenosylmethionine). Your body produces SAM via an S_N2 reaction between ATP and the amino acid methionine:

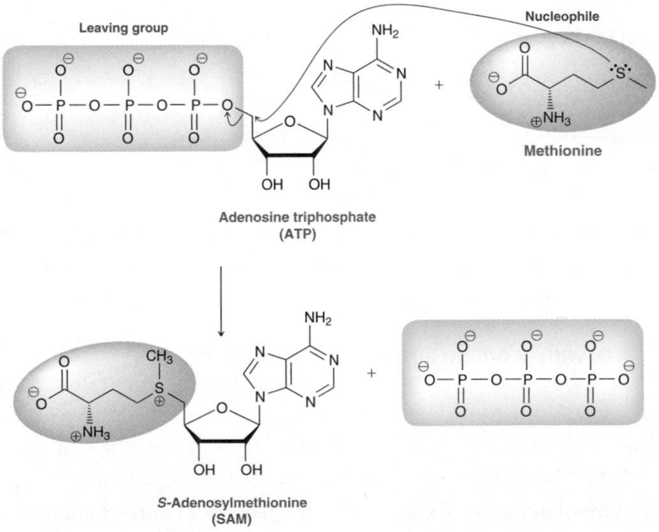

In this reaction, methionine acts as a nucleophile and attacks adenosine triphosphate (ATP), kicking off a triphosphate leaving group. The resulting product, called SAM, is able to function as a methylating agent, very much like CH_3I. Both CH_3I and SAM exhibit a methyl group attached to an excellent leaving group.

Methyl iodide	*S*-Adenosylmethionine (SAM)
H_3C—I **Iodide is a relatively simple leaving group**	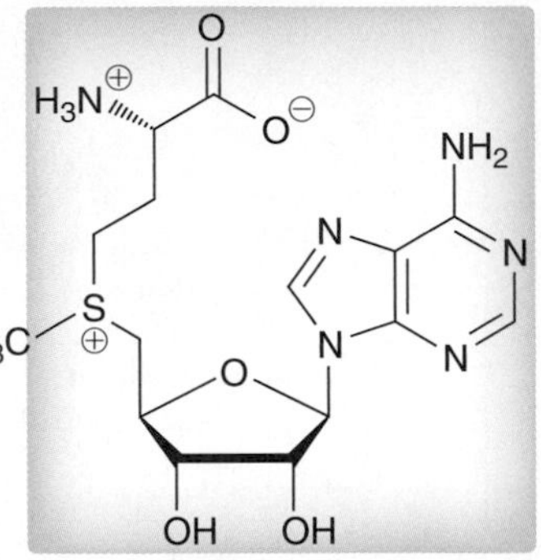 **This leaving group is more complex**

SAM is the biological equivalent of CH_3I. The leaving group is much larger, but SAM functions in the same way as CH_3I. When SAM is attacked by a nucleophile, an excellent leaving group is expelled:

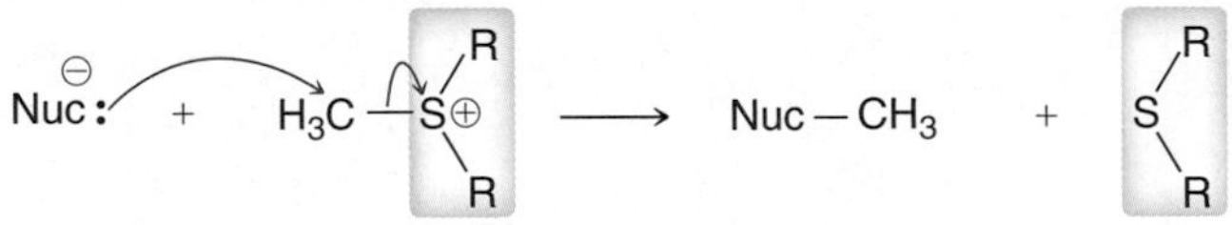

SAM plays a role in the biosynthesis of many compounds, including adrenaline. In response to danger or excitement, adrenaline is produced via a methylation reaction that takes place between noradrenaline and SAM in the adrenal gland:

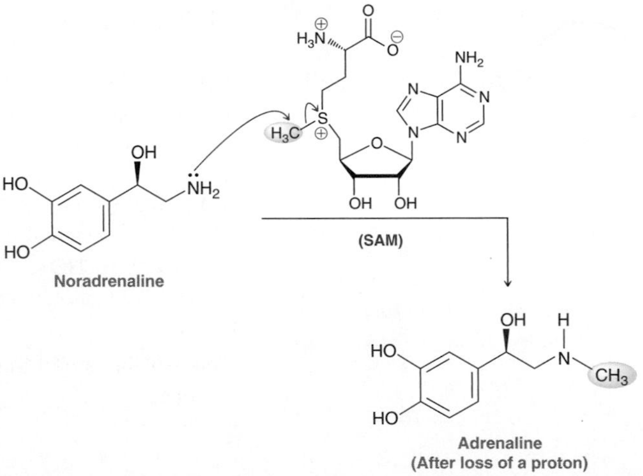

After being released into the bloodstream, adrenaline increases heart rate, elevates sugar levels to provide a boost of energy, and increases levels of oxygen reaching the brain. These physiological responses prepare the body for "fight or flight."

CONCEPTUAL CHECKPOINT

7.12 Nicotine is an addictive compound found in tobacco, and choline is a compound involved in neurotransmission. The biosynthesis of each of these compounds involves the transfer of a methyl group from SAM. Draw a mechanism for both of these transformations:

(a)

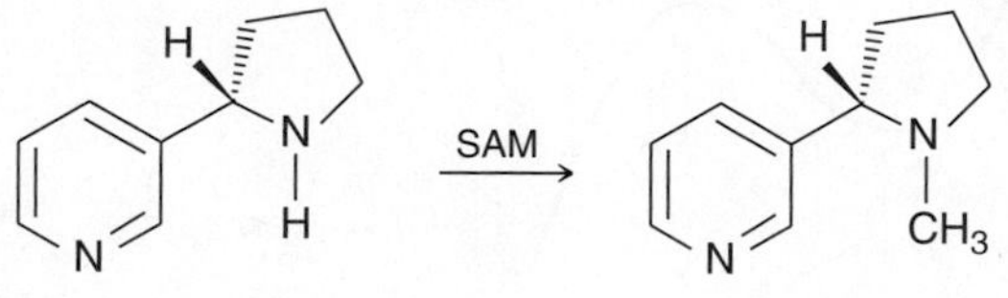

Nicotine

(b)

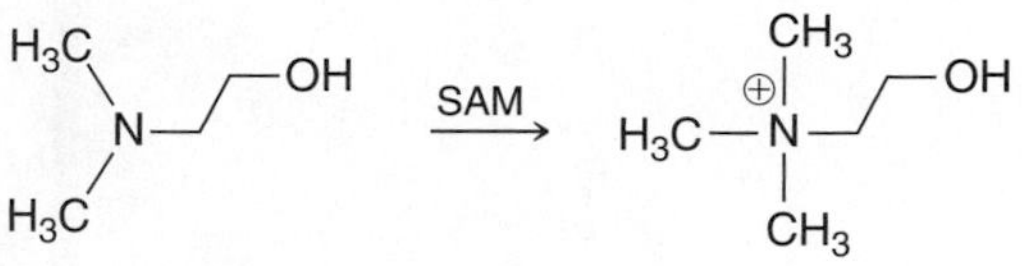

Choline

7.5 The S_N1 Mechanism

The second possible mechanism for a substitution reaction is a stepwise process in which there is (1) loss of the leaving group to form a carbocation intermediate followed by (2) nucleophilic attack on the carbocation intermediate:

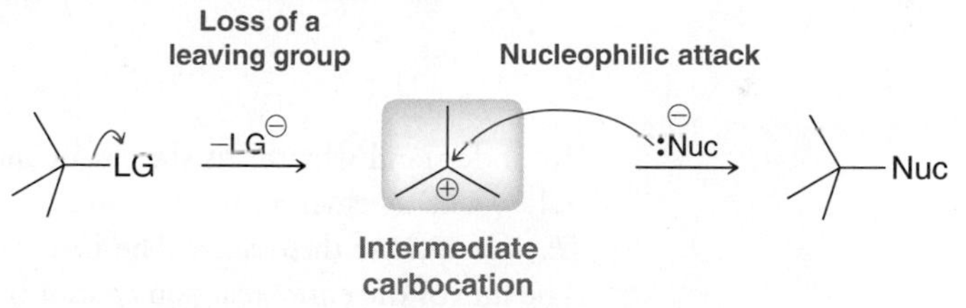

Many reactions appear to follow this stepwise mechanism. Once again, there are several pieces of evidence that support this stepwise mechanism in those cases.

Kinetics

Many substitution reactions do not exhibit second-order kinetics. Consider the following example:

In the reaction above, the rate is dependent only on the concentration of the substrate. The rate equation has the following form:

$$\text{Rate} = k\,[\text{substrate}]$$

Increasing or decreasing the concentration of the nucleophile has no measurable effect on the rate. The rate equation is said to be **first order**, because the rate is linearly dependent on the concentration of only one compound. In such cases, the mechanism must exhibit a slow step in which the nucleophile does not participate. Because that step involves only one chemical entity, it is said to be **unimolecular**. Ingold and Hughes coined the term $\mathbf{S_N1}$ to refer to unimolecular substitution reactions:

When we use the term unimolecular, we don't mean that the nucleophile is completely irrelevant. Clearly, the nucleophile is necessary, or there won't be a reaction. The term *unimolecular* simply describes the fact that only one chemical entity participates in the slowest step of the reaction, and as a result, the rate of the reaction is not affected by how much nucleophile is present.

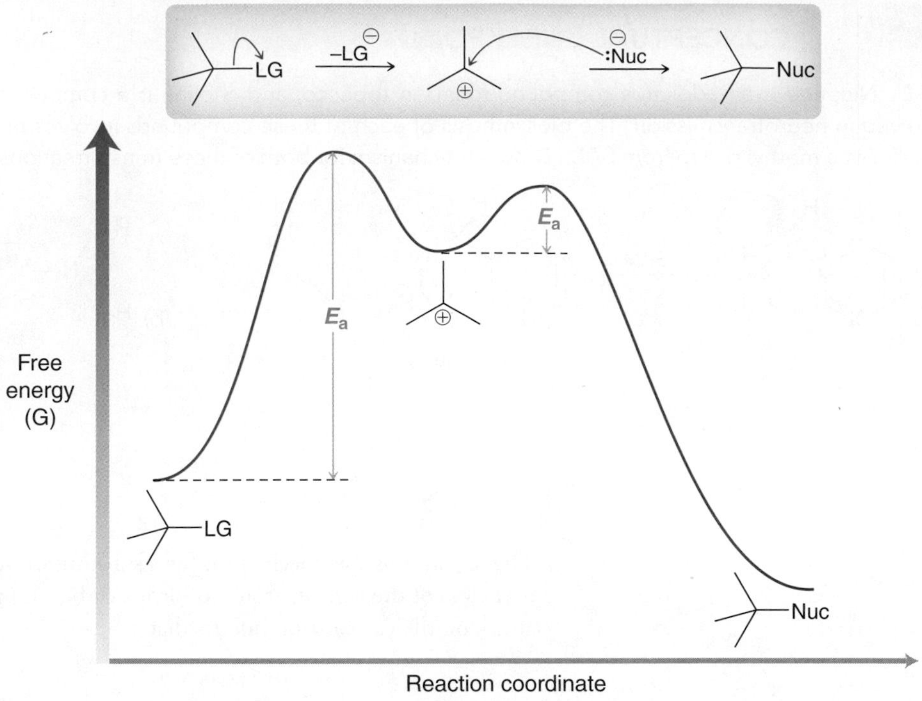

FIGURE 7.12
An energy diagram of an S_N1 process.

To understand why this is the case, consider the energy diagram of an S_N1 mechanism (Figure 7.12). The mechanism has two steps, so we expect two humps. Compare the activation energy (E_a) for each of these steps. The first step has a larger E_a, and therefore, the first step is slower. The rate of the entire reaction cannot be any faster than the rate of the slow step. The slow step is therefore called the **rate-determining step** (RDS). To illustrate this concept, consider the modified hourglass in Figure 7.13. As the sand falls, it passes through two passageways. The first passageway is a narrow opening; it represents the slower step. The second passageway has a larger opening and therefore has no effect on the rate at which the sand falls. If the second passageway is further widened, it will not impact the overall rate in any way. The same idea is true for reactions as well. The rate of an S_N1 process is dependent only on the rate of the slow step, the loss of the leaving group. As a result, the rate of an S_N1 process will only be affected by factors that affect the rate at which the leaving group leaves. Increasing the concentration of the nucleophile has no impact on the rate of the slow step. It is true that the nucleophile must be present in order to obtain the product, but an excess of nucleophile will not speed up the reaction. A unimolecular substitution reaction is therefore consistent with a stepwise mechanism in which the first step is the rate-determining step.

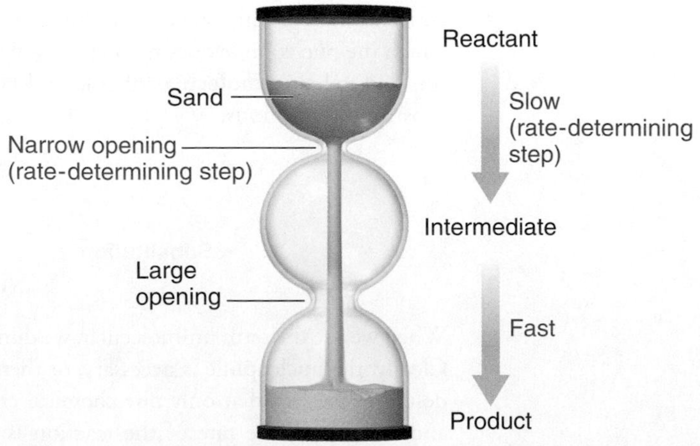

FIGURE 7.13
A modified hourglass with two passageways. The narrow passageway is the rate-determining step.

CONCEPTUAL CHECKPOINT

7.13 The following reaction occurs via an S$_N$1 mechanistic pathway:

$$\text{(CH}_3)_3\text{C}-\text{I} \xrightarrow{\text{NaCl}} \text{(CH}_3)_3\text{C}-\text{Cl} + \text{NaI}$$

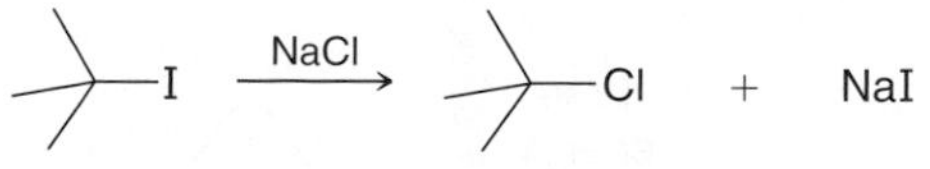

(a) What happens to the rate if the concentration of *tert*-butyl iodide is doubled and the concentration of sodium chloride is tripled? ~~2 times~~

(b) What happens to the rate if the concentration of *tert*-butyl iodide remains the same and the concentration of sodium chloride is doubled? ~~remains same~~

[handwritten margin notes:]
In SN2 rxn depends on nuc. and LG (substrate)
In SN1, It does not depend on nuc above ^. It is only dependent on substrate.

Structure of Substrate

The rate of an S$_N$1 reaction is highly dependent on the nature of the substrate, but the trend is the reverse of the trend we saw for S$_N$2 reactions. With S$_N$1 reactions, tertiary substrates react most quickly, while methyl and primary substrates are mostly unreactive (Figure 7.14). This observation supports a stepwise mechanism (S$_N$1). Why? With S$_N$2 reactions, steric hindrance was the issue because the nucleophile was directly attacking the substrate. In contrast, in S$_N$1 reactions the nucleophile does not attack the substrate directly. Instead, the leaving group leaves first, resulting in the formation of a carbocation, and that step is the rate-determining step. Once the carbocation forms, the nucleophile captures it very quickly. The rate is only dependent on how quickly the leaving group leaves to form a carbocation. Steric hindrance is not at play, because the rate-determining step does not involve nucleophilic attack. The dominant factor now becomes carbocation stability.

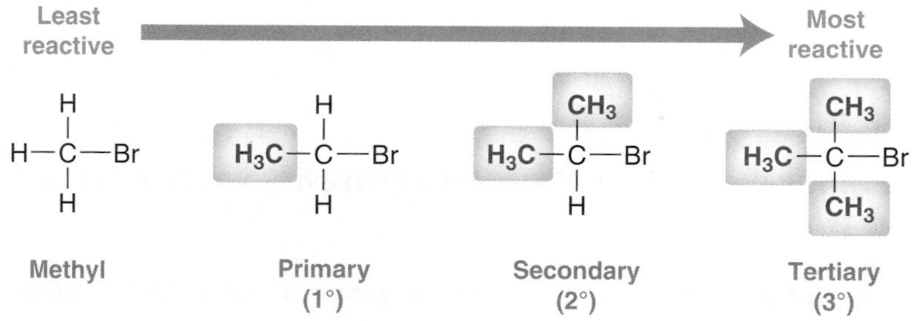

FIGURE 7.14
The relative reactivity of various substrates toward S$_N$1.

Recall that carbocations are stabilized by neighboring alkyl groups (Figure 7.15). Tertiary carbocations are more stable than secondary carbocations, which are more stable than primary carbocations. Therefore, formation of a tertiary carbocation will have a smaller E_a than

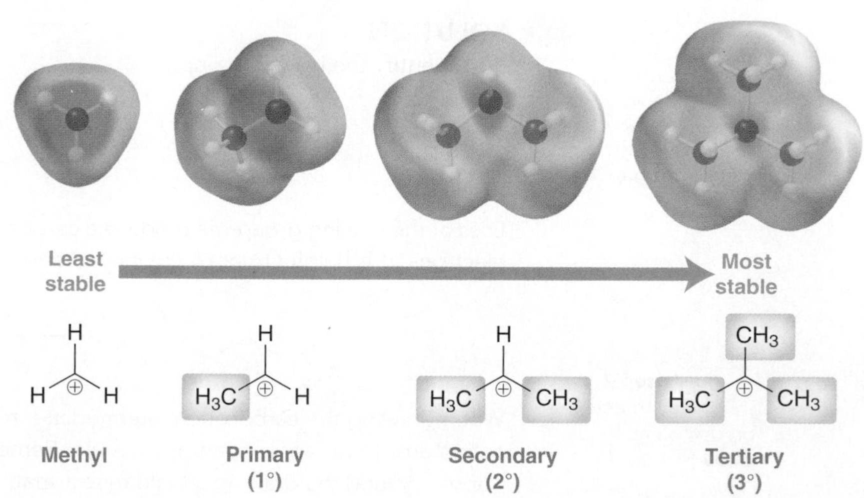

FIGURE 7.15
Electrostatic potential maps of various carbocations. The alkyl groups help spread the positive charge, thereby stabilizing the charge.

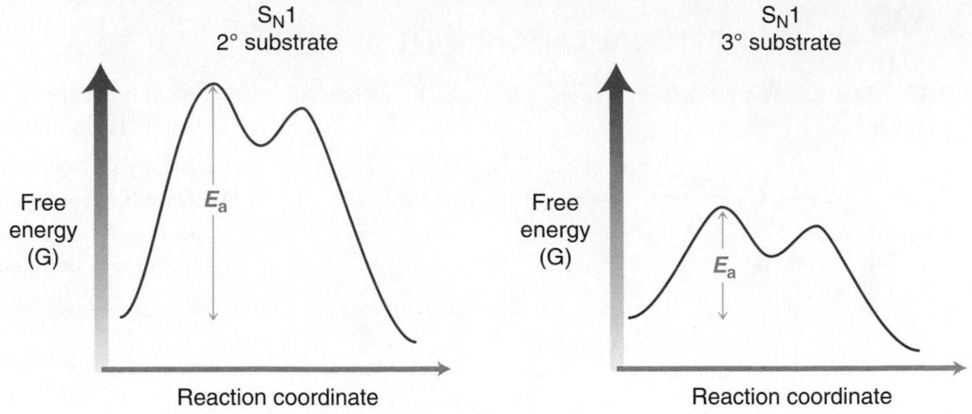

FIGURE 7.16 Energy diagrams comparing S_N1 processes for secondary and tertiary substrates.

formation of a secondary carbocation (Figure 7.16). The larger E_a associated with formation of a secondary carbocation can be explained by the Hammond postulate (Section 6.6). Specifically, the transition state for formation of a tertiary carbocation will be close in energy to a tertiary carbocation, while the transition state for formation of a secondary carbocation will be close in energy to a secondary carbocation. Therefore, formation of a tertiary carbocation will involve a smaller E_a.

The bottom line is that tertiary substrates generally undergo substitution via an S_N1 process, while primary substrates generally undergo substitution via an S_N2 process. Secondary substrates can proceed via either pathway (S_N1 *or* S_N2) depending on other factors, which are discussed later in this chapter.

SKILLBUILDER

7.4 DRAWING THE CARBOCATION INTERMEDIATE OF AN S_N1 PROCESS

LEARN the skill

Draw the carbocation intermediate of the following S_N1 reaction:

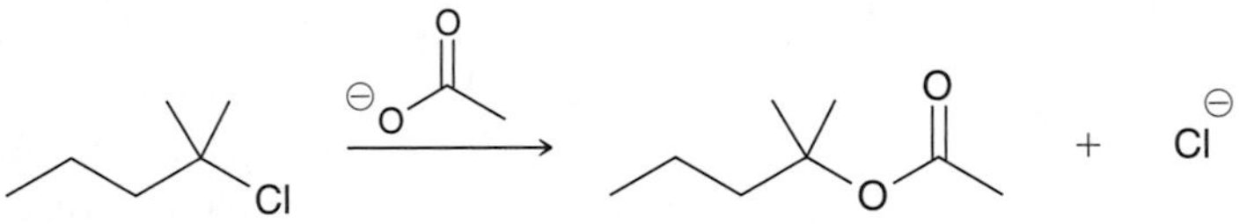

SOLUTION

First identify the leaving group:

Step 1 Identify the leaving group.

Loss of the leaving group will produce a carbocation and a chloride ion. To keep track of the electrons, it is helpful to draw the curved arrow that shows the flow of electrons:

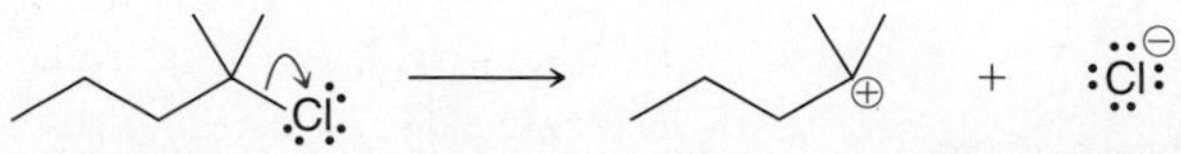

Step 2 Draw all three groups pointing away from each other.

When drawing the carbocation intermediate, make sure that all three groups on the carbocation are drawn as far apart as possible. Remember that a carbocation has trigonal planar geometry, and the drawing should reflect that:

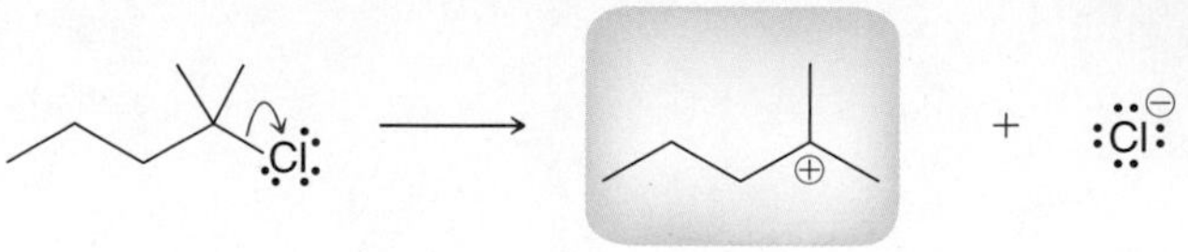

PRACTICE the skill **7.14** Draw the carbocation intermediate generated by each of the following substrates in an S_N1 reaction:

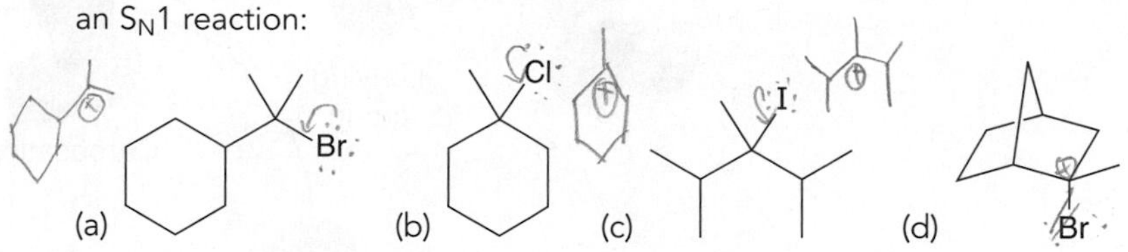

APPLY the skill **7.15** Identify which of the following substrates will undergo an S_N1 reaction more rapidly. Explain your choice.

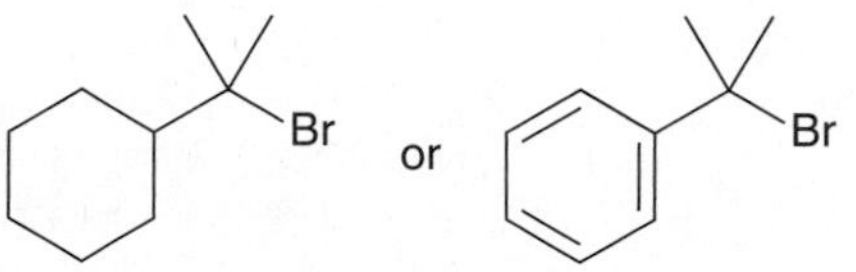

need more **PRACTICE?** **Try Problems 7.50, 7.51**

Stereochemistry of S_N1 Reactions

Recall that S_N2 reactions proceed via an inversion of configuration:

(S)-2-bromobutane
(R)-2-chlorobutane

In contrast, S_N1 reactions involve formation of an intermediate carbocation, which can then be attacked from either side (Figure 7.17), leading to both inversion of configuration and **retention of configuration**.

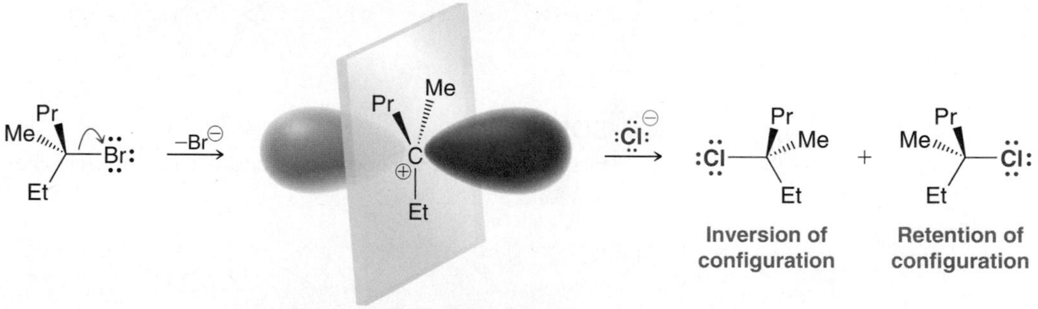
Inversion of configuration
Retention of configuration
A carbocation is planar, and either side of the plane can be attacked by the nucleophile with equal likelihood.

FIGURE 7.17 The intermediate of an S_N1 process is a planar carbocation.

Since the carbocation can be attacked on either side with equal likelihood, we should expect S_N1 reactions to produce a racemic mixture (equal mixture of inversion and retention). In practice, though, S_N1 reactions rarely produce exactly equal amounts of inversion and retention products. There is usually a slight preference for the inversion product. The accepted explanation involves the formation of ion pairs. When the leaving group first leaves, it is initially very close to the intermediate carbocation, forming an intimate ion pair (Figure 7.18). If the nucleophile attacks the carbocation while it is still participating in an ion pair, then the leaving group effectively blocks one face of the carbocation. The other side of the carbocation can experience unhindered attack by a nucleophile. As a result, the nucleophile will attack more often on the side opposite the leaving group, leading to a slight preference for inversion over retention.

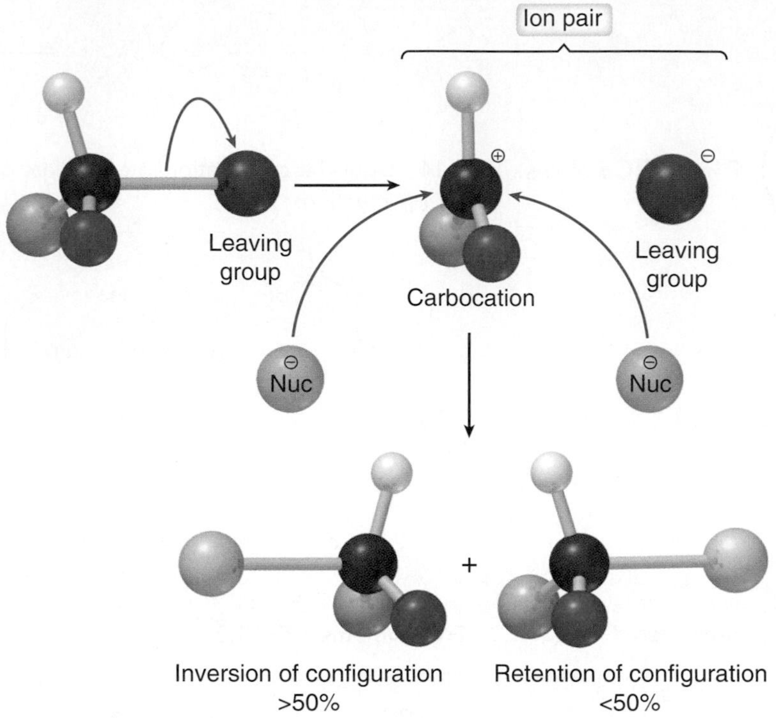

FIGURE 7.18 Loss of a leaving group initially forms an ion pair, which hinders attack on one face of the carbocation.

SKILLBUILDER

7.5 DRAWING THE PRODUCTS OF AN S_N1 PROCESS

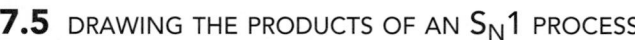

LEARN the skill

Draw the products of the following S_N1 reaction:

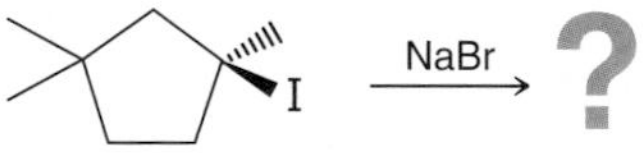

SOLUTION

First identify the leaving group and the nucleophile that will attack once the leaving group has left:

Step 1 Identify the nucleophile and the leaving group.

In an S_N1 process, the leaving group leaves first, generating a carbocation that is then attacked by the nucleophile:

Step 2 Replace the leaving group with the nucleophile.

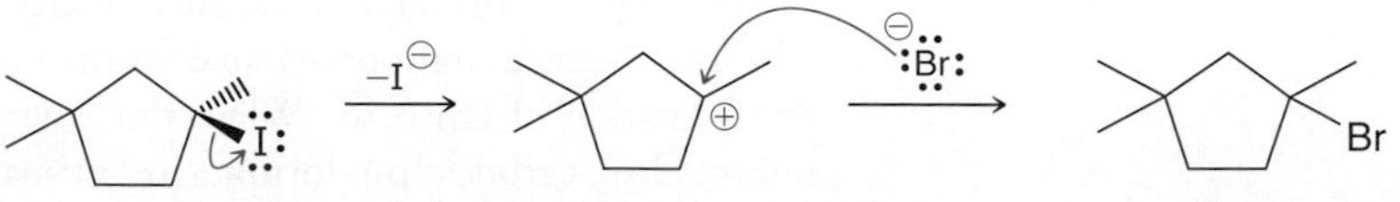

In this example, substitution is taking place at a chirality center, so we must consider the stereochemical outcome. In an S_N1 process, both enantiomers are expected as products, with a slight preference for the enantiomer resulting from inversion of configuration:

Step 3
If the reaction takes place at a chirality center, draw both possible enantiomers.

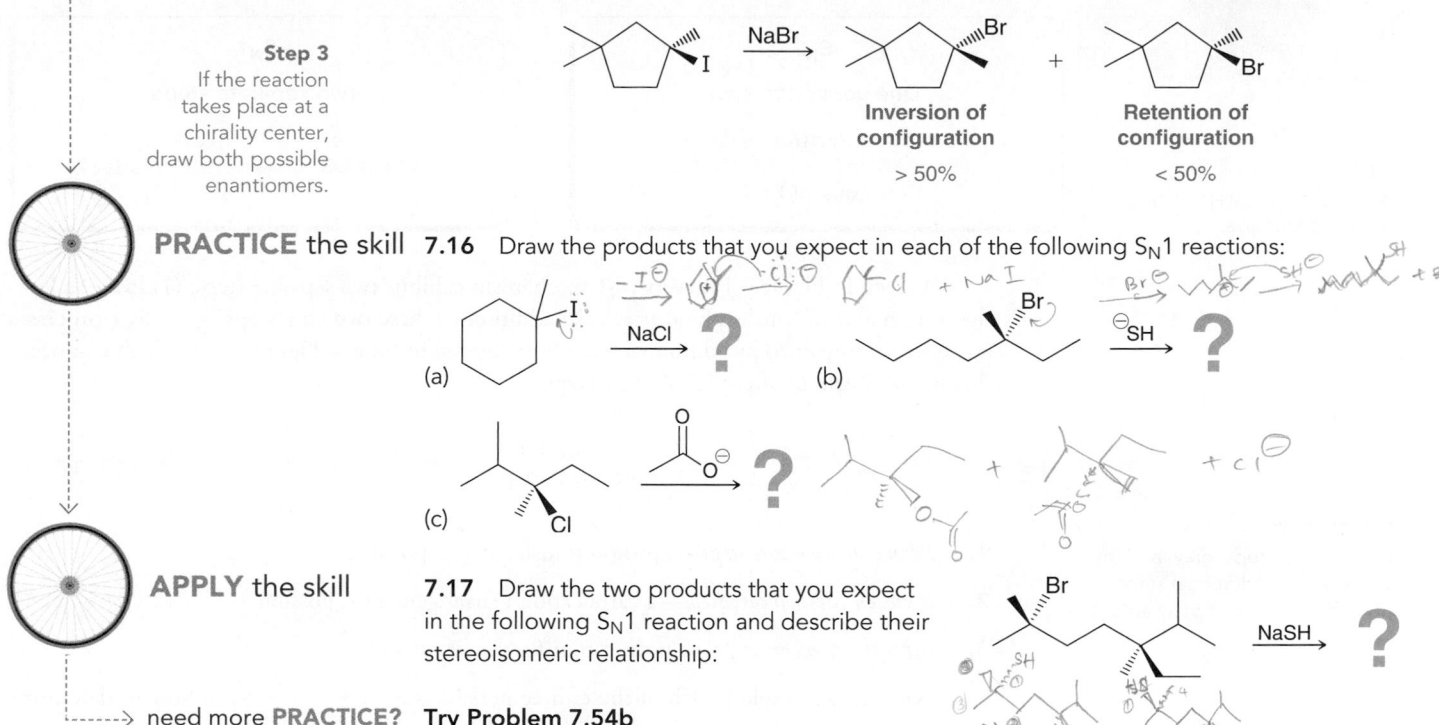

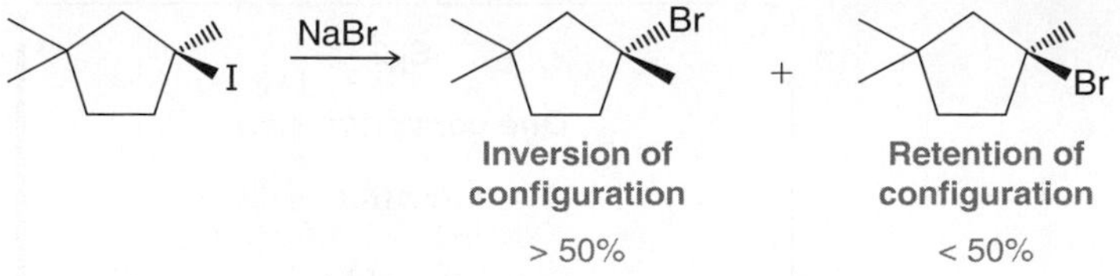

PRACTICE the skill **7.16** Draw the products that you expect in each of the following S_N1 reactions:

(a) 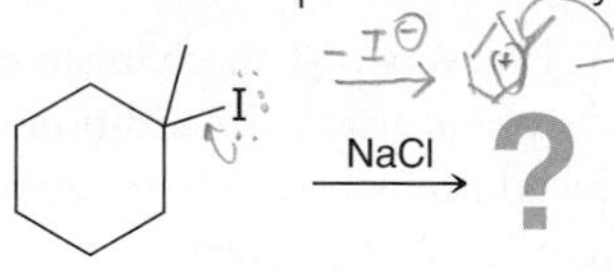

(b) 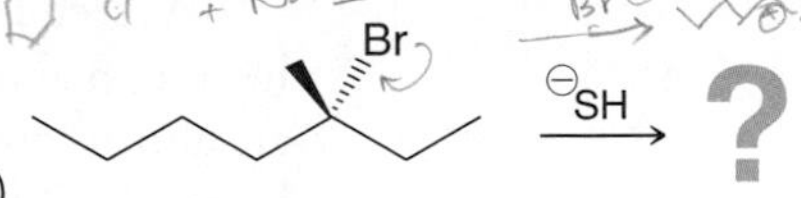

(c) 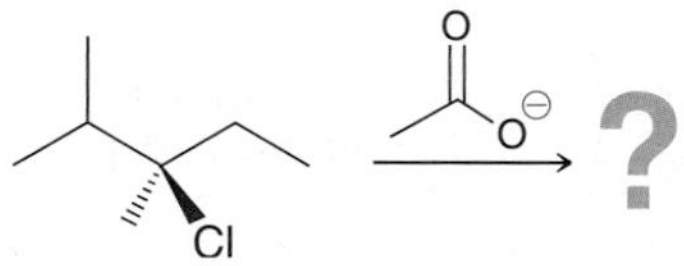

APPLY the skill **7.17** Draw the two products that you expect in the following S_N1 reaction and describe their stereoisomeric relationship:

need more **PRACTICE?** **Try Problem 7.54b**

Let's now summarize the differences that we have seen between S_N2 and S_N1 processes (Table 7.1).

TABLE 7.1 A COMPARISON OF S_N2 AND S_N1 PROCESSES

	S_N2	S_N1
Mechanism	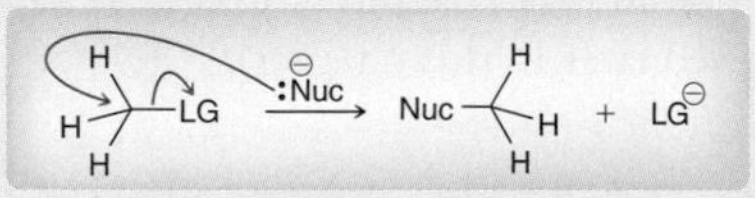	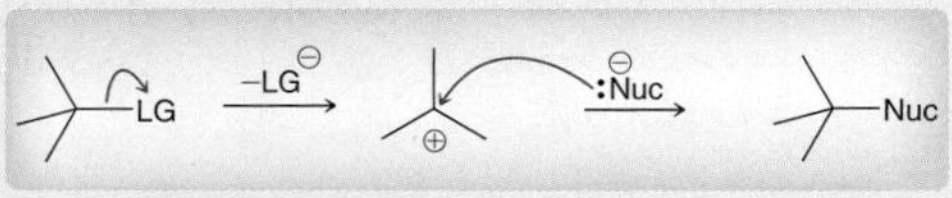
Energy diagram	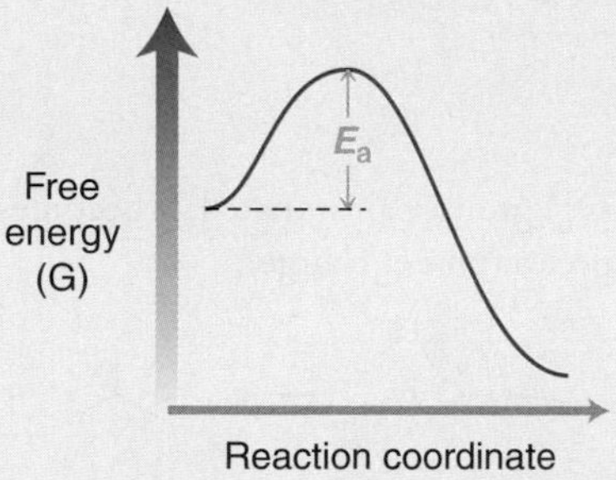	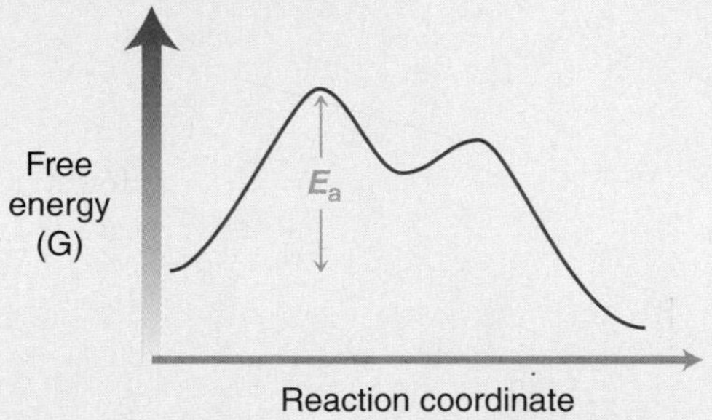
Rate equation	Rate = *k* [substrate] [nucleophile]	Rate = *k* [substrate]
Rate of reaction	Methyl > 1° > 2° > 3°	3° > 2° > 1° > methyl
Stereochemistry	Inversion of configuration	Racemization (with slight preference for inversion due to ion pairs)

7.6 Drawing the Complete Mechanism of an S_N1 Reaction

We have now seen that substitution reactions can occur through either a concerted mechanism (S_N2) or a stepwise mechanism (S_N1) (Figure 7.19). When drawing the mechanism of an S_N2 or S_N1 process, additional mechanistic steps will sometimes be required. In this section, we will focus on the additional steps that can accompany an S_N1 process. Recall from Chapter 6 that ionic mechanisms are constructed using only four different types of arrow-pushing patterns. This will now be important, as all four patterns can play a role in S_N1 processes.

S_N2
One concerted step
Nuc attack
+
Loss of LG
S_N1
Two separate steps
Loss of LG
Nuc attack

FIGURE 7.19
The mechanistic steps in S_N2 and S_N1 processes.

As seen in Figure 7.19, every S_N1 mechanism exhibits two separate steps: (1) loss of a leaving group and (2) nucleophilic attack. In addition to these two core steps, some S_N1 processes are also accompanied by additional steps (highlighted in blue in Figure 7.20), which can occur before, between, or after the two core steps:

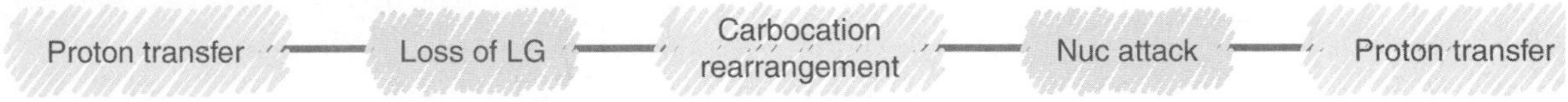
Proton transfer
Loss of LG
Carbocation rearrangement
Nuc attack
Proton transfer

FIGURE 7.20
The two core steps (gray) and the three possible additional steps (blue) that can accompany an S_N1 process.

1. *Before the two core steps*—a proton transfer step is possible.
2. *Between the two core steps*—a carbocation rearrangement is possible.
3. *After the two core steps*—a proton transfer step is possible.

We will now explore each of these three possibilities, and we will learn how to determine whether any of the three additional steps should be included when proposing the mechanism for a transformation that occurs via an S_N1 process.

Proton Transfer at the Beginning of an S_N1 Process

+H⁺
–LG
Nuc attack

Before the two core steps of an S_N1 mechanism, a proton transfer will be necessary whenever the leaving group is an OH group. Hydroxide is a bad leaving group and will not leave by itself (as will be discussed later in this chapter):

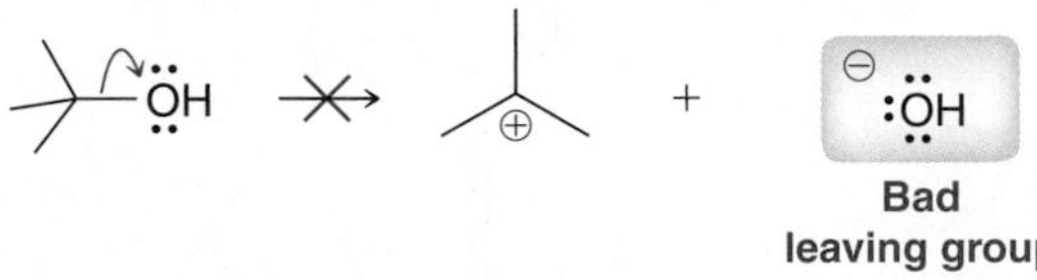

However, once an OH group is protonated, it becomes an excellent leaving group because it can leave as a neutral species (no net charge):

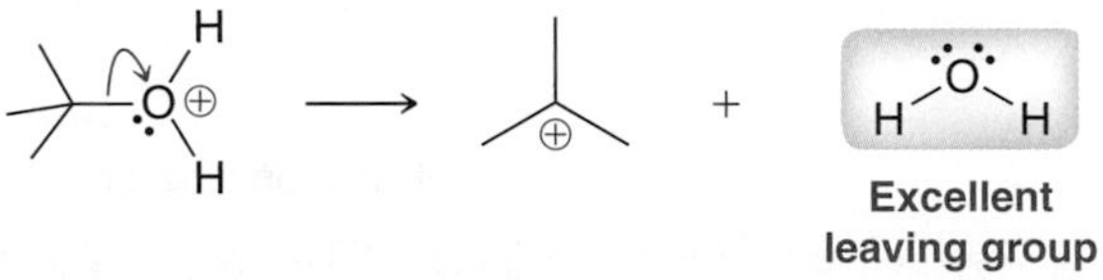
H
O
H
H
O
H
Excellent
leaving group

If a substrate has no leaving group other than an OH group, then acidic conditions will be required in order to perform an S_N1 reaction. In the following example, HCl supplies the H^+ necessary to protonate the OH group as well as the chloride ion that functions as the nucleophile:

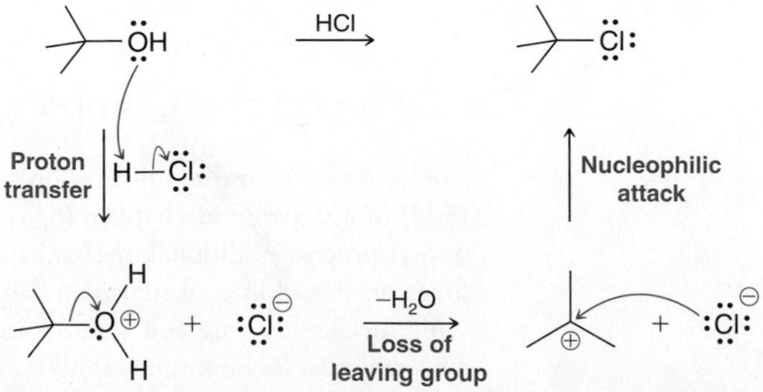

Notice that the two core steps of this S_N1 process are preceded by a proton transfer, giving a total of three mechanistic steps:

+H⁺ — −LG — Nuc attack

CONCEPTUAL CHECKPOINT

7.18 For each of the following substrates, determine whether an S_N1 process will require a proton transfer at the beginning of the mechanism:

(a) 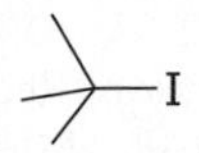

(b) 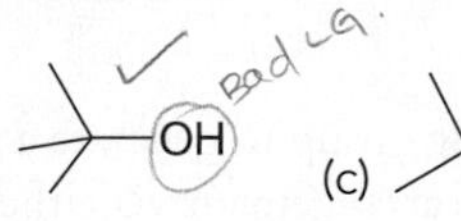

(c) 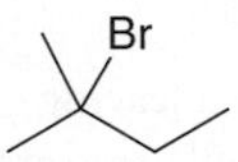

(d) 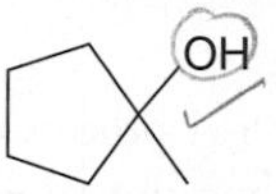

(e) 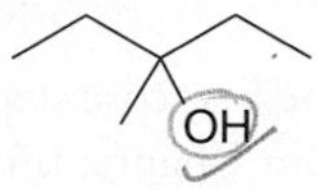

(f) 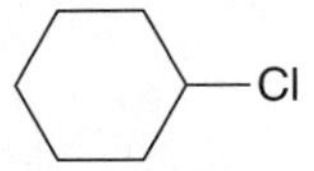

Proton Transfer at the End of an S_N1 Process

−LG — Nuc attack — −H⁺

After the two core steps of an S_N1 mechanism, a proton transfer will be necessary whenever the nucleophile is neutral (not negatively charged). For example:

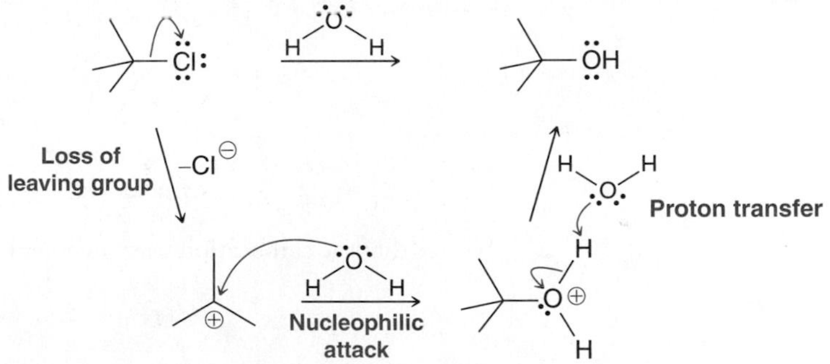

In this case, the nucleophile is water (H_2O), which does not possess a negative charge. In such a case, nucleophilic attack of the carbocation will produce a positively charged species. Removal of the positive charge requires a proton transfer. Notice that the mechanism above has three steps:

−LG — Nuc attack — −H⁺

Any time the attacking nucleophile is neutral, a proton transfer is necessary at the end of the mechanism. Below is one more example. Reactions like this, in which the solvent functions as the nucleophile, are called **solvolysis** reactions.

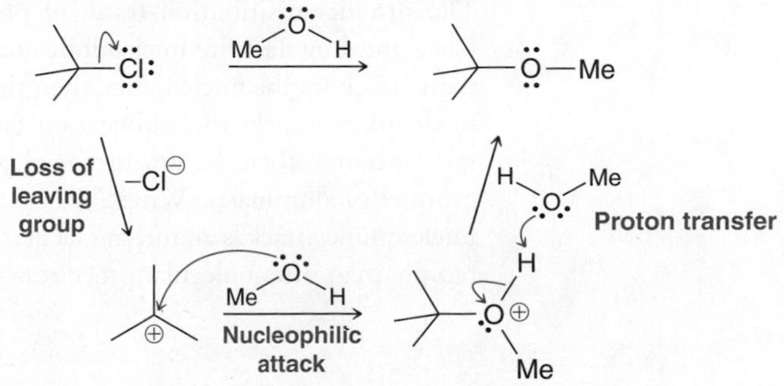

CONCEPTUAL CHECKPOINT

7.19 Will an S_N1 process involving each of the following nucleophiles require a proton transfer at the end of the mechanism?

(a) NaSH (b) H_2S (c) H_2O (d) EtOH (e) NaCN (f) NaCl

(g) $NaNH_2$ (h) NH_3 (i) NaOMe (j) NaOEt (k) MeOH (l) KBr

Carbocation Rearrangements during an S_N1 Process

The first core step of an S_N1 process is loss of a leaving group to generate a carbocation. Recall from Chapter 6 that carbocations are susceptible to rearrangement via either a hydride shift or a methyl shift. Here is an example of an S_N1 mechanism with a carbocation rearrangement:

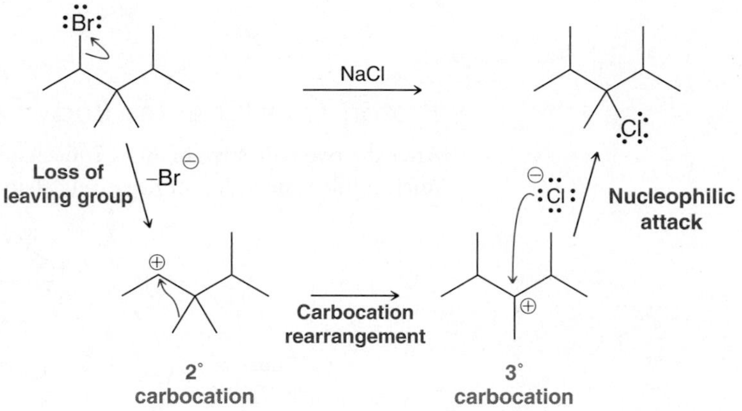

Notice that the carbocation rearrangement occurs between the two core steps of the S_N1 process:

$$-LG \quad — \quad \underset{\text{rearrangement}}{C+} \quad — \quad \text{Nuc attack}$$

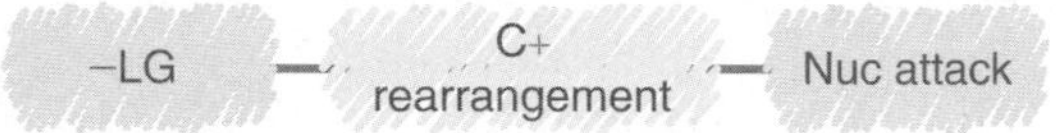

In reactions where a carbocation rearrangement is possible, a mixture of products is generally obtained. The following products are obtained from the above reaction:

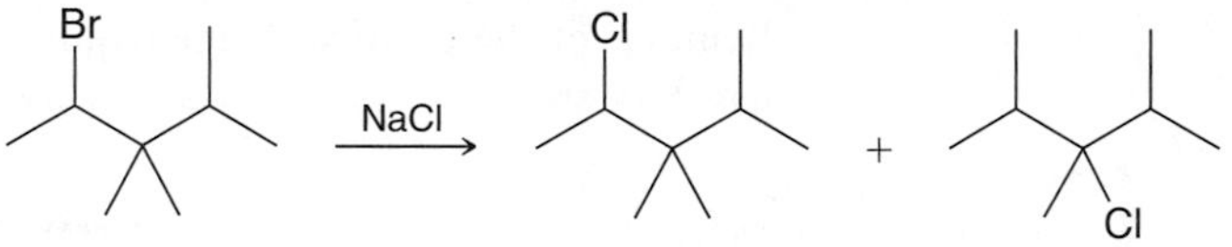

This is the product if the carbocation is captured by the nucleophile *before* rearrangement

This is the product if the carbocation is captured by the nucleophile *after* rearrangement

The product distribution (ratio of products) depends on how fast the rearrangement takes place and how fast the nucleophile attacks the carbocation. If the rearrangement occurs faster than attack by the nucleophile, then the rearranged product will predominate. However, if the nucleophile attacks the carbocation faster than rearrangement (if it attacks before rearrangement occurs), then the unrearranged product will predominate. In most cases, the rearranged product predominates. Why? A carbocation rearrangement is an intramolecular process, while nucleophilic attack is an intermolecular process. In general, intramolecular processes occur more rapidly than intermolecular processes.

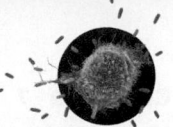

CONCEPTUAL CHECKPOINT

7.20 For each of the following substrates, determine whether an S_N1 process will involve a carbocation rearrangement or not:

(a) 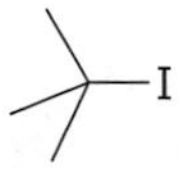

(b) 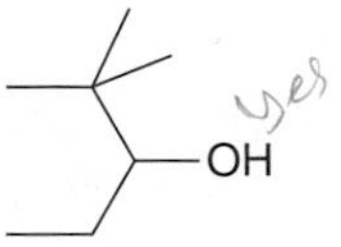

(c) 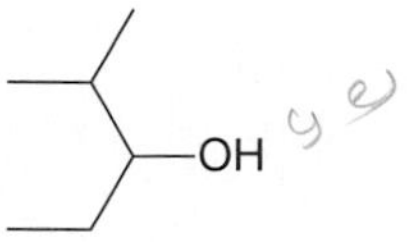

(d) 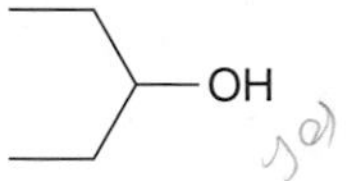

(e) 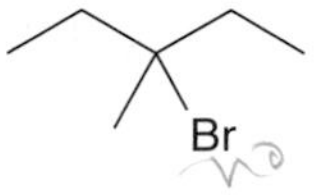

(f) 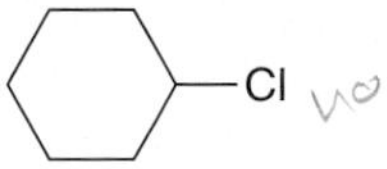

Summary of the S_N1 Process and Its Energy Diagram

We have seen that an S_N1 process has two core steps and can be accompanied by three additional steps, as summarized in Mechanism 7.1.

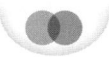

MECHANISM 7.1 THE S_N1 PROCESS

Two core steps

Loss of LG | Nuc attack

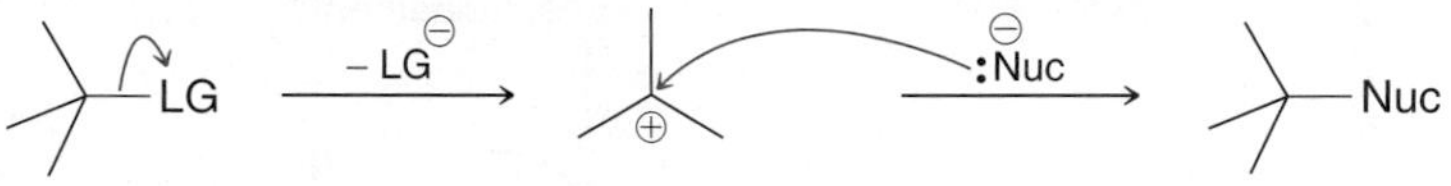

In the first core step of an S_N1 process, the leaving group leaves to form a carbocation

In the second core step of an S_N1 process, a nucleophile attacks the carbocation

Possible additional steps

Proton transfer — Loss of LG — Carbocation rearrangement — Nuc attack — Proton transfer

If the leaving group is an OH group, it must be protonated before it can leave.

If the carbocation initially formed can rearrange to generate a more stable carbocation, then a rearrangement will occur.

If the nucleophile is neutral, a proton transfer is required to remove the positive charge that is generated.

Here is an example of an S_N1 process that is accompanied by all three additional steps:

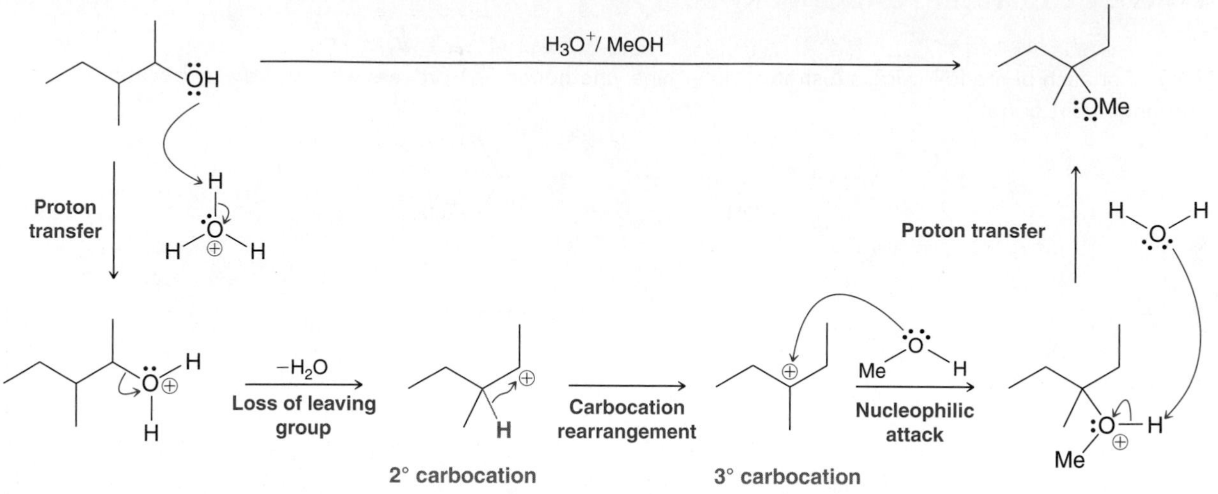

Since this mechanism has five steps, we expect the energy diagram for this reaction to exhibit five humps (Figure 7.21). The number of humps in the energy diagram of an S_N1 process will always be equal to the number of steps in the mechanism. Since the number of steps can range anywhere from two to five, the energy diagram of an S_N1 process can have anywhere from two to five humps. The S_N1 processes encountered most frequently will have two or three steps.

A few aspects of the energy diagram in Figure 7.21 are worth special mention:

- The tertiary carbocation is lower in energy than the secondary carbocation.
- The E_a for the carbocation rearrangement is very small because a carbocation rearrangement is generally a very fast process.
- Oxonium ions (intermediates with a positively charged oxygen atom) are generally lower in energy than carbocations (because the oxygen atom of an oxonium ion has an octet of electrons, while the carbon atom of a carbocation does not have an octet).

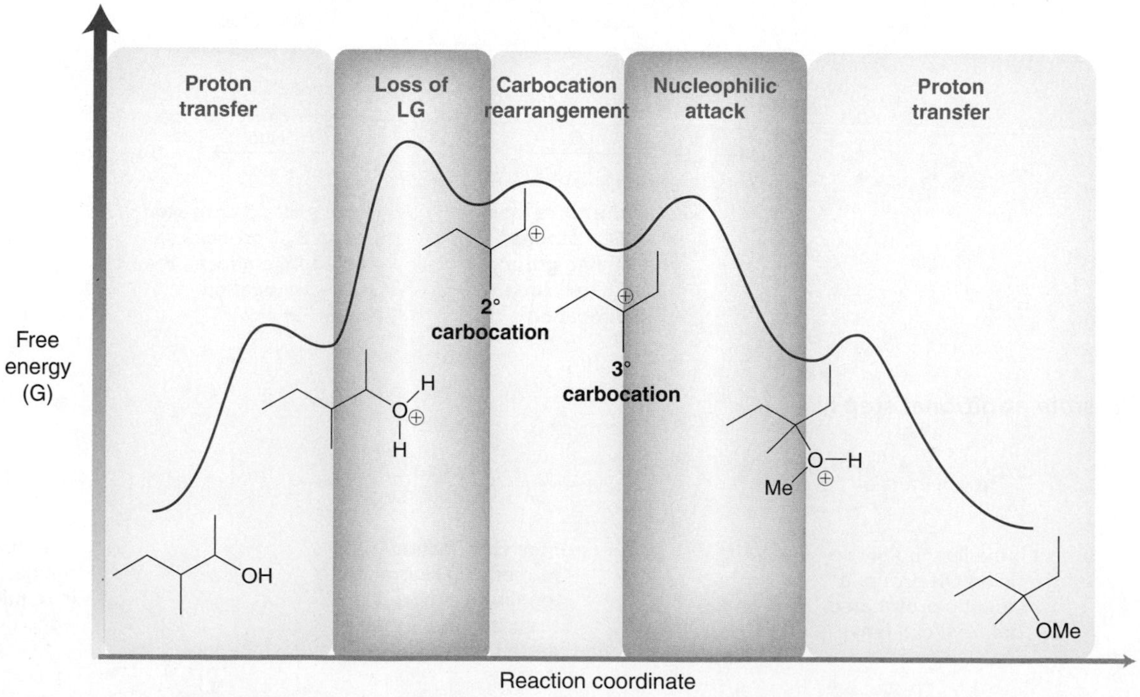

FIGURE 7.21
An energy diagram of an S_N1 process that is accompanied by three additional steps. In total, there are five steps, giving an energy diagram with five humps.

SKILLBUILDER

7.6 DRAWING THE COMPLETE MECHANISM OF AN S_N1 PROCESS

LEARN the skill

Draw the mechanism of the following S_N1 process:

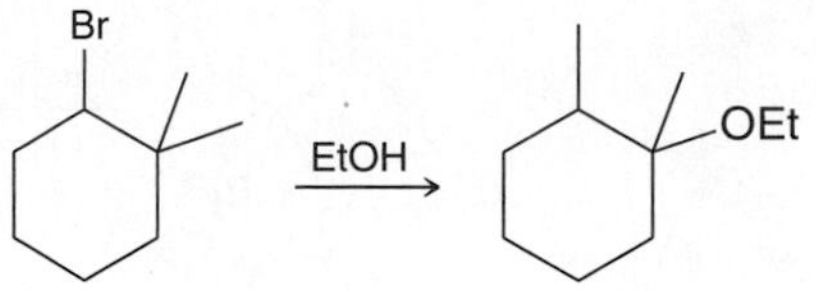

SOLUTION

An S_N1 process must always exhibit two core steps: loss of a leaving group and nucleophilic attack. But we must consider whether any of the other three possible steps will occur:

Proton transfer — Loss of LG — Carbocation rearrangement — Nuc attack — Proton transfer

Does the LG need to be protonated before it can leave?

No. Bromide is a good LG.

Is the nucleophile ultimately positioned at a different location than the leaving group?

Yes. This indicates a carbocation rearrangement.

Is the nucleophile neutral?

Yes. We will therefore need a proton transfer at the end of the mechanism in order to remove the positive charge.

The mechanism will not begin with a proton transfer, but there will be a carbocation rearrangement, and there will be a proton transfer at the end of the mechanism. Therefore, the mechanism will have four steps:

Loss of LG — Carbocation rearrangement — Nuc attack — Proton transfer

Notice that this sequence utilizes each of the four arrow-pushing patterns that are possible for an ionic reaction. To draw these steps, we will rely on the skills we learned in Chapter 6:

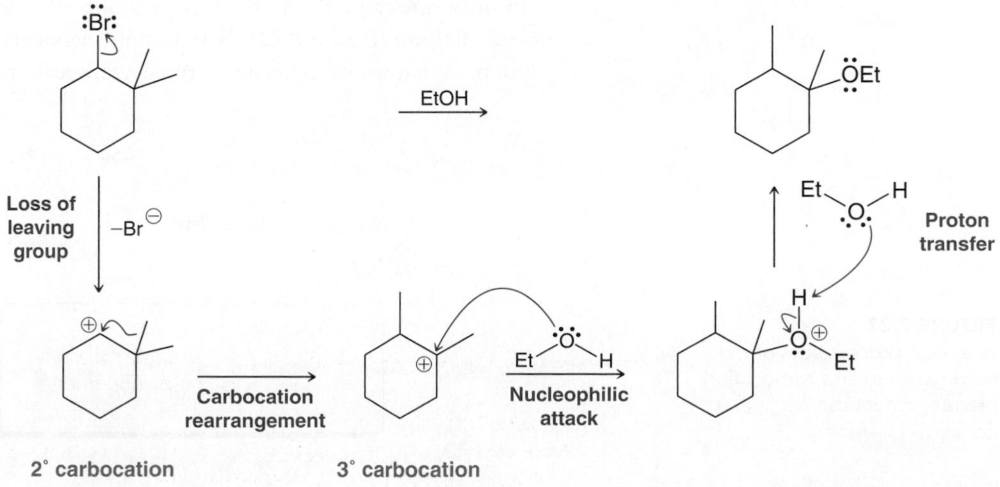

PRACTICE the skill **7.21** Draw the mechanism for each of the following S_N1 processes:

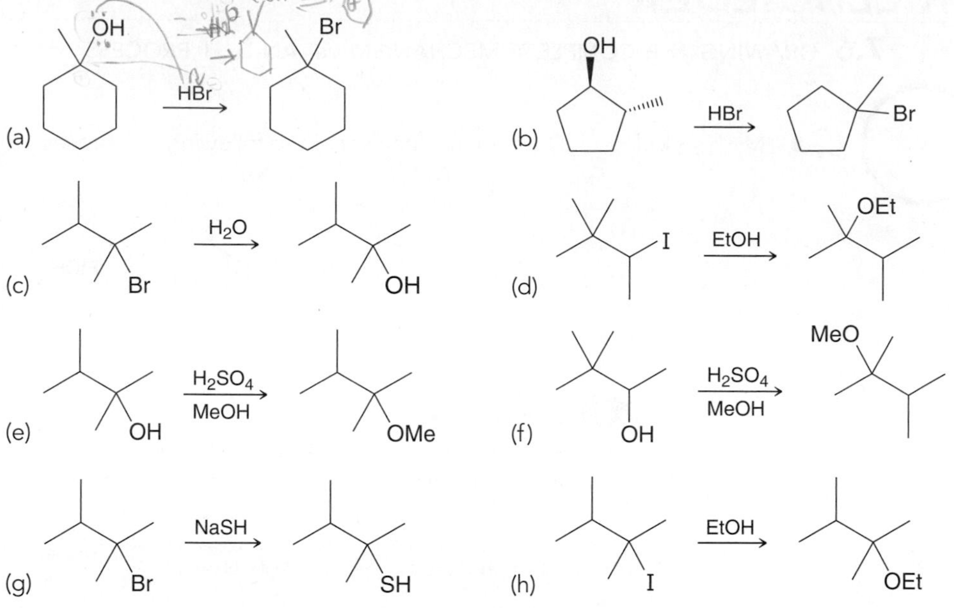

APPLY the skill **7.22** Identify the number of steps (patterns) for the mechanisms in Problems 7.21a–h. For example, the patterns for the first two are:

7.21a: $+H^+$ — $-LG$ — Nuc attack This mechanism exhibits a proton transfer before the two core steps.

7.21b: $+H^+$ — $-LG$ — $C+$ rearrangement — Nuc attack This mechanism exhibits a proton transfer before the two core steps as well as a carbocation rearrangement in between the two core steps.

These patterns are not identical. Draw patterns for the other six problems. Then compare the patterns. There is only one pattern that is repeated in Problem 7.21. Identify the two problems that exhibit the same pattern, and then describe in words why those two reactions are so similar.

7.23 Treatment of (*2R*,*3R*)-3-methyl-2-pentanol with H_3O^+ affords a compound with no chirality centers. Predict the product of this reaction and draw the mechanism of its formation. Use your mechanism to explain how both chirality centers are destroyed.

need more **PRACTICE?** **Try Problems 7.48, 7.49, 7.52, 7.54, 7.65**

In some rare cases, loss of the leaving group and carbocation rearrangement can occur in a concerted fashion (Figure 7.22). For example, neopentyl bromide cannot directly lose its leaving group, as that would generate a primary carbocation, which is too high in energy to form:

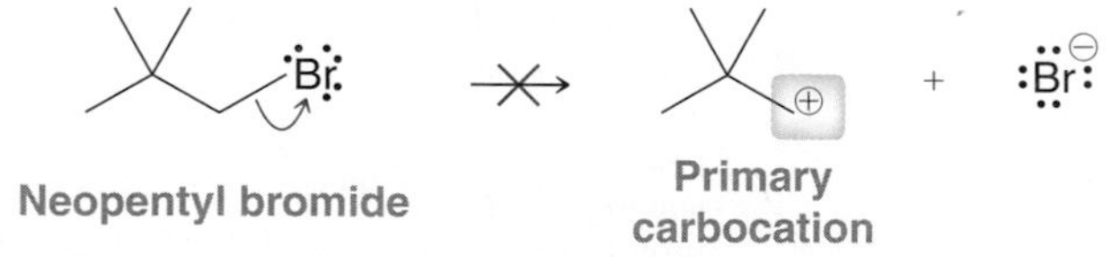

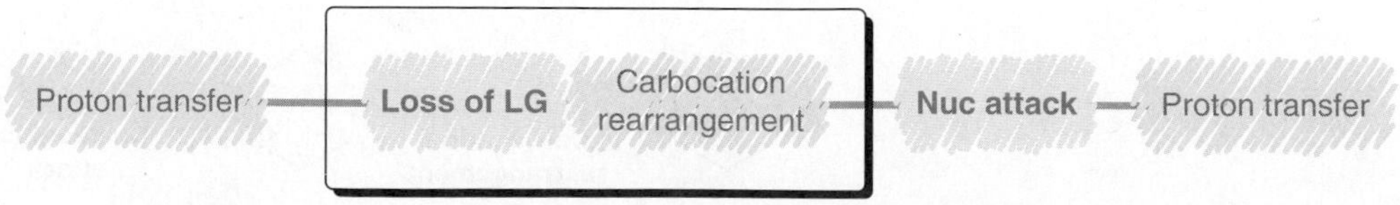

FIGURE 7.22 In an S_N1 process, loss of the leaving group and carbocation rearrangement can occur in a concerted fashion.

But it is possible for the leaving group to leave as a result of a methyl shift:

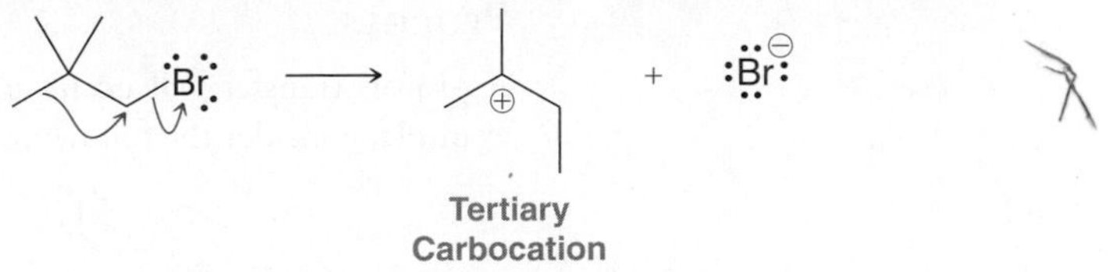

This is essentially a concerted process in which loss of the leaving group occurs simultaneously with a carbocation rearrangement. Examples like this are less common. In the vast majority of cases, each step of an S_N1 process occurs separately.

7.7 Drawing the Complete Mechanism of an S_N2 Reaction

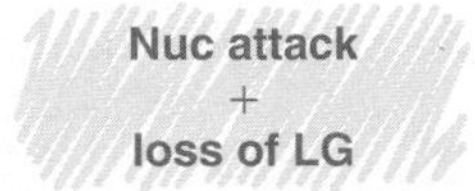

FIGURE 7.23
The one concerted step of an S_N2 process.

In the previous section, we analyzed the additional steps that can accompany an S_N1 process. In this section, we analyze the additional steps that can accompany an S_N2 process. Recall that an S_N2 reaction is a concerted process in which nucleophilic attack and loss of the leaving group occur simultaneously (Figure 7.23). No carbocation is formed, so there can be no carbocation rearrangement. In an S_N2 process, the only two possible additional steps are proton transfers (Figure 7.24). There can be a proton transfer before and/or after the concerted step. Proton transfers will accompany S_N2 processes for the same reasons that they accompany S_N1 processes.

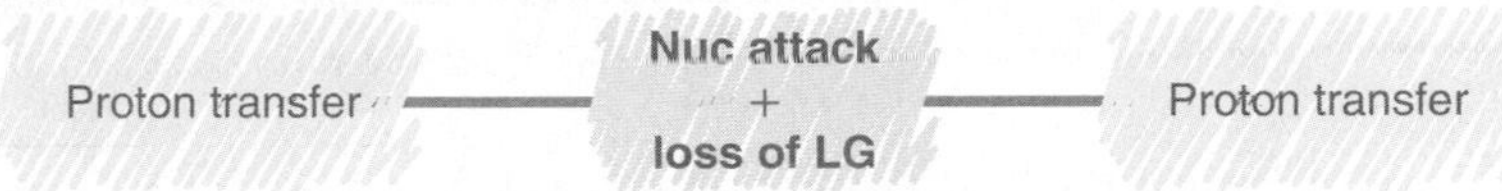

FIGURE 7.24
The concerted step and the two possible additional steps of an S_N2 process.

Specifically, a proton transfer is required at the beginning of a mechanism if the leaving group is an OH group, and it is required at the end of the mechanism if the nucleophile is neutral. Let's see examples of each.

Proton Transfer at the Beginning of an S_N2 Process

A proton transfer is necessary at the beginning of an S_N2 process if the leaving group is an OH group. An example of this reaction is the conversion of methanol to methyl chloride, a reaction first performed by the French chemists Jean-Baptiste Dumas and Eugene Peligot in 1835. This transformation was achieved by boiling a mixture of methanol, sulfuric acid, and sodium chloride.

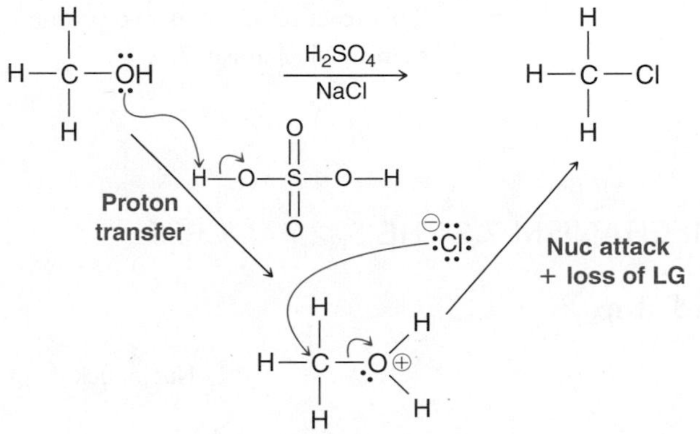

The OH group is first protonated, converting it into a good leaving group, and then the chloride ion attacks in an S_N2 process, displacing the leaving group. Methyl chloride is prepared commercially by a similar process (using HCl as the source of H^+ and Cl^-):

$$CH_3OH \xrightarrow{HCl} CH_3Cl$$

Proton Transfer at the End of an S_N2 Process

S_N2 — $-H^+$

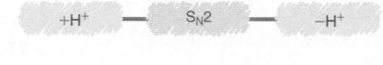

A proton transfer will occur at the end of an S_N2 process if the nucleophile is neutral. For example, consider the following solvolysis reaction:

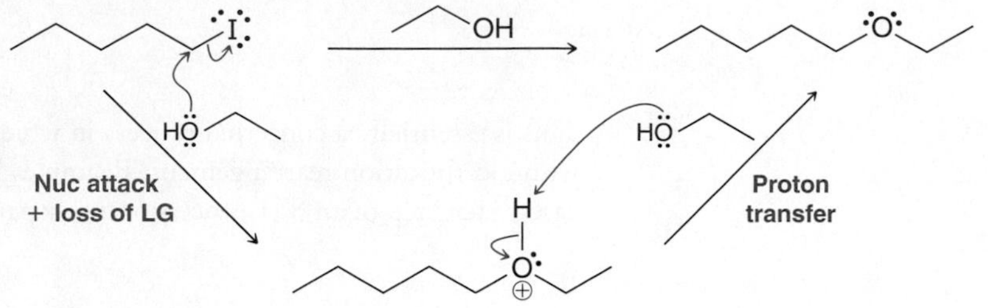

The substrate is primary, and therefore, the reaction must proceed via an S_N2 process. In this case, the solvent (ethanol) is functioning as the nucleophile, so this is a solvolysis reaction. Since the nucleophile is neutral, a proton transfer is required at the end of the mechanism in order to remove the positive charge from the compound.

Proton Transfer Before and After an S_N2 Process

$+H^+$ — S_N2 — $-H^+$

Throughout this course, we will see other examples of S_N2 processes that are accompanied by proton transfers. For example, the following reaction will be explored in Sections 9.16 and 14.10:

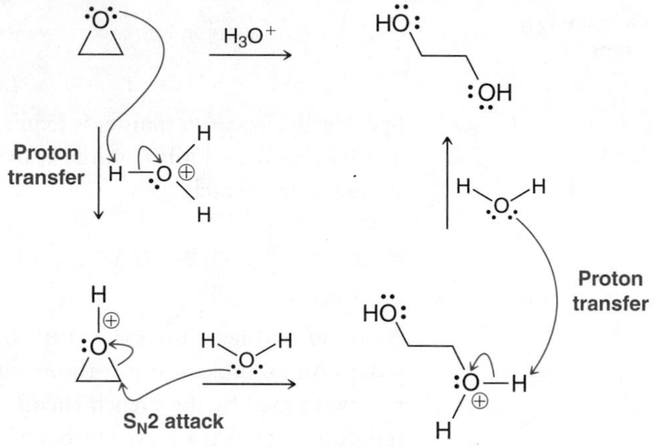

This reaction involves two proton transfer steps—one before and one after the S_N2 attack—as seen in Mechanism 7.2.

MECHANISM 7.2 THE S_N2 PROCESS

One concerted step

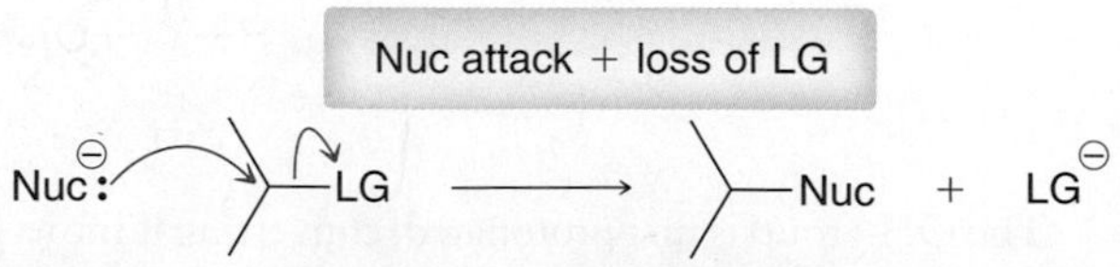

An S_N2 process is comprised of just one concerted step in which the nucleophile attacks with simultaneous loss of the leaving group.

Possible additional steps

If the leaving group is an OH group, it must be protonated before it can leave.

If the nucleophile is neutral, a proton transfer is required to remove the positive charge that is generated.

SKILLBUILDER

7.7 DRAWING THE COMPLETE MECHANISM OF AN S_N2 PROCESS

LEARN the skill

Ethyl bromide was dissolved in water and heated, and the following solvolysis reaction was observed to occur slowly, over a long period of time. Propose a mechanism for this reaction.

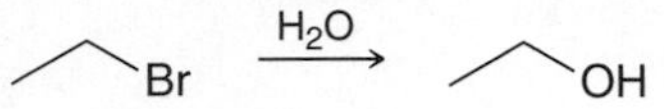

SOLUTION

The substrate is primary, so the reaction must proceed via an S_N2 process, rather than S_N1. An S_N1 mechanism cannot be invoked in this case, because a primary carbocation would be too unstable to form.

In an S_N2 process, there is one concerted step, and we must determine whether a proton transfer will be necessary either before or after the concerted step:

Does the LG need to be protonated first?

No. Bromide is a good LG.

Is the nucleophile neutral?

Yes. We will therefore need a proton transfer at the end of the mechanism in order to remove the positive charge.

The mechanism does not require a proton transfer before the concerted step, and even if it did, the reagents are not acidic and could not donate a proton anyway. In order to have a proton transfer at the beginning of a mechanism, an acid is required to serve as a proton source. At the end of the mechanism, there will be a proton transfer because the attacking nucleophile (H_2O) is neutral. Therefore, the mechanism will have two steps:

The first step involves a simultaneous nucleophilic attack and loss of leaving group, and the second step is a proton transfer:

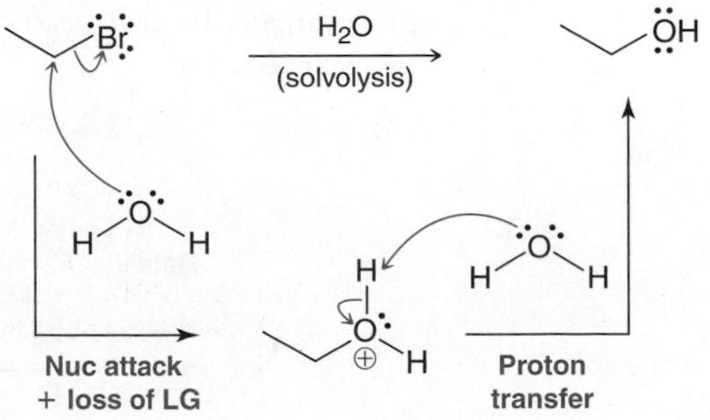

PRACTICE the skill **7.24** Draw the mechanism for each of the following solvolysis reactions:

(a) [structure with Cl] $\xrightarrow[\text{(solvolysis)}]{\text{MeOH}}$ [structure with OMe]

(b) [structure with Br] $\xrightarrow[\text{(solvolysis)}]{\text{EtOH}}$ [structure with O-ethyl ether]

(c) [cyclohexyl structure with I] $\xrightarrow[\text{(solvolysis)}]{\text{H}_2\text{O}}$ [cyclohexyl structure with OH]

(d) [benzyl Cl structure] $\xrightarrow[\text{(solvolysis)}]{}$ [benzyl propyl ether structure]

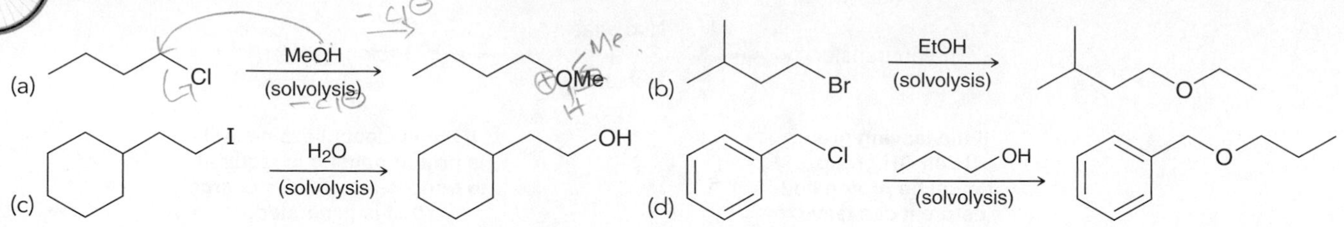

APPLY the skill

7.25 In Chapter 23, we will learn that treatment of ammonia with excess methyl iodide produces a quaternary ammonium salt. This transformation is the result of four sequential S_N2 reactions. Use the tools we have learned in this chapter to draw the mechanism of this transformation. Your mechanism should have seven steps.

$$NH_3 \xrightarrow{\text{Excess MeI}} \underset{\overset{|}{\underset{|}{Me}}}{\overset{\overset{Me}{|}}{Me-\overset{\oplus}{N}-Me}} \quad I^{\ominus}$$

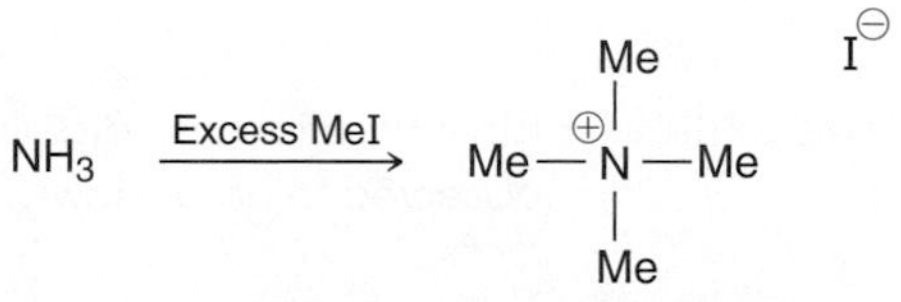

Quaternary ammonium salt

\dashrightarrow need more **PRACTICE?** **Try Problems 7.53, 7.64, 7.66**

7.8 Determining Which Mechanism Predominates

In order to draw the products of a specific substitution reaction, we must first identify the reaction mechanism as either S_N2 or S_N1. This information is important in the following two ways:

- If substitution is taking place at a chirality center, then we must know whether to expect inversion of configuration (S_N2) or racemization (S_N1).
- If the substrate is susceptible to carbocation rearrangement, then we must know whether to expect rearrangement (S_N1) or whether rearrangement is not possible (S_N2).

Four factors have an impact on whether a particular reaction will occur via an S_N2 or an S_N1 mechanism: (1) the substrate, (2) the leaving group, (3) the nucleophile, and (4) the solvent (Figure 7.25). We must learn to look at all four factors, one by one, and to determine whether the factors favor S_N1 or S_N2.

$$R-LG \xrightarrow[\text{Solvent}]{\text{Nuc}} Product(s)$$

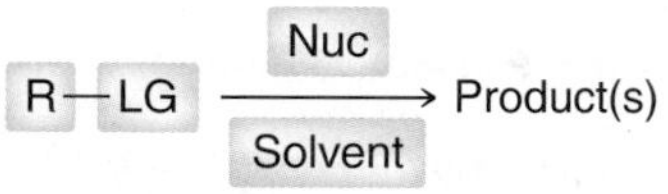

FIGURE 7.25
The four factors that determine which mechanism predominates.

The Substrate

The identity of the substrate is the most important factor in distinguishing between S_N2 and S_N1. Earlier in the chapter, we saw different trends for S_N2 and S_N1 reactions. These trends are compared in the charts in Figure 7.26.

The trend in S_N2 reactions is due to issues of steric hindrance in the transition state, while the trend in S_N1 reactions is due to carbocation stability. The bottom line is that methyl and primary substrates favor S_N2, while tertiary substrates favor S_N1. Secondary substrates can proceed via either mechanism, so a secondary substrate does not indicate which mechanism will predominate. In such a case, you must move on to the next factor, the nucleophile (covered in the next section).

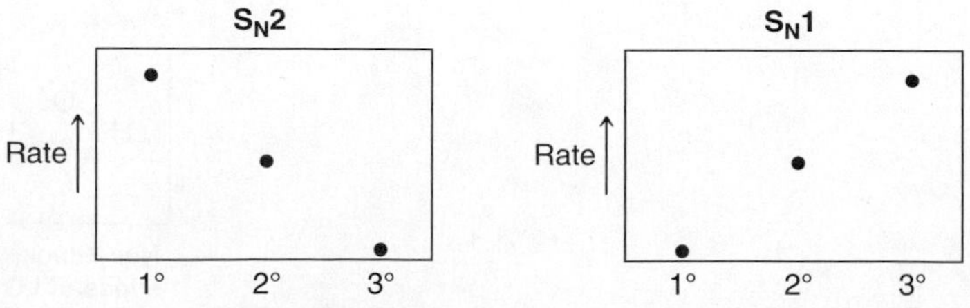

FIGURE 7.26
Substrate effects on the rates of S_N2 and S_N1 processes.

Allylic halides and *benzylic halides* can react either via S_N2 or via S_N1 processes:

Allylic halides **Benzylic halides**

These substrates can react via an S_N2 mechanism because they are relatively unhindered, and they can react via an S_N1 mechanism because loss of a leaving group generates a resonance-stabilized carbocation:

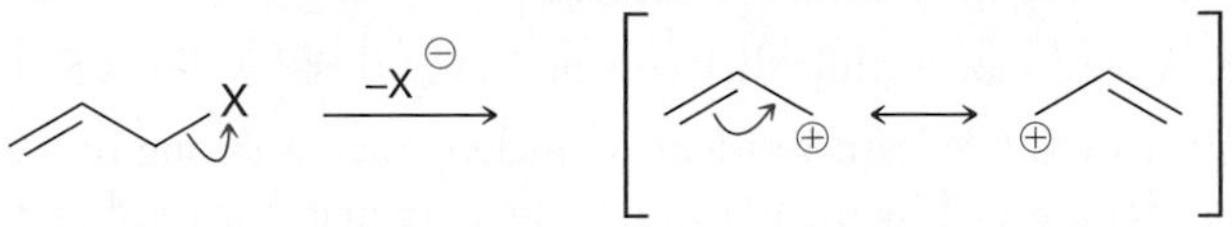

Resonance stabilized

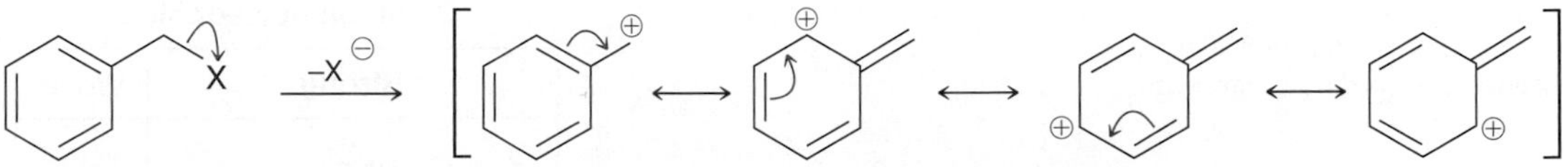

Resonance stabilized

In contrast, *vinyl halides* and *aryl halides* are unreactive in substitution reactions:

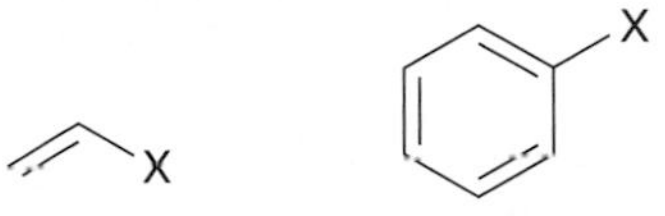

Vinyl halides **Aryl halides**

S_N2 reactions are generally not observed at sp^2-hybridized centers, because back-side attack is sterically encumbered. In addition, vinyl halides and aryl halides are also unreactive toward S_N1, because loss of a leaving group would generate an unstable carbocation:

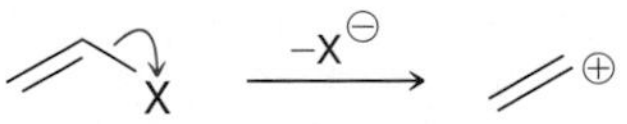

Not stabilized by resonance

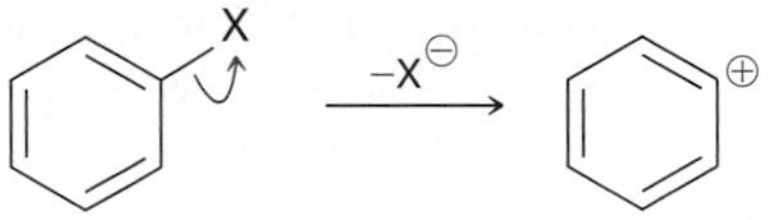

Not stabilized by resonance

To summarize:

- Methyl and primary substrates favor S_N2.
- Tertiary substrates favor S_N1.
- Secondary substrates and allylic and benzylic substrates can react via either mechanism.
- Vinyl and aryl substrates do not react via either mechanism.

CONCEPTUAL CHECKPOINT

7.26 Identify whether each of the following substrates favors S_N2, S_N1, both, or neither:

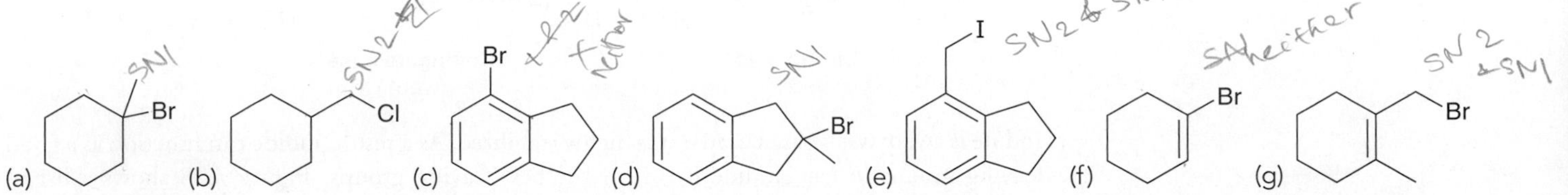

The Nucleophile

Recall that the rate of an S_N2 process is dependent on the concentration of the nucleophile. For the same reason, S_N2 processes are also dependent on the *strength* of the nucleophile. A strong nucleophile will speed up the rate of an S_N2 reaction, while a weak nucleophile will slow down the rate of an S_N2 reaction. In contrast, an S_N1 process is not affected by the concentration or strength of the nucleophile because the nucleophile does not participate in the rate-determining step (remember the hourglass analogy from Section 7.5). In summary, the nucleophile has the following effect on the competition between S_N2 and S_N1:

- A strong nucleophile favors S_N2.
- A weak nucleophile disfavors S_N2 (and thereby allows S_N1 to compete successfully).

We must therefore learn to identify nucleophiles as strong or weak. The strength of a nucleophile is determined by many factors that were first discussed in Section 6.7. Figure 7.27 shows some strong and weak nucleophiles that we will encounter.

FIGURE 7.27 Some common nucleophiles grouped according to strength.

Common nucleophiles

Strong			Weak
$I^{\ominus}$	$HS^{\ominus}$	$HO^{\ominus}$	$F^{\ominus}$
$Br^{\ominus}$	H_2S	$RO^{\ominus}$	H_2O
$Cl^{\ominus}$	RSH	$N{\equiv}C^{\ominus}$	ROH

CONCEPTUAL CHECKPOINT

7.27 Does each of the following nucleophiles favor S_N2 or S_N1?

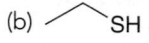

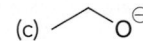

(a) 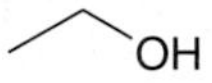 (b) 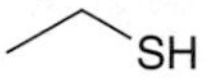 (c) 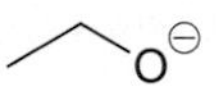 (d) NaOH (e) NaCN

The Leaving Group

Both S_N1 and S_N2 mechanisms are sensitive to the identity of the leaving group. If the leaving group is bad, then neither mechanism can operate, but S_N1 reactions are generally more sensitive to the leaving group than S_N2 reactions. Why? Recall that the rate-determining step of an S_N1 process is loss of a leaving group to form a carbocation and a leaving group:

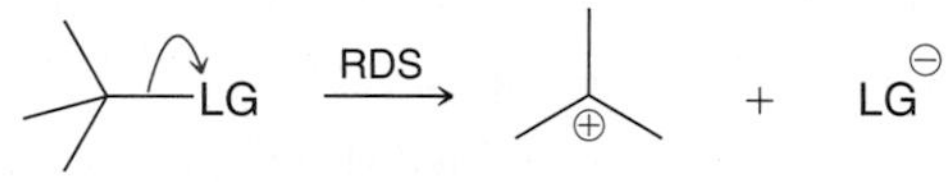

We have already seen that the rate of this step is very sensitive to the stability of the carbocation, but it is also sensitive to the stability of the leaving group. The leaving group must be highly stabilized in order for an S_N1 process to be effective.

What determines the stability of a leaving group? As a general rule, good leaving groups are the conjugate bases of strong acids. For example, iodide (I^-) is the conjugate base of a very strong acid (HI):

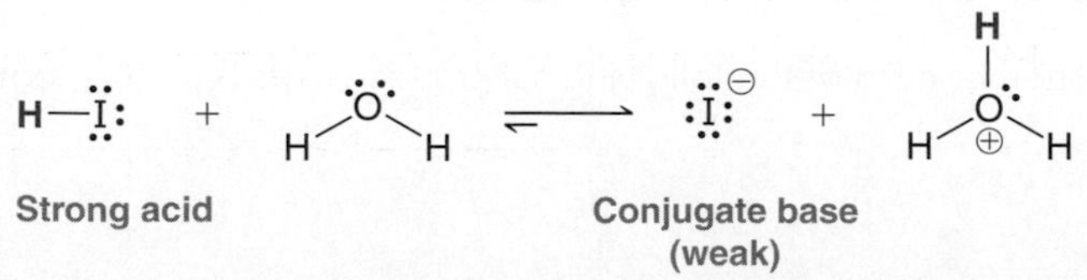

Iodide is a very weak base because it is highly stabilized. As a result, iodide can function as a good leaving group. In fact, iodide is one of the best leaving groups. Figure 7.28 shows a list of good leaving groups, all of which are the conjugate bases of strong acids. In contrast, hydroxide

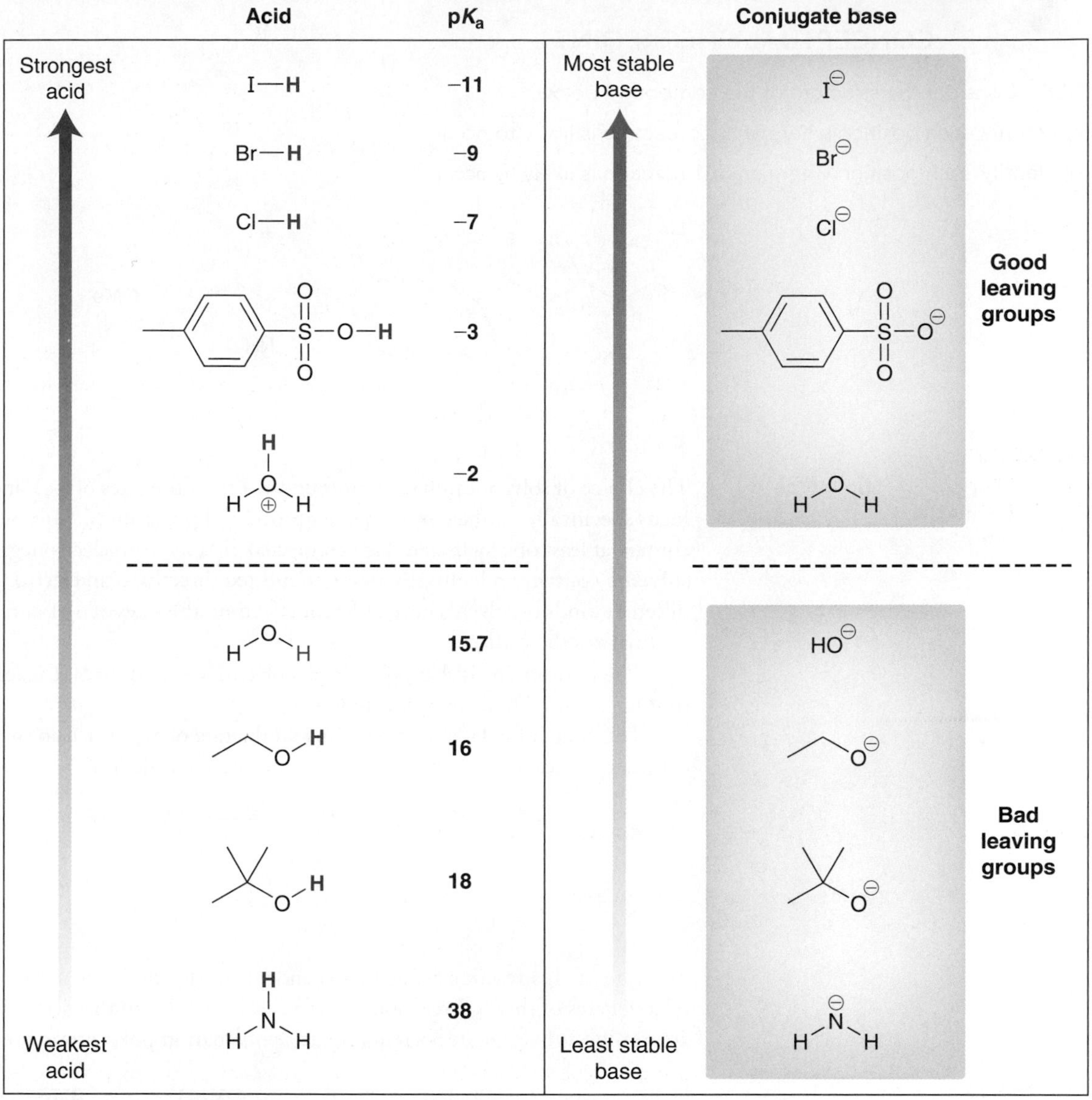

FIGURE 7.28
The conjugate base of a strong acid will generally be a good leaving group. The conjugate base of a weak acid will not be a good leaving group.

is a bad leaving group, because it is not a stabilized base. In fact, hydroxide is a relatively strong base, and therefore, it rarely functions as a leaving group. It is a bad leaving group.

The most commonly used leaving groups are halides and **sulfonate ions** (Figure 7.29). Among the halides, iodide is the best leaving group because it is a weaker base (more stable) than bromide or chloride. Among the sulfonate ions, the best leaving group is the triflate group, but the most commonly used is the **tosylate** group. It is abbreviated as OTs. When you see OTs connected to a compound, you should recognize the presence of a good leaving group.

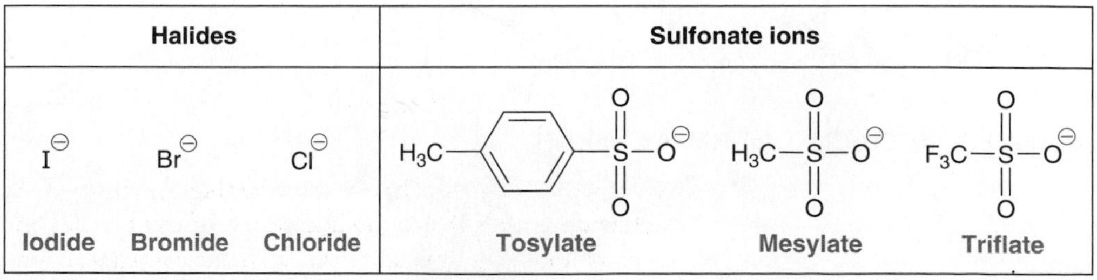

FIGURE 7.29
Common leaving groups.

CONCEPTUAL CHECKPOINT

7.28 Consider the structure of the compound below.

(a) Identify each position where an S_N2 reaction is likely to occur.

(b) Identify each position where an S_N1 reaction is likely to occur.

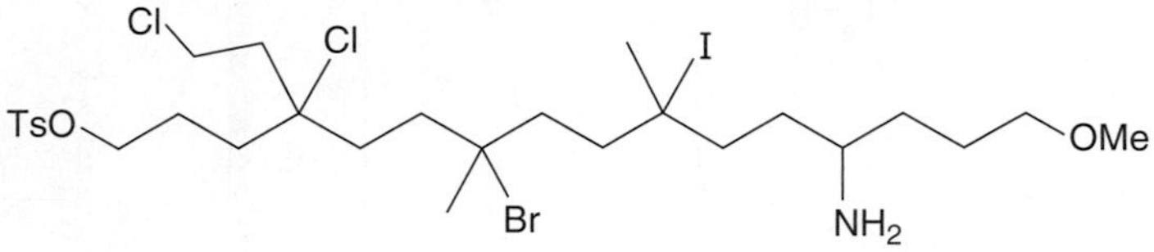

Solvent Effects

The choice of solvent can have a profound effect on the rates of S_N1 and S_N2 reactions. We will focus specifically on the effects of polar protic and polar aprotic solvents. **Polar protic solvents** contain at least one hydrogen atom connected directly to an electronegative atom. **Polar aprotic solvents** contain no hydrogen atoms connected directly to an electronegative atom. These two different kinds of solvents have different effects on the rates of S_N1 and S_N2 processes. Table 7.2 summarizes these effects.

The bottom line is that polar protic solvents are used for S_N1 reactions, while polar aprotic solvents are used to favor S_N2 reactions.

The effect of polar aprotic solvents on the rate of S_N2 reactions is significant. For example, consider the reaction between bromobutane and an azide ion:

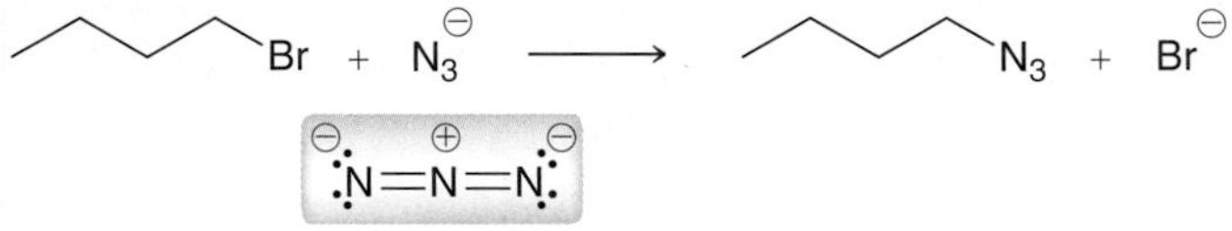

Azide

The rate of this reaction is highly dependent on the choice of solvent. Figure 7.30 shows the relative rates of this S_N2 reaction in various solvents. From these data, we see that S_N2 reactions are significantly faster in polar aprotic solvents than in polar protic solvents.

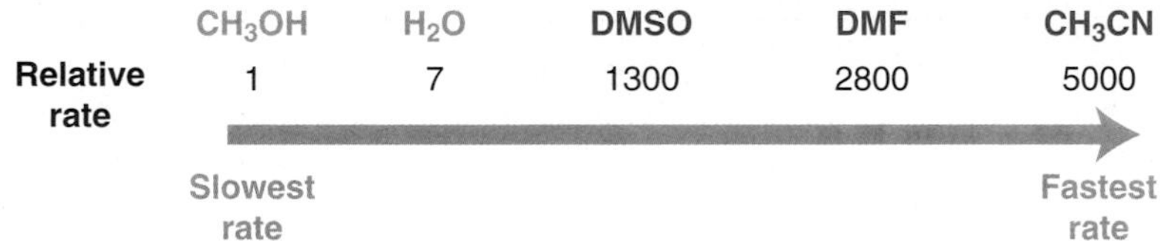

FIGURE 7.30 Relative rates of an S_N2 process in a variety of solvents. Polar protic solvents are shown in blue. Polar aprotic solvents are shown in red.

CONCEPTUAL CHECKPOINT

7.29 Does each of the following solvents favor an S_N2 reaction or an S_N1 reaction? (See Table 7.2)

(a) 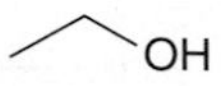

(b) 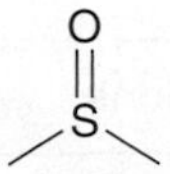

(c) 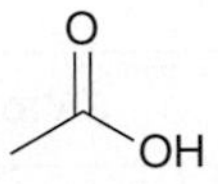

(d) 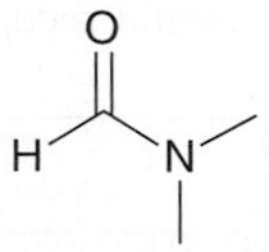

(e) MeOH (f) CH_3CN (g) HMPA (h) NH_3

7.30 When used as a solvent, will acetone favor an S_N2 or an S_N1 mechanism? Explain.

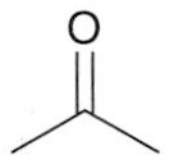

Acetone

TABLE 7.2 THE EFFECTS OF POLAR PROTIC SOLVENTS AND POLAR APROTIC SOLVENTS

	POLAR PROTIC	POLAR APROTIC
Definition	Polar protic solvents contain at least one hydrogen atom connected directly to an electronegative atom.	Polar aprotic solvents contain no hydrogen atoms connected directly to an electronegative atom.
Examples	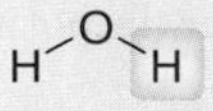 **Water** 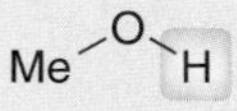 **Methanol** 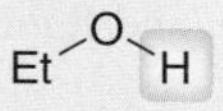 **Ethanol** 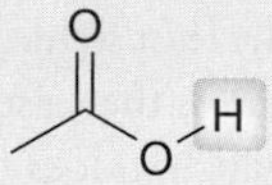 **Acetic acid** 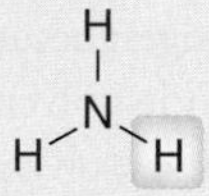 **Ammonia**	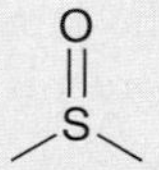 **Dimethylsulfoxide (DMSO)** $H_3C-C\equiv N$ **Acetonitrile** 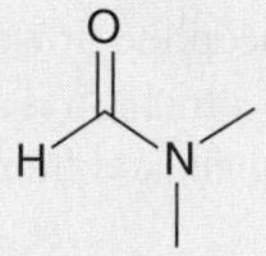 **Dimethylformamide (DMF)** 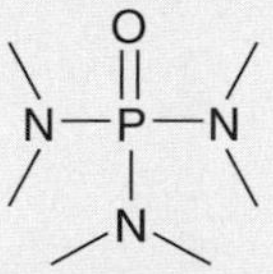 **Hexamethylphosphoramide (HMPA)**
Function	Polar protic solvents stabilize cations and anions. Cations are stabilized by lone pairs from the solvent, while anions are stabilized by H-bonding interactions with the solvent: 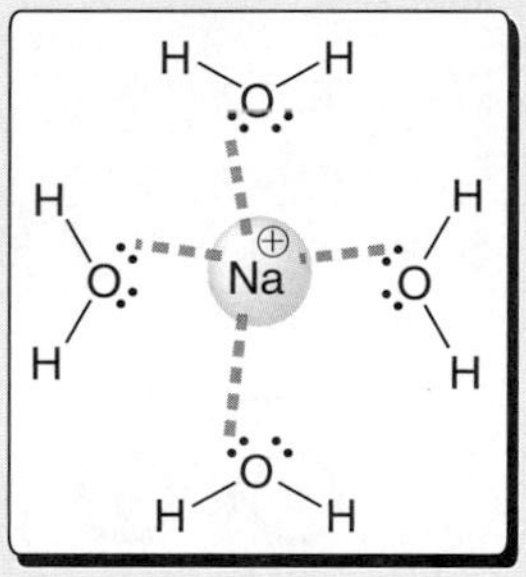 **The lone pairs on the oxygen atoms of H_2O stabilize the cation.** 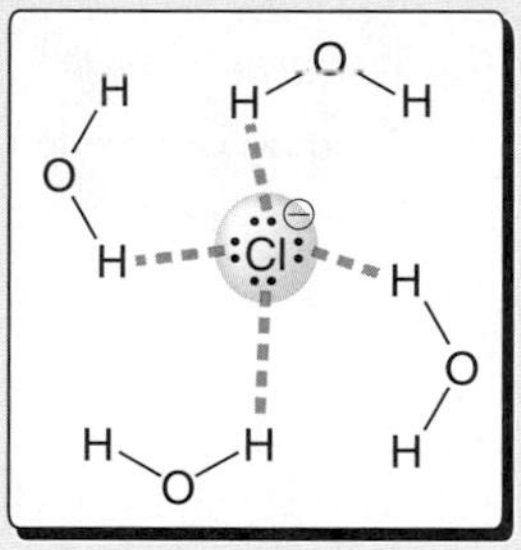 **Hydrogen-bonding interactions stabilize the anion.** As a result, anions and cations are both solvated and surrounded by a solvent shell.	Polar aprotic solvents stabilize cations, but not anions. Cations are stabilized by lone pairs from the solvent, while anions are not stabilized by the solvent: 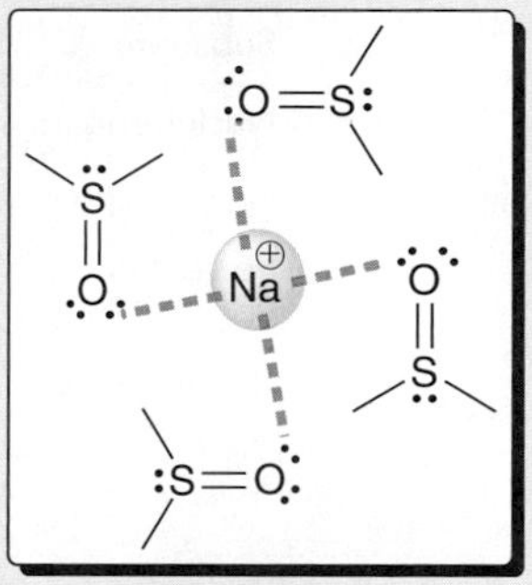 **The lone pairs on the oxygen atoms of DMSO stabilize the cation.**  **The anion is not stabilized by the solvent.** Cations are solvated and surrounded by a solvent shell, but anions are not. As a result, nucleophiles are higher in energy when placed in a polar aprotic solvent.
Effects	Favors S_N1. Polar protic solvents favor S_N1 by stabilizing polar intermediates and transition states: 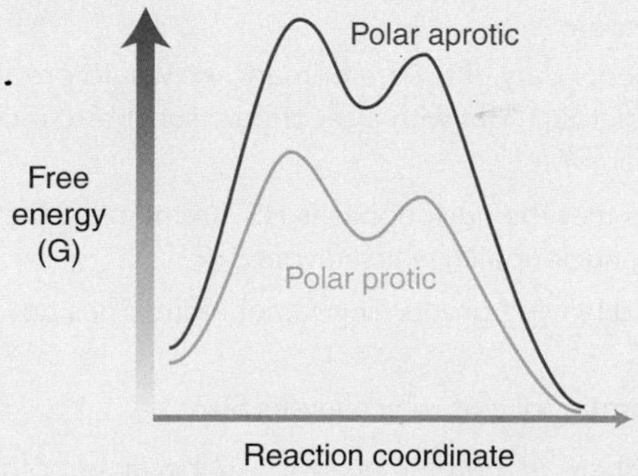	Favors S_N2. Polar aprotic solvents favor S_N2 by raising the energy of the nucleophile, giving a smaller E_a: 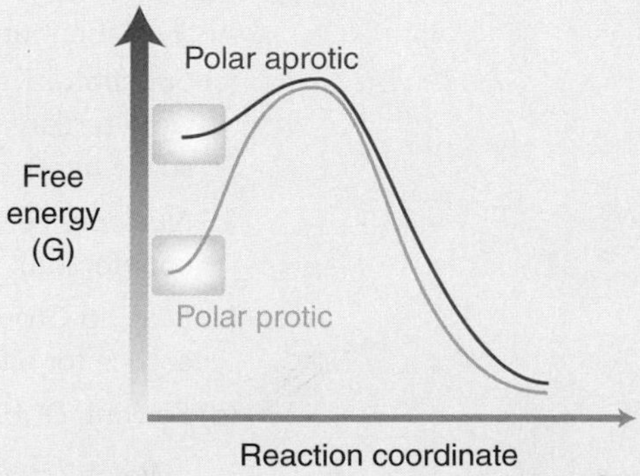

The choice of solvent can also have an impact on the order of reactivity of the halides. If we compare the nucleophilicity of the halides, we find that it is dependent on the solvent. In polar protic solvents, the following order is observed:

$$I^{\ominus} > Br^{\ominus} > Cl^{\ominus} > F^{\ominus}$$

Iodide is the strongest nucleophile, and fluoride is the weakest. However, in polar aprotic solvents, the order is reversed:

$$F^{\ominus} > Cl^{\ominus} > Br^{\ominus} > I^{\ominus}$$

Why the reversal of order? Fluoride is the strongest because it is the least stable anion. In polar protic solvents, fluoride is the most tightly bound to its solvent shell and is the least available to function as a nucleophile (it would have to shed part of its solvent shell, which it does not do often). In such an environment, it is a weak nucleophile. However, when a polar aprotic solvent is used, there is no solvent shell, and fluoride is free to function as a strong nucleophile.

Summary of Factors Affecting S_N2 and S_N1 Mechanisms

Table 7.3 summarizes what we have learned in this section about the four factors that affect S_N2 and S_N1 processes. Now let's get some practice analyzing all four factors:

TABLE 7.3 FACTORS THAT FAVOR S_N2 AND S_N1 PROCESSES

FACTOR	FAVORS S_N2	FAVORS S_N1
Substrate	Methyl or primary	Tertiary
Nucleophile	Strong nucleophile	Weak nucleophile
Leaving group	Good leaving group	Excellent leaving group
Solvent	Polar aprotic	Polar protic

SKILLBUILDER

7.8 DETERMINING WHETHER A REACTION PROCEEDS VIA AN S_N1 OR S_N2 MECHANISM

LEARN the skill

Determine whether the following reaction proceeds via an $S_N1$or an S_N2 mechanism, and then draw the product(s) of the reaction:

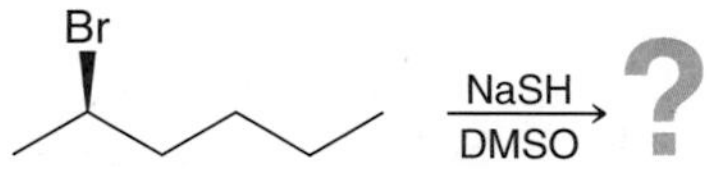

SOLUTION

Analyze the four factors one by one:

(a) *Substrate.* The substrate is secondary. If it were primary, we would predict S_N2, and if it were tertiary, we would predict S_N1. But with a secondary substrate, it could be either, so we move on to the next factor.

(b) *Nucleophile.* NaSH indicates that the nucleophile is HS^- (remember that Na^+ is just the counter ion). HS^- is a strong nucleophile, which favors S_N2.

(c) *Leaving Group.* Br^- is a good leaving group. This factor alone does not indicate a preference for either S_N1 or S_N2.

(d) *Solvent.* DMSO is a polar aprotic solvent, which favors S_N2.

Weighing all four factors, there is a preference for S_N2 because both the nucleophile and the solvent favor S_N2. Therefore, we expect inversion of configuration:

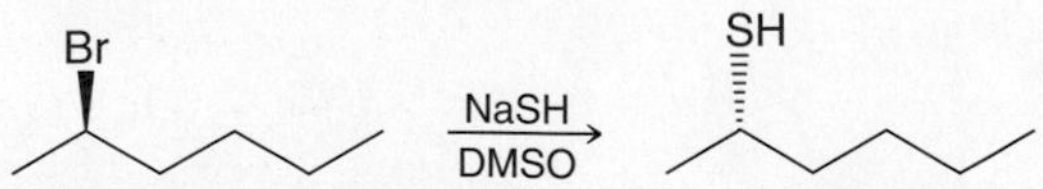

PRACTICE the skill **7.31** Determine whether each of the following reactions proceeds via an S$_N$1 or S$_N$2 mechanism and then draw the product(s) of the reaction:

(a) $\xrightarrow{\text{MeOH}}$ **?**

(b) $\xrightarrow[\text{HMPA}]{\text{Cl}^{\ominus}}$ **?**

(c) $\xrightarrow{\text{H—Br}}$ **?**

(d) $\xrightarrow[\text{DMF}]{\text{NaCN}}$ **?**

(e) $\xrightarrow{\text{H}_2\text{O}}$ **?**

(f) $\xrightarrow[\text{DMSO}]{\text{NaCN}}$ **?**

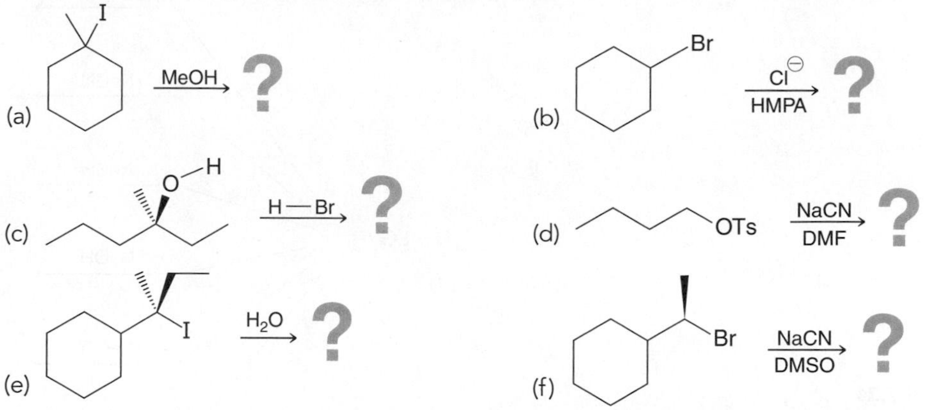

APPLY the skill **7.32** In Chapter 23, we will learn several methods for making primary amines (RNH$_2$). Each of these methods utilizes a different approach for forming the C—N bond. One of these methods, called the Gabriel synthesis, forms the C—N bond by treating potassium phthalimide with an alkyl halide:

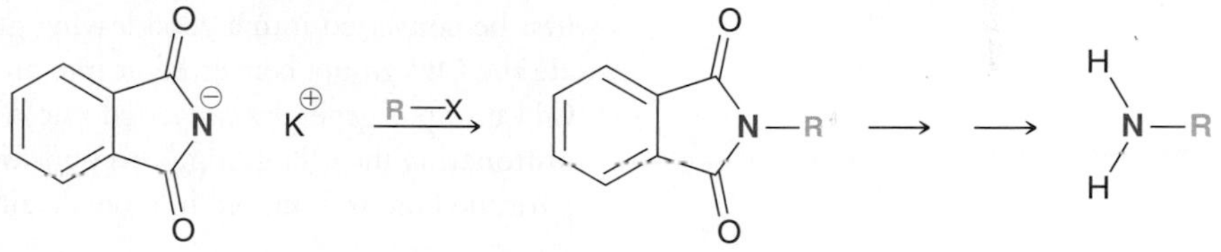

The first step of this process occurs via an S$_N$2 mechanism. Using this information, determine whether the Gabriel synthesis can be used to prepare the following amine. Explain your answer.

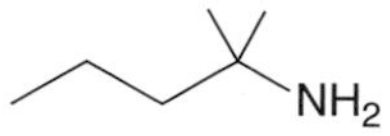

need more **PRACTICE?** **Try Problems 7.37, 7.38, 7.40, 7.41, 7.44, 7.55, 7.57, 7.58**

7.9 Selecting Reagents to Accomplish Functional Group Transformation

As mentioned at the beginning of the chapter, substitution reactions can be utilized to accomplish functional group transformation:

$$ \underset{X}{\bigwedge} \longrightarrow \underset{Y}{\bigwedge} $$

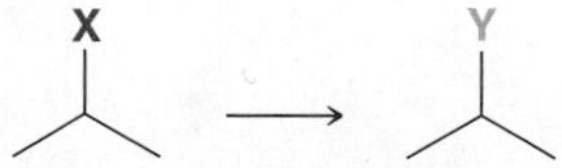

A wide range of nucleophiles can be used, providing a great deal of versatility in the type of products that can be formed with substitution reactions. Figure 7.31 shows some of the types of compounds that can be synthesized using substitution reactions. When selecting reagents for a substitution reaction, remember the following tips:

- *Substrate*. The identity of the substrate indicates which mechanism to use. If the substrate is methyl or primary, the reaction must proceed via an S$_N$2 process. If the substrate is tertiary, the reaction must proceed via an S$_N$1 process. If the substrate is secondary, generally try to use an S$_N$2 process because it avoids the issue of carbocation rearrangement and provides greater control over the stereochemical outcome.

- *Nucleophile and Solvent*. Once you have decided whether you want to use an S$_N$1 or an S$_N$2 mechanism (based on the substrate), make sure to choose a nucleophile and solvent that are consistent with that mechanism. For an S$_N$1 reaction, use a weak nucleophile in a polar protic solvent. For an S$_N$2 reaction, use a strong nucleophile in a polar aprotic solvent.

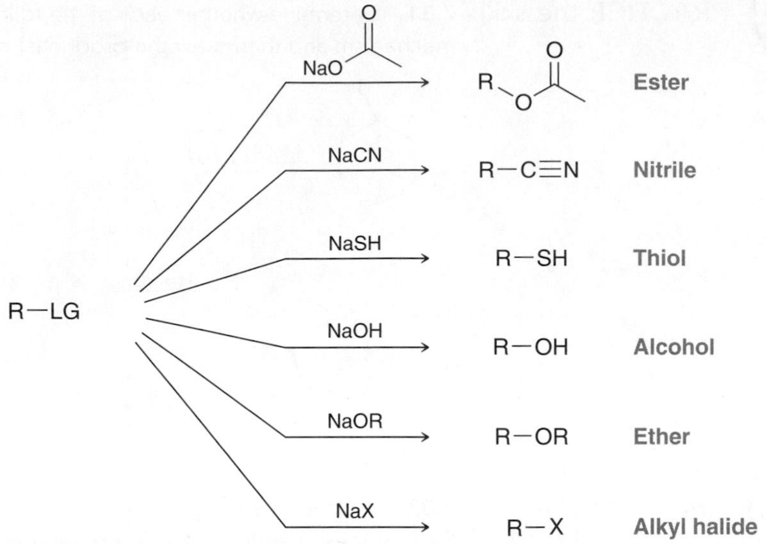

FIGURE 7.31
Various products that can be obtained via substitution reactions.

- *Leaving Group.* Remember that OH is a bad leaving group and will not leave as is. It must first be converted into a good leaving group. In an S_N1 process, use an acid to protonate the OH group, converting it into an excellent leaving group. In an S_N2 reaction, the OH group is generally converted into a tosylate, an excellent leaving group, rather than protonating the OH group. This transformation is accomplished with tosyl chloride and pyridine (and is discussed in more detail in Chapter 13):

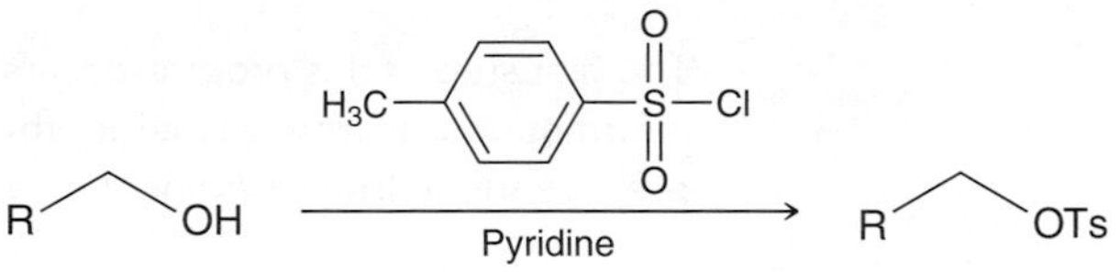

SKILLBUILDER

7.9 IDENTIFYING REAGENTS NECESSARY FOR A SUBSTITUTION REACTION

LEARN the skill

Identify the reagents you would use to accomplish the following transformation:

SOLUTION

First determine which mechanism to use by looking at the substrate:

Step 1 Analyze the substrate and the stereochemical outcome.

(a) *Substrate.* The substrate is secondary, so it could go either way. In general, choose S_N2, because it gives greater control. In this particular case, an S_N2 pathway must be used because the product is formed only through inversion of configuration. Now choose reagents that favor S_N2.

Step 2 Analyze the leaving group.

(b) *Leaving Group.* The OH is a bad leaving group and must be converted into a better leaving group. When performing an S_N2 reaction, the OH should be converted into a tosylate by using TsCl and pyridine:

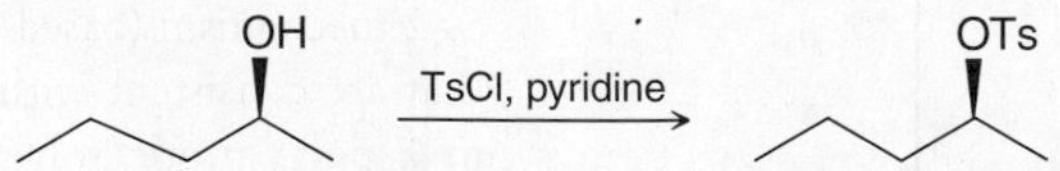

Notice that the stereoisomerism does not change when the OH group is converted into a tosylate group. If the OH is on a wedge, then the tosylate group will also be on a wedge.

Step 3 Choose conditions that favor the required mechanism.

(c) *Nucleophile.* In order to accomplish the desired transformation, the nucleophile needs to be cyanide (CN^-). Cyanide is a strong nucleophile, which supports an S_N2 process.

(d) *Solvent.* In order to favor an S_N2 process, a polar aprotic solvent such as DMSO should be used:

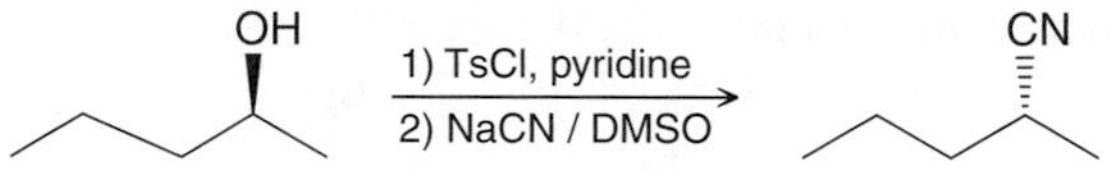

Notice that the conversion of the OH into a tosylate and the S_N2 reaction are two separate synthetic steps, so we place the numbers 1 and 2 before the sets of reagents to indicate that they are separate reactions.

PRACTICE the skill **7.33** Identify the reagents you would use to accomplish each of the following transformations:

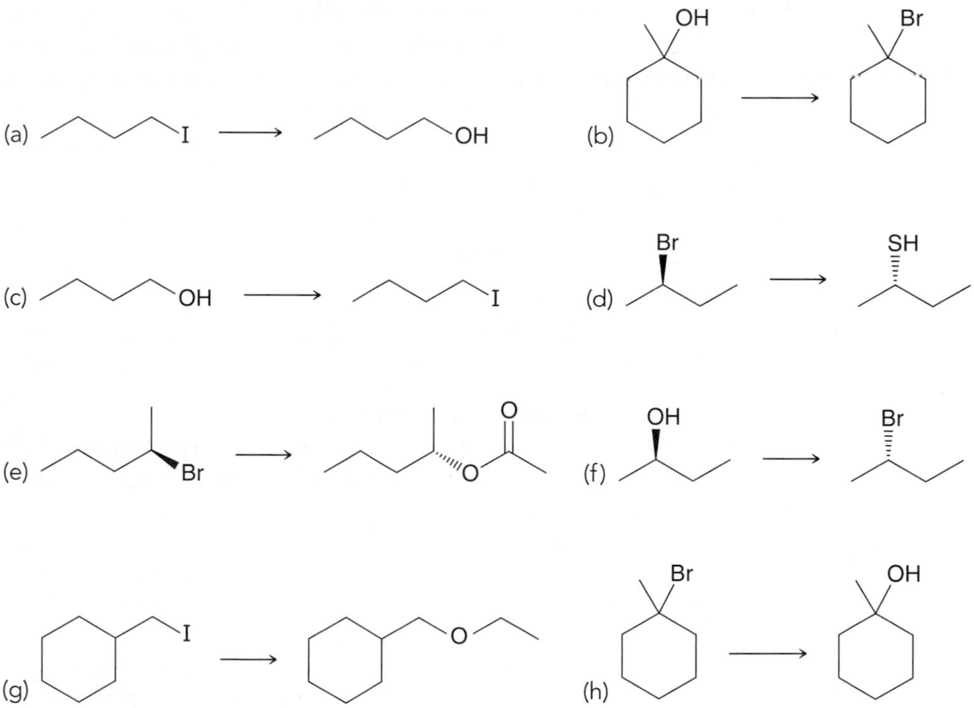

APPLY the skill **7.34** What reagents would you use to accomplish a substitution with retention of configuration; for example:

(*R*)-2-Butanol (*R*)-2-Butanethiol

need more **PRACTICE?** **Try Problems 7.59, 7.60, 7.63**

MEDICALLYSPEAKING

Pharmacology and Drug Design

Pharmacology is the study of how drugs interact with biological systems, including the mechanisms that explain drug action. Pharmacology is a very important field of study because it serves as the basis for the design of new drugs. In this section, we will explore one specific example, the design and development of chlorambucil, an antitumor agent:

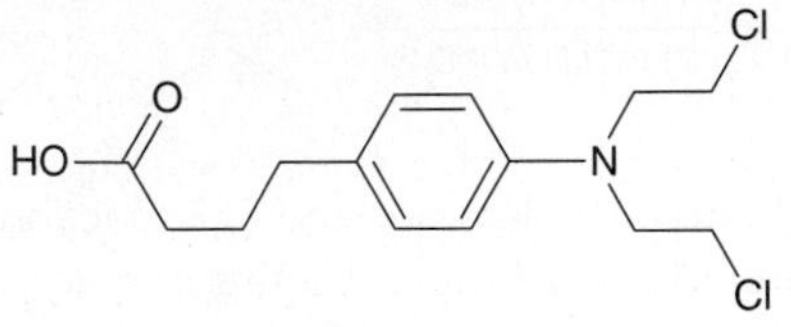

Chlorambucil

Chlorambucil was designed by chemists using principles that we have learned in this and previous chapters. The story of chlorambucil begins with a toxic compound called sulfur mustard.

This compound was first used as a chemical weapon in World War I. It was sprayed as an aerosol mixture with other chemicals and exhibited a characteristic odor similar to that of mustard plants, thus the name *mustard gas.* Sulfur mustard is a powerful alkylating agent. The mechanism of alkylation involves a sequence of two S_N2 reactions:

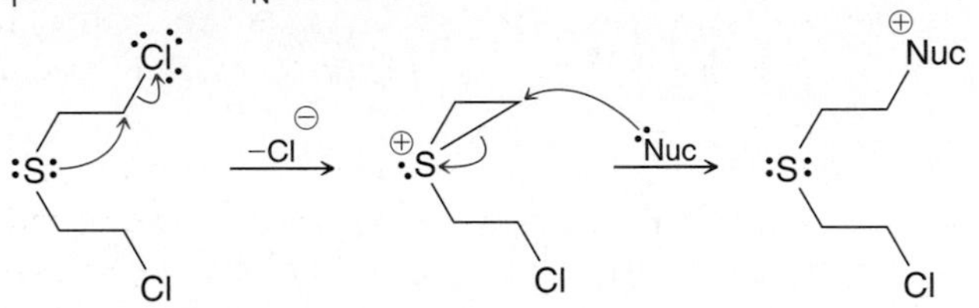

Sulfur mustard

The first substitution reaction is an intramolecular S_N2 process in which a lone pair on the sulfur serves as a nucleophile, expelling chloride as a leaving group. The second reaction is another S_N2 process involving attack of an external nucleophile. The net result is the same as if the nucleophile had attacked directly:

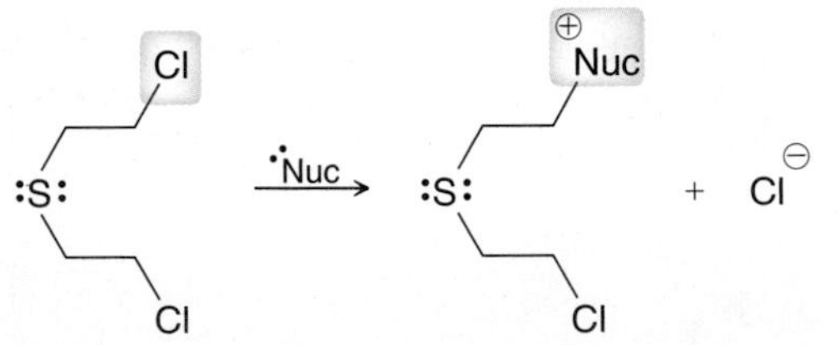

The reaction occurs much more rapidly than a regular S_N2 process on a primary alkyl chloride because the sulfur atom assists in ejecting chloride as a leaving group. The role that sulfur plays in this reaction is called *anchimeric assistance.*

Each molecule of sulfur mustard has two chloride ions and is, therefore, capable of alkylating DNA two times. This causes individual strands of DNA to cross-link (Figure 7.32). Cross-linking of DNA prevents the DNA from replicating and ultimately leads to cell death. The profound impact of sulfur mustard on cell function inspired research on the use of this compound as an antitumor agent. In 1931, sulfur mustard was injected directly into tumors with the intention of stopping tumor growth by interrupting the rapid division of the cancerous cells. Ultimately, sulfur mustard was found to be too toxic for clinical use, and the search began for a similar, less toxic, compound. The first such compound to be produced was a nitrogen analogue called mechlorethamine:

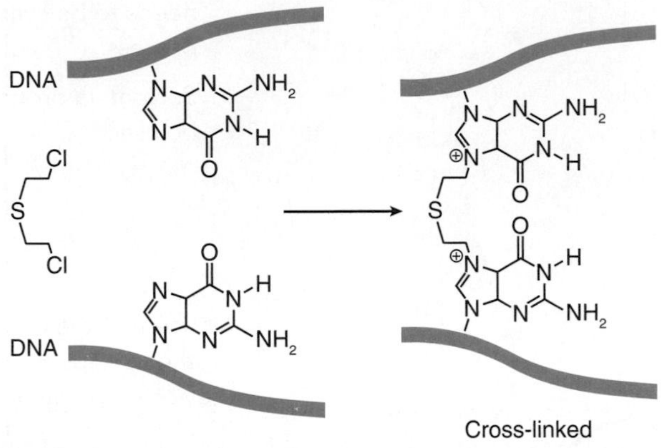

FIGURE 7.32
Sulfur mustard can alkylate two different strands of DNA, causing cross-linking.

Sulfur mustard **Nitrogen mustard (mechlorethamine)**

Mechlorethamine is a "nitrogen mustard" that reacts with nucleophiles in the same way as sulfur mustard, via two successive substitution reactions. The first reaction is an intramolecular S_N2 process involving anchimeric assistance from the nitrogen atom; the second reaction is another S_N2 process involving attack of an external nucleophile:

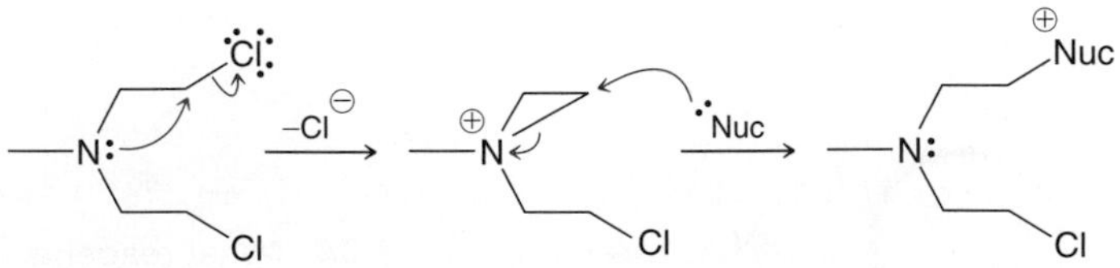

This nitrogen mustard is also capable of alkylating DNA, causing cell death, but is less toxic than sulfur mustard. The discovery of mechlorethamine launched the field of chemotherapy, the use of chemical agents to treat cancer.

Mechlorethamine is still in use today, in combination with other agents, for the treatment of advanced Hodgkin's lymphoma and chronic lymphocytic leukemia (CLL). The use of mechlorethamine is limited, though, by its high rate of reactivity with water. This limitation led to a search for other analogues. Specifically, it was found that replacing the methyl group with an aryl group had the effect of delocalizing the lone pair through resonance, rendering the lone pair less nucleophilic:

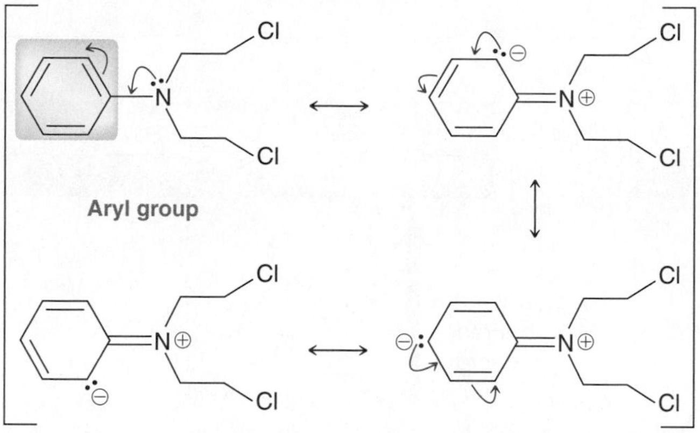

The resonance structures above all exhibit a negative charge on a carbon atom, and therefore, these resonance structures do not contribute very much to the overall resonance hybrid. Nevertheless, they are valid resonance structures, and they do contribute some character. As a result, the lone pair on the nitrogen atom is delocalized (it is spread out over the aryl group) and less nucleophilic. This decreased nucleophilicity is manifested in a slower rate of anchimeric assistance from the nitrogen atom. The compound can still function as an antitumor agent, but its rate of reactivity with water is reduced.

Introduction of the aryl group (in place of the methyl group) might have solved one problem, but it created another problem. Specifically, this new compound was not water soluble, which prevented intravenous administration. This problem was solved by introducing a carboxylate group, which rendered the compound water soluble:

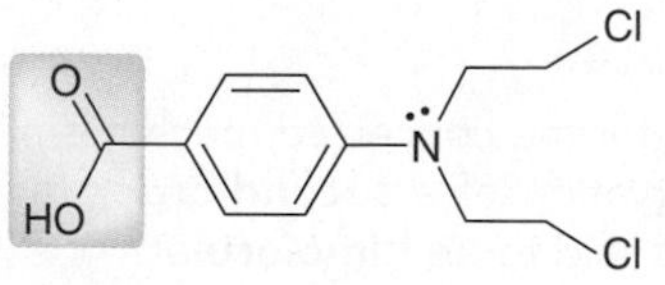

This group renders the compound water soluble

But, once again, solving one problem created another. Now, the lone pair on the nitrogen atom was too delocalized, because of the following resonance structure:

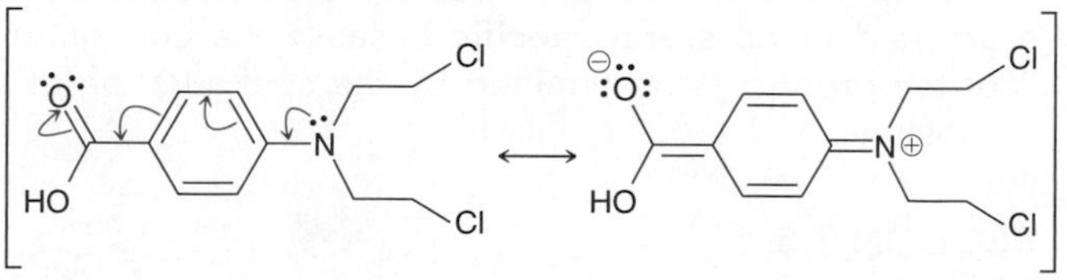

The lone pair was delocalized onto an oxygen atom, and a negative charge on an oxygen atom is much more stable than a negative charge on a carbon atom. The delocalization effect was so pronounced that the reagent no longer functioned as an antitumor agent. The lone pair on the nitrogen atom was not sufficiently nucleophilic to participate in anchimeric assistance at an appreciable rate. Solving all these problems required a way to maintain water solubility without overly stabilizing the lone pair on the nitrogen atom. This was achieved by placing methylene groups (CH_2 groups) between the carboxylate group and the aryl group:

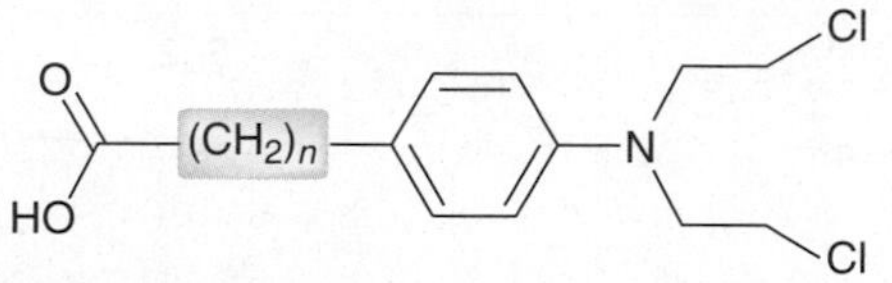

This way, the nitrogen lone pair is no longer participating in resonance with the carboxylate group, but the presence of the carboxylate group is still able to render the compound water soluble. This final change solves all of the problems. In theory, only one methylene group is needed to ensure that the nitrogen lone pair is not overly delocalized by resonance. But in practice, research with various compounds indicated that optimal reactivity was achieved when three methylene groups were placed between the carboxylate group and the aryl group:

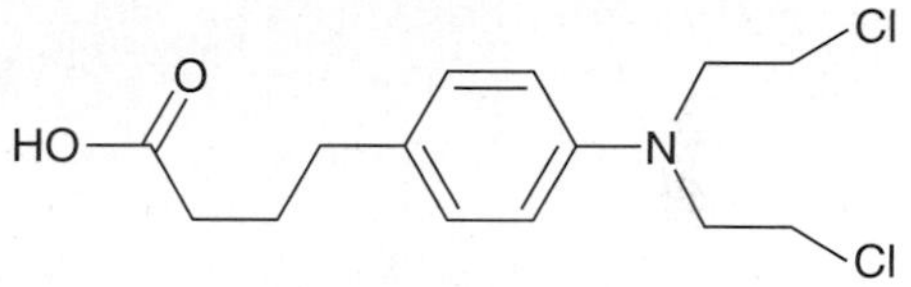

Chlorambucil

The resulting compound, called chlorambucil, was marketed under the trade name Lukeran™ by GlaxoSmithKline. It was mainly used for treatment of CLL, until other, more powerful agents were discovered.

The design and development of chlorambucil is just one example of drug design, but it demonstrates how an understanding of pharmacology, coupled with an understanding of the first principles of organic chemistry, enables chemists to design and create new drugs. Each year, organic chemists and biochemists make enormous strides in the exciting fields of pharmacology and drug design.

CONCEPTUAL CHECKPOINT

7.35 Melphalan is a chemotherapy drug used in the treatment of multiple myeloma and ovarian cancer. Melphalan is an alkylating agent belonging to the nitrogen mustard family. Draw a likely mechanism for the alkylation process that occurs when a nucleophile reacts with melphalan:

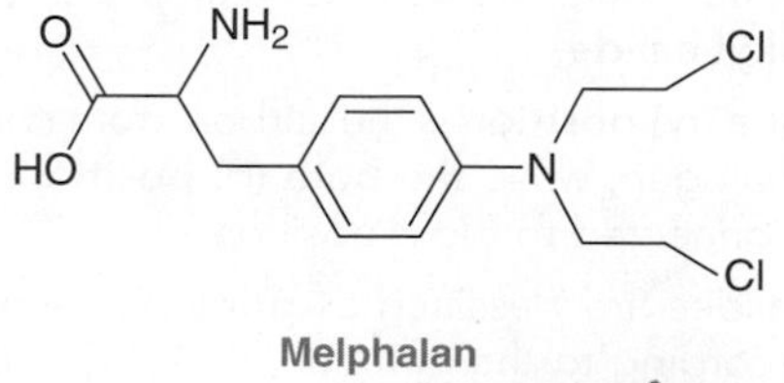

Melphalan

REVIEW OF REACTIONS

	S_N2	S_N1
Mechanism	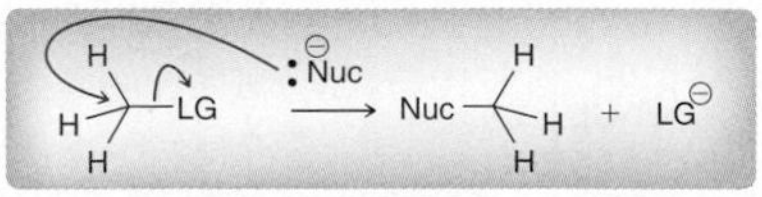	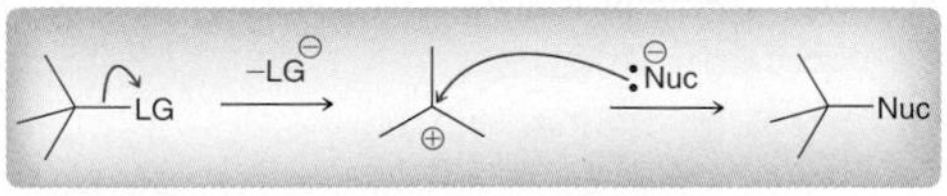
Energy diagram	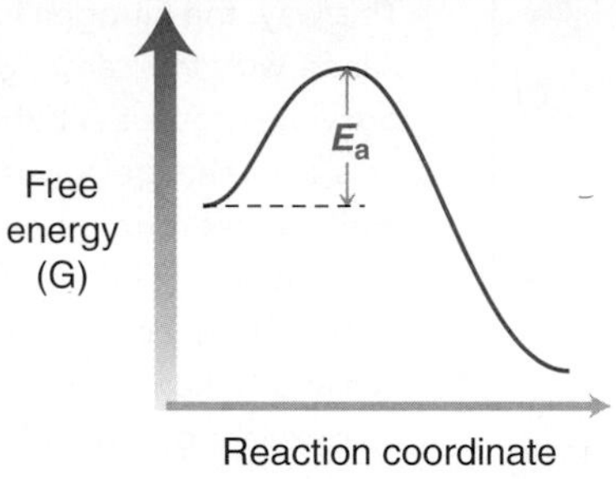	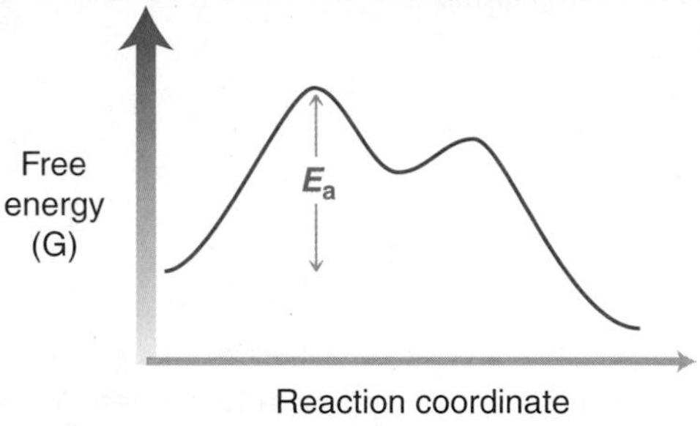
Rate equation	Rate = k [substrate] [nucleophile]	Rate = k [substrate]
Substrate effects	Methyl > 1° > 2° > 3°	3° > 2° > 1° > methyl
Stereochemistry	Inversion of configuration	Racemization (with slight preference for inversion due to the resultant ion pair)

REVIEW OF CONCEPTS AND VOCABULARY

SECTION 7.1

- **Substitution reactions** exchange one functional group for another.
- The electrophile is called the **substrate**, and it must contain a **leaving group**.

SECTION 7.2

- There are two ways to name halogenated organic compounds. The systematic name treats the compound as a **haloalkane**, while the common name treats the compound as an **alkyl halide**.
- The **alpha (α) position** is the carbon atom connected directly to the halogen, while the **beta (β) positions** are the carbon atoms connected to the α position.
- Alkyl halides are classified as **primary**, **secondary**, and **tertiary** according to the number of alkyl groups connected to the α position.

SECTION 7.3

- In a ***concerted process***, nucleophilic attack and loss of the leaving group occur simultaneously.
- In a ***stepwise process***, first the leaving group leaves, and then the nucleophile attacks.

SECTION 7.4

- Evidence for the concerted mechanism, called **S_N2**, includes the observation of a **second-order** rate equation. The S_N2 process is said to be **bimolecular**.
- Methyl halides and primary alkyl halides react most quickly, while tertiary alkyl halides are essentially unreactive toward S_N2.
- When the α position is a chirality center, the reaction proceeds with **inversion of configuration**. The preference for **back-side attack** stems from the need for constructive overlap of orbitals (according to MO theory). S_N2 reactions are said to be **stereospecific** because the configuration of the product is determined by the configuration of the substrate.

SECTION 7.5

- Evidence for the stepwise mechanism, called **S_N1**, includes the observation of a **first-order** rate equation. These reactions are said to be **unimolecular**.
- The first step of an S_N1 process (loss of a leaving group) is the **rate-determining step**.
- The carbocation intermediate can be attacked from either side, leading to **inversion of configuration** and **retention of configuration.** There is usually a slight preference for the inversion product due to the formation of ion pairs.

SECTION 7.6

- A proton transfer is necessary at the beginning of an S_N1 mechanism if the leaving group is an OH group.
- A carbocation rearrangement can take place if it will lead to a more stable carbocation intermediate.
- A proton transfer is necessary at the end of an S_N1 mechanism if the nucleophile is neutral (not negatively charged).
- When the solvent functions as a nucleophile, the reaction is called a **solvolysis** reaction.

SECTION 7.7

- A proton transfer is necessary at the beginning of an S_N2 mechanism if the leaving group is an OH group.
- A proton transfer must occur at the end of an S_N2 mechanism if the nucleophile is neutral.

SECTION 7.8

- There are four factors that impact the competition between the S_N2 mechanism and S_N1: (1) the substrate, (2) the nucleophile, (3) the leaving group, and (4) the solvent.
- The most common leaving groups are halides and **sulfonate ions.** Of the sulfonate ions, the most common is the **tosylate** group.
- **Polar protic** solvents favor S_N1, while **polar aprotic** solvents favor S_N2.

SECTION 7.9

- Depending on the type of nucleophile used, substitution reactions can be used to produce a wide range of different compounds.

KEY TERMINOLOGY

alkyl halide 284
alpha (α) position 284
back-side attack 291
beta (β) position 284
bimolecular 290
first-order 297
haloalkane 284
inversion of configuration 291
leaving group 282
organohalide 284
polar aprotic solvent 318
polar protic solvent 318
primary alkyl halide 284
rate-determining step 298
retention of configuration 301
secondary alkyl halide 284
second-order 290
S_N1 297
S_N2 290
solvolysis 305
stereospecific 291
substitution reaction 282
substrate 282
sulfonate ions 317
tertiary alkyl halide 284
tosylate 317
unimolecular 297

SKILLBUILDER REVIEW

7.1 DRAWING THE CURVED ARROWS OF A SUBSTITUTION REACTION

CONCERTED MECHANISM Two curved arrows drawn in one step. Nucleophilic attack is accompanied by simultaneous loss of a leaving group.

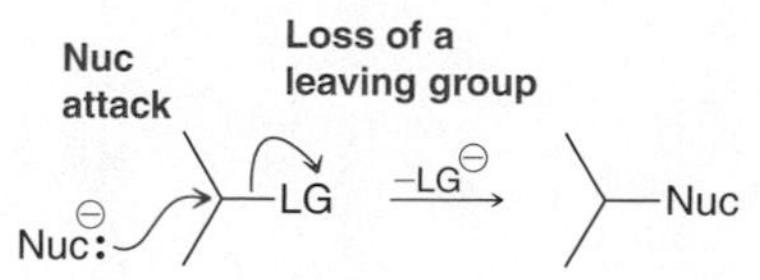

STEPWISE MECHANISM Two curved arrows drawn in two separate steps. Leaving group leaves to form a carbocation intermediate, followed by a nucleophilic attack.

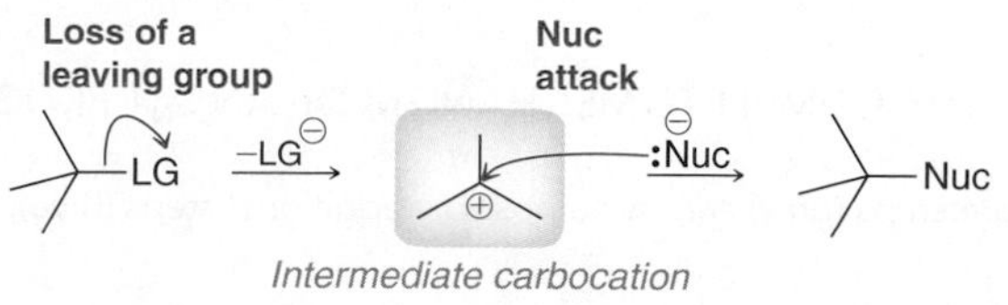

Try Problems 7.2–7.5, 7.64a

7.2 DRAWING THE PRODUCT OF AN S_N2 PROCESS

Replace the LG with the Nuc, and draw inversion of configuration.

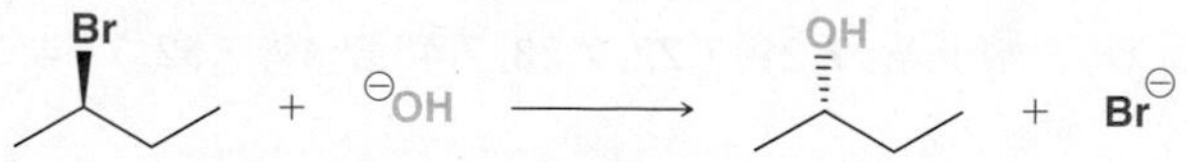

Try Problems 7.7, 7.8, 7.45, 7.56, 7.61

7.3 DRAWING THE TRANSITION STATE OF AN S_N2 PROCESS

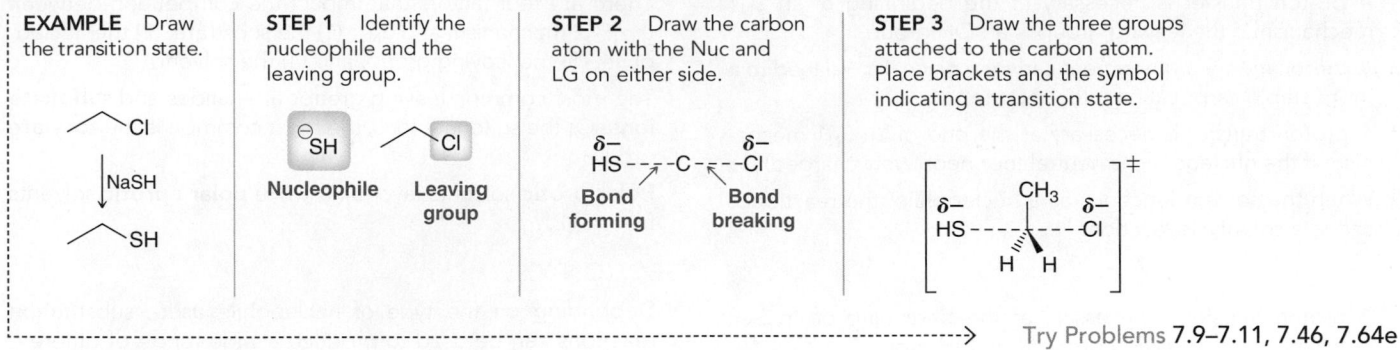

EXAMPLE Draw the transition state.

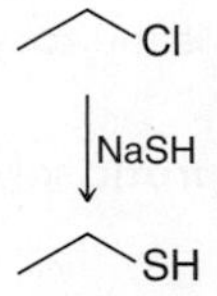

STEP 1 Identify the nucleophile and the leaving group.

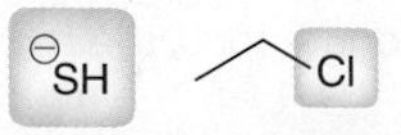

Nucleophile **Leaving group**

STEP 2 Draw the carbon atom with the Nuc and LG on either side.

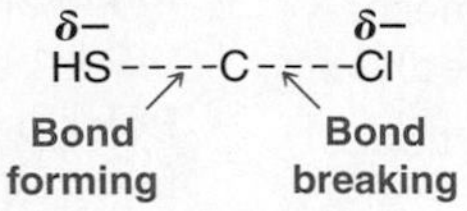

STEP 3 Draw the three groups attached to the carbon atom. Place brackets and the symbol indicating a transition state.

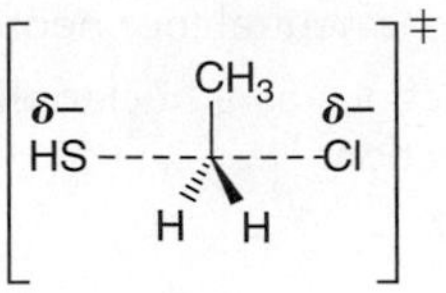

Try Problems **7.9–7.11, 7.46, 7.64e**

7.4 DRAWING THE CARBOCATION INTERMEDIATE OF AN S_N1 PROCESS

STEP 1 Identify the leaving group.

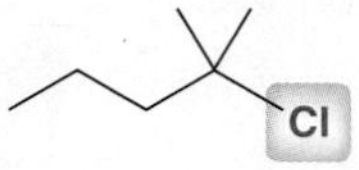

STEP 2 Draw all three groups pointing away from each other.

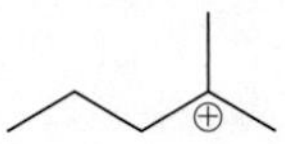

Try Problems **7.14, 7.15, 7.50, 7.51**

7.5 DRAWING THE PRODUCTS OF AN S_N1 PROCESS

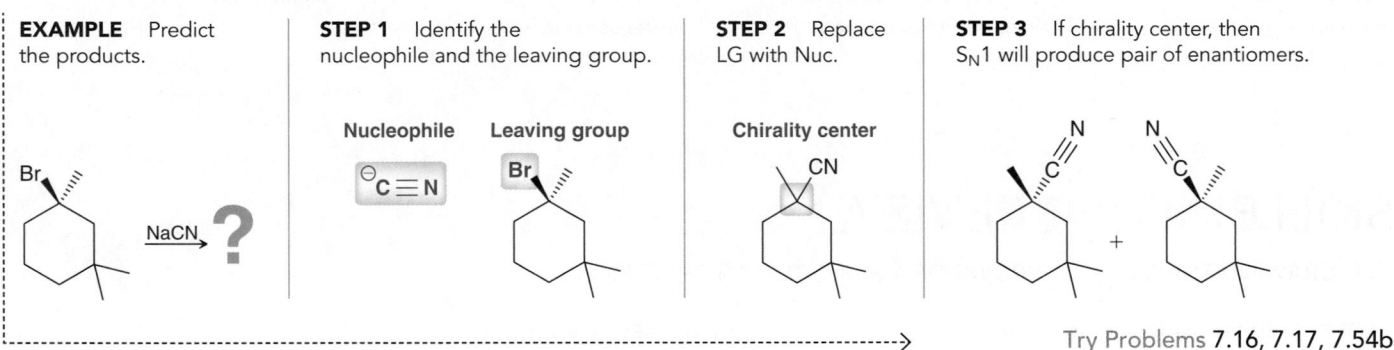

EXAMPLE Predict the products.

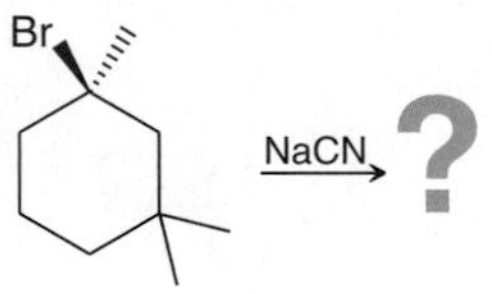

STEP 1 Identify the nucleophile and the leaving group.

Nucleophile **Leaving group**

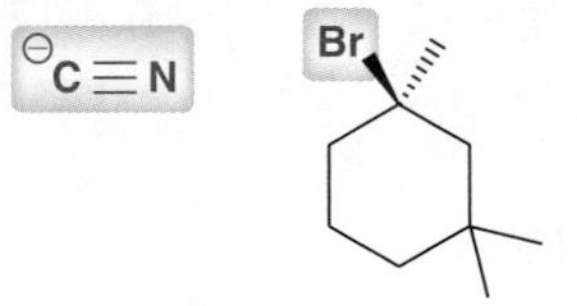

STEP 2 Replace LG with Nuc.

Chirality center

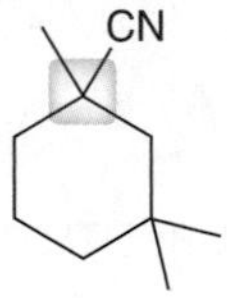

STEP 3 If chirality center, then S_N1 will produce pair of enantiomers.

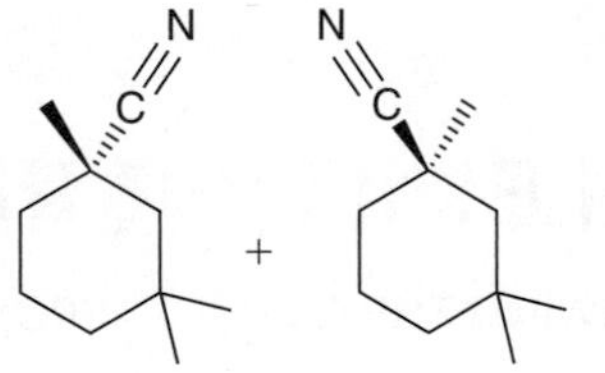

Try Problems **7.16, 7.17, 7.54b**

7.6 DRAWING THE COMPLETE MECHANISM OF AN S_N1 PROCESS

There are two core steps (gray) and three possible additional steps (blue).

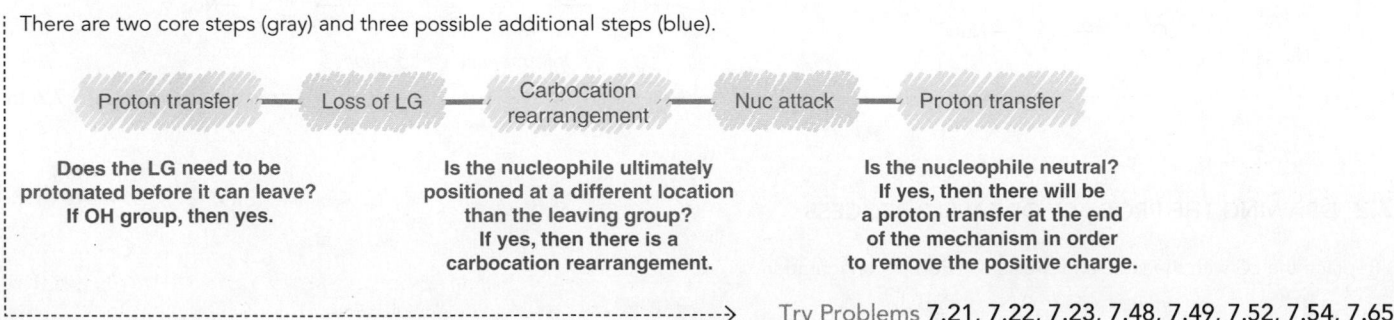

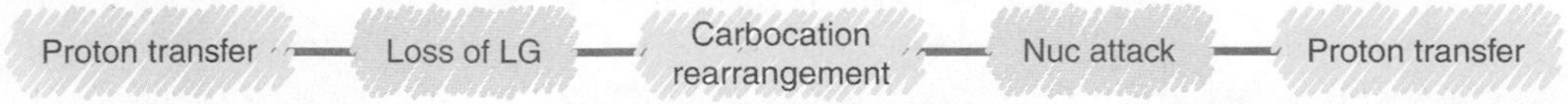

Does the LG need to be protonated before it can leave? If OH group, then yes.

Is the nucleophile ultimately positioned at a different location than the leaving group? If yes, then there is a carbocation rearrangement.

Is the nucleophile neutral? If yes, then there will be a proton transfer at the end of the mechanism in order to remove the positive charge.

Try Problems **7.21, 7.22, 7.23, 7.48, 7.49, 7.52, 7.54, 7.65**

7.7 DRAWING THE COMPLETE MECHANISM OF AN S_N2 PROCESS

There is one core step (concerted) and two possible additional steps.

Proton transfer — Nuc attack + loss of LG — Proton transfer

Does the LG need to be protonated first? If OH group, then yes.

Is the nucleophile neutral? If yes, then there will be a proton transfer at the end of the mechanism in order to remove the positive charge.

Try Problems **7.24, 7.25, 7.53, 7.64, 7.66**

7.8 DETERMINING WHETHER A REACTION PROCEEDS VIA AN S_N1 MECHANISM OR AN S_N2 MECHANISM

RELEVANT FACTORS

	S_N2	S_N1
Substrate	**Methyl or primary**	**Tertiary**
Nuc	**Strong Nuc**	**Weak Nuc**
LG	**Good LG**	**Excellent LG**
Solvent	**Polar aprotic**	**Polar protic**

EXAMPLE

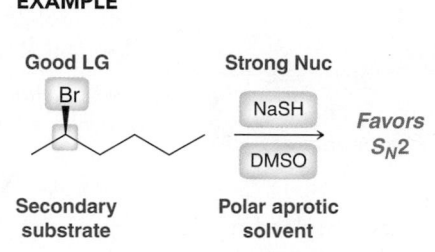

Good LG **Strong Nuc**

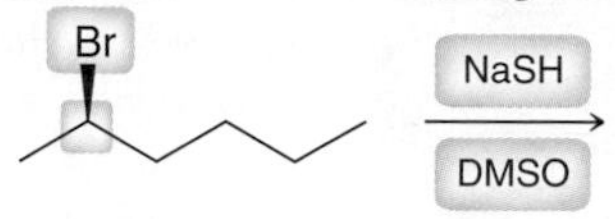

Favors S_N2

Secondary substrate **Polar aprotic solvent**

Try Problems **7.31, 7.32, 7.37, 7.38, 7.40, 7.41, 7.44, 7.55, 7.57, 7.58**

7.9 IDENTIFYING THE REAGENTS NECESSARY FOR A SUBSTITUTION REACTION

EXAMPLE

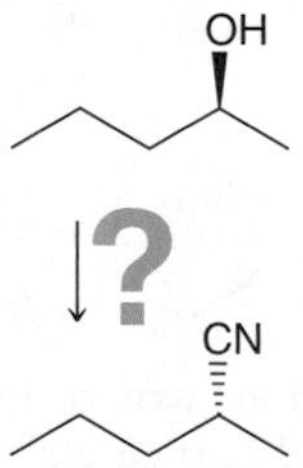

STEP 1 Analyze the substrate and the stereochemistry.

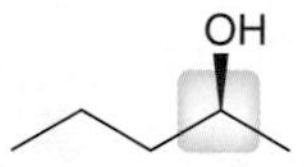

Secondary substrate. Inversion of configuration. Must be S_N2

STEP 2 Analyze the LG.

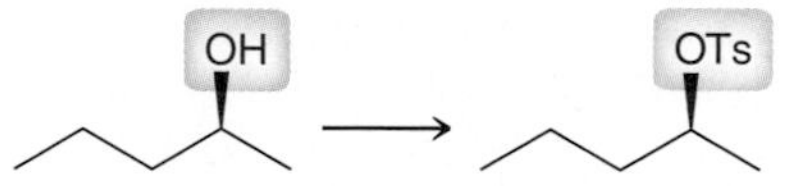

Bad LG. Must convert to tosylate using TsCl and pyridine.

STEP 3 Use conditions that favor S_N2: strong Nuc (NaCN) and a polar aprotic solvent (DMSO).

Reagents:

1. **TsCl, pyridine**
2. **NaCN, DMSO**

Try Problems **7.33, 7.34, 7.59, 7.60, 7.63**

PRACTICE PROBLEMS

Note: Most of the Problems are available within *WileyPLUS*, an online teaching and learning solution.

7.36 List the systematic name and common name for each of the following compounds:

(a) 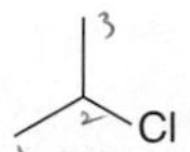

(b) 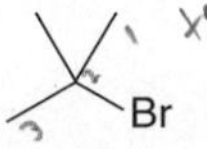

(c) 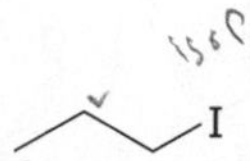

3-chloro propane   2-bromo-2 methyl propane   1-iodopropane

isopropyl chloride ↑ common

(d) 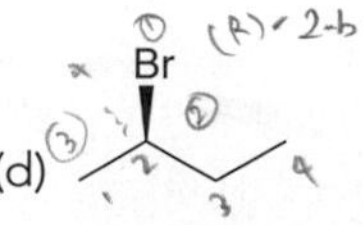

(e) 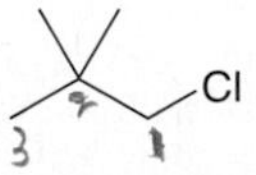

7.37 Draw all isomers of C_4H_9I, and then arrange them in order of increasing reactivity toward an S_N2 reaction.

7.38 For each of the following pairs of compounds, identify which compound would react more rapidly in an S_N2 reaction. Explain your choice in each case.

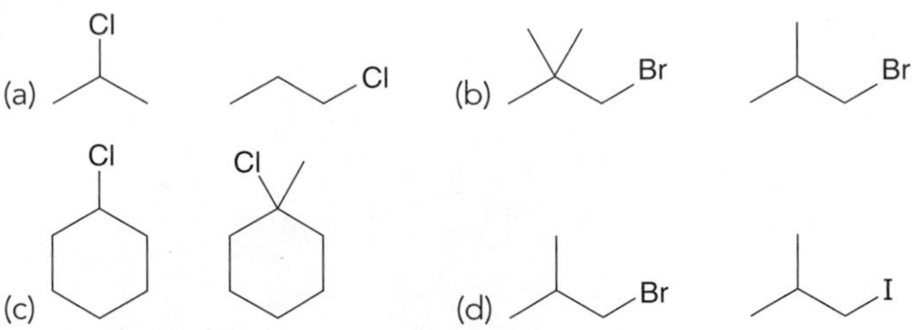

7.39 In Chapter 10, we will see that an acetylide ion (formed by treatment of acetylene with a strong base) can serve as a nucleophile in an S_N2 reaction:

H—C≡C—H $\xrightarrow{\text{Strong base}}$ H—C≡C$^{\ominus}$ $\xrightarrow{\text{R—X}}$ H—C≡C—R

Acetylene **Acetylide ion**

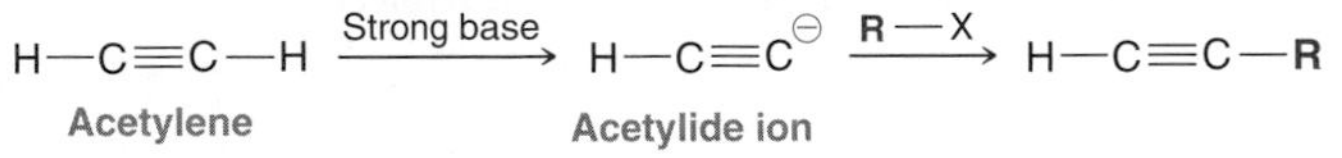

This reaction provides a useful method for making a variety of substituted alkynes. Determine whether this process can be used to make the following alkyne. Explain your answer.

H—C≡C—⟨tert-butyl⟩

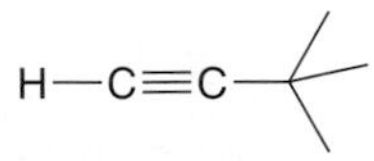

7.40 Identify the stronger nucleophile:

(a) NaSH vs. H_2S

(b) Sodium hydroxide vs. water

(c) Methoxide dissolved in methanol vs. methoxide dissolved in DMSO

7.41 For each pair of the following compounds, identify which compound would react more rapidly in an S_N1 reaction. Explain your choice in each case.

(a) ... (b) ... (c) ... (d) ...

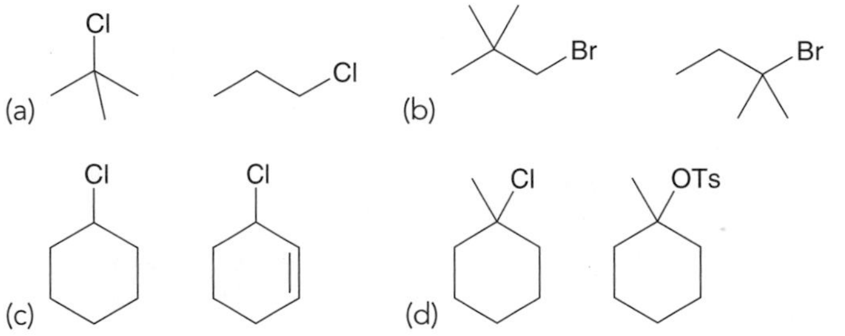

7.42 Consider the following reaction:

[structure with Br] $\xrightarrow[\text{DMSO}]{\text{NaCN}}$ [structure with CN] + NaBr

(a) How would the rate be affected if the concentration of the alkyl halide is doubled?

(b) How would the rate be affected if the concentration of sodium cyanide is doubled?

7.43 Consider the following reaction:

[cyclohexanol] $\xrightarrow{\text{HBr}}$ [cyclohexyl bromide] + H_2O

(a) How would the rate be affected if the concentration of the alcohol is doubled?

(b) How would the rate be affected if the concentration of HBr is doubled?

7.44 Classify each of the following solvents as protic or aprotic:

(a) DMF (c) DMSO (e) Ammonia

(b) Ethanol (d) Water

7.45 Consider the following S_N2 reaction:

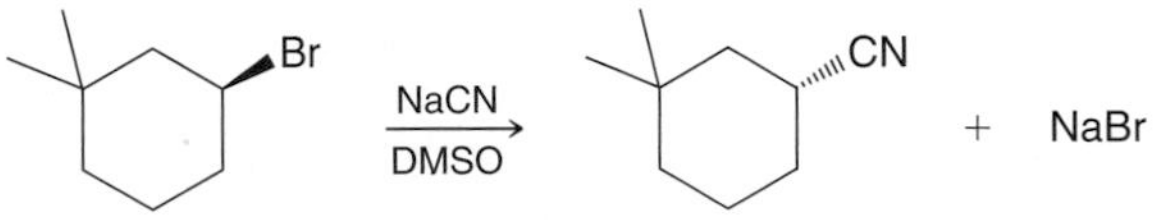

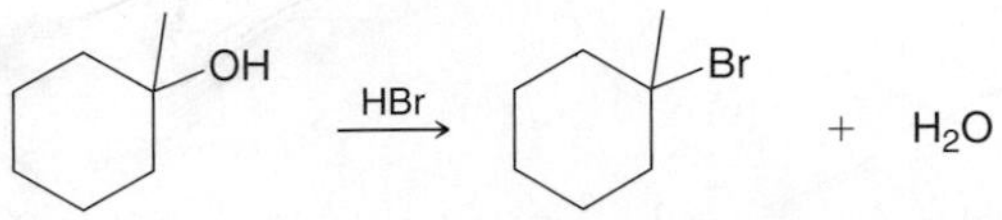

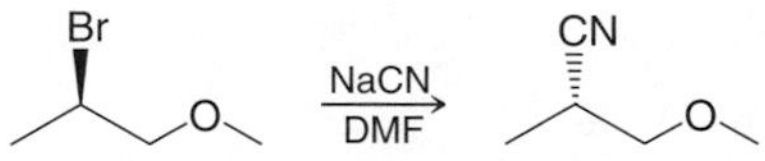

(a) Assign the configuration of the chirality center in the substrate.

(b) Assign the configuration of the chirality center in the product.

(c) Does this S_N2 process proceed with inversion of configuration? Explain.

7.46 Draw the transition state for the reaction between ethyl iodide and sodium acetate

7.47 (S)-2-Iodopentane undergoes racemization in a solution of sodium iodide in DMSO. Explain.

7.48 When the following optically active alcohol is treated with HBr, a racemic mixture of alkyl bromides is obtained:

[structure with OH] $\xrightarrow{\text{HBr}}$ [structure with Br] + H_2O

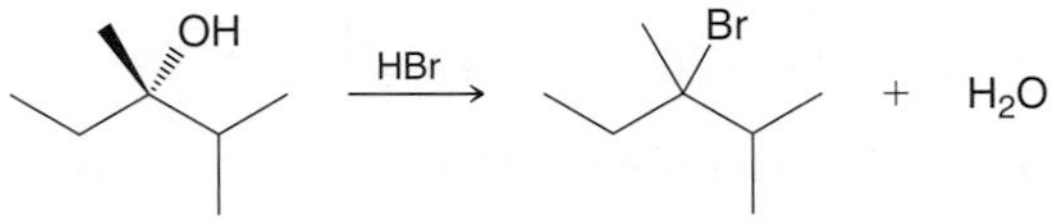

Racemic mixture

Draw the mechanism of the reaction, and explain the stereochemical outcome.

7.49 (R)-2-Pentanol racemizes when placed in dilute sulfuric acid. Draw a mechanism that explains this stereochemical outcome, and draw an energy diagram of the process.

7.50 List the following carbocations in order of increasing stability:

[four carbocation structures]

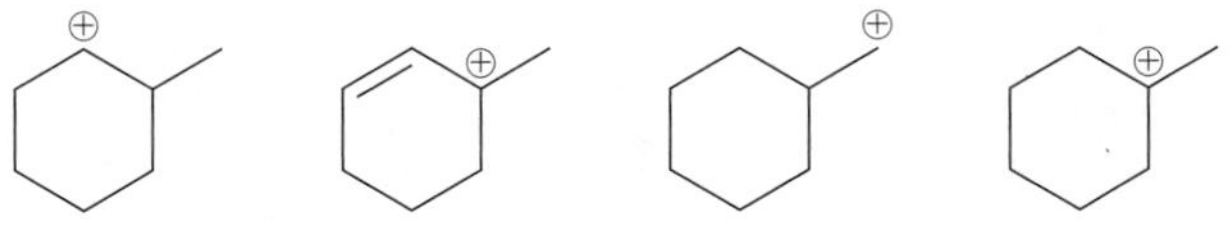

7.51 Draw the carbocation intermediate that would be formed if each of the following substrates would participate in an S_N1 reaction. In each case, identify the carbocation as being primary, secondary, or tertiary.

(a) [structure with Cl] (b) [structure with Br] (c) [structure with I] (d) [structure with Cl]

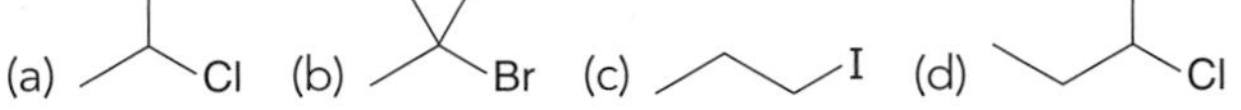

7.52 Propose a mechanism for the following transformation:

[structure with OH] $\xrightarrow{\text{HCl}}$ [structure with Cl] + H_2O

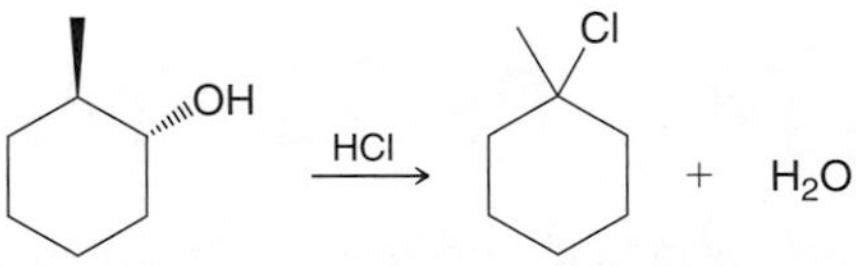

7.53 Draw the mechanism of the following reaction:

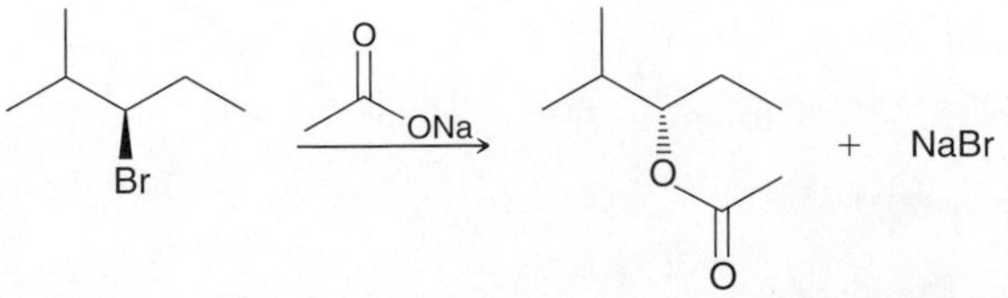

7.54 Each of the following reactions proceeds via an S_N1 mechanism and will have anywhere from two to five steps, as discussed in Section 7.6. Determine the number of steps for each reaction, and then draw the mechanism in each case:

(a) 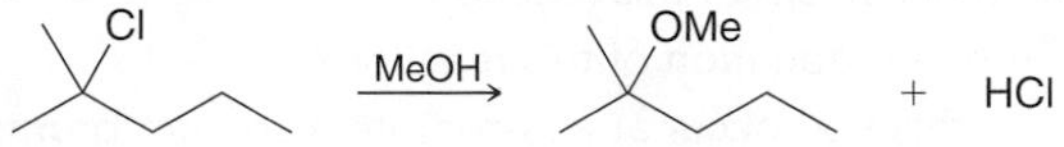

(b) 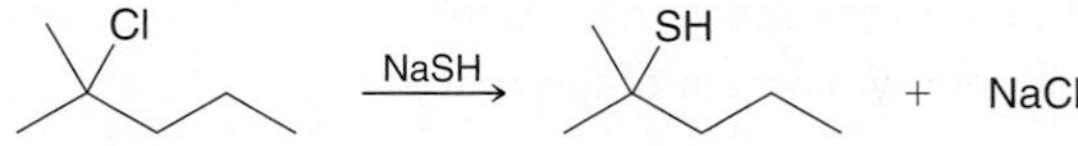

(c) 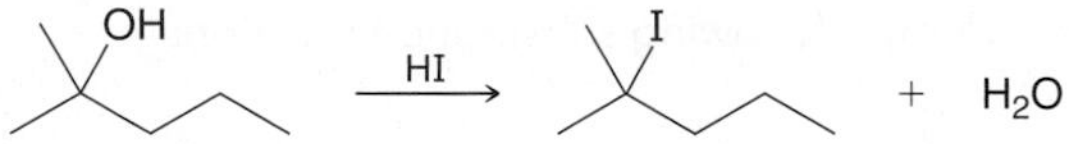

(d) 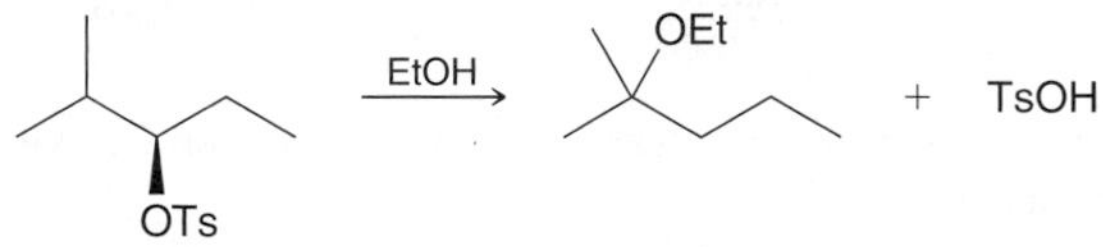

7.55 Identify the product(s) in each of the following reactions:

(a) 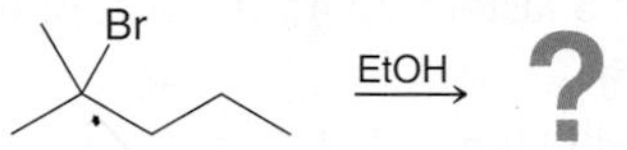

(b) 

(c) 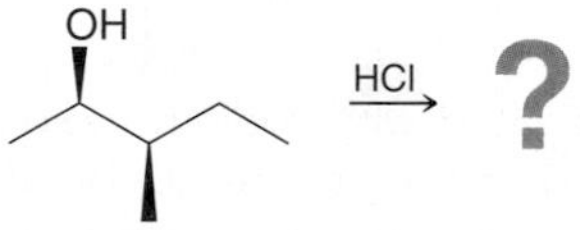

(d) 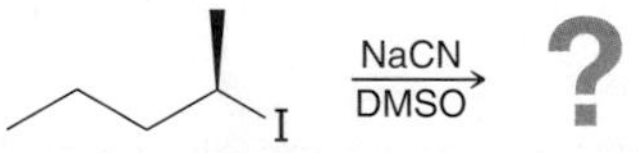

7.56 Identify the product of the following reaction:

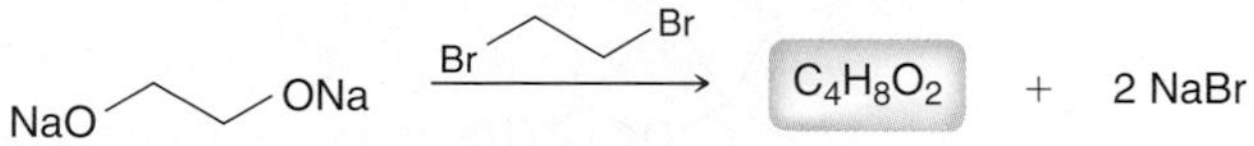

7.57 The following reaction is very slow. Identify the mechanism, and explain why the reaction is so slow.

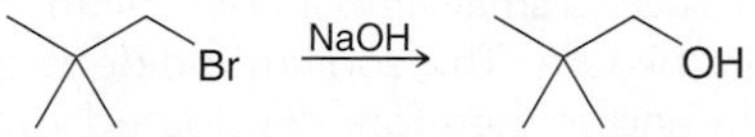

7.58 The following reaction is very slow:

(a) Identify the mechanism.

(b) Explain why the reaction is so slow.

(c) When hydroxide is used instead of water, the reaction is very rapid. Draw the mechanism of this reaction, and explain why it is so fast.

7.59 Identify the reagents you would use to achieve each of the following transformations:

(a) 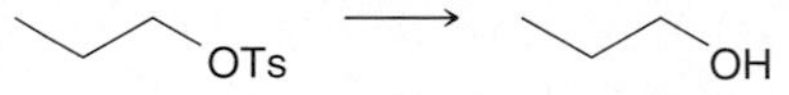

(b) 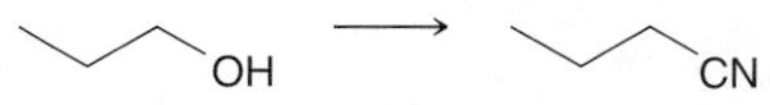

(c) 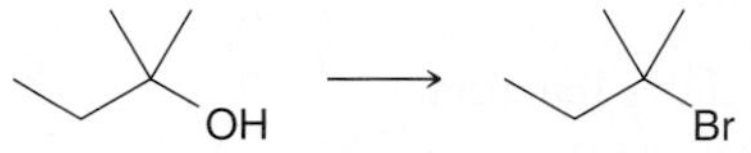

(d) 

(e) 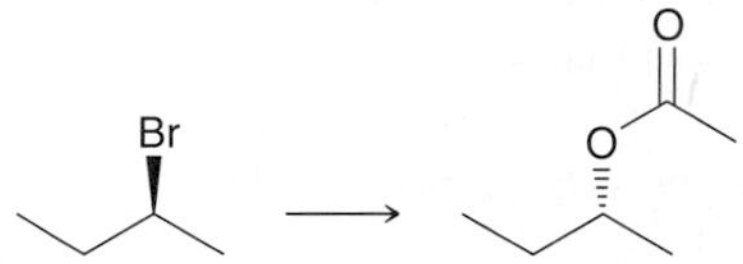

7.60 Each of the following compounds can be prepared with an alkyl iodide and a suitable nucleophile. In each case, identify the alkyl iodide and the nucleophile that you would use.

(a) 

(b) 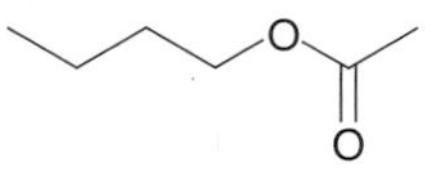

(c) 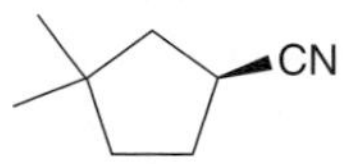

(d) 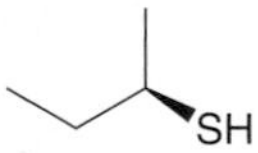

(e) 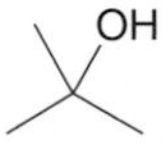

(f) 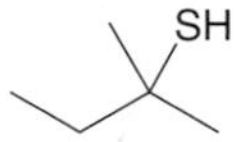

7.61 What products would you expect from the reaction between (*S*)-2-iodobutane and each of the following nucleophiles?

(a) NaSH (b) NaSEt (c) NaCN

7.62 Below are two potential methods for preparing the same ether, but only one of them is successful. Identify the successful aproach and explain your choice.

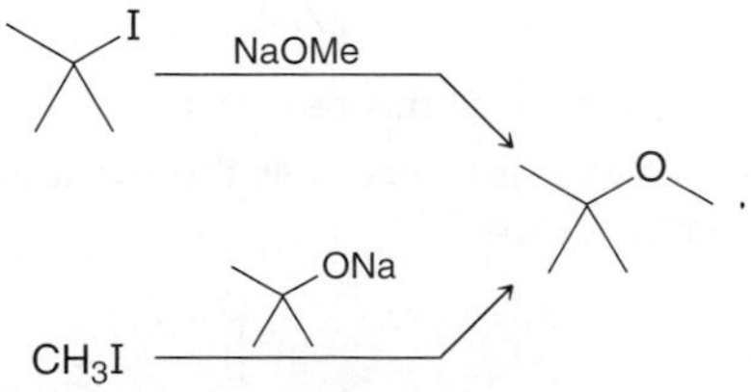

7.63 Identify the reagent you would use to accomplish each of the following transformations:

(a) Cyclobutanol → bromocyclobutane

(b) *tert*-Butanol → *tert*-butyl chloride

(c) Ethyl chloride → ethanol

INTEGRATED PROBLEMS

7.64 Consider the following S_N2 reaction:

(a) Draw the mechanism of this reaction.
(b) What is the rate equation of this reaction?
(c) What would happen to the rate if the solvent is changed from DMSO to ethanol?
(d) Draw an energy diagram of the reaction above.
(e) Draw the transition state of this reaction.

7.65 Consider the following substitution reaction:

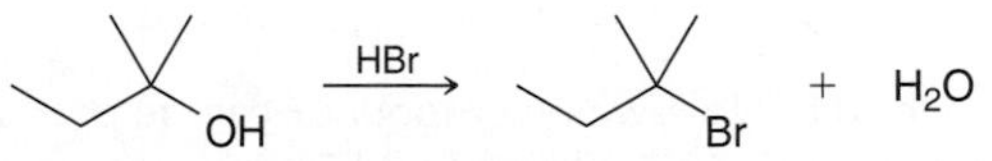

(a) Determine whether this reaction proceeds via an S_N1 or S_N2 process.
(b) Draw the mechanism of this reaction.
(c) What is the rate equation of this reaction?
(d) Would the reaction occur at a faster rate if sodium bromide were added to the reaction mixture?
(e) Draw an energy diagram of this reaction.

7.66 Consider the following substitution reaction:

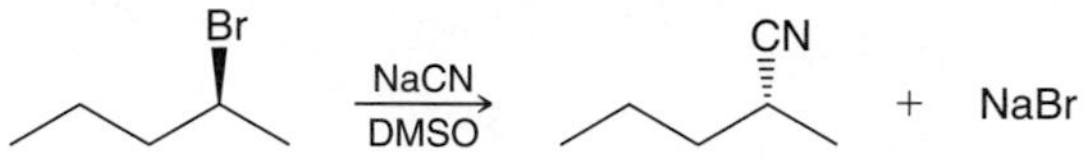

(a) Determine whether this reaction proceeds via an S_N1 or S_N2 process.
(b) Draw the mechanism of this reaction.
(c) What is the rate equation of this reaction?
(d) Would the reaction occur at a faster rate if the concentration of cyanide were doubled?
(e) Draw an energy diagram of the reaction above.

CHALLENGE PROBLEMS

7.67 Propose a mechanism for the following transformation:

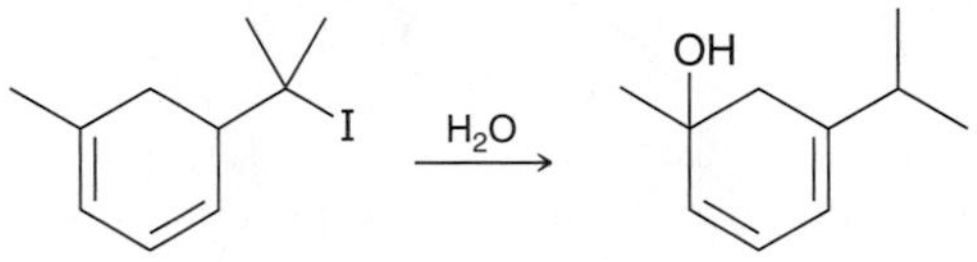

7.68 When the following ester is treated with lithium iodide in DMF, a carboxylate ion is obtained:

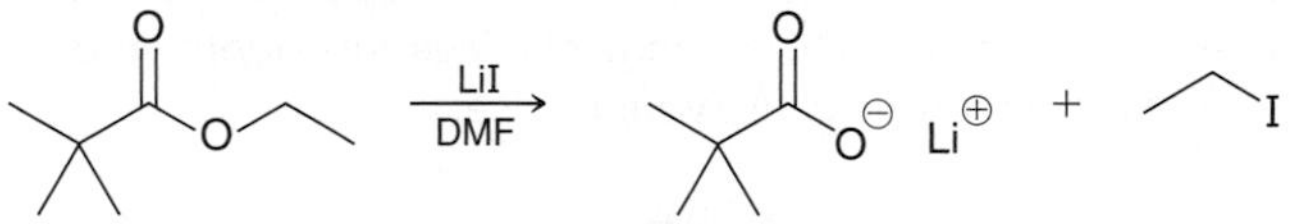

(a) Draw the mechanism of this reaction.
(b) When the methyl ester is used as the substrate, the reaction is 10 times faster:

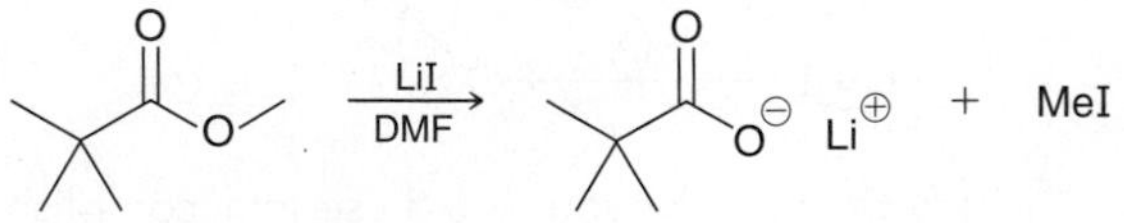

Explain the increase in rate.

7.69 When (1*R*,2*R*)-2-bromocyclohexanol is treated with a strong base, an epoxide (cyclic ether) is formed. Suggest a mechanism for formation of the epoxide:

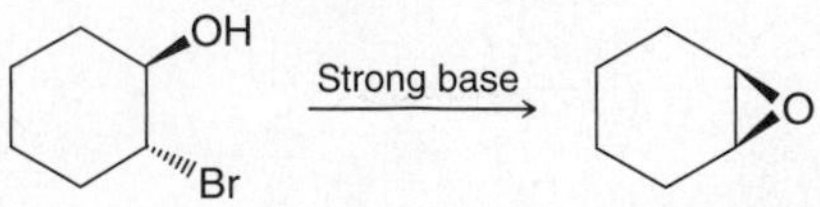

An epoxide

7.70 When butyl bromide is treated with sodium iodide in ethanol, the concentration of iodide quickly decreases but then slowly returns to its original concentration. Identify the major product of the reaction.

7.71 The following compound can react rapidly via an S_N1 process. Explain why this primary substrate will undergo an S_N1 reaction so rapidly.

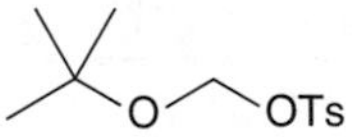

7.72 Consider the reaction below. The rate of this reaction is markedly increased if a small amount of sodium iodide is added to the reaction mixture. The sodium iodide is not consumed by the reaction and is therefore considered to be a catalyst. Explain how the presence of iodide can speed up the rate of the reaction.

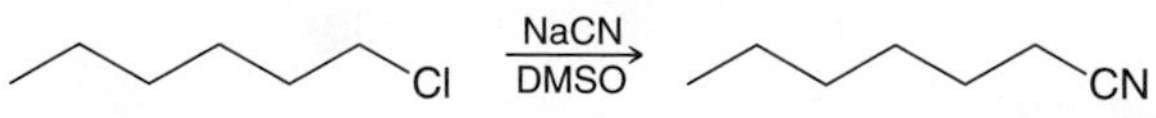

7.73 Propose a mechanism for the following transformation:

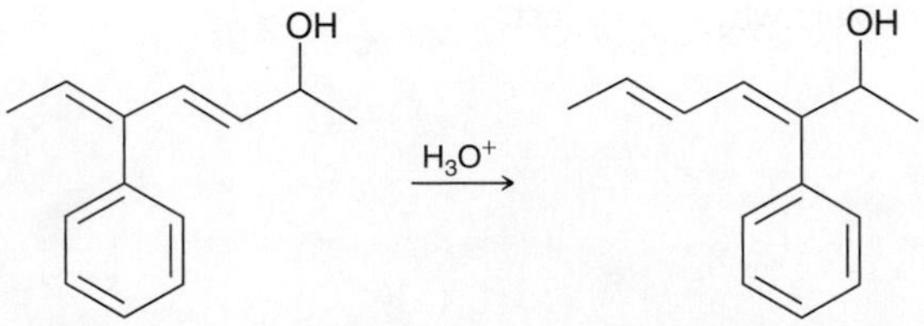

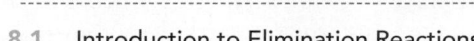

8

Alkenes: Structure and Preparation via Elimination Reactions

DID YOU EVER WONDER...

why fruit ripens more quickly in a paper bag?

The main culprit is a small compound called ethylene ($H_2C{=}CH_2$). At room temperature ethylene is a gas, and it functions as a growth hormone by triggering the ripening process. In a well-ventilated area, a single fruit does not produce enough ethylene to affect nearby fruits significantly. But when fruit are placed in a bag, the concentration of ethylene builds up inside the bag, causing them to ripen more quickly.

Many fruits are picked while still green, transported to stores, and then sprayed with ethylene gas to promote ripening. In some cases, their exposure to ethylene is insufficient, which explains why you might find green bananas in the grocery store. To ripen bananas in your home, simply place them in a paper bag, which subjects the bananas to their own natural ethylene. To speed up the process, place a ripe fruit in the bag as well, because ripe fruits produce more ethylene than unripe fruits. This method will work with any fruit that produces ethylene, including mangos, kiwis, avocados, tomatoes, apples, and pears.

Like ethylene, many naturally occurring compounds contain a carbon-carbon double bond. This chapter will focus on the properties and synthesis of compounds that possess carbon-carbon double bonds.

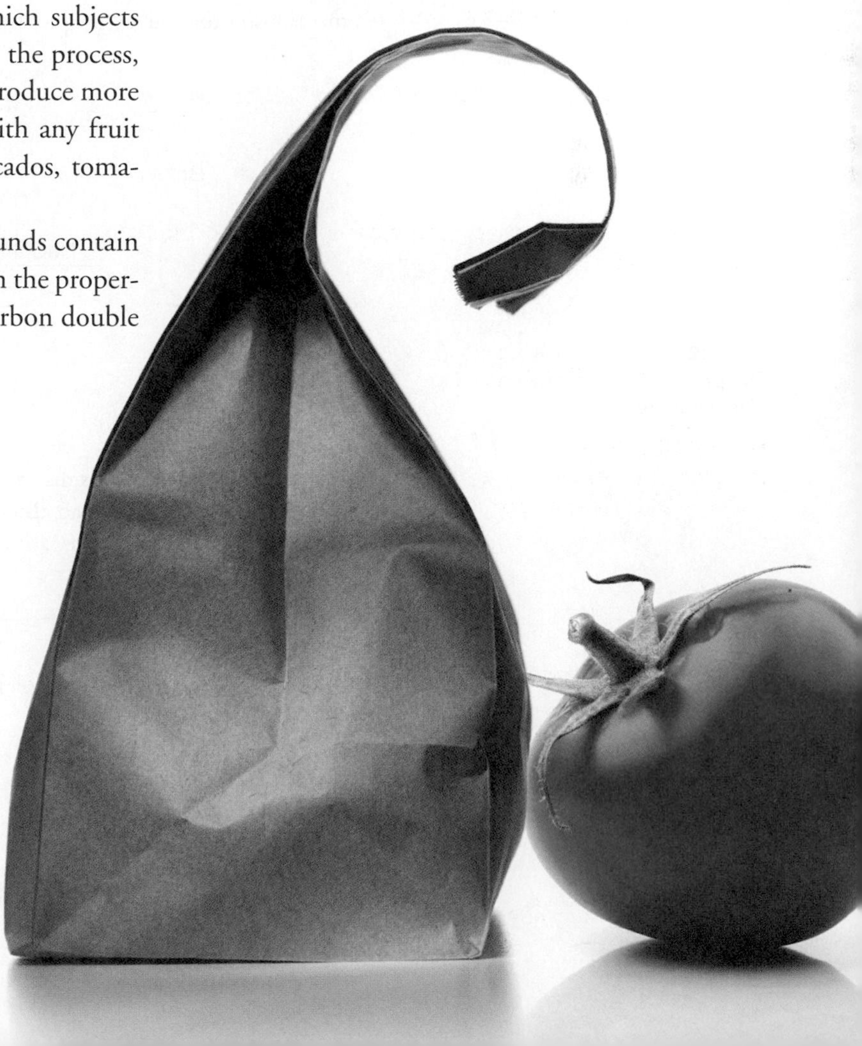

DO YOU REMEMBER?

Before you go on, be sure you understand the following topics.
If necessary, review the suggested sections to prepare for this chapter.

- Kinetics and Energy Diagrams (Sections 6.5, 6.6)
- Arrow Pushing and Carbocation Rearrangements (Sections 6.8 – 6.11)
- Substitution Reactions (Sections 7.3 – 7.8)
- Introduction to Stereoisomerism (Section 5.2)
- Chair Conformations of Cyclohexane (Sections 4.11 – 4.13)

WILEY PLUS Visit www.wileyplus.com to check your understanding and for valuable practice.

8.1 Introduction to Elimination Reactions

In the previous chapter, we saw that a substitution reaction can occur when a compound possesses a leaving group. In this chapter, we will explore another type of reaction, called *elimination*, commonly observed for compounds with leaving groups. Although elimination reactions occur in a variety of different contexts, for now we will examine them as a method for forming alkenes. Consider the difference between substitution and elimination reactions (Figure 8.1). In a substitution reaction, the leaving group is replaced with a nucleophile. In an elimination reaction, a proton from the beta (β) position is removed together with the leaving group, forming a double bond. This type of reaction is called a **beta elimination**, or **1,2-elimination**, and can be accomplished with any good leaving group. Specific classes of beta eliminations were characterized before chemists understood they had a common set of mechanisms. Thus, some types of beta elimination reactions are named on the basis of the leaving group. For example, when the leaving group is specifically a halide, the reaction is also called a **dehydrohalogenation**; when the leaving group is water, the reaction is also called **dehydration**. Regardless of the name, the key is to remember that all of these reactions have a common set of mechanisms that can be used to describe their underlying processes.

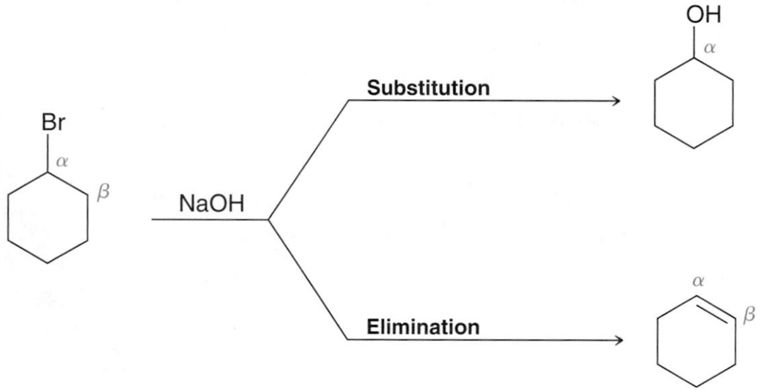

FIGURE 8.1
The products of substitution and elimination reactions.

In the context of this discussion, the product of an elimination reaction possesses a C═C double bond and is called an **alkene**. The chapter will focus on the structure of alkenes and their preparation via beta elimination reactions.

8.2 Alkenes in Nature and in Industry

Alkenes are abundant in nature. Here are just a few examples, all acyclic compounds (compounds that do not contain a ring).

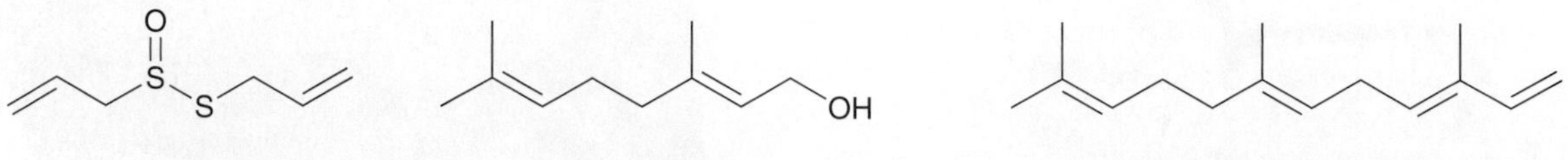

Allicin
Responsible for the odor of garlic

Geraniol
Isolated from roses and used in perfumes

α-Farnesene
Found in the natural waxy coating on apple skins

Nature also produces many cyclic, bicyclic, and polycyclic alkenes:

Limonene
Responsible for the strong smell of oranges

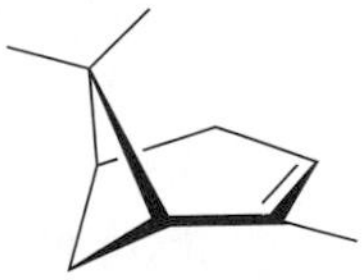

α-Pinene
Isolated from pine resin; a primary constituent of turpentine (paint thinner)

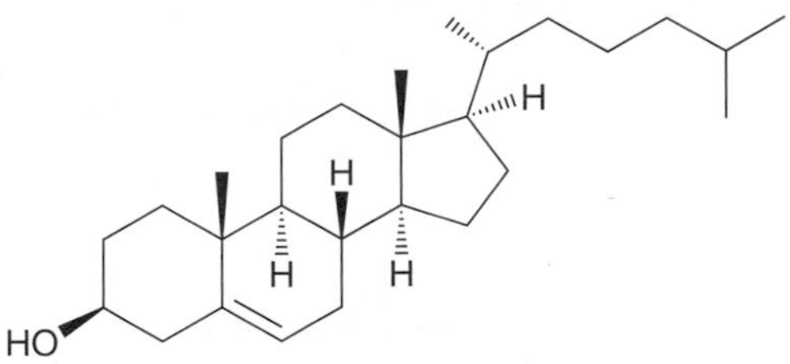

Cholesterol
Produced by all animals; this compound plays a pivotal role in many biological processes

Double bonds are also often found in the structures of pheromones. Recall that pheromones are chemicals used by living organisms to trigger specific behavioral responses in other members of the same species. For example, alarm pheromones are used to signal danger, while sex pheromones are used to attract the opposite sex for mating.

PRACTICALLYSPEAKING)))

Pheromones to Control Insect Populations

The greatest threat to the productivity of apple orchards is an infestation of codling moths. A bad infestation can destroy up to 95% of an apple crop. A female codling moth can lay up to 100 eggs. Once hatched, the larvae dig into the apples, where they are shielded from insecticides. The so-called "worm" in an apple is generally the larva of a codling moth. The main tool for dealing with these pests involves using one of the sex pheromones of the female to disrupt mating:

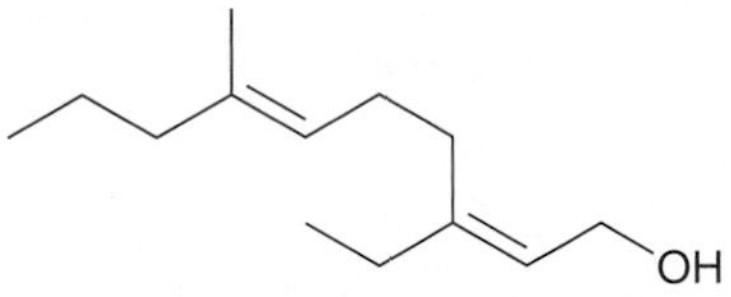

(2*Z*,6*E*)-3-Ethyl-7-methyldeca-2,6-dien-1-ol
A sex pheromone of the codling moth

This compound is easily produced in a laboratory. It can then be sprayed in an apple orchard where its presence interferes with the ability of the male to find the female. The sex pheromone is also used to lure the males into traps, enabling farmers to monitor populations and time the use of insecticides to coincide with the time period during which females are laying eggs, thereby increasing the efficacy of insecticides.

Current research focuses on new compounds that attract both males and females. One such example is ethyl (2*E*,4*Z*)-2,4-decadienoate, known as the pear ester:

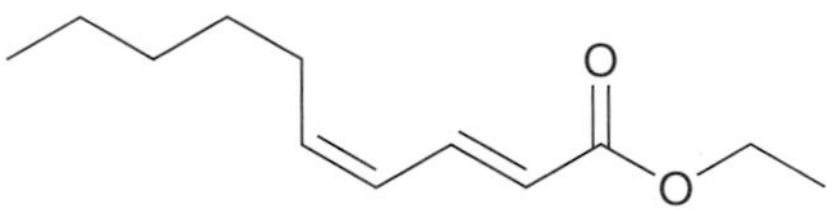

Ethyl (2*E*,4*Z*)-2,4-decadienoate
Pear ester

Researchers at the USDA (U.S. Department of Agriculture) have discovered that the pear ester can be potentially more effective in controlling codling moth populations than other compounds. This compound will potentially allow farmers to target the females and their eggs with greater precision. One of the major advantages of using pheromones as insecticides is that they tend to be less toxic to humans and less harmful to the environment.

Below are several examples of pheromones that contain double bonds:

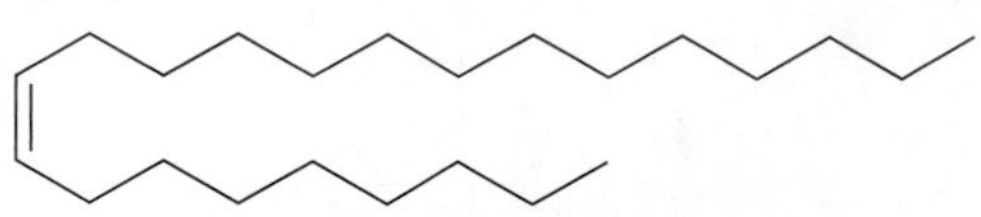

Muscalure
Sex pheromone of the common housefly

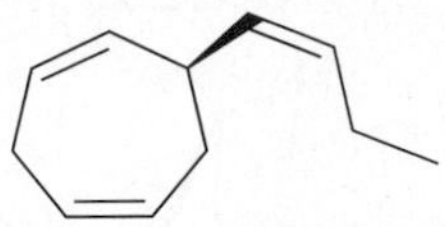

Ectocarpene
A pheromone released by the eggs of the seaweed *Ectocarpus siliculosus* to attract sperm cells

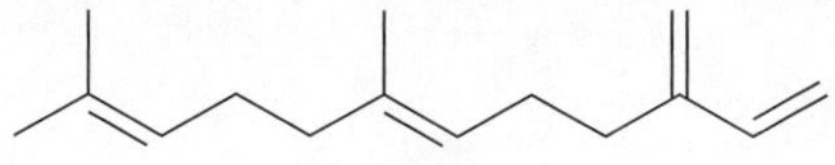

β-Farnesene
An aphid alarm pheromone

Alkenes are also important precursors in the chemical industry. The two most important industrial alkenes, ethylene and propylene, are formed from cracking petroleum, and are used as starting materials for preparing a wide variety of compounds (Figure 8.2):

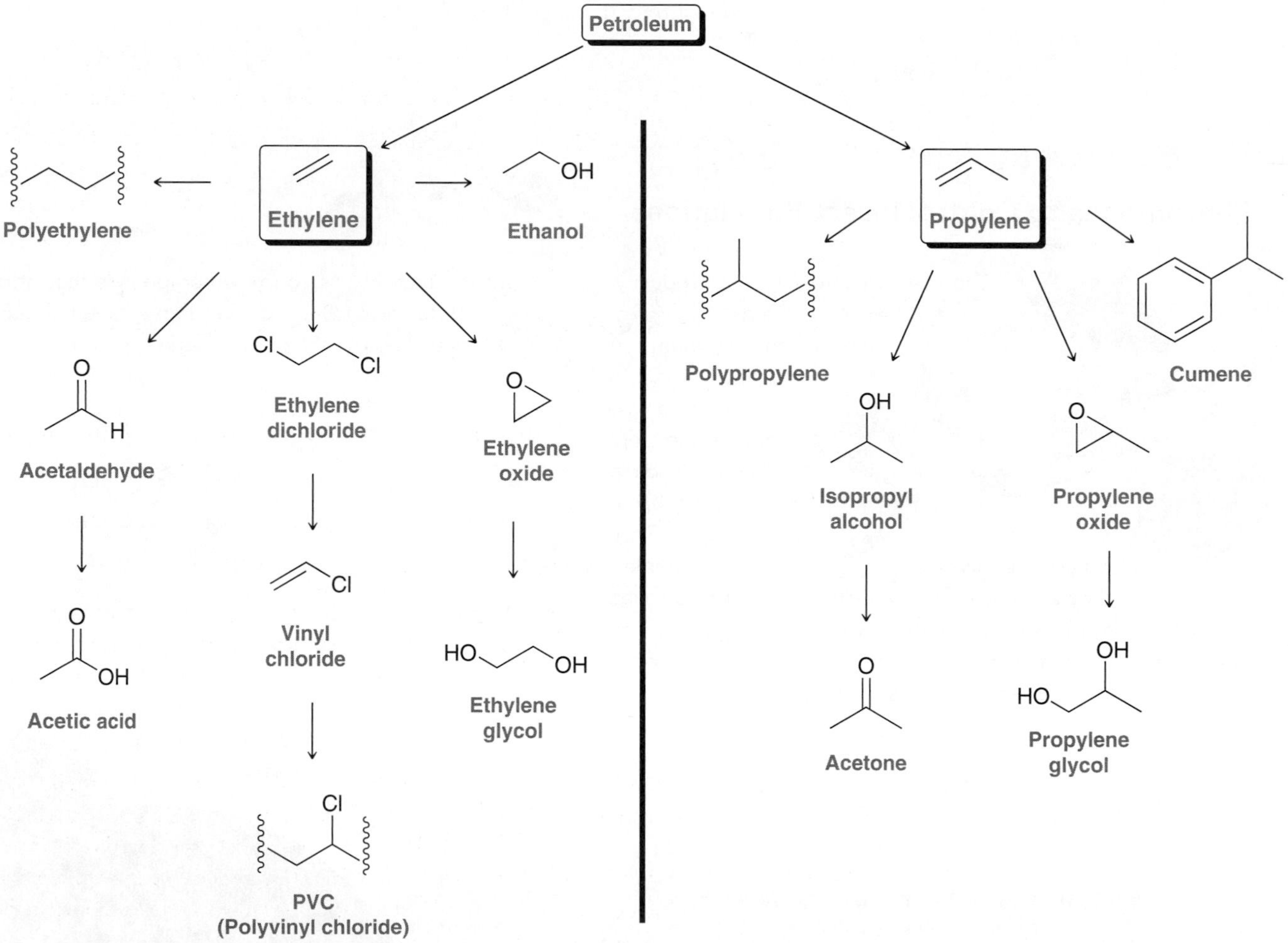

FIGURE 8.2
Industrially important compounds produced from ethylene and propylene.

Each year, over 200 billion pounds of ethylene and 70 billion pounds of propylene are produced globally and used to make many compounds, including those shown in Figure 8.2.

8.3 Nomenclature of Alkenes

Recall from Chapter 4 that naming alk*anes* requires four discrete steps (Section 4.2):

1. Identify the parent.
2. Identify the substituents.
3. Assign a locant to each substituent.
4. Arrange the substituents alphabetically.

Alk*enes* are named using the same four steps, with the following additional guidelines:

When naming the parent, replace the suffix "ane" with "ene" to indicate the presence of a C=C double bond:

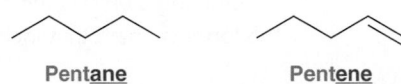

When we learned to name alkanes, we chose the parent by identifying the longest chain. When choosing the parent of an alkene, use the longest chain that includes the π bond:

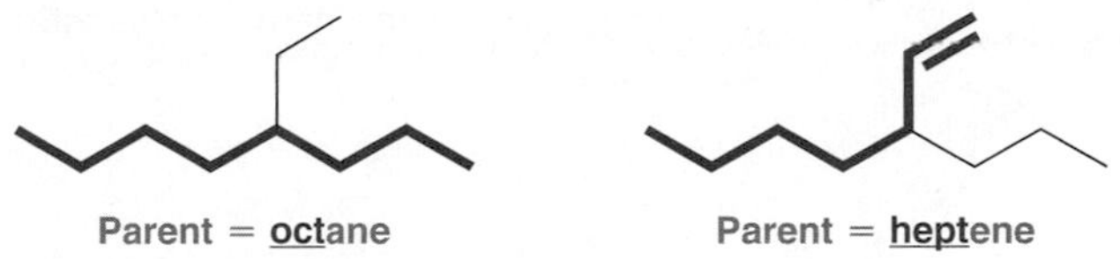

When numbering the parent chain of an alkene, the π bond should receive the lowest number possible despite the presence of alkyl substituents:

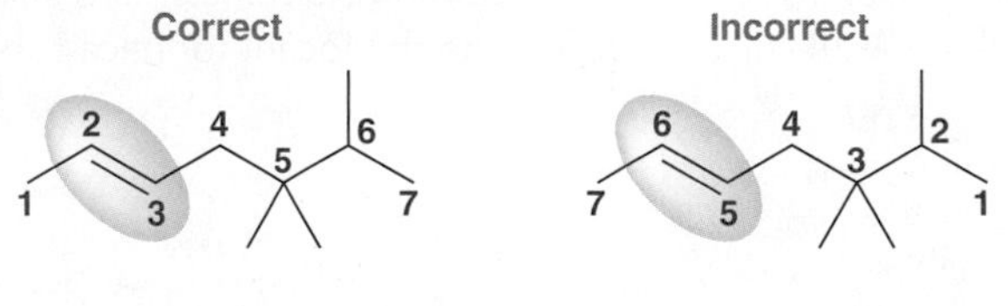

The position of the double bond is indicated using a single locant, rather than two locants. Consider the previous example where the double bond is between C2 and C3 on the parent chain. In this case, the position of the double bond is indicated with number 2. The IUPAC rules published in 1979 dictate that this locant be placed immediately before the parent, while the IUPAC recommendations released in 1993 and 2004 allow for the locant to be placed before the suffix "ene." Both names are acceptable.

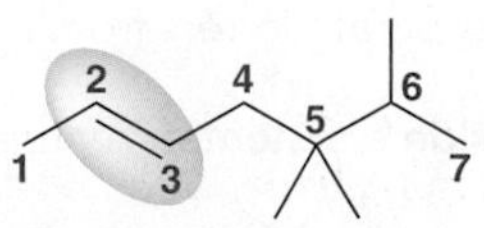

5,5,6-Trimethyl-2-heptene
or
5,5,6-Trimethylhept-2-ene

SKILLBUILDER

8.1 ASSEMBLING THE SYSTEMATIC NAME OF AN ALKENE

LEARN the skill

Provide a systematic name for the following compound:

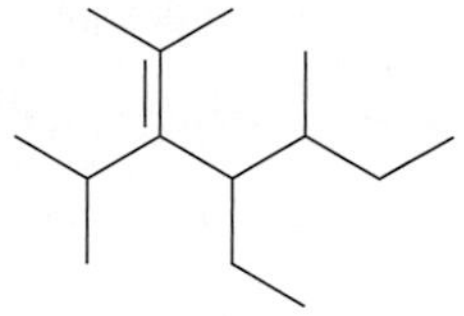

SOLUTION

Assembling a systematic name requires four discrete steps. First, identify the parent. Choose the longest chain that includes the π bond:

STEP 1
Identify the parent.

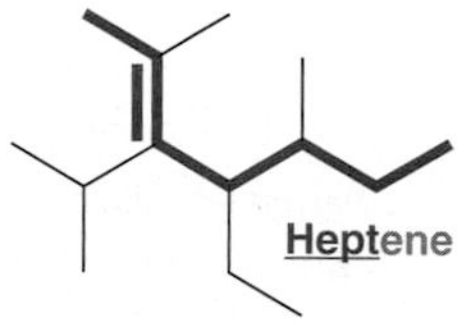

STEP 2
Identify and name the substituents.

Next, identify and name the substituents connected to the parent:

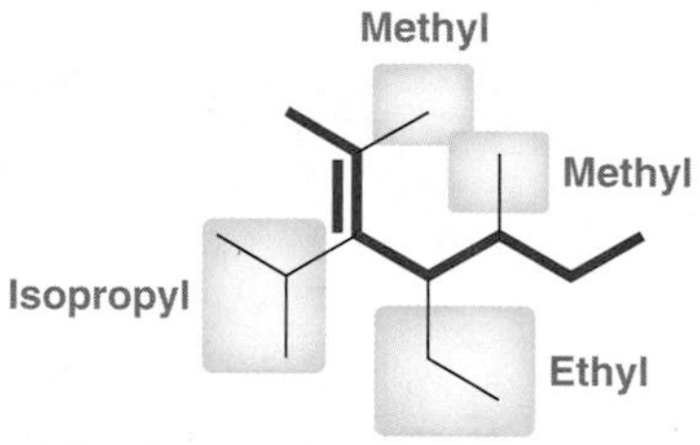

STEP 3
Number the parent chain and assign a locant to each substituent.

STEP 4
Arrange the substituents alphabetically.

LOOKING BACK
Recall that the prefix "iso" is counted as part of the alphabetization scheme, while the prefix "di" is ignored. For a review, see Section 4.2.

Then number the parent chain and assign a locant to each substituent. The parent is numbered starting from the side that is closest to the π bond. This numbering scheme is used to determine the locant for each substituent:

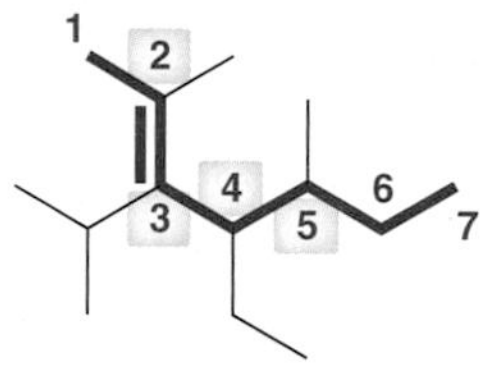

Finally, arrange the substituents alphabetically, making sure to include a locant that identifies the position of the double bond:

4-Ethyl-3-isopropyl-2,5-dimethyl-2-heptene

Notice that ***e***thyl appears first, then ***i***sopropyl, and finally di***m***ethyl. Also, make sure that hyphens separate letters from numbers, while commas separate numbers from each other.

PRACTICE the skill **8.1** Provide a systematic name for each of the following compounds:

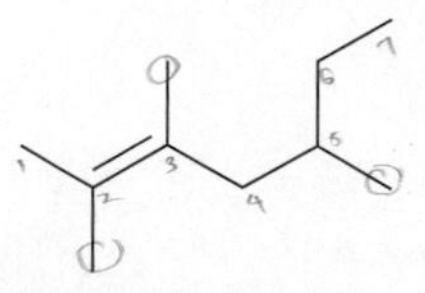
(a)

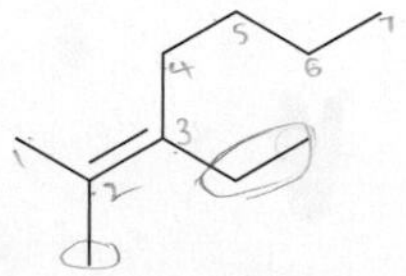
(b)

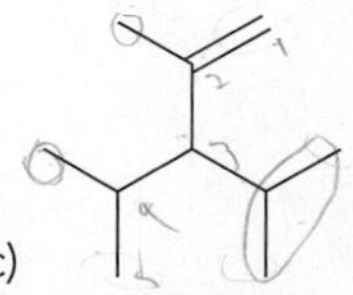
(c)

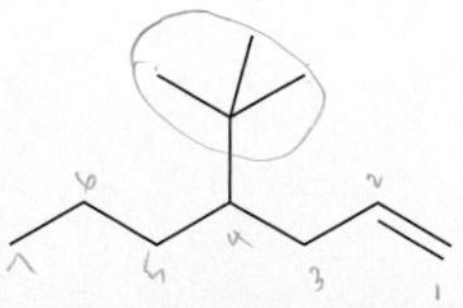
(d)

APPLY the skill

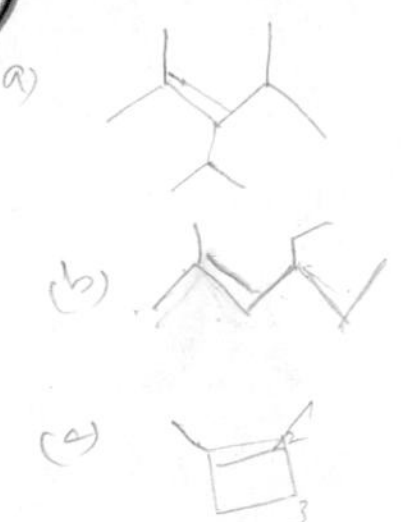

8.2 Draw a bond-line structure for each of the following compounds:

(a) 3-Isopropyl-2,4-dimethyl-2-pentene

(b) 4-Ethyl-2-methyl-2-hexene

(c) 1,2-Dimethylcyclobutene (The name of a cycloalkene will not include a locant to specify the position of the π bond, because, by definition, it is assumed to be between C1 and C2.)

8.3 In Section 4.2, we learned how to name bicyclic compounds. Using those rules, together with the rules discussed in this section, provide a systematic name for the following bicyclic compound:

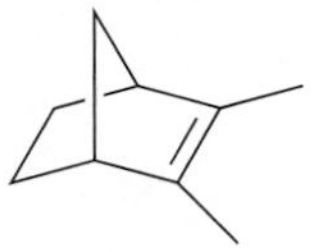

need more **PRACTICE?** **Try Problem 8.50b,c**

IUPAC nomenclature also recognizes common names for many simple alkenes. Here are three such examples:

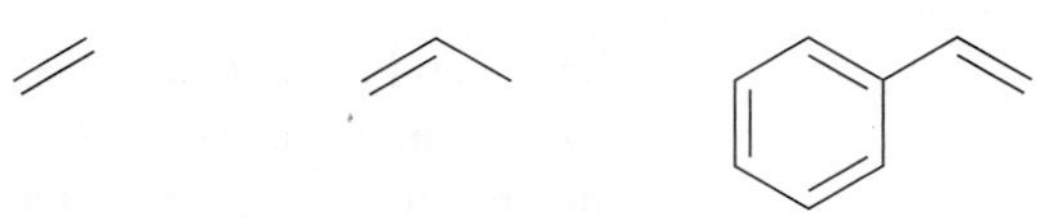

Ethylene **Propylene** **Styrene**

IUPAC nomenclature recognizes common names for the following groups when they appear as substituents in a compound:

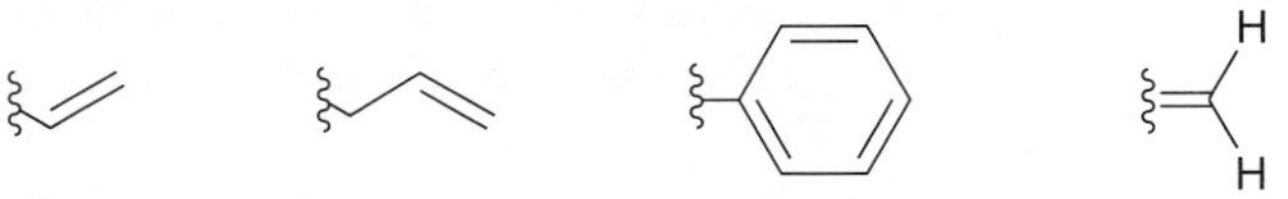

Vinyl **Allyl** **Phenyl** **Methylene**

In addition to their systematic and common names, alkenes are also classified by their **degree of substitution** (Figure 8.3). Do not confuse the word *substitution* with the type of reaction discussed in the previous chapter. The same word has two different meanings. In the previous chapter, the word "substitution" referred to a reaction involving replacement of a leaving group with a nucleophile. In the present context, the word "substitution" refers to the number of alkyl groups connected to a double bond.

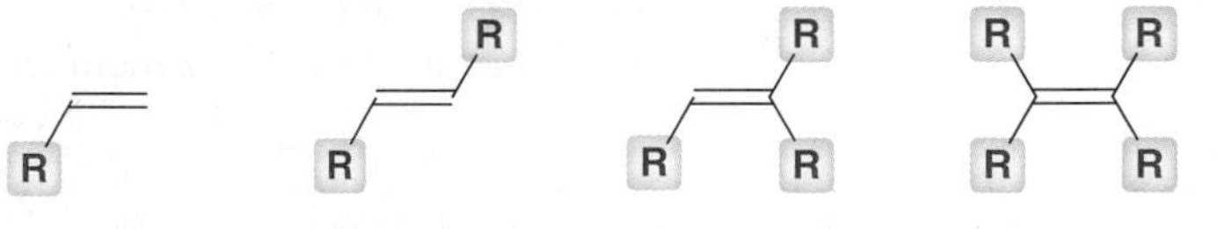

Mono-substituted **Di-substituted** **Tri-substituted** **Tetra-substituted**

FIGURE 8.3
The degree of substitution indicates the number of alkyl groups connected to the double bond.

CONCEPTUAL CHECKPOINT

8.4 Classify each of the following alkenes as monosubstituted, disubstituted, trisubstituted, or tetrasubstituted:

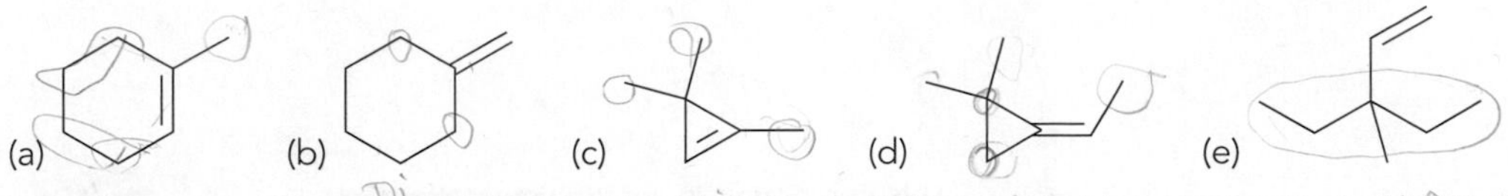

8.4 Stereoisomerism in Alkenes

Using *cis* and *trans* Designations

Recall that a double bond is comprised of a σ bond and a π bond (Figure 8.4).

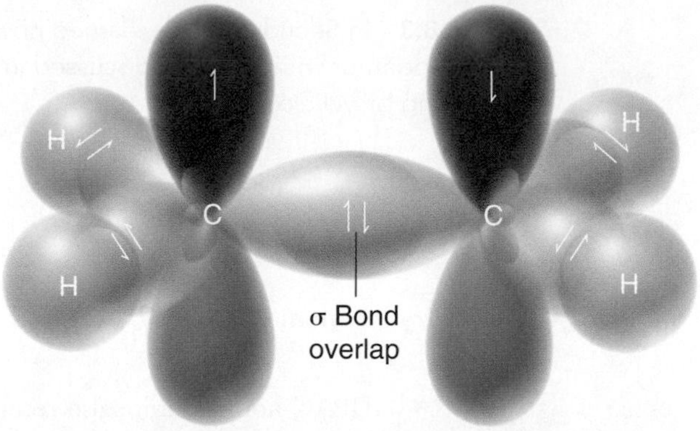

FIGURE 8.4
The σ and π bonds of a C═C double bond.

The σ bond is the result of overlapping sp^2-hybridized orbitals, while the π bond is the result of overlapping *p* orbitals. We have seen that double bonds do not exhibit free rotation at room temperature, giving rise to stereoisomerism:

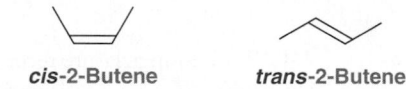

***cis*-2-Butene** ***trans*-2-Butene**

Cycloalkenes comprised of fewer than seven carbon atoms cannot accommodate a *trans* π bond. These rings can only accommodate a π bond in a *cis* configuration:

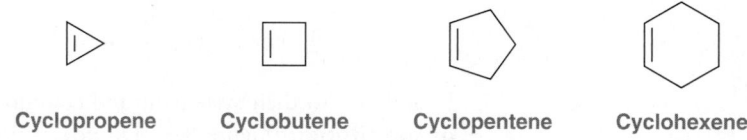

Cyclopropene **Cyclobutene** **Cyclopentene** **Cyclohexene**

In these examples, there is no need to identify the stereoisomerism of the double bond when naming each of these compounds, because the stereoisomerism is inferred. For example, the last compound is called cyclohexene, rather than *cis*-cyclohexene. A seven-membered ring containing a *trans* double bond has been prepared, but this compound (*trans*-cycloheptene) is not stable at room temperature. An eight-membered ring is the smallest ring that can accommodate a *trans* double bond and be stable at room temperature:

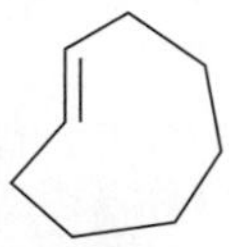

***trans*-Cyclooctene**

This rule also applies to bridged bicyclic compounds and is called **Bredt's rule**. Specifically, Bredt's rule states that it is not possible for a bridgehead carbon of a bicyclic system to possess a C═C double bond if it involves a *trans* π bond being incorporated in a small ring. For example, the following compound is too unstable to form:

This compound is not stable

This compound would require a *trans* double bond in a six-membered ring, highlighted in red The compound is not stable because the geometry of the bridgehead prevents it from maintaining the parallel overlap of *p* orbitals necessary to keep the π bonding intact. As a result, this type of compound is extremely high in energy, and its existence is fleeting.

Bridged bicyclic compounds can only exhibit a double bond at a bridgehead position if one of the rings has at least eight carbon atoms. For example the following compound can maintain parallel overlap of the *p* orbitals, and as a result, it is stable at room temperature:

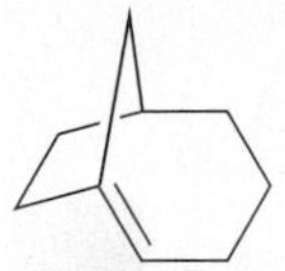

This compound is stable

Using *E* and *Z* Designations

The stereodescriptors *cis* and *trans* can only be used to indicate the relative arrangement of similar groups. When an alkene possesses nonsimilar groups, usage of *cis-trans* terminology would be ambiguous. Consider, for example, the following two compounds:

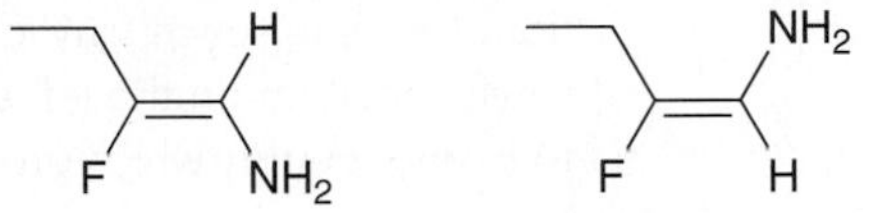

These two compounds are not the same; they are stereoisomers. But which compound should be called *cis* and which should be called *trans*? In situations like this, IUPAC rules provide us with a method for assigning different, unambiguous stereodescriptors. Specifically, we look at the two groups on each vinylic position and choose which of the two groups is assigned the higher priority:

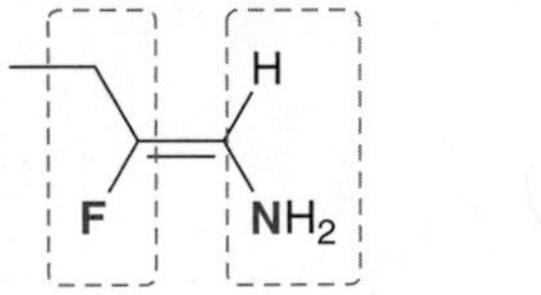

F gets priority over C **N gets priority over H**

In each case, a priority is assigned by the same method used in Section 5.3. Specifically, priority is given to the element with the higher atomic number. In this case, F has priority over C, while N has priority over H. We then compare the position of the higher priority groups. If they are on the same side (as shown above), the configuration is designated with the letter ***Z*** (for the German word *zussamen*, meaning "together"); if they are on opposite sides, the configuration is designated with the letter ***E*** (for the German word *entgegen*, meaning "opposite"):

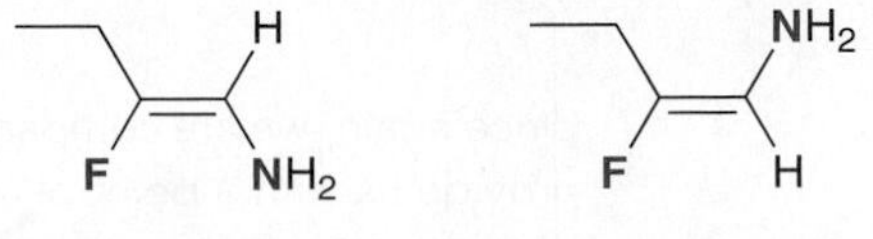

Z ***E***

These examples are fairly straightforward, because the atoms connected directly to the vinylic positions all have different atomic numbers. Other examples may require the comparison of two carbon atoms. In those cases, we will use the same tie-breaking rules that we used when assigning the configurations of chirality centers in Chapter 6. The following SkillBuilder will serve as a reminder of those rules:

SKILLBUILDER

8.2 ASSIGNING THE CONFIGURATION OF A DOUBLE BOND

LEARN the skill

Identify the configuration of the following alkene:

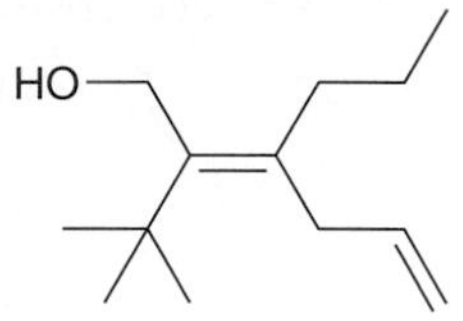

SOLUTION

This compound has two π bonds, but only one of them is stereoisomeric. The π bond shown on the bottom right is not stereoisomeric because it has two hydrogen atoms connected to the same vinylic position.

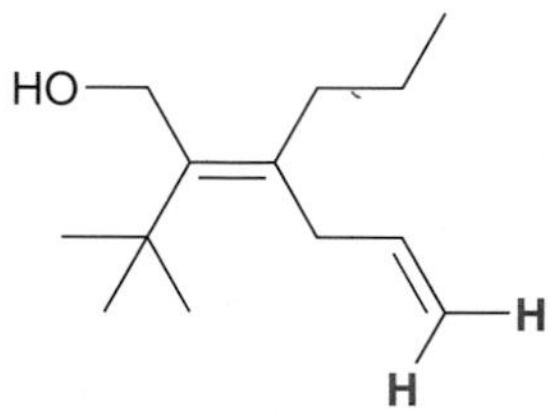

STEP 1
Identify the two groups connected to one vinylic position, and then determine which group has priority.

Let's focus on the other double bond and try to assign its configuration. Consider each vinylic position separately. Let's begin with the vinylic position on the left side. Compare the two groups connected to it, and identify which group should be assigned the higher priority.

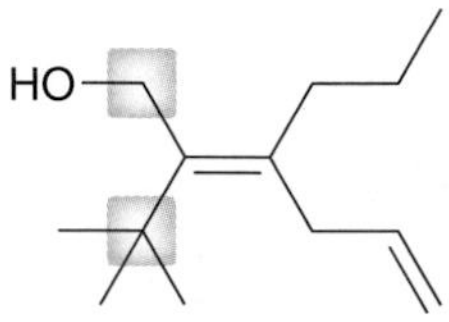

In this case, we are comparing two carbon atoms (which have the same atomic number), so we construct a list for each carbon atom (just as we did in Chapter 6), and look for the first point of difference.

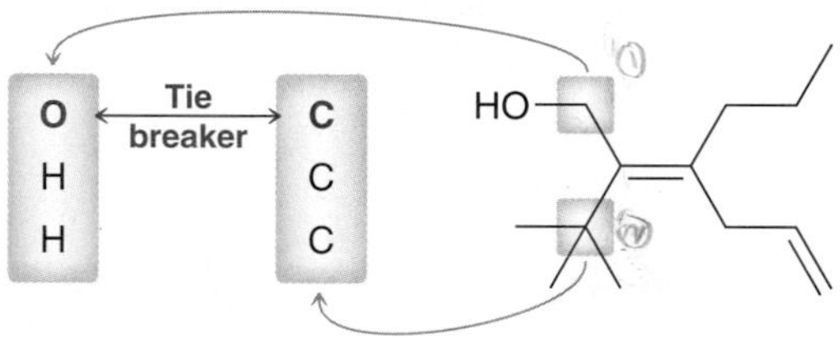

Remember that we do not add atomic numbers of the atoms in each list, but rather, we look for the first point of difference. O has a higher atomic number than C, so the group in the upper left corner is assigned priority over the *tert*-butyl group.

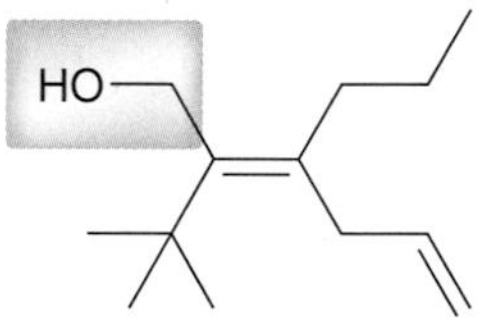

STEP 2
Identify the two groups connected to the other vinylic position, and then determine which group has priority.

Now we turn our attention to the right side of the double bond. Using the same rules, we try to identify which group gets the priority.

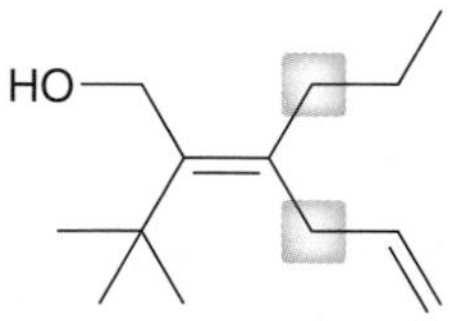

Once again, we are comparing two carbon atoms, but in this case, constructing lists does not provide us with a point of difference. Both lists are identical.

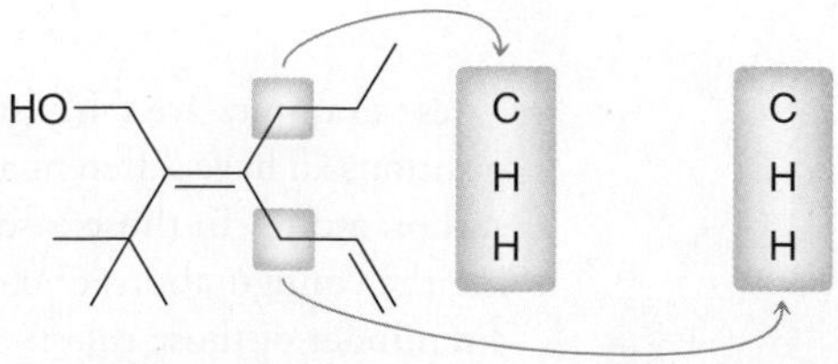

So, we move farther away from the stereoisomeric double bond and again construct lists and look for the first point of difference. Recall that a double bond is treated as two separate single bonds.

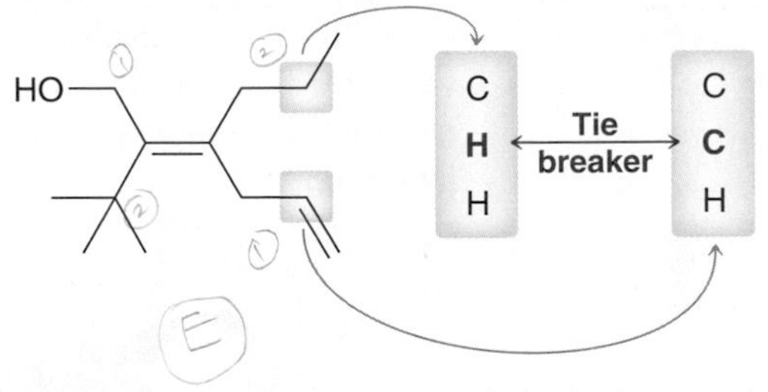

C has a higher atomic number than H, so the group in the bottom right corner is assigned priority over the propyl group.

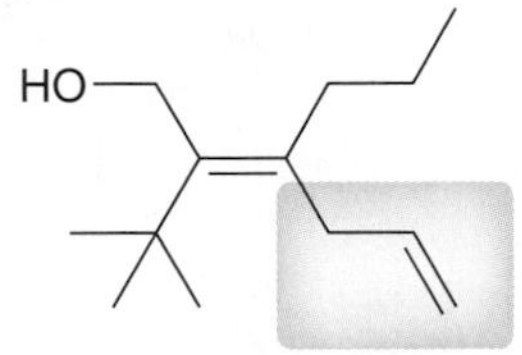

Finally, compare the relative positions of the priority groups. These groups are on opposite sides, so the configuration is *E*.

STEP 3
Determine whether the priorities are on the same side (*Z*) or opposite sides (*E*) of the double bond.

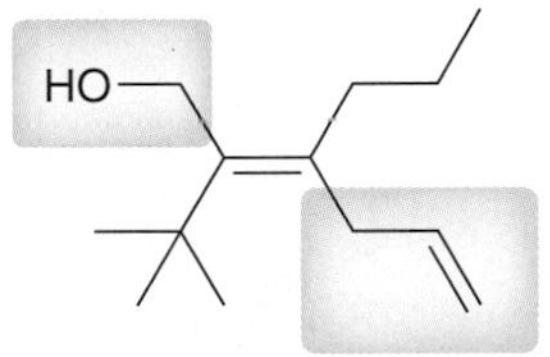

PRACTICE the skill **8.5** For each of the following alkenes, assign the configuration of the double bond as either *E* or *Z*:

(a) 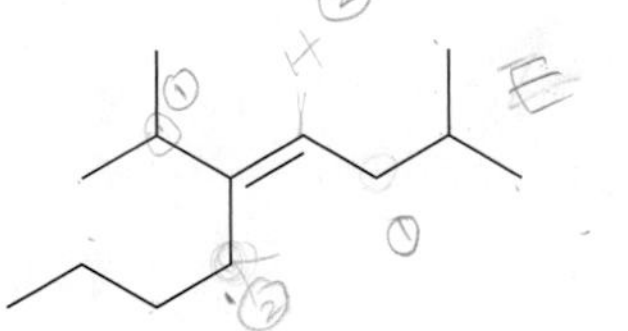

(b) 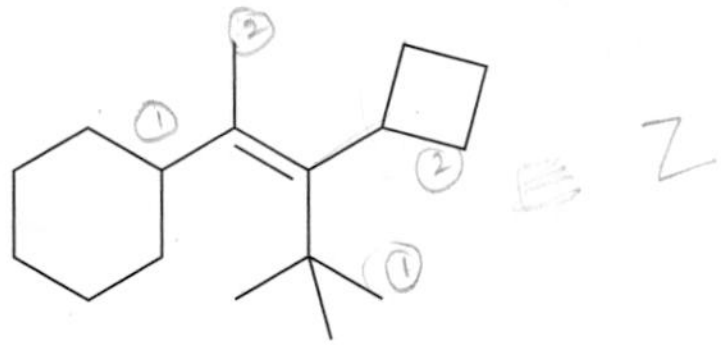

(c) 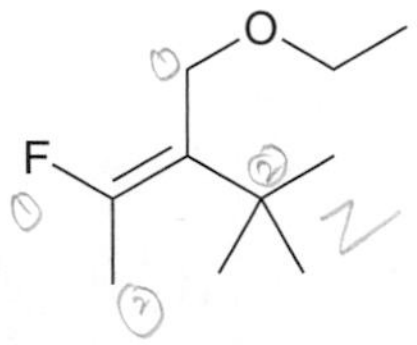

(d) 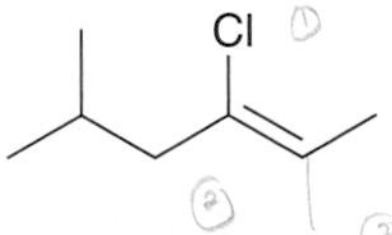

APPLY the skill **8.6** When a double bond has two identical groups (one on each vinylic position), we can use either *cis-trans* terminology or *E-Z* terminology:

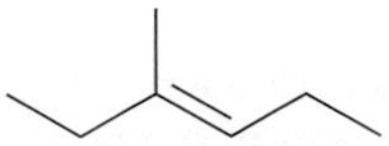

trans or ***E***

The configuration of this double bond can be designated as either *trans* or *E*. Both designations are acceptable. In most cases, *cis* = *Z* and *trans* = *E*. However, there are exceptions. For example:

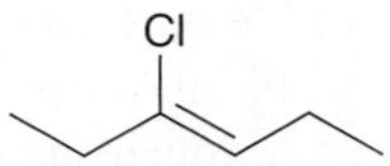

This compound is *trans*, but *Z*. Explain this exception, and then provide two other examples of exceptions.

need more **PRACTICE?** **Try Problem 8.51**

8.5 Alkene Stability

In general, a *cis* alkene will be less stable than its stereoisomeric *trans* alkene. The source of this instability is attributed to steric strain exhibited by the *cis* isomer. This steric strain can be visualized by comparing space-filling models of *cis*-2-butene and *trans*-2-butene (Figure 8.5). The methyl groups are only able to avoid a steric interaction when they occupy a *trans* configuration.

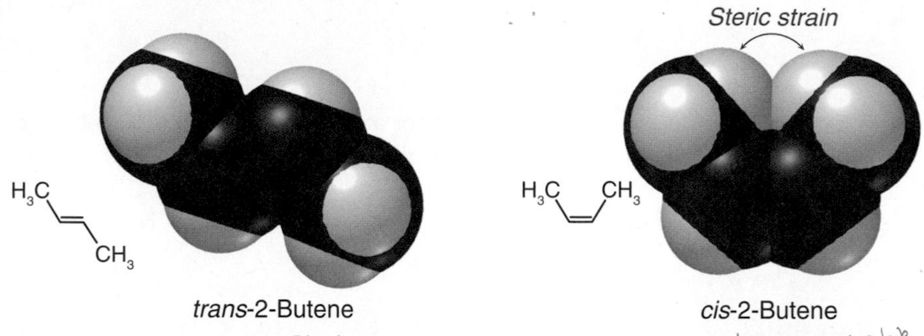

FIGURE 8.5
Space-filling models of the stereoisomers of 2-butene.

trans-2-Butene
more stable

cis-2-Butene
less stable

LOOKING BACK
For a review of heats of combustion, see Section 4.4.

The difference in energy between stereoisomeric alkenes can be quantified by comparing their heats of combustion:

$$\text{(alkene)} \quad + \quad 6 O_2 \quad \longrightarrow \quad 4 CO_2 \quad + \quad 4 H_2O \qquad \Delta H° = -2682 \text{ kJ/mol}$$

$$\text{(alkene)} \quad + \quad 6 O_2 \quad \longrightarrow \quad 4 CO_2 \quad + \quad 4 H_2O \qquad \Delta H° = -2686 \text{ kJ/mol}$$

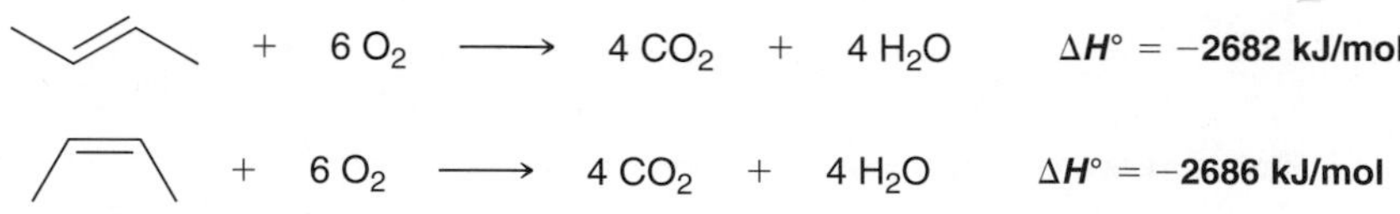

Both reactions yield the same products. Therefore, we can use heats of combustion to compare the relative energy levels of the starting materials (Figure 8.6). This analysis suggests that the *trans* isomer is 4 kJ/mol more stable than the *cis* isomer.

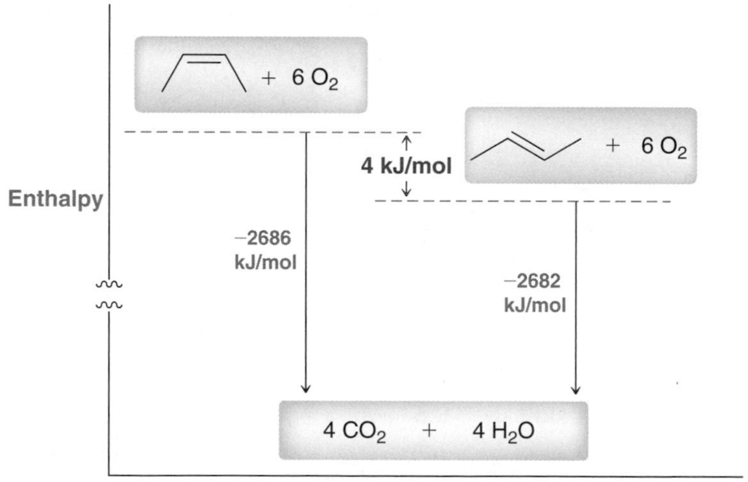

FIGURE 8.6
Heats of combustion of the stereoisomers of 2-butene.

When comparing the stability of alkenes, another factor must be taken into account, in addition to steric effects. We must also consider the degree of substitution. By comparing heats of combustion for isomeric alkenes (all with the same molecular formula, C_6H_{12}), the trend in Figure 8.7 emerges. Alkenes are more stable when they are highly substituted. Tetrasubstituted alkenes are more stable than trisubstituted alkenes. The reason for this trend is not a steric effect (through space), but rather an electronic effect (through bonds). Recall that alkyl groups are electron donating via hyperconjugation (Section 6.11). Specifically, we saw that alkyl groups can donate electron density to the neighboring sp^2-hybridized carbon atom of a carbocation. In a similar way, alkyl groups can also donate electron density to the neighboring sp^2-hybridized carbon atoms of a π bond. The resulting delocalization of electron density is a stabilizing effect.

Increasing stability

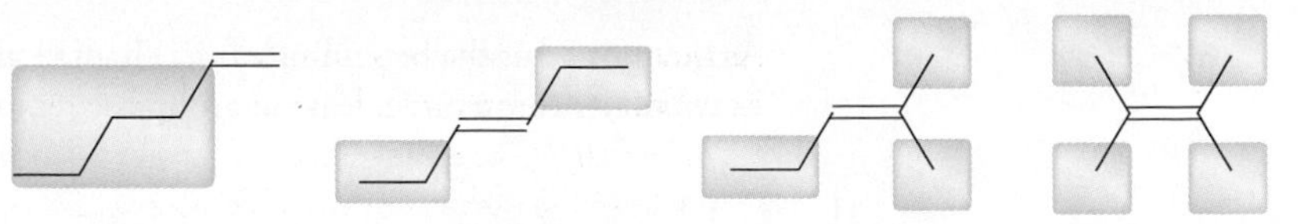

Monosubstituted Disubstituted Trisubstituted Tetrasubstituted

FIGURE 8.7
The relative stability of isomeric alkenes with varying degrees of substitution.

SKILLBUILDER

8.3 COMPARING THE STABILITY OF ISOMERIC ALKENES

LEARN the skill

Arrange the following isomeric alkenes in order of stability:

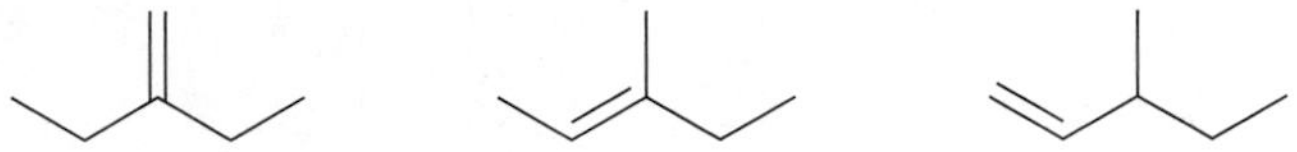

SOLUTION

First identify the degree of substitution for each alkene:

STEP 1
Identify the degree of substitution for each alkene.

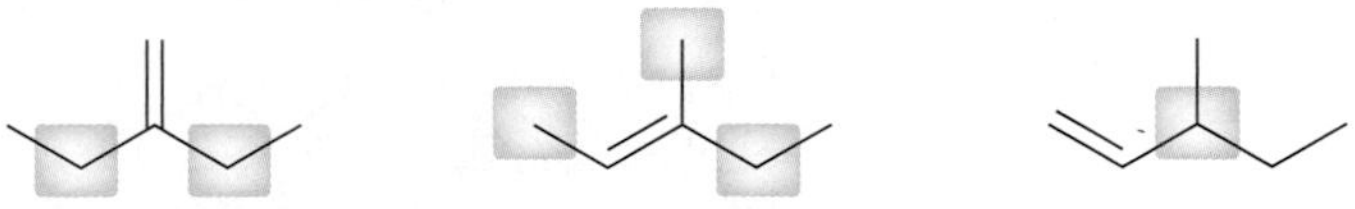

Disubstituted Trisubstituted Monosubstituted

The most highly substituted alkene will be the most stable, and therefore, the following order of stability is expected:

STEP 2
Select the alkene that is more substituted.

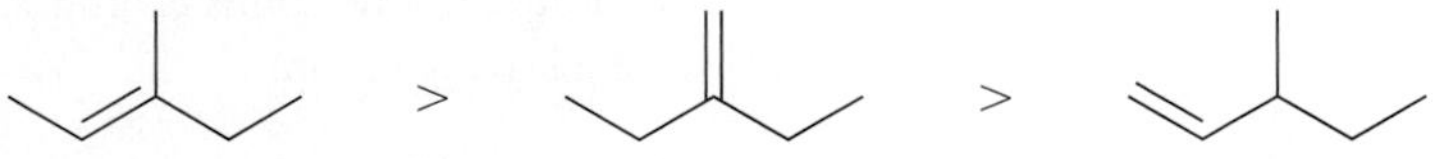

Most stable Least stable

PRACTICE the skill

8.7 Arrange each set of isomeric alkenes in order of stability.

(a)

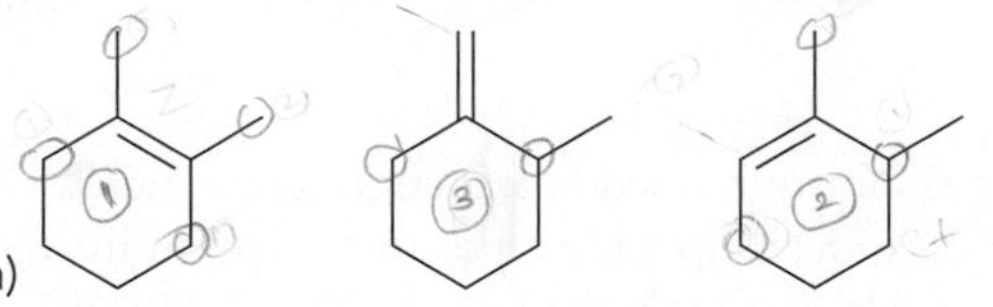

(b)

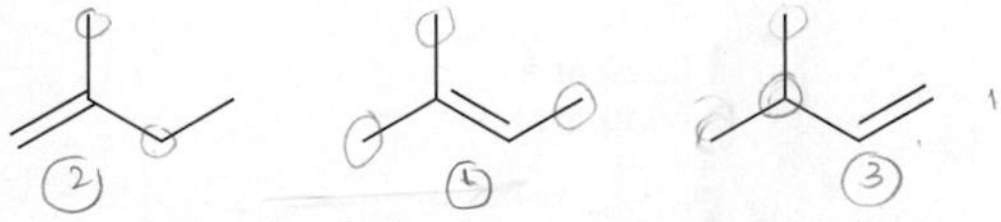

① – most stable
② – middle
③ – least stable

APPLY the skill

8.8 Consider the following two isomeric alkenes. The first isomer is a monosubstituted alkene, while the second isomer is a disubstituted alkene. We might expect the second isomer to be more stable, yet heats of combustion for these two compounds indicate that the first isomer is more stable. Offer an explanation.

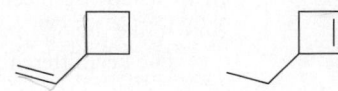

need more **PRACTICE?** **Try Problem 8.53**

8.6 Possible Mechanisms for Elimination

As mentioned at the beginning of the chapter, alkenes can be prepared via elimination reactions, in which a proton and a leaving group are removed to form a π bond:

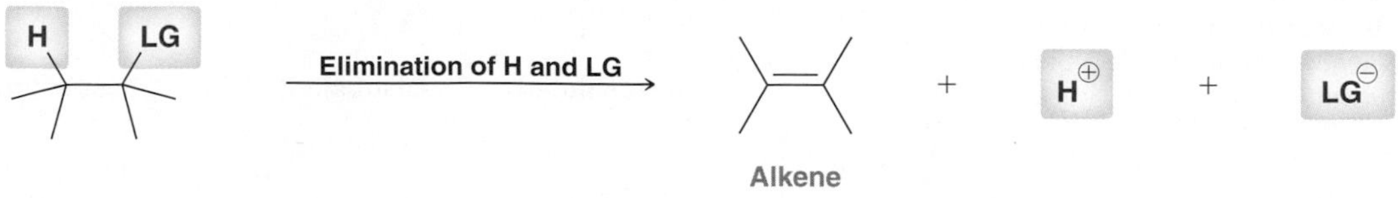

As we begin to explore the possible mechanisms for elimination reactions, recall from Chapter 6 that ionic mechanisms are comprised of only four types of fundamental arrow-pushing patterns (Figure 8.8). All four of these steps will appear in this chapter, so it might be wise to review SkillBuilders 6.3 and 6.4.

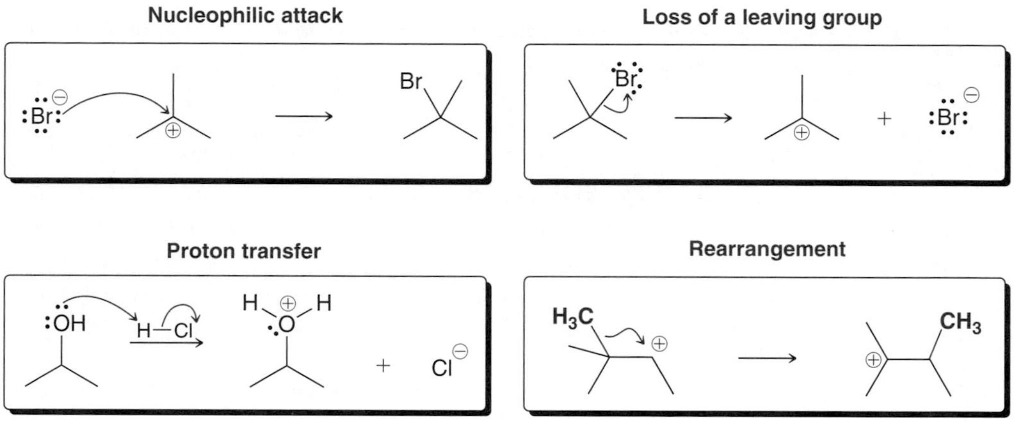

FIGURE 8.8 The four fundamental electron transfer steps for ionic processes.

All elimination reactions exhibit at least two of the four patterns: (1) proton transfer and (2) loss of a leaving group:

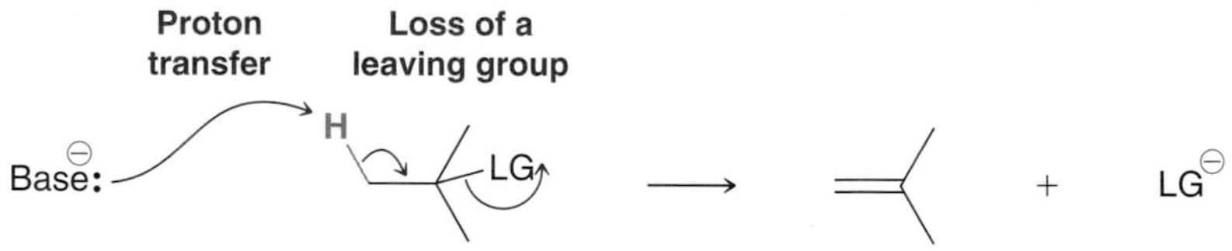

Every elimination reaction features a proton transfer as well as loss of a leaving group. But consider the order of events of these two steps. In the figure above, they occur simultaneously (in a concerted fashion). Alternatively, we can imagine them occurring separately, in a stepwise fashion:

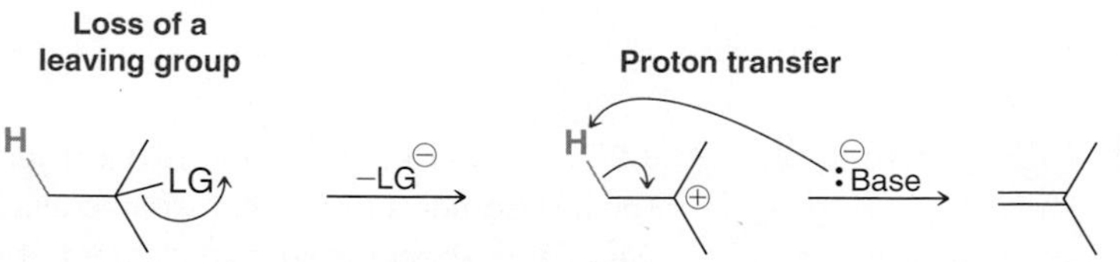

In this stepwise mechanism the leaving group leaves, generating an intermediate carbocation (much like the S_N1 reaction), which is then deprotonated by a base to produce an alkene.

The key difference between these two mechanisms is summarized below:

- In a *concerted process,* a base abstracts a proton and the leaving group leaves simultaneously.
- In a *stepwise process,* first the leaving group leaves, and then the base abstracts a proton.

In this chapter, we will explore these two mechanisms in more detail, and we will see that each mechanism predominates under a different set of conditions. In preparation, we must first practice drawing curved arrows.

SKILLBUILDER

8.4 DRAWING THE CURVED ARROWS OF AN ELIMINATION REACTION

LEARN the skill

Assume that the following elimination reaction proceeds via a concerted process and draw the mechanism:

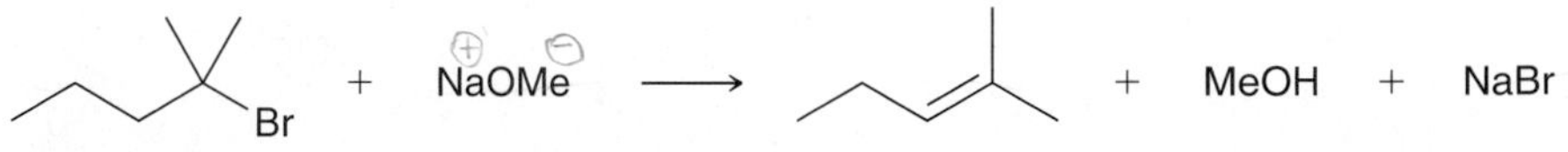

SOLUTION

First identify the base and the substrate. The base in this case is methoxide (remember that Na^+ is the counterion, and its role in the reaction does not concern us in most cases):

Base Substrate

WATCH OUT
When drawing mechanisms, be very precise about the placement of the head and tail of each curved arrow. Students often draw this arrow in the wrong direction.

In a concerted mechanism, the base abstracts a proton and the leaving group leaves simultaneously. This requires a total of three curved arrows. When drawing the first curved arrow, make sure to place the tail on a lone pair of the base, and place the head on the proton that is being removed:

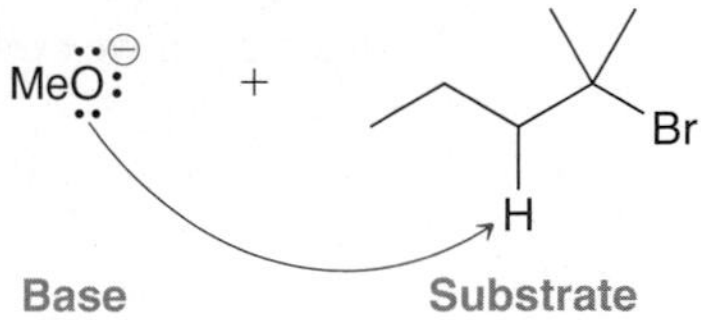

Base Substrate

For the other two curved arrows, one shows the formation of the double bond, and the other shows loss of the leaving group:

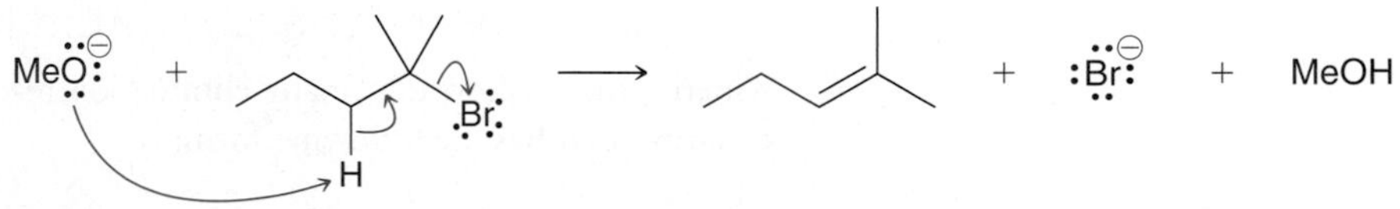

PRACTICE the skill

8.9 For each of the following elimination reactions, assume a concerted process is taking place and draw the mechanism:

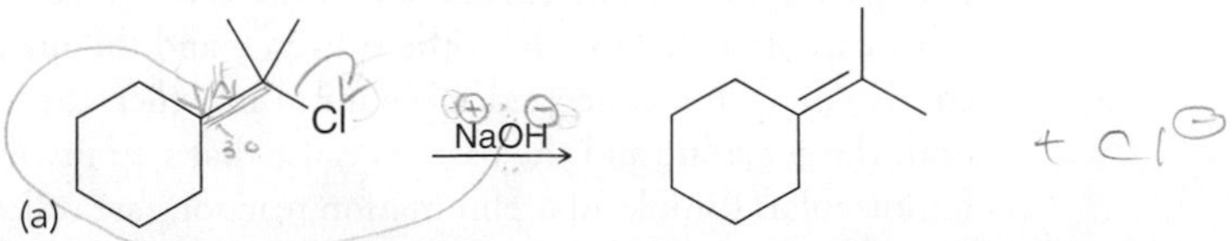

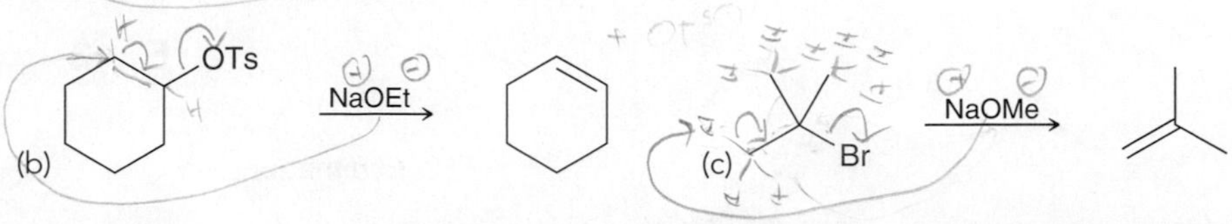

8.10 For each of the following elimination reactions, assume a stepwise process is taking place and draw the mechanism (first loss of the leaving group and then proton transfer):

(a)

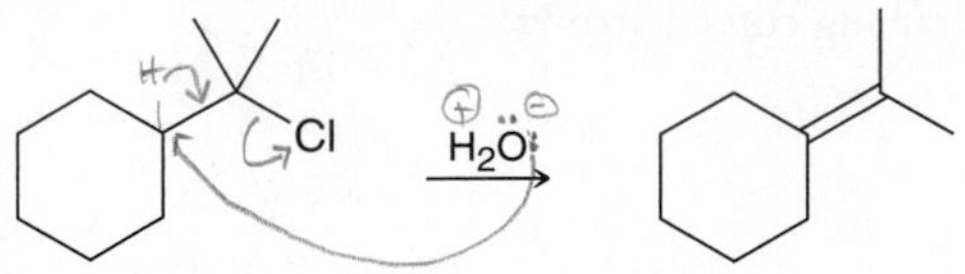

(b)

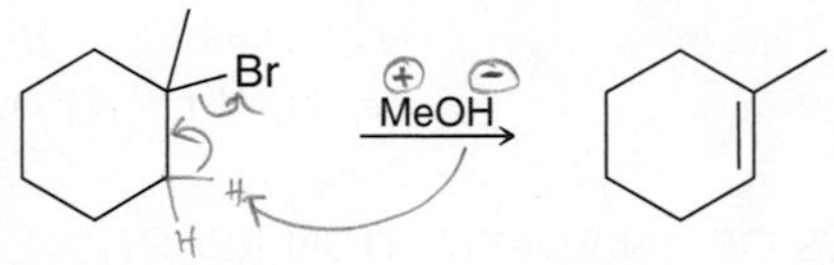

(c)

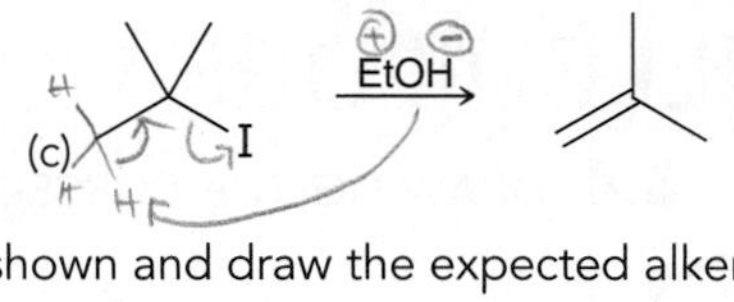

APPLY the skill

8.11 Carefully read the following curved arrows shown and draw the expected alkene that is produced by this elimination reaction. Is this mechanism concerted or stepwise?

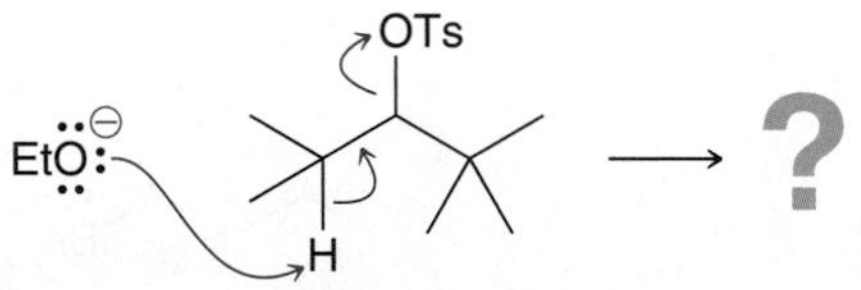

8.12 Draw the intermediate of the following elimination reaction, and then draw the curved arrows for the second step (the reaction between the intermediate and ethanol):

need more **PRACTICE?** **Try Problem 8.69**

8.7 The E2 Mechanism

In this section we explore the concerted mechanism, or E2, pathway:

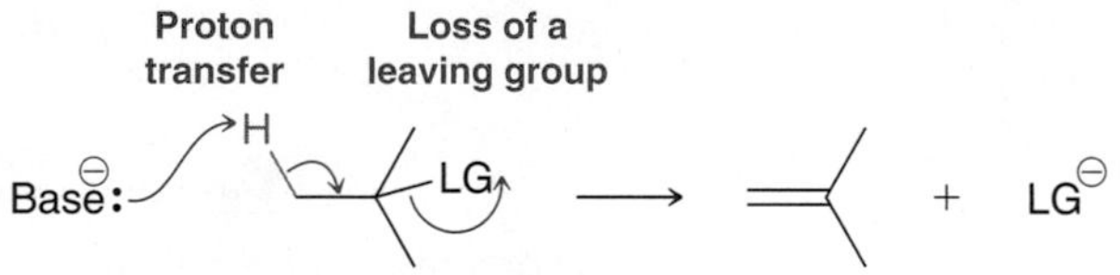

Kinetic Evidence for a Concerted Mechanism

Kinetic studies show that many elimination reactions exhibit second-order kinetics with a rate equation that has the following form:

$$\text{Rate} = k\,[\text{substrate}]\,[\text{base}]$$

Much like the S_N2 reaction, the rate is linearly dependent on the concentrations of two different compounds (the substrate and the base). This observation suggests that the mechanism must exhibit a step in which the substrate and the nucleophile collide with each other. This is consistent with a concerted mechanism in which there is only one mechanistic step involving both the substrate and the base. Because that step involves two chemical entities, it is said to be bimolecular. Bimolecular elimination reactions are called **E2** reactions:

CONCEPTUAL CHECKPOINT

8.13 The following reaction exhibits a second-order rate equation:

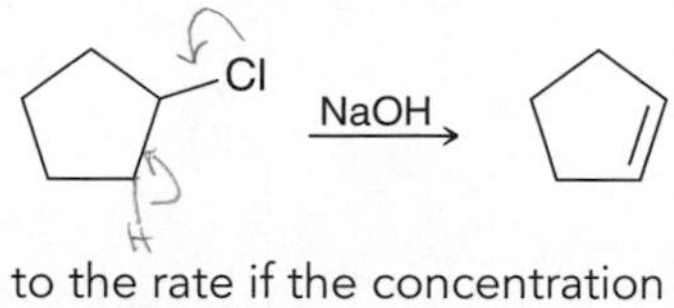

(a) What happens to the rate if the concentration of chlorocyclopentane is tripled and the concentration of sodium hydroxide remains the same? rate is trippled.

(b) What happens to the rate if the concentration of chlorocyclopentane remains the same and the concentration of sodium hydroxide is doubled? rate is two times faster

(c) What happens to the rate if the concentration of chlorocyclopentane is doubled and the concentration of sodium hydroxide is tripled? Six times faster.

* rate dependent on conc. of the two different compounds (substrate and base)

Effect of Substrate

In Chapter 7 we saw that the rate of an S_N2 process with a tertiary substrate is generally so slow that it can be assumed that the substrate is inert under such reaction conditions. It might therefore come as a surprise that tertiary substrates undergo E2 reactions quite rapidly. To explain why tertiary substrates will undergo E2 but not S_N2 reactions, we must recognize that the key difference between substitution and elimination is the role played by the reagent. A substitution reaction occurs when the reagent functions as a nucleophile and attacks an electrophilic position, while an elimination reaction occurs when the reagent functions as a base and abstracts a proton. With a tertiary substrate, steric hindrance prevents the reagent from functioning as a nucleophile at an appreciable rate, but the reagent can still function as a base without encountering much steric hindrance (Figure 8.9).

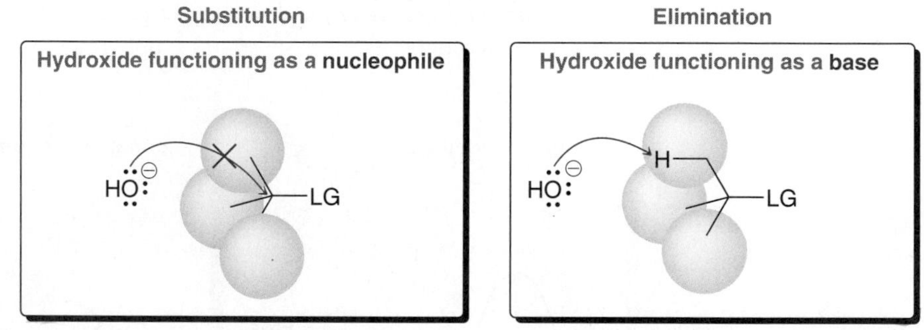

A tertiary substrate is too sterically hindered. The nucleophile cannot penetrate and attack.

A β proton can easily be abstracted even though the substrate is tertiary.

FIGURE 8.9
Steric effects in S_N2 and E2 reactions.

Tertiary substrates react readily in E2 reactions; in fact, they react even more rapidly than primary substrates (Figure 8.10). To understand the reason for this trend, let's look at an energy diagram of an E2 process (Figure 8.11). We focus our attention on the transition state (Figure 8.12). For a tertiary substrate, both R groups (shown in red in Figure 8.12) are alkyl groups. For a primary substrate, both of these R groups are just hydrogen atoms. In the transition state, a C=C double bond is forming. For a tertiary substrate, the transition state exhibits a partial double bond that is more highly substituted, and therefore, the transition state will be lower in energy. Compare the energy diagrams for E2 reactions involving primary, secondary, and

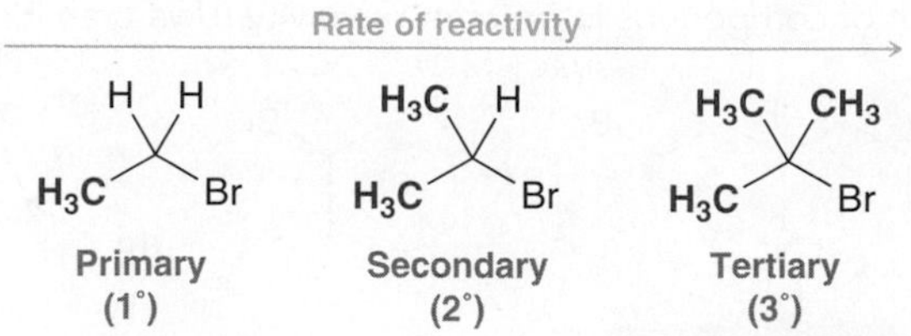

FIGURE 8.10
Relative rates of reactivity for substrates in an E2 reaction.

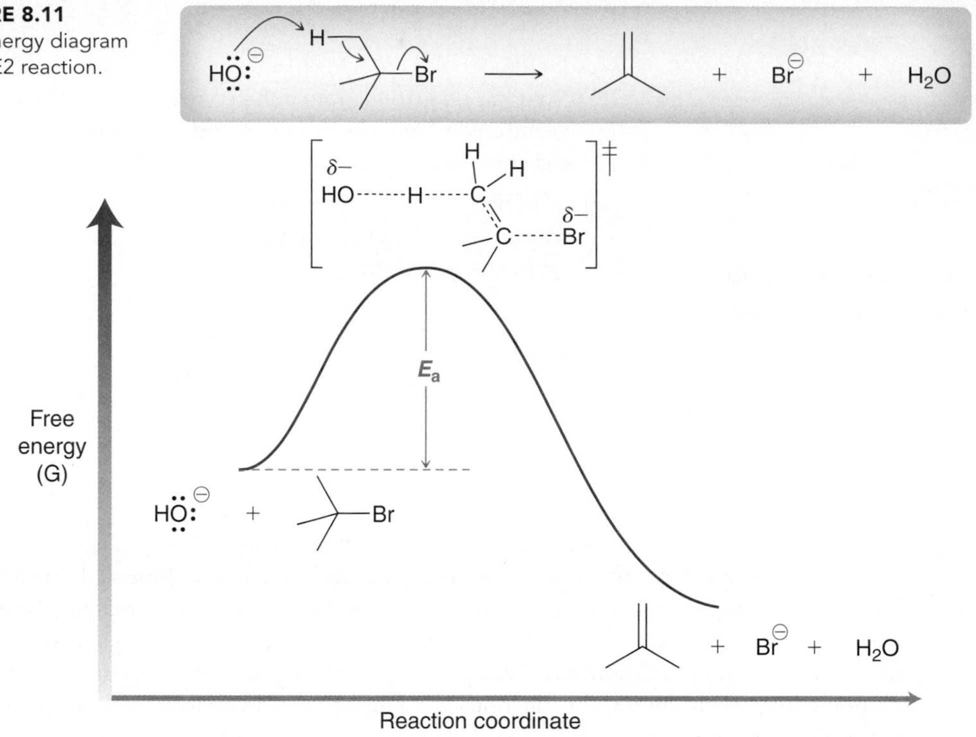

FIGURE 8.11
The energy diagram of an E2 reaction.

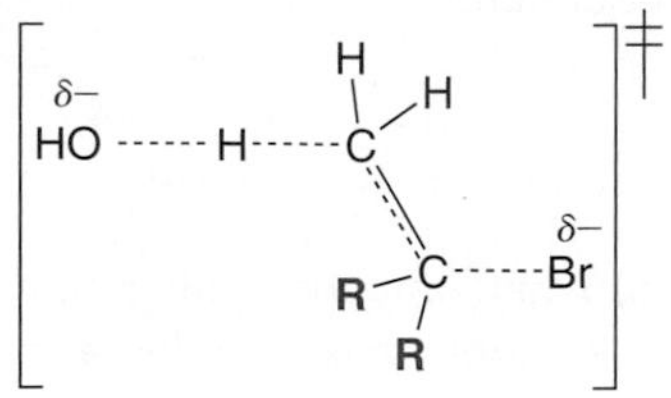

FIGURE 8.12
The transition state of an E2 reaction.

tertiary substrates (Figure 8.13). The transition state is lowest in energy when a tertiary substrate is used, and therefore, the energy of activation is lowest for tertiary substrates. This explains the observation that tertiary substrates react most rapidly in E2 reactions. That doesn't mean that primary substrates are slow to react in E2 reactions. In fact, primary substrates readily undergo E2 reactions. Tertiary substrates simply react more rapidly under the same conditions.

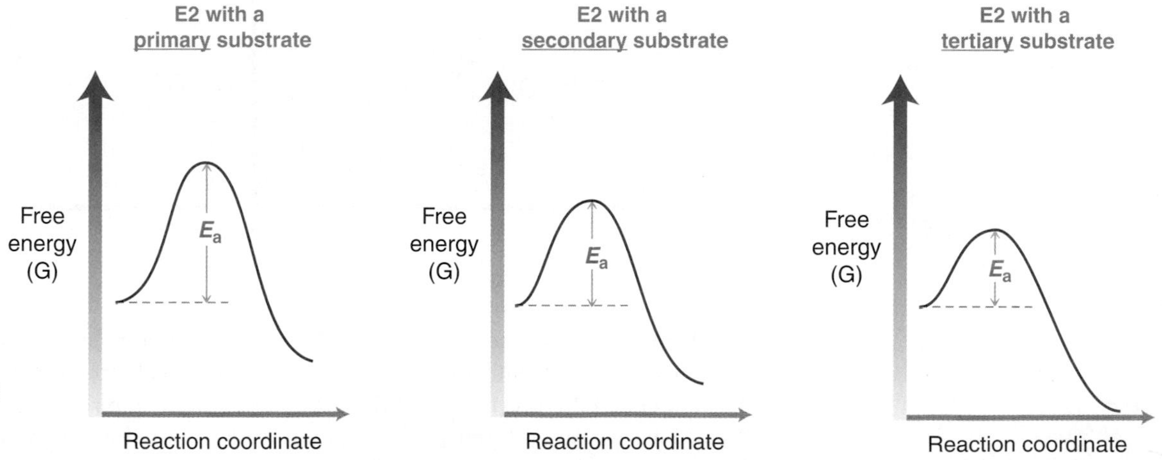

FIGURE 8.13
Energy diagrams for E2 reactions with various substrates.

CONCEPTUAL CHECKPOINT

8.14 Arrange each set of compounds in order of reactivity toward an E2 process:

(a)

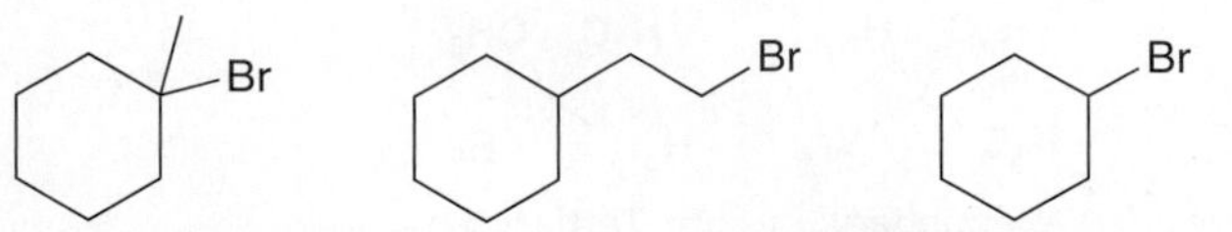

(b)

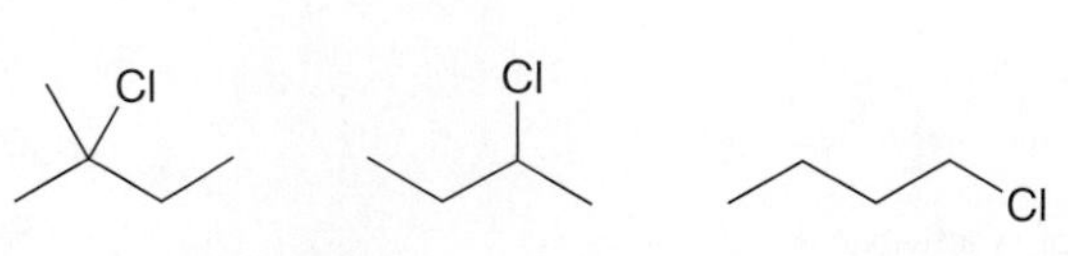

Regioselectivity of E2 Reactions

In many cases, an elimination reaction can produce more than one possible product. In this example, the β positions are not identical, so the double bond can form in two different regions of the molecule. This consideration is an example of **regiochemistry**, and the reaction is said to produce two different regiochemical outcomes. Both products are formed, but the more substituted alkene is generally observed to be the major product. For example:

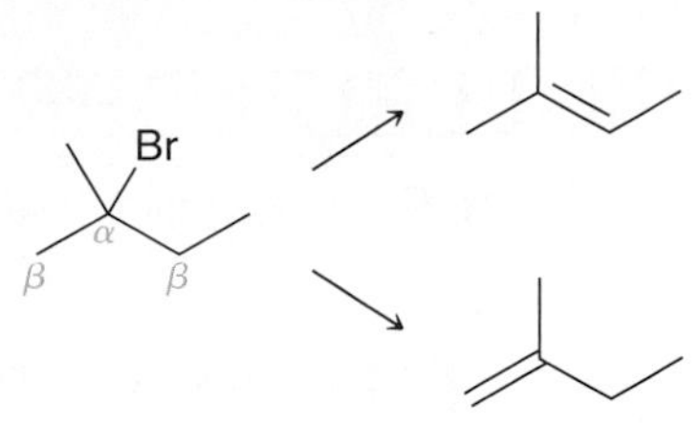

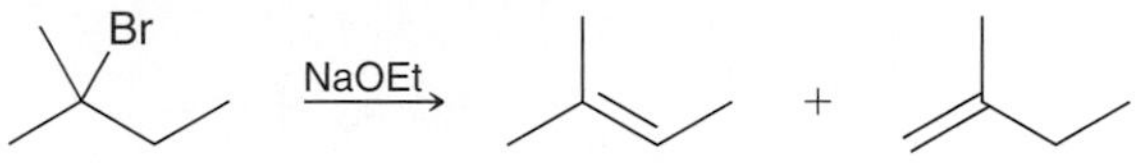

71% 29%

The reaction is said to be **regioselective**. This trend was first observed in 1875 by Russian chemist Alexander M. Zaitsev (University of Kazan), and as a result, the more substituted alkene is called the **Zaitsev product**. However, many exceptions have been observed in which the Zaitsev product is the minor product. For example, when both the substrate and the base are sterically hindered, the less substituted alkene is the major product:

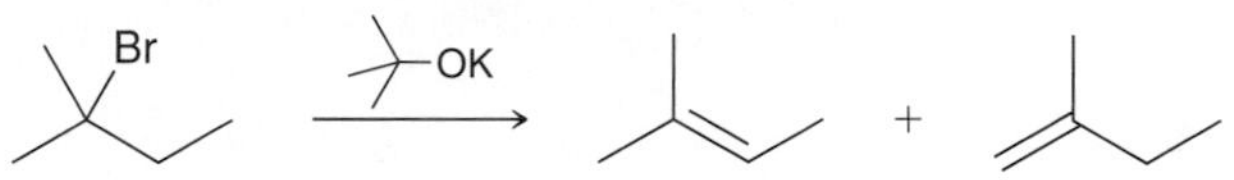

28% 72%

The less substituted alkene is often called the **Hofmann product**. The product distribution (relative ratio of Zaitsev and Hofmann products) is dependent on a number of factors and is often difficult to predict. The choice of base (how sterically hindered the base is) certainly plays an important role. For example, the product distribution for the reaction above is highly dependent on the choice of base (Table 8.1).

TABLE 8.1 PRODUCT DISTRIBUTION OF AN E2 REACTION AS A FUNCTION OF BASE

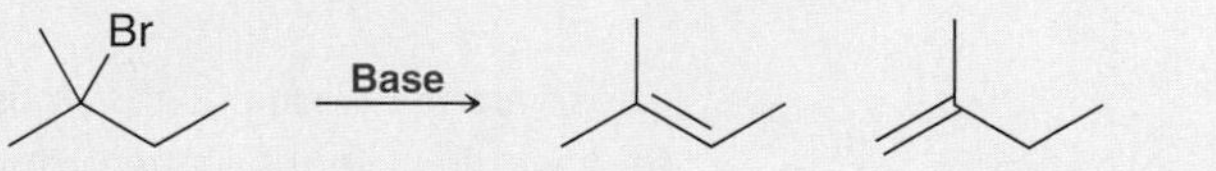

	ZAITSEV	HOFMANN
$EtO^{\ominus}$	71%	29%
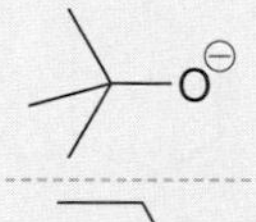	28%	72%
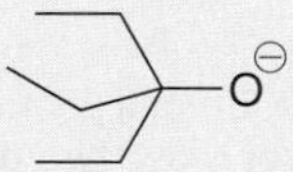	8%	92%

When ethoxide is used, the Zaitsev product is the major product. But when sterically hindered bases are used, the Hofmann product becomes the major product. This case illustrates a critical concept: *The regiochemical outcome of an E2 reaction can often be controlled by carefully choosing the base.* Sterically hindered bases are employed in a variety of reactions, not just elimination, so it is useful to recognize a few sterically hindered bases (Figure 8.14).

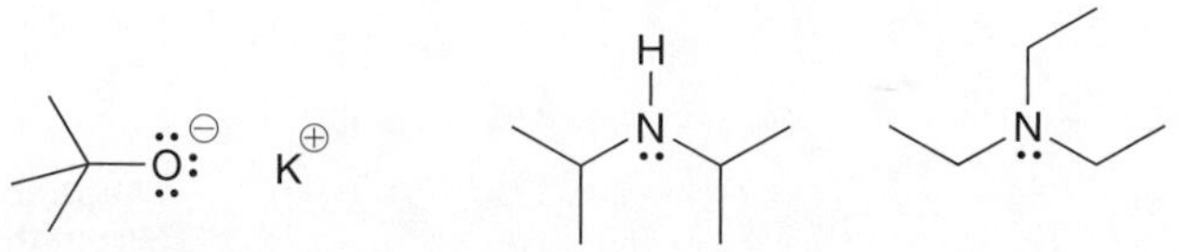

Potassium *tert*-butoxide (*t*-BuOK) **Diisopropylamine** **Triethylamine**

FIGURE 8.14
Commonly used sterically hindered bases.

SKILLBUILDER

8.5 PREDICTING THE REGIOCHEMICAL OUTCOME OF AN E2 REACTION

LEARN the skill

Identify the major and minor products of the following E2 reaction:

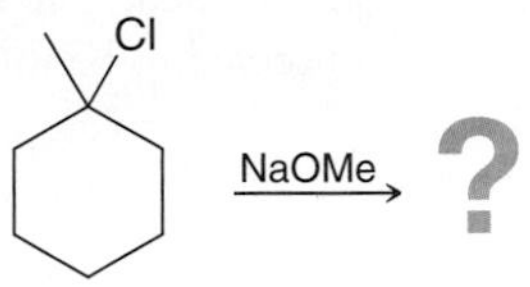

SOLUTION

First identify the α position. This is the position bearing the leaving group:

STEP 1
Identify the α position.

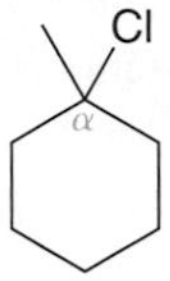

Next, identify all β positions that have protons:

STEP 2
Identify all β positions with protons.

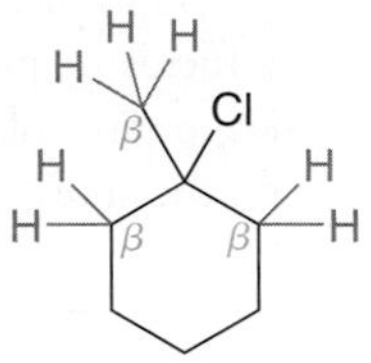

These are the positions that must be explored. Two of these positions are identical, because they would give the same product:

STEP 3
Take notice of β positions that lead to the same product.

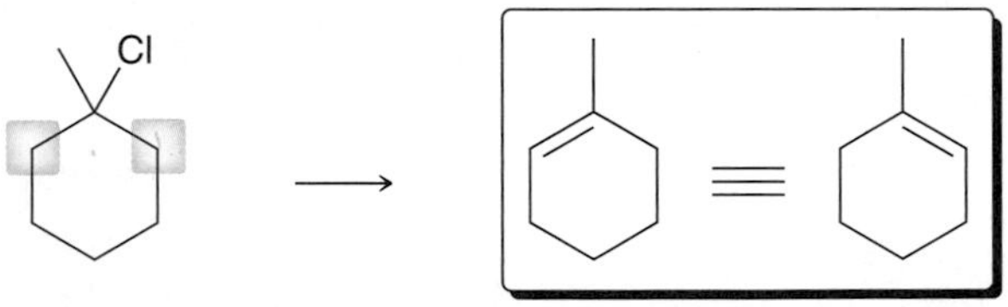

Abstracting a proton from either of these two positions affords the same product

These two drawings represent the same product

Abstracting a proton from the other β position will produce the following alkene:

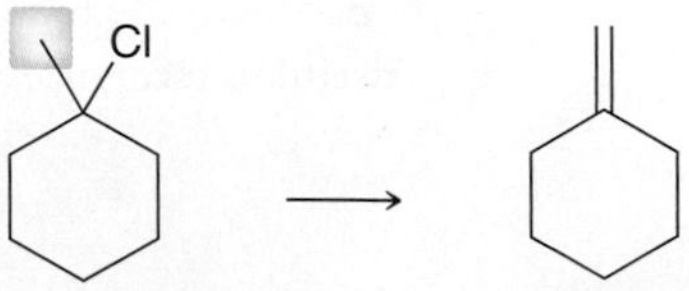

Abstracting a proton from this position...

...affords this product

Therefore, there are two possible products. In order to determine which product predominates, analyze each of the products and determine the degree of substitution:

STEP 4
Remove a proton from each β position and compare the degree of substitution of the resulting alkenes.

Trisubstituted (Zaitsev) **Disubstituted (Hofmann)**

The more substituted alkene is the Zaitsev product, and the less substituted alkene is the Hofmann product. The Zaitsev product is generally the major product, unless a sterically hindered base is used. This example does not employ a sterically hindered base, so we expect that the Zaitsev product will be the major product:

STEP 5
Analyze the base to determine which product predominates.

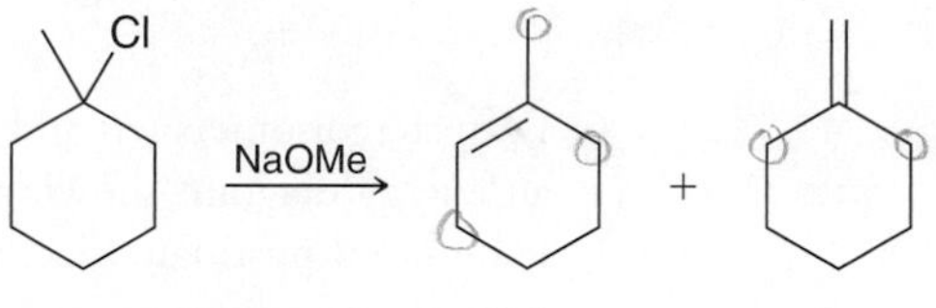

Major **Minor**

PRACTICE the skill

8.15 Identify the major and minor products for each of the following E2 reactions:

(a) 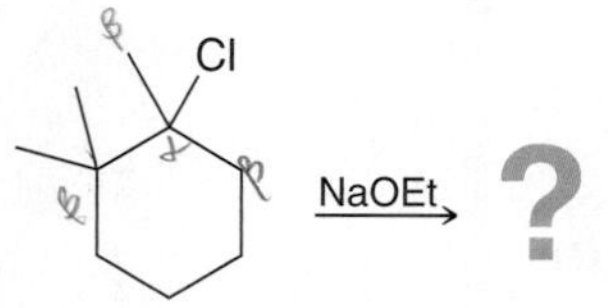

(b) 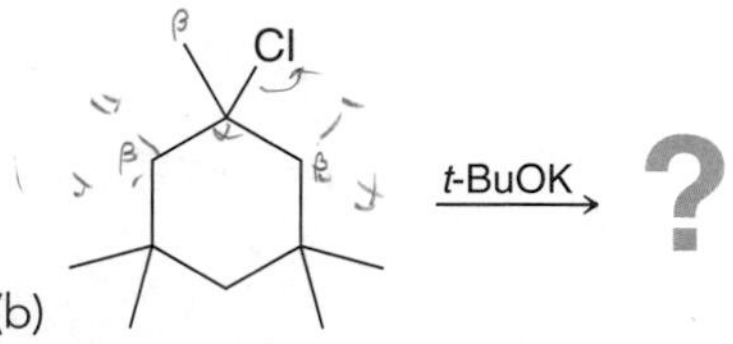

(c) 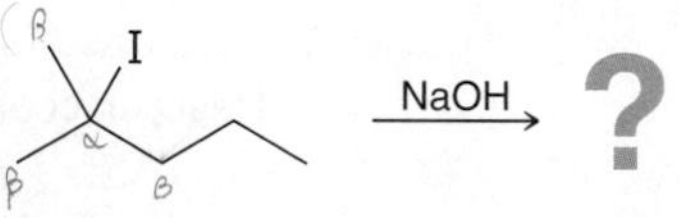

(d) 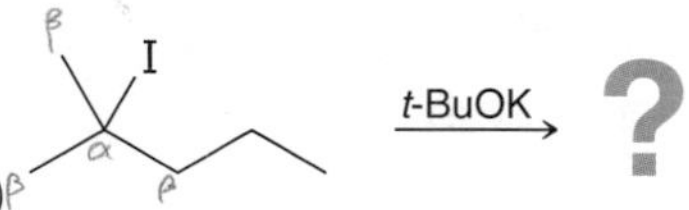

(e) 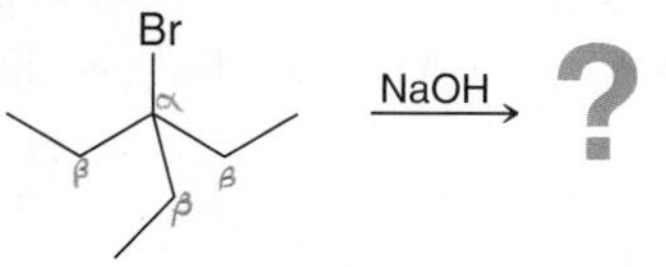

(f) 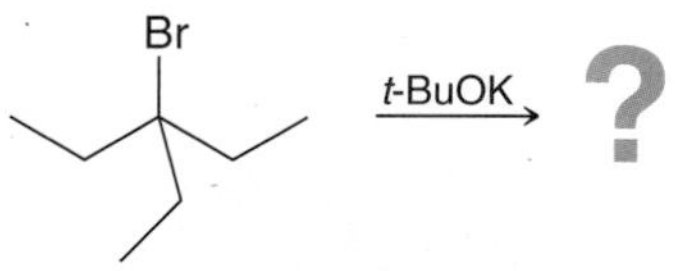

APPLY the skill

8.16 For each of the following reactions, identify whether you would use hydroxide or *tert*-butoxide to accomplish the desired transformation:

(a) 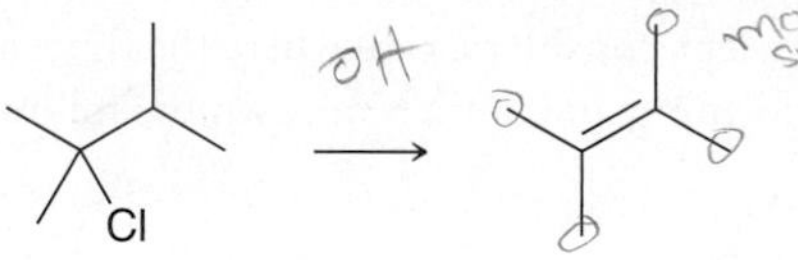

(b) 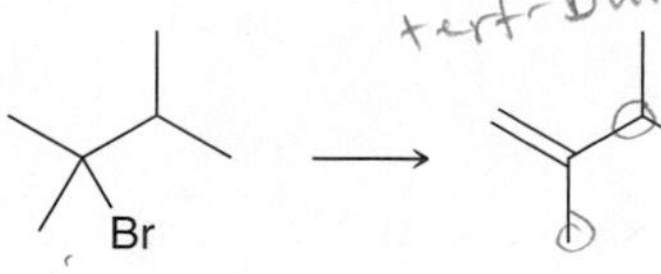

8.17 Show two different methods for preparing each of the following alkenes (one method using a sterically hindered base and the other method using a base that is not sterically hindered):

(a) 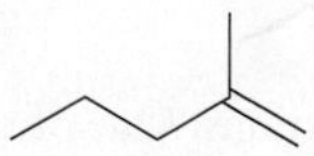

(b) 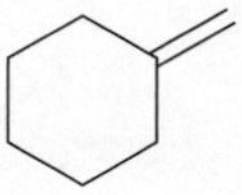

need more **PRACTICE?** **Try Problems 8.60, 8.61b–d, 8.64, 8.66a,d, 8.75**

Stereoselectivity of E2 Reactions

The examples in the previous sections focused on regiochemistry. We will now focus our attention on stereochemistry. For example, consider performing an E2 reaction with 3-bromopentane as the substrate:

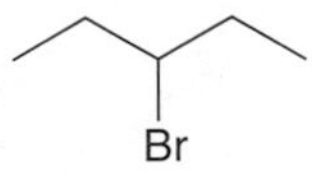

This compound has two identical β positions so regiochemistry is not an issue in this case. Deprotonation of either β position produces the same result. But in this case, stereochemistry is relevant, because two possible stereoisomeric alkenes can be obtained:

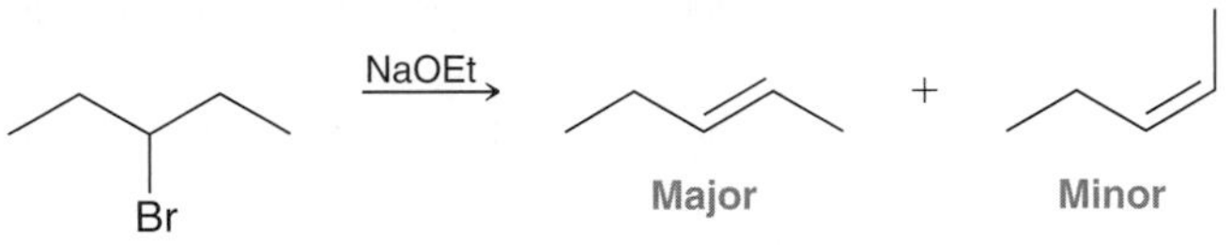

Both stereoisomers (*cis* and *trans*) are produced, but the *trans* product predominates. Consider an energy diagram showing formation of the *cis* and *trans* products (Figure 8.15). Using the Hammond postulate (Section 6.6), we can show that the transition state for formation of the *trans* alkene is more stable than the transition state for formation of the *cis* alkene. This reaction is said to be **stereoselective** because the substrate produces two stereoisomers in unequal amounts.

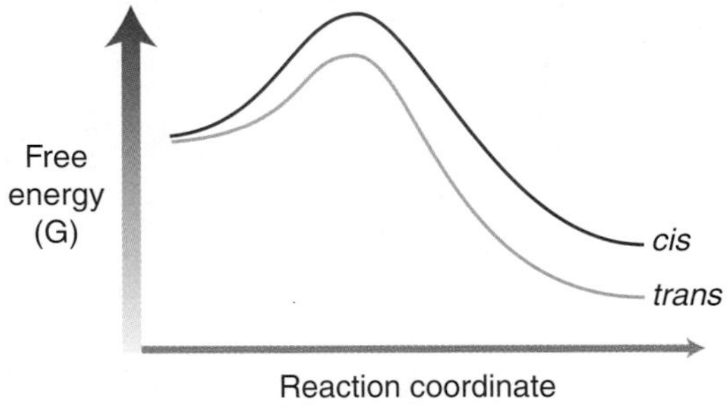

FIGURE 8.15 An energy diagram showing formation of *cis* and *trans* products in an E2 reaction.

Stereospecificity of E2 Reactions

In the previous example, the β position had two different protons:

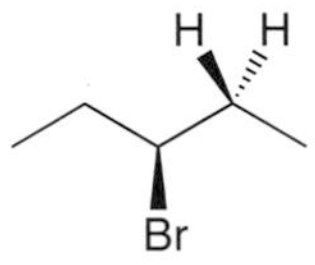

In such a case, both the *cis* and the *trans* isomers were produced, with the *trans* isomer being favored. Now let's consider a case where the β position contains only one proton. For example, consider performing an elimination with the following substrate:

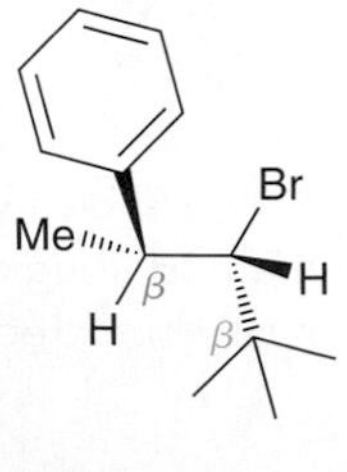

In this example, there are two β positions. One of these positions has no protons at all, and the other position has only one proton. In such a case, a mixture of stereoisomers is not obtained. In this case, there will only be one stereoisomeric product:

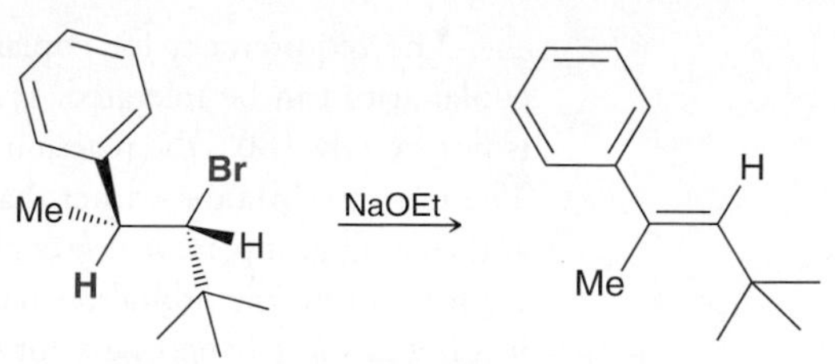

Why is the other possible stereoisomer not obtained? To understand the answer to this question, we must explore the alignment of orbitals in the transition state. In the transition state, a π bond is forming. Recall that a π bond is comprised of overlapping *p* orbitals. Therefore, the transition state must involve the formation of *p* orbitals that are positioned such that they can overlap with each other as they are forming. In order to achieve this kind of orbital overlap, the following four atoms must all lie in the same plane: the proton at the β position, the leaving group, and the two carbon atoms that will ultimately bear the π bond. These four atoms must all be **coplanar**:

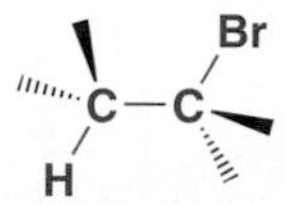

These four atoms (shown in red) must all lie in the same plane

Recall that the C—C single bond is free to rotate before the reaction occurs. If we imagine rotating this bond, we see that there are two ways to achieve a coplanar arrangement:

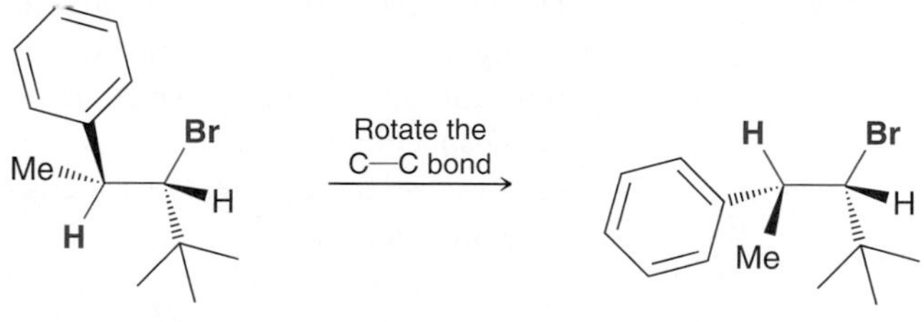

***Anti*-coplanar** ***Syn*-coplanar**

The first conformation is called ***anti*-coplanar**, while the second conformation is called ***syn*-coplanar**. In this context, the terms *anti* and *syn* refer to the relative positions of the proton and the leaving group, which can be seen more clearly with Newman projections (Figure 8.16).

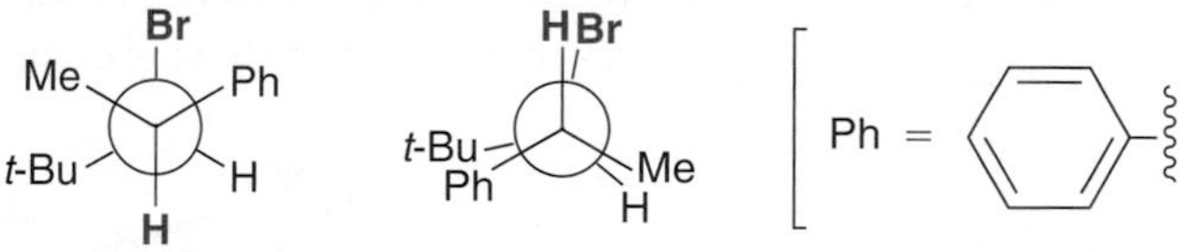

***Anti*-coplanar (staggered)** ***Syn*-coplanar (eclipsed)** **Phenyl**

FIGURE 8.16 Newman projections of *anti* coplanar and *syn*-coplanar conformations.

When viewed in this way, we can see that the *anti*-coplanar conformation is staggered, while the *syn*-coplanar conformation is eclipsed. Elimination via the *syn*-coplanar conformation would involve a transition state of higher energy as a result of the eclipsed geometry. Therefore, elimination occurs more rapidly via the *anti*-coplanar conformation. In fact, in most cases, elimination is observed to occur exclusively via the *anti*-coplanar conformation, which leads to one specific stereoisomeric product:

LOOKING BACK
For practice drawing Newman projections, see Section 4.7.

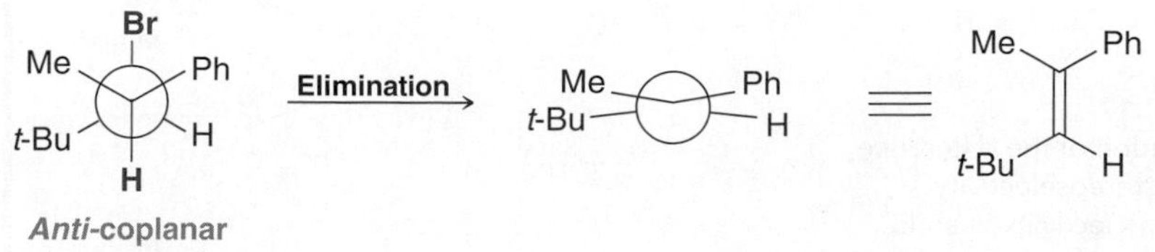

***Anti*-coplanar (staggered)**

The requirement for coplanarity is not entirely absolute. That is, small deviations from coplanarity can be tolerated. If the dihedral angle between the proton and the leaving group is not exactly 180°, the reaction can still proceed as long as the dihedral angle is close to 180°. The term **periplanar** (rather than coplanar) is used to describe a situation in which the proton and leaving group are *nearly* coplanar (for example, a dihedral angle of 178° or 179°). In such a conformation, the orbital overlap is significant enough for an E2 reaction to occur. Therefore, it is not absolutely necessary for the proton and the leaving group to be *anti*-coplanar. Rather, it is sufficient for the proton and the leaving group to be ***anti*-periplanar**. From now on, we will use the term *anti*-periplanar when referring to the stereochemical requirement for an E2 process.

The requirement for an *anti*-periplanar arrangement will determine the stereoisomerism of the product. In other words, *the stereoisomeric product of an E2 process depends on the configuration of the starting alkyl halide*:

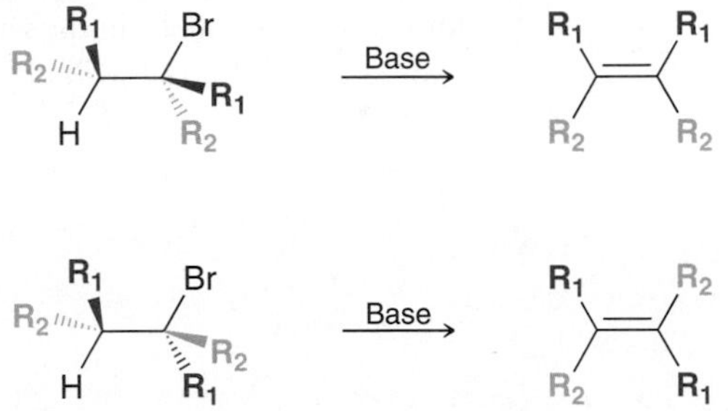

It would be absolutely wrong to say that the product will always be the *trans* isomer. The product obtained depends on the configuration of the starting alkyl halide. The only way to predict the configuration of the resulting alkene is to analyze the substrate carefully, draw a Newman projection, and then determine which stereoisomeric product is obtained. The E2 reaction is said to be **stereospecific**, because the stereoisomerism of the product is dependent on the stereoisomerism of the substrate. The stereospecificity of an E2 reaction is only relevant when the β position has only one proton:

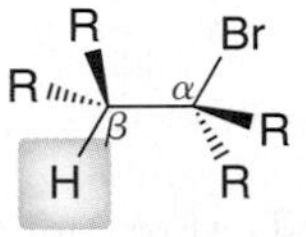

In such a case, the β proton must be arranged *anti*-periplanar to the leaving group in order for the reaction to occur, and that requirement will determine the stereoisomeric product obtained. If, however, the β position has two protons, then either of these two protons can be arranged so that it is *anti*-periplanar to the leaving group. As a result, both stereoisomeric products will be obtained:

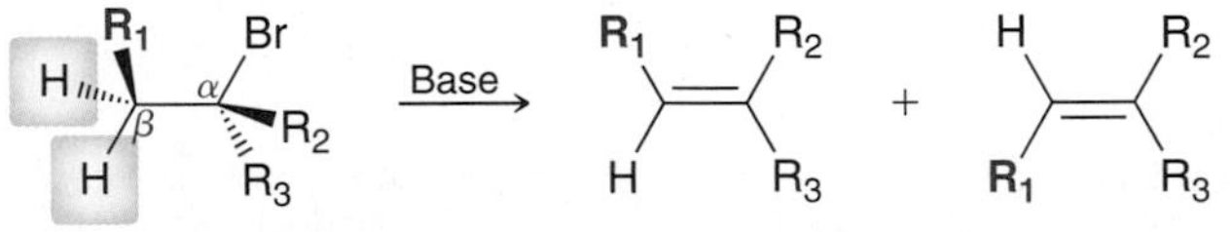

In such a case, the more stable isomeric alkene will predominate. This is an example of stereoselectivity, rather than stereospecificity. The difference between these two terms is often misunderstood, so let's spend a moment on it. The key is to focus on the nature of the substrate (Figure 8.17):

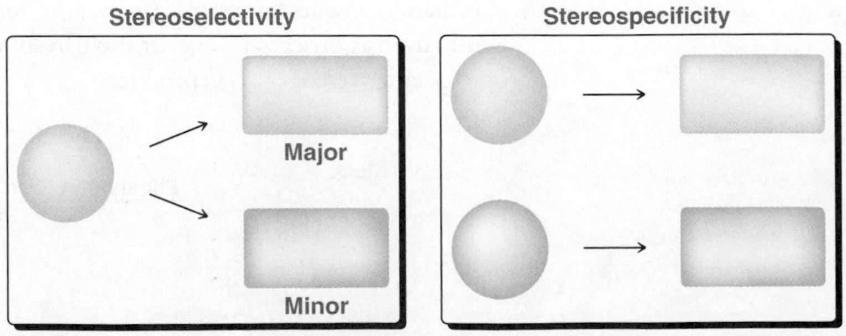

FIGURE 8.17
An illustration of the difference between stereoselectivity and stereospecificity in an E2 reaction.

- In a *stereoselective* E2 reaction: The substrate itself is not necessarily stereoisomeric; nevertheless, this substrate can produce two stereoisomeric products, and it is found that one stereoisomeric product is formed in higher yield.
- In a *stereospecific* E2 reaction: The substrate is stereoisomeric, and the stereochemical outcome is dependent on which stereoisomeric substrate is used.

SKILLBUILDER

8.6 PREDICTING THE STEREOCHEMICAL OUTCOME OF AN E2 REACTION

LEARN the skill

Identify the major and minor products for the E2 reaction that occurs when each of the following substrates is treated with a strong base:

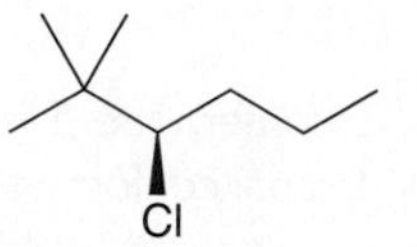
(a)

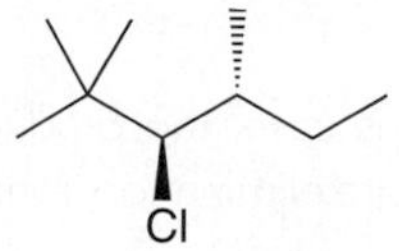
(b)

SOLUTION

(a) In this case, regiochemistry is irrelevant because there is only one β position that bears protons:

STEP 1
Identify all β positions with protons.

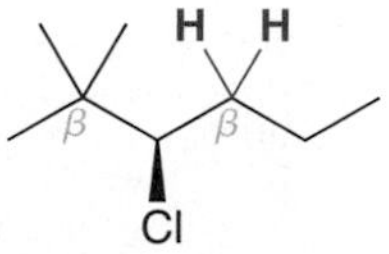

In order to determine if the reaction is stereoselective or stereospecific, count the number of protons at the β position. In this case, the β position has two protons. Therefore, we expect the reaction to be stereoselective. That is, we expect both *cis* and *trans* isomers, with a preference for *trans*:

STEP 2
If the β position has two protons, expect *cis* and *trans* isomers.

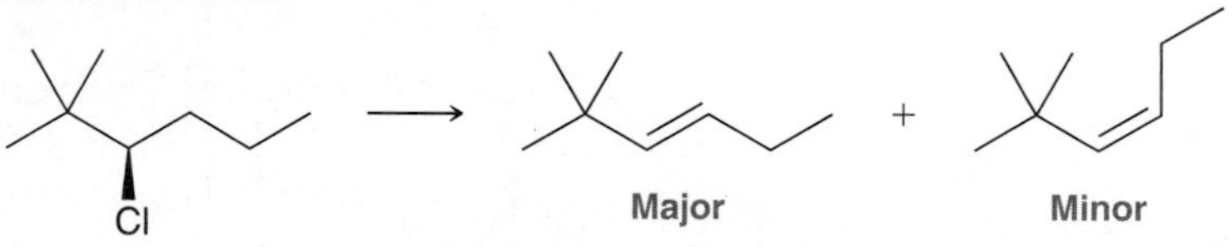

(b) In this case, the regiochemistry is also irrelevant, because there is once again only one β position that has protons:

STEP 1
Identify all β positions with protons.

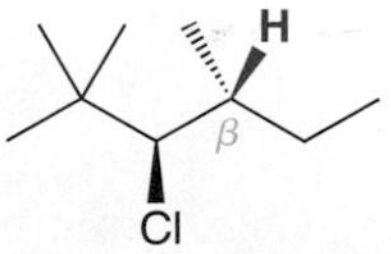

In order to determine if the reaction is stereoselective or stereospecific, count the number of protons at the β position. In this case, the β position has only one proton. Therefore, we expect the reaction to be stereospecific. That is, we expect only one particular stereoisomeric product, not both. To determine which product to expect, begin by drawing the Newman projection:

STEP 2
If the β position has only one proton, draw a Newman projection.

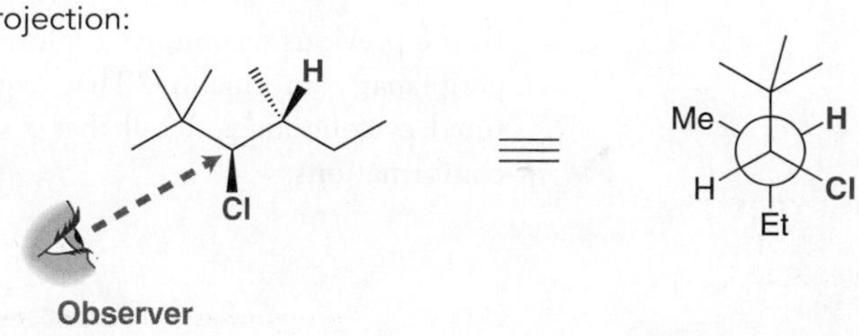

In this conformation, the proton and the leaving group (Cl) are not *anti*-periplanar. In order to achieve the appropriate conformation, rotate the C—C single bond so that the proton and the chlorine atom are *anti*-periplanar:

STEP 3
Rotate the C—C single bond to achieve an *anti*-periplanar conformation.

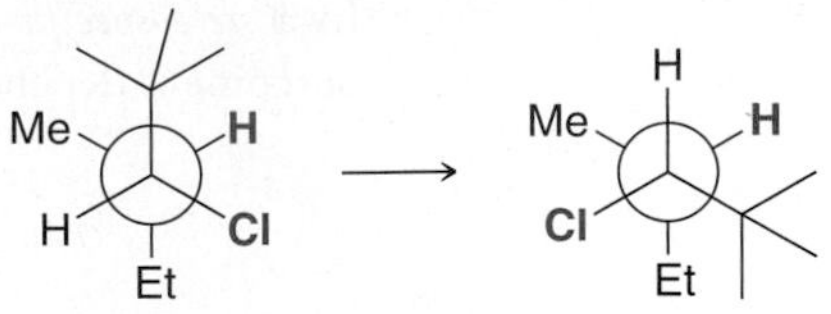

Now use this Newman projection to draw the product:

STEP 4
Use the Newman projection as a guide for drawing the product.

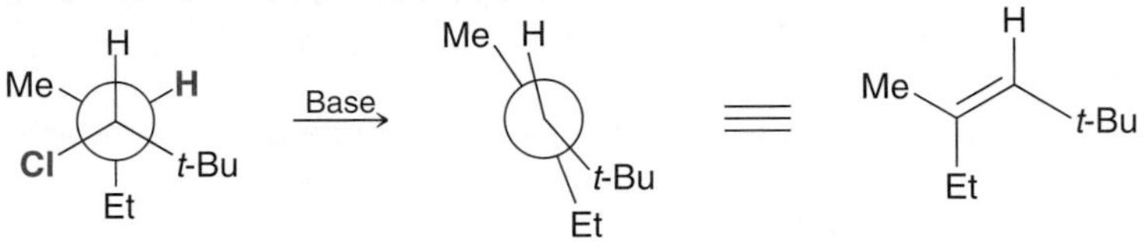

In this case, the product is the *Z* isomer. The *E* isomer is not obtained, because that would require elimination from a *syn*-periplanar conformation, which is eclipsed and too high in energy.

PRACTICE the skill **8.18** Identify the major and minor products for the E2 reaction that occurs when each of the following substrates is treated with a strong base:

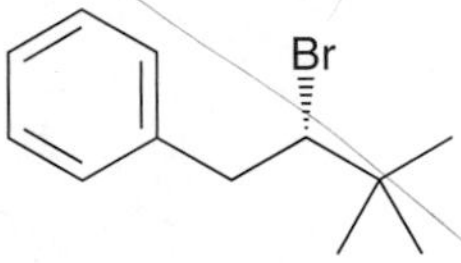

(a)

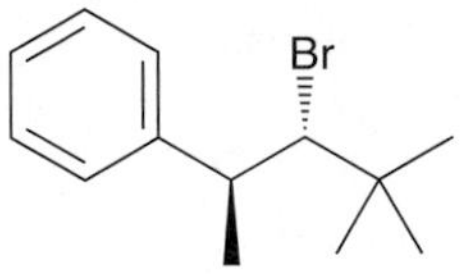

(b)

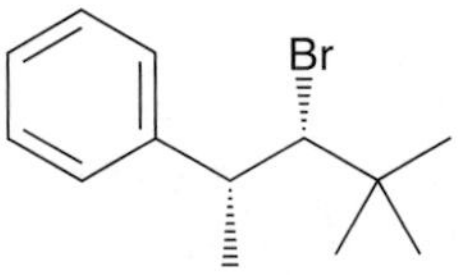

(c)

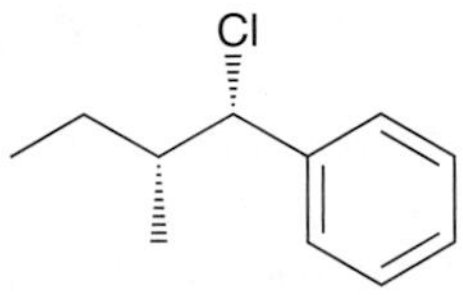

(d)

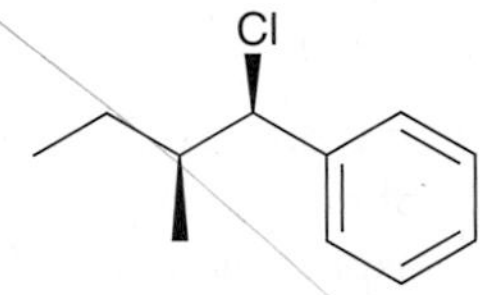

(e)

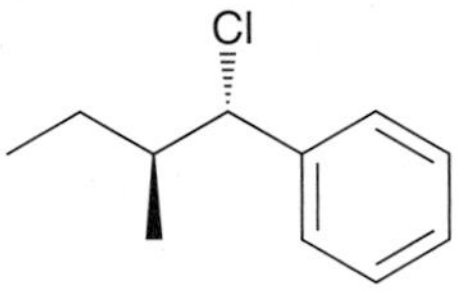

(f)

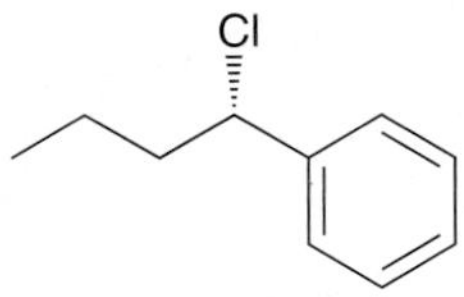

(g)

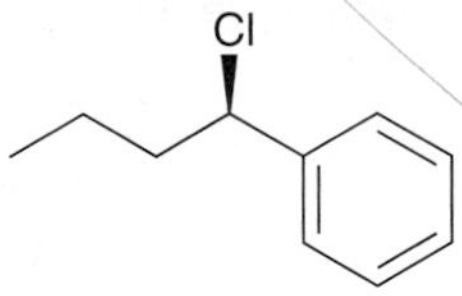

(h)

APPLY the skill **8.19** Identify an alkyl halide that could be used to make the following alkene:

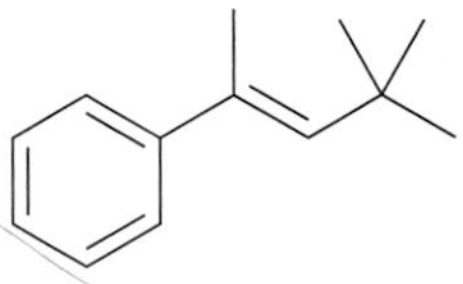

need more **PRACTICE?** **Try Problems 8.56, 8.63, 8.83**

Stereospecificity of E2 Reactions on Substituted Cyclohexanes

In the previous section, we explored the requirement that an E2 reaction proceed via an *anti*-periplanar conformation. That requirement has special significance when dealing with substituted cyclohexanes. Recall that a substituted cyclohexane ring can adopt two different chair conformations:

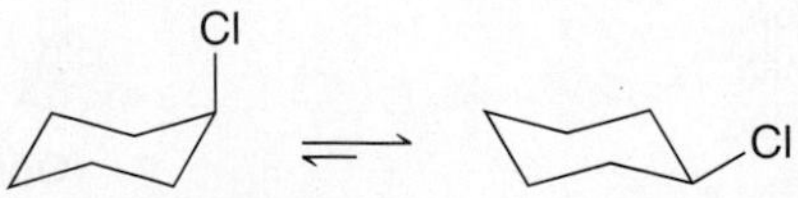

LOOKING BACK
For practice drawing chair conformations, see Section 4.12.

In one chair conformation, the leaving group occupies an axial position. In the other chair conformation, the leaving group occupies an equatorial position. The requirement for an *anti*-periplanar conformation demands that an E2 reaction can only occur from the chair conformation in which the leaving group occupies an axial position. To see this more clearly, consider the Newman projections for each chair conformation:

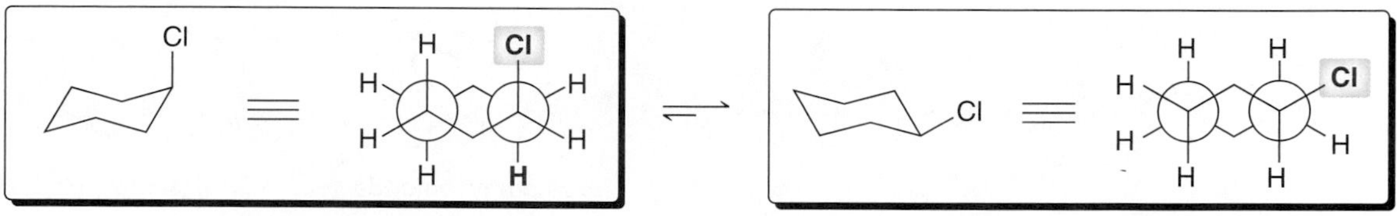

When Cl is axial, it can be *anti*-periplanar with a neighboring hydrogen atom

When Cl is equatorial, it cannot be *anti*-periplanar with any of its neighboring hydrogen atoms

When the leaving group occupies an axial position, it is *anti*-periplanar with a neighboring proton. Yet, when the leaving group occupies an equatorial position, it is not *anti*-periplanar with any neighboring protons. Therefore, on a cyclohexane ring, an E2 reaction only occurs from the chair conformation in which the leaving group is axial. From this, it follows that an E2 reaction can only take place when the leaving group and the proton are on opposite sides of the ring (one on a wedge and the other on a dash):

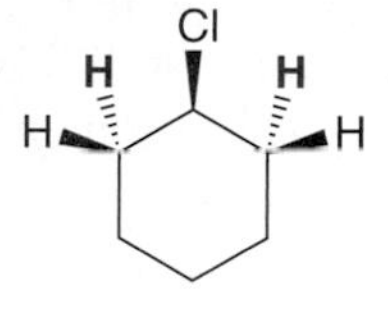

Only the two hydrogen atoms shown in red can participate in an E2 reaction

If there are no such protons, then an E2 reaction cannot occur at an appreciable rate. As an example, consider the following compound:

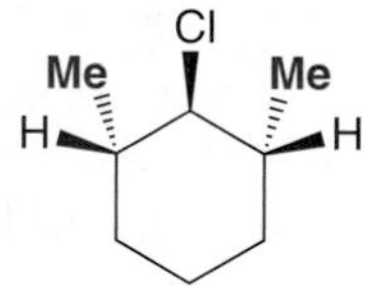

This compound will not undergo an E2 reaction

This compound has two β positions, and each β position bears a proton. But, neither of these protons can be *anti*-periplanar with the leaving group. Since the leaving group is on a wedge, an E2 reaction will only occur if there is a neighboring proton on a dash.

Consider another example:

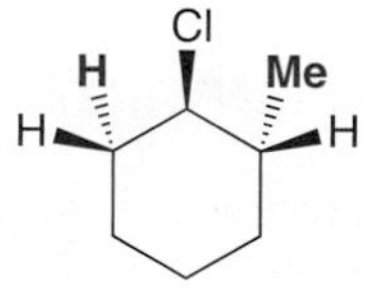

In this case, only one proton can be *anti*-periplanar with the leaving group, so in this case, only one product is obtained:

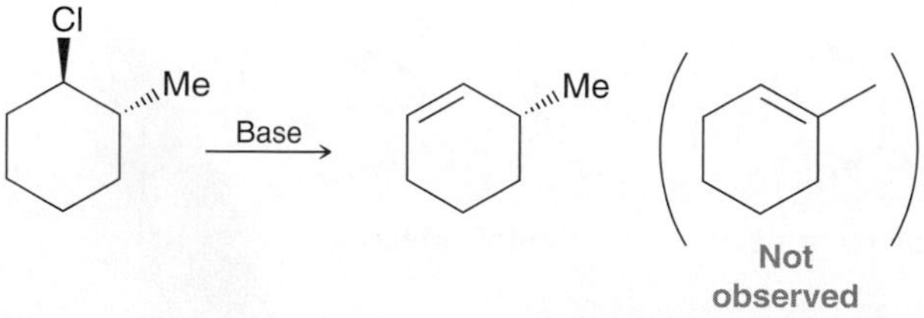

The other product (the Zaitsev product) is not formed in this case, even though it is the more substituted alkene.

For substituted cyclohexanes, the rate of an E2 reaction is greatly affected by the amount of time that the leaving group spends in an axial position. As an example, compare the following two compounds:

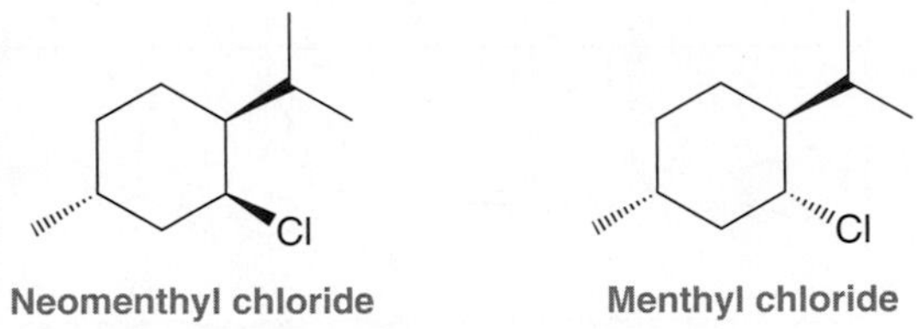

Neomenthyl chloride is 200 times more reactive toward an E2 process. Why? Draw both chair conformations of neomenthyl chloride:

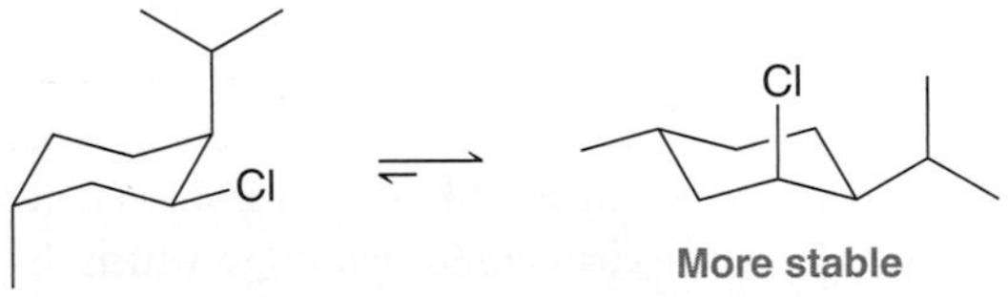

The more stable chair conformation has the larger isopropyl group occupying an equatorial position. In this chair conformation, the chlorine occupies an axial position, which is set up nicely for an E2 elimination. In other words, neomenthyl chloride spends most of its time in the conformation necessary for an E2 process to occur. In contrast, menthyl chloride spends most of its time in the wrong conformation:

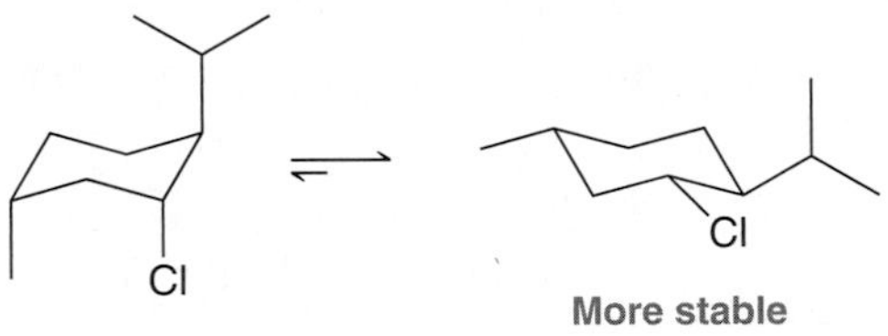

The large isopropyl group occupies an equatorial position in the more stable chair conformation. In this conformation, the leaving group occupies an equatorial position as well, which means that the leaving group spends most of its time in an equatorial position. In this case, an E2 process can only occur from the higher energy chair conformation, whose concentration is small at equilibrium. As a result, menthyl chloride will undergo an E2 reaction more slowly.

CONCEPTUAL CHECKPOINT

8.20 When menthyl chloride is treated with a strong base, only one elimination product is observed. Yet, when neomenthyl chloride is treated with a strong base, two elimination products are observed. Draw the products and explain.

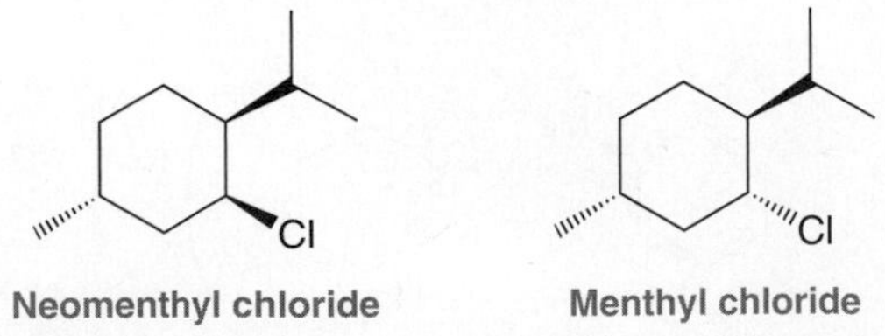

8.21 Predict which of the following two compounds will undergo an E2 reaction more rapidly:

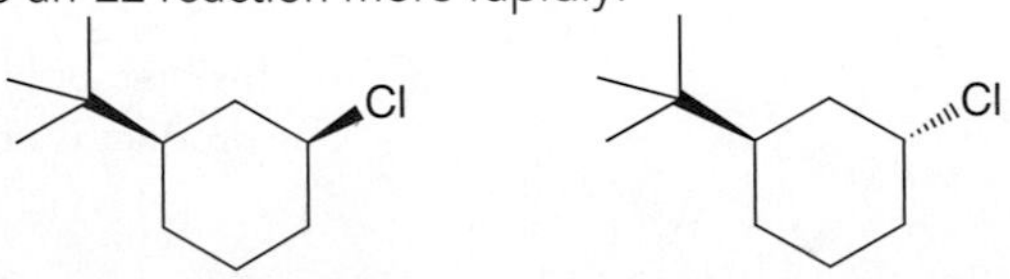

8.8 Drawing the Products of an E2 Reaction

As we have seen, predicting the products of an E2 reaction is often more complex than predicting the products of an S_N2 reaction. With an E2 reaction, two major issues must be considered before drawing the products: regiochemistry (Section 8.4) and stereochemistry (Section 8.5). Both of these issues are illustrated in the following example.

SKILLBUILDER

8.7 DRAWING THE PRODUCTS OF AN E2 REACTION

LEARN the skill

Predict the product(s) of the following reaction:

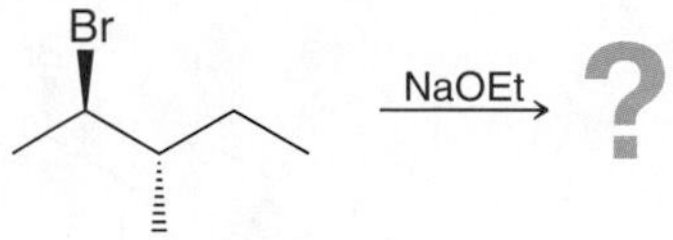

SOLUTION

We must consider both the regiochemistry and the stereochemistry of this reaction. We will start with the regiochemistry. First identify all of the β positions that possess protons:

STEP 1 Assess the regiochemical outcome.

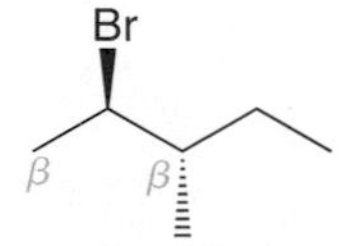

There are two β positions. The base (ethoxide) is not sterically hindered, so we expect the Zaitsev product (the more substituted alkene) as the major product. The less substituted alkene will be the minor product. Next, we must identify the stereochemistry of formation of each of the products. Let's begin with the minor product (the less substituted alkene), because its double bond does not exhibit stereoisomerism:

STEP 2 Assess the stereochemical outcome.

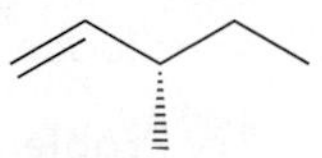

This alkene is neither *E* nor *Z*, so we don't need to worry about stereoisomerism of the double bond in this product. But with the major product we must predict which stereoisomer will be obtained. To do this, draw a Newman projection:

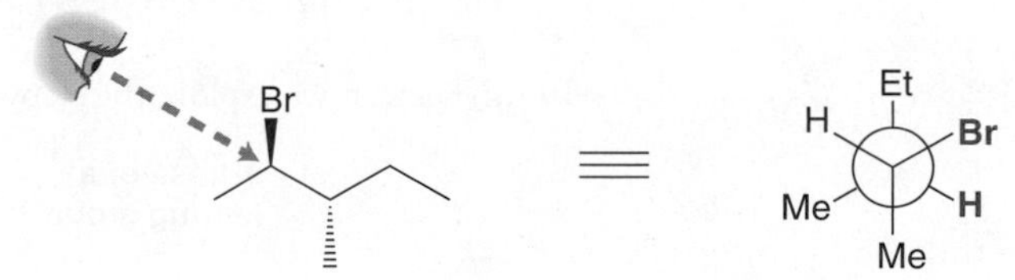

Then rotate the C—C single bond so that the proton and the leaving group are *anti*-coplanar, and draw the product:

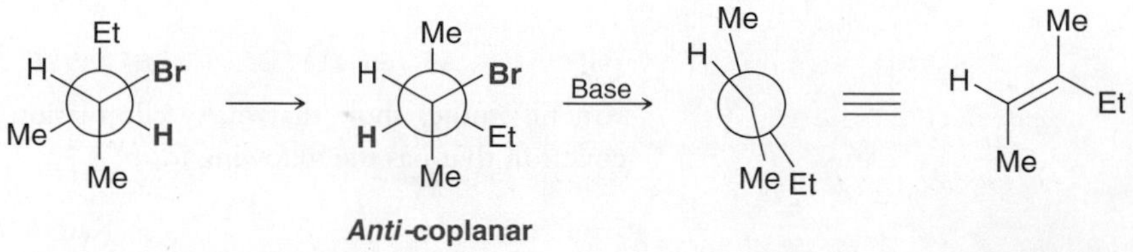

To summarize, we expect the following products:

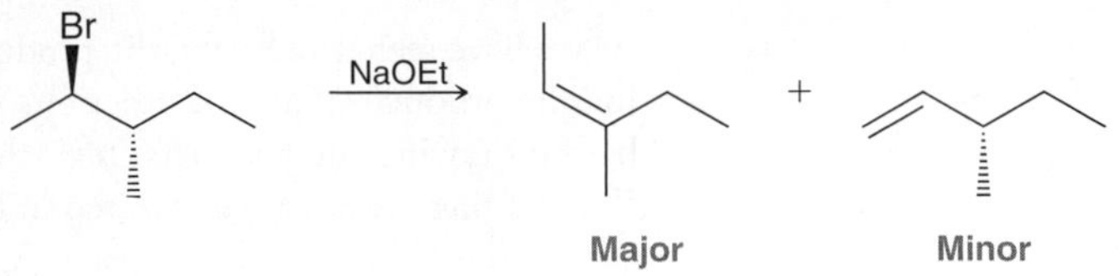

PRACTICE the skill **8.22** Predict the major and minor product for each of the following E2 reactions:

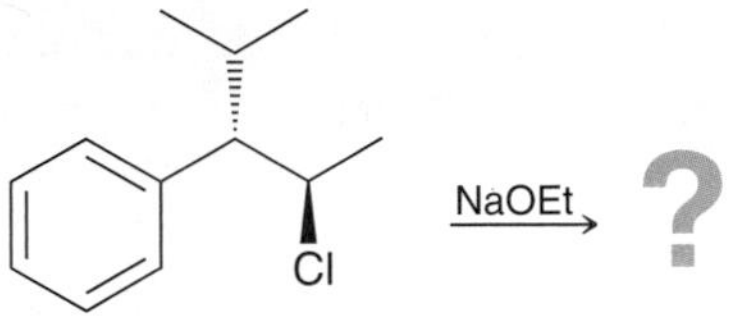

(a)

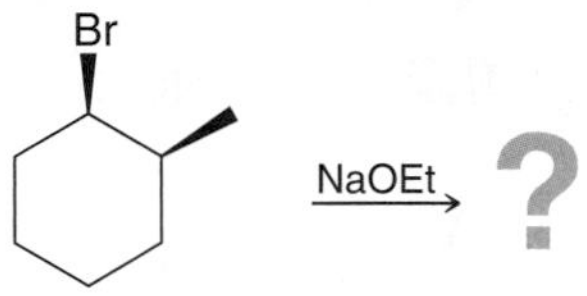

(b)

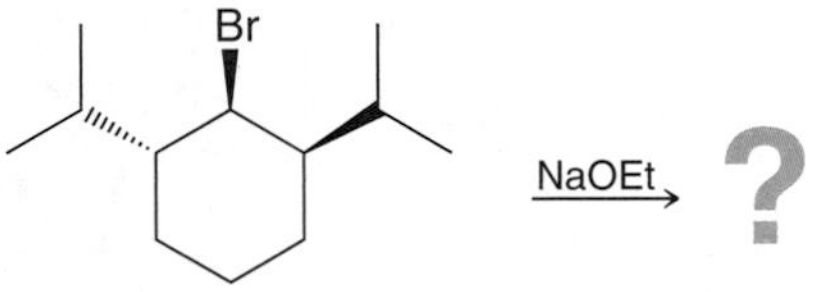

(c)

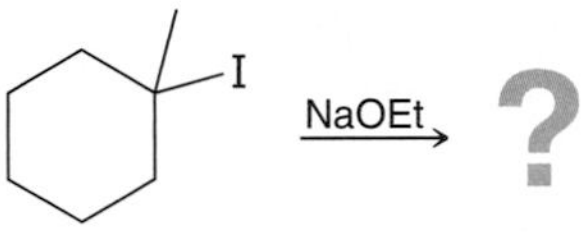

(d)

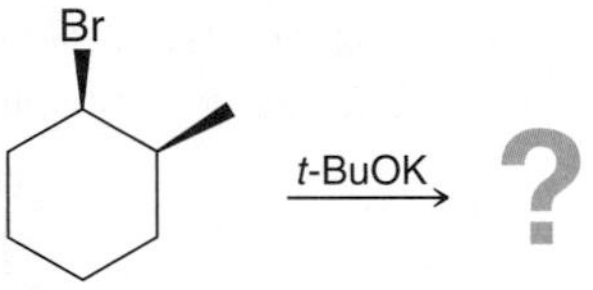

(e)

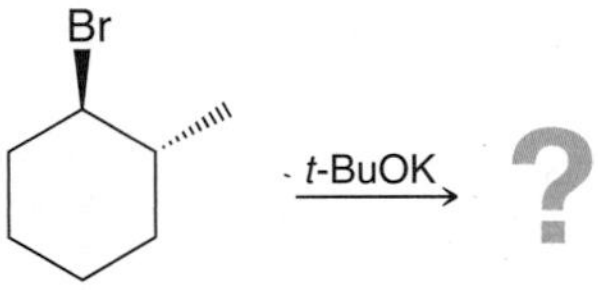

(f)

APPLY the skill

8.23 Draw an alkyl halide that will produce only one alkene upon treatment with a strong base.

8.24 Draw an alkyl halide that will produce exactly two stereoisomeric alkenes upon treatment with a strong base.

8.25 Draw an alkyl halide that will produce two constitutional isomers (but no stereoisomers) upon treatment with a strong base.

need more **PRACTICE?** **Try Problems 8.59–8.61, 8.64, 8.66**

8.9 The E1 Mechanism

In this section we explore the stepwise mechanism, or E1, pathway:

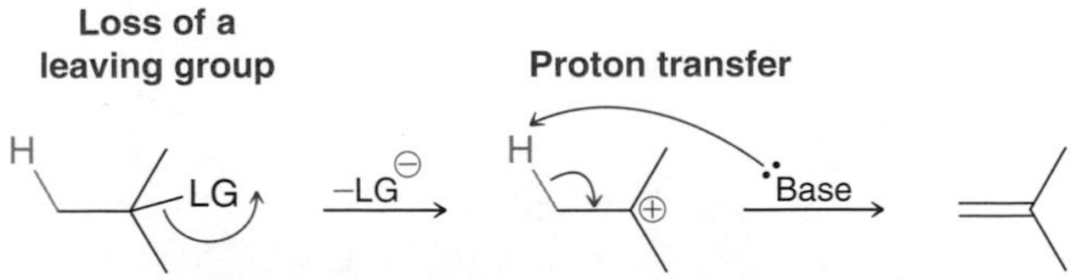

Kinetic Evidence for a Stepwise Mechanism

Kinetic studies show that many elimination reactions exhibit first-order kinetics, with a rate equation that has the following form:

$$\text{Rate} = k\,[\text{substrate}]$$

Much like the S_N1 reaction, the rate is linearly dependent on the concentration of only one compound (the substrate). This observation is consistent with a stepwise mechanism, in which the rate-determining step does not involve the base. The rate-determining step is the first step in the mechanism (loss of the leaving group), just as we saw in the S_N1 reaction. The base does not participate in this step, and therefore, the concentration of the base does not affect the rate. Because this step involves only one chemical entity, it is said to be unimolecular. Unimolecular elimination reactions are called **E1** reactions:

CONCEPTUAL CHECKPOINT

8.26 The following reaction occurs via an E1 mechanistic pathway:

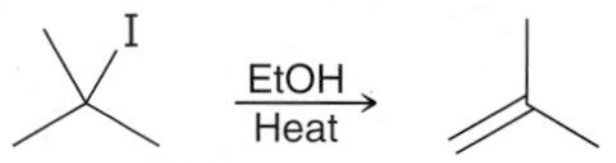

(a) What happens to the rate if the concentration of *tert*-butyl iodide is doubled and the concentration of ethanol is tripled?

(b) What happens to the rate if the concentration of *tert*-butyl iodide remains the same and the concentration of ethanol is doubled?

Effect of Substrate

For E1 reactions, the rate is found to be very sensitive to the nature of the starting alkyl halide, with tertiary halides reacting most readily (Figure 8.18). This trend is identical to the trend we saw for S_N1 reactions, and the reason for the trend is the same as well. A stepwise mechanism involves formation of a carbocation intermediate, and the rate of reaction will be dependent on the stability of the carbocation.

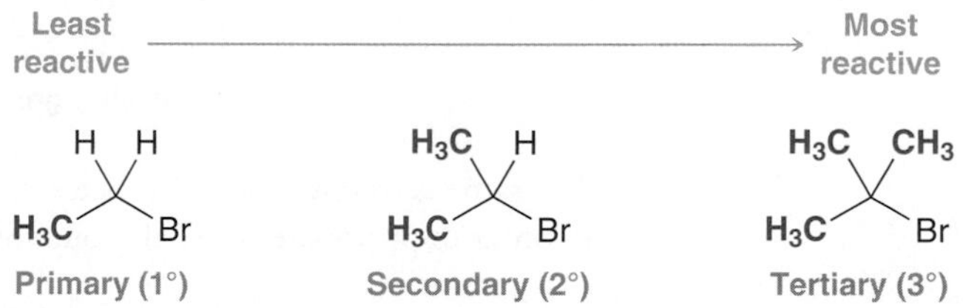

FIGURE 8.18
The rate of reactivity of various substrates in an E1 reaction.

LOOKING BACK
For a review of carbocation stability and hyperconjugation, see Section 6.11.

Recall that tertiary carbocations are more stable than secondary carbocations (Figure 8.19), because alkyl groups are electron donating. Compare the energy diagrams for E1 reactions involving secondary and tertiary substrates (Figure 8.20). Tertiary substrates exhibit a lower energy of activation during an E1 process and therefore react more rapidly. Primary substrates are generally unreactive toward an E1 mechanism, because a primary carbocation is too unstable to form.

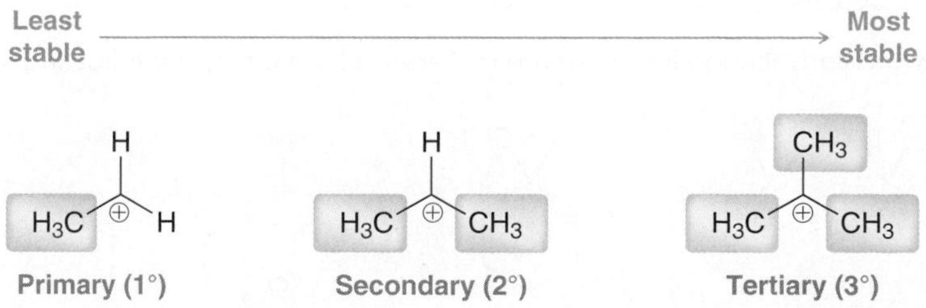

FIGURE 8.19
Relative stability of primary, secondary, and tertiary carbocations.

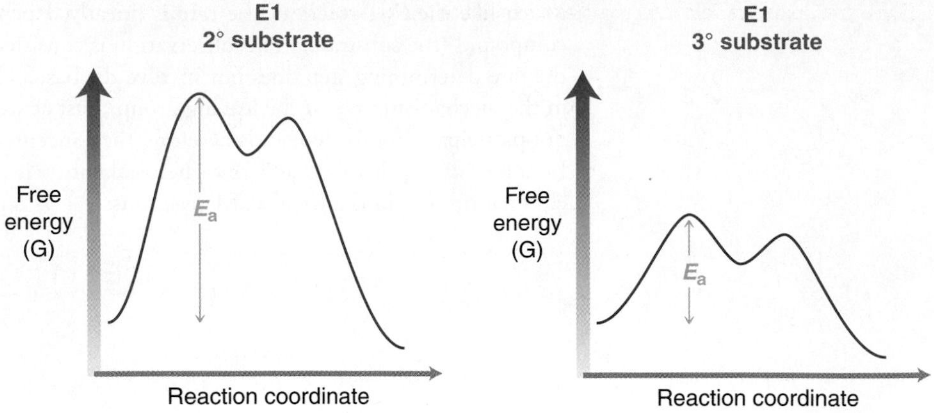

FIGURE 8.20 A comparison of energy diagrams for the E1 reactions of secondary and tertiary substrates.

The first step of an E1 process is identical to the first step of an S_N1 process. In each process, the first step involves loss of the leaving group to form a carbocation intermediate:

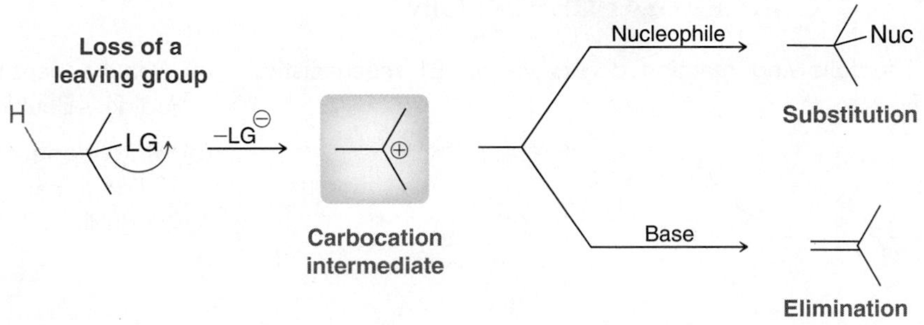

An E1 reaction is generally accompanied by a competing S_N1 reaction, and a mixture of products is generally obtained.

In the previous chapter, we saw that an OH group is a terrible leaving group and that an S_N1 reaction can only occur if the OH group is first protonated to give a better leaving group:

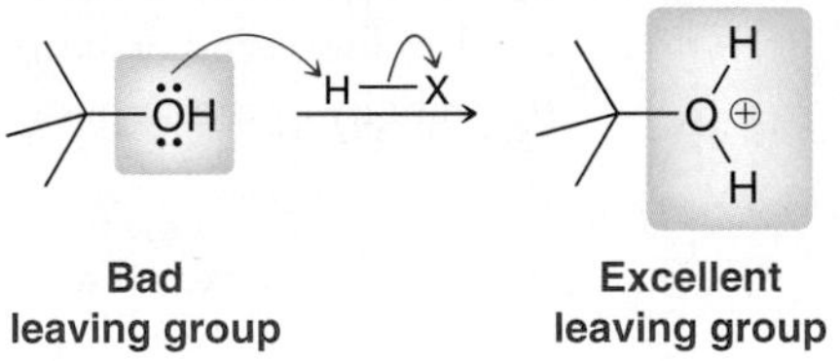

The same is true with an E1 process. If the substrate is an alcohol, a strong acid will be required in order to protonate the OH group. Sulfuric acid is commonly used for this purpose:

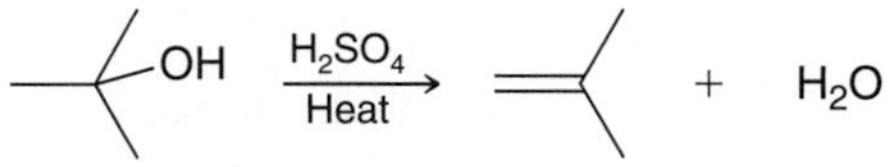

The reaction above involves elimination of water to form an alkene and is therefore called a dehydration reaction.

CONCEPTUAL CHECKPOINT

8.27 Draw the carbocation intermediate generated by each of the following substrates in an E1 reaction:

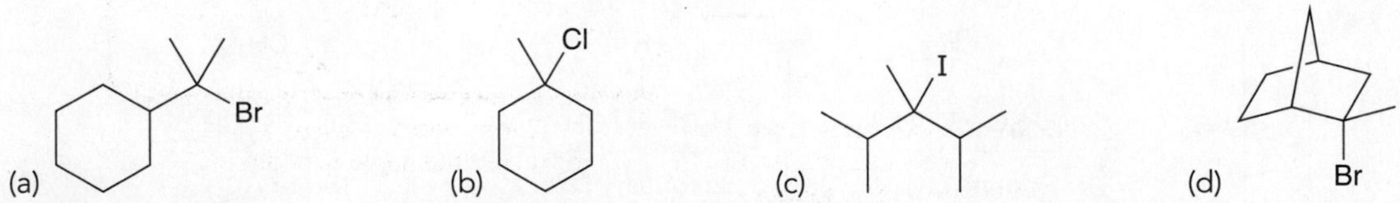

8.28 Draw the carbocation intermediate generated when each of the following substrates is treated with sulfuric acid:

(a) 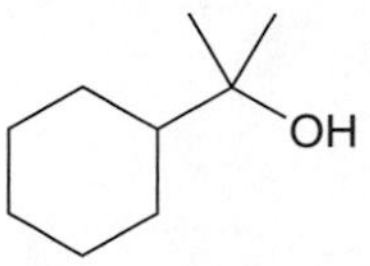

(b) 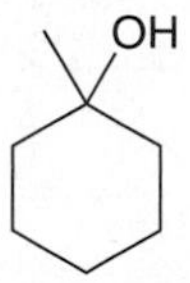

(c) 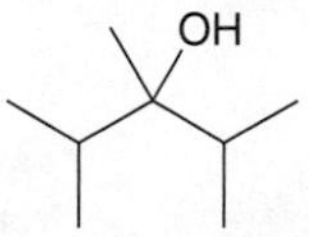

(d) 

Regioselectivity of E1 Reactions

E1 processes exhibit a regiochemical preference for the Zaitsev product, as was observed for E2 reactions. For example:

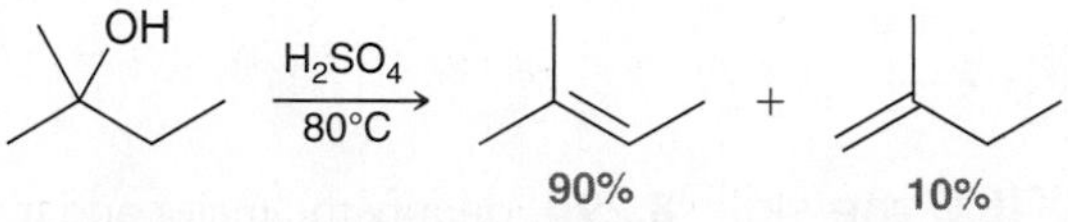

The more substituted alkene (Zaitsev product) is the major product. However, there is one critical difference between the regiochemical outcomes of E1 and E2 reactions. Specifically, we have seen that the regiochemical outcome of an E2 reaction can often be controlled by carefully choosing the base (sterically hindered or not sterically hindered). In contrast, the regiochemical outcome of an E1 process cannot be controlled. The Zaitsev product will generally be obtained.

SKILLBUILDER

8.8 PREDICTING THE REGIOCHEMICAL OUTCOME OF AN E1 REACTION

LEARN the skill

Identify the major and minor products of the following E1 reaction:

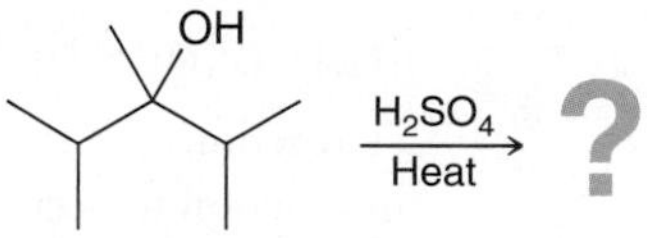

SOLUTION

First identify all β positions that have protons:

STEP 1
Identify all β positions that have protons.

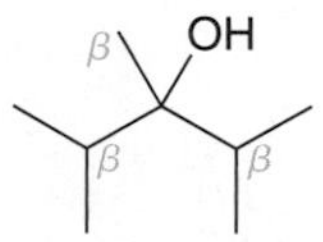

These are the positions that must be explored. Two of these positions are identical, because they would give the same product:

STEP 2
Take notice of the β positions that lead to the same product.

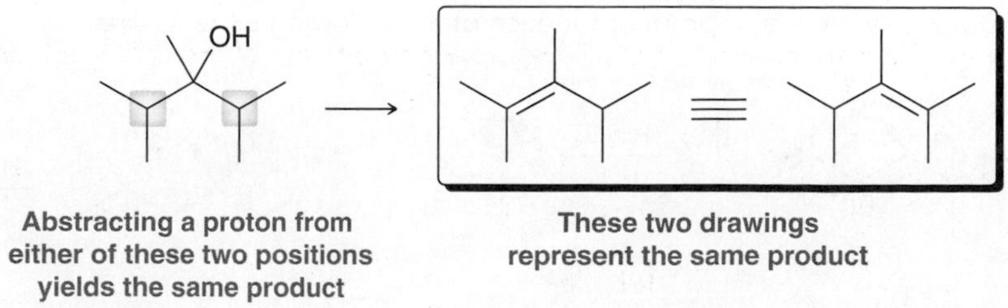

The other β position will produce the following alkene:

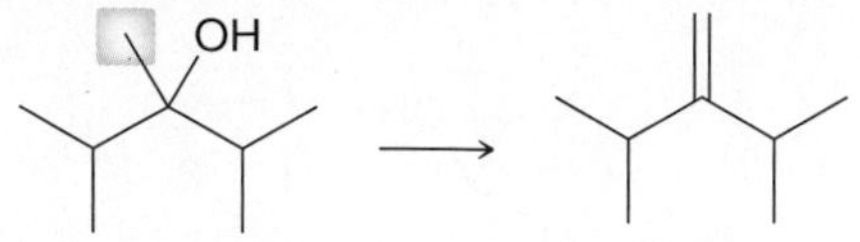

Abstracting a proton from this position... **...yields this product**

Therefore, there are two possible products, and we expect to obtain a mixture of both products. The major product will be the more substituted alkene:

STEP 3
Draw all products and label the more substituted alkene as the major product.

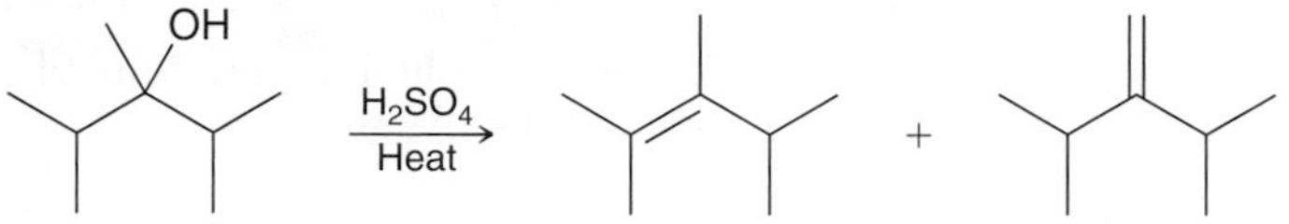

Major **Minor**

PRACTICE the skill **8.29** Identify the major and minor products for each of the following E1 reactions:

(a)
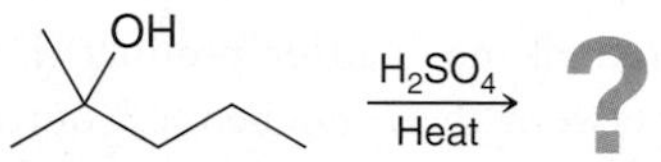

(b)
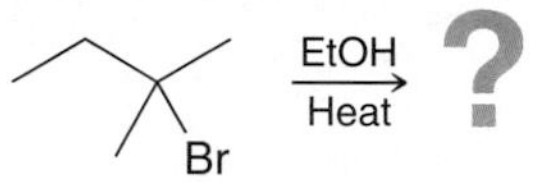

(c)
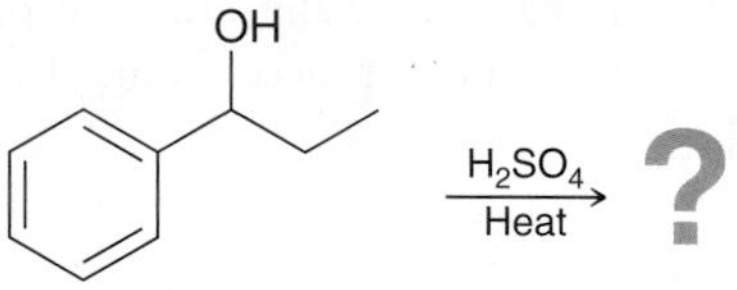

(d)
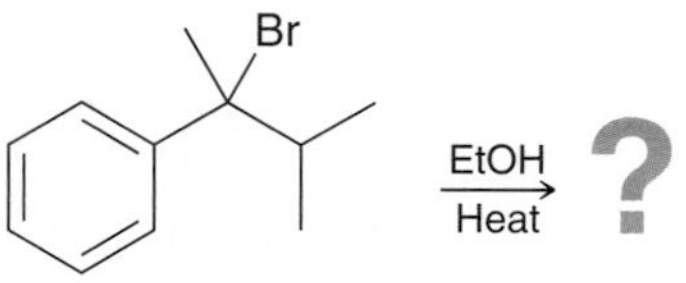

APPLY the skill **8.30** Identify two different starting alcohols that could be used to make 1-methylcyclohexene. Then determine which alcohol would be expected to react more rapidly under acidic conditions. Explain your choice.

need more **PRACTICE?** **Try Problem 8.62**

Stereoselectivity of E1 Reactions

E1 reactions are not stereospecific—that is, they do not require *anti*-periplanarity in order for the reaction to occur. Nevertheless, E1 reactions are stereoselective. When *cis* and *trans* products are possible, we generally observe a preference for formation of the *trans* stereoisomer:

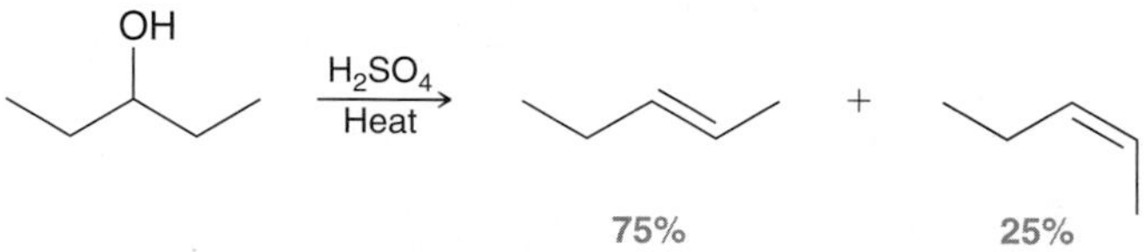

75% **25%**

CONCEPTUAL CHECKPOINT

8.31 Draw only the major product for each of the following E1 reactions:

(a)
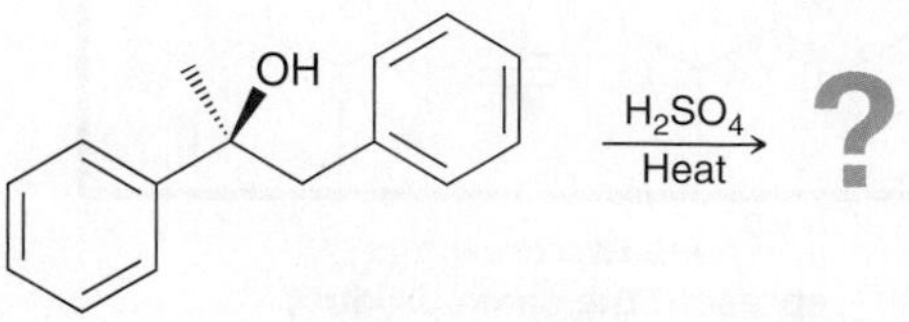

(b)
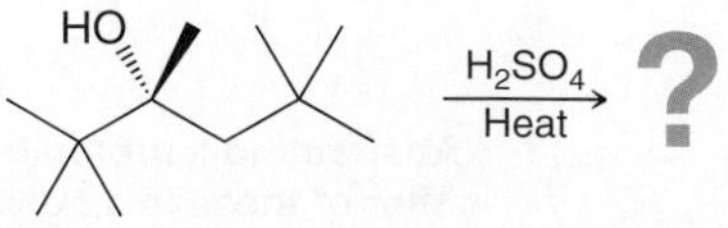

8.10 Drawing the Complete Mechanism of an E1 Process

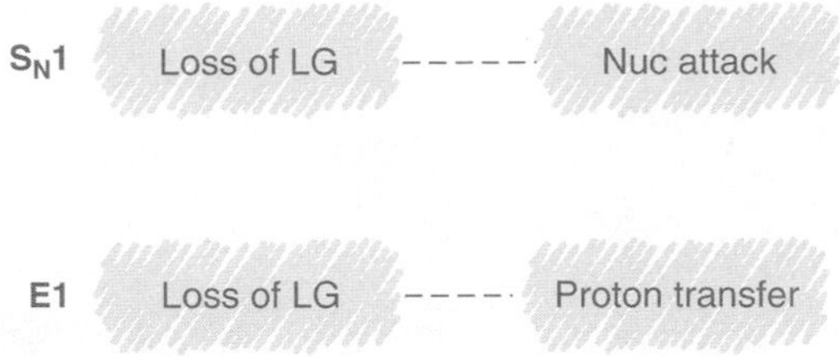

FIGURE 8.21
A comparison of the core steps of S_N1 and E1 processes.

Compare the two core steps of an E1 mechanism with the two core steps of an S_N1 mechanism (Figure 8.21). In Section 7.6, we saw that an S_N1 process can have up to three additional steps (Figure 8.22). Similarly, we might expect that an E1 mechanism could also have up to three additional steps. However, in practice, nothing ever happens after the second core step. An E1 process can only have up to two additional steps (Figure 8.23).

- Proton transfer before the first core step
- Carbocation rearrangement in between the two core steps

Can you explain why a proton transfer is never required at the end of an E1 process? Think about what circumstances require proton transfer at the end of an S_N1 mechanism, and then consider why that doesn't happen in an E1 process.

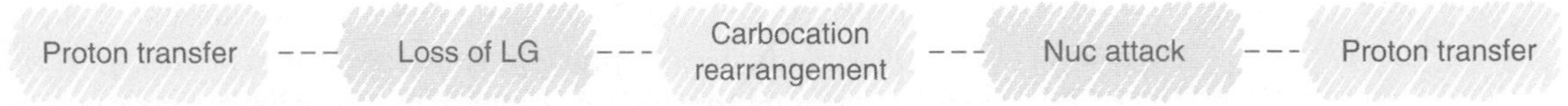

FIGURE 8.22
The core steps (gray) and additional steps (blue) of an S_N1 process.

FIGURE 8.23
The core steps (gray) and additional steps (blue) of an E1 process.

Proton Transfer at the Beginning of an E1 Mechanism

A proton transfer is required at the beginning of an E1 process for the same reason that it was required at the beginning of an S_N1 process. Specifically, a proton transfer is necessary whenever the leaving group is an OH group. Hydroxide is a bad leaving group and will not leave by itself. Therefore, an alcohol will only participate in an E1 reaction if the OH group is first protonated, which means that an acid is needed. Here is an example:

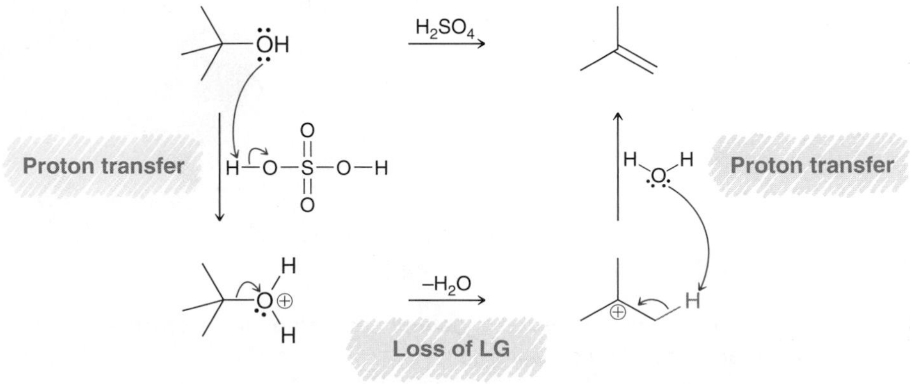

Notice that this mechanism has three steps:

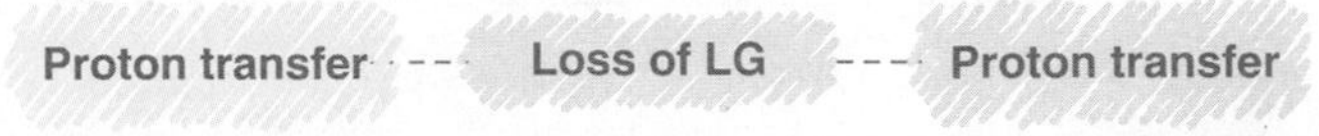

CONCEPTUAL CHECKPOINT

8.32 For each of the following substrates, determine whether an E1 process will require the use of an acid:

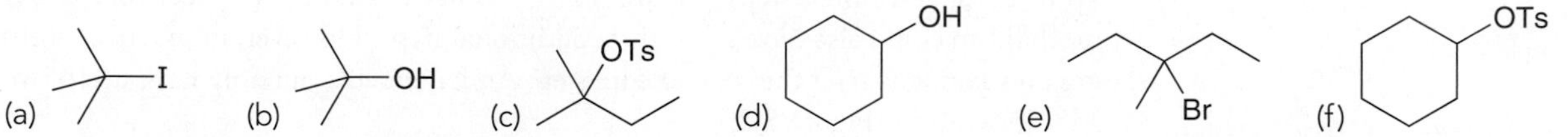

Carbocation Rearrangement During an E1 Mechanism

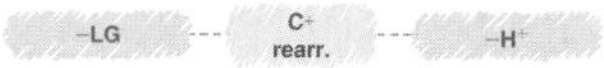

The E1 mechanism involves formation of a carbocation intermediate. Recall from Chapter 6 that carbocations are susceptible to rearrangement via either a hydride shift or a methyl shift. Here is an example of an E1 mechanism that contains a carbocation rearrangement:

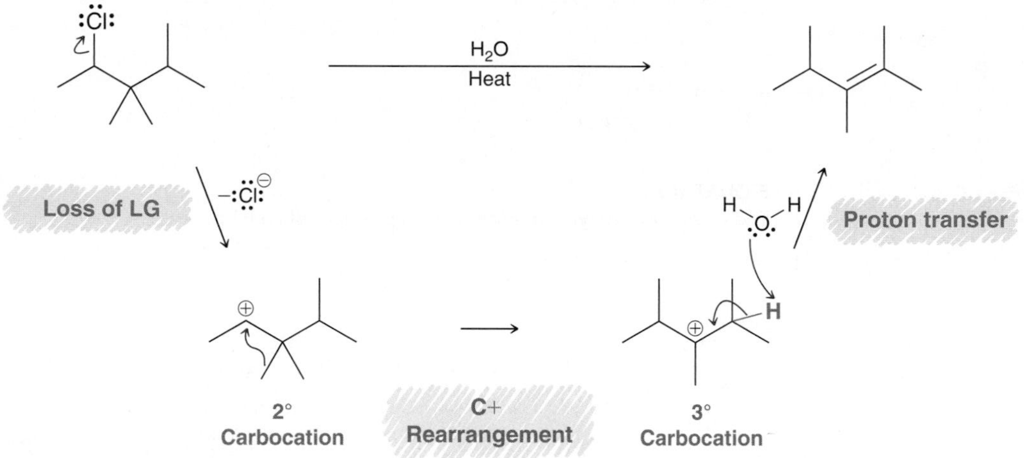

LOOKING BACK
For practice with hydride shifts and methyl shifts, see Section 6.11.

Notice that the carbocation rearrangement occurs between the two core steps of an E1 mechanism:

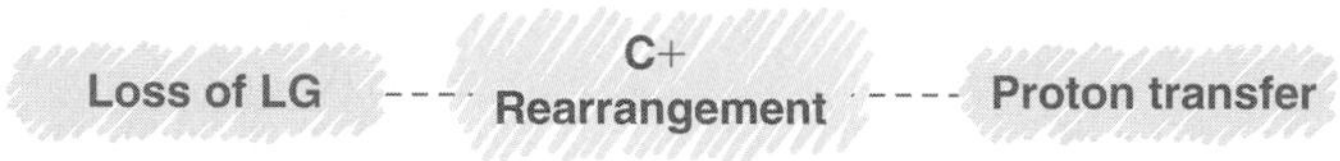

CONCEPTUAL CHECKPOINT

8.33 For each of the following substrates, determine whether an E1 process will involve a carbocation rearrangement or not:

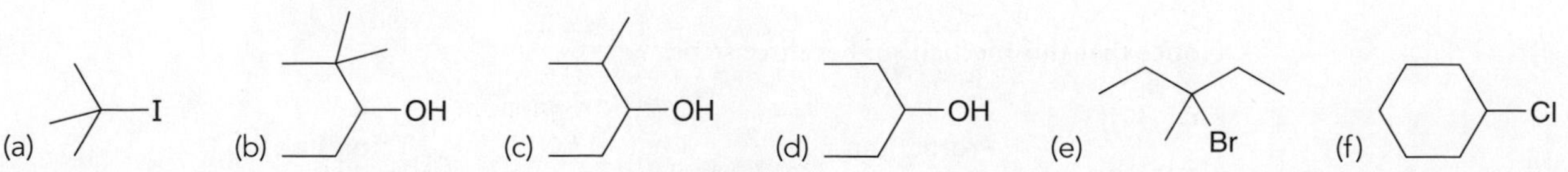

An E1 Mechanism Can Have Many Steps

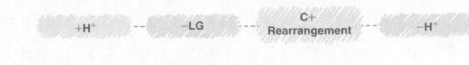

+H⁺ -- –LG --- C+ Rearrangement --- –H⁺

As we have seen, an E1 mechanism must have at least two core steps (loss of leaving group and proton transfer) and can also have up to two additional steps:

1. *Before the two core steps*—It is possible to have a proton transfer.
2. *In between the two core steps*—It is possible to have a carbocation rearrangement.

MECHANISM 8.1 THE E1 MECHANISM

Two core steps

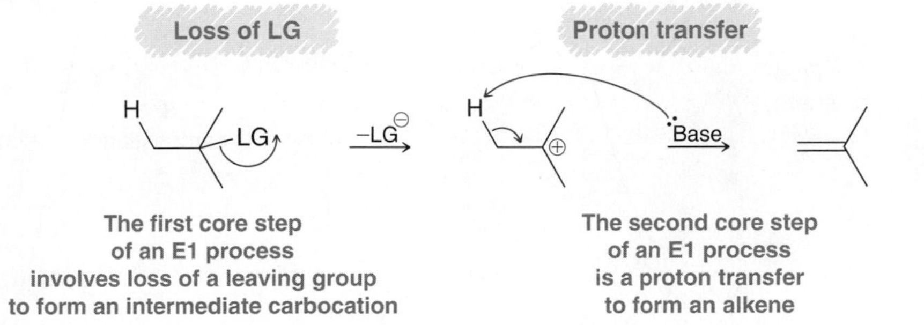

Possible additional steps

Proton transfer ---- **Loss of LG** --- **C+ Rearrangement** ---- **Proton transfer**

If the leaving group is an OH group, it must be protonated before it can leave

If the carbocation initially formed can rearrange to generate a more stable carbocation, then a carbocation rearrangement will occur

An E1 process can have one or both of the additional steps shown in Mechanism 8.1. Here is an example of an E1 process with four steps:

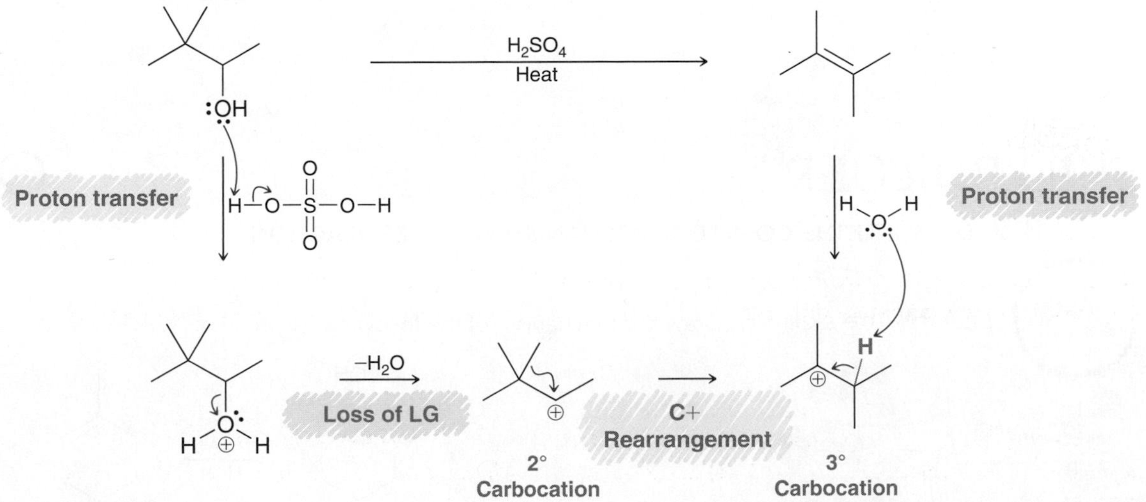

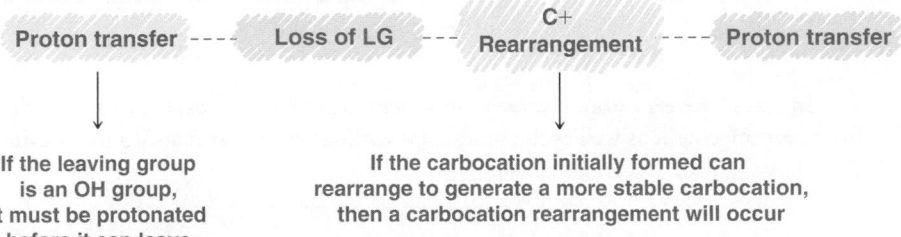

This mechanism has four steps, and therefore, we expect the energy diagram of this reaction to exhibit four humps (Figure 8.24). The number of humps in the energy diagram of an E1 process will always be equal to the number of steps in the mechanism. As we have seen, the number of steps can range from two to four, which means that the energy diagram of an E1 process can have from two to four humps.

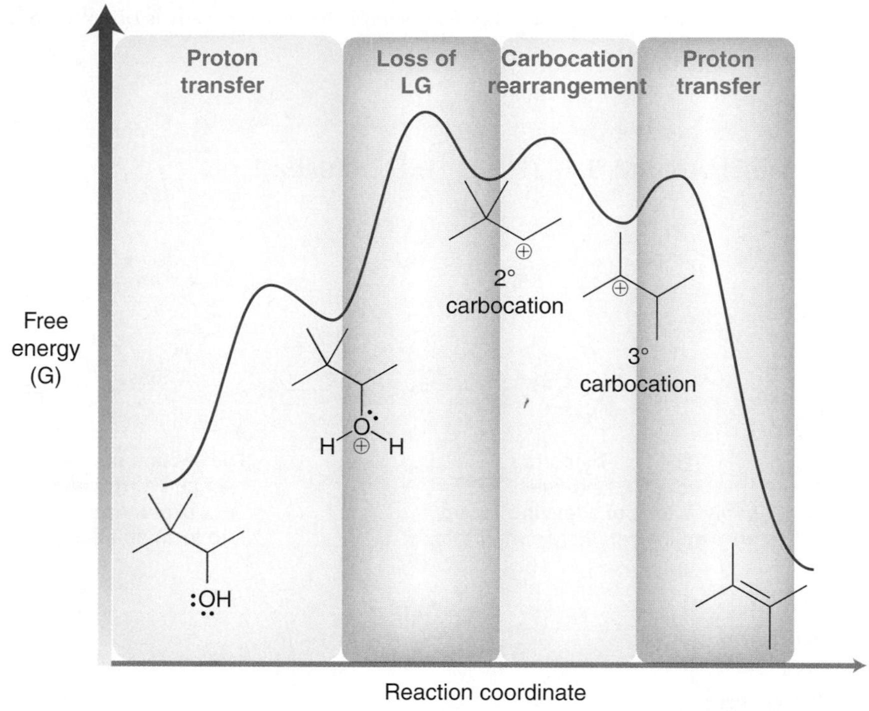

FIGURE 8.24
The energy diagram of an E1 process with four steps.

In cases where a carbocation rearrangement is possible, we expect to obtain the product(s) from rearrangement as well as the product(s) without rearrangement. In the previous example, the following products are obtained:

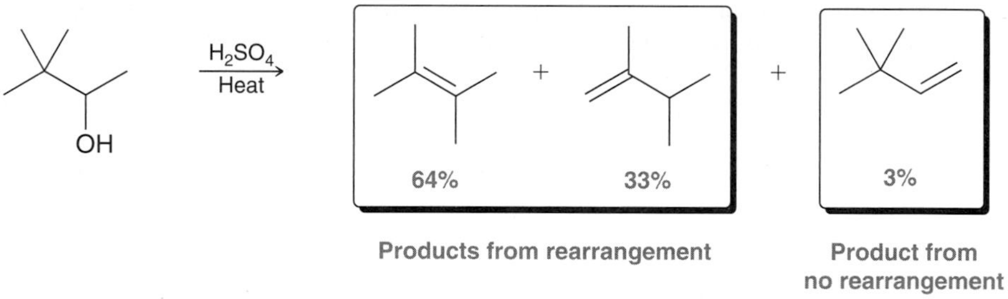

SKILLBUILDER

8.9 DRAWING THE COMPLETE MECHANISM OF AN E1 REACTION

LEARN the skill

Draw the mechanism of the following E1 process:

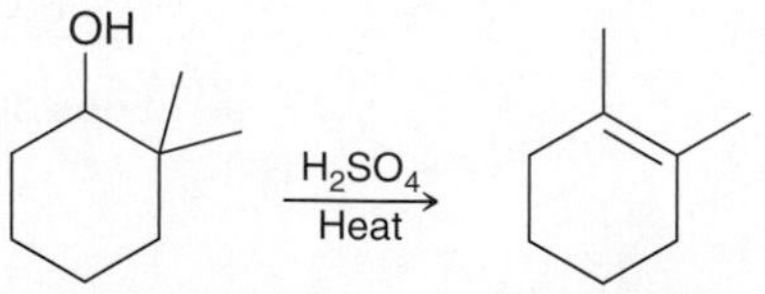

SOLUTION

An E1 process must always involve at least two core steps: loss of a leaving group and proton transfer. But we must consider whether the other two possible steps will occur:

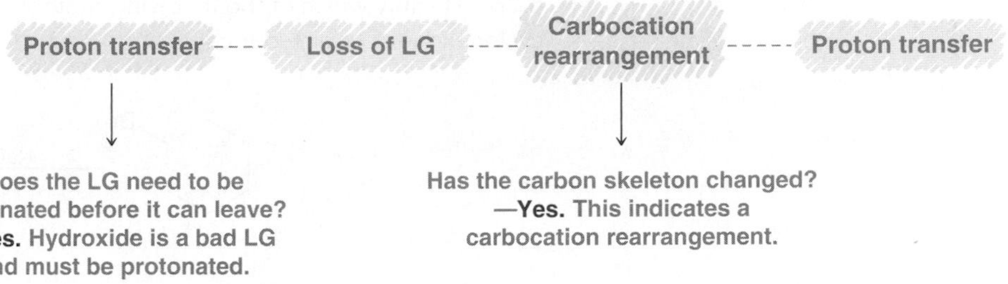

LOOKING BACK
For practice with arrow pushing, see Sections 6.8–6.10).

In this case, the mechanism should exhibit all four steps (the two core steps plus the two additional steps). To draw these steps, we will rely on the skills we learned in Chapter 6:

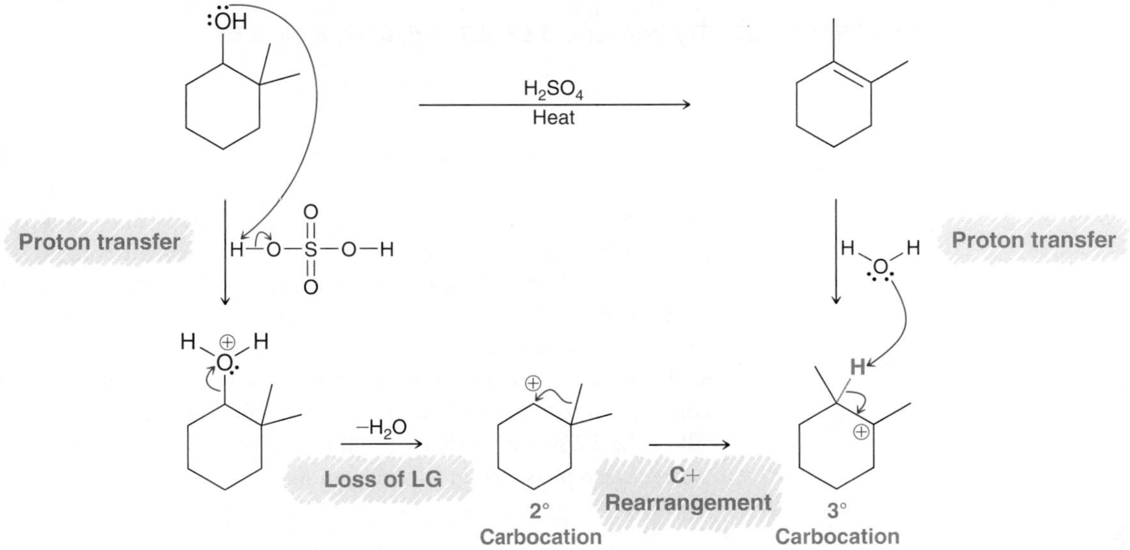

PRACTICE the skill **8.34** Draw a mechanism for each of the following E1 processes:

(a)
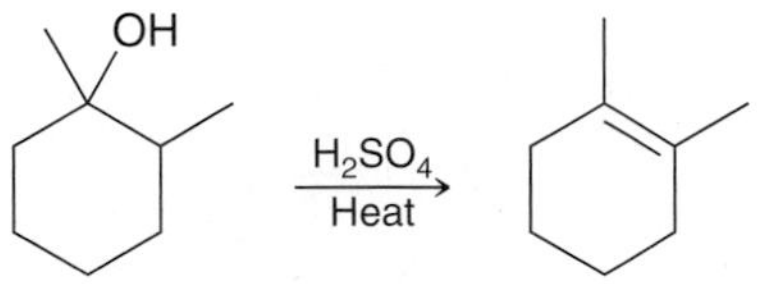

(b)
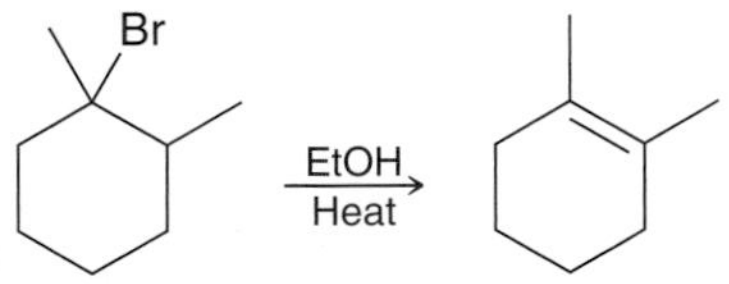

(c)
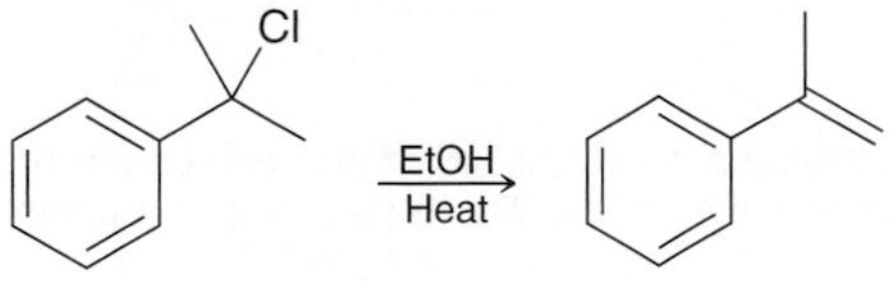

(d)
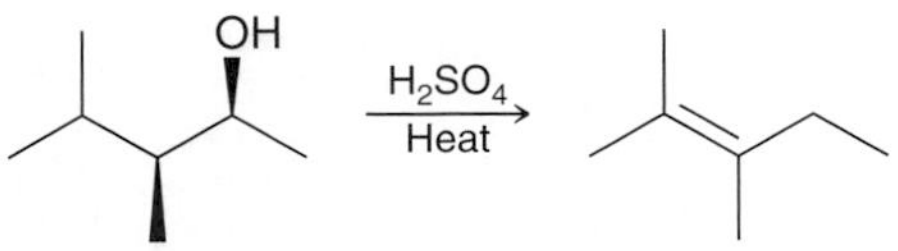

APPLY the skill **8.35** Identify the pattern for each mechanism in Problem 8.34. For example the pattern for Problem 8.34a is:

This mechanism is comprised of a proton transfer followed by the two core steps of an E1 process (loss of leaving group and then proton transfer).

Draw patterns for the other three problems (8.34b–8.34d). Then compare the patterns. Identify which two reactions exhibit the same exact pattern, and then describe in words why those two reactions are so similar.

8.36 Identify which of the following methods is more efficient for producing 3,3-dimethylcyclohexene. Explain your choice.

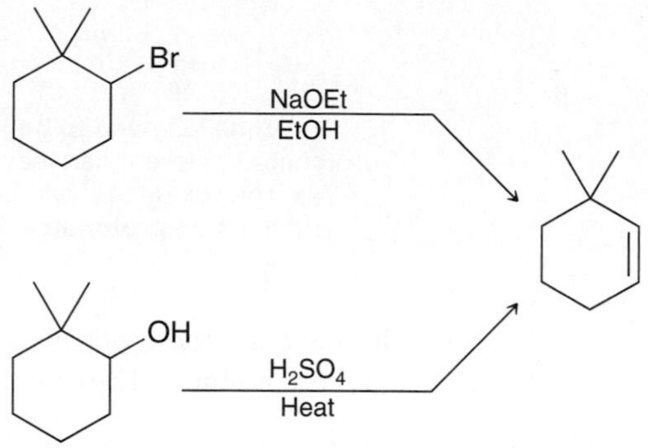

need more **PRACTICE?** **Try Problems 8.68, 8.77a-d, 8.82, 8.84, 8.85**

8.11 Drawing the Complete Mechanism of an E2 Process

In the previous section, we saw that additional steps can accompany an E1 process. In contrast, an E2 process consists of one concerted step and is rarely accompanied by any other steps. A carbocation is never formed, and therefore, there is no possibility for a carbocation rearrangement. In addition, E2 conditions generally require the use of a strong base, and an OH group cannot be protonated under such conditions. It is therefore not common to see an E2 process with a proton transfer at the beginning of the mechanism. It is much more common to see a proton transfer at the beginning of an E1 process (a dehydration reaction). All of the E2 processes that we will encounter in this textbook will be comprised of just one concerted step, as seen in Mechanism 8.2.

MECHANISM 8.2 THE E2 MECHANISM

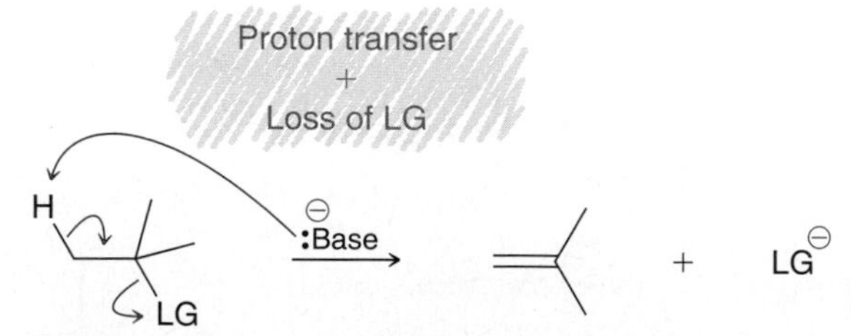

An E2 process is comprised of just one concerted step in which the base abstracts a proton and the leaving group leaves at the same time

The E2 mechanism requires only three curved arrows, all of which must be carefully placed. When drawing the first curved arrow, the tail is placed on a lone pair of the base and the head is drawn pointing to a proton at the β position. When drawing the second curved arrow, the tail is placed on the C—H bond that is breaking, and the head is placed on the bond between the α and β positions. Finally, the third curved arrow is drawn to show the loss of the leaving group.

CONCEPTUAL CHECKPOINT

8.37 Draw the mechanism for each of the following reactions:

(a) 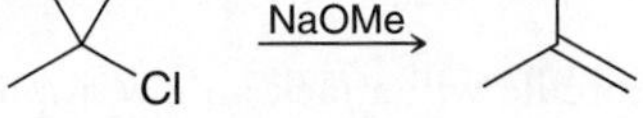

(b) 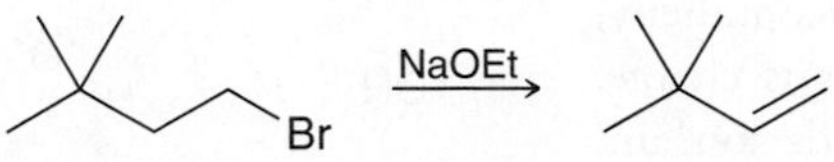

(c) 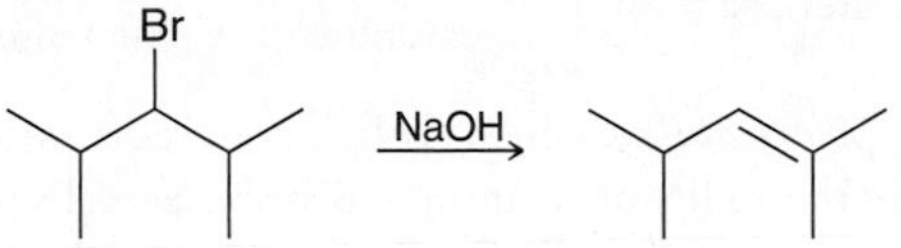

8.38 In the next chapter, we will learn a method for preparing alkynes (compounds containing C≡C triple bonds). In the following reaction, a dihalide (a compound with two halogen atoms) is treated with an excess of strong base (sodium amide), resulting in two successive E2 reactions. Draw the mechanism for this transformation.

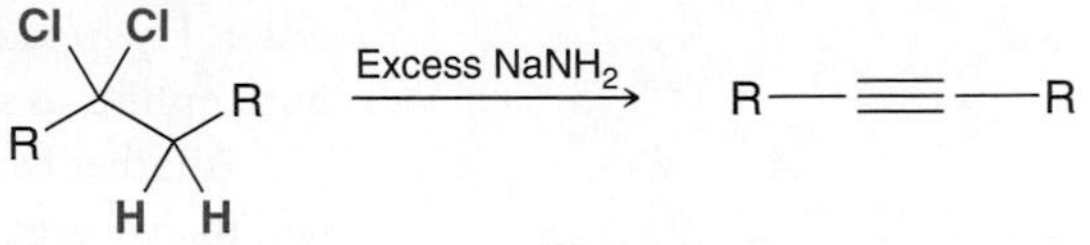

8.12 Substitution vs. Elimination: Identifying the Reagent

Substitution and elimination reactions are almost always in competition with each other. In order to predict the products of a reaction, it is necessary to determine which mechanisms are likely to occur. In some cases, only one mechanism will predominate:

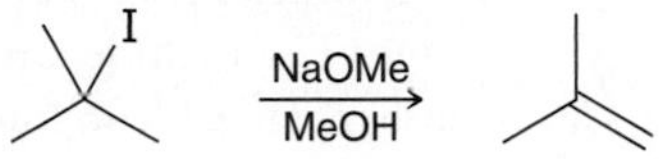

From E2
(only product)

In other cases, two or more mechanisms will compete; for example:

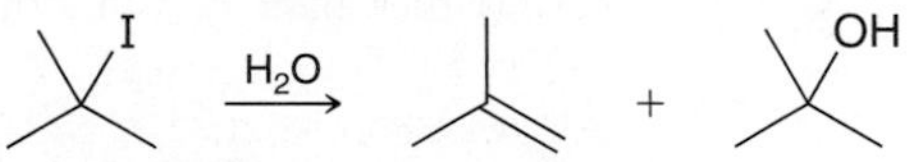

From E1 **From S_N1**

Don't fall into the trap of thinking that there must always be one clear winner. Sometimes there is, but sometimes there are multiple products. The goal is to predict all of the products and to predict which products are major and which are minor. To accomplish this goal, three steps are required:

1. Determine the function of the reagent.
2. Analyze the substrate and determine the expected mechanism(s).
3. Consider any relevant regiochemical and stereochemical requirements.

Each of these three steps will be explored, in detail, in the remainder of this chapter. This section will focus on the skills necessary to achieve the first step, while the second and third steps will be discussed in the following sections. The ultimate goal is proficiency in predicting products, and the skills in this section must be viewed as the first step in achieving that goal.

We have seen earlier in this chapter that the main difference between substitution and elimination is the function of the reagent. A substitution reaction occurs when the reagent functions as a nucleophile, while an elimination reaction occurs when the reagent functions as a base. Accordingly, the first step in any specific case is to determine whether the reagent is a strong nucleophile or a weak nucleophile and whether it is a strong base or a weak base. Nucleophilicity and basicity are not the same concepts, as first described in Section 6.7. Nucleophilicity is a kinetic phenomenon and refers to the rate of reaction, while basicity is a thermodynamic phenomenon and refers to the position of equilibrium. We will soon see that it is possible for a reagent to be a weak nucleophile and a strong base. Similarly, it is possible for a reagent to be a strong nucleophile and a weak base. That is, basicity and nucleophilicity do not always parallel each other. We will now explore nucleophilicity and basicity in more detail in order to develop the skills necessary to identify the expected function of a reagent.

Nucleophilicity

Nucleophilicity refers to the rate at which a particular nucleophile will attack an electrophile. There are many factors that contribute to nucleophilicity, as first described in Section 6.7. One such factor is charge, which can be illustrated by comparing a hydroxide ion and water. Hydroxide is a strong nucleophile, while water is a weak nucleophile, as seen in Section 7.8.

Another factor that impacts nucleophilicity is *polarizability*, which is often even more important than charge. Recall that polarizability describes the ability of an atom to distribute its electron density unevenly as a result of external influences (Section 6.7). Polarizability is directly related to the size of the atom and, more specifically, to the number of electrons that are distant from the nucleus. A sulfur atom is very large and has many electrons that are distant from the nucleus, and it is therefore highly polarizable. Most halogens share this same feature. For this reason, H_2S is a much stronger nucleophile than H_2O:

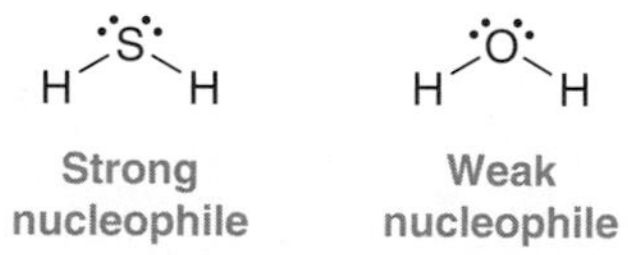

Notice that H_2S lacks a negative charge but is nevertheless a strong nucleophile because it is highly polarizable. The relationship between polarizability and nucleophilicity also explains why the hydride ion (H^-) of NaH does not function as a nucleophile, even though it has a negative charge. The hydride ion is extremely small (hydrogen is the smallest atom), and it is not sufficiently polarizable to function as a nucleophile. We will see, however, that the hydride ion does function as a very strong base. Let's now review the factors that affect basicity.

Basicity

Unlike nucleophilicity, basicity is not a kinetic phenomenon and does not refer to the rate of a process. Rather, basicity is a thermodynamic phenomenon and refers to the position of equilibrium:

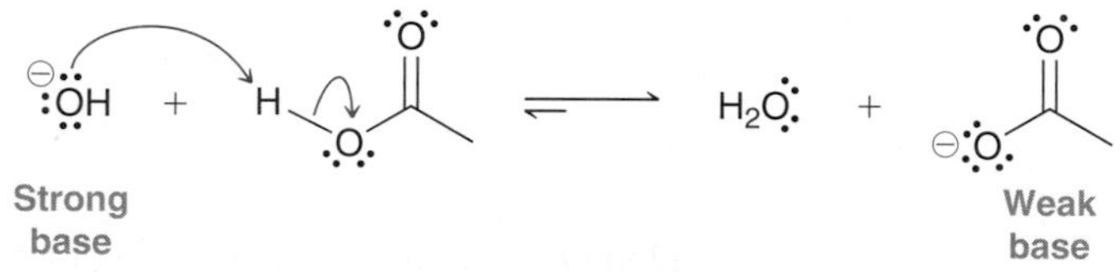

In a proton transfer process, the equilibrium will favor the weaker base.

There are several methods for determining whether a base is strong or weak. Chapter 3 discussed two such methods: a quantitative approach and a qualitative approach. The former requires access to pK_a values. Specifically, the strength of a base is determined by assessing the pK_a of its conjugate acid. For example, a chloride ion is an extremely weak base, because its conjugate acid (HCl) is strongly acidic (p$K_a = -7$). The other approach, the qualitative method, was described in Section 3.4 and involves the use of four factors for determining the relative stability of a base containing a negative charge. For example, a chloride ion has a negative charge on a large, electronegative atom and is therefore highly stabilized (a weak base). Notice that both the quantitative approach and qualitative approach provide the same prediction: that a chloride ion is a weak base.

As another example, consider the bisulfate ion (HSO_4^-). The quantitative approach identifies this ion as a weak base, because its conjugate acid is strongly acidic (p$K_a = -9$). The qualitative approach gives a similar prediction, because the bisulfate ion is highly resonance stabilized:

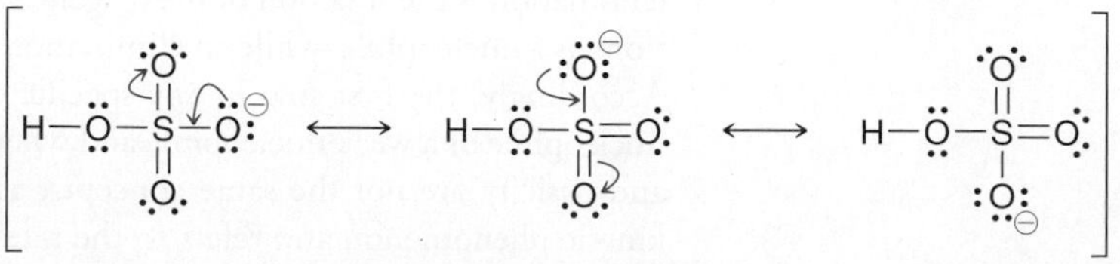

Notice once again that the quantitative approach and qualitative approach are consistent. Either approach can be used. The specific case will usually dictate which method is easier to use.

Nucleophilicity vs. Basicity

After reviewing the main factors contributing to nucleophilicity and basicity, we can now categorize all reagents into four groups (Figure 8.25). Let's quickly review each of these four categories. The first category contains reagents that function only as nucleophiles. That is, they are strong nucleophiles because they are highly polarizable, but they are weak bases because their conjugate acids are fairly acidic. The use of a reagent from this category signifies that a substitution reaction is occurring (not elimination).

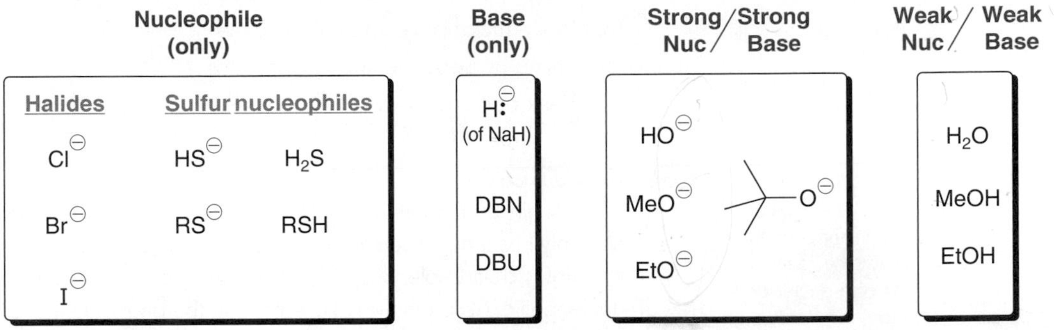

FIGURE 8.25
Classification of common reagents used for substitution/ elimination.

Nucleophile (only)		Base (only)	Strong/Strong Nuc/Base	Weak/Weak Nuc/Base
Halides **Sulfur nucleophiles**		$H:^{\ominus}$ (of NaH)	HO^{\ominus}	H_2O
Cl^{\ominus} HS^{\ominus} H_2S		DBN	MeO^{\ominus} $^{\ominus}O$	MeOH
Br^{\ominus} RS^{\ominus} RSH		DBU	EtO^{\ominus}	EtOH
I^{\ominus}				

LOOKING AHEAD
The hydride ion will be used as a strong base in other applications throughout this book.

The second category contains reagents that function only as bases, not as nucleophiles. The first reagent on this list is the hydride ion, usually shown as NaH, where Na^+ is the counterion. As mentioned earlier, the hydride ion is not a nucleophile because it is not sufficiently polarizable; nevertheless, it is a very strong base because its conjugate acid (H—H) is an extremely weak acid. The use of a hydride ion as the reagent indicates that elimination will occur rather than substitution. There are indeed many reagents, other than hydride, that also function exclusively as bases (and not as nucleophiles). Two commonly used examples are DBN and DBU:

| 1,5-Diazabicyclo[4.3.0]non-5-ene (DBN) | 1,8-Diazabicyclo[5.4.0]undec-7-ene (DBU) |

These two compounds are very similar in structure. When either of these compounds is protonated, the resulting positive charge is resonance stabilized:

Resonance stabilized

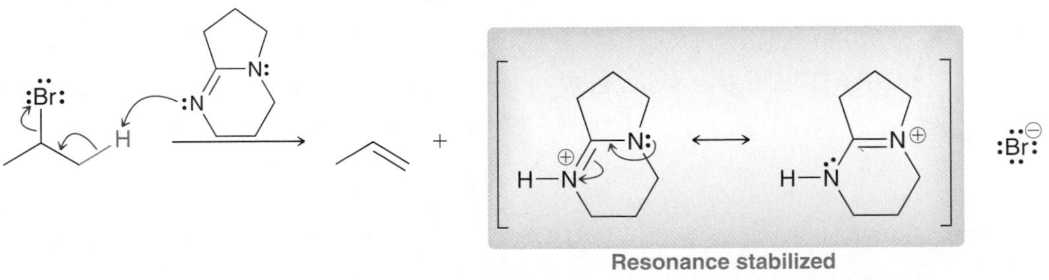

The positive charge is spread over two nitrogen atoms, rather than one. The conjugate acid is stabilized and is therefore not very acidic. As a result, both DBN and DBU are strongly basic. These examples demonstrate that it is possible for a neutral compound to be a strong base.

The third category in Figure 8.25 contains reagents that are both strong nucleophiles and strong bases. These reagents include hydroxide (HO^-) and alkoxide ions (RO^-). Such reagents are generally used for bimolecular processes (S_N2 and E2).

The fourth and final category contains reagents that are weak nucleophiles and weak bases. These reagents include water (H_2O) and alcohols (ROH). Such reagents are generally used for unimolecular processes (S_N1 and E1).

In order to predict the products of a reaction, the first step is determining the identity and nature of the reagent. That is, you must analyze the reagent and determine the category to which it belongs. Let's get some practice with this critical skill.

SKILLBUILDER

8.10 DETERMINING THE FUNCTION OF A REAGENT

LEARN the skill

Consider the phenolate ion. We will explore the reactivity of this ion in Chapter 19. From the following list, identify the category to which the phenolate ion belongs.

(a) Strong nucleophile and weak base
(b) Weak nucleophile and strong base
(c) Strong nucleophile and strong base
(d) Weak nucleophile and weak base

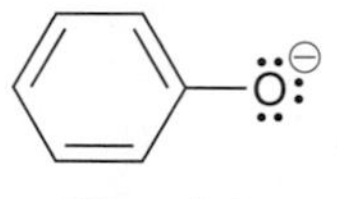
Phenolate

SOLUTION

STEP 1
Determine whether or not the reagent is a strong nucleophile by looking for charge and/or polarizability.

Let's begin by exploring its nucleophilicity. We look for charge and polarizability. The reagent certainly has a negative charge, which should render it a strong nucleophile. The oxygen atom is not highly polarizable, but it is also not small enough of an atom (like hydrogen) to render it non-nucleophilic. Accordingly, we predict that the phenolate ion should be a fairly strong nucleophile. This feature will be an important aspect of its reactivity in Chapter 19.

STEP 2
Determine whether or not the reagent is a strong base using either the quantitative method or the qualitative method.

Next, we explore the basicity of the phenolate ion. There are two methods that can be used, and their results should agree. Using the quantitative approach, we look up the pK_a value of the conjugate acid (called phenol):

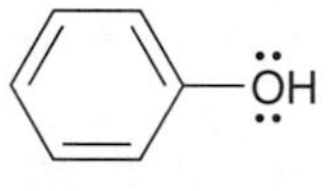
Phenol

Table 3.1 indicates that the pK_a of phenol is 9.9, which is much more acidic than a typical alcohol (ROH usually has a pK_a in the range of 16–18). Since phenol is fairly acidic, we expect the phenolate ion to be a fairly weak base. Alternatively, we could draw the same conclusion by using the qualitative approach. Specifically, we note that the phenolate ion is resonance stabilized:

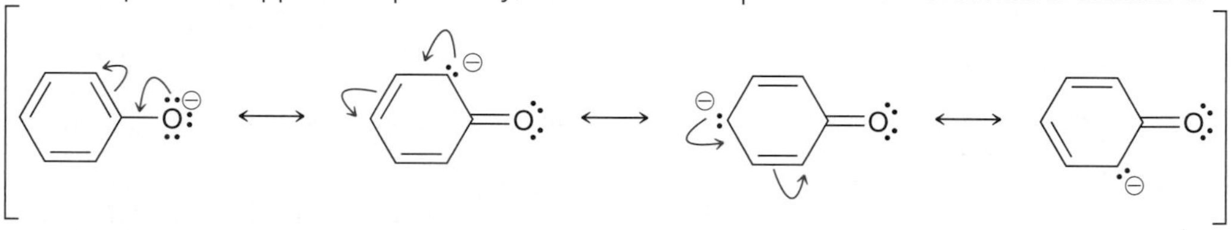

As a result, the phenolate ion is expected to be a weaker base (more stable) than a regular alkoxide ion (RO^-). In conclusion, our analysis predicts that the phenolate ion should be a fairly strong nucleophile and a fairly weak base.

PRACTICE the skill

8.39 Identify whether each of the following reagents would be a strong nucleophile or a weak nucleophile, and also indicate whether it would be a strong base or a weak base:

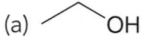

(a) 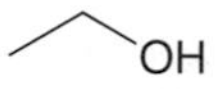

(b) 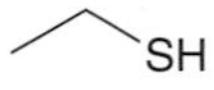

(c) 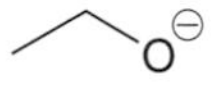

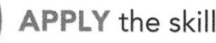

(d) $Br^{\ominus}$
(e) LiOH
(f) MeOH
(g) NaOMe
(h) DBN

APPLY the skill

8.40 We have seen that NaH is a strong base but a weak nucleophile. In contrast, lithium aluminum hydride (LAH) is a reagent that can serve as a source of nucleophilic hydride ion:

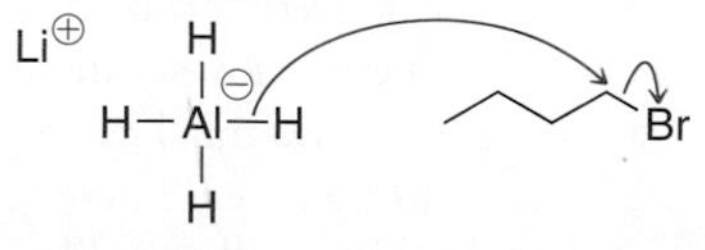

LAH

In this case, LAH functions as a delivery agent of a nucleophilic hydride ion. We will see this reagent in many upcoming chapters. Explain why LAH is capable of functioning as a strong nucleophile while NaH is not.

need more **PRACTICE?** **Try Problem 8.55**

8.13 Substitution vs. Elimination: Identifying the Mechanism(s)

We mentioned that there are three main steps for predicting the products of substitution and elimination reactions. In the previous section, we explored the first step (identifying the function of the reagent). In this section, we now explore the second step of the process in which we analyze the substrate and identify which mechanism(s) operate.

As described in the previous section, there are four categories of reagents. For each category, we must explore the expected outcome with a primary, secondary, or tertiary substrate. Then, we will summarize all of the information in one master flow chart. Let's begin with reagents that function only as nucleophiles.

Possible Outcomes for Reagents That Function Only as Nucleophiles

When the reagent functions exclusively as a nucleophile (and not as a base), only substitution reactions will occur (not elimination), as illustrated in Figure 8.26. The substrate will determine which mechanism operates. S_N2 will predominate for primary substrates, and S_N1 will predominate for tertiary substrates. For secondary substrates, both S_N2 and S_N1 pathways are viable, although S_N2 is generally favored. The rate of an S_N2 process can be further enhanced by using a polar aprotic solvent, as described in Section 7.8.

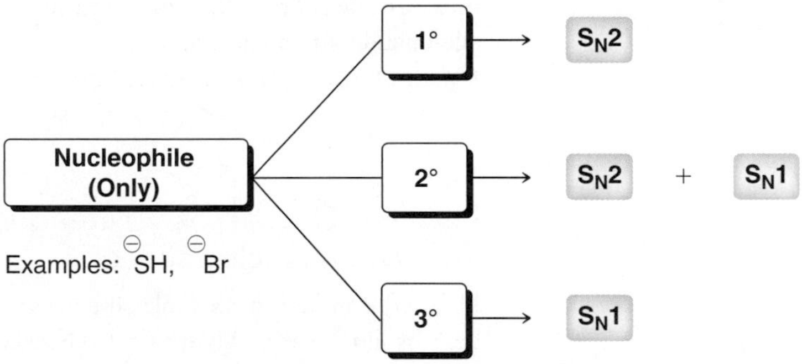

FIGURE 8.26
The outcomes to be expected when a primary, secondary, or tertiary substrate is treated with a reagent that functions only as a nucleophile.

Possible Outcomes for Reagents That Function Only as Bases

When the reagent functions exclusively as a base (and not as a nucleophile), only elimination reactions will occur (not substitution), as illustrated in Figure 8.27. Such reagents are generally strong bases, resulting in an E2 process. This mechanism is not sensitive to steric hindrance and can occur for primary, secondary, or tertiary substrates.

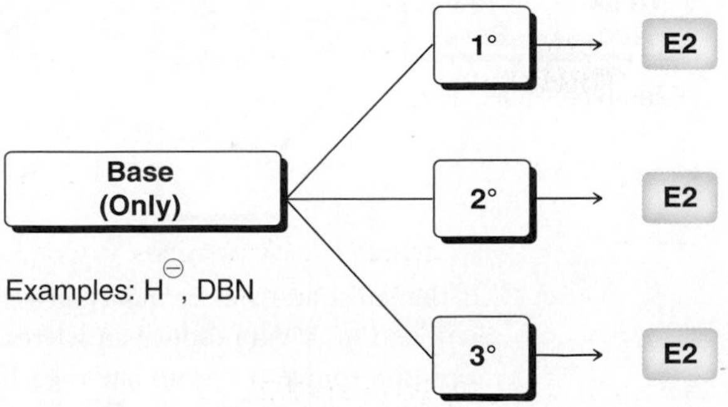

FIGURE 8.27
An E2 process is the expected outcome when a primary, secondary, or tertiary substrate is treated with a reagent that functions only as a base.

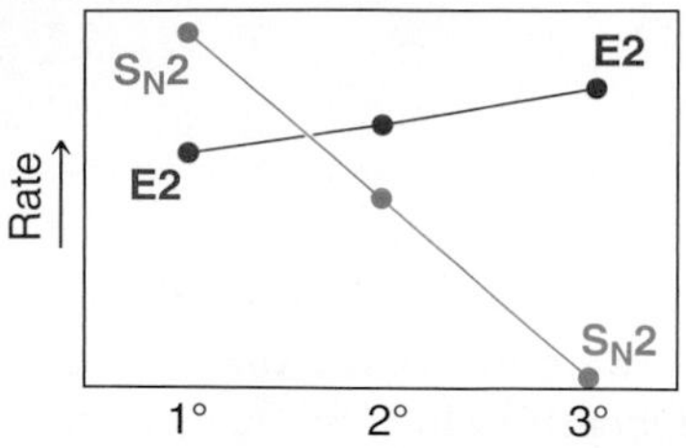

FIGURE 8.28
The effects of substrate on the competition between S_N2 and E2 processes.

Possible Outcomes for Reagents That Are Strong Bases and Strong Nucleophiles

When the reagent is both a strong nucleophile and a strong base, bimolecular mechanisms will dominate (S_N2 and E2), and the substrate plays a critical role in the competition between S_N2 and E2. The nature of this competition is illustrated in Figure 8.28).

The rates of S_N2 and E2 processes are affected differently by the substrate. Notice that the S_N2 process predominates when the substrate is primary, while the E2 process predominates when the substrate is secondary. For tertiary substrates, only E2 is observed, because the S_N2 pathway is too sterically hindered to occur. Recall that the E2 pathway is not sensitive to steric hindrance. This competition between S_N2 and E2 is summarized in Figure 8.29.

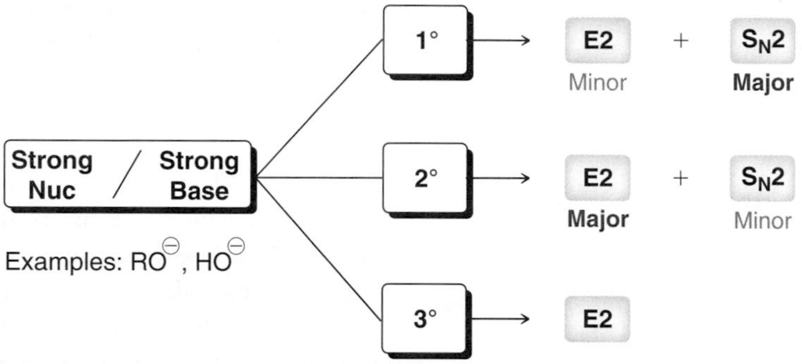

FIGURE 8.29
The outcomes to be expected when a primary, secondary, or tertiary substrate is treated with a reagent that functions both as a strong base and as a strong nucleophile.

Notice that when a primary substrate is treated with an alkoxide ion (RO^-), the S_N2 pathway predominates over the E2 pathway. There is one notable exception to this general rule. Specifically, *tert*-butoxide is a sterically hindered alkoxide, and it favors E2 over S_N2, even when the substrate is primary. This exception is useful because it enables the conversion of a primary alkyl halide into an alkene:

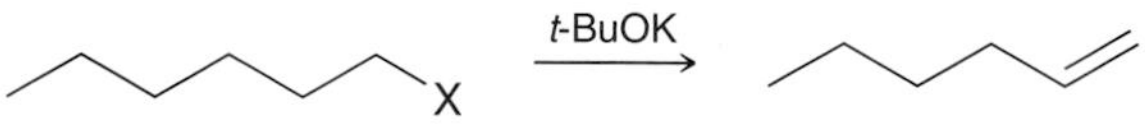

Possible Outcomes for Reagents That Are Weak Bases and Weak Nucleophiles

Now let's consider the possible outcomes when the reagent is both a weak nucleophile and a weak base, as illustrated in Figure 8.30. These conditions are not practical for primary and secondary substrates. Primary substrates generally react slowly with weak nucleophiles and weak bases, and secondary substrates produce a mixture of too many products. It is effective to use a reagent that is both a weak base and a weak nucleophile when the substrate is tertiary. In such a case, unimolecular pathways predominate (S_N1 and E1). A mixture of substitution and elimination products is generally obtained, although the E1 process is often favored at high temperature.

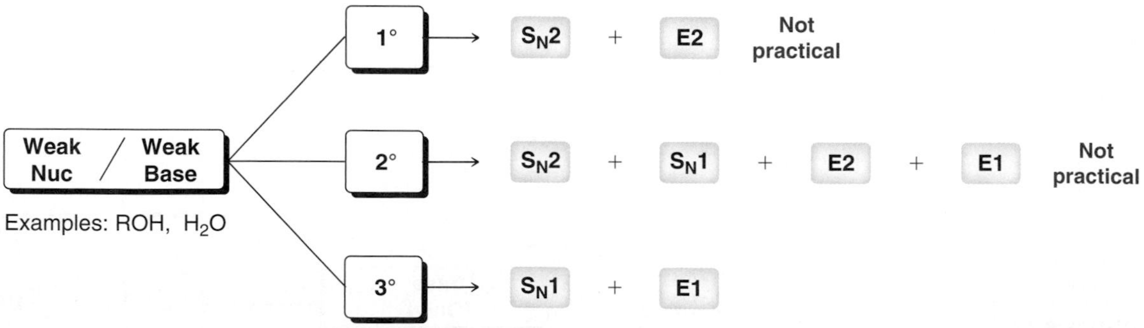

FIGURE 8.30
The outcomes to be expected when a primary, secondary or tertiary substrate is treated with a reagent that functions both as a weak base and as a weak nucleophile.

If the substrate is a secondary or tertiary alcohol, then treatment with concentrated sulfuric acid and heat will produce an E1 reaction (Section 8.9). In such a case, sulfuric acid serves as a proton source (to protonate the OH group), and water serves as a weak base that completes the E1 process.

All of the outcomes previously described are now summarized in one master flow chart (Figure 8.31). It is important to know this flow chart extremely well, but be careful not to memorize it. It is more important to "understand" the reasons for all of these outcomes. A proper understanding will prove to be far more useful on an exam than simply memorizing a set of rules. Figure 8.31 can be used to determine which mechanism(s) operate for any specific case. Let's get some practice.

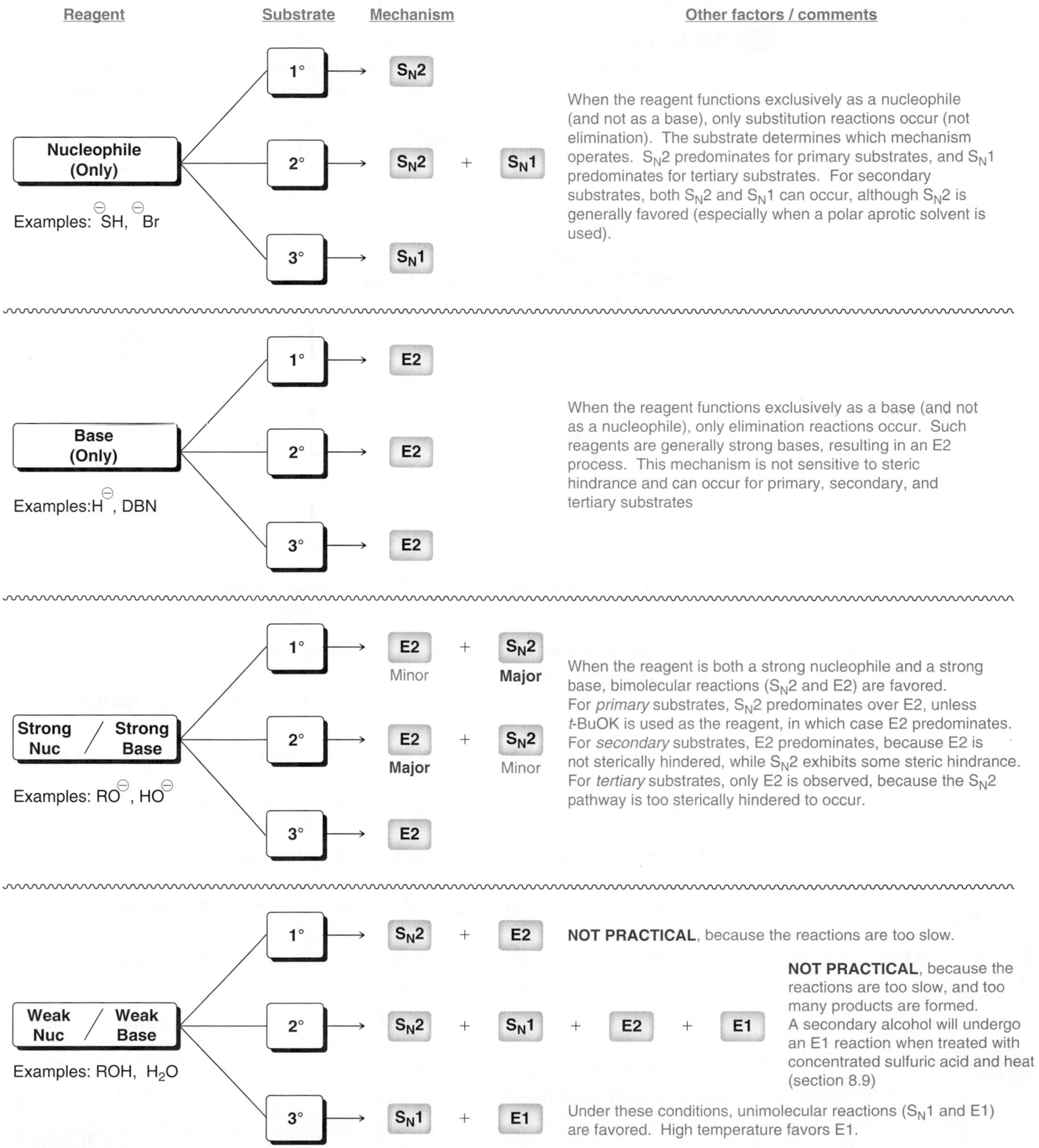

Reagent
Substrate
Mechanism
Other factors / comments
Nucleophile (Only)
Examples: ⊖SH, ⊖Br
1° → SN2
2° → SN2 + SN1
3° → SN1
When the reagent functions exclusively as a nucleophile (and not as a base), only substitution reactions occur (not elimination). The substrate determines which mechanism operates. SN2 predominates for primary substrates, and SN1 predominates for tertiary substrates. For secondary substrates, both SN2 and SN1 can occur, although SN2 is generally favored (especially when a polar aprotic solvent is used).
Base (Only)
Examples: H⊖, DBN
1° → E2
2° → E2
3° → E2
When the reagent functions exclusively as a base (and not as a nucleophile), only elimination reactions occur. Such reagents are generally strong bases, resulting in an E2 process. This mechanism is not sensitive to steric hindrance and can occur for primary, secondary, and tertiary substrates
Strong Nuc / Strong Base
Examples: RO⊖, HO⊖
1° → E2 Minor + SN2 Major
2° → E2 Major + SN2 Minor
3° → E2
When the reagent is both a strong nucleophile and a strong base, bimolecular reactions (SN2 and E2) are favored. For primary substrates, SN2 predominates over E2, unless t-BuOK is used as the reagent, in which case E2 predominates. For secondary substrates, E2 predominates, because E2 is not sterically hindered, while SN2 exhibits some steric hindrance. For tertiary substrates, only E2 is observed, because the SN2 pathway is too sterically hindered to occur.
Weak Nuc / Weak Base
Examples: ROH, H2O
1° → SN2 + E2
NOT PRACTICAL, because the reactions are too slow.
2° → SN2 + SN1 + E2 + E1
NOT PRACTICAL, because the reactions are too slow, and too many products are formed. A secondary alcohol will undergo an E1 reaction when treated with concentrated sulfuric acid and heat (section 8.9)
3° → SN1 + E1
Under these conditions, unimolecular reactions (SN1 and E1) are favored. High temperature favors E1.

FIGURE 8.31
A flow chart for determining the outcome of substitution / elimination reaction.

SKILLBUILDER

8.11 IDENTIFYING THE EXPECTED MECHANISM(S)

LEARN the skill

Identify the mechanism(s) expected to occur when bromocyclohexane is treated with sodium ethoxide.

SOLUTION

STEP 1
Identify the function of the reagent.

First identify the function of the reagent. Using the skills developed in the previous section, we can determine that sodium ethoxide is both a strong nucleophile and a strong base:

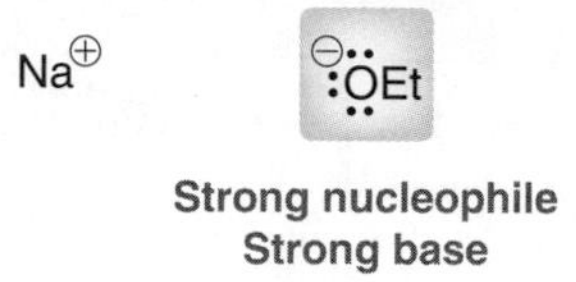

STEP 2
Identify the substrate and determine the mechanisms that operate.

Next, we identify the substrate. Bromocyclohexane is a secondary substrate, and therefore, we expect E2 and S_N2 mechanisms to operate:

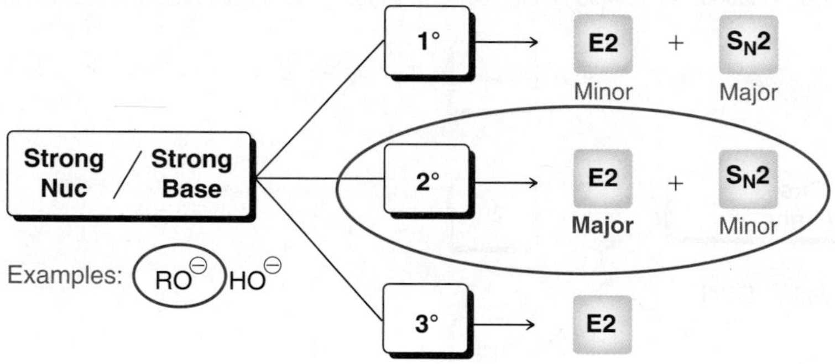

The E2 pathway is expected to provide the major product because the S_N2 pathway is more sensitive to steric hindrance provided by secondary substrates.

PRACTICE the skill

8.41 Identify the mechanism(s) expected to operate when 1-bromobutane is treated with each of the following reagents:

(a) NaOH (b) NaSH (c) *t*-BuOK (d) DBN (e) NaOMe

8.42 Identify the mechanism(s) expected to operate when 2-bromopentane is treated with each of the following reagents:

(a) NaOEt (b) NaI/DMSO (c) DBU (d) NaOH (e) *t*-BuOK

8.43 Identify the mechanism expected to operate when 2-bromo-2-methylpentane is treated with each of the following reagents:

(a) EtOH (b) *t*-BuOK (c) NaI (d) NaOEt (e) NaOH

APPLY the skill

8.44 When 1-chlorobutane is treated with ethanol, neither elimination process (E1 or E2) is observed at an appreciable rate;

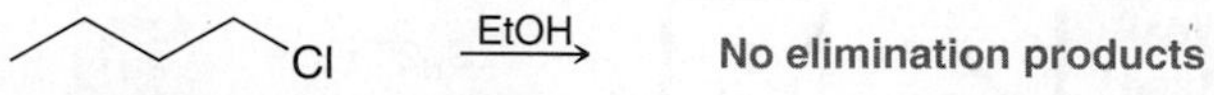

(a) Explain why an E2 reaction does not occur.

(b) Explain why an E1 reaction does not occur.

(c) Modifying the reactants can have a profound effect on the rate of elimination. What modification would you suggest to enhance the rate of an E2 process?

(d) What modification would you suggest to enhance the rate of an E1 process?

8.45 When 2-chloro-1,1,2,3,3-pentamethylcyclohexane is treated with sodium hydroxide, neither E2 nor S_N2 products are formed. Explain.

need more **PRACTICE?** **Try Problems 8.68, 8.69, 8.79a**

8.14 Substitution vs. Elimination: Predicting the Products

We mentioned that predicting the products of substitution and elimination reactions requires three discrete steps:

1. Determine the function of the reagent.
2. Analyze the substrate and determine the expected mechanism(s).
3. Consider any relevant regiochemical and stereochemical requirements.

In the previous two sections, we explored the first two steps of this process. In this last section of the chapter, we explore the third and final step. After determining which mechanism(s) operates, the final step is to consider the regiochemical and stereochemical outcomes for each of the expected mechanisms. Table 8.2 provides a summary of guidelines that must be followed when drawing products. Table 8.2 does not contain any new information. All of the information can be found in this chapter and the previous chapter. The table is meant only as a summary of all of the relevant information, so that it is easily accessible in one location. Let's get some practice applying these guidelines.

TABLE 8.2 GUIDELINES FOR DETERMINING THE REGIOCHEMICAL AND STEREOCHEMICAL OUTCOME OF SUBSTITUTION AND ELIMINATION REACTIONS

	REGIOCHEMICAL OUTCOME	STEREOCHEMICAL OUTCOME
S_N2	The nucleophile attacks the α position, where the leaving group is connected.	The nucleophile replaces the leaving group with inversion of configuration.
S_N1	The nucleophile attacks the carbocation, which is where the leaving group was originally connected, unless a carbocation rearrangement took place.	The nucleophile replaces the leaving group with racemization.
E2	The Zaitsev product is generally favored over the Hofmann product, unless a sterically hindered base is used, in which case the Hofmann product will be favored.	This process is both stereoselective and stereospecific. When applicable, a *trans* disubstituted alkene will be favored over a *cis* disubstituted alkene. When the β position of the substrate has only one proton, the stereoisomeric alkene resulting from *anti*-periplanar elimination will be obtained (exclusively, in most cases).
E1	The Zaitsev product is always favored over the Hofmann product.	The process is stereoselective. When applicable, a *trans* disubstituted alkene will be favored over a *cis* disubstituted alkene.

SKILLBUILDER

8.12 PREDICTING THE PRODUCTS OF SUBSTITUTION AND ELIMINATION REACTIONS

LEARN the skill Predict the product(s) of the following reaction, and identify the major and minor products:

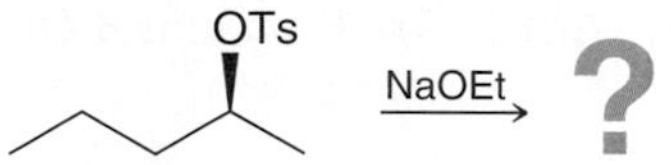

SOLUTION

In order to draw the products, the following three steps must be followed:

1. Determine the function of the reagent.
2. Analyze the substrate and determine the expected mechanism(s).
3. Consider any relevant regiochemical and stereochemical requirements.

In the first step, we analyze the reagent. The ethoxide ion is both a strong base and a strong nucleophile. Next, we analyze the substrate. In this case, the substrate is secondary, so we would expect E2 and S_N2 pathways to compete with each other:

STEP 1 Determine the function of the reagent.

STEP 2 Analyze the substrate and determine the expected mechanisms.

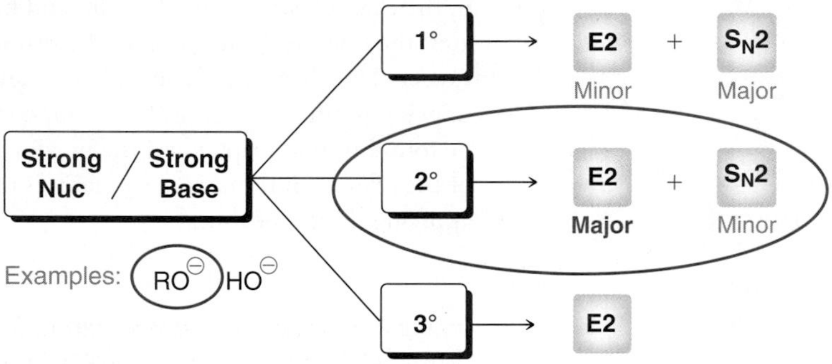

STEP 3 Consider regiochemical and stereochemical requirements.

The E2 pathway would be expected to predominate, because it is less sensitive to steric hinderance than the S_N2 pathway. Therefore, we would expect the major product(s) to be generated via an E2 process and the minor product(s) to be generated via an S_N2 process. In order to draw the products, we must complete the final step. That is, we must consider the regiochemical and stereochemical outcomes for both the E2 and S_N2 processes. Let's begin with the E2 process.

For the regiochemical outcome, we expect the Zaitsev product to be the major product, because the reaction does not utilize a sterically hindered base:

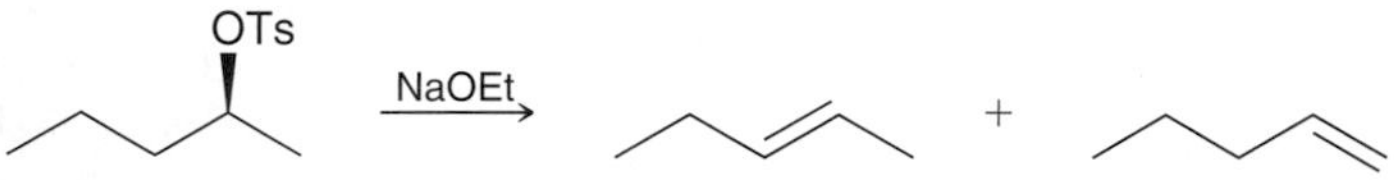

Major **Minor**

Next, look at the stereochemistry. The E2 process is stereoselective, so we expect *cis* and *trans* isomers, with a predominance of the *trans* isomer:

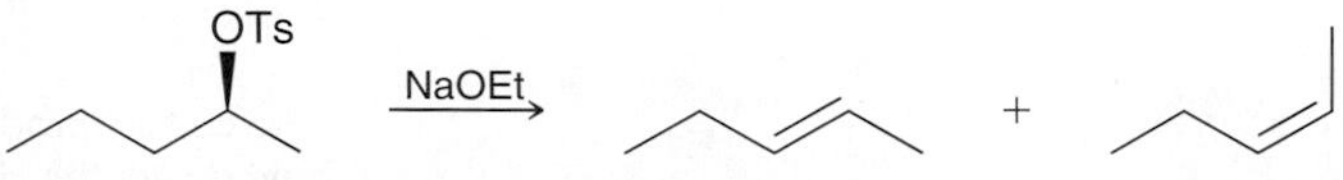

Major **Minor**

The E2 process is also stereospecific, but in this case, the β position has more than one proton, so the stereospecificity of this reaction is not relevant.

Now consider the S_N2 product. This case involves a chirality center, so we expect inversion of configuration:

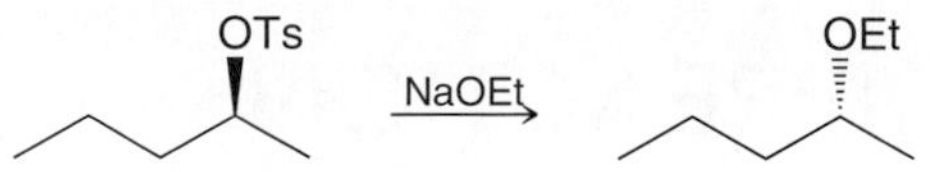

In summary, we expect the following products:

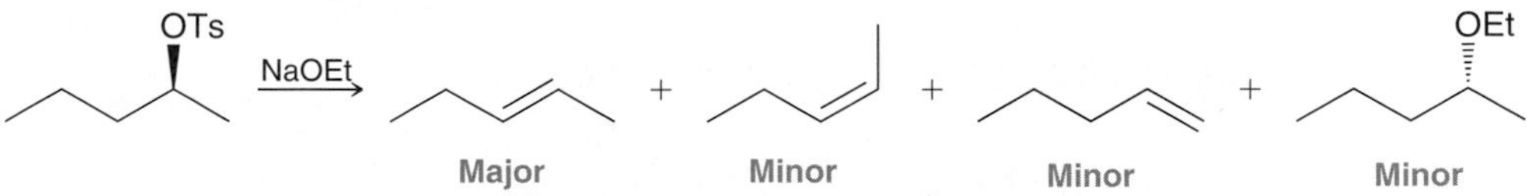

PRACTICE the skill **8.46** Identify the major and minor product(s) that are expected for each of the following reactions:

(a)
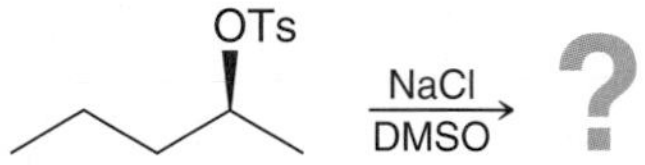

(b)
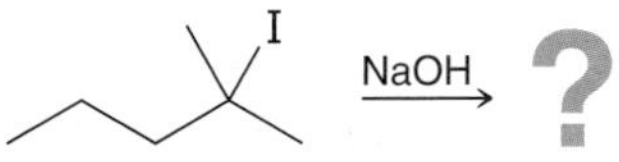

(c)
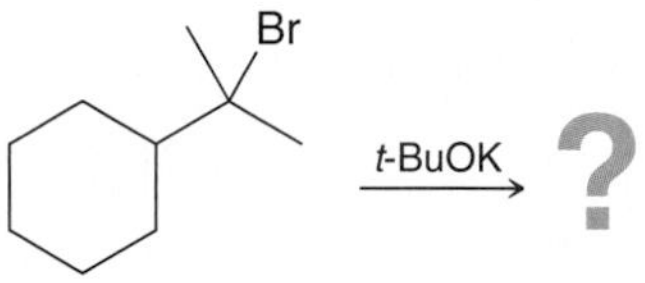

(d)
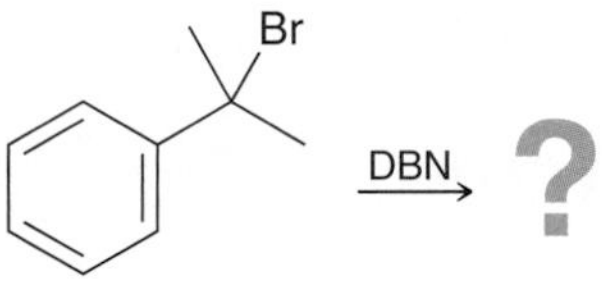

(e)
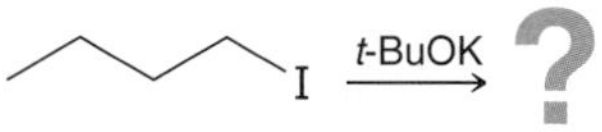

(f)
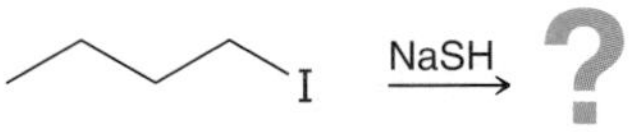

(g)
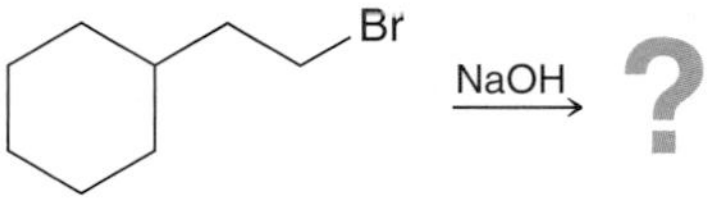

(h)
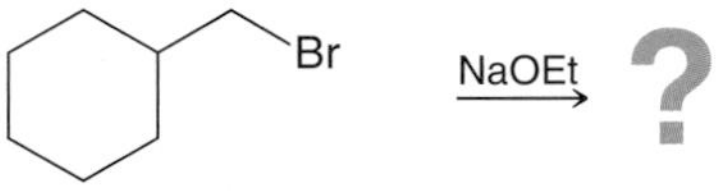

(i)
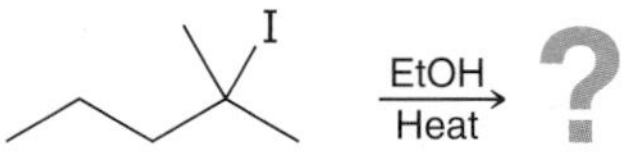

(j)
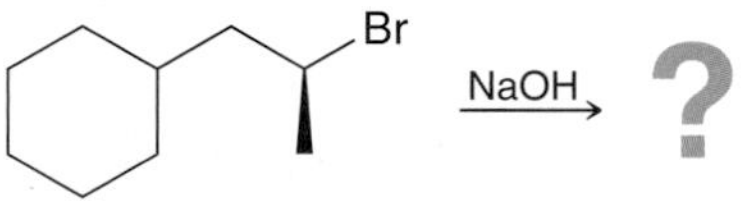

(k)
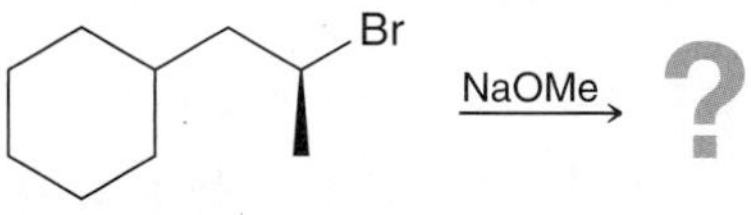

(l)
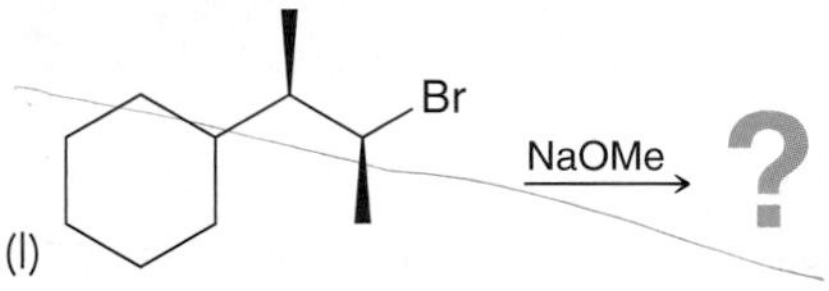

(m)
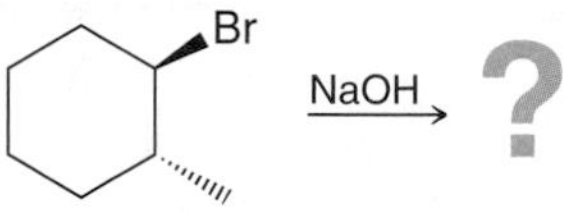

(n)
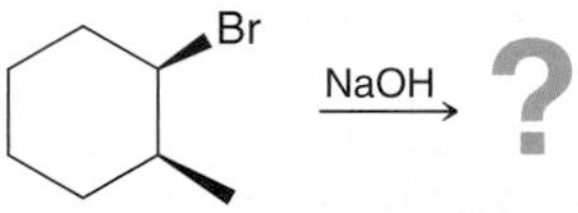

APPLY the skill **8.47** Compound A and compound B are constitutional isomers with molecular formula C_3H_7Cl. When compound A is treated with sodium methoxide, a substitution reaction predominates. When compound B is treated with sodium methoxide, an elimination reaction predominates. Propose structures for compounds A and B.

8.48 Compound A and compound B are constitutional isomers with molecular formula C_4H_9Cl. Treatment of compound A with sodium methoxide gives *trans*-2-butene as the major product, while treatment of compound B with sodium methoxide gives a different disubstituted alkene as the major product.

(a) Draw the structure of compound A.

(b) Propose two structures for compound B.

8.49 An unknown compound with molecular formula $C_6H_{13}Cl$ is treated with sodium ethoxide to produce 2,3-dimethyl-2-butene as the major product. Identify the structure of the unknown compound.

need more **PRACTICE?** **Try Problems 8.78, 8.80**

REVIEW OF REACTIONS

SYNTHETICALLY USEFUL ELIMINATION REACTIONS

DEHYDRATION OF AN ALCOHOL VIA AN E1 PROCESS

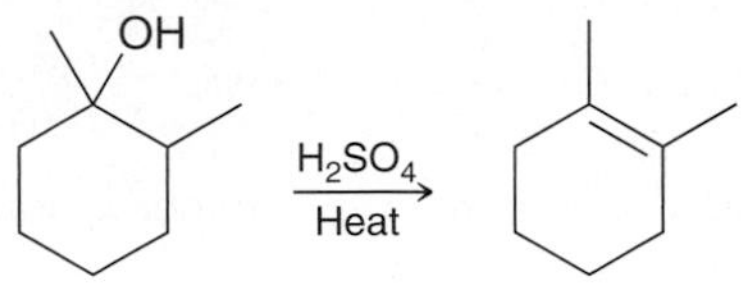

DEHYDROHALOGENATION VIA AN E2 PROCESS—FORMATION OF THE ZAITSEV PRODUCT

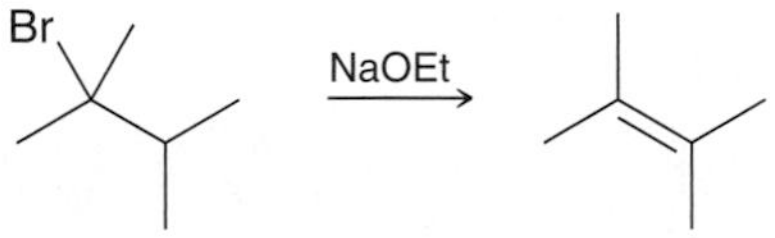

DEHYDROHALOGENATION VIA AN E2 PROCESS—FORMATION OF THE HOFMANN PRODUCT

REVIEW OF CONCEPTS AND VOCABULARY

SECTION 8.1

- **Alkenes** are compounds that possess a C═C double bond and can be prepared via **beta elimination**, or **1,2-elimination.**
- When the leaving group is specifically a halide, the reaction is also called a **dehydrohalogenation.**

SECTION 8.2

- Alkenes are abundant in nature.
- Ethylene and propylene, both formed from cracking petroleum, are used as starting materials for a wide variety of compounds.

SECTION 8.3

- Alkenes are named much like alkanes, with the following additional rules:
  - The suffix "ane" is replaced with "ene."
  - The parent is the longest chain that includes the π bond.
  - The π bond should receive the lowest number possible.
  - The position of the double bond is indicated with a single locant placed either before the parent or before the suffix "ene."
- **Degree of substitution** refers to the number of alkyl groups connected to the double bond.

SECTION 8.4

- Double bonds do not exhibit free rotation at room temperature, giving rise to stereoisomerism.
- An eight-membered ring is the smallest size ring that can accommodate a *trans* double bond. This rule also applies to bridged bicyclic compounds and is called **Bredt's rule.**
- The stereodescriptors *cis* and *trans* can only be used to indicate the relative arrangement of similar groups.
- When an alkene possesses nonsimilar groups, the stereodescriptors *E* and *Z* must be used. ***Z*** indicates priority groups on the same side, while ***E*** indicates priority groups on opposite sides.

SECTION 8.5

- A *trans* alkene is generally more stable than its stereoisomeric *cis* alkene.
- Alkene stability increases with increasing degree of substitution.

SECTION 8.6

- In a *concerted process*, a base abstracts a proton and the leaving group leaves simultaneously.
- In a *stepwise process*, first the leaving group leaves, and then the base abstracts a proton.

SECTION 8.7

- Evidence for the concerted **E2** mechanism includes the observation of a second-order rate equation.
- Tertiary halides react most readily toward E2.
- **Regiochemistry** is an issue in E2 reactions, which are said to be **regioselective**, because the more substituted alkene, called the **Zaitsev product**, is generally the major product.
- When both the substrate and the base are sterically hindered, the less substituted alkene, called the **Hofmann product,** is the major product.
- E2 reactions are **stereospecific**, because they generally occur via the ***anti*-coplanar** conformation, rather than the ***syn*-coplanar** conformation. Small deviations from coplanarity can be tolerated, and it is sufficient for the proton and the leaving group to be ***anti*-periplanar.**
- Substituted cyclohexanes only undergo E2 reactions from the chair conformation in which the leaving group and the proton both occupy axial positions.
- E2 reactions are also **stereoselective**, favoring *trans* disubstituted alkenes over *cis* disubstituted alkenes.

SECTION 8.8

- Regiochemistry and stereochemistry must be considered when predicting the products of an E2 reaction.

SECTION 8.9

- Evidence for the stepwise **E1** mechanism includes the observation of a first-order rate equation.
- Tertiary halides react most readily via an E1 process while primary halides are unreactive toward an E1 process.
- An E1 reaction is generally accompanied by a competing S_N1 reaction, and a mixture of products is generally obtained.
- Elimination of water to form an alkene is called a **dehydration** reaction.
- E1 reactions exhibit a regiochemical preference for the Zaitsev product.
- E1 reactions are not stereospecific, but they are stereoselective.

SECTION 8.10

- If the substrate is an alcohol, a strong acid is required to protonate the OH group.
- During an E1 process, a carbocation rearrangement occurs if it will produce a more stable carbocation intermediate.

SECTION 8.11

- An E2 process is comprised of just one concerted step.

SECTION 8.12

- Strong nucleophiles are compounds that contain a negative charge and/or are polarizable.
- Strong bases are compounds whose conjugate acids are not very acidic.
- All reagents can be classified into four categories: (1) only nucleophiles, (2) only bases, (3) strong nucleophiles and strong bases, and (4) weak nucleophiles and weak bases.

SECTION 8.13

- The reagent and the substrate determine the expected mechanism(s).

SECTION 8.14

- Predicting the products of substitution and elimination reactions requires three discrete steps:
  1. Determine the function of the reagent.
  2. Analyze the substrate and determine the expected mechanism(s).
  3. Consider any relevant regiochemical and stereochemical requirements.

KEY TERMINOLOGY

1,2-elimination 334
alkenes 334
***anti*-coplanar** 355
***anti*-periplanar** 356
beta elimination 334
Bredt's rule 340
coplanar 355
degree of substitution 339
dehydration 334
dehydrohalogenation 334
E 341
E1 363
E2 348
Hofmann product 351
periplanar 356
regiochemistry 351
regioselective 351
stereoselective 359
stereospecific 356
***syn*-coplanar** 355
Z 341
Zaitsev product 351

SKILLBUILDER REVIEW

8.1 ASSEMBLING THE SYSTEMATIC NAME OF AN ALKENE

STEP 1 Identify the parent: choose the longest chain that includes the π bond.

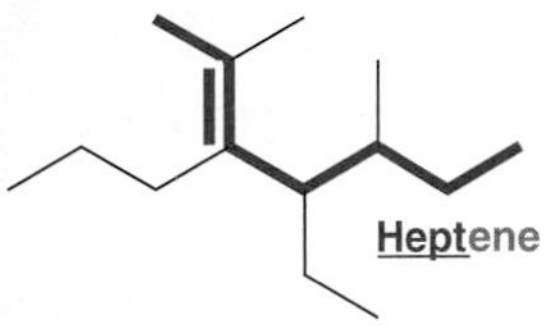

STEP 2 Identify and name the substituents.

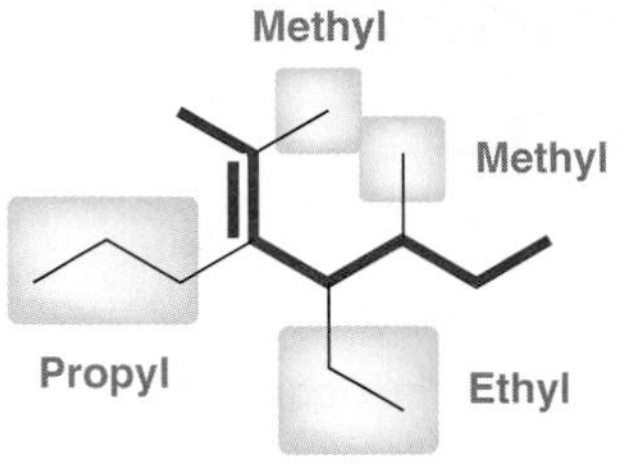

STEP 3 Number the parent chain and assign a locant to each substituent.

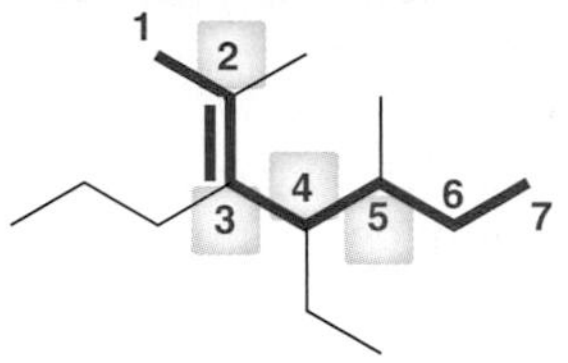

STEP 4 Arrange the substituents alphabetically.

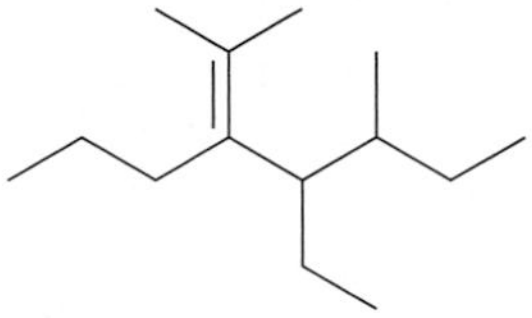

4-Ethyl-2,5-dimethyl-3-propyl-2-heptene

Try Problems **8.1–8.3, 8.50b,c**

8.2 ASSIGNING THE CONFIGURATION OF A DOUBLE BOND

STEP 1 Idenitfy the two groups connected to one vinylic position, and then determine which group has priority.

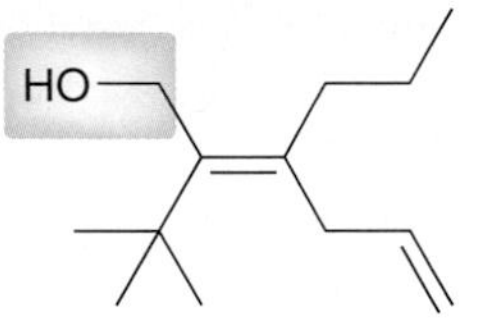

STEP 2 Repeat Step 1 for the other vinylic position, moving away from the double bond and looking for the first point of difference.

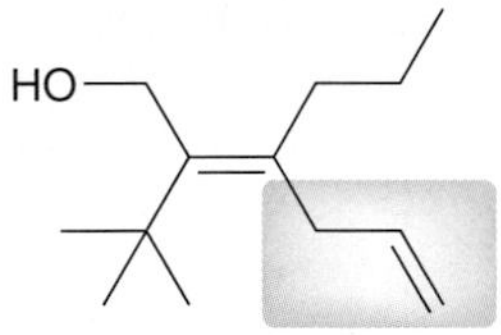

STEP 3 Determine whether the priorities are on the same side (*Z*) or opposite sides (*E*) of the double bond.

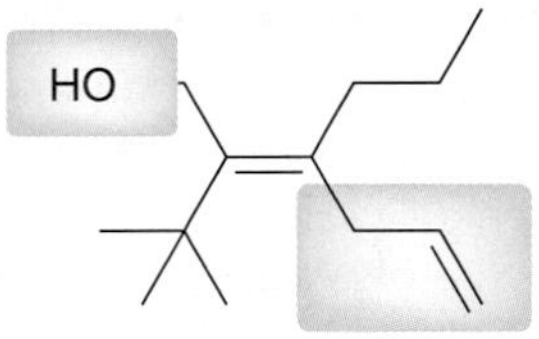

E

Try Problems **8.5, 8.6, 8.51**

8.3 COMPARING THE STABILITY OF ISOMERIC ALKENES

STEP 1 Identify the degree of substitution for each alkene.

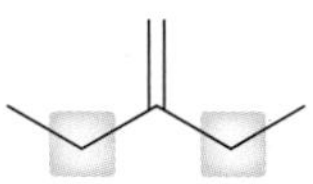

Disubstituted

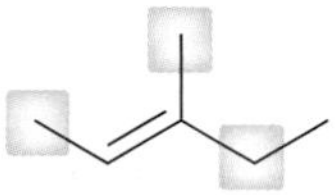

Trisubstituted

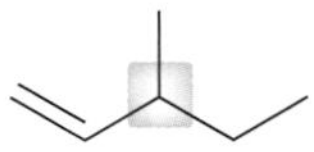

Monosubstituted

STEP 2 Select the alkene that is most substituted.

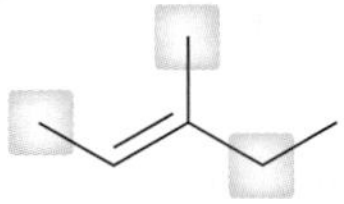

Trisubstituted

Try Problems **8.7, 8.8, 8.53**

8.4 DRAWING THE CURVED ARROWS OF AN ELIMINATION REACTION

CONCERTED MECHANISM Three curved arrows, all drawn in one step. Proton transfer is accompanied by simultaneous loss of a leaving group.

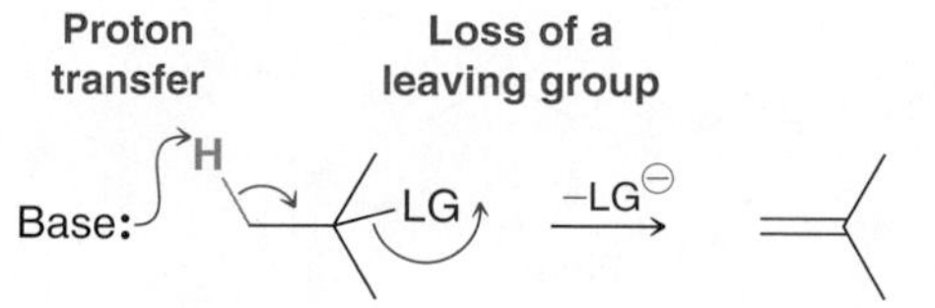

STEPWISE MECHANISM Three curved arrows drawn in two separate steps. Leaving group leaves to form a carbocation intermediate, followed by a proton transfer.

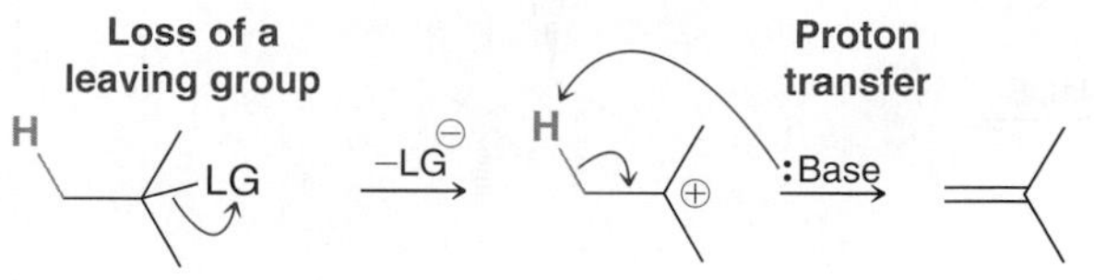

Try Problems 8.9–8.12, 8.69

8.5 PREDICTING THE REGIOCHEMICAL OUTCOME OF AN E2 REACTION

STEP 1 Identify the α position.

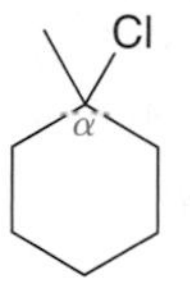

STEP 2 Identify all β positions with protons.

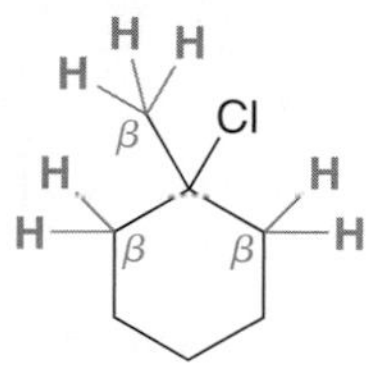

STEP 3 Take notice of β positions that lead to same product.

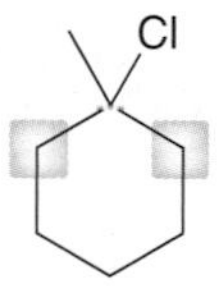

STEP 4 Remove a proton from each β position to draw each product. Then, identify the degree of substitution.

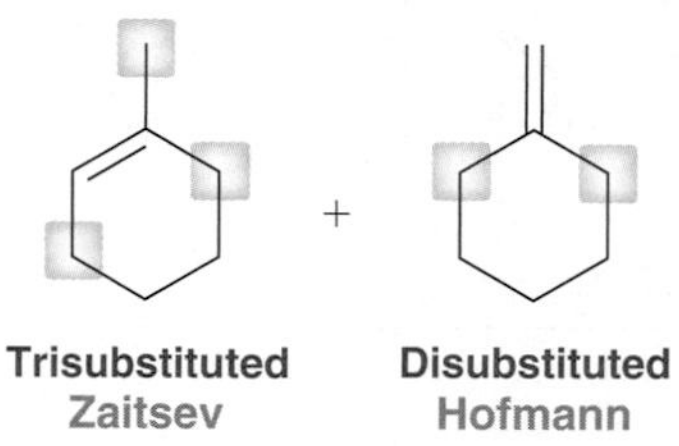

STEP 5 Analyze the base to determine which product predominates.

	Zaitsev	Hofmann
Not sterically hindered	**Major**	Minor
Sterically hindered	Minor	**Major**

Try Problems 8.15–8.17, 8.60, 8.61b–d, 8.64, 8.66a,d, 8.75

8.6 PREDICTING THE STEREOCHEMICAL OUTCOME OF AN E2 REACTION

STEP 1 Identify all β positions with protons.

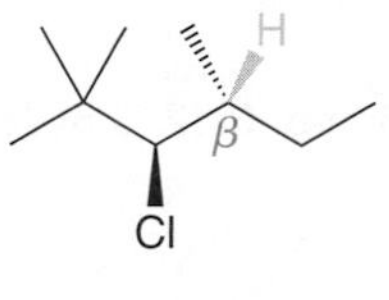

STEP 2 If the β position has two protons, expect *cis* and *trans* isomers. If the β position has only one proton, then draw the Newman projection.

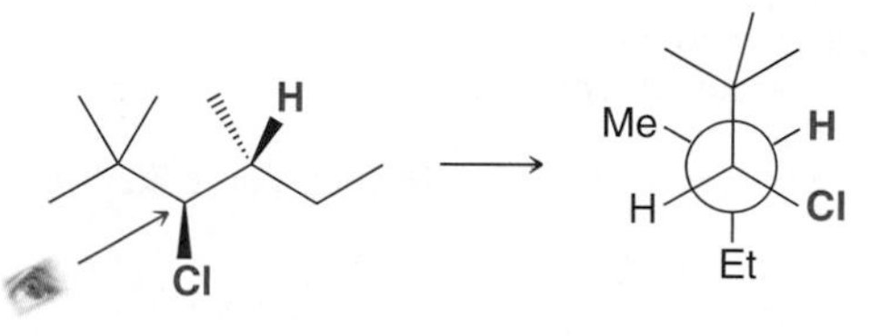

STEP 3 Rotate to achieve *anti*-periplanar conformation.

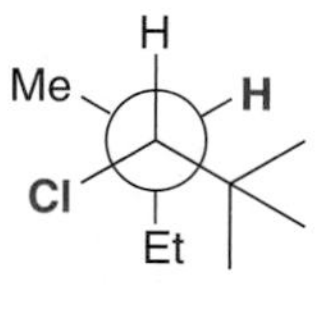

STEP 4 Use Newman projection to draw product.

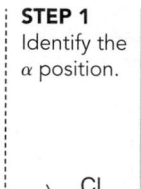

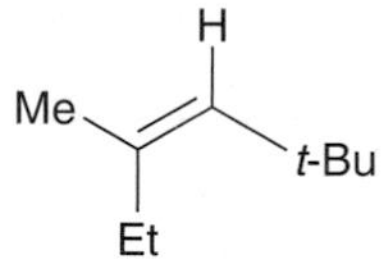

Try Problems 8.18, 8.19, 8.56, 8.63, 8.83

8.7 DRAWING THE PRODUCTS OF AN E2 REACTION

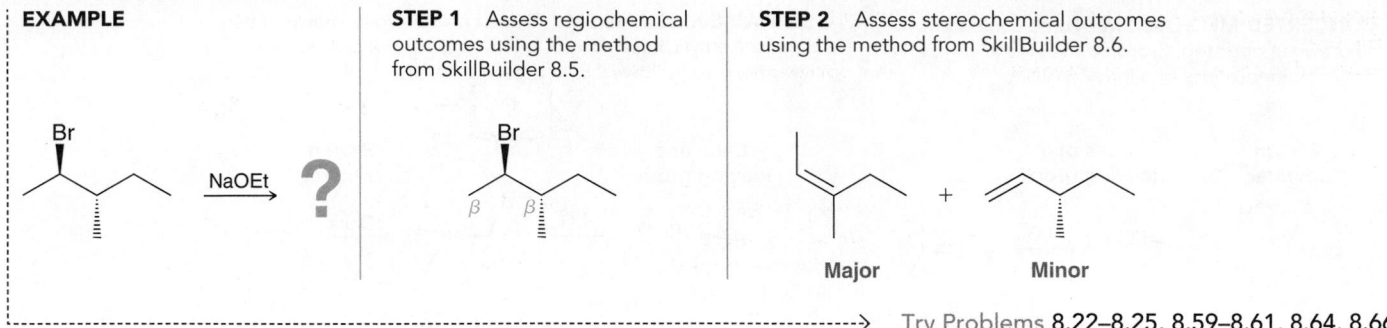

EXAMPLE

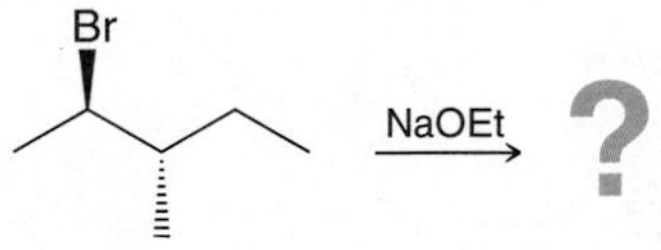

STEP 1 Assess regiochemical outcomes using the method from SkillBuilder 8.5.

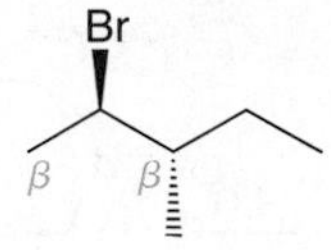

STEP 2 Assess stereochemical outcomes using the method from SkillBuilder 8.6.

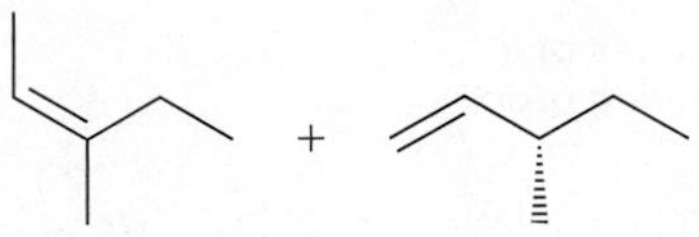

Major **Minor**

Try Problems **8.22–8.25, 8.59–8.61, 8.64, 8.66**

8.8 PREDICTING THE REGIOCHEMICAL OUTCOME OF AN E1 REACTION

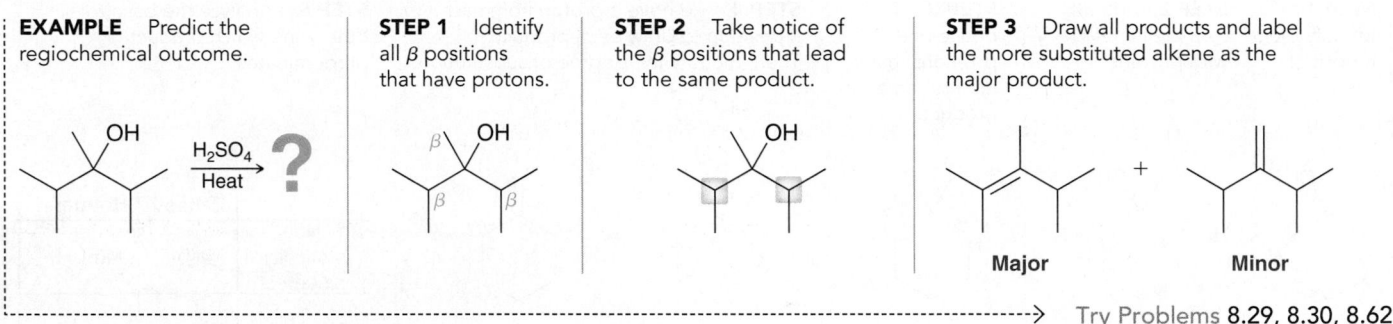

EXAMPLE Predict the regiochemical outcome.

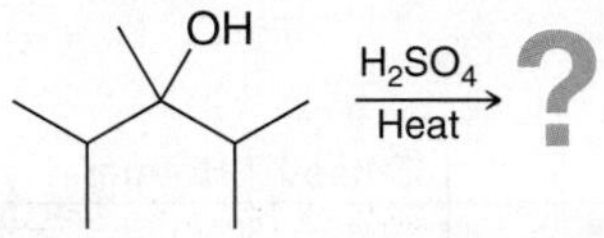

STEP 1 Identify all *β* positions that have protons.

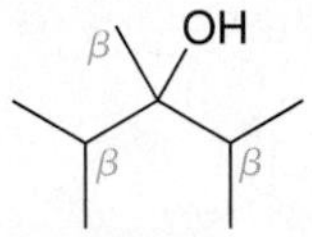

STEP 2 Take notice of the *β* positions that lead to the same product.

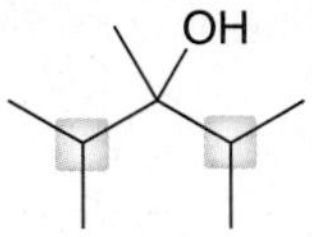

STEP 3 Draw all products and label the more substituted alkene as the major product.

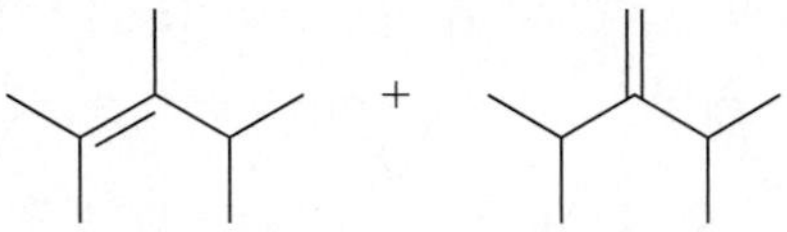

Major **Minor**

Try Problems **8.29, 8.30, 8.62**

8.9 DRAWING THE COMPLETE MECHANISM OF AN E1 REACTION

There are two core steps (gray boxes) and two possible additional steps (pink circles).

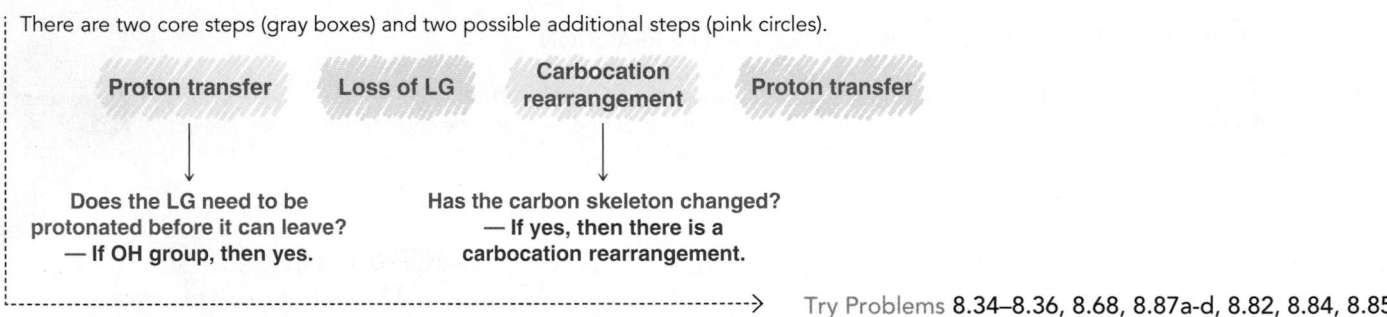

Proton transfer | **Loss of LG** | **Carbocation rearrangement** | **Proton transfer**

Proton transfer ↓ **Does the LG need to be protonated before it can leave? — If OH group, then yes.**

Carbocation rearrangement ↓ **Has the carbon skeleton changed? — If yes, then there is a carbocation rearrangement.**

Try Problems **8.34–8.36, 8.68, 8.87a-d, 8.82, 8.84, 8.85**

8.10 DETERMINING THE FUNCTION OF A REAGENT

Nucleophile (only)

Halides	Sulfur nucleophiles	
$Cl^{\ominus}$	$HS^{\ominus}$	H_2S
$Br^{\ominus}$	$RS^{\ominus}$	RSH
$I^{\ominus}$		

Base (only)

$H{:}^{\ominus}$

DBN

DBU

Strong Nuc / Strong Base

$HO^{\ominus}$

$MeO^{\ominus}$

$EtO^{\ominus}$

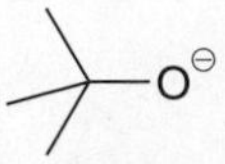

Weak Nuc / Weak Base

H_2O

MeOH

EtOH

Try Problems **8.39, 8.40, 8.55**

8.11 IDENTIFYING THE EXPECTED MECHANISM(S)

EXAMPLE Identify the mechanism(s) expected to occur.

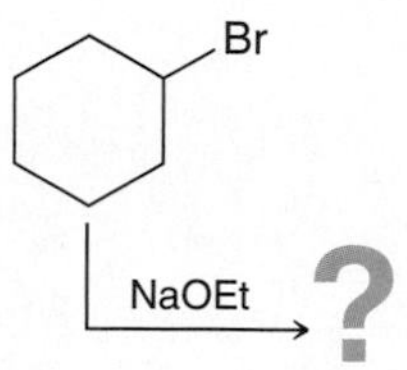

STEP 1 Identify the function of the reagent.

Strong Nuc
Strong base

STEP 2 Identify the substrate, and determine the mechanisms that operate.

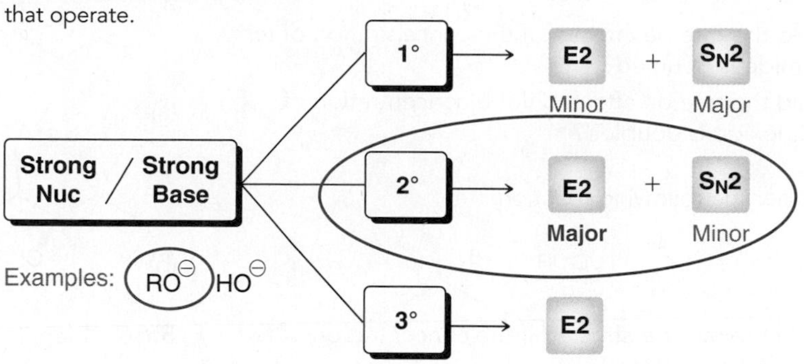

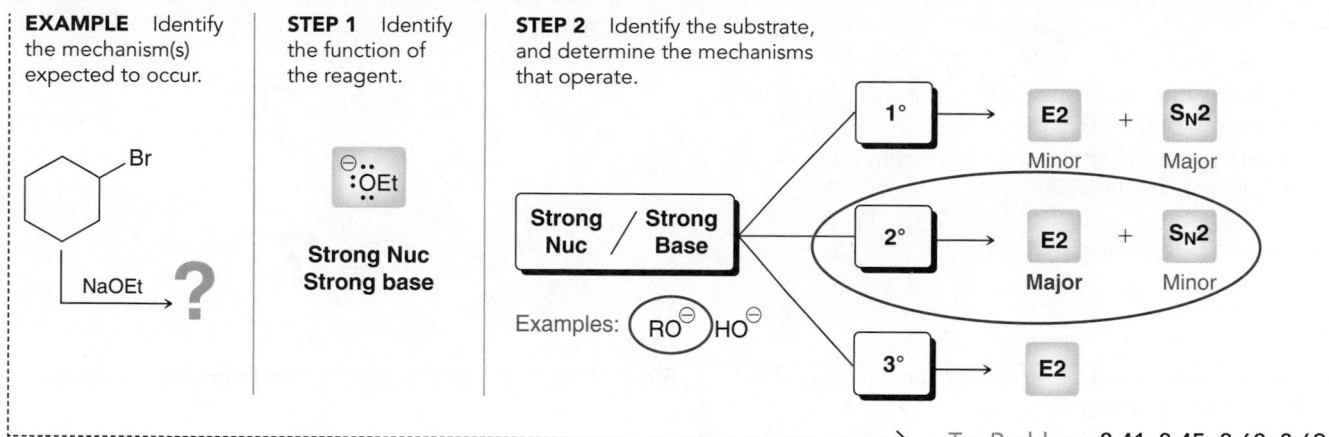

Try Problems **8.41–8.45, 8.68, 8.69, 8.79a**

8.12 PREDICTING THE PRODUCTS OF SUBSTITUTION AND ELIMINATION REACTIONS

STEP 1 Determine the function of the reagent (Section 8.12).

STEP 2 Analyze the substrate and determine the expected mechanism(s) (Section 8.13).

STEP 3 Consider any relevant regiochemical and stereochemical requirements (Section 8.14).

Try Problems **8.46–8.49, 8.78, 8.80**

PRACTICE PROBLEMS

Note: Most of the Problems are available within *WileyPLUS*, an online teaching and learning solution.

8.50 Assign a systematic (IUPAC) name for each of the following compounds:

(a) 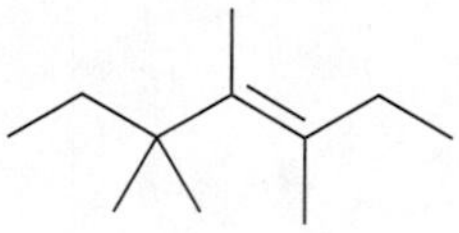 (b) 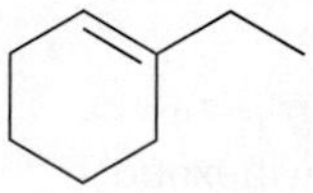 (c) 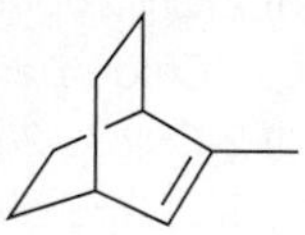

8.51 Using *E-Z* designators, identify the configuration of each C═C double bond in the following compound below:

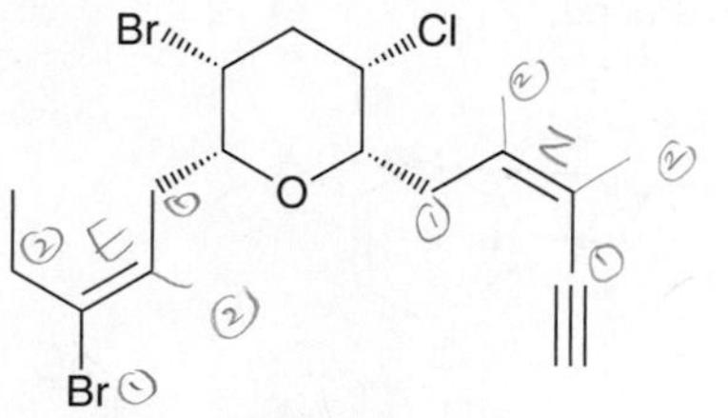

Dactylyne
A natural product
isolated from marine sources

8.52 There are two stereoisomers of 1-*tert*-butyl-4-chlorocyclohexane. One of these isomers reacts with sodium ethoxide in an E2 reaction that is 500 times faster than the reaction of the other isomer. Identify the isomer that reacts faster, and explain the difference in rate for these two isomers.

8.53 Arrange the following alkenes in order of increasing stability:

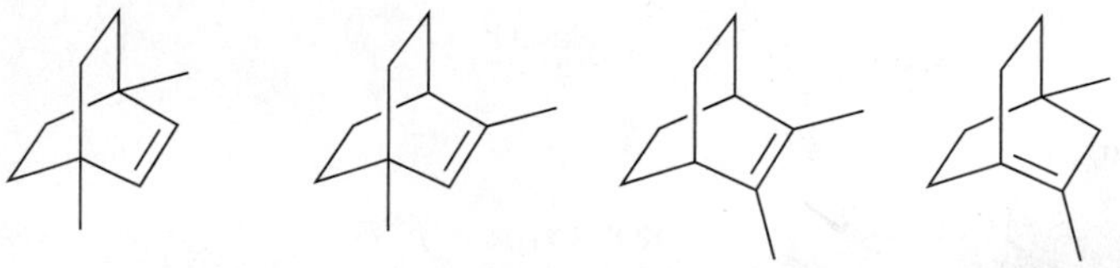

8.54 For each pair of the following compounds identify which compound would react more rapidly in an E1 reaction.

(a) 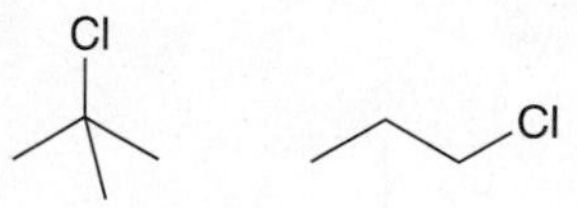

(b) 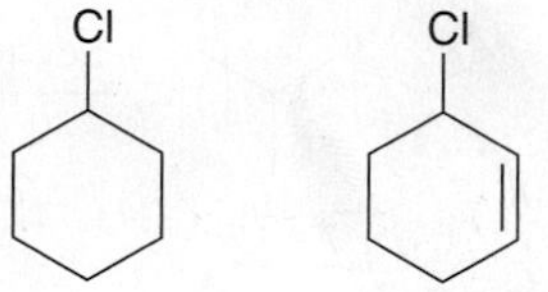

8.55 Identify the stronger base:
(a) NaOH vs. H_2O
(b) Sodium ethoxide vs. ethanol
(c) Ammonia vs. trimethylamine

8.56 (2*S*,3*S*)-2-Bromo-3-phenylbutane undergoes an E2 reaction when treated with a strong base to produce (*E*)-2-phenyl-2-butene. Use Newman projections to explain the stereochemical outcome of this reaction.

8.57 Consider the following reaction:

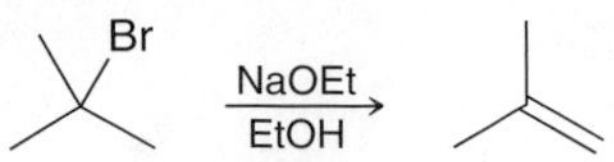

(a) How would the rate be affected if the concentration of *tert*-butyl bromide is doubled? two times faster

(b) How would the rate be affected if the concentration of sodium ethoxide is doubled? two times faster

8.58 Consider the following reaction:

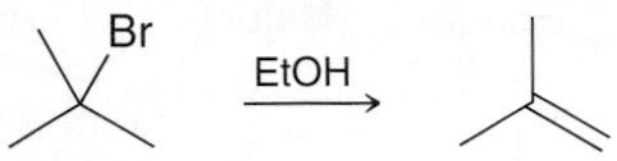

(a) How would the rate be affected if the concentration of *tert*-butyl bromide is doubled?

(b) How would the rate be affected if the concentration of ethanol is doubled?

8.59 When (*R*)-3-bromo-2,3-dimethylpentane is treated with sodium hydroxide, four different alkenes are formed. Draw all four products, and rank them in terms of stability. Which product do you expect to be the major product?

8.60 When 3-bromo-2,4-dimethylpentane is treated with sodium hydroxide, only one alkene is formed. Draw the product and explain why this reaction has only one regiochemical outcome.

8.61 Predict the major product for each of the following E2 reactions:

(a) 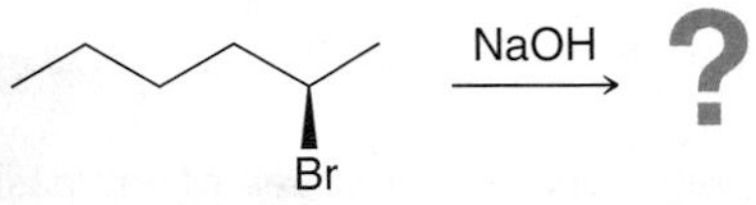

(b) 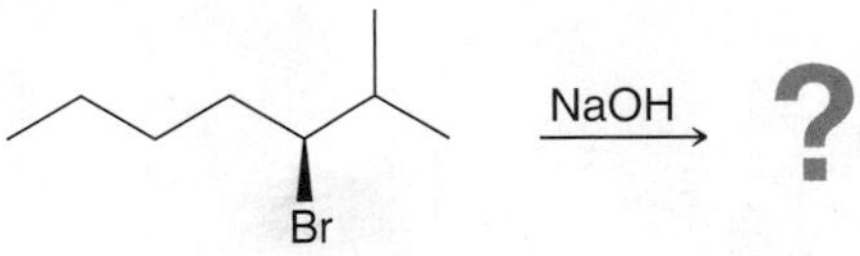

(c) 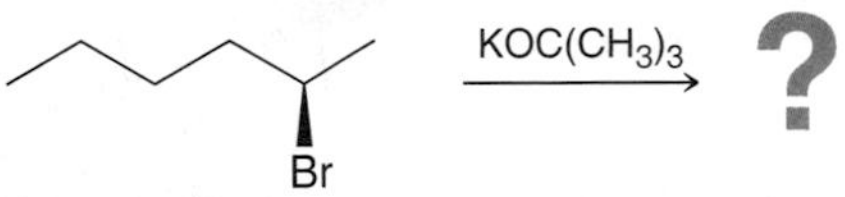

(d) 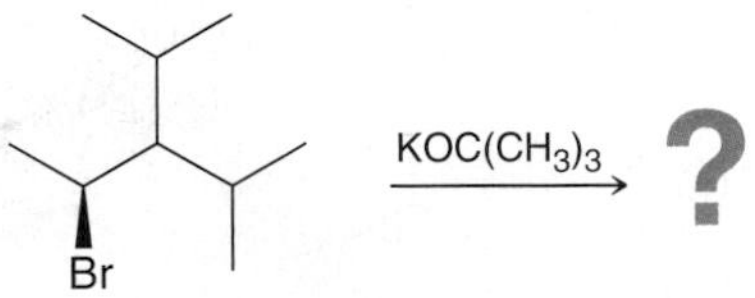

8.62 Predict the major product for each E1 reaction:

(a) 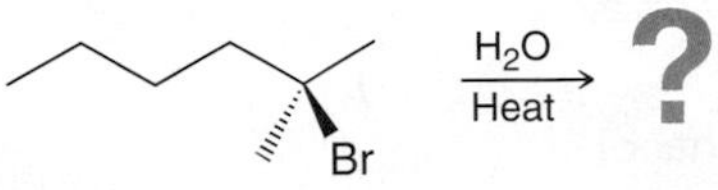

(b) 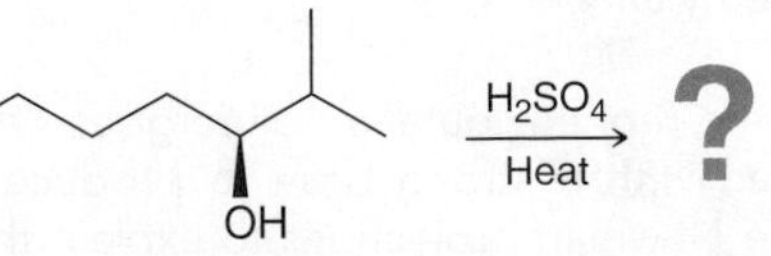

8.63 Predict the stereochemical outcome for each of the following E2 reactions. In each case, draw only the major product of the reaction.

(a) 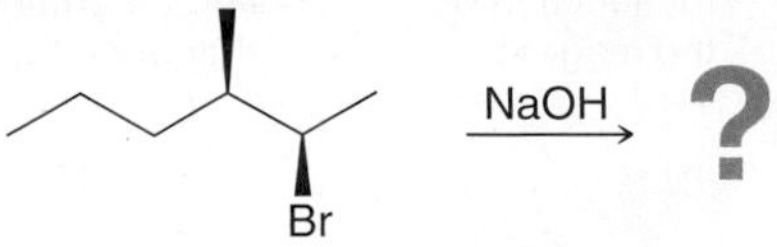

(b) 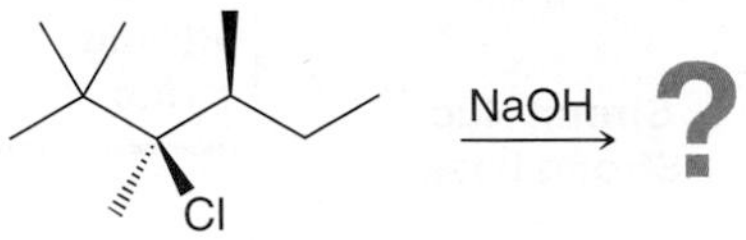

8.64 Identify the sole product of the following reaction:

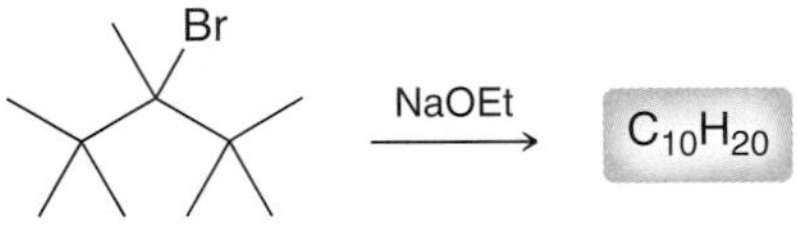

8.65 For each of the following descriptions draw the structure of a compound that fits the description. (*Note*: There are many correct answers for each of these problems.)

(a) An alkyl halide that produces four different alkenes when treated with a strong base

(b) An alkyl halide that produces three different alkenes when treated with a strong base

(c) An alkyl halide that produces two different alkenes when treated with a strong base

(d) An alkyl halide that produces only one alkene when treated with a strong base

8.66 How many different alkenes will be produced when each of the following substrates is treated with a strong base?

(a) 1-Chloropentane

(b) 2-Chloropentane

(c) 3-Chloropentane

(d) 2-Chloro-2-methylpentane

(e) 3-Chloro-3-methylhexane

8.67 In each of the following cases draw the structure of an alkyl halide that will undergo an E2 elimination to yield *only* the indicated alkene.

(a) 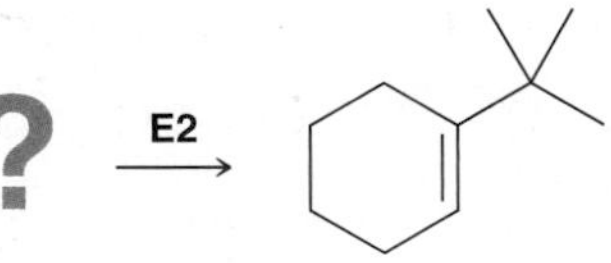

(b) 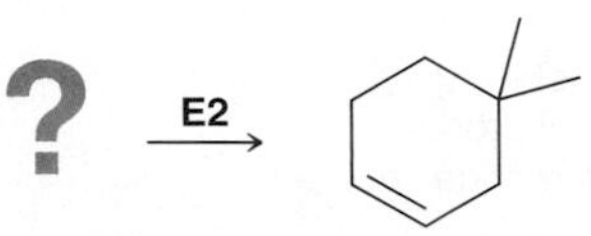

(c) 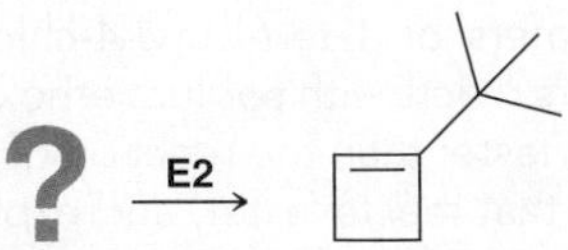

(d) 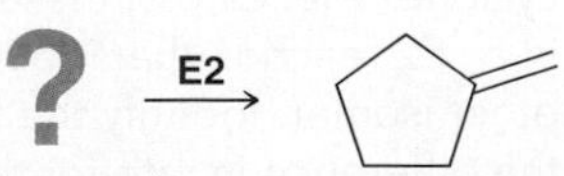

8.68 Consider the following reaction:

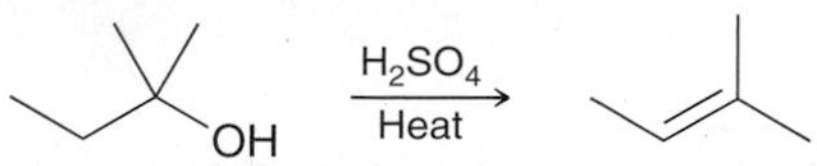

(a) Draw the mechanism of this reaction.
(b) What is the rate equation of this reaction?
(c) Draw an energy diagram of the reaction.

8.69 Draw the carbocation intermediate that would be formed if each of the following substrates participated in a stepwise elimination process (E1). In each case, identify the intermediate carbocation as being primary, secondary, or tertiary. One of the substrates does not undergo an E1 reaction. Identify which one and explain why.

(a) 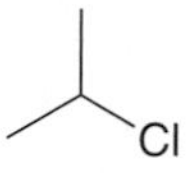

(b) 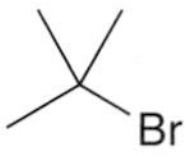

(c) 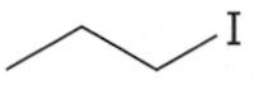

(d) 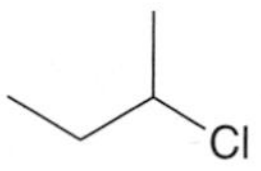

8.70 Draw the transition state for the reaction between *tert*-butyl chloride and sodium hydroxide.

8.71 The following two compounds are both secondary alkyl halides, but they undergo E2 reactions at different rates. The first compound reacts more rapidly than the second compound. Explain.

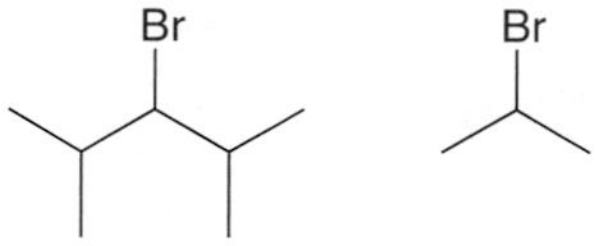

8.72 The following three reactions are similar, differing only in the configuration of the substrate. One of these reactions is very fast, one is very slow, and the other does not occur at all. Identify each reaction and explain your choice.

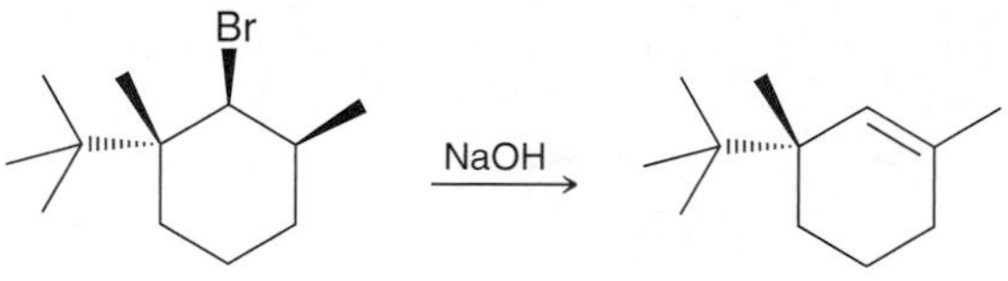

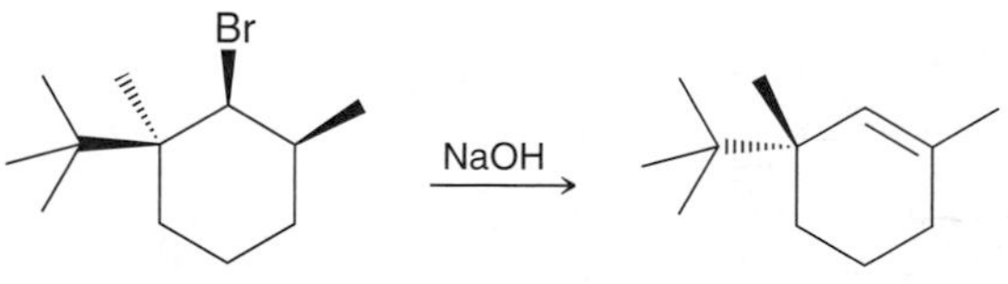

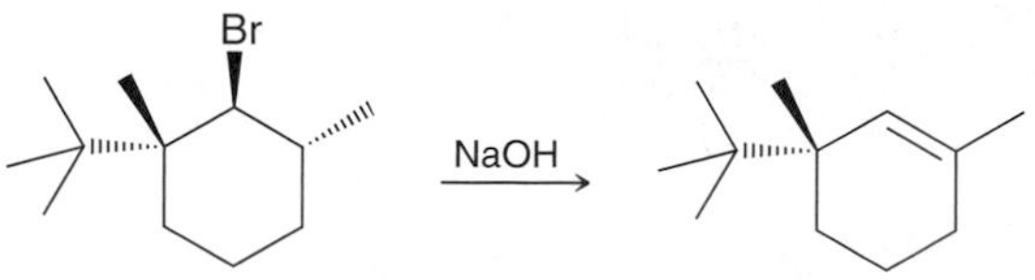

8.73 1-Bromobicyclo[2.2.2]octane does not undergo an E2 reaction when treated with a strong base. Explain why not.

8.74 For each pair of the following compounds identify which compound would react more rapidly in an E2 reaction:

(a) 

(b) 

8.75 Indicate whether you would use sodium ethoxide or potassium *tert*-butoxide to achieve each of the following transformations:

(a) 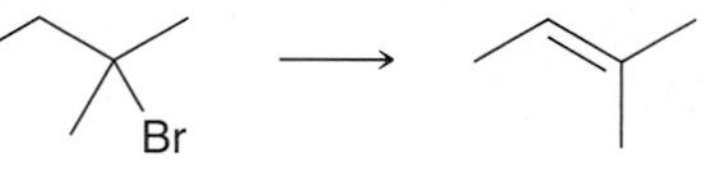

(b) 

(c) 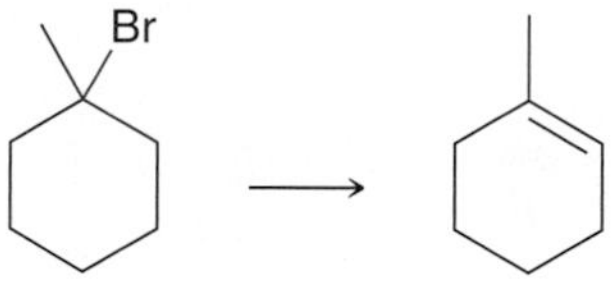

(d) 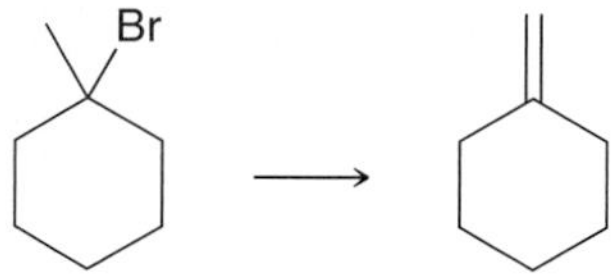

8.76 Explain why the following reaction yields the Hofmann product exclusively (no Zaitsev product at all) even though the base is not sterically hindered:

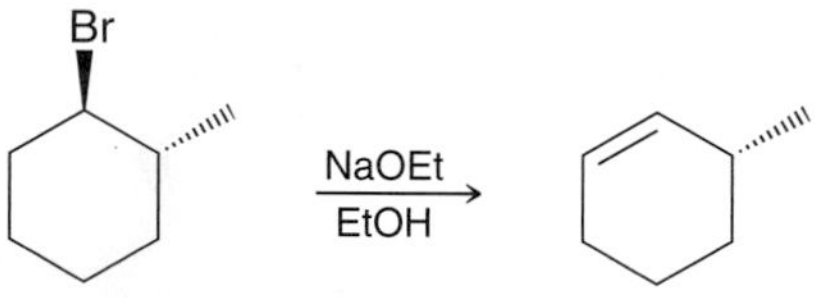

8.77 Propose a mechanism for each of the following transformations:

(a) 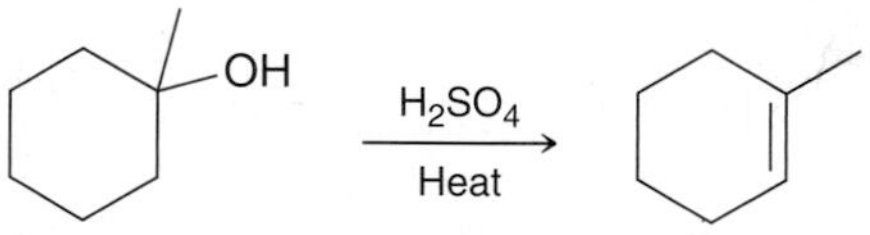

(b) 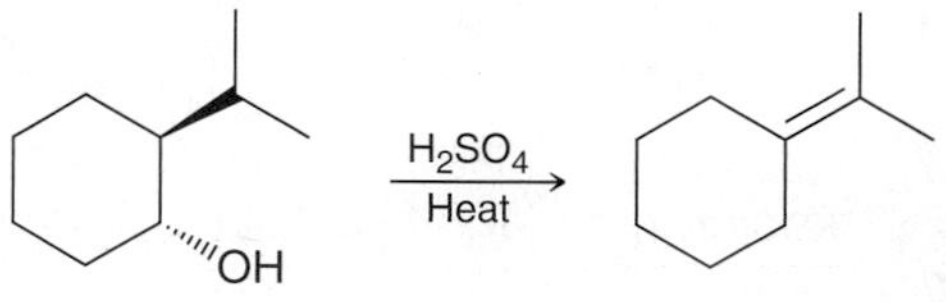

(c) 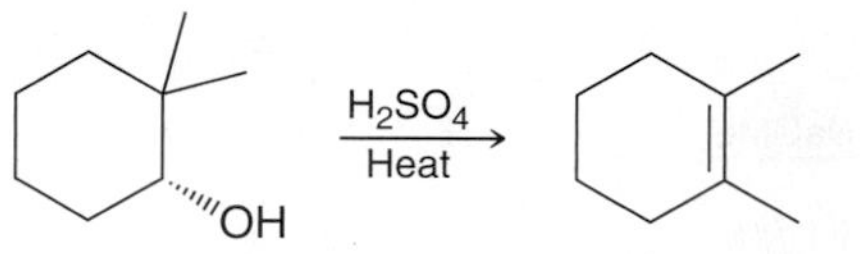

(d) 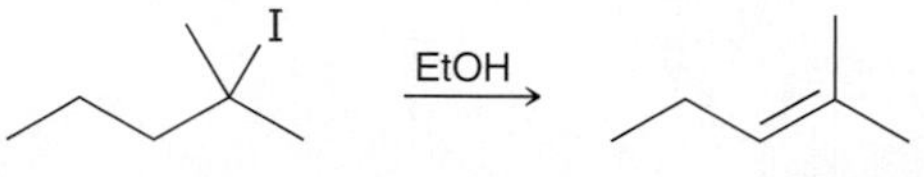

(e) 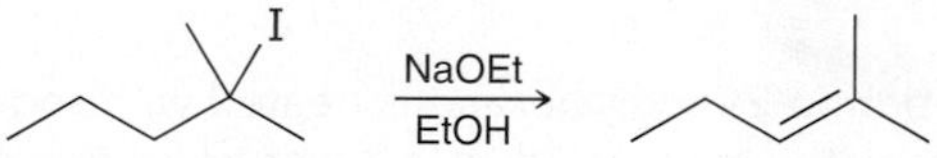

INTEGRATED PROBLEMS

8.78 *Substitution vs. Elimination*: Identify the major and minor product(s) for each of the following reactions:

(a) 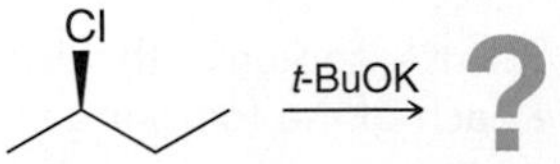

(b) 

(c) 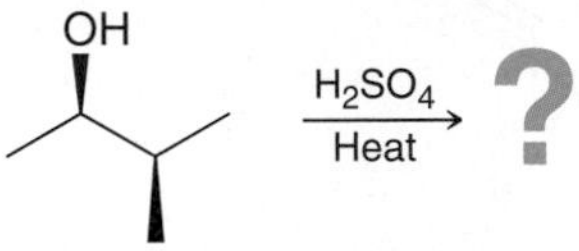

(d) 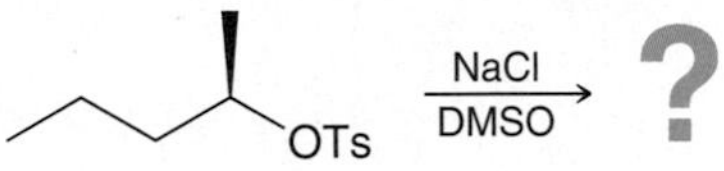

(e) 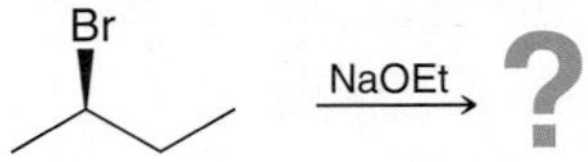

(f) 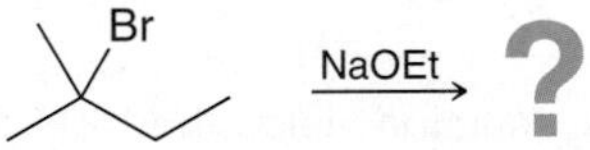

(g) 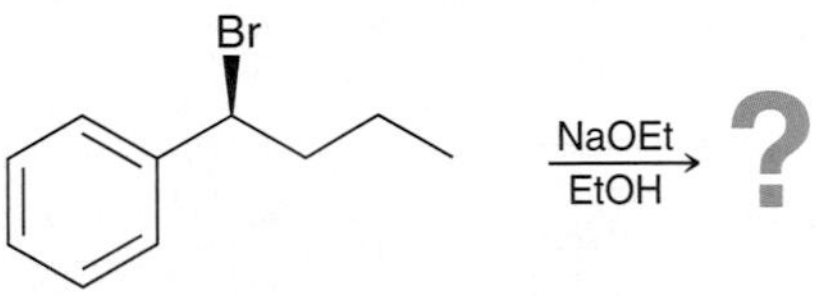

(h) 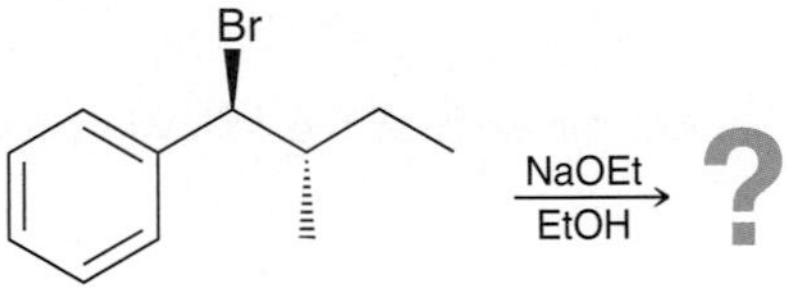

(i) 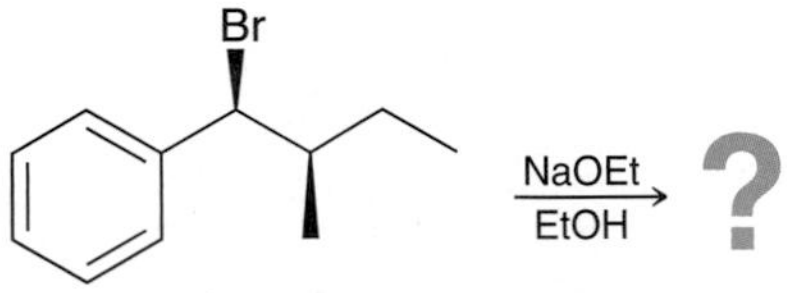

(j) 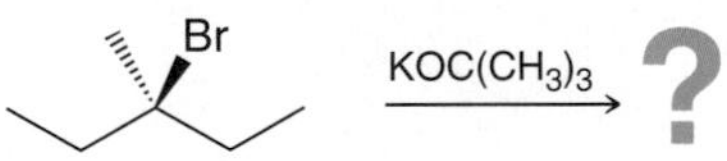

(k) 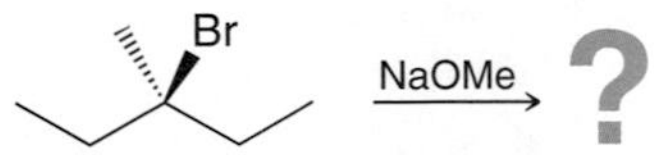

(l) 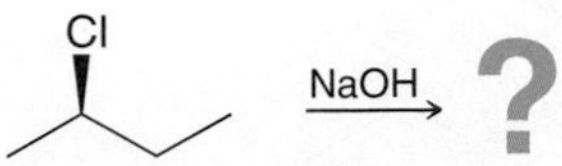

8.79 When 2-bromo-2-methylhexane is treated with sodium ethoxide in ethanol, the major product is 2-methyl-2-hexene.

(a) Draw the mechanism of this reaction.
(b) What is the rate equation of this reaction?
(c) What would happen to the rate if the concentration of base is doubled?
(d) Draw an energy diagram of this reaction.
(e) Draw the transition state of this reaction.

8.80 Predict the major product for each of the following reactions:

(a) 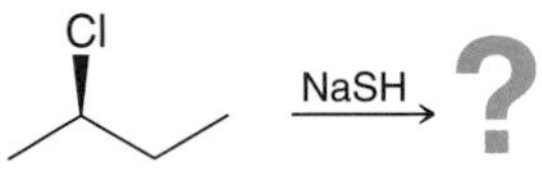

(b) 

(c) 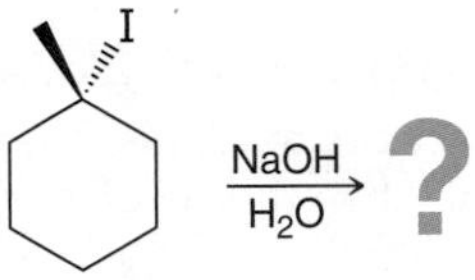

(d) 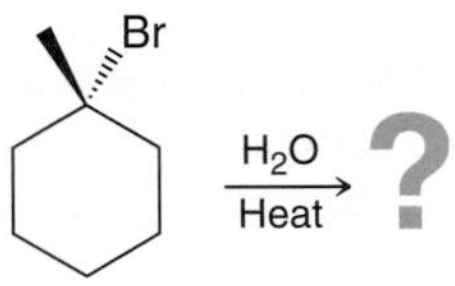

(e) 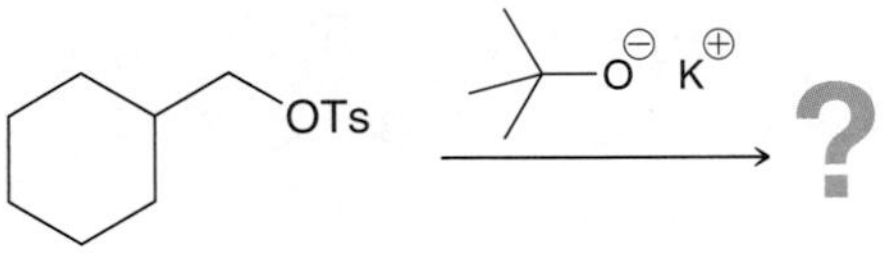

(f) 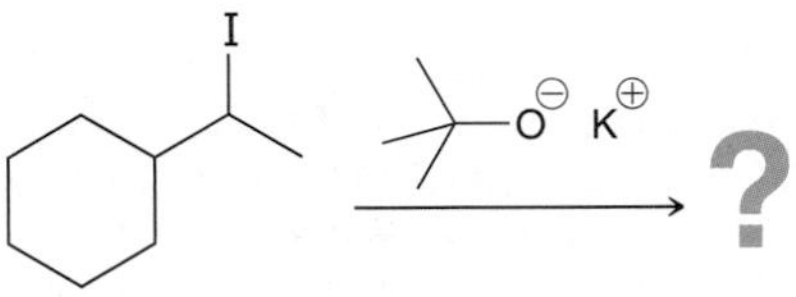

(g) 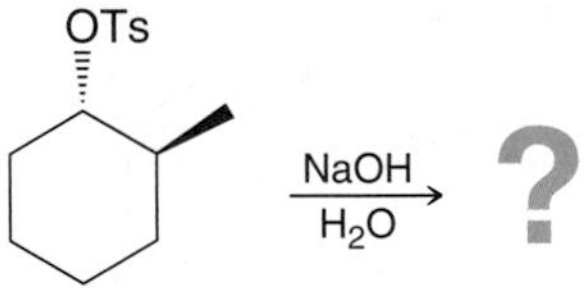

(h) 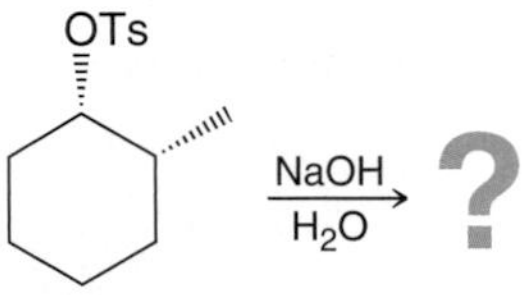

(i) 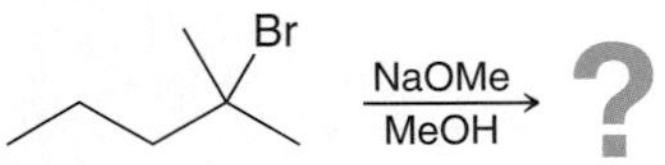

(j) 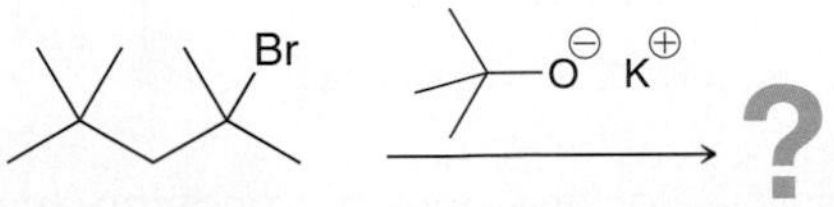

8.81 Draw all constitutional isomers of C_4H_9Br, and then arrange them in order of increasing reactivity toward an E2 reaction.

CHALLENGE PROBLEMS

8.82 Propose a mechanism that explains formation of the following product:

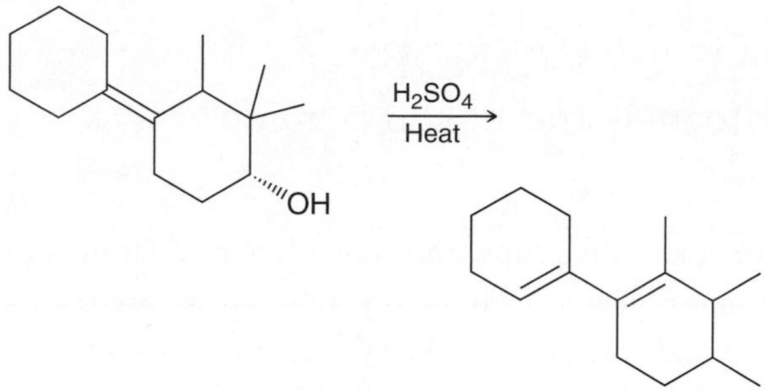

8.83 (*S*)-1-Bromo-1,2-diphenylethane reacts with a strong base to produce *cis*-stilbene and *trans*-stilbene:

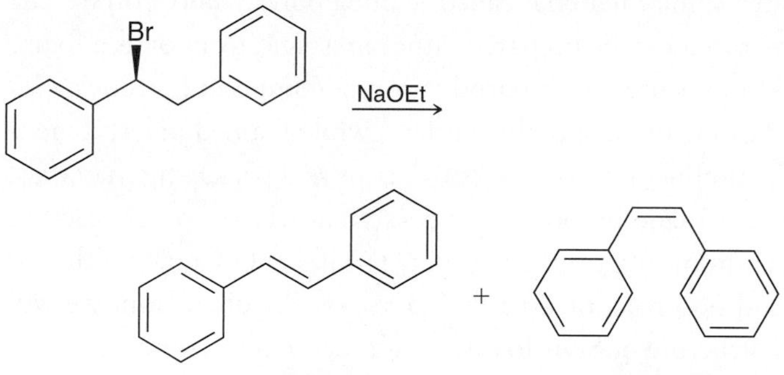

***trans*-Stilbene (major product)** **cis-Stilbene**

(a) This reaction is stereoselective, and the major product is *trans*-stilbene. Explain why the *trans* isomer is the predominant product. To do so, draw the Newman projections that lead to formation of each product and compare their stability.

(b) When (*R*)-1-bromo-1,2-diphenylethane is used as the starting substrate, the stereochemical outcome does not change. That is, *trans*-stilbene is still the major product. Explain.

8.84 Propose a mechanism for the following transformation:

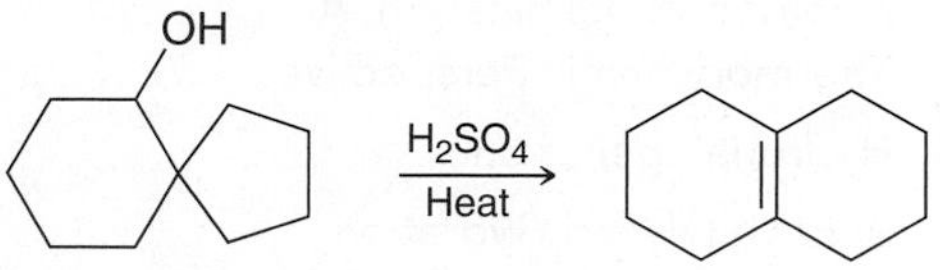

8.85 Propose a mechanism of formation for each of the following products:

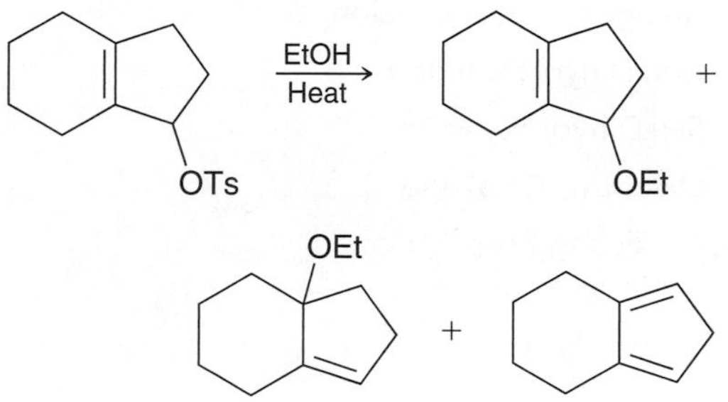

8.86 There are many stereoisomers of 1,2,3,4,5,6-hexachlorocyclohexane. One of those stereoisomers undergoes E2 elimination thousands of times more slowly than the other stereoisomers. Identify which stereoisomer, and explain why it is so slow toward E2.

8.87 Predict which of the following substrates will undergo an E1 reaction more quickly. Explain your choice.

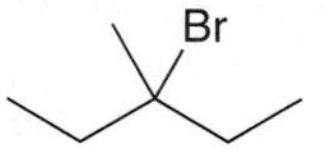

or

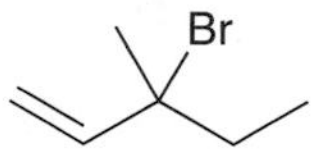

9

- 9.1 Introduction to Addition Reactions
- 9.2 Addition vs. Elimination: A Thermodynamic Perspective
- 9.3 Hydrohalogenation
- 9.4 Acid-Catalyzed Hydration
- 9.5 Oxymercuration-Demercuration
- 9.6 Hydroboration-Oxidation
- 9.7 Catalytic Hydrogenation
- 9.8 Halogenation and Halohydrin Formation
- 9.9 *Anti* Dihydroxylation
- 9.10 *Syn* Dihydroxylation
- 9.11 Oxidative Cleavage
- 9.12 Predicting Products of an Addition Reaction
- 9.13 Synthesis Strategies

Addition Reactions of Alkenes

DID YOU EVER WONDER...
what Styrofoam is and how it is made?

Packing peanuts, coffee cups, and coolers are made from a material that most people erroneously refer to as Styrofoam™. Styrofoam™, a trademark of the Dow Chemical company, is a blue product that is used primarily in the production of building materials for insulating walls, roofs, and foundations. No coffee cups, packing peanuts, or coolers are made from Styrofoam™. Instead, they are made from a similar material called foamed polystyrene. Polystyrene is a polymer that can be prepared either in a rigid form or as a foam. The rigid form is used in the production of computer housings, CD and DVD cases, and disposable cutlery, while foamed polystyrene is used in the production of foam coffee cups and packaging materials. Polystyrene is made by polymerizing styrene via a type of reaction called an addition reaction. This chapter will explore many different types of addition reactions. In the course of our discussion, we will revisit the structure of Styrofoam™ and foamed polystyrene.

DO YOU REMEMBER?

Before you go on, be sure you understand the following topics.
If necessary, review the suggested sections to prepare for this chapter.

- Energy Diagrams (Sections 6.5, 6.6)
- Nucleophiles and Electrophiles (Section 6.7)
- Arrow Pushing and Carbocation Rearrangements (Sections 6.8–6.11)

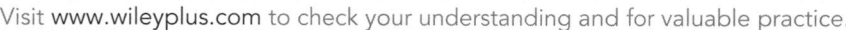

9.1 Introduction to Addition Reactions

In the previous chapter, we learned how to prepare alkenes via elimination reactions. In this chapter, we will explore **addition reactions**, common reactions of alkenes, characterized by the addition of two groups across a double bond. In the process, the pi (π) bond is broken:

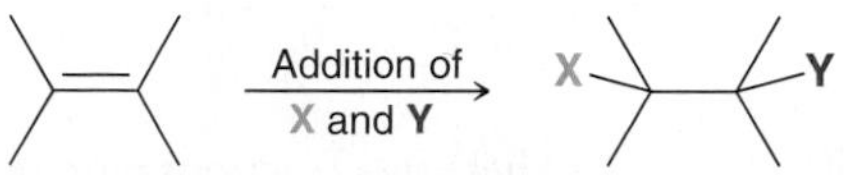

Some addition reactions have special names that indicate the identity of the two groups that were added. Several examples are listed in Table 9.1.

TABLE 9.1 SOME COMMON TYPES OF ADDITION REACTIONS

TYPE OF ADDITION REACTION	NAME	SECTION
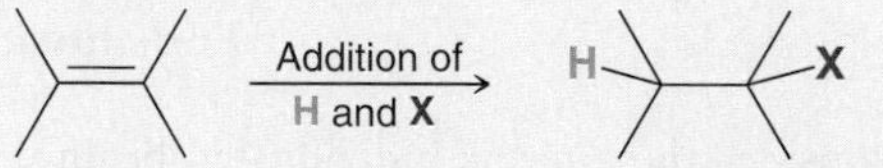	Hydrohalogenation (**X**=Cl, Br, or I)	9.3
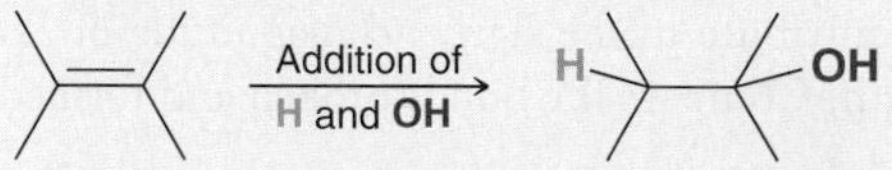	Hydration	9.6
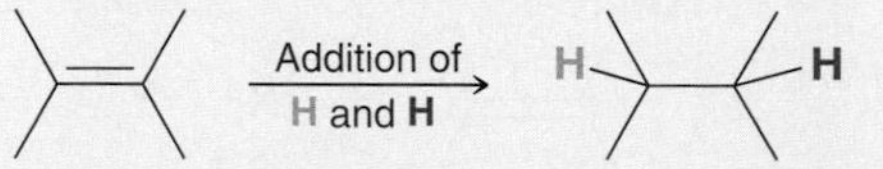	Hydrogenation	9.7
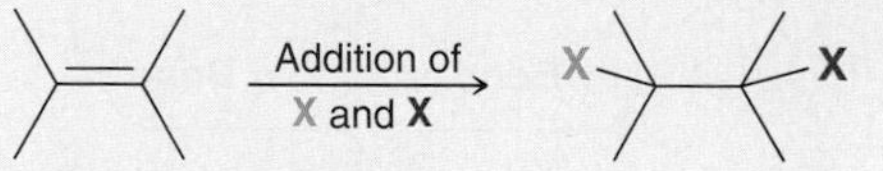	Halogenation (**X**=Cl or Br)	9.8
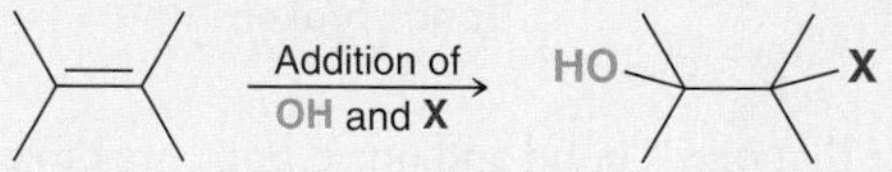	Halohydrin formation (**X**=Cl, Br, or I)	9.8
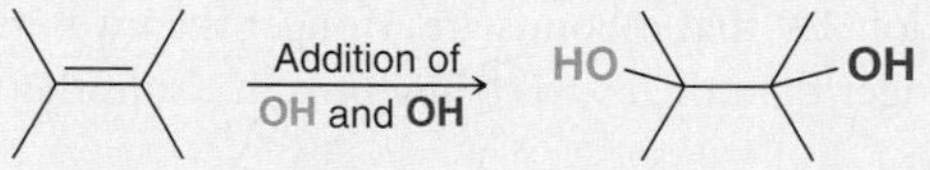	Dihydroxylation	9.9, 9.10

Many different addition reactions are observed for alkenes, enabling them to serve as synthetic precursors for a wide variety of functional groups. The versatility of alkenes can be directly attributed to the reactivity of π bonds, which can function either as weak bases or as weak nucleophiles:

As a base

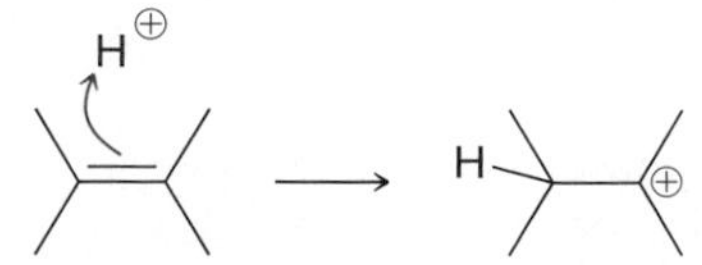

As a nucleophile

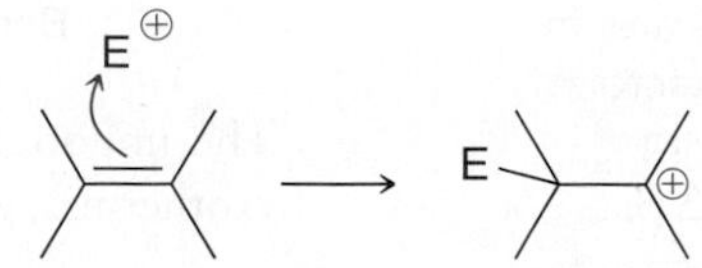

LOOKING BACK
For a review of the subtle differences between basicity and nucleophilicity, see Section 8.12.

The first process illustrates that π bonds can be readily protonated, while the second process illustrates that π bonds can attack electrophilic centers. Both processes will appear many times throughout this chapter.

9.2 Addition vs. Elimination: A Thermodynamic Perspective

In many cases, an addition reaction is simply the reverse of an elimination reaction:

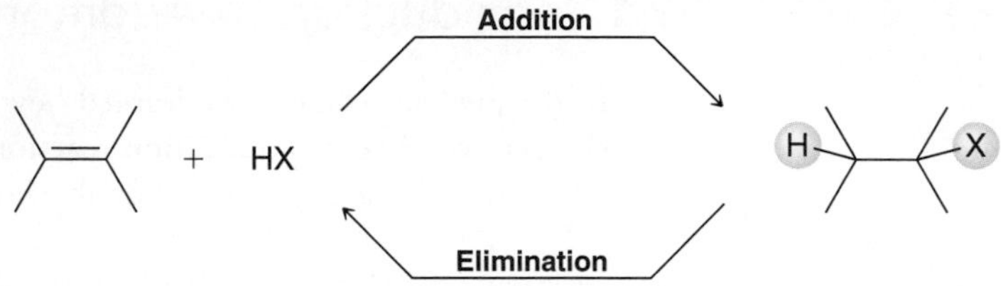

These two reactions represent an equilibrium that is temperature dependent. Addition is favored at low temperature, while elimination is favored at high temperature. To understand the reason for this temperature dependence, recall that the sign of ΔG determines whether the equilibrium favors reactants or products (Section 6.3). ΔG must be negative for the equilibrium to favor products. The sign of ΔG depends on two terms:

$$\Delta G = \underbrace{(\Delta H)}_{\textbf{Enthalpy term}} + \underbrace{(-T\Delta S)}_{\textbf{Entropy term}}$$

Let's consider these terms individually, beginning with the enthalpy term (ΔH). Many factors contribute to the sign and magnitude of ΔH, but the dominant factor is generally bond strength. Compare the bonds broken and formed in an addition reaction:

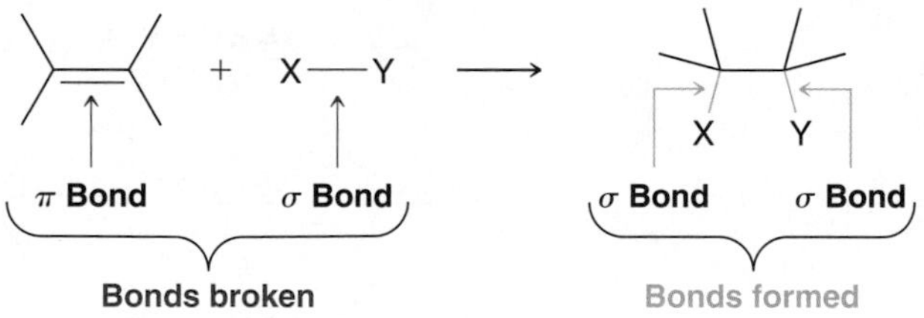

Notice that one π bond and one σ bond are broken, while two σ bonds are formed. Recall from Section 1.9 that σ bonds are stronger than π bonds, and therefore, the bonds being formed are stronger than the bonds being broken. Consider the following example:

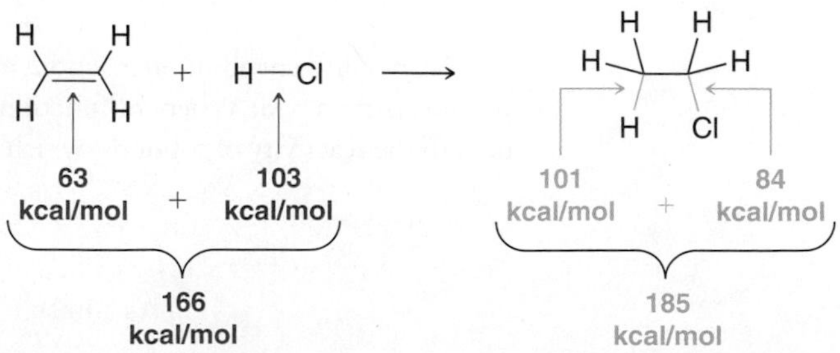

Bonds broken – bonds formed = 166 kcal/mol – 185 kcal/mol = –19 kcal/mol

BY THE WAY
The actual ΔH for this reaction, in the gas phase, has been measured to be –17 kcal/mol, confirming that bond strengths are, in fact, the dominant factor in determining the sign and magnitude of ΔH.

The important feature here is that ΔH has a negative value. In other words, this reaction is exothermic, which is generally the case for addition reactions.

Now let's consider the entropy term $-T\,\Delta S$. This term will always be positive for an addition reaction. Why? In an addition reaction, two molecules are joining together to produce one molecule of product. As described in Section 6.2, this situation represents a decrease in entropy, and ΔS will have a negative value. The temperature component, T (measured in Kelvin), is always positive. As a result, $-T\,\Delta S$ will be positive for addition reactions.

Now let's combine the enthalpy and entropy terms. The enthalpy term is negative and the entropy term is positive, so the sign of ΔG for an addition reaction will be determined by the competition between these two terms:

$$\Delta G = \underbrace{(\Delta H)}_{\substack{\textbf{Enthalpy term} \\ \ominus}} + \underbrace{(-T\,\Delta S)}_{\substack{\textbf{Entropy term} \\ \oplus}}$$

In order for ΔG to be negative, the enthalpy term must be larger than the entropy term. This competition between the enthalpy term and the entropy term is temperature dependent. At low temperature, the entropy term is small, and the enthalpy term dominates. As a result, ΔG will be negative, which means that products are favored over reactants (the equilibrium constant K will be greater than 1). In other words, addition reactions are thermodynamically favorable at low temperature.

However, at high temperature, the entropy term will be large and will dominate the enthalpy term. As a result, ΔG will be positive, which means that reactants will be favored over products (the equilibrium constant K will be less than 1). In other words, the reverse process (elimination) will be thermodynamically favored at high temperature:

BY THE WAY
Not all addition reactions are reversible at high temperature, because in many cases high temperature will cause the reactants and/or products to undergo thermal degradation.

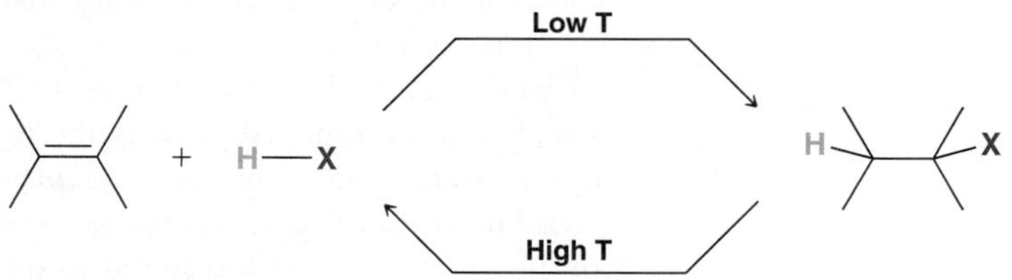

For this reason, the addition reactions discussed in this chapter are generally performed below room temperature.

9.3 Hydrohalogenation

Regioselectivity of Hydrohalogenation

The treatment of alkenes with HX (where X = Cl, Br, or I) results in an addition reaction called **hydrohalogenation**, in which H and X are added across the π bond:

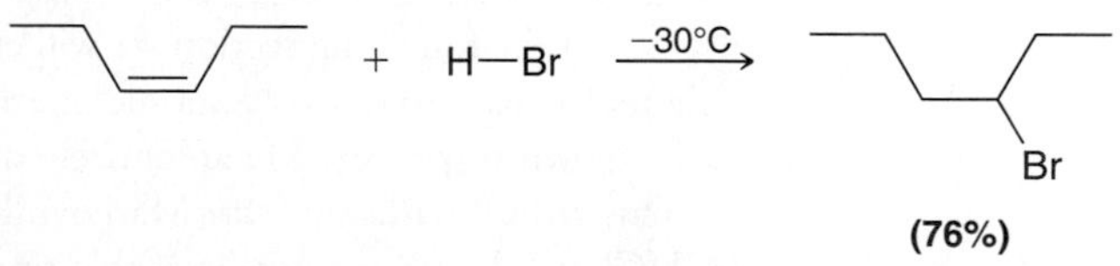

In this example the alkene is symmetrical. However, in cases where the alkene is unsymmetrical, the ultimate placement of H and X must be considered. In the following example, there are two possible vinylic positions where X can be placed:

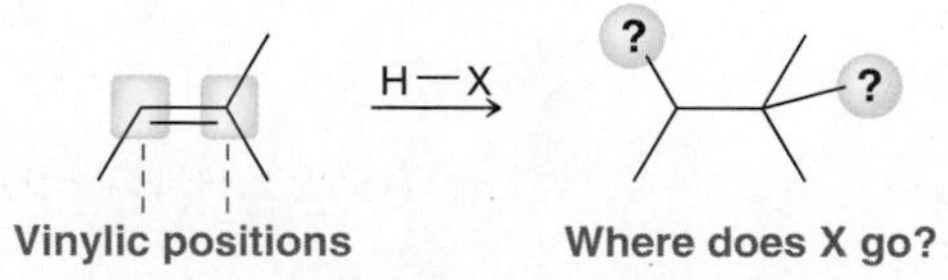

This is an issue of *regiochemistry* that was investigated over a century ago.

In 1869, Vladimir Markovnikov, a Russian chemist, investigated the addition of HBr across many different alkenes, and he noticed that the *H is generally placed at the vinylic position already bearing the larger number of hydrogen atoms*. For example:

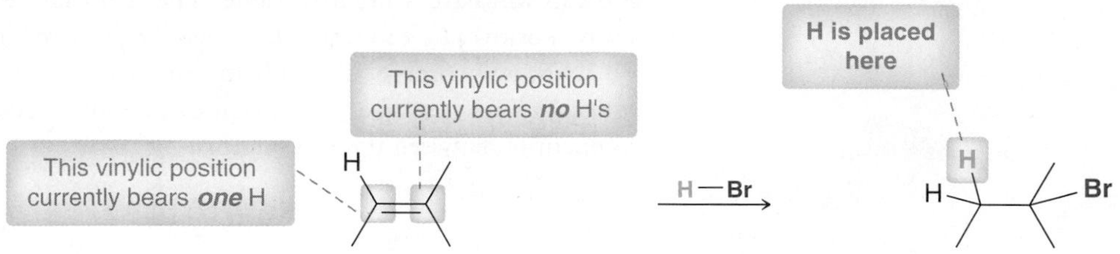

Markovnikov described this regiochemical preference in terms of where the hydrogen is ultimately positioned. However, Markovnikov's observation can alternatively be described in terms of where the halogen (X) is ultimately positioned. Specifically, the *halogen is generally placed at the more substituted position*:

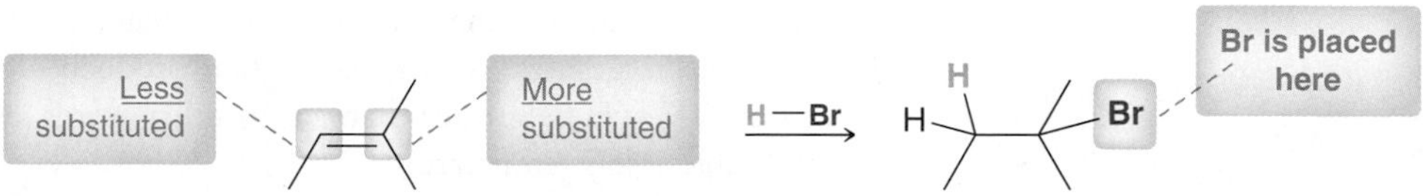

The vinylic position bearing more alkyl groups is more substituted, and that is where the Br is ultimately placed. This regiochemical preference, called **Markovnikov addition,** is also observed for addition reactions involving HCl and HI. Reactions that proceed with a regiochemical preference are said to be **regioselective**.

Interestingly, attempts to repeat Markovnikov's observations occasionally met with failure. In many cases involving the addition of HBr, the observed regioselectivity was, in fact, the opposite of what was expected—that is, the bromine atom would be installed at the less substituted carbon, which came to be known as ***anti*-Markovnikov addition.** These curious observations fueled much speculation over the underlying cause, with some researchers even suggesting that the phases of the moon had an impact on the course of the reaction. Over time, it was realized that purity of reagents was the critical feature. Specifically, Markovnikov addition was observed whenever purified reagents were used, while the use of impure reagents sometimes led to *anti*-Markovnikov addition. Further investigation revealed the identity of the impurity that most greatly affected the regioselectivity of the reaction. It was found that peroxides (ROOR), even in trace amounts, would cause HBr to add across an alkene in an *anti*-Markovnikov fashion:

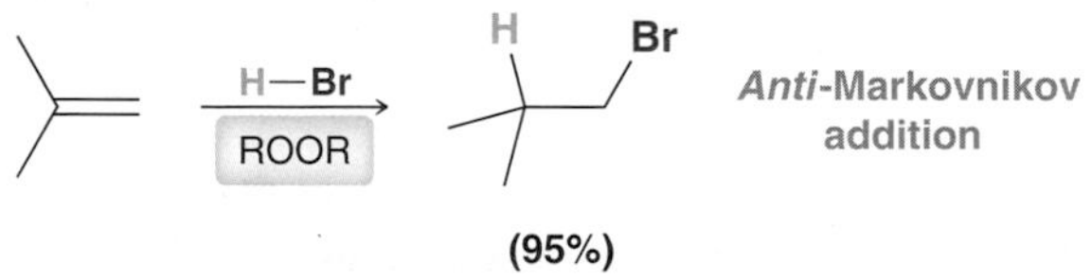

In the following section we will explore Markovnikov addition in more detail and propose a mechanism that involves ionic intermediates. In contrast, *anti*-Markovnikov addition of HBr is known to proceed via an entirely different mechanism, one involving radical intermediates. This radical pathway (*anti*-Markovnikov addition) is efficient for the addition of HBr but not HCl or HI, and the details of that process will be discussed in Section 11.10. For now, we will simply note that the regiochemical outcome of HBr addition can be controlled by choosing whether or not to use peroxides:

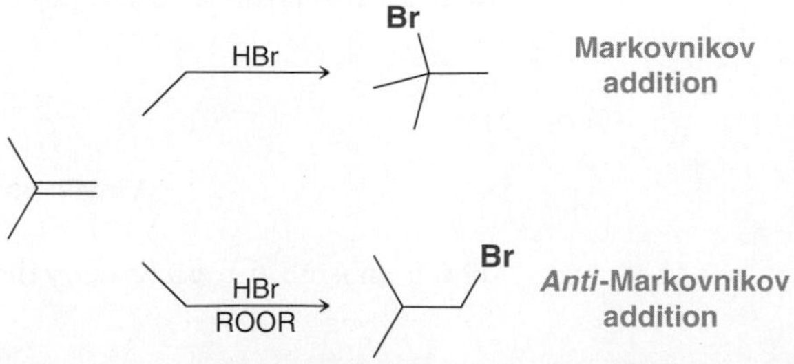

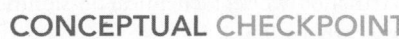

CONCEPTUAL CHECKPOINT

9.1 Draw the expected major product for each of the following reactions:

(a) [structure] $\xrightarrow{\text{HBr}}$ **?**

(b) [structure] $\xrightarrow[\text{ROOR}]{\text{HBr}}$ **?**

(c) [structure] $\xrightarrow{\text{HBr}}$ **?**

(d) [structure] $\xrightarrow{\text{HCl}}$ **?**

(e) [structure] $\xrightarrow{\text{HI}}$ **?**

(f) [structure] $\xrightarrow[\text{ROOR}]{\text{HBr}}$ **?**

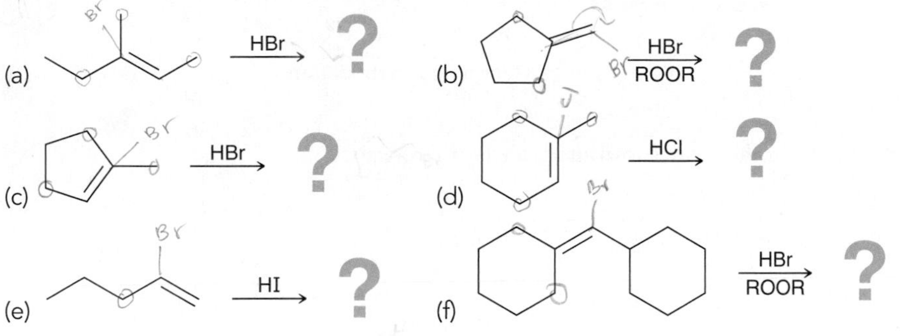

9.2 Identify the reagents that you would use to achieve each of the following transformations:

(a) [structure] $\xrightarrow{\text{HBr}}$ [structure with Br]

(b) [structure] $\xrightarrow[\text{ROOR}]{\text{HBr}}$ [structure with Br]

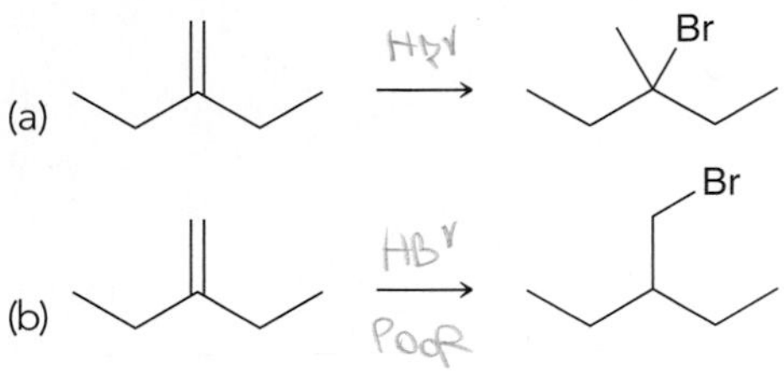

A Mechanism for Hydrohalogenation

Mechanism 9.1 accounts for the Markovnikov addition of HX to alkenes.

MECHANISM 9.1 HYDROHALOGENATION

Proton transfer

The alkene is protonated, forming a carbocation intermediate and a bromide ion

Nucleophilic attack

The bromide ion functions as a nucleophile and attacks the carbocation intermediate

In the first step, the π bond of the alkene is protonated, generating a carbocation intermediate. In the second step, this intermediate is attacked by a bromide ion. Figure 9.1 shows an energy diagram for this two-step process. The observed regioselectivity for this process can be attributed to the first step of the mechanism (proton transfer), which is the rate-determining step because it exhibits a higher transition state energy than the second step of the mechanism.

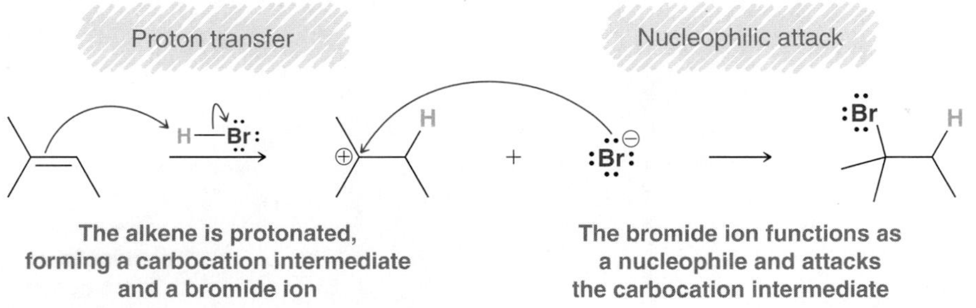

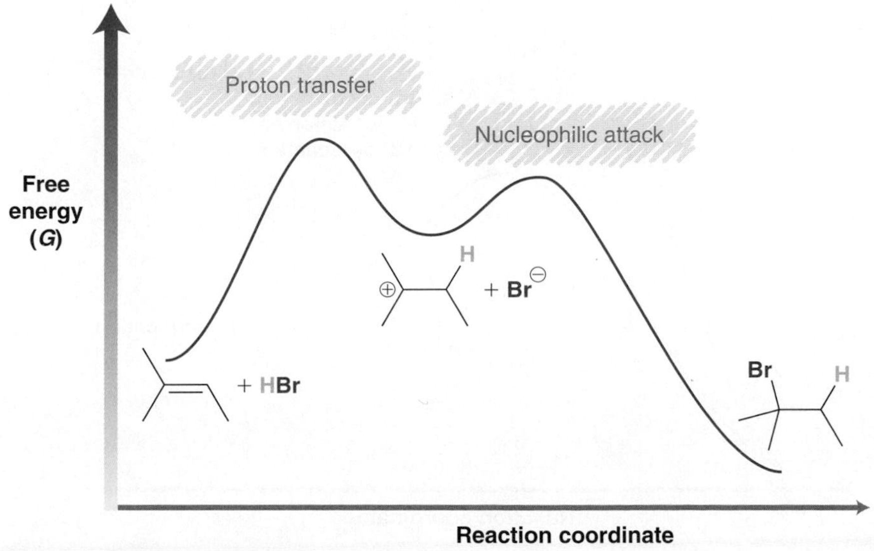

FIGURE 9.1
An energy diagram for the two steps involved in the addition of HBr across an alkene.

In theory, protonation can occur with either of two regiochemical possibilities. It can either occur to form the less substituted, secondary carbocation,

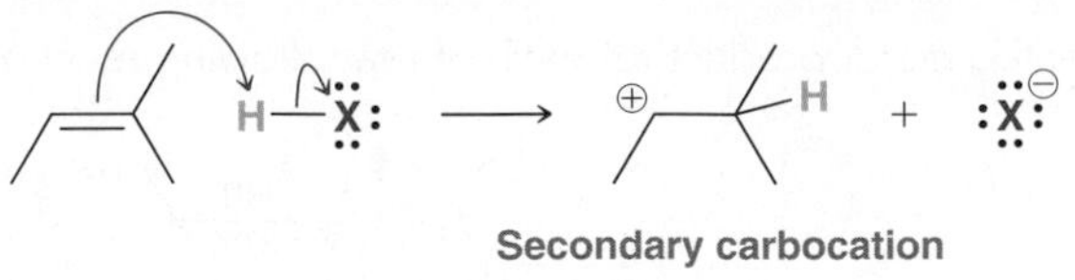

or it can occur to form the more substituted, tertiary carbocation,

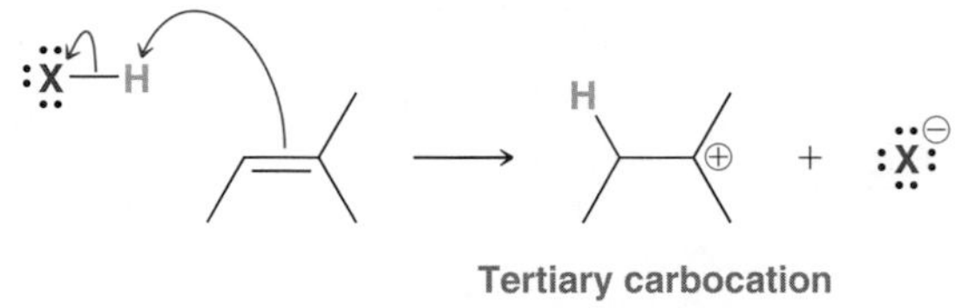

Recall that tertiary carbocations are more stable than secondary carbocations, due to hyperconjugation (Section 6.11). Figure 9.2 shows an energy diagram comparing the two possible pathways for the first step of HX addition. The blue curve represents HX addition proceeding via a secondary carbocation, while the red curve represents HX addition proceeding via a tertiary carbocation. Look closely at the transition states through which the carbocation intermediates are formed. Recall that the Hammond postulate (Section 6.6) suggests that each of these transition states has significant carbocationic character. Therefore, the transition state

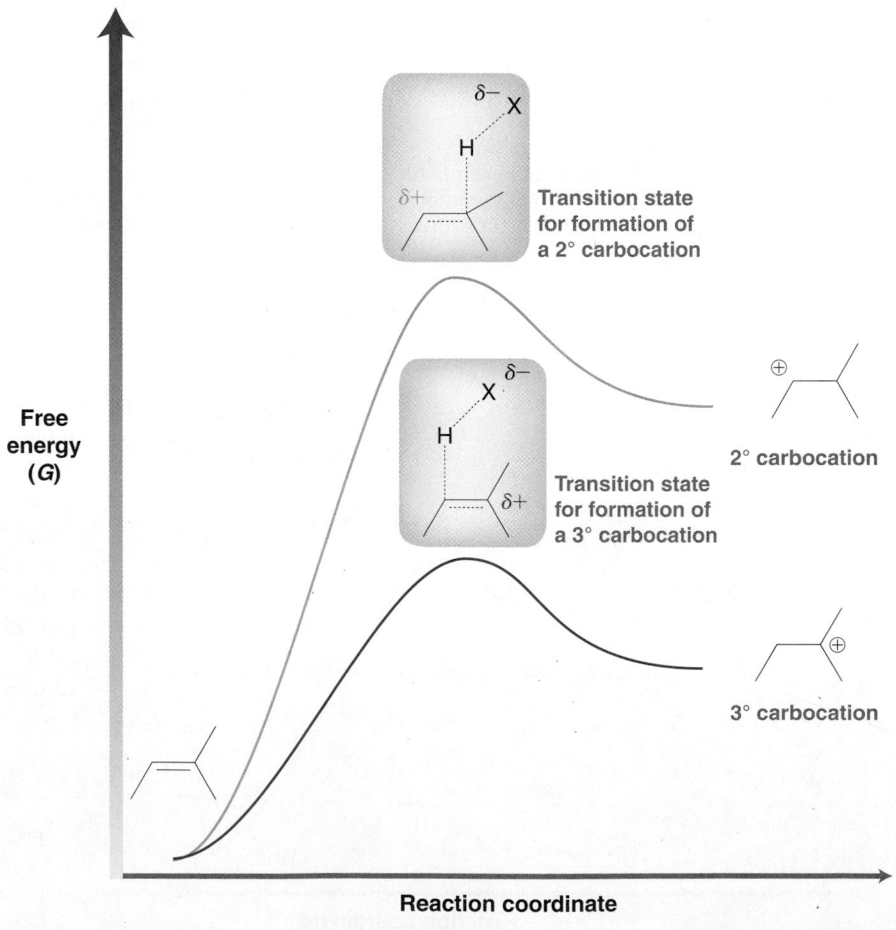

FIGURE 9.2 An energy diagram illustrating the two possible pathways for the first step of hydrohalogenation. One pathway generates a secondary carbocation, while the other pathway generates a tertiary carbocation.

for formation of the tertiary carbocation will be significantly lower in energy than the transition state for formation of the secondary carbocation. The energy barrier for formation of the tertiary carbocation will be smaller than the energy barrier for formation of the secondary carbocation, and as a result, the reaction will proceed more rapidly via the more stable carbocation intermediate. The proposed mechanism therefore provides a theoretical explanation (based on first principles) for the regioselectivity that Markovnikov observed. Specifically, the *regioselectivity of an ionic addition reaction is determined by the preference for the reaction to proceed through the more stable carbocation intermediate.*

SKILLBUILDER

9.1 DRAWING A MECHANISM FOR HYDROHALOGENATION

LEARN the skill

Draw a mechanism for the following transformation:

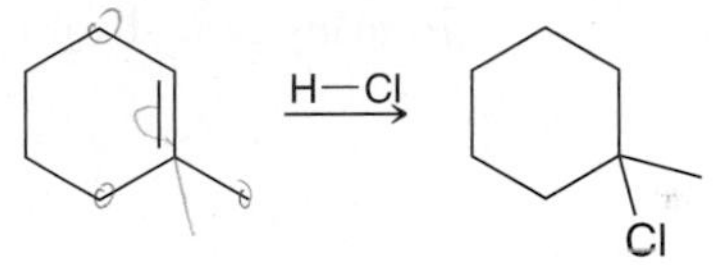

SOLUTION

STEP 1
Using two curved arrows, protonate the alkene to form the more stable carbocation.

WATCH OUT
Take special notice of the arrow with its head placed on the proton; it is a common mistake to draw this arrow in the wrong direction. Remember that curved arrows show the flow of electrons, not atoms.

In this reaction, a hydrogen atom and a halogen are added across an alkene. The halogen (Cl) is ultimately positioned at the more substituted carbon, which verifies that this reaction takes place via an ionic mechanism (Markovnikov addition). The ionic mechanism for hydrohalogenation has two steps: (1) *protonation* of the alkene to form the more stable carbocation and (2) *nucleophilic attack*. Each step must be drawn precisely.

When drawing the first step of the mechanism (protonation), make sure to use two curved arrows:

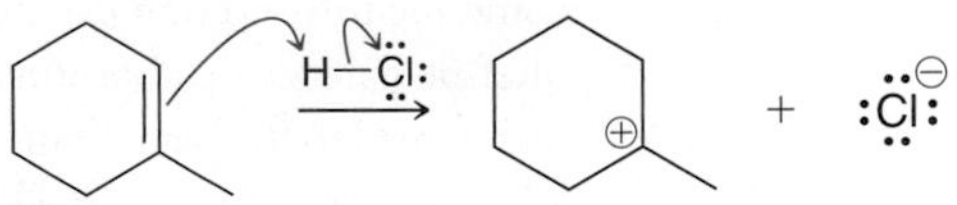

One curved arrow is drawn with its tail on the π bond and its head on the proton. The second curved arrow is drawn with its tail on the H—Cl bond and its head on the Cl.

When drawing the second step of the mechanism (nucleophilic attack of chloride), only one curved arrow is needed. The chloride ion, formed in the previous step, functions as a nucleophile and attacks the carbocation:

STEP 2
Using one curved arrow, draw the halide ion attacking the carbocation.

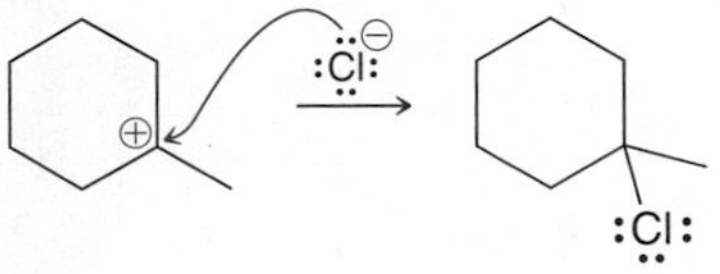

PRACTICE the skill **9.3** Draw the mechanism for each of the following transformations:

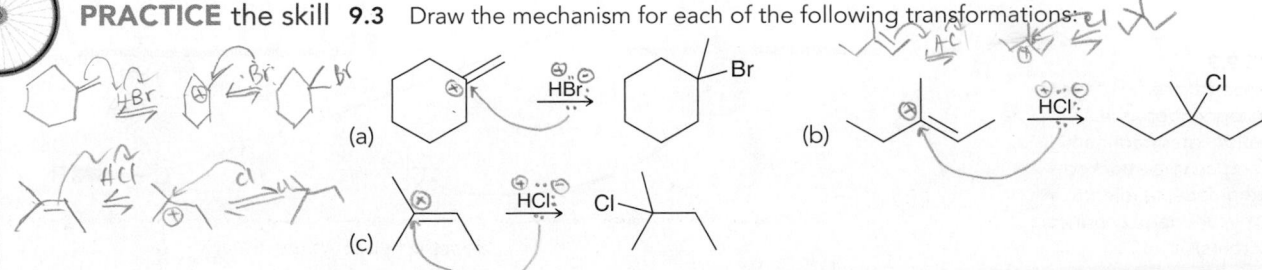

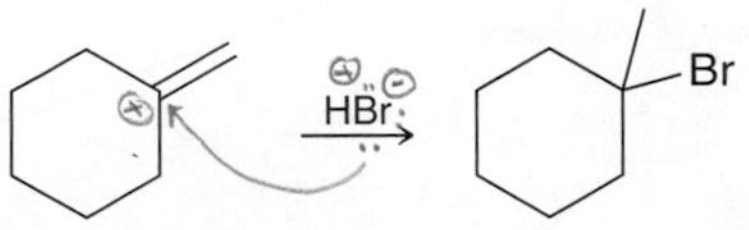

(a)

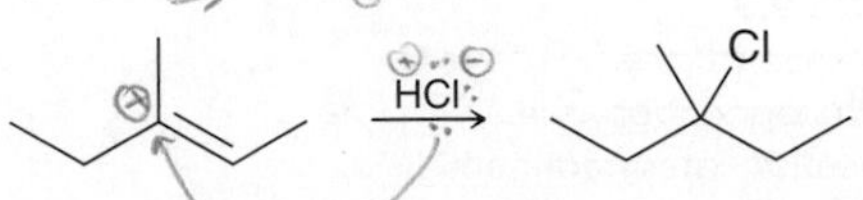

(b)

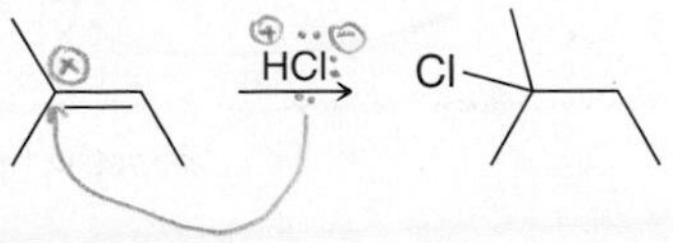

(c)

APPLY the skill

9.4 Draw the intermediate carbocation that is formed when each of the following compounds is treated with HBr:

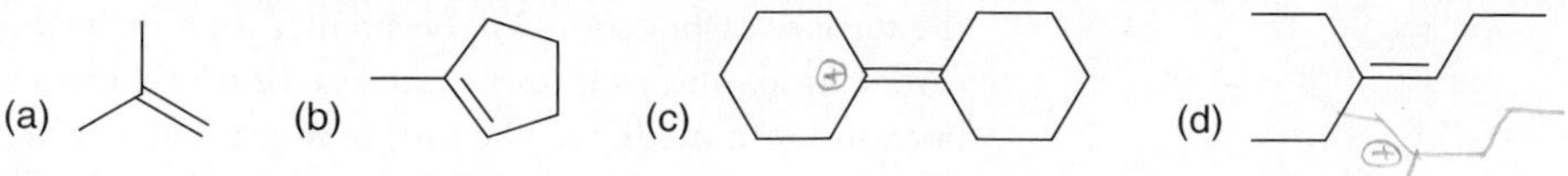

9.5 When 1-methoxy-2-methylpropene is treated with HCl, the major product is 1-chloro-1-methoxy-2-methylpropane. Although this reaction proceeds via an ionic mechanism, the Cl is ultimately positioned at the less substituted carbon. Draw a mechanism that is consistent with this outcome, and then explain why the less substituted carbocation intermediate is more stable in this case.

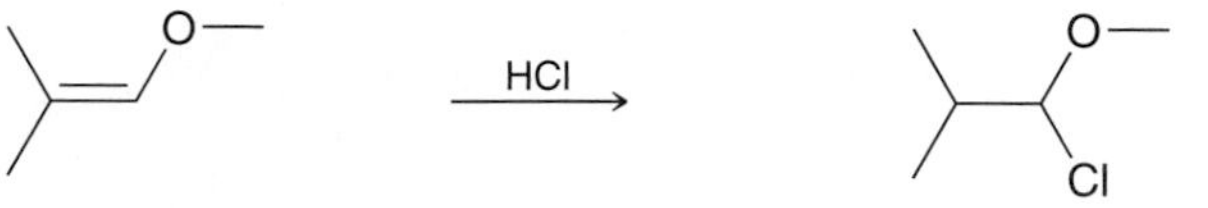

1-Methoxy-2-methylpropene **1-Chloro-1-methoxy-2-methylpropane**

need more **PRACTICE?** **Try Problems 9.51c, 9.64b, 9.69, 9.72**

Stereochemistry of Hydrohalogenation

In many cases, hydrohalogenation involves the formation of a chirality center; for example:

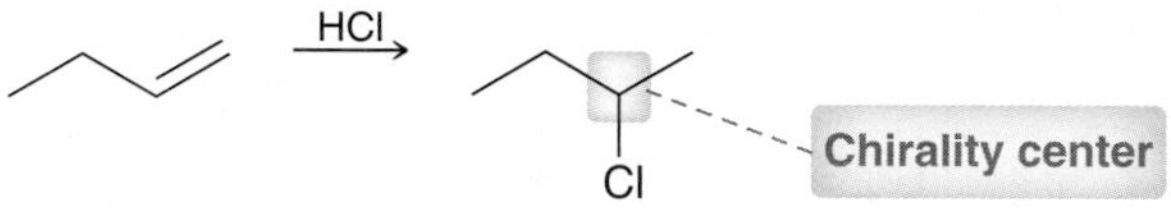

In this reaction, one new chirality center is formed. Therefore, two possible products are expected, representing a pair of enantiomers:

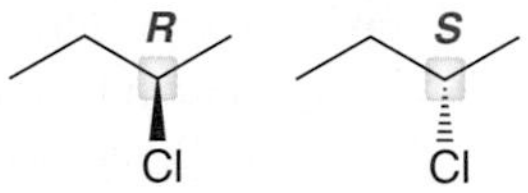

The two enantiomers are produced in equal amounts (a racemic mixture). The stereochemical outcome for this reaction can also be explained by the proposed mechanism, which identifies the key intermediate as a carbocation. Recall that a carbocation is trigonal planar, with an empty *p* orbital orthogonal to the plane. This empty *p* orbital is subject to attack by a nucleophile from either side (the lobe above the plane or the lobe below the plane), as seen in Figure 9.3. Both faces of the plane can be attacked with equal likelihood, and therefore, both enantiomers are produced in equal amounts.

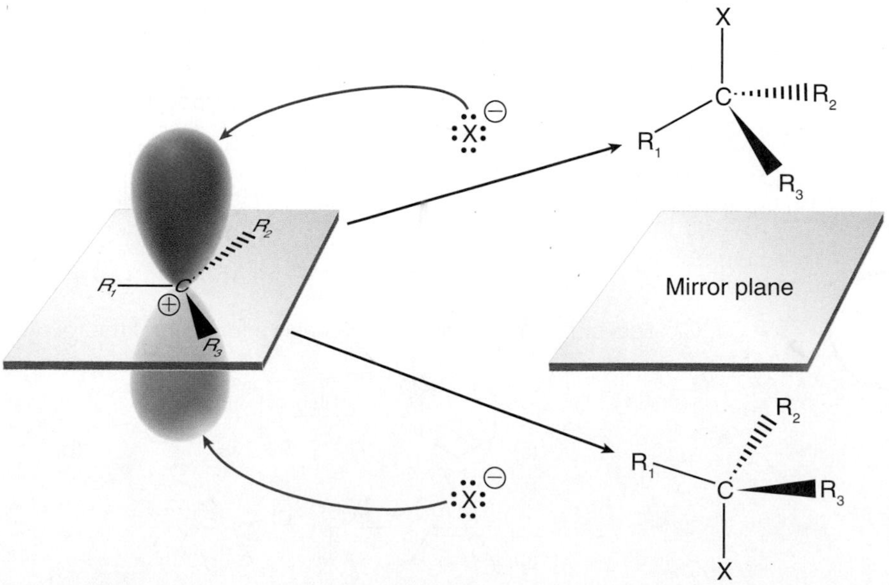

FIGURE 9.3
In the second step of hydrohalogenation, the carbocation intermediate is planar and can be attacked from either face, leading to a pair of mirror-image products (enantiomers).

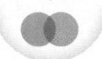

CONCEPTUAL CHECKPOINT

9.6 Predict the products for each of the following reactions. In some cases, the reaction produces a new chirality center, while in other cases, no new chirality center is formed. Consider that when drawing the product(s).

(a) 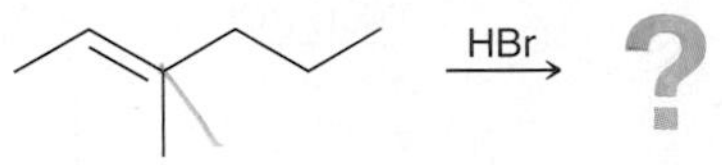

(b) 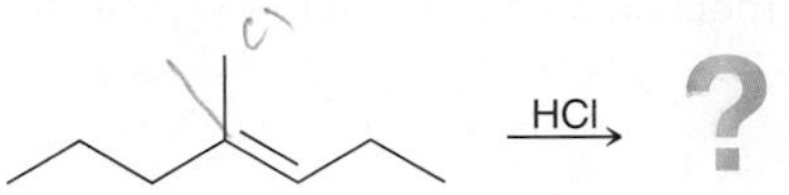

(c) 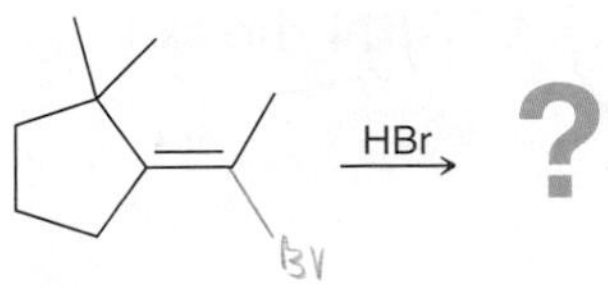

(d) 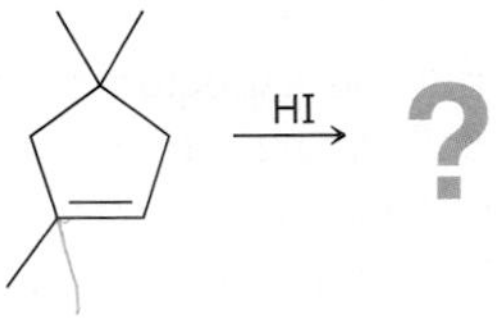

(e) 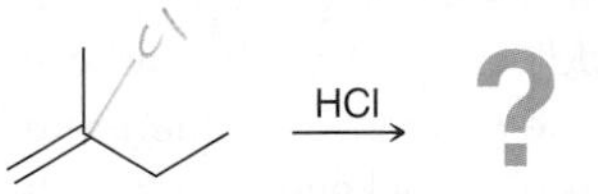

(f) 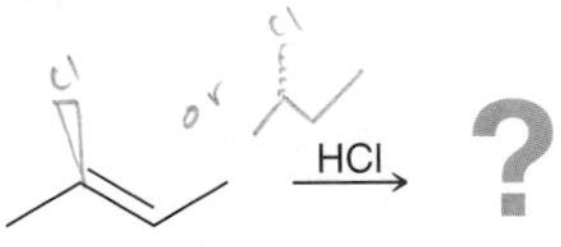

Hydrohalogenation with Carbocation Rearrangements

In Section 6.11, we discussed the stability of carbocations and their ability to rearrange via either a methyl shift or a hydride shift. The problems in that section focused on predicting when and how carbocations rearrange. That skill will be essential now, because the mechanism for HX addition involves formation of an intermediate carbocation. Therefore, HX additions are subject to carbocation rearrangements. Consider the following example, in which, the π bond is protonated to generate the more stable, secondary carbocation, rather than the less stable, primary carbocation:

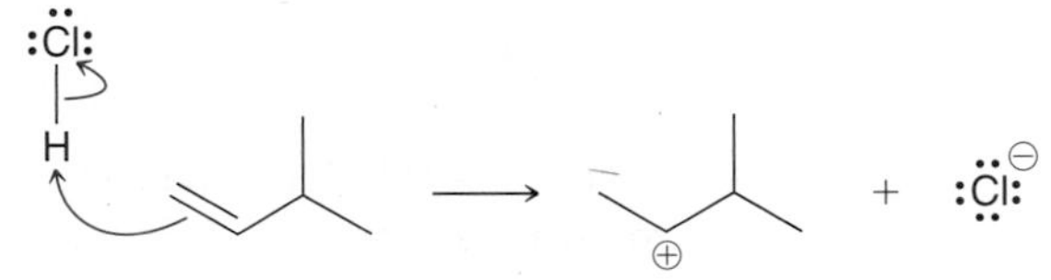

As we might expect, this carbocation can be captured by a chloride ion. However, it is also possible for this secondary carbocation to rearrange before encountering a chloride ion. Specifically, a hydride shift will produce a more stable, tertiary carbocation, which can then be captured by a chloride ion:

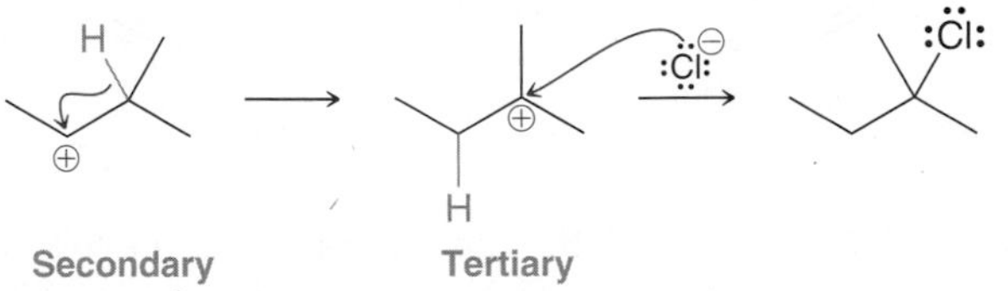

In cases where rearrangements are possible, HX additions produce a mixture of products, including those resulting from a carbocation rearrangement as well as those formed without rearrangement (i.e., if the nucleophile captures the carbocation before rearrangement has a chance to occur):

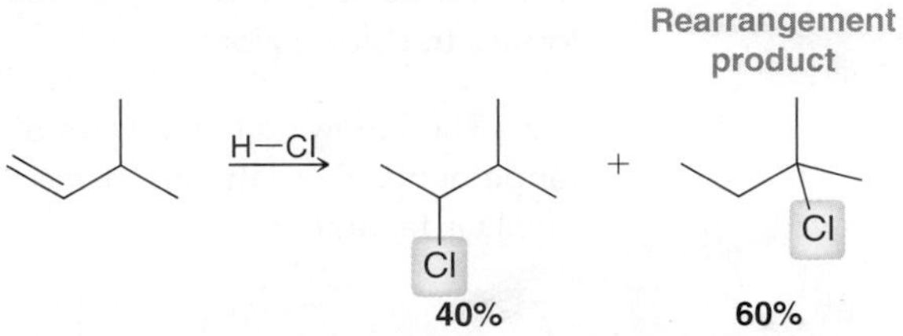

The product ratio shown above (40 : 60) can be moderately affected by altering the concentration of HCl, but a mixture of products is unavoidable. The bottom line is: *When carbocation rearrangements can occur, they do occur.* As a result, HX addition is only synthetically useful in situations where carbocation rearrangements are not possible.

SKILLBUILDER

9.2 DRAWING A MECHANISM FOR A HYDROHALOGENATION PROCESS WITH A CARBOCATION REARRANGEMENT

LEARN the skill

Draw a mechanism for the following transformation:

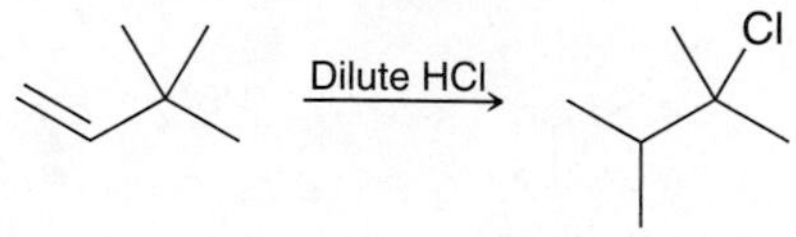

SOLUTION

The reaction is a hydrohalogenation. However, this product is not the expected product from a simple addition. Rather, the carbon skeleton of the product is different from the carbon skeleton of the reactant, and therefore, a carbocation rearrangement is likely.

In the first step, the alkene is protonated to form the more stable carbocation (secondary rather than primary):

STEP 1 Using two curved arrows, protonate the alkene to form the more stable carbocation.

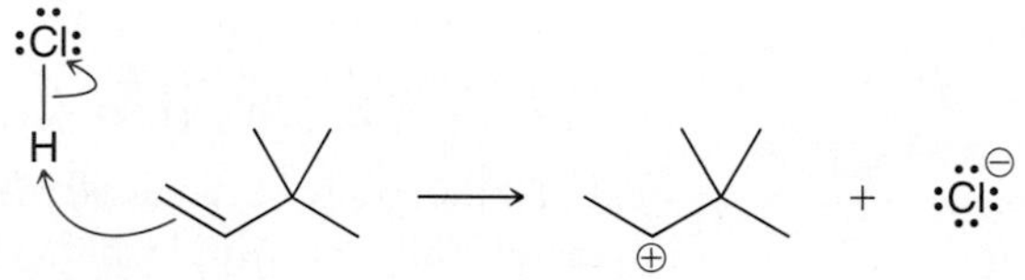

STEP 2 Using one curved arrow, draw the carbocation rearrangement.

Now look to see if a carbocation rearrangement is possible. In this case, a methyl shift can produce a more stable, tertiary carbocation:

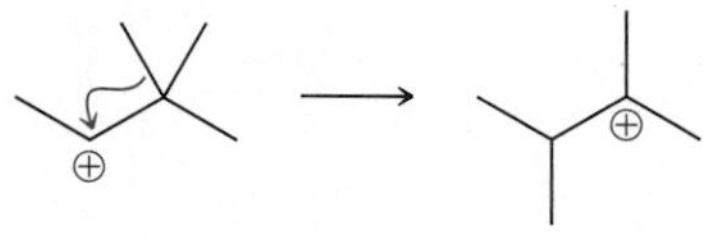

Secondary Tertiary

STEP 3 Using one curved arrow, draw the halide ion attacking the carbocation.

Finally, the chloride ion attacks the tertiary carbocation to give the product:

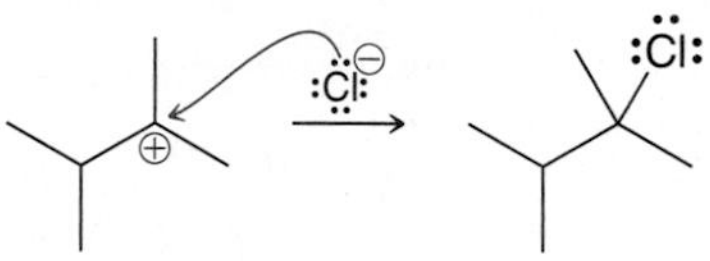

PRACTICE the skill

9.7 Draw a mechanism for each of the following transformations:

(a)
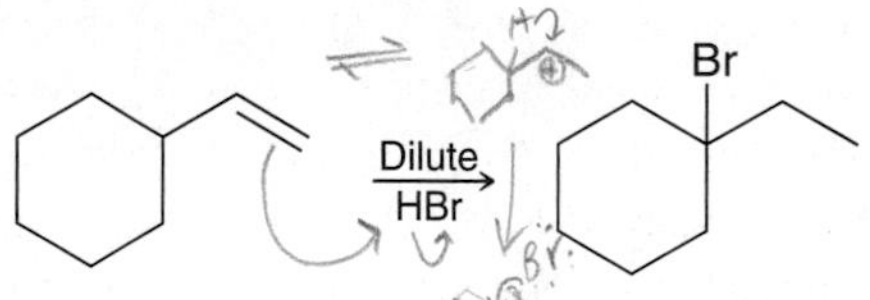

(b)
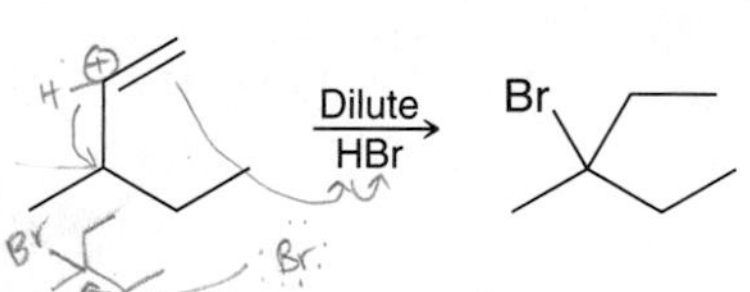

(c)
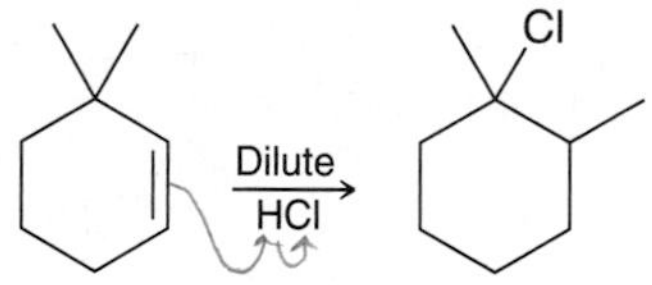

APPLY the skill

9.8 The mechanism of the following transformation involves a carbocation intermediate that rearranges in a way that we have not yet seen. Rather than occurring via a methyl shift or a hydride shift, a carbon atom of the ring migrates, thereby converting a secondary carbocation into a more stable, tertiary carbocation. Using this information, draw the mechanism of the following transformation:

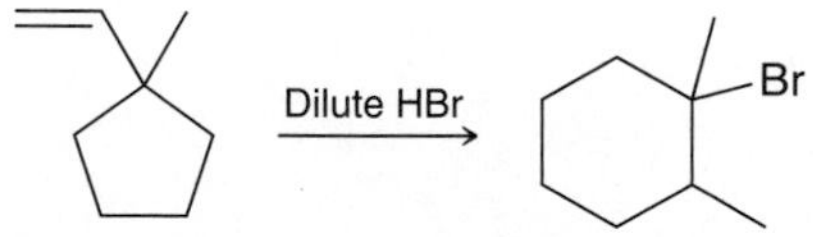

9.9 The following transformation proceeds through two successive carbocation rearrangements. Draw the mechanism, and explain why each of the carbocation rearrangements is favorable:

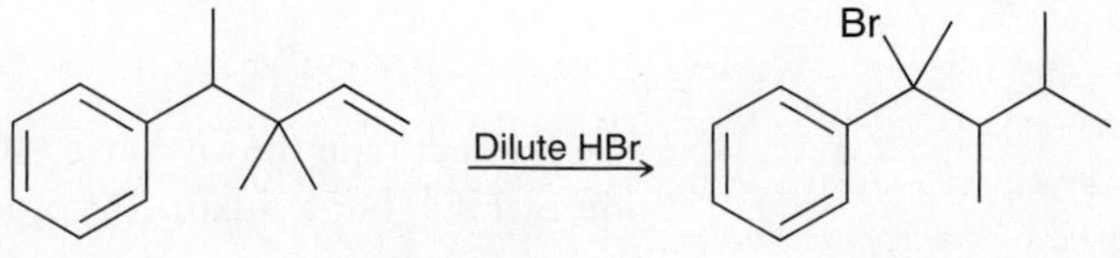

need more **PRACTICE?** **Try Problems 9.51d, 9.76**

PRACTICALLYSPEAKING)))

Cationic Polymerization and Polystyrene

Polymers were first introduced in Chapter 4, and discussions of polymers appear throughout the book. We saw that polymers are large molecules formed by the joining of monomers. More than 200 billion pounds of synthetic organic polymers are produced in the United States every year. They are used in a variety of applications, including automobile tires, carpet fibers, clothing, plumbing pipes, plastic squeeze bottles, cups, plates, cutlery, computers, pens, televisions, radios, CDs and DVDs, paint, and toys. Our society has clearly become dependent on polymers.

Synthetic organic polymers are commonly formed via one of three possible mechanistic pathways: *radical polymerization*, *anionic polymerization*, or *cationic polymerization*. Radical polymerization will be discussed in Chapter 11 and anionic polymerization will be discussed in Chapter 27. In this section, we will briefly introduce cationic polymerization.

The first step of cationic polymerization is similar to the first step of hydrohalogenation. Specifically, an alkene is protonated in the presence of an acid to produce a carbocation intermediate. The most commonly used acid catalyst for cationic polymerization is formed by treating BF_3 with water:

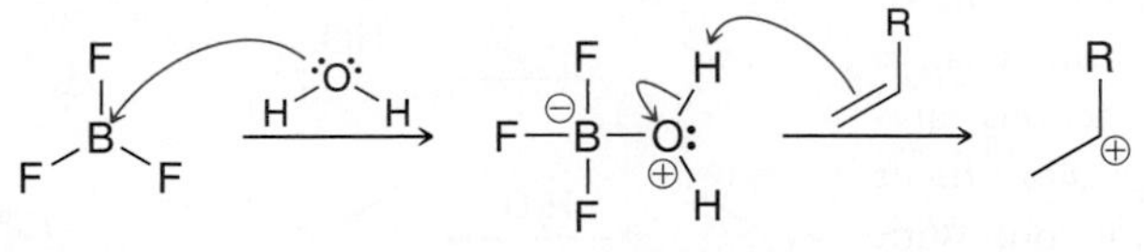

In the absence of halide ions, hydrohalogenation cannot occur. Instead, the carbocation intermediate must react with a nucleophile other than a halide ion. Under these conditions, the only nucleophile present in abundance is the starting alkene, which can attack the carbocation to form a new carbocation. This process repeats itself, adding one monomer at a time; for example:

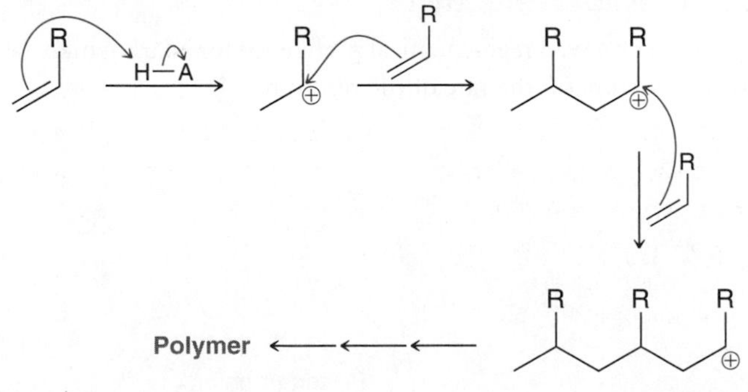

For monomers that undergo cationic polymerization, the process is terminated when the carbocation intermediate is deprotonated by a base or attacked by a nucleophile other than an alkene (such as water).

Many alkenes polymerize readily via cationic polymerization, making possible the production of a wide range of polymers with varying properties and applications. For example, polystyrene (mentioned in the chapter opener) can be prepared via cationic polymerization:

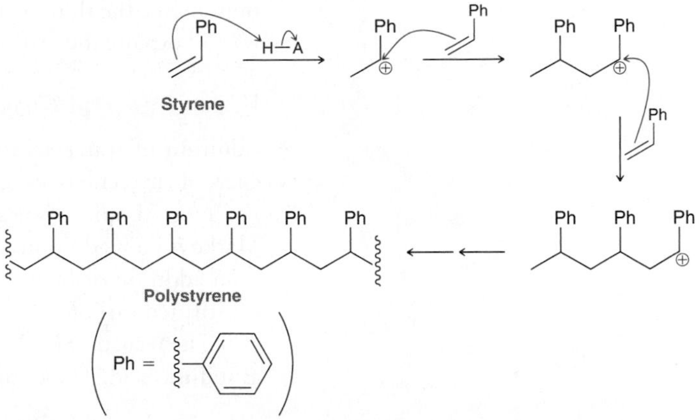

The resulting polymer is a rigid plastic that can be heated, molded into a desired shape, and then cooled again. This feature makes polystyrene ideal in the production of a variety of plastic items, including plastic cutlery, toy pieces, CD jewel cases, computer housings, radios, and televisions. In the early 1940s, an accidental discovery led to the development of Styrofoam™, which is also made from polystyrene. Ray McIntire, working for the Dow Chemical Company, was searching for a polymeric material that could serve as a flexible electrical insulator. His research efforts led to the discovery of a type of foamed polystyrene which is comprised of approximately 95% air and is therefore much lighter (per unit volume) than regular polystyrene. This new material was found to have many useful physical properties. In particular, foamed polystyrene was found to be moisture resistant, unsinkable, and a poor conductor of heat (and therefore an excellent insulator). Dow called its material Styrofoam™ and has continuously improved the method of its production over the last half of a century. Coffee cups, coolers, and packing materials are also made from foamed polystyrene, but the method of preparation is not the same as the process used by Dow, and as a result, the properties are not exactly the same as the properties of Styrofoam™.

Foamed polystyrene is generally prepared by heating polystyrene and then using hot gases, called blowing agents, to expand the polystyrene into a foam. In the past, CFCs (chlorofluorocarbons) were primarily used as blowing agents, but they are no longer in use because of their suspected role in destroying the ozone layer. The alternative blowing agents currently in use are safer for the environment.

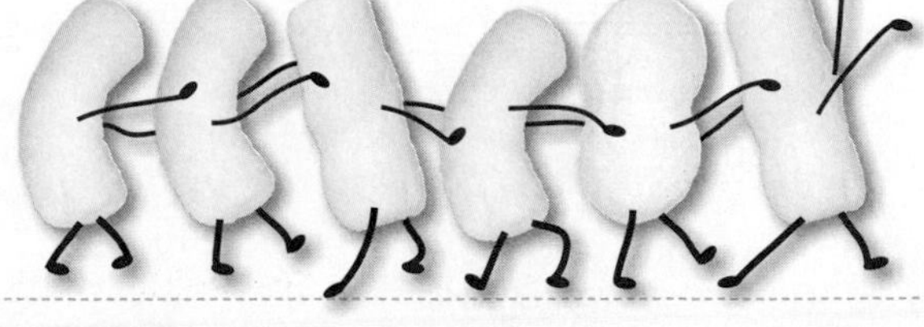

9.4 Acid-Catalyzed Hydration

The next few sections discuss three methods for adding water (H and OH) across a double bond, a process called **hydration**. The first two methods proceed through Markovnikov additions, while the third method proceeds through an *anti*-Markovnikov addition. In this section, we will explore the first of these three reactions.

Experimental Observations

Addition of water across a double bond in the presence of an acid is called **acid-catalyzed hydration**. For most simple alkenes, this reaction proceeds via a Markovnikov addition, as shown here. The net result is an addition of H and OH across the π bond, with the OH group positioned at the more substituted carbon.

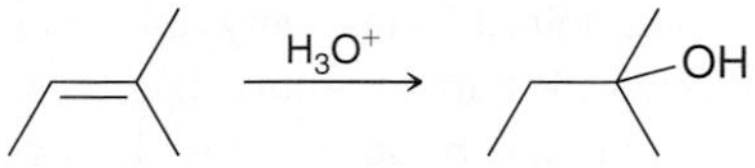

(90%)

The reagent, H_3O^+, represents the presence of both water (H_2O) and an acid source, such as sulfuric acid. These conditions are also commonly shown in the following way:

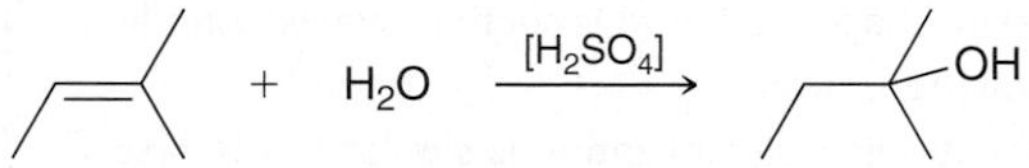

where the brackets indicate that the proton source is not consumed in the reaction. It is a catalyst, and, therefore, this reaction is said to be an *acid-catalyzed hydration*.

The rate of an acid-catalyzed hydration is very much dependent on the structure of the starting alkene. Compare the relative rates of the following three reactions, and analyze the effects of an alkyl substituent on the relative rate of each reaction. With each additional alkyl group, the reaction rate increases by many orders of magnitude. Also take special notice of where the OH is placed in each of the cases above: *at the more substituted position.*

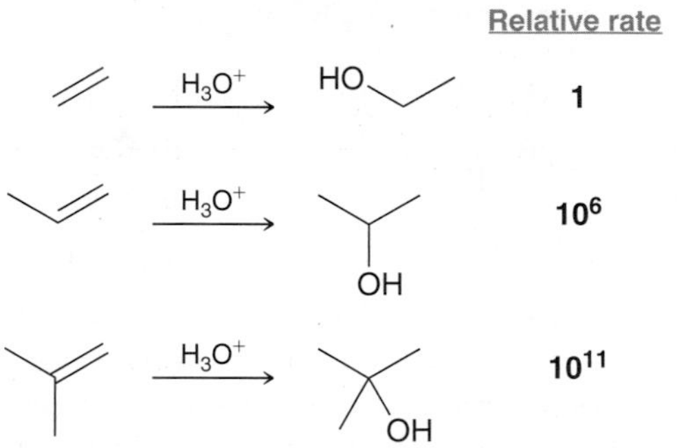

Mechanism and Source of Regioselectivity

Mechanism 9.2 is consistent with the observed regiochemical preference for Markovnikov addition as well as the effect of alkene structure on the rate of the reaction.

MECHANISM 9.2 ACID-CATALYZED HYDRATION

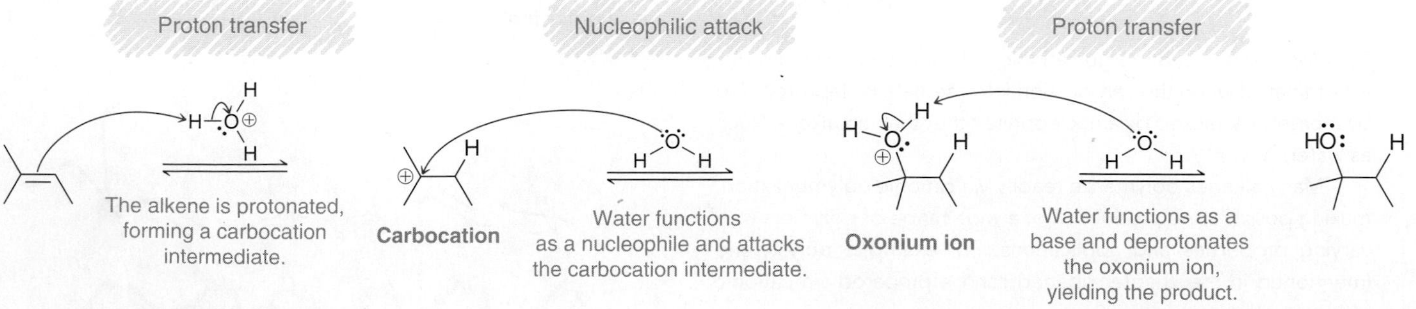

LOOKING AHEAD
A proposed mechanism must always be consistent with the stated conditions, as we see here with the use of H_2O in acidic conditions rather than hydroxide for the last step of the mechanism. This principle will appear many more times throughout this text.

The first two steps of this mechanism are virtually identical to the mechanism we proposed for hydrohalogenation. Specifically, the alkene is first protonated to generate a carbocation intermediate, which is then attacked by a nucleophile. However, in this case, the attacking nucleophile is neutral (H_2O) rather than an anion (X^-), and therefore, a charged intermediate is generated as a result of nucleophilic attack. This intermediate is called an **oxonium ion**, because it bears an oxygen atom with a positive charge. In order to remove this charge and form an electrically neutral product, the mechanism must conclude with a proton transfer. Notice that the base used to deprotonate the oxonium ion is H_2O rather than a hydroxide ion (HO^-). Why? In acidic conditions, the concentration of hydroxide ions is extremely low, but the concentration of H_2O is quite large.

The proposed mechanism is consistent with the experimental observations discussed earlier in this section. The reaction proceeds via a Markovnikov addition, just as we saw for hydrohalogenation, because there is a strong preference for the reaction to proceed via the more stable carbocation intermediate. Similarly, reaction rates for substituted alkenes can be justified by comparing the carbocation intermediates in each case. Reactions that proceed via tertiary carbocations will generally occur more rapidly than reactions that proceed via secondary carbocations.

CONCEPTUAL CHECKPOINT

9.10 In each of the following cases identify the alkene that is expected to be more reactive toward acid-catalyzed hydration.

(a) or *secondary carbocation is more reactive*

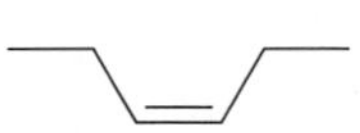

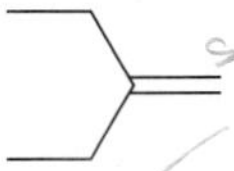

(b) 2-Methyl-2-butene or 3-methyl-1-butene

LOOKING BACK
Acid-catalyzed dehydration was first discussed in Section 8.9.

Controlling the Position of Equilibrium

Carefully examine the proposed mechanism for acid-catalyzed hydration. Notice the equilibrium arrows (\rightleftharpoons rather than \rightarrow). These arrows indicate that the reaction actually goes in both directions. Consider the reverse path (starting from the alcohol and ending with the alkene). That process is an elimination reaction in which an alcohol is converted into an alkene. More specifically, it is an E1 process called acid-catalyzed dehydration. The truth is that most reactions represent equilibrium processes; however, organic chemists generally draw equilibrium arrows only in situations where the equilibrium can be easily manipulated (allowing for control over the product distribution). Acid-catalyzed dehydration is an excellent example of such a reaction.

Earlier in this chapter, thermodynamic arguments were presented to explain why low temperature favors addition, while high temperature favors elimination. But in acid-catalyzed dehydration there is yet another way to easily control the equilibrium. The equilibrium is sensitive not only to temperature but also to the concentration of water that is present. By controlling the amount of water present (using either concentrated acid or dilute acid), one side of the equilibrium can be favored over the other:

$$+ \quad H_2O \quad \xrightleftharpoons[\text{Conc. } H_2SO_4 \;(\text{less } H_2O)]{\text{Dilute } H_2SO_4 \;(\text{more } H_2O)} \quad HO$$

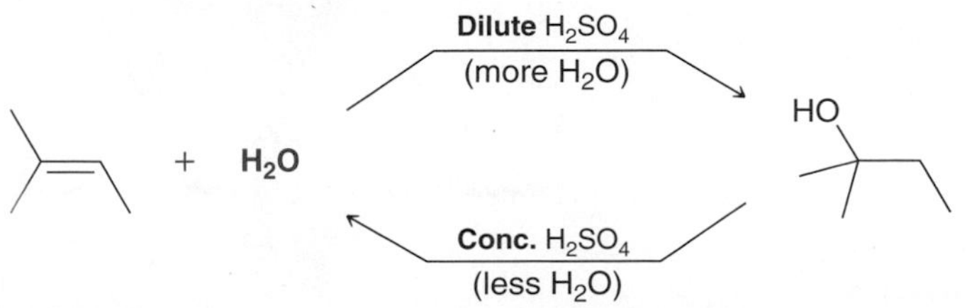

Control over this equilibrium derives from an understanding of Le Châtelier's principle, which states that a system at equilibrium will adjust in order to minimize any stress placed on the system. To understand how this principle applies, consider the above process after equilibrium has been established. Notice that water is on the left side of the reaction. How would the introduction of more water affect the system? The concentrations would no longer be at equilibrium, and the system would have to adjust to reestablish new equilibrium concentrations. The introduction of more water will cause the position of equilibrium to move in such a way that more of the alcohol is produced. Therefore, dilute acid (which is mostly water) is used to convert an alkene into an alcohol. On the flip side, removing water from the system would cause the equilibrium to favor the alkene. Therefore, concentrated acid (very little water) is used to favor formation of the alkene. Alternatively, water can also be removed from the system via a distillation process, which would also favor the alkene.

In summary, the outcome of a reaction can be greatly affected by carefully choosing the reaction conditions and concentrations of reagents.

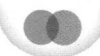

CONCEPTUAL CHECKPOINT

9.11 Identify whether you would use dilute sulfuric acid or concentrated sulfuric acid to achieve each of the following transformations. In each case, explain your choice.

(a)
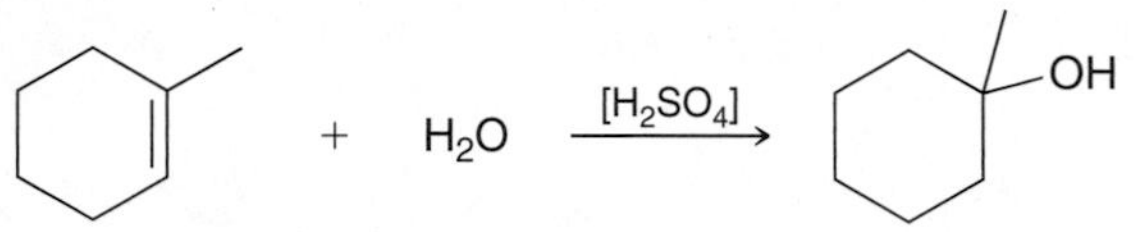

(b)
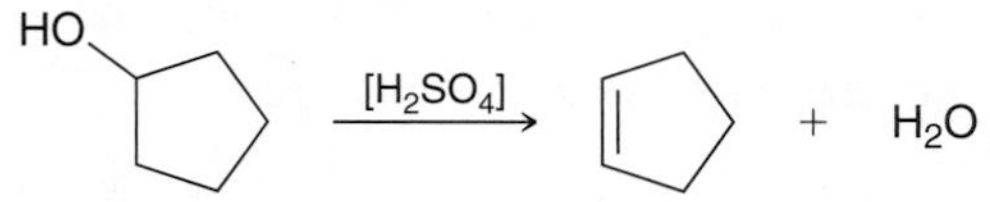

Stereochemistry of Acid-Catalyzed Hydration

The stereochemical outcome of acid-catalyzed hydration is similar to the stereochemical outcome of hydrohalogenation. Once again, the intermediate carbocation can be attacked from either side with equal likelihood (Figure 9.4). Therefore, when a new chirality center is generated, a racemic mixture of enantiomers is expected:

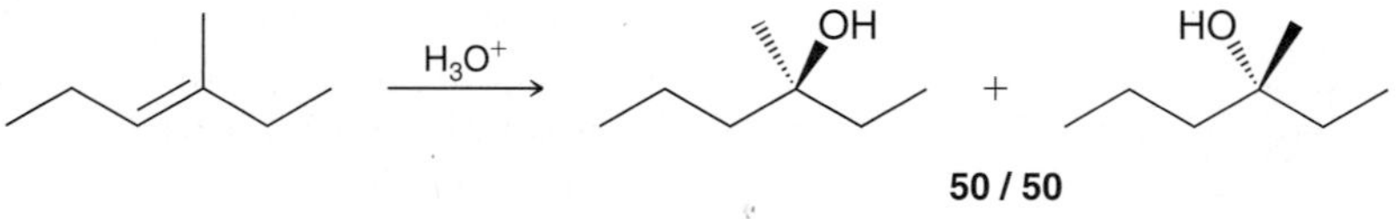

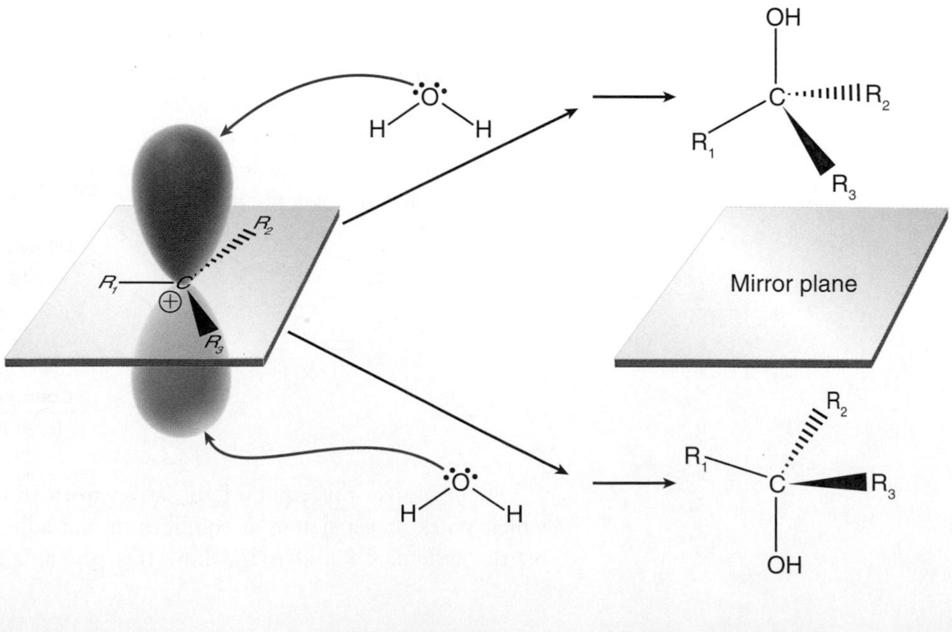

FIGURE 9.4
In the second step of acid-catalyzed hydration, the carbocation intermediate is planar and can be attacked from either face, leading to a pair of mirror-image products (enantiomers).

SKILLBUILDER

9.3 DRAWING A MECHANISM FOR AN ACID-CATALYZED HYDRATION

LEARN the skill

Draw a mechanism for the following transformation:

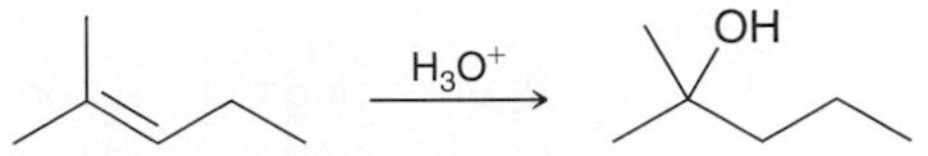

SOLUTION

In this reaction, water is added across an alkene in a Markovnikov fashion under acid-catalyzed conditions. As a result, the OH is positioned at the more substituted carbon. To draw a mechanism for this process, recall that the proposed mechanism for acid-catalyzed hydration has three steps: (1) protonation to give a carbocation, (2) nucleophilic attack of water to give an oxonium ion, and (3) deprotonation to generate a neutral product. When drawing the first step of the mechanism (protonation), make sure to use two curved arrows and make sure to form the more stable carbocation:

STEP 1
Using two curved arrows, protonate the alkene to form the more stable carbocation.

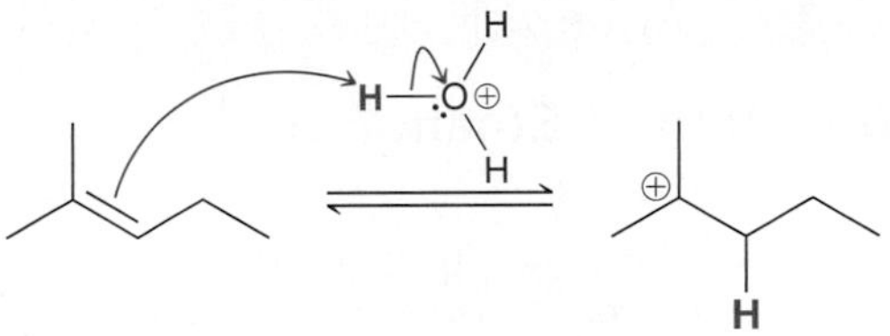

WATCH OUT
Make sure to draw both curved arrows, and be very precise when placing the head and tail of each curved arrow.

One curved arrow is drawn with its tail on the π bond and its head on the proton, while the second curved arrow is drawn with its tail on an O—H bond and its head on the oxygen atom of water. When drawing the second step of the mechanism (nucleophilic attack of water), only one curved arrow is required. The tail should be placed on a lone pair of water, and the head should be placed on the carbocation:

STEP 2
Using one curved arrow, draw a water molecule attacking the carbocation.

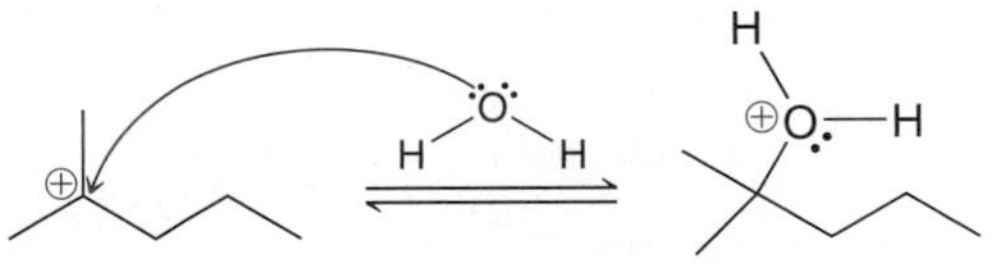

In the final step of the mechanism, water (not hydroxide) functions as a base and abstracts a proton from the oxonium ion. Like all proton transfer steps, this process requires two curved arrows. One curved arrow is drawn with its tail on a lone pair of water and its head on the proton, while the second curved arrow is drawn with its tail on the O—H bond and its head on the oxygen atom:

STEP 3
With two curved arrows, deprotonate the oxonium ion using water as a base.

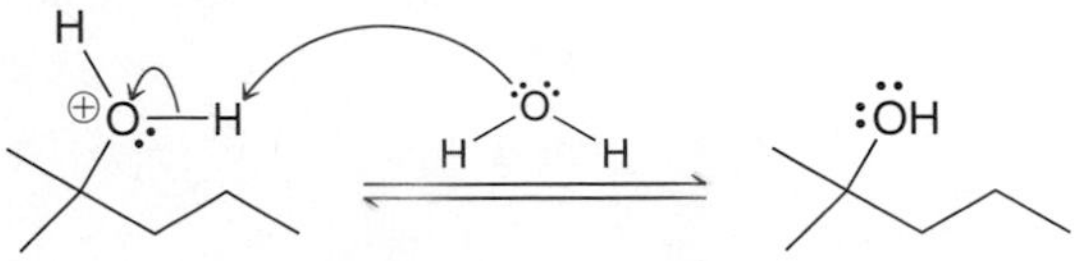

PRACTICE the skill **9.12** Draw a mechanism for each of the following transformations:

(a) 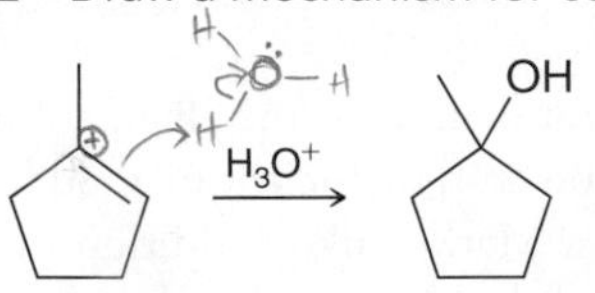

(b) 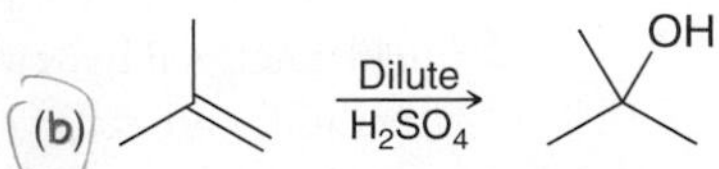

(c) 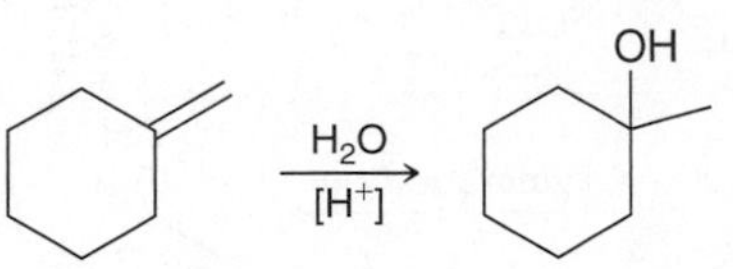

APPLY the skill

9.13 If an alkene is protonated and the solvent is an alcohol rather than water, a reaction takes place that is very similar to acid-catalyzed hydration, but in the second step of the mechanism the alcohol functions as a nucleophile instead of water. Draw a plausible mechanism for the following process:

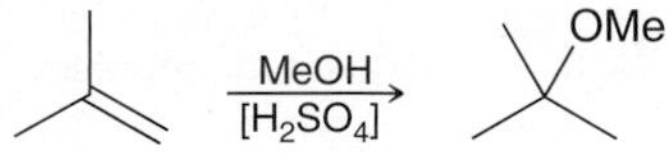

9.14 Using the reaction in the previous problem as a reference, propose a plausible mechanism for the following intramolecular reaction:

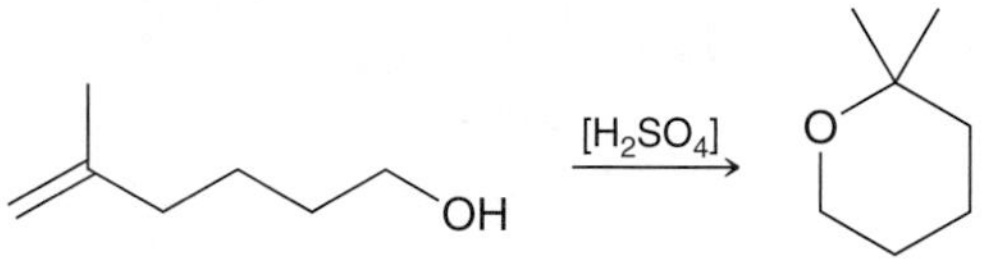

need more **PRACTICE?** **Try Problems 9.51a,b, 9.64a**

PRACTICALLYSPEAKING)))

Industrial Production of Ethanol

Ethanol found in alcoholic beverages is generally produced via the fermentation of sugars by yeast. However, there are many other important industrial uses for ethanol (as a gasoline additive, as a solvent, for perfumes and paints, etc.), and therefore, an alternative, more efficient process for producing pure ethanol is required. This can be accomplished by the acid-catalyzed hydration of ethylene obtained from petroleum:

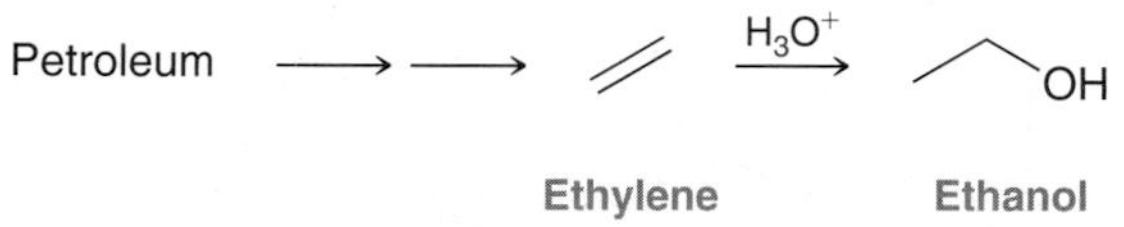

Phosphoric acid is generally employed as the source of protons. The United States produces more than five billion gallons of ethanol each year via this process.

9.5 Oxymercuration-Demercuration

The previous section explored how acid-catalyzed hydration can be used to achieve a Markovnikov addition of water across an alkene. The utility of that process is somewhat diminished by the fact that carbocation rearrangements can produce a mixture of products:

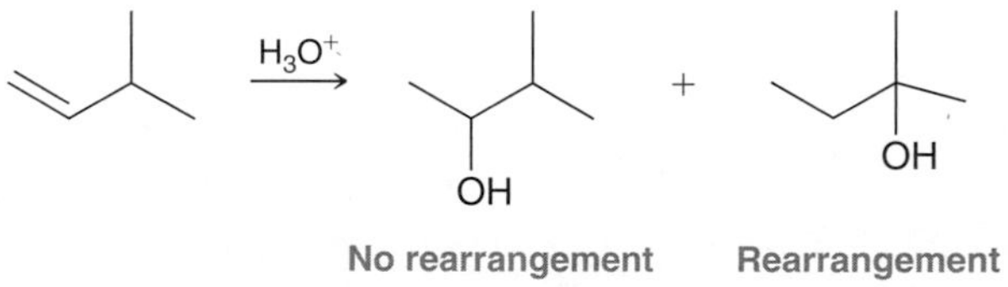

In cases where protonation of the alkene ultimately leads to carbocation rearrangements, acid-catalyzed hydration is an inefficient method for adding water across the alkene. Many other methods can achieve a Markovnikov addition of water across an alkene without carbocation rearrangements. One of the oldest and perhaps best known methods is called **oxymercuration-demercuration**:

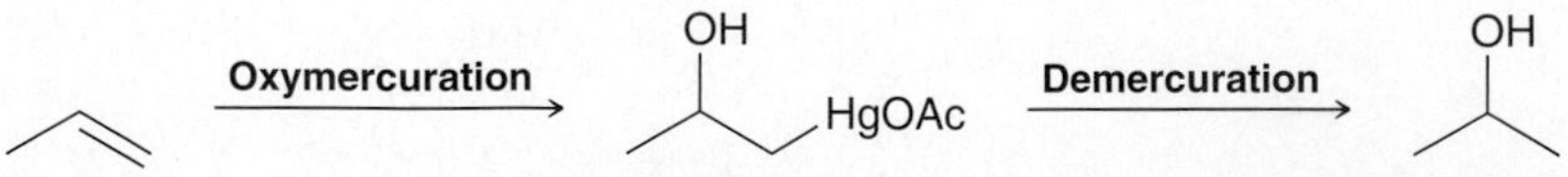

To understand this process, we must explore the reagents employed. The process begins when mercuric acetate, $Hg(OAc)_2$, dissociates to form a mercuric cation:

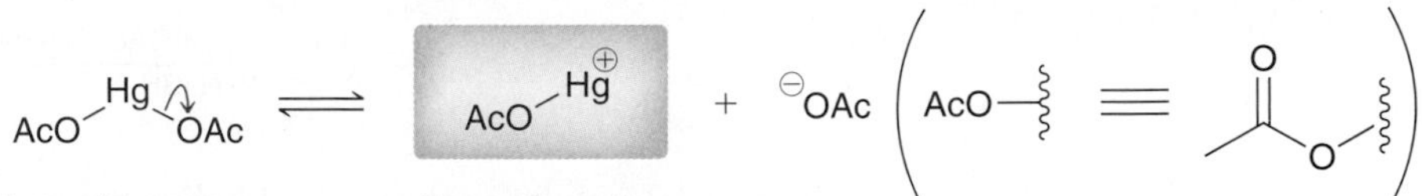

Mercuric acetate   Mercuric cation

This mercuric cation is a powerful electrophile and is subject to attack by a nucleophile, such as the π bond of an alkene. When a π bond attacks a mercuric cation, the nature of the resulting intermediate is quite different from the nature of the intermediate formed when a π bond is simply protonated. Let's compare:

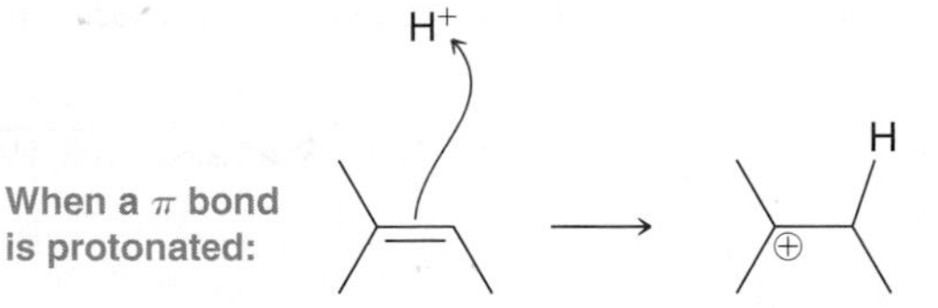

A carbocation

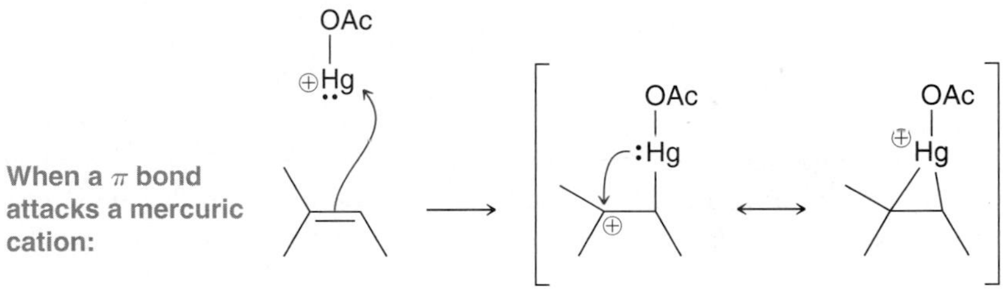

A mercurinium ion

BY THE WAY
As mentioned in Chapter 2, we generally avoid breaking single bonds when drawing resonance structures. However, this is one of the rare exceptions. We will see one other such exception in the next section of this chapter.

When a π bond is protonated, the intermediate formed is simply a carbocation, as we have seen many times in this chapter. In contrast, when a π bond attacks a mercuric cation, the resulting intermediate cannot be considered as a carbocation, because the mercury atom has electrons that can interact with the nearby positive charge to form a bridge. This intermediate, called a **mercurinium ion**, is more adequately described as a hybrid of two resonance structures. A mercurinium ion has some of the character of a carbocation, but it also has some of the character of a bridged, three-membered ring. This dual character can be illustrated with the following drawing:

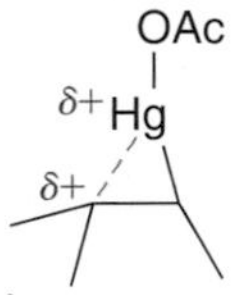

Mercurinium ion

The more substituted carbon atom bears a partial positive charge (δ+), rather than a full positive charge. As a result, this intermediate will not readily undergo carbocation rearrangements, but it is still subject to attack by a nucleophile:

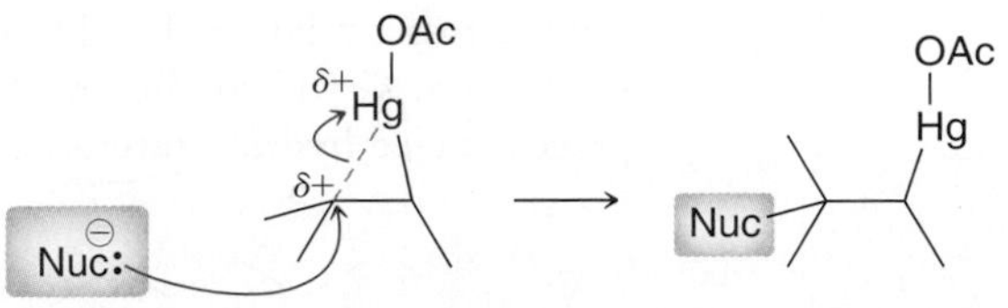

Notice that the attack takes place at the more substituted position, ultimately leading to Markovnikov addition. After attack of the nucleophile, the mercury can be removed through a process called *demercuration*, which is generally accomplished with sodium borohydride. There

LOOKING AHEAD
Radical processes are discussed in more detail in Chapter 11.

is much evidence that demercuration occurs via a radical process. The net result is the addition of H and a nucleophile across an alkene:

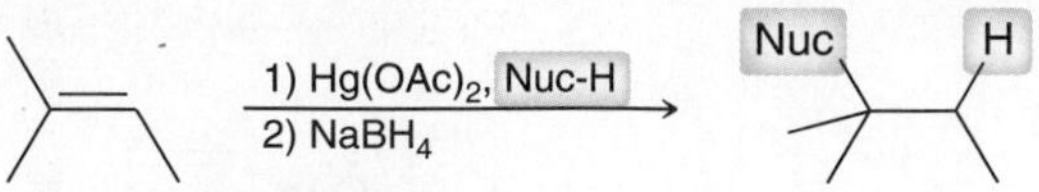

Many nucleophiles can be used, including water:

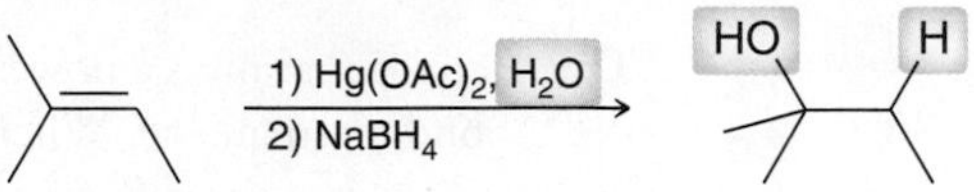

This reaction sequence provides for a two-step process that enables the hydration of an alkene without carbocation rearrangements:

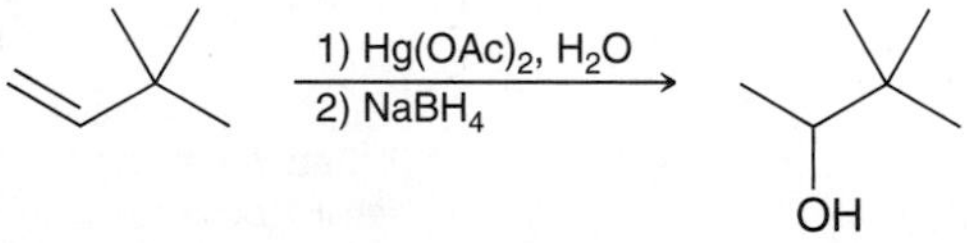

(94%)

No rearrangement

CONCEPTUAL CHECKPOINT

9.15 Predict the product for each reaction, and predict the products if an acid-catalyzed hydration had been performed rather than an oxymercuration-demercuration:

(a) 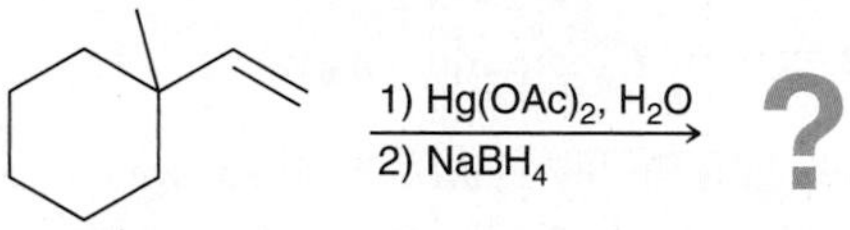

(b) 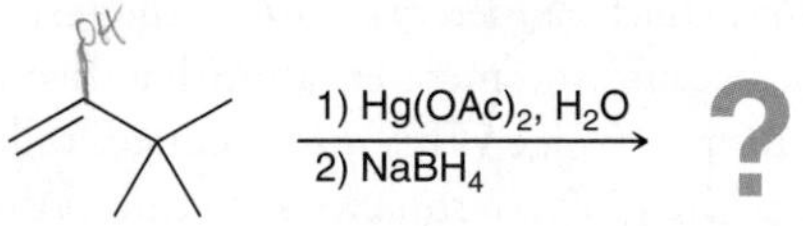

(c) 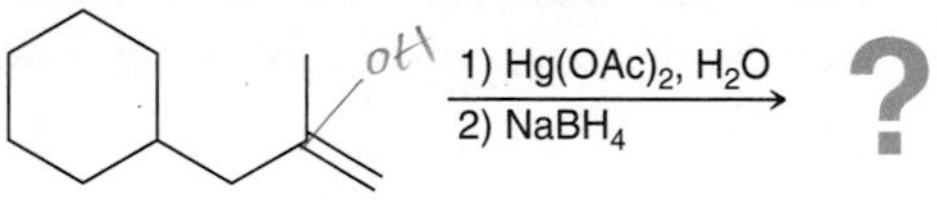

9.16 In the first step of the process (oxymercuration), nucleophiles other than water may be used. Predict the product when each of the following nucleophiles is used instead of water:

(a) 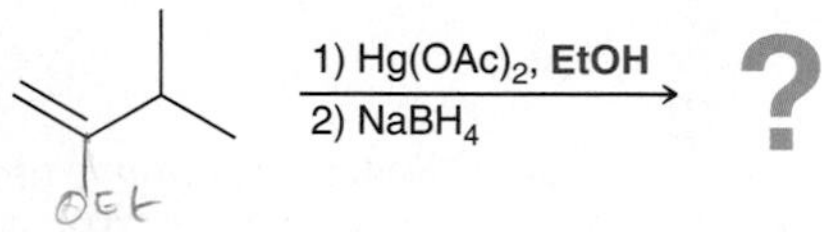

(b) 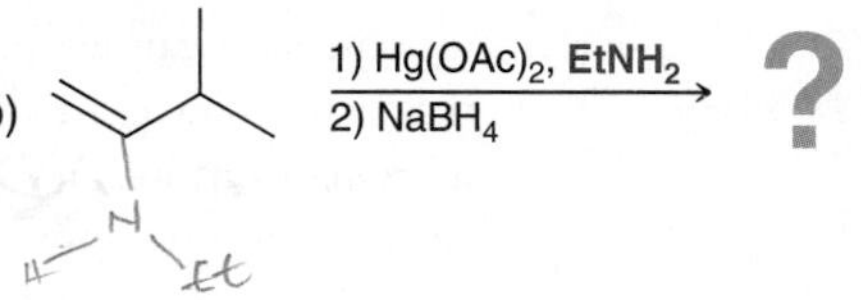

9.6 Hydroboration-Oxidation

An Introduction to Hydroboration-Oxidation

The previous sections covered two different methods for achieving a Markovnikov addition of water across a π bond: (1) acid-catalyzed hydration and (2) oxymercuration-demercuration. In this section, we will explore a method for achieving an *anti*-Markovnikov addition of water. This process, called **hydroboration-oxidation**, places the OH group at the less substituted carbon:

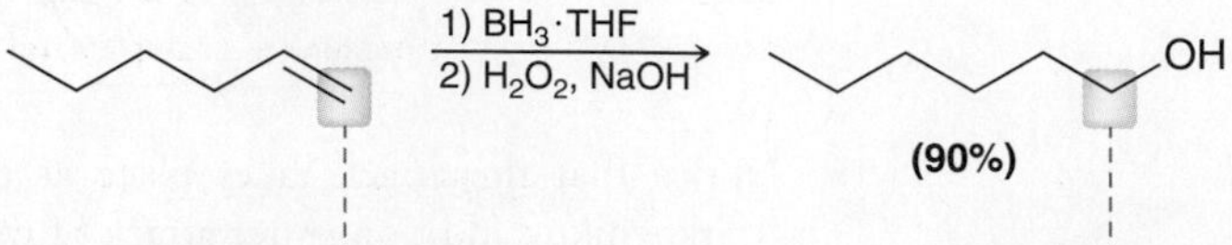

(90%)

Less substituted vinylic position

***Anti*-Markovnikov addition**

The stereochemical outcome of this reaction is also of particular interest. Specifically, when two new chirality centers are formed, the addition of water (H and OH) is observed to occur in a way that places the H and OH on the same face of the π bond:

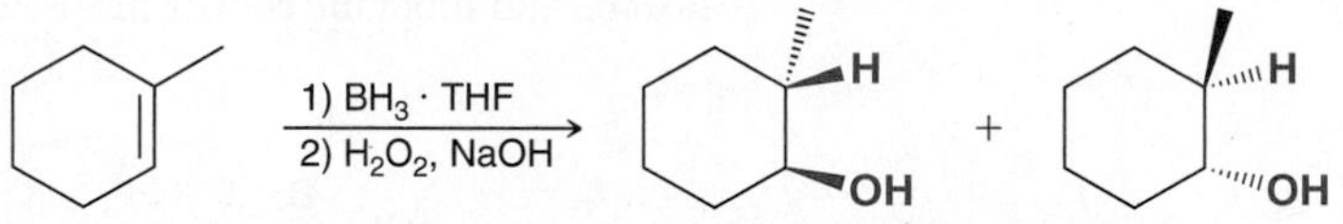

Enantiomers

This mode of addition is called a ***syn* addition**. The reaction is said to be stereospecific because only two of the four possible stereoisomers are formed. That is, the reaction does not produce the two stereoisomers that would result from adding H and OH to opposite sides of the π bond:

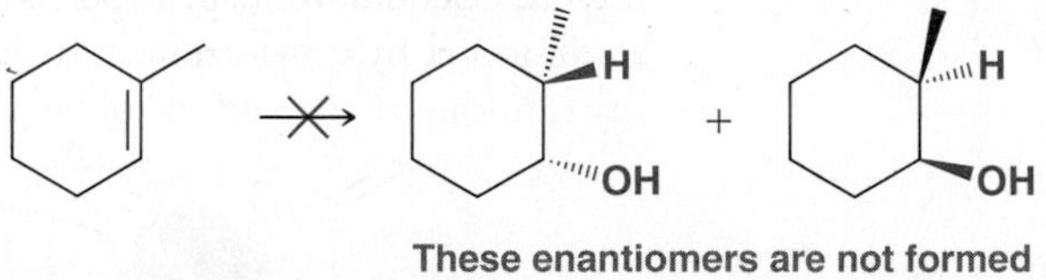

These enantiomers are not formed

Any mechanism that we propose for hydroboration-oxidation must explain both the regioselectivity (*anti*-Markovnikov addition) as well as the stereospecificity (*syn* addition). We will soon propose a mechanism that explains both observations. But first, we must explore the nature of the reagents used for hydroboration-oxidation.

Reagents for Hydroboration-Oxidation

The structure of borane (BH_3) is similar to that of a carbocation, but without the charge:

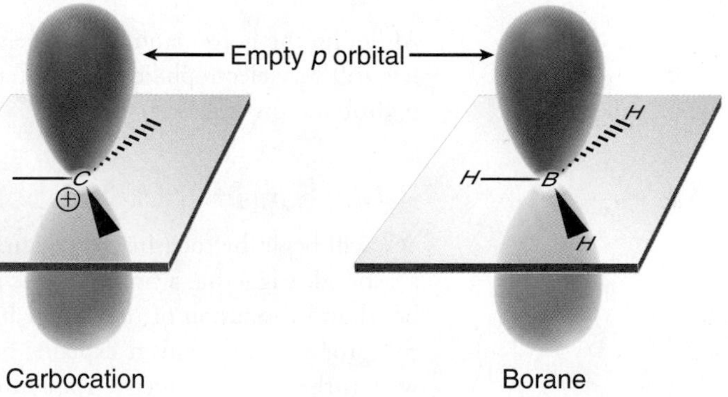

The boron atom lacks an octet of electrons and is therefore very reactive. In fact, one borane molecule will even react with another borane molecule to form a dimeric structure called **diborane**. This dimer is believed to possess a special type of bonding that is unlike anything we have seen thus far. It can be more easily understood by drawing the following resonance structures:

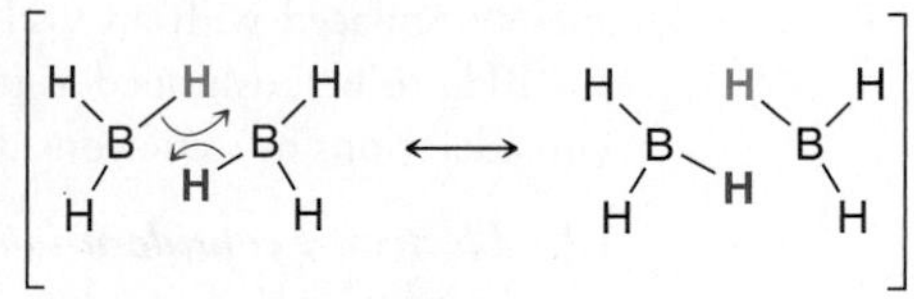

As with the mercurinium ion (Section 9.5), this is another of those rare cases where we break a single bond when drawing the resonance structures. Careful examination of these resonance structures shows that each of the hydrogen atoms, colored in red and blue above, is partially bonded to two boron atoms using a total of two electrons:

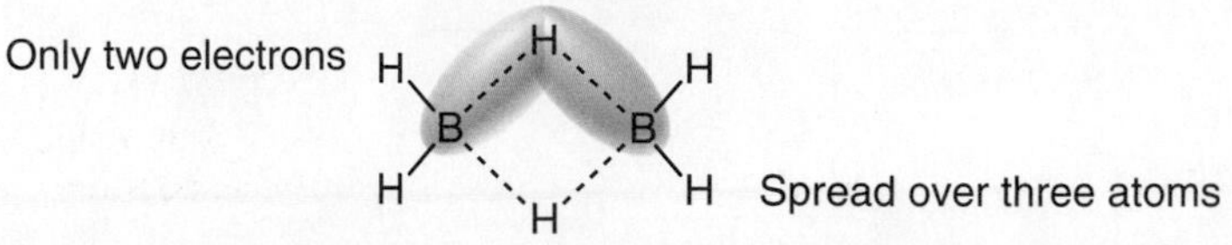

Such bonds are called **three-center, two-electron bonds**. In the world of organic chemistry, there are many other examples of three-center, two-electron bonds; however, we will not encounter many other examples in this text.

Borane and diborane coexist in the following equilibrium:

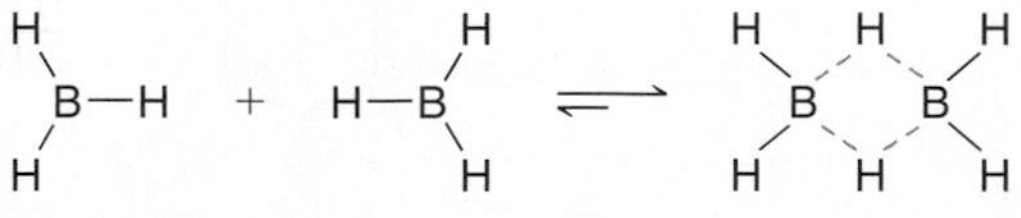

This equilibrium lies very much to the side of diborane (B_2H_6), leaving very little borane (BH_3) present at equilibrium. It is possible to stabilize BH_3, thereby increasing its concentration at equilibrium, by using a solvent such as THF (tetrahydrofuran) which can donate electron density into the empty *p* orbital of boron:

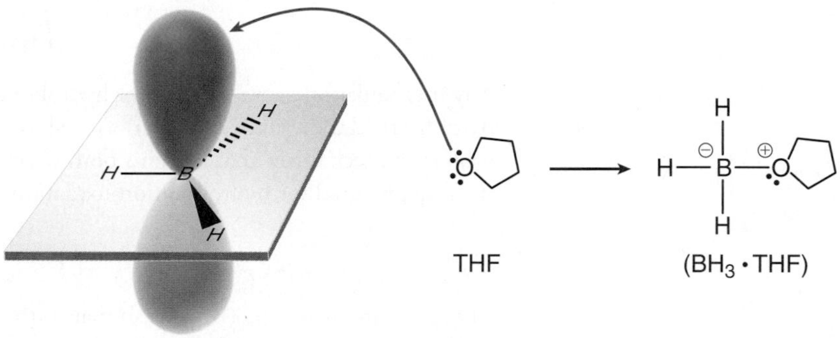

Although the boron atom does receive some electron density from the solvent, it is nevertheless still very electrophilic and subject to attack by the π bond of an alkene. A mechanism for hydroboration-oxidation follows.

A Mechanism for Hydroboration-Oxidation

We will begin by focusing on the first step of hydroboration, in which borane is attacked by a π bond, triggering a simultaneous hydride shift. In other words, formation of the C—BH_2 bond and formation of the C—H bond occur together in a concerted process. This step of the proposed mechanism explains both the regioselectivity (*anti*-Markovnikov addition) as well as the stereospecificity (*syn* addition) for this process. Each of these features will now be discussed in more detail.

Regioselectivity of Hydroboration-Oxidation

As seen in Mechanism 9.3, a BH_2 group is installed at the less substituted carbon and is ultimately replaced with an OH group, leading to the observed regioselectivity. The preference for BH_2 to be positioned at the less substituted carbon can be explained in terms of electronic considerations or steric considerations. Both explanations are presented below:

1. ***Electronic considerations***: In the first step of the proposed mechanism, attack of the π bond triggers a simultaneous hydride shift. However, this process does not have to be perfectly simultaneous. As the π bond attacks the empty *p* orbital of boron, one of the vinylic positions can begin to develop a partial positive charge (δ+). This developing δ+ then triggers a hydride shift:

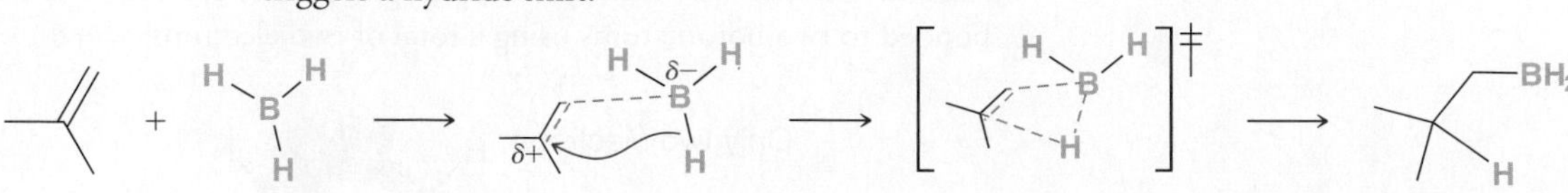

MECHANISM 9.3 HYDROBORATION-OXIDATION

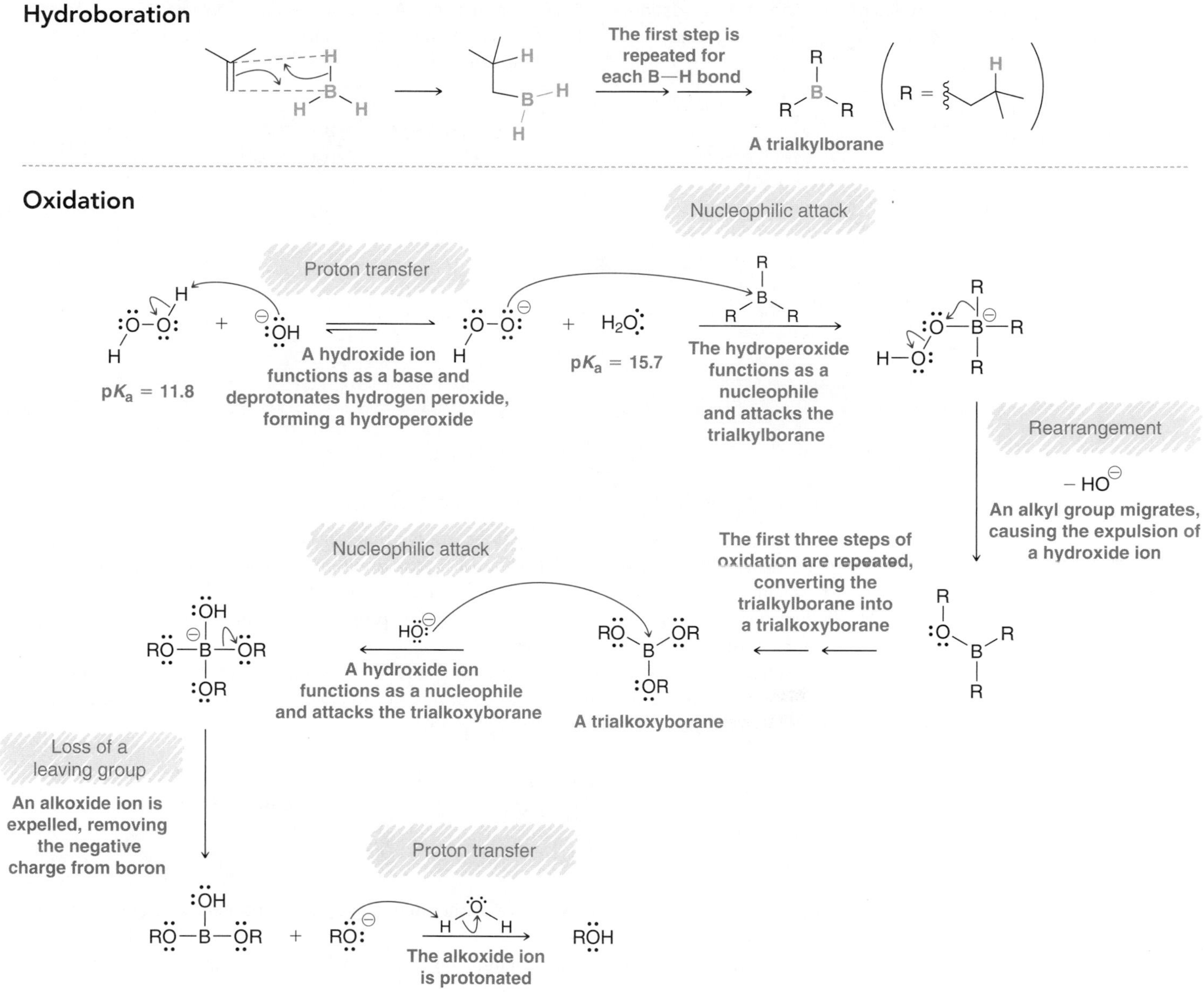

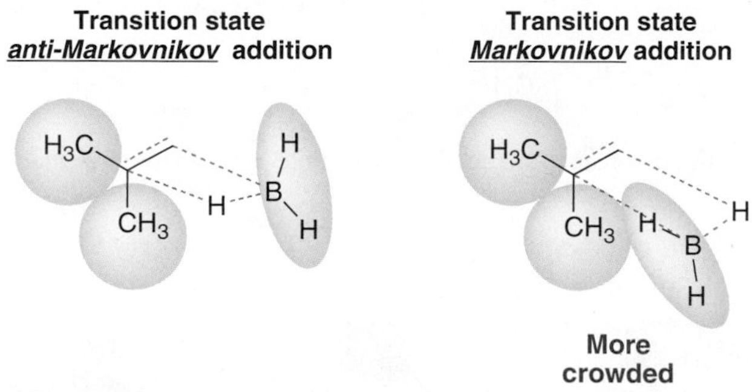

FIGURE 9.5
A comparison of the transition states for hydroboration via Markovnikov addition or anti-Markovnikov addition. The latter will be lower in energy because of decreased steric crowding.

In this way, one of the vinylic carbon atoms develops a partial positive charge when the alkene begins to interact with borane. There will be a preference (as we have seen previously in this chapter) for any positive character to develop at the more substituted carbon. In order to accomplish this, the BH_2 group must be positioned at the less substituted carbon atom.

2. ***Steric considerations***: In the first step of the proposed mechanism, both H and BH_2 are adding across the double bond simultaneously. Since BH_2 is bigger than H, the transition state will be less crowded and lower in energy if the BH_2 group is positioned at the less sterically hindered position (Figure 9.5). It is likely that both electronic and steric factors contribute to the observed regioselectivity for hydroboration-oxidation.

CONCEPTUAL CHECKPOINT

9.17 Below are several examples of hydroboration-oxidation. In each case, consider the expected regioselectivity, and then draw the product:

(a) 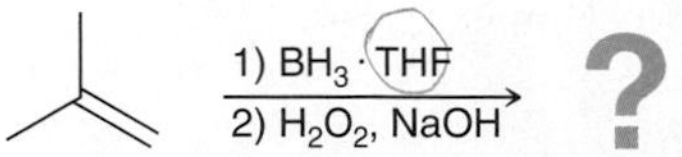

(b) 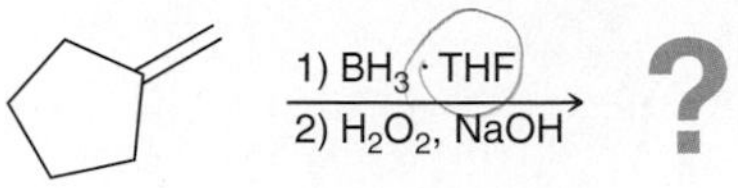

(c) 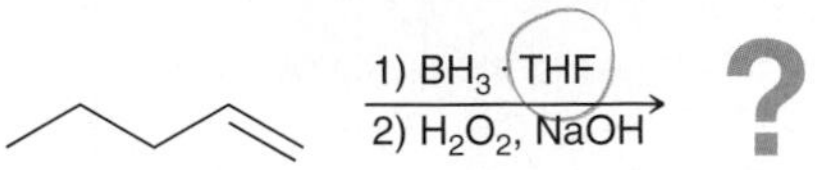

9.18 Compound A has molecular formula C_5H_{10}. Hydroboration-oxidation of compound A produces 2-methylbutan-1-ol. Draw the structure of compound A:

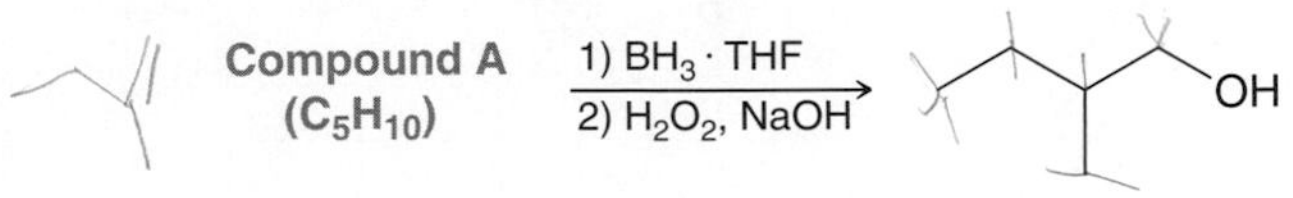

Stereospecificity of Hydroboration-Oxidation

The observed stereospecificity for hydroboration-oxidation is consistent with the first step of the proposed mechanism, in which H and BH_2 are simultaneously added across the π bond of the alkene. The concerted nature of this step requires that both groups add across the same face of the alkene, giving a *syn* addition. In this way, the proposed mechanism explains not only the regiochemistry but also the stereochemistry.

When drawing the products of hydroboration-oxidation, it is essential to consider the number of chirality centers that are created during the process. If no chirality centers are formed, then the stereospecificity of the reaction is not relevant.

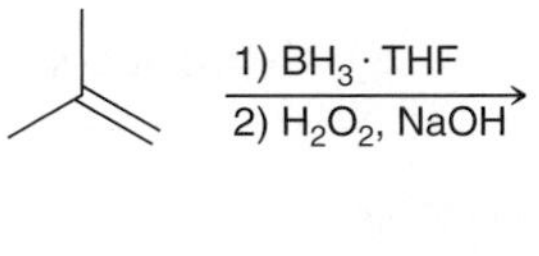

No chirality centers

In this case, only one product is formed, rather than a pair of enantiomers, and the *syn* requirement is irrelevant.

Now consider the stereochemical outcome in a case where one chirality center is formed:

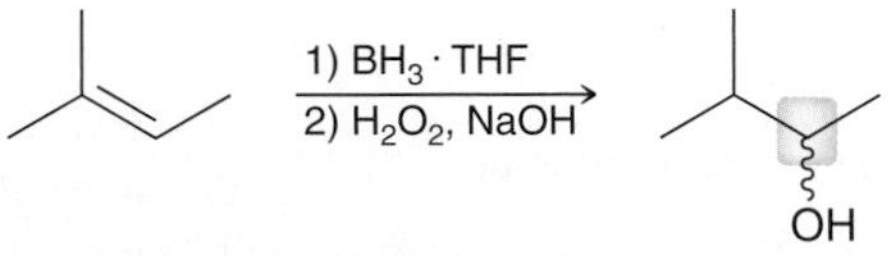

One chirality center

In this case, both enantiomers are obtained, because *syn* addition can take place from either face of the alkene with equal likelihood:

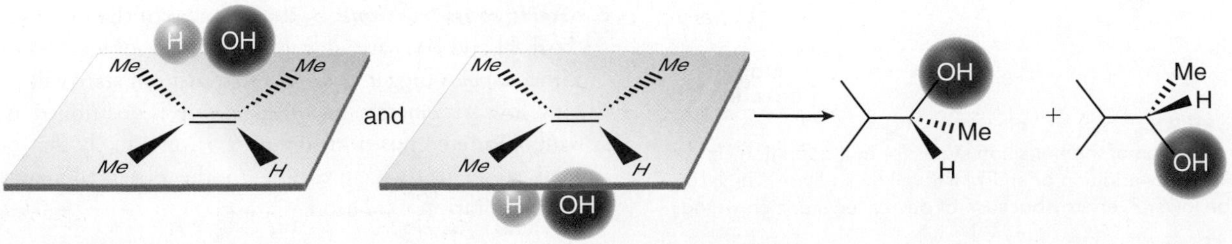

Now consider a case in which two chirality centers are formed:

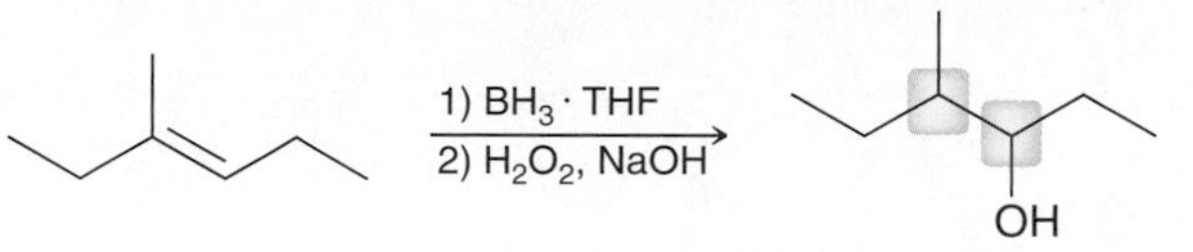

Two chirality centers

In such a case, the requirement for *syn* addition determines which pair of enantiomers is obtained:

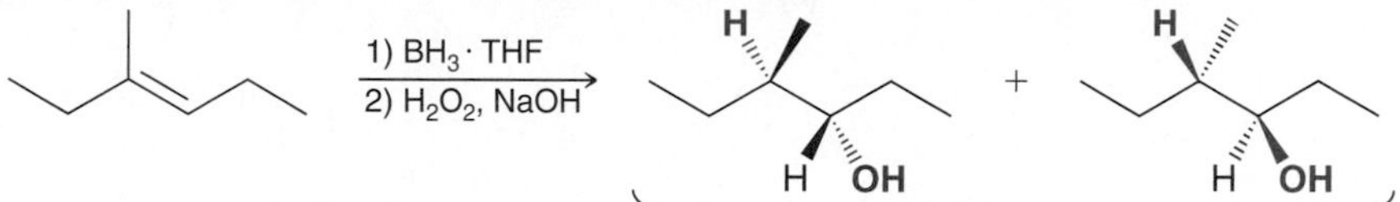

A pair of enantiomers

The other two possible stereoisomers are not obtained.

Under special circumstances, it is possible to achieve *enantioselective addition reactions*, that is, reactions where one enantiomer predominates over the other (and there is an observed % *ee*, or enantiomeric excess). However, we have not yet seen any methods for accomplishing this. We will see one such example later in this chapter, but as of yet, none of the reactions in this chapter have been enantioselective. As a result, all of the reactions presented in this chapter will produce a product mixture that is optically inactive.

SKILLBUILDER

9.4 PREDICTING THE PRODUCTS OF HYDROBORATION-OXIDATION

LEARN the skill

Predict the product(s) for the following transformation:

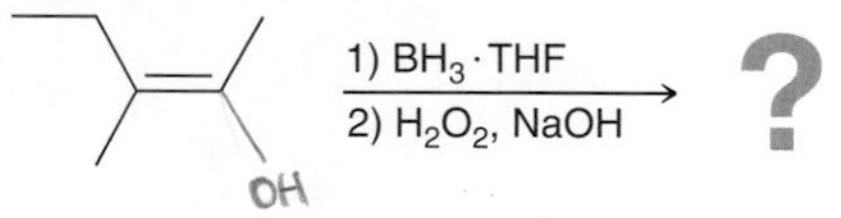

SOLUTION

This process is a hydroboration-oxidation, which will add water across the double bond. The first step is to determine the regiochemical outcome. That is, we must determine where the OH is positioned. Recall that hydroboration-oxidation produces an *anti*-Markovnikov addition, which means that the OH is positioned at the less substituted carbon:

STEP 1
Determine the regiochemical outcome based on the requirement for *anti*-Markovnikov addition.

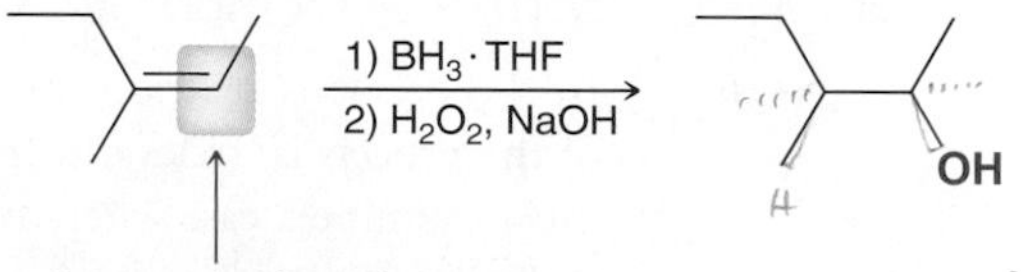

Less substituted

Now we know where to place the OH group, but the drawing is still not complete, because the stereochemical outcome must be indicated. Specifically, we must ask if two new chirality centers are formed in this reaction. In this example, there are, in fact, two new chirality centers being formed. So, the stereospecificity of this process (*syn* addition) is relevant and must be taken into account when drawing the products. The reaction will produce only the pair of enantiomers that results from a *syn* addition:

STEP 2
Determine the stereochemical outcome based on the requirement for *syn* addition.

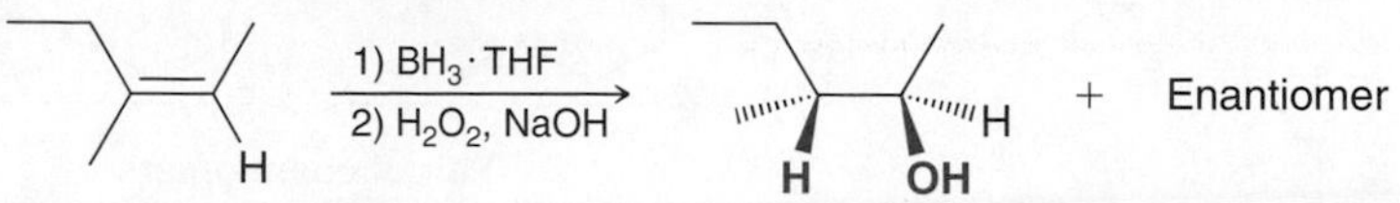

PRACTICE the skill **9.19** Predict the product(s) for each of the following transformations:

(a)
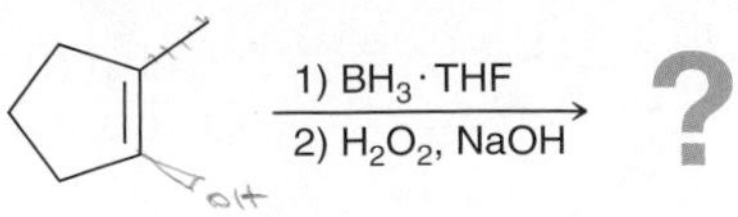

(b)
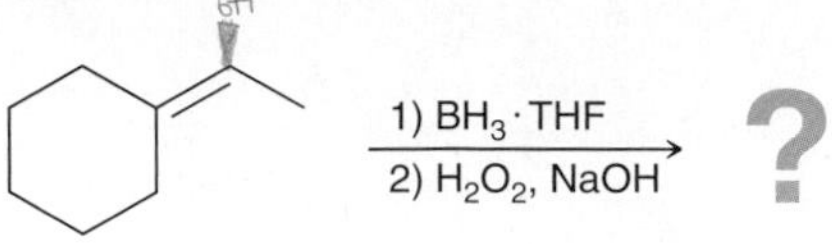

(c)
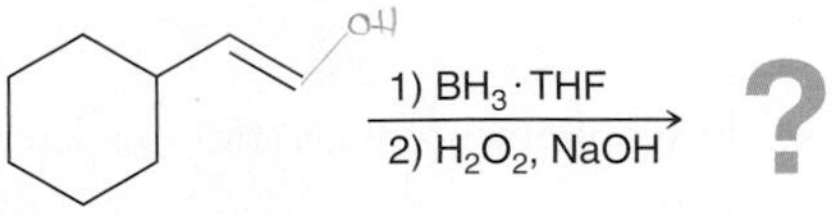

(d)
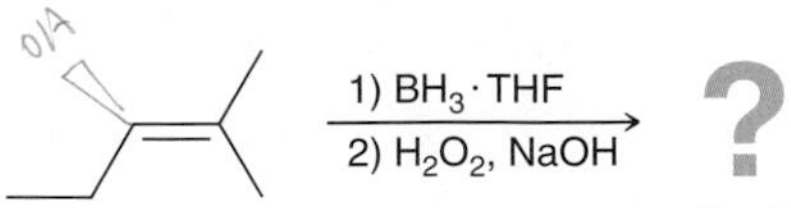

(e)
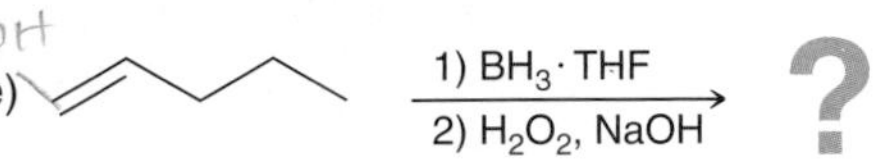

(f)
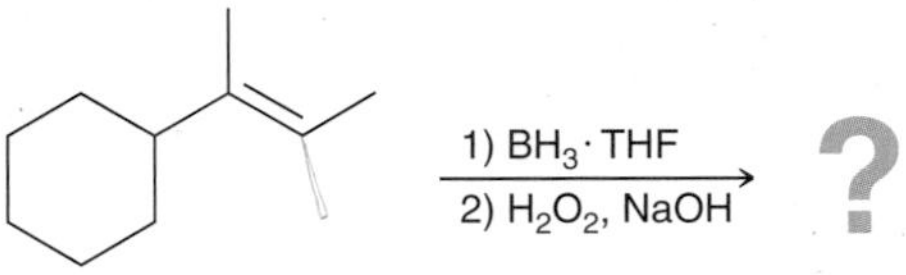

APPLY the skill

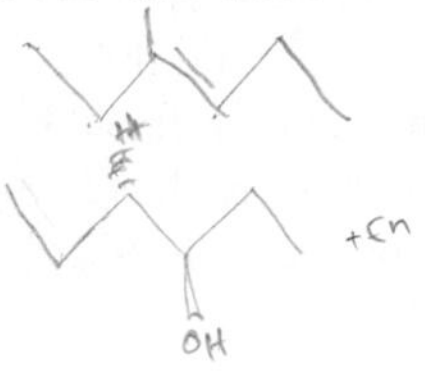

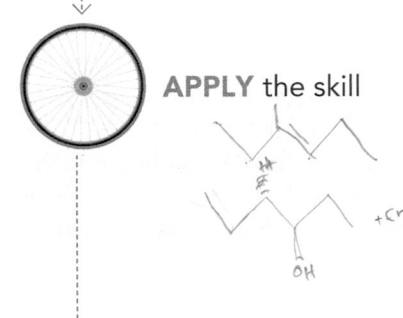

9.20 Draw the product(s) obtained from hydroboration-oxidation of (*E*)-3-methyl-3-hexene.

9.21 The products obtained from hydroboration-oxidation of *cis*-2-butene are identical to the products obtained from hydroboration-oxidation of *trans*-2-butene. Draw the products and explain why the configuration of the starting alkene is not relevant in this case.

9.22 Compound A has molecular formula C_5H_{10}. Hydroboration-oxidation of compound A produces an alcohol with no chirality centers. Draw two possible structures for compound A.

need more **PRACTICE?** **Try Problem 9.66**

9.7 Catalytic Hydrogenation

LOOKING BACK
The hydrogen atoms installed during this process are not explicitly drawn in the product, because the presence of hydrogen atoms can be inferred from bond-line drawings. For more practice "seeing" the hydrogen atoms that are not explicitly drawn, review the exercises in Section 2.2.

Catalytic hydrogenation involves the addition of molecular hydrogen (H_2) across a double bond in the presence of a metal catalyst; for example:

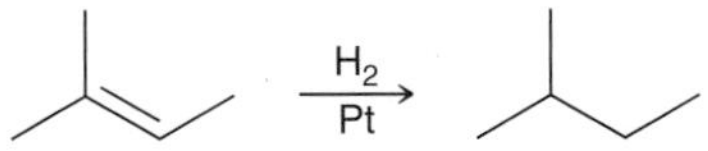

(100%)

The net result of this process is to reduce an alkene to an alkane. Paul Sabatier was the first to demonstrate that catalytic hydrogenation could serve as a general procedure for reducing alkenes, and for his pioneering work, he was a corecipient of the 1912 Nobel Prize in Chemistry.

Stereospecificity of Catalytic Hydrogenation

In the previous reaction there are no chirality centers in the product, so the stereospecificity of the process is irrelevant. In order to explore whether a reaction displays stereospecificity, we must examine a case where two new chirality centers are being formed. For example, consider the following case:

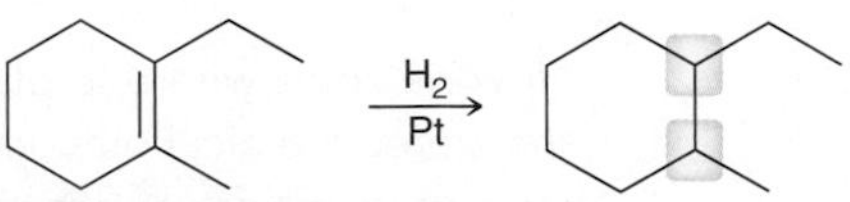

With two chirality centers, there are four possible stereoisomeric products (two pairs of enantiomers):

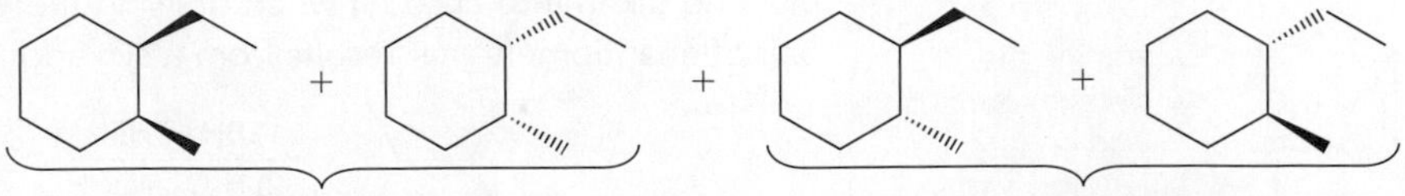

Pair of enantiomers (observed)

Pair of enantiomers (not observed)

However, the reaction does not produce all four products. Only one pair of enantiomers is observed, the pair that results from a *syn* addition. To understand the reason for the observed stereospecificity, we must take a close look at the reagents and their proposed interactions.

The Role of the Catalyst in Catalytic Hydrogenation

Catalytic hydrogenation is accomplished by treating an alkene with H_2 gas and a metal catalyst, often under conditions of high pressure. The role of the catalyst is illustrated by the energy diagram in Figure 9.6. The pathway without the metal catalyst (blue) has a very large energy of activation (E_a), rendering the reaction too slow to be of practical use. The presence of a catalyst provides a pathway (red) with a lower energy of activation, thereby allowing the reaction to occur more rapidly.

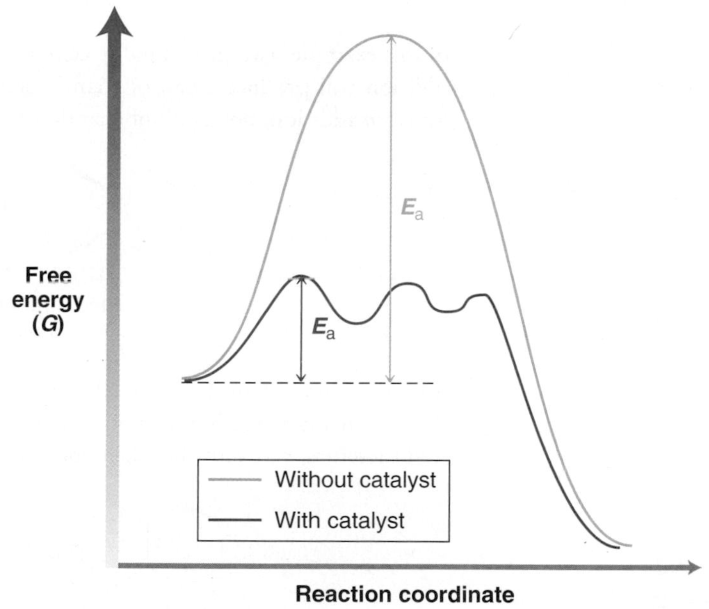

FIGURE 9.6
An energy diagram showing hydrogenation with a catalyst (red) and without a catalyst (blue). The latter has a lower energy of activation and therefore occurs more rapidly.

A variety of metal catalysts can be used, such as Pt, Pd, or Ni. The process is believed to begin when molecular hydrogen (H_2) interacts with the surface of the metal catalyst, effectively breaking the H—H bonds and forming individual hydrogen atoms adsorbed to the surface of the metal. The alkene coordinates with the metal surface, and surface chemistry allows for the reaction between the π bond and two hydrogen atoms, effectively adding H and H across the alkene (Figure 9.7). In this process, both hydrogen atoms add to the same face of the alkene, explaining the observed stereospecificity (*syn* addition).

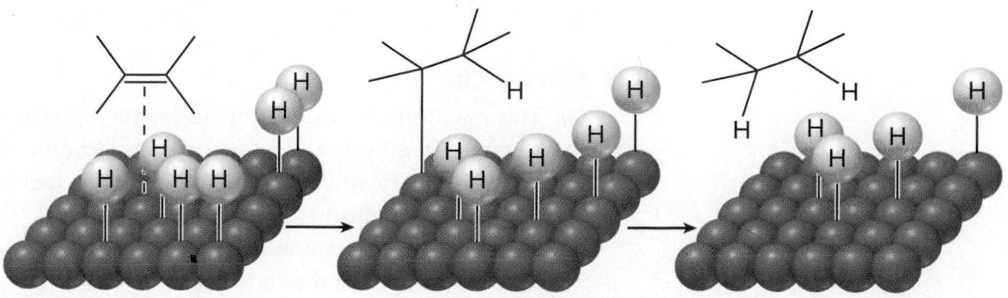

FIGURE 9.7
The addition process takes place on the surface of a metal catalyst.

TABLE 9.2 SUMMARY OF RELATIONSHIP BETWEEN NUMBER OF CHIRALITY CENTERS FORMED AND STEREOCHEMICAL OUTCOME OF CATALYTIC HYDROGENATION

Zero chirality centers	*Syn* requirement is not relevant. Only one product formed.
One chirality center	Both possible enantiomers are formed.
Two chirality centers	The requirement for *syn* addition determines which pair of enantiomers is obtained.

In any given case, *the stereochemical outcome is dependent on the number of chirality centers formed in the process*, as summarized in Table 9.2. Care must be taken when applying the simple paradigm in Table 9.2, because symmetrical alkenes will produce a meso compound rather than a pair of enantiomers. Consider the following case:

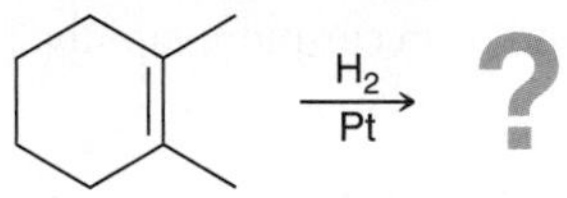

In this example, two new chirality centers are formed, and therefore, we might expect that *syn* addition will produce a pair of enantiomers. However, in this case, there is only one product from *syn* addition, not a pair of enantiomers:

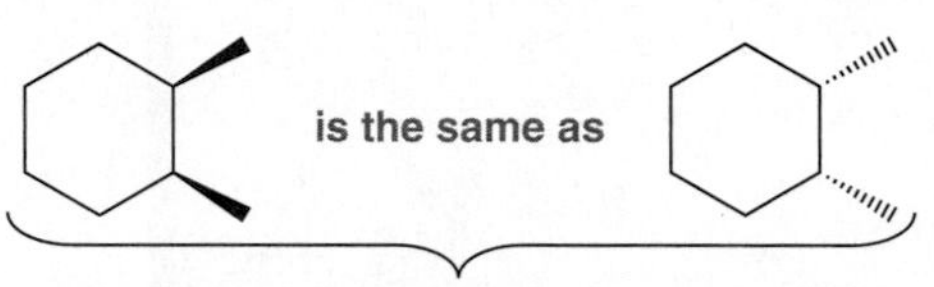

A *meso* compound, by definition, does not have an enantiomer. A *syn* addition on one face of the alkene generates exactly the same compound as a *syn* addition on the other face of the alkene. And therefore, care must be taken not to write "+ Enantiomer" or "+ En" for short:

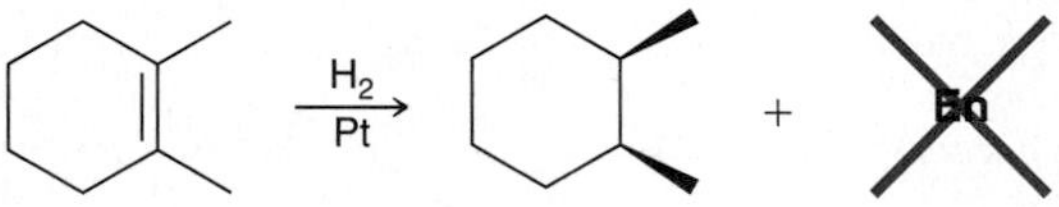

SKILLBUILDER

9.5 PREDICTING THE PRODUCTS OF CATALYTIC HYDROGENATION

LEARN the skill

Predict the products of each of the following reactions:

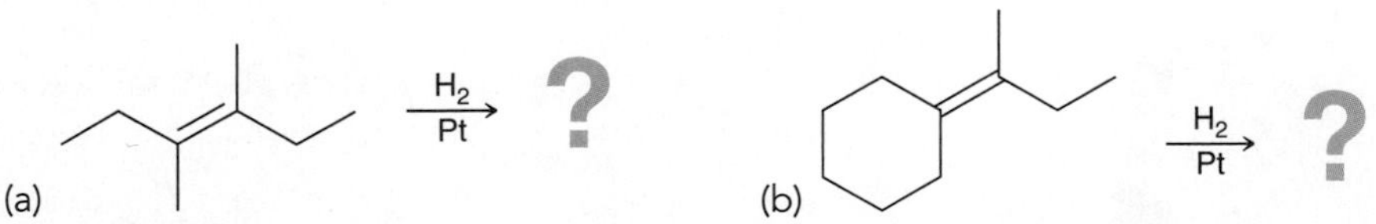

SOLUTION

(a) This reaction is a catalytic hydrogenation process, in which H and H are added across the alkene. Since both groups are identical (H and H), it is not necessary to consider regiochemical issues. However, stereochemistry is a factor that must be considered. In order to properly draw the products, it is first necessary to determine how many chirality centers are formed as a result of this reaction:

STEP 1 Determine the number of chirality centers formed.

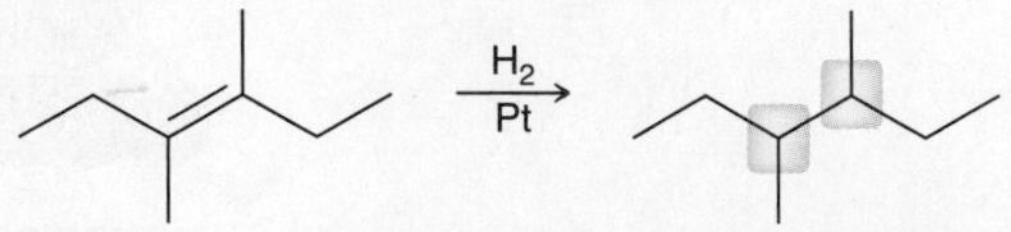

In this case, two chirality centers are formed. Therefore, we expect only the pair of enantiomers that would result from a *syn* addition:

STEP 2
Determine the stereochemical outcome based on the requirement for *syn* addition.

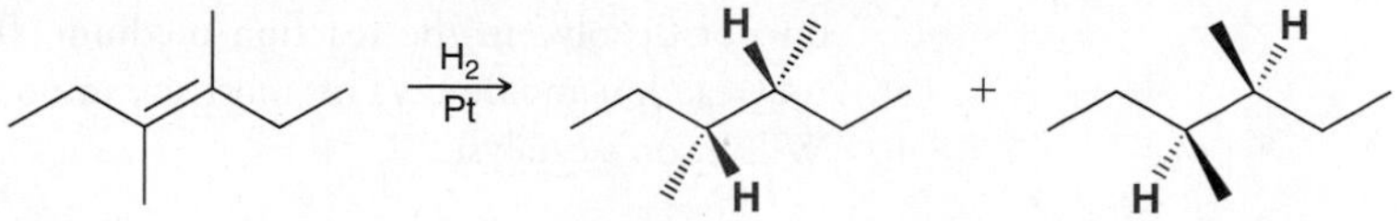

STEP 3
Verify that the products do not represent a single *meso* compound.

As a final check, just make sure that the products do not represent a single meso compound. The compound above lacks an internal plane of symmetry and is not a *meso* compound.

(b) Like the previous example, this reaction is a catalytic hydrogenation process, in which H and H are added across the alkene. In order to properly draw the products, it is first necessary to determine how many chirality centers are formed as a result of this reaction:

LOOKING BACK
For practice identifying *meso* compounds, review the exercises in Section 5.6.

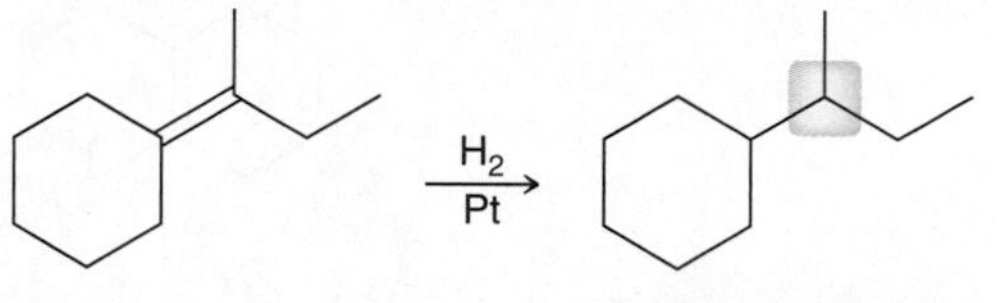

One chirality center

In this case, only one chirality center is formed. Therefore, we expect both possible enantiomers, because *syn* addition can occur from either face of the π bond with equal likelihood:

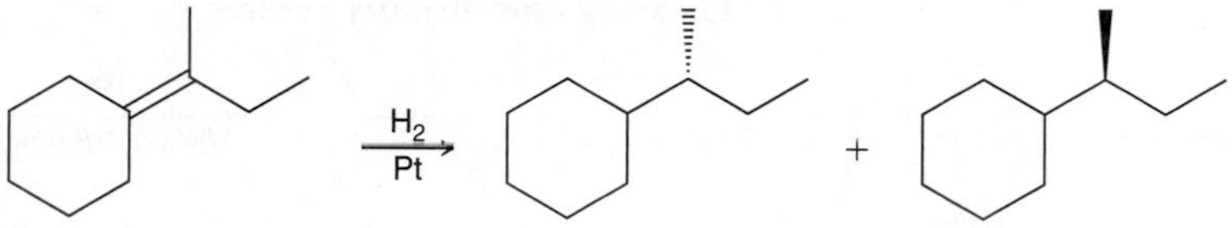

It is impossible for a compound containing exactly one chirality center to be a *meso* compound, so the compounds above are a pair of enantiomers.

PRACTICE the skill **9.23** Predict the product(s) for each of the following reactions:

(a) 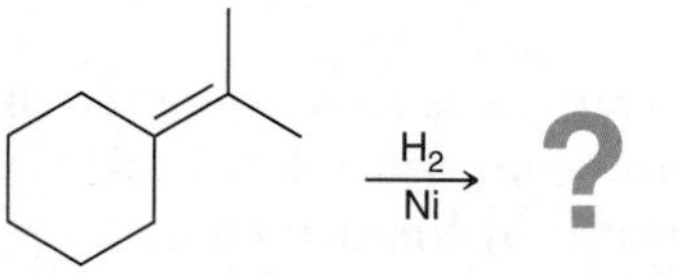

(b) 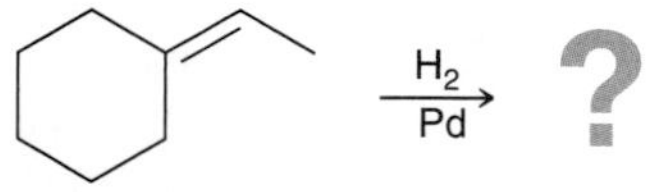

(c) 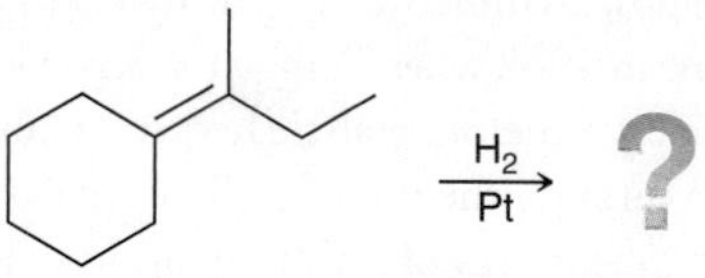

(d) 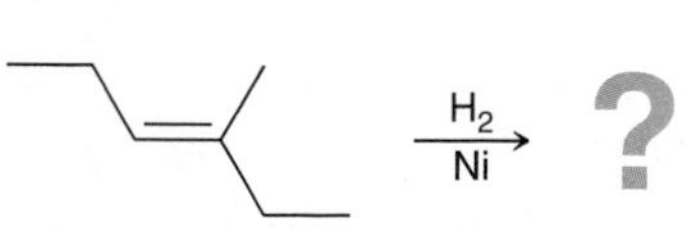

(e) 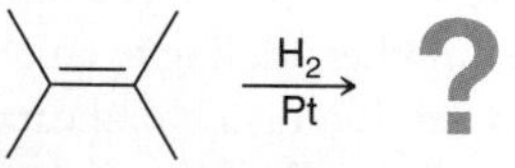

(f) 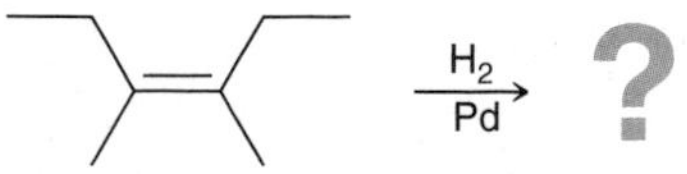

APPLY the skill **9.24** In much the same way that they react with H_2, alkenes also react with D_2 (deuterium is an isotope of hydrogen). Use this information to predict the product(s) of the following reaction:

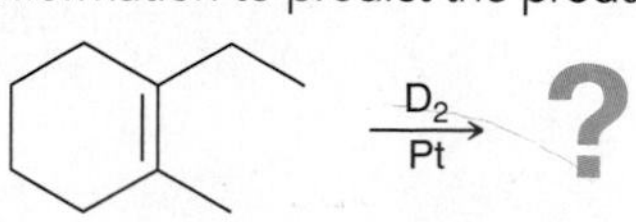

9.25 Compound X has molecular formula C_5H_{10}. In the presence of a metal catalyst, compound X reacts with one equivalent of molecular hydrogen to yield 2-methylbutane.

(a) Suggest three possible structures for compound X.

(b) Hydroboration-oxidation of compound X yields a product with no chirality centers. Identify the structure of compound X.

need more **PRACTICE?** **Try Problems 9.55, 9.75**

Homogeneous Catalysts

The catalysts described so far (Pt, Pd, Ni) are all called **heterogeneous catalysts** because they do not dissolve in the reaction medium. In contrast, **homogeneous catalysts** are soluble in the reaction medium. The most common homogeneous catalyst for hydrogenation is called Wilkinson's catalyst:

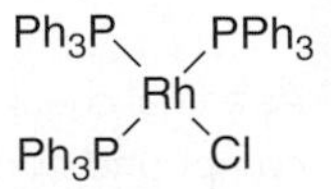

Wilkinson's catalyst

With homogeneous catalysts, a *syn* addition is also observed:

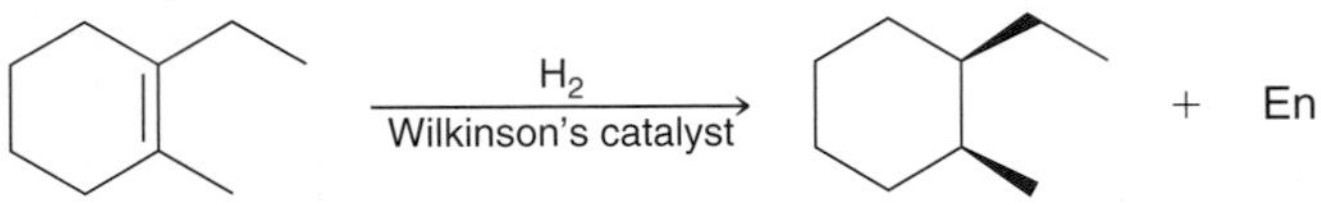

Asymmetric Catalytic Hydrogenation

As seen earlier in this section, when hydrogenation involves the creation of either one or two chirality centers, a pair of enantiomers is expected:

Creating one chirality center:

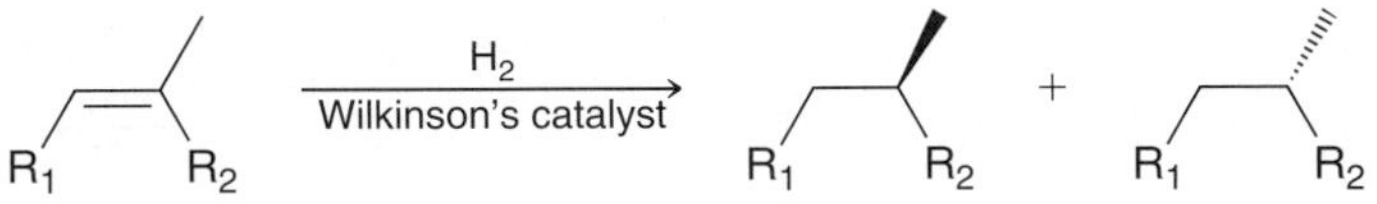

Creating two chirality centers:

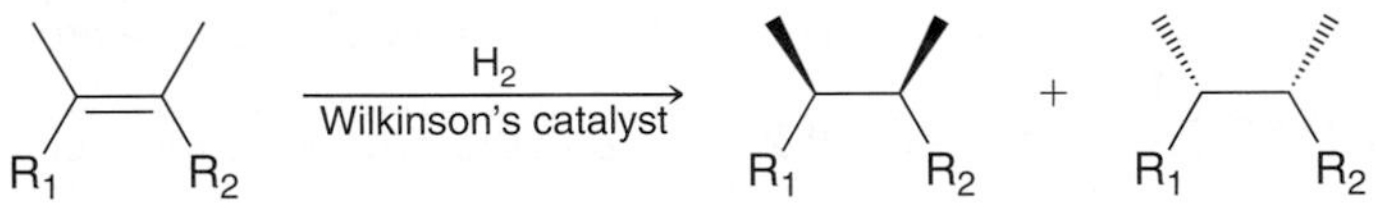

In both reactions a racemic mixture is formed. This begs the obvious question: Is it possible to create only one enantiomer rather than a pair of enantiomers? In other words, is it possible to perform an **asymmetric hydrogenation**?

Before the 1960s, asymmetric catalytic hydrogenation had not been achieved. However, a major breakthrough came in 1968 when William S. Knowles, working for the Monsanto Company, developed a method for asymmetric catalytic hydrogenation. Knowles realized that asymmetric induction might be possible through the use of a chiral catalyst. He reasoned that a chiral catalyst should be capable of lowering the energy of activation for formation of one enantiomer more dramatically than the other enantiomer (Figure 9.8). In this way, a chiral catalyst could theoretically favor the production of one enantiomer over another, leading to an observed enantiomeric excess (*ee*).

Knowles succeeded in developing a chiral catalyst by preparing a cleverly modified version of Wilkinson's catalyst. Recall that Wilkinson's catalyst has three triphenylphosphine ligands:

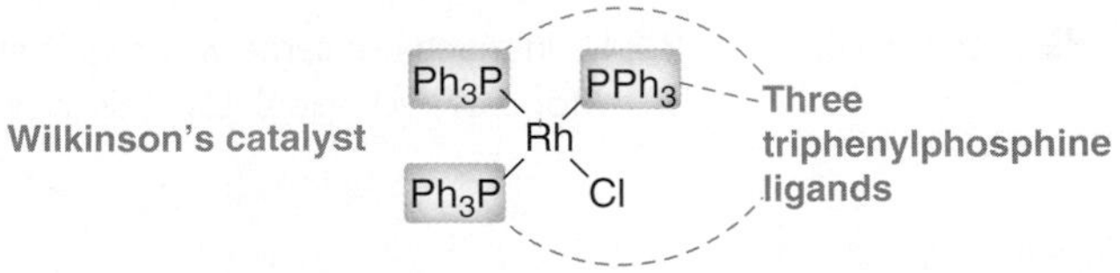

Knowles' idea was to use chiral phosphine ligands, rather than symmetrical phosphine ligands:

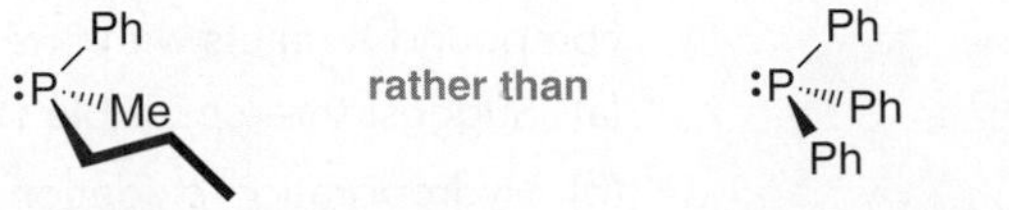

Chiral phosphine ligand

Symmetrical phosphine ligand

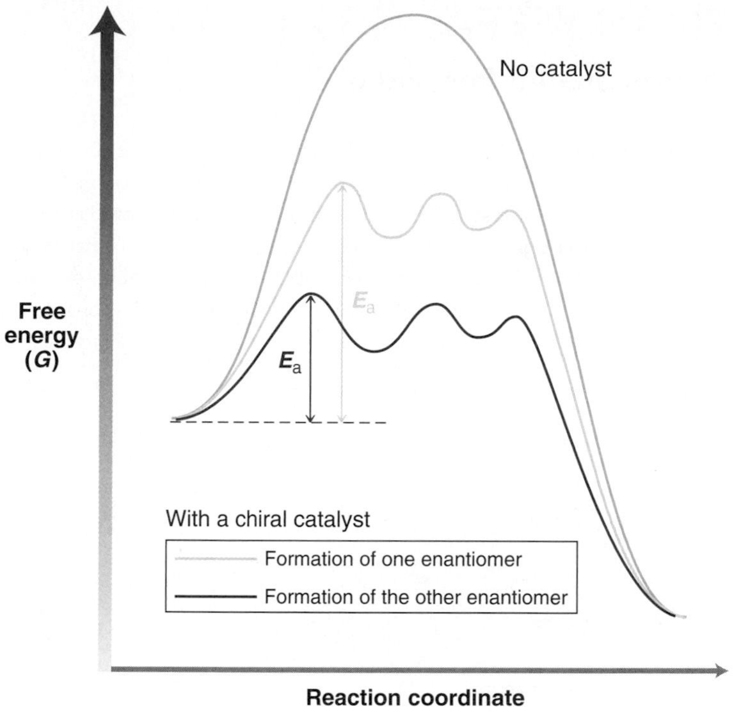

FIGURE 9.8
An energy diagram showing the ability of a chiral catalyst to favor formation of one enantiomer (red pathway) over the other enantiomer (green pathway).

Using chiral phosphine ligands, Knowles prepared a chiral version of Wilkinson's catalyst. He used his chiral catalyst in a hydrogenation reaction and demonstrated a modest enantiomeric excess. He didn't get exclusively one enantiomer, but he got enough of an enantiomeric excess to prove the point—that asymmetric catalytic hydrogenation was in fact possible.

Knowles developed other catalysts capable of much higher enantiomeric excess, and he then set out to use asymmetric catalytic hydrogenation to develop an industrial synthesis of the amino acid L-dopa:

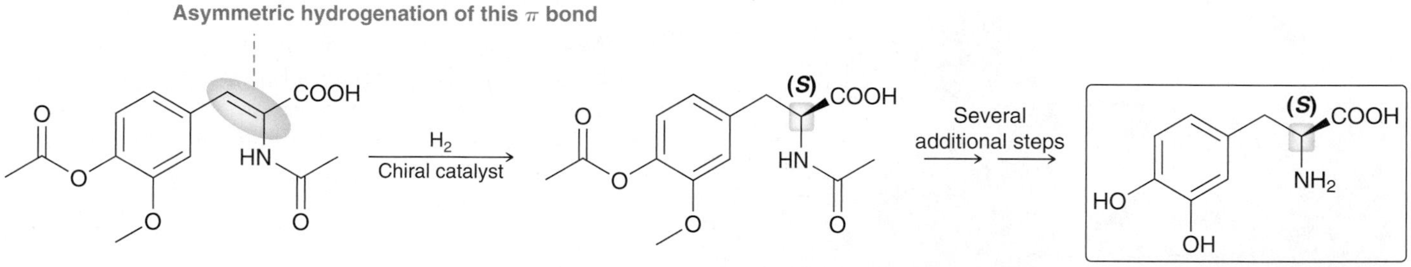

(*S*)-3,4-Dihydroxyphenylalanine
(L-dopa)

LOOKING AHEAD

The L designation describes the configuration of the chirality center present in the compound. This notation will be described in greater detail in Section 25.2.

L-Dopa had been shown to be effective in treating dopamine deficiencies associated with Parkinson's disease (a discovery that earned Arvid Carlsson the 2000 Nobel Prize in Physiology or Medicine). Dopamine is an important neurotransmitter in the brain. A dopamine deficiency cannot be treated by administering dopamine to the patient, because dopamine cannot cross the blood-brain barrier. However, L-dopa can cross this barrier, and it is subsequently converted into dopamine in the central nervous system. This provides a way to increase the levels of dopamine in the brains of patients with Parkinson's disease, thereby providing temporary relief from some of the symptoms associated with the disease. But there is one catch. The enantiomer of L-dopa is believed to be toxic, and therefore, an enantioselective synthesis of L-dopa is required.

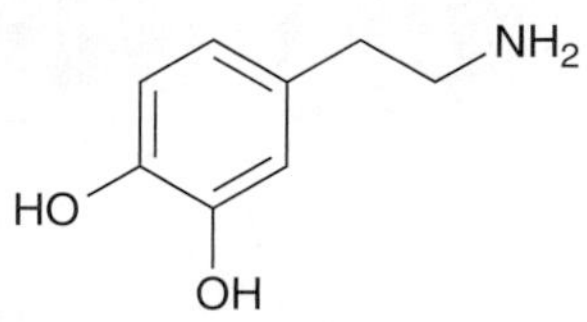

Dopamine

PRACTICALLYSPEAKING)))

Partially hydrogenated fats and oils

Earlier in this chapter, we focused on the role of hydrogenation in the pharmaceutical industry. However, the best known role of hydrogenation is in the food industry, where hydrogenation is used to prepare partially hydrogenated fats and oils.

Naturally occurring fats and oils, such as vegetable oil, are generally mixtures of compounds called *triglycerides*, which contain three long alkyl chains:

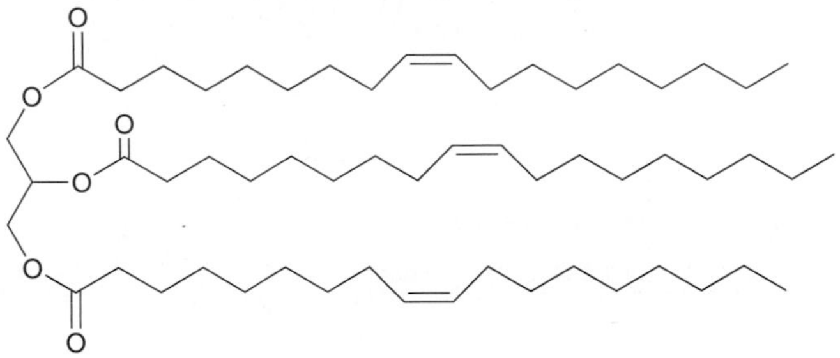

A triglyceride

These important compounds are discussed in more detail in Section 26.3. For now, we focus our attention on the π bonds present in the alkyl chains. Hydrogenation of some of these π bonds alters the physical properties of the oil. For example, cottonseed oil is a liquid at room temperature; however, *partially hydrogenated* cottonseed oil (Crisco®) is a solid at room temperature. This is advantageous because it gives the oil a longer shelf-life. Margarine is prepared in a similar way from a variety of animal and vegetable oils.

Partially hydrogenated oils do have issues, though. There is much evidence that the catalysts present during the hydrogenation process can often isomerize some of the double bonds, producing *trans* double bonds:

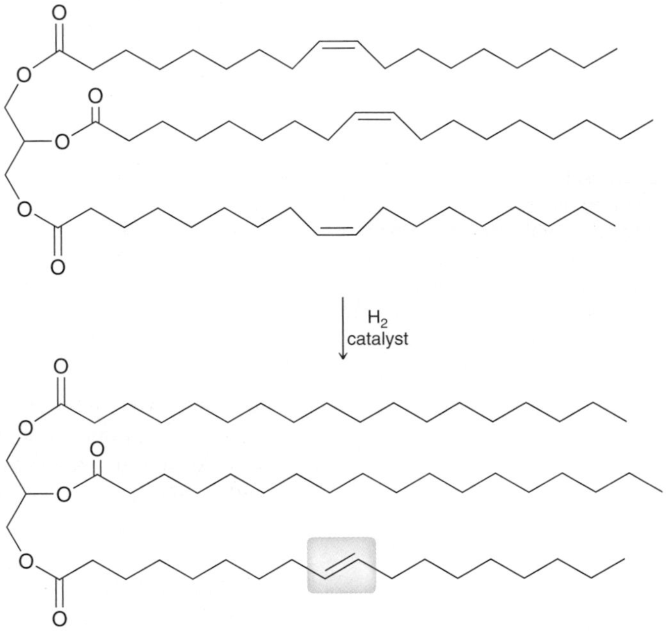

trans

These so-called *trans* fats are believed to cause an increase in LDL (low-density lipoprotein) cholesterol levels, which leads to an increased rate of cardiovascular disease. In response, the food industry is now making an effort to minimize or completely remove the *trans* fats in food products. Current food labels often boast: "0 grams of *trans* fats."

LOOKING AHEAD

For a discussion of the connection between cholesterol and cardiovascular disease, see the Medically Speaking box in Section 26.5.

BY THE WAY

The other half of the 2001 Nobel Prize in Chemistry was awarded to K. Barry Sharpless, for his work on enantioselective synthesis, described in Section 14.9.

Asymmetric catalytic hydrogenation turned out to be an efficient way to prepare L-dopa with high optical purity. For his work, Knowles shared half of the 2001 Nobel Prize in Chemistry, together with Ryoji Noyori (Nagoya University, Japan), who was independently investigating a wide variety of chiral catalysts that could produce an asymmetric catalytic hydrogenation.

Noyori varied both the metal and the ligands attached to the metal, and he was able to create chiral catalysts that achieved enantioselectivity close to 100% *ee*. One example, commonly used in synthesis these days, is based on the chiral ligand called BINAP:

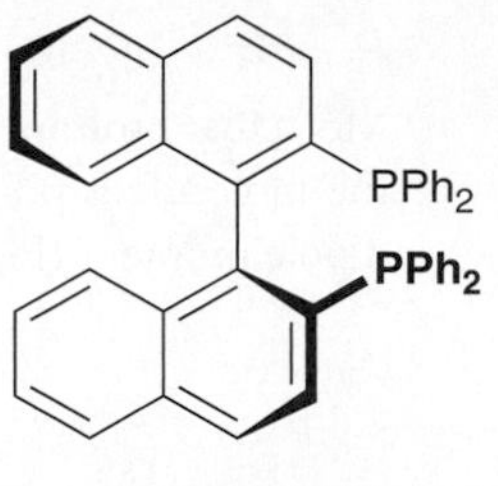

(*S*)-(–)-BINAP
(*S*)-2,2′-Bis(diphenylphosphino)-1,1′-binaphthyl

BINAP does not have a chirality center, but nevertheless, it is a chiral compound because the single bond joining the two ring systems does not experience free rotation (as a result of steric hindrance). BINAP can be used as a chiral ligand to form a complex with ruthenium, producing a chiral catalyst capable of achieving very pronounced enantioselectivity:

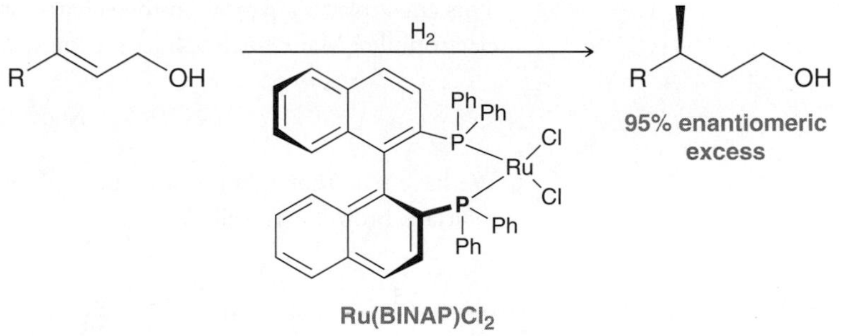

9.8 Halogenation and Halohydrin Formation

Experimental Observations

Halogenation involves the addition of X_2 (either Br_2 or Cl_2) across an alkene. As an example, consider the chlorination of ethylene to produce dichloroethane:

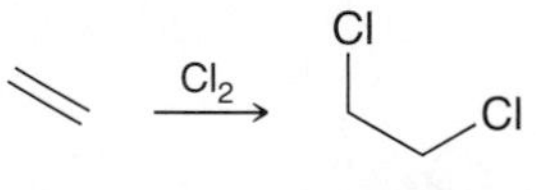

(97%)

This reaction is a key step in the industrial preparation of polyvinylchloride (PVC):

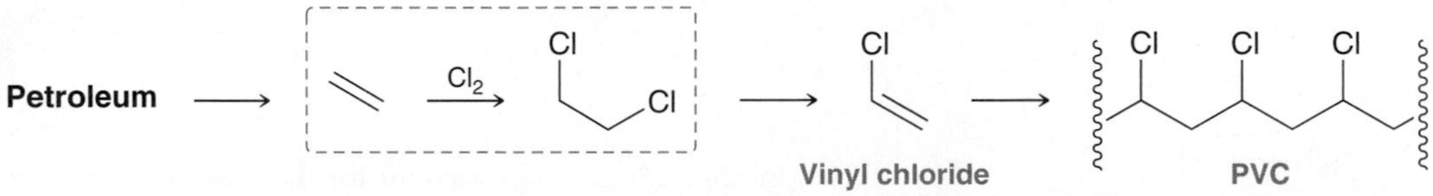

Halogenation of alkenes is only practical for the addition of chlorine or bromine. The reaction with fluorine is too violent, and the reaction with iodine often produces very low yields.

The stereospecificity of halogenation reactions can be explored in a case where two new chirality centers are formed. For example, consider the products that are formed when cyclopentene is treated with molecular bromine (Br_2):

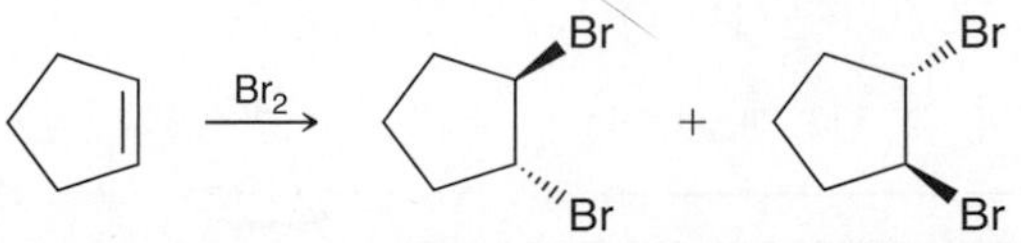

WATCH OUT
The products of this reaction are a pair of enantiomers. They are not the same compound. Students are often confused with this particular example. For a review of enantiomers, go to Section 5.5.

Notice that addition occurs in a way that places the two halogen atoms on opposite sides of the π bond. This mode of addition is called an ***anti* addition**. *For most simple alkenes, halogenation appears to proceed primarily via an anti addition.* Any proposed mechanism must be consistent with this observation.

A Mechanism for Halogenation

Molecular bromine is a nonpolar compound, because the Br—Br bond is covalent. Nevertheless, the molecule is polarizable, and the proximity of a nucleophile can cause a temporary, induced dipole moment (Figure 9.9).

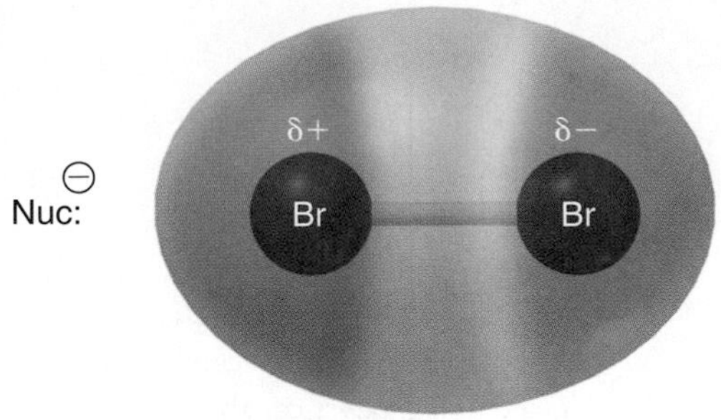

δ+
δ−
Nuc:
Br
Br

FIGURE 9.9 An electrostatic potential map showing the temporary induced dipole moment of molecular bromine when it is in the vicinity of a nucleophile.

This effect places a partial positive charge on one of the bromine atoms, rendering the molecule electrophilic. Many nucleophiles are known to react with bromine:

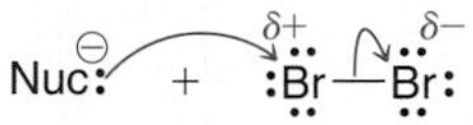

We have seen that π bonds are nucleophilic, and therefore, it is reasonable to expect an alkene to attack bromine as well:

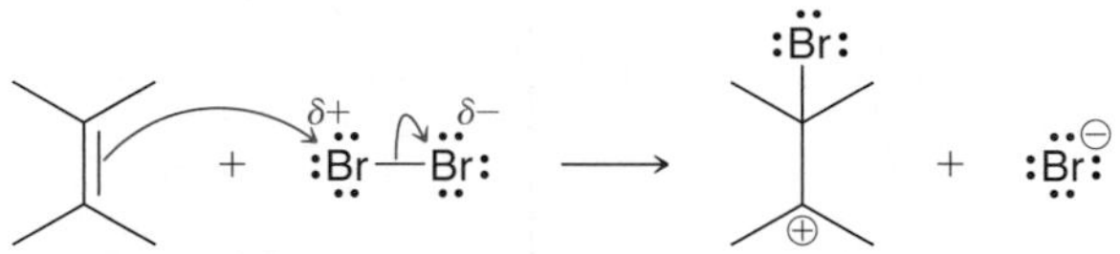

Although this step seems plausible, there is a fatal flaw in this proposal. Specifically, the production of a free carbocation is inconsistent with the observed *anti* stereospecificity of halogenation. If a free carbocation were produced in the process, then both *syn* and *anti* addition would be expected to occur, because the carbocation could be attacked from either side:

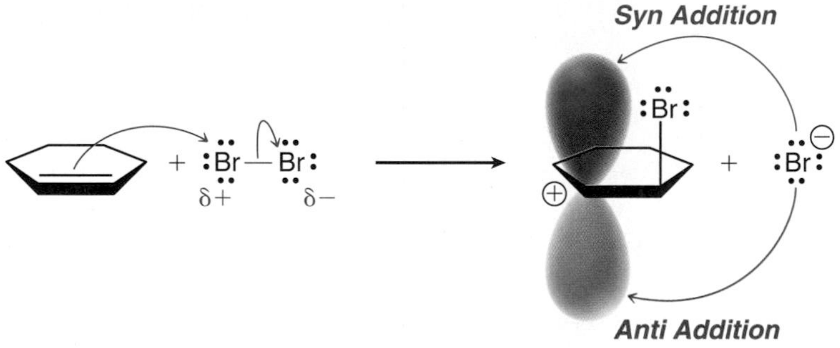

Syn Addition
Anti Addition

This mechanism does not account for the observed *anti* stereospecificity of halogenation. The modified mechanism shown in Mechanism 9.4 is consistent with an *anti* addition:

MECHANISM 9.4 HALOGENATION

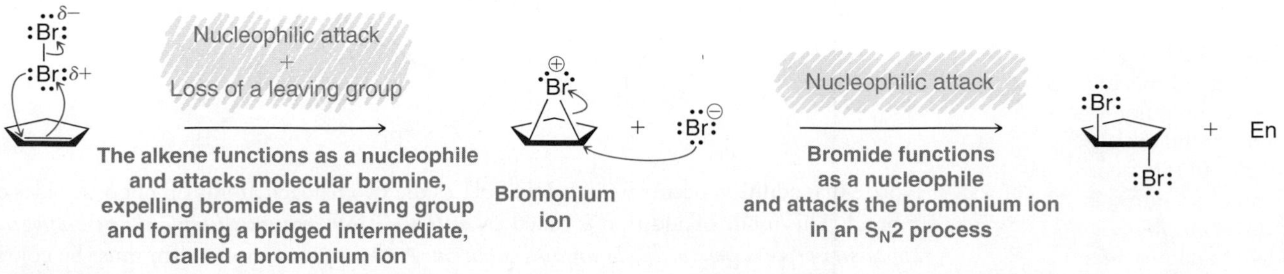

Nucleophilic attack
+
Loss of a leaving group
The alkene functions as a nucleophile and attacks molecular bromine, expelling bromide as a leaving group and forming a bridged intermediate, called a bromonium ion
Bromonium ion
Nucleophilic attack
Bromide functions as a nucleophile and attacks the bromonium ion in an S_N2 process
En

In this mechanism, an additional curved arrow has been introduced, forming a bridged intermediate rather than a free carbocation. This bridged intermediate, called a **bromonium ion**, is similar in structure and reactivity to the mercurinium ion discussed in Section 9.5. Compare their structures:

Bromonium ion Mercurinium ion

In the second step of the proposed mechanism, the bromonium ion is attacked by the bromide ion that was produced in the first step. This step is an S_N2 process and must therefore proceed via a back-side attack (as seen in Section 7.4). The requirement for back-side attack explains the observed stereochemical requirement for *anti* addition.

The stereochemical outcome for halogenation reactions is dependent on the configuration of the starting alkene. For example, *cis*-2-butene will yield different products than *trans*-2-butene:

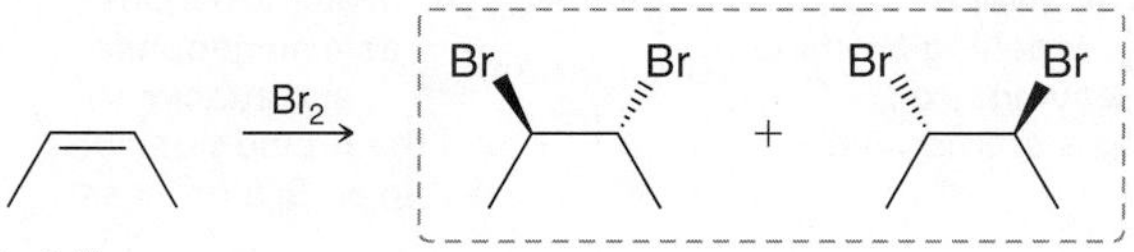

cis-2-Butene

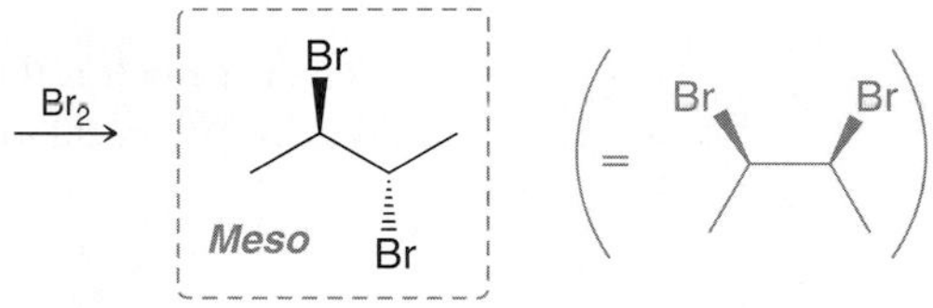

trans-2-Butene

Anti addition across *cis*-2-butene leads to a pair of enantiomers, while *anti* addition across *trans*-2-butene leads to a *meso* compound. These examples illustrate that *the configuration of the starting alkene determines the configuration of the product for halogenation reactions.*

CONCEPTUAL CHECKPOINT

9.26 Predict the major product(s) for each of the following reactions:

(a) 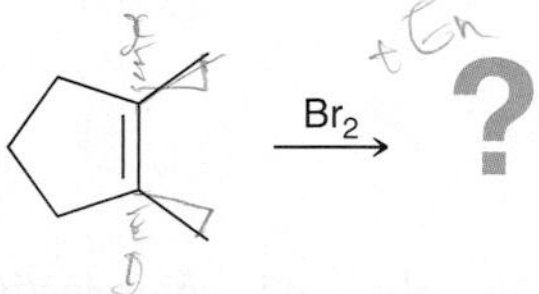

(b) 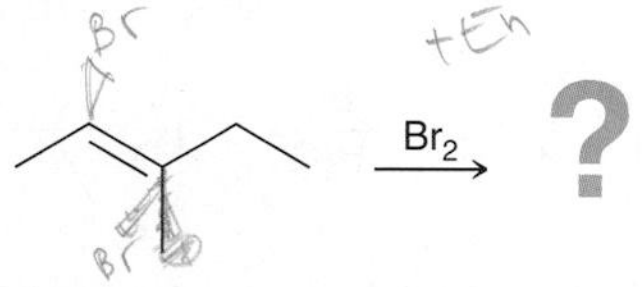

(c) 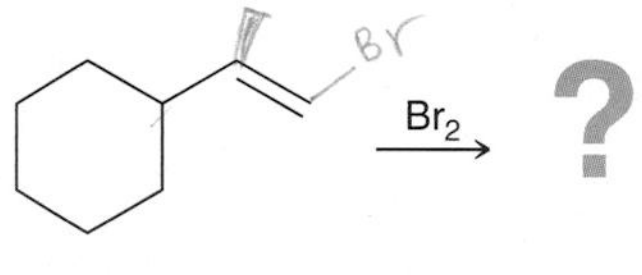

(d) 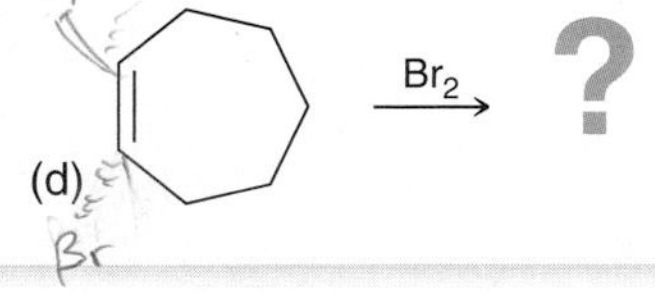

Halohydrin Formation

When bromination occurs in a non-nucleophilic solvent, such as $CHCl_3$, the result is the addition of Br_2 across the π bond (as seen in the previous sections). However, when the reaction is performed in the presence of water, the bromonium ion that is initially formed can be captured by a water molecule, rather than bromide:

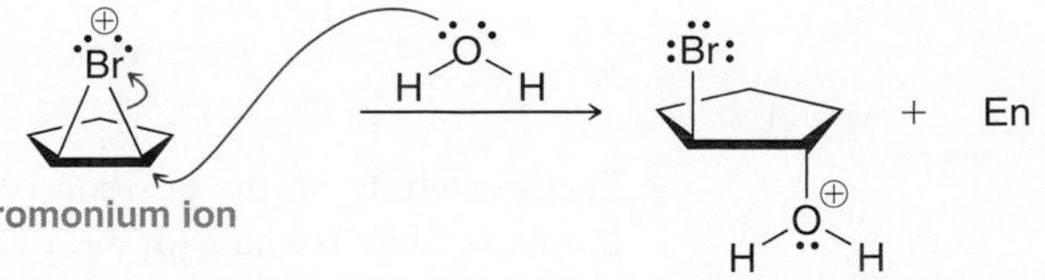

The intermediate bromonium ion is a high-energy intermediate and will react with any nucleophile that it encounters. When water is the solvent, it is more likely that the bromonium ion will be captured by a water molecule before having a chance to react with a bromide ion (although some dibromide product is likely to be formed as well). The resulting oxonium ion is then deprotonated to give the product (Mechanism 9.5).

MECHANISM 9.5 HALOHYDRIN FORMATION

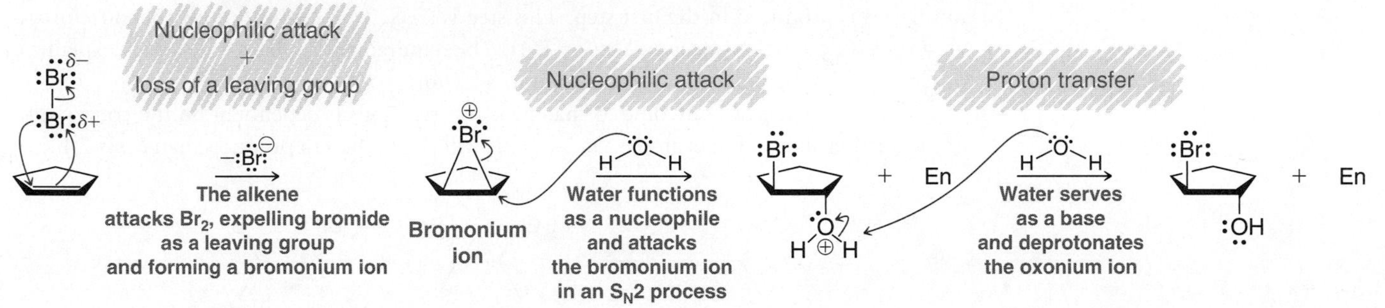
Nucleophilic attack + loss of a leaving group
Nucleophilic attack
Proton transfer
The alkene attacks Br_2, expelling bromide as a leaving group and forming a bromonium ion
Bromonium ion
Water functions as a nucleophile and attacks the bromonium ion in an S_N2 process
+ En
Water serves as a base and deprotonates the oxonium ion
+ En

The net result is the addition of Br and OH across the alkene. The product is called a **bromohydrin**. When chlorine is used in the presence of water, the product is called a **chlorohydrin**:

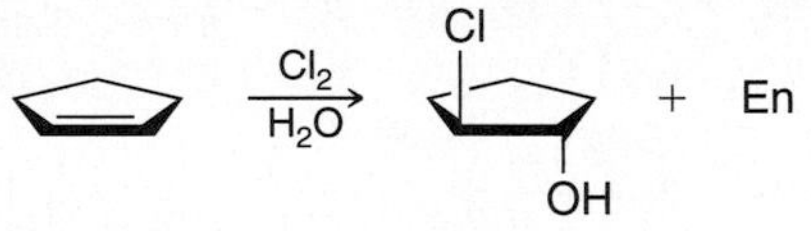
Cl_2
H_2O
Cl
OH
+ En

A chlorohydrin

These reactions are generally referred to as **halohydrin formation**.

Regiochemistry of Halohydrin Formation

In most cases, halohydrin formation is observed to be a regioselective process. Specifically, the OH is generally positioned at the more substituted position:

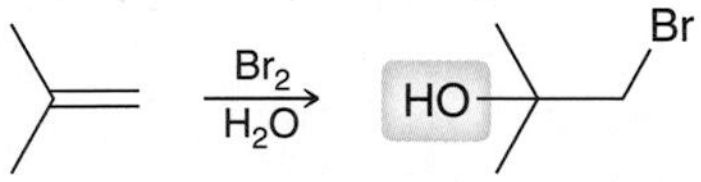
Br_2
H_2O
HO
Br

The proposed mechanism for halohydrin formation can justify the observed regioselectivity. Recall that in the second step of the mechanism the bromonium ion is captured by a water molecule:

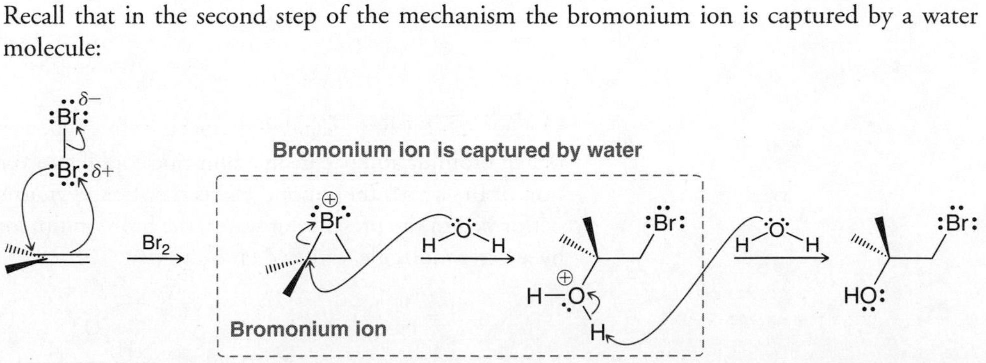
Br_2
Bromonium ion is captured by water
Bromonium ion

Focus carefully on the position of the positive charge throughout the reaction. Think of the positive charge as a hole (or more accurately, a site of electron deficiency) that is passed from one place to another. It begins on the bromine atom and ends up on the oxygen atom. In order to do so, the positive charge must pass through a carbon atom in the transition state:

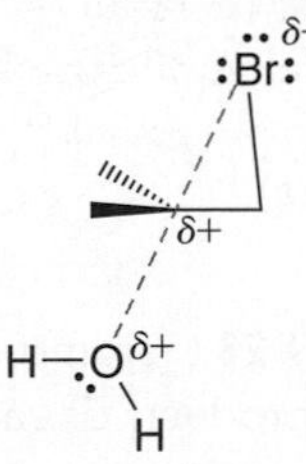

In other words, the transition state for this step will bear partial carbocationic character. This explains why the water molecule is observed to attack the more substituted carbon. The more substituted carbon is more capable of stabilizing the partial positive charge in the transition state. As a result, the transition state will be lower in energy when the attack takes place at the more substituted carbon atom. The proposed mechanism is therefore consistent with the observed regioselectivity of halohydrin formation.

SKILLBUILDER

9.6 PREDICTING THE PRODUCTS OF HALOHYDRIN FORMATION

LEARN the skill

Predict the major product(s) for the following reaction:

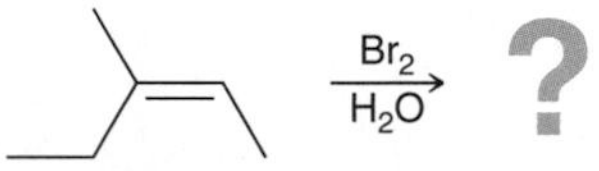

SOLUTION

The presence of water indicates halohydrin formation (addition of Br and OH). The first step is to identify the regiochemical outcome. Recall that the OH group is expected to be positioned at the more substituted carbon:

STEP 1
Determine the regiochemical outcome. The OH group should be placed at the more substituted carbon.

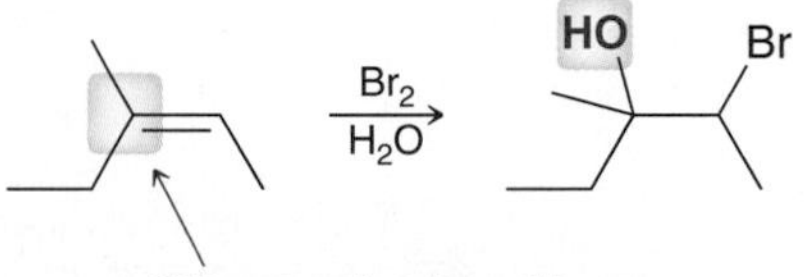

The next step is to identify the stereochemical outcome. In this case, two new chirality centers are formed, so we expect only the pair of enantiomers that would result from *anti* addition. That is, the OH and Br will be installed on opposite sides of the π bond:

STEP 2
Determine the stereochemical outcome based on the requirement for *anti* addition.

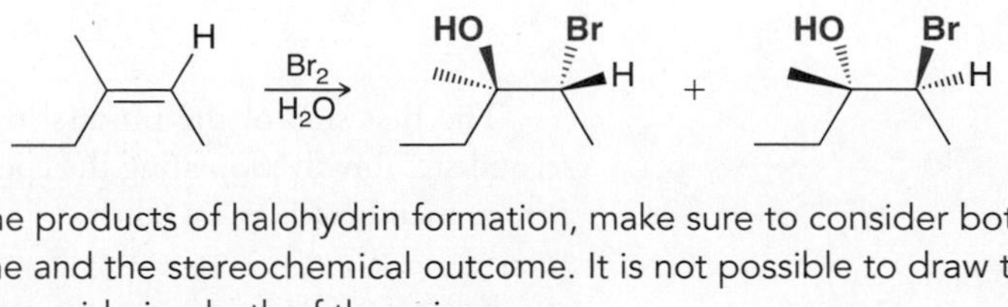

When drawing the products of halohydrin formation, make sure to consider both the regiochemical outcome and the stereochemical outcome. It is not possible to draw the products correctly without considering both of these issues.

PRACTICE the skill **9.27** Predict the major product(s) that are expected when each of the following alkenes is treated with Br_2/H_2O:

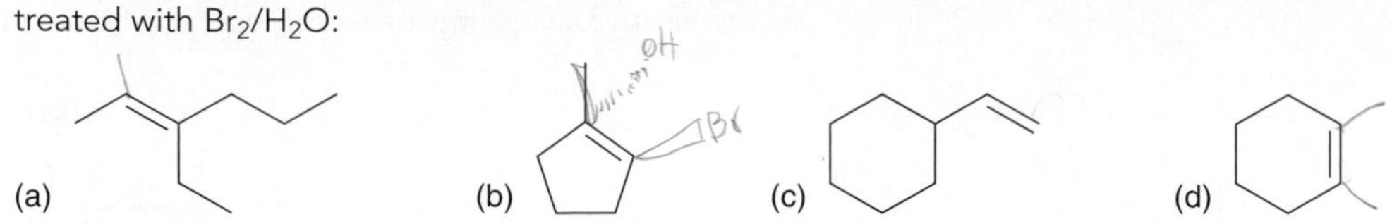

APPLY the skill **9.28** Bromonium ions can be captured by nucleophiles other than water. Predict the products of each of the following reactions:

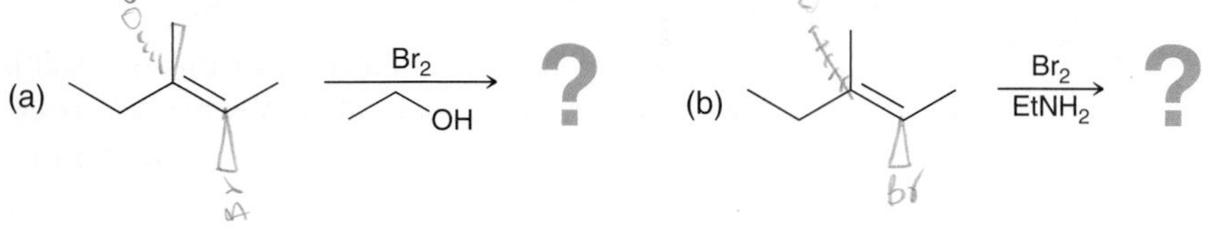

9.29 When *trans*-1-phenylpropene is treated with bromine, some *syn* addition is observed. Explain why the presence of a phenyl group causes a loss of stereospecificity.

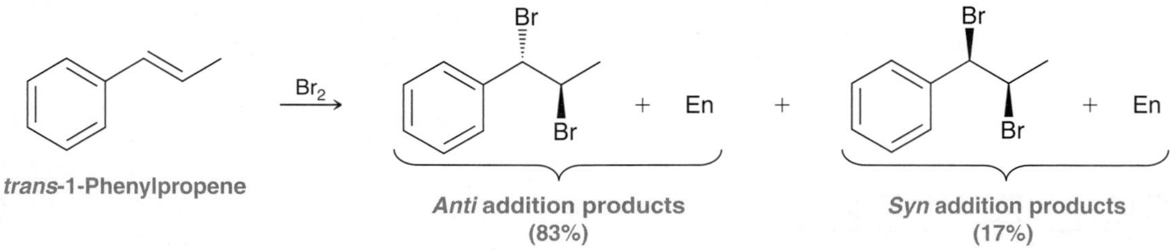

need more **PRACTICE?** **Try Problem 9.73**

9.9 *Anti* Dihydroxylation

As mentioned in the introductory section of this chapter, **dihydroxylation** reactions are characterized by the addition of OH and OH across an alkene. As an example, consider the dihydroxylation of ethylene to produce ethylene glycol:

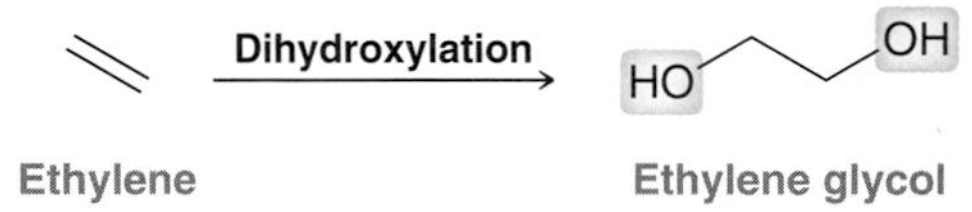

There are a number of reagents well suited to carry out this transformation. Some reagents provide for an *anti* dihydroxylation, while others provide for a *syn* dihydroxylation. In this section, we will explore a two-step procedure for achieving *anti* dihydroxylation:

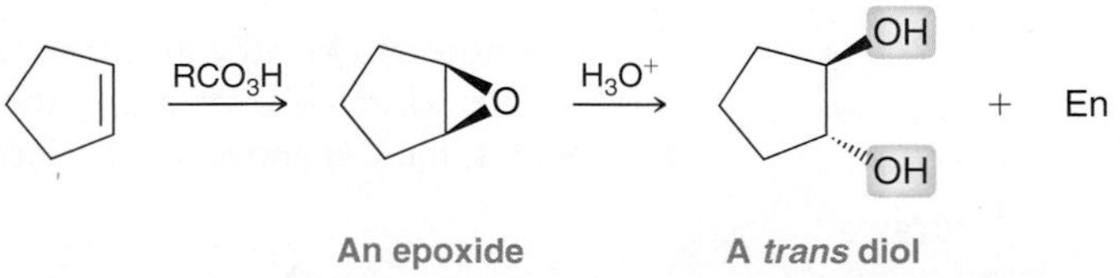

The first step of the process involves conversion of the alkene into an epoxide, and the second step involves opening the epoxide to form a *trans* diol (Mechanism 9.6). An **epoxide** is a three-membered, cyclic ether.

MECHANISM 9.6 *ANTI* DIHYDROXYLATION

Formation of an epoxide

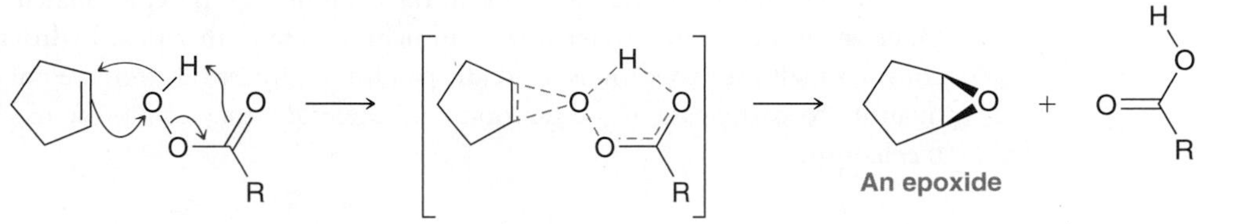
An epoxide
Transition state

Acid-catalyzed opening of an epoxide

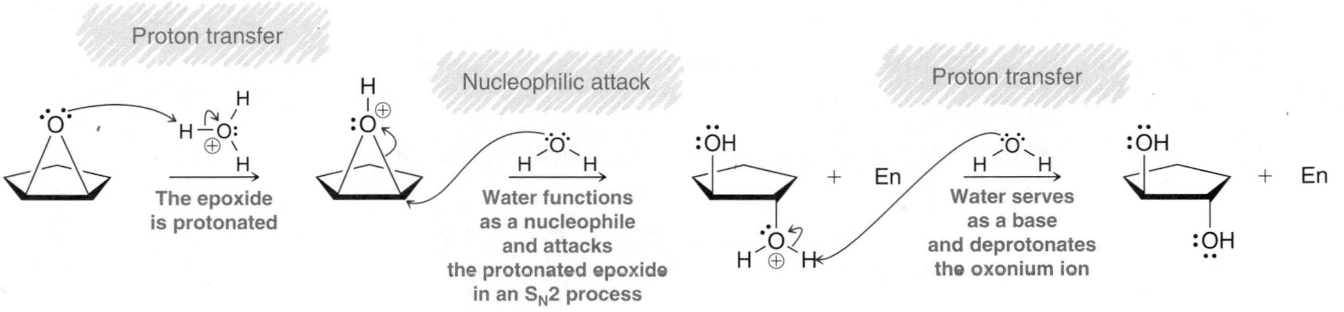
Proton transfer
The epoxide is protonated
Nucleophilic attack
Water functions as a nucleophile and attacks the protonated epoxide in an S_N2 process
Proton transfer
Water serves as a base and deprotonates the oxonium ion
En

In the first part of the process, a peroxy acid (RCO_3H) reacts with the alkene to form an epoxide. Peroxy acids resemble carboxylic acids in structure, possessing just one additional oxygen atom. Two common peroxy acids are shown below:

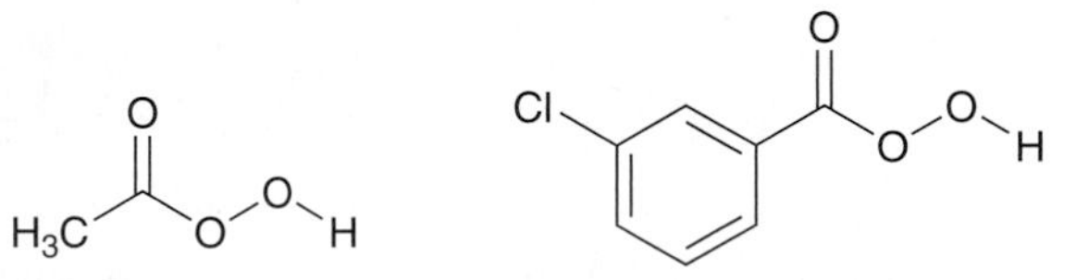

Peroxyacetic acid **_meta_-Chloroperoxybenzoic acid (MCPBA)**

Peroxy acids are strong oxidizing agents and are capable of delivering an oxygen atom to an alkene in a single step. The product is an epoxide.

Once the epoxide has been formed, it can then be opened with water under either acid-catalyzed or base-catalyzed conditions. Both sets of conditions are explored and compared in more detail in Section 14.10. For now, we will explore only the acid-catalyzed opening of epoxides, as seen in Mechanism 9.6. Under these conditions, the epoxide is first protonated to produce an intermediate that is very similar to a bromonium or mercurinium ion. All three cases involve a three-membered ring bearing a positive charge:

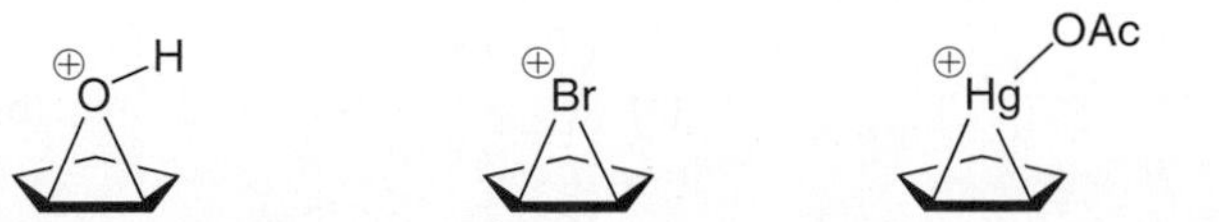

A protonated epoxide **A bromonium ion** **A mercurinium ion**

LOOKING AHEAD

You might be wondering why a weak nucleophile can participate in an S_N2 process at a secondary substrate. In fact, this reaction can occur even if the center being attacked is tertiary. This will be explained in Section 14.10.

We have seen that bromonium and mercurinium ions can be attacked by water from the back side. In much the same way, a protonated epoxide can also be attacked by water from the back side, as seen in Mechanism 9.6. The necessity for back-side attack (S_N2) explains the observed stereochemical preference for *anti* addition.

In the final step of the mechanism, the oxonium ion is deprotonated to yield a *trans* diol. Once again, notice that water is used to deprotonate (rather than hydroxide) in order to stay consistent with the conditions. In acidic conditions, hydroxide ions are not present in sufficient quantity to participate in the reaction, and therefore, they cannot be used when drawing the mechanism.

SKILLBUILDER

9.7 DRAWING THE PRODUCTS OF *ANTI* DIHYDROXYLATION

LEARN the skill

Predict the major product(s) for each of the following reactions:

(a) 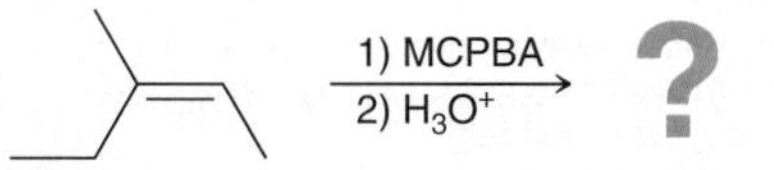

(b) 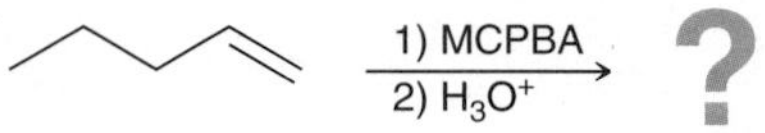

SOLUTION

STEP 1 Determine the number of chirality centers formed.

STEP 2 Determine the stereochemical outcome based on the requirement for *anti* addition.

(a) These reagents achieve dihydroxylation, which means that OH and OH will add across the alkene. Since the two groups are identical (OH and OH), we don't need to consider the regiochemistry of this process. However, the stereochemistry must be considered. Begin by determining the number of chirality centers formed. In this case, two new chirality centers are formed, so we expect only the pair of enantiomers that result from an *anti* addition. That is, the OH groups will be added to opposite sides of the π bond:

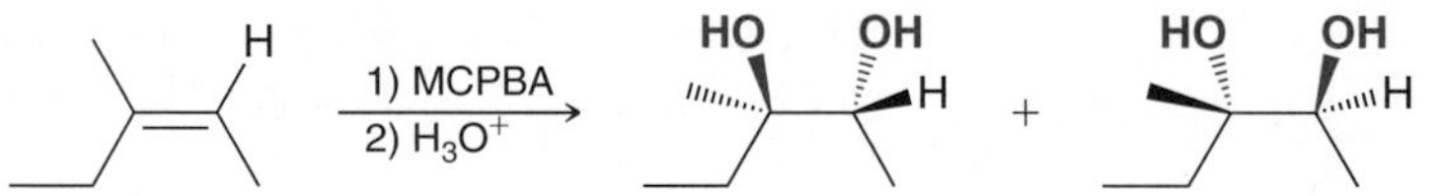

(b) In this example, only one new chirality center is formed. It is true that the reaction proceeds through an *anti* addition of OH and OH. However, with only one chirality center in the product, the preference for *anti* addition becomes irrelevant. Both possible enantiomers are formed:

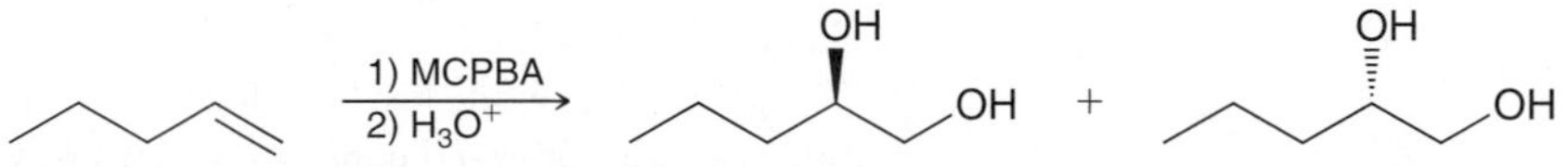

PRACTICE the skill

9.30 Predict the products that are expected when each of the following alkenes is treated with a peroxy acid (such as MCPBA) followed by aqueous acid:

(a) 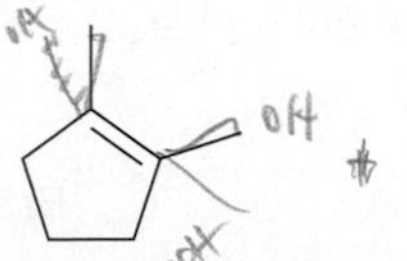

(b) 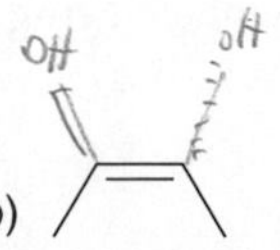

(c) 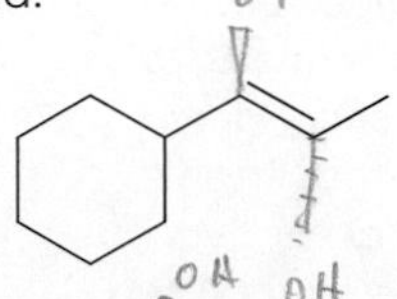

(d) 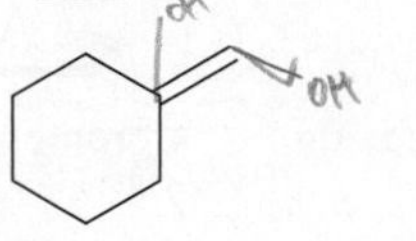

(e) 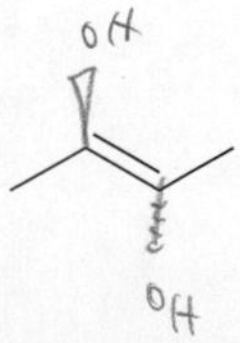

(f) 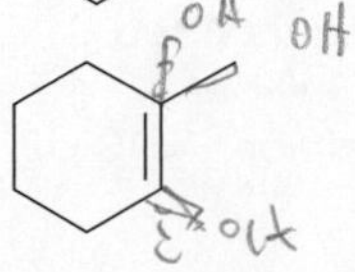

APPLY the skill

9.31 Under acid-catalyzed conditions, epoxides can be opened by a variety of nucleophiles other than water, such as alcohols. In such a case, the nucleophile will generally attack at the more substituted position. Using this information, predict the products for each of the following reactions:

(a)
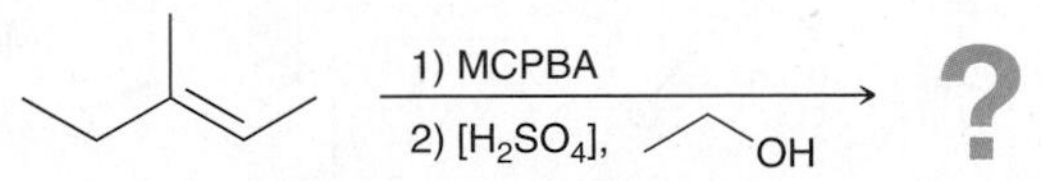

(b)
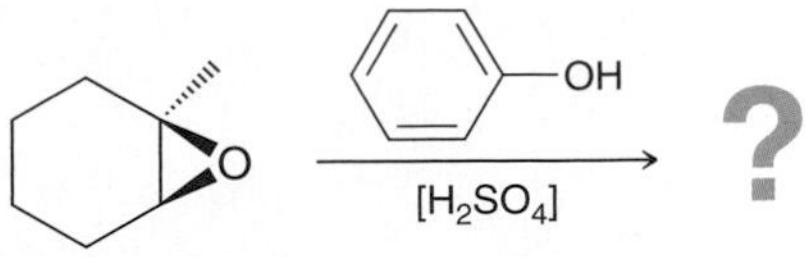

9.32 Compound A and compound B both have molecular formula C_6H_{12}. Both compounds produce epoxides when treated with MCPBA.

(a) The epoxide resulting from compound A was treated with aqueous acid (H_3O^+) and the resulting diol had no chirality centers. Propose two possible structures for compound A.

(b) The epoxide resulting from compound B was treated with H_3O^+ and the resulting diol was a *meso* compound. Draw the structure of compound B.

need more **PRACTICE?** **Try Problem 9.65d**

9.10 *Syn* Dihydroxylation

The previous section described a method for *anti* dihydroxylation of alkenes. In this section, we will explore two different sets of reagents that can accomplish *syn* dihydroxylation of alkenes. When an alkene is treated with osmium tetroxide (OsO_4), a cyclic osmate ester is produced:

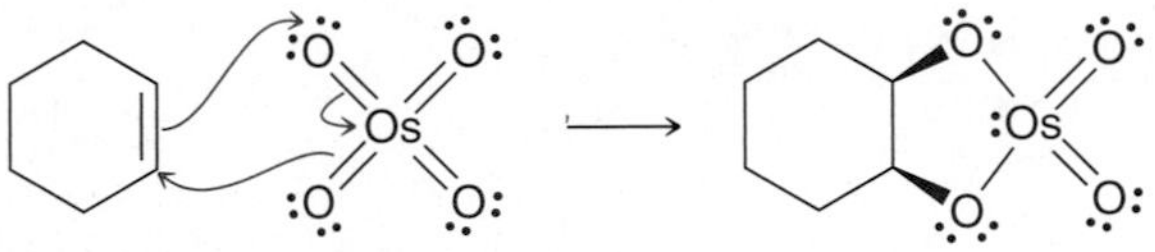

A cyclic osmate ester

Osmium tetroxide adds across the alkene in a concerted process. In other words, both oxygen atoms attach to the alkene simultaneously. This effectively adds two groups across the same face of the alkene; hence, the *syn* addition. The cyclic osmate ester that results can be isolated and then treated with either aqueous sodium sulfite (Na_2SO_3) or sodium bisulfite ($NaHSO_3$) to produce a diol:

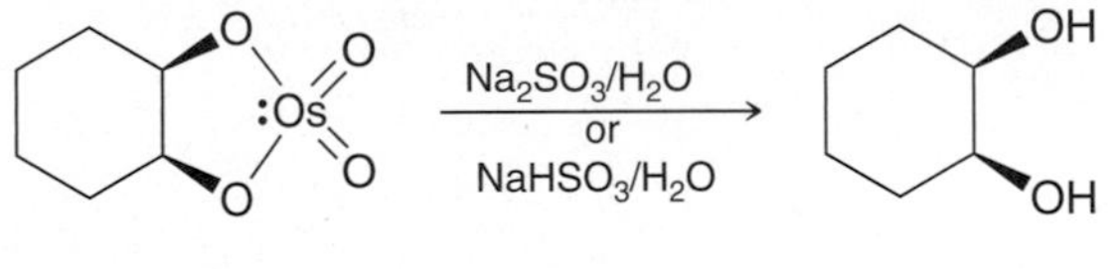

This method can be used to convert alkenes into diols with fairly high yields, but there are several disadvantages. In particular, OsO_4 is expensive and toxic. To deal with these issues, several methods have been developed that use a co-oxidant that serves to regenerate OsO_4 as it is consumed during the reaction. In this way, OsO_4 functions as a catalyst so that even small quantities can produce large quantities of the diol. Typical co-oxidants include *N*-methylmorpholine *N*-oxide (NMO) and *tert*-butyl hydroperoxide:

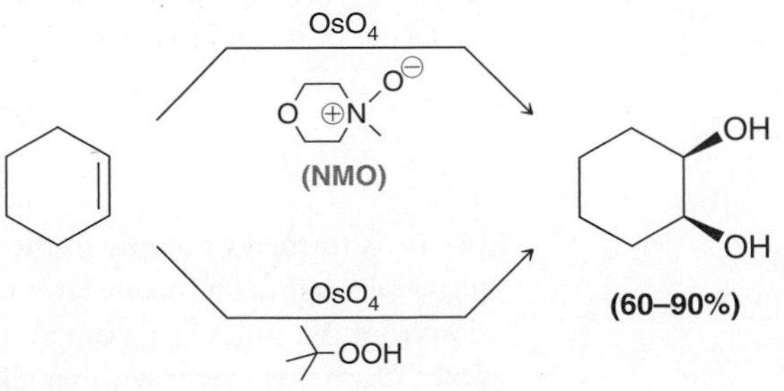

A different, although mechanistically similar, method for achieving *syn* dihydroxylation involves treatment of alkenes with cold potassium permanganate under basic conditions:

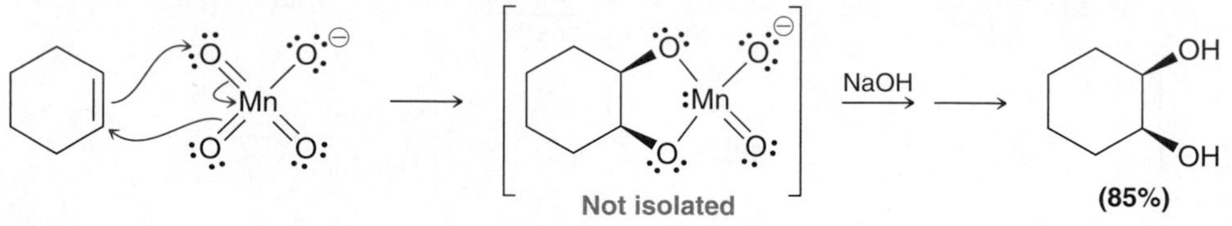

Once again, a concerted process adds both oxygen atoms simultaneously across the double bond. Notice the similarity between the mechanisms of these two methods (OsO_4 vs. $KMnO_4$).

Potassium permanganate is fairly inexpensive; however, it is a very strong oxidizing agent, and it often causes further oxidation of the diol. Therefore, synthetic organic chemists often choose to use OsO_4 together with a co-oxidant to achieve *syn* dihydroxylation.

CONCEPTUAL CHECKPOINT

9.33 Predict the product(s) for each of the following reactions. In each case, make sure to consider the number of chirality centers being formed.

(a) $\xrightarrow[\text{NMO}]{\text{OsO}_4 \text{ (catalytic)}}$ **?**

(b) $\xrightarrow[\text{2) NaHSO}_3 / \text{H}_2\text{O}]{\text{1) OsO}_4}$ **?**

(c) $\xrightarrow[\text{Cold}]{\text{KMnO}_4, \text{NaOH}}$ **?**

(d) $\xrightarrow[\text{Cold}]{\text{KMnO}_4, \text{NaOH}}$ **?**

(e) $\xrightarrow[\text{OOH, NaOH}]{\text{OsO}_4 \text{ (catalytic)}}$ **?**

(f) $\xrightarrow[\text{NMO}]{\text{OsO}_4 \text{ (catalytic)}}$ **?**

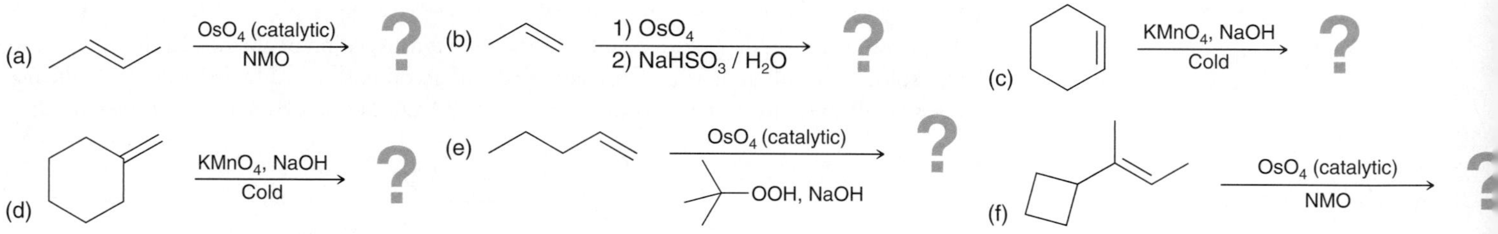

9.11 Oxidative Cleavage

There are many reagents that will add across an alkene and completely cleave the C=C bond. In this section, we will explore one such reaction, called **ozonolysis**. Consider the following example:

$$\xrightarrow[\text{2) DMS}]{\text{1) O}_3}$$

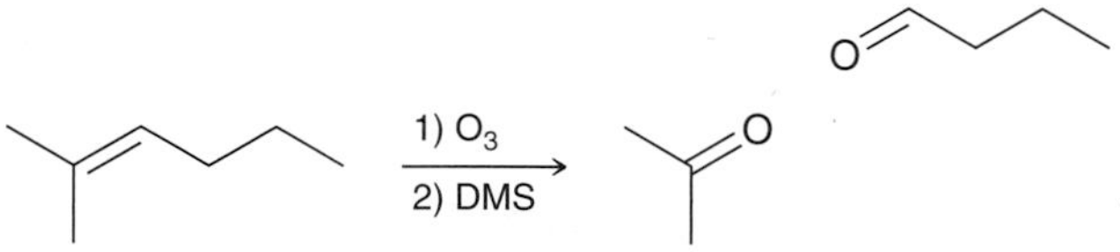

Notice that the C=C bond is completely split apart to form two C=O double bonds. Therefore, issues of stereochemistry and regiochemistry become irrelevant. In order to understand how this reaction occurs, we must first explore the structure of ozone.

Ozone is a compound with the following resonance structures:

LOOKING AHEAD
The role of ozone in atmospheric chemistry is discussed in Section 11.8.

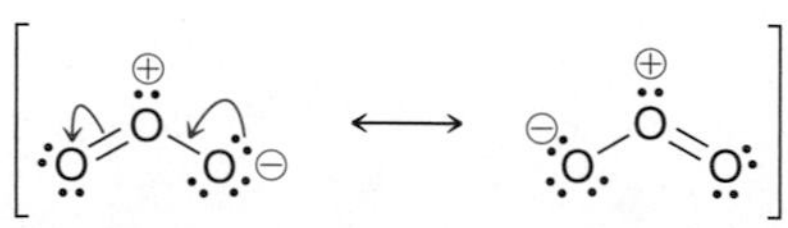

Ozone is formed primarily in the upper atmosphere where oxygen gas (O_2) is bombarded with ultraviolet light. The ozone layer in our atmosphere serves to protect us from harmful UV radiation from the sun. Ozone can also be prepared in the laboratory, where it can serve a useful purpose. Ozone will react with an alkene to produce an initial, primary ozonide (or molozonide), which undergoes rearrangement to produce a more stable ozonide:

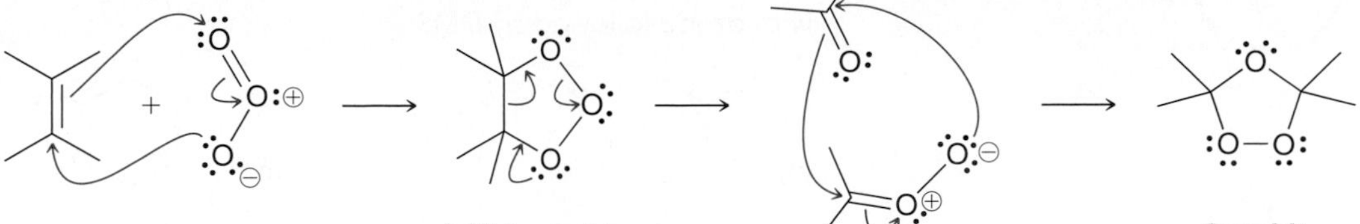

Initial ozonide (molozonide) | Ozonide

When treated with a mild reducing agent, the ozonide is converted into products:

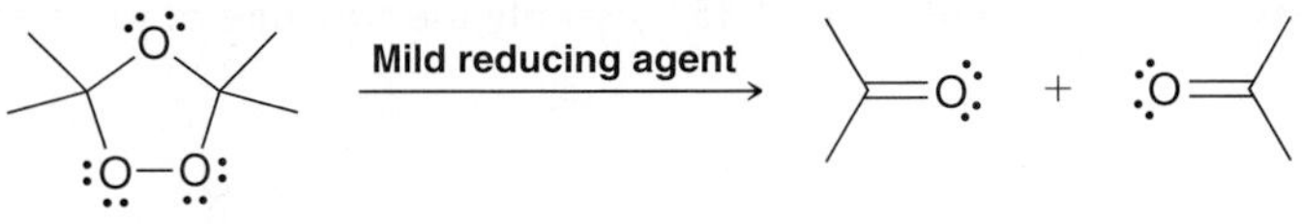

Ozonide

Common examples of reducing agents include dimethyl sulfide (DMS) or Zn/H_2O:

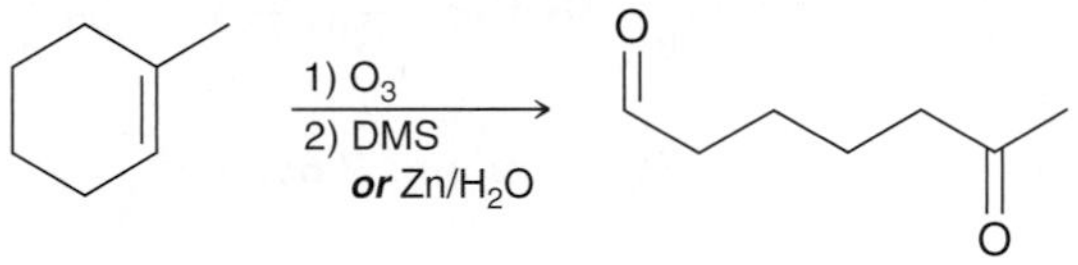

The following worked example illustrates a simple method for drawing the products of ozonolysis.

SKILLBUILDER

9.8 PREDICTING THE PRODUCTS OF OZONOLYSIS

LEARN the skill

Predict the products of the following reaction:

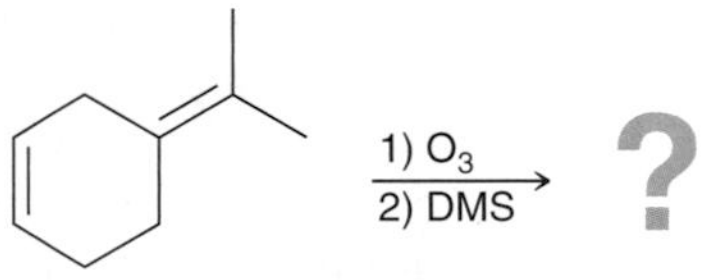

SOLUTION

There are two C=C double bonds in this compound. Begin by redrawing the compound with the C=C double bonds longer than normal:

STEP 1
Redraw the C=C double bonds longer than normal.

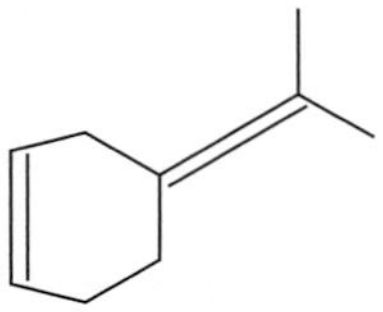

Then, erase the center of each C=C double bond and place two oxygen atoms in the space:

STEP 2
Erase the center of each C=C double bond and place two oxygen atoms in the space.

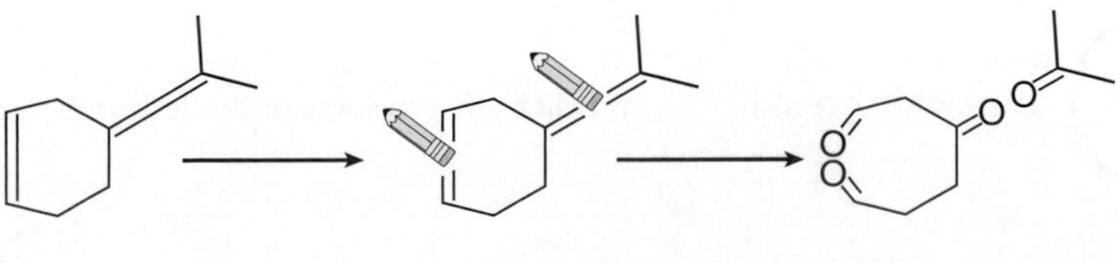

This simple procedure can be used for quickly drawing the products of any ozonolysis reaction.

PRACTICE the skill **9.34** Predict the products that are expected when each of the following alkenes is treated with ozone followed by DMS:

(a)

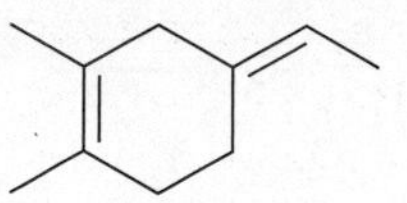

(b)

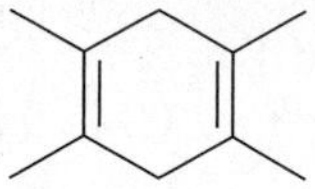

(c)

(d) (e) (f)

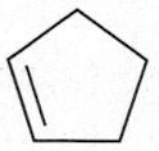

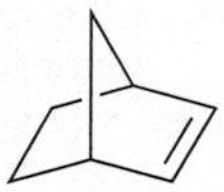

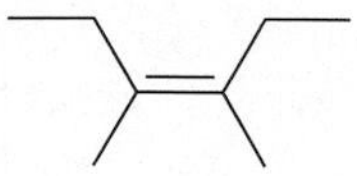

APPLY the skill **9.35** Identify the structure of the starting alkene in each of the following cases:

(a) C_8H_{14} $\xrightarrow[\text{2) DMS}]{\text{1) O}_3}$

(b) $C_{10}H_{16}$ $\xrightarrow[\text{2) DMS}]{\text{1) O}_3}$

(c) $C_{10}H_{16}$ $\xrightarrow[\text{2) DMS}]{\text{1) O}_3}$

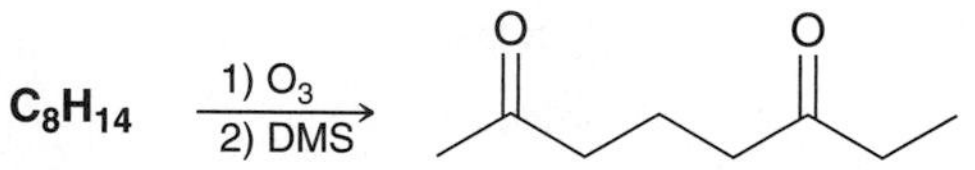

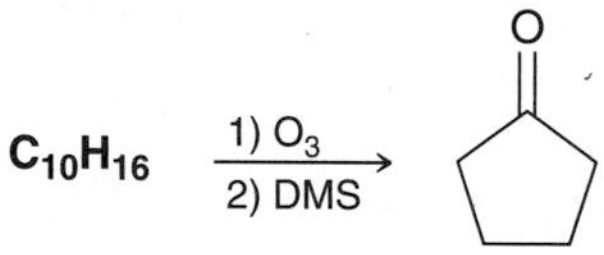

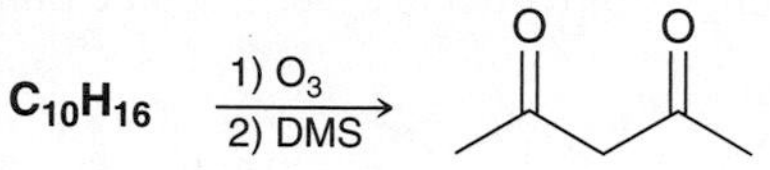

need more **PRACTICE?** Try Problems 9.68, 9.77

9.12 Predicting the Products of an Addition Reaction

Many addition reactions have been covered in this chapter, and in each case, there are several factors to take into account when predicting the products of a reaction. Let's now summarize the factors that are common to all addition reactions: In order to predict products properly, the following three questions must be considered:

1. What are the identities of the groups being added across the double bond?

2. What is the expected regioselectivity (Markovnikov or *anti*-Markovnikov addition)?

3. What is the expected stereospecificity (*syn* or *anti* addition)?

Answering all three questions requires a careful analysis of both the starting alkene and the reagents employed. It is absolutely essential to recognize reagents, and while that might sound like it involves a lot of memorization, it actually does not. By understanding the proposed mechanism for each reaction, you will intuitively understand all three pieces of information for each reaction. Remember that a proposed mechanism must explain the experimental observations. Therefore, the mechanism for each reaction can serve as the key to remembering the three pieces of information listed above.

SKILLBUILDER

9.9 PREDICTING THE PRODUCTS OF AN ADDITION REACTION

LEARN the skill Predict the products of the following reaction:

$\xrightarrow[\text{2) H}_2\text{O}_2\text{, NaOH}]{\text{1) BH}_3 \cdot \text{THF}}$ **?**

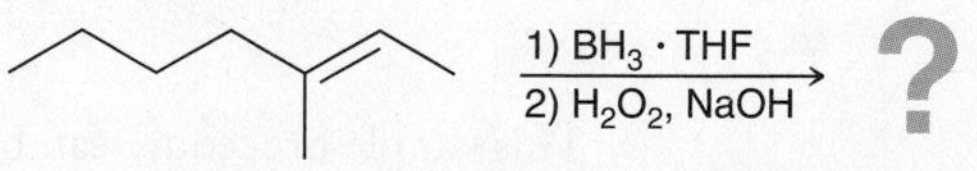

SOLUTION

In order to predict the products, the following three questions must be answered:

STEP 1
Identify the two groups being added across the π bond.

1. ***Which two groups are being added across the double bond?*** To answer this question, it is necessary to recognize that these reagents achieve hydroboration-oxidation, which adds H and OH across a π bond. It would be impossible to solve this problem without being able to recognize the reagents. But even with that recognition, there are still two more questions that must be answered in order to arrive at the correct answer.

2. ***What is the expected regioselectivity (Markovnikov or anti-Markovnikov)?*** We have seen that hydroboration-oxidation is an *anti*-Markovnikov process. Think about the mechanism of this process, and recall that there were two different explanations for the observed regioselectivity, one based on a steric argument and the other based on an electronic argument. An *anti*-Markovnikov addition means that the OH group is placed at the less substituted carbon:

STEP 2
Identify the expected regioselectivity.

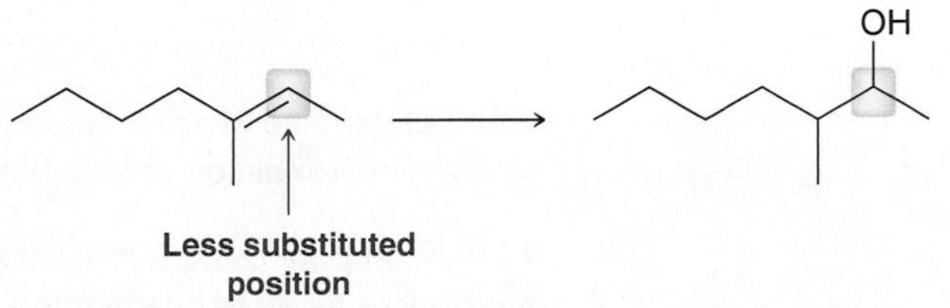

STEP 3
Identify the expected stereospecificity.

3. ***What is the expected stereospecificity (syn or anti)?*** We have seen that hydroboration-oxidation produces a *syn* addition. Think about the mechanism of this process, and recall the reason for *syn* addition. In the first step, BH_2 and H are added to the same face of the alkene in a concerted process. To determine if this *syn* requirement is even relevant in this case, we must analyze how many chirality centers are being formed in the process:

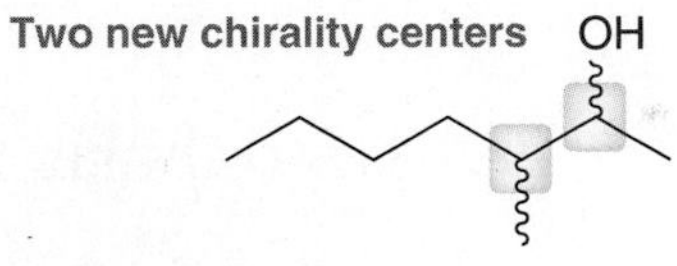

Two new chirality centers are formed, and therefore, the requirement for *syn* addition is relevant. The reaction is expected to produce only the pair of enantiomers that would result from a *syn* addition:

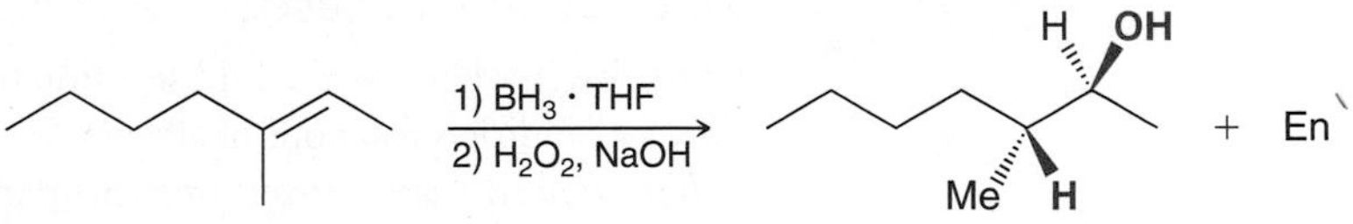

PRACTICE the skill 9.36 Predict the products of each of the following reactions:

(a)
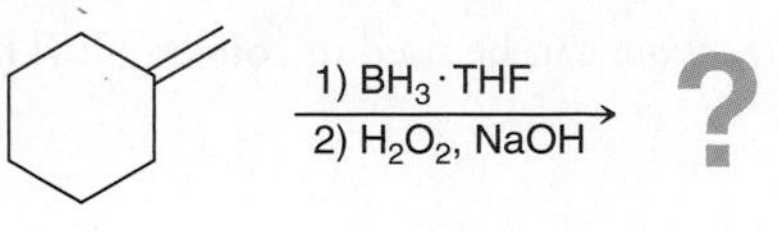

(b)
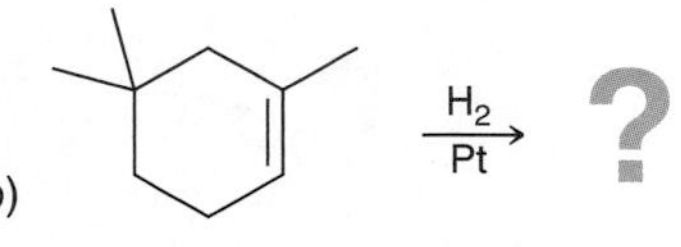

(c)
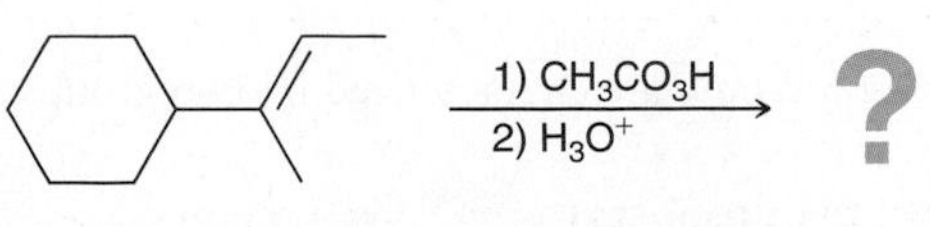

(d)
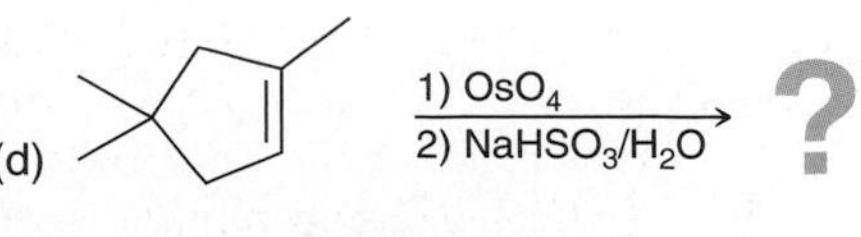

(e)
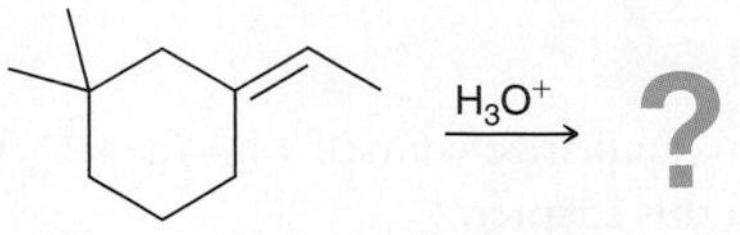

(f)
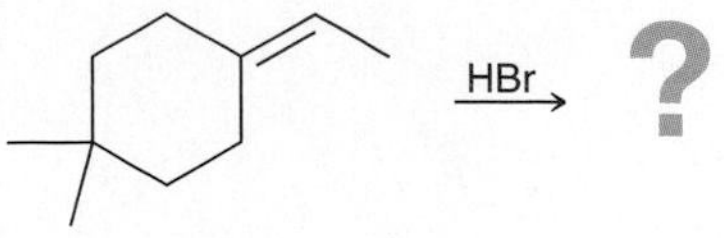

(g) 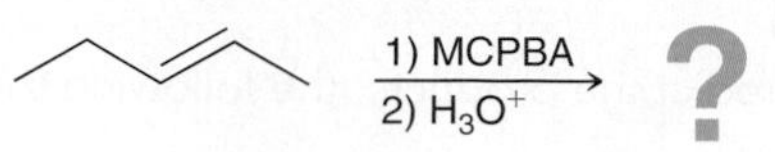

(h) 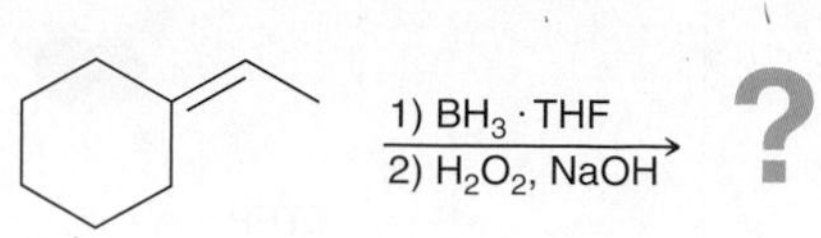

(i) 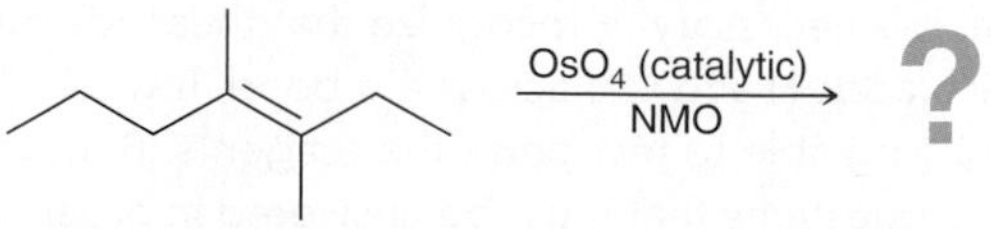

APPLY the skill

9.37 *Syn* dihydroxylation of the compound below yields two products. Draw both products and describe their stereoisomeric relationship (i.e., are they enantiomers or diastereomers?):

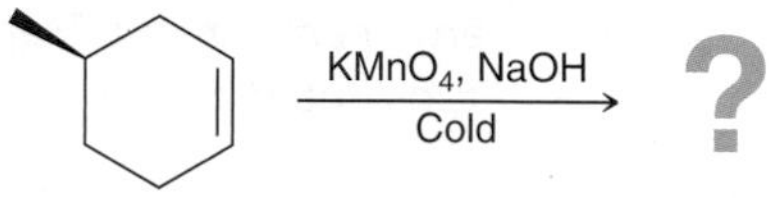

9.38 Determine whether *syn* dihydroxylation of *trans*-2-butene will yield the same products as *anti* dihydroxylation of *cis*-2-butene. Draw the products in each case and compare them.

9.39 Compound A has molecular formula C_5H_{10}. Hydroboration-oxidation of compound A produces a pair of enantiomers, compounds B and C. When treated with HBr, compound A is converted into compound D, which is a tertiary alkyl bromide. When treated with O_3 followed by DMS, compound A is converted into compounds E and F. Compound E has three carbon atoms, while compound F has only two carbon atoms. Identify the structures of compounds A, B, C, D, E, and F.

need more **PRACTICE?** **Try Problems 9.49, 9.50, 9.65, 9.73**

9.13 Synthesis Strategies

In order to begin practicing synthesis problems, it is essential to master all of the individual reactions covered thus far. We will begin with one-step synthesis problems and then progress to cover multistep problems.

One-Step Syntheses

Until now, we have covered substitution reactions (S_N1 and S_N2), elimination reactions (E1 and E2), and addition reactions of alkenes. Let's quickly review what these reactions can accomplish. *Substitution reactions* convert one group into another:

Elimination reactions can be used to convert alkyl halides or alcohols into alkenes:

Addition reactions are characterized by two groups adding across a double bond:

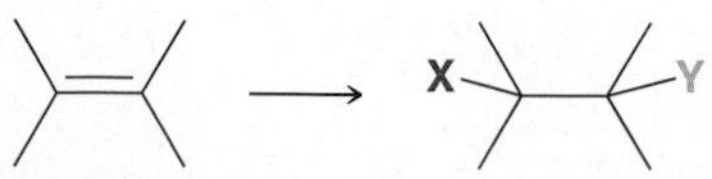

It is essential to familiarize yourself with the reagents employed for each type of addition reaction covered in this chapter.

SKILLBUILDER

9.10 PROPOSING A ONE-STEP SYNTHESIS

LEARN the skill

Identify the reagents that you would use to accomplish the following transformation:

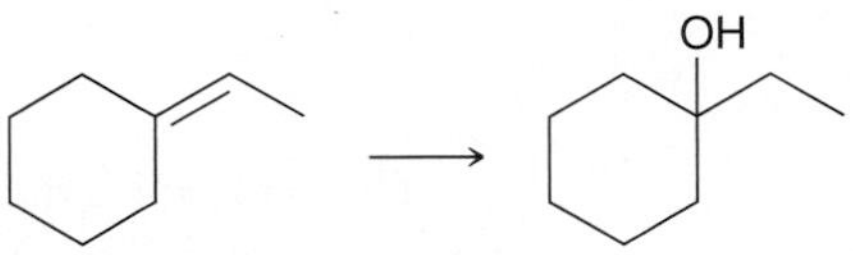

STEP 1
Identify the two groups being added across the π bond.

STEP 2
Identify the regioselectivity.

STEP 3
Identify the stereospecificity.

STEP 4
Identify reagents that will achieve the details described in the first three steps.

SOLUTION

We approach this problem using the same three questions developed in the previous section:

1. Which two groups are being added across the double bond?—H and OH.
2. What is the regioselectivity?—Markovnikov addition.
3. What is the stereospecificity?—Not relevant (no chirality centers formed).

This transformation requires reagents that will give a Markovnikov addition of H and OH. This can be accomplished in either of two ways: acid-catalyzed hydration or oxymercuration-demercuration. Acid-catalyzed hydration might be more simple in this case, because there is no chance of a carbocation rearrangement:

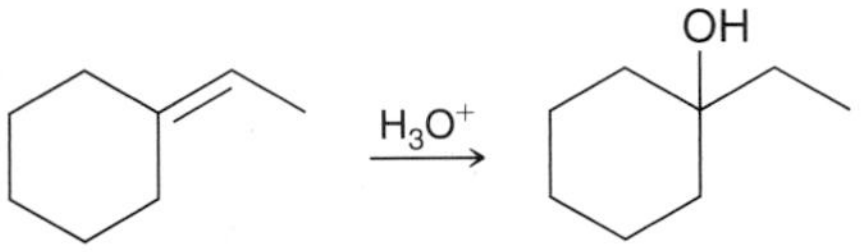

If rearrangement were possible, then oxymercuration-demercuration would have been the preferred route.

PRACTICE the skill

9.40 Identify the reagents that you would use to accomplish each of the following transformations:

(a)
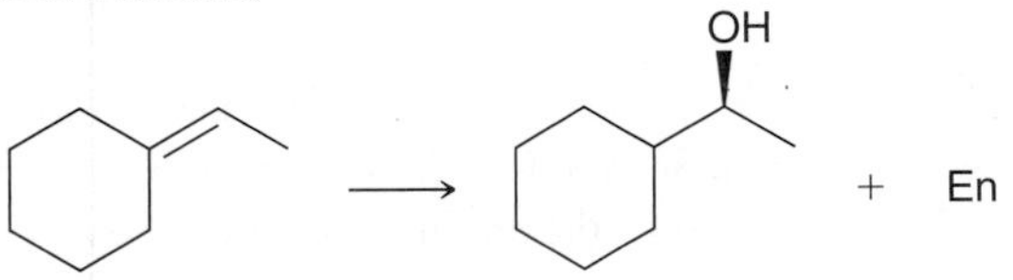

(b)
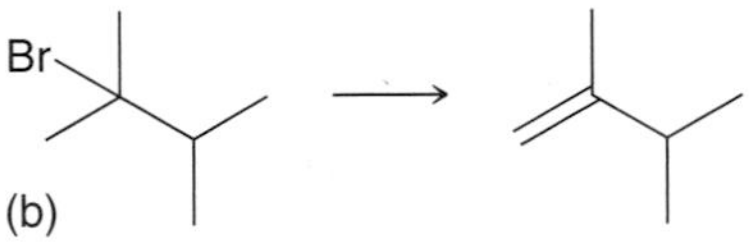

(c)
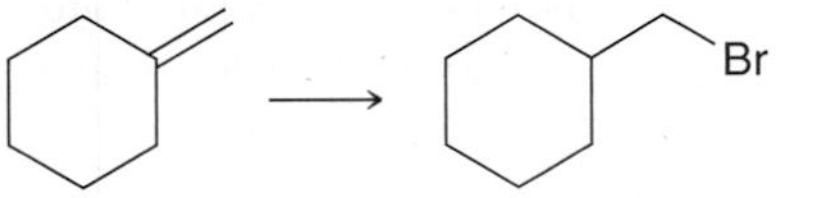

(d)

(e)
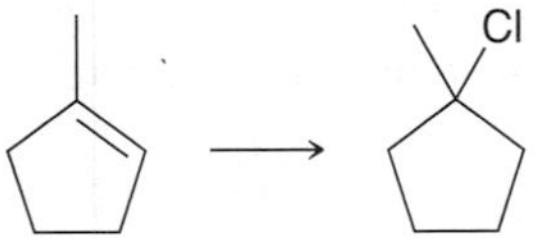

(f)
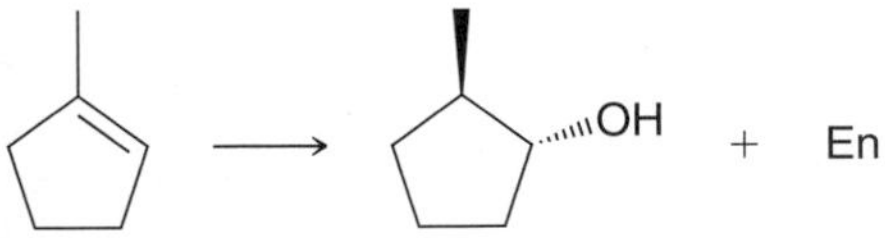

(g)
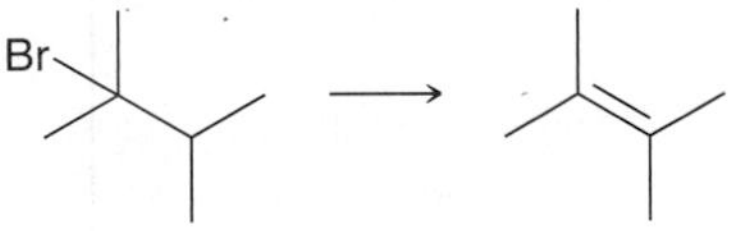

(h)
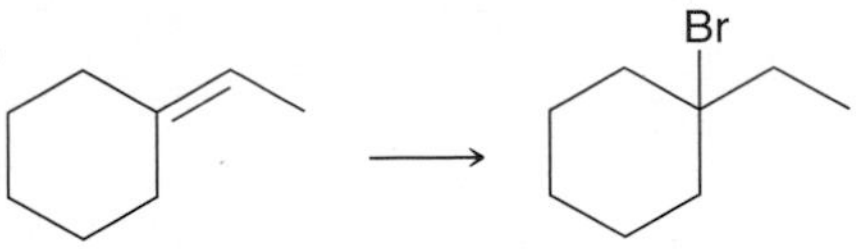

APPLY the skill

9.41 Identify what reagents you would use to achieve each transformation:

(a) Conversion of 2-methyl-2-butene into a secondary alkyl halide
(b) Conversion of 2-methyl-2-butene into a tertiary alkyl halide
(c) Conversion of *cis*-2-butene into a *meso* diol
(d) Conversion of *cis*-2-butene into enantiomeric diols

need more **PRACTICE?** **Try Problems 9.60, 9.62, 9.70, 9.71**

Changing the Position of a Leaving Group

Now let's get some practice combining the reactions covered thus far. As an example, consider the following transformation:

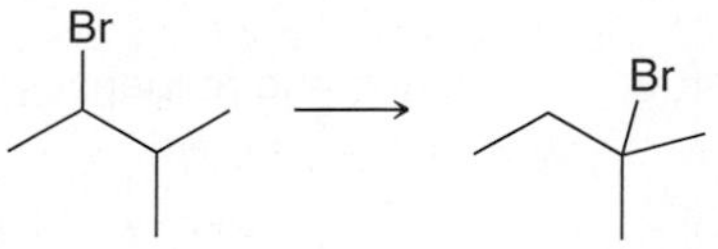

The net result is the change in position of the Br atom. How can this type of transformation be achieved? This chapter did not present a one-step method for moving the location of a bromine atom. However, this transformation can be accomplished in two steps: an elimination reaction followed by an addition reaction:

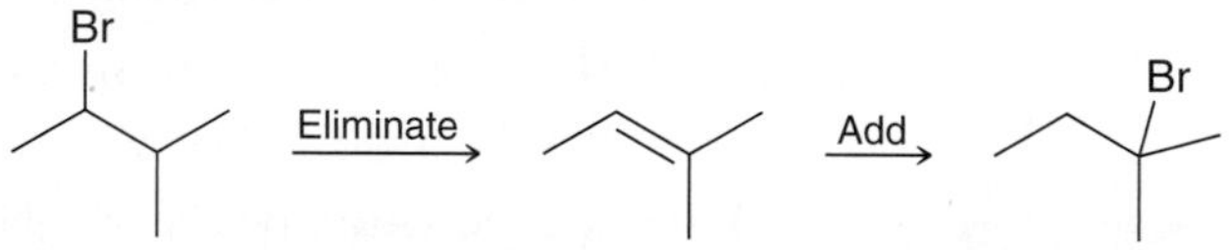

When performing this two-step sequence, there are a few important issues to keep in mind. In the first step (elimination), the product can be the more substituted alkene (Zaitsev product) or the less substituted alkene (Hofmann product):

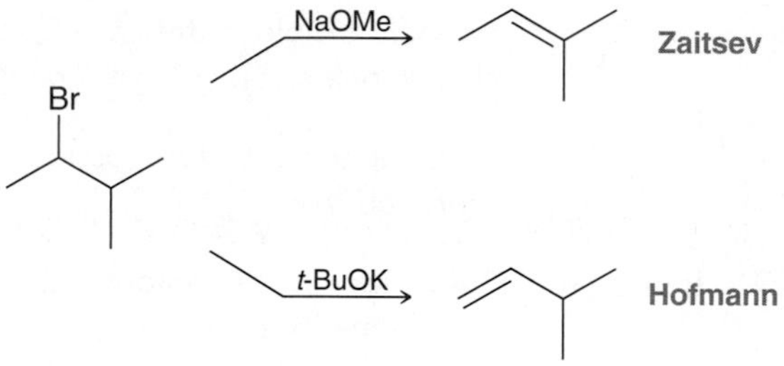

Note that a careful choice of reagents makes it possible to control the regiochemical outcome, as we first saw in Section 8.7. With a strong base, such as sodium methoxide (NaOMe) or sodium ethoxide (NaOEt), the product is the more substituted alkene. With a strong, sterically hindered base, such as potassium *tert*-butoxide (*t*-BuOK), the product is the less substituted alkene. After forming the double bond, the regiochemical outcome of the next step (addition of HBr) can also be controlled by the reagents used. HBr produces a Markovnikov addition while HBr/ROOR produces an *anti*-Markovnikov addition.

When changing the location of a leaving group, make sure to remember that OH is a terrible leaving group. Let's see what to do when dealing with an OH group. As an example, consider how the following transformation might be achieved:

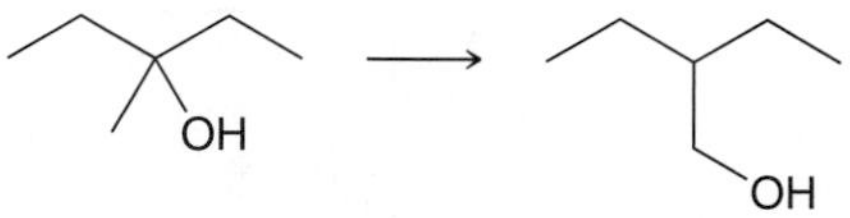

Our strategy would suggest the following steps:

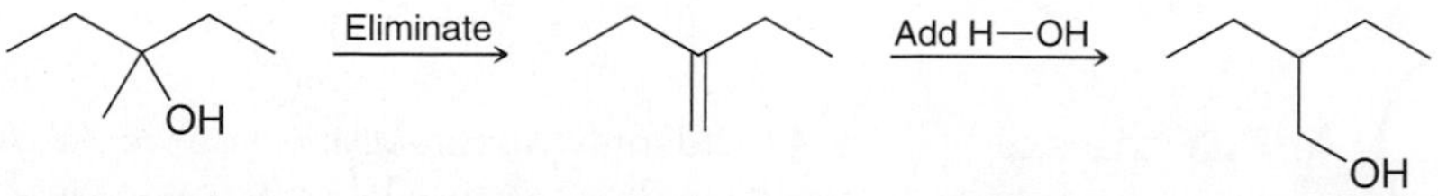

However, this sequence presents a serious obstacle in that the first step is an elimination reaction in which OH would be the leaving group. It is possible to protonate an OH group using concentrated acid, which converts a bad leaving group into an excellent leaving group (see Section 7.6). However, that reaction cannot be used here because it is an E1 process, which always yields the Zaitsev product, not the Hofmann product. The regiochemical outcome of an E1 process cannot be controlled. The regiochemical outcome of an E2 process can be controlled, but an E2 process cannot be used in this example, because OH is a bad leaving group. It is not possible to protonate the OH group with a strong acid and then use a strong base to achieve an E2 reaction, because when mixed together, a strong base and a strong acid will simply neutralize each other. The question remains: How can an E2 process be performed when the desired leaving group is an OH group?

In Chapter 7, we explored a method that would allow an E2 process in this example, which maintains control over the regiochemical outcome. The OH group can first be converted into a tosylate, which is a much better leaving group than OH (for a review of tosylates, see Section 7.8). After the OH is converted into a tosylate, the strategy outlined above can be followed. Specifically, a strong, sterically hindered base is used for the elimination reaction, followed by *anti*-Markovnikov addition of H—OH:

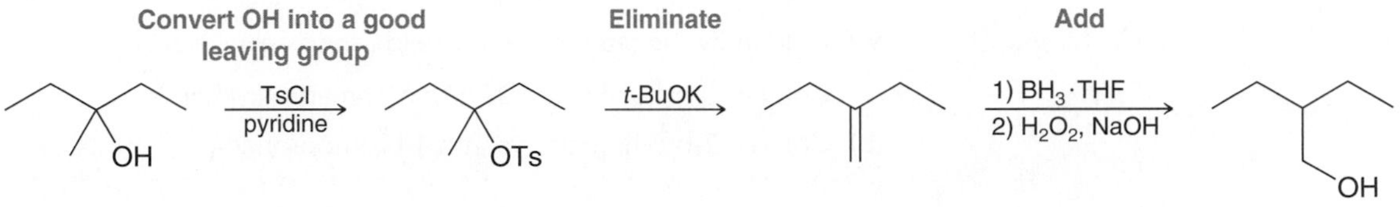

SKILLBUILDER

9.11 CHANGING THE POSITION OF A LEAVING GROUP

LEARN the skill

Identify the reagents you would use to accomplish the following transformation:

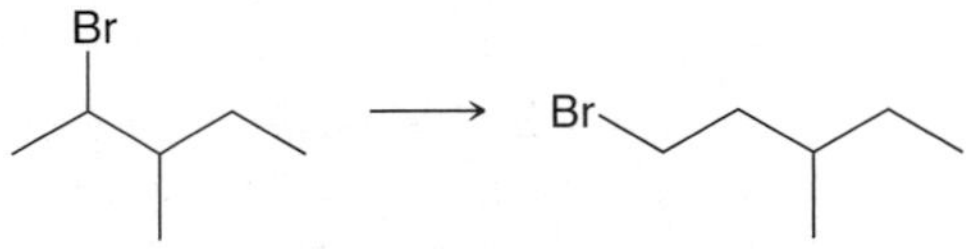

SOLUTION

This problem shows the Br changing its position. This can be accomplished with a two-step process: (1) elimination to form a double bond followed by (2) addition across that double bond:

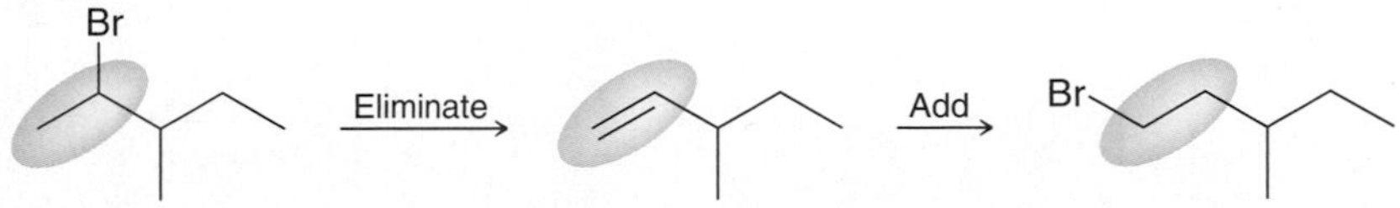

STEP 1
Control the regiochemical outcome of elimination by choosing the appropriate base.

STEP 2
Control the regiochemical outcome of addition by choosing the appropriate reagents.

Care must be taken to control the regiochemical outcome in each of these steps. In the elimination step, the product with the less substituted double bond (i.e., the Hofmann product) is desired, and therefore, a sterically hindered base must be used. In the addition step, the Br must be positioned at the less substituted carbon (*anti*-Markovnikov addition), and therefore, we must use HBr with peroxides. This gives the following overall synthesis:

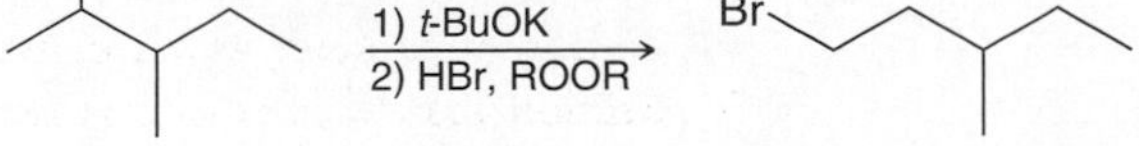

PRACTICE the skill **9.42** Identify the reagents you would use to accomplish each of the following transformations:

(a) 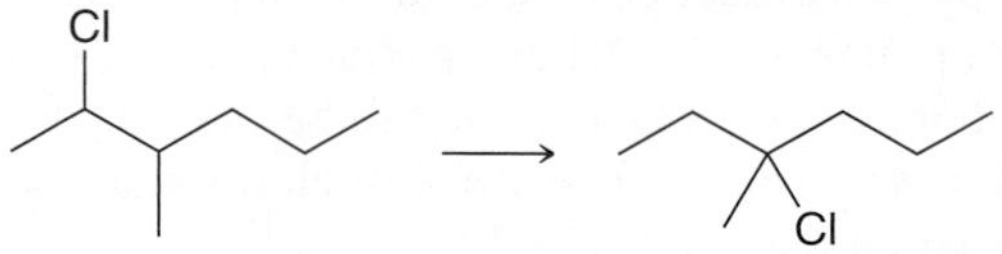

(b) 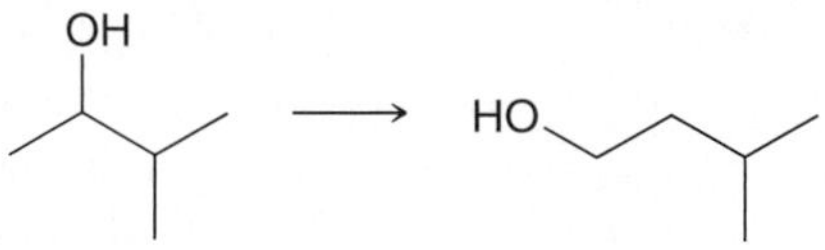

(c) 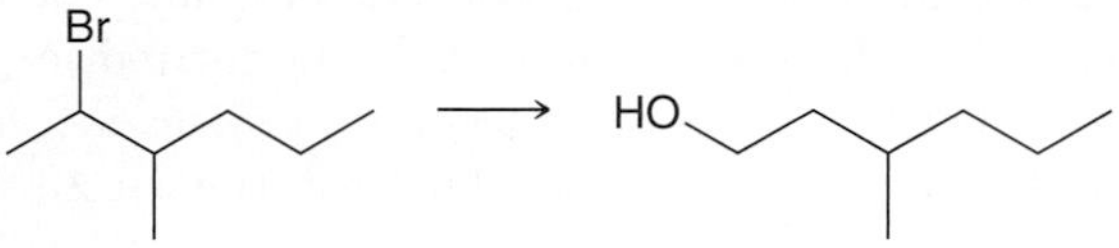

(d) 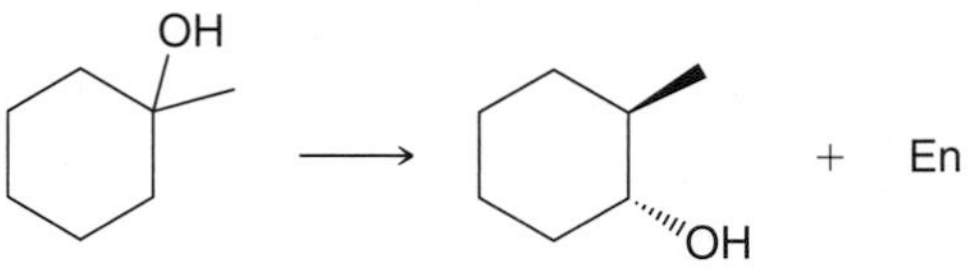

APPLY the skill

9.43 Identify the reagents you would use to achieve each of the following transformations:

(a) Convert *tert*-butyl bromide into a primary alkyl halide

(b) Convert 2-bromopropane into 1-bromopropane

9.44 Identify the reagents you would use to accomplish the following transformation

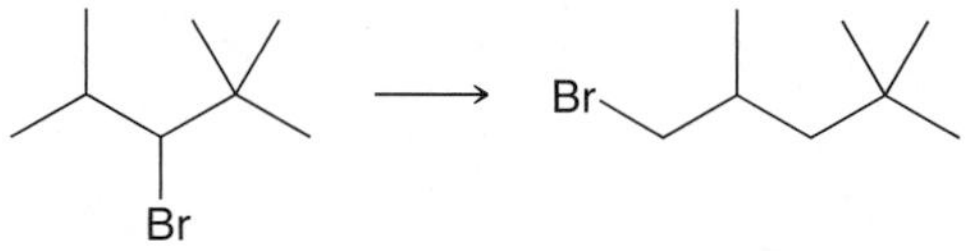

need more **PRACTICE?** **Try Problems 9.57, 9.58b**

Changing the Position of a π Bond

In the previous section, we combined two reactions in one synthetic strategy—*eliminate and then add*—which enabled us to change the location of a halogen or hydroxyl group. Now, let's focus on another type of strategy—*add and then eliminate*:

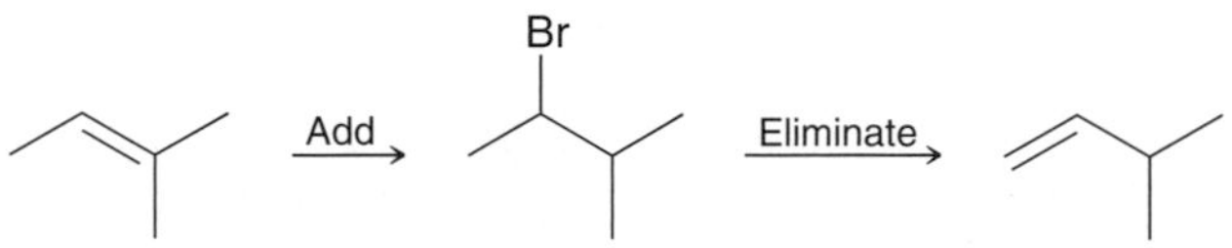

This two-step sequence makes it possible to change the position of a π bond. When using this strategy, the regioselectivity of each step can be carefully controlled. In the first step (addition), Markovnikov addition is achieved by using HBr, while *anti*-Markovnikov addition is achieved by using HBr with peroxides. In the second step (elimination), the Zaitsev product can be obtained by using a strong base, while the Hofmann product can be obtained by using a strong, sterically hindered base.

SKILLBUILDER

9.12 CHANGING THE POSITION OF A π BOND

LEARN the skill

Identify the reagents you would use to accomplish the following transformation:

SOLUTION

STEP 1
Control the regiochemical outcome of addition by choosing the appropriate reagents.

STEP 2
Control the regiochemical outcome of elimination by choosing the appropriate base.

This example involves moving the position of a π bond. We have not seen a way to accomplish this transformation in one step. However, it can be accomplished in two steps—addition followed by elimination:

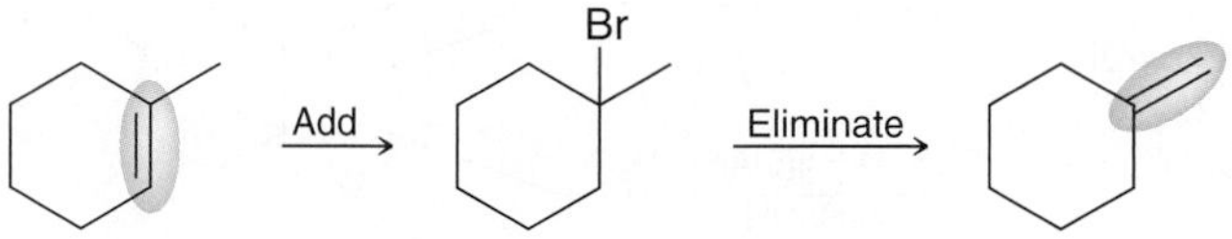

In the first step, a Markovnikov addition is required (Br must be placed at the more substituted carbon), which can be accomplished by using HBr. The second step requires an elimination to give the Hofmann product, which can be accomplished with a sterically hindered base, such as *tert*-butoxide. Therefore, the overall synthesis is:

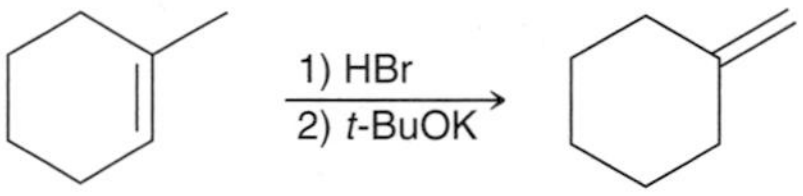

PRACTICE the skill

9.45 Identify the reagents you would use to accomplish each of the following transformations:

(a) 

(b) 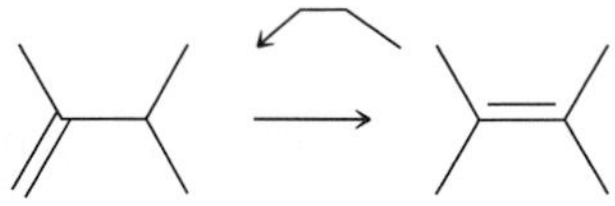

APPLY the skill

9.46 Identify the reagents you would use to accomplish each of the following transformations:

(a) Convert 2-methyl-2-butene into a monosubstituted alkene

(b) Convert 2,3-dimethyl-1-hexene into a tetrasubstituted alkene

9.47 Identify the reagents you would use to accomplish each of the following transformations:

(a) 

(b) 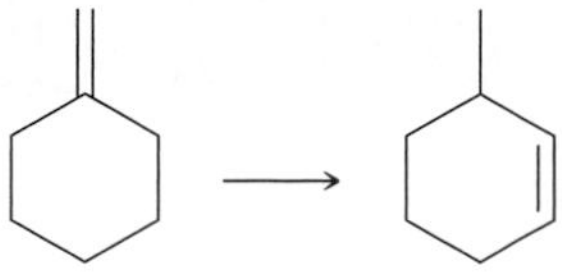

need more **PRACTICE?** **Try Problems 9.53, 9.58a**

REVIEW OF REACTIONS

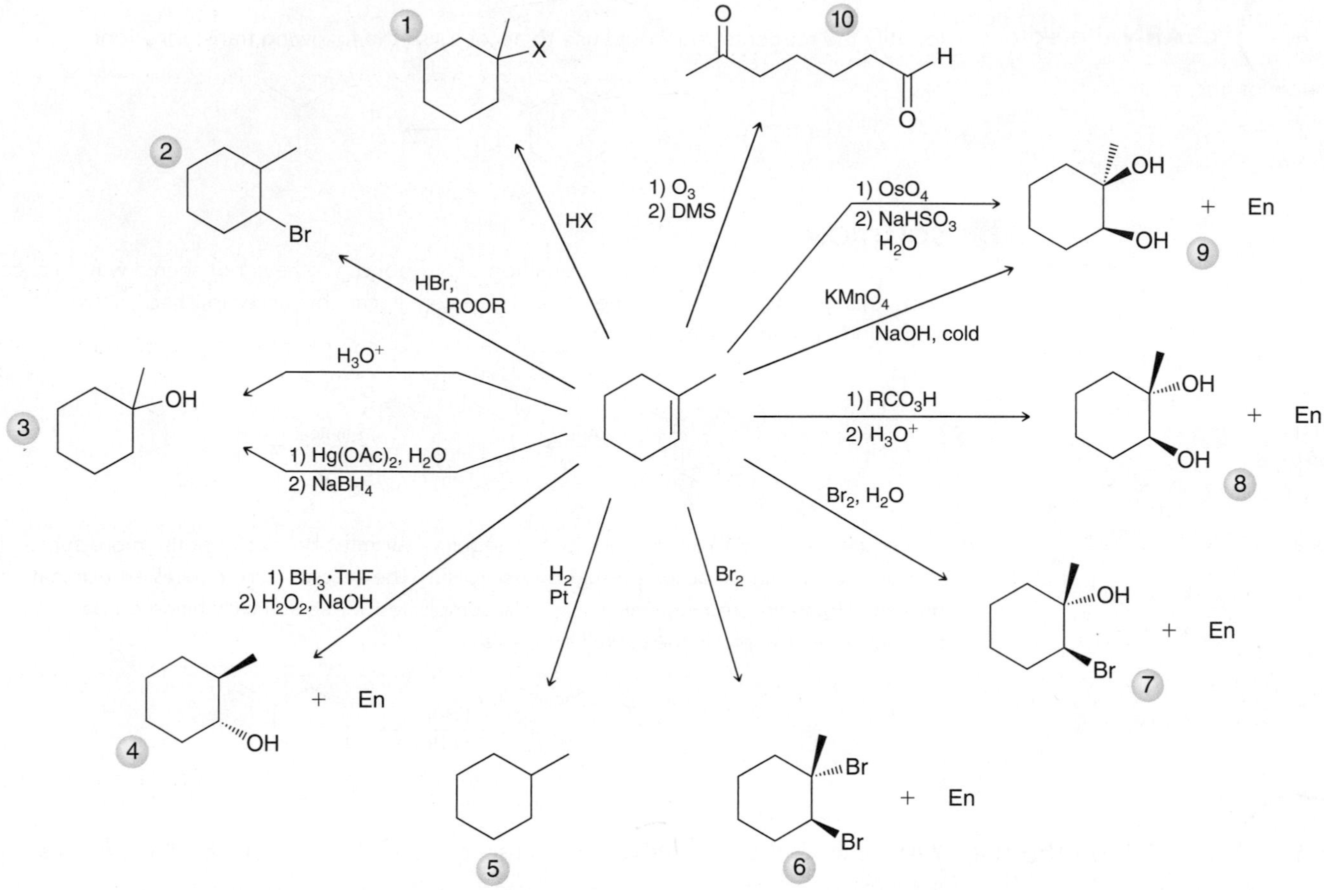

1. Hydrohalogenation (Markovnikov)
2. Hydrohalogenation (*anti*-Markovnikov)
3. Acid-catalyzed hydration and oxymercuration-demercuration
4. Hydroboration-oxidation
5. Hydrogenation
6. Bromination
7. Halohydrin formation
8. *Anti* dihydroxylation
9. *Syn* dihydroxylation
10. Ozonolysis

REVIEW OF CONCEPTS AND VOCABULARY

SECTION 9.1

- **Addition reactions** are characterized by the addition of two groups across a double bond.

SECTION 9.2

- Addition reactions are thermodynamically favorable at low temperature and disfavored at high temperature.

SECTION 9.3

- **Hydrohalogenation** reactions are characterized by the addition of H and X across a π bond.
- For unsymmetrical alkenes, the placement of the halogen represents an issue of regiochemistry. Hydrohalogenation reactions are **regioselective**, because the halogen is generally placed at the more substituted position, called **Markovnikov addition.**

- In the presence of peroxides, addition of HBr proceeds via an ***anti*-Markovnikov addition.**
- The regioselectivity of an ionic addition reaction is determined by the preference for the reaction to proceed through the more stable carbocation intermediate.
- When one new chirality center is formed, a racemic mixture of enantiomers is obtained.
- Hydrohalogenation reactions are only efficient when carbocation rearrangements are not possible.

SECTION 9.4

- Addition of water (H and OH) across a double bond is called **hydration**.
- Addition of water in the presence of an acid is called **acid-catalyzed hydration**, which generally proceeds via a Markovnikov addition.
- Acid-catalyzed hydration proceeds via a carbocation intermediate, which is attacked by water to produce an **oxonium ion**, followed by deprotonation.
- Acid-catalyzed hydration is inefficient when carbocation rearrangements are possible.
- Dilute acid favors formation of the alcohol, while concentrated acid favors the alkene.
- When generating a new chirality center, a racemic mixture of enantiomers is expected.

SECTION 9.5

- **Oxymercuration-demercuration** achieves hydration of an alkene without carbocation rearrangements.
- The reaction is believed to proceed via a bridged intermediate called a **mercurinium ion**.

SECTION 9.6

- **Hydroboration-oxidation** can be used to achieve an *anti*-Markovnikov addition of water across an alkene. The reaction is stereospecific and proceeds via a ***syn* addition**.
- Borane (BH_3) exists in equilibrium with its dimer, **diborane**, which exhibits **three-center, two-electron bonds**.
- Hydroboration proceeds via a concerted process, in which borane is attacked by a π bond, triggering a *simultaneous* hydride shift.

SECTION 9.7

- **Catalytic hydrogenation** involves the addition of H_2 across an alkene in the presence of a metal catalyst.
- The reaction proceeds via a *syn* addition.
- **Heterogeneous catalysts** are not soluble in the reaction medium, while **homogeneous catalysts** are.
- **Asymmetric hydrogenation** can be achieved with a chiral catalyst.

SECTION 9.8

- **Halogenation** involves the addition of X_2 (either Br_2 or Cl_2) across an alkene.
- Bromination proceeds via a bridged intermediate, called a **bromonium ion**, which is opened by an S_N2 process that produces an ***anti* addition**.
- In the presence of water, the product is a **bromohydrin** or a **chlorohydrin**, and the reaction is called **halohydrin formation**.

SECTION 9.9

- **Dihydroxylation** reactions are characterized by the addition of OH and OH across an alkene.
- A two-step procedure for *anti* dihydroxylation involves conversion of an alkene to an **epoxide**, followed by acid-catalyzed ring opening.

SECTION 9.10

- *Syn* dihydroxylation can be achieved with osmium tetroxide or potassium permanganate.

SECTION 9.11

- **Ozonolysis** can be used to cleave a double bond and produce two carbonyl groups.

SECTION 9.12

- In order to predict products properly, the following three questions must be considered:
  - What is the identity of the groups being added across the double bond?
  - What is the expected regioselectivity (Markovnikov or *anti*-Markovnikov addition)?
  - What is the expected stereospecificity (*syn* or *anti* addition)?

SECTION 9.13

- The position of a leaving group can be changed via elimination followed by addition.
- The position of a π bond can be changed via addition followed by elimination.

KEY TERMINOLOGY

acid-catalyzed hydration 406
addition reactions 395
***anti* addition** 425
***anti*-Markovnikov addition** 398
asymmetric hydrogenation 422
bromohydrin 428
bromonium ion 427
catalytic hydrogenation 418
chlorohydrin 428
diborane 413
dihydroxylation 430
epoxide 430
halogenation 425
halohydrin formation 428
heterogeneous catalysts 422
homogeneous catalysts 422
hydration 406
hydroboration-oxidation 412
hydrohalogenation 397
Markovnikov addition 398
mercurinium ion 411
oxonium ion 407
oxymercuration-demercuration 410
ozonolysis 434
regioselective 398
***syn* addition** 413
three-center, two-electron bonds 414

SKILLBUILDER REVIEW

9.1 DRAWING A MECHANISM FOR HYDROHALOGENATION

STEP 1 Using two curved arrows, protonate the alkene to form the more stable carbocation.

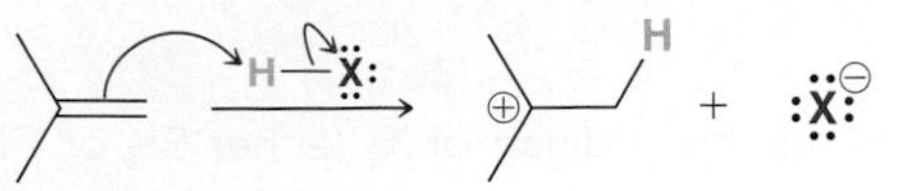

STEP 2 Using one curved arrow, draw the halide ion attacking the carbocation.

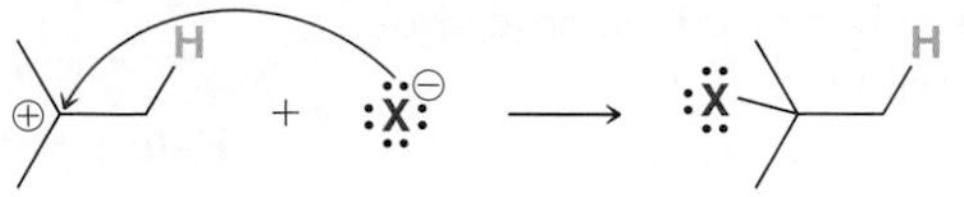

Try Problems 9.3–9.5, 9.51c, 9.64b, 9.69, 9.72

9.2 DRAWING A MECHANISM FOR HYDROHALOGENATION WITH A CARBOCATION REARRANGEMENT

STEP 1 Using two curved arrows, protonate the alkene to form the more stable carbocation.

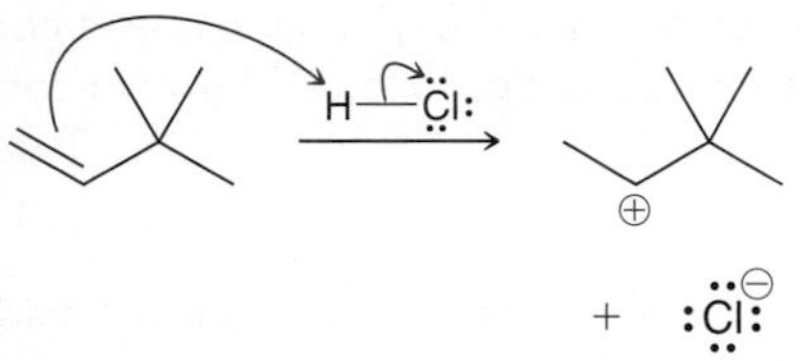

STEP 2 Using one curved arrow, draw a carbocation rearrangement that forms a more stable carbocation, via either a hydride shift or a methyl shift.

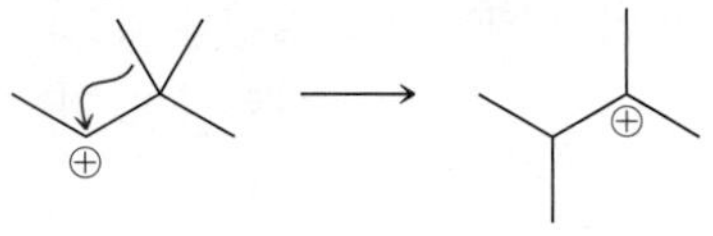

Secondary **Tertiary**

STEP 3 Using one curved arrow, draw the halide ion attacking the carbocation.

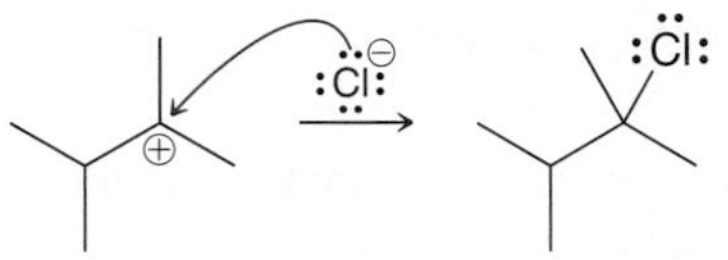

Try Problems 9.7–9.9, 9.51d, 9.76

9.3 DRAWING A MECHANISM FOR AN ACID-CATALYZED HYDRATION

STEP 1 Using two curved arrows, protonate the alkene to form the more stable carbocation.

STEP 2 Using one curved arrow, draw water attacking the carbocation.

STEP 3 Using two curved arrows, deprotonate the oxonium ion using water as a base.

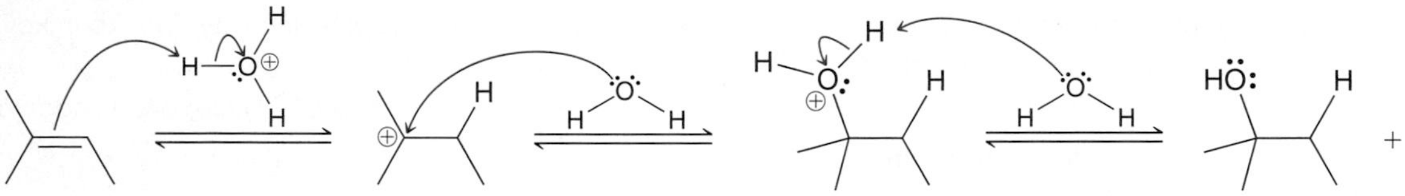

Try Problems 9.12–9.14, 9.51a,b, 9.64a

9.4 PREDICTING THE PRODUCTS OF HYDROBORATION-OXIDATION

STEP 1 Determine the regiochemical outcome based on the requirement for *anti*-Markovnikov addition.

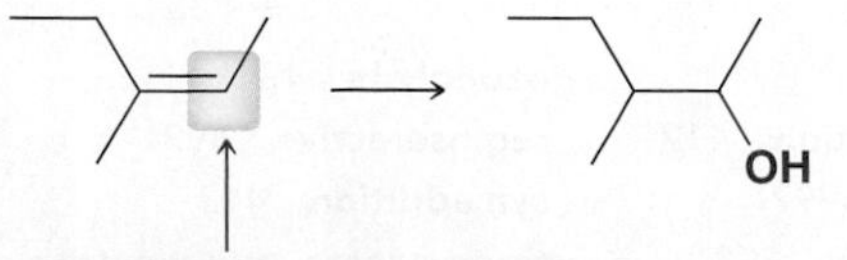

Less substituted

STEP 2 Determine the stereochemical outcome based on the requirement for *syn* addition.

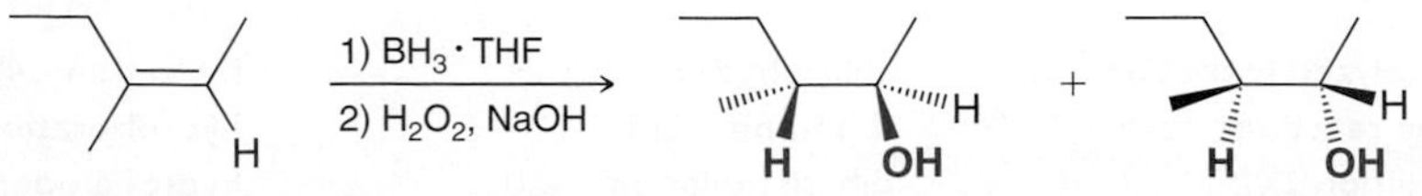

Enantiomers

Try Problems 9.19–9.22, 9.66

9.5 PREDICTING THE PRODUCTS OF CATALYTIC HYDROGENATION

STEP 1 Determine the number of chirality centers formed in the process.

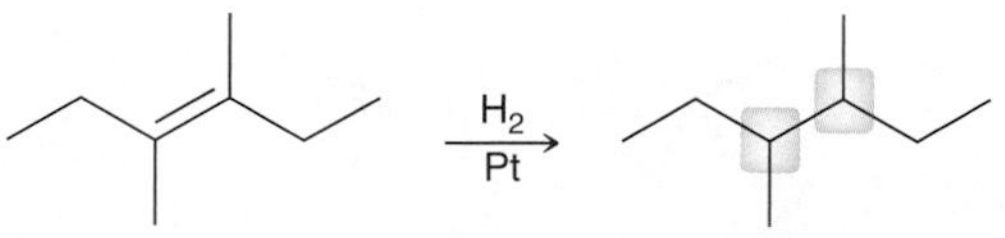

Two stereocenters

STEP 2 Determine the stereochemical outcome based on the requirement for *syn* addition.

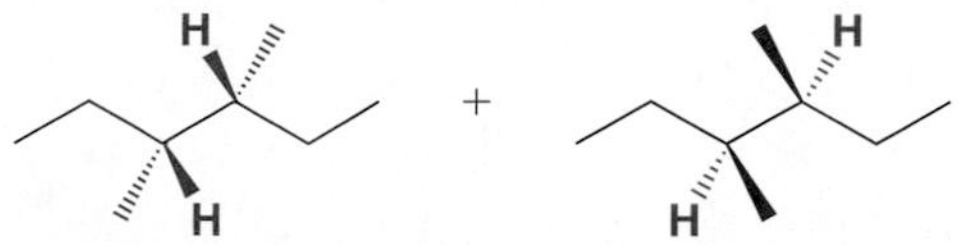

Enantiomers

STEP 3 Verify that the products do not represent a single *meso* compound.

Try Problems 9.23–9.25, 9.55, 9.75

9.6 PREDICTING THE PRODUCTS OF HALOHYDRIN FORMATION

STEP 1 Determine the regiochemical outcome. The OH group should be placed at the more substituted position.

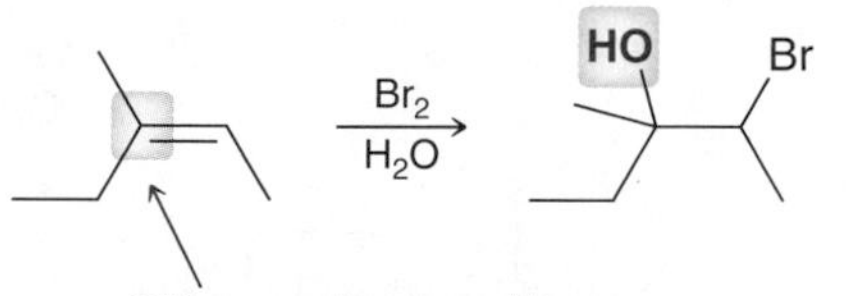

OH group Is placed here at the more substituted position

STEP 2 Determine the stereochemical outcome based on the requirement for *anti* addition.

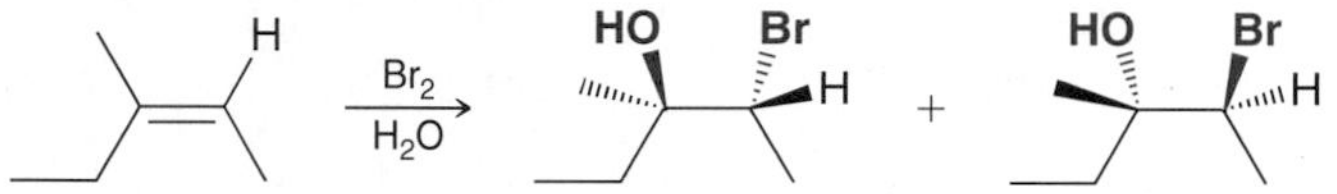

Enantiomers

Try Problems 9.27–9.29, 9.73

9.7 DRAWING THE PRODUCTS OF *ANTI* DIHYDROXYLATION

STEP 1 Determine the number of chirality centers formed in the process.

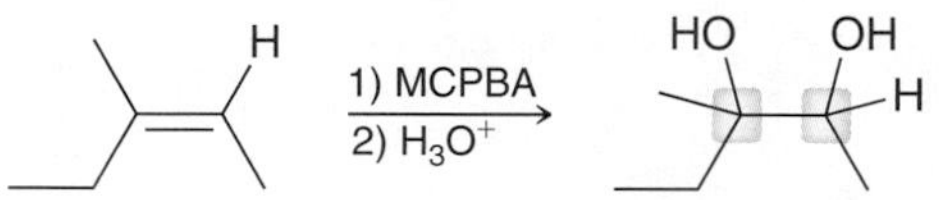

Two stereocenters

STEP 2 Determine the stereochemical outcome based on the requirement for *anti* addition.

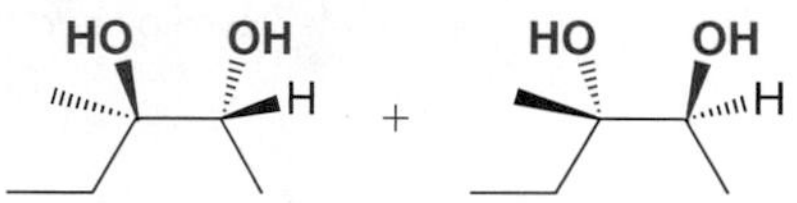

Enantiomers

Try Problems 9.30 – 9.32, 9.65d

9.8 PREDICTING THE PRODUCTS OF OZONOLYSIS

STEP 1 Redraw the compound with the C=C double bonds longer than normal.

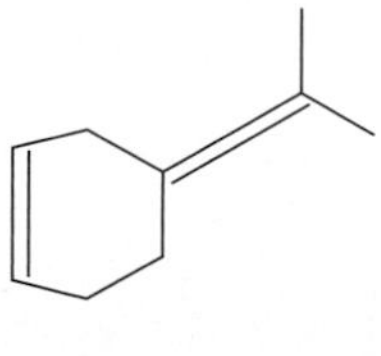

STEP 2 Erase the center of each C=C double bond and place two oxygen atoms in the space.

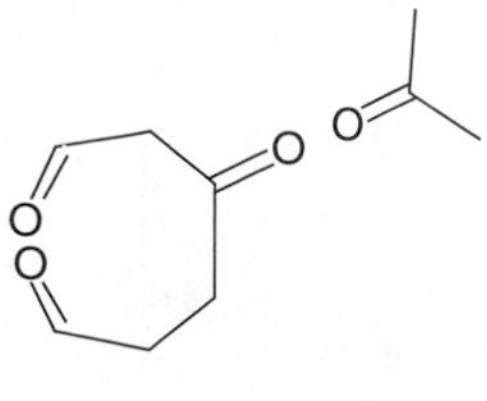

Try Problems 9.34, 9.35, 9.68, 9.77

9.9 PREDICTING THE PRODUCTS OF AN ADDITION REACTION

STEP 1 Identify the two groups being added across the π bond.

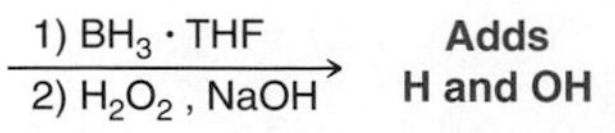

STEP 2 Identify the expected regioselectivity.

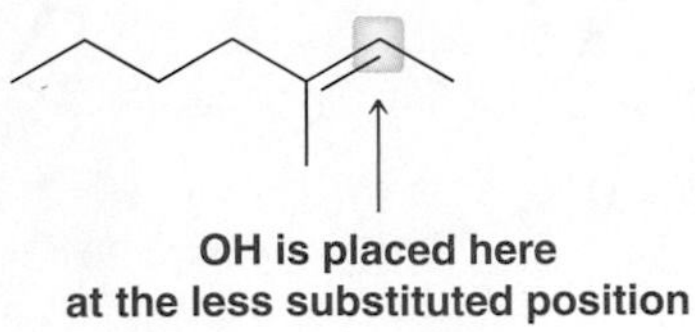

STEP 3 Identify the expected stereospecificity.

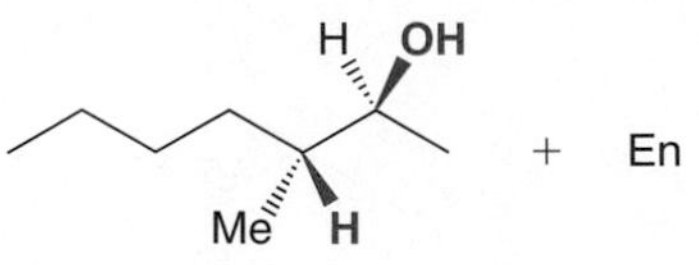

Try Problems 9.3–9.39, 9.49, 9.50, 9.65, 9.73

9.10 PROPOSING A ONE-STEP SYNTHESIS

STEP 1 Identify which two groups were added across the π bond.

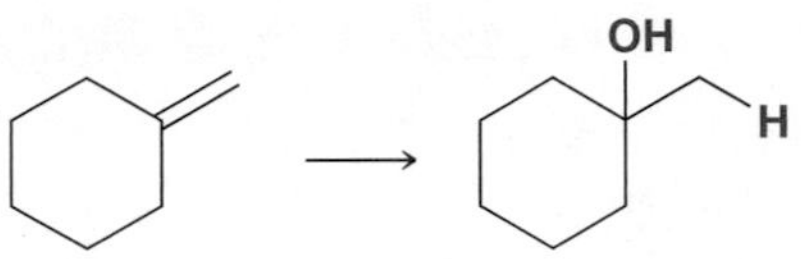

STEP 2 Identify the regioselectivity.

STEP 3 Identify the stereospecificity.

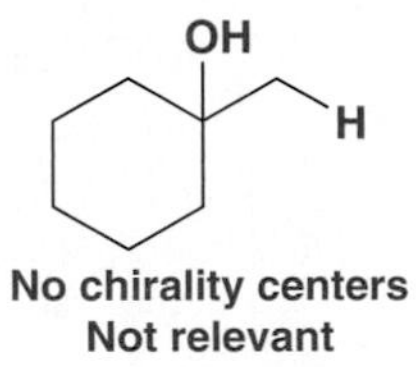

STEP 4 Identify reagents that will achieve the details specified in the first three steps.

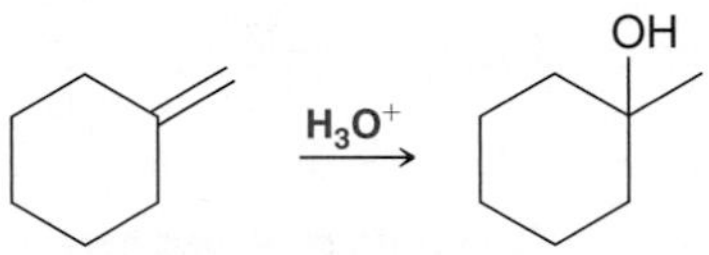

Try Problems 9.40, 9.41, 9.60, 9.62, 9.70, 9.71

9.11 CHANGING THE POSITION OF A LEAVING GROUP

STEP 1 Control the regiochemical outcome of the first step by choosing the appropriate base.

STEP 2 Control the regiochemical outcome of the second step by choosing the reagents to achieve either Markovnikov or *anti*-Markovnikov addition.

Try Problems 9.42–9.44, 9.57, 9.58b

9.12 CHANGING THE POSITION OF A π BOND

STEP 1 Control the regiochemical outcome of the first step by choosing the reagents to achieve either Markovnikov or *anti*-Markovnikov addition.

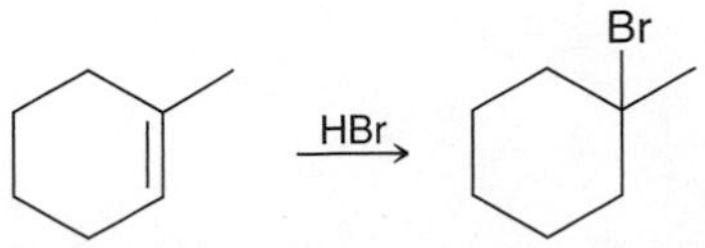

STEP 2 Control the regiochemical outcome of the second step by choosing the appropriate base.

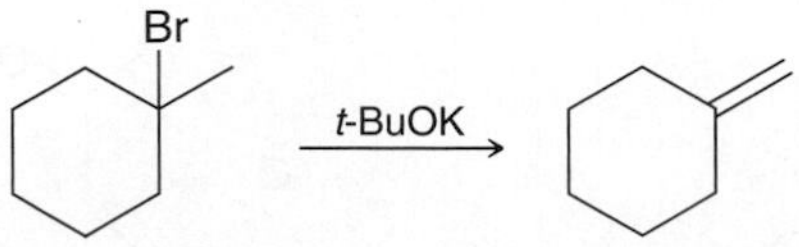

Try Problems 9.45–9.47, 9.53, 9.58

PRACTICE PROBLEMS

Note: Most of the Problems are available within *WileyPLUS*, an online teaching and learning solution.

9.48 At high temperatures, alkanes can undergo *dehydrogenation* to produce alkenes. For example:

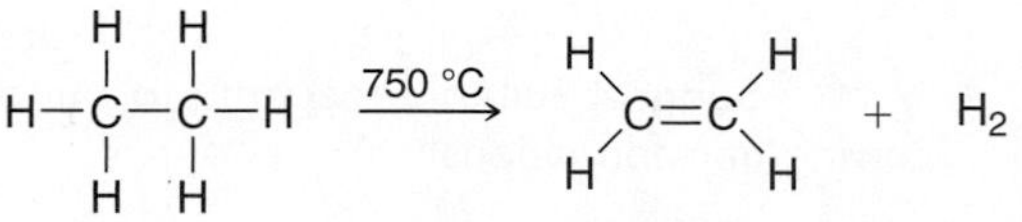

Ethane **Ethylene** **Hydrogen gas**

This reaction is used industrially to prepare ethylene while simultaneously serving as a source of hydrogen gas. Explain why dehydrogenation only works at high temperatures.

9.49 Predict the major product(s) for each of the following reactions:

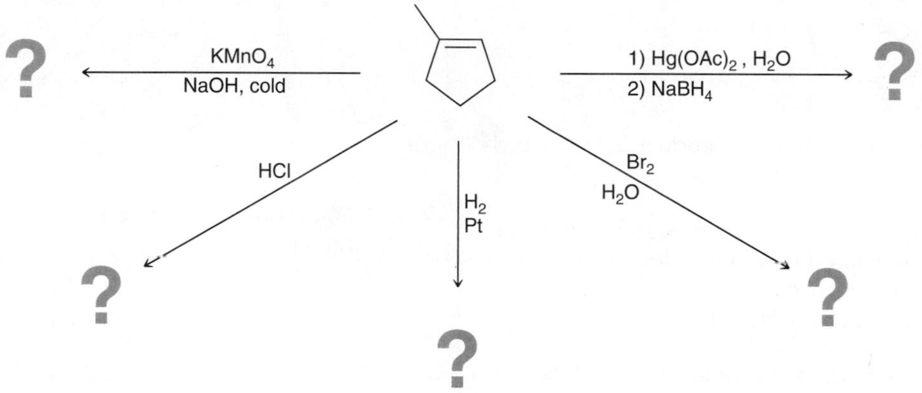

9.50 Predict the major product(s) for each of the following reactions:

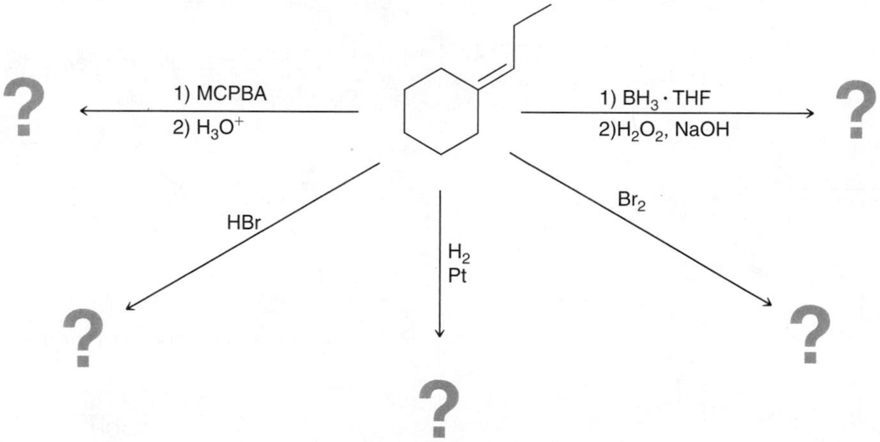

9.51 Propose a mechanism for each of the following reactions:

(a) 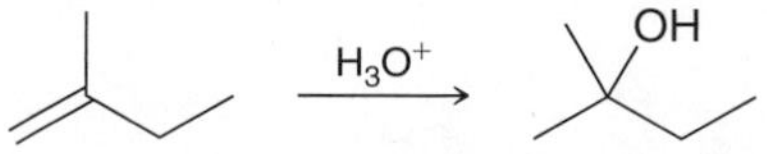

(b) 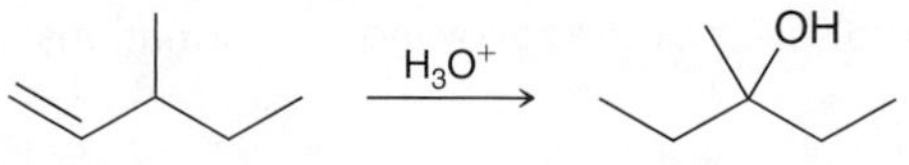

(c) 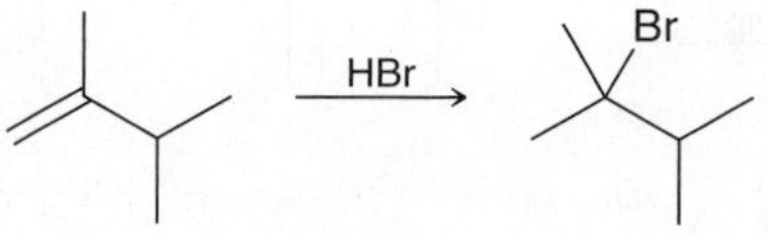

(d) 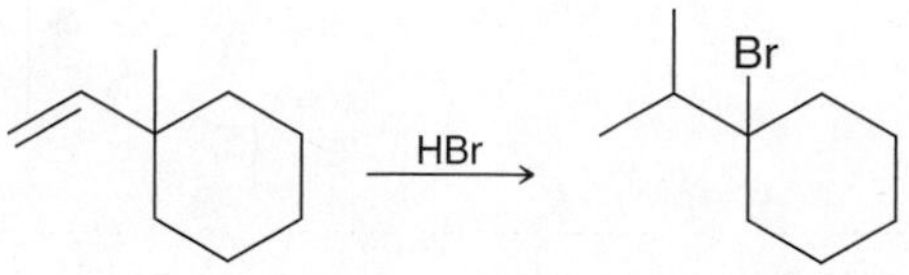

9.52 Compound A reacts with one equivalent of H_2 in the presence of a catalyst to give methylcyclohexane. Compound A can be formed upon treatment of 1-bromo-1-methylcyclohexane with sodium methoxide. What is the structure of compound A?

9.53 Suggest an efficient synthesis for each of the following transformations:

(a)
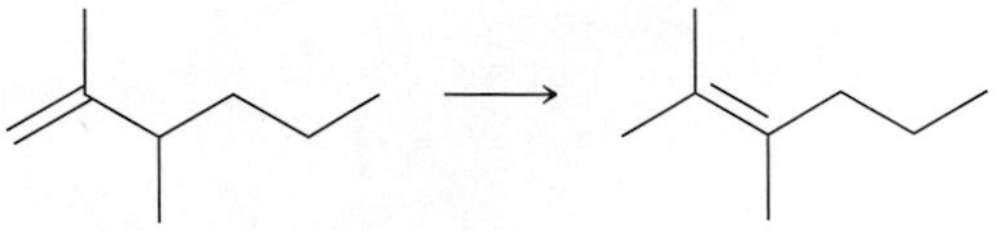

(b)

9.54 Suggest an efficient synthesis for the following transformation:

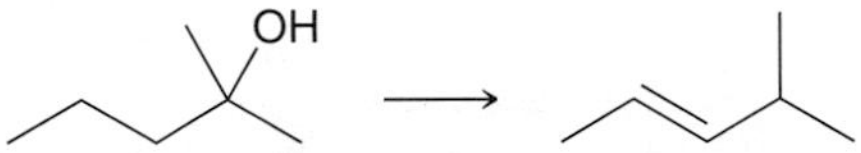

9.55 How many different alkenes will produce 2,4-dimethylpentane upon hydrogenation? Draw them.

9.56 Compound A is an alkene that was treated with ozone (followed by DMS) to yield only 4-heptanone. Identify the major product that is expected when compound A is treated with MCPBA followed by aqueous acid (H_3O^+).

9.57 Suggest an efficient synthesis for each of the following transformations:

(a)
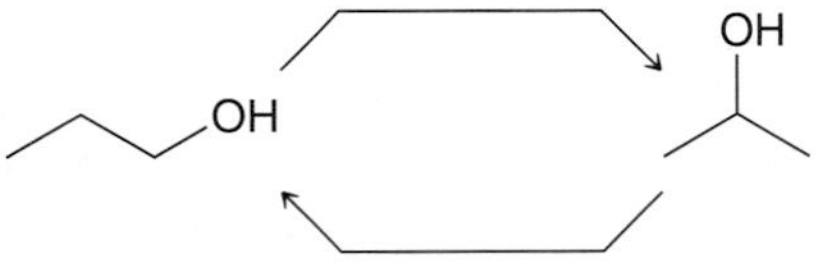

(b)
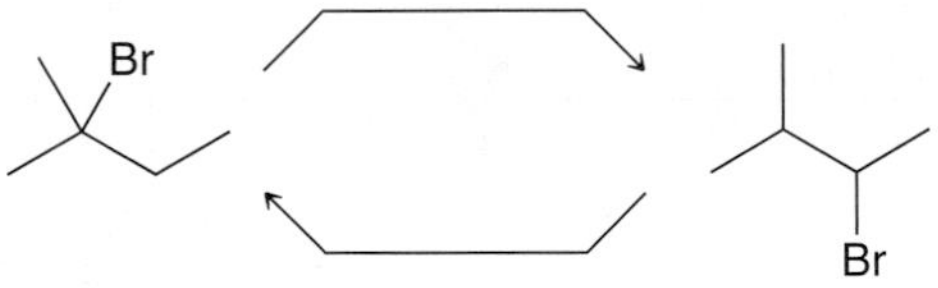

(c)
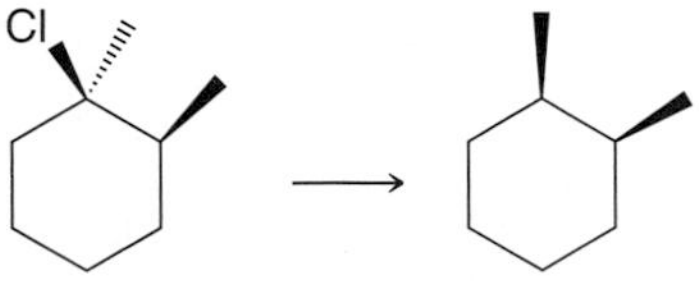

(d)
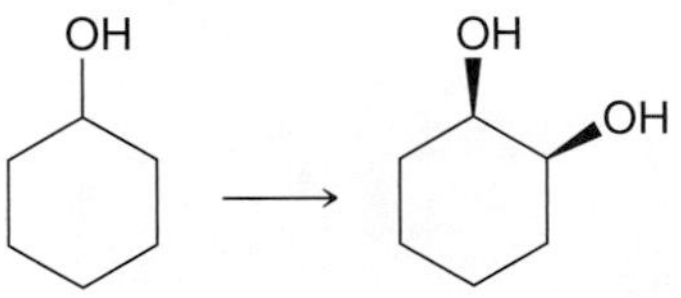

9.58 Suggest an efficient synthesis for each of the following transformations:

(a)
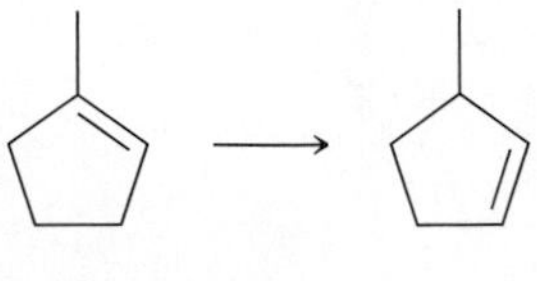

(b)
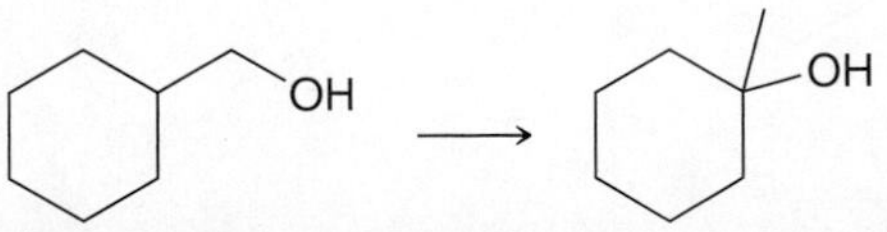

9.59 Compound A has molecular formula $C_7H_{15}Br$. Treatment of compound A with sodium ethoxide yields only one elimination product (compound B) and no substitution products. When compound B is treated with dilute sulfuric acid, compound C is obtained, which has molecular formula $C_7H_{16}O$. Draw the structures of compounds A, B, and C.

9.60 Suggest suitable reagents to perform each of the following transformations:

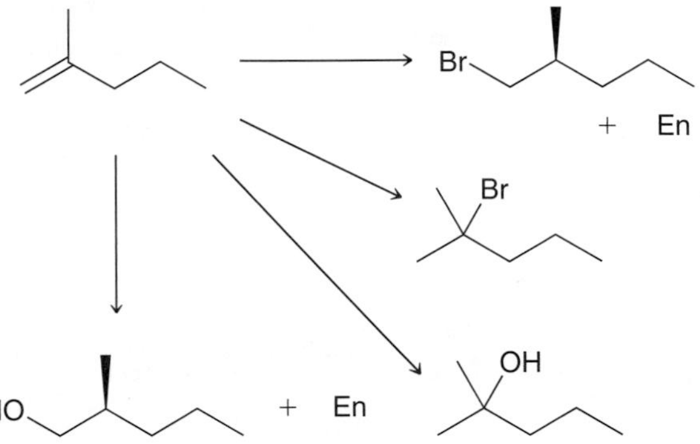

9.61 (*R*)-Limonene is found in many citrus fruits, including oranges and lemons:

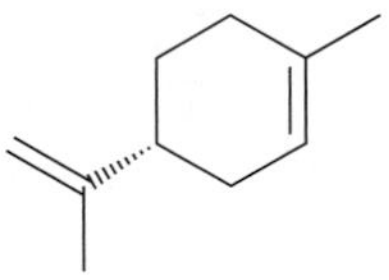

Draw the structures and identify the relationship of the two products obtained when (*R*)-limonene is treated with excess hydrogen in the presence of a catalyst.

9.62 Suggest suitable reagents to perform the following transformation:

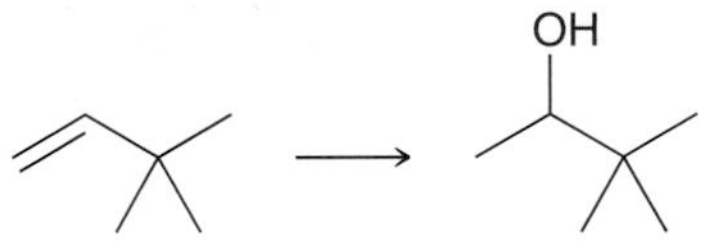

Racemic

9.63 Propose a mechanism for the following transformation:

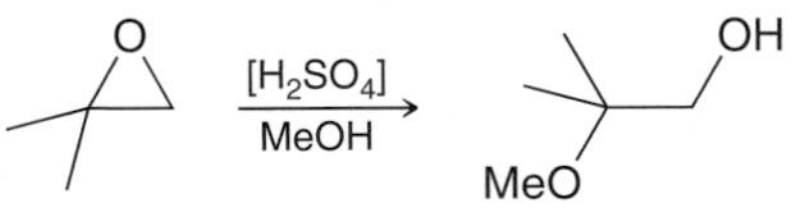

9.64 Propose a mechanism for each of the following transformations:

(a)
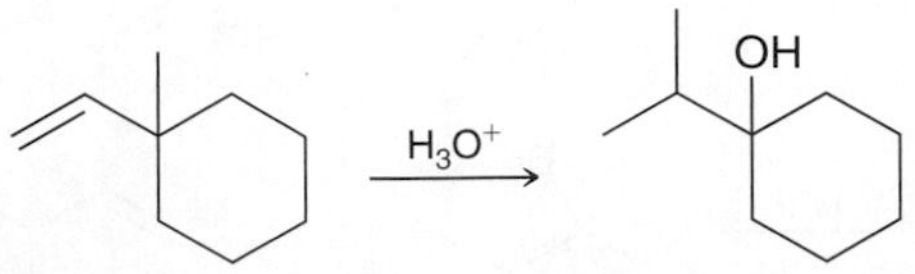

(b)
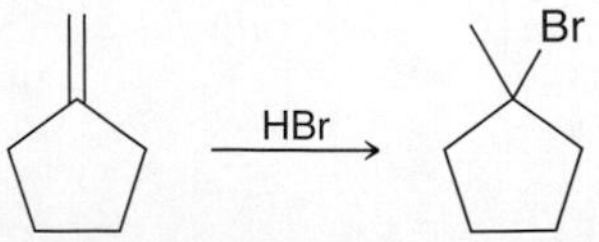

9.65 Predict the major product(s) for each of the following reactions:

(a) 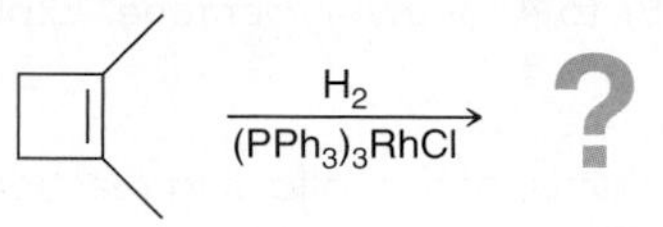

(b) 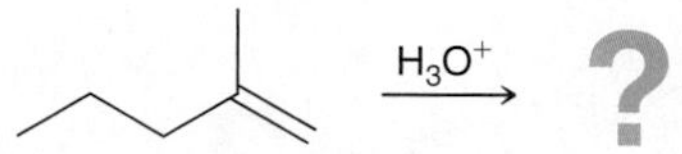

(c) 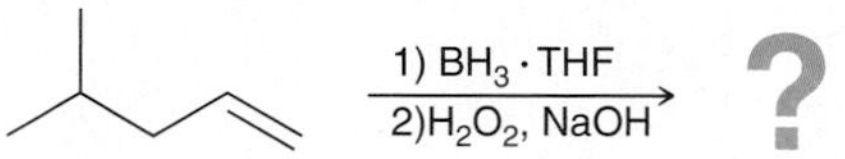

(d) 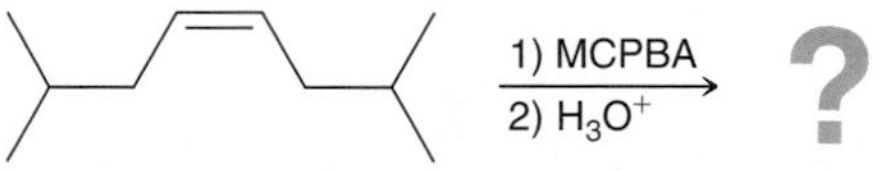

9.66 Explain why each of the following alcohols cannot be prepared via hydroboration-oxidation:

(a) 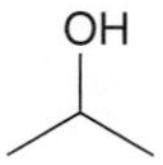

(b) 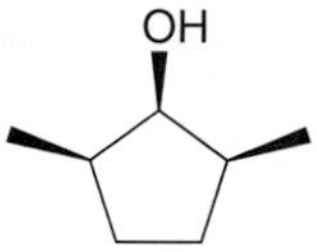

(c) 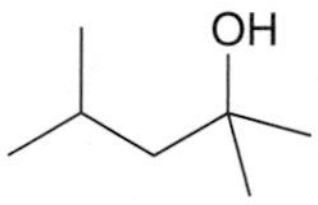

9.67 Compound X is treated with Br_2 to yield *meso*-2,3-dibromobutane. What is the structure of compound X?

9.68 Identify the alkene that would yield the following products via ozonolysis:

(a) 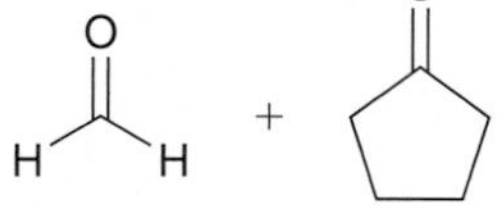
(b) 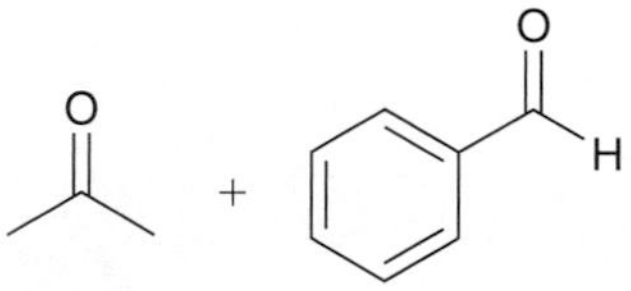
(c) 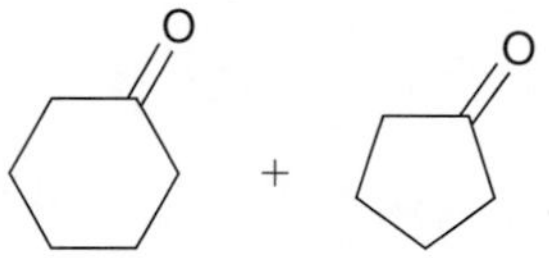
(d) 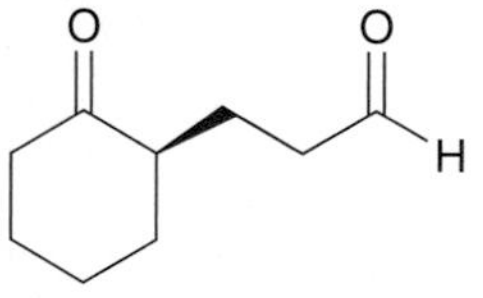

9.69 The following reaction is observed to be regioselective. Draw a mechanism for the reaction, and explain the source of regioselectivity in this case:

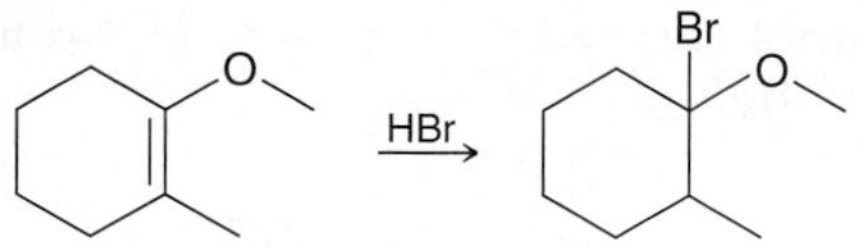

9.70 Identify the reagents you would use to accomplish each of the following transformations:

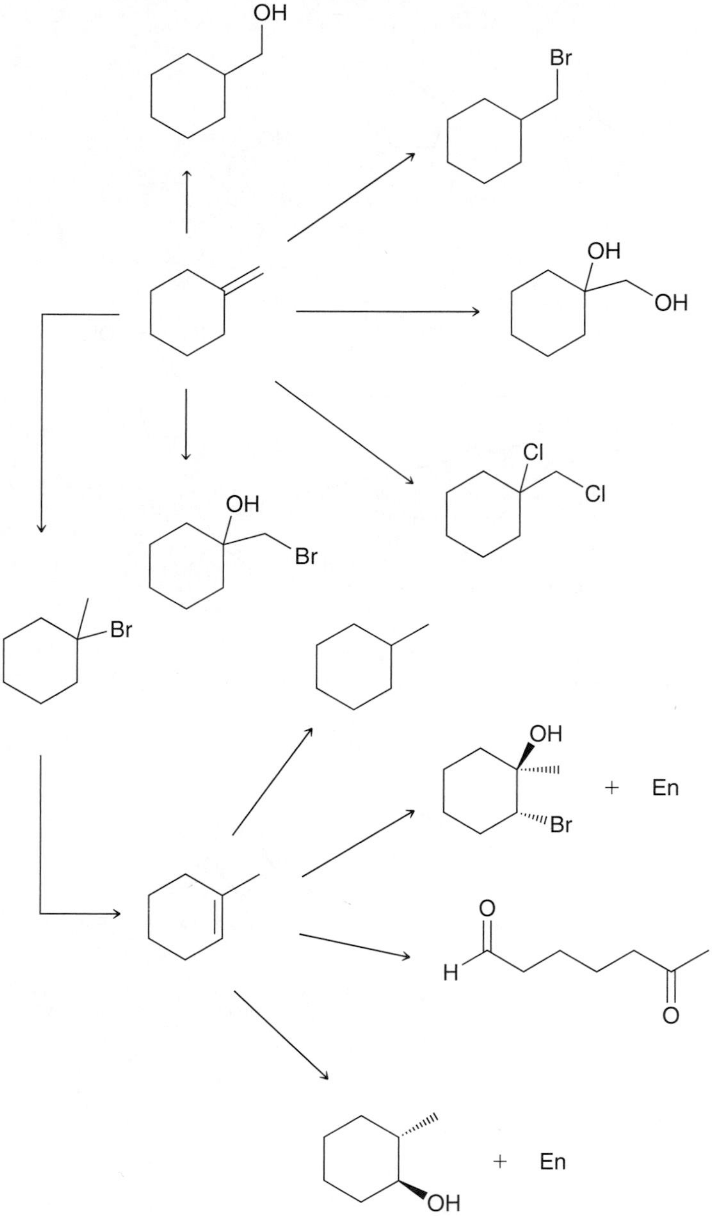

9.71 Identify the reagents you would use to accomplish each of the following transformations:

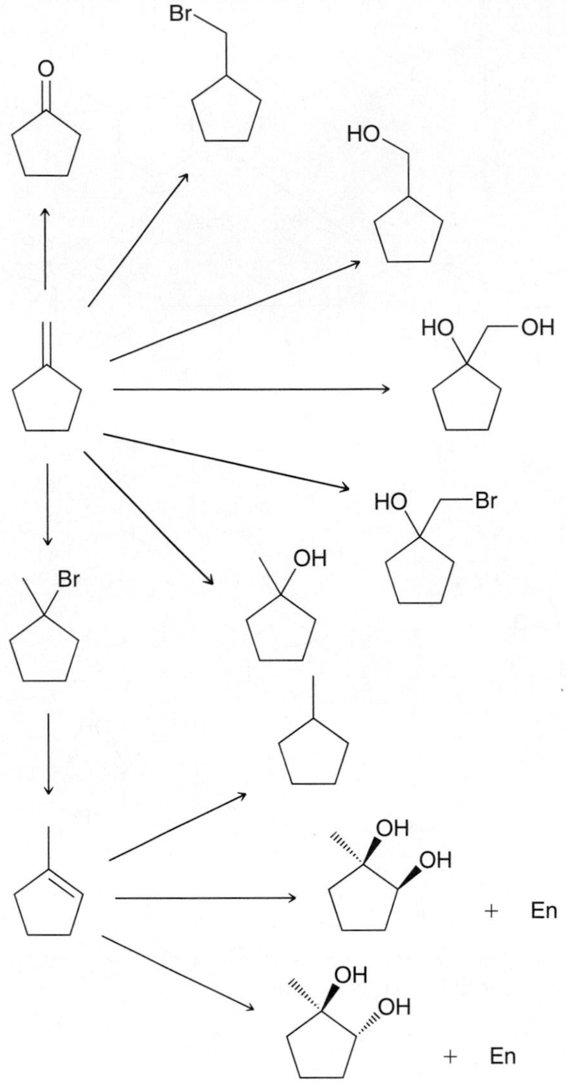

9.72 Identify which of the following two reactions you would expect to occur more rapidly: (1) addition of HBr to 2-methyl-2-pentene or (2) addition of HBr to 4-methyl-1-pentene. Explain your choice.

9.73 Predict the major product(s) of the following reaction:

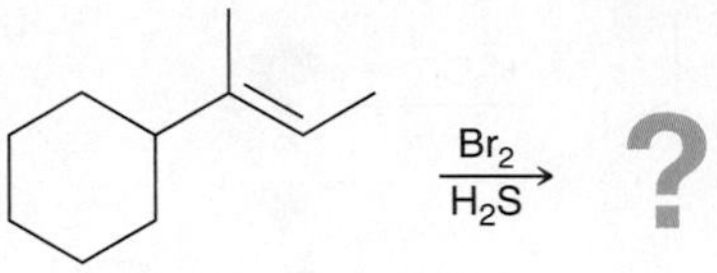

INTEGRATED PROBLEMS

9.74 Suggest an efficient synthesis for the following transformation:

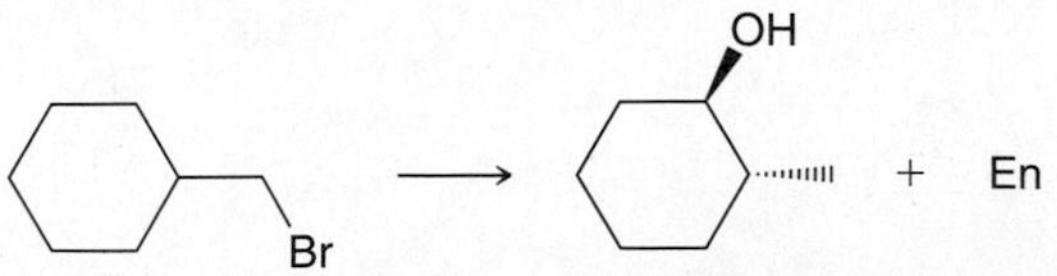

9.75 Compound X has molecular formula C_7H_{14}. Hydrogenation of compound X produces 2,4-dimethylpentane. Hydroboration-oxidation of compound X produces a racemic mixture of 2,4-dimethylpentan-1-ol (shown below). Predict the major product(s) obtained when compound X is treated with aqueous acid (H_3O^+).

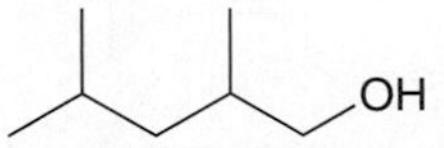

2,4-Dimethyl-1-pentanol

9.76 When (*R*)-2-chloro-3-methylbutane is treated with potassium *tert*-butoxide, a monosubstituted alkene is obtained. When this alkene is treated with HBr, a mixture of products is obtained. Draw all of the expected products.

9.77 Compound Y has molecular formula C_7H_{12}. Hydrogenation of compound Y produces methylcyclohexane. Treatment of compound Y with HBr in the presence of peroxides produces the following compound:

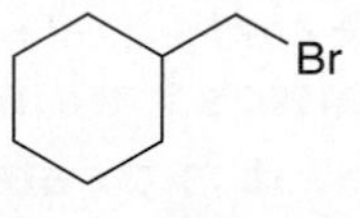

Predict the products when compound Y undergoes ozonolysis.

9.78 Muscalure is the sex pheromone of the common housefly and has the molecular formula $C_{23}H_{46}$. When treated with O_3 followed by DMS, the following two compounds are produced. Draw two possible structures for muscalure.

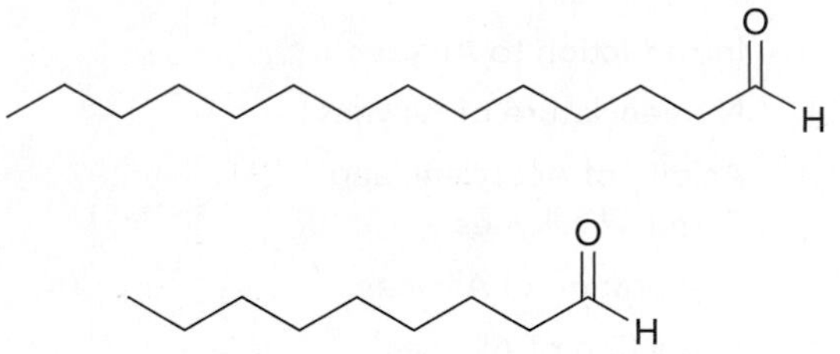

CHALLENGE PROBLEMS

9.79 Propose a plausible mechanism for each of the following reactions:

(a)

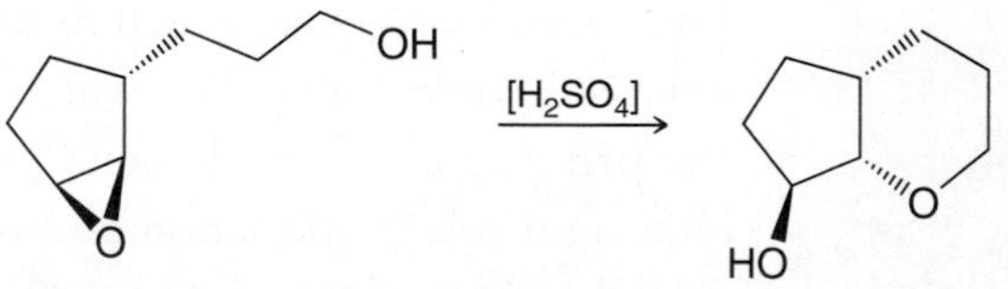

(b)

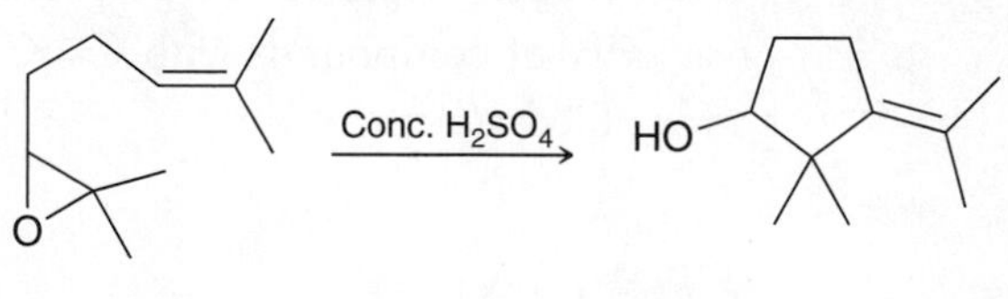

9.80 Suggest an efficient synthesis for the following transformation:

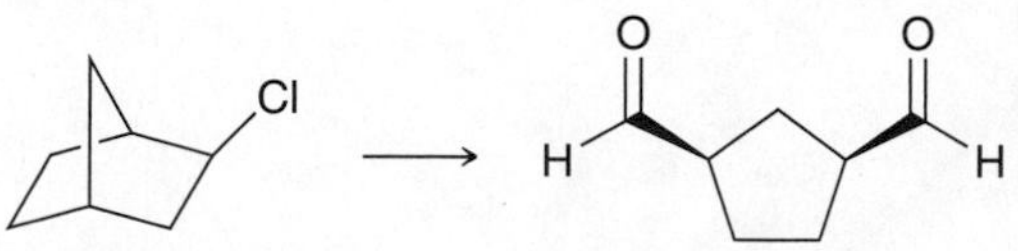

9.81 Propose a plausible mechanism for each of the following reactions:

(a)

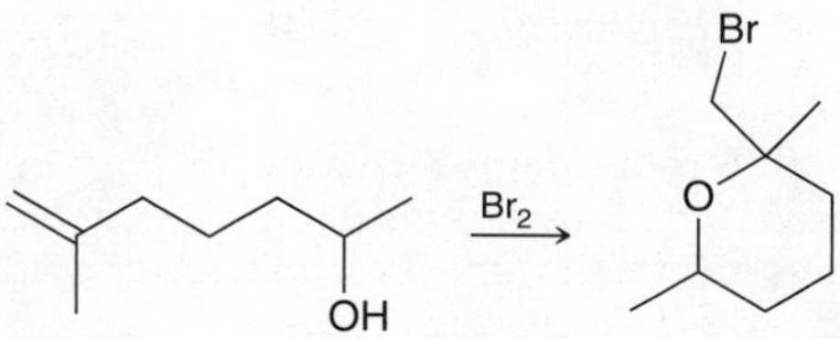

(b)

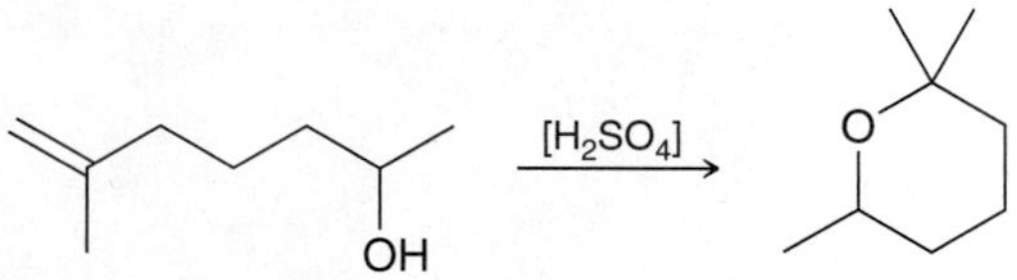

9.82 Propose a plausible mechanism for the following process, called iodolactonization:

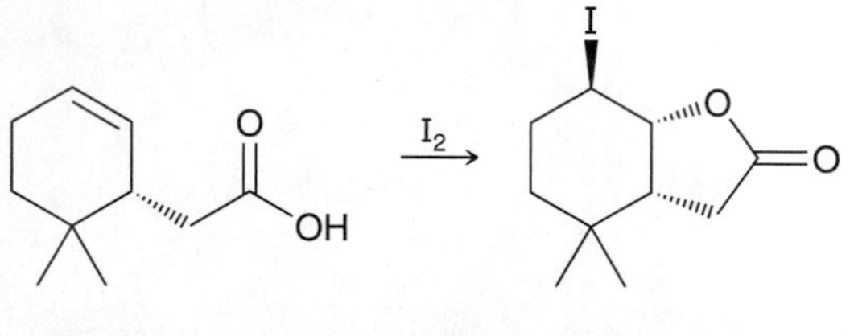

9.83 When 3-bromocyclopentene is treated with HBr, the observed product is a racemic mixture of *trans*-1,2-dibromocyclopentane. None of the corresponding *cis*-dibromide is observed. Propose a mechanism that accounts for the observed stereochemical outcome:

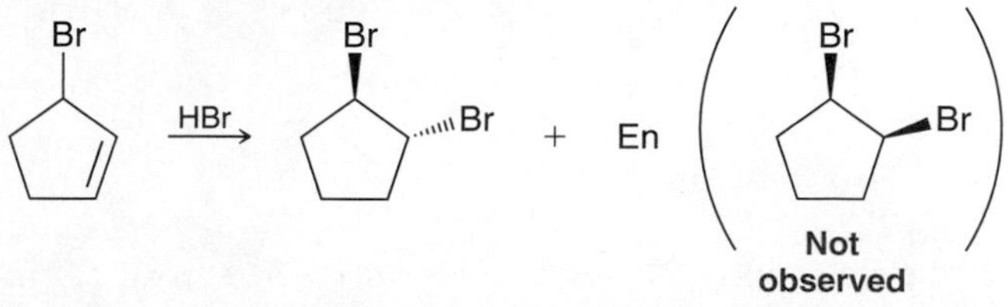

10 Alkynes

DID YOU EVER WONDER…
what causes Parkinson's disease and how it is treated?

Parkinson's disease is a motor system disorder that affects an estimated 3% of the U.S population over the age of 60. The main symptoms of Parkinson's disease include trembling and stiffness of the limbs, slowness of movement, and impaired balance. Parkinson's disease is a neurodegenerative disease, because the symptoms are caused by the degeneration of neurons (brain cells). The neurons most affected by the disease are located in a region of the brain called the *substantia nigra.* When these neurons die, they cease to produce dopamine, the neurotransmitter used by the brain to regulate voluntary movement. The symptoms described above begin to appear when 50–80% of dopamine-producing neurons have died. There is no known cure for the disease, which is progressive. However, the symptoms can be treated through a variety of methods. One such method utilizes a drug called selegiline, whose molecular structure contains a C≡C triple bond. We will see that the presence of the triple bond plays an important role in the action of this drug. This chapter explores the properties and reactivity of compounds with C≡C triple bonds called *alkynes*.

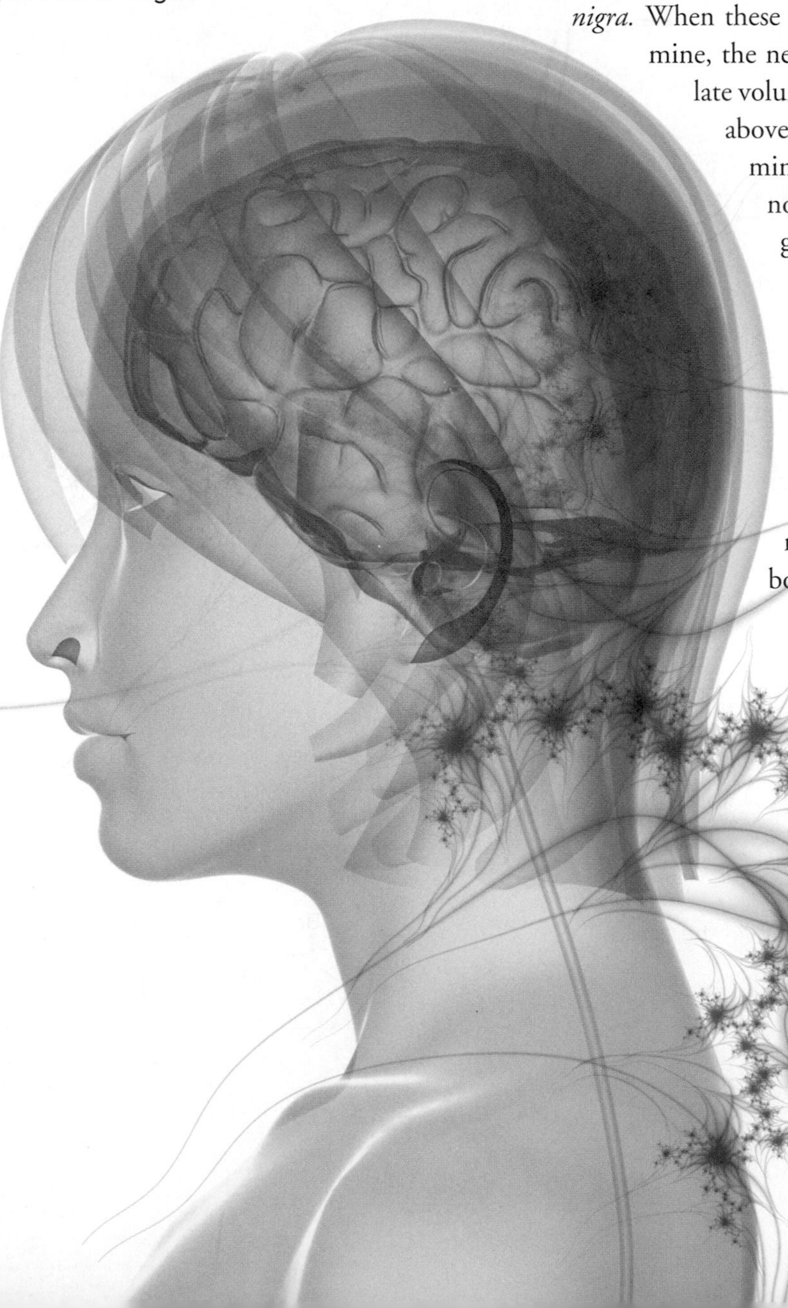

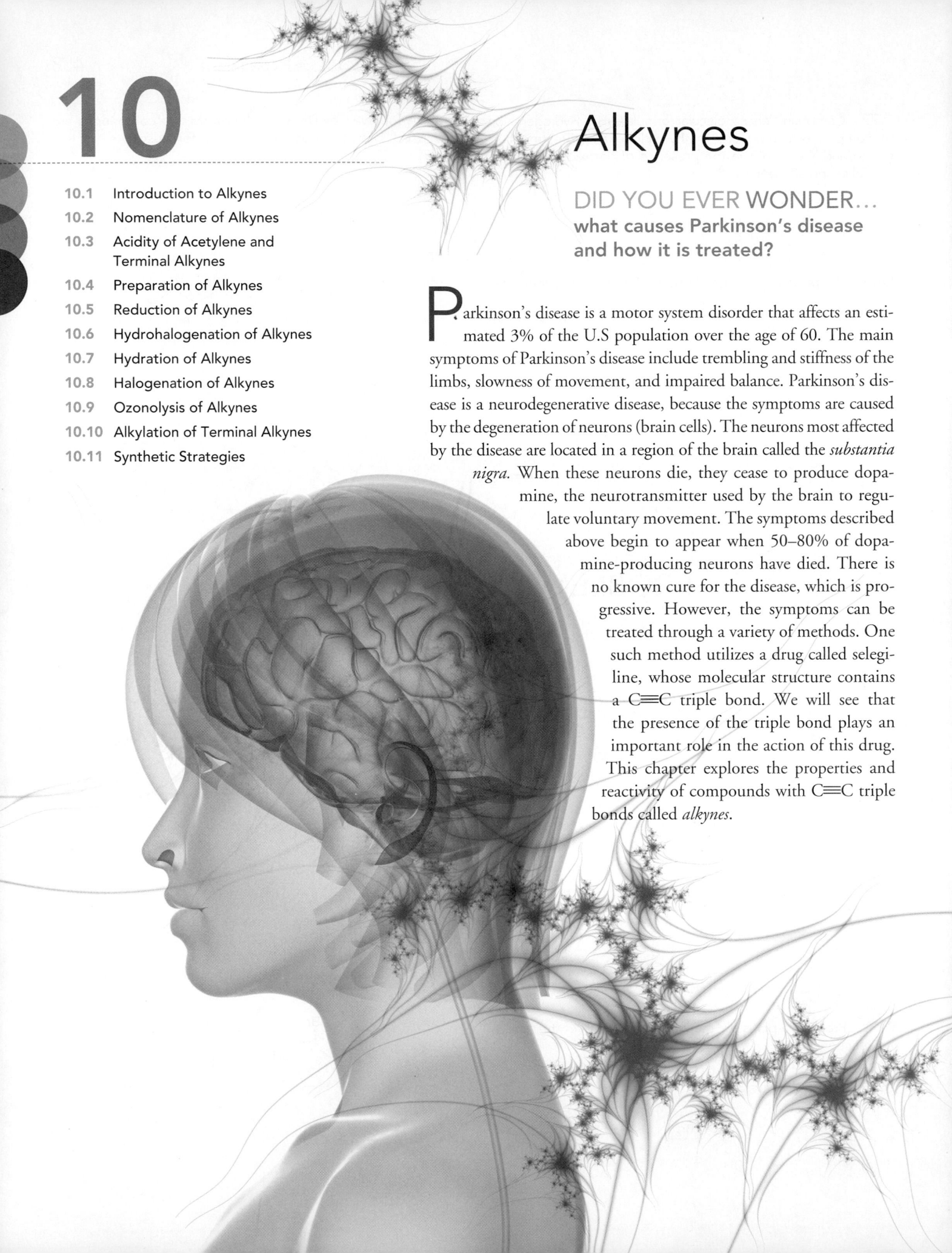

DO YOU REMEMBER?

Before you go on, be sure you understand the following topics.
If necessary, review the suggested sections to prepare for this chapter.

- Brønsted-Lowry Acidity (Sections 3.4, 3.5)
- Nucleophiles and Electrophiles (Section 6.7)
- Energy Diagrams (Sections 6.5, 6.6)
- Arrow Pushing (Sections 6.8–6.10)

WILEY PLUS Visit www.wileyplus.com to check your understanding and for valuable practice.

10.1 Introduction to Alkynes

Structure and Geometry of Alkynes

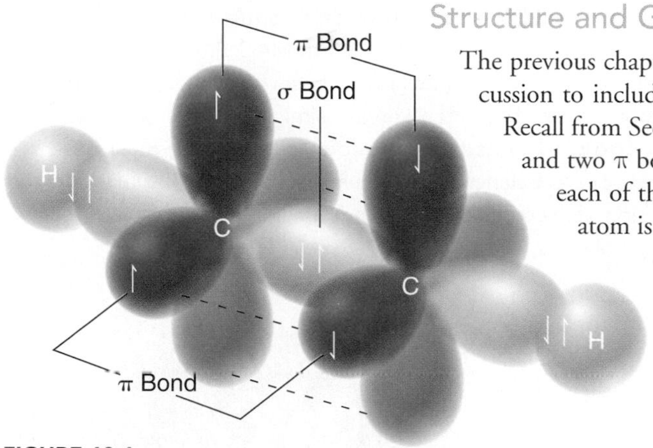

FIGURE 10.1 The atomic orbitals used to form a triple bond. The σ bond is formed from the overlap of two hybridized orbitals, while each of the two π bonds is formed from overlapping *p* orbitals.

The previous chapter explored the reactivity of alkenes. In this chapter, we expand that discussion to include the reactivity of **alkynes**, compounds containing a C≡C triple bond. Recall from Section 1.9 that a triple bond is comprised of three separate bonds: a σ bond and two π bonds. The σ bond results from the overlap of *sp*-hybridized orbitals, while each of the π bonds results from overlapping *p* orbitals (Figure 10.1). Each carbon atom is *sp*-hybridized and exhibits linear geometry. An electrostatic potential map of acetylene (H—C≡C—H) reveals a cylindrical region of high electron density encircling the triple bond (Figure 10.2). This region (shown in red) explains why alkynes react with compounds that bear a region of low electron density. As a result, alkynes are similar to alkenes in their ability to function either as bases or as nucleophiles. We will see examples of both behaviors in this chapter.

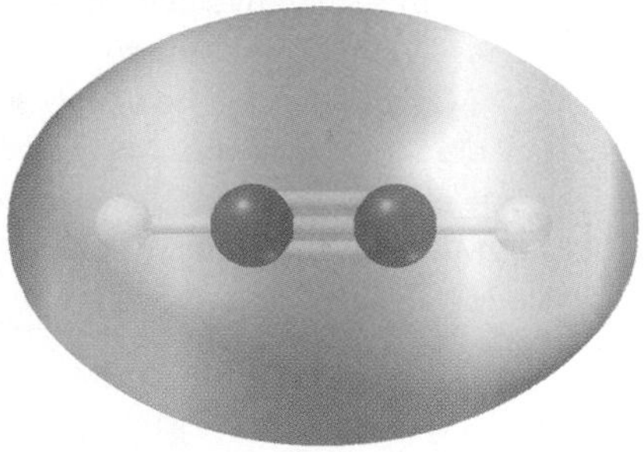

FIGURE 10.2 An electrostatic potential map of acetylene, indicating a cylindrical region of high electron density (red).

Alkynes in Industry and Nature

The simplest alkyne, acetylene (H—C≡C—H), is a colorless gas that undergoes combustion to produce a high-temperature flame (2800°C) and is used as a fuel for welding torches. Acetylene is also used as a precursor for the preparation of higher alkynes, as we will see in Section 10.10.

Alkynes are less common in nature than alkenes, although more than 1000 different alkynes have been isolated from natural sources. One such example is histrionicotoxin, which is one of the many toxins secreted by the South American frog, *Dendrobates histrionicus*, as a defense against predators. For many centuries, South American tribes have extracted the mixture of toxins from the frog's skin and placed it on the tips of arrows to produce poison arrows.

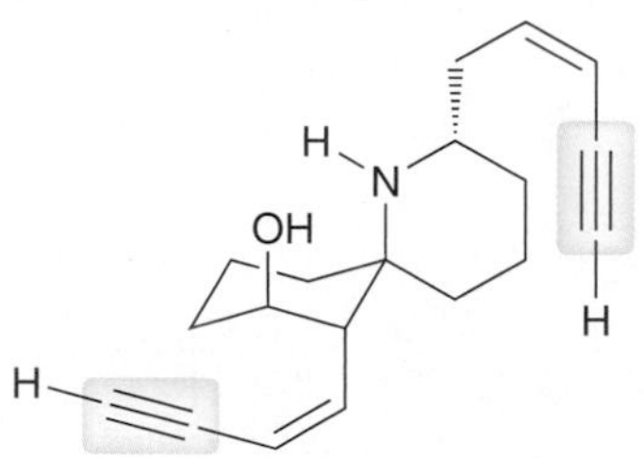

Histrionicotoxin

MEDICALLYSPEAKING)))

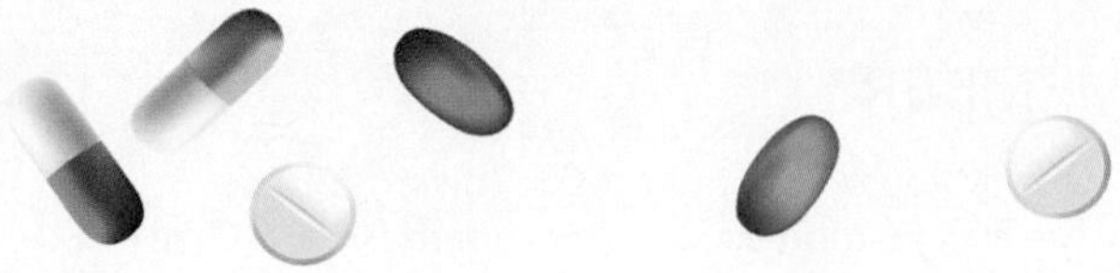

The Role of Molecular Rigidity

As mentioned in the chapter opening, Parkinson's disease is a neurodegenerative disorder marked by a decreased production of dopamine in the brain. Since dopamine regulates motor function, the decrease in dopamine leads to impaired motor control. Although there is no cure for Parkinson's disease, the symptoms can be treated by a variety of methods. The most effective course of treatment is to administer a drug called L-dopa, which is converted into dopamine in the brain:

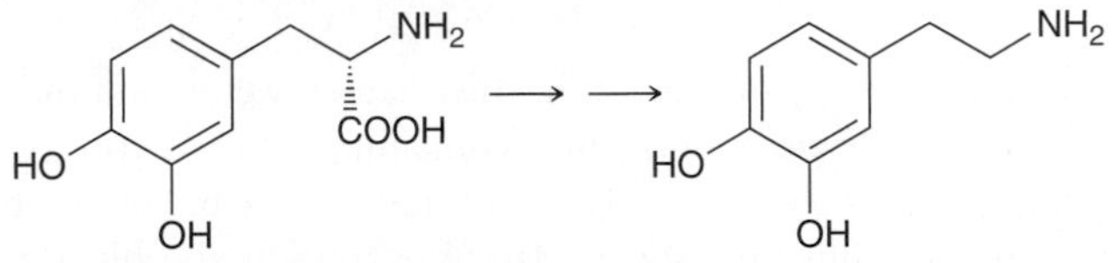

(*S*)-3,4-Dihydroxyphenylalanine (L-dopa) **Dopamine**

This method was described previously in Section 9.7 and is effective because it replenishes the supply of dopamine in the brain. Another method for increasing dopamine levels is to slow the rate at which dopamine is removed from the brain. Dopamine is primarily metabolized under the influence of an enzyme (a biological catalyst), called monoamine oxidase B (MAO B). Any drug that inactivates this enzyme will effectively slow the rate at which dopamine is metabolized, thereby slowing the rate at which dopamine levels decrease in the brain. Unfortunately, a closely related enzyme, called MAO A, is used for the metabolism of other compounds, and any drug that inactivates MAO A leads to significant cardiovascular side effects. Therefore, the selective inactivation of MAO B (but not MAO A) is required. The first selective MAO B inactivator, called selegiline, was approved by the FDA in 1989 for the treatment of Parkinson's disease:

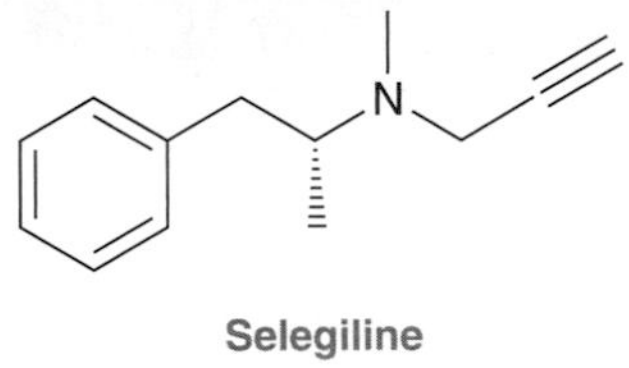

Selegiline

Selegiline, sold under the trade name Eldepryl™, is often prescribed in combination with L-dopa. The combination of these two drugs offers a more effective method for combating the diminishing supply of dopamine in the brain.

Notice that the structure of selegiline exhibits a C≡C triple bond, which serves an important function. Specifically, its linear geometry imparts structural rigidity to the compound. The aromatic ring on the other side of the compound also imparts structural rigidity, and both of these structural subunits enable the compound to selectively bind to MAO B, thereby causing its inactivation. Triple bonds appear in several other FDA-approved drugs, where they often serve a similar function. In order for a drug to bind to its target receptor effectively, it must have the appropriate balance of structural rigidity and flexibility. In the design of new drugs, triple bonds are sometimes used to achieve that balance.

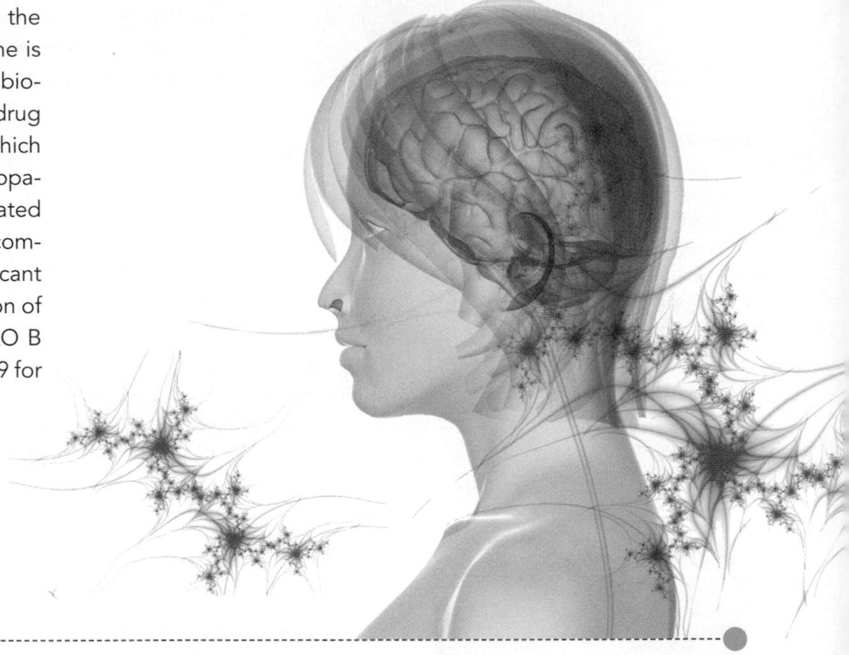

In addition to the alkynes found in nature, many synthetic alkynes (prepared in the laboratory) are of particular interest. One such example is ethynylestradiol, found in many birth control formulations. This synthetic oral contraceptive elevates hormone levels in women and prevents ovulation. The presence of the triple bond renders this compound a more potent contraceptive than its natural analogue, which lacks a triple bond. The effect of the triple bond is attributed to the additional structural rigidity that it imparts to the compound (as described in the box on molecular rigidity).

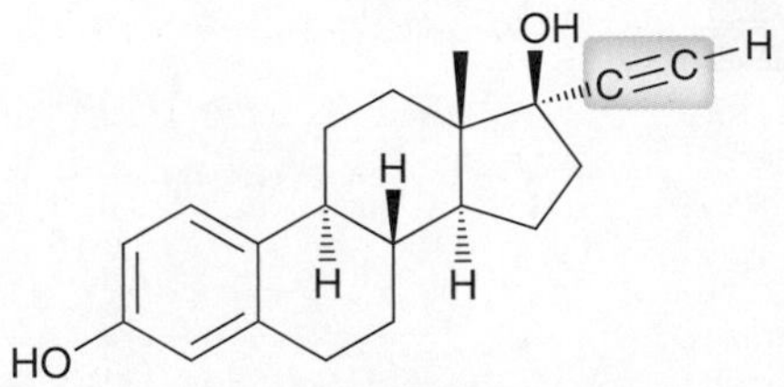

Ethynylestradiol

PRACTICALLYSPEAKING)))

Conducting Organic Polymers

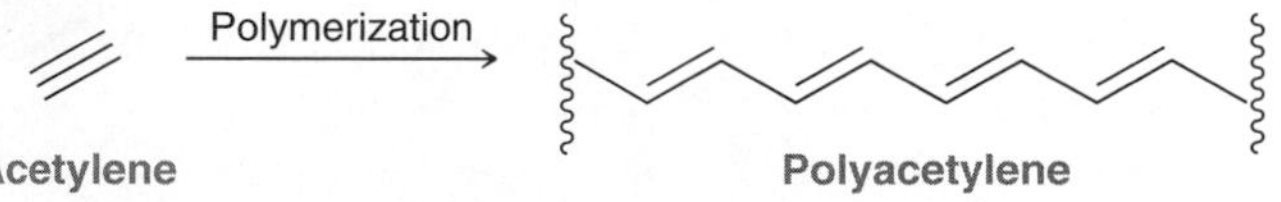

Polyacetylene, which can be prepared via the polymerization of gaseous acetylene, was the first known example of an organic polymer capable of conducting electricity.

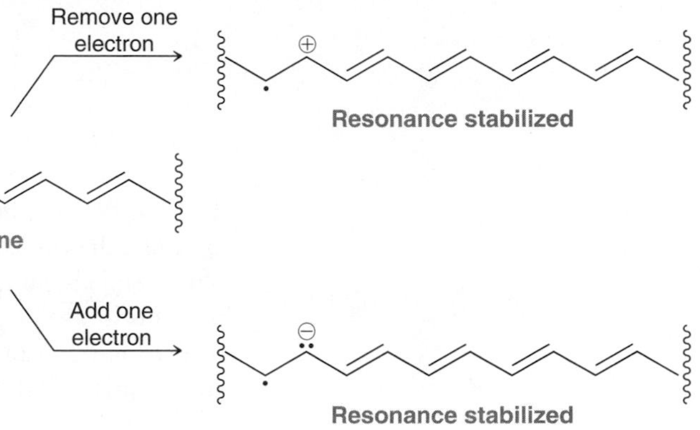

In practice, polyacetylene only poorly conducts electricity. However, when the polymer is prepared as a cation or anion (a process called *doping*), it can conduct electricity almost as well as a copper wire. Both the cationic polymer and the anionic polymer are resonance stabilized, and they each conduct electricity very efficiently. This discovery effectively opened the doorway to the exciting field of conducting organic polymers, and the 2000 Nobel Prize in Chemistry was awarded to its discoverers: Alan Heeger, Alan MacDiarmid, and Hideki Shirakawa. Polyacetylene is used in the packaging materials for computer parts, due to its ability to dissipate static charges that could damage sensitive circuitry.

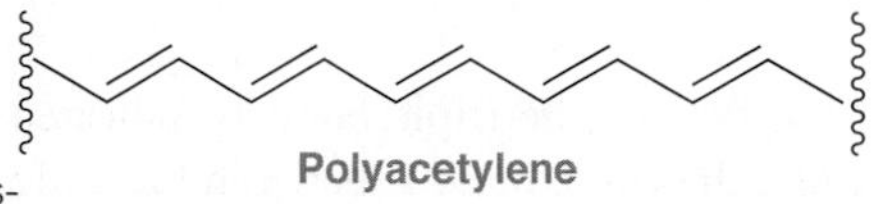

Polyacetylene

Polyacetylene itself has limited application because of its sensitivity to air and moisture, but many other conducting polymers have been developed. Consider the structure of poly(*p*-phenylene vinylene), or PPV, which is a conducting polymer:

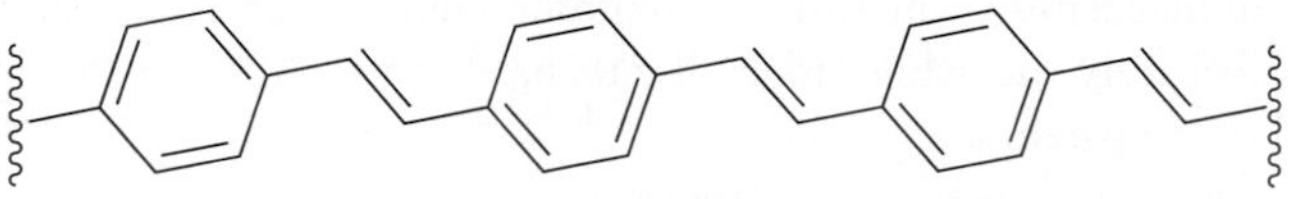

Poly(*p*-phenylene vinylene)

Conducting polymers such as PPV are sometimes used in LED (light emitting diode) displays. When subjected to an electric field, LEDs emit light, a phenomenon called electroluminescence. Organic LED systems, or OLEDs, are generally less effective than inorganic LED systems, although many useful OLEDs have been developed over the past several decades. They are used in cell phone displays, digital cameras, and the displays of MP3 players. LEDs are used in a wide range of applications, including traffic lights, flat-panel TV displays, and alarm clock displays.

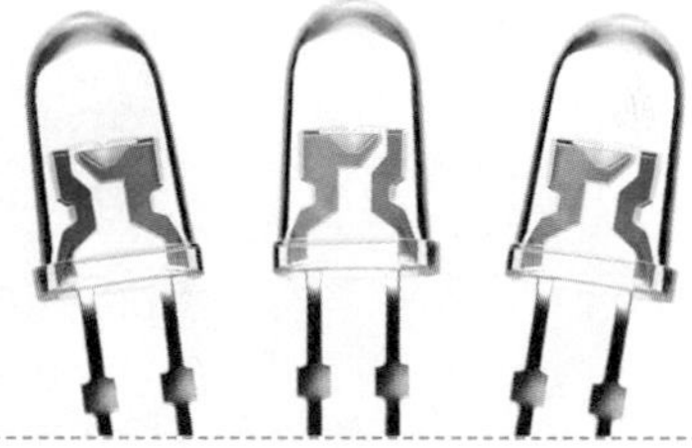

10.2 Nomenclature of Alkynes

Recall from Sections 4.2 and 8.3 that naming alkanes and alkenes requires four discrete steps:

1. Identify and name the parent.
2. Identify and name the substituents.
3. Number the parent chain and assign a locant to each substituent.
4. Assemble the substituents alphabetically.

Alkynes are named using the same four steps, with the following additional rules:
When naming the parent, the suffix "yne" is used to indicate the presence of a C≡C triple bond:

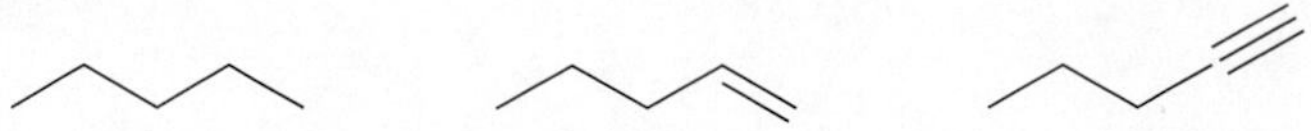

Pent<u>ane</u> **Pent<u>ene</u>** **Pent<u>yne</u>**

The parent of an alkyne is the longest chain *that includes the C≡C triple bond:*

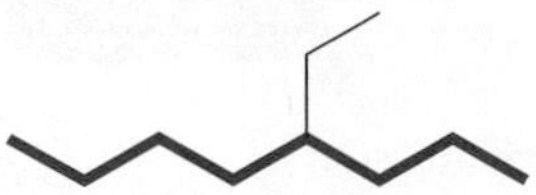

Parent = octane

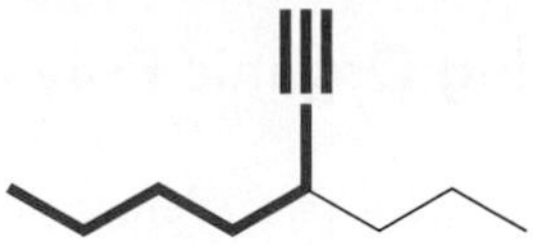

Parent = heptyne

When numbering the parent chain of an alkyne, the triple bond should receive the lowest number possible, despite the presence of alkyl substituents:

Correct

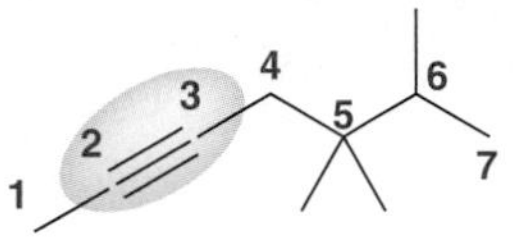

Incorrect

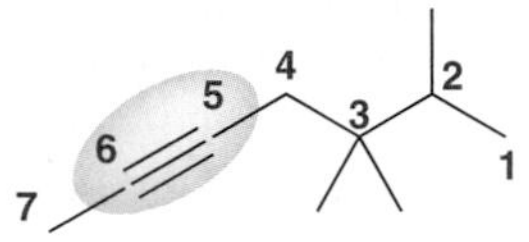

The position of the triple bond is indicated using a single locant, not two. In the previous example, the triple bond is between C2 and C3 on the parent chain. In this case, the position of the triple bond is indicated with the number 2. The IUPAC rules published in 1979 dictate that this locant be placed immediately before the parent, while the IUPAC recommendations released in 1993 and 2004 allow for the locant to be placed before the suffix "yne". Both names are acceptable IUPAC names.

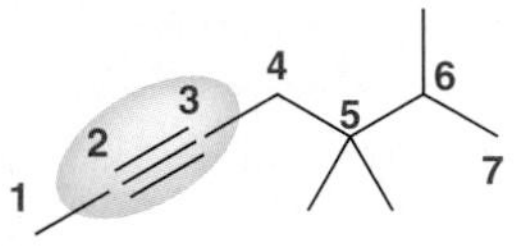

5,5,6-Trimethyl-2-heptyne
or
5,5,6-Trimethylhept-2-yne

In addition to IUPAC nomenclature, chemists use common names for many alkynes. Ethyne (H—C≡C—H) is called acetylene, while larger alkynes have common names that identify the alkyl groups attached to the parent acetylene:

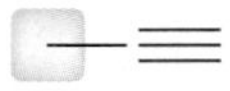

Methylacetylene

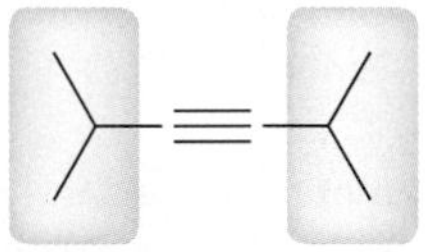

Diisopropylacetylene

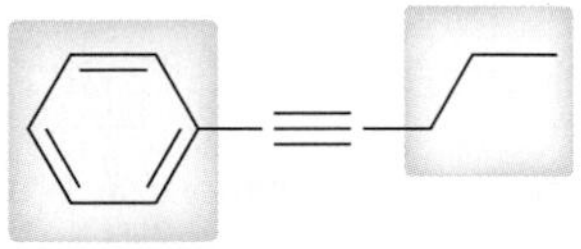

Phenylpropylacetylene

Notice that the first example is monosubstituted; it possesses only one alkyl group. Monosubstituted acetylenes are called **terminal alkynes**, while disubstituted acetylenes are called **internal alkynes**. This distinction will be important in the upcoming sections of this chapter.

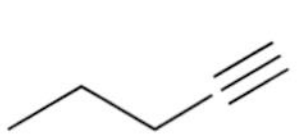

Terminal

Internal

SKILLBUILDER

10.1 ASSEMBLING THE SYSTEMATIC NAME OF AN ALKYNE

LEARN the skill

Provide a systematic name for the following compound:

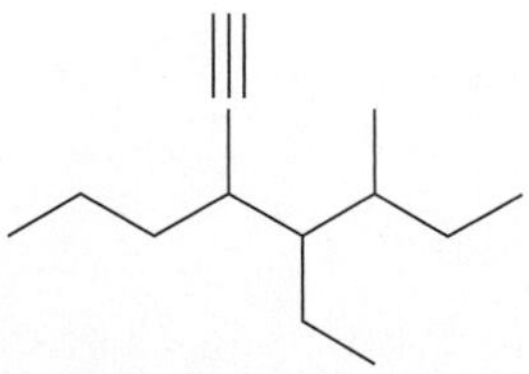

SOLUTION

Assembling a systematic name requires four discrete steps:

STEP 1
Identify the parent.

Begin by identifying the parent. Choose the longest chain that includes the triple bond:

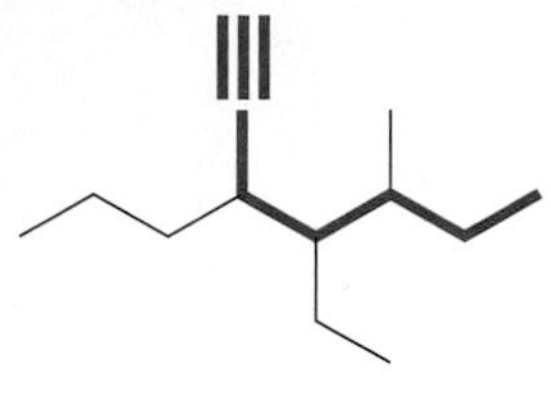

Heptyne

STEP 2
Identify and name the substituents.

Then, identify and name the substituents:

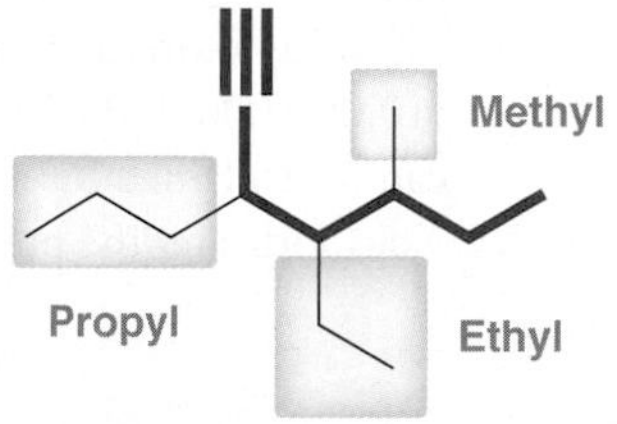

STEP 3
Number the parent chain and assign a locant to each substituent.

Next, number the parent chain and assign a locant to each substituent:

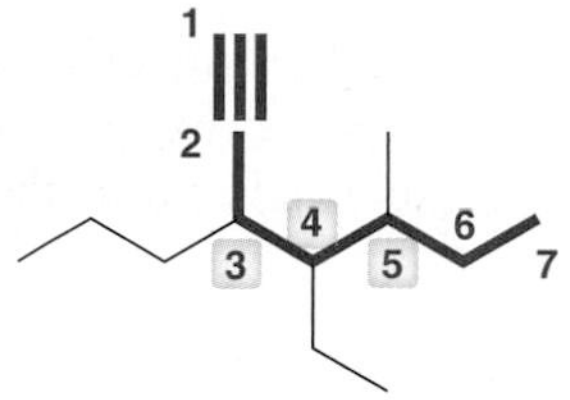

STEP 4
Assemble the substituents alphabetically.

Finally, assemble the substituents alphabetically, making sure to include a locant that identifies the position of the triple bond:

4-ethyl-5-methyl-3-propyl-1-heptyne

Make sure that hyphens separate letters from numbers, while commas separate numbers from each other.

PRACTICE the skill **10.1** Provide a systematic name for each of the following compounds:

(a)  (b) 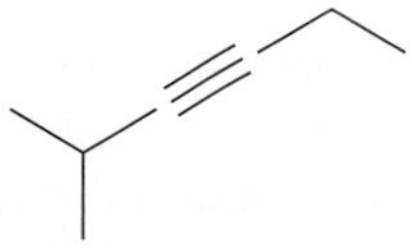 (c) 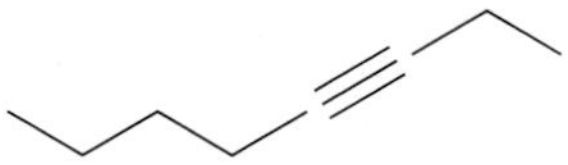 (d) 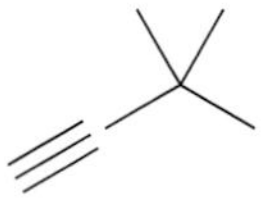

APPLY the skill

10.2 Draw a bond-line structure for each of the following compounds:

(a) 4,4-Dimethyl-2-pentyne (b) 5-Ethyl-2,5-dimethyl-3-heptyne

10.3 Although triple bonds have linear geometry, they do have some flexibility and can be incorporated in large rings. Cycloalkynes containing more than eight carbon atoms have been isolated and are stable at room temperature. When naming cycloalkynes that lack any other functional groups, the triple bond does not require a locant, because it is assumed to be between C1 and C2. Draw the structure of (*R*)-3-methylcyclononyne.

10.4 Compound A is a terminal alkyne with molecular formula C_6H_{10}. Draw all four possible constitutional isomers that are possible structures for compound A, and provide a systematic name for each.

need more **PRACTICE?** **Try Problems 10.35, 10.36**

10.3 Acidity of Acetylene and Terminal Alkynes

Compare the pK_a values for ethane, ethylene, and acetylene:

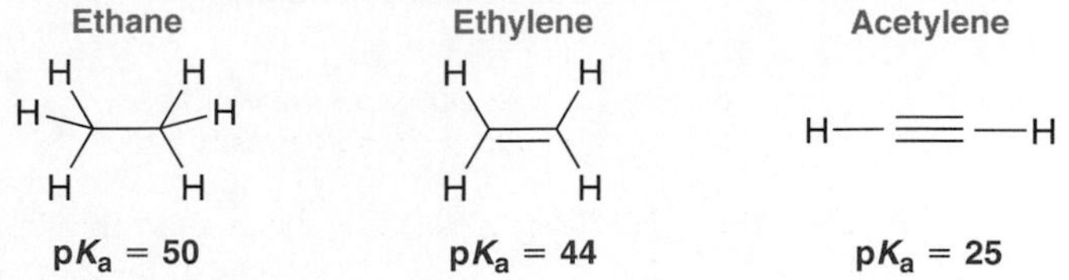

Recall that a lower pK_a corresponds to a greater acidity. Therefore, acetylene ($pK_a = 25$) is significantly more acidic than ethane or ethylene. To be precise, acetylene is 19 orders of magnitude (10,000,000,000,000,000,000 times) more acidic than ethylene. The relative acidity of acetylene can be explained by exploring the stability of its conjugate base, called an **acetylide ion** (the suffix "ide" indicates the presence of a negative charge):

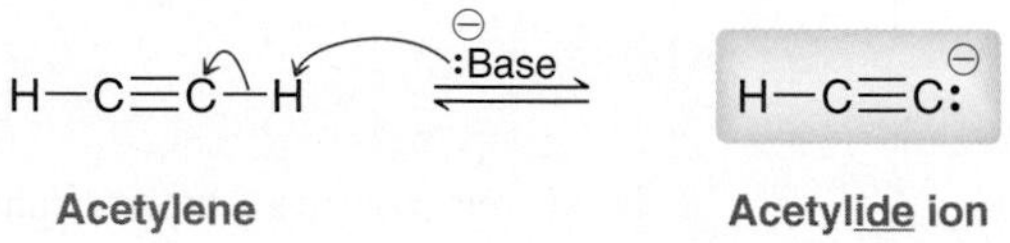

LOOKING BACK
This effect was first discussed in Section 3.4.

The stability of an acetylide ion can be rationalized using hybridization theory, in which the negative charge is considered to be associated with a lone pair that occupies an *sp*-hybridized orbital. Compare the conjugate bases for ethane, ethylene, and acetylene (Figure 10.3).

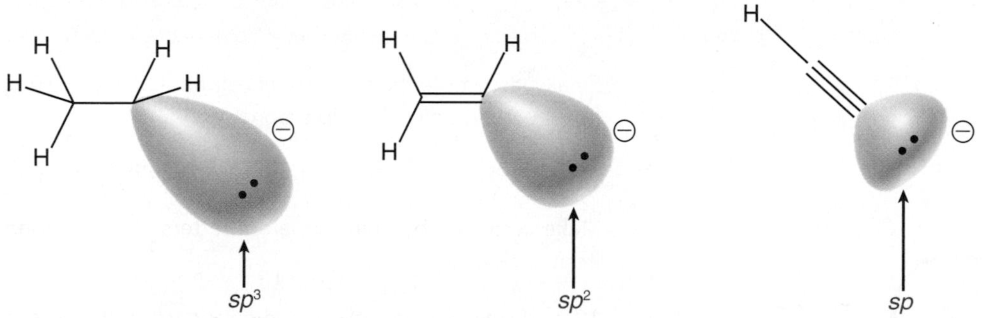

FIGURE 10.3 The conjugate base of ethane exhibits a lone pair in an sp^3-hybridized orbital. The conjugate base of ethylene has the lone pair in an sp^2-hybridized orbital, and the conjugate base of acetylene has the lone pair in an *sp*-hybridized orbital.

In an *sp*-hybridized orbital, the electron density is closer to the positively charged nucleus and is therefore more stable.

Now let's consider the equilibrium that is established when a strong base is used to deprotonate acetylene. Recall that the equilibrium of an acid-base reaction will always favor formation of the weaker acid and weaker base. For example, consider the equilibrium established when an amide ion (H_2N^-) is used as a base to deprotonate acetylene:

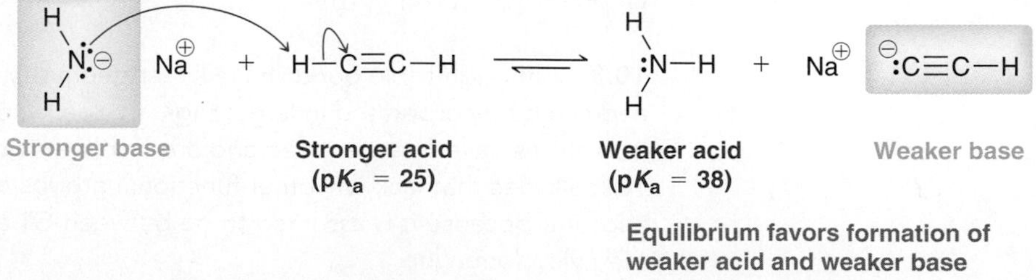

In this case, the equilibrium favors formation of the acetylide ion, because it is more stable (a weaker base) than the amide ion. In contrast, consider what happens when a hydroxide ion is used as the base:

$$\text{HÖ:}^{\ominus} \text{Na}^{\oplus} \;+\; \text{H}-\text{C}\equiv\text{C}-\text{H} \;\rightleftharpoons\; \text{H}-\overset{\cdot\cdot}{\text{O}}-\text{H} \;+\; \text{Na}^{\oplus}\,{}^{\ominus}\text{:C}\equiv\text{C}-\text{H}$$

Weaker base	Weaker acid	Stronger acid	Stronger base
	(pK_a = 25)	(pK_a = 15.7)	

**Equilibrium favors the
weaker acid and weaker base**

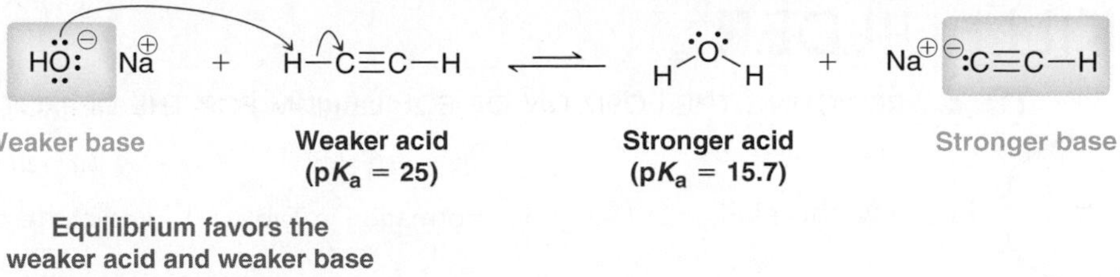

In this case, the equilibrium does not favor formation of the acetylide ion, because the acetylide ion is less stable (a stronger base) than the hydroxide ion. Therefore, hydroxide is not sufficiently basic to produce a significant amount of the acetylide ion. That is, hydroxide cannot be used to deprotonate acetylene.

Much like acetylene, terminal alkynes are also acidic and can be deprotonated with a suitable base:

$$\text{R}-\text{C}\equiv\text{C}-\text{H} \quad {}^{\ominus}\text{:Base} \;\rightleftharpoons\; \text{R}-\text{C}\equiv\text{C:}^{\ominus}$$

An alkyne	An alkynide ion

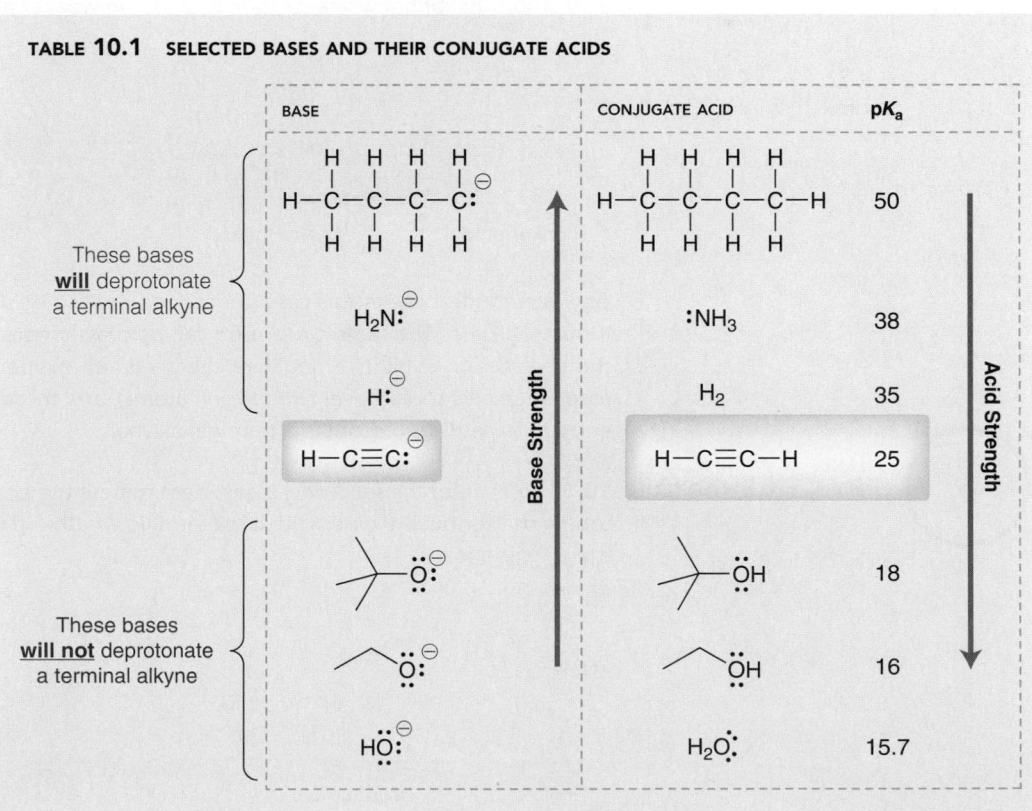

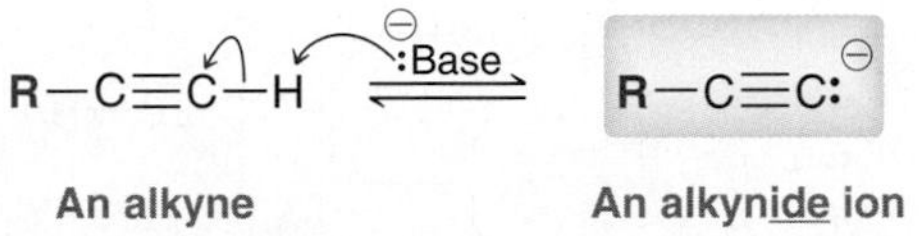

The conjugate base of a terminal alkyne, called an **alkynide ion**, can only be formed with a sufficiently strong base. Sodium hydroxide (NaOH) is not a suitable base for this purpose, but sodium amide (NaNH$_2$) can be used. There are several bases that can be used to deprotonate acetylene or terminal alkynes, as seen in Table 10.1. The three bases shown on the top left of the chart are all strong enough to deprotonate a terminal alkyne, and all three are commonly used to do so. Notice the position of the negative charge in each of these cases (N$^-$, H$^-$, or C$^-$). In contrast, the three bases on the bottom left of the chart all have the negative charge on an oxygen atom, which is not strong enough to deprotonate a terminal alkyne.

TABLE 10.1 SELECTED BASES AND THEIR CONJUGATE ACIDS

	BASE	CONJUGATE ACID	pK$_a$
These bases **will** deprotonate a terminal alkyne	H–C–C–C–C:$^{\ominus}$	H–C–C–C–C–H	50
	H$_2$N:$^{\ominus}$	:NH$_3$	38
	H:$^{\ominus}$	H$_2$	35
	H–C≡C:$^{\ominus}$	H–C≡C–H	25
These bases **will not** deprotonate a terminal alkyne	\searrowÖ:$^{\ominus}$	\searrowÖH	18
	\diagdownÖ:$^{\ominus}$	\diagdownÖH	16
	HÖ:$^{\ominus}$	H$_2$Ö:	15.7

SKILLBUILDER

10.2 PREDICTING THE POSITION OF EQUILIBRIUM FOR THE DEPROTONATION OF A TERMINAL ALKYNE

LEARN the skill

Using the information in Table 3.1, predict the position for the following equilibrium:

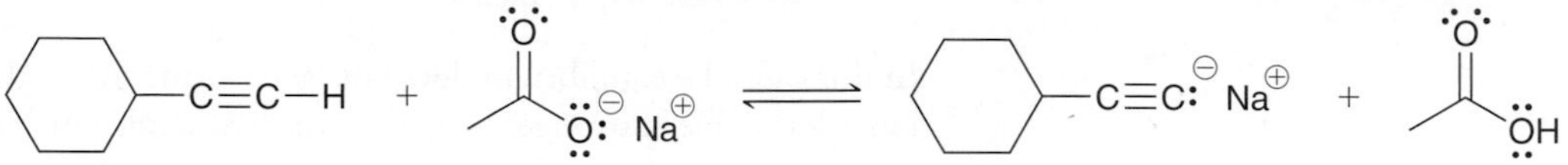

SOLUTION

Begin by identifying the acid and base on each side of the equilibrium:

STEP 1
Identify the acid and base on each side of the equilibrium.

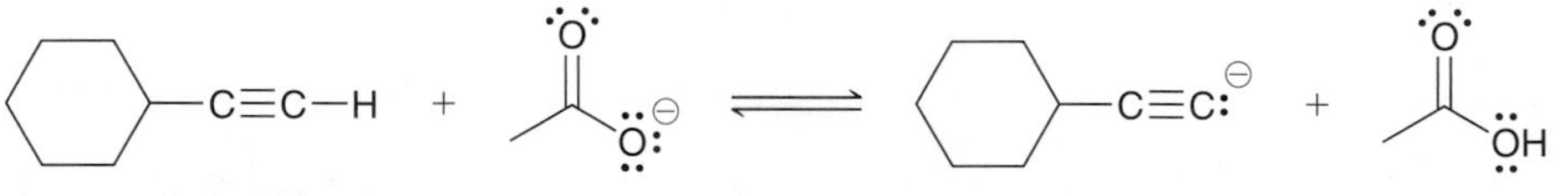

Acid | Base | Base | **Acid**

BY THE WAY
Na^+ is simply the counterion for each base and can be ignored in most cases.

Next, compare the pK_a values of the two acids in order to determine which one is the weaker acid. These values can be found in Table 3.1:

STEP 2
Determine which acid is weaker.

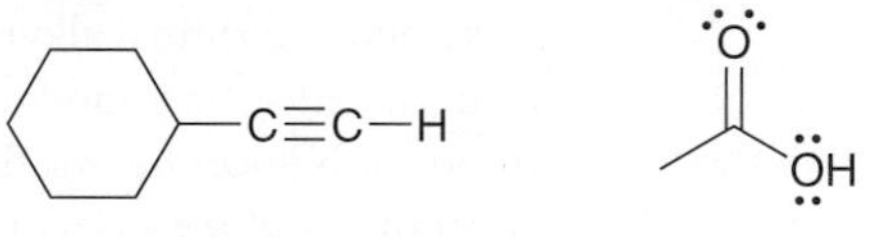

(pK_a ~ 25) **(pK_a = 4.75)**

STEP 3
Identify the position of equilibrium.

Recall from Section 3.3 that a higher pK_a indicates a weaker acid, so the alkyne is the weaker acid. The equilibrium will favor formation of the weaker acid and weaker base:

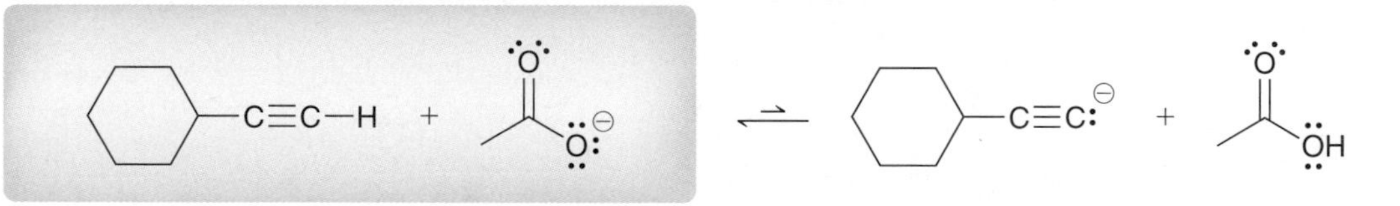

Weaker acid | Weaker base | Stronger base | **Stronger acid**

As a result, the base in this case (an acetate ion) is not sufficiently strong to deprotonate a terminal alkyne. The same conclusion can be drawn more quickly by simply recognizing that the acetate ion exhibits a negative charge on an oxygen atom (actually, it is stabilized by resonance and spread over two oxygen atoms), and therefore, it cannot possibly be a strong enough base to deprotonate a terminal alkyne.

PRACTICE the skill

10.5 In each of the following cases, determine if the base is sufficiently strong to deprotonate the terminal alkyne. That is, determine whether the equilibrium favors formation of the alkynide ion.

(a) 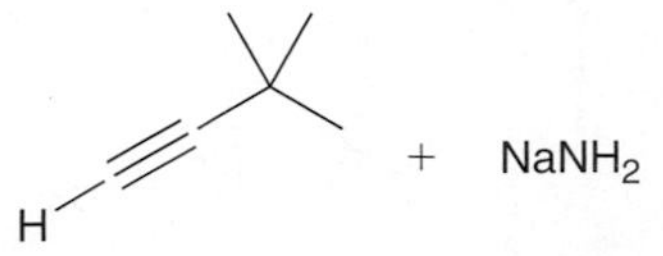

(b) 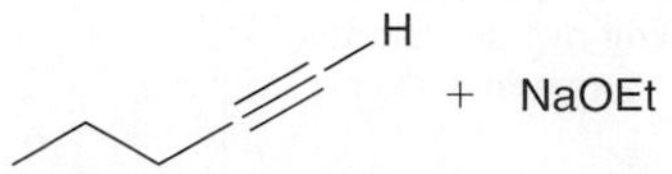

(c) 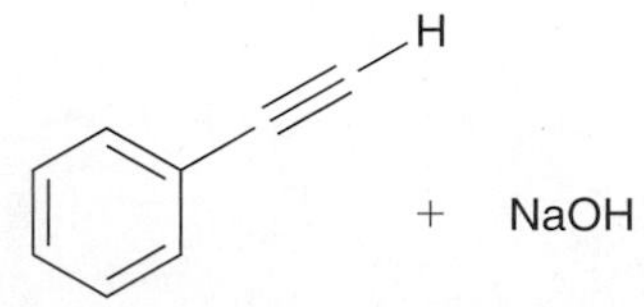

(d) 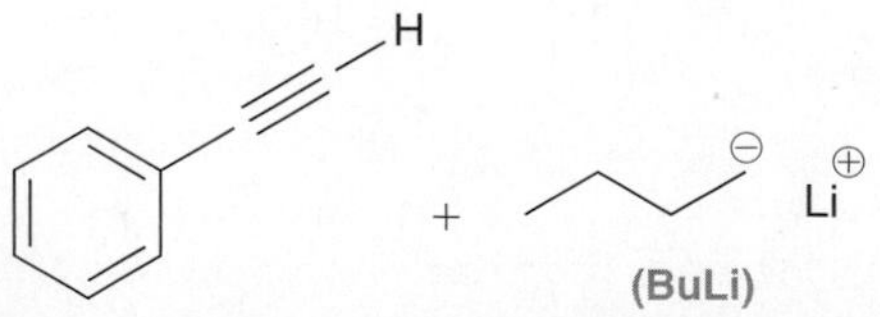

(e) 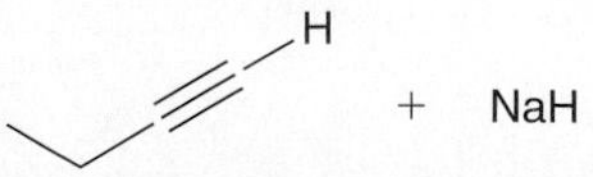

(f) 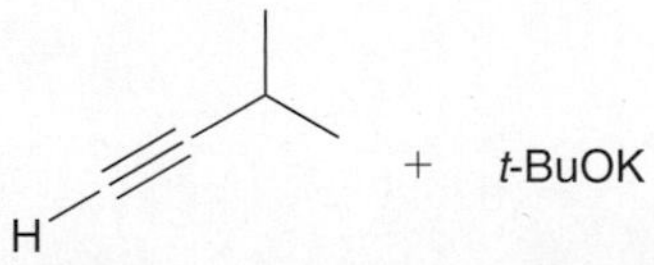

APPLY the skill

10.6 The pK_a of CH_3NH_2 is 40, while the pK_a of HCN is 9.

(a) Explain this difference in acidity.

(b) Can the cyanide anion (the conjugate base of HCN) be used as a base to deprotonate a terminal alkyne? Explain.

need more **PRACTICE?** **Try Problems 10.39, 10.42**

10.4 Preparation of Alkynes

Just as alkenes can be prepared from alkyl halides, alkynes can be prepared from alkyl dihalides:

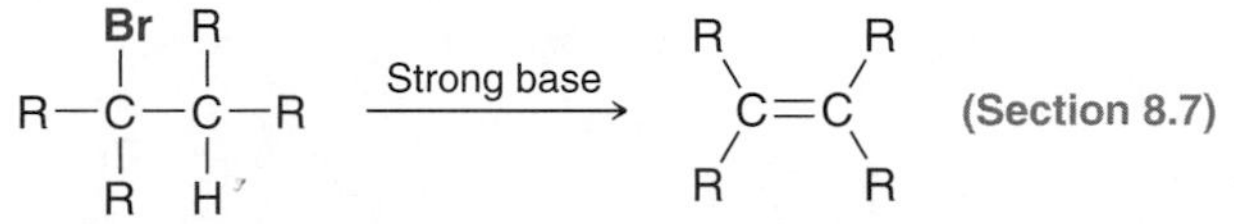

An alkyl halide

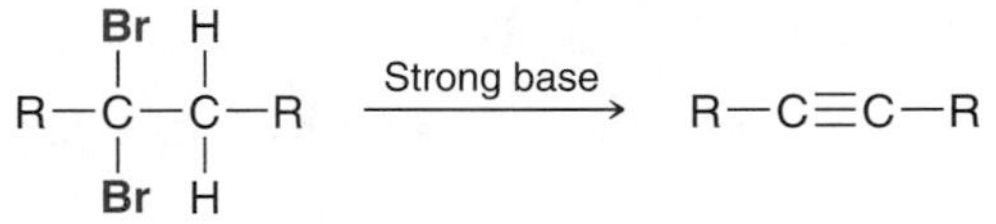

An alkyl dihalide

An alkyl dihalide has two leaving groups, and the transformation is accomplished via two successive elimination (E2) reactions:

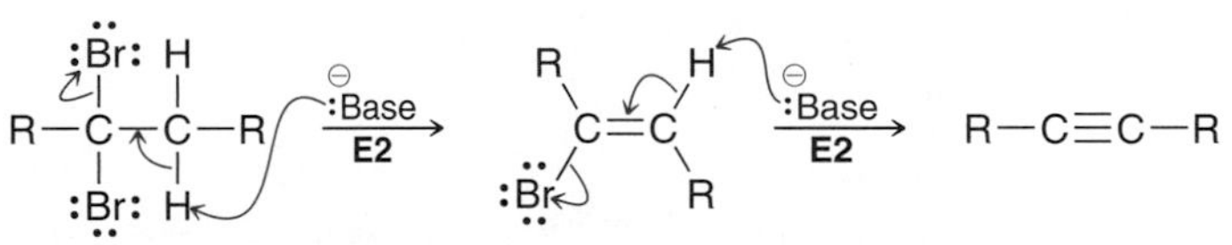

In this example, the dihalide used is a **geminal** dihalide, which means that both halogens are connected to the same carbon atom. Alternatively, alkynes can also be prepared from **vicinal** dihalides, in which the two halogens are connected to adjacent carbon atoms:

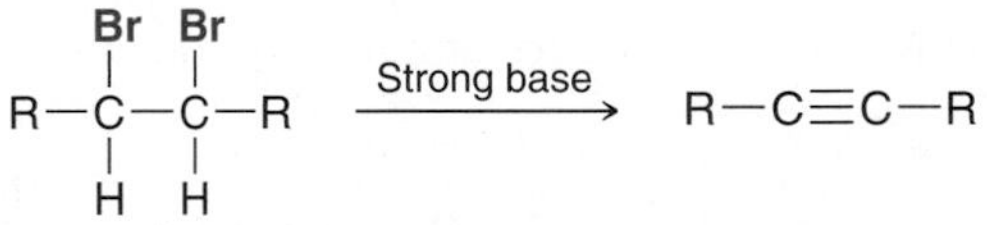

Whether the starting dihalide is geminal or vicinal, the alkyne is obtained as the result of two successive elimination reactions. The first elimination can be readily accomplished using many different bases, but the second elimination requires a very strong base. Sodium amide ($NaNH_2$), dissolved in liquid ammonia (NH_3), is a suitable base for achieving two successive elimination reactions in a single reaction vessel. This method is used most frequently for the preparation of terminal alkynes, because the strongly basic conditions favor production of an alkynide ion, which serves as a driving force for the overall process:

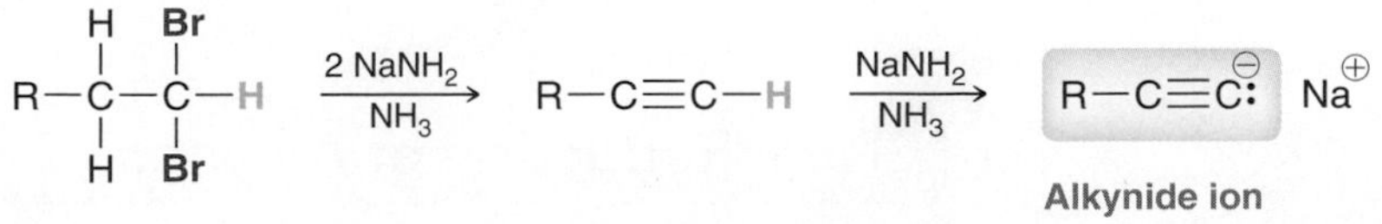

Alkynide ion

In total, three equivalents of the amide ion are required: two equivalents for the two E2 reactions, and one equivalent to deprotonate the terminal alkyne and form the alkynide ion. After the alkynide ion has formed and the reaction is complete, a proton source can be introduced into the reaction vessel, thereby protonating the alkynide ion to regenerate the terminal alkyne:

$$R-C{\equiv}C{:}^{\ominus} \ Na^{\oplus} \ + \ H{-}\overset{..}{\underset{..}{O}}{-}H \ \rightleftharpoons \ R-C{\equiv}C-H \ + \ Na^{\oplus} \ {:}\overset{..}{\underset{..}{O}}H^{\ominus}$$

| Stronger base | Stronger acid (pK_a = 15.7) | Weaker acid (pK_a = 25) | Weaker base |

Equilibrium favors formation of weaker acid and weaker base

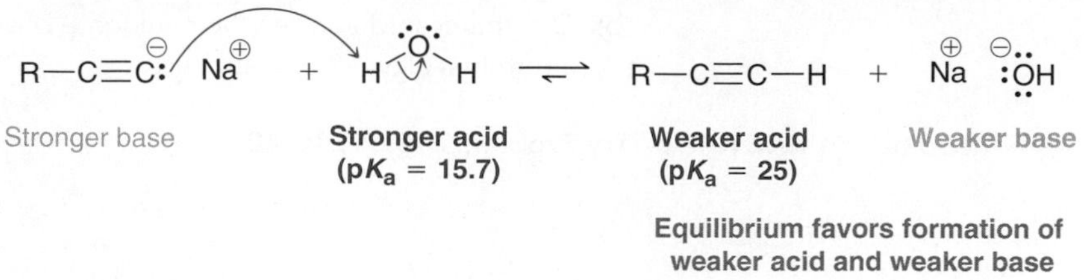

Comparison of the pK_a values indicates that water is a sufficient proton source.

In summary, a terminal alkyne can be prepared by treating a dihalide with excess (xs) sodium amide followed by water:

1) xs NaNH$_2$/NH$_3$
2) H$_2$O

(60%)

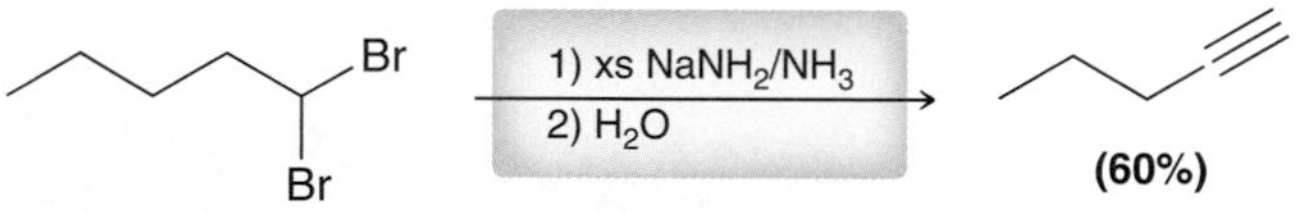

CONCEPTUAL CHECKPOINT

10.7 For each of the following transformations, predict the major product, and draw a mechanism for its formation:

(a)
1) xs NaNH$_2$/NH$_3$
2) H$_2$O
?

(b)
1) xs NaNH$_2$/NH$_3$
2) H$_2$O
?

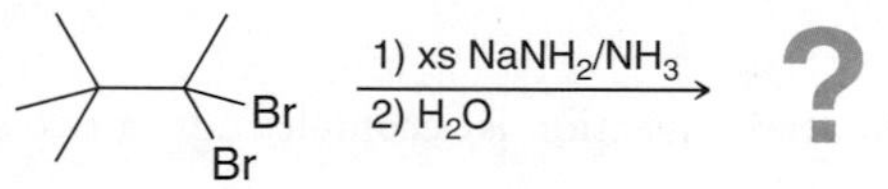

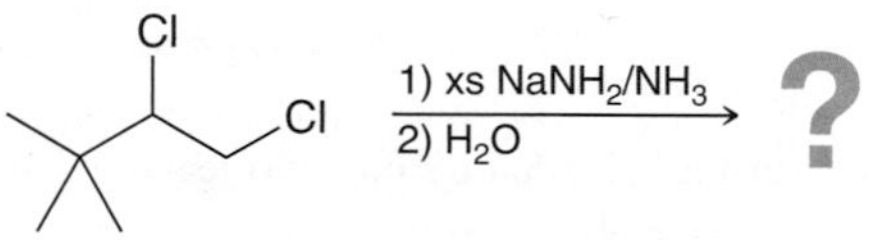

10.8 When 3,3-dichloropentane is treated with excess sodium amide in liquid ammonia, the initial product is 2-pentyne:

xs NaNH$_2$ →

2-Pentyne

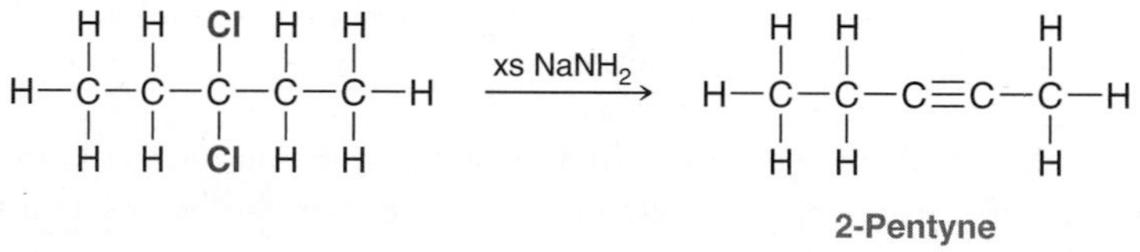

However, under these conditions, this internal alkyne quickly isomerizes to form a terminal alkyne that is subsequently deprotonated to form an alkynide ion:

2-Pentyne ⇌ (xs NaNH$_2$) **1-Pentyne** ⇌ (xs NaNH$_2$) **Alkynide ion**

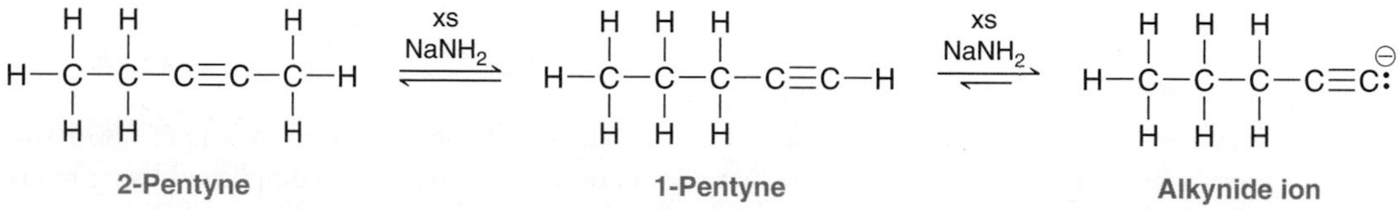

The isomerization process is believed to occur via a mechanism with the following four steps: (1) deprotonate, (2) protonate, (3) deprotonate, and (4) protonate. Using these four steps as a guide, try to draw the mechanism for isomerization using resonance structures whenever possible. Explain why the equilibrium favors formation of the terminal alkyne.

10.5 Reduction of Alkynes

Catalytic Hydrogenation

Alkynes undergo many of the same addition reactions as alkenes. For example, alkynes will undergo catalytic hydrogenation just as alkenes do:

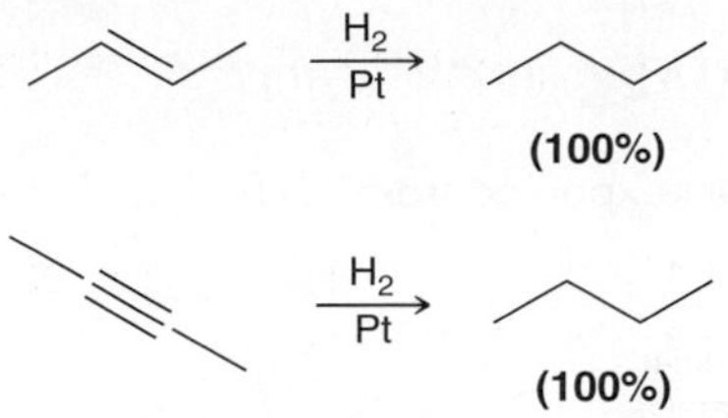

In the process, the alkyne consumes two equivalents of molecular hydrogen:

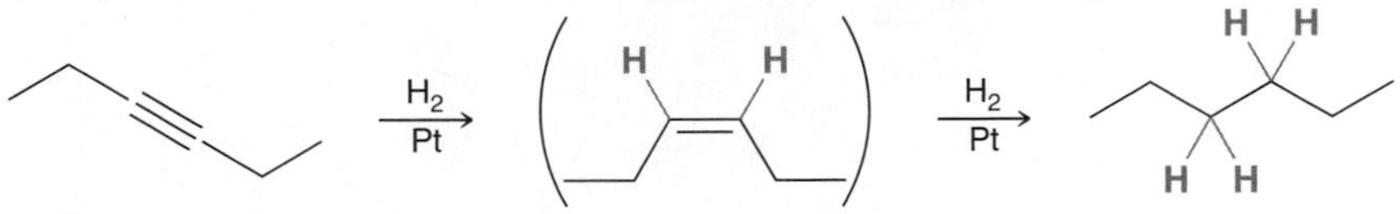

Under these conditions, the *cis* alkene is difficult to isolate because it is even more reactive toward further hydrogenation than the starting alkyne. The obvious question then is whether it is possible to add just one equivalent of hydrogen to form the alkene. With the catalysts we have seen thus far (Pt, Pd, or Ni), this is difficult to achieve. However, with a partially deactivated catalyst, called a **poisoned catalyst**, it is possible to convert an alkyne into a *cis* alkene (without further reduction):

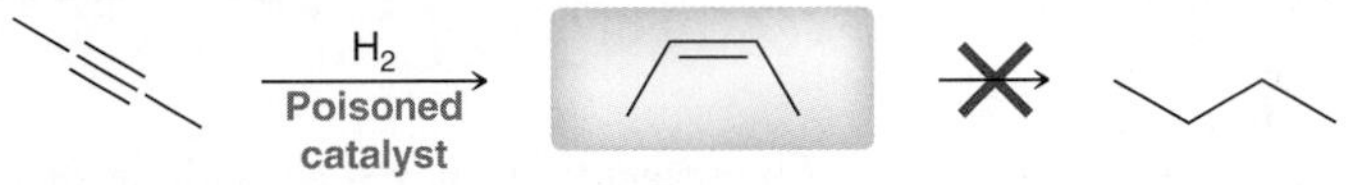

There are many poisoned catalysts. One common example is called *Lindlar's catalyst*:

Lindlar's catalyst = , Pd / BaSO$_4$, CH$_3$OH

Quinoline

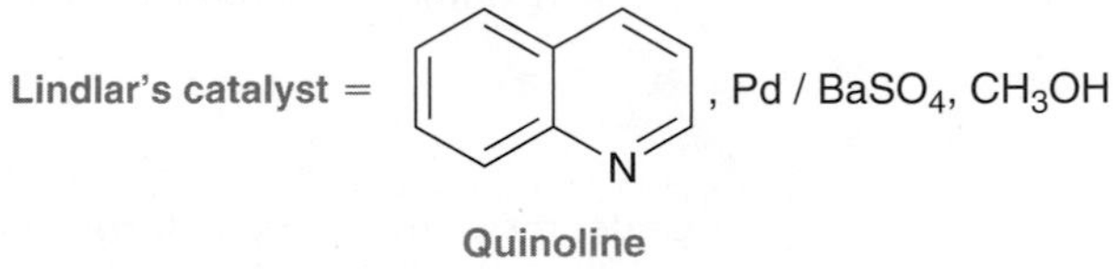

Another common example is a nickel-boron complex (Ni$_2$B), which is often called the P-2 catalyst. A poisoned catalyst will catalyze the conversion of an alkyne into a *cis* alkene, but it will not catalyze the subsequent reduction to form the alkane (Figure 10.4). Therefore, a poisoned catalyst can be used to convert an alkyne into a *cis* alkene.

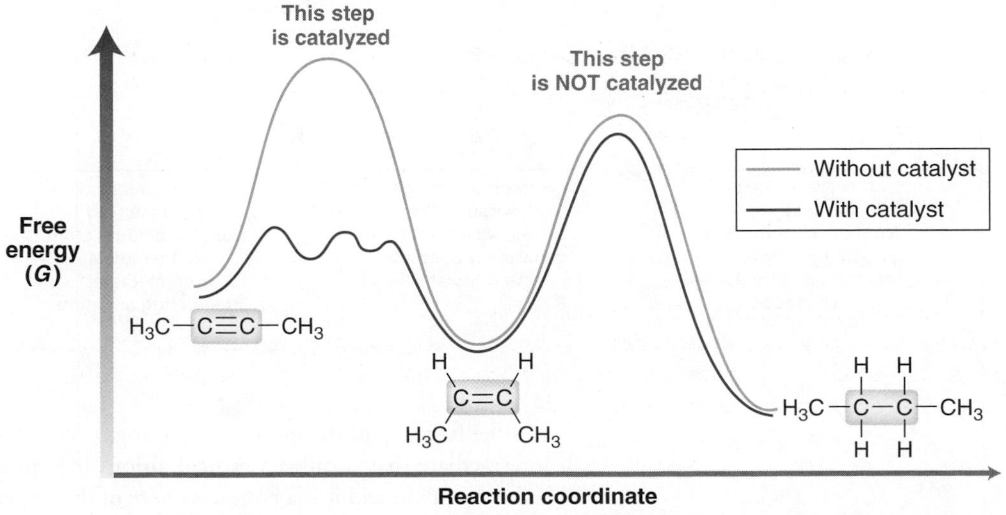

FIGURE 10.4
An energy diagram showing the effect of a poisoned catalyst. Hydrogenation of the alkyne is catalyzed, but subsequent hydrogenation of the alkene is not catalyzed.

This process does not produce any *trans* alkene. The stereochemical outcome of alkyne hydrogenation can be rationalized in the same way that we rationalized the outcome of alkene hydrogenation (Section 9.7). Both hydrogen atoms are added to the same face of the alkene (*syn* addition) to give the *cis* alkene as the major product.

CONCEPTUAL CHECKPOINT

10.9 Draw the major product expected from each of the following reactions:

(a)

$$\xrightarrow[\text{catalyst}]{\text{H}_2 \quad \text{Lindlar's}} \quad \textbf{?}$$

$$\xrightarrow[\text{Pt}]{\text{H}_2} \quad \textbf{?}$$

(b)

$$\xrightarrow[\text{Ni}_2\text{B}]{\text{H}_2} \quad \textbf{?}$$

$$\xrightarrow[\text{Ni}]{\text{H}_2} \quad \textbf{?}$$

WATCH OUT

These reagents should not be confused with sodium amide (NaNH$_2$), which we saw earlier in this chapter. NaNH$_2$ is a source of NH$_2^-$, a very strong base. In contrast, the reagents employed here (Na, NH$_3$) represent a source of electrons.

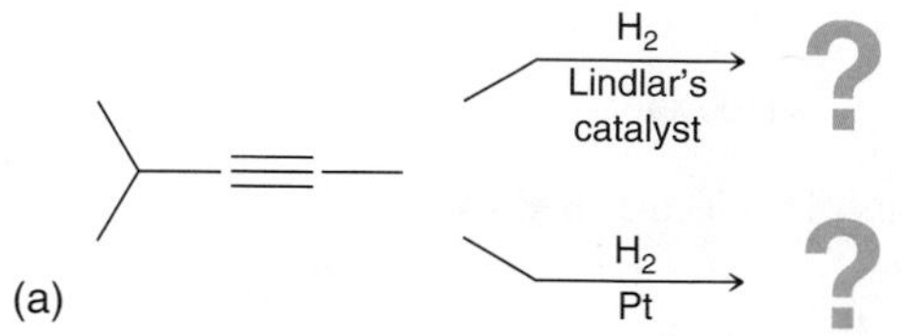

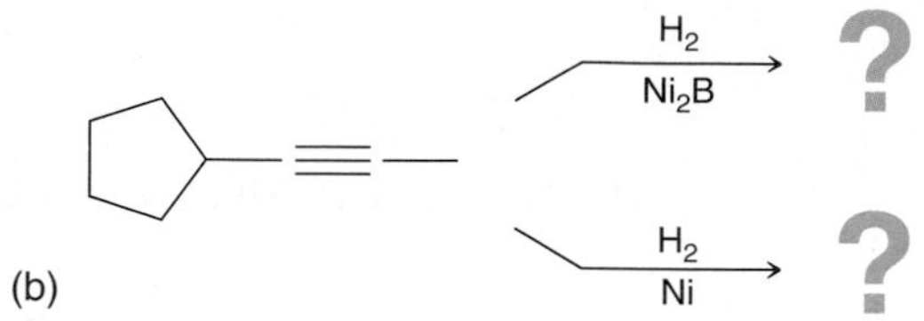

Dissolving Metal Reduction

In the previous section, we explored the conditions that enable the reduction of an alkyne to a *cis* alkene. Alkynes can also be reduced to *trans* alkenes via an entirely different reaction called **dissolving metal reduction**:

$$-\!\!\equiv\!\!- \quad \xrightarrow[\text{NH}_3\,(l)]{\text{Na}} \quad \diagdown\!\!=\!\!\diagup \quad \textbf{(80\%)}$$

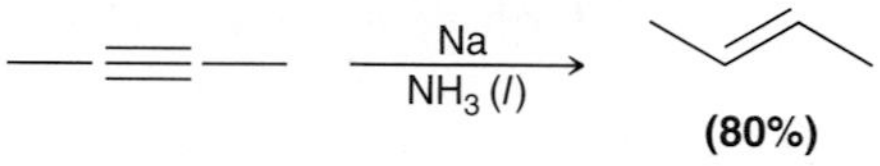

The reagents employed are sodium metal (Na) in liquid ammonia (NH$_3$). Ammonia has a very low boiling point (-33 °C), so use of these reagents requires low temperature. When dissolved in liquid ammonia, sodium atoms serve as a source of electrons:

$$\text{Na}^{\bullet} \longrightarrow \text{Na}^{+} + e^{\ominus}$$

Under these conditions, reduction of alkynes is believed to proceed via Mechanism 10.1.

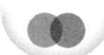

MECHANISM 10.1 DISSOLVING METAL REDUCTIONS

Nucleophilic attack	Proton transfer	Nucleophilic attack	Proton transfer
A single electron is transferred from the sodium atom to the alkyne, generating a radical anion intermediate	Ammonia donates a proton to the radical anion, generating a radical intermediate	A single electron is transferred from the sodium atom to the radical intermediate, generating an anion	Ammonia donates a proton to the anion, generating a *trans* alkene

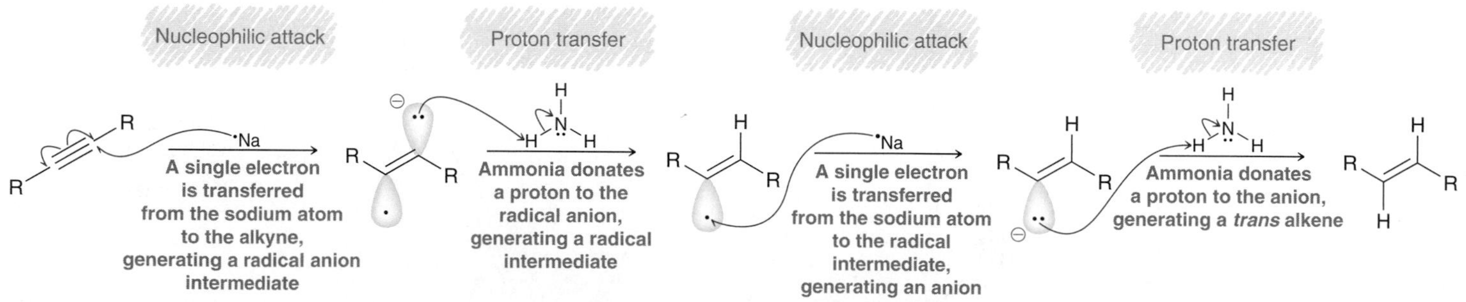

In the first step of the mechanism, a single electron is transferred to the alkyne, generating an intermediate that is called a **radical anion**. It is an anion because of the charge associated with the lone pair, and it is a radical because of the unpaired electron:

Anion

Radical

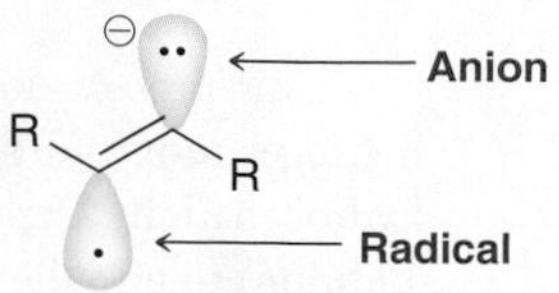

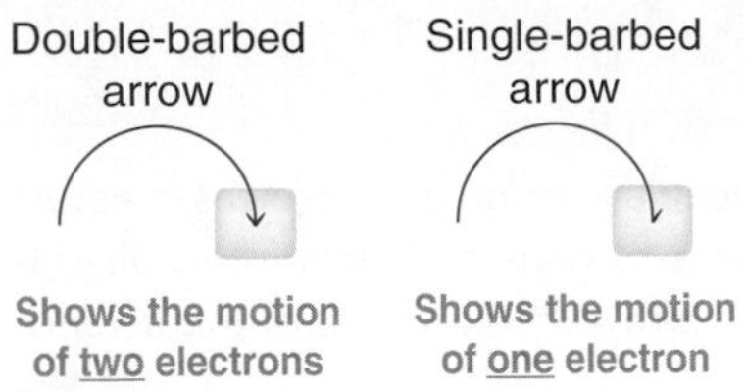

FIGURE 10.5 Arrows used in ionic mechanisms (double barbed) and radical mechanisms (single barbed).

Radicals and their chemistry are discussed in more detail in Chapter 11. For now, we will just point out that mechanistic steps involving radicals utilize single-barbed curved arrows, often called *fishhook arrows*, rather than double-barbed curved arrows (Figure 10.5). Single-barbed curved arrows indicate the movement of one electron, while double-barbed arrows indicate the movement of two electrons. Notice the use of single-barbed curved arrows in the first step of the mechanism to form the intermediate radical anion. The nature of this intermediate explains the stereochemical preference for formation of a *trans* alkene. Specifically, the intermediate achieves a lower energy state when the paired and unpaired electrons are positioned as far apart as possible, minimizing their repulsion.

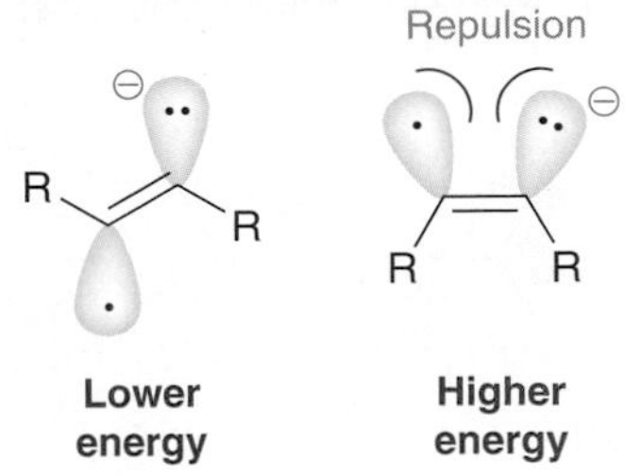

The reaction proceeds more rapidly through the lower energy intermediate, which is then protonated by ammonia under these conditions. In the remaining two steps of the mechanism, another electron is transferred, followed by one more proton transfer. The mechanism therefore is comprised of the following four steps: (1) electron transfer, (2) proton transfer, (3) electron transfer, and (4) proton transfer. That is, the net addition of molecular hydrogen (H_2) is achieved via the installation of two electrons and two protons in the following order: e^-, H^+, e^-, H^+. The net result is an *anti* addition of two hydrogen atoms across the alkyne:

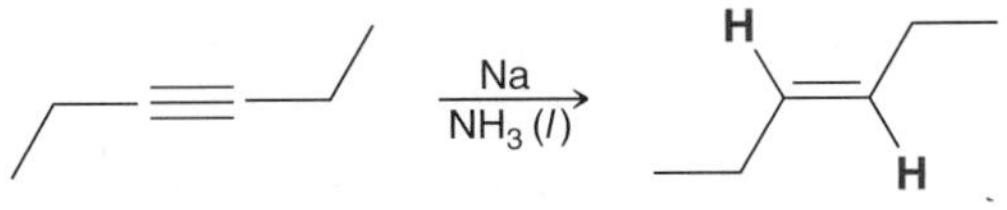

CONCEPTUAL CHECKPOINT

10.10 Draw the major product expected when each of the following alkynes is treated with sodium in liquid ammonia:

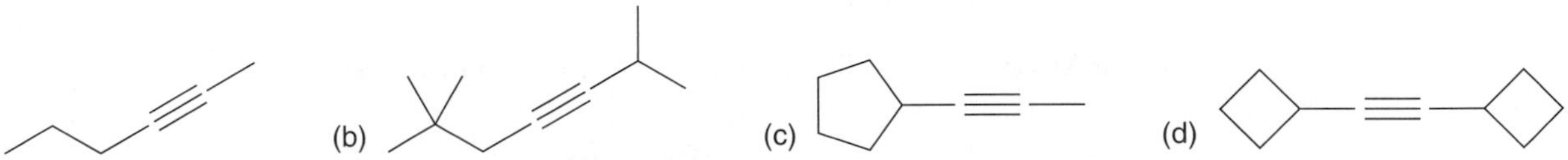

Catalytic Hydrogenation vs. Dissolving Metal Reduction

Figure 10.6 summarizes the various methods we have seen for reducing an alkyne. This diagram illustrates how the outcome of alkyne reduction can be controlled by a careful choice of reagents:

- To produce an *alkane*, an alkyne can be treated with H_2 in the presence of a metal catalyst, such as Pt, Pd, or Ni.
- To produce a *cis alkene*, an alkyne can be treated with H_2 in the presence of a poisoned catalyst, such as Lindlar's catalyst or Ni_2B.
- To produce a *trans alkene*, an alkyne can be treated with sodium in liquid ammonia.

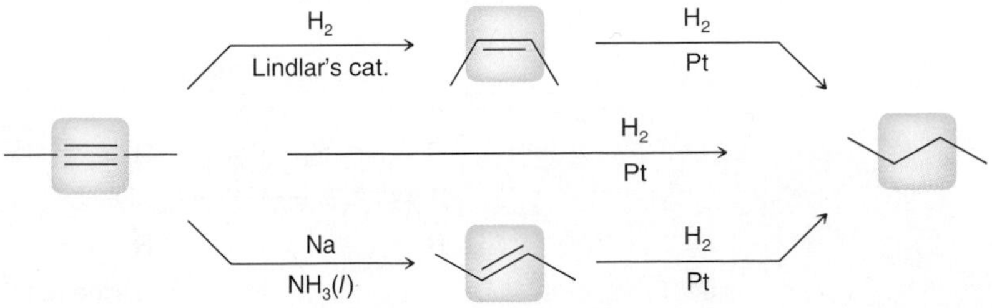

FIGURE 10.6 A summary of the various reagents that can be used to reduce an alkyne.

CONCEPTUAL CHECKPOINT

10.11 Identify the reagents you would use to achieve each of the following transformations:

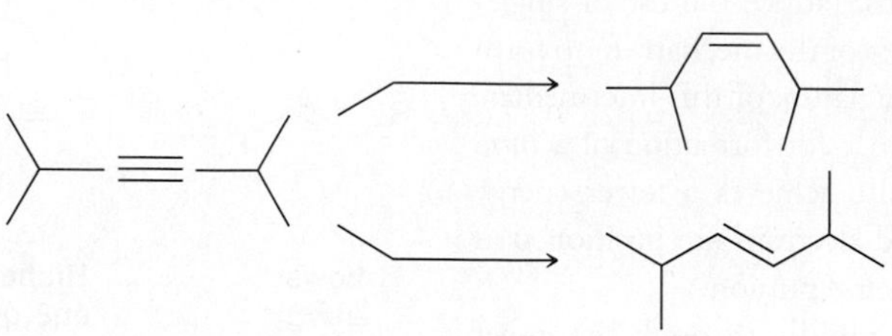

(a)

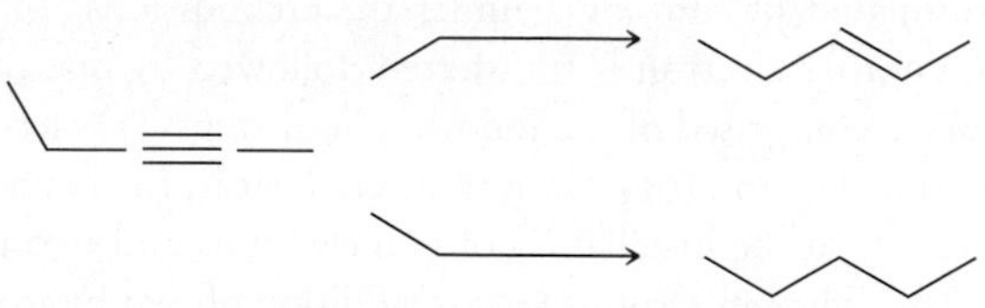

(b)

10.12 An alkyne with molecular formula C_5H_8 was treated with sodium in liquid ammonia to give a disubstituted alkene with molecular formula C_5H_{10}. Draw the structure of the alkene.

10.6 Hydrohalogenation of Alkynes

Experimental Observations

In the previous chapter, we saw that alkenes will react with HX via a Markovnikov addition, thereby installing a halogen at the more substituted position:

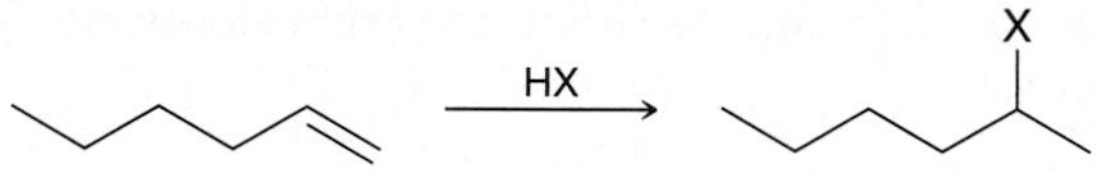

A similar Markovnikov addition is observed when alkynes are treated with HX:

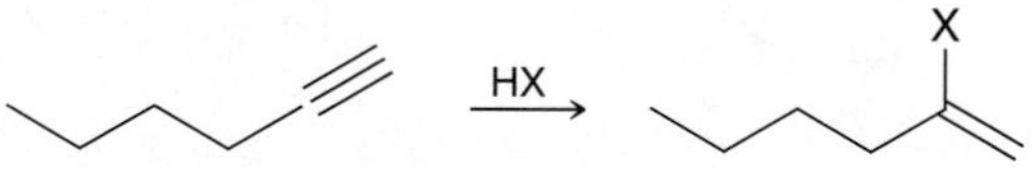

(60–80%)

Once again, the halogen is installed at the more substituted position.

When the starting alkyne is treated with excess HX, two successive addition reactions occur, producing a geminal dihalide:

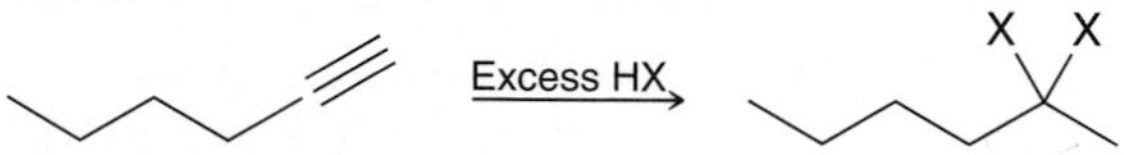

A Mechanism for Hydrohalogenation

In Section 9.3, we proposed the following two-step mechanism for HX addition to alkenes: (1) protonation of the alkene to form the more stable carbocation intermediate followed by (2) nucleophilic attack:

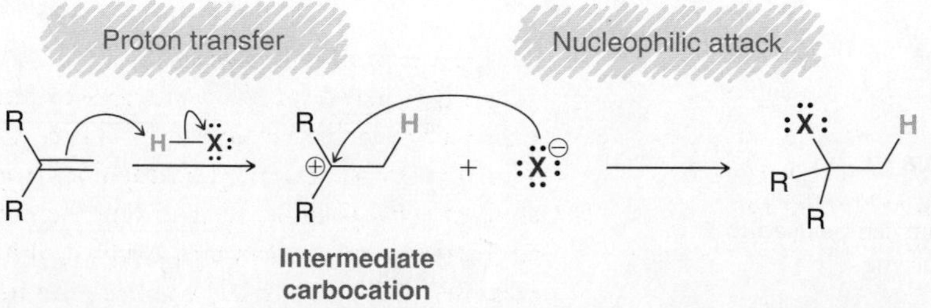

A similar mechanism can be proposed for addition of HX to a triple bond: (1) protonation to form a carbocation followed by (2) nucleophilic attack:

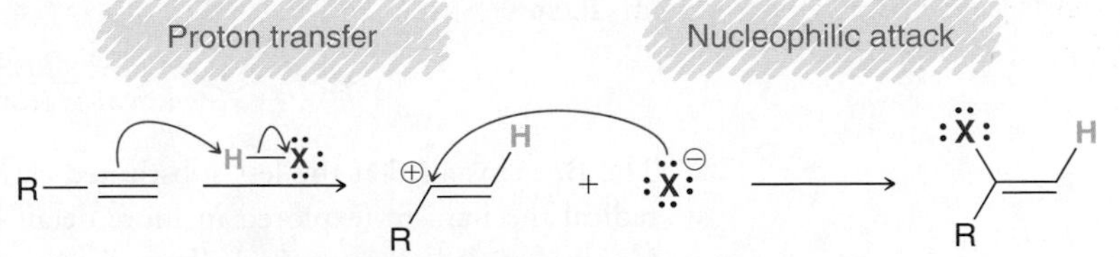

Vinylic carbocation

This proposed mechanism invokes an intermediate **vinylic carbocation** (*vinyl* = a carbon atom bearing a double bond) and can successfully explain the observed regioselectivity. Specifically, the reaction is expected to proceed via the more stable, secondary vinylic carbocation, rather than via the less stable, primary vinylic carbocation:

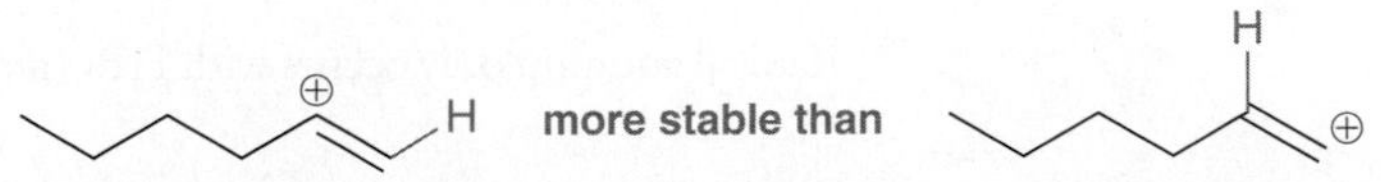

Secondary vinylic **Primary vinylic**

Unfortunately, this proposed mechanism is not consistent with all of the experimental observations. Most notably, studies in the gas phase indicate that vinylic carbocations are not particularly stable. Secondary vinylic carbocations are believed to be similar in energy to regular primary carbocations. Accordingly, we would expect HX addition of alkynes to be significantly slower than HX addition of alkenes. A difference in rate is, in fact, observed, but this difference is not as large as expected—HX addition to alkynes is only slightly slower than HX addition to alkenes. Accordingly, other mechanisms have been proposed that avoid formation of a vinylic carbocation. For example, it is possible that the alkyne interacts with two molecules of HX simultaneously. This process is said to be **termolecular** (involving three molecules) and proceeds through the following transition state:

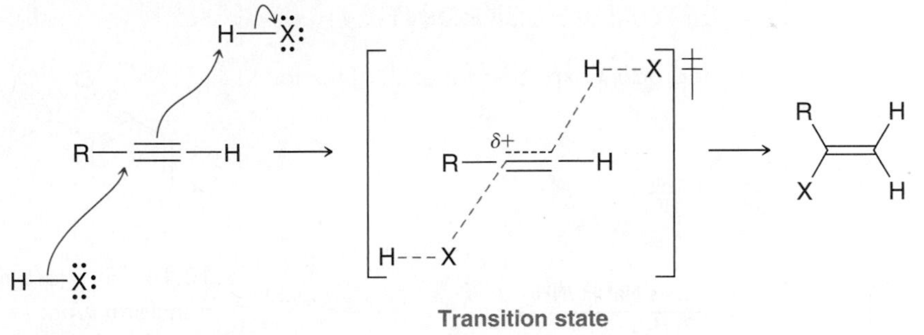

Transition state

This one-step mechanism avoids the formation of a vinylic carbocation but still invokes a transition that exhibits some partial carbocationic character (notice the δ+ shown in the transition state). The development of a partial positive charge can effectively explain the observed regioselectivity, because the transition state will be lower in energy when this partial positive charge forms at the more substituted position. This more complex mechanism is supported in many cases by kinetic studies in which the rate expression is found to be overall third order:

$$\text{Rate} = k\,[\text{alkyne}]\,[\text{HX}]^2$$

LOOKING BACK
Recall from Section 6.5 that "third order" means that the sum of the exponents is 3 for the rate expression of the rate-determining step.

This rate expression is consistent with the termolecular process proposed above.

It is believed that in most cases the addition of HX to alkynes probably occurs through a variety of mechanistic pathways, all occurring at the same time and competing with each other. The vinylic carbocation probably does play some limited role, but it cannot, by itself, explain all of the observations. In several of the mechanisms throughout this chapter, we may invoke a vinylic carbocation as an intermediate, although it should be understood that other, more complex mechanisms are likely operating simultaneously.

Radical Addition of HBr

Recall that in the presence of peroxides, HBr undergoes an *anti*-Markovnikov addition across an alkene:

The Br is installed at the less substituted carbon, and the reaction is believed to proceed via a radical mechanism (explored in more detail in Section 11.10). A similar reaction is observed for alkynes. When a terminal alkyne is treated with HBr in the presence of peroxides, an *anti*-Markovnikov addition is observed. The Br is installed at the terminal position, producing a mixture of *E* and *Z* isomers:

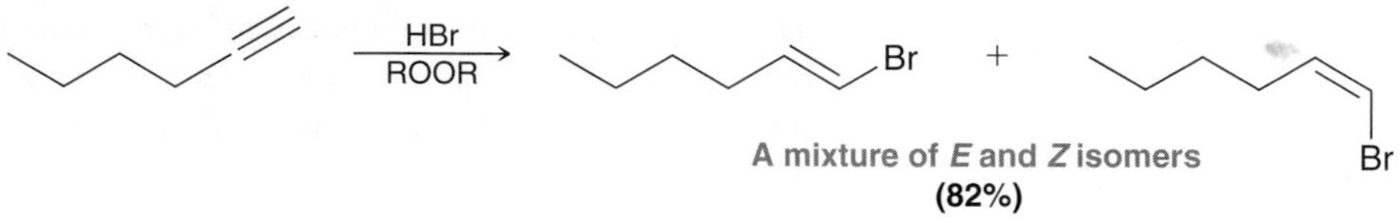

A mixture of *E* and *Z* isomers
(82%)

Radical addition only occurs with HBr (not with HCl or HI), as will be explained in Section 11.10.

Interconversion of Dihalides and Alkynes

The reactions discussed thus far enable the interconversion between dihalides and terminal alkynes:

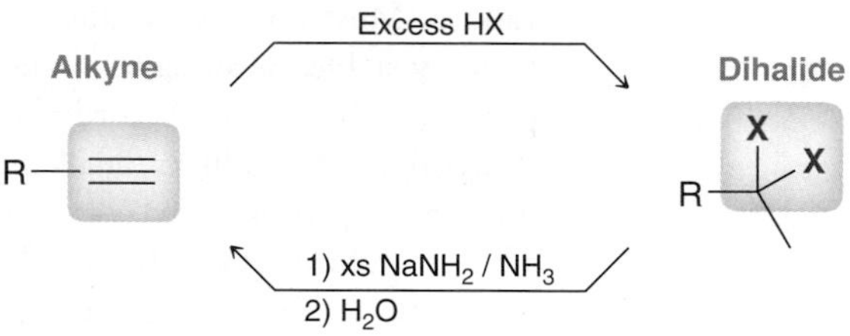

CONCEPTUAL CHECKPOINT

10.13 Predict the major product expected for each of the following reactions:

(a)
$\xrightarrow[\text{HCl}]{\text{xs}}$?

(b)
$\xrightarrow[\text{2) H}_2\text{O}]{\text{1) xs NaNH}_2/\text{NH}_3}$?

(c)
$\xrightarrow[\text{HBr}]{\text{xs}}$?

(d)
$\xrightarrow[\text{2) H}_2\text{O}]{\text{1) xs NaNH}_2/\text{NH}_3}$?

(e)
$\xrightarrow[\substack{\text{2) H}_2\text{O} \\ \text{3) HBr, ROOR}}]{\text{1) xs NaNH}_2/\text{NH}_3}$?

(f)
$\xrightarrow[\substack{\text{2) H}_2\text{O} \\ \text{3) xs HBr}}]{\text{1) xs NaNH}_2/\text{NH}_3}$?

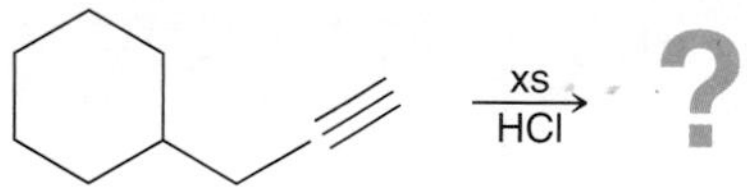

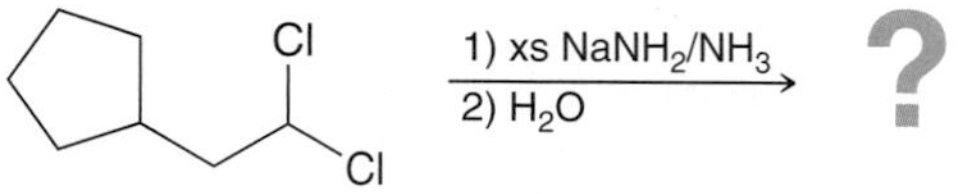

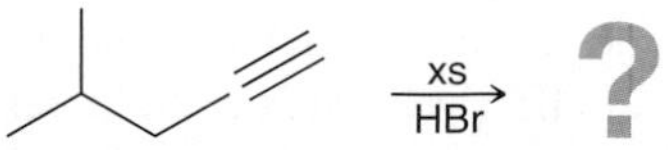

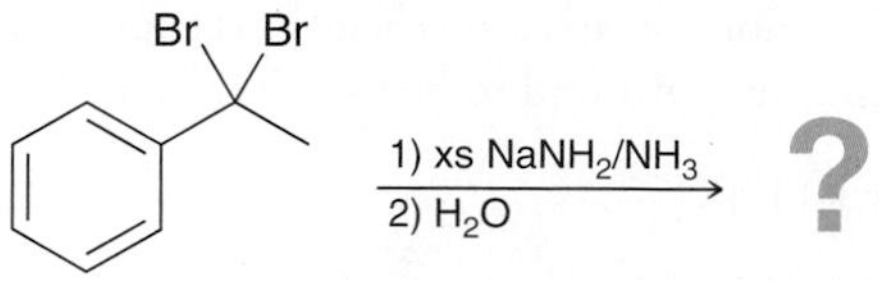

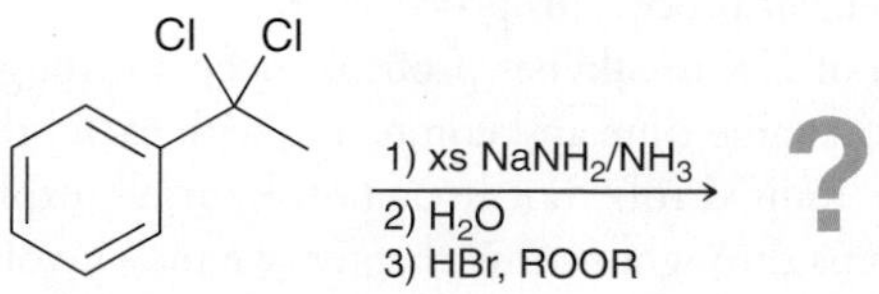

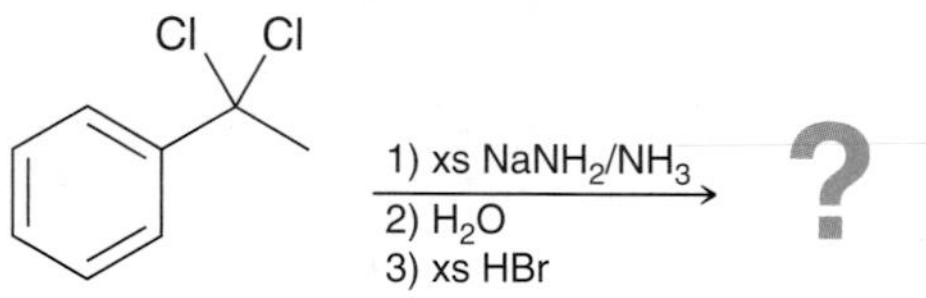

10.14 Suggest reagents that would achieve the following transformation:

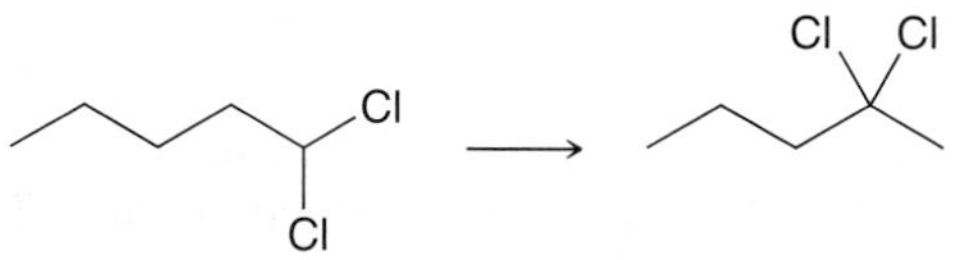

10.15 An alkyne with molecular formula C_5H_8 is treated with excess HBr, and two different products are obtained, each of which has molecular formula $C_5H_{10}Br_2$.

(a) Identify the starting alkyne.

(b) Identify the two products.

10.7 Hydration of Alkynes

Acid-Catalyzed Hydration of Alkynes

In the previous chapter, we saw that alkenes will undergo acid-catalyzed hydration when treated with aqueous acid (H_3O^+). The reaction proceeds via a Markovnikov addition, thereby installing a hydroxyl group at the more substituted position:

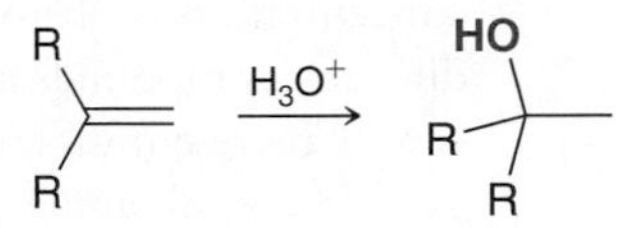

Alkynes are also observed to undergo acid-catalyzed hydration, but the reaction is slower than the corresponding reaction with alkenes. As noted earlier in this chapter, the difference in rate is attributed to the high-energy, vinylic carbocation intermediate that is formed when an alkyne is protonated. The rate of alkyne hydration is markedly enhanced in the presence of mercuric sulfate ($HgSO_4$), which catalyzes the reaction:

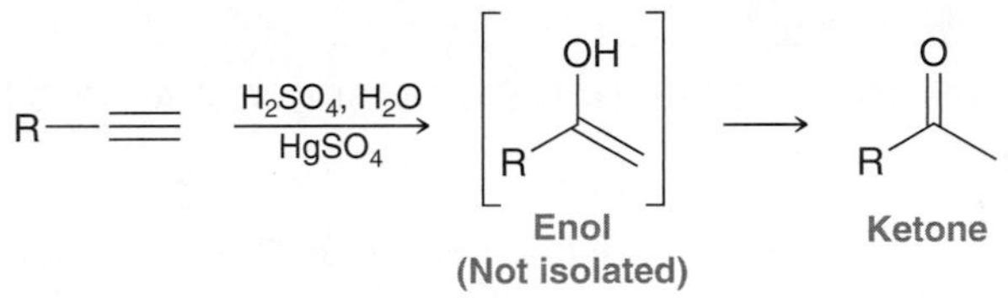

The initial product of this reaction has a double bond (***en***) and an OH group (***ol***) and is therefore called an **enol**. But the enol cannot be isolated because it is rapidly converted into a ketone. The conversion of an enol into a ketone will appear again in many subsequent chapters and therefore warrants further discussion. Acid-catalyzed conversion of an enol to a ketone occurs via two steps (Mechanism 10.2).

MECHANISM 10.2 ACID-CATALYZED TAUTOMERIZATION

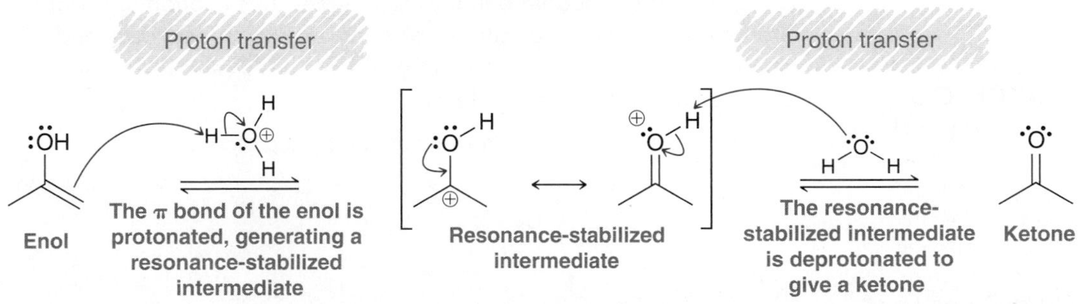

The π bond of the enol is first protonated, generating a resonance-stabilized intermediate, which is then deprotonated to give the ketone. Notice that both steps of this mechanism are proton transfers. The result of this process is the migration of a proton from one location to another, accompanied by a change in location of the π bond:

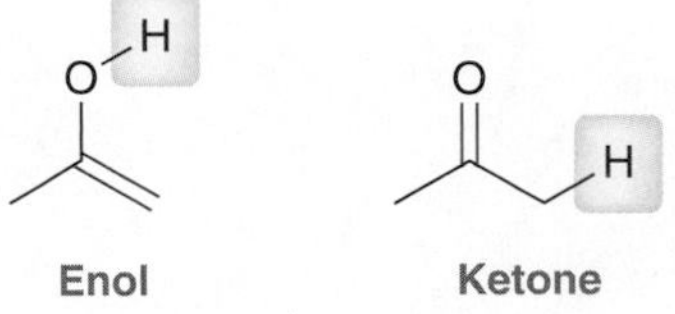

The enol and ketone are said to be **tautomers**, which are constitutional isomers that rapidly interconvert via the migration of a proton. The interconversion between an enol and a ketone is called **keto-enol tautomerization**. Tautomerization is an equilibrium process, which means that the equilibrium will establish specific concentrations for both the enol and the ketone. Generally, the ketone is highly favored, and the concentration of enol will be quite small. Be very careful not to confuse tautomers with resonance structures. Tautomers are constitutional isomers that exist in equilibrium with one another. Once the equilibrium has been reached, the concentrations of ketone and enol can be measured. In contrast, resonance structures are not different compounds and they are not in equilibrium with one another. Resonance structures simply represent different drawings of one compound.

LOOKING BACK
For a review of resonance structures, go to Section 2.7.

Keto-enol tautomerization is an equilibrium process that is catalyzed by even trace amounts of acid (or base). Glassware that is scrupulously cleaned will still have trace amounts of acid or base adsorbed to its surface. As a result, it is extremely difficult to prevent a keto-enol tautomerization from reaching equilibrium.

SKILLBUILDER

10.3 DRAWING THE MECHANISM OF ACID-CATALYZED KETO-ENOL TAUTOMERIZATION

LEARN the skill

Under normal conditions, 1-cyclohexenol cannot be isolated or stored in a bottle, because it undergoes rapid tautomerization to yield cyclohexanone. Draw a mechanism for this tautomerization:

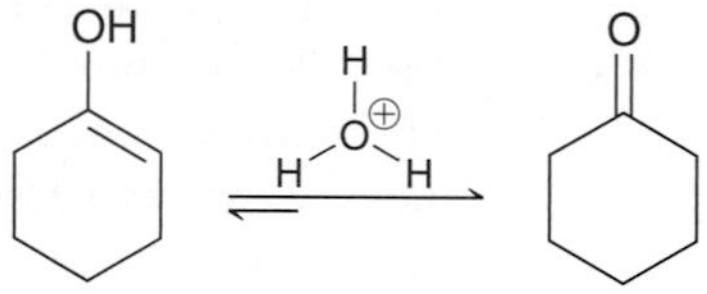

SOLUTION

The mechanism of a keto-enol tautomerization has two steps: (1) protonate and then (2) deprotonate. To draw this mechanism, it is essential to remember where to protonate in the first step. There are two places where protonation could possibly occur: the OH group or the double bond. In fact, under these conditions, both sites are reversibly protonated (in acidic conditions, protons are transferred back and forth, wherever possible). In order to choose where to protonate, let's carefully consider the positions of the protons involved in this transformation:

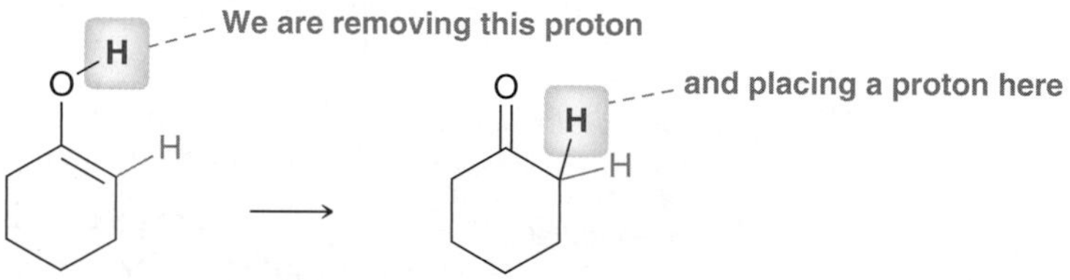

WATCH OUT
Do not give the OH another proton, as that pathway will not quickly lead to the ketone. Students who attempt to protonate the OH group invariably continue by losing water as a leaving group to form a vinylic carbocation. That pathway is too high in energy and simply does not lead to the product.

In acidic conditions, we must protonate first and only then deprotonate. In order to accomplish the transformation above, the double bond must be protonated, rather than the OH group. It might be tempting to protonate the OH group first (and this is a very common mistake), but look carefully at the OH group in the enol. The OH must lose its proton in the course of the reaction (to form the ketone). Make sure to begin by protonating the π bond rather than the OH group:

STEP 1
Protonate the π bond of the enol.

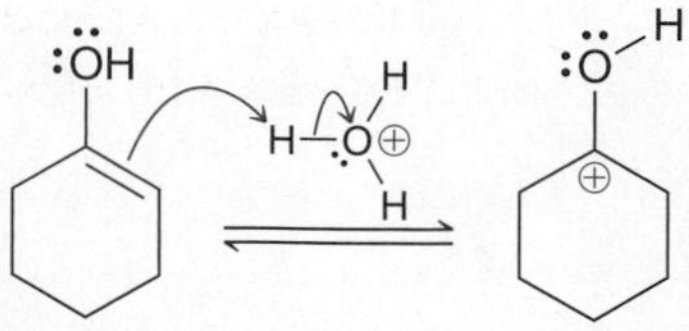

As is the case for all proton transfer steps, two curved arrows are required. Be sure to draw both of them.

Next, draw the resonance structure of the intermediate formed when the enol is protonated:

STEP 2
Draw the resonance structures of the intermediate.

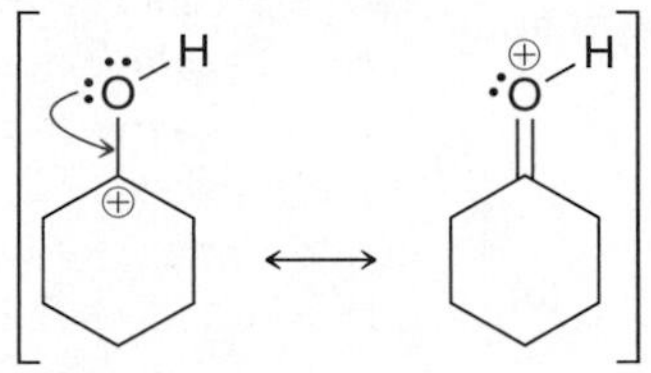

Finally, remove a proton to form the ketone. Once again, two curved arrows are required:

STEP 3
Remove a proton to form the ketone.

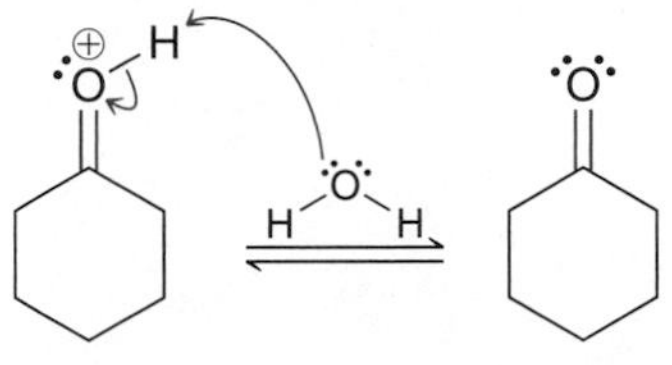

PRACTICE the skill **10.16** The following enols cannot be isolated. They rapidly tautomerize to produce ketones. In each case, draw the expected ketone, and show a mechanism for its formation under acid-catalyzed conditions (H_3O^+).

(a) 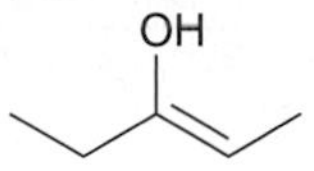 (b) 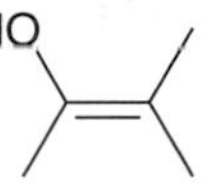 (c) 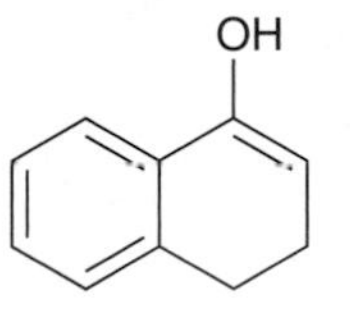 (d) 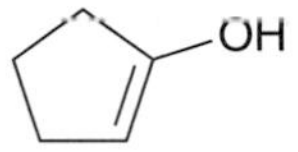

APPLY the skill **10.17** Consider the following equilibrium:

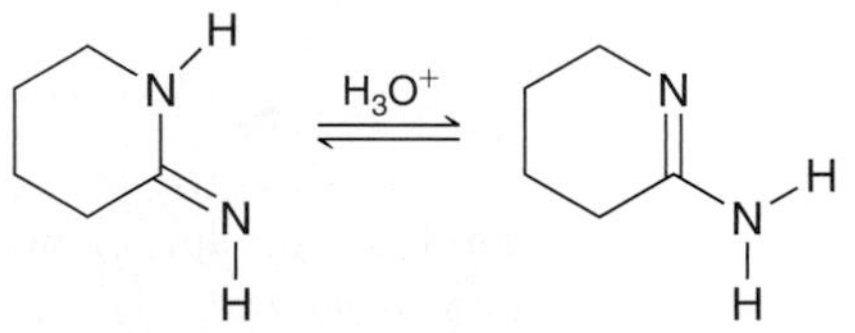

These constitutional isomers rapidly interconvert in the presence of even trace amounts of acid and are therefore said to be tautomers of each other. Draw a mechanism for this transformation.

need more **PRACTICE?** **Try Problems 10.49b, 10.63, 10.64, 10.67**

We have seen that even trace amounts of acid will catalyze a keto-enol tautomerization. Therefore, the enol initially produced by hydration of an alkyne will immediately tautomerize to form a ketone. When asked to predict the product of alkyne hydration, do not make the mistake of drawing an enol. Simply draw the ketone.

Acid-catalyzed hydration of unsymmetrical internal alkynes yields a mixture of ketones:

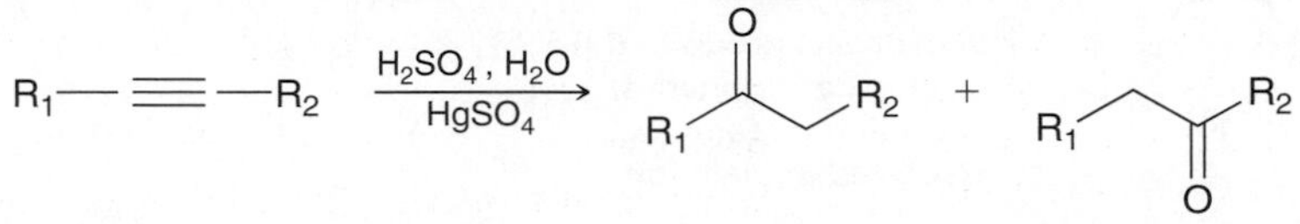

The lack of regiochemical control renders this process significantly less useful. It is most often used for the hydration of a terminal alkyne, which generates a methyl ketone as the product:

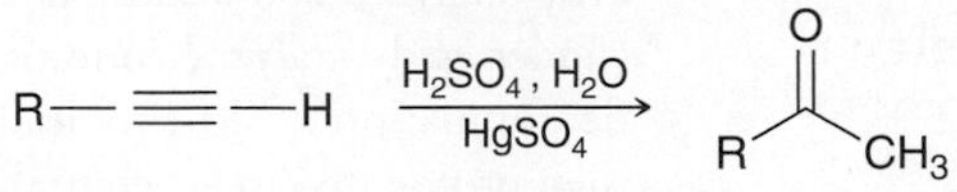

CONCEPTUAL CHECKPOINT

10.18 Draw the major product(s) expected when each of the following alkynes is treated with aqueous acid in the presence of mercuric sulfate ($HgSO_4$):

(a) 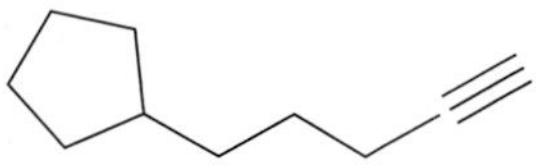

(b) 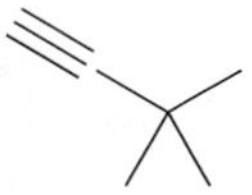

(c) 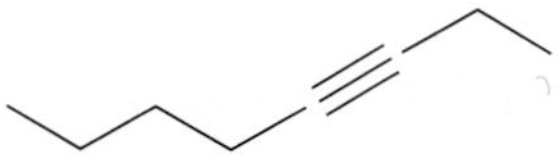

(d) 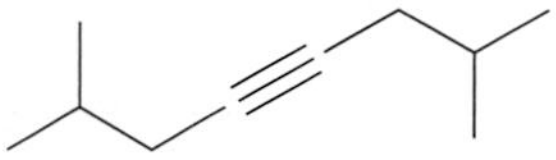

(e) 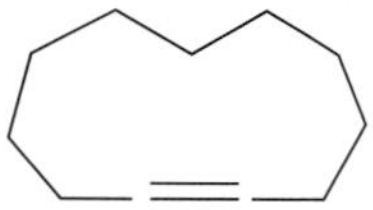

10.19 Identify the alkyne you would use to prepare each of the following ketones via acid-catalyzed hydration:

(a) 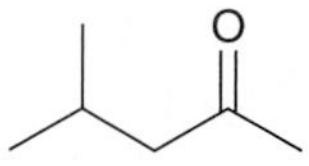

(b) 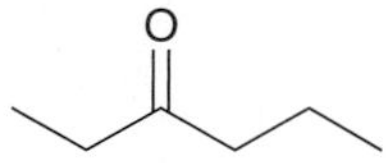

(c) 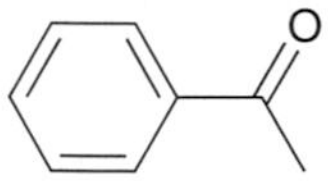

Hydroboration-Oxidation of Alkynes

In the previous chapter, we saw that alkenes will undergo hydroboration-oxidation (Section 9.6). The reaction proceeds via an *anti*-Markovnikov addition, thereby installing a hydroxyl group at the less substituted position:

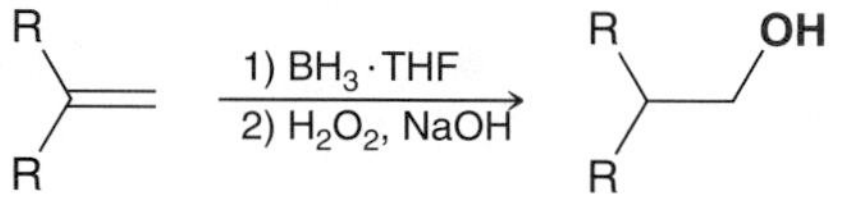

Alkynes are also observed to undergo a similar process:

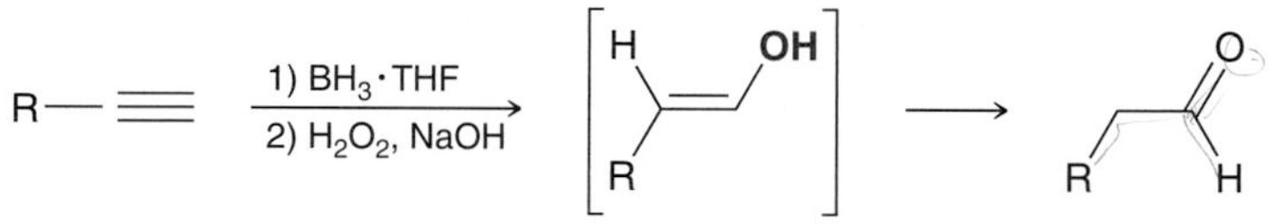

The initial product of this reaction is an enol that cannot be isolated because it is rapidly converted into an aldehyde via tautomerization. As we saw in the previous section, tautomerization cannot be prevented, and it is catalyzed by either acid or base. In this case, basic conditions are employed, so the tautomerization process occurs via a base-catalyzed mechanism (Mechanism 10.3).

MECHANISM 10.3 BASE-CATALYZED TAUTOMERIZATION

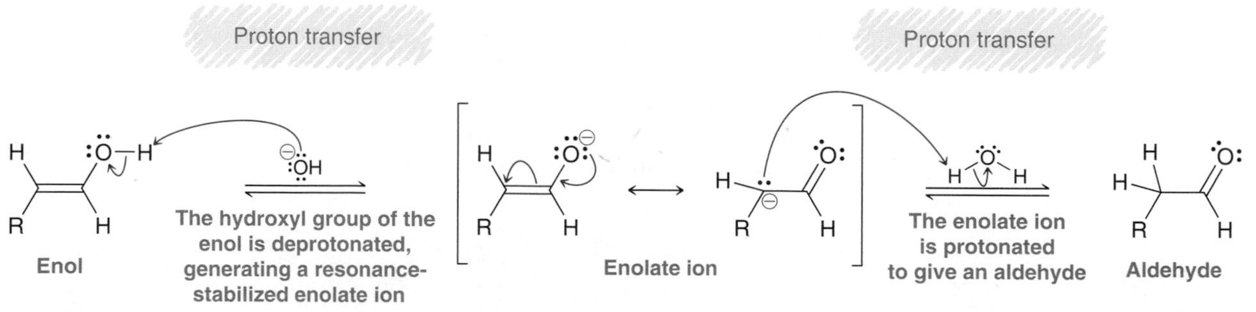

LOOKING AHEAD
Enolates are very useful intermediates, and we will explore their chemistry in Chapter 22.

Notice that the order of events under base-catalyzed conditions is the reverse of the order under acid-catalyzed conditions. That is, the enol is first deprotonated, and only then is it protonated (under acid-catalyzed conditions, the first step is to protonate the enol). In basic conditions, deprotonation of the enol leads to a resonance-stabilized anion called an **enolate ion**, which is then protonated to generate the aldehyde.

Hydroboration-oxidation of alkynes is believed to proceed via a mechanism that is similar to the mechanism invoked for hydroboration-oxidation of alkenes (Mechanism 9.3). Specifically, borane adds to the alkyne in a concerted process that gives an *anti*-Markovnikov addition. There is, however, one critical difference. Unlike an alkene, which only possesses one π bond, an alkyne possesses two π bonds. As a result, two molecules of BH_3 can add across the alkyne. To prevent the second addition, a dialkyl borane (R_2BH) is employed instead of BH_3. The two alkyl groups provide steric hindrance that prevents the second addition. Two commonly used dialkyl boranes are disiamylborane and 9-BBN:

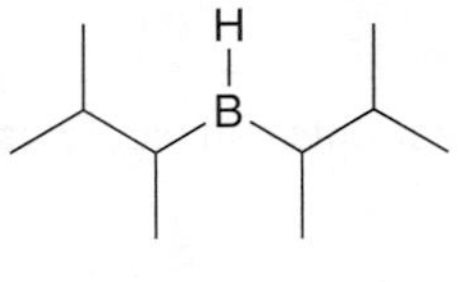

Disiamylborane

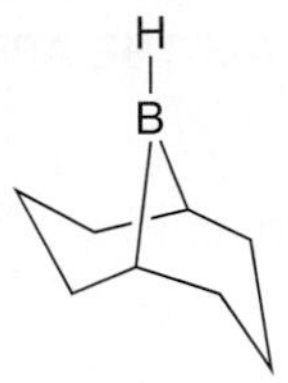

9-BBN
(9-Borabicyclo[3.3.1]nonane)

With these modified borane reagents, hydroboration-oxidation is an efficient method for converting a terminal alkyne into an aldehyde:

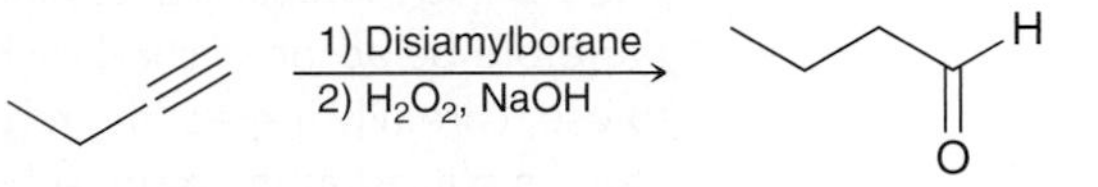

Notice that the ultimate product of this reaction sequence is an aldehyde rather than a ketone.

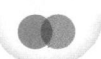

CONCEPTUAL CHECKPOINT

10.20 Draw the major product for each of the following reactions:

(a) 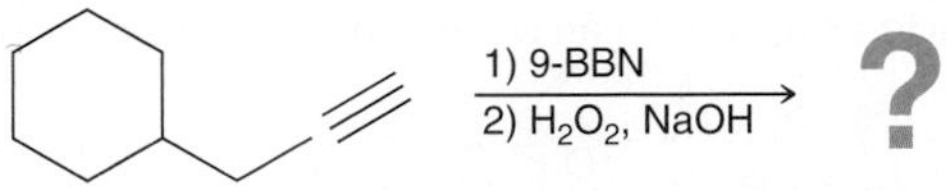

(b) 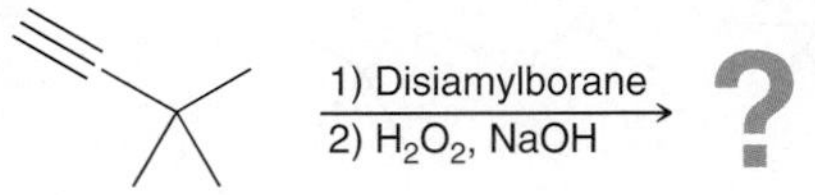

(c) 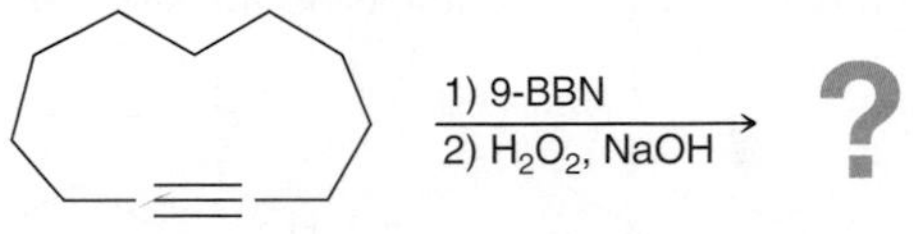

10.21 Identify the alkyne you would use to prepare each of the following compounds via hydroboration-oxidation:

(a) 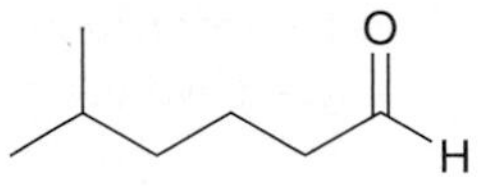

(b) 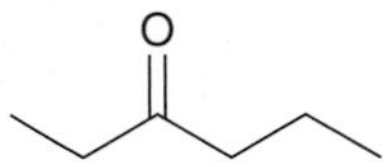

(c) 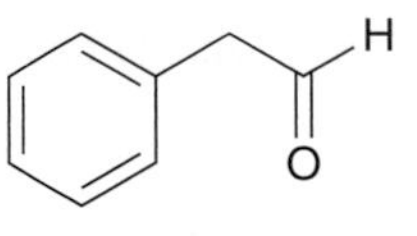

Controlling the Regiochemistry of Alkyne Hydration

In the previous sections, we explored two methods for the hydration of a terminal alkyne:

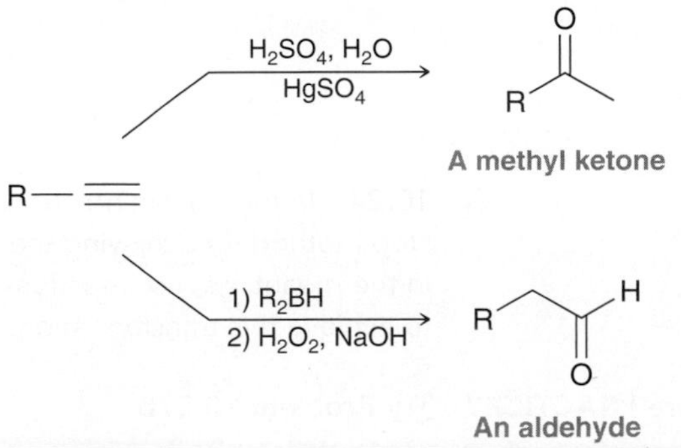

Acid-catalyzed hydration of a terminal alkyne produces a methyl ketone, while hydroboration-oxidation produces an aldehyde. In other words, the regiochemical outcome of alkyne hydration can be controlled by the choice of reagents. Let's get some practice determining which reagents to use.

SKILLBUILDER

10.4 CHOOSING THE APPROPRIATE REAGENTS FOR THE HYDRATION OF AN ALKYNE

LEARN the skill

Identify the reagents you would use to carry out the following transformation:

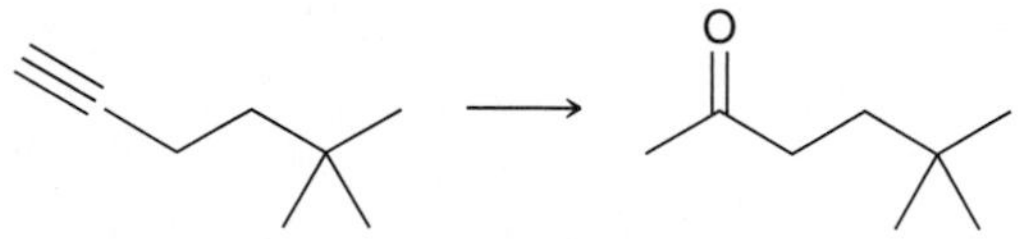

SOLUTION

The starting material is a terminal alkyne and the product has a C=O bond. This reaction can therefore be accomplished via hydration of the starting alkyne. To determine what reagents to use, carefully inspect the regiochemical outcome of the process. Specifically, the oxygen atom is placed at the more substituted position to produce a methyl ketone:

STEP 1 Identify the regiochemical outcome.

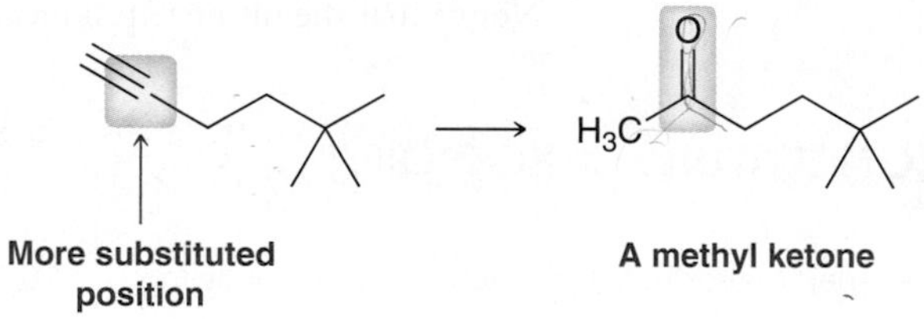

This transformation requires a Markovnikov addition, which can be accomplished via an acid-catalyzed hydration:

STEP 2 Identify the reagents that achieve the desired regiochemical outcome.

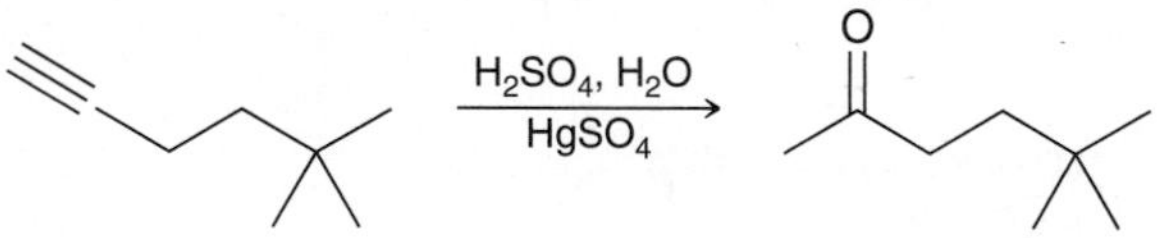

PRACTICE the skill

10.22 Identify the reagents you would use to carry out each of the following transformations:

(a) 

(b) 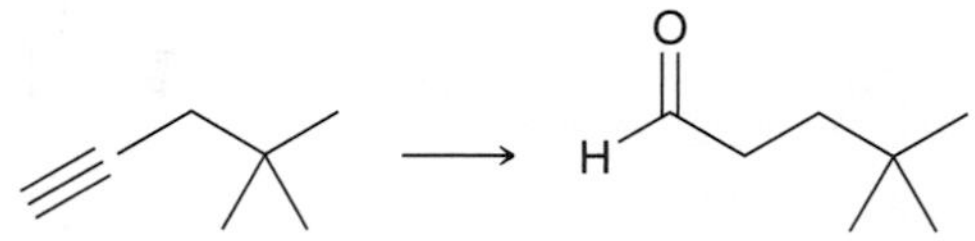

APPLY the skill

10.23 Identify the reagents you would use to carry out each of the following transformations:

(a) 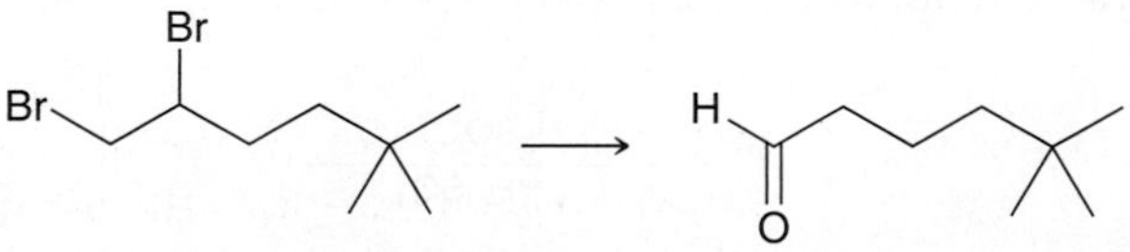

(b) 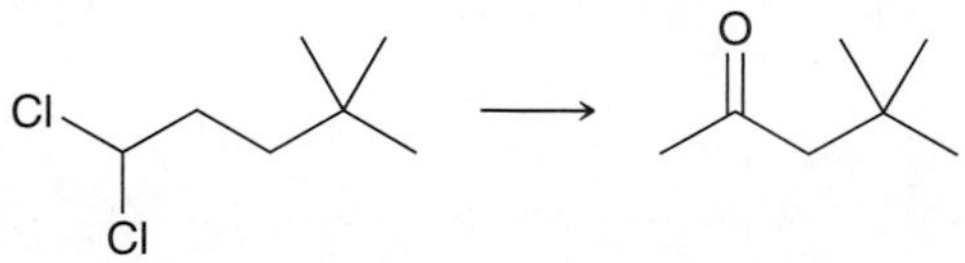

10.24 In the upcoming chapters, we will learn a two-step method for achieving the following transformation. In the meantime, use reactions we have already learned to achieve this transformation:

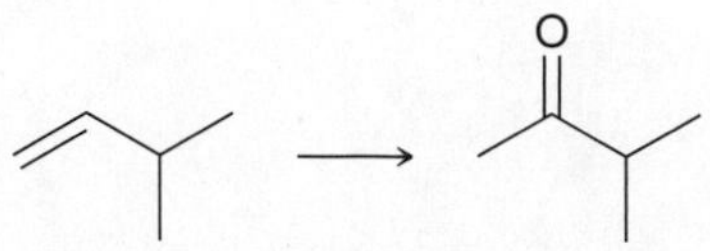

need more **PRACTICE?** **Try Problem 10.57b**

10.8 Halogenation of Alkynes

In the previous chapter, we saw that alkenes will react with Br_2 or Cl_2 to produce a dihalide. In much the same way, alkynes are also observed to undergo halogenation. The one major difference is that alkynes have two π bonds rather than one and can, therefore, add two equivalents of the halogen to form a tetrahalide:

$$R-C{\equiv}C-R \xrightarrow[\text{CCl}_4]{\text{excess X}_2} R-\overset{\overset{\displaystyle X}{|}}{\underset{\underset{\displaystyle X}{|}}{C}}-\overset{\overset{\displaystyle X}{|}}{\underset{\underset{\displaystyle X}{|}}{C}}-R$$

(X = Cl or Br)

(60–70%)

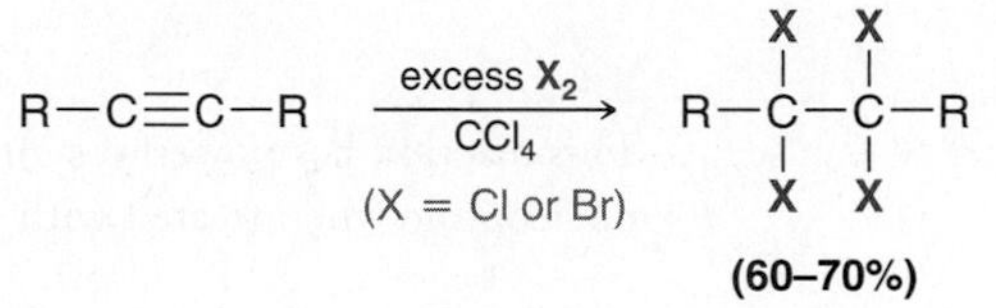

In some cases, it is possible to add just one equivalent of halogen to produce a dihalide. Such a reaction generally proceeds via an *anti* addition (just as we saw with alkenes), producing the *E* isomer as the major product:

$$R-{\equiv}-R \xrightarrow[\text{CCl}_4]{\text{X}_2\text{ (one equivalent)}}$$

Major + **Minor**

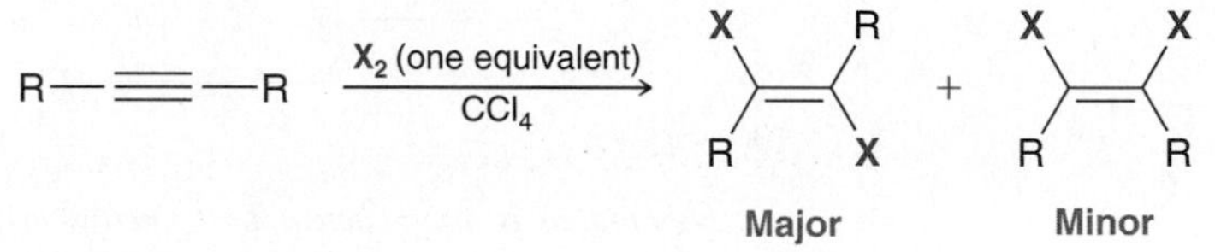

The mechanism of alkyne halogenation is not entirely understood.

10.9 Ozonolysis of Alkynes

When treated with ozone followed by water, alkynes undergo oxidative cleavage to produce carboxylic acids:

$$R_1-C{\equiv}C-R_2 \xrightarrow[\text{2) H}_2\text{O}]{\text{1) O}_3} R_1-\overset{\overset{\displaystyle O}{\|}}{C}-OH + HO-\overset{\overset{\displaystyle O}{\|}}{C}-R_2$$

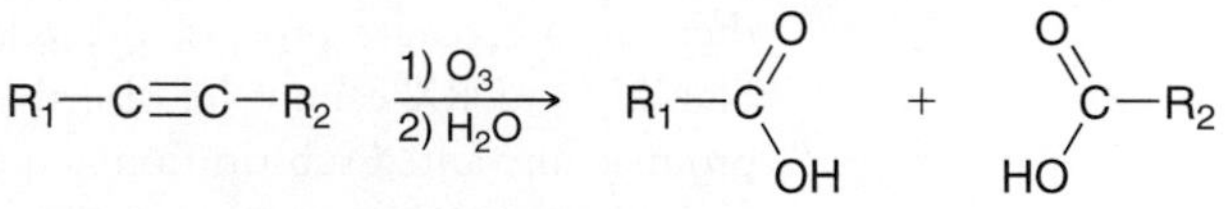

When a terminal alkyne undergoes oxidative cleavage, the terminal side is converted into carbon dioxide:

$$R-C{\equiv}C-H \xrightarrow[\text{2) H}_2\text{O}]{\text{1) O}_3} R-\overset{\overset{\displaystyle O}{\|}}{C}-OH + \overset{\overset{\displaystyle O}{\|}}{\underset{\underset{\displaystyle O}{\|}}{C}}$$

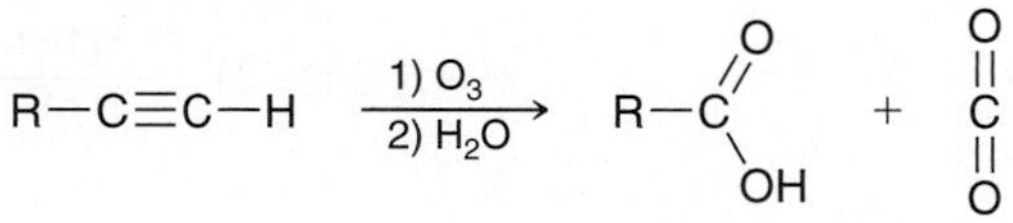

Decades ago, chemists used oxidative cleavage to help with structural determinations. An unknown alkyne would be treated with ozone followed by water, and the resulting carboxylic acids would be identified. This technique allowed chemists to identify the location of a triple bond in an unknown alkyne. However, the advent of spectroscopic methods (Chapters 15 and 16) has rendered this process obsolete as a tool for structural determination.

CONCEPTUAL CHECKPOINT

10.25 Draw the major products that are expected when each of the following alkynes is treated with O_3 followed by H_2O:

(a)

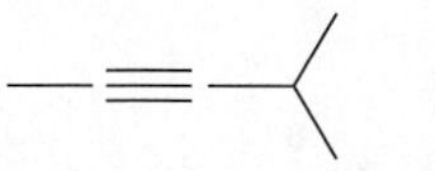

(b)

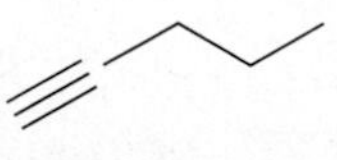

(c)

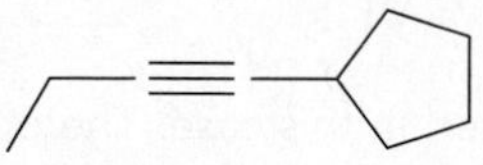

(d)

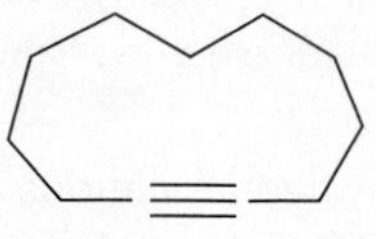

10.26 An alkyne with molecular formula C_6H_{10} was treated with ozone followed by water to produce only one type of carboxylic acid. Draw the structure of the starting alkyne and the product of ozonolysis.

10.27 An alkyne with molecular formula C_4H_6 was treated with ozone followed by water to produce a carboxylic acid and carbon dioxide. Draw the expected product when the alkyne is treated with aqueous acid in the presence of mercuric sulfate.

10.10 Alkylation of Terminal Alkynes

In Section 10.3, we saw that a terminal alkyne can be deprotonated in the presence of a sufficiently strong base, such as sodium amide ($NaNH_2$):

$$R-C\equiv C-H \xrightarrow{\ominus\text{:}NH_2} R-C\equiv C\text{:}^{\ominus}$$

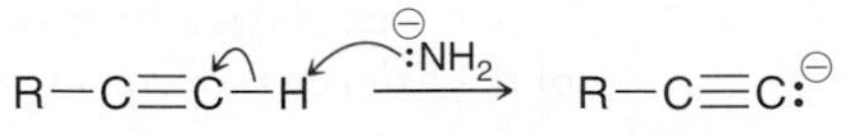

Alkynide ion

This reaction has powerful synthetic utility, because the resulting alkynide ion can function as a nucleophile when treated with an alkyl halide:

$$R-C\equiv C\text{:}^{\ominus} \xrightarrow{R-X} R-C\equiv C-R$$

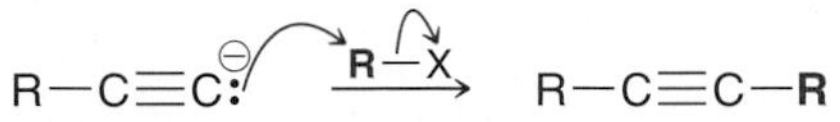

This transformation proceeds via an S_N2 reaction and provides a method to install an alkyl group on a terminal alkyne. This process is called **alkylation**, and it is achieved in just two steps; for example:

$$\xrightarrow[\text{2)} \diagup\diagdown I]{\text{1) NaNH}_2}$$

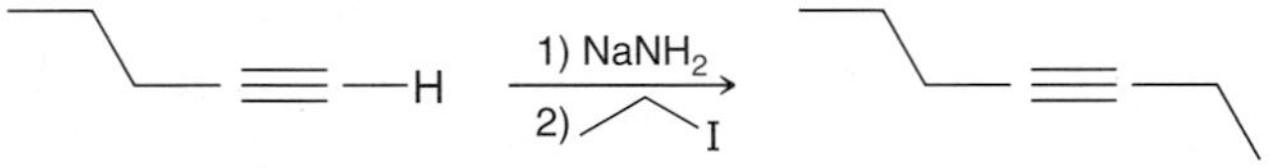

This process is only efficient with methyl or primary alkyl halides. When secondary or tertiary alkyl halides are used, the alkynide ion functions primarily as a base and elimination products are obtained. This observation is consistent with the pattern we saw in Section 8.14 (substitution vs. elimination).

Acetylene possesses two terminal protons (one on either side) and can therefore undergo alkylation twice:

$$H-C\equiv C-H \xrightarrow[\text{2) RX}]{\text{1) NaNH}_2} R-C\equiv C-H \xrightarrow[\text{2) RX}]{\text{1) NaNH}_2} R-C\equiv C-R$$

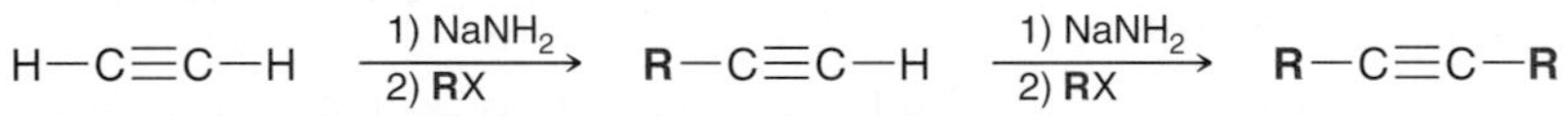

Notice that two separate alkylations are required. One side of acetylene is first alkylated, and then, in a separate process, the other side is alkylated. This repetition is required because $NaNH_2$ and RX cannot be placed into the reaction flask at the same time. Doing so would produce unwanted substitution and elimination products resulting from the reactions between $NaNH_2$ and RX. It might seem burdensome to require two separate alkylation processes, but this requirement does provide additional synthetic utility. Specifically, it enables the installation of two different alkyl groups; for example:

$$H-C\equiv C-H \xrightarrow[\text{2) EtI}]{\text{1) NaNH}_2} Et-C\equiv C-H \xrightarrow[\text{2) MeI}]{\text{1) NaNH}_2} Et-C\equiv C-Me$$

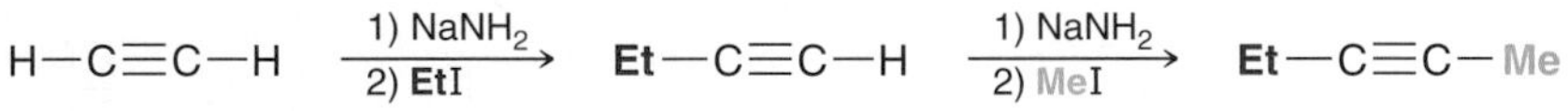

SKILLBUILDER

10.5 ALKYLATING TERMINAL ALKYNES

LEARN the skill

Identify the reagents necessary to convert acetylene into 7-methyl-3-octyne.

SOLUTION

Whenever confronted with a problem that is written only in words, with no corresponding structures, the first step will always be to draw the structures described in the problem. In this case, the task is to identify the reagents necessary to accomplish the following transformation:

STEP 1
Draw the structures described in the problem.

$$H-\equiv-H \longrightarrow$$

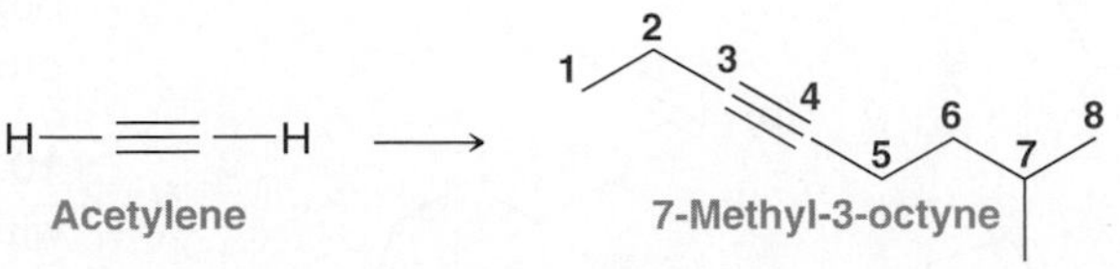

Acetylene

7-Methyl-3-octyne

When drawn out, it becomes clear that two alkylations are required in this case. The order of events (which alkyl group is installed first) is not important. Just make sure to install each alkyl group separately. The first alkylation is accomplished by treating acetylene with $NaNH_2$ followed by an alkyl halide:

STEP 2
Install the first alkyl group.

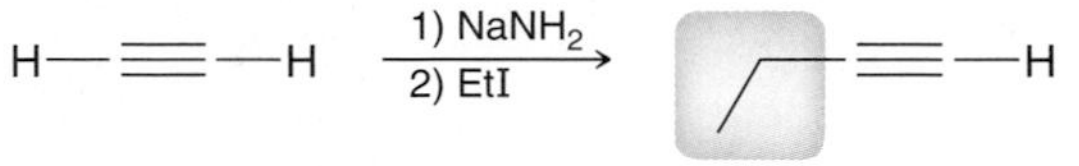

The terminal alkyne is then alkylated once again using $NaNH_2$ followed by the appropriate alkyl halide:

STEP 3
Install the second alkyl group.

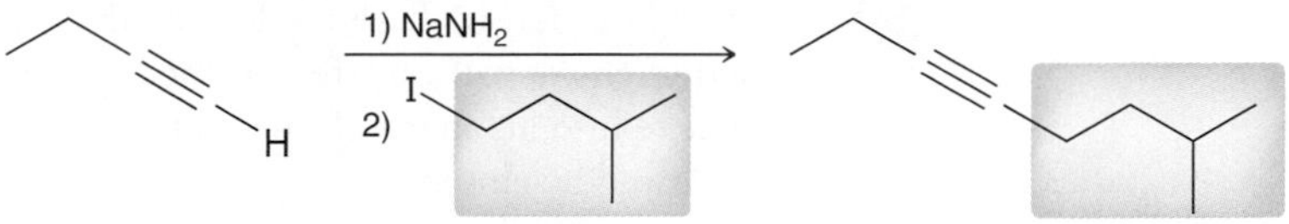

The overall synthesis therefore has four steps:

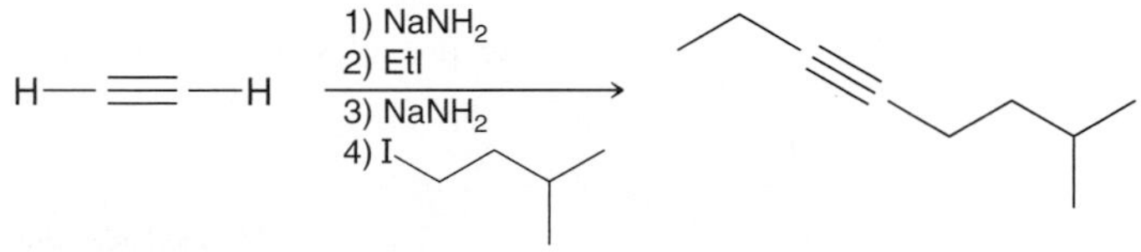

Both alkyl halides are primary, so the process would be expected to be efficient.

PRACTICE the skill **10.28** Starting with acetylene, show the reagents you would use to prepare the following compounds:

(a) 1-Butyne (b) 2-Butyne (c) 3-Hexyne (d) 2-Hexyne (e) 1-Hexyne
(f) 2-Heptyne (g) 3-Heptyne (h) 2-Octyne (i) 2-Pentyne

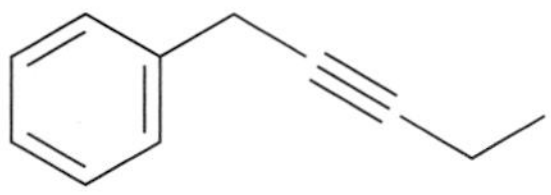
(j)

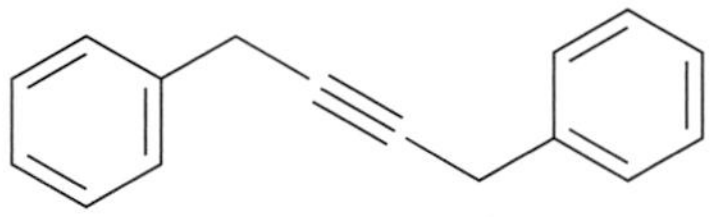
(k)

APPLY the skill **10.29** Preparation of 2,2-dimethyl-3-octyne cannot be achieved via alkylation of acetylene. Explain.

10.30 A terminal alkyne was treated with $NaNH_2$ followed by propyl iodide. The resulting internal alkyne was treated with ozone followed by water, giving only one type of carboxylic acid. Provide a systematic, IUPAC name for the internal alkyne.

10.31 Using acetylene as your only source of carbon atoms, outline a synthesis for 3-hexyne.

need more **PRACTICE?** **Try Problems 10.40f, 10.46, 10.50, 10.59**

10.11 Synthesis Strategies

Earlier in this chapter, we saw how to control the reduction of an alkyne. In particular, a triple bond can be converted into a double bond (either *cis* or *trans*), or into a single bond:

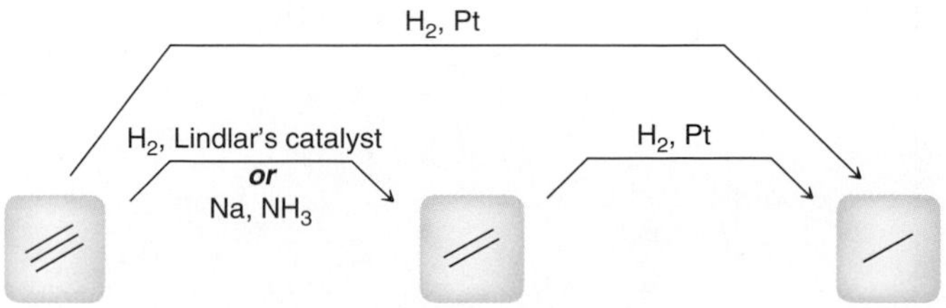

Alkyne **Alkene** **Alkane**

Now let's consider how to go in the other direction, that is, to convert a single bond or double bond into a triple bond:

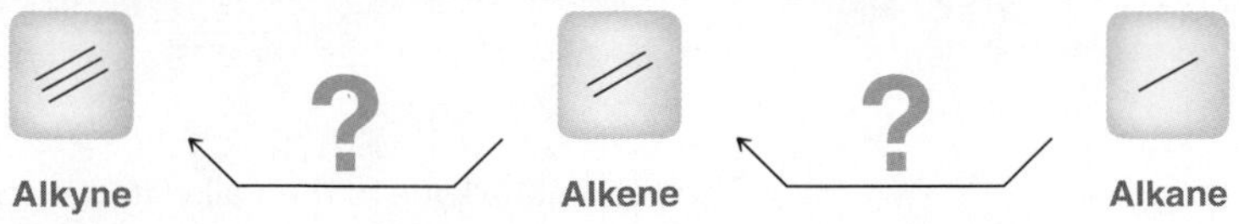

Thus far, we have not learned the reactions necessary to convert an alkane into an alkene, although we will explore a method in Chapter 11. We did, however, learn how to convert an alkene into an alkyne. Specifically, an alkene can be converted into an alkyne by bromination followed by elimination:

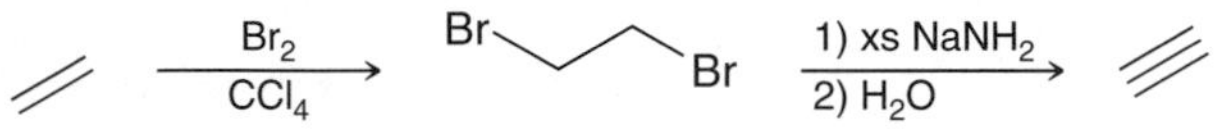

This now provides us with the flexibility to interconvert single, double, and triple bonds (Figure 10.7).

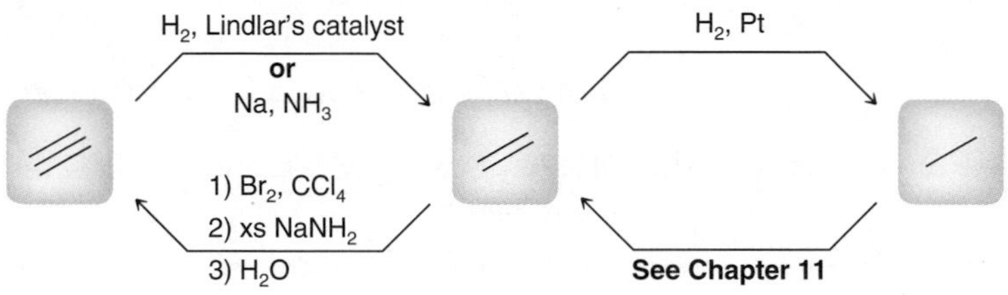

FIGURE 10.7
A summary of the reagents that can be used to interconvert alkanes, alkenes and alkynes.

SKILLBUILDER

10.6 INTERCONVERTING ALKANES, ALKENES, AND ALKYNES

LEARN the skill

Propose a plausible synthesis for the following transformation:

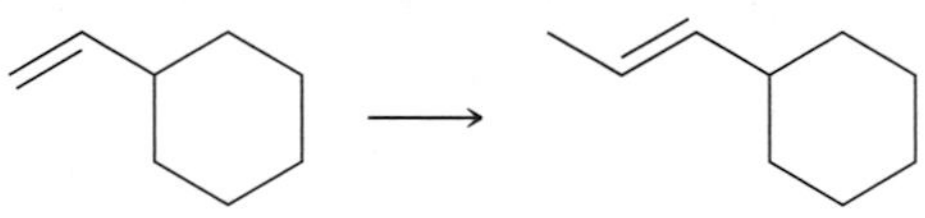

SOLUTION

This problem requires the installation of a methyl group at a vinylic position (i.e., alkylation of an alkene). We have not yet seen a direct way to alkylate an alkene. However, we did learn how to alkylate an alkyne. The ability to interconvert alkenes and alkynes therefore enables us to achieve the desired transformation. Specifically, the alkene can first be converted into an alkyne, which can be readily alkylated. Then, after the alkylation, the alkyne can be converted back into an alkene:

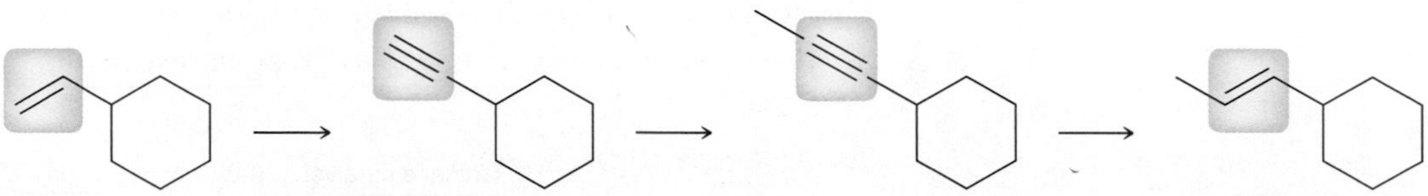

The first part of this strategy (conversion of the alkene into an alkyne) can be accomplished by brominating the alkene to give a dibromide followed by elimination with excess $NaNH_2$. The resulting alkyne can be purified and isolated and then alkylated by treating with $NaNH_2$ followed by MeI. Finally, a dissolving metal reduction will convert the alkyne into a *trans* alkene:

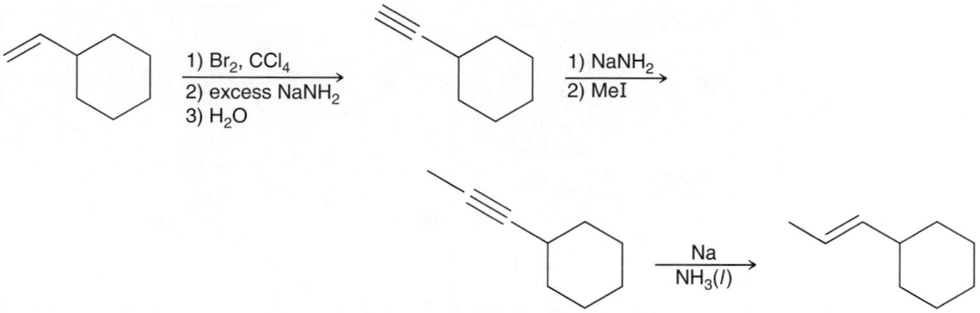

PRACTICE the skill **10.32** Propose a plausible synthesis for each of the following transformations:

(a) 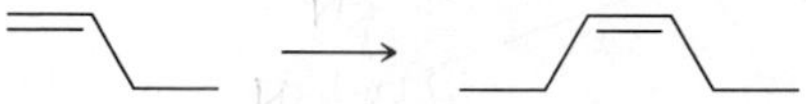

(b) 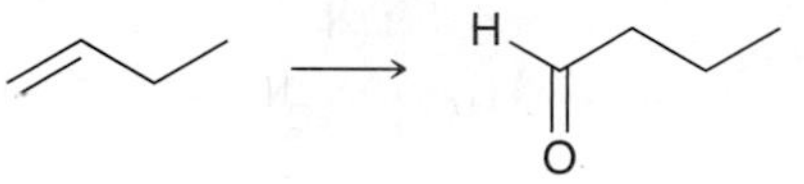

(c) 

(d) 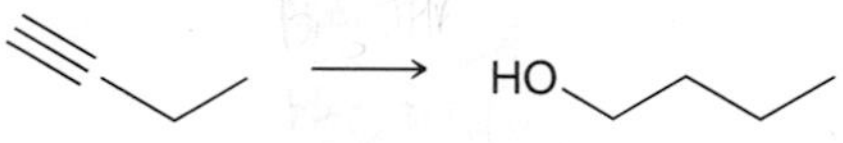

(e) 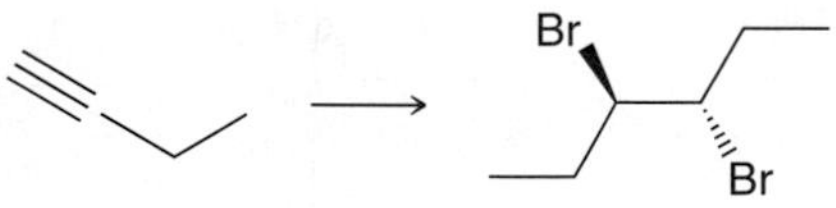

(f) 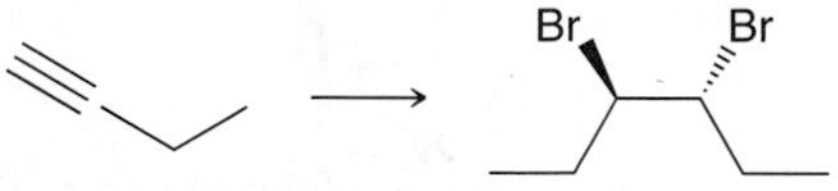

\+ En

APPLY the skill **10.33** Identify the reagents you would use to achieve each of the following transformations:

(a) Conversion of all of the carbon atoms in bromoethane into CO_2 gas

(b) Conversion of all of the carbon atoms in 2-bromopropane into acetic acid (CH_3COOH)

10.34 Using ethylene ($H_2C{=}CH_2$) as your only source of carbon atoms, outline a synthesis for 3-hexanone.

need more **PRACTICE?** **Try Problems 10.57, 10.60**

REVIEW OF REACTIONS

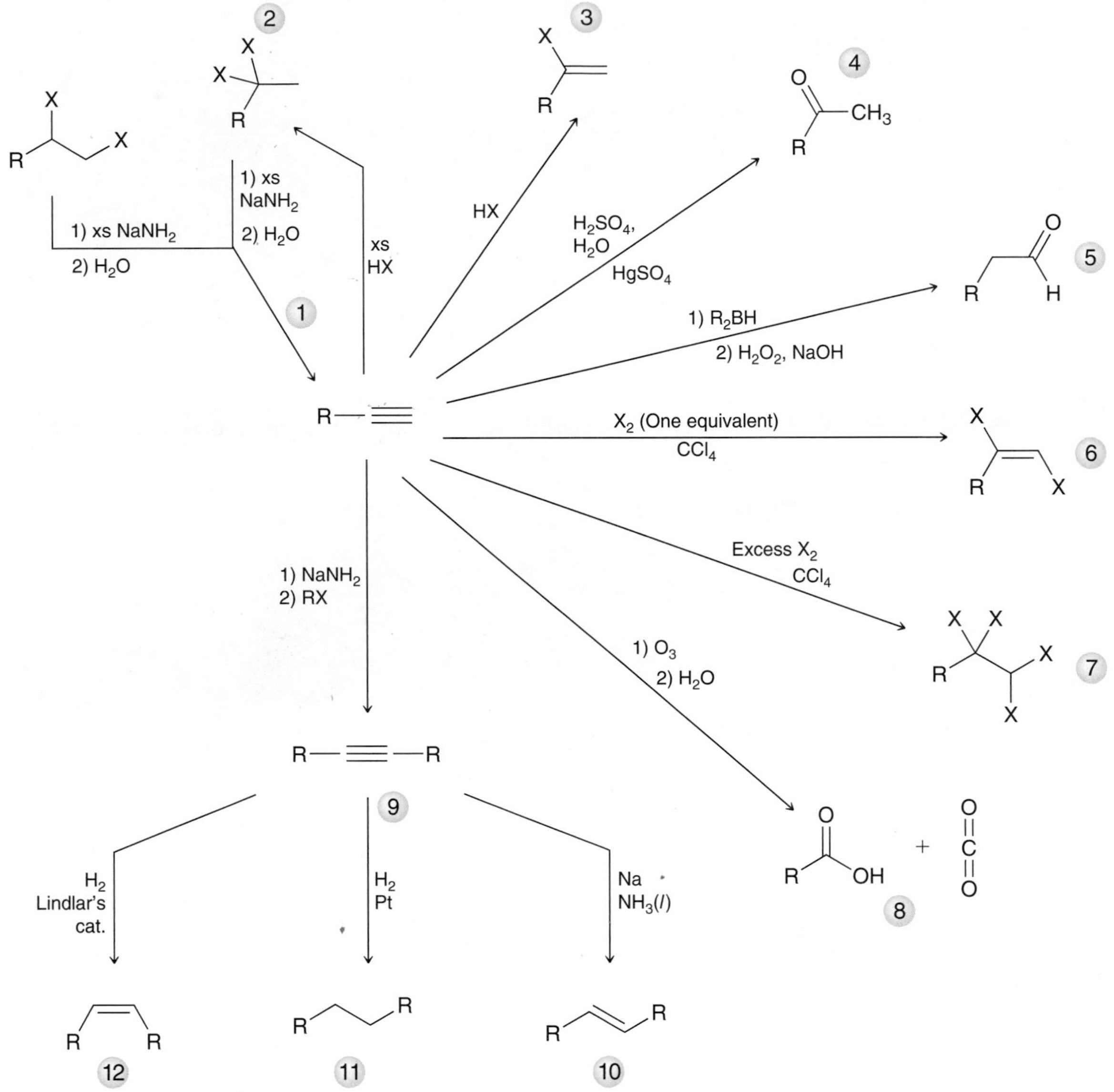

1. Elimination
2. Hydrohalogenation (two equivalents)
3. Hydrohalogenation (one equivalent)
4. Acid-catalyzed hydration
5. Hydroboration-oxidation
6. Halogenation (one equivalent)
7. Halogenation (two equivalents)
8. Ozonolysis
9. Alkylation
10. Dissolving metal reduction
11. Hydrogenation
12. Hydrogenation with a poisoned catalyst

REVIEW OF CONCEPTS AND VOCABULARY

SECTION 10.1

- A triple bond is comprised of three separate bonds: one σ bond and two π bonds.
- Alkynes exhibit linear geometry and can function either as bases or as nucleophiles.

SECTION 10.2

- Alkynes are named much like alkanes, with the following additional rules:
- The suffix "ane" is replaced with "yne."
- The parent is the longest chain that includes the C≡C triple bond.
- The triple bond should receive the lowest number possible.
- The position of the triple bond is indicated with a single locant placed either before the parent or the suffix.
- Monosubstituted acetylenes are **terminal alkynes**, while disubstituted acetylenes are **internal alkynes**.

SECTION 10.3

- The conjugate base of acetylene, called an **acetylide ion**, is relatively stabilized because the lone pair occupies an *sp*-hybridized orbital.
- The conjugate base of a terminal alkyne, called an **alkynide ion**, can only be formed with a sufficiently strong base, such as $NaNH_2$.

SECTION 10.4

- Alkynes can be prepared from either **geminal** or **vicinal** dihalides via two successive E2 reactions.

SECTION 10.5

- Catalytic hydrogenation of an alkyne yields an alkane.
- Catalytic hydrogenation in the presence of a poisoned catalyst (Lindlar's catalyst or Ni_2B) yields a *cis* alkene.
- A **dissolving metal reduction** will convert an alkyne into a *trans* alkene. The reaction involves an intermediate **radical anion** and employs *fishhook arrows*, which indicate the movement of only one electron.

SECTION 10.6

- Alkynes react with HX via a Markovnikov addition.
- One possible mechanism for the hydrohalogenation of alkynes involves a **vinylic carbocation**, while another possible mechanism is **termolecular**.
- Addition of HX to alkynes probably occurs through a variety of mechanistic pathways all of which are occurring at the same time and competing with each other.
- Treatment of a terminal alkyne with HBr and peroxides gives an *anti*-Markovnikov addition of HBr.

SECTION 10.7

- Acid-catalyzed hydration of alkynes is catalyzed by mercuric sulfate ($HgSO_4$) to produce an **enol** that cannot be isolated because it is rapidly converted into a ketone.
- Enols and ketones are **tautomers**, which are constitutional isomers that rapidly interconvert via the migration of a proton.
- The interconversion between an enol and a ketone is called **keto-enol tautomerization** and is catalyzed by trace amounts of acid or base.
- Hydroboration-oxidation of a terminal alkyne proceeds via an *anti*-Markovnikov addition to produce an enol that is rapidly converted into an aldehyde via tautomerization.
- In basic conditions, tautomerization proceeds via a resonance-stabilized anion called an **enolate ion**.

SECTION 10.8

- Alkynes can undergo halogenation to form a tetrahalide.

SECTION 10.9

- When treated with ozone followed by water, internal alkynes undergo oxidative cleavage to produce carboxylic acids.
- When a terminal alkyne undergoes oxidative cleavage, the terminal side is converted into carbon dioxide.

SECTION 10.10

- Alkynide ions undergo **alkylation** when treated with an alkyl halide (methyl or primary).
- Acetylene possesses two terminal protons and can undergo two separate alkylations.

SECTION 10.11

- An alkene can be converted into an alkyne via bromination followed by elimination with excess $NaNH_2$.

KEY TERMINOLOGY

acetylide ion 460
alkylation 478
alkyne 455
alkynide ion 461
dissolving metal reduction 466
enol 471
enolate ion 474
geminal 473
internal alkynes 458
keto-enol tautomerization 472
poisoned catalyst 465
radical anion 466
tautomers 472
terminal alkynes 458
termolecular 469
vicinal 463
vinylic carbocation 469

SKILLBUILDER REVIEW

10.1 ASSEMBLING THE SYSTEMATIC NAME OF AN ALKYNE

STEP 1 Identify the parent: choose the longest chain that includes the triple bond.

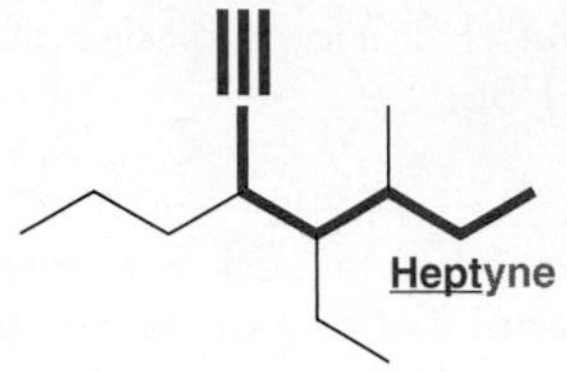

STEP 2 Identify and name substituents.

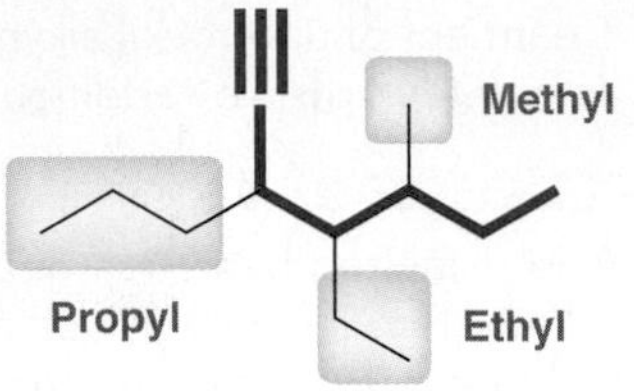

STEP 3 Number the parent chain and assign a locant to each substituent.

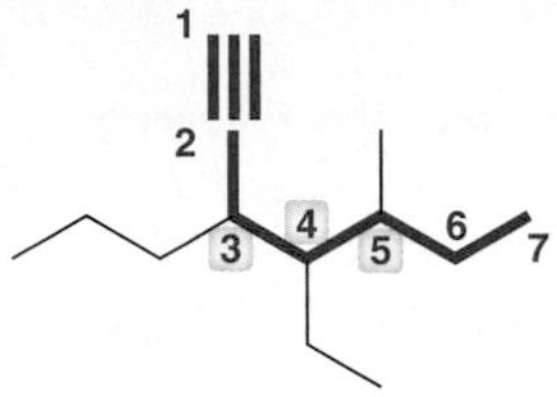

STEP 4 Assemble the substituents alphabetically.

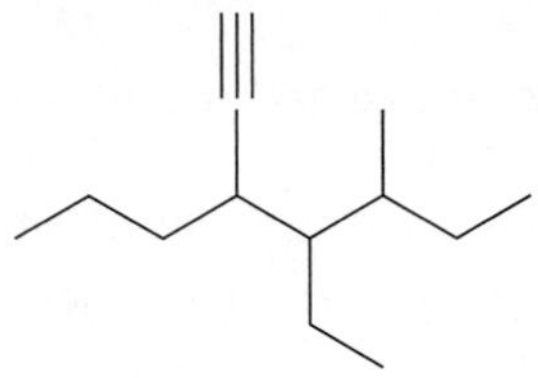

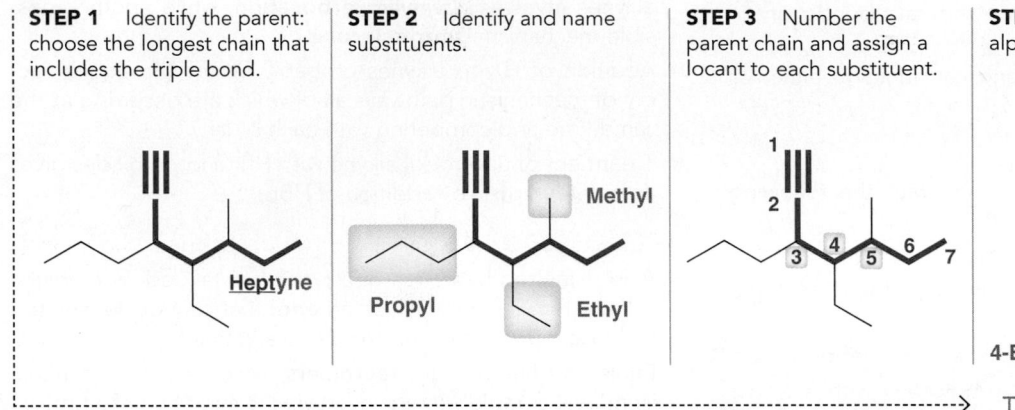

4-Ethyl-5-methyl-3-propyl-1-heptyne

Try Problems 10.1–10.4, 10.35, 10.36

10.2 PREDICTING THE POSITION OF EQUILIBRIUM FOR THE DEPROTONATION OF A TERMINAL ALKYNE

STEP 1 Identify the acid and base on each side of the equilibrium.

$$\underset{\textbf{Acid}}{R{-}C{\equiv}C{-}H} + \underset{\textbf{Base}}{:\ddot{O}H^{\ominus}} \rightleftharpoons \underset{\textbf{Base}}{R{-}C{\equiv}C:^{\ominus}} + \underset{\textbf{Acid}}{H_2\ddot{O}:}$$

STEP 2 Compare the two acids and determine which is the weaker acid (higher pK_a).

$$\underset{(\mathbf{p}K_a \sim 25)}{R{-}C{\equiv}C{-}H} \qquad \underset{(\mathbf{p}K_a = 15.7)}{H_2O}$$

STEP 3 Equilibrium favors weaker acid and weaker base. In this case, the equilibrium favors the left side.

Try Problems 10.5, 10.6, 10.39, 10.42

10.3 DRAWING THE MECHANISM OF ACID-CATALYZED KETO-ENOL TAUTOMERIZATION

STEP 1 Protonate the π bond of the enol (do not protonate the hydroxyl group).

STEP 2 Draw the resonance structure of the intermediate.

STEP 3 Remove a proton to form the ketone.

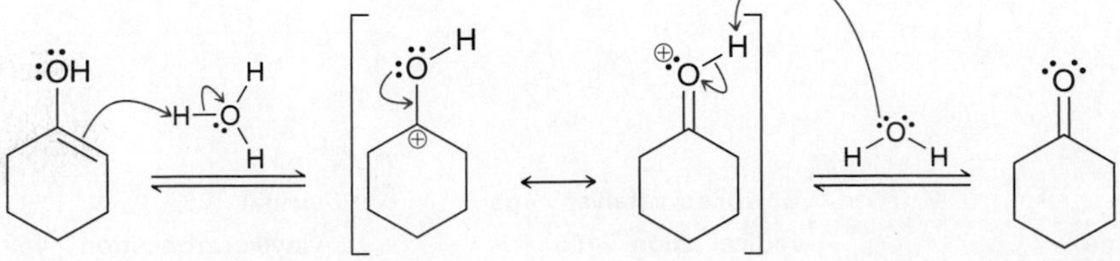

Try Problems 10.16, 10.17, 10.49b, 10.63, 10.64, 10.67

10.4 CHOOSING THE APPROPRIATE REAGENTS FOR THE HYDRATION OF AN ALKYNE

STEP 1 Identify the regiochemical outcome.

STEP 2 Choose the reagents that achieve that outcome.

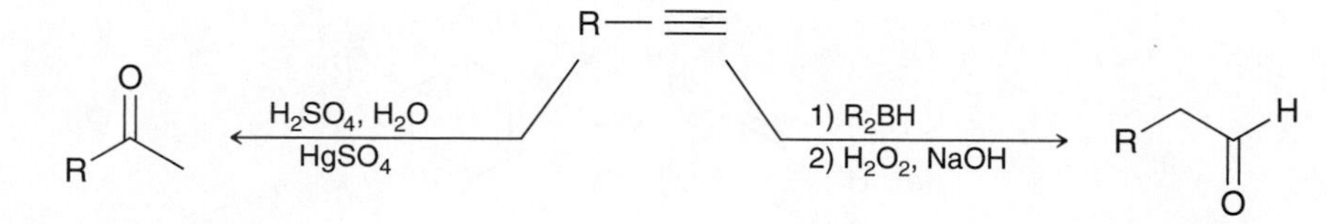

A methyl ketone **An aldehyde**

Try Problems 10.22–10.24, 10.57b

10.5 ALKYLATING TERMINAL ALKYNES

STEP 1 Draw the structures described in the problem.

STEP 2 Install the first alkyl group using sodium amide, followed by the appropriate alkyl halide.

STEP 3 Install the second alkyl group (if necessary).

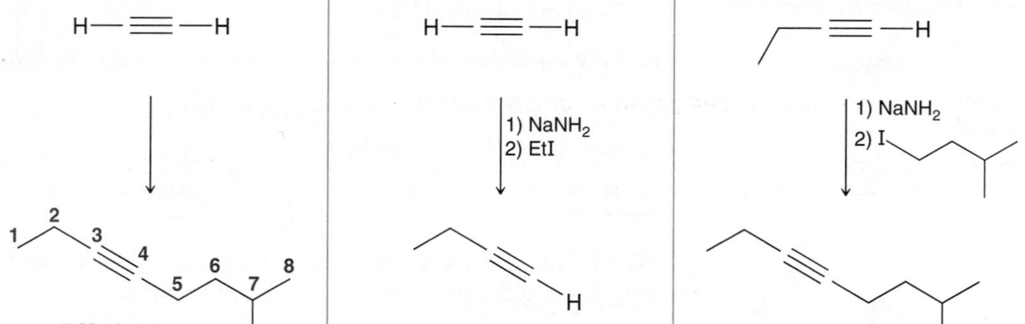

7-Methyl-3-octyne

Try Problems 10.28–10.31, 10.40f, 10.46, 10.50, 10.59

10.6 INTERCONVERTING ALKANES, ALKENES, AND ALKYNES

Reagents for interconversion.

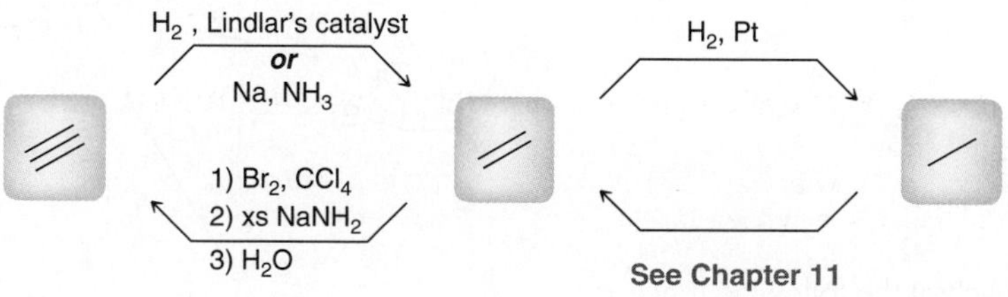

Example:

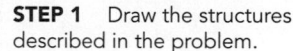

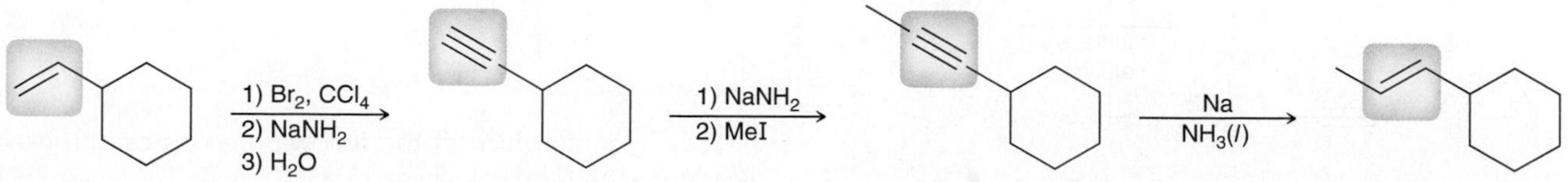

Try Problems 10.32–10.34, 10.57, 10.60

PRACTICE PROBLEMS

Note: Most of the Problems are available within WileyPLUS, an online teaching and learning solution.

10.35 Provide a systematic name for each of the following compounds:

(a) 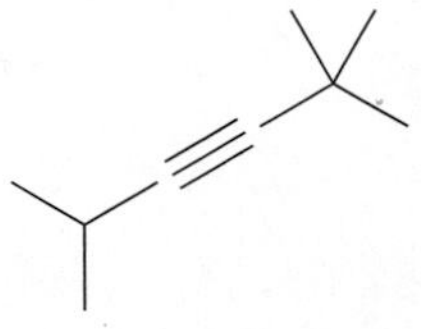

(b) 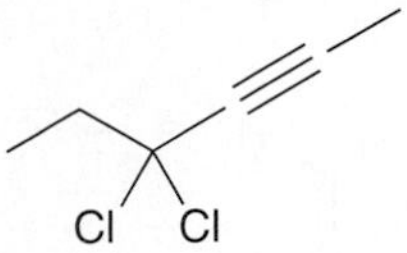

(c) 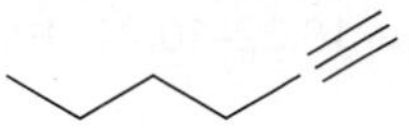

(d) 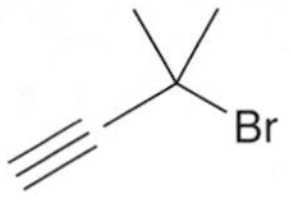

10.36 Draw a bond-line structure for each of the following compounds:

(a) 2-Heptyne

(b) 2,2-Dimethyl-4-octyne

(c) 3,3-Diethylcyclodecyne

10.37 Predict the products for each of the following reactions:

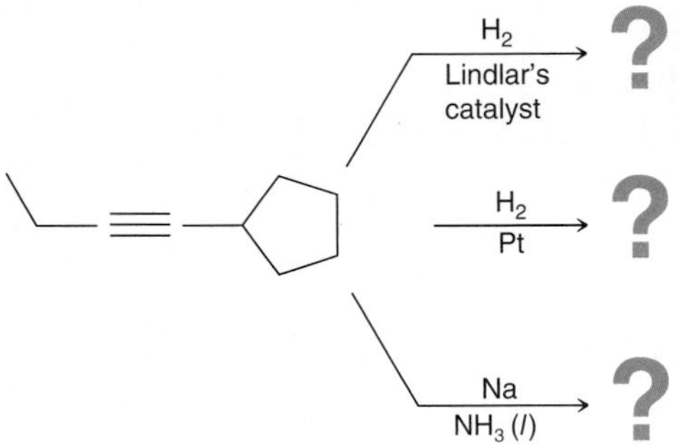

(a)

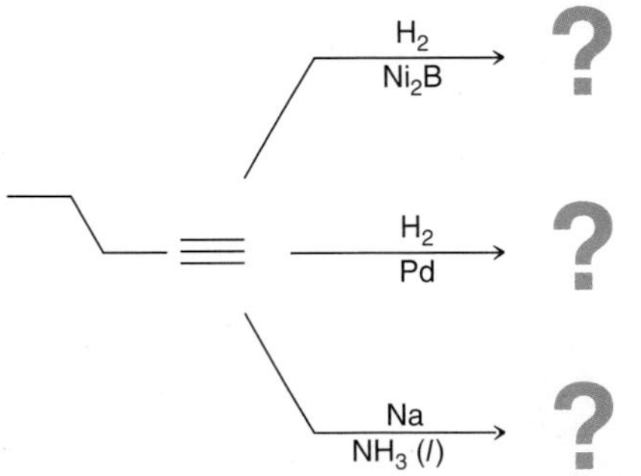

(b)

10.38 Predict the products for each of the following reactions:

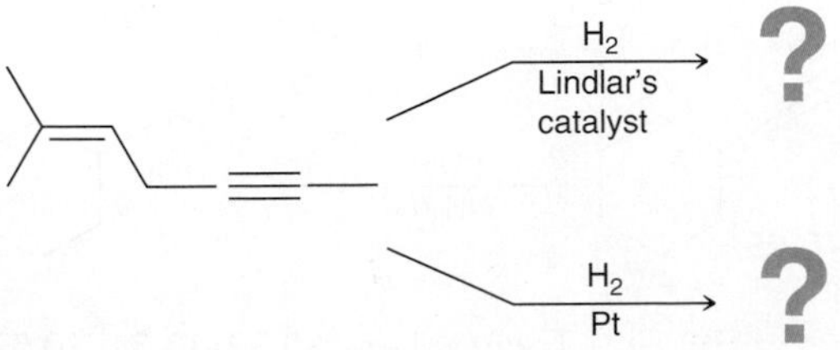

10.39 Draw the products of each of the following acid-base reactions, and then predict the position of equilibrium in each case:

(a) 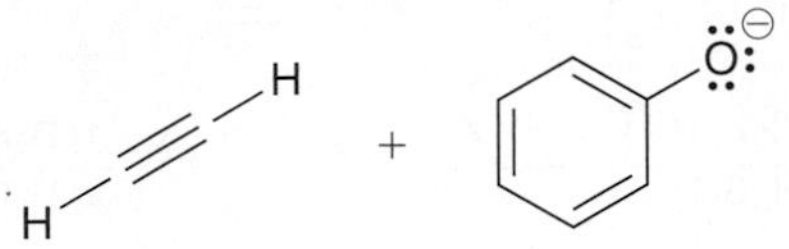

(b) 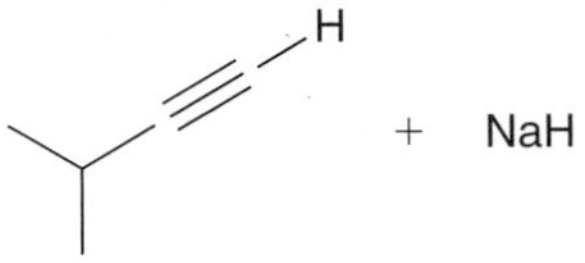

10.40 Predict the products obtained when 1-pentyne reacts with each of the following reagents:

(a) H_2SO_4, H_2O, $HgSO_4$

(b) 9-BBN followed by H_2O_2, NaOH

(c) Two equivalents of HBr

(d) One equivalent of HCl

(e) Two equivalents of Br_2 in CCl_4

(f) $NaNH_2$ in NH_3 followed by MeI

(g) H_2, Pt

10.41 Identify the reagents you would use to achieve each of the following transformations:

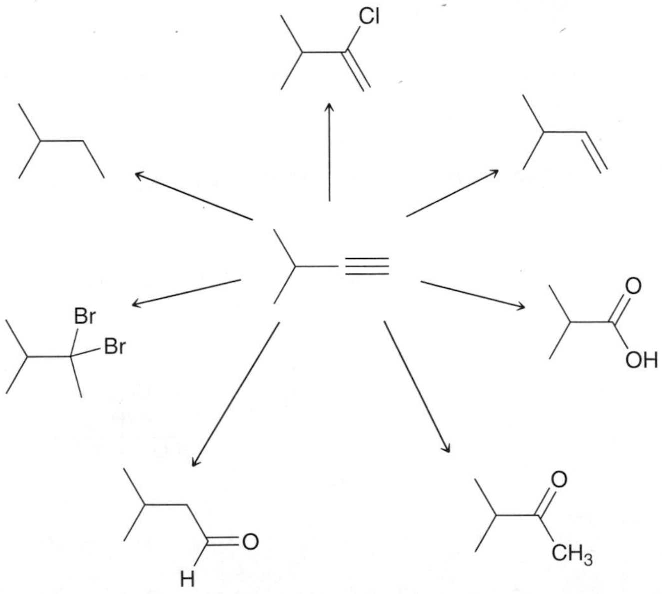

10.42 Identify which of the following bases can be used to deprotonate a terminal alkyne:

(a) $NaOCH_3$ (b) NaH (c) BuLi (d) NaOH (e) $NaNH_2$

10.43 Identify which of the following compounds represent a pair of keto-enol tautomers:

(a)
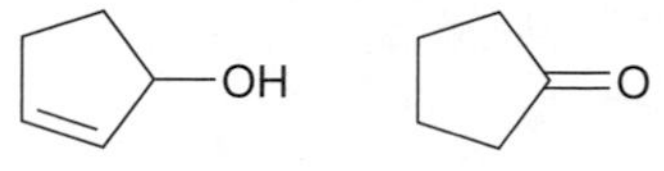

(b)
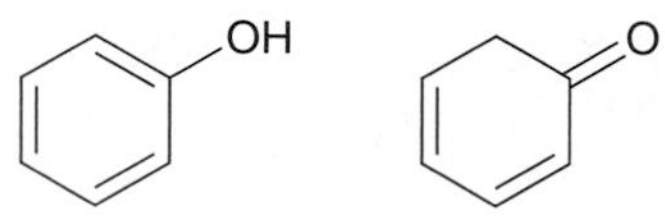

(c)
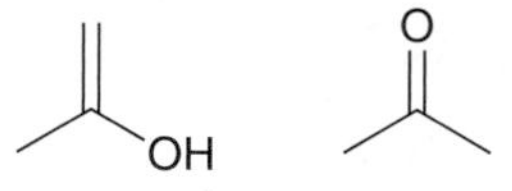

(d)
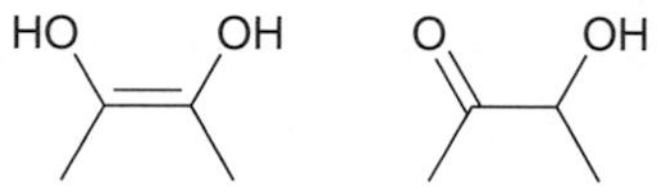

10.44 Draw the enol tautomer of each of the following ketones:

(a)
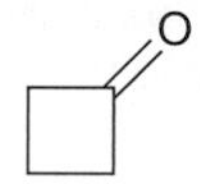

(b)
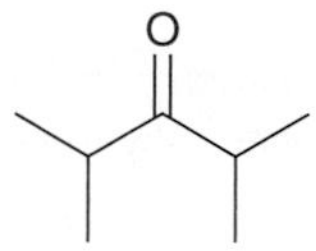

(c)
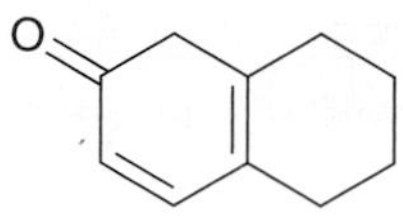

10.45 Oleic acid and elaidic acid are isomeric alkenes:

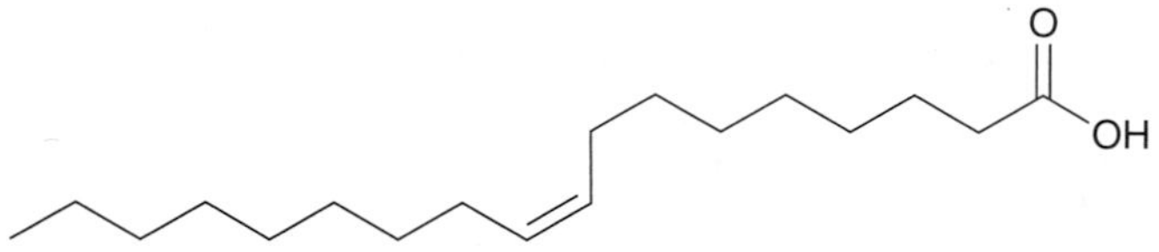

Oleic acid

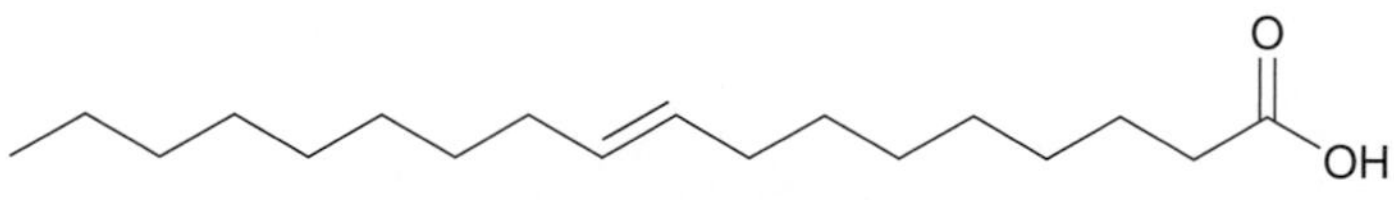

Elaidic acid

Oleic acid, a major component of butter fat, is a colorless liquid at room temperature. Elaidic acid, a major component of partially hydrogenated vegetable oils, is a white solid at room temperature. Oleic acid and elaidic acid can both be prepared in the laboratory by reduction of an alkyne called stearolic acid. Draw the structure of stearolic acid, and identify the reagents you would use to convert stearolic acid into oleic acid or elaidic acid.

10.46 Predict the final product(s) for each sequence of reactions:

(a)
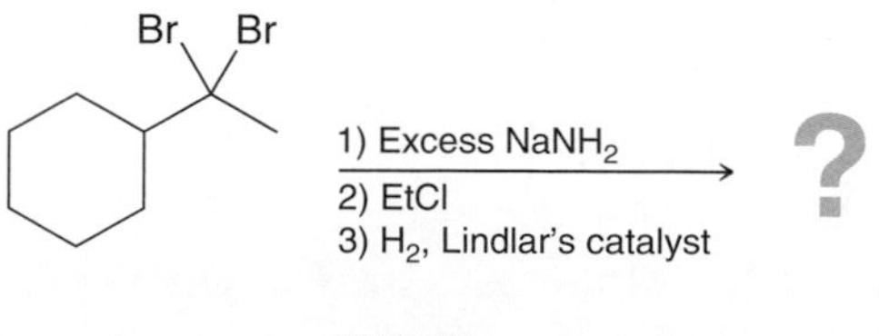

(b)
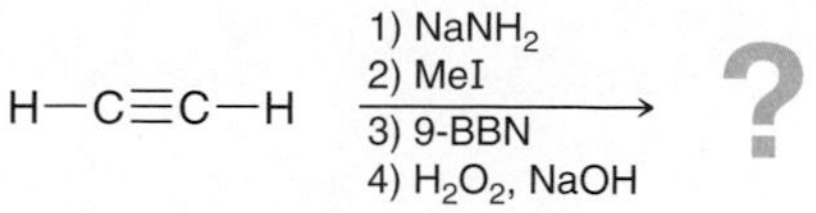

(c)
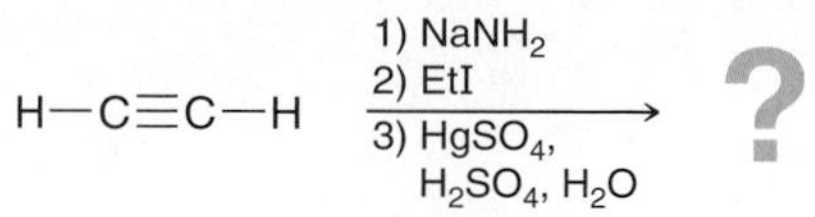

(d)
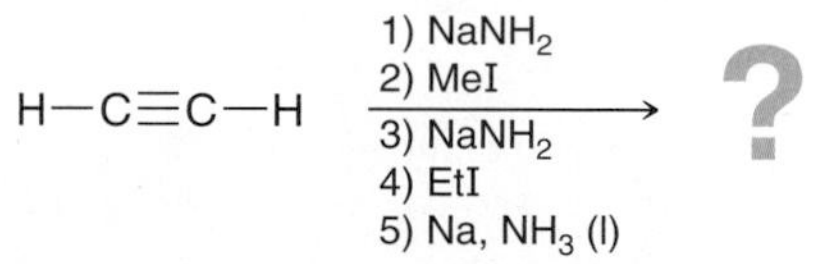

10.47 When (*R*)-4-bromohept-2-yne is treated with H_2 in the presence of Pt, the product is optically inactive. Yet, when (*R*)-4-bromohex-2-yne is treated with the same conditions, the product is optically active. Explain.

10.48 Draw the structure of an alkyne that can be converted into 3-ethylpentane upon hydrogenation. Provide a systematic name for this compound.

10.49 Propose a mechanism for each of the following transformations:

(a)
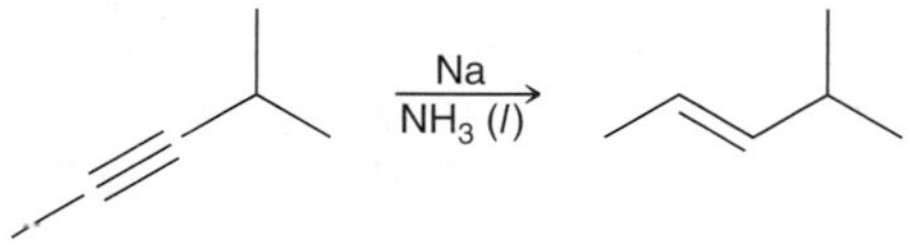

(b)
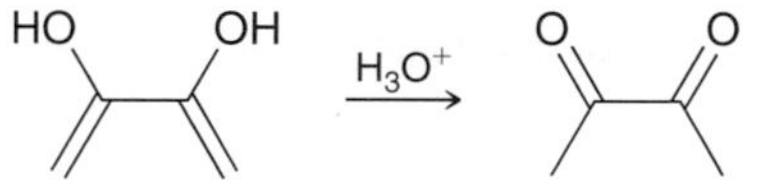

10.50 Draw the expected product of each of the following reactions, showing stereochemistry where appropriate:

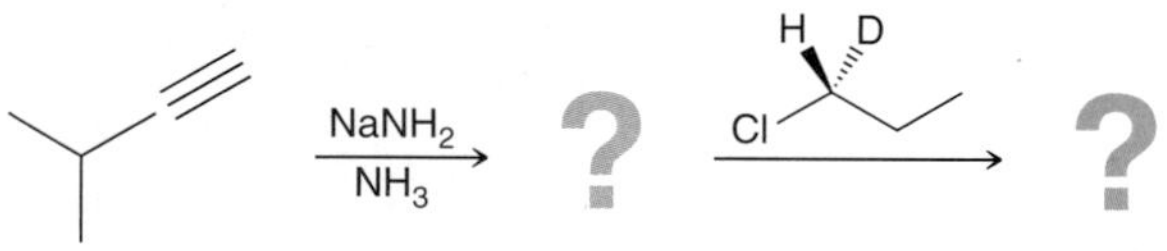

10.51 Compound A is an alkyne that reacts with two equivalents of H_2 in the presence of Pd to give 2,4,6-trimethyloctane.

(a) Draw the structure of compound A.

(b) How many chirality centers are present in compound A?

(c) Identify the locants for the methyl groups in compound A. Explain why the locants are not 2, 4, and 6 as seen in the product of hydrogenation.

10.52 Compound A has molecular formula C_7H_{12}. Hydrogenation of compound A produces 2-methylhexane. Hydroboration-oxidation of compound A produces an aldehyde. Draw the structure of compound A, and draw the structure of the aldehyde produced upon hydroboration-oxidation of compound A.

10.53 Propose a plausible synthesis for each of the following transformations:

(a)
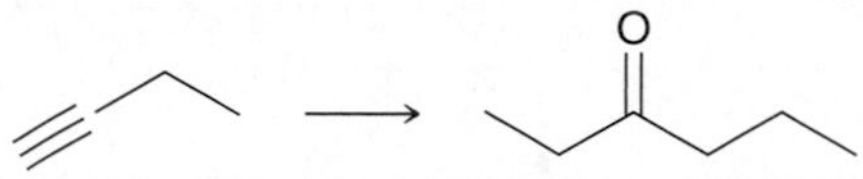

(b) 

(c) 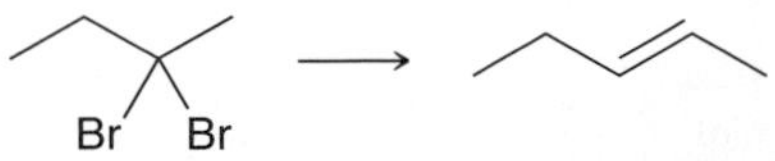

(d) 

(e) 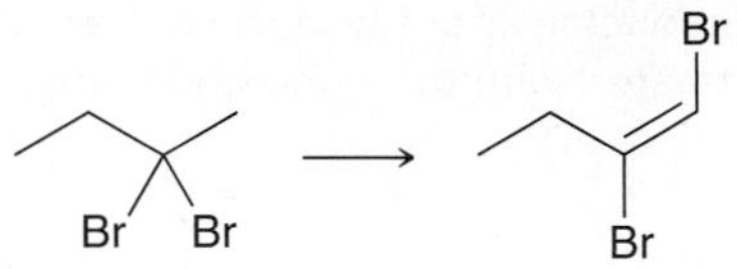

(f) 

10.54 1,2-Dichloropentane reacts with excess sodium amide in liquid ammonia to produce compound X. Compound X undergoes acid-catalyzed hydration to produce a ketone. Draw the structure of the ketone produced upon hydration of compound X.

10.55 An unknown alkyne is treated with ozone (followed by hydrolysis) to yield acetic acid and carbon dioxide. What is the structure of the alkyne?

10.56 Compound A is an alkyne with molecular formula C_5H_8. When treated with aqueous sulfuric acid and mercuric sulfate, two different products with molecular formula $C_5H_{10}O$ are obtained in equal amounts. Draw the structure of compound A, and draw the two products obtained.

10.57 Propose a plausible synthesis for each of the following transformations:

(a) 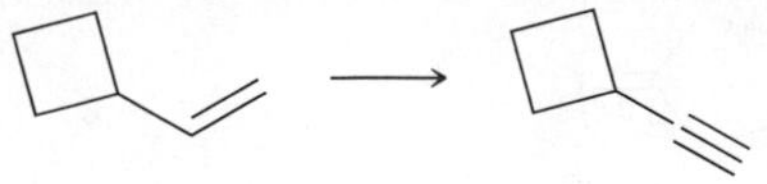

(b) 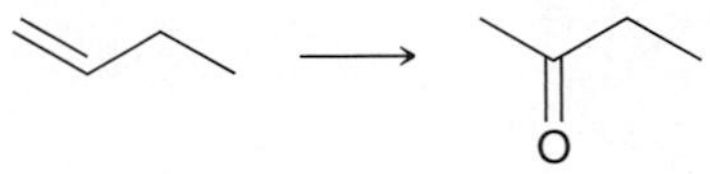

(c) 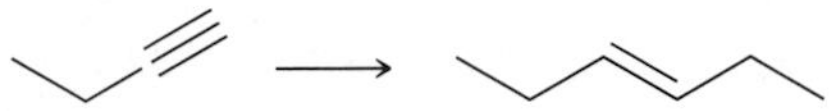

(d) 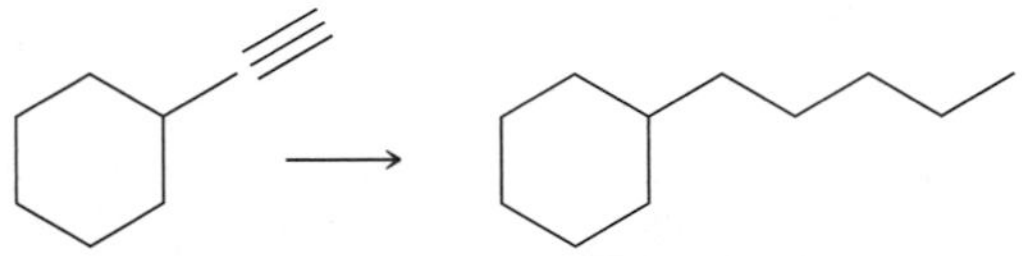

10.58 Draw the structure of each possible dichloride that can be used to prepare the following alkyne via elimination:

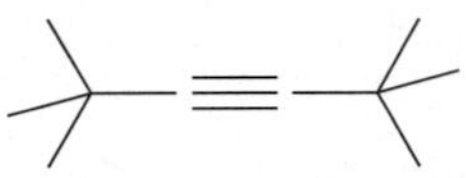

10.59 Draw the structures of compounds **A** to **D**:

A (C_6H_{12}) $\xrightarrow{Br_2}$ **B** $\xrightarrow[2)\ H_2O]{1)\ \text{Excess } NaNH_2}$ **C** $\xrightarrow{NaNH_2}$

D 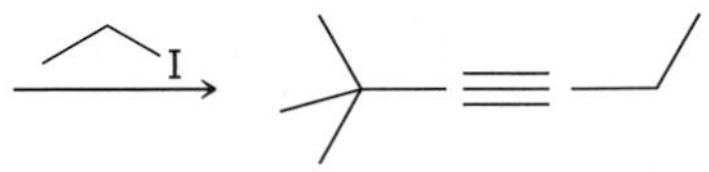

INTEGRATED PROBLEMS

10.60 Identify the reagents necessary to achieve each transformation below. In each case, you will need to use at least one reaction from this chapter and at least one reaction from the previous chapter. The essence of each problem is to choose reagents that will achieve the desired stereochemical outcome:

(a) 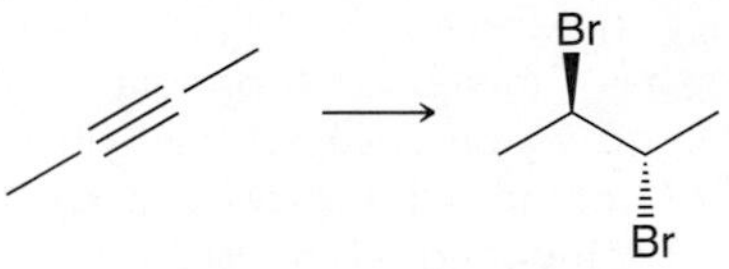

(b) 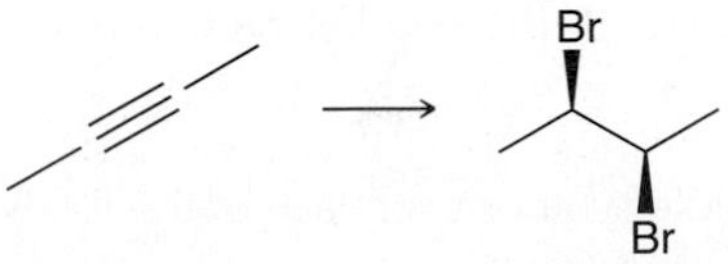 + En

(c) 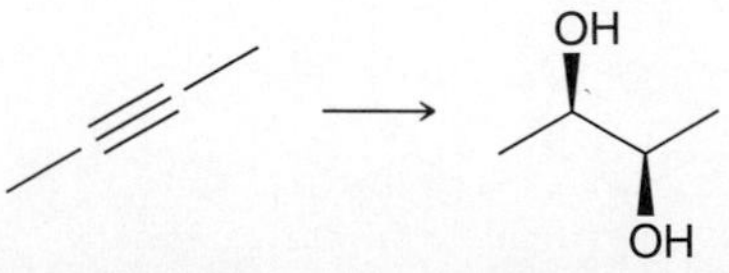 + En

(d) 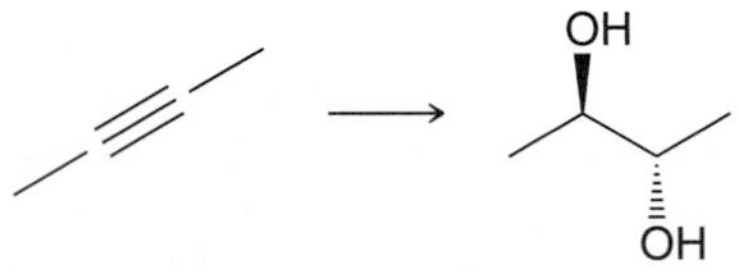

(e) 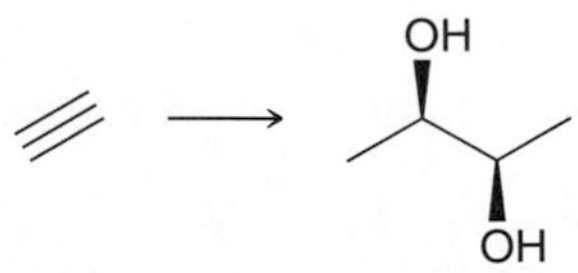 + En

(f) 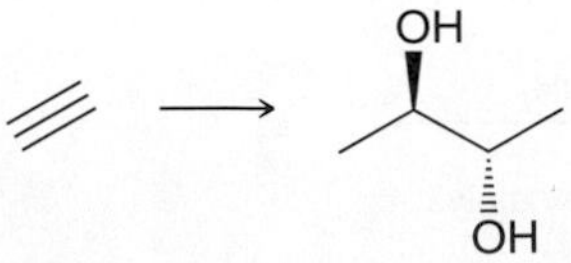

10.61 Identify the reagents necessary to achieve each transformation below:

(a) 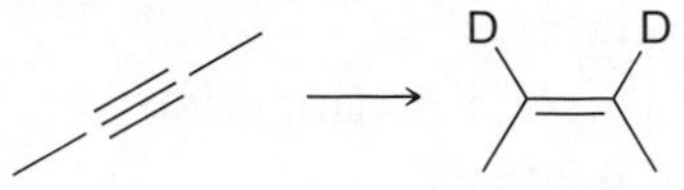

(b) 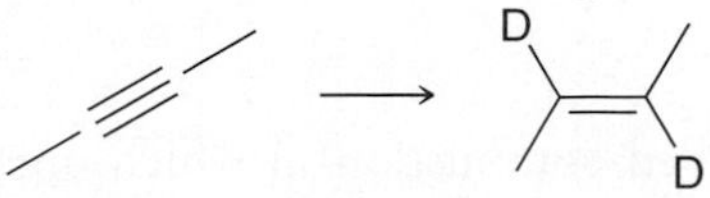

(c) 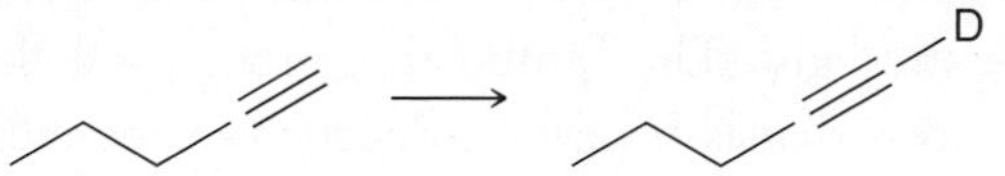

10.62 The following reaction does not produce the desired product, but does produce a product that is a constitutional isomer of the desired product. Draw the product that is obtained, and propose a mechanism for its formation:

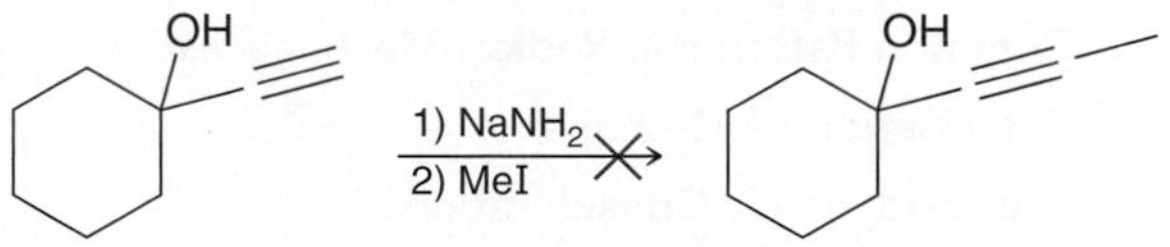

CHALLENGE PROBLEMS

10.63 Propose a plausible mechanism for the following transformation:

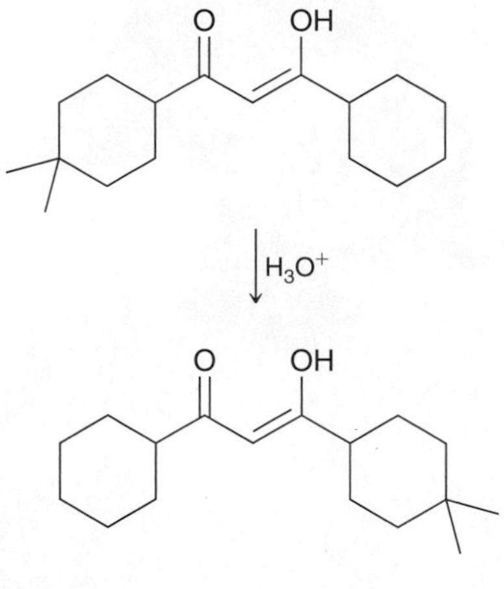

10.64 Propose a plausible mechanism for the following tautomerization process:

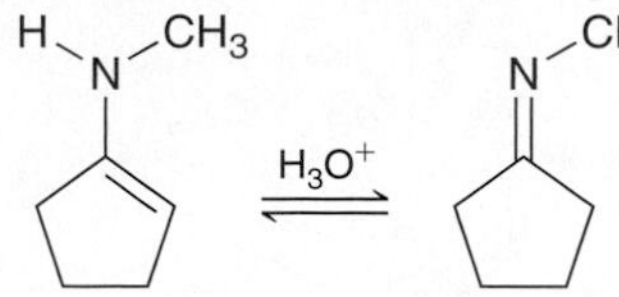

10.65 Using acetylene and methyl bromide as your only sources of carbon atoms, propose a synthesis for each of the following compounds:

(a) 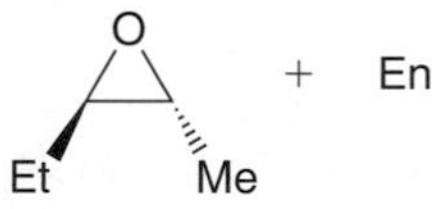

(b) 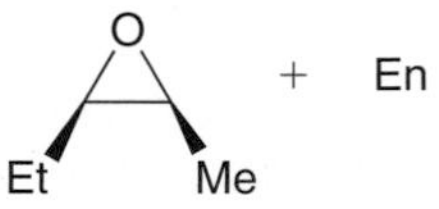

10.66 Propose a plausible mechanism for the following transformation:

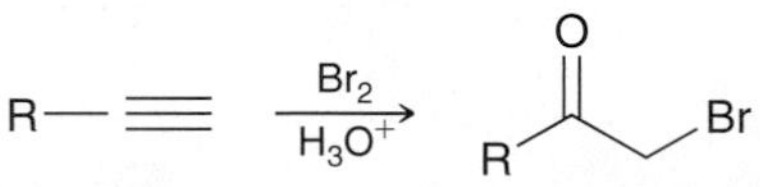

10.67 Propose a plausible mechanism for the following transformation:

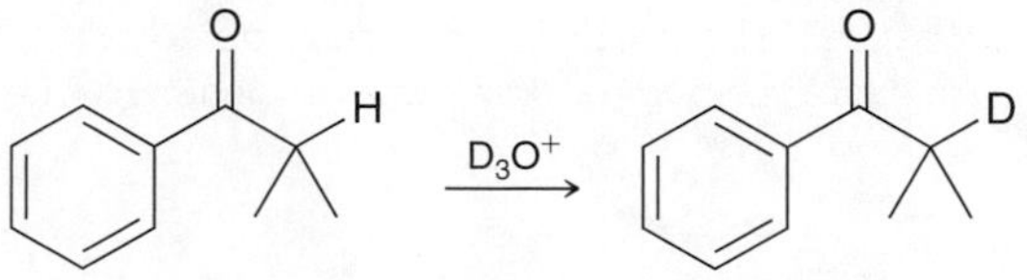

11 Radical Reactions

DID YOU EVER WONDER…

how certain chemicals are able to extinguish fires more successfully than water?

Fire is a chemical reaction called combustion in which organic compounds are converted into CO_2 and water together with the liberation of heat and light. The combustion process is a chain (self-perpetuating) process that is believed to occur via free radical intermediates. Understanding the nature of these free radicals is the key to understanding how fires can be extinguished most effectively.

This chapter will focus on radicals. We will learn about their structure and reactivity, and we will explore some of the important roles that radicals play in the food and chemical industries and in our overall health. We will also return to the topic of fire to explain how specially designed chemicals are able to destroy the radical intermediates in a fire, thereby stopping the combustion process and extinguishing the fire.

DO YOU REMEMBER?

Before you go on, be sure you understand the following topics.
If necessary, review the suggested sections to prepare for this chapter.

- Enthalpy (Section 6.1)
- Entropy (Section 6.2)
- Gibbs Free Energy (Section 6.3)
- Reading Energy Diagrams (Section 6.6)

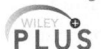

11.1 Radicals

Introduction to Radicals

In Section 6.1, we mentioned that a bond can be broken in two different ways: heterolytic bond cleavage forms ions, while homolytic bond cleavage forms radicals (Figure 11.1). Until now, we have focused mostly on ionic reactions—that is, we have been exploring mechanisms that involve ions. This chapter will focus exclusively on radicals. Look carefully at the curved arrows

FIGURE 11.1 An illustration of the difference between homolytic and heterolytic bond cleavage.

used in the two processes above. Specifically, an ionic process employs double-barbed curved arrows, while a radical process employs single-barbed arrows (Figure 11.2). A double-barbed curved arrow represents the motion of two electrons, while a single-barbed arrow represents the motion of only one electron. Single-barbed arrows are called **fishhook arrows** because of their appearance. This chapter will use fishhook arrows exclusively.

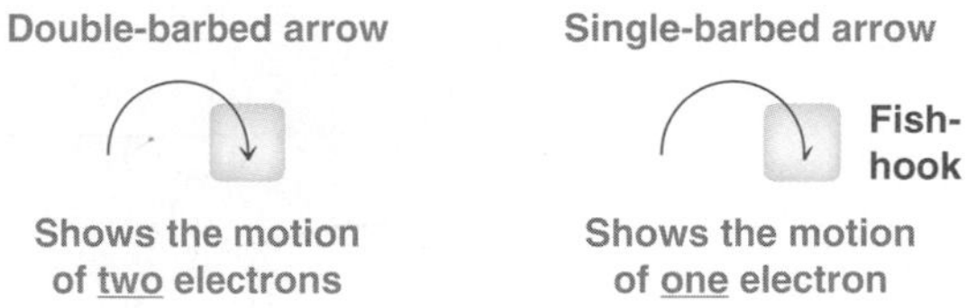

FIGURE 11.2 Arrows used in ionic mechanisms are double-barbed, while arrows used in radical mechanisms are single-barbed.

Structure and Geometry of Radicals

In order to understand the structure and geometry of a radical, we must quickly review the structures of carbocations and carbanions (Figure 11.3). A carbocation is sp^2 hybridized and

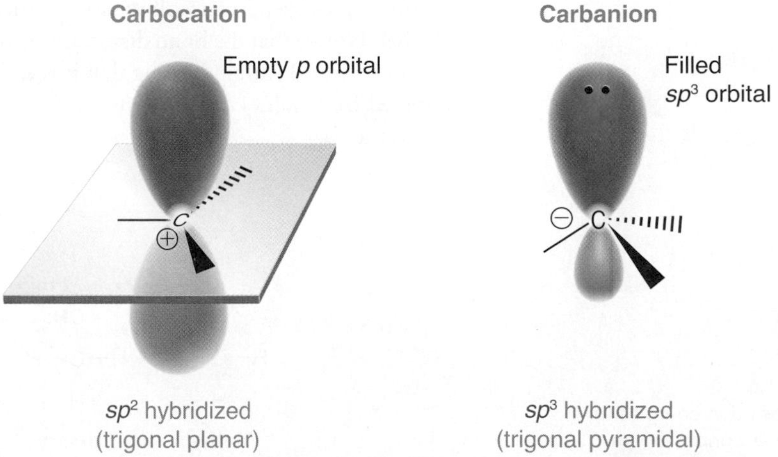

FIGURE 11.3 A comparison of the geometries of a carbocation and a carbanion.

trigonal planar, while a carbanion is sp^3 hybridized and trigonal pyramidal. The difference in geometry results from the difference in the number of nonbonding electrons. A carbocation has zero nonbonding electrons, while a carbanion has two nonbonding electrons. A carbon radical is between these two cases, because it has one nonbonding electron. It therefore might seem reasonable to expect that the geometry of a carbon radical would be somewhere in between trigonal planar and trigonal pyramidal. Experiments suggest that carbon radicals either are trigonal planar or exhibit a very shallow pyramid (nearly planar) with a very low barrier to inversion (Figure 11.4). Either way, carbon radicals can be treated as trigonal planar entities. This geometry will have significance later in the chapter when we deal with the stereochemical outcomes of radical reactions.

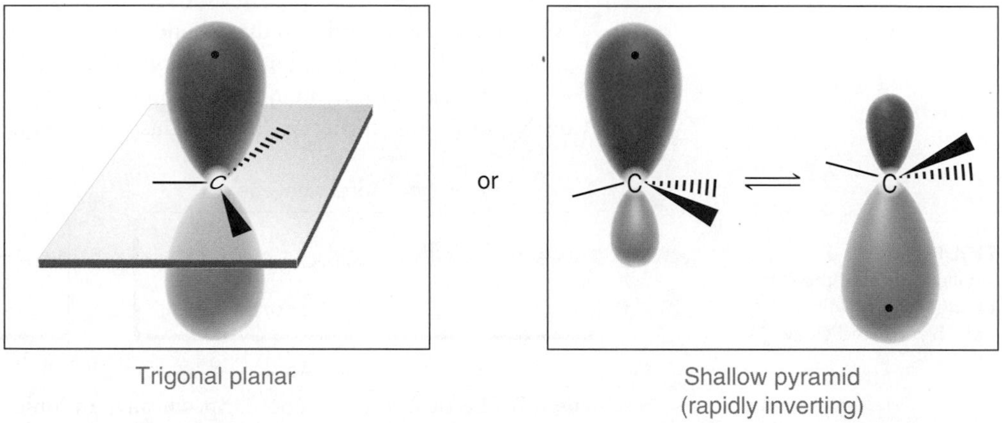

FIGURE 11.4 The geometry of a carbon radical.

LOOKING BACK
For a review of carbocation stability and hyperconjugation, see Section 6.8.

The order of stability for radicals follows the same trend exhibited by carbocations in which tertiary radicals are more stable than secondary radicals, which in turn are more stable than primary radicals (Figure 11.5). The explanation for this trend is similar to the explanation for

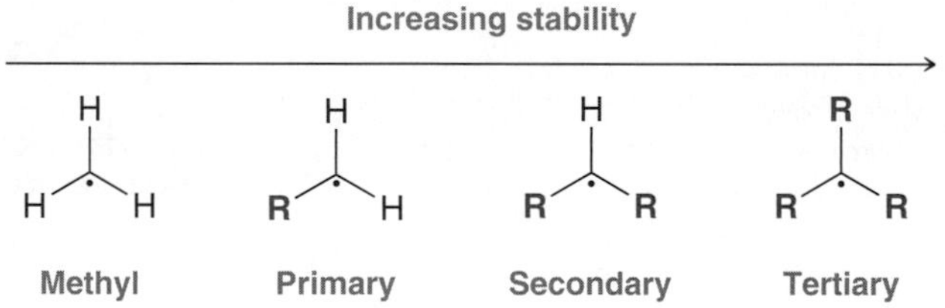

FIGURE 11.5 The order of stability for carbon radicals.

carbocation stability. Specifically, alkyl groups are electron donating (because of hyperconjugation), and they are able to donate electron density into the orbital containing the unpaired electron. This trend in stability is supported by a comparison of bond dissociation energies (BDEs). Notice that the bond dissociation energy is lowest for a C—H bond involving a tertiary carbon (Figure 11.6), indicating that it is easiest to cleave this C—H bond homolytically. These observed BDE values suggest that a tertiary radical is about 8 kJ/mol more stable than a secondary radical.

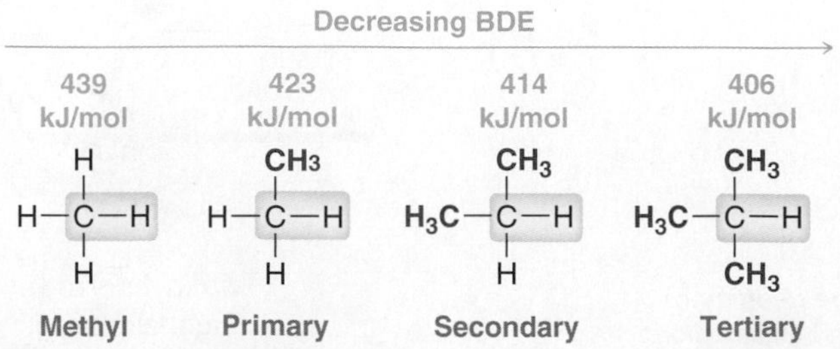

FIGURE 11.6 Bond dissociation energies for various C—H bonds.

CONCEPTUAL CHECKPOINT

11.1 Rank the following radicals in order of stability:

(a) 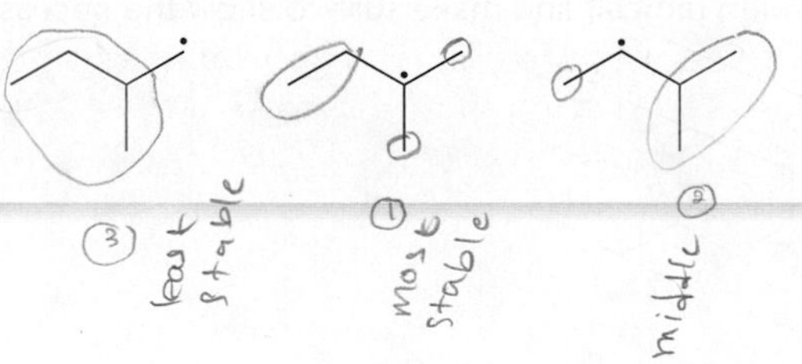

(b) 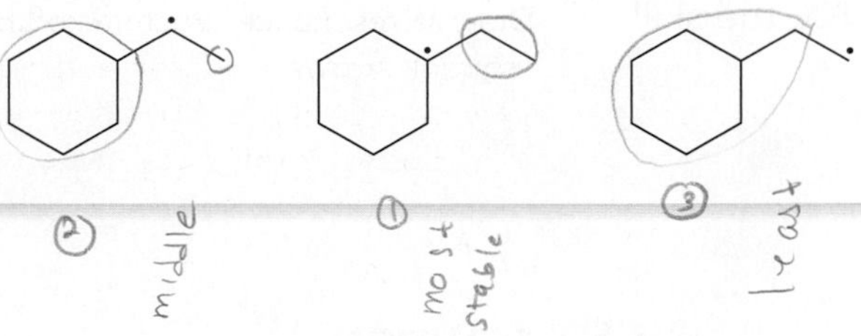

Resonance Structures of Radicals

In Chapter 6, we saw five patterns for drawing resonance structures. There are also several patterns for drawing resonance structures of radicals. However, the vast majority of situations require only one pattern, which is characterized by an unpaired electron next to a π bond, in an allylic position.

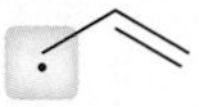

Allylic position

In such a case, the unpaired electron is resonance stabilized, and three fishhook arrows are used to draw the resonance structure.

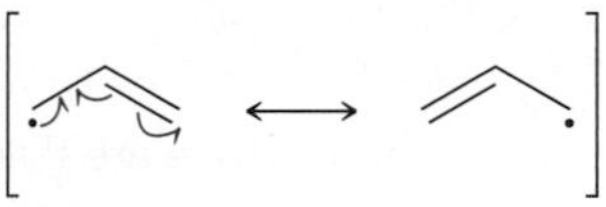

Benzylic positions also exhibit the same pattern, with several resonance structures contributing to the overall resonance hybrid.

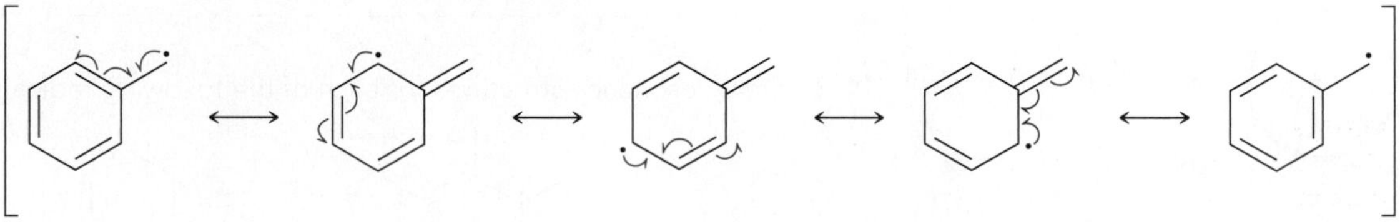

Resonance-stabilized radicals are even more stable than tertiary radicals, as can be seen by comparing the BDE values in Figure 11.7. A C—H bond at a benzylic or allylic position is more easily cleaved than a C—H bond at a tertiary position. These observed BDE values suggest that resonance-stabilized radicals are about 40–50 kJ/mol more stable than tertiary radicals. This is a large difference when compared with the 8 kJ difference between secondary and tertiary radicals.

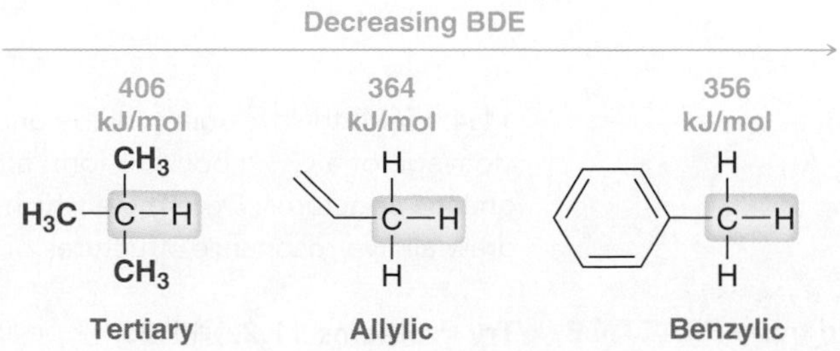

FIGURE 11.7 Bond dissociation energies for various C—H bonds.

SKILLBUILDER

11.1 DRAWING RESONANCE STRUCTURES OF RADICALS

LEARN the skill

Draw all resonance structures of the following radical, and make sure to show the necessary fishhook arrows:

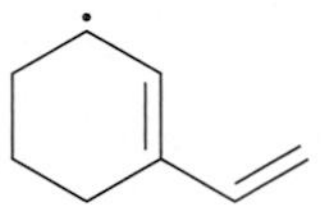

SOLUTION

STEP 1
Look for an unpaired electron next to a π bond.

STEP 2
Draw three fishhook arrows, and then draw the corresponding resonance structure.

When drawing resonance structures of a radical, look for an unpaired electron next to a π bond. In this case, the unpaired electron is in fact located at an allylic position, so we draw the following three fishhook arrows:

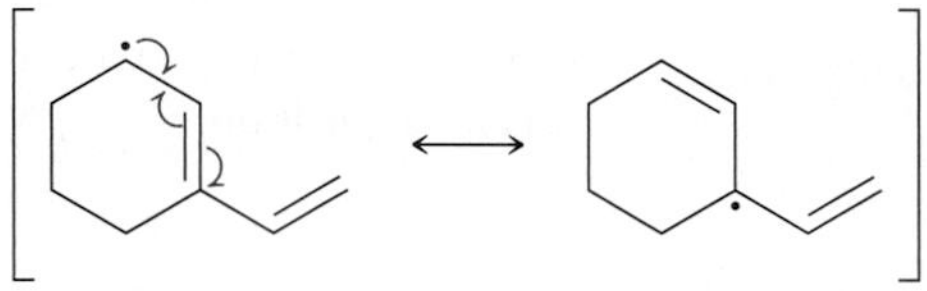

The new resonance structure exhibits an unpaired electron that is allylic to another π bond, so again, we draw three fishhook arrows to arrive at another resonance structure:

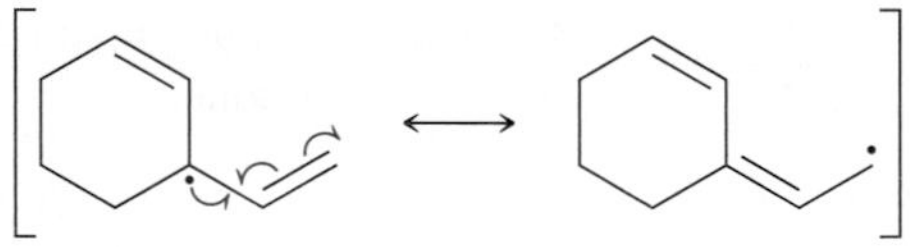

In total, there are three resonance structures:

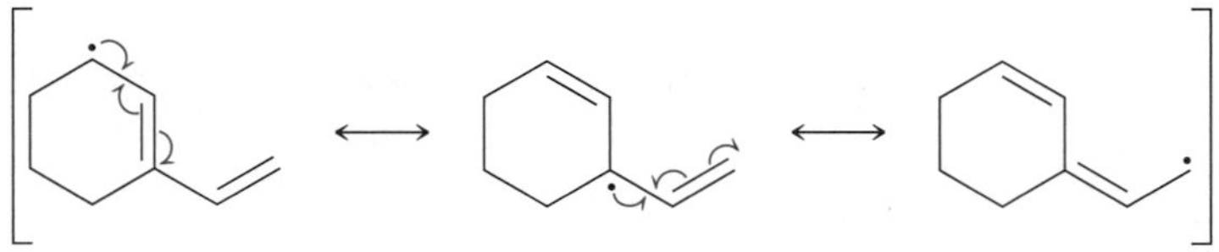

PRACTICE the skill

11.2 Draw resonance structures for each of the following radicals:

(a) 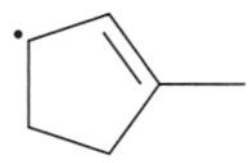

(b) 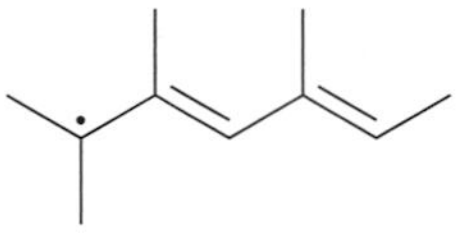

(c) 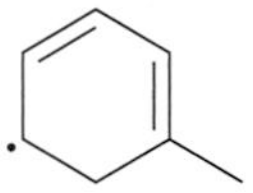

(d) 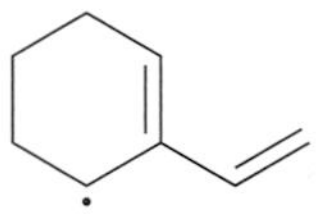

APPLY the skill

11.3 The triphenylmethyl radical was the first radical to be observed. Draw all resonance structures of this radical, and explain why this radical is unusually stable:

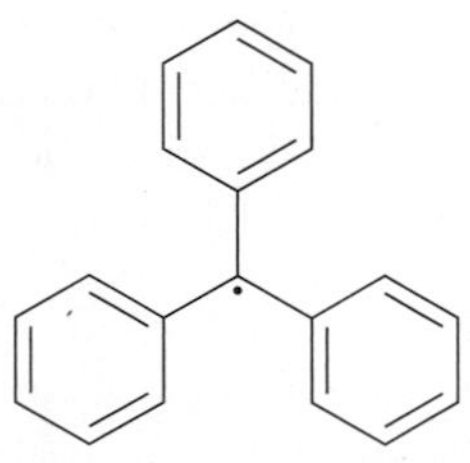

11.4 5-Methylcyclopentadiene undergoes homolytic bond cleavage of a C—H bond to form a radical that exhibits five resonance structures. Determine which hydrogen is abstracted and draw all five resonance structures of the resulting radical.

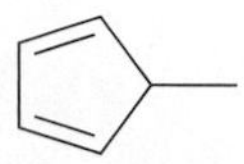

5-Methylcyclopentadiene

need more **PRACTICE?** **Try Problems 11.22, 11.25**

In this section, we have seen that allylic and benzylic radicals are stabilized by resonance. When analyzing the stability of a radical, make sure not to confuse allylic and vinylic positions:

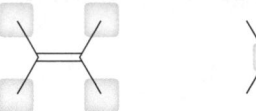

Allylic Positions
Vinylic Positions

A vinylic radical is not resonance stabilized and does not have a resonance structure:

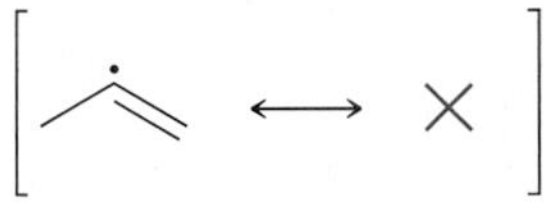

In fact, a vinylic radical is even less stable than a primary radical. This can be seen by comparing BDE values (Figure 11.8). A C—H bond at a vinylic position requires even more energy to cleave than a C—H bond at a primary position. These observed BDE values suggest that a vinylic radical is more than 40 kJ/mol higher in energy than a primary radical (a very large difference).

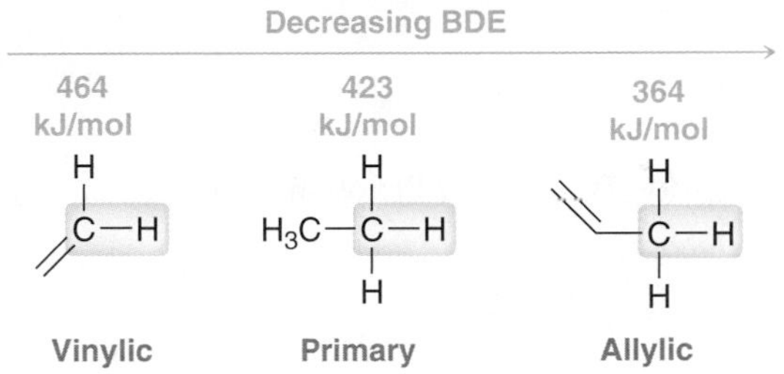
Decreasing BDE
464 kJ/mol
423 kJ/mol
364 kJ/mol
Vinylic
Primary
Allylic

FIGURE 11.8
Bond dissociation energies for various C—H bonds.

SKILLBUILDER

11.2 IDENTIFYING THE WEAKEST C—H BOND IN A COMPOUND

LEARN the skill

Identify the weakest C—H bond in the following compound:

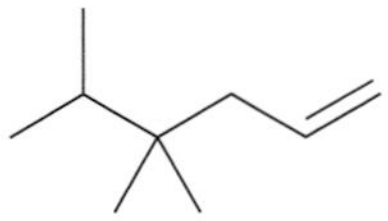

SOLUTION

Imagine homolytically breaking each different type of C—H bond in the compound. This would produce the following radicals:

STEP 1
Consider all possible radicals that would result from homolytic cleavage of a C—H bond.

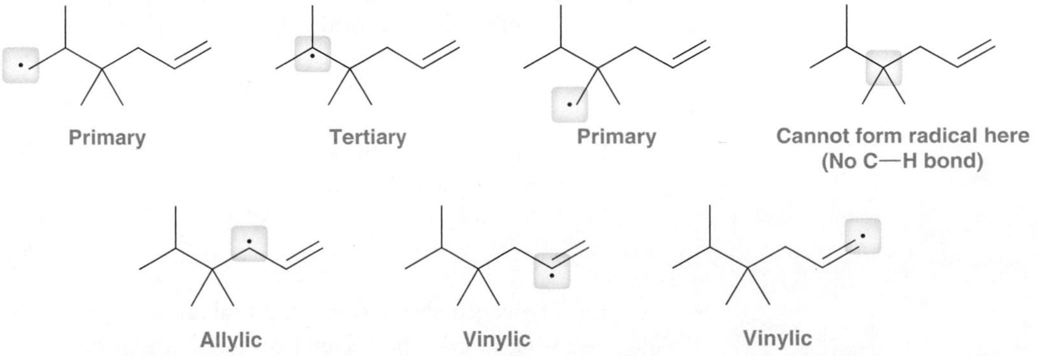
Primary
Tertiary
Primary
Cannot form radical here (No C—H bond)
Allylic
Vinylic
Vinylic

STEP 2
Identify the most stable radical.

Of all the possibilities, the allylic radical will be the most stable radical. The weakest C—H bond is the one that leads to the most stable radical. Therefore, the C—H bond at the allylic position is the weakest C—H bond. It is the easiest bond to break homolytically:

STEP 3
Determine which bond is the weakest.

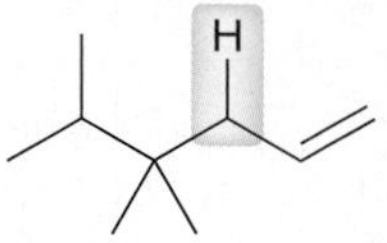

PRACTICE the skill

11.5 Identify the weakest C—H bond in each of the following compounds:

(a) 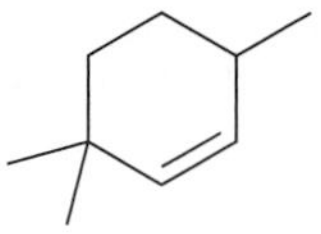 (b) 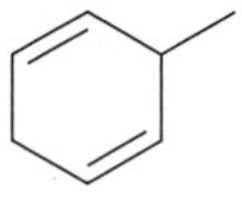 (c) 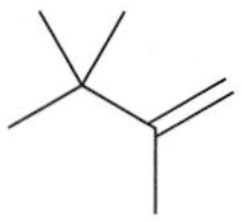 (d) 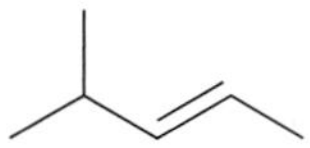

APPLY the skill

11.6 The C—H bonds shown in red exhibit very similar BDEs, because homolytic cleavage of either bond results in a resonance-stabilized radical. Nevertheless, one of these C—H bonds is weaker than the other. Identify the weaker bond, and explain your choice:

need more **PRACTICE?** **Try Problem 11.23**

11.2 Common Patterns in Radical Mechanisms

In Chapter 6, we saw that ionic mechanisms are comprised of only four different kinds of arrow-pushing patterns (nucleophilic attack, loss of a leaving group, proton transfer, and rearrangement). Similarly, radical mechanisms are also comprised of only a few different kinds of arrow-pushing patterns, although these patterns are very different from the patterns in ionic mechanisms. For example, carbocations can undergo rearrangement (as seen in Section 6.11), but radicals are not observed to undergo rearrangement:

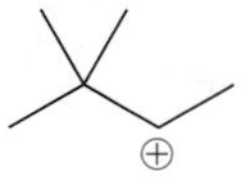

This carbocation will rearrange to produce a more stable tertiary carbocation

This radical will not rearrange to produce a more stable tertiary radical

There are six different kinds of arrow-pushing patterns that comprise radical reactions. We will now explore these patterns one by one:

1. **Homolytic cleavage**: Homolytic cleavage requires a large input of energy. This energy can be supplied in the form of heat (Δ) or light (*hν*).

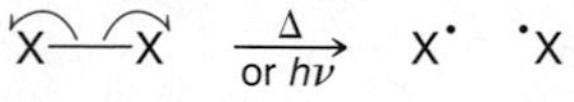

2. **Addition to a π bond**: A radical adds to a π bond, thereby destroying the π bond and generating a new radical.

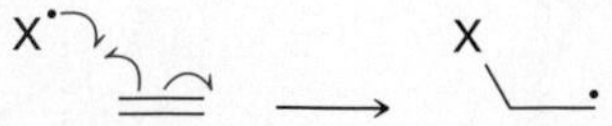

3. **Hydrogen abstraction**: A radical can abstract a hydrogen atom from a compound, generating a new radical. Don't confuse this step with a proton transfer, which is an ionic step.

In a proton transfer, only the nucleus of the hydrogen atom (a proton, H^+) is being transferred. Here, the entire hydrogen atom (proton and electron, H^{\bullet}) is being transferred from one location to another.

$$X^{\bullet} \quad H-R \longrightarrow X-H \quad {}^{\bullet}R$$

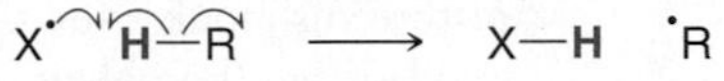

4. **Halogen abstraction**: A radical can abstract a halogen atom, generating a new radical. This step is similar to hydrogen abstraction, only a halogen atom is being abstracted instead of a hydrogen atom.

$$R^{\bullet} \quad X-X \longrightarrow R-X \quad {}^{\bullet}X$$

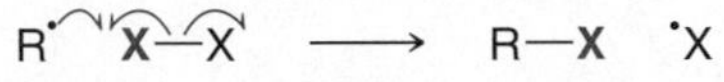

5. **Elimination**: The position bearing the unpaired electron is called the alpha position. In an elimination step, a double bond forms between the alpha (α) and beta (β) positions. As a result, a single bond at the β position is cleaved, causing the compound to fragment into two pieces.

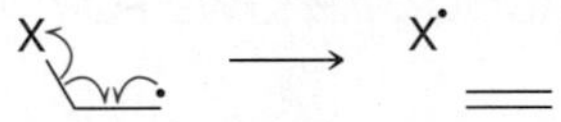

6. **Coupling**: Two radicals join together and form a bond.

$$X^{\bullet} \quad {}^{\bullet}X \longrightarrow X-X$$

Figure 11.9 summarizes the six common steps in radical mechanisms. This might seem like a lot to remember. It is therefore helpful to group them. For example, notice that the first step (homolytic cleavage) and the sixth step (coupling) are just the reverse of each other. Homolytic cleavage creates radicals, while coupling destroys them:

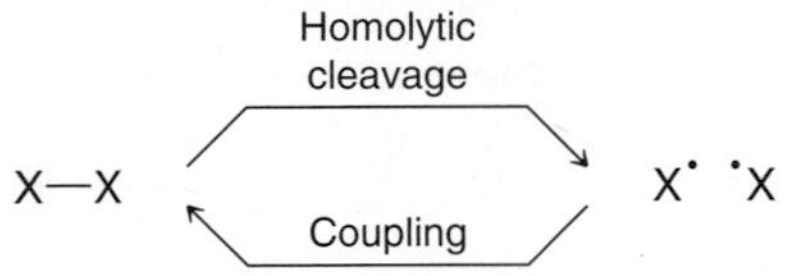

The second and fifth steps shown in Figure 11.9 are also the reverse of each other. Addition to a π bond is just the reverse of elimination:

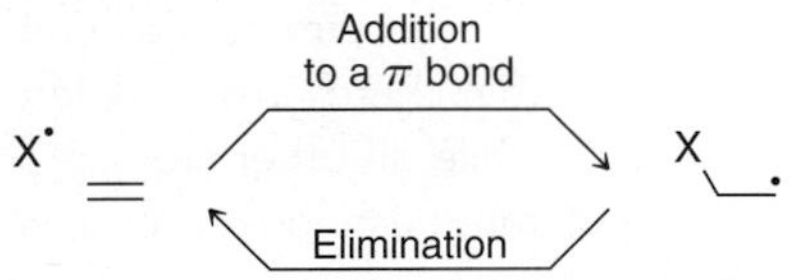

That leaves only two more steps, both of which are called abstractions (hydrogen abstraction and halogen abstraction). These two steps are not the reverse of each other. The reverse of a hydrogen abstraction is simply a hydrogen abstraction.

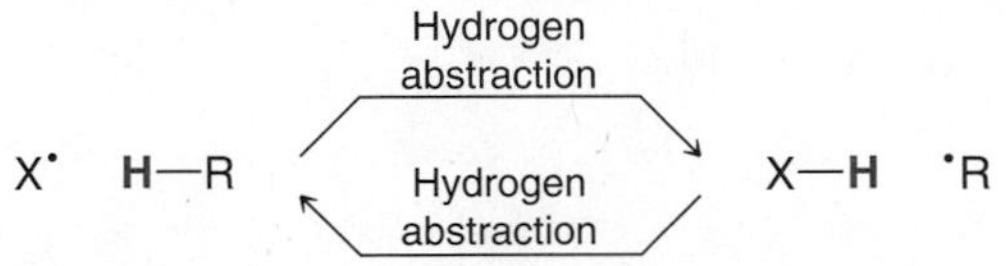

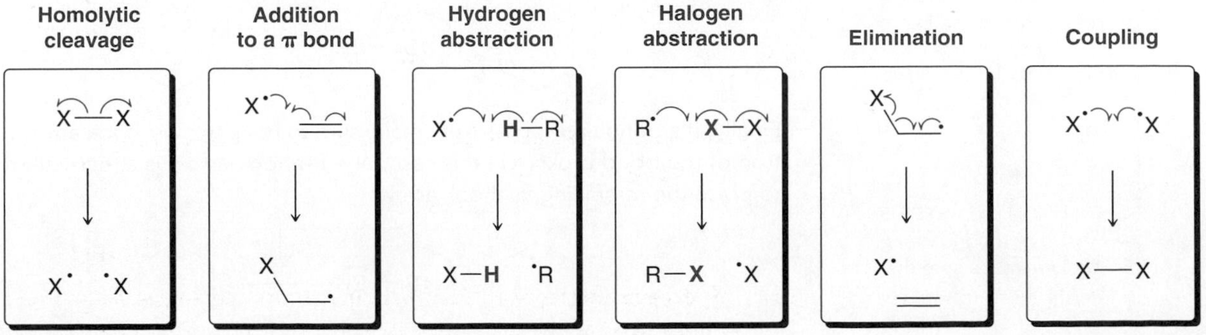

FIGURE 11.9
Six steps common in radical mechanisms.

The same is true for halogen abstraction. That is, the reverse of a halogen abstraction is simply a halogen abstraction.

When drawing a radical mechanism, remember that every step will generally have either two or three fishhook arrows. Take another close look at the six steps shown in Figure 11.9. The first and last step (homolytic cleavage and coupling) each require exactly two fishhook arrows, because two electrons are moving in each case. In contrast, all other steps have three electrons moving and therefore require three fishhook arrows. In any step, the number of fishhook arrows must correspond to the number of electrons moving. Let's get some practice drawing fishhook arrows.

SKILLBUILDER

11.3 DRAWING FISHHOOK ARROWS FOR A RADICAL PROCESS

LEARN the skill

Draw the appropriate fishhook arrows for the following radical process:

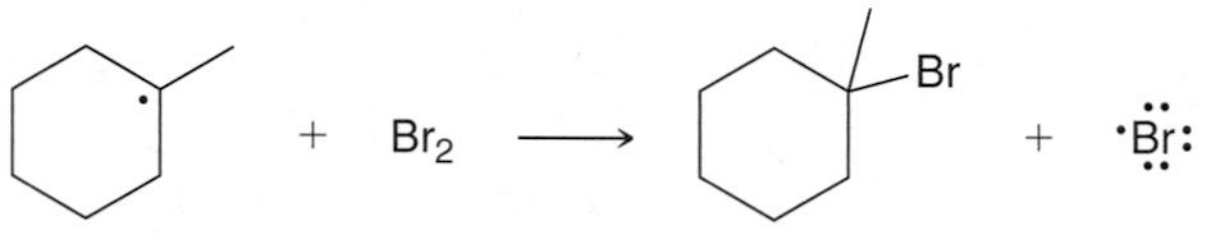

SOLUTION

The following procedure can be used for all six types of radical steps (not just the one in this problem). First identify the type of process taking place. A radical is reacting with Br_2, and the result is the transfer of a bromine atom:

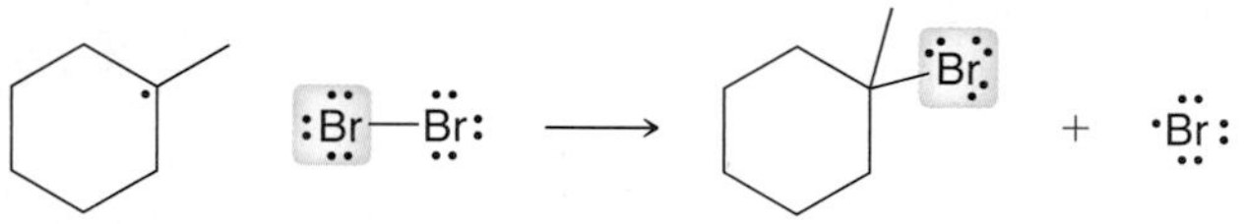

STEP 1 Identify the process.

Therefore, this step is a halogen abstraction.

STEP 2 Determine the number of fishhook arrows required.

Determine the number of fishhook arrows that must be used (either two or three). Recall that there are six different kinds of steps. Cleavage and coupling require two fishhooks, while all other steps require three fishhook arrows. This process (halogenation) must have *three* fishhook arrows, which means that three electrons are moving. Each of the fishhooks will have its tail placed on one of the electrons that is moving.

STEP 3 Identify any bonds being broken or formed.

Next, identify any bonds being broken or formed:

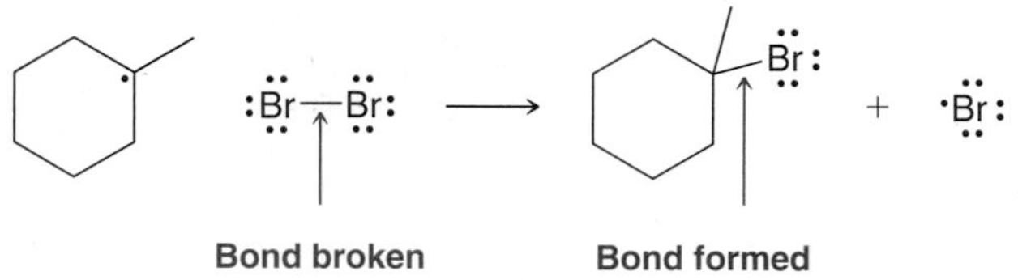

STEP 4 For a bond being formed, draw two fishhook arrows.

If a bond is being formed, draw the two fishhook arrows that show bond formation:

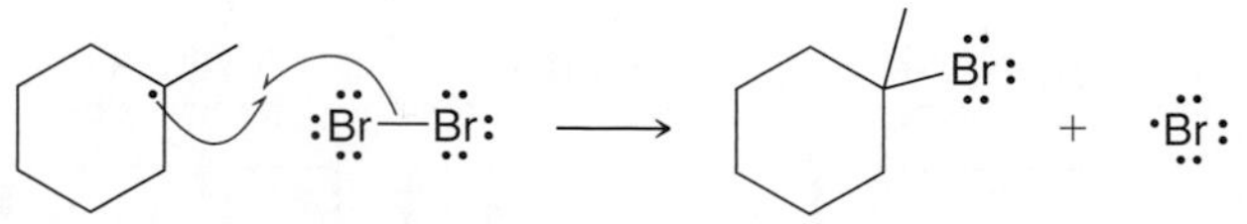

STEP 5 For a bond being broken, make sure that two fishhook arrows are shown.

Finally, if a bond is being broken, make sure to have two fishhook arrows, one for each electron of the bond broken. In this case, one fishhook arrow is already there, so we only need to place the remaining fishhook arrow:

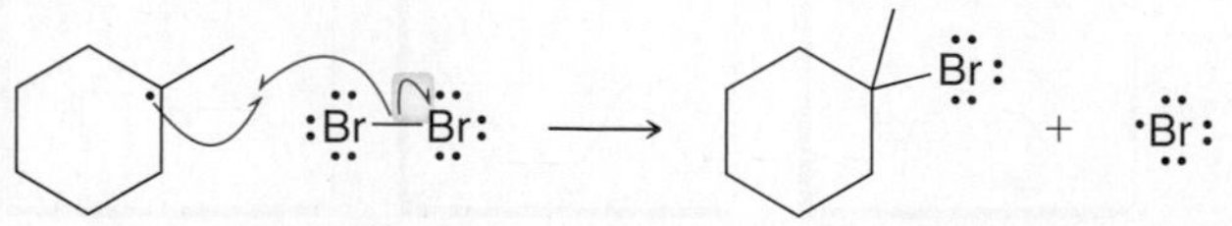

PRACTICE the skill **11.7** For each of the following reactions identify the type of radical process involved, and draw the appropriate fishhook arrows:

(a) 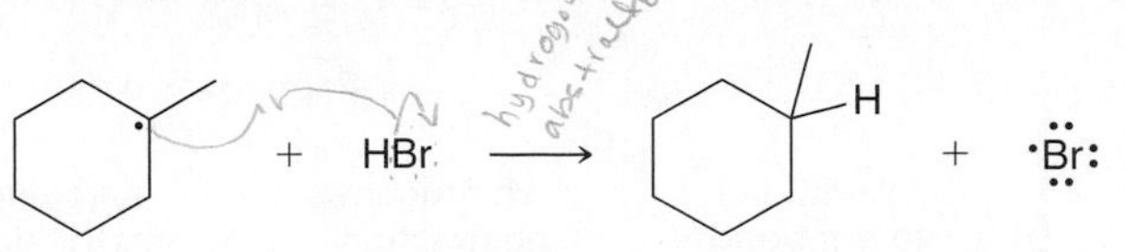

(b) 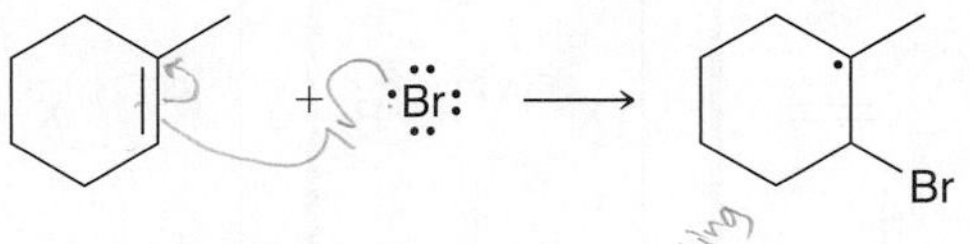

(c) 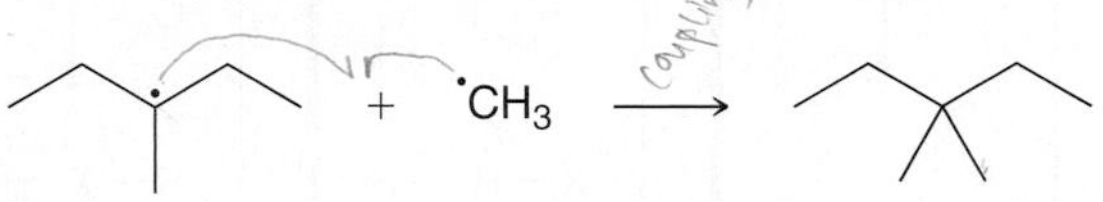

(d) 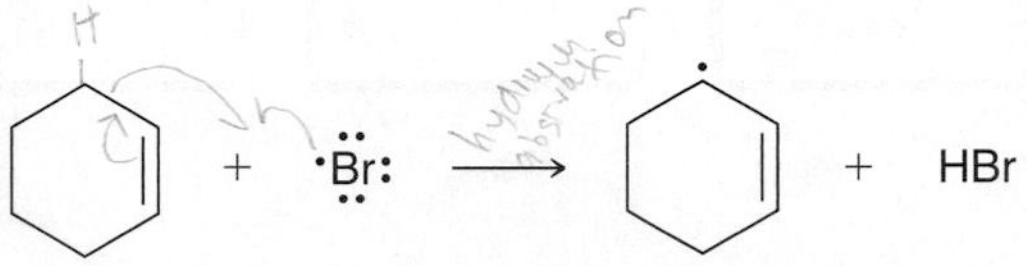

(e) 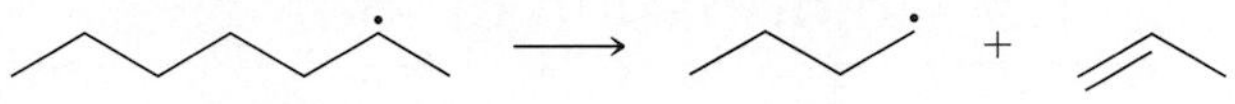

(f) 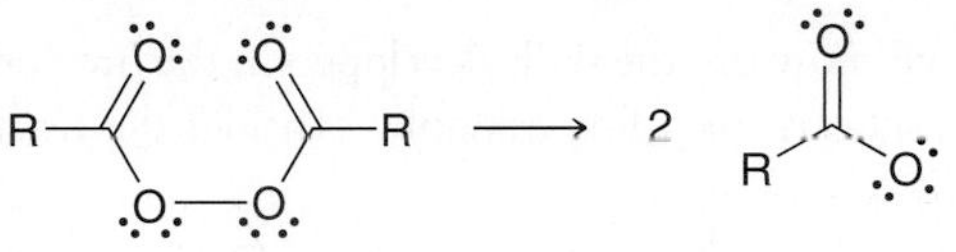

APPLY the skill **11.8** Draw a mechanism for the following intramolecular process:

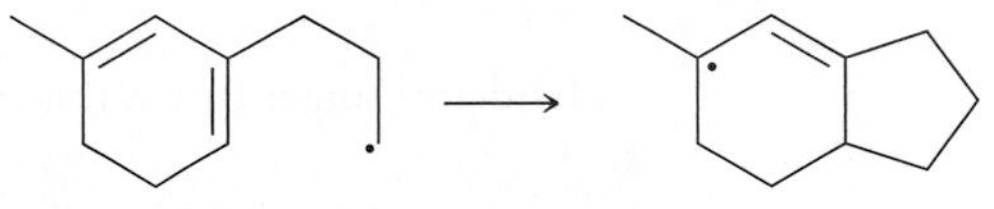

11.9 The triphenylmethyl radical reacts with itself to form the following dimer:

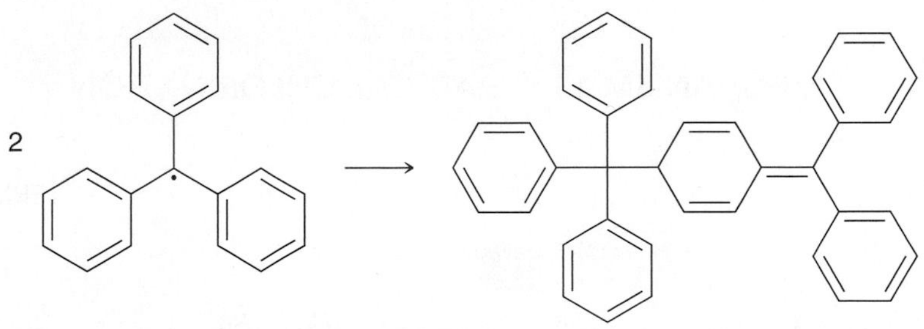

Identify the type of radical process taking place, and draw the appropriate fishhook arrows.

(Hint: It will help to draw just a few resonance structures of the triphenylmethyl radical.)

need more **PRACTICE?** **Try Problem 11.48**

Each of the six patterns shown in Figure 11.9 can be placed in one of three categories based on the fate of the radicals (Figure 11.10). **Initiation** is when radicals are created, while **termination** is when two radicals annihilate each other by forming a bond. The other four steps are generally **propagation** steps, in which the location of the unpaired electron is moved from place to place. Later in this chapter, we will encounter situations that will require us to refine our definitions for initiation, propagation, and termination, but for now, these simplistic (and incomplete) definitions provided will be helpful to us as we begin exploring radical mechanisms in the upcoming sections.

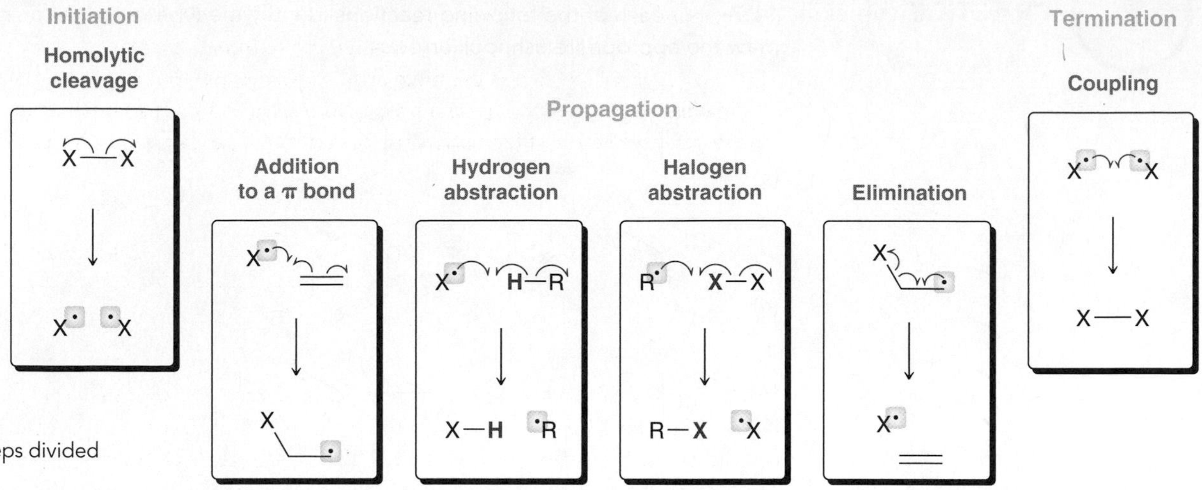

FIGURE 11.10 The six common steps divided into categories.

11.3 Chlorination of Methane

A Radical Mechanism for Chlorination of Methane

We will now use the skills developed in the previous section to explore the mechanisms of radical reactions. As a first example, consider the reaction between methane and chlorine to form methyl chloride:

$$\underset{\text{Methane}}{CH_4} \xrightarrow[h\nu]{Cl_2} \underset{\text{Methyl chloride}}{CH_3Cl} + HCl$$

Evidence suggests that this reaction proceeds via a radical mechanism (Mechanism 11.1).

MECHANISM 11.1 RADICAL CHLORINATION

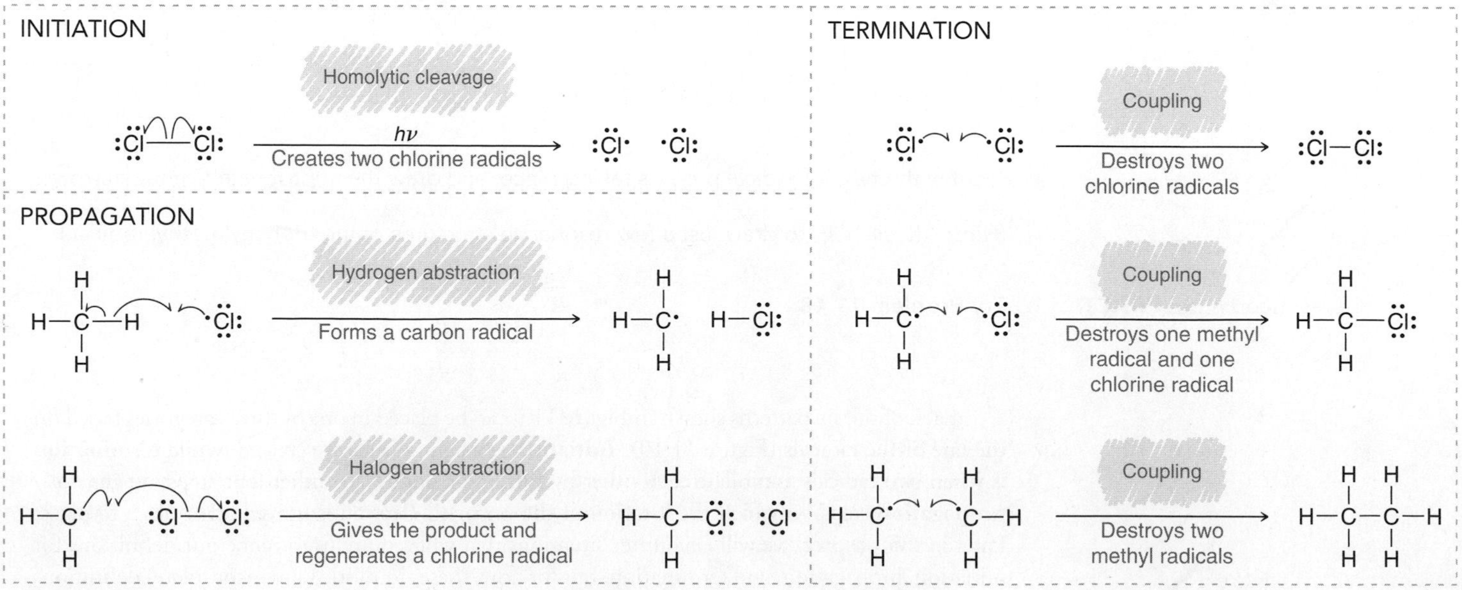

The reaction mechanism for radical chlorination is divided into three distinct stages. The initiation stage involves creation of radicals, while the termination stage involves destruction of radicals. The propagation stage is the most important, because the propagation steps are responsible for the observed reaction. The first propagation step is a hydrogen abstraction, and the second is a halogen abstraction. Notice that the sum of these propagation steps gives the net reaction:

Hydrogen abstraction	CH_4	+	~~$\cdot Cl$~~	⟶	~~$\cdot CH_3$~~	+	HCl
Halogen abstraction	~~$\cdot CH_3$~~	+	Cl_2	⟶	CH_3Cl	+	~~$\cdot Cl$~~
Net reaction	CH_4	+	Cl_2	⟶	CH_3Cl	+	HCl

This now provides us with a new, refined definition for propagation steps. Specifically, the sum of the propagation steps gives the net chemical reaction. All other steps must be either initiation or termination, not propagation. We will revisit this refined definition later in the chapter.

There is one more critical aspect of the propagation steps shown above. Notice that the first propagation step *consumes* a chlorine radical, while the second propagation step *regenerates* a chlorine radical. In this way, one chlorine radical can ultimately cause thousands of molecules of methane to be converted into chloromethane (assuming enough Cl_2 is present). Therefore, the reaction is called a **chain reaction**.

When excess Cl_2 is present, polychlorination is observed:

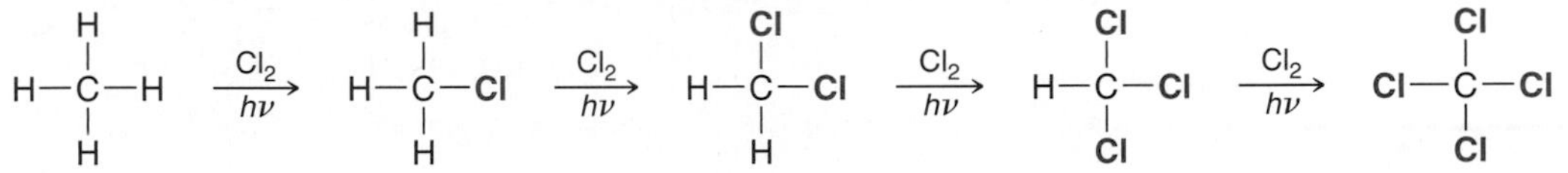

Methyl chloride **Methylene chloride** **Chloroform** **Carbon tetrachloride**

The initial product, methyl chloride (CH_3Cl), is even more reactive toward radical halogenation than methane. As methyl chloride is formed, it reacts with chlorine to produce methylene chloride. The process continues until carbon tetrachloride is produced. In order to produce methyl chloride as the major product (monohalogenation), it is necessary to use an excess of methane and a small amount of Cl_2. Unless otherwise indicated, the conditions of a halogenation reaction are generally controlled so as to produce monohalogenation.

SKILLBUILDER

11.4 DRAWING A MECHANISM FOR RADICAL HALOGENATION

LEARN the skill

Draw the mechanism of the radical chlorination of methyl chloride to produce methylene chloride:

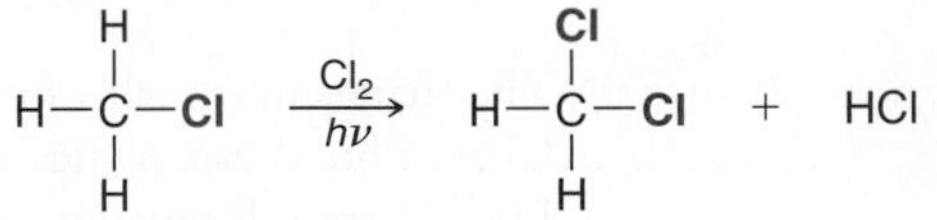

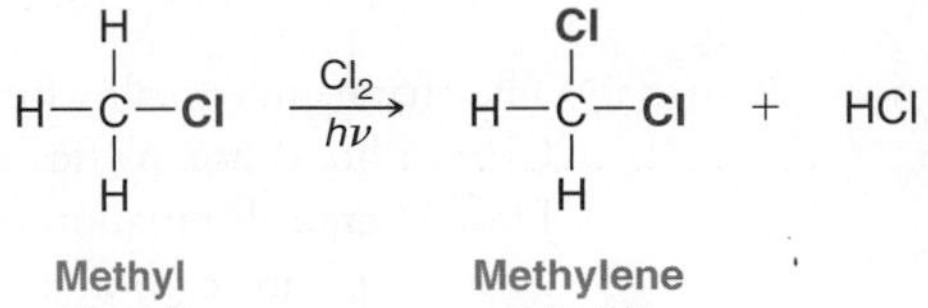

Methyl chloride **Methylene chloride**

SOLUTION

The mechanism will have three distinct stages. The first stage is initiation, which creates chlorine radicals. This initiation step involves homolytic bond cleavage and should employ only two fishhook arrows:

STEP 1
Draw the initiation step(s).

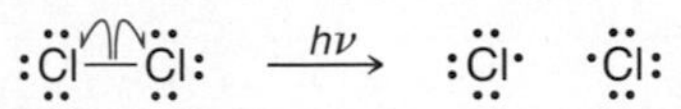

The next stage involves the propagation steps. There are two propagation steps: hydrogen abstraction to remove the hydrogen atom followed by halogen abstraction to attach the chlorine atom. Each of these steps should have three fishhook arrows:

STEP 2
Draw the propagation steps.

Hydrogen abstraction

Halogen abstraction

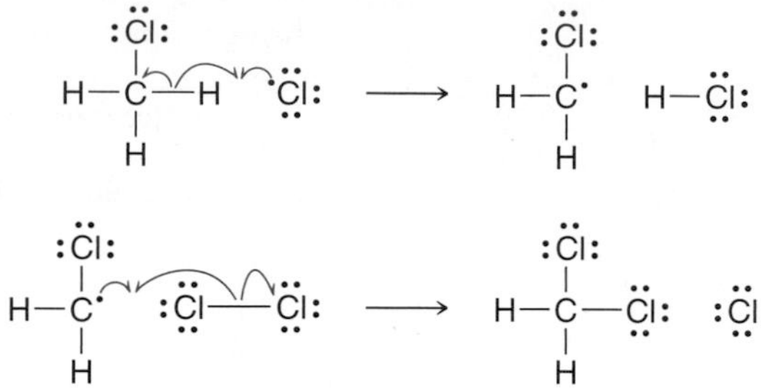

These two abstractions together represent the core of the reaction. They show how the product is formed.

Termination ends the process. There are a number of possible termination steps. When drawing a mechanism for a radical reaction, it is generally not necessary to draw all possible termination steps unless specifically asked to do so. It is sufficient to draw the one termination step that also produces the desired product:

STEP 3
Draw at least one termination step.

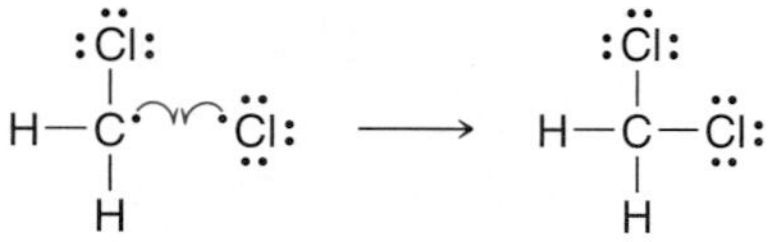

PRACTICE the skill **11.10** Draw a mechanism for each of the following processes:

(a) Chlorination of methylene chloride to produce chloroform

(b) Chlorination of chloroform to produce carbon tetrachloride

(c) Chlorination of ethane to produce ethyl chloride

(d) Chlorination of 1,1,1-trichloroethane to produce 1,1,1,2-tetrachloroethane

(e) Chlorination of 2,2-dichloropropane to produce 1,2,2-trichloropropane

APPLY the skill **11.11** In practice, the chlorination of methane often produces many by-products. For example, ethyl chloride is obtained in small quantities. Can you suggest a mechanism for the formation of ethyl chloride?

Radical Initiators

Energy is required to initiate a radical chain reaction:

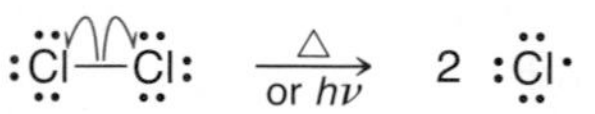

$$:\overset{..}{\underset{..}{Cl}}-\overset{..}{\underset{..}{Cl}}: \quad \xrightarrow[\text{or } h\nu]{\triangle} \quad 2 \; :\overset{..}{\underset{..}{Cl}}\cdot \qquad \Delta H° = 243 \text{ kJ/mol}$$

The change in enthalpy for homolytic bond cleavage of Cl_2 is 243 kJ/mol. In order to break the Cl—Cl bond homolytically, this energy must be supplied in the form of either heat or light. Photochemical initiation requires the use of UV light, while thermal initiation requires a very high temperature (several hundred degrees Celsius) in order to produce the requisite energy necessary to cause homolytic bond cleavage of a Cl—Cl bond. Achieving thermal initiation at more moderate temperatures requires the use of a **radical initiator,** a compound with a weak bond that undergoes homolytic bond cleavage with greater ease. For example, consider the homolytic bond cleavage of **peroxides,** which are compounds containing an O—O bond:

$$RO\overset{..}{\underset{..}{-}}OR \quad \xrightarrow[\text{or } h\nu]{\triangle} \quad 2 \; R\overset{..}{\underset{..}{O}}\cdot \qquad \Delta H° = 159 \text{ kJ/mol}$$

A peroxide

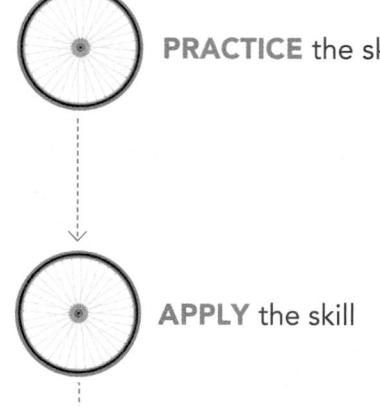

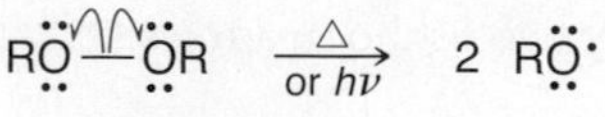

Notice that this process requires less energy, because the O—O bond is weaker than the Cl—Cl bond. This process can be achieved at 80°C, and the radicals produced (RO•) can start the chain process.

Acyl peroxides are often used as radical initiators, because the O—O bond is especially weak:

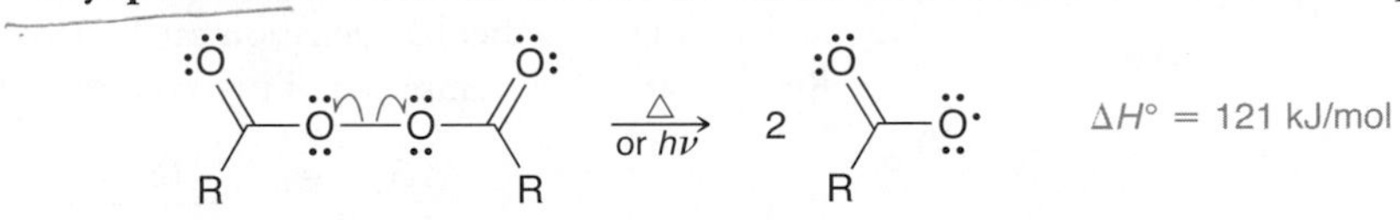

An acyl peroxide

The energy required to homolytically cleave this bond is only 121 kJ/mol. This bond has a particularly small BDE because the radical produced is resonance stabilized:

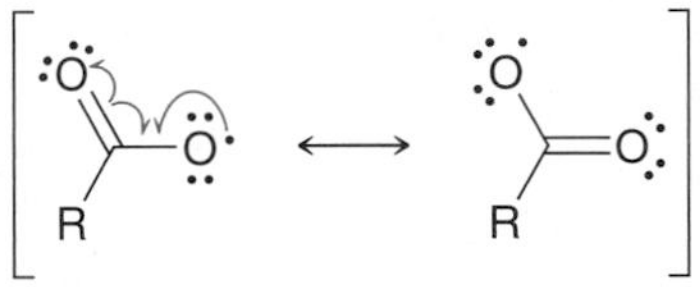

Radical Inhibitors

While some compounds, like peroxides, help initiate radical reactions, compounds called radical inhibitors have the opposite effect. A **radical inhibitor** is a compound that prevents a chain process from either getting started or continuing. Radical inhibitors effectively destroy radicals and are therefore also called radical scavengers. For example, molecular oxygen (O_2) is a diradical:

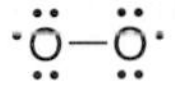

Molecular oxygen is able to couple with other radicals, thereby destroying them. Each molecule of oxygen is capable of destroying two radicals. As a result, radical chain reactions generally won't occur rapidly until all of the available oxygen has been consumed.

Another example of a radical inhibitor is hydroquinone. When hydroquinone meets a radical, a hydrogen abstraction can take place, generating a resonance-stabilized radical that is less reactive than the original radical:

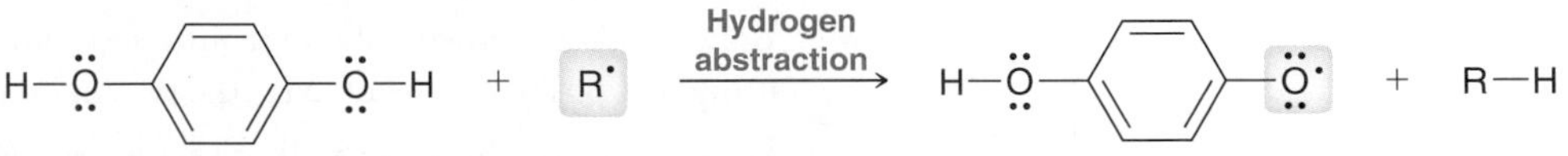

Hydroquinone   Resonance stabilized

This resonance-stabilized radical can then destroy another radical via another hydrogen abstraction to form benzoquinone:

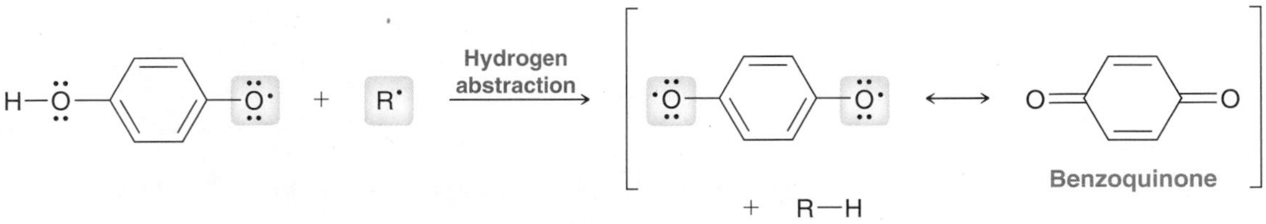

In Section 11.9, we will discuss the role of radical inhibitors in biological processes.

11.4 Thermodynamic Considerations for Halogenation Reactions

In the previous section, we explored the proposed mechanism for the chlorination of methane. We will now explore whether this reaction can be accomplished with other halogens as well. Is it possible to achieve a radical fluorination, bromination, or iodination? To answer this question, we must explore some thermodynamic aspects of halogenation.

In Section 6.3, we saw that ΔG (the change in Gibbs free energy) ultimately determines whether or not a reaction can occur. In order for a reaction to favor products over starting materials, the reaction must exhibit a negative ΔG. If ΔG is positive, starting materials will be favored, and the reaction will not produce the desired products. We will now use this information to determine whether halogenation can be accomplished with halogens other than chlorine.

Recall that ΔG is comprised of two components—enthalpy and entropy:

$$\Delta G = \underbrace{(\Delta H)}_{\text{Enthalpy term}} + \underbrace{(-T\,\Delta S)}_{\text{Entropy term}}$$

For halogenation of an alkane, the entropy component is negligible because two molecules of starting material are converted into two molecules of products:

$$\underbrace{CH_4 + X_2}_{\text{Two molecules}} \xrightarrow{h\nu} \underbrace{CH_3X + HX}_{\text{Two molecules}}$$

Since the change in entropy of a halogenation reaction is negligible, we can assess the value of ΔG by analyzing the enthalpy term alone:

$$\Delta G \approx \Delta H$$

The enthalpy term is determined by a number of factors, but the most important factor is bond strength. We can estimate ΔH by comparing the energy of the bonds broken and the energy of bonds formed:

$$\underbrace{H_3C{-}H + X{-}X}_{\text{Bonds broken}} \xrightarrow{h\nu} \underbrace{H_3C{-}X + H{-}X}_{\text{Bonds formed}}$$

To make the comparison, we look at bond dissociation energies. The relevant values are shown in Table 11.1. Using these values, we can estimate the sign of ΔH (positive or negative) associated with each halogenation reaction. Remember that:

- Bonds broken (shown in blue) require an input of energy in order to be broken, contributing to a positive value of ΔH (the system increases in energy).
- Bonds formed (shown in red) release energy when they are formed, contributing to a negative value of ΔH (the system decreases in energy).

TABLE 11.1 BOND DISSOCIATION ENERGIES OF RELEVANT BONDS (kJ/mol)

	$H_3C{-}H$	$X{-}X$	$H_3C{-}X$	$H{-}X$
F	435	159	456	569
Cl	435	243	351	431
Br	435	193	293	368
I	435	151	234	297

The data in Table 11.1 can be used to make the following predictions:

$$CH_4 + F_2 \xrightarrow{h\nu} CH_3F + HF \qquad \Delta H^\circ = -431 \text{ kJ/mol}$$

$$CH_4 + Cl_2 \xrightarrow{h\nu} CH_3Cl + HCl \qquad \Delta H^\circ = -104 \text{ kJ/mol}$$

$$CH_4 + Br_2 \xrightarrow{h\nu} CH_3Br + HBr \qquad \Delta H^\circ = -33 \text{ kJ/mol}$$

$$CH_4 + I_2 \xrightarrow{h\nu} CH_3I + HI \qquad \Delta H^\circ = +55 \text{ kJ/mol}$$

All of the processes above have a negative value for ΔH and are exothermic except for iodination. Iodination of methane has a positive ΔH, which means that ΔG will also be positive for that reaction. As a result, iodination is not thermodynamically favorable, and the reaction simply does not occur. The other halogenation reactions are all thermodynamically favorable, but fluorination is so exothermic that the reaction is too violent to be of practical use. Therefore, only chlorination and bromination are practical in the laboratory.

When this type of thermodynamic analysis is applied to alkanes other than methane, similar results are obtained. For example, ethane will undergo both radical chlorination and radical bromination:

$$\text{H}_3\text{C}-\text{CH}_3 + \text{X}_2 \xrightarrow[\text{or ROOR, } \triangle]{h\nu} \text{H}_3\text{C}-\text{CH}_2-\text{X} + \text{HX} \qquad \textbf{(X = Br or Cl)}$$

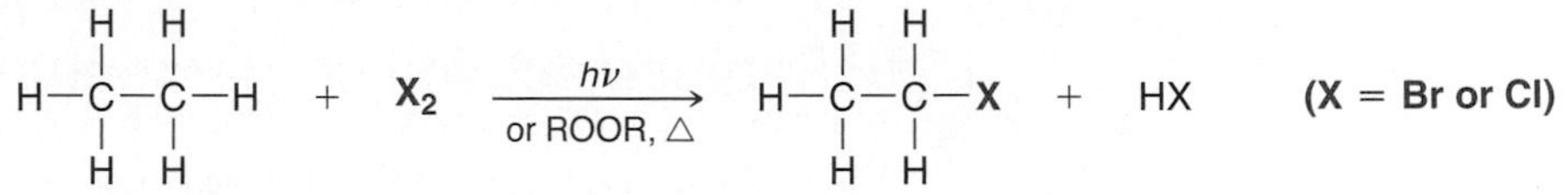

Radical fluorination of ethane is too violent to be practical, and radical iodination of ethane does not occur.

A comparison of chlorination and bromination reveals that bromination is generally a much slower process than chlorination. To understand why, we must take a closer look at the individual propagation steps. We will compare ΔH of each propagation step for the chlorination and bromination of ethane:

	ΔH (kJ/mol)	
	X = Cl	**X = Br**
First propagation step: hydrogen abstraction	21	+126
Second propagation step: halogen abstraction	−96	−176
Net reaction:	−117	−50
	kJ/mol	kJ/mol

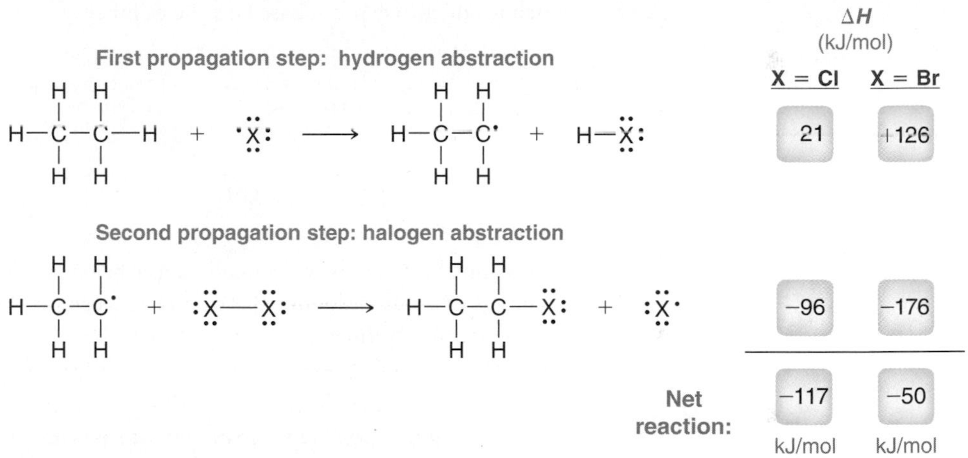

In each case, the net reaction is exothermic. The estimated values for ΔH for chlorination and bromination are -117 and -50 kJ/mol, respectively. Therefore, both of these reactions are thermodynamically favorable. But notice that the first step of bromination is an endothermic process (ΔH has a positive value). This endothermic step does not prevent bromination from occurring, because the net reaction is still exothermic. However, the endothermic first step does greatly affect the rate of the reaction. Compare the energy diagrams for the chlorination and bromination of ethane (Figure 11.11). Each energy diagram displays both propagation steps.

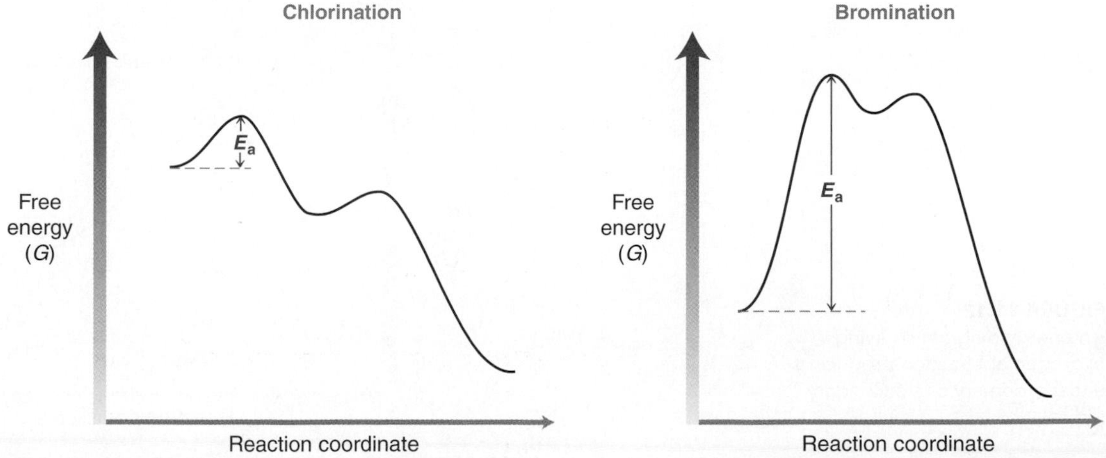

FIGURE 11.11
Energy diagrams for the two propagation steps of radical chlorination and radical bromination of ethane.

In each case, the first propagation step (hydrogen abstraction) is the rate-determining step. For chlorination, the rate-determining step is exothermic, and the energy of activation (E_a) is relatively small. In contrast, the rate-determining step of the bromination process is endothermic, and the energy of activation (E_a) is relatively large. As a result, bromination occurs more slowly than chlorination.

According to this analysis, it might seem to be a disadvantage that the first step of bromination is endothermic. We will see in the next section that this endothermic step actually makes bromination more useful (albeit slower) than chlorination.

11.5 Regioselectivity of Halogenation

When propane undergoes radical halogenation, regiochemistry becomes an issue. There are two possible products, and both are formed.

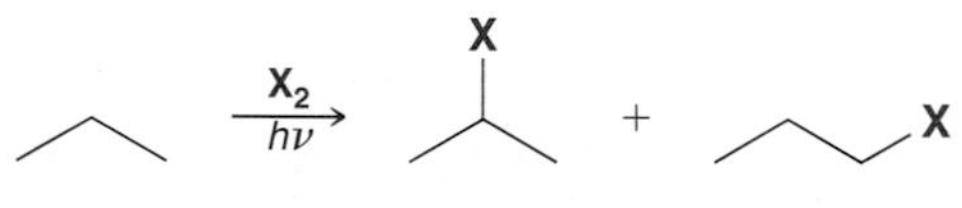

Statistically, we should be able to predict what the product distribution should be (how much of each product to expect), based on the number of each type of hydrogen atom:

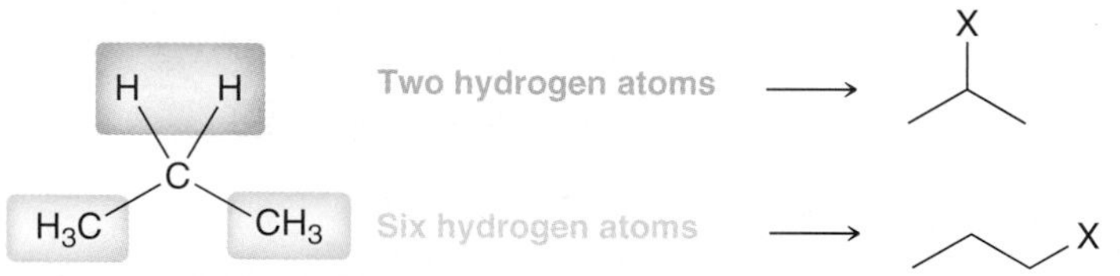

Based on this analysis, we should expect halogenation to occur at the primary position three times as often as it does at the secondary position. However, these expectations are not supported by observations.

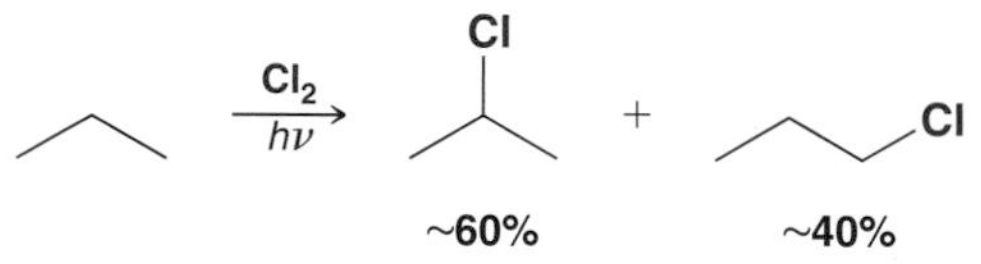

These results show that halogenation occurs at the secondary position more readily than statistics alone would suggest. Why? Recall that the rate-determining step is the first propagation step—hydrogen abstraction. We therefore focus our attention on that step. Specifically, compare the transition state for abstraction at the primary position with the transition state for abstraction at the secondary position (Figure 11.12). Compare the energy levels of the transition states highlighted in green in Figure 11.12. In each transition state, an unpaired electron (radical) is developing on a carbon atom. The developing radical is more stable at the secondary position

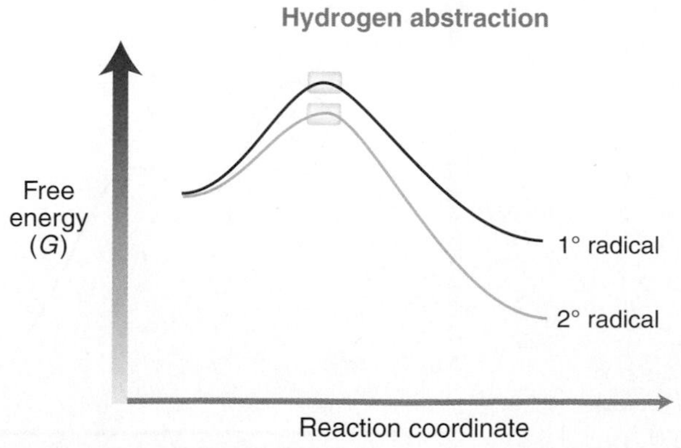

FIGURE 11.12
An energy diagram showing hydrogen abstraction producing either a primary or a secondary radical.

than at the primary position. As a result, E_a is lower for hydrogen abstraction at the secondary position, and the reaction occurs more rapidly at that site. This explains the observation that chlorination at the secondary position occurs more readily than is predicted by statistics alone.

When propane undergoes radical bromination (rather than chlorination), the following product distribution is observed:

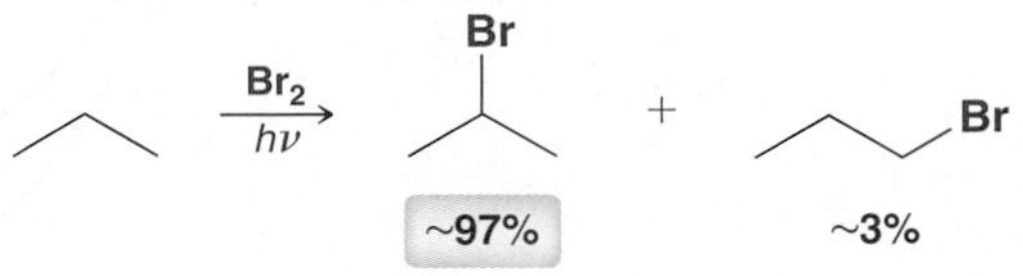

These results indicate that the tendency for bromination at the secondary position is much more pronounced, with 97% of the product occurring from reaction at the secondary position. Bromination is said to be more *selective* than chlorination. To understand the reason for this selectivity, recall that the rate-determining step of chlorination is exothermic, while the rate-determining step of bromination is endothermic. In the previous section, we used that fact to explain why bromination is slower than chlorination. Now, we will use that fact to explain why bromination is more selective than chlorination.

Recall the Hammond postulate (Section 6.6), which describes the nature of the transition state in each case. For the chlorination process, the rate-determining step is exothermic, and therefore, the energy and structure of the transition state more closely resemble reactants than products (Figure 11.13).

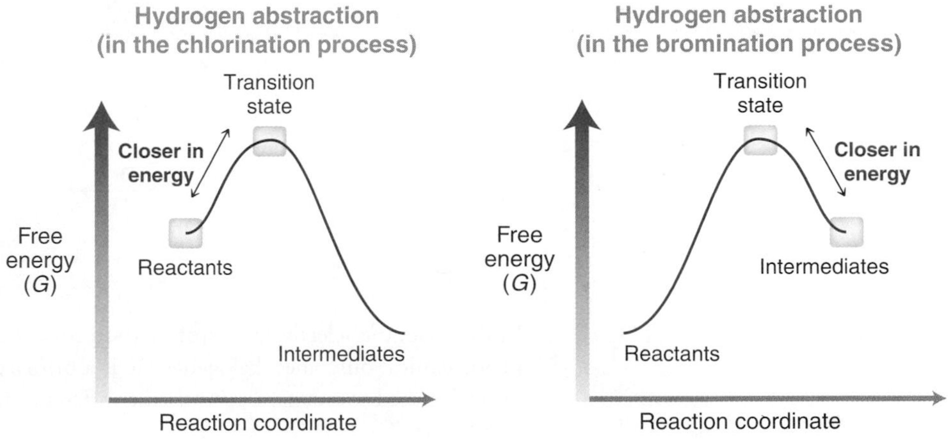

FIGURE 11.13
Use of the Hammond postulate to compare the transition states in the first propagation step of radical chlorination and radical bromination.

That means that the C—H bond has only begun to break, and the carbon atom has very little radical character. In the case of bromination, the rate-determining step is endothermic, and therefore, the energy and structure of the transition state more closely resemble intermediates than reactants (Figure 11.13). As a result, the C—H bond is almost completely broken in the transition state, and the carbon atom has significant radical character (Figure 11.14). In both cases, the carbon atom has partial radical character (δ•), but during chlorination, the amount of radical character is small. During bromination, the amount of radical character is much larger. As a result, the transition state will be more sensitive to the nature of the substrate during bro-

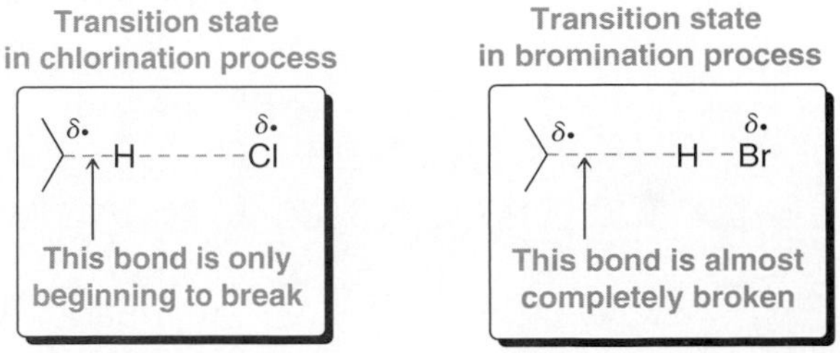

FIGURE 11.14
A comparison of the transition states during the first propagation step of radical chlorination and radical bromination.

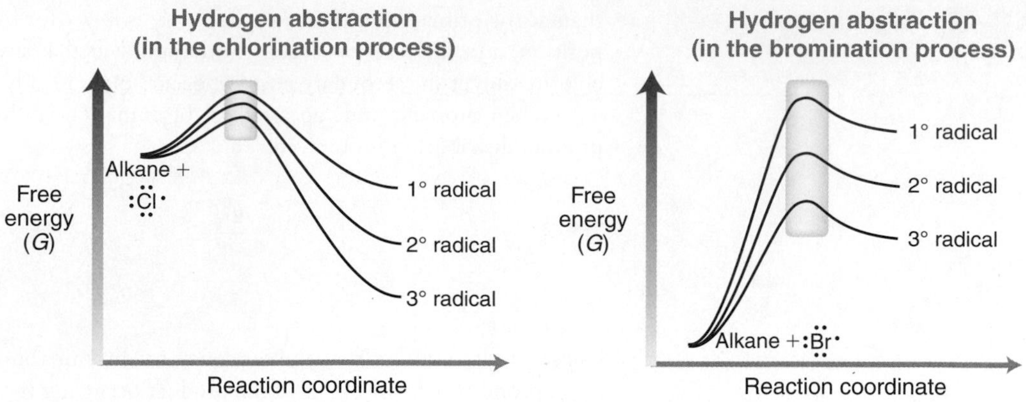

FIGURE 11.15
Energy diagrams for the first propagation step of chlorination and bromination, comparing the difference in transition states leading to primary, secondary, or tertiary radicals.

mination. In the energy diagram for the chlorination process, there is only a small difference in energy between the transition states leading to primary, secondary, or tertiary radicals (Figure 11.15, left). In contrast, the energy diagram for the bromination process shows a large difference in energy between the transition states leading to primary, secondary, or tertiary radicals (Figure 11.15, right). As a result, bromination shows greater selectivity than chlorination.

Here is another example that illustrates the difference in selectivity between chlorination and bromination:

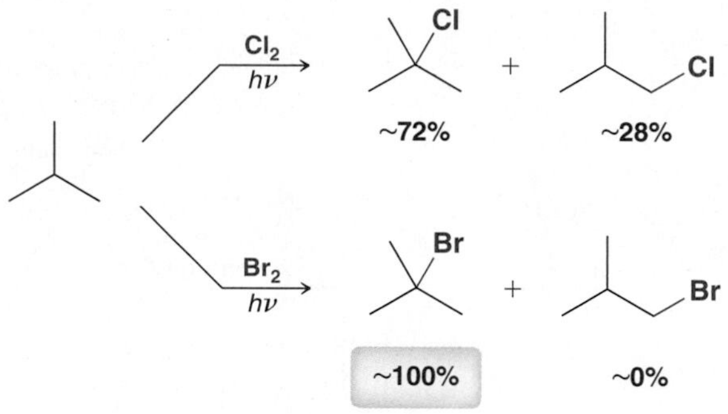

In this case, the selectivity of bromination is extremely pronounced because there are only two regiochemical outcomes: halogenation at a primary position or halogenation at a tertiary position. For the bromination process, the difference in energy of the transition states will be very significant.

Table 11.2 summarizes the relative selectivity of fluorination, chlorination, and bromination. Fluorine only shows a very small selectivity for tertiary over primary halogenation. Fluorine is the most reactive of all the halogens, and it is therefore the least selective. In contrast, bromine shows a very large selectivity for halogenation at the tertiary position (1600 times greater than at the primary position). Bromine is less reactive than fluorine and chlorine, and it is therefore the most selective. The relationship between reactivity and selectivity is a general trend that we will observe many times throughout this course (not just with radical reactions). Specifically, there is an inverse relationship between reactivity and selectivity: *Reagents that are the least reactive will generally be the most selective.*

TABLE 11.2 THE RELATIVE SELECTIVITY OF FLUORINATION, CHLORINATION, AND BROMINATION

	PRIMARY	SECONDARY	TERTIARY
F	1	1.2	1.4
Cl	1	3.9	5.1
Br	1	82	1600

SKILLBUILDER

11.5 PREDICTING THE REGIOCHEMISTRY OF RADICAL BROMINATION

LEARN the skill

Predict the major product obtained upon radical bromination of 2,2,4-trimethylpentane:

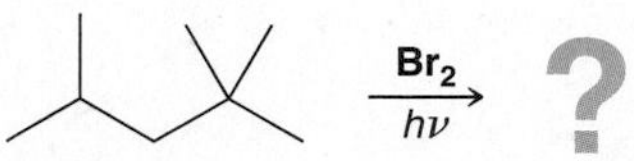

SOLUTION

Analyze all possible positions that can possibly undergo bromination, and identify each position as primary, secondary, or tertiary:

STEPS 1 AND 2 Identify all possible positions and determine the most stable location for a radical.

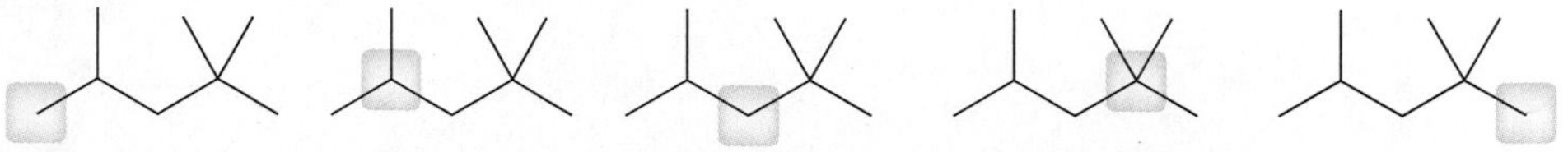

Make sure to avoid selecting a quaternary position, as quaternary positions do not possess a C—H bond. Bromination cannot occur at a quaternary site. The major product is expected to result from bromination at the tertiary position:

STEP 3 Draw the product of bromination at the location of the most stable radical.

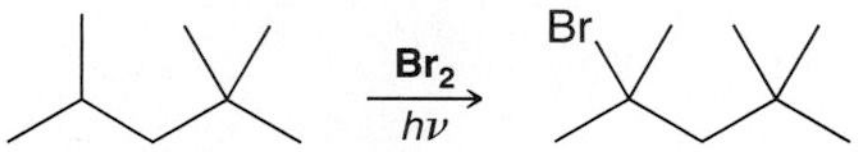

Major

PRACTICE the skill

11.12 Predict the major product obtained upon radical bromination of each of the following compounds:

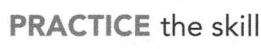

(a) 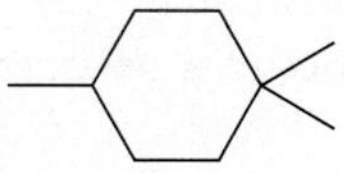

(b) 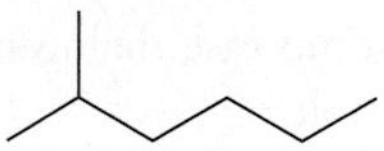

(c) 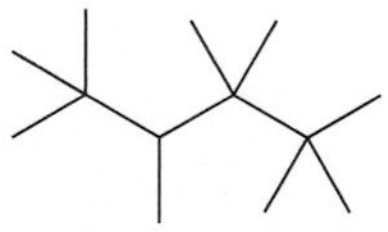

APPLY the skill

11.13 Compound A has molecular formula C_5H_{12} and undergoes monochlorination to produce four different constitutional isomers.

(a) Draw the structure of compound A.

(b) Draw all four monochlorination products.

(c) If compound A undergoes monobromination (instead of monochlorination), one product predominates. Draw that product.

need more **PRACTICE?** **Try Problems 11.27, 11.33a-c, f, 11.38, 11.40, 11.44**

11.6 Stereochemistry of Halogenation

We now focus our attention on the stereochemical outcome of radical halogenation. We will investigate two different situations: (1) halogenation that creates a new chirality center and (2) halogenation that occurs at an existing chirality center.

Halogenation That Creates a New Chirality Center

When butane undergoes radical chlorination, two constitutional isomers are obtained:

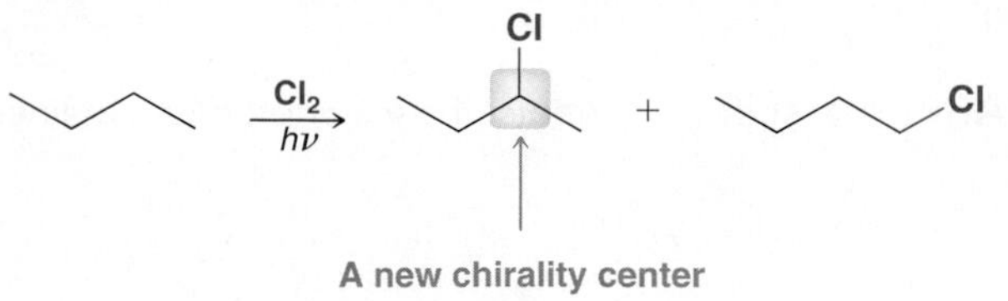

The products are 2-chlorobutane and 1-chlorobutane. The former has a new chirality center that was created by the reaction. In this case, a racemic mixture of 2-chlorobutane is obtained. Why? Consider the structure of the radical intermediate:

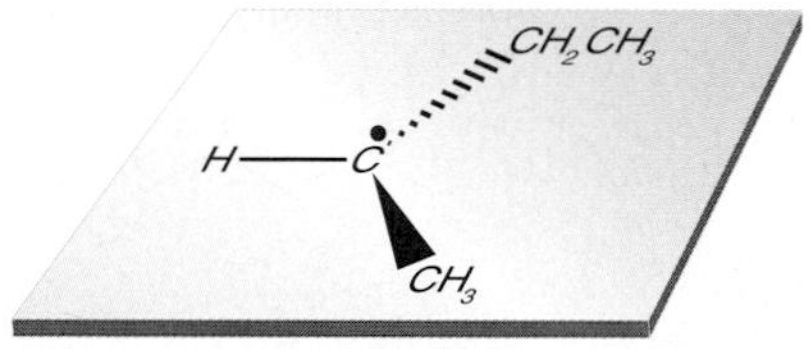

Trigonal planar

At the beginning of this chapter, we saw that a carbon radical either is trigonal planar or is a shallow pyramid that is rapidly inverting. Either way, it can be treated as trigonal planar. We therefore expect that halogen abstraction can occur on either face of the plane with equal likelihood, leading to a racemic mixture of 2-chlorobutane:

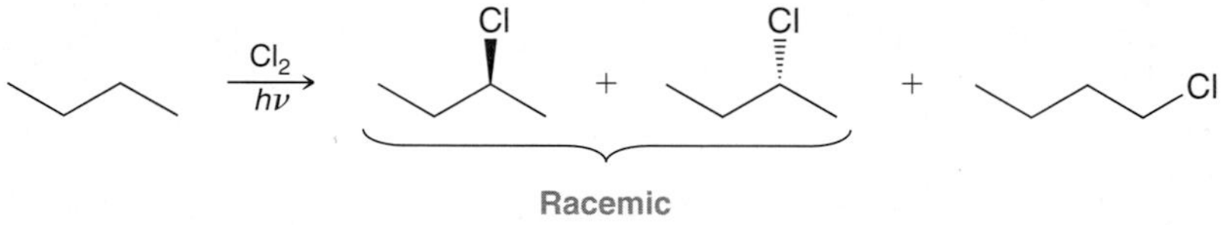

Halogenation at an Existing Chirality Center

In some case, halogenation will occur at an existing chirality center. Consider the following example:

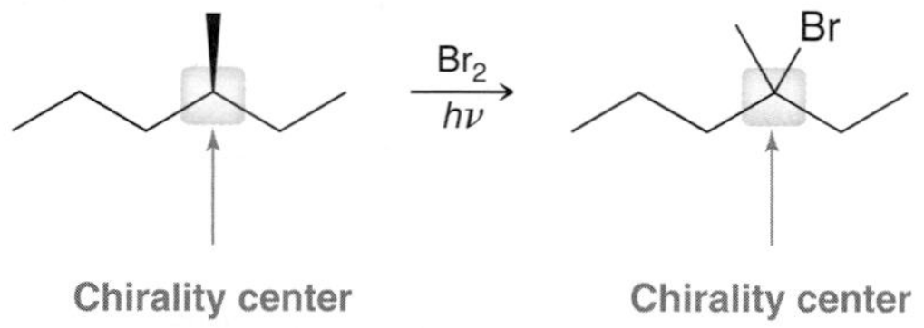

We expect bromination to occur at the tertiary position, and this position is already a chirality center before the reaction takes place. In such a case, what happens to the chirality center? Once again, we expect a racemic product, regardless of the configuration of the starting material. The first propagation step is removal of a hydrogen atom from the chirality center to form a radical intermediate that can be treated as planar.

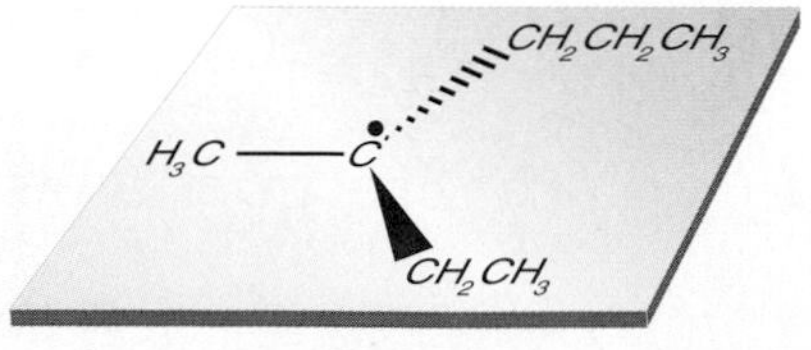

Trigonal planar

At this point, the configuration of the starting alkane has been lost. The second propagation step can now occur on either face of the plane with equal likelihood, leading to a racemic mixture:

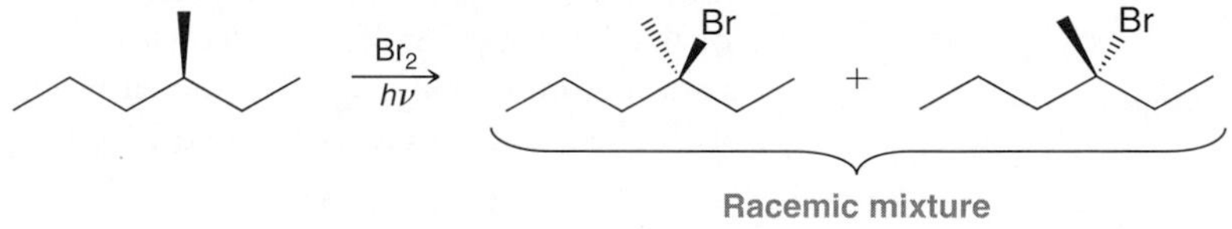

SKILLBUILDER

11.6 PREDICTING THE STEREOCHEMICAL OUTCOME OF RADICAL BROMINATION

LEARN the skill

Predict the stereochemical outcome of radical bromination of the following alkane:

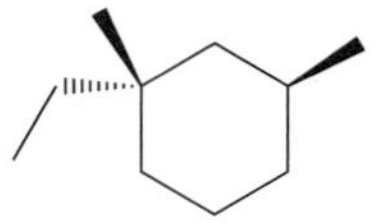

SOLUTION

First identify the regiochemical outcome—that is, identify the position that will undergo bromination. In this example, there is only one tertiary position:

STEP 1
Identify the location where bromination will occur.

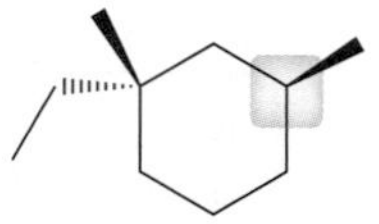

STEP 2
If the location of bromination is a chirality center, or will become a chirality center, draw both possible stereoisomers.

This position is an existing chirality center, so we expect loss of configuration to produce both possible stereoisomers:

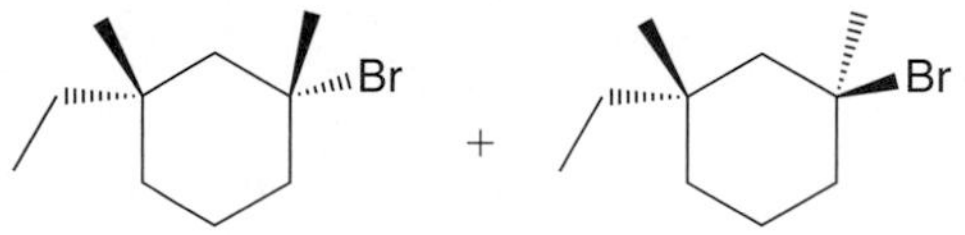

Notice that the other chirality center was unaffected by the reaction. The products of this reaction are not enantiomers and cannot be called a racemic mixture. In this case, the products are diastereomers.

PRACTICE the skill

11.14 Predict the stereochemical outcome of radical bromination of the following alkanes:

(a) 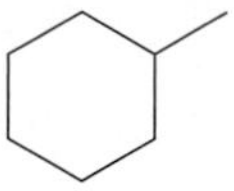 (b) 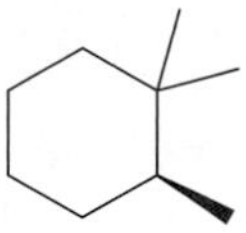 (c)  (d) 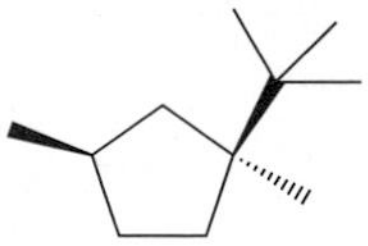

APPLY the skill

11.15 Compound A has molecular formula $C_5H_{11}Br$. When compound A is treated with bromine in the presence of UV light, the major product is 2,2-dibromopentane. Treatment of compound A with NaSH (a strong nucleophile) produces a compound with one chirality center having the *R* configuration. What is the structure of compound A?

need more **PRACTICE?** **Try Problems 11.35, 11.41**

11.7 Allylic Bromination

Until now, we have focused on reactions of alkanes. Now let's consider the radical halogenation of alkenes. For example, consider what outcome you might expect when cyclohexene undergoes radical bromination. Begin by comparing all C—H bonds to identify the bond that is most easily broken. Specifically, compare the BDE for each type of C—H bond in cyclohexene.

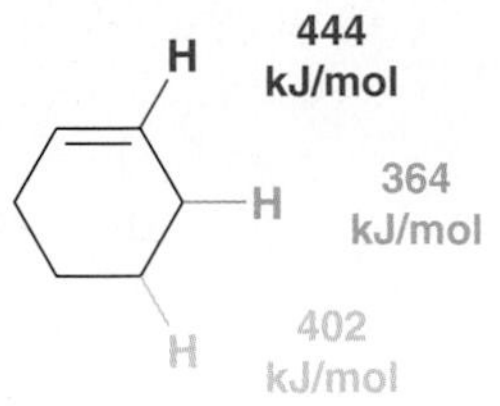

Of the three different types of C—H bonds in cyclohexene, the C—H bond at an allylic site has the lowest BDE, because hydrogen abstraction at that site generates a resonance-stabilized allylic radical:

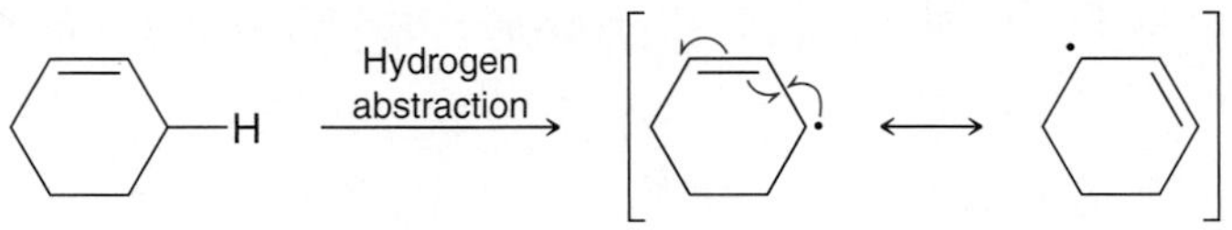

Therefore, we expect bromination of cyclohexene to produce the allylic bromide:

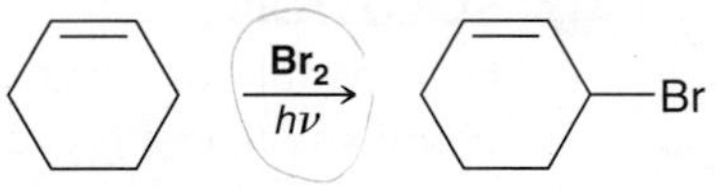

This reaction is called an **allylic bromination**, and it suffers from one serious flaw. When the reaction is performed with Br_2 as the reagent, a competition occurs between allylic bromination and ionic addition of bromine across the π bond (as we saw in Section 9.8):

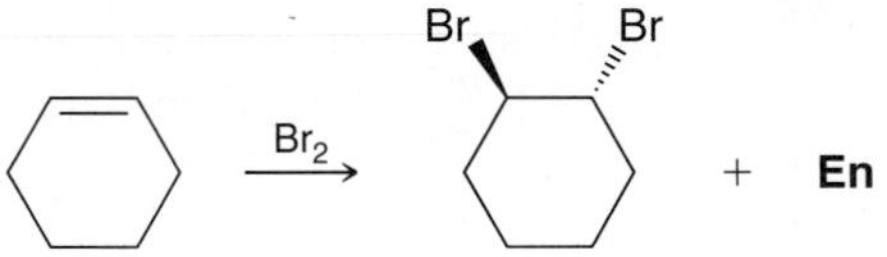

To avoid this competitive reaction, the concentration of bromine (Br_2) must be kept as low as possible throughout the reaction. This can be accomplished by using ***N*-bromosuccinimide** (NBS) as a reagent instead of Br_2. NBS is an alternate source of the bromine radical:

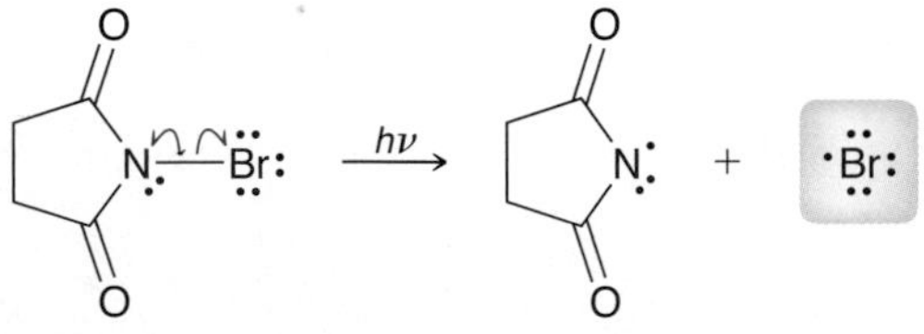

N-Bromosuccinimide (NBS)

Stabilized by resonance

The N—Br bond is weak and is easily cleaved to produce a bromine radical, which achieves the first propagation step:

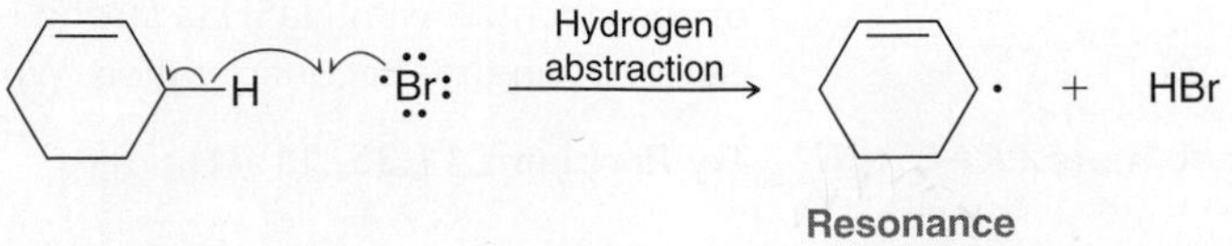

Resonance stabilized

The HBr produced in this step then reacts with NBS in an ionic reaction that produces Br_2. This Br_2 is then used in the second propagation step to form the product:

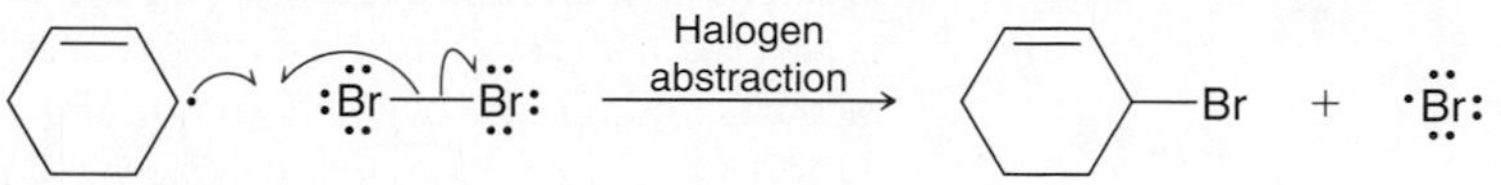

Throughout the process, concentrations of HBr and Br_2 are kept at a minimum. Under these conditions, ionic addition of Br_2 does not successfully compete with radical bromination.

With many alkenes, a mixture of products is obtained, because the allylic radical initially formed is resonance stabilized and can undergo halogen abstraction at either site:

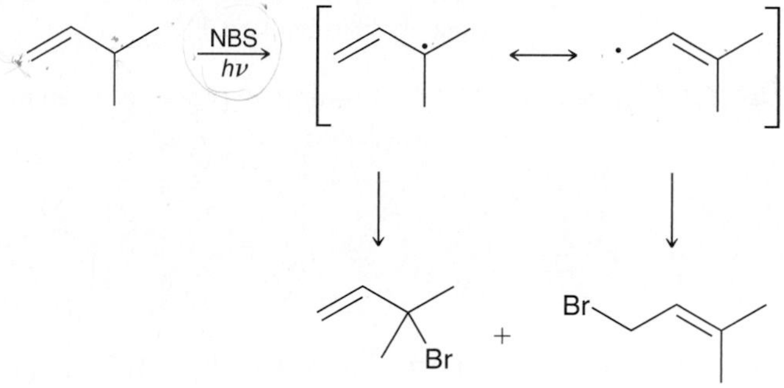

SKILLBUILDER

11.7 PREDICTING THE PRODUCTS OF ALLYLIC BROMINATION

LEARN the skill

Predict the products that are obtained when methylenecyclohexane is treated with NBS and irradiated with UV light:

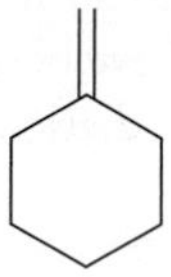

SOLUTION

First identify any allylic positions:

STEP 1
Identify the allylic position.

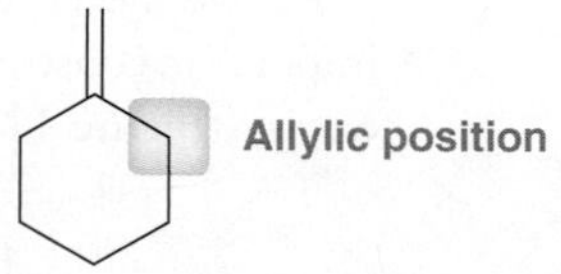

In this case, there is only one unique allylic position, because the allylic position on the other side of the compound is identical (it will lead to the same products).

Next, remove a hydrogen atom from the allylic position, and draw the resonance structures of the resulting allylic radical:

STEP 2
Remove a hydrogen atom and draw resonance structures.

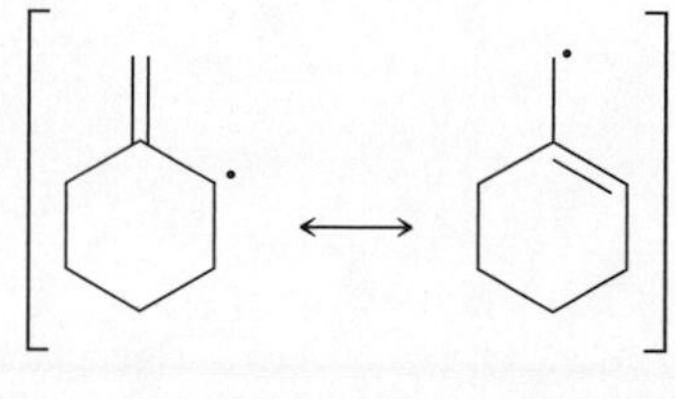

STEP 3
Place a bromine at each location that bears an unpaired electron.

Finally, use these resonance structures to determine the products of the second propagation step (halogen abstraction). Simply place a bromine at the position of the unpaired electron in each resonance structure, giving the following products:

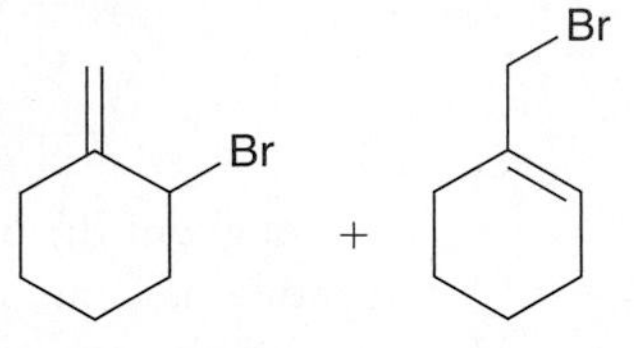

(Racemic mixture)

The first product is expected to be obtained as a racemic mixture of enantiomers, as described in Section 11.6.

PRACTICE the skill

11.16 Predict the products when each of the following compounds is treated with NBS and irradiated with UV light:

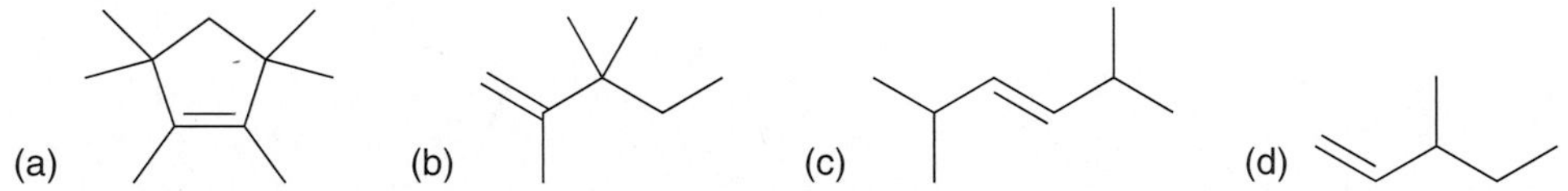

APPLY the skill

11.17 When 2-methyl-2-butene is treated with NBS and irradiated with UV light, five different monobromination products are obtained, one of which is a racemic mixture of enantiomers. Draw all five monobromination products and identify the product that is obtained as a racemic mixture.

need more **PRACTICE?** **Try Problems 11.26, 11.28, 11.33d, e, 11.36, 11.39**

11.8 Atmospheric Chemistry and the Ozone Layer

Ozone (O_3) is constantly produced and destroyed in the stratosphere, and its presence plays a vital role in shielding us from harmful UV radiation emitted by the sun. It is believed that life could not have flourished on land without this protective ozone layer, and instead, life would have been restricted to the depths of the ocean. The ability of ozone to protect us from harmful radiation is believed to result from the following mechanism:

$$O_3 \xrightarrow{h\nu} O_2 + \cdot\ddot{\underset{\cdot\cdot}{O}}\cdot$$

$$O_2 + \cdot\ddot{\underset{\cdot\cdot}{O}}\cdot \longrightarrow O_3 + \text{heat}$$

In the first step above, ozone absorbs UV light and splits into two pieces. In the second step, these two pieces recombine to release energy. There is no net chemical change, but there is one important consequence of this process: Harmful UV light is converted into another form of energy (Figure 11.16). This process exemplifies the role of entropy in nature. Light and heat

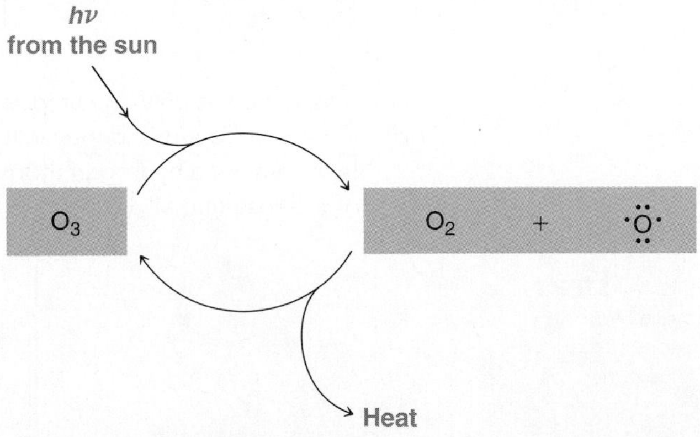

FIGURE 11.16
An illustration of the ability of stratospheric ozone to convert light into heat.

are both forms of energy, but heat is a more disordered form of energy. The driving force for the conversion of light into heat is an increase in entropy. Ozone is simply the vehicle by which ordered energy (light) is converted into disordered energy (heat).

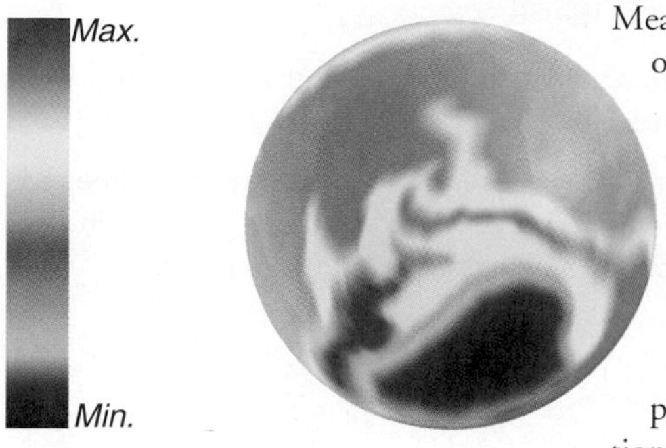

FIGURE 11.17
An illustration of the hole in the ozone layer over Antarctica.

Measurements over the last several decades have indicated a rapid decrease of stratospheric ozone. This decrease has been most drastic over Antarctica, where the ozone layer is now almost completely absent (Figure 11.17). Stratospheric ozone over the rest of the planet has decreased at a rate of about 6% each year. Each year, we are exposed to increasing levels of harmful UV radiation that are linked with skin cancer and other health issues. While many factors contribute to ozone depletion, it is believed that the main culprits are compounds called **chlorofluorocarbons** (**CFCs**). CFCs are compounds containing only carbon, chlorine, and fluorine. In the past they were heavily used for a wide variety of commercial applications, including as refrigerants, propellants, in the production of foam insulation, as fire-fighting materials, and many other useful applications. They were sold under the trade name "**Freons.**"

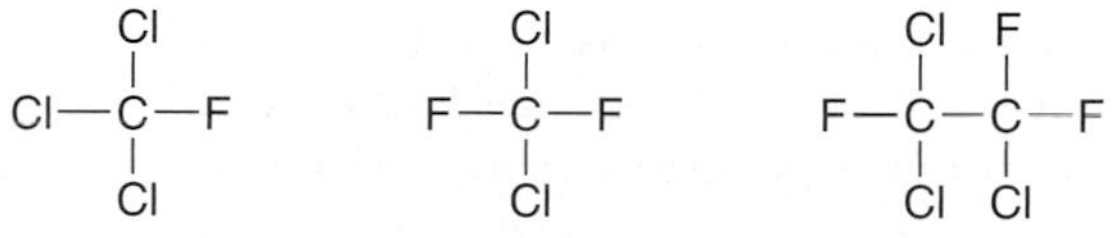

CFC-11 **(Freon 11)** **CFC-12** **(Freon 12)** **CFC-113** **(Freon 113)**

The harmful effects of CFCs on the ozone layer were elucidated in the late 1960s and early 1970s by Mario Molina, Frank Rowland, and Paul Crutzen, who shared the 1995 Nobel Prize in Chemistry for their work. CFCs are stable compounds that do not undergo chemical change until they reach the stratosphere. In the stratosphere, they interact with high-energy UV light and undergo homolytic cleavage, forming chlorine radicals. These radicals are then believed to destroy ozone by the following mechanism:

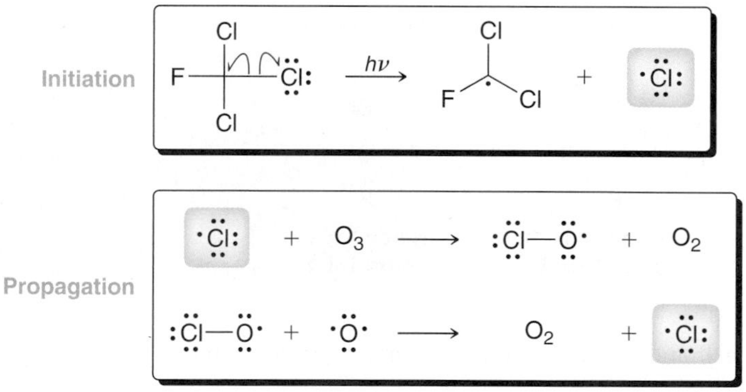

The second propagation step regenerates a chlorine radical, which continues the chain reaction. In this way, each CFC molecule can destroy thousands of ozone molecules. Awareness of the effect of CFCs led to a ban on their production in most countries as of January 1, 1996, as part of a global treaty called the Montreal Protocol on Substances that Deplete the Ozone Layer. As a result, much research has been directed toward finding suitable substitutes, such as the following two categories of compounds.

Hydrochlorofluorocarbons (HCFCs) are compounds that possess at least one C—H bond. Below are a few examples:

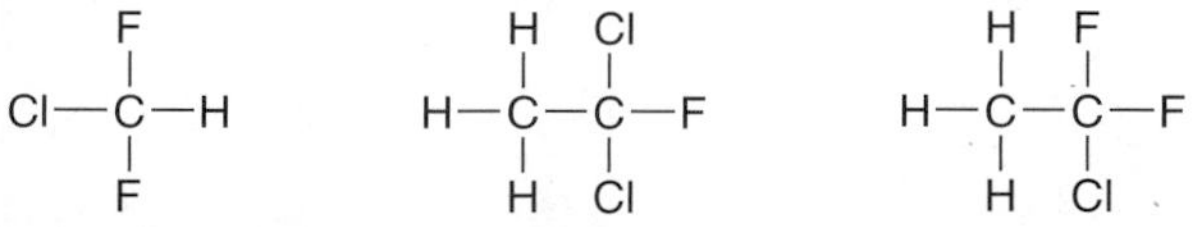

HCFC-22 **HCFC-141b** **HCFC-142b**

These compounds are believed to be less destructive to the ozone layer, because the C—H bond allows them to decompose before they reach the stratosphere. HCFC-22 and HCFC-141b have replaced CFC-11 in the production of foam insulation.

PRACTICALLY SPEAKING)))

Fighting Fires with Chemicals

As mentioned in the chapter-opening paragraphs, the combustion process is believed to involve free radicals. When subjected to excessive heat, the single bonds in organic compounds undergo homolytic bond cleavage to produce radicals, which then react with molecular oxygen in a coupling reaction. A series of radical chain reactions continue until most of the C—C and C—H bonds have been broken to produce CO_2 and H_2O. In order for this chain process to continue, a fire needs three essential ingredients: fuel (such as wood, which is comprised of organic compounds), oxygen, and heat. In order to extinguish a fire, we must deprive the fire of at least one of these three ingredients. Alternatively, we can find a way to stop the radical chain reaction by destroying the radical intermediates.

Many reagents can be used to extinguish a fire. Common examples found in many small fire extinguishers are CO_2, water, and argon. A sudden discharge of CO_2 or argon gas will deprive a fire of oxygen, while water deprives a fire of heat (by absorbing the heat in order to evaporate or boil).

The most powerful reagents for extinguishing fires are called halons, because they are organic compounds containing halogen atoms. These compounds are generally either CFCs or BFCs (bromofluorocarbons). Here are just two examples of halons that have been extensively used as fire suppression agents:

<div align="center">

Cl
|
F—C—Br
|
F

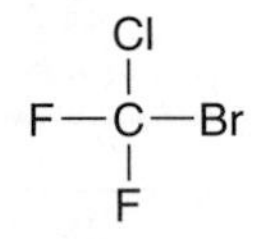

Halon 1211
(Freon 12B1)

Br
|
F—C—F
|
F

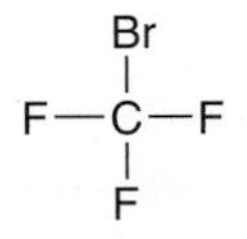

Halon 1301
(Freon 13B1)

</div>

Halons are extremely effective, because they fight fire in three different ways:

1. Halons are gases, so a sudden discharge of a halon gas will deprive the fire of oxygen.

2. Halons absorb heat to undergo homolytic bond cleavage:

<div align="center">

Cl
|
F—C—Br: →(heat)→ F—C• + •Br:
|
F

</div>

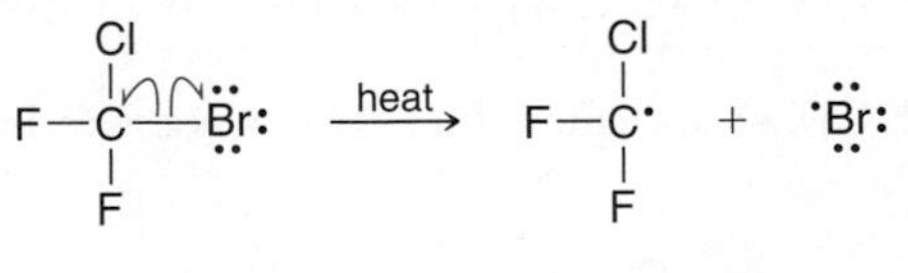

By absorbing heat, they deprive the fire of one of its essential ingredients.

3. Homolytic bond cleavage results in the formation of free radicals, which can then couple with the radicals participating in the chain reaction, thereby terminating the process:

<div align="center">

Cl
|
F—C• •R → F—C—R
| |
F F

</div>

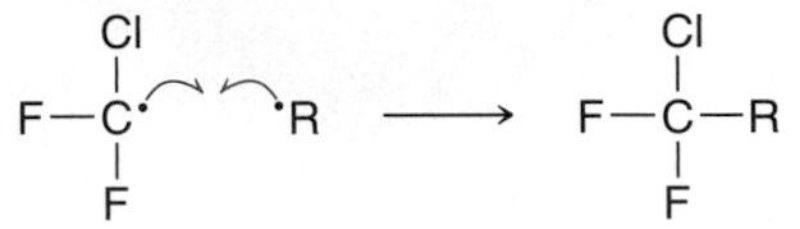

Halons are therefore able to speed up the rate of termination steps so that they compete with propagation steps.

For all of these reasons halons are extremely effective as firefighting agents and were heavily used over the last three decades. Halons also have the added benefit of not leaving behind any residue after being used, rendering them particularly useful for fighting fires involving sensitive electronic equipment or documents. Unfortunately, halons have been shown to contribute to ozone depletion, and their production is now banned by the Montreal Protocol. The ban only applies to production, and it is still permitted to use existing stockpiles of halon gases. These stockpiles have been designated for use in special situations involving sensitive equipment, such as on planes or in control rooms. For all other situations, halon gases have been replaced with alternative gases that do not contribute to ozone depletion but are less effective as firefighting agents. One such example is called FM-200:

<div align="center">

F H F
| | |
F—C—C—C—F
| | |
F F F

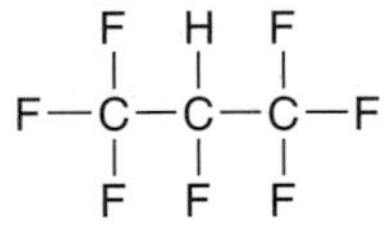

FM-200

</div>

Hydrofluorocarbons (HFCs) are compounds that contain only carbon, fluorine, and hydrogen (no chlorine). Below are a few examples:

$$
\begin{array}{ccc}
\text{HFC-32} & \text{HFC-125} & \text{HFC-134a}
\end{array}
$$

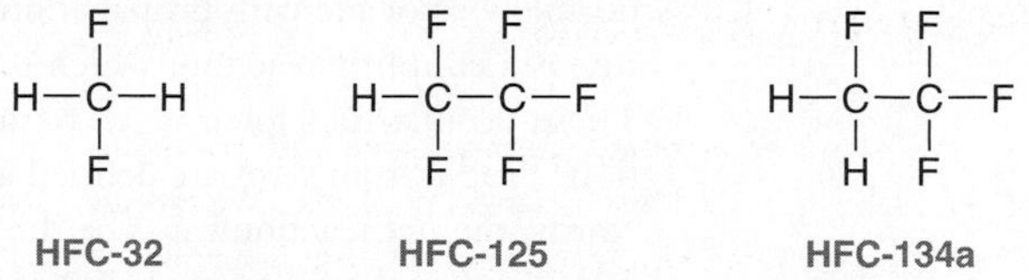

HFCs are not believed to contribute to ozone depletion, as they do not contain chlorine and therefore do not produce chlorine radicals. Nevertheless, they are believed to contribute to global warming. HFC-134a has replaced CFC-12 in refrigeration systems and medical aerosols.

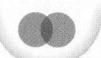

CONCEPTUAL CHECKPOINT

11.18 Most supersonic planes produce exhaust of hot gases containing many compounds, including nitric oxide (NO). Nitric oxide is a radical that is believed to play a role in ozone depletion. Propose propagation steps that show how nitric oxide can destroy ozone in a chain process.

11.9 Autooxidation and Antioxidants

Autooxidation

In the presence of atmospheric oxygen, organic compounds are known to undergo a slow oxidation process called **autooxidation**. For example, consider the reaction of cumene with oxygen to form a **hydroperoxide** (ROOH):

Cumene Cumene hydroperoxide

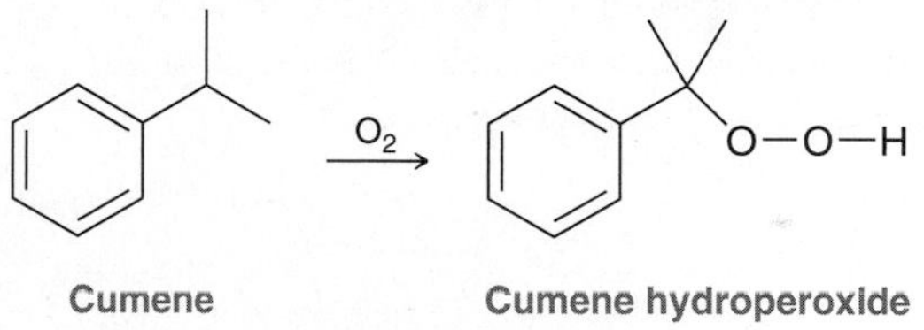

Autooxidation is believed to proceed via Mechanism 11.2.

MECHANISM 11.2 AUTOOXIDATION

INITIATION

Hydrogen abstraction

$$R-H \xrightarrow[\text{Forms a carbon radical}]{\cdot \text{Initiator}} R\cdot$$

PROPAGATION

Coupling

$$R\cdot \quad \cdot\ddot{O}-\ddot{O}\cdot \xrightarrow[\substack{\text{A carbon radical couples} \\ \text{with molecular oxygen}}]{} R-\ddot{O}-\ddot{O}\cdot$$

Hydrogen abstraction

$$R-\ddot{O}-\ddot{O}\cdot \quad H-R \xrightarrow[\substack{\text{Gives the product and} \\ \text{regenerates a carbon radical}}]{} R-\ddot{O}-\ddot{O}H \quad R\cdot$$

TERMINATION

Coupling

$$R\cdot \quad \cdot R \xrightarrow[\substack{\text{Destroys two} \\ \text{carbon radicals}}]{} R-R$$

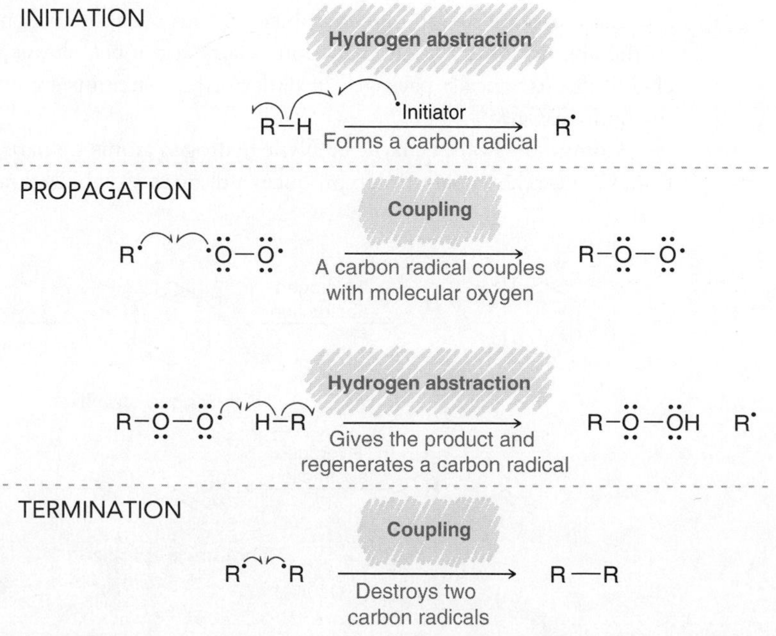

The initiation step might seem abnormal at first glance, because it is not a homolytic bond cleavage. The initiation step in this case is a hydrogen abstraction, which is a step that we normally associate with propagation. Moreover, look closely at the first propagation step. That step is a coupling reaction, which is a step that we have associated with termination until now. These peculiarities force us to refine our definitions of initiation, propagation, and termination. Propagation steps are defined as the steps that produce the net chemical reaction. In other words, the net reaction must be the sum of the propagation steps. In this case, the net reaction is the sum of the following two steps:

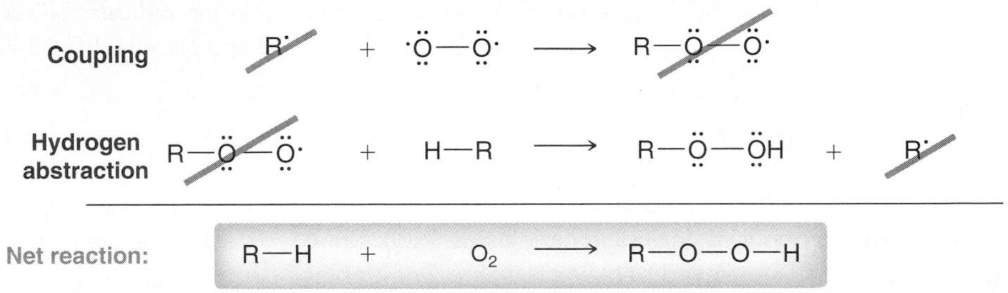

These two steps are therefore the propagation steps, and any other steps must be classified as either initiation or termination steps.

Many organic compounds, such as ethers, are particularly susceptible to autooxidation. For example, consider the reaction between diethyl ether and oxygen to form a hydroperoxide:

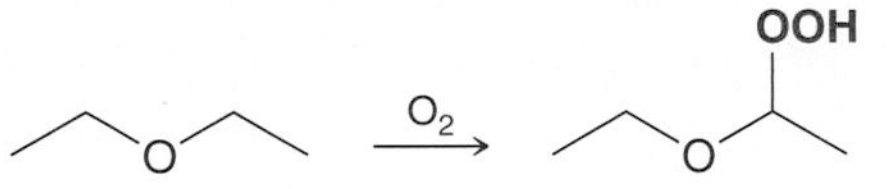

Diethyl ether A hydroperoxide

This reaction is slow, but old bottles of ether will invariably contain a small concentration of hydroperoxides, rendering the solvent very dangerous to use. Hydroperoxides are unstable and decompose violently when heated. Many laboratory explosions have been caused by distillation of ether that was contaminated with hydroperoxides. For this reason, ethers used in the laboratory must be dated and used in a timely fashion.

Most organic compounds are subject to autooxidation, which is a process initiated by light. In the absence of light, autooxidation occurs at a much slower rate. For this reason, organic chemicals are typically packaged in dark bottles. Vitamins are often sold in brown bottles for the same reason.

Compounds with benzylic or allylic hydrogen atoms are particularly sensitive to autooxidation, because the initiation step produces a resonance-stabilized radical:

Resonance stabilized

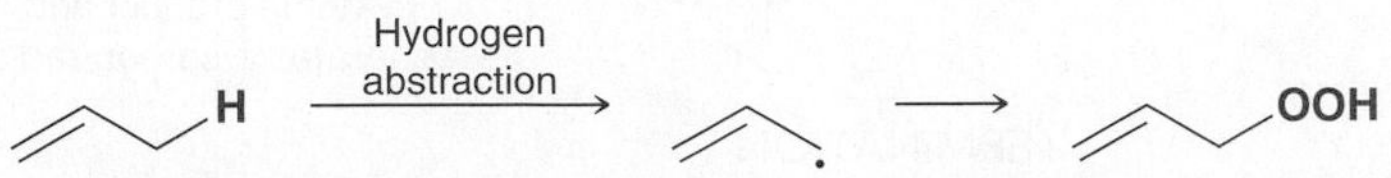

Resonance stabilized

Antioxidants as Food Additives

Naturally occurring fats and oils, such as vegetable oil, are generally mixtures of compounds called triglycerides, which contain three long alkyl chains:

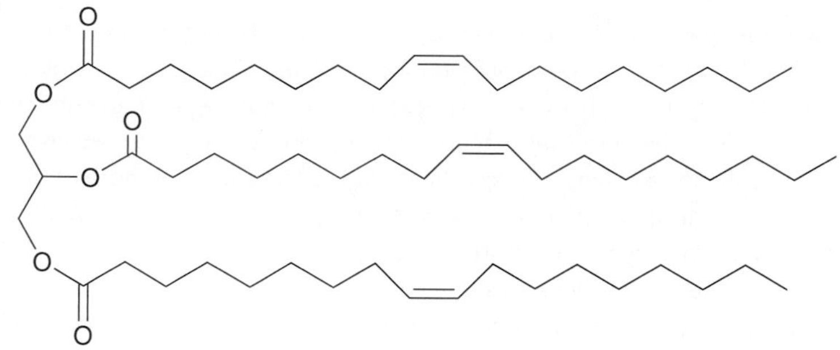

A triglyceride

The alkyl chains generally contain double bonds, which render them susceptible to autooxidation, specifically at the allylic positions where hydrogen abstraction can occur more readily:

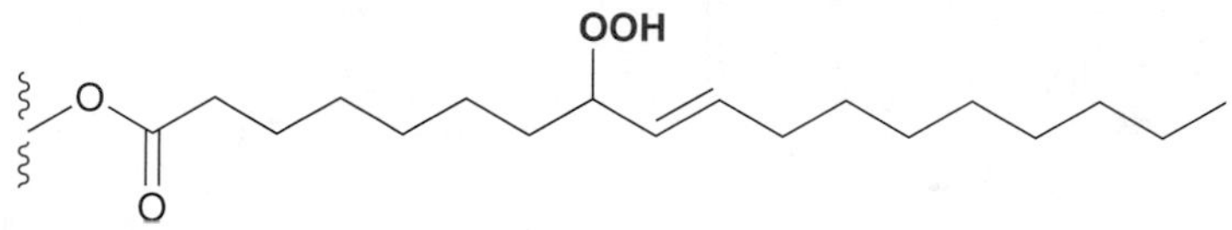

The resulting hydroperoxides contribute to the rancid smell that develops over time in foods containing unsaturated oils. Moreover, hydroperoxides are also toxic. Food products with unsaturated oils therefore have a short shelf life unless radical inhibitors are used to slow the autooxidation process. Many radical inhibitors are used as food preservatives, including BHT and BHA:

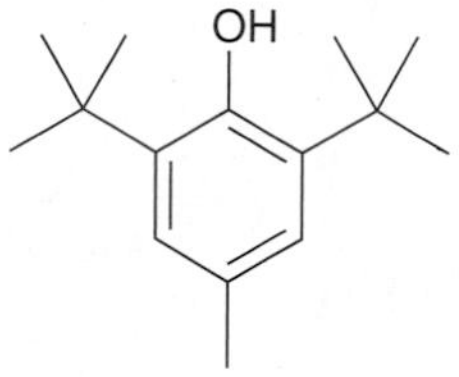

Butylated hydroxytoluene (BHT)

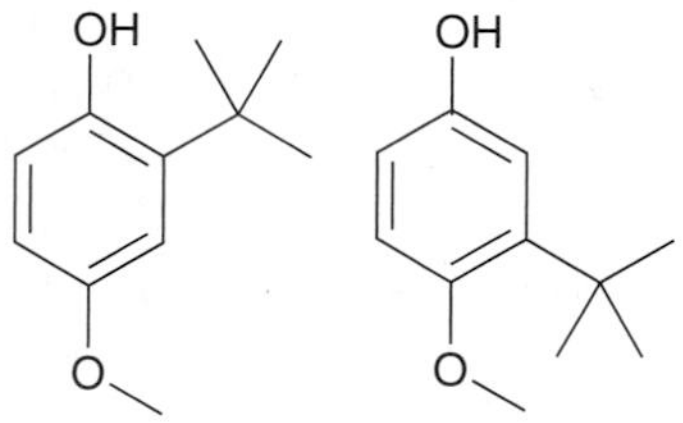

Butylated hydroxyanisole (BHA)

BHA is a mixture of constitutional isomers. These compounds function as radical inhibitors, because they react with radicals to generate resonance-stabilized radicals:

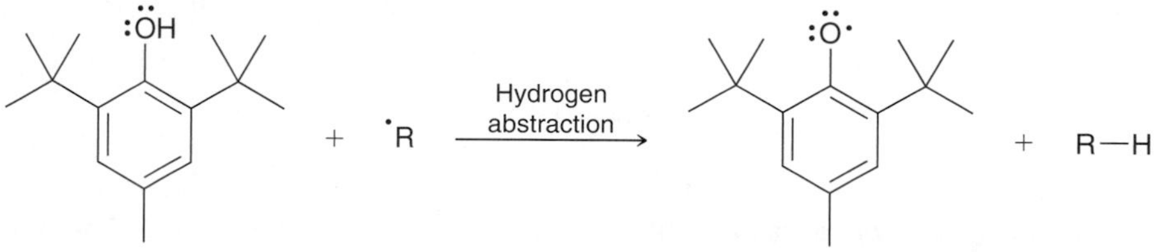

Resonance-stabilized

The *tert*-butyl groups provide steric hindrance, which further reduces the reactivity of the stabilized radical. BHT and BHA effectively scavenge and destroy radicals. They are called **antioxidants** because one molecule of a radical scavenger can prevent the autooxidation of thousands of oil molecules by not allowing the chain process to begin.

MEDICALLYSPEAKING)))

Why Is an Overdose of Acetaminophen Fatal?

Tylenol (acetaminophen) can be very helpful in alleviating pain, but it is well known that an overdose of Tylenol can be fatal. The source of this toxicity (in high doses) involves radical chemistry. Our bodies utilize many radical reactions for a wide variety of functions, but these reactions are controlled and localized. If uncontrolled, free radicals are very dangerous and are capable of damaging DNA and enzymes, ultimately leading to cell death. Free radicals are regularly produced as by-products of metabolism, and our bodies utilize a variety of compounds to destroy these radicals. One such example is glutathione, which bears a mercapto group (SH), highlighted in red:

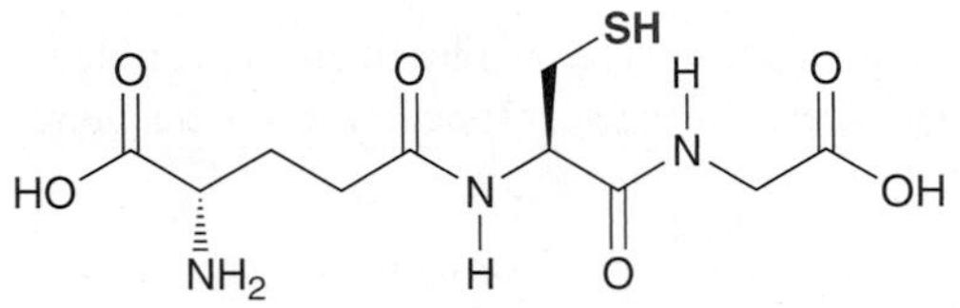

Glutathione

Glutathione, often abbreviated GSH, is produced in the liver and plays many key roles, including its ability to function as a radical scavenger. The SH bond is particularly susceptible to hydrogen abstraction, which produces a stabilized radical on a sulfur atom:

| GS—H | + | •R | ⟶ | GS• | + | H—R |

Glutathione **Glutathione radical**

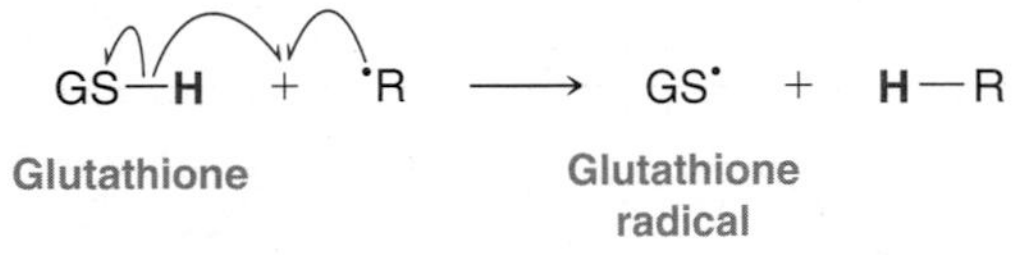

The original radical is destroyed and replaced by a glutathione radical. Two glutathione radicals, formed in this way, can then couple with each other to produce a compound called glutathione disulfide, abbreviated GSSG, which is then ultimately converted back into glutathione:

GS• + •SG ⟶ GS—SG

Glutathione disulfide

↓ reduction

2 GS—H

Glutathione

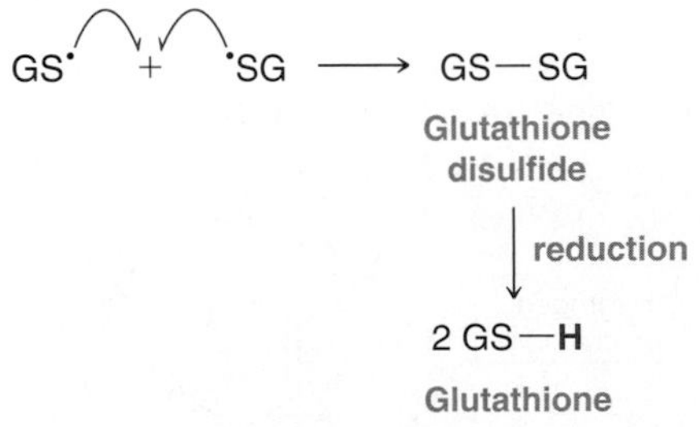

The important function of glutathione as a radical scavenger is compromised by an overdose of acetaminophen (Tylenol).

Acetaminophen is metabolized in the liver via a process that consumes glutathione, thereby causing a temporary reduction in glutathione levels. For people with healthy livers, the body is able to replenish the supply of glutathione quickly by biosynthesizing more of this important compound. In this way, the concentration of glutathione never reaches dangerously low levels. However, an overdose of acetaminophen can cause a temporary depletion of glutathione. During that time, free radicals are uncontrolled and cause a host of problems that lead to irreversible liver damage. If untreated, an overdose of acetaminophen can lead to liver failure and death within a few days. Early intervention can prevent irreversible liver damage and consists of treatment with *N*-acetylcysteine (NAC). NAC acts as an antidote for an acetaminophen overdose by delivering to the liver a high concentration of cysteine:

N-acetylcysteine (NAC) **Cysteine**

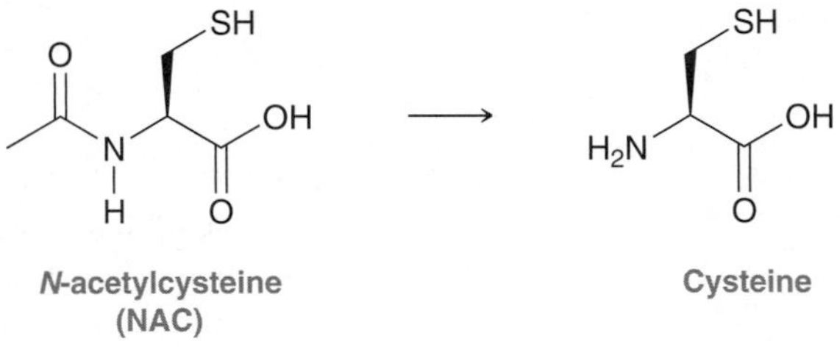

Glutathione is biosynthesized in the liver from cysteine, glycine, and glutamic acid:

Glutathione

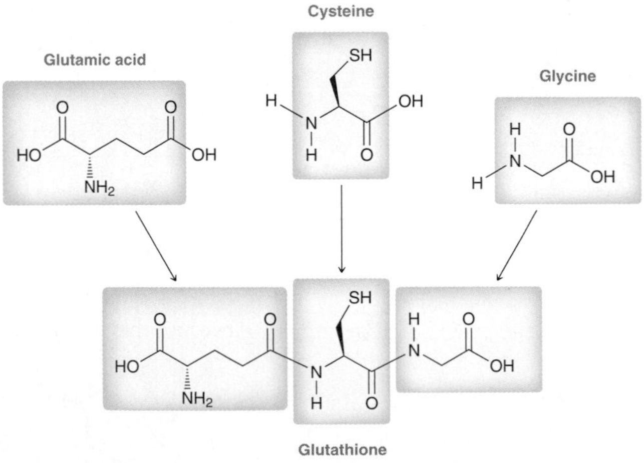

Glycine and glutamic acid are abundant, while cysteine is the limiting reagent. By supplying the liver with excess cysteine, the body is able to produce glutathione rapidly and bring the glutathione levels back up to healthy concentrations.

LOOKING AHEAD
The structure of cell membranes is discussed in more detail in Section 26.5.

Naturally Occurring Antioxidants

Nature employs many of its own natural antioxidants to prevent the oxidation of cell membranes and protect a variety of biologically important compounds. Examples of natural antioxidants include vitamin E and vitamin C:

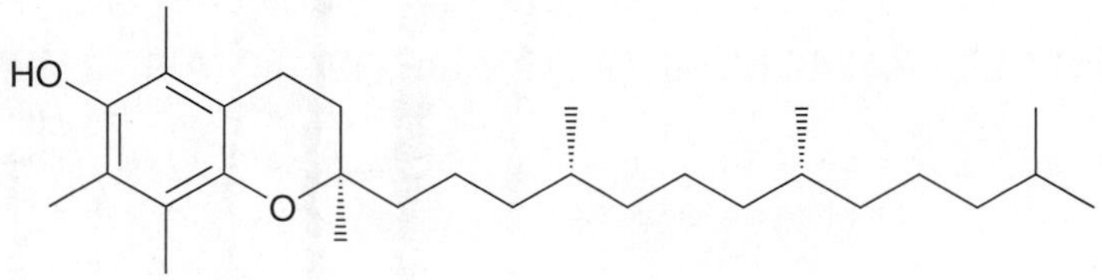

Vitamin E

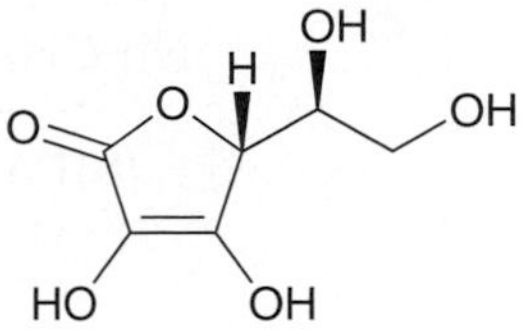

Vitamin C

Vitamin E has a long alkyl chain and is therefore hydrophobic, making it capable of reaching the hydrophobic regions of cell membranes. Vitamin C is a small molecule with multiple OH groups, rendering it water soluble. It functions as an antioxidant in hydrophilic regions, such as in blood.

Each of these compounds can destroy a reactive radical by undergoing hydrogen abstraction, transferring a hydrogen atom to the radical, thereby generating a more stable, less reactive radical. A connection has been suggested between the aging process and the natural oxidation processes that occur in the body. This suggestion has led to the widespread use of antioxidant products that include compounds such as vitamin E, despite the lack of evidence for a measurable connection between these products and the rate of aging.

CONCEPTUAL CHECKPOINT

11.19 Compare the structure of vitamin E with the structures of BHT and BHA, and then determine which hydrogen atom is most easily abstracted from vitamin E.

11.10 Radical Addition of HBr: *Anti*-Markovnikov Addition

Regiochemical Observations

In Chapter 9, we saw that an alkene will react with HBr via an ionic addition that installs a bromine atom at the more substituted position (Markovnikov addition):

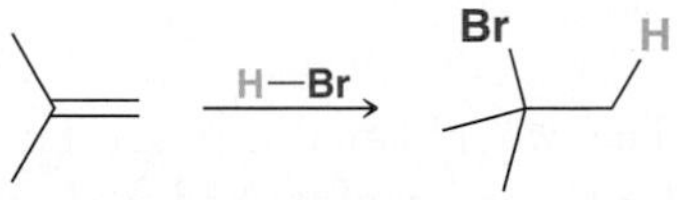

Markovnikov addition

Recall that purity of reagents was found to be a critical feature of HBr additions. With impure reagents the reaction proceeds via *anti-Markovnikov addition*, where the halogen appears in the product at the less substituted position:

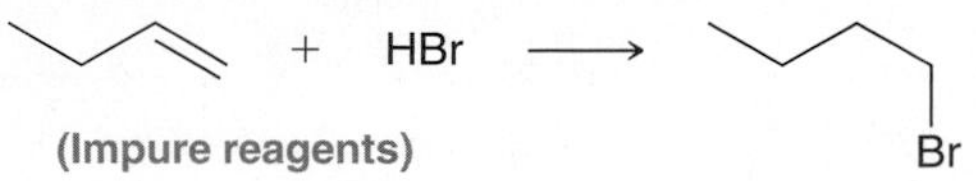

Further investigation revealed that even trace amounts of peroxides (ROOR) can cause this regiochemical preference for *anti*-Markovnikov addition:

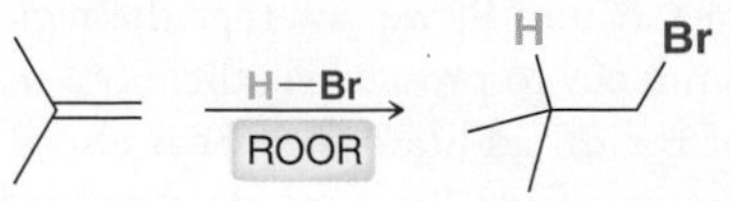

(95%)

***Anti*-Markovnikov addition**

A Mechanism for *Anti*-Markovnikov Addition of HX

Anti-Markovnikov addition of HBr in the presence of peroxides can be explained by invoking a
radical mechanism (Mechanism 11.3).

MECHANISM 11.3 RADICAL ADDITION OF HBr TO AN ALKENE

INITIATION

Homolytic cleavage

$$RO\overset{}{-}OR \xrightarrow[\text{Creates two alkoxy radicals}]{\text{Heat}} RO\cdot \quad \cdot OR \qquad \Delta H° = +151 \text{ kJ/mol}$$

Hydrogen abstraction

$$RO\cdot \quad H\overset{}{-}Br: \xrightarrow[\text{Forms a bromine radical}]{} ROH \quad + \quad \cdot Br: \qquad \Delta H° = -63 \text{ kJ/mol}$$

PROPAGATION

:Br·

Addition

The bromine radical adds
to the π bond giving the
more stable carbon radical

:Br

$\Delta H° = -13$ kJ/mol

:Br

Hydrogen abstraction

H—Br:

Gives the product
and regenerates
a bromine radical

:Br

—H + :Br·

$\Delta H° = -25$ kJ/mol

TERMINATION

Coupling

$$:Br\cdot \quad \cdot Br: \xrightarrow[\substack{\text{Destroys two} \\ \text{bromine radicals}}]{} :Br\text{—}Br:$$

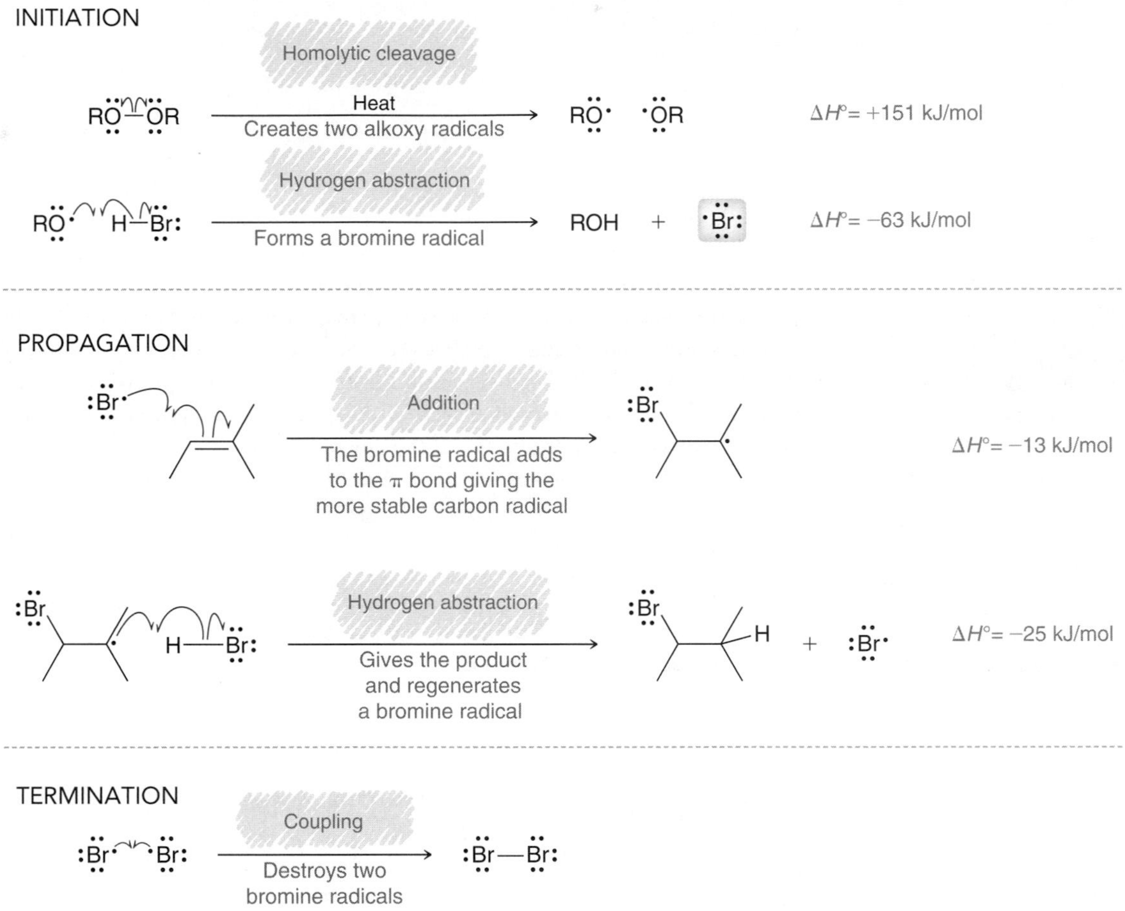

The two initiation steps generate a bromine radical. The two propagation steps are responsible
for formation of the observed product, so we must focus on these two steps in order to understand the regiochemical preference for *anti*-Markovnikov addition. Notice that the intermediate
formed in the first propagation step is a tertiary carbon radical, rather than a secondary radical:

| 3° radical | 2° radical |
| more stable, favored | less stable, <u>not formed</u> |

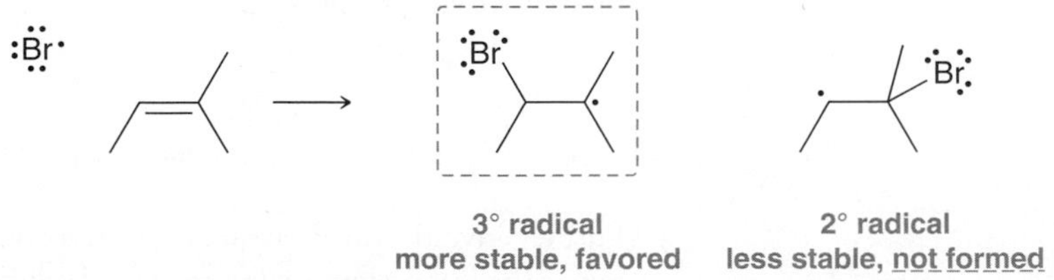

In Chapter 9, we saw that the regiochemistry of ionic addition of HBr is determined by the
tendency to proceed via the more stable carbocation intermediate. Similarly, the regiochemistry
of radical addition of HBr is also determined by the tendency to proceed via the more stable
intermediate. But here, the intermediate is a radical, rather than a carbocation. To see this more

clearly, compare the intermediate involved in an ionic mechanism with the intermediate of a radical mechanism:

Ionic mechanism　　　　Radical mechanism

Tertiary carbocation intermediate　　　　Tertiary radical intermediate

Each reaction proceeds via the lowest energy pathway available—via either a tertiary carbocation or a tertiary radical. But take special notice of the fundamental difference. In the ionic mechanism, the alkene reacts with the proton first, while in the radical mechanism, the alkene reacts with the bromine first. Therefore:

- An ionic mechanism results in a Markovnikov addition.
- A radical mechanism results in an *anti*-Markovnikov addition.

In each case, the regioselectivity is based on the tendency to proceed via the most stable possible intermediate.

Thermodynamic Considerations

Markovnikov addition can be accomplished with HCl, HBr, or HI. In contrast, *anti*-Markovnikov addition can only be accomplished with HBr. The radical pathway is not thermodynamically favorable for addition of either HCl or HI. To understand the reason for this, we must explore the thermodynamics of each step of the radical mechanism.

In Section 6.3 (and earlier in this chapter), we saw that the sign of ΔG must be negative in order for a reaction to be spontaneous. Recall that the two components of ΔG are enthalpy and entropy:

$$\Delta G = \underbrace{(\Delta H)}_{\text{Enthalpy term}} + \underbrace{(-T\,\Delta S)}_{\text{Entropy term}}$$

In order to assess the sign of ΔG for any process, we must evaluate the signs of both the enthalpy and entropy terms. At the beginning of this chapter, we posed this type of thermodynamic argument to explore the halogenation of alkanes. In this section, we will explore each propagation step of the radical addition mechanism separately.

Let's begin with the first propagation step. In the following chart the enthalpy and entropy terms are evaluated individually for the cases of HCl, HBr, and HI:

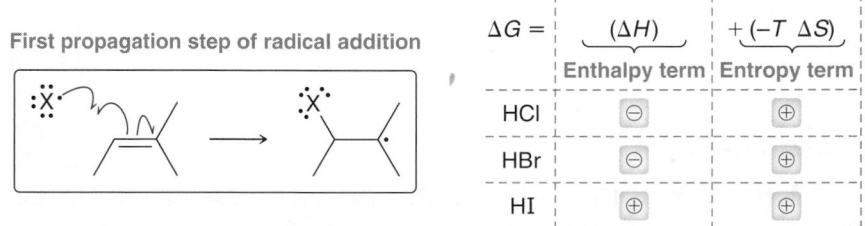

First propagation step of radical addition

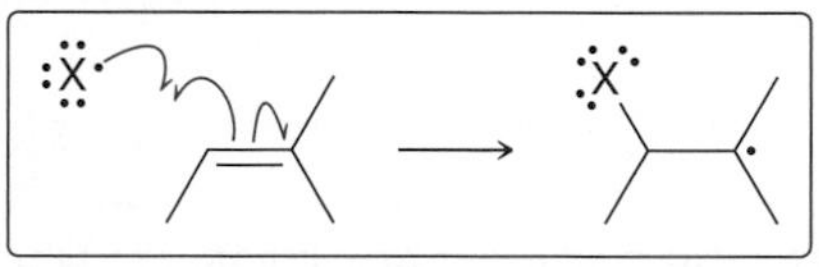

$\Delta G =$	(ΔH) Enthalpy term	$+ (-T\,\Delta S)$ Entropy term
HCl	⊖	⊕
HBr	⊖	⊕
HI	⊕	⊕

For the cases of HCl and HBr, the sign of ΔG is determined by a competition between the enthalpy and entropy terms. At high temperature, the entropy term will dominate, and ΔG will be positive. At low temperature, the enthalpy term will dominate, and ΔG will be negative. Therefore, the process will be thermodynamically favorable at low temperatures.

The case of HI is fundamentally different. Look at the previous chart and notice that the enthalpy term is positive in the case of HI (the step is endothermic). The enthalpy and entropy terms are not in competition here. Regardless of temperature ΔG will be positive. Therefore, radical addition of HI (*anti*-Markovnikov addition) is not observed at all.

Now let's perform the same type of analysis for the second propagation step in the radical mechanism:

Second propagation step of radical addition

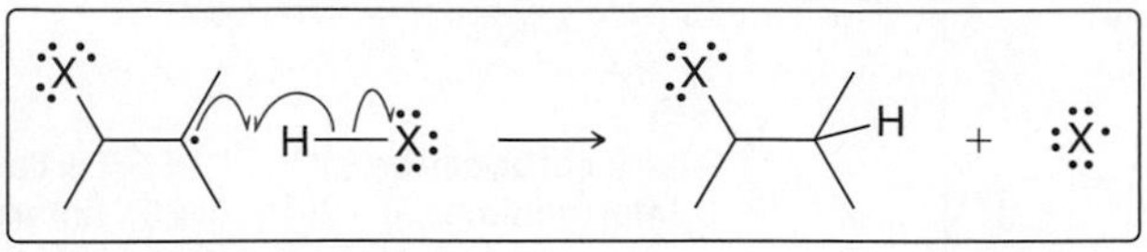

$\Delta G =$	(ΔH) Enthalpy term	$+(-T\,\Delta S)$ Entropy term
HCl	⊕	~0
HBr	⊖	~0
HI	⊖	~0

In each of the cases above, the second term ($-T\Delta S$) is insignificant. Why? In each case, *two* chemical entities are reacting with each other to produce *two* new chemical entities. Based on this, we might expect $\Delta S = 0$. However, there is still some small value of ΔS in each case, due to changes in vibrational degrees of freedom. This effect will be very small though, and as a result, ΔS will be close to zero. Therefore, the sign of ΔH will be the dominant factor in determining the sign of ΔG in each case. In the case of HCl, ΔH has a positive value. Therefore, it will be difficult to find a temperature at which the second term ($-T\Delta S$) can overcome the ΔH term. As a result, radical addition of HCl is not an effective process.

To summarize our analysis for radical addition of HX across a π bond: The first propagation step cannot occur in the case of HI, and the second propagation step is unlikely to occur in the case of HCl. Only in the case of HBr are both propagation steps thermodynamically favorable.

Stereochemistry for Radical Addition of HBr

For some alkenes, radical addition of HBr results in the formation of one new chirality center:

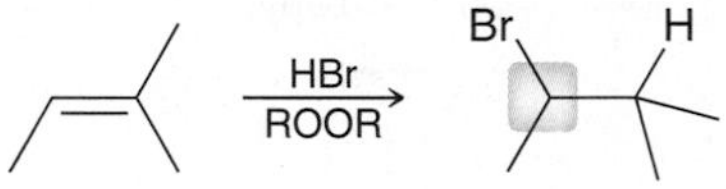

In cases like this, there is no reason to expect one enantiomer to prevail over the other, as the first propagation step can occur from either face of the alkene. Therefore, the reaction will produce a racemic mixture of enantiomers:

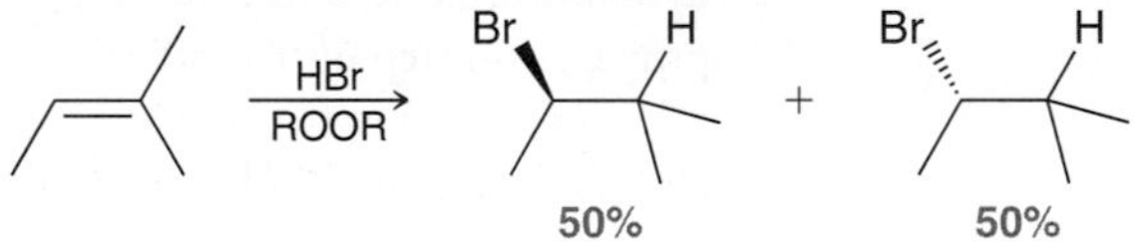

SKILLBUILDER

11.8 PREDICTING THE PRODUCTS FOR RADICAL ADDITION OF HBr

LEARN the skill

Predict the products for the following reaction:

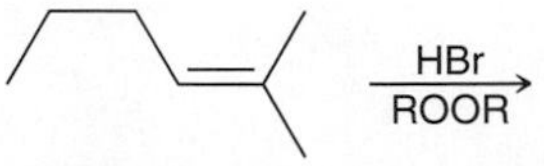

SOLUTION

In Section 9.12, we saw that predicting the products of an addition reaction requires the following three questions:

STEP 1 Identify the two groups being added across the double bond.

1. Which two groups are being added across the double bond? HBr indicates an addition of H and Br across the double bond.

2. What is the expected regioselectivity (Markovnikov or *anti*-Markovnikov)? The presence of peroxides indicates that the reaction proceeds via an *anti*-Markovnikov addition. That is, the Br is expected to be placed at the less substituted carbon:

STEP 2
Identify the expected regioselectivity.

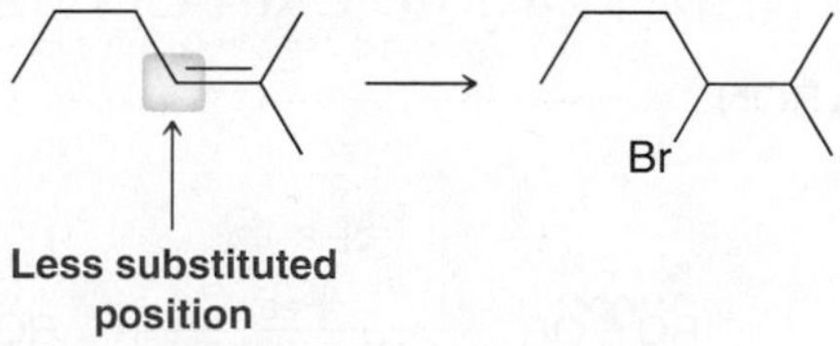

3. What is the expected stereospecificity? In this case, one new chirality center is created, which results in a racemic mixture of the two possible enantiomers:

STEP 3
Identify the expected stereospecificity.

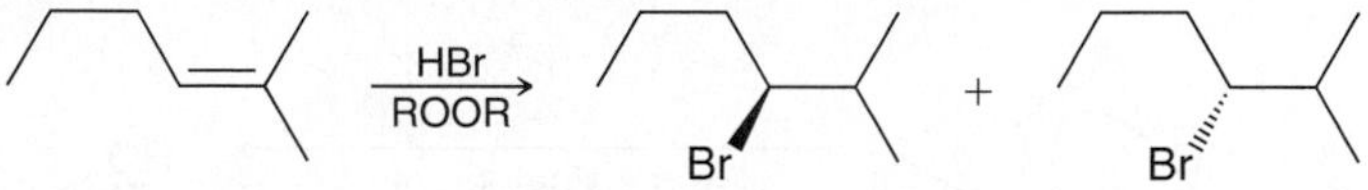

PRACTICE the skill

11.20 Predict the products for each reaction. In each case, be sure to consider whether a chirality center is being generated and then draw all expected stereoisomers.

(a) 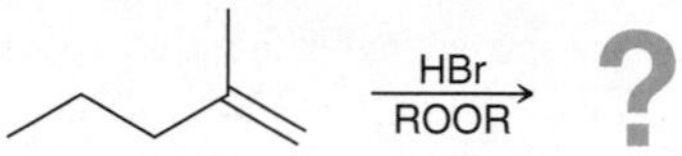

(b) 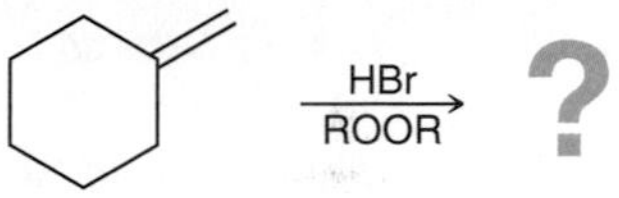

(c) 

(d) 

(e) 

(f) 

APPLY the skill

11.21 The initiation step for radical addition of HBr is highly endothermic:

$$\mathrm{R\ddot{\underset{\cdot\cdot}{O}}{-}\ddot{\underset{\cdot\cdot}{O}}R} \xrightarrow[\text{or heat}]{h\nu} \mathrm{R\ddot{\underset{\cdot\cdot}{O}}\cdot\quad \cdot\ddot{\underset{\cdot\cdot}{O}}R} \qquad \Delta H° = +151 \text{ kJ/mol}$$

(a) Explain how this step can be thermodynamically favorable at high temperature even though it is endothermic.

(b) Explain why this step is not thermodynamically favorable at low temperature.

11.11 Radical Polymerization

Radical Polymerization of Ethylene

In Chapter 9 we explored polymerization that occurs via an ionic mechanism. In this section, we will see that polymerization can also occur via a radical process. Consider, for example, the radical polymerization of ethylene to form polyethylene:

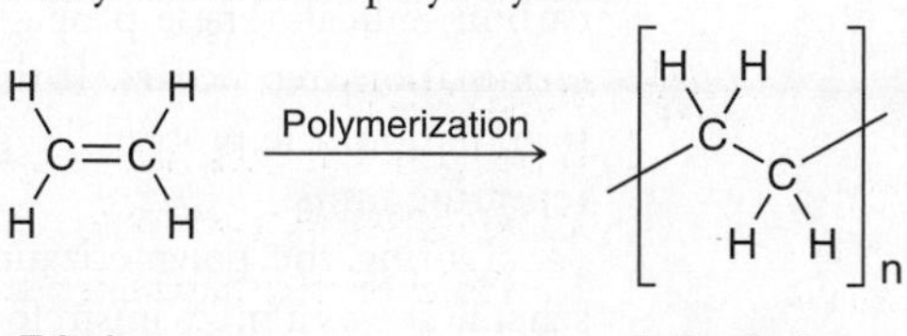

Ethylene **Polyethylene**

This polymerization can occur via a radical mechanism (Mechanism 11.4).

MECHANISM 11.4 RADICAL POLYMERIZATION

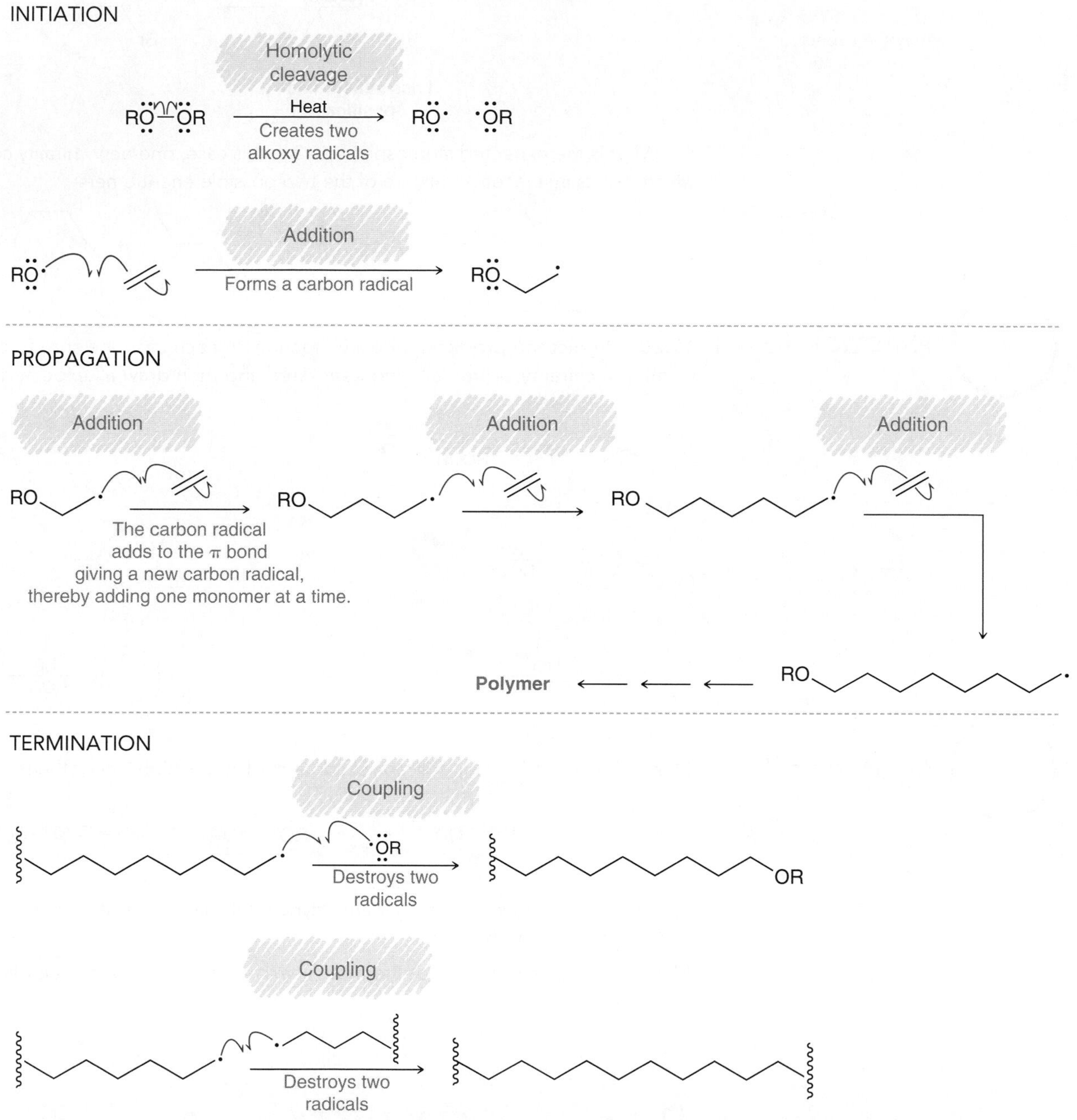
INITIATION
Homolytic cleavage
Heat
Creates two alkoxy radicals
Addition
Forms a carbon radical
PROPAGATION
Addition
The carbon radical adds to the π bond giving a new carbon radical, thereby adding one monomer at a time.
Addition
Addition
Polymer
TERMINATION
Coupling
Destroys two radicals
Coupling
Destroys two radicals

When ethylene is heated in the presence of peroxides, two initiation steps produce a carbon radical. This carbon radical then attacks another molecule of ethylene, generating a new carbon radical. These propagation steps continue, adding one monomer at a time, until a termination step occurs. If the conditions are carefully controlled to favor propagation over termination, the process can generate very large molecules of polyethylene with over 10,000 repeating units.

During the polymerization process, **chain branching** inevitably occurs. The following scheme shows a mechanism for the formation of chain branches, which begins with a hydrogen abstraction that occurs on an existing chain:

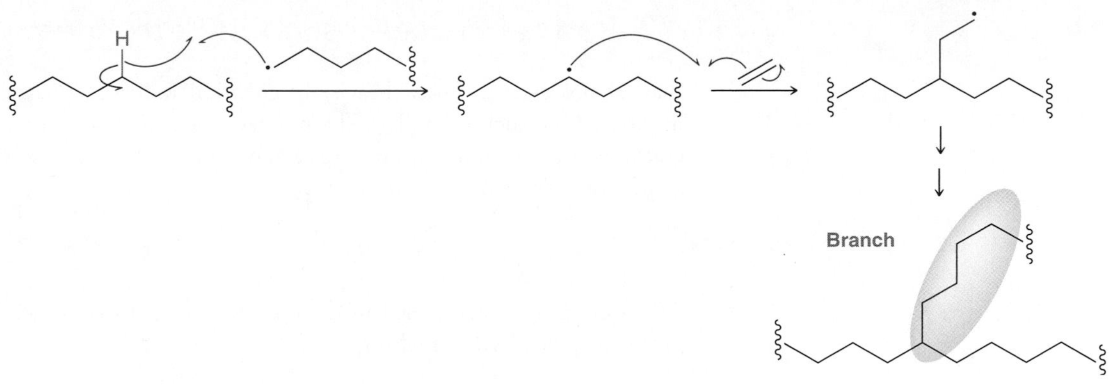

The extent of chain branching determines the physical properties of the resulting polymer. For example, a flexible plastic squeeze bottle and its relatively inflexible cap are both made from polyethylene, but the former has extensive branching while the latter has relatively little. These processes will be discussed in more detail in Chapter 27, and we will see that the extent of branching can be controlled with specialized catalysts, although a small amount of chain branching is generally unavoidable.

Radical Polymerization of a Substituted Ethylene

A substituted ethylene (ethylene that bears a substituent) will generally undergo radical polymerization to produce a polymer with the following structure:

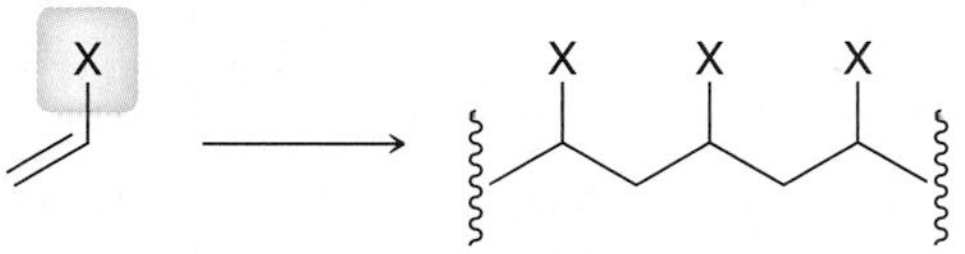

For example, vinyl chloride is polymerized to form **polyvinyl chloride (PVC)**, which is a very hard polymer used to make plumbing pipes. PVC can be softened if polymerization takes place in the presence of compounds called plasticizers, as we will explore in more detail in Chapter 27. PVC made with plasticizers is a bit more flexible (although durable and strong) and is used for a wide variety of purposes, such as garden hoses, plastic raincoats, and shower curtains. Table 11.3 shows a number of common polymers produced via a radical process.

TABLE 11.3 COMMON POLYMERS FORMED FROM ETHYLENE AND SUBSTITUTED ETHYLENE

MONOMER	POLYMER	APPLICATION	MONOMER	POLYMER	APPLICATION
 **Ethylene**	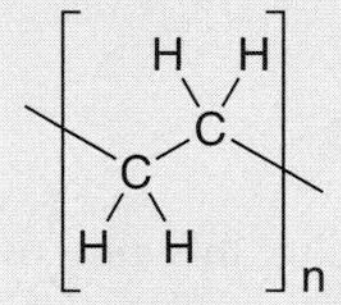 **Polyethylene**	Plastic bottles and packaging material, insulation, Tupperware	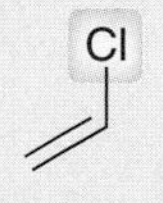 **Vinyl chloride**	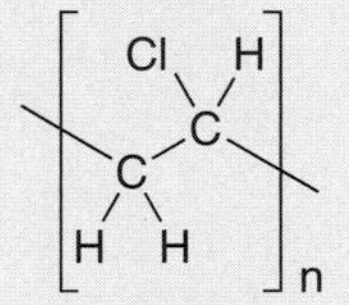 **Poly(vinyl chloride)**	PVC piping for plumbing, CDs, garden hoses, raincoats, shower curtains
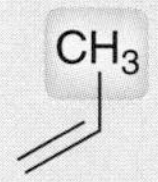 **Propylene**	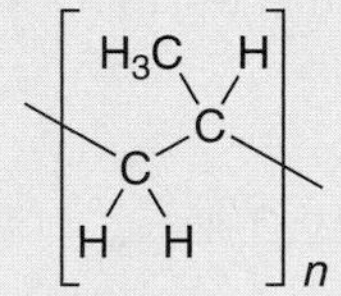 **Polypropylene**	Carpet fibers, appliances, car tires	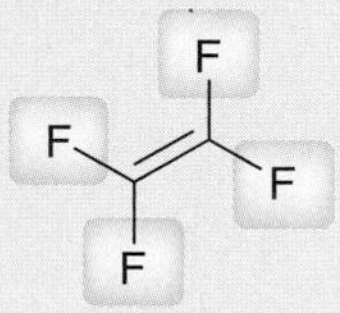 **Tetrafluoroethylene**	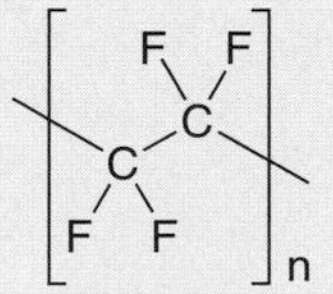 **Teflon**	Nonstick coating for frying pans
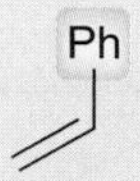 **Styrene**	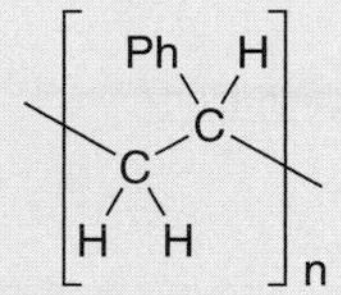 **Polystyrene**	Televisions, radios, Styrofoam			

11.12 Radical Processes in the Petrochemical Industry

Radical processes are used heavily in chemical industries, particularly in the petrochemical industry. One such example is the process called *cracking*. In Chapter 4, we mentioned that cracking of petroleum converts large alkanes into smaller alkanes that are more suitable for use as gasoline. Cracking is a radical process:

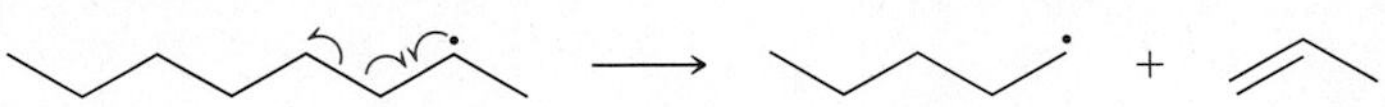

When cracking is performed in the presence of hydrogen gas, alkanes are produced and the process is called **hydrocracking**:

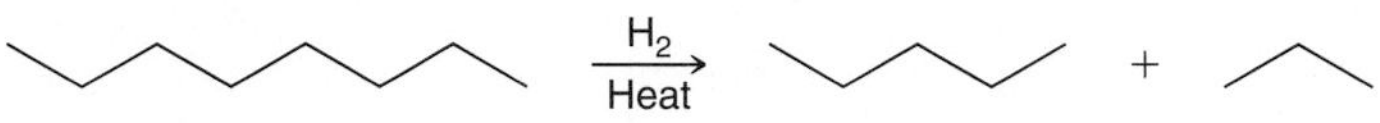

Reforming, also mentioned in Chapter 4, is a process that causes straight-chain alkanes to become more highly branched, and this process also involves radical intermediates.

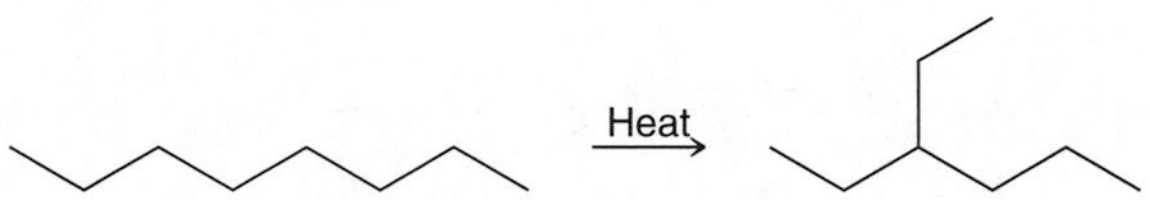

11.13 Halogenation as a Synthetic Technique

In this chapter, we have seen that radical chlorination and radical bromination are both thermodynamically favorable processes. Bromination is slower but is more selective than chlorination. Both reactions are used in synthesis. When the starting compound has only one kind of hydrogen (all hydrogen atoms are equivalent), chlorination can be accomplished without having to worry about regiochemical outcomes:

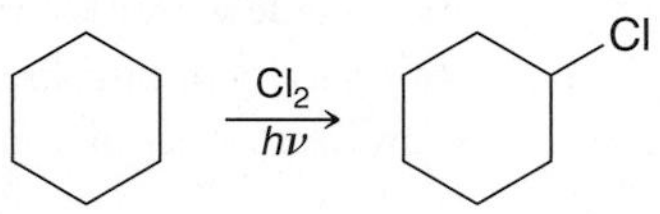

When different types of hydrogen atoms are present in the compound, it is best to use bromination in order to control the regiochemical outcome and avoid a mixture of products:

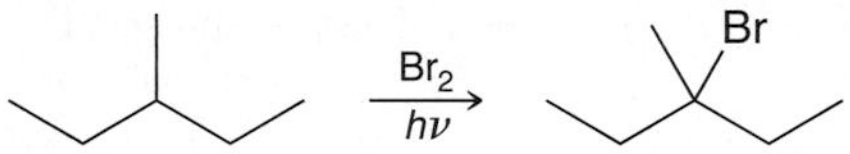

In truth, even radical bromination has very limited utility in synthesis. Its greatest utility is to serve as a method for introducing a functional group into an alkane. When the starting material is an alkane, there is very little that can be done other than radical halogenation. By introducing a functional group into the compound, the door is opened for a wide variety of reactions:

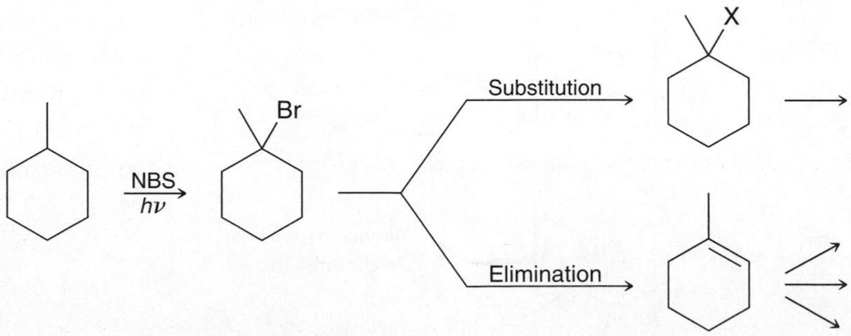

Chapter 12 is devoted entirely to synthesis techniques, and we will revisit the role of radical halogenation in designing syntheses.

REVIEW OF REACTIONS

SYNTHETICALLY USEFUL RADICAL REACTIONS

BROMINATION OF ALKANES

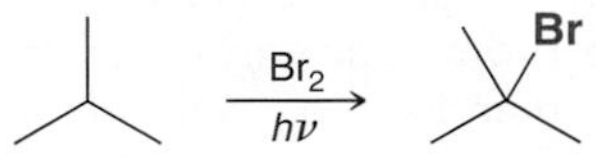

ANTI-MARKOVNIKOV ADDITION OF HBr TO ALKENES

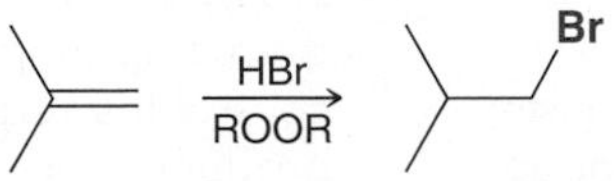

ALLYLIC BROMINATION

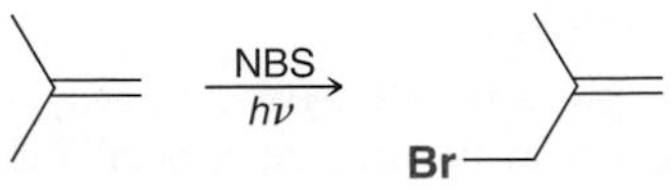

REVIEW OF CONCEPTS AND VOCABULARY

SECTION 11.1

- Radical mechanisms utilize **fishhook arrows**, each of which represents the flow of only one electron.
- The order of stability for radicals follows the same trend exhibited by carbocations.
- Allylic and benzylic radicals are resonance stabilized. Vinylic radicals are not.

SECTION 11.2

- Radical mechanisms are characterized by six different kinds of steps: (1) **homolytic cleavage**, (2) **addition to a π bond**, (3) **hydrogen abstraction**, (4) **halogen abstraction**, (5) **elimination**, and (6) **coupling.**
- Every step in a radical mechanism can be classified as **initiation**, **propagation**, or **termination**.

SECTION 11.3

- Methane reacts with chlorine via a radical mechanism.
- The sum of the two propagation steps gives the net chemical reaction. These steps together represent a **chain reaction.**
- A **radical initiator** is a compound with a weak bond that readily undergoes homolytic bond cleavage. Examples include **peroxides** and **acyl peroxides**.
- A **radical inhibitor**, also called a radical scavenger, is a compound that prevents a chain process from either getting started or continuing. Examples include molecular oxygen and hydroquinone.

SECTION 11.4

- Only radical chlorination or radical bromination have practical use in the laboratory.
- Bromination is generally a much slower process than chlorination.

SECTION 11.5

- Halogenation occurs more readily at substituted positions. Bromination is more selective than chlorination.
- In general, there is an inverse relationship between reactivity and selectivity.

SECTION 11.6

- When a new chirality center is created during a radical halogenation process, a racemic mixture is obtained.
- When a halogenation reaction takes place at a chirality center, a racemic mixture is obtained regardless of the configuration of the starting material.

SECTION 11.7

- Alkenes can undergo **allylic bromination**, in which bromination occurs at the allylic position.
- To avoid a competing ionic addition reaction, ***N*-bromosuccinimide** (NBS) can be used instead of Br_2.

SECTION 11.8

- Ozone is produced and destroyed by a radical process that shields the earth's surface from harmful UV radiation.
- A rapid decrease of stratospheric ozone is attributed to the use of **CFCs**, or **chlorofluorocarbons**, sold under the trade name **Freons.**
- A ban on CFCs prompted a search for viable substitutes, such as **hydrochlorofluorocarbons (HCFCs)** and **hydrofluorocarbons (HFCs).**

SECTION 11.9

- Organic compounds undergo oxidation in the presence of atmospheric oxygen to produce **hydroperoxides**. This process, called **autooxidation**, is believed to proceed via a radical mechanism.
- **Antioxidants**, such as BHT and BHA, are used as food preservatives to prevent autooxidation of unsaturated oils.
- Natural antioxidants prevent the oxidation of cell membranes and protect a variety of biologically important compounds. Vitamins E and C are natural antioxidants.

SECTION 11.10

- Alkenes will react with HBr in the presence of peroxides to produce a radical addition reaction.

SECTION 11.11

- Polymerization of ethylene via a radical process generally involves **chain branching**.
- When vinyl chloride is polymerized, **polyvinyl chloride (PVC)** is obtained.

SECTION 11.12

- Radical processes are used heavily in the chemical industry, particularly in the petrochemical industry. Examples include cracking and reforming. When cracking is performed in the presence of hydrogen, it is called **hydrocracking**.

SECTION 11.13

- Radical halogenation provides a method for introducing functionality into an alkane.
- When the starting compound has only one kind of hydrogen, chlorination can be used.
- When different types of hydrogen atoms are present in the compound, it is best to use bromination in order to control the regiochemical outcome and avoid a mixture of products.

KEY TERMINOLOGY

acyl peroxides 503
addition to π bond 496
allylic bromination 512
antioxidants 519
autooxidation 517
CFCs 515
chain branching 526
chain reaction 501
coupling 497
elimination 497
fishhook arrows 491
Freon 515
halogen abstraction 497
HCFCs 515
HFCs 517
homolytic cleavage 496
hydrocracking 528
hydrogen abstraction 496
hydroperoxide 517
initiation 499
***N*-bromosuccinimide** 512
peroxides 502
propagation 499
PVC 527
radical inhibitor 503
radical initiator 502
termination 499

SKILLBUILDER REVIEW

11.1 DRAWING RESONANCE STRUCTURES OF RADICALS

STEP 1 Look for an unpaired electron next to a π bond.

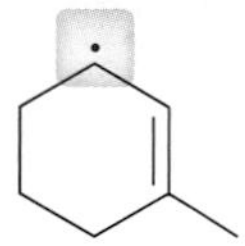

STEP 2 Draw three fishhook arrows, and then draw the corresponding resonance structure.

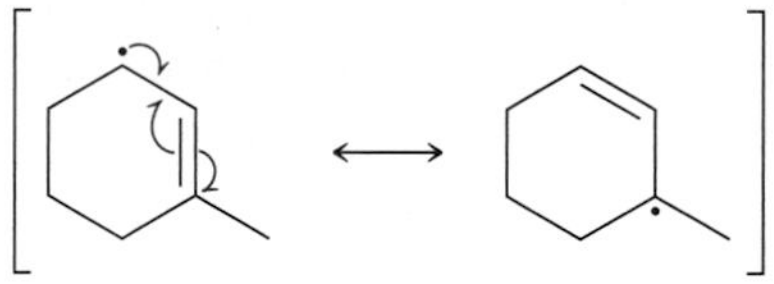

Try Problems 11.2–11.4, 11.22, 11.25

11.2 IDENTIFYING THE WEAKEST C—H BOND IN A COMPOUND

STEP 1 Consider all possible radicals that would result from homolytic cleavage of a C—H bond.

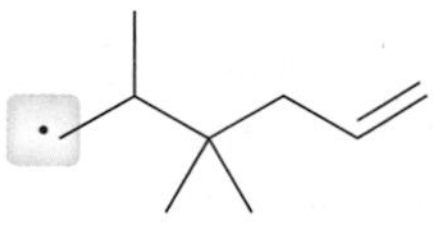

Primary

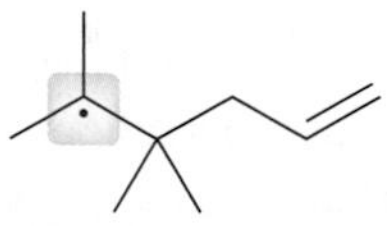

Tertiary

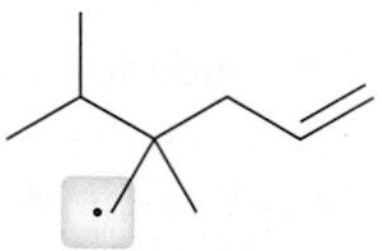

Primary

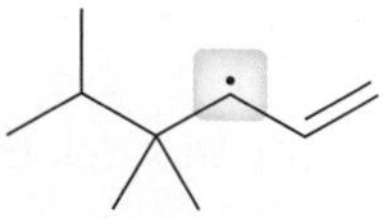

Allylic

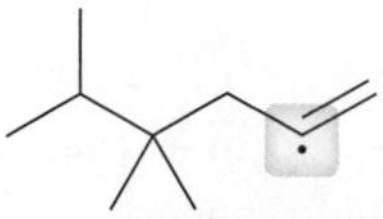

Vinylic

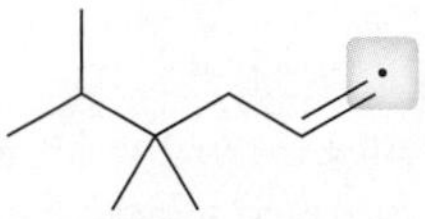

Vinylic

STEP 2 Identify the most stable radical.

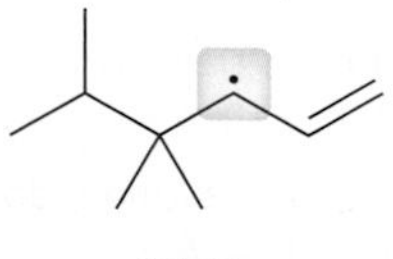

Allylic

STEP 3 The most stable radical indicates which bond is the weakest.

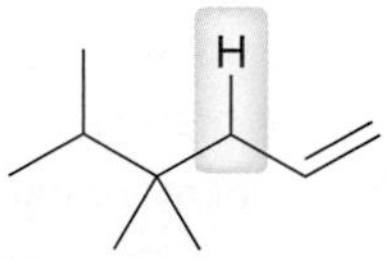

Weakest bond

Try Problems 11.5, 11.6, 11.23

11.3 IDENTIFYING A RADICAL PATTERN AND DRAWING FISHHOOK ARROWS

There are six characteristic patterns to recognize.

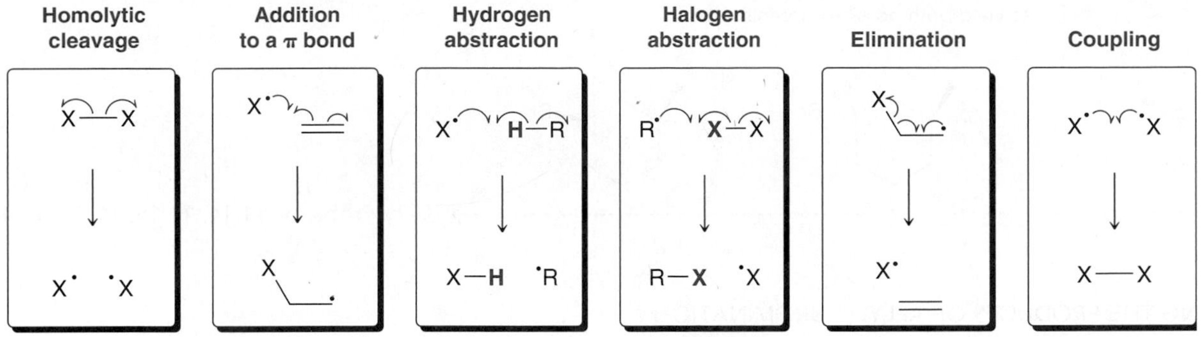

STEP 1–3 1. Identify the type of process—halogen abstraction.
2. Determine the number of fishhook arrows required—three.
3. Identify any bonds being broken or formed.

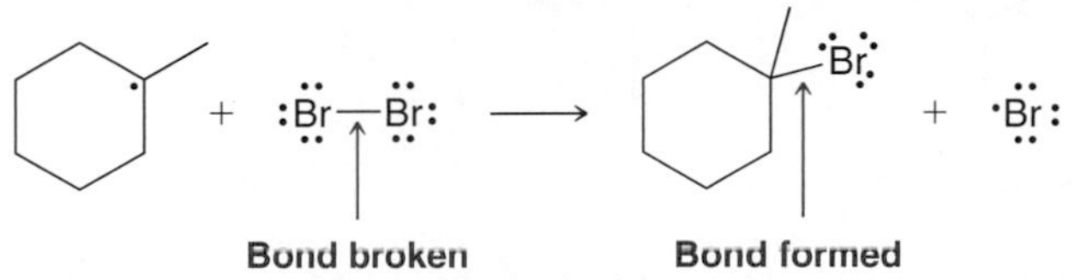

STEP 4 For a bond formed, draw two fishhook arrows.

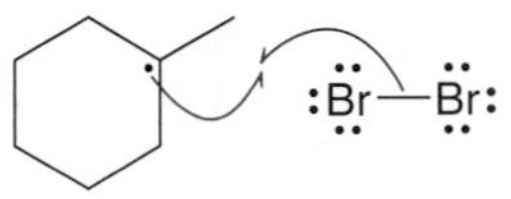

STEP 5 For a bond broken, make sure to have both fishhook arrows.

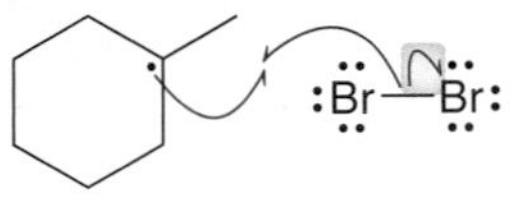

Try Problems 11.7–11.9, 11.48

11.4 DRAWING A MECHANISM FOR RADICAL HALOGENATION

Initiation Homolytic bond cleavage creates radicals, thereby initiating a radical process.

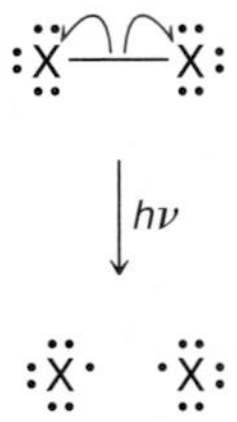

Propagation • Hydrogen abstraction forms a radical on a carbon atom.
• Halogen abstraction forms the product and regenerates a halogen radical.

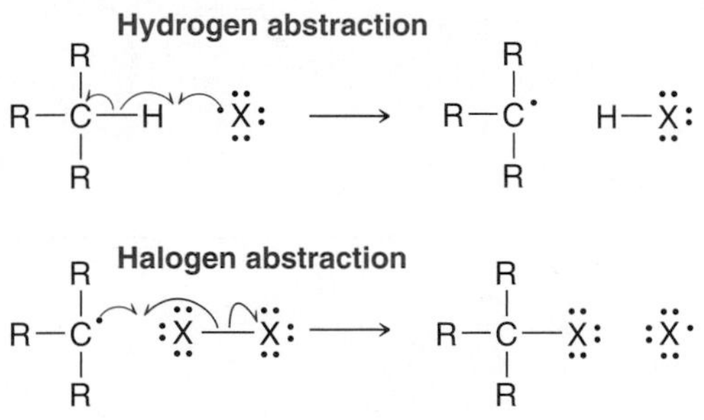

Termination Coupling can give the product, but this step destroys radicals.

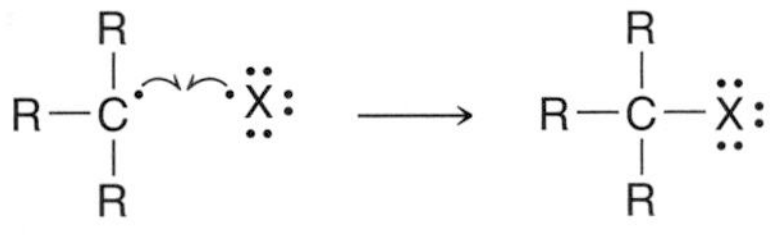

Try Problems 11.10, 11.11

11.5 PREDICTING THE REGIOCHEMISTRY OF RADICAL BROMINATION

STEP 1 Identify all possible positions.

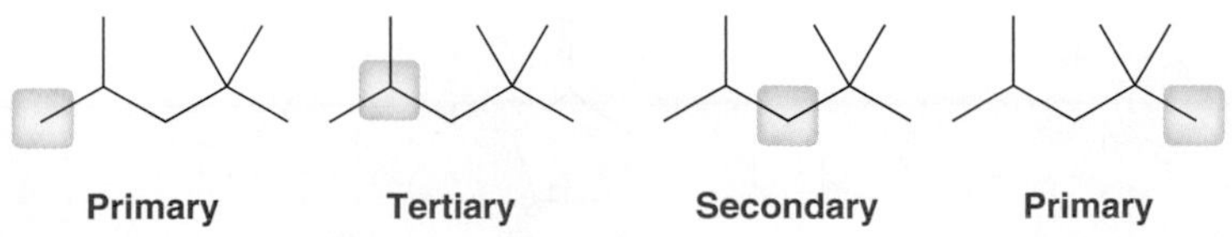

STEP 2 Identify the most stable location for a radical.

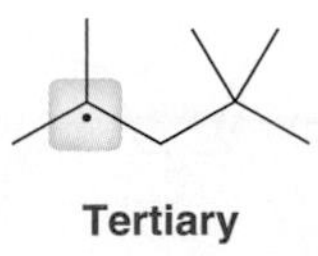

STEP 3 Bromination occurs at the location of the most stable radical.

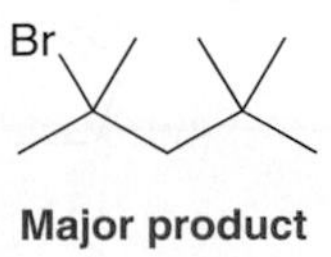

Try Problems 11.12, 11.13, 11.27, 11.33a-c, f, 11.38, 11.40, 11.44

11.6 PREDICTING THE STEREOCHEMICAL OUTCOME OF RADICAL BROMINATION

STEP 1 First identify the location where bromination will occur.

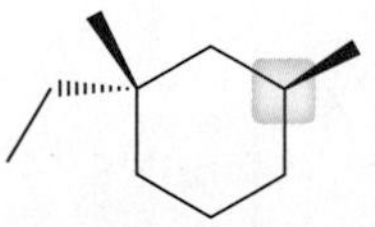

STEP 2 If this position is a chirality center, or if it will become a chirality center, expect a loss of configuration to yield both possible stereoisomers.

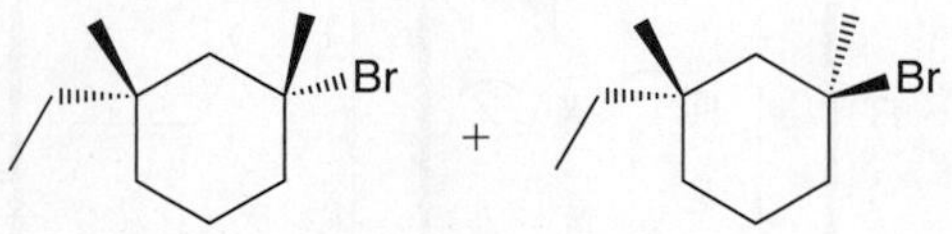

Try Problems 11.14, 11.15, 11.35, 11.41

11.7 PREDICTING THE PRODUCTS OF ALLYLIC BROMINATION

STEP 1 Identify the allylic position.

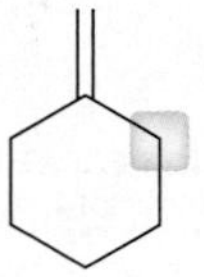

STEP 2 Remove a hydrogen atom and draw resonance structures.

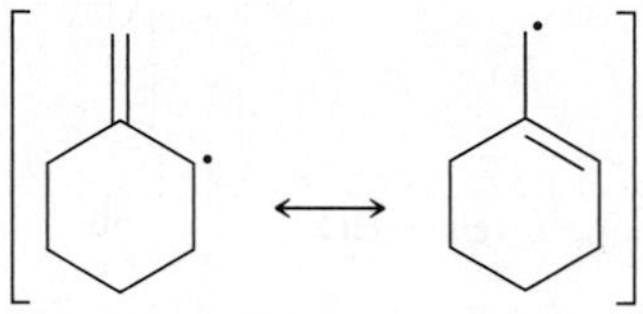

STEP 3 Place a bromine at each location that bears an unpaired electron.

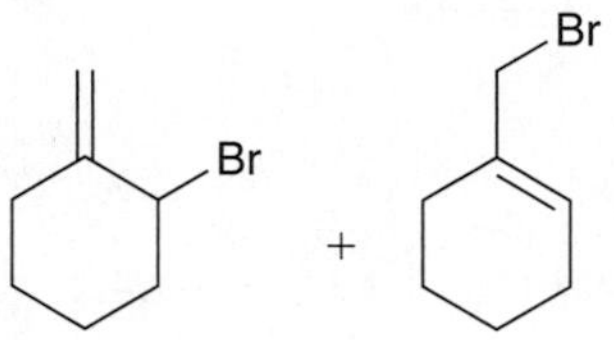

Racemic

Try Problems 11.16, 11.17, 11.26, 11.28, 11.33d, e, 11.36, 11.39

11.8 PREDICTING THE PRODUCTS FOR RADICAL ADDITION OF HBr

STEP 1 Identify the two groups being added across the double bond.

$$\xrightarrow[\text{ROOR}]{\text{HBr}} \quad \textbf{Adds H and Br}$$

STEP 2 Identify the expected regioselectivity.

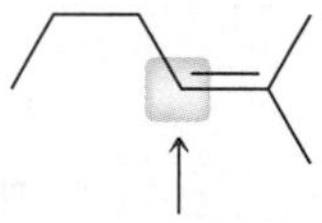

Br is placed here at the less substituted position

STEP 3 Identify the expected stereospecificity.

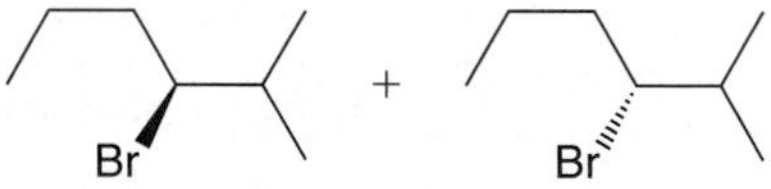

Try Problems 11.20, 11.21

PRACTICE PROBLEMS

Note: Most of the Problems are available within *WileyPLUS*, an online teaching and learning solution.

11.22 Draw all resonance structures for each of the following radicals:

(a)

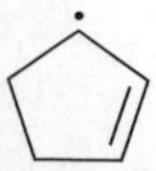

(b)

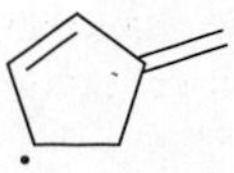

(c)

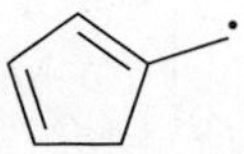

(d)

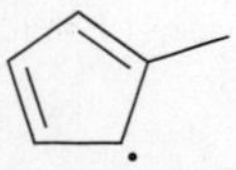

(e)

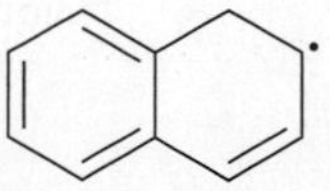

11.23 Consider all of the different C—H bonds in cyclopentene, and rank them in order of increasing bond strength:

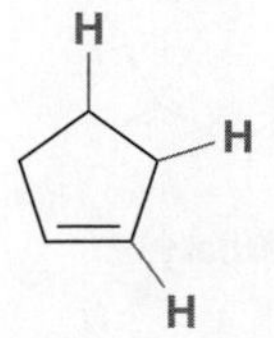

11.24 Rank each group of radicals in order of increasing stability:

(a) 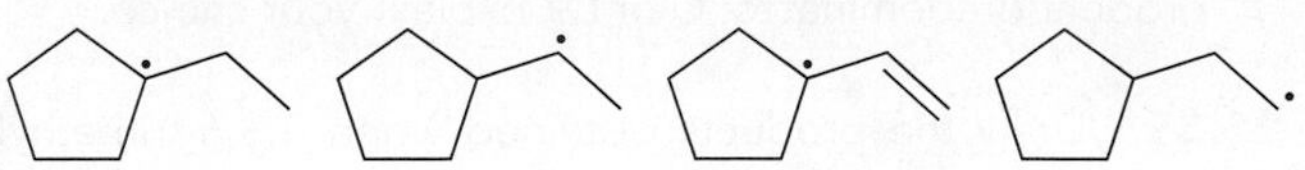

(b) 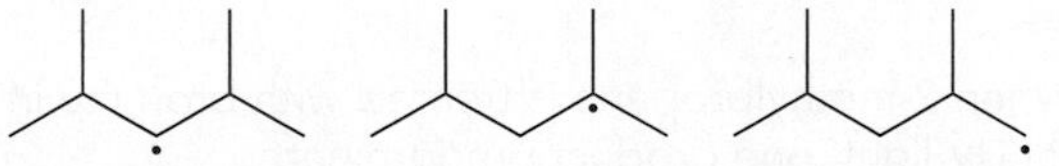

11.25 Draw all resonance structures of the radical produced when a hydrogen atom is abstracted from the OH group in BHT:

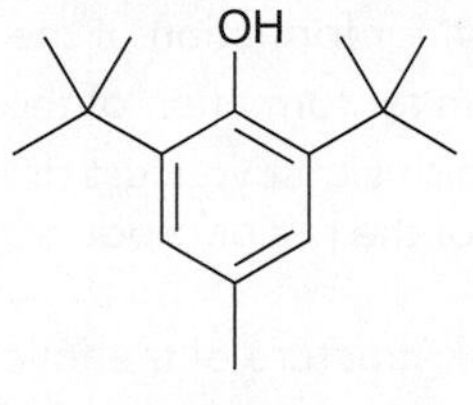

Butylated **h**ydroxy**t**oluene
(BHT)

11.26 When isopropylbenzene (cumene) is treated with NBS and irradiated with UV light, only one product is obtained. Propose a mechanism, and explain why only one product is formed.

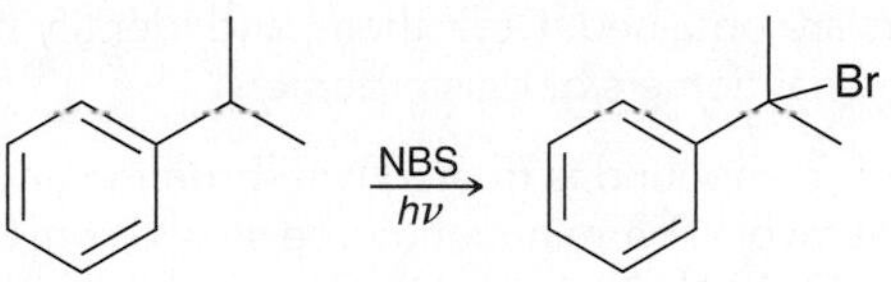

11.27 There are three constitutional isomers with the molecular formula C_5H_{12}. Chlorination of one of these isomers yields only one product. Identify the isomer, and draw the product of chlorination.

11.28 When ethylbenzene is treated with NBS and irradiated with UV light, two stereoisomeric compounds are obtained in equal amounts. Draw the products and explain why they are obtained in equal amounts.

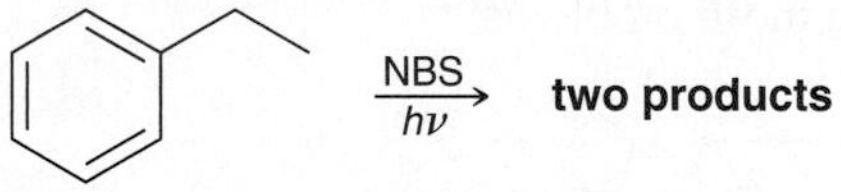

11.29 AIBN is an azo compound (a compound with a N═N double bond) that is often used as a radical initiator. Upon heating, AIBN liberates nitrogen gas to produce two identical radicals:

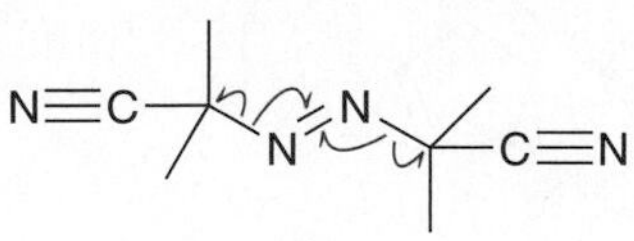

AIBN

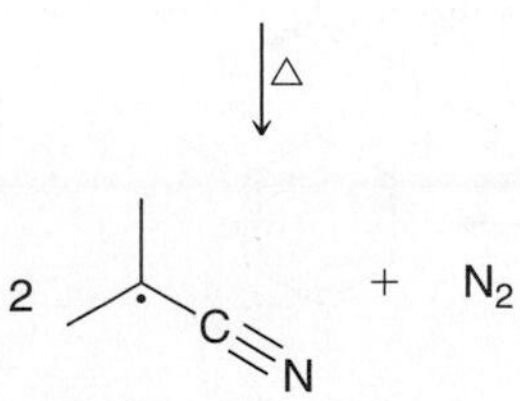

(a) Give two reasons why these radicals are so stable.

(b) Explain why the following azo compound is not useful as a radical initiator:

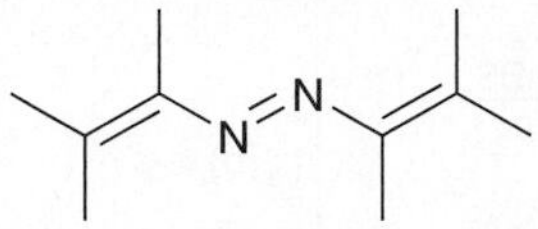

11.30 Triphenylmethane readily undergoes autooxidation to produce a hydroperoxide:

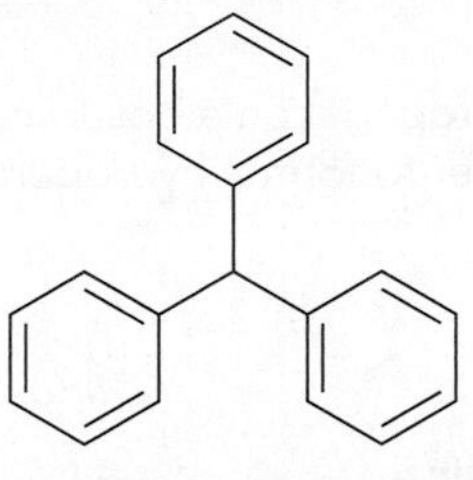

Triphenylmethane

(a) Draw the expected hydroperoxide.

(b) Explain why triphenylmethane is so susceptible to autooxidation.

(c) In the presence of phenol, triphenylmethane undergoes autooxidation at a much slower rate. Explain this observation.

11.31 For each of the products shown in the following reaction, propose a mechanism that explains its formation:

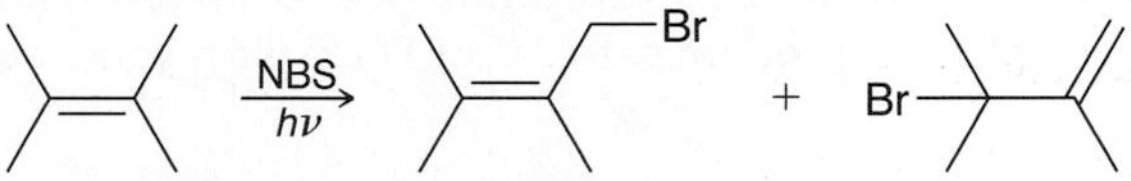

11.32 Abstraction of a hydrogen atom from 3-ethylpentane can yield three different radicals, depending on which hydrogen atom is abstracted. Draw all three radicals and rank them in order of increasing stability.

11.33 Identify the major product(s) for each of the following reactions. If any of the reactions do not yield a product, indicate "no reaction."

(a) 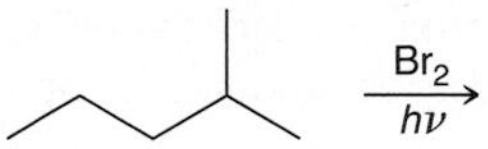

(b) 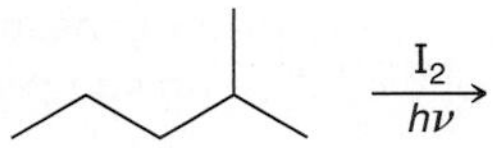

(c) 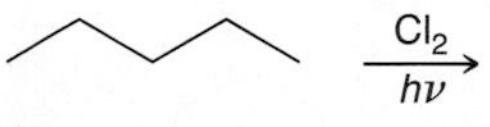

(d) 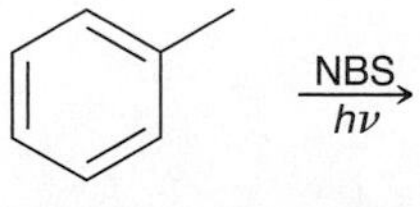

(e) 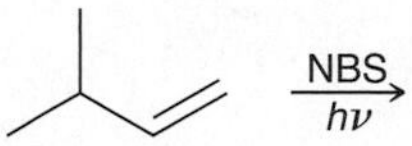

(f) 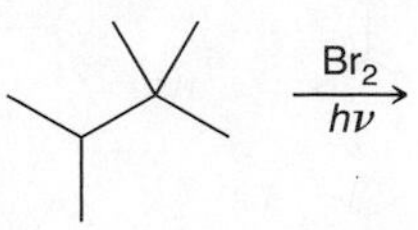

11.34 Chlorination of (*S*)-2-chloropentane produces a mixture of isomers with molecular formula $C_5H_{10}Cl_2$. How many isomers are obtained (consider stereoisomers as separate products)? Draw all of the products.

11.35 Predict the major product(s) obtained upon bromination of (*S*)-3-methylhexane.

11.36 Identify all products expected for each of the following reactions. Take stereochemistry into account, and draw expected stereoisomer(s), if any:

(a) 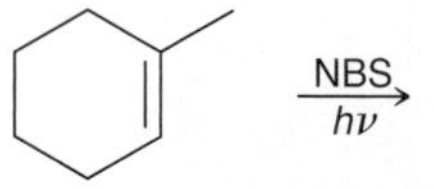

(b) 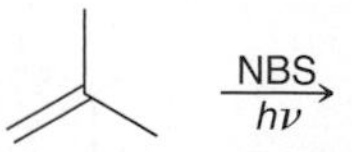

(c) 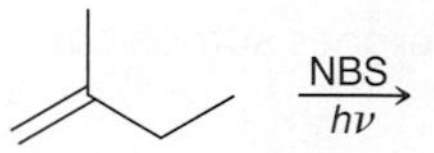

(d) 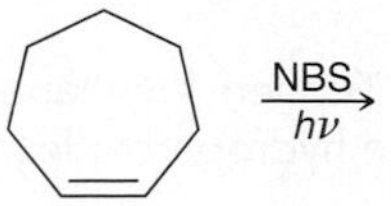

11.37 Draw the propagation steps that achieve the autooxidation of diethyl ether to form a hydroperoxide:

Diethyl ether **A hydroperoxide**

11.38 Compound A has molecular formula C_5H_{12}, and monobromination of compound A produces only compound B. When compound B is treated with a strong base, a mixture is obtained containing compound C and compound D. Using this information, answer the following questions:

(a) Draw the structure of compound A.

(b) Draw the structure of compound B.

(c) Draw the structures of compounds C and D.

(d) When compound B is treated with potassium *tert*-butoxide, which product predominates: C or D? Explain your choice.

(e) When compound B is treated with sodium ethoxide, which product predominates: C or D? Explain your choice.

11.39 Draw the products obtained when 3,3,6-trimethylcyclohexene is treated with NBS and irradiated with UV light.

11.40 When 2-methylpropane is treated with bromine in the presence of UV light, one product predominates.

(a) Identify the structure of the product.

(b) Draw the structure of the expected minor product.

(c) Draw a mechanism for formation of the major product.

(d) Draw a mechanism for formation of the minor product.

(e) Using the mechanisms that you just drew, explain why we expect very little of the minor product.

11.41 Consider the structure of the following compound:

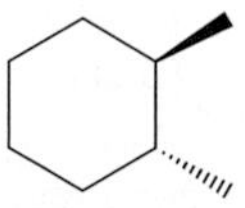

(a) When this compound is treated with bromine under conditions that favor monobromination, two stereoisomeric products are obtained. Draw them, and identify whether they are enantiomers or diastereomers.

(b) When this compound is treated with bromine under conditions that favor dibromination, three stereoisomeric products are obtained. Draw them, and explain why there are only three products and not four.

INTEGRATED PROBLEMS

11.42 The rate at which two methyl radicals couple to form ethane is significantly faster than the rate at which two *tert*-butyl radicals couple. Offer two explanations for this observation.

11.43 How many constitutional isomers are obtained when each of the following compounds undergoes monochlorination?

(a) 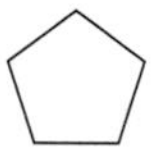

(b) 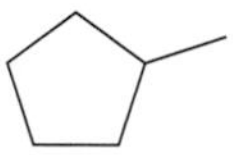

(c) 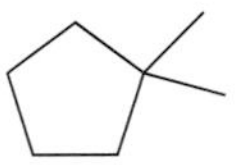

(d) 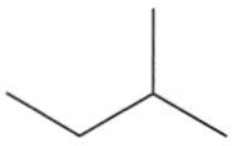

(e) 

(f) 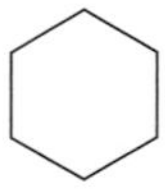

(g) 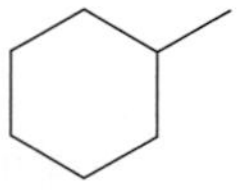

(h) 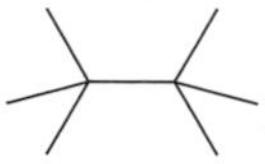

(i) 

(j) 

11.44 Using acetylene and 2-methylpropane as your only sources of carbon atoms, propose a plausible synthesis for 4-methyl-2-pentanone. You will need to utilize many reactions from previous chapters.

11.45 Propose an efficient synthesis for each of the following transformations. You might find it useful to review Section 11.13 before doing these problems.

(a) 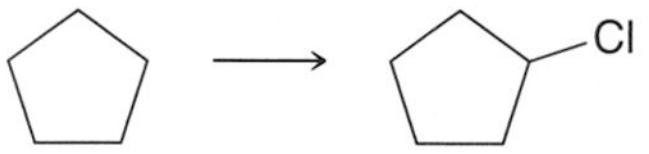

(b) 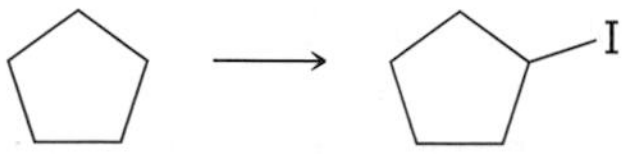

(c) 

(d) 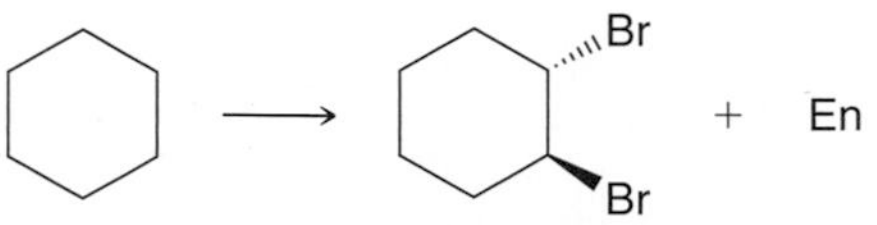

(e) 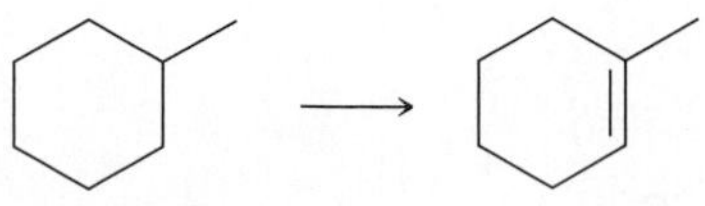

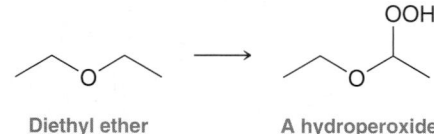

CHALLENGE PROBLEMS

11.46 Consider the following two compounds. Monochlorination of one of these compounds produces twice as many stereoisomeric products as the other. Draw the products in each case, and identify which compound yields more products upon chlorination.

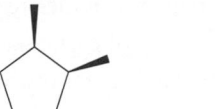

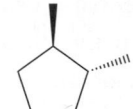

11.47 When butane reacts with Br_2 in the presence of Cl_2, both brominated and chlorinated products are obtained. Under such conditions, the usual selectivity of bromination is not observed. In other words, the ratio of 2-bromobutane to 1-bromobutane is very similar to the ratio of 2-chlorobutane to 1-chlorobutane. Can you offer an explanation as to why we do not observe the normal selectivity expected for bromination?

11.48 When an acyl peroxide undergoes homolytic bond cleavage, the radicals produced can liberate carbon dioxide to form alkyl radicals:

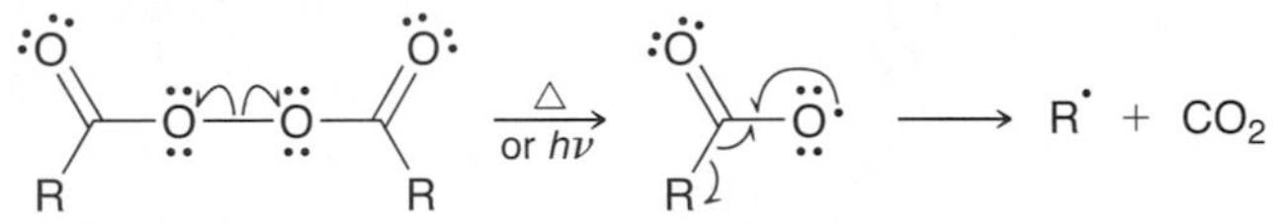

An acyl peroxide

Using this information, provide a mechanism of formation for each of the following products:

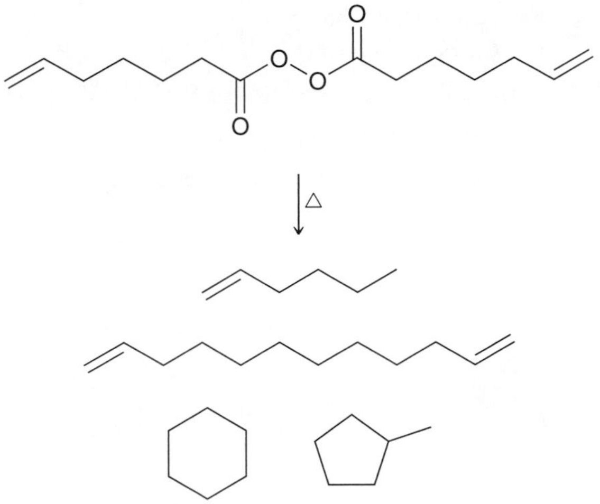

12 Synthesis

DID YOU EVER WONDER...

what vitamins are and why we need them?

Vitamins are essential nutrients that our bodies require in order to function properly, and a deficiency of particular vitamins can lead to diseases, many of which can be fatal. Later in this chapter, we will learn more about the discovery of vitamins, and we will see that the laboratory synthesis of one particular vitamin represented a landmark event in the history of synthetic organic chemistry. This chapter serves as a brief introduction to organic synthesis.

Until this point in the text, we have only seen a limited number of reactions (a few dozen, at most). In this chapter, our modest repertoire of reactions will allow us to develop a methodical, step-by-step process for proposing syntheses. We will begin with one-step synthesis problems and then progress toward more challenging multistep problems. The goal of this chapter is to develop the fundamental skills required for proposing a synthesis.

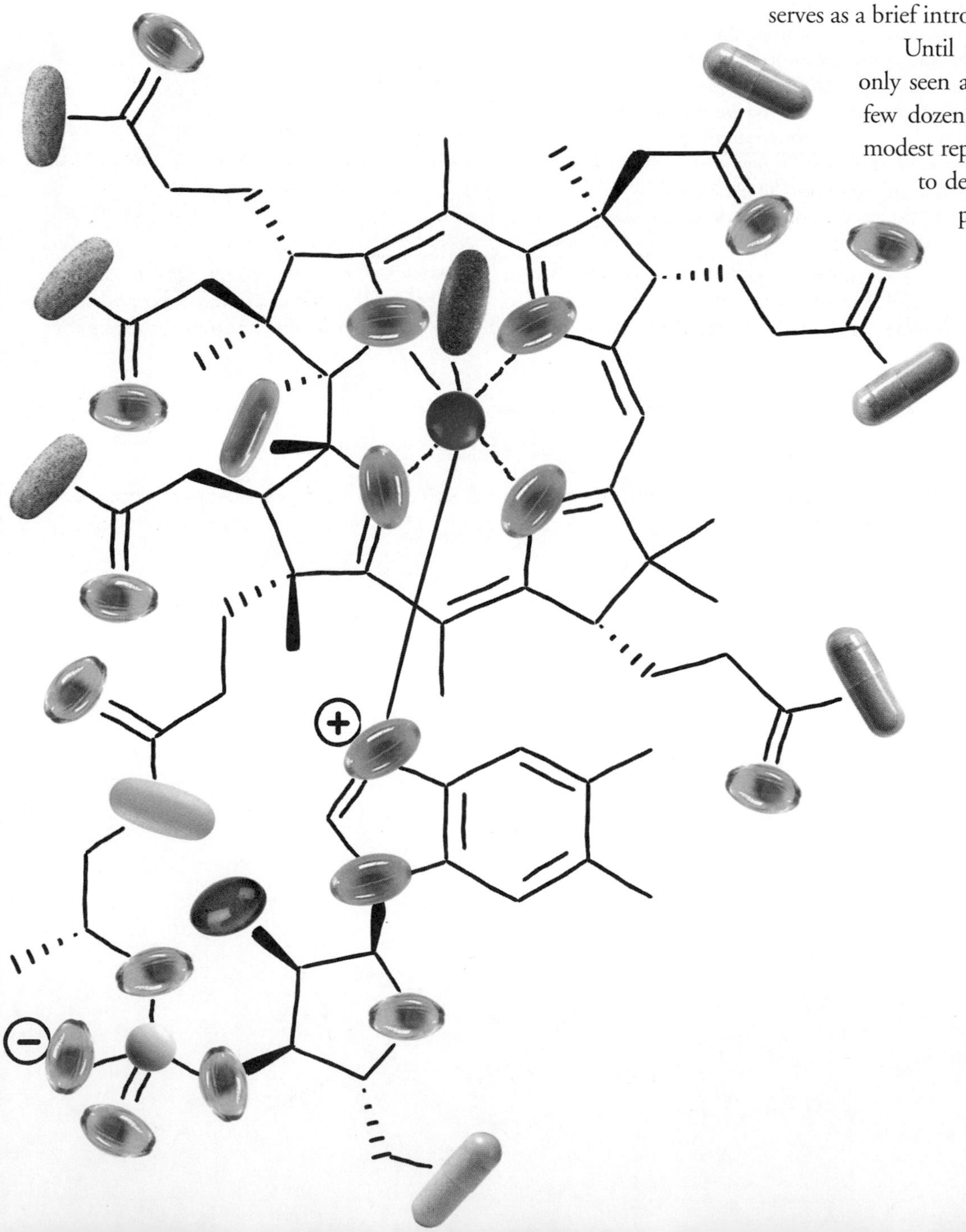

DO YOU REMEMBER?

Before you go on, be sure you understand the following topics.
If necessary, review the suggested sections to prepare for this chapter.

- Selecting Reagents to Accomplish Functional Group Transformation (Section 7.9)
- Synthesis Strategies for Alkenes and Alkynes (Sections 9.13, 10.11)
- Substitution vs. Elimination (Section 8.13)

WILEY PLUS Visit www.wileyplus.com to check your understanding and for valuable practice.

12.1 One-Step Syntheses

The most straightforward synthesis problems are the ones that can be solved in just one step. For example, consider the following:

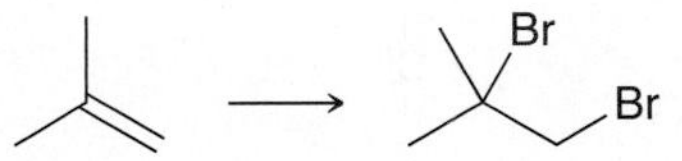

This transformation can be accomplished by treating the alkene with Br_2 in an inert solvent, such as CCl_4. Other synthesis problems might require more than a single step, and those problems will be more challenging. Before approaching multistep synthesis problems, it is absolutely essential to become comfortable with one-step syntheses. In other words, it is critical to achieve mastery over all reagents described in the previous chapters. If you can't identify the reagents necessary for a one-step synthesis problem, then certainly you will be unable to solve more complex problems. The following exercises represent a broad review of the reactions in previous chapters. These exercises are designed to help you identify which reagents are still not at the forefront of your consciousness.

CONCEPTUAL CHECKPOINT

12.1 Identify the reagents necessary to accomplish each of the transformations shown below. If you are having trouble, the reagents for these transformations appear on page 444, but you should first try to identify the reagents yourself without help:

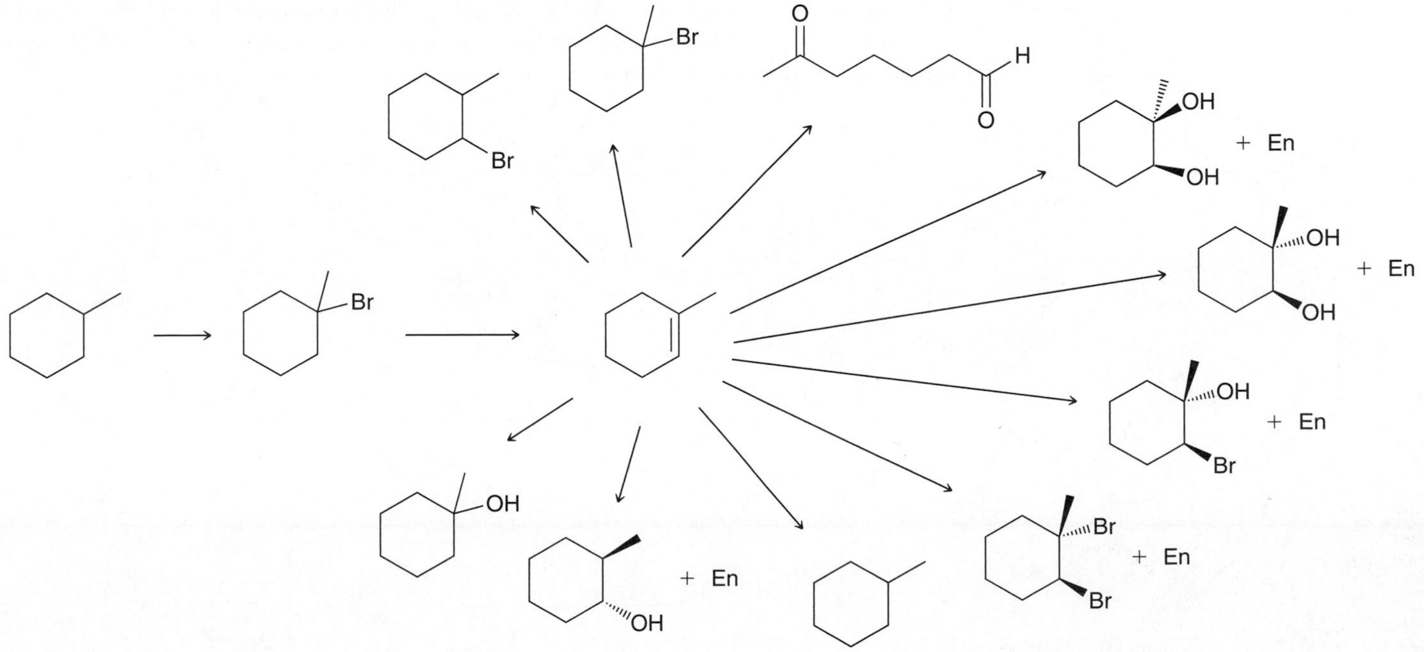

12.2 Identify the reagents necessary to accomplish each of the following transformations. If you are having trouble, the reagents for these transformations appear on page 482, but you should first try to identify the reagents yourself without help:

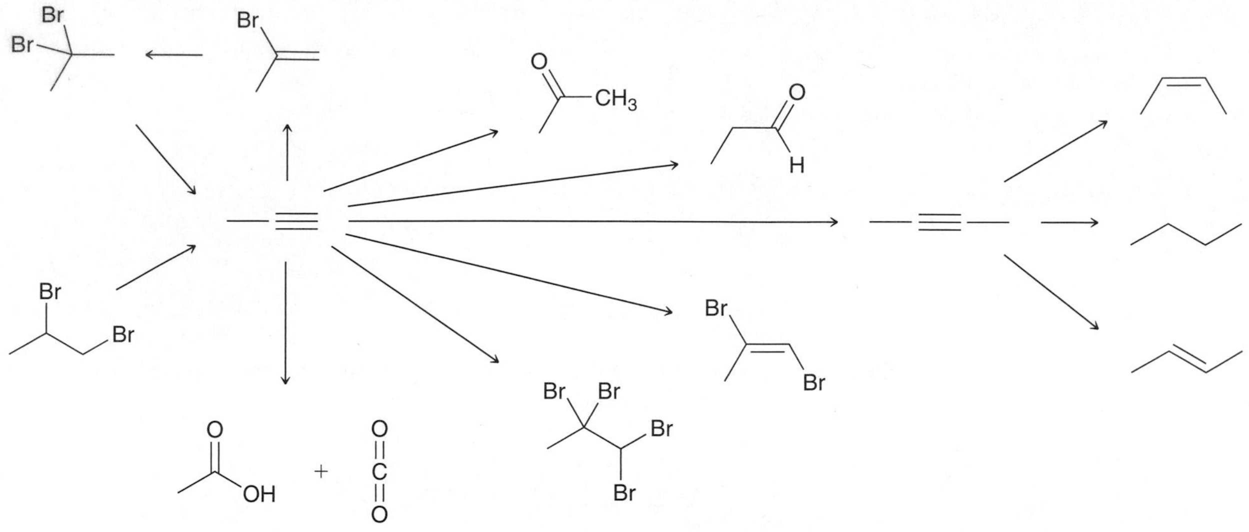

12.2 Functional Group Transformations

In the previous few chapters, we developed several synthesis strategies that enable us to move the location of a functional group or change its identity. Let's briefly review these techniques, as they will be extremely helpful when solving multistep synthesis problems.

In Chapter 9, we developed a technique for changing the position of a halogen by performing an elimination reaction followed by an addition reaction. For example:

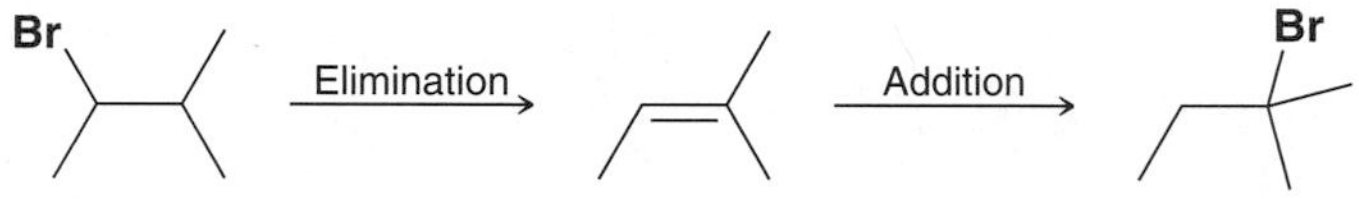

In this two-step process, the halogen is removed and then reinstalled at a different location. The regiochemical outcome of each step must be carefully controlled. The choice of base in the elimination step determines whether the more substituted or the less substituted alkene is formed. In the addition step, the decision whether or not to use peroxides will determine whether a Markovnikov addition or an *anti*-Markovnikov addition occurs.

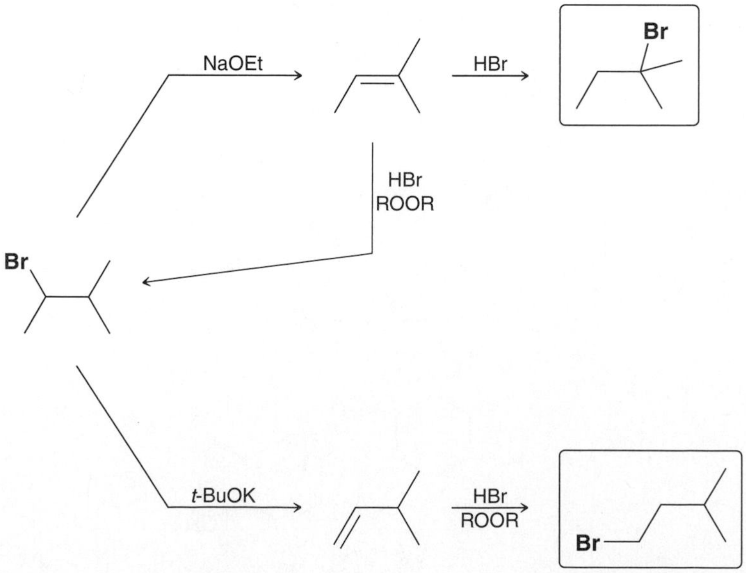

As we saw in Chapter 9, this technique must be slightly modified when the functional group is a hydroxyl group (OH). In such a case, the hydroxyl group must first be converted into a tosylate (a better leaving group), and only then can the technique be employed (elimination followed by addition):

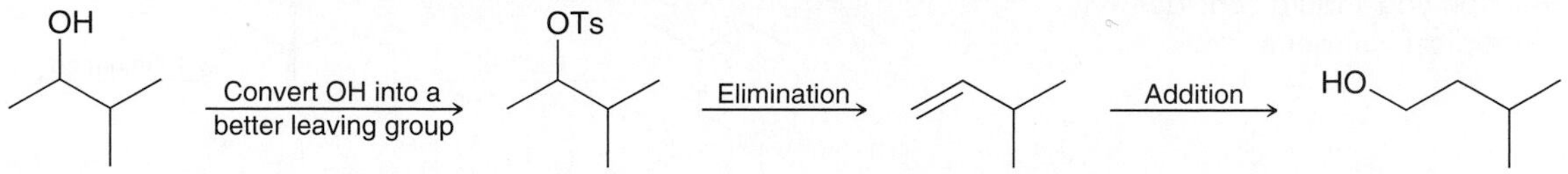

After converting the hydroxyl group into a tosylate, the regiochemical outcome for elimination and addition can be carefully controlled, as summarized below:

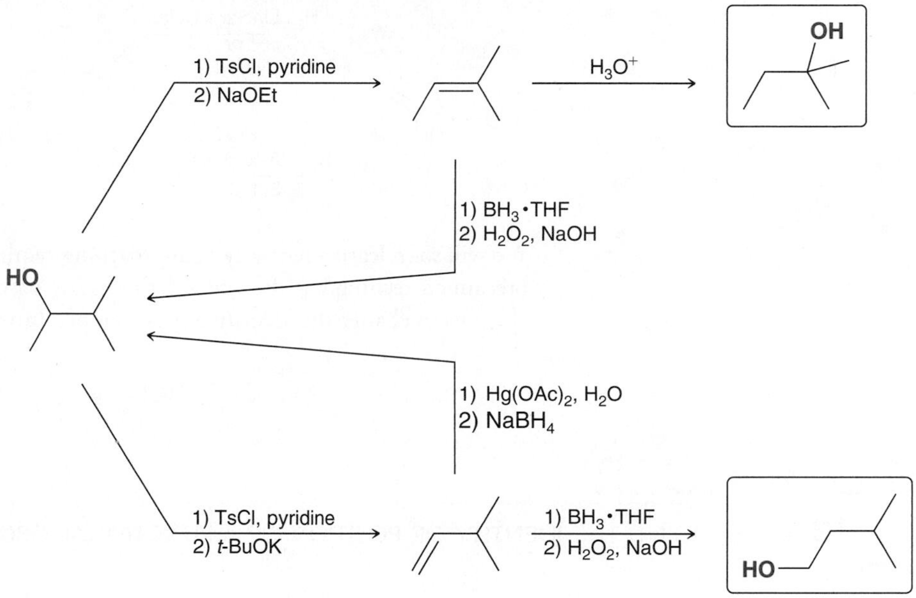

In Chapter 9, we also developed a two-step technique for moving the position of a double bond. For example:

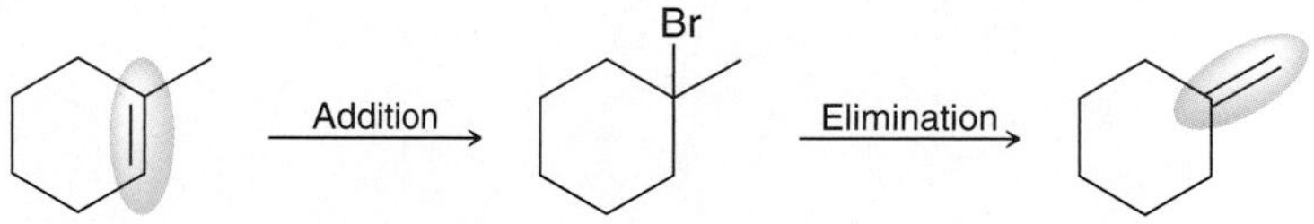

Once again, the regiochemical outcome of each step can be controlled by choice of reagents, as summarized below:

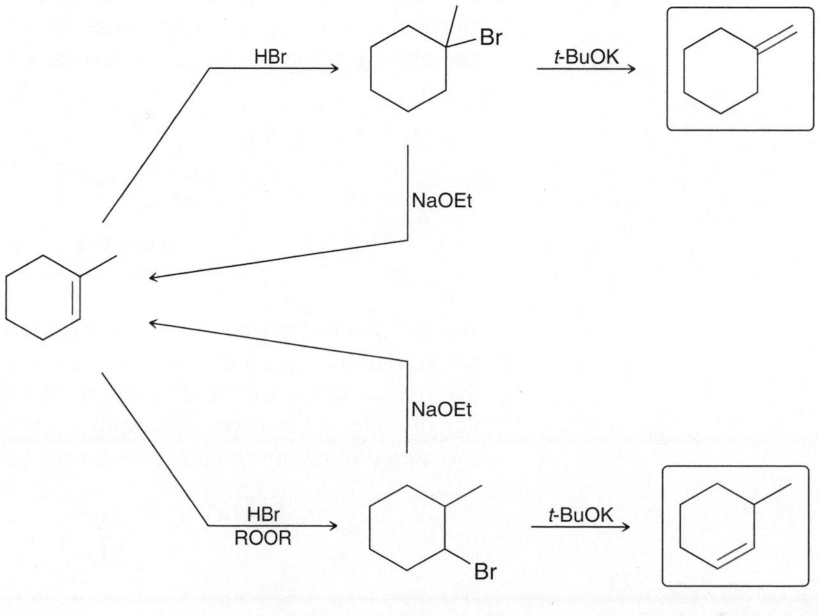

BY THE WAY
If your starting material is an alkane, the only useful reaction you should consider is a radical halogenation.

In Chapter 11, we developed one other important technique: installing functionality in a compound with no functional groups:

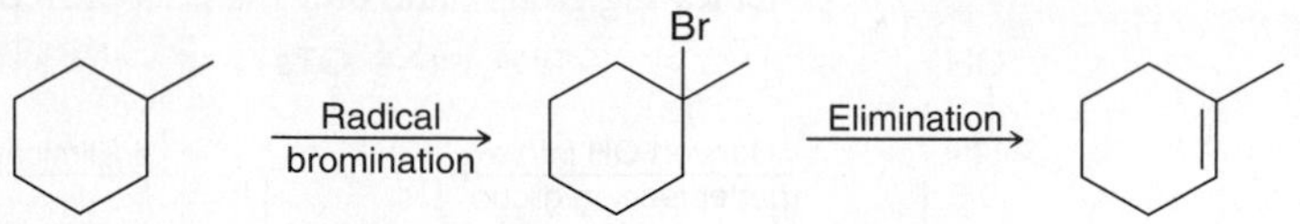

This procedure, together with the other reactions covered in the previous chapters, enables the interconversion between single, double, and triple bonds:

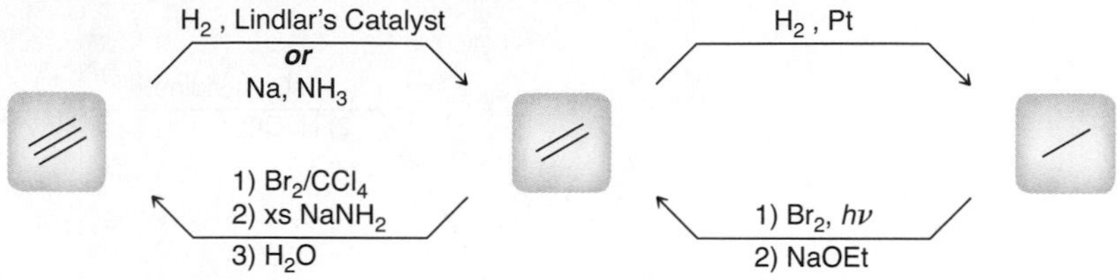

We will soon learn a new way of approaching synthesis problems (rather than relying on a few, precanned techniques). For now, let's ensure mastery over the reactions and techniques that allow us to change the identity or position of a functional group.

SKILLBUILDER

12.1 CHANGING THE IDENTITY OR POSITION OF A FUNCTIONAL GROUP

LEARN the skill

Propose a plausible synthesis for the following transformation:

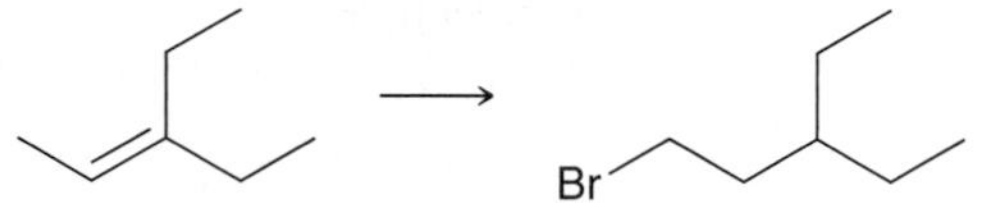

SOLUTION

We begin by analyzing the identity and location of the functional groups in both the starting material and the product. The identity of the functional group has certainly changed (an alkene has been converted into an alkyl halide), and the position of the functional group has also changed. This can be seen more easily by numbering the parent chain:

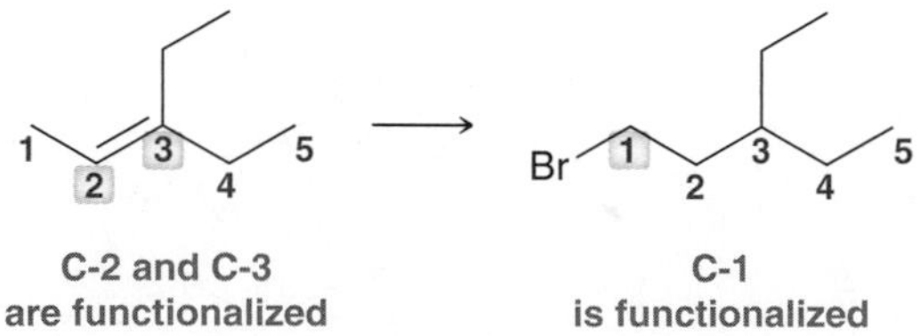

The C-2 and C-3 positions are functionalized (have a functional group) in the starting material, but the C-1 position is functionalized in the product. Therefore, both the location and the identity of the functional group must be changed. That is, we must find a way to functionalize the C-1 position by moving the existing functional group. We have already seen a two-step method for changing the location of a π bond:

This strategy would achieve the desired goal of functionalizing the C-1 position, and this type of transformation can be accomplished via addition followed by elimination. In order to move the position of the π bond, the reagents must be chosen carefully for each step of the process. Specifically, an *anti*-Markovnikov addition must be followed by a Hofmann elimination:

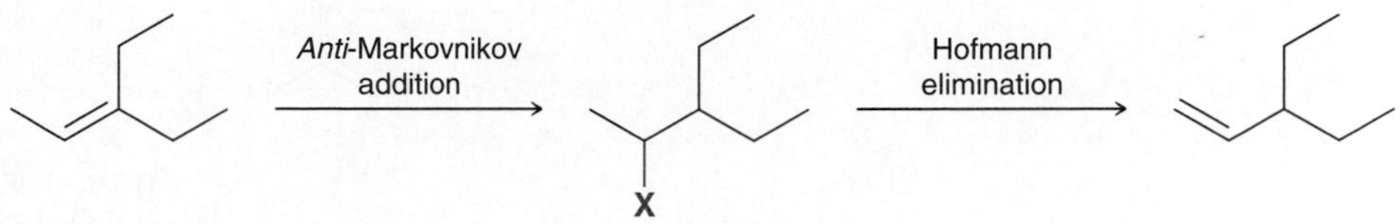

For the first step of this process, we have learned only two reactions that proceed via *anti*-Markovnikov addition: (a) either addition of HBr with peroxides or (b) hydroboration-oxidation:

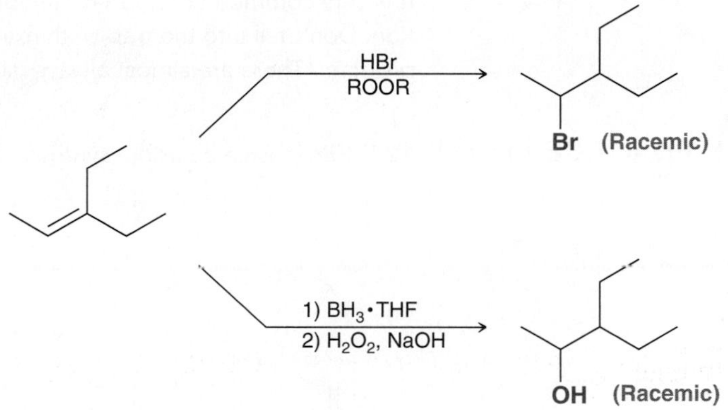

Both reactions will produce an *anti*-Markovnikov addition. However, when hydroboration oxidation is used, the resulting hydroxyl group must then be converted to a tosylate prior to the Hofmann elimination process:

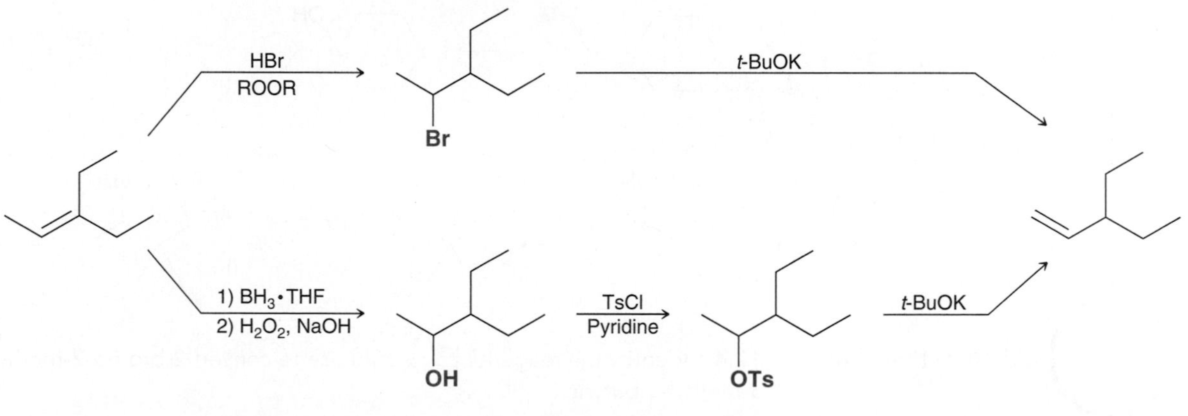

Both routes will work, although the first route might be more efficient, because it requires fewer steps.

Now that the C-1 position has been functionalized, the last step is to install a bromine atom at the C-1 position. This can be accomplished with an *anti*-Markovnikov addition of HBr:

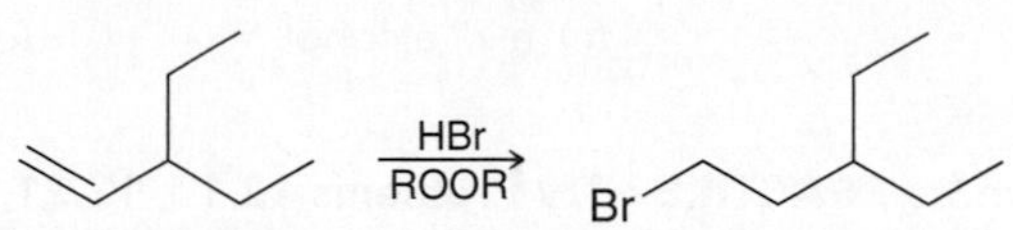

In summary, there are two plausible routes to achieve the desired synthetic transformation:

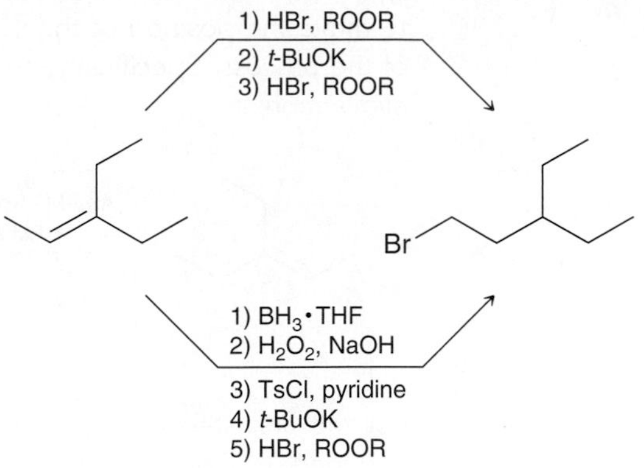

It is very common to find that multiple routes can be used to achieve a desired transformation. Don't fall into the trap of thinking that there is only one correct solution to a synthesis problem. There are almost always multiple pathways that are feasible.

PRACTICE the skill **12.3** Propose a plausible synthesis for each of the following transformations:

(a) 

(b) 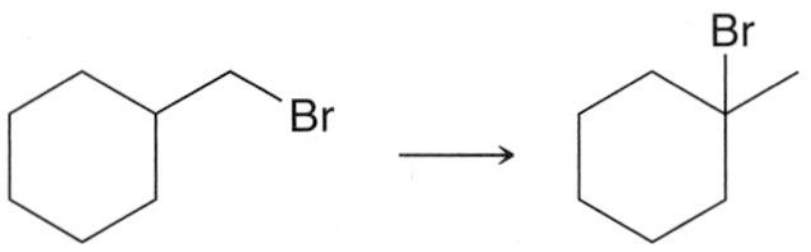

(c) 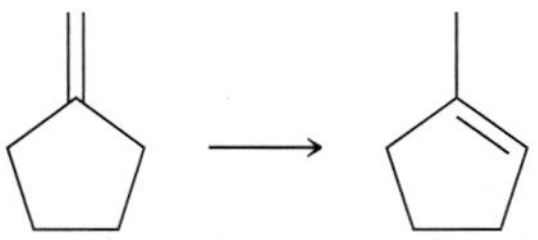

(d) 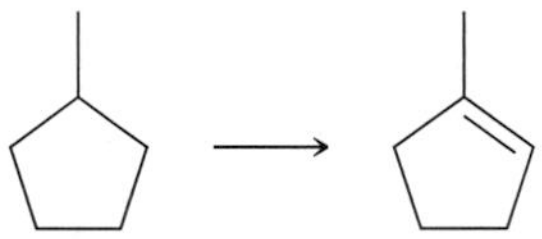

(e) 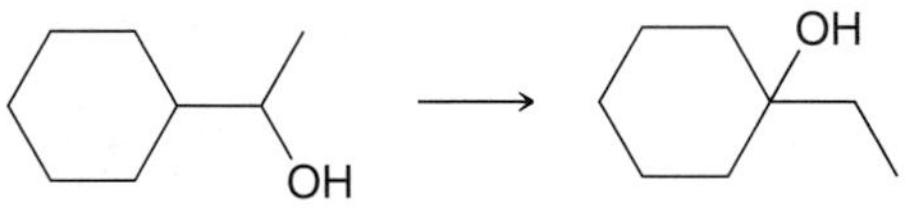

(f) 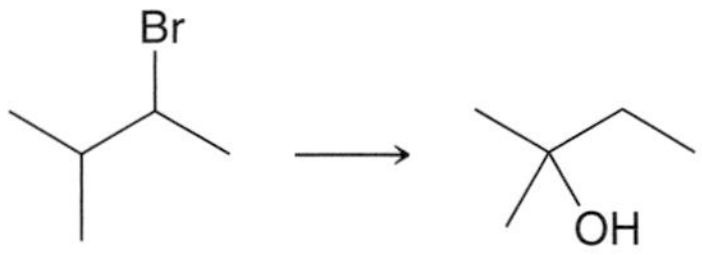

(g) 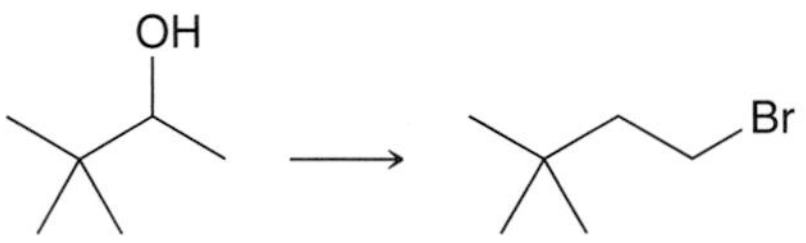

(h) 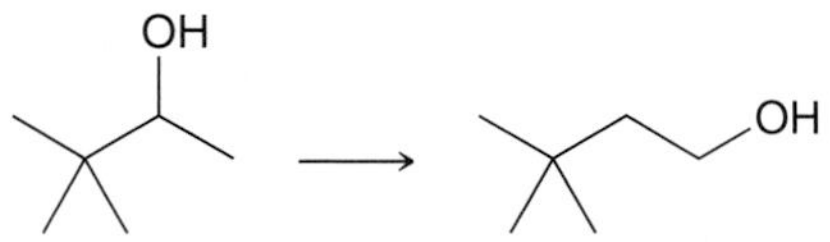

APPLY the skill

12.4 Identify the reagents you would use to convert 2-bromo-2-methylbutane into 3-methyl-1-butyne.

12.5 Identify the reagents you would use to convert 1-pentene into a geminal dibromide ("geminal" indicates that both bromine atoms are connected to the same carbon atom).

12.6 Identify the reagents you would use to convert methylcyclohexane into each of the following:

(a) a 3° alkyl halide (b) a trisubstituted alkene

(c) a 2° alcohol (d) 3-methylcyclohexene

need more **PRACTICE?** **Try Problems 12.17, 12.21, 12.22**

12.3 Reactions That Change the Carbon Skeleton

In all of the problems in the previous section, the functional group changed its identity or location, but the carbon skeleton always remained the same. In this section, we will focus on examples in which the carbon skeleton changes. In some cases, the number of carbon atoms in the skeleton increases, and in other cases, the number of carbon atoms decreases.

If the size of the carbon skeleton increases, then a C—C bond-forming reaction is required. Thus far, we have only learned one reaction that can be used to introduce an alkyl group onto an existing carbon skeleton. Alkylation of a terminal alkyne (Section 10.10) will increase the size of a carbon skeleton:

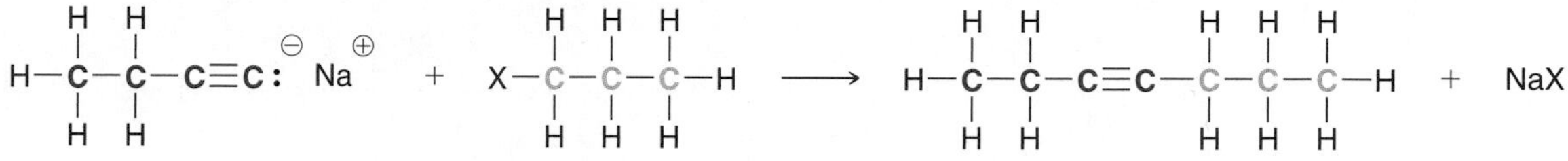

Four carbon atoms Three carbon atoms Seven carbon atoms

Over time, we will see many other C—C bond-forming reactions, but for now, the knowledge that we have only seen one such reaction should greatly simplify the problems in this section, enabling a smooth transition into the world of synthetic organic chemistry.

If the size of the carbon skeleton decreases, then a C—C bond-breaking reaction, called **bond cleavage**, is required. Once again, we have only seen one such reaction. Ozonolysis of an alkene (or alkyne) achieves bond cleavage at the location of the π bond:

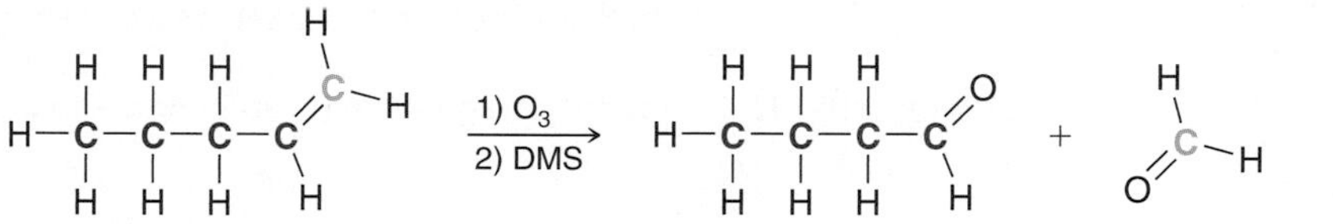

Five carbon atoms Four carbon atoms One carbon atom

Over time, we will see other reactions that involve C—C bond cleavage. For now, the knowledge that we have only seen one such reaction should greatly simplify the problems in this section.

SKILLBUILDER

12.2 CHANGING THE CARBON SKELETON

LEARN the skill

Identify reagents that can be used to achieve the following transformation:

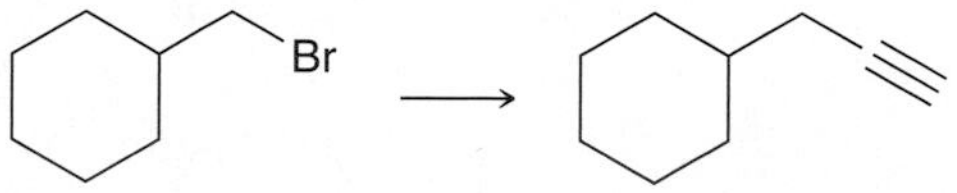

SOLUTION

Count the carbon atoms in the starting material and in the desired product. There are seven carbon atoms in the starting material, and there are nine carbon atoms in the product. Therefore two carbon atoms must be installed. We have only learned one reaction capable of installing two carbon atoms on an existing carbon skeleton. This process requires the use of an alkynide ion and an alkyl halide:

$$R-C\equiv C:^{\ominus}\ Na^{\oplus}\quad R-X\longrightarrow R-C\equiv C-R\ +\ NaX$$

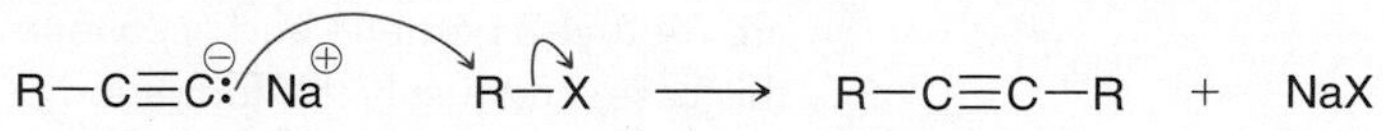

Alkynide **Alkyl halide**

Until now, this reaction has always been viewed *from the perspective of the alkyne*. That is, the alkyne is the starting material, and the alkyl halide is used as a reagent in the second step of the process, thereby achieving the installation of an alkyl group (R):

$$H-C\equiv C-H \xrightarrow[\text{2) } \boxed{R-X}]{\text{1) NaNH}_2} H-C\equiv C-\boxed{R}$$

Starting material

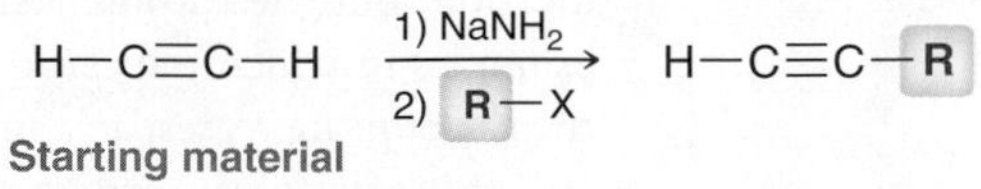

Alternatively, this reaction can be viewed from the perspective of the alkyl halide. That is, the alkyl halide is the starting material, and an alkynide ion is used to achieve the installation of a triple bond onto an existing carbon skeleton:

$$R-X \xrightarrow{Na^{\oplus} \ {}^{\ominus}:C\equiv C-H} R-C\equiv C-H$$

Starting material

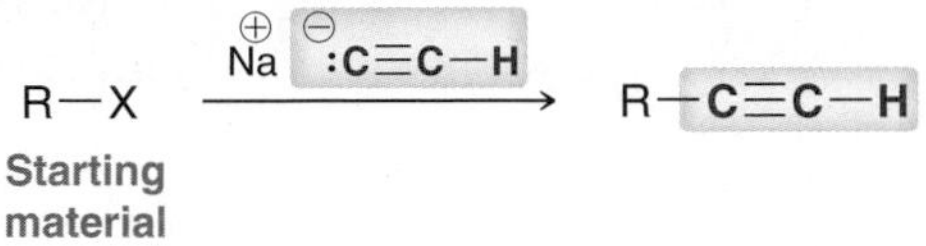

When viewed in this way, the alkylation process represents a technique for introducing an acetylenic group. This is exactly what is needed to solve this problem:

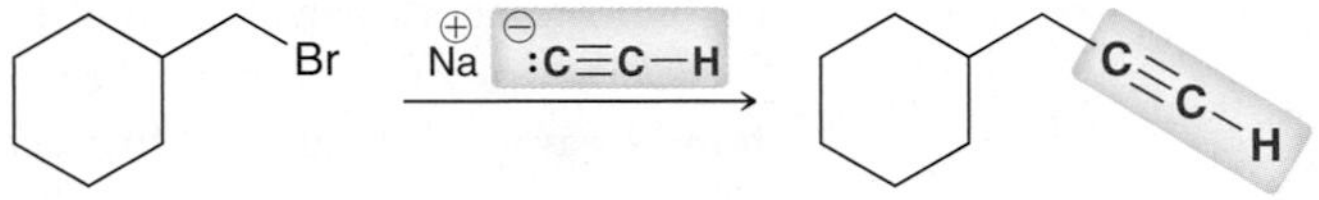

In fact, this one step provides the answer. This is just a one-step synthesis problem.

PRACTICE the skill 12.7 Identify reagents that can be used to achieve each of the following transformations:

(a)

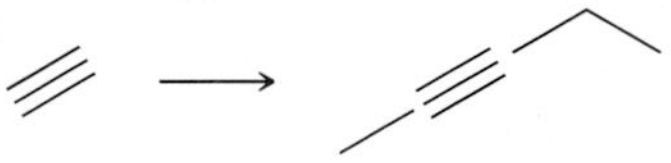

(b)

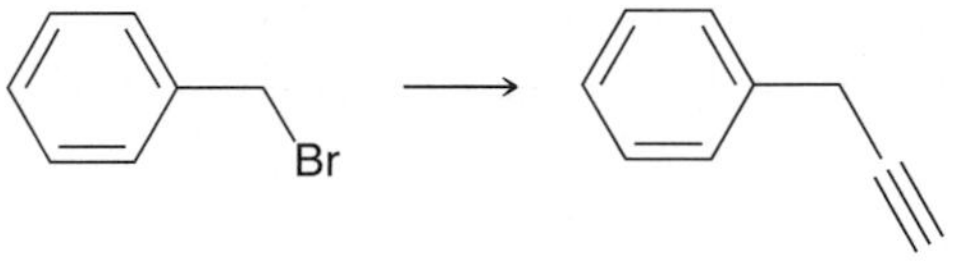

(c)

APPLY the skill

12.8 Propose a plausible synthesis for each of the following transformations:

(a)

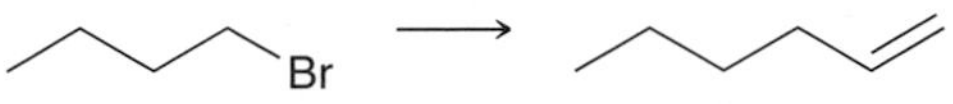

(b)

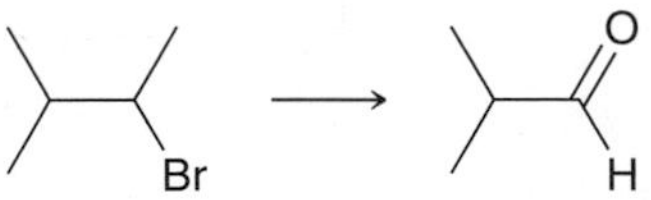

(c)

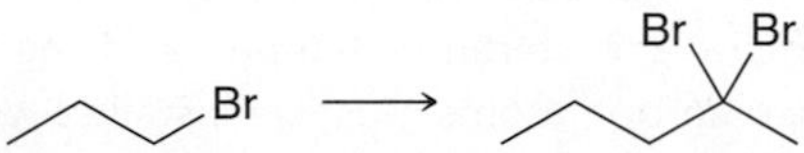

12.9 When 3-bromo-3-ethylpentane is treated with sodium acetylide, the major products are 3-ethyl-2-pentene and acetylene. Explain why the carbon skeleton does not change in this case, and justify the formation of the observed products.

······> need more **PRACTICE?** Try Problems **12.18, 12.19, 12.20, 12.23, 12.26**

MEDICALLYSPEAKING

Vitamins

Vitamins are compounds that our bodies require for normal functioning and must be obtained from food. An inadequate intake of certain vitamins causes specific diseases. This phenomenon had been observed long before the exact role of vitamins was understood. For example, sailors who remained at sea for extended periods would suffer from a disease called scurvy, characterized by the loss of teeth, swollen limbs, and bruising. If left untreated, the disease would be fatal. In 1747, a British naval physician named James Lind demonstrated that the effects of scurvy could be reversed by eating oranges and lemons. It was recognized that oranges and lemons must contain some "factor" that our bodies require, and lime juice became a normal part of a sailor's diet. For this reason, British sailors were called "Limeys."

Other studies revealed that a variety of foods contained mysterious "growth factors." For example, Gowland Hopkins (University of Cambridge) conducted a series of experiments in which he controlled the dietary intake of rats. He fed them carbohydrates, proteins and fats (see Chapters 24, 25 and 26). This mixture was insufficient to sustain the rats, and it was necessary to add a few drops of milk to their dietary intake to achieve sustainable growth. These experiments demonstrated that milk contained some unknown ingredient, or factor, necessary for promoting proper growth.

Similar observations were made by a Dutch physician, Christiaan Eijkman, in the Dutch colonies in Indonesia. While investigating the cause for a massive outbreak of beriberi, a disease characterized by paralysis, he noted that the hens in the laboratory began to exhibit a form of paralysis as well. He discovered that they were being fed rice from which the fibrous husk had been removed (called polished rice). When he fed the hens raw whole rice, their condition improved drastically. It was therefore realized that the fibrous husk of rice contained some vital growth factor. In 1912, Polish biochemist Casimir Funk isolated the active compound from rice husks. Careful studies revealed that the structure contained an amino group, and therefore belongs to a class of compounds called amines (see Chapter 23):

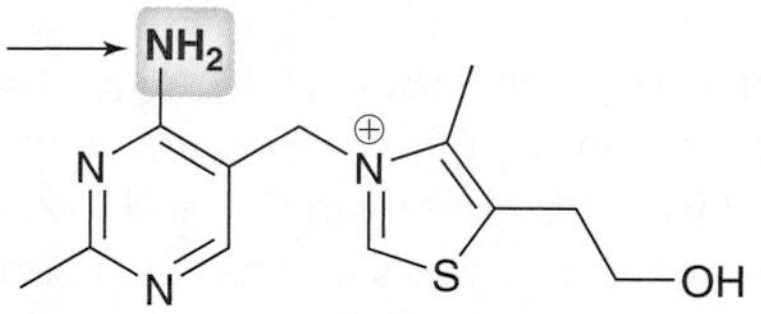

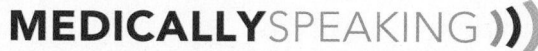

Thiamine (vitamin B_1)

It was originally believed that all vital growth factors were amines, so they were called vitamins (a combination of *vital* and *amine*). Further research demonstrated that most vitamins lack an amino group, yet the term "vitamins" persisted. For their discovery of the role vitamins play in nutrition, Eijkman and Hopkins were awarded the 1929 Nobel Prize in Physiology or Medicine.

Vitamins are grouped into families based on their behavior (rather than their structure); each family is designated with a letter. For example, vitamin B is a family of many compounds, each designated with a letter and a number (B_1, B_2, B_3, etc.). The table shows representative vitamins from several families:

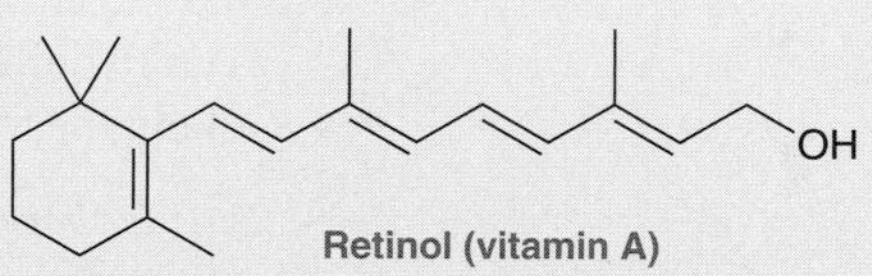

Retinol (vitamin A)

Sources: Milk, eggs, fruit, vegetables, and fish
Deficiency disease: Night-blindness (see Section 17.13)

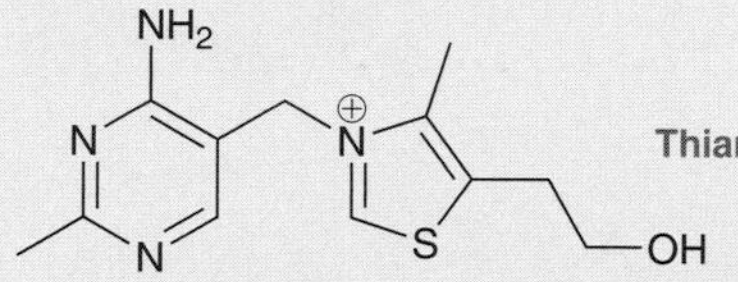

Thiamine (vitamin B_1)

Sources: Liver, potatoes, whole grains, and legumes
Deficiency disease: Beriberi

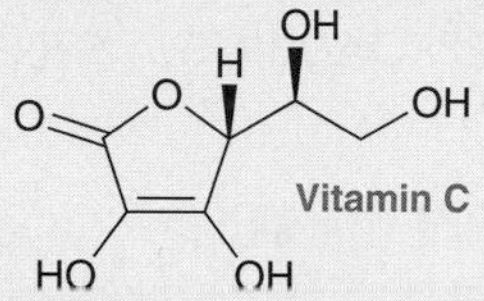

Vitamin C

Sources: Citrus fruits, bell peppers, tomatoes, and broccoli
Deficiency disease: Scurvy

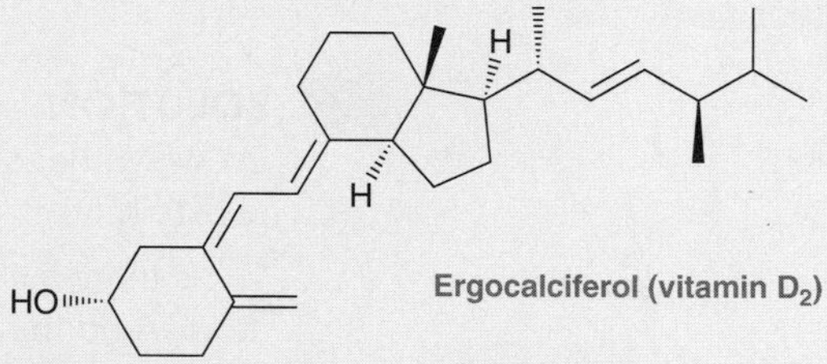

Ergocalciferol (vitamin D_2)

Sources: Fish, produced by the body when exposed to sunlight
Deficiency disease: Rickets (see Section 17.10)

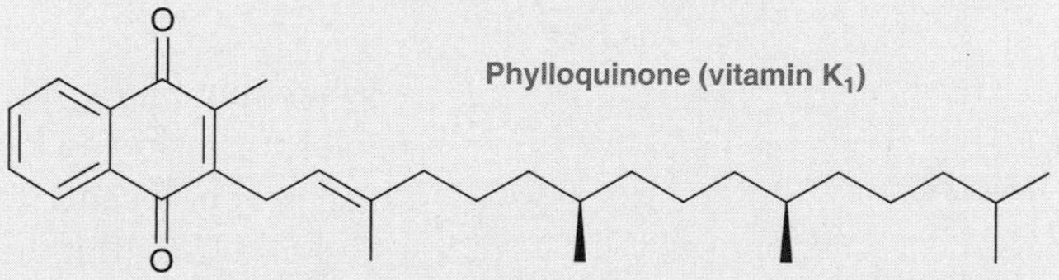

Phylloquinone (vitamin K_1)

Sources: Soybean oil, green vegetables, and lettuce
Deficiency disease: Hemorrhaging (internal bleeding)

Early studies revealed the structure of vitamin C, which enabled chemists to devise a synthetic strategy for preparing vitamin C in the laboratory. Success led to the preparation of vitamin C on an industrial scale. Such early successes encouraged the notion that chemists would soon elucidate the structures of all vitamins and devise methods for their preparation in the laboratory. The synthesis of vitamin B_{12}, however, would prove to be more complex, as we will see in the next Medically Speaking box.

12.4 How to Approach a Synthesis Problem

In the previous two sections, we covered two critical skills: (1) functional group transformations and (2) changing the carbon skeleton. In this section, we will explore synthesis problems that require both skills. From this point forward, every synthesis problem should be approached by asking the following two questions:

1. ***Is there a change in the carbon skeleton?*** Compare the starting material with the product to determine if the carbon skeleton is gaining or losing carbon atoms.
2. ***Is there a change in the identity or location of the functional group?*** Is one functional group converted into another, and does the position of functionality change?

The following example demonstrates how these two questions should be applied.

SKILLBUILDER

12.3 APPROACHING A SYNTHESIS PROBLEM

LEARN the skill

Propose a plausible synthesis for the following transformation:

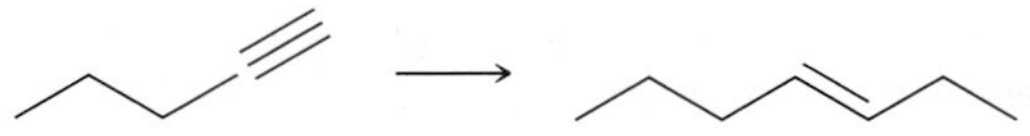

SOLUTION

Every synthesis problem should always be analyzed through the lens of the following two questions:

1. *Is there a change in the carbon skeleton?* The starting compound has five carbon atoms, and the product has seven carbon atoms. This transformation therefore requires the installation of two carbon atoms:

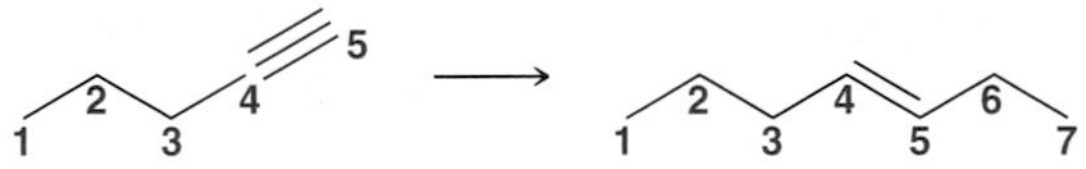

When numbering the carbon atoms, as shown above, it is not necessary to follow IUPAC rules for assigning locants. If we were naming this compound, we would be compelled to use proper locants (the triple bond would be between C-1 and C-2, rather than between C-4 and C-5). But the numbers here are only tools and can be used in whatever way is easiest for you to count.

2. *Is there a change in the identity or location of the functional group?* Certainly, the identity of the functional group has changed (a triple bond has been converted into a double bond), but consider the location of the functional group:

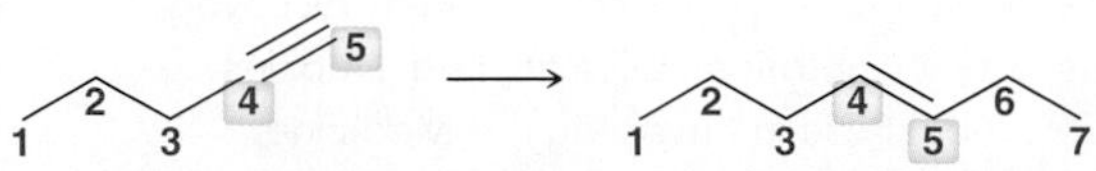

Once again, these numbers are not IUPAC numbers; rather, they are tools that help us identify that the double bond in the product occupies the same position as the triple bond in the starting material. In this case, strict adherence to the IUPAC numbering system would have been confusing and misleading.

By asking both questions, the following two tasks have been identified: (1) two carbon atoms must be installed and (2) the triple bond must be converted into a double bond in its current location. For each of these tasks, we must determine what reagents to use:

1. What reagents will add two carbon atoms to a skeleton?
2. What reagents will convert a triple bond into a *trans* double bond?

Two new carbon atoms can be introduced via alkylation of the starting alkyne:

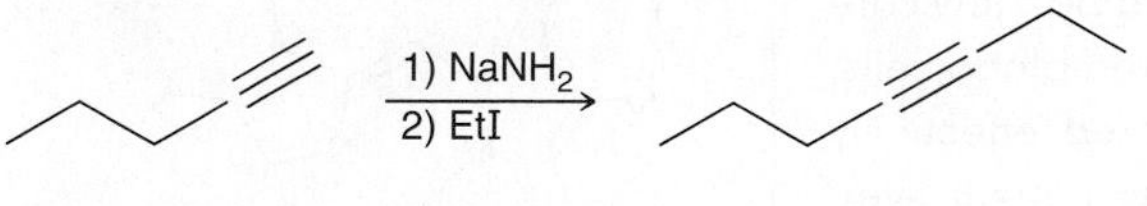

Now that the correct carbon skeleton has been established, reduction of the triple bond can be accomplished via a dissolving metal reduction to afford the *trans* alkene:

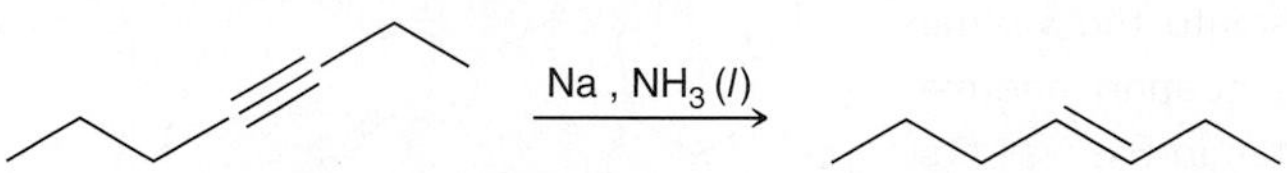

The solution to this problem requires two steps: (1) alkylation of the alkyne followed by (2) conversion of the triple bond into a double bond. Notice the order of events. If the triple bond had first been converted into a double bond, the alkylation process would not work. Only a terminal triple bond can be alkylated, not a terminal double bond.

PRACTICE the skill **12.10** Identify reagents that can be used to achieve each of the following transformations:

(a) 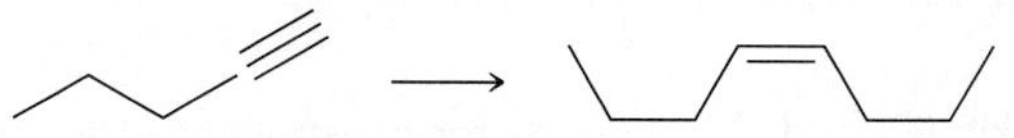

(b) 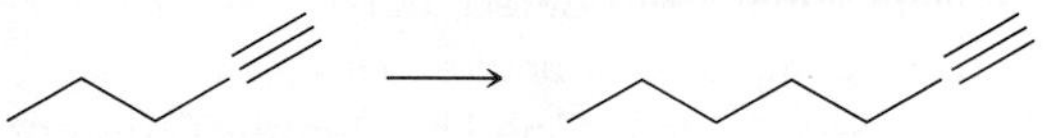

(c) 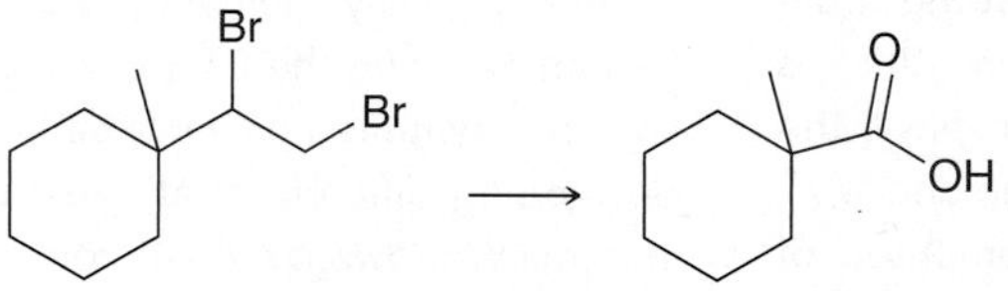

(d) 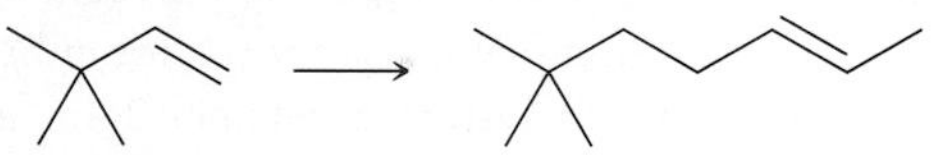

(e) 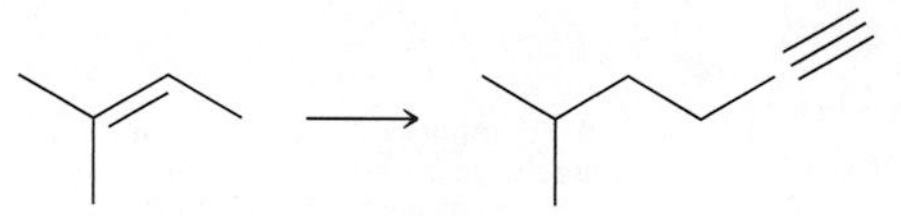

(f) 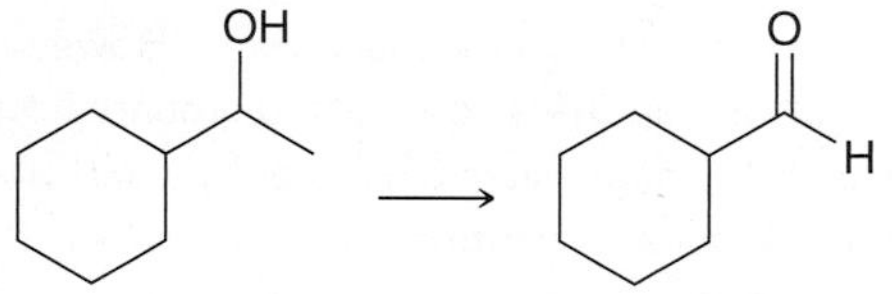

APPLY the skill **12.11** Propose a plausible synthesis for the following transformation (many steps are required):

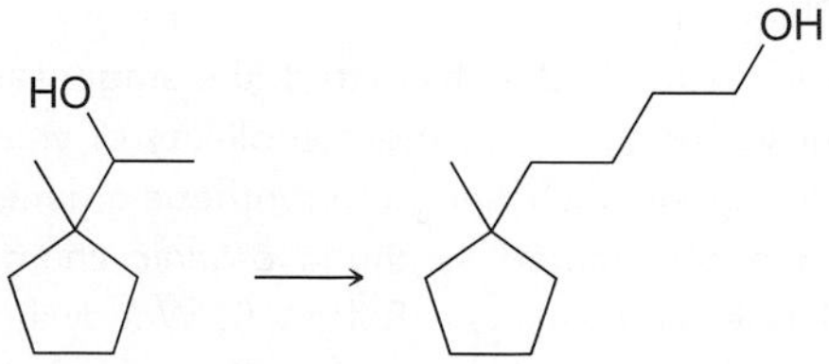

12.12 Propose a plausible synthesis for the following transformation, in which the carbon skeleton is increased by only one carbon atom:

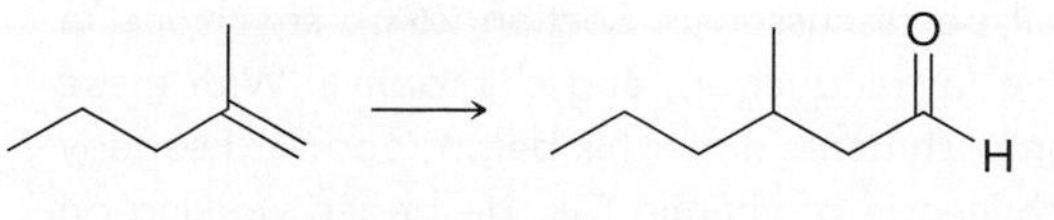

need more **PRACTICE?** **Try Problems 12.19–12.26**

MEDICALLYSPEAKING

The Total Synthesis of Vitamin B_{12}

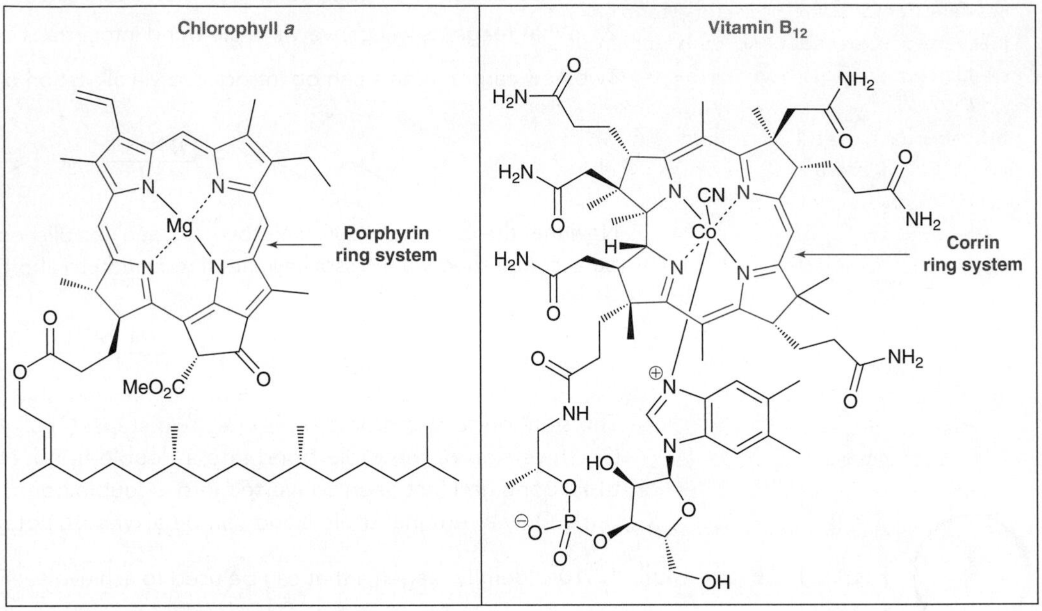

The story of vitamin B_{12} began when physicians recognized that anemia, a fatal disease caused by a low concentration of red blood cells, could be treated effectively by feeding the patient liver. This finding sparked a race to extract the compounds in liver and isolate the vitamin capable of treating anemia. In 1947, vitamin B_{12} was first isolated and purified as deep red crystals by Ed Rickes (a scientist working at the Merck chemical company). Efforts then focused on determining the structure. Some structural features were initially elucidated, but the complete structure was successfully determined by Dorothy Crowfoot Hodgkin (Oxford University) using X-ray crystallography. She found that the structure of vitamin B_{12} is built upon a corrin ring system, which is similar to the porphyrin ring system present in chlorophyll, the green pigment that plants use for photosynthesis (see above).

The corrin ring system of vitamin B_{12} is also comprised of four heterocycles (rings containing a heteroatom, such as nitrogen) joined together in a macrocycle. However, the corrin ring system of vitamin B_{12} is constructed around a central cobalt atom instead of the magnesium of chlorophyll, and vitamin B_{12} contains many more chirality centers than chlorophyll. The determination of this complex structure pushed the boundaries of X-ray crystallography, which until then had never been used to elucidate such a complex structure. Hodgkin was a pioneer in the field of X-ray crystallography and identified the structures of many important biochemical compounds. For her efforts, she was awarded the 1964 Nobel Prize in Chemistry.

With the structure of vitamin B_{12} elucidated, the stage was set for its total synthesis. At the time, the complexity of vitamin B_{12} represented the greatest challenge to synthetic organic chemists, and a couple of talented synthetic organic chemists viewed the challenge as irresistible. Robert B. Woodward (Harvard University) had already established himself as a leading player in the field of organic chemistry with his successful total synthesis of many important natural products, including quinine (used in the treatment of malaria), cholesterol and cortisone (steroids that will be discussed in Section 26.6), strychnine (a poison), reserpine (a tranquilizer), and chlorophyll. With these impressive accomplishments under his belt, Woodward eagerly embraced the challenge of vitamin B_{12}. He began working on methods for constructing the corrin ring system as well as the stereochemically demanding side chain. Meanwhile, Albert Eschenmoser (at the ETH in Zurich, Switzerland) was also working on a synthesis of the vitamin. However, the two men were developing different strategies for constructing the corrin ring system. Woodward's A→B route involved forming the macrocycle between rings A and B, while Eschemoser's A→D route involved forming the macrocycle between rings A and D:

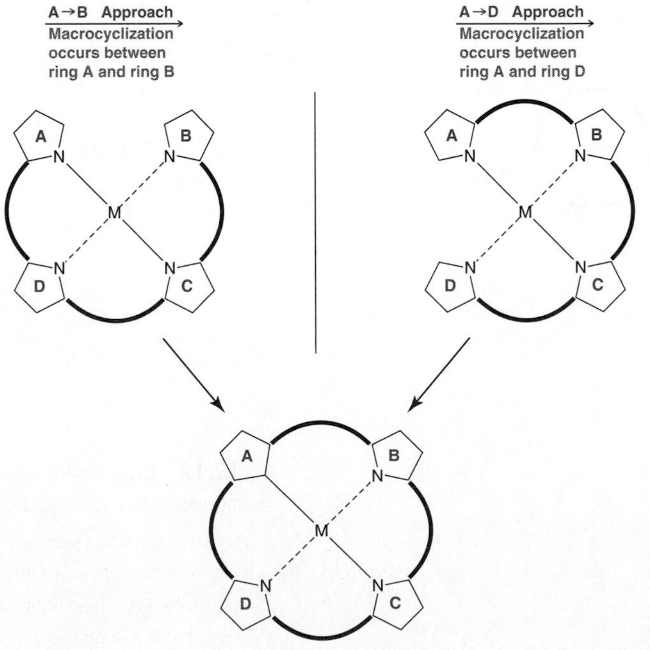

During the development of each pathway, unexpected obstacles emerged that required the development of new strategies and techniques. For example, Eschenmoser had successfully demonstrated a method for coupling the heterocycles, enabling construction of a simple corrin system (without the large, bulky side groups found in vitamin B_{12}); however, this approach failed to couple the heterocycles that needed to be joined to form the corrin ring system of vitamin B_{12}. This failure was attributed to the steric bulk of the substituents on each heterocycle. To circumvent the problem, Eschenmoser developed an ingenious method by which he tethered two rings together with a temporary sulfur bridge. By doing so, the coupling process became an *intra*molecular process rather than an *inter*molecular process (see above).

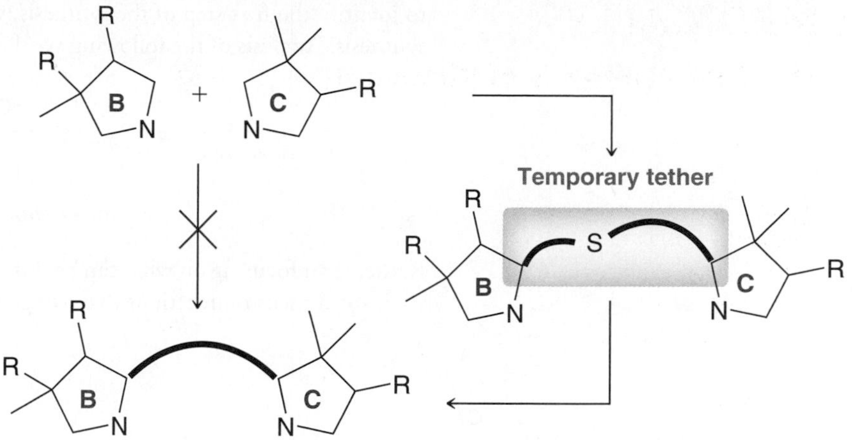

The sulfur bridge was readily expelled during the coupling process, and the overall process became known as "sulfide contraction." This is just one example of the creative solutions that synthetic chemists must develop when a planned synthetic route fails. A number of obstacles still faced both Woodward and Eschenmoser, and a partnership was forged in 1965 to tackle the problem together. In fact, that was the same year that Woodward was awarded the Nobel Prize in Chemistry for his contributions to the field of synthetic organic chemistry.

Woodward and Eschenmoser continued to work together for another seven years, often spending an entire year optimizing the conditions for an individual step. The intense effort, which involved nearly 100 graduate students working for a decade, would ultimately be rewarded. Woodward's team completed the assembly of the stereochemically demanding side chain, and the two groups combined some of the best methods and practices developed during studies aimed at construction of the corrin system. The pieces were finally joined together, and the synthesis was completed to produce vitamin B_{12} in 1972. This landmark event represents one of the greatest achievements in the history of synthetic organic chemistry, and it demonstrated that organic chemists could prepare any compound, regardless of complexity, given enough time.

During his journey toward the total synthesis of vitamin B_{12}, Woodward encountered a class of reactions that were known to proceed with unexplained stereochemical outcomes. Together with his colleague Roald Hoffmann, he developed a theory and a set of rules that would successfully explain the stereochemical outcomes of an entire area of organic chemistry called pericyclic reactions. This class of reactions will be covered in Chapter 17 together with the so-called Woodward–Hoffmann rules used to describe and predict the stereochemical outcomes for these reactions. The development of these rules led to another Nobel Prize in 1981, for which Woodward would have been a co-recipient had he not died two years earlier.

The story of vitamin B_{12} is a wonderful example of how organic chemistry progresses. During the total synthesis of a structurally complex compound, there is inevitably a point at which the planned route fails, requiring a creative method for circumventing the obstacle. In this way, new ideas and techniques are constantly developed. In the decades since the total synthesis of vitamin B_{12}, thousands of synthetic targets, most of them pharmaceuticals, have been constructed. New techniques, reagents, and principles constantly emerge form these endeavors. With time, synthetic targets are getting more and more complex, and the leaders in the field of organic chemistry are constantly pushing the boundaries of synthetic organic chemistry, which continues to evolve on a daily basis.

12.5 Retrosynthetic Analysis

As we progress through the course and increase our repertoire of reactions, synthesis problems will become increasingly more challenging. To meet this challenge, a modified approach will be necessary. The same two fundamental questions (as described in the previous section) will continue to serve as a starting point for analyzing all synthesis problems, but instead of trying to identify the first step of the synthesis, we will begin by trying to identify the last step of the synthesis. Analysis of the following synthesis problem will illustrate this process:

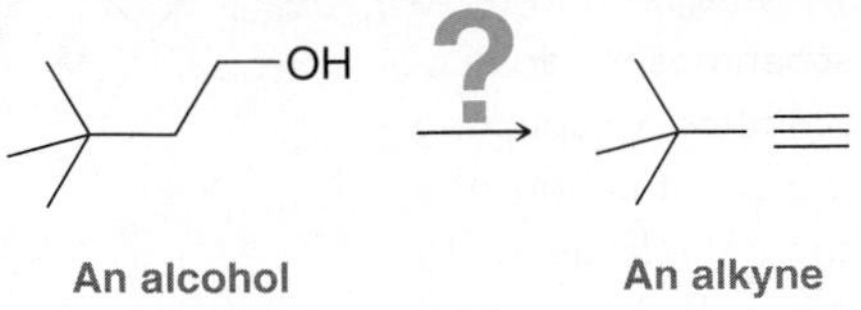

Rather than focusing on what can be done with an alcohol that will ultimately lead to an alkyne, we instead focus on reactions that can generate an alkyne:

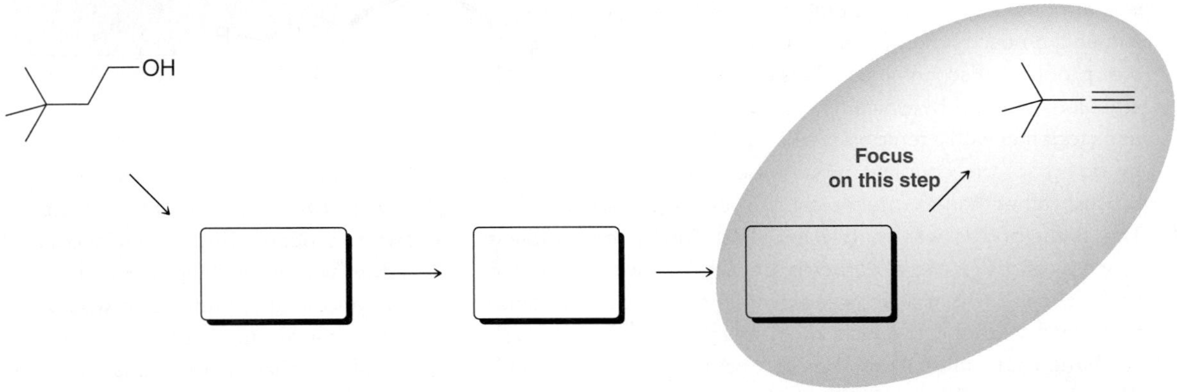

In this way, we work backward until arriving at the starting material. Chemists have intuitively used this approach for many years, but E. J. Corey (Harvard University) was the first to develop a systematic set of principles for application of this approach, which he called **retrosynthetic analysis**. Let's use a retrosynthetic analysis to solve the problem above.

We must always begin by determining whether there is a change in the carbon skeleton or in the identity or location of the functional group. In this case, both the starting material and the product contain six carbon atoms, and the carbon skeleton is not changing in this instance. However, there is a change in the functional group. Specifically, an alcohol is converted into an alkyne but remains in the same location in the skeleton. We have not learned a way to do this in just one step. In fact, using reactions covered so far, this transformation cannot be accomplished even in two steps. So we approach this problem *backward* and ask: "How are triple bonds made?" We have only covered one way to make a triple bond. Specifically, a dihalide undergoes two successive E2 eliminations in the presence of excess $NaNH_2$ (Section 10.4). Any one of the following three dihalides could be used to form the desired alkyne:

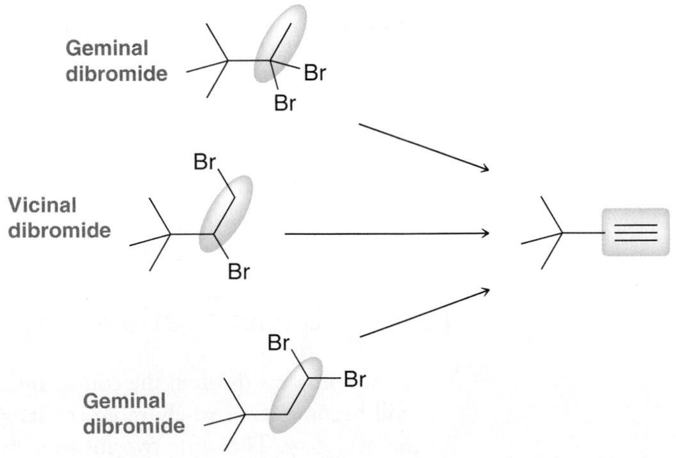

The geminal dibromides can be ruled out, because we only saw one way to make a geminal dihalide—and that was starting from an alkyne. We certainly do not want to start with an alkyne in order to produce the very same alkyne:

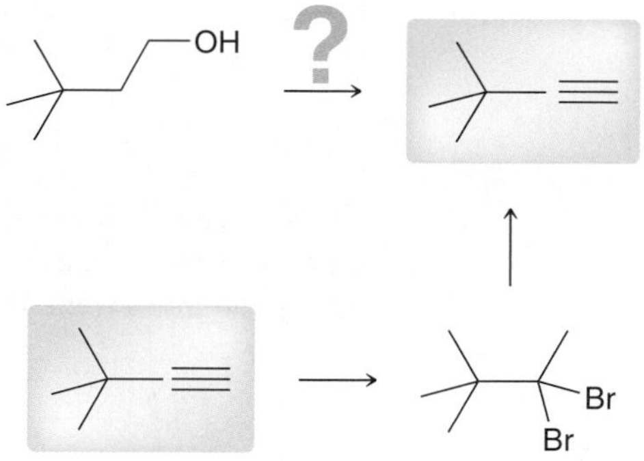

Therefore, the last step of our synthesis must be formation of the alkyne from a *vicinal* dihalide:

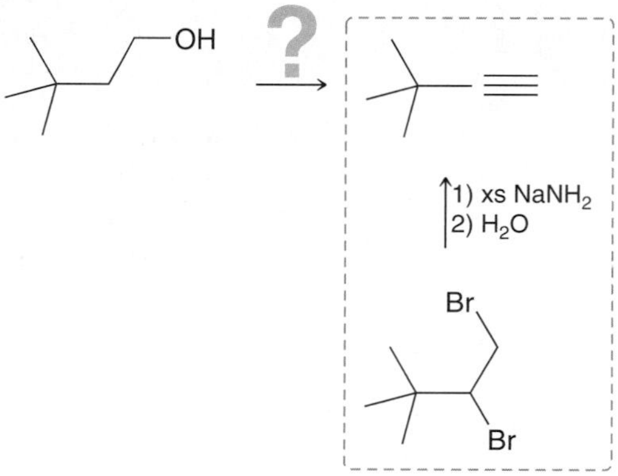

A special retrosynthetic arrow is used by chemists to indicate this type of "backward" thinking:

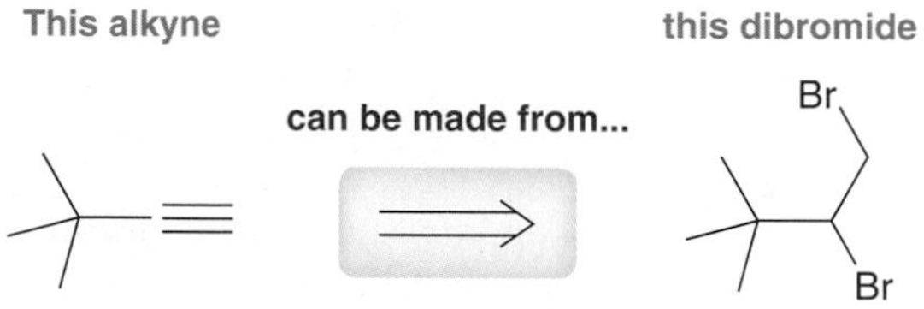

Don't be confused by this retrosynthetic arrow. It indicates a hypothetical synthetic pathway *thinking backward* from the product (alkyne). In other words, the previous figure should be read as: "In the last step of our synthesis, the alkyne *can be made from* a vicinal dibromide."

Now let's try to go backward one more step. We have learned only one way to make a vicinal dihalide, starting with an alkene:

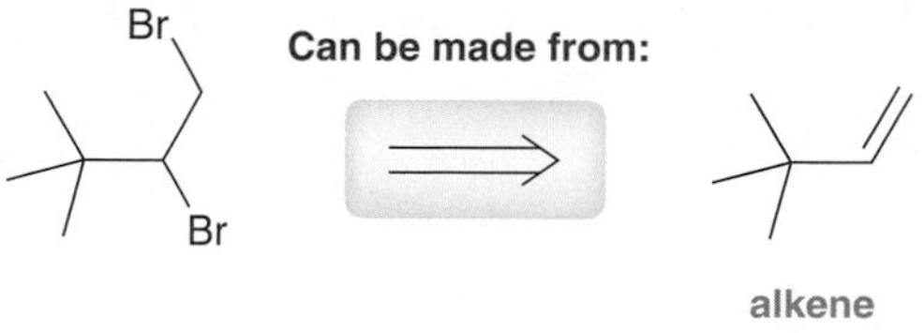

Notice again the retrosynthetic arrow. The figure indicates that the vicinal dibromide can be made from an alkene. In other words, the alkene can be used as a precursor to prepare the desired dibromide.

Therefore, our retrosynthetic analysis, so far, looks like this:

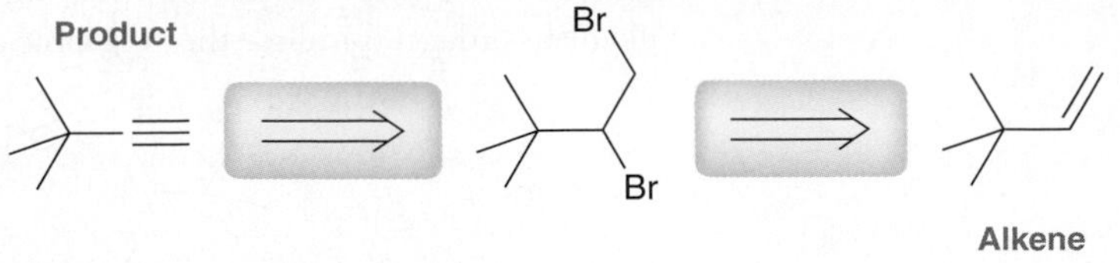

This scheme indicates that the product (alkyne) can be prepared from the alkene. This sequence of events represents one of the strategies discussed earlier in this chapter—converting a double bond into a triple bond:

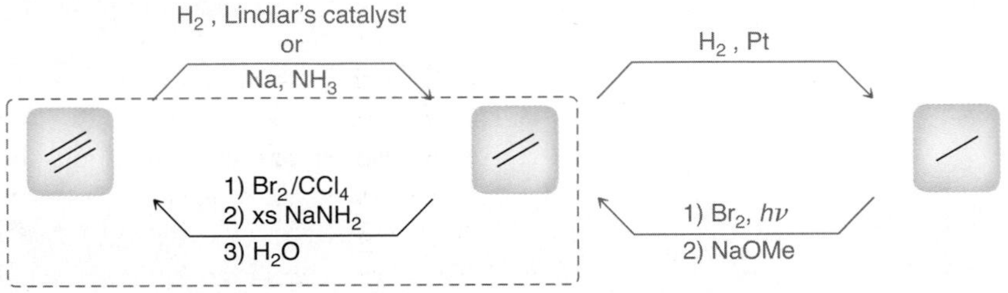

In order to complete the synthesis, the starting material must be converted into the alkene. At this point, we can think forward, in an attempt to converge with the pathway revealed by the retrosynthetic analysis:

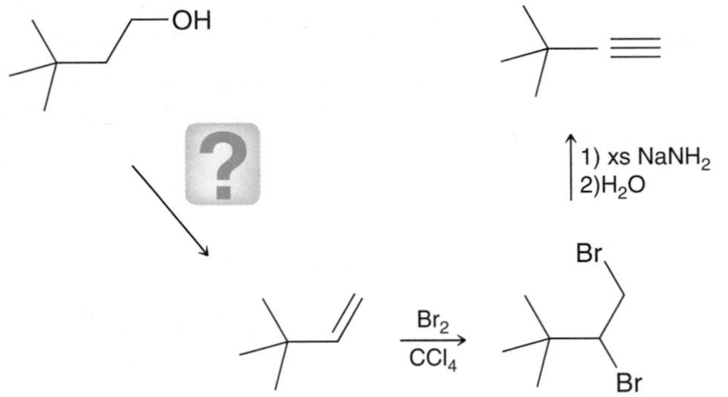

This step can be accomplished with an E2 elimination. Just remember that the hydroxyl group must first be converted into a tosylate (a better leaving group). Then, an E2 elimination will create an alkene, which bridges the gap between the starting material and the product:

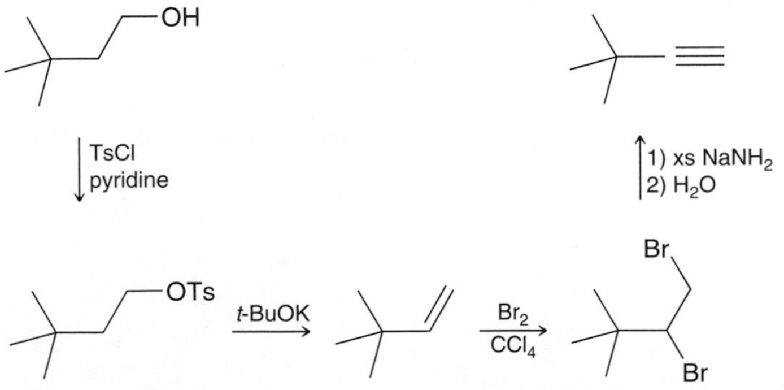

The synthesis seems complete. However, before recording the answer, it is always helpful to review all of the proposed steps and make sure that the regiochemistry and stereochemistry of each step will lead to the desired product *as a major product*. It would be inefficient to involve any steps that would rely on the formation of a minor product. We should only use steps that produce the desired product as the major product. After reviewing every step of the proposed synthesis, the answer is recorded like this:

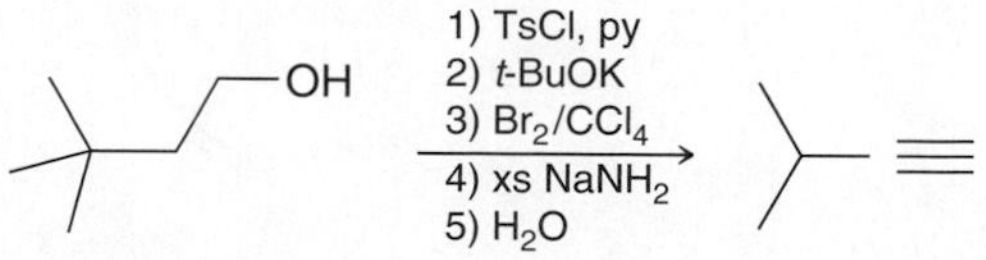

SKILLBUILDER

12.4 RETROSYNTHETIC ANALYSIS

LEARN the skill

Propose a plausible synthesis for each of the following transformations:

(a) 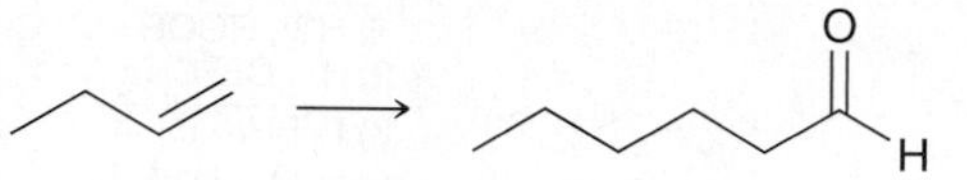 (b) 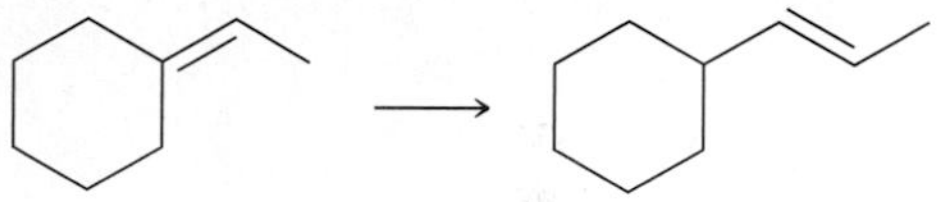

SOLUTION

(a) First determine whether there is a change in the carbon skeleton. The starting compound has four carbon atoms while the product has six. This transformation therefore requires the installation of two carbon atoms:

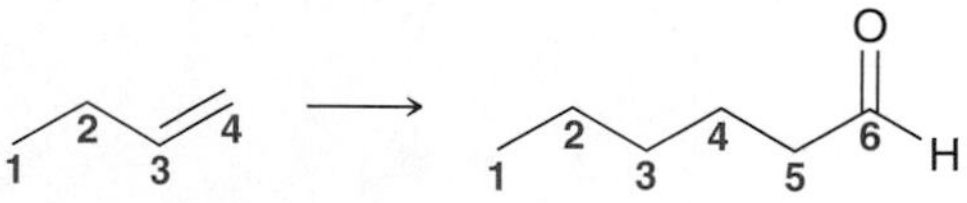

As we have seen many times in this chapter, the only method we have learned for achieving this transformation is the alkylation of a terminal alkyne. Our synthesis must therefore involve an alkylation step. This should be taken into account when performing a retrosynthetic analysis.

Second, we need to determine whether there is a change in the identity or location of the functional group. In this case, the functional group has changed both its identity and its location. The functional group in the product is an aldehyde moiety. As seen in Section 10.7, an aldehyde can be made via hydroboration-oxidation of an alkyne. As shown with the retrosynthetic arrow, a terminal alkyne can be converted into an aldehyde.

Can be made from:

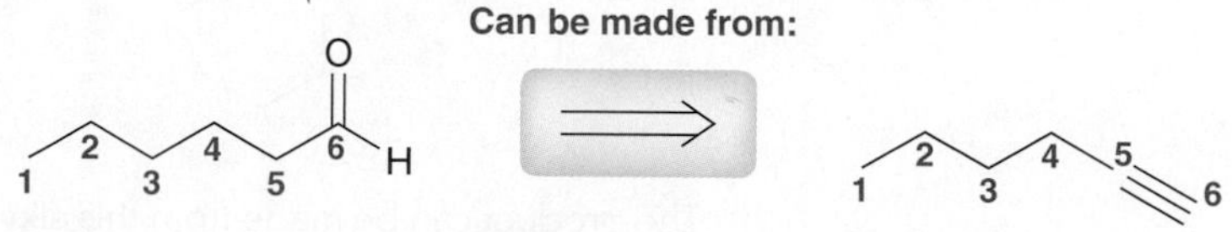

Continuing with a retrosynthetic analysis, we must consider the step that might be used to produce the alkyne. Recall that our synthesis must contain a step involving the alkylation of an alkyne. We therefore propose the following retrosynthetic step:

Can be made from:

This step accomplishes the necessary installation of two carbon atoms.

To complete the synthesis, we just need to bridge the gap between the starting material and the alkyl halide:

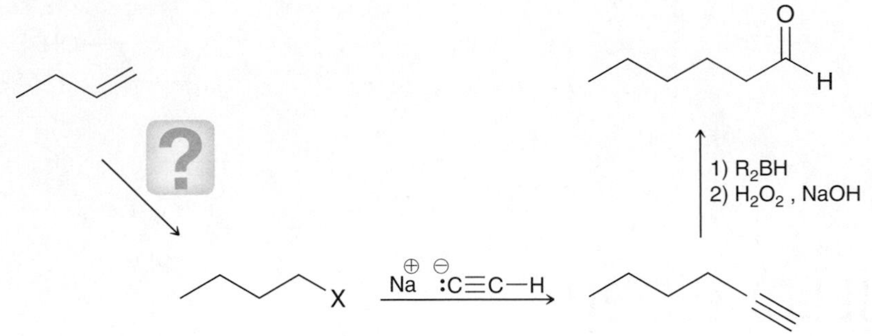

Working forward now, it is apparent that the necessary transformation can be achieved with an *anti*-Markovnikov addition, which can be accomplished with HBr in the presence of peroxides. In summary, we propose the following synthesis.

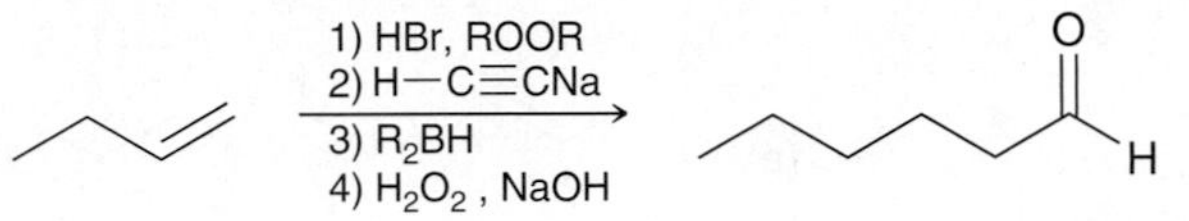

(b) Always begin by determining whether there is a change in the carbon skeleton. The starting compound has a two carbon chain connected to the ring, while the product has a three-carbon chain connected to the ring:

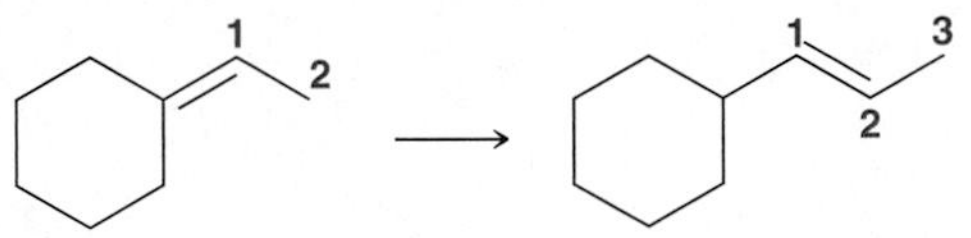

This transformation therefore requires the installation of one carbon atom.

Next determine whether there is a change in the identity or location of the functional group. In this case the identity of the functional group does not change, but its location does change.

In order to achieve the necessary change in the carbon skeleton, our synthesis must involve an alkyne that will be alkylated and then converted into the product. Working backward, we focus on the alkyne:

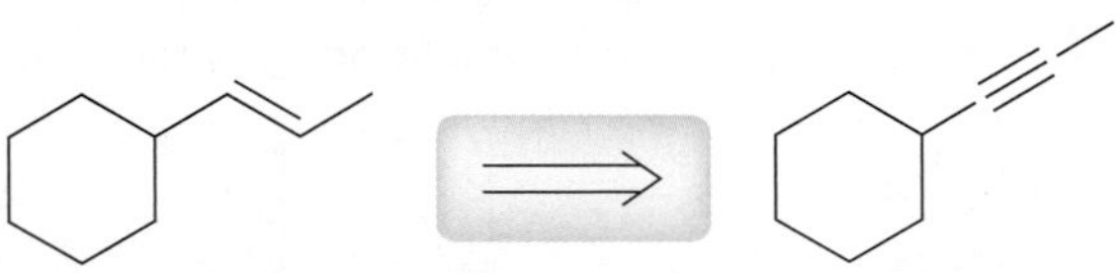

The product can be made from this alkyne via a dissolving metal reduction. Recall the reason for going through the alkyne in the first place: to introduce a carbon atom via alkylation. The last two steps of our synthesis can therefore be written in the following way, using retrosynthetic arrows:

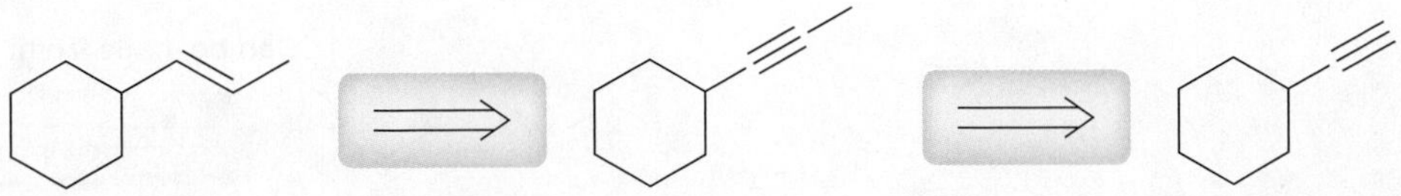

Now we must bridge the gap:

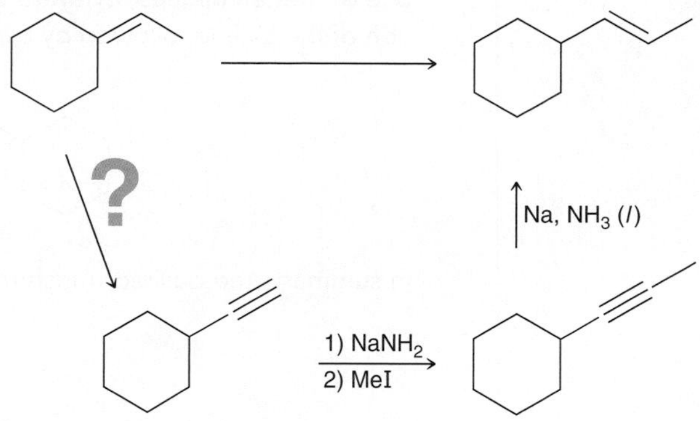

To bridge the gap, a double bond must be converted into a triple bond and its position must be moved. We cannot convert it into a triple bond in its current location—that would give a pentavalent carbon, which is impossible:

WATCH OUT
Never draw a pentavalent carbon because that would violate the octet rule. Recall that carbon only has four orbitals with which to form bonds, and as a result, it can never form more than four bonds.

We must first move the position of the double bond and only then convert the double bond into a triple bond. The location of the double bond can be moved using the technique discussed earlier in the chapter—addition followed by elimination:

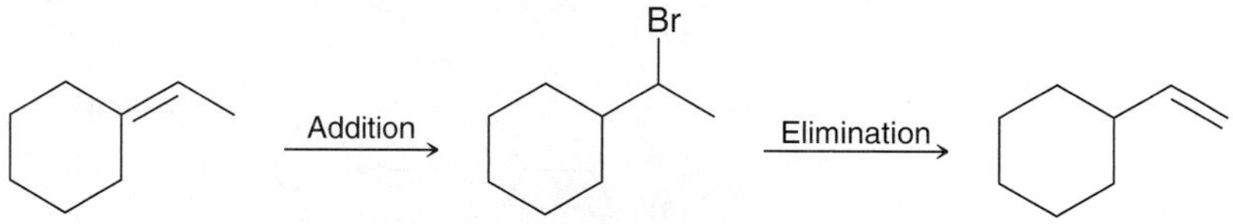

Take special note of the regiochemistry in each case. In the first step, an *anti*-Markovnikov addition of HBr is needed, so peroxides are required. Then, in the second step, the Hofmann product (the less substituted alkene) is needed, so a sterically hindered base is required:

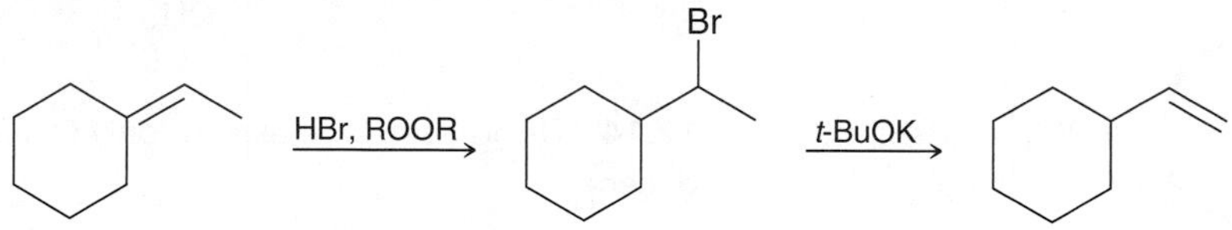

Let's summarize what we have so far, working forward.

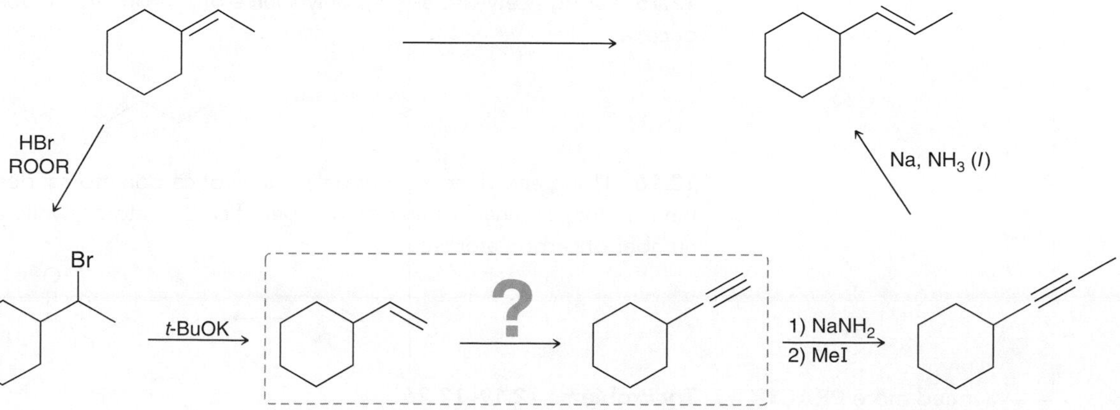

To bridge the gap, a double bond must be converted into a triple bond. Once again, this is one of the techniques reviewed earlier in this chapter. It is accomplished through bromination of the alkene followed by two successive elimination reactions to produce an alkyne.

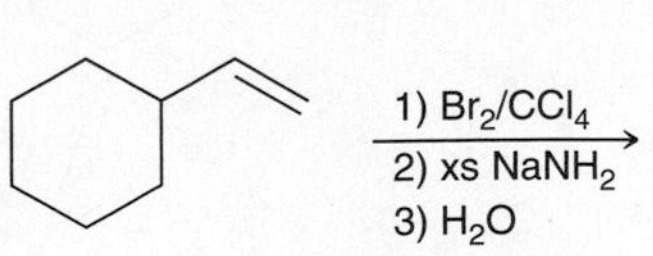

In summary, the desired transformation can be achieved with the following reagents:

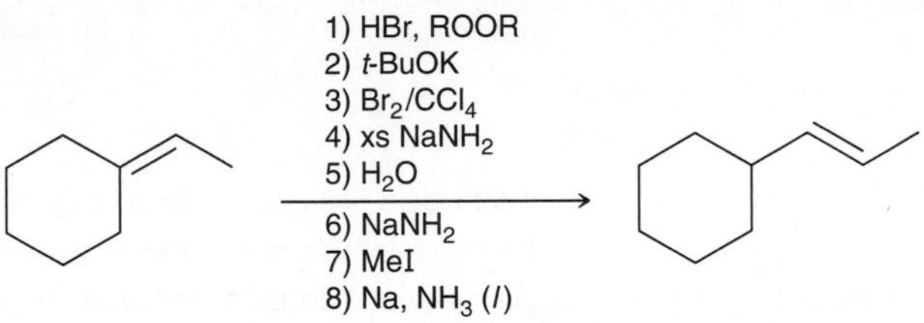

PRACTICE the skill **12.13** Propose a plausible synthesis for each of the following transformations:

(a) 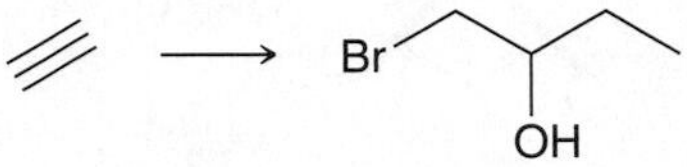

(b) 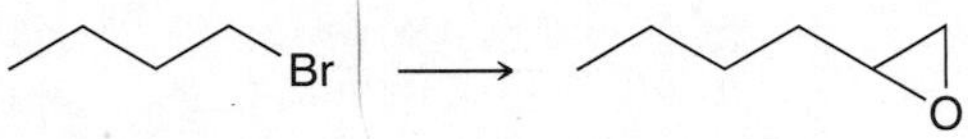

(c) 

(d) 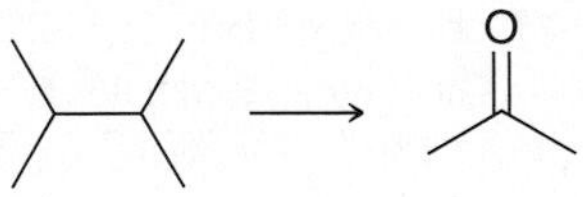

(e) 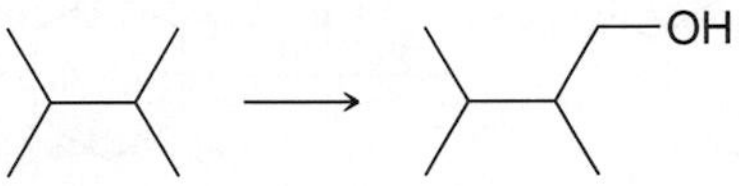

(f) 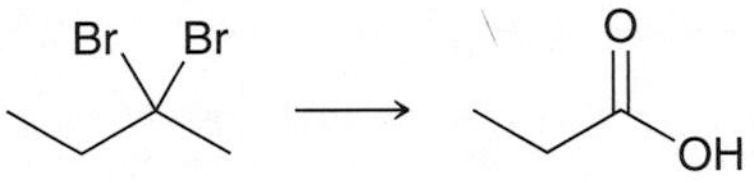

(g) 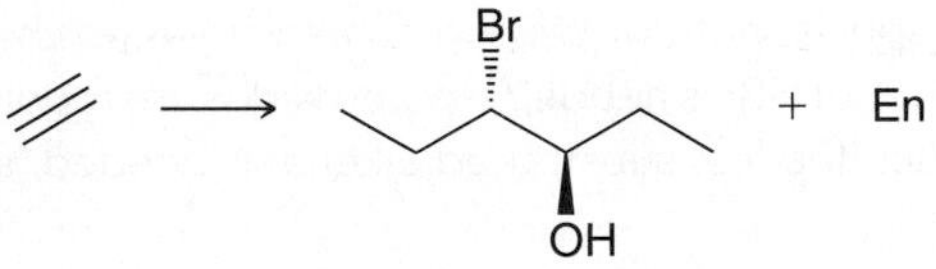

(h) 

APPLY the skill

12.14 Using acetylene as your only source of carbon atoms, design a synthesis of *trans*-5-decene:

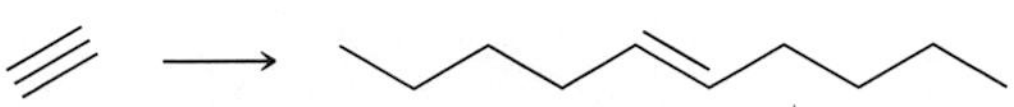

12.15 Using acetylene as your only source of carbon atoms, design a synthesis of *cis*-3-decene:

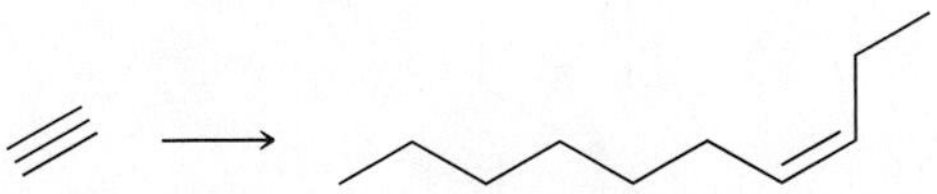

12.16 Using acetylene as your only source of carbon atoms, design a synthesis of pentanal (note: pentanal has an odd number of carbon atoms, while acetylene has an even number of carbon atoms):

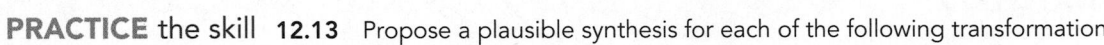

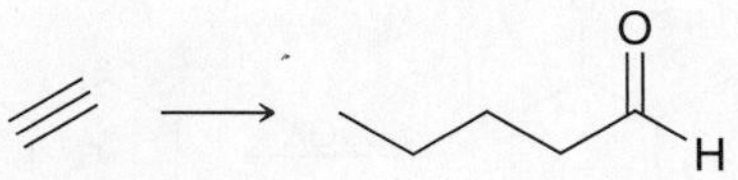

need more **PRACTICE?** **Try Problems 12.19–12.26**

PRACTICALLYSPEAKING

Retrosynthetic Analysis

Thus far, we have only explored a limited number of reactions, and the complexity of synthesis problems is therefore limited. At this stage, it might be difficult to see why retrosynthetic analysis is so critical. However, as we expand our repertoire of reactions, we will begin to explore the synthesis of more sophisticated structures, and the need for retrosynthetic analysis will become more apparent.

E. J. Corey first developed his retrosynthetic methodology as a faculty member at the University of Illinois. During his sabbatical year, which he spent at Harvard University, he shared his ideas with Robert Woodward (the very same Woodward who developed a total synthesis of vitamin B_{12}, described in the previous Medically Speaking box). Corey demonstrated his ideas by presenting a retrosynthetic analysis of longifolene, a natural compound that represented a challenging target because of its unique tricyclic structure. Corey identified the following strategic bond of longifolene for disconnection:

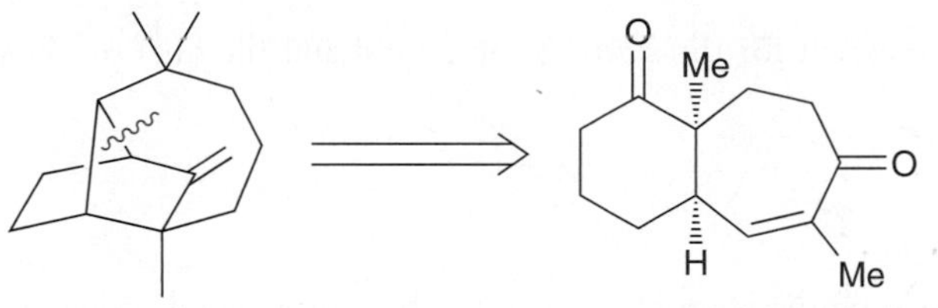

The functional groups in the precursor serve as handles enabling formation of the strategic bond (utilizing a reaction that we will explore in Chapter 22). Longifolene represented an excellent example of the power of Corey's retrosynthetic disconnection approach. For most of the bonds in longifolene, disconnection would produce a complex tricyclic precursor. That is, there are not many strategic bonds that can be disconnected to provide a simpler precursor. Woodward was impressed by Corey's insights and immediately recognized that Corey would soon be another leader in the field. This likely played a large role in the offer that Corey subsequently received to join the chemistry department at Harvard as a professor.

Corey developed his ideas further and established a set of rules and principles for proposing retrosynthetic analysis. He also spent much time creating computer programs to assist chemists in performing retrosynthetic analysis. It is clear that computers cannot replace the chemist altogether, as the chemist must provide the insight and creativity in determining which strategic bonds to disconnect. In addition, the chemist must exhibit creativity when a planned route fails. As discussed in the previous box, this is how new ideas, reagents, and techniques are developed. Corey used his approach for the total synthesis of dozens of compounds. In the process, he developed many new reagents, strategies, and reactions that now bear his name. Corey was an instrumental player in developing the field of synthetic organic chemistry. Retrosynthetic analysis is now taught to all students of organic chemistry. Similarly, nearly all reports of total synthesis in the chemical literature begin with a retrosynthetic analysis. For his contributions to the development of synthetic organic chemistry, Corey was awarded the Nobel Prize in Chemistry in 1990.

12.6 Practical Tips for Increasing Proficiency

Organizing a Synthetic "Toolbox" of Reactions

All of the reactions in this course will collectively represent your "toolbox" for proposing syntheses. It will be very helpful to prepare two lists that parallel the two questions that must be considered in every synthesis problem (*change in the carbon skeleton and change in the functional group*). The first list should contain C—C bond-forming reactions and C—C bond-breaking reactions. At this point, this list is very small. We have only seen one of each: alkylation of alkynes for C—C bond forming and ozonolysis for C—C bond breaking. As the text progresses, more reactions will be added to the list, which will remain relatively small but very powerful. The second list should contain functional group transformations, and it will be longer.

As we move through the text, both lists will grow. For solving synthesis problems, it will be helpful to have the reactions categorized in this way in your mind.

Creating Your Own Synthesis Problems

A helpful method for practicing synthesis strategies is to construct your own problems. The process of designing problems often uncovers patterns and new ways of thinking about the reactions. Begin by choosing a starting compound. To illustrate this, let's begin by choosing a simple starting compound, such as acetylene. Then choose a reaction expected for a triple bond, perhaps an alkylation:

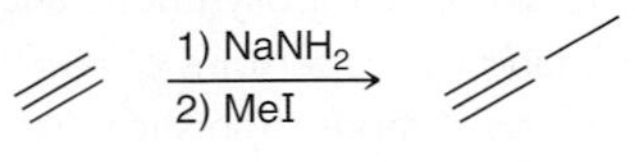

Next, choose another reaction, perhaps another alkylation:

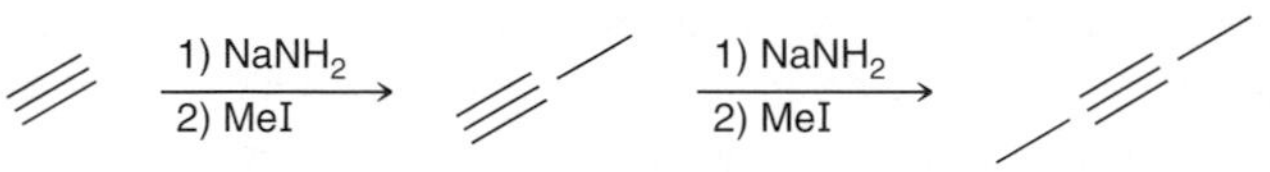

Then, treat the alkyne with another reagent. Look at the list of reactions of alkynes and choose one; perhaps hydrogenation with a poisoned catalyst:

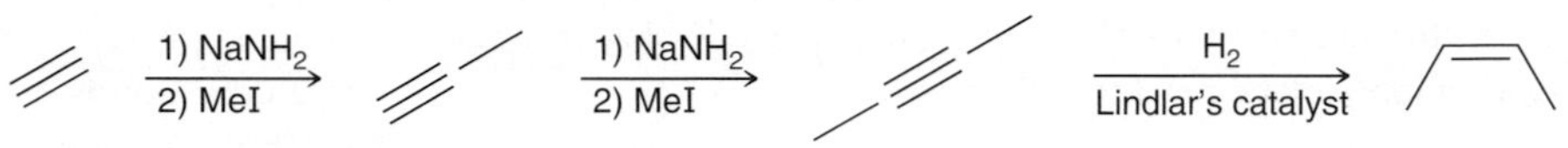

Finally, simply erase everything except for the starting compound and the final product. The result is a synthesis problem:

Once you have created your own synthesis problem, it might be a really good problem, but it won't be helpful for you to solve it. You already know the answer! Nevertheless, the process of creating your own synthesis problems will be very helpful to you in sharpening your synthesis skills.

Once you have created several of your own problems, try to find a study partner who is also willing to create several problems. Each of you can swap problems, try to solve each other's problems, and then get back together to discuss the answers. You are likely to find that exercise to be very rewarding.

Multiple Correct Answers

Remember that most synthesis problems will have numerous correct answers. As an example, *anti*-Markovnikov hydration of an alkene can be achieved through either of the two possible routes:

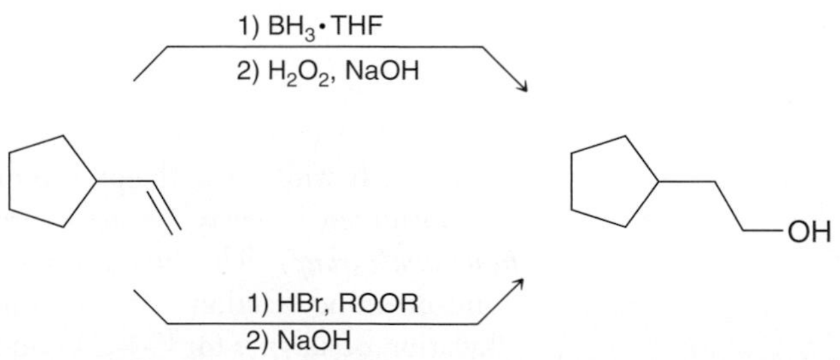

The first pathway represents hydroboration-oxidation of the alkene to achieve *anti*-Markovnikov addition of water. The second pathway represents a radical (*anti*-Markovnikov) addition of HBr followed by an S_N2 reaction to replace the halogen with a hydroxyl group. Each of these pathways represents a valid synthesis. As we learn more reactions, it will become more common to encounter synthesis problems with several perfectly correct answers. The goal should always be efficiency. A 3-step synthesis will generally be more efficient than a 10-step synthesis.

REVIEW OF CONCEPTS AND VOCABULARY

SECTION 12.1

- It is essential to achieve mastery over all reagents described in the previous chapters.

SECTION 12.2

- The position of a halogen can be moved by performing elimination followed by addition.
- The position of a π bond can be moved by performing addition followed by elimination.
- An alkane can be functionalized via radical bromination.

SECTION 12.3

- If the size of the carbon skeleton increases, then a C—C bond-forming reaction is required.
- If the size of the carbon skeleton decreases, then a C—C bond-breaking reaction, called **bond cleavage**, is required.

SECTION 12.4

- Every synthesis problem should be approached by asking the following two questions:
  - Is there any change in the carbon skeleton?
  - Is there any change in the identity or location of the functional group?

SECTION 12.5

- In a **retrosynthetic analysis**, the last step of the synthetic route is first established, and the remaining steps are determined, working backward from the product.

KEY TERMINOLOGY

bond cleavage 543

retrosynthetic analysis 550

SKILLBUILDER REVIEW

12.1 CHANGING THE IDENTITY OR POSITION OF A FUNCTIONAL GROUP

Change the position of a halogen.

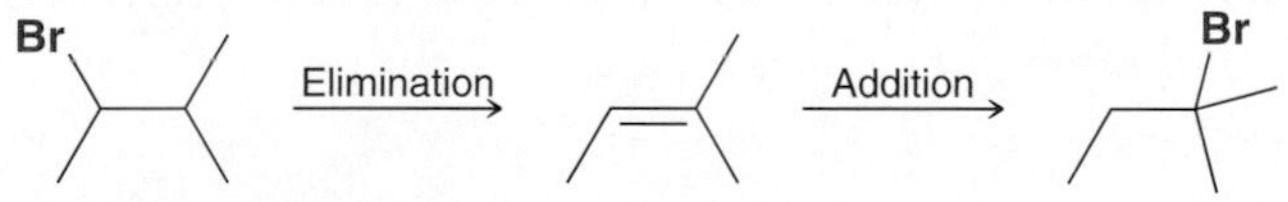

Change the position of a π bond.

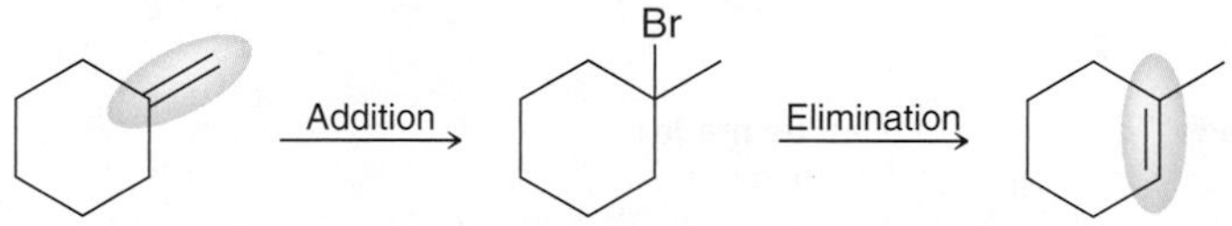

Install a functional group.

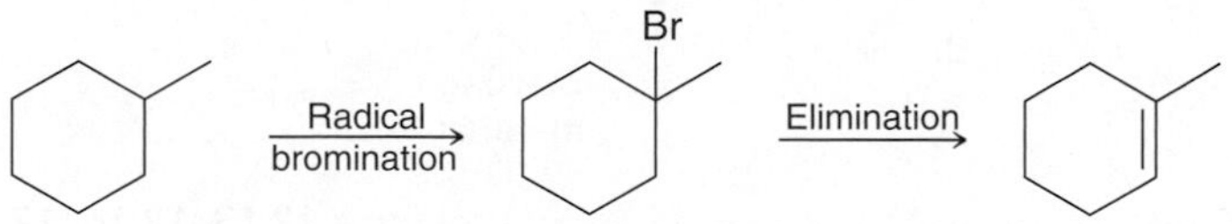

Interconvert between single, double, and triple bonds.

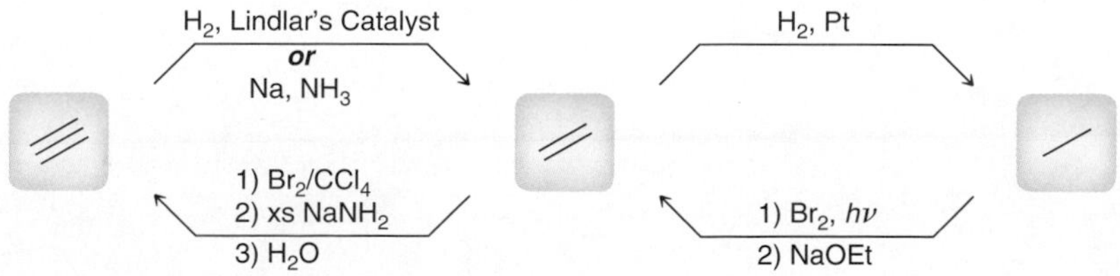

Try Problems 12.3–12.6, 12.17, 12.21, 12.22

12.2 CHANGING THE CARBON SKELETON

C—C bond formation.

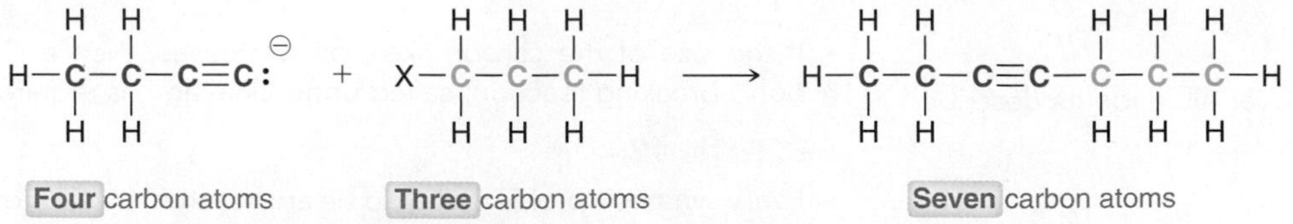

C—C bond cleavage.

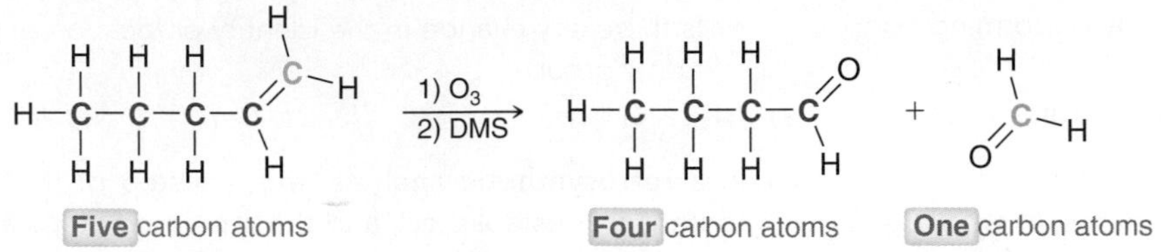

Try Problems 12.7–12.9, 12.18, 12.19, 12.20, 12.23, 12.26

12.3 APPROACHING A SYNTHESIS PROBLEM BY ASKING TWO QUESTIONS

1. Is there any change in the carbon skeleton?

Compare the starting material with the product to detemine if the carbon skeleton is gaining or losing any carbon atoms.

2. Is there any change in the identity or location of the functional group?

In other words, is one functional group converted into another, and does the position of functionality change?

Try Problems 12.10–12.12, 12.19–12.26

12.4 RETROSYNTHETIC ANALYSIS

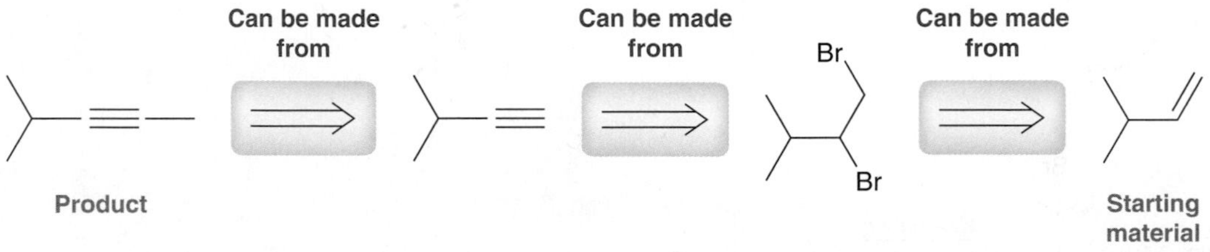

Try Problems 12.13–12.16, 12.19–12.26

PRACTICE PROBLEMS

Note: Most of the Problems are available within WileyPLUS, an online teaching and learning solution.

12.17 Identify the reagents necessary to achieve each of the following transformations:

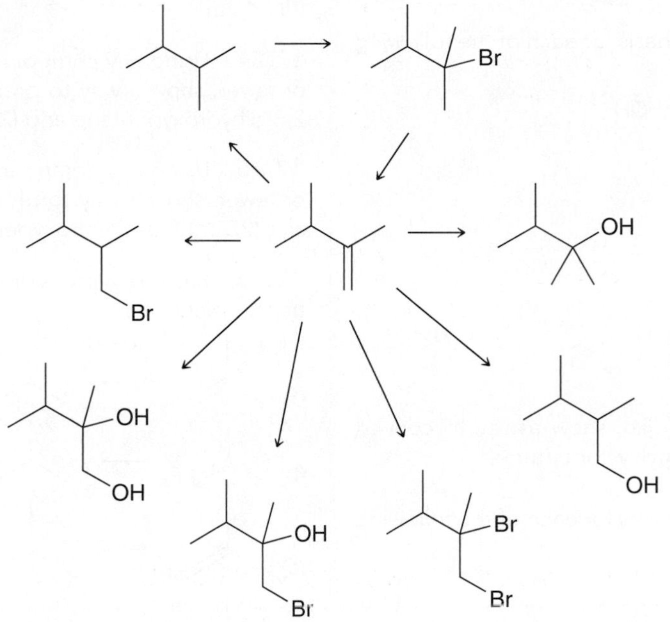

12.18 Identify the reagents necessary to achieve each of the following transformations:

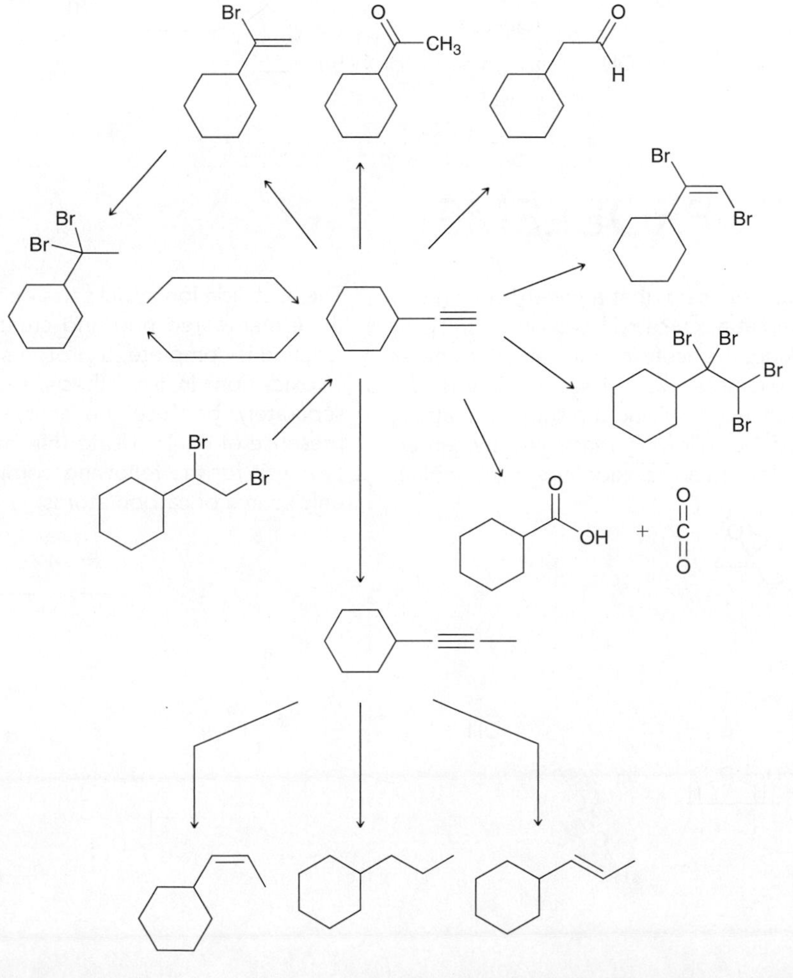

12.19 Using acetylene as your only source of carbon atoms, identify a synthetic route for the production of 2-bromobutane.

12.20 Using acetylene as your only source of carbon atoms, identify a synthetic route for the production of 1-bromobutane.

12.21 Propose a plausible synthesis for each of the following transformations:

(a)

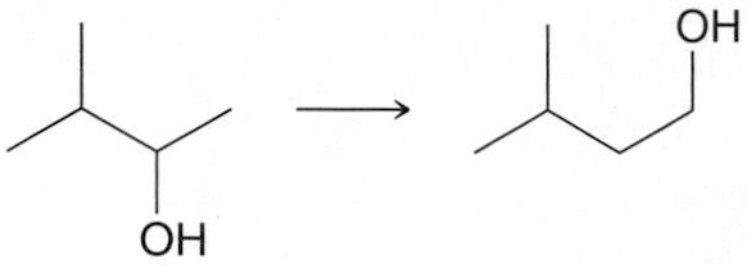

(b)

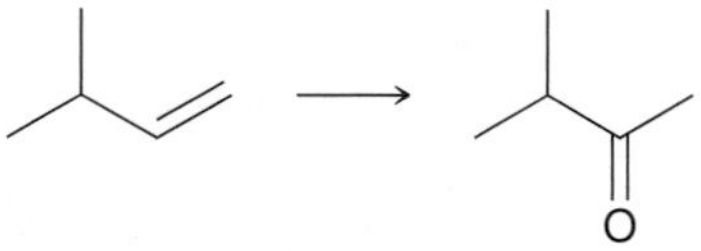

12.22 Using any reagents you like, show a way to convert 1-methylcyclopentene into 3-methylcyclopentene.

12.23 Propose a plausible synthesis for each of the following transformations:

(a)

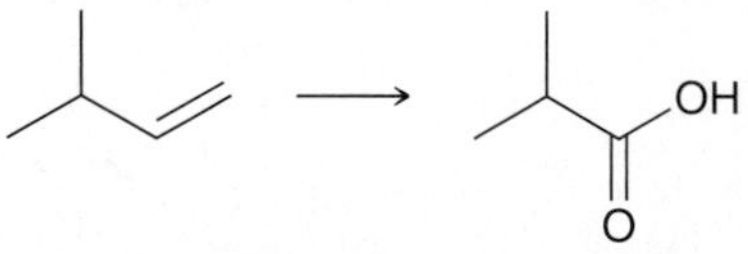

(b)

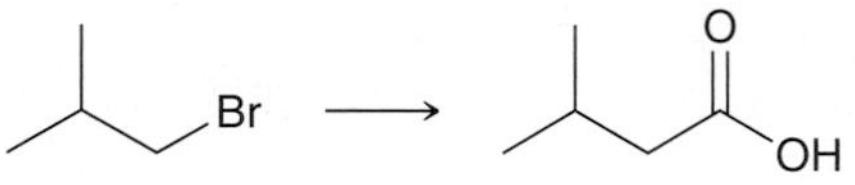

(c)

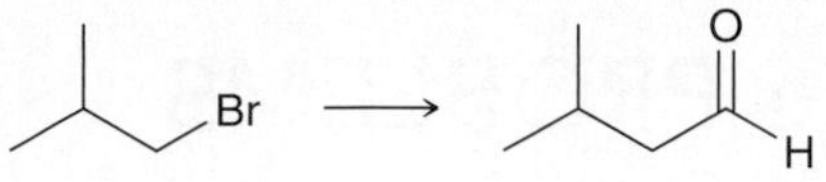

(d)

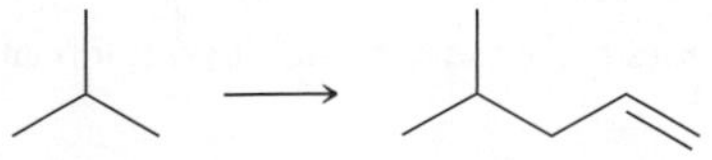

12.24 Using any compounds that contain two carbon atoms or fewer, show a way to prepare a racemic mixture of (2*R*,3*R*)-2,3-dihydroxypentane and (2*S*,3*S*)-2,3-dihydroxypentane.

12.25 Using any compounds that contain two carbon atoms or fewer, show a way to prepare a racemic mixture of (2*R*,3*S*)- and (2*S*,3*R*)-2,3-dihydroxypentane.

12.26 Propose a plausible synthesis for each of the following transformations:

(a)

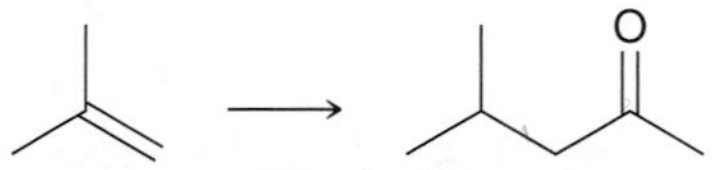

(b)

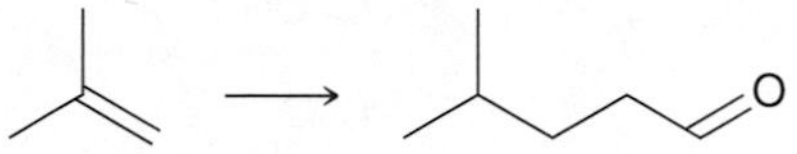

(c)

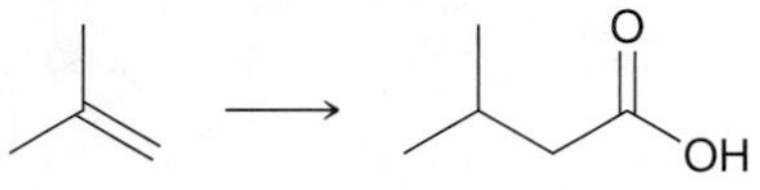

(d)

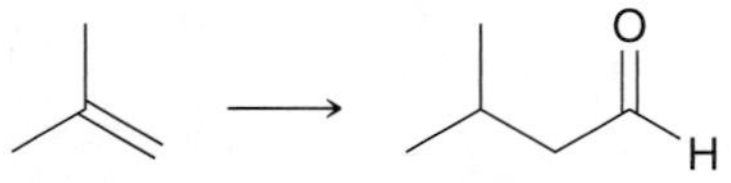

(e)

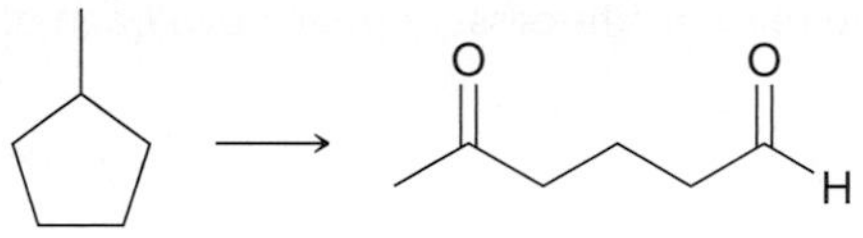

INTEGRATED PROBLEMS

12.27 In this chapter, we have seen that an acetylide ion can function as a nucleophile and attack an alkyl halide in an S_N2 process. More generally speaking, the acetylide ion can attack other electrophiles as well. For example, we will see in Chapter 14 that epoxides function as electrophiles and are subject to attack by a nucleophile. Consider the following reaction between an acetylide ion (the nucleophile) and an epoxide (the electrophile):

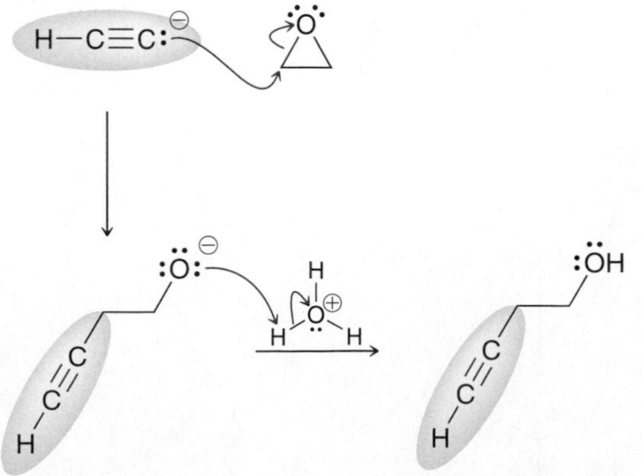

The acetylide ion attacks the epoxide, opening up the strained, three-membered ring and creating an alkoxide ion. After the reaction is complete, a proton source is used to protonate the alkoxide ion. In a synthesis, these two steps must be shown separately, because the acetylide ion will not survive in the presence of H_3O^+. Using this information, propose a plausible synthesis for the following compound using acetylene as your only source of carbon atoms:

12.28 In the previous problem, we saw that an acetylide ion can attack a variety of electrophiles. In Chapter 20, we will see that a C=O bond can also function as an electrophile. Consider the following reaction between an acetylide ion (the nucleophile) and a ketone (the electrophile):

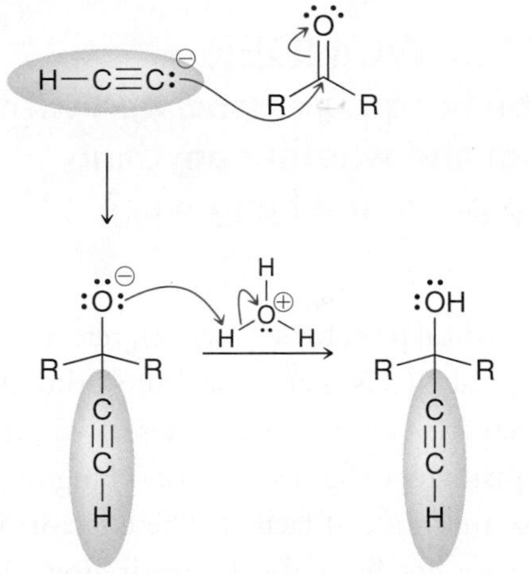

The acetylide ion attacks the ketone, generating an alkoxide ion. After the reaction is complete, a proton source is used to protonate the alkoxide ion. In a synthesis, these two steps must be shown separately, because the acetylide ion will not survive in the presence of H_3O^+. Using this information, propose a plausible synthesis for allyl alcohol, using acetylene as your only source of carbon atoms:

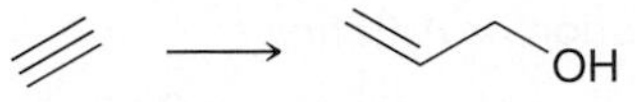

CHALLENGE PROBLEMS

12.29 Identify the reagents you might use to achieve each of the following transformations:

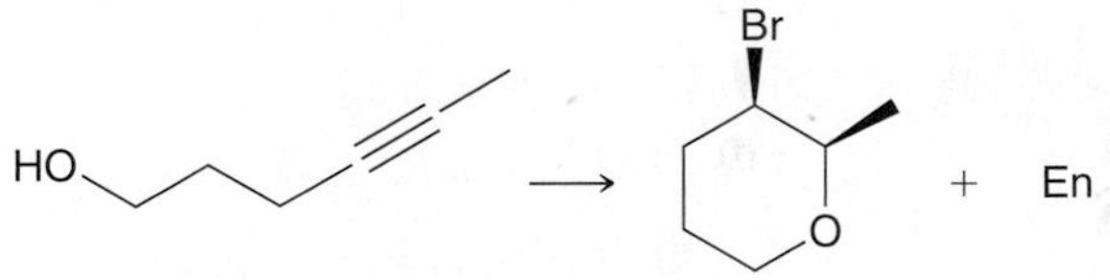

12.30 Starting with acetylene as your only source of carbon atoms, identify how you would prepare each member of the following homologous series of aldehydes:

(a) Ethanal (b) Propanal
(c) Butanal (d) Pentanal

12.31 Starting with acetylene as your only source of carbon atoms, propose a plausible synthesis for 1,4-dioxane:

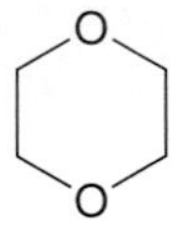

1,4-Dioxane

13 Alcohols and Phenols

DID YOU EVER WONDER...

what causes the hangover associated with drinking alcohol and whether anything can be done to prevent a hangover?

Veisalgia, the medical term for a hangover, refers to the unpleasant physiological effects that result from drinking too much alcohol. These effects include headache, nausea, vomiting, fatigue, and a heightened sensitivity to light and noise. Hangovers are caused by a multitude of factors. These factors include (but are not limited to) dehydration caused by the stimulation of urine production, the loss of vitamin B, and the production of acetaldehyde in the body. Acetaldehyde is a product of the oxidation of ethanol. Oxidation is just one of the many reactions that alcohols undergo. In this chapter, we will learn about alcohols and their reactions. Then, we will revisit the topic of acetaldehyde production in the body, and we will see if anything can be done to prevent a hangover.

DO YOU REMEMBER?

Before you go on, be sure you understand the following topics.
If necessary, review the suggested sections to prepare for this chapter.

- Brønsted-Lowry Acidity (Sections 3.3-3.4)
- Mechanisms and Curved Arrows (Sections 6.8-6.11)
- E2 and E1 Reactions (Sections 8.6-8.11)
- Designating the Configuration of a Chirality Center (Section 5.3)
- S_N2 and S_N1 Reactions (Sections 7.4-7.8)

WILEY PLUS Visit www.wileyplus.com to check your understanding and for valuable practice.

13.1 Structure and Properties of Alcohols

Alcohols are compounds that possess a **hydroxyl group** (OH) and are characterized by names ending in "ol":

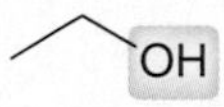

Ethanol

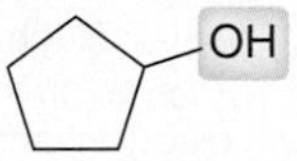

Cyclopentanol

A vast number of naturally occurring compounds contain the hydroxyl group. Here are just a few examples.

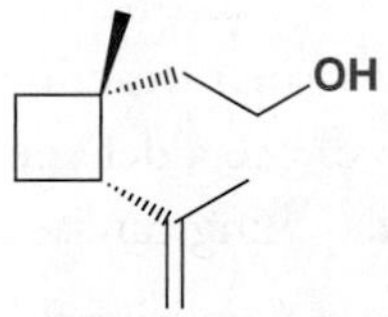

Grandisol
The sex pheromone of the male boll weevil

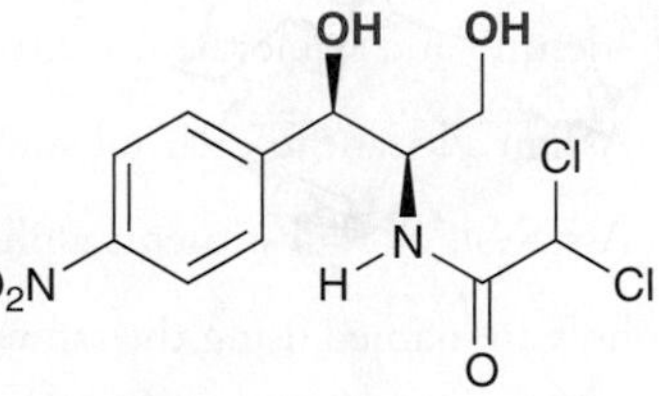

Chloramphenicol
An antibiotic isolated from the *Streptomyces venezuelae* bacterium. Potent against typhoid fever

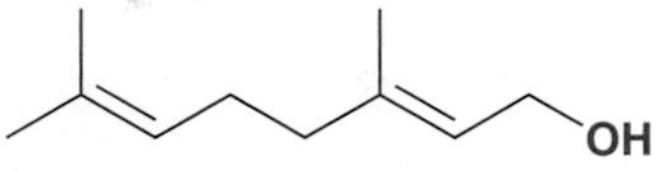

Geraniol
Isolated from roses and geraniums. Used in perfumes

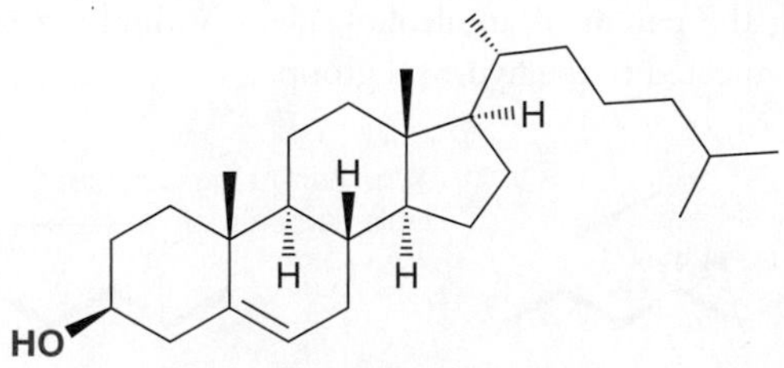

Cholesterol
Plays a vital role in the biosynthesis of many steroids

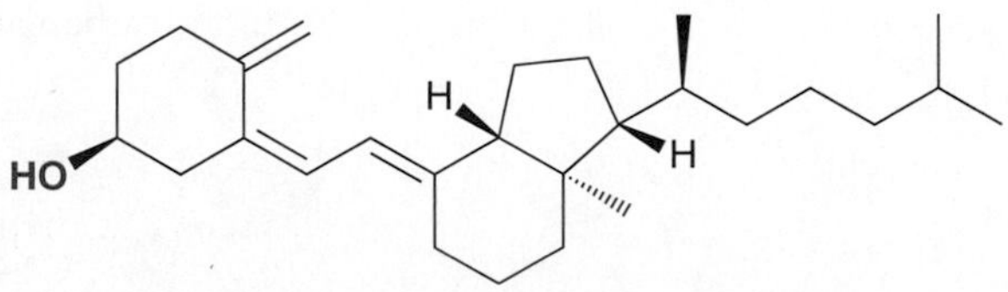

Cholecalciferol (vitamin D_3)
Regulates calcium levels and helps to form and maintain strong bones

Phenol is a special kind of alcohol. It is comprised of a hydroxyl group attached directly to a phenyl ring. Substituted phenols are extremely common in nature and exhibit a wide variety of properties and functions, as seen in the following examples.

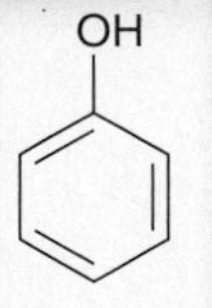

Phenol

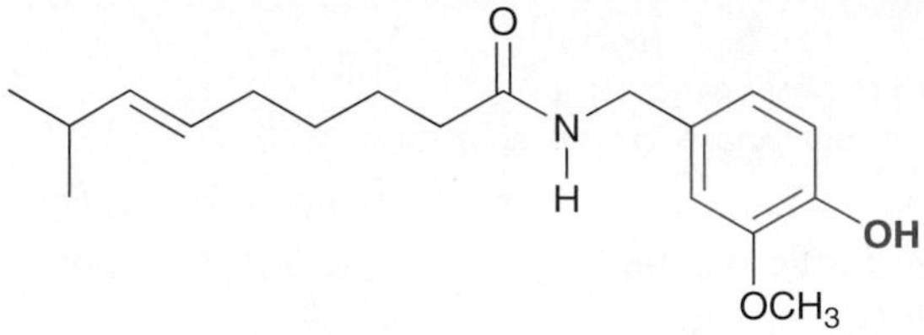

Capsaicin
The compound responsible for the spicy hot flavor of chili peppers

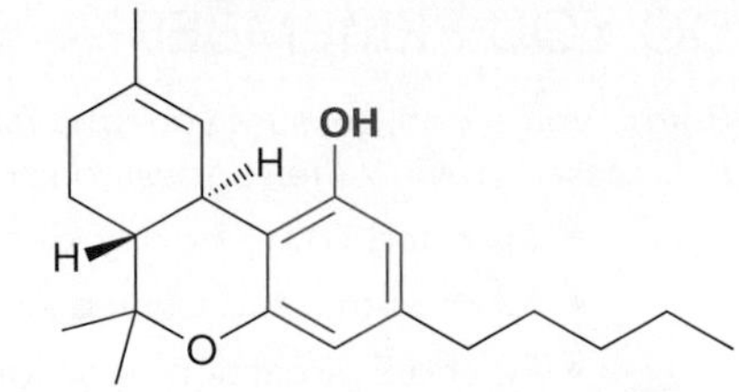

Tetrahydrocannabinol (THC)
The psychoactive drug found in marijuana (cannabis)

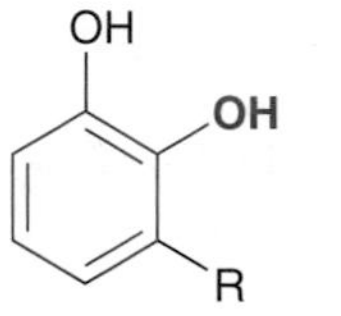

Urushiols
Present in the leaves of poison ivy and poison oak. Cause skin irritation

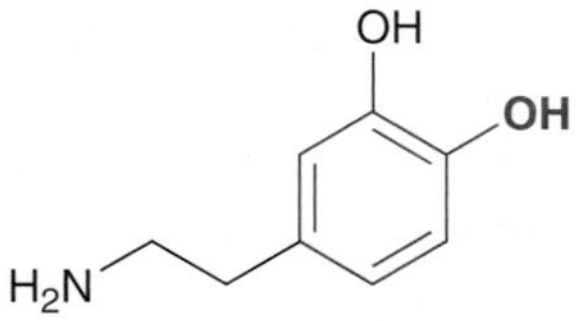

Dopamine
A neurotransmitter that is deficient in Parkinson's disease

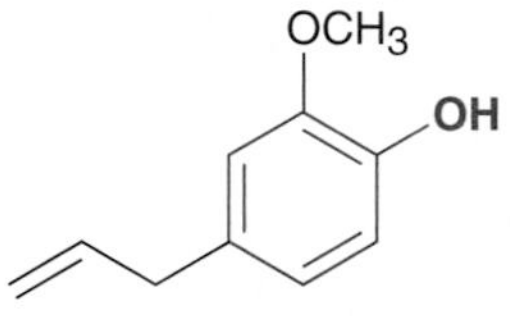

Eugenol
Isolated from cloves and used in perfumes and as a flavor additive

Nomenclature

Recall that four discrete steps are required to name alkanes, alkenes, and alkynes.

1. Identify and name the parent.
2. Identify and name the substituents.
3. Assign a locant to each substituent.
4. Assemble the substituents alphabetically.

Alcohols are named using the same four steps and applying the following rules.

- When naming the parent, replace the suffix "e" with "ol" to indicate the presence of a hydroxyl group:

Pentan<u>e</u>

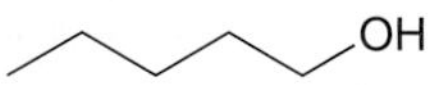

Petan<u>ol</u>

- When choosing the parent of an alcohol, identify the longest chain that includes the carbon atom connected to the hydroxyl group.

Parent = <u>oct</u>ane

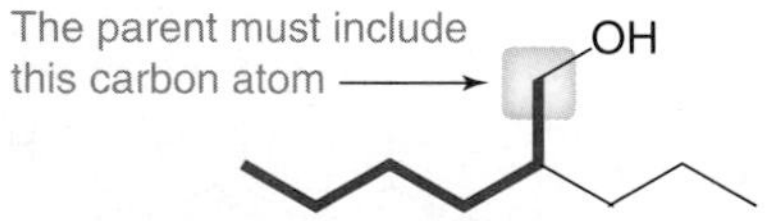

Parent = <u>hex</u>anol

- When numbering the parent chain of an alcohol, the hydroxyl group should receive the lowest number possible, despite the presence of alkyl substituents or π bonds.

Correct

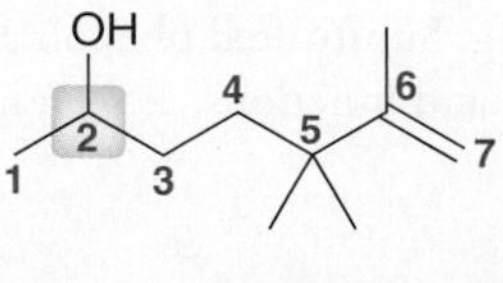

Incorrect

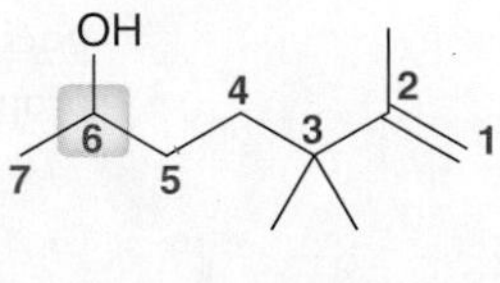

- The position of the hydroxyl group is indicated using a locant. The IUPAC rules published in 1979 dictate that this locant be placed immediately before the parent, while the IUPAC recommendations released in 1993 and 2004 allow for the locant to be placed before the suffix "ol." Both names are acceptable IUPAC names.

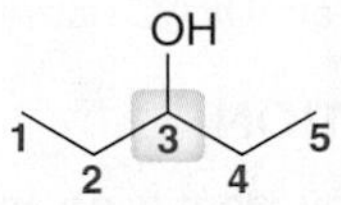

3-Pentanol
or
Pentan-3-ol

- When a chirality center is present, the configuration must be indicated at the beginning of the name; for example:

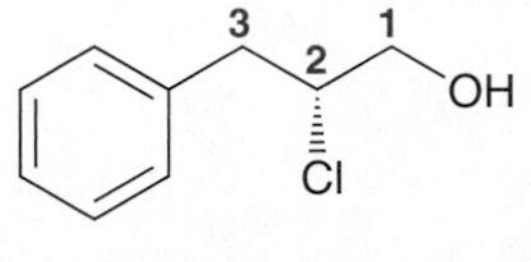

(*R*)-2-Chloro-3-phenyl-1-propanol

- Cyclic alcohols are numbered starting at the position bearing the hydroxyl group, so there is no need to indicate the position of the hydroxyl group; it is understood to be at C–1.

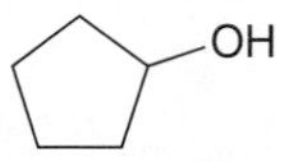

Cyclopentanol

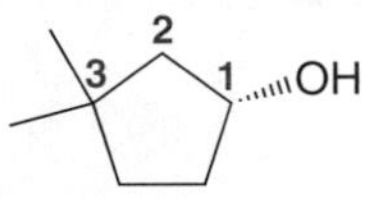

(*R*)-3,3-Dimethylcyclopentanol

IUPAC nomenclature recognizes the common names of many alcohols, such as the following three examples:

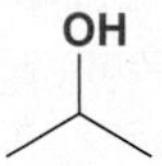

Isopropyl alcohol (2-propanol)

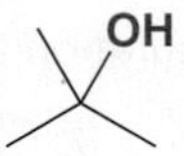

***tert*-Butyl alcohol (2-methyl-2-propanol)**

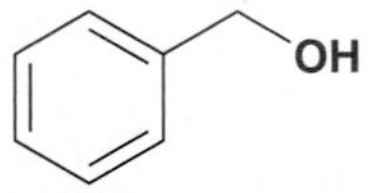

Benzyl alcohol (phenylmethanol)

Alcohols are also designated as primary, secondary, or tertiary, depending on the number of alkyl groups attached directly to the alpha (α) position (the carbon atom bearing the hydroxyl group).

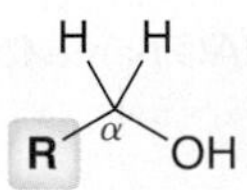

Primary

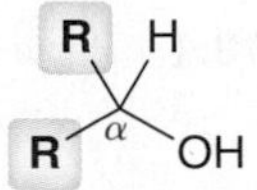

Secondary

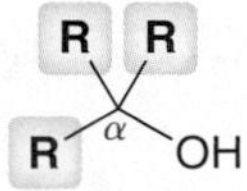

Tertiary

The word "phenol" is used to describe a specific compound (hydroxybenzene) but is also used as the parent name when substituents are attached.

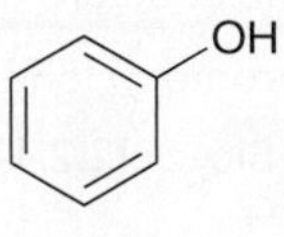

Phenol

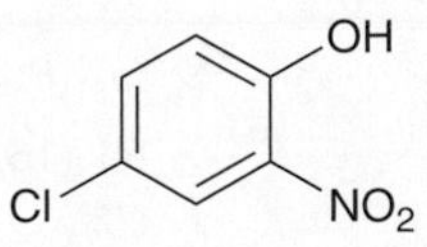

4-Chloro-2-nitrophenol

SKILLBUILDER

13.1 NAMING AN ALCOHOL

LEARN the skill

Assign an IUPAC name for the following alcohol.

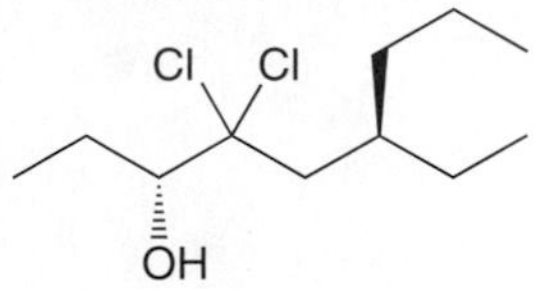

SOLUTION

Begin by identifying and naming the parent. Choose the longest chain that includes the carbon atom connected to the hydroxyl group, and then number the chain to give the hydroxyl group the lowest number possible.

STEP 1
Identify and name the parent.

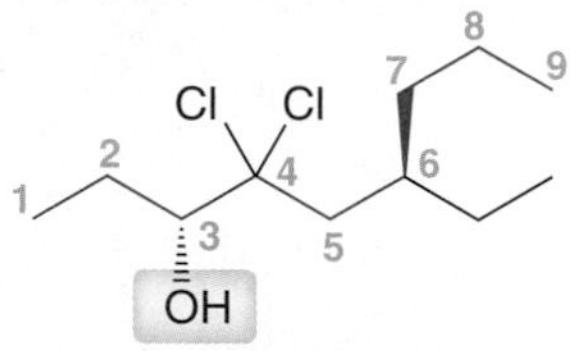

3-Nonanol

Then identify the substituents and assign locants.

STEPS 2 AND 3
Identify the substituents and assign locants.

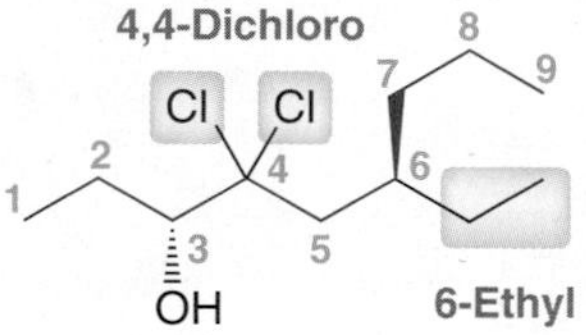

STEP 4
Assemble the substituents alphabetically.

Next, assemble the substituents alphabetically: 4,4-Dichloro-6-ethyl-3-nonanol

Before concluding, we must always check to see if there are any chirality centers This compound has two chirality centers. Using the skills from Section 5.3, we can assign the *R* configuration to both chirality centers.

STEP 5
Assign the configuration of any chirality center.

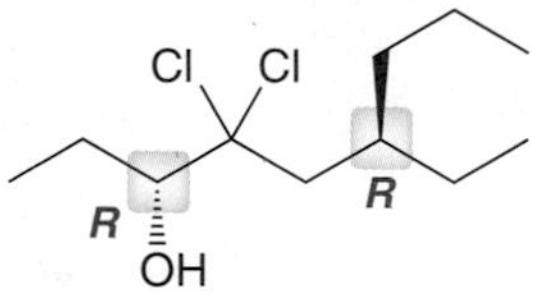

Therefore, the complete name is (3*R*,6*R*)-4,4-Dichloro-6-ethyl-3-nonanol.

PRACTICE the skill

13.1 Provide an IUPAC name for each of the following alcohols:

(a) 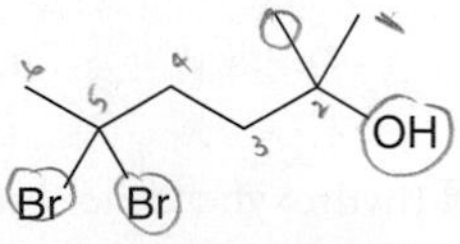

(b) 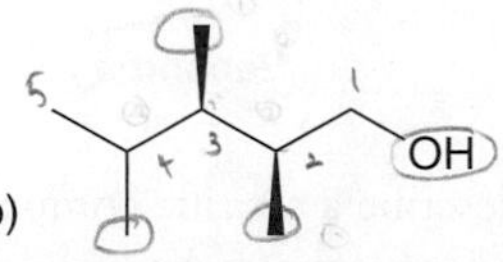

(c) 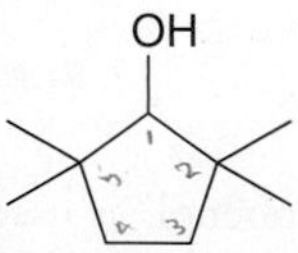

(d) 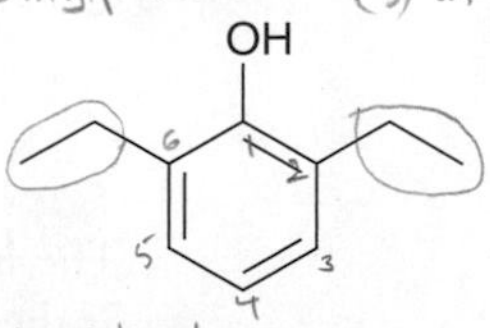

(e) 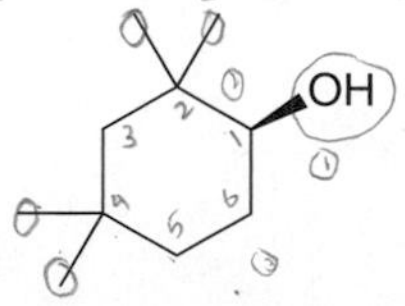

APPLY the skill

13.2 Draw the structure of each of the following compounds:

(a) (*R*)-3,3-Dibromocyclohexanol (b) (*S*)-2,3-Dimethyl-3-pentanol

(c) (1*S*,2*S*,4*R*)-Bicyclo[2.2.1]heptan-2-ol

need more **PRACTICE?** **Try Problems 13.30–13.31a-d,f, 13.32**

Commercially Important Alcohols

Methanol (CH_3OH) is the simplest alcohol. It is toxic, and ingestion can cause blindness and death, even in small quantities. Methanol can be obtained from heating wood in the absence of air and has therefore been called "wood alcohol." Industrially, methanol is prepared by the reaction between carbon dioxide (CO_2) and hydrogen gas (H_2) in the presence of suitable catalysts. Each year, the United States produces approximately 2 billion gallons of methanol, which is used as a solvent and as a precursor in the production of other commercially important compounds.

Methanol can also be used as a fuel to power combustion engines. In the second lap of the 1964 Indianapolis 500, a bad accident involving seven cars resulted in a large fire that claimed the lives of two drivers. That accident prompted the decision that all race cars switch from gasoline-powered engines to methanol-powered engines, because methanol fires produce no smoke and are more easily extinguished. In 2006, the Indianapolis 500 switched the choice of fuels for the race cars again, replacing methanol with ethanol.

Ethanol (CH_3CH_2OH), also called grain alcohol, is obtained from the fermentation of grains or fruits, a process that has been widely used for thousands of years. Industrially, ethanol is prepared via the acid-catalyzed hydration of ethylene. Each year, the United States produces approximately 5 billion gallons of ethanol, used as a solvent and as a precursor for the production of other commercially important compounds. Ethanol that is suitable for drinking is highly taxed by most governments. To avoid these taxes, industrial-grade ethanol is contaminated with small quantities of toxic compounds (such as methanol) that render the mixture unfit for human consumption. The resulting solution is called "denatured alcohol."

Isopropanol, also called rubbing alcohol, is prepared industrially via the acid-catalyzed hydration of propylene. Isopropanol has antibacterial properties and is used as a local antiseptic. Sterilizing pads typically contain a solution of isopropanol in water. Isopropanol is also used as an industrial solvent and as a gasoline additive.

Physical Properties of Alcohols

The physical properties of alcohols are quite different from the physical properties of alkanes or alkyl halides. For example, compare the boiling points for ethane, chloroethane, and ethanol.

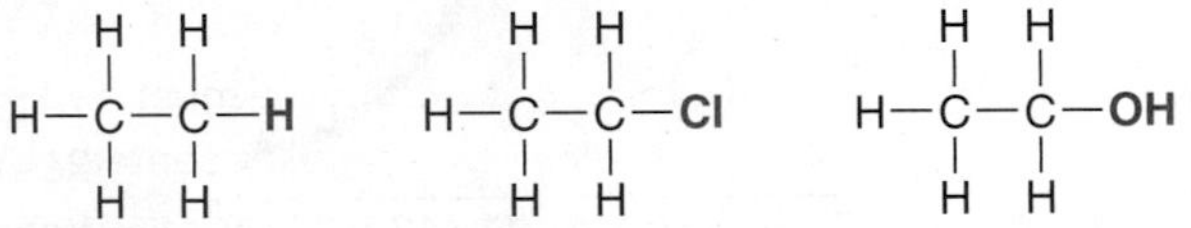

Ethane bp = −89°C | **Chloroethane** bp = 12°C | **Ethanol** bp = 78°C

The boiling point of ethanol is much higher than the other two compounds as a result of the hydrogen-bonding interactions that occur between molecules of ethanol.

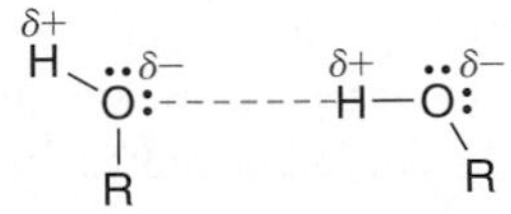

These interactions are fairly strong intermolecular forces, and they are also critical in understanding how alcohols interact with water. For example, methanol is **miscible** with water, which means that methanol can be mixed with water in any proportion (they will never separate into two layers like a mixture of water and oil). However, not all alcohols are miscible with water. To understand why, we must realize that every alcohol has two regions. The **hydrophobic** region does *not* interact well with water, while the **hydrophilic** region *does* interact with water via hydrogen bonding. Figure 13.1 shows the hydrophobic and hydrophilic regions of methanol and octanol. In the case of methanol, the hydrophobic end of the molecule is fairly small. This

FIGURE 13.1
The hydrophobic and hydrophilic regions of methanol and octanol.

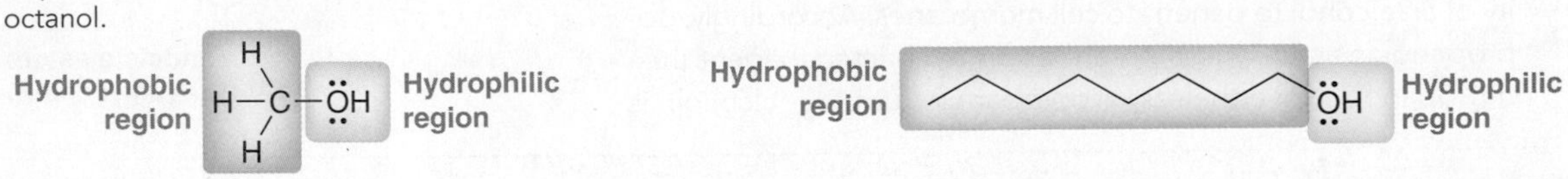

is true even of ethanol and propanol, but it is not true of butanol. The hydrophobic end of the butanol molecule is large enough to prevent miscibility. Water can still be mixed with butanol, but not in all proportions. In other words, butanol is considered to be soluble in water, rather than miscible. The term **soluble** means that only a certain volume of butanol will dissolve in a specified amount of water at room temperature.

As the size of the hydrophobic region increases, solubility in water decreases. For example, octanol exhibits extremely low solubility in water at room temperature. Alcohols with more than eight carbon atoms, such as nonanol and decanol, are considered to be insoluble in water.

MEDICALLYSPEAKING

Chain Length as a Factor in Drug Design

Primary alcohols (methanol, ethanol, propanol, butanol, etc.) exhibit antibacterial properties. Research indicates that the antibacterial potency of primary alcohols increases with increasing molecular weight, and this trend continues up to an alkyl chain length of eight carbon atoms (octanol). Beyond eight carbon atoms, the potency decreases. That is, nonanol is less potent than octanol, and dodecanol (12 carbon atoms) has very little potency.

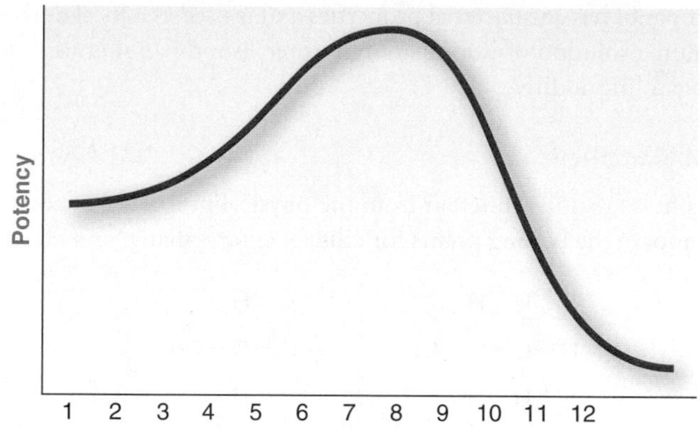

Two trends explain the observations:

- An alcohol with a larger alkyl chain (hydrophobic region) exhibits a greater ability to penetrate microbial membranes, which are composed of molecules with hydrophobic regions. According to this trend, potency should continue to increase with increased alkyl chain length, even beyond eight carbon atoms.
- A compound with a larger alkyl chain exhibits lower solubility in water, decreasing its ability to be transported through aqueous media. This trend explains why the potency of alcohols decreases steeply as the alkyl chain becomes larger than eight carbon atoms. A larger alcohol simply cannot reach its destination and therefore has low potency.

The balance between these two trends is achieved for octanol, which has the highest antibacterial potency of the primary alcohols.

Studies also show that chain branching decreases the ability of an alcohol to penetrate cell membranes. Accordingly, isopropanol is actually less potent as an antibacterial agent than *n*-propanol. Nevertheless, isopropanol (rubbing alcohol) is used because it is less expensive to produce than *n*-propanol, and the difference in antibacterial potency does not justify the added expense of production.

Many other antibacterial agents are specifically designed with alkyl chains that enable them to penetrate cell membranes. The design of these agents is optimized by carefully striking the right balance between the two trends previously discussed. Various chain lengths are tested to determine optimal potency. In most cases, the optimal chain length is found to be between five and nine carbon atoms. Consider, for example, the structure of resorcinol.

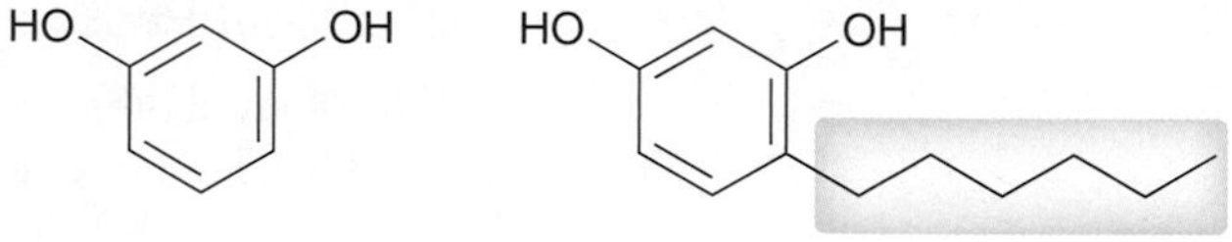

Resorcinol Hexylresorcinol

Resorcinol is a weak antiseptic (antimicrobial agent) used in the treatment of skin conditions such as eczema and psoriasis. Placing an alkyl chain on the ring increases its potency as an antiseptic. Studies indicate that the optimal potency is achieved with a six-carbon chain length. Hexylresorcinol exhibits bactericidal and fungicidal properties and is used in many throat lozenges.

CONCEPTUAL CHECKPOINT

13.3 Mandelate esters exhibit spasmolytic activity (they act as muscle relaxants). The nature of the alkyl group (R) greatly affects potency. Research indicates that the optimal potency is achieved when R is a nine-carbon chain (a nonyl group). Explain why nonyl mandelate is more potent than either octyl mandelate or decyl mandelate.

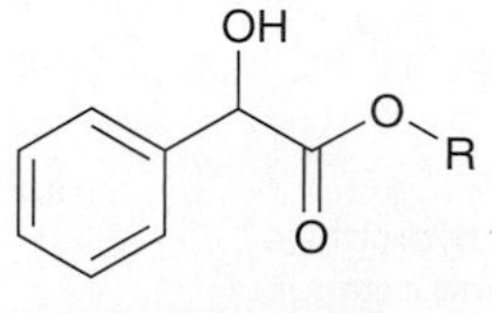

Mandelate esters (R = alkyl chain)

13.2 Acidity of Alcohols and Phenols

Acidity of the Hydroxyl Functional Group

LOOKING BACK
The factors affecting the stability of a negative charge were first discussed in Section 3.4.

As we learned in Chapter 3, the acidity of a compound can be qualitatively evaluated by analyzing the stability of its conjugate base:

$$R-\ddot{O}-H \xrightarrow{-H^+} R-\ddot{O}\colon^{\ominus}$$

To evaluate the acidity ...deprotonate... ...and assess the stability
of this compound... of the conjugate base
 (an alkoxide ion)

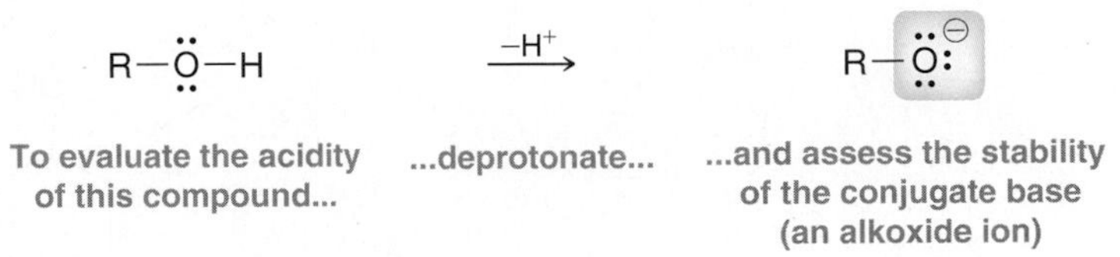

LOOKING BACK
Recall that a strong acid will have a low pK_a value. To review the relationship between pK_a and acidity, see SkillBuilder 3.2.

The conjugate base of an alcohol is called an **alkoxide** ion, and it exhibits a negative charge on an oxygen atom. A negative charge on an oxygen atom is more stable than a negative charge on a carbon or nitrogen atom but less stable than a negative charge on a halogen, X (Figure 13.2).

Increasing stability →

$$R\colon^{\ominus} \qquad \underset{H}{\overset{R}{\underset{|}{N}}}\colon^{\ominus} \qquad R-\ddot{O}\colon^{\ominus} \qquad \colon\ddot{X}\colon^{\ominus}$$

Least Most
stable stable

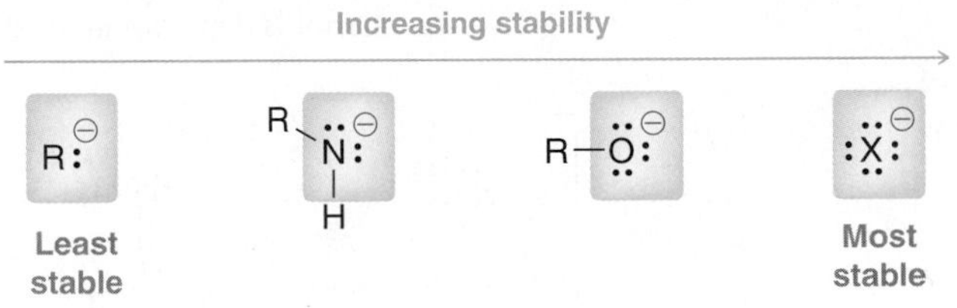

FIGURE 13.2
The relative stability of various anions.

Therefore, alcohols are more acidic than amines and alkanes but less acidic than hydrogen halides (Figure 13.3). The pK_a for most alcohols falls in the range of 15–18.

Increasing acidity →

R—H	R—$\ddot{N}H_2$	R—$\ddot{O}H$	$\colon\ddot{X}$—H
pK_a between 45 and 50	pK_a between 35 and 40	**pK_a between 15 and 18**	pK_a between −10 and 3

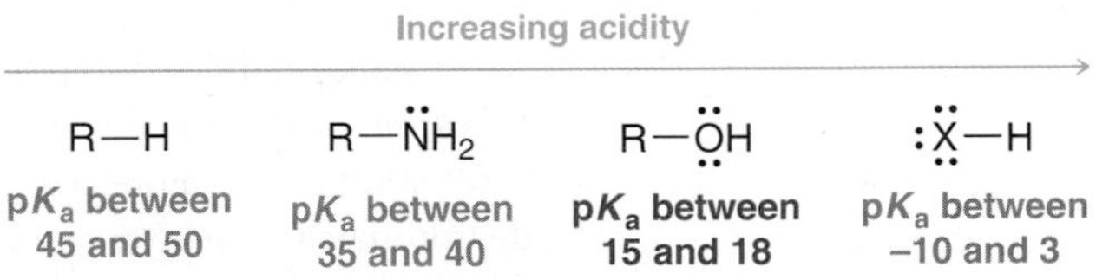

FIGURE 13.3
The relative acidity of alkanes, amines, alcohols, and hydrogen halides.

Reagents for Deprotonating an Alcohol

There are two common ways to deprotonate an alcohol, forming an alkoxide ion.

1. A strong base can be used to deprotonate the alcohol. A commonly used base is sodium hydride (NaH), because hydride (H⁻) deprotonates the alcohol to generate hydrogen gas, which bubbles out of solution:

$$\text{Ethanol} + \overset{\oplus}{Na} \colon H^{\ominus} \longrightarrow \text{Sodium ethoxide} + H_2\uparrow$$

Ethanol Sodium hydride Sodium ethoxide Hydrogen gas

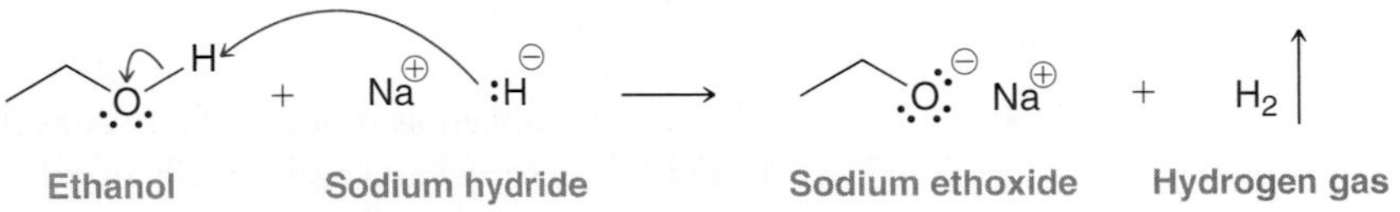

2. Alternatively, it is often more practical to use Li, Na, or K. These metals react with the alcohol to liberate hydrogen gas, producing the alkoxide ion.

$$\sim\!\!O\text{--}H \xrightarrow{Na} \sim\!\!O^{\ominus}\ \overset{\oplus}{Na} + \tfrac{1}{2}H_2\uparrow$$

CONCEPTUAL CHECKPOINT

13.4 Draw the alkoxide formed in each of the following cases:

(a) Na → ? (b) NaH → ? (c) Li → ? (d) NaH → ?

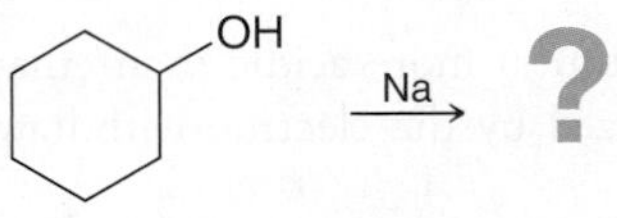

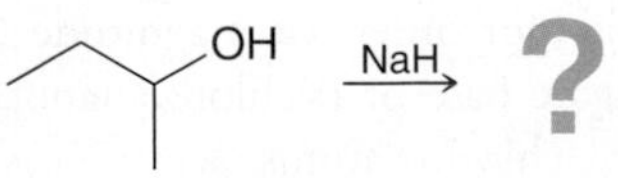

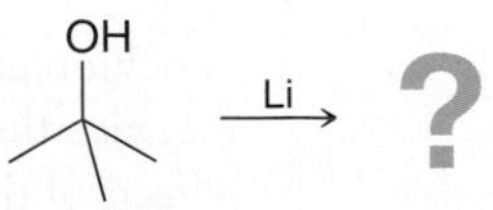

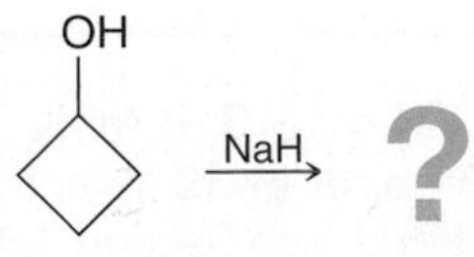

Factors Affecting the Acidity of Alcohols and Phenols

How can we predict which, of a number of alcohols, is more acidic? In this section, we will explore three factors for comparing the acidity of alcohols.

1. *Resonance.* One of the most significant factors affecting the acidity of alcohols is resonance. As a striking example, compare the pK_a values of cyclohexanol and phenol:

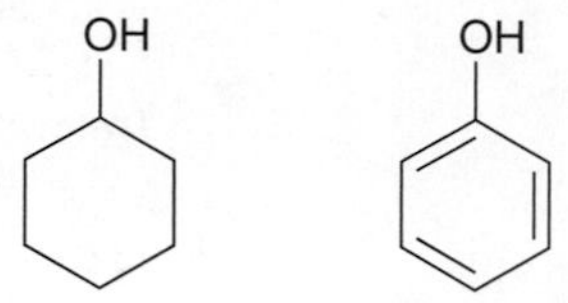

Cyclohexanol (pK_a = 18) **Phenol (pK_a = 10)**

When phenol is deprotonated, the conjugate base is stabilized by resonance.

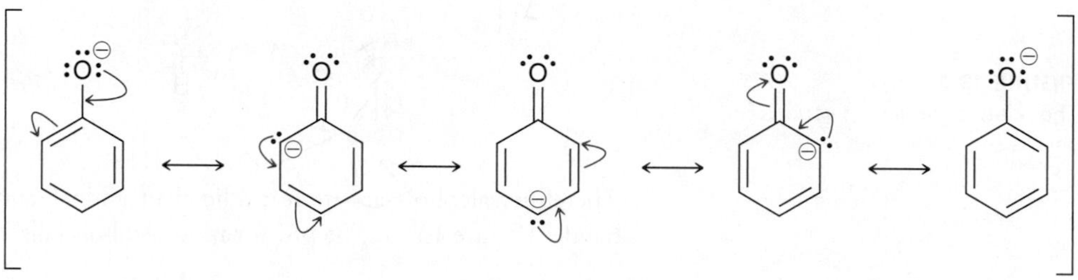

This resonance-stabilized anion is called a **phenolate**, or a **phenoxide** ion. Resonance stabilization of the phenoxide ion explains why phenol is eight orders of magnitude (100,000,000 times) more acidic than cyclohexanol. As a result, phenol does not need to be deprotonated with a very strong base like sodium hydride. Instead, it can be deprotonated by hydroxide.

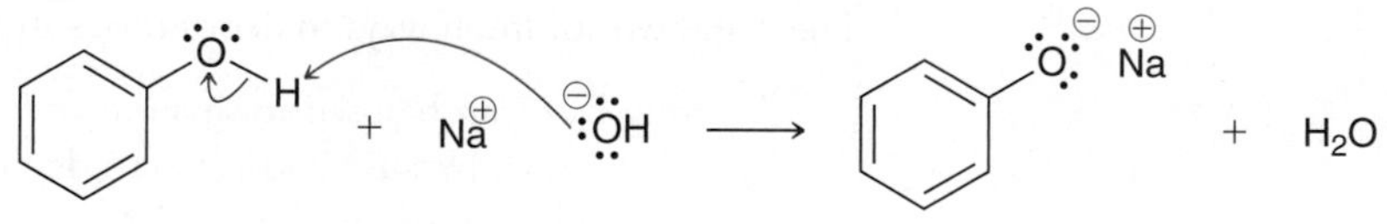

(pK_a = 10) **(pK_a = 15.7)**

The acidity of phenols is one of the reasons that phenols are a special category of alcohols. Later in this chapter and again in Chapter 19 we will see other reasons why phenols belong to a class of their own.

2. *Induction.* Another factor in comparing the acidity of alcohols is induction. As an example, compare the pK_a values of ethanol and trichloroethanol.

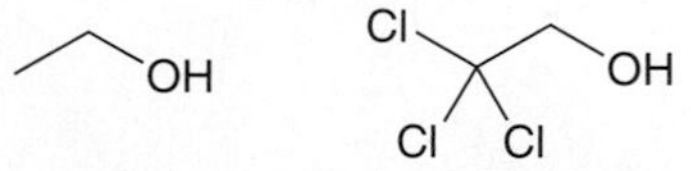

Ethanol (pK_a = 16) **Trichloroethanol (pK_a = 12.2)**

LOOKING BACK
For a review of inductive effects, see Section 3.4.

Trichloroethanol is four orders of magnitude (10,000 times) more acidic than ethanol, because the conjugate base of trichloroethanol is stabilized by the electron-withdrawing effects of the nearby chlorine atoms.

3. *Solvation effects.* To explore the effect of alkyl branching, compare the acidity of ethanol and *tert*-butanol.

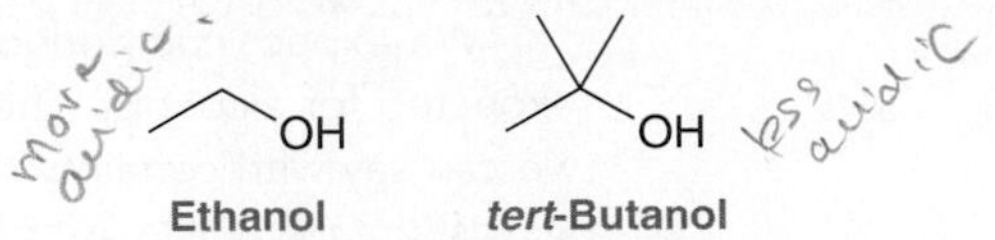

Ethanol ($pK_a = 16$) *tert*-Butanol ($pK_a = 18$)

The pK_a values indicate that *tert*-butanol is less acidic than ethanol, by two orders of magnitude. This difference in acidity is best explained by a steric effect. The ethoxide ion is not sterically hindered and is therefore easily solvated (stabilized) by the solvent, while *tert*-butoxide is sterically hindered and is less easily solvated (Figure 13.4). The conjugate base of *tert*-butanol is less stabilized than the conjugate base of ethanol, rendering *tert*-butanol less acidic.

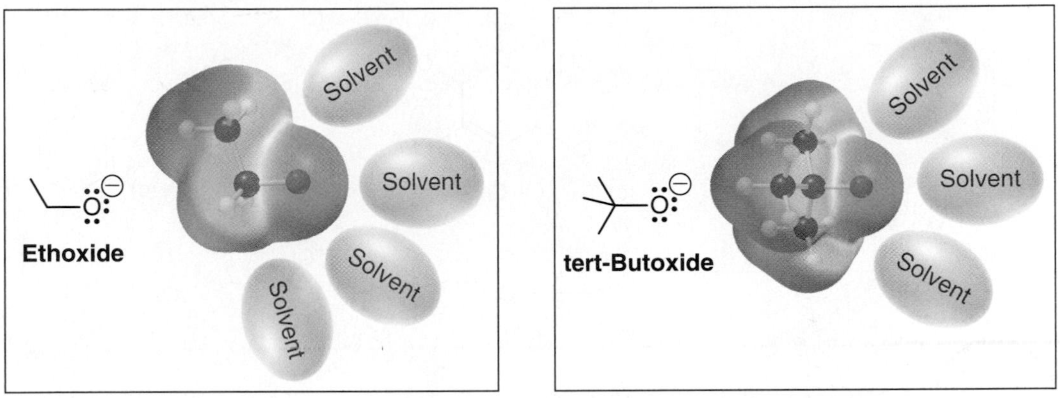

FIGURE 13.4 An ethoxide ion is stabilized by the solvent to a greater extent than *tert*-butoxide is stabilized by the solvent.

SKILLBUILDER

13.2 COMPARING THE ACIDITY OF ALCOHOLS

LEARN the skill

Identify which of the following compounds is expected to be more acidic.

Compound A Compound B

SOLUTION

Begin by drawing the conjugate base of each, and then compare the stability of those conjugate bases.

Conjugate base of compound A Conjugate base of compound B

The conjugate base of compound **B** is not resonance stabilized, but the conjugate base of compound **A** is resonance stabilized.

Conjugate base of compound A

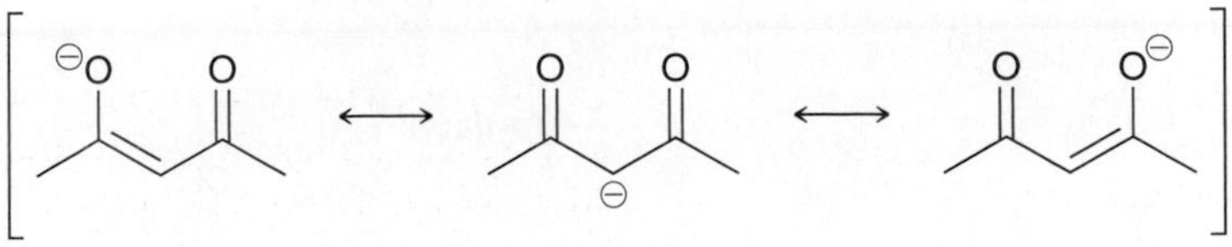

The conjugate base of compound **A** will be more stable than the conjugate base of compound **B**. Therefore, compound **A** will be more acidic.

We expect compound **B** to have a pK_a somewhere in the range of 15–18 (the range expected for alcohols). The pK_a of compound **A** will be more difficult to predict. However, we can say with certainty that it will be lower (more acidic) than a regular alcohol. In other words, the pK_a value will be lower than 15.

PRACTICE the skill **13.5** For each of the following pairs of alcohols, identify the one that is more acidic, and explain your choice:

(a)

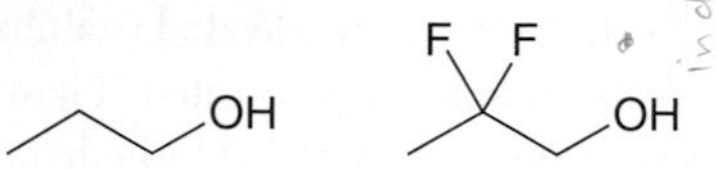

(b)

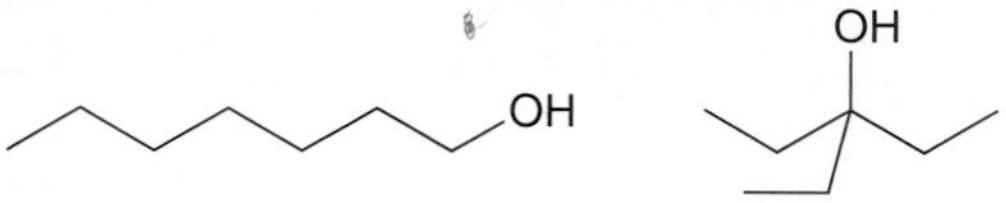

(c)

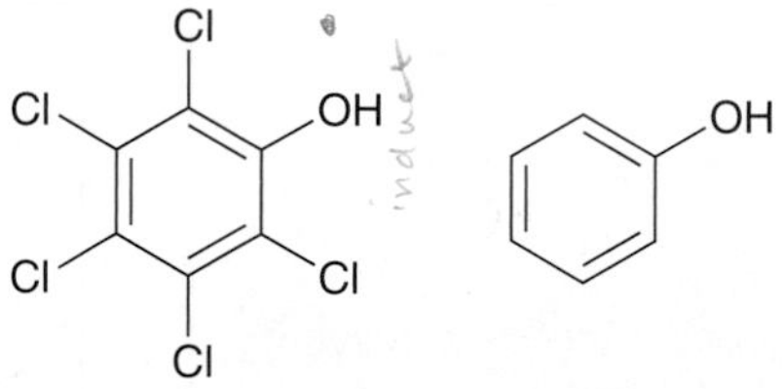

(d)

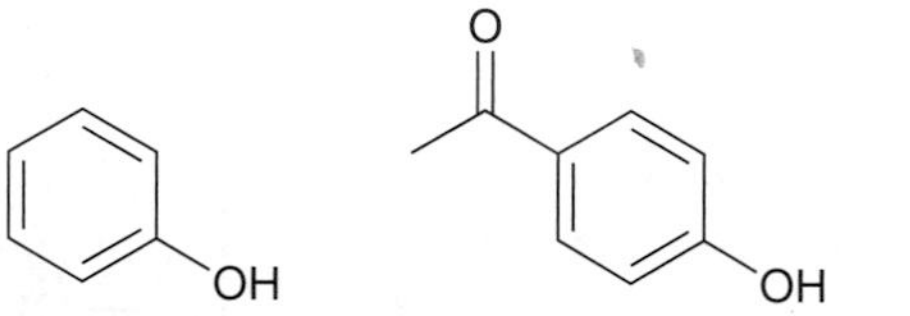

(e)

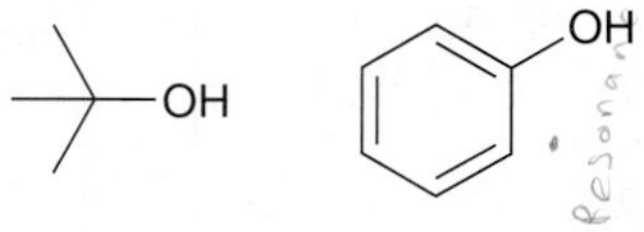

APPLY the skill **13.6** Consider the structures of 2-nitrophenol and 3-nitrophenol. These compounds have very different pK_a values. Predict which one has the lower pK_a, and explain why. (**Hint:** In order to solve this problem, you must draw the structure of each nitro group.)

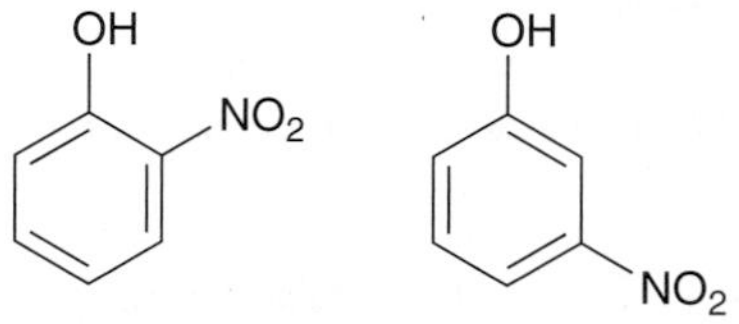

2-Nitrophenol **3-Nitrophenol**

need more **PRACTICE?** **Try Problems 13.33, 13.34**

13.3 Preparation of Alcohols via Substitution or Addition

Substitution Reactions

As we saw in Chapter 7, alcohols can be prepared by substitution reactions in which a leaving group is replaced by a hydroxyl group.

$$R{-}\mathbf{X} \longrightarrow R{-}\mathbf{OH}$$

A primary substrate will require S_N2 conditions (a strong nucleophile), while a tertiary substrate will require S_N1 conditions (a weak nucleophile).

Primary:

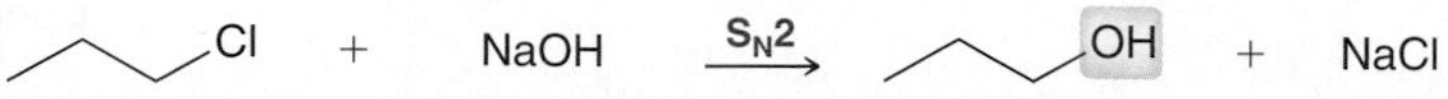

Tertiary:

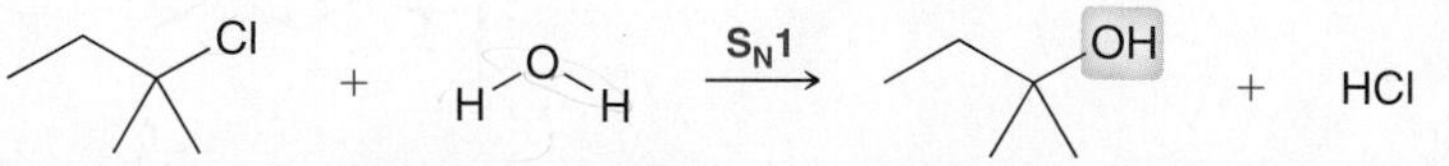

LOOKING BACK
For a review of the factors that affect substitution vs. elimination, see Section 8.14.

With a secondary substrate, neither S_N2 nor S_N1 are particularly effective for preparing a secondary alcohol. Under S_N1 conditions, the reaction is generally too slow, while S_N2 conditions (use of hydroxide as the nucleophile) will generally favor elimination over substitution.

Addition Reactions

In Chapter 9, we learned several addition reactions that produce alcohols.

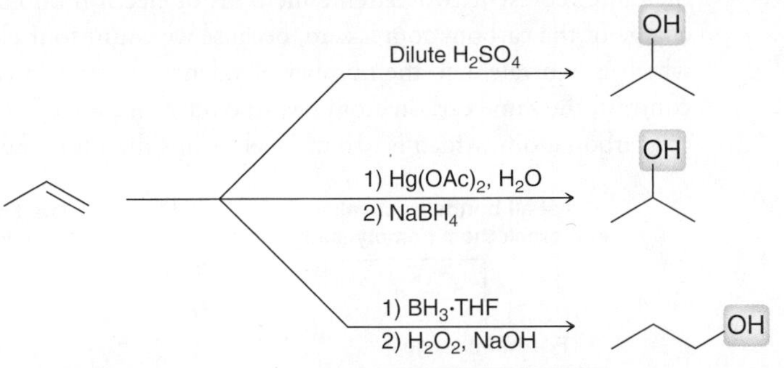

Acid-catalyzed hydration (Section 9.4)

Oxymercuration-demercuration (Section 9.5)

Hydroboration-oxidation (Section 9.6)

Acid-catalyzed hydration proceeds with Markovnikov addition (Section 9.4). That is, the hydroxyl group is positioned at the more substituted carbon. It is a useful method if the substrate is not susceptible to carbocation rearrangements (Section 6.11). In a case where the substrate can possibly rearrange, oxymercuration-demercuration can be employed. This approach also proceeds via Markovnikov addition, but it does not involve carbocation rearrangements. Hydroboration-oxidation is used to achieve an *anti*-Markovnikov addition of water.

CONCEPTUAL CHECKPOINT

13.7 Identify the reagents that you would use to accomplish each of the following transformations:

(a) 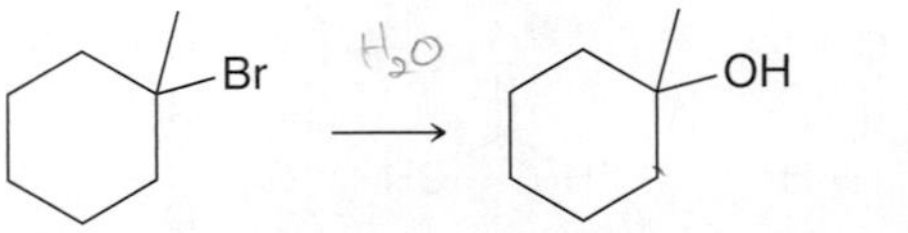

(b) 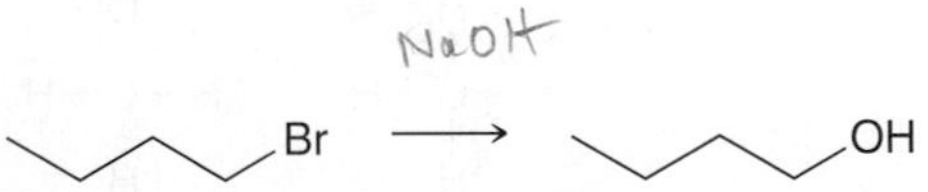

(c) 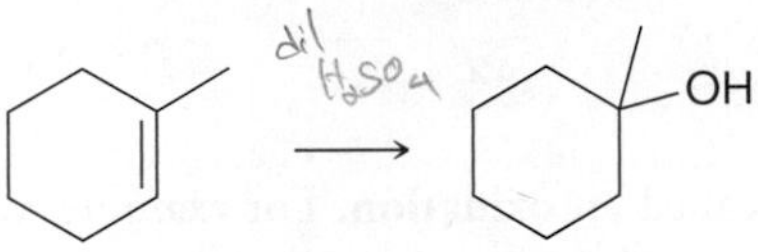

(d) 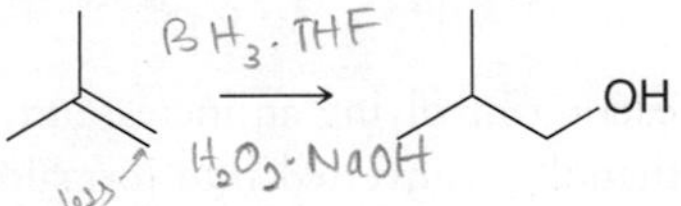

(e) 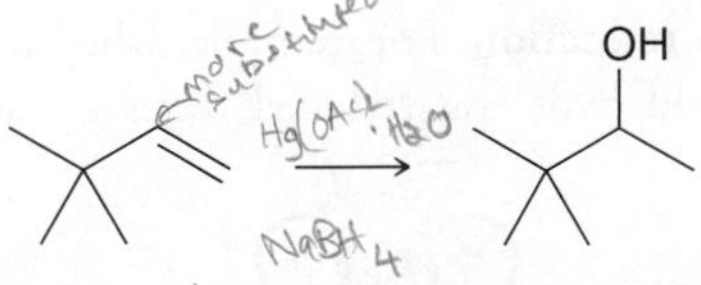

(f) 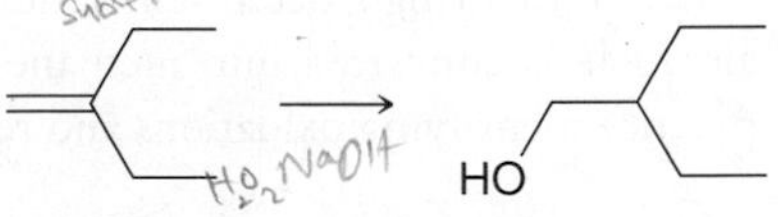

13.8 Identify reagents that can be used to achieve each of the following transformations:

(a) To convert 1-hexene into a primary alcohol

(b) To convert 3,3-dimethyl-1-hexene into a secondary alcohol

(c) To convert 2-methyl-1-hexene into a tertiary alcohol

13.4 Preparation of Alcohols via Reduction

In this section, we will explore a new method for preparing alcohols. This method involves a change in **oxidation state**, so let's spend a few moments to understand oxidation states and their relationship with formal charges.

Oxidation States

An oxidation state refers to a method of electron bookkeeping. Chapter 1 developed a different method for electron bookkeeping called formal charge. To calculate the formal charge on an atom, we treat all bonds to that atom as if they were purely *covalent*, and we break them *homolytically*. For calculating oxidation states, we will take another extreme approach. We treat all bonds as if they were purely *ionic*, and we break them *heterolytically*, giving each pair of electrons to the more electronegative atom in each case. Formal charges and oxidation states therefore represent two extreme methods of electron bookkeeping. In Figure 13.5, the formal charge of the carbon atom is zero, because we count four electrons on the central carbon atom, which is equivalent to the number of valence electrons a carbon atom is supposed to have. In contrast, the same carbon atom has an oxidation state of −2, because we count six electrons on the carbon atom, which is two more electrons than it is supposed to have.

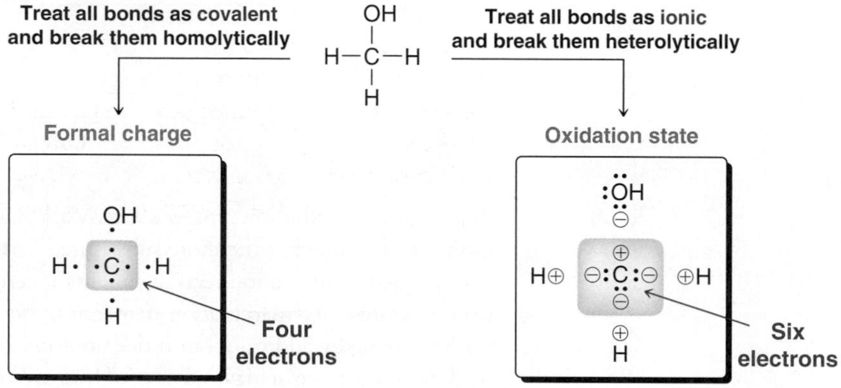

FIGURE 13.5 Two different electron bookkeeping methods: formal charge vs. oxidation state.

A carbon atom with four bonds will always have no formal charge, but its oxidation state can range anywhere from −4 to +4.

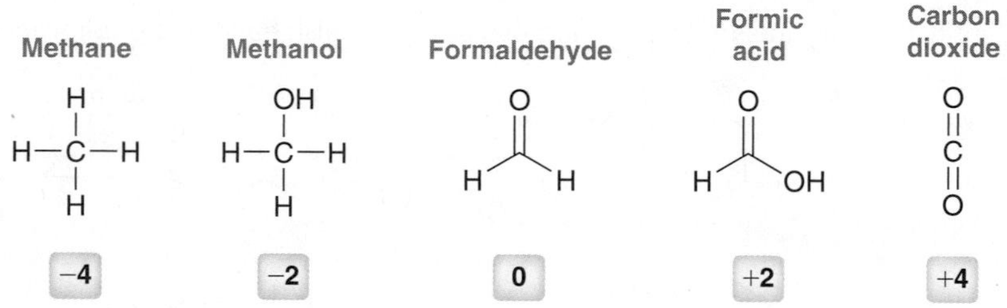

A reaction involving an increase in oxidation state is called an **oxidation**. For example, when methanol is converted into formaldehyde, we say that methanol was oxidized. In contrast, a reaction involving a decrease in oxidation state is called a **reduction**. For example, when formaldehyde is converted into methanol, we say that formaldehyde was reduced. Let's get some practice identifying oxidations and reductions.

SKILLBUILDER

13.3 IDENTIFYING OXIDATION AND REDUCTION REACTIONS

LEARN the skill

In the following transformation, identify whether the compound has been oxidized, reduced, or neither.

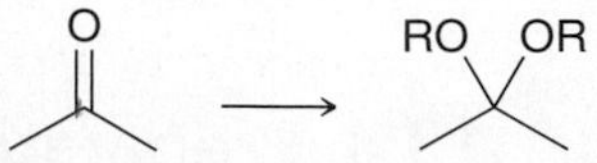

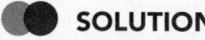

SOLUTION

Focus on the carbon atom where a change has occurred, and determine whether the oxidation state has changed as a result of the transformation. Let's begin with the starting material.

STEP 1
Determine the oxidation state of the starting material.

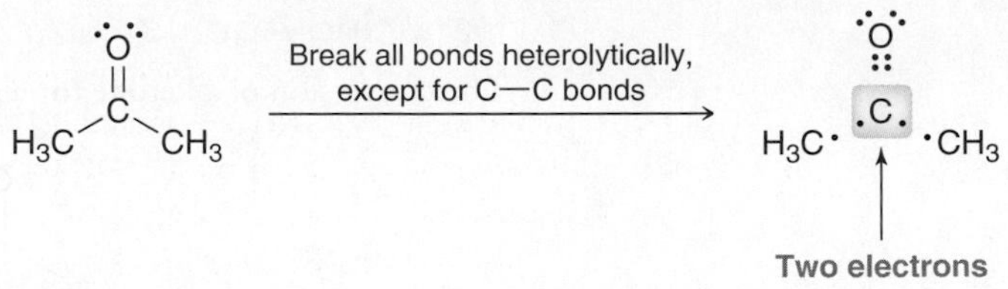
Break all bonds heterolytically, except for C—C bonds
Two electrons

Each C—O bond is broken heterolytically, giving all four electrons of the C=O bond to the oxygen atom. Each C—C bond cannot be broken heterolytically, because C and C have the same electronegativity. For each C—C bond, just divide the electrons between the two carbon atoms, breaking the bond homolytically. This leaves a total of two electrons on the central carbon atom. Compare this number with the number of valence electrons that a carbon atom is supposed to have (four). The carbon atom in this example is missing two electrons. Therefore, it has an oxidation state of +2.

STEP 2
Determine the oxidation state of the product.

STEP 3
Determine if there has been a change in oxidation state.

Now analyze the carbon atom in the product. The same result is obtained: an oxidation state of +2. This should make sense, because the reaction has simply exchanged one C=O bond for two C—O single bonds. The oxidation state of the carbon atom has not changed during the reaction, and therefore the starting material has been neither oxidized nor reduced.

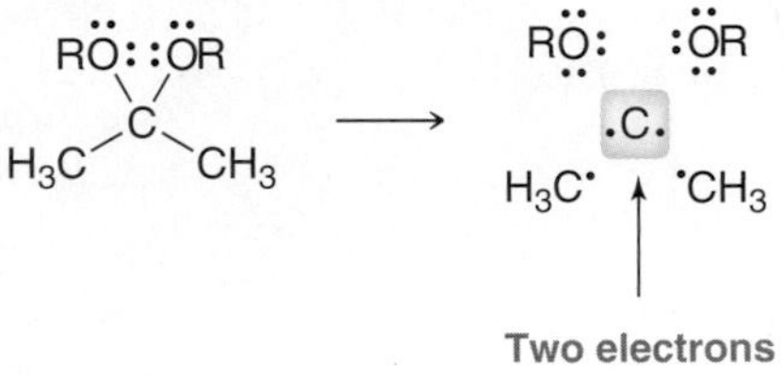
Two electrons

PRACTICE the skill

13.9 In each of the following transformations, identify whether the starting material has been oxidized, reduced, or neither. Try to determine the answer without calculating oxidation states, and then use the calculations to see if your intuition was correct.

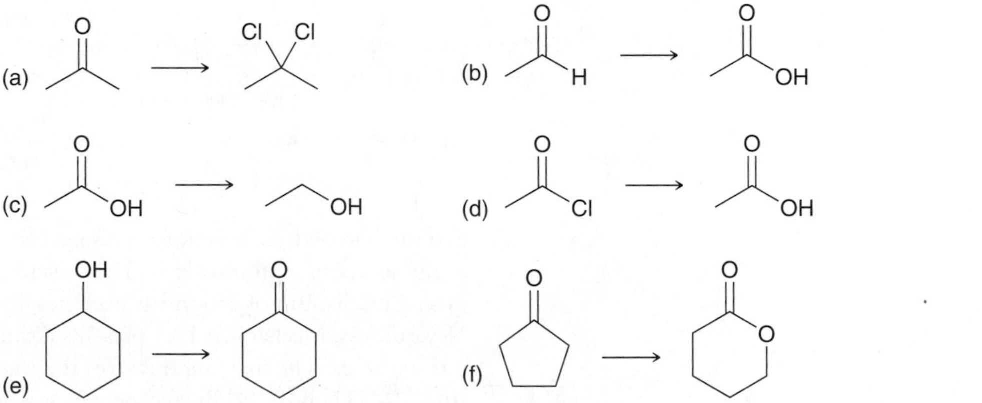

APPLY the skill

13.10 In Chapter 9, we learned about addition of water across a π bond. Identify whether the alkene has been oxidized, reduced, or neither.

(Hint: First look at each carbon atom separately, and then look at the net change for the alkene as a whole.)

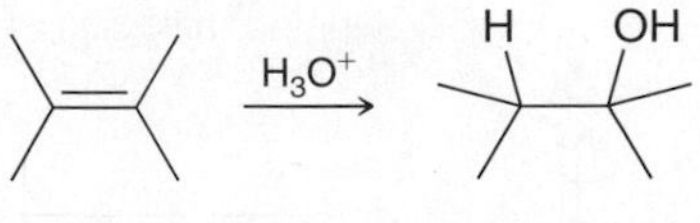

13.11 In the following reaction, determine whether the alkyne has been oxidized, reduced, or neither. Using the answer from the previous problem, try to determine the answer without calculating oxidation states, and then use the calculations to see if your intuition was correct.

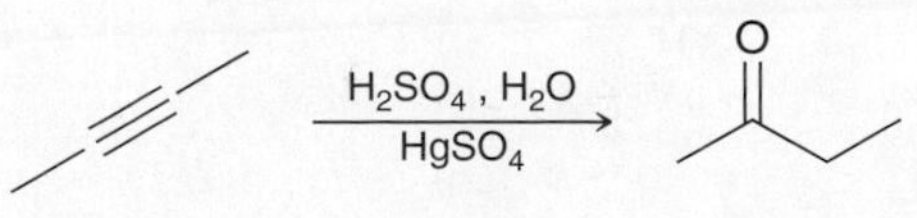

need more **PRACTICE?** **Try Problem 13.62**

Reducing Agents

The conversion of a ketone (or aldehyde) to an alcohol is a reduction.

The reaction requires a **reducing agent,** which is itself oxidized as a result of the reaction. In this section, we will explore three reducing agents that can be used to convert a ketone or aldehyde to an alcohol:

1. In Chapter 9, we learned that an alkene can undergo hydrogenation in the presence of a metal catalyst such as platinum, palladium, or nickel. A similar reaction can occur for ketones or aldehydes, although more forcing conditions are generally required (higher temperature and pressure). This process works well for both ketones and aldehydes, often with fairly good yields.

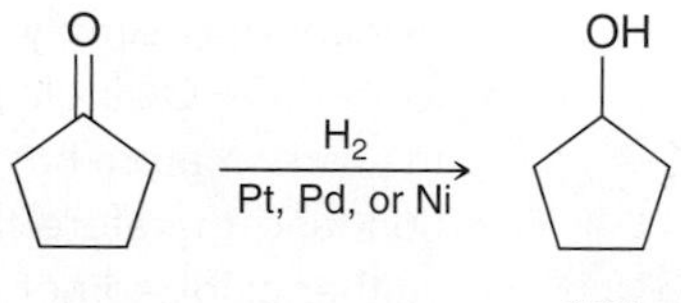

2. Sodium borohydride ($NaBH_4$) is another common reducing agent that can be used to reduce ketones or aldehydes.

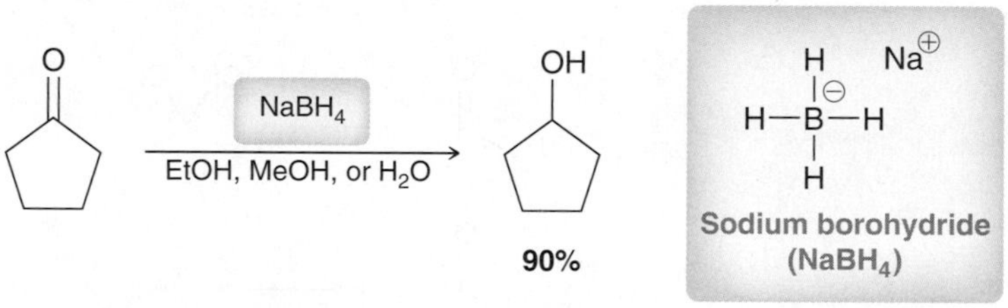

Sodium borohydride functions as a source of hydride ($H{:}^-$) and the solvent functions as the source of a proton (H^+). The solvent can be ethanol, methanol, or water. The precise mechanism of action has been heavily investigated and is somewhat complex. Nevertheless, Mechanism 13.1 presents a simplified version that will be sufficient for our purposes. The first step involves the transfer of hydride to the **carbonyl group** (the C=O bond), and the second step is a proton transfer.

MECHANISM 13.1 REDUCTION OF A KETONE OR ALDEHYDE WITH $NaBH_4$

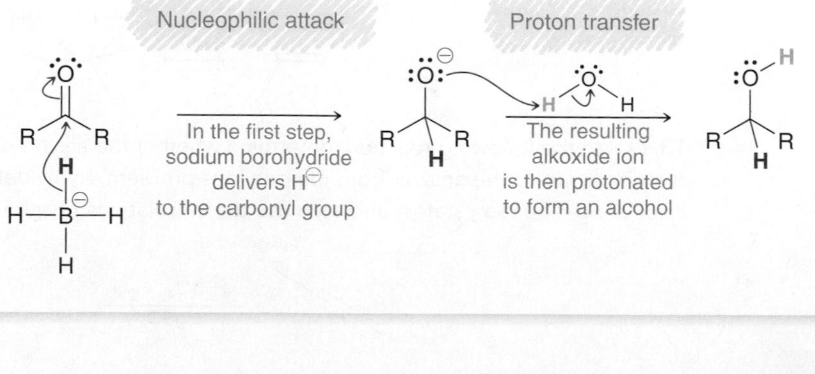

LOOKING BACK
For a review of the factors affecting nucleophilicity, see Section 6.7.

Hydride (H:$^-$), by itself, is not a good nucleophile because it is not polarizable. As a result, the reaction above cannot be achieved by using NaH (sodium hydride). NaH only functions as a base, not as a nucleophile. But $NaBH_4$ does function as a nucleophile. Specifically, $NaBH_4$ functions as a *delivery agent* of nucleophilic H:$^-$.

3. Lithium aluminum hydride (LAH) is another common reducing agent, and its structure is very similar to $NaBH_4$.

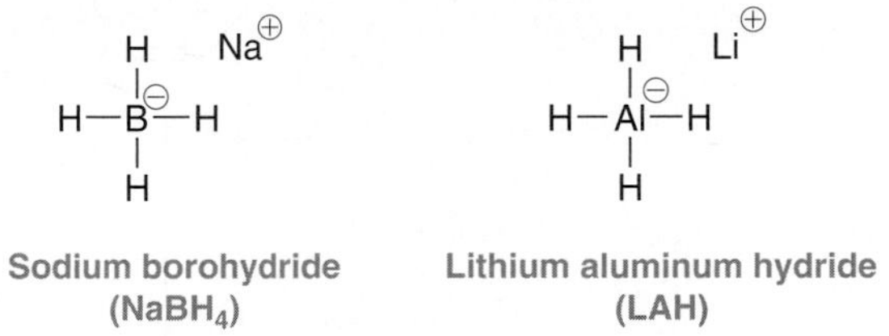

Lithium aluminum hydride is commonly abbreviated either as LAH or as $LiAlH_4$. It is also a delivery agent of H:$^-$, but it is a much stronger reagent. It reacts violently with water, and therefore, a protic solvent cannot be present together with LAH in the reaction flask. First the ketone or aldehyde is treated with LAH, and then, in a separate step, the proton source is added to the reaction flask. Water (H_2O) can serve as a proton source, although H_3O^+ can also be used as a proton source:

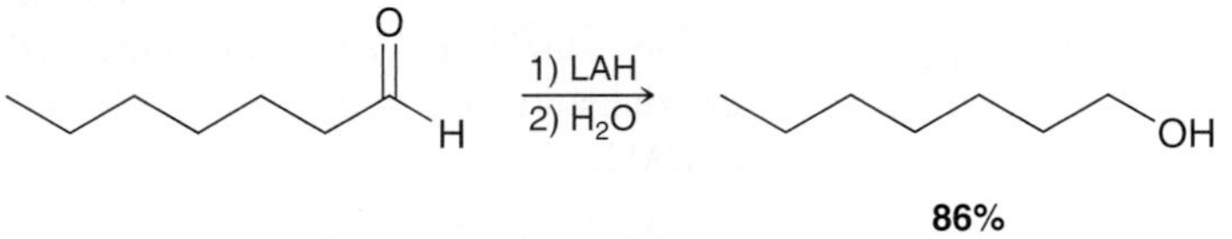
1) LAH
2) H₂O
86%

Notice that LAH and the proton source are listed as two separate steps. The mechanism of reduction with LAH (Mechanism 13.2) is similar to the mechanism of reduction with $NaBH_4$.

MECHANISM 13.2 REDUCTION OF A KETONE OR ALDEHYDE WITH LAH

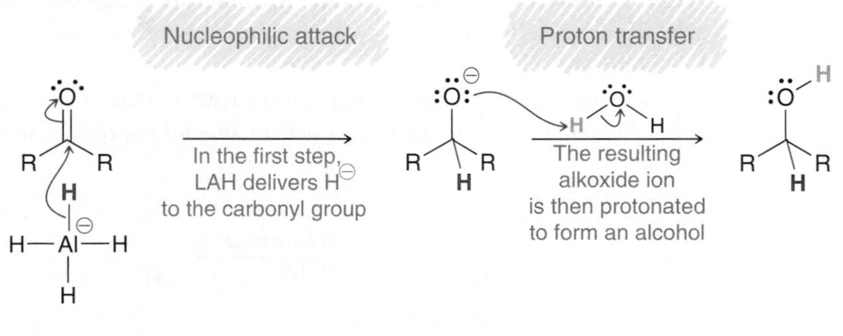
Nucleophilic attack
Proton transfer
In the first step, LAH delivers H⁻ to the carbonyl group
The resulting alkoxide ion is then protonated to form an alcohol

This simplified mechanism does not take into account many important observations, such as the role of the lithium cation (Li^+). However, a full treatment of the mechanism of hydride reducing agents is beyond the scope of this text, and this simplified version will suffice.

Both $NaBH_4$ and LAH will reduce ketones or aldehydes. These hydride delivery agents offer one significant advantage over catalytic hydrogenation in that they can selectively reduce a carbonyl group in the presence of a C=C bond. Consider the following example.

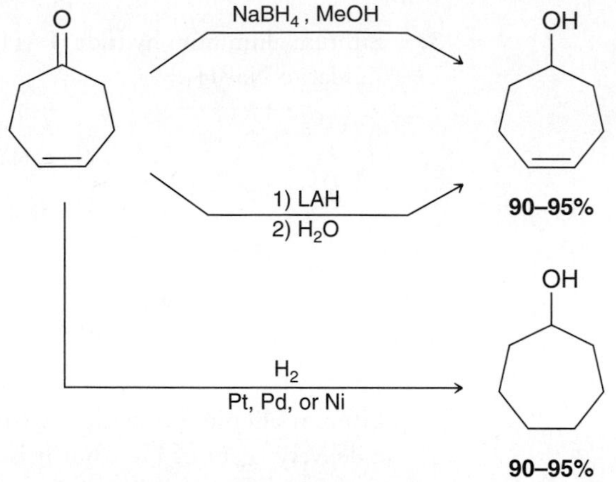

When treated with a hydride reducing agent, only the carbonyl group is reduced. In contrast, catalytic hydrogenation will also reduce the C=C bond under the conditions required to reduce the carbonyl group (high temperature and pressure). For this reason, hydride reducing agents such as $NaBH_4$ and LAH are generally preferred over catalytic hydrogenation. Many hydride reducing reagents are commercially available. Some are even more reactive than LAH and others are even milder than $NaBH_4$. Many of these reagents are derivatives of $NaBH_4$ and LAH.

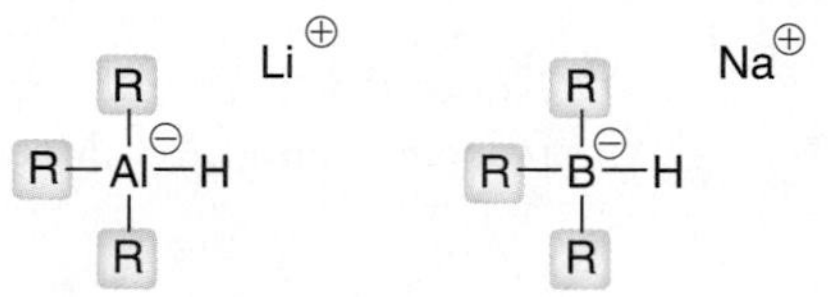

Each R group can be an alkyl group, a cyano group, an alkoxy group, or any one of a number of groups. By carefully choosing the three R groups to be either electron donating or electron withdrawing, it is possible to modify the reactivity of the hydride reagent. Hundreds of different hydride reducing agents are available, and each has its own selectivity and advantages. For now, we will simply focus our attention on the differences between LAH and $NaBH_4$.

BY THE WAY
Although $NaBH_4$ does not reduce esters under mild conditions, it is observed that $NaBH_4$ will sometimes reduce esters when more forcing conditions are employed, such as high temperature.

We mentioned that LAH is much more reactive than $NaBH_4$. As a result, LAH is less selective. LAH will react with a carboxylic acid or an ester to produce an alcohol, but $NaBH_4$ will not.

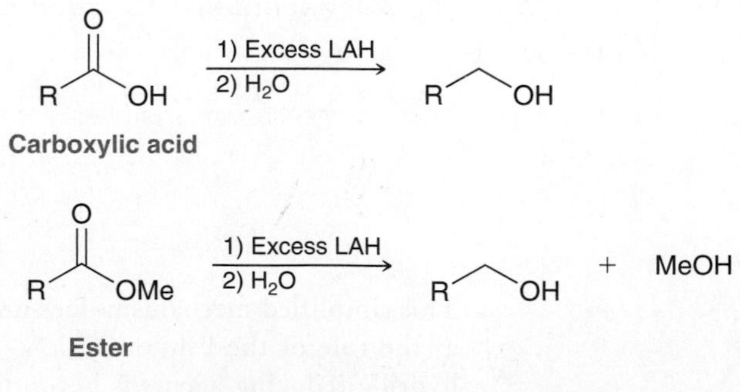

The mechanism for reduction of the ester involves the transfer of hydride two times (Mechanism 13.3).

MECHANISM 13.3 REDUCTION OF AN ESTER WITH LAH

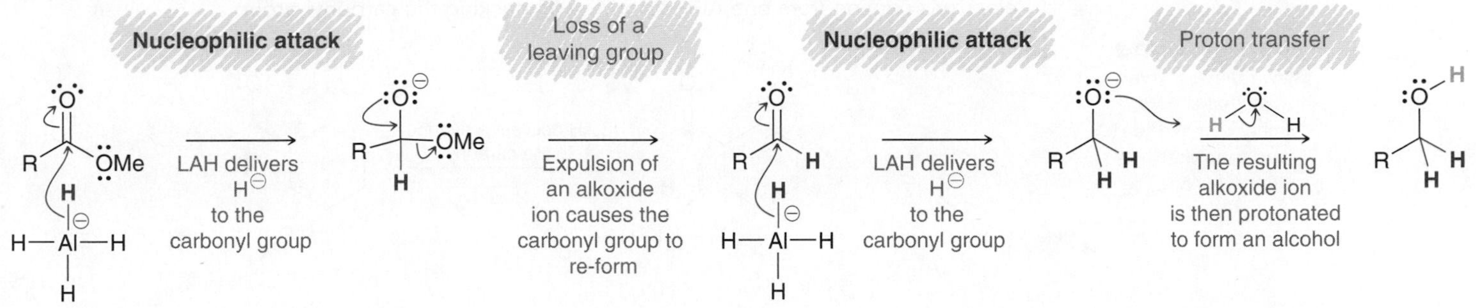
Nucleophilic attack
Loss of a leaving group
Nucleophilic attack
Proton transfer
LAH delivers H⊖ to the carbonyl group
Expulsion of an alkoxide ion causes the carbonyl group to re-form
LAH delivers H⊖ to the carbonyl group
The resulting alkoxide ion is then protonated to form an alcohol

LAH delivers hydride to the carbonyl group, but then loss of a leaving group causes the carbonyl group to re-form. In the presence of LAH, the newly formed carbonyl group can be attacked again by hydride. The leaving group in the second step of the mechanism is a methoxide ion, which is generally not a good leaving group. For example, methoxide does not function as a leaving group in E2 or S_N2 reactions. The reason that it can function as a leaving group in this case stems from the nature of the intermediate after the first attack of hydride. That intermediate is high in energy and already exhibits a negatively charged oxygen atom. This intermediate is therefore capable of ejecting methoxide, because such a step is not uphill in energy.

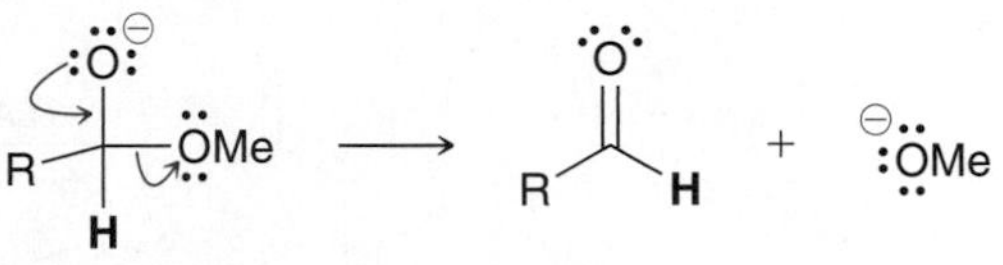

High-energy intermediate

In Chapter 21, we will discuss this reaction in greater detail as well as the mechanism for the reaction of a carboxylic acid with LAH.

SKILLBUILDER

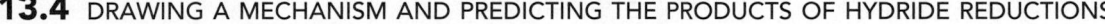

13.4 DRAWING A MECHANISM AND PREDICTING THE PRODUCTS OF HYDRIDE REDUCTIONS

LEARN the skill

Draw the mechanism and predict the product of the following reaction.

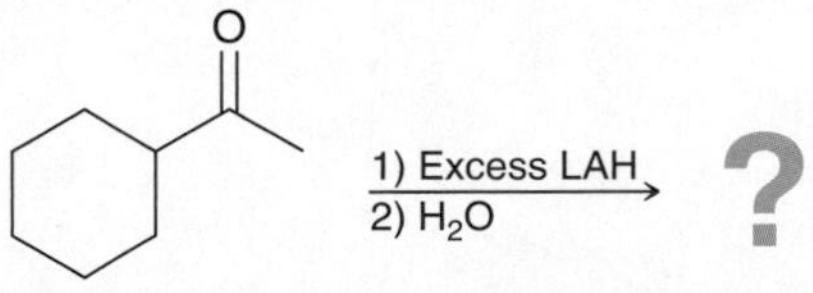
1) Excess LAH
2) H2O
?

SOLUTION

LAH is a hydride reducing agent, and the starting material is a ketone. When hydride reducing agents react with a ketone or aldehyde, the mechanism consists of a nucleophilic attack followed by proton transfer. In the first step, draw the structure of LAH, and show a hydride being delivered to the carbonyl group. Do not simply draw H:⁻ as the reagent, because H:⁻ is not nucleophilic by itself. It must be delivered by the delivery agent (LAH). Draw the complete structure of LAH (showing all bonds), and then draw a curved arrow that shows the electrons coming from one Al—H bond and attacking the carbonyl group.

STEPS 1 AND 2
Draw the two curved arrows that show the delivery of a hydride, and then draw the resulting alkoxide ion.

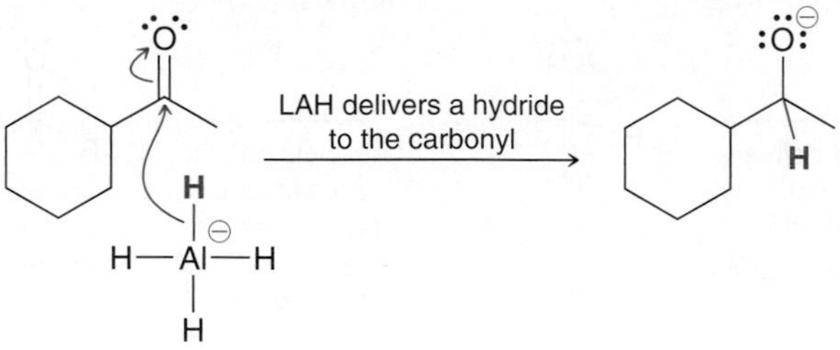

In the second step of the mechanism, the alkoxide ion is protonated by a proton source, in this case, water.

STEP 3
Draw the two curved arrows showing the alkoxide ion being protonated by the proton source.

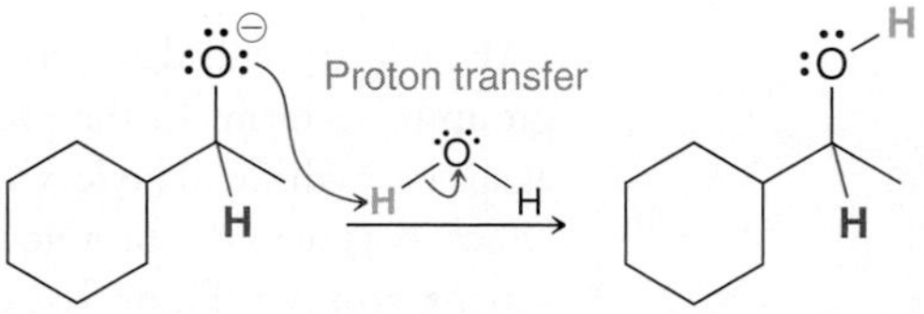

The product is a secondary alcohol, which is what we expect from reduction of a ketone.

PRACTICE the skill **13.12** Draw a mechanism and predict the major product for each reaction.

(a)
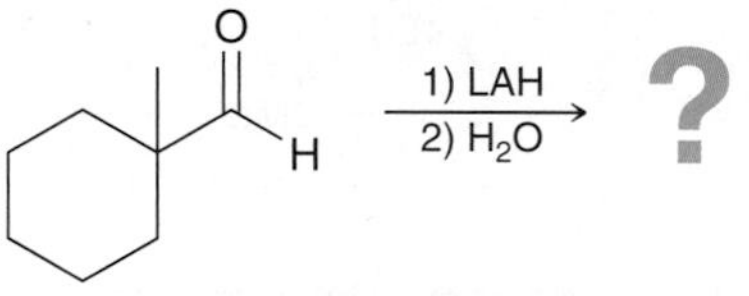

(b)
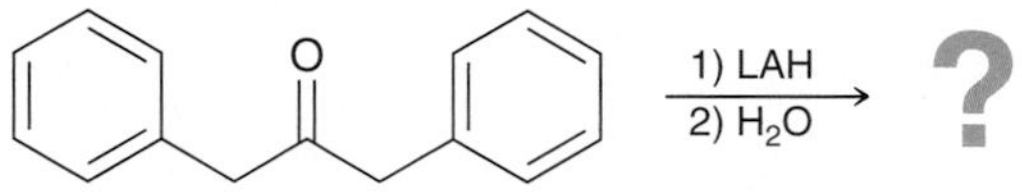

(c)
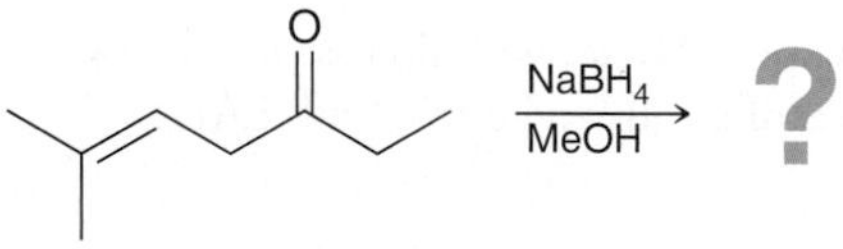

(d)
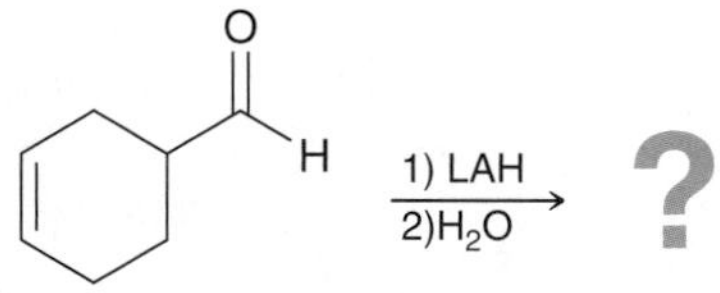

(e)
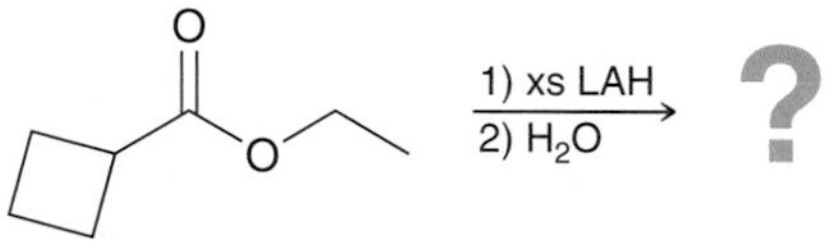

(f)
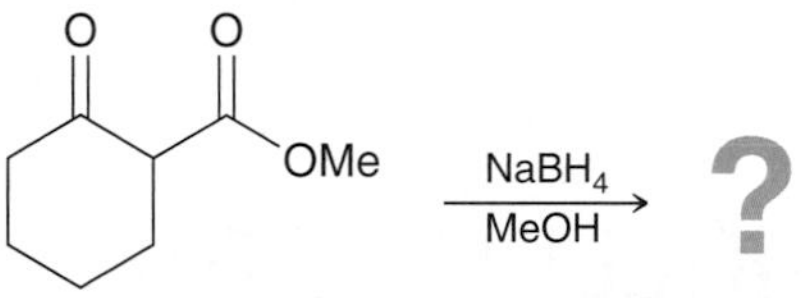

APPLY the skill **13.13** Draw a mechanism and predict the major product of the following reaction.

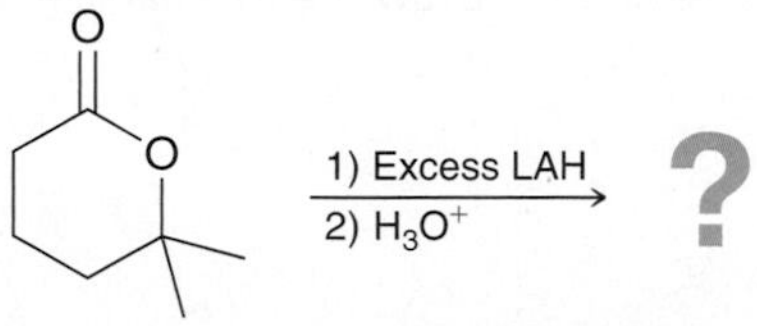

need more **PRACTICE?** **Try Problems 13.46, 13.47c, 13.48e,f, 13.60**

13.5 Preparation of Diols

Diols are compounds with two hydroxyl groups, and the following additional rules are used to name them:

1. The positions of both hydroxyl groups are identified with numbers placed before the parent.
2. The suffix "diol" is added to the end of the name:

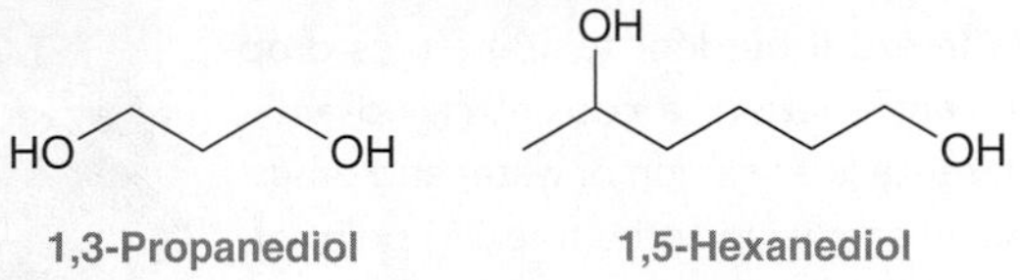

Notice that an "e" appears in between the parent and the suffix. In a regular alcohol, the "e" is dropped (i.e., propanol or hexanol). A few simple diols have common names that are accepted by IUPAC nomenclature.

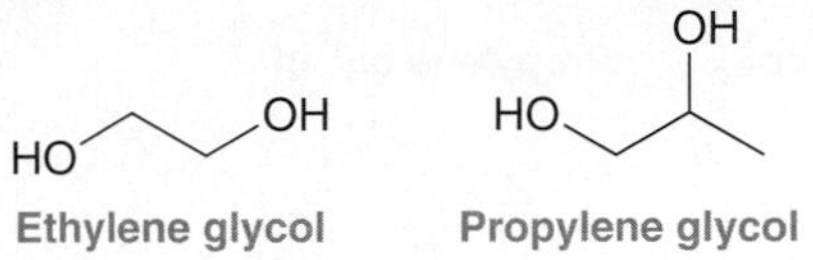

The term glycol indicates the presence of two hydroxyl groups. Diols can be prepared from diketones via reduction using any of the reducing agents that we have seen.

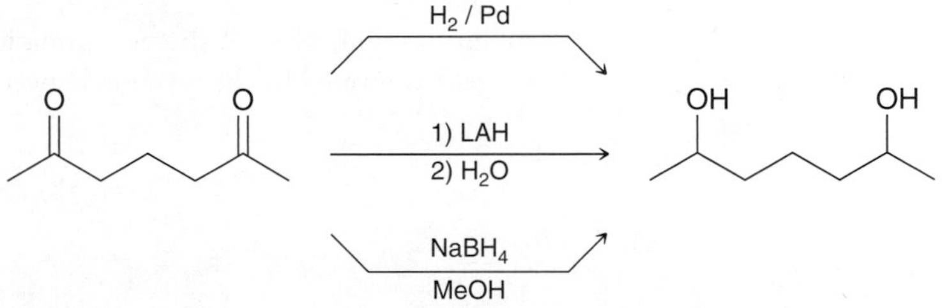

Alternatively, diols can be made via dihydroxylation of an alkene. In Chapter 9, we explored reagents for achieving either *syn* or *anti* dihydroxylation.

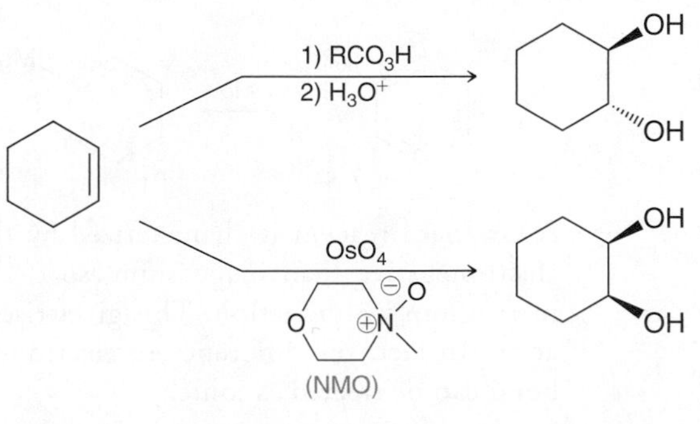

Anti **dihydroxylation (Section 9.9)**

Syn **dihydroxylation (Section 9.10)**

PRACTICALLYSPEAKING)))

Antifreeze

Automobiles are powered by internal combustion engines, and various parts of the engines can become very hot. To prevent damage caused by overheating, a liquid coolant is used to transfer some of the heat away from the sensitive engine parts. In most climates, pure water is unsuitable as a coolant because it can freeze if outdoor temperatures drop below 0°C. To prevent freezing, a coolant called antifreeze is used. Antifreeze is a solution of water and other compounds that significantly lower the freezing point of the mixture. Ethylene glycol and propylene glycol are the two most commonly used compounds for this purpose.

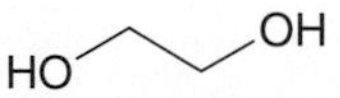

Ethylene glycol

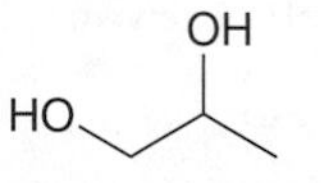

Propylene glycol

13.6 Preparation of Alcohols via Grignard Reagents

In this section, we will discuss formation of alcohols using Grignard reagents. A **Grignard reagent** is formed by the reaction between an alkyl halide and magnesium.

$$\mathrm{R{-}X} \xrightarrow{\mathbf{Mg}} \mathrm{R{-}\mathbf{Mg}{-}X}$$

Grignard reagent

These reagents are named after the French chemist Victor Grignard, who demonstrated their utility in preparing alcohols. For his achievements, he was awarded the 1912 Nobel Prize in Chemistry. Below are a couple of specific examples of Grignard reagents.

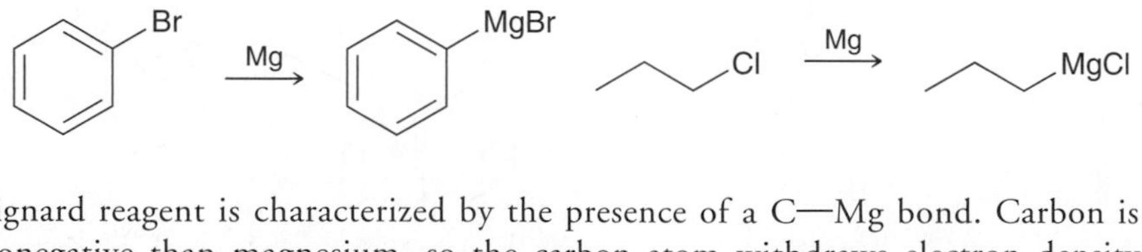

A Grignard reagent is characterized by the presence of a C—Mg bond. Carbon is more electronegative than magnesium, so the carbon atom withdraws electron density from magnesium via induction. This gives rise to a partial negative charge (δ−) on the carbon atom. In fact, the difference in electronegativity between C and Mg is so large that the bond can be treated as ionic.

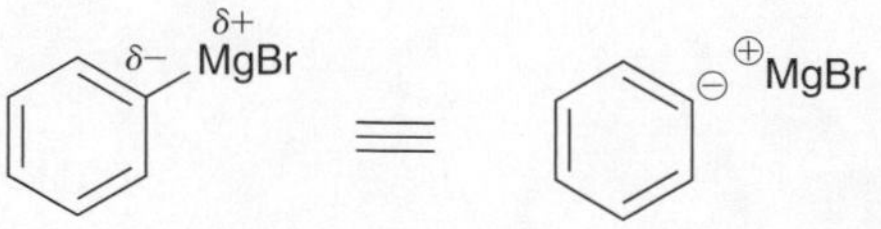

Grignard reagents are carbon nucleophiles capable of attacking a wide range of electrophiles, including the carbonyl group of ketones or aldehydes (Mechanism 13.4).

MECHANISM 13.4 THE REACTION BETWEEN A GRIGNARD REAGENT AND A KETONE OR ALDEHYDE

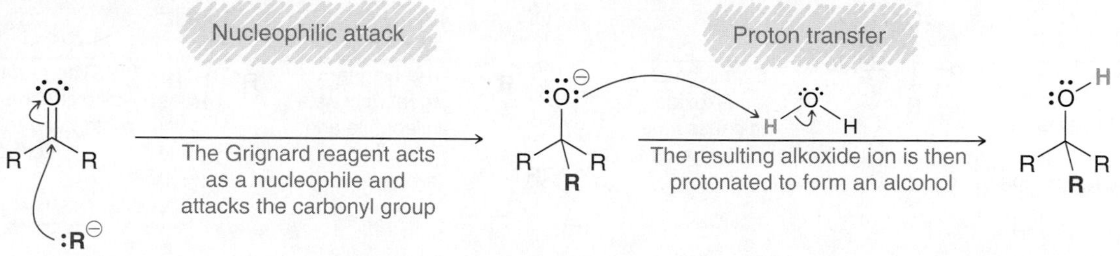

The product is an alcohol, and the mechanism is similar to the mechanism of reduction via hydride reagents (LAH or $NaBH_4$). In fact, the reaction here is also a reduction, but it involves introduction of an R group.

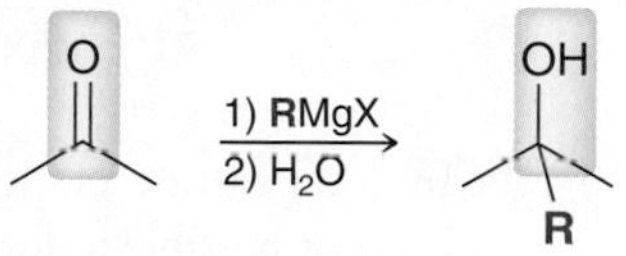

Notice that water is added to the reaction flask in a separate step (similar to reduction with LAH). Water cannot be present together with the Grignard reagent, because the Grignard reagent is also a strong base and will deprotonate water.

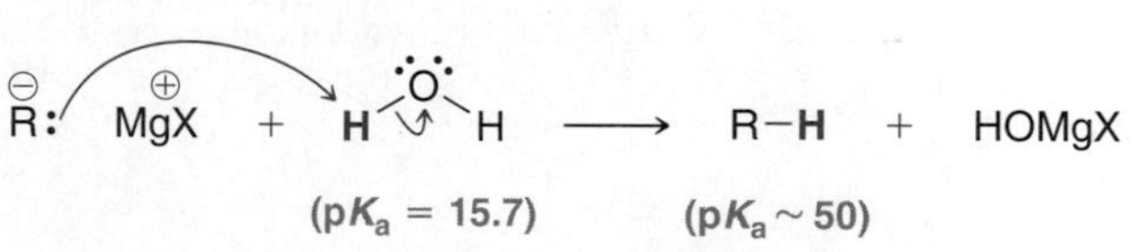

The difference in pK_a values is so vast that the reaction is essentially irreversible. Every water molecule present in the reaction flask will destroy one molecule of Grignard reagent. Grignard reagents will even react with the moisture in the air, so care must be taken to use conditions that scrupulously avoid the presence of water. After the Grignard reagent has attacked the ketone, then water can be added to protonate the alkoxide.

A Grignard reagent will react with a ketone or an aldehyde to produce an alcohol.

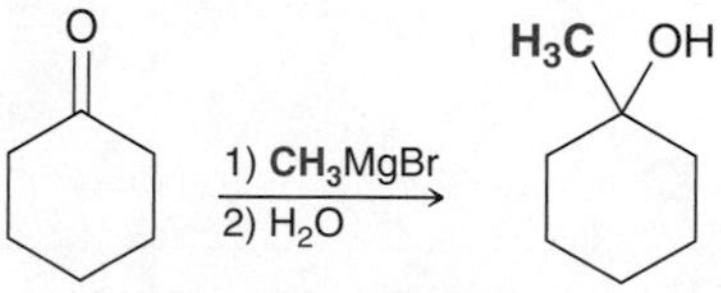

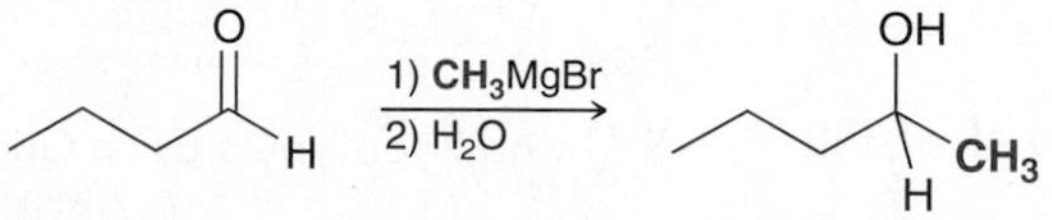

Grignard reagents also react with esters to produce alcohols, with introduction of two R groups.

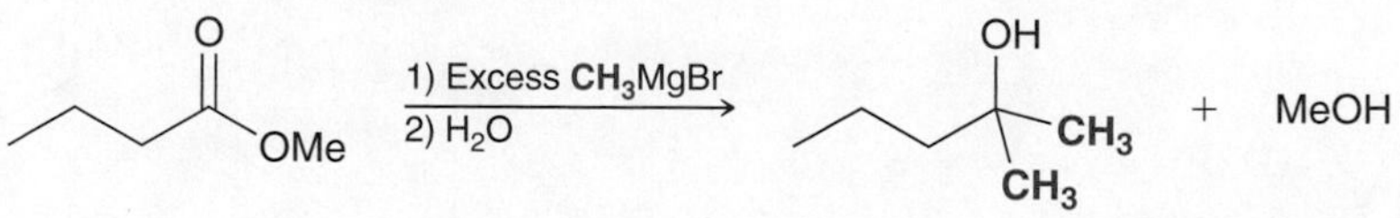

The mechanism of this process (Mechanism 13.5) is similar to the mechanism for reduction of an ester with LAH.

MECHANISM 13.5 THE REACTION BETWEEN A GRIGNARD REAGENT AND AN ESTER

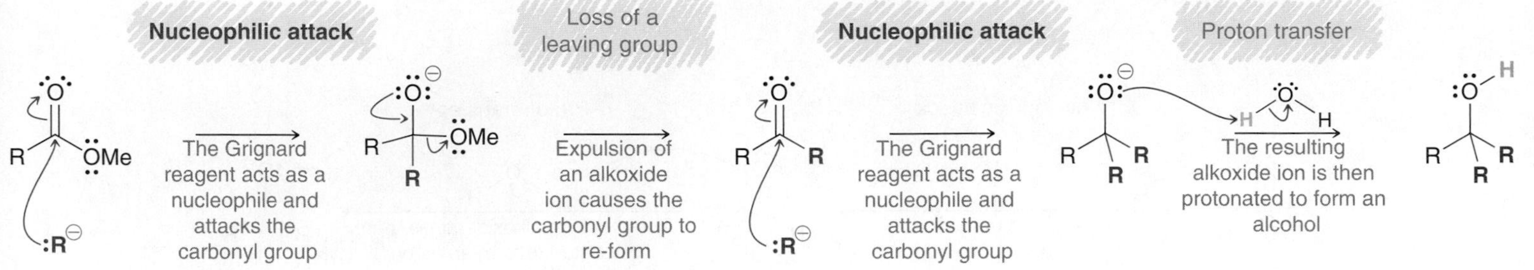

Nucleophilic attack
The Grignard reagent acts as a nucleophile and attacks the carbonyl group
Loss of a leaving group
Expulsion of an alkoxide ion causes the carbonyl group to re-form
Nucleophilic attack
The Grignard reagent acts as a nucleophile and attacks the carbonyl group
Proton transfer
The resulting alkoxide ion is then protonated to form an alcohol

We will explore this reaction and others like it in much more detail in Chapter 21. A Grignard reagent will not attack the carbonyl group of a carboxylic acid. Instead, the Grignard reagent will simply function as a base to deprotonate the carboxylic acid.

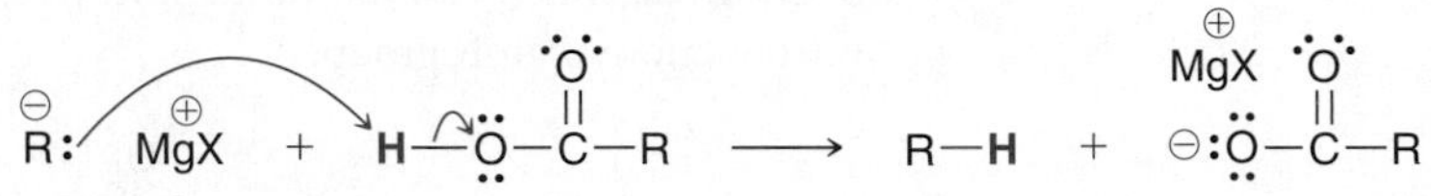

In other words, a Grignard reagent is incompatible with a carboxylic acid. For similar reasons, it is not possible to form a Grignard reagent in the presence of even mildly acidic protons, such as the proton of a hydroxyl group.

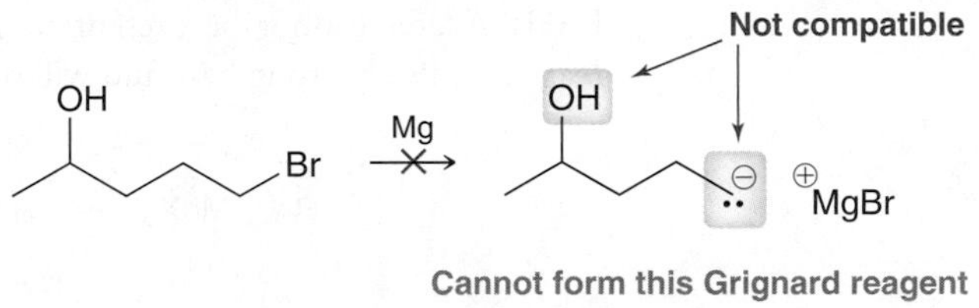

Not compatible
Cannot form this Grignard reagent

This Grignard reagent cannot be formed, as it will simply attack itself to produce an alkoxide. In the next section, we will learn how to circumvent this problem. But first, let's get some practice preparing alcohols via Grignard reactions.

SKILLBUILDER

13.5 PREPARING AN ALCOHOL VIA A GRIGNARD REACTION

LEARN the skill

Show how you would use a Grignard reaction to prepare the following compound.

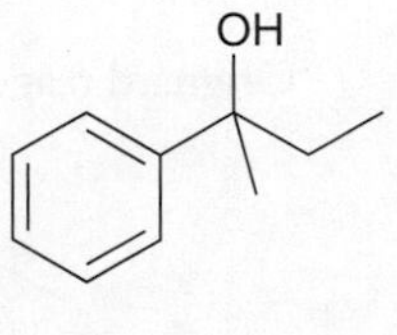

SOLUTION

First identify the carbon atom attached directly to the hydroxyl group (the α position):

STEP 1
Identify the α position.

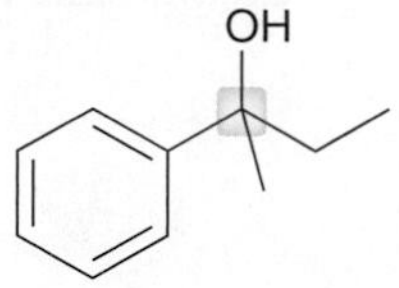

STEP 2
Identify the three groups connected to the α position.

Next, identify all groups connected to this position. There are three: phenyl, methyl, and ethyl. We would need to start with a ketone in which two of the groups are already present, and we introduce the third group as a Grignard reagent. Here are all three possibilities.

STEP 3
Show how each group could have been installed via a Grignard reaction.

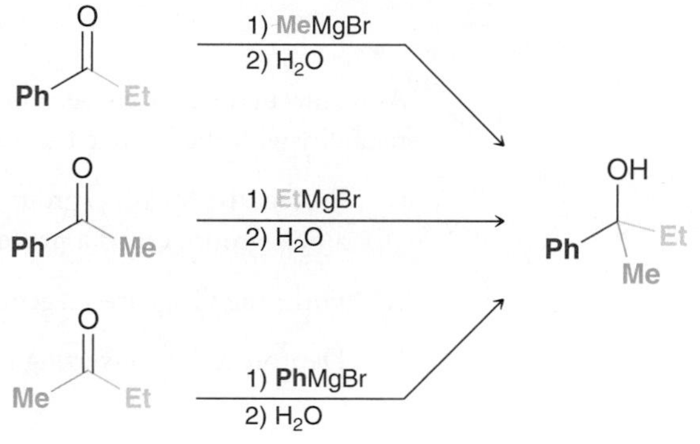

This problem illustrates an important point—that there are three perfectly correct answers to this problem. In fact, synthesis problems will rarely have only one solution. More often, synthesis problems will have multiple correct solutions.

PRACTICE the skill

13.14 Show how you would use a Grignard reaction to prepare each compound below.

(a) 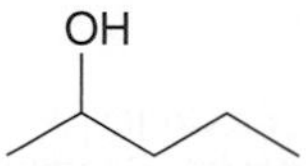

(b) 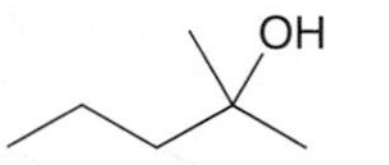

(c) 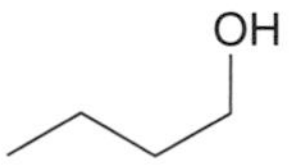

(d) 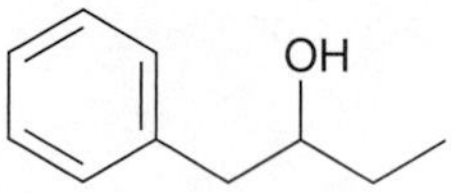

(e) 

(f) 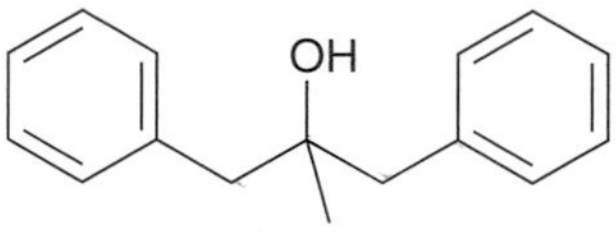

APPLY the skill

13.15 Two of the compounds from Problem 13.14 can be prepared from the reaction between a Grignard reagent and an ester. Identify those two compounds, and explain why the other four compounds cannot be prepared from an ester.

13.16 Three of the compounds from Problem 13.14 can be prepared from the reaction between a hydride reducing agent ($NaBH_4$ or LAH) and a ketone or aldehyde. Identify those three compounds, and explain why the other three compounds cannot be prepared via a hydride reducing agent.

13.17 Draw the mechanism and predict the product of the following reaction. In this case, H_3O^+ must be used as a proton source instead of water. Explain why.

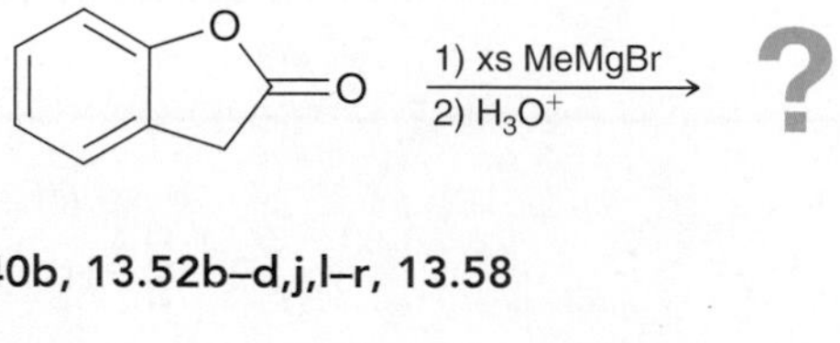

need more **PRACTICE?** **Try Problems 13.38, 13.40b, 13.52b–d,j,l–r, 13.58**

13.7 Protection of Alcohols

Consider the following transformation.

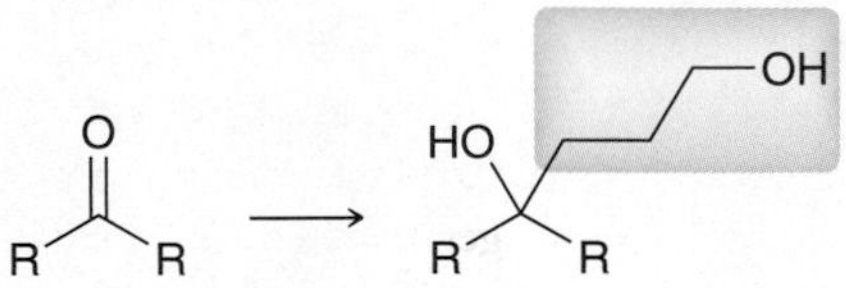

To achieve this transformation via a Grignard reaction, the following Grignard reagent would be required:

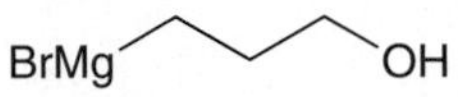

As we saw in the previous section, it is not possible to form this Grignard reagent, because of incompatibility with the hydroxyl group. To circumvent this problem, we employ a three-step process.

1. Protect the hydroxyl group by removing its proton and converting the hydroxyl group into a new group, called a **protecting group**, that is compatible with a Grignard reagent.
2. Form the Grignard reagent and perform the desired Grignard reaction.
3. Deprotect, by converting the protecting group back into a hydroxyl group.

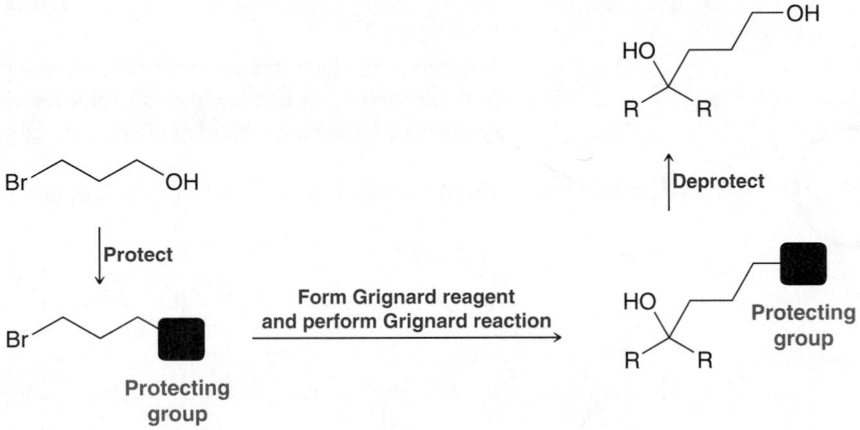

The protecting group enables us to perform the desired Grignard reaction. One such example of a protecting group is the conversion of the hydroxyl group into a trimethylsilyl ether.

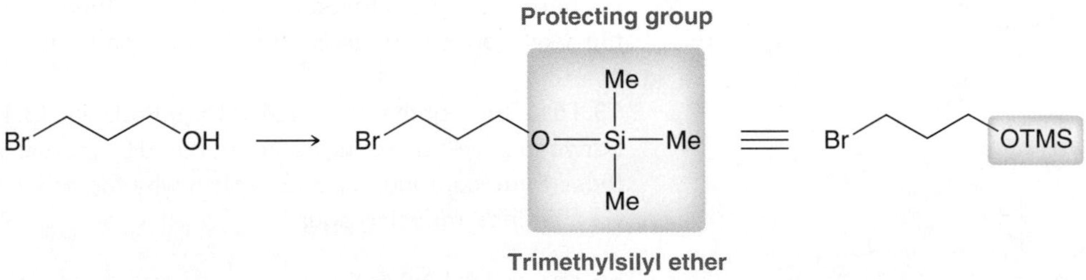

This protecting group is abbreviated as OTMS. The trimethylsilyl ether is formed via the reaction between an alcohol and trimethylsilyl chloride, abbreviated TMSCl.

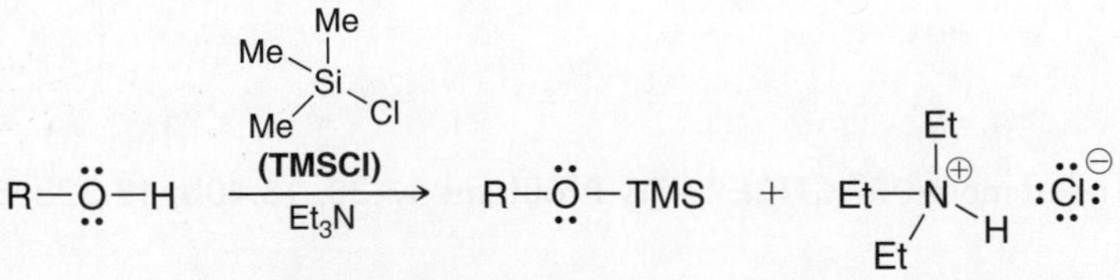

This reaction is believed to proceed via an S_N2-like process (called S_N2-Si), in which the hydroxyl group functions as a nucleophile to attack the silicon atom and a chloride ion is expelled as a leaving group. A base, such as triethylamine, is then used to remove the proton connected to the oxygen atom. Notice that the first step involves an S_N2-like process occurring at a tertiary substrate. This should seem surprising because in Chapter 7 we learned that a nucleophile cannot effectively attack a sterically hindered substrate. This case is different because the electrophilic center is a silicon atom rather than a carbon atom. Bonds to silicon atoms are typically much longer than bonds to carbon atoms, and this longer bond length opens up the back side for attack. To visualize this, compare the space-filling models of *tert*-butyl chloride and trimethylsilyl chloride (Figure 13.6).

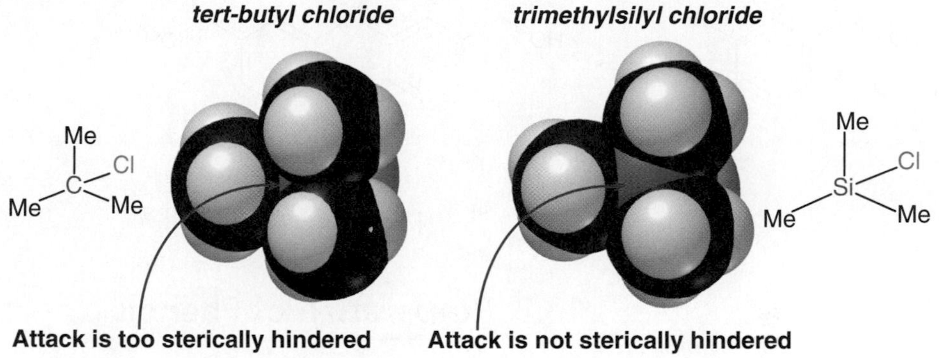

FIGURE 13.6
Space-filling models of *tert*-butyl chloride and trimethylsilyl chloride. The latter is not sterically hindered and is susceptible to attack by a nucleophile.

tert-butyl chloride **trimethylsilyl chloride**

Attack is too sterically hindered **Attack is not sterically hindered**

After the desired Grignard reaction has been performed, the trimethylsilyl group can be removed easily with either H_3O^+ or fluoride ion.

$$R-O-TMS \xrightarrow[or\ F^{\ominus}]{H_3O^{\oplus}} R-OH$$

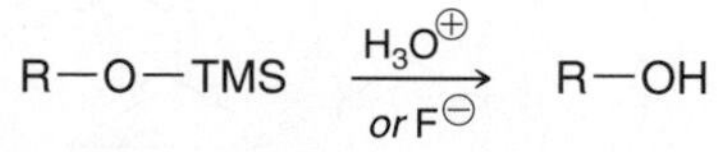

A commonly used source of fluoride ion is tetrabutylammonium fluoride (TBAF).

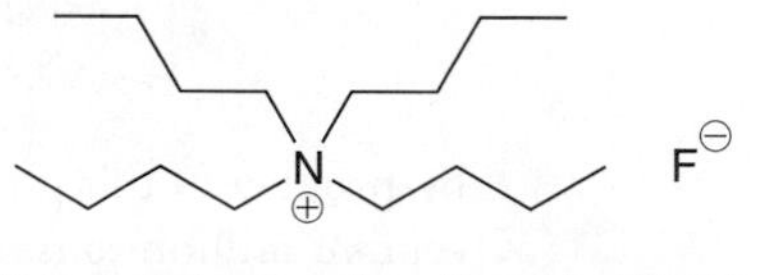

Tetrabutylammonium fluoride (TBAF)

The overall process is shown below.

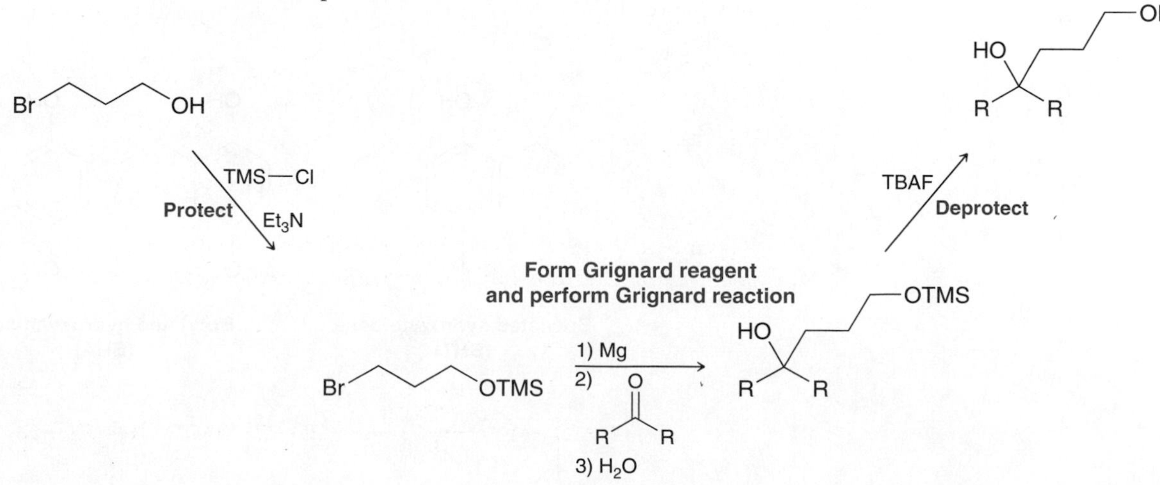

CONCEPTUAL CHECKPOINT

13.18 Identify the reagents you would use to achieve the following transformations:

(a)

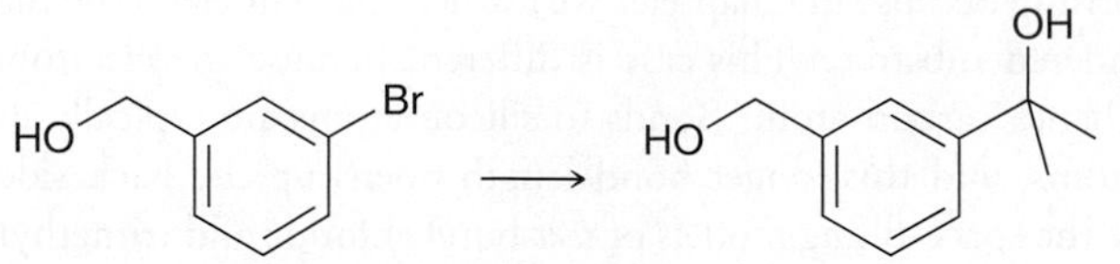

(b)

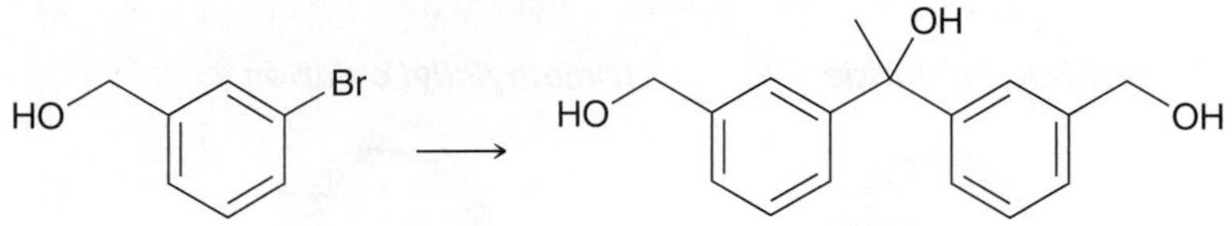

13.8 Preparation of Phenols

Phenol is prepared industrially via a multistep process involving the formation and oxidation of cumene.

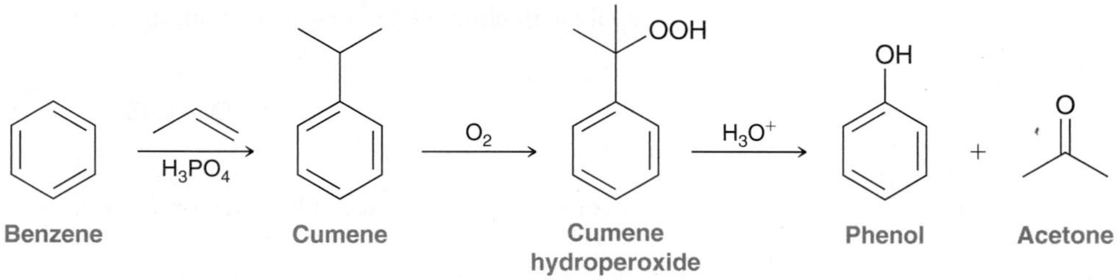

A by-product of this process is acetone, which is also a commercially important compound. Over two million tons of phenol is produced each year in the United States. Phenol is used as a precursor in the synthesis of a wide variety of pharmaceuticals and other commercially useful compounds, including bakelite (a synthetic polymer made from phenol and formaldehyde), adhesives for plywood, and antioxidant food additives (BHT and BHA, discussed in Chapter 11).

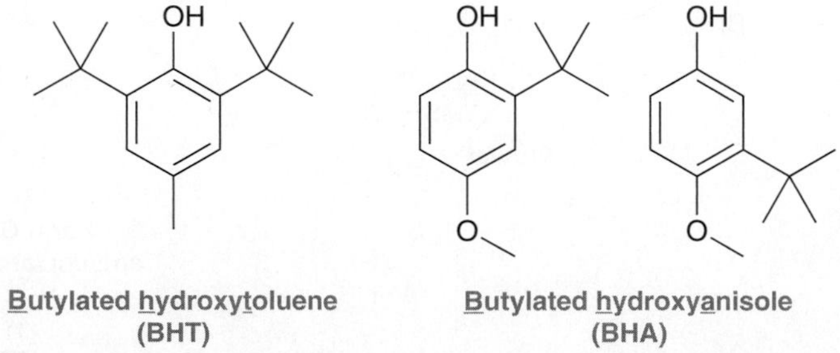

MEDICALLYSPEAKING)))

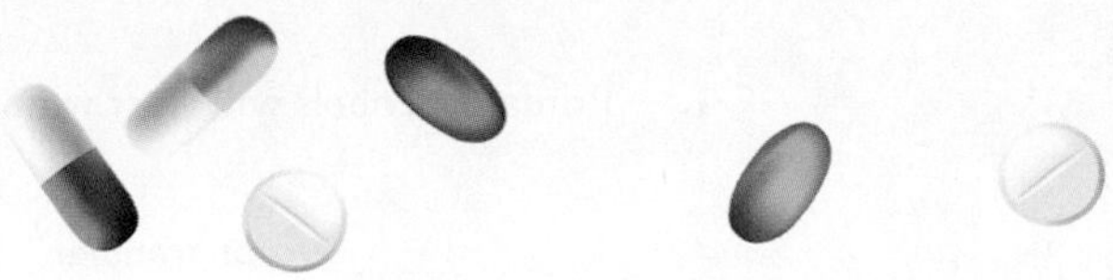

Phenols as Antifungal Agents

Many phenols and their derivatives exhibit topical antifungal properties. These agents are believed to interfere with the cell membrane function of fungi. The following are just a few examples.

These compounds, as well as many others, are used in the treatment of athlete's foot, jock itch, and ringworm. Tolnaftate is the active ingredient in Tinactin™, Odor Eaters™, and Desenex™.

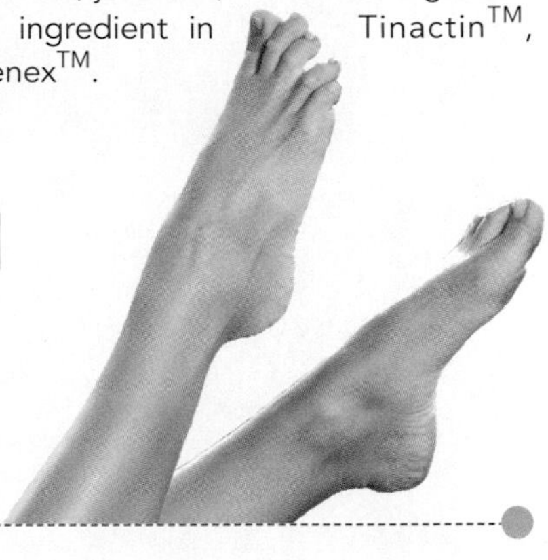

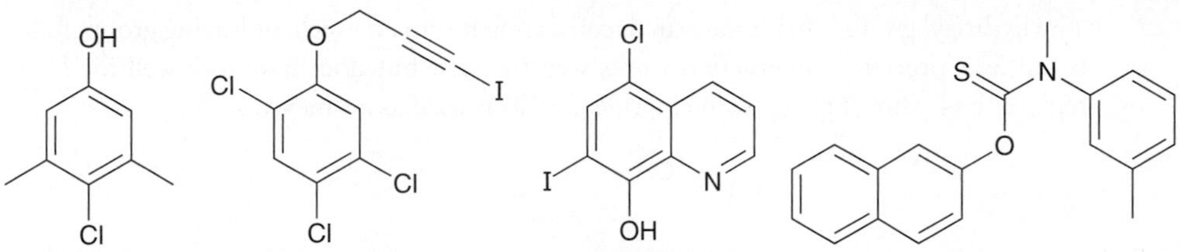

para-Chloro-*meta*-xylenol   Haloprogin   Clioquinol   Tolnaftate

13.9 Reactions of Alcohols: Substitution and Elimination

S_N1 Reactions with Alcohols

As seen in Section 7.5, tertiary alcohols will undergo a substitution reaction when treated with a hydrogen halide.

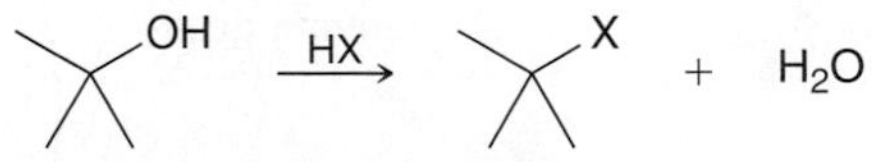

We saw that this reaction proceeds via an S_N1 mechanism.

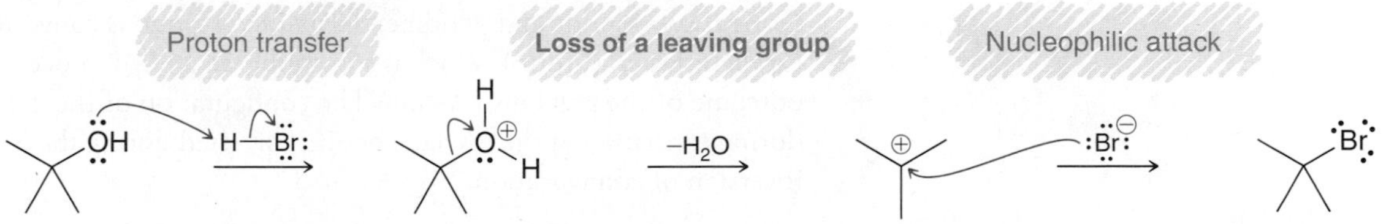

Recall that an S_N1 mechanism has two core steps (loss of leaving group and nucleophilic attack). When the starting material is an alcohol, we saw that an additional step is required in order to protonate the hydroxyl group first. This reaction proceeds via a carbocation intermediate and is therefore most appropriate for tertiary alcohols. Secondary alcohols undergo S_N1 more slowly, and primary alcohols will not undergo S_N1 at an appreciable rate. When dealing with a primary alcohol, an S_N2 pathway is required in order to convert an alcohol into an alkyl halide.

S_N2 Reactions with Alcohols

Primary and secondary alcohols will undergo substitution reactions with a variety of reagents, all of which proceed via an S_N2 process. In this section, we will explore three such reactions that all employ an S_N2 process.

1. Primary alcohols will react with HBr via an S_N2 process.

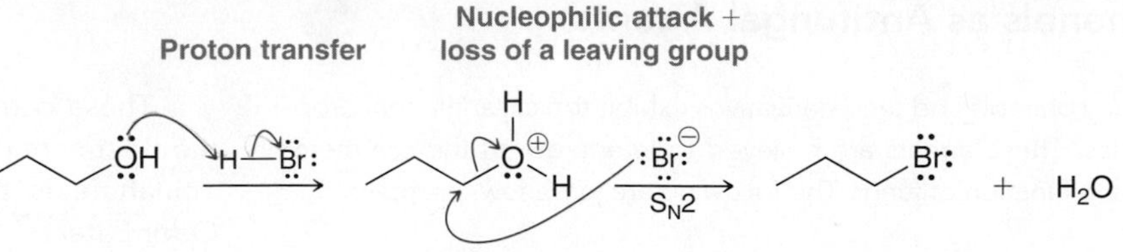

The hydroxyl group is first protonated, converting it into an excellent leaving group, followed by an S_N2 process. This reaction works well for HBr but does not work well for HCl. To replace the hydroxyl group with chloride, $ZnCl_2$ is used as a catalyst.

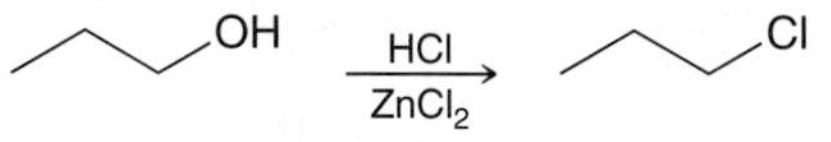

The catalyst is a Lewis acid that converts the hydroxyl group into a better leaving group.

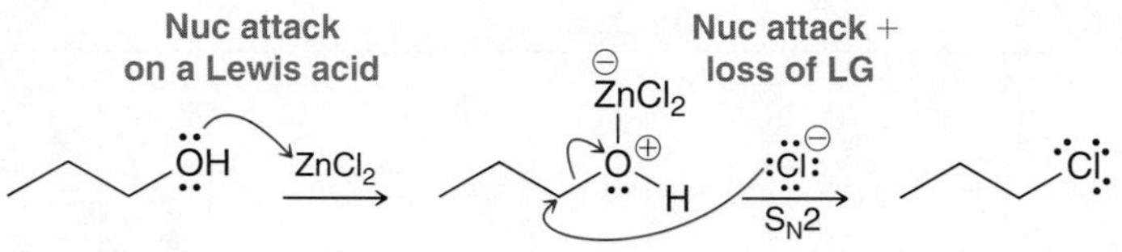

2. As seen in Section 7.8, an alcohol can be converted into a tosylate, followed by nucleophilic attack.

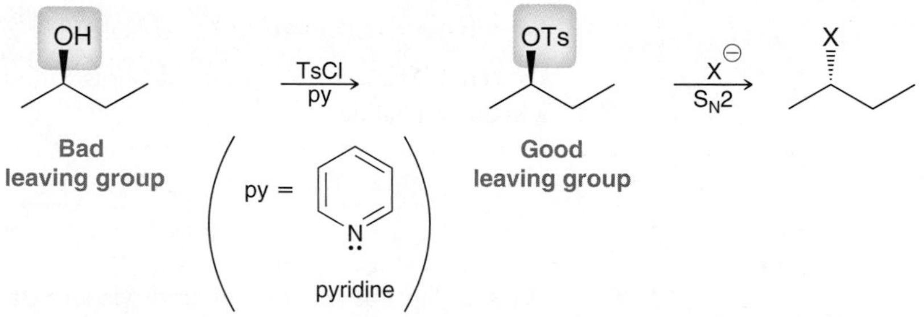

Using tosyl chloride and pyridine, the hydroxyl group is converted into a tosylate group (an excellent leaving group), which is susceptible to an S_N2 process. Notice the stereochemical outcome of the previous reaction. The configuration of the chirality center is not inverted during formation of the tosylate, but it is inverted during the S_N2 process. The net result is inversion of configuration.

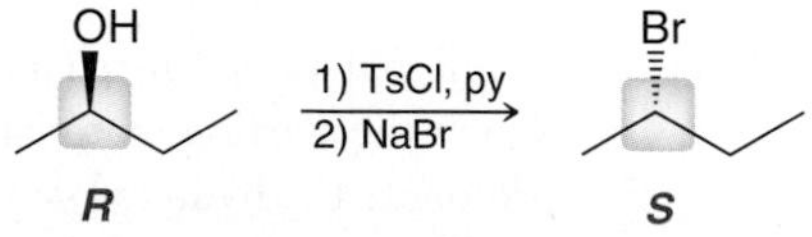

3. Primary and secondary alcohols react with $SOCl_2$ or PBr_3 via an S_N2 process.

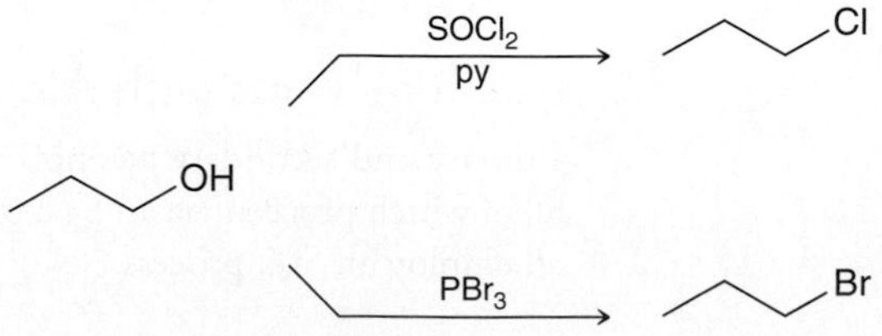

The mechanisms for these two pathways are very similar. The first few steps convert a bad leaving group into a good leaving group, then the halide attacks in an S_N2 process (Mechanism 13.6).

MECHANISM 13.6 THE REACTION BETWEEN $SOCl_2$ AND AN ALCOHOL

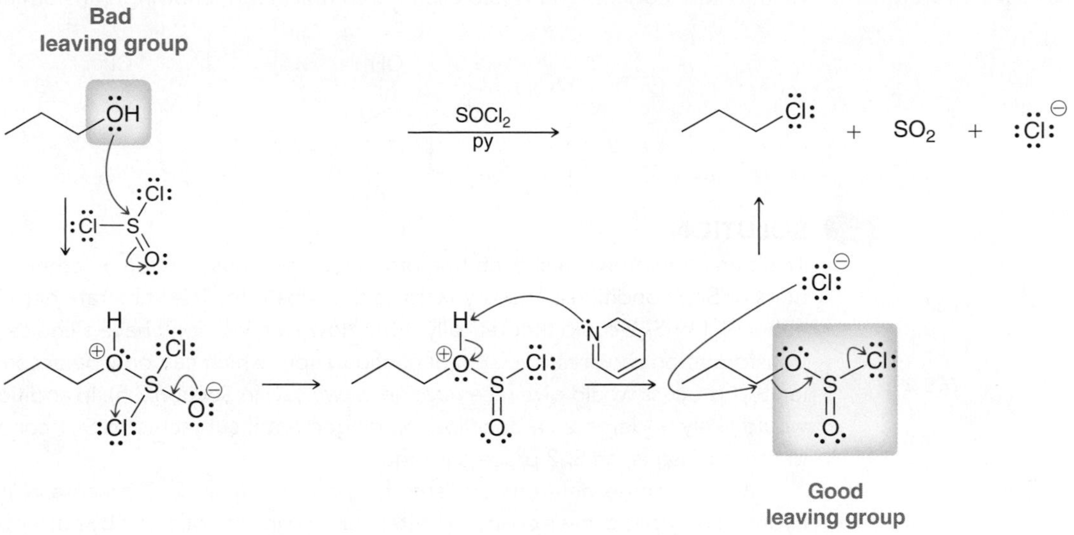

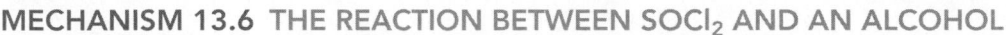

The reaction mechanism for PBr_3 has similar characteristics, including conversion of the hydroxyl group into a better leaving group followed by nucleophilic attack (Mechanism 13.7).

MECHANISM 13.7 THE REACTION BETWEEN PBr_3 AND AN ALCOHOL

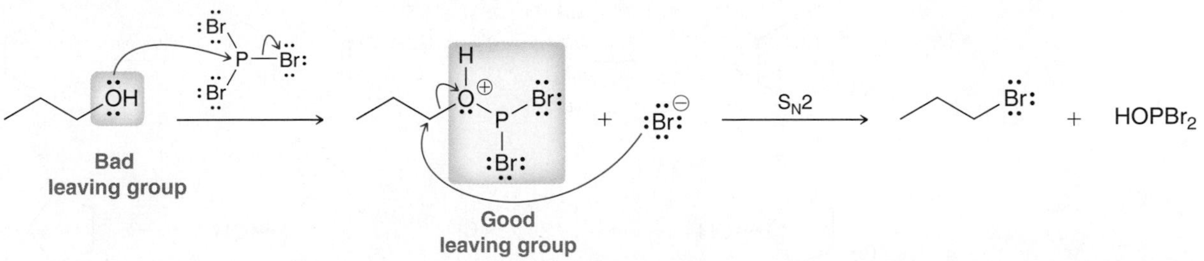

Notice the similarity among all of the S_N2 processes that we have seen in this section. All involve the conversion of the hydroxyl group into a better leaving group followed by nucleophilic attack. If any of these reactions occurs at a chirality center, inversion of configuration is to be expected.

SKILLBUILDER

13.6 PROPOSING REAGENTS FOR THE CONVERSION OF AN ALCOHOL INTO AN ALKYL HALIDE

LEARN the skill

Identify the reagents you would use to accomplish the following transformation.

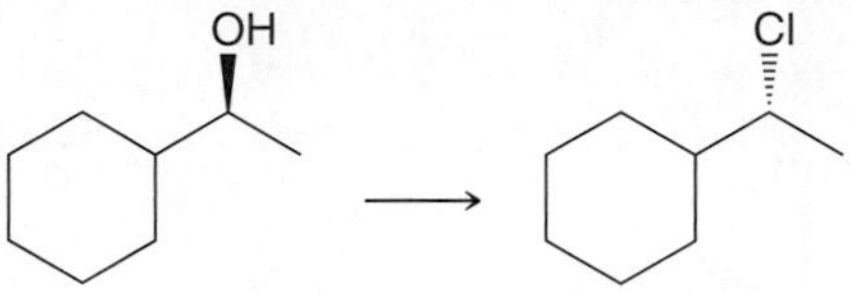

SOLUTION

STEP 1 Analyze the substrate.

STEP 2 Analyze the stereochemical outcome and determine whether S_N2 is required.

This transformation is a substitution process, so we must decide whether to use S_N1 conditions or S_N2 conditions. The key factor is the substrate. The substrate here is secondary, so either S_N1 or S_N2 could theoretically work. However, we don't have a choice in this case. The transformation requires inversion of configuration, which can only be accomplished via S_N2 (an S_N1 process would give racemization, as we saw in Section 7.5). In addition, the substrate would likely undergo a carbocation rearrangement if subjected to S_N1 conditions. For both of these reasons, an S_N2 process is required.

We saw three different reagents for performing an S_N2 process. With all three, the hydroxyl group is converted into a better leaving group, followed by nucleophilic attack. Any of the three methods can be used.

STEP 3 Identify reagents that achieve the process determined in step 2.

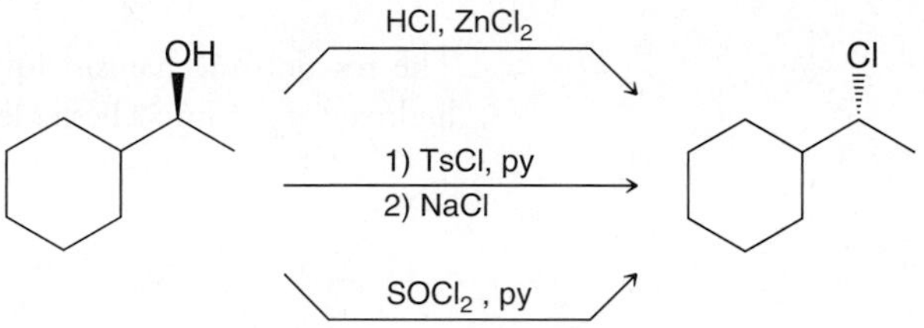

PRACTICE the skill

13.19 Identify the reagents you would use to accomplish each of the following transformations.

(a) 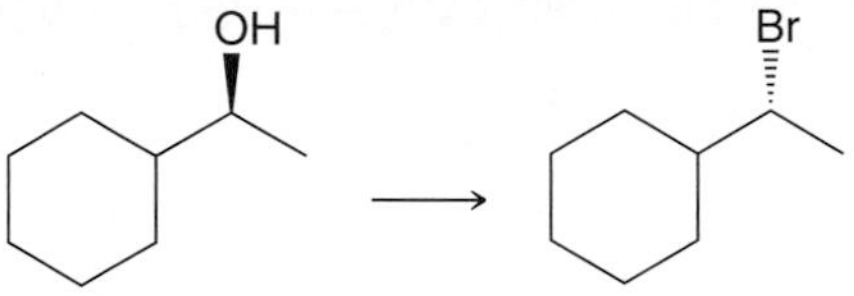 (b) 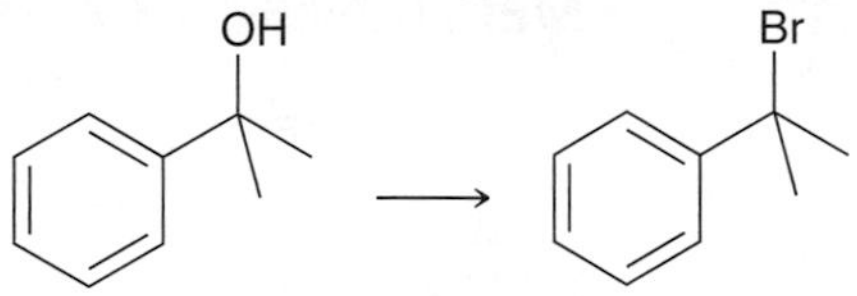

(c) 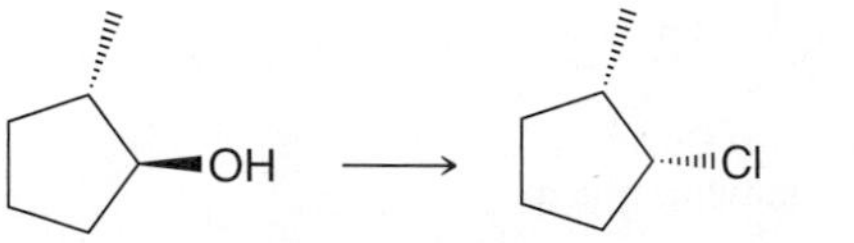 (d) 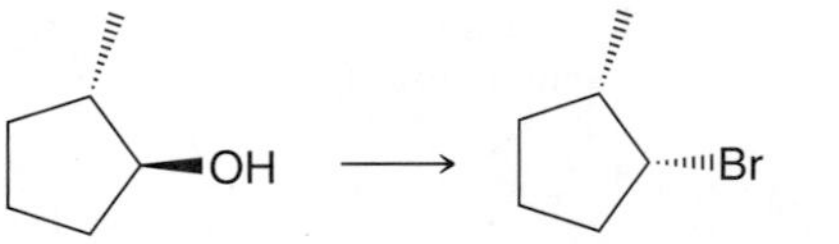

(e)  (f) 

APPLY the skill

13.20 Identify the reagents you would use to accomplish the following transformation.

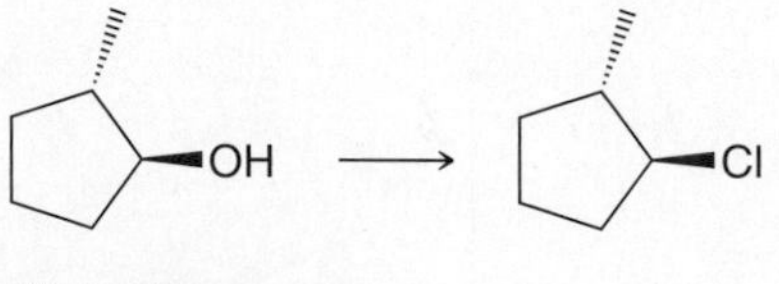

need more **PRACTICE?** **Try Problems 13.35a–c, 13.44f, 13.52r**

PRACTICALLYSPEAKING)))

Drug Metabolism

Drug metabolism refers to the set of reactions by which drugs are converted into other compounds that are either used by the body or excreted. Our bodies dispose of the medications we ingest by a variety of metabolic pathways.

One common metabolic pathway is called glucuronic acid conjugation, or simply glucuronidation. This process is very similar to all of the S_N2 processes that we investigated in the previous section. Specifically, a bad leaving group (hydroxyl) is first converted into a good leaving group, followed by nucleophilic attack. Glucuronidation exhibits the same two key steps.

1. Formation of UDPGA (uridine-5′-diphospho-α-D-glucuronic acid) from glucose involves conversion of a bad leaving group into a good leaving group.

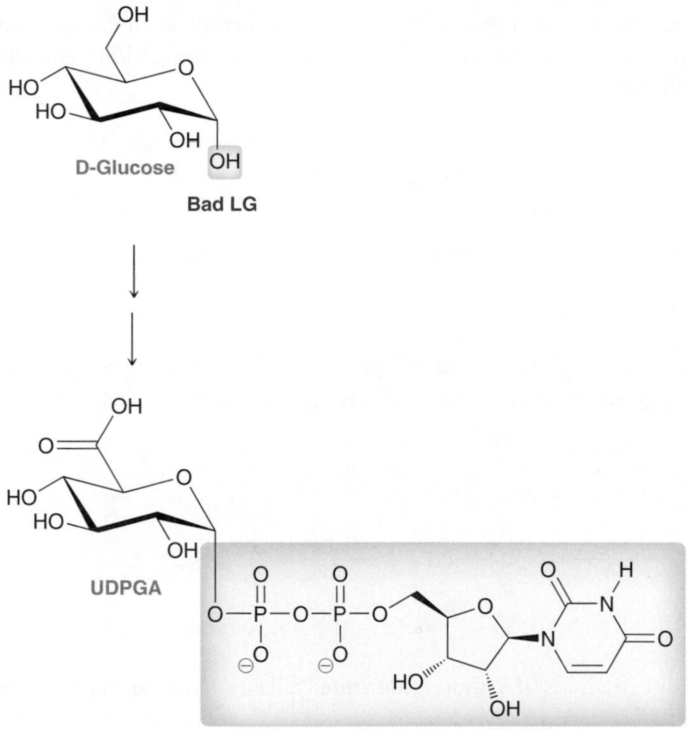

Good LG (called UDP)

UDPGA is a compound with a very good leaving group. This large leaving group is called UDP. In this transformation, one of the hydroxyl groups in glucose has been converted into a good leaving group.

2. Next, an S_N2 process occurs in which the drug being metabolized (such as an alcohol) attacks UDPGA, expelling the good leaving group.

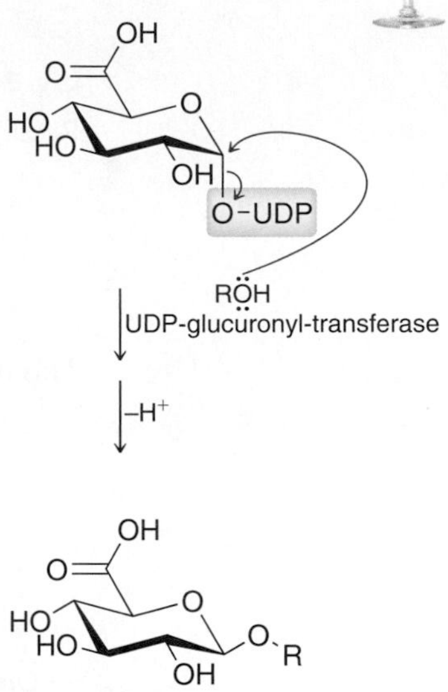

***β*-Glucuronide**

This S_N2 process requires an enzyme (a biological catalyst) called UDP-glucuronyl transferase. The reaction proceeds via inversion of configuration (as expected of an S_N2 process) to produce a β-glucuronide, which is highly water soluble and is readily excreted in the urine.

Many functional groups undergo glucuronidation, but alcohols and phenols are the most common classes of compounds that undergo this metabolic pathway. For example, morphine, acetaminophen, and chloramphenicol are all metabolized via glucuronidation:

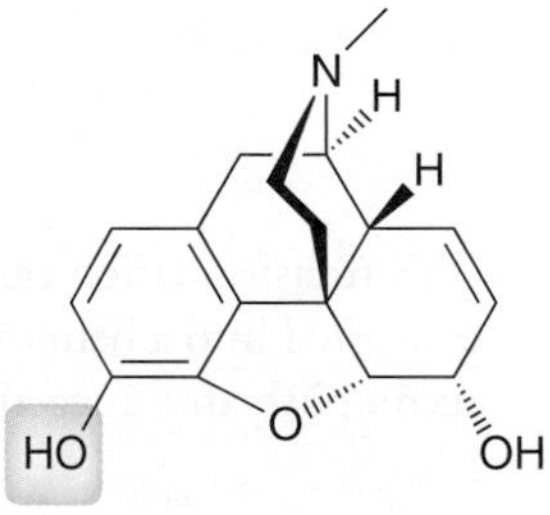

Morphine
An opiate analgesic used to treat severe pain

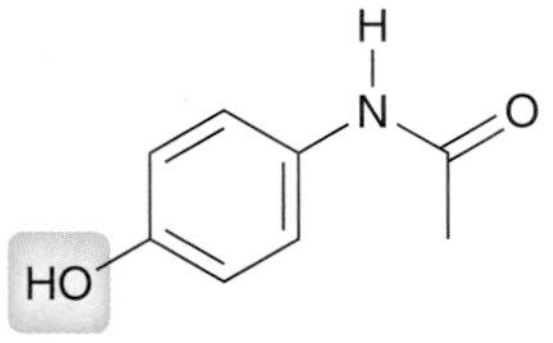

Acetaminophen
An analgesic (pain-relieving) and antipyretic (fever-reducing) agent, sold under the trade name Tylenol™

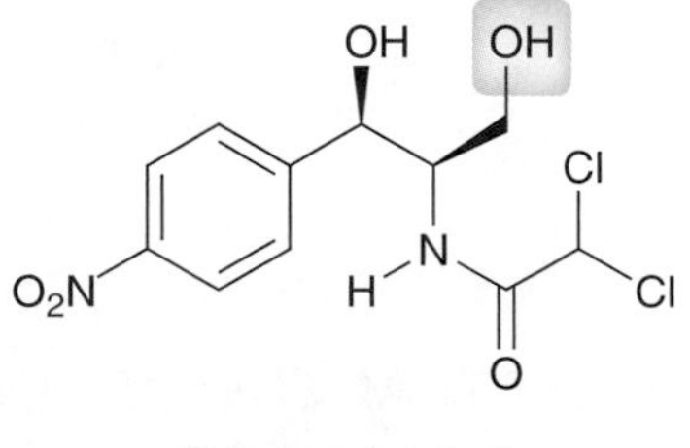

Chloramphenicol
An antibiotic used in eye drops to treat bacterial conjuctivitis

In each of these three compounds, the highlighted hydroxyl group attacks UDPGA. Glucuronidation is the main metabolic pathway by which these drugs are eliminated from the body.

E1 and E2 Reactions with Alcohols

Recall from Section 8.9 that alcohols undergo elimination reactions in acidic conditions.

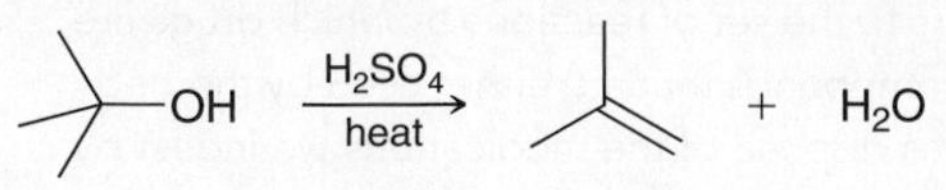

This transformation follows an E1 mechanism:

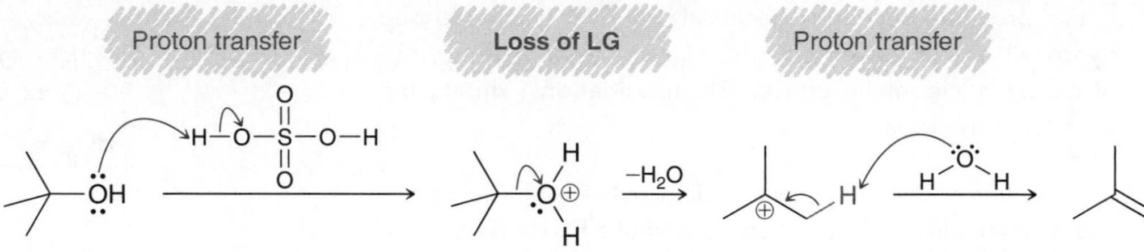

Recall that the two core steps of an E1 mechanism are loss of a leaving group followed by a proton transfer. However, when the starting material is an alcohol, an additional step is first required in order to protonate the hydroxyl group. This reaction proceeds via a carbocation intermediate and is therefore best for tertiary alcohols. Also recall that elimination generally favors the more substituted alkene.

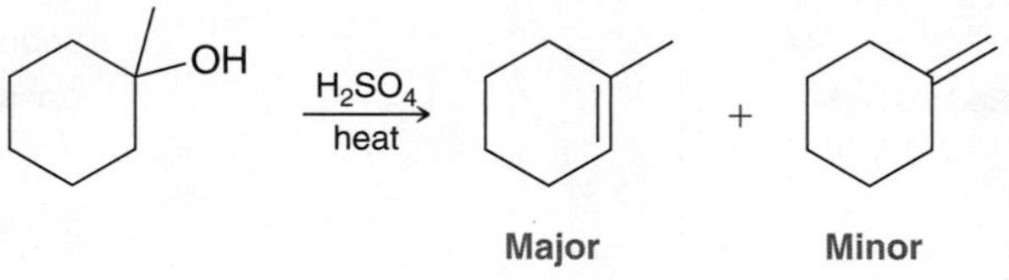

This transformation can also be accomplished via an E2 pathway if the hydroxyl group is first converted into a better leaving group, such as a tosylate. A strong base can then be employed to accomplish an E2 reaction.

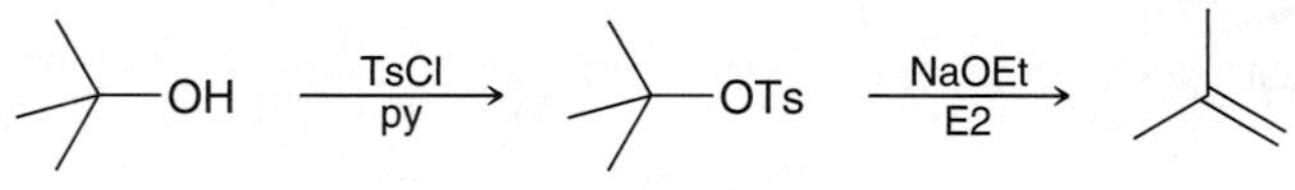

The E2 process also generally produces the more substituted alkene, and no carbocation rearrangements are observed in E2 processes.

CONCEPTUAL CHECKPOINT

13.21 Predict the products for each of the following transformations.

(a)

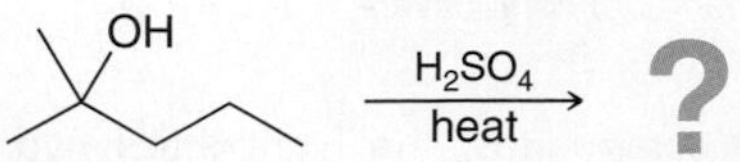

(b)

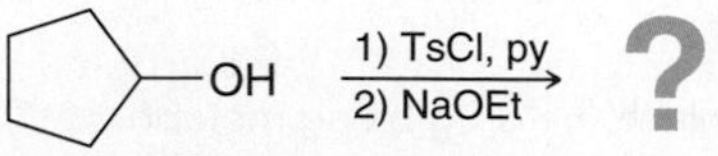

13.10 Reactions of Alcohols: Oxidation

In Section 13.4, we saw that alcohols can be formed via a reduction process. In this section, we will explore the reverse process, called **oxidation**, which involves an increase in oxidation state.

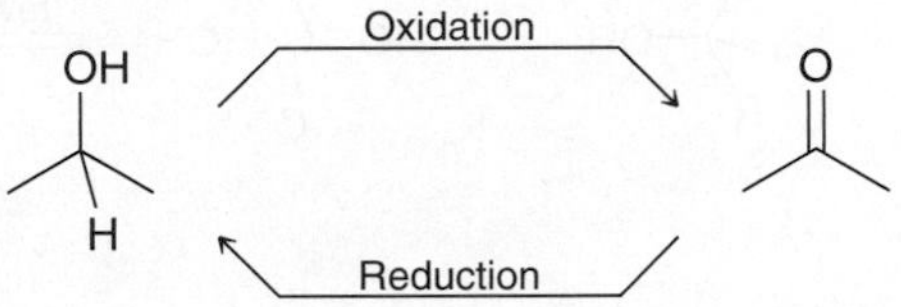

The outcome of an oxidation process depends on whether the starting alcohol is primary, secondary, or tertiary. Let's first consider the oxidation of a primary alcohol.

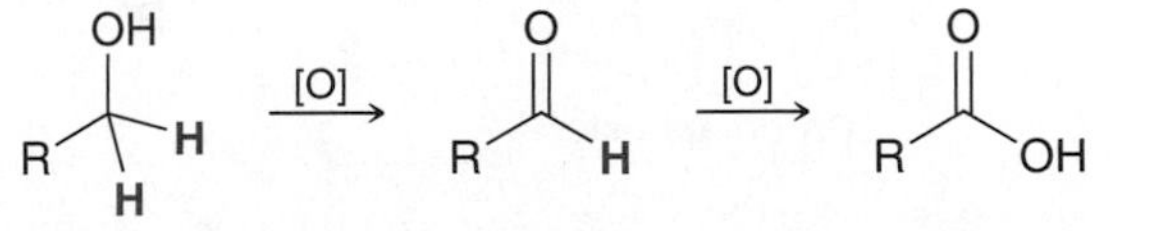

Primary alcohol **Aldehyde** **Carboxylic acid**

Notice that a primary alcohol has two protons at the α position (the carbon atom bearing the hydroxyl group). As a result, primary alcohols can be oxidized twice. The first oxidation produces an aldehyde, and then oxidation of the aldehyde produces a carboxylic acid.

Secondary alcohols only have one proton at the α position so they can only be oxidized once, forming a ketone.

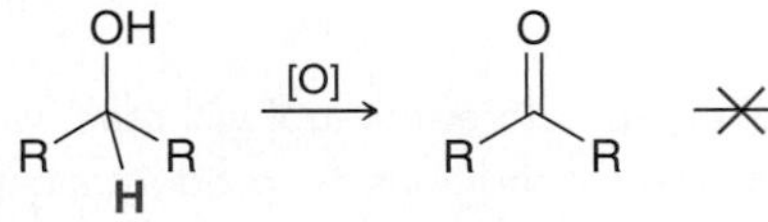

Secondary alcohol **Ketone**

LOOKING AHEAD
For an exception to this general rule, see Section 20.11, where we will learn about a special oxidizing reagent that can oxidize a ketone to form an ester.

Generally speaking, the ketone is not further oxidized. Tertiary alcohols do not have any protons at the α position, and as a result, they generally do not undergo oxidation:

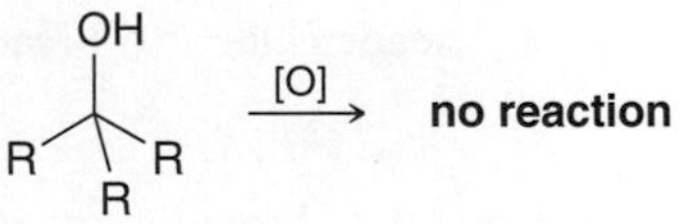

A large number of reagents are available for oxidizing primary and secondary alcohols. The most common oxidizing reagent is chromic acid (H_2CrO_4), which can be formed either from chromium trioxide (CrO_3) or from sodium dichromate ($Na_2Cr_2O_7$) in aqueous acidic solution.

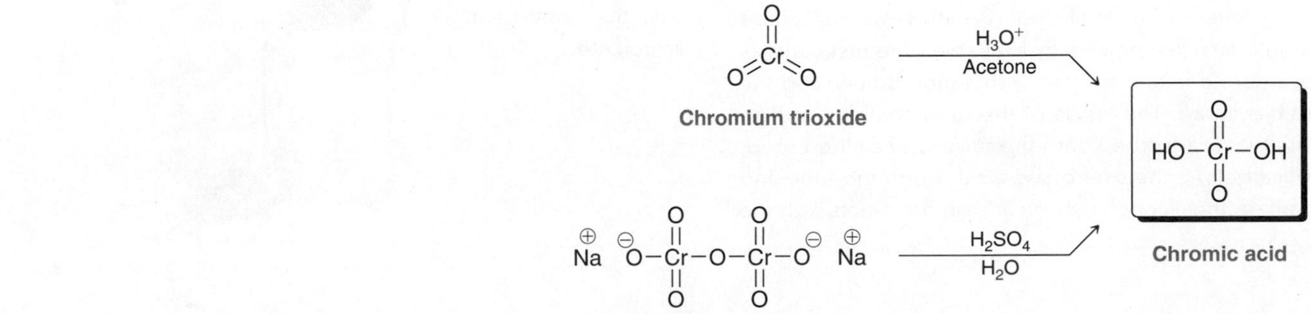

The mechanism of oxidation with chromic acid has two main steps (Mechanism 13.8). The first step involves formation of a chromate ester, and the second step is an E2 process to form a carbon-oxygen π bond (rather than a carbon-carbon π bond).

MECHANISM 13.8 OXIDATION OF AN ALCOHOL WITH CHROMIC ACID

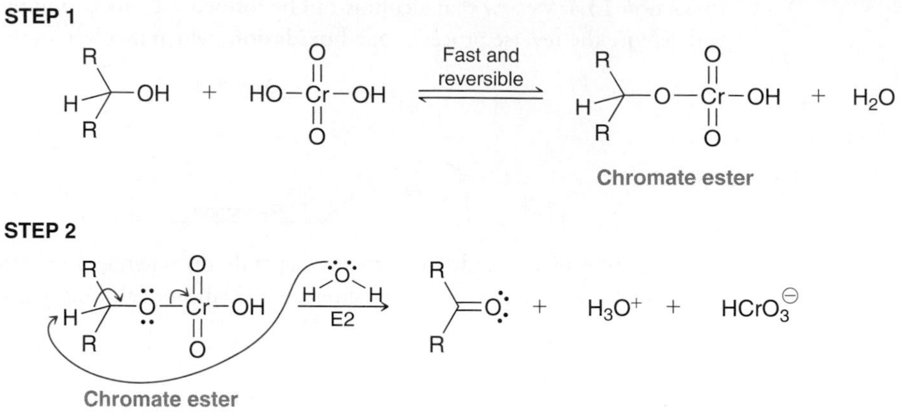
STEP 1
Fast and reversible
Chromate ester
STEP 2
E2
Chromate ester

PRACTICALLYSPEAKING)))

Breath Tests to Measure Blood Alcohol Level

Ethanol is a primary alcohol. Therefore, ethanol will react with potassium dichromate in acidic conditions to produce acetic acid.

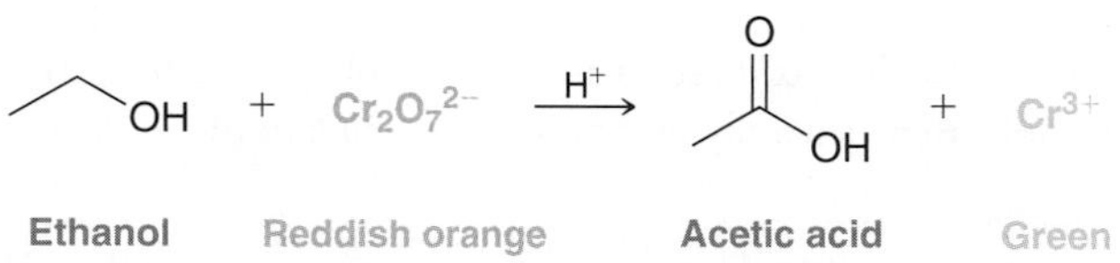
Ethanol
Reddish orange
Acetic acid
Green

In the process, ethanol is oxidized, and the chromium reagent is reduced. The new chromium compound is a different color, and therefore, the progress of the reaction can be monitored by the color change from reddish-orange to green. This reaction formed the basis for many of the early breath tests that assessed the level of blood alcohol. Breath tests that utilize this reaction are still commercially available. The test consists of a tube, where one end is fitted with a mouthpiece and the other end has a bag. The inside of the tube contains sodium dichromate that is adsorbed onto the surface of an inert solid, such as silica gel. As the user blows air through the tube and fills up the bag, the alcohol in the user's breath reacts with the oxidizing agent in the tube, and a color change occurs. The extent of the color change gives an indication of the blood alcohol content.

The Breathalyzer™ is based on the same idea, but it is more accurate. A measured volume of breath is bubbled through an aqueous acidic solution of potassium dichromate, and the change in color is measured with an ultraviolet-visible (UV-VIS) spectrophotometer (Section 17.11). Contrary to popular belief, it is not possible to beat the test by sucking a breath mint or rinsing with a breath freshener. These techniques might fool a person, but they won't fool the potassium dichromate.

When a primary alcohol is oxidized with chromic acid, a carboxylic acid is obtained. It is generally difficult to control the reaction to produce the aldehyde.

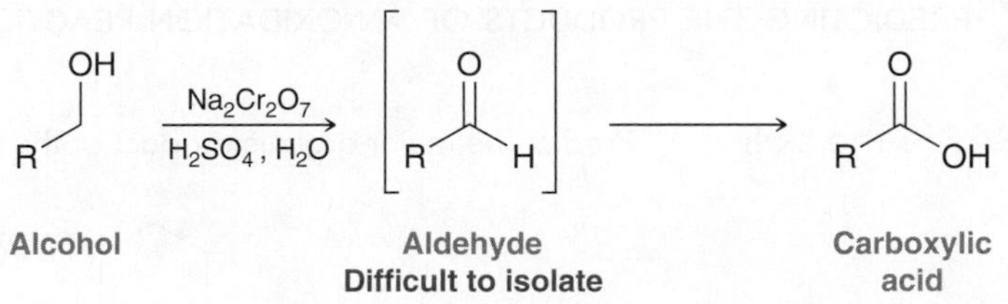

In order to produce the aldehyde as the final product, it is necessary to use a more selective oxidizing reagent, one that will react with the alcohol but will not react with the aldehyde. Many such reagents are available, including pyridinium chlorochromate (PCC). PCC is formed from the reaction between pyridine, chromium trioxide, and hydrochloric acid.

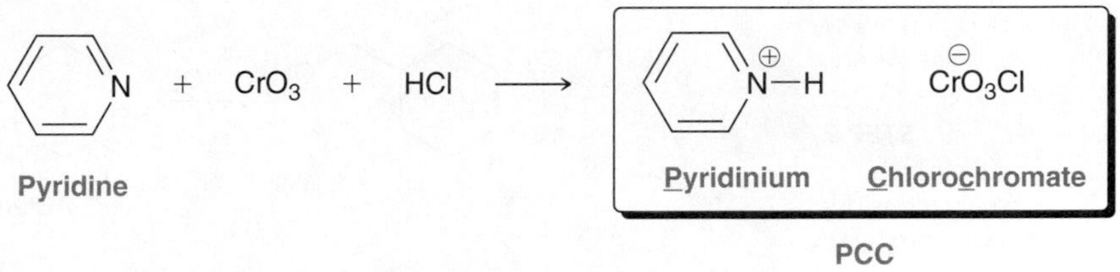

When PCC is used as the oxidizing agent, an aldehyde is produced as the major product. Under these conditions, the aldehyde is not further oxidized to the carboxylic acid.

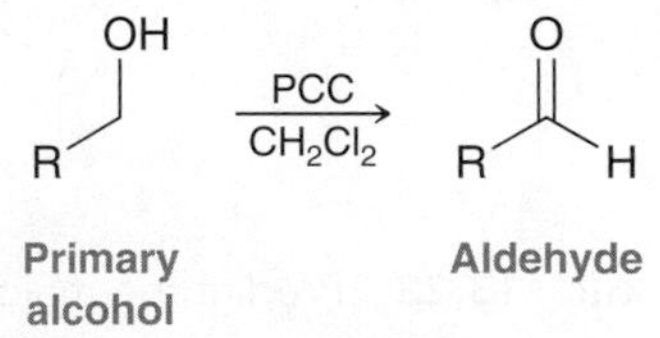

Methylene chloride (CH_2Cl_2) is typically the solvent used when PCC is employed.

Secondary alcohols are oxidized only once to form a ketone, which is stable under oxidizing conditions. Therefore, a secondary alcohol can be oxidized either with chromic acid or with PCC.

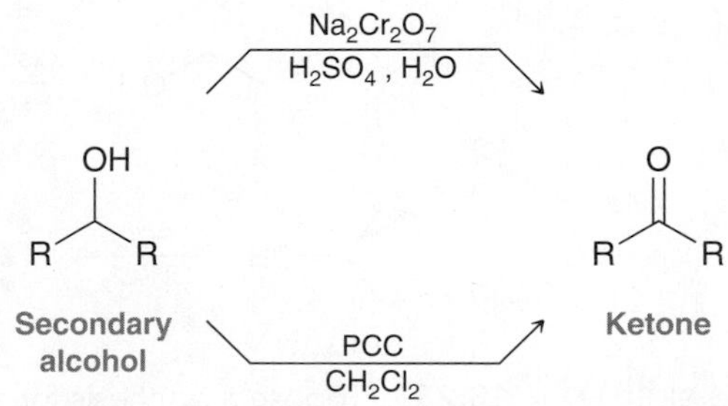

Sodium dichromate is less expensive, but PCC is more gentle and often preferred if other sensitive functional groups are present in the compound.

SKILLBUILDER

13.7 PREDICTING THE PRODUCTS OF AN OXIDATION REACTION

LEARN the skill

Predict the major organic product of the following reaction.

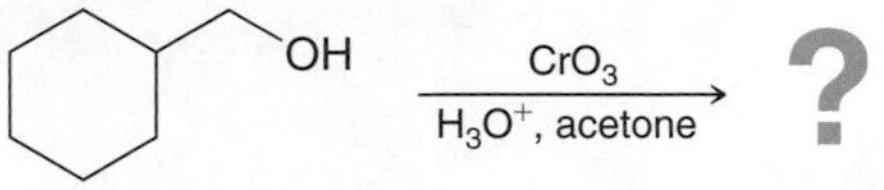

SOLUTION

In order to predict the product, identify whether the alcohol is primary or secondary. In this case, the alcohol is primary, and therefore, we must decide whether the process will produce an aldehyde or a carboxylic acid:

STEP 1
Identify whether the alcohol is primary or secondary.

STEP 2
If the alcohol is primary, consider whether an aldehyde or a carboxylic acid is obtained.

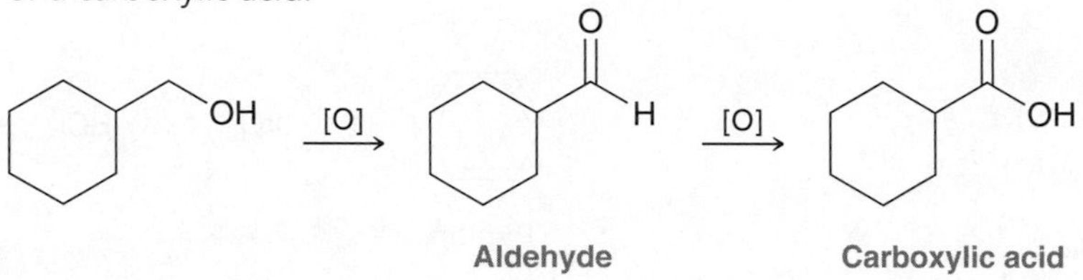

The product is determined by the choice of reagent. Chromium trioxide in acidic conditions forms chromic acid, which will oxidize the alcohol twice to give the carboxylic acid. In order to stop at the aldehyde, PCC would be used instead.

STEP 3
Draw the product.

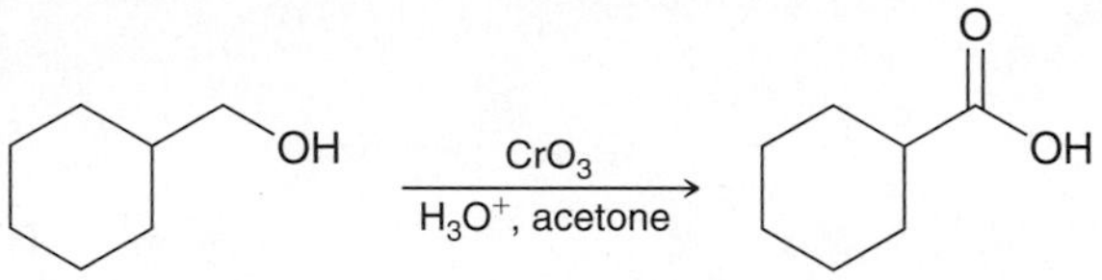

PRACTICE the skill

13.22 Predict the major organic product for each of the following reactions.

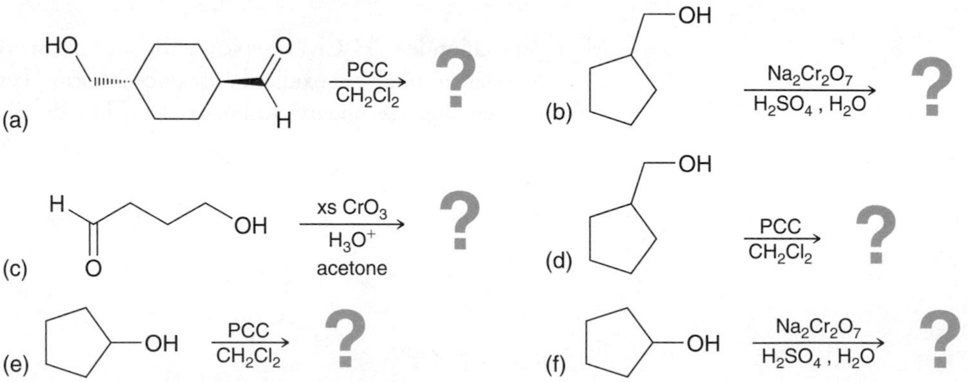

APPLY the skill

13.23 Propose a synthesis for each of the following transformations.

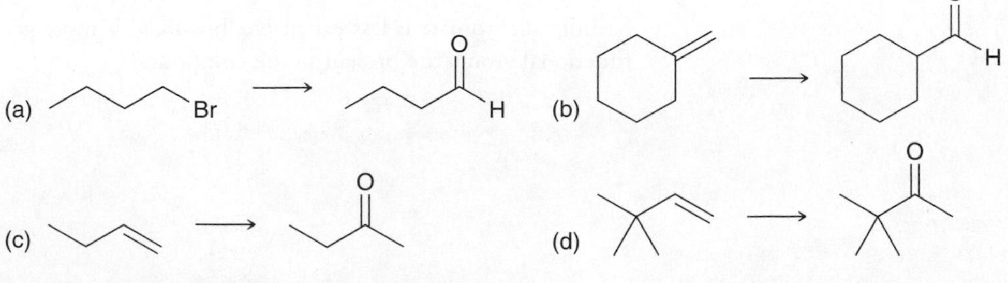

need more **PRACTICE?** **Try Problems 13.35e–f, 13.37, 13.48**

13.11 Biological Redox Reactions

In this chapter, we have seen several reducing agents and several oxidizing reagents used by chemists in the laboratory. Nature employs its own reducing and oxidizing agents, although they are generally much more complex in structure and much more selective in their reactivity. One important biological reducing agent is NADH, which is a molecule comprised of many parts, shown in Figure 13.7.

BY THE WAY
NADH is much less reactive than $NaBH_4$ or LAH and requires a catalyst to do its job. Nature's catalysts are called enzymes (Section 25.8.)

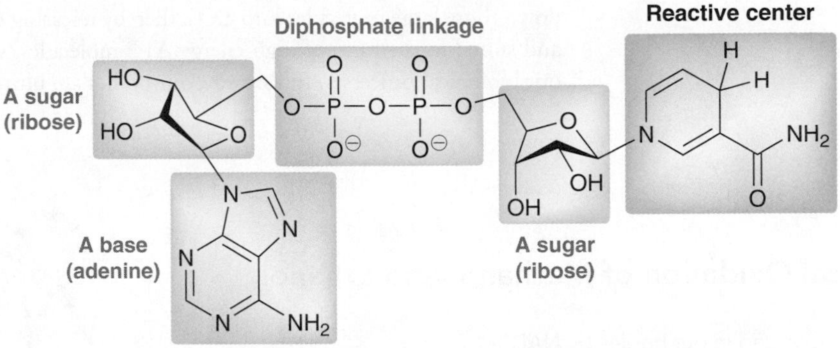

FIGURE 13.7
The structure of NADH, a biological reducing agent.

The reactive center in NADH (highlighted in orange) functions as a hydride delivery agent (very much like $NaBH_4$ or LAH) and can reduce ketones or aldehydes to form alcohols. NADH acts as a reducing agent, and in the process, it is oxidized. The oxidized form is called NAD^+.

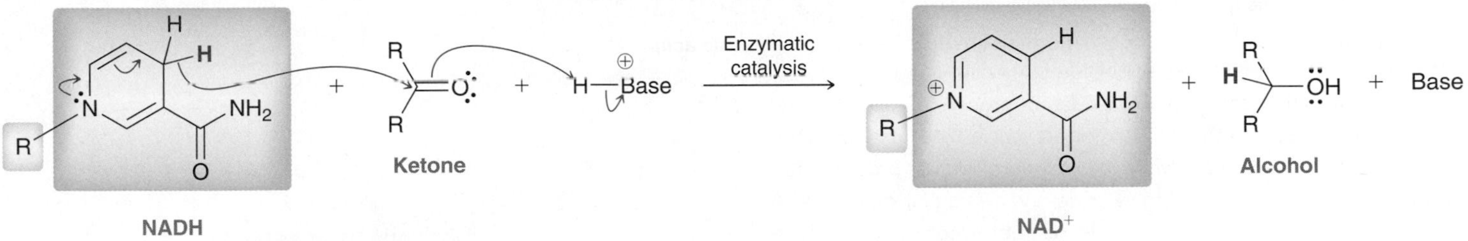

The reverse process can also occur. That is, NAD^+ can act as an oxidizing agent by accepting a hydride from an alcohol, and in the process, NAD^+ is reduced to give NADH.

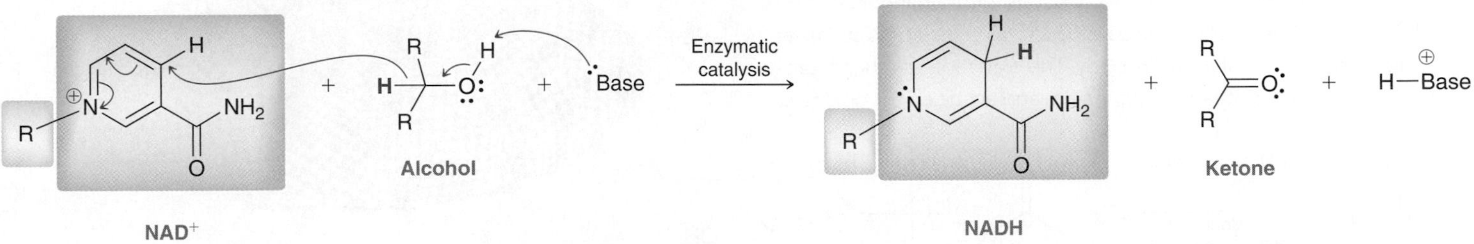

BY THE WAY
To remember which is the reducing agent and which is the oxidizing agent, simply remember that NADH has an H at the end of its name, because it can function as a delivery agent of hydride.

NAD^+ and NADH are present in all living cells and function in a wide variety of redox reactions. NADH is a reducing agent, and NAD^+ is an oxidizing agent. Two important biological processes demonstrate the critical role of NADH and NAD^+ in biological systems (Figure 13.8). The *citric acid cycle* is part of the process by which food is metabolized. This process involves the conversion of NAD^+ to NADH. Another important biological process is the conversion of ADP (adenosine diphosphate) to ATP (adenosine triphosphate), called *ATP synthesis*, which involves the conversion of NADH to NAD^+. ATP is a molecule that is higher in energy than ADP, and the conversion of ADP to ATP is the way that our bodies store the energy that is obtained from metabolism of the food we eat. The energy stored in ATP can be released by conversion back into ADP, and that stored energy is used to power all of our life functions.

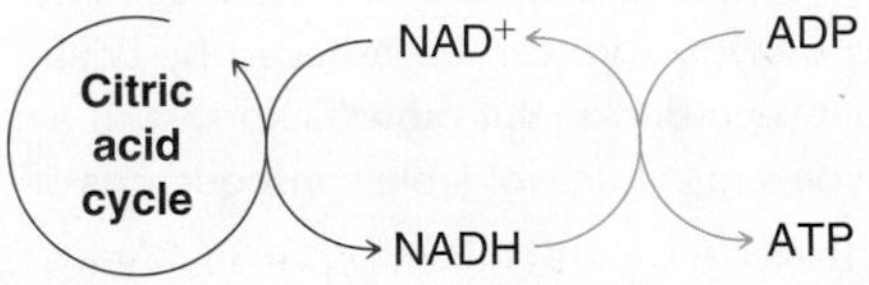

FIGURE 13.8
NADH and NAD^+ play an important role in the citric acid cycle as well as in ATP synthesis.

The citric acid cycle and ATP synthesis are linked into one story. That story can be summarized like this: Energy from the sun is absorbed by vegetation and is used to convert CO_2 molecules into larger organic compounds (a process called carbon dioxide fixing). These organic compounds have bonds (such as C—C and C—H) that are higher in energy than the C=O bonds in CO_2. Therefore, energy will be released if the organic compounds are converted back into CO_2 (this is precisely why the combustion of gasoline releases energy). Humans eat the vegetation (or we eat the animals that ate the vegetation), and our bodies utilize a series of chemical reactions that break down those large molecules into CO_2, thereby releasing energy. That energy is once again captured and stored in the form of high-energy ATP molecules, which are then used to provide energy for our biological processes. In this way, our bodies are ultimately powered by energy from the sun.

PRACTICALLYSPEAKING)))

Biological Oxidation of Methanol and Ethanol

Methanol is oxidized in our bodies by NAD^+.

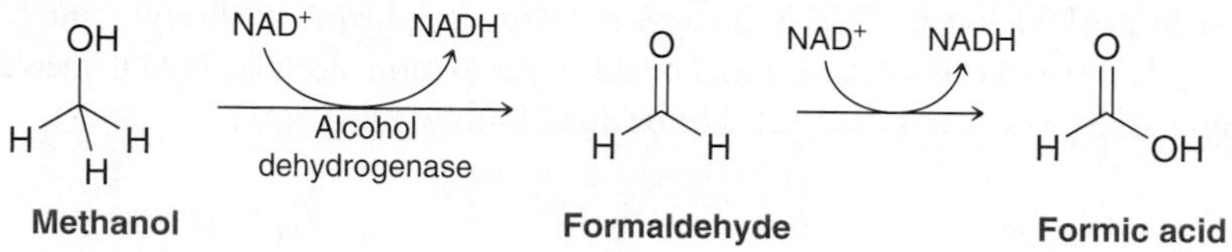

The enzyme that catalyzes this oxidation process is called alcohol dehydrogenase. Methanol is oxidized twice. The first oxidation produces formaldehyde, while the second oxidation produces formic acid. Formic acid is highly toxic, even in small quantities. A buildup of formic acid in the eyes leads to blindness, and a buildup of formic acid in other organs leads to organ failure and death.

A methanol overdose is typically treated by administering ethanol to the patient. Ethanol undergoes oxidation more rapidly than methanol, and the resulting decreased rate of methanol oxidation allows other metabolic pathways (such as glucuronidation) to remove the methanol from the body. The oxidation of ethanol produces acetic acid instead of formic acid, and acetic acid is not toxic.

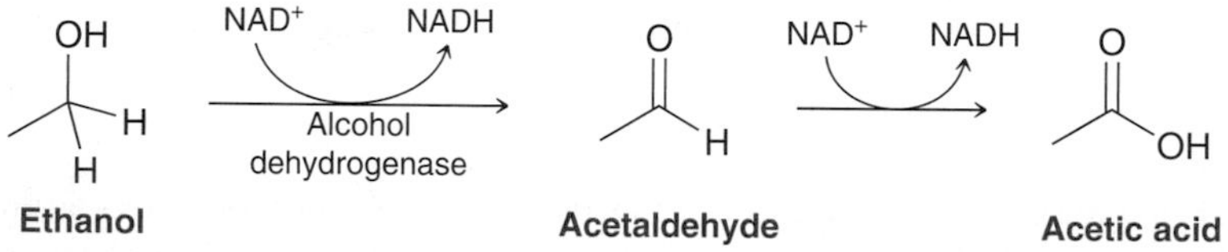

Ethanol is a primary alcohol and is oxidized twice. The first oxidation produces acetaldehyde, while the second oxidation produces acetic acid. Acetic acid can be used by the body for a variety of functions, but acetaldehyde is less useful. When a person drinks large quantities of ethanol (binge drinking), the concentration of acetaldehyde temporarily builds up. A high concentration of acetaldehyde causes nausea, vomiting, and other nasty symptoms.

We mentioned in the chapter opener that the effects of a hangover are caused by a variety of factors. The impact of some of those factors, such as dehydration, can be reduced by drinking a glass of water in between drinks. But other factors, such as the buildup of acetaldehyde, are the unavoidable consequences of binge drinking. The only way to avoid high concentrations of acetaldehyde is to drink small quantities of alcohol over a long period of time. Binge drinking will always produce the unpleasant effects of a hangover. There are many products on the market that claim to prevent hangovers, but there is little scientific evidence that any of these products are effective. The only way to prevent a hangover is to drink responsibly. That is, to drink small quantities of alcohol over a long period of time, together with plenty of water.

In addition to the unpleasant effects of a hangover, binge drinking can also potentially cause a host of more serious health issues, such as an irregular heart rhythm or acute pancreatitis (inflammation of the pancreas), both of which can be life threatening.

13.12 Oxidation of Phenol

In the previous two sections, we discussed reactions that achieve the oxidation of primary and secondary alcohols. In this section, we consider the oxidation of phenol. Based on our discussion thus far, we might expect that phenol would not readily undergo oxidation because it lacks a proton at the α position, much like a tertiary alcohol.

3° alcohol Phenol

Nevertheless, phenol is observed to undergo oxidation even more readily than primary and secondary alcohols. The product is benzoquinone.

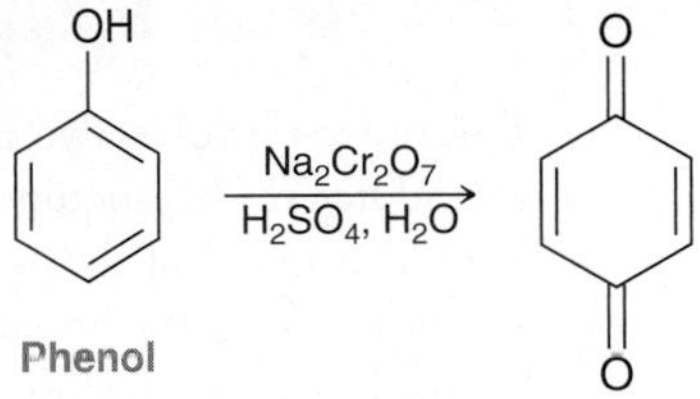

Phenol Benzoquinone

Quinones are important because they are readily converted to hydroquinones.

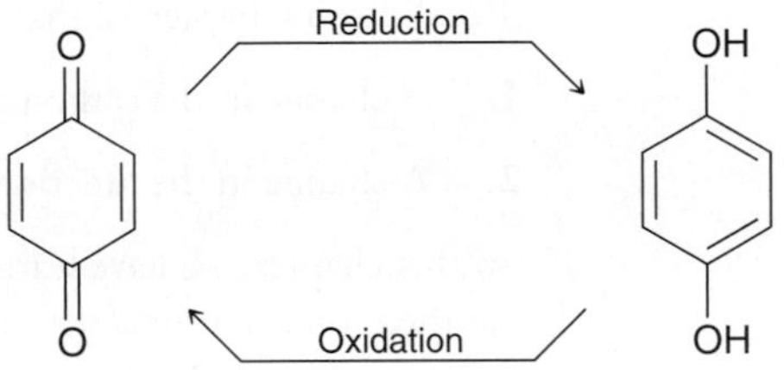

Benzoquinone Hydroquinone

The reversibility of this process is critical for **cellular respiration**, a process by which molecular oxygen is used to convert food into CO_2, water, and energy. Key players in this process are a group of quinones called ubiquinones.

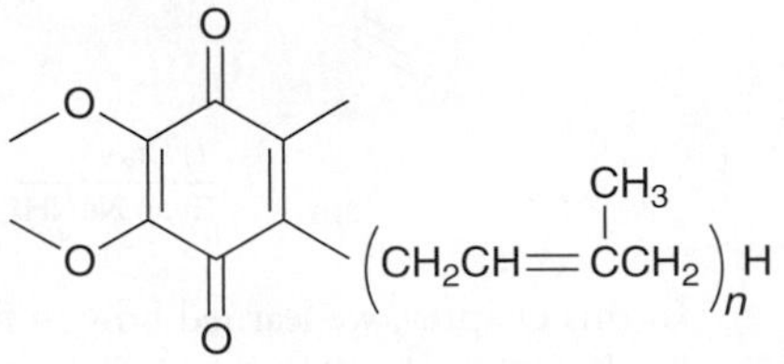

n = 6–10

Ubiquinones

These compounds are called ubiquinones because they are ubiquitous in nature—that is, they are found in all cells. The redox properties of ubiquinones are utilized to convert molecular oxygen into water.

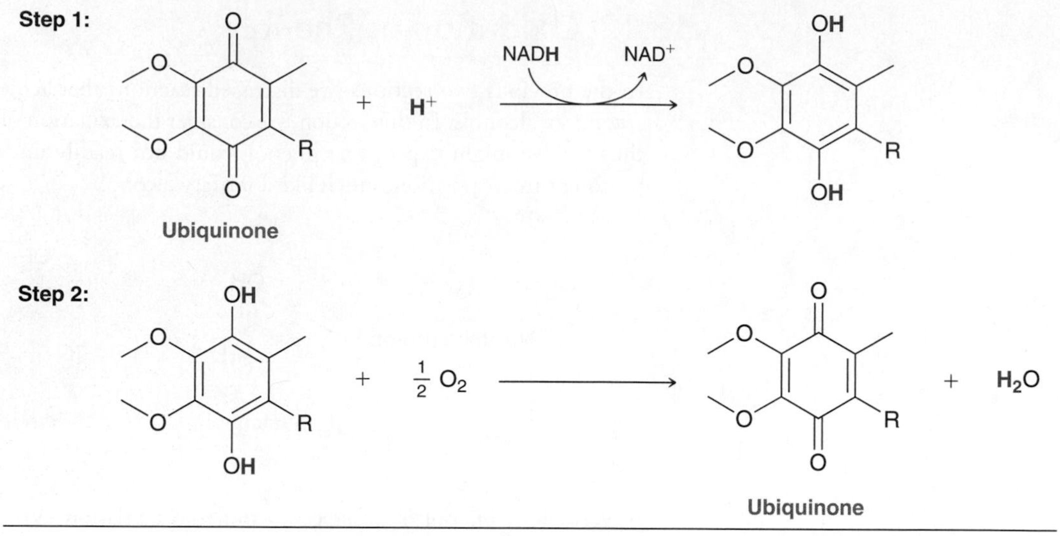

Net reaction: $NADH + \frac{1}{2}O_2 + H^+ \longrightarrow NAD^+ + H_2O$

This process involves two steps. In the first step, a ubiquinone is reduced to a hydroquinone. In the second step the hydroquinone is oxidized to regenerate the ubiquinone. In this way, the ubiquinone is not consumed by the process. It is a catalyst for the conversion of molecular oxygen into water, a critical step in the breakdown of food molecules to release the energy stored in their chemical bonds. British biochemist Peter Mitchell was awarded the 1978 Nobel Prize in Chemistry for the discovery of the role that ubiquinones play in energy production (ATP synthesis).

13.13 Synthesis Strategies

Recall from Chapter 12 that there are two issues to consider when proposing a synthesis:

1. A change in the carbon skeleton
2. A change in the functional group

In this chapter, we have learned valuable skills for dealing with both of these issues. Let's focus on them one at a time.

Functional Group Interconversion

In Chapter 12, we saw how to interconvert triple bonds, double bonds, and single bonds (Figure 13.9).

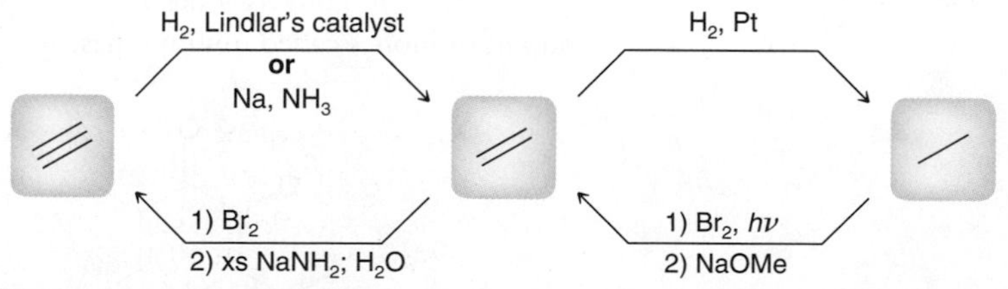

FIGURE 13.9
A map showing the reactions that enable the interconversion between alkanes, alkenes, and alkynes.

In this chapter, we learned how to interconvert between ketones and secondary alcohols. This level of control will be very important later, as we will see many reactions involving compounds containing carbonyl groups.

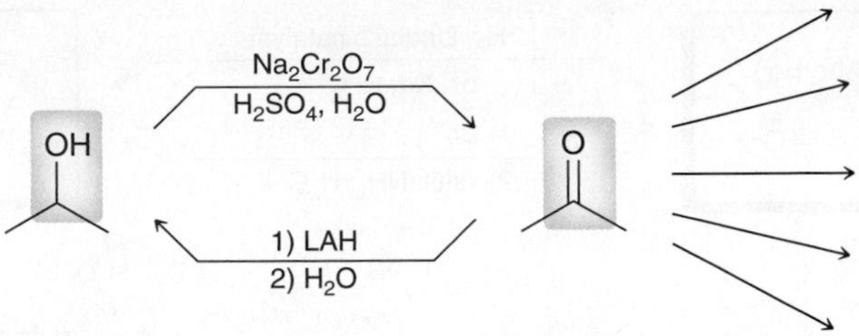

Primary alcohols and aldehydes can also be interconverted. PCC is used in place of chromic acid to convert an alcohol into an aldehyde.

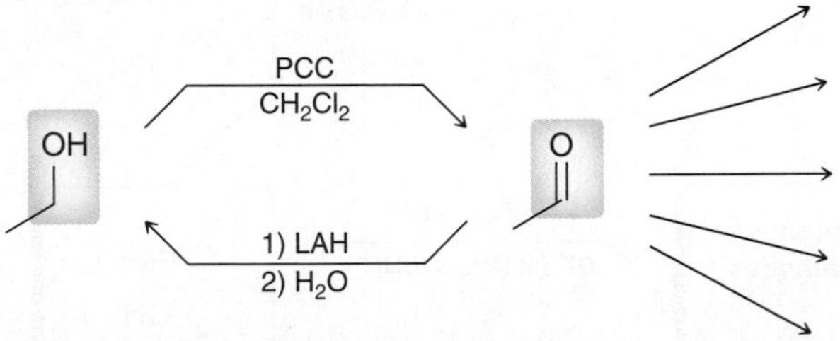

Let's now combine some important reactions we have seen in the previous few chapters into one map that will help us organize reactions that interconvert functional groups. We will focus on the six functional groups highlighted in Figure 13.10. These six functional groups have been separated into categories based on oxidation state. Conversion of a functional group from one category to another category is either a reduction or an oxidation. For example, converting an alkanc into an alkyl halide is an oxidation reaction. In contrast, interconverting functional groups within a category is neither oxidation nor reduction. For example, conversion of an alcohol into an alkyl halide does not involve a change in oxidation state.

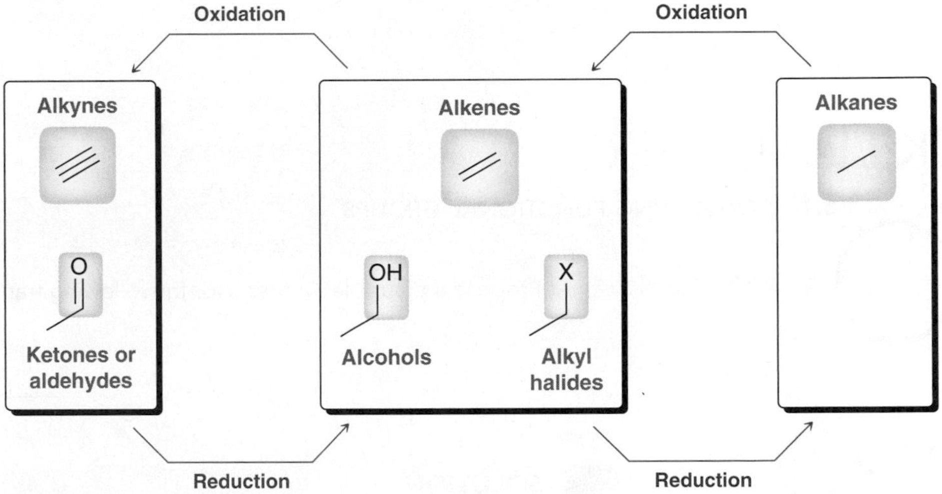

FIGURE 13.10
A diagram showing six different functional groups and their relative oxidation states. The interconversion between alkenes, alcohols, and alkyl halides does not constitute oxidation or reduction.

We have seen many reagents that allow us to interconvert functional groups, both within a category and across categories (oxidation/reduction). Those reagents are shown in Figure 13.11. It is worth your time to study Figure 13.11 carefully, as it summarizes many of the key reactions that we have seen until now. Note that each functional group can be converted into any one of the other functional groups. Some conversions can be done in one step, such as the conversion of an alcohol into a ketone. Other conversions require multiple steps, such as the conversion of an alkane into a ketone. Having a clear mental picture of this chart will be extremely helpful when solving synthesis problems. Let's see an example.

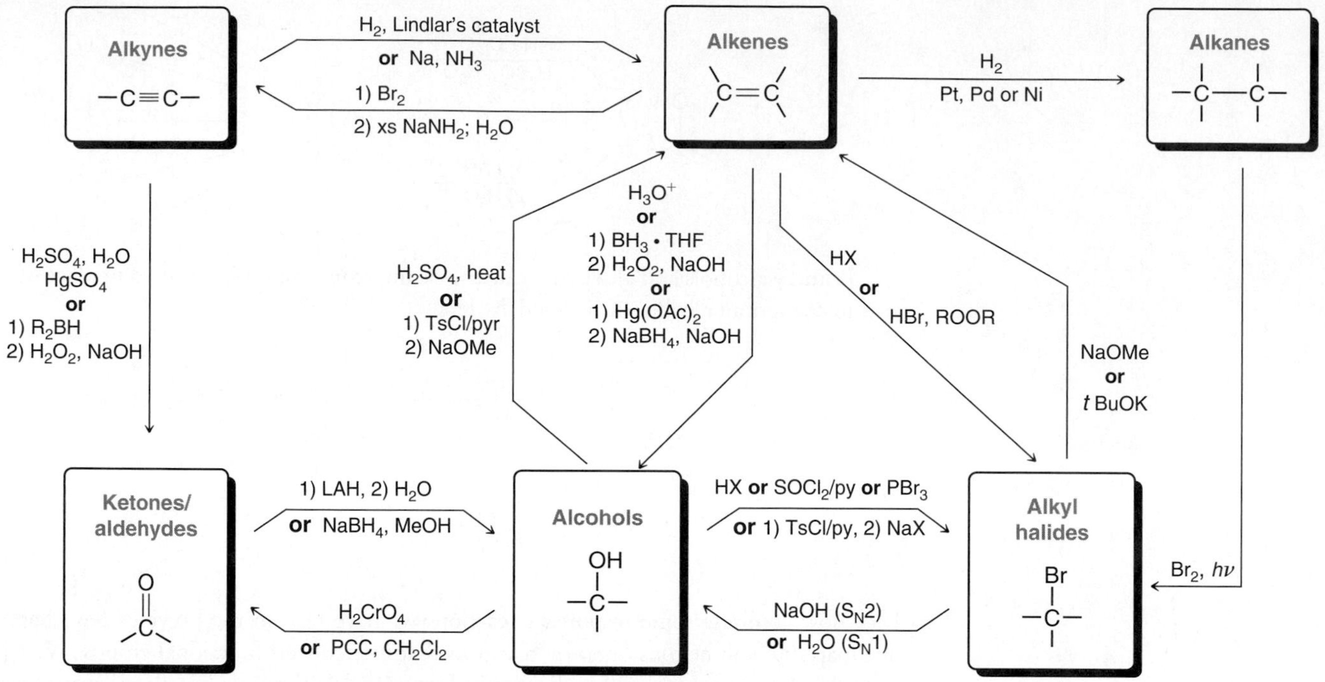

FIGURE 13.11
A map of reactions for interconversion between alkanes, alkenes, alkynes, alkyl halides, alcohols, and ketones/aldehydes.

SKILLBUILDER

13.8 CONVERTING FUNCTIONAL GROUPS

LEARN the skill

Propose a plausible synthesis for the following transformation.

SOLUTION

In this example, the reactant and product have the same carbon skeleton and differ only in the identity of the functional group. The reactant has a carbon-carbon double bond, while the product has a carbon-oxygen double bond (a carbonyl group). It is always preferable to achieve a transformation in the fewest number of steps possible, so we first consider whether there is a single step that will achieve this transformation. As seen in Figure 13.11, we have not yet discussed a one-step method for achieving the desired transformation. Therefore, we must consider whether the reactant can be converted into the product using a two-step synthesis. Figure 13.11 shows two possible routes.

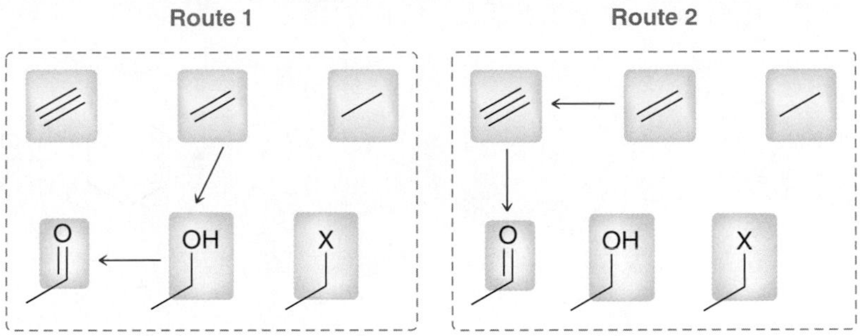

The first route passes through the alcohol, and the second route passes through an alkyne. Therefore, there are (at least) two perfectly acceptable answers to this problem.

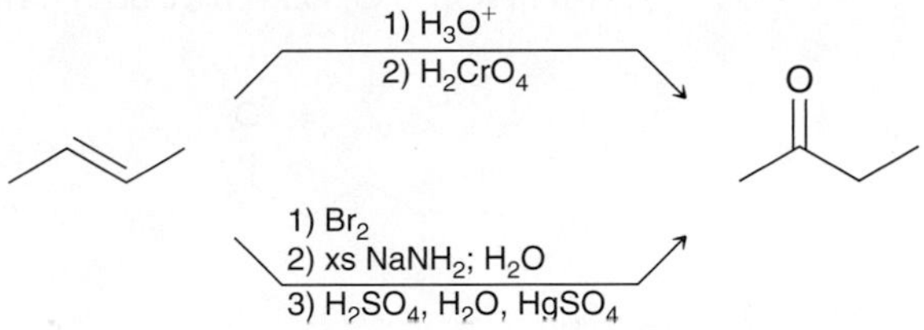

PRACTICE the skill **13.24** Propose a plausible synthesis for each of the following transformations.

(a) 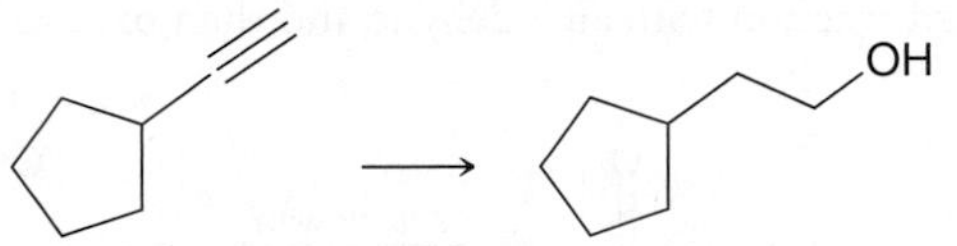

(b) 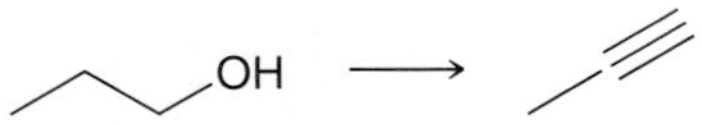

(c) 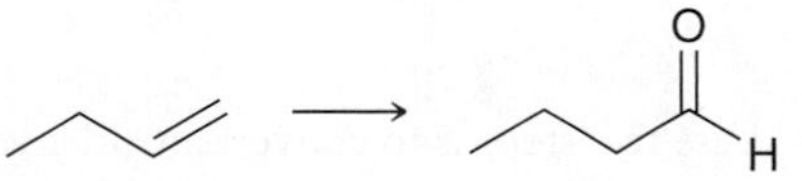

(d) 

(e) 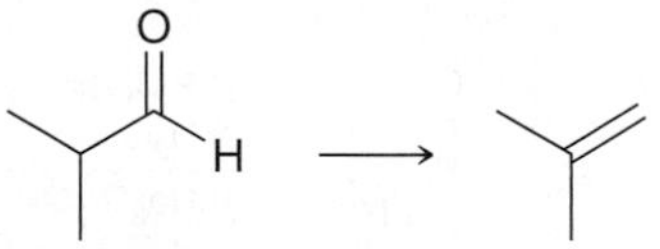

(f) 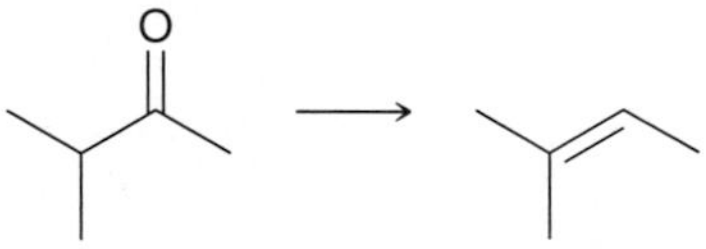

APPLY the skill

13.25 Show at least two different methods for preparing 1-methylcyclohexene from 1-methylcyclohexanol.

13.26 Using any reagents of your choosing, show how you would convert *tert*-butyl alcohol into 2-methyl-1-propanol

need more **PRACTICE?** **Try Problems 13.35, 13.39, 13.48, 13.51**

C—C Bond Formation

In this chapter, we learned a new way to form C—C bonds using the reaction between a Grignard reagent and a ketone or aldehyde.

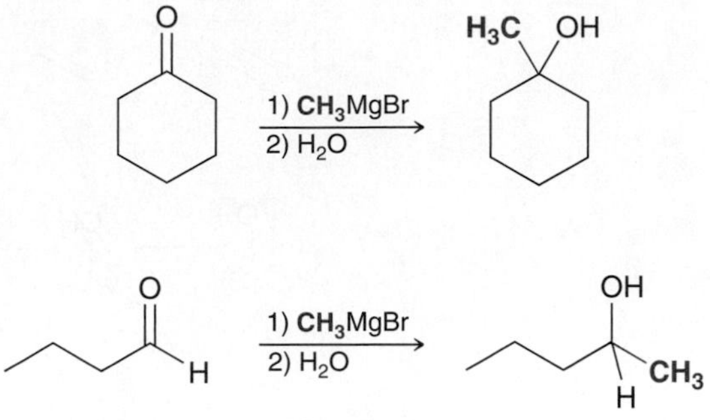

We also saw that a Grignard reagent will attack an ester twice to form an alcohol. In this reaction, two new C—C bonds have been formed.

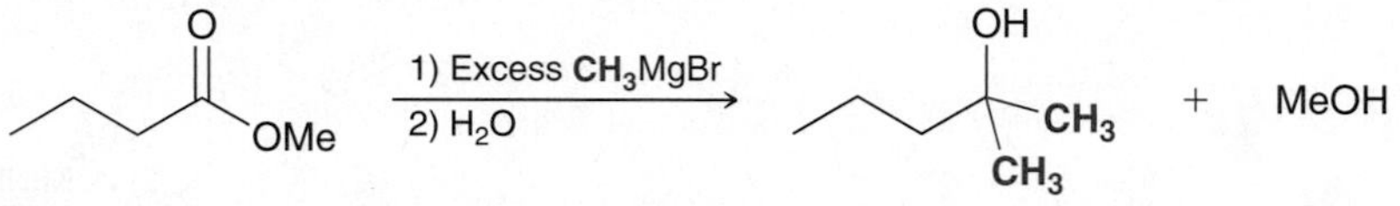

A Grignard reaction accomplishes more than just C—C bond formation; it also reduces the carbonyl group to an alcohol in the process. Now consider the following two steps: Suppose we perform a Grignard reaction with an aldehyde and then oxidize back up to a ketone.

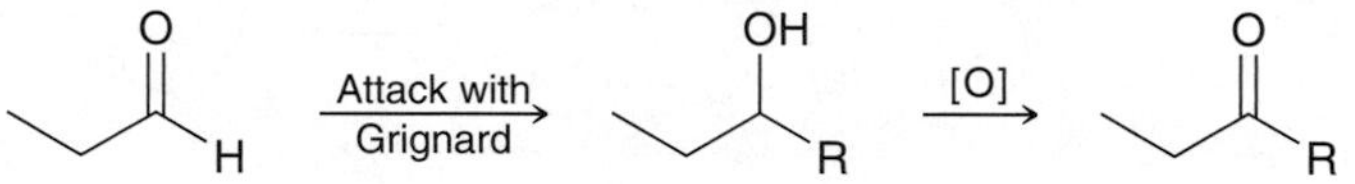

The net result of these two steps is to convert an aldehyde into a ketone. This is very helpful, because we have not yet seen a direct way to convert aldehydes into ketones.

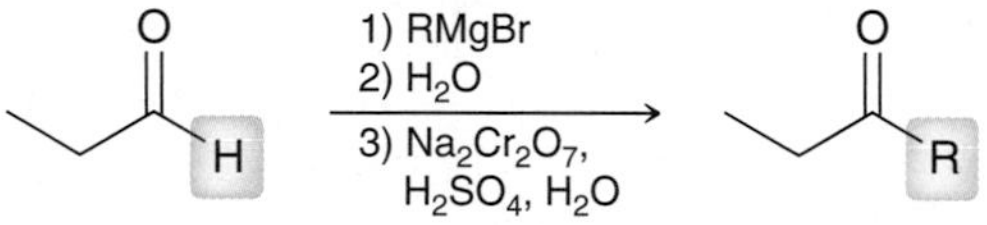

CONCEPTUAL CHECKPOINT

13.27 Identify the reagents that you would need to accomplish each of the following transformations.

(a)

(b)
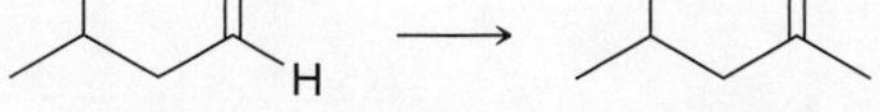

Functional Group Transformations and C—C Bond Formation

In this final section of the chapter, we will combine the skills from the previous two sections and work on synthesis problems that involve functional group interconversion and C—C bond formation. Recall from Chapter 12 that it is very helpful to approach a synthesis problem working forward as well as working backward (retrosynthetic analysis). We will use all of these skills in the following example.

SKILLBUILDER

13.9 PROPOSING A SYNTHESIS

LEARN the skill

Propose a plausible synthesis for the following transformation.

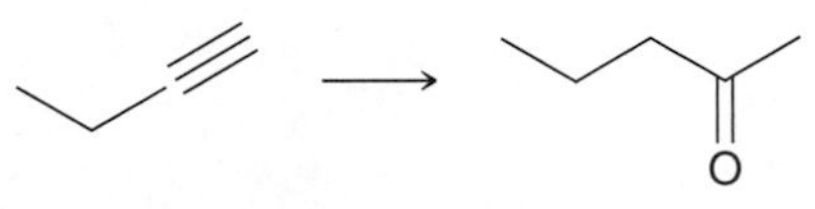

SOLUTION

Always approach a synthesis problem by initially asking two questions.

1. Is there a change in the carbon skeleton? Yes, the carbon skeleton is increasing in size by one carbon atom.

2. Is there a change in the functional groups? Yes, the starting material has a triple bond, and the product has a carbonyl group.

To introduce one carbon atom, we will need to use a reaction that forms a C—C bond. So far, we have learned to form a C—C bond by alkylation of an alkyne (Section 10.10) and by using a Grignard reagent to attack a carbonyl group. Since our starting material is a terminal alkyne, let's first try alkylating the alkyne and see where that takes us.

Our first proposed strategy is to alkylate the alkyne and then hydrate the triple bond.

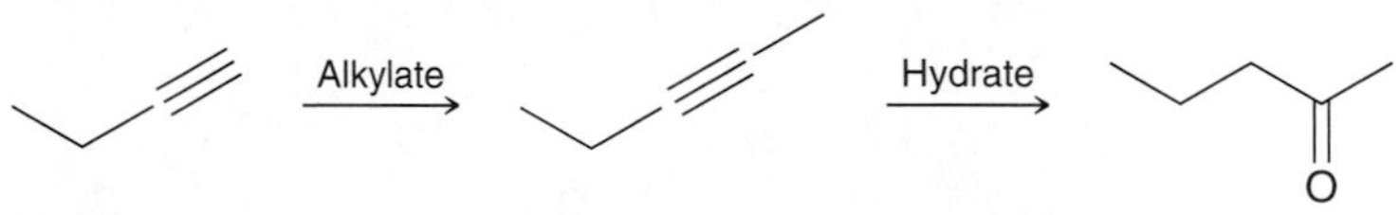

It is always critical to analyze any proposed strategy and make sure that the regiochemistry and stereochemistry of each step are correct. This particular strategy does not have any issues of stereochemistry (there are no chirality centers or stereoisomeric alkenes). However, the regiochemistry is problematic. There is a serious flaw with the second step (hydration of the alkyne to give a ketone). Specifically, it is not possible to control the regiochemical outcome of that reaction. That step will likely produce a mixture of two possible ketones (2-pentanone and 3-pentanone). This represents a critical flaw. We cannot propose a synthesis in which we have no control over the regiochemistry of one of the steps. All of our steps should lead to the desired product as the major product.

In the failed strategy above, we attempted to form the C—C bond via alkylation of an alkyne. So, let's now focus on our second method of forming a C—C bond, using a Grignard reagent to attack a carbonyl group. In other words, the last two steps of our synthesis could look like this.

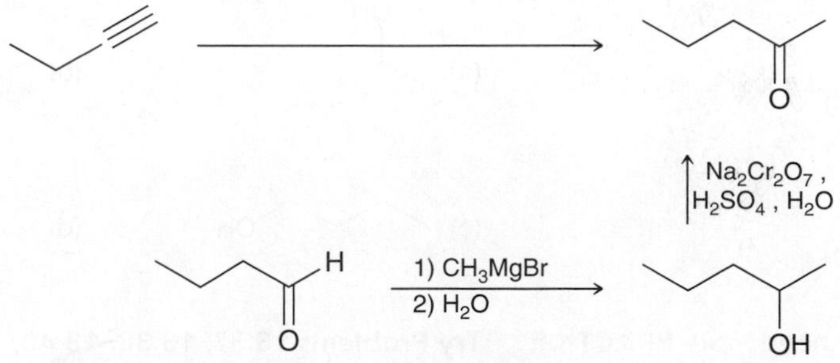

We are now using a retrosynthetic approach to this problem, after the forward approach proved to be unfruitful. Specifically, we have recognized that the final product can be made from an aldehyde via a two-step process. Now we just need to bridge the gap between the starting material and the aldehyde, which can be accomplished via hydroboration-oxidation.

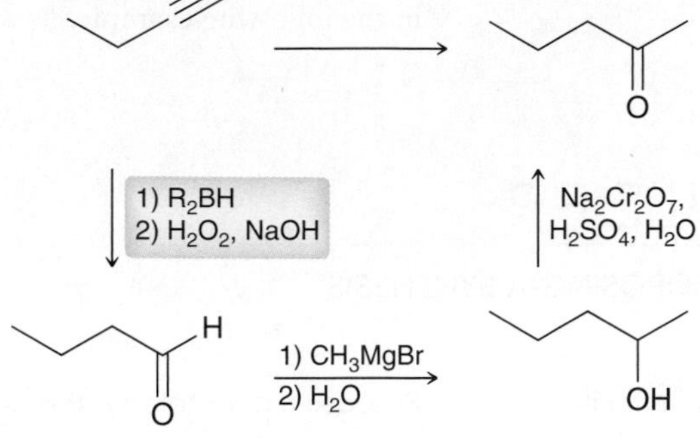

This proposed strategy exhibits the correct regiochemistry for each step. Our answer is:

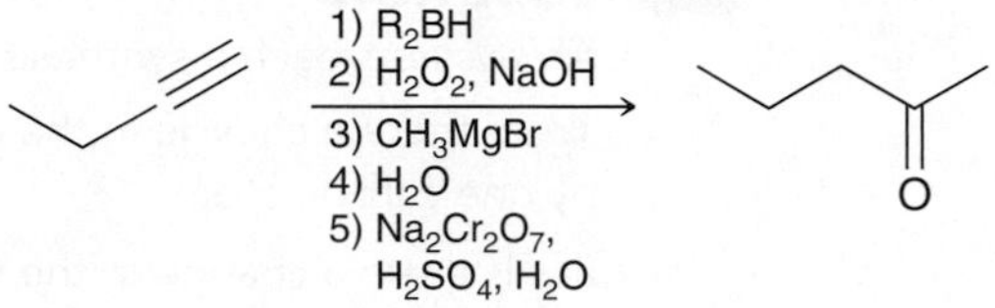

PRACTICE the skill **13.28** Propose a plausible synthesis for each of the following transformations.

(a) 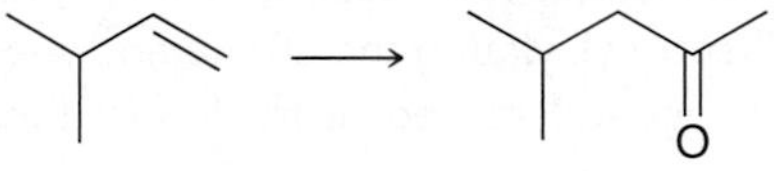

(b) 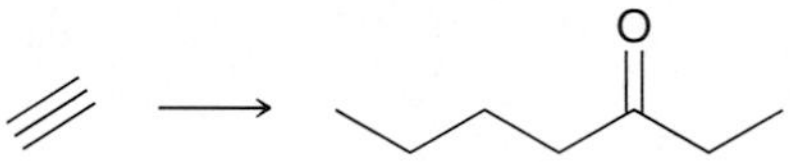

(c) 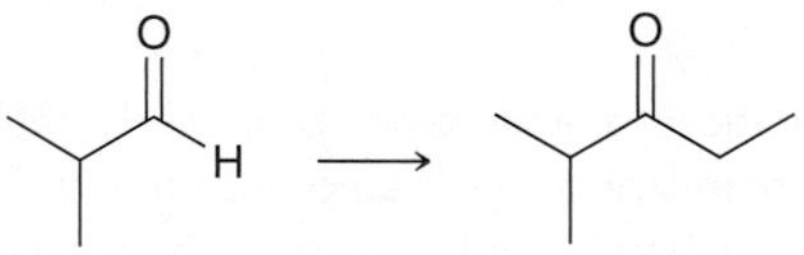

(d) 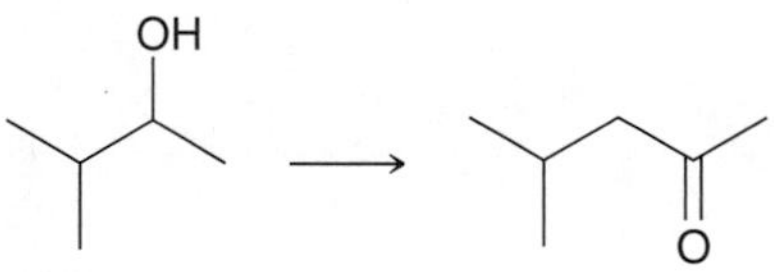

(e) 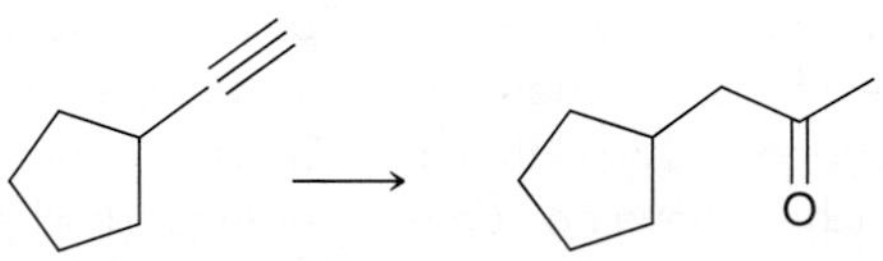

(f) 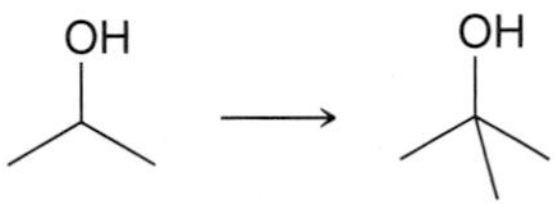

APPLY the skill **13.29** Using any compounds that have no more than *two* carbon atoms, identify a method for preparing each of the following compounds:

(a) 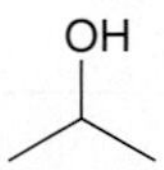

(b) 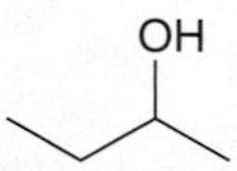

(c) 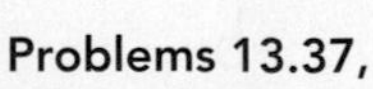

(d) 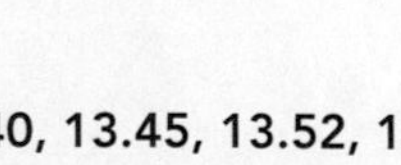

need more **PRACTICE?** **Try Problems 13.37, 13.38, 13.40, 13.45, 13.52, 13.59**

REVIEW OF REACTIONS

PREPARATION OF ALKOXIDES

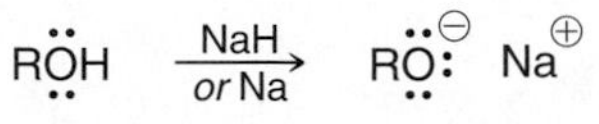

$$\text{R}\overset{..}{\underset{..}{\text{O}}}\text{H} \xrightarrow[\text{or Na}]{\text{NaH}} \text{R}\overset{..}{\underset{..}{\text{O}}}{:}^{\ominus} \ \text{Na}^{\oplus}$$

PREPARATION OF ALCOHOLS VIA REDUCTION

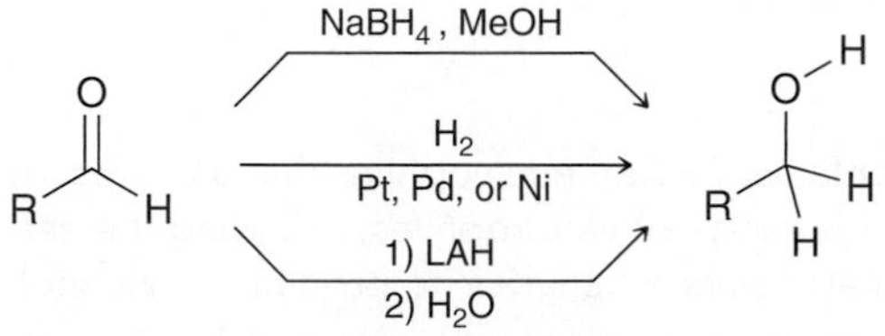

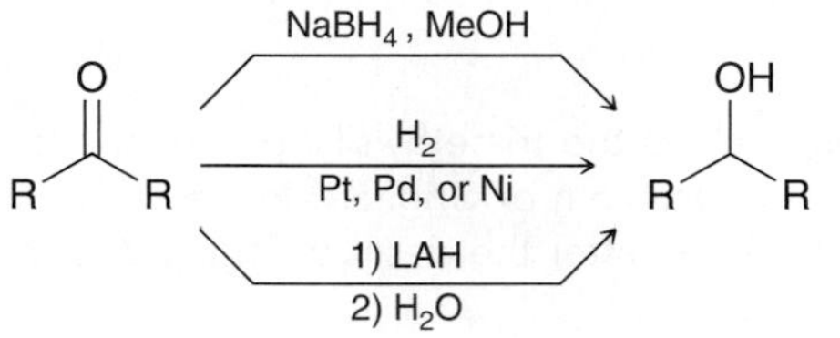

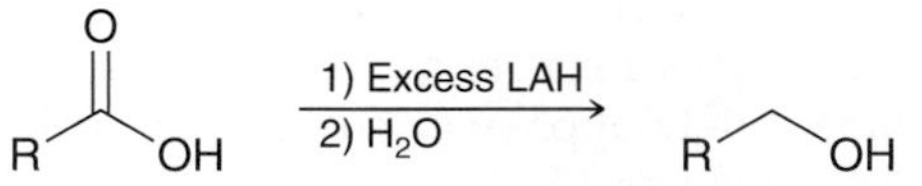

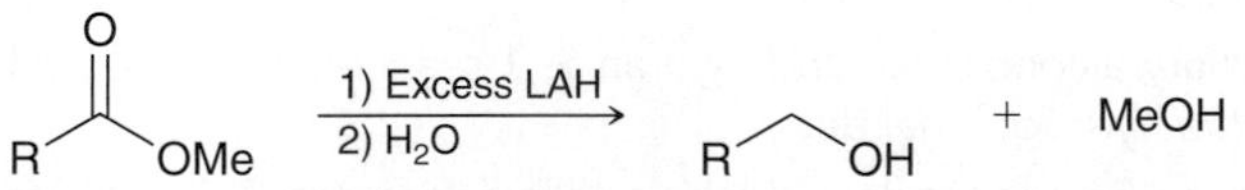

PREPARATION OF ALCOHOLS VIA GRIGNARD REAGENTS

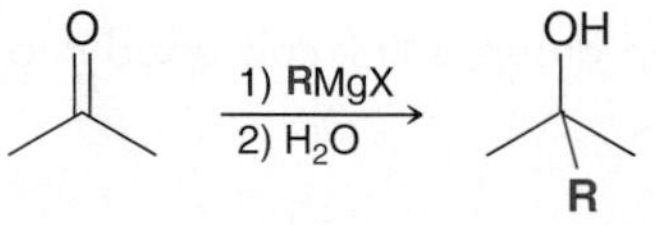

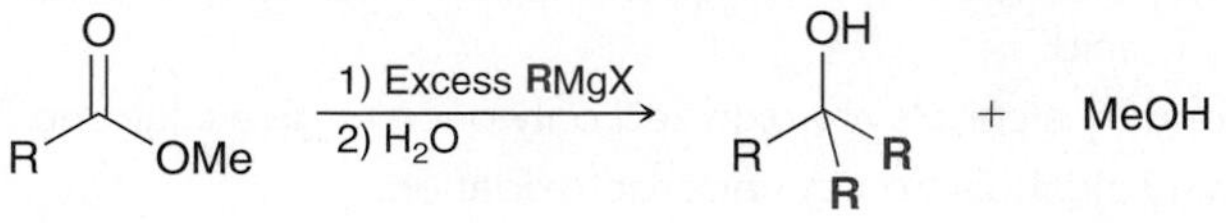

PROTECTION AND DEPROTECTION OF ALCOHOLS

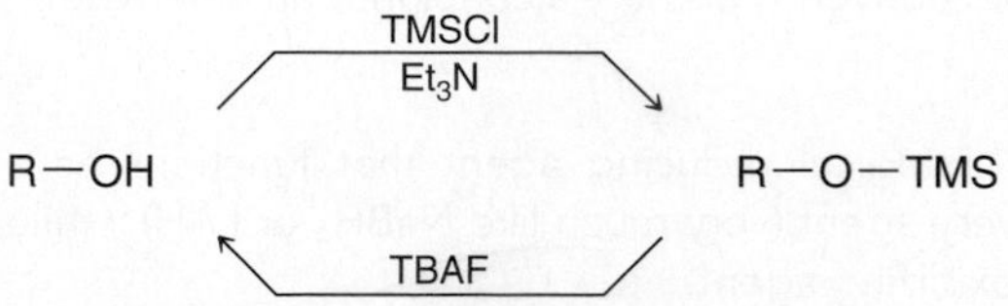

S$_N$1 REACTIONS WITH ALCOHOLS

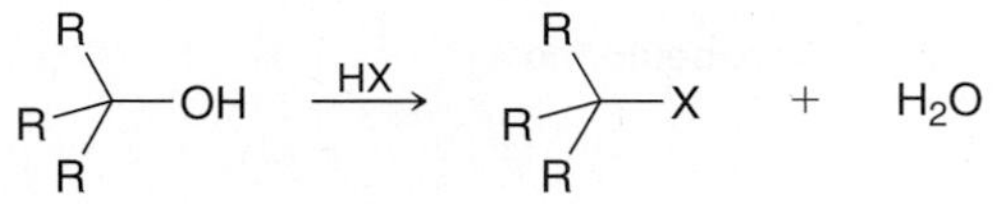

S$_N$2 REACTIONS WITH ALCOHOLS

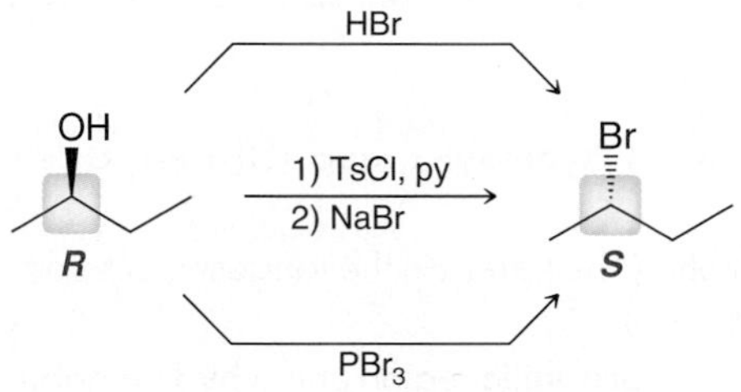

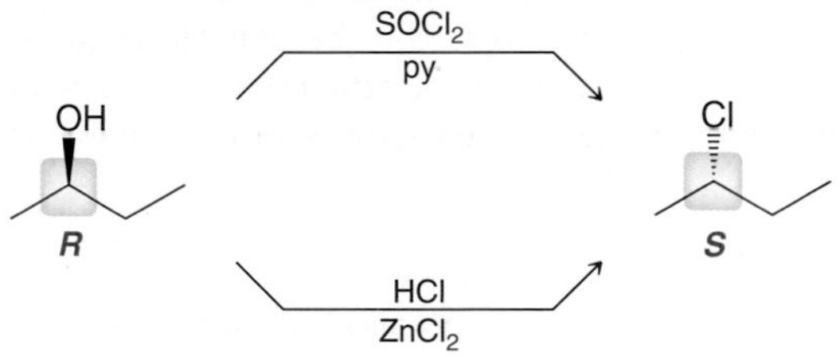

E1 AND E2 REACTIONS WITH ALCOHOLS

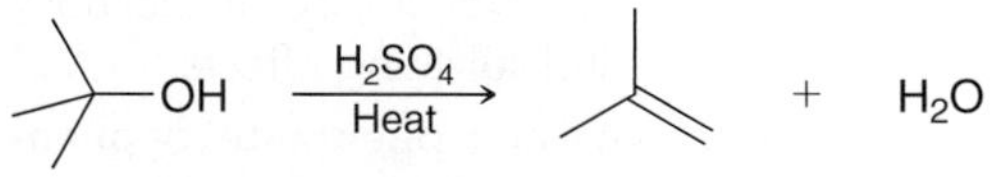

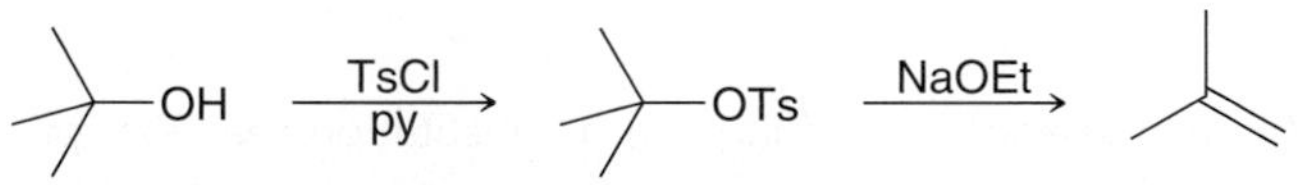

OXIDATION OF ALCOHOLS AND PHENOLS

With Chromic Acid

Secondary alcohol → Ketone

Primary alcohol → Carboxylic acid

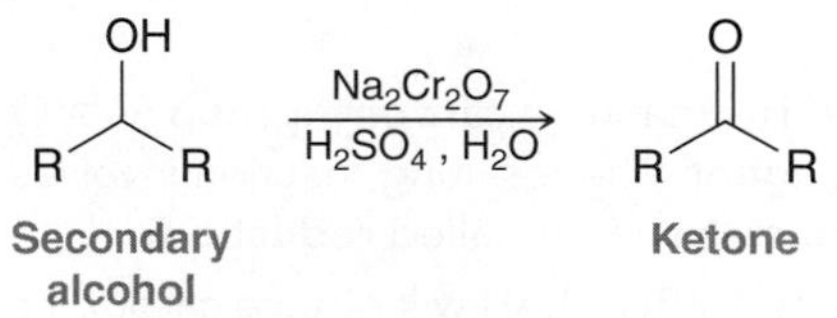

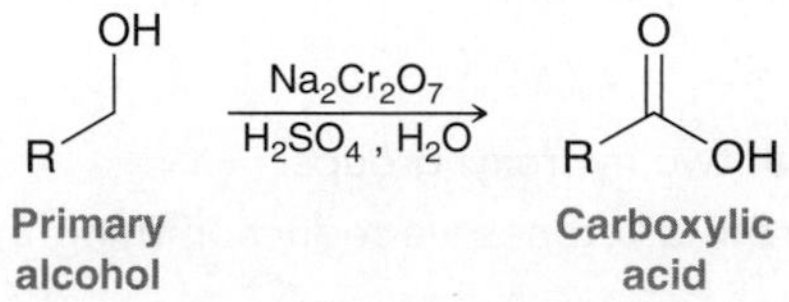

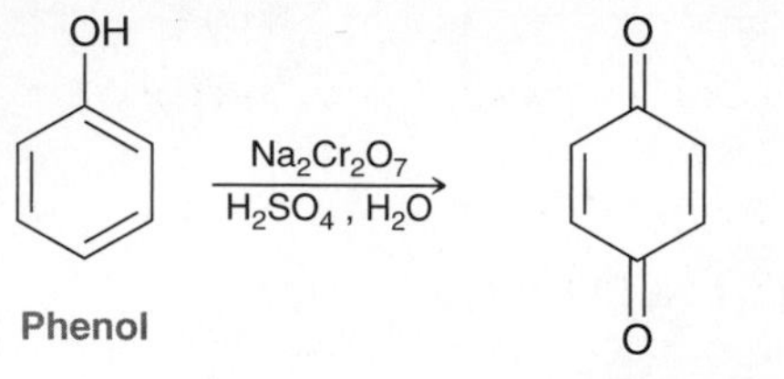

With PCC

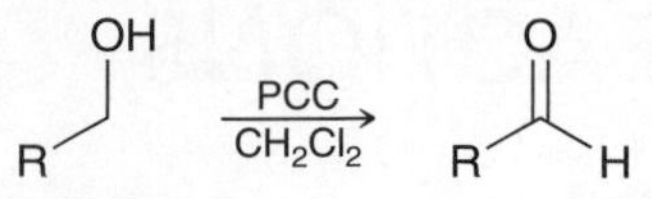

REVIEW OF CONCEPTS AND VOCABULARY

SECTION 13.1

- Compounds that have a **hydroxyl group** (OH) are called **alcohols**.
- When naming an alcohol, the parent is the longest chain containing the hydroxyl group.
- All alcohols possess a **hydrophilic** region and **a hydrophobic** region. Small alcohols (methanol, ethanol, propanol) are **miscible** with water. A substance is said to be **soluble** in water when only a certain volume of the substance will dissolve in a specified amount of water at room temperature. Butanol is soluble in water.

SECTION 13.2

- The conjugate base of an alcohol is called an **alkoxide** ion.
- The pK_a for most alcohols falls in the range of 15–18.
- Alcohols are commonly deprotonated with either sodium hydride (NaH) or an alkali metal (Na, Li, or K).
- Several factors determine the relative acidity of alcohols, including resonance, induction, and solvating effects.
- The conjugate base of phenol is called a **phenolate**, or **phenoxide** ion.

SECTION 13.3

- When preparing an alcohol via a substitution reaction, primary substrates will require S_N2 conditions, while tertiary substrates will require S_N1 conditions.
- Addition reactions that will produce alcohols include acid-catalyzed hydration, oxymercuration-demercuration, and hydroboration-oxidation.

SECTION 13.4

- Alcohols can be formed by treating a **carbonyl group** (C=O bond) with a **reducing agent**. The resulting reaction involves a decrease in **oxidation state** and is called **reduction.**
- LAH is more reactive than $NaBH_4$. LAH will reduce carboxylic acids and esters, while $NaBH_4$ will not.

SECTION 13.5

- **Diols** are compounds with two hydroxyl groups.
- Diols can be prepared from diketones via reduction using a reducing agent.
- Diols can also be made via *syn* dihydroxylation or *anti* dihydroxylation of an alkene.

SECTION 13.6

- **Grignard reagents** are carbon nucleophiles that are capable of attacking a wide range of electrophiles, including the carbonyl group of ketones or aldehydes, to produce an alcohol.
- Grignard reagents also react with esters to produce alcohols with introduction of two R groups.

SECTION 13.7

- **Protecting groups**, such as the trimethylsilyl group, can be used to circumvent the problem of Grignard incompatibility and can be easily removed after the desired Grignard reaction has been performed.

SECTION 13.8

- Phenol, also called hydroxybenzene, is used as a precursor in the synthesis of a wide variety of pharmaceuticals and other commercially useful compounds.

SECTION 13.9

- Tertiary alcohols will undergo an S_N1 reaction when treated with a hydrogen halide.
- Primary and secondary alcohols will undergo an S_N2 process when treated with HX, $SOCl_2$, or PBr_3 or when the hydroxyl group is converted into a tosylate group followed by nucleophilic attack.
- Tertiary alcohols undergo E1 elimination when treated with sulfuric acid.
- For an E2 process, the hydroxyl group is first converted into a tosylate or an alkyl halide.

SECTION 13.10

- Primary alcohols can undergo **oxidation** twice to give a carboxylic acid.
- Secondary alcohols are oxidized only once to give a ketone
- Tertiary alcohols do not undergo oxidation.
- The most common oxidizing reagent is chromic acid (H_2CrO_4), which can be formed either from chromium trioxide (CrO_3) or from sodium dichromate ($Na_2Cr_2O_7$) in aqueous acidic solution.
- PCC is used to convert a primary alcohol into an aldehyde.

SECTION 13.11

- NADH is a biological reducing agent that functions as a hydride delivery agent (very much like $NaBH_4$ or LAH), while NAD^+ is an oxidizing agent.

- NADH and NAD$^+$ play critical roles in biological systems. Examples include the **citric acid cycle** and **ATP** synthesis.

SECTION 13.12

- Phenols undergo oxidation to quinones. Quinones are biologically important because their redox properties play a significant role in **cellular respiration**.

SECTION 13.13

- There are two key issues to consider when proposing a synthesis:
  1. A change in the carbon skeleton
  2. A change in the functional group

KEY TERMINOLOGY

alcohols 565
alkoxide 571
carbonyl group 578
cellular respiration 603
diol 583
Grignard reagent 584
hydrophilic 569
hydrophobic 569
hydroxyl group 565
miscible 569
oxidation 576
oxidation state 575
phenolate 572
phenoxide 572
protecting group 588
reducing agents 578
reduction 576
soluble 570

SKILLBUILDER REVIEW

13.1 NAMING AN ALCOHOL

STEP 1 Choose the longest chain containing the OH group, and number the chain starting from the end closest to the OH group.

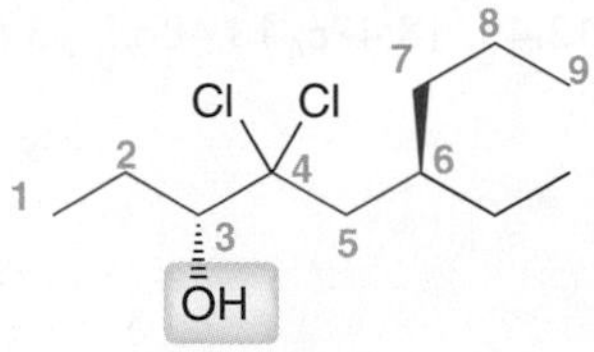

3-Nonanol

STEPS 2 AND 3 Identify the substituents and assign locants.

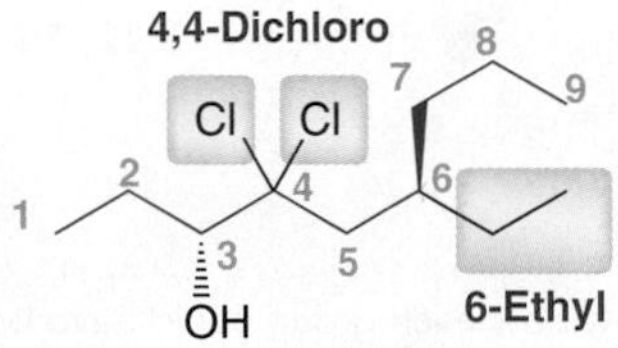

STEP 4 Assemble the substituents alphabetically.

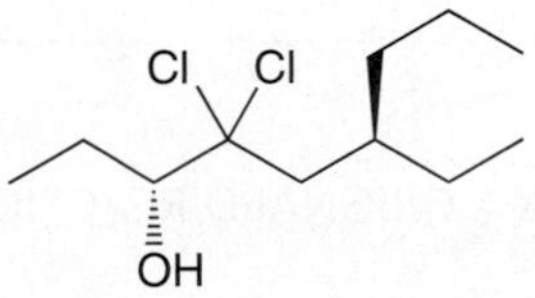

4,4-Dichloro - 6-ethyl-3-nonanol

STEP 5 Assign the configuration of any chirality center.

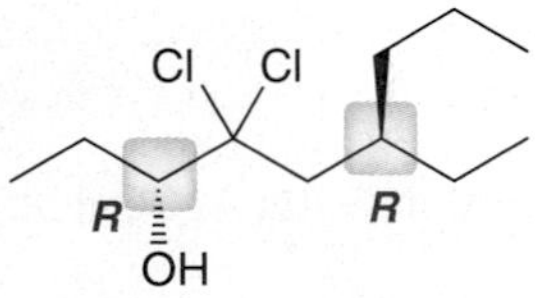

(3*R*,6*R*)-4,4-dichloro- 6-ethyl-3-nonanol

Try Problems 13.1, 13.2, 13.30, 13.31a-d,f, 13.32

13.2 COMPARING THE ACIDITY OF ALCOHOLS

Look for **resonance effects**; for example:

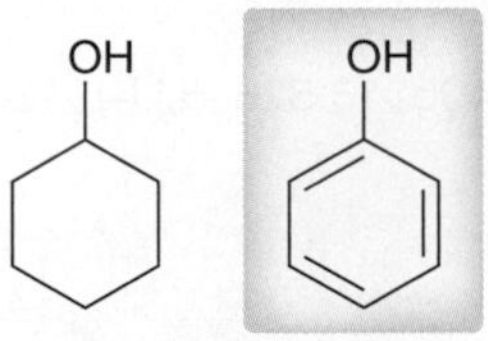

More acidic

Look for **inductive effects**; for example:

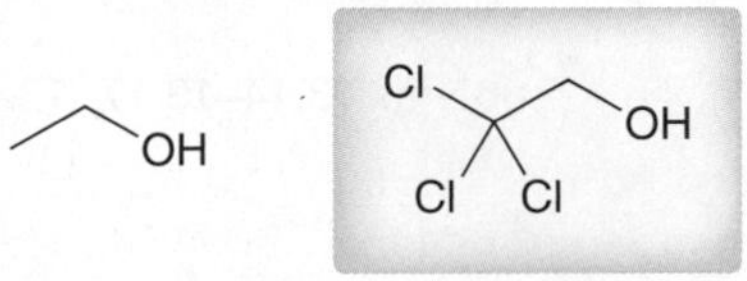

More acidic

Look for **solvating effects**; for example:

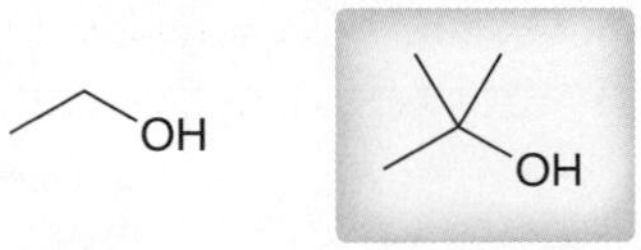

Less acidic

Try Problems 13.5, 13.6, 13.33, 13.34

13.3 IDENTIFYING OXIDATION AND REDUCTION REACTIONS

EXAMPLE Determine whether starting material has been oxidized, reduced, or neither.

STEP 1 Determine oxidation state of starting material. Break all bonds heterolytically, except for C—C bonds.

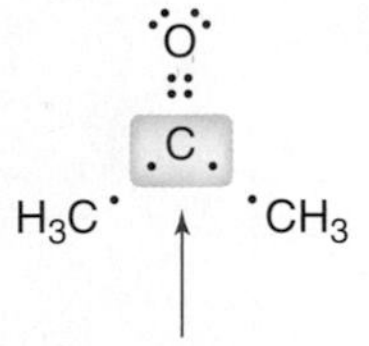

Two electrons, but carbon should have four. This carbon is missing two electrons

Oxidation state = +2

STEP 2 Determine oxidation state of product. Break all bonds heterolytically, except for C—C bonds.

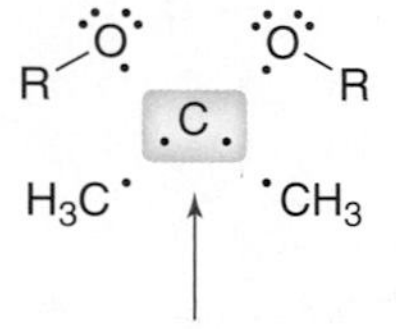

Two electrons, but carbon should have four. This carbon is missing two electrons

Oxidation state = +2

STEP 3 Determine if there has been a change in oxidation state.

Increase = oxidation
Decrease = reduction
No change = neither

+2 ⟶ **+2**

This example is neither an oxidation nor a reduction

Try Problems **13.9–13.11, 13.62**

13.4 DRAWING A MECHANISM AND PREDICTING THE PRODUCTS OF HYDRIDE REDUCTIONS

STEP 1 Draw the complete structure of LAH,and draw two curved arrows that show the delivery of hydride to the carbonyl group.

STEP 2 Draw the alkoxide intermediate.

STEP 3 Draw two curved arrows showing the alkoxide intermediate being protonated by the proton source.

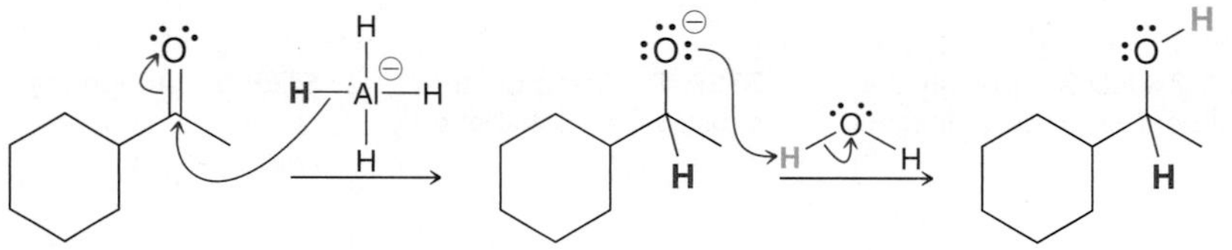

Try Problems **13.12, 13.13, 13.46, 13.47c, 13.48e,f, 13.60**

13.5 PREPARING AN ALCOHOL VIA A GRIGNARD REACTION

STEP 1 Identify the alpha position.

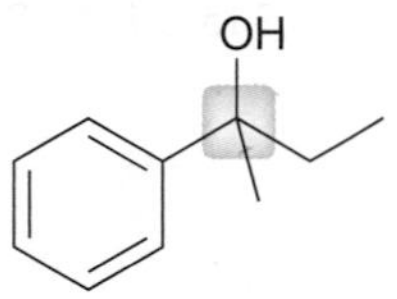

STEP 2 Identify the three groups connected to the alpha position.

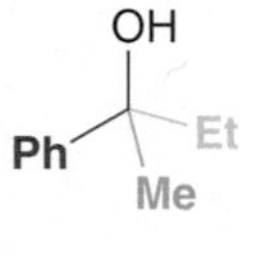

STEP 3 Show how each group could have been installed via a Grignard reaction.

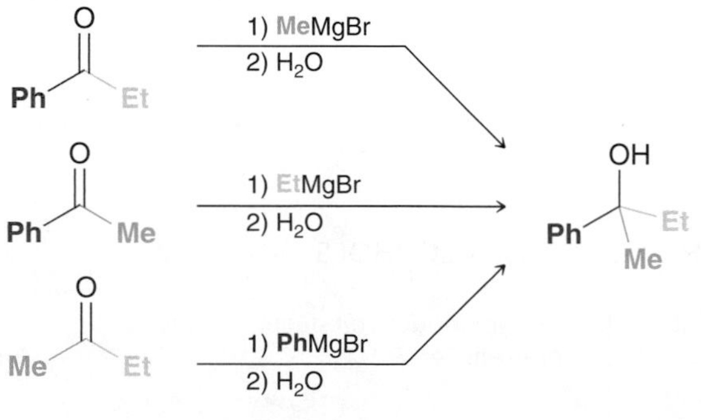

Try Problems **13.14–13.17, 13.38, 13.40b, 13.52b–d,j,l–r, 13.58**

13.6 PROPOSING REAGENTS FOR THE CONVERSION OF AN ALCOHOL INTO AN ALKYL HALIDE

EXAMPLE Identify the necessary reagents.

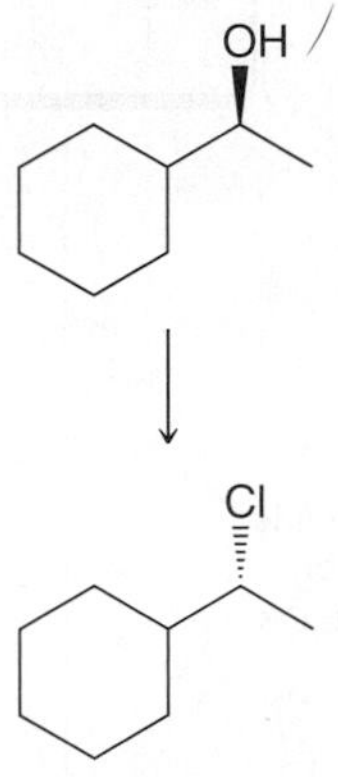

STEP 1 Analyze the substrate:

Primary = S_N2
Tertiary = S_N1

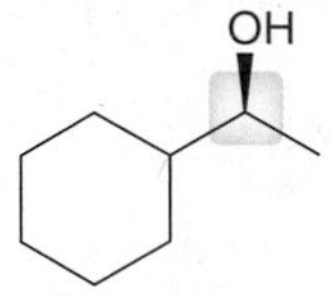

Substrate is secondary

STEP 2 Analyze the stereochemistry: inversion = S_N2.

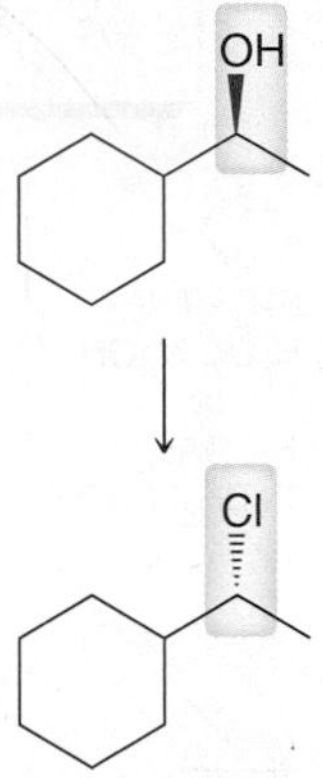

STEP 3 Reaction must occur via S_N2 so use reagents that favor S_N2.

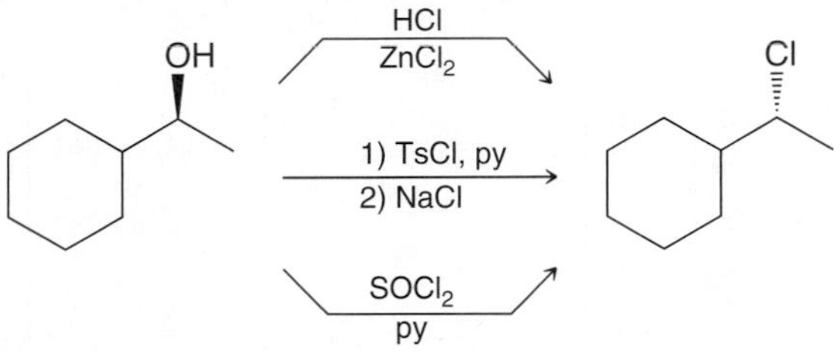

Try Problems 13.19, 13.20, 13.35a–c, 13.44f, 13.52r

13.7 PREDICTING THE PRODUCTS OF AN OXIDATION REACTION

EXAMPLE

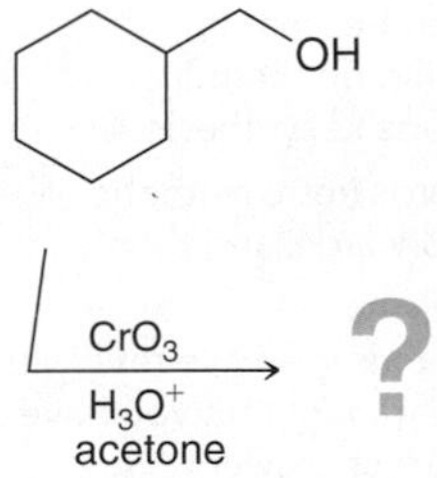

STEP 1 Identify whether the alcohol is primary or secondary.

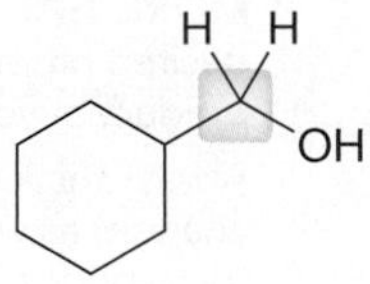

Primary

STEP 2 A primary alcohol can be oxidized either to an aldehyde or to a carboxylic acid, depending on the reagents.

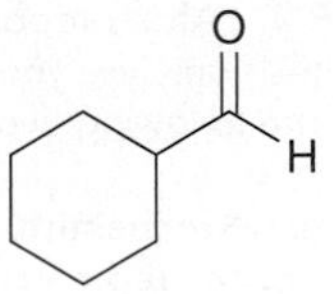

Aldehyde

PCC

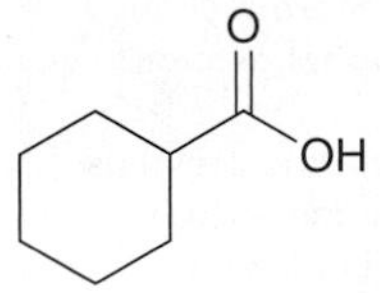

Carboxylic acid

Chromic acid

STEP 3 Analyze reagents. PCC is used to form the aldehyde. Chromic acid is used to form the carboxylic acid.

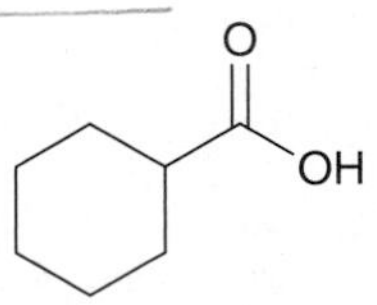

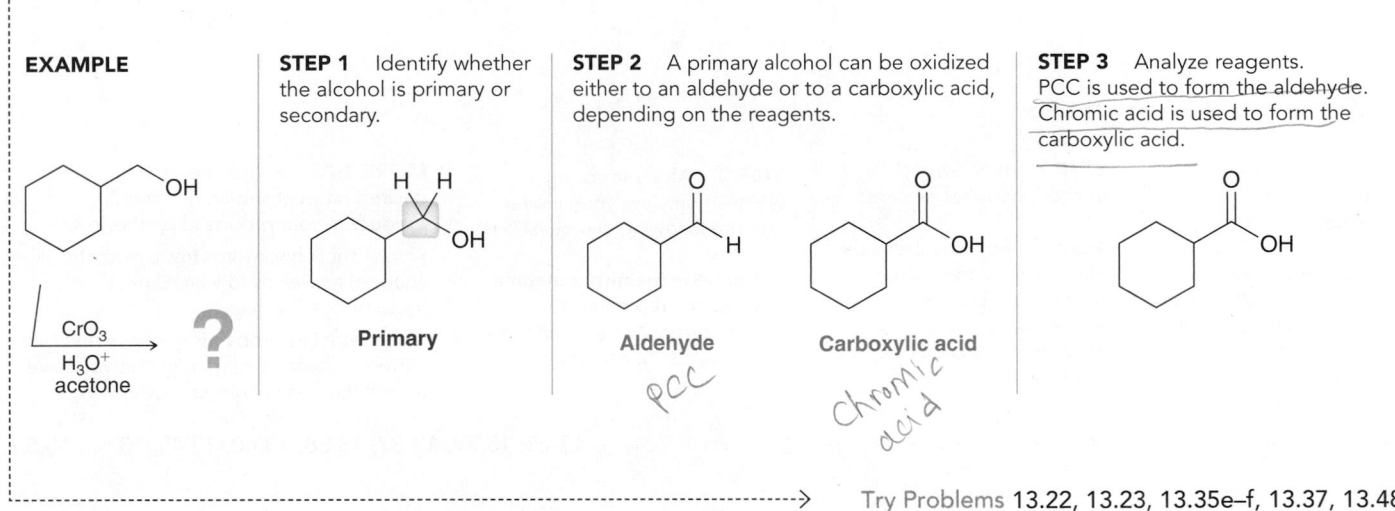

Try Problems 13.22, 13.23, 13.35e–f, 13.37, 13.48

13.8 CONVERTING FUNCTIONAL GROUPS

Functional groups can be interconverted using the following reagents, which should be committed to memory.

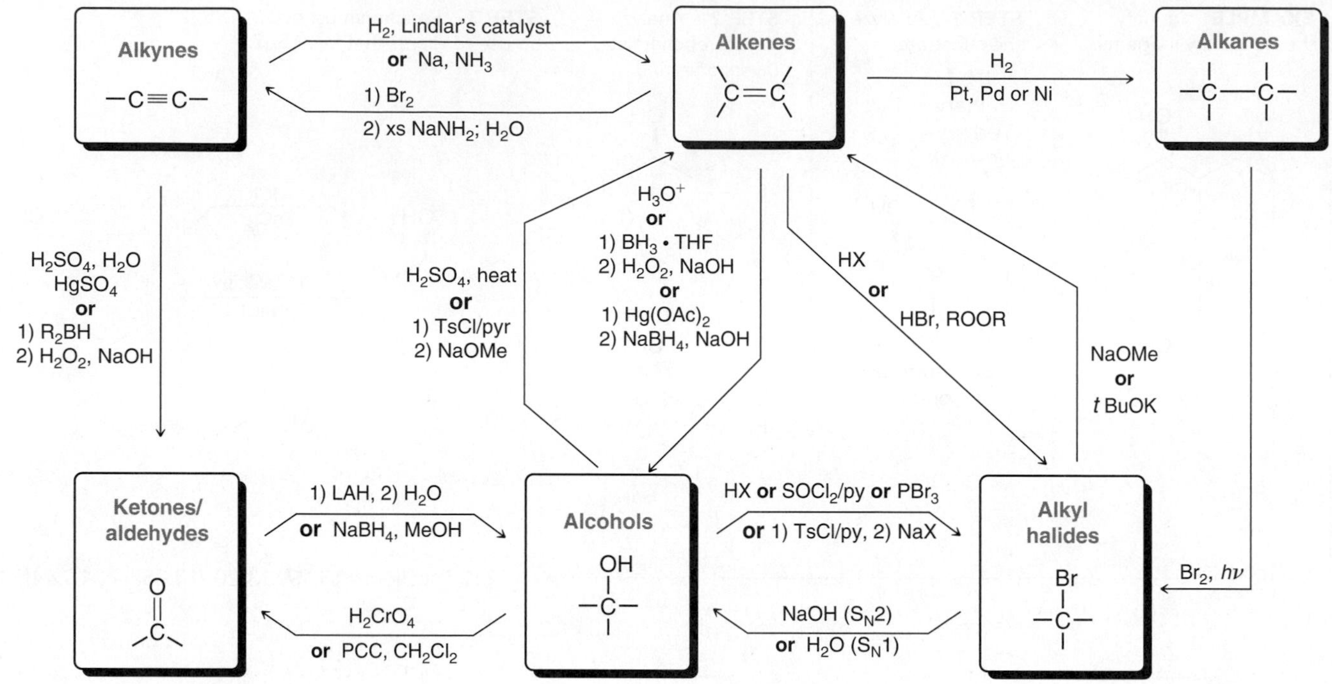

Try Problems 13.24–13.26, 13.39, 13.48, 13.51

13.9 PROPOSING A SYNTHESIS

STEP 1 **Is there a change in the carbon skeleton?**

Keep track of all the C—C bond-forming reactions that you have learned until now.

STEP 2 **Is there a change in the functional groups?**

The chart from the previous SkillBuilder summarizes many of the important functional group interconversions that we have seen.

STEP 3 After proposing a synthesis, analyze your answer with the following two questions:

- **Is the regiochemical outcome of each step correct?**
- **Is the stereochemical outcome of each step correct?**

MORE TIPS Remember that the desired product should be the major product of your proposed synthesis.

Always think backwards (retrosynthetic analysis) as well as forwards, and then try to bridge the gap.

Most synthesis problems will have multiple correct answers. Do not feel that you have to find the "one" correct answer.

Try Problems 13.28, 13.29, 13.37, 13.38, 13.40, 13.45, 13.52, 13.59

PRACTICE PROBLEMS

Note: Most of the Problems are available within *WileyPLUS*, an online teaching and learning solution.

13.30 Assign an IUPAC name for each of the following compounds:

(a) 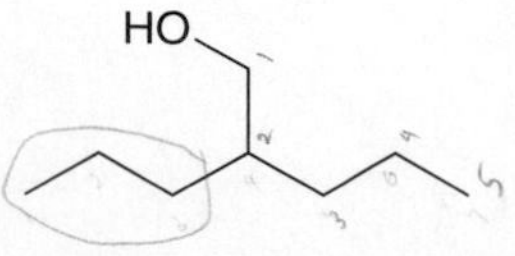

(b) 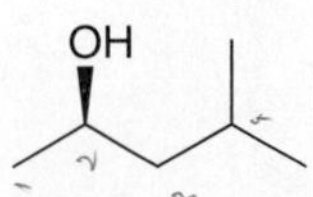

(c) 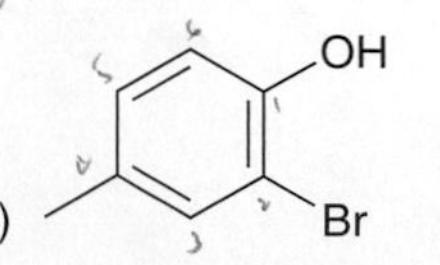

(d) 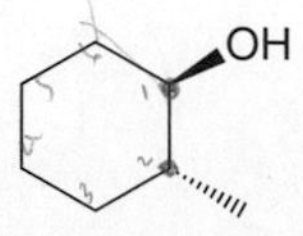

13.31 Draw the structure of each compound:

(a) *cis*-1,2-Cyclohexanediol (b) Isobutanol

(c) 2,4,6-Trinitrophenol
(d) (*R*)-2,2-Dimethyl-3-heptanol
(e) Ethylene glycol
(f) (*S*)-2-Methyl-1-butanol

13.32 Draw and name all constitutionally isomeric alcohols with molecular formula $C_4H_{10}O$.

13.33 Rank each set of alcohols below in order of increasing acidity.

(a) 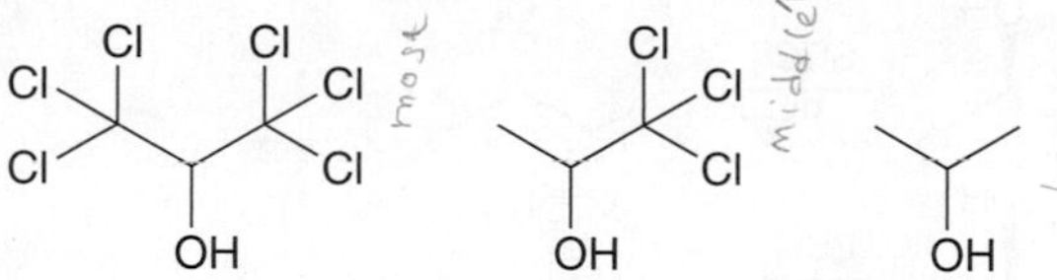

(b) 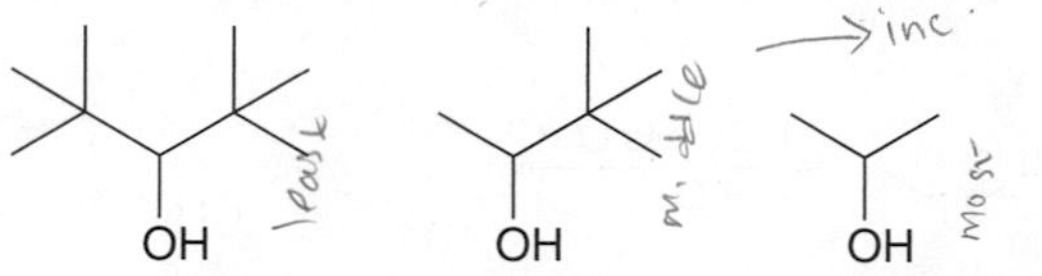

(c) 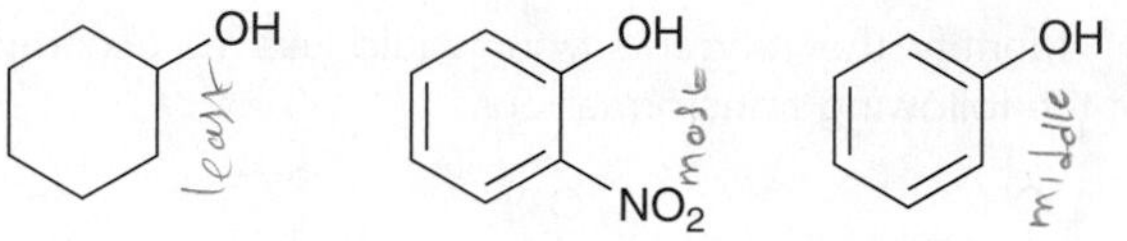

13.34 Draw resonance structures for each of the following anions.

(a) 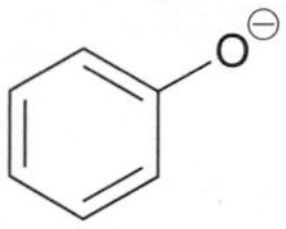 (b) 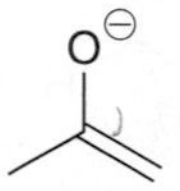 (c) 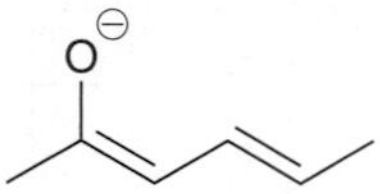

13.35 Predict the major product of the reaction between 1-butanol and:

(a) PBr_3
(b) $SOCl_2$, py
(c) HCl, $ZnCl_2$
(d) H_2SO_4, heat
(e) PCC, CH_2Cl_2
(f) $Na_2Cr_2O_7$, H_2SO_4, H_2O
(g) Li
(h) NaH
(i) TMSCl, Et_3N
(j) TsCl, pyridine
(k) Na
(l) Potassium *tert*-butoxide

13.36 Acid-catalyzed hydration of 3,3-dimethyl-1-butene produces 2,3-dimethyl-2-butanol. Show a mechanism for this reaction.

13.37 Starting with 1-butanol, show the reagents you would use to prepare each of the following compounds.

(a) 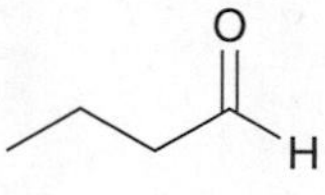 (b) 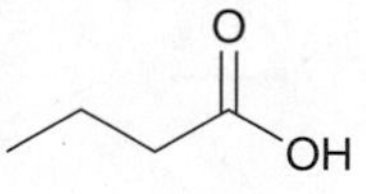

(c) 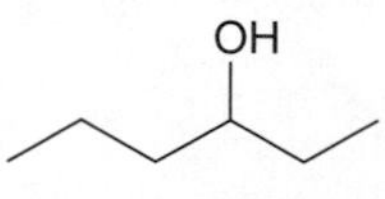 (d) 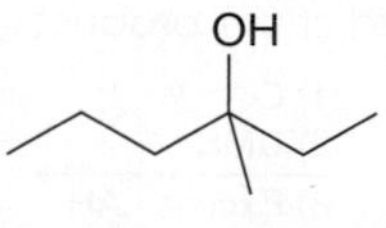

(e) 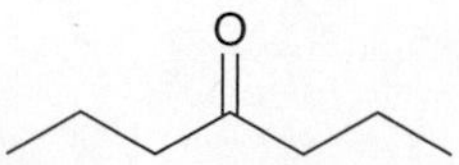

13.38 Using a Grignard reaction, show how you could prepare each of the following alcohols.

(a) 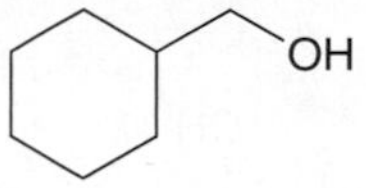 (b) 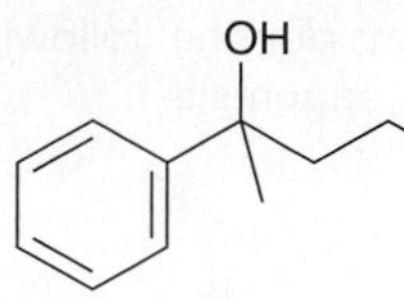

(c) 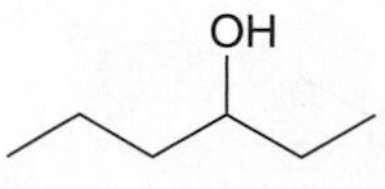 (d) 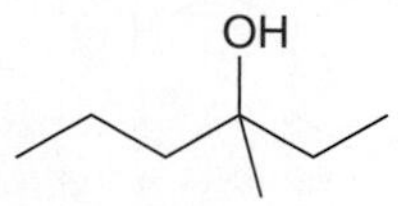

13.39 Each of the following alcohols can be prepared via reduction of a ketone or aldehyde. In each case, identify the aldehyde or ketone that would be required.

(a) 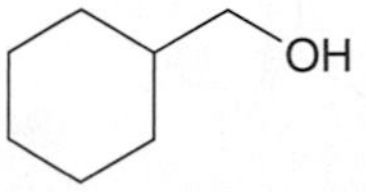 (b) 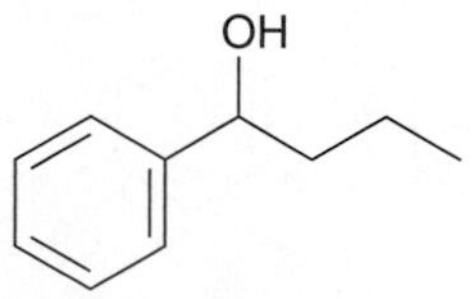

(c) 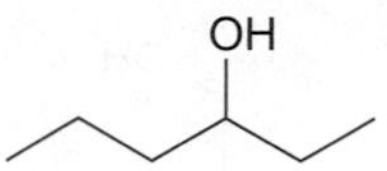 (d) 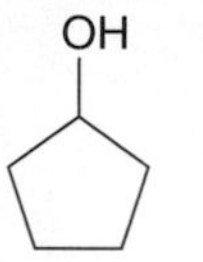

13.40 What reagents would you use to perform each of the following transformations?

(a) 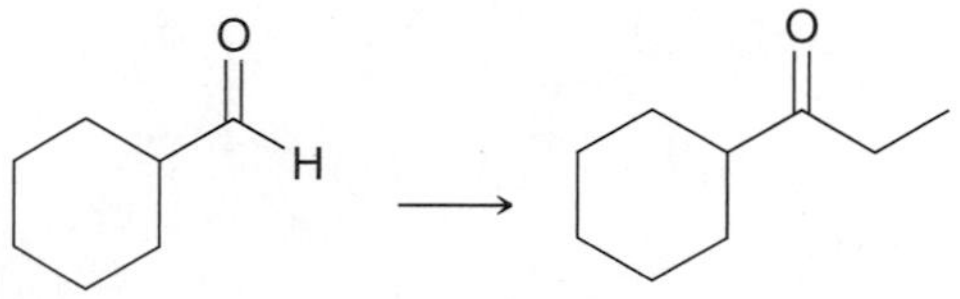

(b) 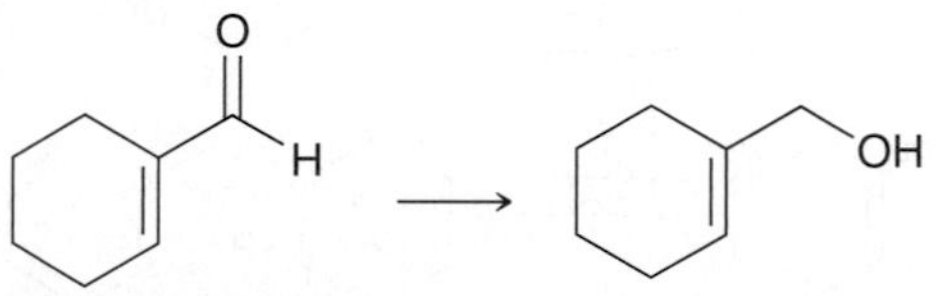

13.41 Propose a mechanism for the following reaction.

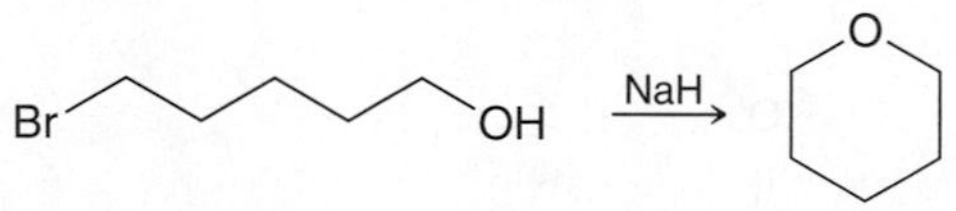

13.42 Acid-catalyzed hydration of 1-methylcyclohexene yields two alcohols. The major product does not undergo oxidation, while the minor product will undergo oxidation. Explain.

13.43 Consider the following sequence of reactions, and identify the structures of compounds **A**, **B**, and **C**.

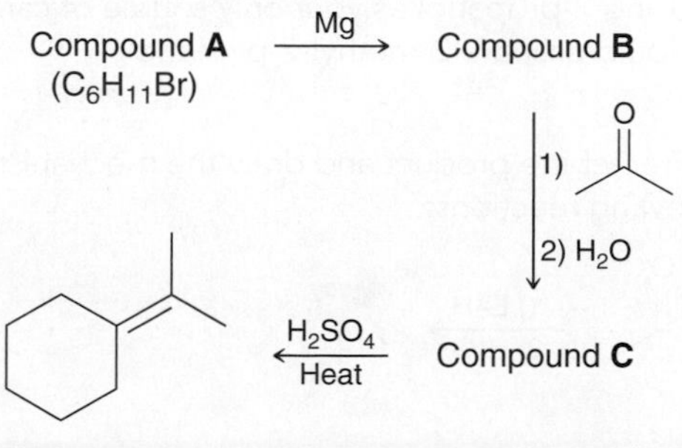

13.44 Consider the following sequence of reactions, and identify the reagents a–h.

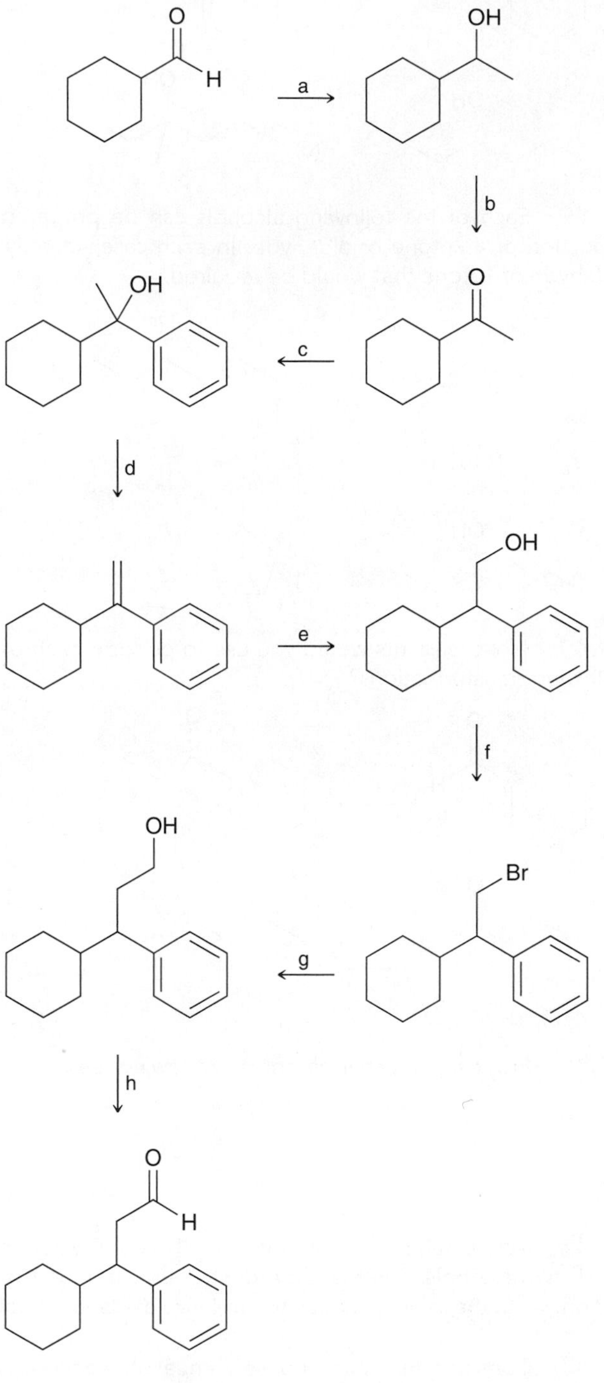

13.45 Using 2-propanol as your only source of carbon, show how you would prepare 2-methyl-2-pentanol.

13.46 Predict the product and draw the mechanism for each of the following reactions:

(a)
$$\xrightarrow[\text{2) H}_2\text{O}]{\text{1) LAH}}$$ **?**

(b)
$$\xrightarrow[\text{2) H}_2\text{O}]{\text{1) LAH}}$$ **?**

(c)
$$\xrightarrow[\text{MeOH}]{\text{NaBH}_4}$$ **?**

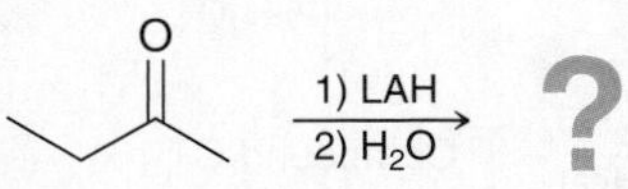

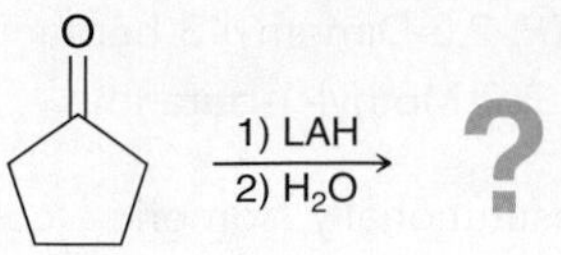

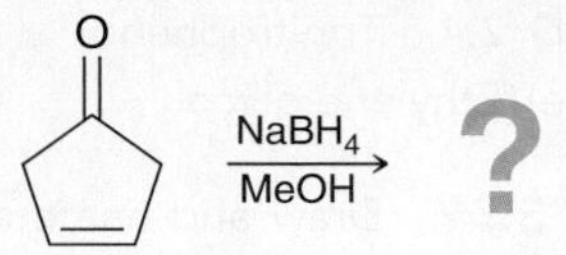

13.47 Draw the mechanism for each of the following reactions.

(a)
$$\xrightarrow[\text{py}]{\text{SOCl}_2}$$

(b)
$$\xrightarrow{\text{PBr}_3}$$

(c)
$$\xrightarrow[\text{2) H}_2\text{O}]{\text{1) Excess LAH}}$$ + CH_3OH

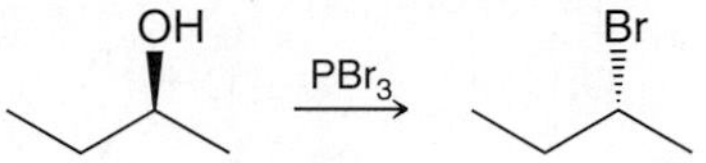

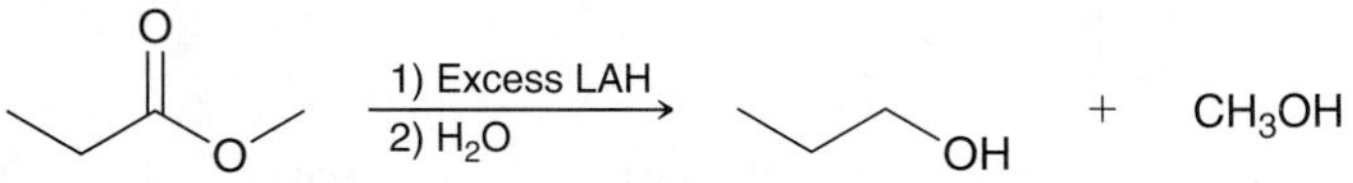

13.48 Identify the reagents you would use to accomplish each of the following transformations.

(a)

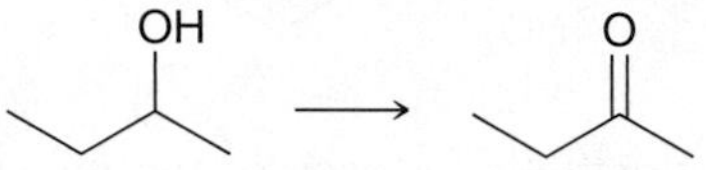

(b)

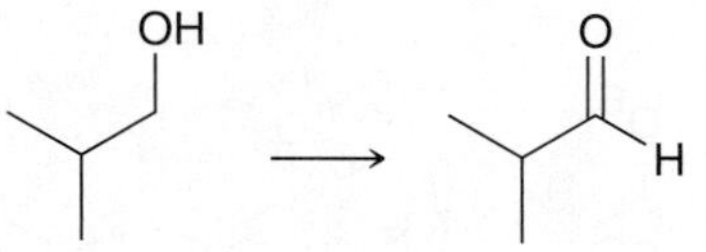

(c)

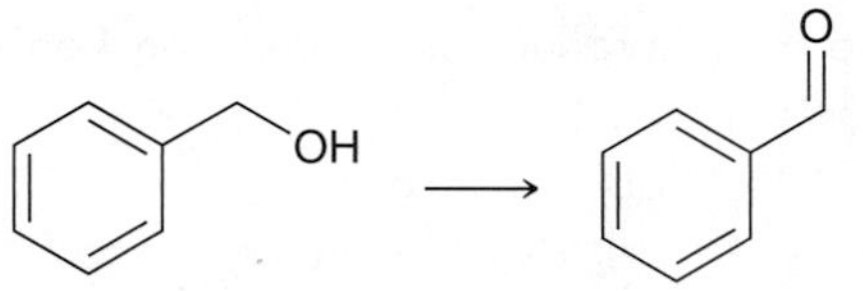

(d)

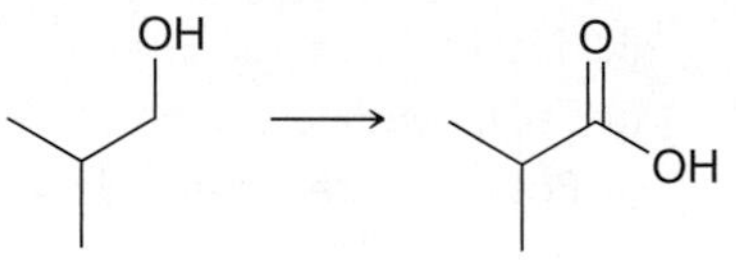

(e)

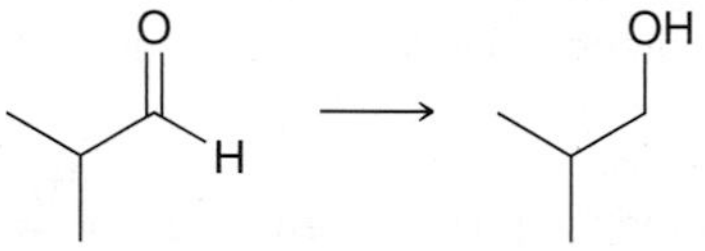

(f)

13.49 Predict the products for each of the following:

(a)
$$\xrightarrow{\begin{array}{l}\text{1) O}_3\\\text{2) DMS}\\\text{3) Excess LAH}\\\text{4) H}_2\text{O}\end{array}}$$ **?**

(b)
$$\xrightarrow{\begin{array}{l}\text{1) O}_3\\\text{2) DMS}\\\text{3) Excess LAH}\\\text{4) H}_2\text{O}\end{array}}$$ **?**

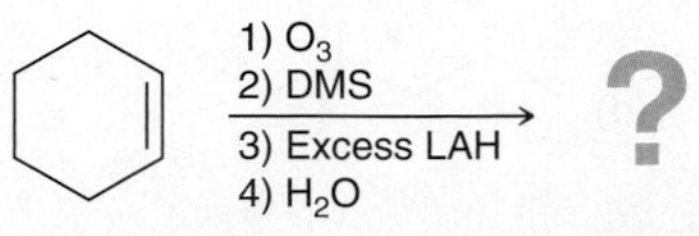

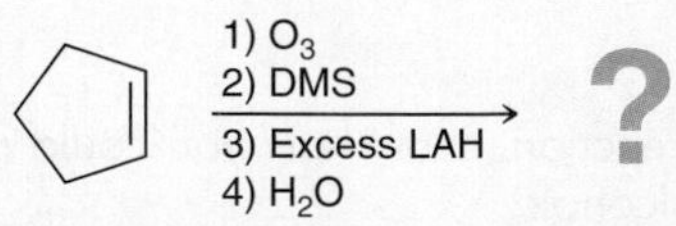

(c) 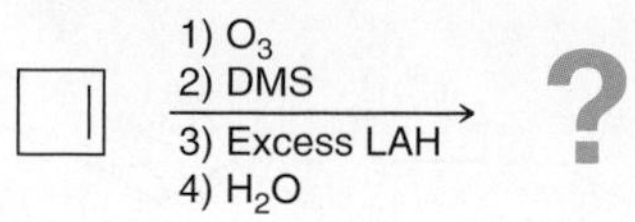

(d) 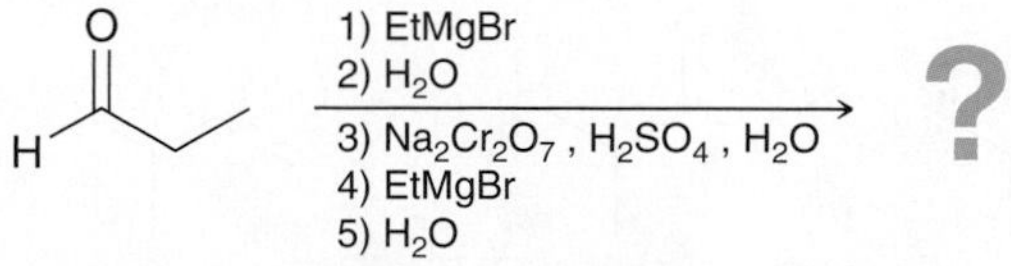

(e) 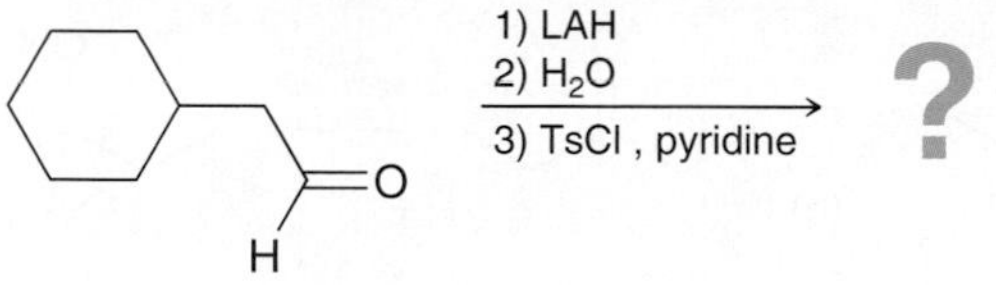

(f) 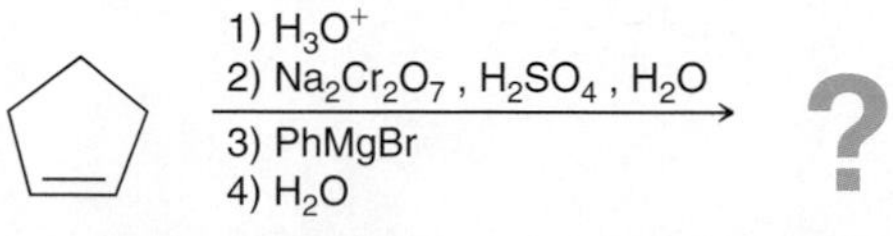

13.50 Propose a plausible mechanism for each of the following transformations.

(a) 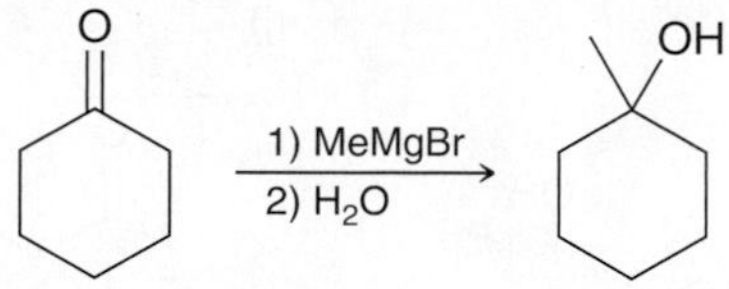

(b) 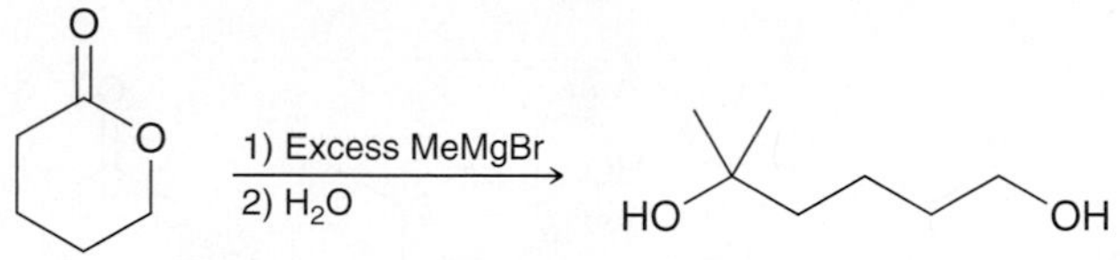

INTEGRATED PROBLEMS

13.51 Identify the reagents that would be necessary to accomplish each of the transformations shown here:

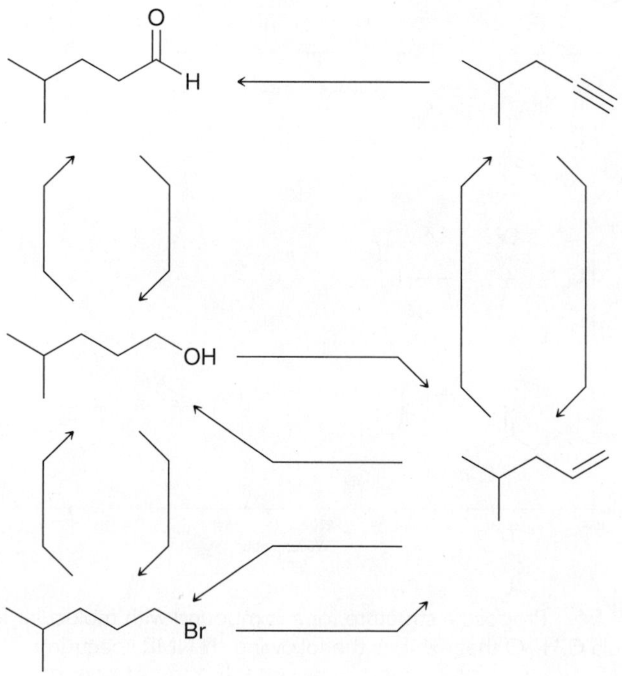

13.52 Propose a plausible synthesis for each of the following transformations:

(a) 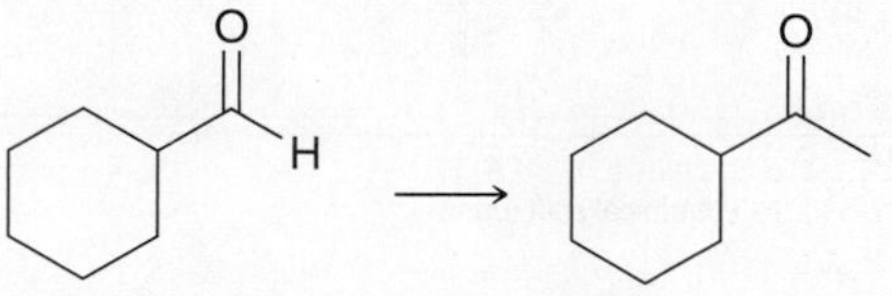

(b) 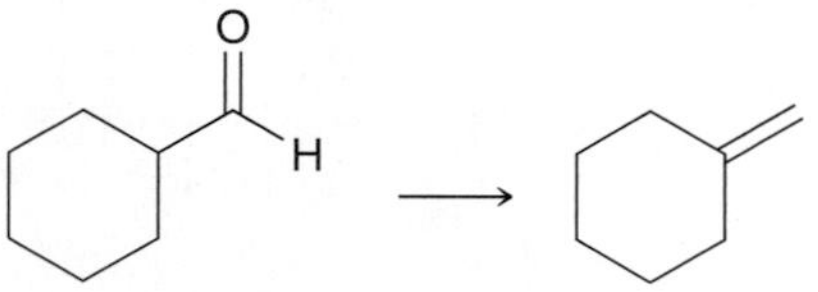

(c) 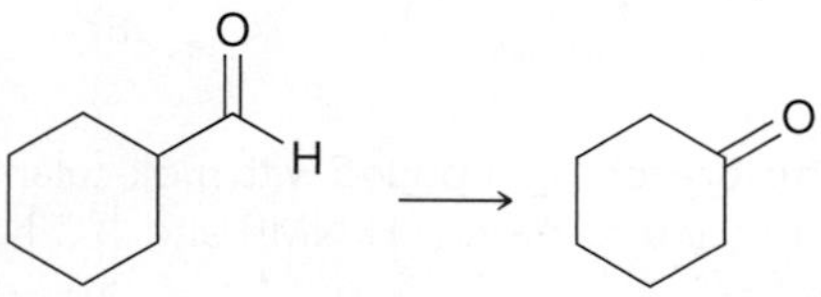

(d) 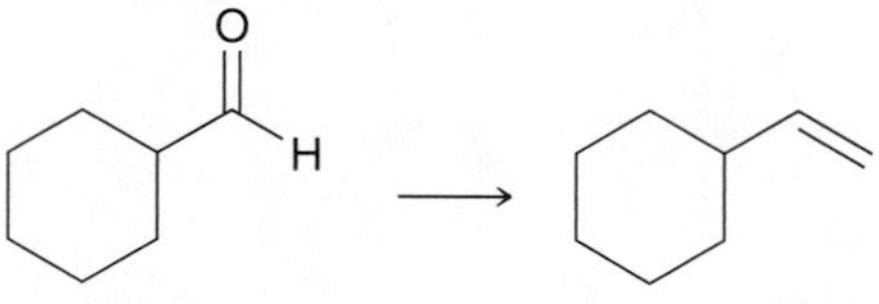

(e) 

(f) 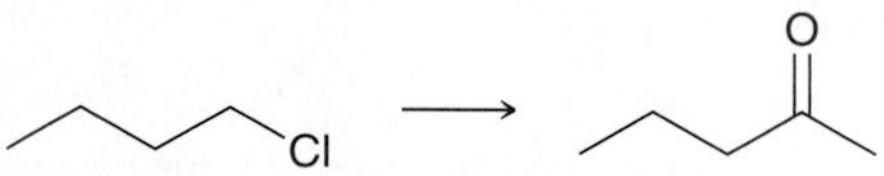

(g) 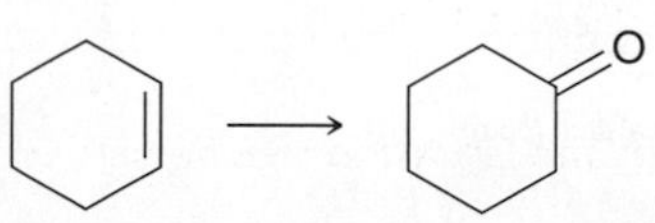

(h) 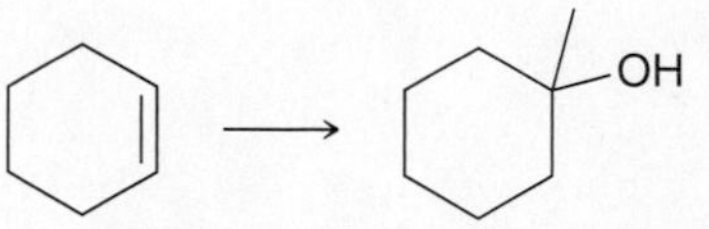

(i) 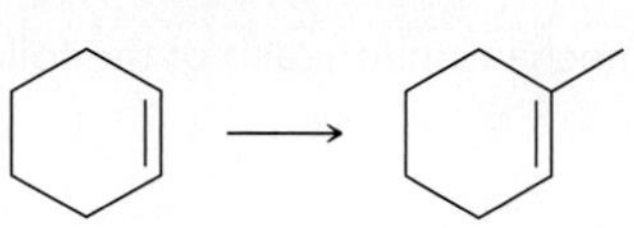

(j) 

(k) 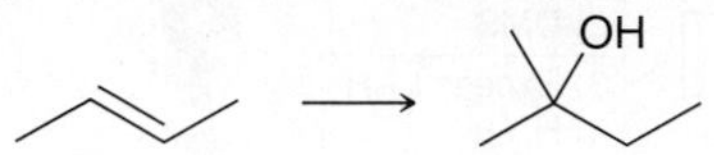

(l) 

(m) 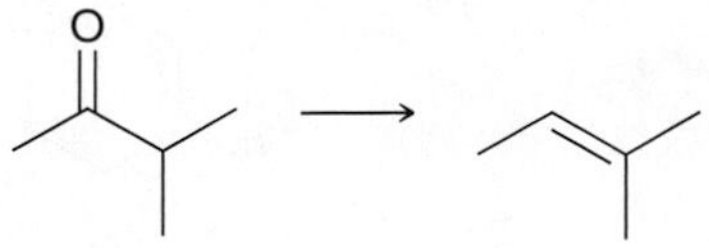

(n) 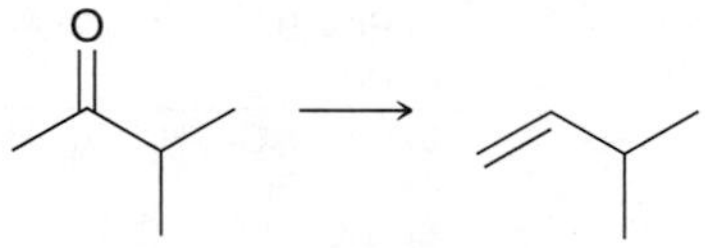

(o) 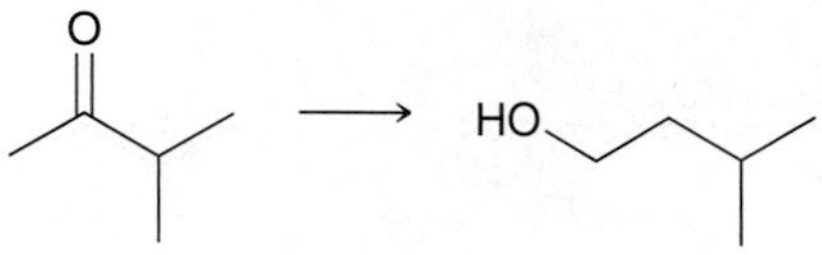

(p) 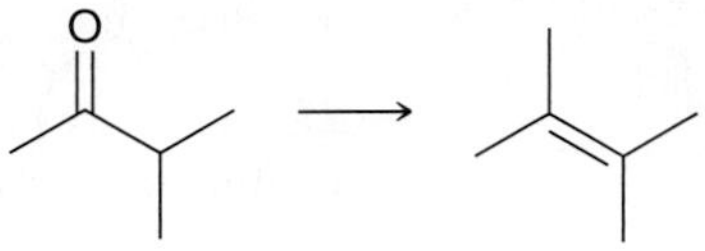

(q) 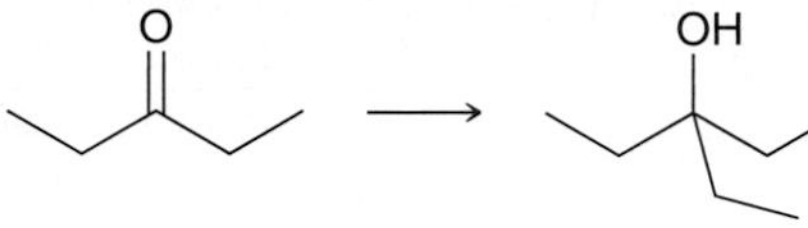

(r) 

(s) 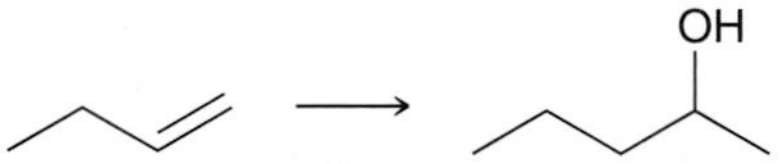

Problems 13.53–13.56 are intended for students who have already covered spectroscopy (Chapters 15 and 16).

13.53 Propose a structure for a compound with molecular formula $C_{10}H_{14}O$ that exhibits the following ^1H NMR spectrum.

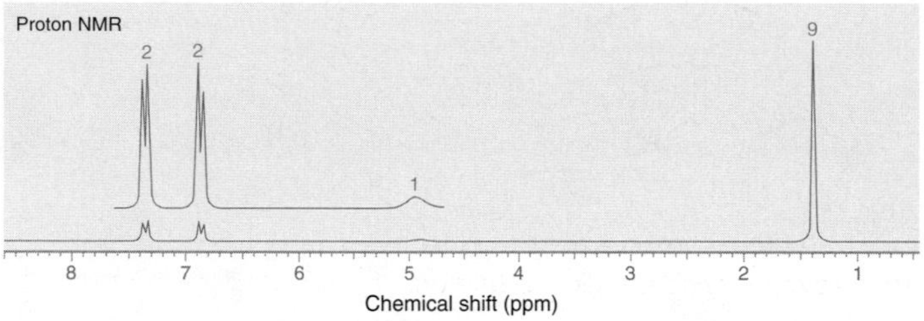

13.54 Propose a structure for a compound with molecular formula C_3H_8O that exhibits the following ^1H NMR and ^{13}C NMR spectra:

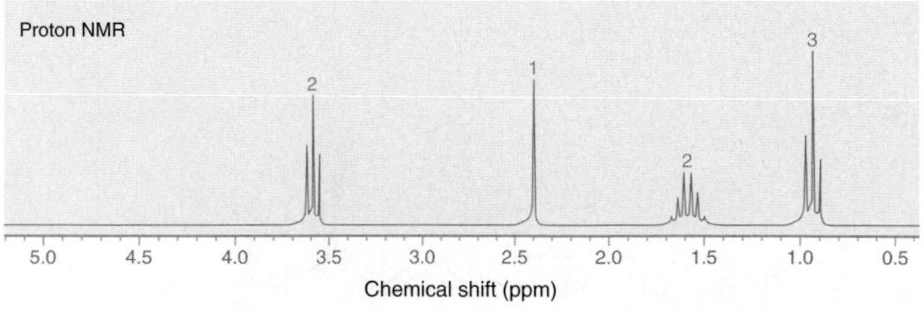

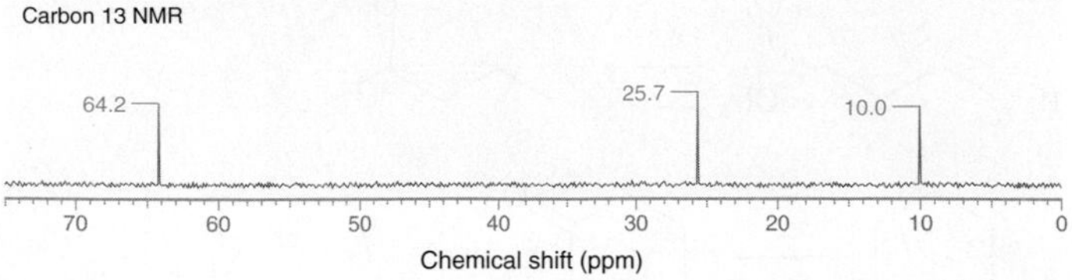

13.55 Propose two possible structures for a compound with molecular formula $C_5H_{12}O$ that exhibits the following ^{13}C NMR and IR spectra:

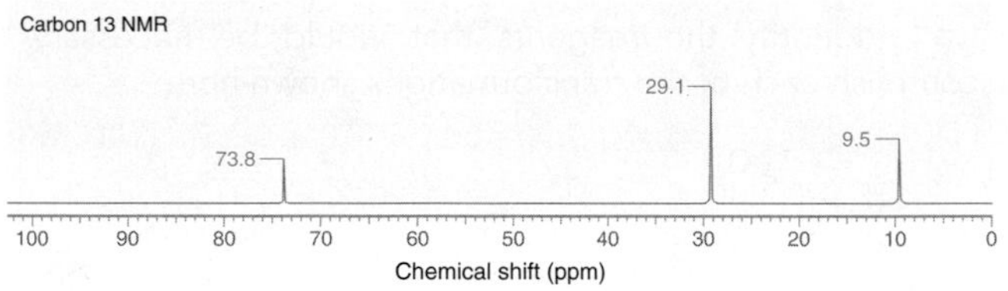

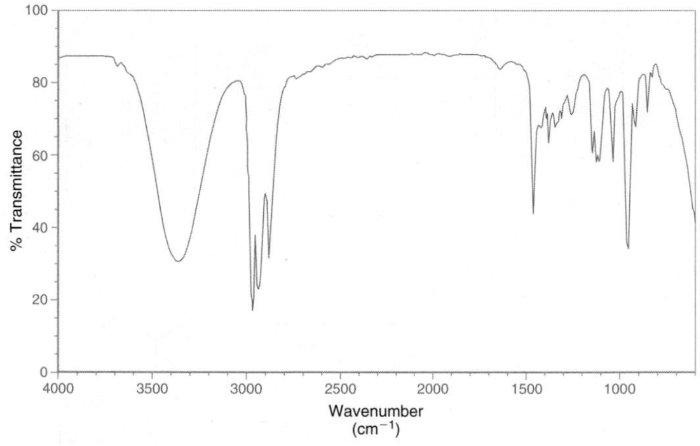

13.56 Propose a structure for a compound with molecular formula $C_8H_{10}O$ that exhibits the following ^1H NMR spectrum:

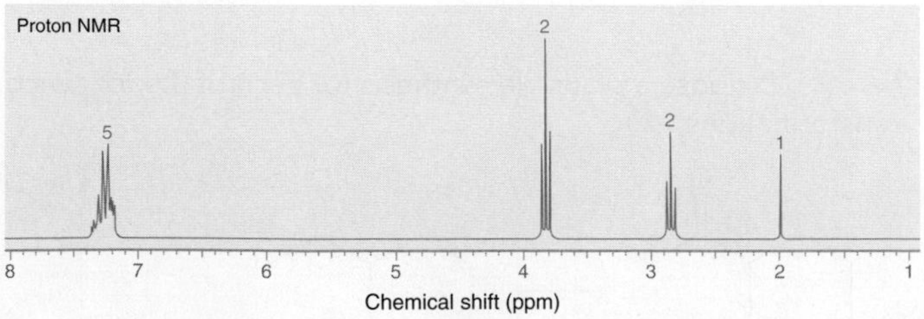

CHALLENGE PROBLEMS

13.57 Propose a mechanism for the following transformation:

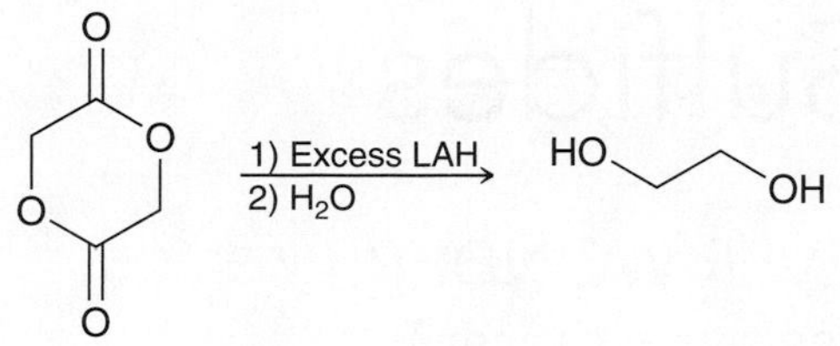

13.58 Propose a mechanism for the following transformation:

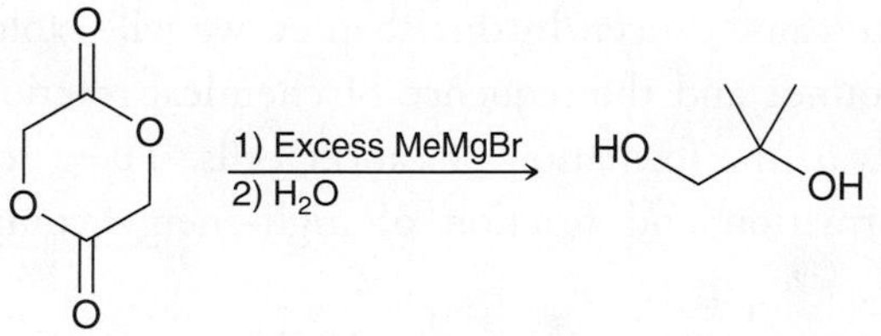

13.59 Show the reagents you would use to achieve the following transformation:

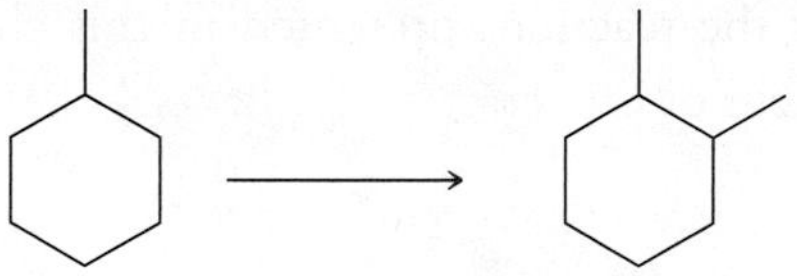

13.60 Propose a mechanism for the following transformation:

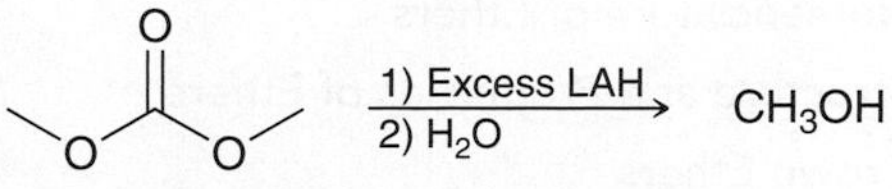

13.61 A carbocation is resonance stabilized when it is adjacent to an oxygen atom:

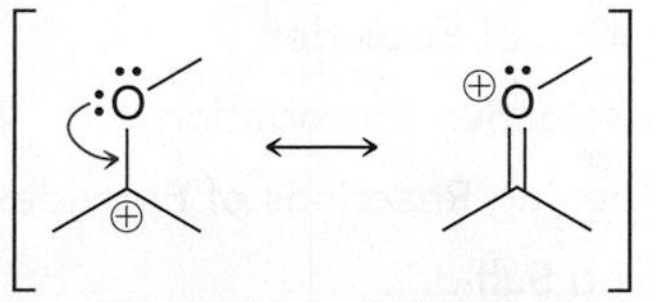

Resonance stabilized

Such a carbocation is even more stable than a tertiary carbocation. Using this information, propose a mechanism for the following transformation exhibited by a diol. This reaction is called a pinacol rearrangement:

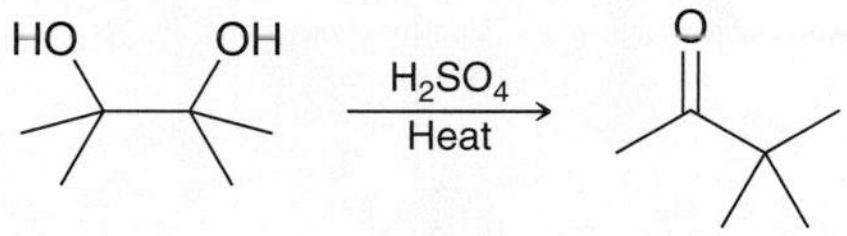

13.62 Determine whether the pinacol rearrangement, shown in the previous problem, is a reduction, an oxidation, or neither.

14

Ethers and Epoxides; Thiols and Sulfides

DID YOU EVER WONDER...

how cigarettes cause cancer?

Cigarette smoke contains many compounds that have been shown to cause cancer. In this chapter, we will explore one of those compounds and the sequence of chemical reactions that ultimately leads to the formation of cancer cells. These reactions involve the formation and reaction of high-energy compounds called epoxides.

Epoxides are a special category of ethers, and these compounds comprise the main topic of this chapter. We will learn about the properties and reactions of ethers, epoxides, and other related compounds. Then, we will return to the subject of this chapter opener to learn how the reactions presented in this chapter play a role in the development of cancer.

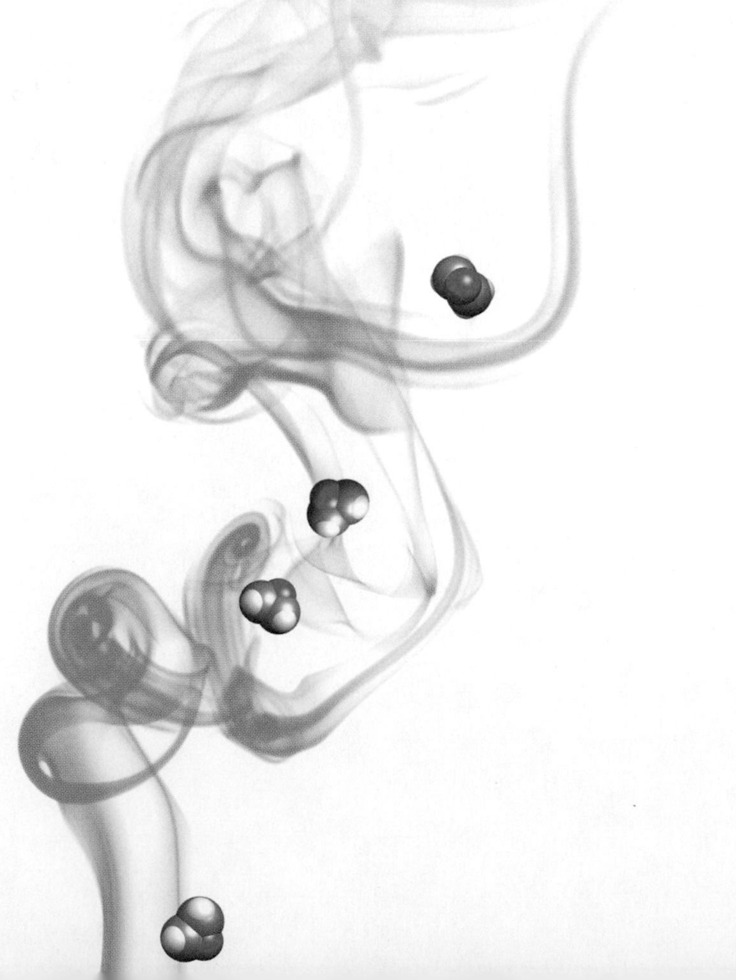

DO YOU REMEMBER?

Before you go on, be sure you understand the following topics.
If necessary, review the suggested sections to prepare for this chapter.

- Reading Energy Diagrams (Section 6.6)
- S_N2 Reactions (Sections 7.4, 7.7)
- Halohydrin Formation (Section 9.8)
- Mechanisms and Curved Arrows (Sections 6.8-6.11)
- Oxymercuration-Demercuration (Section 9.5)

14.1 Introduction to Ethers

Ethers are compounds that exhibit an oxygen atom bonded to two R groups, where each R group can be an alkyl, aryl, or vinyl group.

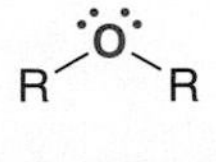

An ether

The ether moiety is a common structural feature of many natural compounds, for example:

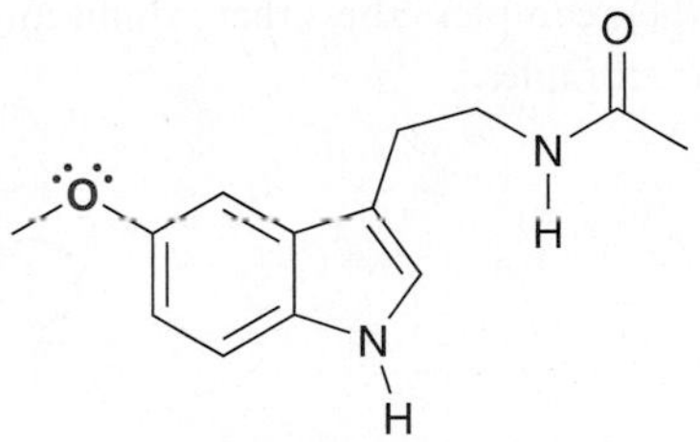

Melatonin
A hormone that is believed to regulate the sleep cycle

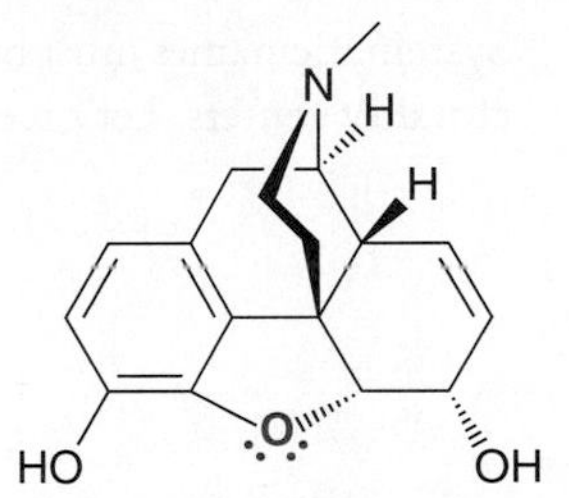

Morphine
An opiate analgesic used to treat severe pain

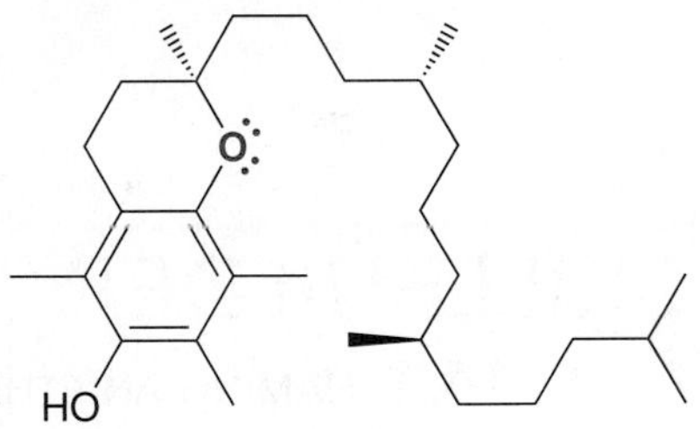

Vitamin E
An antioxidant

Many pharmaceuticals also exhibit an ether moiety, for example:

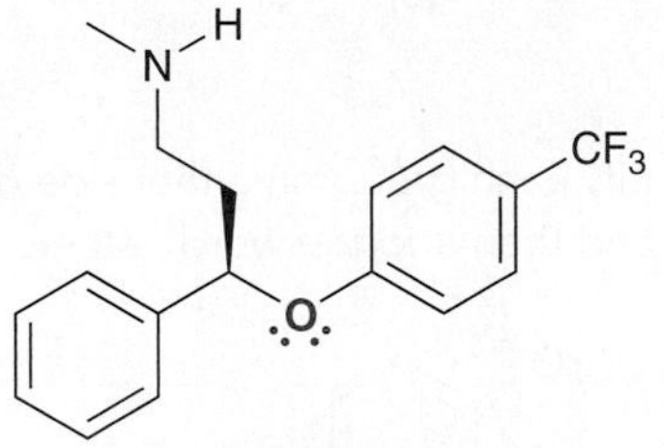

(*R*)-Fluoxetine
A powerful antidepressant sold under the trade name Prozac®

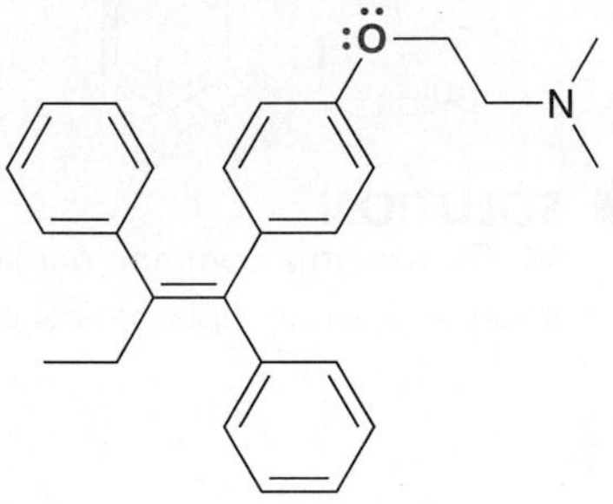

Tamoxifen
Inhibits the growth of some breast tumors

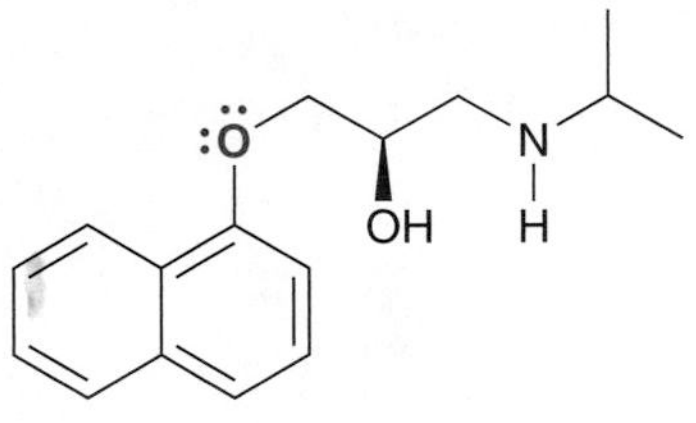

Propanolol
Used in the treatment of high blood pressure

14.2 Nomenclature of Ethers

IUPAC rules allow two different methods for naming ethers.

1. A common name is constructed by identifying each R group, arranging them in alphabetical order, and then adding the word "ether", for example:

Ethyl Methyl

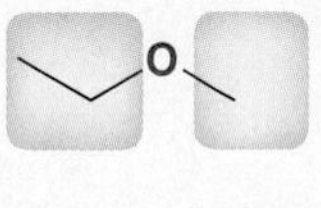

Ethyl methyl ether

***tert*-Butyl Methyl**

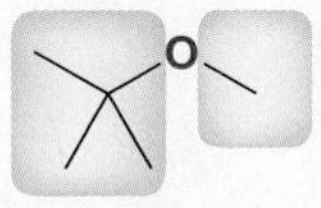

***tert*-Butyl methyl ether**

In these examples, the oxygen atom is connected to two different alkyl groups. Such compounds are called **unsymmetrical ethers.** When the two alkyl groups are identical, the compound is called a **symmetrical ether** and is named as a *di*alkyl ether.

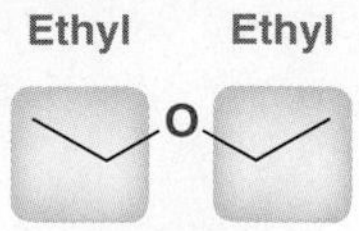

Diethyl ether
(ethyl ether)

2. A systematic name is constructed by choosing the larger group to be the parent alkane and naming the smaller group as an **alkoxy** substituent.

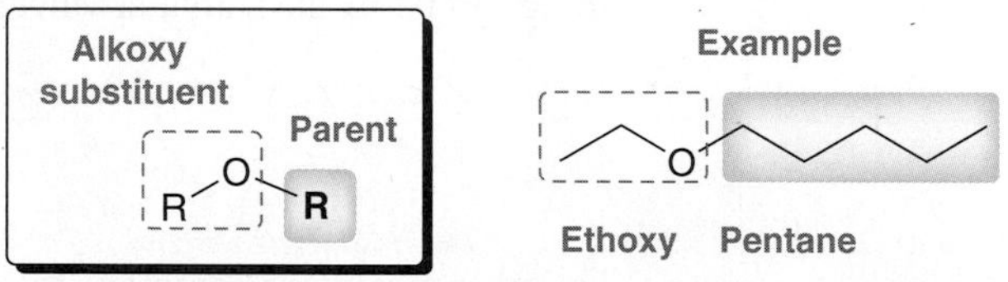

Systematic names must be used for complex ethers that exhibit multiple substituents and/or chirality centers. Let's see some examples.

SKILLBUILDER

14.1 NAMING AN ETHER

LEARN the skill

Name the following compounds:

SOLUTION

(a) To assign a common name, identify each group on either side of the oxygen atom, arrange them in alphabetical order, and then add the word "ether."

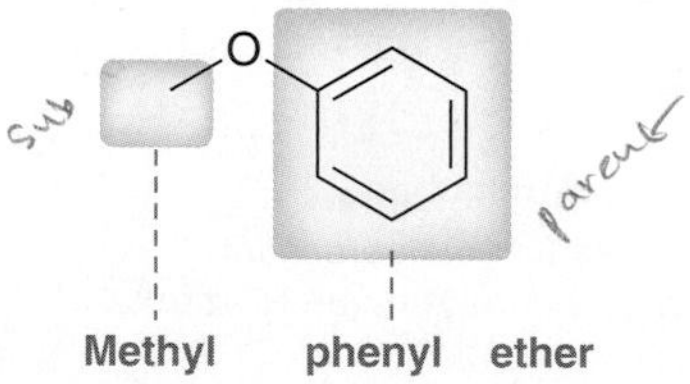

To assign a systematic name, choose the more complex (larger) group as the parent, and name the smaller group as an alkoxy substituent.

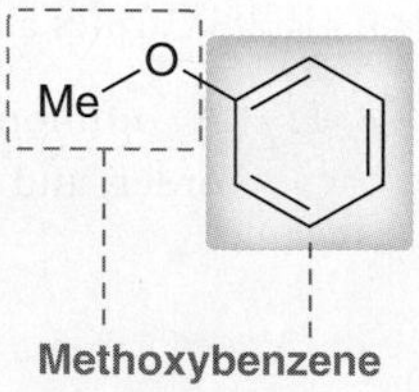

This compound therefore can be called methyl phenyl ether or methoxybenzene. Both names are accepted by IUPAC rules.

(b) The second compound is more complex. It has a chirality center and several substituents. Therefore, it will not have a common name. To assign a systematic name, begin by choosing the more complex group as the parent.

Parent **Substituent**

ethoxy

ethoxy

1,1 dichloro cyclopentane.

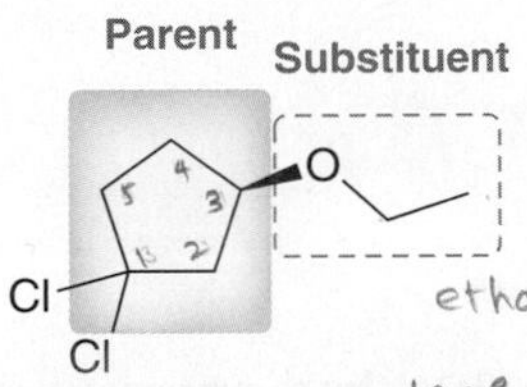

The cyclopentane ring becomes the parent, and the ethoxy group is listed as one of the three substituents on the cyclopentane ring. Locants are then assigned so as to give the lowest possible numbers to all three substituents (1,1,3 rather than 1,3,3):

Parent

1,1 - dichloro-3 -ethoxycyclopentane

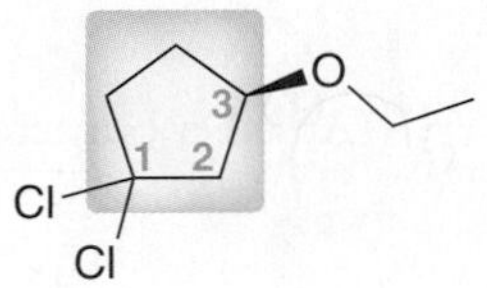

The configuration of the chirality center is identified at the beginning of the name:

(*R*)-1,1-Dichloro-3-ethoxycyclopentane

PRACTICE the skill **14.1** Provide an IUPAC name for each of the following compounds.

1,3-dichloro benzene

ethoxy *propane* *2-chloropropane* *ethoxy* *CI* *CI*

(a) *2-exthoxy propane* (b) *CI* *ethoxy* (c) *ethoxy*

(S)- 2-chloro-1-ethoxypropane ✓ *2,4-dichloro-1-ethoxybenzene* ✓

parent cyclohexanol

OH *OEt*

(d) (e) *ethoxy*

(1R), (2R)- 2-ethoxy eeyclohexanol *1-ethoxy cyclohexane* ✓

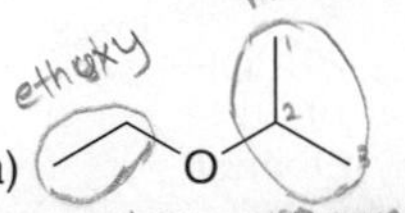

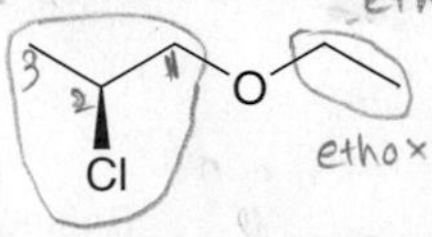

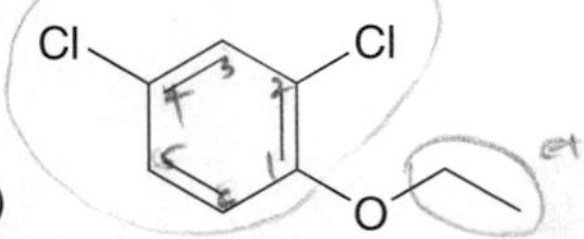

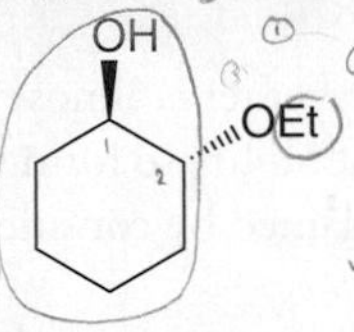

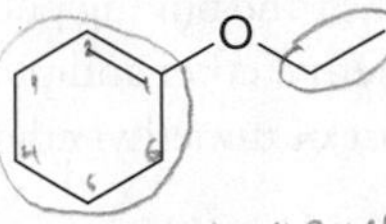

APPLY the skill **14.2** Draw the structure of each of the following compounds.

(a) (*R*)-2-Ethoxy-1,1-dimethylcyclobutane

(b) Cyclopropyl isopropyl ether

14.3 There are six ethers with molecular formula $C_5H_{12}O$ that are constitutional isomers.

(a) Draw all six constitutional isomers.

(b) Provide a systematic name for each of the six compounds.

(c) Provide a common name for each of the six compounds.

(d) Only one of these compounds has a chirality center. Identify that compound.

need more **PRACTICE?** **Try Problems 14.30, 14.32**

14.3 Structure and Properties of Ethers

The geometry of an oxygen atom is similar for water, alcohols, and ethers. In all three cases, the oxygen atom is sp^3 hybridized, and the orbitals are arranged in a nearly tetrahedral shape. The exact bond angle depends on the groups attached to the oxygen atom, with ethers having the largest bond angles.

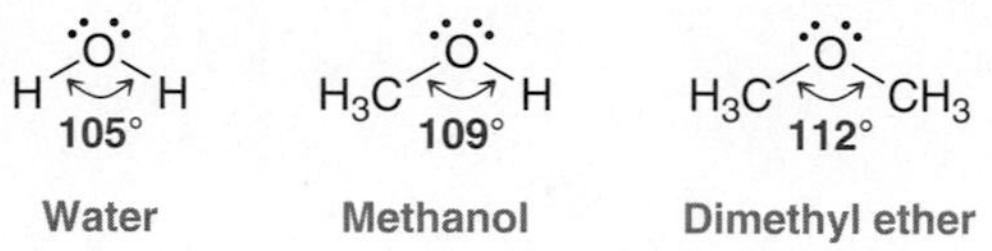

In the previous chapter, we saw that alcohols have relatively high boiling points due to the effects of intermolecular hydrogen bonding.

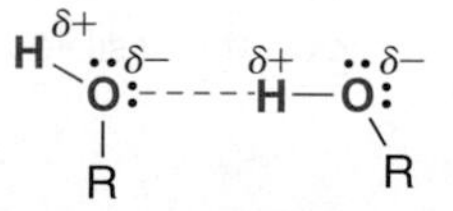

An ether can act as a hydrogen bond acceptor and can interact with the proton of an alcohol.

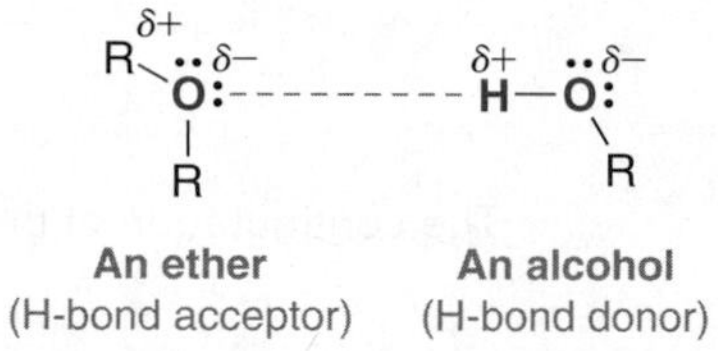

However, ethers cannot function as hydrogen bond donors, and therefore, ethers cannot form hydrogen bonds with each other. As a result, the boiling points of ethers are significantly lower than their isomeric alcohols.

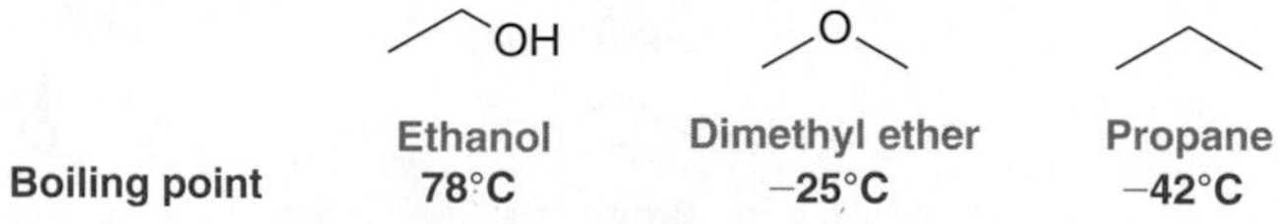

In fact, the boiling point of dimethyl ether is almost as low as the boiling point of propane. Both dimethyl ether and propane lack the ability to form hydrogen bonds. The slightly higher boiling point of dimethyl ether can be explained by considering the net dipole moment.

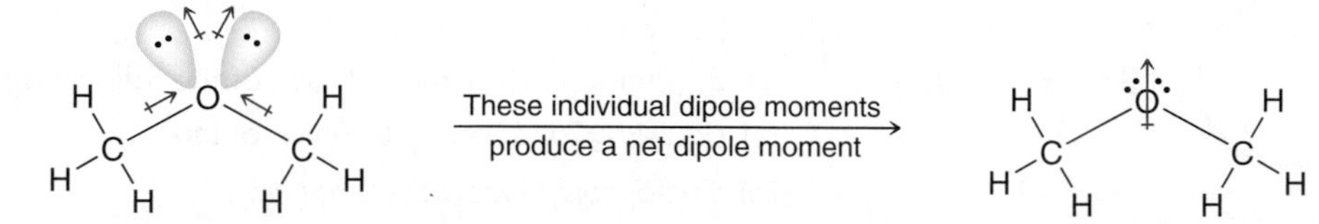

Ethers therefore exhibit dipole-dipole interactions, which slightly elevate the boiling point relative to propane. Ethers with larger alkyl groups have higher boiling points due to London dispersion forces between the alkyl groups on different molecules. This trend is significant.

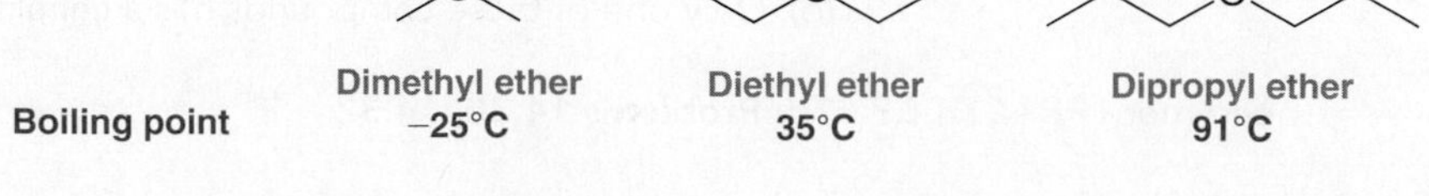

Ethers are often used as solvents for organic reactions, because they are fairly unreactive, they dissolve a wide variety of organic compounds, and their low boiling points allow them to be readily evaporated after a reaction is complete. Below are three common solvents.

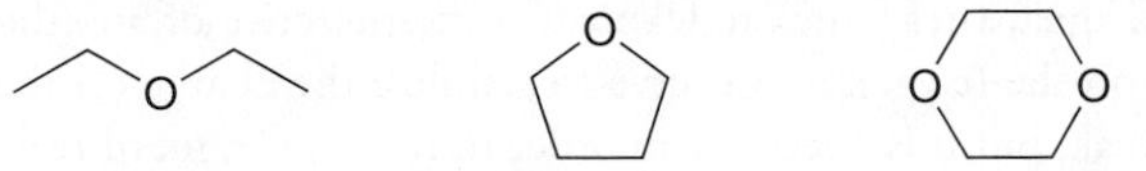

Diethyl ether **Tetrahydrofuran** **1,4-Dioxane**

MEDICALLY SPEAKING

Ethers as Inhalation Anesthetics

Diethyl ether was once used as an inhalation anesthetic, but the side-effects were unpleasant, and the recovery was often accompanied by nausea and vomiting. Diethyl ether was eventually replaced by halogenated ethers, such as the ones shown below.

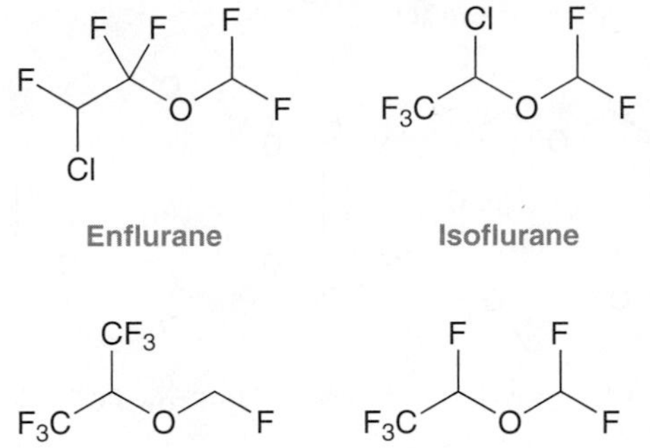

Enflurane **Isoflurane** **Sevoflurane** **Desflurane**

Enflurane was introduced in the mid-1970s and was eventually replaced by isoflurane. The use of isoflurane is now also declining as the newer generation ethers (sevoflurane and desflurane) are being more heavily used.

Inhalation anesthetics are introduced into the body via the lungs and distributed by the circulatory system. They specifically target the nerve endings in the brain. Nerve endings, which are separated by a synaptic gap, transmit signals across the gap by means of small organic compounds called neurotransmitters (shown as blue balls in the following figure).

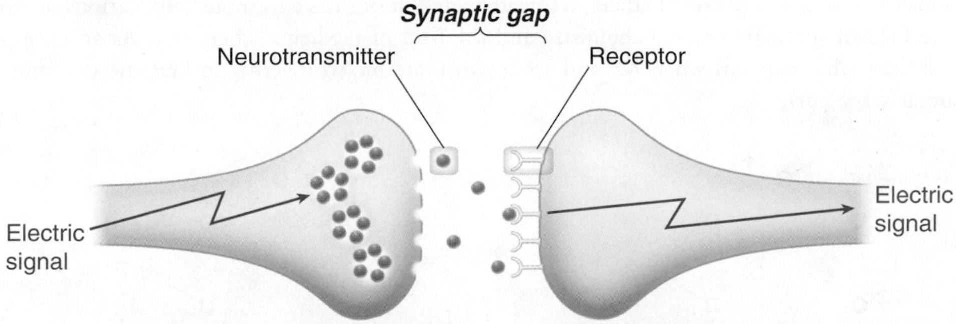

A change in ionic conductance (electrical signal) causes the presynaptic cell to release neurotransmitters, which travel across the synaptic gap until they reach the receptors at the postsynaptic cell. When the neurotransmitters bind to the receptors, a change in conductance is triggered once again. In this way, a signal either can be relayed across the synaptic gap or can be stopped, depending on whether the neurotransmitters are allowed to do their job. Several factors are involved that either can inhibit or increase the function of the neurotransmitters. By controlling whether signals are sent or stopped at each synaptic gap, the nervous system is able to control the various systems in the body (similar to the way a computer uses zeros and ones to perform all of its functions).

Inhalation anesthetics disrupt the normal synaptic transmission process. Many mechanisms of action for anesthetics have been suggested, including the following:

1. Interfering with the release of neurotransmitters from the presynaptic nerve cell
2. Interfering with the binding of the neurotransmitters at the postsynaptic receptors
3. Affecting the ionic conductance (the electrical signal that causes neurotransmission)
4. Affecting reuptake of neurotransmitters into the presynaptic cell

The main mechanism of action is likely to be a combination of many of these factors.

14.4 Crown Ethers

BY THE WAY
Although, like ethers, alcohols (ROH) also have an oxygen atom with lone pairs, they cannot be used to stabilize Grignard reagents, because alcohols possess acidic protons. As we saw in Section 13.6, Grignard reagents cannot be prepared in the presence of acidic protons.

Ethers can interact with metals that have either a full positive charge or a partial positive charge. For example, Grignard reagents are formed in the presence of an ether, such as diethyl ether. The lone pairs on the oxygen atom serve to stabilize the charge on the magnesium atom. The interaction is weak, but it is necessary in order to form a Grignard reagent.

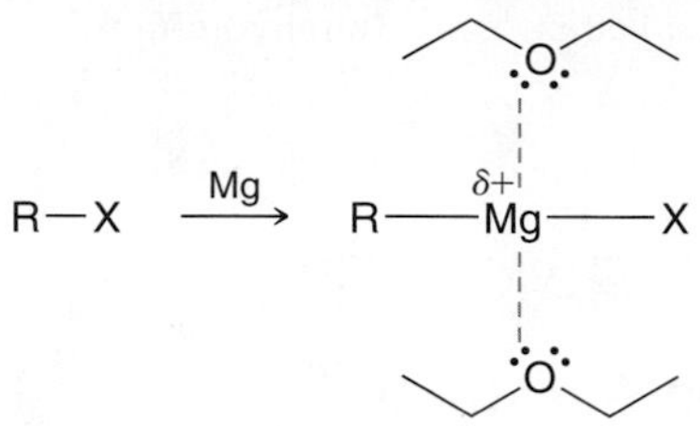

Charles J. Pedersen, working for Du Pont, discovered that the interaction between ethers and metal ions is significantly stronger for compounds with multiple ether moieties. Such compounds are called **polyethers**. Pedersen prepared and investigated the properties of many cyclic polyethers, such as the following examples. Pedersen called them **crown ethers** because their molecular models resemble crowns.

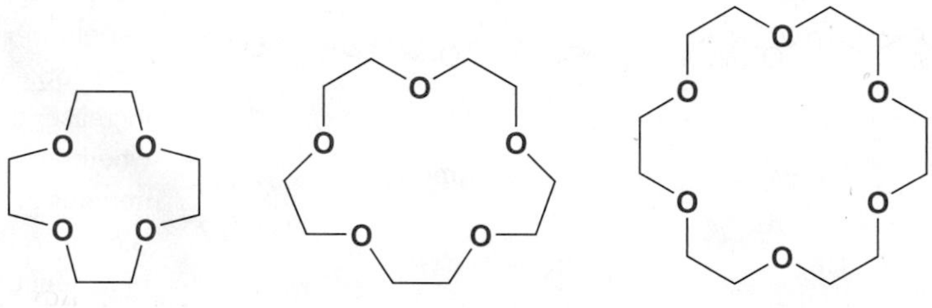

12-Crown-4 15-Crown-5 18-Crown-6

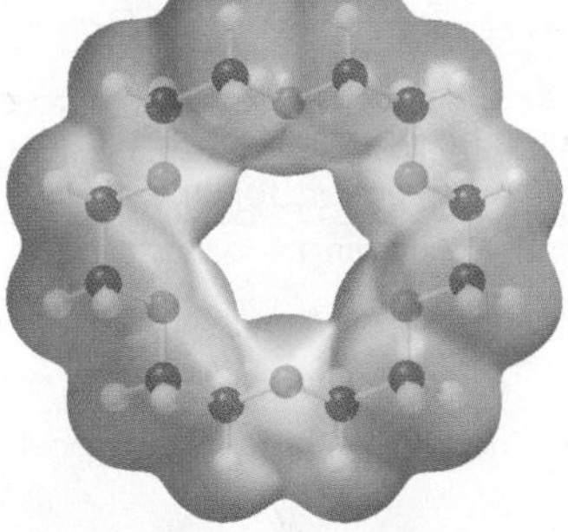

FIGURE 14.1a
An electrostatic potential map of 18-crown-6 shows the oxygen atoms facing the inside of the internal cavity.

FIGURE 14.1b
A space-filling model of 18-crown-6 shows that a potassium cation can fit nicely inside the internal cavity.

These compounds contain multiple oxygen atoms and are therefore capable of binding more tightly to metal ions. Systematic nomenclature for these compounds can be complex, so Pedersen developed a simple method for naming them. He used the formula X-crown-Y, where X indicates the total number of atoms in the ring and Y represents the number of oxygen atoms. For example, 18-crown-6 is an 18-membered ring in which 6 of the 18 atoms are oxygen atoms.

The unique properties of these compounds derive from the size of their internal cavities. For example, the internal cavity of 18-crown-6 comfortably hosts a potassium cation (K^+). In the electrostatic potential map in Figure 14.1a, it is clear that the oxygen atoms all face toward the inside of the cavity, where they can bind to the metal cation. The space-filling model in Figure 14.1b shows how a potassium cation fits perfectly into the internal cavity. Once inside the cavity, the entire complex has an outer surface that resembles a hydrocarbon, rendering the complex soluble in organic solvents. In this way, 18-crown-6 is capable of solvating potassium ions in organic solvents. Normally, the metal cation by itself would not be soluble in a nonpolar solvent. The ability of crown ethers to solvate metal cations has enormous implications, in both the field of synthetic organic chemistry and the field of medicinal chemistry. As an example, consider what happens when KF and 18-crown-6 are mixed together in benzene (a common organic solvent).

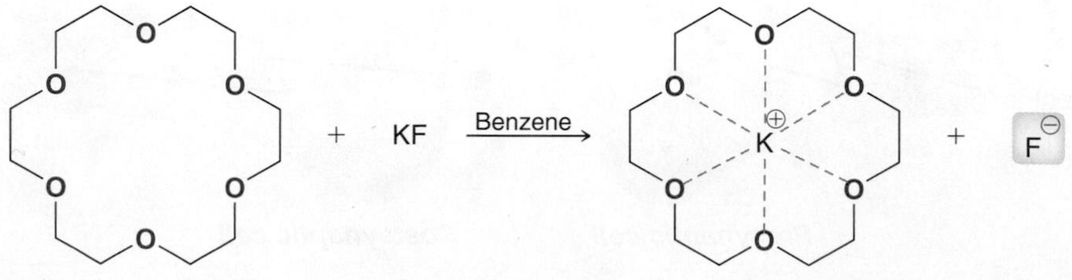

Without the crown ether, KF would simply not dissolve in benzene. The presence of 18-crown-6 generates a complex that dissolves in benzene. The result is a solution containing fluoride ions, which enables us to perform substitution reactions with F^- as a nucleophile. Generally, it is too difficult to use F^- as a nucleophile, because it will usually interact too strongly with the polar solvents in which it dissolves. The strong interaction between fluoride ions and polar solvents makes it difficult for F^- to become "free" to serve as a nucleophile. However, the use of 18-crown-6 allows the creation of free fluoride ions in a nonpolar solvent, making substitution reactions possible. For example:

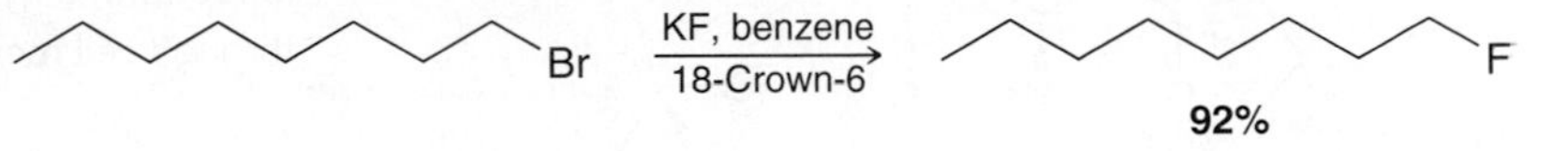

Another example is the ability of 18-crown-6 to dissolve potassium permanganate ($KMnO_4$) in benzene. Such a solution is very useful for performing a wide variety of oxidation reactions.

Other metal cations can be solvated by other crown ethers. For example, a lithium ion is solvated by 12-crown-4, and a sodium ion is solvated by 15-crown-5.

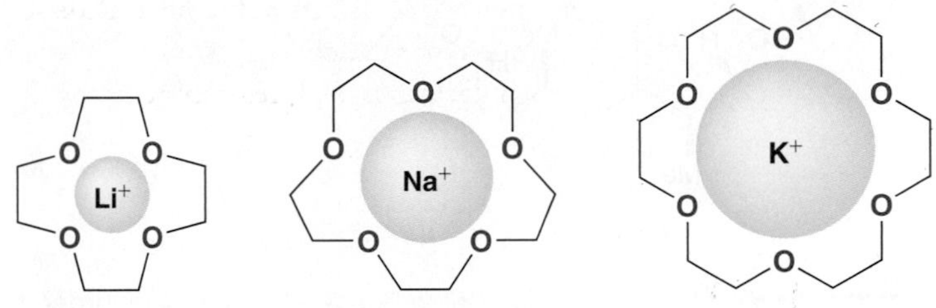

12-Crown-4
Solvates Li$^+$

15-Crown-5
Solvates Na$^+$

18-Crown-6
Solvates K$^+$

The discovery of these compounds led to a whole new field of chemistry, called *host-guest chemistry*. For his contribution, Pedersen shared the 1987 Nobel Prize in Chemistry together with Donald Cram and Jean-Marie Lehn, who were also pioneers in the field of host-guest chemistry.

CONCEPTUAL CHECKPOINT

14.4 Identify the missing reagent needed to achieve the following transformations:

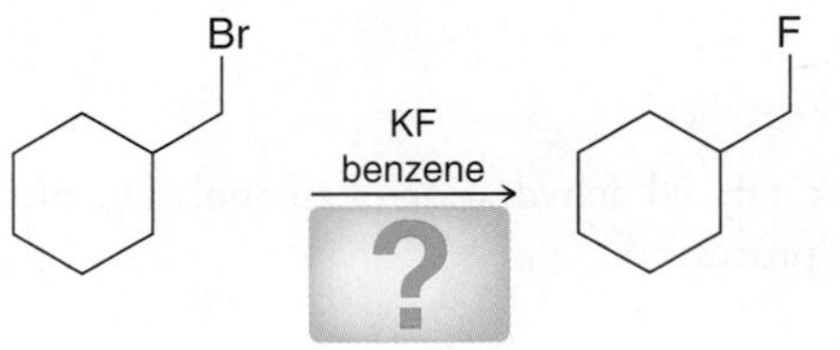

(a)

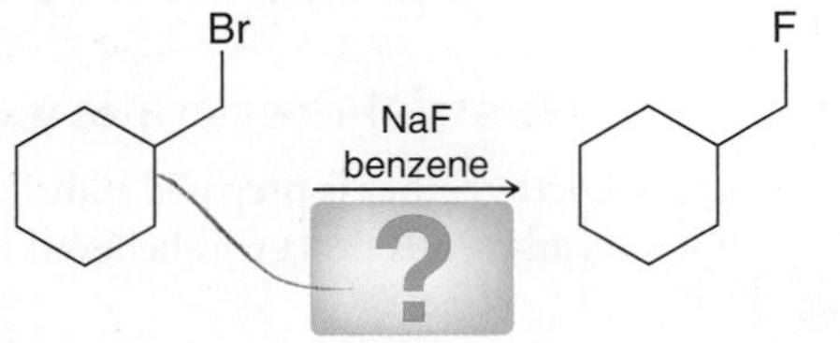

(b)

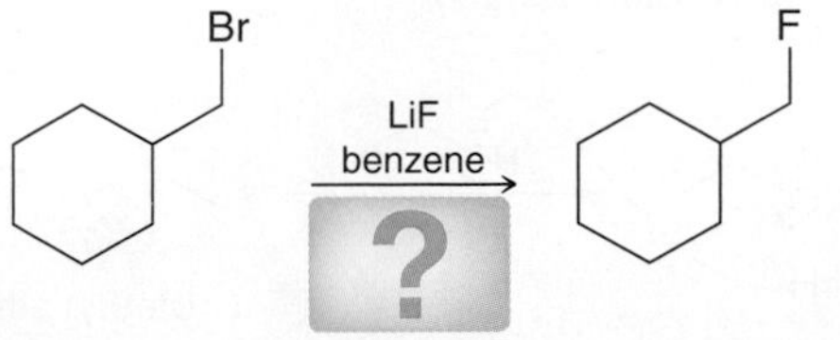

(c)

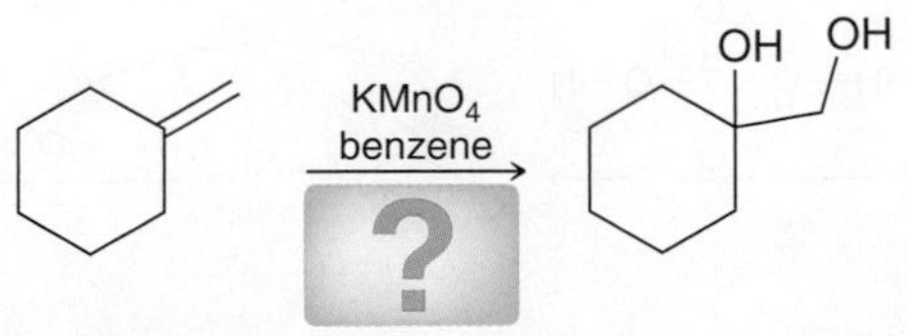

(d)

MEDICALLYSPEAKING

Polyether Antibiotics

Some antibiotics function very much like crown ethers. For example, consider the structures of nonactin and monensin.

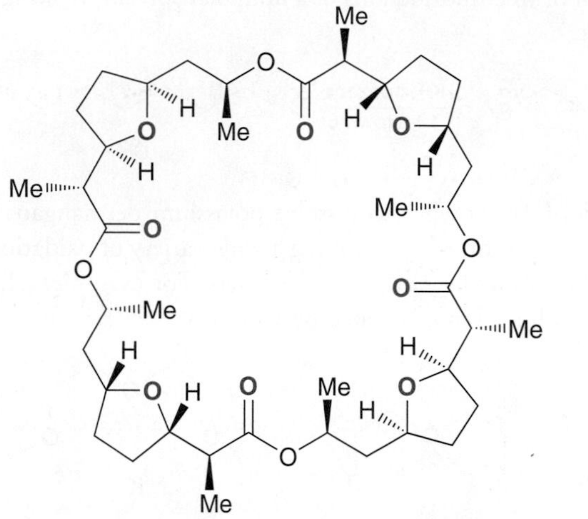

Nonactin

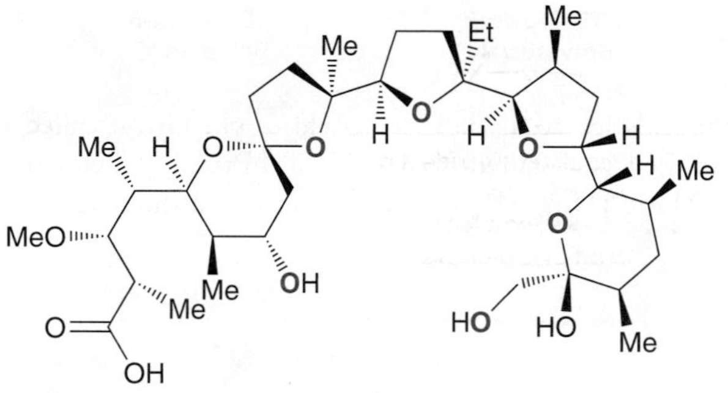

Monensin

These compounds are polyethers and therefore are capable of serving as hosts for metal cations, much like crown ethers. These polyethers are called *ionophores* because the internal cavity is capable of binding a metal ion. The outside surface of the ionophore is hydrocarbon-like (or *lipophilic*), allowing it to pass through cell membranes readily.

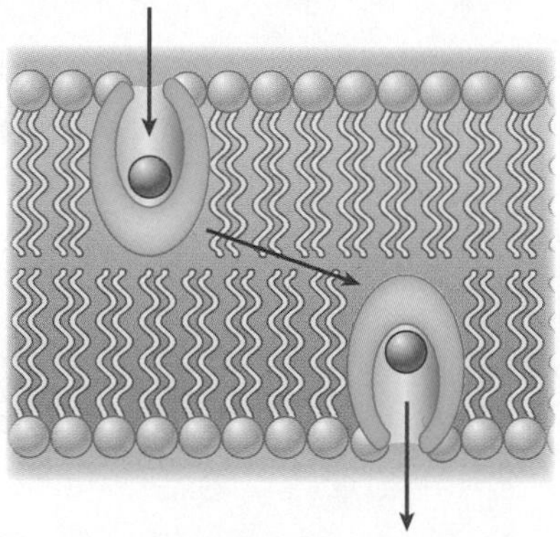

In order to function properly, cells must maintain a gradient between the concentration of sodium and potassium ions inside and outside the cell. That gradient is established because ions are not free to pass through the cell membrane, except through special ion channels where K^+ ions are pumped into the cell and Na^+ ions are pumped out of the cell. Ionophores effectively render the cell membrane permeable to these ions. The ionophores serve as hosts to the ions, carrying them across the cell membrane and destroying the necessary concentration gradient. In this way, ionophores interrupt cell function, thereby killing bacteria. Many new ionophores are currently under investigation as new potential antibiotics.

14.5 Preparation of Ethers

Industrial Preparation of Diethyl Ether

Diethyl ether is prepared industrially via the acid-catalyzed dehydration of ethanol. The mechanism of this process is believed to involve an S_N2 process.

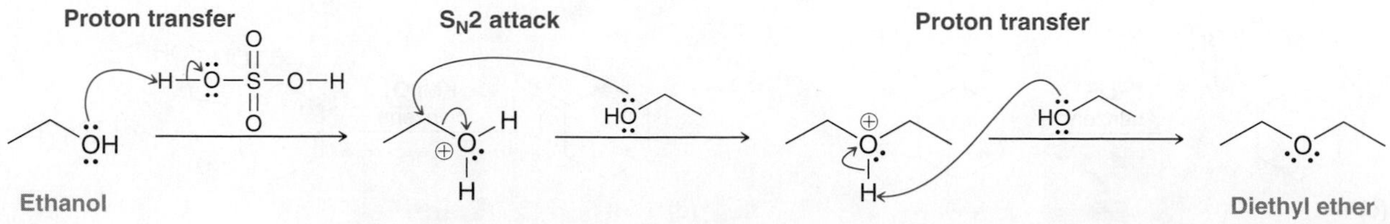

A molecule of ethanol is protonated and then attacked by another molecule of ethanol in an S_N2 process. As a final step, deprotonation generates the product. Notice that a proton is used in the first step of the mechanism, and then another proton is liberated in the last step of the mechanism. The acid is therefore a catalyst (not consumed by the reaction) that enables the S_N2 process to proceed.

This process has many limitations. For example, it only works well for primary alcohols (since it proceeds via an S_N2 pathway), and it produces symmetrical ethers. As a result, this process for preparing ethers is too limited to be of any practical value for organic synthesis.

Williamson Ether Synthesis

Ethers can be readily prepared via a two-step process called a **Williamson ether synthesis.**

$$R-OH \xrightarrow[\text{2) RX}]{\text{1) NaH}} R-O-R$$

We learned both of these steps in the previous chapter. In the first step, the alcohol is deprotonated to form an alkoxide ion. In the second step, the alkoxide ion functions as a nucleophile in an S_N2 reaction (Mechanism 14.1).

MECHANISM 14.1 THE WILLIAMSON ETHER SYNTHESIS

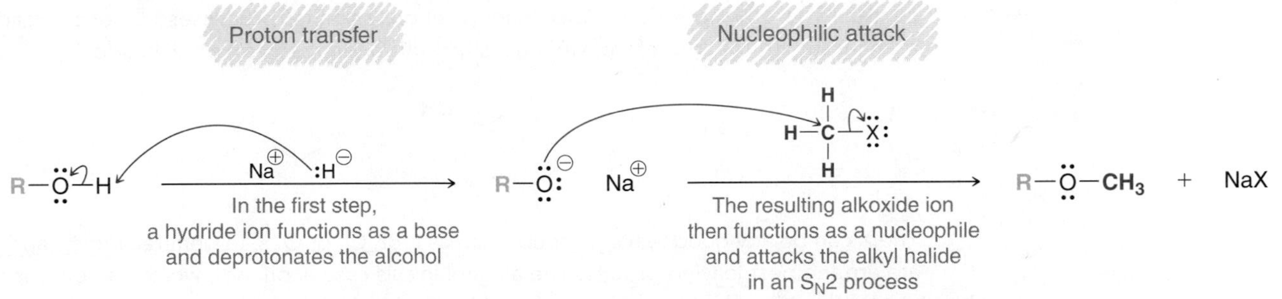
Proton transfer
Nucleophilic attack
In the first step, a hydride ion functions as a base and deprotonates the alcohol
The resulting alkoxide ion then functions as a nucleophile and attacks the alkyl halide in an SN2 process

BY THE WAY

The *tert*-butyl group is alphabetized by the letter "b" rather than "t," and therefore, the *tert*-butyl group precedes the methyl group in the name. This compound is commonly called MTBE, which is an acronym of the incorrect name.

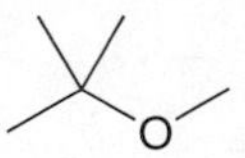

***tert*-Butyl methyl ether (MTBE)**

This process is named after Alexander Williamson, a British scientist who first demonstrated this method in 1850 as a way of preparing diethyl ether. Since the second step is an S_N2 process, steric effects must be considered. Specifically, the process works best when methyl or primary alkyl halides are used. Secondary alkyl halides are less efficient because elimination is favored over substitution and tertiary alkyl halides cannot be used. This limitation must be taken into account when choosing which C—O bond to form. For example, consider the structure of *tert*-butyl methyl ether. MTBE was used heavily as a gasoline additive until concerns emerged that it might contribute to groundwater contamination. As a result, its use has declined in recent years. There are two possible routes to consider in the preparation of MTBE, but only one is efficient.

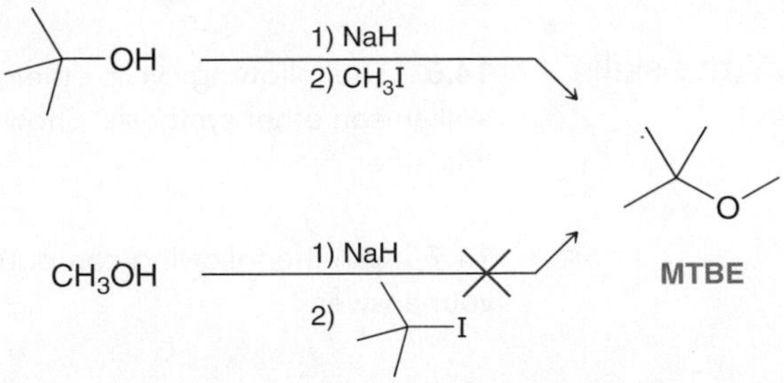
OH
1) NaH
2) CH3I
CH3OH
1) NaH
2)
I
MTBE

The first route is efficient because it employs a methyl halide, which is a suitable substrate for an S_N2 process. The second route does not work because it employs a tertiary alkyl halide, which will undergo elimination rather than substitution.

SKILLBUILDER

14.2 PREPARING AN ETHER VIA A WILLIAMSON ETHER SYNTHESIS

LEARN the skill

Show the reagents you would use to prepare the following ether via a Williamson ether synthesis.

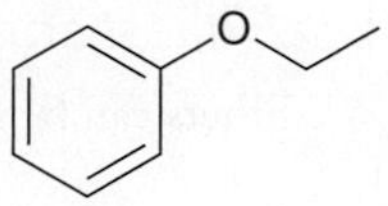

SOLUTION

In order to prepare this ether via a Williamson ether synthesis, we must decide which starting alcohol and which starting alkyl halide to use. To make the proper choice, we analyze the positions on either side of the oxygen atom:

STEP 1
Classify the groups on either side of the oxygen.

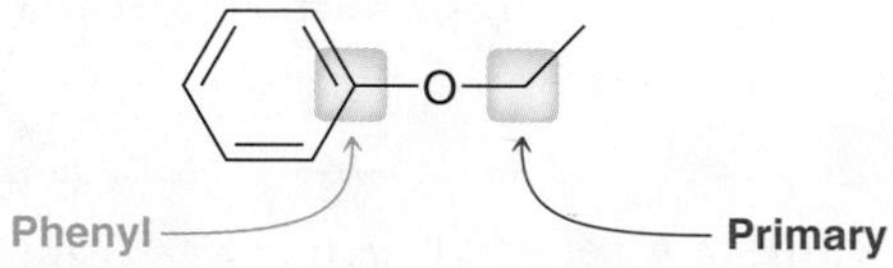

STEP 2
Determine which side is more capable of serving as a substrate in an S_N2 reaction.

The phenyl position is *sp*^{2} hybridized, and S_N2 processes do not occur at *sp*^{2}-hybridized centers. The other position is a primary position, and S_N2 processes can occur readily at primary substrates. Therefore, we must start with phenol and an ethyl halide.

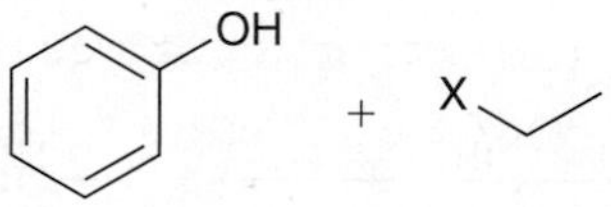

LOOKING BACK
For a review of leaving groups, see Section 7.8.

The X can be any good leaving group, such as I, Br, Cl, or OTs. In general, iodide and tosylate are the best leaving groups. The alcohol in this case is phenol, which can be deprotonated with sodium hydroxide (as we saw in Section 13.2). We therefore propose the following synthesis.

STEP 3
Use a base to deprotonate the alcohol and then introduce the alkyl halide.

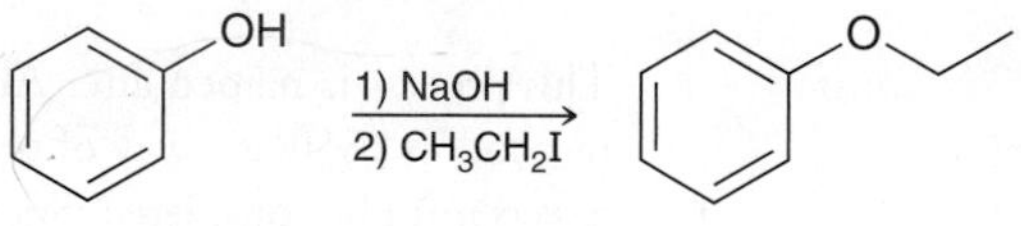

PRACTICE the skill

14.5 Show what reagents you would use to prepare each of the following ethers via a Williamson ether synthesis, and explain your reasoning.

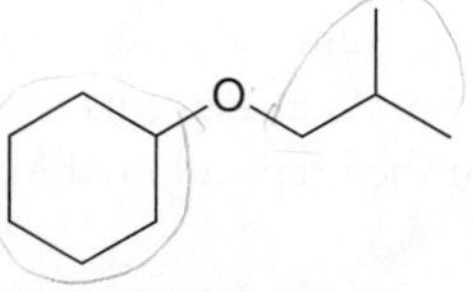
(a)

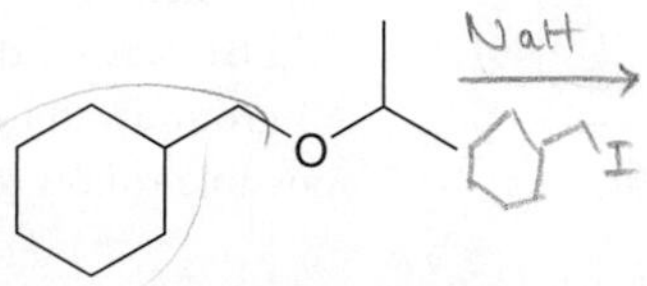
(b)

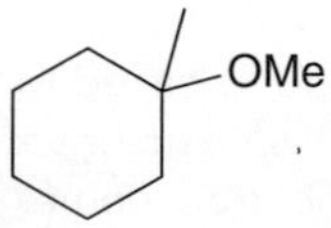
(c)

APPLY the skill

14.6 The following cyclic ether can be prepared via an intramolecular Williamson ether synthesis. Show what reagents you would use to make this ether.

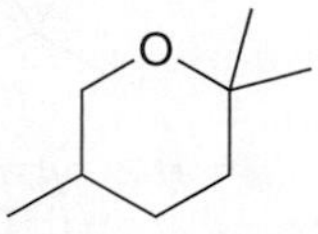

14.7 Can the following compound be prepared via a Williamson ether synthesis? Explain your answer.

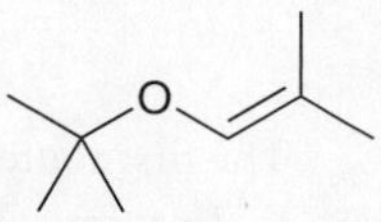

need more **PRACTICE?** **Try Problems 14.33a,c, 14.37, 14.40, 14.42d, 14.43b**

Alkoxymercuration-Demercuration

Recall from Section 9.5 that alcohols can be prepared from alkenes via a process called oxymercuration-demercuration.

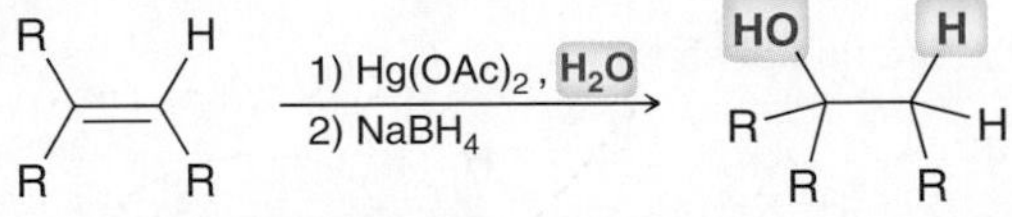

The net result is a Markovnikov addition of water (H and OH) across an alkene. That is, the hydroxyl group is ultimately placed at the more substituted position. The mechanism for this process was discussed in Section 9.3.

If an alcohol (ROH) is used in place of water, then the result is a Markovnikov addition of the alcohol (RO and H) across the alkene. This process is called **alkoxymercuration-demercuration**, and it can be used as another way of preparing ethers.

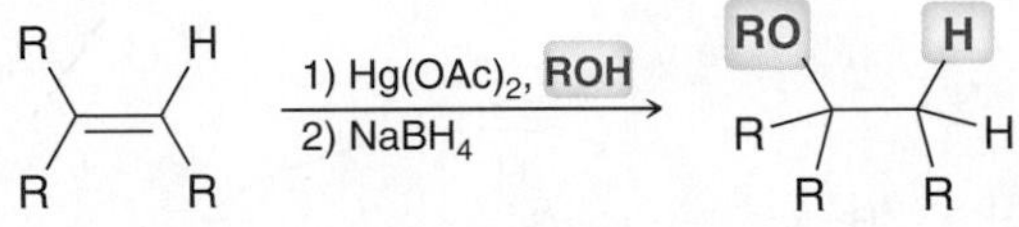

CONCEPTUAL CHECKPOINT

14.8 Show what reagents you would use to prepare each of the following ethers via an alkoxymercuration-demercuration.

(a)

(b)
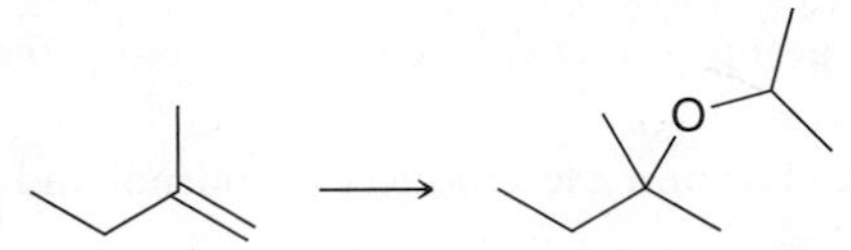

(c)
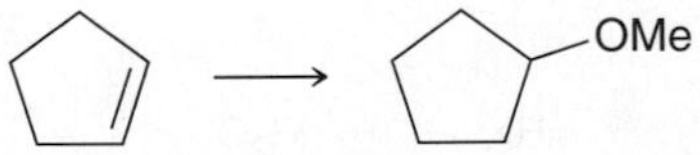

(d)
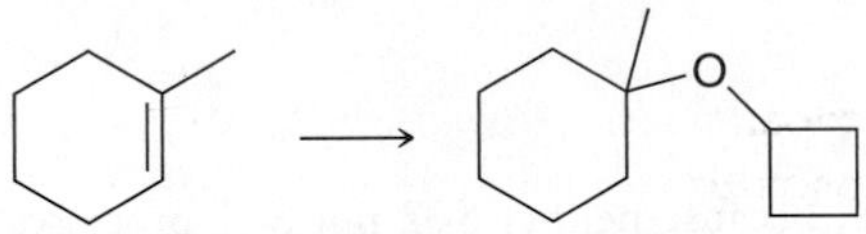

14.9 How would you use an alkoxymercuration-demercuration to prepare dicyclopentyl ether using cyclopentene as your only source of carbon?

14.10 Show how you would use an alkoxymercuration-demercuration to prepare isopropyl propyl ether using propene as your only source of carbon and any other reagents of your choosing.

14.6 Reactions of Ethers

Ethers are generally unreactive under basic or mildly acidic conditions. As a result, they are an ideal choice as solvents for many reactions. Nevertheless, ethers are not completely unreactive, and two reactions of ethers will be explored in this section.

Acidic Cleavage

When heated with a concentrated solution of a strong acid, an ether will undergo **acidic cleavage**, in which the ether is converted into two alkyl halides.

$$R{-}O{-}R \xrightarrow[\text{heat}]{\text{excess HX}} R{-}X + R{-}X + H_2O$$

This process involves two substitution reactions (Mechanism 14.2).

MECHANISM 14.2 ACIDIC CLEAVAGE OF AN ETHER

FORMATION OF FIRST ALKYL HALIDE

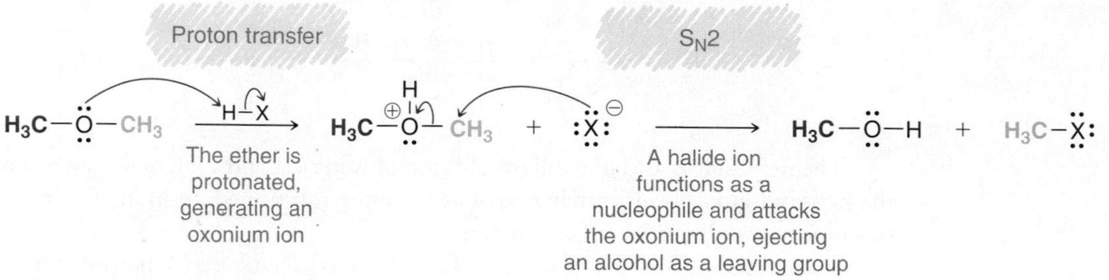
Proton transfer
S_N2
The ether is protonated, generating an oxonium ion
A halide ion functions as a nucleophile and attacks the oxonium ion, ejecting an alcohol as a leaving group

FORMATION OF SECOND ALKYL HALIDE

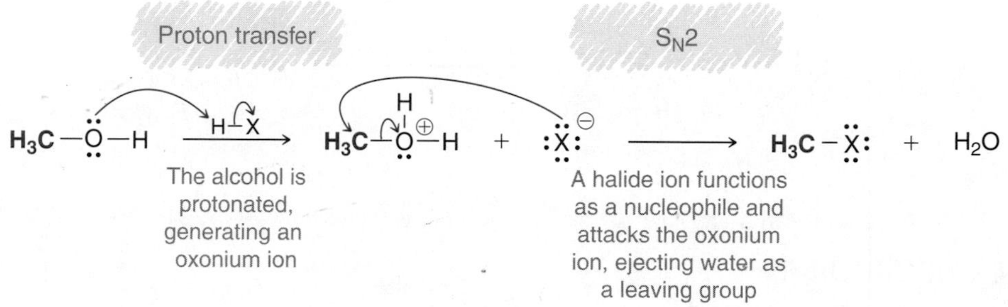
Proton transfer
S_N2
The alcohol is protonated, generating an oxonium ion
A halide ion functions as a nucleophile and attacks the oxonium ion, ejecting water as a leaving group

The formation of the first alkyl halide begins with protonation of the ether to form a good leaving group, followed by an S_N2 process in which a halide ion functions as a nucleophile and attacks the protonated ether. The second alkyl halide is then formed with the same two steps—protonation followed by an S_N2 attack. If either R group is tertiary, then substitution is more likely to proceed via an S_N1 process rather than S_N2.

When a phenyl ether is cleaved under acidic conditions, the products are phenol and an alkyl halide.

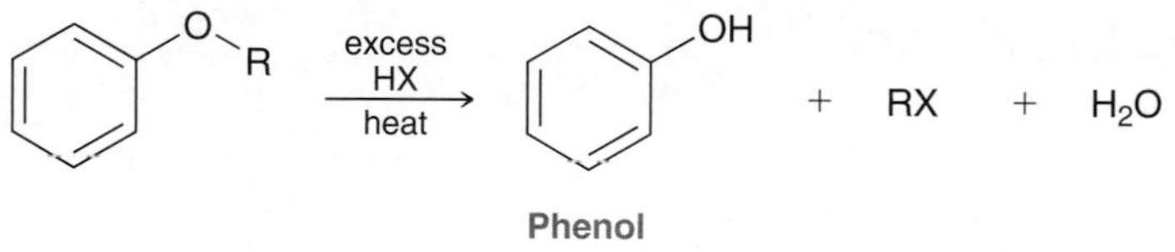
excess HX
heat
RX
H_2O
Phenol

LOOKING BACK
For a review of the relative nucleophilicity of the halide ions, see Section 7.8.

The phenol is not further converted into a halide, because neither S_N2 nor S_N1 processes are efficient at sp^2-hybridized centers.

Both HI and HBr can be used to cleave ethers. HCl is less efficient, and HF does not cause acidic cleavage of ethers. This reactivity is a result of the relative nucleophilicity of the halide ions.

CONCEPTUAL CHECKPOINT

14.11 Predict the products for each of the following reactions:

(a)
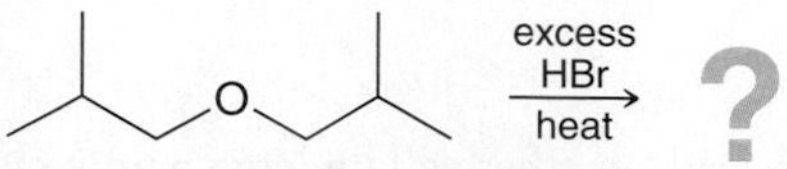
excess HBr
heat
?

(b)
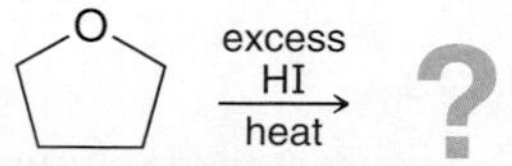
excess HI
heat
?

(c)
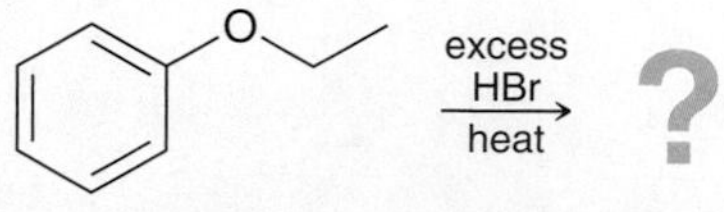
excess HBr
heat
?

(d)
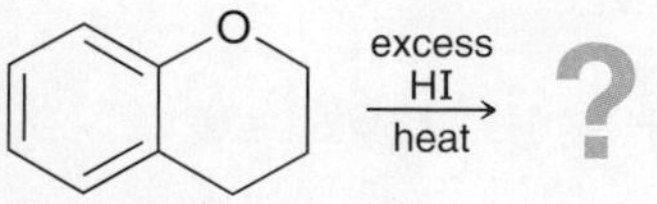
excess HI
heat
?

(e)
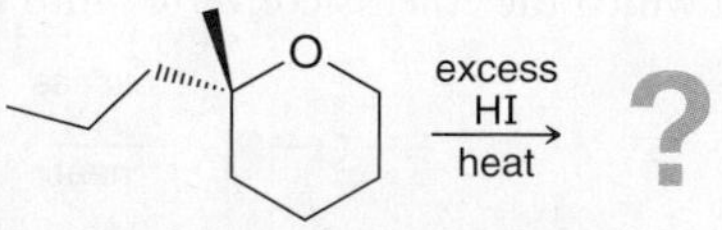
excess HI
heat
?

(f)
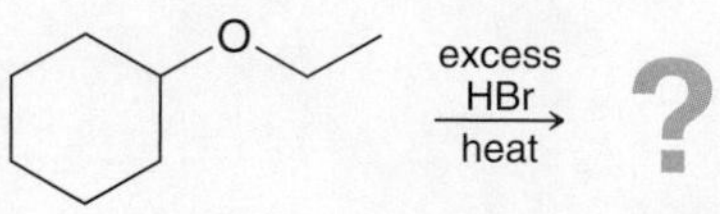
excess HBr
heat
?

Autooxidation

Recall from Section 11.9 that ethers undergo autooxidation in the presence of atmospheric oxygen to form hydroperoxides:

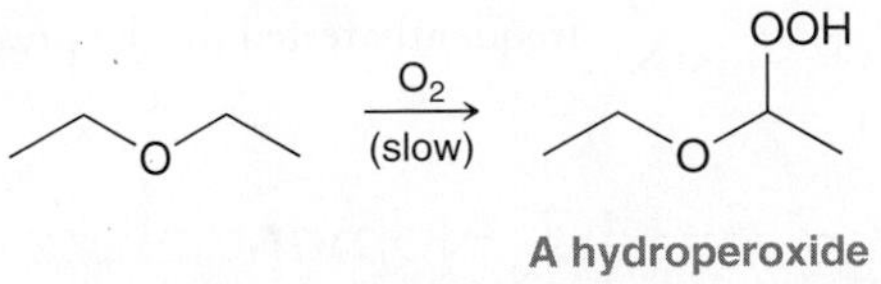

This process occurs via a radical mechanism which is initiated by a hydrogen abstraction (Mechanism 14.3).

MECHANISM 14.3 AUTOOXIDATION OF ETHERS

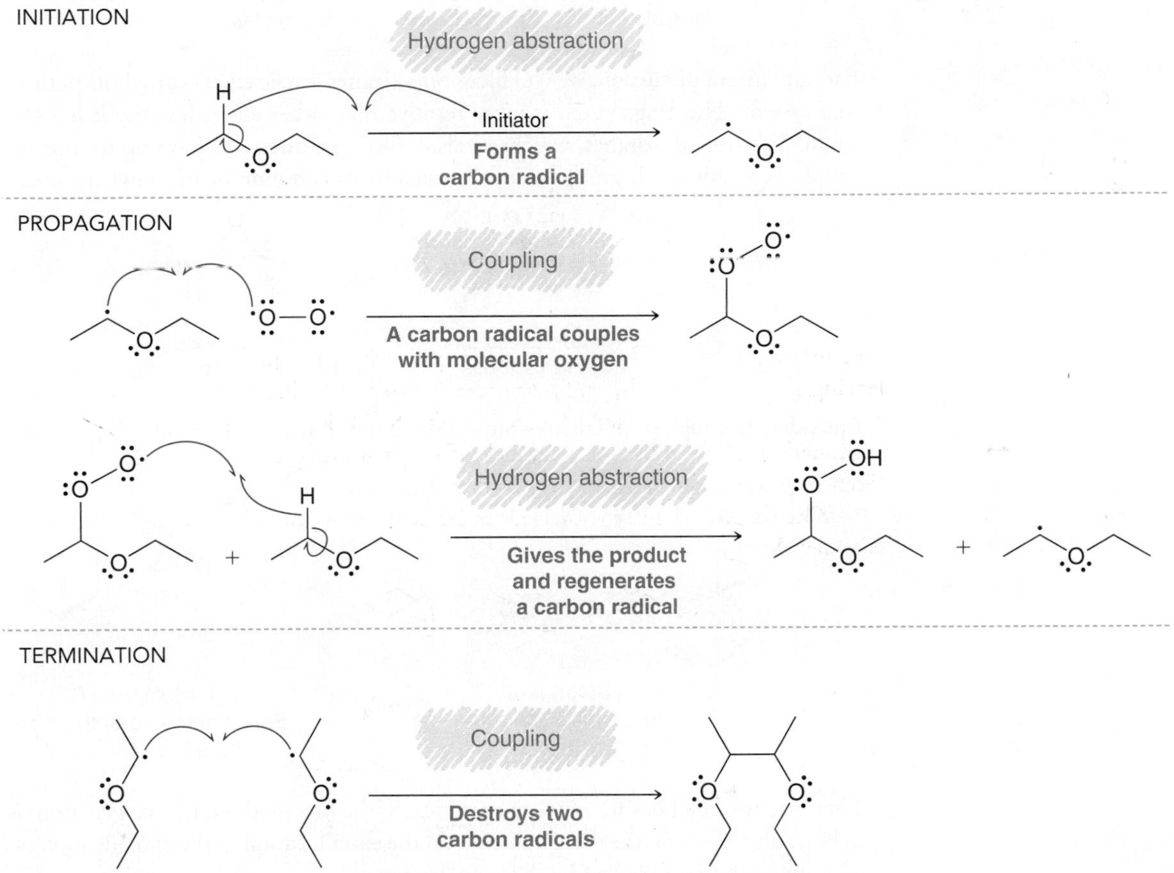

As with all radical mechanisms, the net reaction is the sum of the propagation steps:

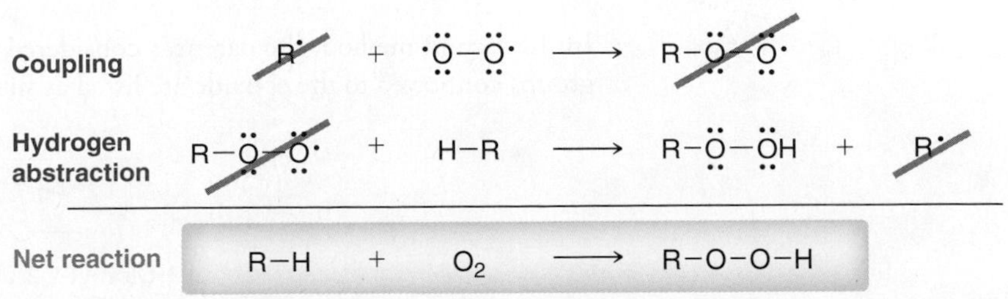

The reaction is slow, but old bottles of ether will invariably contain a small concentration of hydroperoxides, rendering the solvent very dangerous to use. Hydroperoxides are unstable and decompose violently when heated. Many laboratory explosions have been caused by distillation of ethers that were contaminated with hydroperoxides. Ethers used in the laboratory must be frequently tested for the presence of hydroperoxides and purified prior to use.

14.7 Nomenclature of Epoxides

Cyclic ethers are compounds that contain an oxygen atom incorporated in a ring. Special parent names are used to indicate the ring size.

Oxirane ring system | Oxetane ring system | Oxolane ring system | Oxane ring system

BY THE WAY
The common name "ethylene oxide" denotes the fact that it is produced from ethylene.

For our current discussion, we will focus on **oxiranes,** cyclic ethers containing a three-membered ring system. This ring system is more reactive than other ethers because it has significant ring strain. Substituted oxiranes, which are also called **epoxides**, can have up to four R groups. The simplest epoxide (no R groups) is often called by its common name, ethylene oxide.

A substituted oxirane (an epoxide)

Ethylene oxide (the simplest epoxide)

Epoxides, although strained, are commonly found in nature. The following are two examples.

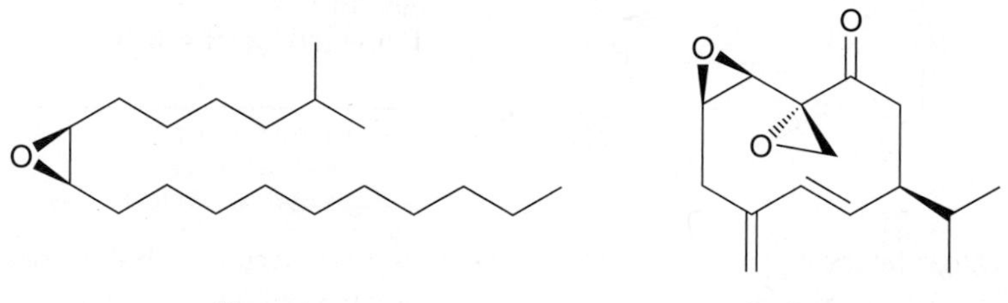

Disparlure
Sex pheromone of the female gypsy moth

Periplanone B
Sex pheromone of the female American cockroach

There are two methods for naming epoxides. In the first method, the oxygen atom is considered to be a substituent on the parent chain, and the exact location of the epoxide moiety is identified with two numbers followed by the term "epoxy."

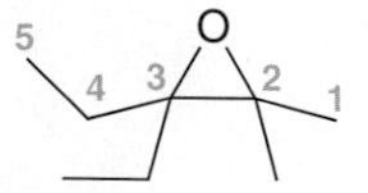

3-Ethyl-2-methyl-2,3-epoxypentane

In the second method, the parent is considered to be the epoxide (parent = oxirane), and any groups connected to the epoxide are listed as substituents.

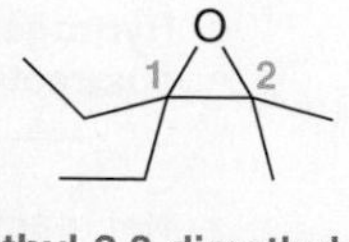

1,1-Diethyl-2,2-dimethyloxirane

CONCEPTUAL CHECKPOINT

14.12 Assign a name for each of the following compounds.

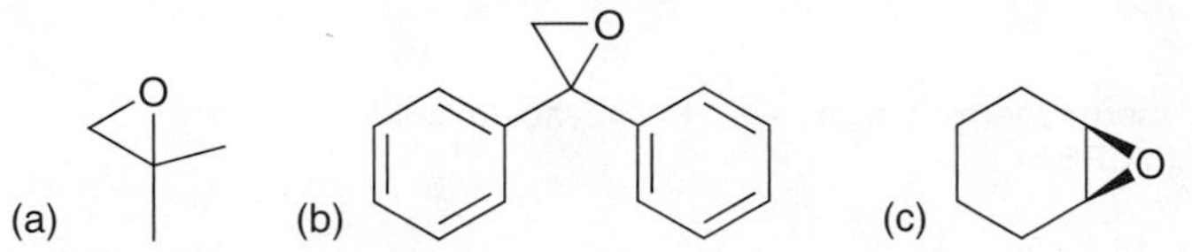

14.13 Assign a name for each of the following compounds. Be sure to assign the configuration of each chirality center and indicate the configuration(s) at the beginning of the name.

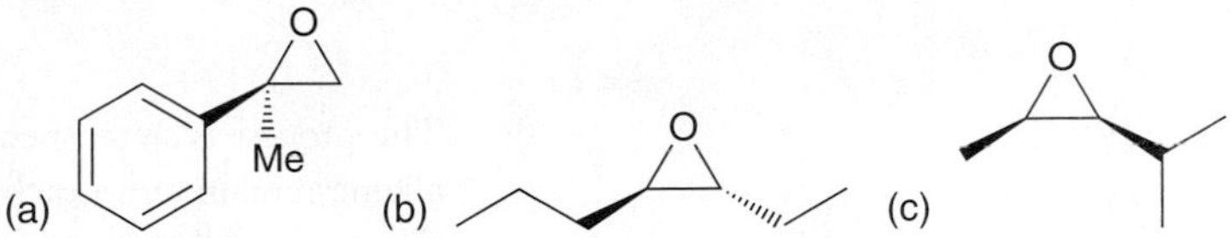

MEDICALLYSPEAKING

Epothilones as Novel Anticancer Agents

Epothilones are a class of novel compounds first isolated from the bacterium *Sorangium cellulosum* in Southern Africa.

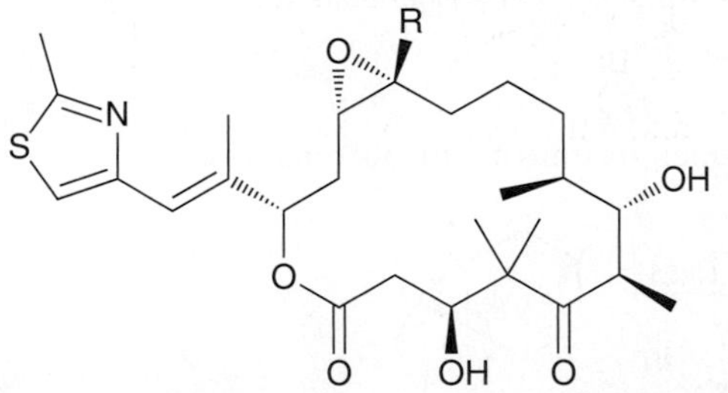

Epothilone A (R = H)
Epothilone B (R = Me)

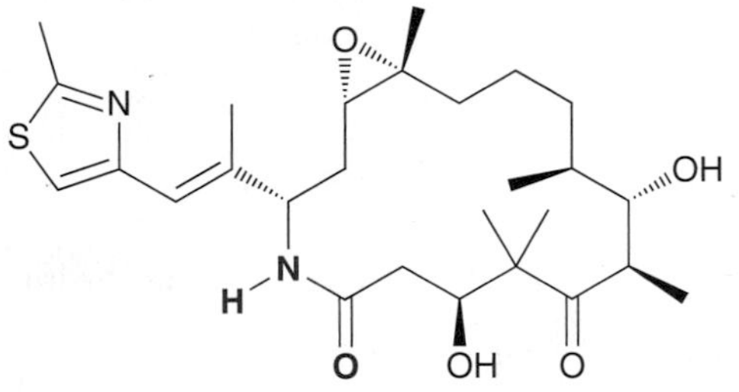

Ixabepilone

The discovery of the antitumor behavior of these naturally occurring epoxides led to a search for related compounds that might exhibit enhanced potency and selectivity. In October 2007, the FDA approved one such derivative, called ixabepilone, for treatment of advanced breast cancer. Ixabepilone is an analog of epothilone B, in which the ester linkage is replaced with an amide linkage, highlighted in red.

Ixabepilone is currently being marketed by Bristol-Myers Squibb under the trade name Ixempra. Several other analogs of the epothilones are currently undergoing clinical trials for treatment of many different forms of cancer. The next decade is likely to witness several epothilone analogs emerge as new anticancer agents.

14.8 Preparation of Epoxides

Preparation with Peroxy Acids

Recall from Section 9.9 that alkenes can be converted into epoxides upon treatment with peroxy acids (see Mechanism 9.6).

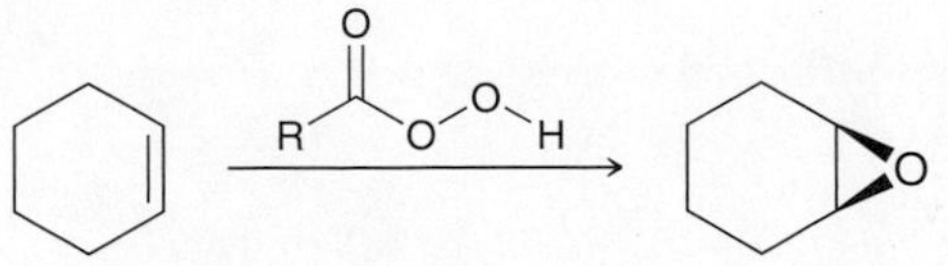

Any peroxy acid can be used, although MCPBA and peroxyacetic acid are the most common:

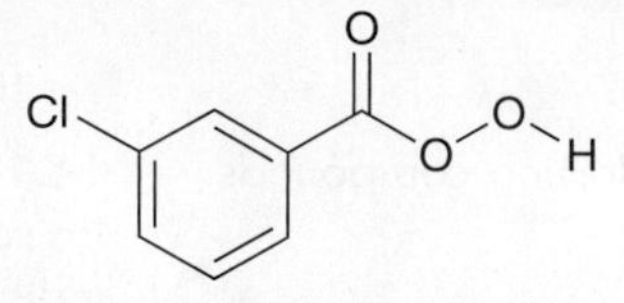

***meta*-Chloroperoxybenzoic acid (MCPBA)**

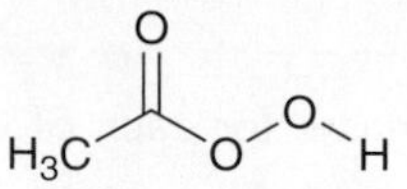

Peroxyacetic acid

The process is stereospecific. Specifically, substituents that are *cis* to each other in the starting alkene remain *cis* to each other in the epoxide, and substituents that are *trans* to each other in the starting alkene remain *trans* to each other in the epoxide:

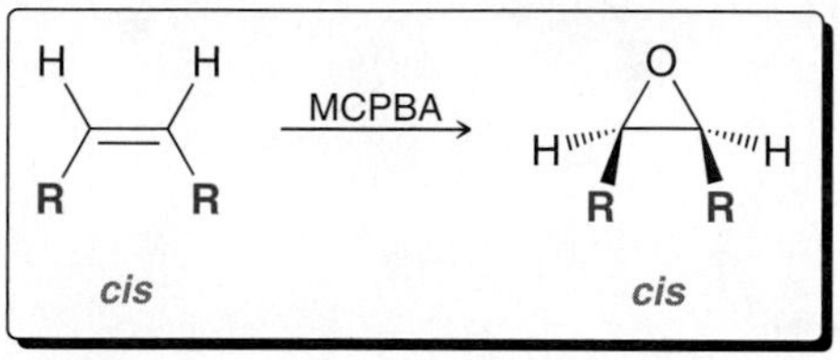
MCPBA
cis
cis

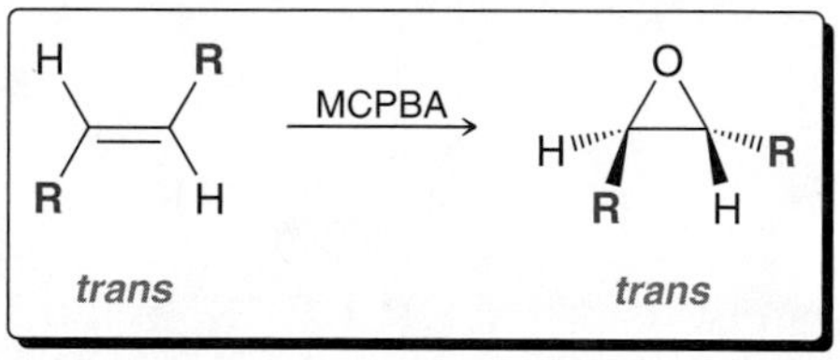
MCPBA
trans
trans

Preparation from Halohydrins

Recall from Section 9.8 that alkenes can be converted into halohydrins when treated with a halogen in the presence of water (see Mechanism 9.5).

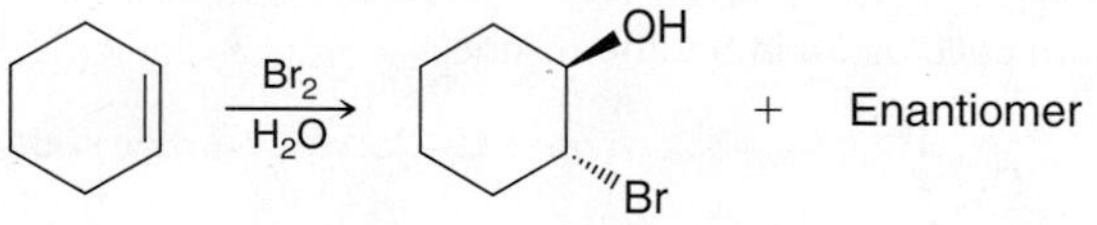
Br2
H2O
OH
Br
+ Enantiomer

Halohydrins can be converted into epoxides upon treatment with a strong base:

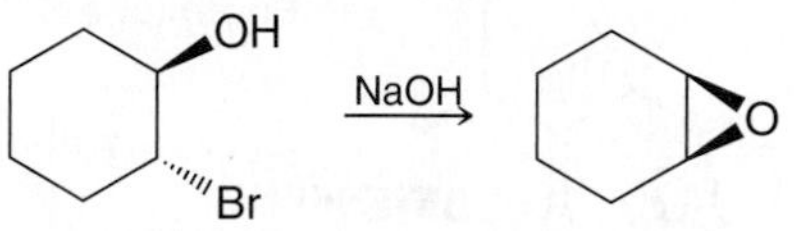
OH
Br
NaOH

The process is achieved via an intramolecular Williamson ether synthesis. An alkoxide ion is formed, which then functions as a nucleophile in an intramolecular S_N2-like process (Mechanism 14.4).

MECHANISM 14.4 EPOXIDE FORMATION FROM HALOHYDRINS

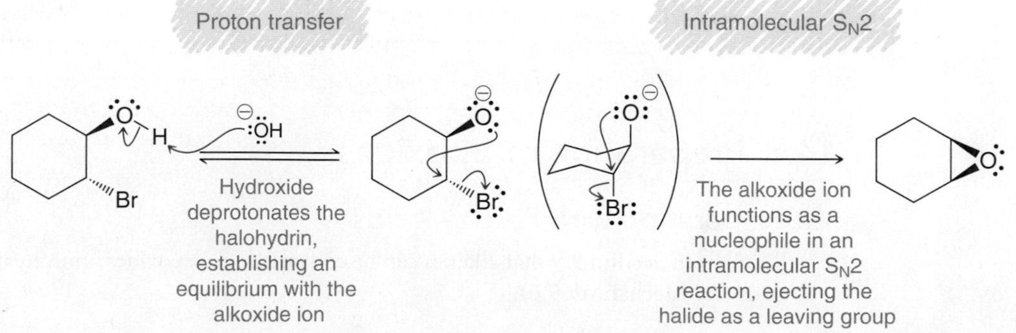
Proton transfer
Intramolecular SN2
Hydroxide deprotonates the halohydrin, establishing an equilibrium with the alkoxide ion
The alkoxide ion functions as a nucleophile in an intramolecular SN2 reaction, ejecting the halide as a leaving group

This provides us with another way of forming an epoxide from an alkene.

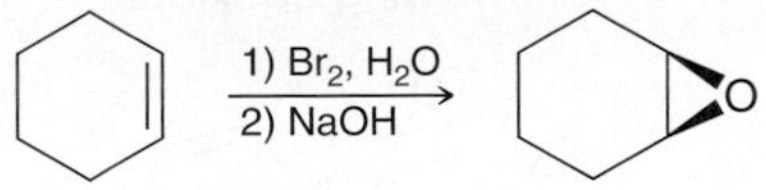

The overall stereochemical outcome is the same as direct epoxidation with MCPBA. That is, substituents that are *cis* to each other in the starting alkene remain *cis* to each other in the epoxide, and substituents that are *trans* to each other in the starting alkene remain *trans* to each other in the epoxide.

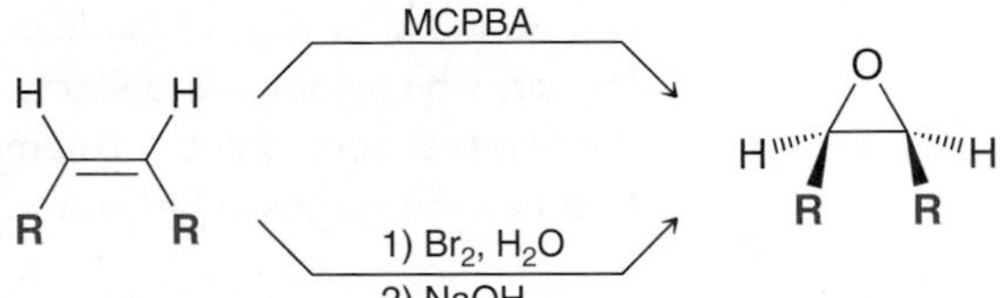

cis *cis*

SKILLBUILDER

14.3 PREPARING EPOXIDES

LEARN the skill

Show what reagents you would use to prepare the following epoxide.

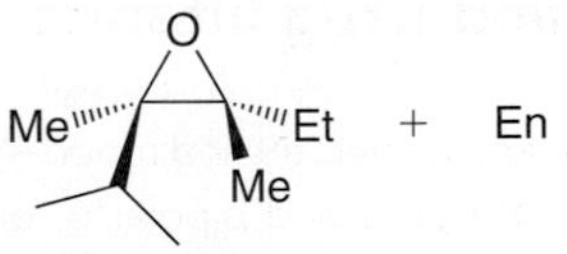

SOLUTION

Begin by analyzing all four groups connected to the epoxide.

STEP 1
Identify the four groups attached to the epoxide.

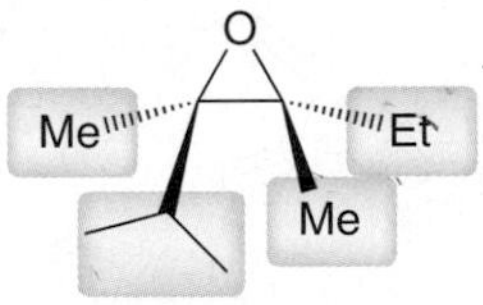

STEP 2
Identify the relative configuration of the four groups in the starting alkene.

The starting alkene must contain these four groups. Look carefully at the relative configuration of these groups. The methyl groups are *trans* to each other in the epoxide, which means that they must have been *trans* to each other in the starting alkene. To convert this alkene into the epoxide, we can use either of the following acceptable methods.

STEP 3
Use either of the following two methods to prepare the epoxide from the starting alkene.

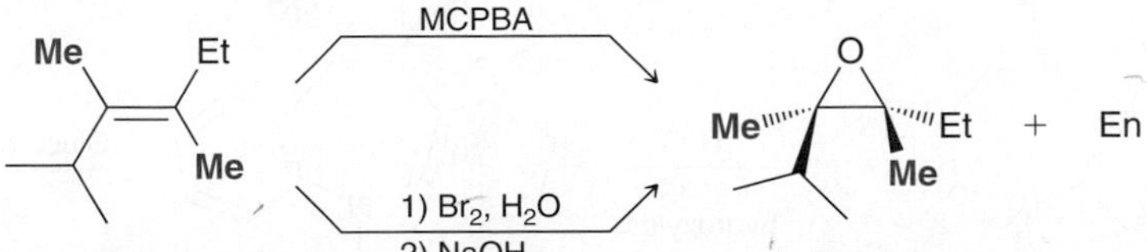

trans *trans*

PRACTICE the skill **14.14** Identify the reactants you would use to form a racemic mixture of each of the following epoxides:

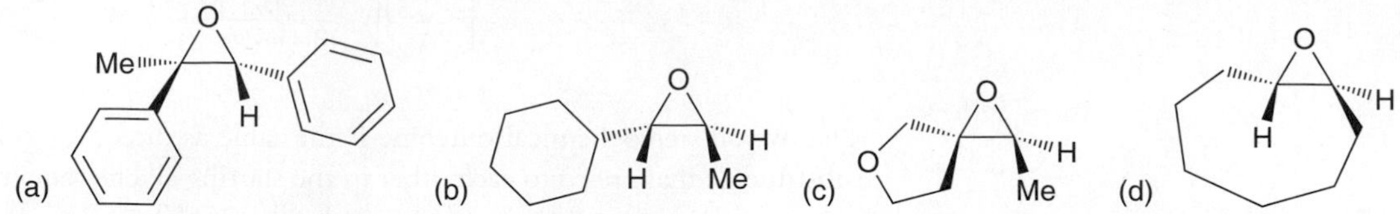

APPLY the skill **14.15** Consider the following two compounds. When treated with NaOH, one of these compounds forms an epoxide quite rapidly, while the other forms an epoxide very slowly. Identify which compound reacts more rapidly and explain the difference in rate between the two reactions. (Hint: You may find it helpful to review the conformations of substituted cyclohexanes in Section 4.13.)

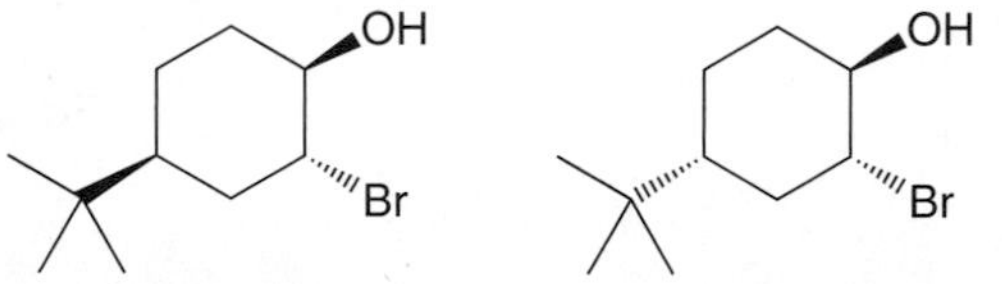

Compound A **Compound B**

need more **PRACTICE?** **Try Problems 14.39, 14.51t**

MEDICALLYSPEAKING
Active Metabolites and Drug Interactions

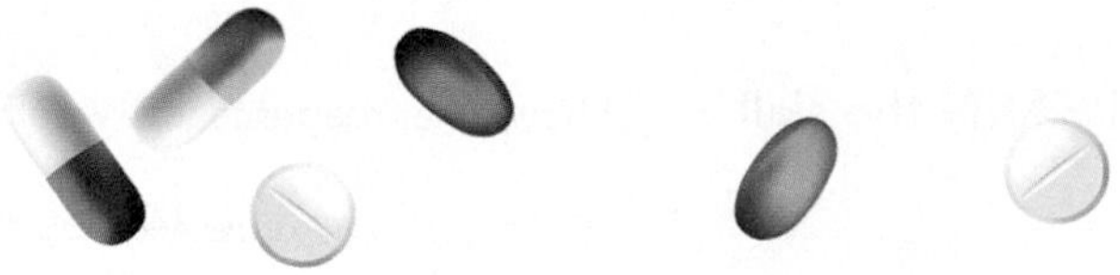

Carbamazepine (Tegretol) is an anticonvulsant and mood-stabilizing drug used in the treatment of epilepsy and bipolar disorder. It is also used to treat ADD (attention-deficit disorder). It is metabolized in the liver to produce an epoxide.

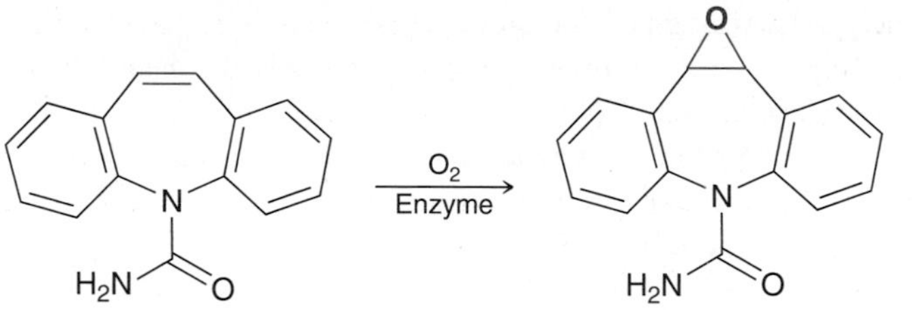

Carbamazepine **Carbamazepine-10,11-epoxide**

This epoxide is further metabolized by the enzyme epoxide hydroxylase to form a *trans* diol, which undergoes glucuronidation to produce a water-soluble adduct that can be excreted in the urine.

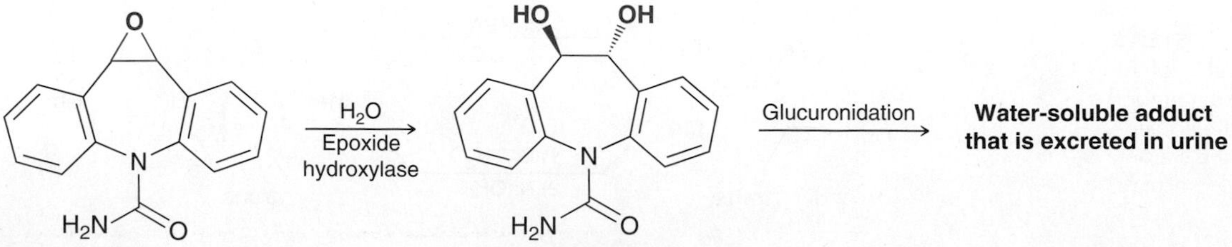

Carbamazepine-10,11-epoxide ***trans*-10,11-Dihydroxy-carbamazepine**

The epoxide metabolite is believed to exhibit activity similar to that of the parent compound and therefore contributes substantially to the overall therapeutic effects of carbamazepine. This fact must be taken into consideration when a patient is taking other medications. For instance, the antibiotic clarithromycin has been found to inhibit the action of the enzyme epoxide hydroxylase. This causes the concentration of the epoxide to be higher than normal, increasing the potency of carbamazepine. Before a physician prescribes carbamazepine for a patient, potential drug interactions must be taken into account. This is an example of one important factor that practicing physicians must consider—specifically, the effect that one drug can have on the potency of another drug.

14.9 Enantioselective Epoxidation

When forming an epoxide that is chiral, both of the previous methods will provide a racemic mixture.

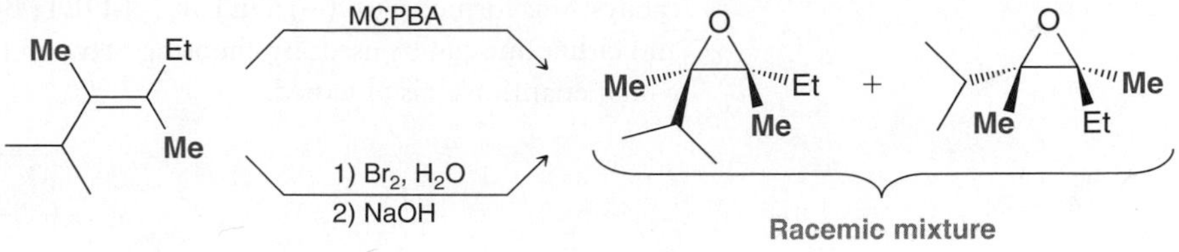

That is, the two enantiomers are formed in equal amounts, because the epoxide can be formed on either face of the alkene with equal likelihood.

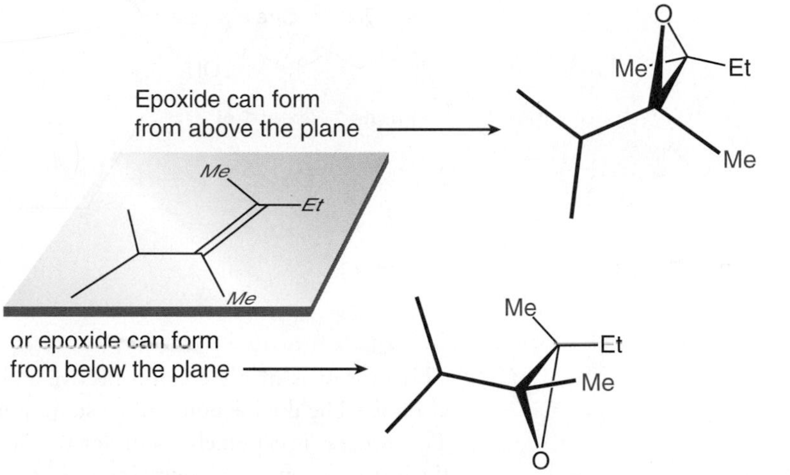

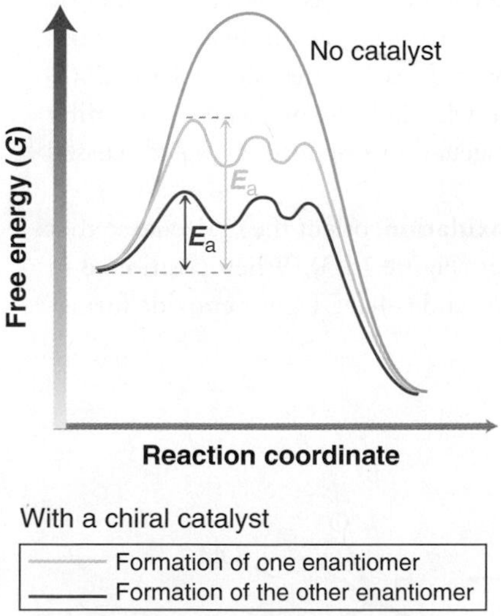

FIGURE 14.2
An energy diagram that depicts the effect of a chiral catalyst. The formation of one enantiomer is more effectively catalyzed than the other.

If only one enantiomer is desired, then the methods we have learned for preparing epoxides will be inefficient, as half of the product is unusable and must be separated from the desired product. To favor formation of just one enantiomer, we must somehow favor epoxidation at one face of the alkene. K. Barry Sharpless, currently at the Scripps Research Institute, recognized that this could be accomplished with a chiral catalyst. He reasoned that a chiral catalyst could, in theory, create a chiral environment that would favor epoxidation at one face of the alkene. Specifically, a chiral catalyst can lower the energy of activation for formation of one enantiomer more dramatically than the other enantiomer (Figure 14.2). In this way, a chiral catalyst favors the production of one enantiomer over the other, leading to an observed enantiomeric excess (*ee*). Sharpless succeeded in developing such a catalyst for the *enantioselective epoxidation* of allylic alcohols. An allylic alcohol is an alkene in which a hydroxyl group is attached to an allylic position. Recall that the allylic position is the position next to a C=C bond.

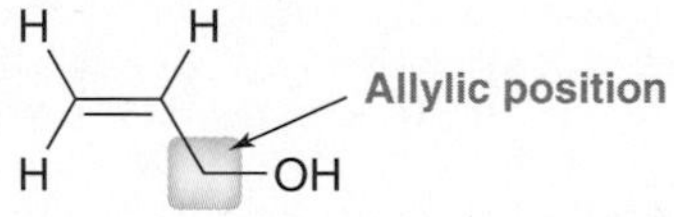

Sharpless' catalyst is comprised of titanium tetraisopropoxide and one enantiomer of diethyl tartrate (DET).

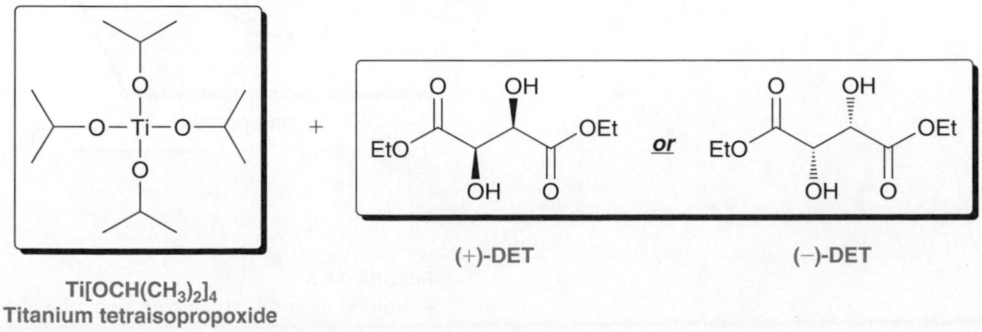

Titanium tetraisopropoxide forms a chiral complex with either (+)-DET or (–)-DET, and this complex serves as the chiral catalyst. In the presence of such a catalyst, an oxidizing agent such as *tert*-butyl hydroperoxide (ROOH, where R = *tert*-butyl) can be employed to convert the alkene into an epoxide. The stereochemical outcome of the reaction depends on whether the chiral catalyst was formed with (+)-DET or (–)-DET. Both enantiomers of DET are readily available, and either one can be used. By choosing between (+)-DET or (–)-DET, it is possible to control which enantiomer is obtained.

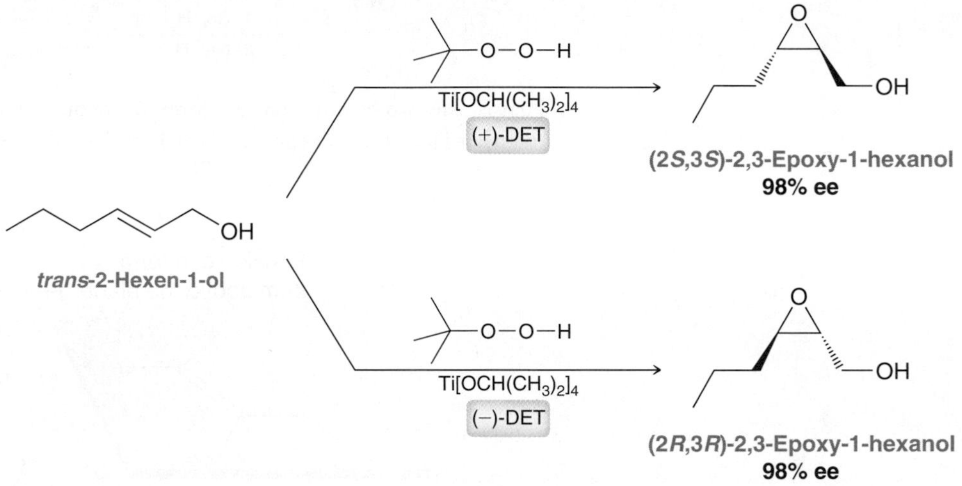

This process is highly enantioselective and is extremely successful for a wide range of allylic alcohols. The double bond in the starting material can be mono-, di-, tri-, or tetrasubstituted. This process is extremely useful for the practicing synthetic organic chemist because it allows the introduction of a chirality center with enantioselectivity. Since Sharpless pioneered the field of enantioselective epoxidation, many more reagents have been developed for the asymmetric epoxidation of other alkenes that do not require the presence of an allylic hydroxyl group. Sharpless was instrumental in opening an important door and was a corecipient of the 2001 Nobel Prize in Chemistry (the other corecipients were Knowles and Noyori, who used similar reasoning to develop chiral catalysts for asymmetric hydrogenation reactions, as was discussed in Section 9.7).

To predict the product of a **Sharpless asymmetric epoxidation**, orient the molecule so that the allylic hydroxyl group appears in the upper right corner (Figure 14.3). When positioned in this way, (+)-DET gives epoxide formation above the plane, and (–)-DET gives epoxide formation below the plane.

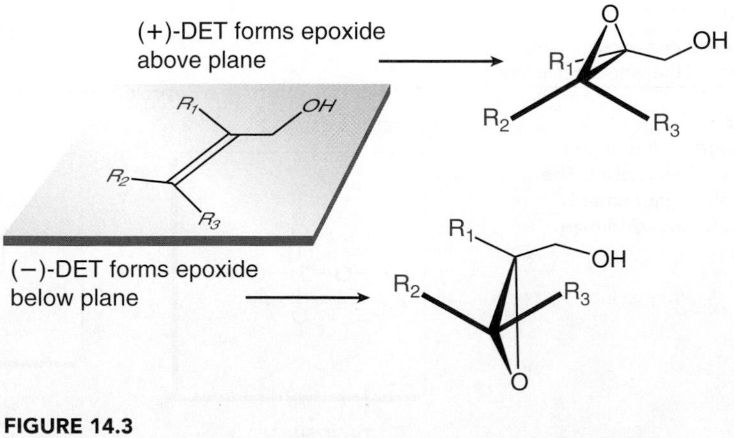

FIGURE 14.3
A method for predicting the product of a Sharpless epoxidation.

CONCEPTUAL CHECKPOINT

14.16 Predict the products for each of the following reactions:

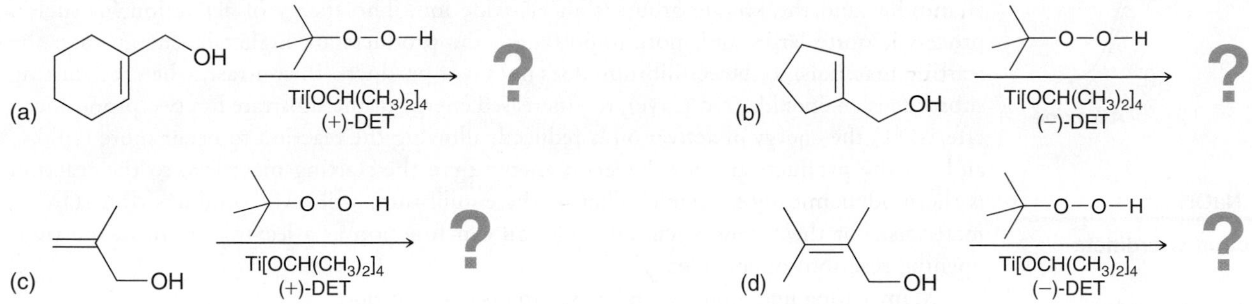

14.10 Ring-Opening Reactions of Epoxides

Epoxides have significant ring strain, and as a result, they exhibit unique reactivity. Specifically, epoxides undergo reactions in which the ring is opened, which alleviates the strain. In this section, we will see that epoxides can be opened under conditions involving a strong nucleophile or under acid-catalyzed conditions.

Reactions of Epoxides with Strong Nucleophiles

When an epoxide is subjected to attack by a strong nucleophile, a **ring-opening reaction** occurs. For example, consider the opening of ethylene oxide by a hydroxide ion.

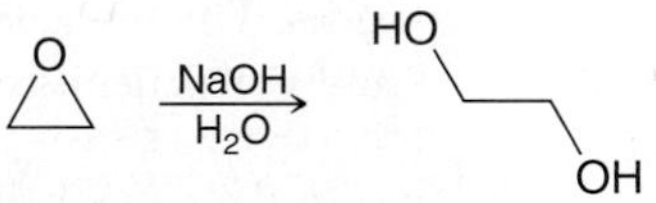

The transformation involves two mechanistic steps (Mechanism 14.5).

MECHANISM 14.5 EPOXIDE RING OPENING WITH A STRONG NUCLEOPHILE

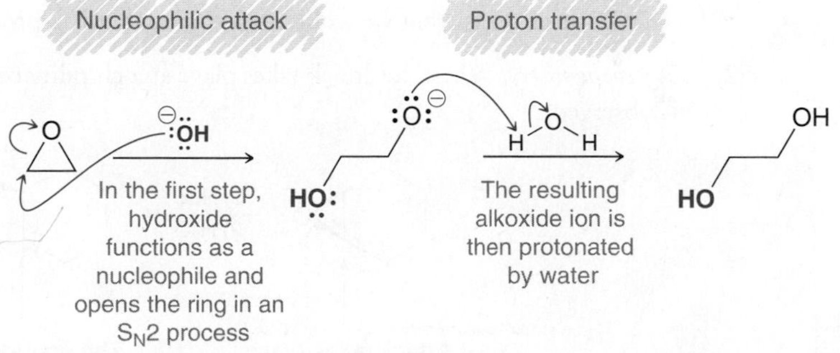

The first step of the mechanism is an S_N2 process, involving an alkoxide ion functioning as a leaving group. Although we learned in Chapter 7 that alkoxide ions do not function as leaving groups in S_N2 reactions, the exception here can be explained by focusing on the

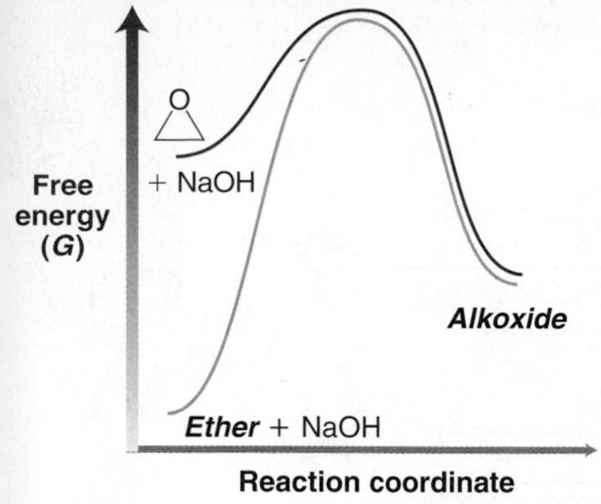

FIGURE 14.4
An energy diagram showing the effect of using a high-energy substrate in an S_N2 reaction.

substrate. In this case, the substrate is an epoxide that exhibits significant ring strain and is therefore higher in energy than the substrates we encountered when we first learned about S_N2 reactions. The effects of a high-energy substrate are illustrated in the energy diagram in Figure 14.4. The blue curve represents a hypothetical S_N2 process in which the substrate is an ether and the leaving group is an alkoxide ion. The energy of activation for such a process is quite large, and more importantly, the products are higher in energy than the starting materials, so the equilibrium does not favor products. In contrast, when the starting substrate is an epoxide (red curve), the increased energy of the substrate has two pronounced effects: (1) the energy of activation is reduced, allowing the reaction to occur more rapidly, and (2) the products are now lower in energy than the starting materials, so the reaction is thermodynamically favorable. That is, the equilibrium will favor products over starting materials. For these reasons, an alkoxide ion can function as a leaving group in the ring-opening reactions of epoxides.

Many strong nucleophiles can be used to open an epoxide.

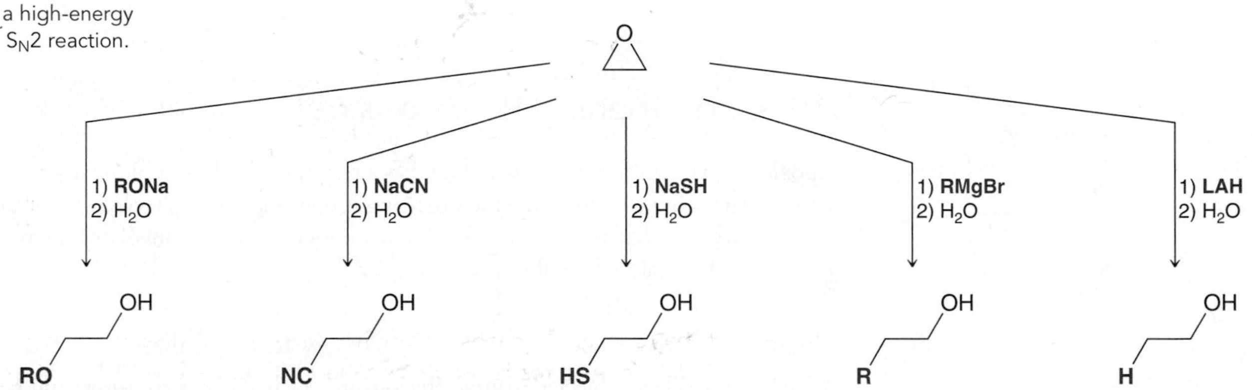

All of these nucleophiles are reagents that we have previously encountered, and they can all open epoxides. These reactions exhibit two important features that must be considered, regiochemistry and stereochemistry.

1. *Regiochemistry*. When the starting epoxide is unsymmetrical, the nucleophile attacks at the less substituted (less hindered) position.

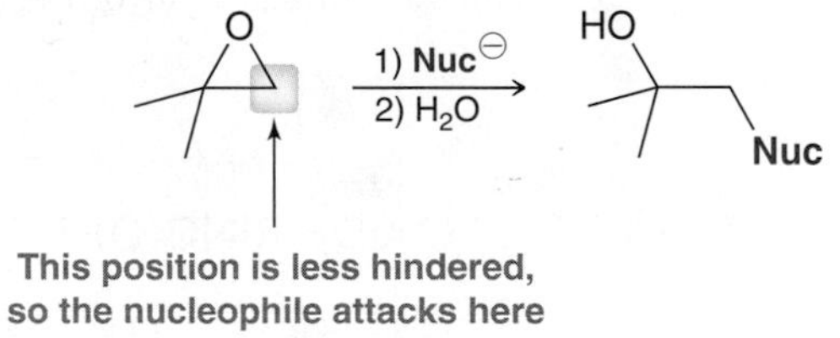

This steric effect is what we would expect from an S_N2 process.

2. *Stereochemistry*. When the attack takes place at a chirality center, inversion of configuration is observed.

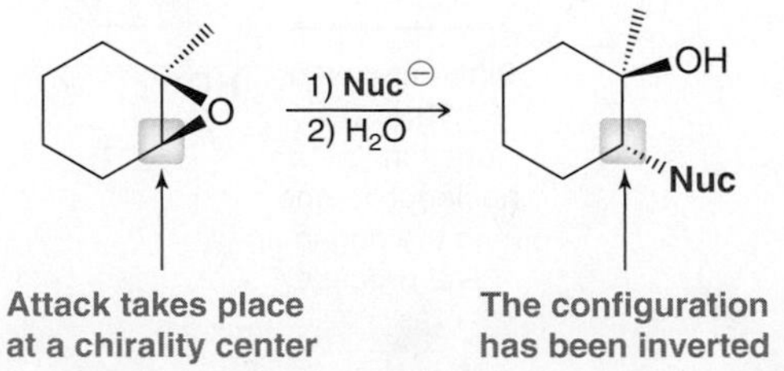

This result is also expected for an S_N2 process as a consequence of the requirement for backside attack of the nucleophile. Notice that the configuration of the other chirality center is not affected by the process. Only the center being attacked undergoes an inversion of configuration.

SKILLBUILDER

14.4 DRAWING THE MECHANISM AND PREDICTING THE PRODUCT OF THE REACTION BETWEEN A STRONG NUCLEOPHILE AND AN EPOXIDE

LEARN the skill

Predict the product of the following reaction, and draw the likely mechanism for its formation.

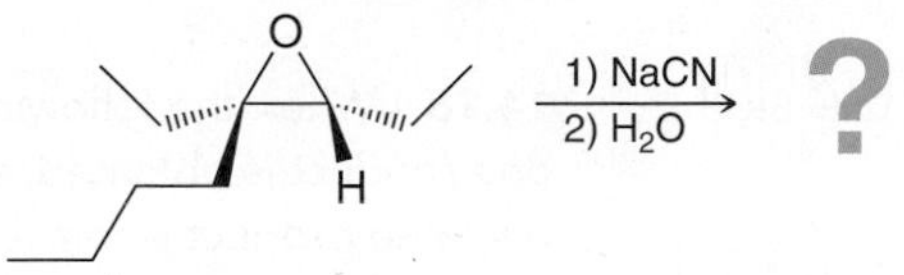

SOLUTION

Cyanide is a strong nucleophile, so we expect a ring-opening reaction. In order to draw the product, we must consider the regiochemistry and stereochemistry of the reaction.

STEP 1
Identify the regiochemistry by selecting the less hindered position as the site of nucleophilic attack.

1. To predict the regiochemistry, identify the less substituted (less hindered) position.

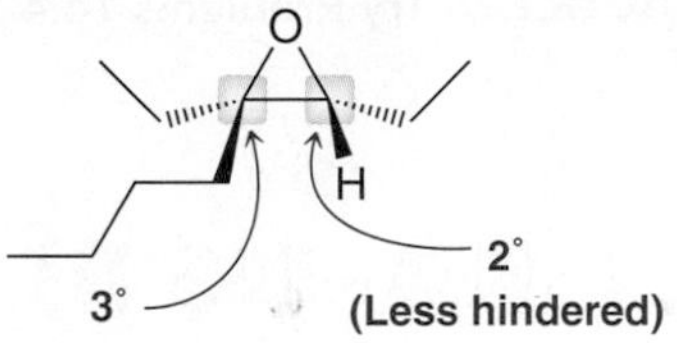

Recall that the reaction proceeds via an S_N2 process and is therefore highly sensitive to steric hindrance. As a result, we expect the nucleophile to attack the less hindered, secondary position, rather than the more hindered, tertiary position.

STEP 2
Identify the stereochemistry by determining whether the nucleophile attacks a chirality center. If so, expect inversion of configuration.

2. Next identify the stereochemical outcome. Look at the center being attacked and determine if it is a chirality center. In this case, it is a chirality center. An S_N2 process is expected to proceed via back-side attack to give inversion of configuration at that center.

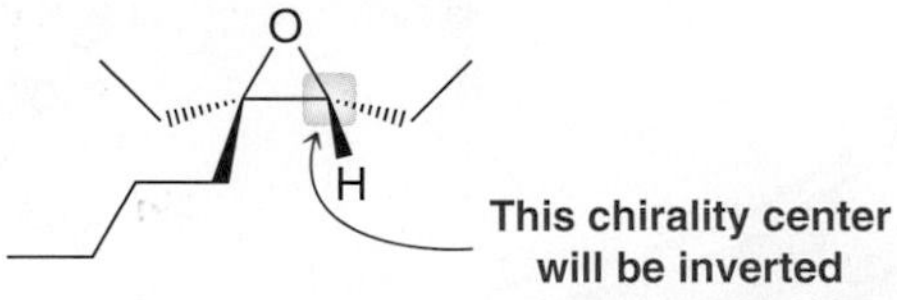

Now that we have predicted the regiochemical and stereochemical outcomes, we are ready to draw the mechanism. There are two separate steps: attack of the nucleophile to open the ring, followed by protonation of the alkoxide.

STEP 3
Draw both steps of the mechanism.

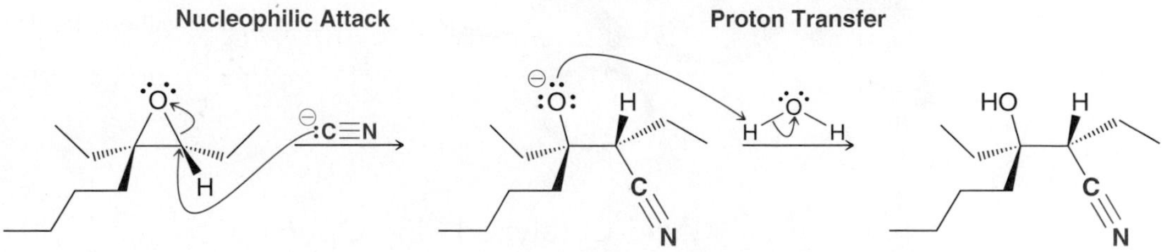

PRACTICE the skill

14.17 For each of the following reactions, predict the product and draw the mechanism of its formation.

(a) 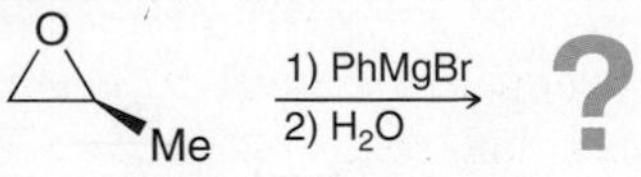

(b) 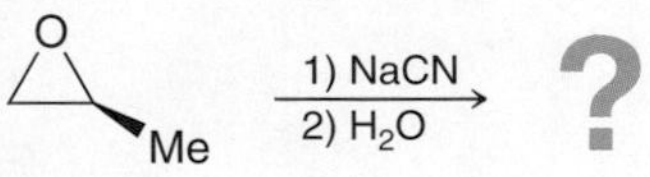

(c) 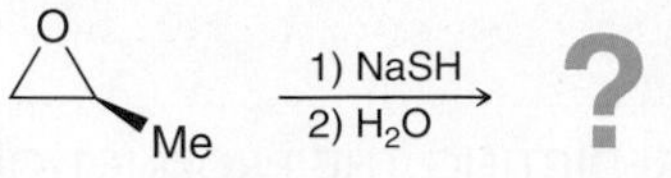

(d) 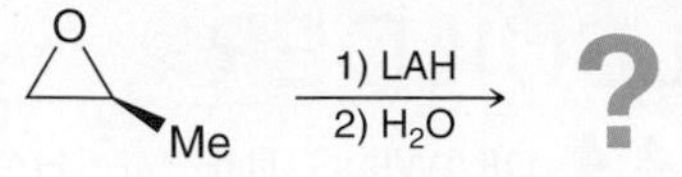

(e) 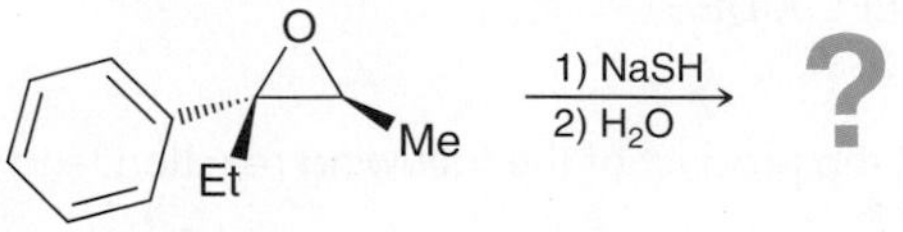

(f) 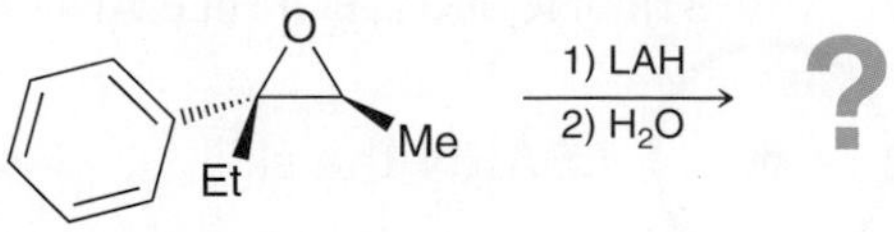

APPLY the skill

14.18 When the following chiral epoxide is treated with aqueous sodium hydroxide, only one product is obtained, and that product is achiral. Draw the product and explain why only one product is formed.

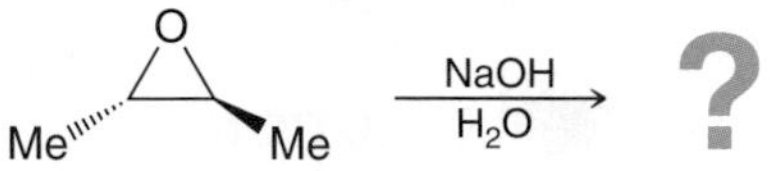

14.19 When *meso*-2,3-epoxybutane is treated with aqueous sodium hydroxide, two products are obtained. Draw both products and describe their relationship.

need more **PRACTICE?** **Try Problems 14.42a, 14.42c, 14.42e–h, 14.43a, c**

PRACTICALLYSPEAKING

Ethylene Oxide as a Sterilizing Agent for Sensitive Medical Equipment

Ethylene oxide is a colorless, flammable gas that is often used to sterilize temperature-sensitive medical equipment. The gas easily diffuses through porous materials and effectively kills all forms of microorganisms, even at room temperature. The mechanism of action likely involves a functional group in DNA attacking the ring and causing a ring opening of the epoxide, effectively alkylating that site.

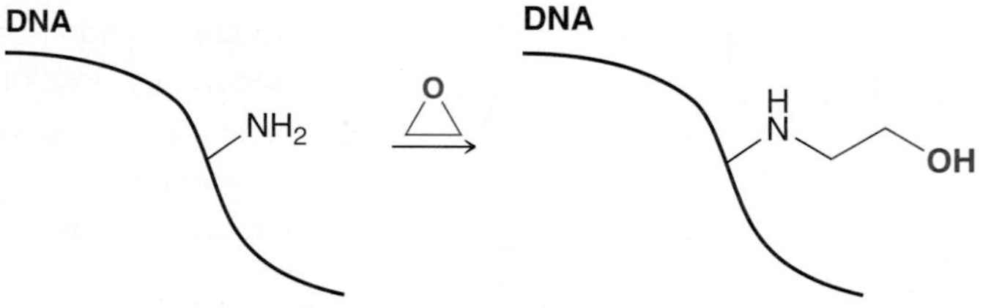

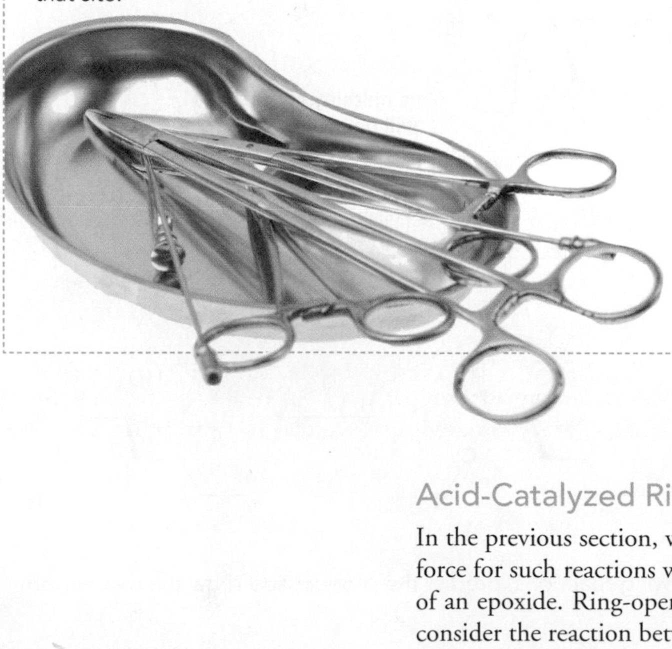

This alkylation process interferes with the normal function of DNA, thereby killing the microorganisms. The use of pure ethylene oxide presents a hazard, because it mixes with atmospheric oxygen and becomes susceptible to explosion. This problem is circumvented by using a mixture of ethylene oxide and carbon dioxide, which is no longer explosive. Such mixtures are sold commercially for the sterilization of medical equipment and agricultural grains. One such mixture is called Carboxide and is comprised of 10% ethylene oxide and 90% CO_2. Carboxide can be exposed to air without the danger of explosion.

Acid-Catalyzed Ring Opening

In the previous section, we saw the reactions of epoxides with strong nucleophiles. The driving force for such reactions was the removal of ring strain associated with the three-membered ring of an epoxide. Ring-opening reactions can also occur under acidic conditions. As an example, consider the reaction between ethylene oxide and HX.

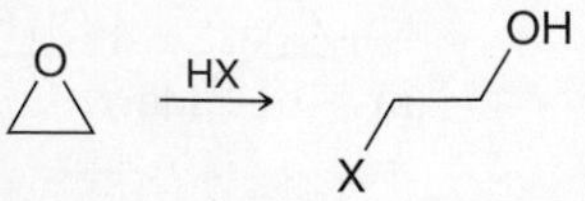

This transformation involves two mechanistic steps (Mechanism 14.6).

MECHANISM 14.6 ACID-CATALYZED RING OPENING OF AN EPOXIDE

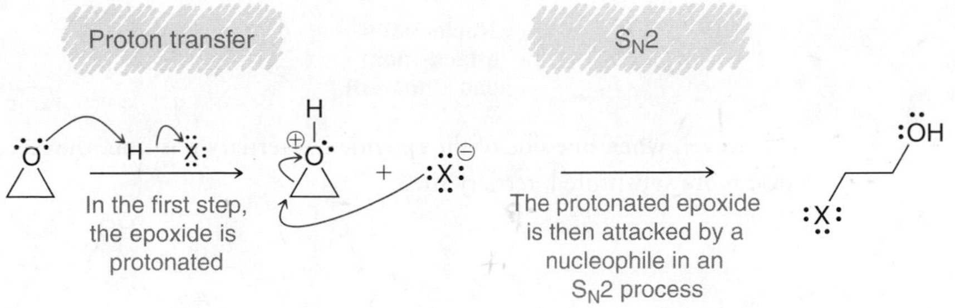

Proton transfer
S_N2
In the first step, the epoxide is protonated
The protonated epoxide is then attacked by a nucleophile in an S_N2 process

The first step is a proton transfer, and the second step is nucleophilic attack (S_N2) by a halide ion. This reaction can be accomplished with HCl, HBr, or HI. Other nucleophiles such as water or an alcohol can also open an epoxide ring under acidic conditions. A small amount of acid (often sulfuric acid) is used to catalyze the reaction.

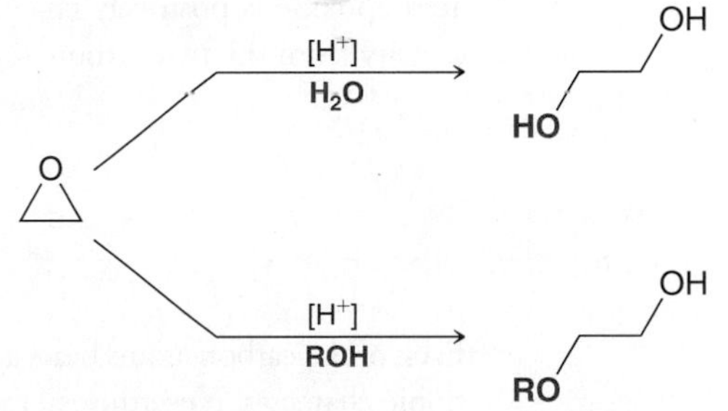

[H⁺]
H_2O
ROH

The brackets around the H^+ indicate that the acid functions as a catalyst. In each of the reactions above, the mechanism involves a proton transfer as the final step of the mechanism.

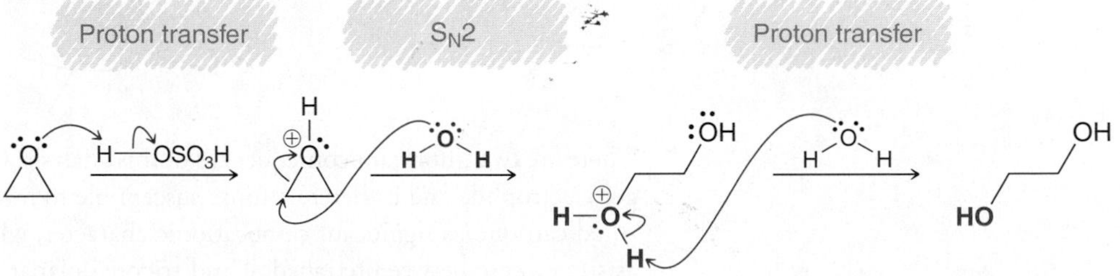

Proton transfer
S_N2
Proton transfer

The first two steps are analogous to the two steps in Mechanism 14.6. The additional proton transfer step at the end of the mechanism is required to remove the charge formed after the attack of a neutral nucleophile. The process above is used for the mass production of ethylene glycol.

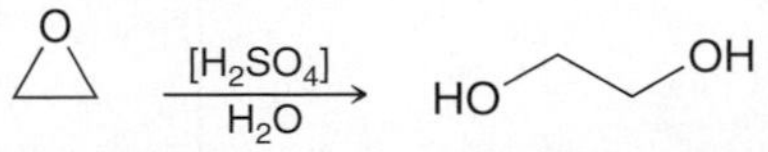

$[H_2SO_4]$
H_2O

Ethylene oxide **Ethylene glycol**

Each year, over three million tons of ethylene glycol are produced in the United States via the acid-catalyzed ring opening of ethylene oxide. Most of the ethylene glycol is used as antifreeze.

As we saw in the previous section, there are two important features of ring-opening reactions: the regiochemical outcome and the stereochemical outcome. We'll begin with regiochemistry.

When the starting epoxide is unsymmetrical, the regiochemical outcome depends on the nature of the epoxide. If one side is primary and the other side is secondary, then attack takes place at the less hindered primary position, just as we would expect for an S_N2 process.

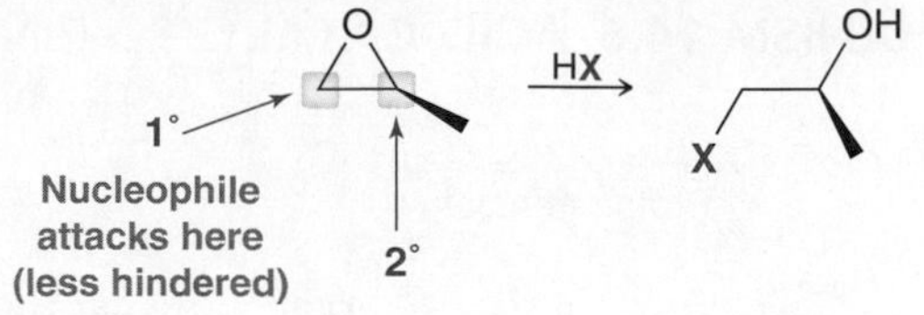

However, when one side of the epoxide is a tertiary position, the reaction is observed to occur at the more substituted, tertiary site.

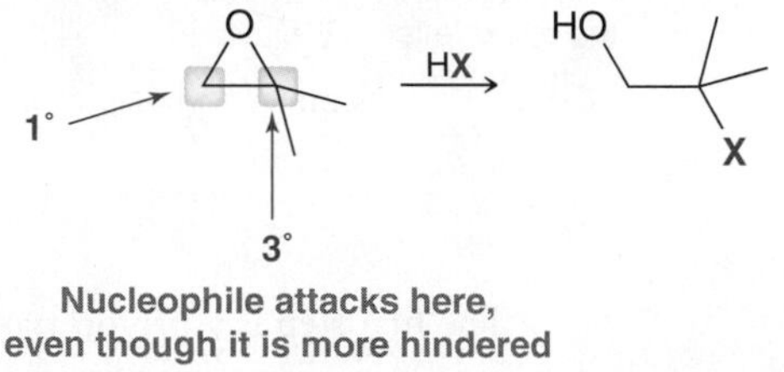

Why should this be the case? It is true that the primary site is less hindered, but there is a factor that is even more dominant than steric hindrance. That factor is an **electronic effect**. A protonated epoxide is positively charged, and the positively charged oxygen atom withdraws electron density from the two carbon atoms of the epoxide.

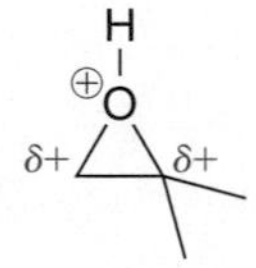

Each of the carbon atoms bears a partial positive charge (δ+). That is, they both have partial carbocationic character. Nevertheless, these two carbon atoms are not equivalent in their ability to support a partial positive charge. The tertiary position is significantly better at supporting a partial positive charge, so the tertiary position has significantly more partial carbocationic character than the primary position. The protonated epoxide is therefore more accurately drawn in the following way.

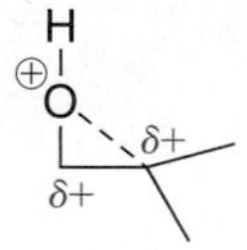

There are two important consequences of this analysis: (1) the more substituted carbon is a stronger electrophile and is therefore more susceptible to nucleophilic attack and (2) the more substituted carbon has significant carbocationic character, which means that its geometry is described as somewhere between tetrahedral and trigonal planar, allowing nucleophilic attack to occur at that position even though it is tertiary.

To summarize, the regiochemical outcome of acid-catalyzed ring opening depends on the nature of the epoxide.

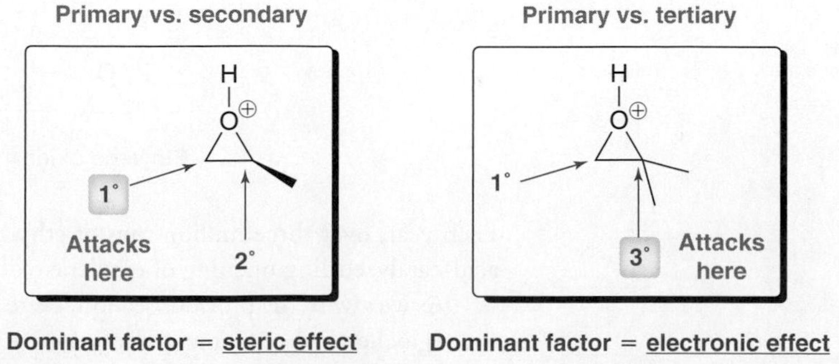

There are two factors competing to control the regiochemistry: electronic effects vs. steric effects. The former favors attack at the more substituted position, while the latter favors attack at the less substituted position. To determine which factor is dominant, we must analyze the epoxide. When the epoxide possesses a tertiary position, the electronic effect will be dominant. When the epoxide possesses only primary and secondary positions, the steric effect will be dominant. The regiochemistry of acid-catalyzed ring opening is just one example where steric effects and electronic effects compete. As we progress through the course, we will see other examples of electronic vs. steric effects.

In the previous section (ring opening with strong nucleophiles), the regiochemistry was more straightforward, because electronics was not a factor at all. The epoxide was attacked by a nucleophile before being protonated, so the epoxide did not bear a positive charge when it was attacked. In such a case, steric hindrance was the only consideration.

Now let's turn our attention to the stereochemistry of ring-opening reactions under acid-catalyzed conditions. When the attack takes place at a chirality center, inversion of configuration is observed. This result is consistent with an S_N2-like process involving back-side attack of the nucleophile.

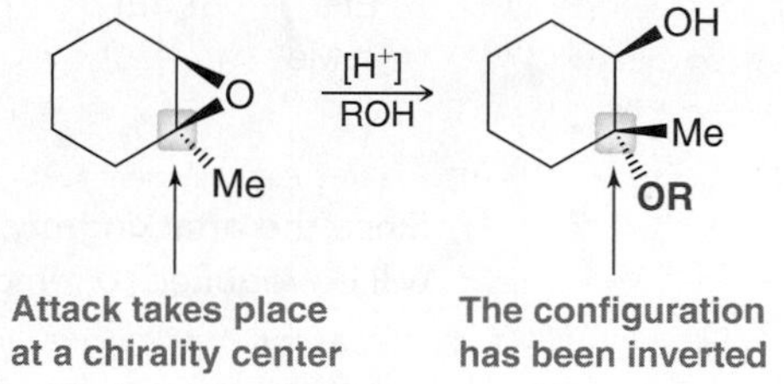

SKILLBUILDER

14.5 DRAWING THE MECHANISM AND PREDICTING THE PRODUCT OF ACID-CATALYZED RING OPENING

LEARN the skill

Predict the product of the reaction below, and draw the likely mechanism for its formation:

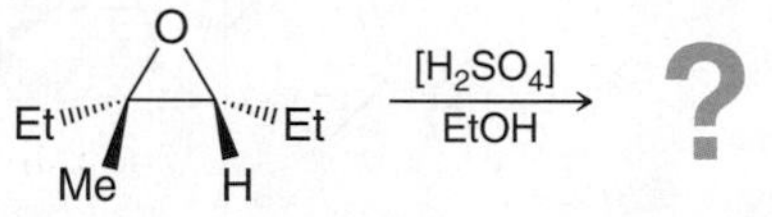

SOLUTION

The presence of sulfuric acid indicates that the epoxide is opened under acid-catalyzed conditions. In order to draw the product, we must consider the regiochemistry and stereochemistry of the reaction.

STEP 1 Identify the regiochemistry by determining whether steric or electronic effects will dominate.

1. To predict the regiochemical outcome, analyze the epoxide.

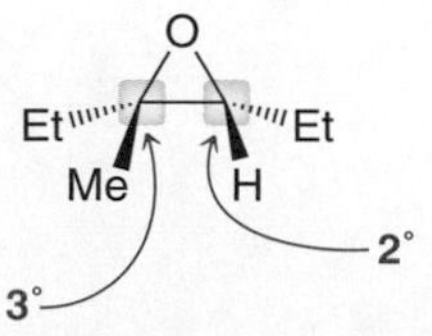

One side is tertiary, so we expect that electronic effects will control the regiochemistry, and we predict that nucleophilic attack will occur at the more substituted, tertiary position.

STEP 2
Identify the stereochemistry by determining if the nucleophile attacks a chirality center. If so, expect inversion.

2. Next identify the stereochemical outcome. Look at the center being attacked and determine if it is a chirality center. In this case, it is a chirality center, so we expect back-side attack of the nucleophile to invert the configuration of that center.

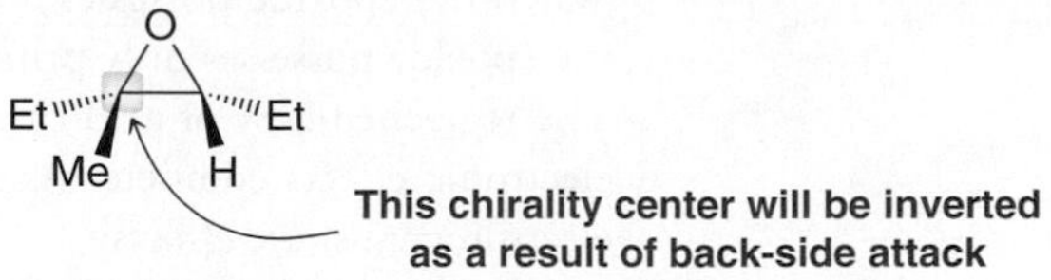

Now that we have predicted the regiochemical and stereochemical outcomes, we are ready to draw the mechanism. The epoxide is first protonated and then attacked by a nucleophile (EtOH).

STEP 3
Draw all three steps of the mechanism.

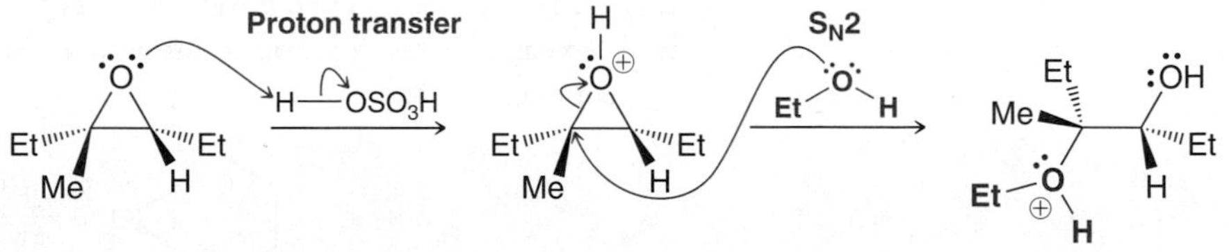

Since the attacking nucleophile was neutral (EtOH), an additional proton transfer step will be required to remove the charge and generate the final product.

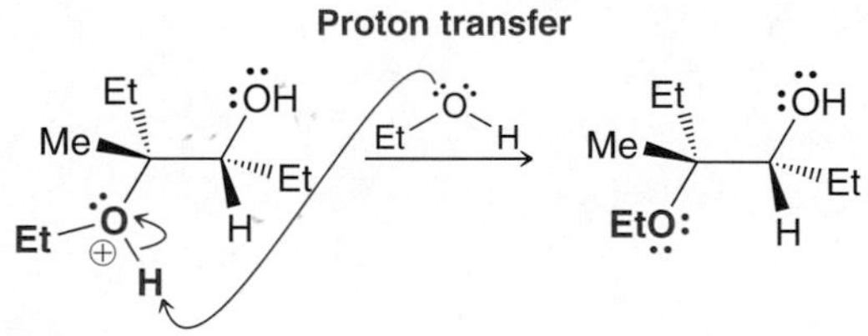

PRACTICE the skill **14.20** For each reaction, predict the product and draw the mechanism of its formation.

(a)
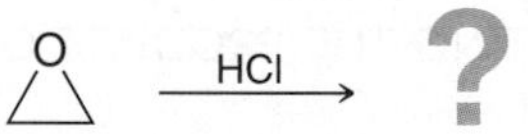

(b)
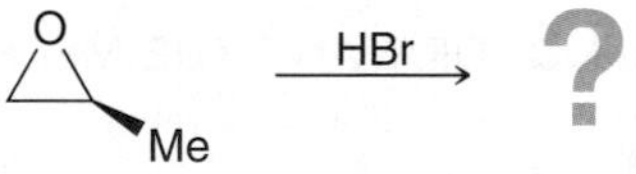

(c)
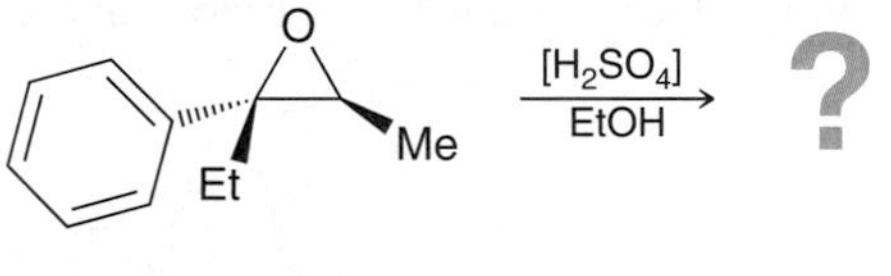

(d)
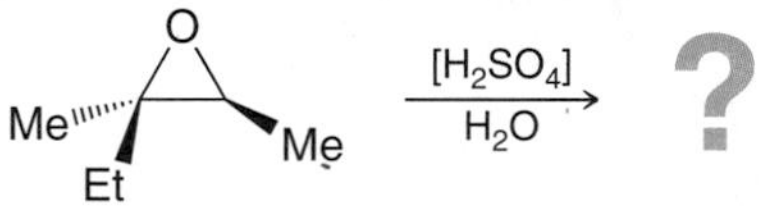

(e)
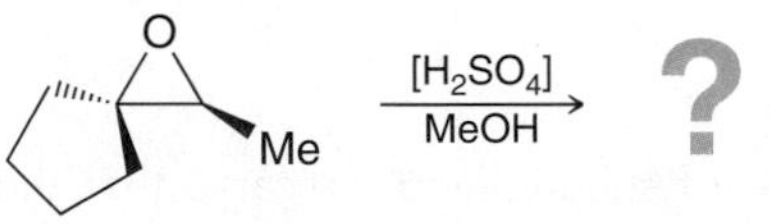

(f)
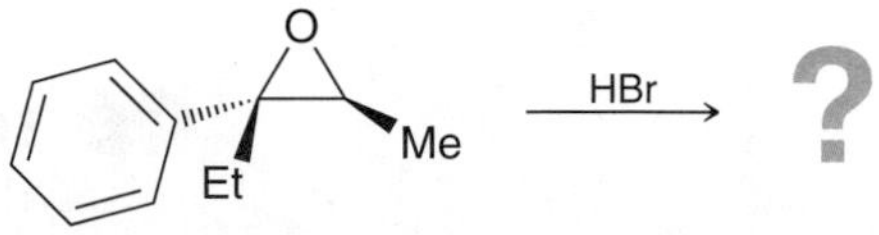

APPLY the skill **14.21** Propose a mechanism for the following transformation.

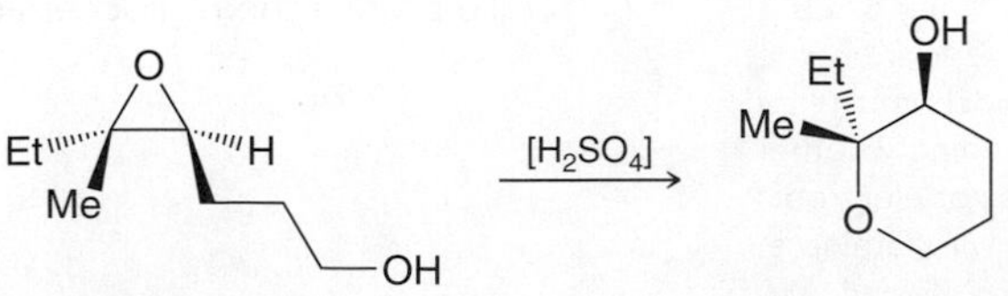

need more **PRACTICE?** **Try Problems 14.43d, 14.49, 14.50**

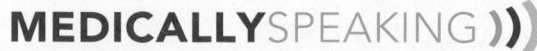

MEDICALLYSPEAKING

Cigarette Smoke and Carcinogenic Epoxides

As we have seen many times in previous chapters, the combustion of organic materials should produce CO_2 and water. But combustion rarely produces only these two products. Usually, incomplete combustion produces organic compounds that account for the smoke that is observed to emanate from a fire. One of those compounds is called benzo[*a*]pyrene. This highly carcinogenic compound is produced from the burning of organic materials, such as gasoline, wood, and cigarettes. In recent years, extensive research has elucidated the likely mechanistic pathway for the carcinogenicity of this compound, and it has been shown that benzo[*a*]pyrene itself is not the compound that causes cancer. Rather, it is one of the metabolites (one of the compounds produced during the metabolism of benzo[*a*]pyrene) that is a highly reactive intermediate capable of alkylating DNA and thereby interfering with the normal function of DNA.

When benzo[*a*]pyrene is metabolized, an initial epoxide is formed (called an arene oxide) which is then opened by water to give a diol.

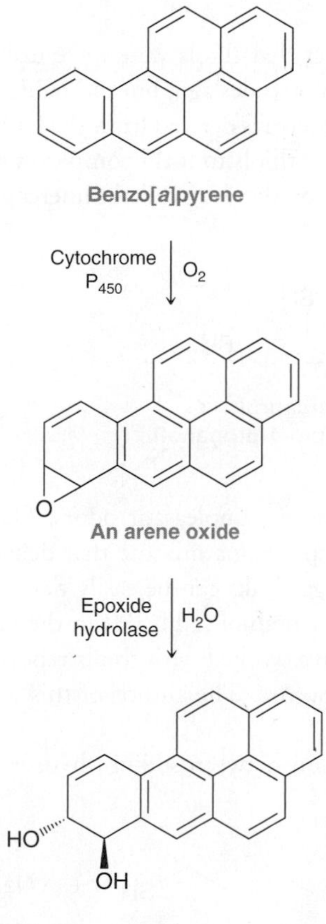

These two steps should be very familiar to you. The first step is epoxide formation, and the second step is a ring-opening reaction in the presence of a catalyst to form a diol. This diol can then undergo epoxidation another time to give the following diol epoxide.

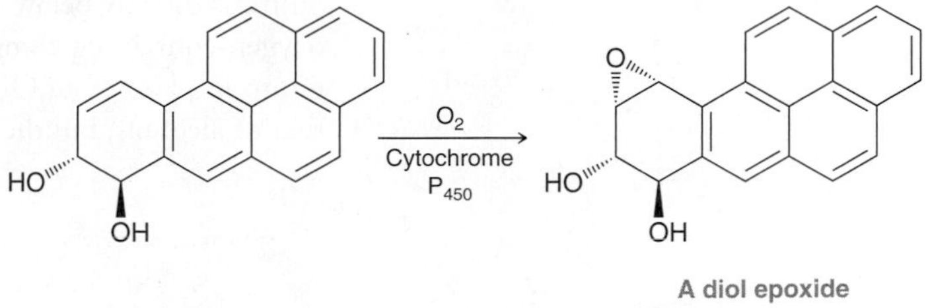

This diol epoxide is the carcinogenic metabolite and is capable of alkylating DNA. Studies show that the amino group of deoxyguanosine (in DNA) attacks the epoxide. This reaction changes the structure of DNA and causes genetic code alterations that ultimately lead to the formation of cancer cells.

Benzo[*a*]pyrene is not the only carcinogenic compound in cigarette smoke. Throughout this book, we will see several other carcinogenic compounds that are also present in cigarette smoke.

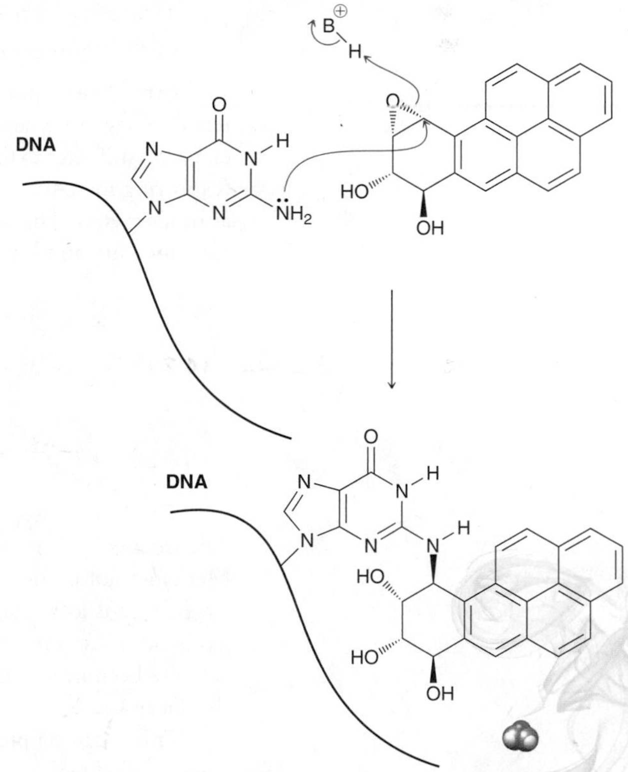

14.11 Thiols and Sulfides

Thiols

Sulfur is directly below oxygen in the periodic table (in the same column), and therefore, many oxygen-containing compounds have sulfur analogs. Sulfur analogs of alcohols contain an SH group in place of an OH group and are called **thiols**. The nomenclature of thiols is similar to that of alcohols, but the suffix of the name is "thiol" instead of "ol":

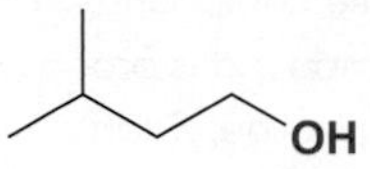

3-Methyl-1-butanol

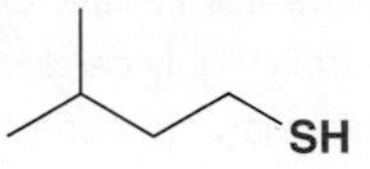

3-Methyl-1-butanethiol

Notice that the "e" is kept before the suffix "thiol." When another functional group is present in the compound, the SH group is named as a substituent and is called a **mercapto group**:

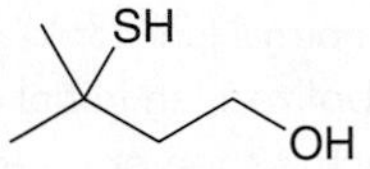

3-Mercapto-3-methyl-1-butanol

The name "mercapto" is derived from the fact that thiols were once called mercaptans. This terminology was abandoned by IUPAC several decades ago, but old habits die hard, and many chemists still refer to thiols as mercaptans. The term is derived from the Latin *mercurium captans* (capturing mercury) and describes the ability of thiols to form complexes with mercury as well as other metals. This ability is put to good use by the drug called dimercaprol, which is used to treat mercury and lead poisoning.

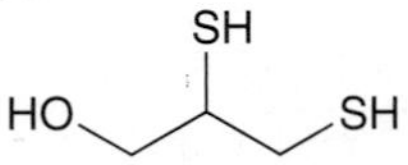

Dimercaprol
(2,3-dimercapto-1-propanol)

Thiols are most notorious for their pungent, unpleasant odors. Skunks use thiols as a defense mechanism to ward off predators by spraying a mixture that delivers a mighty stench. Methanethiol is added to natural gas so that gas leaks can be easily detected. If you have ever smelled a gas leak, you were smelling the methanethiol (CH_3SH) in the natural gas, as natural gas is odorless. Surprisingly, scientists who have worked with thiols report that the nasty odor actually becomes pleasant after prolonged exposure. The author of this textbook can attest to this fact.

Thiols can be prepared via an S_N2 reaction between sodium hydrosulfide (NaSH) and a suitable alkyl halide; for example:

This reaction can occur even at secondary substrates without competing E2 reactions, because the hydrosulfide ion (HS^-) is an excellent nucleophile and a poor base. When this nucleophile attacks a chirality center, inversion of configuration is observed.

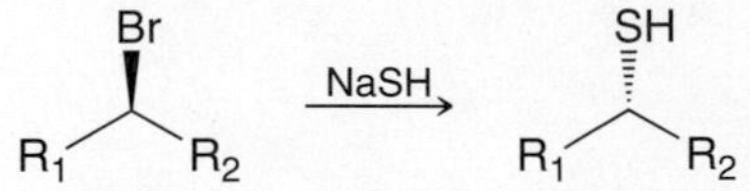

CONCEPTUAL CHECKPOINT

14.22 What reagents would you use to prepare each of the following thiols:

(a)

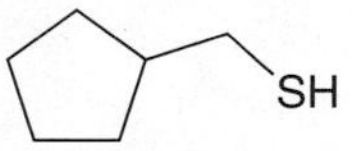

(b)

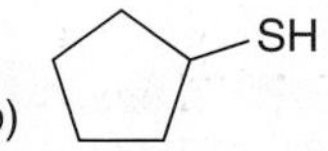

(c)

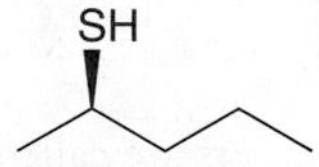

Thiols easily undergo oxidation to produce **disulfides**.

$$\text{SH} + \text{HS} \xrightarrow{\text{NaOH/H}_2\text{O, Br}_2} \text{S}\!-\!\text{S}$$

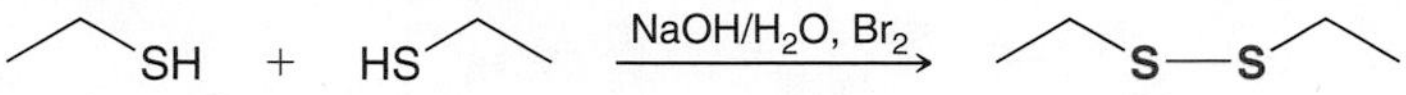

A disulfide

The conversion of thiols into disulfides requires an oxidizing reagent, such as bromine in aqueous hydroxide. The process begins with deprotonation of the thiol to generate a **thiolate** ion (Mechanism 14.7).

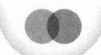

MECHANISM 14.7 OXIDATION OF THIOLS

DEPROTONATION OF THE THIOL

Proton transfer

$$R\!-\!\overset{..}{\underset{..}{S}}\!-\!H \; + \; \overset{\oplus}{Na} \quad \overset{\ominus}{:}\!\overset{..}{O}H \; \rightleftharpoons \; R\!-\!\overset{..}{\underset{..}{S}}\!:\!\overset{\ominus}{} \quad \overset{\oplus}{Na} \; + \; H_2\overset{..}{\underset{..}{O}}\!:$$

| **Thiol** | **Hydroxide** | In the first step, the thiol is deprotonated to form a thiolate ion | **Thiolate ion** | **Water** |

Thiol
(pK_a = 10.5)
(Stronger acid)

Water
(pK_a = 15.7)
(Weaker acid)

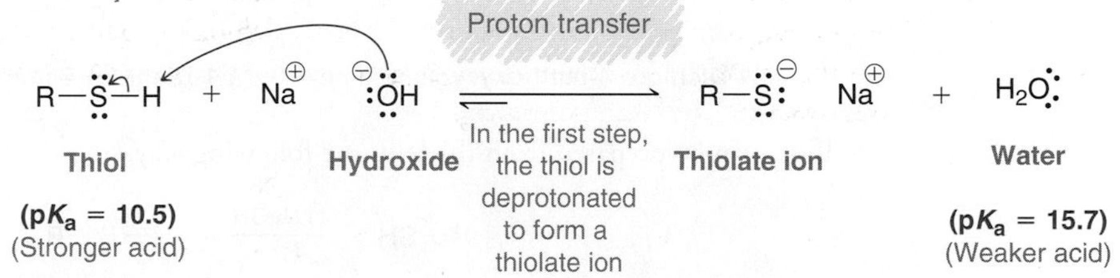

FORMATION OF THE DISULFIDE

S_N2 S_N2

$$R\!-\!\overset{..}{\underset{..}{S}}\!:\!\overset{\ominus}{} \quad \overset{\oplus}{Na} \xrightarrow{\quad :\!\overset{..}{Br}\!-\!\overset{..}{Br}\!: \quad} R\!-\!\overset{..}{\underset{..}{S}}\!-\!\overset{..}{Br}\!: \xrightarrow{\quad \overset{\oplus}{Na} \;\; \overset{\ominus}{:}\!\overset{..}{S}\!-\!R \quad} R\underset{\overset{..}{\underset{..}{S}}}{\diagdown}\!\!\diagup R \; + \; NaBr$$

A thiolate ion functions as a nucleophile and attacks molecular bromine
+ NaBr

Another thiolate functions as a nucleophile in a second S_N2 process

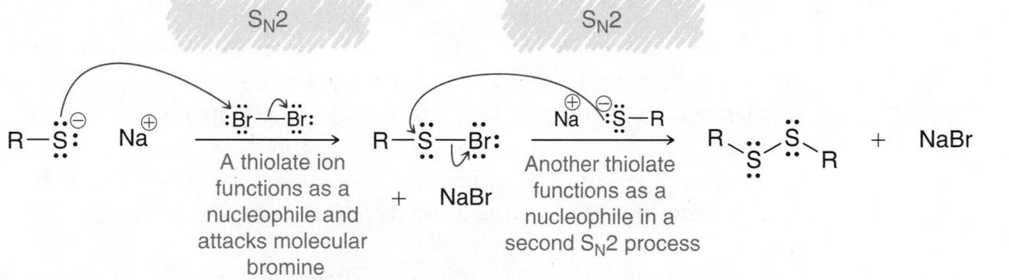

Hydroxide is a strong enough base that the equilibrium favors formation of the thiolate ion. This thiolate ion is an excellent nucleophile and can attack molecular bromine in an S_N2 process. A second S_N2 process then produces the disulfide.

There are many oxidizing agents that can be used to convert thiols into disulfides. In fact, the reaction is accomplished with so much ease that atmospheric oxygen can function as an oxidizing agent to produce disulfides. Catalysts can be used to speed up the process. Disulfides are also easily reduced back to thiols when treated with a reducing agent, such as HCl in the presence of zinc.

$$\text{S}\!-\!\text{S} \xrightarrow[\text{[Reduction]}]{\text{HCl, Zn}} \text{SH} + \text{HS}$$

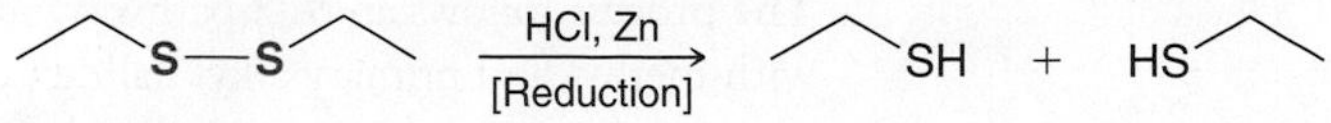

The ease of interconversion between thiols and disulfides is attributed to the nature of the S—S bond. It has a bond strength of approximately 220 kJ/mol, which is about half the strength of many other covalent bonds. The interconversion between thiols and disulfides is extremely important in determining the shape of many biologically active compounds, as we will explore in Section 25.4.

Sulfides

The sulfur analogs of ethers are called **sulfides,** or thioethers.

An ether | A sulfide (thioether)

Nomenclature of sulfides is similar to that of ethers. Common names are assigned using the suffix "sulfide" instead of "ether."

Diethyl **ether** | Diethyl **sulfide**

More complex sulfides are named systematically, much the way ethers are named, with the alkoxy group being replaced by an **alkylthio group**.

1,1-Dichloro-4-**methoxy**cyclohexane | 1,1-Dichloro-4-**(methylthio)**cyclohexane

Sulfides can be prepared from thiols in the following way.

$$R{-}SH \xrightarrow[\text{2) RX}]{\text{1) NaOH}} R{-}S{-}R$$

This process is essentially the sulfur analog of the Williamson ether synthesis (Mechanism 14.8).

MECHANISM 14.8 PREPARATION OF SULFIDES FROM THIOLS

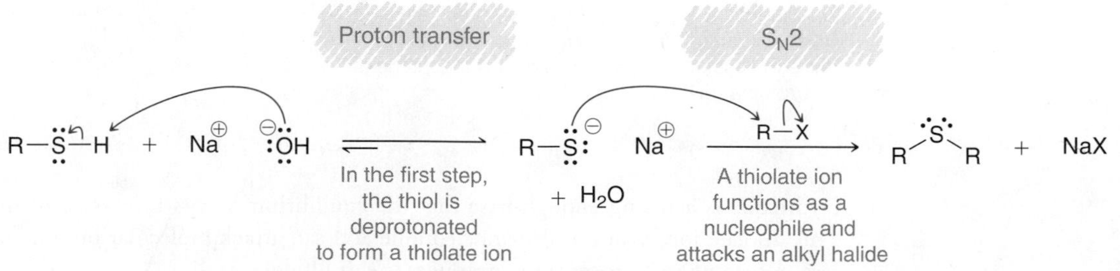

In the first step, hydroxide is used to deprotonate the thiol and produce a thiolate ion. The thiolate ion then functions as a nucleophile and attacks an alkyl halide, producing the sulfide. The process follows an S_N2 pathway, so the regular restrictions apply. The reaction works well with methyl and primary alkyl halides, can often be accomplished with secondary alkyl halides, and does not work for tertiary alkyl halides.

Since sulfides are structurally similar to ethers, we might expect sulfides to be as unreactive as ethers, but this is not the case. Sulfides undergo several important reactions.

1. Sulfides will attack alkyl halides in an S_N2 process.

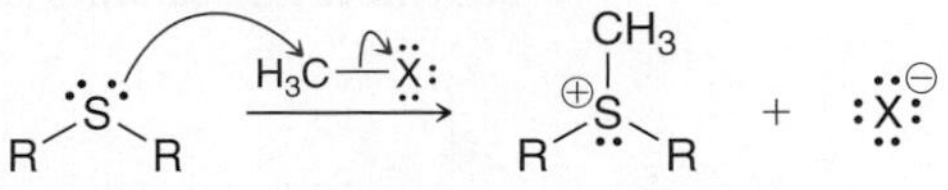

The product of this step is a powerful alkylating agent, because it is capable of transferring a methyl group to a nucleophile.

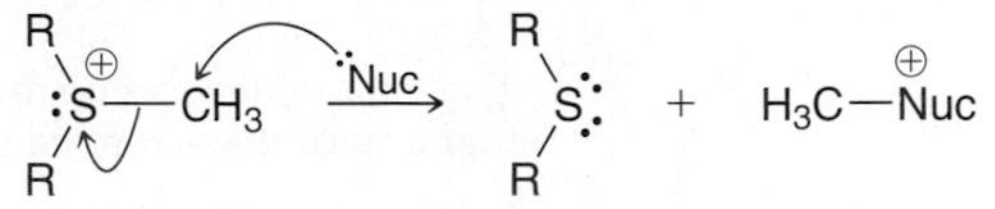

LOOKING BACK
For a review of alkylation reactions with SAM, see Section 7.4.

We have already seen one such example of this process in Chapter 7. Recall that SAM is a biological methylating agent.

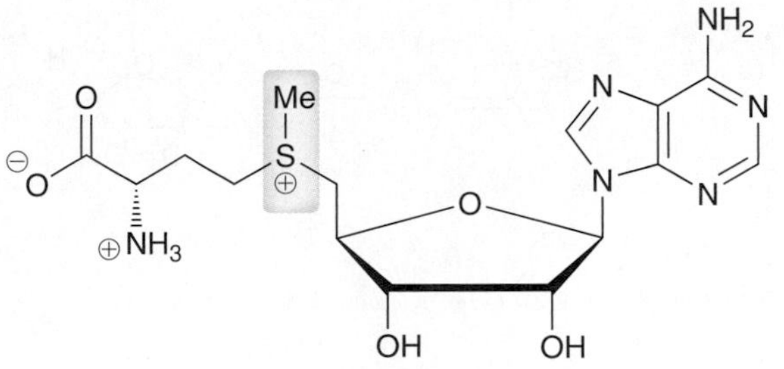

***S*-Adenosylmethionine (SAM)**

2. Sulfides also undergo oxidation to give **sulfoxides** and then **sulfones**.

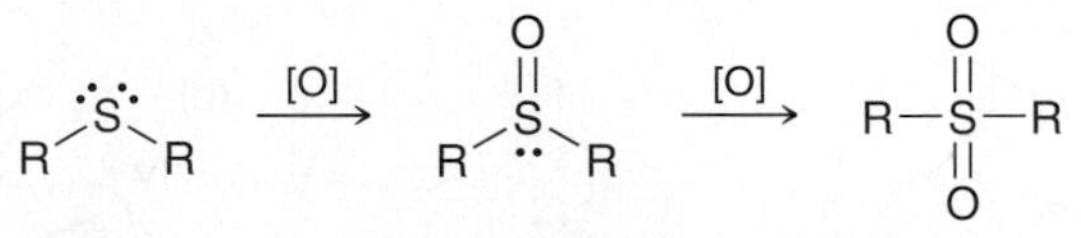

The initial product is a sulfoxide. If the oxidizing agent is strong enough and present in excess, then the sulfoxide is oxidized further to give the sulfone. For good yields of the sulfoxide without further oxidation to the sulfone, it is necessary to use an oxidizing reagent that will not oxidize the sulfoxide. Many such reagents are available, including sodium *meta*-periodiate, $NaIO_4$.

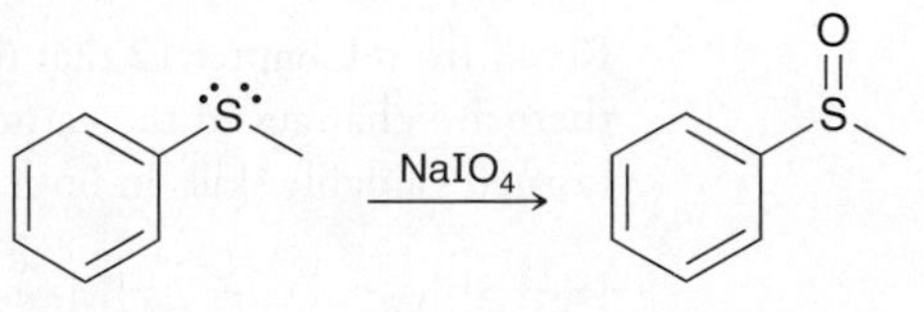

If the sulfone is the desired product, then two equivalents of hydrogen peroxide are used.

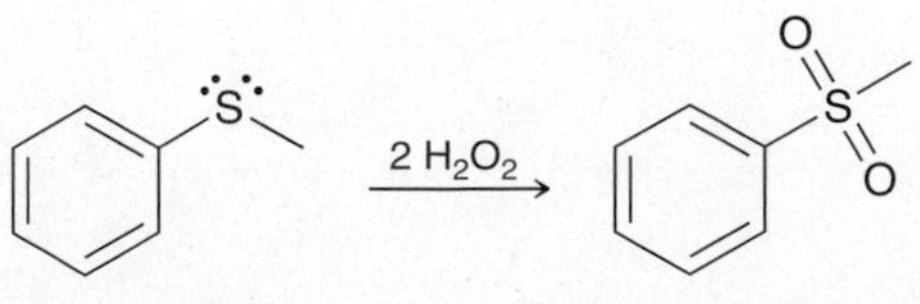

The S=O bonds in sulfoxides and sulfones actually have little double-bond character. The $3p$ orbital of a sulfur atom is much larger than the $2p$ orbital of an oxygen, and therefore, the orbital overlap of the π bond between these two atoms is not effective. Consequently, sulfoxides and sulfones are often drawn with each S—O bond as a single bond.

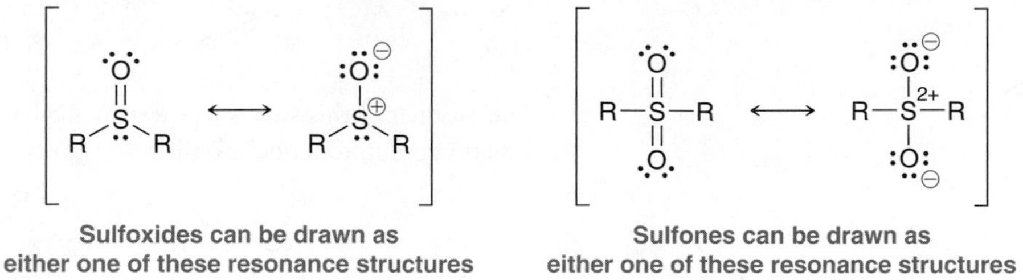
Sulfoxides can be drawn as
either one of these resonance structures
Sulfones can be drawn as
either one of these resonance structures

The ease with which sulfides are oxidized renders them ideal reducing agents in a wide variety of applications. For example, recall that DMS (dimethyl sulfide) is used as a reducing agent in ozonolysis (Section 9.11). The by-product is dimethyl sulfoxide (DMSO).

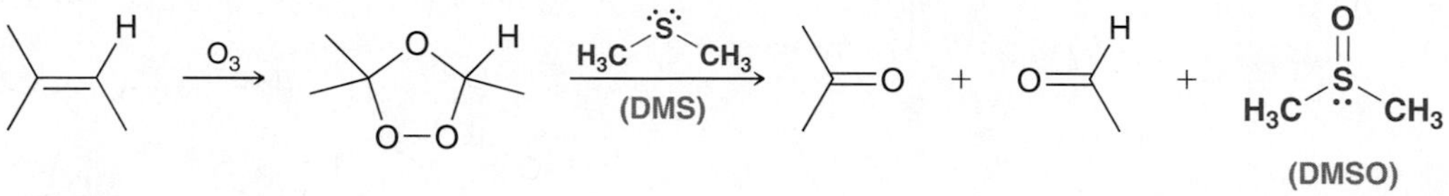

CONCEPTUAL CHECKPOINT

14.23 Predict the products for each of the following reactions.

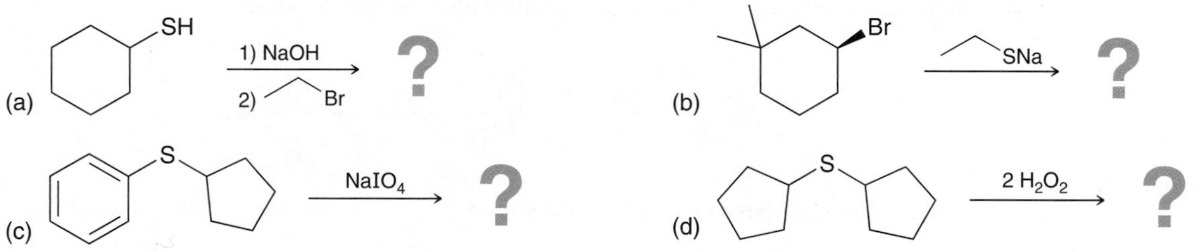

14.12 Synthesis Strategies Involving Epoxides

Recall from Chapter 12 that the two issues to consider when proposing a synthesis are whether there are changes in the carbon skeleton or in the functional group. In this chapter, we have learned valuable skills in both categories. Let's focus on them one at a time.

Installing Two Adjacent Functional Groups

The most useful synthetic techniques that we learned in this chapter involve epoxides. We saw several ways to form epoxides and many reagents that open epoxides. Note that opening an epoxide provides two functional groups on adjacent carbon atoms.

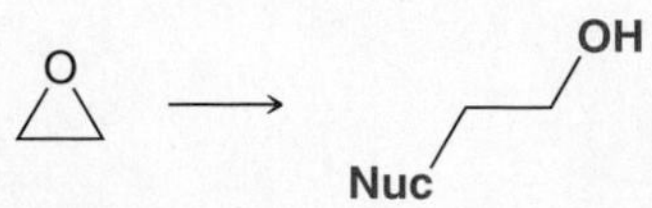

Whenever you see two adjacent functional groups, you should think of epoxides. Let's see an example.

SKILLBUILDER

14.6 INSTALLING TWO ADJACENT FUNCTIONAL GROUPS

LEARN the skill

Propose a synthesis for the following transformation:

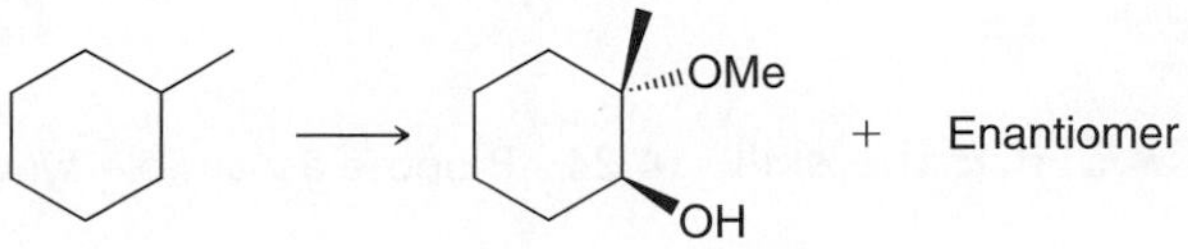

SOLUTION

Always approach a synthesis problem by initially asking two questions.

1. Is there a change in the carbon skeleton? No, the carbon skeleton is not changing.
2. Is there a change in the functional groups? Yes, the starting material has no functional groups, and the product has two adjacent functional groups.

The answers to these questions dictate what must be done. Specifically, we must install two adjacent functional groups. This suggests that we consider using a ring-opening reaction of an epoxide. Using a retrosynthetic analysis, we draw the epoxide that would be necessary.

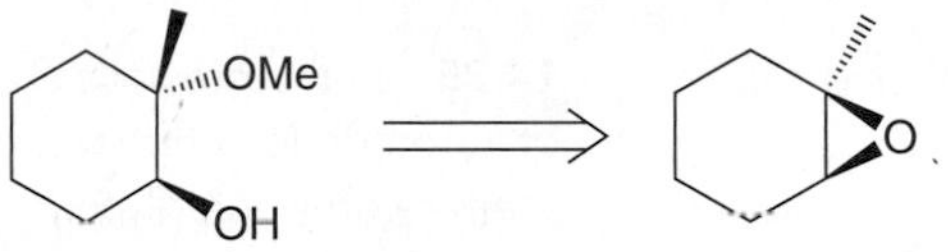

The regiochemistry of this step requires that the methoxy group must be placed at the more substituted position. This dictates that the epoxide must be opened under acidic conditions to ensure that the nucleophile (MeOH) attacks at the more substituted position.

Our next step is to determine how to make the epoxide. We have seen a couple of ways to make epoxides, both of which start with an alkene.

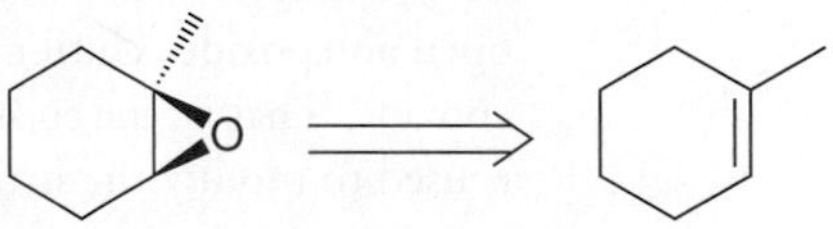

At this point, we can start working forward, focusing on converting the starting material into the desired alkene. The starting material has no functional groups, and we have seen only one method for introducing a functional group into an alkane. Specifically, we must employ a radical bromination.

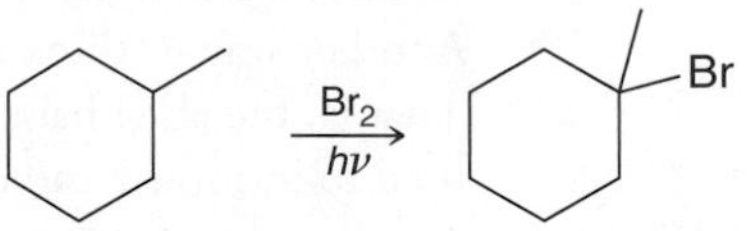

At this point, we just need to bridge the gap.

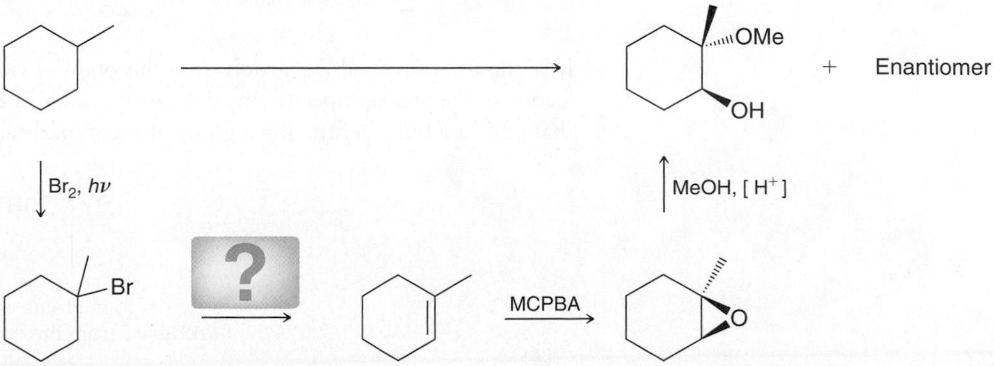

A strong base, such as ethoxide, will produce an elimination reaction that will form the desired alkene. So, our proposed synthesis is:

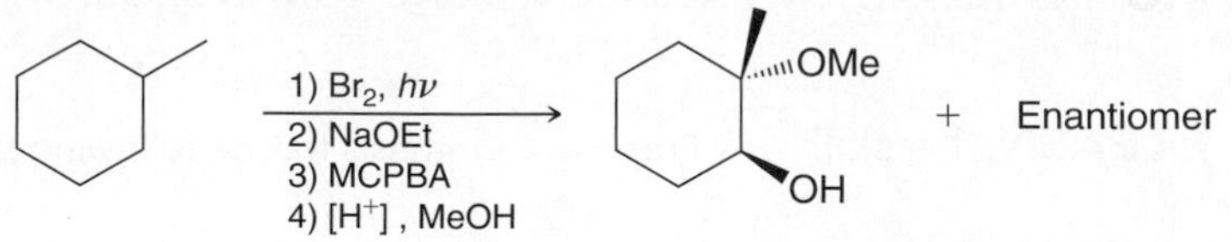

PRACTICE the skill **14.24** Propose a plausible synthesis for each of the following transformations.

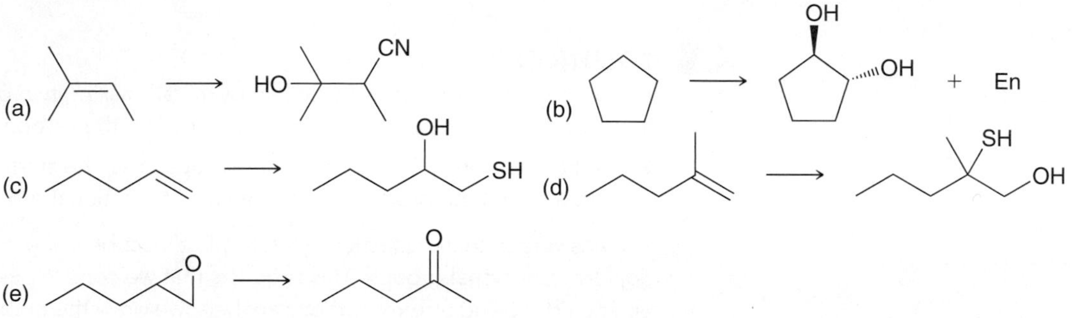

APPLY the skill **14.25** In the previous chapter, we saw that ethylene glycol is one of the main components of automobile antifreeze. Using iodoethane as your starting material, show how you could prepare ethylene glycol.

need more **PRACTICE?** **Try Problem 14.51**

Grignard Reagents: Controlling the Location of the Resulting Functional Group

In this chapter, we have seen a new way to make a C—C bond by using a Grignard reagent to open an epoxide. Until now, we have thought about this reaction from the point of view of the epoxide. That is, the epoxide is considered to be the starting material, and the Grignard reagent is used to modify the structure of the starting material. For example:

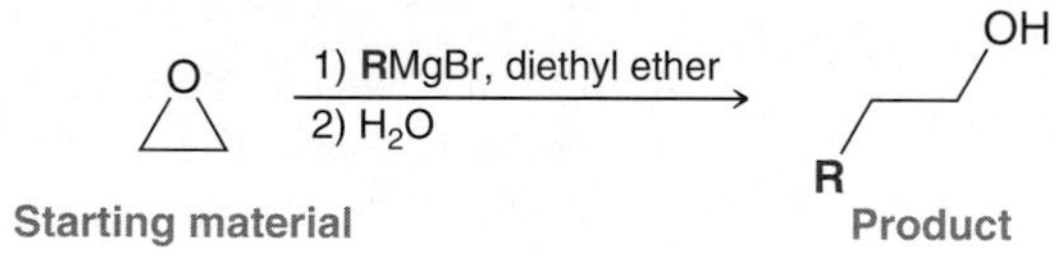

In this reaction, the epoxide is opened and an alkyl group (R) is introduced into the structure. Another way to think about this type of reaction is from the point of view of the alkyl halide. That is, the alkyl halide can be considered to be the starting material, and the epoxide is then used to introduce carbon atoms into the structure of the alkyl halide.

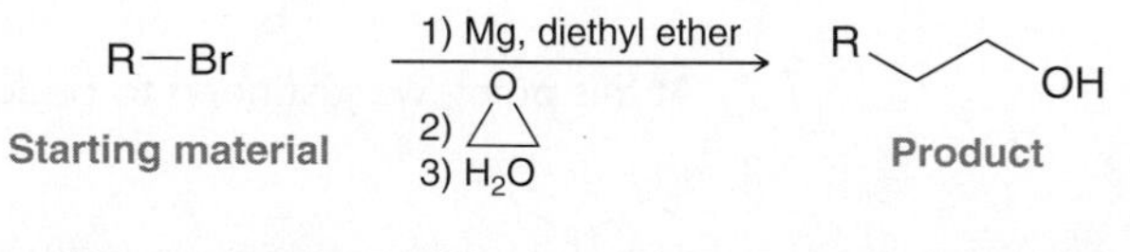

It is important to see this reaction from this point of view as well, as it highlights an important feature of the process. Specifically, this process can be used to introduce a chain of carbon atoms that possess a built-in functional group at the second carbon atom.

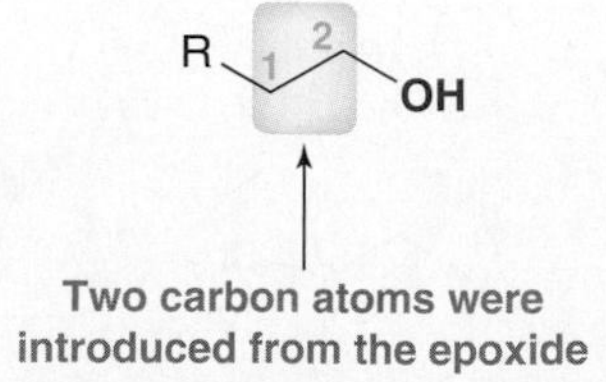

Notice the position of the functional group. It is on the second carbon atom of the newly installed chain. That is an extremely important feature, because we obtain a different outcome when a Grignard reagent attacks a ketone or aldehyde (Section 13.6). Specifically, the functional group appears on the first carbon atom of the new chain that was introduced.

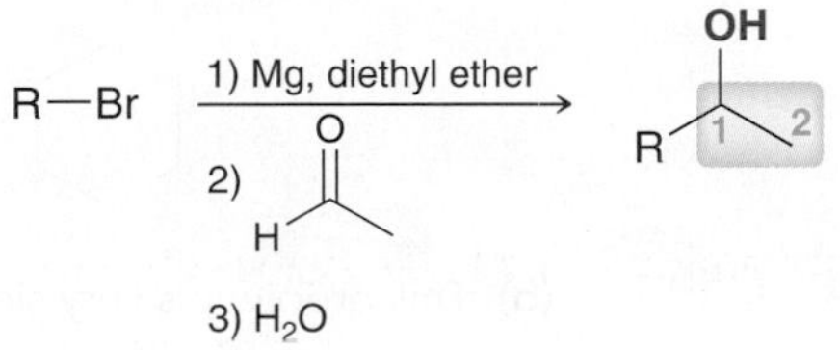

Notice the difference. You must train your eyes to look at the precise location of a functional group when it appears on a newly introduced alkyl fragment. Let's see a specific example of this.

SKILLBUILDER

14.7 CHOOSING THE APPROPRIATE GRIGNARD REACTION

LEARN the skill

Propose a synthesis for each of the following two transformations.

(a) 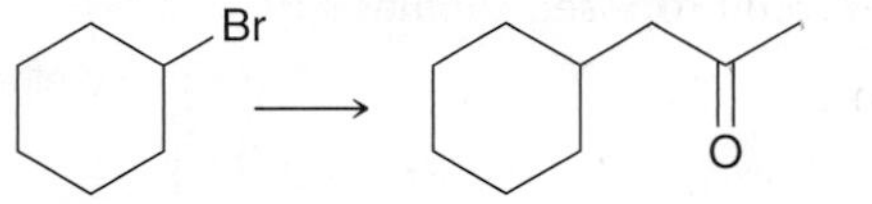

(b) 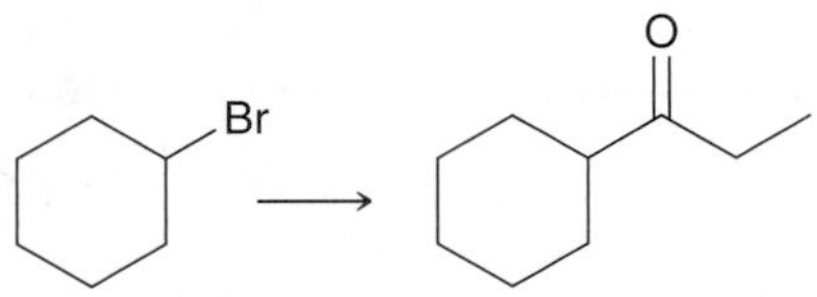

SOLUTION

(a) Always approach a synthesis problem by initially asking whether there are changes in the carbon skeleton or in the functional groups. In this case, there are changes both in the carbon skeleton and in the functional group. A three-carbon chain is introduced, and both the identity and location of the functional group have changed.

From the answers to these two questions we know that we must introduce three carbon atoms and a functional group (C═O). It would be inefficient to think of these two tasks as separate. If we first attach a three-carbon chain and only then think about how to introduce the C═O bond in exactly the correct position, we will find it very difficult to install the carbonyl group. It is more efficient to introduce the three carbon atoms in such a way that a functional group is already in the correct location. So let's analyze the precise location.

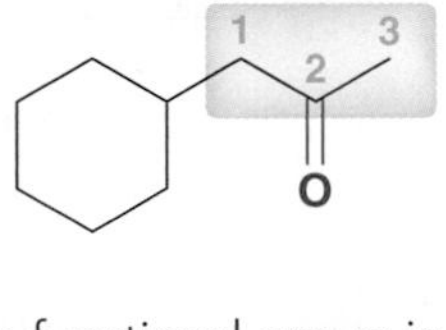

If we look at the three carbon atoms that are being introduced, the functional group is on the second carbon atom. That should signal an epoxide opening.

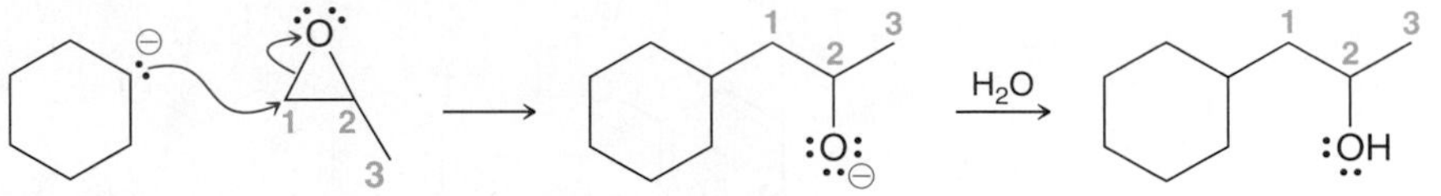

This one step introduces the three-carbon chain and simultaneously places a functional group in the proper location. Granted, the functional group is a hydroxyl group, rather than a carbonyl group. But once the hydroxyl group is in the right location, it is easy enough to perform an oxidation and form the carbonyl group.

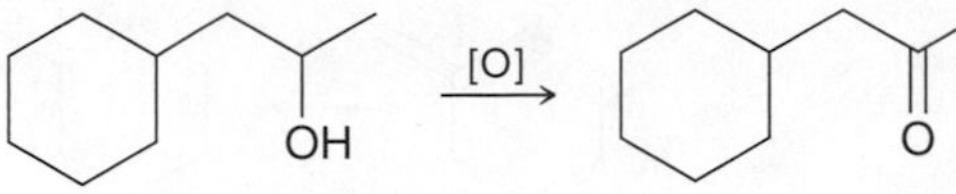

Always remember that functional groups can be easily interconverted. The important consideration is how to place a functional group in the desired location. In summary, our proposed synthesis is:

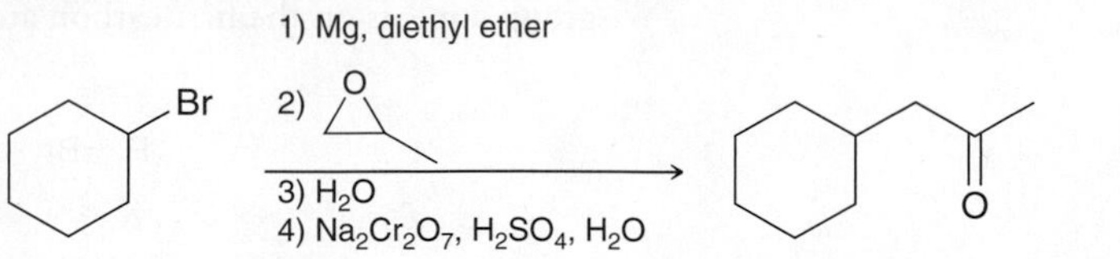

(b) This problem is very similar to the previous problem. Once again, we are introducing three carbon atoms, but this time, the functional group is located at the first carbon atom on the chain that was introduced.

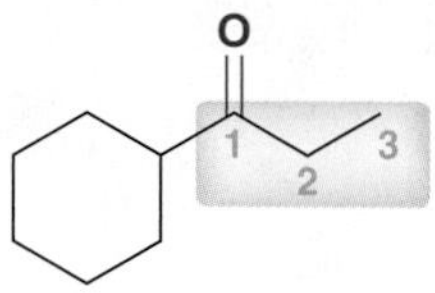

This means that the compound cannot be formed using an epoxide ring opening. In this case, the Grignard reagent must attack an aldehyde rather than an epoxide.

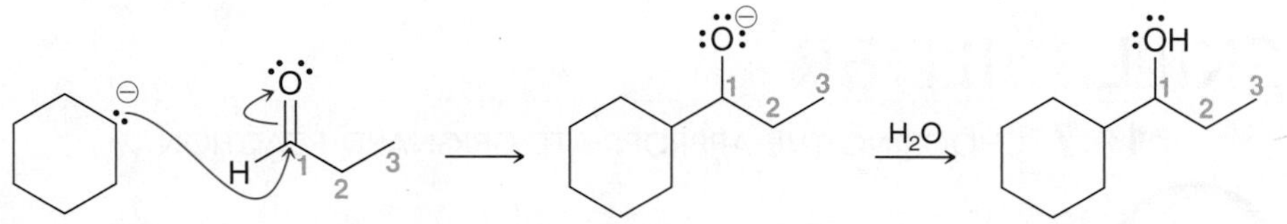

This sequence places the functional group in the desired location. It is then easy enough to oxidize the alcohol to form the desired ketone, as in the previous problem. In summary, our proposed synthesis is:

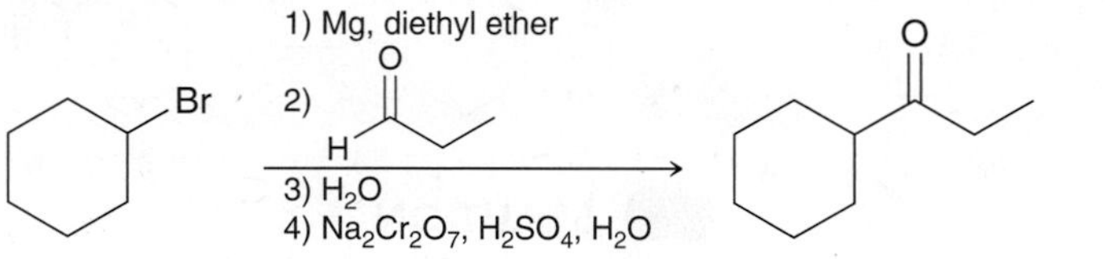

Notice the difference between the two syntheses above. Both involve introduction of carbon atoms via a Grignard reagent, but the precise location of the functional group dictates whether the Grignard reagent should attack an epoxide or an aldehyde.

PRACTICE the skill **14.26** Propose a plausible synthesis for each of the following transformations.

(a)
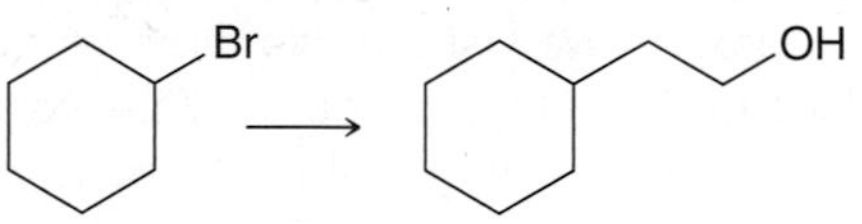

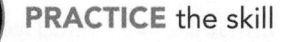

(b)
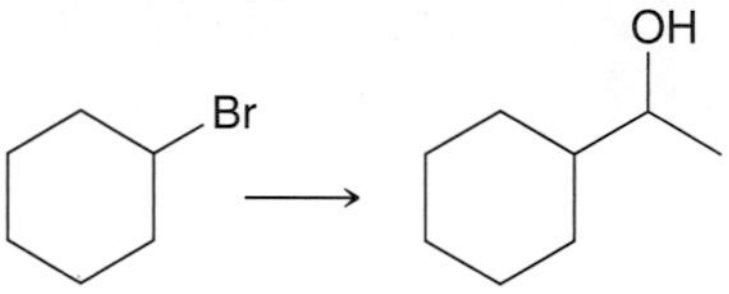

(c)
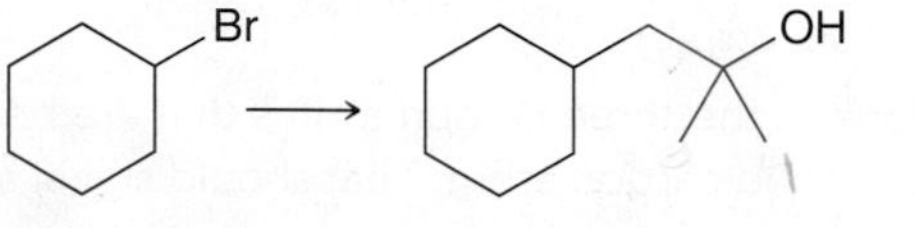

(d)
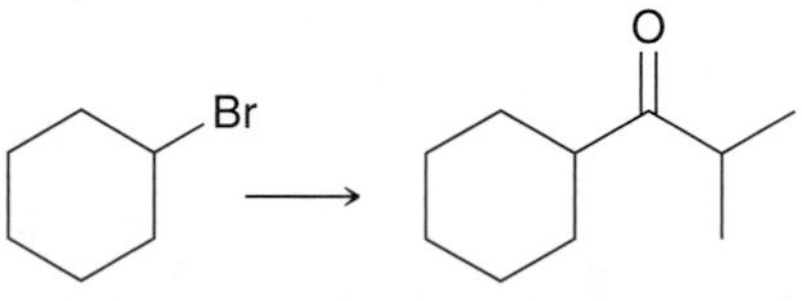

(e)
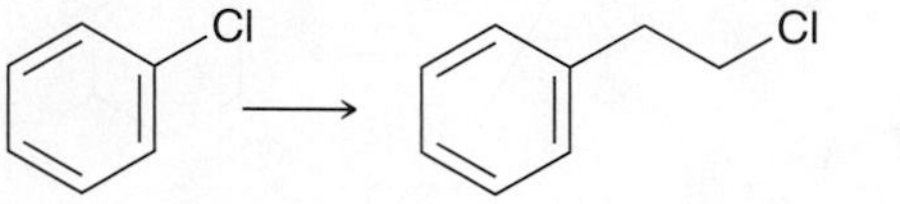

(f)
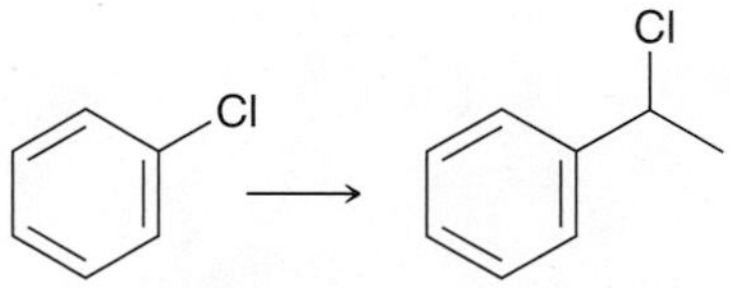

(g)
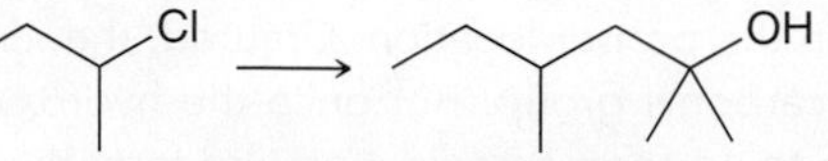

(h)
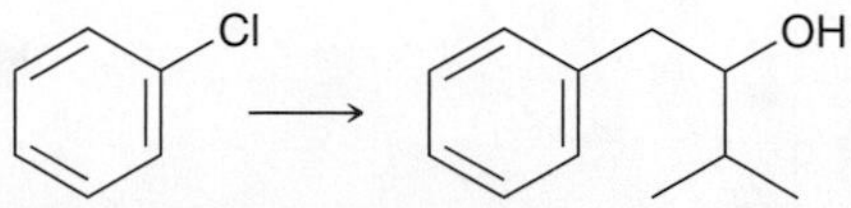

(i)
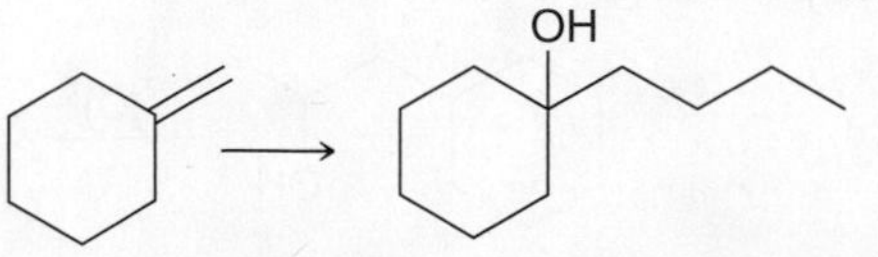

APPLY the skill

14.27 Propose a plausible synthesis for 1,4-dioxane using acetylene as your only source of carbon atoms.

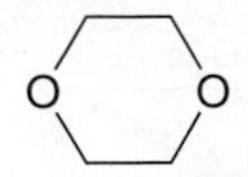

1,4-Dioxane

14.28 Dimethoxyethane (DME) is a polar aprotic solvent often used for S_N2 reactions. Propose a plausible synthesis for DME using acetylene and methyl iodide as your only sources of carbon atoms.

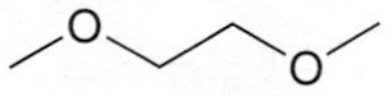

Dimethoxyethane

14.29 Using compounds that possess no more than two carbon atoms, propose a plausible synthesis for the following compound.

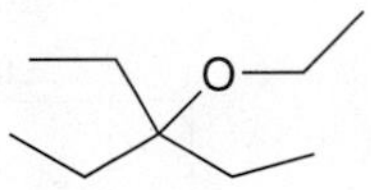

----> need more **PRACTICE?** Try Problem **14.51**

REVIEW OF REACTIONS

PREPARATION OF ETHERS

Williamson Ether Synthesis

$$R-OH \xrightarrow[\text{2) RX}]{\text{1) NaH}} R-O-R$$

Alkoxymercuration-Demercuration

$$\xrightarrow[\text{2) NaBH}_4]{\text{1) Hg(OAc)}_2, \textbf{ROH}}$$

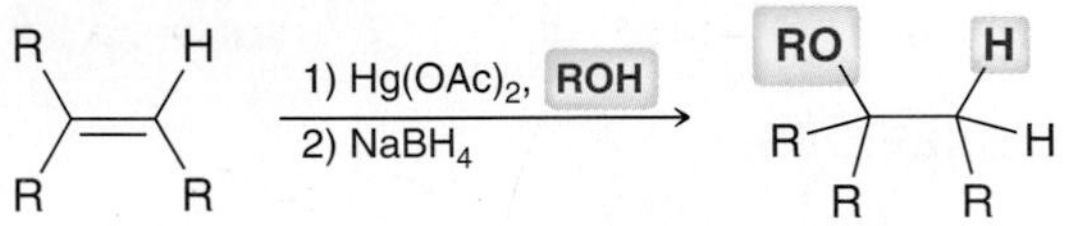

REACTIONS OF ETHERS

Acidic Cleavage

$$R-O-R \xrightarrow[\text{heat}]{\substack{\text{excess} \\ \text{HX}}} R-X \ + \ R-X \ + \ H_2O$$

$$\text{Ph}-O-R \xrightarrow[\text{heat}]{\substack{\text{excess} \\ \text{HX}}} \text{Ph}-OH \ + \ R-X$$

Autooxidation

$$\xrightarrow[\text{(slow)}]{O_2}$$

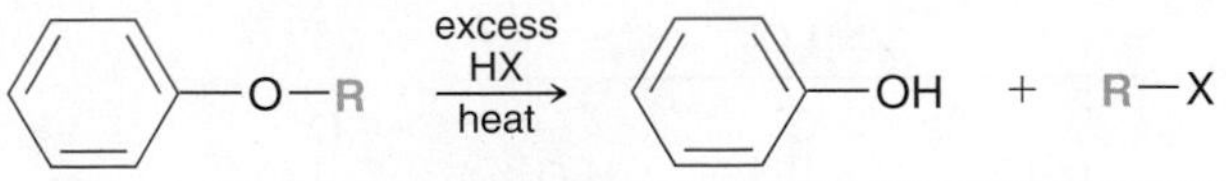

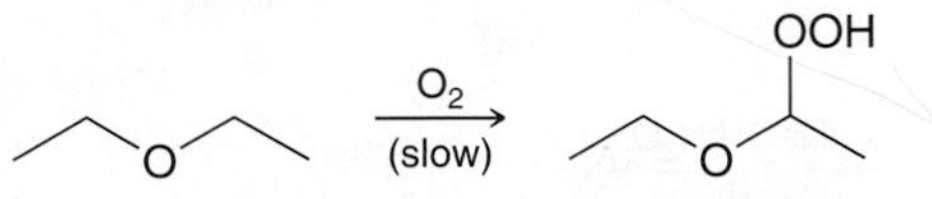

A hydroperoxide

PREPARATION OF EPOXIDES

MCPBA

1) Br₂, H₂O
2) NaOH

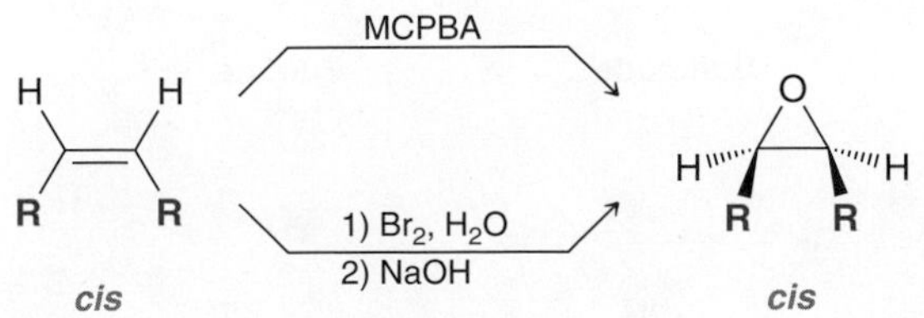

cis *cis*

ENANTIOSELECTIVE EPOXIDATION

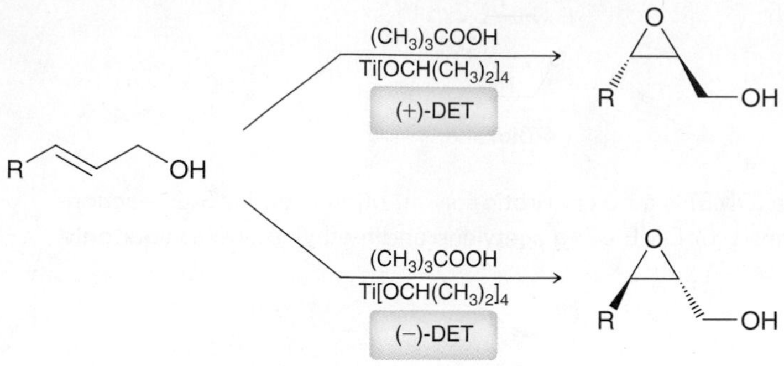

RING-OPENING REACTIONS OF EPOXIDES

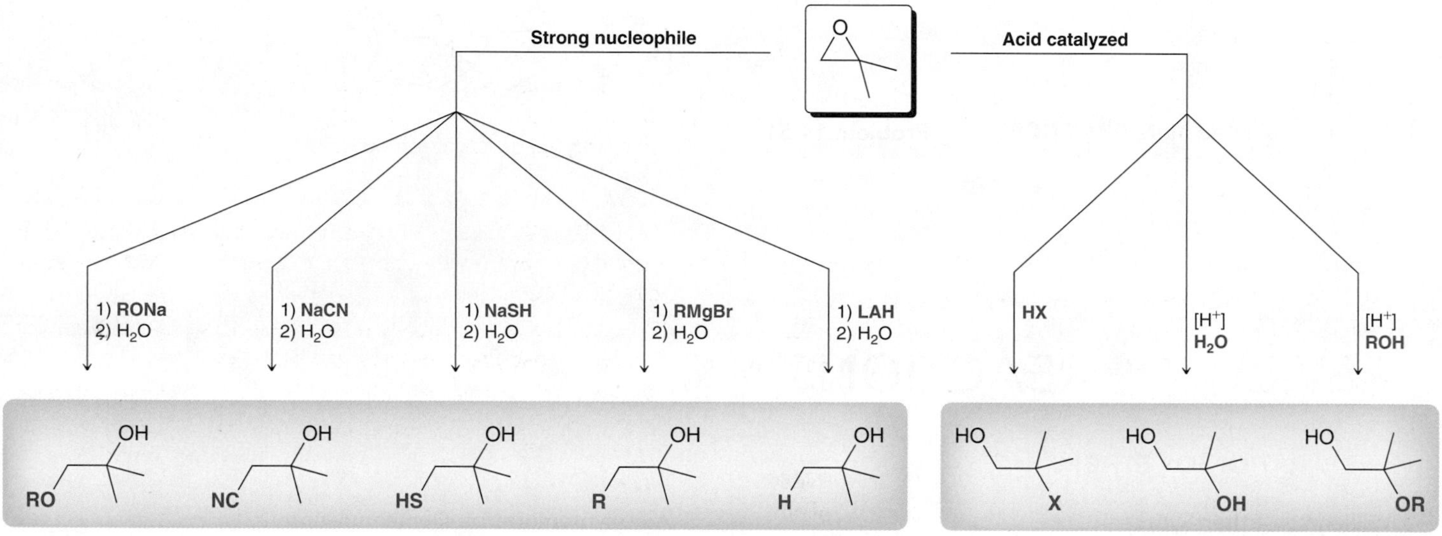

THIOLS AND SULFIDES

Thiols

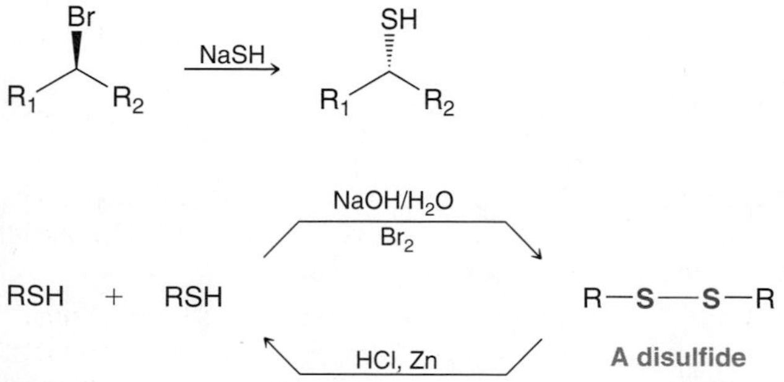

Sulfides

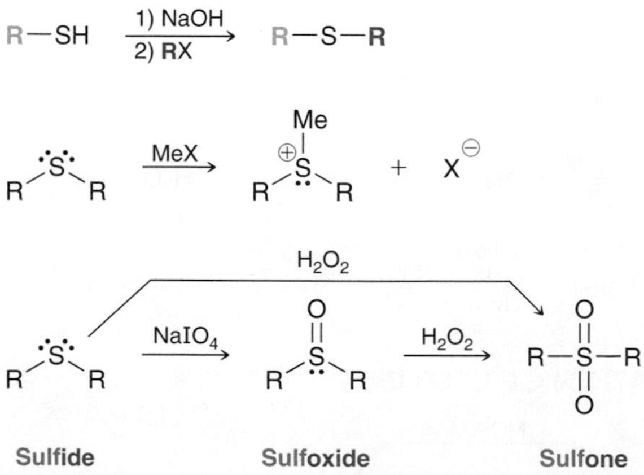

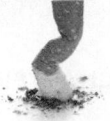

REVIEW OF CONCEPTS AND VOCABULARY

SECTION 14.1

- **Ethers** are compounds that have an oxygen atom bonded to two groups, which can be alkyl, aryl, or vinyl groups.
- The ether moiety is a common structural feature of many natural compounds and pharmaceuticals.

SECTION 14.2

- **Unsymmetrical ethers** have two different alkyl groups, while **symmetrical ethers** have two identical groups.
- The common name of an ether is constructed by assigning a name to each R group, arranging them in alphabetical order, and then adding the word "ether."
- The systematic name of an ether is constructed by choosing the larger group to be the parent alkane and naming the smaller group as an **alkoxy** substituent.

SECTION 14.3

- Ethers of low molecular weight have low boiling points, while ethers with larger alkyl groups have higher boiling points due to London dispersion forces between the alkyl groups.
- Ethers are often used as solvents for organic reactions.

SECTION 14.4

- The interaction between ethers and metal ions is very strong for **polyethers,** compounds with multiple ether moieties.
- Cyclic polyethers, or **crown ethers**, are capable of solvating metal ions in organic (nonpolar) solvents.

SECTION 14.5

- Ethers can be readily prepared from the reaction between an alkoxide ion and an alkyl halide, a process called a **Williamson ether synthesis**. This process works best for methyl or primary alkyl halides. Secondary alkyl halides are significantly less efficient, and tertiary alkyl halides cannot be used.
- Ethers can be prepared from alkenes via **alkoxymercuration-demercuration**, which results in a Markovnikov addition of RO and H across an alkene.

SECTION 14.6

- When treated with a strong acid, an ether will undergo **acidic cleavage** in which it is converted into two alkyl halides.
- When a phenyl ether is cleaved under acidic conditions, the products are phenol and an alkyl halide.
- Ethers undergo autooxidation in the presence of atmospheric oxygen to form hydroperoxides.

SECTION 14.7

- A three-membered cyclic ether is called **oxirane.** It possesses significant ring strain and is therefore more reactive than other ethers.
- Substituted oxiranes are also called **epoxides**, which are named in either of two ways:
  - The oxygen atom is considered to be a substituent on the parent chain, and the exact location of the epoxide moiety is identified with two numbers followed by the term "epoxy."
  - The parent is considered to be the epoxide (parent = oxirane), and any groups connected to the epoxide are listed as substituents.

SECTION 14.8

- Alkenes can be converted into epoxides by treatment with peroxy acids or via halohydrin formation and subsequent epoxidation. Both procedures are stereospecific.
- Substituents that are *cis* to each other in the starting alkene remain *cis* to each other in the epoxide, and substituents that are *trans* to each other in the starting alkene remain *trans* to each other in the epoxide.

SECTION 14.9

- Chiral catalysts can be used to achieve the enantioselective epoxidation of allylic alcohols.
- In a **Sharpless asymmetric epoxidation**, the catalyst favors the production of one enantiomer over the other, leading to an observed enantiomeric excess.

SECTION 14.10

- Epoxides will undergo **ring-opening reactions** either (1) in conditions involving a strong nucleophile or (2) under acid-catalyzed conditions.
- When a strong nucleophile is used, the nucleophile attacks at the less substituted (less hindered) position.
- Under acid-catalyzed conditions, the regiochemical outcome is dependent on the nature of the epoxide and is explained in terms of a competition between **electronic effects** and **steric effects**.
- The stereochemical outcome involves inversion of configuration under all conditions.

SECTION 14.11

- Sulfur analogs of alcohols contain an SH group rather than an OH group and are called **thiols**.
- When another functional group is present in the compound, the SH group is named as a substituent and is called a **mercapto group.**
- Thiols can be prepared via an S_N2 reaction between sodium hydrosulfide (NaSH) and a suitable alkyl halide.
- Thiols easily undergo oxidation to produce **disulfides**, and disulfides are also easily reduced back to thiols when treated with a reducing agent.
- The sulfur analogs of ethers (thioethers) are called **sulfides**. The nomenclature of sulfides is similar to that of ethers. Common names are assigned using the suffix "sulfide" instead of "ether." More complex sulfides are named systematically, much the way ethers are named, with the alkoxy group being replaced by an **alkylthio group.**
- Sulfides can be prepared from thiols in a process that is essentially the sulfur analog of the Williamson ether synthesis, involving a **thiolate** ion, rather than an alkoxide.
- Sulfides will attack alkyl halides to produce alkylating agents, such as the biological alkylating agent SAM.

- Sulfides undergo oxidation to give **sulfoxides** and then **sulfones.**

SECTION 14.12

- Ring opening of an epoxide produces a compound with two functional groups on adjacent carbon atoms. Whenever you see two adjacent functional groups, you should think of epoxides.
- When a Grignard reagent reacts with an epoxide, a C—C bond is formed. This reaction can be used to introduce a chain of carbon atoms that possess a built-in functional group at the second carbon atom.
- In contrast, when a Grignard reagent attacks a ketone or aldehyde, the functional group appears on the first carbon atom that was introduced.
- You must train your eyes to look at the precise location of a functional group when it appears on a newly introduced alkyl fragment.

KEY TERMINOLOGY

SKILLBUILDER REVIEW

14.1 NAMING AN ETHER

COMMON NAME
Treat both sides as substituents, and list them alphabetically.

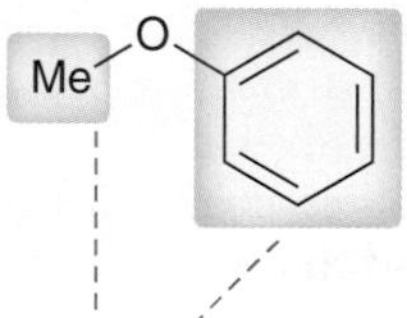

Methyl phenyl ether

SYSTEMATIC NAME
1) Choose the parent (the more complex side).
2) Identify all substituents (including alkoxy group).
3) Assign locants to give lowest number to alkoxy group.
4) Assemble substituents alphabetically with locants.
5) Assign the configuration of any chirality centers.

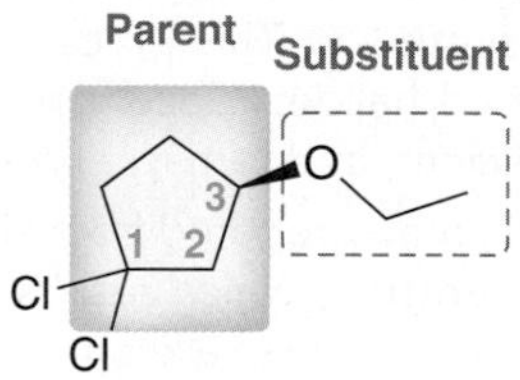

(*R*)-1,1-Dichloro-3-ethoxycyclopentane

Try Problems 14.1–14.3, 14.30, 14.32

14.2 PREPARING AN ETHER VIA A WILLIAMSON ETHER SYNTHESIS

STEP 1 Identify the two groups on either side of the oxygen atom.

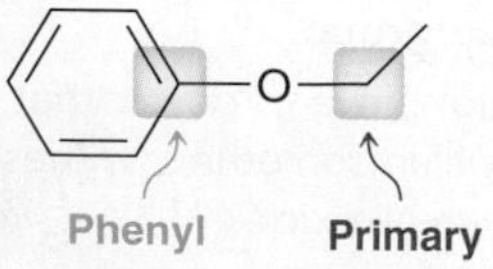

STEP 2 Determine which side is more capable of serving as a substrate in an S_N2 reaction.

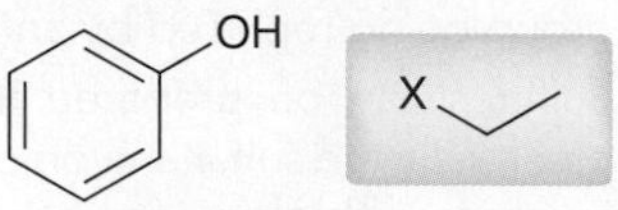

STEP 3 Use a base to deprotonate the alcohol, and then identify the alkyl halide.

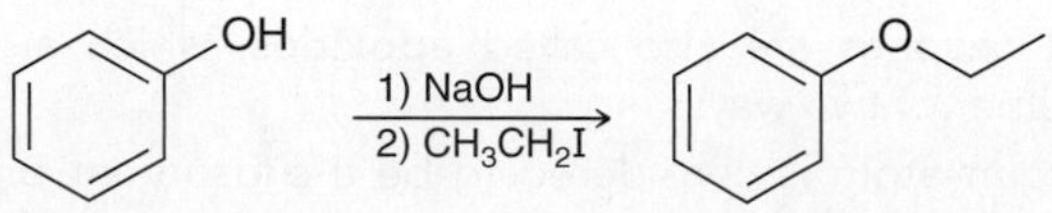

Try Problems 14.5–14.7, 14.33a,c, 14.37, 14.40, 14.42d, 14.43b

14.3 PREPARING EPOXIDES

STEP 1 Identify the four groups attached to the epoxide.

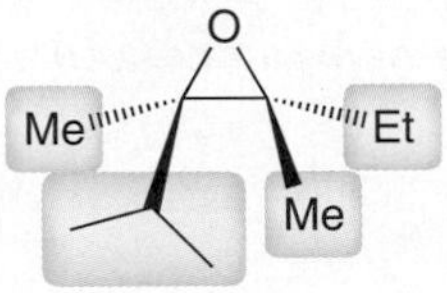

STEP 2 Identify the relative configuration of the four groups, and draw the starting alkene.

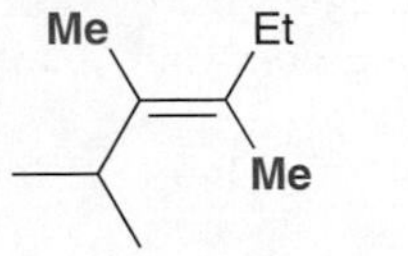

STEP 3 Use either of the folloiwing two methods to prepare the epoxide from the starting alkene.

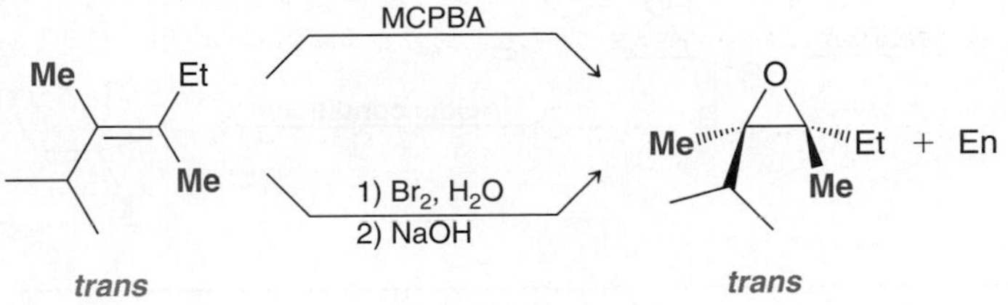

Try Problems 14.14, 14.15, 14.39, 14.51t

14.4 DRAWING THE MECHANISM AND PREDICTING THE PRODUCT OF THE REACTION BETWEEN A STRONG NUCLEOPHILE AND AN EPOXIDE

EXAMPLE Predict the product, and propose a mechanism for its formation.

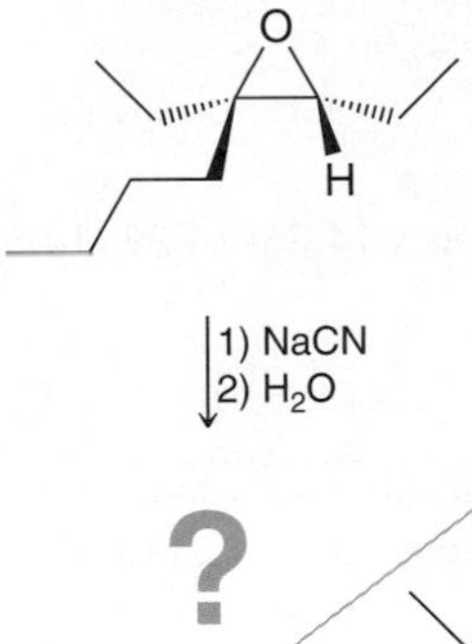

STEP 1 Identify the regiochemistry by selecting the less-hindered position as the site of nucleophilic attack.

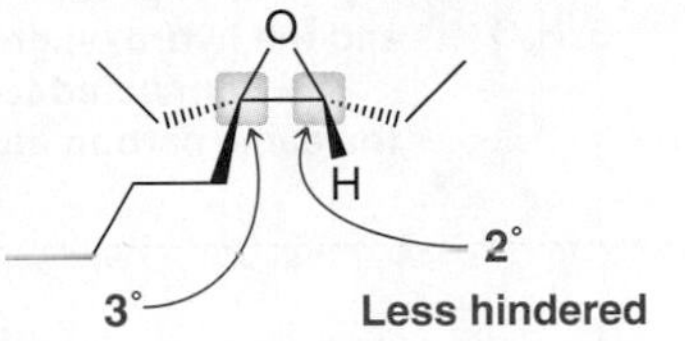

STEP 2 Identify the stereochemistry by determining whether the nucleophile attacks a chirality center. If so, expect inversion of configuration.

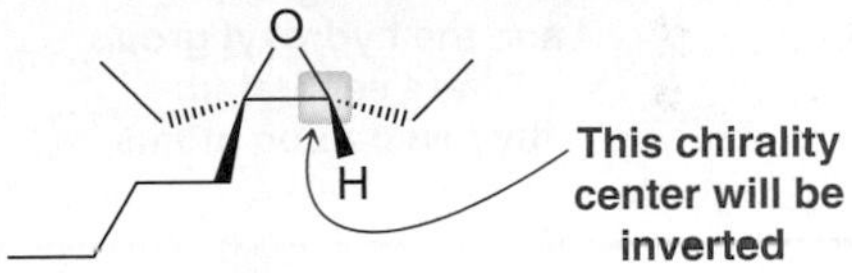

STEP 3 Draw both steps of the mechanism:

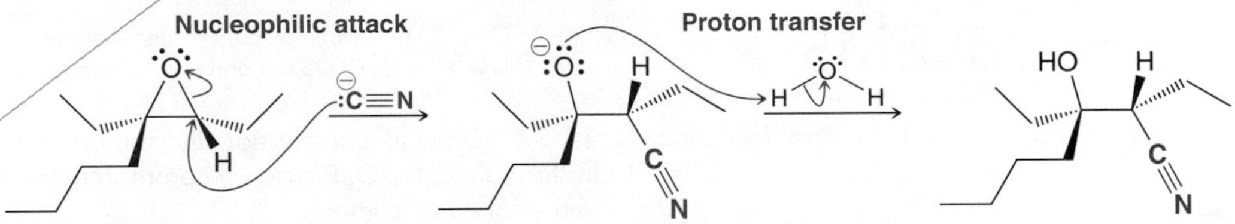

Try Problems 14.17–14.19, 14.42a, c, e–h, 14.43

14.5 DRAWING THE MECHANISM AND PREDICTING THE PRODUCT OF ACID-CATALYZED RING OPENING

EXAMPLE Predict the product, and propose a mechanism for its formation.

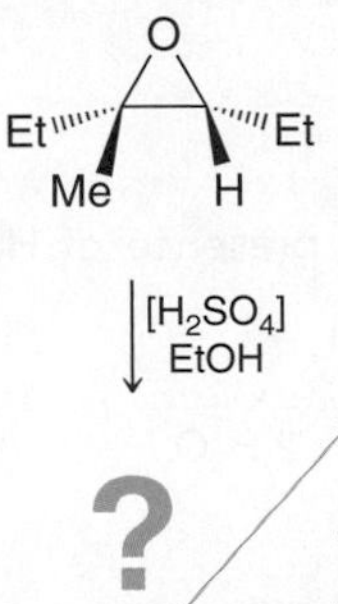

STEP 1 Identify the regiochemistry: determine whether steric or electronic effects will dominate.

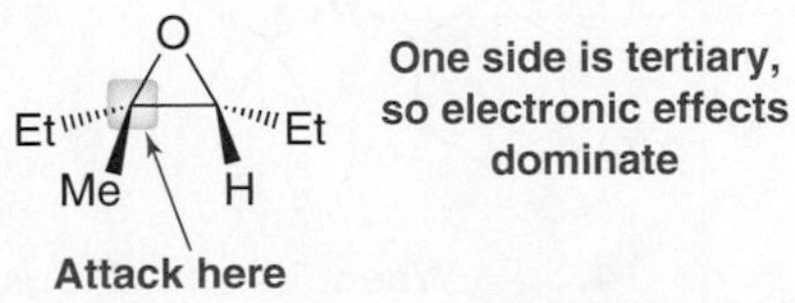

STEP 2 Identify the stereochemistry: if nucleophile attacks a chirality center, expect inversion of configuration.

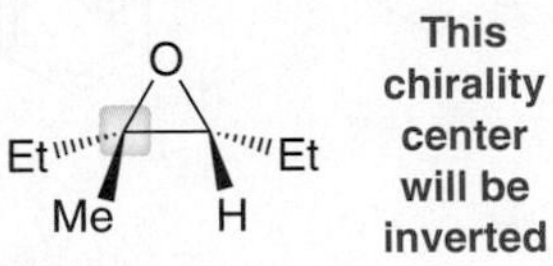

STEP 3 Draw all three steps of the mechanism: (1) proton transfer, (2) nucleophilic attack, and (3) proton transfer.

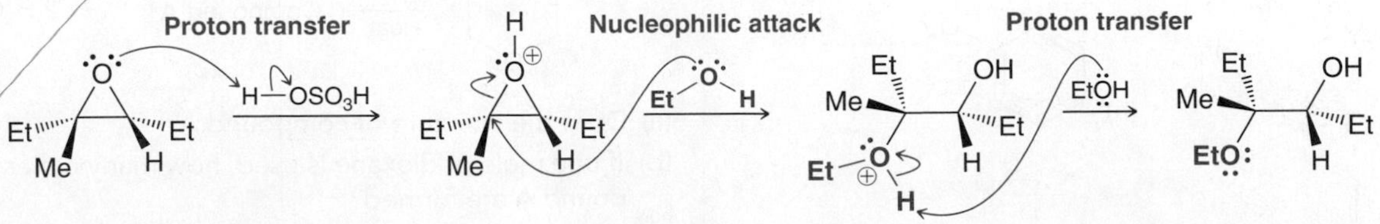

Try Problems 14.20, 14.21, 14.43d, 14.49, 14.50

14.6 INSTALLING TWO ADJACENT FUNCTIONAL GROUPS

Convert an alkene into an epoxide. | Open the epoxide with regiochemical control.

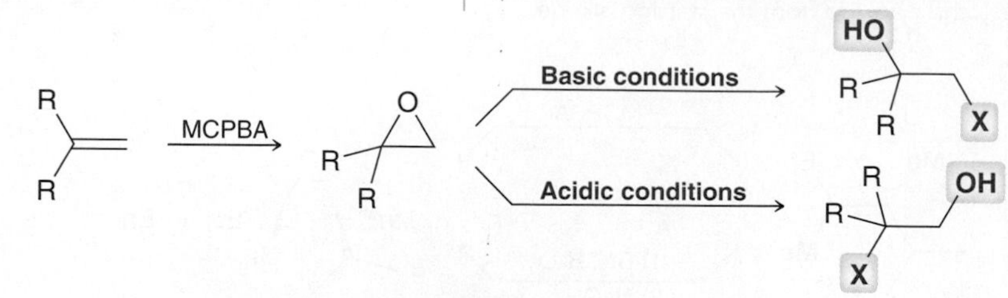

\dashrightarrow Try Problems **14.24, 14.25, 14.51**

14.7 CHOOSING THE APPROPRIATE GRIGNARD REACTION

Grignard reagent attacking an epoxide:

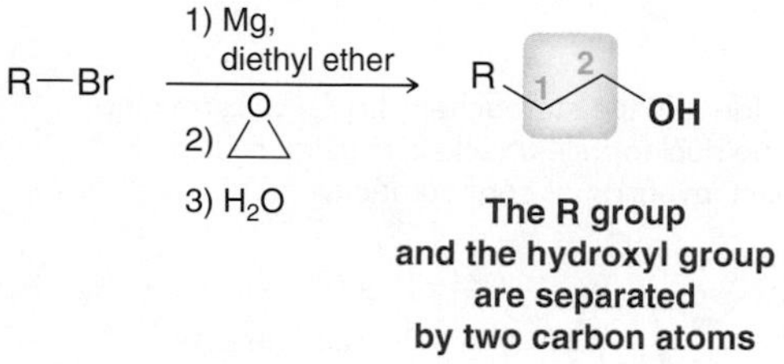

R—Br
1) Mg, diethyl ether
2) [epoxide]
3) H$_2$O

The R group and the hydroxyl group are separated by two carbon atoms

Grignard reagent attacking an aldehyde or ketone:

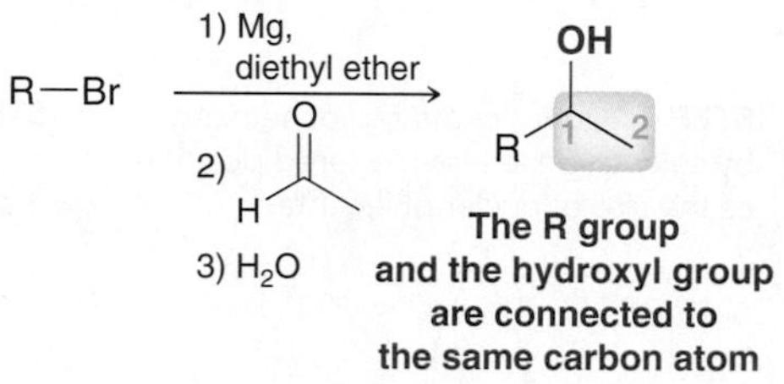

R—Br
1) Mg, diethyl ether
2) [aldehyde]
3) H$_2$O

The R group and the hydroxyl group are connected to the same carbon atom

\dashrightarrow Try Problems **14.26–14.29, 14.51**

PRACTICE PROBLEMS

Note: Most of the Problems are available within *WileyPLUS*, an online teaching and learning solution.

14.30 Assign an IUPAC name for each of the following compounds.

(a)

(b)

SH
(c)

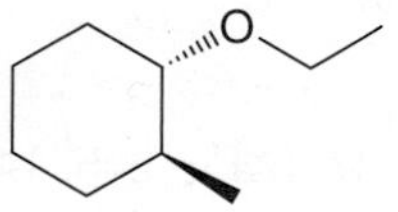

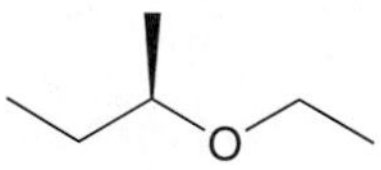

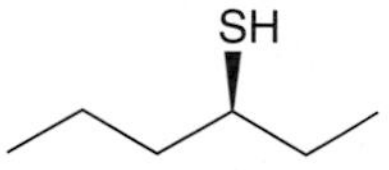

(d)

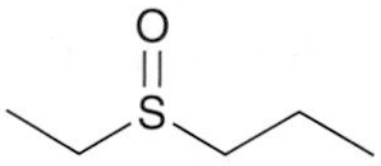

(e)

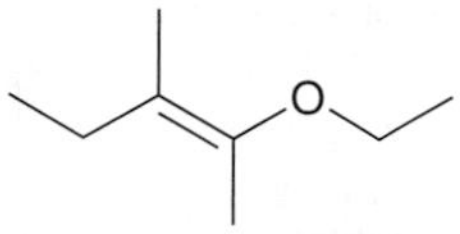

(f)

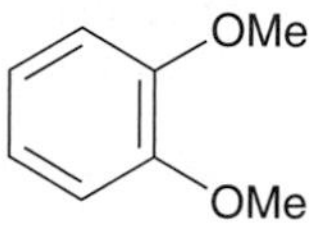

(g)

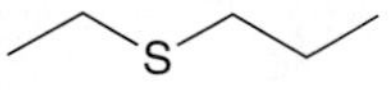

14.31 Predict the products that are expected when each of the following compounds is heated with concentrated HBr.

(a)

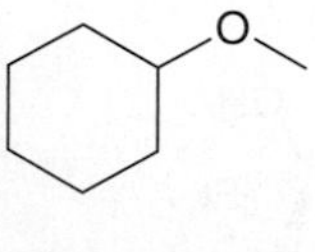

(b)

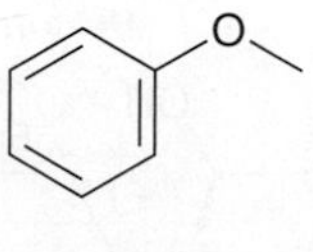

(c)

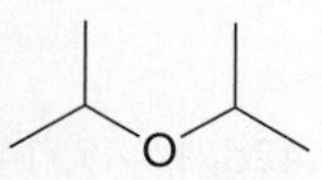

(d)

14.32 Draw all constitutionally isomeric ethers with molecular formula C$_4$H$_{10}$O. Provide a common name and a systematic name for each isomer.

14.33 Starting with cyclohexene and using any other reagents of your choice, show how you would prepare each of the following compounds.

(a)

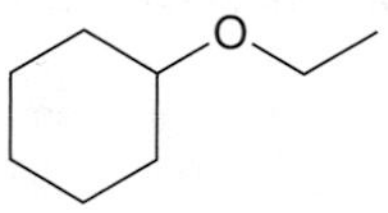

(b)

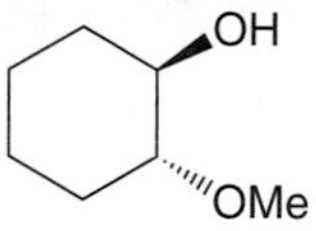

(c)

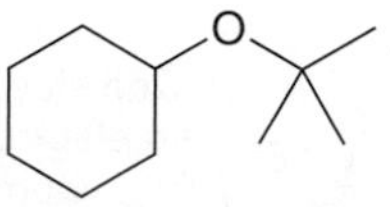

14.34 When 1,4-dioxane is heated in the presence of HI, compound **A** is obtained:

[1,4-dioxane] $\xrightarrow[\text{Heat}]{\substack{\text{excess} \\ \text{HI}}}$ Compound **A** + 2 H$_2$O

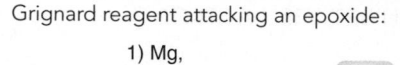

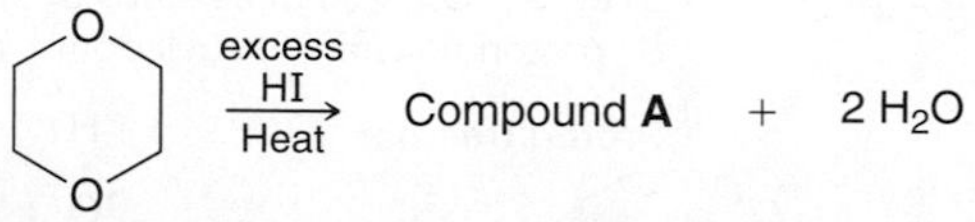

(a) Draw the structure of compound **A**.

(b) If one mole of dioxane is used, how many moles of compound **A** are formed?

(c) Show a plausible mechanism for the conversion of dioxane into compound **A**.

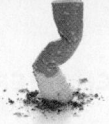

14.35 Tetrahydrofuran (THF) can be formed by treating 1,4-butanediol with sulfuric acid. Propose a mechanism for this transformation.

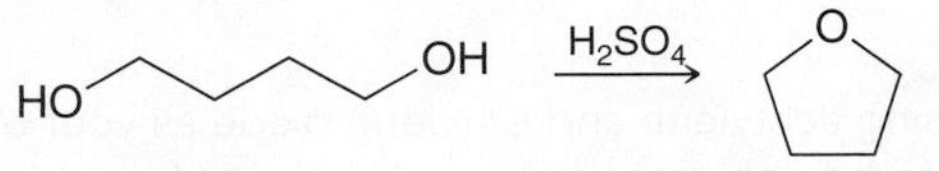

1,4-Butanediol **Tetrahydrofuran (THF)**

14.36 When ethylene glycol is treated with sulfuric acid, 1,4-dioxane is obtained. Propose a mechanism for this transformation:

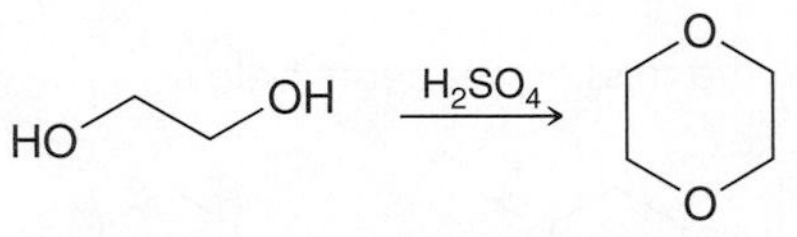

Ethylene glycol **1,4-Dioxane**

14.37 The Williamson ether synthesis cannot be used to prepare *tert*-butyl phenyl ether.

(a) Explain why this method cannot be used in this case.

(b) Suggest an alternative method for preparing *tert*-butyl phenyl ether.

14.38 Methylmagnesium bromide reacts rapidly with ethylene oxide, it reacts slowly with oxetane, and it does not react at all with tetrahydrofuran.

Ethylene oxide

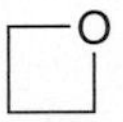

Oxetane

Tetrahydrofuran (THF)

Explain this difference in reactivity.

14.39 Identify the reagents necessary to accomplish each of the following transformations.

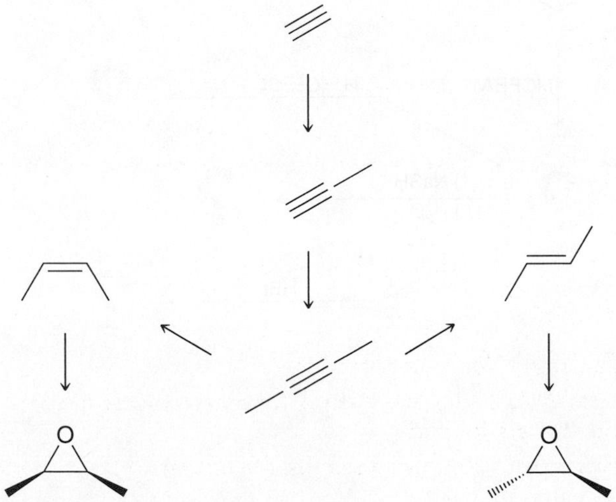

14.40 When 5-bromo-2,2-dimethyl-1-pentanol is treated with sodium hydride, a compound with molecular formula $C_7H_{14}O$ is obtained. Identify the structure of this compound.

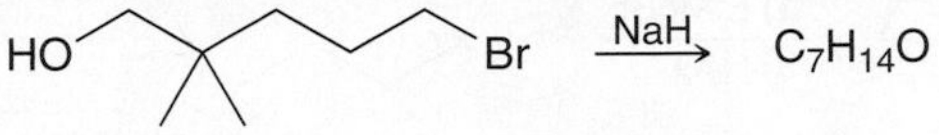

14.41 Problem 14.39 outlines a general method for the preparation of *cis*- or *trans*-disubstituted epoxides. Using that method, identify what reagents you would use to prepare a racemic mixture of each of the following epoxides from acetylene.

(a) 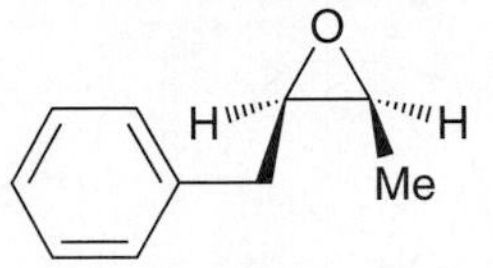

(b) 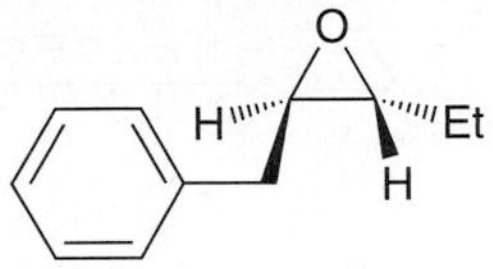

(c) 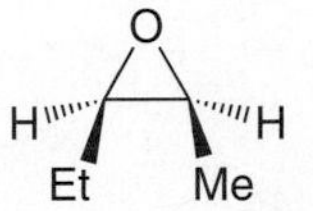

(d) 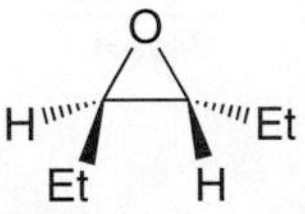

14.42 Predict the products for each of the following.

(a) 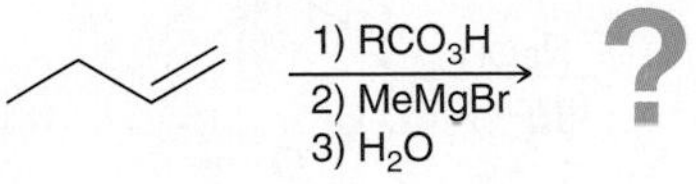

(b) 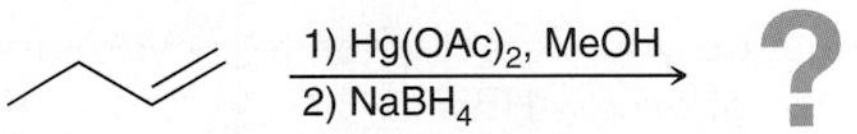

(c) 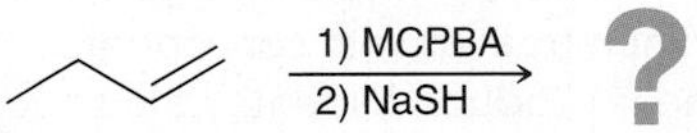

(d) 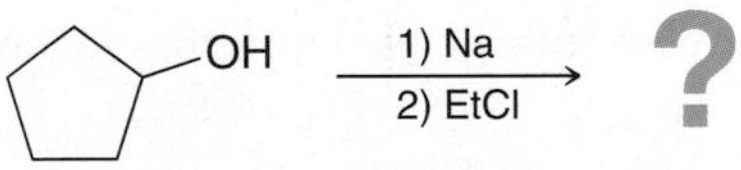

(e) 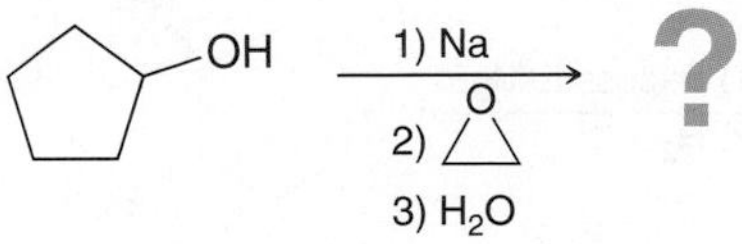

(f) 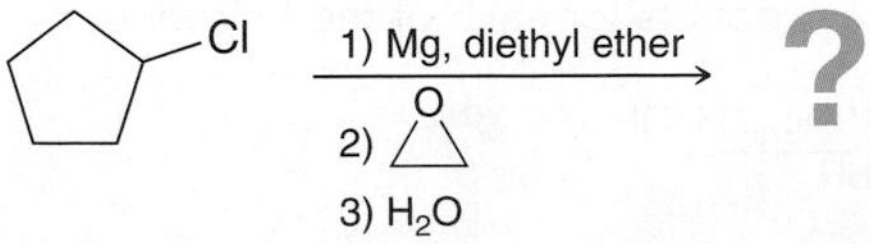

(g) 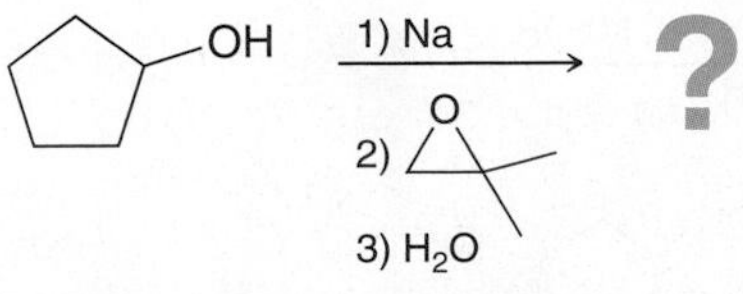

(h) 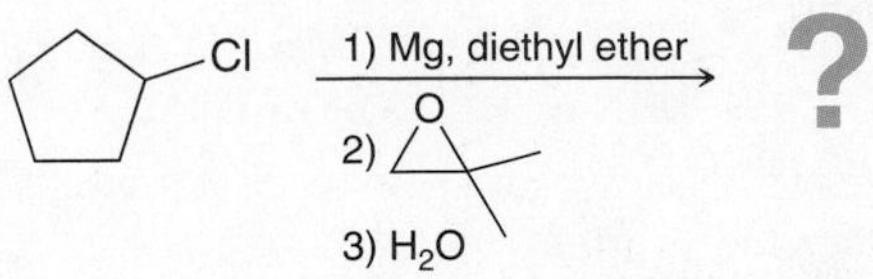

14.43 Propose a plausible mechanism for each of the following transformations.

(a) 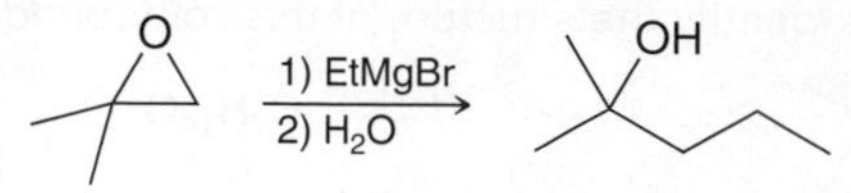

(b) 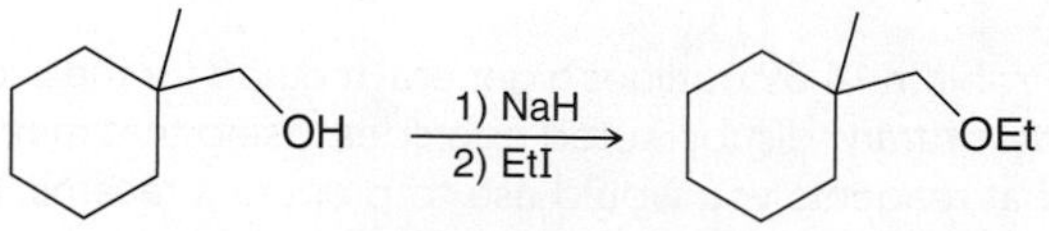

(c) 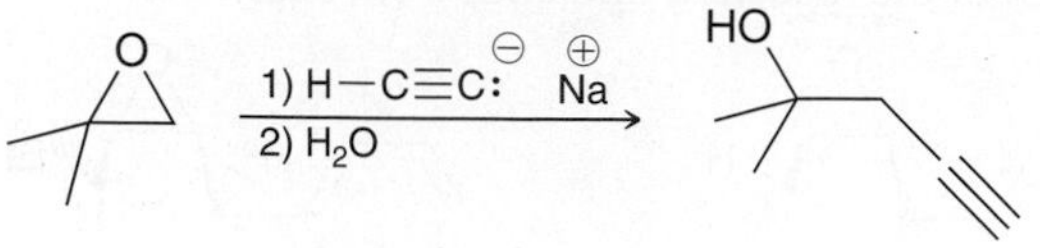

(d) 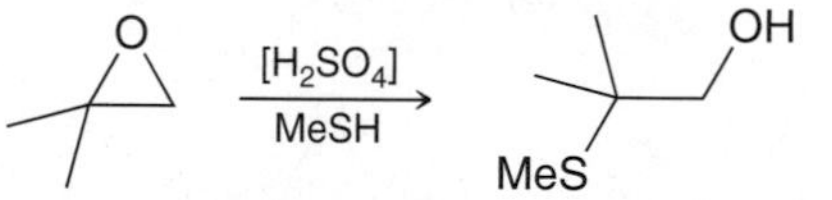

(e) 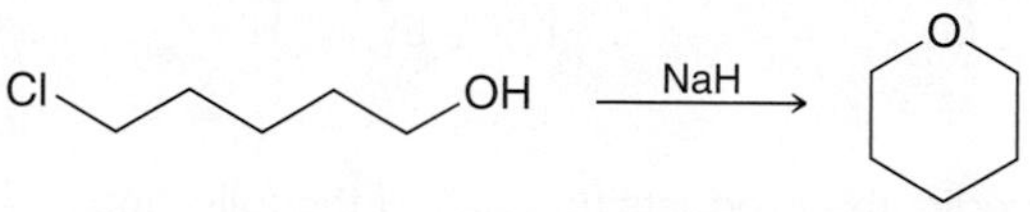

(f) 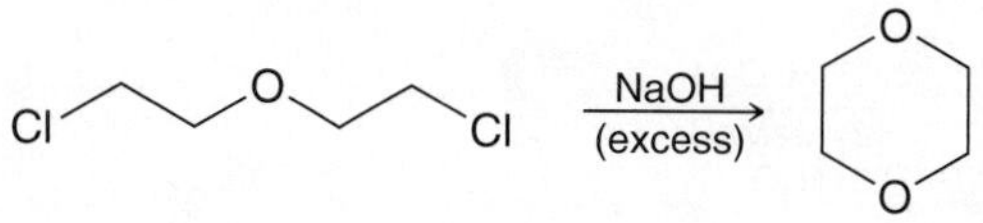

14.44 What product do you expect when tetrahydrofuran is heated in the presence of excess HBr?

14.45 Compound **B** has molecular formula $C_6H_{10}O$ and does not possess any π bonds. When treated with concentrated HBr, *cis*-1,4-dibromocyclohexane is produced. Identify the structure of compound **B**.

14.46 Propose a stepwise mechanism for the following transformation.

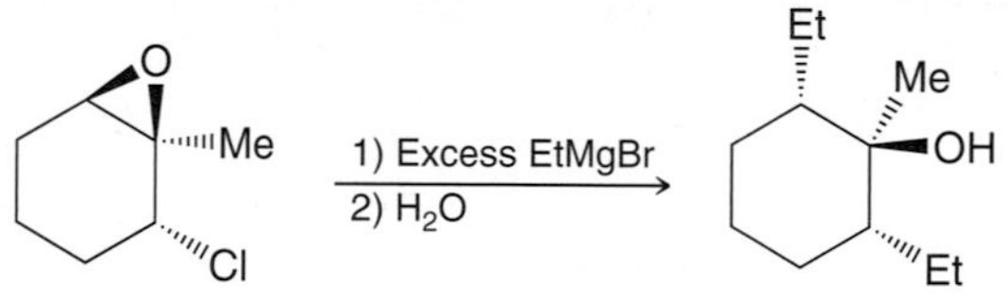

14.47 Predict the products for each of the following:

(a) 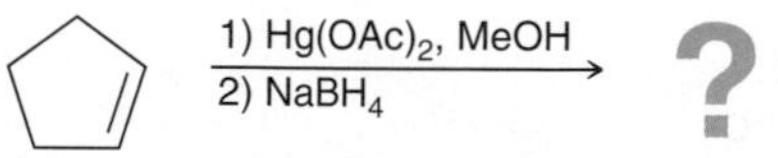

(b) 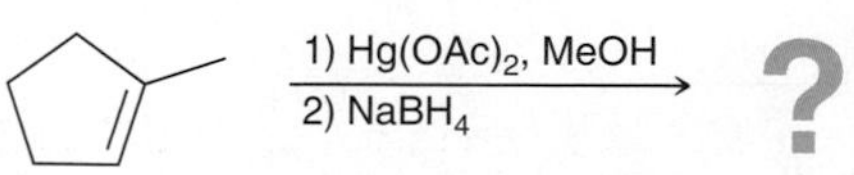

(c) 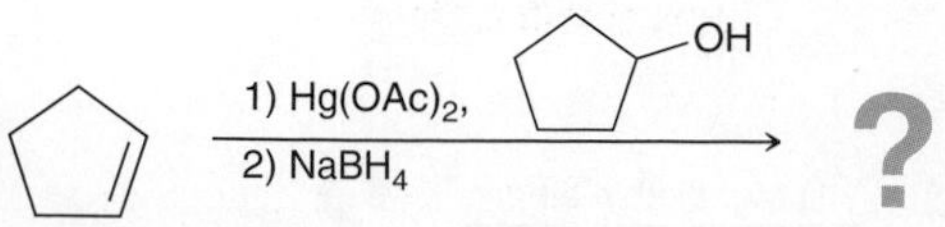

(d) 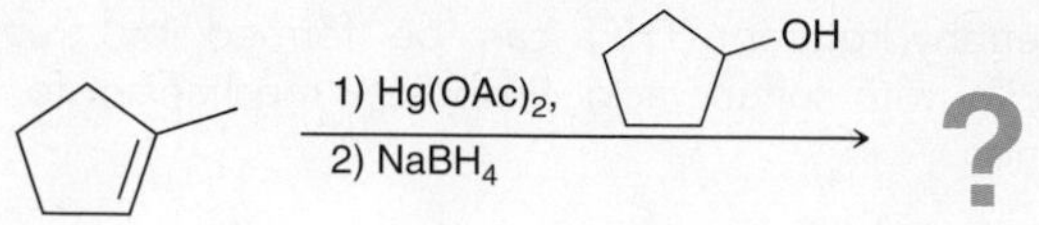

14.48 Using acetylene and ethylene oxide as your only sources of carbon atoms, propose a synthesis for each of the following compounds.

(a) 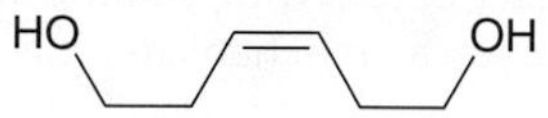

(b) 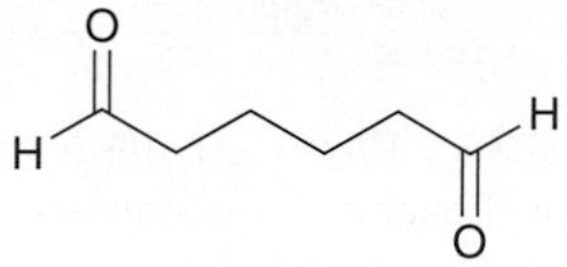

14.49 Fill in the missing reagents below.

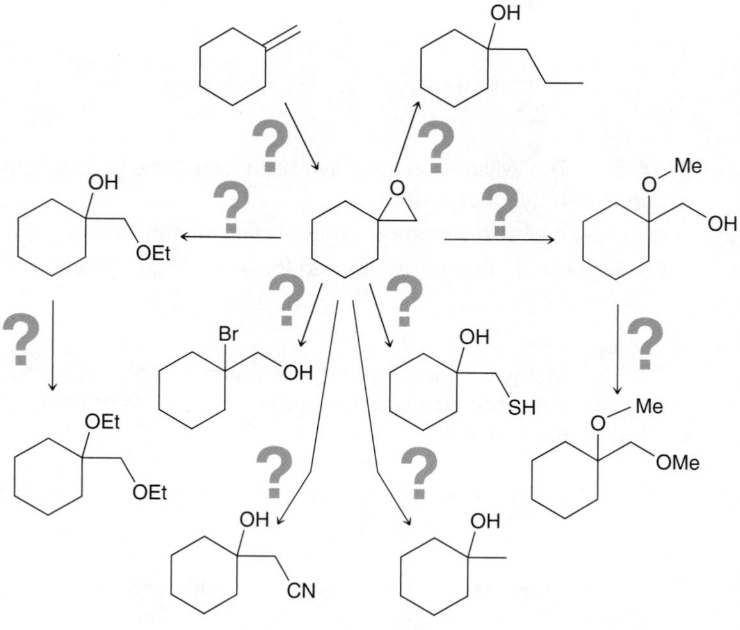

14.50 Fill in the missing products below.

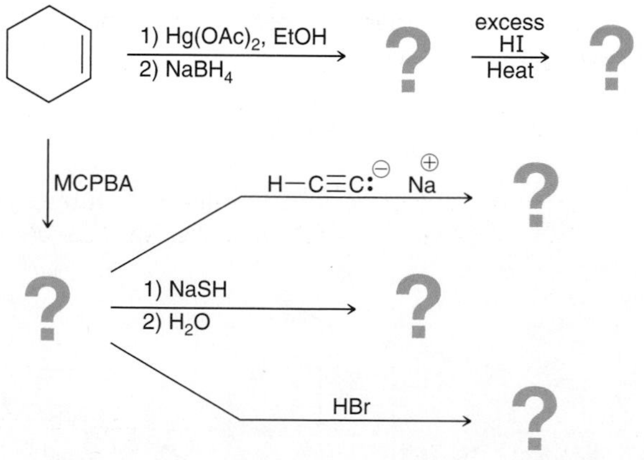

INTEGRATED PROBLEMS

14.51 Propose a plausible synthesis for each transformation.

(a) 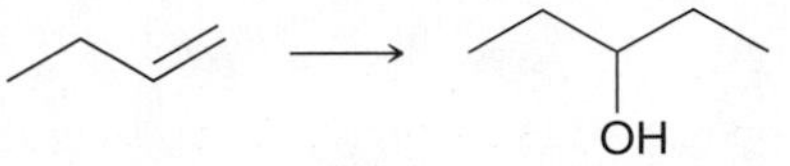

(b) 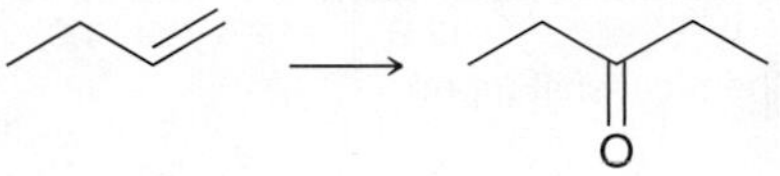

(c) 

(d) 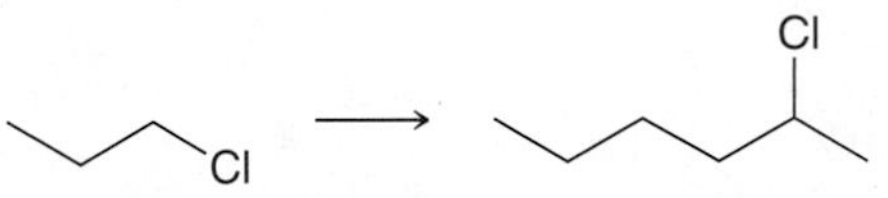

(e) 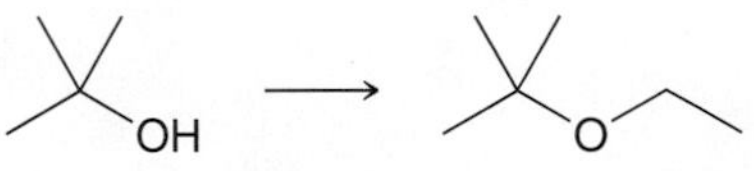

(f) 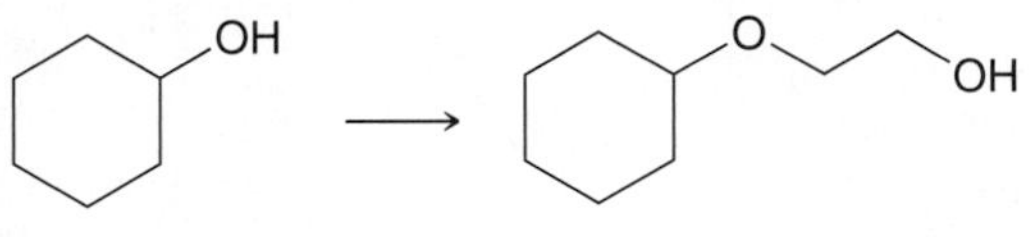

(g) 

(h) 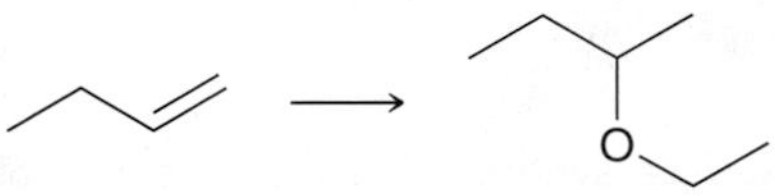

(i) 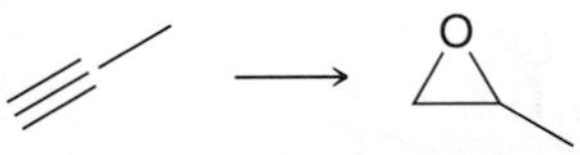

(j) 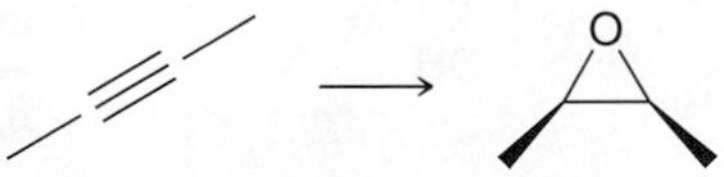

(k) 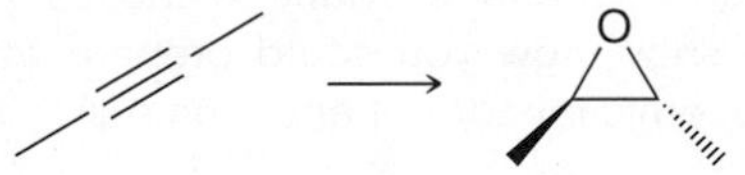 + En

(l) 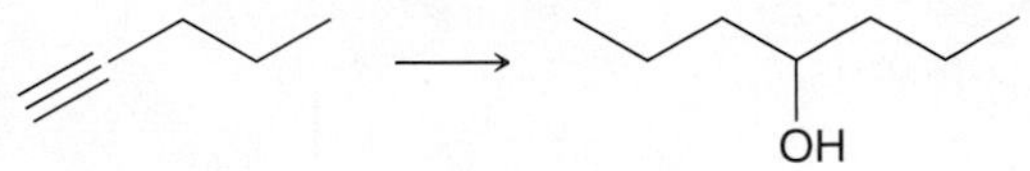

(m) 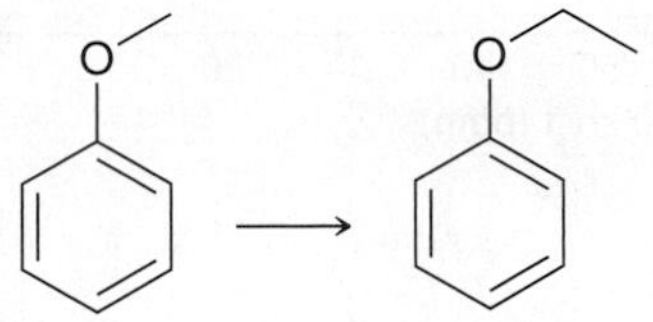

(n) 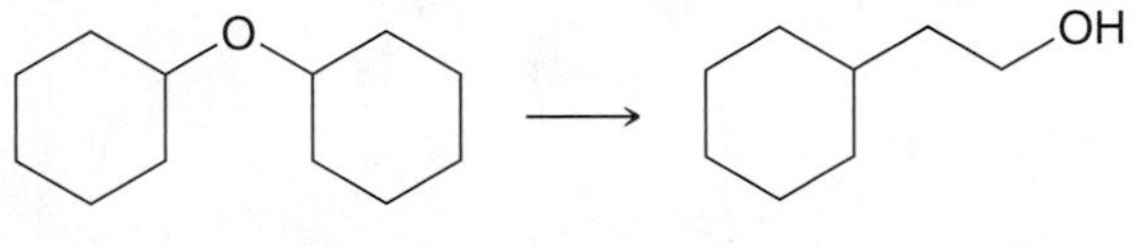

(o) 

(p) 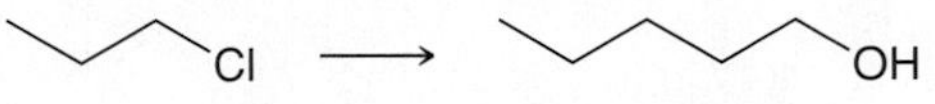

(q) 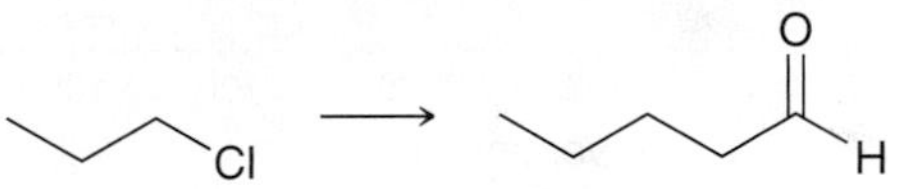

(r) 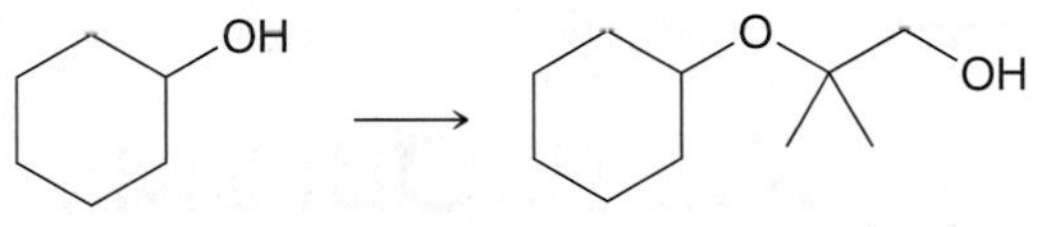

(s) 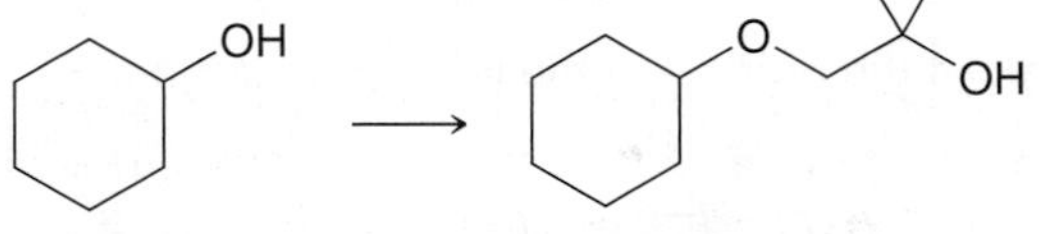

(t) 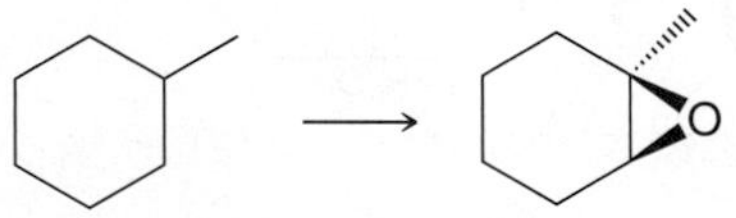 + En

(u) 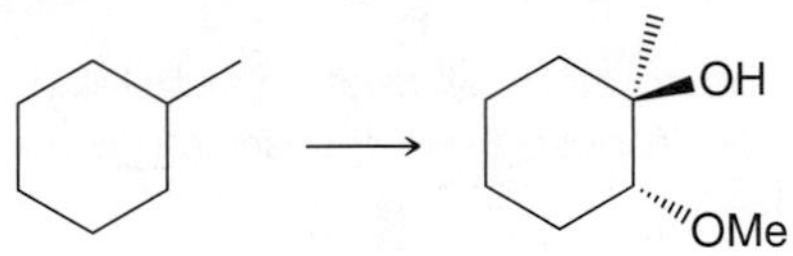 + En

Problems 14.52-14.55 are intended for students who have already covered spectroscopy (Chapters 15 and 16).

14.52 Propose a structure for an ether with molecular formula C_7H_8O that exhibits the following ^{13}C NMR spectrum.

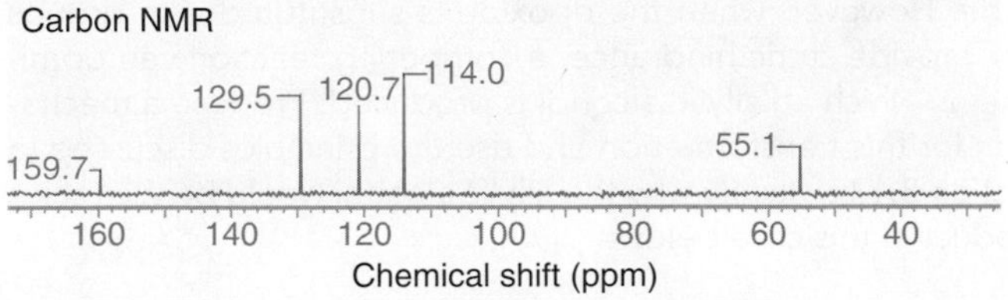

14.53 Propose a structure for a compound with molecular formula $C_8H_{18}O$ that exhibits the following ^1H NMR and ^{13}C NMR spectra.

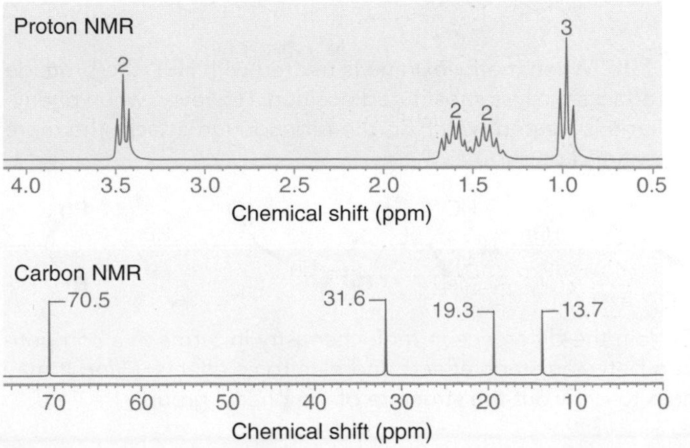

14.54 Propose a structure for a compound with molecular formula C_4H_8O that exhibits the following ^{13}C NMR and FTIR spectra.

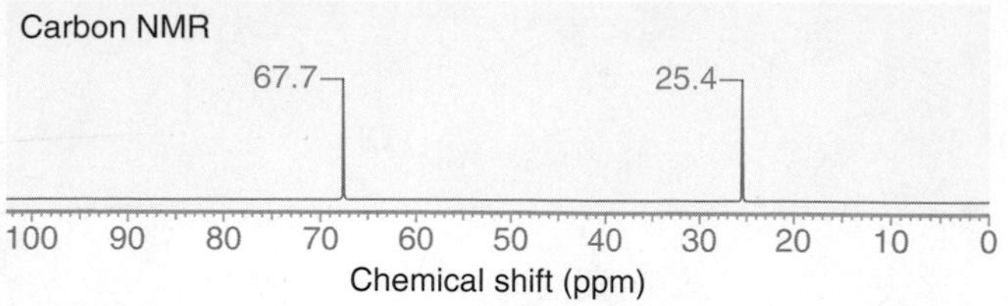

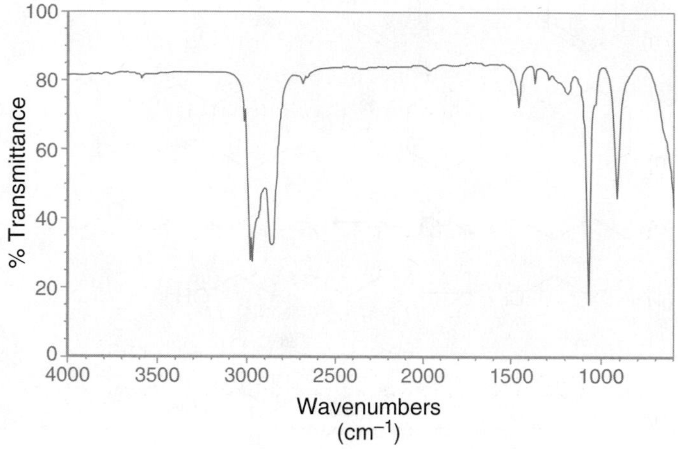

14.55 Propose a structure for a compound with molecular formula $C_4H_{10}O$ that exhibits the following ^1H NMR spectrum.

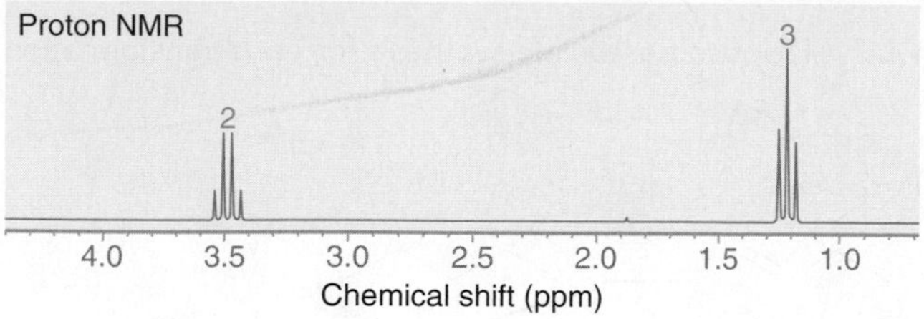

CHALLENGE PROBLEMS

14.56 Predict the product of the following reaction.

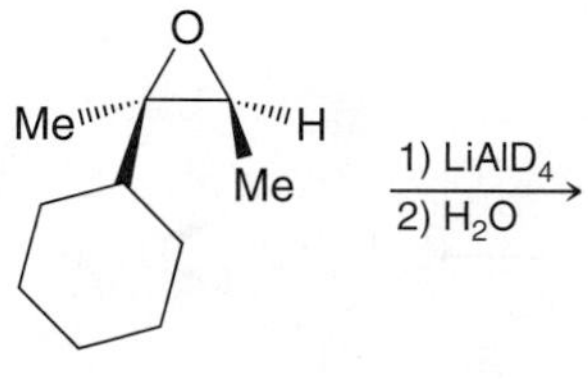

14.57 Epoxides can be formed by treating α-haloketones with sodium borohydride. Propose a mechanism for formation of the following epoxide.

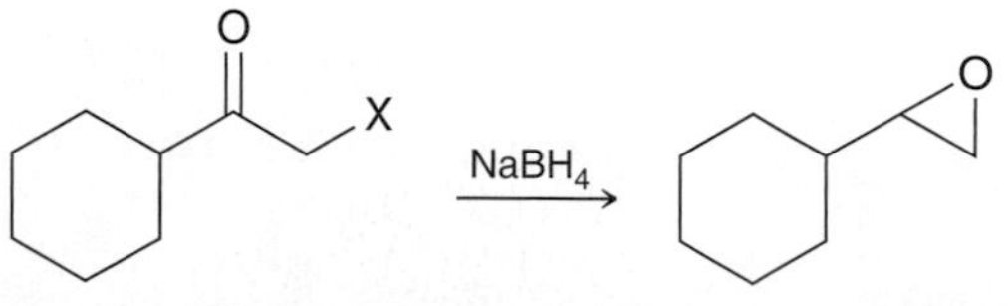

14.58 When methyloxirane is treated with HBr, the bromide ion attacks the less substituted position. However, when phenyloxirane is treated with HBr, the bromide ion attacks the more substituted position.

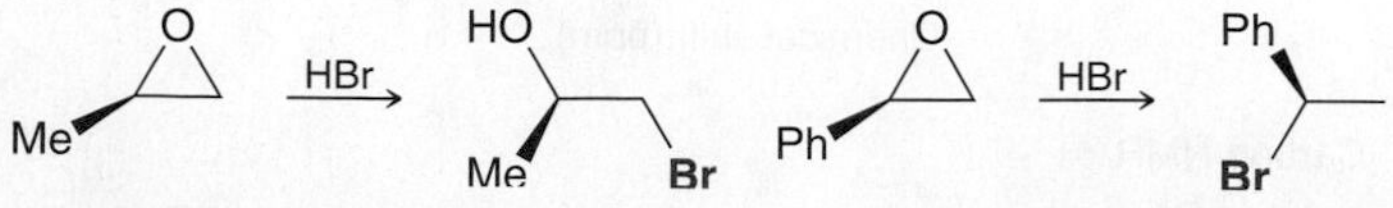

Explain the difference in regiochemistry in terms of a competition between steric effects and electronic effects. (*Hint*: It may help to draw out the structure of the phenyl group).

14.59 Propose a plausible synthesis for the following transformation.

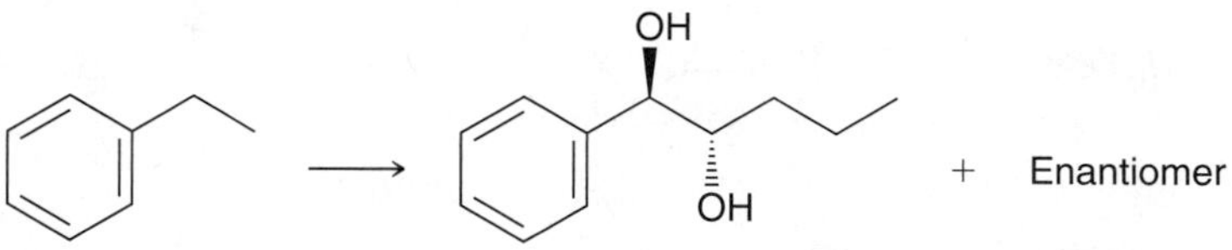

14.60 Using bromobenzene and ethylene oxide as your only sources of carbon, show how you could prepare *trans*-1,2-diphenyloxirane (a racemic mixture of enantiomers).

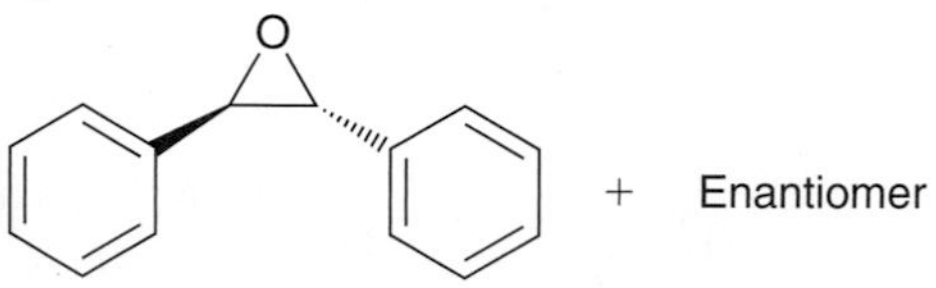

14.61 The S_N2 reaction between a Grignard reagent and an epoxide works reasonably well when the epoxide is ethylene oxide. However, when the epoxide is substituted with groups that provide steric hindrance, a competing reaction can dominate, in which an allylic alcohol is produced. Propose a mechanism for this transformation and use the principles discussed in Section 8.13 to justify why the allylic alcohol would be the major product in this case below.

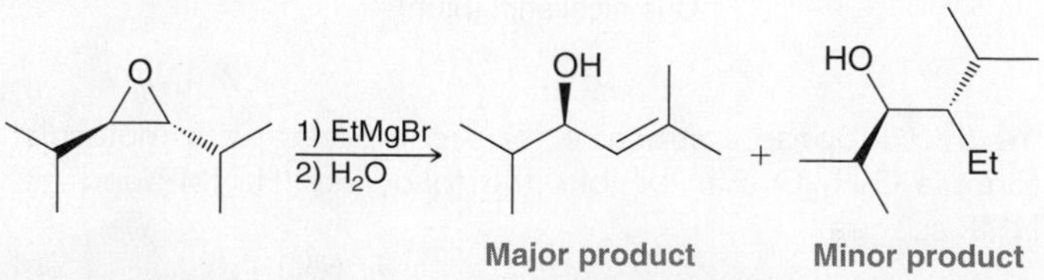

15 Infrared Spectroscopy and Mass Spectrometry

DID YOU EVER WONDER…

how a microwave oven works?

Microwave ovens make use of *the interaction between light and matter*, the study of which is called **spectroscopy**. This chapter will serve as an introduction to spectroscopy and its relevance to the field of organic chemistry. After briefly reviewing how light and matter interact with each other (and how microwave ovens work), we will spend most of the chapter learning two very important spectroscopic techniques that are frequently used by organic chemists to verify the structures of compounds. Less than one hundred years ago, structural determination was a difficult and time-consuming task. It was not uncommon for a chemist to spend years determining the structure of an unknown compound. The advent of modern spectroscopic techniques has completely transformed the field of chemistry, and structures can now be determined in several minutes. In this chapter, we will begin to appreciate how spectroscopy can be used to determine the structures of unknown compounds.

DO YOU REMEMBER?

Before you go on, be sure you understand the following topics.
If necessary, review the suggested sections to prepare for this chapter:

- Quantum Mechanics (Section 1.6)
- Hydrogen Bonding (Section 1.12)
- Drawing Resonance Structures (Sections 2.7–2.10)
- Carbocation Stability (Section 6.11)

WILEY PLUS Visit www.wileyplus.com to check your understanding and for valuable practice.

15.1 Introduction to Spectroscopy

In order to understand how spectroscopy is used for structure determination, we must first review some of the basic features of light and matter.

The Nature of Light

Electromagnetic radiation (light) exhibits both wave-like properties and particle-like properties. Consequently, electromagnetic radiation can be viewed as a wave or as a particle. When viewed as a wave, electromagnetic radiation consists of perpendicular oscillating, electric and magnetic fields (Figure 15.1).

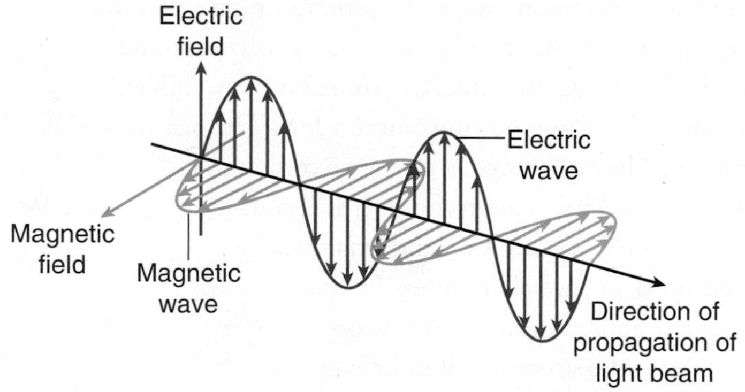

FIGURE 15.1 The perpendicular, oscillating, electric and magnetic fields associated with electromagnetic radiation.

The **wavelength** describes the distance between adjacent peaks of an oscillating field, while the **frequency** describes the number of wavelengths that pass a particular point in space per unit time. Accordingly, a long wavelength corresponds with a small frequency, and a short wavelength corresponds with a large frequency. This inverse relationship is summarized in the following equation where frequency (ν) and wavelength (λ) are inversely proportional. The constant of proportionality is the speed of light (c):

$$\nu = \frac{c}{\lambda}$$

When viewed as a particle, electromagnetic radiation consists of packets of energy, called **photons**. The energy of each photon is directly proportional to its frequency,

$$E = h\nu$$

where h is Planck's constant ($h = 6.626 \times 10^{-34}$ J·s). The range of all possible frequencies is known as the **electromagnetic spectrum**, which is arbitrarily divided into several regions by wavelength (Figure 15.2). Different regions of the electromagnetic spectrum are used to probe different aspects of molecular structure (Table 15.1). In this chapter, we will learn about the information obtained when IR radiation interacts with a compound. NMR and UV-VIS spectroscopy will be discussed in subsequent chapters.

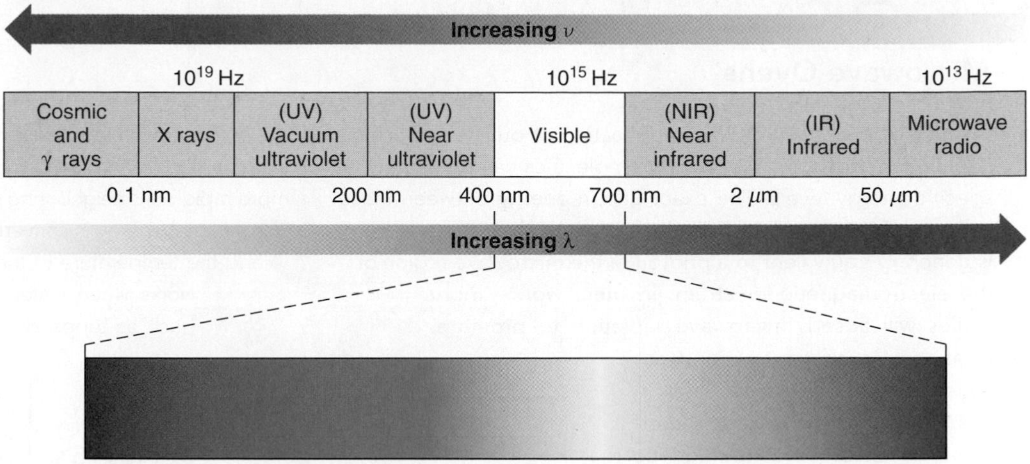

FIGURE 15.2
The electromagnetic spectrum.

TABLE 15.1 SOME COMMON FORMS OF SPECTROSCOPY AND THEIR USES

TYPE OF SPECTROSCOPY	REGION OF ELECTROMAGNETIC SPECTRUM	INFORMATION OBTAINED
Nuclear magnetic resonance (NMR) spectroscopy	Radio waves	The specific arrangement of all carbon and hydrogen atoms in the compound
IR spectroscopy	Infrared	The functional groups present in the compound
UV-VIS spectroscopy	Visible and ultraviolet	Any conjugated π system present in the compound

The Nature of Matter

Matter, like electromagnetic radiation, also exhibits both wave-like properties and particle-like properties. In Section 1.6 we saw that quantum mechanics describes the wave-like properties of matter. According to the principles of quantum mechanics, the energy of a molecule is "quantized." To understand what this means, we will compare the rotation of a molecule with the rotation of an automobile tire connected directly to a motor. When the motor is turned on, the tire begins to rotate and a measuring device is used to determine the rate of rotation. Our ability to control the rate of rotation appears to be limited only by the precision of the motor and the sensitivity of the measuring device. It would be inconceivable to suggest that a measurable rotation rate would be unattainable by the laws of physics. If we want to spin the tire at exactly 60.06251 revolutions per second, there is nothing preventing us from doing so. In contrast, molecules behave differently. The rotation of a molecule appears to be restricted to specific energy levels. That is, a molecule can only rotate at specific rates, which are defined by the nature of the molecule. Other rotation rates are simply not allowed by the laws of physics. The rotational energy of a molecule is said to be quantized.

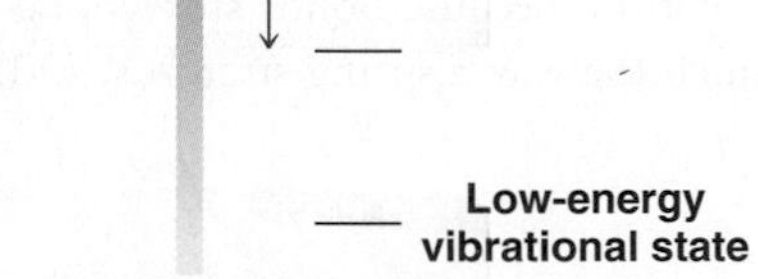

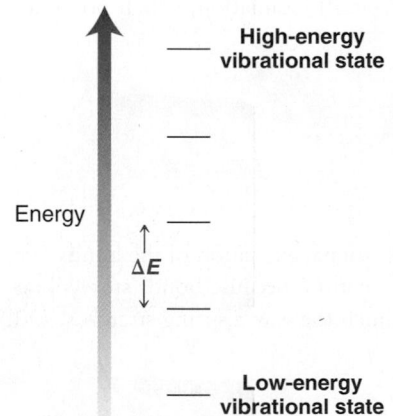

FIGURE 15.3
An energy diagram showing the energy gap between allowed vibrational states.

Molecules can store energy in a variety of ways. They rotate in space, their bonds vibrate like springs, their electrons can occupy a number of possible molecular orbitals, and so on. Each of these forms of energy is quantized. For example, a bond in a molecule can only vibrate at specific energy levels (Figure 15.3). The horizontal lines in the diagram represent allowed vibrational energy levels for a particular bond. The bond is restricted to these energy levels and cannot vibrate with an energy that is in between the allowed levels. The difference in energy (ΔE) between allowed energy levels is determined by the nature of the bond.

The Interaction between Light and Matter

In the previous section, we saw that the vibrational energy levels for a particular bond are separated from each other by an energy gap (ΔE). If a photon of light possesses exactly this amount of energy, the bond can absorb the photon to promote a **vibrational excitation**. The energy of the photon is temporarily stored as vibrational energy until it is released back into the environment, usually in the form of heat. Each form of spectroscopy uses a different region

PRACTICALLY SPEAKING)))

Microwave Ovens

Microwave ovens are a direct application of quantum mechanics and spectroscopy. We saw that molecules can only rotate at specific energy levels. The exact gap in energy between these levels (ΔE) is dependent on the nature of the molecule but is generally equivalent to a photon in the microwave region of the electromagnetic spectrum. In other words, molecules will absorb microwave radiation to promote rotational excitations. For reasons that we will discuss later in this chapter, the ability of a molecule to absorb electromagnetic radiation depends very much on the presence of a permanent dipole moment. Water molecules, in particular, possess a strong dipole moment and are therefore extremely efficient at absorbing microwave radiation.

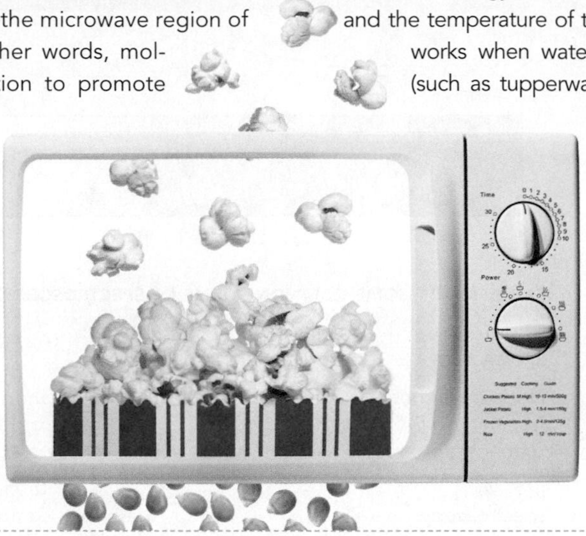

When frozen food is bombarded with microwave radiation, the water molecules in the food absorb the energy and begin to rotate more rapidly. As neighboring molecules collide with each other, the rotational energy is converted into translational energy (or heat), and the temperature of the food rises quickly. This process only works when water molecules are present. Plastic items (such as tupperware) generally do not get hot, because they are composed of long polymers that cannot freely rotate.

In addition to causing rotational excitation, microwave radiation can also cause electronic excitation in metals. That is, electrons can be promoted to higher energy orbitals. This type of electronic excitation accounts for the sparks that can be observed when metal objects are placed in a microwave oven.

of the electromagnetic spectrum and involves a different kind of excitation. In this chapter, we will focus mainly on the interaction between molecules and IR radiation, which promotes vibrational excitations of the bonds in a molecule.

15.2 IR Spectroscopy

Vibrational Excitation

In the previous section, we saw that IR radiation causes vibrational excitation of the bonds in a molecule. There are many different kinds of vibrational excitation, because bonds store vibrational energy a number of ways. Bonds can **stretch**, very much the way a spring stretches, and bonds can **bend** in a number of ways.

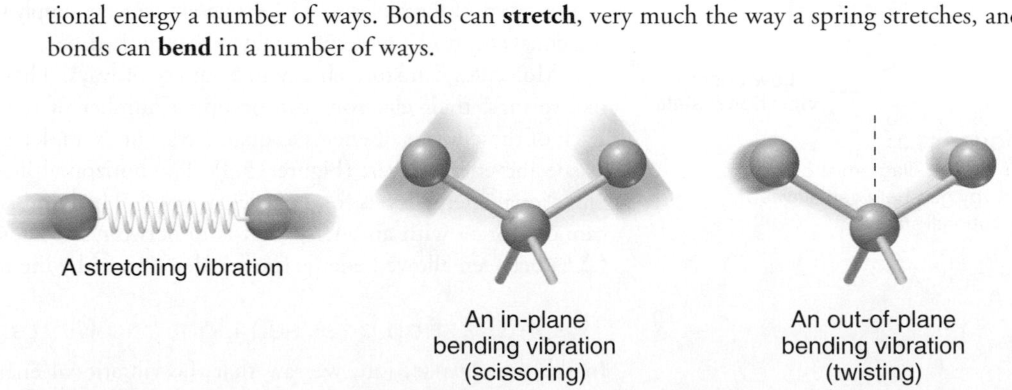

There are many other types of bending vibrations, but we will focus most of our attention in this chapter on stretching vibrations. Bending will only be mentioned briefly in this chapter.

MEDICALLY SPEAKING

IR Thermal Imaging for Cancer Detection

We have seen that vibrating bonds will absorb IR radiation to achieve a higher vibrational state, but it is worthwhile to note that the reverse process also occurs in nature. That is, vibrating bonds can be demoted to a *lower* vibrational state by *emitting* IR radiation. In this way, warm objects are able to release some of their energy in the form of IR radiation. This idea has found application in a wide variety of products. For example, some varieties of night vision goggles are able to produce an image in the total absence of visible light by detecting the IR radiation emitted by warm objects. Warmer areas emit more IR radiation, providing the contrast necessary to construct an image.

More recently, thermal imaging has found application in the detection of breast cancer. When a patient is exposed to cold air, the nervous system responds by reducing the flow of blood to the surface, thereby conserving heat. Cancerous cells and the surrounding tissue are less affected by this process and remain warm for a longer period of time. The contrast between warm and cool regions in the body can then be measured using IR sensors that detect the amount of IR radiation being emitted from different regions. This thermal imaging technique allows for the early detection of breast cancer. Various regions are color coded to indicate temperature, with the red areas being the warmest. The image shows the detection of a breast tumor.

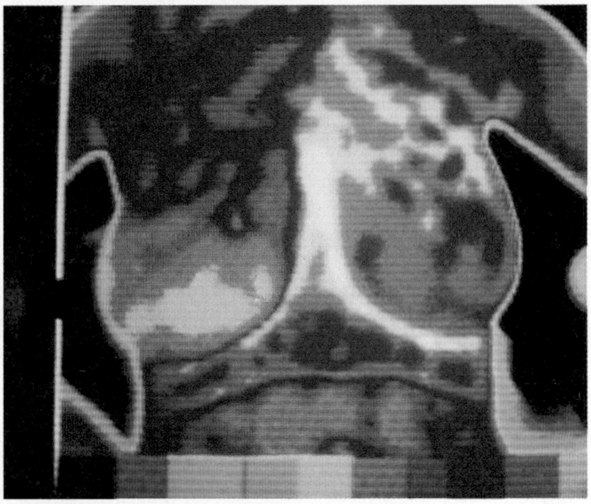

Identification of Functional Groups with IR Spectroscopy

For each and every bond in a molecule, the energy gap between vibrational states is very much dependent on the nature of the bond. For example, the energy gap for a C—H bond is much larger than the energy gap for a C—O bond (Figure 15.4).

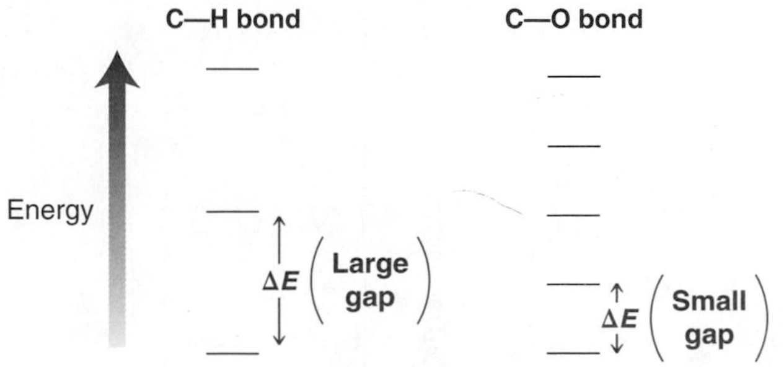

FIGURE 15.4 The energy gap between vibrational energy levels is dependent on the nature of the bond.

Both bonds will absorb IR radiation but the C—H bond will absorb a higher energy photon. In fact, each type of bond will absorb a characteristic frequency, allowing us to determine which types of bonds are present in a compound. We simply irradiate the compound with all frequencies of IR radiation and then detect which frequencies were absorbed. For example, a compound containing an O—H bond will absorb a frequency of IR radiation characteristic of the O—H bond. In this way IR spectroscopy can be used to identify the presence of functional groups in a compound.

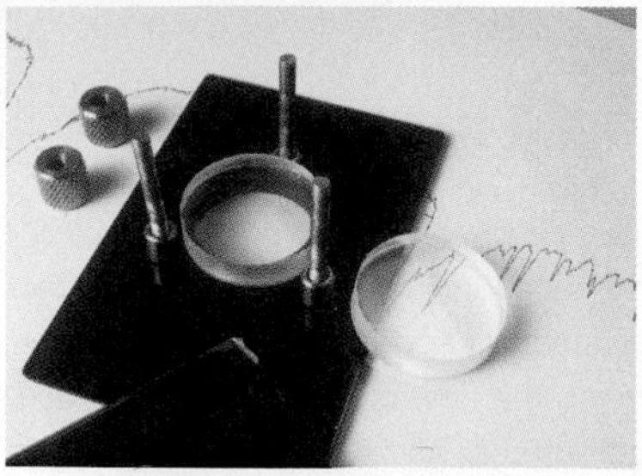

A salt plate used for IR spectroscopy. These plates are moisture sensitive and must be stored in a moisture-free environment.

IR Spectrometer

In an IR spectrometer, a sample is irradiated with frequencies of IR radiation, and the frequencies that pass through (that are not absorbed by the sample) are detected. A plot is then constructed showing which frequencies were absorbed by the sample. The most commonly used type of spectrometer, called a Fourier transform (FT-IR) spectrometer, irradiates the sample with all frequencies simultaneously and then utilizes a mathematical operation called a Fourier transform to determine which frequencies passed through the sample.

Several techniques are used for preparing a sample for IR spectroscopy. The most common method involves the use of salt plates. These expensive plates are made from sodium chloride and are used because they are transparent to IR radiation. If the compound under investigation is a liquid at room temperature, a drop of the sample is sandwiched in between two salt plates and is called a *neat* sample. If the compound is a solid at room temperature, it can be dissolved in a suitable solvent and placed in between two salt plates. Alternatively, insoluble compounds can be mixed with powdered KBr and then pressed into a thin, transparent film, called a KBr pellet. All of these sampling techniques are commonly used for IR spectroscopy.

The General Shape of an IR Absorbance Spectrum

An IR spectrometer measures the percent transmittance as a function of frequency. This plot is called an **absorption spectrum** (Figure 15.5). All the signals, called absorption bands, point down on an IR spectrum. The location of each signal on the spectrum can be specified either by the corresponding *wavelength* or by the corresponding *frequency* of radiation that was absorbed. Several decades ago, signals were reported by their wavelengths (measured in micrometers, or microns). Currently, the location of each signal is more often reported in terms of a frequency-related unit, called **wavenumber ($\tilde{\nu}$).** The wavenumber is simply the frequency of light divided by a constant (the speed of light, *c*):

$$\tilde{\nu} = \frac{\nu}{c}$$

The units of wavenumber are inverse centimeters (cm^{-1}), and the values range from 400 to 4000 cm^{-1}. All of the spectra in this chapter will be reported in wavenumber rather than wavelength, to be consistent with common practice. Don't confuse the terms wavenumber and wavelength. Wavenumber is proportional to frequency, and therefore, a larger wavenumber represents higher energy. Signals that appear on the left side of the spectrum correspond with higher energy radiation, while signals on the right side of the spectrum correspond with lower energy radiation.

Every signal in an IR spectrum has three characteristics: wavenumber, intensity, and shape. We will now explore each of these three characteristics, starting with wavenumber.

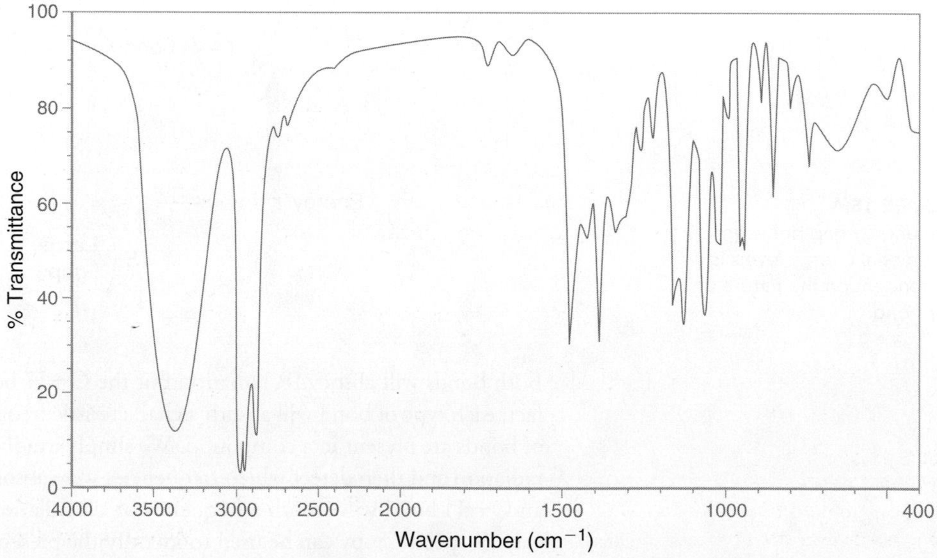

FIGURE 15.5
An example of an IR absorbance spectrum.

15.3 Signal Characteristics: Wavenumber

Hooke's Law

For every bond, the wavenumber of absorption associated with bond stretching is dependent on two factors: (1) bond strength and (2) masses of the atoms sharing the bond. The impact of these two factors can be rationalized when we treat a bond as if it were a vibrating spring connecting two weights.

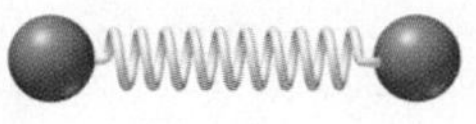

Using this analogy, we can construct the following equation, derived from Hooke's law, which enables us to approximate the frequency of vibration for a bond between two atoms of mass m_1 and m_2:

$$\tilde{\nu} = \left(\frac{1}{2\pi c}\right)\left(\frac{f}{m_{\text{red}}}\right)^{\frac{1}{2}}$$

force constant (bond strength)

$$\textbf{reduced mass} = \left(\frac{m_1 m_2}{m_1 + m_2}\right)$$

In this equation, f is the force constant of the spring, which represents the bond strength of the bond, and m_{red} is the *reduced mass* of the system. Use of the reduced mass in this equation allows us to treat the two atoms as one system. Notice that m_{red} appears in the denominator. This means that smaller atoms give bonds that vibrate at higher frequencies, thereby corresponding to a higher wavenumber of absorption. For example, compare the following bonds. The C—H bond involves the smallest atom (H) and therefore appears at the highest wavenumber.

C—H	C—D	C—O	C—Cl
~3000 cm^{-1}	~2200 cm^{-1}	~1100 cm^{-1}	~700 cm^{-1}

While m_{red} appears in the denominator of the equation, the force constant (f) appears in the numerator. This means that stronger bonds will vibrate at higher frequencies, thereby corresponding to a higher wavenumber of absorption. For example, compare the following bonds. The C≡N bond is the strongest of the three bonds and therefore appears at the highest wavenumber.

C≡N	C=N	C—N
~2200 cm^{-1}	~1600 cm^{-1}	~1100 cm^{-1}

Using the two trends shown, we can understand why different types of bonds will appear in different regions of an IR spectrum (Figure 15.6). Single bonds (except for X—H bonds) appear on the right side of the spectrum (below 1500 cm^{-1}) because single bonds are generally the weakest bonds. Double bonds appear at higher wavenumber (1600–1850 cm^{-1}) because they are stronger than single bonds, while triple bonds appear at even higher wavenumber (2100–2300 cm^{-1}) because they are even stronger than double bonds. And finally, the left side of the spectrum contains signals produced by X—H bonds (such as C—H, O—H, or N—H), all of which stretch at a high wavenumber because hydrogen has the smallest mass.

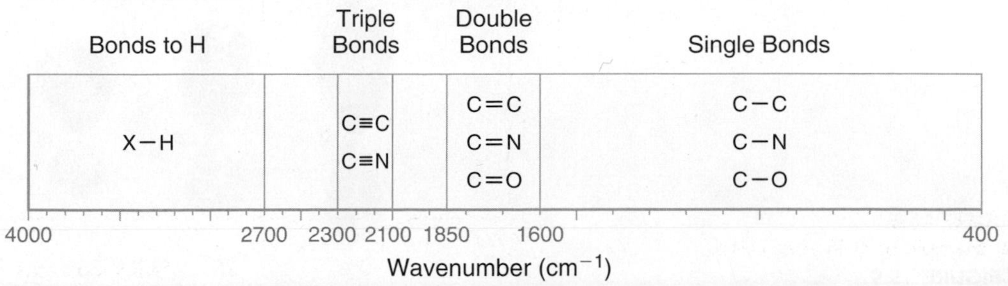

FIGURE 15.6 An IR absorbance spectrum divided into regions based on bond strength and atomic mass.

IR spectra can be divided into two main regions (Figure 15.7). The **diagnostic region** generally has fewer peaks and provides the clearest information. This region contains all signals that arise from double bonds, triple bonds, and X—H bonds. The **fingerprint region** contains signals resulting from the vibrational excitation of most single bonds (stretching and bending). This region generally contains many signals and is more difficult to analyze. What appears like a C—C stretch might in fact be another bond that is bending. This region is called the fingerprint region because each compound has a unique pattern of signals in this region, much the way each person has a unique fingerprint. To illustrate this point, compare the spectra for 2-butanol and 2-propanol (Figure 15.8). The diagnostic regions of these compounds are virtually indistinguishable (they both contain characteristic C—H and O—H signals), but the fingerprint regions of these compounds are very different.

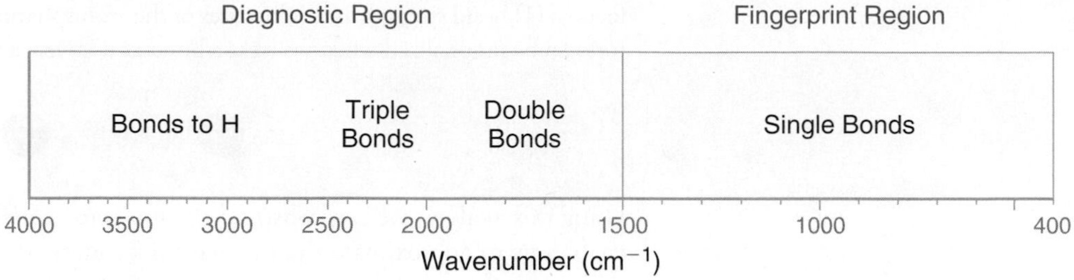

FIGURE 15.7
The diagnostic and fingerprint regions of an IR spectrum.

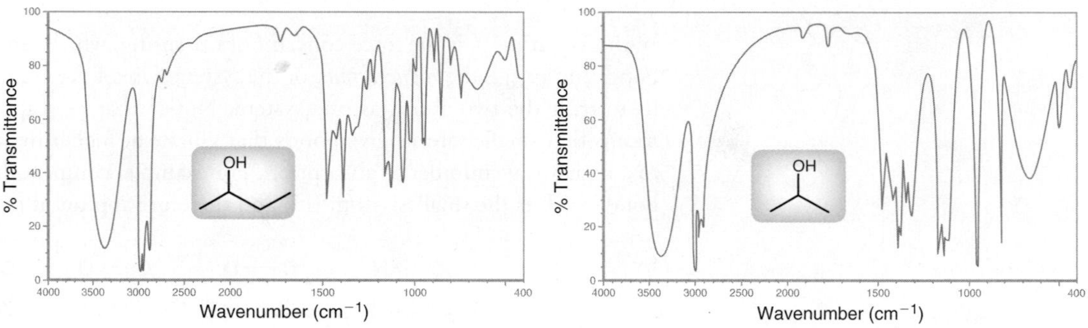

FIGURE 15.8
IR spectra for 2-butanol and 2-propanol.

Effect of Hybridization States on Wavenumber of Absorption

In the previous section, we saw that bonds to hydrogen (such as C—H bonds) appear on the left side of an IR spectrum (high wavenumber). We will now compare various kinds of C—H bonds.

The wavenumber of absorption for a C—H bond is very much dependent on the hybridization state of the carbon atom. Compare the following three C—H bonds.

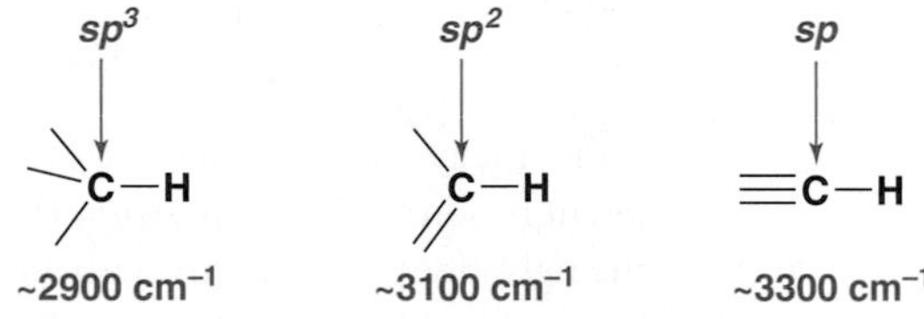

Of the three bonds shown, the C_{sp}—H bond produces the highest energy signal (~3300 cm^{-1}), while a C_{sp^3}—H bond produces the lowest energy signal (~2900 cm^{-1}). To understand this trend, we must revisit the shapes of the hybridized atomic orbitals (Figure 15.9). As illustrated,

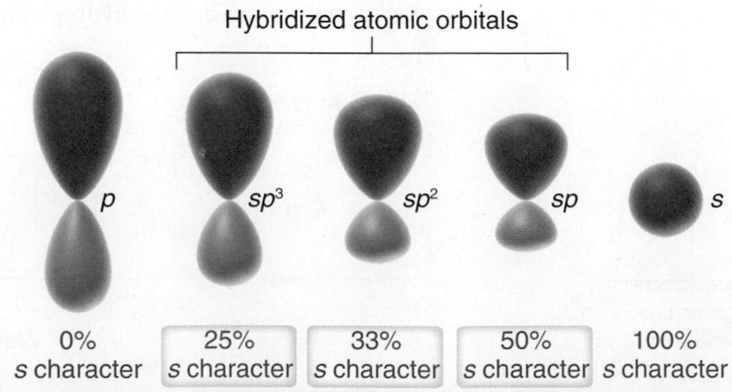

FIGURE 15.9
The shapes of atomic orbitals.

sp orbitals have more *s* character than the other hybridized atomic orbitals and therefore, *sp* orbitals more closely resemble *s* orbitals. Compare the shapes of the hybridized atomic orbitals, and note that the electron density of an *sp* orbital is closest to the nucleus (much like an *s* orbital). As a result, a C_{sp}—H bond will be shorter than other C—H bonds. Compare the bond lengths of the following three C—H bonds.

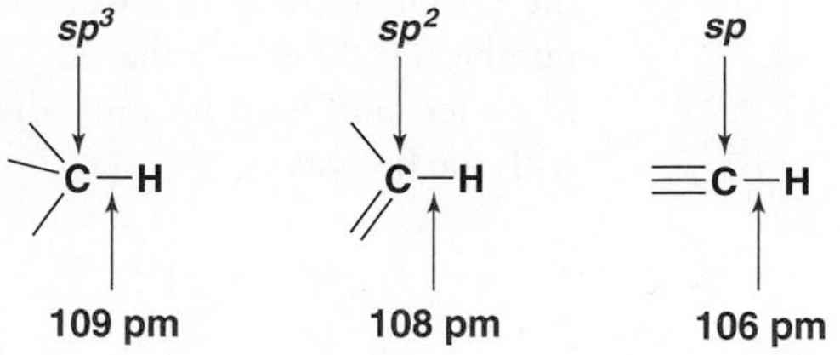

The C_{sp}—H bond has the shortest bond length and is therefore the strongest bond. The C_{sp^3}—H bond has the longest bond length and is therefore the weakest bond. Compare the spectra of an alkane, an alkene, and an alkyne (Figure 15.10). In each case, we draw a line at 3000 cm^{-1}.

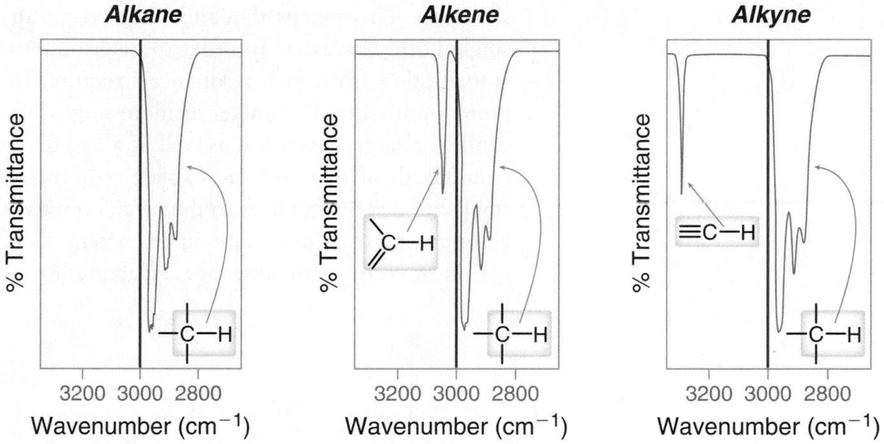

FIGURE 15.10
The relevant portions of the IR spectra for an alkane, an alkene, and an alkyne.

All three spectra have signals to the right of the line, resulting from C_{sp^3}—H bonds. The key is to look for any signals to the left of the line. An alkane does not have a signal to the left of 3000 cm^{-1}. An alkene has a signal at 3100 cm^{-1}, and an alkyne has a signal at 3300 cm^{-1}. But be careful—the absence of a signal to the left of 3000 cm^{-1} does *not* necessarily indicate the absence of a double bond or triple bond in the compound. Tetrasubstituted double bonds do not possess any C_{sp^2}—H bonds, and internal triple bonds also do not possess any C_{sp}—H bonds.

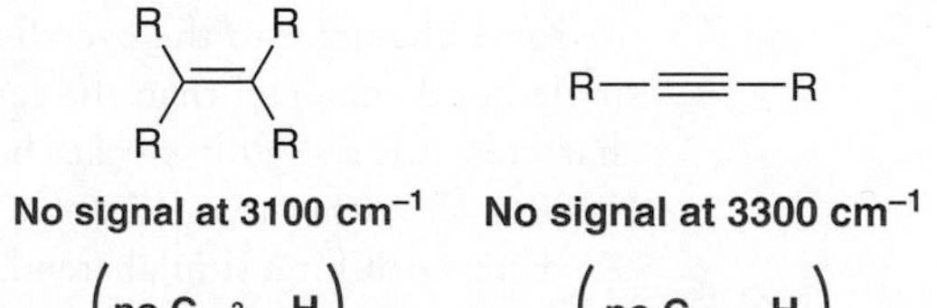

Effect of Resonance on Wavenumber of Absorption

We will now consider the effects of resonance on the wavenumber of absorption. As an illustration, compare the carbonyl groups (C=O bonds) in the following two compounds.

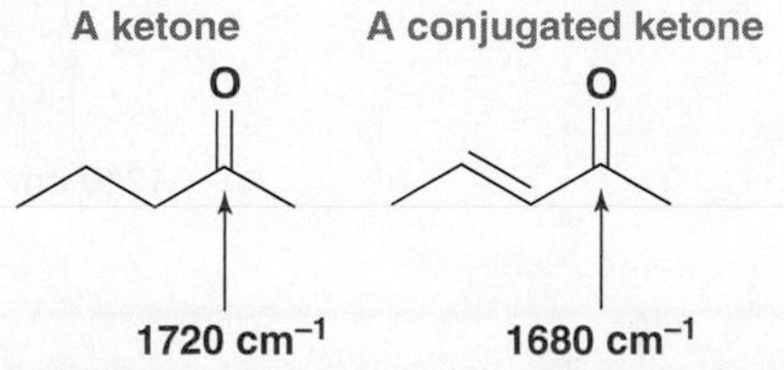

The second compound is called an unsaturated, conjugated ketone. It is *unsaturated* because of the presence of a C=C bond (we will see more on this terminology later in this chapter), and it is **conjugated** because the π bonds are separated from each other by exactly one single bond. We will explore conjugated π systems in more detail in Chapter 17. For now, we will just analyze the effect of conjugation on the IR absorption of the carbonyl group. As shown, the carbonyl group of an unsaturated, conjugated ketone produces a signal at lower wavenumber (1680 cm^{-1}) than the carbonyl group of a saturated ketone (1720 cm^{-1}). In order to understand why, we must draw the resonance structures for each compound. Let's begin with the ketone.

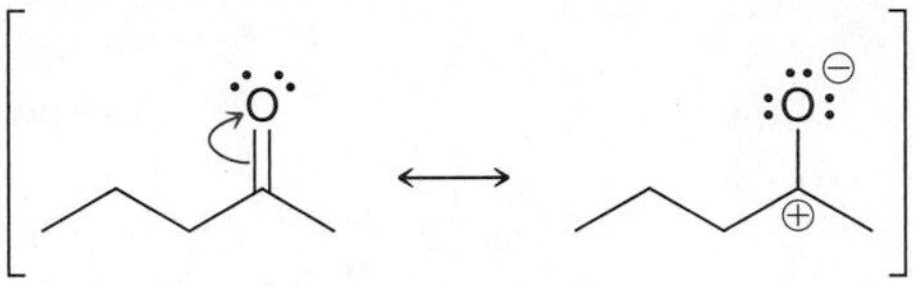

Ketones have two significant resonance structures. The carbonyl group is drawn as a double bond in the first resonance structure, and it is drawn as a single bond in the second resonance structure. This means that the carbonyl group has some double-bond character and some single-bond character. In order to determine the nature of this bond, we must consider the contribution from each resonance structure. In other words, does the carbonyl group have more double-bond character or more single-bond character? The second resonance structure exhibits charge separation as well as a carbon atom (C+) that has less than an octet of electrons. Both of these reasons explain why the second resonance structure contributes only a small amount of character to the overall resonance hybrid. Therefore, the carbonyl group of a ketone has mostly double-bond character.

Now consider the resonance structures for a conjugated, unsaturated ketone.

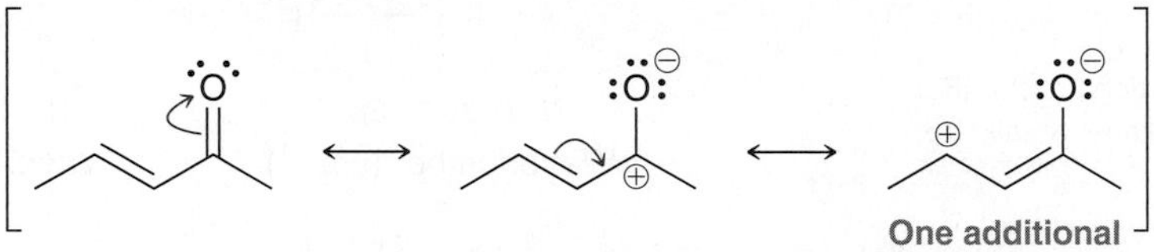

Conjugated, unsaturated ketones have three resonance structures rather than two. In the third resonance structure, the carbonyl group is drawn as a single bond. Once again, this resonance structure exhibits charge separation as well as a carbon atom (C+) with less than an octet of electrons. As a result, this resonance structure also contributes only a small amount of character to the overall resonance hybrid. Nevertheless, this third resonance structure does contribute some character to the overall resonance hybrid, giving this carbonyl group slightly more single-bond character than the carbonyl group of a saturated ketone. With more single-bond character, it is a slightly weaker bond and therefore produces a signal at a lower wavenumber (1680 cm^{-1} rather than 1720 cm^{-1}).

Esters exhibit a similar trend. An ester typically produces a signal at around 1740 cm^{-1}, but conjugated, unsaturated esters produce lower energy signals, usually around 1710 cm^{-1}. Once again, the carbonyl group of a conjugated, unsaturated ester is a weaker bond, due to resonance.

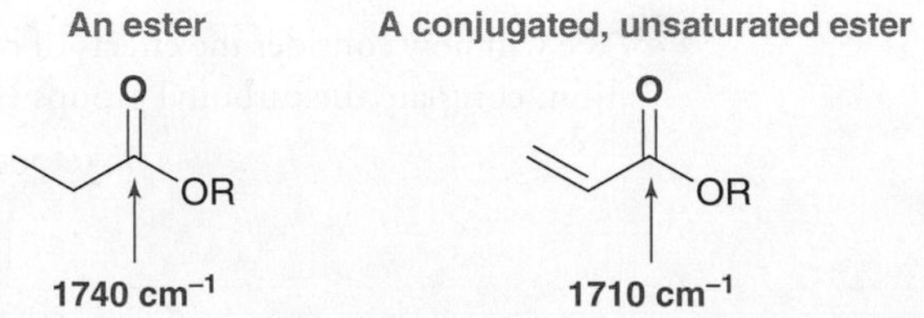

CONCEPTUAL CHECKPOINT

15.1 For each of the following compounds, rank the highlighted bonds in terms of increasing wavenumber:

(a) 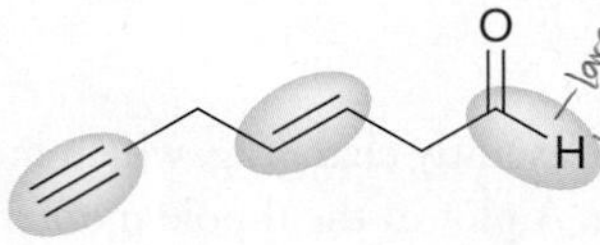

(b) 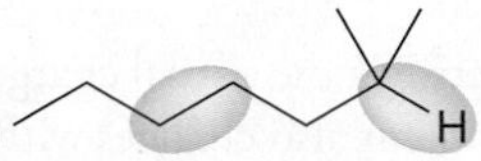

15.2 For each of the following compounds, determine whether or not you would expect its IR spectrum to exhibit a signal to the left of 3000 cm^{-1}.

(a) 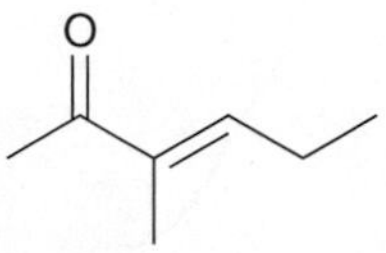

(b) 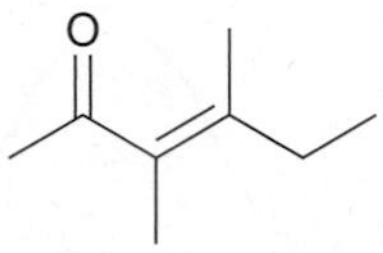

(c) 

(d) 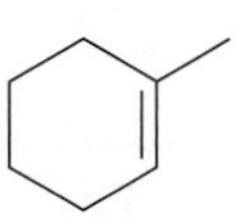

(e) 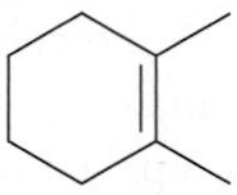

(f) 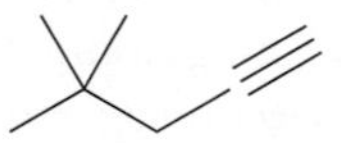

15.3 Each of the following compounds contains two carbonyl groups. Identify which carbonyl group will exhibit a signal at lower wavenumber.

(a) 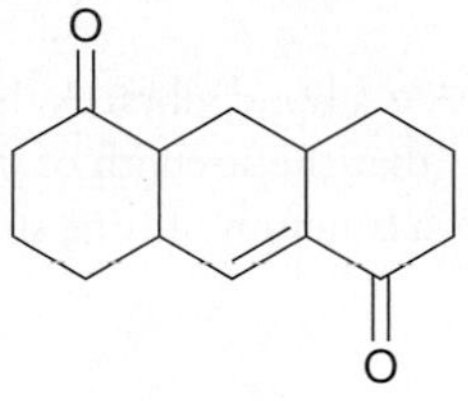

(b) 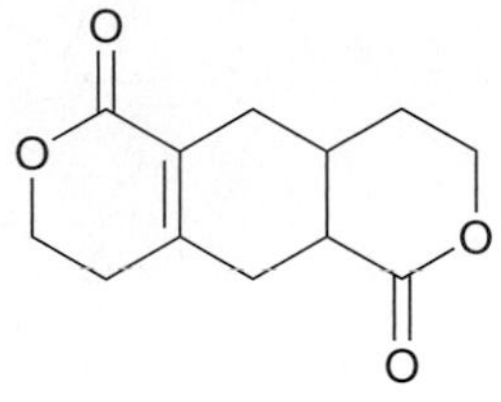

(c) 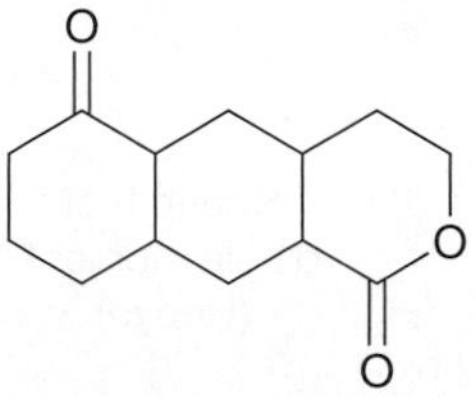

15.4 Compare the wavenumber of absorption for the following two C=C bonds. Use resonance structures to explain why the C=C bond in the conjugated compound produces a signal at lower wavenumber.

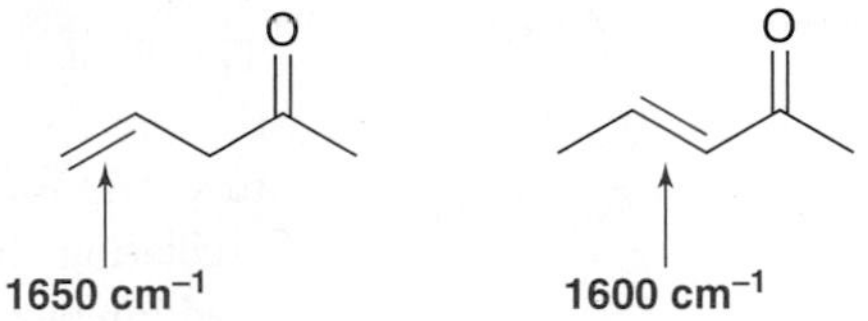

15.4 Signal Characteristics: Intensity

In an IR spectrum, some signals will be very strong in comparison with other signals on the same spectrum (Figure 15.11). That is, some bonds absorb IR radiation very efficiently, while other bonds are less efficient at absorbing IR radiation. To understand the reason for this, we must

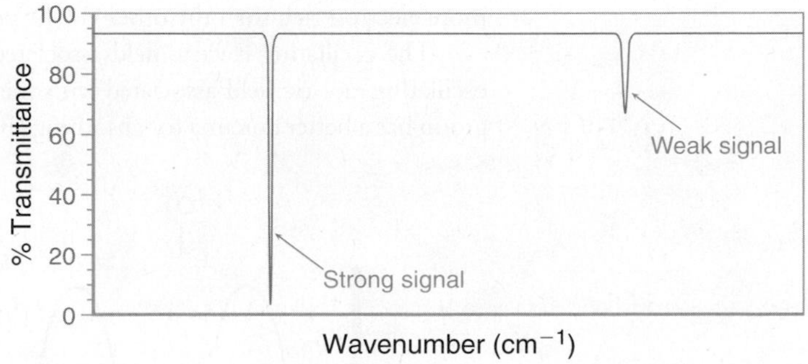

FIGURE 15.11 A comparison of a strong signal and weak signal in an IR spectrum.

consider how the strength of a dipole moment changes when a bond vibrates. Recall that the dipole moment (μ) of a bond is defined by the following equation.

$$\mu = e \times d$$

where *e* is the strength of the partial charges (δ+ and δ–) and *d* is the distance separating them. Now consider what happens to the strength of a dipole moment as a bond stretches.

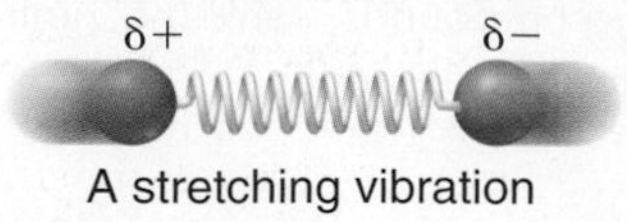

As a bond vibrates, the distance between the partial charges is constantly changing, which means that the strength of the dipole moment also changes with time. A plot of the dipole moment as a function of time shows an oscillating dipole moment (Figure 15.12).

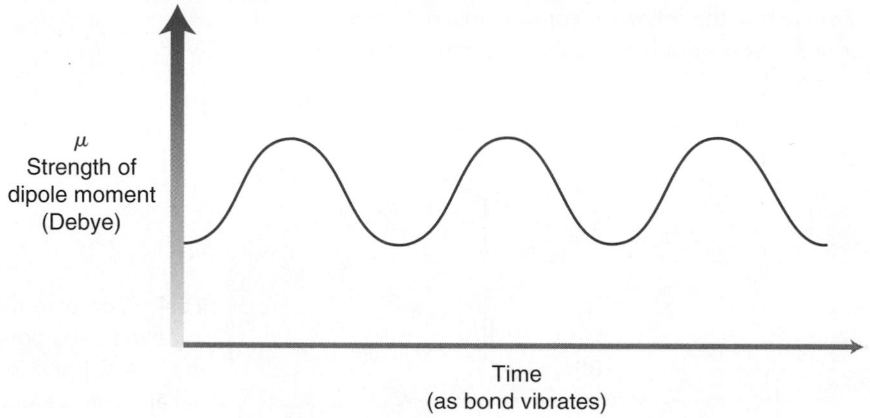

FIGURE 15.12 For a vibrating bond, the strength of the dipole moment oscillates as a function of time.

The dipole moment is an electric field surrounding the bond. So as the dipole moment oscillates, the bond is essentially surrounded by an oscillating electric field, which serves as an antenna (so to speak) for absorbing IR radiation. Since electromagnetic radiation itself is comprised of an oscillating electric field, the bond can absorb a photon because the bond's oscillating electric field interacts with the oscillating electric field of the IR radiation.

The efficiency of a bond at absorbing IR radiation therefore depends on the strength of the dipole moment. For example, compare the following two highlighted bonds:

Each of these bonds has a measurable dipole moment, but they differ significantly in strength. Let's first analyze the carbonyl group (C=O bond). Due to resonance and induction, the carbon atom bears a large partial positive charge, and the oxygen atom bears a large partial negative charge. The carbonyl group therefore has a large dipole moment. Now let's analyze the C=C bond. One vinylic position is connected to electron-donating alkyl groups, while the other vinylic position is connected to hydrogen atoms. As a result, the vinylic position bearing two alkyl groups is slightly more electron rich than the other vinylic position, producing a small dipole moment.

The oscillating electric field associated with the carbonyl group is much stronger than the oscillating electric field associated with the C=C bond (Figure 15.13). Therefore, the carbonyl group has a better antenna for absorbing IR radiation. That is, the carbonyl group is more efficient

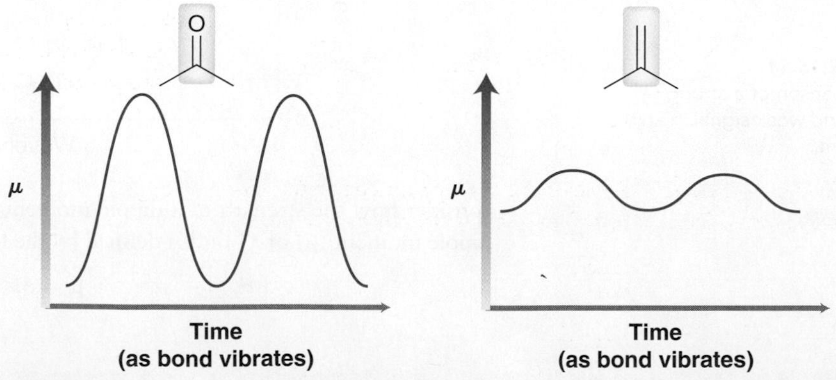

FIGURE 15.13 A comparison of the oscillating electric fields for a C=O bond and a C=C bond.

at absorbing IR radiation, producing a stronger signal (Figure 15.14). Carbonyl groups often produce the strongest signals in an IR spectrum, while C=C bonds often produce fairly weak signals. In fact, some alkenes do not even produce any C=C signal at all. For example, consider the IR spectrum of 2,3-dimethyl-2-butene (Figure 15.15). This alkene is symmetrical. That is, both vinylic positions are electronically identical, and the bond has no dipole moment at all. As the C=C bond vibrates, there is no change in dipole moment, which means that the bond does not have an antenna for absorbing IR radiation. In other words, symmetrical C=C bonds are completely inefficient at absorbing IR radiation, and no signal is observed. The same is true for symmetrical C≡C bonds.

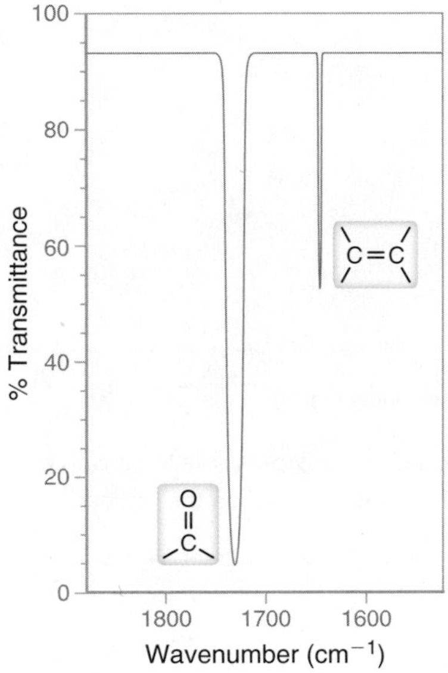

FIGURE 15.14
A comparison of the intensities of the signals arising from a C=O bond and a C=C bond.

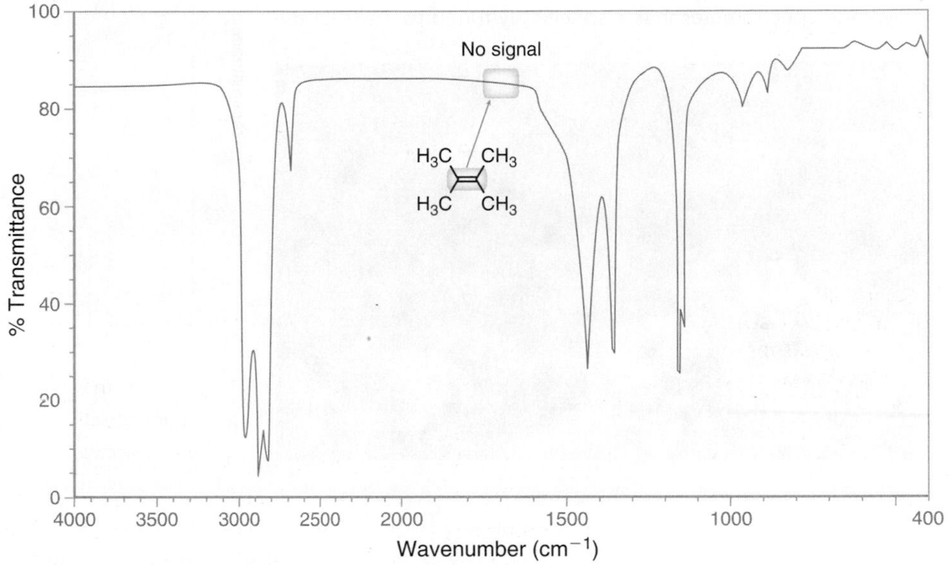

FIGURE 15.15
An IR spectrum of 2,3-dimethyl-2-butene.

LOOKING AHEAD
You might be wondering why there is more than one signal for these C—H bonds. This question will be answered in the next section.

There is one other factor that can contribute significantly to the intensity of signals in an IR spectrum. Consider the group of signals appearing just below 3000 cm^{-1} in Figure 15.15. These signals are associated with the stretching of the C—H bonds in the compound. The intensity of these signals derives from the number of C—H bonds giving rise to the signals. In fact, the signals just below 3000 cm^{-1} are typically among the strongest signals in an IR spectrum.

CONCEPTUAL CHECKPOINT

15.5 For each pair of compounds, determine which C=C bond will produce a stronger signal.

(a) 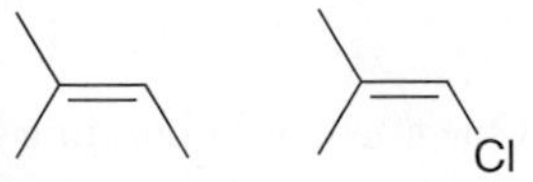

(b) 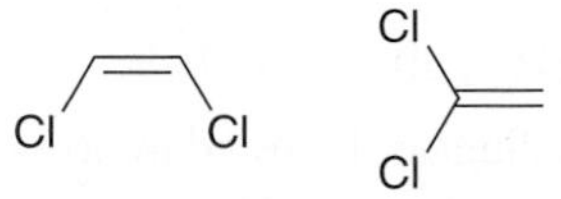

15.6 The C=C bond in 2-cyclohexenone produces an unusually strong signal. Explain using resonance structures.

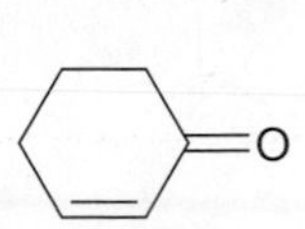

15.7 *trans*-2-Butene does not exhibit a signal in the double-bond region of the spectrum (1600–1850 cm^{-1}); however, IR spectroscopy is still helpful in identifying the presence of the double bond. Identify the other signal that would indicate the presence of a C=C bond.

PRACTICALLYSPEAKING

IR Spectroscopy for Testing Blood-Alcohol Levels

In the previous section, we saw that the intensity of a signal in an IR spectrum is dependent on how efficiently a bond absorbs IR radiation. The intensity of the signal is also dependent on other factors, such as the concentration of the sample being analyzed. This fact is used by law enforcement officials to measure blood alcohol levels accurately. The Intoxilyzer is the most current and accurate device for measuring blood alcohol levels. It is essentially an IR spectrometer that is specifically tuned to measure the intensity of the signals for the C—H bonds in ethanol. For example, the Intoxilyzer 5000 measures the intensity of absorption at the five frequencies illustrated.

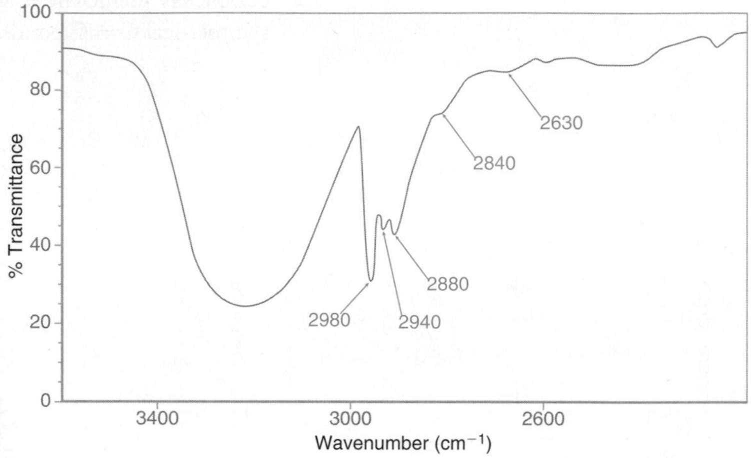

A breath sample is analyzed, and the device is able to calculate the concentration of ethanol in the sample based on the intensities of these signals.

15.5 Signal Characteristics: Shape

In this section, we will explore some of the factors that affect the shape of a signal. The shape of the signals in an IR spectrum varies—some are very broad while other signals are very narrow (Figure 15.16).

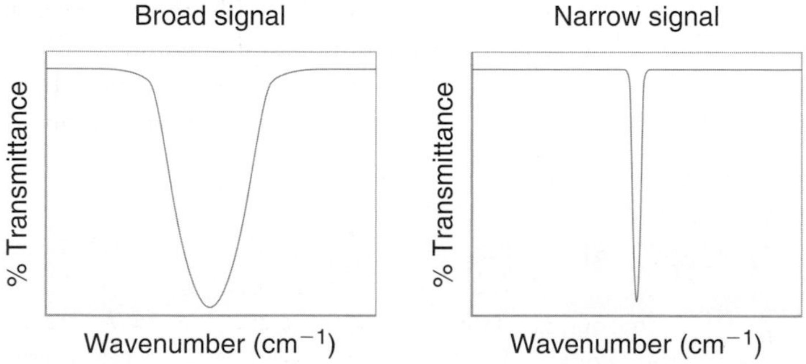

FIGURE 15.16 A comparison of a broad signal and narrow signal in an IR spectrum.

Effects of Hydrogen Bonding

Alcohols exhibit hydrogen bonding, as discussed in Section 13.1. One effect of H bonding is to weaken the existing O—H bond.

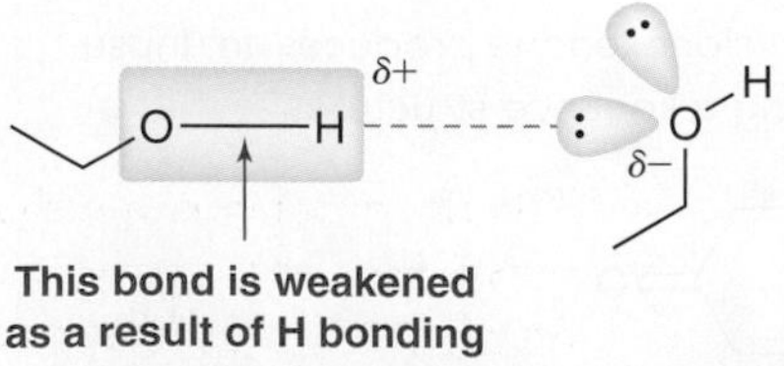

Concentrated alcohols give rise to broad signals because of this weakening effect. At any given moment in time, the O—H bond in each molecule is weakened to a different extent. As a result, the O—H bonds do not have a consistent bond strength, but rather, there is a *distribution* of bond strength. That is, some molecules are barely participating in H bonding, while others are participating in H bonding to varying degrees. The result is a broad signal.

The shape of an OH signal is different when the alcohol is diluted in a solvent that cannot form hydrogen bonds with the alcohol. In such an environment, it is likely that the O—H bonds will not participate in an H-bonding interaction. The result is a narrow signal. When the solution is neither very concentrated nor very dilute, two signals are observed. The molecules that are not participating in H bonding will give rise to a narrow signal, while the molecules participating in H bonding will give rise to a broad signal. As an example, consider the spectrum of 2-butanol in which both signals can be observed (Figure 15.17). When O—H bonds do not participate in H bonding, they generally produce a signal at approximately 3600 cm^{-1}. That signal can be seen in the spectrum. When O—H bonds participate in H bonding, they generally produce a broad signal between 3200 and 3600 cm^{-1}. That signal can also be seen in the spectrum. Depending on the conditions, an alcohol will give a broad signal, or a narrow signal, or both.

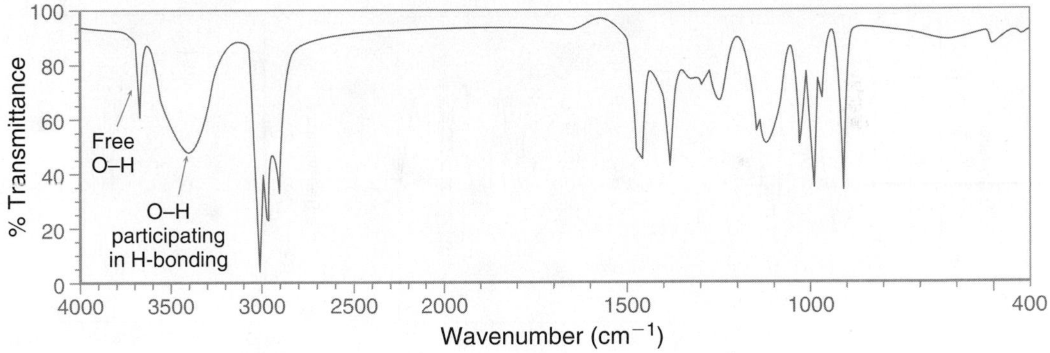

FIGURE 15.17
The IR spectrum of 2-butanol.

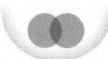

CONCEPTUAL CHECKPOINT

15.8 As explained previously, the concentration of an alcohol can be selected such that both a broad signal and a narrow signal appear simultaneously. In such cases, the broad signal is always to the *right* of the narrow signal, never to the left. Explain.

Carboxylic acids exhibit similar behavior, only more pronounced. For example, consider the spectrum of butyric acid (Figure 15.18). Notice the very broad signal on the left side of the spectrum, extending from 2200 to 3600 cm^{-1}. This signal is so broad that it extends over the usual C—H signals that appear around 2900 cm^{-1}. This very broad signal, characteristic of carboxylic acids, is a result of H bonding. The effect is more pronounced than alcohols, because molecules of the carboxylic acid can form two hydrogen-bonding interactions, resulting in a dimer.

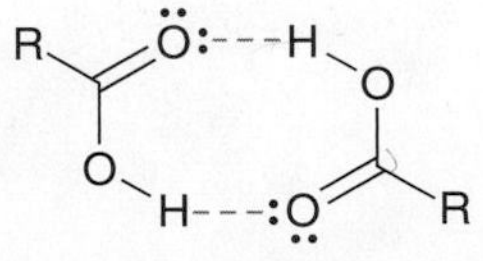

The IR spectrum of a carboxylic acid is easy to recognize, because of the characteristic broad signal that covers nearly one third of the spectrum. This broad signal is also accompanied by a broad C═O signal just above 1700 cm^{-1}.

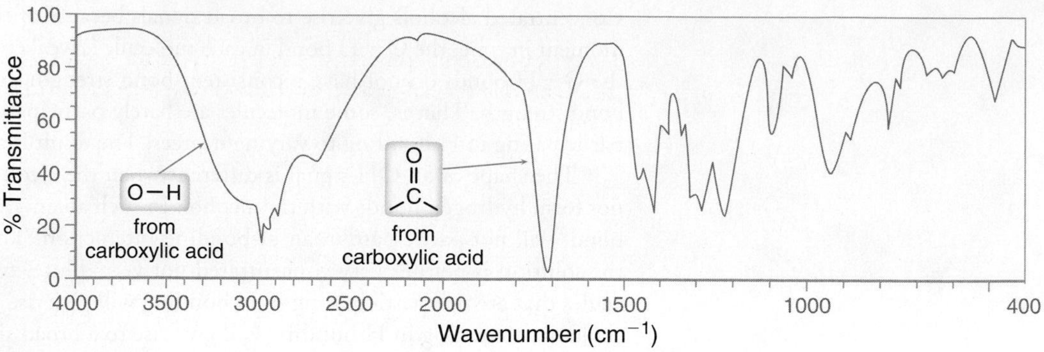

FIGURE 15.18
The IR spectrum of butyric acid.

CONCEPTUAL CHECKPOINT

15.9 For each of the following IR spectra, identify whether it is consistent with the structure of an alcohol, a carboxylic acid, or neither.

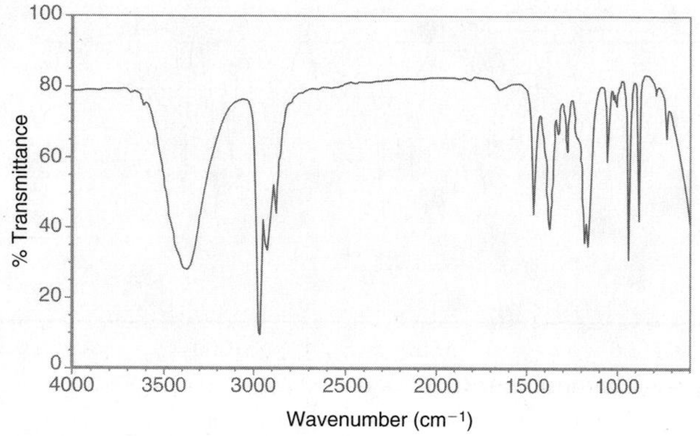

(a)

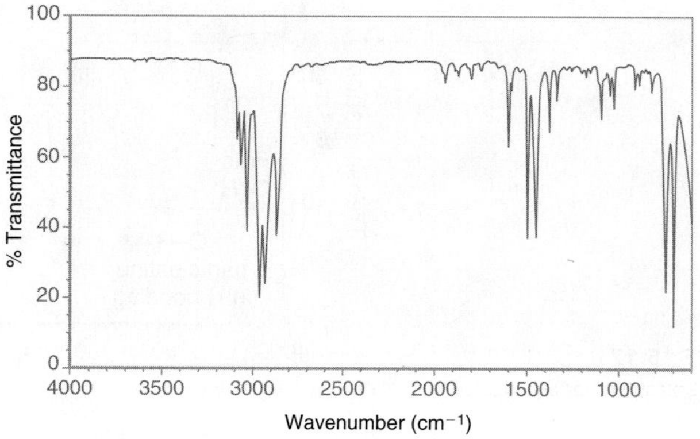

(b)

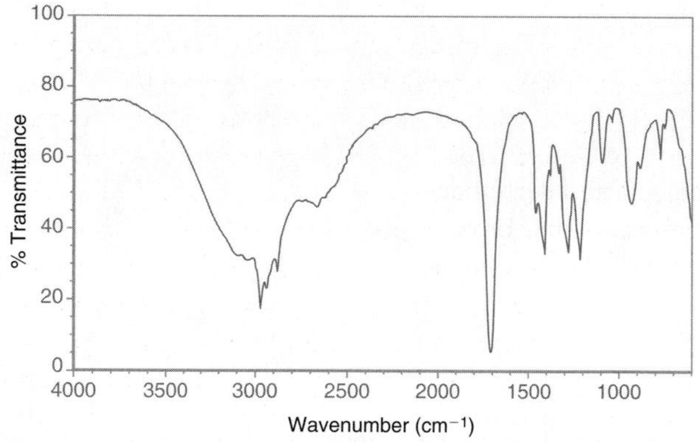

(c)

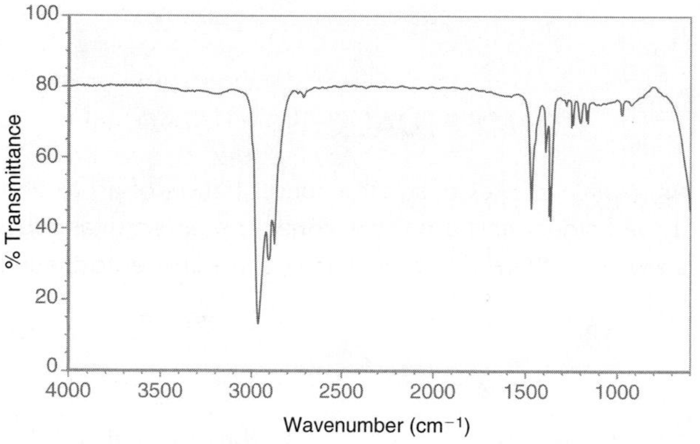

(d)

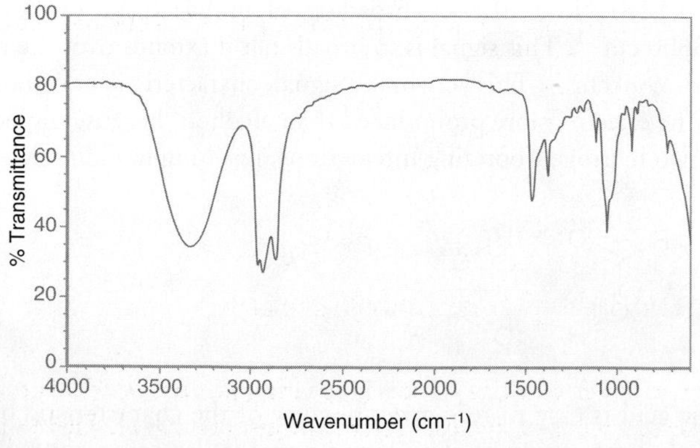

(e)

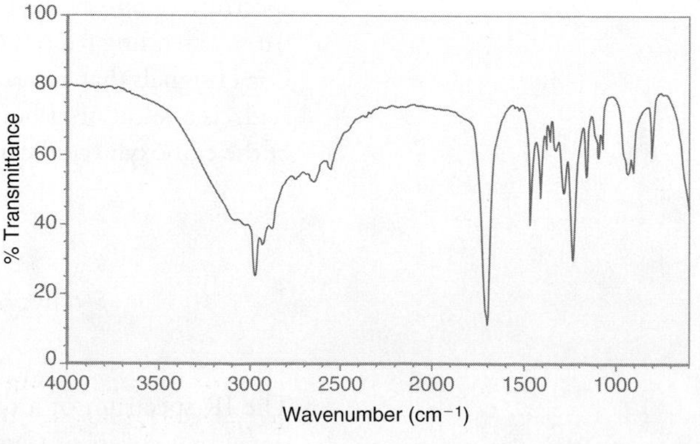

(f)

Amines: Symmetrical vs. Asymmetrical

There is one other important factor in addition to H bonding that affects the shape of a signal. Consider the difference in shape of the N—H signals for primary and secondary amines (Figure 15.19).

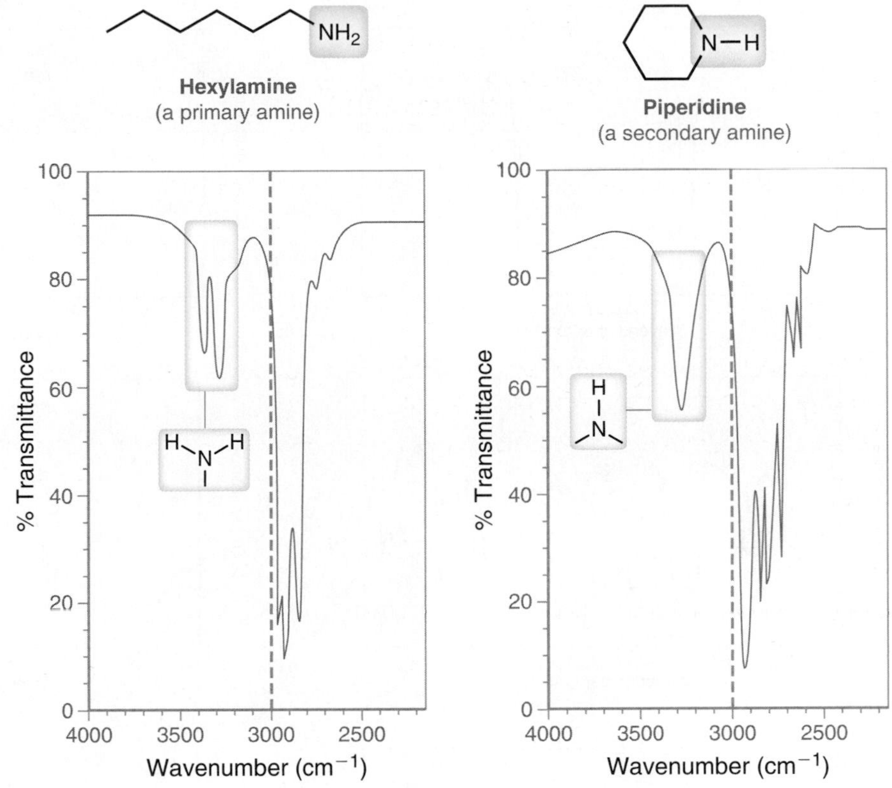

FIGURE 15.19 The difference in shape of the signals for a primary amine and a secondary amine.

The primary amine exhibits two signals: one at 3350 cm^{-1} and the other at 3450 cm^{-1}. In contrast, the secondary amine exhibits only one signal. It might be tempting to explain this by arguing that each N—H bond gives rise to a signal, and therefore a primary amine gives two signals because it has two N—H bonds. Unfortunately, the explanation is not quite that simple. In fact, both N—H bonds of a single molecule will together produce only one signal. The reason for the appearance of two signals is more accurately explained by considering the two possible ways in which the entire NH_2 group can vibrate. The N—H bonds can be stretching in phase with each other, called **symmetric stretching**, or they can be stretching out of phase with each other, called **asymmetric stretching**. At any given moment in time, approximately half of the molecules will be vibrating symmetrically, while the other half will be vibrating asymmetrically. The molecules vibrating symmetrically will absorb a particular frequency of IR radiation to promote a vibrational excitation, while the molecules vibrating asymmetrically will absorb a different frequency. In other words, one of the signals is produced by half of the molecules, and the other signal is produced by the other half of the molecules.

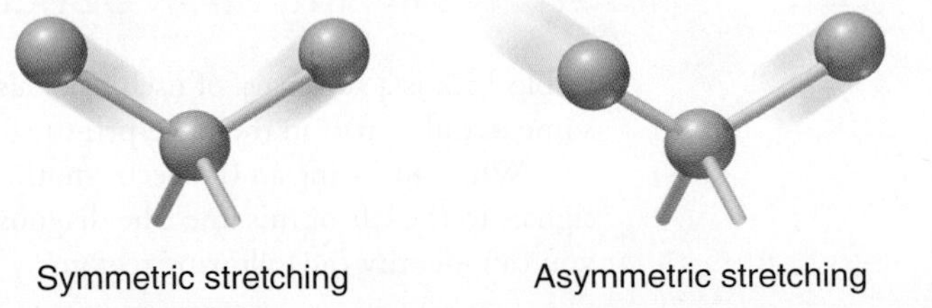

For a similar reason, the C—H bonds of a CH_3 group (appearing just below 3000 cm^{-1} in an IR spectrum) generally give rise to a series of signals, rather than just one signal. These signals arise from the various ways in which a CH_3 group can be excited.

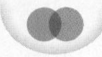

CONCEPTUAL CHECKPOINT

15.10 For each of the following IR spectra, determine whether it is consistent with the structure of a ketone, an alcohol, a carboxylic acid, a primary amine, or a secondary amine.

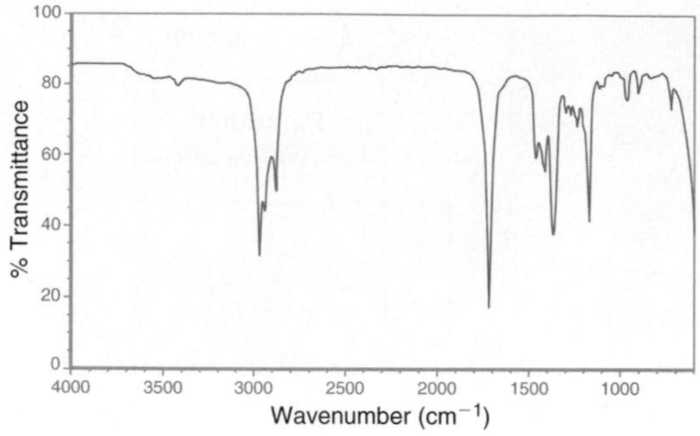

(a)

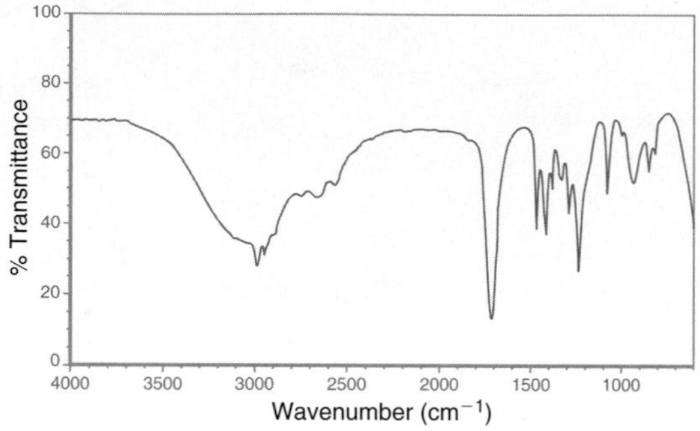

(b)

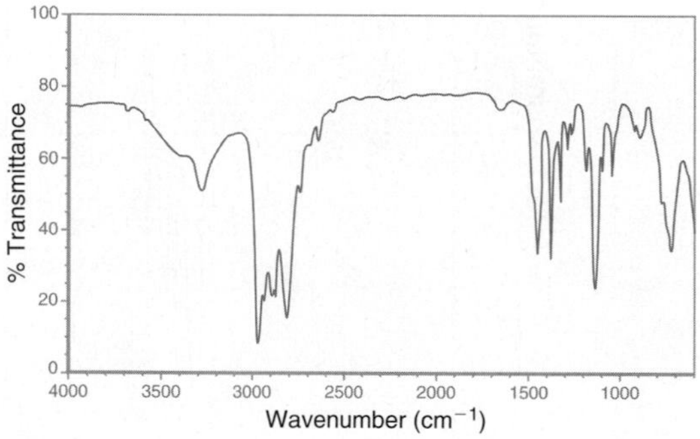

(c)

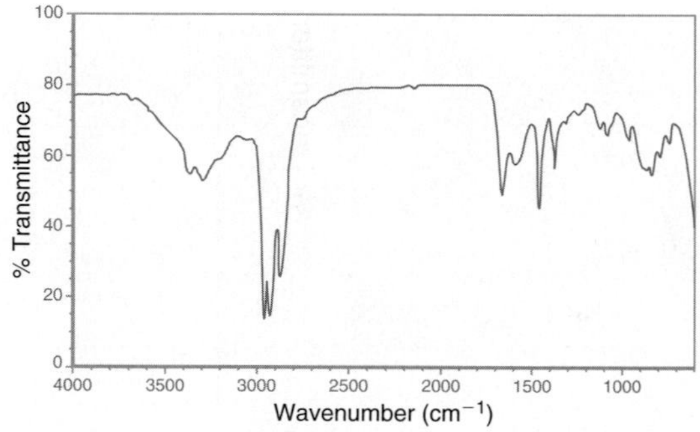

(d)

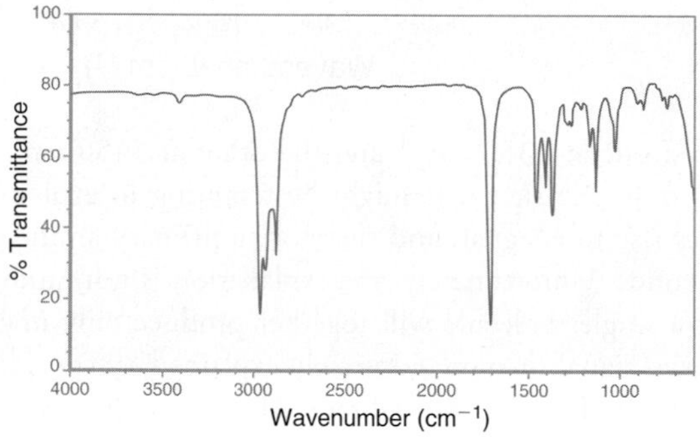

(e)

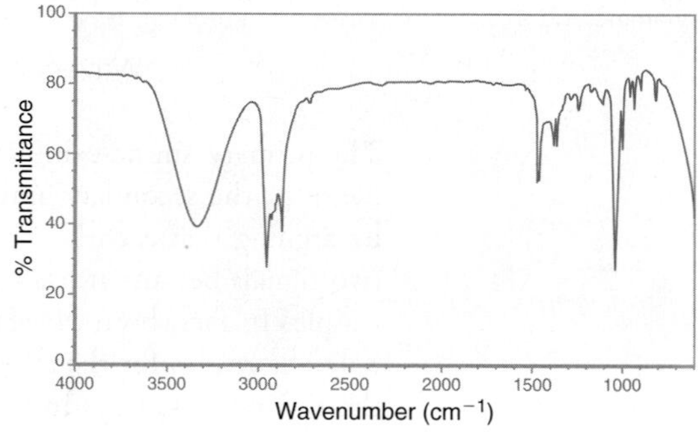

(f)

15.11 Carefully consider the structure of 2,3-dimethyl-2-butene. There are twelve C_{sp^3}—H bonds, and they are all identical. Nevertheless, there is more than one signal just to the right of 3000 cm^{-1} in the IR spectrum of this compound. Can you offer an explanation?

15.6 Analyzing an IR Spectrum

Table 15.2 is a summary of useful signals in the diagnostic region of an IR spectrum as well as some useful signals in the fingerprint region.

When analyzing an IR spectrum, the first step is to draw a line at 1500 cm^{-1}. Focus on any signals to the left of this line (the diagnostic region). In doing so, it will be extremely helpful if you can identify the following regions:

- Double bonds: 1600–1850 cm^{-1}
- Triple bonds: 2100–2300 cm^{-1}
- X—H bonds: 2700–4000 cm^{-1}

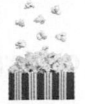

TABLE 15.2 IMPORTANT SIGNALS IN IR SPECTROSCOPY

USEFUL SIGNALS IN THE DIAGNOSTIC REGION

STRUCTURAL UNIT	FREQUENCY (cm^{-1})	
Single Bonds (X—H)		
—O—H	3200–3600	
$RC(=O)$—O—H	2200–3600	
N—H	3350–3500	
≡C—H	~3300	
=C—H	3000–3100	
—C—H	2850–3000	
$O=$C—H	2750–2850	Discussed in Chapter 20
Triple Bonds		
—C≡C—	2100–2200	
—C≡N	2200–2300	

STRUCTURAL UNIT	FREQUENCY (cm^{-1})	
Double Bonds		
Cl–C=O	1750–1850	Discussed in Chapter 20
RO–C(R)=O	1700–1750	Discussed in Chapter 20
HO–C(R)=O	1700–1750	Discussed in Chapter 20
R–C(R)=O	1680–1750	Discussed in Chapter 20
H_2N–C(R)=O	1650–1700	Discussed in Chapter 20
C=C	1600–1700	
benzene ring	1450–1600 1650–2000	Discussed in Chapter 18

USEFUL SIGNALS IN THE FINGERPRINT REGION

STRUCTURAL UNIT	FREQUENCY (cm^{-1})	
—C—Cl	600–800	
—C—Br	500–600	
—C—O—	1000–1100	
$O=$C—O—	1250–1350	
—C—N	1000–1200	
R–CH=CH_2	900–920 980–1000	*(bending)*
R_2C=CH_2	880–900	*(bending)*
R–CH(CH_3)$_2$	1370, 1380	*(bending)*
R–C(CH_3)$_3$	1370, 1390	*(bending)*

Remember that each signal appearing in the diagnostic region will have three characteristics (wavenumber, intensity, and shape). Make sure to analyze all three characteristics.

When looking for X—H bonds, draw a line at 3000 cm^{-1} and look for signals that appear to the left of the line (Figure 15.20).

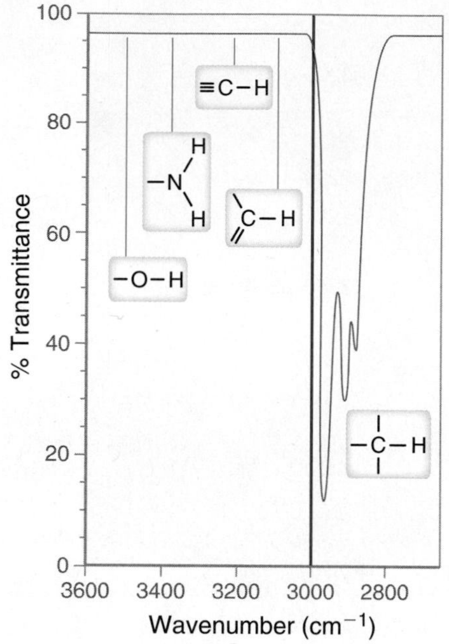

FIGURE 15.20
The location of signals arising from several different X—H bonds.

SKILLBUILDER

15.1 ANALYZING AN IR SPECTRUM

LEARN the skill

A compound with molecular formula $C_6H_{10}O$ gives the following IR spectrum.

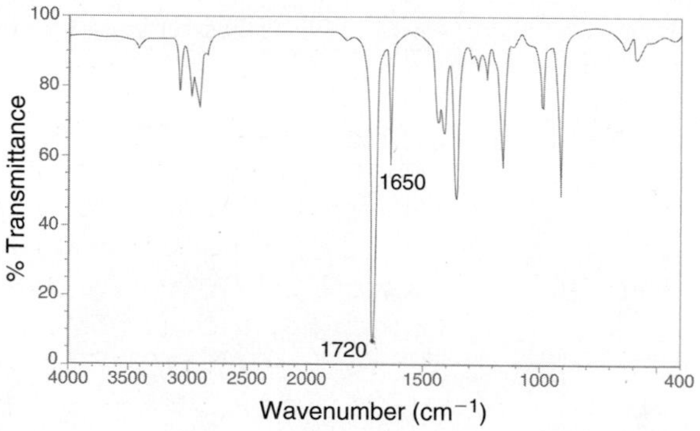

Identify the structure below that is most consistent with the spectrum:

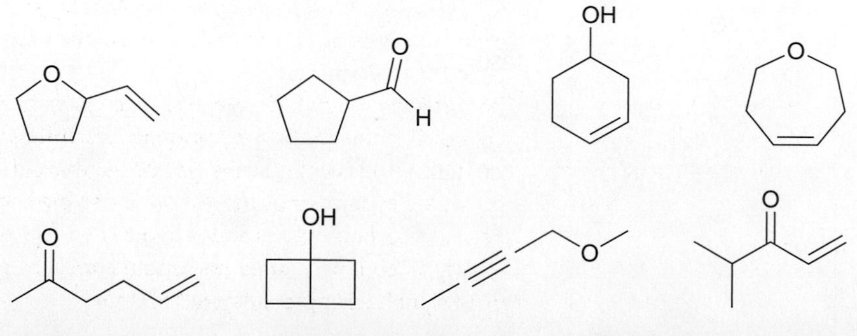

STEPS 1 AND 2
Look for double bonds (1600–1850 cm^{-1}) and triple bonds (2100–2300 cm^{-1})

SOLUTION

Draw a line at 1500 cm^{-1}, and focus on the diagnostic region (to the left of the line). Start by looking at the double-bond region and the triple-bond region.

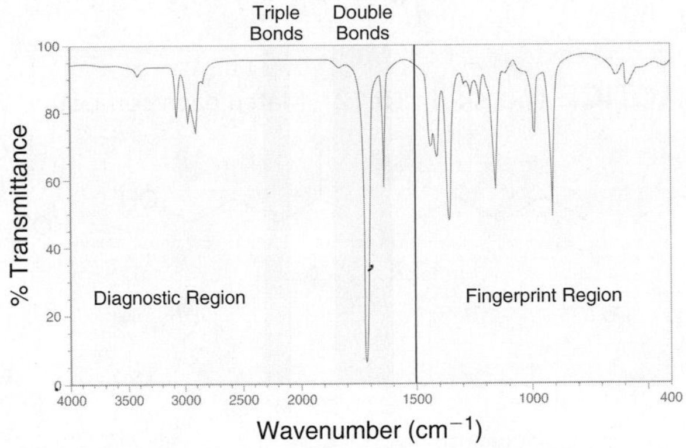

STEP 3
Look for X—H bonds. Draw a line at 3000 cm^{-1}.

There are no signals in the triple-bond region, but there are two signals in the double-bond region. The signal at 1650 cm^{-1} is narrow and weak, consistent with a C═C bond. The signal at 1720 cm^{-1} is broad and strong, consistent with a C═O bond.

Next, look for X—H bonds. Draw a line at 3000 cm^{-1}, and identify if there are any signals to the left of this line.

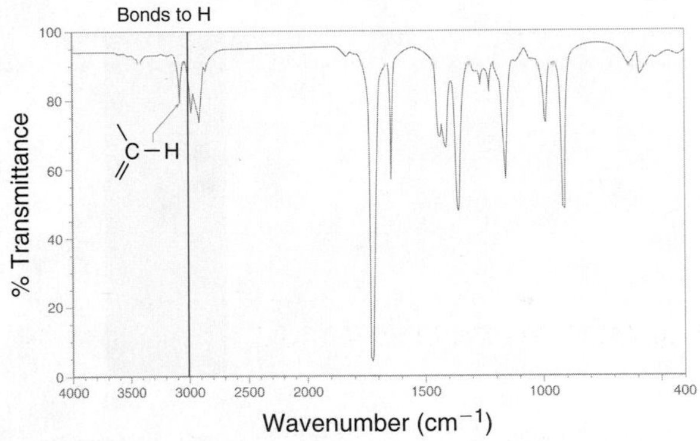

This spectrum exhibits one signal just above 3000 cm^{-1}, indicating a vinylic C—H bond.

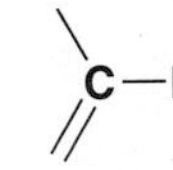

The identification of a vinylic C—H bond is consistent with the observed C═C signal present in the double-bond region (1650 cm^{-1}). There are no other signals above 3000 cm^{-1}, so the compound does not possess any OH or NH bonds.

The little bump between 3400 and 3500 cm^{-1} is not strong enough to be considered a signal. These bumps are often observed in the spectra of compounds containing a C═O bond. The bump occurs at exactly twice the wavenumber of the C═O signal and is called an *overtone* of the C═O signal.

The diagnostic region provides the information necessary to solve this problem. Specifically, the compound must have the following bonds: C═C, C═O, and vinylic C—H. Among the possible choices, there are only two compounds that have these features.

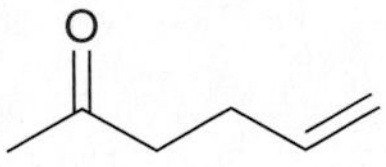

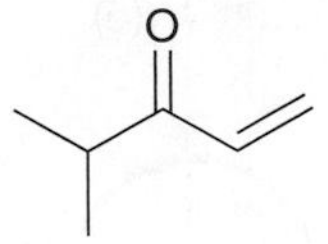

To distinguish between these two possibilities, notice that the second compound is conjugated, while the first compound is *not* conjugated (the π bonds are separated by more than one single bond). Recall that ketones produce signals at approximately 1720 cm^{-1}, while conjugated ketones produce signals at approximately 1680 cm^{-1}.

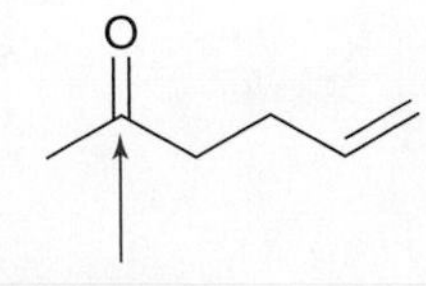

1720 cm^{-1}

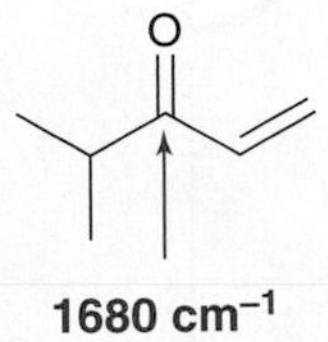

1680 cm^{-1}

In the spectrum provided, the C=O signal appears at 1720 cm^{-1}, indicating that it is not conjugated. The spectrum is therefore consistent with the following compound.

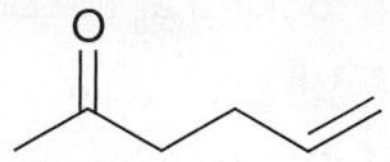

PRACTICE the skill **15.12** Match each compound with the appropriate IR spectrum:

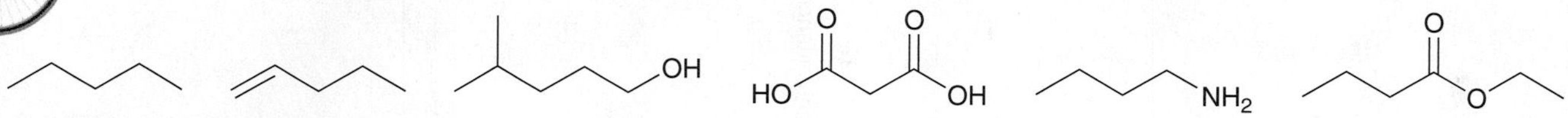

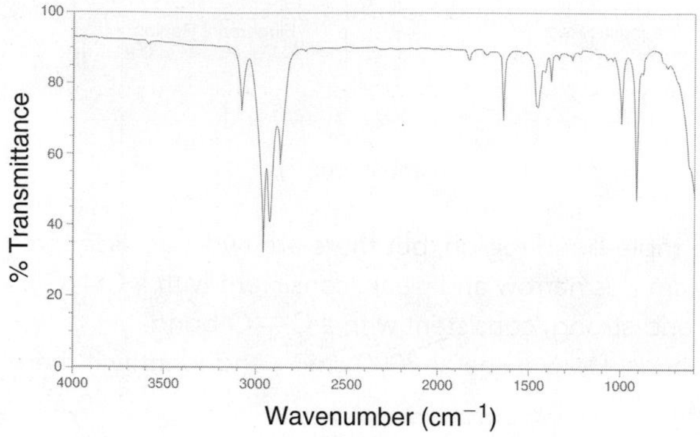

(a)

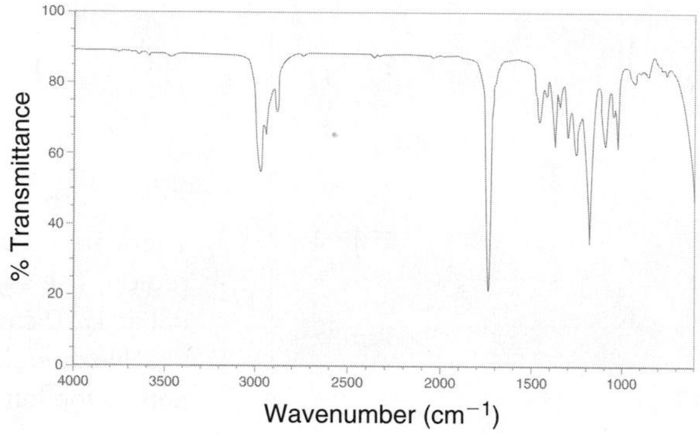

(b)

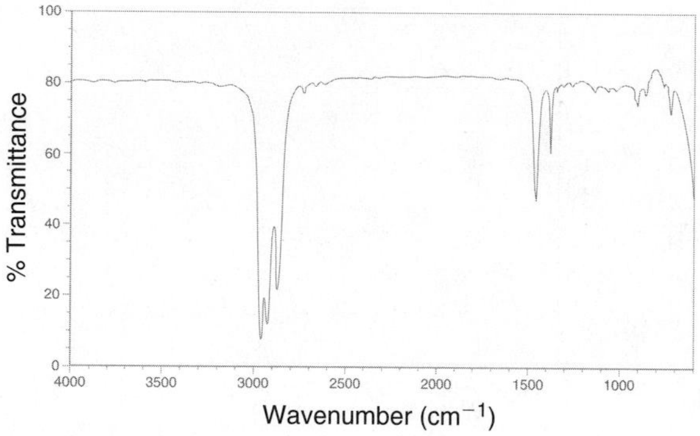

(c)

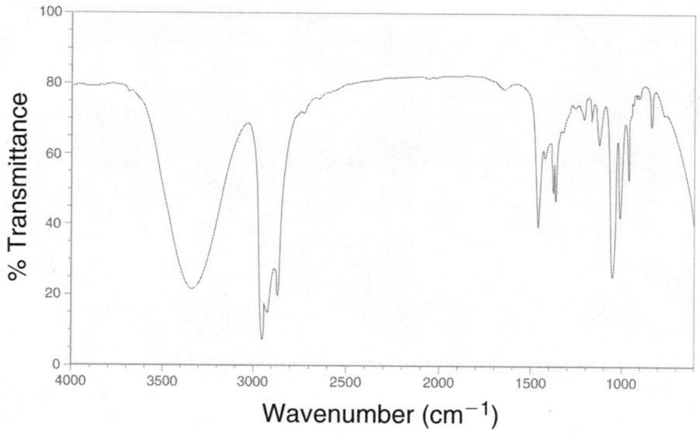

(d)

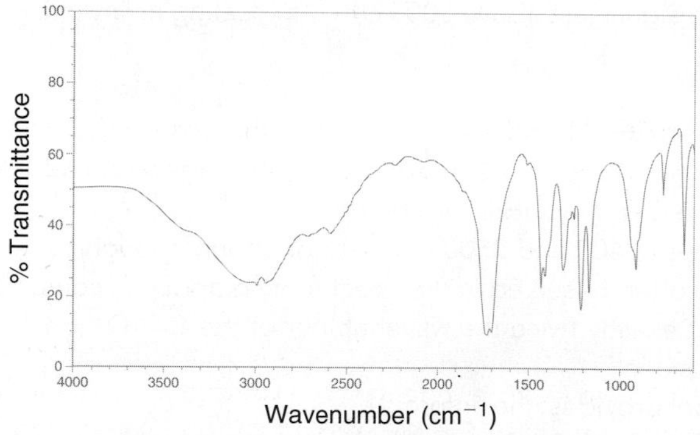

(e)

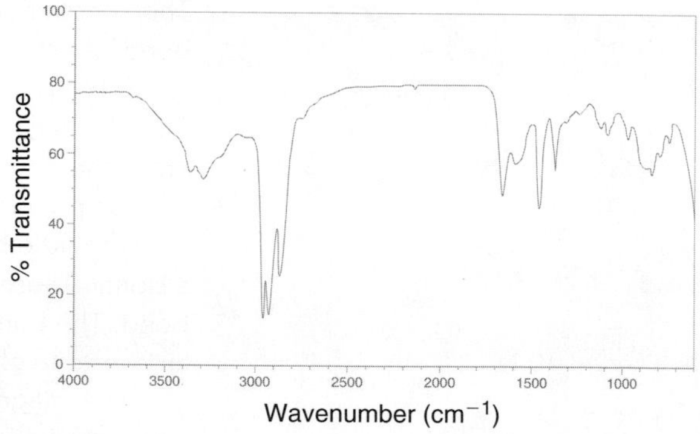

(f)

APPLY the skill

15.13 Chrysanthemic acid is isolated from chrysanthemum flowers. The IR spectrum of chrysanthemic acid exhibits five signals above 1500 cm^{-1}. Identify the source for each of these signals.

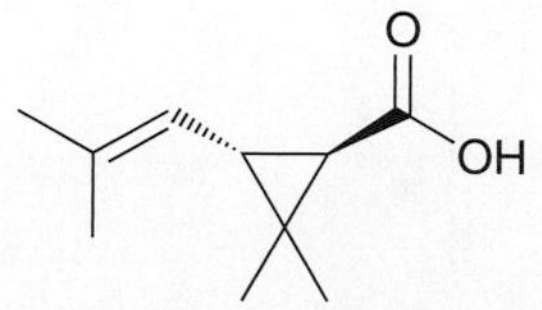

(+)-*trans*-Chrysanthemic acid

need more **PRACTICE?** **Try Problems 15.42, 15.57**

15.7 Using IR Spectroscopy to Distinguish between Two Compounds

IR spectroscopy can be used to differentiate between two compounds. This technique is often extremely helpful when performing a reaction in which a functional group is transformed into a different functional group. For example, consider the following reaction.

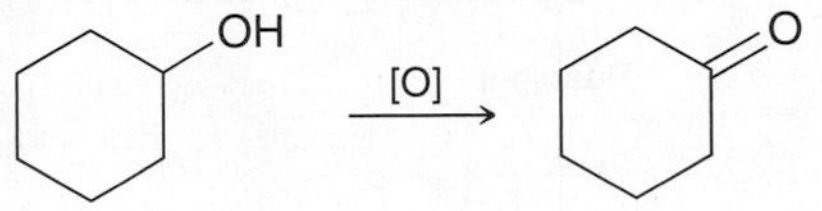

After running the reaction, it is possible to verify formation of the product by looking for the absence of an O—H signal and the presence of a C=O signal. In order to distinguish two compounds using IR spectroscopy, you must know what to look for in the spectrum.

SKILLBUILDER

15.2 DISTINGUISHING BETWEEN TWO COMPOUNDS USING IR SPECTROSCOPY

LEARN the skill

Identify how IR spectroscopy might be used to monitor the progress of the following reaction.

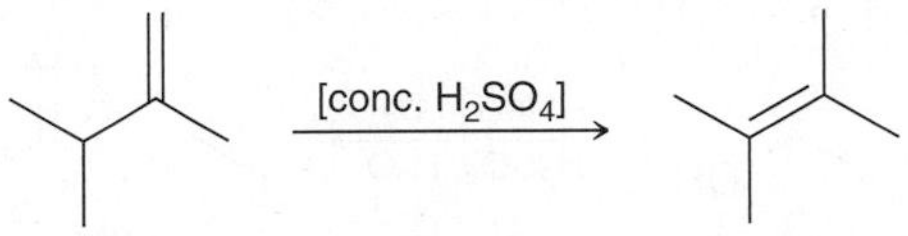

SOLUTION

Carefully inspect both compounds (reactant and product), and think about what signals each compound would produce in its IR spectrum. It might be helpful to work your way through the various regions of the spectrum and determine if there would be a difference in each region.

Let's begin with the double-bond region (1600–1850 cm^{-1}). Each compound has a C=C bond, but consider the intensity of the expected signals. The reactant has an unsymmetrical C=C bond, while the product has a symmetrical C=C bond.

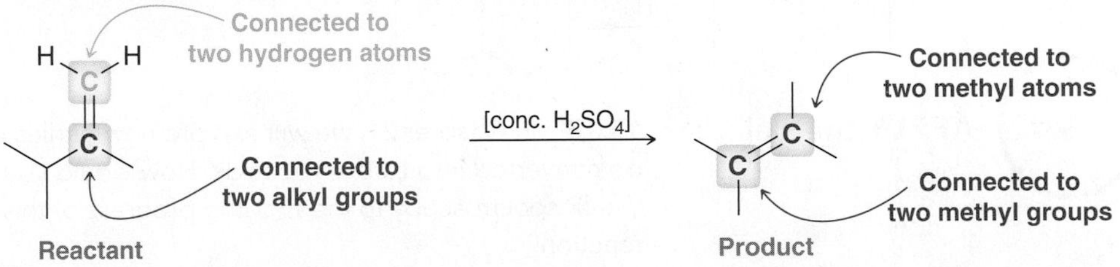

The reactant has a weak dipole moment because each vinylic carbon atom is in a slightly different electronic environment. A weak signal is therefore expected at 1650 cm^{-1}. In contrast, each vinylic position of the product is connected to two methyl groups, so there is no dipole moment associated with this bond. It is completely inefficient at absorbing IR radiation, and therefore, it will not produce a signal at 1650 cm^{-1}.

This difference can be used to monitor the progress of the reaction. A drop of the reaction mixture can be removed from the reaction flask at periodic intervals and analyzed in an IR spectrum. As the product forms, the signal at 1650 cm^{-1} should vanish.

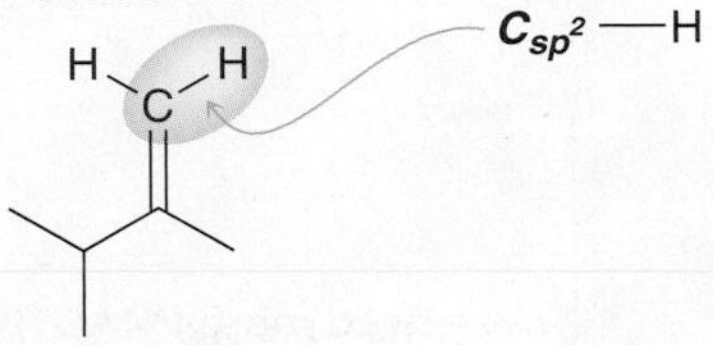

Now let's consider the region containing signals from X—H bonds (2700–4000 cm^{-1}). Neither of the compounds has an O—H or N—H bond. Both compounds have C_{sp^3}—H bonds, but only the reactant has C_{sp^2}—H bonds.

The product does not have this type of C—H bond. We therefore expect that the reactant will produce a signal at 3100 cm^{-1} while the product will not. Once again, this difference can be used to monitor the progress of the reaction. As the product forms, the signal at 3100 cm^{-1} should vanish.

Our analysis has produced two different ways to distinguish between these compounds. Following are the actual IR spectra of these two compounds with the relevant regions highlighted.

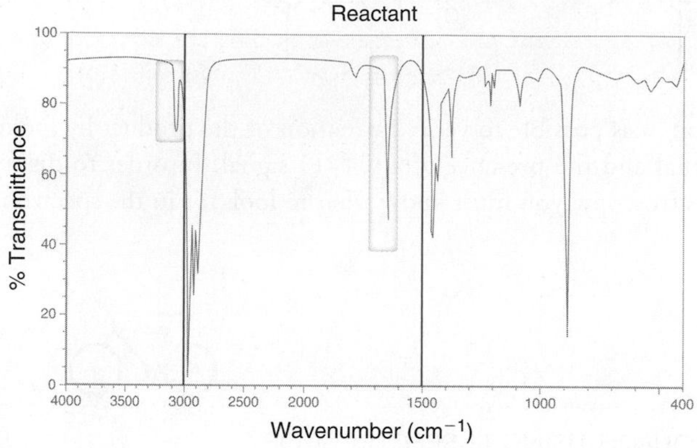

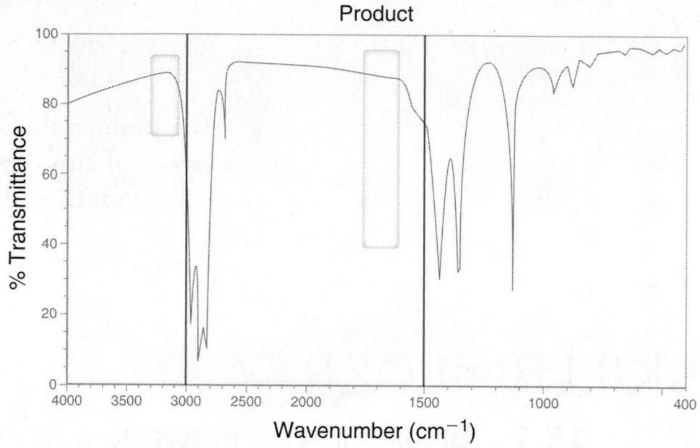

PRACTICE the skill **15.14** Describe how IR spectroscopy might be used to monitor the progress of each of the following reactions.

(a) (structure) $\xrightarrow[\text{H}_2\text{SO}_4,\ \text{H}_2\text{O}]{\text{Na}_2\text{Cr}_2\text{O}_7}$ (structure)

(b) (structure) $\xrightarrow{\text{CH}_3\text{I}}$ (structure)

(c) (structure) $\xrightarrow[\text{catalyst}]{\text{H}_2 \ \text{Lindlar's}}$ (structure)

(d) (structure) $\xrightarrow[\text{NH}_3]{\text{Na}}$ (structure)

(e) (structure) $\xrightarrow[\text{Ni}]{\text{H}_2}$ (structure)

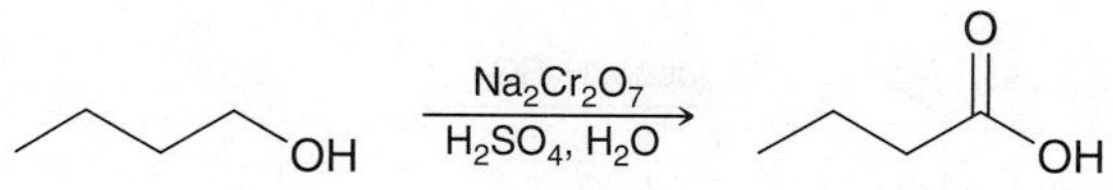

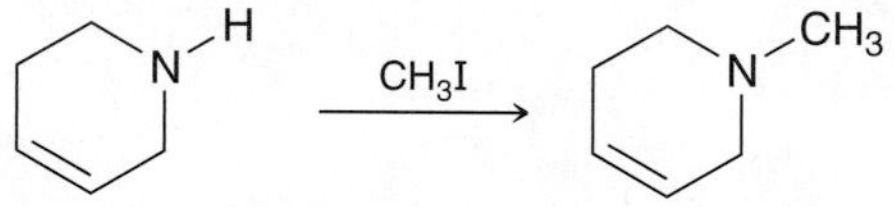

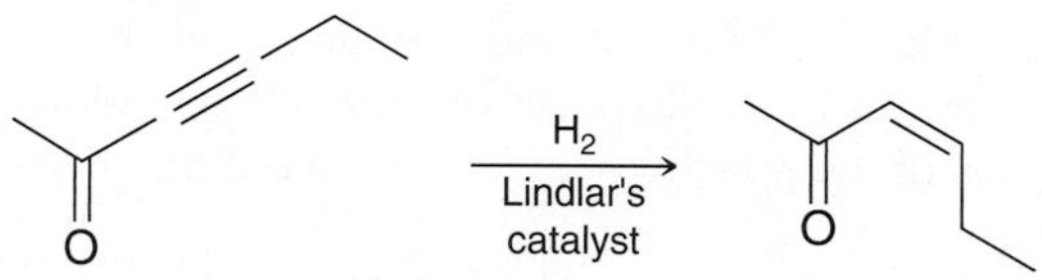

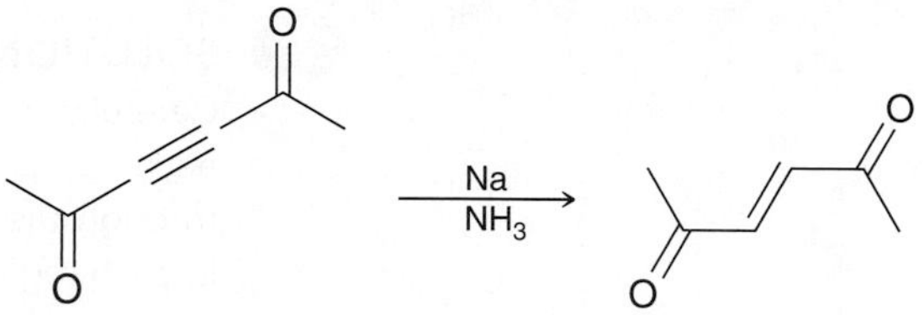

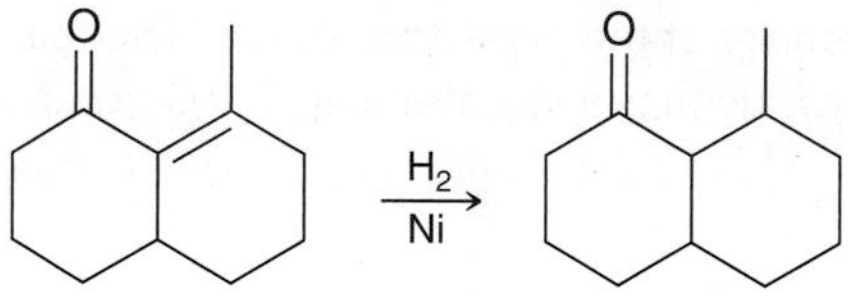

APPLY the skill **15.15** In Chapter 21, we will explore how nitriles can be converted into carboxylic acids. How would you use IR spectroscopy to monitor the progress of this reaction?

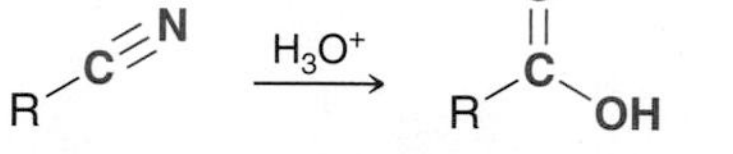

A nitrile **A carboxylic acid**

15.16 1-Butyne was treated with sodium amide followed by ethyl iodide. An IR spectrum of the product was acquired and then compared with the spectrum of the starting alkyne. A signal at 2200 cm^{-1} was observed in the IR spectrum of the starting alkyne, but the same signal was not observed in the IR spectrum of the product. Explain this observation.

15.17 Cyclopentanone was treated with lithium aluminum hydride followed by H$_3$O$^+$. Explain what you would look for in the IR spectrum of the product to verify that the expected reaction had occurred. Identify which signal should be present and which signal should be absent.

15.18 When 1-chlorobutane is treated with sodium hydroxide, two products are formed. Identify the two products, and explain how these products could be distinguished using IR spectroscopy.

need more **PRACTICE?** **Try Problems 15.38–15.41, 15.48d**

15.8 Introduction to Mass Spectrometry

In the beginning of this chapter, we defined spectroscopy as the study of the interaction between matter and electromagnetic radiation. In contrast, **mass spectrometry** is the study of the interaction between matter and an energy source other than electromagnetic radiation. Mass spectrometry is used primarily to determine the molecular weight and molecular formula of a compound.

In a **mass spectrometer**, a compound is first vaporized and converted into ions, which are then separated and detected. The most common ionization technique involves bombarding the compound with high-energy electrons. These electrons carry an extraordinary amount of energy, usually around 1600 kcal/mol, or 70 electron volts (eV). When a high–energy electron strikes the molecule, it causes one of the electrons in the molecule to be ejected. This technique, called **electron impact ionization** (**EI**), generates a high-energy intermediate that is both a radical and a cation.

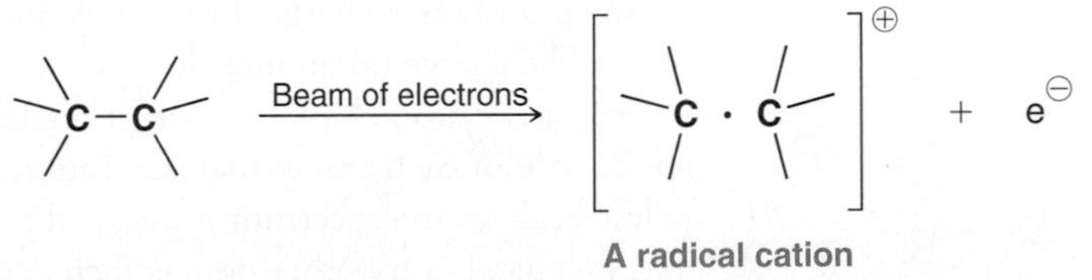

It is a radical because it has an unpaired electron, and it is a cation because it bears a positive charge as a result of losing an electron. The mass of the ejected electron is negligible compared to the mass of the molecule, so the mass of the radical cation is essentially equivalent to the mass of the original molecule. This radical cation, symbolized by $(M)^{+\bullet}$, is called the **molecular ion**, or the **parent ion**. The molecular ion is often very unstable and is susceptible to **fragmentation**, which generates two distinct fragments. Most commonly, one fragment carries the unpaired electron, while the other fragment carries the charge.

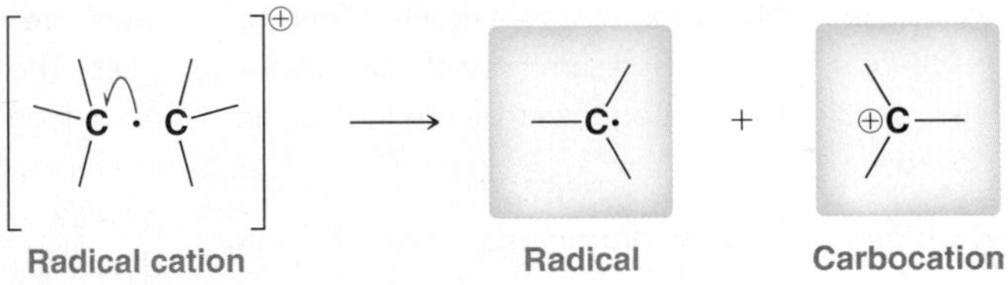

In this way, the ionization process generates many different cations—the molecular ion as well as many different carbocation fragments. All of these ions are accelerated and then sent through a magnetic field, where they are deflected in curved paths (Figure 15.21).

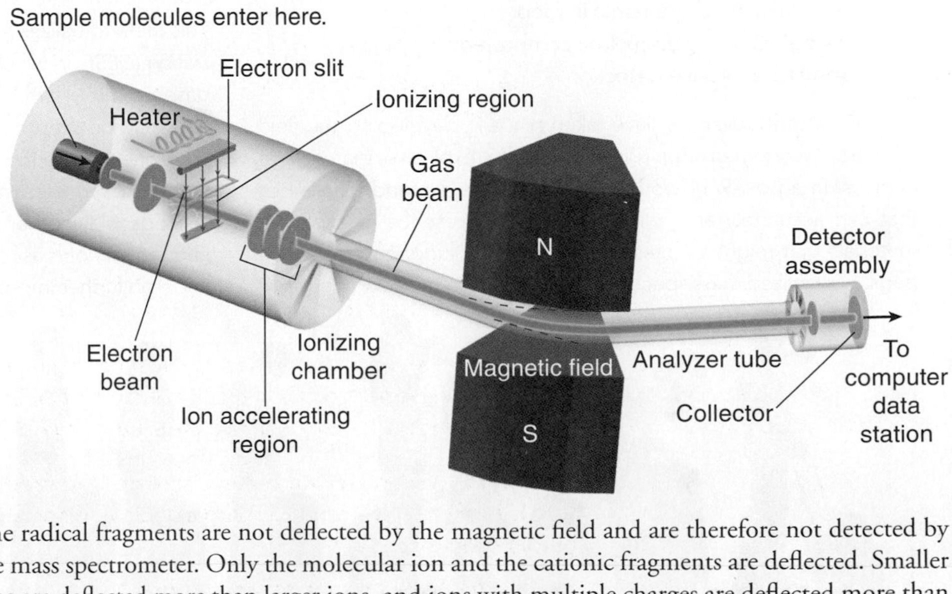

FIGURE 15.21
A schematic of a mass spectrometer.

The radical fragments are not deflected by the magnetic field and are therefore not detected by the mass spectrometer. Only the molecular ion and the cationic fragments are deflected. Smaller ions are deflected more than larger ions, and ions with multiple charges are deflected more than

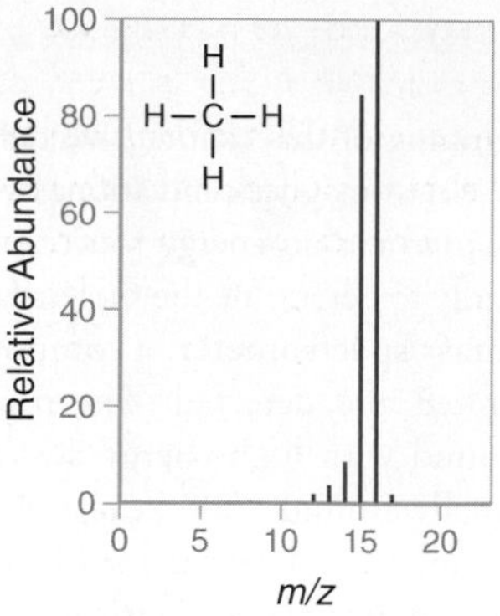

(a)

(b)

MASS SPECTRUM DATA	
m/z	RELATIVE HEIGHT (%)
12	1.0
13	3.9
14	9.2
15	85.0
16	100 (base peak)
17	1.1

FIGURE 15.22
A mass spectrum of methane.

ions with only one charge. In this way, the cations are separated by their **mass-to-charge ratio** (m/z). The charge (z) on most ions is +1, and therefore, m/z is effectively a measure of the mass (m) of each cation. A plot is then generated, called a **mass spectrum**, which shows the relative abundance of each cation that was detected. Figure 15.22 is a mass spectrum of methane. The tallest peak in the spectrum is assigned a relative value of 100% and is called the **base peak**. The height of every other peak is then described relative to the height of the base peak. In the case of methane, the molecular ion peak is the base peak, but this is not always the case for other compounds. In the mass spectra of larger compounds, it is very common for one of the

PRACTICALLYSPEAKING)))

Mass Spectrometry for Detecting Explosives

Mass spectrometry is an incredibly important tool that has found a broad array of applications. Following is a summary of some major applications of mass spectrometry, categorized by the industry in which they are applied:

Pharmaceutical: drug discovery, drug metabolism, reaction monitoring
Biotechnology: amino acid sequencing, analysis of macromolecules
Clinical: neonatal screening, hemoglobin analysis
Environmental: drug testing, water quality testing, contamination level measurements in food
Geological: evaluation of oil compositions
Forensic: explosive detection

Significant advances have taken place in developing the field of mass spectrometry for use in detecting explosive materials at airports. In a post-9/11 world, there is a great demand for devices that can accurately and reliably detect the presence of explosive materials that might be present in a traveler's luggage or carry-on bags. Specialized mass spectrometers, called ion mobility spectrometers, are now being utilized at hundreds of major airports in U.S. cities. These devices collect chemicals from the surface of a traveler's luggage, subject these chemicals to a process that converts them into ions, and then measures the speed of these ions as they pass through an electric field. These spectrometers are designed to detect the presence of any ions that move at a speed consistent with known explosive materials. Several recent advances have made this possible, and new techniques are constantly being developed as this is an area of significant ongoing research. One such technique, developed by researchers at Purdue University, enables the acquisition of chemicals from the surface of luggage in just a few seconds. This method utilizes a technique called *desorption electrospray ionization (DESI)*, in which the surface of the suitcase is sprayed with a gaseous mixture that dislodges any residual explosive compounds that may be present on the surface of the suitcase as a result of someone loading the suitcase with explosives. The gaseous mixture is then sucked into a mass spectrometer where it can be analyzed in seconds. Many other advances in this field are emerging each year, and the role of mass spectrometry in explosive detection is likely to undergo further improvement in the years to come.

fragments to produce the tallest peak. In such a case, the molecular ion peak is not the base peak. We will see examples of this in the upcoming sections.

In the spectrum of methane, the peaks below 16 are formed by fragmentation of the molecular ion. Methane is a small molecule, and there are very few ways for the molecular ion to fragment. There are only C—H bonds, so fragmentation of methane simply involves the loss of hydrogen atoms.

$$\left[\begin{array}{c} H \\ | \\ H-C \cdot H \\ | \\ H \end{array} \right]^{\oplus} \longrightarrow \begin{array}{c} H \\ | \\ H-C-H \\ \oplus \end{array} + \; H^{\bullet}$$

Molecular ion Fragment
m/z = 16 *m/z* = 15

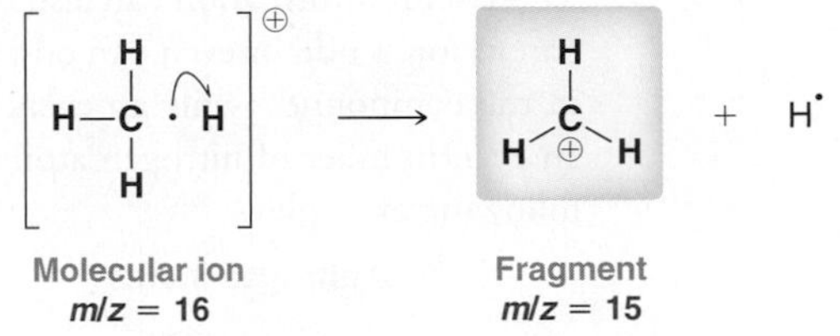

The loss of a hydrogen atom produces a carbocation fragment with *m/z* = 15. The mass spectrum of methane (Figure 15.22) indicates that this fragment is almost as abundant as the molecular ion itself. This carbocation can then lose another hydrogen atom, resulting in a new fragment with *m/z* = 14, although this new fragment is not very abundant, as can be seen on the spectrum. This process can continue until all four hydrogen atoms have been lost, giving rise to a series of peaks with *m/z* values ranging from 12 to 15.

We have explained all but one of the peaks in the mass spectrum of methane. Notice that there is a peak *m/z* = 17. This peak, called the (M+1)$^{+\bullet}$ peak, will be discussed in more detail in Section 15.10.

15.9 Analyzing the (M)$^{+\bullet}$ Peak

For some compounds, the (M)$^{+\bullet}$ is the base peak. Consider, for example, the mass spectrum of benzene (Figure 15.23). Clearly, the molecular ion (*m/z* = 78) is not very susceptible to fragmentation, as it is the most abundant ion to pass through the spectrometer. This is generally not the case. Most compounds will easily fragment, and the (M)$^{+\bullet}$ peak will not be the most abundant ion.

For example, consider the mass spectrum of pentane (Figure 15.24). In the mass spectrum of pentane, the (M)$^{+\bullet}$ peak (at *m/z* = 72) is very small indeed. In this case, the fragment at *m/z* = 43 is the most abundant ion, and that peak is therefore called the base peak and assigned a value of

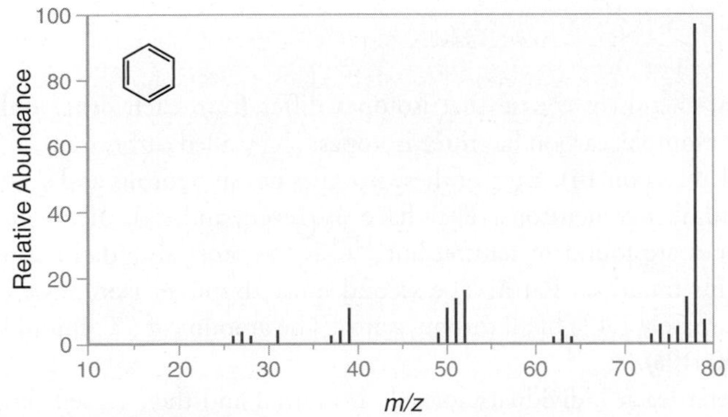

FIGURE 15.23
A mass spectrum of benzene.

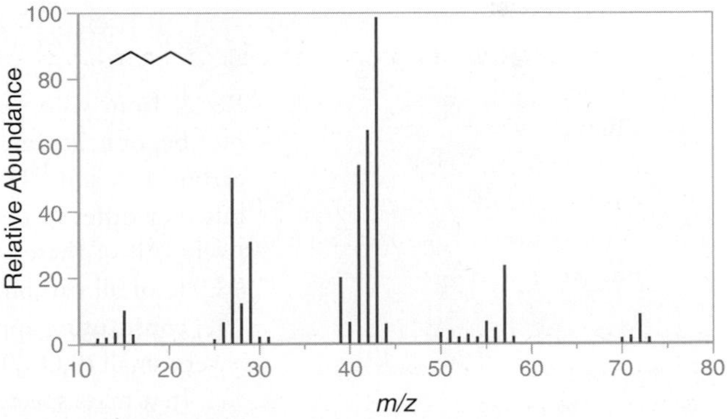

FIGURE 15.24
A mass spectrum of pentane.

100%. In some cases, it is possible for the (M)$^{+\bullet}$ peak to be entirely absent, if it is particularly susceptible to fragmentation. In such cases, there are more gentle methods of ionization (other than EI) that allow the parent ion to survive long enough to pass through the spectrometer. We will briefly review one of those methods later in this chapter.

When analyzing a mass spectrum, the first step is to look for the (M)$^{+\bullet}$ peak, because it indicates the molecular weight of the molecule. This technique can be used to distinguish compounds. For example, compare the molecular weights of pentane and of 1-pentene.

Pentane 1-Pentene
(MW = **72**) (MW = **70**)

Pentane has 5 carbon atoms ($5 \times 12 = 60$) and 12 hydrogen atoms ($12 \times 1 = 12$) and therefore a molecular weight of 72. In contrast, 1-pentene only has 10 hydrogen atoms and therefore has a molecular weight of 70. As a result, we expect the mass spectrum of pentane to exhibit a molecular ion peak at 72, while the mass spectrum of 1-pentene should exhibit a molecular ion peak at 70.

Useful information can also be obtained by analyzing whether the molecular weight of the parent ion is odd or even. An odd molecular weight indicates an odd number of nitrogen atoms in the compound, while an even molecular weight indicates either the absence of nitrogen or an even number of nitrogen atoms. This is called the **nitrogen rule**, and it is illustrated in the following examples.

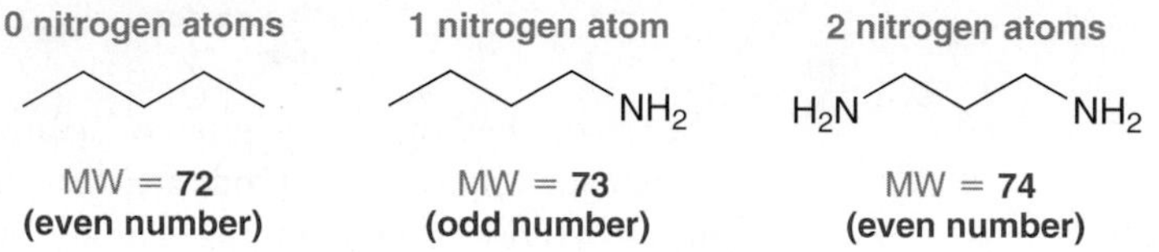

CONCEPTUAL CHECKPOINT

15.19 How would you distinguish between each pair of compounds using mass spectrometry?

(a) 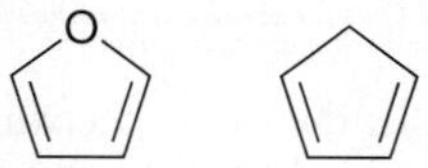

(b) 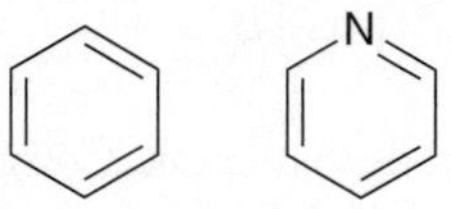

15.20 For each of the following compounds, use the nitrogen rule to determine whether the molecular weight should be even or odd. Then calculate the expected *m/z* value for the molecular ion.

(a) 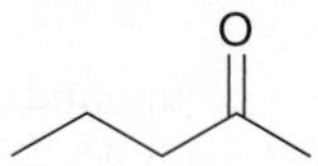

(b) 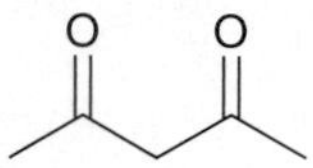

(c) 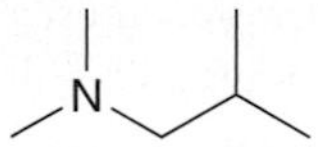

(d) 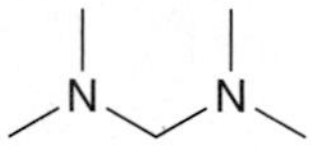

15.10 Analyzing the $(M+1)^{+\bullet}$ Peak

Recall from your general chemistry course that isotopes differ from each other only in the number of neutrons. For example, carbon has three isotopes: ^{12}C (called carbon 12), ^{13}C (called carbon 13), and ^{14}C (called carbon 14). Each of these isotopes has six protons and six electrons, but they differ in their number of neutrons. They have six, seven, and eight neutrons, respectively. All of these isotopes are found in nature, but ^{12}C is the most abundant, constituting 98.9% of all carbon atoms found on Earth. The second most abundant isotope of carbon is ^{13}C, constituting approximately 1.1% of all carbon atoms. The amount of ^{14}C found in nature is very small (0.0000000001%).

In a mass spectrometer, each individual molecule is ionized and then passed through the magnetic field. When methane is analyzed, 98.9% of the molecular ions will contain a ^{12}C atom, while only 1.1% will contain a ^{13}C atom. The latter group of molecular ions are responsible for the observed peak at $(M+1)^{+\bullet}$. The relative height of this peak is approximately 1.1% as tall as $(M)^{+\bullet}$ peak, just as expected. Larger compounds, containing more carbon atoms, will have a larger $(M+1)^{+\bullet}$ peak. For example, decane has 10 carbon atoms in its structure, so the chances that a molecule of decane will possess one ^{13}C atom are 10 times greater than the chances that a molecule of methane will possess a ^{13}C atom. Consequently, the $(M+1)^{+\bullet}$ peak in the mass spectrum of decane is 11% as tall as the molecular ion peak ($10 \times 1.1\%$). Similarly, the $(M+1)^{+\bullet}$ peak in the mass spectrum of icosane ($C_{20}H_{42}$) is 22% as tall as the molecular ion peak ($20 \times 1.1\%$) (Figure 15.25). Isotopes of other elements also contribute to the $(M+1)^{+\bullet}$ peak, but ^{13}C is the greatest contributor, and therefore, it is generally possible to determine the number of carbon atoms in an unknown compound by comparing the relative heights of the $(M)^{+\bullet}$

peak and the $(M+1)^{+\bullet}$ peak. This piece of information can be very helpful in determining the molecular formula. The following exercise illustrates this technique.

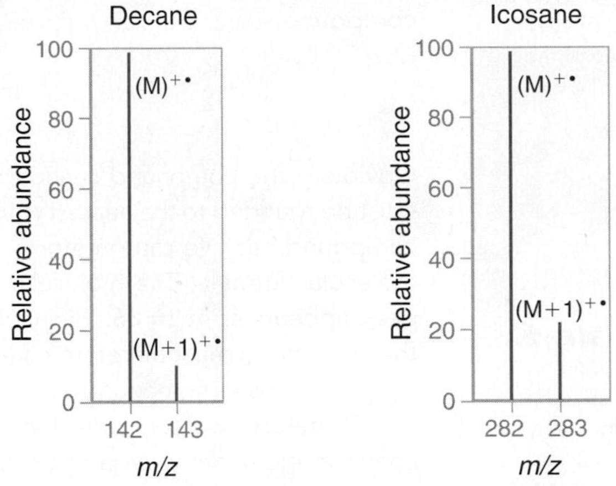

FIGURE 15.25
The relative heights of the $(M)^{+\bullet}$ and $(M+1)^{+\bullet}$ peaks for decane and icosane.

SKILLBUILDER

15.3 USING THE RELATIVE ABUNDANCE OF THE $(M+1)^{+\bullet}$ PEAK TO PROPOSE A MOLECULAR FORMULA

LEARN the skill

Below is the mass spectrum as well as the tabulated mass spectrum data for an unknown compound. Propose a molecular formula for this compound.

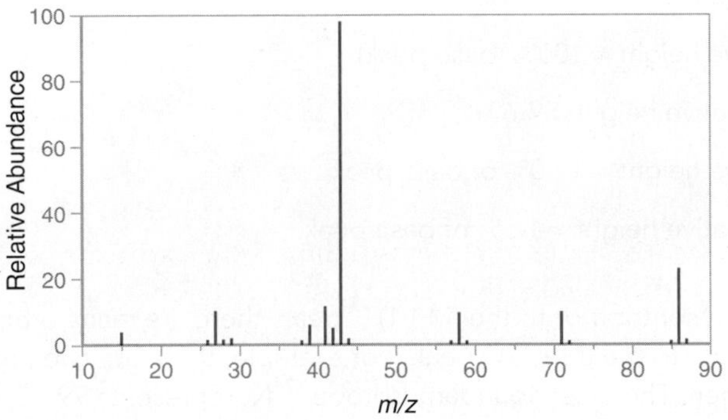

MASS SPECTRUM DATA

m/z	RELATIVE HEIGHT (%)	*m/z*	RELATIVE HEIGHT (%)
15	4.8	42	4.0
26	1.3	43	100 (base peak)
27	10.5	44	2.3
28	1.3	58	10.3
29	1.9	71	11.0
38	1.2	86	20.9 ($M^{+\bullet}$)
39	6.3	87	1.2
41	11.9		

SOLUTION

To solve this problem, it is not necessary to visually inspect the spectrum. The data alone are sufficient. It is important to become accustomed to interpreting data even when the spectrum itself is not provided, much the way pilots learn to fly planes at night using instrument readings.

Begin by identifying the molecular ion peak. The data indicate that this peak occurs at $m/z = 86$. Now compare the relative abundance of this peak and the $(M+1)^{+\bullet}$ peak (which appears at $m/z = 87$). The relative height of the $(M+1)^{+\bullet}$ peak is 1.2%, but be careful here. In this case, the molecular ion is not the tallest peak. The tallest peak (base peak) appears at $m/z = 43$. The data indicate that the $(M+1)^{+\bullet}$ peak is 1.2% as tall as the base peak. But we need to know how the $(M+1)^{+\bullet}$ peak compares to the molecular ion peak. To do this, we take the relative height of the $(M+1)^{+\bullet}$ peak, divide by the relative height of the $(M)^{+\bullet}$ peak, and then multiply by 100%.

$$\frac{1.2\%}{20.9\%} \times 100\% = 5.7\%$$

STEP 1
Determine the number of carbon atoms in the compound by analyzing the relative abundance of the M+1 peak.

In other words, the $(M+1)^{+\bullet}$ peak is 5.7% as tall as the $(M)^{+\bullet}$ peak.

Recall that each carbon atom in the compound contributes 1.1% to the height of the $(M+1)^{+\bullet}$ peak, so we divide by 1.1% to determine the number of carbon atoms in the compound.

$$\text{Number of C} = \frac{5.7\%}{1.1\%} = 5.2$$

STEP 2
Analyze the mass of the molecular ion to determine if any heteroatoms are present.

Obviously the compound cannot have a fractional number of carbon atoms, so the value must be rounded to the nearest whole number, or 5. This analysis suggests that the unknown compound has five carbon atoms. This information is incredibly useful in determining the molecular formula. The molecular weight is known to be 86, because the molecular ion peak appears at $m/z = 86$. Five carbon atoms equals $5 \times 12 = 60$, so the other elements in the compound must therefore give a total of $86–60 = 26$. The molecular formula cannot be C_5H_{26}, because a compound with five carbon atoms cannot have that many hydrogen atoms.

Therefore, we conclude that there must be another element present. The two most common elements in organic chemistry (other than C and H) are nitrogen and oxygen. It cannot be a nitrogen atom, because that would give an odd molecular weight (remember the nitrogen rule). So, we try oxygen. This gives the following possible molecular formula:

$$C_5H_{10}O$$

PRACTICE the skill **15.21** Propose a molecular formula for a compound that exhibits the following peaks in its mass spectrum.

(a) $(M)^{+\bullet}$ at $m/z = 72$, relative height = 38.3% of base peak

$(M+1)^{+\bullet}$ at $m/z = 73$, relative height = 1.8% of base peak

(b) $(M)^{+\bullet}$ at $m/z = 68$, relative height = 100% (base peak)

$(M+1)^{+\bullet}$ at $m/z = 69$, relative height = 4.3%

(c) $(M)^{+\bullet}$ at $m/z = 54$, relative height = 100% (base peak)

$(M+1)^{+\bullet}$ at $m/z = 55$, relative height = 4.6%

(d) $(M)^{+\bullet}$ at $m/z = 96$, relative height = 19.0% of base peak

$(M+1)^{+\bullet}$ at $m/z = 97$, relative height = 1.5% of base peak

APPLY the skill **15.22** While ^{13}C is the main contributor to the $(M+1)^{+\bullet}$ peak, there are many other elements that can also contribute to the $(M+1)^{+\bullet}$ peak. For example, there are two naturally occurring isotopes of nitrogen. The most abundant isotope, ^{14}N, represents 99.63% of all nitrogen atoms on Earth. The other isotope, ^{15}N, represents 0.37% of all nitrogen atoms. In a compound with molecular formula $C_8H_{11}N_3$, if the molecular ion peak has a relative abundance of 24.5%, then what percent abundance do you expect for the $(M+1)^{+\bullet}$ peak?

need more **PRACTICE?** **Try Problems 15.45, 15.47**

15.11 Analyzing the $(M+2)^{+\bullet}$ Peak

Most elements have only one dominant isotope. For example, the dominant isotope of hydrogen is ^{1}H, while ^{2}H (deuterium) and ^{3}H (tritium) represent only a small fraction of all hydrogen atoms. Similarly, the dominant isotope of carbon is ^{12}C, while ^{13}C and ^{14}C represent only a small fraction of all carbon atoms. In contrast, chlorine has two major isotopes. One isotope of chlorine, ^{35}Cl, represents 75.8% of all chlorine atoms; the other isotope of chlorine, ^{37}Cl, represents 24.2% of all chlorine atoms. As a result, compounds that contain a chlorine atom will give a characteristically strong $(M+2)^{+\bullet}$ peak. For example, consider the mass spectrum of chlorobenzene (Figure 15.26). The molecular ion appears at $m/z = 112$. The peak at (M+2) ($m/z = 114$) is approximately one-third the height of the parent peak. This pattern is characteristic of compounds containing a chlorine atom.

Compounds containing bromine also give a characteristic pattern. Bromine has two isotopes, ^{79}Br and ^{81}Br, that are almost equally abundant in nature (50.7% and 49.3%, respectively). Compounds containing bromine will therefore have a characteristic peak at (M+2) that is approximately the same height as the molecular ion peak. For example, consider the mass spectrum of bromobenzene (Figure 15.27).

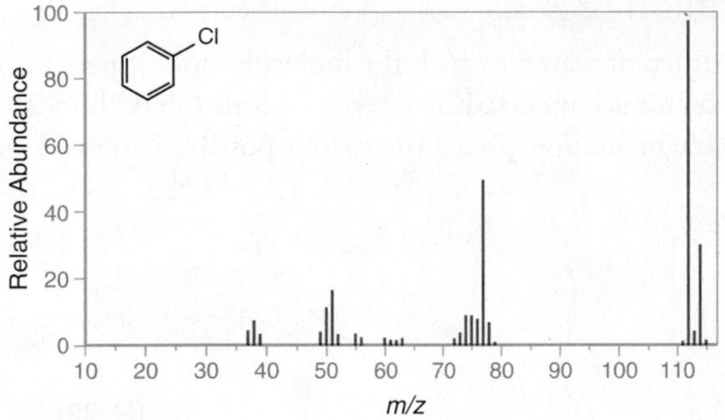

FIGURE 15.26
A mass spectrum of chlorobenzene.

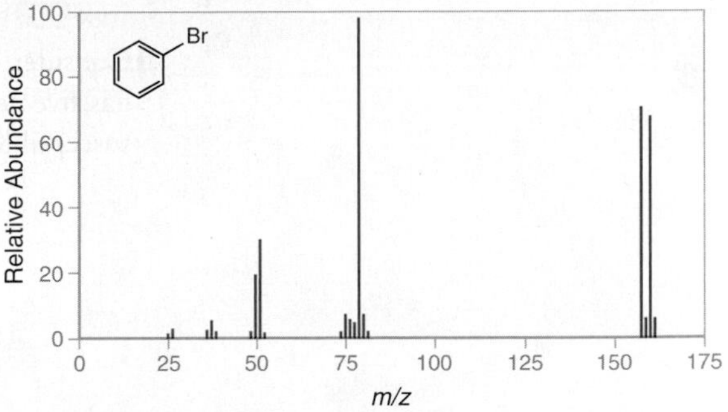

FIGURE 15.27
A mass spectrum of bromobenzene.

The presence of chlorine or bromine in a compound is readily identified by analyzing the height of the $(M+2)^{+\bullet}$ peak and comparing it with the height of the $(M)^{+\bullet}$ peak.

CONCEPTUAL CHECKPOINT

15.23 In the mass spectrum of bromobenzene (Figure 15.27), the base peak appears at $m/z = 77$.

(a) Does this fragment contain Br? Explain your reasoning.

(b) Draw the cationic fragment that represents the base peak.

15.24 Below are mass spectra for four different compounds. Identify whether each of these compounds contains a bromine atom, a chlorine atom, or neither.

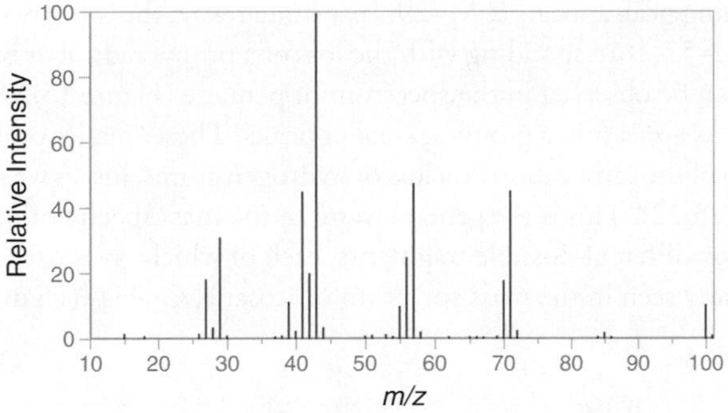

(a)

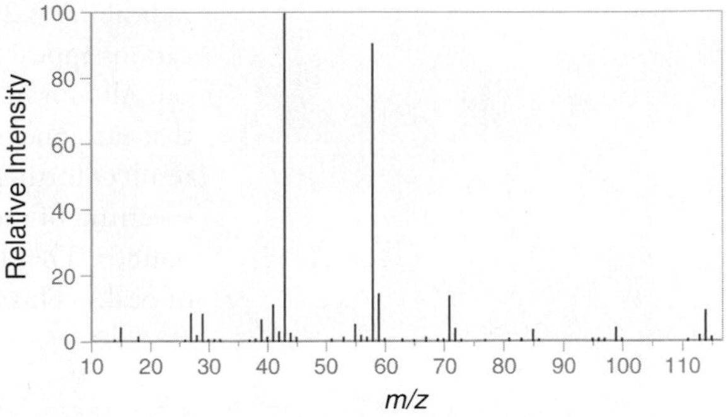

(b)

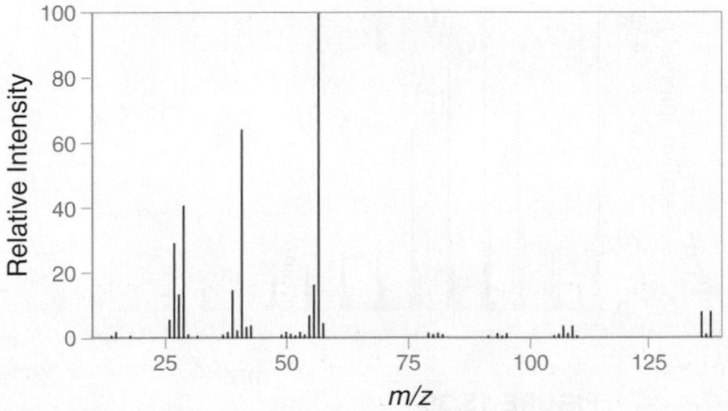

(c)

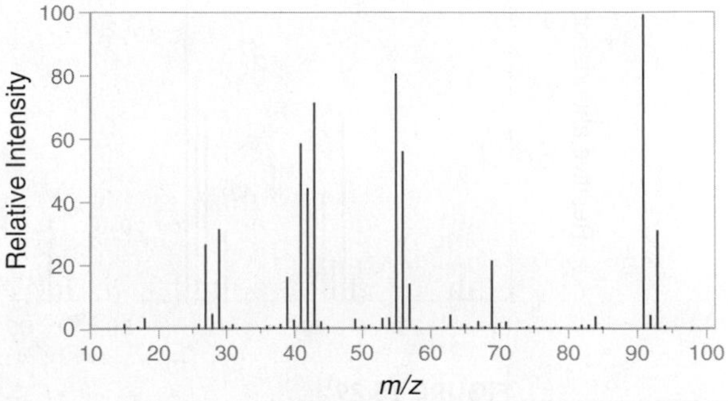

(d)

15.12 Analyzing the Fragments

The majority of the peaks in a mass spectrum are produced from fragmentation of the molecular ion. In this section, we will explore the characteristic fragmentation patterns for a number of compounds. These patterns are often helpful in identifying certain structural features of a compound, but it is generally not possible to use the fragmentation patterns to determine the entire structure of the compound.

Fragmentation of Alkanes

Consider the different ways in which the molecular ion of pentane can fragment. Pentane has five carbon atoms connected by a series of four C—C bonds. Each of these bonds is susceptible to fragmentation, giving rise to four possible cations (Figure 15.28). Remember

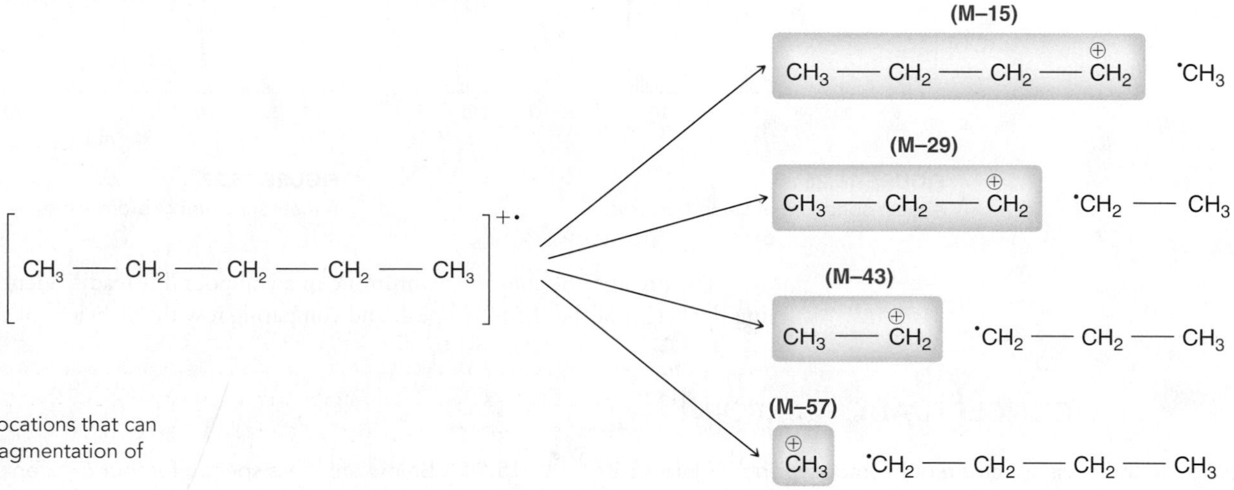

FIGURE 15.28
The various carbocations that can result from the fragmentation of pentane.

that a mass spectrometer does not detect the radical fragments; it only detects the ions. The first cation shown is formed from loss of a methyl radical. The methyl radical has a mass of 15, so the resulting cation appears as a peak at M–15. The second cation is formed from the loss of an ethyl radical (mass 29), so the resulting peak appears at M–29. In a similar way, the other two possible cations appear at M–43 and M–57, corresponding with the loss of a propyl radical or butyl radical. All four of these cations can be observed in the spectrum of pentane (Figure 15.29). Notice that each one of these four peaks appears in a group of smaller peaks. These smaller peaks are the result of further fragmentation of the carbocations via loss of hydrogen atoms, just as we saw in the spectrum of methane (Figure 15.22). This is the general trend in the mass spectra of most compounds. That is, there are many different possible fragments, each of which gives rise to a group of peaks. This can be very clearly seen in the mass spectrum of icosane, $C_{20}H_{42}$ (Figure 15.30).

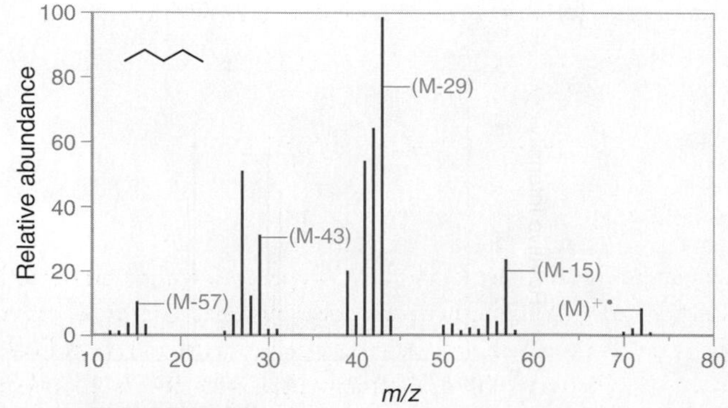

FIGURE 15.29
A mass spectrum of pentane.

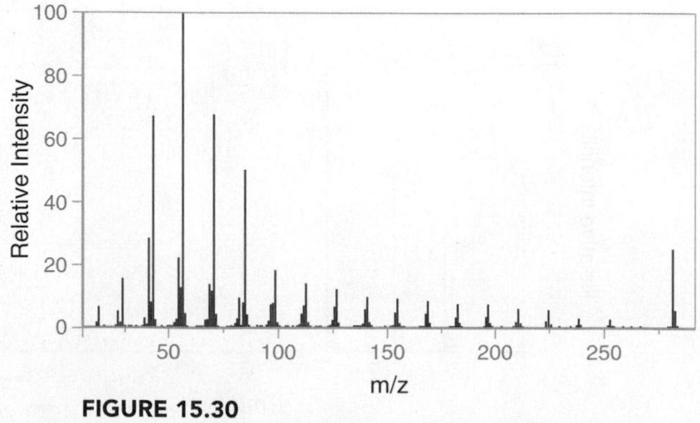

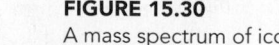

FIGURE 15.30
A mass spectrum of icosane.

LOOKING BACK
For a review of carbocation stability, see Section 6.11.

Icosane is a straight-chain hydrocarbon consisting of 20 carbon atoms connected by 19 C—C bonds. Each of these bonds is susceptible to fragmentation to produce a cation that is detected by the mass spectrometer. Each one of these 19 possible carbocations appears in the spectrum as a group of peaks (except for the M–15 peak, which has a very small relative abundance).

When analyzing a fragment peak, the key is to look at its proximity to the molecular ion peak. For example, a signal at M–15 indicates the loss of a methyl group, and a signal at M–29 indicates the loss of an ethyl group. The likelihood of fragmentation increases with the stability of the carbocation formed as well as the stability of the radical that is ejected.

For example, look again at the various possible carbocations formed by fragmentation of pentane (Figure 15.28). The carbocation corresponding to M–57 is a methyl carbocation, which is less stable than the other possible carbocations (all of which are primary). That explains why the peak at M–57 is a fairly small peak in the spectrum. In general, fragmentation will occur in all possible locations but will typically favor the formation of the most stable carbocation.

As another example, consider the most likely fragmentation of the following molecular ion.

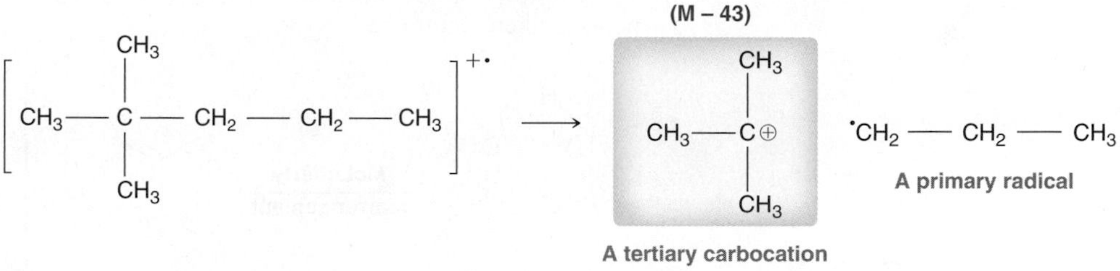

The most abundant peak in the spectrum of this compound is expected to be at M–43, corresponding with formation of a tertiary carbocation, via loss of a propyl radical. A tertiary carbocation can also be produced via loss of a methyl radical (from the left side of the molecular ion above); however, a methyl radical is less stable than a primary radical. Certainly, all possible fragmentations are observed under the high-energy conditions employed, but the most abundant peak will generally result from formation of the most stable carbocation via expulsion of the most stable possible radical. Therefore, it is generally possible to predict the location of the most abundant peak that is expected in the mass spectrum of a simple alkane and branched alkane.

Fragmentation of Alcohols

Alcohols exhibit two common fragmentation patterns: alpha cleavage and dehydration. During alpha cleavage, a bond to the alpha position of the alcohol is cleaved to form a resonance-stabilized cation and a radical.

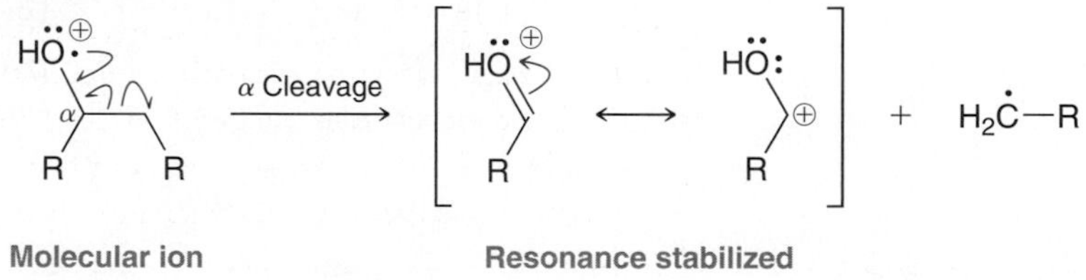

Alternatively, alcohols can undergo dehydration via loss of a water molecule.

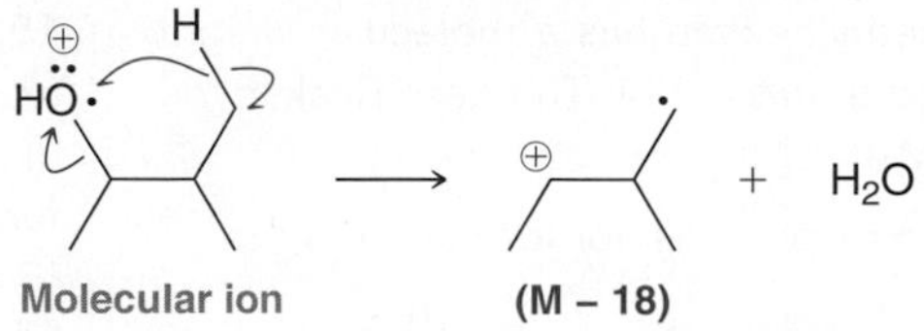

This fragmentation results from an intramolecular elimination process, which can occur because of the high energy of the parent ion. This fragmentation does not eject a radical, but instead, it ejects a neutral molecule (water). Loss of a neutral molecule produces a new radical cation, which produces a peak at M–18 (because the molecular weight of water is 18). The water molecule itself does not appear on the spectrum, because the spectrometer only detects ions. A signal at M-18 is therefore characteristic of alcohols.

Fragmentation of Amines

Much like alcohols, amines are also observed to undergo alpha cleavage to generate a resonance-stabilized cation and a radical.

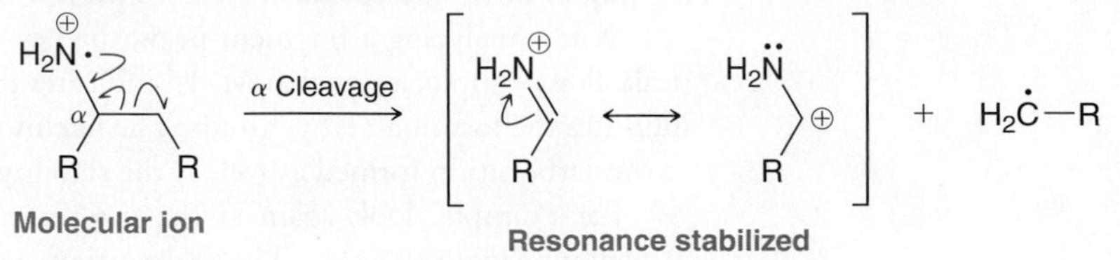

Fragmentation of Ketones and Aldehydes

Aldehydes and ketones containing a hydrogen atom at the gamma position generally undergo a characteristic fragmentation called the *McLafferty rearrangement*, which results in the loss of a neutral alkene fragment.

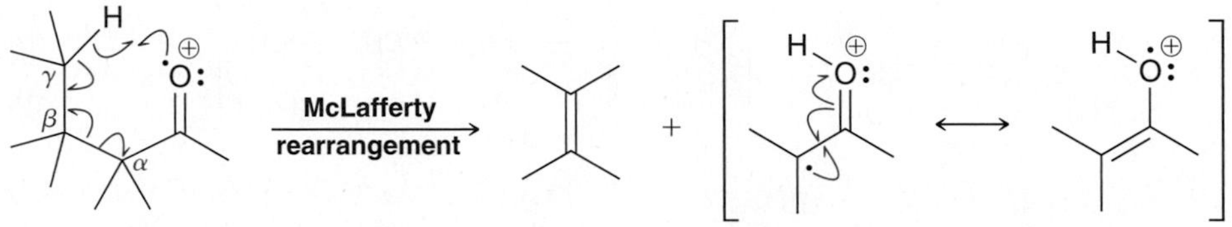

The alkene fragment has an even mass. In contrast, most radical fragments encountered thus far have an odd mass (M–15 for loss of a methyl group, M–29 for loss of an ethyl group, M–43 for loss of a propyl group, etc.). Therefore, the mass spectrum of a ketone or aldehyde often contains a peak at M–*x*, where *x* is an even number. Some of the most common fragments are listed in Table 15.3.

TABLE 15.3 COMMON FRAGMENTS IN MASS SPECTROMETRY

M–15	Loss of a methyl radical
M–29	Loss of an ethyl radical
M–43	Loss of a propyl radical
M–57	Loss of a butyl radical
M–18	Loss of water (from an alcohol)
M–x (where x = even number)	McLafferty rearrangement (ketone or aldehyde)

CONCEPTUAL CHECKPOINT

15.25 Although 2,2-dimethylhexane has a molecular weight of 114, no peak is observed at $m/z = 114$. The base peak in the mass spectrum occurs at M–57.

(a) Draw the fragmentation responsible for formation of the M–57 ion.

(b) Explain why this cation is the most abundant ion to pass through the spectrometer.

(c) Explain why no molecular ions survive long enough to be detected.

(d) Can you offer an explanation as to why the M–15 peak is not the base peak?

15.26 Identify two peaks that are expected to appear in the mass spectrum of 3-pentanol. For each peak, identify the fragment associated with the peak, and show a mechanism for its formation.

15.27 Identify the expected base peak in the mass spectrum of 2,2,3-trimethylbutane. Draw the fragment associated with this peak, and explain why the base peak results from this fragment.

15.28 The following are mass spectra for the constitutional isomers ethylcyclohexane and 1,1-dimethylcyclohexane. Based on likely fragmentation patterns, match the compound with its spectrum.

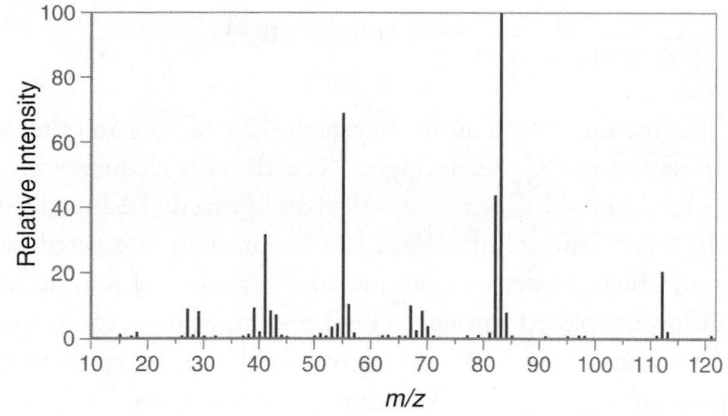

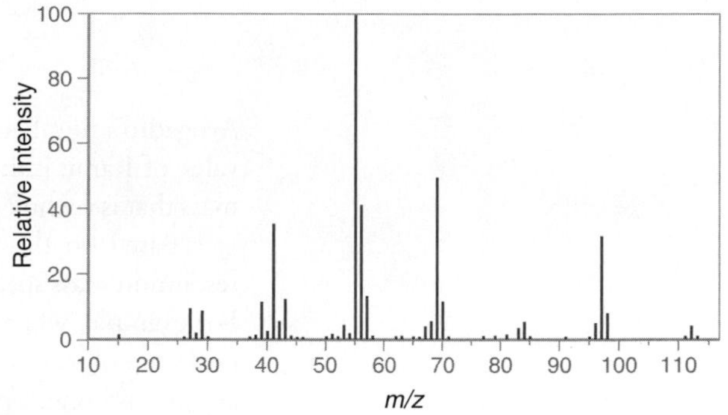

15.13 High-Resolution Mass Spectrometry

High-resolution mass spectrometry involves the use of a detector that can measure *m/z* values to four decimal places. This technique allows for the determination of the molecular formula of an unknown compound. In order to analyze the data obtained from high-resolution mass spectrometry, we must first review some background information. Specifically, we must discuss why atomic masses are not whole numbers (despite the fact that we have treated them as such until now).

The mass of an atom is approximately equal to the sum of its protons and neutrons, because the mass of the electrons are negligible compared to the protons and neutrons. Originally, the mass of a proton was considered to be the same as the mass of a neutron, resulting in the *relative* atomic weights given in Table 15.4. With this simple model, the atomic weights appear as whole numbers. These are the numbers we have used in this chapter so far, and they are a good approximation. But they are not accurate for two reasons:

TABLE 15.4 RELATIVE ATOMIC WEIGHTS OF SEVERAL ELEMENTS

ELEMENT	NUMBER OF PROTONS	NUMBER OF NEUTRONS	RELATIVE ATOMIC WEIGHT
H	1	0	1
He	2	2	4
C	6	6	12
N	7	7	14
O	8	8	16

1. Protons do not have exactly the same mass as neutrons:

$$\text{one proton} = 1.6726 \times 10^{-24}\ \text{g}$$
$$\text{one neutron} = 1.6749 \times 10^{-24}\ \text{g}$$

As a result, a helium atom (which has two protons and two neutrons) is not exactly four times the weight of a hydrogen atom (which has one proton).

2. Protons repel each other but will nevertheless bind together in the nuclei of atoms under the influence of the strong nuclear force. These protons possess an enormous amount of potential energy, which is achieved at the expense of some mass. According to Einstein's famous equation ($E = mc^2$), matter and energy are interconvertible. When protons bind together to form the nucleus of an atom, some of their mass is converted into potential energy. As a result, two bound protons will have less mass than two individual protons. This explains why carbon, which has six protons and six neutrons, is less than six times the mass of a deuterium atom, which has one proton and one neutron.

For the two reasons given, atomic masses are not whole numbers. For reporting atomic mass, chemists have defined the **atomic mass unit (amu)** as being equivalent to 1 g divided by Avogadro's number:

$$1 \text{ amu} = \frac{1 \text{ g}}{6.02214 \times 10^{23}} = 1.6605 \times 10^{-24} \text{ g}$$

Avogadro's number is defined as the number of atoms in exactly 12 g of ^{12}C. In other words, the value of 1 amu is defined with respect to ^{12}C. Accordingly, ^{12}C is the only element with an atomic mass that is a whole number. One atom of ^{12}C has an atomic mass of exactly 12 amu, by definition.

Based on this definition, other isotopes of carbon can be precisely measured using high-resolution mass spectrometry, in which the detector can measure *m/z* values to four decimal places. For example, when carbon atoms are passed through a high-resolution mass spectrometer, three peaks are observed: one peak representing the ^{12}C atoms, one smaller peak representing the ^{13}C atoms, and one very small peak representing the ^{14}C atoms. By defining the ^{12}C peak as exactly 12.0000 amu, the other two peaks are measured at $m/z = 13.0034$ amu and $m/z = 14.0032$ amu. In this way, the isotopes of each element can be accurately "weighed," giving the values in Table 15.5.

TABLE 15.5 RELATIVE ATOMIC MASS AND ABUNDANCE OF SEVERAL ELEMENTS

ISOTOPE	RELATIVE ATOMIC MASS (amu)	ABUNDANCE IN NATURE	ISOTOPE	RELATIVE ATOMIC MASS (amu)	ABUNDANCE IN NATURE
^{1}H	1.0078	99.99%	^{16}O	15.9949	99.76%
^{2}H	2.0141	0.01%	^{17}O	16.9991	0.04%
^{3}H	3.0161	<0.01%	^{18}O	17.9992	0.20%
^{12}C	12.0000	98.93%	^{35}Cl	34.9689	75.78%
^{13}C	13.0034	1.07%	^{37}Cl	36.9659	24.22%
^{14}C	14.0032	<0.01%	^{79}Br	78.9183	50.69%
^{14}N	14.0031	99.63%	^{81}Br	80.9163	49.31%
^{15}N	15.0001	0.37%			

Note: Data obtained from the National Institute of Standards and Technology (NIST).

The values that typically appear on a periodic table are the weighted averages for each element, or **standard atomic weight**, which takes into account isotopic abundance. For example, the standard atomic weight of carbon is 12.011 amu (rather than 12.0000 amu), which is a weighted average of the various isotopes based on their abundance in nature. With high-resolution mass spectrometry, the values on the periodic table are irrelevant, because each molecular ion (or cationic fragment) passes through the spectrometer individually and strikes the detector in a particular location that is dependent on which isotopes are present in that specific molecule. The majority of the ions that strike the detector will be comprised of the most abundant isotopes (^{1}H, ^{12}C, ^{14}N, and ^{16}O). Therefore, for purposes of interpreting the data from high-resolution mass spectrometry, the values in Table 15.5 must be used, rather than the values on the periodic table.

Now we are ready to see how high-resolution mass spectrometry can reveal the molecular formula of an unknown compound. As an example, compare the following two compounds.

C_5H_8O (MW = **84**) C_6H_{12} (MW = **84**)

Both of these compounds have the same molecular weight when rounded to the nearest whole number. The low-resolution mass spectra of each compound is therefore expected to give a molecular ion peak at $m/z = 84$. However, with high-resolution mass spectrometry, the molecular ion peaks of these compounds can be differentiated. The calculations for each compound are as follows.

$$C_5H_8O = (5 \times 12.0000) + (8 \times 1.0078) + (1 \times 15.9949) = \mathbf{84.0573} \text{ amu}$$
$$C_6H_{12} = (6 \times 12.0000) + (12 \times 1.0078) = \mathbf{84.0936} \text{ amu}$$

When measured to four decimal places, the mass of these compounds are in fact different, and this difference is detectable. In a high-resolution mass spectrometer, the molecular ion for the first compound will appear near $m/z = 84.0573$, while the molecular ion of the second compound will appear near $m/z = 84.0936$. In this way, the molecular formula of an unknown compound can be determined via high-resolution mass spectrometry. The mass of the molecular ion is measured accurately, and a simple computer program can then calculate the correct molecular formula. The computer program is not entirely necessary, because published data tables can be cross-referenced to find the matching molecular formula.

CONCEPTUAL CHECKPOINT

15.29 How would you distinguish between each pair of compounds using high-resolution mass spectrometry?

(a)

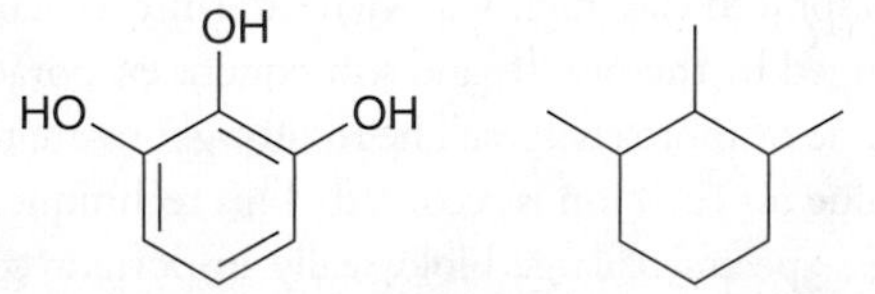

(b)

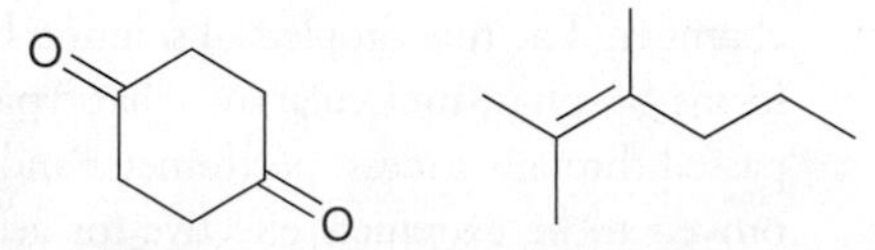

15.30 How would you distinguish between each pair of compounds in Problem 15.29 using IR spectroscopy?

15.14 Gas Chromatography–Mass Spectrometry

Mass spectrometry is ideally suited for analyzing pure compounds. However, when dealing with a mixture that contains several compounds, the compounds must first be separated from each other and then individually injected into the mass spectrometer to yield different spectra. This process has been greatly simplified by the advent of the **gas chromatograph-mass spectrometer** (Figure 15.31), often called "GC mass spec." This powerful diagnostic tool combines a gas

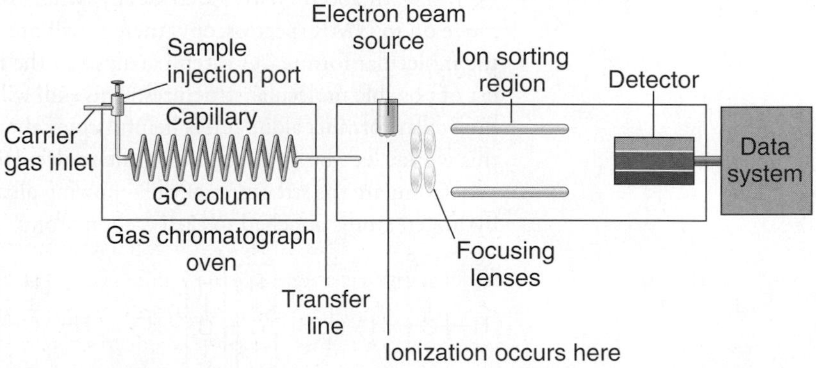

FIGURE 15.31
A schematic of a gas chromatograph–mass spectrometer.

chromatograph with a mass spectrometer. First, the mixture of compounds is separated in the gas chromatograph and then each compound is analyzed sequentially by the mass spectrometer. The gas chromatograph is comprised of a long, narrow tube containing a viscous, high–boiling liquid on a solid support, called the **stationary phase**. The tube is contained in an oven, allowing the temperature to be controlled. A syringe is used to inject the sample into the gas chromatograph where it is vaporized, mixed with an inert gas, and then carried through the tube. The various compounds in the mixture travel through the stationary phase at different rates based on their boiling points and their affinity for the stationary phase. Each compound in the mixture generally exhibits a unique **retention time**, which is the amount of time required for it to exit from the gas chromatograph. In this way, the compounds are separated from each other based on their different retention times. A plot, called a **chromatogram**, identifies the retention time of each compound in the mixture. The chromatogram in Figure 15.32 shows five different compounds exiting the gas chromatograph at different times. Each of these compounds is then passed through the mass spectrometer where its molecular weight can be measured.

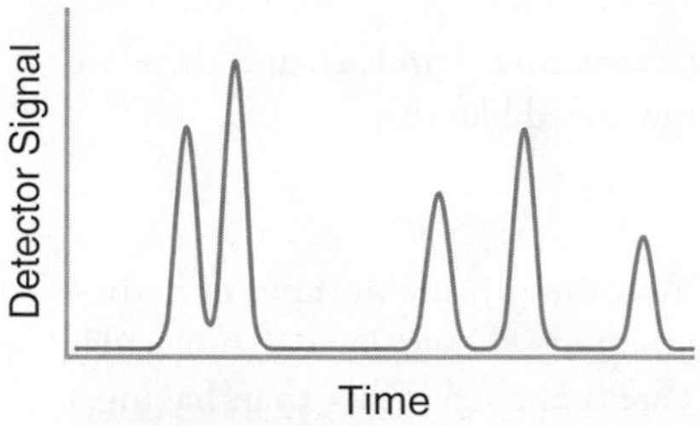

FIGURE 15.32
An example of a chromatogram, showing five different compounds, each with a unique retention time.

GC-MS analysis is often used for drug screening, which can be performed on a urine sample. The organic compounds present in the urine are first extracted and then injected into the gas chromatograph. Each specific drug has a unique retention time, and its identity can be verified via mass spectrometry. This technique is used to test for marijuana, illegal steroids, and many other illegal drugs.

15.15 Mass Spectrometry of Large Biomolecules

Until about 30 years ago, mass spectrometry was limited to compounds with molecular weights under 1000 amu. Compounds with higher molecular weights could not be vaporized without undergoing decomposition. Over the last 30 years, a number of new techniques have emerged, expanding the scope of mass spectrometry. One such technique, called **electrospray ionization (ESI)**, is used most often for large molecules such as proteins and nucleic acids. In this technique, the compound is first dissolved in a solvent and then sprayed via a high-voltage needle into a vacuum chamber. The tiny droplets of solution become charged by the needle, and subsequent evaporation forms gas-phase molecular ions that typically carry one or more charges. The resulting ions are then passed through a mass spectrometer and the *m/z* value for each ion is recorded. This technique has proven to be extremely effective for acquiring mass spectra of large biologically important compounds, particularly because the molecular ions formed generally do not undergo fragmentation. For the development of the ESI technique, Dr. John Fenn (Yale University) shared the 2002 Nobel Prize in Chemistry.

15.16 Hydrogen Deficiency Index: Degrees of Unsaturation

This chapter explained how mass spectrometry can be used to ascertain the molecular formula of an unknown compound. For example, it is possible to determine that the molecular formula of a compound is $C_6H_{12}O$, but that is still not enough information to draw the structure of the compound. There are many constitutional isomers of $C_6H_{12}O$. IR spectroscopy can tell us whether or not a double bond is present and whether or not an O—H bond is present. But, once again, that is not enough information to draw the structure of the compound. In Chapter 16, we will learn about NMR spectroscopy, which provides even more information. But before we move on to NMR spectroscopy, there is still one piece of information that can be gleaned from the molecular formula. A careful analysis of the molecular formula can often provide us with a list of possible molecular structures. This skill will be important in the next chapter, because the molecular formula alone offers helpful clues about the structure of the compound. To see how this works, let's begin by analyzing the molecular formula of several alkanes.

Compare the structures of the following alkanes, paying special attention to the number of hydrogen atoms attached to each carbon atom.

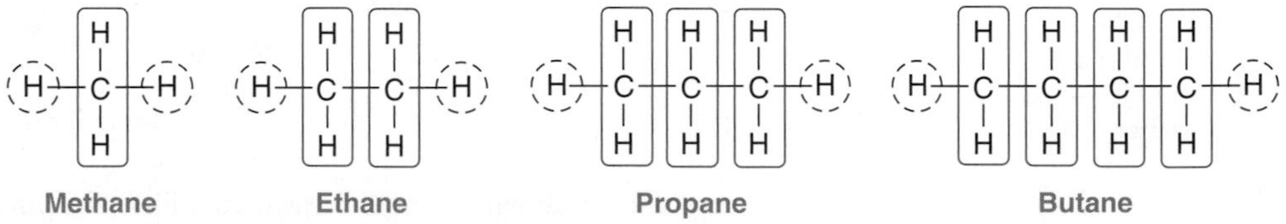

In each case there are two hydrogen atoms on the ends of the structures (circled), and there are two hydrogen atoms on every carbon atom. This can be summarized like this:

$$H—(CH_2)_n—H$$

where n is the number of carbon atoms in the compound. Accordingly, the number of hydrogen atoms will be $2n+2$. In other words, all of the compounds shown have a molecular formula of C_nH_{2n+2}. This is true even for compounds that are branched rather than having a straight chain.

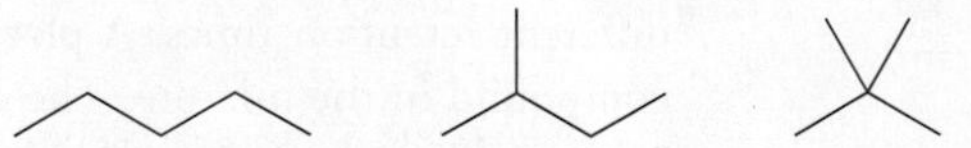

C_5H_{12} C_5H_{12} C_5H_{12}

These compounds are said to be saturated because they possess the maximum number of hydrogen atoms possible, relative to the number of carbon atoms present.

A compound with a π bond (a double or triple bond) will have fewer than the maximum number of hydrogen atoms. Such compounds are said to be unsaturated.

C_5H_{10} C_5H_8

A compound containing a ring will also have fewer than the maximum number of hydrogen atoms, just like a compound with a double bond. For example, compare the structures of 1-hexene and cyclohexane.

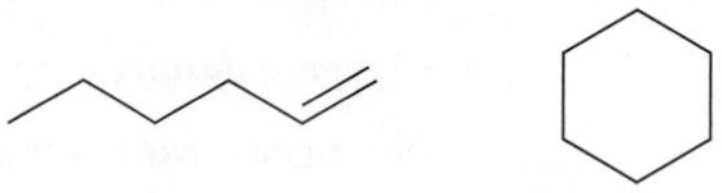

1-Hexene **(C_6H_{12})** **Cyclohexane** **(C_6H_{12})**

Both compounds have molecular formula C_6H_{12} because both are "missing" two hydrogen atoms [6 carbon atoms require $(2 \times 6) + 2 = 14$ hydrogen atoms]. Each of these compounds is said to have one **degree of unsaturation**. The **hydrogen deficiency index (HDI)** is a measure of the number of degrees of unsaturation. A compound is said to have one degree of unsaturation for every two hydrogen atoms that are missing. For example, a compound with molecular formula C_4H_6 is missing four hydrogen atoms (if saturated, it would be C_4H_{10}), so it has two degrees of unsaturation (HDI = 2).

There are several ways for a compound to possess two degrees of unsaturation: two double bonds, two rings, one double bond and one ring, or one triple bond. Let's explore all of these possibilities for C_4H_6 (Figure 15.33).

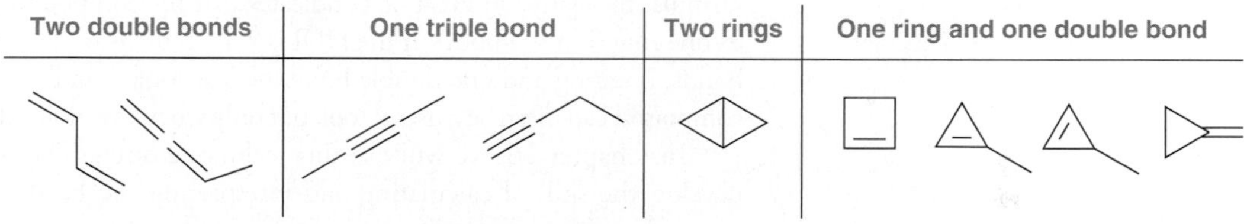

Two double bonds
One triple bond
Two rings
One ring and one double bond

FIGURE 15.33 All possible constitutional isomers of C_4H_6.

With this in mind, let's expand our skills set. Let's explore how to calculate the HDI when other elements are present in the molecular formula.

1. ***Halogens:*** Compare the following two compounds. Notice that chlorine takes the place of a hydrogen atom. Therefore, for purposes of calculating the HDI, treat a halogen as if it were a hydrogen atom. For example, C_4H_9Cl should have the same HDI as C_4H_{10}.

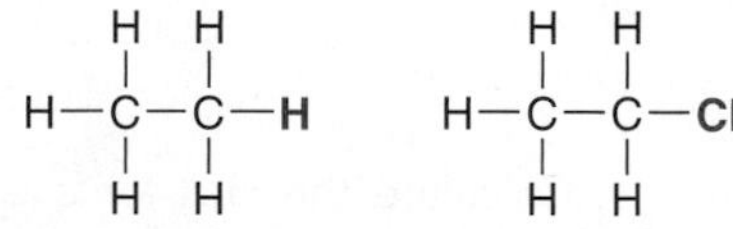

Ethane **Chloroethane**

2. ***Oxygen:*** Compare the following two compounds. Notice that the presence of the oxygen atom does not affect the expected number of hydrogen atoms. Therefore, whenever an oxygen atom appears in the molecular formula, it should be ignored for purposes of calculating the HDI. For example, C_4H_8O should have the same HDI as C_4H_8.

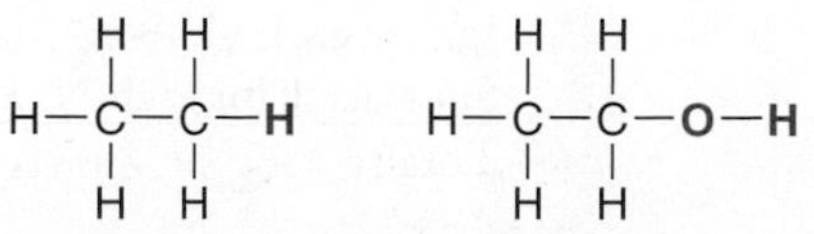

Ethane **Ethanol**

3. ***Nitrogen:*** Compare the following two compounds. Notice that the presence of a nitrogen atom changes the number of expected hydrogen atoms. It gives one more hydrogen atom

than would be expected. Therefore, whenever a nitrogen atom appears in the molecular formula, one hydrogen atom must be subtracted from the molecular formula. For example, C_4H_9N should have the same HDI as C_4H_8.

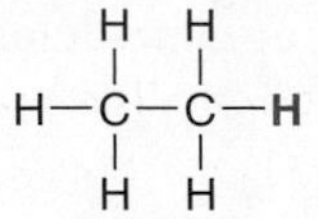

Ethane

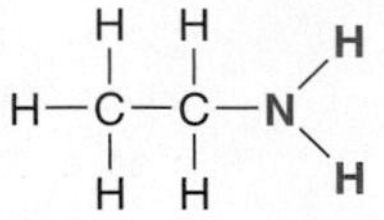

Ethyl Amine

In summary:

- Halogens: ***Add*** one H for each halogen.
- Oxygen: ***Ignore***.
- Nitrogen: ***Subtract*** one H for each N.

These rules will enable you to determine the HDI for most simple compounds. Alternatively, the following formula can be used,

$$\mathrm{HDI} = \frac{1}{2}(2C + 2 + N - H - X)$$

where C is the number of carbon atoms, N is the number of nitrogen atoms, H is the number of hydrogen atoms, and X is the number of halogens. This formula will work for all compounds containing C, H, N, O, and X.

Calculating the HDI is particularly helpful, because it provides clues about the structural features of the compound. For example, an HDI of zero indicates that the compound cannot have any rings or π bonds. That is extremely useful information when trying to determine the structure of a compound, and it is information that is easily obtained by simply analyzing the molecular formula. Similarly, an HDI of 1 indicates that the compound must have either one double bond ***or*** one ring (but not both). If the HDI is 2, then there are a few possibilities: two rings, two double bonds, one ring and one double bond, or one triple bond. Analysis of the HDI for an unknown compound can often be a useful tool, but only when the molecular formula is known with certainty.

In Chapter 16, we will use this technique often. The following exercises are designed to develop the skill of calculating and interpreting the HDI of an unknown compound whose molecular formula is known.

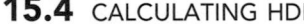

SKILLBUILDER

15.4 CALCULATING HDI

LEARN the skill

Calculate the HDI for a compound with molecular formula $C_4H_8ClNO_2$, and identify the structural information provided by the HDI.

SOLUTION

The calculation is:

STEP 1 Identify the molecular formula of a hydrocarbon (a compound with only C and H) that will have the same HDI.

Number of H's:	8
Add 1 for each Cl:	+1
Ignore each O:	0
Subtract 1 for each N:	−1
Total:	8

This compound will have the same HDI as a compound with molecular formula C_4H_8. To be fully saturated, four carbon atoms would require $(4 \times 2) + 2 = 10$ H's. According to our calculation, two hydrogen atoms are missing, and therefore, this compound has one degree of unsaturation: HDI = 1.

STEP 2
Determine how many hydrogen atoms are missing and assign an HDI.

Alternatively, the following formula can be used.

$$\text{HDI} = \frac{1}{2}(2C + 2 + N - H - X) = \frac{1}{2}(8 + 2 + 1 - 8 - 1) = \frac{2}{2} = 1$$

With one degree of unsaturation, the compound must contain either one ring or one double bond, but not both. The compound cannot have a triple bond, as this would require two degrees of unsaturation.

PRACTICE the skill

15.31 Calculate the degree of unsaturation for each of the following molecular formulas.

(a) C_6H_{10} (b) $C_5H_{10}O_2$ (c) C_5H_9N (d) C_3H_5ClO (e) $C_{10}H_{20}$
(f) $C_4H_6Br_2$ (g) C_6H_6 (h) C_2Cl_6 (i) $C_2H_4O_2$ (j) $C_{100}H_{200}Cl_2O_{16}$

15.32 Identify which two compounds shown here have the same degree of unsaturation.

C_3H_8O $C_3H_5ClO_2$ $C_3H_5NO_2$ C_3H_6

APPLY the skill

15.33 Propose all possible structures for a compound with molecular formula C_4H_8O that exhibits a signal at 1720 cm^{-1} in its IR spectrum.

15.34 Propose all possible structures for a compound with molecular formula C_4H_8O that exhibits a broad signal between 3200 and 3600 cm^{-1} in its IR spectrum and does not contain any signals between 1600 and 1850 cm^{-1}.

15.35 Propose the only possible structure for a compound with molecular formula C_4H_6 that exhibits a signal at 2200 cm^{-1} in its IR spectrum.

15.36 What structural features are present in a compound with molecular formula $C_{10}H_{20}O$?

15.37 Each pair of compounds below has the same number of carbon atoms. Without counting the hydrogen atoms, determine whether the pair of compounds have the same molecular formula. In each case, simply determine the HDI of each compound to make your decision. Then, count the number of hydrogen atoms to see if your analysis was correct.

(a) 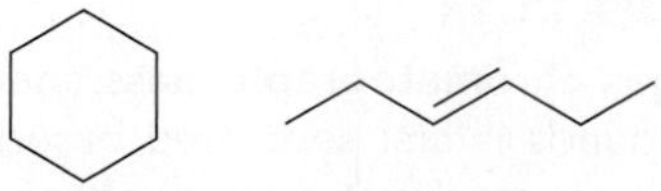 (b) 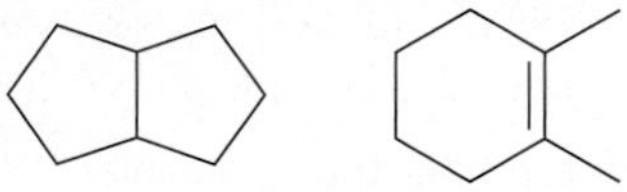 (c) 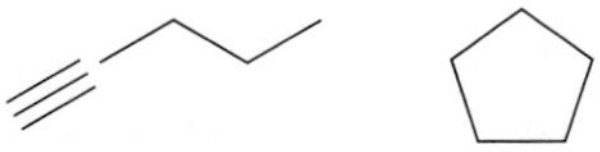

need more **PRACTICE?** ***Try Problems 15.48b, 15.54–15.55***

REVIEW OF CONCEPTS AND VOCABULARY

SECTION 15.1

- **Spectroscopy** is the study of the interaction between light and matter.
- The **wavelength** of light describes the distance between adjacent peaks of an oscillating field, while the **frequency** describes the number of wavelengths that pass a particular point in space per unit time.
- The energy of each **photon** is determined by its frequency. The range of all possible frequencies is known as the **electromagnetic spectrum**.
- Molecules can store energy in a variety of ways. Each of these forms of energy is quantized.
- The difference in energy (ΔE) between vibrational energy levels is determined by the nature of the bond. If a photon of light possesses exactly this amount of energy, the bond can absorb the photon to promote a **vibrational excitation**.

SECTION 15.2

- There are many different kinds of vibrational excitation, including **stretching** and **bending**.
- IR spectroscopy can be used to identify which functional groups are present in a compound.
- An **absorption spectrum** is a plot that measures the percent transmittance as a function of frequency.
- The location of each signal in an IR spectrum is reported in terms of a frequency-related unit called **wavenumber**.

SECTION 15.3

- The wavenumber of each signal is determined primarily by bond strength and the masses of the atoms sharing the bond.
- The **diagnostic region** contains all signals that arise from double bonds, triple bonds, and X—H bonds.
- The **fingerprint region** contains signals resulting from the vibrational excitation of most single bonds (stretching and bending).
- A **conjugated** C=O bond will produce a lower energy signal than an unconjugated C=O bond.

SECTION 15.4

- The intensity of a signal is dependent on the dipole moment of the bond giving rise to the signal.
- C=O bonds produce strong signals in an IR spectrum, while C=C bonds often produce fairly weak signals.
- Symmetrical C=C bonds do not produce signals. The same is true for symmetrical triple bonds.

SECTION 15.5

- Concentrated alcohols give rise to broad signals, while dilute alcohols give rise to narrow signals.
- Primary amines exhibit two signals resulting from **symmetric stretching** and **asymmetric stretching**.

SECTIONS 15.6 AND 15.7

- When analyzing an IR spectrum, look for double bonds, triple bonds, and X—H bonds.
- Every signal has three characteristics—wavenumber, intensity, and shape. Analyze all three.
- When looking for X—H bonds, identify whether any signals appear to the left of a line drawn at 3000 cm^{-1}.

SECTION 15.8

- **Mass spectrometry** is used to determine the molecular weight and molecular formula of a compound.
- In a **mass spectrometer**, a compound is converted into ions, which are then separated by a magnetic field.
- **Electron impact ionization** (**EI**) involves bombarding the compound with high-energy electrons, generating a radical cation that is symbolized by $(M)^{+\bullet}$ and is called the **molecular ion**, or the **parent ion**.
- The molecular ion is often very unstable and susceptible to **fragmentation**.
- Only the molecular ion and the cationic fragments are deflected, and they are then separated by their **mass-to-charge ratio** (*m/z*).
- A **mass spectrum** shows the relative abundance of each cation detected.
- The tallest peak in a mass spectrum is assigned a relative value of 100% and is called the **base peak**.

SECTION 15.9

- The $(M)^{+\bullet}$ peak may be weak or entirely absent if it is particularly susceptible to fragmentation.
- The $(M)^{+\bullet}$ peak indicates the molecular weight of the molecule.
- According to the **nitrogen rule**, an odd molecular weight indicates an odd number of nitrogen atoms, while an even molecular weight indicates either the absence of nitrogen or an even number of nitrogen atoms.

SECTION 15.10

- The relative heights of the $(M)^{+\bullet}$ peak and the $(M+1)^{+\bullet}$ peak indicate the number of carbon atoms.

SECTION 15.11

- For compounds containing a bromine atom, the $(M+2)^{+\bullet}$ peak will be the same height as the $(M)^{+\bullet}$ peak.
- For compounds containing a chlorine atom, the $(M+2)^{+\bullet}$ peak will be one-third as tall as the $(M)^{+\bullet}$ peak.

SECTION 15.12

- A signal at M–15 indicates the loss of a methyl group; a signal at M–29 indicates the loss of an ethyl group.
- The likelihood of fragmentation increases with the stability of the carbocation formed.

SECTION 15.13

- The molecular formula of a compound can be determined with **high-resolution mass spectrometry**.
- The **atomic mass unit** (**amu**) is equivalent to 1 g divided by Avogadro's number.
- The **standard atomic weight** is a weighted average that takes into account relative isotopic abundance.

SECTION 15.14

- In a **gas chromatograph–mass spectrometer**, a mixture of compounds is first separated based on their boiling points and affinity for the **stationary phase**. Each compound is then analyzed individually.
- Each compound in the mixture generally exhibits a unique **retention time**, which is plotted on a **chromatogram.**

SECTION 15.15

- **Electrospray ionization (ESI)** is used most often for large molecules such as proteins and nucleic acids.

SECTION 15.16

- Saturated alkanes have a molecular formula of the form C_nH_{2n+2}.
- A compound possessing a π bond is **unsaturated**.
- Each double bond and each ring represents one **degree of unsaturation**.
- The **hydrogen deficiency index** (**HDI**) is a measure of the number of degrees of unsaturation.

KEY TERMINOLOGY

absorption spectrum 676
asymmetric stretching 687
atomic mass unit (amu) 706
base peak 696
bending 674
chromatogram 707
conjugated 680
degree of unsaturation 709
diagnostic region 678
electromagnetic spectrum 672
electron impact ionization (EI) 695
electrospray ionization (ESI) 708
fingerprint region 678
fragmentation 695
frequency 672
GC-MS 707
high resolution mass spectrometry 705
hydrogen deficiency index (HDI) 709
mass spectrometer 695
mass spectrometry 695
mass spectrum 696
mass-to-charge ratio (*m/z*) 696
molecular ion 695
nitrogen rule 698
parent ion 695
photons 672
retention time 707
spectroscopy 671
standard atomic weight 706
stationary phase 707
stretching (of bond) 674
symmetric stretching 687
vibrational excitation 673
wavelength 672
wavenumber 676

SKILLBUILDER REVIEW

15.1 ANALYZING AN IR SPECTRUM

STEP 1 Look for double bonds between 1600 and 1850 cm^{-1}.

Guidelines:

- C=O bonds produce strong signals.
- C=C bonds generally produce weak signals. Symmetrical C=C bonds do not appear at all.
- The exact position of a signal indicates subtle factors that affect bond stiffness, such as resonance.

STEP 2 Look for triple bonds between 2100 and 2300 cm^{-1}.

Guidelines:

- Symmetrical triple bonds do not produce signals.

STEP 3 Look for X—H bonds between 2750 and 4000 cm^{-1}.

Guidelines:

- Draw a line at 3000, and look for vinylic or acetylenic C—H bonds to the left of the line.
- The shape of an O—H signal is affected by concentration (due to H bonding).
- Primary amines exhibit two N—H signals (symmetric and asymetric stretching).

Try Problems **15.12, 15.13, 15.42, 15.57**

15.2 DISTINGUISHING BETWEEN TWO COMPOUNDS USING IR SPECTROSCOPY

STEP 1 Work methodically through the expected IR spectrum of each compound.

STEP 2 Determine if any signals will be present in one compound but absent in the other.

STEP 3 For each expected signal, compare for any possible differences in wavenumber, intensity, or shape.

Try Problems **15.14–15.18, 15.38–15.41, 15.48d**

15.3 USING THE RELATIVE ABUNDANCE OF THE $(M+1)^{+\bullet}$ PEAK TO PROPOSE A MOLECULAR FORMULA

STEP 1 Determine the number of carbon atoms in the compound by analyzing the relative abundance of the M + 1 peak:

$$\frac{\left(\dfrac{\text{Abundance of } (M+1)^{+\bullet} \text{ peak}}{\text{Abundance of } (M)^{+\bullet} \text{ peak}}\right) \times 100\%}{1.1\ \%}$$

STEP 2 Analyze the mass of the molecular ion to determine if any heteroatoms (such as oxygen or nitrogen) are present.

Try Problems **15.21, 15.22, 15.45, 15.47**

15.4 CALCULATING HDI

STEP 1 Rewrite the molecular formula "as if" the compound had no elements other than C and H, using the following rules:

- Add one H for each halogen.
- Ignore all oxygen atoms.
- Subtract one H for each nitrogen.

STEP 2 Determine whether any H's are missing. Every two H's represents one degree of unsaturation:

$$C_4H_9Cl \longrightarrow C_4H_{10} \longrightarrow \text{HDI} = 0$$

$$C_4H_8O \longrightarrow C_4H_8 \longrightarrow \text{HDI} = 1$$

$$C_4H_9N \longrightarrow C_4H_8 \longrightarrow \text{HDI} = 1$$

Try Problems **15.31–15.37, 15.48b, 15.54, 15.55**

PRACTICE PROBLEMS

15.38 All of the following compounds absorb IR radiation in the range between 1600 and 1850 cm^{-1}. In each case, identify the specific bond(s) responsible for the absorption(s), and predict the approximate wavenumber of absorption for each of those bonds.

(a) 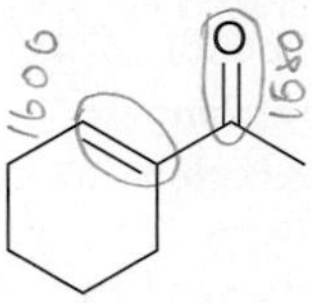
(b) 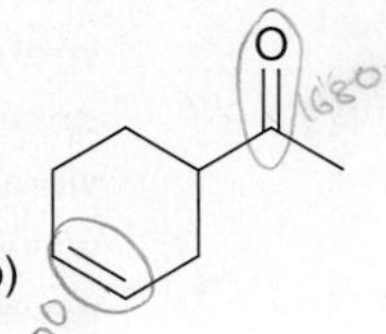
(c) 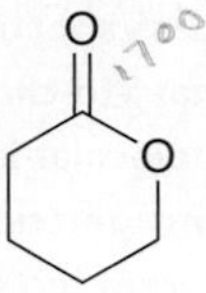
(d) 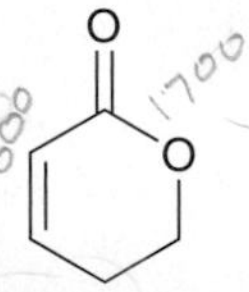
(e) 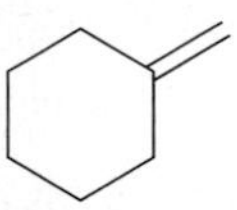

15.39 Rank each of the bonds identified in order of increasing wavenumber.

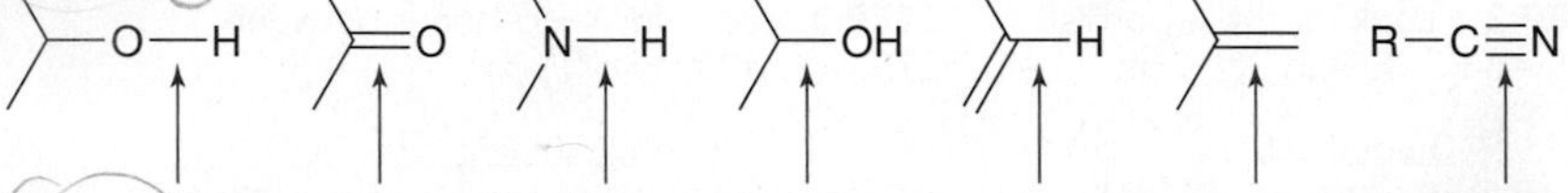

15.40 Identify the signals you would expect in the diagnostic region of the IR spectrum for each of the following compounds.

(a) 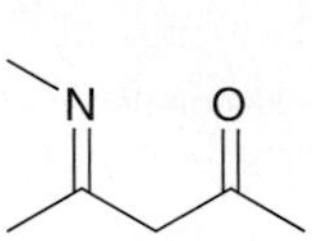
(b) 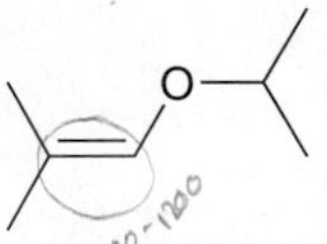
(c) 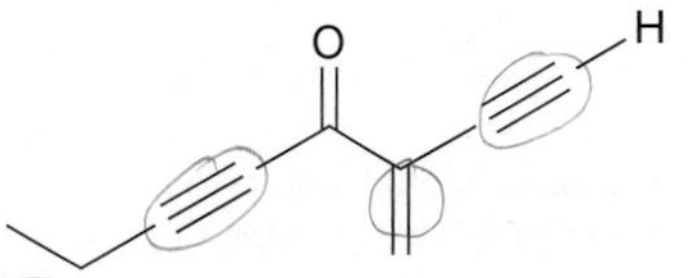
(d) 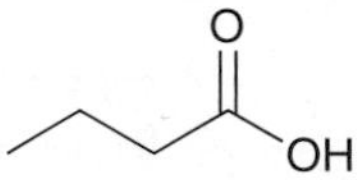

15.41 Identify how IR spectroscopy might be used to monitor the progress of each of the following reactions.

(a) 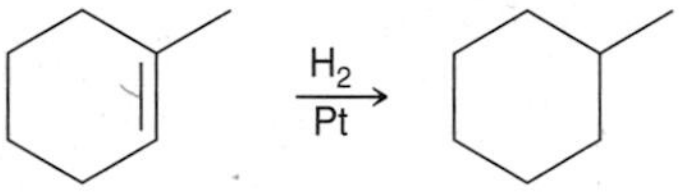
(b) 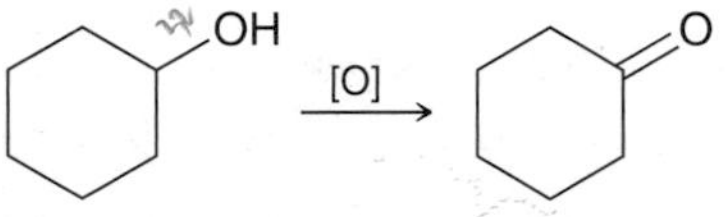
(c) 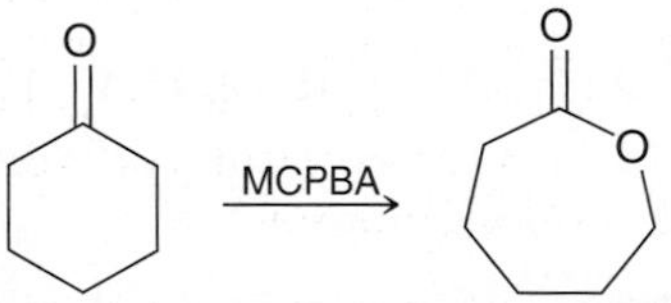
(d) 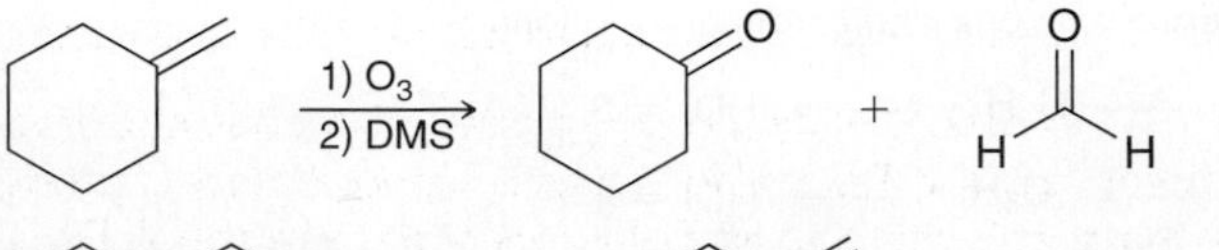
(e) 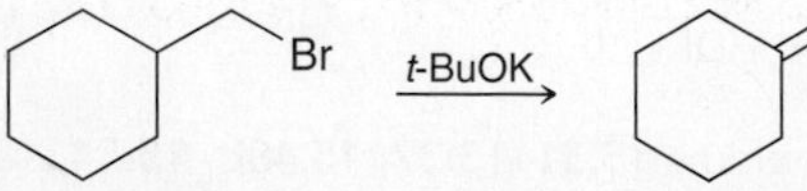

15.42 Identify the characteristic signals that you would expect in the diagnostic region of an IR spectrum of each of the following compounds.

(a) 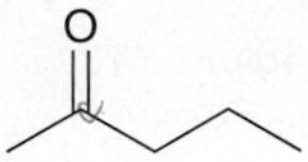
(b) 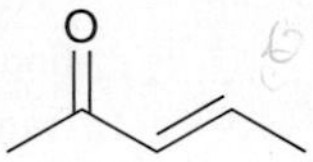
(c) 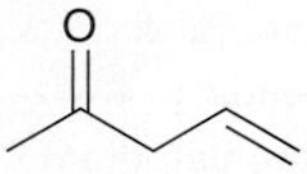
(d) 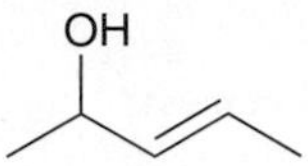
(e) 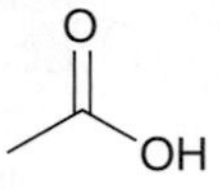
(f) 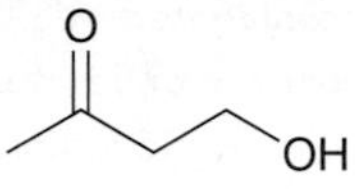

15.43 Identify the molecular formula for each of the following compounds, and then predict the mass of the expected molecular ion in the mass spectrum of each compound.

(a) 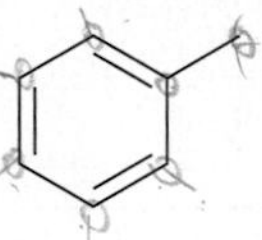
(b) 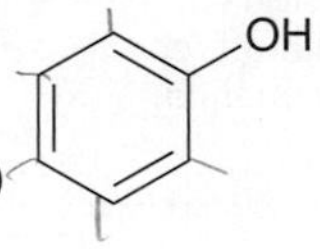
(c) 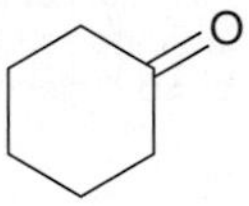
(d) 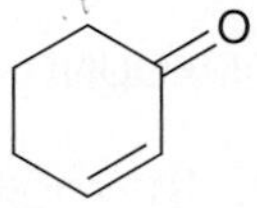
(e) 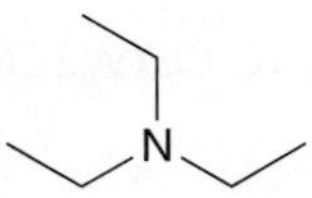

15.44 Propose a molecular formula for a compound that has one degree of unsaturation and a mass spectrum that displays a molecular ion signal at $m/z = 86$.

15.45 The mass spectrum of an unknown hydrocarbon exhibits an (M+1) peak that is 10% as tall as the molecular ion peak. Identify the number of carbon atoms in the unknown compound.

15.46 Match each compound with the appropriate spectrum.

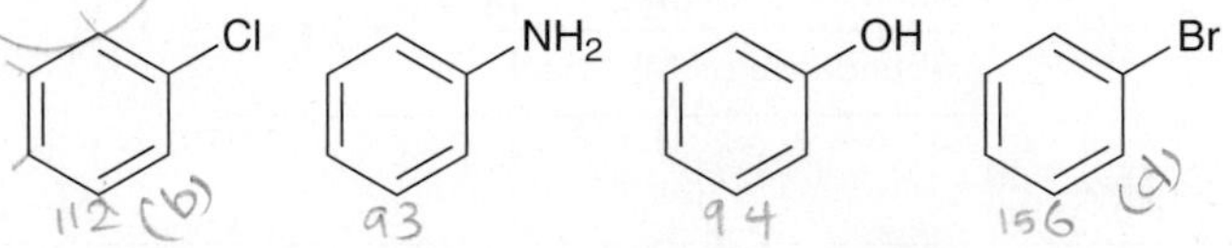

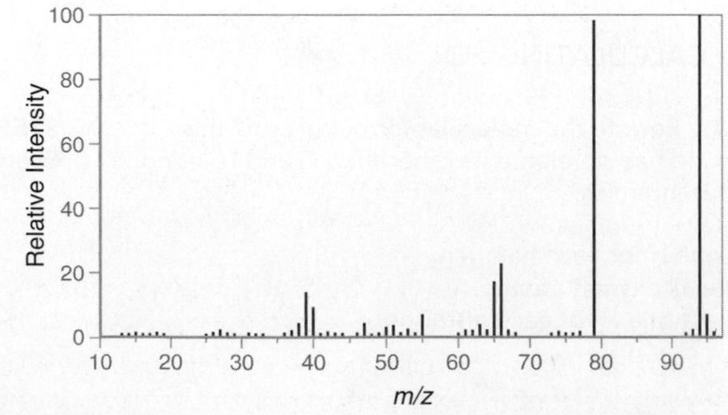

(a)

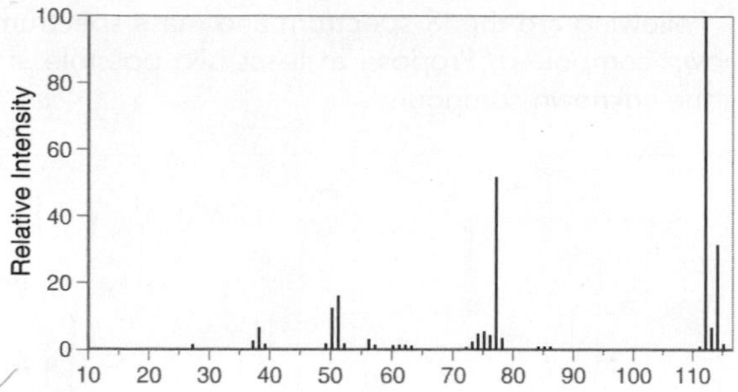

(b)

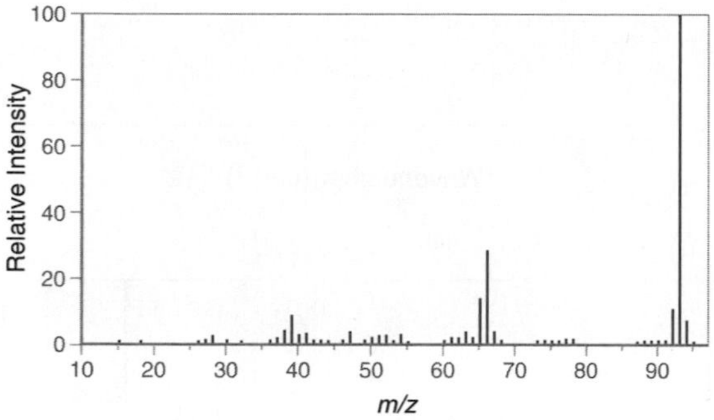

(c)

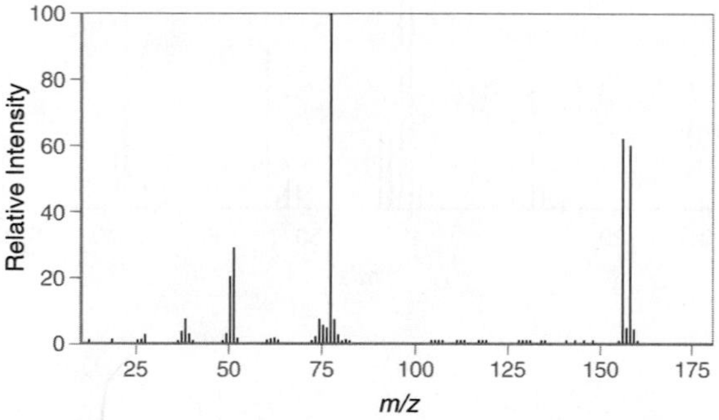

(d)

15.47 The sex attractant of the codling moth gives an IR spectrum with a broad signal between 3200 and 3600 cm^{-1} and two signals between 1600 and 1700 cm^{-1}. In the mass spectrum of this compound, the molecular ion peak appears at $m/z = 196$, and the relative abundance of the molecular ion and the (M+1) peak are 27.2% and 3.9%, respectively.

(a) What functional groups are present in this compound?
(b) How many carbon atoms are present in the compound?
(c) Based on the information given, propose a molecular formula for the compound.

15.48 Compare the structures of cyclohexane and 2-methyl-2-pentene.

(a) What is the molecular formula of each compound?
(b) What is the HDI of each compound?
(c) Can high-resolution mass spectrometry be used to distinguish between these compounds? Explain.
(d) How would you differentiate between these two compounds using IR spectroscopy?

15.49 The mass spectrum of 1-ethyl-1-methylcyclohexane shows many fragments, with two in very large abundance. One appears at $m/z = 111$ and the other appears at $m/z = 97$. Identify the structure of each of these fragments.

15.50 The mass spectrum of 2-bromopentane shows many fragments.

(a) One fragment appears at M–79. Would you expect a signal at M–77 that is equal in height to the M–79 peak? Explain.
(b) A fragment appears at M–15. Would you expect a signal at M–13 that is equal in height to the M–15 peak? Explain.
(c) One fragment appears at M–29. Would you expect a signal at M–27 that is equal in height to the M–29 peak? Explain.

15.51 When treated with a strong base, 2-bromo-2,3-dimethylbutane will undergo an elimination reaction to produce two products. The choice of base (ethoxide vs. *tert*-butoxide) will determine which of the two products predominates. Draw both products and determine how you could distinguish between them using IR spectroscopy.

15.52 Propose a molecular formula that fits the following data.

(a) A hydrocarbon (C_xH_y) with a molecular ion peak at $m/z = 66$
(b) A compound that absorbs IR radiation at 1720 cm^{-1} and exhibits a molecular ion peak at $m/z = 70$

15.53 The following mass spectrum is for octane.

(a) Which peak represents the molecular ion?
(b) Which peak is the base peak?
(c) Draw the structure of the fragment that produces the base peak.

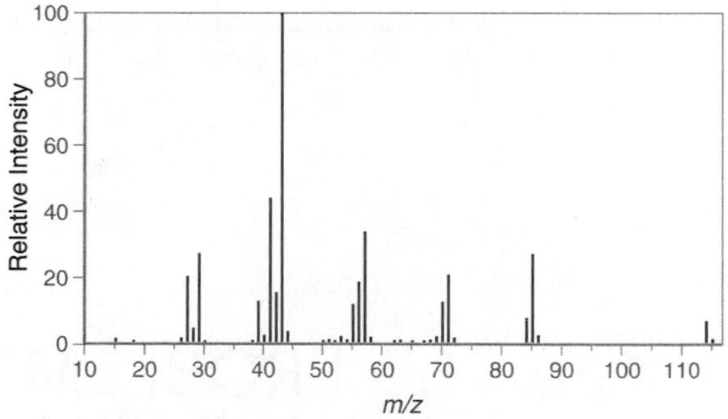

15.54 Calculate the HDI for each molecular formula.

(a) C_4H_6 (b) C_5H_8 (c) $C_{40}H_{78}$ (d) $C_{72}H_{74}$
(e) $C_6H_6O_2$ (f) $C_7H_9NO_2$ (g) $C_8H_{10}N_2O$ (h) $C_5H_7Cl_3$
(i) C_6H_5Br (j) $C_6H_{12}O_6$

15.55 Propose two possible structures for a compound with molecular formula C_5H_8 that produces an IR signal at 3300 cm^{-1}.

15.56 Limonene is a hydrocarbon found in the peels of lemons and contributes significantly to the smell of lemons. Limonene has a molecular ion peak at $m/z = 136$ in its mass spectrum, and it has two double bonds and one ring in its structure. What is the molecular formula of limonene?

15.57 Explain how you would use IR spectroscopy to distinguish between *trans*-3-hexene and 2,3-dimethyl-2-butene.

15.58 A dilute solution of 1,3-pentanediol does not produce the characteristic IR signal for a dilute alcohol. Rather, it produces a signal that is characteristic of a concentrated alcohol. Explain.

15.59 Following are the IR spectrum and mass spectrum of an unknown compound. Propose at least two possible structures for the unknown compound.

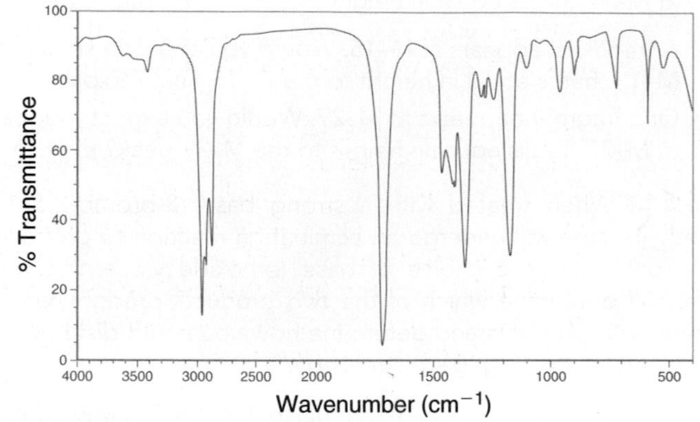

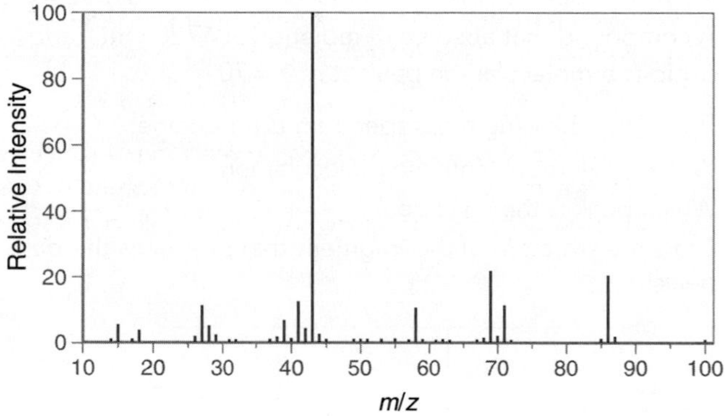

15.60 Following are the IR spectrum and mass spectrum of an unknown compound. Propose at least two possible structures for the unknown compound.

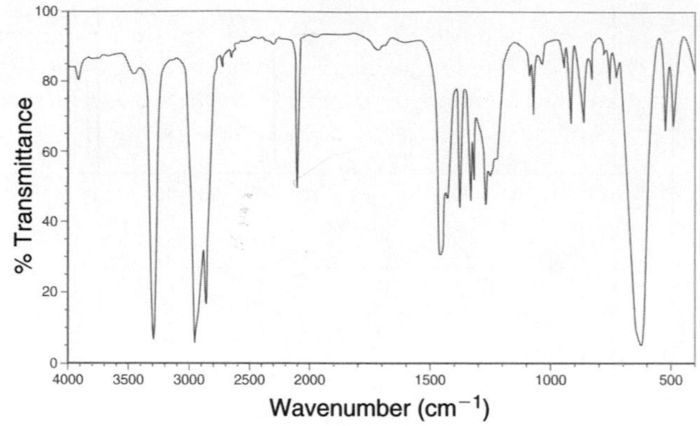

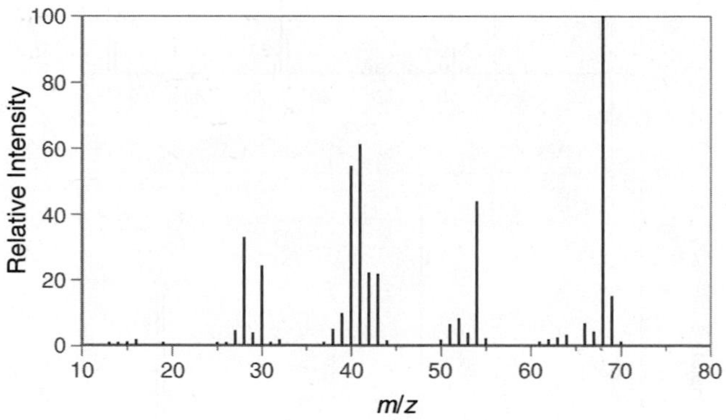

INTEGRATED PROBLEMS

15.61 Consider the following sequence of reactions:

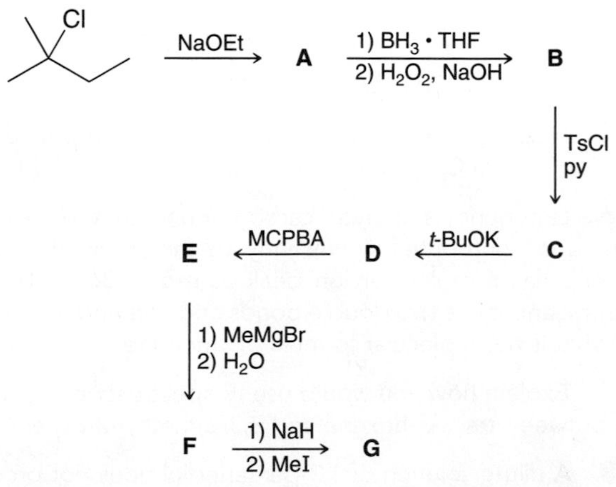

(a) Explain how you could use IR spectroscopy to differentiate between compounds **F** and **G**.

(b) Explain how you could use IR spectroscopy to differentiate between compounds **D** and **E**.

(c) If you wanted to distinguish between compounds **B** and **F**, would it be more suitable to use IR spectroscopy or mass spectrometry? Explain.

(d) Would mass spectrometry be helpful for distinguishing between compounds **A** and **D**? Explain.

15.62 There are five constitutional isomers with molecular formula C_4H_8. One of the isomers exhibits a particularly strong signal at M–15 in its mass spectrum. Identify this isomer, and explain why the signal at M–15 is so strong.

15.63 There are four isomers with molecular formula C_4H_9Cl. Only one of these isomers (compound **A**) has a chirality center. When compound **A** is treated with sodium ethoxide, three products are formed: compounds **B**, **C**, and **D**. Compounds **B** and **C** are diastereomers, with compound **B** being the less stable diastereomer. Do you expect compound **D** to exhibit a signal at approximately 1650 cm^{-1} in its IR spectrum? Explain.

CHALLENGE PROBLEMS

15.64 Chloramphenicol is an antibiotic isolated from the *Streptomyces venezuelae* bacterium. Predict the expected isotope pattern in the mass spectrum of this compound (the relative heights of the molecular ion peak and surrounding peaks).

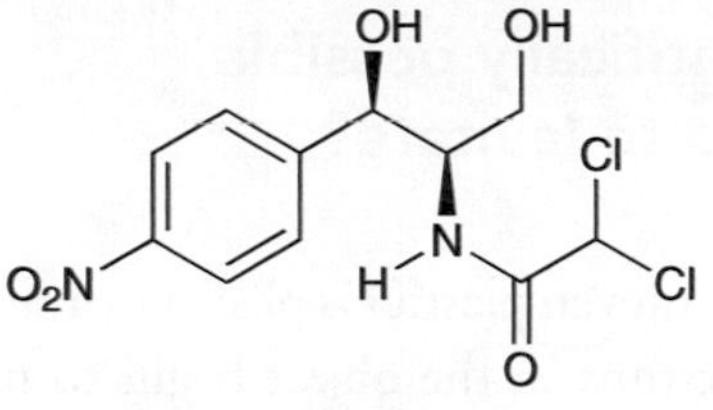

Chloramphenicol

15.65 Ephedrine is a bronchodilator and decongestant obtained from the Chinese plant *Ephedra sinica*. A concentrated solution of ephedrine gives an IR spectrum with a broad signal between 3200 and 3600 cm^{-1}. An IR spectrum of a dilute solution of ephedrine is very similar to the IR spectrum of the concentrated solution. That is, the broad signal between 3200 and 3600 cm^{-1} is not transformed into narrow signals, as we might expect for a compound containing an OH group. Explain.

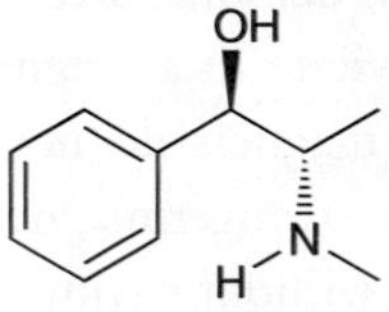

Ephedrine

15.66 Predict the expected isotope pattern in the mass spectrum of a compound with molecular formula $C_{90}H_{180}Br_2$.

15.67 Esters contain two C—O bonds and therefore will produce two separate stretching signals in the fingerprint region of an IR spectrum. One of these signals typically appears at approximately 1000 cm^{-1}, while the other appears at approximately 1300 cm^{-1}. Predict which of the two C—O bonds produces the higher energy signal. Using principles that we learned in this chapter (factors that affect the wavenumber of absorption), offer two separate explanations for your choice.

15.68 Treating 1,2-cyclohexanediol with concentrated sulfuric acid yields a product with molecular formula $C_6H_{10}O$. An IR spectrum of the product exhibits a strong signal at 1720 cm^{-1}. Identify the structure of the product, and show a mechanism for its formation.

15.69 Draw the expected isotope pattern that would be observed in the mass spectrum of CH_2BrCl. In other words, predict the relative heights of the peaks at M, M + 2, and M + 4.

16

Nuclear Magnetic Resonance Spectroscopy

DID YOU EVER WONDER . . .

whether it is scientifically possible for a human being to levitate?

When any object (even plastic) is placed in a strong magnetic field, the electrons in the object begin to move in a way that produces a small magnetic field that opposes the external magnetic field. As a result, the object repels, and is repelled by, the external magnetic field. This weak effect, called diamagnetism, will cause objects to levitate in a strong magnetic field. Larger objects require a larger external magnetic field to induce levitation, but in theory, any object will levitate if it is placed in a strong enough magnetic field.

The strongest magnetic fields currently available are capable of causing small objects to levitate. In 1997, researchers at Radboud University (in the Netherlands) demonstrated that hazelnuts, strawberries, and even frogs will levitate in a magnetic field of 16 tesla. Unfortunately, our strongest magnets do not yet allow a human being to levitate. But in theory, stronger magnets would enable even people to enjoy weightlessness without having to travel to space.

In this chapter, we will see the role that diamagnetism plays in nuclear magnetic resonance (NMR) spectroscopy, which provides more structural information than any other form of spectroscopy. We will also learn how NMR spectroscopy is used as a powerful tool for structure determination.

DO YOU REMEMBER?

Before you go on, be sure you understand the following topics.
If necessary, review the suggested sections to prepare for this chapter:

- Quantum Mechanics (Section 1.6)
- Stereoisomeric Relationships: Enantiomers and Diastereomers (Section 5.5)
- Symmetry and Chirality (Section 5.6)
- Introduction to Spectroscopy (Section 15.1)
- Hydrogen Deficiency Index (Section 15.16)

WILEY PLUS Visit www.wileyplus.com to check your understanding and for valuable practice.

16.1 Introduction to NMR Spectroscopy

Nuclear magnetic resonance (NMR) spectroscopy is arguably the most powerful and broadly applicable technique for structure determination available to organic chemists. It provides the most information about molecular structure, and in some cases, the structure of a compound can be determined using only NMR spectroscopy. In practice, the structures of complicated molecules are determined through a combination of techniques that include NMR and IR spectroscopy and mass spectrometry.

NMR spectroscopy involves the study of the interaction between electromagnetic radiation and the nuclei of atoms. A wide variety of nuclei can be studied using NMR spectroscopy, including ^{1}H, ^{13}C, ^{15}N, ^{19}F, and ^{31}P. In practice, ^{1}H NMR spectroscopy and ^{13}C NMR spectroscopy are used most often by organic chemists, because hydrogen and carbon are the primary constituents of organic compounds. Analysis of an NMR spectrum provides information about how the individual carbon and hydrogen atoms are connected to each other in a molecule. This information enables us to determine the carbon-hydrogen framework of a compound, much the way puzzle pieces can be assembled to form a picture.

A nucleus with an odd number of protons and/or an odd number of neutrons possesses a quantum mechanical property called *nuclear spin,* and it can be probed by an NMR spectrometer. Consider the nucleus of a hydrogen atom, which consists of just one proton and therefore has a spin. A spinning proton can be viewed as a rotating sphere of charge, which generates a magnetic field, called a **magnetic moment**. The magnetic moment of a spinning proton is similar to the magnetic field produced by a bar magnet (Figure 16.1). The nucleus of a ^{12}C atom has an even number of

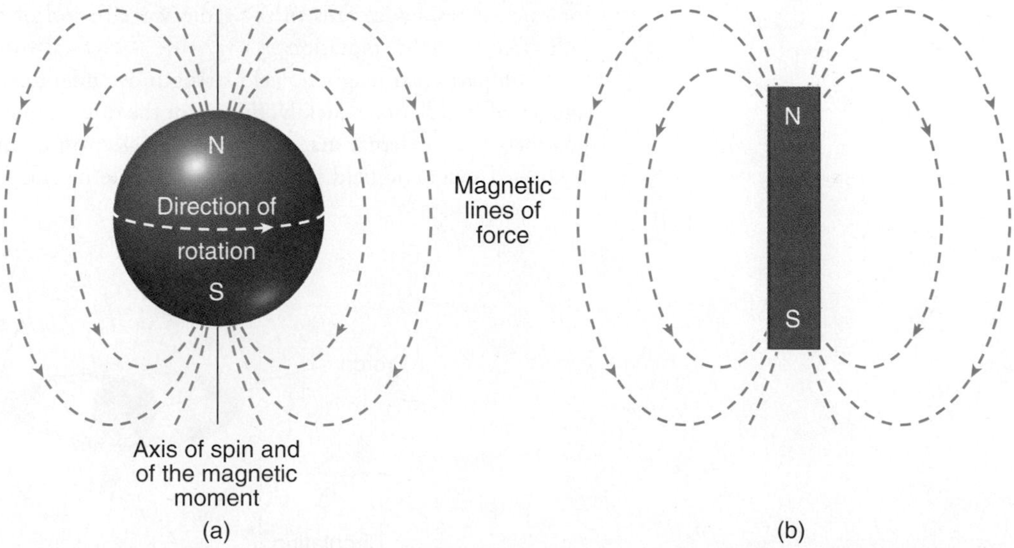

FIGURE 16.1
(a) The magnetic moment of a spinning proton.
(b) The magnetic field of a bar magnet.

protons and an even number of neutrons and therefore does not possess this property. In contrast, the nucleus of ^{13}C has an odd number of neutrons and therefore exhibits spin. Our study of NMR spectroscopy will begin with ^{1}H NMR spectroscopy and will conclude with ^{13}C NMR spectroscopy.

When the nucleus of a hydrogen atom (a proton) is subjected to an external magnetic field, the interaction between the magnetic moment and the magnetic field is quantized, and the magnetic moment must align either with the field or against the field (Figure 16.2). A proton aligned with the field is said to occupy the alpha (α) spin state, while a proton aligned against the field is

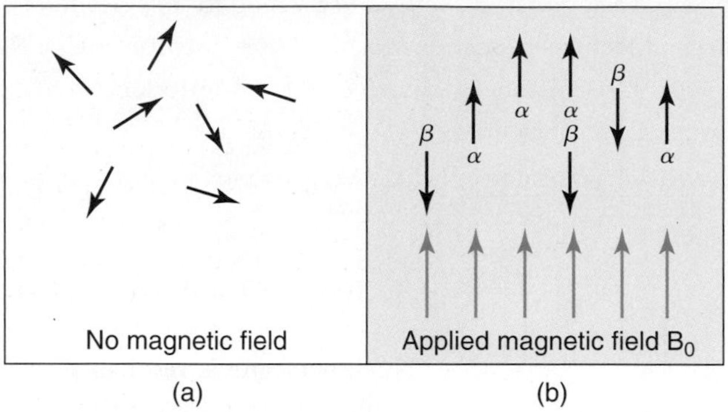

FIGURE 16.2
The orientation of the magnetic moments of protons in
(a) the absence of an external magnetic field or
(b) the presence of an external magnetic field.

said to occupy the beta (β) spin state. The two spin states are not equivalent in energy, and there is a quantifiable difference in energy (ΔE) between them (Figure 16.3).

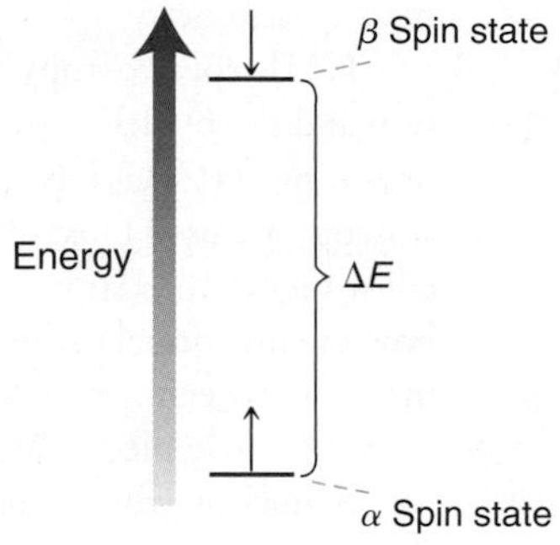

FIGURE 16.3
The energy gap between the alpha and beta spin states as the result of an applied, external magnetic field.

WATCH OUT
Don't confuse the term "resonance" used here with resonance structures of compounds. These are two different concepts.

When a nucleus occupying the α spin state is subjected to electromagnetic radiation, an absorption can take place if the energy of the photon is equivalent to the energy gap between the spin states. The absorption causes the nucleus to *flip* to the β spin state, and the nucleus is said to be in **resonance** with the external magnetic field; thus the term *nuclear magnetic resonance*. When a strong magnetic field is employed, the frequency of radiation typically required for nuclear resonance falls in the radio wave region of the electromagnetic spectrum [called radio frequency (rf) radiation].

At a particular magnetic field strength, we might expect all nuclei to absorb the same frequency of rf radiation. Luckily, this is not the case, as nuclei are surrounded by electrons. In the presence of an external magnetic field, the electron density circulates, which produces a local (induced) magnetic field that opposes the external magnetic field (Figure 16.4). This effect,

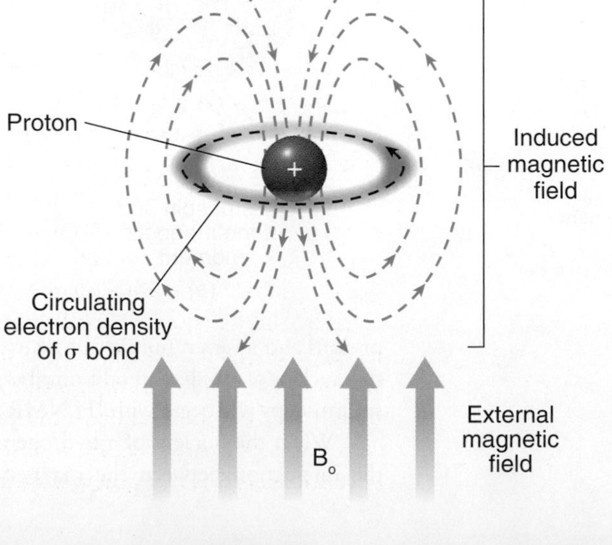

FIGURE 16.4
The induced magnetic field that is generated as a result of the motion of electrons that are surrounding the proton.

called **diamagnetism**, was discussed in the chapter opener. All materials possess diamagnetic properties, because all materials contain electrons. This effect is extremely important for NMR spectroscopy. Without this effect, all protons would absorb the same frequency of rf radiation, and NMR spectroscopy would not provide us with any useful information.

When electron density circulates around a proton, the induced magnetic field has a small but important effect on the proton. The proton is now subjected to two magnetic fields—the strong, external magnetic field and the weak induced magnetic field established by the circulating electron density. The proton therefore experiences a net magnetic field strength that is slightly smaller than the external magnetic field. The proton is said to be **shielded** by the electrons.

Not all protons occupy identical electronic environments. Some protons are surrounded by more electron density and are more shielded, while other protons are surrounded by less electron density and are less shielded, or **deshielded**. As a result, protons in different electronic environments will exhibit a different energy gap between the α and β spin states and will therefore absorb different frequencies of rf radiation. This allows us to probe the electronic environment of each hydrogen atom in a molecule.

16.2 Acquiring a ^{1}H NMR Spectrum

Magnetic Field Strength

We have seen that NMR spectroscopy requires a strong magnetic field as well as a source of rf radiation. The magnetic field establishes an energy gap (ΔE) between spin states, which enables the nuclei to absorb rf radiation. The magnitude of this energy gap depends on the strength of the imposed external magnetic field (Figure 16.5). The energy gap increases with increasing magnetic field strength.

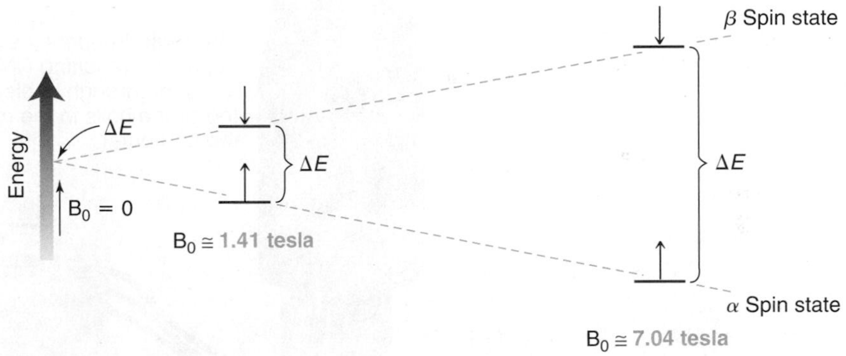

FIGURE 16.5 The relationship between the strength of the magnetic field and the energy gap between the alpha and beta spin states.

With a magnetic field strength of 1.41 tesla, all of the protons in organic compounds will resonate over a narrow range of frequencies near 60 MHz (60,000,000 Hz). If, however, a magnetic field strength of 7.04 tesla is employed, all of the protons in organic compounds will resonate over a range of frequencies near 300 MHz (300,000,000 Hz). In other words, the strength of the magnetic field determines the range of frequencies that must be used. An NMR spectrometer that utilizes a magnetic field strength of 7.04 tesla is said to have an **operating frequency** of 300 MHz. Later in this chapter we will see the advantage of using an NMR spectrometer with a higher operating frequency.

BY THE WAY
Magnetic field strengths are often measured in gauss; 10,000 gauss = 1 tesla.

The strong magnetic fields employed in NMR spectroscopy are produced by passing a current of electrons through a loop composed of superconducting materials. These materials offer very little resistance to the electric current, allowing for large magnetic fields to be produced. The superconducting materials only maintain their properties at extremely low temperatures (just a few degrees above absolute zero) and must therefore be kept in a very low temperature container. Most spectrometers employ three chambers to achieve this low temperature. The innermost chamber contains liquid helium, which has a boiling point of 4.3 K. Surrounding the liquid helium is a chamber containing liquid nitrogen (boiling point 77 K). The third and outermost chamber is a vacuum that minimizes heat transfer from the environment. Any heat developed in the system is transferred from the liquid helium to the liquid nitrogen. In this way, the innermost chamber can be maintained at 4 degrees above absolute zero. This extremely cold environment enables the use of superconductors that generate the large magnetic fields required in NMR spectroscopy.

NMR Spectrometers

The previous generation of spectrometers, called **continuous-wave (CW) spectrometers**, held the magnetic field constant and slowly swept through a range of rf frequencies, monitoring which frequencies were absorbed. Alternatively, these devices could produce the same result by holding the frequency of rf radiation constant and slowly increasing the magnetic field strength, while monitoring which field strengths produced a signal. Both techniques work, but CW spectrometers are rarely used anymore. They have been replaced by pulsed **Fourier-transform NMR (FT-NMR)**.

In an FT-NMR spectrometer (Figure 16.6), the sample is irradiated with a short pulse that covers the entire range of relevant rf frequencies. All protons are excited simultaneously and then begin to return (or decay) to their original spin states. As each type of proton decays, it releases energy in a particular way, generating an electrical impulse in a receiver coil. The receiver coil records a complex signal, called a **free induction decay** (FID), which is a combination of all of the electrical impulses generated by each type of proton. The FID is then converted into a spectrum via a mathematical technique called a Fourier transform. Since each FID is acquired in 1–2 seconds, it is possible to acquire hundreds of FIDs in just a few minutes, and the FIDs can be averaged. Later in the chapter, we will see that signal averaging is the only practical way to produce a ^{13}C NMR spectrum.

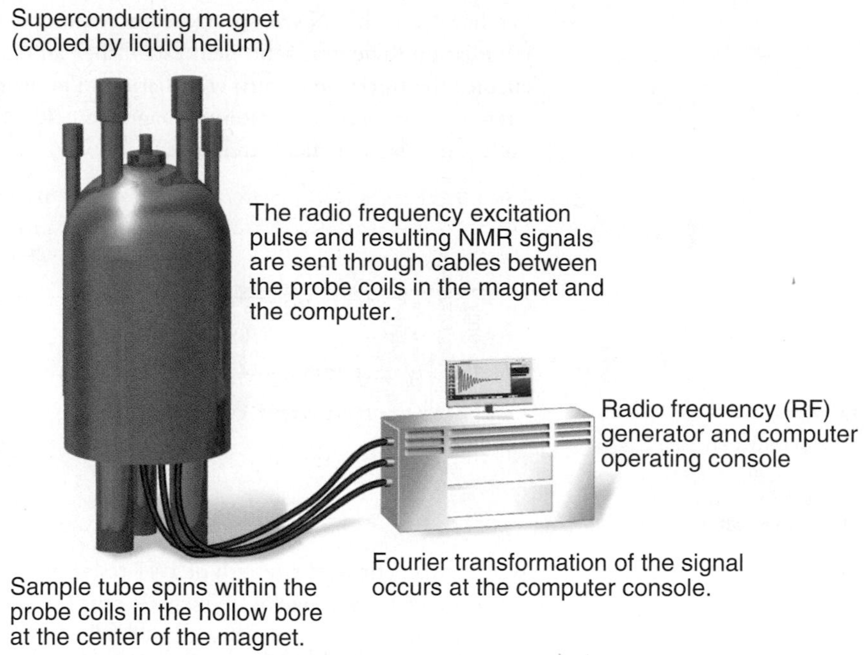
Superconducting magnet
(cooled by liquid helium)
The radio frequency excitation pulse and resulting NMR signals are sent through cables between the probe coils in the magnet and the computer.
Radio frequency (RF) generator and computer operating console
Fourier transformation of the signal occurs at the computer console.
Sample tube spins within the probe coils in the hollow bore at the center of the magnet.

FIGURE 16.6
An FT-NMR spectrometer.

Preparing the Sample

In order to acquire a ^{1}H NMR (called a "proton NMR") spectrum of a compound, the compound must first be dissolved in a solvent and placed in a narrow glass tube, which is then inserted into the NMR spectrometer. If the solvent itself has protons, the spectrum will be saturated with signals from the solvent, rendering it unreadable. As a result, solvents without protons must be used. Although there are several solvents that lack protons, such as CCl_4, these solvents do not dissolve all compounds. In practice, deuterated solvents are generally used.

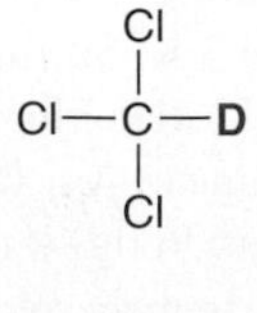
Chloroform-d

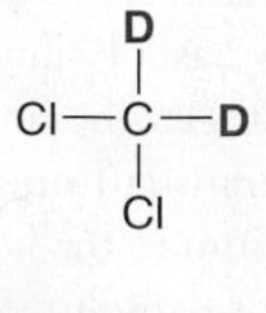
Methylene chloride-d_2

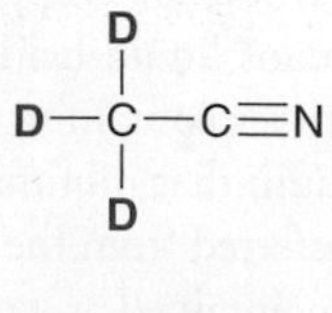
Acetonitrile-d_3

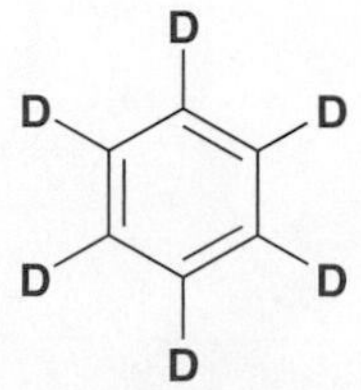
Benzene-d_6

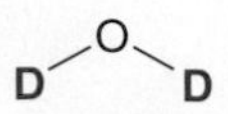
Deuterium oxide

The nuclei of deuterium also exhibit nuclear spin, and therefore also resonate, but they absorb rf radiation over a very different range of frequencies than do protons. In an NMR spectrometer, a very narrow range of frequencies is used, covering just the frequencies absorbed by protons. For example, a 300-MHz spectrometer will use a pulse that consists only of frequencies between 300,000,000 and 300,005,000 Hz. The frequencies required for deuterium resonance do not fall in this range, so the deuterium atoms are invisible to the NMR spectrometer. All of the solvents shown above are routinely used, and many other deuterated solvents are also available commercially, although quite expensive.

16.3 Characteristics of a ^{1}H NMR Spectrum

The spectrum produced by ^{1}H NMR spectroscopy is generally rich with information that can be interpreted to determine a molecular structure. Consider the following ^{1}H NMR spectrum.

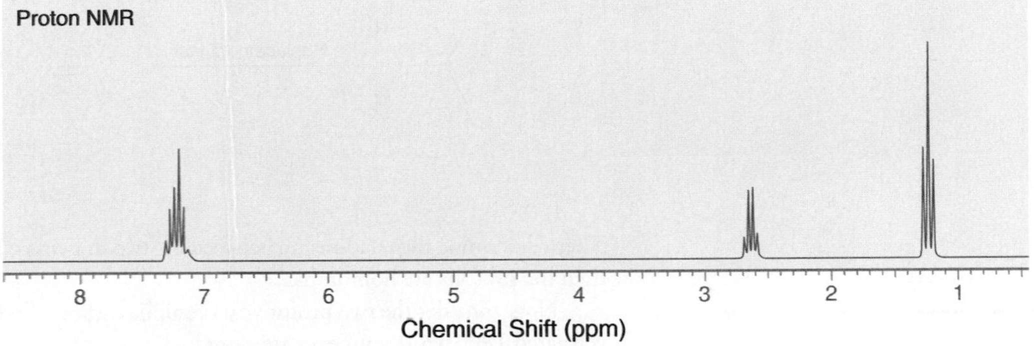

The first valuable piece of information is the number of signals. This spectrum appears to have three different signals. In Section 16.4, we will learn how to interpret this piece of information. In addition, each signal has three important characteristics:

1. The *location* of each signal indicates the electronic environment of the protons giving rise to the signal.
2. The *area* under each signal indicates the number of protons giving rise to the signal.
3. The *shape* of the signal indicates the number of neighboring protons.

We will discuss these characteristics in Sections 16.5–16.7.

16.4 Number of Signals

Chemical Equivalence

LOOKING BACK
For a review of symmetry operations, see Section 5.6.

The number of signals in a ^{1}H NMR spectrum indicates the number of different kinds of protons (protons in different electronic environments). Protons that occupy identical electronic environments are called **chemically equivalent**, and they will produce only one signal. Two protons are chemically equivalent if they can be interchanged via a symmetry operation—either rotation or reflection. The following examples will illustrate the relationship between symmetry and chemical equivalence. Let's begin by considering the two protons on the middle carbon of propane. Imagine that this molecule is rotated 180° about the following axis while your eyes are closed.

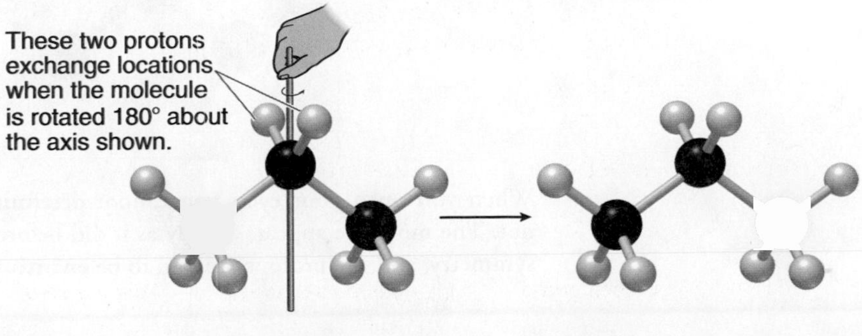

When you open your eyes, you cannot determine whether the molecule was rotated or not. From your point of view, the molecule appears exactly as it did before rotation, and it therefore has an axis of symmetry. The two protons on the middle carbon of propane are interchangeable by rotational symmetry and are therefore said to be **homotopic**. Homotopic protons are chemically equivalent. Below are other examples of homotopic protons.

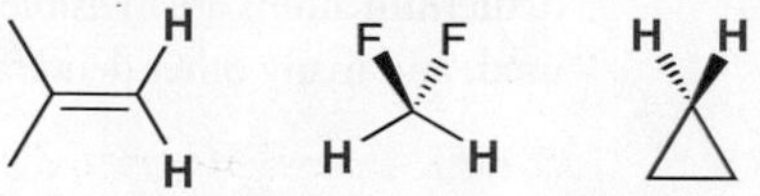

In each of these examples, the two identified protons are homotopic, because they can be interchanged by rotational symmetry. (Can you identify the axis of rotation in each compound above?) If you are having trouble seeing axes of symmetry, there is a simple method, called the **replacement test**, that will allow you to verify whether or not two protons are homotopic. Draw the compound two times, each time replacing one of the protons with deuterium; for example:

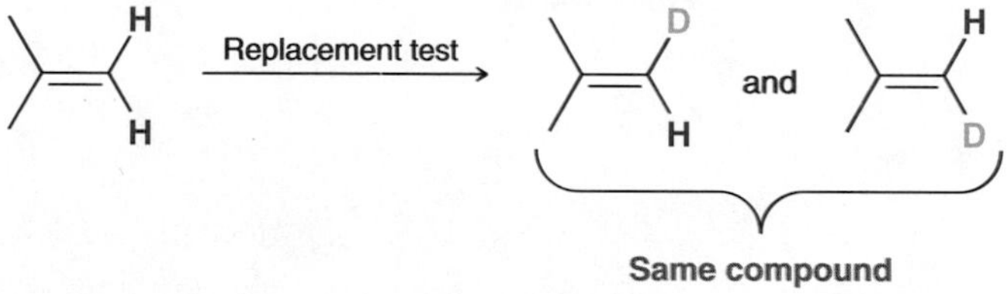
Replacement test
and
Same compound

Then, determine the relationship between the two drawings. If they represent the same compound, then the protons are homotopic.

Now consider the two protons on the alpha carbon of ethanol, and imagine that this molecule is rotated 180° while your eyes are closed.

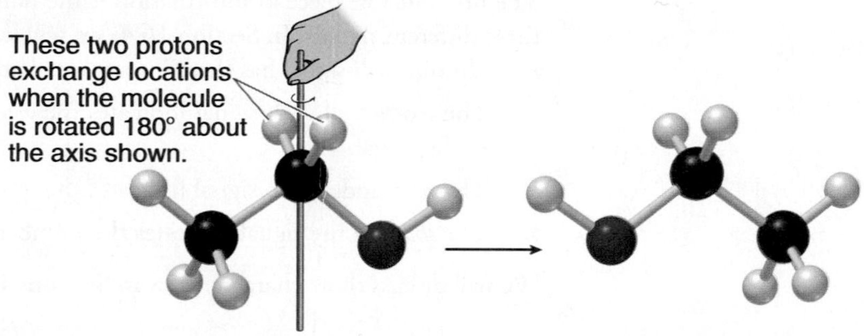
These two protons exchange locations when the molecule is rotated 180° about the axis shown.

When you open your eyes, you *will* be able to determine that the molecule has been rotated. Specifically, the OH is now on the left side. That is, the two protons on the alpha carbon of ethanol are not interchangeable by rotational symmetry. These protons are therefore not homotopic. Nevertheless, they can be interchanged by reflectional symmetry. Imagine that the molecule is reflected about the plane of the page while your eyes are closed.

These two protons exchange locations when the molecule is reflected through the plane shown

When you open your eyes, you cannot determine whether the molecule was reflected or not. The molecule appears exactly as it did before reflection. In this case, there is a plane of symmetry, and the protons are said to be **enantiotopic**. Enantiotopic protons are chemically equivalent, because they are interchangeable by reflectional symmetry. Here are two other examples of enantiotopic protons.

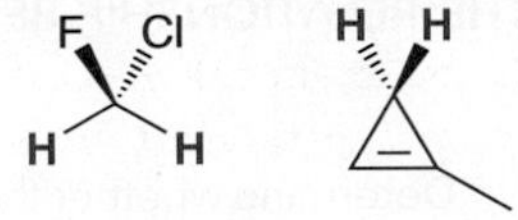

In each of these examples the two highlighted protons are enantiotopic, because they are interchangeable by reflectional symmetry. (Can you see the plane of symmetry in each compound?) If you are having trouble seeing planes of symmetry, you can resort once again to the replacement test. Simply draw the compound twice, each time replacing one of the protons with deuterium. Then determine the relationship between the two drawings. If they are enantiomers, then the protons are enantiotopic.

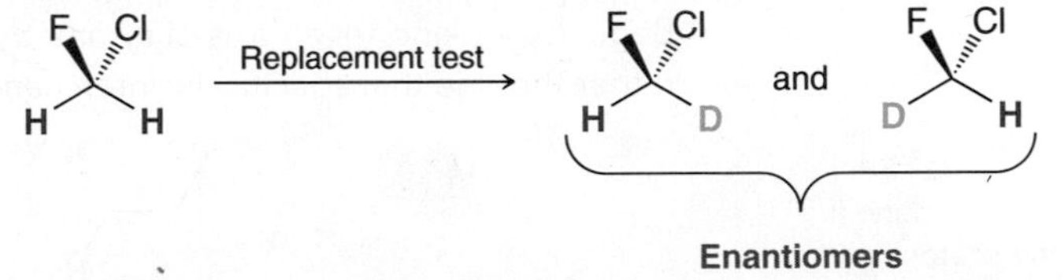

When determining the relationship between two protons, always look for rotational symmetry first. Figure 16.7 indicates how to determine the relationship between two protons. First determine if there is an axis of symmetry that interchanges the protons. If there is, then the protons are homotopic, whether or not there is a plane of symmetry. If the protons cannot be interchanged by rotation, then look for reflectional symmetry. If there is a plane of symmetry, then the protons are enantiotopic.

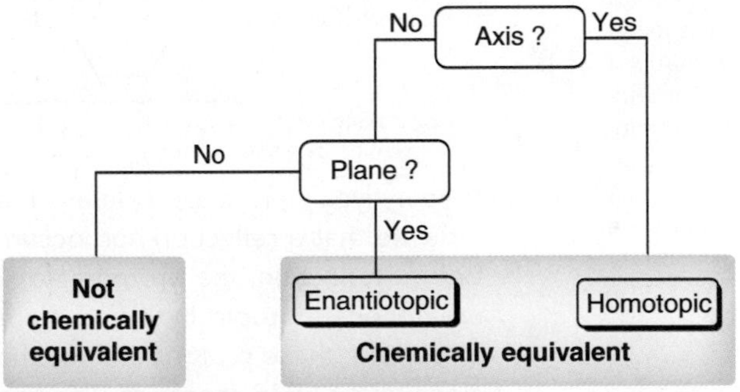

FIGURE 16.7 A flow chart showing the process for determining whether two protons are chemically equivalent.

If two protons are neither homotopic nor enantiotopic, then they are not chemically equivalent. As an example, consider the protons on C3 of (*R*)-2-butanol.

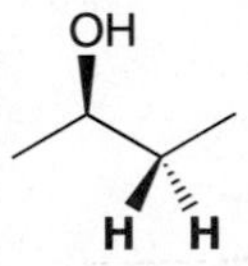

These protons *cannot* be interchanged by either rotational symmetry or reflectional symmetry. Therefore, these two protons are not chemically equivalent. In this case, the replacement test produces diastereomers.

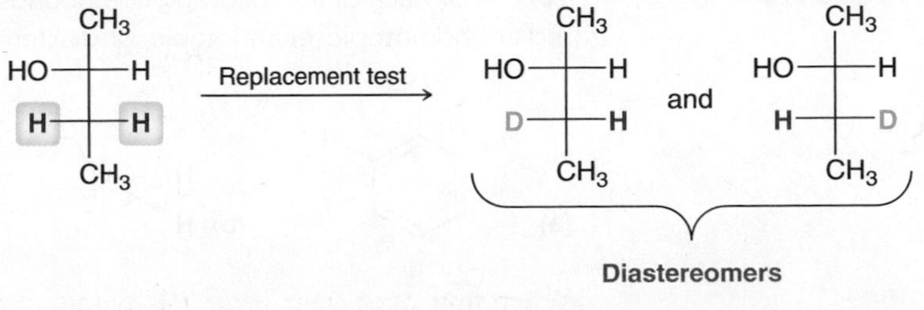

These protons are therefore said to be **diastereotopic**. They are not chemically equivalent, because they cannot be interchanged by symmetry.

SKILLBUILDER

16.1 DETERMINING THE RELATIONSHIP BETWEEN TWO PROTONS IN A COMPOUND

LEARN the skill

Determine whether the two protons shown in red are homotopic, enantiotopic, diastereotopic, or simply not related at all.

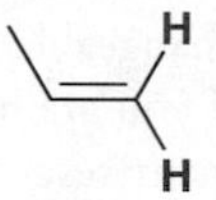

SOLUTION

Begin by looking for an axis of symmetry. Do not look for any and all axes of symmetry, but rather the one that specifically interchanges the two protons of interest.

STEP 1 Determine if the protons are interchangeable by rotational symmetry.

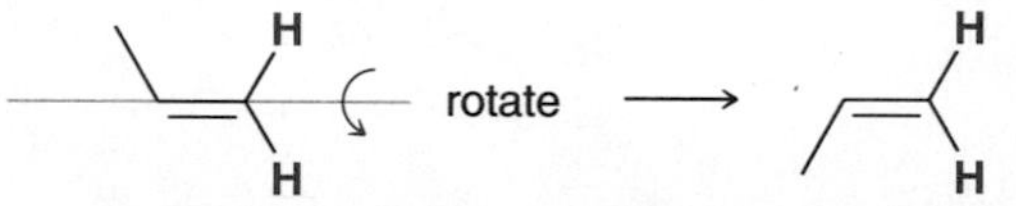

If your eyes were closed during the operation, the position of the methyl group would indicate that the rotation had occurred. Since the molecule is not drawn exactly as it was before rotation, the protons are not interchangeable by rotation. They are not homotopic.

STEP 2 If the protons are not interchangeable by rotational symmetry, determine if they are interchangeable by reflectional symmetry.

Next, look for a plane of symmetry. Once again, do not look for any and all planes of symmetry, but rather the one that specifically interchanges the two protons of interest.

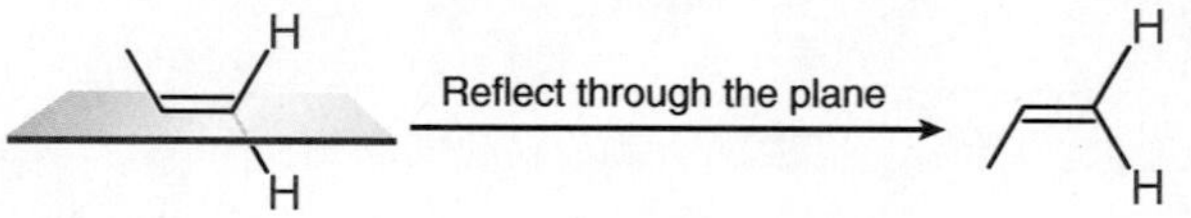

If your eyes were closed during the operation, the position of the methyl group would indicate that a reflection had occurred. Since the molecule is not drawn exactly as it was before reflection, the protons are not interchangeable by reflection, and therefore, they are *not* enantiotopic. In this problem, the protons are neither homotopic nor enantiotopic. Therefore, these protons are not chemically equivalent. They will each produce a different signal in an NMR spectrum. The two protons must either be diastereotopic or simply not related to each other at all. To determine if they are diastereotopic, do the replacement test.

STEP 3 If the protons cannot be interchanged via a symmetry operation, then they are not chemically equivalent.

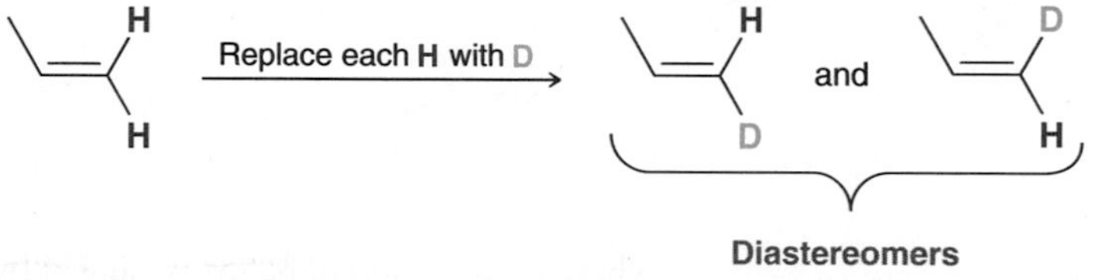

The two compounds are diastereomers, and therefore, the protons are diastereotopic.

PRACTICE the skill

16.1 For each of the following compounds, determine whether the two protons shown in red are homotopic, enantiotopic, or diastereotopic:

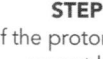

(a) 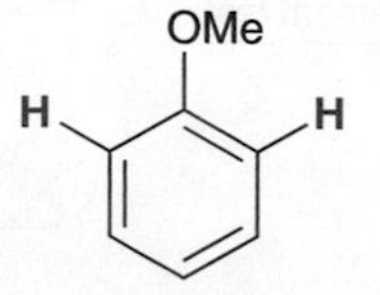

(b) 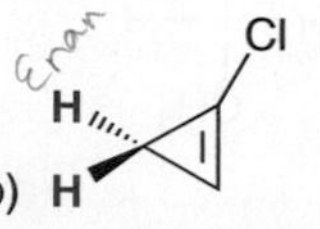

(c) 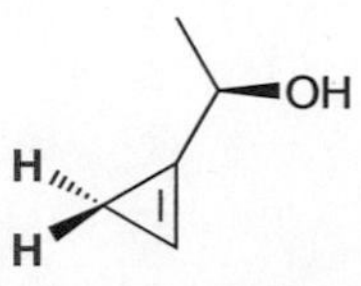

(d) 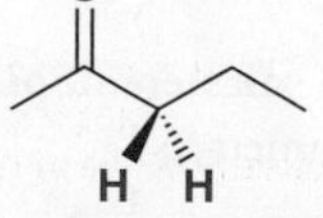

(e) 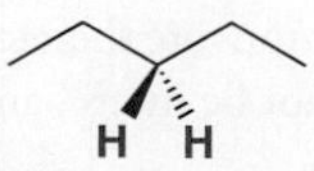

APPLY the skill

16.2 Butane (C_4H_{10}) exhibits only two different kinds of protons, shown here in red and blue.

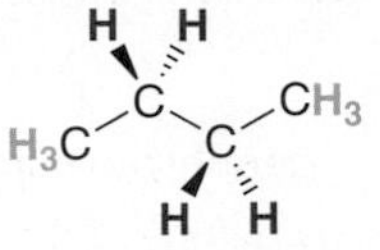
H H
H3C C C CH3
H H

(a) Explain why all four protons shown in red are chemically equivalent.

(b) Explain why all six protons shown in blue are chemically equivalent.

(c) How many different kinds of protons are present in pentane?

(d) How many different kinds of protons are present in hexane?

(e) How many different kinds of protons are present in 1-chlorohexane?

16.3 Identify the structure of a compound with molecular formula C_5H_{12} that exhibits only one kind of proton. That is, all 12 protons are chemically equivalent.

need more **PRACTICE?** **Try Problem 16.40**

In order to predict the number of expected signals in the ^{1}H NMR spectrum of a compound, it is not necessary to compare all of the protons and drive yourself crazy looking for axes and planes of symmetry. In general, it is possible to determine the number of expected signals for a compound using a few simple rules:

- The three protons of a CH_3 group are always chemically equivalent. Example:

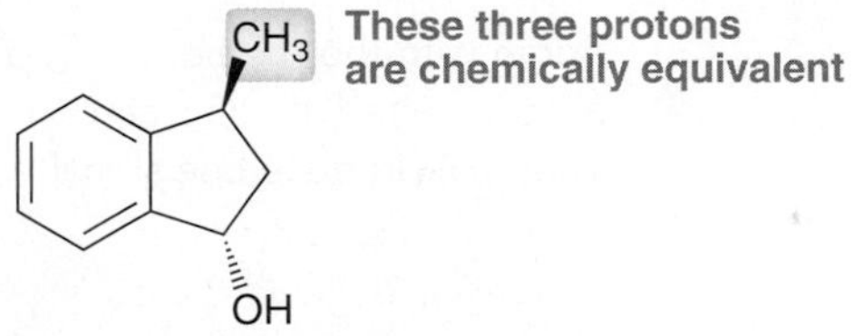
CH3
These three protons are chemically equivalent
OH

- The two protons of a CH_2 group will generally be chemically equivalent if the compound has no chirality centers. If the compound has a chirality center, then the protons of a CH_2 group will generally not be chemically equivalent. Examples:

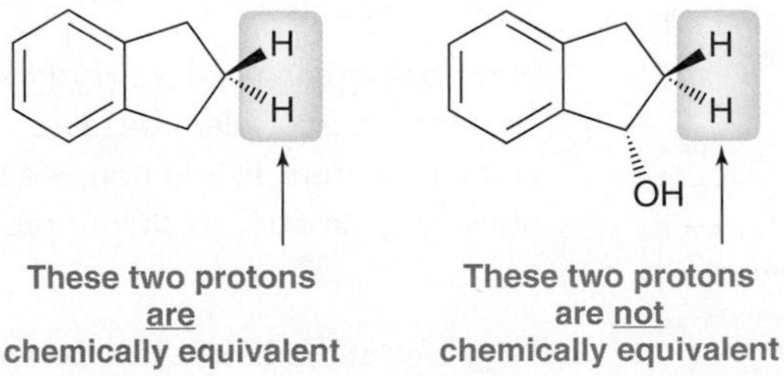
H
H
These two protons are chemically equivalent
H
H
OH
These two protons are not chemically equivalent

- Two CH_2 groups will be equivalent to each other (giving four equivalent protons) if the CH_2 groups can be interchanged by either rotation or reflection. Example:

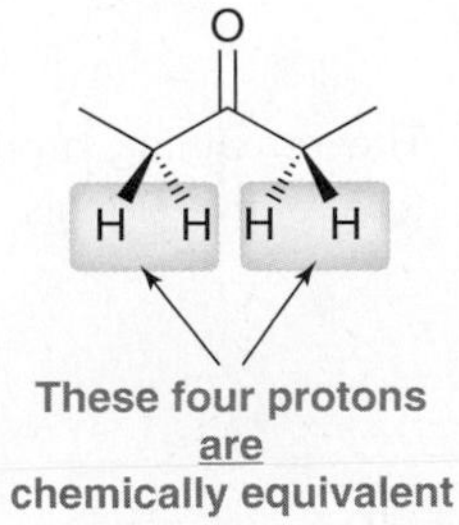
O
H H H H
These four protons are chemically equivalent

SKILLBUILDER

16.2 IDENTIFYING THE NUMBER OF EXPECTED SIGNALS IN A ^{1}H NMR SPECTRUM

LEARN the skill

Identify the number of signals expected in the ^{1}H NMR spectrum of the following compound.

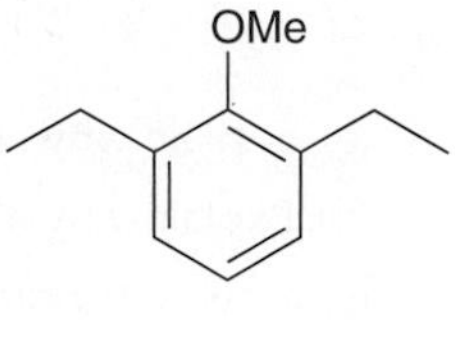

SOLUTION

When looking for symmetry, don't be confused by the position of the double bonds in the aromatic ring. Recall that we can draw the following two resonance structures.

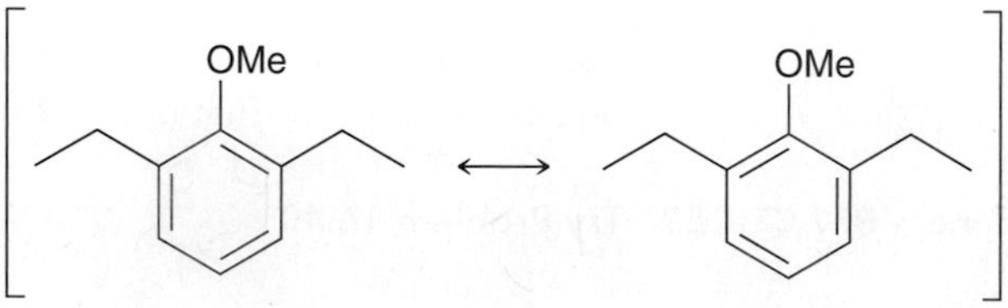

Neither resonance structure is more correct than the other. For purposes of looking for symmetry, it will be less confusing to draw the compound like this:

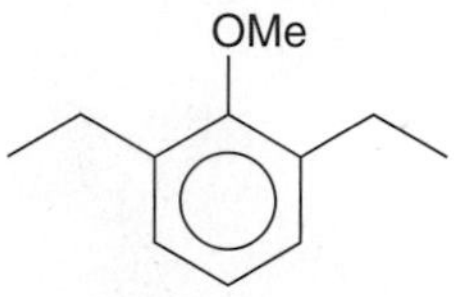

Begin with the methoxy group (OCH_3). These three protons are all connected to one carbon atom and are therefore equivalent. They will produce one signal.

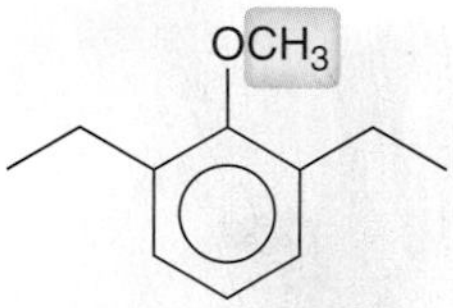

Now look at the other remaining CH_3 groups in the compound. These two CH_3 groups are interchangeable by symmetry, which means that all six protons give rise to one signal.

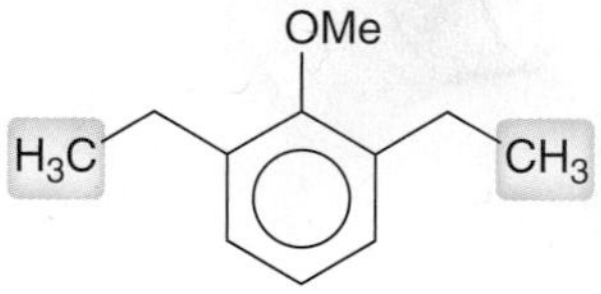

Now look at each of the CH_2 groups. For each CH_2 group, the two protons are equivalent because the compound does not have any chirality centers. In addition, the two CH_2 groups are interchangeable by symmetry, so all four protons give rise to one signal.

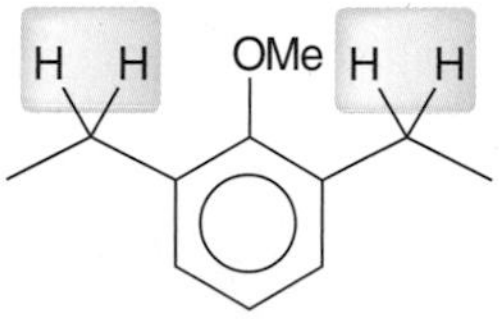

Now look at the protons on the aromatic ring. Two of them are interchangeable by symmetry, giving rise to one signal.

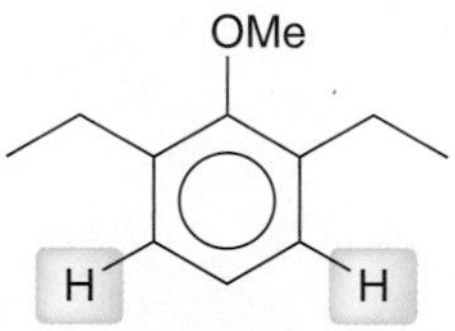

The remaining aromatic proton gives one more signal, giving a total of five signals.

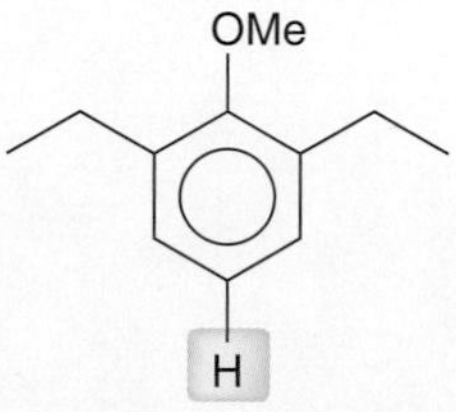

PRACTICE the skill **16.4** Identify the number of signals expected in the ^1H NMR spectrum of each of the following compounds.

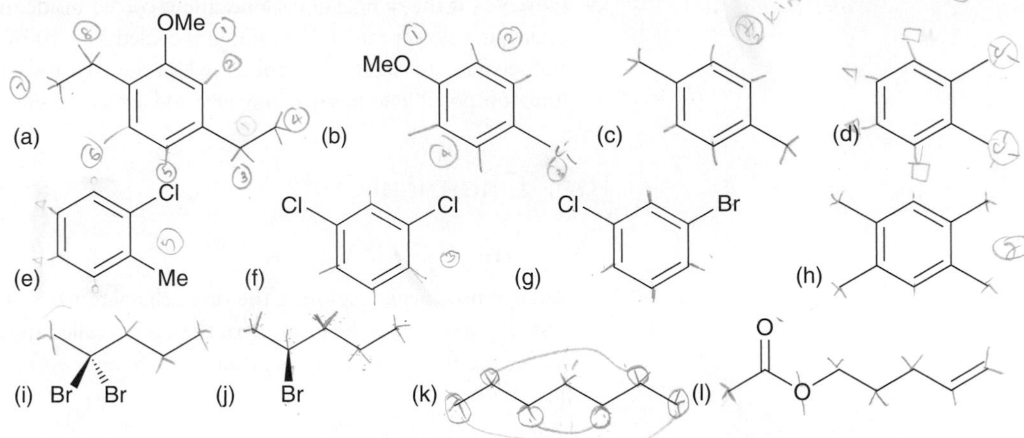

APPLY the skill **16.5** We saw a general rule that the two protons of a CH_2 group will be chemically equivalent if there are no chirality centers in the compound. An example of an exception is 3-bromopentane. This compound does not possess a chirality center. Nevertheless, the two highlighted protons are not chemically equivalent. Explain.

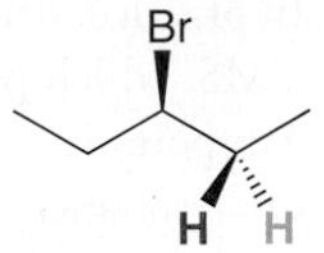

16.6 Identify the structure of a compound with molecular formula C_9H_{20} that exhibits four CH_2 groups, all of which are chemically equivalent. How many total signals would you expect in the ^1H NMR spectrum of this compound?

need more **PRACTICE?** **Try Problems 16.34, 16.42a, 16.44, 16.48**

Variable-Temperature NMR

In Chapter 4, we learned that the most stable conformation of cyclohexane is a chair conformation in which six of the protons occupy axial positions and six occupy equatorial positions.

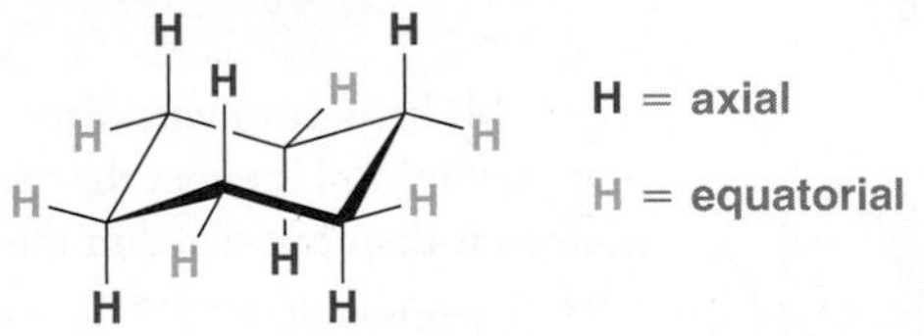

Axial protons (shown in red) occupy a different electronic environment than equatorial protons (shown in blue). That is, axial and equatorial protons are not chemically equivalent, because they cannot be interchanged by a symmetry operation. Consequently, we might expect two signals in the ^1H NMR spectrum: one for the six axial protons and one for the six equatorial protons. Yet, the ^1H NMR spectrum of cyclohexane only exhibits one signal. Why?

The observation can be explained by considering the rapid rate at which ring flipping occurs at room temperature.

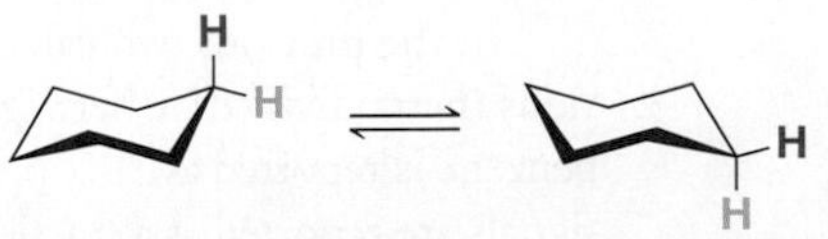

The NMR spectrometer is analogous to a camera with a slow shutter speed that produces a blurred picture when a fast-moving object is photographed. The NMR spectrometer is too slow to acquire a spectrum of a single chair conformation. While the spectrometer is acquiring the spectrum, the ring is flipping rapidly between the two chair conformations, producing a blurry picture. The spectrometer only "sees" the average electronic environment of the protons, and only one signal is observed. However, if the sample of cyclohexane is cooled inside the spectrometer, the ring-flipping process occurs at a slower rate. If the sample is cooled to −100°C, ring flipping occurs at a very slow rate, and separate signals are in fact observed for the axial and equatorial protons. By varying the temperature, it is possible to measure the rates and activation energies of many rapid processes.

16.5 Chemical Shift

Chemical Shift Values

We will now begin exploring the three characteristics of every signal in an NMR spectrum. The first characteristic is the location of the signal, called its **chemical shift (δ)**, which is defined relative to the frequency of absorption of a reference compound, tetramethylsilane (TMS).

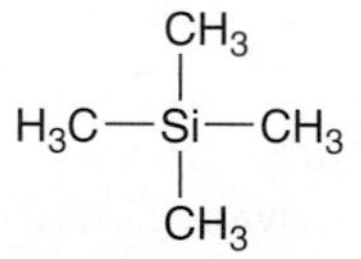

TMS

In practice, deuterated solvents used for NMR spectroscopy typically contain a small amount of TMS, which produces a signal at a lower frequency than the signals produced by most organic compounds. The frequency of each signal is then described as the difference (in hertz) between the resonance frequency of the proton being observed and that of TMS, divided by the operating frequency of the spectrometer.

$$\delta = \frac{\text{observed shift from TMS in hertz} \times 10^6}{\text{operating frequency of the instrument in hertz}}$$

For example, when benzene is analyzed using an NMR spectrometer operating at 300 MHz, the protons of benzene absorb a frequency of rf radiation that is 2181 Hz larger than the frequency of absorption of TMS. The chemical shift of these protons is then calculated in the following way:

$$\delta = \frac{2181\ \text{Hz} \times 10^6}{300 \times 10^6\ \text{Hz}} = 7.27$$

If a 60 MHz spectrometer is used instead, the protons of benzene absorb a frequency of rf radiation that is 436 Hz larger than the frequency of absorption of TMS. The chemical shift of these protons is then calculated in the following way:

$$\delta = \frac{436\ \text{Hz} \times 10^6}{60 \times 10^6\ \text{Hz}} = 7.27$$

Notice that the chemical shift of the protons is a constant, regardless of the operating frequency of the spectrometer. That is precisely why chemical shifts have been defined in relative terms, rather than absolute terms (hertz). If signals were reported in hertz (the precise frequency of rf radiation absorbed), then the frequency of absorption would be dependent on the strength of the magnetic field and would not be a constant.

In the previous two calculations, notice that the value obtained does not possess any dimensions (hertz divided by hertz gives a dimensionless number). The chemical shift for the protons of benzene is reported as 7.27 ppm (parts per million), which are dimensionless units indicating that signals are reported as a fraction of the operating frequency of the spectrometer. For most organic compounds, the signals produced will fall in a range between 0 and 12 ppm. In rare cases, it is possible to observe a signal occurring at a chemical shift below 0 ppm, which results from a proton that absorbs a lower frequency than TMS. Most protons in organic compounds absorb a higher frequency than TMS, so most chemical shifts that we encounter will be positive numbers.

The left side of an NMR spectrum is described as **downfield**, and the right side of the spectrum is described as **upfield**.

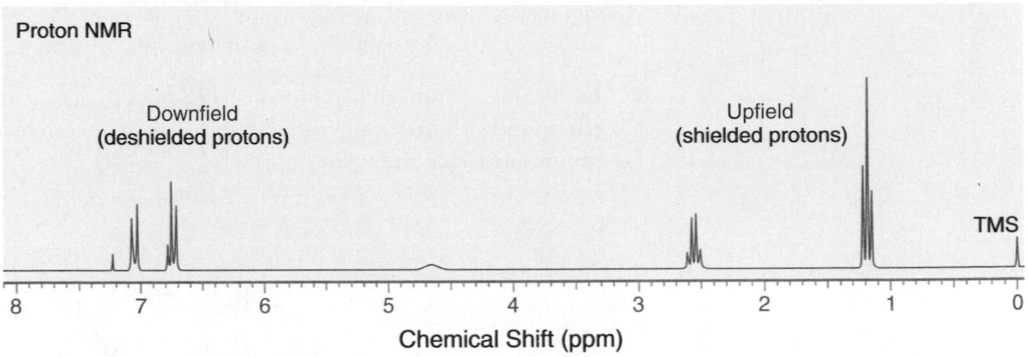

These terms are historical artifacts that reflect the way spectra were once acquired. As mentioned earlier in this chapter, continuous-wave spectrometers would hold the frequency of rf radiation constant and slowly increase the magnetic field strength while monitoring which field strengths produced a signal. Signals on the left side of the spectrum were produced at lower field strength (downfield), while signals on the right side of the spectrum were produced at higher field strength (upfield). With the advent of FT-NMR spectrometers, spectra are no longer acquired in this fashion. In modern spectrometers, the magnetic field strength is held constant while the sample is irradiated with a short pulse that covers the entire range of relevant rf frequencies. Accordingly, signals on the left side of the spectrum (downfield) are "high-frequency signals" because they result from deshielded protons that absorb higher frequencies of rf radiation. In contrast, signals on the right side of the spectrum (upfield) are "low-frequency signals" because they result from shielded protons that absorb lower frequencies of rf radiation. Despite the advent of modern spectrometers, the older terms "downfield" and "upfield" are still commonly used to describe the position of a signal in an NMR spectrum.

LOOKING BACK
Recall that a deshielded proton experiences a stronger magnetic field strength, thereby creating a larger difference in energy between the alpha and beta spin states (Figure 16.5). A shielded proton experiences a weaker magnetic field strength, thereby creating a smaller difference in energy between the alpha and beta spin states.

Protons of alkanes produce upfield signals (generally between 1 and 2 ppm). We will now explore some of the effects that can push a signal downfield.

Inductive Effects

Recall from Section 1.11 that electronegative atoms, such as halogens, withdraw electron density from neighboring atoms (Figure 16.8). This inductive effect causes the protons of the methyl group to be deshielded (surrounded by less electron density), and as a result, the

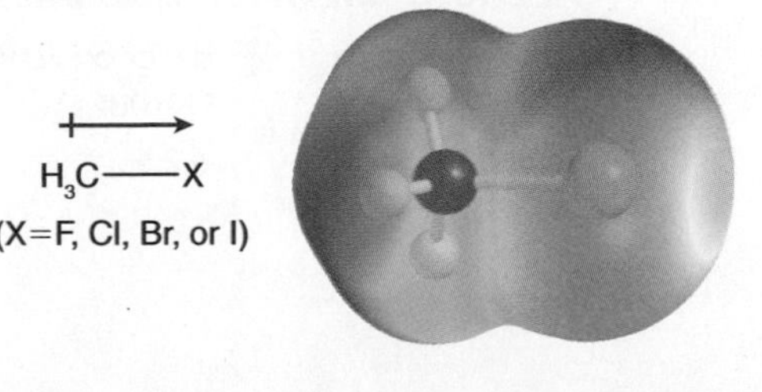

FIGURE 16.8
The inductive effect of an electronegative atom causes nearby protons to be deshielded.

signal produced by these protons is shifted downfield—that is, the signal appears at a higher chemical shift than the protons of an alkane. The strength of this effect depends on the electronegativity of the halogen. Compare the chemical shifts of the protons in the following compounds.

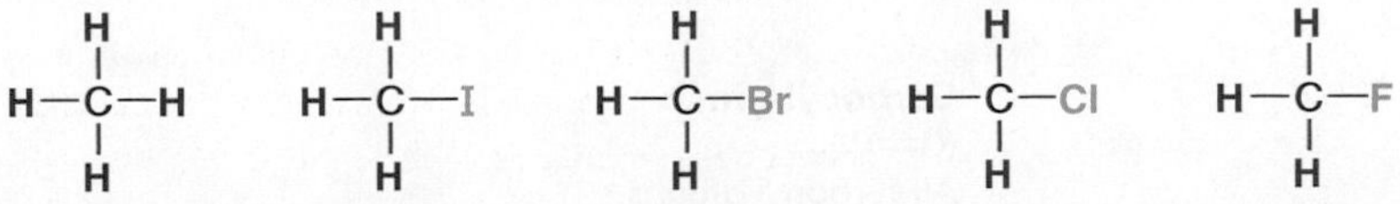

1.0 ppm **2.2 ppm** **2.7 ppm** **3.1 ppm** **4.3 ppm**

Fluorine is the most electronegative element and therefore produces the strongest effect. When multiple halogens are present, the effect is additive, as can be seen when comparing the following compounds.

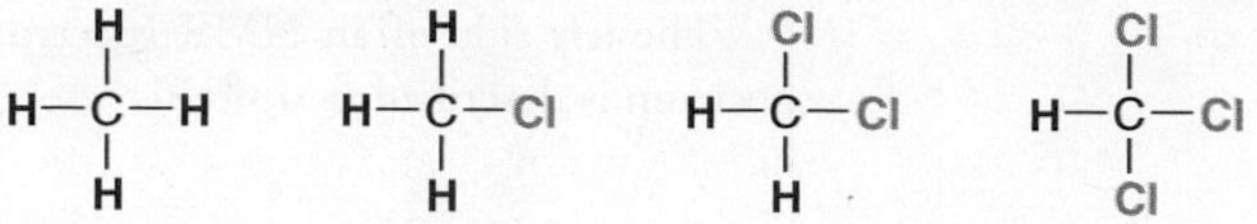

1.0 ppm 3.1 ppm 5.3 ppm 7.3 ppm

Each chlorine atom adds approximately 2 ppm to the chemical shift of the signal. The inductive effect tapers off drastically with distance, as can be seen by comparing the chemical shifts of the protons in 1-chloropropane.

1.6 ppm

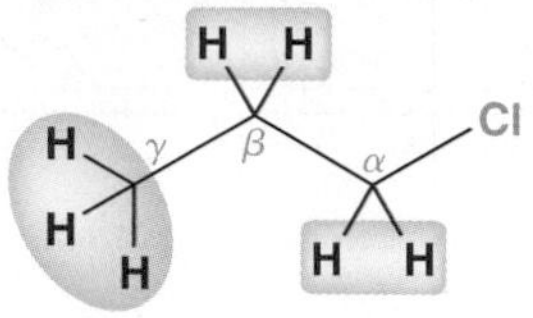

0.9 ppm 3.3 ppm

The effect is most significant for the protons at the alpha position. The protons at the beta position are only slightly affected, and the protons at the gamma position are virtually unaffected by the presence of the chlorine atom.

By committing a few numbers to memory, it is possible to predict the chemical shifts for the protons in a wide variety of compounds, including alcohols, ethers, ketones, esters, and carboxylic acids. The following numbers are used as benchmark values:

Methyl Methylene Methine

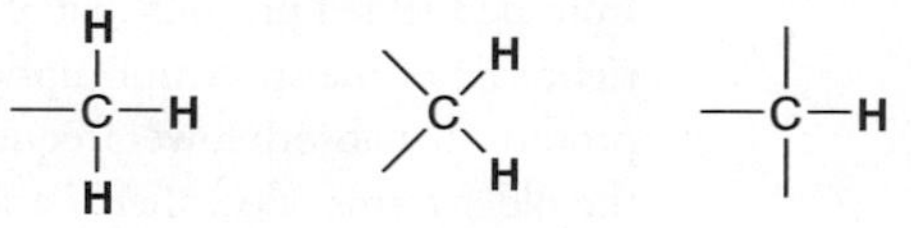

~ 0.9 ppm ~ 1.2 ppm ~ 1.7 ppm

These are the expected chemical shifts for protons that lack neighboring electronegative atoms. In the absence of inductive effects, a methyl group (CH_3) will produce a signal near 0.9 ppm, a **methylene** group (CH_2) will produce a signal near 1.2 ppm, and a **methine** group (CH) will produce a signal near 1.7 ppm. These benchmark values are then modified by the presence of neighboring functional groups. Table 16.1 shows the effect of a few functional groups on the chemical shifts of

TABLE 16.1 THE EFFECT OF NEIGHBORING FUNCTIONAL GROUPS ON CHEMICAL SHIFT

FUNCTIONAL GROUP	EFFECT ON ALPHA PROTONS	EXAMPLE
Oxygen of an alcohol or ether	+ 2.5	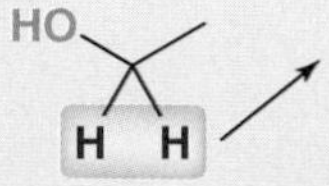 Methylene group (CH_2) = 1.2 ppm; Next to oxygen = +2.5 ppm; 3.7 ppm; Actual chemical shift = 3.7 ppm
Oxygen of an ester	+3	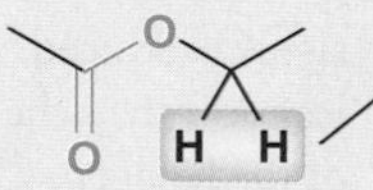 Methylene group (CH_2) = 1.2 ppm; Next to oxygen = +3.0 ppm; 4.2 ppm; Actual chemical shift = 4.1 ppm
Carbonyl group (C=O) All carbonyl groups, including ketones, aldehydes, esters, etc.	+1	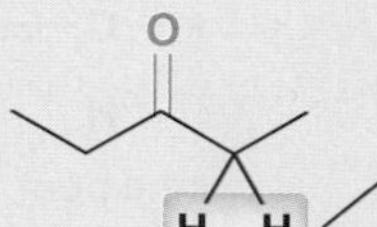 Methylene group (CH_2) = 1.2 ppm; Next to oxygen = +1.0 ppm; 2.2 ppm; Actual chemical shift = 2.4 ppm

alpha protons. The effect on beta protons is generally about one-fifth of the effect on the alpha protons. For example, in an alcohol, the presence of an oxygen atom adds +2.5 ppm to the chemical shift of the alpha protons but adds only +0.5 ppm to the beta protons. Similarly, a carbonyl group adds +1 ppm to the chemical shift of the alpha protons but only +0.2 to the beta protons.

The three benchmark values, together with the three values shown in Table 16.1, enable us to predict the chemical shifts for the protons in a wide variety of compounds, as illustrated in the following exercise.

SKILLBUILDER

16.3 PREDICTING CHEMICAL SHIFTS

LEARN the skill

Predict the chemical shifts for the signals in the ^1H NMR spectrum of the following compound.

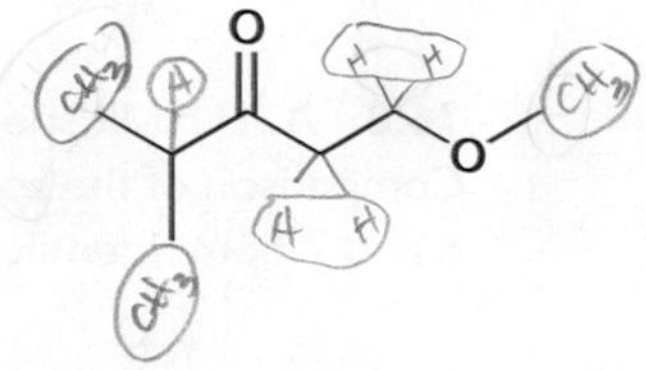

SOLUTION

First determine the total number of expected signals. In this compound, there are five different kinds of protons, giving rise to five distinct signals. For each type of signal, identify whether it represents methyl (0.9 ppm), methylene (1.2 ppm), or methine (1.7 ppm) groups.

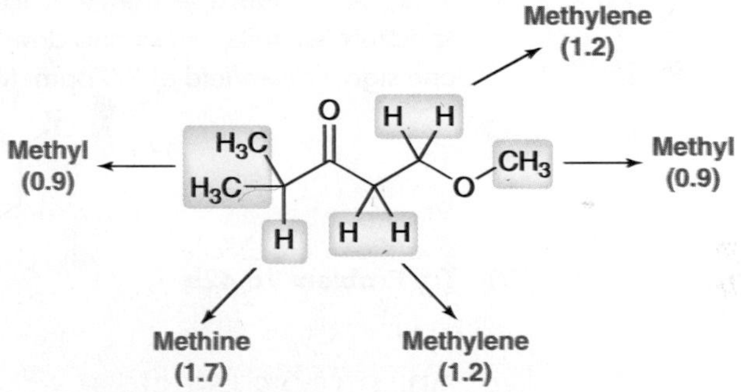

Finally, modify each of these numbers based on proximity to the oxygen and the carbonyl group.

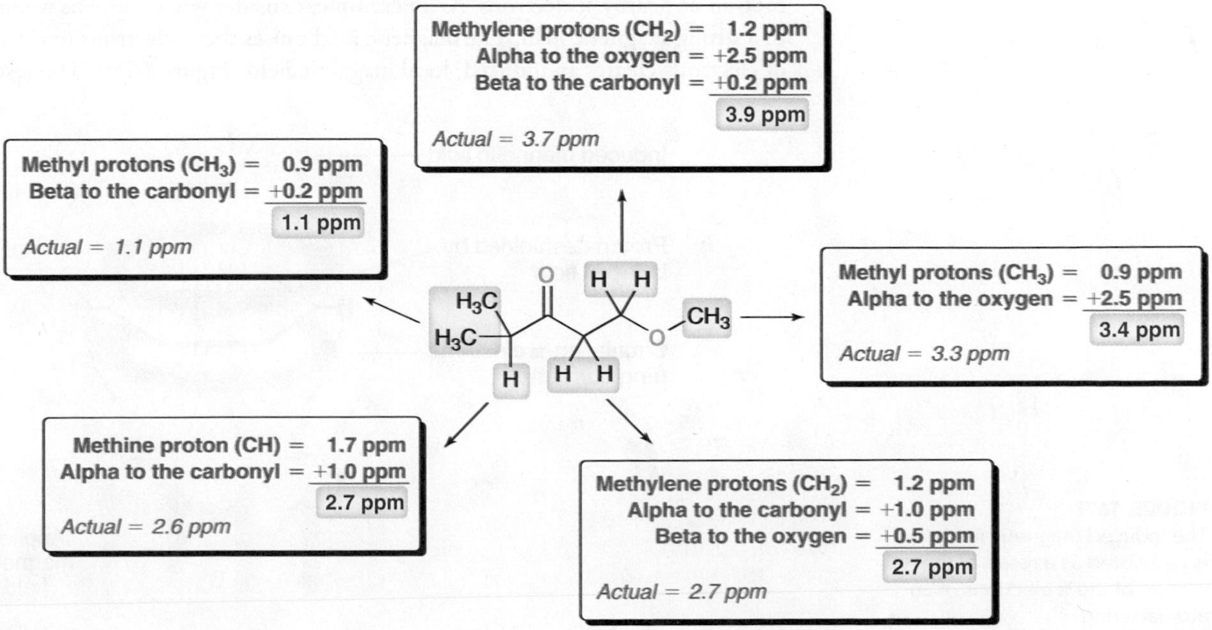

These values are only estimates, and the actual chemical shifts might differ slightly from the predicted values. The actual values are also shown, and they are indeed very close to the estimated values.

PRACTICE the skill **16.7** Predict the chemical shifts for the signals in the ^1H NMR spectrum of each of the following compounds:

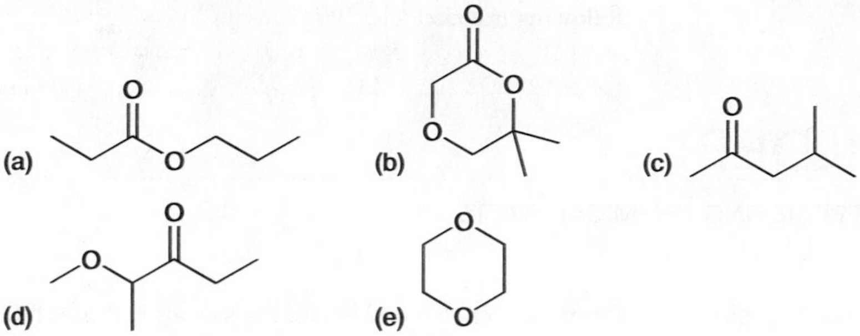

APPLY the skill **16.8** A ^1H NMR spectrum was acquired for each of the following constitutional isomers. Comparison of the spectra reveals that only one of these spectra exhibits a signal between 6 and 7 ppm. Identify the structure that corresponds with this spectrum.

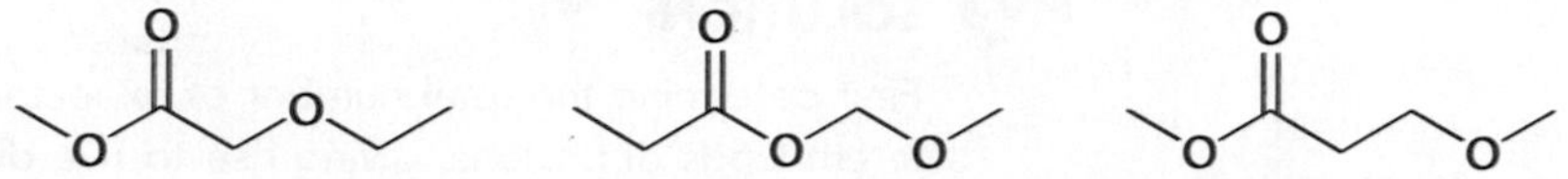

16.9 A ^1H NMR spectrum was acquired for each of the following two compounds. One spectrum exhibits two signals downfield of 2.0 ppm, while the other spectrum exhibits only one signal downfield of 2.0 ppm. Match each spectrum with its corresponding compound.

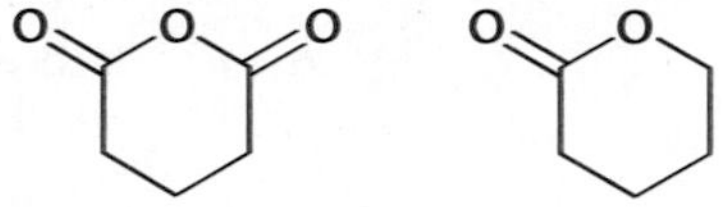

need more **PRACTICE?** **Try Problem 16.42b**

Anisotropic Effects

The chemical shift of a proton is also sensitive to diamagnetic effects that result from the motion of nearby π electrons. As an example, consider what happens when benzene is placed in a strong magnetic field. The magnetic field causes the π electrons to circulate, and this flow of electrons creates an induced, local magnetic field (Figure 16.9). The result is **diamagnetic**

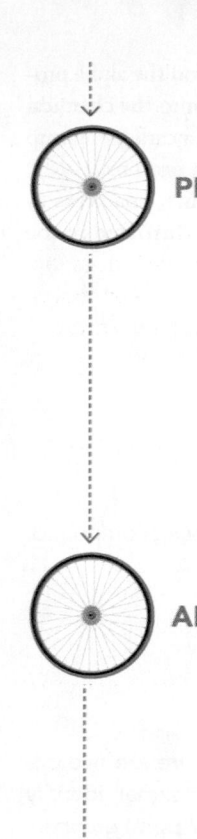

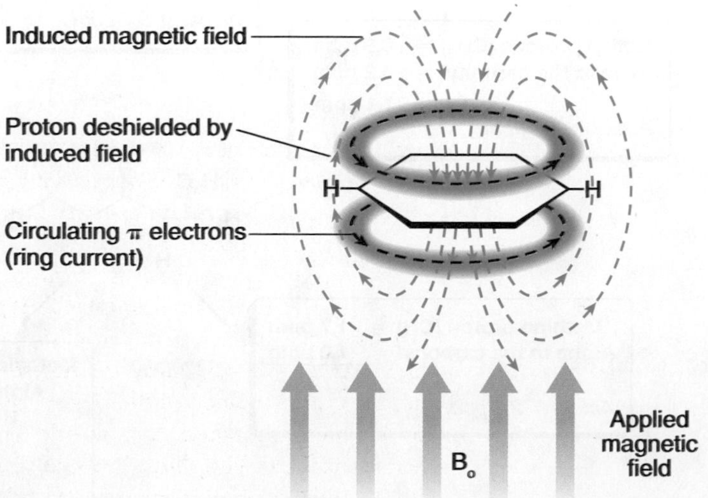

FIGURE 16.9 The induced magnetic field that is generated as a result of the motion of the π electrons of an aromatic ring.

anisotropy, which means that different regions of space are characterized by different magnetic field strengths. Locations inside the ring are characterized by a local magnetic field that opposes the external field, while locations outside the ring are characterized by a local magnetic field that adds to the external field. The protons connected to the ring are permanently positioned outside of the ring, and as a result, they experience a stronger magnetic field. These protons experience the external magnetic field plus the local magnetic field. The effect is similar to a deshielding effect, and therefore, the protons are shifted downfield. Aromatic protons produce a signal in the neighborhood of 7 ppm (sometimes just above 7, sometimes just below 7) in an NMR spectrum. For example, consider the spectrum of ethylbenzene.

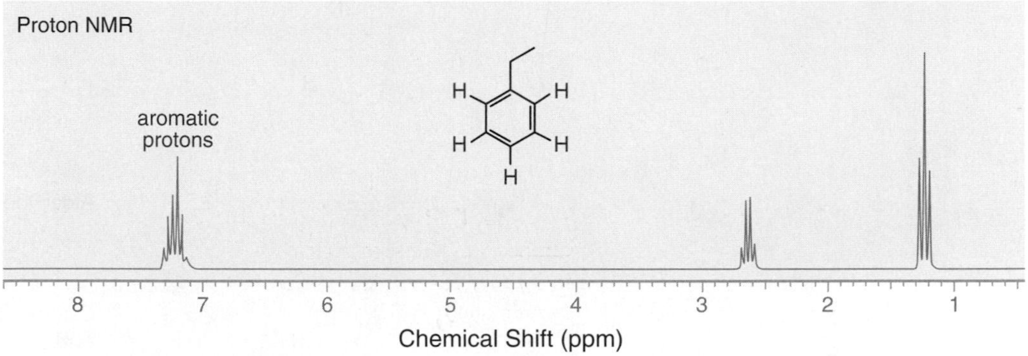

Ethylbenzene has three different kinds of aromatic protons (can you identify them in the structure?), producing three (complex) overlapping signals just above 7 ppm. A complex signal around 7 ppm is characteristic of compounds with aromatic protons.

The methylene group (CH_2) in ethylbenzene produces a signal at 2.6 ppm, rather than the expected benchmark value of 1.2 ppm. These protons have been shifted downfield because of their position in the local magnetic field. They are not shifted as much as the aromatic protons themselves, because the methylene protons are farther away from the ring, where the induced, local magnetic field is weaker.

A similar effect is observed for [14] annulene.

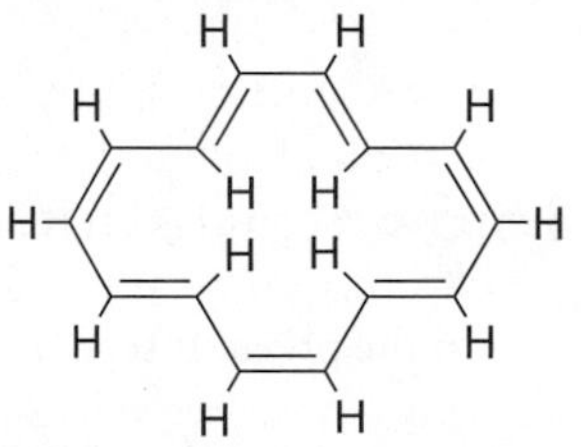

[14] Annulene

The protons outside of the ring produce signals at approximately 8 ppm, but in this case, there are four protons positioned inside the ring, where the local magnetic field opposes the external magnetic field. These protons experience the external magnetic field minus the local magnetic field. The effect is similar to a shielding effect, because the protons experience a weaker magnetic field, and therefore, the protons are shifted upfield. This effect is quite strong, producing a signal at -1 ppm (even further upfield than TMS).

All π bonds exhibit a similar anisotropic effect. That is, π electrons circulate under the influence of an external magentic field, generating a local magnetic field. For each type of π bond, the precise location of the nearby protons determines their chemical shift. For example, aldehydic protons produce characteristic signals at approximately 10 ppm. Table 16.2 summarizes important chemical shifts. It would be wise to become familiar with these numbers, as they will be required in order to interpret ^1H NMR spectra.

TABLE 16.2 CHEMICAL SHIFTS FOR PROTONS IN DIFFERENT ELECTRONIC ENVIRONMENTS

TYPE OF PROTON		CHEMICAL SHIFT (δ)	TYPE OF PROTON		CHEMICAL SHIFT (δ)
Methyl	R—CH_3	~0.9	Alkyl halide	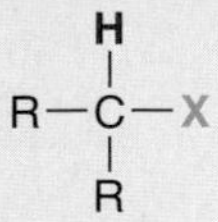	2–4
Methylene	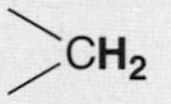	~1.2	Alcohol	R—O—H	2–5
Methine	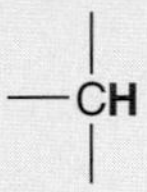	~1.7	Vinylic	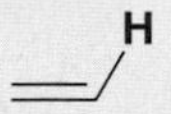	4.5–6.5
Allylic	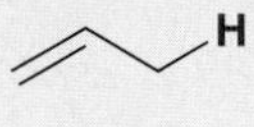	~2	Aryl	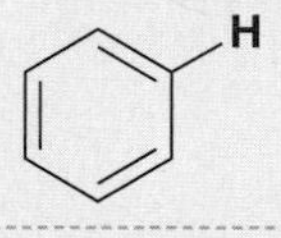	6.5–8
Alkynyl	R—≡—H	~2.5	Aldehyde	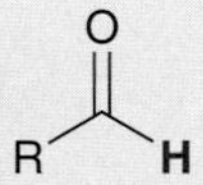	~10
Aromatic methyl	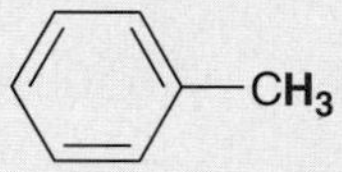	~2.5	Carboxylic acid	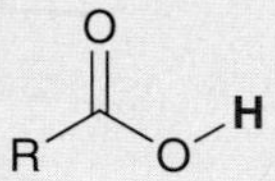	~12

CONCEPTUAL CHECKPOINT

16.10 For each of the following compounds, identify the expected chemical shift for each type of proton:

(a) 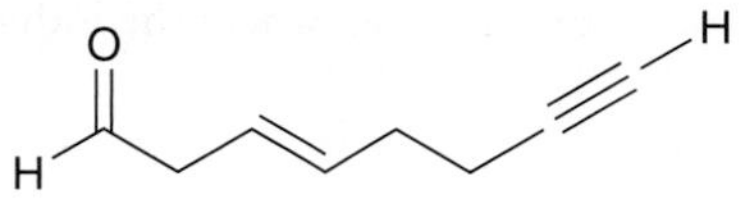
(b) 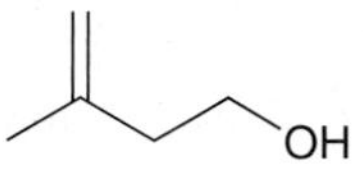
(c) 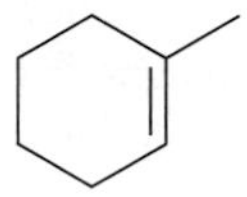
(d) 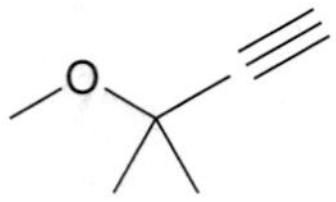

16.6 Integration

In the previous section, we learned about the first characteristic of every signal, chemical shift. In this section, we will explore the second characteristic, **integration**, or the area under each signal. This value indicates the number of protons giving rise to the signal. After acquiring a spectrum, the computer calculates the area under each signal and then displays this area as a numerical value placed either above or below the signal.

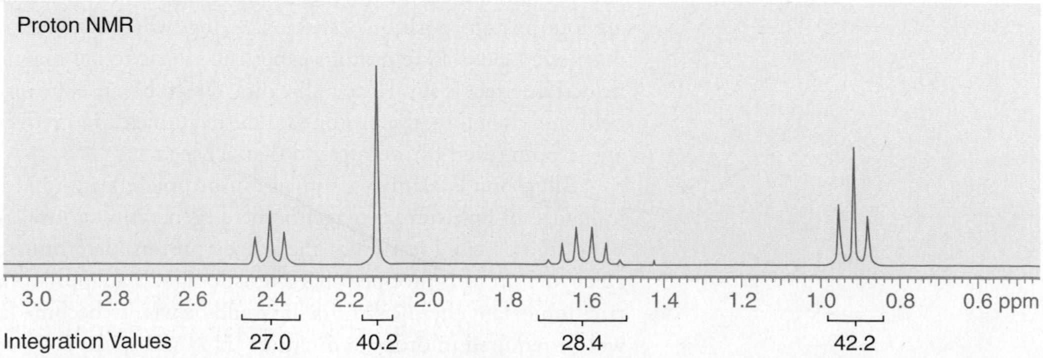

These numbers only have meaning when compared to each other. In order to convert these numbers into useful information, choose the smallest number (27.0 in this case), and then divide all integration values by this number.

$$\frac{27.0}{27.0} = 1 \qquad \frac{40.2}{27.0} = 1.49 \qquad \frac{28.4}{27.0} = 1.05 \qquad \frac{42.2}{27.0} = 1.56$$

These numbers provide the *relative number*, or ratio, of protons giving rise to each signal. This means that a signal with an integration of 1.5 involves one and a half times as many protons as a signal with an integration of 1. In order to arrive at whole numbers (there is no such thing as half a proton), multiply all the numbers by 2, giving the same ratio now expressed in whole numbers, 2 : 3 : 2 : 3. In other words, the signal at 2.4 ppm represents two equivalent protons, and the signal at 2.1 ppm represents three equivalent protons.

Integration is often represented by **step curves**:

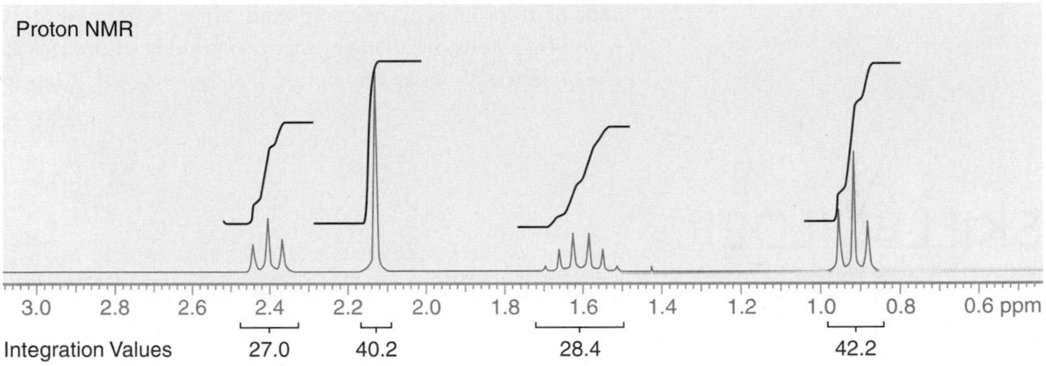

The height of each step curve represents the area under the signal. In this case, a comparison of the heights of the four step curves reveals a ratio of 2 : 3 : 2 : 3.

When interpreting integration values, don't forget that the numbers are only relative. To illustrate this point, consider the structure of *tert*-butyl methyl ether (MTBE).

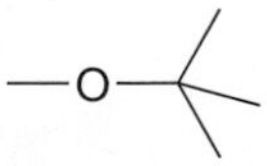

MTBE

MTBE has two kinds of protons (the methyl group and the *tert*-butyl group) and will produce two signals in its ^{1}H NMR spectrum. The computer analyzes the area under each signal and gives numbers that allow us to calculate a ratio of 1 : 3. This ratio only indicates the relative number of protons giving rise to each signal, not the exact number of protons. In this case, the exact numbers are 3 (for the methyl group) and 9 (for the *tert*-butyl group). When analyzing the NMR spectrum of an unknown compound, the molecular formula provides extremely useful information because it enables us to determine the exact number of protons giving rise to each signal. If we were analyzing the spectrum of MTBE, the molecular formula ($C_5H_{12}O$) would indicate that the compound has a total of 12 protons. This information then allows us to determine that the ratio of 1 : 3 must correspond with 3 protons and 9 protons, in order to give a total of 12 protons.

When analyzing an NMR spectrum of an unknown compound, we must also consider the impact of symmetry on integration values. For example, consider the structure of 3-pentanone.

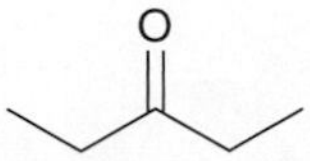

3-Pentanone

This compound has only two kinds of protons, because the methylene groups are equivalent to each other, and the methyl groups are equivalent to each other. The ^{1}H NMR spectrum is therefore expected to exhibit only two signals.

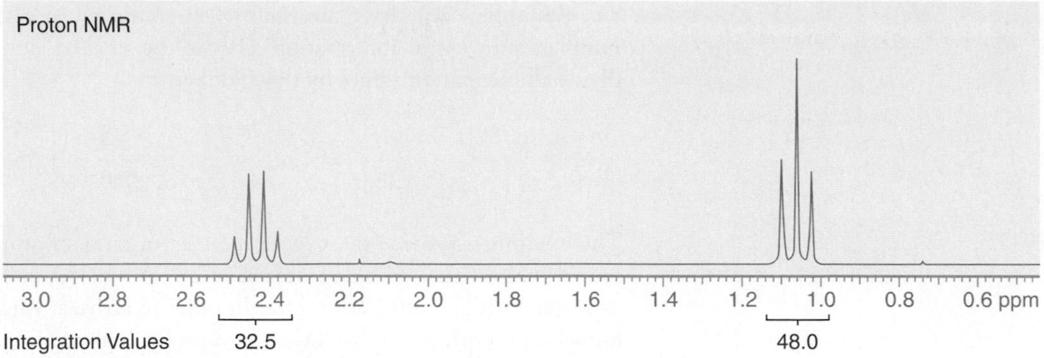

Compare the relative integration values: 32.5 and 48.0. These values give a ratio of 2:3, but again the values 2 and 3 are just relative numbers. They actually represent 4 protons and 6 protons. This can be determined by inspecting the molecular formula ($C_5H_{10}O$), which indicates a total of 10 protons in the compound. Since the ratio of protons is 2:3, this ratio must represent 4 and 6 protons, in order for the total number of protons to be 10. This analysis indicates that the molecule possesses symmetry.

SKILLBUILDER

16.4 DETERMINING THE NUMBER OF PROTONS GIVING RISE TO A SIGNAL

LEARN the skill

A compound with molecular formula $C_5H_{10}O_2$ has the following ^1H NMR spectrum.

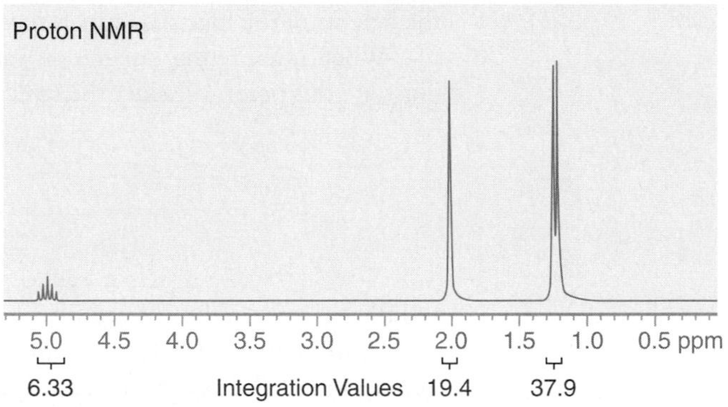

Determine the number of protons giving rise to each signal.

STEP 1
Compare the relative integration values and choose the lowest number.

SOLUTION

The spectrum exhibits three signals.

Begin by comparing the relative integration values: 6.33, 19.4, and 37.9. Divide each of these three numbers by the smallest number (6.33).

$$\frac{6.33}{6.33} = 1 \qquad \frac{19.4}{6.33} = 3.06 \qquad \frac{37.9}{6.33} = 5.99$$

STEP 2
Divide all the integration values by the number from Step 1, which gives the ratio of protons.

This gives a ratio of 1:3:6, but these are just relative numbers. To determine the exact number of protons giving rise to each signal, look at the molecular formula, which indicates a total of ten protons in the compound.

Therefore, the numbers 1:3:6 are not only relative values, but they are also the exact values. Exact integration values can be illustrated in the following way:

STEP 3
Identify the number of protons in the compound (from the molecular formula) and then adjust the relative integration values so that the sum total equals the number of protons in the compound.

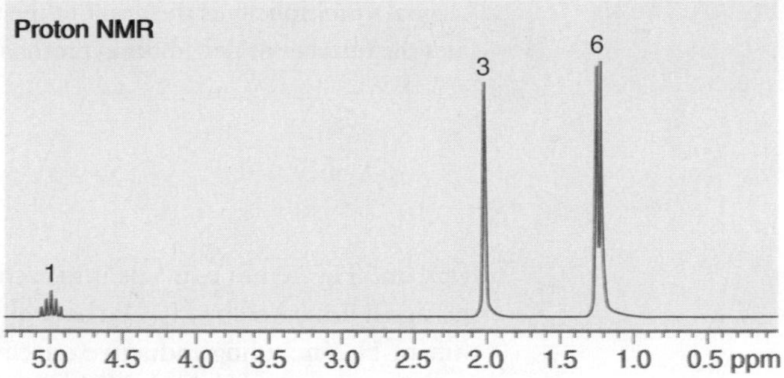

PRACTICE the skill

16.11 A compound with molecular formula $C_5H_{10}O_2$ has the following NMR spectrum. Determine the number of protons giving rise to each signal.

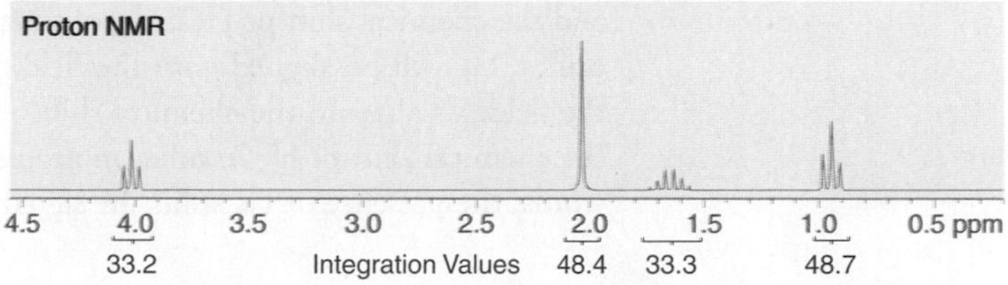

16.12 A compound with molecular formula $C_{10}H_{10}O$ has the following NMR spectrum. Determine the number of protons giving rise to each signal.

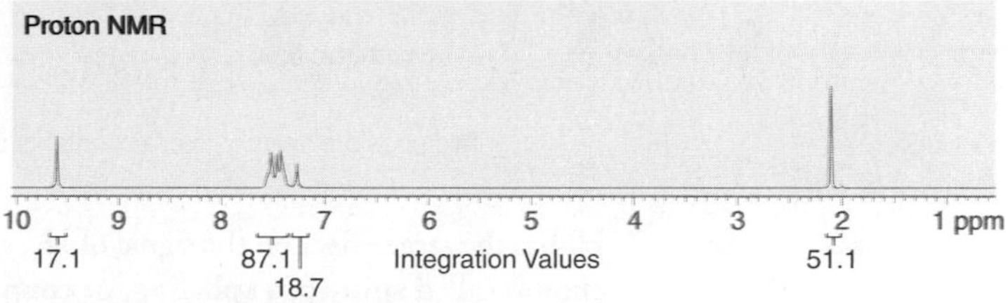

16.13 A compound with molecular formula $C_4H_6O_2$ has the following NMR spectrum. Determine the number of protons giving rise to each signal.

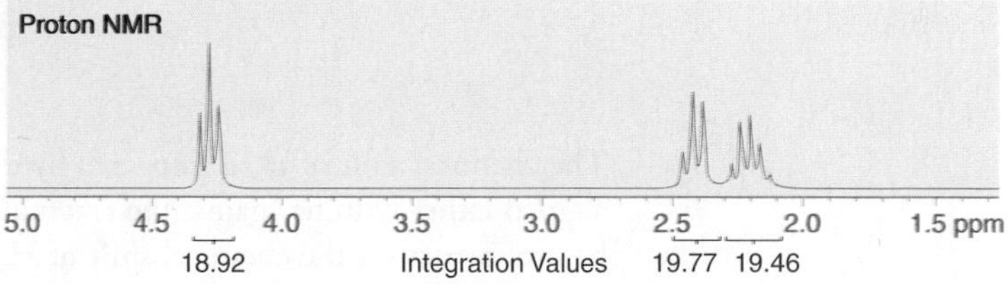

APPLY the skill

16.14 The ^{1}H NMR spectrum of a compound with molecular formula $C_7H_{15}Cl$ exhibits two signals with relative integration 2:3. Propose a structure for this compound.

need more **PRACTICE?** **Try Problem 16.54**

16.7 Multiplicity

Coupling

The third, and final, characteristic of each signal is its **multiplicity**, which is defined by the number of peaks in the signal. A **singlet** has one peak, a **doublet** has two peaks, a **triplet** has three peaks, a **quartet** has four peaks, a **quintet** has five peaks, and so on.

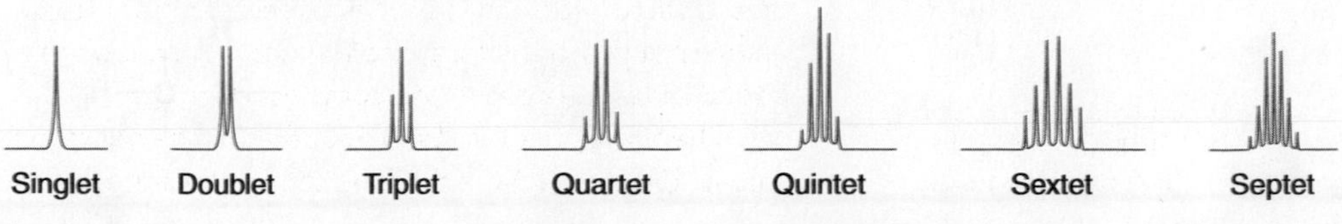

A signal's multiplicity is the result of the magnetic effects of neighboring protons and therefore indicates the number of neighboring protons. To illustrate this concept, consider the following example.

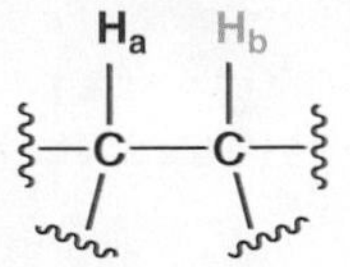

If H_a and H_b are not equivalent to each other, they will produce different signals. Let's focus on the signal produced by H_a. We have already seen a variety of factors that will affect the chemical shift of H_a, including inductive effects and diamagnetic anisotropy effects. All of these effects modify the magnetic field felt by H_a, thereby affecting the resonance frequency of H_a. The chemical shift of H_a is also impacted by the presence of H_b, because H_b has a magnetic moment that can either be aligned with or against the external magnetic field. H_b is like a tiny magnet, and the chemical shift of H_a is dependent on the alignment of this tiny magnet. In some molecules, H_b will be aligned with the field, while in other molecules, H_b will be aligned against the field. As a result, the chemical shift of H_a in some molecules will be slightly different than the chemical shift of H_a in other molecules, resulting in the appearance of two peaks. In other words, the presence of H_b splits the signal for H_a into a doublet (Figure 16.10).

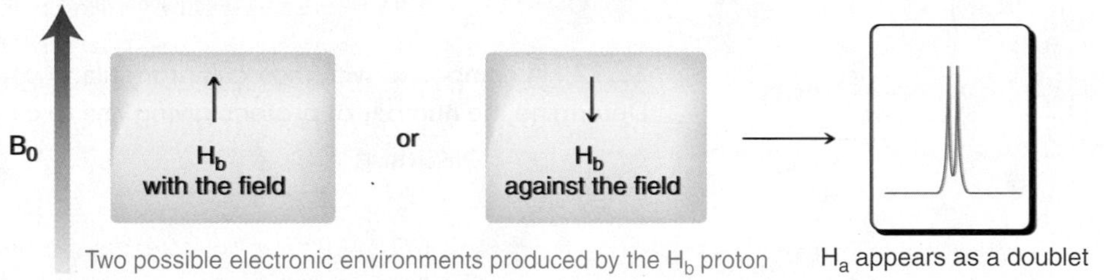

FIGURE 16.10
The source of a doublet.

H_a has the same effect on the signal of H_b, splitting the signal for H_b into a doublet. This phenomenon is called **spin-spin splitting**, or **coupling**

Now consider a scenario in which H_a has two neighboring protons.

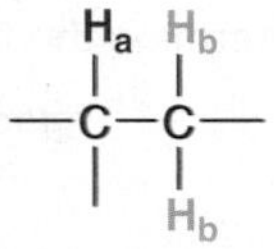

The chemical shift of H_a is impacted by the presence of both H_b protons, each of which can be aligned either with or against the external field. Once again, each H_b is like a tiny magnet and has an impact on the chemical shift of H_a. In each molecule, H_a can find itself in one of three possible electronic environments, resulting in a triplet (Figure 16.11). If each peak of the triplet is separately integrated, a ratio of 1 : 2 : 1 is observed, consistent with statistical expectations.

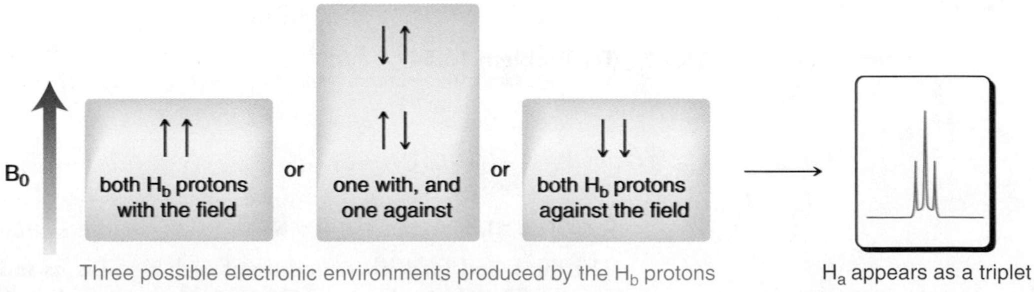

FIGURE 16.11
The source of a triplet.

Now consider a scenario in which H_a has three neighbors.

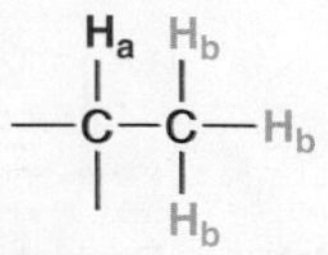

The chemical shift of H_a is impacted by the presence of all three H_b protons, each of which can be aligned either with the field or against the field. Once again, each H_b is like a tiny magnet and has an impact on the chemical shift of H_a. In each molecule, H_a can find itself in one of four possible electronic environments, resulting in a quartet (Figure 16.12). If each peak of the quartet is integrated separately, a ratio of 1:3:3:1 is observed, consistent with statistical expectations.

FIGURE 16.12
The source of a quartet.

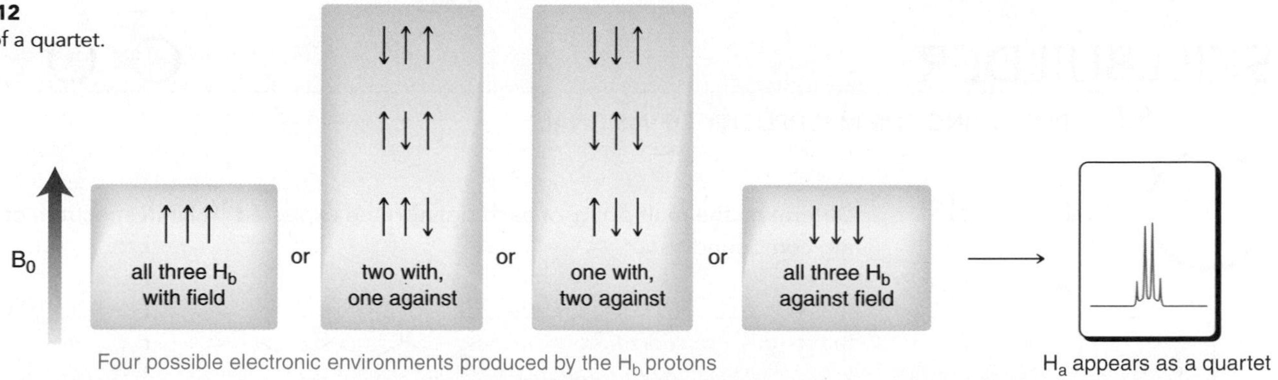

LOOKING AHEAD

The $n+1$ rule only applies when all of the neighboring protons are chemically equivalent to each other. When the neighbors are not all chemically equivalent, the observed splitting will be more complex, as we will see at the end of this section.

Table 16.3 summarizes the splitting patterns and peak intensities for signals that result from coupling with neighboring protons. When analyzing this information, a pattern emerges. Specifically, if n is the number of neighboring protons, then the multiplicity will be $n+1$. Extending this rule, a proton with six neighbors ($n=6$) will be split into a septet (7 peaks, or $n+1$). This observation is called **the n+1 rule**.

TABLE 16.3 MULTIPLICITY INDICATES THE NUMBER OF NEIGHBORING PROTONS

NUMBER OF NEIGHBORS	MULTIPLICITY	RELATIVE INTENSITIES OF INDIVIDUAL PEAKS
1	Doublet	1:1
2	Triplet	1:2:1
3	Quartet	1:3:3:1
4	Quintet	1:4: 6:4:1
5	Sextet	1:5:10:10:5:1
6	Septet	1:6:15:20:15:6:1

There are two major factors that determine whether or not splitting occurs:

1. Equivalent protons do not split each other. Consider the two methylene groups in 1,2-dichloroethane. All four protons are chemically equivalent, and therefore, they do not split each other. In order for splitting to occur, the neighboring protons must be different than the protons producing the signal.

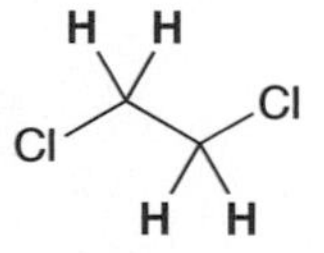

Four equivalent protons
No splitting

2. Splitting is most commonly observed when protons are separated by either two or three σ bonds; that is, when the protons are either diastereotopic protons on the same carbon atom (geminal) or when they are connected to adjacent carbon atoms (vicinal).

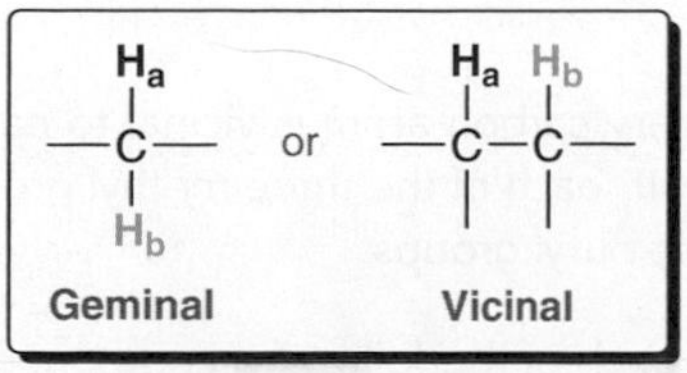

Splitting is observed

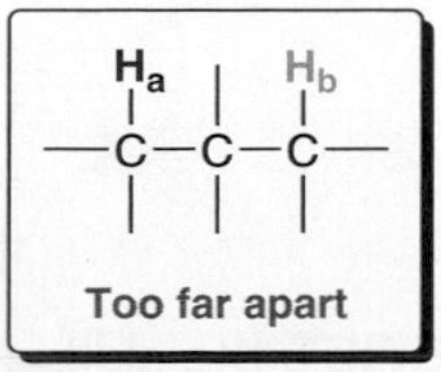

Splitting is generally not observed

When two protons are separated by more than three sigma bonds, splitting is generally not observed. Such long-range splitting is only observed in rigid molecules, such as bicyclic compounds, or in molecules that contain rigid structural moieties, such as allylic systems. For purposes of this introductory treatment of NMR spectroscopy, we will avoid examples that exhibit long-range coupling.

SKILLBUILDER

16.5 PREDICTING THE MULTIPLICITY OF A SIGNAL

LEARN the skill

Determine the multiplicity of each signal in the expected ^{1}H NMR spectrum of the following compound.

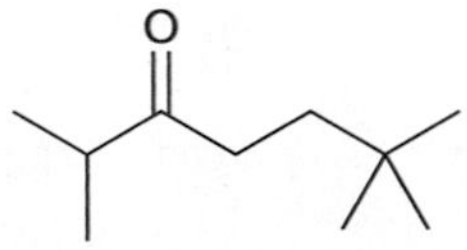

SOLUTION

Begin by identifying the different kinds of protons. That is, determine the number of expected signals.

STEP 1
Identify all of the different kinds of protons.

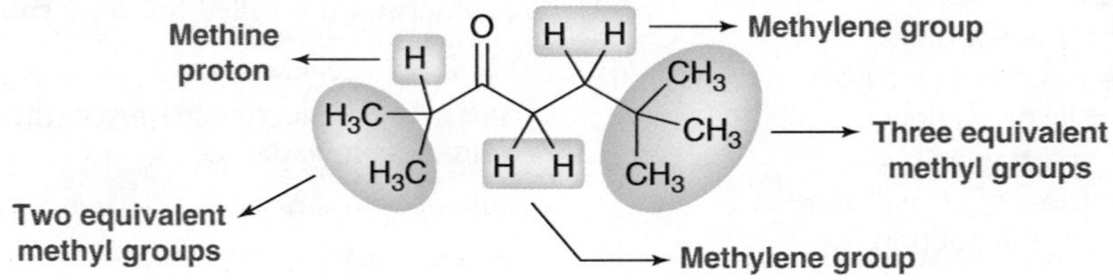

This compound is expected to produce five signals in its ^{1}H NMR spectrum. Now let's analyze each signal using the $n+1$ rule.

STEP 2
For each kind of proton, identify the number of neighbors (n). The multiplicity will follow the $n+1$ rule.

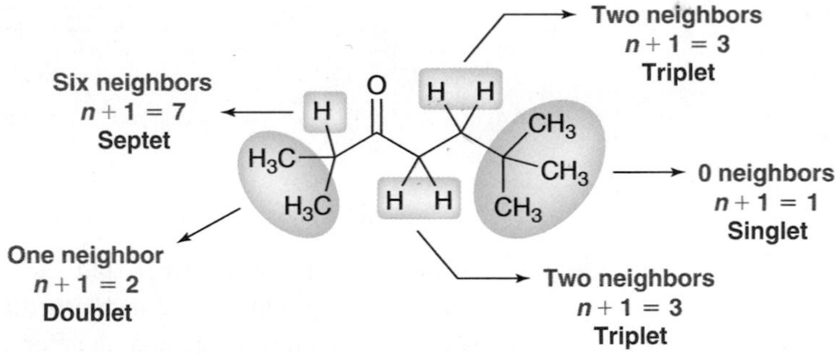

Notice that the *tert*-butyl group (on the right side of the molecule) appears as a singlet, because the following carbon atom has no protons.

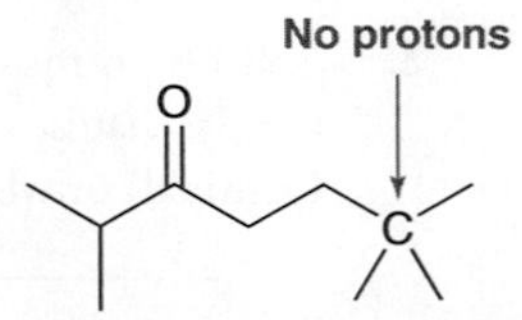

This quaternary carbon atom is vicinal to each of the three methyl groups connected to it, and as a result, each of the three methyl groups has no neighboring protons. This is characteristic of *tert*-butyl groups.

PRACTICE the skill **16.15** For each of the following compounds, determine the multiplicity of each signal in the expected ^{1}H NMR spectrum:

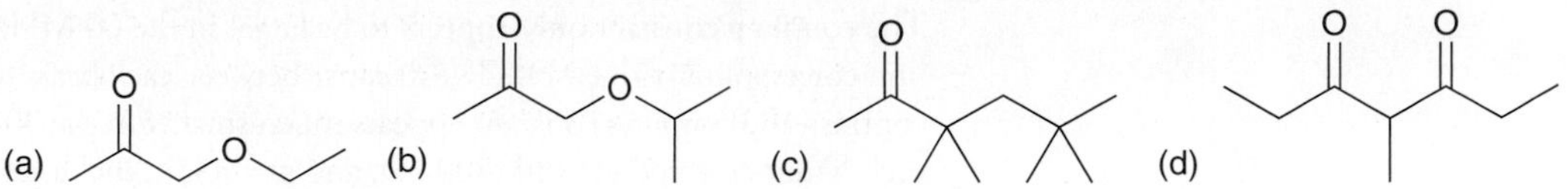

APPLY the skill

16.16 Propose the structure for a compound that lacks a methyl group but nevertheless exhibits a quartet in its ^1H NMR spectrum.

need more **PRACTICE?** **Try Problem 16.37**

Coupling Constant

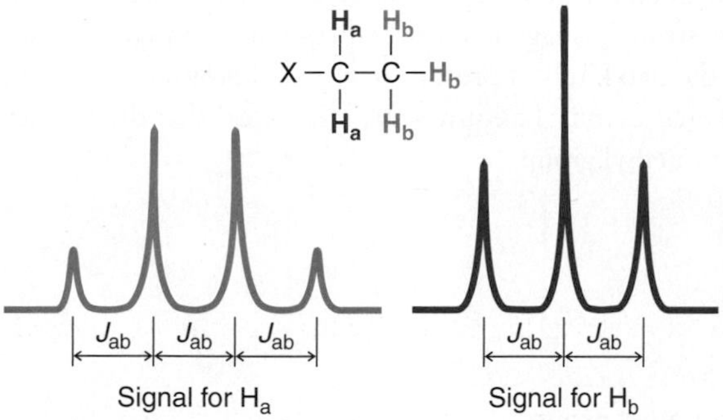

FIGURE 16.13
The coupling constant (J_{ab}) is the same for both signals because these signals are splitting each other.

When signal splitting occurs, the distance between the individual peaks of a signal is called the **coupling constant**, or ***J* value**, and is measured in hertz. Neighboring protons always split each other with equivalent *J* values. For example, consider the two kinds of protons in an ethyl group (Figure 16.13). The H_a signal is split into a quartet under the influence of its three neighbors, while the H_b signal is split into a triplet under the influence of its two neighbors. H_a and H_b are said to be coupled to each other. The coupling constant J_{ab} is the same in both signals. *J* values can range anywhere from 0 to 20 Hz, depending on the type of protons involved, and are independent of the operating frequency of the spectrometer. For example, if J_{ab} is measured to be 7.3 Hz on one spectrometer, the value does not change when the spectrum is acquired on a different spectrometer that uses a stronger magnetic field.

As a result, NMR spectrometers with higher operating frequencies provide better resolution. As an example, compare the two spectra of ethyl chloroacetate in Figure 16.14. The first spectrum was acquired on a 60 MHz NMR spectrometer, and the second spectrum was acquired on a 300 MHz NMR spectrometer. In each spectrum, the coupling constant (J_{ab}) is approximately 7 Hz.

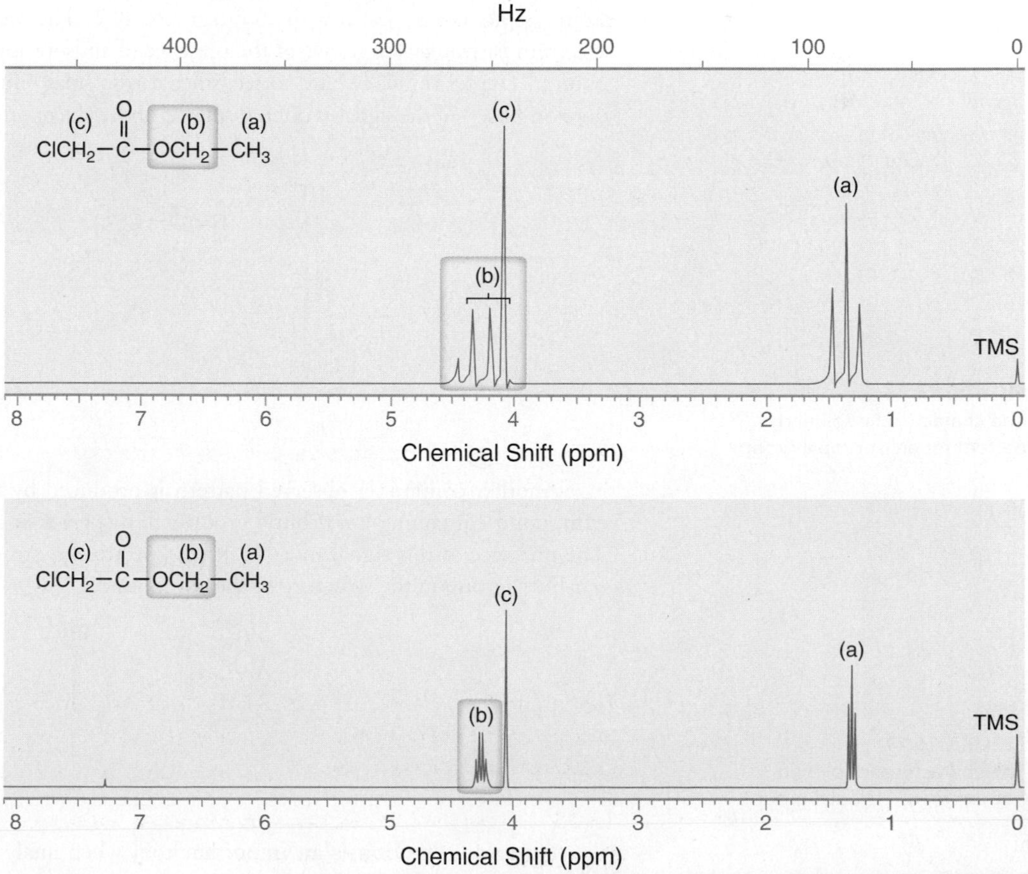

FIGURE 16.14
The 60 MHz ^1H NMR spectrum (top) of ethyl chloroacetate and the 300 MHz ^1H NMR spectrum (bottom) of ethyl chloroacetate. The 300 MHz spectrum has a better resolution and avoids overlapping signals.

The coupling constant only appears to be larger in the 60 MHz ^{1}H NMR spectrum, because each δ unit corresponds with 60 Hz. The distance between each peak (7 Hz) is more than 10% of a δ unit. In contrast, the coupling constant appears much smaller in the 300 MHz ^{1}H NMR spectrum, because each δ unit corresponds with 300 Hz, and as a result, the distance between each peak (7 Hz) is only 2% of a δ unit. This example illustrates why spectrometers with higher operating frequencies provide better resolution and avoid overlapping signals. For this reason, 60 MHz NMR spectrometers are rarely used for routine research. They have been widely replaced with 300 and 500 MHz instruments.

Pattern Recognition

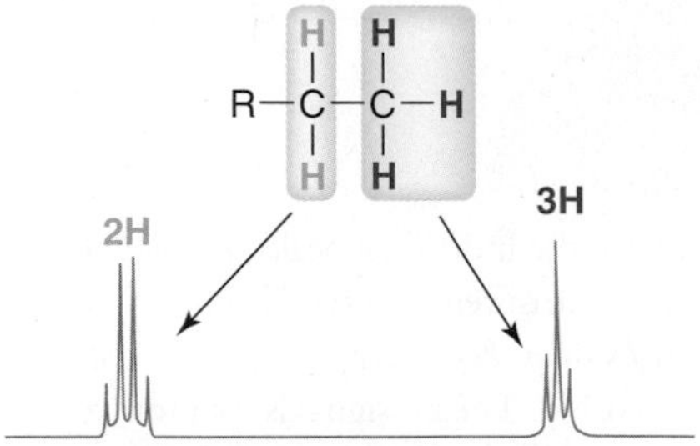

FIGURE 16.15
The characteristic splitting pattern for an ethyl group.

Specific splitting patterns are commonly seen in ^{1}H NMR spectra, and recognizing these patterns allows for a more efficient analysis. For example, consider the splitting pattern produced by an ethyl group (Figure 16.15). A compound containing an ethyl group will display a triplet with an integration of 3, upfield from a quartet with an integration of 2 in its ^{1}H NMR spectrum. The presence of these signals in a spectrum is strongly suggestive of the presence of an ethyl group in the structure of the compound. Since the two kinds of protons of an ethyl group are splitting each other, the *J* values for the triplet and quartet must be equivalent. If it is clear that the *J* values are different, then the two signals are not an ethyl group.

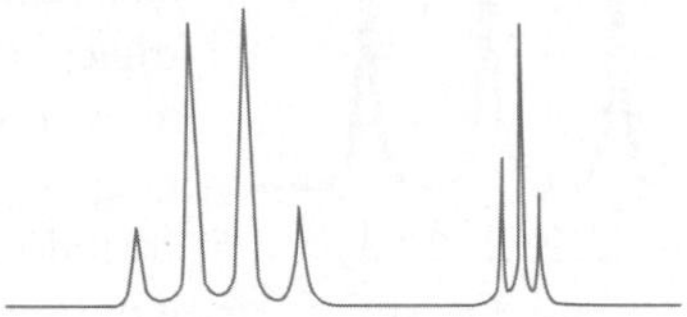

not an ethyl group

In this image, the quartet has a larger *J* value than the triplet, so these two signals are not splitting each other. In a situation like this, you would need to look at the rest of the NMR spectrum and find the signals that are splitting each other.

Another commonly observed splitting pattern is produced by isopropyl groups (Figure 16.16). A compound containing an isopropyl group will display a doublet with an integration of 6, upfield from a septet (seven peaks) with an integration of 1. The presence of these signals in a ^{1}H NMR spectrum is strongly suggestive of the presence of an isopropyl group in the structure of the compound. A septet is usually hard to see, since it is so small (integration of 1), so an enlarged reproduction (inset) of the signal is often displayed above the original signal (as shown in Figure 16.16).

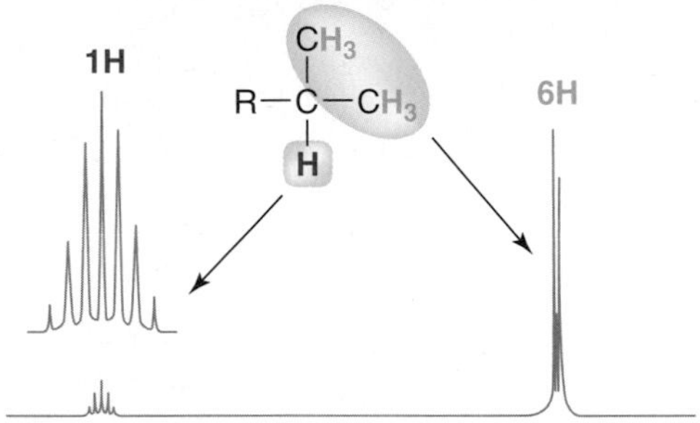

FIGURE 16.16
The characteristic splitting pattern for an isopropyl group.

Another commonly observed pattern is produced by *tert*-butyl groups (Figure 16.17). A compound containing a *tert*-butyl group will display a singlet with a relative integration of 9. The presence of this signal in a ^{1}H NMR spectrum is strongly suggestive of the presence of a *tert*-butyl group in the structure of the compound.

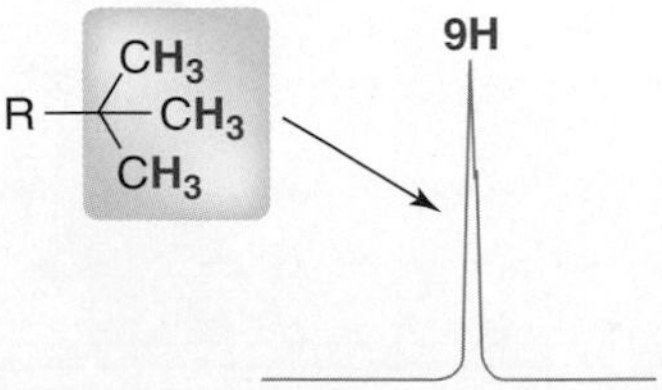

FIGURE 16.17
The characteristic splitting pattern for a *tert*-butyl group.

Pattern recognition is an important tool when analyzing NMR spectra, so let's quickly review the patterns we have seen (Figure 16.18).

Ethyl

Isopropyl

***tert*-butyl**

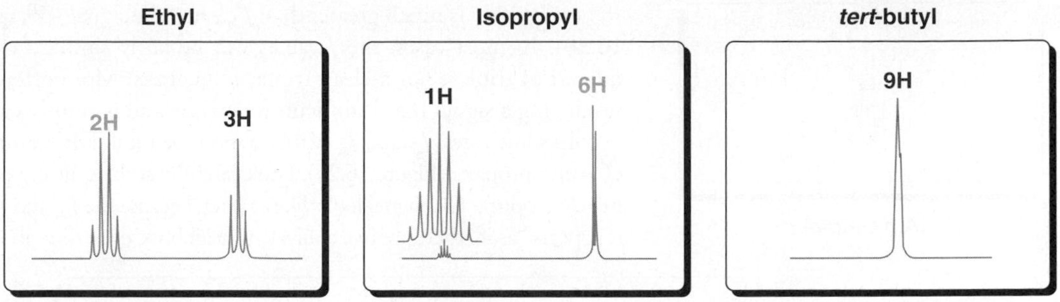

FIGURE 16.18
The characteristic splitting patterns for ethyl, isopropyl, and *tert*-butyl groups.

CONCEPTUAL CHECKPOINT

16.17 Below are NMR spectra of several compounds. Identify whether these compounds are likely to contain ethyl, isopropyl, and/or *tert*-butyl groups:

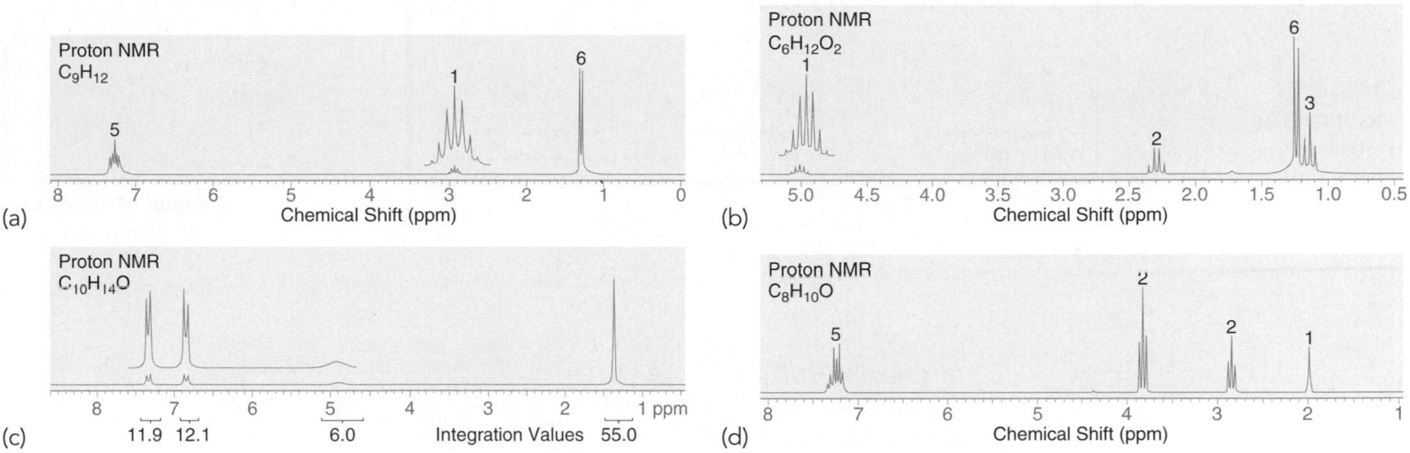

Complex Splitting

Complex splitting occurs when a proton has two different kinds of neighboring protons. For example:

$$\overset{\displaystyle H_a \quad H_b \quad H_c}{\underset{\displaystyle H_a \quad H_b \quad H_c}{H_a\!-\!C\!-\!C\!-\!C\!-\!X}}$$

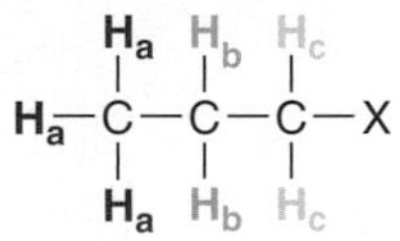

Consider the expected splitting pattern for H_b in this example. The signal for H_b is being split into a quartet because of the nearby H_a protons, and it is being split into a triplet because of the nearby H_c protons. The signal will therefore be comprised of 12 peaks (4×3). The appearance of the signal will depend greatly on the J values. If J_{ab} is much greater than J_{bc}, then the signal will appear as a quartet of triplets. This is illustrated in the splitting tree shown in Figure 16.19.

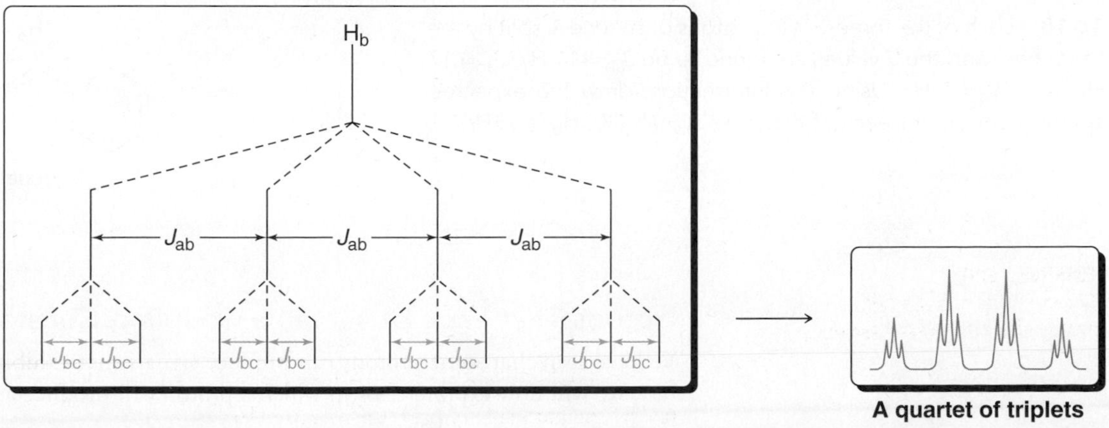

FIGURE 16.19
A splitting tree showing how a quartet of triplets is formed.

A quartet of triplets

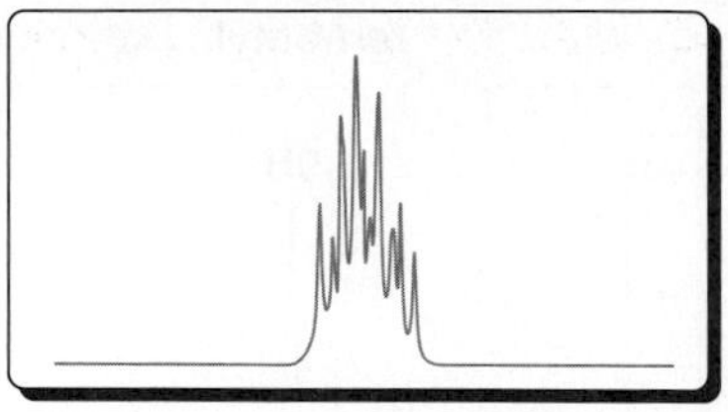

A multiplet

If, however, J_{bc} is much greater than J_{ab}, then the signal will appear as a triplet of quartets (Figure 16.20). In most cases, the J values will be fairly similar, and we will observe neither a clean quartet of triplets nor a clean triplet of quartets. More often, several of the peaks will overlap, producing a signal that is difficult to analyze and is simply called a **multiplet**.

In some cases, J_{ab} and J_{ac} will be almost identical. For example, consider the ^{1}H NMR spectrum of 1-nitropropane (Figure 16.21). Look carefully at the splitting pattern of the H_b protons (at approximately 2 ppm). This signal looks like a sextet, because the J_{ab} and J_{bc} are so close in value. In such a case, it appears "as if" there are five equivalent neighbors, even though all five protons are not equivalent.

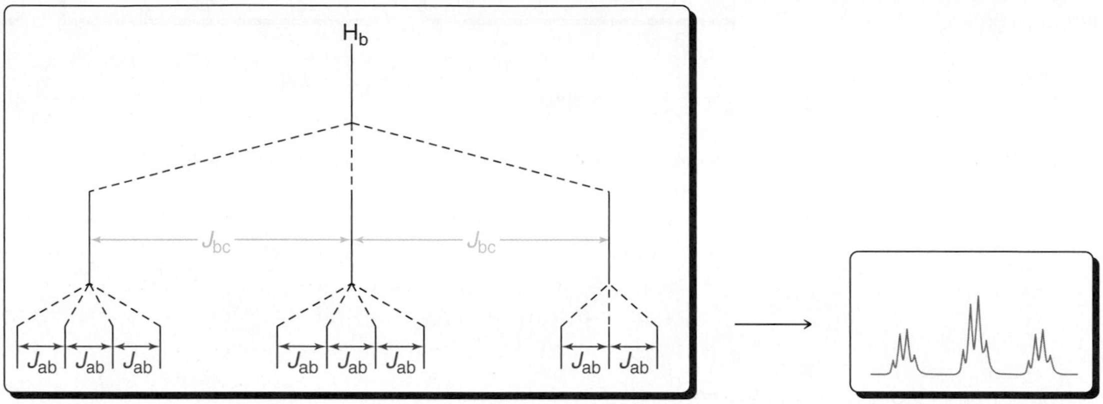

FIGURE 16.20
A splitting tree showing how a triplet of quartets is formed.

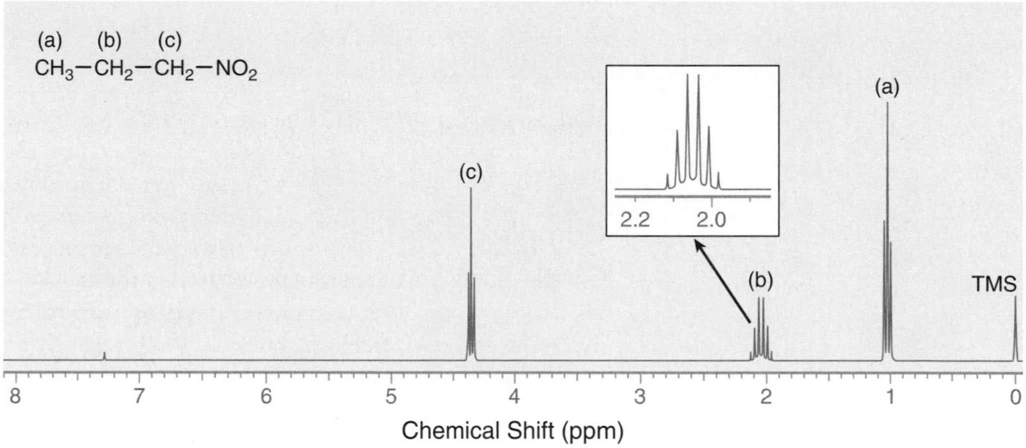

FIGURE 16.21
The ^{1}H NMR spectrum of 1-nitropropane, showing an "apparent" sextet for H_b.

CONCEPTUAL CHECKPOINT

16.18 Each of the three vinylic protons of styrene is split by the other two, and the J values are found to be $J_{ab} = 11$ Hz, $J_{ac} = 17$ Hz, and $J_{bc} = 1$ Hz. Using this information, draw the expected splitting pattern for each of the three signals (H_a, H_b, and H_c).

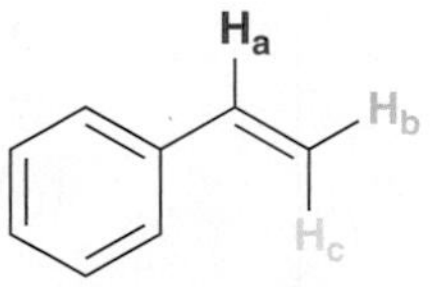

Styrene

Protons That Lack Observable Coupling Constants

We have seen that some protons can produce signals that exhibit complex splitting. In contrast, we will now explore cases in which a proton can produce a signal for which splitting is not observed despite the presence of neighboring protons. Consider the ¹H NMR spectrum of ethanol:

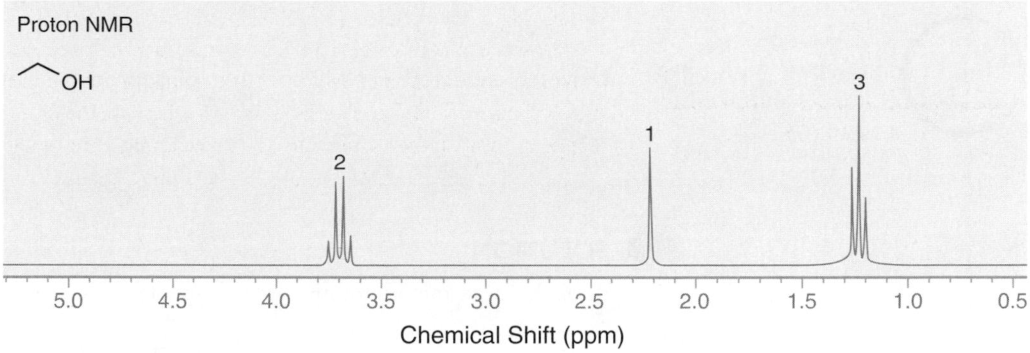

As expected, the spectrum exhibits the characteristic signals of an ethyl group. In addition, another signal is observed at 2.2 ppm, representing the hydroxyl proton (OH). Hydroxyl protons typically produce a signal between 2 and 5 ppm, and it is often difficult to predict exactly where that signal will appear. In this NMR spectrum notice that the hydroxyl proton is not split into a triplet from the neighboring methylene group. Generally, no splitting is observed across the oxygen of an alcohol, because proton exchange is a very rapid process that is catalyzed by trace amounts of acid or base.

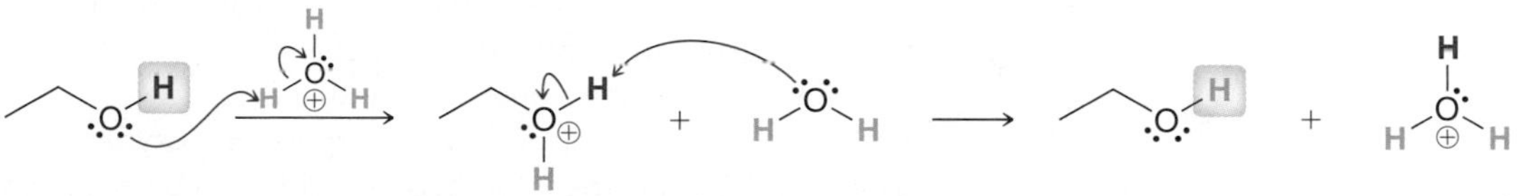

Hydroxyl protons are said to be **labile**, because of the rapid rate at which they are exchanged. This proton transfer process occurs at a faster rate than the timescale of an NMR spectrometer, producing a blurring effect that averages out any possible splitting effect. It is possible to slow down the rate of proton transfer by scrupulously removing the trace amounts of acid and base dissolved in ethanol. Such purified ethanol does in fact exhibit splitting across the oxygen atom, and the signal at 2.2 ppm is observed to be a triplet.

There is one other common example of neighboring protons that often do not produce observable splitting. Aldehydic protons, which generally produce signals near 10 ppm, will often couple only weakly with their neighbors (i.e., a very small *J* value).

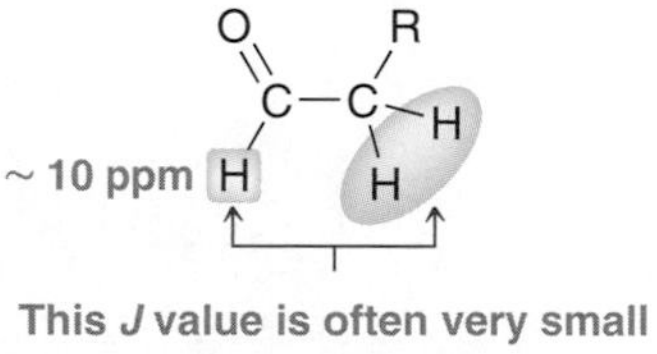

Depending on the size of the *J* value, splitting may or may not be noticeable. If the *J* value is too small, then the signal near 10 ppm will be appear to be a singlet, despite the presence of neighboring protons.

16.8 Drawing the Expected ¹H NMR Spectrum of a Compound

In the previous sections, we explored the three characteristics of every signal (chemical shift, multiplicity, and integration). In this section, we will apply the concepts and skills developed in the previous sections, and we will practice drawing the expected ¹H NMR spectrum of a compound. The following exercise illustrates the procedure.

SKILLBUILDER

16.6 DRAWING THE EXPECTED ^{1}H NMR SPECTRUM OF A COMPOUND

LEARN the skill

Draw the expected ^{1}H NMR spectrum of isopropyl acetate.

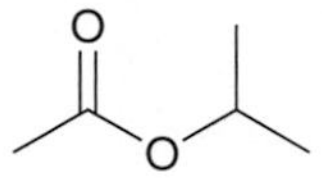

SOLUTION

Begin by determining the number of signals:

STEP 1
Determine the number of signals.

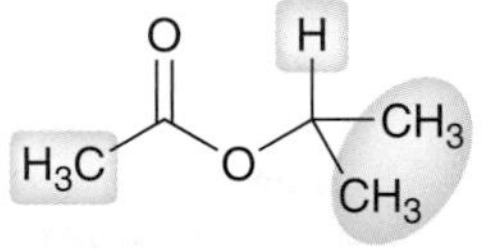

STEPS 2–4
Predict the chemical shift, integration and multiplicity of each signal.

STEP 5
Draw each signal.

This compound is expected to produce three signals in its ^{1}H NMR spectrum. For each signal, analyze all three characteristics methodically. Let's begin with the methyl group on the left side of the molecule. A methyl group is expected to produce a signal at 0.9 ppm, and the neighboring carbonyl adds +1, so we expect the signal to appear at approximately 1.9 ppm. The integration should be 3, because there are three protons. The multiplicity should be a singlet, because there are no neighbors.

Now consider the signal of the methine proton on the right. The benchmark value for a methine proton is 1.7 ppm, and the neighboring oxygen adds +3, so we expect the signal to appear at 4.7 ppm. The integration should be 1, because there is one proton. The multiplicity should be a septet, because there are six neighbors.

The last signal is from the two methyl groups on the right. Methyl groups have a benchmark value of 0.9 ppm, and the distant oxygen adds +0.6, so we expect a signal at approximately 1.5 ppm. The integration should be 6 because there are six protons, and the multiplicity should be a doublet because there is only one neighbor. This information enables us to draw the expected ^{1}H NMR spectrum. On a separate piece of paper, try to draw the spectrum. Then compare your drawn spectrum to the actual spectrum below.

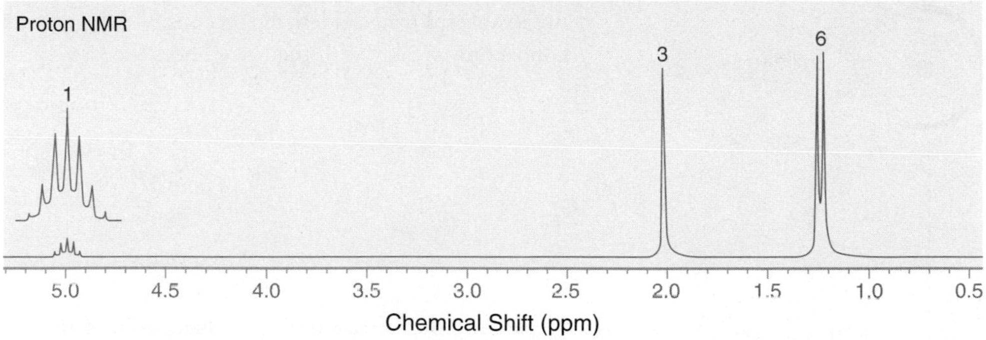

Note that our predicted chemical shifts are only estimates and should be treated as such. The actual chemical shifts will usually be fairly close to the predicted value.

PRACTICE the skill **16.19** Draw the expected ^{1}H NMR spectrum for each of the following compounds:

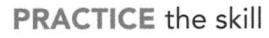

(a) 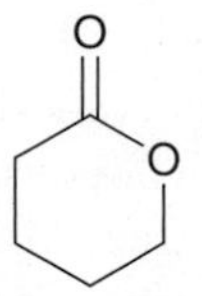
(b) 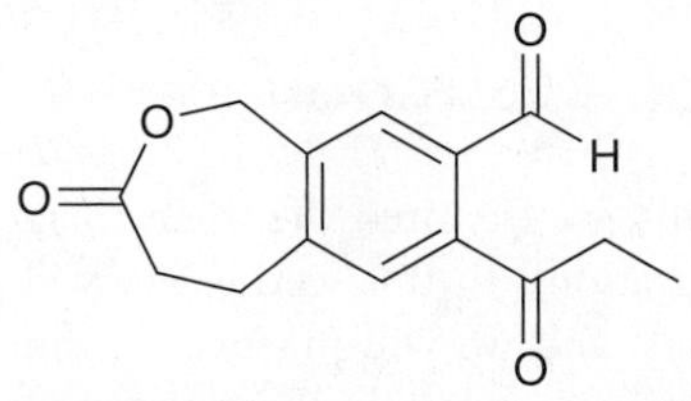
(c) 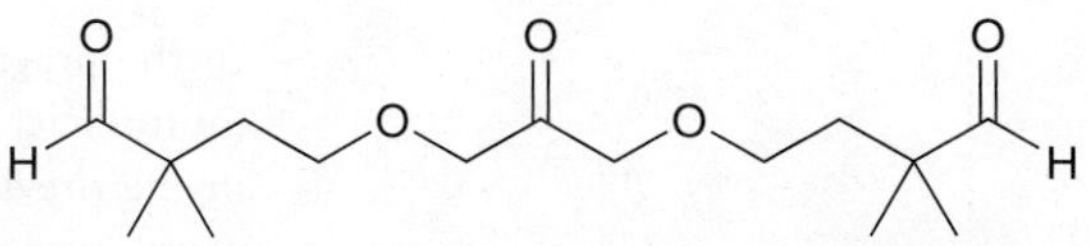

APPLY the skill

16.20 The ^{1}H NMR spectrum of a compound with molecular formula $C_7H_{14}O_3$ exhibits only three signals, and all three signals appear above 2 ppm (downfield of 2 ppm) on the spectrum. Propose a structure for this compound.

need more **PRACTICE?** **Try Problem 16.41**

16.9 Using ^{1}H NMR Spectroscopy to Distinguish between Compounds

NMR spectroscopy is a powerful tool for distinguishing compounds from each other. For example, consider the following three constitutional isomers, which produce different numbers of signals in their ^{1}H NMR spectra.

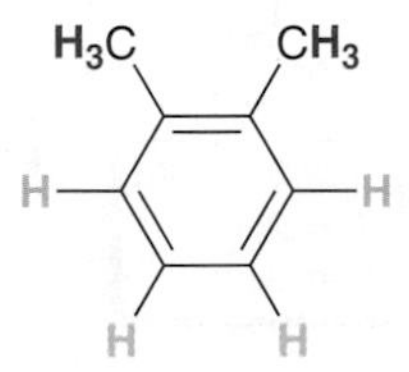

Three signals

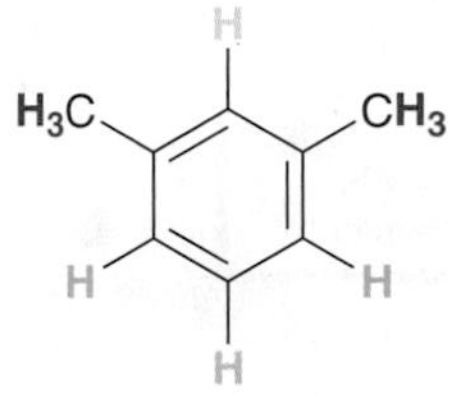

Four signals

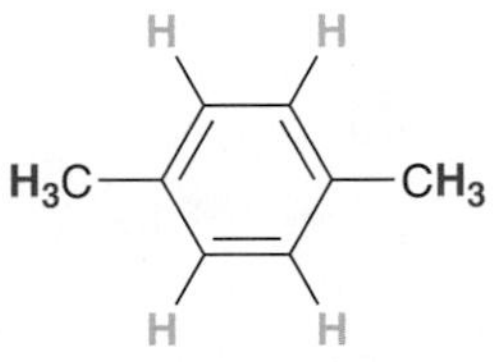

Two signals

These examples illustrate how similar compounds can have vastly different NMR spectra. Even compounds that produce the same number of signals can often be easily distinguished with NMR spectroscopy. You just need to identify one signal that would be different in the two spectra. The following exercise illustrates this point.

SKILLBUILDER

16.7 USING ^{1}H NMR SPECTROSCOPY TO DISTINGUISH BETWEEN COMPOUNDS

LEARN the skill

How would you use ^{1}H NMR spectroscopy to distinguish between the following compounds?

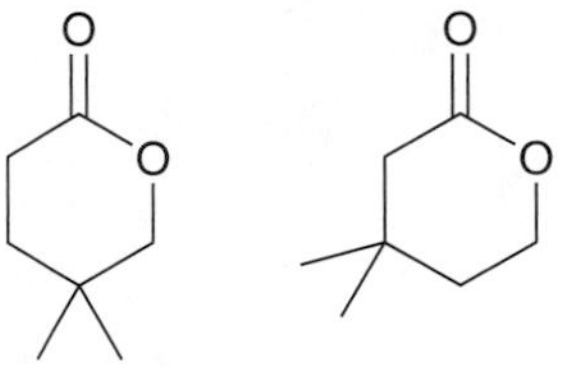

SOLUTION

First look to see if the two compounds will produce a different number of signals. For each of the structures, we expect four signals.

STEP 1
Identify the number of signals that each compound will produce.

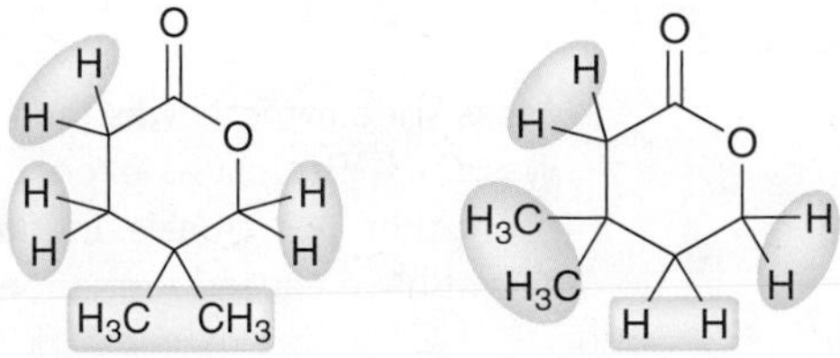

Nevertheless, there are many other ways to distinguish these two compounds. In order to see all of the possible ways, let's methodically analyze all three characteristics for each signal:

STEP 2
Determine the expected chemical shift, integration and multiplicity of each signal in both compounds.

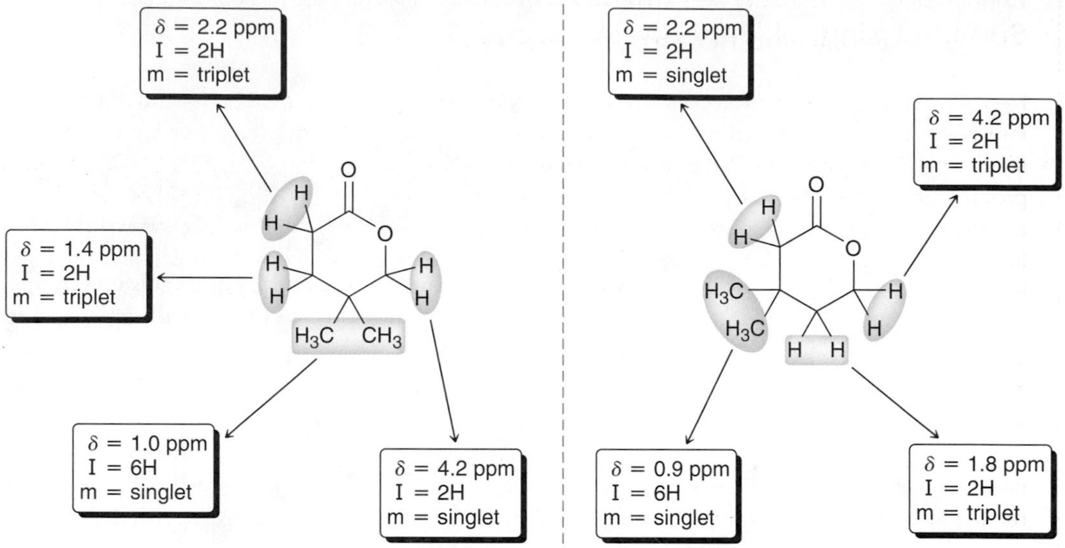

STEP 3
Look for differences in the chemical shifts, multiplicities, or integration values of the signals.

Now we can compare the data and look for differences. If we look at the chemical shifts for all four signals, we find that both compounds produce signals in similar locations. However, the nature of these signals is different. For example, both compounds will produce signals at 4.2 ppm, but the first compound produces a singlet, while the second compound produces a triplet. If we look at 2.2 ppm, we find that the first compound should have a triplet at that location, while the second compound should have a singlet.

Here is the bottom line: If you are able to predict all three characteristics of every signal, then you should be able to find several differences, even for compounds that are very similar in structure. At first, it might be helpful to actually write out all three characteristics of every signal and then compare the data. But after working several problems, you should find that you can identify differences without having to write out all of the data.

PRACTICE the skill **16.21** How would you use ^{1}H NMR spectroscopy to distinguish between the following compounds?

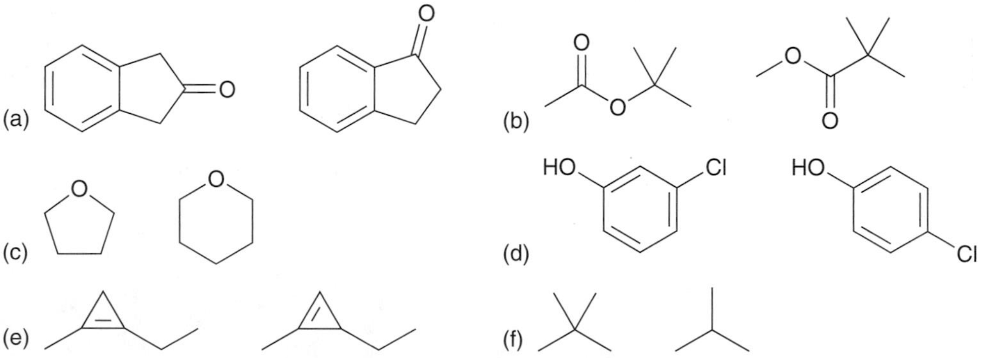

APPLY the skill **16.22** A chemist was attempting to achieve the following transformation:

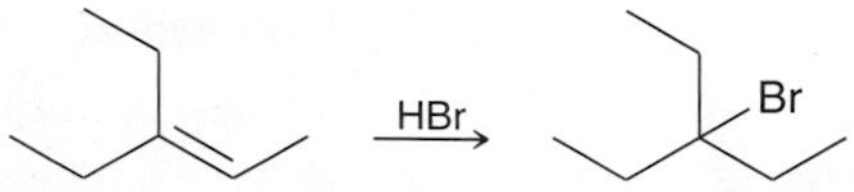

Mass spectrometry was then used to verify that the molecular formula of the major product was $C_7H_{15}Br$, just as expected. However, the ^{1}H NMR spectrum of the product exhibited more than two signals. Identify what product was actually obtained. Can you suggest what might have caused formation of this product?

need more **PRACTICE?** **Try Problems 16.38, 16.47**

MEDICALLYSPEAKING

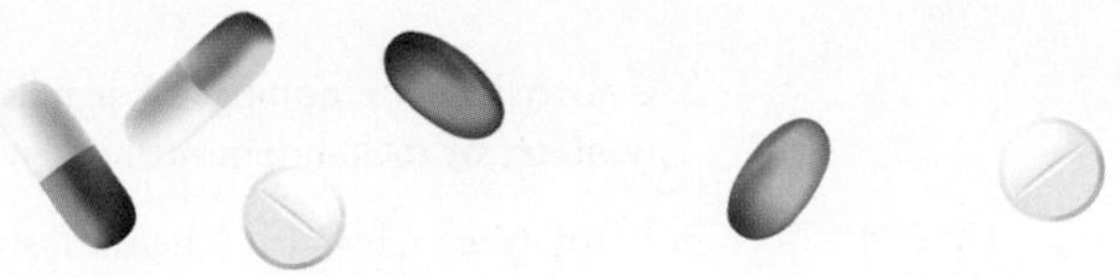

Detection of Impurities in Heparin Sodium Using ^{1}H NMR Spectroscopy

Heparin sodium is an anticoagulant that is derived from pig intestines. It has been used since the 1930s for the prevention of life-threatening blood clots. Because heparin is an animal-derived product, a number of steps are used in the manufacturing process to ensure that infectious agents and impurities are removed. In January 2008, the FDA announced a recall for certain lots of heparin sodium due to an increased number of reported adverse reactions (>350) in association with heparin administration. Some of the adverse events reported were serious in nature, including allergic or hypersensitivity-type reactions, with symptoms of oral swelling, shortness of breath, and severe hypotension.

During the course of the FDA investigation, contaminants similar in structure to heparin were found in some of the recalled products. The contaminants were found in significant quantities (accounting for 5–20% of the total mass of each sample). In order to ensure a safe supply of this life-saving medication, the FDA released information on two analytical methods for detection of the contaminants.

One of these methods utilizes ^{1}H NMR spectroscopy to differentiate products with and without the contaminants. In pure heparin sodium (no contaminants), a singlet peak appears upfield of 2 ppm without any additional features between 2.0 and 3.0 ppm. In heparin sodium products containing the contaminants, two additional peaks appear near 2 ppm. For future production of heparin, the FDA has recommended that manufacturers use ^{1}H NMR spectroscopy in conjunction with other conventional methods to screen for impurities prior to the release of the final products into the market.

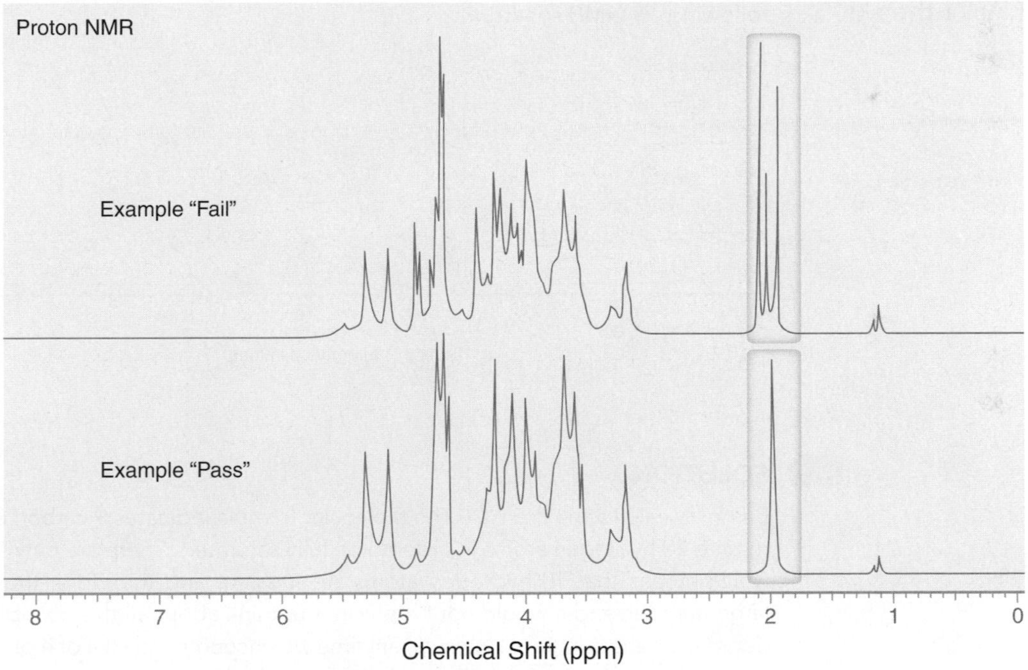

16.10 Analyzing a ^{1}H NMR Spectrum

In this section, we will practice analyzing and interpreting NMR spectra, a process that involves four discrete steps:

LOOKING BACK
For a review of how to calculate and interpret the HDI of a compound, see Section 15.16.

1. Always begin by inspecting the molecular formula (if it is given), as it provides useful information. Specifically, calculating the hydrogen deficiency index (HDI) can provide important clues about the structure of the compound. An HDI of zero indicates that the compound does not possess any rings or π bonds. An HDI of 1 indicates that the compound has either one ring or one π bond. The larger the HDI, the less useful it is. However, an HDI of 4 or more should indicate the likely presence of an aromatic ring:

Four degrees of unsaturation (HDI = 4)

2. Consider the number of signals and integration of each signal (gives clues about the symmetry of the compound).
3. Analyze each signal (chemical shift, integration, and multiplicity), and then draw fragments consistent with each signal. These fragments become our puzzle pieces that must be assembled to produce a molecular structure.
4. Assemble the fragments into a molecular structure.

The following exercise illustrates how this is done.

SKILLBUILDER

16.8 ANALYZING A ^{1}H NMR SPECTRUM AND PROPOSING THE STRUCTURE OF A COMPOUND

LEARN the skill

Identify the structure of a compound with molecular formula $C_9H_{10}O$ that exhibits the following ^{1}H NMR spectrum.

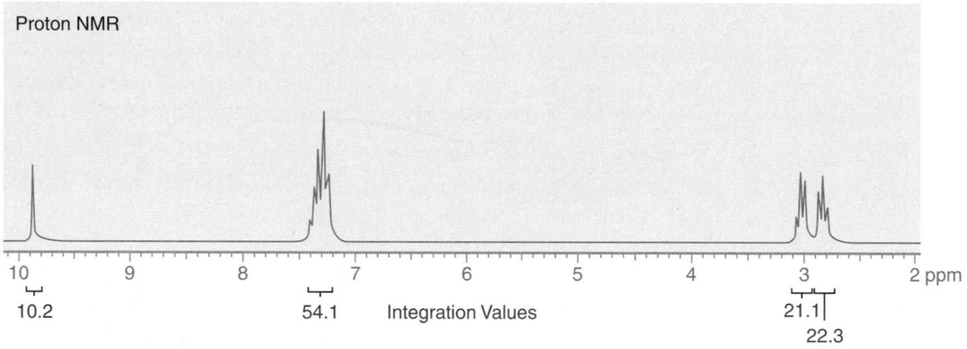

SOLUTION

STEP 1
Using the molecular formula, calculate and interpret the HDI.

Begin by calculating the HDI. The molecular formula indicates 9 carbon atoms, which would require 20 hydrogen atoms in order to be fully saturated. There are only 10 hydrogen atoms, which means that 10 hydrogen atoms are missing, and therefore, the HDI is 5. This is a large number, and it would not be efficient to think about all the possible ways to have five degrees of unsaturation. However, anytime we encounter an HDI of 4 or more, we should be on the lookout for an aromatic ring. We must keep this in mind when analyzing the spectrum. We should expect to see an aromatic ring (HDI = 4) plus one other degree of unsaturation (either a ring or a double bond).

STEP 2
Consider the number of signals and integration values of each signal.

Next, consider the number of signals and the integration value for each signal. Be on the lookout for integration values that would suggest the presence of symmetry elements. For example, a signal with an integration of 4 would suggest two equivalent CH_2 groups.

In this spectrum, we see four signals. In order to analyze the integration of each signal, we must first divide by the lowest number (10.2).

$$\frac{10.2}{10.2} = 1 \qquad \frac{54.1}{10.2} = 5.30 \qquad \frac{21.1}{10.2} = 2.07 \qquad \frac{22.3}{10.2} = 2.19$$

The ratio is approximately 1:5:2:2. Now look at the molecular formula. There are 10 protons in the compound, so the relative integration values represent the actual number of protons giving rise to each signal.

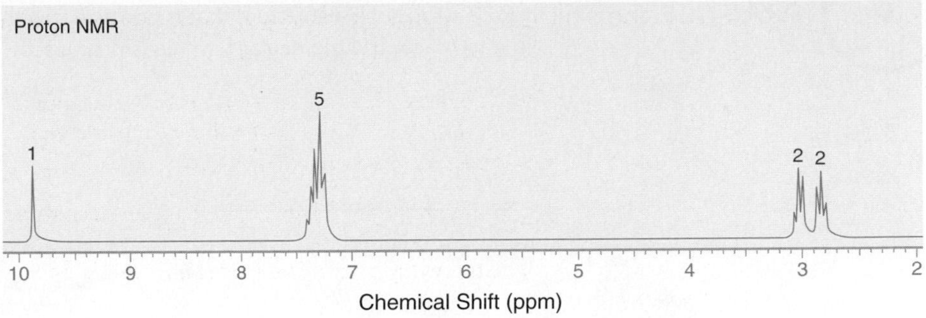

STEP 3
Analyze the chemical shift, multiplicity, and integration of each signal, and then draw a fragment consistent with each signal.

Now analyze each signal. Starting upfield, there are two triplets skewed toward each other, each with an integration of 2. This suggests that there are two adjacent methylene groups.

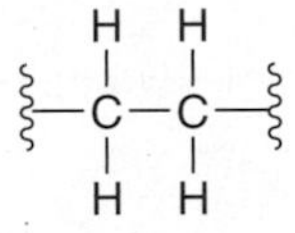

These signals do not appear at 1.2, where methylene groups are expected, so one or more factors are shifting these signals downfield. Our proposed structure must take this into account.

Moving downfield through the spectrum, the next signal appears just above 7 ppm, characteristic of aromatic protons (just as we suspected after analyzing the HDI). The multiplicity of aromatic protons only rarely gives useful information. More often, a messy multiplet of overlapping signals is observed. But the integration value gives valuable information. Specifically, there are five aromatic protons, which means that the aromatic ring is monosubstituted.

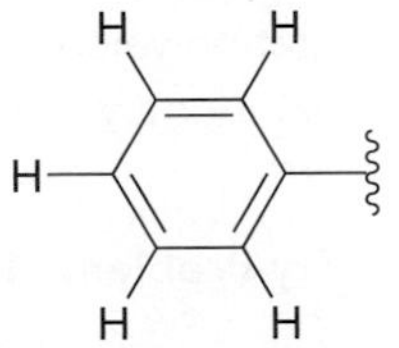

Five aromatic protons

LOOKING BACK
Recall from Section 16.7 that aldehydic protons often appear as singlets despite the presence of neighboring protons.

Next, move on to the last signal, which is a singlet at 10 ppm with an integration of 1. This is suggestive of an aldehydic proton. Our analysis has produced the following fragments.

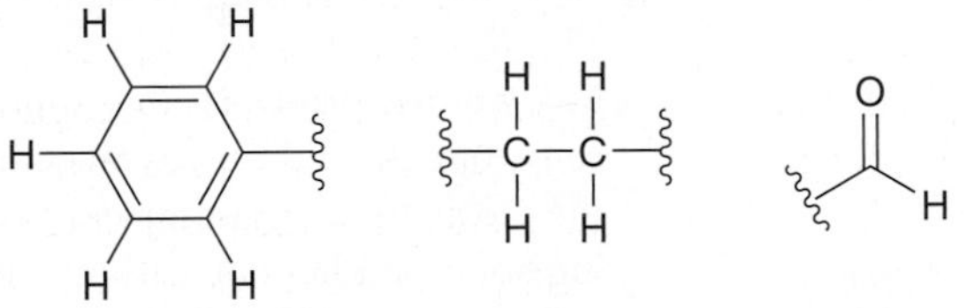

STEP 4
Assemble the fragments.

The final step is to assemble these fragments. Fortunately, there is only one way to assemble these three puzzle pieces.

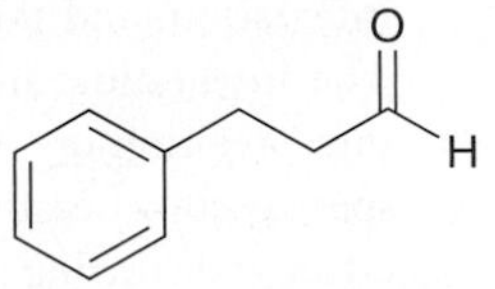

We mentioned before that each methylene group is being shifted downfield by one or more factors. Our proposed structure explains the observed chemical shifts. In particular, one methylene group is shifted significantly by the carbonyl group and slightly by the aromatic ring. The other methylene group is being shifted significantly by the aromatic ring and slightly by the carbonyl group.

PRACTICE the skill **16.23** Propose a structure that is consistent with each of the following ^1H NMR spectra. In each case, the molecular formula is provided.

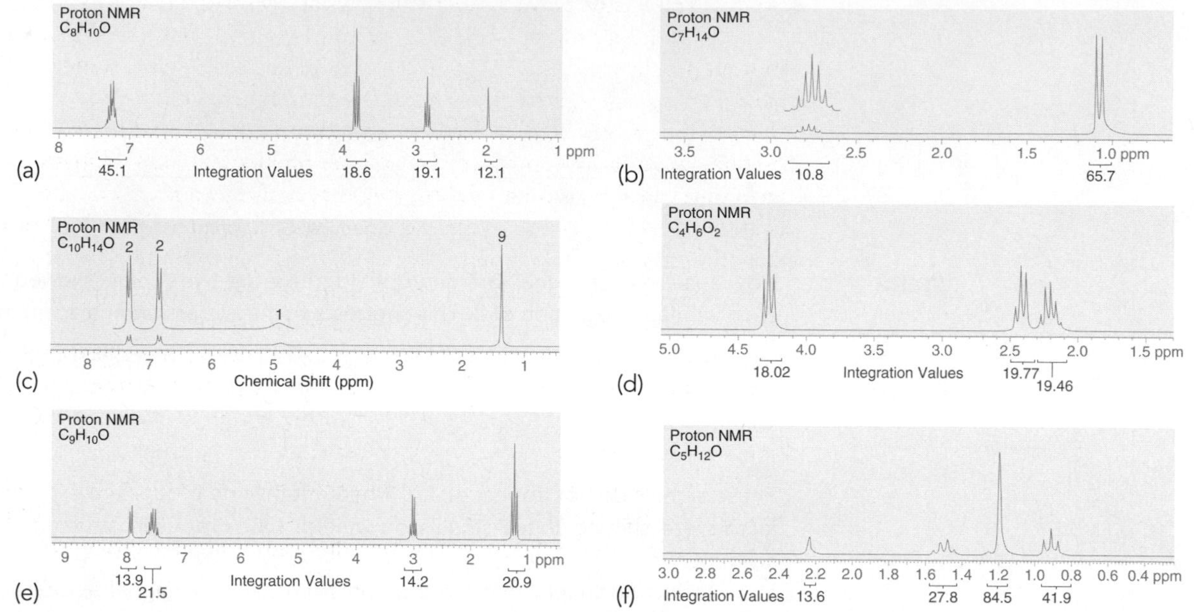

APPLY the skill **16.24** A compound with molecular formula $C_{10}H_{10}O_4$ produces a ^1H NMR spectrum that exhibits only two signals, both singlets. One signal appears at 3.9 ppm with a relative integration value of 79. The other signal appears at 8.1 ppm with a relative integration value of 52. Identify the structure of this compound.

need more **PRACTICE?** **Try Problems 16.54, 16.57**

16.11 Acquiring a ^{13}C NMR Spectrum

LOOKING BACK
Recall from Section 16.1 that ^{12}C does not have a nuclear spin and therefore cannot be probed with NMR spectroscopy.

Many of the principles that apply to ^1H NMR spectroscopy also apply to ^{13}C NMR spectroscopy, but there are a few major differences, and we will focus on those. For example, ^1H is the most abundant isotope of hydrogen, but ^{13}C is only a minor isotope of carbon, representing about 1.1% of all carbon atoms found in nature. As a result, only one in every hundred carbon atoms will resonate, which demands the use of a sensitive receiver coil for ^{13}C NMR spectroscopy.

In ^1H NMR spectroscopy, we saw that each signal has three characteristics (chemical shift, integration, and multiplicity). In ^{13}C NMR spectroscopy, only the chemical shift is important. The integration and multiplicity of ^{13}C signals are not reported, which greatly simplifies the interpretation of ^{13}C NMR spectra. Integration values are not routinely calculated in ^{13}C NMR spectroscopy because the pulse technique employed by FT NMR spectrometers has the undesired effect of distorting the integration values, rendering them useless in most cases. Multiplicity is also not a common characteristic of ^{13}C NMR spectra.

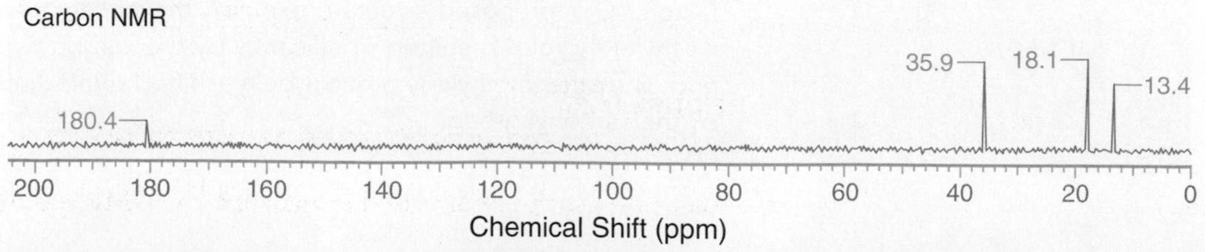

Notice that all of the signals are recorded as singlets. There are several good reasons for this. First, no splitting is observed between neighboring carbon atoms because of the low abundance of ^{13}C. The likelihood of a compound having two neighboring ^{13}C atoms is quite small, so ^{13}C -^{13}C splitting is not observed. In contrast, ^{13}C -^{1}H splitting does occur, and it creates significant problems. The signal of each ^{13}C atom nucleus is split not only by the protons directly connected to it (separated by only one sigma bond) but also by the protons that are two or three sigma bonds removed. This leads to very complex splitting patterns, and signals overlap to produce an unreadable spectrum. To solve the problem, all ^{13}C -^{1}H splitting is suppressed with a technique called **broadband decoupling**, which uses two rf transmitters. The first transmitter provides brief pulses in the range of frequencies that cause ^{13}C nuclei to resonate, while the second transmitter continuously irradiates the sample with the range of frequencies that cause all ^{1}H nuclei to resonate. This second rf source effectively decouples the ^{1}H nuclei from the ^{13}C nuclei, causing all of the ^{13}C signals to collapse to singlets.

The advantage of broadband decoupling comes at the expense of useful information that would otherwise be obtained from spin-spin coupling. A technique called **off-resonance decoupling** allows us to retrieve some of this information. With this technique, only the one-bond couplings are observed, so CH_3 groups appear as quartets, CH_2 groups appear as triplets, CH groups appear as doublets, and quaternary carbon atoms appear as singlets. Nonetheless, off-resonance decoupling is rarely used because it often produces overlapping peaks that are difficult to interpret. The desired information can be obtained using newer techniques, one of which is described in Section 16.13.

16.12 Chemical Shifts in ^{13}C NMR Spectroscopy

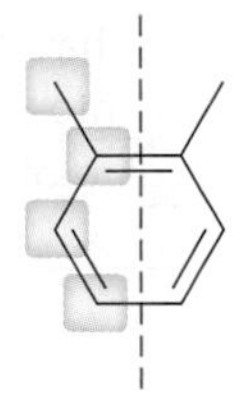
Four signals

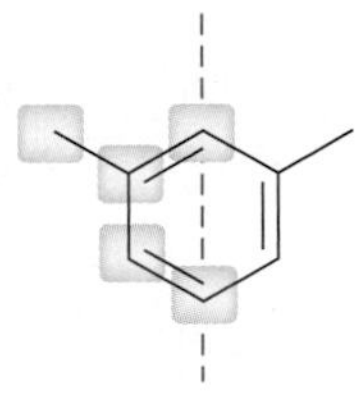
Five signals

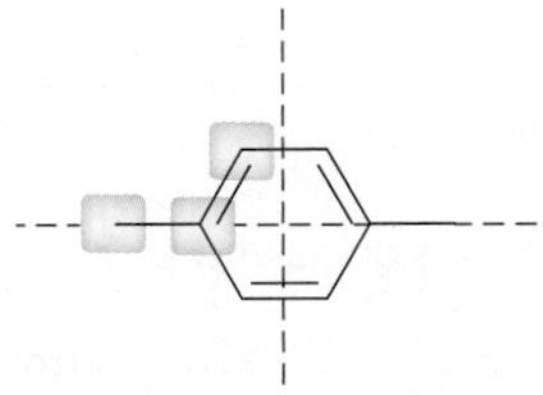
Three signals

The range of rf frequencies in ^{13}C spectroscopy is different from that used in ^{1}H NMR spectroscopy, because ^{13}C atoms resonate over a different frequency range. Just as in ^{1}H NMR spectroscopy, the position of each signal is defined relative to the frequency of absorption of a reference compound, TMS. With this definition, the chemical shift of each ^{13}C atom is constant, regardless of the operating frequency of the spectrometer. In ^{13}C NMR spectroscopy, chemical shift values typically range from 0 to 220 ppm.

The number of signals in a ^{13}C NMR spectrum represents the number of carbon atoms in different electronic environments (not interchangeable by symmetry). Carbon atoms that are interchangeable by a symmetry operation (either rotation or reflection) will only produce one signal. To illustrate this point, consider the compounds in the margin. Each compound has eight carbon atoms but does not produce eight signals. The unique carbon atoms in each compound are highlighted. Each carbon atom that is not highlighted is equivalent to one of the highlighted carbon atoms.

The location of each signal is dependent on shielding and deshielding effects, just as we saw in ^{1}H NMR spectroscopy. Figure 16.22 shows the chemical shifts of several important types of carbon atoms.

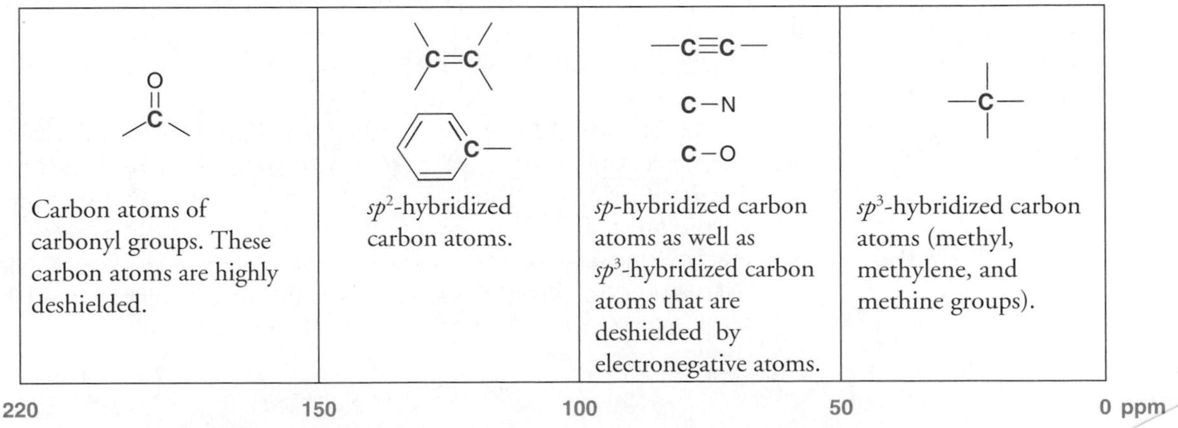

Carbon atoms of carbonyl groups. These carbon atoms are highly deshielded.
sp²-hybridized carbon atoms.
sp-hybridized carbon atoms as well as sp³-hybridized carbon atoms that are deshielded by electronegative atoms.
sp³-hybridized carbon atoms (methyl, methylene, and methine groups).
220
150
100
50
0 ppm

FIGURE 16.22
The general location of signals produced by different types of carbon atoms.

Let's now use this information to analyze a ^{13}C NMR spectrum.

SKILLBUILDER

16.9 PREDICTING THE NUMBER OF SIGNALS AND APPROXIMATE LOCATION OF EACH SIGNAL IN A ^{13}C NMR SPECTRUM

LEARN the skill

Consider the following compound.

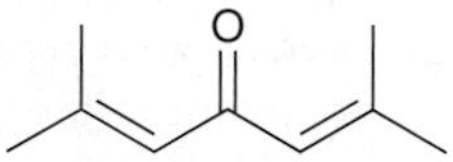

Predict the number of signals and the location of each signal in a ^{13}C NMR spectrum of this compound.

STEP 1
Look to see if any of the carbon atoms are interchangeable by either rotational or reflectional symmetry, and determine the number of expected signals.

SOLUTION

Begin by determining the number of expected signals. The compound has nine carbon atoms, but we must look to see if any of these carbon atoms are interchangeable by a symmetry operation. In this case, there is both a plane and an axis of symmetry. As a result, we expect only five signals in the ^{13}C NMR spectrum.

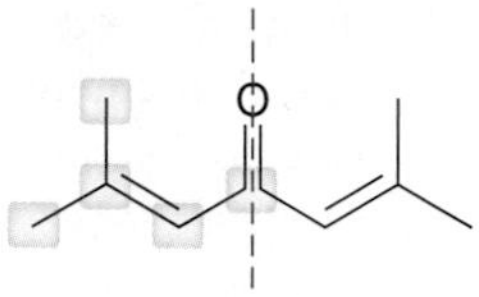

STEP 2
Predict the expected region in which each signal will appear, based on hybridization states and deshielding effects.

In general, when two methyl groups are attached to the same carbon atom, they will be equivalent. This is not the case in this example, because one methyl group is *cis* to the carbonyl group, while the other methyl group is *trans* to the carbonyl group.

The expected chemical shifts are shown below, categorized according to the region of the spectrum in which each signal is expected to appear:

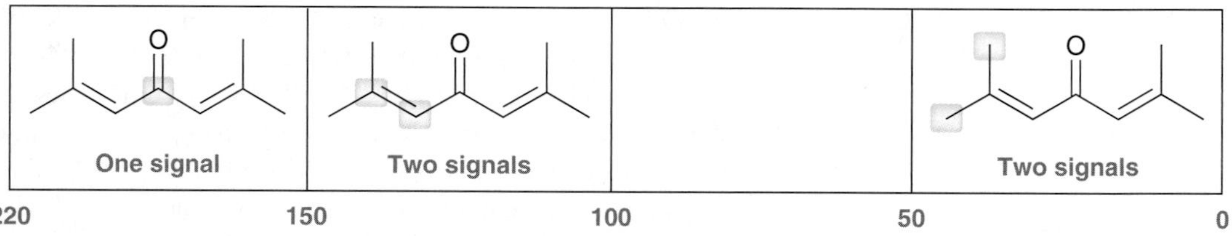

PRACTICE the skill

16.25 For each of the following compounds, predict the number of signals and location of each signal in a ^{13}C NMR spectrum:

(a) 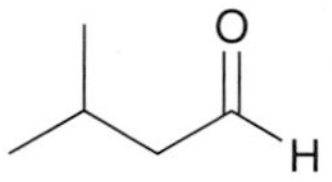 (b) 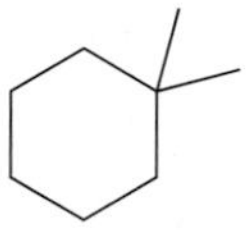 (c) 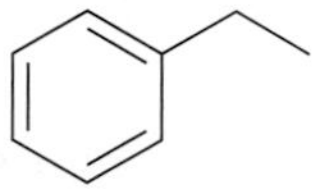 (d) 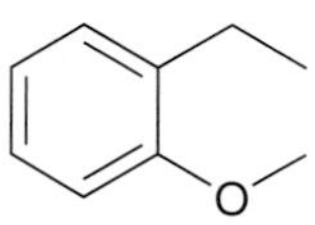 (e) 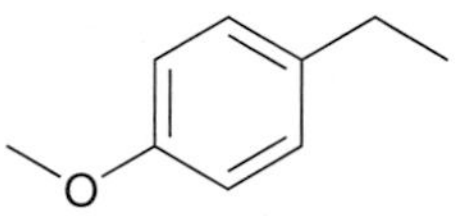

(f) 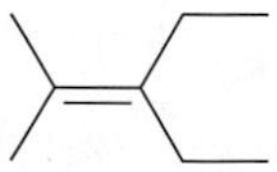 (g) 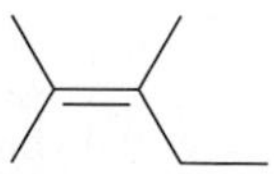 (h) 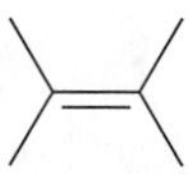 (i) 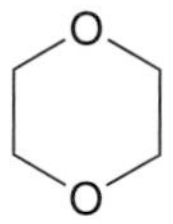 (j) 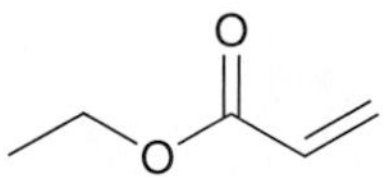

APPLY the skill

16.26 Compare the following two constitutional isomers. The ^{13}C NMR spectrum of the first compound exhibits five signals, while the second compound exhibits six signals. Explain.

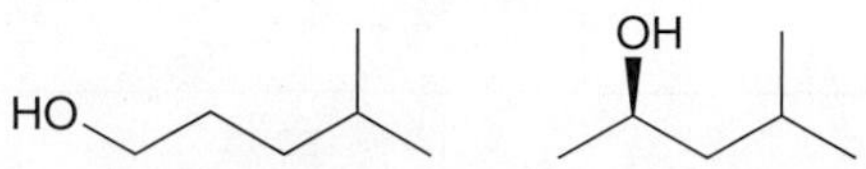

16.27 Draw the structure of a compound with molecular formula C_8H_{10} that exhibits five signals in its ^{13}C NMR spectrum, four of which appear between 100 and 150 ppm.

need more **PRACTICE?** **Try Problems 16.35, 16.38, 16.46**

16.13 DEPT ^{13}C NMR Spectroscopy

As mentioned in Section 16.11, a broadband-decoupled ^{13}C spectrum does not provide information regarding the number of protons attached to each carbon atom in a compound. This information can be obtained through a variety of recently developed techniques, one of which is called **distortionless enhancement by polarization transfer (DEPT)**. DEPT ^{13}C NMR spectroscopy utilizes two rf radiation emitters and relies on the fact that the intensity of each particular signal will respond to different pulse sequences in a predictable fashion, depending on the number of protons attached. This technique involves the acquisition of several spectra. First, a regular broadband-decoupled ^{13}C spectrum is acquired, indicating the chemical shifts associated with all carbon atoms in the compound. Then a special pulse sequence is utilized to produce a spectrum called a DEPT-90, in which only signals from CH groups appear. This spectrum does not show any signals resulting from CH_3 groups, CH_2 groups, or quaternary carbon atoms (C with no protons). Then, a different pulse sequence is employed to generate a spectrum, called a DEPT-135, in which CH_3 groups and CH groups appear as positive signals, CH_2 groups appear as negative signals (pointing down), and quaternary carbon atoms do not appear.

By comparing all of the spectra, it is possible to identify each signal in the broadband-decoupled spectrum as arising from either a CH_3 group, a CH_2 group, a CH group, or a quaternary carbon atom. This information is summarized in Table 16.4. Notice that each type of group exhibits a different absorption pattern when all three spectra are compared. For example, only CH groups give positive signals in all three spectra, while CH_2 groups are the only groups that give negative signals in the DEPT-135 spectrum. This technique therefore produces a series of spectra that collectively contain all of the information in an off-resonance decoupled spectrum, but without the disadvantage of overlapping signals. The following example illustrates how DEPT spectra can be interpreted.

TABLE 16.4 SIGNAL PATTERNS FOR DEPT ^{13}C SPECTROSCOPY

	CH_3	CH_2	CH	C
BROADBAND DECOUPLED	⅄	⅄	⅄	⅄
DEPT -90	—	—	⅄	—
DEPT -135	⅄	Y	⅄	—

SKILLBUILDER

16.10 DETERMINING MOLECULAR STRUCTURE USING DEPT ^{13}C NMR SPECTROSCOPY

LEARN the skill

Determine the structure of an alcohol with molecular formula $C_4H_{10}O$ that exhibits the following ^{13}C NMR spectra.

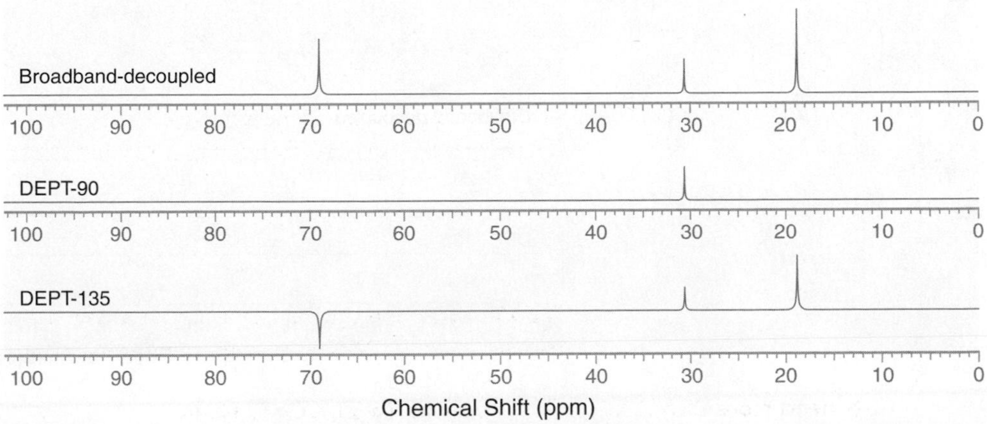

SOLUTION

Whenever the molecular formula is known, the first step is always to calculate the HDI and determine the information that it provides. In this case, there are zero degrees of unsaturation, which means that the compound does not have any rings or π bonds. Next, we focus on the number of signals in the broadband-decoupled spectrum. There are only three signals, but the molecular formula indicates that the compound has four carbon atoms. Therefore, we conclude that one of the signals must represent two chemically equivalent carbon atoms. Now we are ready to analyze each individual signal in the spectra. Using the information in Table 16.4, we can determine the following information:

- The signal at approximately 69 ppm is a CH_2 group (signal is negative in the DEPT-135).
- The signal at approximately 30 ppm is a CH group (signal is positive in all spectra).
- The signal at approximately 19 ppm is a CH_3 group (signal is positive in the broadband-decoupled spectrum, absent in DEPT-90, and positive in DEPT-135).

We can now record this information on the broadband-decoupled spectrum to aid in our analysis.

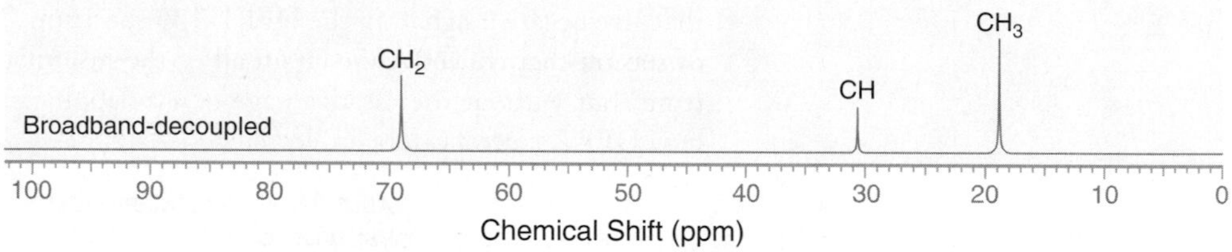

In order to determine which signal represents two carbon atoms, we notice that the molecular formula indicates that the structure must have 10 protons. So far, we have only accounted for 7 of the protons (CH_2+CH+CH_3+OH=7 protons). We need to account for 3 more protons. Therefore, we can conclude that the CH_3 signal must represent 2 equivalent carbon atoms.

We can now analyze the chemical shifts of each signal. The CH_2 signal is more downfield than the others (at 70 ppm), so this signal must represent the carbon atom attached directly to the oxygen. The signal at 30 ppm is also shifted slightly downfield (relative to the signal at 19 ppm), and therefore we expect this signal to represent the carbon atom that is beta to the OH group. Finally, the signal at 19 ppm represents two equivalent methyl groups, giving the following structure.

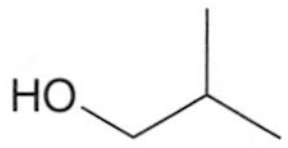

PRACTICE the skill **16.28** Determine the structure of a compound with molecular formula $C_5H_{10}O$ that exhibits the following broadband-decoupled and DEPT-135 spectra. The DEPT-90 spectrum has no signals.

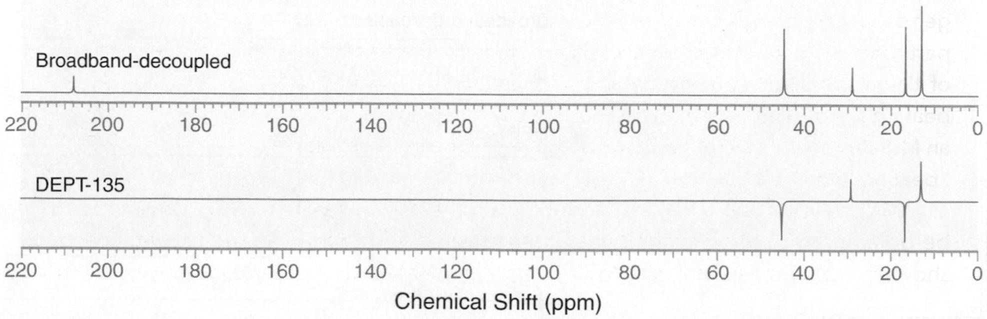

16.29 Determine the structure of an alcohol with molecular formula $C_5H_{12}O$ that exhibits the following signals in its ^{13}C NMR spectra:

(a) Broadband decoupled: 73.8 δ, 29.1 δ, and 9.5 δ

(b) DEPT-90: 73.8 δ

(c) DEPT-135: positive signals at 73.8 δ and 9.5 δ; negative signal at 29.1 δ

APPLY the skill

16.30 A compound with molecular formula $C_7H_{14}O$ exhibits the following ^{13}C NMR spectra:

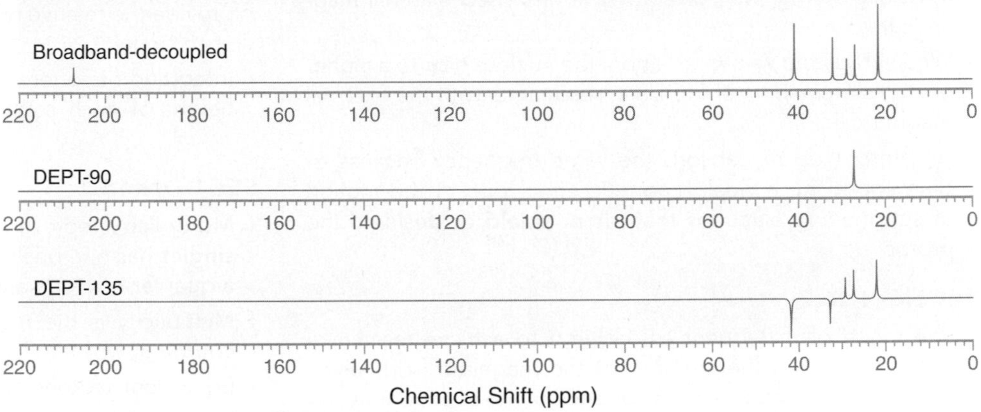

Several structures are consistent with these spectra. To determine which structure is correct, a ^{1}H NMR spectrum was acquired which exhibits six signals. One of those signals is a singlet at 1.9 ppm with an integration of 3, and another of the signals is a doublet at 0.9 ppm with an integration of 6. Using this information, identify the correct structure of the compound.

need more **PRACTICE?** **Try Problems 16.60, 16.61**

MEDICALLYSPEAKING

Magnetic Resonance Imaging (MRI)

Magnetic resonance imaging has developed into an invaluable medical diagnostic tool for its ability to take pictures of internal organs. MRI devices are essentially NMR spectrometers that are large enough to accommodate a human being. In the presence of a strong magnetic field, the subject is irradiated with rf waves that cover the range of resonance frequencies for ^{1}H nuclei (protons). Protons are abundant in the human body where there is water or fat. When these protons are excited, the signals produced are monitored and recorded. Unlike NMR spectrometers, MRI devices use sophisticated techniques that analyze the three-dimensional location of each signal as well as its intensity. The intensity of each signal is dependent on the density of hydrogen atoms and the nature of their electronic environment in that particular location. This allows the identification of different types of tissues and can even be used to monitor motion, such as a beating heart. Many different videos of a beating heart taken by an MRI device are readily available online (try a Google search for "beating heart MRI").

MRI devices are able to produce pictures that cannot be obtained with other techniques. The following MRI image shows a ruptured disc pressing on a patient's spinal cord.

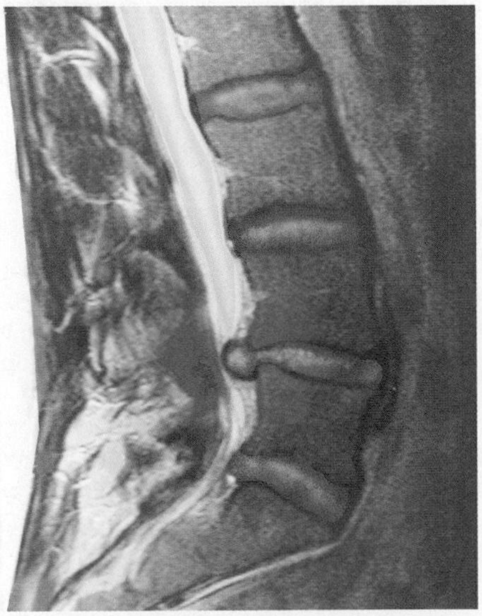

One of the most important features of MRI devices is the low risk to the patient. The magnetic field does not cause any known health concerns, and rf radiation is completely harmless.

REVIEW OF CONCEPTS AND VOCABULARY

SECTION 16.1

- **Nuclear magnetic resonance** (NMR) spectroscopy provides information about how the individual carbon and hydrogen atoms in a molecule are connected to each other.
- A spinning proton generates a **magnetic moment**, which must align either with or against an imposed external magnetic field.
- When irradiated with rf radiation, the nucleus flips to a higher energy state and is said to be in **resonance** with the external magnetic field.
- All protons do not absorb the same frequency because of **diamagnetism**, a weak magnetic effect due to the motion of surrounding electrons that either **shield** or **deshield** the proton.

SECTION 16.2

- The range of rf frequencies required to achieve resonance is dependent on the strength of the magnetic field, which determines the **operating frequency** of the spectrometer.
- **Continuous wave spectrometers** have been replaced by **FT-NMR** spectrometers in which the sample is irradiated with a short pulse that covers the entire range of relevant rf frequencies, and a **free induction decay** (FID) is recorded and then converted into a spectrum.
- Deuterated solvents containing a small amount of TMS are generally used for acquiring NMR spectra.

SECTION 16.3

- In a ^{1}H NMR spectrum, each signal has three important characteristics: location, area, and shape.

SECTION 16.4

- **Chemically equivalent** protons occupy identical electronic environments and produce only one signal.
- When two protons are interchangeable by rotational symmetry, the protons are said to be **homotopic**.
- When two protons are interchangeable by reflectional symmetry, the protons are said to be **enantiotopic**.
- If two protons are neither homotopic nor enantiotopic, then they are not chemically equivalent.
- The **replacement test** is a simple technique for verifying the relationship between protons. If the replacement test produces diastereomers, then the protons are **diastereotopic**.

SECTION 16.5

- **Chemical shift** (δ) is defined as the difference (in hertz) between the resonance frequency of the proton being observed and that of TMS divided by the operating frequency of the spectrometer.
- The left side of an NMR spectrum is described as **downfield**, and the right side is described as **upfield**.
- In the absence of inductive effects, a methyl group (CH_3) will produce a signal near 0.9 ppm, a **methylene group** (CH_2) will produce a signal near 1.2 ppm, and a **methine group** (CH) will produce a signal near 1.7 ppm. The presence of nearby groups increases these values somewhat predictably.
- **Diamagnetic anisotropy** causes the aromatic protons of benzene to be strongly deshielded.

SECTION 16.6

- The **integration**, or area under each signal, indicates the number of protons giving rise to the signal. These numbers provide the relative number, or ratio, of protons giving rise to each signal.
- Integration is often represented by **step curves**, where the height of each step curve represents the area under the signal.

SECTION 16.7

- **Multiplicity** represents the number of peaks in a signal. A **singlet** has one peak, a **doublet** has two, a **triplet** has three, a **quartet** has four, and a **quintet** has five.
- Multiplicity is the result of **spin-spin splitting**, also called **coupling**, which follows the ***n*+1 rule**.
- Equivalent protons do not split each other.
- In order for protons to split each other, they must be separated by no more than three sigma bonds.
- When signal splitting occurs, the distance between the individual peaks of a signal is called the **coupling constant**, or ***J* value**, and is measured in hertz.
- Generally, no splitting is observed across the oxygen of an alcohol, because hydroxyl protons are **labile**.
- Complex splitting occurs when a proton has two different kinds of neighbors, often producing a **multiplet**.

SECTION 16.10

- Analyzing a ^{1}H NMR spectrum involves four discrete steps:
  - Calculate the HDI.
  - Consider the number of signals and the integration of each signal.
  - Analyze each signal (δ, m, I), and then draw fragments consistent with each signal.
  - Assemble the fragments.

SECTION 16.11

- ^{13}C is an isotope of carbon, representing 1.1% of all carbon atoms.
- In ^{13}C NMR spectroscopy, only the chemical shift is important.
- All ^{13}C-^{1}H splitting is suppressed with a technique called **broadband decoupling**, causing all of the ^{13}C signals to collapse to singlets.
- **Off-resonance decoupling** allows the retrieval of some of the coupling information that was lost, but this technique is rarely used because it often produces overlapping peaks that are difficult to interpret.

SECTION 16.12

- The position of each signal in a ^{13}C NMR spectrum is defined relative to the frequency of absorption of a reference compound, tetramethylsilane (TMS).
- In ^{13}C NMR spectroscopy, chemical shift values typically range from 0 to 220 ppm.

- The number of signals in a ^{13}C NMR spectrum represents the number of carbon atoms in different electronic environments (carbons that are not interchangeable by symmetry). Carbon atoms that are interchangeable by a symmetry operation (either rotation or reflection) will only produce one signal.
- The location of a signal is dependent on a number of factors, including hybridization state and shielding effects.

SECTION 16.13

- **Distortionless enhancement by polarization transfer (DEPT)** involves the acquisition of several spectra and allows the determination of the number of protons attached to each carbon atom.

KEY TERMINOLOGY

broadband decoupling 755
chemical shift (δ) 730
chemically equivalent 723
continuous-wave (CW) spectrometer 722
coupling 740
coupling constant 743
DEPT ^{13}C NMR 757
deshielding 721
diamagnetic anisotropy 734
diamagnetism 721
diastereotopic 725
doublet 739
downfield 731
enantiotopic 724
free induction decay 722
FT-NMR 722
homotopic 724
integration 736
***J* value** 743
labile 747
magnetic moment 719
methine group 732
methylene group 732
multiplet 746
multiplicity 739
***n* +1 rule** 741
nuclear magnetic resonance (NMR) 719
off-resonance decoupling 755
operating frequency 721
quartet 739
quintet 739
replacement test 724
resonance 720
shielding 721
singlet 739
spin-spin splitting 740
step curves 737
triplet 739
upfield 731

SKILLBUILDER REVIEW

16.1 DETERMINING THE RELATIONSHIP BETWEEN TWO PROTONS IN A COMPOUND

METHOD 1: LOOKING FOR SYMMETRY

STEP 1 Determine if the protons are interchangeable by rotational symmetry.

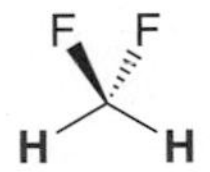

Homotopic
Chemically equivalent

STEP 2 If there is no rotational symmetry, look for a plane of symmetry:

Enantiotopic
Chemically equivalent

STEP 3 If the protons cannot be interchanged via a symmetry operation:

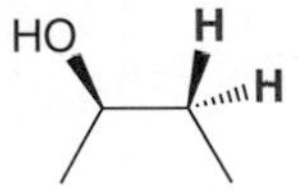

Diastereotopic
Not chemically equivalent

METHOD 2: REPLACEMENT TEST

Replace each proton with a deuterium and compare the resulting structures.

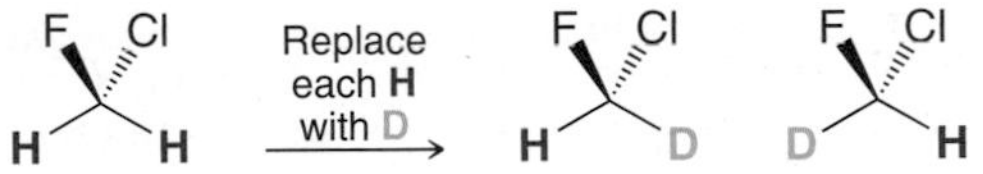
Replace each H with D

Enantiomers
Therefore, protons are enantiotopic.
Chemically equivalent

Try Problems **16.1–16.3, 16.40**

16.2 IDENTIFYING THE NUMBER OF EXPECTED SIGNALS IN A ^{1}H NMR SPECTRUM

Count the number of different protons, using the following rules of thumb:

The three protons of a methyl group are always equivalent.

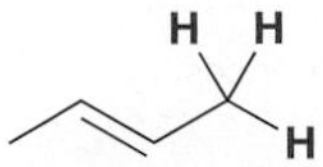

Chemically equivalent

The two protons of a methylene group (CH_2) are generally equivalent if the compound has no chirality centers.

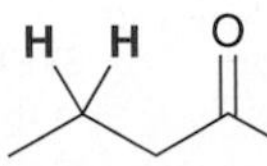

Chemically equivalent

The two protons of a methylene group (CH_2) are generally not equivalent if the compound has a chirality center.

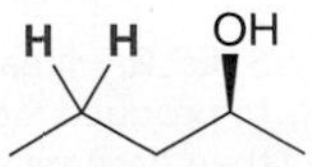

Not chemically equivalent

Two methylene groups will be equivalent if they can be interchanged by either rotation or reflection.

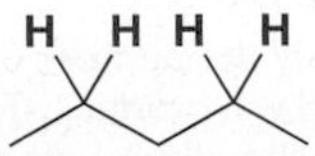

Chemically equivalent

Try Problems **16.4–16.6, 16.34, 16.42a, 16.44, 16.48**

16.3 PREDICTING CHEMICAL SHIFTS

Benchmark values

Methyl	Methylene	Methine
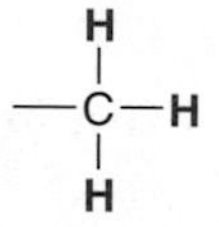	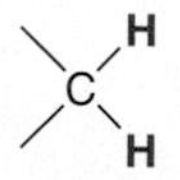	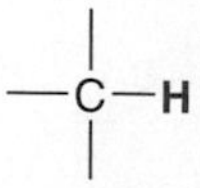
~ 0.9 ppm	~ 1.2 ppm	~ 1.7 ppm

Deviations from benchmark values

Effect on alpha proton

+ 2.5 **An oxygen of an alcohol or ether**

Example:

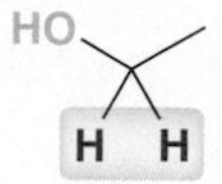

1.2
+ 2.5
= 3.7 ppm

+ 3.0 **An oxygen of an ester or acid**

Example:

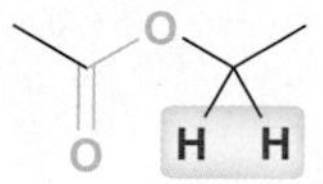

1.2
+ 3.0
= 4.2 ppm

+ 1.0 **A carbonyl group**

Example:

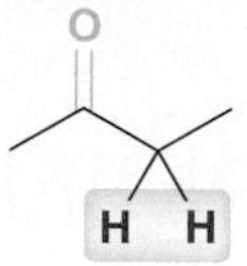

1.2
+ 1.0
= 2.2 ppm

Try Problems 16.7–16.9, 16.42b

16.4 DETERMINING THE NUMBER OF PROTONS GIVING RISE TO A SIGNAL

STEP 1 Compare the relative integration values, and choose the lowest number.

STEP 2 Divide all integration values by the number from step 1,which gives the ratio of protons.

STEP 3 Identify the number of protons in the compound (from the molecular formula) and then adjust the relative integration values so that the sum of the integrals equals the number of protons in the compound.

Try Problems 16.11–16.14, 16.54

16.5 PREDICTING THE MULTIPLICITY OF A SIGNAL

STEP 1 Identify all of the different kinds of protons.

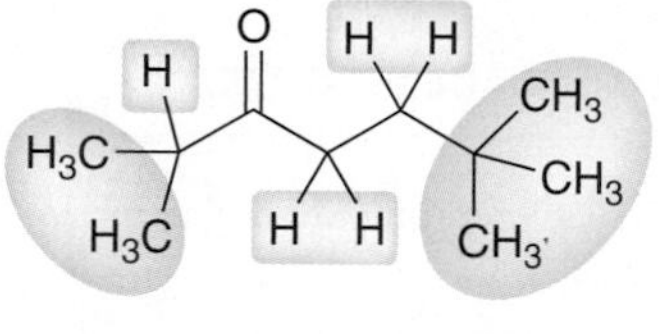

Five kinds of protons

STEP 2 For each kind of proton, identify the number of neighbors (N). The multiplicity will be $n + 1$.

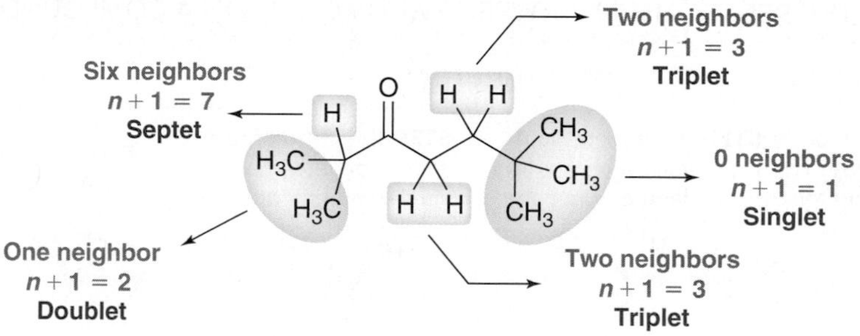

Try Problems 16.15, 16.16, 16.37

16.6 DRAWING THE EXPECTED ^{1}H NMR SPECTRUM OF A COMPOUND

STEP 1 Identify the number of signals.

STEP 2 Predict the chemical shift of each signal.

STEP 3 Determine the integration of each signal by counting the number of protons giving rise to each signal.

STEP 4 Predict the multiplicity of each signal.

STEP 5 Draw each signal.

Try Problems 16.19, 16.20, 16.41

16.7 USING ^{1}H NMR SPECTROSCOPY TO DISTINGUISH BETWEEN COMPOUNDS

STEP 1 Identify the number of signals that each compound will produce. The simplest way to distinguish compounds is if they are expected to produce a different number of signals.

STEP 2 If each compound is expected to produce the same number of signals, then determine the chemical shift, multiplicity, and integration of each signal in both compounds.

STEP 3 Look for differences in the chemical shifts, multiplicities, or integration values of the expected signals.

Try Problems 16.21, 16.22, 16.38, 16.47

16.8 ANALYZING A ^{1}H NMR SPECTRUM AND PROPOSING THE STRUCTURE OF A COMPOUND

STEP 1 Use the molecular formula to determine the HDI. An HDI of 4 or more indicates the possibility of an aromatic ring.

STEP 2 Consider the number of signals and integration of each signal (gives clues about the symmetry of the compound).

STEP 3 Analyze each signal (δ, m, I), and then draw fragments consistent with each signal. These fragments become our puzzle pieces that must be assembled to produce a molecular structure.

STEP 4 Assemble the fragments.

Try Problems 16.23, 16.24, 16.54, 16.57

16.9 PREDICTING THE NUMBER OF SIGNALS AND APPROXIMATE LOCATION OF EACH SIGNAL IN A ^{13}C NMR SPECTRUM

STEP 1 Determine the number of expected signals. Look to see if any of the carbon atoms are interchangeable by either rotational or reflectional symmetry.

STEP 2 Predict the expected region in which each signal will appear, based on hybridization states and deshielding effects.

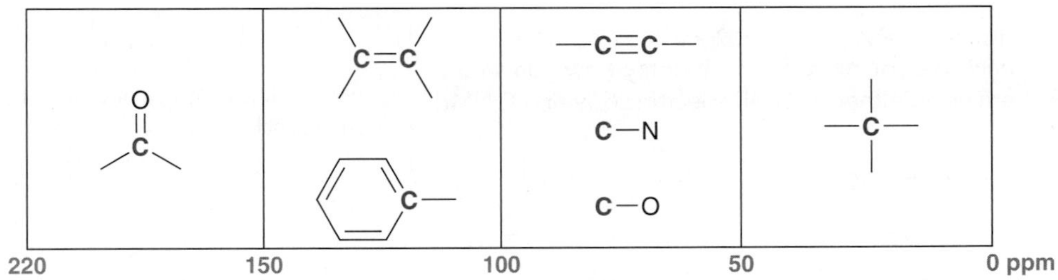

Try Problems 16.25–16.27, 16.35, 16.38, 16.46

16.10 DETERMINING MOLECULAR STRUCTURE USING DEPT ^{13}C NMR SPECTROSCOPY

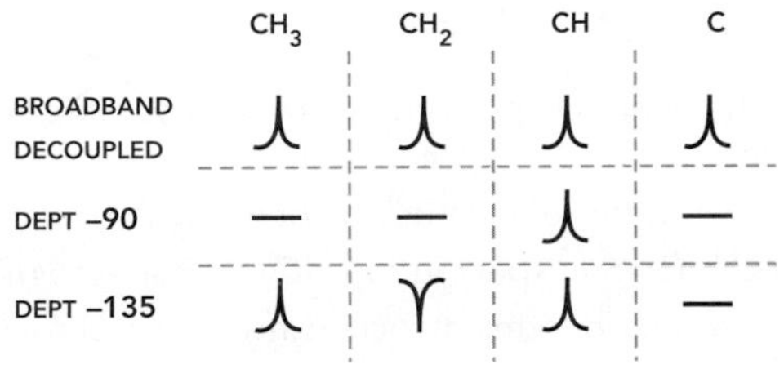

Try Problems 16.28–16.30, 16.60, 16.61

PRACTICE PROBLEMS

Note: Most of the Problems are available within *WileyPLUS*, an online teaching and learning solution.

16.31 Each of the following compounds exhibits a ^{1}H NMR spectrum with only one signal. Deduce the structure of each compound:

(a) C_5H_{10} (b) $C_5H_8Cl_4$ (c) $C_{12}H_{18}$

16.32 A compound with molecular formula $C_{12}H_{24}$ exhibits a ^{1}H NMR spectrum with only one signal and a ^{13}C NMR spectrum with two signals. Deduce the structure of this compound.

16.33 A compound with molecular formula $C_{17}H_{36}$ exhibits a ^{1}H NMR spectrum with only one signal. How many signals would you expect in the ^{13}C NMR spectrum of this compound?

16.34 How many signals would you expect in the ^{1}H NMR spectrum of each of the following compounds:

(a) 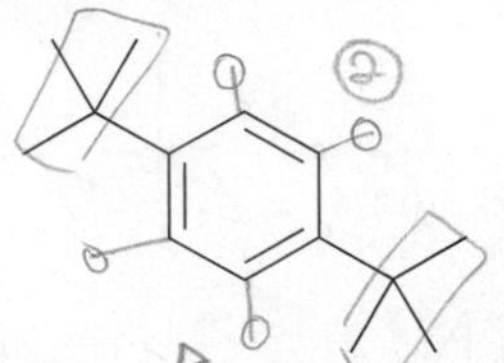

(b) 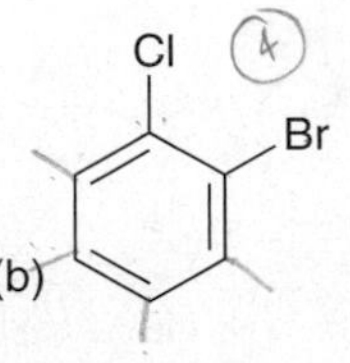

(c) 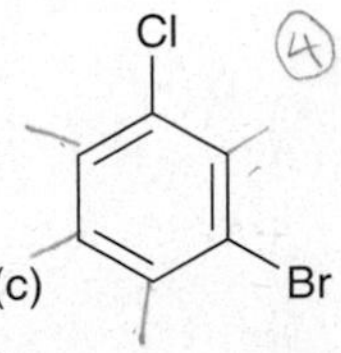

(d) 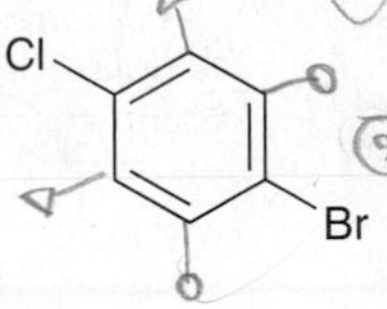

(e) 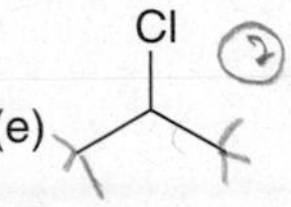

(f) 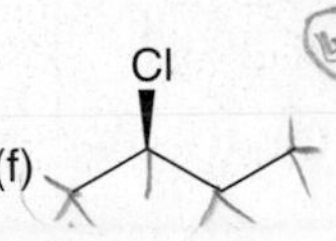

16.35 How many signals would you expect in the ^{13}C NMR spectrum of each of the compounds in Problem 16.34?

16.36 How would you distinguish between the following compounds using ^{13}C NMR spectroscopy?

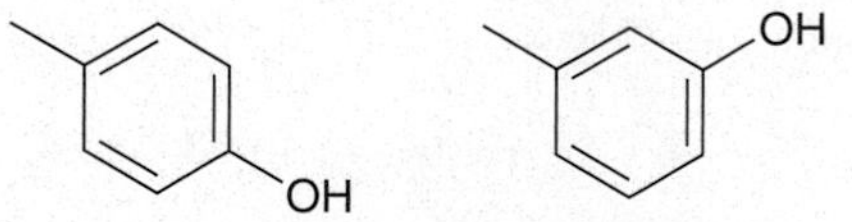

16.37 Predict the multiplicity of each signal in the ^{1}H NMR spectrum of the following compound:

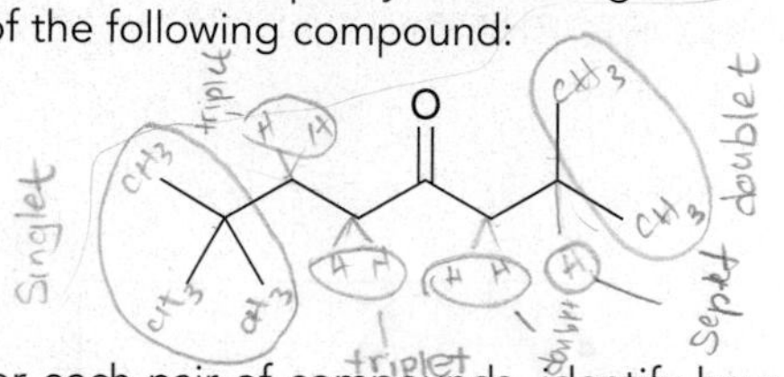

16.38 For each pair of compounds, identify how you would distinguish them using either ^{1}H NMR spectroscopy or ^{13}C NMR spectroscopy:

(a) 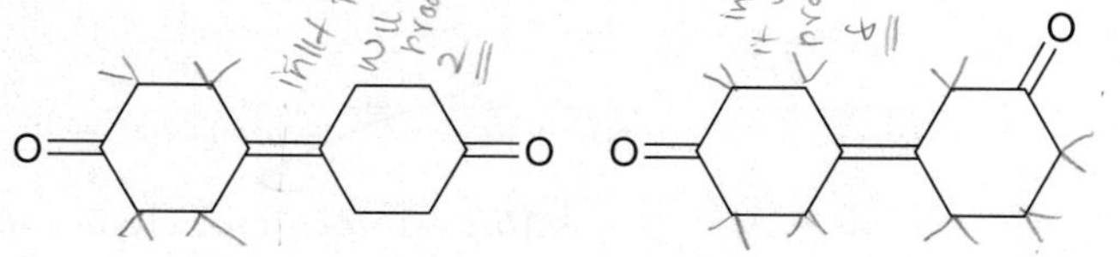

(b) 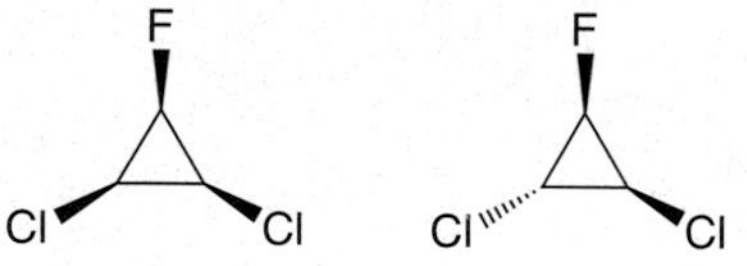

(c) 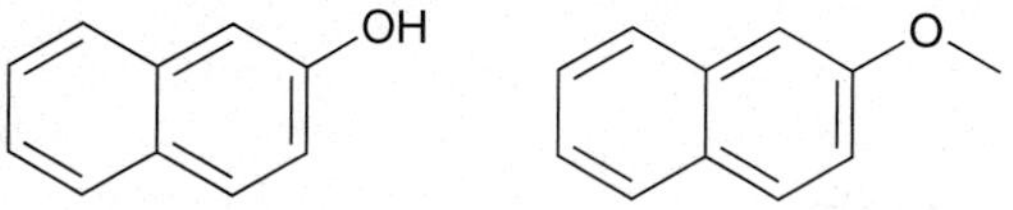

(d) 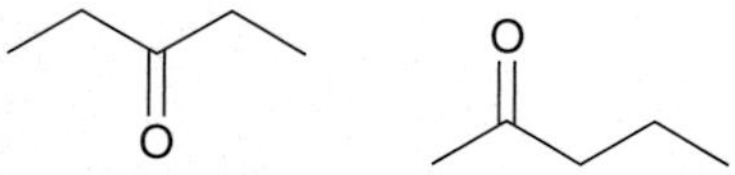

16.39 A compound with molecular formula C_8H_{18} exhibits a ^{1}H NMR spectrum with only one signal. How many signals would you expect in the ^{13}C NMR spectrum of this compound?

16.40 For each of the following compounds, compare the two indicated protons and determine whether they are enantiotopic, homotopic, or diastereotopic:

(a) 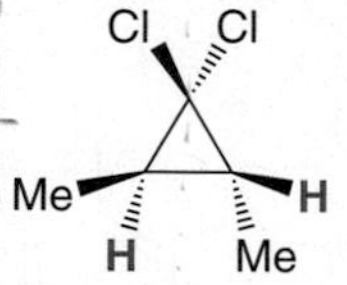
(b) 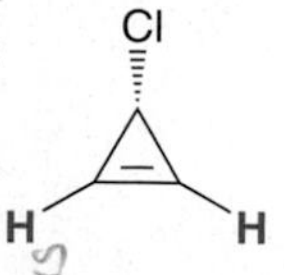
(c) 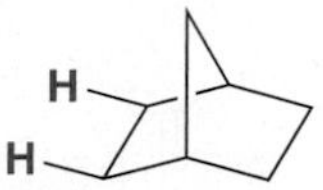

(d) 
(e) 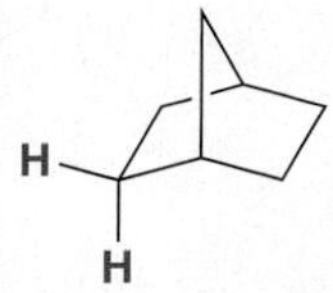
(f) 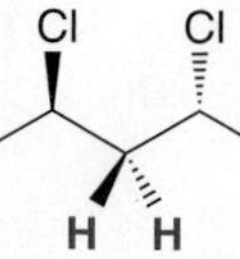

(g) 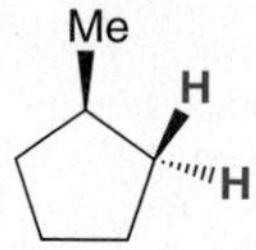
(h) 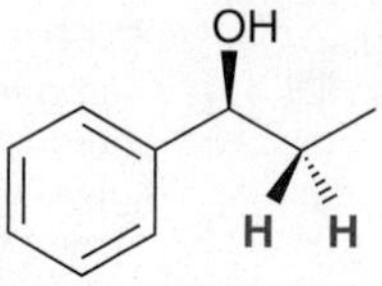
(i) 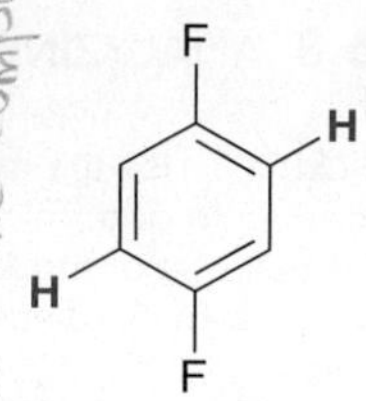

(j) 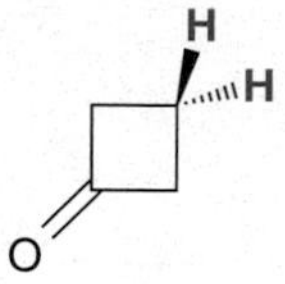
(k) 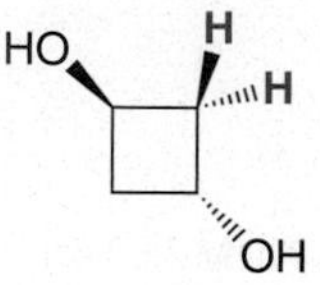
(l) 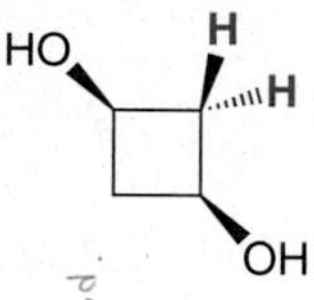

(m) 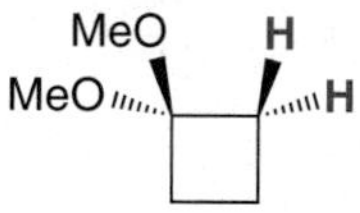
(n) 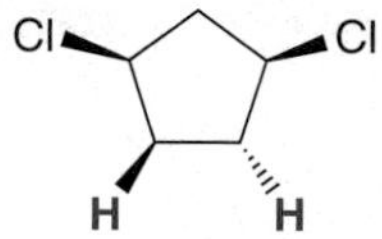
(o) 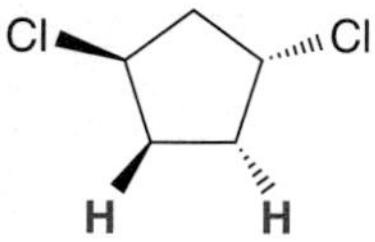

16.41 Draw the expected ^{1}H NMR spectrum of the following compound:

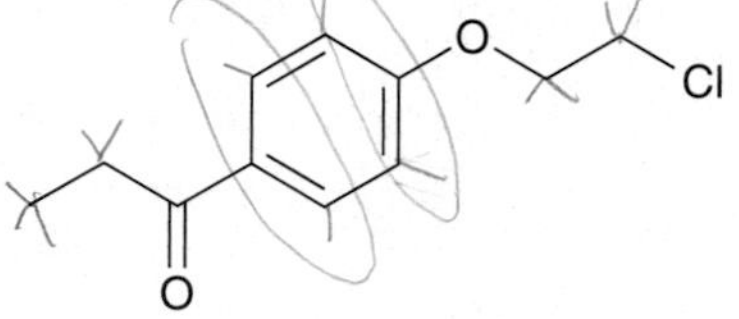

16.42 Consider the following compound:

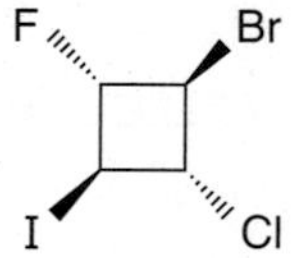

(a) How many signals do you expect in the ^{1}H NMR spectrum of this compound?

(b) Rank the protons in terms of increasing chemical shift.

(c) How many signals do you expect in the ^{13}C NMR spectrum?

(d) Rank the carbon atoms in terms of increasing chemical shift.

16.43 A compound with molecular formula C_9H_{18} exhibits a ^{1}H NMR spectrum with only one signal and a ^{13}C NMR spectrum with two signals. Deduce the structure of this compound.

16.44 How many signals do you expect in the ^{1}H NMR spectrum of each of the following compounds:

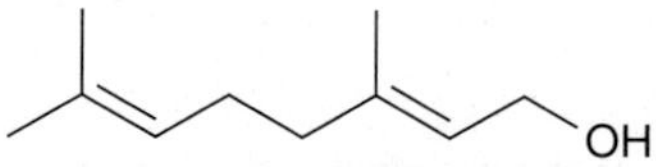

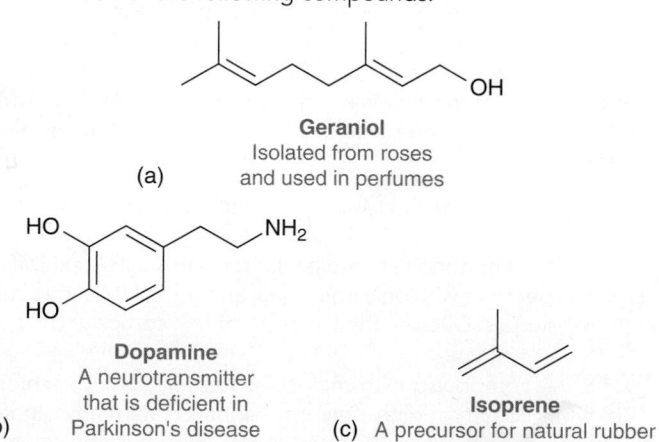

(a) **Geraniol** Isolated from roses and used in perfumes

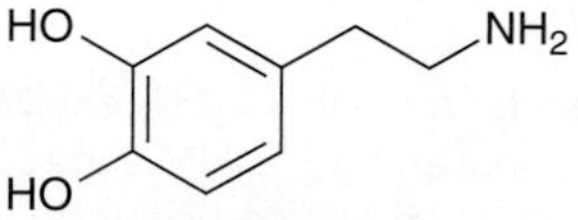

(b) **Dopamine** A neurotransmitter that is deficient in Parkinson's disease

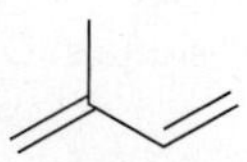

(c) **Isoprene** A precursor for natural rubber

16.45 Rank the signals of the following compound in terms of increasing chemical shift. Identify the proton(s) giving rise to each signal:

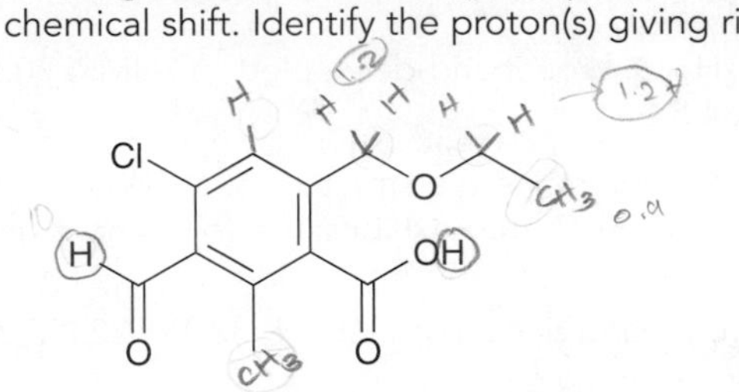

16.46 Predict the expected number of signals in the ^{13}C NMR spectrum of each of the following compounds. For each signal, identify where you expect it to appear in the ^{13}C NMR spectrum:

(a) 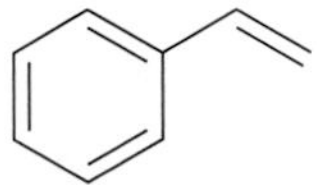

(b) 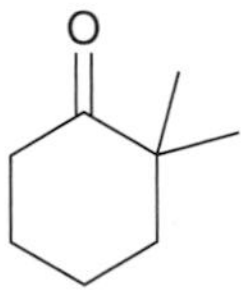

(c) 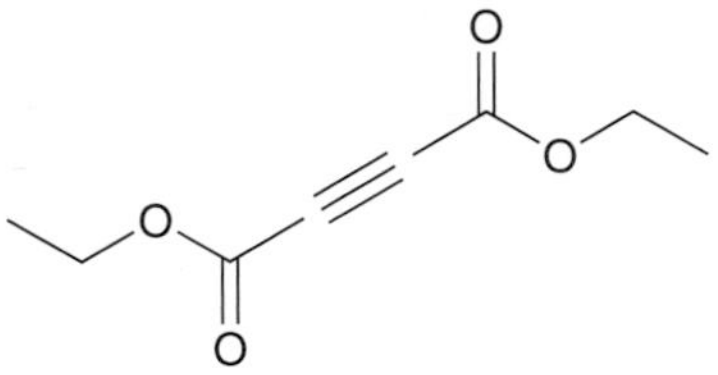

16.47 When 1-methylcyclohexene is treated with HCl, a Markovnikov addition is observed. How would you use ^{1}H NMR spectroscopy to determine that the major product is indeed the Markovnikov product?

16.48 How many signals will be expected in the ^{1}H NMR spectrum of each of the following compounds?

(a) 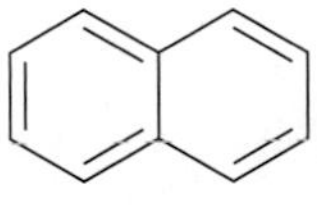

(b) 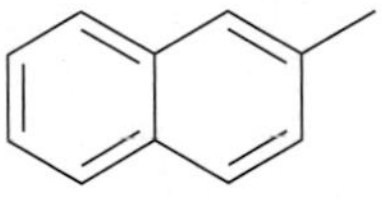

(c) 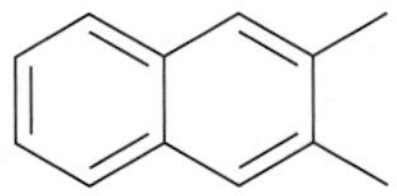

(d) 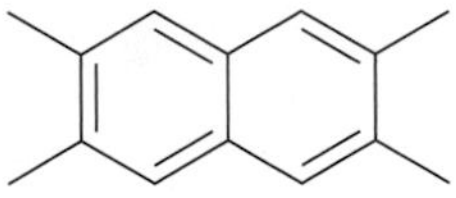

(e) 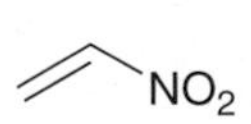

(f) 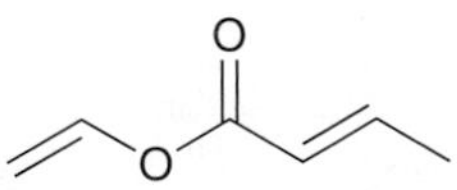

(g) 

(h) 

16.49 Compare the structures of ethylene, acetylene, and benzene. Each of these compounds produces only one signal in its ^{1}H NMR spectrum. Arrange these signals in order of increasing chemical shift.

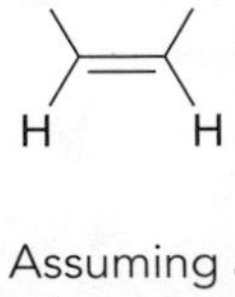

16.50 Assuming a 300 MHz instrument is used, calculate the difference between the frequency of absorption (in hertz) of TMS and the frequency of absorption of a proton with a δ value of 1.2 ppm.

16.51 A compound with molecular formula $C_{13}H_{28}$ exhibits a ^{1}H NMR spectrum with two signals: a septet with an integration of 1 and a doublet with an integration of 6. Deduce the structure of this compound.

16.52 A compound with molecular formula C_8H_{10} produces three signals in its ^{13}C NMR spectrum and only two signals in its ^{1}H NMR spectrum. Deduce the structure of the compound.

16.53 A compound with molecular formula C_3H_8O produces a broad signal between 3200 and 3600 cm^{-1} in its IR spectrum and produces two signals in its ^{13}C NMR spectrum. Deduce the structure of the compound.

16.54 A compound with molecular formula $C_4H_6O_4$ produces a broad signal between 2500 and 3600 cm^{-1} in its IR spectrum and produces two signals in its ^{1}H NMR spectrum (a singlet at 12.1 ppm with a relative integration of 1 and a singlet at 2.4 ppm with a relative integration of 2). Deduce the structure of the compound.

16.55 Propose the structure of a compound that exhibits the following ^{1}H NMR data:

(a) $C_5H_{10}O$
- 1.09 δ (6H, doublet)
- 2.12 δ (3H, singlet)
- 2.58 δ (1H, septet)

(b) $C_5H_{12}O$
- 0.91 δ (3H, triplet)
- 1.19 δ (6H, singlet)
- 1.50 δ (2H, quartet)
- 2.24 δ (1H, singlet)

(c) $C_4H_{10}O$
- 0.90 δ (6H, doublet)
- 1.76 δ (1H, multiplet)
- 3.38 δ (2H, doublet)
- 3.92 δ (1H, singlet)

(d) $C_4H_8O_2$
- 1.21 δ (6H, doublet)
- 2.59 δ (1H, septet)
- 11.38 δ (1H, singlet)

16.56 A compound with molecular formula $C_8H_{10}O$ produces six signals in its ^{13}C NMR spectrum and exhibits the following ^{1}H NMR spectrum. Deduce the structure of the compound.

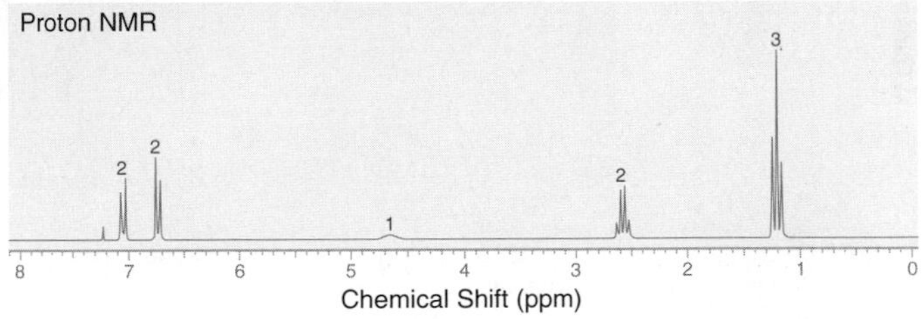

16.57 Deduce the structure of a compound with molecular formula C_9H_{12} that produces the following ^{1}H NMR spectrum:

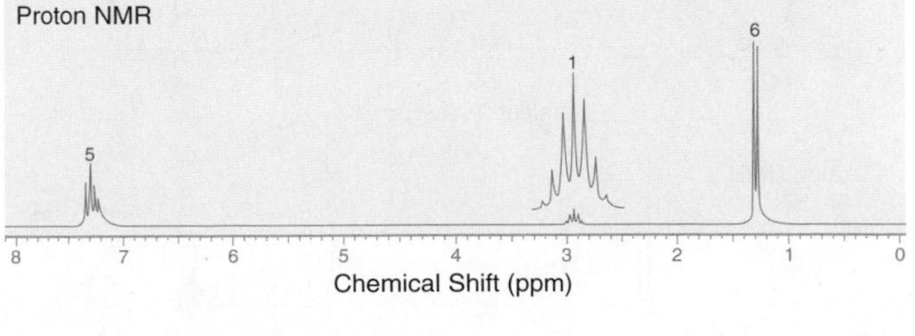

16.58 Deduce the structure of a compound with molecular formula $C_9H_{10}O_2$ that produces the following ^1H NMR spectrum and ^{13}C NMR spectrum:

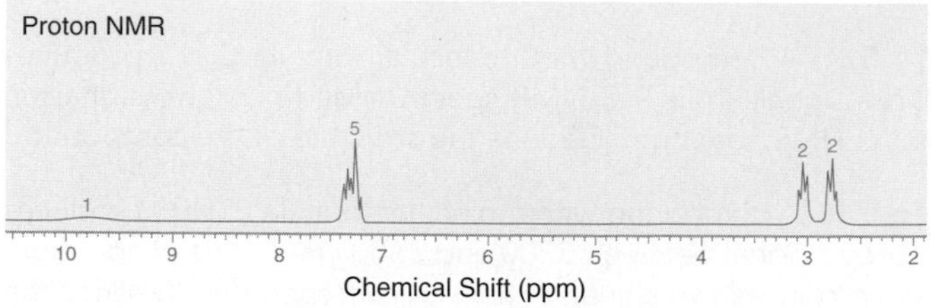

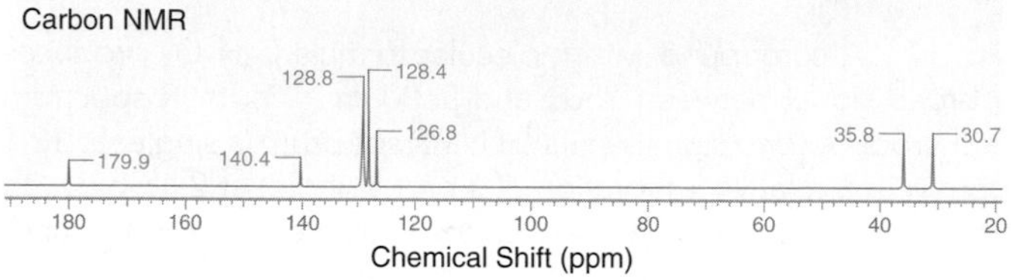

16.59 Propose the structure of a compound consistent with the following data:

(a) $C_5H_{10}O$, broadband-decoupled ^{13}C NMR: 7.1, 34.6, 210.5 δ

(b) $C_6H_{10}O$, broadband-decoupled ^{13}C NMR: 70.8, 116.2, 134.8 δ

16.60 Determine the structure of an alcohol with molecular formula $C_4H_{10}O$ that exhibits the following signals in its ^{13}C NMR spectra:

(a) Broadband decoupled: 69.3 δ, 32.1 δ, 22.8 δ, and 10.0 δ

(b) DEPT-90: 69.3 δ

(c) DEPT-135: positive signals at 69.3 δ, 22.8 δ, and 10.0 δ; negative signal at 32.1 δ

16.61 Determine the structure of an alcohol with molecular formula $C_6H_{14}O$ that exhibits the following DEPT-135 spectrum:

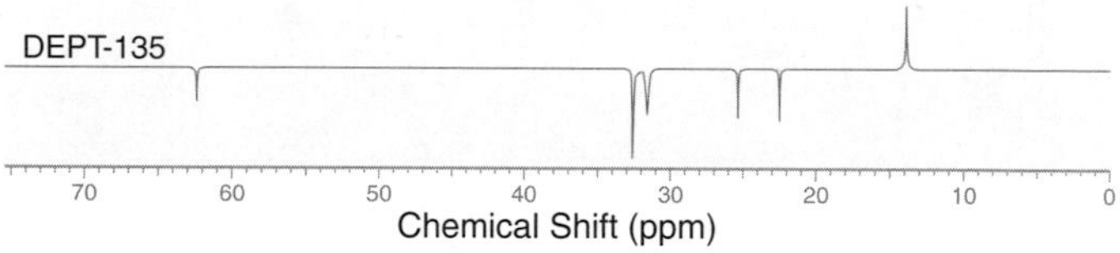

INTEGRATED PROBLEMS

Note: Most of the Problems are available within *WileyPLUS*, an online teaching and learning solution.

16.62 Deduce the structure of a compound with molecular formula $C_6H_{14}O_2$ that exhibits the following IR, ^1H NMR, and ^{13}C NMR spectra.

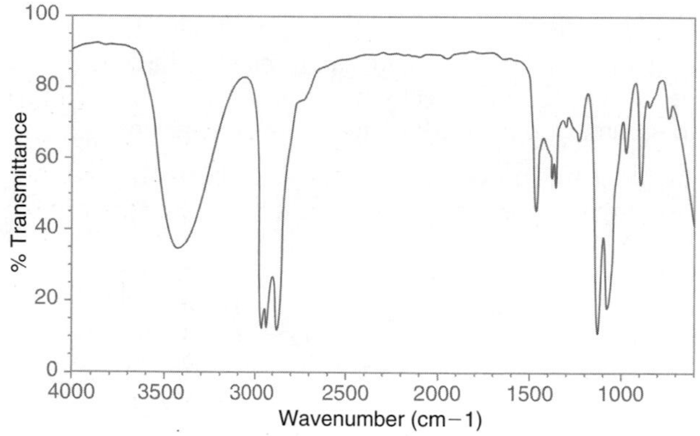

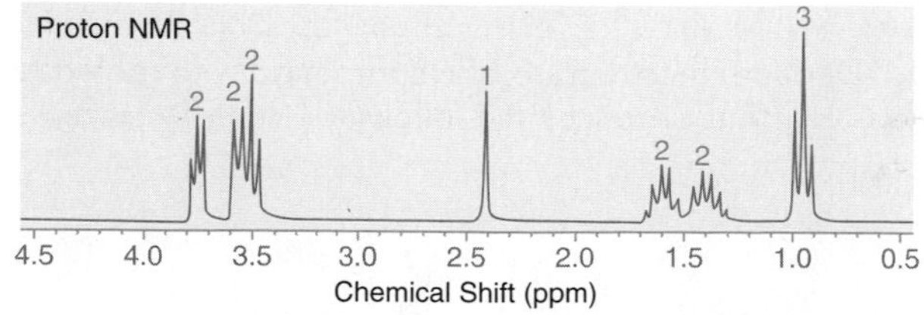

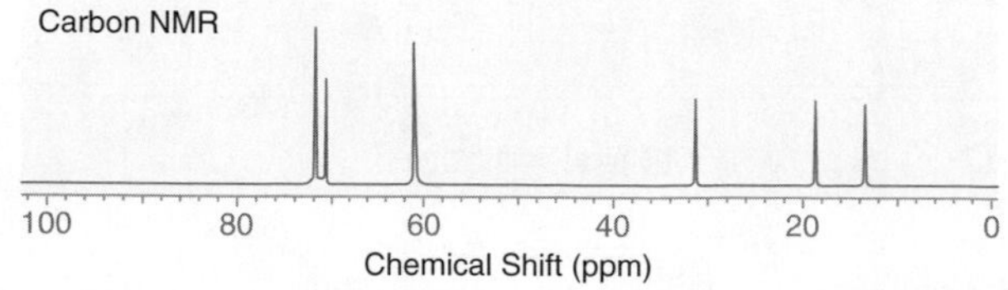

16.63 Deduce the structure of a compound with molecular formula $C_8H_{10}O$ that exhibits the following IR, ^1H NMR, and ^{13}C NMR spectra.

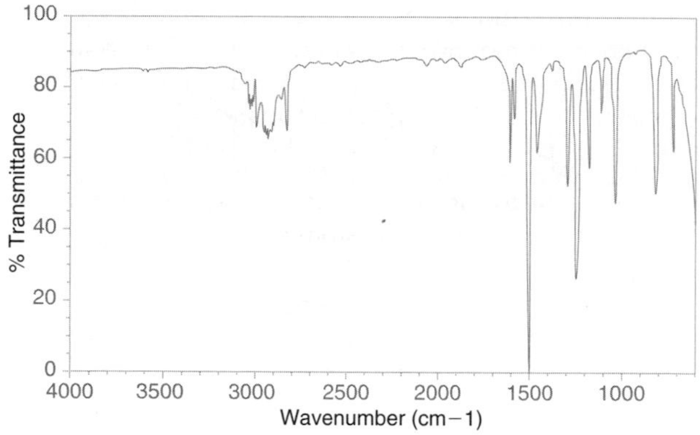

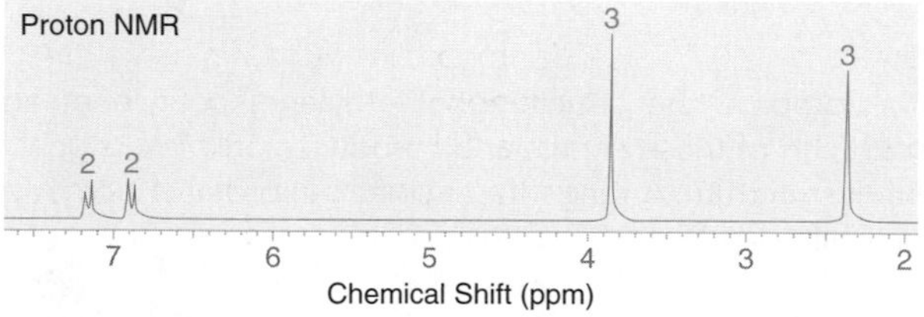

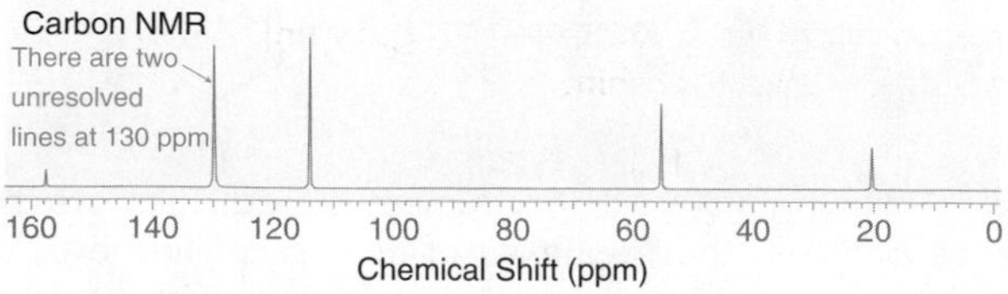

16.64 Deduce the structure of a compound with molecular formula $C_5H_{10}O$ that exhibits the following IR, ^1H NMR, and ^{13}C NMR spectra. Data from the mass spectrum are also provided.

Mass Spec. Data

m/z	relative abund.
15	23
26	20
27	61
29	92
30	20
31	45
39	47
41	100
45	10
57	82
58	86
86	12

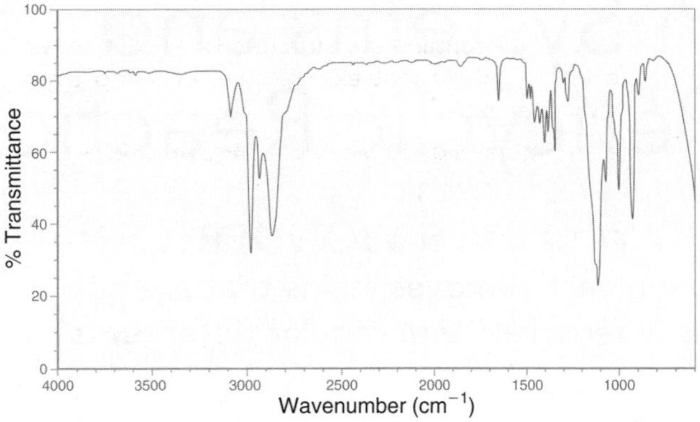

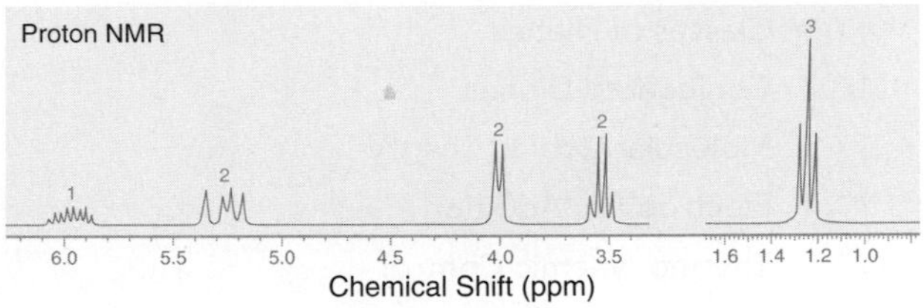

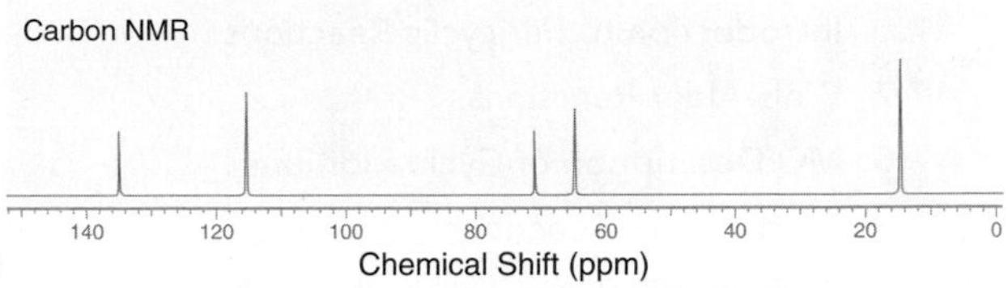

CHALLENGE PROBLEMS

16.65 Consider the structure of *N,N*-dimethylformamide (DMF):

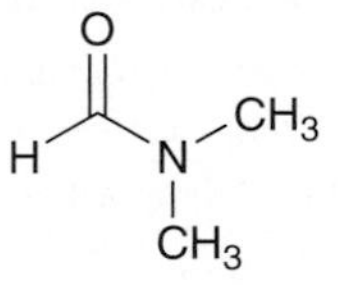

We might expect the two methyl groups to be equivalent; however, both the proton and carbon NMR spectra of DMF show two separate signals for the methyl groups. Propose an explanation for the nonequivalence of the methyl groups. Would you expect the signals to collapse into one signal at high temperature?

(*Hint*: You may want to turn to Section 2.12 and consider the hybridization state of the nitrogen atom.)

16.66 Consider the structure of phenol:

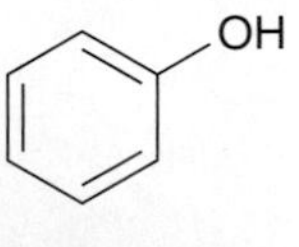

The chemical shift of the hydroxyl proton is found to be sensitive to the concentration of phenol. In a concentrated solution, the hydroxyl proton produces a signal at 7.5 ppm. However, in a dilute solution, the hydroxyl proton produces a signal at 4.5 ppm. Can you offer an explanation for this difference?

16.67 Consider the two methyl groups shown in the following compound. Explain why the methyl group on the right side appears at lower chemical shift.

1.0 ppm 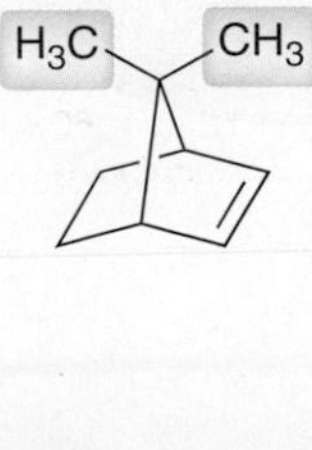 **0.8 ppm**

16.68 Compare the chemical shifts of the carbon atoms in the ^{13}C NMR spectra of the following compounds:

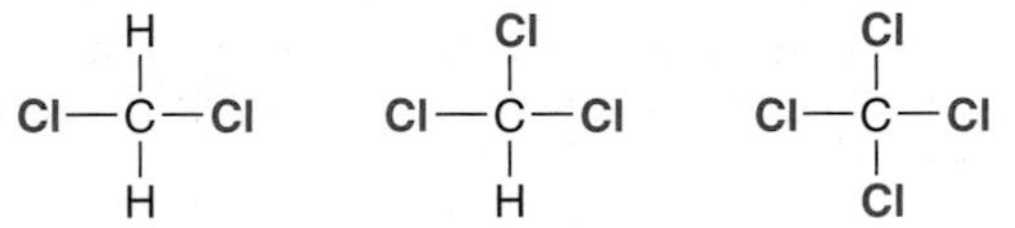

53.5 ppm **77.2 ppm** **96.1 ppm**

This is exactly what we would expect. Each chlorine atom withdraws electron density via induction, which deshields the carbon atom. The greater the number of chlorine atoms, the stronger the deshielding effect. However, when we compare the bromine analogues of the above compounds, we see the opposite trend:

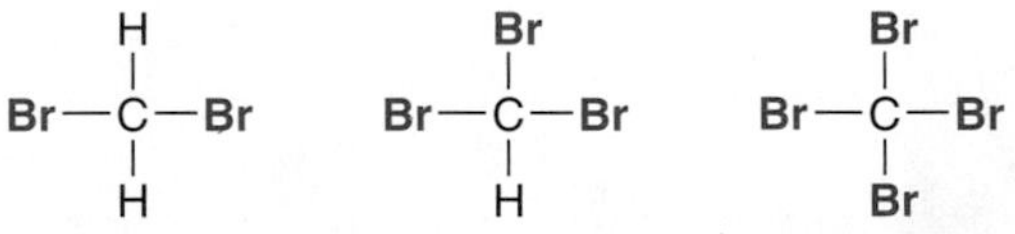

21 ppm **12 ppm** **–28.5 ppm**

Can you offer an explanation for this anomalous trend? Consider the fact that a bromine atom is significantly larger than a chlorine atom and its electron density is spread over a larger volume of space.

17 Conjugated Pi Systems and Pericyclic Reactions

DID YOU EVER WONDER...
how bleach removes stains that are not easily removed with regular detergents?

Most bleaching agents do not actually remove stains, but instead, they react with the colored compounds in the stain to produce colorless compounds. The stain is still there, but it has been rendered "invisible." When illuminated with ultraviolet light, the invisible stain can be seen once again. In this chapter, we will learn about the structural features that cause organic compounds to absorb visible light. Specifically, we will see that color is a consequence of special conjugated π systems. In this chapter, we will learn about the structure, properties, and reactions of conjugated π systems.

DO YOU REMEMBER?

Before you go on, be sure you understand the following topics.
If necessary, review the suggested sections to prepare for this chapter:

- MO Theory and Pi Bonds (Sections 1.8, 1.9)
- Enthalpy, Entropy and Gibbs Free Energy (Sections 6.1–6.3)
- Conformations (Sections 4.7, 4.8)
- Equilibria, Kinetics, and Energy Diagrams (Sections 6.4–6.6)

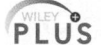

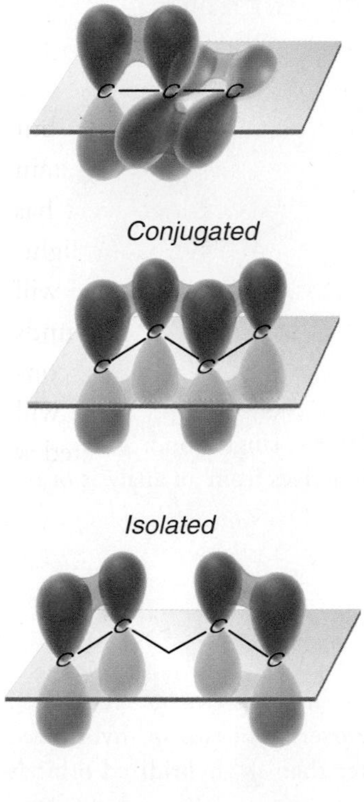

FIGURE 17.1
The relative spatial arrangement of the *p* orbitals in cumulated, conjugated, and isolated dienes.

17.1 Classes of Dienes

Dienes are compounds that possess two C=C bonds. Depending on the proximity of the π bonds, they are classified as cumulated, conjugated, or isolated.

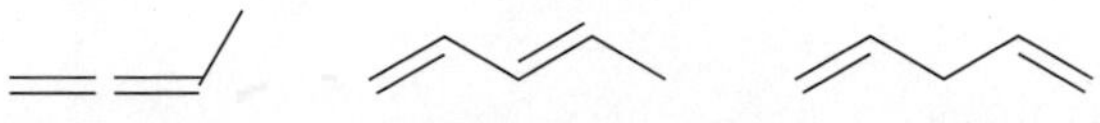

- In **cumulated dienes**, also called *allenes*, the π bonds are adjacent.
- In **conjugated dienes**, the π bonds are separated by exactly one single bond.
- In **isolated dienes**, the π bonds are separated by two or more single bonds.

The proximity of the π bonds is critical to understanding the structure and reactivity of a diene. In particular, consider the spatial arrangement of *p* orbitals in each of the dienes in Figure 17.1. Conjugated dienes represent a special category, because they contain one continuous system of overlapping *p* orbitals—that is, the two π bonds together comprise one functional group composed of four overlapping *p* orbitals. As such, conjugated dienes exhibit special properties and reactivity.

A C=C π bond can also be conjugated with other types of π bonds, such as carbonyl groups.

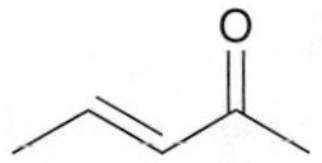

A conjugated enone

This chapter will focus on the properties and reactivity of conjugated π systems.

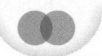

CONCEPTUAL CHECKPOINT

17.1 For each of the following compounds, identify whether each C=C π system is cumulated, conjugated, or isolated:

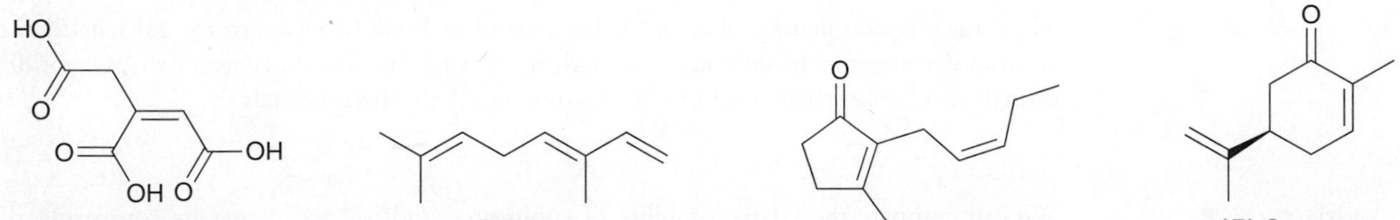

(a) ***cis*-Aconitic acid** Plays a role in the citric acid cycle (b) **Ocimene** Present in the essential oils of many plants (c) ***cis*-Jasmone** A consituent of many perfumes (d) **(*R*)-Carvone** Responsible for the flavor of spearmint

17.2 Conjugated Dienes

Preparation

Conjugated dienes can be prepared from allylic halides via an elimination process.

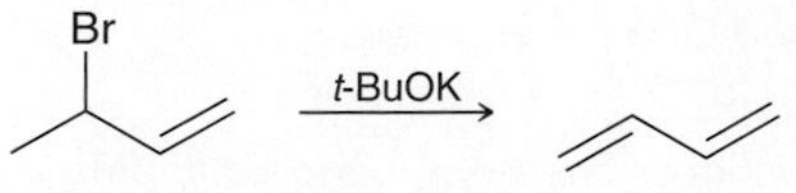

A sterically hindered base, such as potassium *tert*-butoxide, is used to prevent the competing S_N2 reaction from occurring. Conjugated dienes can also be formed from dihalides via two successive elimination reactions.

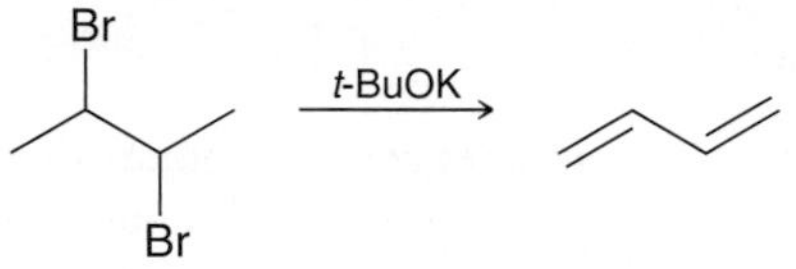

Bond Lengths

Compare the bond lengths of the following single bonds:

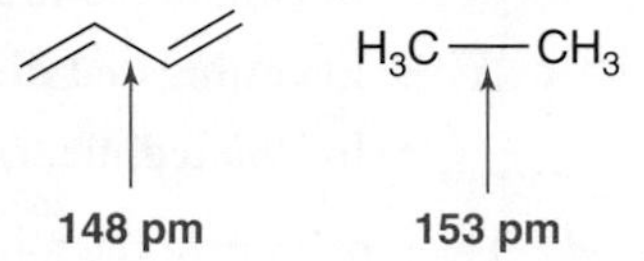

The single bond in a conjugated diene is shorter than a typical C—C single bond. There are a couple of ways to explain this difference. The simpler approach derives from an analysis of the hybridization states involved:

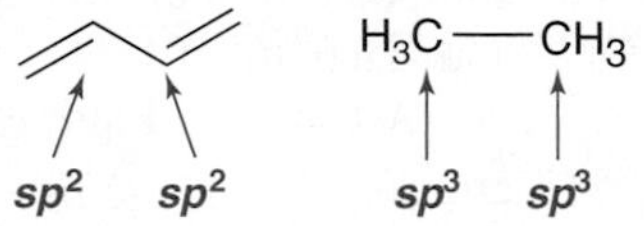

The C—C bond of a conjugated diene is formed from the overlap of two sp^2-hybridized orbitals, while the C—C bond in ethane is formed from the overlap of two sp^3-hybridized orbitals. Orbitals that are sp^2-hybridized have more *s* character than sp^3-hybridized orbitals (Figure 17.2).

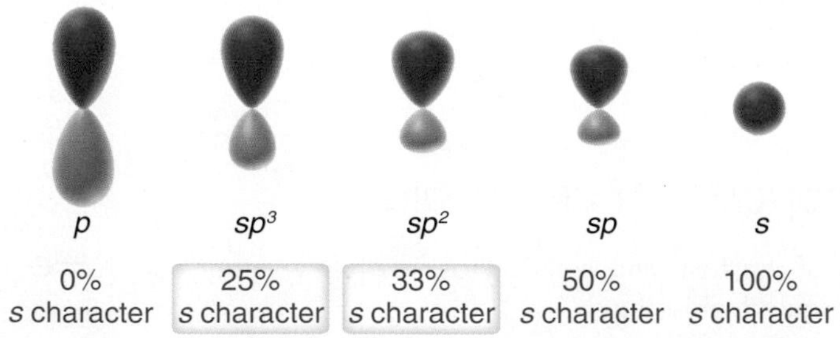

FIGURE 17.2
The relative amount of "*s* character" for atomic orbitals.

Thus, the electron density of an sp^2-hybridized orbital will be closer to the nucleus than the electron density of an sp^3-hybridized orbital, and as a result, a bond between two sp^2-hybridized orbitals will be shorter than a bond between two sp^3-hybridized orbitals.

Stability

We can compare the relative stability of conjugated and isolated dienes by comparing their heats of hydrogenation. For example, compare the heats of hydrogenation of 1-butene and 1,3-butadiene (Figure 17.3). We might expect that the heat of hydrogenation for 2 mol of 1-butene would be the same as the heat of hydrogenation for 1 mol of 1,3-butadiene. After all, both reactions produce butane as the sole product. Experiments reveal, however, that the heat

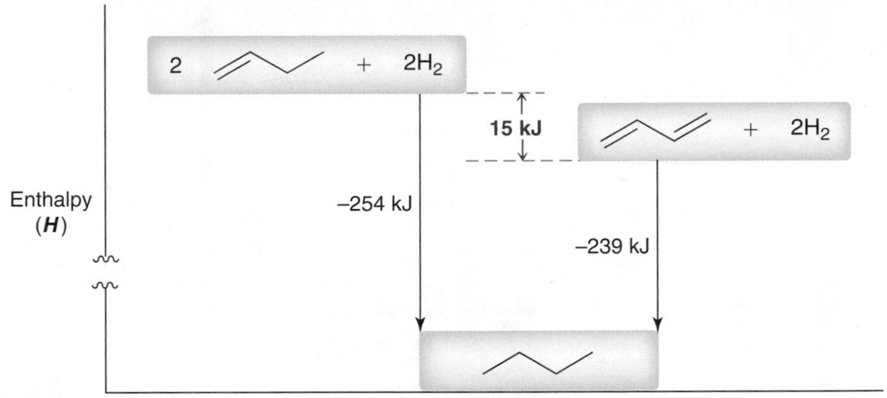

FIGURE 17.3
A comparison of the heats of hydrogenation of isolated and conjugated π bonds.

of hydrogenation for the conjugated diene is less than expected. These results suggest that conjugated double bonds are more stable than isolated double bonds, and we can determine that the **stabilization energy** associated with a conjugated diene is approximately 15 kJ/mol. The source of this stabilization energy will be discussed in Section 17.3.

CONCEPTUAL CHECKPOINT

17.2 Identify reagents necessary to convert cyclohexane into 1,3-cylohexadiene: Notice that the starting material has no leaving groups and cannot simply be treated with a strong base. You must first introduce functionality into the starting material (for help with introducing functional groups, see Section 12.2).

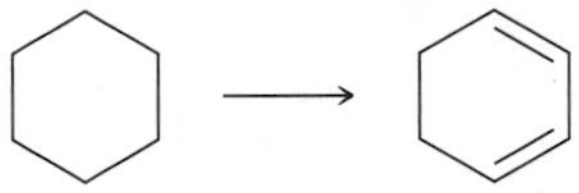

17.3 Rank the three C—C bonds of 1-butene in terms of bond length (from shortest to longest).

17.4 Compare the following three isomeric dienes:

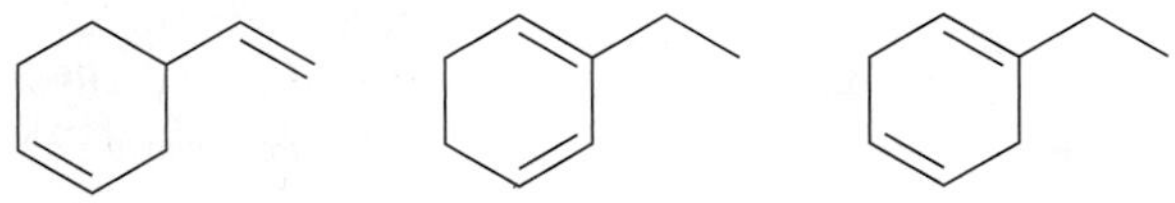

(a) Which compound will liberate the least heat upon hydrogenation with 2 mol of hydrogen gas? Why?

(b) Which compound will liberate the most heat upon hydrogenation with 2 mol of hydrogen gas? Why? (For help, see Section 8.5).

17.5 Identify the most stable compound:

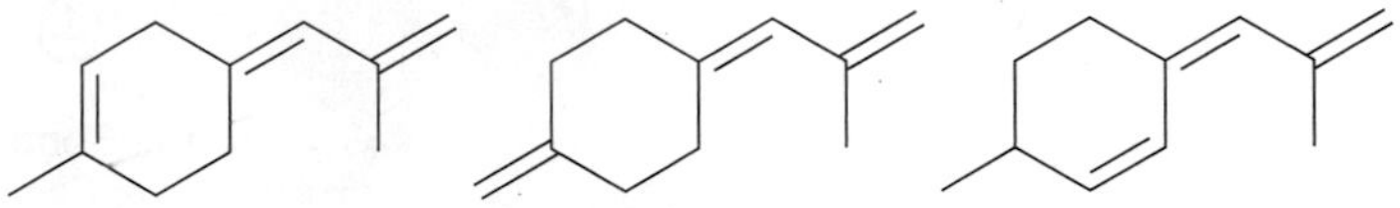

Conformations of 1,3-Butadiene

At room temperature, 1,3-butadiene experiences free rotation about the C2—C3 bond, giving rise to two important conformers.

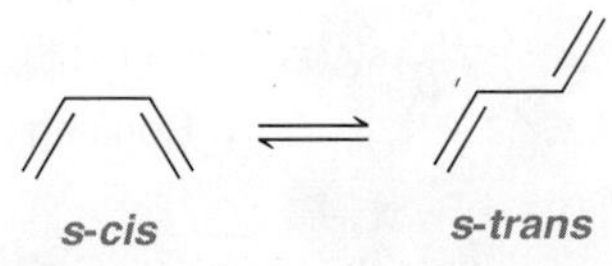

In the ***s-cis*** conformer, the disposition of the two π bonds with regard to the connecting single bond is *cis-like* (a dihedral angle of 0°). In the ***s-trans*** conformer, the disposition of the two π bonds with regard to the connecting single bond is *trans-like* (a dihedral angle of 180°). In each of these conformers, the *p* orbitals effectively overlap to give a continuous, conjugated π system (Figure 17.4).

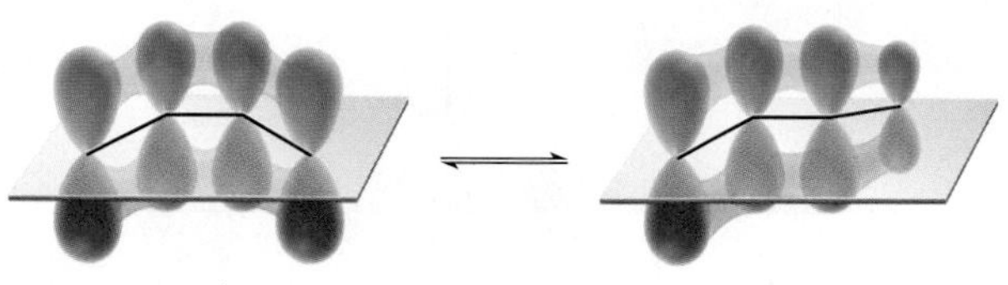

FIGURE 17.4
The *p* orbitals of 1,3-butadiene effectively overlap in both the *s-cis* and *s-trans* conformers.

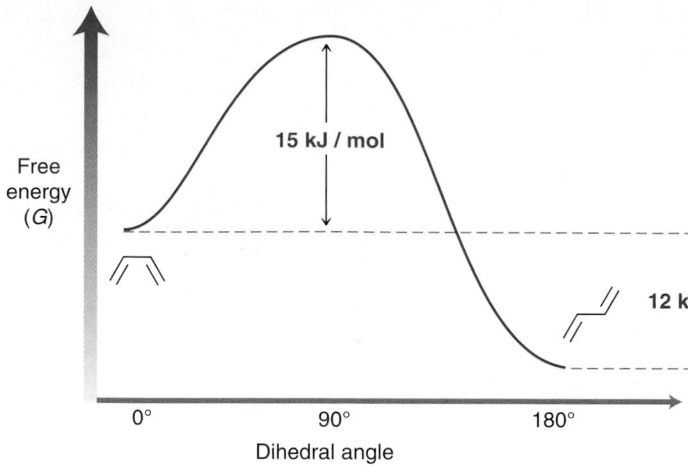

FIGURE 17.5
An energy diagram that illustrates the equilibrium between the s-cis and s-trans conformations.

At room temperature, these two conformers rapidly interconvert, establishing an equilibrium (Figure 17.5). The activation energy for conversion from the *s-cis* to the *s-trans* conformer is approximately 15 kJ/mol, which is equivalent to the stabilization energy associated with conjugated double bonds. In other words, the stabilizing effect of conjugation is completely destroyed when the C2—C3 bond is rotated by 90°. In such a high-energy conformation, the two π bonds do not effectively overlap, so they are essentially equivalent to isolated double bonds (Figure 17.6).

Due to steric factors, the *s-trans* conformer is significantly lower in energy than the *s-cis* conformer, and as a result, the equilibrium favors the *s-trans* conformer. At any given time, approximately 98% of the molecules assume an *s-trans* conformation, while only 2% of the molecules assume an *s-cis* conformation.

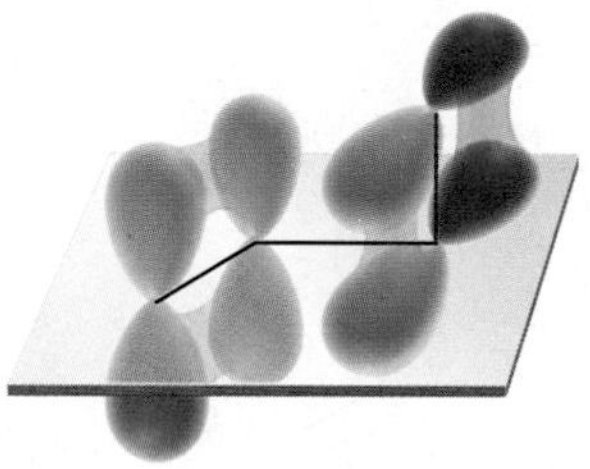

FIGURE 17.6
The *p* orbitals of 1,3-butadiene do not effectively overlap when the C2—C3 bond is rotated by 90°.

17.3 Molecular Orbital Theory

Throughout this chapter, we will invoke molecular orbital (MO) theory to explain the reactivity and properties of conjugated π systems. This section is designed to review and introduce some of the most basic features of MO theory.

Recall that a π bond is constructed from overlapping *p* orbitals (Figure 1.29). According to MO theory, these two atomic *p* orbitals are mathematically combined to produce two new orbitals, called *molecular orbitals*. There is a profound difference between an atomic orbital and a molecular orbital. Specifically, electrons that occupy atomic orbitals are associated with an individual atom, while electrons that occupy a molecular orbital are associated with the entire molecule. In Section 1.9, we discussed the molecular orbitals associated with the π bond of ethylene (Figure 17.7). According to MO theory, the two atomic *p* orbitals are replaced by two molecular orbitals—a bonding MO and an antibonding MO. The antibonding MO exhibits a node (shown in Figure 17.7) and is therefore higher in energy than the bonding MO. The π bond results from the ability of the π electrons to achieve a lower energy state by occupying the bonding MO.

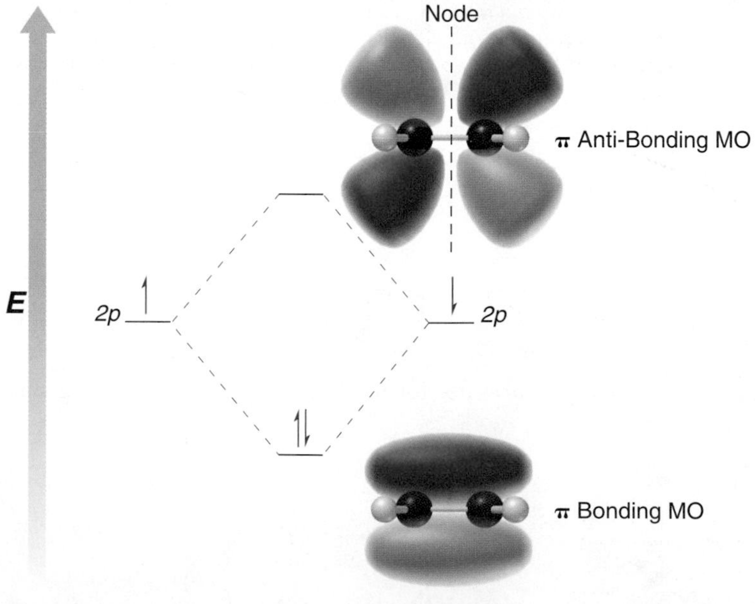

FIGURE 17.7
An energy diagram showing images of the bonding and antibonding MOs associated with the π bond in ethylene.

Molecular Orbitals of Butadiene

A conjugated diene, such as 1,3-butadiene, is comprised of four overlapping *p* orbitals. According to MO theory, these four atomic *p* orbitals are mathematically combined to produce four molecular orbitals (Figure 17.8). The lowest energy MO (ψ_1) has no nodes. The next MO (ψ_2) has one node, and each higher energy MO has one additional node. These four MOs are often represented with the shorthand drawings in Figure 17.9. Although these drawings resemble *p* orbitals, they are actually used to represent *molecular* orbitals, rather than *atomic* orbitals. These drawings are just a quick, shorthand method for drawing the phases and nodes of the MOs from Figure 17.8. We will use this method several times in this chapter, so it is important to understand that these drawings are merely simplified representations of molecular orbitals.

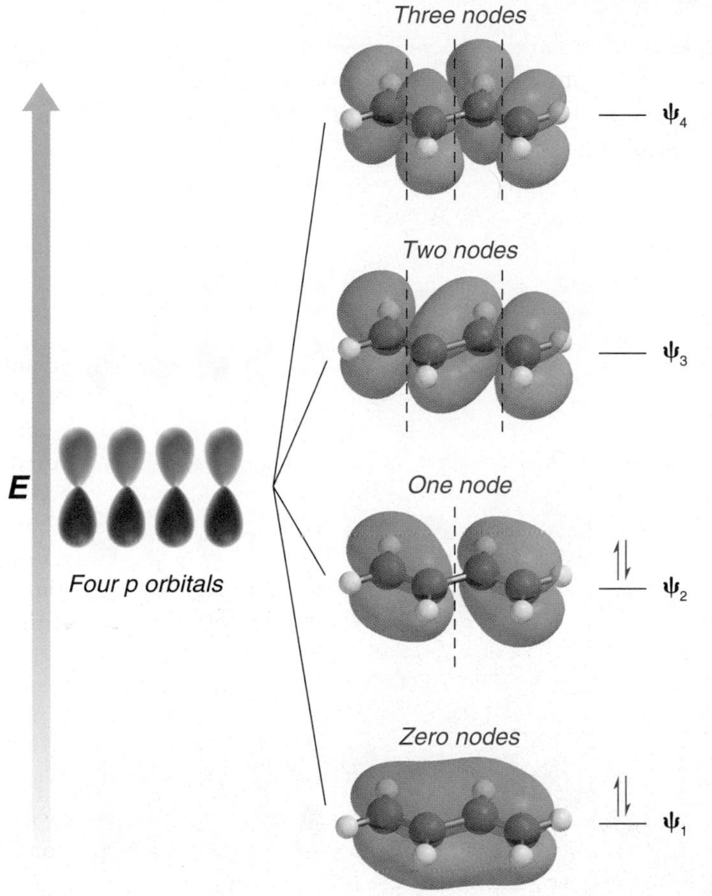

FIGURE 17.8
The MOs of 1,3-butadiene.

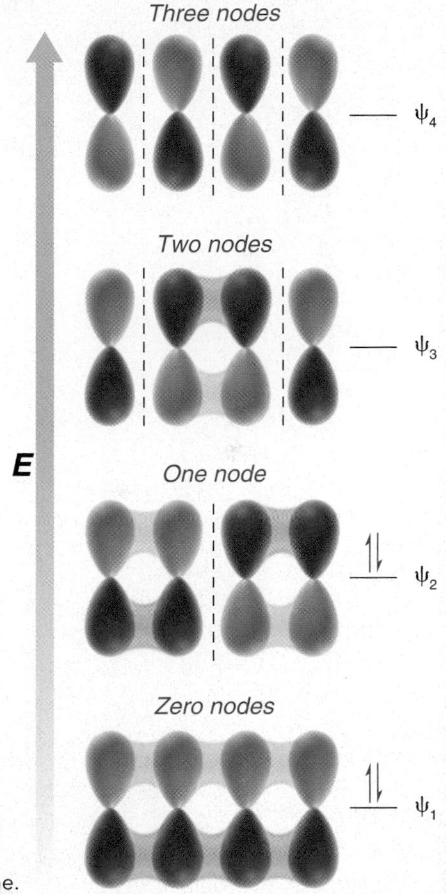

FIGURE 17.9
A shorthand method for representing the MOs of 1,3-butadiene.

MO theory provides an explanation for the stability associated with conjugated dienes, as discussed in Section 17.2. As seen in Figure 17.9, the four π electrons of a conjugated diene occupy the two lowest energy MOs (the bonding MOs). So let's focus our attention on those two MOs. The lowest energy MO (ψ_1) exhibits double-bond character at C2—C3, while the second MO (ψ_2) does not (Figure 17.10). All four π electrons occupy these two MOs (ψ_1 and ψ_2).

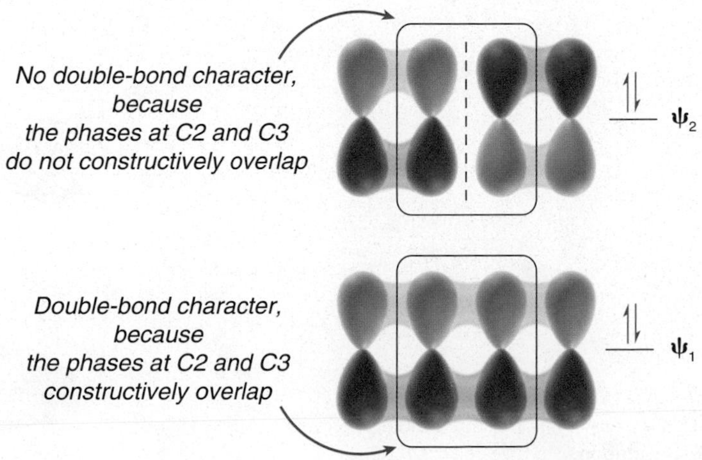

FIGURE 17.10
The occupied MOs of 1,3-butadiene.

Therefore, the C2—C3 bond has some double-bond character and some single-bond character. MO theory therefore provides us with an alternate way of explaining the short bond distance of the C2—C3 bond. In addition, MO theory also explains the stabilization energy associated with a conjugated diene. Specifically, the two electrons that occupy ψ_1 are delocalized over four carbon atoms. This delocalization accounts for the observed stabilization energy discussed in the previous section.

Molecular Orbitals of Hexatriene

A conjugated triene, such as 1,3,5-hexatriene, is comprised of six overlapping *p* orbitals. According to MO theory, these six atomic *p* orbitals are mathematically combined to produce six molecular orbitals (Figure 17.11). In this case, the six π electrons occupy the three MOs that are lowest in energy (the bonding MOs). Of these three MOs, the highest in energy is called the

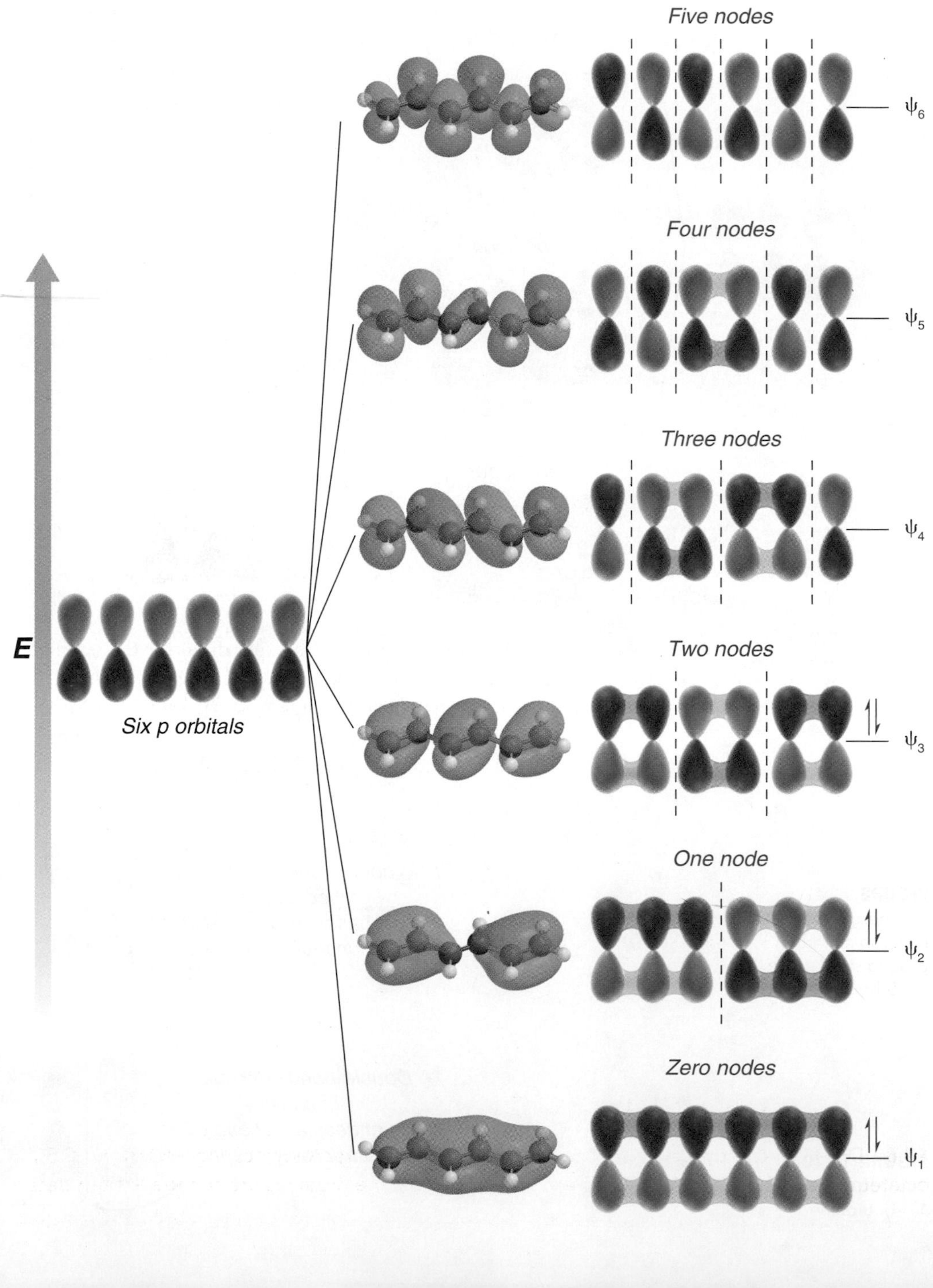

FIGURE 17.11
The MOs of 1,3,5-hexatriene.

*H*ighest *O*ccupied *M*olecular *O*rbital, or **HOMO** for short. Of the three unoccupied MOs (the antibonding MOs), the lowest in energy is called the *L*owest *U*noccupied *M*olecular *O*rbital, or **LUMO** for short.

For any conjugated polyene, the HOMO and the LUMO are the most important MOs to consider and are called the **frontier orbitals**. The HOMO contains the highest energy π electrons, which are most readily available to participate in a reaction, and the LUMO is the lowest energy MO that is capable of accepting electron density. Throughout this chapter, we will explore the reactivity of conjugated polyenes by focusing on their frontier orbitals, the HOMO and LUMO. This approach, called **frontier orbital theory**, was first advanced in1954 by Kenichi Fukui (Kyoto University, Japan), a corecipient of the Nobel Prize in Chemistry in 1981.

Conjugated π systems are capable of interacting with light, a phenomenon that will be discussed in greater detail at the end of this chapter. Under the right conditions, a π electron in the HOMO can absorb a photon of light bearing the appropriate energy necessary to promote the electron to the LUMO. For example, consider the photochemical excitation of hexatriene (Figure 17.12). In the ground state of hexatriene (prior to excitation), the HOMO is ψ_3, but in the **excited state**, the HOMO is ψ_4. Excitation causes a change in the identities of the frontier orbitals. The ability of light to affect the frontier orbitals will be important in Section 17.9, when we will discuss reactions induced by light, called **photochemical reactions**.

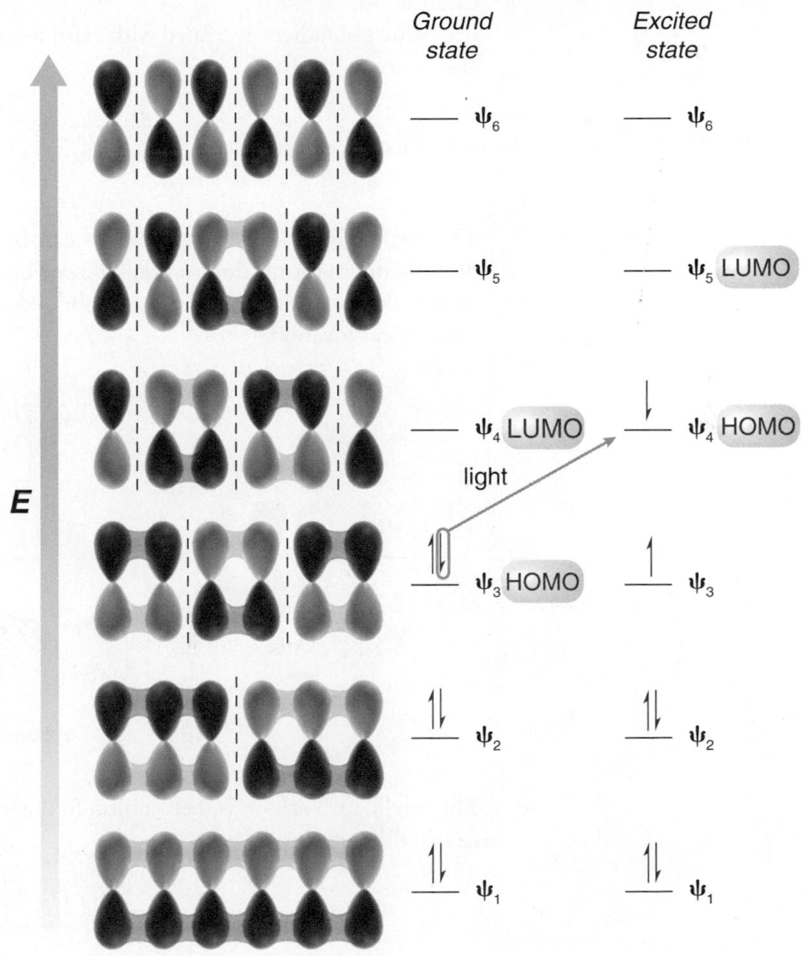

FIGURE 17.12 An energy diagram showing the electron configuration of the ground state and excited state of 1,3,5-hexatriene.

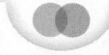

CONCEPTUAL CHECKPOINT

17.6 Draw an energy diagram showing the relative energy levels of the MOs for 1,3,5,7-octatetraene and identify the HOMO and LUMO for both the ground state and the excited state.

17.4 Electrophilic Addition

Addition of HX Across 1,3-Butadiene

In Section 9.3, we learned about the addition of HX across an alkene.

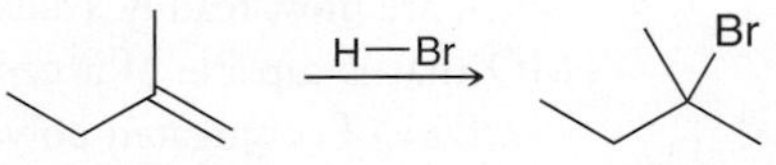

We saw that the reaction proceeds via Markovnikov addition, that is, the bromine is positioned at the more substituted carbon. This regiochemical outcome was explained with the following proposed mechanism, comprised of two steps:

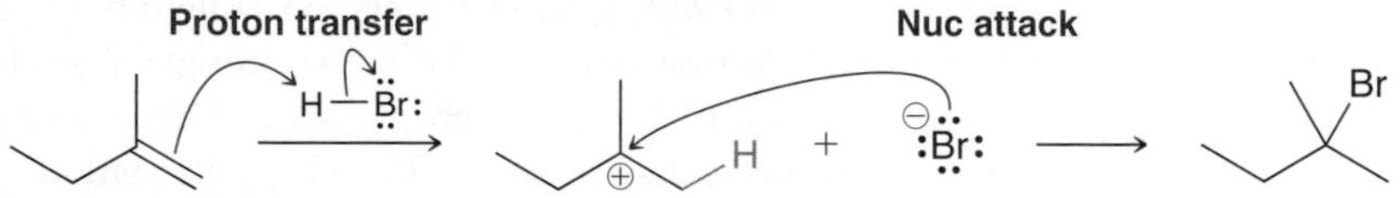

The first step of the mechanism controls the regiochemical outcome. Specifically, protonation occurs so as to create the more stable, tertiary carbocation, rather than the less stable, primary carbocation.

When butadiene is treated with HBr, a similar process takes place, but two major products are observed.

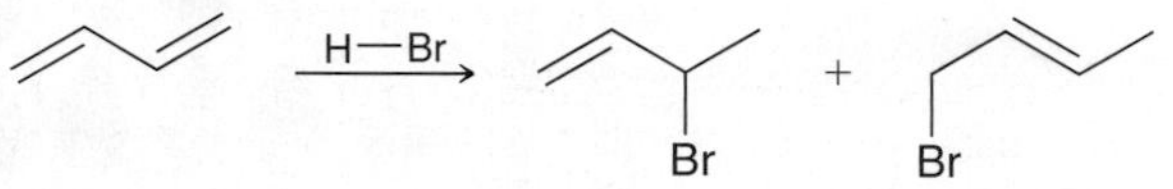

The formation of these two products can be explained with a similar two-step mechanism: protonation to form a carbocation followed by nucleophilic attack. In the first step, protonation creates the more stable, resonance-stabilized, allylic carbocation, rather than an unstabilized primary carbocation.

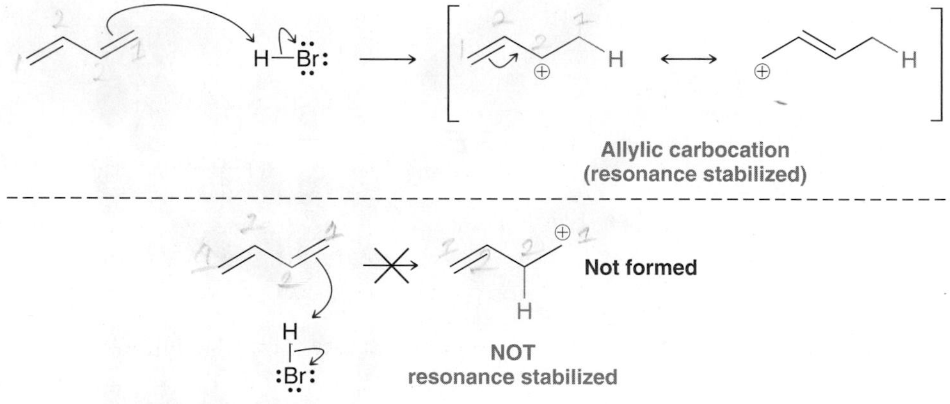

This allylic carbocation is then subject to nucleophilic attack in either of two positions, leading to two different products.

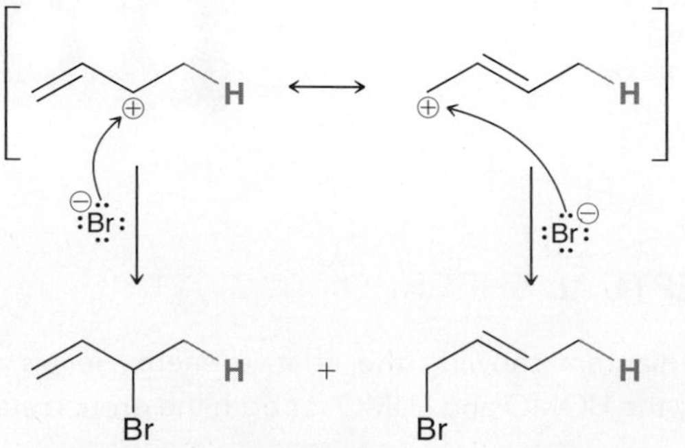

These compounds are said to be the products of **1,2-addition** and **1,4-addition**, respectively. This terminology derives from the fact that the starting diene contains a π system spread over four atoms, and the positions of H and Br are either at C1 and C2 or at C1 and C4. The products are called the **1,2-adduct** and the **1,4-adduct**, respectively.

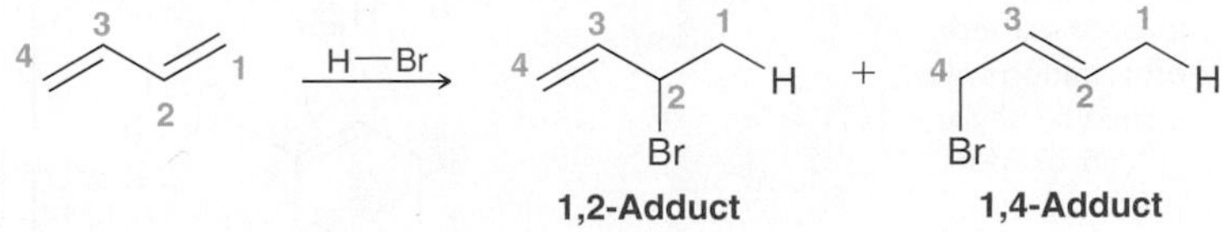

The exact product distribution (the ratio of products) is temperature-dependent. This temperature dependence will be explored in detail in the next section. For now, let's practice drawing both products and showing the mechanism of their formation.

SKILLBUILDER

17.1 PROPOSING THE MECHANISM AND PREDICTING THE PRODUCTS OF ELECTROPHILIC ADDITION TO CONJUGATED DIENES

LEARN the skill

Predict the products of the following reaction, and propose a mechanism that explains the formation of each product.

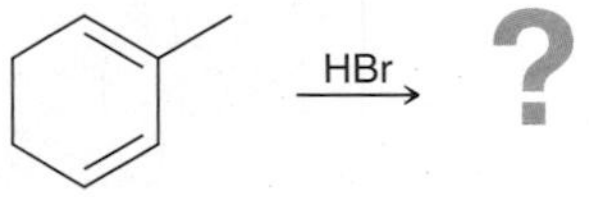

SOLUTION

This problem involves the addition of HBr across a conjugated π system. We therefore expect the mechanism to exhibit two steps: (1) proton transfer and (2) nucleophilic attack. In order to predict the products properly, we must explore the mechanism and identify the carbocation that is formed. In other words, we must carefully consider the regiochemistry of the protonation step. In this case, there are four distinct locations where the proton can be placed.

STEP 1
Identify the possible locations where protonation can occur.

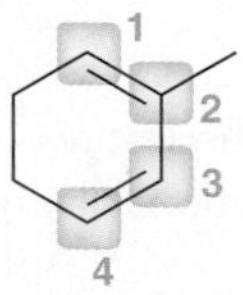

Protonation at either C2 or C3 will not produce an allylic carbocation.

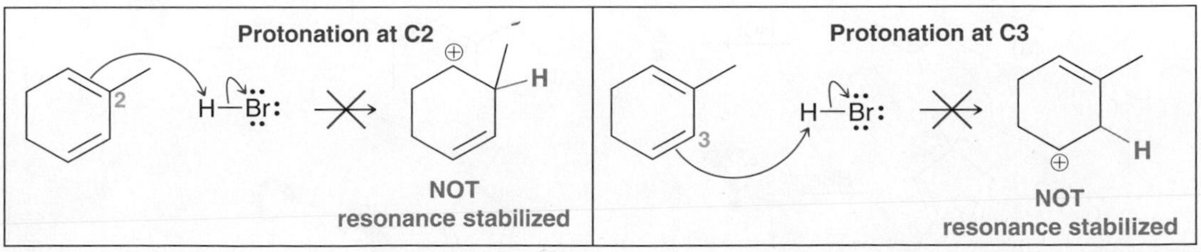

In either case, protonation produces a secondary carbocation that is not resonance stabilized. Only protonation at C1 or C4 will produce an allylic carbocation:

STEP 2
Determine the locations where protonation will produce an allylic carbocation.

Protonation at C1

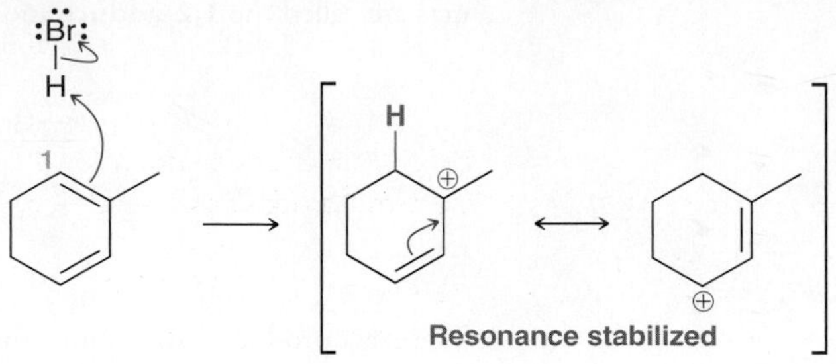

Protonation at C4

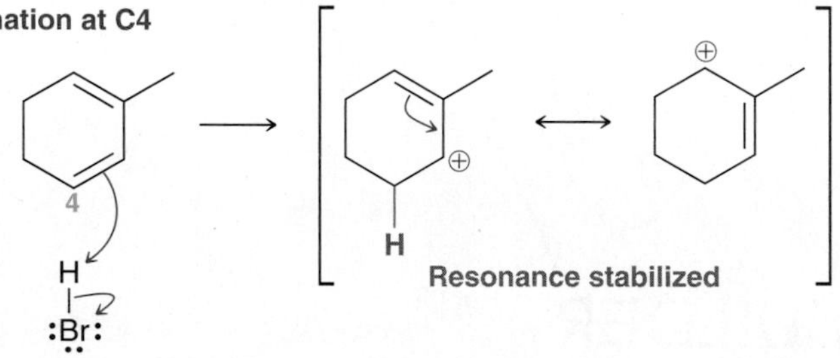

In this example, there are two different allylic carbocation intermediates that could be formed, each of which is resonance stabilized. In order to draw all possible products, consider the consequence of nucleophilic attack at each possible electrophilic position in each of the allylic carbocations:

STEP 3
Draw a nucleophilic attack occurring at each possible electrophilic position.

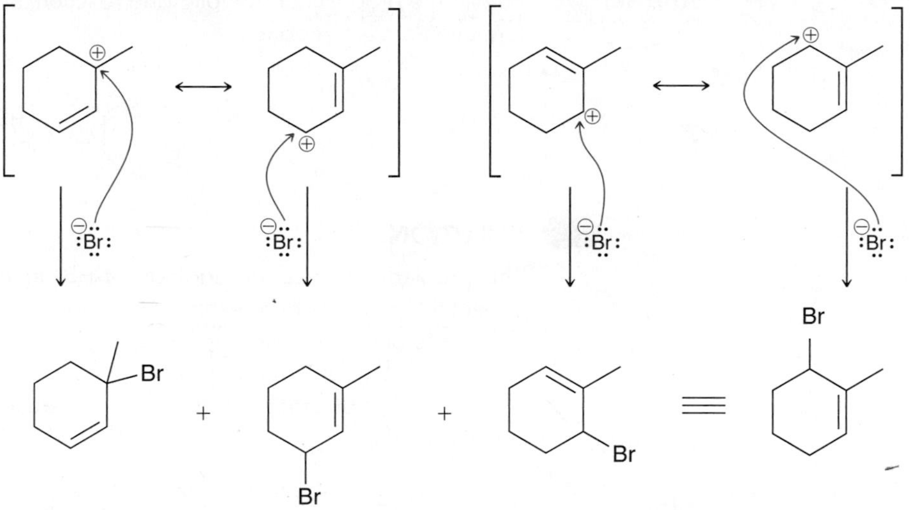

The last two drawings represent the same compound, giving a total of three expected constitutional isomers.

PRACTICE the skill **17.7** Predict the products for each of the following reactions, and propose a mechanism that explains the formation of each product:

(a) 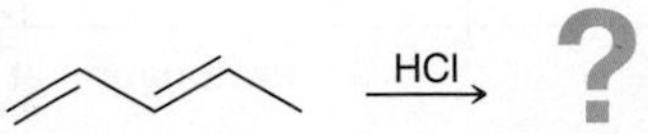

(b) 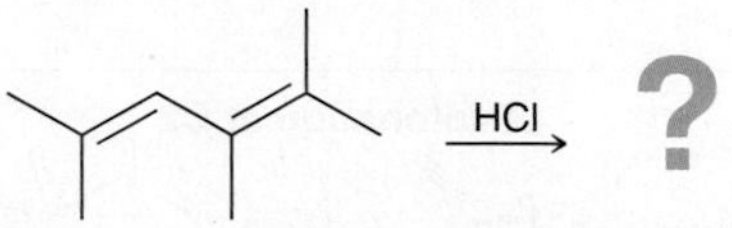

(c) 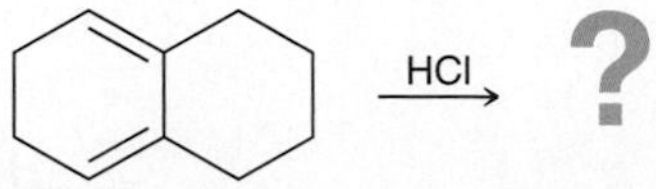

(d) 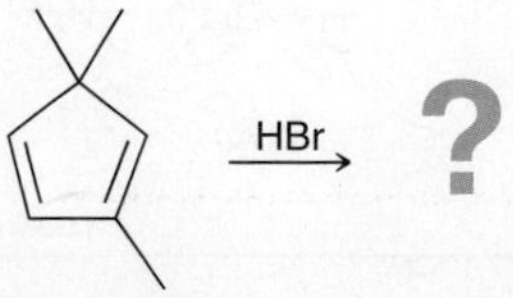

(e) 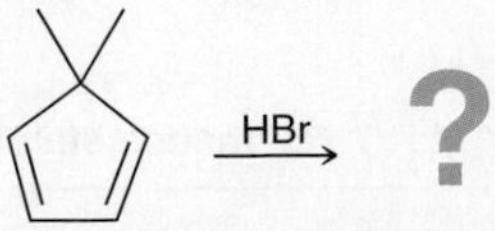

(f) 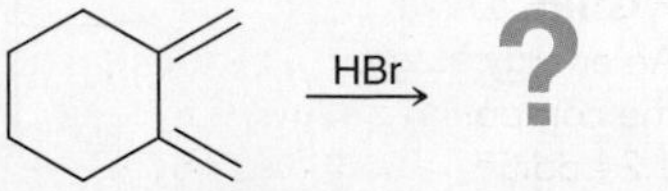

APPLY the skill

17.8 Consider the following two dienes. When treated with HBr, one of these dienes yields four products, while the other diene yields only two products. Explain.

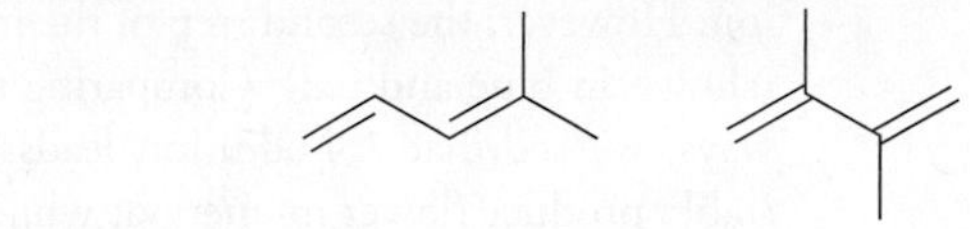

17.9 Propose the structure of a conjugated diene for which 1,2-addition of HBr produces the same product as 1,4-addition of HBr.

need more **PRACTICE?** **Try Problems 17.35, 17.38**

Addition of Br_2 Across 1,3-Butadiene

Many other electrophiles will also add across conjugated π systems to produce mixtures of 1,2- and 1,4-adducts. For example, bromine will add across 1,3-butadiene to produce the following mixture of products.

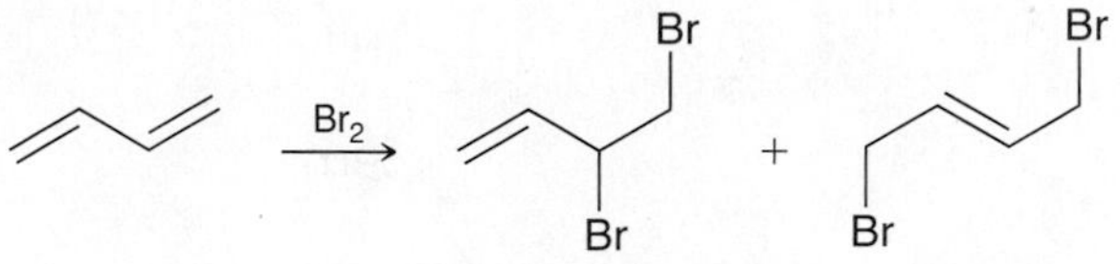

1,2-Adduct **1,4-Adduct**

17.5 Thermodynamic Control vs. Kinetic Control

In the previous section, we saw that conjugated π systems will undergo 1,2-addition as well as 1,4-addition. The exact ratio of products is highly dependent on the temperature at which the reaction is performed.

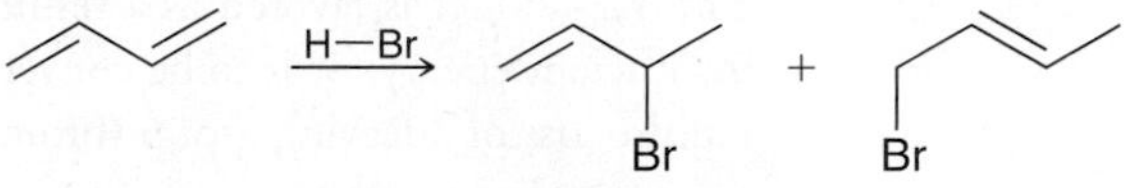

	1,2-Adduct	1,4-Adduct
At 0° C	71%	29%
At 40° C	15%	85%

When butadiene is treated with HBr at low temperature (0°C), the 1,2-adduct is favored. However, when the same reaction is performed at elevated temperature (40°C), the 1,4-adduct is favored. To understand the role temperature plays in this competition, we must look carefully at an energy diagram showing the formation of both products (Figure 17.13).

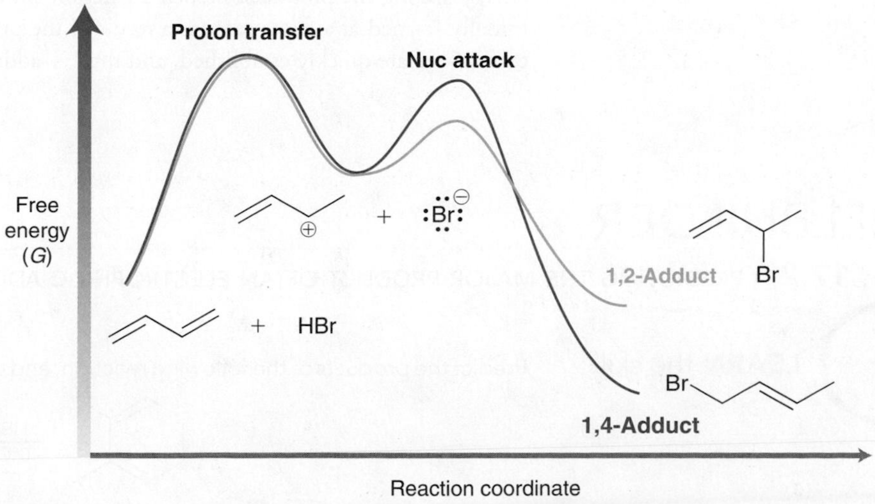

FIGURE 17.13
An energy diagram illustrating the competing pathways for 1,2-addition and 1,4-addition.

The first step of the mechanism is identical for both 1,2-addition and 1,4-addition, namely, the conjugated diene is protonated to give a resonance-stabilized, allylic carbocation and a bromide ion. However, the second step of the mechanism can occur via either of two competing pathways (shown in blue and red). Comparing these pathways, we see that 1,4-addition leads to a more stable product (lower in energy), while 1,2-addition occurs more rapidly (lower energy of activation). The 1,4-adduct is lower in energy because it exhibits a more substituted double bond:

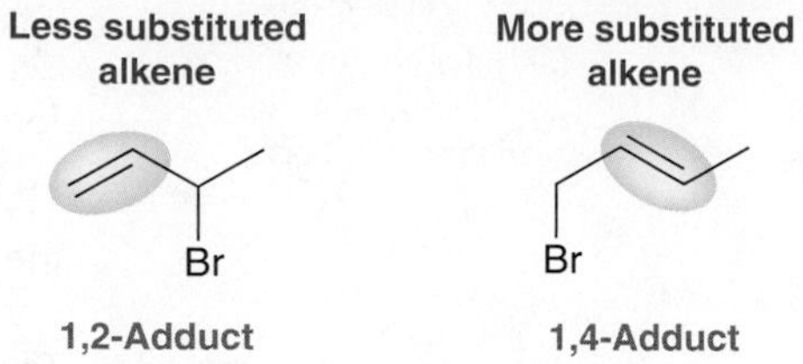

LOOKING BACK
For a review of alkene stability, see Section 8.5.

The 1,2-adduct is believed to form more rapidly as a result of a *proximity effect*. Specifically, the carbocation and the bromide ion are initially very close to each other immediately after their formation in the first step of the mechanism. The bromide ion is simply closer in proximity to C2 than C4, so attack at C2 occurs more rapidly.

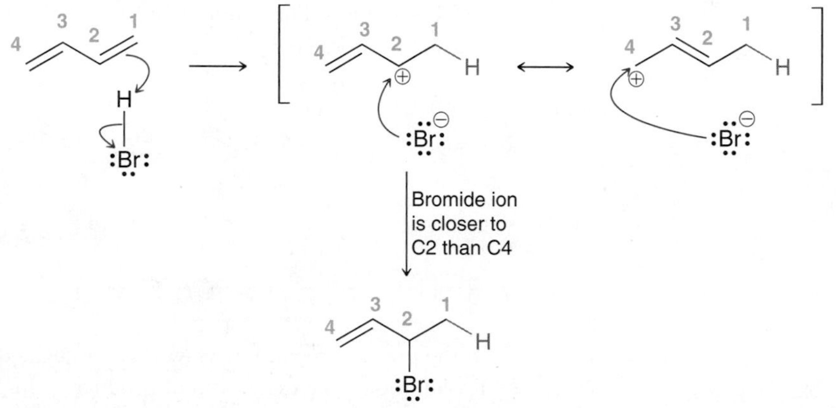

The 1,2-adduct is favored as a result of the proximity effect, and at low temperatures, there is insufficient energy for it to be converted back into the allylic carbocation. Such a process would require loss of a leaving group (bromide), which is simply too slow at low temperatures. Under these conditions, the competing reaction pathways (1,2-addition and 1,4-addition) are both practically irreversible. The 1,2-addition pathway occurs more rapidly and therefore generates the 1,2-adduct as the major product. The reaction is said to be under **kinetic control**, which means that the product distribution is determined by the *relative rates* at which the products are formed.

At elevated temperatures, the two competing pathways are no longer irreversible. The products have sufficient energy to lose a leaving group, reforming the allylic carbocation intermediate. Under such conditions, an equilibrium is established, and the product distribution will depend only on the relative energy levels of the two products. The product that is lower in energy (the 1,4-adduct) will predominate. The reaction is said to be under **thermodynamic control,** which means that the ratio of products is determined solely by the distribution of energy among the products. If such a reaction is monitored, it is found that the 1,2-adduct is initially formed at a faster rate (as a result of the proximity effect). However, equilibrium concentrations are quickly established, and the 1,4-adduct ultimately predominates.

SKILLBUILDER

17.2 PREDICTING THE MAJOR PRODUCT OF AN ELECTROPHILIC ADDITION TO CONJUGATED DIENES

LEARN the skill

Predict the products of the following reaction, and determine which product will predominate.

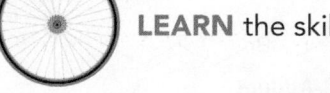

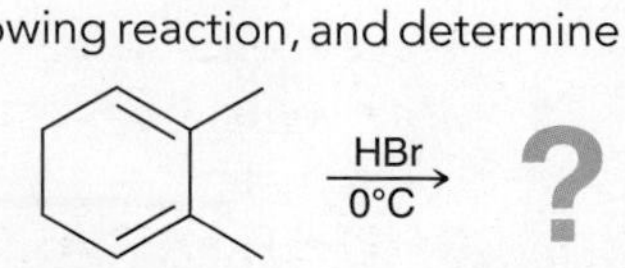

SOLUTION

STEP 1
Identify the possible locations where protonation can occur.

First let's predict all of the products. This problem involves the addition of HBr across a conjugated π system. We therefore expect the mechanism to exhibit two steps: (1) proton transfer and (2) nucleophilic attack. In order to predict the products properly, we must explore the mechanism and identify the carbocation that is formed. Begin by carefully considering the regiochemistry of the protonation step. In this case, it might seem that there are four possibilities.

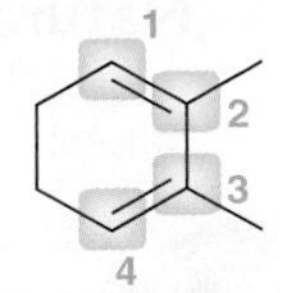

But this compound is symmetrical, with C1 being equivalent to C4, and C2 being equivalent to C3. Therefore, there are only two distinct possibilities for the protonation step: at C1 or at C2. The resonance-stabilized, allylic carbocation is only formed via protonation at C1:

STEP 2
Draw the structure of the allylic carbocation that is expected to form.

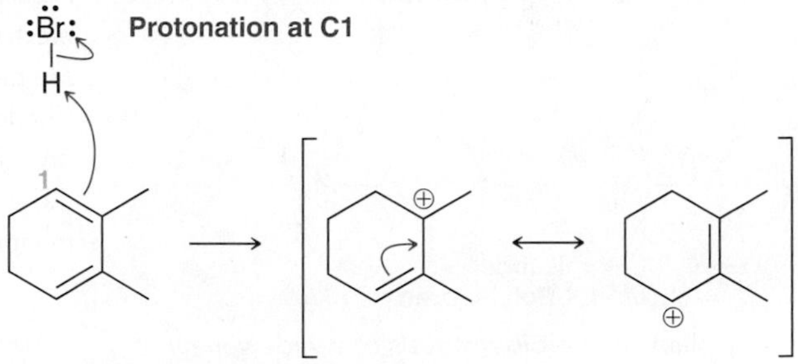

STEP 3
Attack each electrophilic position with a bromide ion and draw the possible products.

This allylic carbocation can then be attacked at either of two positions, leading to 1,2- or 1,4-addition. To predict which product predominates, we must determine which adduct is the kinetic product and which is the thermodynamic product. The 1,2-adduct is expected to be the kinetic product due to the proximity effect, while the 1,4-adduct exhibits a more highly substituted (tetrasubstituted) π bond and is therefore the thermodynamic product.

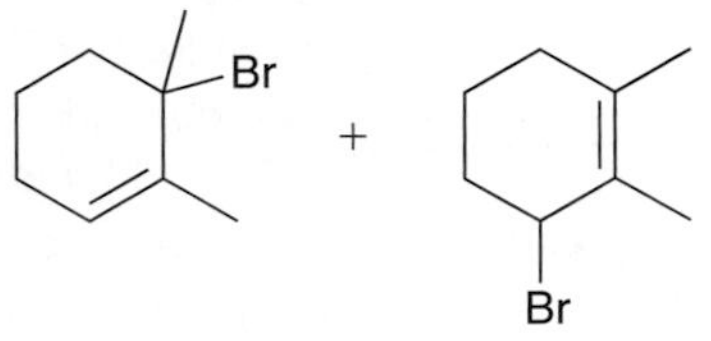

1,2-Adduct **1,4-Adduct**

STEP 4
Identify the kinetic and thermodynamic products, and then choose which product predominates based on the temperature of the reaction.

Finally, look carefully at the temperature indicated. In this case, a low temperature is used, so the reaction is under kinetic control. Therefore, in this case, the 1,2-adduct is expected to be favored.

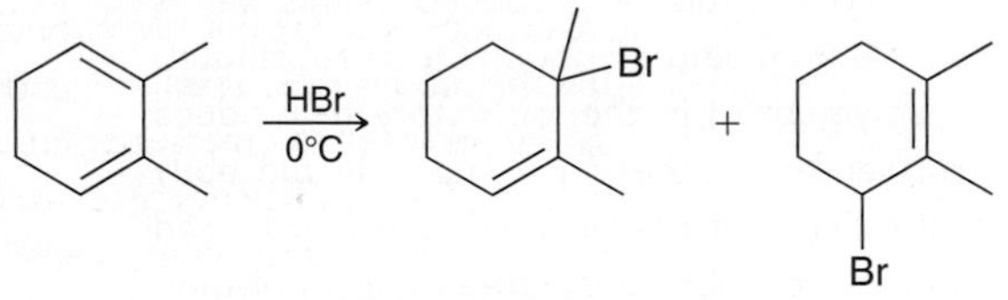

Major **Minor**

PRACTICE the skill **17.10** Predict the products for each of the following reactions, and in each case, determine which product will predominate:

(a) 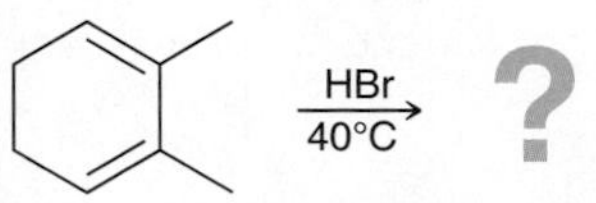

(b) 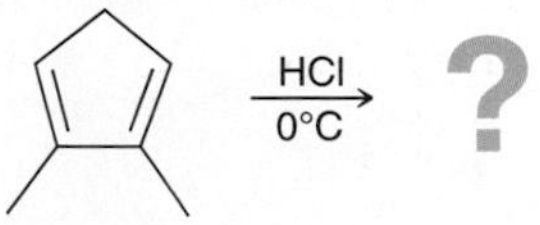

(c) 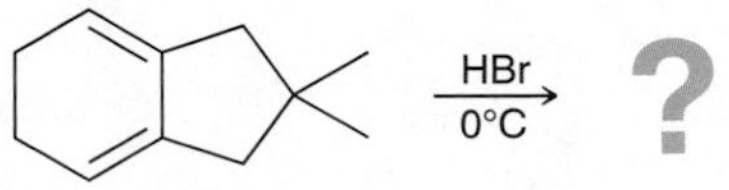

APPLY the skill **17.11** When 1,4-dimethylcyclohepta-1,3-diene is treated with HBr at elevated temperature, the 1,2-adduct predominates, rather than the 1,4-adduct. Explain this result.

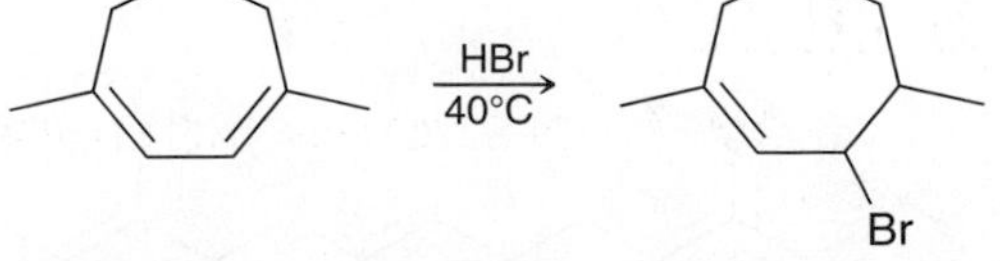

17.12 When 1,3-cyclopentadiene is treated with HBr, the temperature does not have an influence on the identity of the major product. Explain.

need more **PRACTICE?** **Try Problems 17.36, 17.37, 17.39**

PRACTICALLYSPEAKING)))

Natural and Synthetic Rubbers

Natural rubber is produced from the polymerization of isoprene:

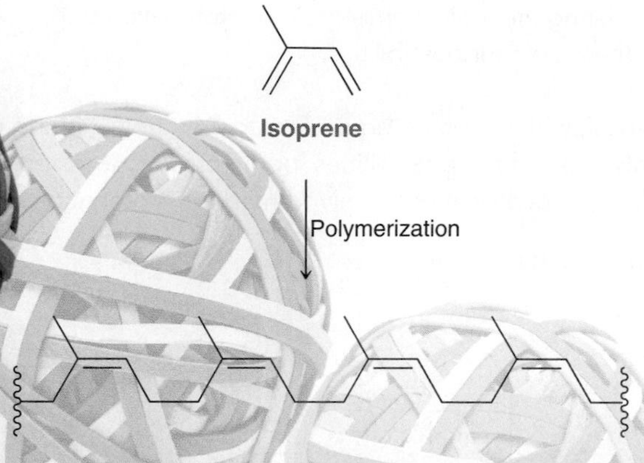

Rubber
(*cis*-1,4-Polyisoprene)

Isoprene is used by plants in the biosynthesis of a wide variety of compounds. The importance of isoprene as a chemical precursor will be discussed in Section 26.7. During the polymerization of isoprene to form rubber, the monomers are joined by the head-to-tail linkage of C1 and C4 positions. The polymerization of isoprene is therefore a special case of 1,4-addition.

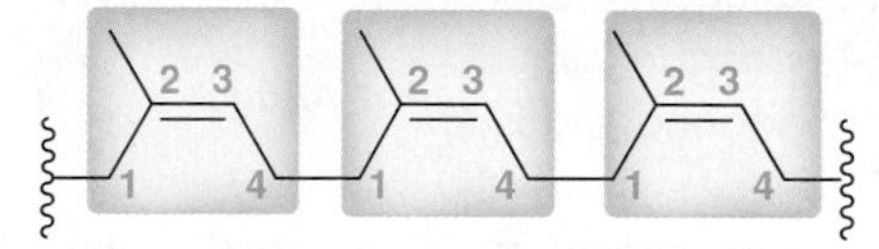

Rubber can also be made in the laboratory from polymerizing isoprene in the presence of suitable catalysts that favor formation of the *cis* polymer. Rubber produced in this way is virtually indistinguishable from natural rubber. Other substituted dienes can also be polymerized in the laboratory to produce a wide variety of rubberlike, synthetic polymers. In the early 1930s, chemists at the Du Pont chemical company produced a commercially important polymer via the free-radical polymerization of chloroprene. The resulting polymer is sold under the trade name Neoprene.

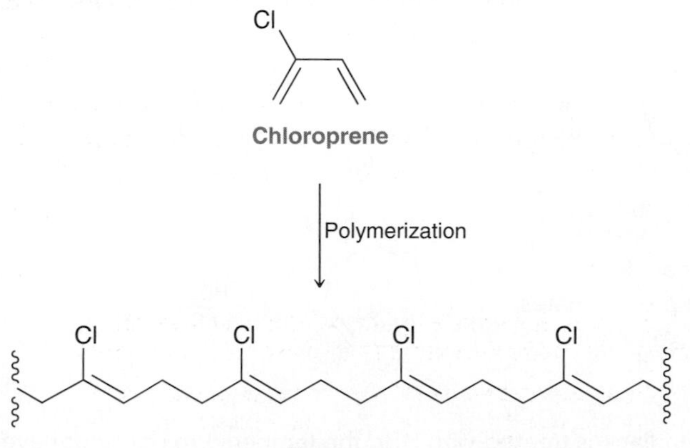

Neoprene
(*trans*-1,4-Polychloroprene)

Neoprene is used in many applications, including insulation for electrical wiring, fan belts for automobiles, and scuba diving wetsuits.

When natural or synthetic rubber is stretched, neighboring polymer chains slide past one another and ultimately snap back into their original positions when the external force is removed. The ability of a polymer to return to its original shape after being stretched is called *elasticity*, and polymers that exhibit this quality are called *elastomers*. The elasticity of natural rubber is restricted to a specific temperature range, which limits its potential applications. In 1839, Charles Goodyear discovered that polymerization of isoprene in the presence of sulfur, a process called *vulcanization*, vastly improves the elasticity of natural rubber. Vulcanization is the result of the formation of disulfide linkages between neighboring polymeric chains:

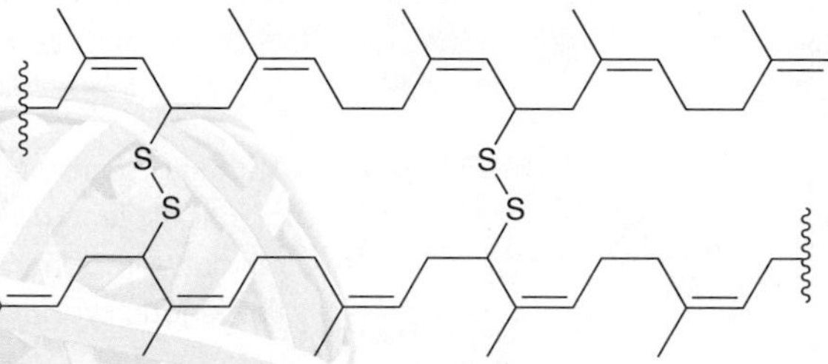

Vulcanized rubber

The sulfide linkages greatly improved the temperature range of rubber's elasticity. Vulcanized rubber maintains its elasticity even at high temperatures, because the disulfide linkages help snap the chains back into their original shape after the external force is removed. The vulcanized elastomer produced in greatest quantity is styrene-butadiene rubber (SBR). SBR is commercially prepared from styrene and butadiene via a free-radical polymerization process. It is called a copolymer, because it is made from two different monomers:

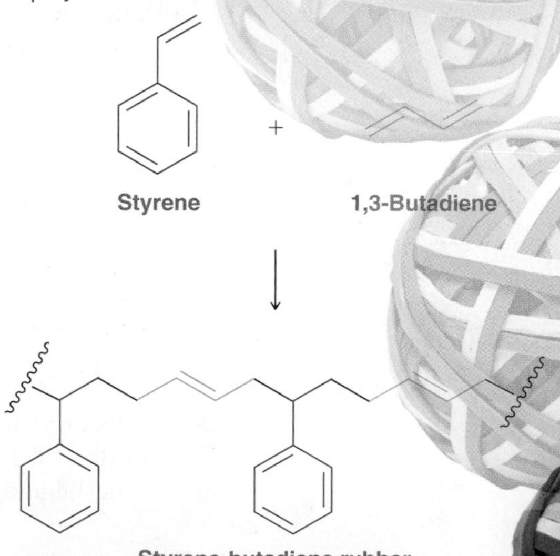

Styrene-butadiene rubber

In the United States, approximately one billion pounds of vulcanized SBR is produced annually for the production of automobile tires.

CONCEPTUAL CHECKPOINT

17.13 What monomer should be used to make each of the following polymers:

(a) 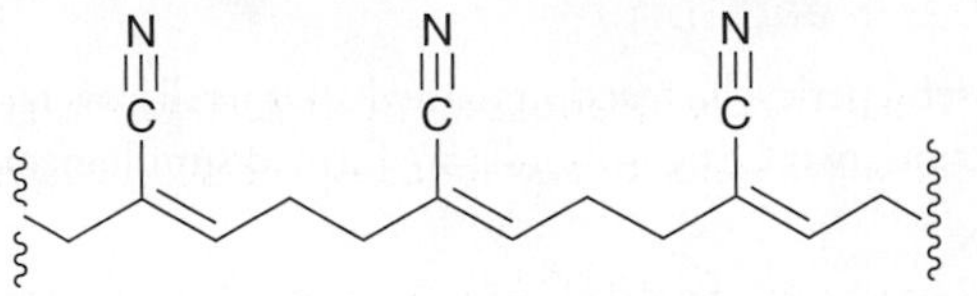

(b) 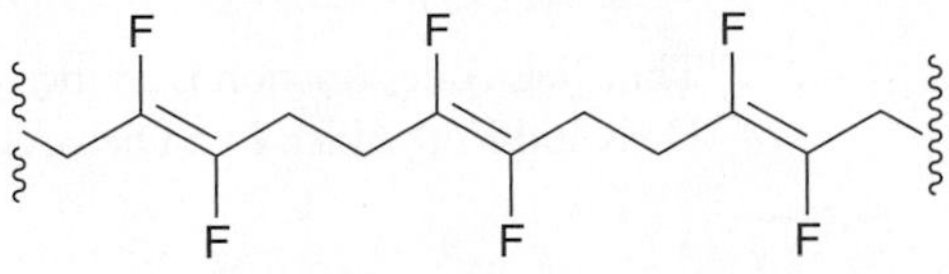

17.6 An Introduction to Pericyclic Reactions

Most organic reactions, both in the laboratory and in living organisms, proceed either via ionic intermediates or radical intermediates. There is, however, a third category of organic reactions, called **pericyclic reactions**, which do not involve either ionic or radical intermediates. These reactions can be classified into three major groups: **cycloaddition reactions**, **electrocyclic reactions**, and **sigmatropic rearrangements** (Figure 17.14).

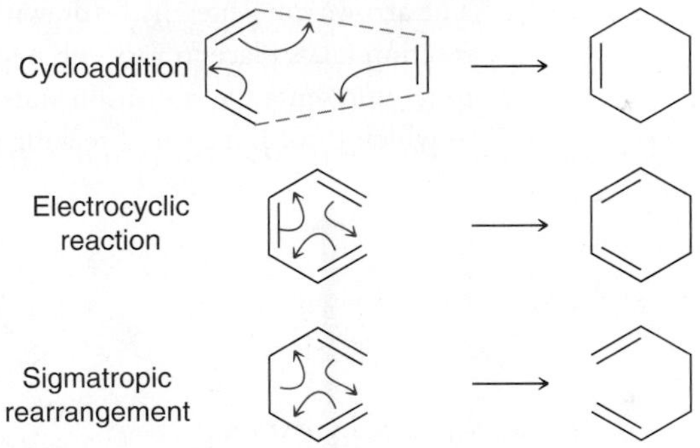

FIGURE 17.14
Examples of the three major classes of pericyclic reactions.

Each of these reactions has the following features, characteristic of pericyclic reactions:

- The reaction proceeds via a concerted process, which means that all changes in bonding occur in a single step. As a result, the reaction mechanism has no intermediates.
- The reaction involves a ring of electrons moving around in a closed loop.
- The reaction occurs through a cyclic transition state.
- The polarity of the solvent generally does not have a large impact on the rate or yield of the reaction, suggesting that the transition state bears very little (if any) partial charge.

The three major types of pericyclic reactions differ in terms of the number and type of bonds that are broken and formed (Table 17.1). In a cycloaddition reaction, two π bonds are converted into two σ bonds, resulting in the addition of two reactants to form a ring. In an electrocyclic reaction, one π bond is converted into a σ bond, which effectively joins the ends of one reactant to form a ring. In a sigmatropic rearrangement, one σ bond is formed at the expense of another, and the π bonds change location. The following sections will explore each of the three categories of pericyclic reactions, beginning with cycloadditions.

TABLE 17.1 A COMPARISON OF THE NUMBER OF BONDS BROKEN AND FORMED IN EACH OF THE THREE MAJOR TYPES OF PERICYCLIC REACTIONS

	CHANGE IN THE NUMBER OF σ BONDS	CHANGE IN THE NUMBER OF π BONDS
Cycloaddition	+2	−2
Electrocyclic	+1	−1
Sigmatropic	0	0

17.7 Diels-Alder Reactions

BY THE WAY
Diels and Alder were awarded the 1950 Nobel Prize in Chemistry for their discovery of this reaction.

The Mechanism of a Diels-Alder Reaction

The Diels-Alder reaction is an incredibly useful pericyclic reaction named after its discoverers, Otto Diels and Kurt Alder. In a Diels-Alder reaction, two C—C σ bonds are formed simultaneously:

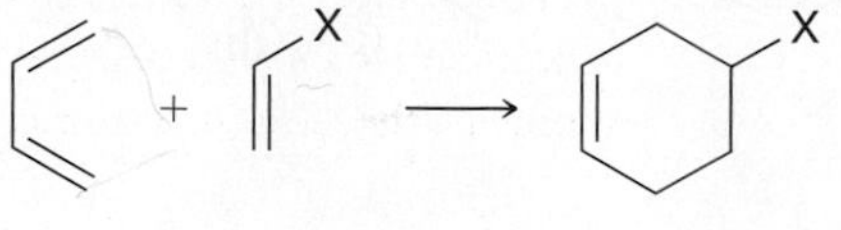

In the process, a ring is also formed, and the process is therefore a cycloaddition. Specifically, the Diels-Alder reaction is called a **[4+2] cycloaddition** because the reaction takes place between two different π systems, one of which is associated with four atoms, while the other is associated with two atoms. The product of a Diels-Alder reaction is always a substituted cyclohexene. As is the case for all pericyclic reactions, the Diels-Alder reaction is a concerted process.

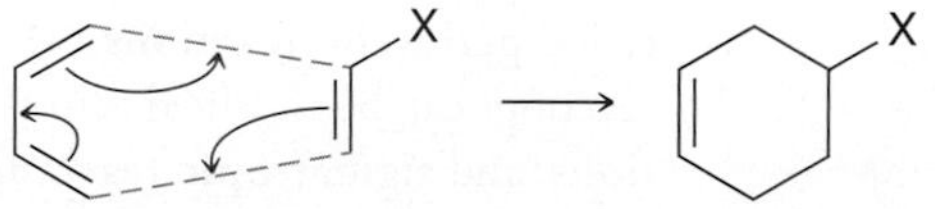

The arrows can generally be drawn in either a clockwise or a counterclockwise fashion. Since the reaction takes place in just one step, the energy diagram has just one peak, where the top of the peak represents the transition state (Figure 17.15). The transition state is a six-membered ring in which three bonds are breaking and three bonds are forming simultaneously.

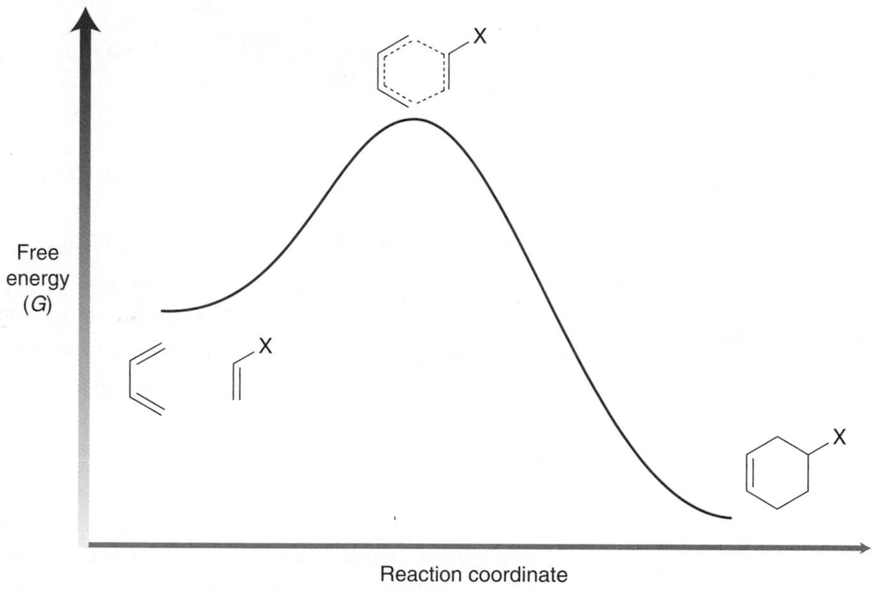

FIGURE 17.15
An energy diagram of a Diels-Alder reaction.

Thermodynamic Considerations

Moderate temperatures favor the formation of products in a Diels-Alder reaction, but very high temperatures (over 200°C) tend to disfavor product formation. In fact, in many cases, high temperatures can be used to achieve the reverse of a Diels-Alder reaction, called a **retro Diels-Alder reaction.**

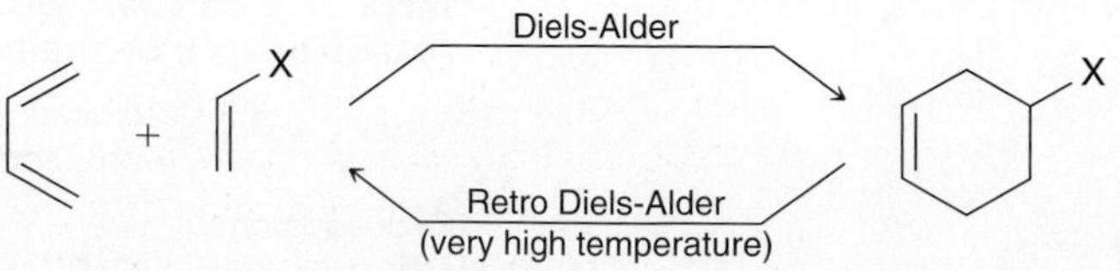

To understand why high temperatures favor ring opening rather than ring formation, recall from Section 6.3 that the sign of ΔG determines whether the equilibrium favors reactants or products. The sign of ΔG must be negative for the equilibrium to favor products. To determine the sign of ΔG, there are two terms that must be considered.

$$\Delta G = \underbrace{(\Delta H)}_{\text{Enthalpy term}} + \underbrace{(-T\Delta S)}_{\text{Entropy term}}$$

Let's consider each term individually, beginning with the enthalpy term. There are many factors that contribute to the sign and magnitude of ΔH, but the dominant factor is generally bond strength. We compare the bond strengths of the bonds broken and the bonds formed in a Diels-Alder reaction.

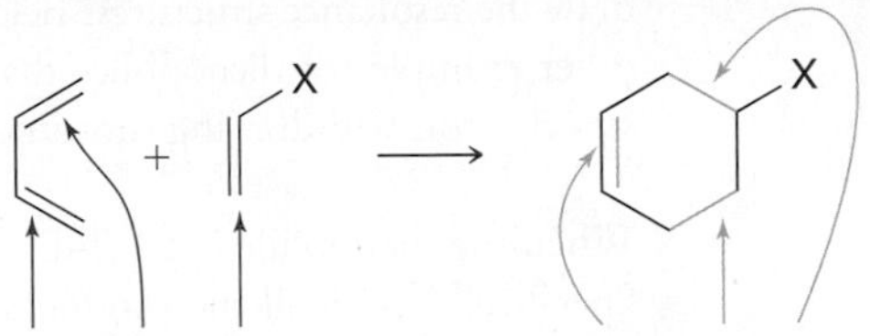

Bonds broken
3 π bonds

Bonds formed
1 π bond and 2 σ bonds

Three π bonds are broken and replaced by one π bond and two σ bonds. We saw in Chapter 1 that σ bonds are stronger than π bonds, and as a result, ΔH for this process has a *negative* value so the reaction is exothermic.

Now let's consider the second term, the entropy term $-T\,\Delta S$. This term will always be positive for a Diels-Alder reaction. Why? There are two main reasons: (1) two molecules are joining together to produce one molecule of product and (2) a ring is being formed. As described in Section 6.3, each of these factors represents a decrease in entropy, contributing to a negative value for ΔS. Temperature (measured in kelvin) is always positive, and therefore, $-T\,\Delta S$ will be positive.

Now let's combine both terms (enthalpy and entropy). The sign of ΔG for a Diels-Alder reaction will be determined by the competition between these two terms:

$$\Delta G = \underbrace{(\Delta H)}_{\substack{\text{Enthalpy term} \\ \ominus}} + \underbrace{(-T\Delta S)}_{\substack{\text{Entropy term} \\ \oplus}}$$

In order for ΔG to be negative, the enthalpy term must be larger than the entropy term, which is temperature dependent. At moderate temperatures, the entropy term is small, and the enthalpy term dominates. As a result, ΔG will be negative, which means that products will be favored over reactants (the equilibrium constant K will be greater than 1). In other words, Diels-Alder reactions are thermodynamically favorable at moderate temperatures.

However, at very high temperatures, the entropy term will be large and will dominate the enthalpy term. As a result, ΔG will be positive, which means that reactants will be favored over products (the equilibrium constant K will be less than 1). In other words, the reverse reaction (a retro Diels-Alder) will be thermodynamically favored at high temperature.

In summary, Diels-Alder reactions are generally performed at moderate temperatures, usually between room temperature and 200°C, depending on the specific case.

The Dienophile

The starting materials for a Diels-Alder reaction are a diene and a compound that reacts with the diene, called the **dienophile.**

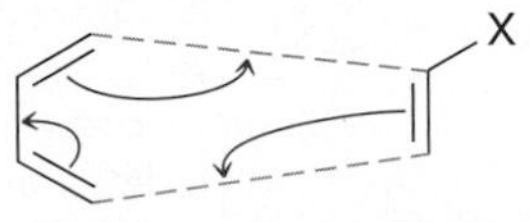

Diene **Dienophile**

We will begin our discussion with the dienophile. When the dienophile does not contain any substituents, the reaction exhibits a large activation energy and proceeds slowly. If the temperature is raised to overcome the energy barrier, the starting materials are favored over the products, and the resulting yield is low:

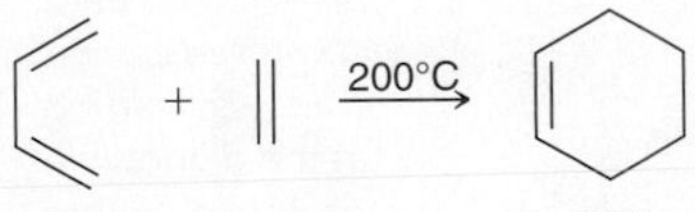

20%

A Diels-Alder reaction will proceed more rapidly and with a much higher yield when the dienophile has an electron-withdrawing substituent such as a carbonyl group:

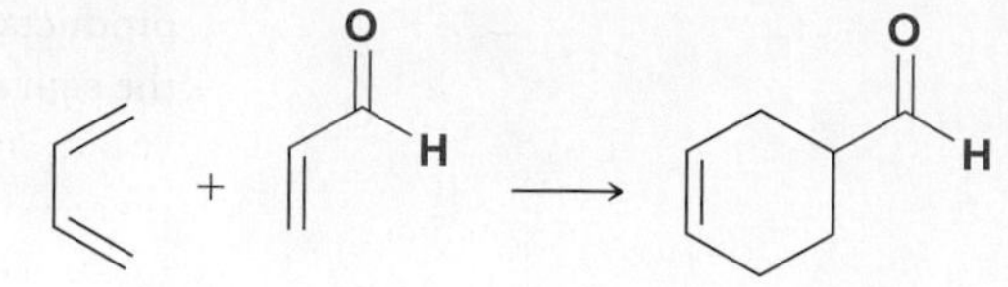

100%

The carbonyl group is electron withdrawing because of resonance. Can you draw the resonance structures? Below are other examples of dienophiles that possess electron-withdrawing substituents.

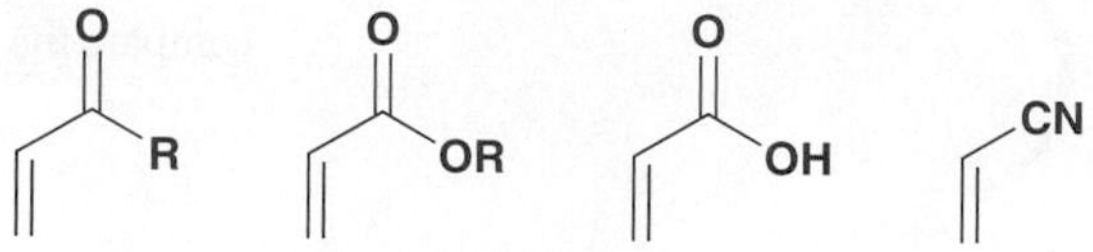

When the dienophile is a 1,2-disubstituted alkene, the reaction proceeds with stereospecificity. Specifically, a *cis* alkene produces a *cis* disubstituted ring, and a *trans* alkene produces a *trans* disubstituted ring.

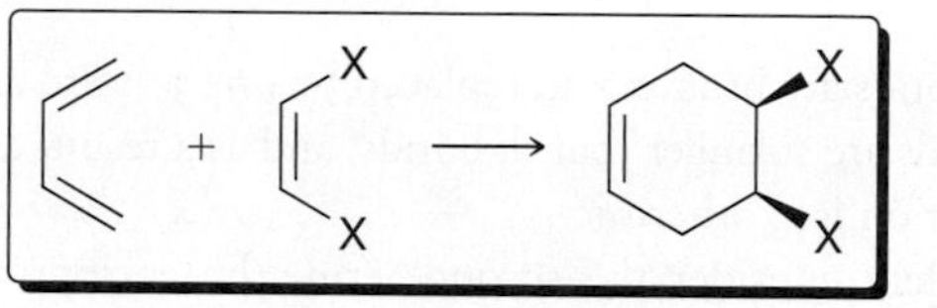

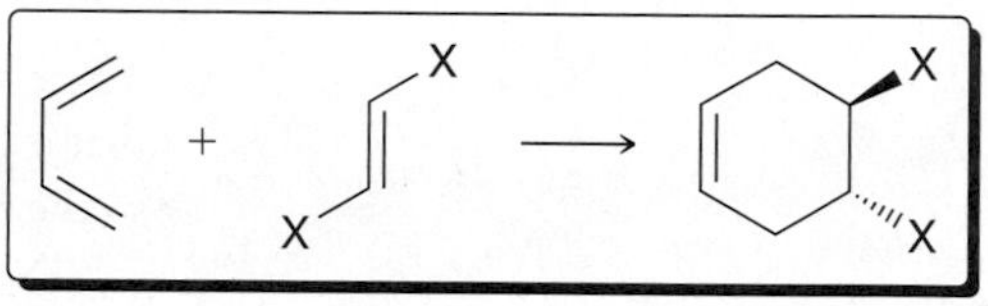

A triple bond can also function as a dienophile, in which case the product is a ring with two double bonds (a 1,4-cyclohexadiene).

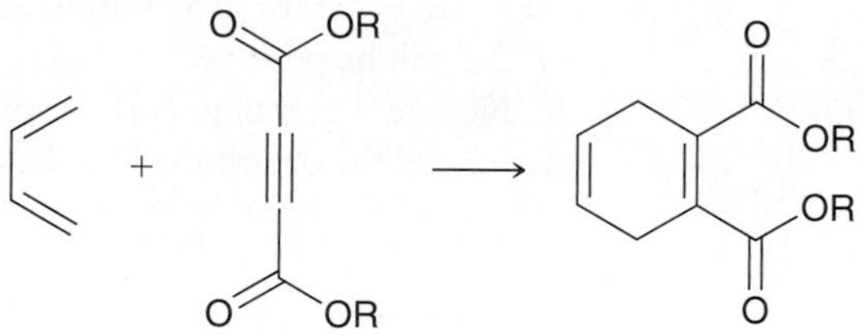

SKILLBUILDER

17.3 PREDICTING THE PRODUCT OF A DIELS-ALDER REACTION

LEARN the skill

Predict the product of the following reaction:

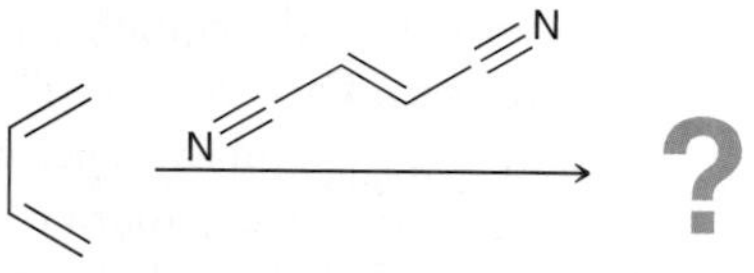

STEP 1
Redraw and line up the ends of the diene with the dienophile.

BY THE WAY
It is not necessary to draw the dotted lines between the diene and the dienophile, but you might find it helpful.

SOLUTION

First redraw the diene and the dienophile so that the ends of the diene are close to the dienophile:

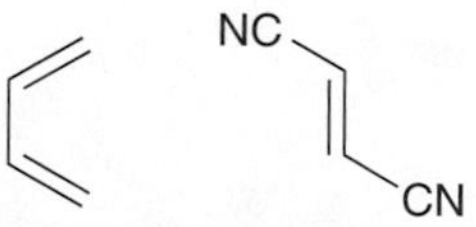

STEP 2
Draw three curved arrows, starting at the dienophile and going either clockwise or counterclockwise.

Next, draw three curved arrows that move around in a circle. Place the tail of the first curved arrow on the dienophile, and then continue to draw all three arrows in a circle, either clockwise or counterclockwise:

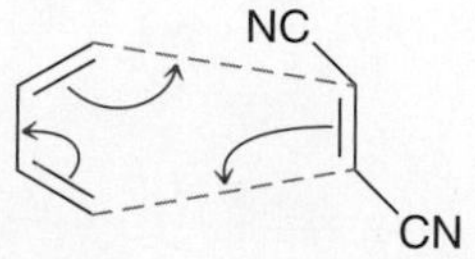

STEP 3
Draw the product with the correct stereochemical outcome.

Finally, draw the product with the correct stereochemical outcome. In this case, the starting dienophile is a *trans*-disubstituted alkene, so we expect the cyano groups to be *trans* to each other in the product:

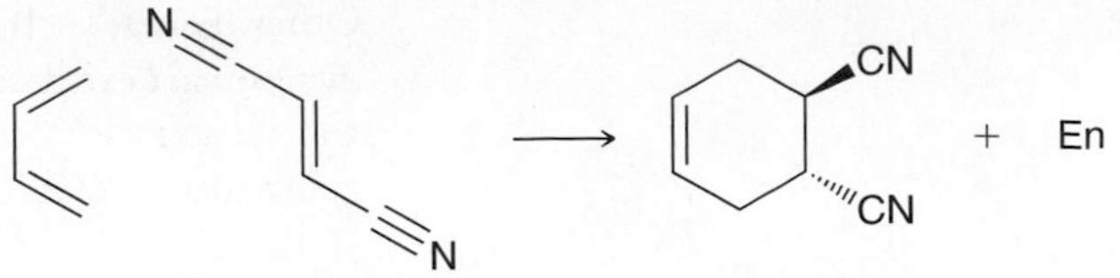

PRACTICE the skill **17.14** Predict the products for each of the following reactions:

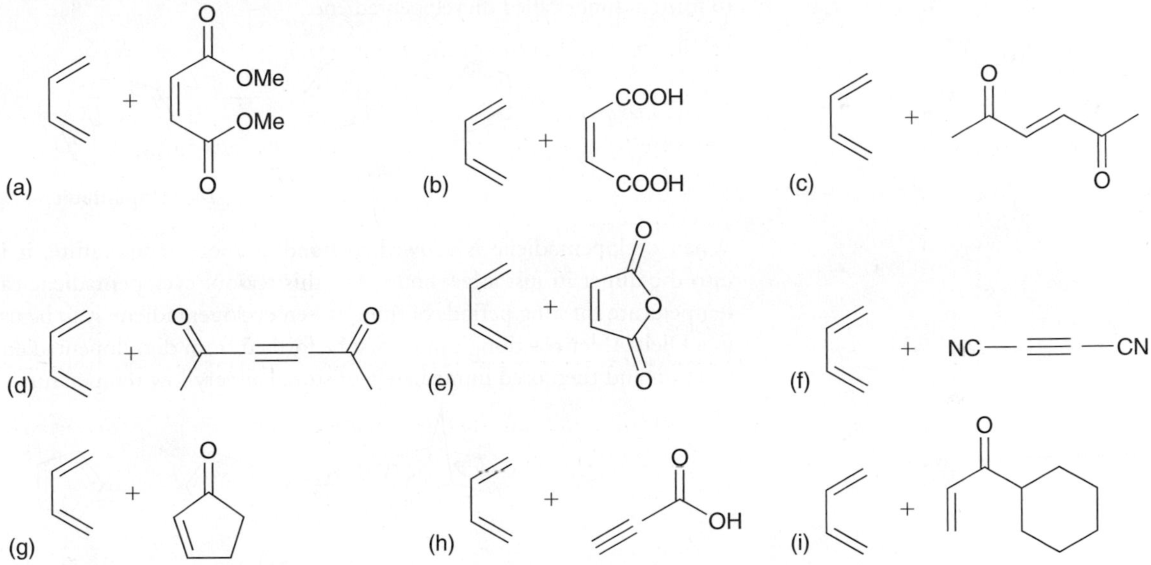

APPLY the skill **17.15** Predict the product(s) obtained when benzoquinone is treated with excess butadiene:

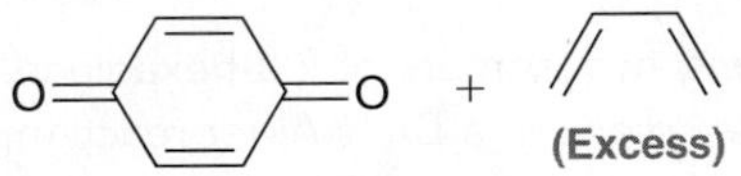

17.16 Propose a mechanism for the following transformation:

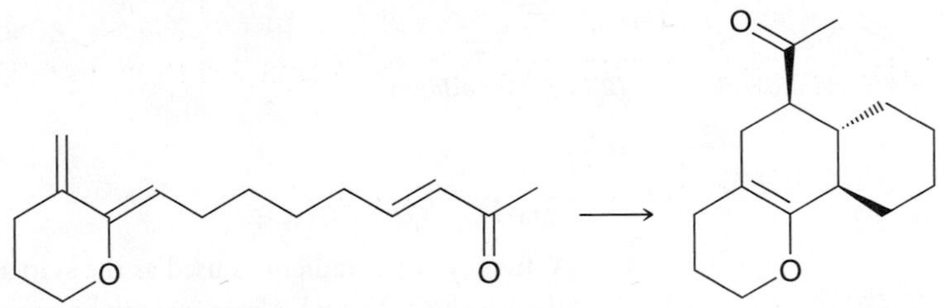

need more **PRACTICE?** **Try Problems 17.44, 17.46**

The Diene

Recall that 1,3-butadiene exists as an equilibrium between the *s-cis* conformation and the *s-trans* conformation.

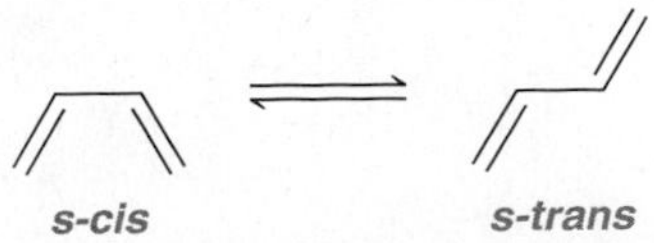

The Diels-Alder reaction only occurs when the diene is in an *s-cis* conformation. When the compound is in an *s-trans* conformation, the ends of the diene are too far apart to react with the dienophile. Some dienes are incapable of adopting an *s-cis* conformation, and are therefore unreactive toward a Diels-Alder reaction; for example:

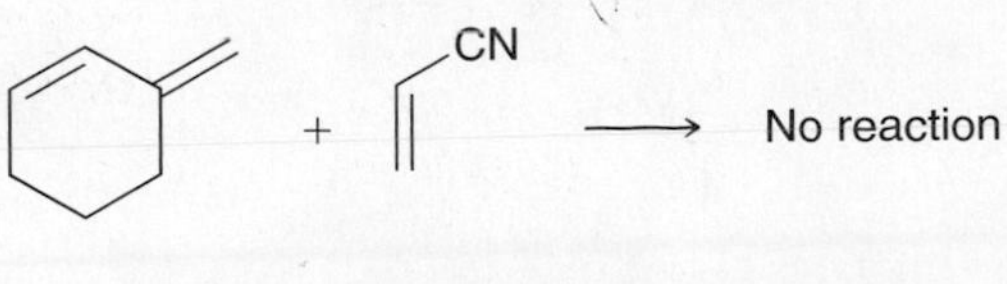

Other dienes, such as cyclopentadiene, are permanently locked in an *s-cis* conformation. Such dienes react extremely rapidly in Diels-Alder reactions.

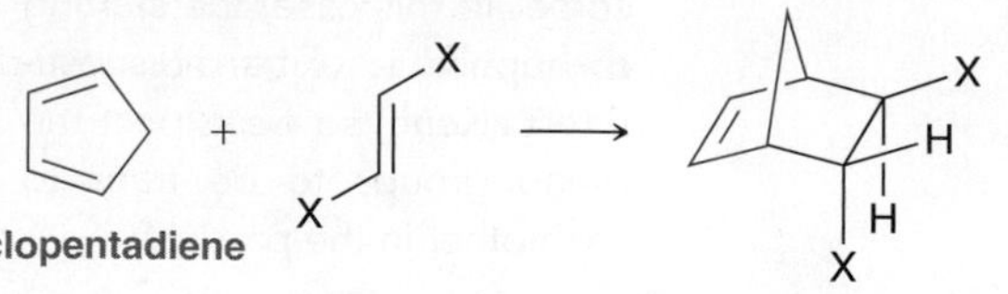

Cyclopentadiene

In fact, cyclopentadiene is so reactive toward Diels-Alder reactions that it even reacts with itself to form a dimer called dicyclopentadiene.

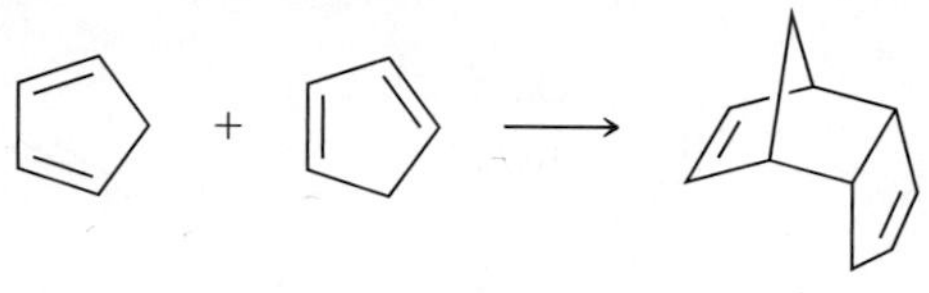

Dicyclopentadiene

When cyclopentadiene is allowed to stand at room temperature, it is completely converted into the dimer in just a few hours. For this reason, cyclopentadiene cannot be stored at room temperature for long periods of time. When cyclopentadiene is to be used as a starting material in a Diels-Alder reaction, it must first be formed from dicyclopentadiene via a retro Diels-Alder reaction and then used immediately or stored at very low temperature.

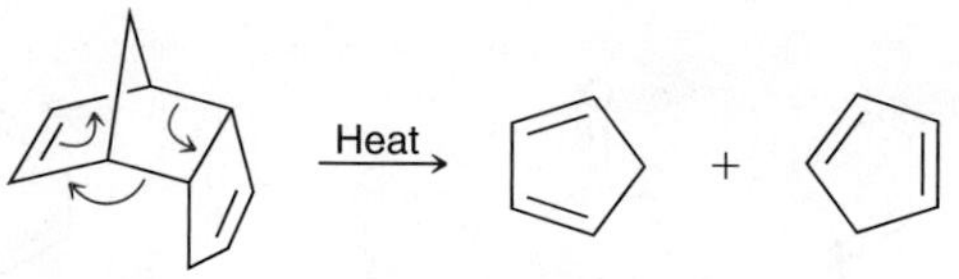

CONCEPTUAL CHECKPOINT

17.17 Consider the following two isomers of 2,4-hexadiene. One isomer reacts rapidly as a diene in a Diels-Alder reaction, and the other does not. Identify which isomer is more reactive, and explain your choice.

(2*E*,4*E*)-Hexadiene **(2*Z*,4*Z*)-Hexadiene**

17.18 Rank the following dienes in terms of reactivity in Diels-Alder reactions (from least reactive to most reactive):

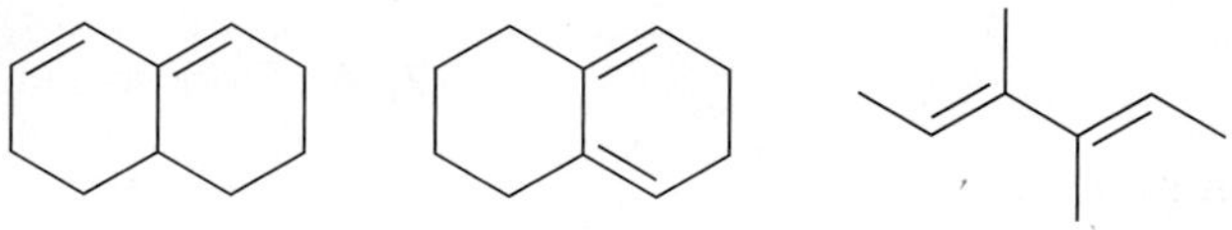

Endo Preference

When cyclopentadiene is used as the starting diene, a bridged bicyclic compound is obtained as the product. In such a case, we might expect to obtain the following two products.

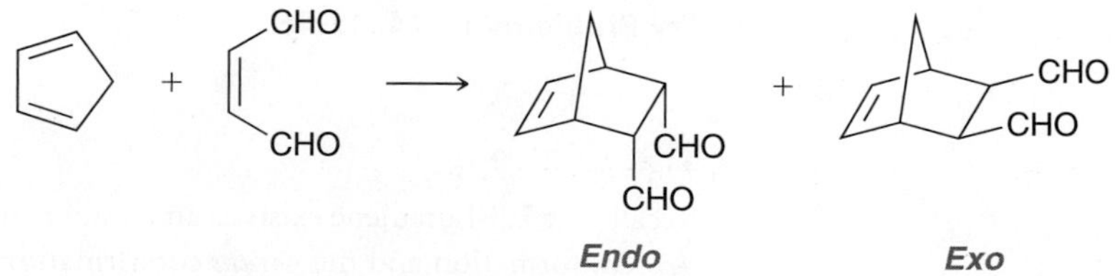

Endo ***Exo***

In one cycloadduct, the electron-withdrawing substituents occupy *endo* positions, and in the other cycloadduct, the substituents occupy *exo* positions. The **endo** positions are syn to the larger bridge, and the **exo** positions are *anti* to the larger bridge.

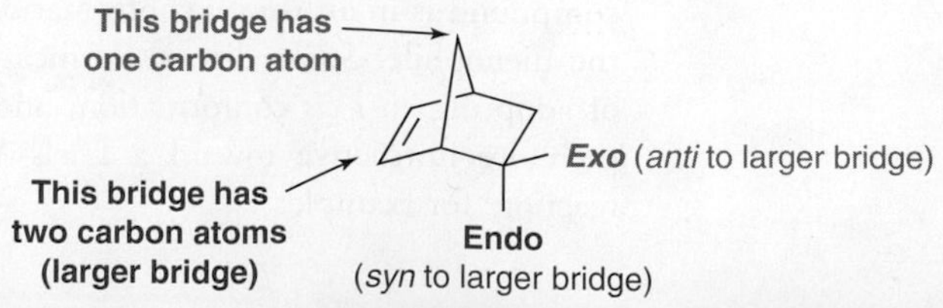

In general, the endo cycloadduct is highly favored over the exo cycloadduct. In many cases, the *endo* product is the exclusive product. The explanation for this preference is based on an analysis of the transition states leading to the *endo* and *exo* products. During formation of the *endo* product, a favorable interaction exists between the electron-withdrawing substituents and the developing π bond.

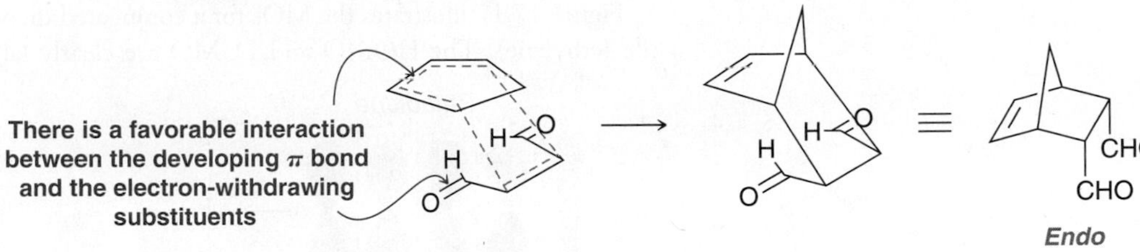

The transition state leading to the exo product does not exhibit this favorable interaction:

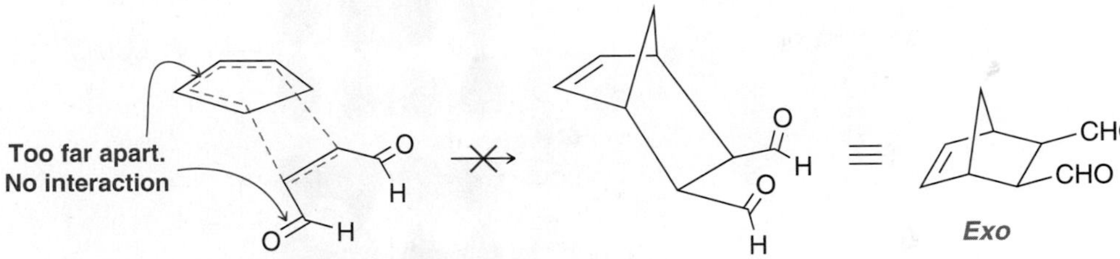

The transition state leading to formation of the endo product is lower in energy than the transition state leading to formation of the exo product (Figure 17.16). As a result, the endo product is formed more rapidly.

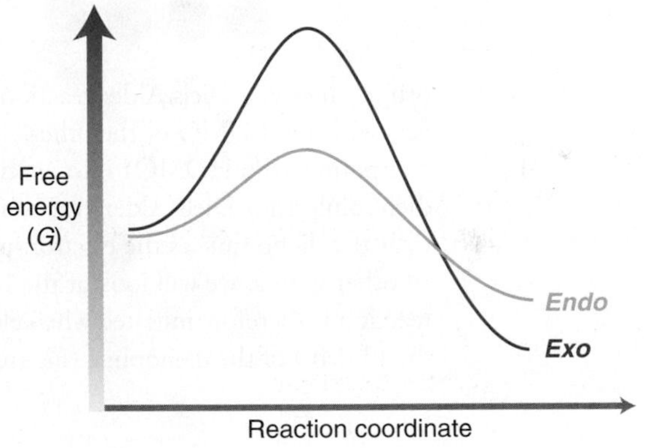

FIGURE 17.16
An energy diagram illustrating the formation of endo products (blue) and exo products (red).

CONCEPTUAL CHECKPOINT

17.19 Predict the products for each of the following reactions:

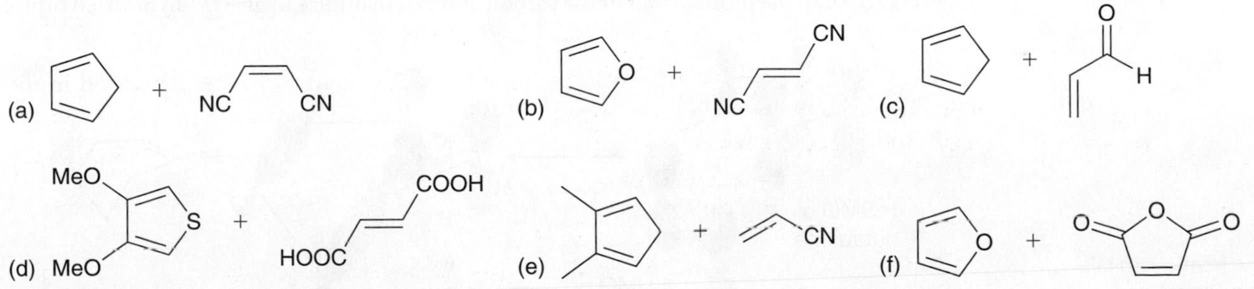

17.8 MO Description of Cycloadditions

We now turn our attention to the MOs involved in a Diels-Alder reaction. By studying the interactions between the relevant MOs we will gain insight into the reaction, and we will be able to explore the feasibility of reactions similar to Diels-Alder reactions.

Figure 17.17 illustrates the MOs for a conjugated diene (1,3-butadiene) and a simple dienophile (ethylene). The HOMO and LUMO are clearly labeled for both. According to frontier orbital theory, a Diels-Alder reaction is accomplished when the HOMO of one compound interacts with the LUMO of the other. That is, the electron density flows from a filled orbital of one compound (the HOMO) into an empty orbital of another compound (the LUMO). Since the dienophile in a Diels-Alder reaction generally has an electron-withdrawing substituent, we will treat the dienophile as the electron-poor species (the empty orbital) that accepts electron density. In other words, we will look at the LUMO of the dienophile and the HOMO of the diene. The reaction is therefore initiated when electron density is transferred from the HOMO of the diene to the LUMO of the dienophile (Figure 17.18).

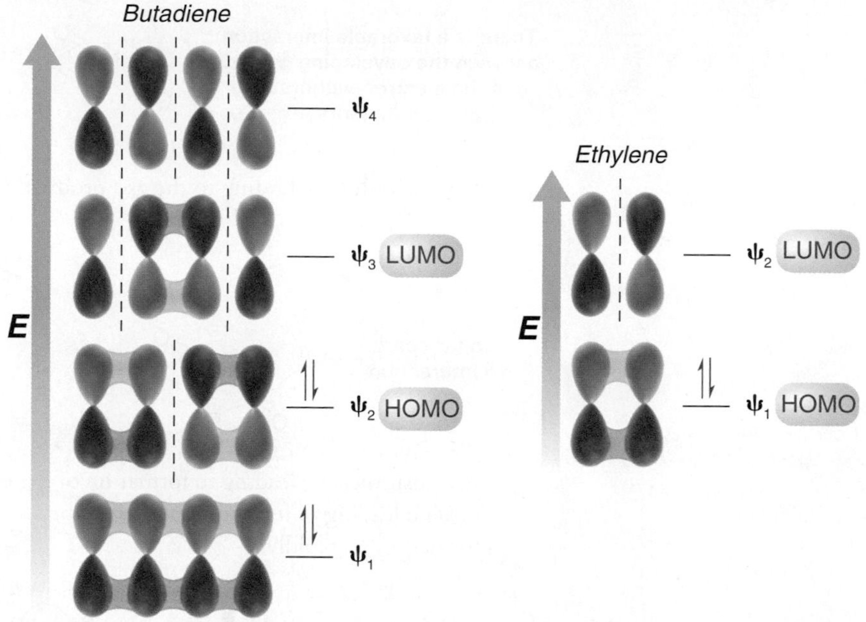

FIGURE 17.17
The MOs of butadiene and ethylene.

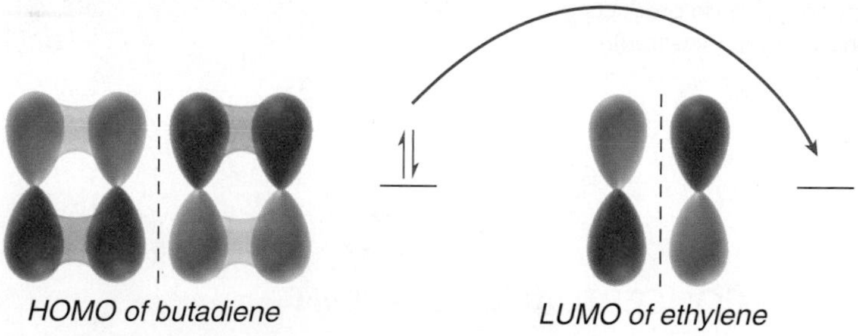

FIGURE 17.18
In a Diels-Alder reaction, the electron density flows from the HOMO of the diene to the LUMO of the dienophile.

When we align these frontier orbitals, we find that the phases of the MOs overlap nicely (Figure 17.19). In the process, four of the carbon atoms rehybridize to give sp^3-hybridized orbit-

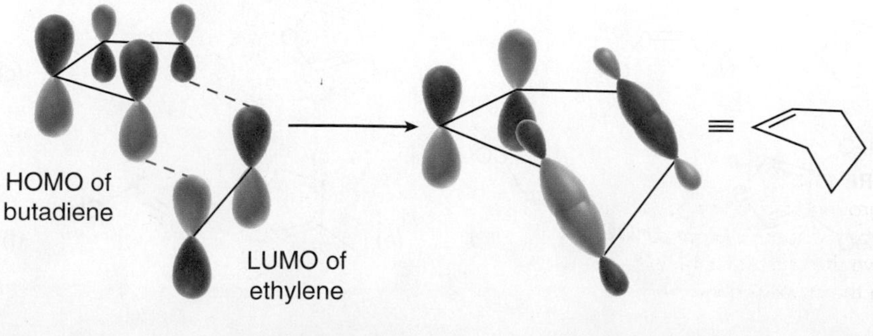

FIGURE 17.19
The phases of the frontier orbitals align properly in a Diels-Alder reaction.

als that form σ bonds. In order for this to occur, the phases of the MOs must overlap, that is, the phases must be symmetric. This requirement, called **conservation of orbital symmetry**, was first described by R. B. Woodward and Roald Hoffman (both at Harvard University) in 1965. In the Diels-Alder reaction, orbital symmetry is indeed conserved, and the reaction is therefore called a **symmetry-allowed** process.

Now let's use the same approach (analyzing the frontier orbitals) to determine whether the following reaction is possible.

Ethylene **Ethylene** **Cyclobutane**

Like the Diels-Alder reaction, this reaction is also a cycloaddition. Specifically, it is called a [2+2] cycloaddition, because the reaction takes place between two different π systems, each of which is associated with two atoms. To determine if this reaction is feasible, we once again look at the frontier orbitals—the HOMO of one compound and the LUMO of the other (Figure 17.20).

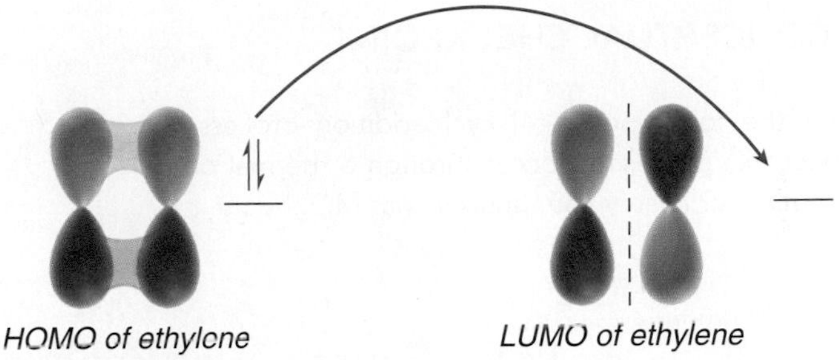

FIGURE 17.20 In a [2+2] cycloaddition, the electron density must flow from the HOMO of one compound to the LUMO of the other compound.

When we try to align these frontier orbitals, we find that the phases of the MOs do not overlap (Figure 17.21). This reaction is therefore said to be **symmetry-forbidden**, and the reaction does not occur. A [2+2] cycloaddition can only be performed with photochemical excitation.

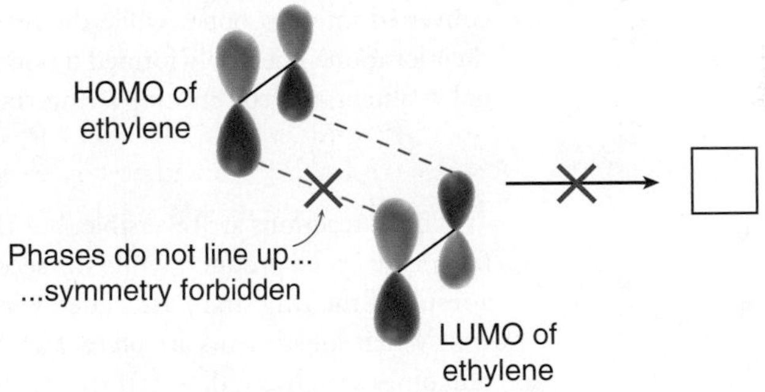

FIGURE 17.21 The phases of the frontier orbitals do not align properly in a [2+2] cycloaddition.

When one of the compounds is subjected to UV light, it can absorb the light to promote a π electron to the next higher energy level (Figure 17.22). In this excited state, the HOMO is now

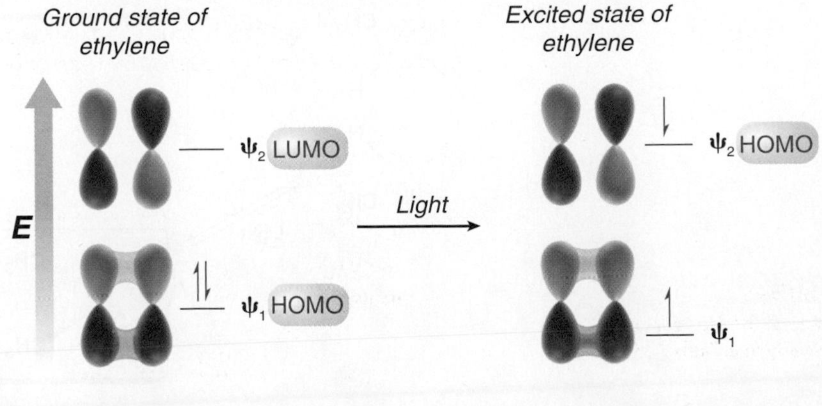

FIGURE 17.22 The ground state of ethylene absorbs a photon of light to achieve the excited state, in which the HOMO is ψ_2.

considered to be ψ_2 rather than ψ_1. The HOMO of the excited state can now interact with the LUMO of a ground-state molecule, and the reaction is symmetry allowed (Figure 17.23).

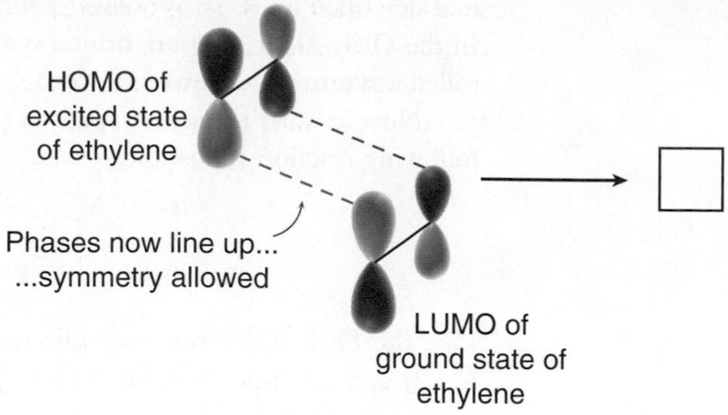

FIGURE 17.23
The phases of the frontier orbitals align properly in a photochemical [2+2] cycloaddition.

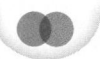

CONCEPTUAL CHECKPOINT

17.20 Consider the following [4+4] cycloaddition process. Would you expect this process to occur through a thermal or photochemical pathway? Justify your answer with MO theory.

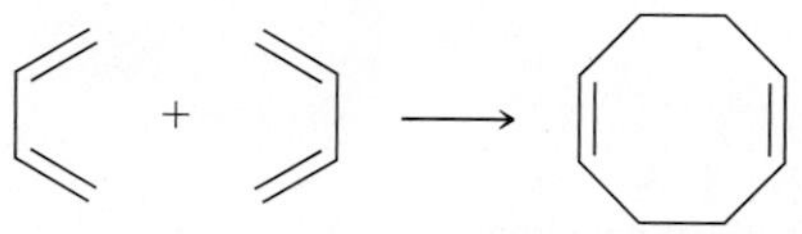

17.9 Electrocyclic Reactions

An Introduction to Electrocyclic Reactions

An electrocyclic reaction is a pericyclic process in which a conjugated polyene undergoes cyclization. In the process, one π bond is converted into a σ bond, while the remaining π bonds all change their location. The newly formed σ bond joins the ends of the original π system, thereby creating a ring. Two examples are shown.

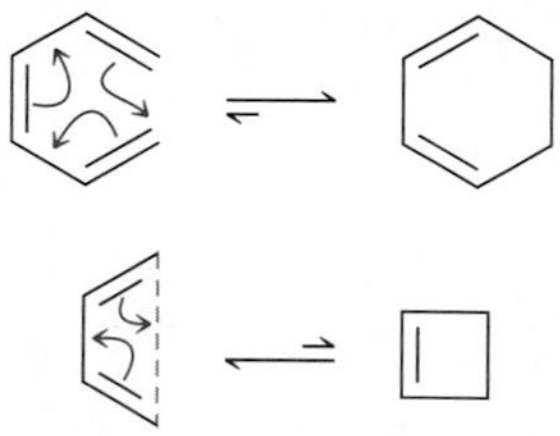

Both reactions are reversible, but the position of equilibrium is different. The first example favors the cyclic product, while the second example disfavors formation of the cyclic product as a result of the ring strain associated with a four-membered ring.

When substituents are present at the termini of the π system, the following stereochemical outcomes are observed.

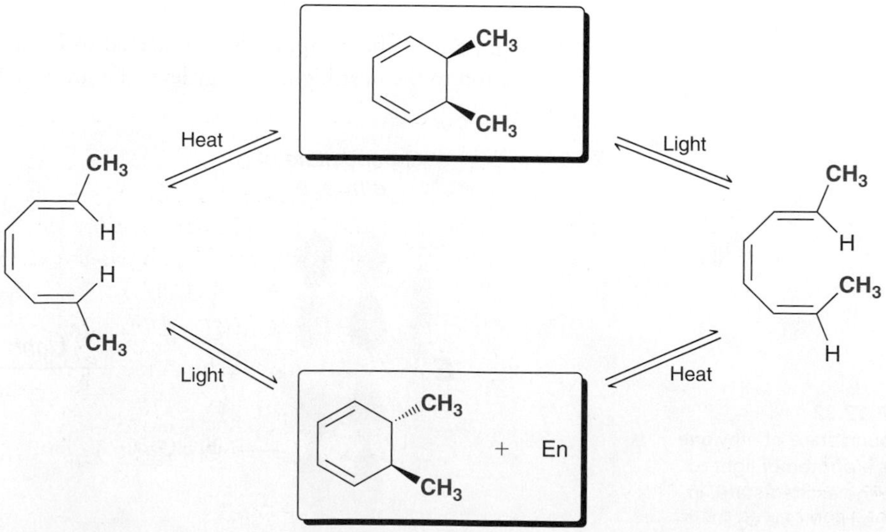

Note that the configuration of the product is dependent not only on the configuration of the reactant but also on the conditions of ring closure. That is, a different outcome is observed when the reaction is performed under thermal conditions (using heat) or under photochemical conditions (using UV light).

The choice of whether to use thermal or photochemical conditions also has an impact on the stereochemical outcome of electrocyclic reactions with four π electrons:

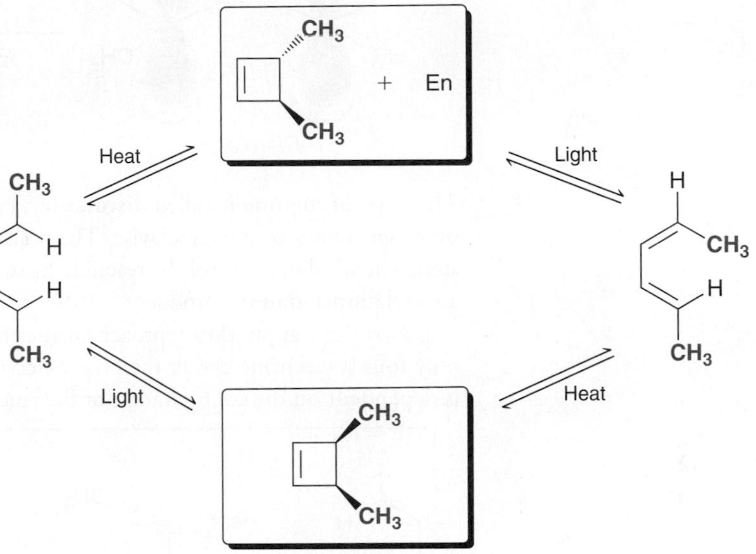

These stereochemical observations puzzled chemists for many years, until Woodward and Hoffmann developed their theory describing conservation of orbital symmetry. This single theory is capable of explaining all of the observations. We will first apply this theory to explain the stereochemical outcome of electrocyclic reactions taking place under thermal conditions, and then we will explore electrocyclic reactions taking place under photochemical conditions.

The Stereochemistry of Thermal Electrocyclic Reactions

Despite their name, thermal electrocyclic reactions do not necessarily require the use of elevated temperature. They can, in fact, take place at or even below room temperature. In many cases, the heat available at room temperature is sufficient for a thermal electrocyclic process to occur.

Under thermal conditions, the configuration of the reactant determines the configuration of the product.

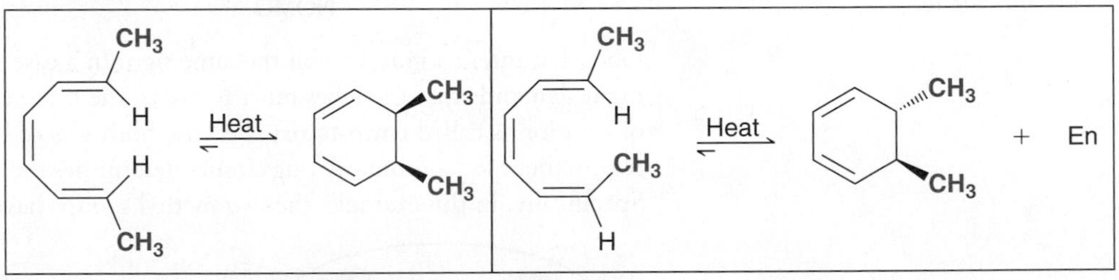

To explain these stereochemical outcomes, we focus on the symmetry of the HOMO for a π system with three conjugated π bonds. As seen in Section 17.3, the HOMO of a conjugated triene has two nodes and can be drawn using our shorthand method (Figure 17.24). Focus, in particular, on the signs (illustrated with red and blue) of the outermost lobes, because these are

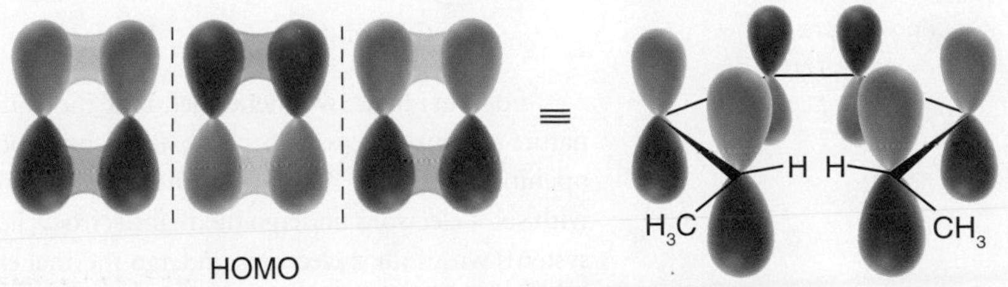

FIGURE 17.24
A shorthand method for depicting the HOMO of a conjugated triene.

the lobes that will participate in formation of the new σ bond. In order to form a bond, the lobes that interact with each other must exhibit the same sign. This requirement demands that the lobes must rotate in the manner illustrated in Figure 17.25.

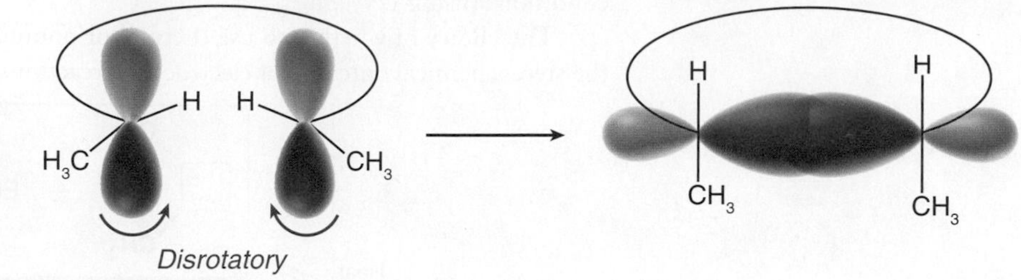

FIGURE 17.25 An illustration of disrotatory ring closure.

This type of rotation is called **disrotatory**, because one set of lobes rotates clockwise, while the other set rotates counterclockwise. The requirement for disrotatory ring closure determines the stereochemical outcome of the reaction. Specifically, in this example, the two methyl groups have a *cis* relationship in the product.

Now let's apply this approach to thermal electrocyclic reactions of π systems containing only four π electrons, rather than six π electrons. Once again, the configuration of the product is dependent on the configuration of the reactant.

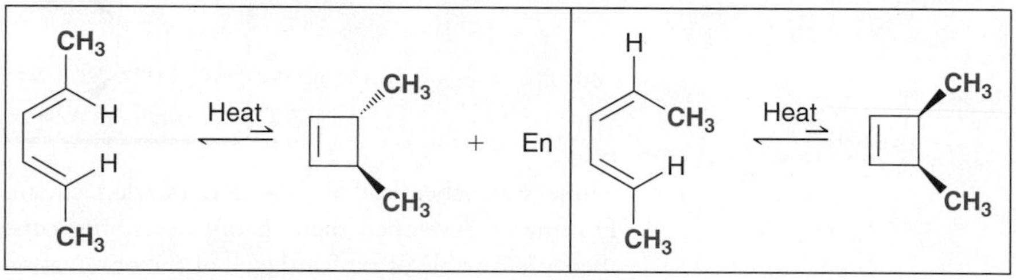

To account for these results, we once again look at the outermost lobes of the HOMO. As seen in Section 17.3, the HOMO of a π system containing four π electrons has one node and can be drawn using our shorthand method (Figure 17.26). In order to form a bond, recall that the

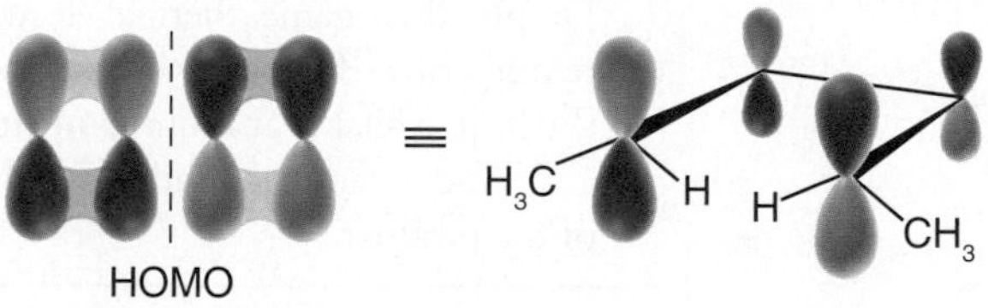

FIGURE 17.26 A shorthand method for depicting the HOMO of a conjugated diene.

lobes that interact must exhibit the same sign. In a case with only four π electrons, this requirement demands that the lobes must rotate in the manner illustrated in Figure 17.27. This type of rotation is called **conrotatory**, because both sets of lobes must rotate in the same way. The requirement for conrotatory ring closure determines the stereochemical outcome of the reaction. Specifically, in this example, the two methyl groups have a *trans* relationship in the product.

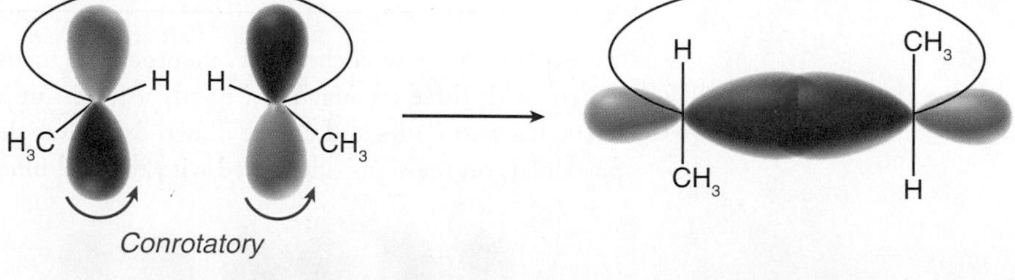

FIGURE 17.27 A conrotatory ring closure.

For butadiene, we mentioned that the equilibrium favors the open chain, so the conrotatory nature of this process can be observed by monitoring the stereochemical outcome for the ring-opening reaction of 3,4-disubstituted cyclobutene (Figure 17.28). In summary, conjugated systems with six π electrons undergo thermal electrocyclic reactions in a disrotatory fashion, while conjugated systems with four π electrons undergo thermal electrocyclic reactions in a conrotatory fashion.

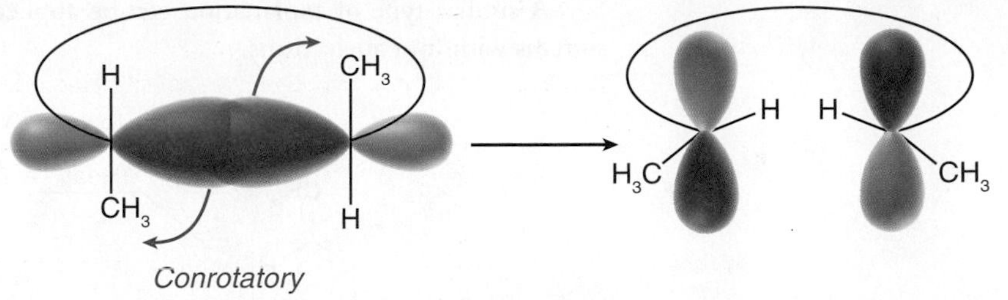

FIGURE 17.28
A conrotatory ring opening.

CONCEPTUAL CHECKPOINT

17.21 Predict the major product(s) for each of the following thermal electrocyclic reactions:

(a) 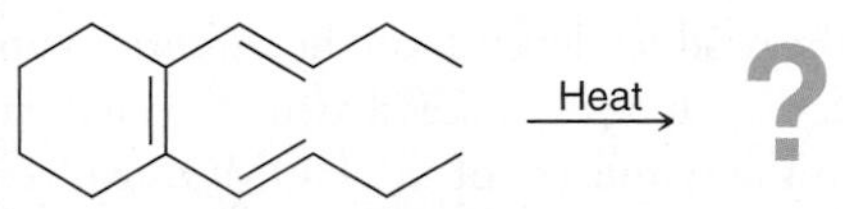

(b) 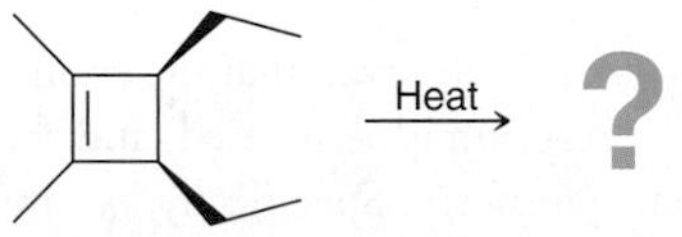

(c) 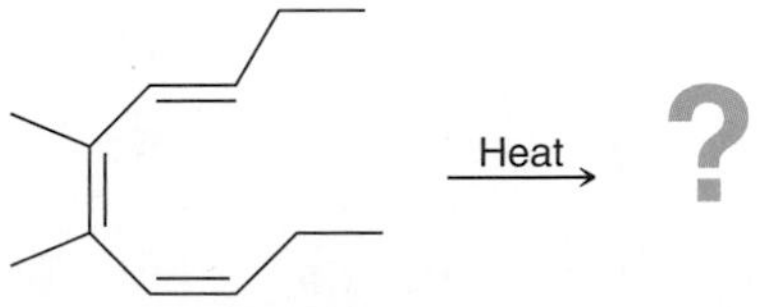

Stereochemistry of Photochemical Electrocyclic Reactions

We have seen that the stereochemical outcome of an electrocyclic reaction depends on whether the reaction is performed under thermal conditions or under photochemical conditions.

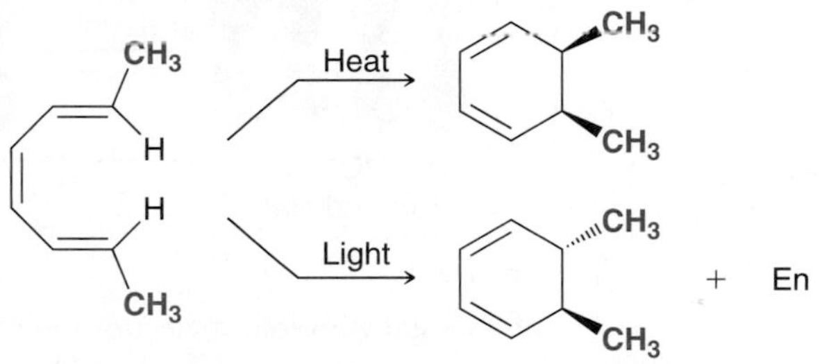

In this example, the configuration of the product depends on the conditions under which the reaction is performed. Once again, we can explain these observations with the theory of conservation of orbital symmetry. Specifically, we must focus on the symmetry of the HOMO involved in the reaction. Recall from Section 17.3 that excitation of an electron redefines the identity of the HOMO (Figure 17.29). Focusing on the outer lobes, we find that photochemical ring closure of

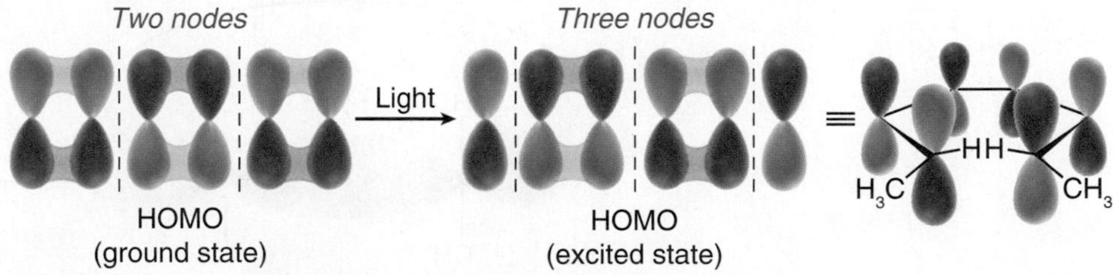

FIGURE 17.29
A shorthand method for depicting the HOMO of the excited state of a conjugated triene.

a system with six π electrons must occur in a conrotatory fashion (Figure 17.30), as opposed to the disrotatory ring closure observed under thermal conditions. The requirement for conrotatory ring closure correctly explains the observed stereochemical outcome for this reaction.

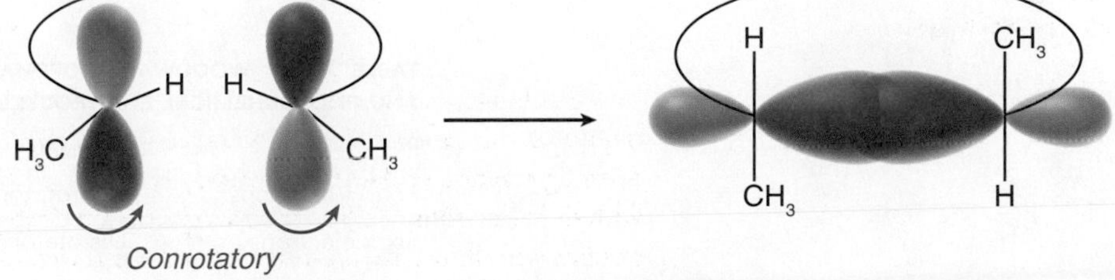

FIGURE 17.30
A system with six π electrons will undergo conrotatory ring closure under photochemical conditions.

A similar type of explanation can be applied to photochemical electrocyclic reactions of systems with four π electrons.

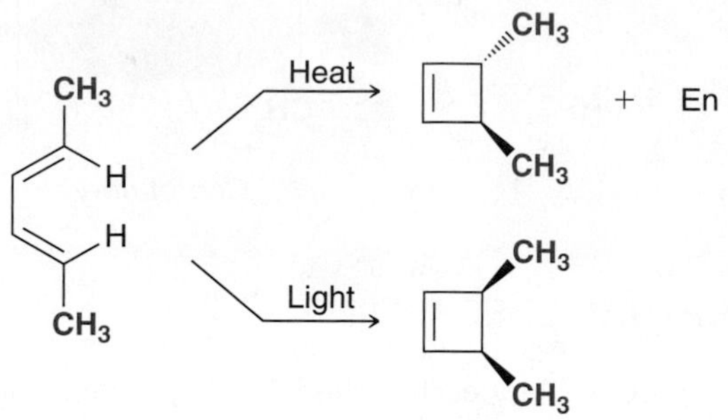

Once again, it is observed that the configuration of the product depends on the conditions under which the reaction is performed, and these observations can be explained based on conservation of orbital symmetry. Specifically, we must focus on the symmetry of the HOMO involved in the reaction. Recall from Section 17.3 that excitation of an electron redefines the identity of the HOMO (Figure 17.31). Focusing on the outer lobes, we find that photochemical ring closure

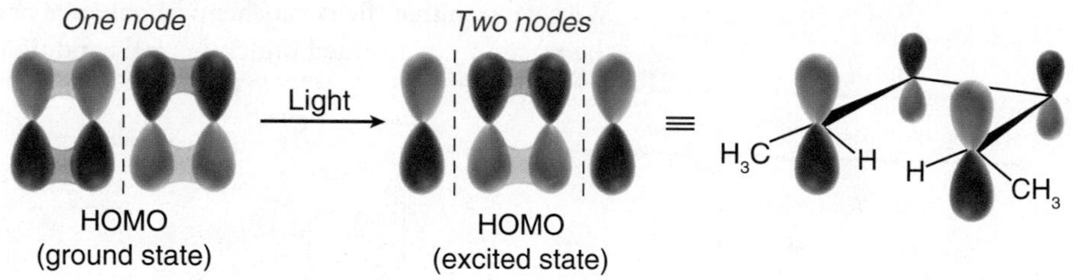

FIGURE 17.31
A shorthand method for depicting the HOMO of the excited state of a conjugated diene.

of a system with four π electrons must occur in a disrotatory fashion, as opposed to the conrotatory ring closure observed under thermal conditions. For butadiene, the equilibrium favors the open chain, so the disrotatory nature of this process can be observed by monitoring the stereochemical outcome for the ring-opening reaction of 3,4-disubstituted cyclobutene under photochemical conditions (Figure 17.32). Table 17.2 summarizes the Woodward-Hoffmann rules for common electrocyclic reactions.

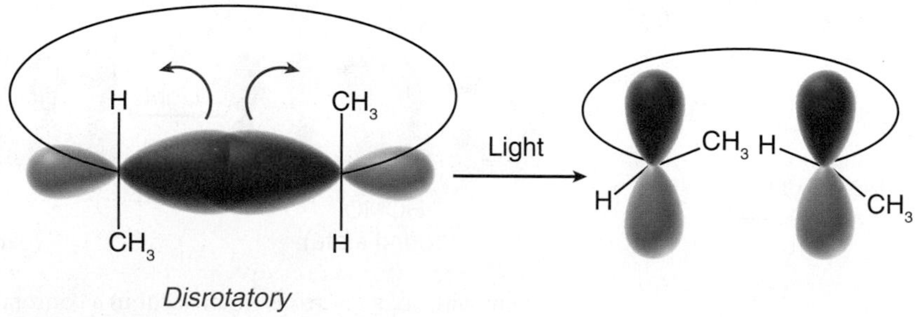

FIGURE 17.32
A disrotatory ring opening.

TABLE 17.2 WOODWARD-HOFFMANN RULES FOR THERMAL AND PHOTOCHEMICAL ELECTROCYCLIC REACTIONS

	THERMAL	PHOTOCHEMICAL
Four π electrons	Conrotatory	Disrotatory
Six π electrons	Disrotatory	Conrotatory

SKILLBUILDER

17.4 PREDICTING THE PRODUCT OF AN ELECTROCYCLIC REACTION

LEARN the skill

Predict the product of the following electrocyclic reaction.

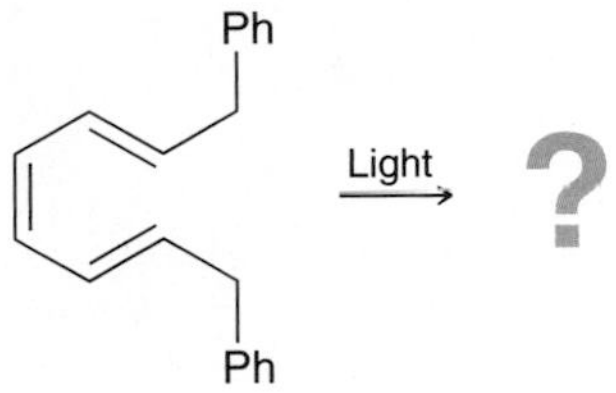

STEP 1
Count the number of π electrons.

SOLUTION

We begin by counting the number of π electrons. In this case, there are six π electrons (the π electrons of the phenyl groups are not participating in the reaction).

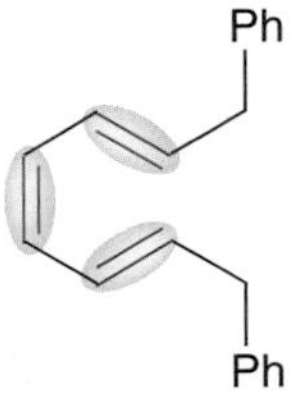

6 π electrons

STEP 2
Determine whether the reaction is conrotatory or disrotatory.

Next, we must analyze the reaction conditions and determine whether the reaction occurs in a conrotatory or disrotatory fashion. As shown in Table 17.2, for a system with six π electrons, photochemical conditions cause a conrotatory ring closure. This information is used to determine whether the substituents are *cis* or *trans* to each other in the product. In this case, the substituents are *trans* to each other in the product.

STEP 3
Determine whether the substituents are *cis* or *trans* to each other in the product.

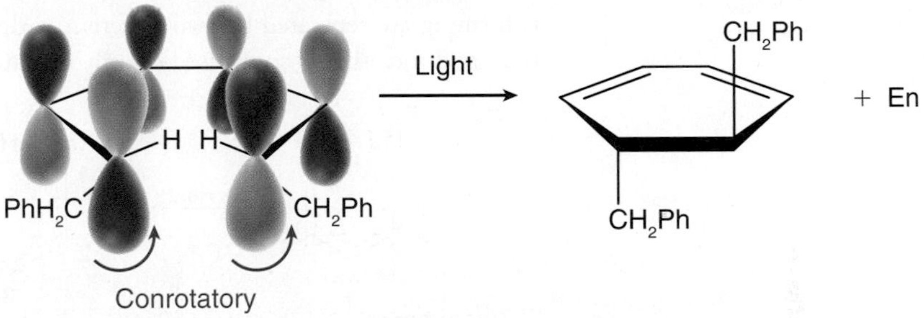

PRACTICE the skill **17.22** Predict the product for each of the following reactions:

(a) 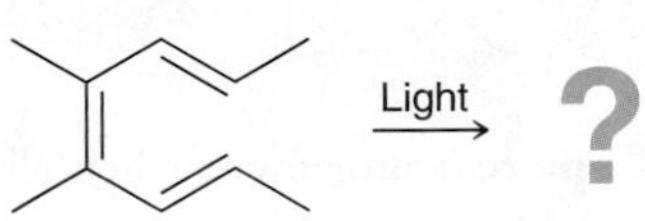

(b) 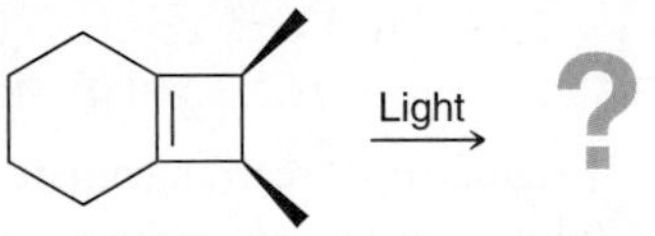

(c) 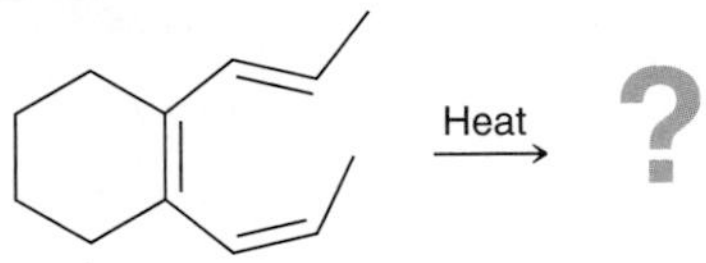

APPLY the skill

17.23 Draw the structure of *cis*-3,4-diethylcyclobutene.

(a) Conrotatory ring opening produces only one product. Draw the product and determine whether its formation is best achieved under thermal conditions or photochemical conditions.

(b) Disrotatory ring opening can produce two possible products, but only one of these products is obtained in practice. Identify the product that is formed, and explain why the other possible product is not formed.

17.24 Identify whether the product obtained from each of the following reactions is a *meso* compound or a pair of enantiomers:

(a) Irradiation of (2*E*,4*Z*,6*Z*)-4,5-dimethyl-2,4,6-octatriene with UV light

(b) Subjecting (2*E*,4*Z*,6*Z*)- 4,5-dimethyl-2,4,6-octatriene to elevated temperature

(c) Irradiation of (2*E*,4*Z*,6*E*)-4,5-dimethyl-2,4,6-octatriene with UV light

need more **PRACTICE?** **Try Problems 17.55, 17.56**

17.10 Sigmatropic Rearrangements

An Introduction to Sigmatropic Rearrangements

A sigmatropic rearrangement is a pericyclic reaction in which one σ bond is formed at the expense of another. In the process, the π bonds change their location. The term *sigmatropic* comes from the Greek word, tropos, which means "change." A sigmatropic rearrangement is therefore a reaction in which a σ bond has undergone a change (in its location).

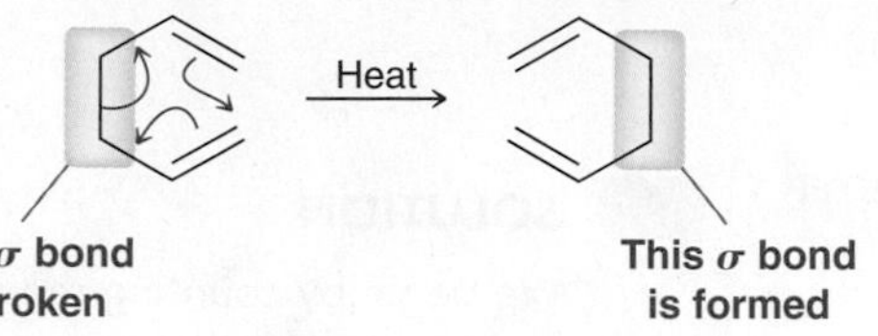

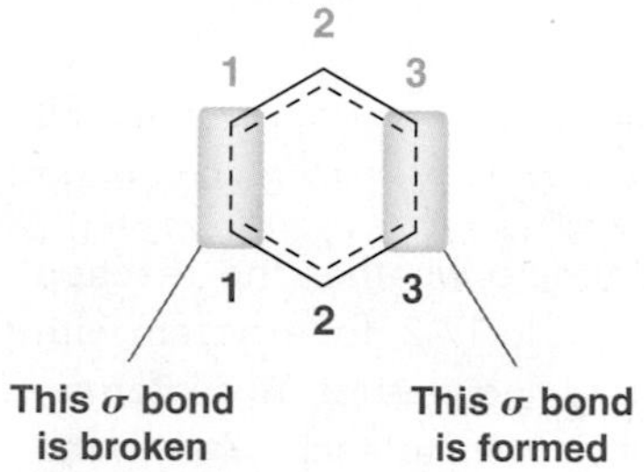

FIGURE 17.33
The transition state for a [3,3] sigmatropic rearrangement.

There are many different types of sigmatropic rearrangements. The previous example is called a [3,3] sigmatropic rearrangement. This notation is different than any other notation used thus far in this book. The two numbers in the brackets indicate the number of atoms separating the bond that is forming and the bond that is breaking in the transition state (Figure 17.33). Notice that the transition state is cyclic, which is a characteristic feature of all pericyclic reactions. In the transition state, the bond that is breaking and the bond that is forming are separated by two different pathways, each of which is comprised of three atoms (one path is labeled in red and the other in blue). Therefore, it is called a [3,3] sigmatropic rearrangement.

The following is an example of a [1,5] sigmatropic rearrangement, also sometimes referred to as a [1,5] hydrogen shift. In the transition state, the bond that is breaking and the bond that is forming are separated by two different pathways: one is comprised of five atoms (labeled in red), and the other is comprised of only one atom (labeled in blue).

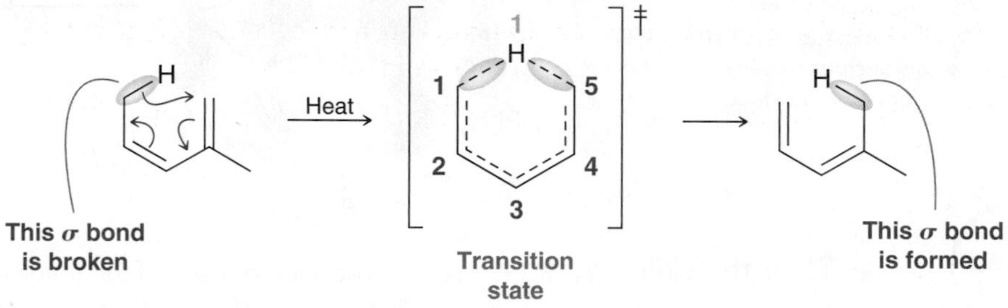

The Cope Rearrangement

A [3,3] sigmatropic rearrangement is called a **Cope rearrangement** when all six atoms of the cyclic transition state are carbon atoms.

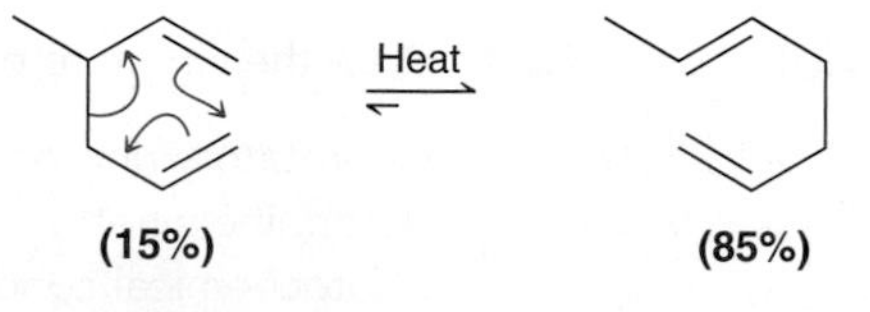

The equilibrium for a Cope rearrangement generally favors formation of the more substituted alkene. In the example above, both π bonds of the reactant are monosubstituted, but in the product, one of the π bonds is disubstituted. For this reason, the equilibrium for the reaction favors the product.

The Claisen Rearrangement

The oxygen analogue of a Cope rearrangement is called a **Claisen rearrangement**.

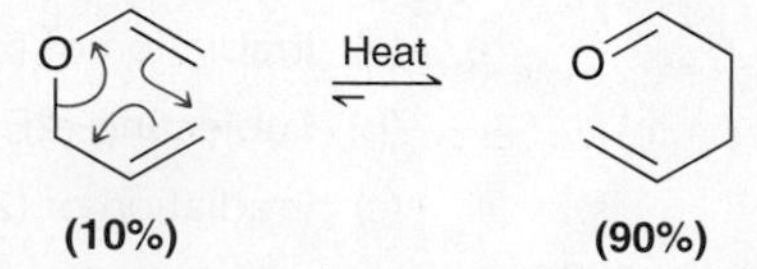

The Claisen rearrangement is a [3,3] sigmatropic rearrangement and is commonly observed for allylic vinylic ethers.

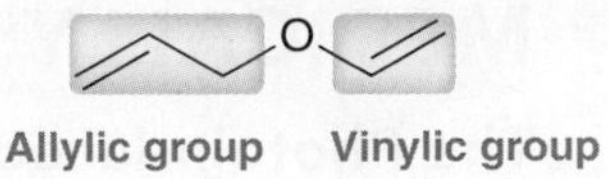

The equilibrium greatly favors the product because of formation of a C=O bond, which is thermodynamically more stable (lower in energy) than a C=C bond. The Claisen rearrangement is also observed for allylic aryl ethers.

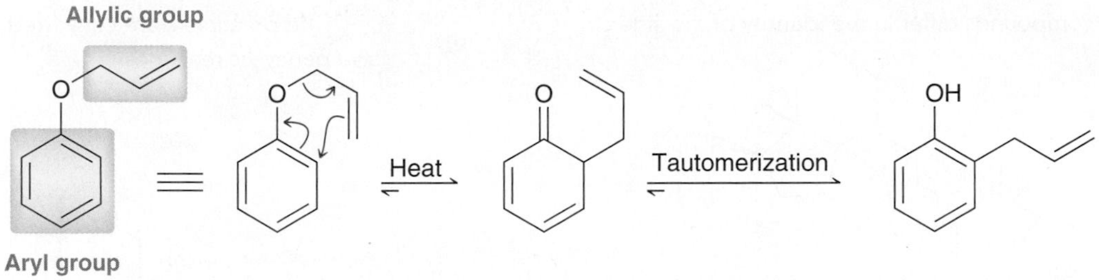

Aryl groups possess an aromatic ring (a six-membered ring drawn with alternating double and single bonds), which is particularly stable, as we will see in the next chapter. In the case of allylic aryl ethers, the Claisen rearrangement initially destroys the aromatic ring but is rapidly followed by a spontaneous tautomerization process that regenerates the aromatic ring. In this tautomerization process, the conversion of a ketone into an enol is a testament to the stability of the aromatic ring system. Aromatic compounds, as well as their stability and reactivity, will be discussed in detail in the next two chapters.

CONCEPTUAL CHECKPOINT

17.25 For each of the following reactions, use brackets and two numbers to identify the type of sigmatropic rearrangement taking place:

(a)

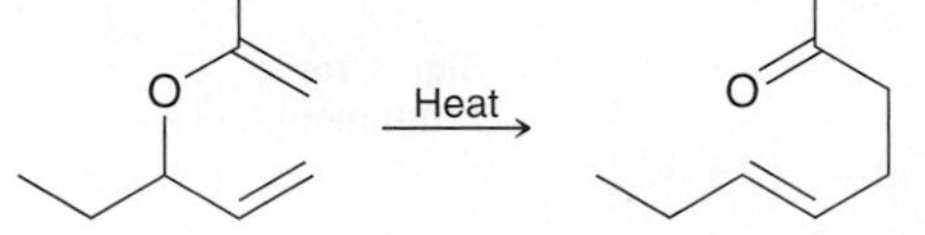

(b)

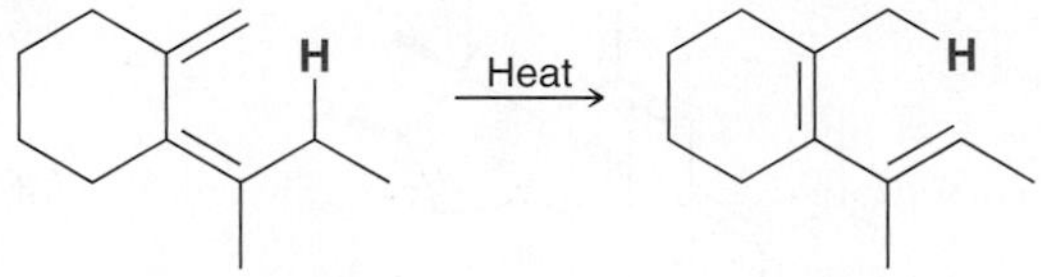

17.26 Consider the structure of *cis*-1,2-divinylcyclopropane:

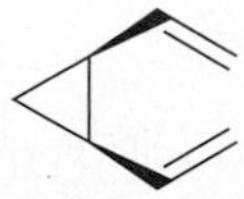

This compound is stable at low temperature but rearranges at room temperature to produce 1,4-cycloheptadiene.

(a) Draw a mechanism for this transformation.

(b) Using brackets and numbers, identify the type of sigmatropic rearrangement taking place in this case.

(c) This reaction proceeds to completion, which means that the concentration of the reactant is insignificant once the equilibrium has been established. Explain why this reaction proceeds to completion.

17.27 Predict the product for each of the following reactions.

(a)

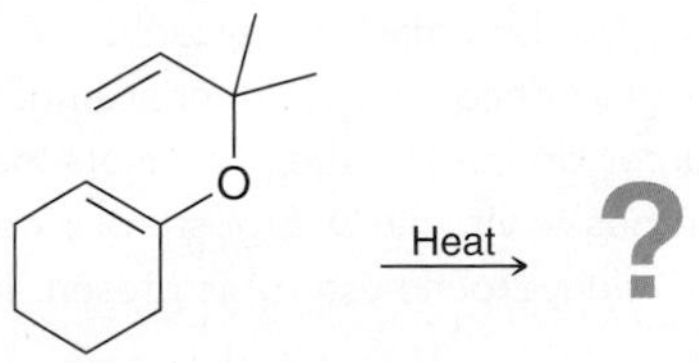

(b)

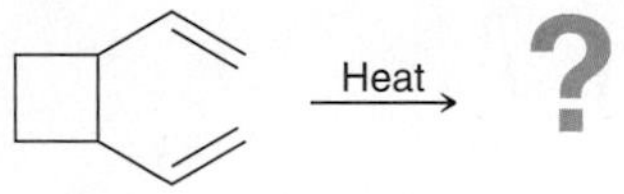

(c)

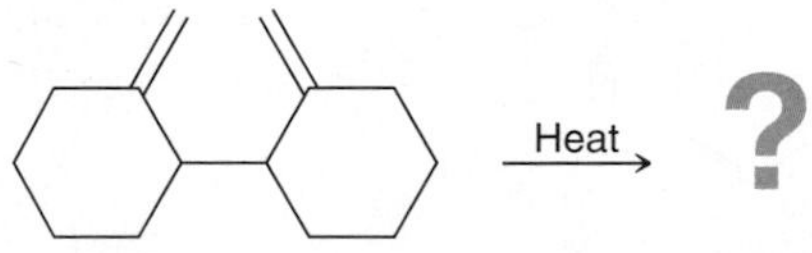

(d)

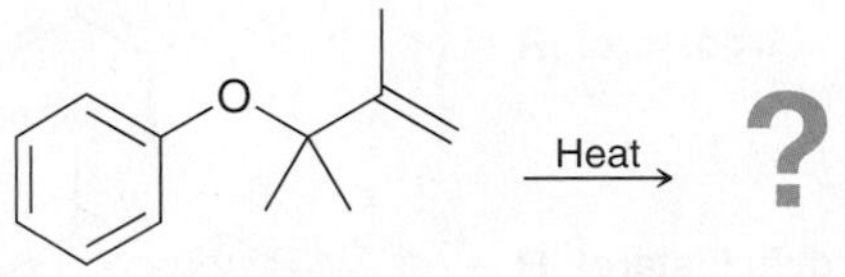

17.28 Draw a plausible mechanism for the following transformation:

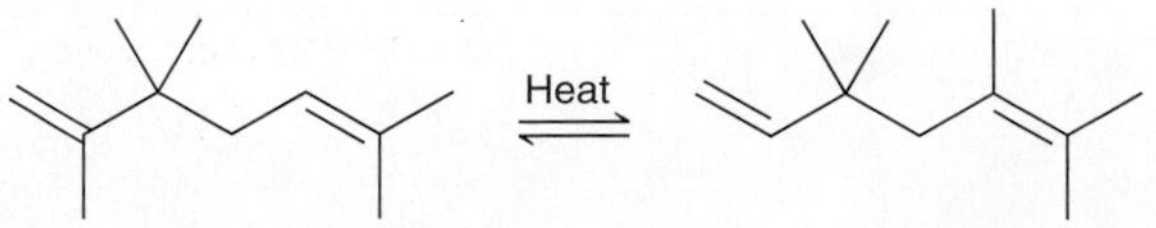

MEDICALLYSPEAKING

The Photoinduced Biosynthesis of Vitamin D

Vitamin D is the general name for two related compounds called cholecalciferol (vitamin D_3) and ergocalciferol (vitamin D_2). These two compounds differ in the identity of the side chain, R:

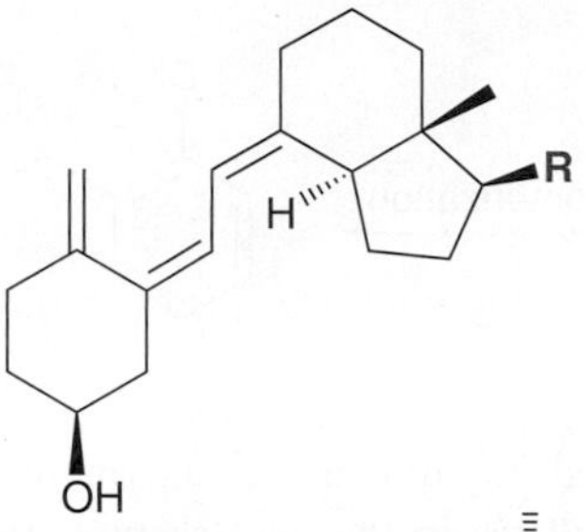

Ergocalciferol (vitamin D_2) R =

Cholecalciferol (vitamin D_3) R =

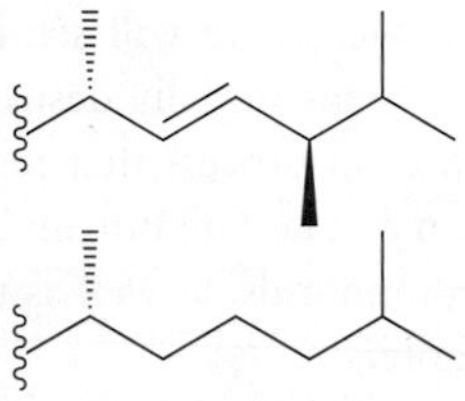

Vitamin D serves an important function in that it increases the body's ability to metabolize the calcium present in the foods we eat. Calcium is essential because it is used in the formation of bones and teeth. A deficiency of vitamin D can cause the childhood disease called rickets, which is characterized by abnormal bone growth.

Although most food does not contain substantial amounts of vitamin D, some food sources do contain the precursors that the body needs in order to synthesize vitamin D. Ergosterol is present in many vegetables, and 7-dehydrocholesterol is present in fish and dairy products.

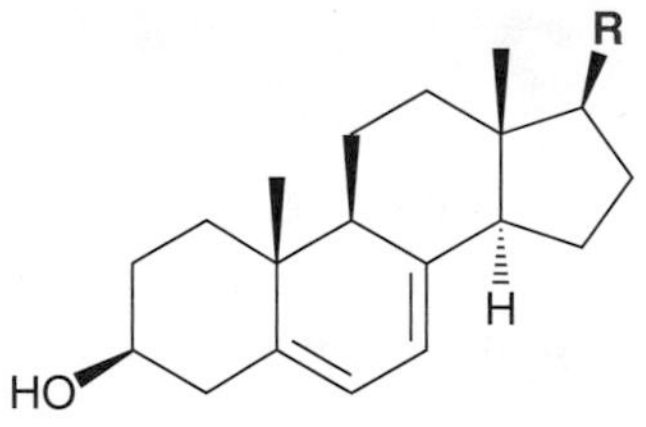

Ergosterol R =

7-Dehydrocholesterol R =

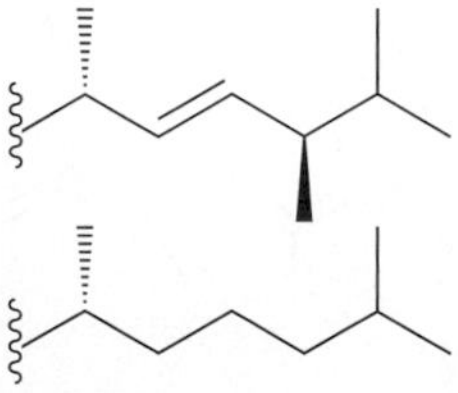

When the skin is exposed to sunlight, these precursors, located just beneath the skin, are converted into ergocalciferol and cholecalciferol, respectively. For this reason, vitamin D is commonly called the "sunshine vitamin."

The precursors are converted into vitamin D via two successive pericyclic reactions.

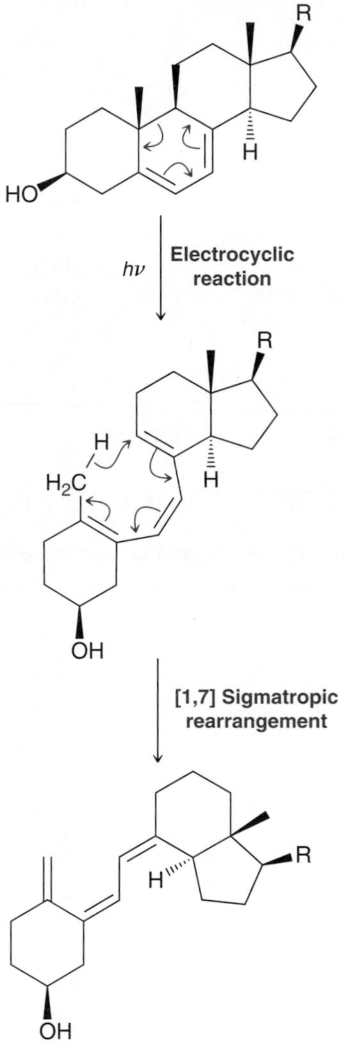

The first step is an electrocyclic ring-opening reaction, and the second step is a [1,7] sigmatropic rearrangement. This sequence of steps not only occurs in our bodies but is also utilized commercially to enrich milk with vitamin D_3. All milk sold in the United States is irradiated with UV light, which converts the 7-dehydrocholesterol present in the milk into vitamin D_3.

17.11 UV-Vis Spectroscopy

The Information in a UV-Vis Spectrum

As seen in the previous two chapters, spectroscopy allows us to probe molecular structure by studying the interaction between matter and electromagnetic radiation. Recall that the frequency of light determines the energy of a photon, and the range of all possible frequencies is known as the *electromagnetic spectrum* (Figure 15.2). In this chapter, we will focus on the interaction between matter and the UV-Vis region of the electromagnetic spectrum. Specifically, organic compounds with conjugated π systems absorb UV or visible light to promote an electronic excitation. That is, a π electron absorbs the photon and is promoted to a higher energy level. As an example, consider what happens when butadiene is irradiated with UV light (Figure 17.34).

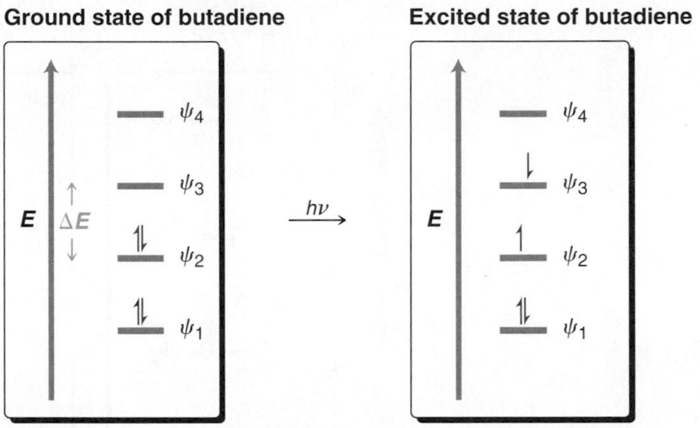

FIGURE 17.34 The ground state of butadiene can absorb a photon of UV light to produce the excited state, in which an electron has been promoted to a higher energy MO.

The energy of the photon ($h\nu$) is absorbed by a π electron in the HOMO (ψ_2) and is consequently promoted to a higher energy MO (ψ_3). In order for this excitation to occur, the photon must possess the right amount of energy—the energy of the photon must be equivalent to the energy gap between the two MOs (ψ_2 and ψ_3). For most organic compounds containing conjugated π systems, the relevant energy gap corresponds with UV or visible light. This excitation is called a $\pi \rightarrow \pi^*$ transition (pronounced "pi to pi star"). The excited electron then returns to its original MO, releasing the energy in the form of heat or light. In this way, compounds with conjugated π systems will absorb UV or visible light.

A standard UV-Vis spectrophotometer irradiates a sample with wavelengths of light ranging from 200 to 800 nm. The light is first split into two beams. One beam passes through a cuvette (a small glass container) containing an organic compound dissolved in some solvent, while the other beam (the reference beam) is passed through a cuvette containing only the solvent. The spectrophotometer then compares the intensities of the beams at each particular wavelength, and the results are plotted on a chart that shows **absorbance** as a function of wavelength. Absorbance is defined as

$$A = \log \frac{I_0}{I}$$

where I_0 is the intensity of the reference beam and I is the intensity of the sample beam. The plot that is generated is called a UV-Vis **absorption spectrum**. As an example, consider the absorption spectrum of butadiene (Figure 17.35). For our purposes, the most important feature of the

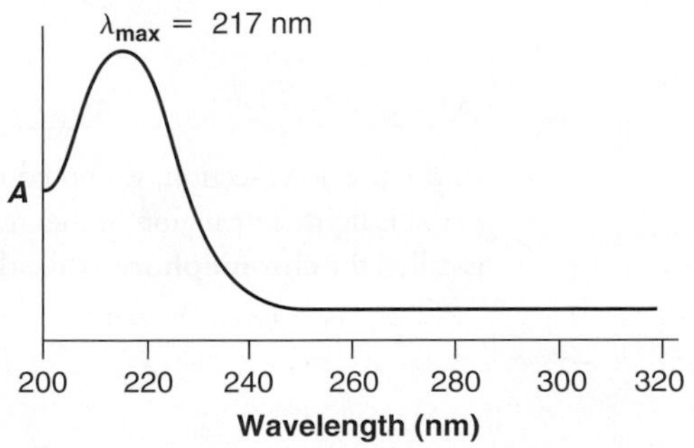

FIGURE 17.35 An absorbance spectrum of butadiene.

absorption spectrum is the λ_{max} (pronounced **lambda max**), which indicates the wavelength of maximum absorption. For butadiene, λ_{max} is 217 nm. The amount of UV light absorbed at the λ_{max} for any compound is described by the **molar absorptivity** (ε) and is expressed by the following equation, called **Beer's law**:

$$\varepsilon = \frac{A}{C \times l}$$

where A is the absorbance, C is the concentration of the solution (mol/L), and l is the sample path length (the width of the cuvette) measured in centimeters. The molar absorptivity (ε) is a physical characteristic of the particular compound being investigated and is typically in the range of 0–15,000.

The value of λ_{max} for a particular compound is highly dependent on the extent of conjugation. To illustrate this point, compare the MOs of butadiene, hexatriene, and octatetraene (Figure 17.36).

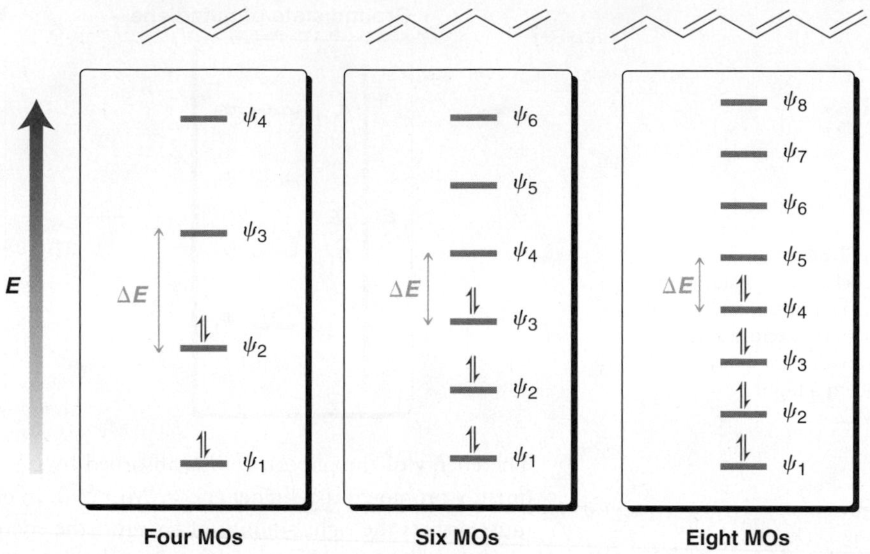

FIGURE 17.36
An energy diagram showing the relative energy levels for MOs for various conjugated π systems.

Compounds that are more highly conjugated will have more MOs and will exhibit smaller gaps between MOs. Compare the energy gaps (ΔE) for each of the three compounds and notice the trend. Compounds with more highly conjugated systems require less energy to promote an electronic excitation. Recall that the energy of a photon is directly proportional to the frequency of the light but inversely proportional to the wavelength. As a result, a longer wavelength corresponds with less energy. We therefore find that compounds with a greater extent of conjugation will have a longer λ_{max} (Table 17.3). The λ_{max} of a compound therefore indicates the extent of conjugation present in the compound. Each additional conjugated double bond adds between 30 and 40 nm.

TABLE 17.3 THE WAVELENGTH OF MAXIMUM ABSORPTION FOR A SIMPLE DIENE, TRIENE, AND TETRAENE

COMPOUND	λ_{MAX}
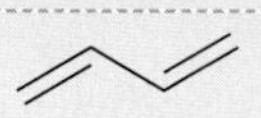	217 nm
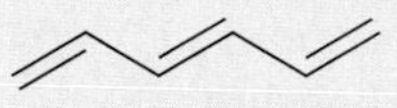	258 nm
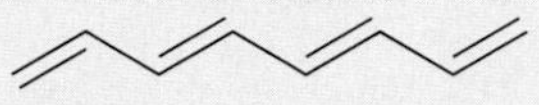	290 nm

Woodward-Fieser Rules

In the previous section, we noted that a compound containing a conjugated π system will absorb UV-Vis light. The region of the molecule responsible for the absorption (the conjugated π system) is called the **chromophore**, while the groups attached to the chromophore are called **auxochromes.**

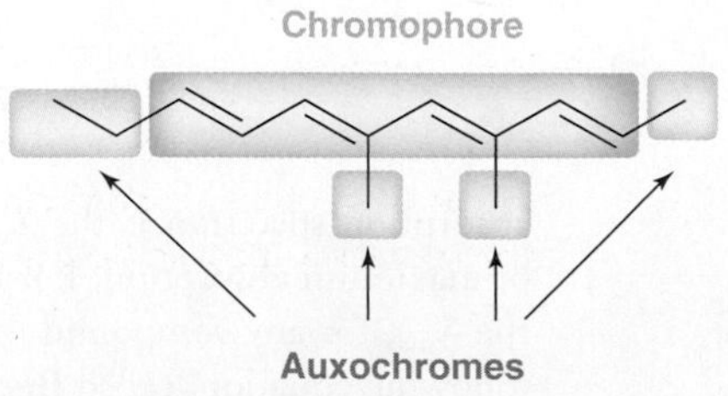

Auxochromes can also have a substantial effect on the value of λ_{max}. **Woodward-Fieser rules** can be used to make simple predictions (Table 17.4). The full list of Woodward-Fieser rules is quite long, but the few rules in Table 17.4 allow us to make simple predictions. Note that Woodward-Fieser rules only provide us with an estimate for λ_{max}. In some cases, the estimate will be very close to the observed value, and in other cases, there will be a small discrepancy. The rules are only meant as a rough guide for making meaningful predictions, and they do not work well for compounds that contain more than six double bonds in conjugation.

TABLE 17.4 WOODWARD–FIESER RULES FOR PREDICTING λ_{MAX} FOR CONJUGATED π SYSTEMS

FEATURE TO LOOK FOR:	λ_{MAX}	EXAMPLE	CALCULATION
Conjugated diene	Base value = 217	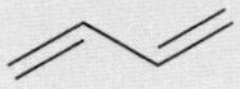	217 nm
Each additional double bond	+30	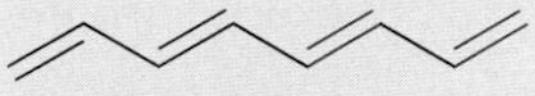 **Two additional double bonds**	Base: 217 +2 × 30 277 nm (observed = 290 nm)
Each auxochromic alkyl group	+5	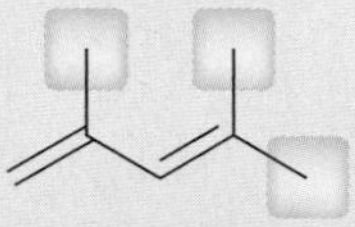 **Three alkyl groups connected to chromophore**	Base: 217 +3 × 5 232 nm (observed = 232 nm)
Each exocyclic double bond (a double bond where one vinylic position is part of a ring and the other vinylic position is outside the ring)	+5	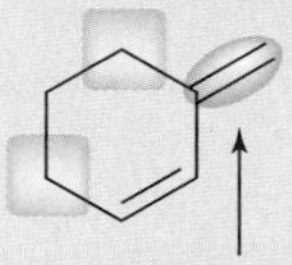 **Exocyclic**	Base: 217 +2 × 5 alkyl groups +5 exocyclic double bond 232 nm (observed = 230 nm)
Homoannular diene—both double bonds are contained in one ring, so the diene moiety is locked in an *s-cis* conformation	+39	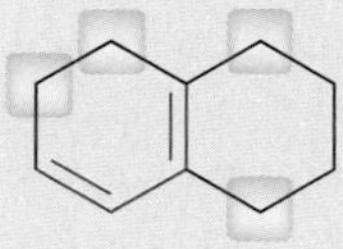	Base: 217 +4 × 5 alkyl groups +39 homoannular diene 276 nm (observed = 269 nm)

SKILLBUILDER

17.5 USING WOODWARD-FIESER RULES TO ESTIMATE λ_{MAX}

LEARN the skill

Use Woodward-Fieser rules to estimate the expected λ_{max} for the following compound:

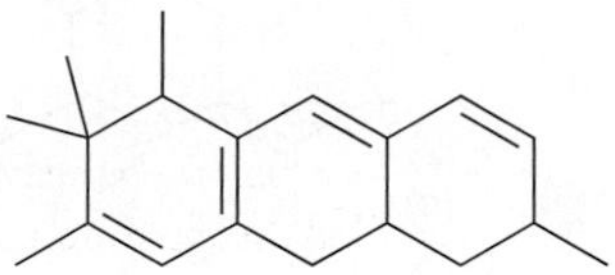

SOLUTION

STEP 1
Count the number of conjugated double bonds.

First count the number of conjugated double bonds. This compound has four double bonds in conjugation (highlighted in red). Two of them count toward the base value of 217 nm, and the other two double bonds will each add +30.

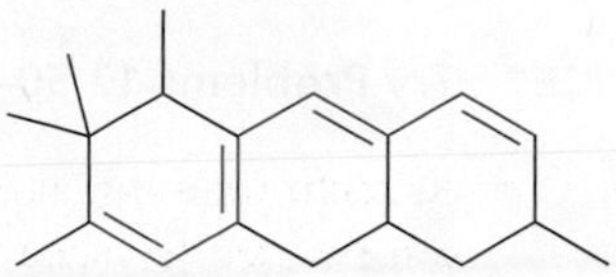

Next, look for any auxochromic alkyl groups. These are the carbon atoms connected directly to the chromophore. This compound has six auxochromic alkyl groups, each of which adds +5, giving another +30.

STEP 2
Count the number of auxochromic alkyl groups.

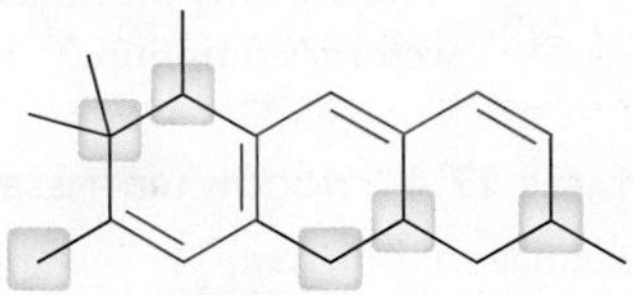

Next look for exocyclic double bonds. In this case, the double bond highlighted in green is exocyclic to the ring highlighted in red. This adds another +5.

STEP 3
Count the number of exocyclic double bonds.

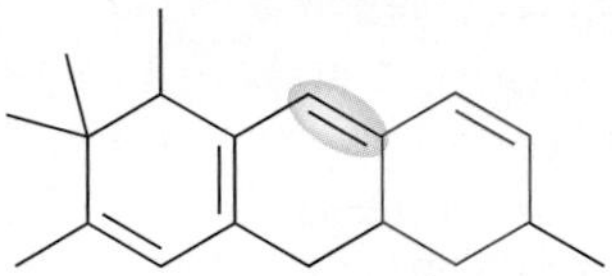

Finally, look for a homoannular diene. The first ring (on the left) exhibits two double bonds in the same ring, which adds +39. The second ring also exhibits two double bonds in the same ring, which adds another +39.

STEP 4
Look for homoannular dienes.

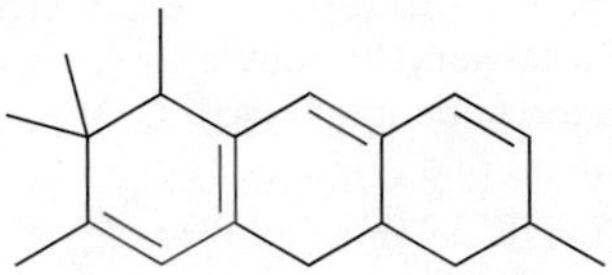

We predict this compound will have a λ_{max} somewhere in the neighborhood of 390 nm.

STEP 5
Add up all factors.

Base = 217
Additional double bonds (2) = +60
Auxochromic alkyl groups (6) = +30
Exocyclic double bond (1) = +5
Homoannular dienes (2) = +78
Total = 390 nm

PRACTICE the skill

17.29 Use Woodward-Fieser rules to estimate the expected λ_{max} for each of the following compounds:

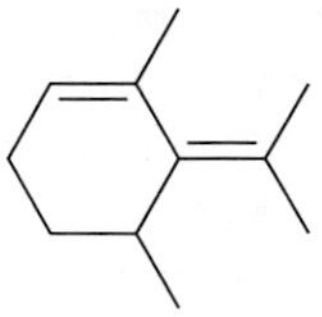
(a)

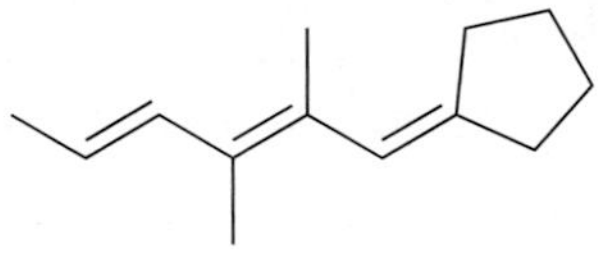
(b)

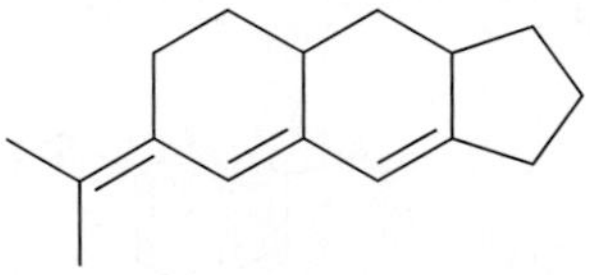
(c)

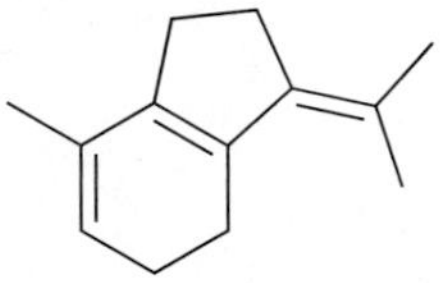
(d)

APPLY the skill

17.30 Identify which of the following compounds is expected to have a larger λ_{max}:

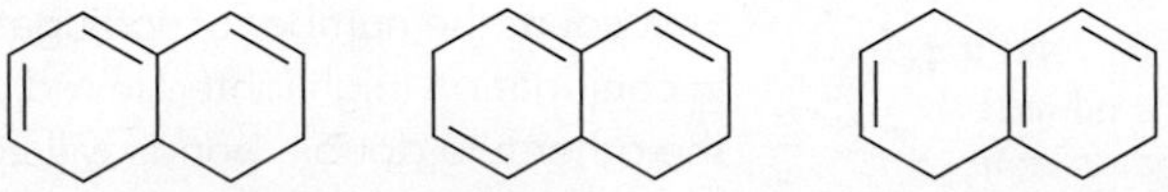

need more **PRACTICE?** **Try Problems 17.50–17.52, 17.60c**

PRACTICALLYSPEAKING

Sunscreens

In the United States, societal attitudes toward suntans have changed many times over the last two centuries. Before the industrial revolution, a suntan was considered a sign of lower class. Wealthy people stayed in the shade, while manual laborers worked in the sun. After the industrial revolution, manual laborers began to work indoors, and this distinction disappeared. In fact, societal attitudes took a 180° turn when in the 1920s sunlight was considered to be a cure-all for everything from acne to tuberculosis. By the 1940s, suntan lotions first appeared on the market, although the function of these products was not to block out the sun but to aid in the tanning process. In the 1960s, sun lamps were invented, so that people could tan in the winter.

In the last several decades, societal attitudes have shifted once again, with the discovery of the link between sunlight and skin cancer. Harmful UV radiation from the sun is mostly screened out by the ozone layer, although some harmful UV rays do make it to the earth's surface. There are two types of damaging UV light, called UV-A (315–400 nm) and UV-B (280–315 nm). UV-B is higher energy and more damaging, but UV-A is also dangerous, as it can penetrate the skin more deeply. It has been estimated that 1 in every 10 Americans will develop skin cancer, and it is believed that the situation will worsen over the coming decades, as a result of the slow destruction of the ozone layer by CFCs (Section 11.8). That means that there is a 10% chance that you will develop skin cancer at some point in your life unless you take precautions.

The mechanism by which UV light damages DNA has been extensively studied, and many pathways are responsible. One such pathway involves a thymine base of DNA absorbing UV light to give an excited state that is capable of undergoing a [2+2] cycloaddition with a neighboring thymine base.

In Section 17.8, we saw that [2+2] cycloadditions become symmetry-allowed in the presence of UV light, and this is just a real-world example of that process. This reaction changes the structure of DNA and causes genetic code alterations that ultimately lead to the development of cancer cells.

Sunscreens are used to prevent UV light from causing damage to the structure of DNA. There are two basic types of sunscreens: inorganic and organic. Inorganic sunscreens (such as titanium dioxide or zinc oxide) reflect and scatter UV light. Organic sunscreens are conjugated π systems that absorb UV light and release the absorbed energy in the form of heat. The following compounds are examples of common organic sunscreens.

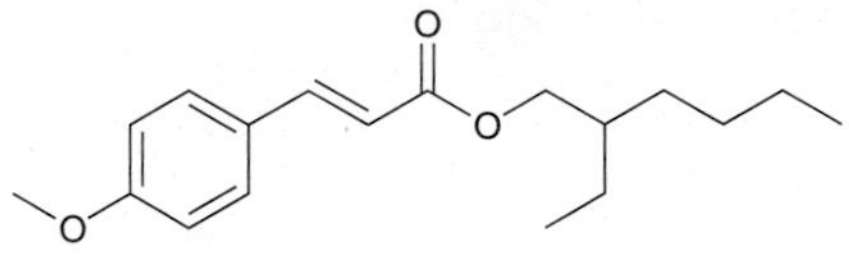

Octyl Methoxycinnamate

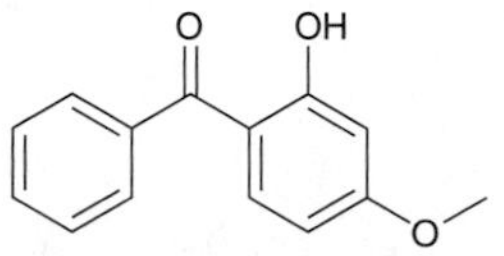

Oxybenzone

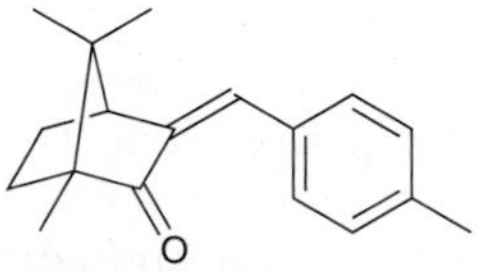

4-Methylbenzylidene camphor

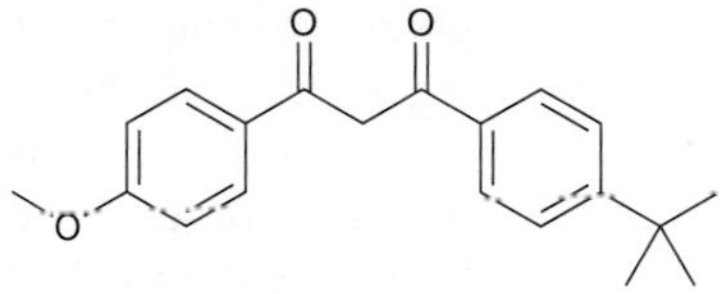

Avobenzone

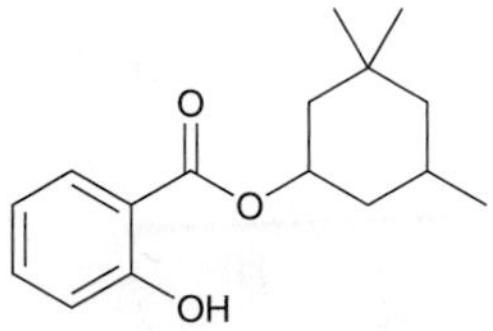

Homosalate

All of these compounds possess conjugated π systems. Both organic and inorganic sunscreens provide protection from UV-B radiation. Some sunscreens, although not all, also provide additional protection from UV-A radiation. Many commercially available sunscreens contain mixtures of both organic and inorganic compounds, thereby achieving high SPF (sun protection factor). However the SPF rating only indicates the strength of protection from UV-B, not UV-A. For products sold in the United States, there is no clear way to compare the abilities of sunscreens to filter UV-A radiation other than reading the list of ingredients. Among the organic sunscreens approved in the United States, it is believed that avobenzone offers the best protection against UV-A radiation.

Despite the established link between sunlight and skin cancer, many people continue to view suntans as sexy. This dangerous attitude will likely become less popular as the public becomes more aware of the dangers involved. As evidence of rising public awareness, annual sales of sunscreens rise each year in the United States and currently exceed $100 million.

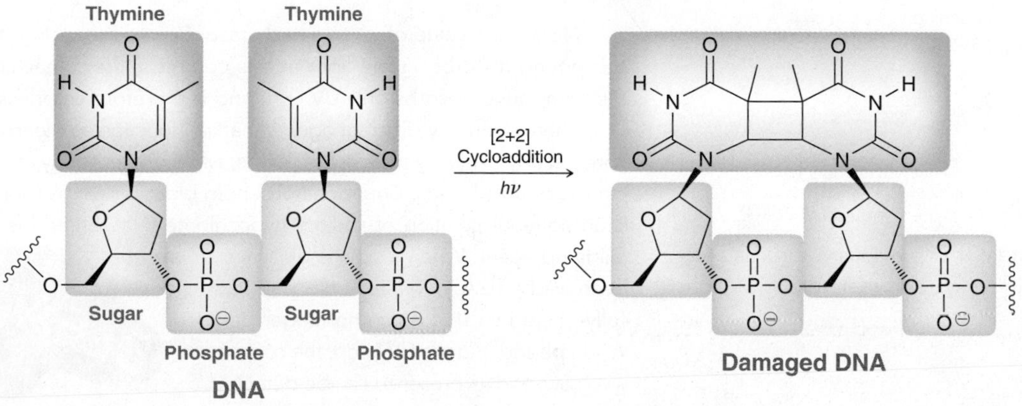

BY THE WAY

What happened to indigo? You probably have learned at some point that the rainbow is composed of the colors ROYGBIV (red, orange, etc.). The "I" stands for indigo, but you might notice that indigo is absent from our color wheel. The truth is that indigo is not worth mentioning (it is intermediate between blue and violet). Indigo is mentioned as one of the colors of the rainbow, because Sir Isaac Newton, who described them, believed in metaphysics and felt that the number 7 was more spiritually meaningful than 6. Thus, the "seven" colors of the rainbow.

17.12 Color

The visible region of the electromagnetic spectrum has wavelengths of 400–700 nm. When a compound possesses a highly conjugated π system, it is possible for the λ_{max} to be above 400 nm. Such a compound will absorb visible light, rather than UV light, and will therefore be colored. Consider the following two examples.

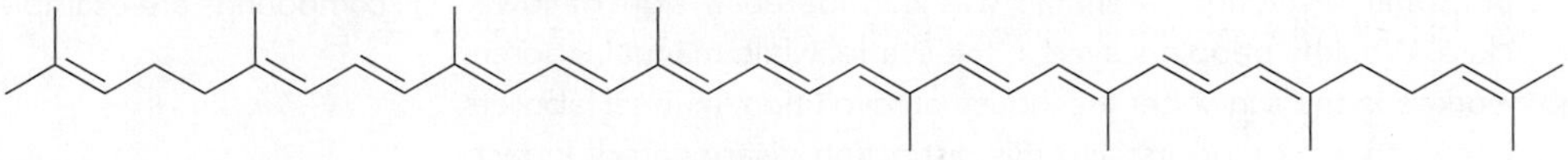

Lycopene

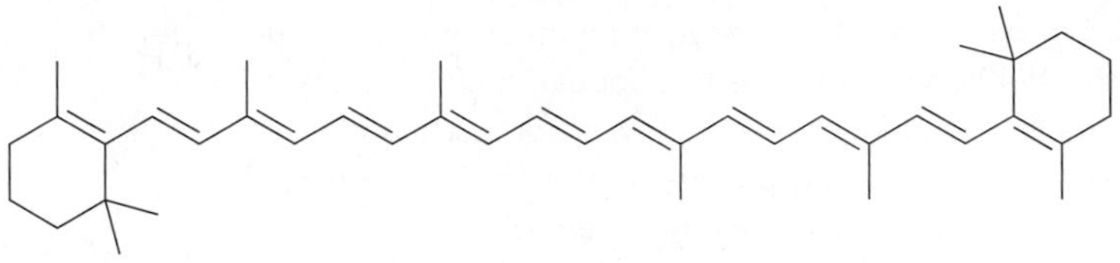

***β*-Carotene**

Lycopene is responsible for the red color of tomatos; β-carotene is responsible for the orange color of carrots. Both compounds are colored because they possess highly conjugated π systems.

White light is composed of all visible wavelengths (400–700 nm) and can be divided into groups of complementary colors, which appear opposite from each other in the color wheel in Figure 17.37. Each pair of complementary colors (such as red and green) can be thought of as canceling each other (in a way) to produce white light. A compound will be colored if it absorbs a specific color more strongly than its complementary color. For example, β-carotene absorbs light of 455 nm (blue light) but reflects orange light. The compound therefore appears to be orange.

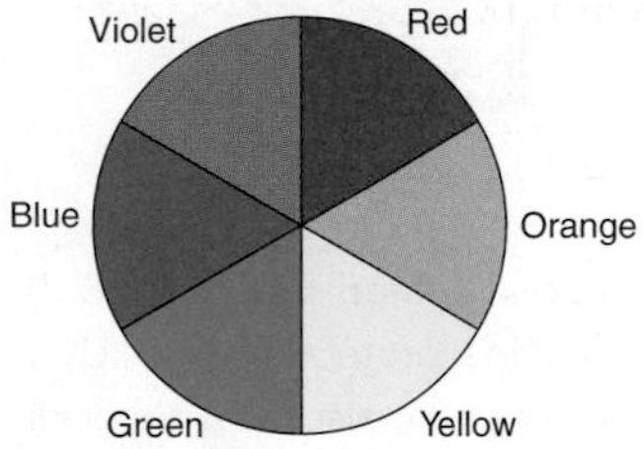

FIGURE 17.37 The color wheel, showing pairs of complementary colors.

PRACTICALLYSPEAKING

Bleach

Now that we understand the origin of color in organic compounds, we are finally able to explore how bleaches function. Most bleaching agents react with conjugated π systems, disrupting the extended conjugation:

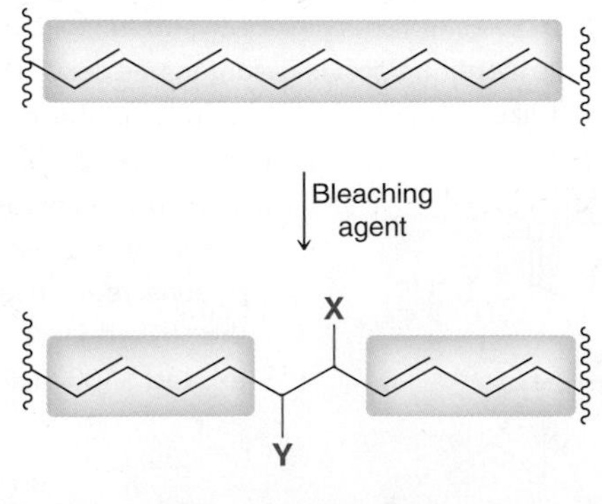

In this way, a conjugated π system that previously absorbed visible light is converted into smaller π systems that absorb UV light instead.

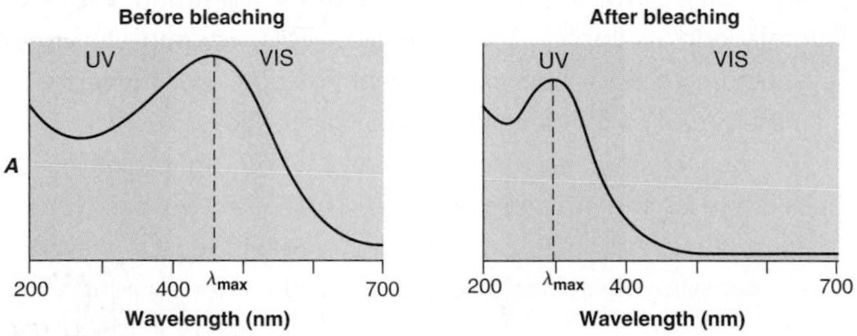

Notice the value of λ_{max} in each case. Before bleaching, the compound absorbs visible light and is colored. After bleaching, the compound absorbs only UV light and is therefore colorless.

There are many different agents that will function as bleaches. Some act by oxidizing the double bonds, while others act by reducing the double bonds. Common household bleach (such as Clorox) is an aqueous solution of sodium hypochlorite (NaOCl) and is an oxidizing agent. When a stain is bleached, it has not been washed away. Rather, it has just been chemically altered so that it can no longer be seen. When placed under a UV lamp, the compound will glow, and its presence can be detected:

CONCEPTUAL CHECKPOINT

17.31 Use the color wheel in Figure 17.37.

(a) Identify the color of a compound that absorbs orange light.

(b) Identify the color of a compound that absorbs blue-green light.

(c) Identify the color of a compound that absorbs orange-yellow light.

17.13 Chemistry of Vision

The chemical reactions responsible for vision have been the subject of much investigation over the last 60 years.

Two types of light-sensitive cells function as photoreceptors: *rods* and *cones*. Rods do not detect color, and they function as the dominant light receptors in dim light. Cones contain the pigments necessary for color vision, but they only function as the dominant receptor cells in bright light. Humans possess both rods and cones, but some species possess only one of the two cell types. Pigeons, for example, have only cones. As a result, they see quite well in bright light, but they are blind at night. Owls, on the other hand, have only rods. Owls see well at night but are colorblind. In this section, we will focus solely on the chemistry of rods.

The light-sensitive compound in rods is called *rhodopsin*. In 1952, Nobel Laureate George Wald (Harvard University) and his co-workers showed that the chromophore in rhodopsin is the conjugated polyunsaturated system of 11-*cis*-retinal. Rhodopsin is produced by a chemical reaction between 11-*cis*-retinal and a protein called opsin.

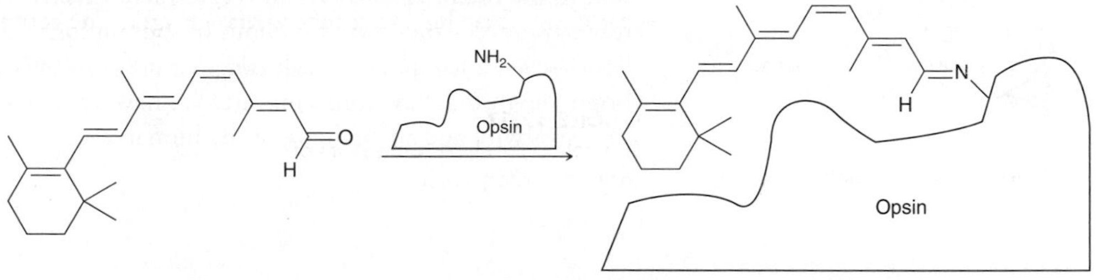

11-*cis*-Retinal **Rhodopsin**

The conjugated π system fits precisely into an internal cavity of opsin and absorbs light over a broad region of the visible spectrum (400–600 nm). Sources of 11-*cis*-retinal include vitamin A and β-carotene. A vitamin A deficiency causes night blindness, while a diet rich in β-carotene can improve vision.

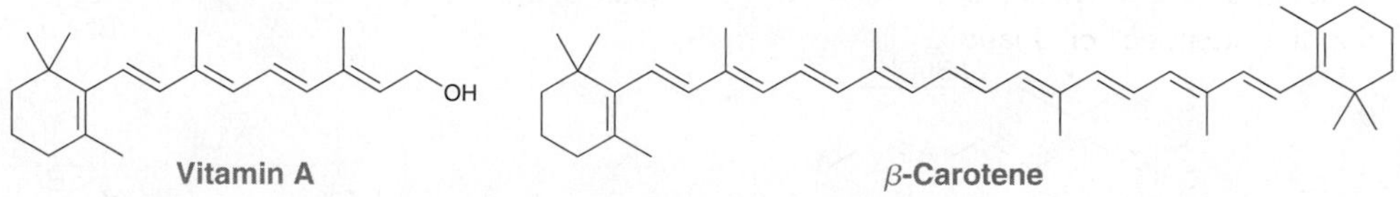

Vitamin A **β-Carotene**

Wald demonstrated that rhodopsin made from 11-*cis*-retinal gave the same photoproduct as rhodopsin made from 9-*cis*-retinal. He concluded that the primary photochemical reaction must be a *cis-trans* isomerization.

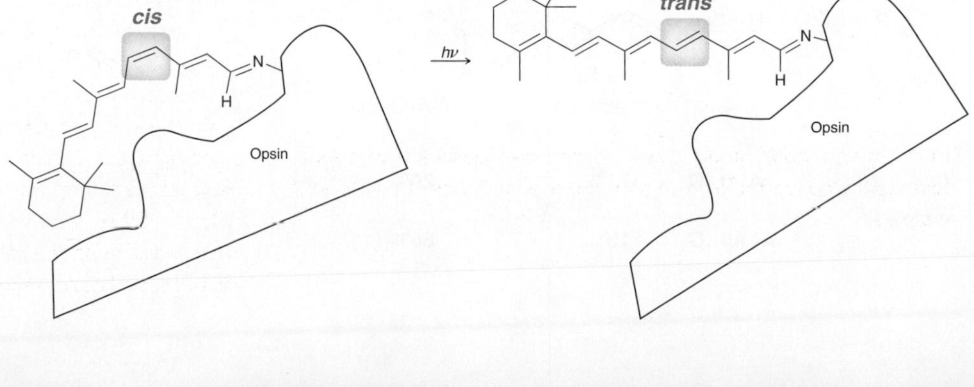

Further evidence supporting this photoisomerization step came from Koji Nakanishi's work at Columbia University. Nakanishi's group synthesized an analogue of 11-*cis*-retinal, in which the *cis* double bond was incorporated in a seven-membered ring, thereby preventing it from isomerizing to the *trans* configuration.

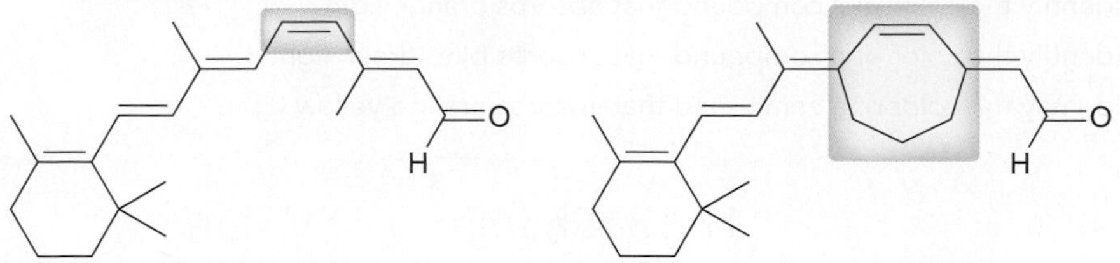

11-*cis*-Retinal **Nakanishi's analogue**

Recall from Section 8.4 that rings of fewer than eight carbon atoms cannot accommodate a *trans* π bond at room temperature. When Nakanishi's analogue binds with opsin, the product is similar to rhodopsin but is unable to isomerize under the influence of light. When Nakanishi's analogue was administered to rats, their vision was severely impaired. Based on these and other convincing pieces of evidence, it is widely accepted that the first step in the chemistry of vision is a photoisomerization reaction.

When rhodopsin is excited to the high-energy all-*trans* isomer, the resulting change in rhodopsin's shape initiates a cascade of enzymatic reactions that causes the release of calcium ions. These ions block channels that normally allow the passage of billions of sodium ions per second. This regular flow of sodium ions is called the *dark current,* and it is reduced when calcium ions block the channels. This reduction in current culminates in a nerve impulse being sent to the brain. Human vision is extremely sensitive, because the absorption of a single photon can prevent the flow of millions of sodium ions. Our eyes are capable of sensing very dim light—even a few photons will trigger a nerve impulse. We can also adjust our eyes to a very bright environment within a minute. We have yet to create a photographic system that parallels the sensitivity and adaptability of the human eye.

REVIEW OF REACTIONS

PREPARATION OF DIENES

From Allylic Halides

From Dihalides

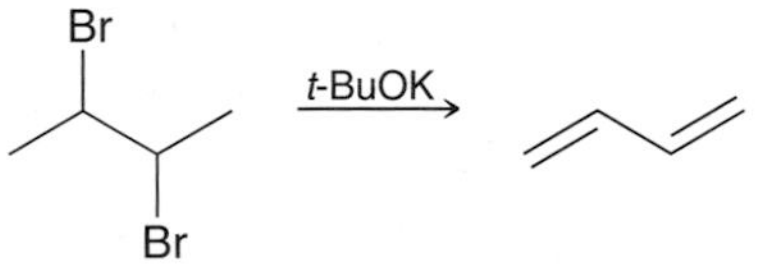

ELECTROPHILIC ADDITION

With HBr

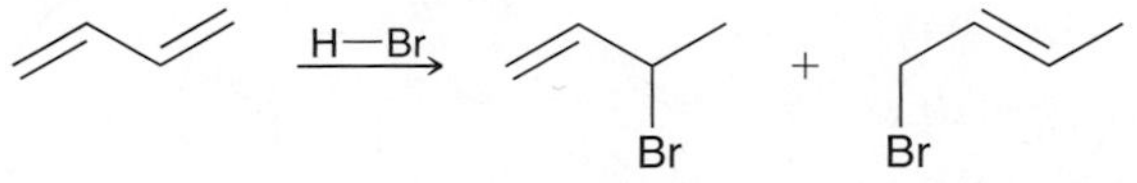

	1,2-Adduct	1,4-Adduct
At 0 °C	71%	29%
At 40 °C	15%	85%

With Br_2

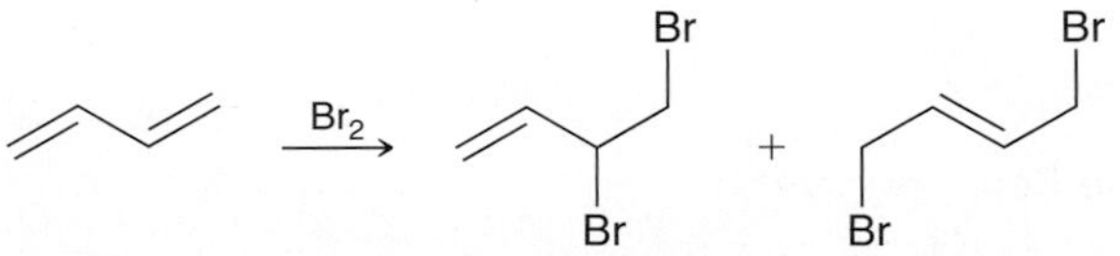

1,2-Adduct **1,4-Adduct**

DIELS-ALDER REACTION

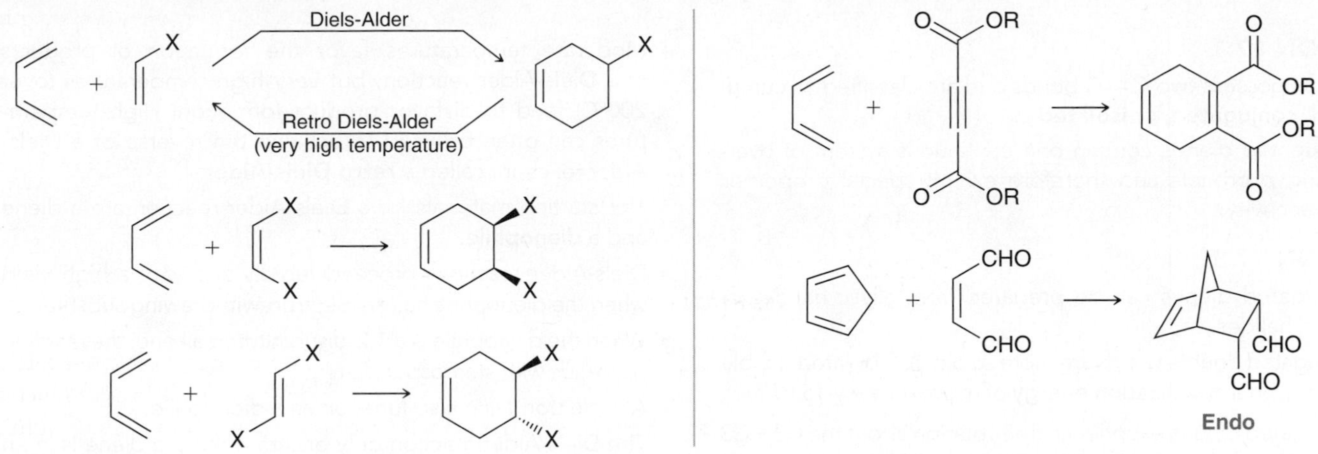

ELECTROCYCLIC REACTIONS

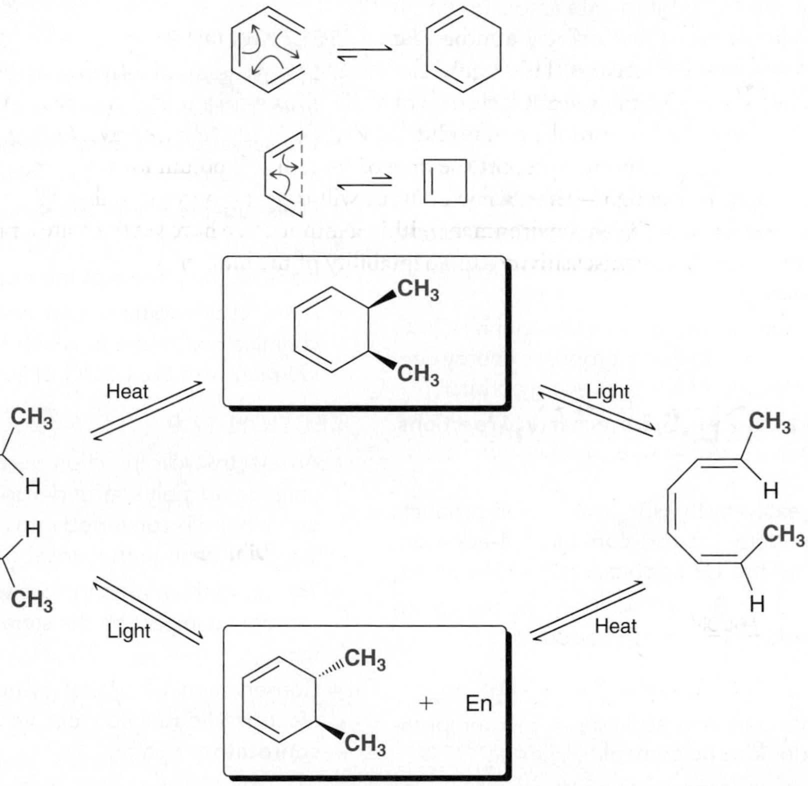

SIGMATROPIC REARRANGEMENTS

Cope Rearrangement

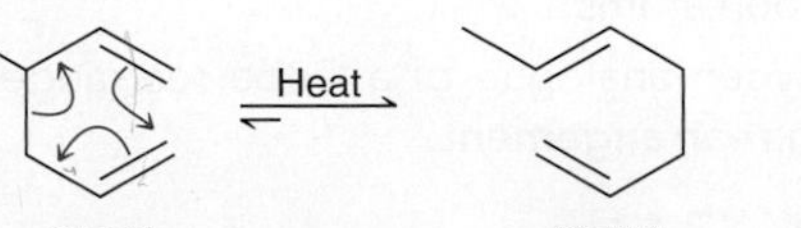

(15%) (85%)

Claisen Rearrangement

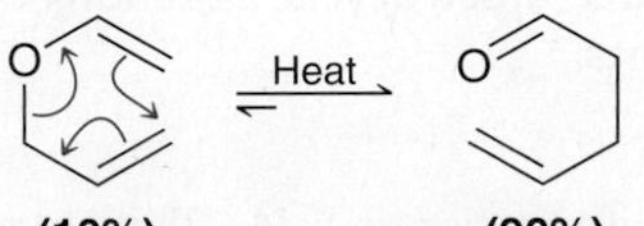

(10%) (90%)

REVIEW OF CONCEPTS AND VOCABULARY

SECTION 17.1

- **Dienes** possess two C=C bonds and are classified as **cumulated**, **conjugated**, or **isolated**.
- Conjugated dienes contain one continuous system of overlapping *p* orbitals and therefore exhibit special properties and reactivity.

SECTION 17.2

- Conjugated dienes can be prepared from allylic halides or from dihalides.
- Conjugated double bonds are more stable than isolated double bonds, with a **stabilization energy** of approximately 15 kJ/mol.
- Conjugated dienes experience free-rotation about the C2—C3 bond, giving rise to two important conformations: ***s-cis*** and ***s-trans***. The *s-trans* conformation is lower in energy.

SECTION 17.3

- Electrons that occupy atomic orbitals are associated with an individual atom, while electrons that occupy a molecular orbital (MO) are associated with the entire molecule.
- The *H*ighest *O*ccupied *M*olecular *O*rbital, or **HOMO**, contains the π electrons most readily available to participate in a reaction.
- The *L*owest *U*noccupied *M*olecular *O*rbital, or **LUMO**, is the lowest energy MO that is capable of accepting electron density.
- The HOMO and LUMO are referred to as **frontier orbitals**, and the reactivity of conjugated polyenes can be explained with **frontier orbital theory.**
- An **excited state** is produced when a π electron in the HOMO absorbs a photon of light bearing the appropriate energy necessary to promote the electron to a higher energy orbital.
- Reactions induced by light are called **photochemical reactions**.

SECTION 17.4

- When butadiene is treated with HBr, two major products are observed, resulting from **1,2-addition** and **1,4-addition**. These products are called the **1,2-adduct** and the **1,4-adduct.**
- Other electrophiles, such as Br_2, will also add across conjugated π systems to produce 1,2- and 1,4-adducts.

SECTION 17.5

- Conjugated dienes that undergo addition at low temperature are said to be under **kinetic control.**
- Conjugated dienes that undergo addition at elevated temperature are said to be under **thermodynamic control**.

SECTION 17.6

- **Pericyclic reactions** proceed via a concerted process with a cyclic transition state, and they are classified as **cycloaddition reactions**, **electrocyclic reactions**, and **sigmatropic rearrangements.**

SECTION 17.7

- The Diels-Alder reaction is a **[4+2] cycloaddition** in which two C—C bonds are formed simultaneously.
- The product of a Diels-Alder reaction is always a substituted cyclohexene.
- Moderate temperatures favor the formation of products in a Diels-Alder reaction, but very high temperatures (over 200°C) tend to disfavor product formation. High temperatures can often be used to achieve the reverse of a Diels-Alder reaction, called a **retro Diels-Alder**.
- The starting materials for a Diels-Alder reaction are a diene and a **dienophile**.
- Diels-Alder reactions proceed rapidly and with a high yield when the dienophile has an electron-withdrawing substituent.
- When the dienophile is a 1,2-disubstituted alkene, the reaction proceeds with stereospecificity.
- A triple bond can also function as a dienophile.
- The Diels-Alder reaction only occurs when the diene is in an *s-cis* conformation.
- When cyclopentadiene is used as the starting diene, a bridged bicyclic compound is obtained, and the **endo** cycloadduct is favored over the **exo** cycloadduct.

SECTION 17.8

- In a Diels-Alder reaction, the electron density is transferred from the HOMO of the diene to the LUMO of the dienophile.
- **Conservation of orbital symmetry** requires that MOs have the same phases in order to effectively overlap during a reaction.
- Diels-Alder reactions are **symmetry allowed.**
- Thermal [2+2] cycloadditions are **symmetry forbidden**, because the phases of the frontier orbitals do not overlap.
- [2+2] cycloadditions can only be performed with photochemical excitation in which the HOMO of an excited state interacts with the LUMO of a ground-state molecule.

SECTION 17.9

- An electrocyclic reaction is a pericyclic process in which a conjugated polyene undergoes cyclization. In the process, one π bond is converted into a σ bond, while the remaining π bonds all change their locations.
- The use of thermal conditions vs. photochemical conditions has a profound impact on the stereochemical outcome of the reaction.
- Conservation of orbital symmetry determines whether an electrocyclic reaction occurs in a **disrotatory** fashion or a **conrotatory** fashion.

SECTION 17.10

- A sigmatropic rearrangement is a pericyclic reaction in which one σ bond is formed at the expense of another.
- A [3,3] sigmatropic rearrangement is called a **Cope rearrangement** when all six atoms of the cyclic transition state are carbon atoms.
- The oxygen analogue of a Cope rearrangement is called a **Claisen rearrangement.**

SECTION 17.11

- Compounds that possess a conjugated π system will absorb UV or visible light to promote an electronic excitation called a $\pi \rightarrow \pi^*$ transition.

- A standard UV-VIS spectrophotometer will irradiate a sample with UV and visible light and will generate an **absorption spectrum**, which plots **absorbance** as a function of wavelength.
- The most important feature of the absorption spectrum is the λ_{max} (pronounced **lambda max**), which indicates the wavelength of maximum absorption.
- The amount of UV light absorbed at the λ_{max} for any compound is described by the **molar absorptivity** (ε) and is related to absorbance by an equation called **Beer's law.**
- Compounds with a greater extent of conjugation will have a longer λ_{max}.
- The region of the molecule responsible for the absorption (the conjugated π system) is called the **chromophore**, while the groups attached to the chromophore are called **auxochromes**.
- The λ_{max} for a simple compound can be predicted with **Woodward-Fieser rules.**

SECTION 17.12

- When a compound exhibits a λ_{max} above 400 nm, the compound will absorb visible light, rather than UV light. Compounds that absorb light in this range will be colored.

SECTION 17.13

- The light-sensitive compound in rods is called rhodopsin.
- When rhodopsin absorbs light, a *cis-trans* isomerization occurs to give the all-*trans* isomer. The resulting change in rhodopsin's shape initiates a cascade of enzymatic reactions that culminate with a nerve impulse being sent to the brain.

KEY TERMINOLOGY

[4+2] cycloaddition 784
1,2-addition 777
1,4-addition 777
1,2 adduct 777
1,4-adduct 777
absorbance 801
absorption spectrum 801
auxochrome 802
Beer's law 801
chromophore 802
Claisen rearrangement 798
conjugated diene 769
conrotatory 794
conservation of orbital symmetry 791
Cope rearrangement 798
cumulated diene 769
cycloaddition reactions 783
diene 769
dienophile 785
disrotatory 794
electrocyclic reactions 783
endo 788
excited state 775
exo 788
frontier orbitals 775
frontier orbital theory 775
HOMO 775
isolated diene 769
kinetic control 780
lambda max 801
LUMO 775
molar absorptivity 801
pericylic reactions 783
photochemical reaction 775
retro Diels-Alder reaction 784
s-cis 771
sigmatropic rearrangements 783
stabilization energy 771
s-trans 771
symmetry allowed 791
symmetry forbidden 791
thermodynamic control 780
Woodward-Fieser rules 802

SKILLBUILDER REVIEW

17.1 PROPOSING THE MECHANISM AND PREDICTING THE PRODUCTS OF ELECTROPHILIC ADDITION TO CONJUGATED DIENES

EXAMPLE Predict the products and propose a mechanism.

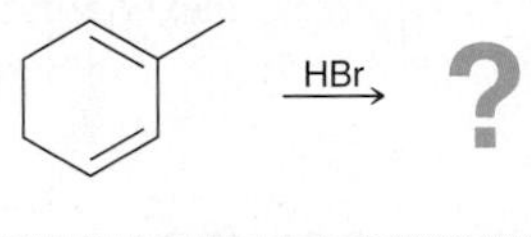

STEP 1 Identify the possible locations where protonation can occur.

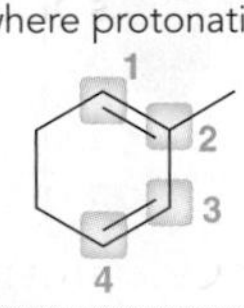

STEP 2 Determine the location where protonation will produce an allylic carbocation.

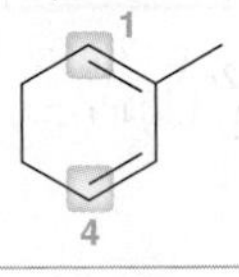

STEP 3 Draw both resonance structures of each possible allylic carbocation, and then draw a nucleophilic attack occurring at each possible electrophilic position.

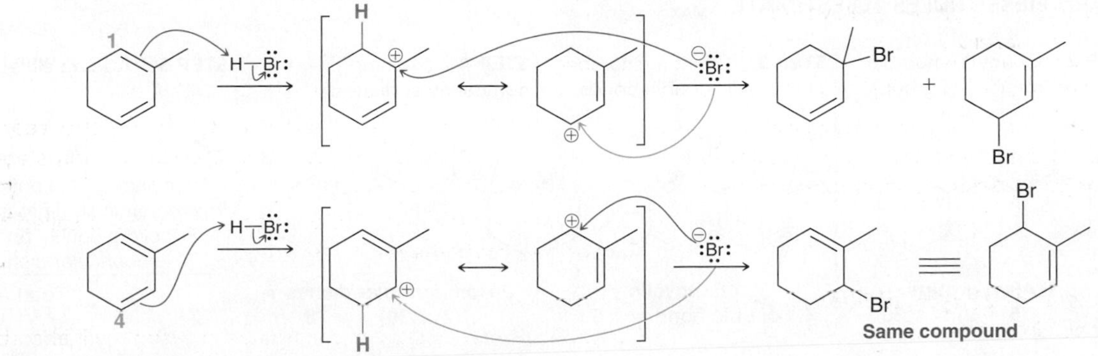

Try Problems 17.7–17.9, 17.35, 17.38

17.2 PREDICTING THE MAJOR PRODUCT OF AN ELECTROPHILIC ADDITION TO CONJUGATED DIENES

EXAMPLE

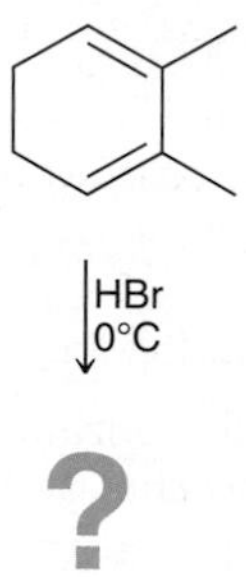

STEP 1 Identify the possible locations for protonation.

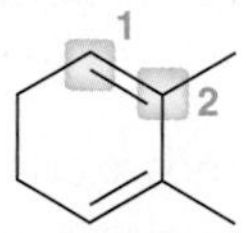

Only two unique positions

STEP 2 Draw the resonance structures of the allylic carbocation that is expected to form.

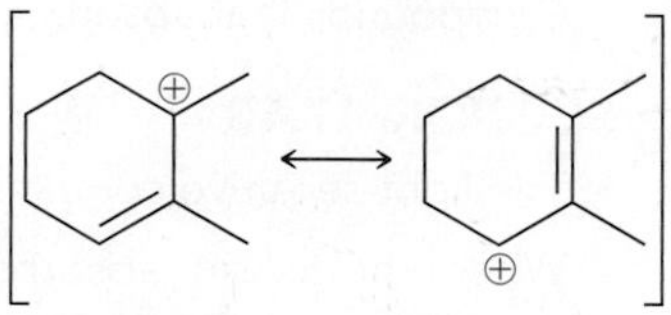

Allylic carbocation

STEP 3 Attack each position with bromide, and draw both possible products.

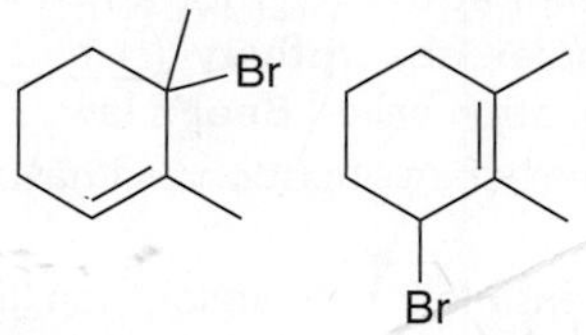

1,2-Adduct **1,4-Adduct**

STEP 4 Identify the kinetic and thermodynamic products, and then choose which product predominates based on the temperature of the reaction.

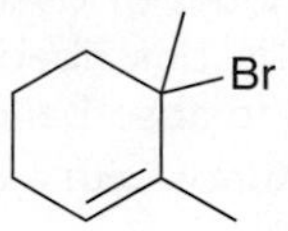

The 1,2-adduct is the kinetic product and is favored at low temperature

Try Problems 17.10–17.12, 17.36, 17.37, 17.39

17.3 PREDICTING THE PRODUCT OF A DIELS-ALDER REACTION

EXAMPLE Predict the product:

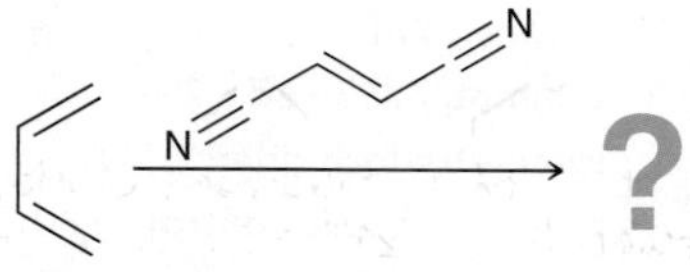

STEP 1 Redraw, and line up the ends of the diene with the dienophile.

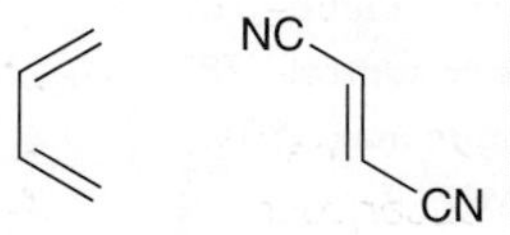

STEP 2 Draw three curved arrows, starting at dienophile, and moving either clockwise or counterclockwise.

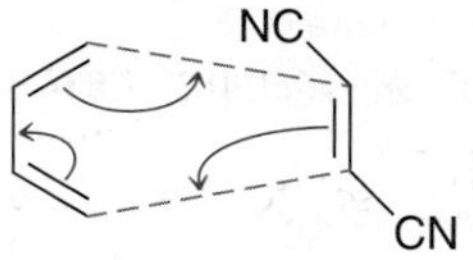

STEP 3 Draw the product with the correct stereochemical outcome.

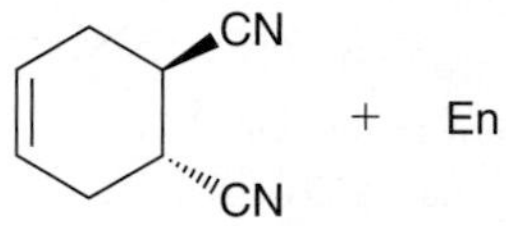

Try Problems 17.14–17.16, 17.44, 17.46

17.4 PREDICTING THE PRODUCT OF AN ELECTROCYCLIC REACTION

STEP 1 Count the number of π electrons.

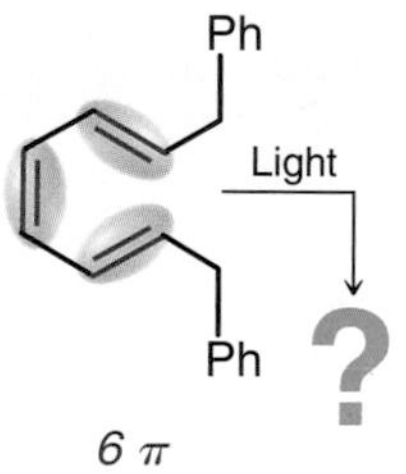

6 π Electrons

STEP 2 Determine whether the reaction is conrotatory or disrotatory.

	Heat	Light
4π	Con	Dis
6π	Dis	Con

6π, Light = Conrotatory

STEP 3 Determine whether the substituents are *cis* or *trans* to each other in the product.

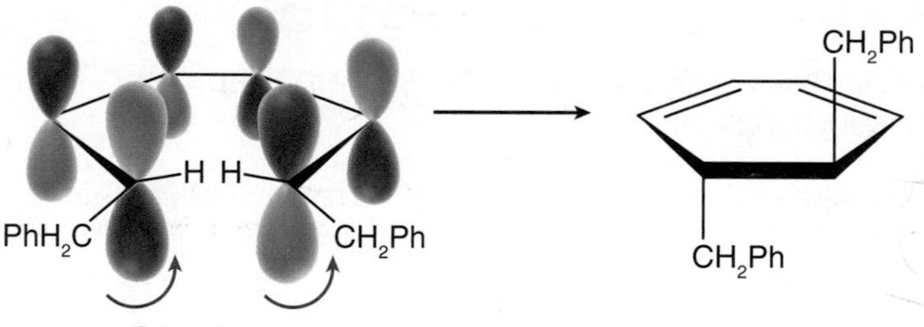

Try Problems 17.21–17.24, 17.55, 17.56

17.5 USING WOODWARD-FIESER RULES TO ESTIMATE λ_{MAX}

STEP 1 Count the number of double bonds.

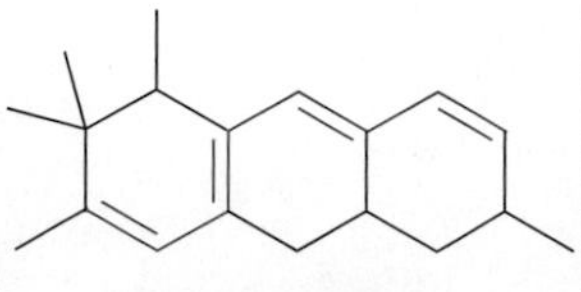

4 Double bonds = 217 + 30 + 30

STEP 2 Count the number of auxochromic alkyl groups.

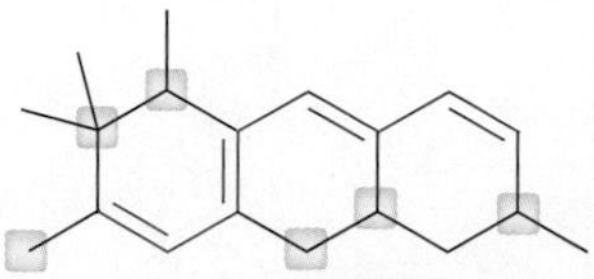

6 Alkyl groups = + (6 × 5) = + 30

STEP 3 Count the number of exocyclic double bonds.

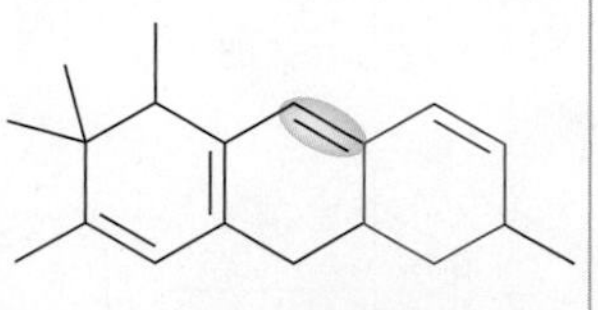

1 Exocyclic double bond = + 5

STEP 4 Look for homoannular dienes.

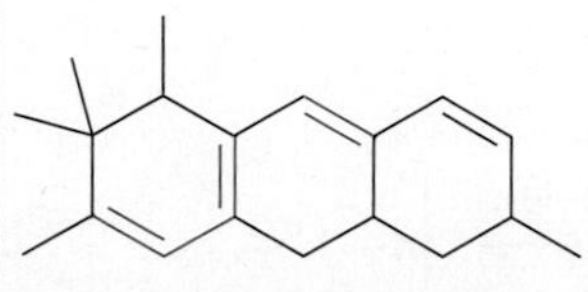

2 Homoannular dienes = (2 × 39) = + 78

STEP 5 Add up all factors:

Base value	= 217
Additional double bonds	= +60
Auxochromic alkyl groups	= +30
Exocyclic double bond	= +5
Homoannular dienes	= +78
Total	**= 390 nm**

Try Problems 17.29, 17.30, 17.50–17.52, 17.60c

PRACTICE PROBLEMS

Note: Most of the Problems are available within WileyPLUS, an online teaching and learning solution.

17.32 Draw the structure of each of the following compounds:

(a) 1,4-Cyclohexadiene (b) 1,3-Cyclohexadiene
(c) (*Z*)-1,3-Pentadiene (d) (2*Z*,4*E*)-Hepta-2,4-diene
(e) 2,3-Dimethyl-1,3-butadiene

17.33 Circle each compound that has a conjugated π system:

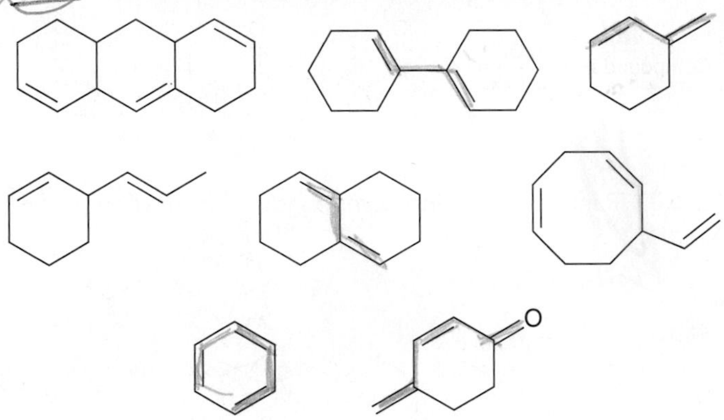

17.34 For each of the following pairs of compounds determine whether the two compounds would be at equilibrium at room temperature or whether they could be isolated from one another:

(a) 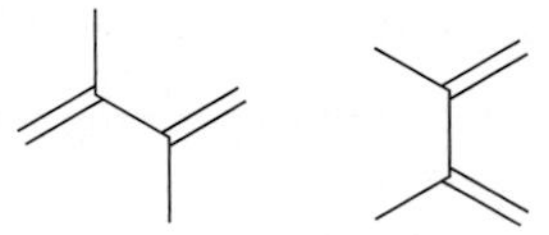

(b) 

(c) 

17.35 Identify the structure of the conjugated diene that will react with one equivalent of HBr to yield a racemic mixture of 3-bromocyclohexene.

17.36 Draw the major product expected when 1,3-butadiene is treated with one equivalent of HBr at 0°C, and show the mechanism of its formation.

17.37 Draw the major product expected when 1,3-butadiene is treated with one equivalent of HBr at 40°C, and show the mechanism of its formation.

17.38 Draw all four products obtained when 2-ethyl-3-methyl-1,3-cyclohexadiene is treated with HBr at room temperature, and show the mechanism of their formation.

17.39 After performing the reaction from Problem 17.38, the reaction flask is heated to 40°C and two of the products become the major products. When the flask is then cooled to 0°C, no change occurs in the product distribution. Explain why an increase in temperature causes a change in the product distribution, and explain why a subsequent reduction in temperature has no effect.

17.40 Each of the following compounds does not participate as a diene in a Diels-Alder reaction. Explain why in each case.

(a) 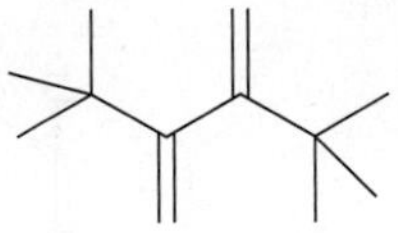

(b) 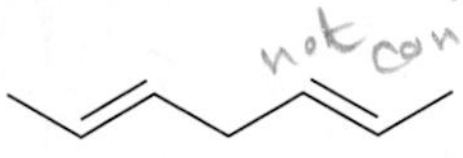

(c) 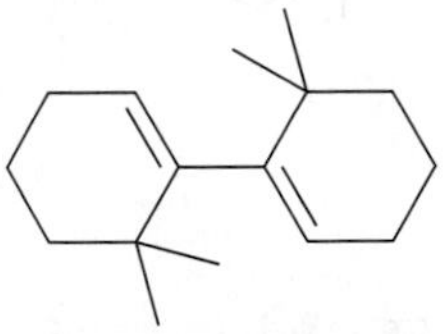

(d) 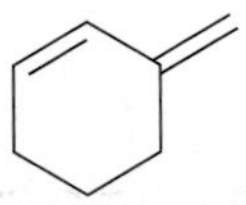

17.41 Rank the following dienophiles (from least reactive to most reactive) in terms of reactivity in a Diels-Alder reaction:

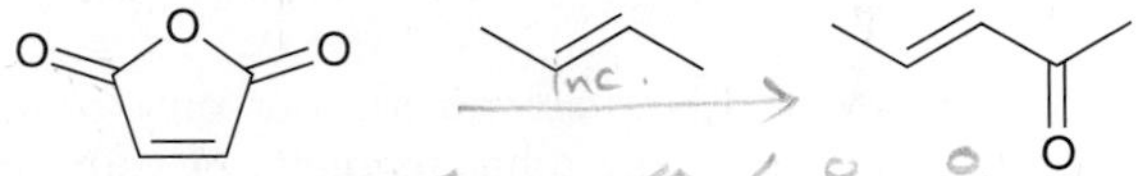

17.42 The absorbance spectrum of 1,3-butadiene displays an absorption in the UV region (λ_{max} = 217 nm), while the absorption spectrum of 1,2-butadiene does not display a similar absorption. Explain.

17.43 Predict the products for each of the following Diels-Alder reactions:

(a) 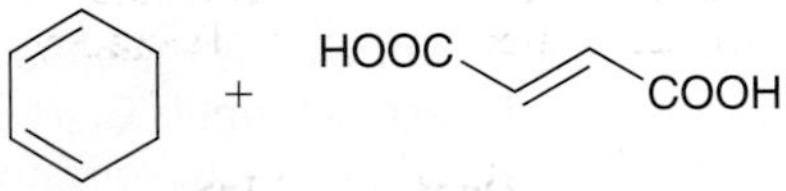

(b) 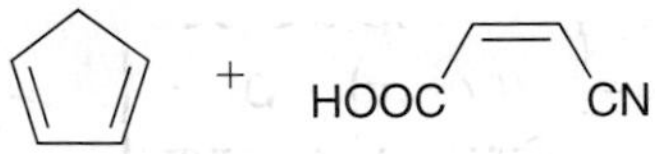

(c) 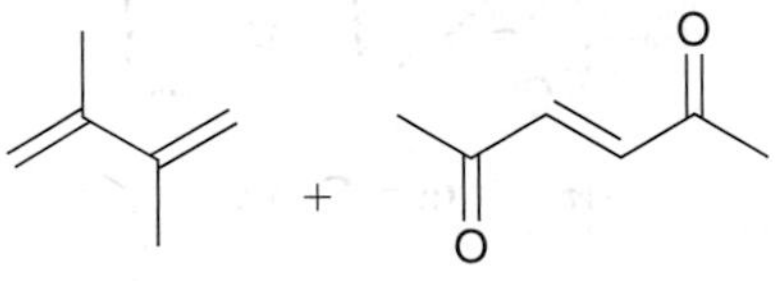

(d) 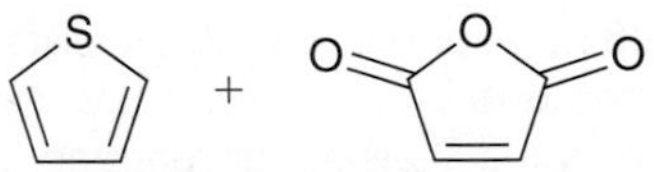

(e) 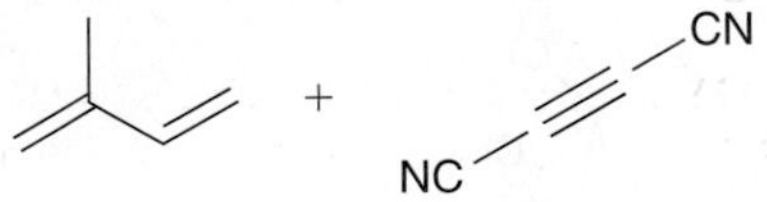

(f) 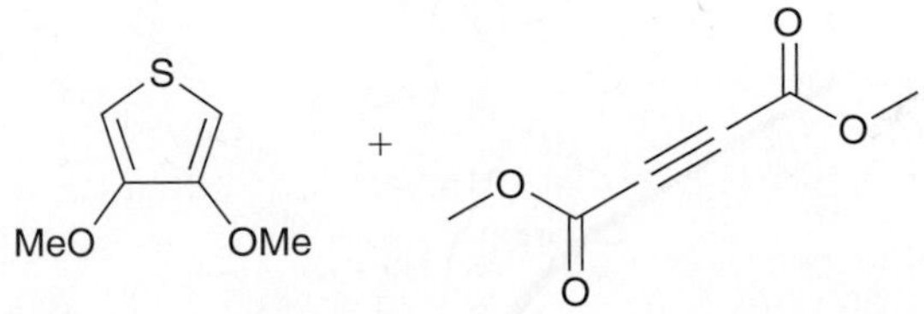

17.44 Identify the reagents you would use to prepare each of the following compounds via a Diels-Alder reaction:

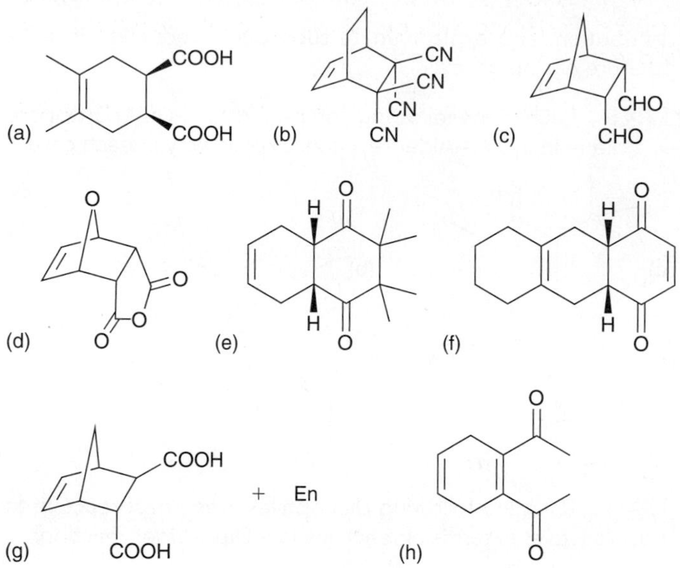

17.45 Starting with 1,3-butadiene as your only source of carbon atoms, and using any other reagents of your choice, design a synthesis of the following compound:

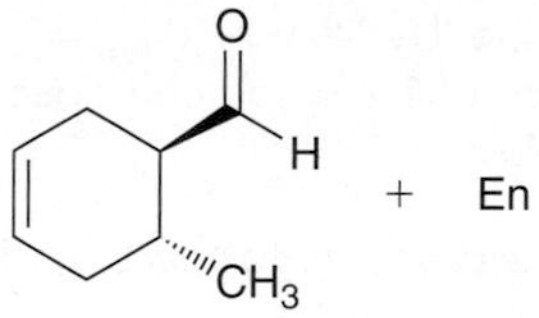

17.46 The following triene reacts with excess maleic anhydride to produce a compound with molecular formula $C_{14}H_{12}O_6$. Draw the structure of this product (ignoring stereochemistry).

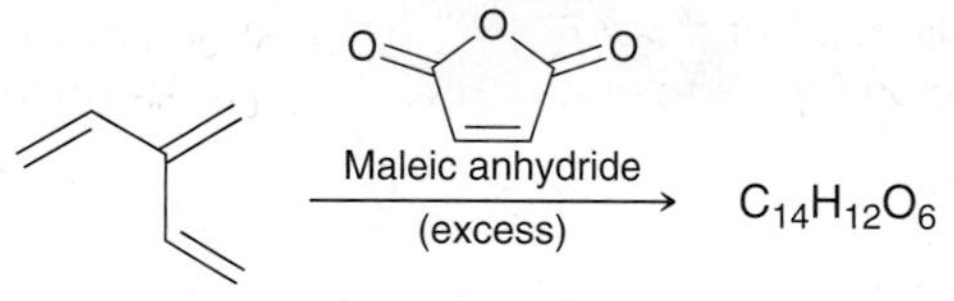

17.47 Chlordane is a powerful insecticide that was used in the United States during the second half of the twentieth century. Its use was discontinued in1988 in recognition of its persistence and accumulation in the environment. Identify the reagents you would use to prepare chlordane via a Diels-Alder reaction.

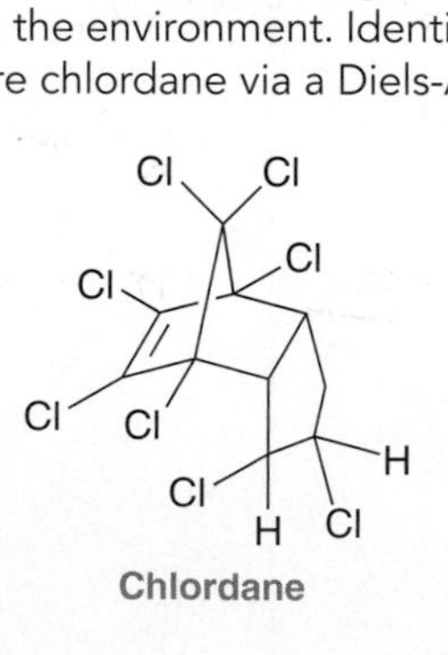

Chlordane

17.48 Cylopentadiene reacts very rapidly in Diels-Alder reactions. In contrast, 1,3-cyclohexadiene reacts more slowly and 1,3-cycloheptadiene is practically unreactive. Can you offer an explanation for this trend?

17.49 Compound **A** (C_7H_{10}) exhibits a λ_{max} of 230 nm in its UV absorption spectrum. Upon hydrogenation with a metal catalyst, compound **A** will react with two equivalents of hydrogen gas. Ozonolysis of compound **A** yields the following two compounds. Identify two possible structures for compound **A**.

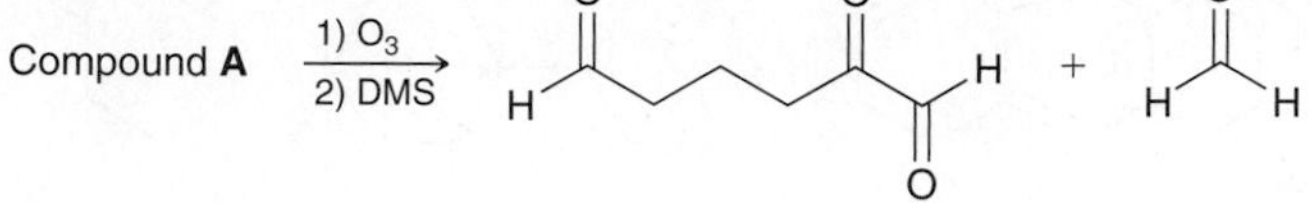

17.50 Rank the following compounds in order of increasing λ_{max}:

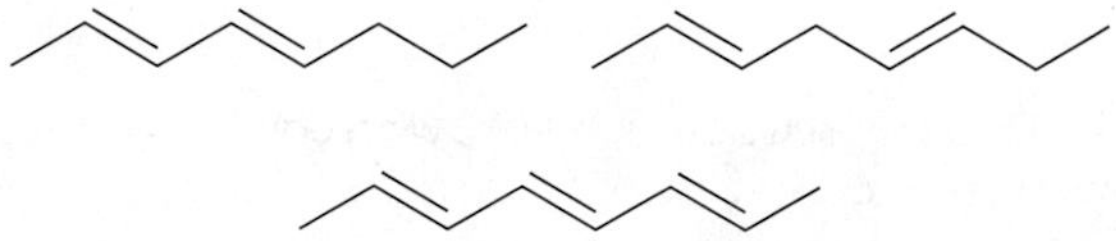

17.51 Which of the following compounds below do you expect to have a longer λ_{max}?

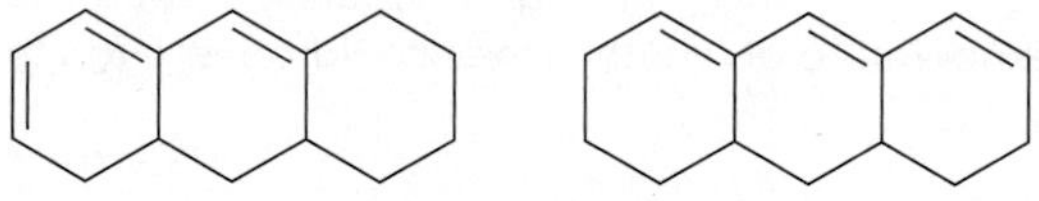

17.52 Predict the expected λ_{max} of the following compound:

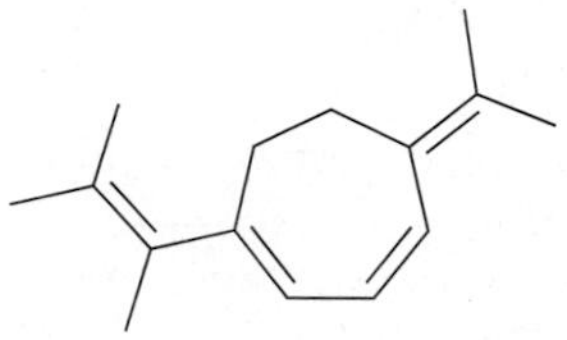

17.53 When 5-deutero-5-methyl-1,3-cyclopentadiene is warmed to room temperature, it rapidly rearranges, giving an equilibrium mixture containing the original compound as well as two others. Propose a plausible mechanism for the formation of these other two compounds.

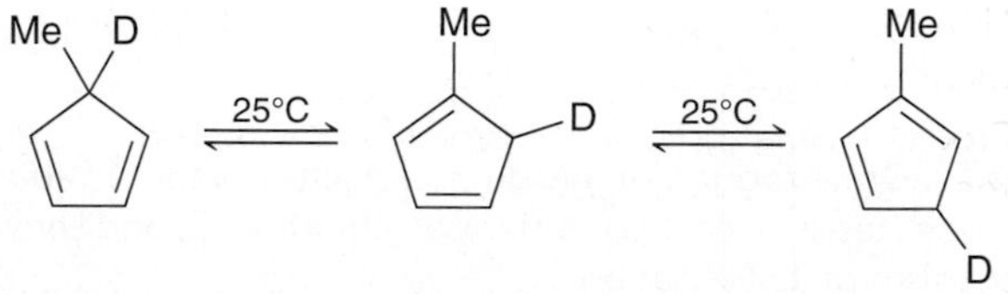

17.54 Propose a plausible mechanism for the following transformation:

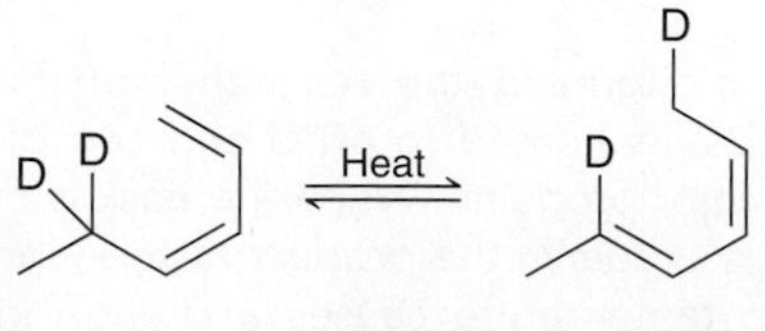

17.55 Propose a method for achieving the following transformation:

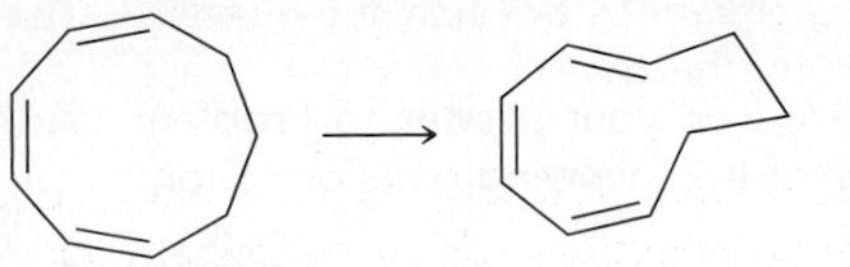

17.56 Predict the product for each of the following electrocyclic reactions:

(a) 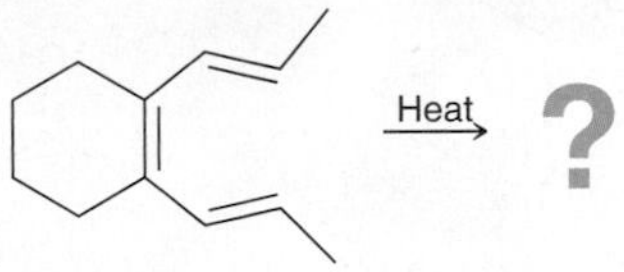

(b) 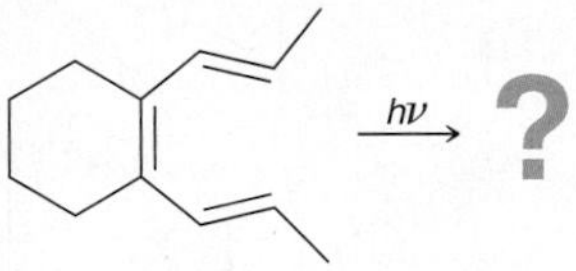

(c) 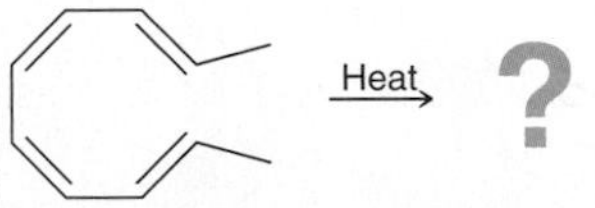

(d) 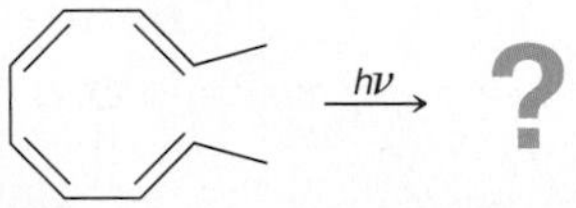

17.57 Predict which side of the following equilibrium is favored, and explain your choice.

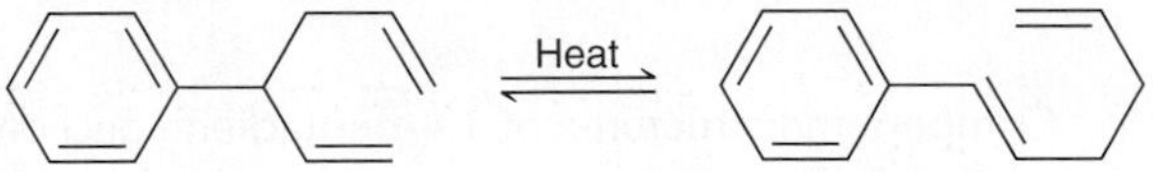

17.58 When *trans*-3,4-dimethylcyclobutene is heated, conrotatory ring opening can produce two different products, yet only one is formed. Draw both products, identify which product is formed, and then explain why the other product is not formed.

17.59 Predict the product for each of the following reactions:

(a) 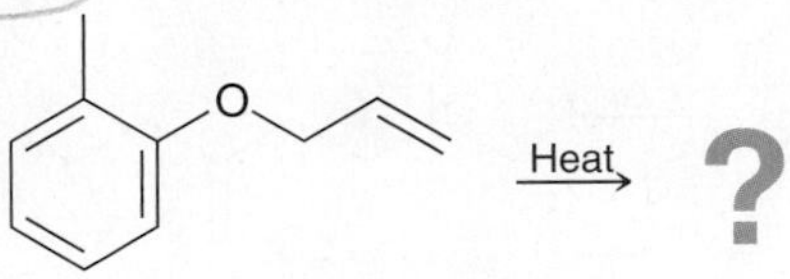

(b) 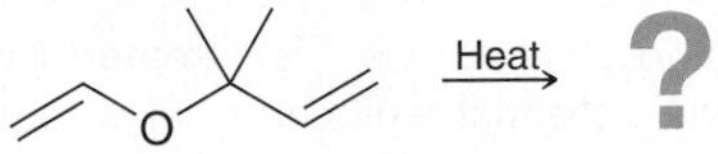

(c) 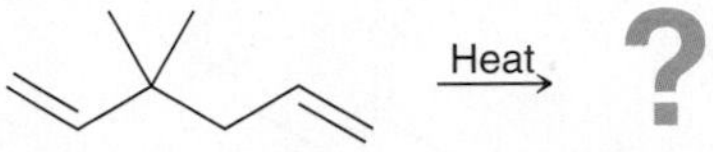

INTEGRATED PROBLEMS

17.60 α-Terpinene is a pleasant-smelling compound present in the essential oil of marjoram, a perennial herb. Upon hydrogenation with a metal catalyst, α-terpinene reacts with two equivalents of hydrogen gas to produce 1-isopropyl-4-methylcyclohexane. Ozonolysis of α-terpinene yields the following two compounds:

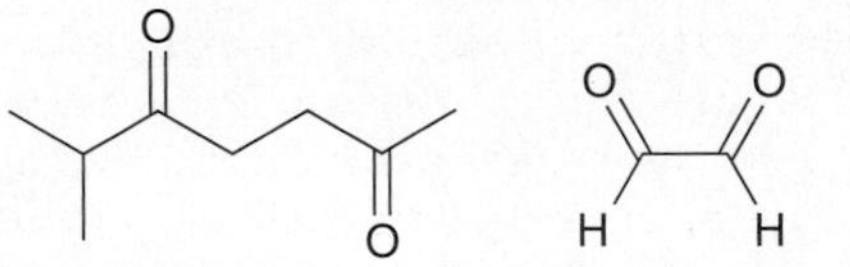

(a) How many double bonds are present in α-terpinene?
(b) Identify the structure of α-terpinene.
(c) Predict the wavelength of maximum absorption (λ_{max}) in the absorption spectrum of α-terpinene.

17.61 Draw all possible conjugated dienes with molecular formula C_6H_{10}, taking special care not to draw the same compound twice.

17.62 Treatment of 1,2-dibromocycloheptane with excess potassium *tert*-butoxide yields a product that absorbs UV light. Identify the product.

17.63 In each of the following pairs of compounds identify the compound that liberates the most heat upon hydrogenation.

(a) 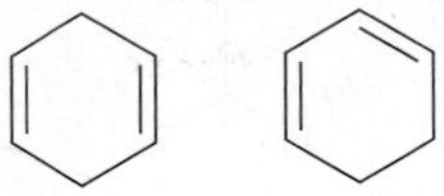

(b) 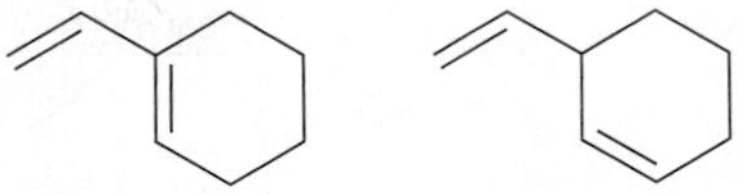

17.64 Would you expect nitroethylene to be more or less reactive than ethylene in a Diels-Alder reaction? (*Hint*: Draw the resonance structures of nitroethylene.)

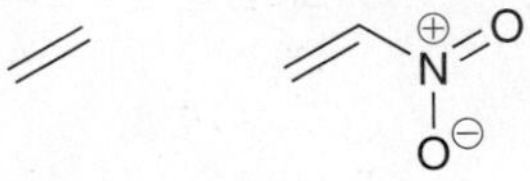

Ethylene **Nitroethylene**

CHALLENGE PROBLEMS

17.65 Propose a plausible mechanism for the following transformatio. (*Hint:* Only two sequential pericyclic reactions are required.)

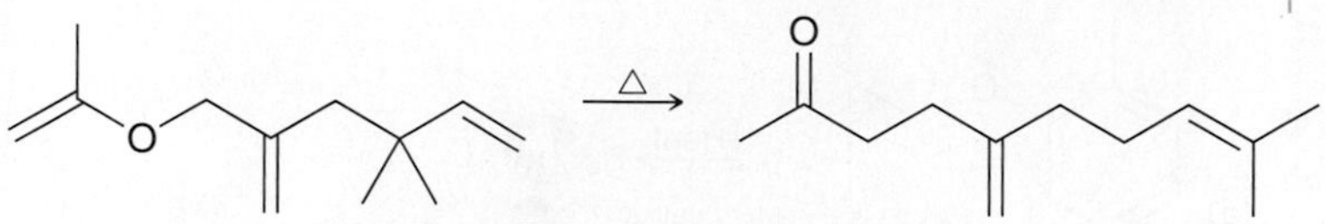

17.66 When 2-methoxy-1,3-butadiene is treated with 3-buten-2-one, there are two potential regiochemical outcomes for the Diels-Alder reaction that occurs:

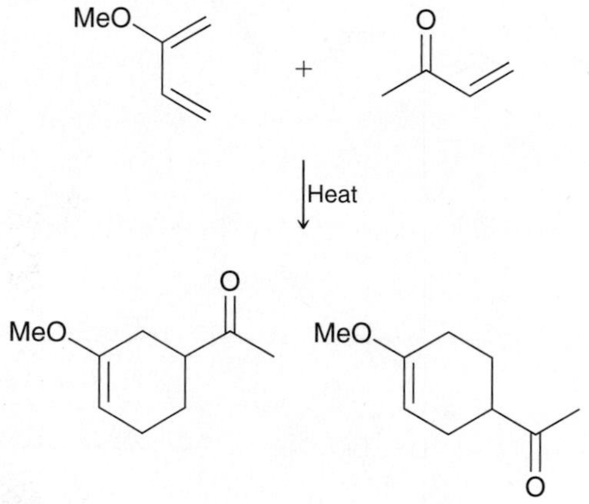

Nevertheless, only one of these products is obtained. Draw resonance structures for the diene and the dienophile, and use these resonance structures to make an educated guess regarding which product is formed.

17.67 Propose a mechanism for the following transformation:

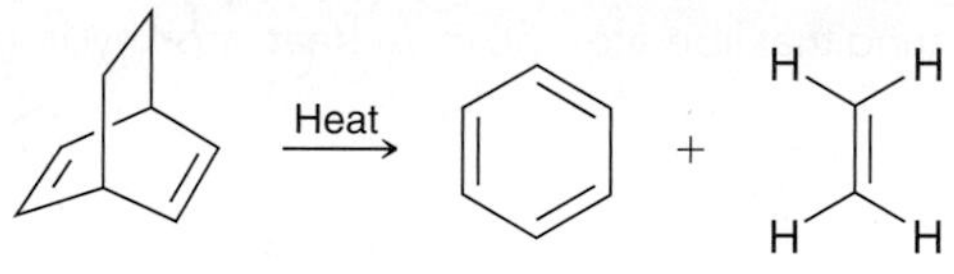

17.68 Based on your answer to Problem 17.67, propose a mechanism for the following transformation:

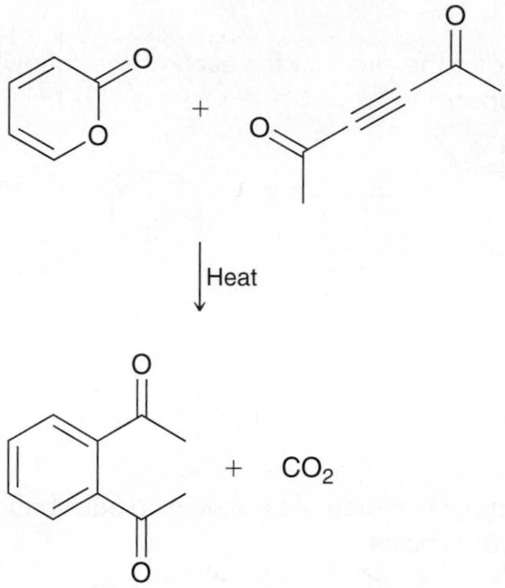

17.69 Compare the structures of 1,4-pentadiene and divinyl amine:

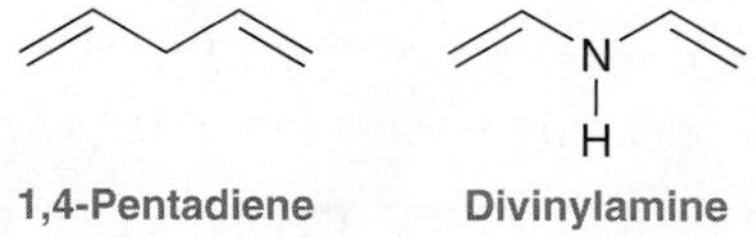

The first compound does not absorb UV light in the region between 200 and 400 nm. The second compound does absorb light above 200 nm. Using this information, identify the hybridization state of the nitrogen atom in divinylamine, and justify your answer.

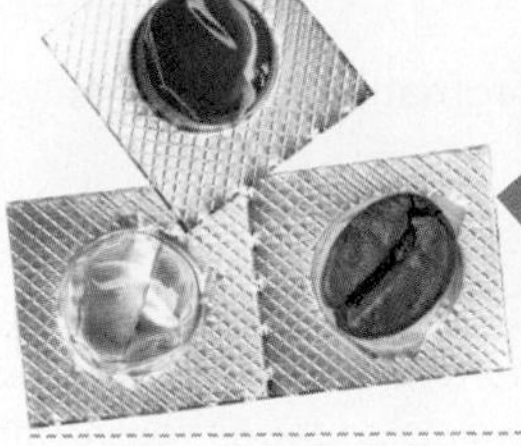

Aromatic Compounds

18

DID YOU EVER WONDER...

what the difference is between antihistamines that cause drowsiness and those that don't?

In this chapter, we expand our discussion of conjugated π systems and explore compounds that exhibit conjugated π bonds enclosed in a ring. For example, benzene and its derivatives belong to a class of compounds called *arenes*.

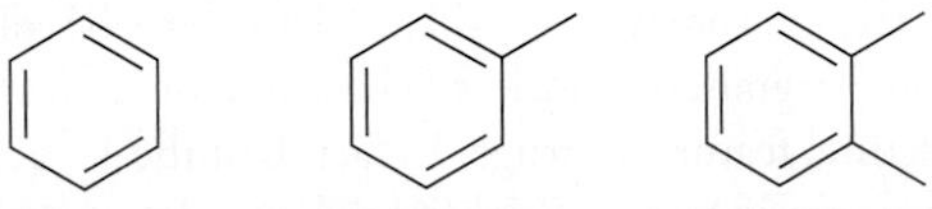

Benzene **Toluene** ***ortho*-Xylene**

As noted in earlier chapters, these compounds are said to be aromatic. This chapter will explore the unusual stability of aromatic rings as well as their properties and reactions. We will learn the criteria for aromaticity and then apply those criteria to identify other ring systems that are also aromatic. With this knowledge, we will be ready to explore the important role that aromatic rings play in the function of antihistamines.

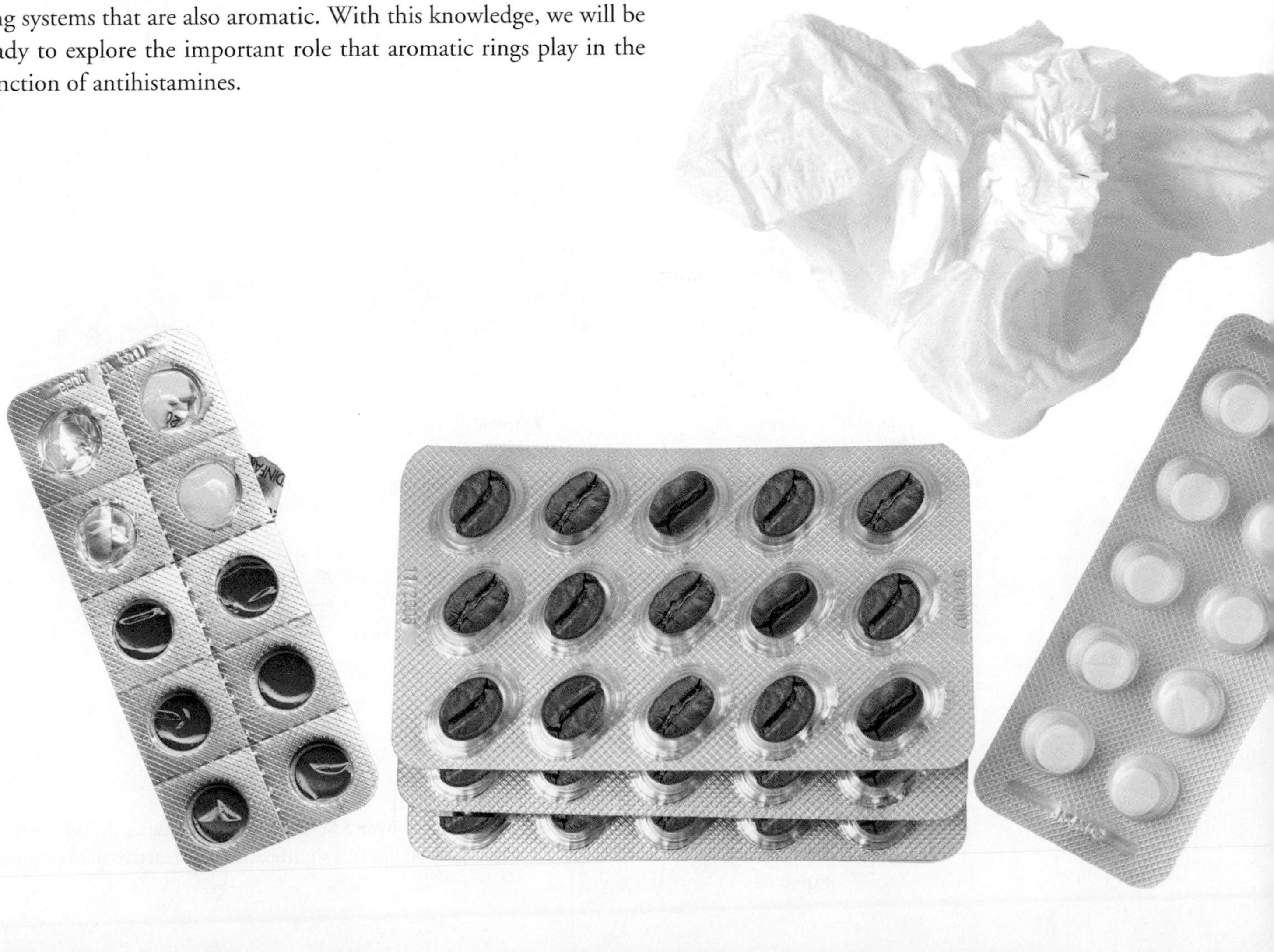

DO YOU REMEMBER?

Before you go on, be sure you understand the following topics.
If necessary, review the suggested sections to prepare for this chapter:

- Molecular Orbital Theory (Section 1.8)
- Dissolving Metal Reduction (Section 10.5)
- Delocalized and Localized Lone Pairs (Section 2.12)
- MO Description of Conjugated Dienes (Section 17.3)

WILEY PLUS Visit www.wileyplus.com to check your understanding and for valuable practice.

18.1 Introduction to Aromatic Compounds

Many derivatives of benzene were originally isolated from the fragrant balsams obtained from trees and plants, and these compounds were therefore described as being **aromatic** in reference to their pleasant odors. Over time, chemists discovered that many derivatives of benzene are, in fact, odorless. Nevertheless, the term "aromatic" is still used to describe all derivatives of benzene, regardless of whether they are fragrant or odorless. The aromatic moiety—the benzene ring—is a particularly common structural feature in drugs. In fact, 8 of the 10 best-selling drugs in 2007 contained the benzene-like aromatic moiety (highlighted in red).

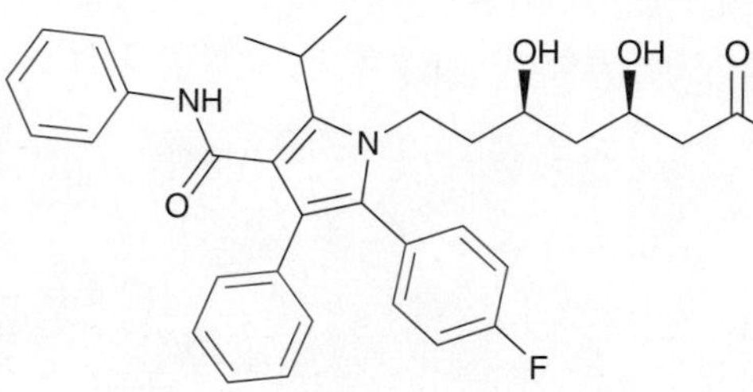

Lipitor™
(atorvastatin)
Lowers cholesterol levels and reduces risk of heart attack and stroke

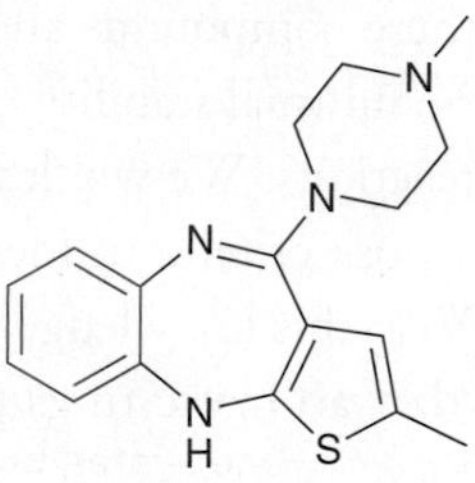

Zyprexa™
(olanzapine)
An antipsychotic used in the treatment of schizophrenia and bipolar disorder

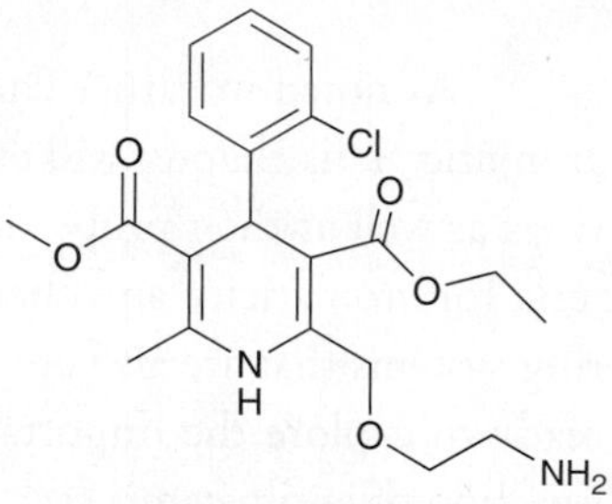

Norvasc™
(amlodipine)
Used in the treatment of angina and high blood pressure

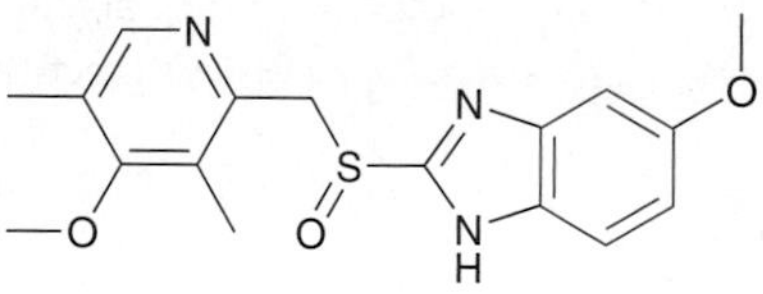

Nexium™
(omeprazole)
A proton-pump inhibitor used in the treatment of ulcers and acid reflux

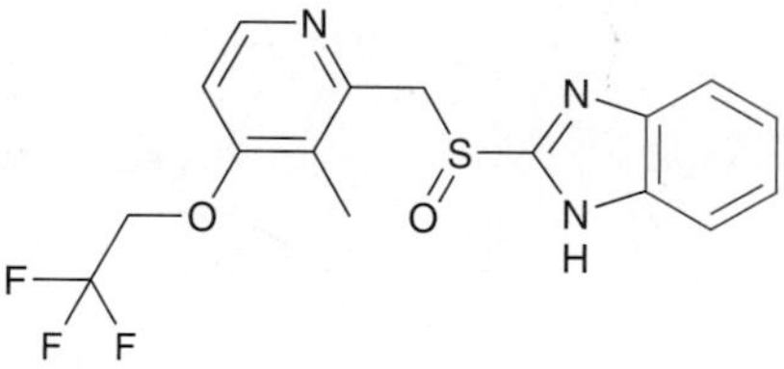

Prevacid™
(lansoprazole)
A proton-pump inhibitor used in the treatment of ulcers and acid reflux

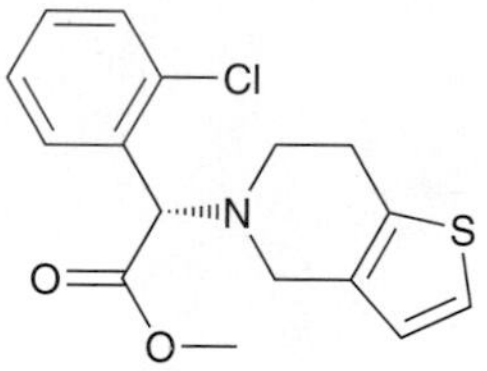

Plavix™
(clopidogrel)
An antiplatelet agent (prevents formation of blood clots) used in the treatment of coronary artery disease

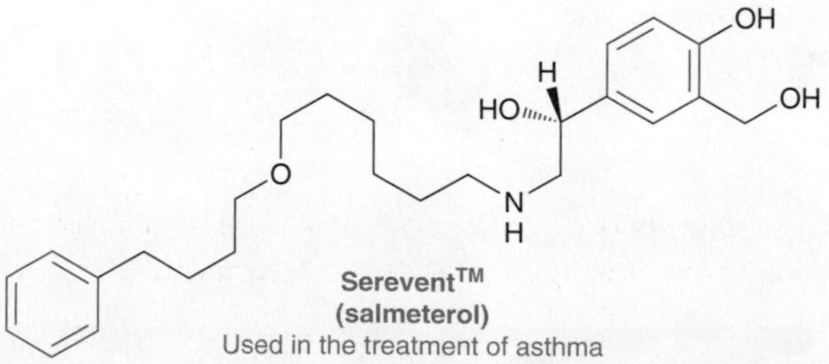

Serevent™
(salmeterol)
Used in the treatment of asthma

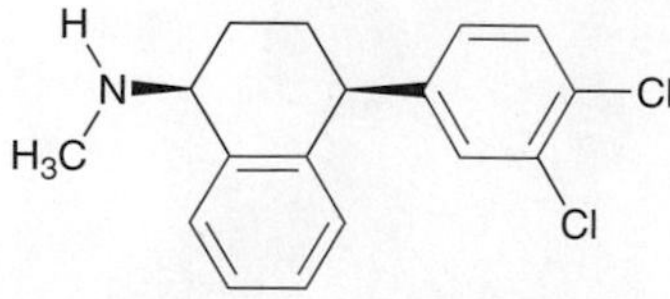

Zoloft™
(sertraline)
Used in the treatment of depression

In 2007, these eight drugs collectively generated over \$38 billion in global sales. In this chapter, we will see that aromatic rings are particularly stable and significantly less reactive than originally expected.

PRACTICALLYSPEAKING)))

What Is Coal?

It is estimated that 25% of the world's recoverable coal (coal that can be readily mined) is located in the United States. As such, it is believed that the United States has enough natural resources to meet all of its energy needs for at least 200 years. To provide some additional perspective, consider the fact that these natural resources contain more than double the energy contained in all of the combined oil reserves in the Middle East.

It is believed that the earth's coal deposits were formed as organic matter from prehistoric plant life was subjected to high pressures and temperatures over long periods of time. The molecular structure of coal involves haphazard arrangements of aromatic moieties (highlighted in orange boxes), connected to each other by nonaromatic links (shown in green boxes):

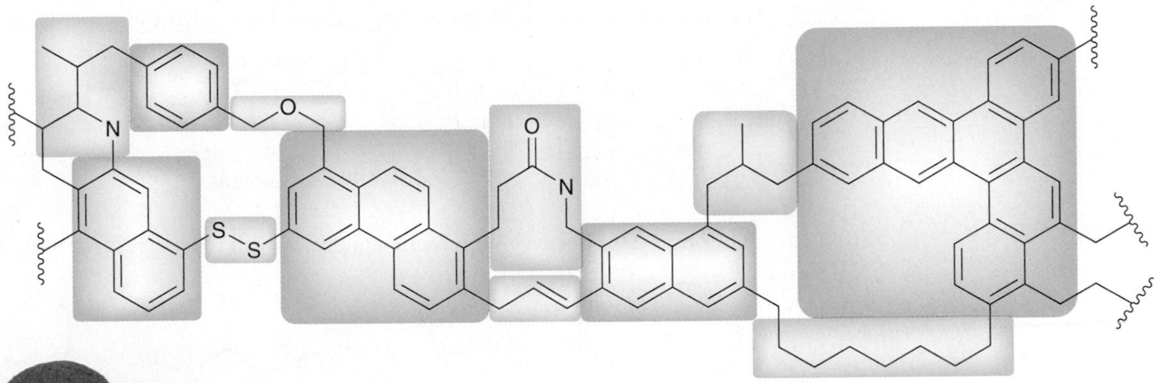

When heated in the presence of oxygen, a combustion process takes place, producing carbon dioxide, water, and many other products. During the combustion process, the C—C and C—H bonds are broken and replaced by lower energy C=O bonds, thereby releasing energy that can be used to generate electricity. In fact, more than half of the electricity consumed in the United States is produced from the combustion of coal.

When coal is heated to about 1000°C in the absence of oxygen, a mixture of compounds, called *coal tar*, is released. Distillation of coal tar yields many aromatic compounds.

Although aromatic compounds can be obtained from coal, the primary source of aromatic compounds is from petroleum, using a process called *reforming* (Section 4.5).

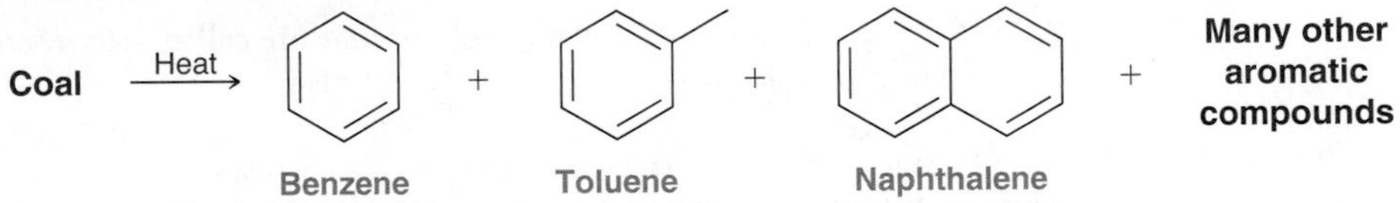

Benzene (b.p. = 80°C) **Toluene** (b.p. = 110°C) **Naphthalene** (b.p. = 218°C)

18.2 Nomenclature of Benzene Derivatives

Monosubstituted Derivatives of Benzene

Monosubstituted derivatives of benzene are named systematically using benzene as the parent and listing the substituent as a prefix. Below are several examples.

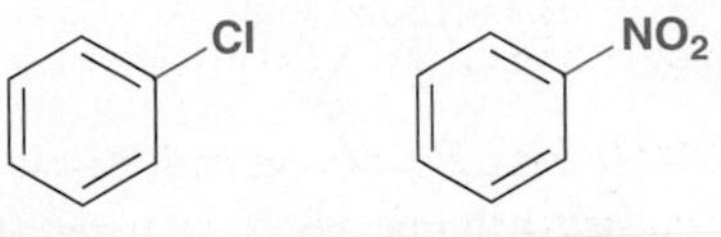

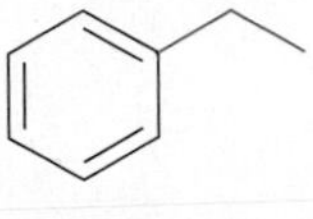

Chlorobenzene **Nitrobenzene** **Ethylbenzene**

The following are some monosubstituted aromatic compounds that have common names accepted by IUPAC. You must commit these names to memory, as they will be used extensively throughout the remaining chapters.

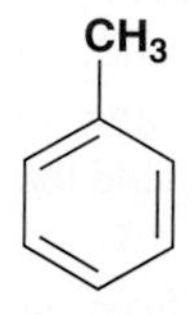
Toluene

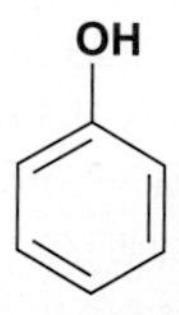
Phenol

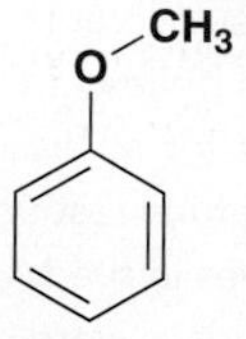
Anisole

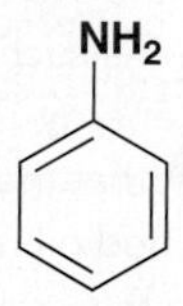
Aniline

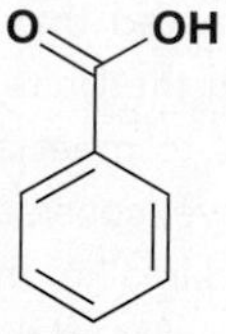
Benzoic acid

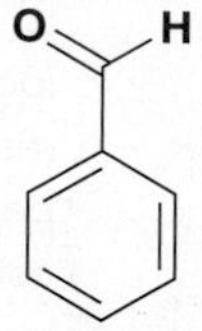
Benzaldehyde

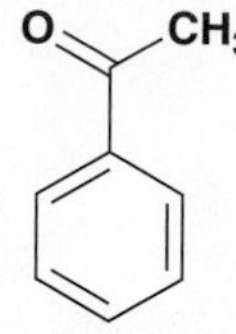
Acetophenone

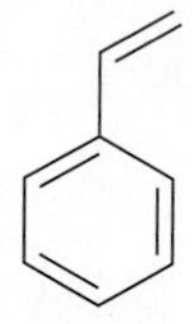
Styrene

If the substituent is larger than the benzene ring (i.e., if the substituent has more than six carbon atoms), then the benzene ring is treated as a substituent and is called a **phenyl group**.

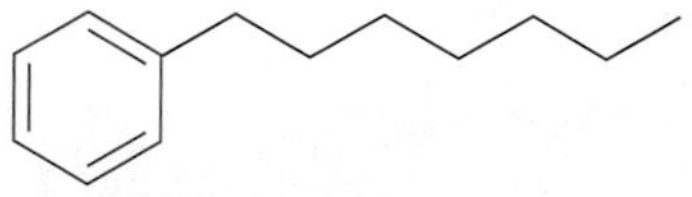
1-Phenylheptane

The presence of phenyl groups is often indicated with the letters Ph or with the Greek letter phi (ϕ).

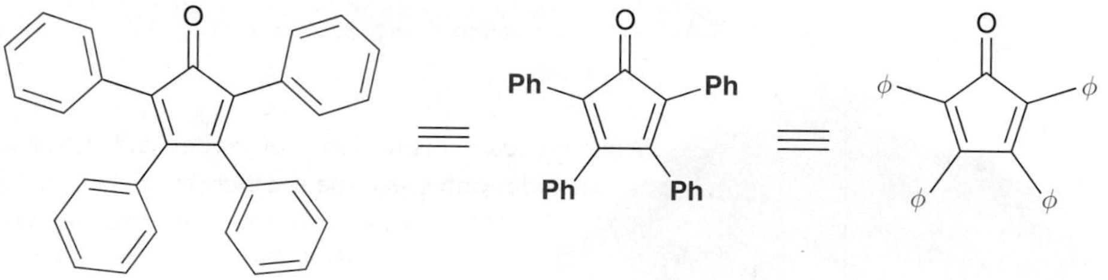
Tetraphenylcyclopentadienone

Phenyl groups bearing substituents are sometimes indicated with the letters Ar, indicating the presence of an aromatic ring.

Disubstituted Derivatives of Benzene

Dimethyl derivatives of benzene are called xylene, and there are three constitutionally isomeric xylenes.

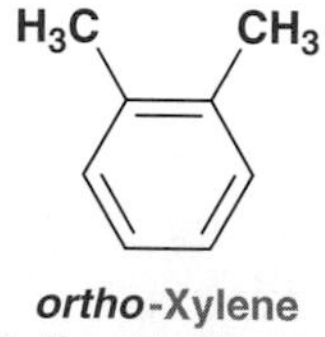
ortho-Xylene
(1,2-dimethylbenzene)

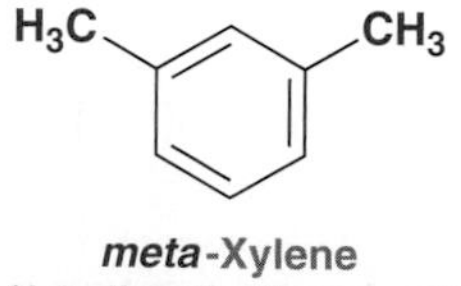
meta-Xylene
(1,3-dimethylbenzene)

para-Xylene
(1,4-dimethylbenzene)

These isomers differ from each other in the relative positions of the methyl groups and can be named in two ways: (1) using the descriptors ***ortho***, ***meta***, and ***para*** or (2) using locants (i.e., 1,3 is the same as *meta*). Both methods can be used when the parent is a common name:

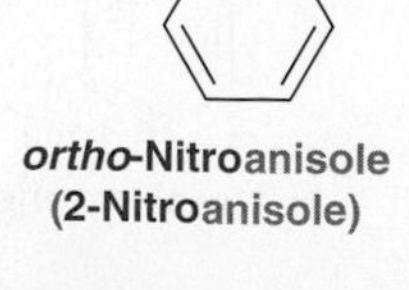
ortho-Nitroanisole
(2-Nitroanisole)

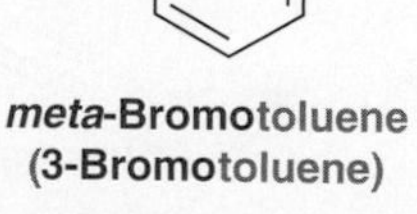
meta-Bromotoluene
(3-Bromotoluene)

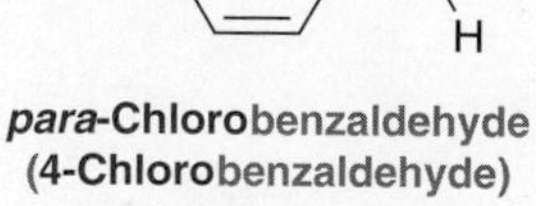
para-Chlorobenzaldehyde
(4-Chlorobenzaldehyde)

Polysubstituted Derivatives of Benzene

The descriptors *ortho*, *meta*, and *para* cannot be used when naming an aromatic ring bearing three or more substituents. In such a case, locants are required. That is, each substituent is designated with a number to indicate its location on the ring.

When naming a polysubstituted benzene ring, we will follow the same four-step process used for naming alkanes, alkenes, alkynes, and alcohols.

1. Identify and name the parent.
2. Identify and name the substituents.
3. Assign a locant to each substituent.
4. Arrange the substituents alphabetically.

When identifying the parent, it is acceptable (and common practice) to choose a common name. Consider the following example.

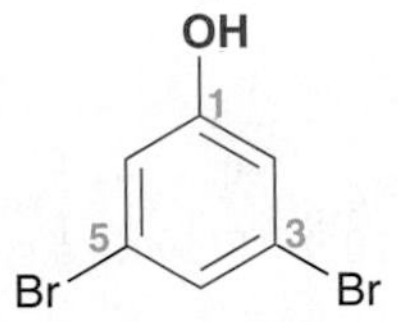

3,5-Dibromophenol

This compound could certainly be named as a trisubstituted benzene ring. However, it is much more efficient to name the parent as phenol rather than benzene and to list the two bromine atoms as substituents. The choice to name this compound as a phenol dictates that the carbon atom connected to the OH group must receive the lowest locant (number 1). When a choice exists, place numbers so that the second substituent receives the lower possible number.

Correct

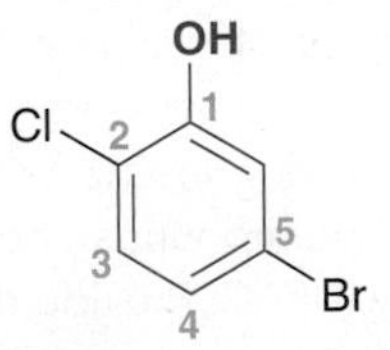

5-Bromo-2-chlorophenol

Incorrect

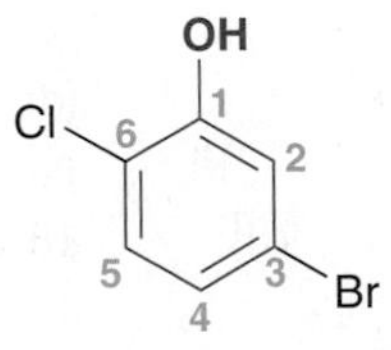

3-Bromo-6-chlorophenol

When arranging the name, make sure to alphabetize the substituents. In the example above, *b*romo is listed before *c*hloro.

SKILLBUILDER

18.1 NAMING A POLYSUBSTITUTED BENZENE

LEARN the skill

Provide a systematic name for TNT, a well-known explosive with the following molecular structure.

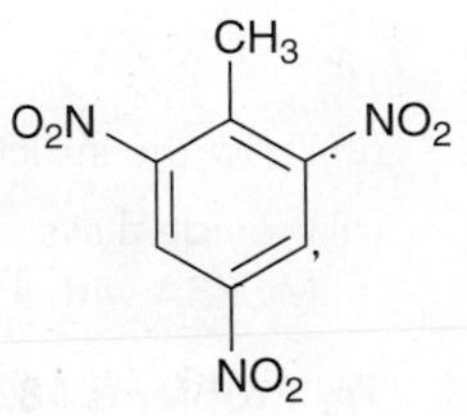

SOLUTION

STEP 1 Identify and name the parent.

Apply the four-step method. First, identify and name the parent. In this case, we choose toluene as the parent (shown in red).

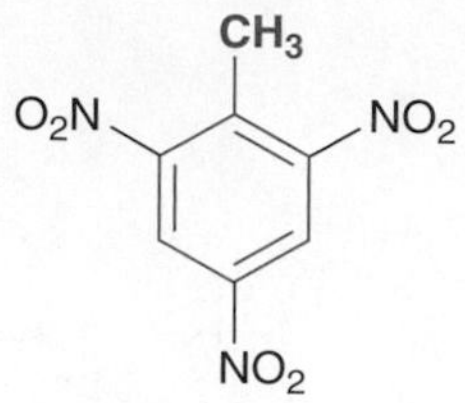

STEPS 2 AND 3 Identify the substituents and assign locants.

Next, identify the substituents and assign locants. The decision to name this compound as a trisubstituted toluene (rather than a tetrasubstituted benzene) dictates that the carbon atom bearing the methyl group be assigned the lowest locant (number 1). In this case, the remaining locants can be assigned either clockwise or counterclockwise because the result is the same.

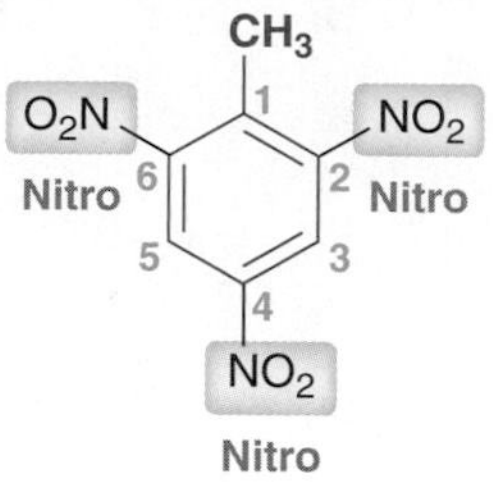

STEP 4 Arrange the substituents alphabetically.

Finally, arrange the substituents alphabetically. In this case, there are three nitro groups. Make sure to indicate all three numbers in the name.

2,4,6-Trinitrotoluene

PRACTICE the skill **18.1** Provide a systematic name for each of the following compounds.

(a) 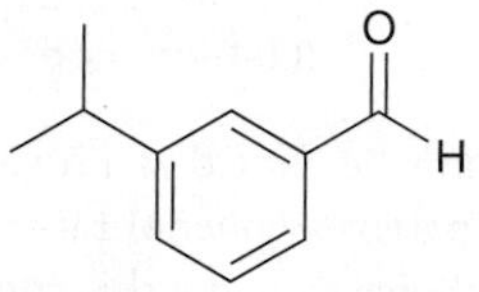

(b) 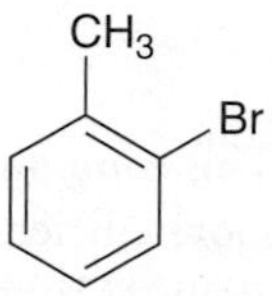

(c) 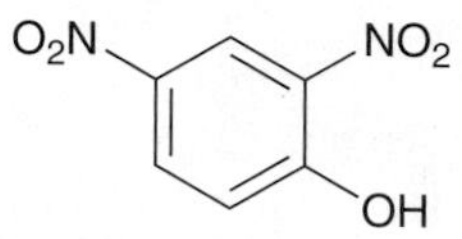

(d) 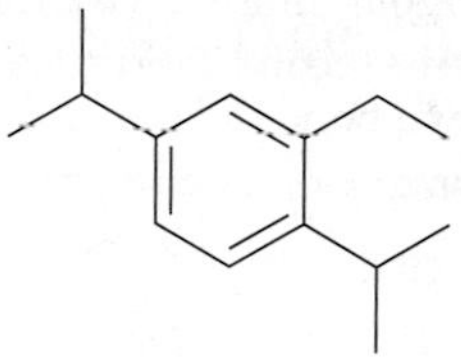

(e) 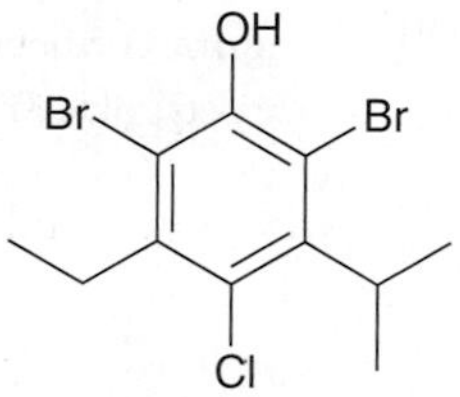

APPLY the skill

18.2 Aromatic compounds often have multiple names that are all accepted by IUPAC. Provide three different systematic (IUPAC) names for the following compound.

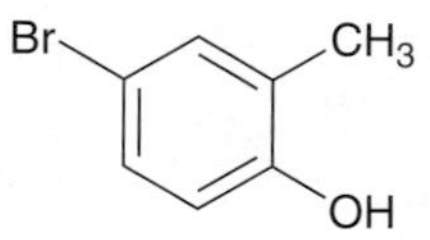

18.3 For each of the following compounds, draw its structure.

(a) 2,6-Dibromo-4-chloroanisole (b) *meta*-Nitrophenol

18.4 Provide at least five different acceptable IUPAC names for the following compound.

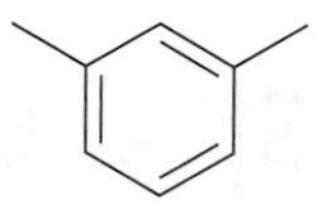

18.5 In Chapter 9, we saw that *meta*-chloroperoxybenzoic acid (MCPBA) is a peroxy acid commonly used to convert alkenes into epoxides. Recall that peroxy acids have the following structure:

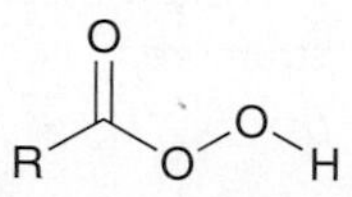

(a) Draw the structure of MCPBA.

(b) Provide a systematic name for the compound formed by replacing the chlorine atom in MCPBA with a methyl group.

need more **PRACTICE?** **Try Problems 18.28, 18.29, 18.33**

18.3 Structure of Benzene

In 1825, Michael Faraday isolated benzene from the oily residue left by illuminating gas in London street lamps. Further investigation showed that the molecular formula of this compound was C_6H_6: a hydrocarbon comprised of six carbon atoms and six hydrogen atoms.

In 1866, August Kekulé used his recently published structural theory of matter to propose a structure for benzene. Specifically, he proposed a ring comprised of alternating double and single bonds.

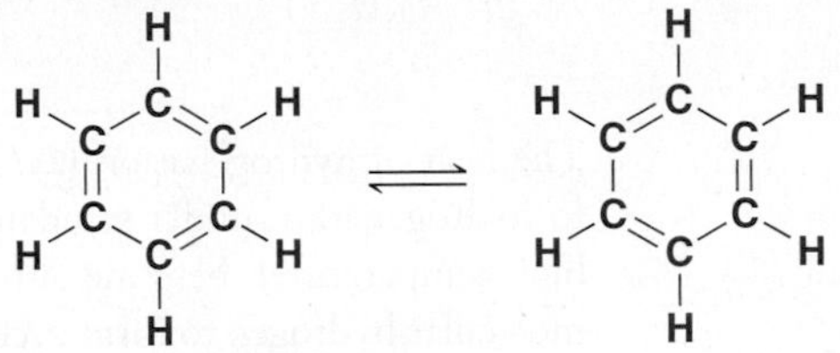

Kekulé described the exchange of double and single bonds to be an equilibrium process. Over time, this view was refined by the advent of resonance theory and molecular orbital concepts of delocalization. The two drawings above are now viewed as resonance structures, not as an equilibrium process.

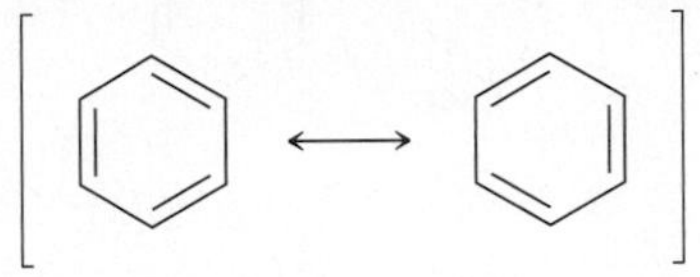

Recall that resonance does not describe the motion of electrons, but rather, resonance is the way that chemists deal with the inadequacy of bond-line drawings. Specifically, each drawing alone is inadequate to describe the structure of benzene. The problem is that each C—C bond is neither a single bond nor a double bond, nor is it vibrating back and forth between these two states. Instead, each C—C bond has a bond order of 1.5, exactly midway between a single bond and a double bond. To avoid drawing resonance structures, benzene is often drawn like this:

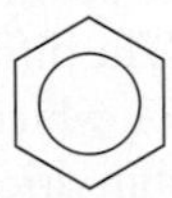

This type of drawing appears often in the literature and will likely appear on the chalkboard of your organic chemistry lecture hall. Nevertheless, these drawings should be avoided when proposing reaction mechanisms, which require scrupulous bookkeeping of electrons. Throughout the remainder of this textbook, we will draw benzene rings as alternating single and double bonds (Kekulé structures).

18.4 Stability of Benzene

Evidence for Unusual Stability

There is much evidence that the aromatic moiety is particularly stable, much more so than expected. Below, we will explore two pieces of evidence for this stability.

Recall from Chapter 9 that alkenes readily undergo addition reactions, such as the addition of bromine to form a dihalide.

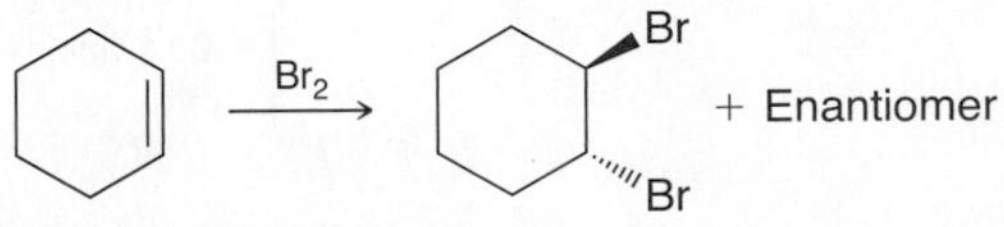

We might therefore expect benzene to undergo a similar reaction, perhaps even three times. But in fact, no reaction is observed.

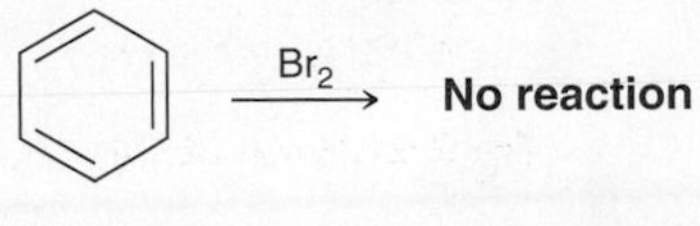

The aromatic moiety apparently exhibits a special stability that would be lost if an addition reaction were to occur. The stability of the aromatic moiety is also observed when comparing heats of hydrogenation for several similar compounds. Recall that hydrogenation of a π bond proceeds in the presence of a metal catalyst.

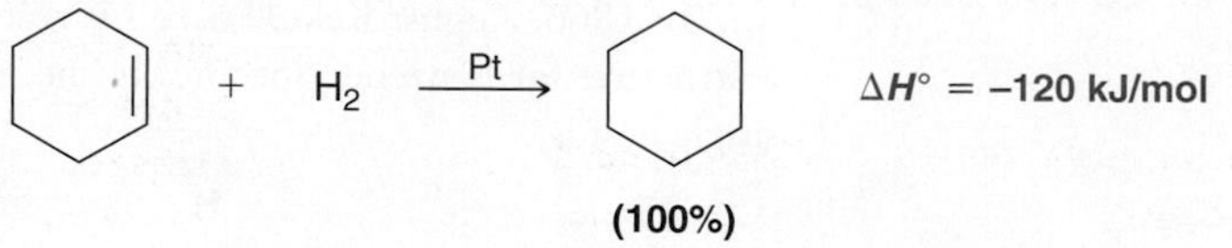

The heat of hydrogenation (ΔH) for this reaction is −120 kJ/mol. Benzene is generally stable to hydrogenation under standard conditions, but under forcing conditions (high pressure and high temperature), benzene also undergoes hydrogenation and reacts with three equivalents of molecular hydrogen to form cyclohexane. Therefore, we might expect $\Delta H = -360$ kJ/mol. But, in fact, the heat of hydrogenation for hydrogenation of benzene is only −208 kJ/mol.

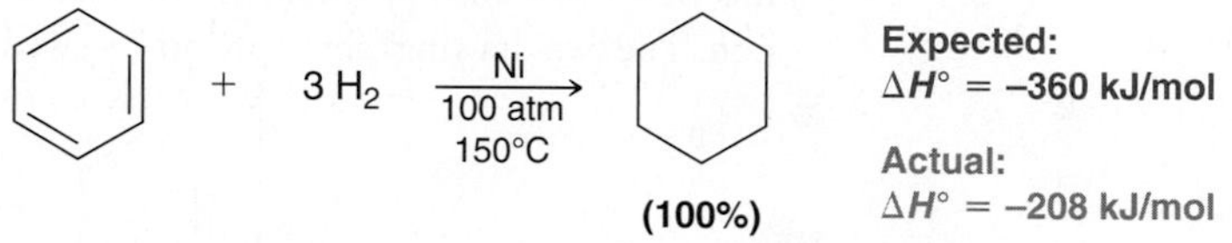

In other words, benzene is much more stable than expected. This information is graphically represented on the energy diagram in Figure 18.1. This energy diagram shows the relative heats of hydrogenation for cyclohexene, 1,3-cyclohexadiene, and benzene. Notice that the heat of hydrogenation of 1,3-cyclohexadiene is not quite twice the value of the heat of hydrogenation of cyclohexene. This is attributed to conjugation. Extending this logic to an imaginary cyclohexatriene molecule, we would expect that the heat of hydrogenation would be just below −360 kJ/mol. In fact, the heat of hydrogenation for benzene is only −208 kJ/mol. The difference between the expected value (−360) and the observed value (−208) is 152 kJ/mol, which is called the **stabilization energy** of benzene. This value represents the amount of stabilization associated with aromaticity.

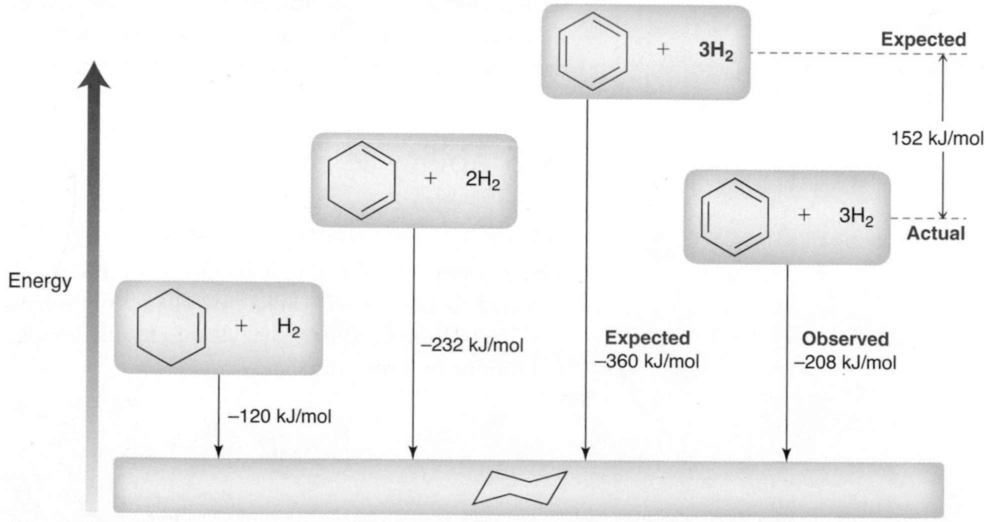

FIGURE 18.1
An energy diagram comparing the heats of hydrogenation for cyclohexene, cyclohexadiene, and benzene.

CONCEPTUAL CHECKPOINT

18.6 Compound **A** has molecular formula C_8H_8. When treated with excess Br_2, compound **A** is converted into compound **B**, with molecular formula $C_8H_8Br_2$. Identify the structures of compounds **A** and **B**.

$$\text{Compound A } (C_8H_8) \longrightarrow \text{Compound B } (C_8H_8Br_2)$$

18.7 In some circumstances, dehydrogenation is observed. Dehydrogenation involves the loss of two hydrogen atoms (the reverse of hydrogenation). Analyze each of the following dehydrogenation reactions and then use the information in Figure 18.1 to predict whether each transformation will be downhill in energy (negative value for ΔH) or uphill in energy (positive value for ΔH).

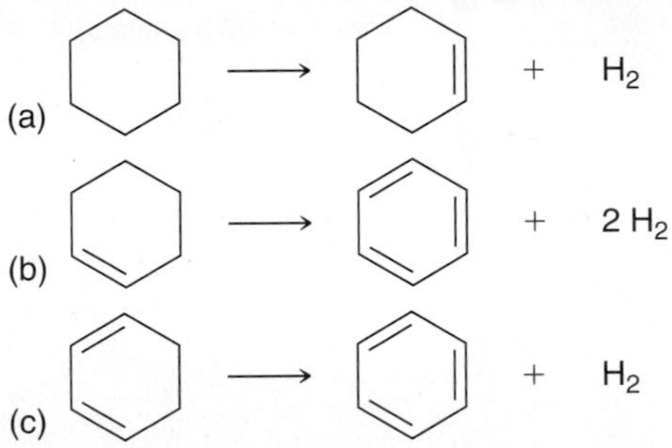

Source of Stability

In order to explain the stability of benzene, we will need to invoke MO theory. Recall that, according to MO theory, a π bond results when two atomic *p* orbitals overlap to form two new orbitals, called molecular orbitals (Figure 18.2). The lower energy MO is the bonding MO, while the higher energy MO is the antibonding MO. Both π electrons occupy the bonding MO and thereby achieve a lower energy state.

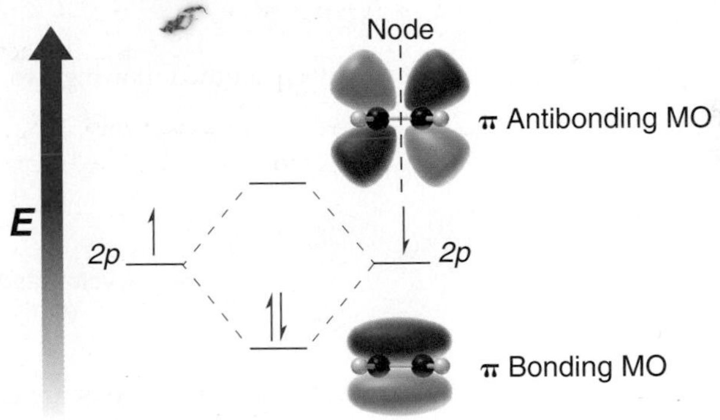

FIGURE 18.2
The molecular orbitals of a π bond.

Now let's explore how MO theory describes the nature of benzene, which is comprised of six overlapping *p* orbitals (Figure 18.3). According to MO theory, these six atomic *p* orbitals are replaced with six molecular orbitals (Figure 18.4). The precise shape and energy level of each MO are determined by sophisticated mathematics that is beyond the scope of our current discussion. For now, let's focus on the important concepts that are illustrated by the energy diagram in Figure 18.4.

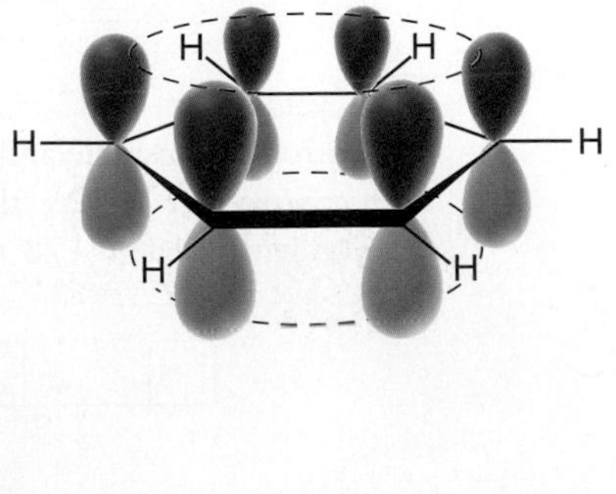

FIGURE 18.3
The *p* orbitals of benzene overlap continuously around the ring.

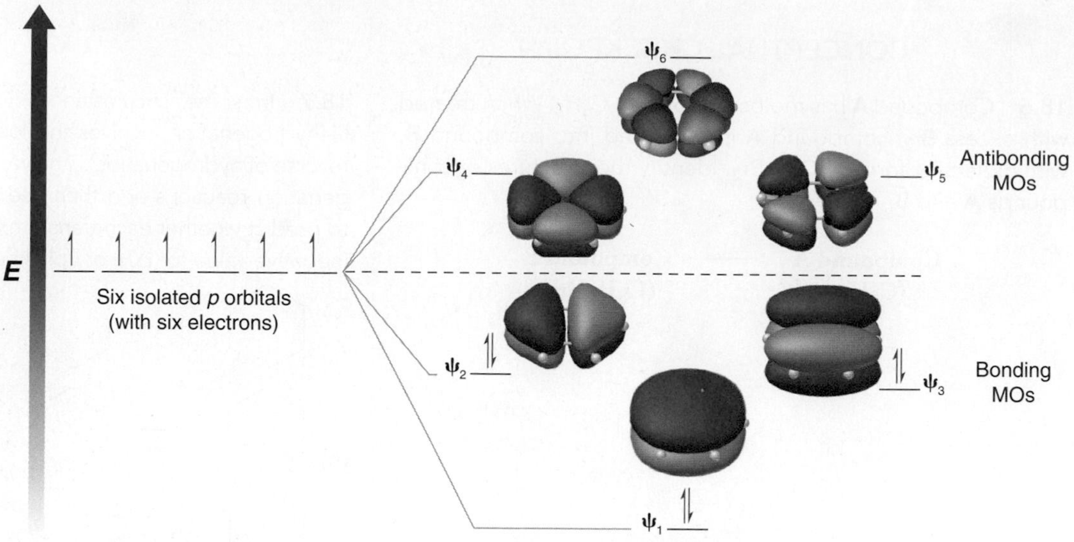

FIGURE 18.4 The molecular orbitals of benzene.

- There are six molecular orbitals, each of which is associated with the entire molecule (rather than being associated with any specific bond).
- Three of the six MOs (those below the dashed line) are bonding MOs, while the other three MOs (those above the dashed line) are antibonding MOs.
- Since each MO can contain two electrons, the three bonding MOs can collectively accommodate up to six π electrons.
- By occupying the bonding MOs, all six electrons achieve a lower energy state and are said to be delocalized. This is the source of the stabilization energy associated with benzene.

Hückel's Rule

We might expect the following two compounds to exhibit aromatic stabilization like benzene.

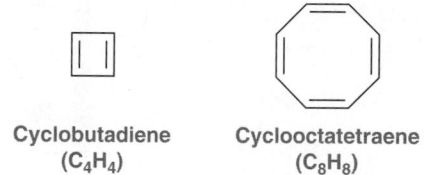

BY THE WAY
Willstätter was awarded the 1915 Nobel Prize in Chemistry for his research on elucidating the structure of chlorophyll.

After all, they are similar to benzene in that each compound is comprised of a ring of alternating single and double bonds. Cyclooctatetraene (C_8H_8) was first isolated by Richard Willstätter in 1911. The reactivity of this compound suggests that cyclooctatetraene does not exhibit the same stability exhibited by benzene. For example, the compound readily undergoes addition reactions, such as bromination.

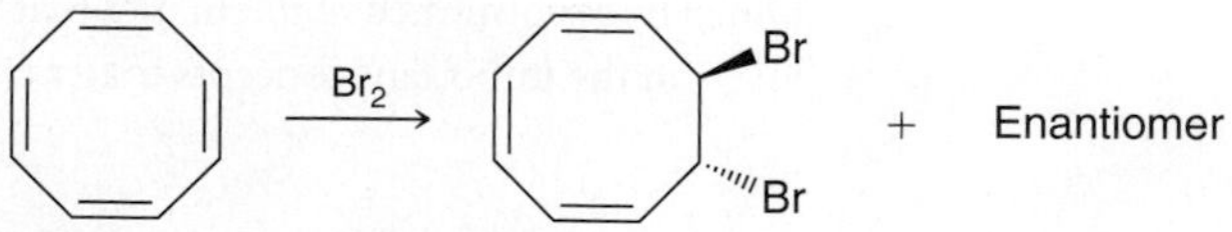

Cyclobutadiene (C_4H_4) also does not exhibit aromatic stability. In fact, quite the opposite. It is extremely unstable and resisted all attempts to prepare it until the second half of the twentieth century (see the Practically Speaking box that follows). Cyclobutadiene is so unstable that it reacts with itself at −78°C in a Diels-Alder reaction.

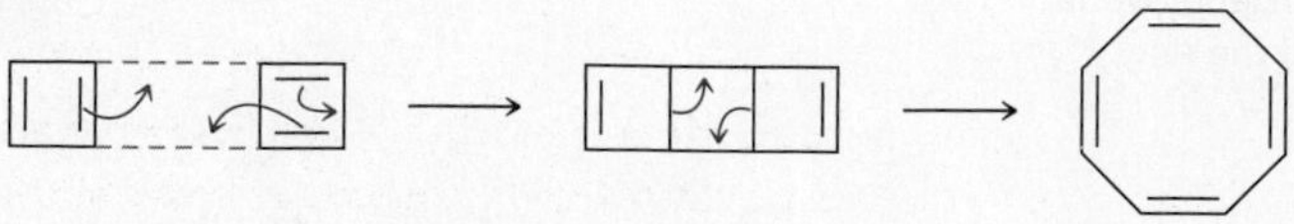

The Diels-Alder reaction generates an initial tricyclic product that is also unstable and rapidly rearranges to form cyclooctatetraene.

These observations indicate that the presence of a fully conjugated ring of π electrons is not the sole requirement for aromaticity; the number of π electrons in the ring is also important. Specifically, we have seen that an odd number of electron pairs is required for aromaticity.

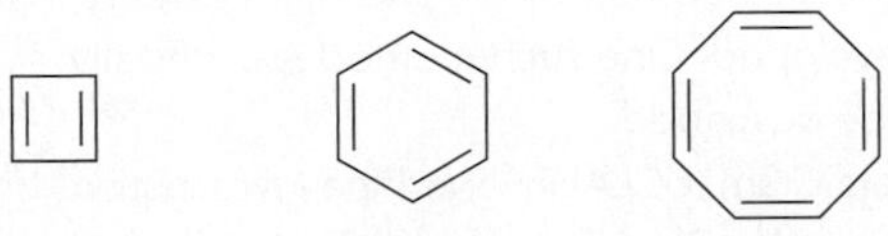

2 pairs of π electrons — 3 pairs of π electrons — 4 pairs of π electrons

Benzene has a total of six π electrons, or *three* electron pairs. With an odd number of electron pairs, benzene is aromatic. In contrast, cyclobutadiene and cyclooctatetraene each have an even number of electron pairs (two pairs and four pairs, respectively) and do not exhibit aromatic stabilization.

The requirement for an odd number of electron pairs is called **Hückel's rule**. Specifically, a compound can only be aromatic if the number of π electrons in the ring is 2, 6, 10, 14, 18, and so on. This series of numbers can be expressed mathematically as $4n+2$, where n is a whole number. In the upcoming section, we will use MO theory to explain why aromatic stabilization requires an odd number of electron pairs ($4n+2$ electrons).

CONCEPTUAL CHECKPOINT

18.8 Predict whether each of the following compounds should be aromatic.

(a)
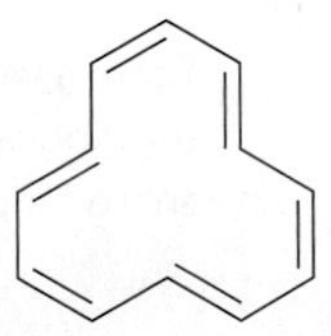

(b)
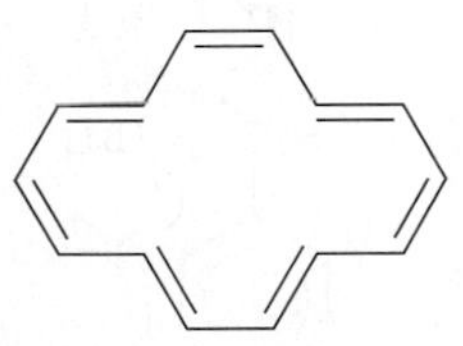

(c)

MO Theory and Frost Circles

In this section, we will use MO theory to explain the source of Hückel's rule for planar, conjugated ring systems. We already analyzed the MOs of benzene (Figure 18.4) to explain its stability. Let's now turn our attention to the MOs of cyclobutadiene (C_4H_4) and cyclooctatetraene (C_8H_8) to see how MO theory can explain the observed instability of these compounds. We will start with square cyclobutadiene, which is a system constructed from the overlap of four atomic p orbitals (Figure 18.5).

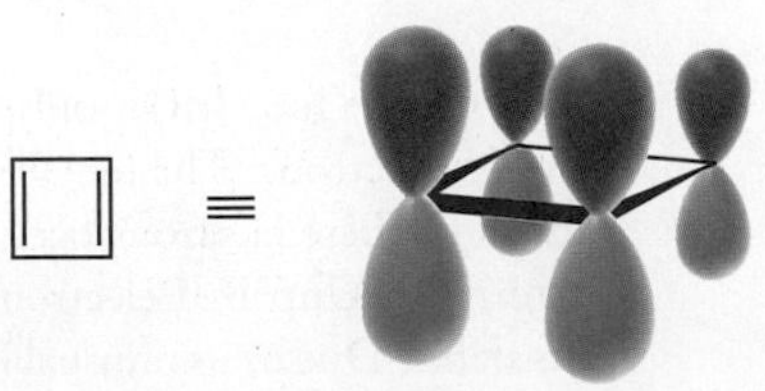

FIGURE 18.5 The p orbitals that are used in cyclobutadiene.

According to MO theory, these four atomic orbitals are replaced with four molecular orbitals, with the relative energy levels shown in Figure 18.6.

PRACTICALLYSPEAKING

Molecular Cages

Early attempts at preparing cyclobutadiene were met with failure again and again. In the second half of the twentieth century, several methods were developed. One such method is especially fascinating and will now be described.

The research of Donald Cram (UCLA) involved the investigation of cagelike molecules. Specifically, Cram prepared molecules shaped like hemispheres and then linked them together to form a cage:

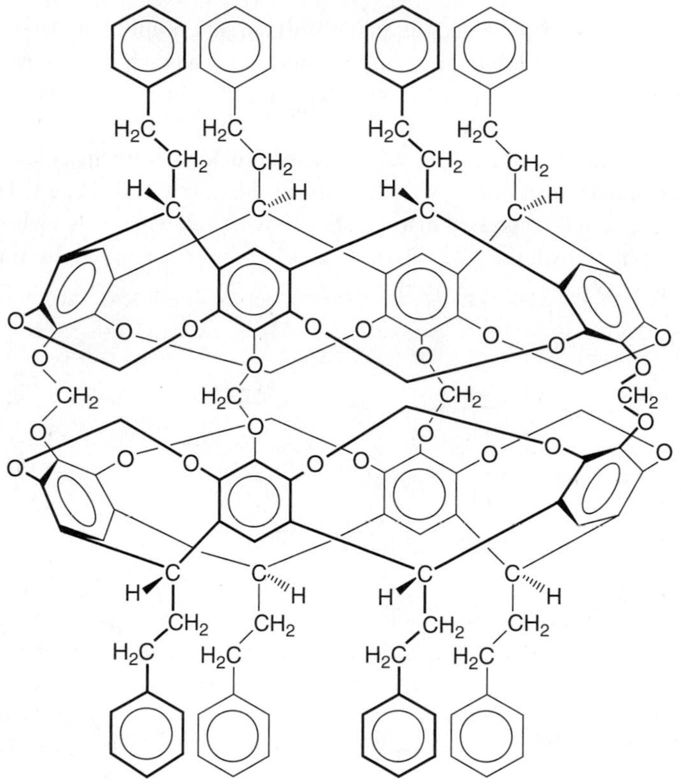

Cram then used these cages to trap α-pyrone inside:

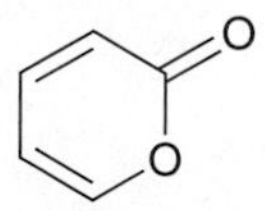

α-Pyrone

When irradiated with UV light, α-pyrone was known to liberate CO_2 to form cyclobutadiene:

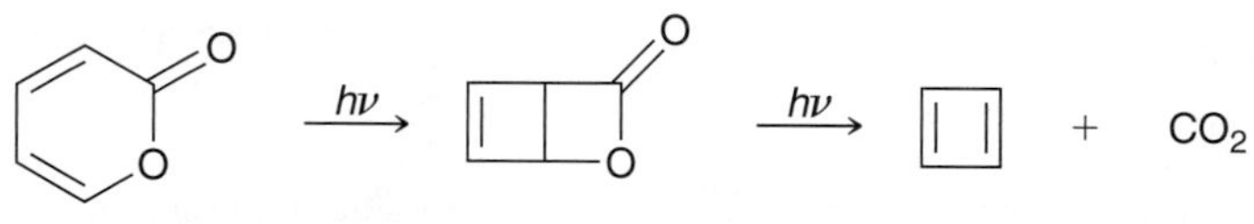

The CO_2 molecules generated by this process were then able to escape through the spaces between the linkers holding the hemispheres together. However, cyclobutadiene is too large to escape through the holes in the cage. As a result, a single cyclobutadiene molecule would be trapped and unable to react with other molecules of cyclobutadiene. This extremely clever method for imprisoning cyclobutadiene worked beautifully and enabled its spectroscopic analysis. The many studies that followed provided strong evidence that cyclobutadiene rapidly equilibrates between two rectangular forms:

Cram's work with cagelike molecules, called *carcerands*, opened the doorway to a new field of chemistry. For his contributions, he shared a portion of the 1987 Nobel Prize in Chemistry.

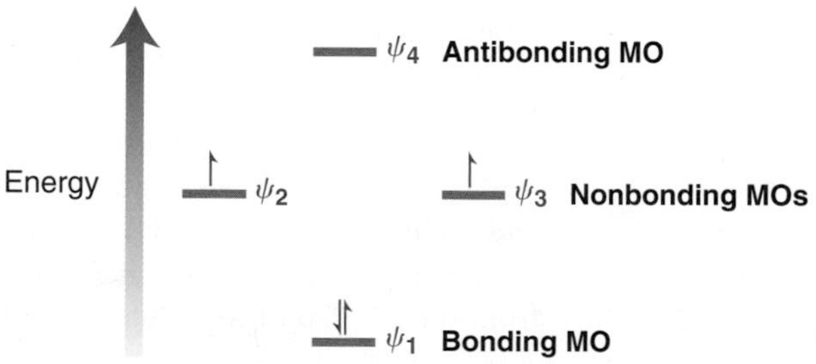

FIGURE 18.6 The relative energy levels of the molecular orbitals of cyclobutadiene.

Of the four MOs, only one MO is a bonding MO, and it can accommodate no more than two π electrons. The remaining two π electrons must go into the degenerate nonbonding MOs. Each of these electrons occupies a different MO (according to Hund's rule), so each electron is unpaired. Unpaired electrons are very reactive, which explains why square cyclobutadiene is so unstable. Due to its unusual instability, this compound is said to be **antiaromatic**. Cyclobutadiene can alleviate some if its instability by adopting a rectangular shape (as described in the previous Practically Speaking box), in which the compound functions as two isolated π bonds, rather than a diradical. Nevertheless, cyclobutadiene is still extremely unstable and highly reactive.

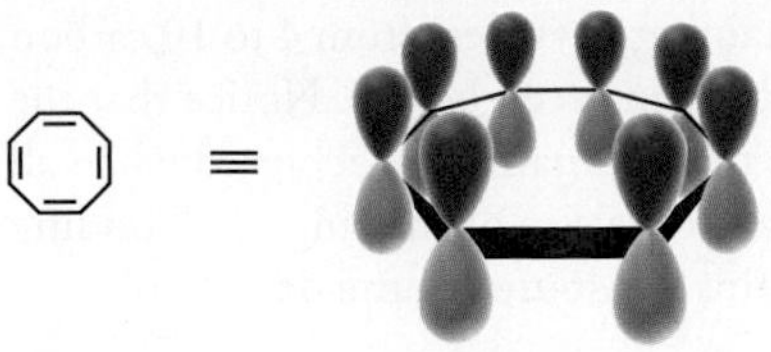

FIGURE 18.7
In a planar conformation, the *p* orbitals of cyclooctatetraene overlap continuously around the ring.

Now consider how MO theory treats planar cyclooctatetraene, which is a system constructed from the overlap of eight atomic *p* orbitals (Figure 18.7). According to MO theory, these eight atomic orbitals are replaced with eight molecular orbitals, with the relative energy levels shown in Figure 18.8. When we fill these MOs with eight π electrons, we encounter a situation similar to that which we encountered with square cyclobutadiene—specifically, two of the electrons are unpaired and occupy nonbonding orbitals. Once again, this electronic configuration leads to remarkable instability, and the compound should be antiaromatic. But the compound can avoid this unnecessary instability by assuming a tub shape (Figure 18.9). In this conformation, the eight *p* orbitals no longer overlap as one continuous system, and the energy diagram in Figure 18.8 (of the eight MOs) simply does not apply. Rather, there are four isolated π bonds, each of which contains two electrons in a bonding MO (as shown in Figure 18.2).

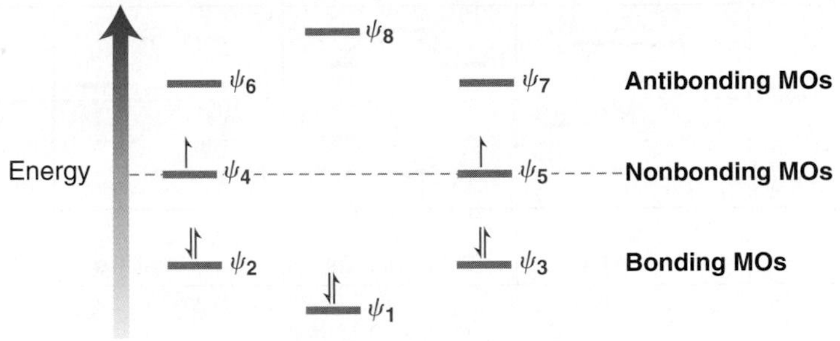

FIGURE 18.8
The relative energy levels of the molecular orbitals of cyclooctatetraene.

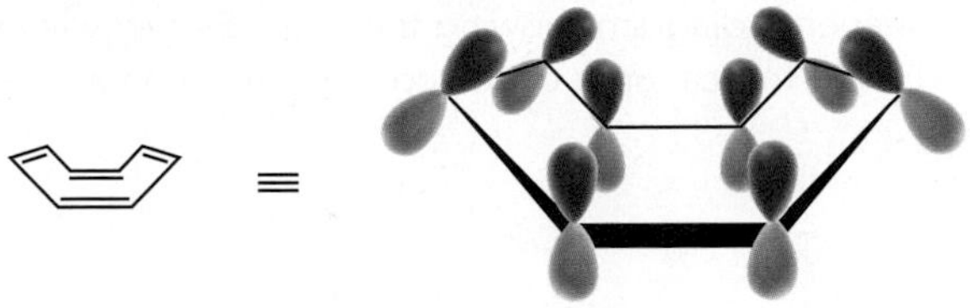

FIGURE 18.9
The tub-shaped structure of cyclooctatetraene.

The explanation above is predicated on our ability to draw the relative energy levels of the MOs, and as we already mentioned, that ability requires sophisticated mathematics that is beyond the scope of our course. Without getting heavily into the mathematics of MO theory, there is a simple method that will enable us to predict and draw the relative energy levels for any conjugated ring system. This method is illustrated with Figure 18.10, in which we analyze a seven-membered ring comprised of continuous, overlapping *p* orbitals.

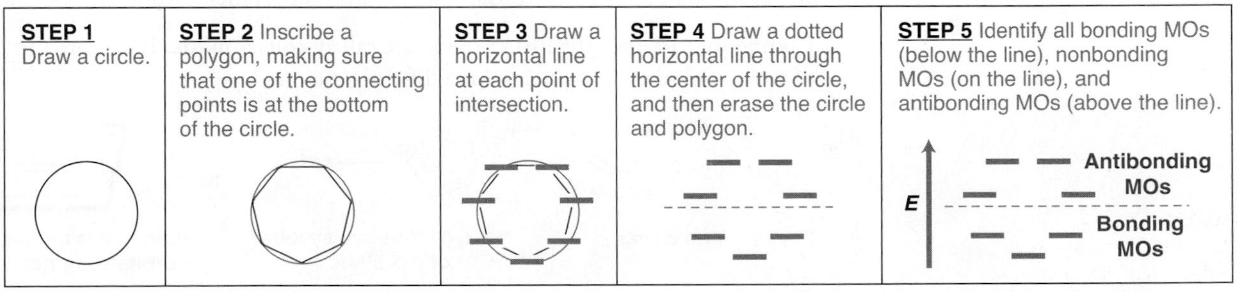

STEP 1 Draw a circle.
STEP 2 Inscribe a polygon, making sure that one of the connecting points is at the bottom of the circle.
STEP 3 Draw a horizontal line at each point of intersection.
STEP 4 Draw a dotted horizontal line through the center of the circle, and then erase the circle and polygon.
STEP 5 Identify all bonding MOs (below the line), nonbonding MOs (on the line), and antibonding MOs (above the line).
E
Antibonding MOs
Bonding MOs

FIGURE 18.10
A method for determining the relative energy levels of the molecular orbitals for a ring system comprised of continuous, overlapping *p* orbitals.

Begin by drawing a circle and inscribing a seven-membered polygon inside the circle, with one of the connecting points of the polygon at the bottom of the circle. Each location where the polygon touches the circle represents an energy level. All energy levels in the bottom half of the circle are the bonding MOs, and all energy levels in the top half of the circle are antibonding MOs. This method is extremely helpful, because the relative distances between the energy levels do in fact represent the relative difference in energy between the MOs. This method was first developed by Arthur Frost (Northwestern University), and it is referred to as a **Frost circle**.

If we accept that Frost circles accurately predict the relative energy levels of the MOs in a conjugated ring system, then we can use Frost circles to gain a better understanding of the

$4n+2$ rule. Consider the Frost circles for ring systems containing anywhere from 4 to 10 carbon atoms (Figure 18.11). Look at the bonding MOs (highlighted in green boxes). Notice that the number of bonding MOs is always odd (1, 3, or 5). Recall that aromaticity is achieved when all of the π electrons are paired in bonding MOs. Depending on the size of the ring, the bonding MOs will accommodate 2, 6, or 10 π electrons. This explains the source of the $4n+2$ rule.

FIGURE 18.11
Frost circles for different-size ring systems.

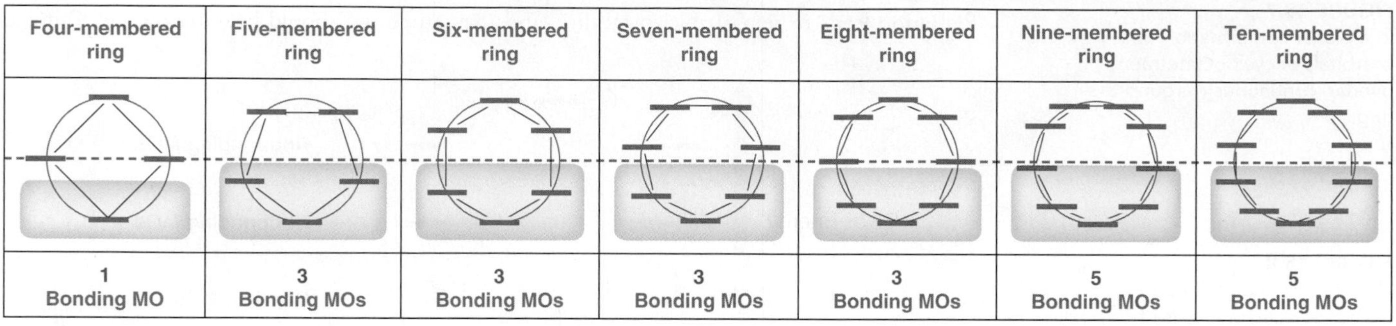
Four-membered ring
Five-membered ring
Six-membered ring
Seven-membered ring
Eight-membered ring
Nine-membered ring
Ten-membered ring
1 Bonding MO
3 Bonding MOs
3 Bonding MOs
3 Bonding MOs
3 Bonding MOs
5 Bonding MOs
5 Bonding MOs

CONCEPTUAL CHECKPOINT

18.9 The cyclopropenyl cation has a three-membered ring that contains a continuous system of overlapping *p* orbitals. This system contains a total of two π electrons. Using a Frost circle, draw an energy diagram showing the relative energy levels of all three MOs, and then predict whether this cation is expected to exhibit aromatic stabilization.

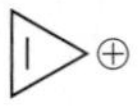

18.5 Aromatic Compounds Other than Benzene

The Criteria for Aromaticity

Benzene is not the only compound that exhibits aromatic stabilization. A compound will be aromatic if it satisfies the following two criteria.

1. The compound must contain a ring comprised of continuously overlapping *p* orbitals.
2. The number of π electrons in the ring must be a Hückel number.

Compounds that fail the first criterion are called **nonaromatic**. Below are three examples, each of which fails the first criterion for a different reason.

Not a ring

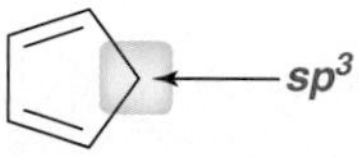

Not a continuous system of *p* orbitals

Molecule is not planar, so the *p* orbitals are not overlapping

The first compound (1,3,5-hexatriene) is not aromatic, because it does not possess a ring. The second compound is not aromatic because the ring is not a continuous system of *p* orbitals (there is an intervening sp^3-hybridized carbon atom). The third compound is not planar, and the *p* orbitals do not effectively overlap.

Compounds that satisfy the first criterion but have $4n$ electrons (rather than $4n+2$) are antiaromatic. In practice, there are very few examples of antiaromatic compounds, because most compounds can change their geometry to avoid being antiaromatic, just as we saw in the case of cyclooctatetraene.

In the following sections, we will explore several examples of compounds that meet both criteria and are therefore aromatic.

Annulenes

Annulenes are compounds consisting of a single ring containing a fully conjugated π system.

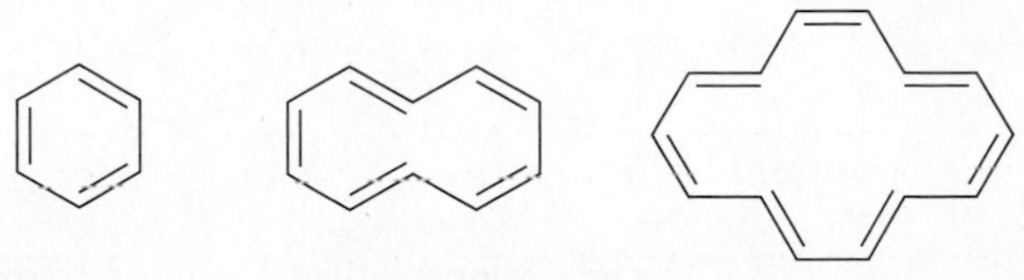

[6]Annulene [10]Annulene [14]Annulene

[6]Annulene is benzene, which is aromatic. We might also expect [10]annulene to be aromatic since it is a ring with a continuous system of *p* orbitals and it has a Hückel number of π electrons. However, the hydrogen atoms positioned inside the ring (shown in red) experience steric hindrance that forces the compound out of planarity:

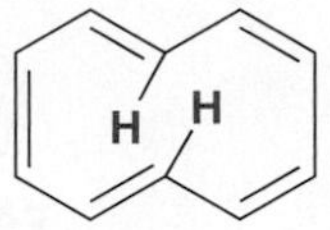
H H

Since the molecule cannot adopt a planar conformation, the *p* orbitals cannot continuously overlap with each other to form one system, and as a result, [10]annulene does not meet the first criterion for aromaticity. It is nonaromatic. In a similar way, [14]annulene is also somewhat destabilized by steric hindrance, and is also nonplanar. In contrast, [18]annulene is aromatic, becauses it satisfies both criteria for aromaticity.

CONCEPTUAL CHECKPOINT

18.10 Predict whether the following compound will be aromatic, nonaromatic, or antiaromatic. Explain your reasoning.

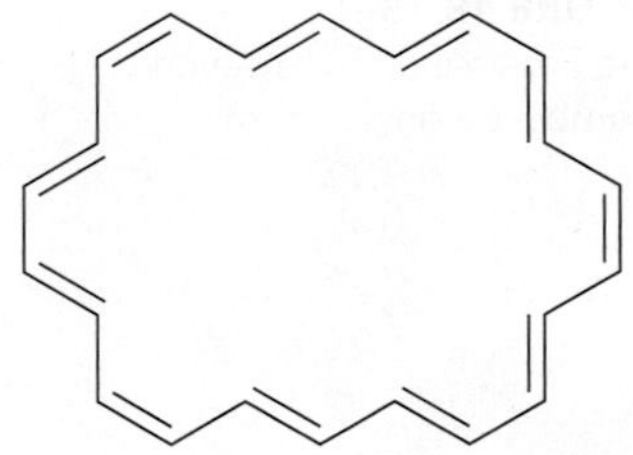

Aromatic Ions

Previously, we used MO theory and Frost circles to explain the requirement for a Hückel number of π electrons. Let's now take a closer look at the Frost circle of a five-membered ring (Figure 18.12). A five-membered ring will be aromatic if it contains six π electrons (a Hückel number). In order

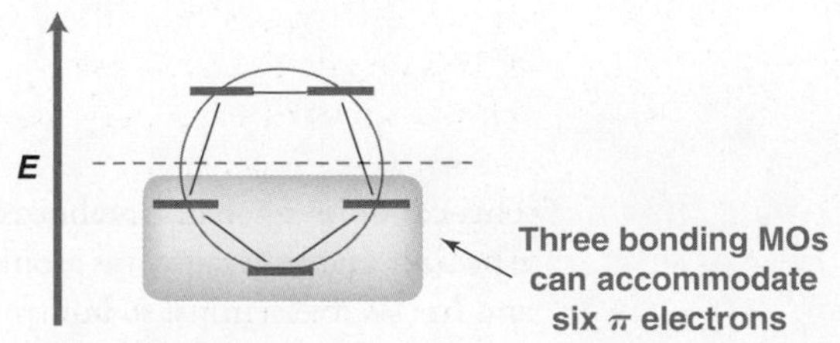
E
Three bonding MOs can accommodate six π electrons

FIGURE 18.12 The Frost circle for a five-membered ring system.

to have six π electrons, one of the carbon atoms must possess two electrons (a carbanion). The resulting anion, called the cyclopentadienyl anion, has five resonance structures.

The cyclopentadienyl anion is resonance stabilized, but that alone does not explain the observed stability. This anion is especially stable because it is aromatic. The lone pair occupies a *p* orbital (Section 2.12), and as a result, the cyclopentadienyl anion has a continuous system of overlapping *p* orbitals. With six π electrons, this anion fulfills both criteria for aromaticity. The stabilization energy of the cyclopentadienyl anion explains the remarkable acidity of cyclopentadiene.

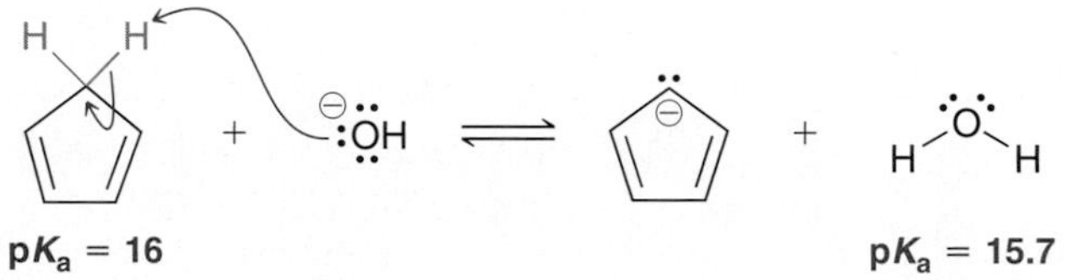

Notice that the pK_a of cyclopentadiene is very similar to the pK_a of water (around 16), which is very unusual for a hydrocarbon (the pK_a of cyclopentane is >50). The acidity of cyclopentadiene is attributed to the stability of its conjugate base, which is aromatic.

Now let's explore the Frost circle of a seven-membered ring (Figure 18.13). Once again, exactly six π electrons are necessary to achieve aromaticity.

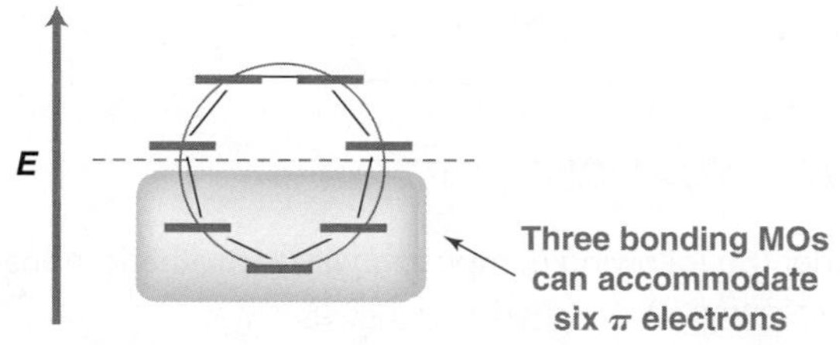

FIGURE 18.13
The Frost circle for a seven-membered ring system.

In order to have six π electrons in a seven-membered ring, one of the carbon atoms must possess an empty *p* orbital (a carbocation). The resulting ion is called the tropylium cation, and it has seven resonance structures.

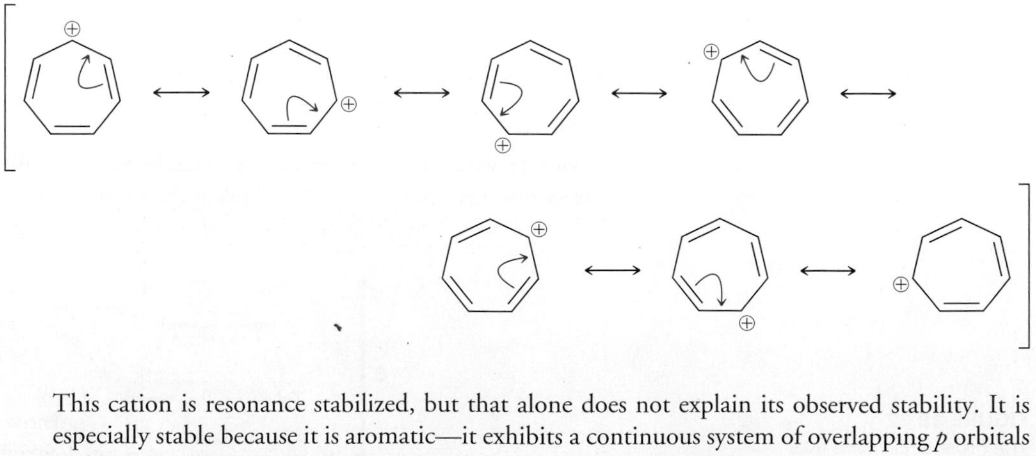

This cation is resonance stabilized, but that alone does not explain its observed stability. It is especially stable because it is aromatic—it exhibits a continuous system of overlapping *p* orbitals and has six π electrons, so both criteria for aromaticity are satisfied.

SKILLBUILDER

18.2 DETERMINING WHETHER A COMPOUND IS AROMATIC, NONAROMATIC, OR ANTIAROMATIC

LEARN the skill

Determine whether the following compound is aromatic, nonaromatic, or antiaromatic.

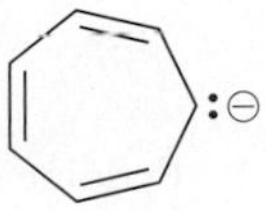

SOLUTION

To determine if this anion is aromatic, we must ask two questions:

1. Does the compound contain a ring comprised of continuously overlapping *p* orbitals?
2. Is there a Hückel number of π electrons in the ring?

The answer to the first question appears to be yes, that is, the lone pair can occupy a *p* orbital, providing for continuous overlap of *p* orbitals around the ring. However, when we try to answer the second question, we discover that this anion would have eight π electrons, which would render the compound antiaromatic. As such, the geometry of the anion will change to avoid the instability associated with antiaromaticity. Specifically, the lone pair can occupy an sp^3-hybridized orbital (rather than a *p* orbital), so the first criterion is no longer met. This renders the compound nonaromatic.

PRACTICE the skill

18.11 Determine whether each of following ions is aromatic, nonaromatic, or antiaromatic.

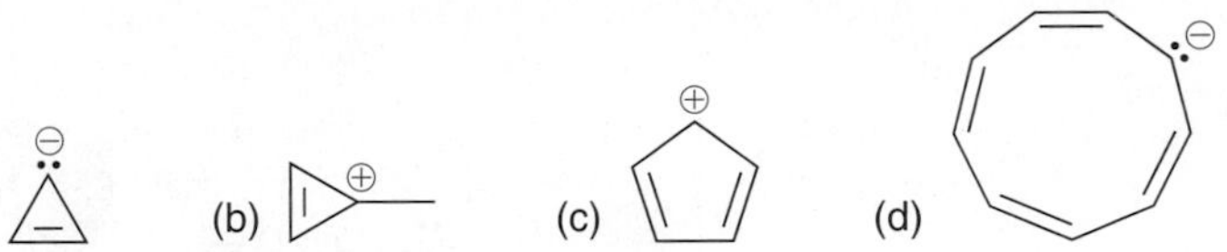

APPLY the skill

18.12 Explain the vast difference in pK_a values for the following two apparently similar compounds.

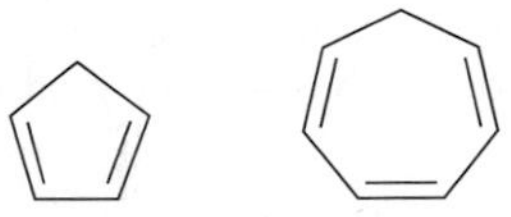

pK_a = 16 pK_a = 36

18.13 Predict which compound will react more readily in an S_N1 process, and explain your choice.

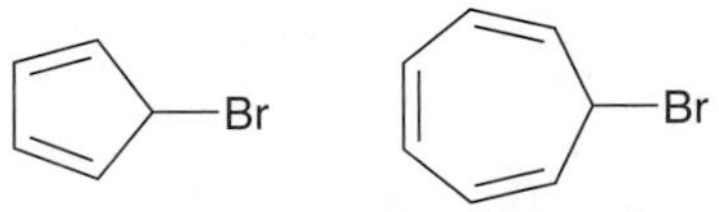

18.14 Identify which of the following compounds is more acidic and explain your choice.

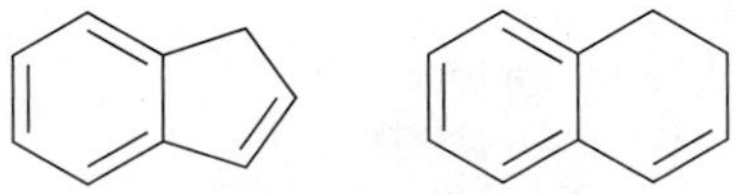

need more **PRACTICE?** **Try Problems 18.34a,c,e, 18.36e, 18.52**

Aromatic Heterocycles

Cyclic compounds containing heteroatoms (such as S, N, or O) are called **heterocycles.** Below are two examples of nitrogen-containing heterocycles.

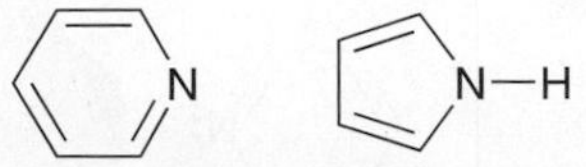

Pyridine **Pyrrole**

Both of these compounds are aromatic, but for very different reasons. We will discuss them first separately, and then compare them to each other.

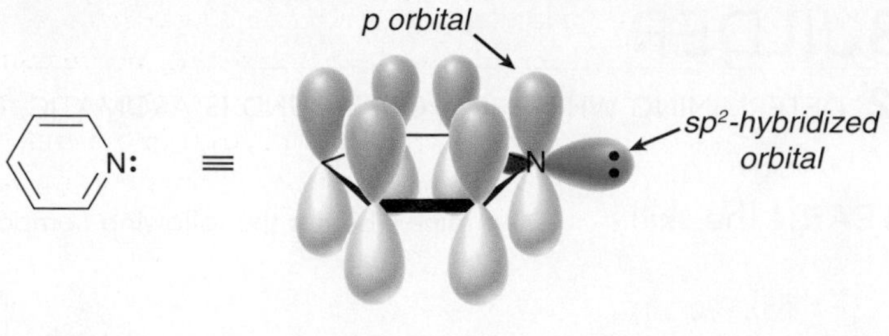

FIGURE 18.14
The atomic *p* orbitals of pyridine that overlap to give an aromatic system. The lone pair on nitrogen occupies an sp^2-hybridized orbital.

Pyridine exhibits a continuous system of overlapping *p* orbitals (Figure 18.14) and therefore satisfies the first criterion for aromaticity. The nitrogen atom is sp^2 hybridized, and the lone pair on the nitrogen atom occupies an sp^2-hybridized orbital, which is pointing away from the ring. This lone pair is not part of the conjugated system and therefore is not included when we count the number of π electrons. In this case, there are six π electrons, so the compound is aromatic. Since the lone pair of pyridine does not participate in resonance or aromaticity, it is free to function as a base:

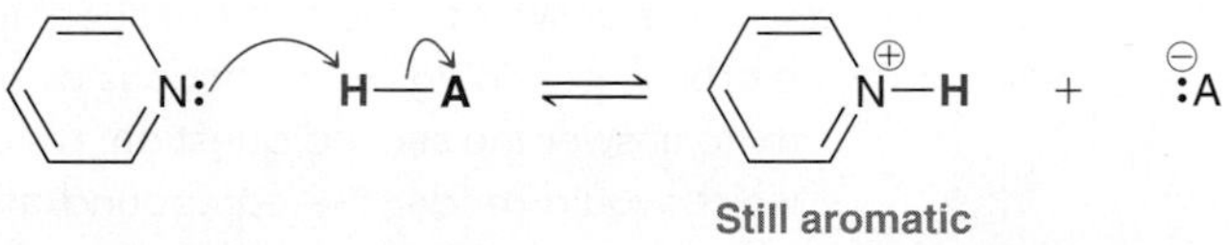

Pyridine can function as a base because protonation of the nitrogen atom does not destroy aromaticity.

Now let's analyze the structure of pyrrole. Once again, the nitrogen atom is sp^2 hybridized, but in this case, the lone pair occupies a *p* orbital (Figure 18.15). In order for the pyrrole ring

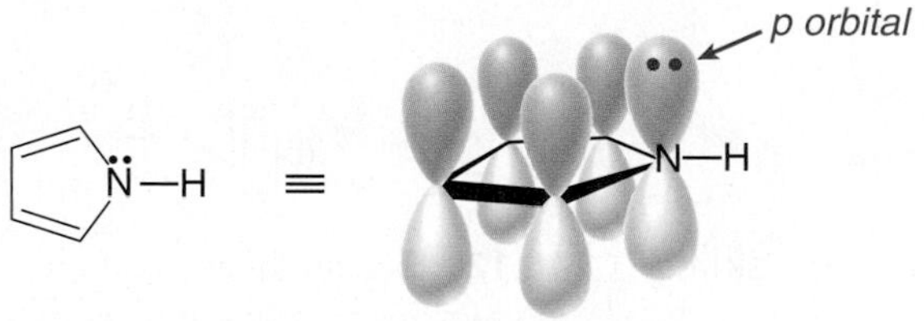

FIGURE 18.15
The atomic *p* orbitals of pyrrole that overlap to give an aromatic system. The lone pair on nitrogen occupies one of the *p* orbitals.

to achieve a continuous system of overlapping *p* orbitals, the lone pair of the nitrogen atom must occupy a *p* orbital. With six π electrons (four from the π bonds and two from the lone pair), this compound is aromatic. In this case, the lone pair is crucial in establishing aromaticity. The delocalized nature of the lone pair explains why pyrrole does not readily function as a base. Protonation of the lone pair on nitrogen effectively destroys the *p* orbital overlap, thereby destroying aromaticity.

LOOKING BACK
Recall from Section 2.12 that a lone pair occupies a *p* orbital if it participates in resonance.

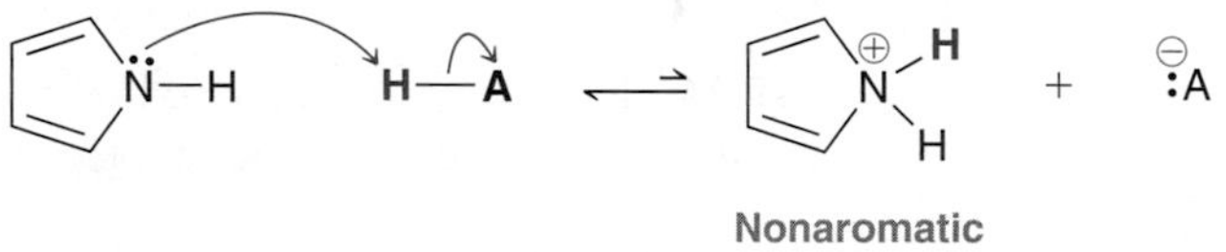

The difference between pyridine and pyrrole can be seen in their electrostatic potential maps (Figure 18.16). In general, it is not safe to compare electrostatic potential maps throughout this book, because the color scale on the maps might differ. However, in this case, the same color scale was used to generate both electrostatic potential maps, which allows us to compare

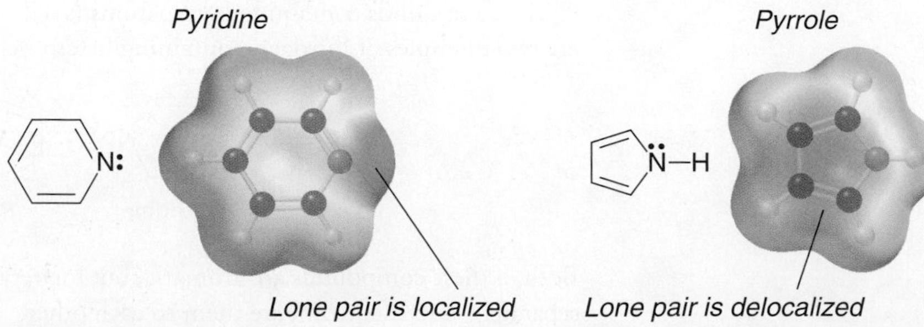

FIGURE 18.16
Electrostatic potential maps of pyridine and pyrrole.

the location of electron density in the two compounds. The map of pyridine shows a high concentration of electron density on the nitrogen atom, which represents the localized lone pair. That lone pair is not part of the aromatic system and is therefore free to function as a base. In contrast, the map of pyrrole indicates that the lone pair has been delocalized into the ring, and the nitrogen atom does not represent a site of high electron density.

SKILLBUILDER

18.3 DETERMINING WHETHER A LONE PAIR PARTICIPATES IN AROMATICITY

LEARN the skill

Histamine is responsible for many physiological responses and is known to mediate the onset of allergic reactions.

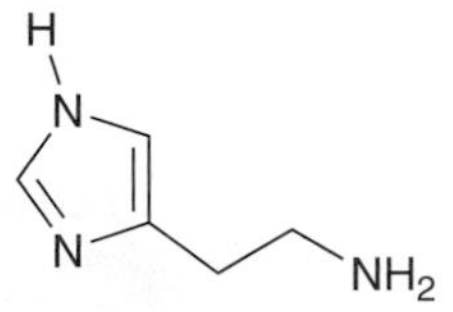

Histamine

Histamine has three lone pairs. Determine which lone pair(s) participate in aromaticity.

SOLUTION

The lone pair on the side chain is certainly not participating in aromaticity, because it is not situated in a ring. Let's focus on the two nitrogen atoms in the ring, beginning with the nitrogen in the upper left corner.

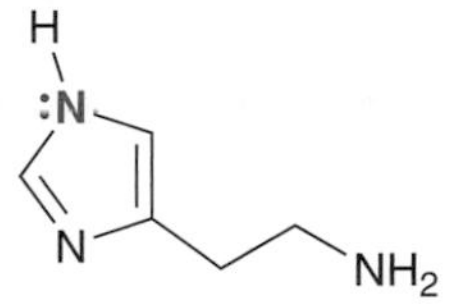

This lone pair must occupy a *p* orbital in order to have a continuous system of overlapping *p* orbitals. If this nitrogen were sp^3 hybridized, then the compound could not be aromatic. This lone pair is part of the aromatic system. Now look at the rest of the ring, and count the π electrons:

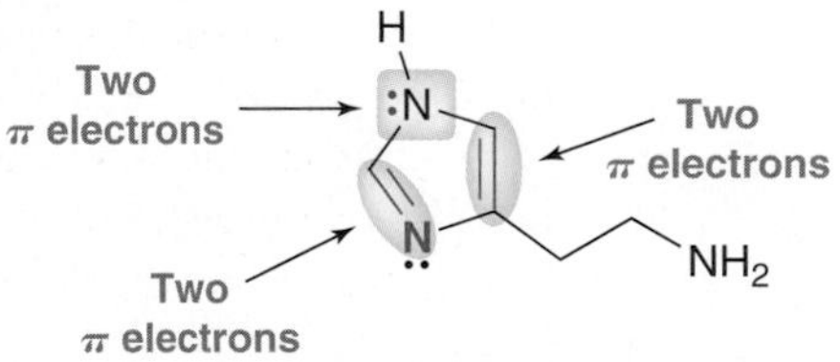

There are six π electrons, without including the lone pair on the other nitrogen atom.

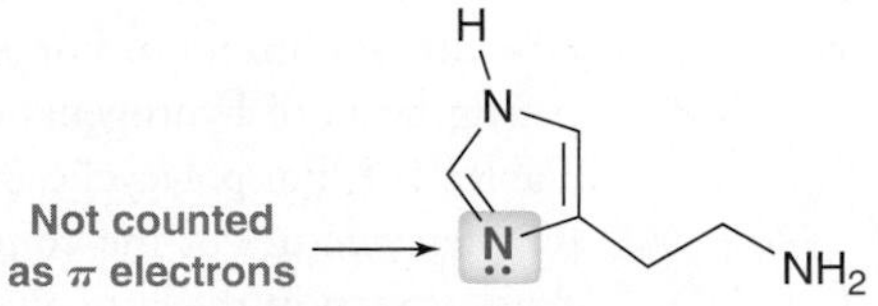

This lone pair cannot be included in the count, because each atom in the ring can only have one *p* orbital overlapping with the *p* orbitals of the ring. In this case, the nitrogen atom already has a *p* orbital from the double bond, and therefore, the lone pair cannot also occupy a *p* orbital.

PRACTICE the skill **18.15** For each of the following compounds determine which (if any) lone pairs are participating in aromaticity.

(a) 

(b) 

(c) 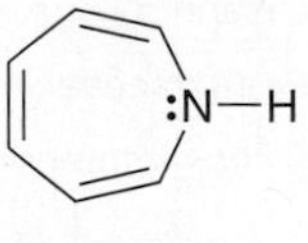

(d) 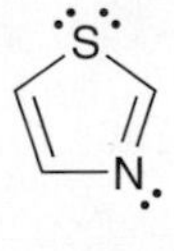

(e) 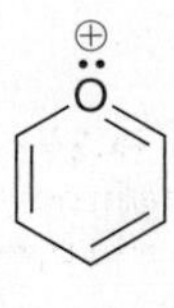

(f) 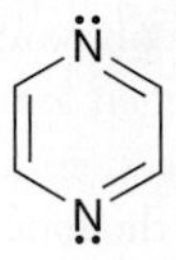 (g) 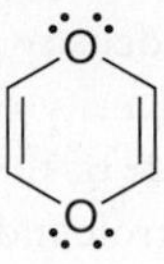 (h) 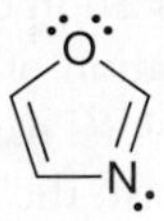

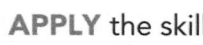

APPLY the skill

18.16 Go to the beginning of Section 18.1 where 8 best-selling drugs were shown. Review the structures of those compounds, and identify all of the aromatic rings that are not already highlighted in red.

18.17 Identify which compound is expected to have a lower pK_a. Justify your choice.

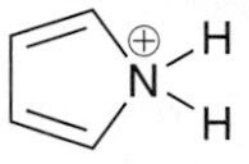 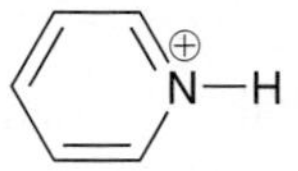

need more **PRACTICE?** **Try Problems 18.34b,d, 18.36c,d, 18.38, 18.41, 18.44, 18,62, 18.64**

Polycyclic Aromatic Compounds

Hückel's rule ($4n+2$) can only be applied to compounds that exhibit a single ring of overlapping *p* orbitals (monocyclic compounds). Nevertheless, many **polycyclic aromatic hydrocarbons (PAHs)** are known to be stable and have been thoroughly studied:

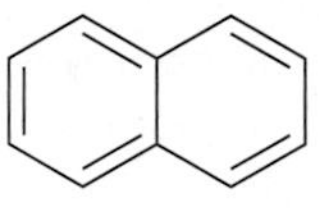

Naphthalene

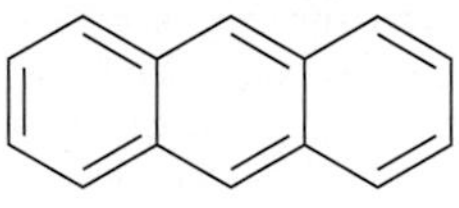

Anthracene

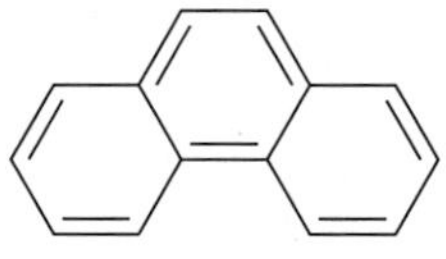

Phenanthrene

These compounds are comprised of fused benzene-like rings, and they exhibit significant aromatic stabilization. For each compound, the stabilization energy can be measured by comparing heats of hydrogenation, just as we did with benzene. These values are summarized in Table 18.1. For polycyclic aromatic hydrocarbons, the stabilization energy *per ring* is dependent on the structure of the compound and is generally less than the stabilization energy of benzene, as seen in the last column of Table 18.1. In addition, the individual rings of a polycyclic aromatic hydrocarbon will often exhibit different levels of stabilization. For example, the middle ring of anthracene is known to be more reactive than either of the outer rings.

TABLE 18.1 STABILIZATION ENERGY FOR A FEW POLYCYCLIC AROMATIC HYDROCARBONS

COMPOUND	STABILIZATION ENERGY (KJ/MOL)	AVERAGE STABILIZATION ENERGY PER RING (KJ/MOL)
Benzene	152	152
Naphthalene	255	128
Anthracene	347	116
Phenanthrene	381	127

MEDICALLYSPEAKING)))

The Development of Nonsedating Antihistamines

Histamine (as seen in SkillBuilder 18.3) is an aromatic compound that plays many roles in the biological processes of mammals. It is involved in the mediation of allergic reactions and the regulation of gastric acid secretion in the stomach. Extensive pharmacological studies reveal the existence of at least three different types of histamine receptors, called H_1, H_2, and H_3 receptors. The binding of histamine to H_1 receptors is responsible for triggering allergic reactions. Compounds that compete with histamine to bind with H_1 receptors (without triggering the allergic reaction) are called H_1 antagonists and belong to a class of compounds called antihistamines. Examples of H_1 antagonists include Benadryl™ and Chlortrimeton™.

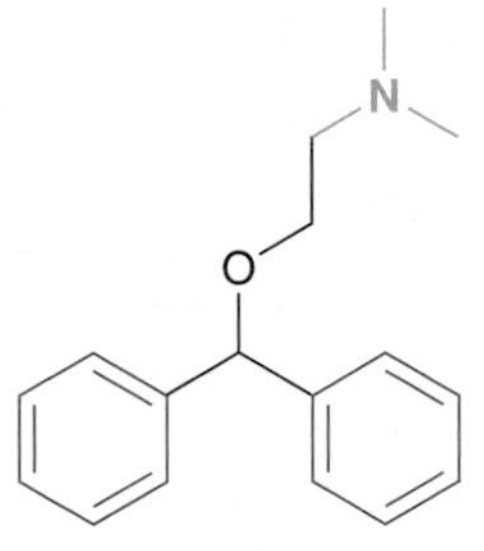

Benadryl™ (diphenhydramine)

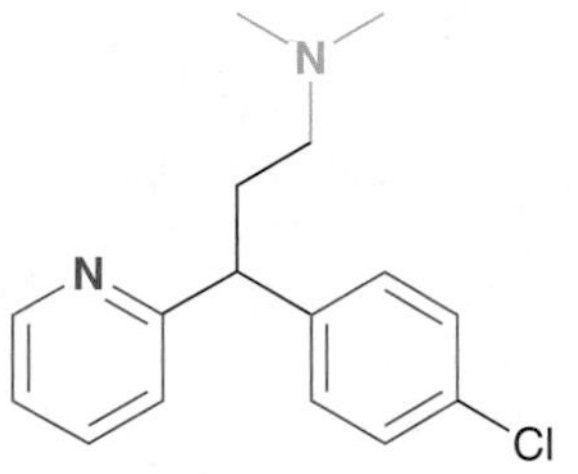

Chlortrimeton™ (chlorpheniramine)

Notice the structural similarities between these compounds. Both compounds exhibit two aromatic moieties, highlighted in red, as well as a tertiary amine (a nitrogen atom connected to three alkyl groups), highlighted in blue. Extensive research indicates that these structural features are necessary in order for a drug to exhibit H_1 antagonism. Studies also show that the two aromatic rings must be in close proximity (separated by either one or two carbon atoms). To understand this requirement, consider the structure of a simple compound that bears two phenyl groups separated by one carbon atom.

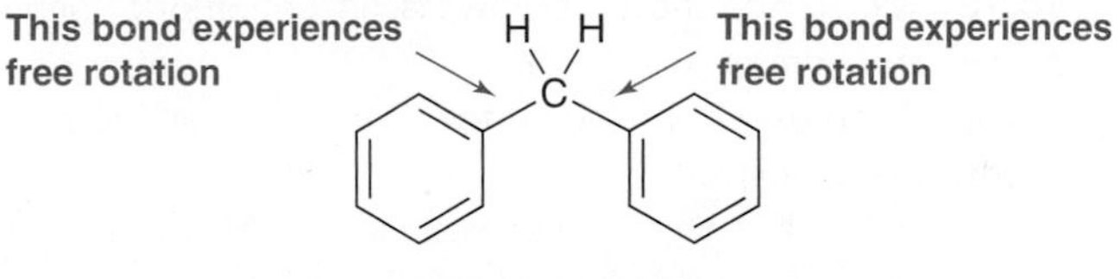

Diphenylmethane

The single bonds experience free rotation at room temperature, which enables the molecule to adopt different conformations. When the two rings are coplanar (when they lie in the same plane), there is significant steric hindrance between two of the aromatic hydrogen atoms.

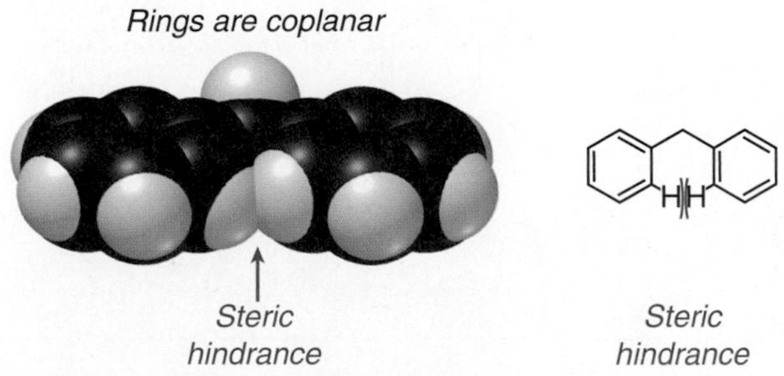

As a result, this conformation is very high in energy. The molecule can alleviate most of this strain by adopting a conformation in which the two rings are not coplanar. This noncoplanar conformation resembles a twisted waffle maker.

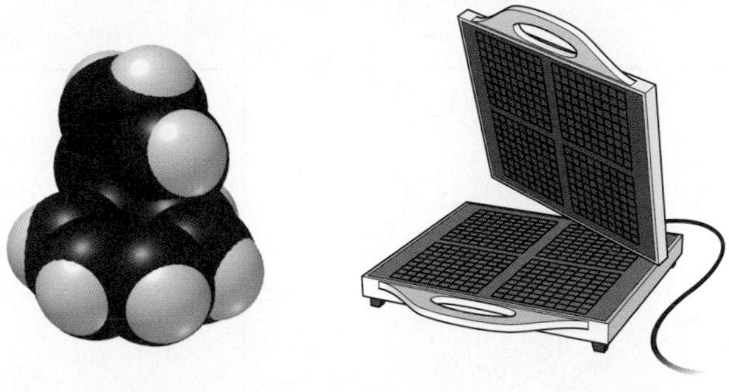

Rings are not coplanar Resembles a waffle maker

This conformation is much lower in energy than the coplanar conformation. As a result, a molecule that contains two closely situated aromatic rings will spend most of its time in a noncoplanar conformation.

Antihistamines possess this "waffle maker" structural feature. For example, the structure of Benadryl™ exhibits two aromatic rings, and these rings adopt a noncoplanar conformation. This feature is important, because it enables the molecule to experience significant binding with the H_1 receptor.

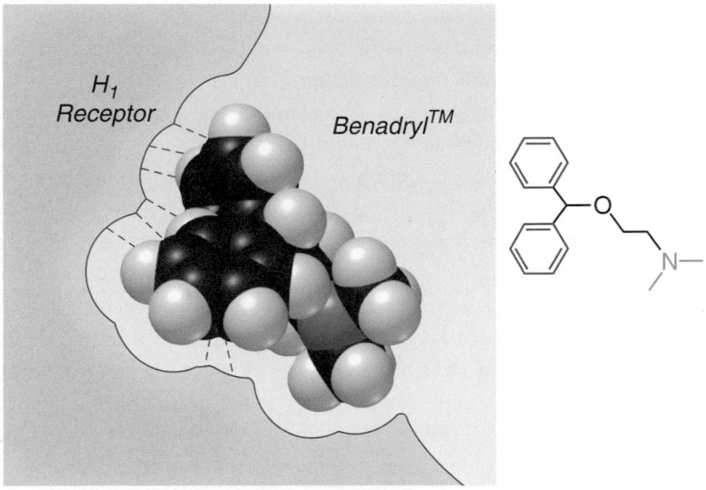

In a noncoplanar conformation, the two aromatic moieties achieve a combined binding force of approximately 40 kJ/mol (equivalent in strength to two hydrogen bonds). The nitrogen atom (shown in blue) also binds to the receptor. That interaction will be discussed in more detail in Chapter 23.

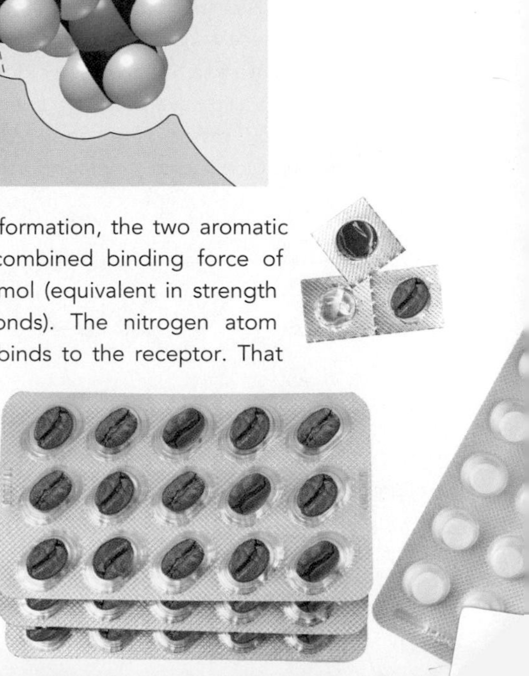

(Continued on the next page).

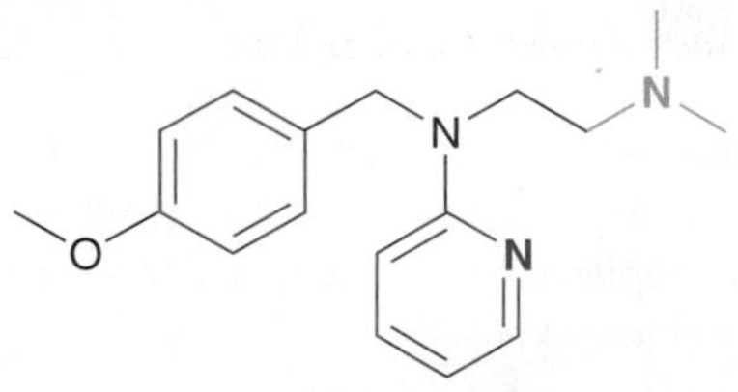
Pyrilamine

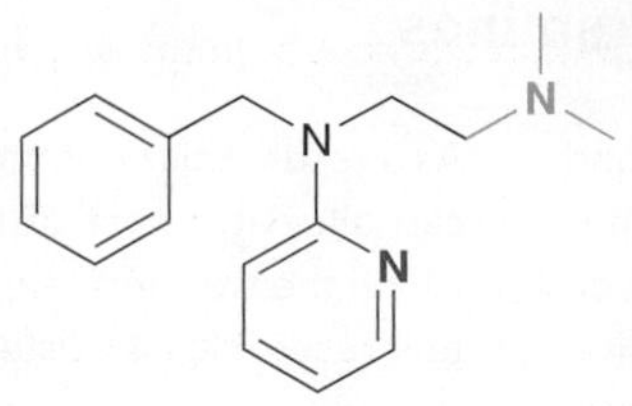
Tripelennamine

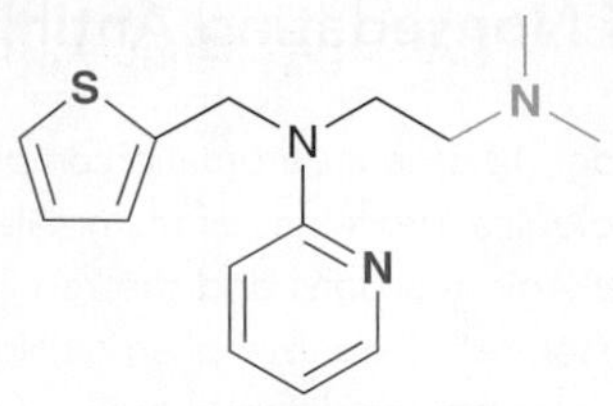
Methapyrilene

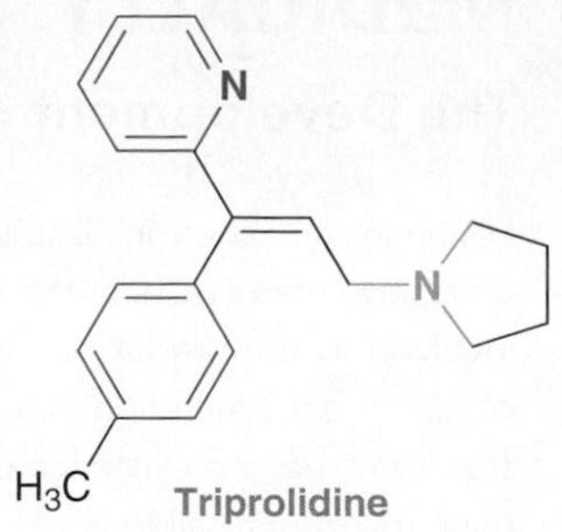
Triprolidine

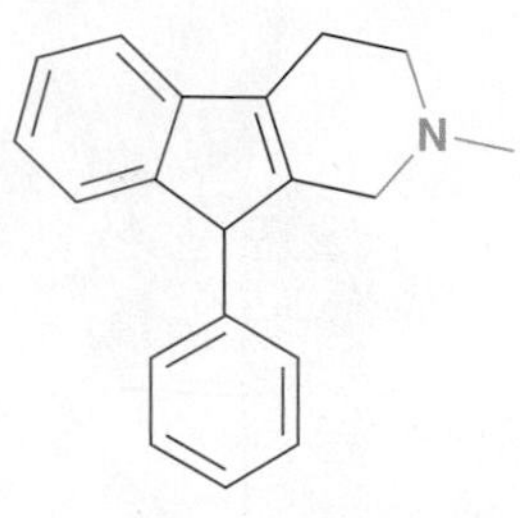
Phenindamine

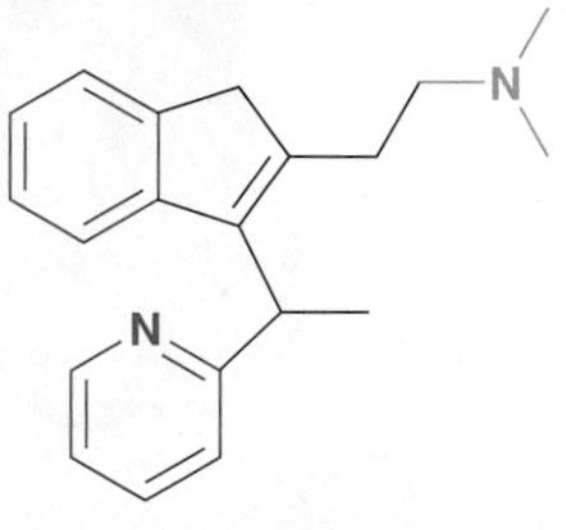
Dimethindene

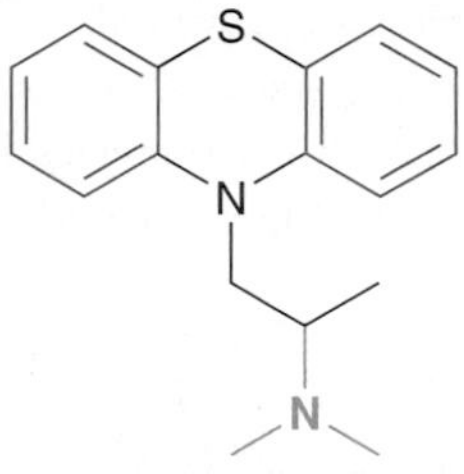
Promethazine

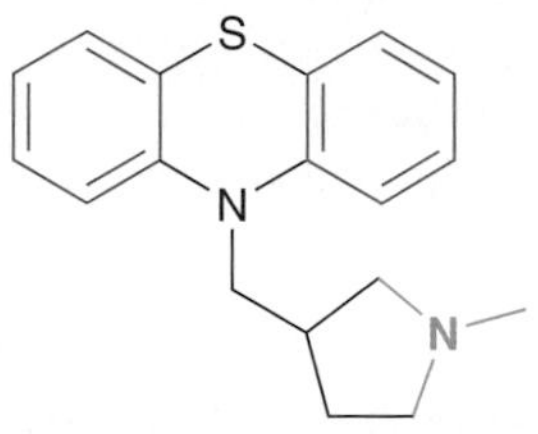
Methdilazine

The table shows a sampling of antihistamines. Notice that each structure contains two aromatic moieties in close proximity (red) as well as the tertiary amine (blue).

In addition to their ability to bind with the H_1 receptors, these antihistamines also bind with receptors in the central nervous system, causing undesired side effects such as sedation (drowsiness). All of these compounds, called *first-generation antihistamines,* cause these undesired side effects. The sedative effect of first-generation antihistamines has been attributed to their ability to cross the blood-brain barrier, allowing them to interact with receptors in the central nervous system.

One of the goals of antihistamine research over the last two decades was the development of antihistamines that could bind with H_1 receptors but could not readily cross the blood-brain barrier and therefore could not reach the receptors that trigger sedation. Extensive research has led to the development of several new drugs, called *second-generation antihistamines,* which are nonsedating. As we might expect, these drugs contain the pharmacophore necessary for H_1 antagonism (two aromatic rings in close proximity and a tertiary amine). However, second-generation antihistamines also have polar functional groups that prevent the compound from crossing the nonpolar environment of the blood-brain barrier. One such example is fexofenadine.

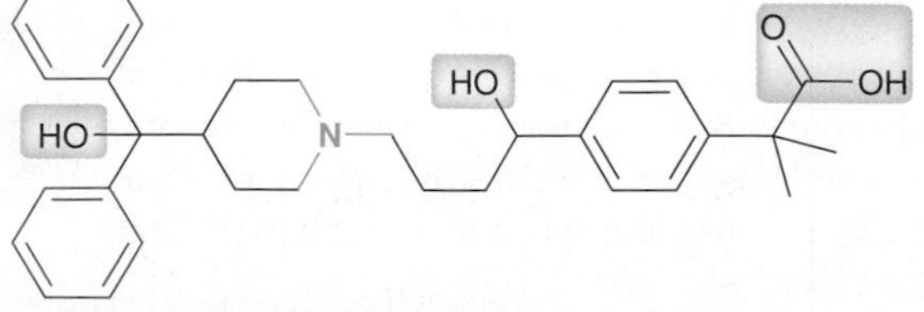
Fexofenadine

This compound contains several polar groups (highlighted in green boxes). In addition, the COOH group is deprotonated under physiological pH to produce a carboxylate anion (COO^-). This ionic group, together with the polar hydroxyl groups, prevents the compound from crossing the blood-brain barrier and from reaching receptors in the central nervous system. As a result, fexofenadine does not produce the sedative effects that are typical of first-generation antihistamines. Many other nonsedating antihistamines have also been developed, such as loratidine, sold under the trade name Claritin™. Apparently, the carbamate group (highlighted in green) is sufficiently polar to reduce the ability of this compound to cross the blood-brain barrier.

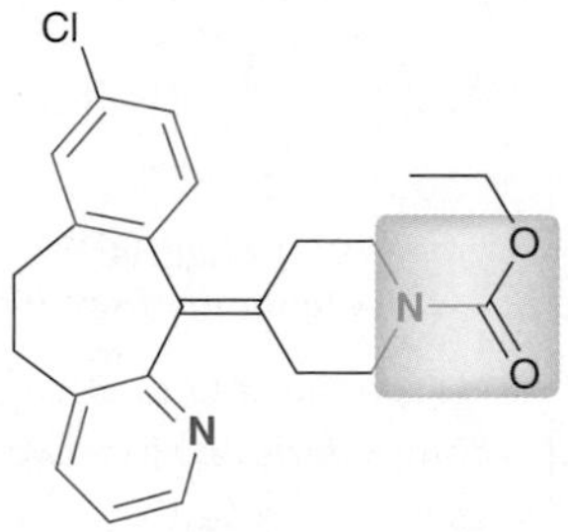
Claritin™ (loratadine)

CONCEPTUAL CHECKPOINT

18.18 Meclizine, shown below, is an antiemetic (prevents nausea and vomiting).

a) Would you expect meclizine to be an antihistamine as well? Justify your answer.

b) This drug is known to cause sedation. Describe the source of the sedative properties of meclizine.

c) Can you suggest a structural modification that could possibly modify the sedative properties of meclizine?

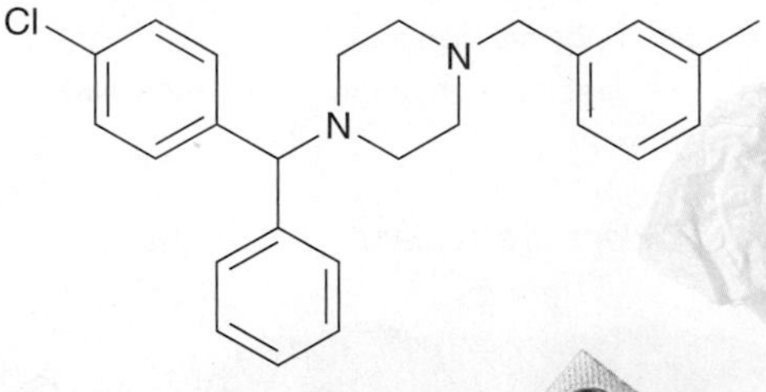
Meclizine

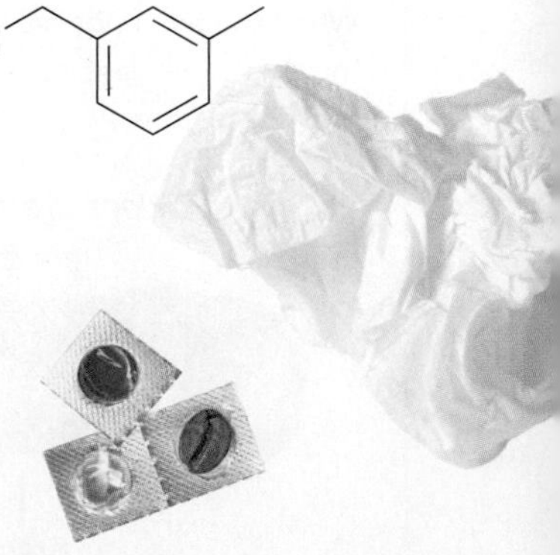

18.6 Reactions at the Benzylic Position

Any carbon atom attached directly to a benzene ring is called a **benzylic position**:

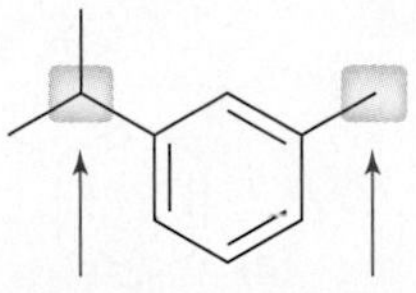

In the following sections, we will explore reactions that can occur at the benzylic position.

Oxidation

Recall from Section 13.10 that chromic acid (H_2CrO_4) is a strong oxidizing agent used to oxidize primary or secondary alcohols. Chromic acid does not readily react with benzene or with alkanes:

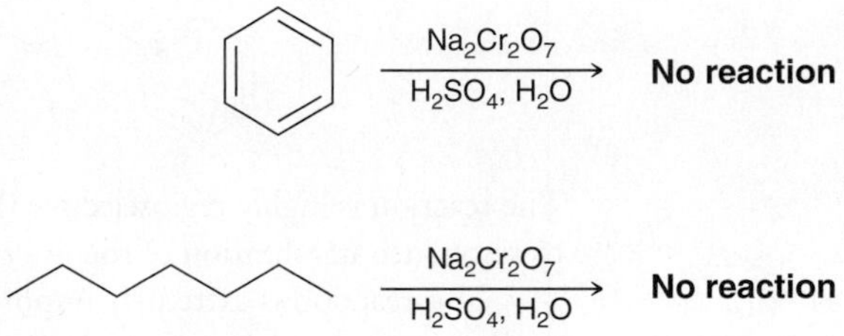

Interestingly, however, alkylbenzenes are readily oxidized by chromic acid. The oxidation takes place selectively at the benzylic position:

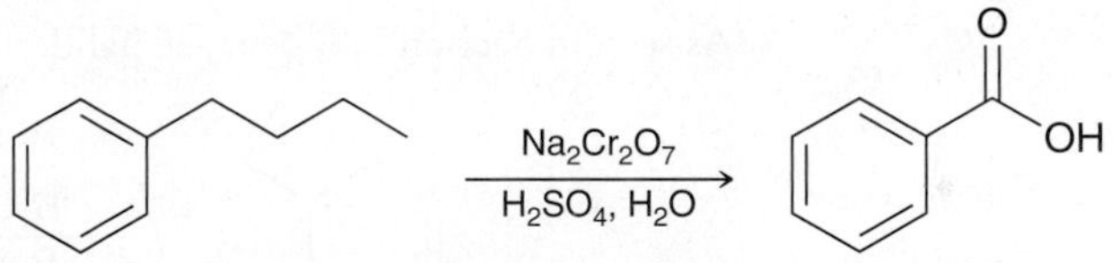

Although the aromatic moiety itself is stable in the presence of chromic acid, the benzylic position is particularly susceptible to oxidation. Notice that the alkyl group is entirely excised, leaving only the benzylic carbon atom behind. The product is benzoic acid, irrespective of the identity of the alkyl group. The only condition is that the benzylic position must have at least one proton. If the benzylic position lacks a proton, then oxidation does not occur.

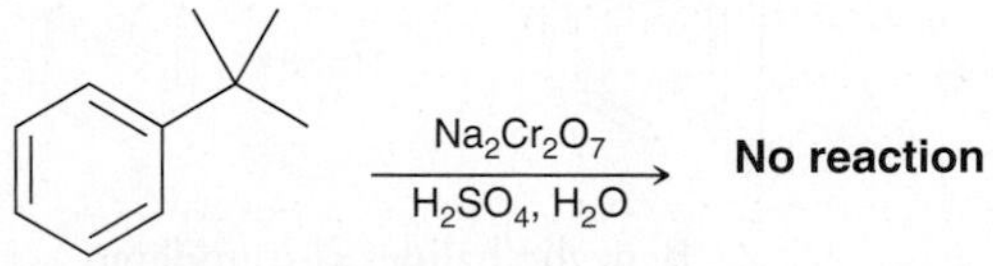

Oxidation of a proton-bearing benzylic position can also be accomplished with other reagents, including potassium permanganate ($KMnO_4$).

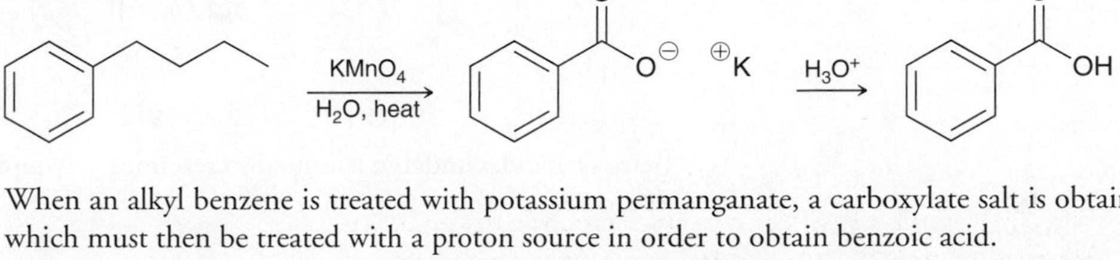

When an alkyl benzene is treated with potassium permanganate, a carboxylate salt is obtained, which must then be treated with a proton source in order to obtain benzoic acid.

CONCEPTUAL CHECKPOINT

18.19 Draw the expected product when each of the following compounds is treated with chromic acid.

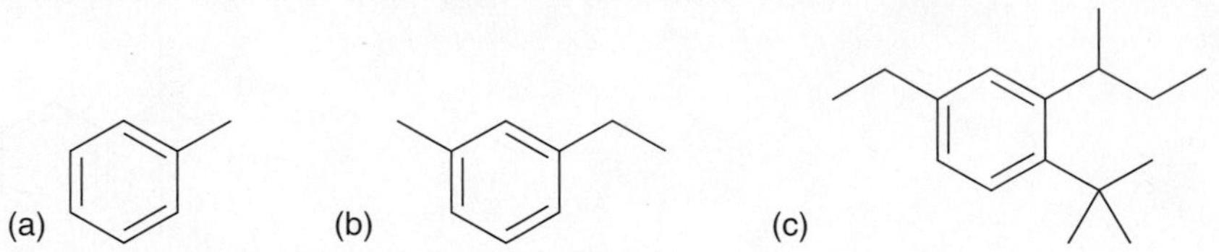

Free-Radical Bromination

In Section 11.7 we explored the free-radical bromination of allylic positions. Similarly, free-radical bromination also occurs readily at benzylic positions.

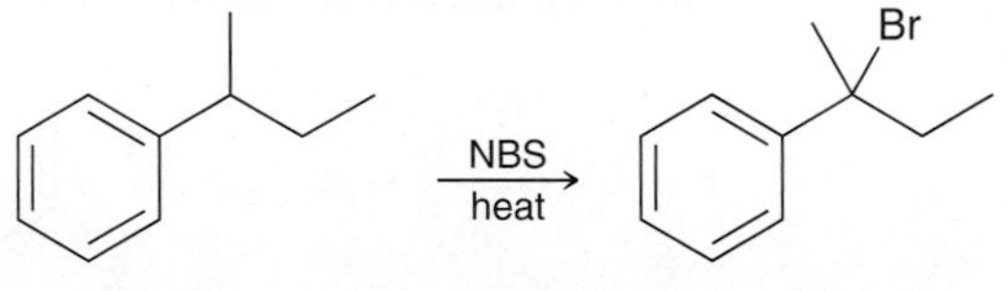

The reaction is highly regioselective (bromination occurs primarily at the benzylic position) due to resonance stabilization of the intermediate benzylic radical.

This reaction is extremely important, because it enables us to introduce a functional group at the benzylic position. Once introduced, that functional group can then be exchanged for a different group, as seen in the next section.

Substitution Reactions of Benzylic Halides

As seen in Section 7.8, benzylic halides undergo S_N1 reactions very rapidly.

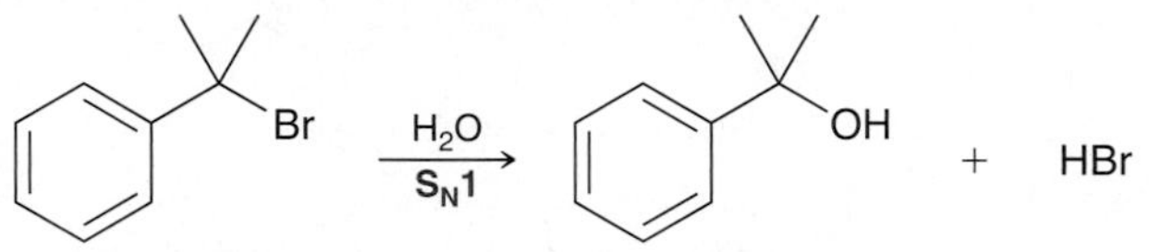

The relative ease of the reaction is attributed to the stability of the carbocation intermediate. Specifically, a benzylic carbocation is resonance stabilized.

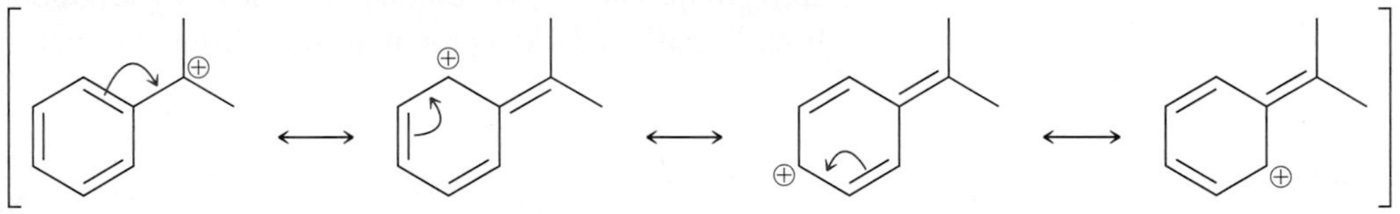

Benzylic halides also undergo S_N2 reactions very rapidly, provided that they are not sterically hindered.

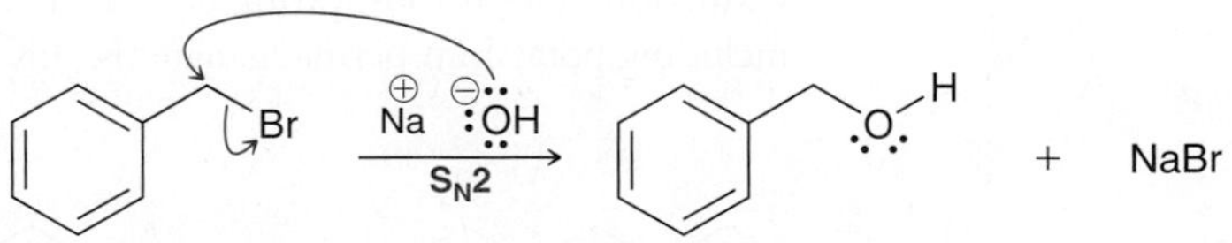

Elimination Reactions of Benzylic Halides

Benzylic halides undergo elimination reactions very rapidly. For example, the following E1 reaction occurs readily.

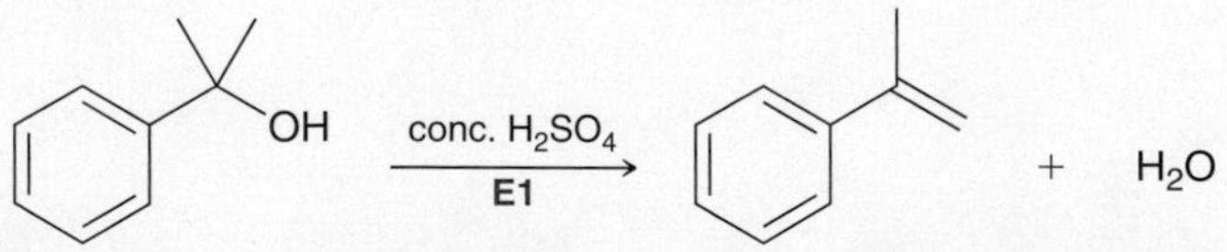

The relative ease of the reaction is attributed to the stability of the carbocation intermediate (a benzylic carbocation), just as we saw with S_N1 reactions.

Benzylic halides also undergo E2 reactions very rapidly.

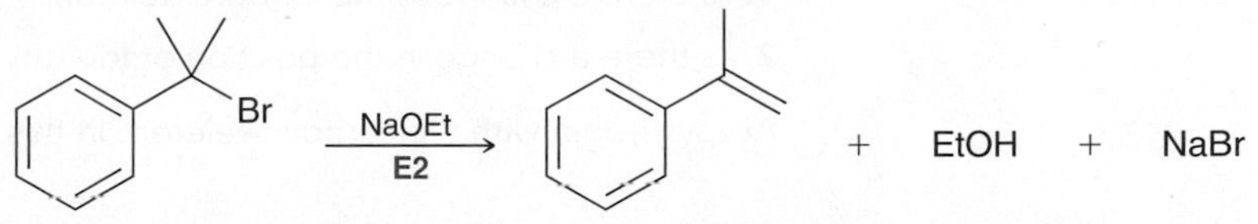

The relative ease of the reaction is attributed to the low energy of the transition state as a result of conjugation between the forming double bond and the aromatic ring.

Summary of Reactions at the Benzylic Position

Figure 18.17 is a summary of reactions that can occur at the benzylic position. Note that there are two ways to introduce functionality at a benzylic position: (1) oxidation and (2) radical bromination. The first method produces benzoic acid, while the second produces a benzylic halide that can undergo substitution or elimination.

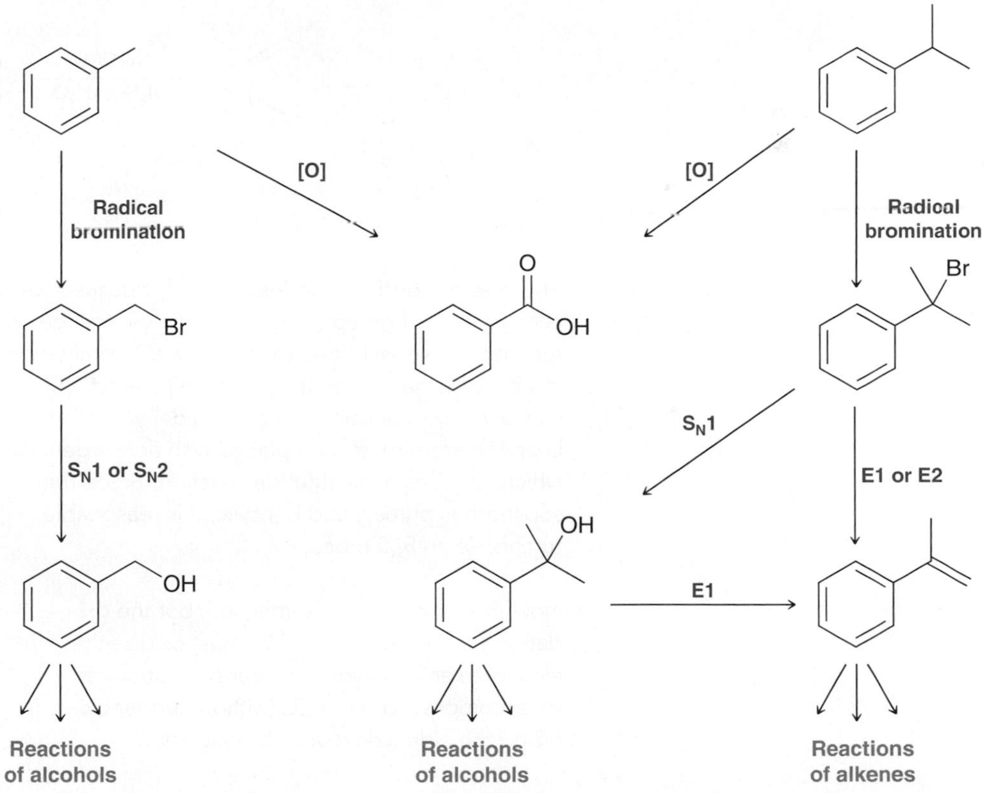

FIGURE 18.17
Reactions that occur at the benzylic position.

SKILLBUILDER

18.4 MANIPULATING THE SIDE CHAIN OF AN AROMATIC COMPOUND

LEARN the skill

Propose a plausible synthesis for the following transformation:

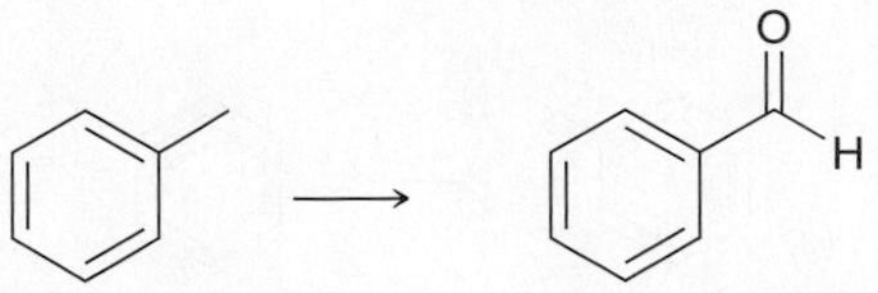

SOLUTION

Recall from Chapter 12 the two questions to ask when approaching a synthesis problem:

1. Is there a change in the carbon skeleton?

2. Is there a change in the position or identity of the functional groups?

Always begin with the carbon skeleton. In this case, there is no change:

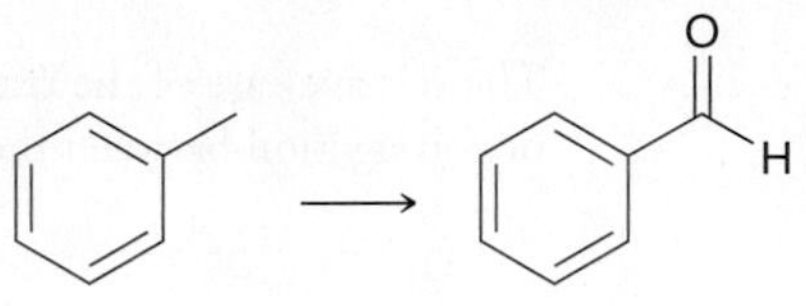

Now focus on the change in functional groups. In this example, the reactant does not have a functional group at the benzylic position, while the product does. Therefore, we must first introduce functionality at the benzylic position, and there are two ways to do that.

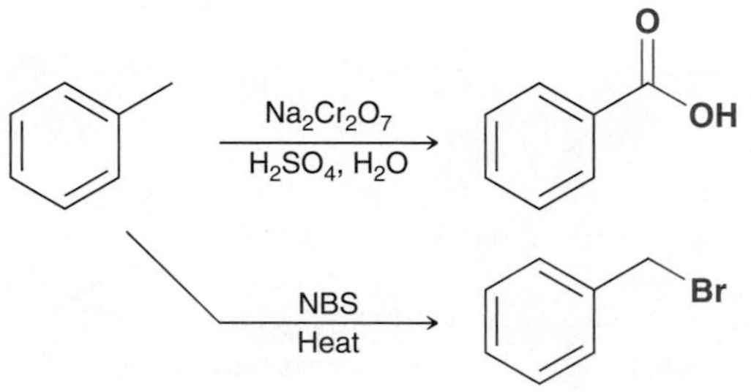

The first method will be less helpful, because we do not yet know how to manipulate a carboxylic acid group (COOH). That topic will be covered extensively in Chapter 21. In the meantime, we will have to use radical bromination to introduce functionality. The resulting benzylic halide can then be manipulated. But can a Br be converted into an aldehyde? The bromine atom must be replaced with an oxygen, which requires a substitution reaction. Since the substrate is primary and benzylic, it is reasonable to choose an S_N2 process.

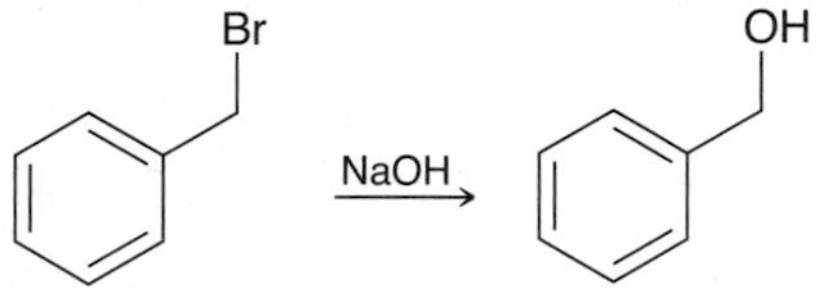

Now the oxygen atom is attached, but the oxidation state is not correct. We must oxidize the alcohol to an aldehyde. This transformation can be accomplished with PCC (without further oxidizing the aldehyde to a carboxylic acid).

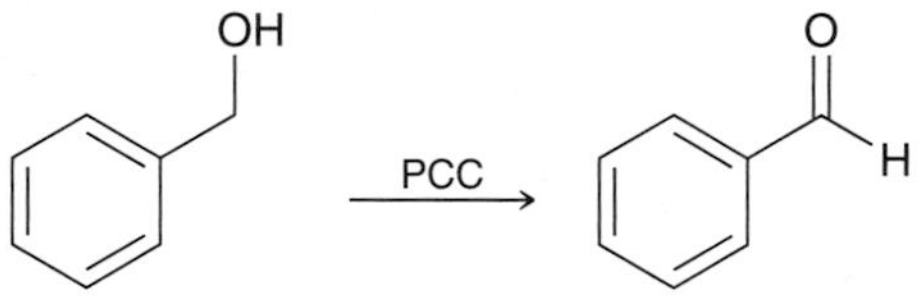

In summary, our synthesis has just a few steps:

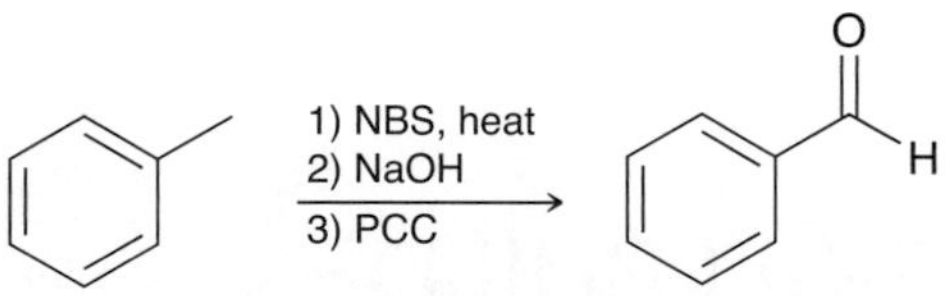

PRACTICE the skill **18.20** Propose a plausible synthesis for each of the following transformations.

(a) 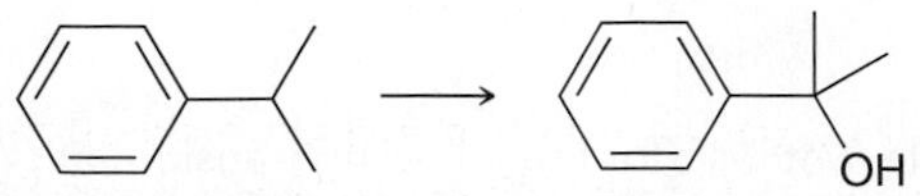

(b) 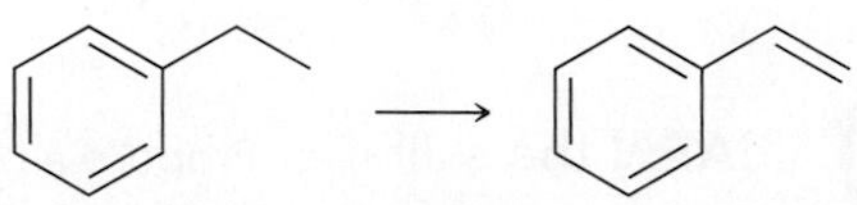

(c) 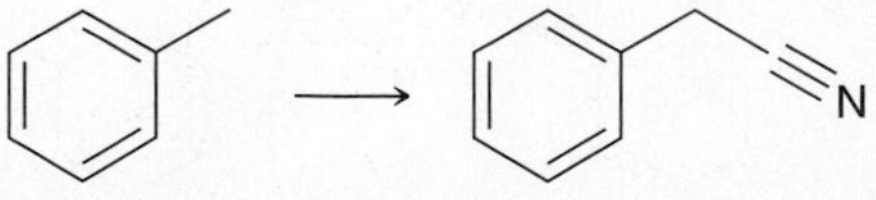

(d) 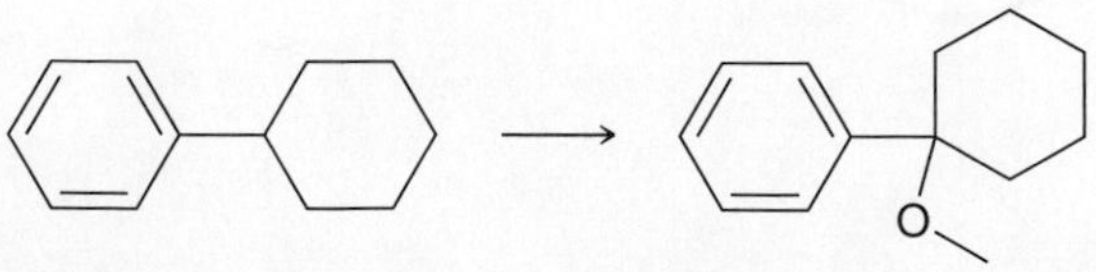

APPLY the skill

18.21 Using toluene as your only source of carbon atoms, show how you would prepare the following compound.

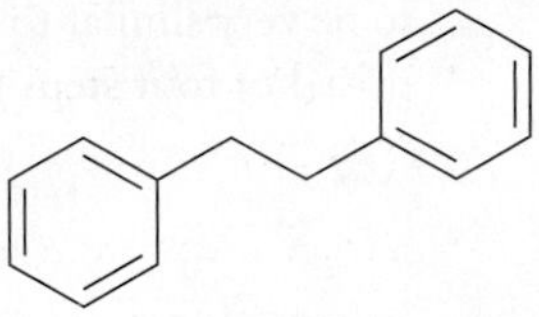

18.22 Starting with isopropyl benzene, propose a synthesis for acetophenone. (*Hint*: Make sure to count the number of carbon atoms in the starting material and product.)

18.23 Propose a plausible synthesis for the following transformation.

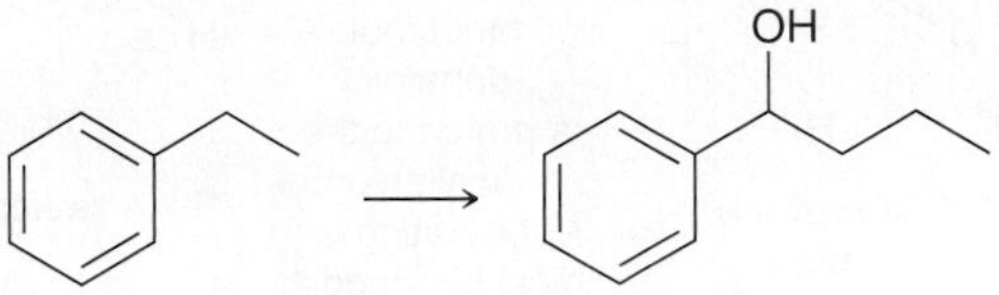

need more **PRACTICE?** **Try Problems 18.47, 18.56**

18.7 Reduction of the Aromatic Moiety

Hydrogenation

Recall from earlier in this chapter that under forcing conditions benzene will react with three equivalents of molecular hydrogen to produce cyclohexane.

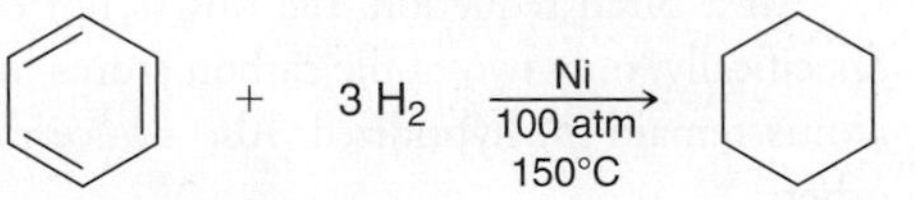

$$+ \quad 3\ H_2 \xrightarrow[\substack{100\ \text{atm} \\ 150°C}]{Ni}$$ $\Delta H° = -208$ kJ/mol

With some catalysts and under certain conditions, it is possible to selectively hydrogenate a vinyl group in the presence of an aromatic ring.

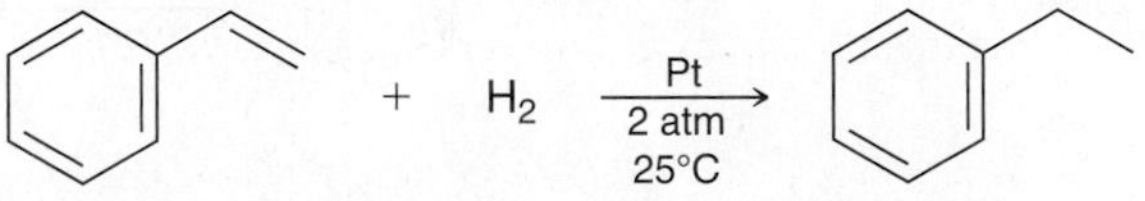

$$+ \quad H_2 \xrightarrow[\substack{2\ \text{atm} \\ 25°C}]{Pt}$$ $\Delta H° = -117$ kJ/mol

(100%)

The vinyl group is reduced, but the aromatic moiety is not. Notice that ΔH for this reaction is only slightly lower than what we expect for a double bond (-120 kJ/mol), which can be attributed to the fact that the double bond is conjugated in this case.

Birch Reduction

Recall from Section 10.5 that alkynes can be reduced via the dissolving metal reduction. Benzene can also be reduced under similar conditions to give 1,4-cyclohexadiene.

$$\xrightarrow[\text{NH}_3]{\text{Na, CH}_3\text{OH}}$$

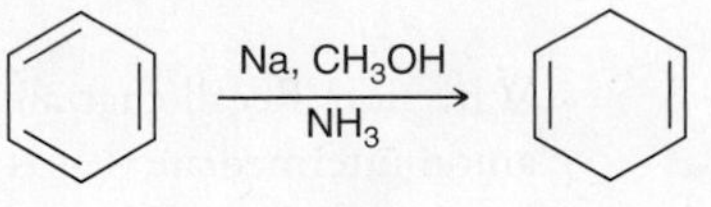

This reaction is called the **Birch reduction**, named after the Australian chemist Arthur Birch, who systematically explored the details of this reaction. The mechanism, which is believed to be very similar to the mechanism of alkyne reduction via a dissolving metal reduction, is comprised of four steps (Mechanism 18.1).

MECHANISM 18.1 THE BIRCH REDUCTION

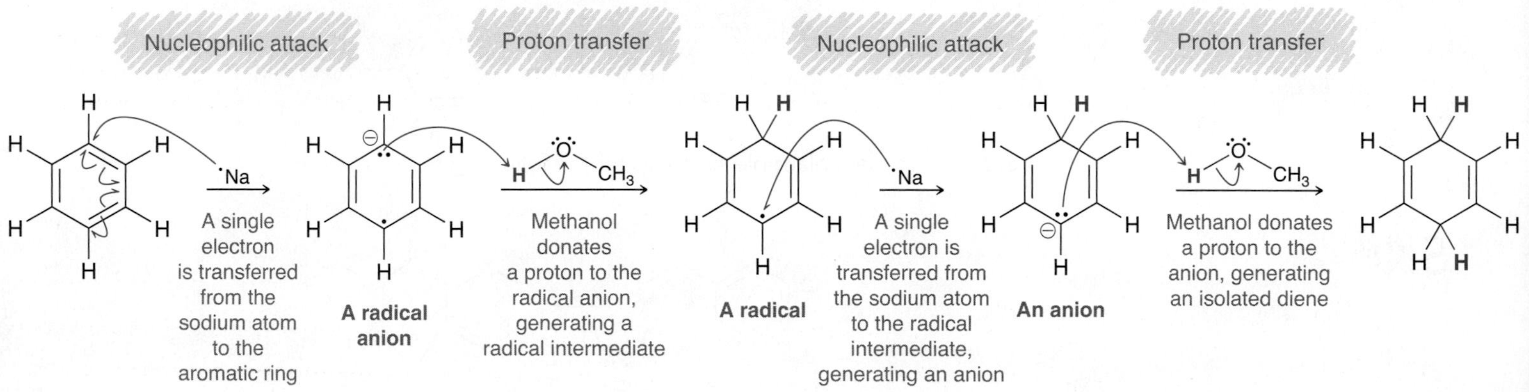
Nucleophilic attack
Proton transfer
Nucleophilic attack
Proton transfer
A single electron is transferred from the sodium atom to the aromatic ring
A radical anion
Methanol donates a proton to the radical anion, generating a radical intermediate
A radical
A single electron is transferred from the sodium atom to the radical intermediate, generating an anion
An anion
Methanol donates a proton to the anion, generating an isolated diene

In step 1, a single electron is transferred to the aromatic ring, giving a radical anion, which is then protonated in step 2. Steps 3 and 4 are repeats of steps 1 and 2, with the transfer of an electron followed by protonation. In order to remember the mechanism of this reaction, it might be helpful to summarize the steps as (1) electron, (2) proton, (3) electron, and (4) proton.

In a Birch reduction, the ring is not completely reduced, since two double bonds remain. Specifically, only two of the carbon atoms in the ring are actually reduced. The other four carbon atoms remain sp^2 hybridized. Also notice that the two reduced carbon atoms are opposite each other:

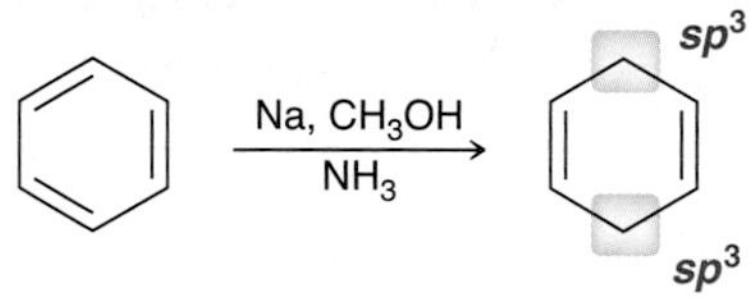
Na, CH3OH
NH3
sp3
sp3

The product is a nonconjugated diene, rather than a conjugated diene.

When an alkyl benzene is treated with Birch conditions, the carbon atom connected to the alkyl group is not reduced:

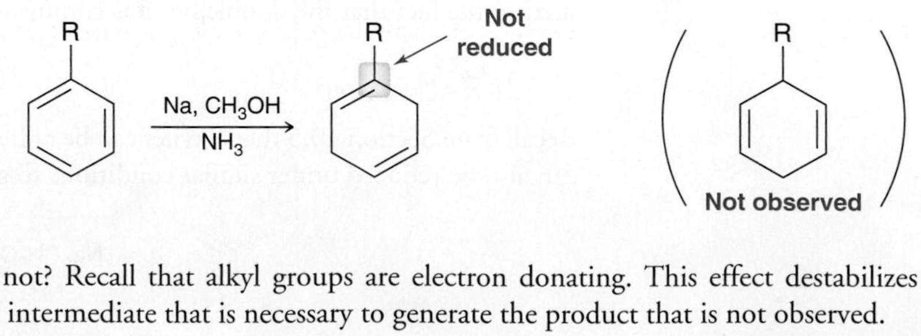
R
Na, CH3OH
NH3
Not reduced
Not observed

Why not? Recall that alkyl groups are electron donating. This effect destabilizes the radical anion intermediate that is necessary to generate the product that is not observed.

When electron-withdrawing groups are used, a different regiochemical outcome is observed. For example, consider the structure of acetophenone, which has a carbonyl group (C=O bond) next to the aromatic ring:

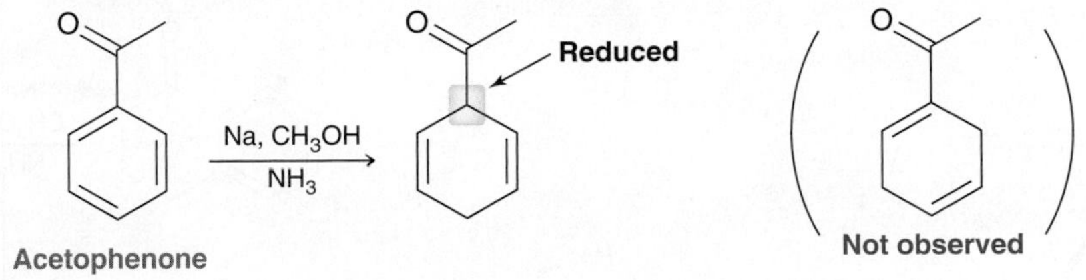

The carbonyl group is electron withdrawing via resonance. (Can you draw the resonance structures of acetophenone and explain why the carbonyl group is electron withdrawing?) This electron-withdrawing effect stabilizes the intermediate that is necessary to generate the observed product.

SKILLBUILDER

18.5 PREDICTING THE PRODUCT OF A BIRCH REDUCTION

LEARN the skill

Predict the major product obtained when the following compound is treated with Birch conditions.

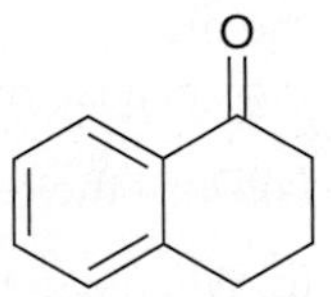

SOLUTION

Determine whether the groups attached to the aromatic ring are electron donating or electron withdrawing.

In this case, there are two groups. The carbonyl group is electron withdrawing, while the alkyl group is electron donating.

STEP 1
Determine whether each substituent is electron donating or electron withdrawing.

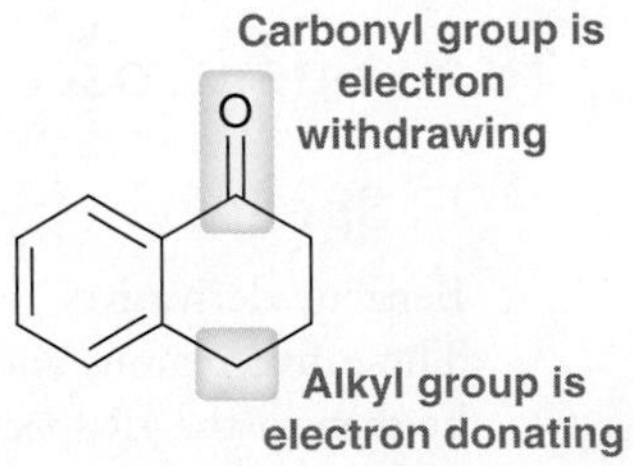

STEP 2
Identify which carbon atoms are reduced.

We therefore predict that the carbon atom next to the carbonyl group will be reduced, while the carbon atom next to the alkyl group will not be reduced.

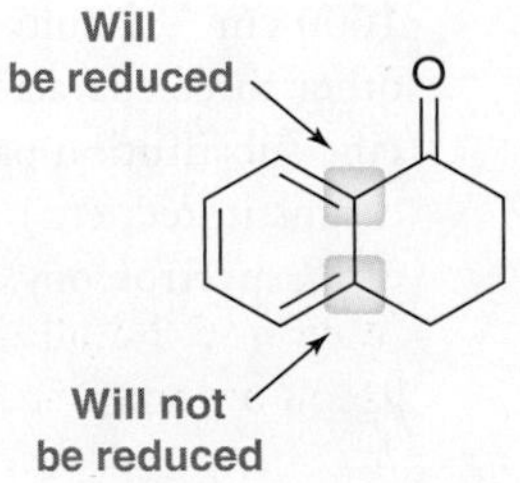

STEP 3
Draw the product, making sure that the two positions not reduced are 1,4 to each other.

Remember that only two positions are reduced in a Birch reaction, and these two positions must be on opposite sides of the ring (1,4 to each other). This requirement, together with the information above, dictates the following outcome.

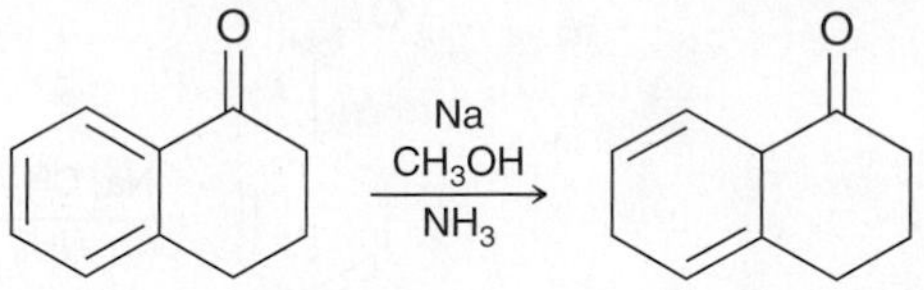

PRACTICE the skill **18.24** Predict the major product obtained when each of the following compounds is treated with Birch conditions.

(a) 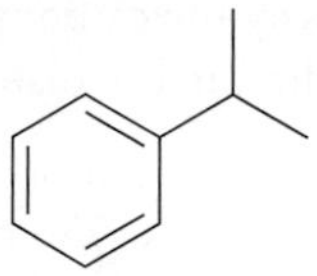

(b) 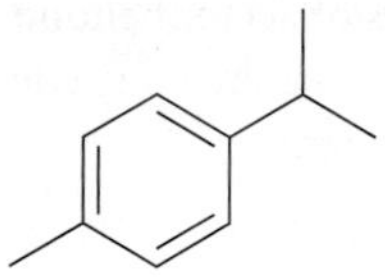

(c) 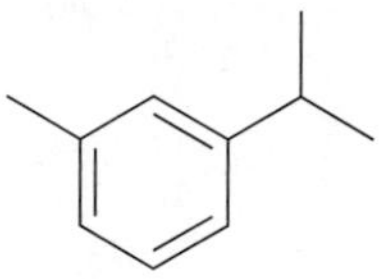

(d) 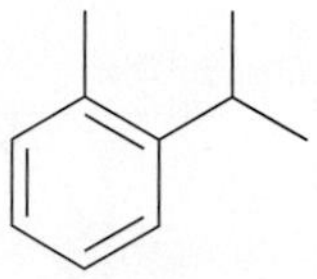

(e) 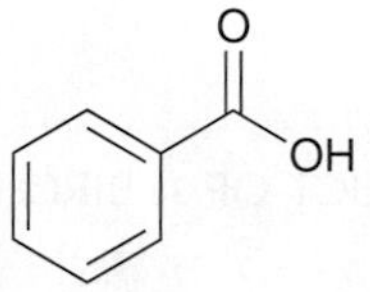

(f) 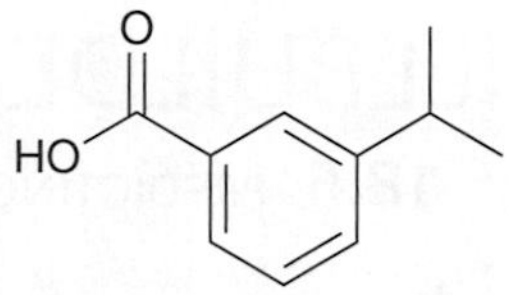

APPLY the skill **18.25** Consider the structure of anisole (also called methoxybenzene). In the next chapter, we will discuss whether a methoxy group is electron donating or electron withdrawing. We will see that there is a competition between two factors. Specifically, we will see that the methoxy group is electron withdrawing because of induction, but it is electron donating because of resonance.

(a) Draw the resonance structures of anisole.

(b) Whenever resonance and induction compete with each other, resonance is generally the dominant factor. Using this information, predict the regiochemical outcome of a Birch reduction of anisole.

need more **PRACTICE?** **Try Problems 18.49, 18.50**

18.8 Spectroscopy of Aromatic Compounds

IR Spectroscopy

Benzene derivatives generally produce signals in five characteristic regions of an IR spectrum. These five regions and the vibrations associated with them are listed in Table 18.2 and can be seen in the IR spectrum of ethylbenzene (Figure 18.18). Notice that the signals just above 3000 cm^{-1}, corresponding with C_{sp^2}—H stretching, appear on the shoulder of the signals for all other C—H stretching (just below 3000 cm^{-1}). This is often the case, and these signal(s) can be identified by drawing a line at 3000 cm^{-1} and looking for any signals to the left of that line (Section 15.3). Aromatic compounds also produce a series of signals between 1450 and 1600 cm^{-1}, resulting from C═C stretching and ring vibrations. The pattern of signals in the other three characteristic regions (as seen in Table 18.2) can often be used to identify the specific substitution pattern of the aromatic ring (i.e., monosubstituted, *ortho* disubstituted, *meta* disubstituted, etc.), although this level of analysis will not be discussed in our current treatment of IR spectroscopy.

TABLE 18.2 CHARACTERISTIC SIGNALS IN THE IR SPECTRA OF AROMATIC COMPOUNDS

ABSORPTION	FEATURE	COMMENTS
3000–3100 cm^{-1}	C_{sp^2}—H stretching	One or more signals just above 3000 cm^{-1}. Intensity is generally weak or medium.
1700–2000 cm^{-1}	Combination bands and overtones	A group of very weak signals
1450–1650 cm^{-1}	C=C stretching and ring vibrations	Generally three signals (medium intensity) at around 1450, 1500, and 1600 cm^{-1}
1000–1275 cm^{-1}	C—H bending (in plane)	Several signals of strong intensity
690–900 cm^{-1}	C—H bending (out of plane)	One or two strong signals

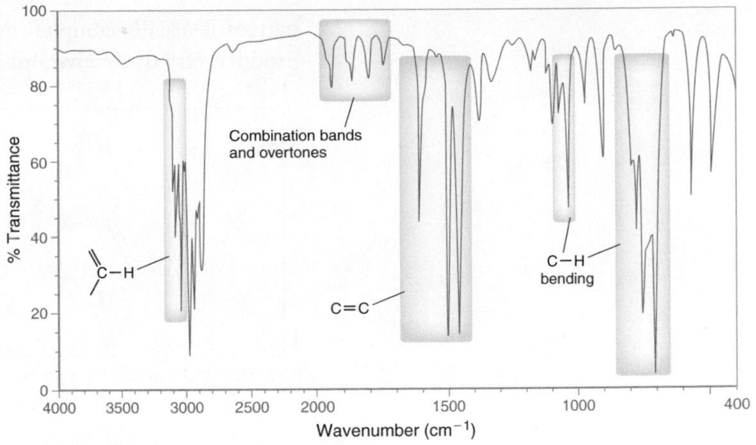

FIGURE 18.18
An IR spectrum of ethylbenzene indicating the five regions of absorption, characteristic of aromatic compounds.

LOOKING BACK
To see a picture of the shielding and deshielding regions established by an aromatic ring, see Figure 16.9.

^{1}H NMR Spectroscopy

In Section 16.5, we first discussed the anisotropic effects of an aromatic ring. Specifically, the motion of the π electrons generates a local magnetic field that effectively deshields the protons connected directly to the ring.

The signals from these protons typically appear between 6.5 and 8 ppm. For example, consider the ^{1}H NMR spectrum of ethylbenzene (Figure 18.19). The presence of a multiplet near 7 ppm is one of the best ways to verify the presence of an aromatic ring. Notice that the deshielding effects of the aromatic ring are felt most strongly for protons connected directly to the ring. Protons in benzylic positions are farther removed from the ring and are deshielded to a lesser extent. These protons generally produce signals between 2 and 3 ppm. Protons that are even farther removed from the aromatic ring exhibit almost no shielding effects.

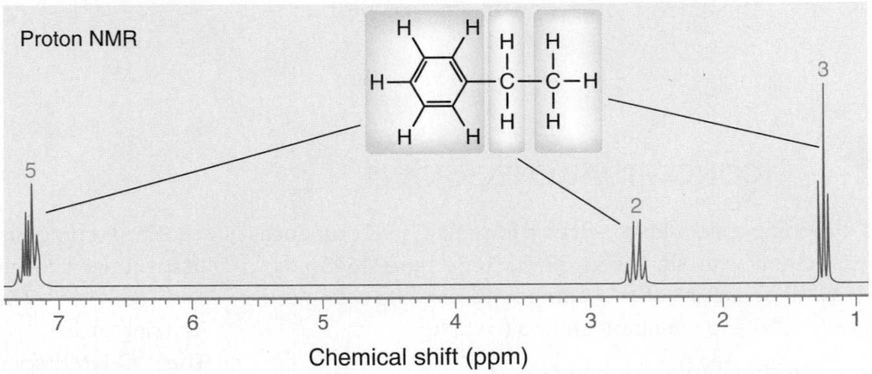

FIGURE 18.19
A ^{1}H NMR spectrum of ethylbenzene showing that the deshielding effects of the aromatic ring are strongly dependent on the proximity of the proton to the ring.

The integration value of the multiplet near 7 ppm is very useful information because it indicates the extent of substitution of the aromatic ring (monosubstituted, disubstituted, trisubstituted, etc.). An integration of 5 is indicative of a monosubstituted ring, an integration of 4 is indicative of a disubstituted ring, and so forth. The splitting pattern of this multiplet is generally too complex to analyze, because the rigid planar structure of benzene causes long-range coupling between all of the different aromatic protons (even non-neighbors). Therefore, the substitution pattern is usually complex, except in the following two *para*-disubstituted cases, both of which produce very distinctive and simple patterns.

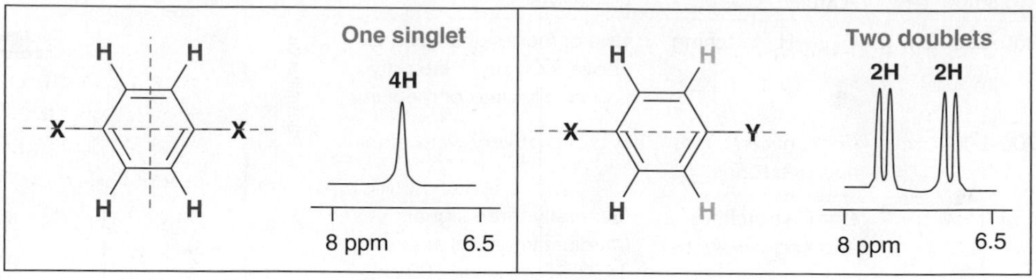

^{13}C NMR Spectroscopy

As first mentioned in Section 16.12, the carbon atoms of aromatic rings typically produce signals in the range of 100–150 ppm in a ^{13}C NMR spectrum. The number of signals is very helpful in determining the specific substitution pattern for substituted aromatic rings. Several common substitution patterns are shown below.

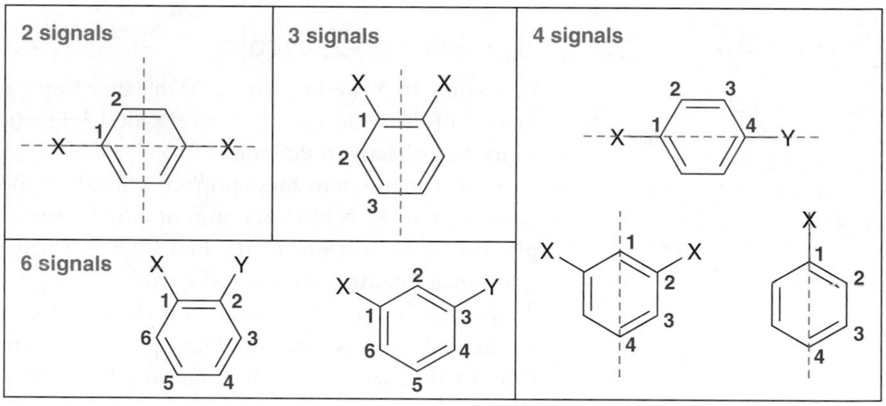

The number of signals in the region of 150–200 signals can therefore provide valuable information.

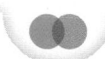

CONCEPTUAL CHECKPOINT

18.26 A compound with molecular formula C_8H_8O produces an IR spectrum with signals at 3063, 1686, and 1646 cm^{-1}. The ^1H NMR spectrum of this compound exhibits a singlet at 2.6 ppm (I = 3H) and a multiplet at 7.5 (I = 5H).

(a) Draw the structure of this compound.

(b) What is the common name of this compound?

(c) When this compound is treated with Na, CH_3OH, and NH_3, a reduction takes place, giving a new compound with molecular formula $C_8H_{10}O$. Draw the product of this reaction.

18.27 A compound with molecular formula C_8H_{10} produces an IR spectrum with many signals, including 3108, 3066, 3050, 3018, and 1608 cm^{-1}. The ^1H NMR spectrum of this compound exhibits a singlet at 2.2 ppm (I = 6H) and a multiplet at 7.1 ppm (I = 4H). The ^{13}C NMR spectrum of this compound exhibits signals at 19.7, 125.9, 129.6, and 136.4 ppm.

(a) Draw the structure of this compound.

(b) What is the common name of this compound?

(c) Treating this compound with chromic acid yields a product with molecular formula $C_8H_6O_4$. Draw the product of this reaction.

PRACTICALLYSPEAKING)))

Buckyballs and Nanotubes

Until the mid-1980s, only two forms of elemental carbon were known—diamond and graphite:

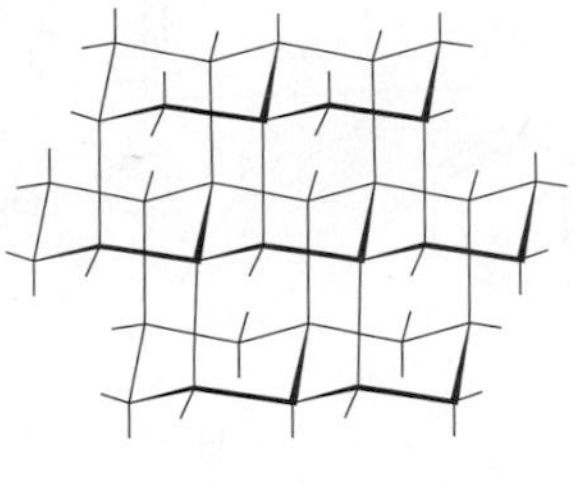

Diamond

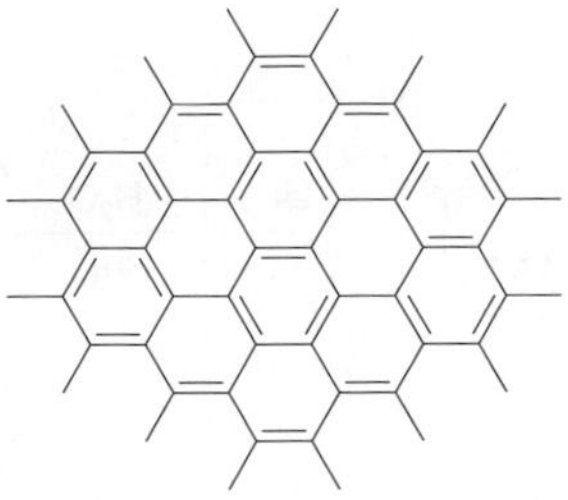

Graphite

Diamond is a three-dimensional lattice of interlocking chair conformations. Graphite is comprised of flat sheets of interlocking benzene rings. These sheets adhere to one another due to van der Waals interaction, but they can readily slide past each other, which makes graphite an excellent lubricant.

In 1985, a new form of elemental carbon was discovered, very much by accident. Harold Kroto (University of Sussex) was investigating how certain organic compounds are formed in space. While visiting with Robert Curl and Richard Smalley (Rice University), they discussed a way to re-create the type of chemical transformations that might occur in stars that are carbon rich. Smalley had developed a method for laser-induced evaporation of metals, and the group of scientists agreed to apply this method to graphite in the hopes of creating polyacetylenic compounds:

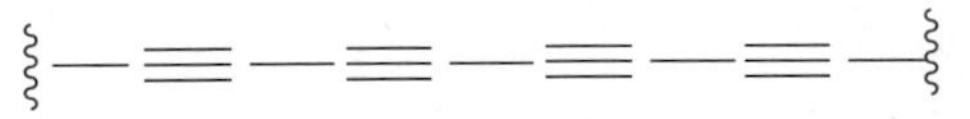

The compounds generated by this procedure were then analyzed by mass spectrometry, and to everyone's surprise, certain conditions would reliably produce a compound with molecular formula C_{60}. They theorized that the structure of C_{60} is based on alternating five- and six-membered rings. To illustrate this, imagine a five-membered ring completely surrounded by fused benzene rings:

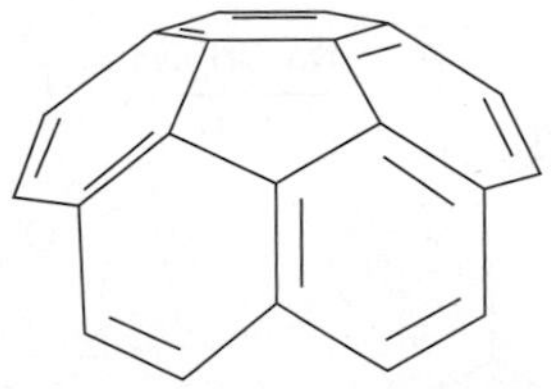

This group of atoms provides for a natural curvature. Extending this bonding pattern provides the possibility of a spherical molecule. Based on this reasoning, Kroto, Curl, and Smalley theorized that C_{60} is comprised of fused rings (20 hexagons and 12 pentagons), resembling the pattern of the seams on a soccer ball.

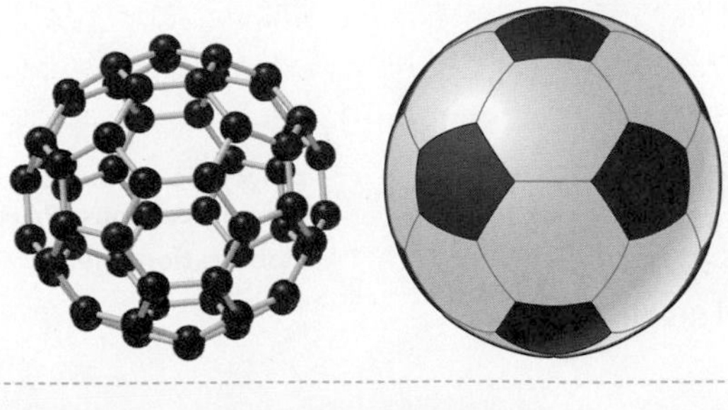

They called this unusual compound buckminsterfullerene (or just fullerene, for short), named after American architect R. Buckminster Fuller, who was famous for building geodesic domes. This compound is also commonly referred to as a *buckyball.*

Shortly after the discovery of C_{60}, methods were developed for preparing C_{60} in larger quantities, enabling its spectroscopic analysis as well as the exploration of its chemistry. A ^{13}C NMR spectrum of C_{60} shows a single peak at 143 ppm, because all 60 carbon atoms are sp^2 hybridized and chemically equivalent.

Since buckyballs are comprised of interlocking aromatic rings, we might expect C_{60} to function as if it were one large aromatic compound. However, this is not the case, because the curvature of the sphere prevents all of the *p* orbitals from overlapping with each other, so the first criterion for aromaticity is not met. This explains why C_{60} does not exhibit the same stability as benzene. Specifically, C_{60} readily undergoes addition reactions, much like alkenes.

Since 1990, chemists have been preparing and exploring larger fullerenes and their interesting chemistry. For example, C_{60} can be prepared under conditions in which ions can be trapped inside the center of the sphere. Such compounds are superconductors at low temperature and offer the potential of many exciting applications. These compounds are also being explored as novel drug delivery systems. Tubular fullerenes have also been prepared:

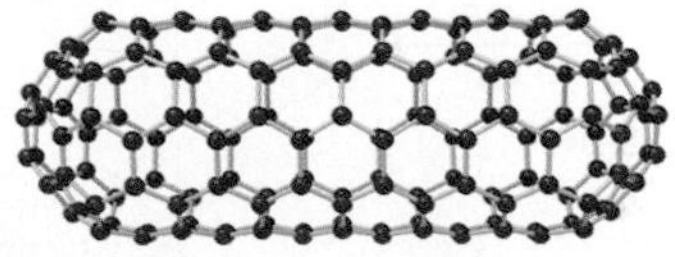

These compounds, which are called *nanotubes*, can be thought of as a rolled-up sheet of graphite capped on either end by half of a buckyball. Nanotubes have many potential applications. They can be spun into fibers that are stronger and ligher than steel, and they can also be made to carry electrical currents more efficiently than metals. The next several decades are likely to see many exciting applications of buckyballs and nanotubes. For their discovery of fullerenes, Kroto, Curl, and Smalley were awarded the 1996 Nobel Prize in Chemistry.

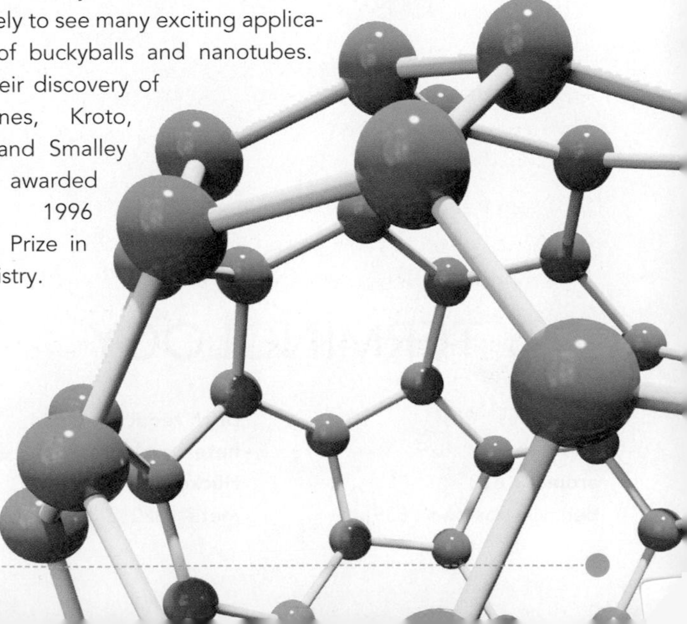

REVIEW OF REACTIONS

REACTIONS AT THE BENZYLIC POSITION

Oxidation

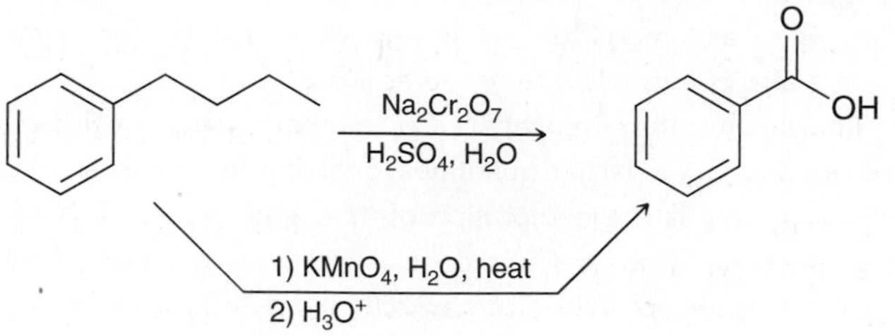

Free-Radical Bromination

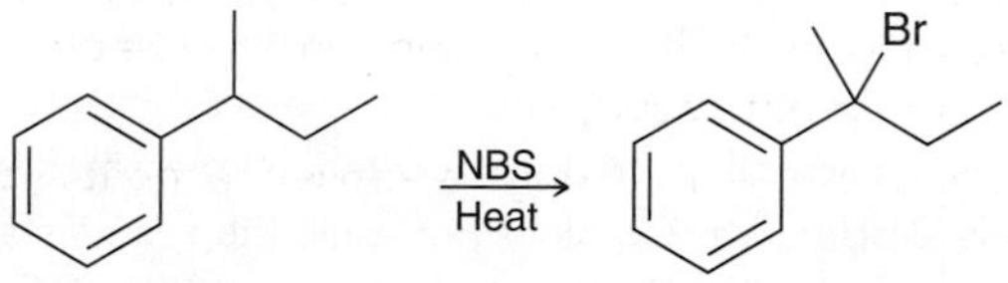

Elimination Reactions

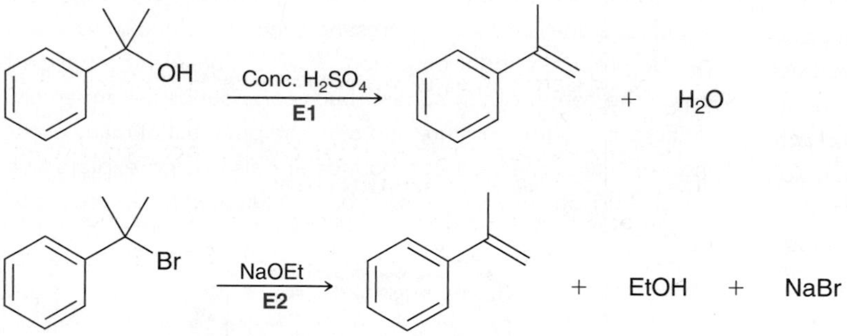

Substitution Reactions

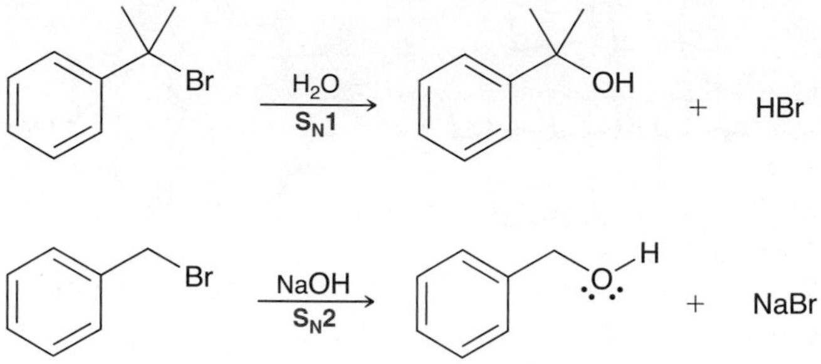

REDUCTION

Catalytic Hydrogenation

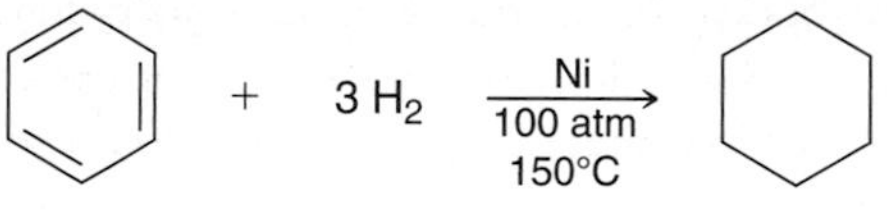

Birch Reduction

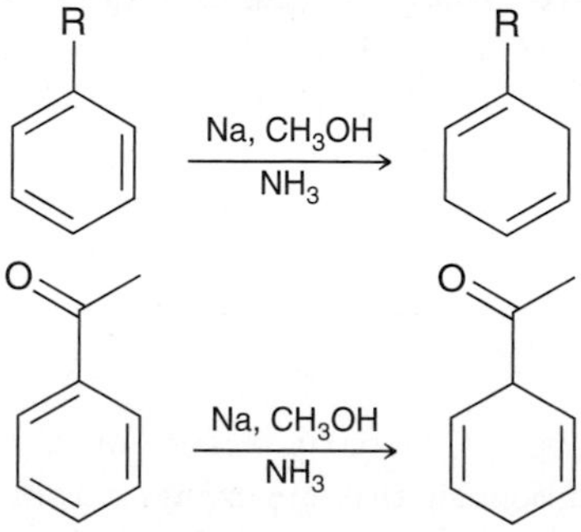

KEY TERMINOLOGY

annulenes 831
antiaromatic 829
aromatic 818
benzylic position 839
Birch reduction 844
heterocycles 833
Hückel's rule 827
meta 820
nonaromatic 830
ortho 820
para 820
phenyl group 820
polycyclic aromatic hydrocarbons (PAHs) 836
stabilization energy 824

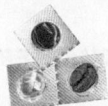

REVIEW OF CONCEPTS AND VOCABULARY

SECTION 18.1

- Derivatives of benzene are called **aromatic compounds**, regardless of whether they are fragrant or odorless.

SECTION 18.2

- Monosubstituted derivatives of benzene are named systematically using benzene as the parent and listing the substituent as a prefix.
- IUPAC also accepts many common names for monosubstituted benzenes.
- When a benzene ring is a substituent, it is called a **phenyl group**.
- Disubstituted derivatives of benzene can be differentiated by the use of the descriptors ***ortho***, ***meta***, and ***para*** or by the use of locants.
- Polysubstituted derivatives of benzene are named using locants. Common names can be used as parents.

SECTION 18.3

- Benzene is comprised of a ring of six identical C—C bonds, each of which has a bond order of 1.5.
- No single Lewis structure adequately describes the structure of benzene. Resonance structures are required.

SECTION 18.4

- Benzene exhibits unusual stability. It does not react with bromine in an addition reaction.
- The **stabilization energy** of benzene can be measured by comparing heats of hydrogenation.
- The stability of benzene can be explained with MO theory. The six π electrons all occupy bonding MOs.
- The presence of a fully conjugated ring of π electrons is not the sole requirement for aromaticity. The requirement for an odd number of electron pairs is called **Hückel's rule**.
- Cyclobutadiene is **antiaromatic**; cyclooctatetraene adopts a tub-shaped conformation and is nonaromatic.
- **Frost circles** accurately predict the relative energy levels of the MOs in a conjugated ring system.

SECTION 18.5

- A compound is aromatic if it contains a ring comprised of continuously overlapping *p* orbitals and if it has a Hückel number of π electrons in the ring.
- Compounds that fail the first criterion are called **nonaromatic**.
- Compounds that satisfy the first criterion but have $4n$ electrons (rather than $4n+2$) are antiaromatic.
- **Annulenes** are compounds consisting of a single ring containing a fully conjugated π system. Due to steric hindrance, [10]annulene and [14]annulene do not meet the first criterion and are nonaromatic.
- The cyclopentadienyl anion exhibits aromatic stabilization, as does the tropylium cation.
- Cyclic compounds containing hetereoatoms, such as S, N, and O, are called **heterocycles**.
- The lone pair in pyridine is localized and does not participate in resonance, while the lone pair in pyrrole is delocalized and participates in aromaticity.
- Hückel's rule ($4n+2$) can only be applied to monocyclic compounds.
- Many **polycyclic aromatic hydrocarbons (PAHs)** are known to be stable.

SECTION 18.6

- Any carbon atom attached directly to a benzene ring is called a **benzylic position**.
- Alkylbenzenes are oxidized at the benzylic position by chromic acid or potassium permanganate.
- Free-radical bromination occurs readily at benzylic positions.
- Benzylic halides readily undergo S_N1, S_N2, E1, and E2 reactions.

SECTION 18.7

- Under certain conditions, a vinyl group can be selectively hydrogenated in the presence of an aromatic ring.
- In a **Birch reduction**, the aromatic moiety is reduced to give a nonconjugated diene. The carbon atom connected to an alkyl group is not reduced, while the carbon atom connected to an electron-withdrawing group is reduced.

SECTION 18.8

- Aromatic compounds generally produce IR signals in five distinctive regions of the IR spectrum.
- Benzylic protons produce ^1H NMR signals between 2 and 3 ppm, while aromatic protons produce a characteristic signal (usually a multiplet) around 7 ppm.
- The sp^2-hybridized carbon atoms of an aromatic ring produce ^{13}C NMR signals between 100 and 150 ppm.

SKILLBUILDER REVIEW

18.1 NAMING A POLYSUBSTITUTED BENZENE

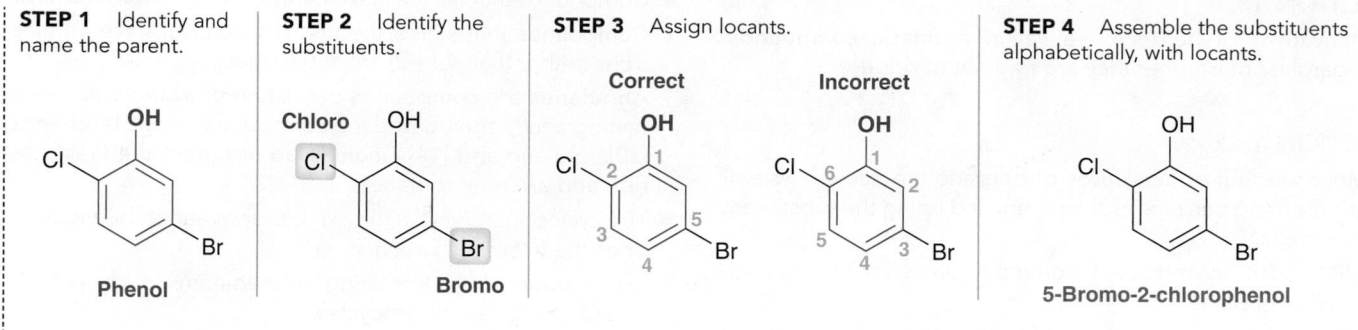

STEP 1 Identify and name the parent.

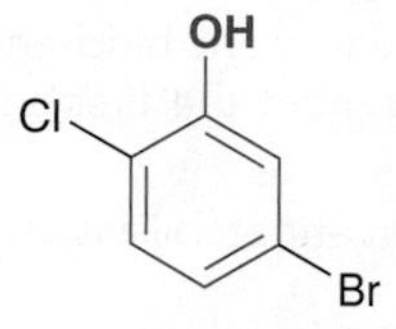

Phenol

STEP 2 Identify the substituents.

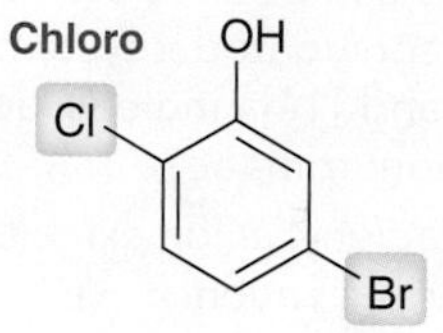

Bromo

STEP 3 Assign locants.

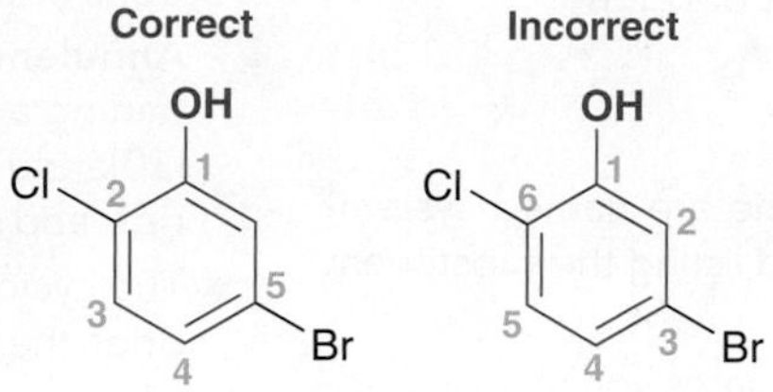

STEP 4 Assemble the substituents alphabetically, with locants.

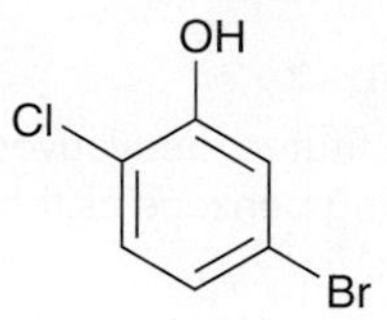

5-Bromo-2-chlorophenol

Try Problems **18.1–18.5, 18.28, 18.29, 18.33**

18.2 DETERMINING WHETHER A COMPOUND IS AROMATIC, NONAROMATIC, OR ANTIAROMATIC

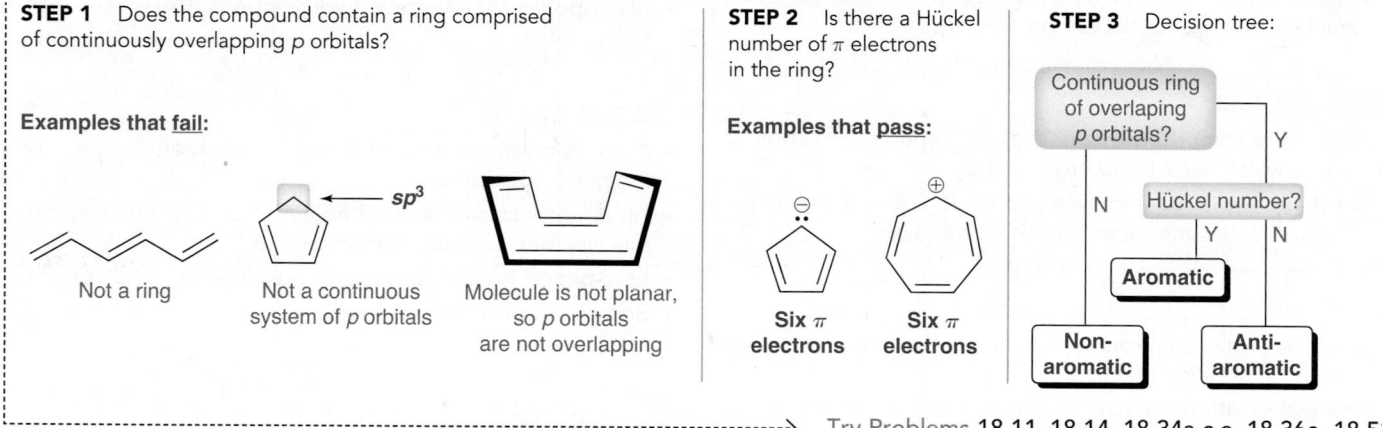

STEP 1 Does the compound contain a ring comprised of continuously overlapping *p* orbitals?

Examples that fail:

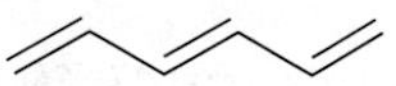

Not a ring

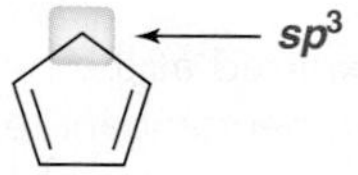

Not a continuous system of *p* orbitals

Molecule is not planar, so *p* orbitals are not overlapping

STEP 2 Is there a Hückel number of π electrons in the ring?

Examples that pass:

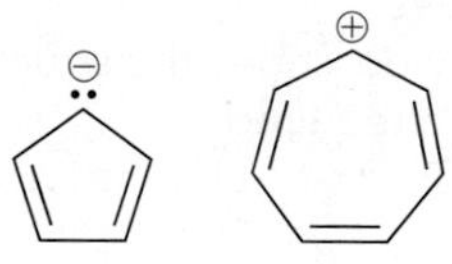

Six π electrons **Six π electrons**

STEP 3 Decision tree:

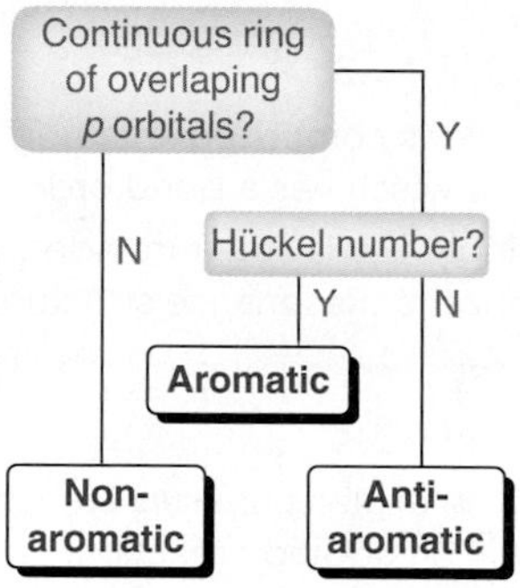

Try Problems **18.11–18.14, 18.34a,c,e, 18.36e, 18.52**

18.3 DETERMINING WHETHER A LONE PAIR PARTICIPATES IN AROMATICITY

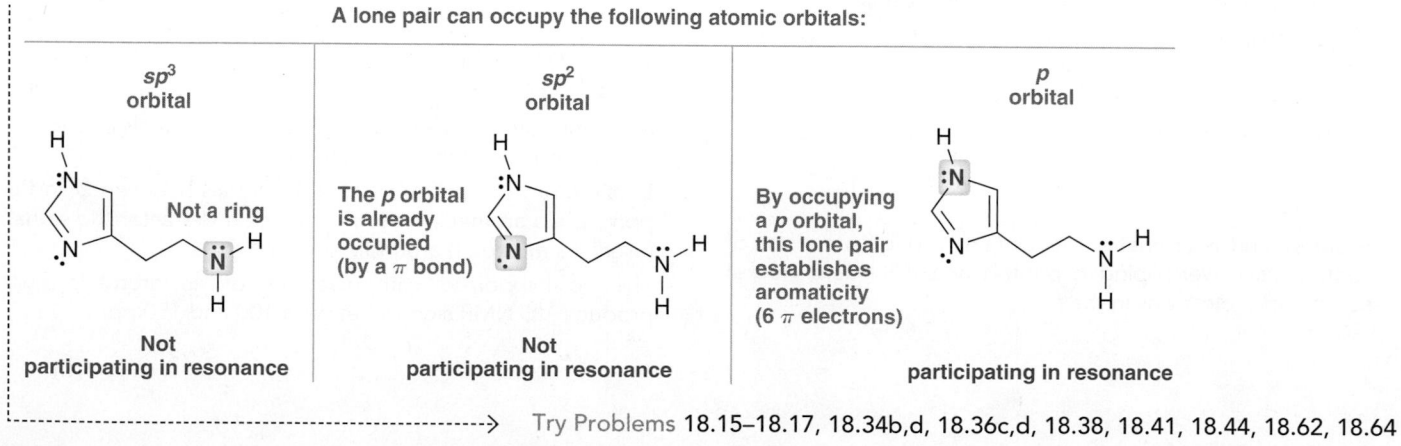

A lone pair can occupy the following atomic orbitals:

***sp*^{3} orbital**	***sp*^{2} orbital**	***p* orbital**
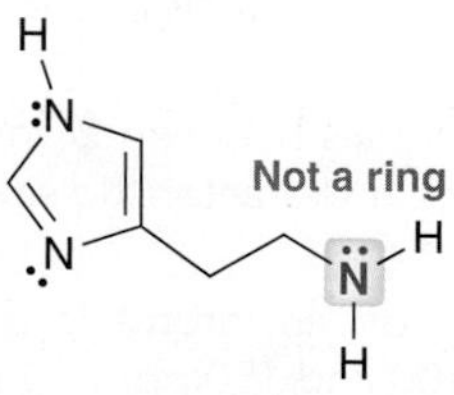	**The *p* orbital is already occupied (by a π bond)** 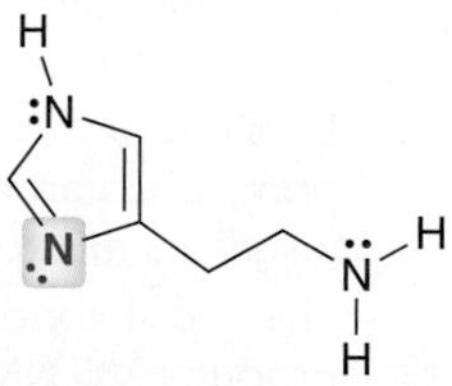	**By occupying a *p* orbital, this lone pair establishes aromaticity (6 π electrons)** 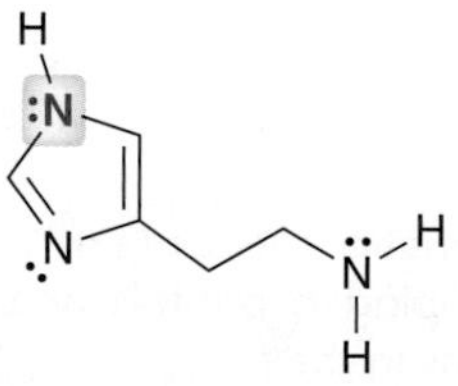
Not participating in resonance	**Not participating in resonance**	**participating in resonance**

Try Problems **18.15–18.17, 18.34b,d, 18.36c,d, 18.38, 18.41, 18.44, 18.62, 18.64**

18.4 MANIPULATING THE SIDE CHAIN OF AN AROMATIC COMPOUND

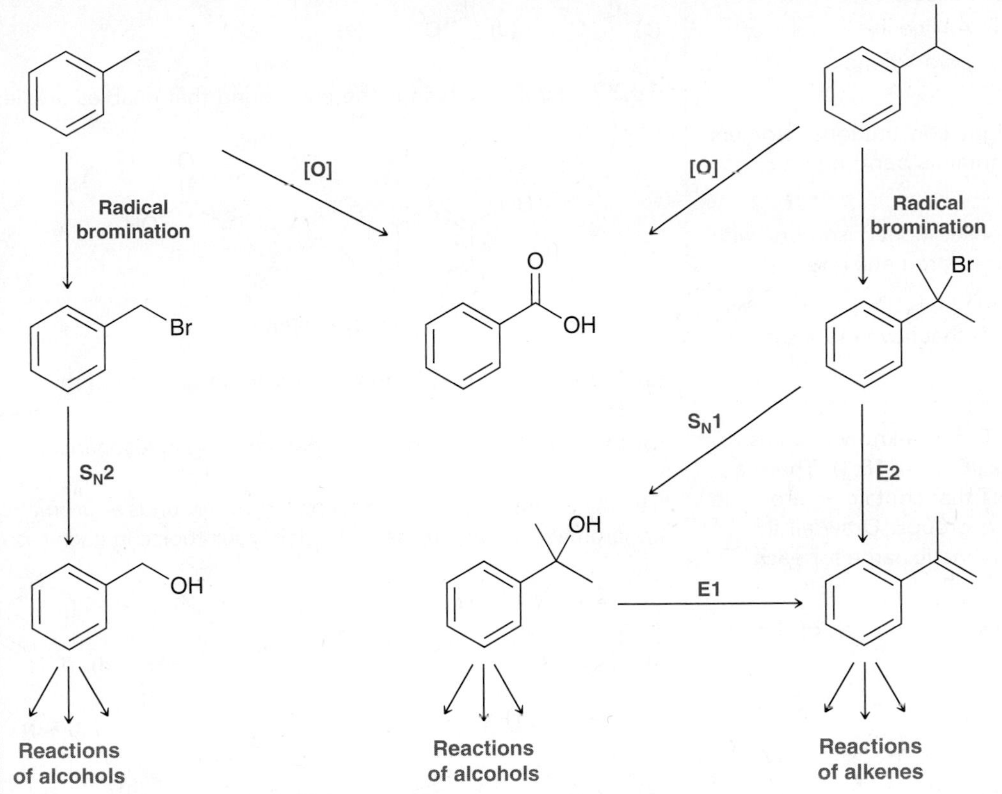

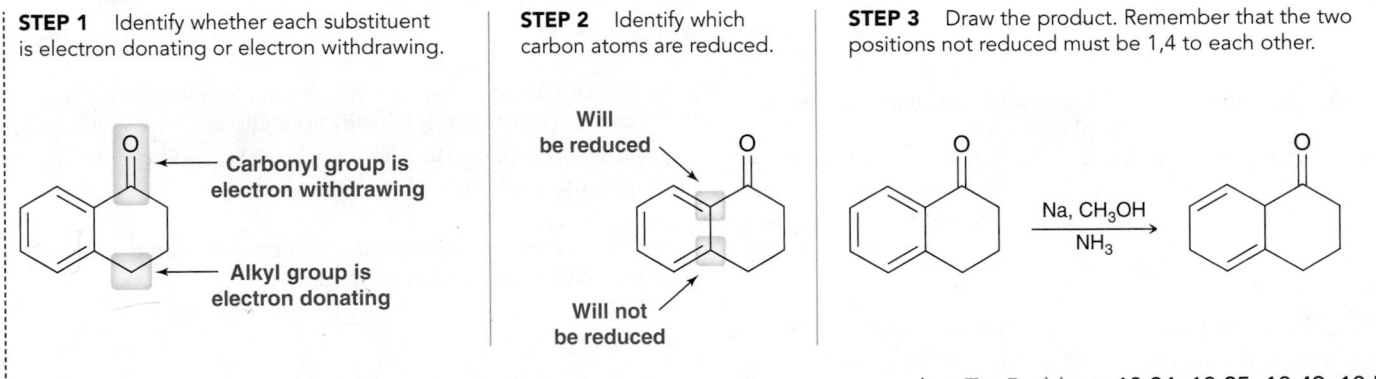

Try Problems 18.20–18.23, 18.47, 18.56

18.5 PREDICTING THE PRODUCT OF A BIRCH REDUCTION

STEP 1 Identify whether each substituent is electron donating or electron withdrawing.

Carbonyl group is electron withdrawing

Alkyl group is electron donating

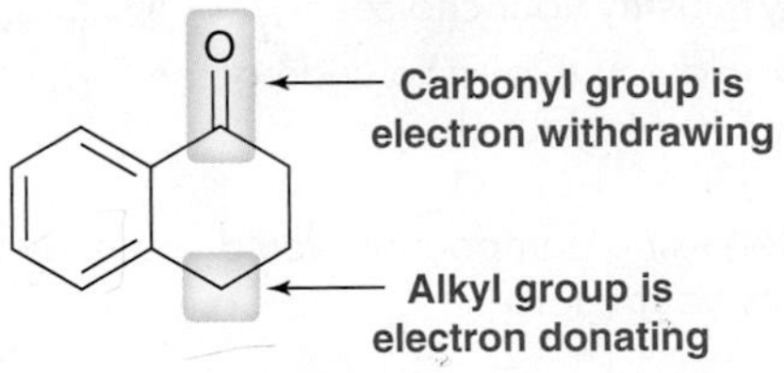

STEP 2 Identify which carbon atoms are reduced.

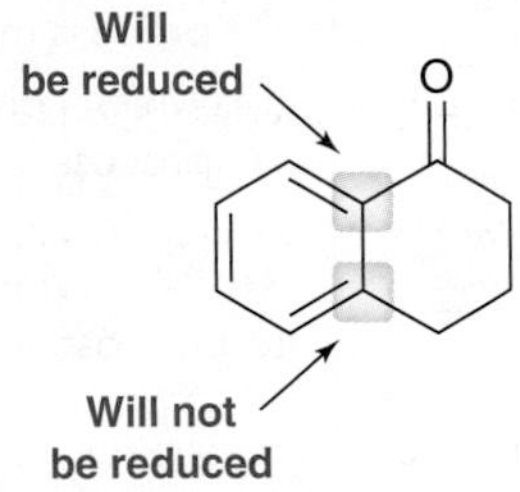

STEP 3 Draw the product. Remember that the two positions not reduced must be 1,4 to each other.

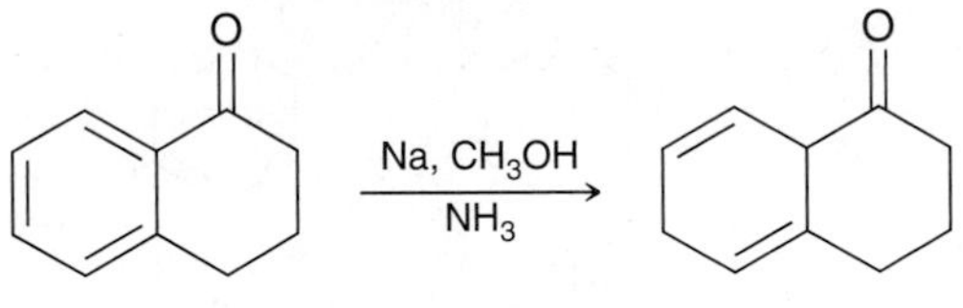

Try Problems 18.24, 18.25, 18.49, 18.50

PRACTICE PROBLEMS

Note: Most of the Problems are available within *WileyPLUS*, an online teaching and learning solution.

18.28 Provide a systematic name for each of the following compounds.

(a) para-ethylbenzoic acid

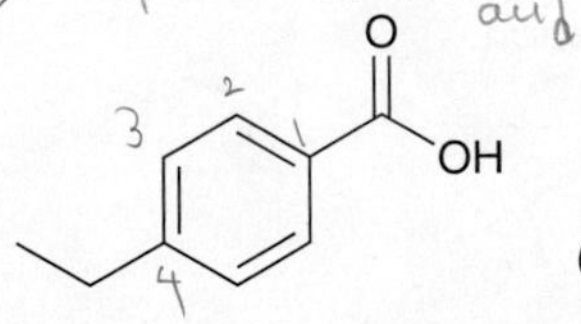

(b)

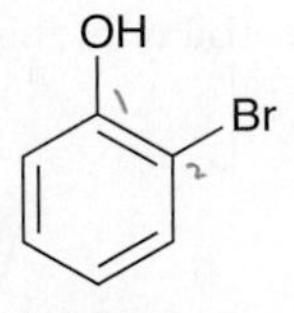

2 bromo phenol

(c) 2 chloro-4-nitrophenol

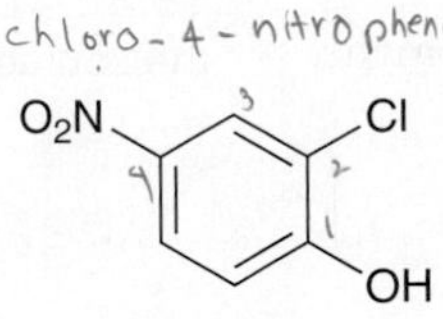

(d)

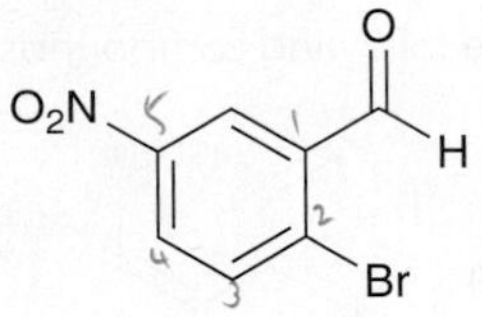

2-bromo-5-nitro benzaldehyde

(e) 1,4 diisopropyl benzene

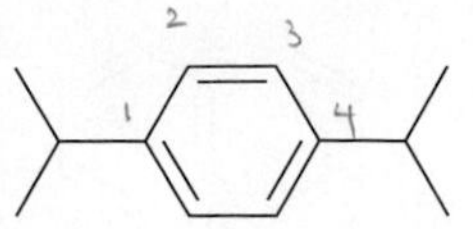

18.29 Draw a structure for each of the following compounds.

(a) *ortho*-Dichlorobenzene (b) Anisole
(c) *meta*-Nitrotoluene (d) Aniline
(e) 2,4,6-Tribromophenol (f) *para*-Xylene

18.30 Draw structures for the eight constitutional isomers with molecular formula C_9H_{12} that contain a benzene ring.

18.31 Draw structures for all constitutional isomers with molecular formula C_8H_{10} that contain an aromatic ring.

18.32 Draw all aromatic compounds that have molecular formula C_8H_9Cl.

18.33 The systematic name of TNT, a well-known explosive, is 2,4,6-trinitrotoluene (as seen in SkillBuilder 18.1). There are only five constitutional isomers of TNT that contain an aromatic ring, a methyl group, and three nitro groups. Draw all five of these compounds, and provide a systematic name for each.

18.34 Identify the number of π electrons in each of the following compounds.

(a) 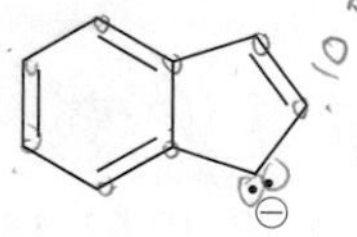
(b) 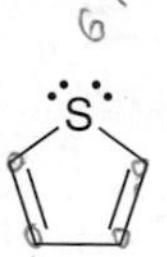
(c) 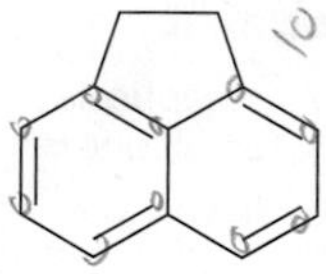
(d) 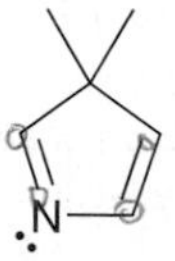
(e) 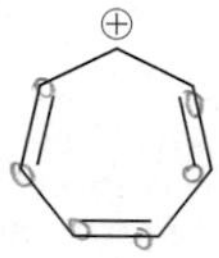

18.35 Students often confuse cyclohexane and benzene:

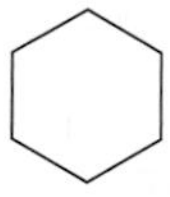
Cyclohexane

Benzene

In fact, these compounds have different properties, different geometry, and different reactivity. Each of these compounds also has a unique set of terminology. For each of the following terms, identify whether it is used in reference to benzene or to cyclohexane:

(a) *meta* (b) Frost circle (c) sp^2
(d) Chair (e) *ortho* (f) sp^3
(g) Resonance (h) π Electrons (i) *para*
(j) Ring flip (k) Boat

18.36 Identify which of the following compounds are aromatic.

(a) 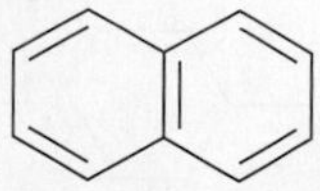
(b) 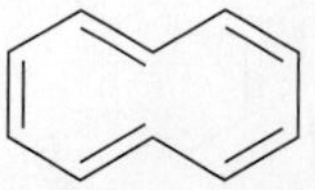
(c) 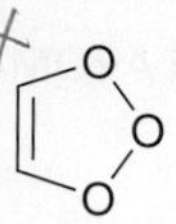
(d) 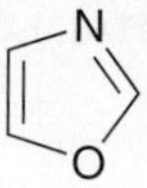
(e) 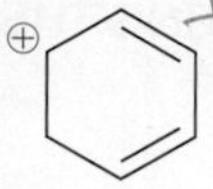

18.37 Firefly luciferin is the compound that enables fireflies to glow.

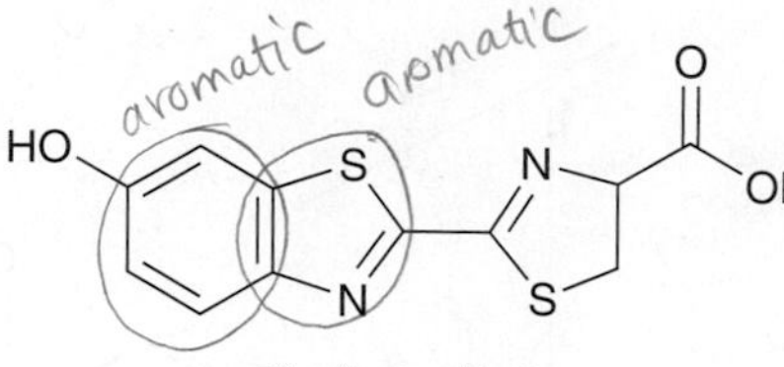

Firefly luciferin

(a) The structure exhibits three rings. Identify which of the rings are aromatic.
(b) Identify which lone pairs are participating in resonance.

18.38 Identify each of the following compounds as aromatic, nonaromatic, or antiaromatic. Explain your choice in each case.

(a) 
(b) 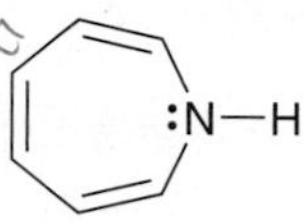
(c) 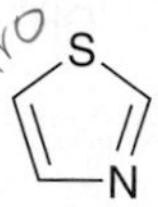
(d) 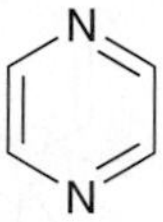
(e) 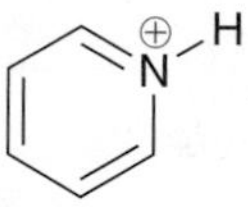
(f) 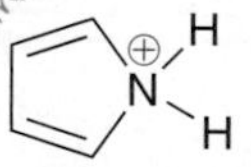
(g) 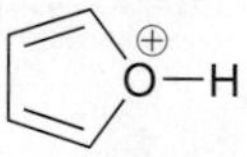
(h) 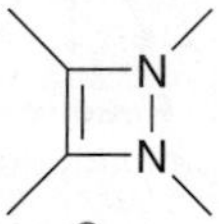

18.39 Consider the structures of the following alkyl chlorides:

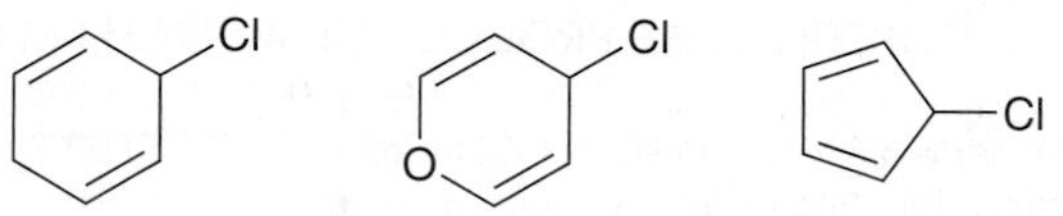

(a) Which compound would you expect to undergo an S_N1 process most readily? Justify your choice.
(b) Which compound would you expect to undergo an S_N1 process least readily? Justify your choice.

18.40 Which of the following compounds would you expect to be most acidic? Justify your choice.

18.41 Identify which of the following compounds is expected to be a stronger base. Justify your choice.

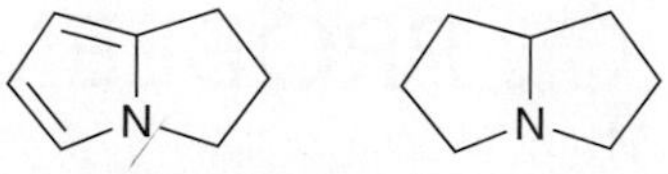

18.42 Draw a Frost circle for the following cation, and explain the source of instability of this cation.

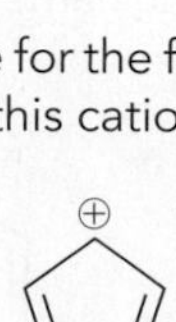

18.43 Do you expect the following dianion to exhibit aromatic stabilization? Explain.

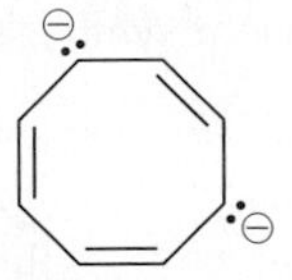

18.44 Would you expect the following compound to be aromatic? Justify your answer.

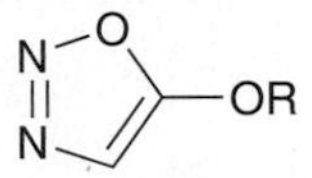

18.45 Diphenylmethane exhibits two aromatic rings, which achieve coplanarity in the highest energy conformation. Explain.

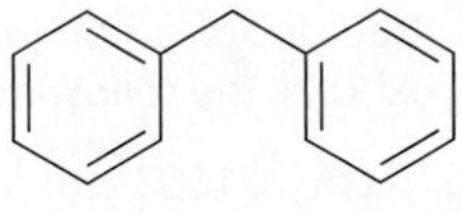

Diphenylmethane

18.46 The following two drawings are resonance structures of one compound:

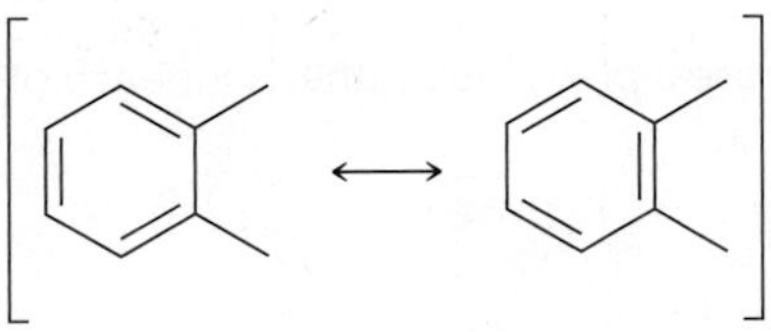

But the following two drawings are not resonance structures:

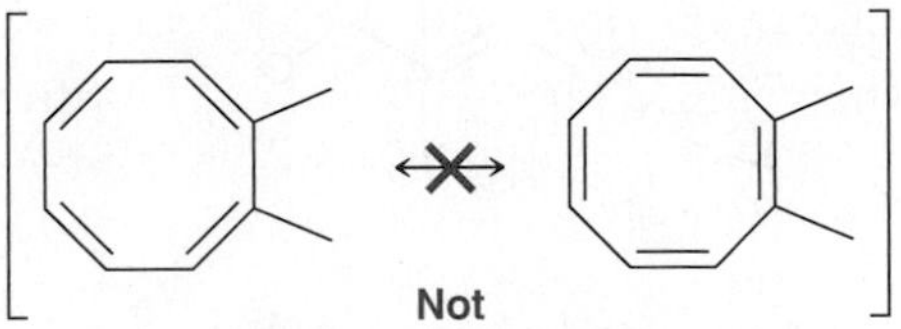

Not resonance structures

They are, in fact, two different compounds. Explain.

18.47 Predict the major product of the following reactions.

(a)
$$\xrightarrow[\text{Heat or light}]{\text{NBS}}$$?

(b)
$$\xrightarrow[\text{H}_2\text{SO}_4, \text{H}_2\text{O}]{\text{Na}_2\text{Cr}_2\text{O}_7}$$?

(c)
$$\xrightarrow[\text{Heat}]{\text{H}_2\text{SO}_4}$$?

(d)
$$\xrightarrow{\text{NaOEt}}$$?

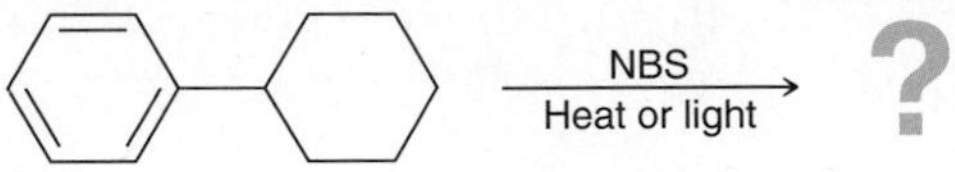

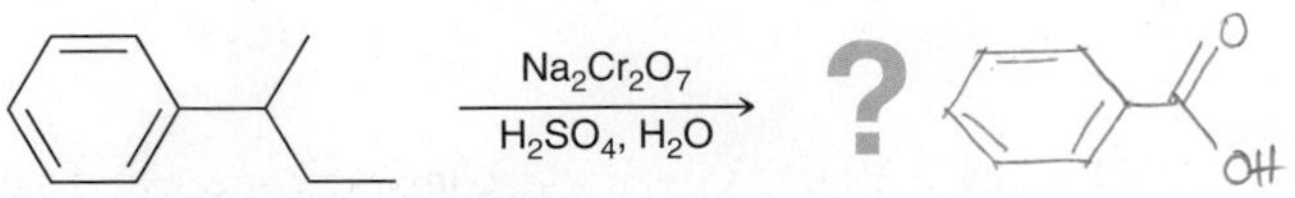

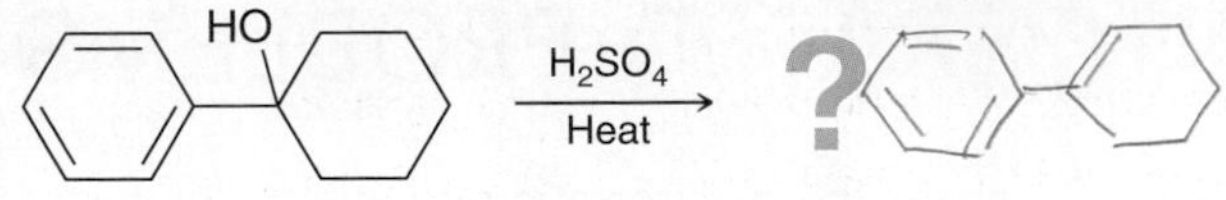

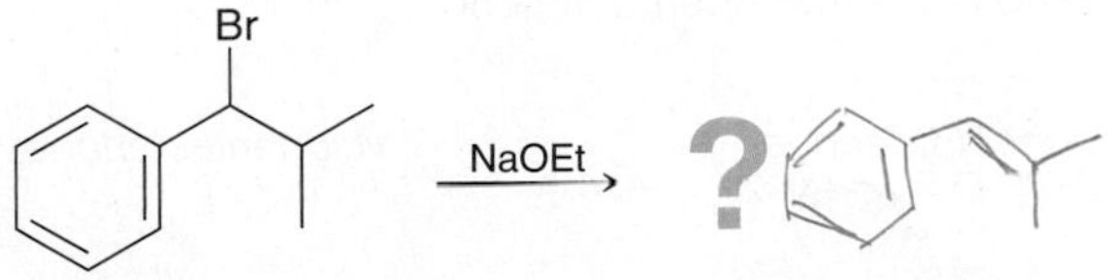

18.48 How many signals do you expect in the ^{13}C NMR spectrum of each of the following compounds?

(a) 6

(b) 5

(c) 3

(d) 9

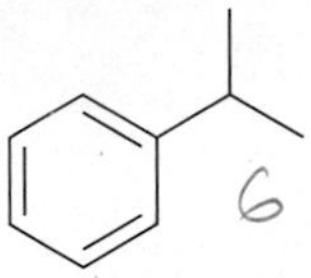

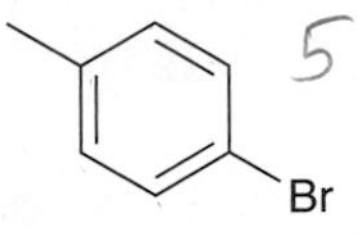

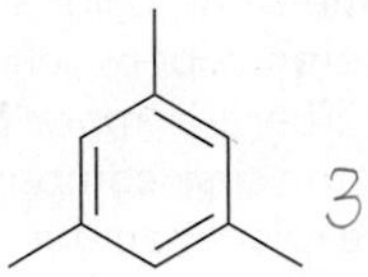

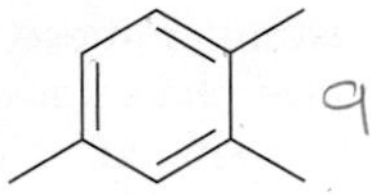

18.49 Predict the product of the following reaction, and propose a mechanism for its formation.

$$\xrightarrow[\text{NH}_3]{\text{Na, CH}_3\text{OH}}$$?

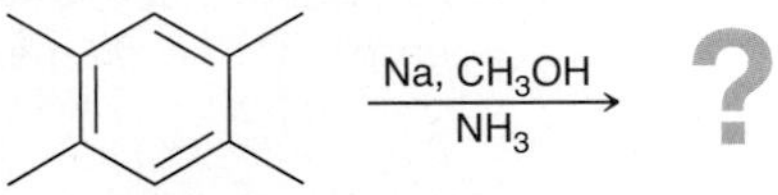

18.50 One of the constitutional isomers of xylene was treated with sodium, methanol, and ammonia to yield a product that exhibited five signals in its ^{13}C NMR spectrum. Identify which constitutional isomer of xylene was used as the starting material.

18.51 Consider the following two compounds. How would you distinguish between them using:

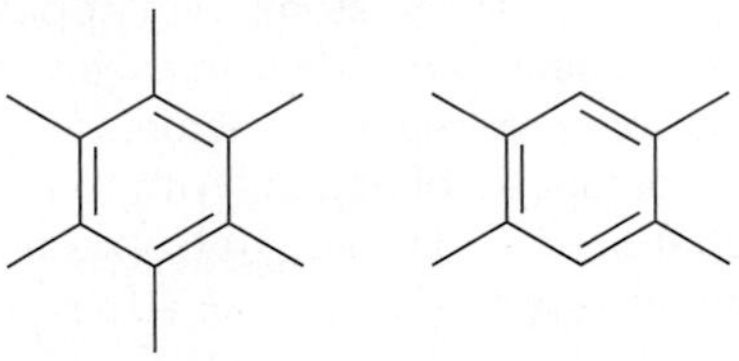

(a) IR spectroscopy?
(b) ^1H NMR spectroscopy?
(c) ^{13}C NMR spectroscopy?

18.52 Explain how the following two compounds can have the same conjugate base. Is this conjugate base aromatic?

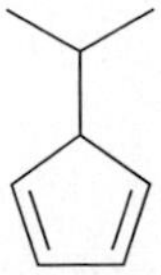

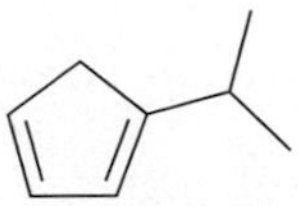

INTEGRATED PROBLEMS

18.53 Compare the electrostatic potential maps for cycloheptatrienone and cyclopentadienone.

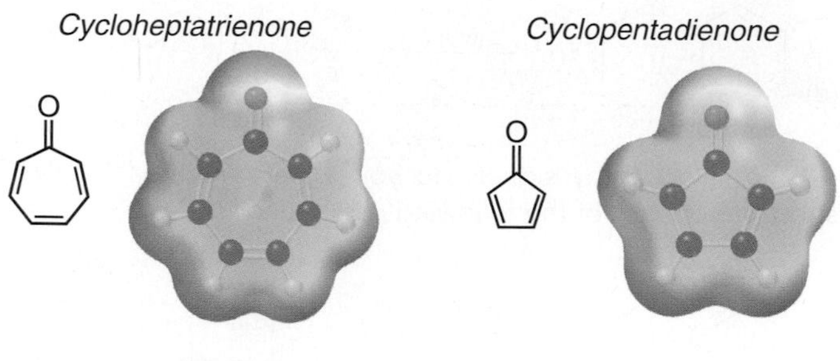

Both of these maps were created using the same color scale so they can be compared. Notice the difference between the oxygen atoms in these two compounds. There is more partial negative character on the oxygen in the first compound (cycloheptatrienone). Can you offer an explanation for this difference?

18.54 Azulene exhibits an appreciable dipole moment, and an electrostatic potential map indicates that the five-membered ring is electron rich (at the expense of the seven-membered ring).

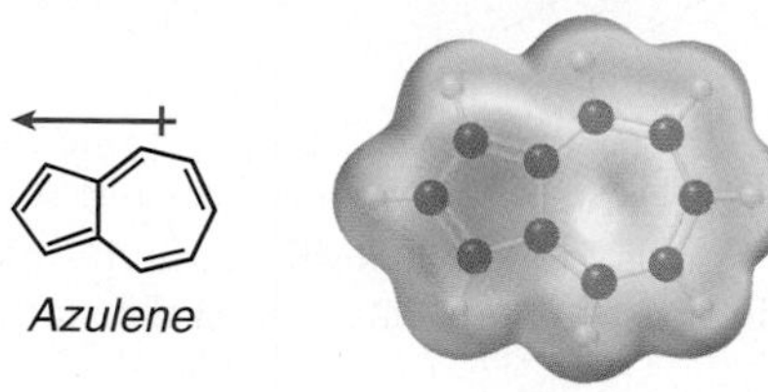

(a) In Chapter 2, we saw that a resonance structure will be insignificant if it has carbon atoms with opposite charges (C− and C+). Azulene represents an exception to this rule, because some resonance structures (with C− and C+) exhibit aromatic stabilization. With this in mind, draw structures of azulene and use them to explain the observed dipole moment.

(b) Based on your explanation, determine which compound is expected to exhibit a larger dipole moment.

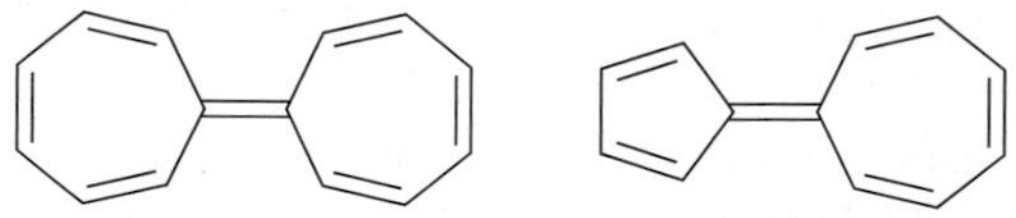

18.55 Propose an efficient synthesis for the following transformation.

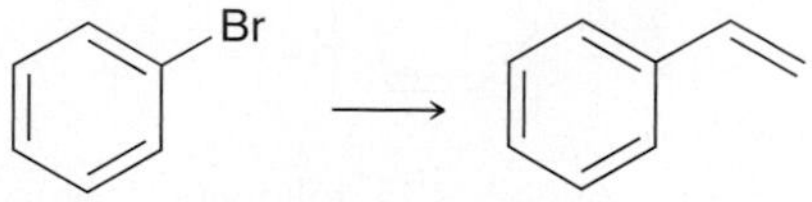

18.56 Propose a plausible mechanism for the following transformation.

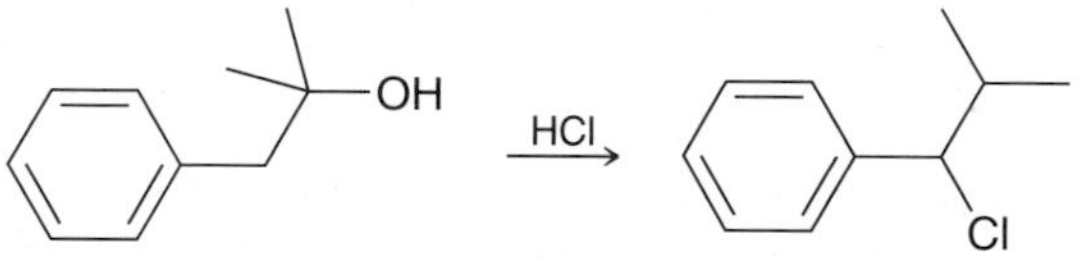

18.57 Identify the structure of a compound with molecular formula $C_9H_{10}O_2$ that exhibits the following spectral data.

(a) IR: 3005 cm^{-1}, 1676 cm^{-1}, 1603 cm^{-1}

(b) ^1H NMR: 2.6 ppm (singlet, I = 3H), 3.9 ppm (singlet, I = 3H), 6.9 ppm (doublet, I = 2H), 7.9 ppm (doublet, I = 2H)

(c) ^{13}C NMR: 26.2, 55.4, 113.7, 130.3, 130.5, 163.5, 196.6 ppm

18.58 Propose a plausible synthesis for each of the following transformations.

(a) 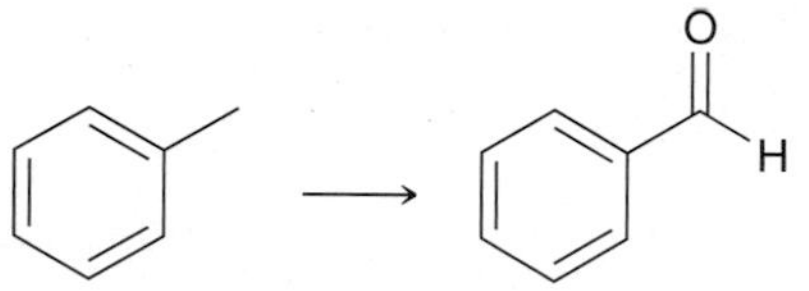

(b) 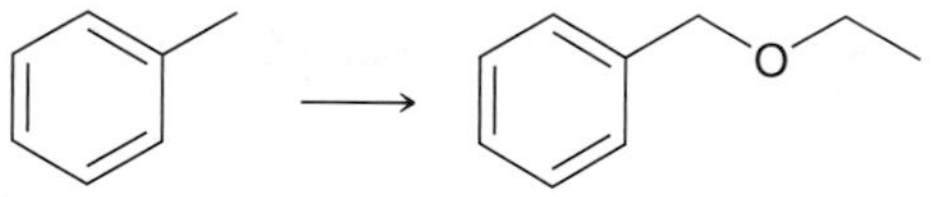

(c) 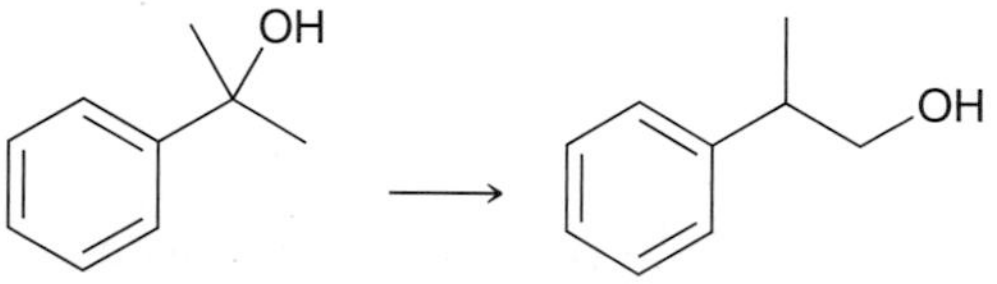

(d) 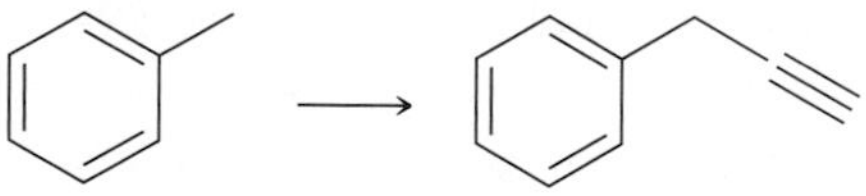

18.59 A compound with molecular formula $C_{11}H_{14}O_2$ exhibits the following spectra (^1H NMR, ^{13}C NMR, and IR). Identify the structure of this compound.

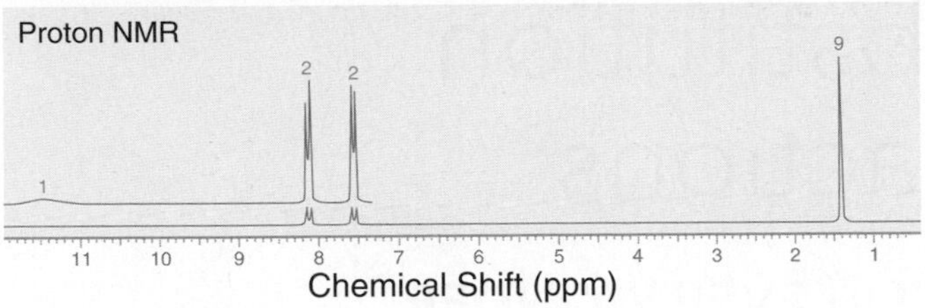

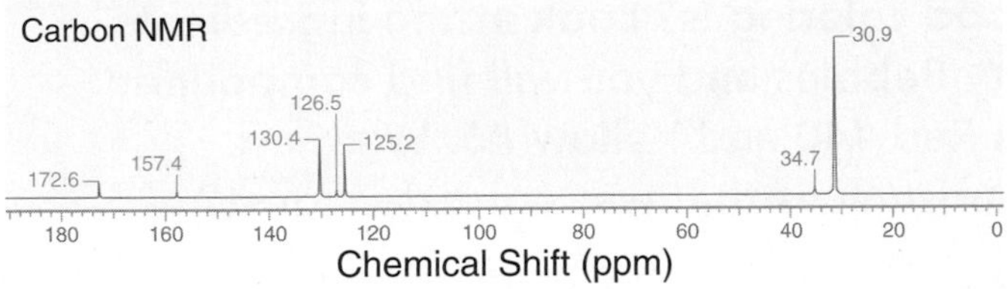

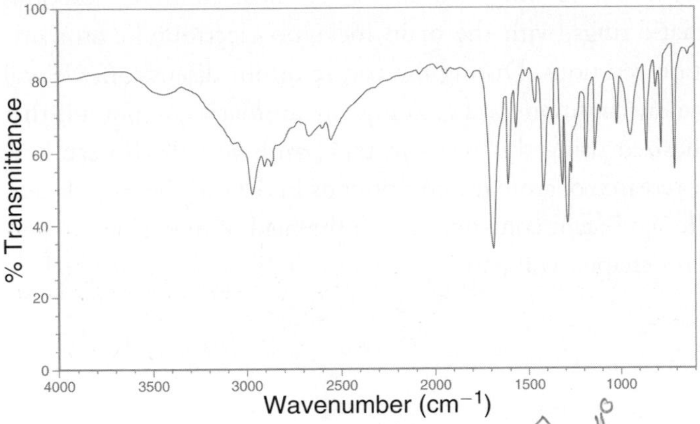

18.60 A compound with molecular formula $C_9H_{10}O$ exhibits the following spectra (^1H NMR, ^{13}C NMR, and IR). Identify the structure of this compound.

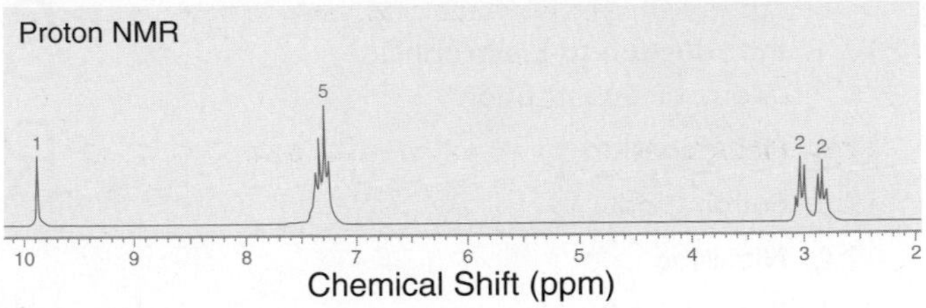

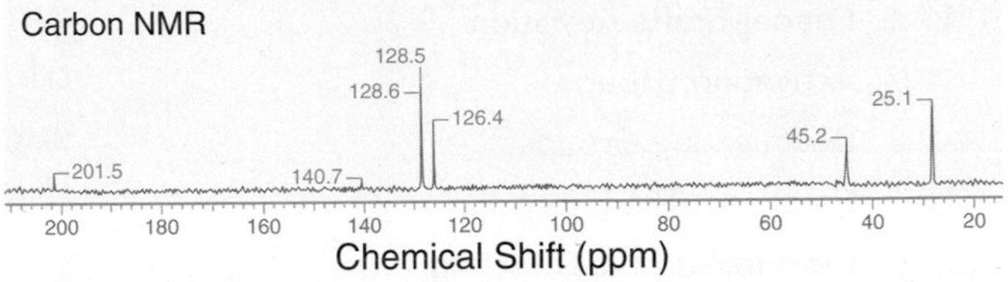

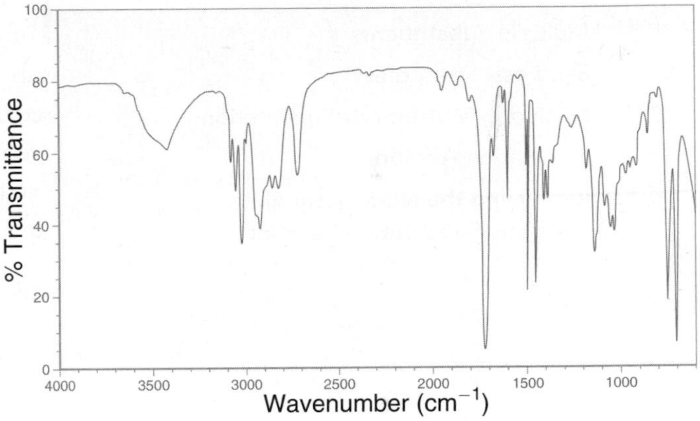

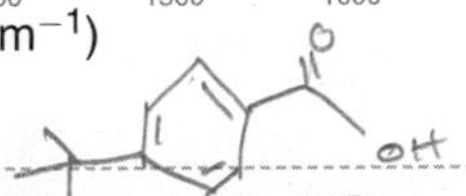

CHALLENGE PROBLEMS

18.61 Below are two hypothetical compounds.

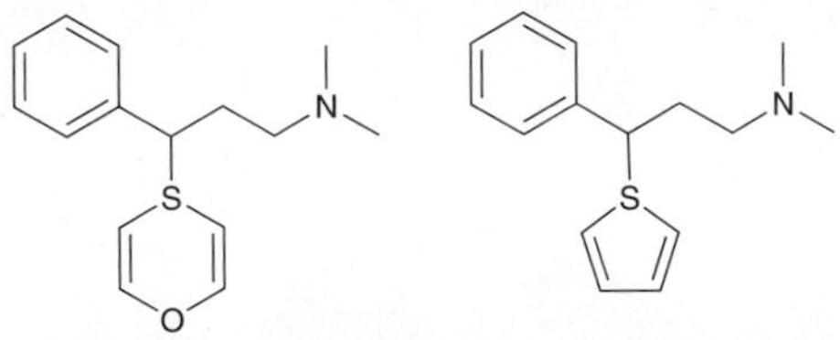

(a) Which compound would you expect to hold greater promise as a potential antihistamine? Explain your choice.

(b) Do you expect the compound you chose (in part a) to exhibit sedative properties? Explain your reasoning.

18.62 Would you expect the following compound to be aromatic? Explain your answer.

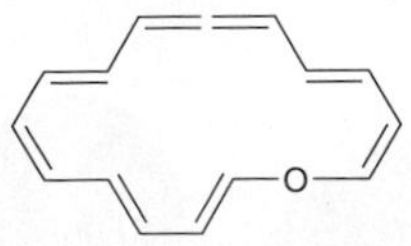

18.63 Compounds **A**, **B**, **C**, and **D** are constitutionally isomeric, aromatic compounds with molecular formula C_8H_{10}. Deduce the structure of compound **D** using the following clues.

- The ^1H NMR spectrum of compound **A** exhibits two upfield signals as well as a multiplet near 7 ppm (with I=5).
- The ^{13}C NMR spectrum of compound **B** exhibits four signals.
- The ^{13}C NMR spectrum of compound **C** exhibits only three signals.

18.64 The following two compounds each exhibit two heteroatoms (one nitrogen atom and one oxygen atom).

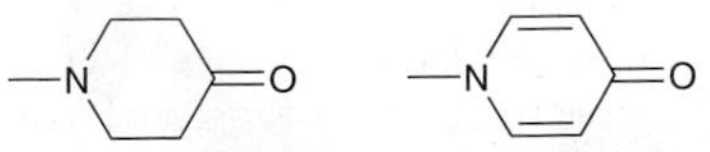

Compound A **Compound B**

In compound **A**, the lone pair on the nitrogen atom is more likely to function as a base. However, in compound **B**, the lone pair on the oxygen atom is more likely to function as a base. Explain this difference.

18.65 Propose a plausible synthesis for the following transformation.

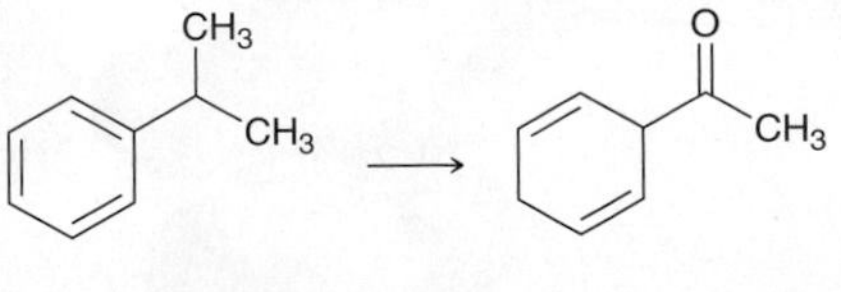

18.66 Using toluene and acetylene as your only sources of carbon atoms, show how you would prepare the following compound.

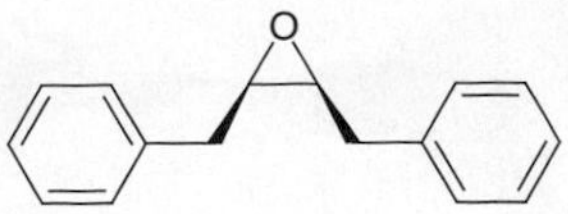

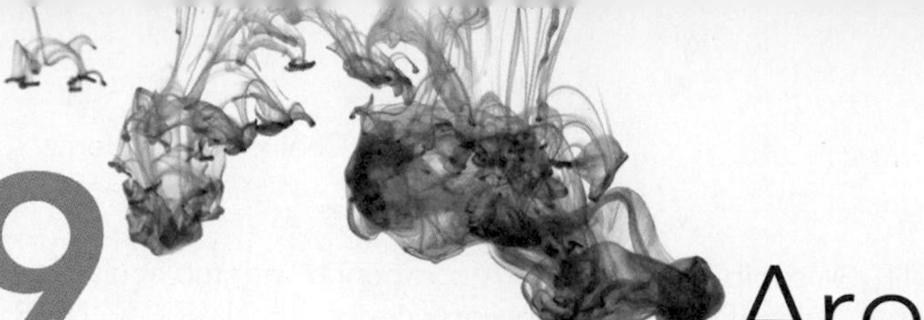

19 Aromatic Substitution Reactions

DID YOU EVER WONDER...

what food coloring is? Look at the ingredients of Fruity Pebbles and you will find compounds such as Red #40 and Yellow #6. What are these substances that we regularly ingest?

In this chapter, we will learn about the most common reactions of aromatic rings, with the main focus on electrophilic aromatic substitution reactions. During the course of our discussion, we will see that many common food colorings are aromatic compounds that are synthesized using this reaction type, and we will also see how extensive research of aromatic compounds in the early twentieth century made significant contributions to the field of medicine.

This chapter will paint Fruity Pebbles in a whole new light.

DO YOU REMEMBER?

Before you go on, be sure you understand the following topics.
If necessary, review the suggested sections to prepare for this chapter:

- Resonance Structures (Sections 2.7–2.11)
- Lewis Acids (Section 3.9)
- Retrosynthetic Analysis (Section 12.5)
- Delocalized Lone Pairs (Section 2.12)
- Reading Energy Diagrams (Section 6.6)
- Aromaticity and Nomenclature of Aromatic Compounds (Sections 18.1–18.4)

WILEY PLUS Visit www.wileyplus.com to check your understanding and for valuable practice.

19.1 Introduction to Electrophilic Aromatic Substitution

In the previous chapter, we explored the remarkable stability of benzene. Specifically, we saw that while alkenes undergo an addition reaction when treated with bromine, benzene is inert under the same conditions.

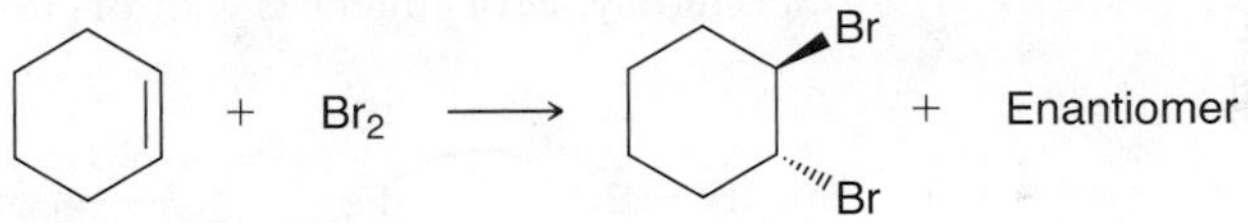

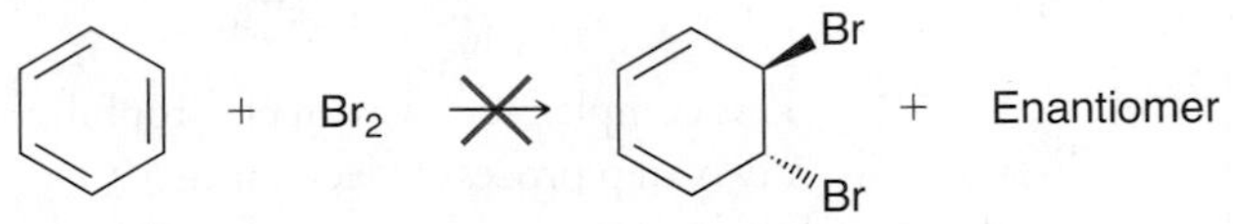

Curiously though, when Fe (iron) is introduced into the mixture, a reaction does in fact take place, although the product is not what we might have expected.

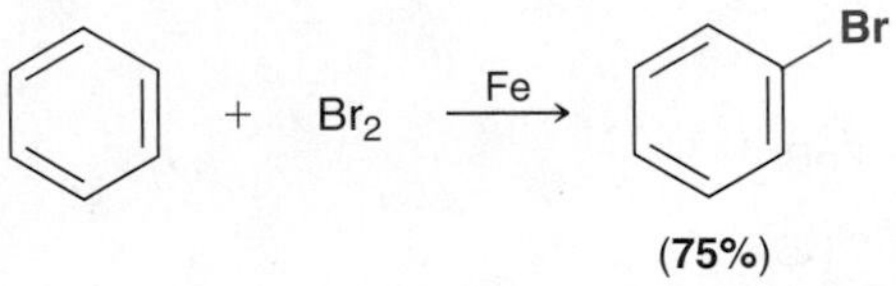

Rather than an addition reaction taking place, the observed reaction is **electrophilic aromatic substitution** reaction in which one of the aromatic protons is replaced by an electrophile, and the aromatic moiety is preserved. In this chapter, we will see many other groups that can also be installed on an aromatic ring via an electrophilic aromatic substitution reaction.

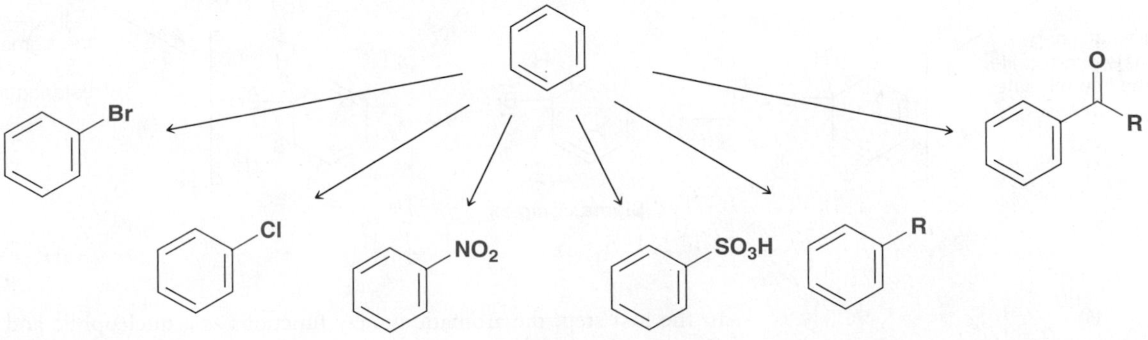

19.2 Halogenation

Recall from Section 9.8 that during bromination of an alkene, Br_2 functions as an electrophile.

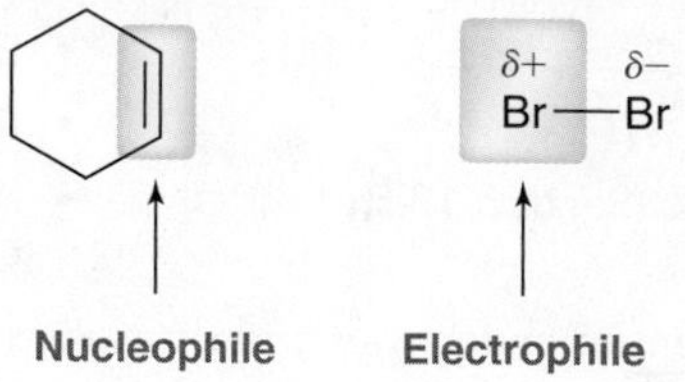

δ+ δ−
Br—Br
Nucleophile
Electrophile

As it approaches the π electron cloud of the alkene, Br_2 becomes temporarily polarized, rendering one of the bromine atoms electrophilic (δ+). This bromine atom becomes sufficiently electrophilic to react with the alkene but is not sufficiently electrophilic to react with benzene. The presence of iron (Fe) in the reaction mixture enhances the electrophilicity of this bromine atom. To understand how iron accomplishes this task, we must recognize that iron itself is not the real catalyst. Rather, it first reacts with Br_2 to generate iron tribromide ($FeBr_3$).

$$2\,Fe + 3\,Br_2 \longrightarrow 2\,FeBr_3$$

Iron tribromide, a Lewis acid, is the real catalyst in the reaction between benzene and bromine. Specifically, $FeBr_3$ interacts with Br_2 to form a complex which reacts as if it were Br^+.

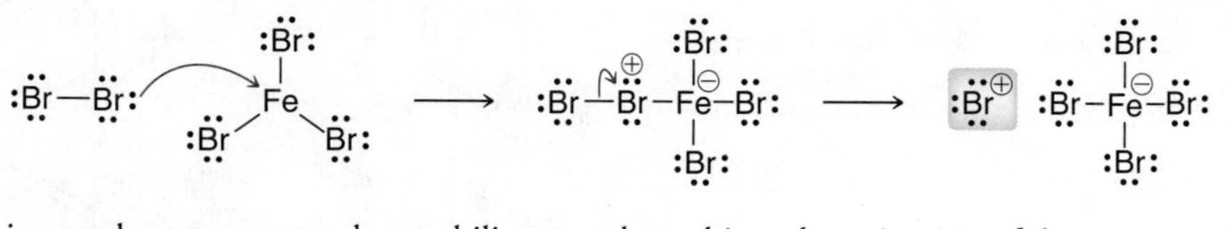

This complex serves as an electrophilic agent that achieves bromination of the aromatic ring via a two-step process (Mechanism 19.1).

MECHANISM 19.1 BROMINATION OF BENZENE

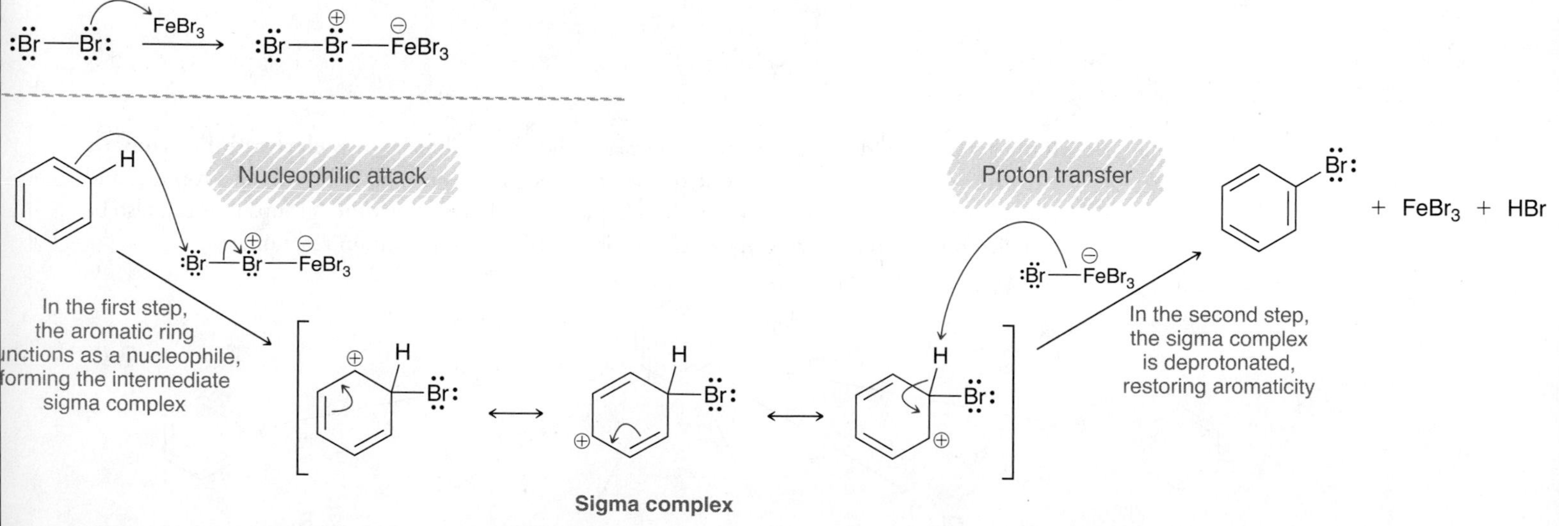

FeBr3
Nucleophilic attack
Proton transfer
In the first step, the aromatic ring functions as a nucleophile, forming the intermediate sigma complex
In the second step, the sigma complex is deprotonated, restoring aromaticity
+ FeBr3 + HBr
Sigma complex

In the first step, the aromatic moiety functions as a nucleophile and attacks the electrophilic agent, generating a positively charged intermediate called a **sigma complex**, or **arenium ion**, which is resonance stabilized. This step requires an input of energy because it involves the temporary loss of aromatic stabilization. The loss of stabilization occurs because the sigma complex is not aromatic— it does not possess a continuous system of overlapping *p* orbitals.

In the second step of the mechanism, the sigma complex is then deprotonated, thereby restoring aromaticity and regenerating the Lewis acid ($FeBr_3$). Notice that the Lewis acid is ultimately not consumed by the reaction and is, therefore, a catalyst. Aluminum tribromide ($AlBr_3$) is another common Lewis acid that can serve as a suitable alternative to $FeBr_3$.

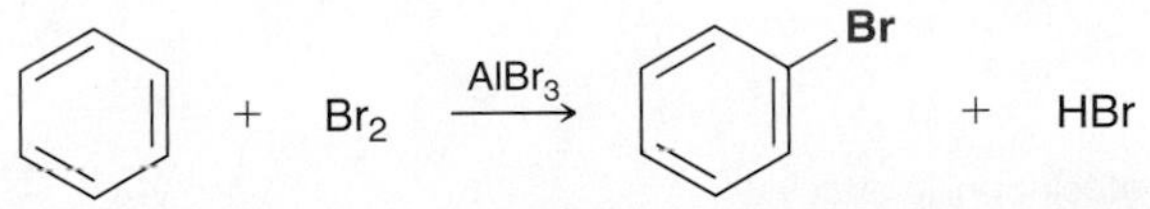

The observed substitution reaction is not accompanied by the formation of any addition products. Addition is not observed because it would involve a permanent loss of aromaticity, which is thermodynamically unfavorable (Figure 19.1). Notice that, overall, substitution is an exergonic

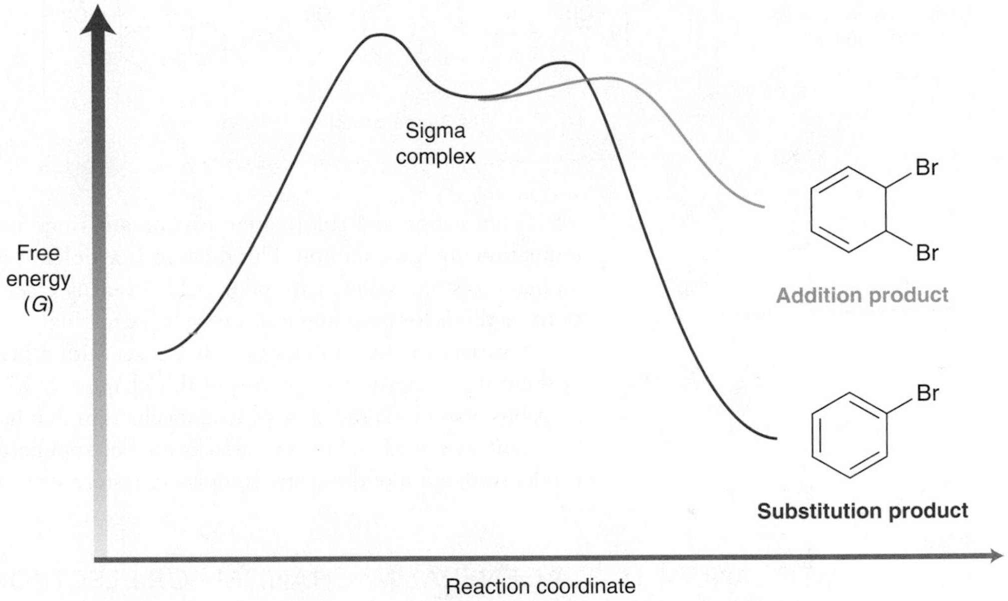

FIGURE 19.1
An energy diagram comparing substitution and addition pathways for benzene.

process (downhill in energy), while addition is an endergonic process (uphill in energy). For this reason, only substitution is observed.

A similar reaction occurs when chlorine is used rather than bromine. Chlorination of benzene is accomplished with a suitable Lewis acid, such as aluminum trichloride.

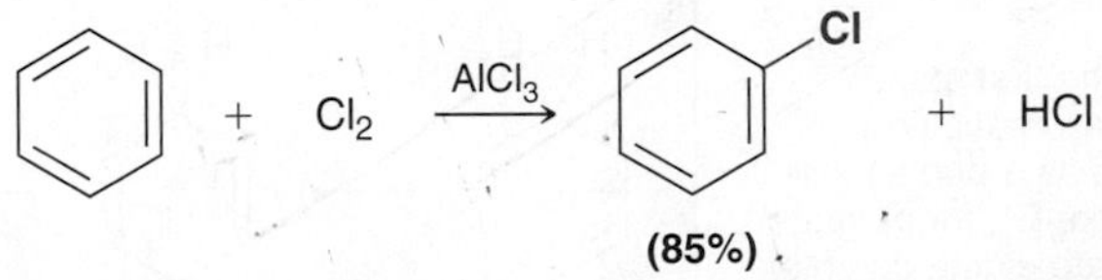

Chlorine reacts with $AlCl_3$ to form a complex, which reacts as if it were Cl^+.

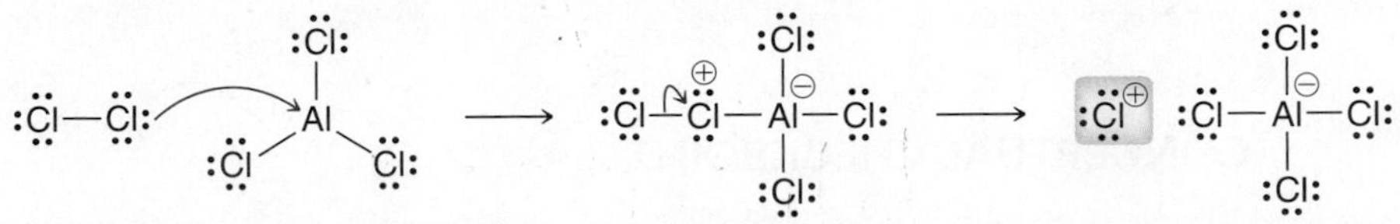

This complex is the electrophilic agent that achieves chlorination of the aromatic ring, as illustrated in Mechanism 19.2. This mechanism is directly analogous to the mechanism of bromination and involves the same two steps. In the first step, the aromatic moiety functions as a nucleophile and attacks the electrophilic agent, generating a sigma complex. Then, in the second step, the sigma complex is deprotonated, thereby restoring aromaticity and regenerating the Lewis acid ($AlCl_3$).

MECHANISM 19.2 CHLORINATION OF BENZENE

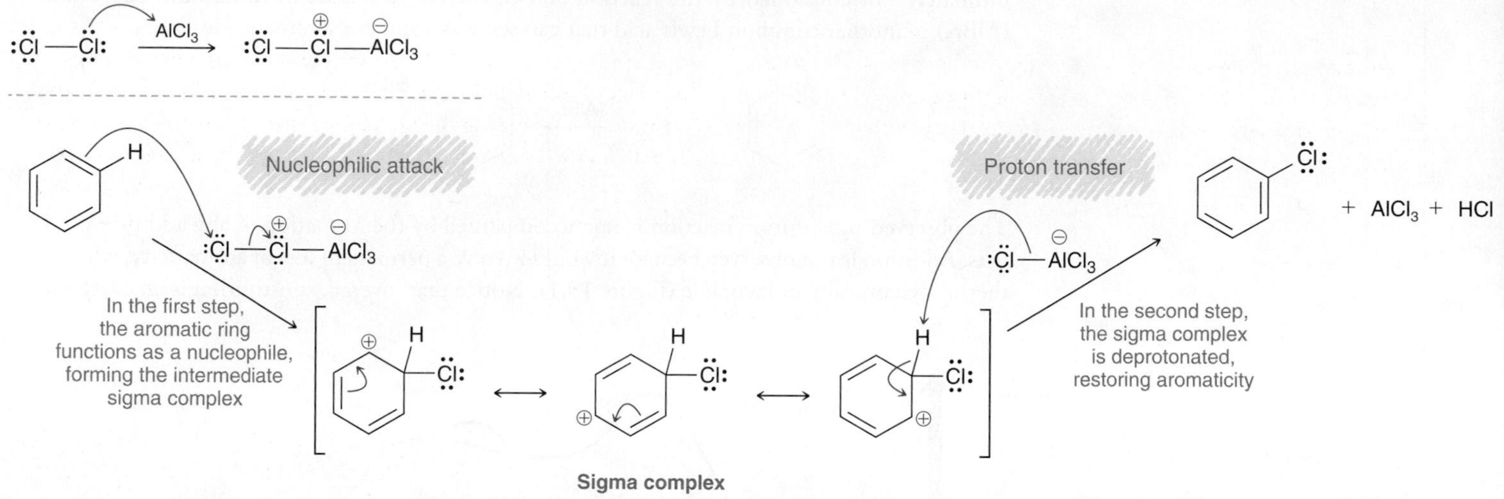

AlCl3
Nucleophilic attack
In the first step, the aromatic ring functions as a nucleophile, forming the intermediate sigma complex
Sigma complex
Proton transfer
In the second step, the sigma complex is deprotonated, restoring aromaticity
+ AlCl3 + HCl

Bromination and chlorination of aromatic rings are easily achieved, but fluorination and iodination are less common. Fluorination is a violent process and is difficult to control, while iodination is often slow, with poor yields in many cases. In Chapter 23, we will see more efficient methods for installing F or I on a benzene ring.

A variety of electrophiles (E^+) will react with a benzene ring, and we will explore many of them in the upcoming sections of this chapter. It will be helpful to realize that all of these reactions operate via the same general mechanism that has only two steps: (1) the aromatic ring functions as a nucleophile and attacks an electrophile to form a sigma complex followed by (2) deprotonation of the sigma complex to restore aromaticity (Mechanism 19.3).

MECHANISM 19.3 A GENERAL MECHANISM FOR ELECTROPHILIC AROMATIC SUBSTITUTION

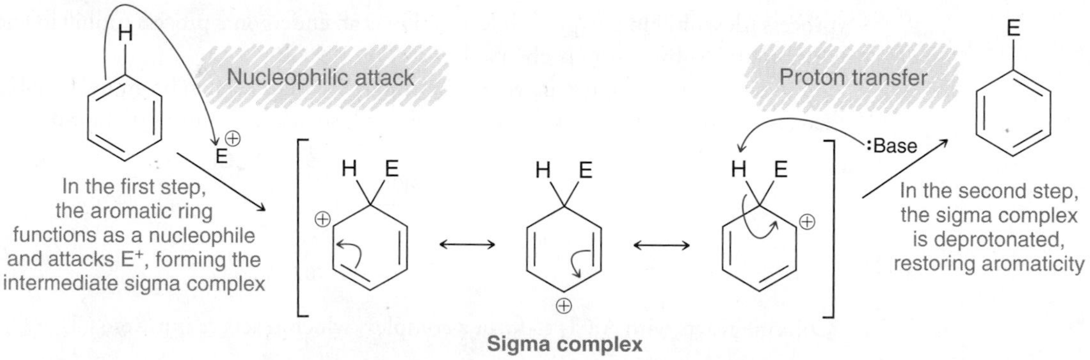

Nucleophilic attack
In the first step, the aromatic ring functions as a nucleophile and attacks E+, forming the intermediate sigma complex
Sigma complex
Proton transfer
:Base
In the second step, the sigma complex is deprotonated, restoring aromaticity

CONCEPTUAL CHECKPOINT

19.1 When benzene is treated with I_2 in the presence of $CuCl_2$, iodination of the ring is achieved with modest yields. It is believed that $CuCl_2$ interacts with I_2 to generate I^+, which is an excellent electrophile. The aromatic ring then reacts with I^+ in an electrophilic aromatic substitution reaction. Draw the mechanism of the reaction between benzene and I^+. Make sure that your mechanism has two steps, and make sure to draw all of the resonance structures of the sigma complex.

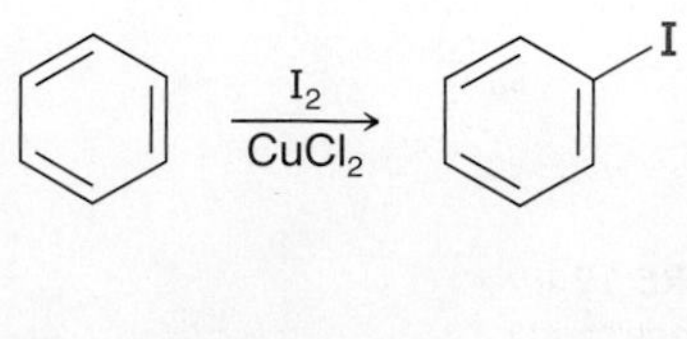

I2
CuCl2

MEDICALLYSPEAKING))

Halogenation in Drug Design

Aromatic halogenation is a common technique used in drug design, as chemists attempt to modify the structure of a known drug to produce new drugs with enhanced properties. For example, consider the following three compounds:

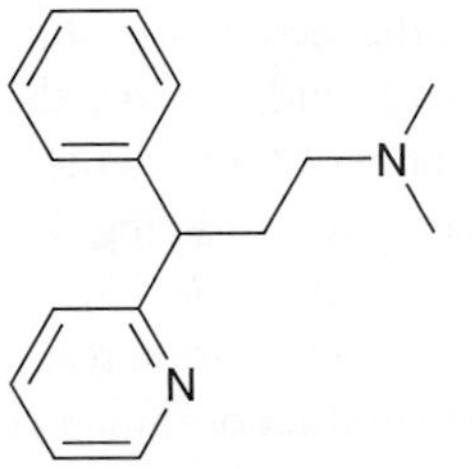

Pheniramine

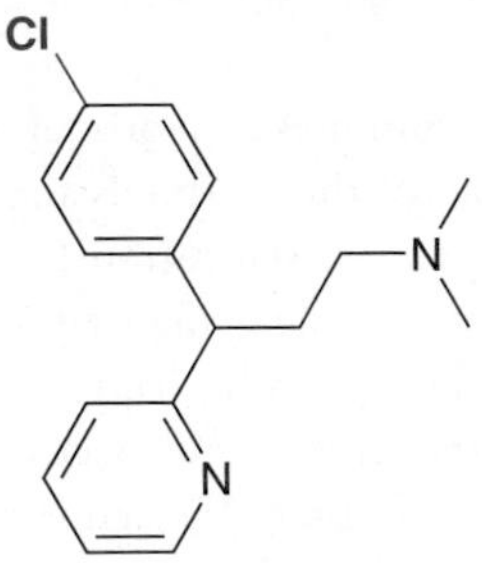

Chlorpheniramine

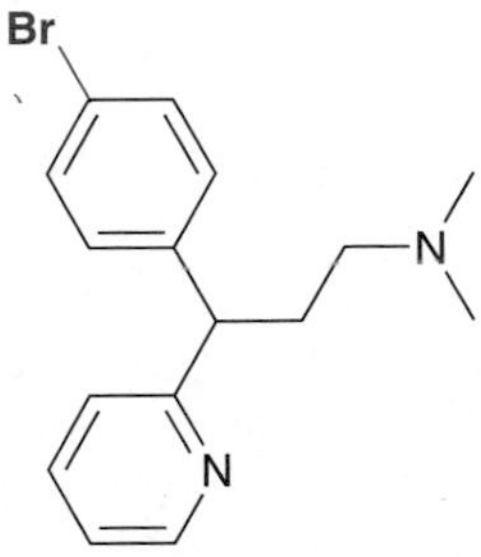

Brompheniramine

Pheniramine is an antihistamine (for a discussion on antihistamines, see Section 18.5). When a chlorine atom is installed in the *para* position of one of the rings, a new compound called chlorpheniramine is obtained. Chlorpheniramine is 10 times more potent than pheniramine and is marketed under the trade name Chlortrimeton™. When a bromine atom is installed instead of chlorine, brompheniramine is obtained, which is marketed under the trade name Dimetane™. This compound is one of the active ingredients in Dimetapp™. It is similar in potency to Chlortrimeton, but its effects last almost twice as long.

Halogenation is a critical process in the design of many other types of drugs as well. For example, consider the antifungal agents chlotrimazole and econazole:

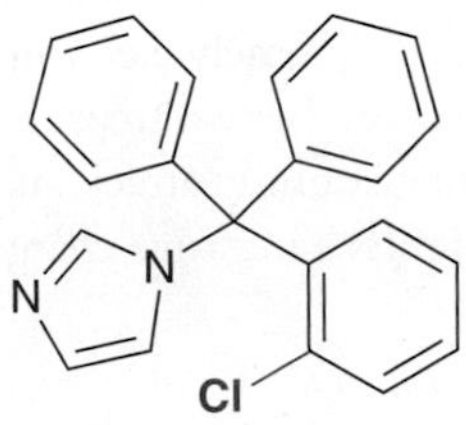

Chlotrimazole

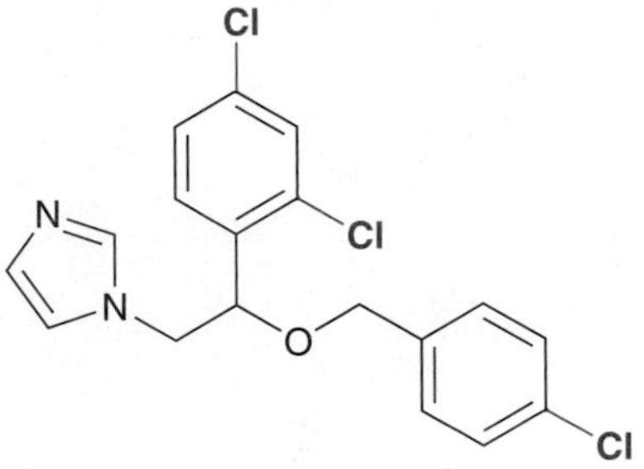

Econazole

Both of these compounds are examples of azole antifungal agents (azole is a five-membered aromatic ring containing two nitrogen atoms). Azole antifungal agents typically contain two or three additional aromatic rings, at least one of which is substituted with a halogen. Structure-activity studies have revealed that the presence of a halogen is critical for drug activity. Chlotrimazole is marketed under the trade name Lotrimin™, and econazole is marketed under the trade name Spectazole™. Notice that in both of these compounds the halogens are positioned in the *ortho* and *para* positions. In the later sections of this chapter, we will see why the *ortho* and *para* positions are more easily halogenated.

19.3 Sulfonation

When benzene is treated with fuming sulfuric acid, a **sulfonation** reaction occurs and benzenesulfonic acid is obtained.

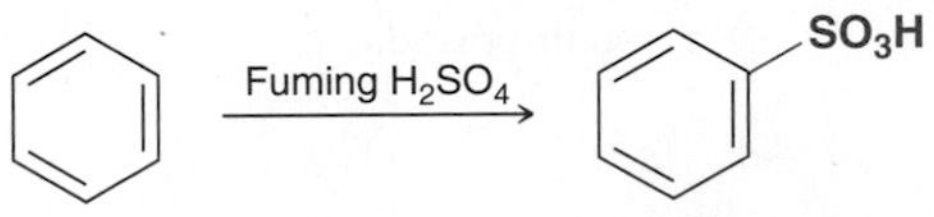

(95%)

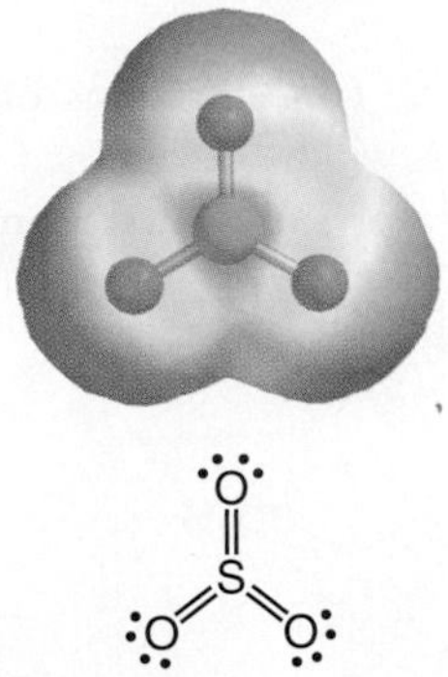

FIGURE 19.2
An electrostatic potential map of sulfur trioxide.

Fuming sulfuric acid is a mixture of H_2SO_4 and SO_3 gas. Sulfur trioxide (SO_3) is a very powerful electrophile, as can be seen in the electrostatic potential map in Figure 19.2. This image illustrates that the sulfur atom is a site of low electron density (an electrophilic center). To understand the reason for this, we must explore the nature of S=O double bonds. Recall that a C=C double bond is formed from the overlap of *p* orbitals. The S=O double bond is also formed from the overlap of *p* orbitals, but the overlap is less efficient because the *p* orbitals are different sizes (Figure 19.3). The sulfur atom uses a 3*p* orbital (sulfur is in the third row of the

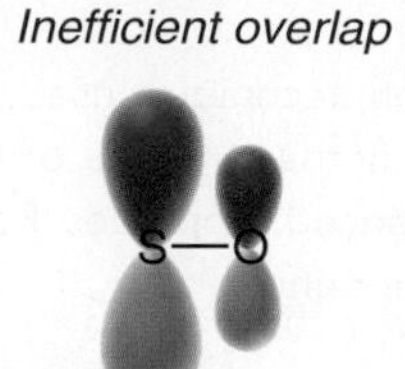
Inefficient overlap
S—O

FIGURE 19.3 The *p* orbitals involved in an S=O bond.

periodic table), while the oxygen atom uses a 2*p* orbital (oxygen is in the second row of the periodic table). The inefficient overlap of these orbitals suggests that we should consider the bond to be a single bond that exhibits charge separation (S^+ and O^-), rather than a double bond. Each of the S—O bonds in sulfur trioxide is highly polarized in this way, rendering the sulfur atom extremely electron poor and sufficiently electrophilic to react with benzene. The reaction involves the two steps that are characteristic of all electrophilic aromatic substitution reactions—a nucleophilic attack and a proton transfer (Mechanism 19.4). The product of these two steps exhibits a negative charge and is protonated in the presence of sulfuric acid.

MECHANISM 19.4 SULFONATION OF BENZENE

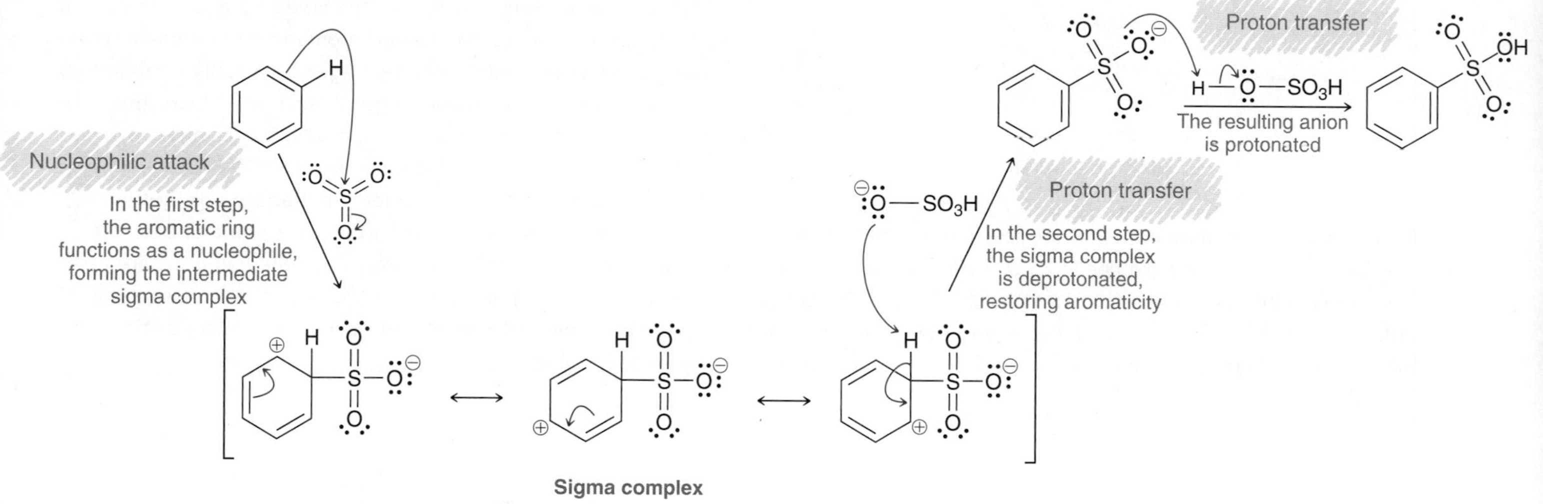
Nucleophilic attack
In the first step, the aromatic ring functions as a nucleophile, forming the intermediate sigma complex
Sigma complex
Proton transfer
In the second step, the sigma complex is deprotonated, restoring aromaticity
Proton transfer
H—O—SO3H
The resulting anion is protonated

The reaction between benzene and SO_3 is highly sensitive to the concentrations of the reagents and is, therefore, reversible. The reversibility of this process will be reexamined later in this chapter and will also be utilized heavily in the synthesis of polysubstituted aromatic compounds.

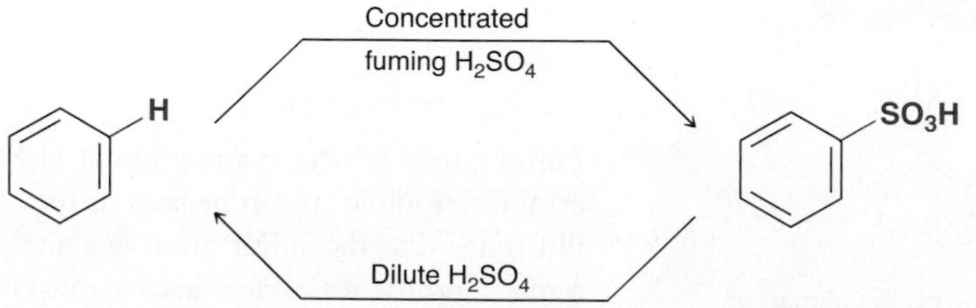
Concentrated
fuming H2SO4
Dilute H2SO4
H
SO3H

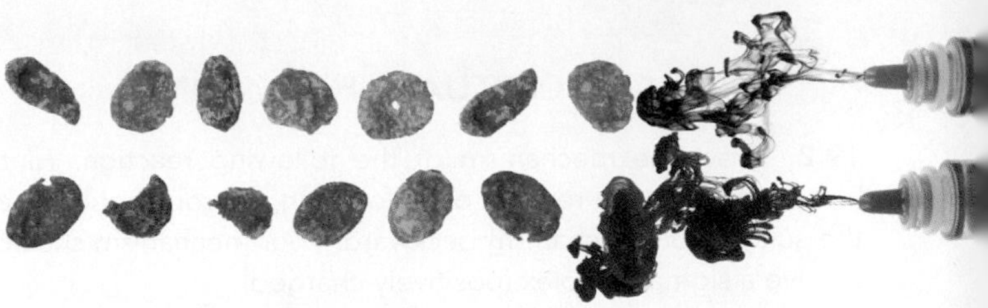

PRACTICALLYSPEAKING)))

What Are Those Colors in Fruity Pebbles?

In the mid-nineteenth century, it was discovered that two aromatic moieties could be joined by an azo group (—N=N—) in a process called *azo coupling*:

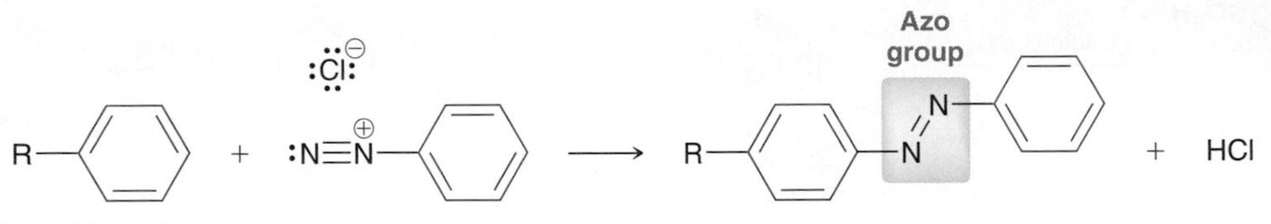

This process, which will be explored in more detail in Chapter 23, is believed to occur via an electrophilic aromatic substitution reaction:

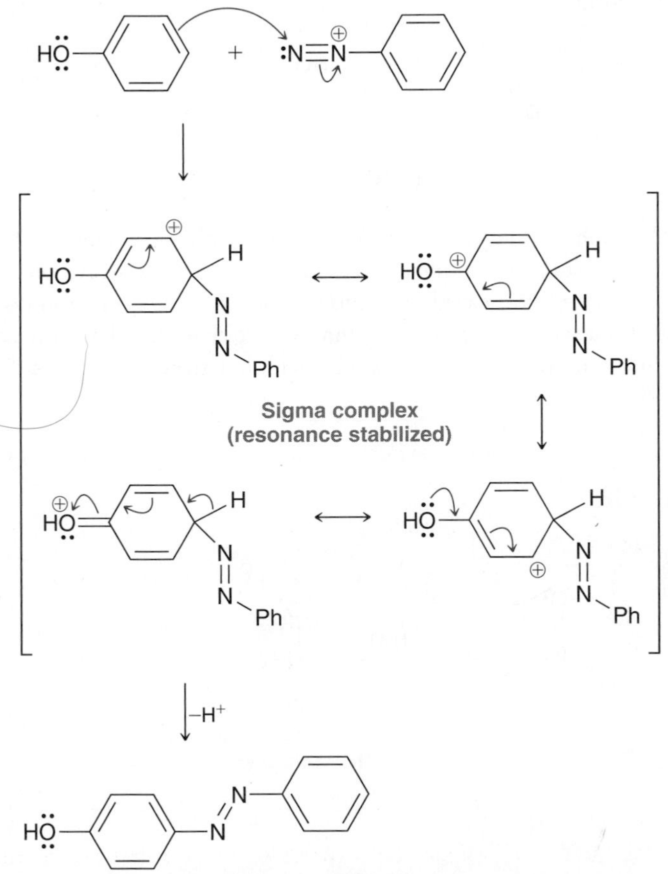

The resulting compound exhibits extended conjugation and is therefore colored (for more on the source of color, see Section 17.12).

By structurally modifying the starting materials (by placing substituents on the aromatic rings prior to azo coupling), a variety of products can be made, each exhibiting a unique color. Due to the variety of colors that can be prepared, a significant amount of research was aimed at designing compounds that could serve as fabric dyes. These compounds, called azo dyes, were produced in large quantities, and by the late nineteenth century, there was a very large market for them. While other types of dyes have been discovered since then, azo dyes still represent more than 50% of the synthetic dye market.

Among many other applications, azo dyes are currently used in paints, cosmetics, and food. Food colorings are regulated by the FDA (Food and Drug Administration) and include compounds such as Red #40 and Yellow #6, both of which are used in Fruity Pebbles:

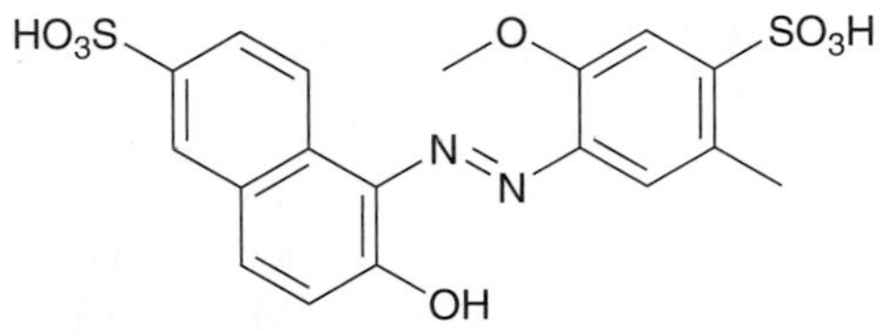

Red #40

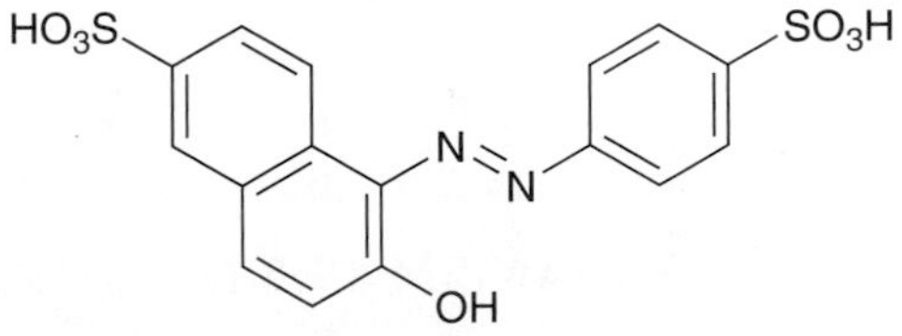

Yellow #6

Notice the presence of sulfonic acid groups (—SO_3H) in both of these compounds. These groups are necessary because they are easily deprotonated to give anions, rendering them water soluble. The sulfonic acid groups are introduced via the sulfonation process that we learned in this section.

Many azo dyes also contain nitro groups, such as Orange #1:

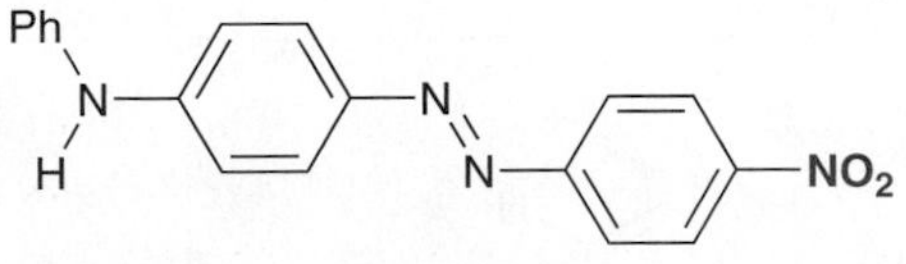

Orange #1

In the following section, we will learn how to install a nitro group on an aromatic ring.

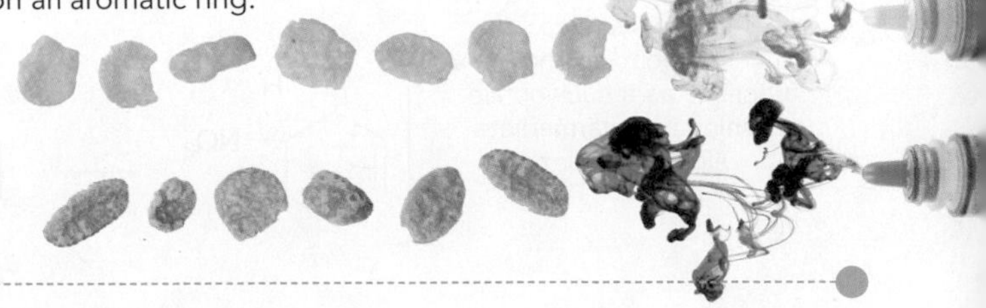

CONCEPTUAL CHECKPOINT

19.2 Draw the mechanism of the following reaction. *Hint:* This reaction is the reverse of sulfonation, so you should read the sulfonation mechanism backward. Your mechanism should involve a sigma complex (positively charged).

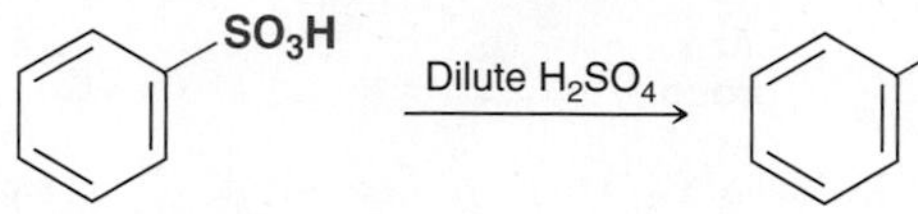
SO_3H
Dilute H_2SO_4
H

19.3 When benzene is treated with D_2SO_4, a deuterium atom replaces one of the hydrogen atoms. Propose a mechanism for this reaction. Once again, make sure that your mechanism involves a sigma complex.

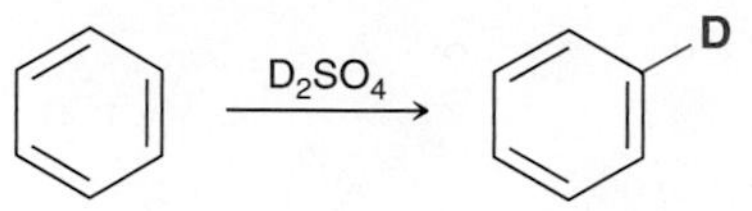
D_2SO_4
D

19.4 Nitration

When benzene is treated with a mixture of nitric acid and sulfuric acid, a **nitration** reaction occurs in which nitrobenzene is formed.

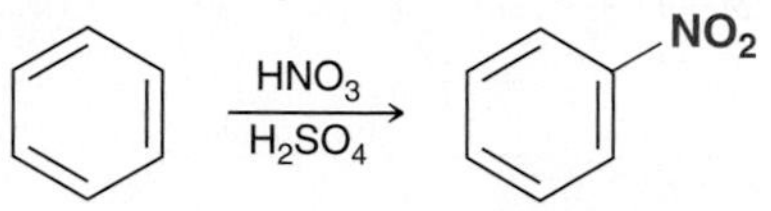
HNO_3
H_2SO_4
NO_2

(95%)

This reaction proceeds via an electrophilic aromatic substitution in which a **nitronium ion** (NO_2^+) is believed to be the electrophile. This strong electrophile is formed from the acid-base reaction that takes place between HNO_3 and H_2SO_4. Nitric acid functions as a base to accept a proton from sulfuric acid, followed by loss of water to produce a nitronium ion (Mechanism 19.5). It might seem strange that nitric acid functions as a base rather than an acid, but remember that acidity

MECHANISM 19.5 NITRATION OF BENZENE

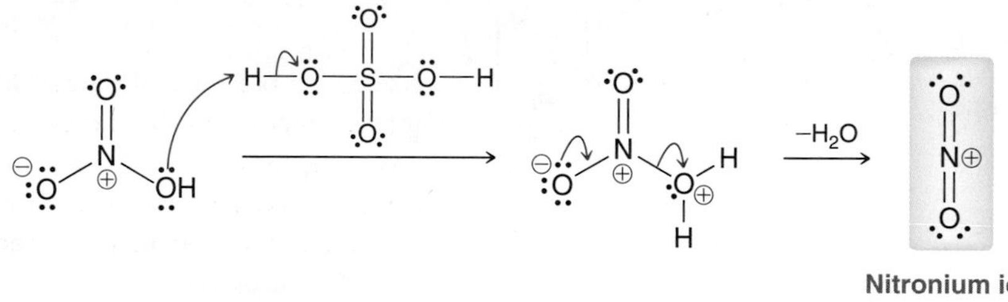
$-H_2O$
Nitronium ion

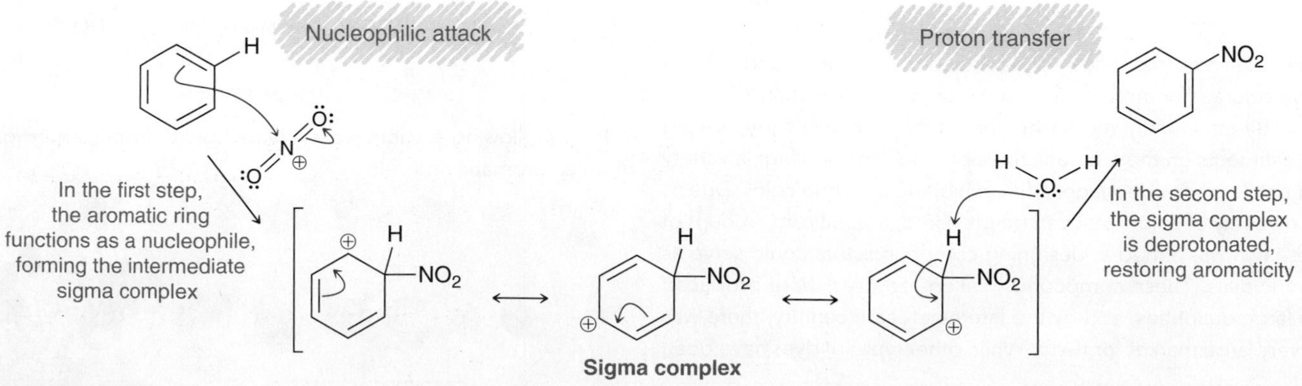
Nucleophilic attack
In the first step, the aromatic ring functions as a nucleophile, forming the intermediate sigma complex
Proton transfer
In the second step, the sigma complex is deprotonated, restoring aromaticity
Sigma complex

is relative. Sulfuric acid is a much stronger acid than nitric acid, and it will protonate nitric acid when mixed together. The resulting nitronium ion then serves as an electrophile in an electrophilic aromatic substitution reaction.

This method can be used to install a nitro group on an aromatic ring. Once on the ring, the nitro group can be reduced to give an amino group (NH_2).

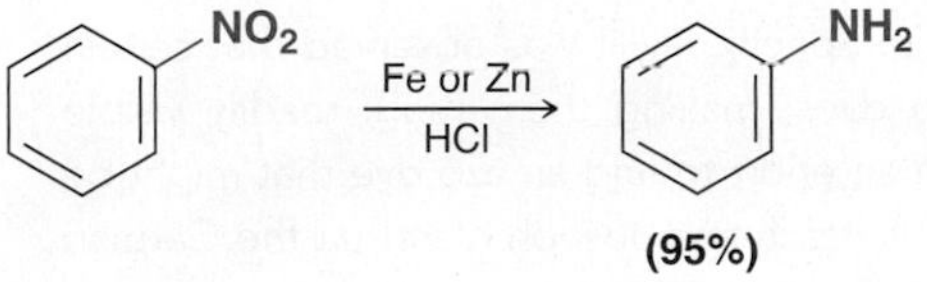

This provides us with a two-step method for installing an amino group on an aromatic ring: (1) nitration, followed by (2) reduction of the nitro group.

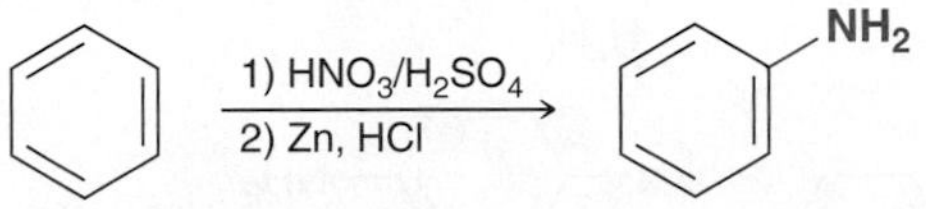

CONCEPTUAL CHECKPOINT

19.4 Draw the mechanism of the following reaction, and make sure to draw all three resonance structures of the sigma complex.

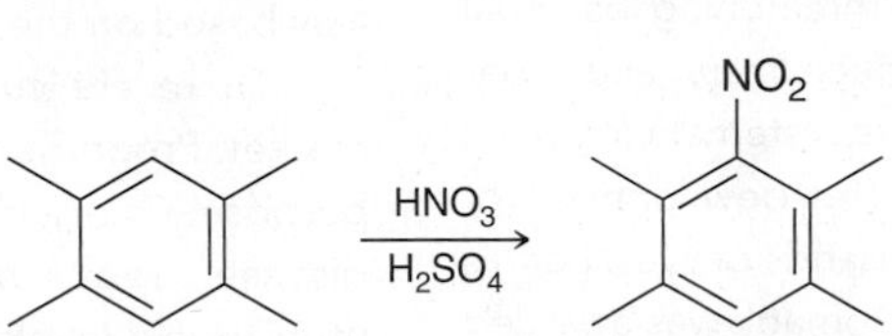

19.5 Friedel-Crafts Alkylation

In the previous sections, we have seen that a variety of electrophiles (Br^+, Cl^+, SO_3, and $NO_2{}^+$) will react with benzene in an electrophilic aromatic substitution reaction. In this section and the next, we will explore electrophiles in which the electrophilic center is a carbon atom.

The **Friedel–Crafts alkylation**, discovered by Charles Friedel and James Crafts in 1877, makes possible the installation of an alkyl group on an aromatic ring.

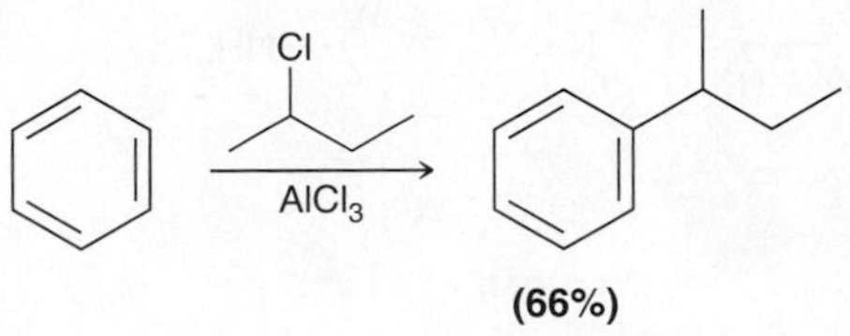

Although an alkyl halide such as 2-chlorobutane is, by itself, electrophilic, it is not sufficiently electrophilic to react with benzene. However, in the presence of a Lewis acid, such as aluminum trichloride, the alkyl halide is converted into a carbocation.

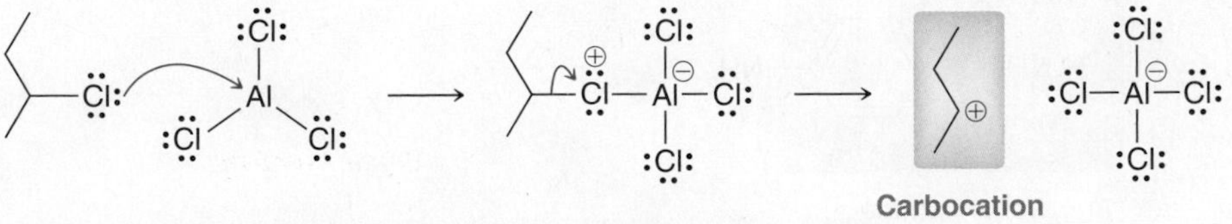

MEDICALLYSPEAKING)))
The Discovery of Prodrugs

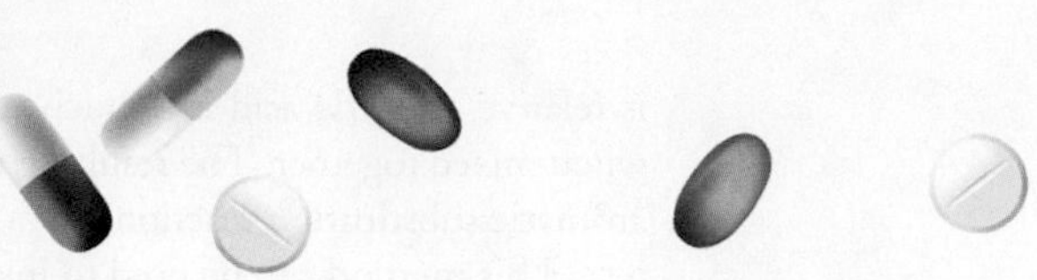

We have seen that azo dyes are used in a variety of applications. One such application led to a discovery that had a profound impact on the field of medicine. Specifically, it was observed that certain bacteria absorbed azo dyes, making them more readily visible under a microscope. In an effort to find an azo dye that might be toxic to bacteria, Fritz Mietzsch and Joseph Klarer (at the German dye company, I.G. Farbenindustrie) began cataloguing azo dyes for possible antibacterial properties. A physician named Gerhard Domagk evaluated the dyes for potential activity, which led to the discovery of the potent antibacterial properties of prontosil.

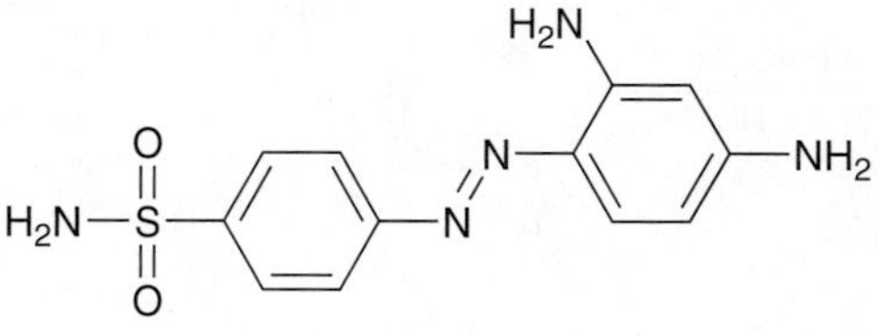

Prontosil

Domagk was able to demonstrate that prontosil cured streptococcal infections in mice. In 1933, physicians began using prontosil in human patients suffering from life-threatening bacterial infections. The success of this drug was extraordinary, and prontosil staked its claim as the first drug that was systematically used for the treatment of bacterial infections. The development of prontosil has been credited with saving thousands of lives. For his pioneering work that led to this discovery, Domagk was awarded the 1939 Nobel Prize in Physiology or Medicine.

Prontosil exhibited one very curious property that intrigued scientists. Specifically, it was found to be totally inactive against bacteria *in vitro* (literally "in glass," in bacterial cultures grown in glass dishes). Its antibacterial properties were only observed *in vivo* (literally "in life", when administered to living creatures, such as mice and humans). These observations inspired much research on the activity of prontosil, and in 1935, it was found that prontosil is metabolized in the body to produce a compound called sulfanilamide.

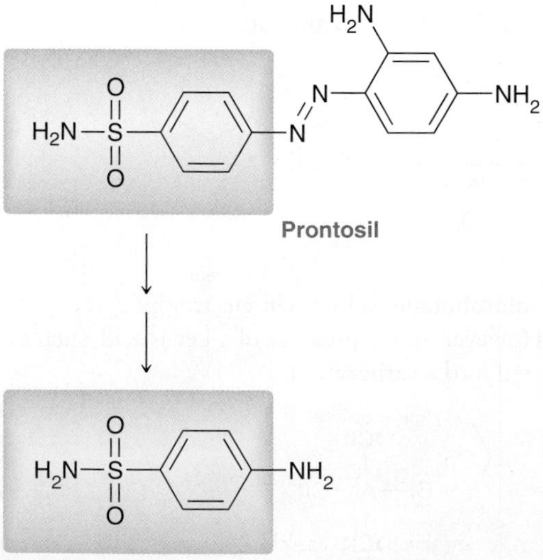

Sulfanilamide

Sulfanilamide was determined to be the active drug, as it interferes with bacterial cell growth. In a glass dish, prontosil is not converted into sulfanilamide, explaining why the antibacterial properties were only observed *in vivo*. This discovery ushered in the era of prodrugs. *Prodrugs* are pharmacologically inactive compounds that are converted by the body into active compounds. This discovery led scientists to direct their research in new directions. They began designing new potential drugs based on structural modifications to sulfanilamide rather than prontosil. Extensive research was directed at making these sulfanilamide analogues, called sulfonamides. By 1948, over 5000 sulfonamides were created, of which more than 20 were ultimately used in clinical practice.

The emergence of bacterial strains resistant to sulfanilamide, together with the advent of penicillins (discussed in Chapter 23), rendered most sulfonamides obsolete. Some sulfonamides are still used today to treat specific bacterial infection in patients with AIDS, as well as a few other applications. Despite their small role in current practice, sulfonamides occupy a unique role in history, because their development was based on the discovery of the first known prodrug.

There are currently a large number of prodrugs on the market. Prodrugs are often designed intentionally for a specific purpose. One such drug is used in the treatment of Parkinson's disease. The symptoms of Parkinson's disease are attributed to low levels of dopamine in a specific part of the brain. Administering dopamine to a patient does not raise the concentration of dopamine in the brain, because the compound does not readily cross the blood-brain barrier. Instead, the prodrug L-dopa is used because it readily crosses the blood-brain barrier, after which it undergoes decarboxylation to produce the needed dopamine.

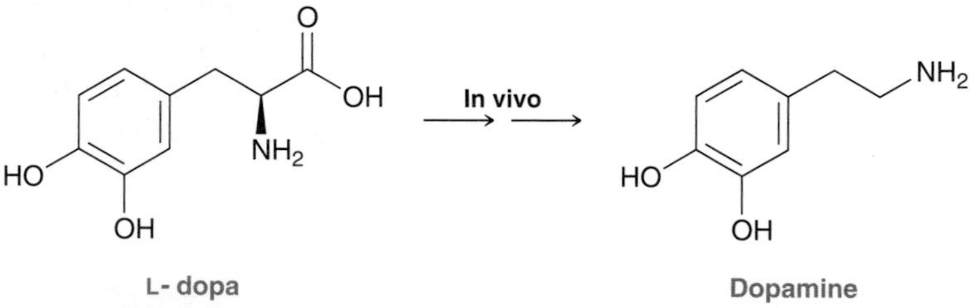

There are many different varieties and classes of prodrugs, and a thorough treatment is beyond the scope of our discussion. The discovery of prodrugs was an extremely important achievement in the development of medicinal chemistry, and it all started with a careful analysis of azo dyes, just like the ones found in Fruity Pebbles.

The catalyst functions exactly as expected (compare the role of $AlCl_3$ here to the role that it plays in Section 19.2). The result here is the formation of a carbocation, which is an excellent electrophile and is capable of reacting with benzene in an electrophilic aromatic substitution reaction (Mechanism 19.6).

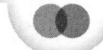

MECHANISM 19.6 FRIEDEL-CRAFTS ALKYLATION

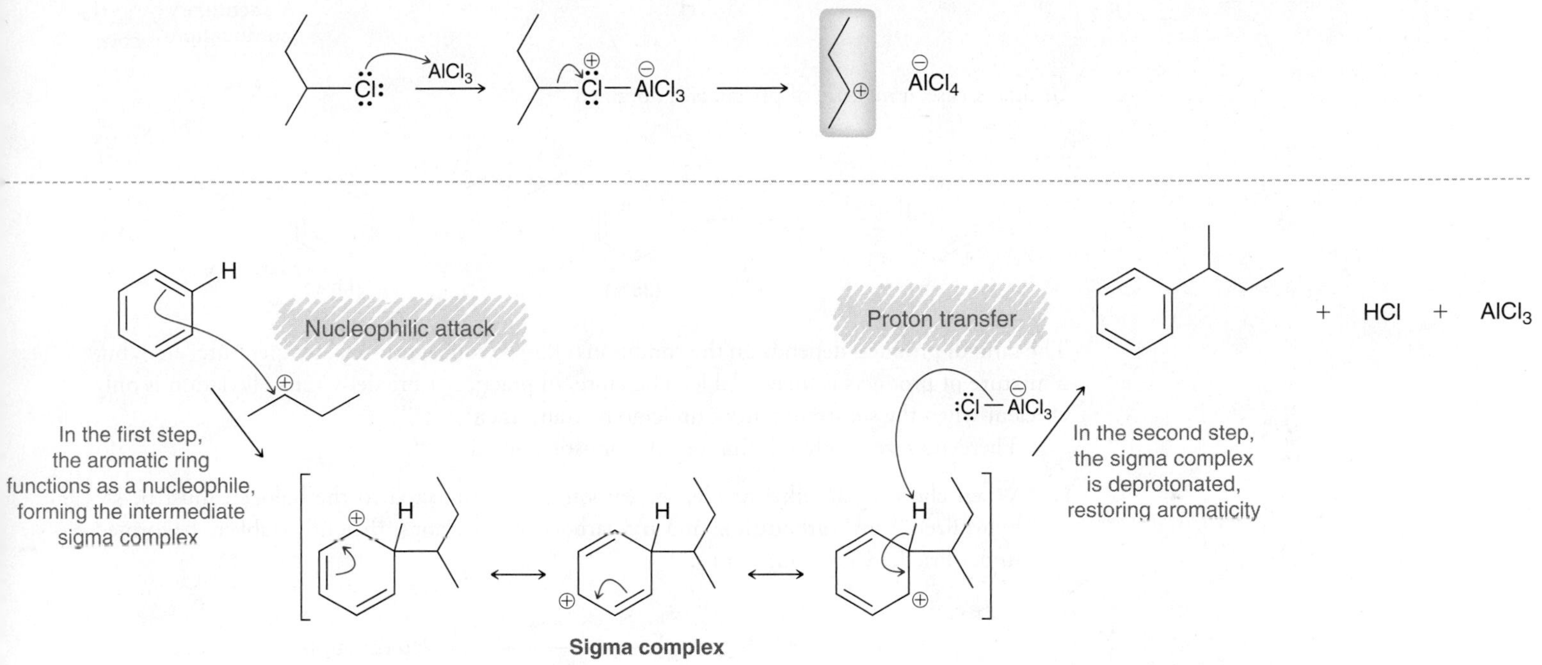

AlCl3
Cl—AlCl3
AlCl4
Nucleophilic attack
Proton transfer
+ HCl + AlCl3
:Cl—AlCl3
In the first step, the aromatic ring functions as a nucleophile, forming the intermediate sigma complex
In the second step, the sigma complex is deprotonated, restoring aromaticity
Sigma complex

After formation of the carbocation, two steps are involved in this electrophilic aromatic substitution. In the first step, the aromatic ring functions as a nucleophile and attacks the carbocation, forming a sigma complex. The sigma complex is then deprotonated to restore aromaticity.

Many different alkyl halides can be used in a Friedel-Crafts alkylation. Secondary and tertiary halides are readily converted into carbocations in the presence of $AlCl_3$. Primary alkyl halides are not converted into carbocations, since primary carbocations are extremely high in energy. Nevertheless, a Friedel-Crafts alkylation is indeed observed when benzene is treated with ethyl chloride in the presence of $AlCl_3$.

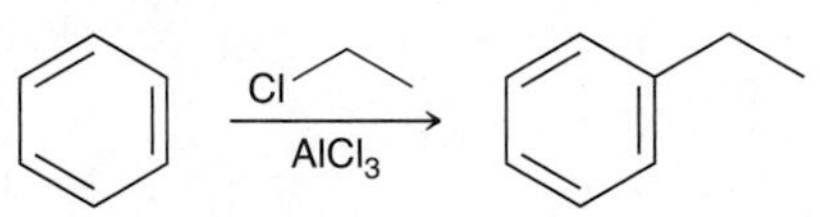

Cl
AlCl3

In this case, the electrophilic agent is presumed to be a complex between ethyl chloride and $AlCl_3$.

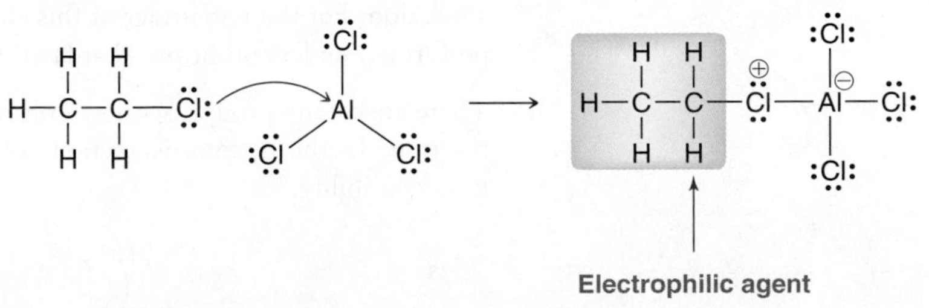

Electrophilic agent

This complex can be attacked by an aromatic ring, much like we saw during chlorination (Mechanism 19.2). Although a Friedel-Crafts alkylation is effective when ethyl chloride is

LOOKING BACK
For a review of carbocation rearrangements, see Section 6.11.

used, most other primary alkyl halides cannot be used effectively, because their complexes with $AlCl_3$ readily undergo rearrangement to form secondary or tertiary carbocations. For example, when 1-chlorobutane is treated with aluminum trichloride, a secondary carbocation is formed via a hydride shift.

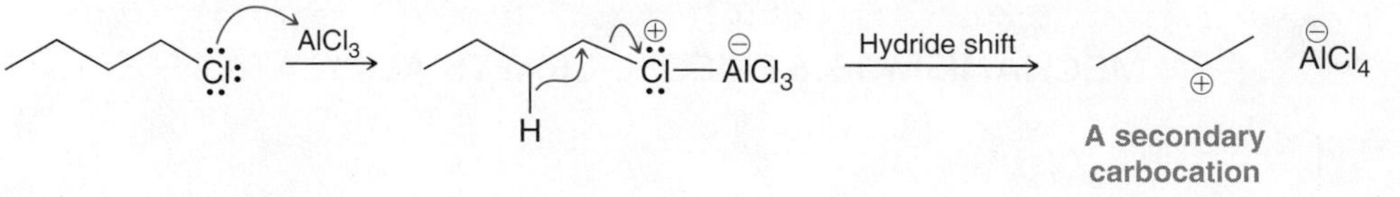

In such a case, a mixture of products is obtained.

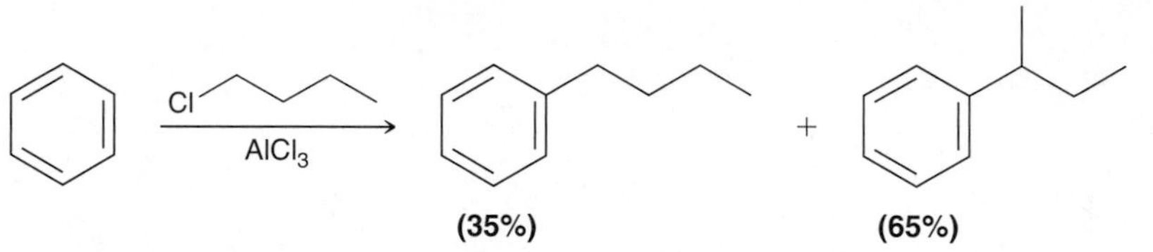

The ratio of products depends on the conditions chosen (concentrations, temperature, etc.), but a mixture of products is unavoidable. Therefore, in practice, a Friedel-Crafts alkylation is only efficient when the substrate cannot undergo rearrangement.

There are several other limitations that must be noted.

1. When choosing an alkyl halide, the carbon atom connected to the halogen must be sp^3 hybridized. Vinyl carbocations and aryl carbocations are not sufficiently stable to be formed under Friedel-Crafts conditions.

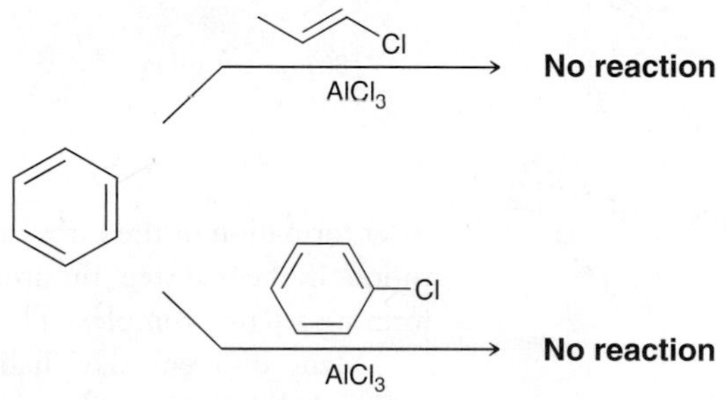

2. Installation of an alkyl group activates the ring toward further alkylation (for reasons that we will explore in the upcoming sections of this chapter). Therefore, polyalkylations often occur.

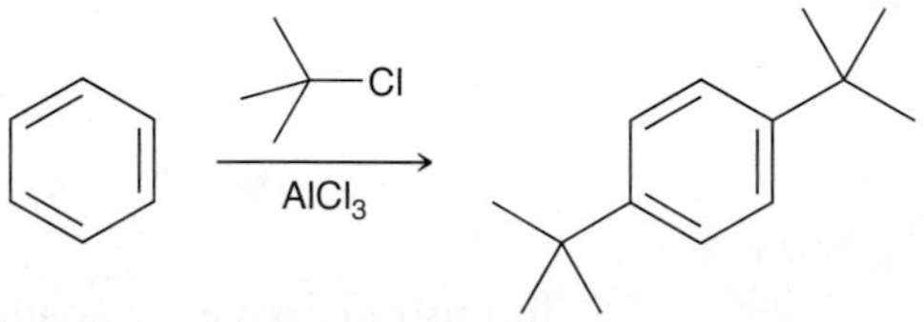

This problem can generally be avoided by choosing reaction conditions that favor monoalkylation. For the remainder of this chapter, assume that all Friedel-Crafts alkylations are performed under conditions that favor monoalkylation, unless otherwise specified.

3. There are certain groups, such as a nitro group, that are incompatible with a Friedel-Crafts reaction. In the upcoming sections of this chapter, we will explore the reason for this incompatibility.

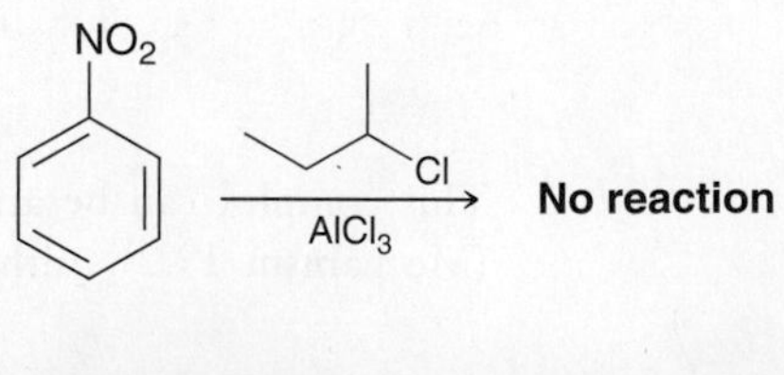

CONCEPTUAL CHECKPOINT

19.5 Predict the expected product(s) when benzene is treated with each of the following alkyl halides in the presence of AlCl₃. In each case, assume conditions have been controlled to favor monoalkylation.

(a)

(b)

(c)

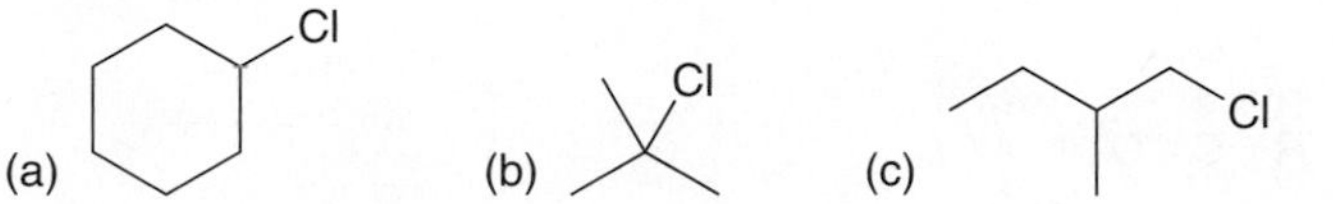

19.6 Draw the mechanism of the following reaction, which involves two consecutive Friedel-Crafts alkylations. When drawing the mechanism, do not try to draw the two alkylations as occurring simultaneously (such a mechanism would have too many curved arrows and too many simultaneous charges). First draw the steps that install one alkyl group, and then draw the steps that install the second alkyl group.

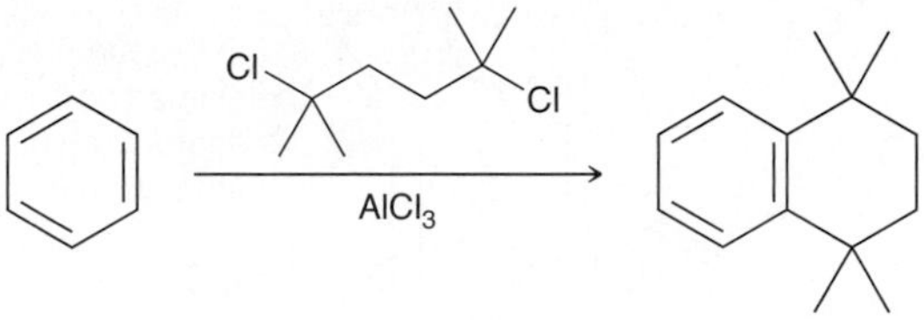

19.7 A Friedel-Crafts alkylation is an electrophilic aromatic substitution in which the electrophile (E⁺) is a carbocation. In previous chapters, we have seen other methods of forming carbocations, such as protonation of an alkene using a strong acid. The resulting carbocation can also be attacked by a benzene ring, resulting in alkylation of the aromatic ring. With this in mind, draw a mechanism for the following transformation:

$$\xrightarrow{\text{H}_2\text{SO}_4}$$

(68%)

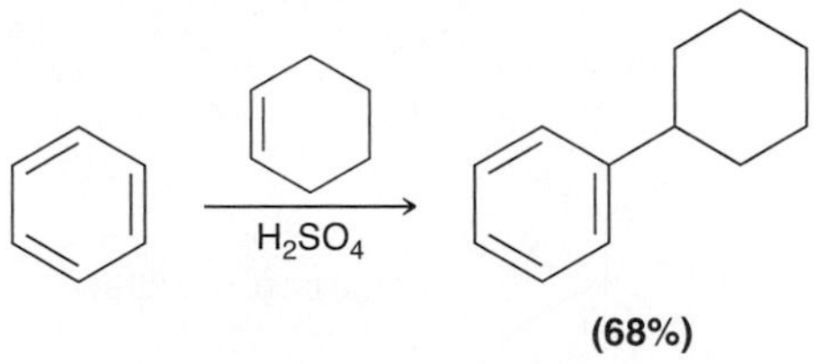

19.6 Friedel-Crafts Acylation

In the previous section, we learned how to install an alkyl group on an aromatic ring. A similar method can be used to install an **acyl group**. The difference between an alkyl group and an acyl group is shown below.

Alkyl group

Acyl group

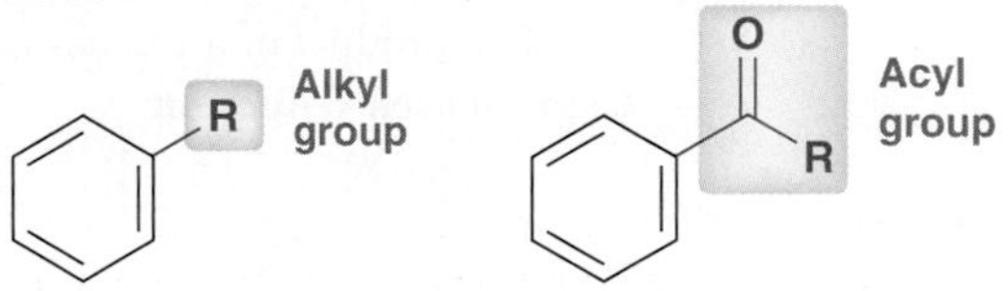

A reaction that installs an acyl group is called an *acylation*.

$$\xrightarrow[\text{AlCl}_3]{}$$

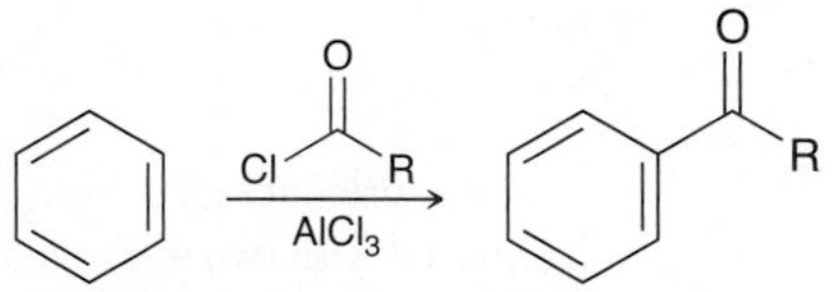

In a **Friedel-Crafts acylation**, the mechanism is very similar to the alkylation process discussed in the previous section. An acyl chloride is treated with a Lewis acid to form a cationic species, called an **acylium ion**.

Acylium ion

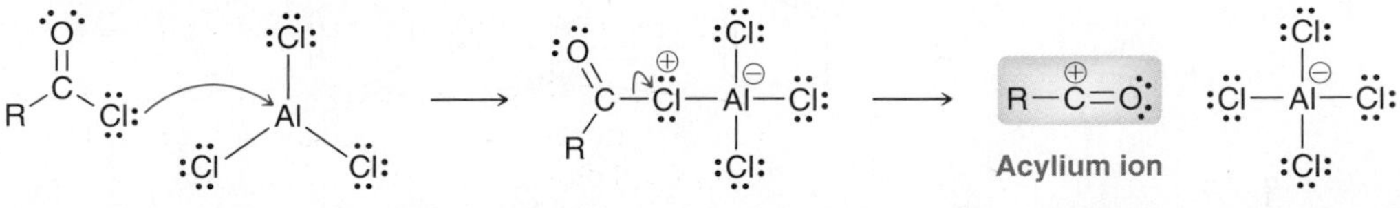

An acylium ion is resonance stabilized and is therefore not susceptible to rearrangement.

$$\left[\overset{\oplus}{R-C}=\overset{..}{\overset{..}{O}:} \longleftrightarrow R-C\equiv\overset{\oplus}{O}: \right]$$

Resonance stabilized

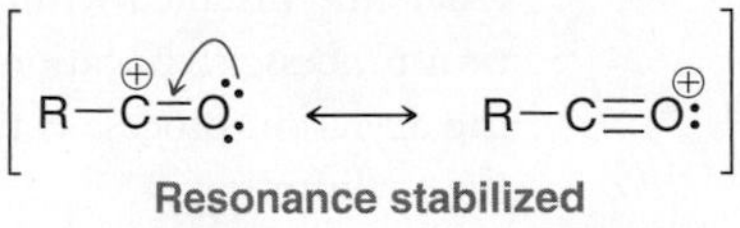

Rearrangement does not occur because a carbocation rearrangement would result in loss of resonance stabilization (an endergonic process). The acylium ion is an excellent electrophile, producing an electrophilic aromatic substitution reaction (Mechanism 19.7).

MECHANISM 19.7 FRIEDEL-CRAFTS ACYLATION

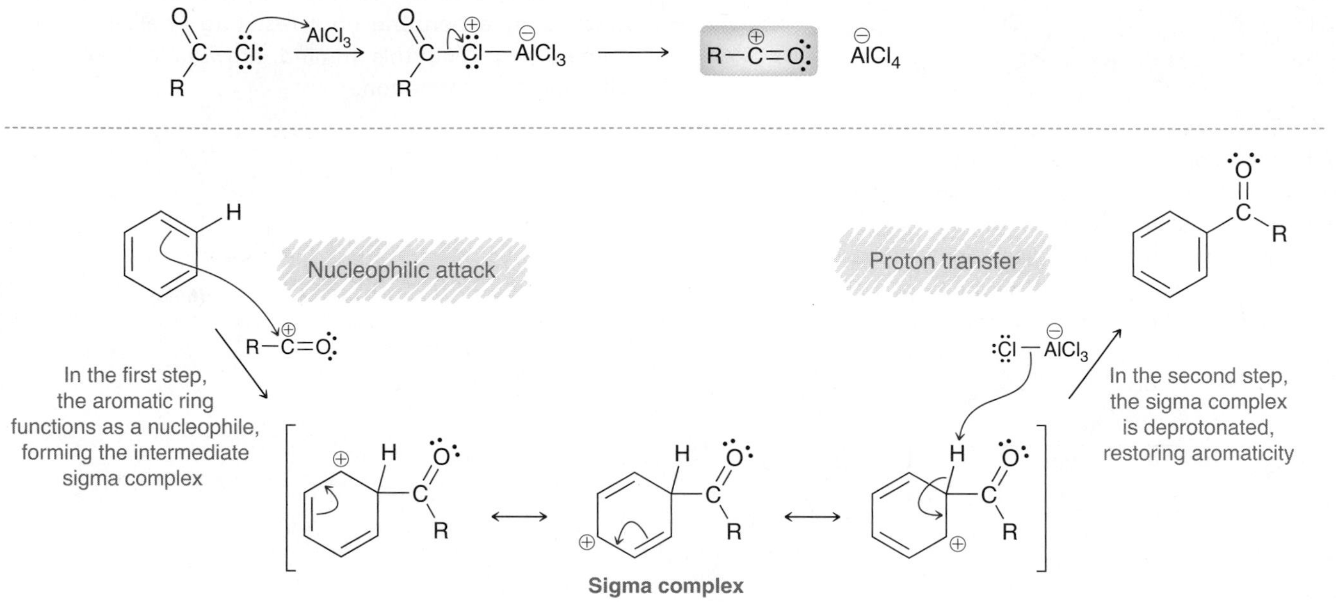

AlCl$_3$
Nucleophilic attack
Proton transfer
In the first step, the aromatic ring functions as a nucleophile, forming the intermediate sigma complex
In the second step, the sigma complex is deprotonated, restoring aromaticity
Sigma complex

The acylium ion is attacked by the benzene ring to produce an intermediate sigma complex, which is then deprotonated to restore aromaticity.

The product of a Friedel-Crafts acylation is an aryl ketone, which can be reduced using a **Clemmensen reduction**.

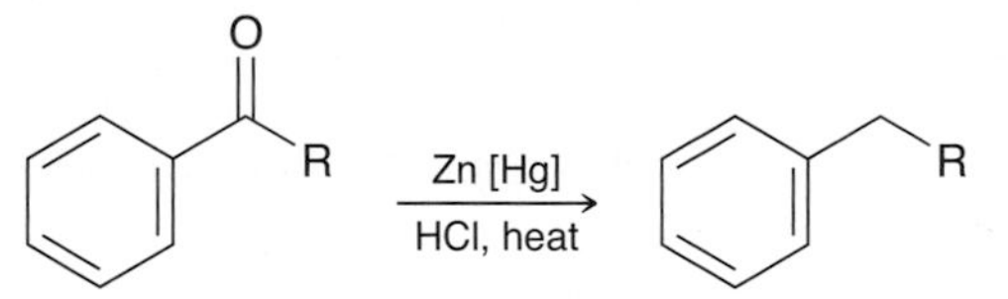

Zn [Hg]
HCl, heat

In the presence of a zinc amalgam and HCl, the carbonyl group is completely reduced and replaced with two hydrogen atoms. When a Friedel-Crafts acylation is followed by a Clemmensen reduction, the net result is the installation of an alkyl group.

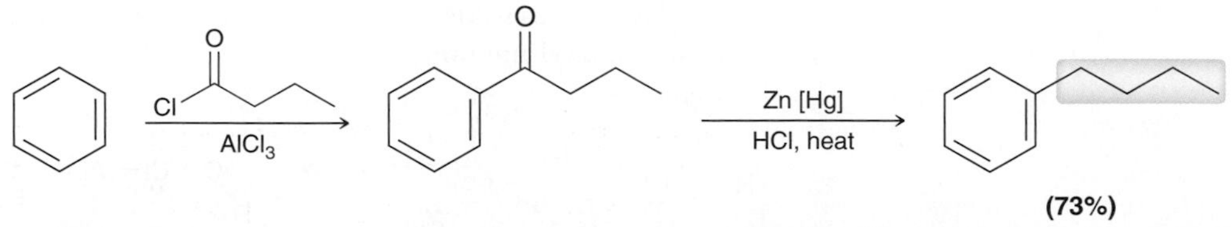

AlCl$_3$
Zn [Hg]
HCl, heat
(73%)

This two-step process is a useful synthetic method for installing alkyl groups that cannot be efficiently installed with a direct alkylation process. If the product is made with a direct alkylation process, carbocation rearrangements would give a mixture of products. The advantage of the acylation process is that carbocation rearrangements are avoided because of the stability of the acylium ion.

Polyacylation is not observed, because introduction of an acyl group deactivates the ring toward further acylation. This will be explained in more detail in the upcoming sections.

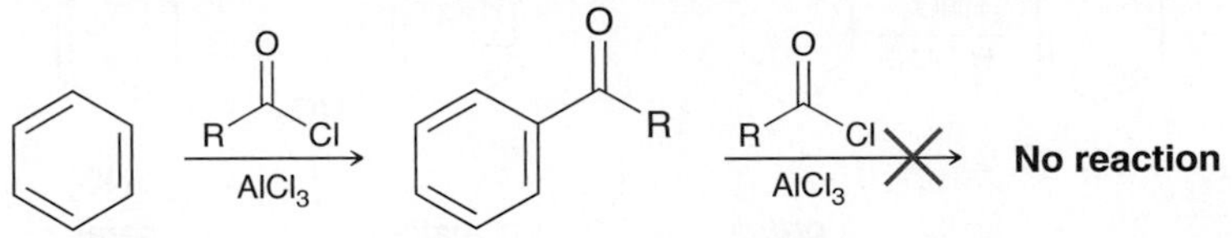

CONCEPTUAL CHECKPOINT

19.8 Identify whether each of the following compounds can be made using a direct Friedel-Crafts alkylation or whether it is necessary to perform an acylation followed by a Clemmensen reduction to avoid carbocation rearrangements:

(a) 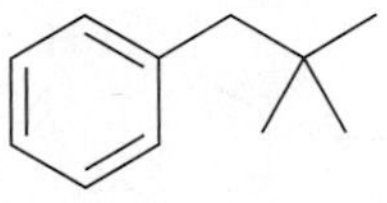

(b) 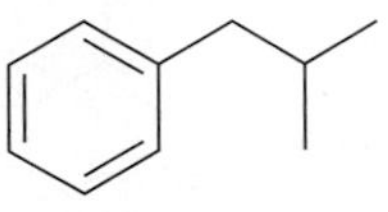

(c) 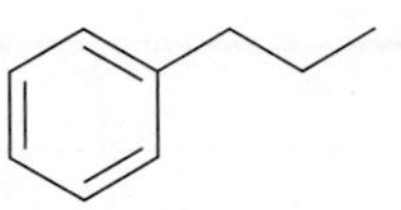

(d) 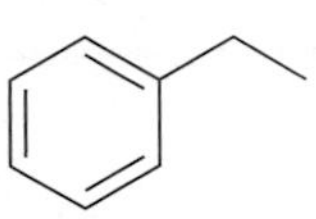

19.9 The following compound cannot be made with either a Friedel-Crafts alkylation or acylation. Explain.

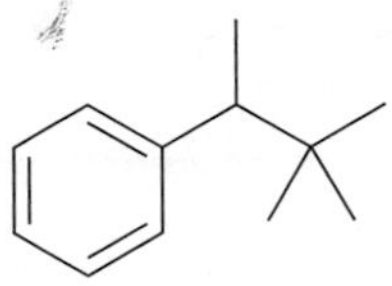

19.10 A Friedel-Crafts acylation is an electrophilic aromatic substitution in which the electrophile (E^+) is an acylium ion. There are other methods of forming acylium ions, such as treatment of an anhydride with a Lewis acid. The resulting acylium ion can also be attacked by a benzene ring, resulting in acylation of the aromatic ring. With this in mind, draw the mechanism of the following transformation:

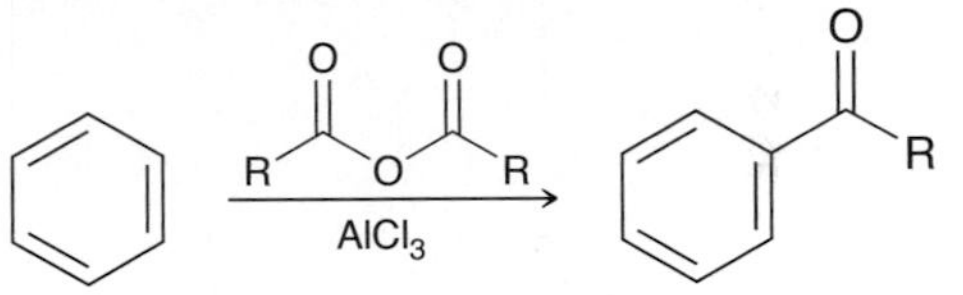

19.7 Activating Groups

Nitration of Toluene

Thus far, we have dealt only with reactions of benzene. We now expand our discussion to include reactions of aromatic compounds that already possess substituents, such as toluene.

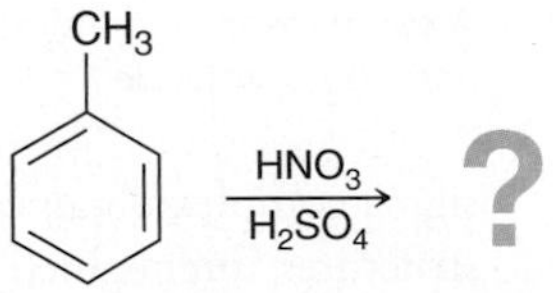

In this nitration reaction, the presence of the methyl group raises two issues: (1) the effect of the methyl group on the rate of reaction and (2) the effect of the methyl group on the regiochemical outcome of the reaction. Let's begin with the rate of reaction.

Toluene undergoes nitration approximately 25 times faster than benzene. In other words, the methyl group is said to **activate** the aromatic ring. Why? Recall that alkyl groups are electron donating because of hyperconjugation (see Section 6.11). As a result, a methyl group donates electron density to the ring, thereby stabilizing the positively charged sigma complex and lowering the energy of activation for its formation.

Now let's focus on the regiochemical outcome. The nitro group could be installed *ortho*, *meta*, or *para* to the methyl group, but the three possible products are not obtained in equal amounts. As seen in Figure 19.4, the *ortho* and *para* products predominate, while very little *meta*

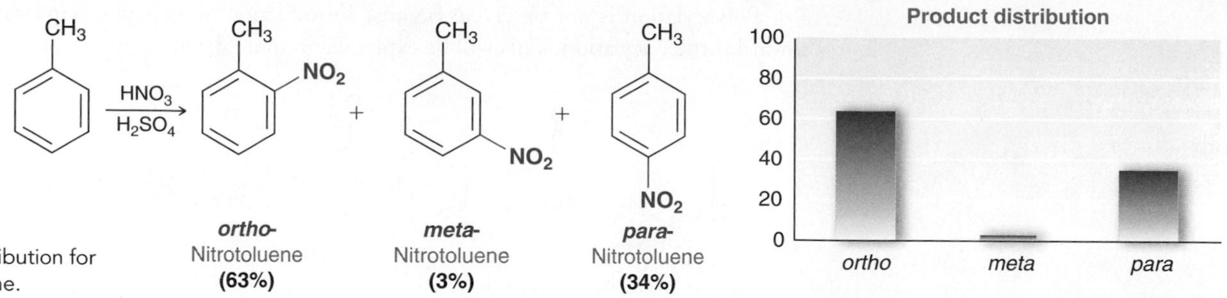

FIGURE 19.4
The product distribution for nitration of toluene.

product is obtained. To explain this observation, we must compare the stability of the sigma complex formed for *ortho* attack, *meta* attack, and *para* attack (Figure 19.5). Notice that the

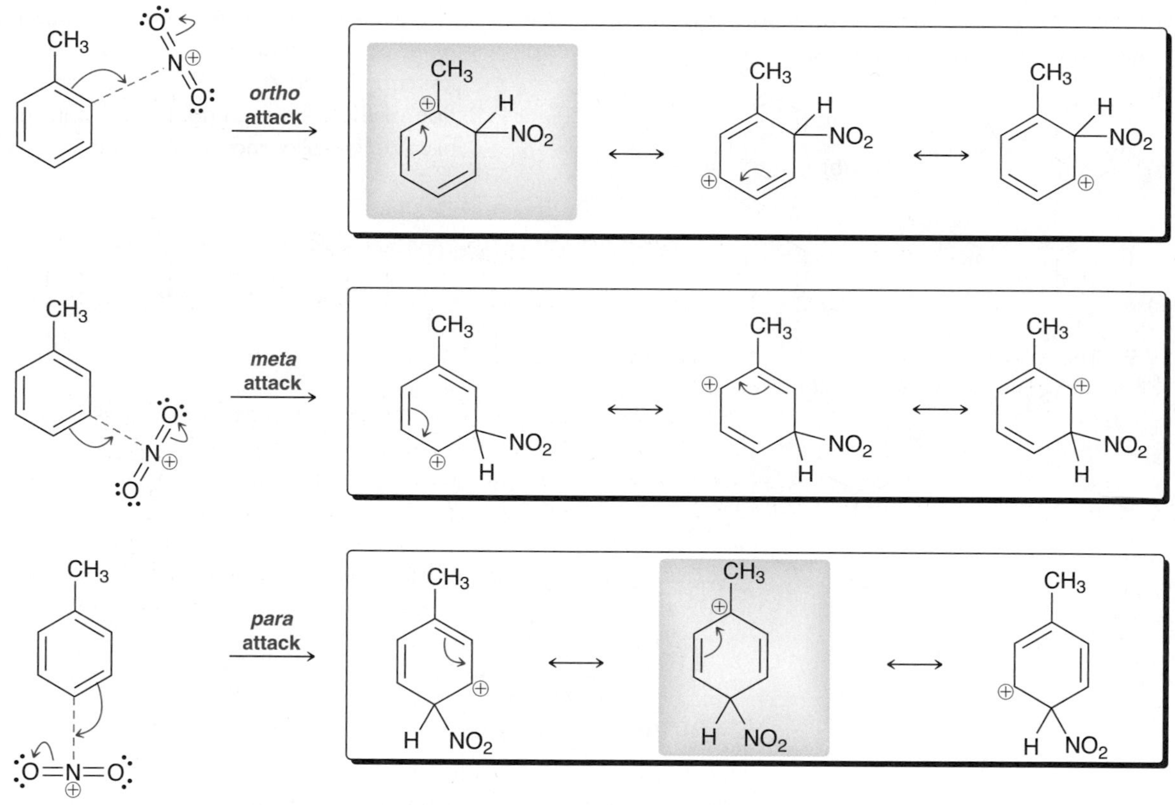

FIGURE 19.5
A comparison of the sigma complexes formed for each of the possible regiochemical outcomes for the nitration of toluene.

BY THE WAY
The exact ratio of *ortho* and *para* products is often sensitive to the conditions employed, such as the choice of solvent. In some cases, nitration of toluene has been observed to favor the *para* product over the *ortho* product.

sigma complex obtained from *ortho* attack has additional stability because one of the resonance structures (highlighted in green) exhibits the positive charge directly adjacent to the electron-donating alkyl group. Similarly, the sigma complex obtained from *para* attack also exhibits this additional stability. The sigma complex obtained from *meta* attack does not exhibit this stability and is, therefore, higher in energy. The relative energy of each sigma complex is best visualized by comparing energy diagrams for *ortho* attack, *meta* attack, and *para* attack (Figure 19.6, as seen on the next page). Notice that the intermediate sigma complex formed from *meta* attack is the highest in energy and therefore requires the largest energy of activation (E_a). This explains why the *meta* product is only obtained in very small amounts. The products of the reaction are generated from *ortho* attack and *para* attack, both of which involve a lower E_a. If we compare E_a for *ortho* attack and *para* attack, we see that *ortho* attack involves a slightly higher E_a than *para* attack, because the methyl group and the nitro group are in close proximity and exhibit a small amount of steric hindrance. As a result, we might have expected the *para* product to be the major product. In fact, the *ortho* product predominates in this case for statistical reasons—there are two *ortho* positions and only one *para* position.

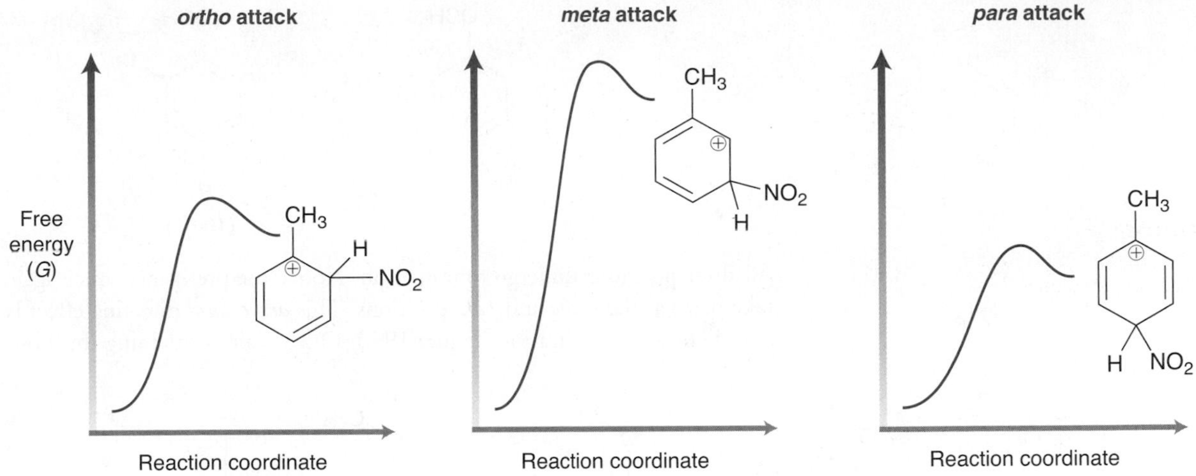

FIGURE 19.6
Energy diagrams comparing the relative energy levels of the possible sigma complexes that could be formed during nitration of toluene. The differences in energy between these three pathways have been slightly exaggerated for clarity of presentation.

Comparison of the energy diagrams in Figure 19.6 provides an explanation for the observation that the methyl group is an ***ortho-para* director**. In other words, the presence of the methyl group directs the incoming nitro group into the *ortho* and *para* positions.

Nitration of Anisole

In the previous section, we saw that the presence of a methyl group activates the ring toward electrophilic aromatic substitution. There are many different groups that activate a ring, with some more activating than others. For example, a methoxy group is a more powerful activator than a methyl group, and methoxybenzene (anisole) undergoes nitration 400 times faster than toluene. To understand why a methoxy group is more activating than a methyl group, we must explore the electronic effects of a methoxy group connected to an aromatic ring.

A methoxy group is inductively electron withdrawing, because oxygen is more electronegative than carbon.

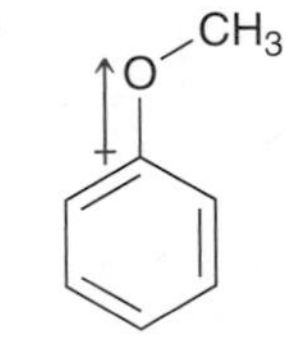

From this point of view, the methoxy group withdraws electron density from the aromatic ring. However, we see a different picture when we draw the resonance structures of anisole.

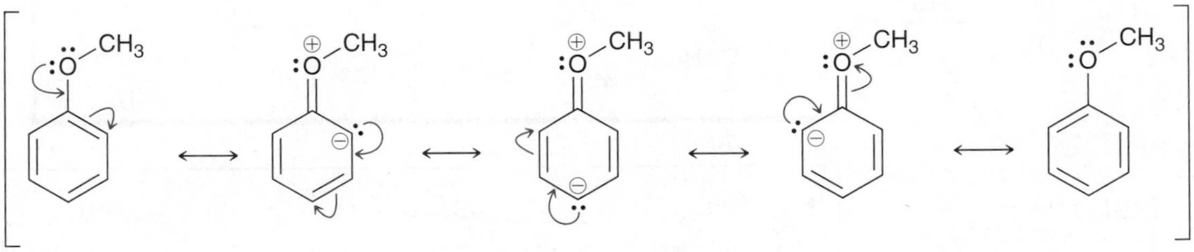

Three of the resonance structures exhibit a negative charge in the ring. This suggests that the methoxy group *donates* electron density to the ring. Clearly, there is a competition here between induction and resonance. Induction suggests that the methoxy group is electron withdrawing, while resonance suggests that the methoxy group is electron donating. Whenever resonance and induction compete with each other, resonance is generally the dominant factor and vastly overshadows any inductive effects. Therefore, the net effect of the methoxy group is to donate electron density to the ring. This effect stabilizes the positively charged sigma complex and lowers the energy of activation for its formation. In fact, the ring is so activated that treatment with bromine *without* a Lewis acid gives a trisubstituted product.

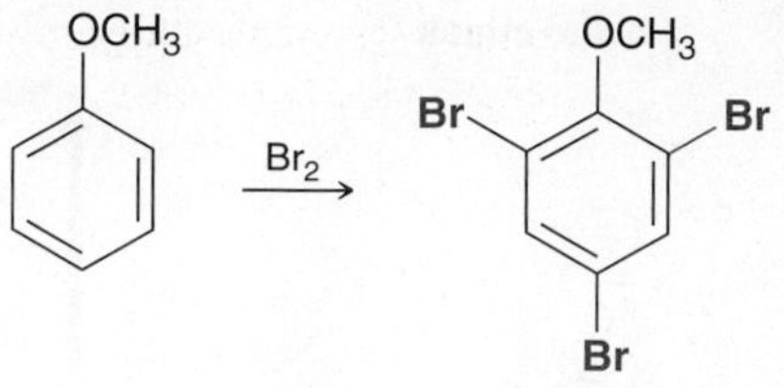

(100%)

All three positions undergo bromination. Notice the preference, once again, for the reaction to take place at the *ortho* and *para* positions. This *ortho-para* directing effect is also observed when anisole undergoes nitration (Figure 19.7). Once again, explaining this observation requires that

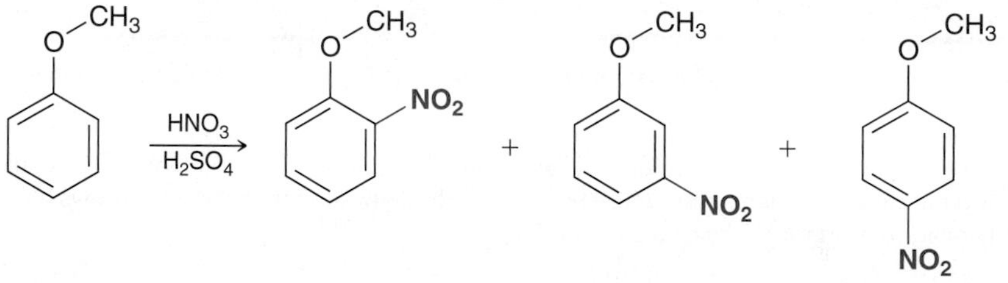

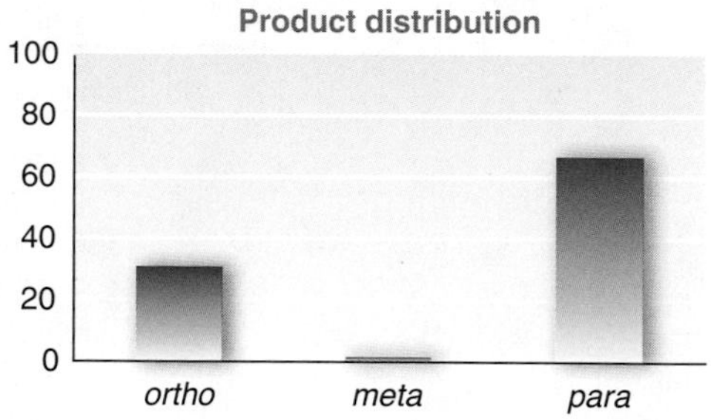

FIGURE 19.7
The product distribution for nitration of anisole.

***ortho*-** Nitroanisole **(31%)**

***meta*-** Nitroanisole **(2%)**

***para*-** Nitroanisole **(67%)**

we compare the stability of the sigma complexes formed for *ortho*, *meta*, and *para* attack (Figure 19.8). Notice that the sigma complex obtained from *ortho* attack has one additional resonance structure (highlighted in green), which stabilizes the sigma complex. Similarly, the sigma complex

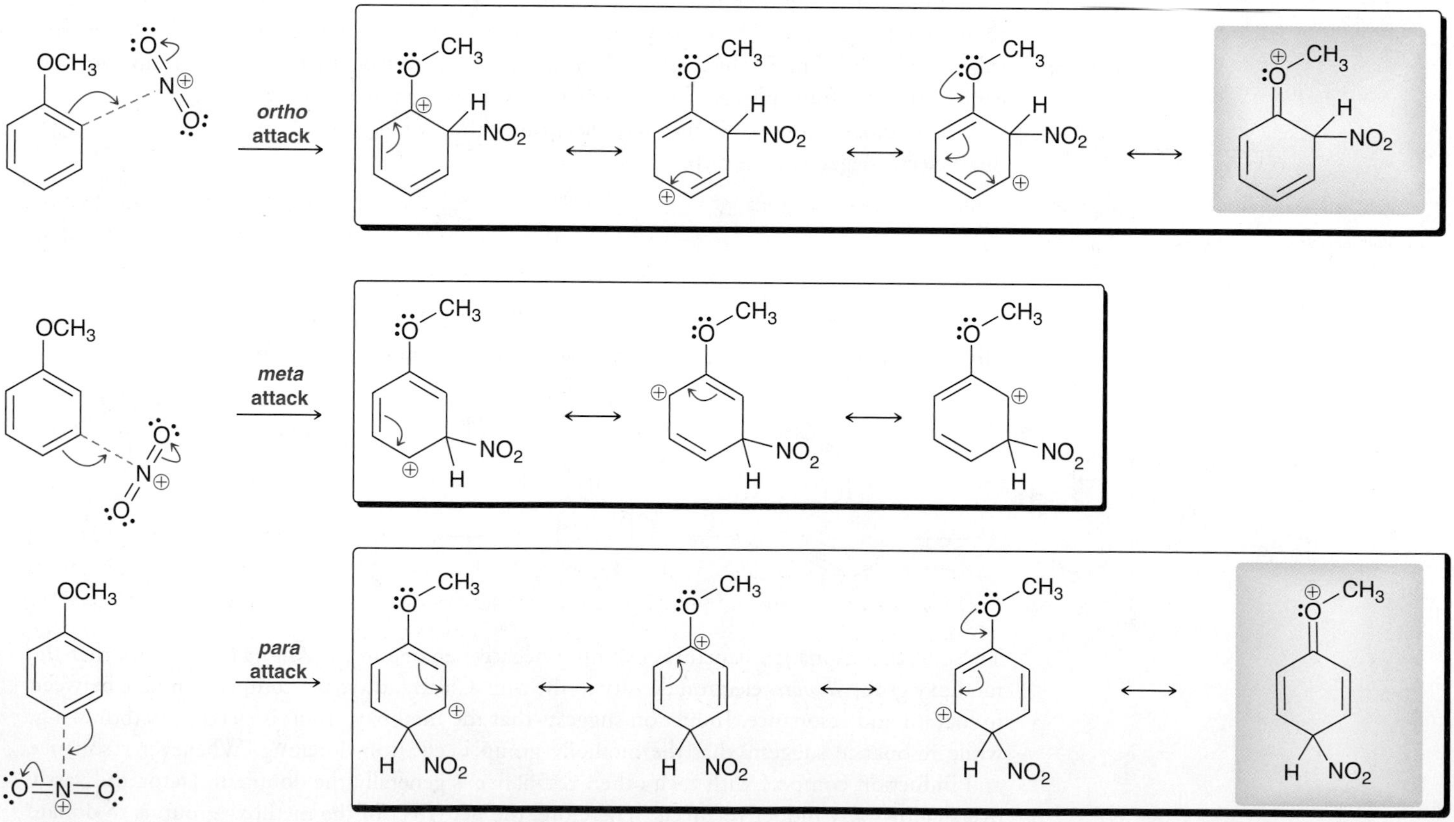

FIGURE 19.8
A comparison of the sigma complexes formed for each of the possible regiochemical outcomes for the nitration of anisole.

obtained from *para* attack also exhibits this additional stability. The sigma complex obtained from *meta* attack does not exhibit this extra stability and is, therefore, higher in energy. Figure 19.9

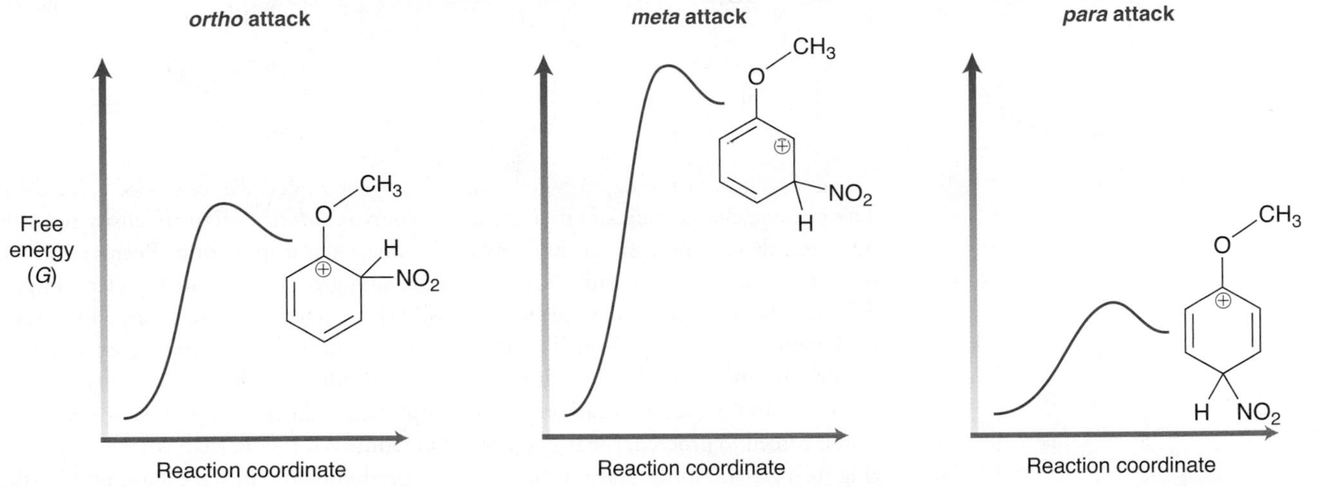

FIGURE 19.9 Energy diagrams comparing the relative energy levels of the sigma complexes that could be formed during nitration of anisole. The differences in energy between these three pathways have been slightly exaggerated for clarity of presentation.

shows a comparison of the energy diagrams for *ortho*, *meta*, and *para* attack. Notice that the intermediate sigma complex formed from *meta* attack is the highest in energy and therefore requires the largest energy of activation (E_a). This explains why the *meta* product is only obtained in very small amounts. The products of the reaction are generated from *ortho* attack and *para* attack, both of which involve a lower E_a.

In the nitration of anisole, the *para* product is favored over the *ortho* product despite the fact that there are two *ortho* positions. Several factors contribute to this observation. One factor is most certainly a steric consideration. That is, the sigma complex resulting from *ortho* attack exhibits more steric hindrance and is higher in energy than the sigma complex resulting from *para* attack.

In summary, we have seen that both a methyl group and a methoxy group activate the ring and are *ortho-para* directors. This is in fact a general rule that will be used extensively throughout the rest of this chapter: *All activators are ortho-para directors.*

CONCEPTUAL CHECKPOINT

19.11 Draw the two major products obtained when toluene undergoes monobromination.

19.12 When ethoxybenzene is treated with a mixture of nitric acid and sulfuric acid, two products are obtained each of which has the molecular formula $C_8H_9NO_3$.

(a) Draw the structure of each product.

(b) Propose a mechanism of formation for the major product.

19.8 Deactivating Groups

In the previous section, we saw that certain groups will activate the ring toward electrophilic aromatic substitution. In this section, we explore the effects of a nitro group, which **deactivates** the ring toward electrophilic aromatic substitution. To understand why the nitro group deactivates the ring, we must explore the electronic effects of a nitro group connected to an aromatic ring.

A nitro group is inductively electron withdrawing, because a positively charged nitrogen atom is extremely electronegative.

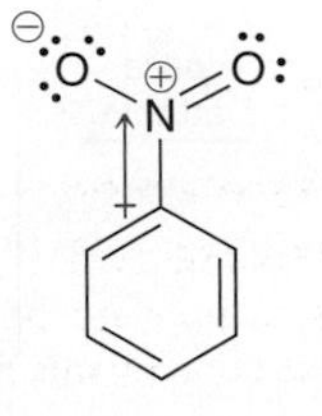

Now let's consider resonance. Many of the resonance structures exhibit a positive charge in the ring.

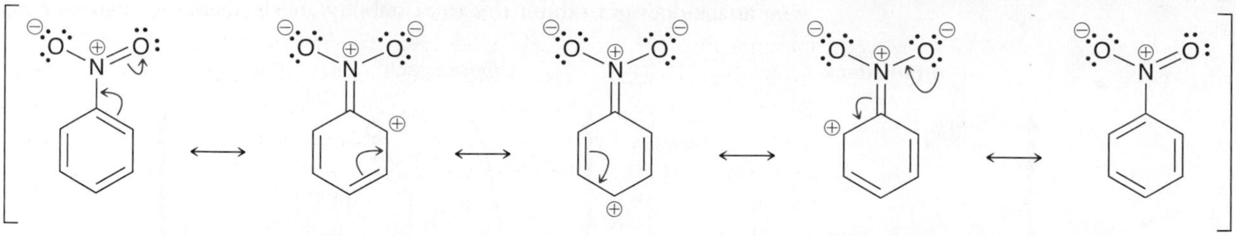

The positive charge indicates that the nitro group *withdraws* electron density from the ring. In this case, there is no competition between resonance and induction. Both factors suggest that the nitro group is a powerful electron-withdrawing group. By removing electron density from the ring, the nitro group destabilizes the positively charged sigma complex and raises the energy of activation for its formation. This effect is quite significant and can be observed by comparing rates of nitration. Specifically, nitrobenzene is 100,000 times less reactive than benzene toward nitration, and the reaction can only be accomplished at an elevated temperature. When the reaction is forced to proceed, the regiochemical outcome is different than what we have seen thus far (Figure 19.10). Notice that the *meta* product predominates, in stark contrast with the previous

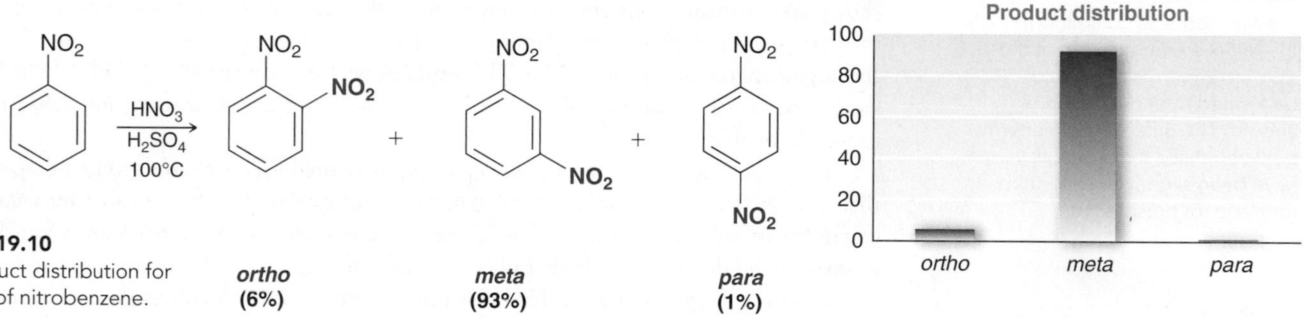

FIGURE 19.10
The product distribution for nitration of nitrobenzene.

examples in which *ortho* and *para* products predominated. To explain this observation, we must compare the stability of the sigma complexes formed for *ortho*, *meta*, and *para* attack (Figure 19.11). Notice that the sigma complex obtained from *ortho* attack exhibits

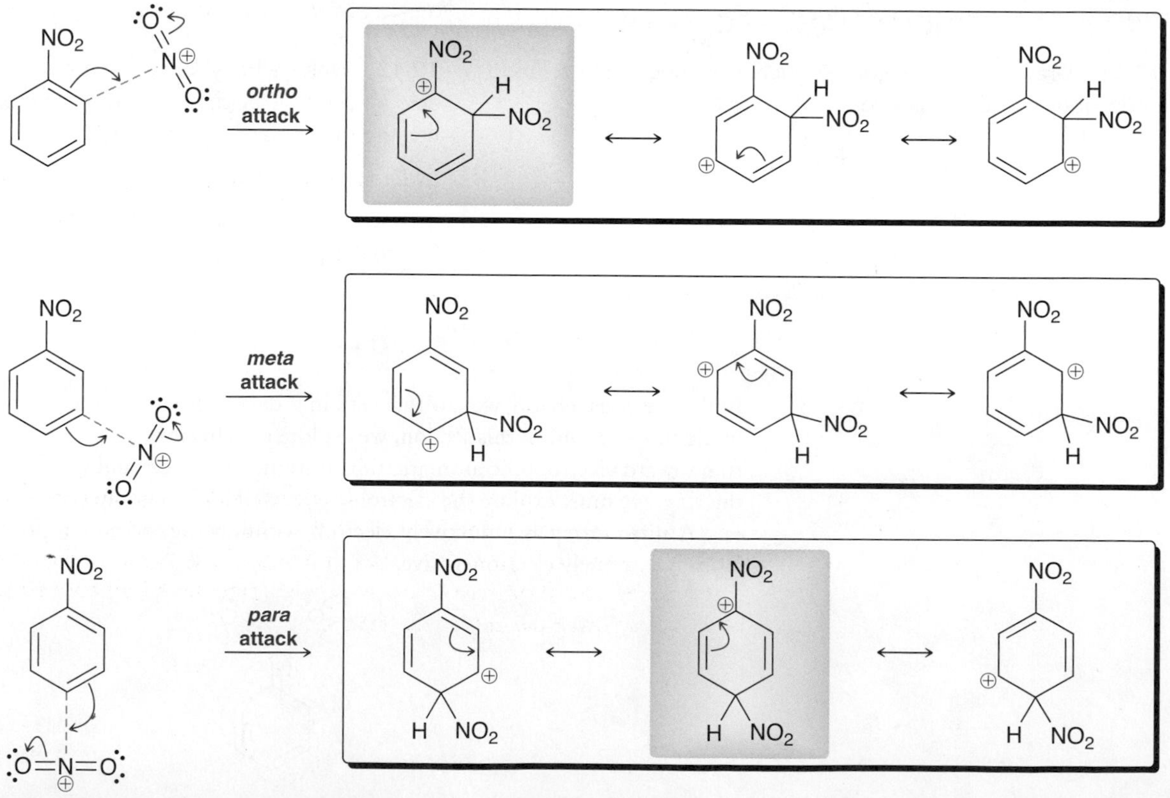

FIGURE 19.11
A comparison of the sigma complexes formed for each of the possible regiochemical outcomes for the nitration of nitrobenzene.

instability because one of the resonance structures (highlighted in green) has a positive charge directly adjacent to an electronegative, positively charged nitrogen atom. Similarly, the sigma complex obtained from *para* attack also exhibits this instability. The sigma complex obtained from *meta* attack does not exhibit this instability and is therefore lower in energy. Compare the energy diagrams for *ortho*, *meta*, and *para* attack (Figure 19.12). In summary, we have seen that a nitro group deactivates the ring and is a ***meta* director**. This is in fact a general rule that will be used extensively throughout the rest of this chapter: *Most deactivators are meta directors.*

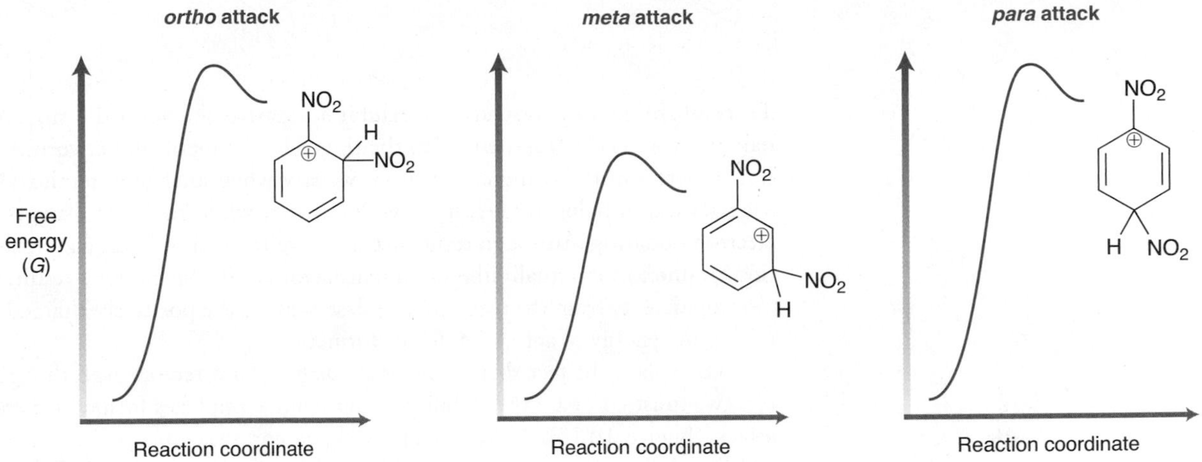

FIGURE 19.12
Energy diagrams comparing the relative energy levels of the sigma complexes that could be formed during nitration of nitrobenzene. The differences in energy between these three pathways have been slightly exaggerated for clarity of presentation.

CONCEPTUAL CHECKPOINT

19.13 When 1,3-dinitrobenzene is treated with nitric acid and sulfuric acid at elevated temperature, the product is 1,3,5-trinitrobenzene. Explain the regiochemical outcome of this reaction. In other words, explain why nitration takes place at the C5 position. Make sure to draw the sigma complex for each possible pathway and to compare the relative stability of each sigma complex.

19.9 Halogens: The Exception

In the previous sections, we have seen that activators are *ortho-para* directors and deactivators are *meta* directors.

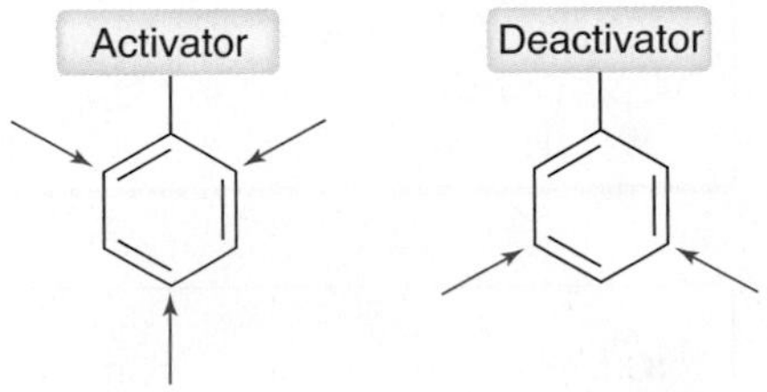

There is one important exception to these general rules—the halogens (F, Cl, Br, or I). Halogens are *ortho-para* directors despite the fact that they are deactivators. To rationalize this curious exception, we must explore the electronic effects of a halogen connected to an aromatic ring. As we have seen several times, it is necessary to consider both inductive effects and resonance. Let's begin by exploring the inductive effects.

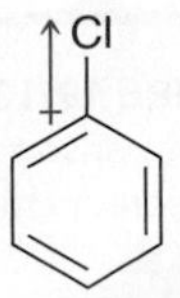

Halogens are fairly electronegative (more so than carbon) and are therefore inductively electron *withdrawing*. When we draw the resonance structures, a different picture emerges.

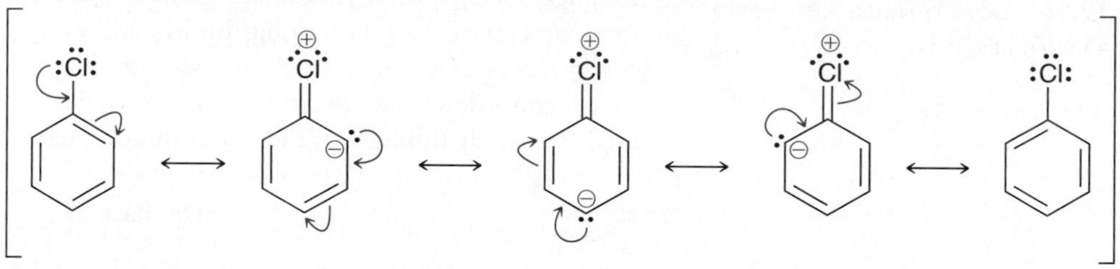

Three of the resonance structures exhibit a negative charge in the ring. This suggests that a halogen *donates* electron density to the ring. The competition between resonance and induction is very similar to the competition we saw when analyzing methoxybenzene. Induction suggests that a halogen is electron withdrawing, while resonance suggests that a halogen is electron donating. Although resonance is generally the dominant factor, this case is the exception. Induction is actually the dominant factor for halogens. As a result, halogens withdraw electron density from the ring, thereby destabilizing the positively charged sigma complex and raising the energy of activation for its formation.

To explain the fact that halogens are *ortho-para* directors even though they are deactivators, we must compare the stability of the sigma complexes formed for *ortho*, *meta*, and *para* attack (Figure 19.13). Notice that the sigma complex obtained from *ortho* attack has one additional resonance structure (highlighted in green), which stabilizes the sigma complex. Similarly, the sigma complex obtained from *para* attack also exhibits this additional stability, but the sigma complex from *meta* attack does not exhibit this extra stability, and therefore it is higher in energy. For this reason, halogens are *ortho-para* directors, despite the fact that they are deactivators.

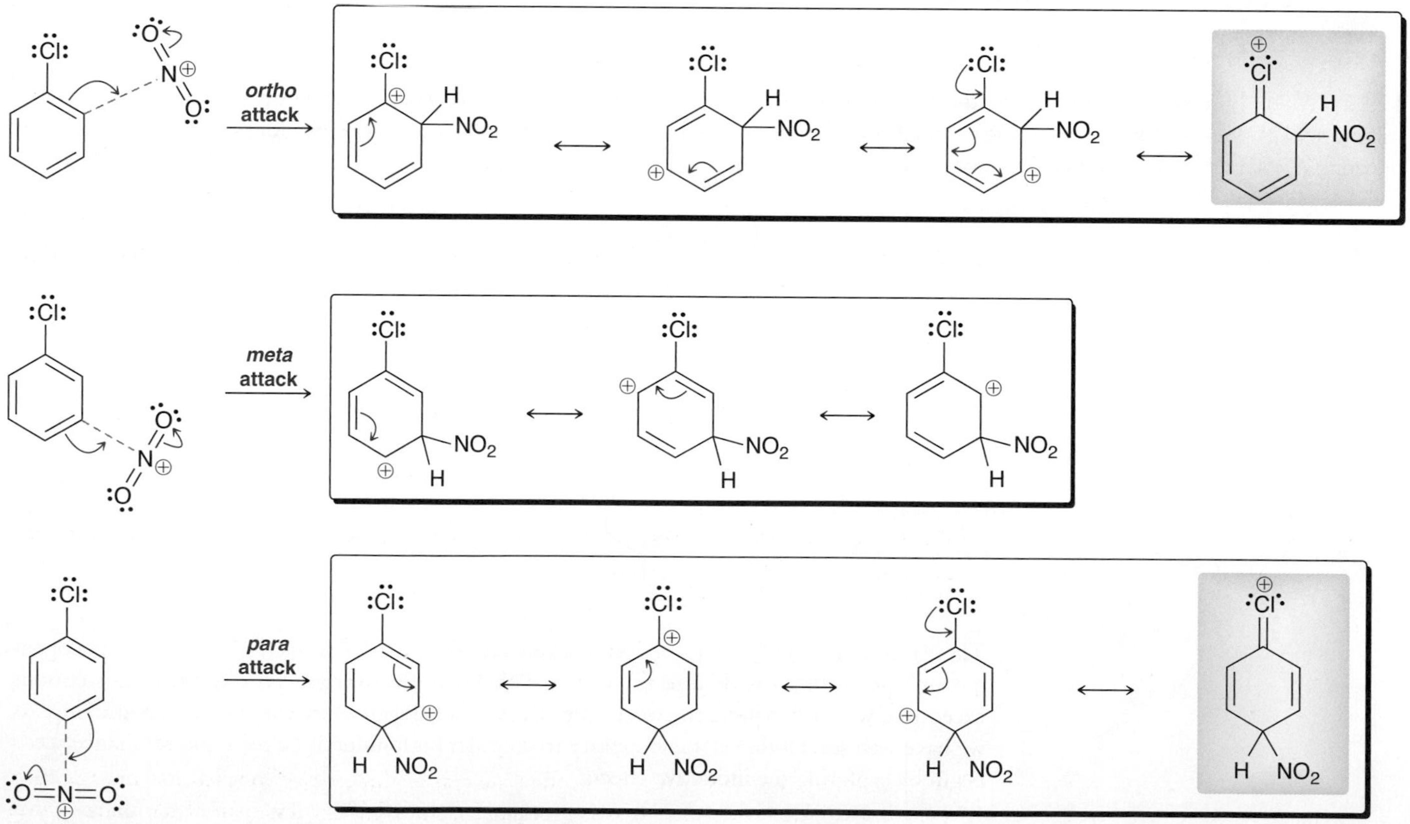

FIGURE 19.13
A comparison of the sigma complexes formed for each of the possible regiochemical outcomes for the nitration of chlorobenzene.

CONCEPTUAL CHECKPOINT

19.14 Does chlorination of chlorobenzene require the use of a Lewis acid? Explain why or why not?

19.15 Predict and explain the regiochemical outcome for chlorination of bromobenzene.

19.10 Determining the Directing Effects of a Substituent

The previous sections focused on the directing effects of a few specific groups (methyl, methoxy, nitro, and halogens). In this section, we will learn how to predict the directing effects for any substituent. That skill will prove to be essential in subsequent sections that deal with synthesis.

Both activators and deactivators can be classified as strong, moderate, and weak. Each of these categories is described below, followed by a summary chart of all six categories. As discussed in the preceding sections, the activators are *ortho-para* directors, while the deactivators, except for the halogens, are meta directors.

Activators

Strong activators are characterized by the presence of a lone pair immediately adjacent to the aromatic ring.

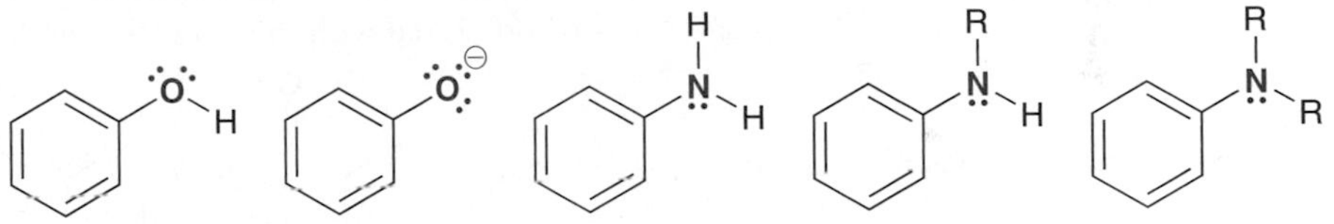

All of these groups exhibit a lone pair that is delocalized into the ring, as can be seen in their resonance structures. For example, phenol has the following resonance structures:

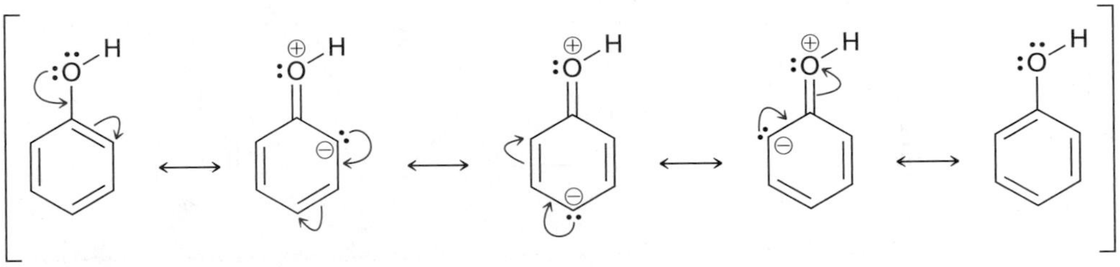

Many of these resonance structures have a negative charge in the ring, indicating that the OH group is donating electron density into the ring. This electron-donating effect strongly activates the ring.

Moderate activators exhibit a lone pair that is already delocalized outside of the ring.

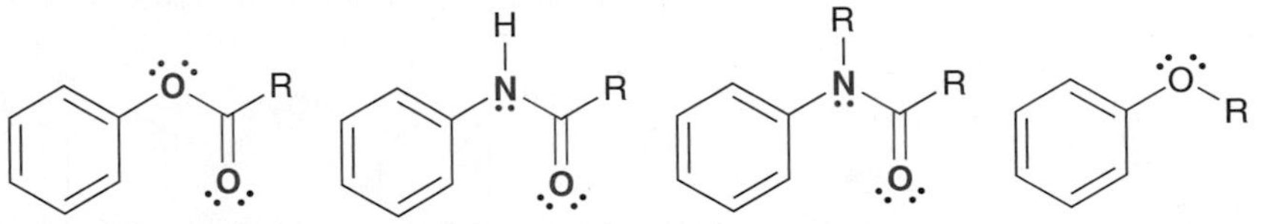

In the first three compounds, there is a lone pair next to the ring, but that lone pair is participating in resonance outside of the ring.

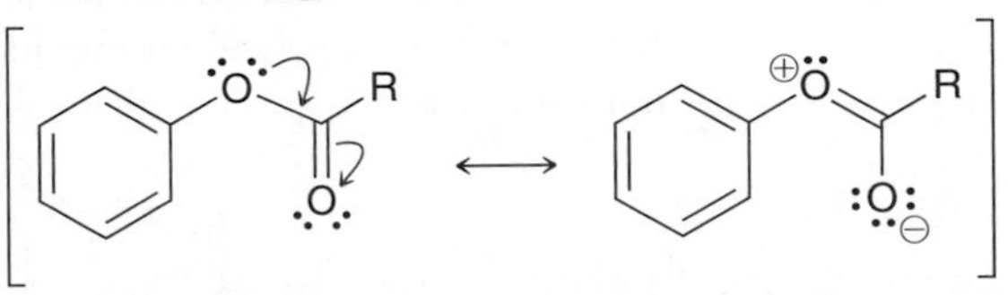

This effect diminishes the capability of the lone pair to donate electron density into the ring. These groups are activating, but they are moderate activators. The lone pair of the alkoxy group (OR) is not participating in resonance outside of the ring, and we might therefore expect that it would be a strong activator. Nevertheless, alkoxy groups belong to the class of moderate activators. You can think of alkoxy groups as an exception.

Alkyl groups are **weak activators**, because they donate electron density by the relatively weak effect of hyperconjugation (as described in Section 6.11).

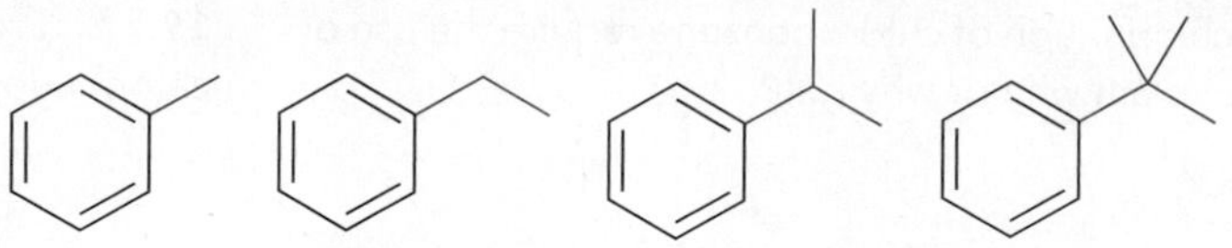

We will now turn our attention to deactivators, starting with weak deactivators and progressing to strong deactivators.

Deactivators

As we have already seen, halogens deactivate a benzene ring:

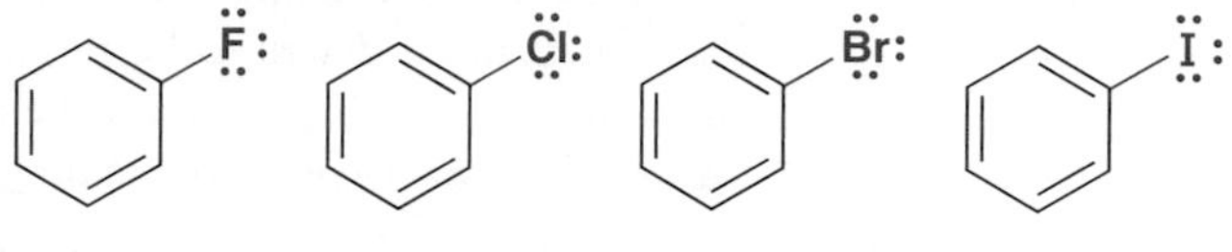

We have seen that the electronic effects of halogens are determined by the delicate competition between resonance and induction, with induction emerging as the dominant effect. As a result, halogens are **weak deactivators**.

Moderate deactivators are groups that exhibit a π bond to an electronegative atom, where the π bond is conjugated with the aromatic ring. Below are several examples.

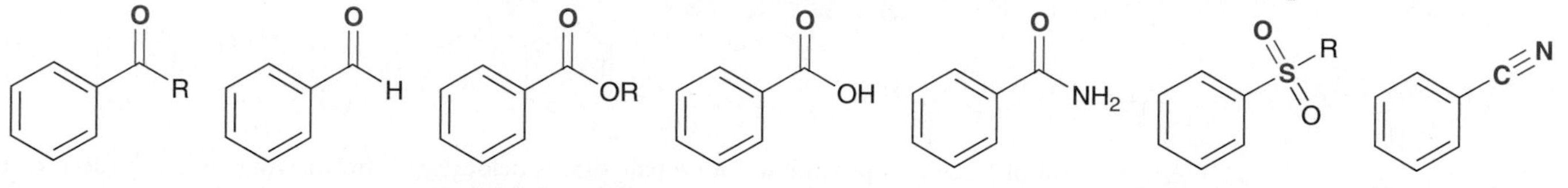

Each of these groups withdraws electron density from the ring via resonance. For example.

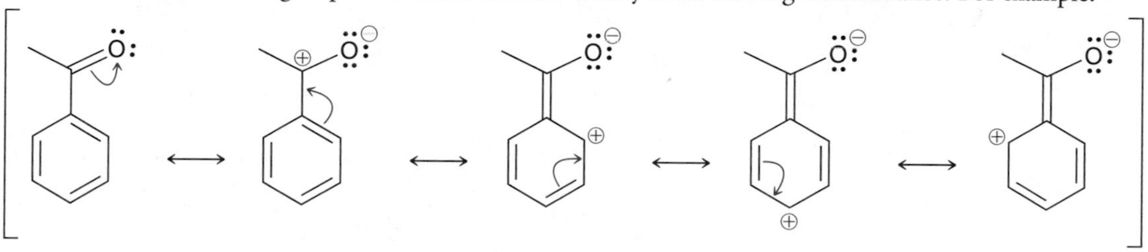

Three of the resonance structures have a positive charge in the ring, indicating that the group is withdrawing electron density from the ring. This electron-withdrawing effect moderately deactivates the ring.

There are only a few common substituents that are **strong deactivators**.

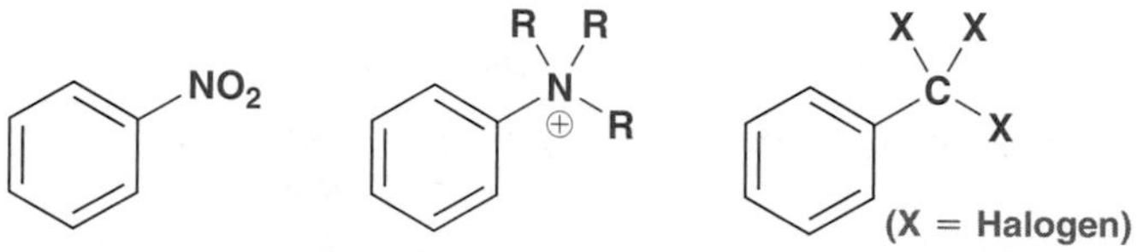

The nitro group is a strong deactivator because of resonance and induction. The other two groups are strong deactivators because of powerful inductive effects. A positively charged nitrogen atom is extremely electronegative, and CX_3 has three electron-withdrawing halogens. Do not confuse a CX_3 group with a halogen (X).

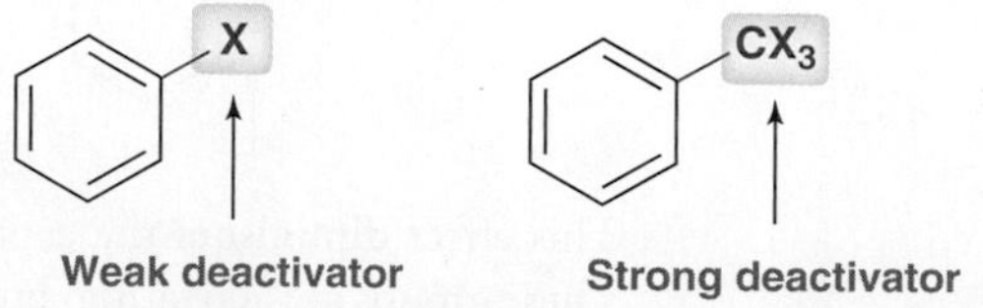

Table 19.1 summarizes the six categories of activators and deactivators. Notice the unique position of the halogens. In general, activators are *ortho-para* directors, while deactivators are *meta* directors, but halogens are the exception.

TABLE 19.1 A LIST OF ACTIVATORS AND DEACTIVATORS BY CATEGORY.

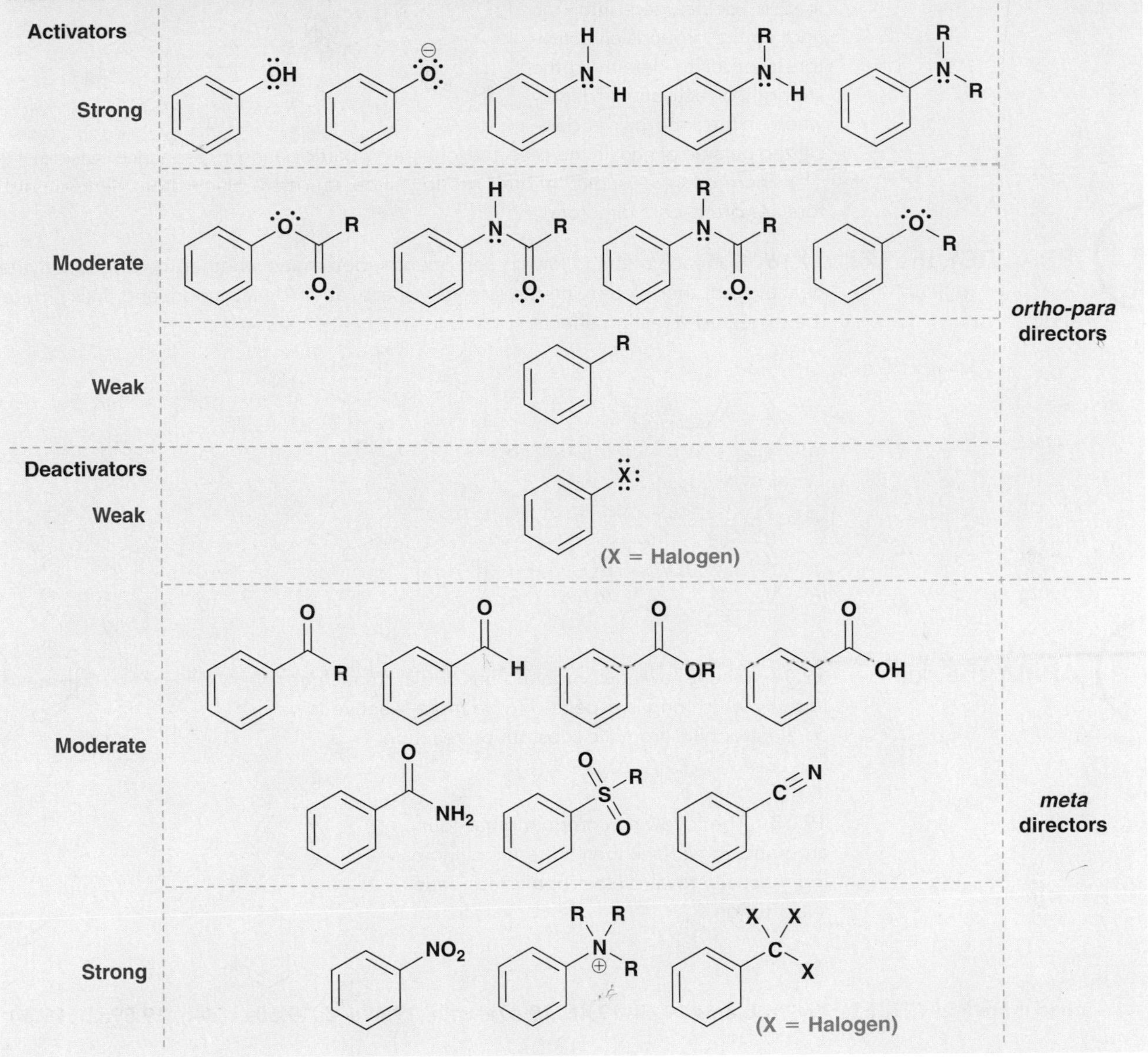

SKILLBUILDER

19.1 IDENTIFYING THE EFFECTS OF A SUBSTITUENT

LEARN the skill

Consider the following monosubstitued aromatic ring. Determine whether the group on the aromatic ring is an activator or a deactivator. Then determine the strength of activation/deactivation (i.e., is it strong, moderate, or weak). Finally, determine the directing effects of the group.

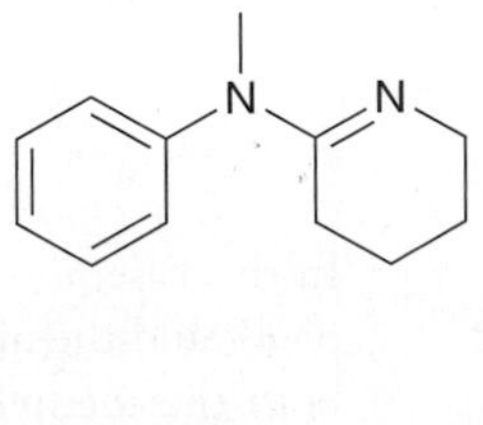

SOLUTION

First look for a lone pair immediately adjacent to the ring.

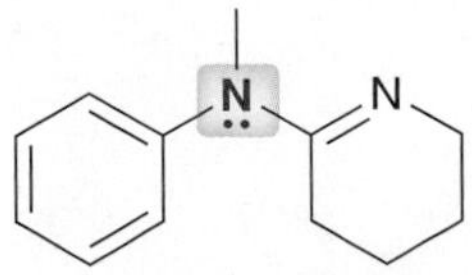

In this case, there is a lone pair that is delocalized into the ring, so the group is an activator. In order to determine the strength of activation, identify whether the lone pair is delocalized outside of ring. In this case, the lone pair is participating in resonance outside of the ring.

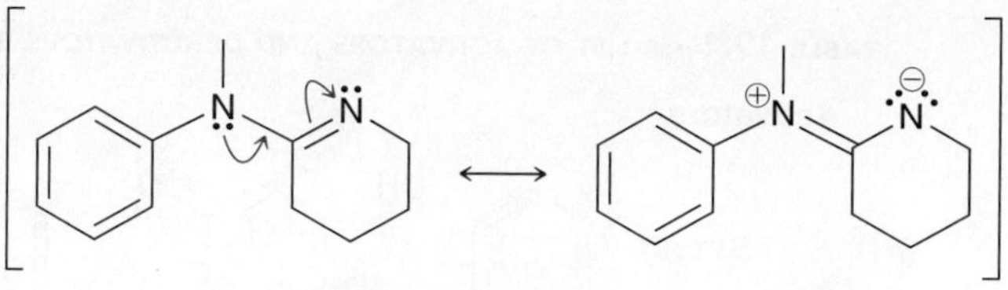

Therefore, we predict that this group will be a moderate activator. All moderate activators are *ortho-para* directors.

PRACTICE the skill **19.16** For each of the following compounds, determine whether the ring is activated or deactivated, then determine the strength of activation/deactivation, and finally, determine the expected directing effects.

(a) 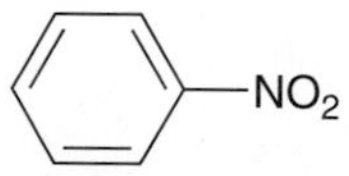

(b) 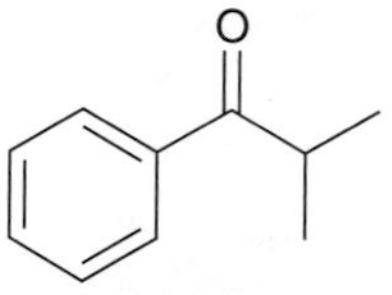

(c) 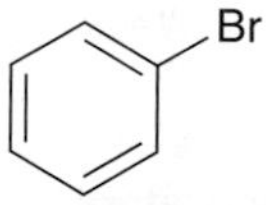

(d) 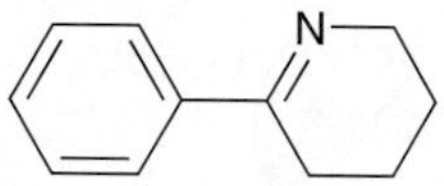

(e) 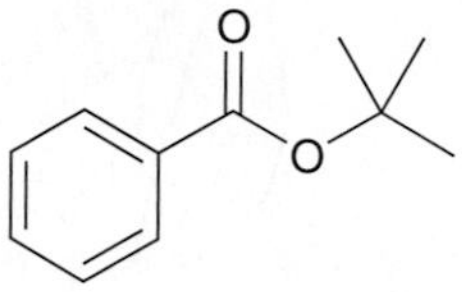

(f) 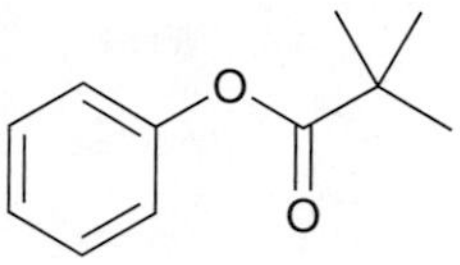

APPLY the skill

19.17 The following compound has two aromatic rings. Identify which ring is expected to be more reactive toward an electrophilic aromatic substitution reaction.

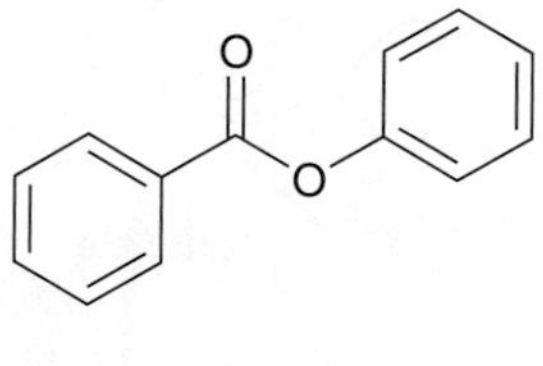

19.18 The following compound has four aromatic rings. Rank them in terms of increasing reactivity toward electrophilic aromatic substitution.

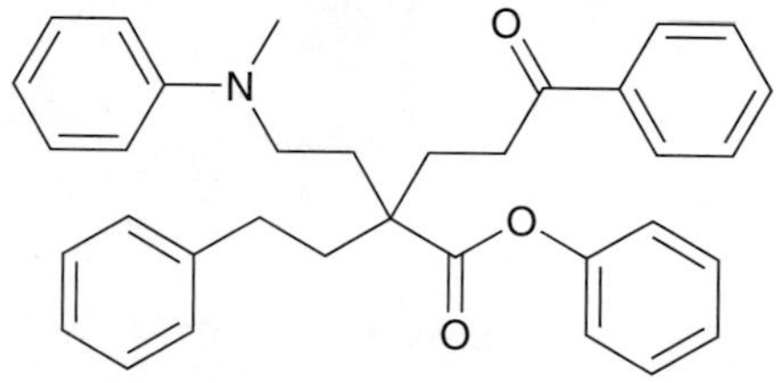

need more **PRACTICE?** **Try Problems 19.44–19.46, 19.47a–c,f,h, 19.49a–d, 19.50a,b,d–g, 19.59a,b, 19.64, 19.66**

19.11 Multiple Substituents

Directing Effects

We will now explore directing effects when multiple substituents are present on a ring. In some cases, the directing effects of all substituents reinforce each other, for example:

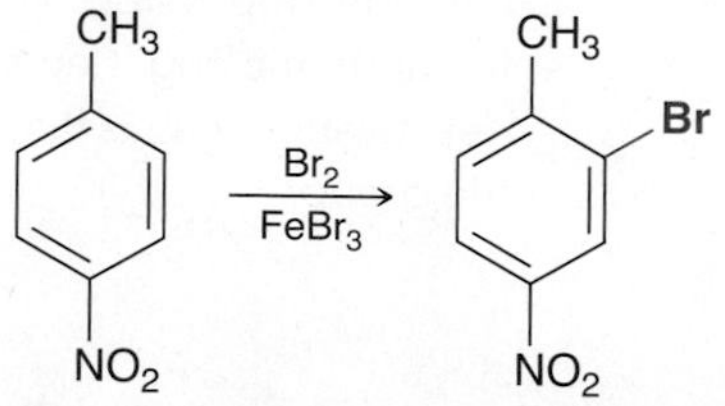

In this case, the methyl group directs to the *ortho* positions (the *para* position is already occupied), and the nitro group directs to the positions that are *meta* to the nitro group. In this case, both the methyl group and the nitro group direct to the same two locations. Since the two locations are identical (by symmetry), only one product is obtained.

In other cases, the directing effects of the various substituents may compete with each other. In such cases, the more powerful activating group dominates the directing effects.

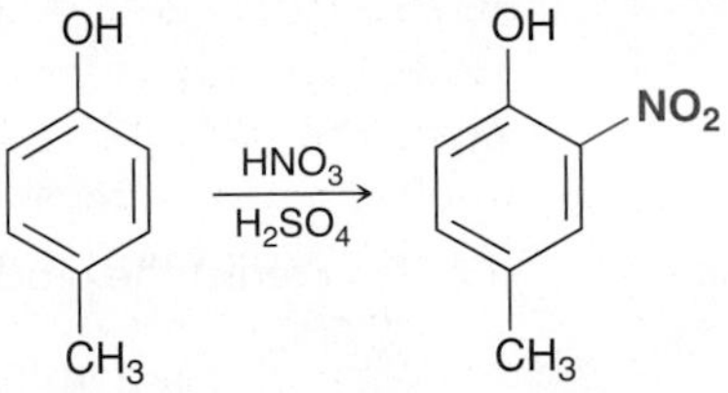

In this case, there are two substituents on the ring: an OH group (strong activator) and a methyl group (weak activator). The strong activator dominates, so the incoming nitro group is installed at a position that is *ortho* to the strong activator (the *para* position is already occupied).

SKILLBUILDER

19.2 IDENTIFYING DIRECTING EFFECTS FOR DISUBSTITUTED AND POLYSUBSTITUTED BENZENE RINGS

LEARN the skill

Identify the position that is most likely to undergo an electrophilic aromatic substitution reaction.

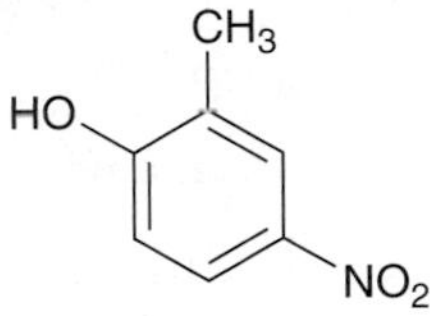

SOLUTION

STEP 1 Identify the nature of each group.

Begin by identifying the effect of each group on the aromatic ring.

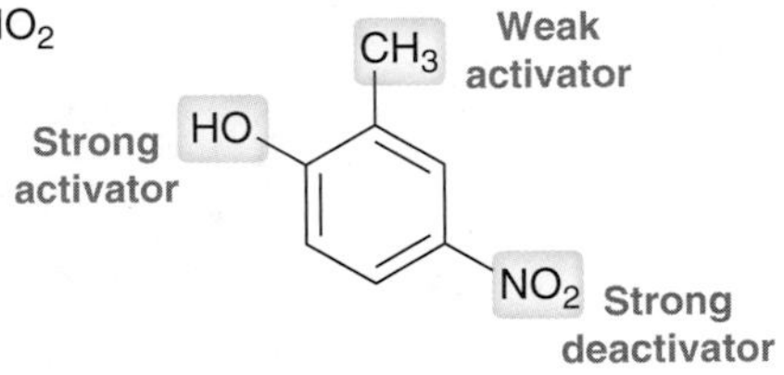

STEP 2 Select the most powerful activator and identify the positions that are *ortho* or *para* to that group.

The most powerful activator will dominate the directing effects. In this case, the OH group is the strongest activator. Now consider the positions that are *ortho* and *para* to the OH group.

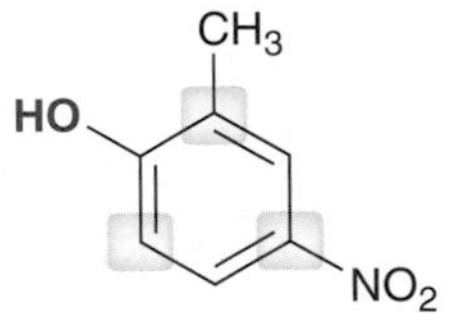

STEP 3 Identify the unoccupied positions.

Two of these positions are already occupied. Only one position remains. We therefore predict that this position is most likely to undergo an electrophilic aromatic substitution reaction.

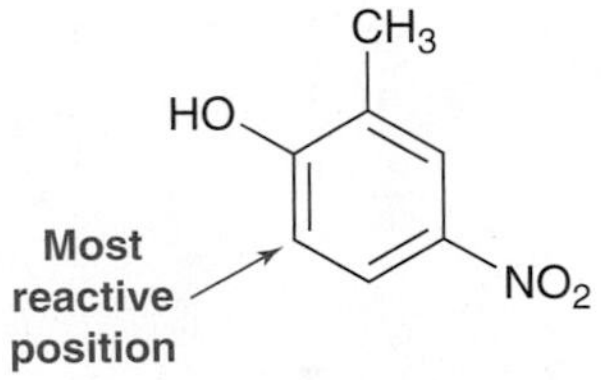

PRACTICE the skill **19.19** For each compound below, identify which position(s) is/are most likely to undergo an electrophilic aromatic substitution reaction.

(a)
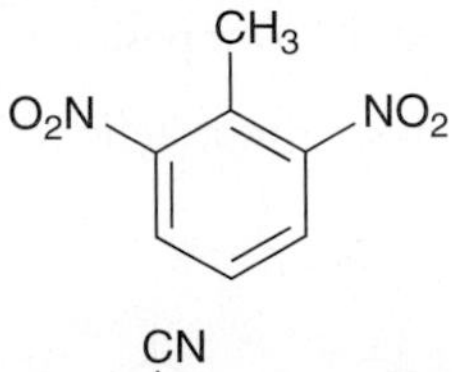

(b)
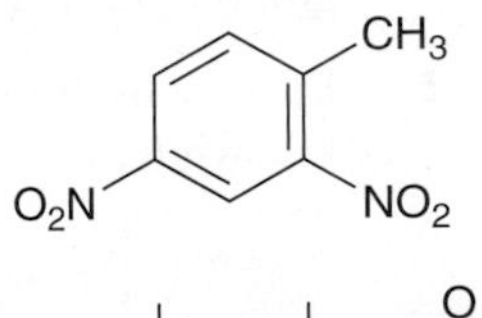

(c)
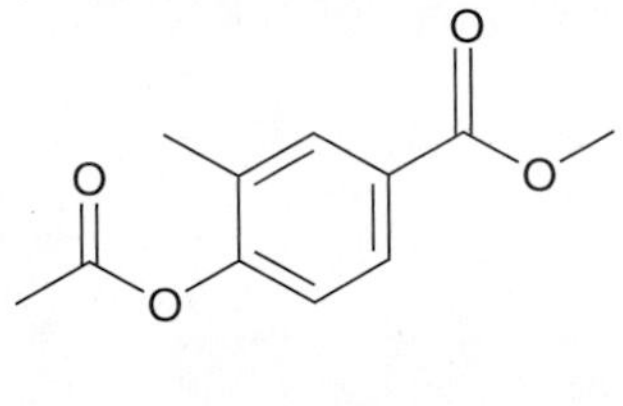

(d)
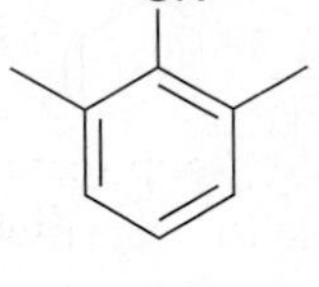

(e)
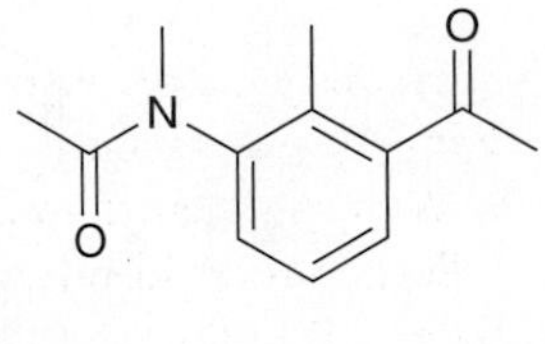

(f)
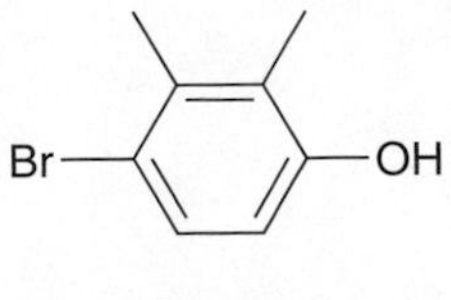

(g) 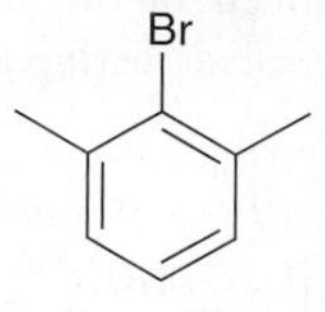

(h) 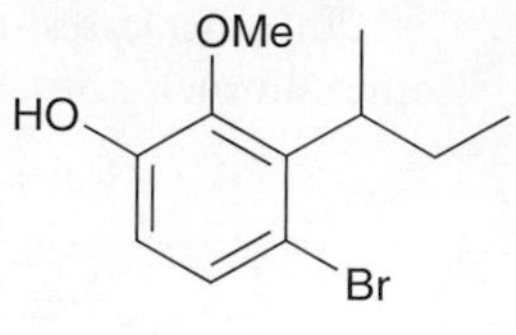

(i) 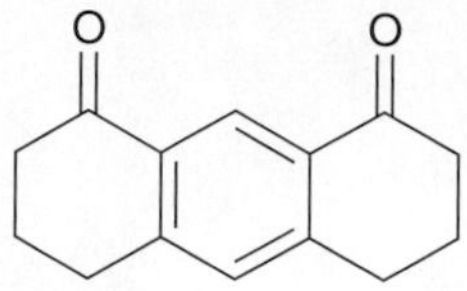

APPLY the skill

19.20 Predict the product(s) for each of the following reactions:

(a) 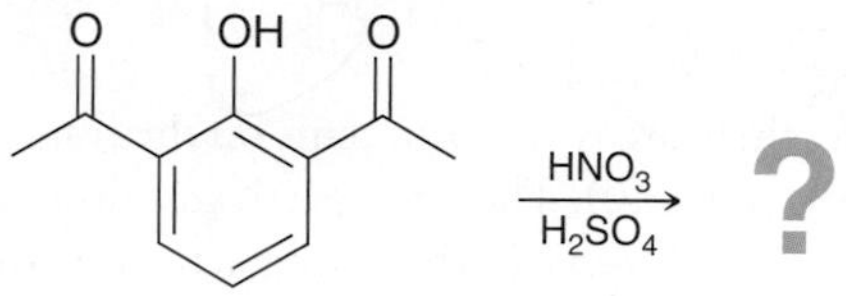

(b) 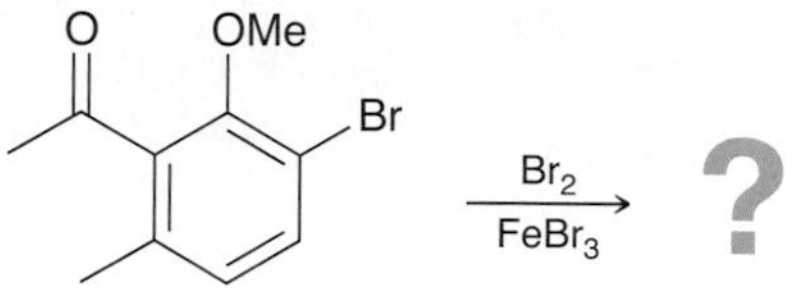

(c) 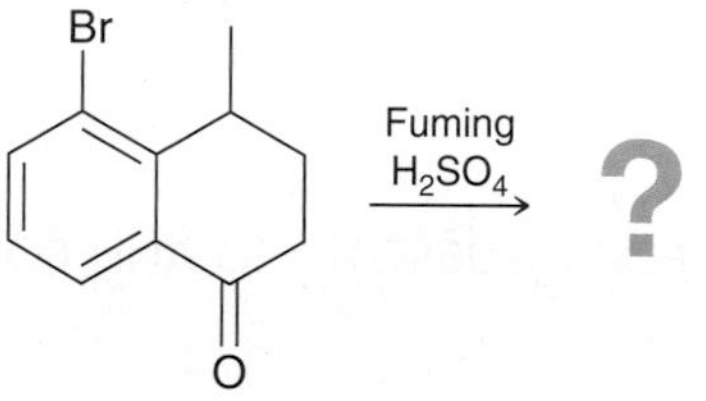

19.21 When 2,4-dibromo-3-methyltoluene is treated with bromine in the presence of iron (Fe), a compound with molecular formula $C_8H_7Br_3$ is obtained. Identify the structure of this product.

need more **PRACTICE?** **Try Problems 19.47e, 19.50k, 19.56a,b, 19.59, 19.69**

Steric effects

In many cases, steric effects can play an important role in determining product distribution. Let's begin with a simple case in which only one substituent is on the ring.

When an *ortho-para* director is present on the ring, it is difficult to predict the exact ratio of *ortho* and *para* products. Nevertheless, the following guidelines are helpful in most cases:

1. For most monosubstituted aromatic rings, the *para* product generally dominates over the *ortho* product as a result of steric considerations.

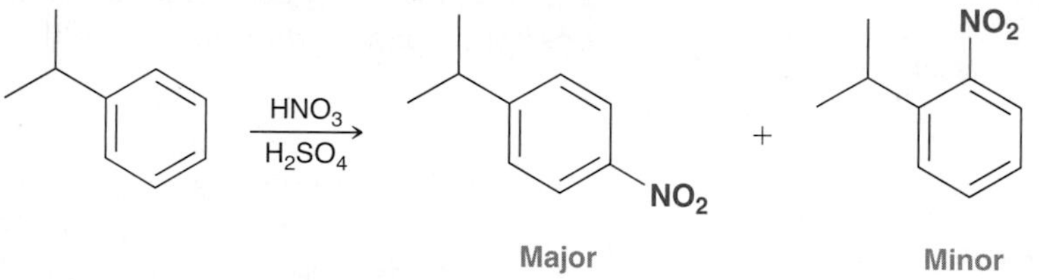

Steric hindrance reduces the likelihood of attack at the *ortho* position, and as a result, the *para* product is the major product. A notable exception is toluene (methylbenzene), for which the ratio of *ortho* and *para* products is sensitive to the conditions employed, such as the choice of solvent. In some cases, the *para* product is favored; in others, the *ortho* product is favored; Therefore, it is generally not wise to utilize the directing effects of a methyl group to favor a reaction at the *para* position over the *ortho* position.

2. For 1,4-disubstituted aromatic rings, steric effects again play a significant role. Consider the following case:

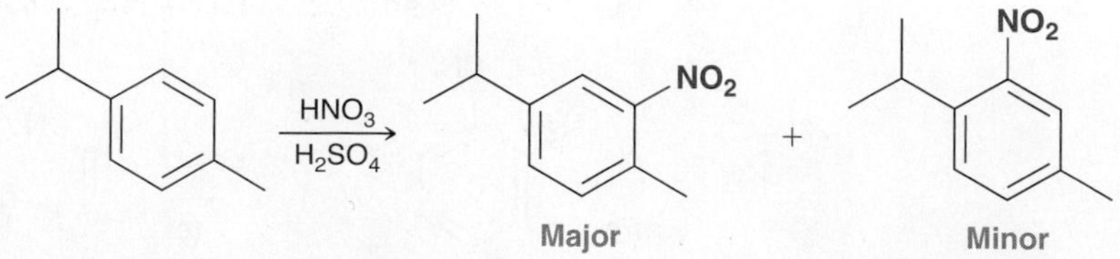

The regiochemical outcome of this reaction is controlled by sterics. Nitration is more likely to occur at the site that is less sterically hindered (*ortho* to the methyl group).

3. For 1,3-disubstituted aromatic rings, it is extremely unlikely that substitution will occur at the position between the two substituents. That position is the most sterically hindered position on the ring, and a reaction generally does not take place at that position.

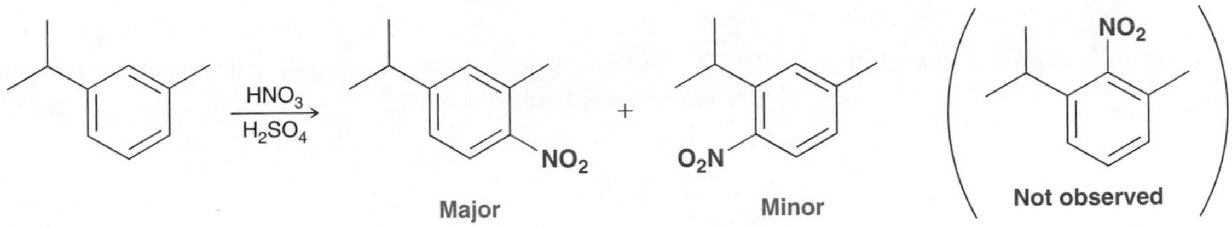

Using the previous three guidelines, let's get some practice predicting the product distribution in cases where steric effects control the outcome.

SKILLBUILDER

19.3 IDENTIFYING STERIC EFFECTS FOR DISUBSTITUTED AND POLYSUBSTITUTED BENZENE RINGS

LEARN the skill

Determine the position that is most likely to be the site of an electrophilic aromatic substitution reaction.

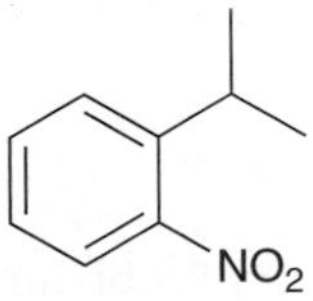

SOLUTION

STEP 1 Identify the nature of each group.

Begin by identifying the effect of each group on the aromatic ring.

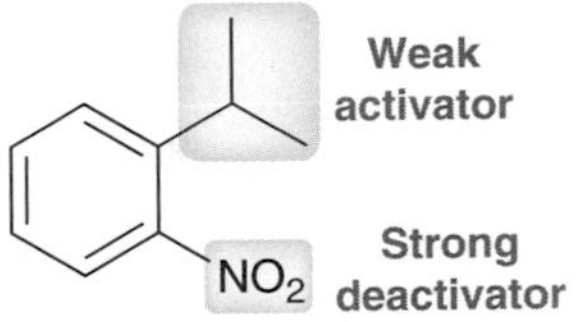

STEP 2 Select the most powerful activator and identify the positions that are *ortho* or *para* to that group.

In this case, a weak activator is competing with a strong deactivator for directing effects. Recall that the more powerful activator controls the directing effects, so the isopropyl group determines the outcome in this case. The isopropyl group is *ortho-para* directing, so we must consider the positions that are *ortho* and *para* to the isopropyl group.

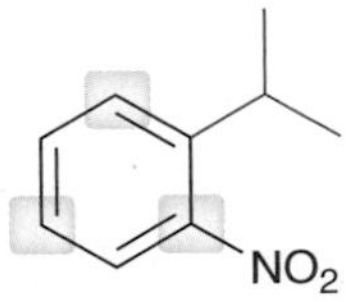

STEP 3 Identify the unoccupied positions that are the least sterically hindered.

One of these locations is already occupied, leaving two choices. The *ortho* position is sterically hindered, while the *para* position is sterically unencumbered. In this case, we expect substitution to take place primarily at the *para* position and to a lesser extent at the *ortho* position.

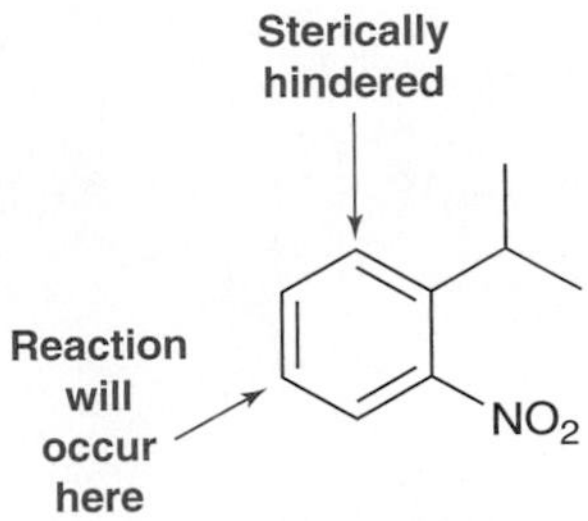

PRACTICE the skill

19.22 For each of the following compounds, determine the position that is most likely to be the site of an electrophilic aromatic substitution reaction:

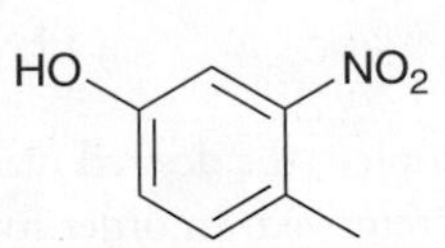

(a)

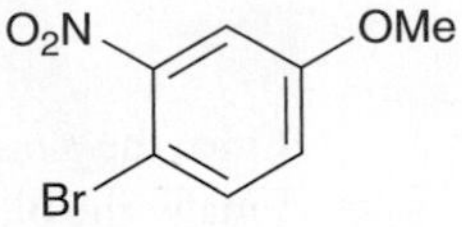

(b)

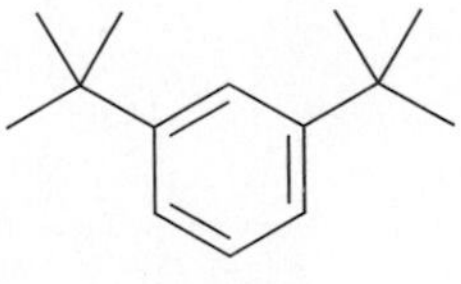
(c)

(d) 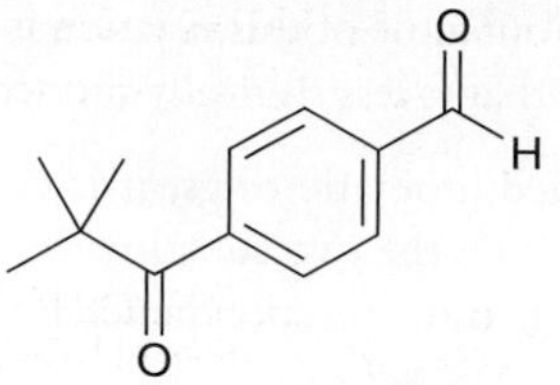

(e) 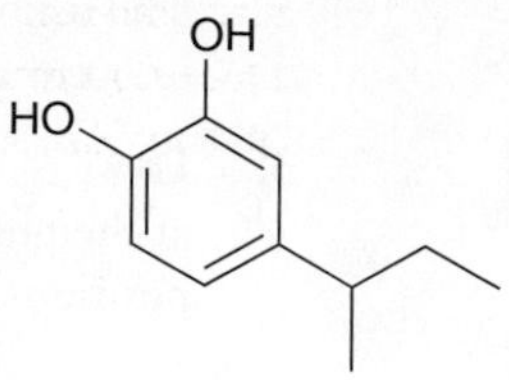

APPLY the skill

19.23 The following compound is highly activated, but nevertheless undergoes bromination very slowly. Explain.

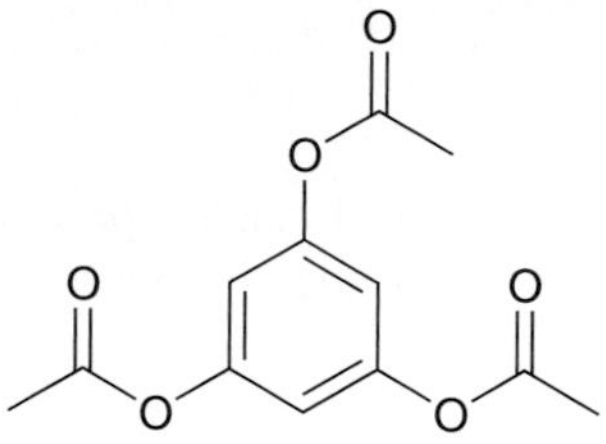

19.24 When the following compound is treated with Br_2 in the presence of a Lewis acid, one product predominates. Determine the structure of that product.

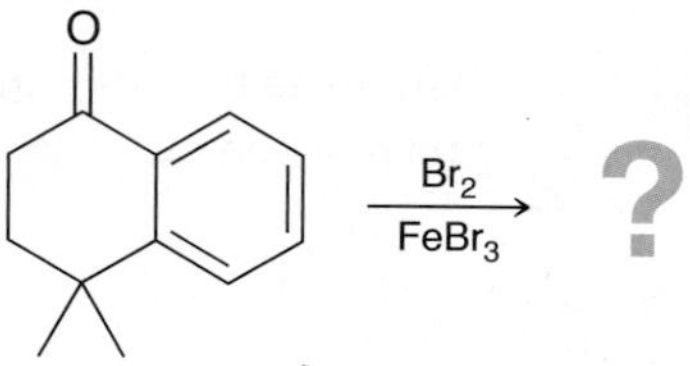

need more **PRACTICE?** **Try Problems 19.59, 19.63, 19.69**

Blocking Groups

Consider how the following transformation might be achieved:

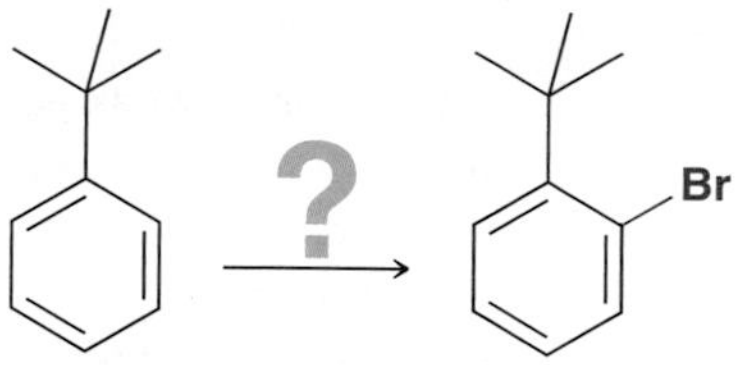

Direct bromination of *tert*-butylbenzene produces the *para* product as the major product, while the desired *ortho* product is the minor product. In such a situation, a **blocking group** can be used to direct the bromination toward the *ortho* position. In this case, the blocking group is first installed at the *para* position.

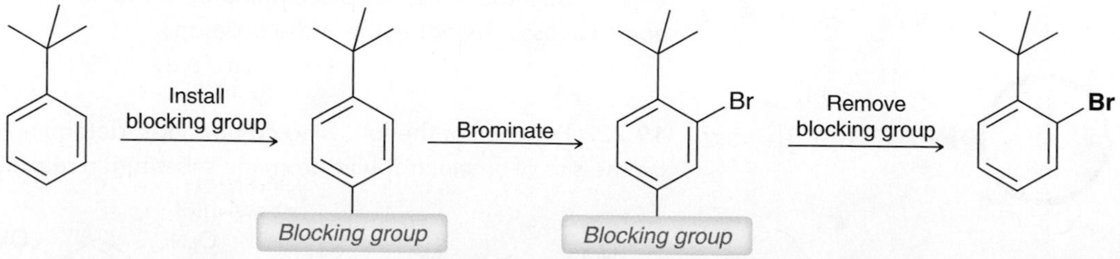

Once the *para* position is occupied, the desired reaction is forced to occur at the *ortho* position. Finally the blocking group is removed. In order for a group to function as a blocking group, it must be easily removable after the desired reaction has been achieved. There are many different blocking groups that can be used. Sulfonation is commonly used for this purpose, because the sulfonation process is reversible.

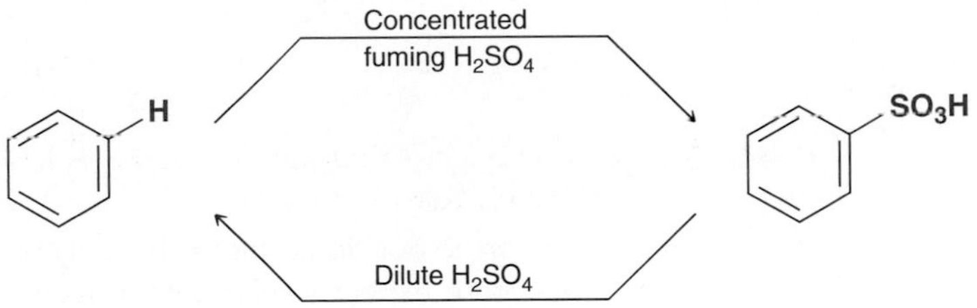

Sulfonation provides a valuable blocking technique that enables us to achieve the desired transformation.

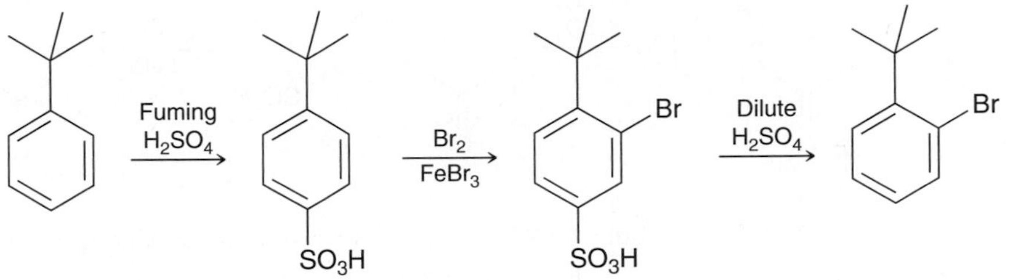

SKILLBUILDER

19.4 USING BLOCKING GROUPS TO CONTROL THE REGIOCHEMICAL OUTCOME OF AN ELECTROPHILIC AROMATIC SUBSTITUTION REACTION

LEARN the skill

Identify whether a blocking group is necessary to accomplish the following transformation:

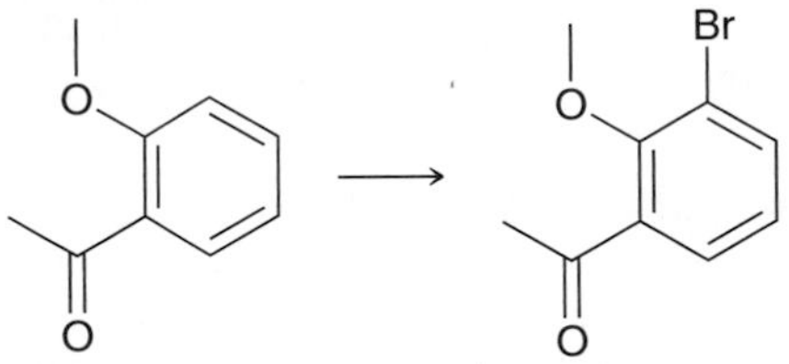

SOLUTION

Analyze the starting material. The two substituents are the methoxy group, which is a moderate activator, and the acyl group, which is a moderate deactivator.

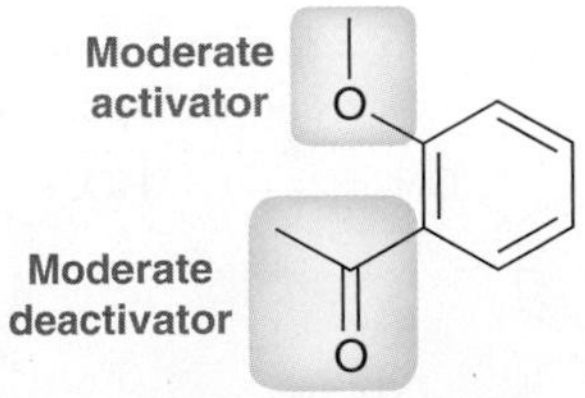

In this case, the methoxy group controls the directing effects, and therefore, the reactive centers are the unoccupied *ortho* and *para* positions.

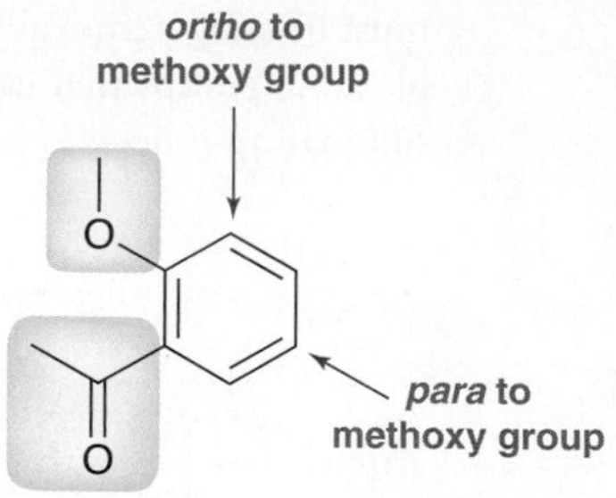

The *ortho* position is more sterically hindered, while the *para* position is not hindered. We therefore expect a substitution reaction to take place at the position that is *para* to the methoxy group. If we want to place a group in the *ortho* position, a blocking group would be required.

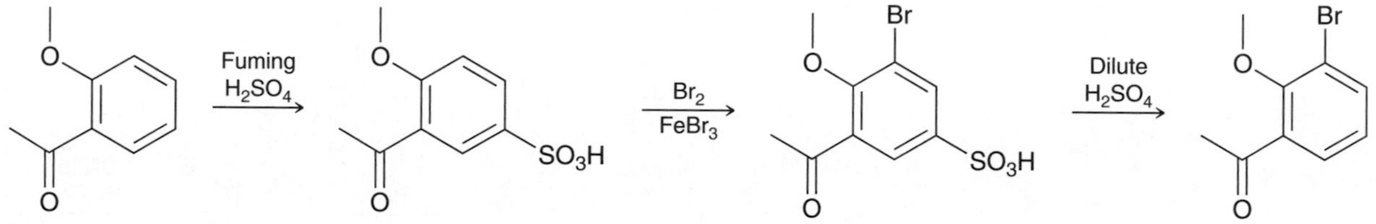

PRACTICE the skill **19.25** Determine whether a blocking group is necessary to accomplish each of the following transformations.

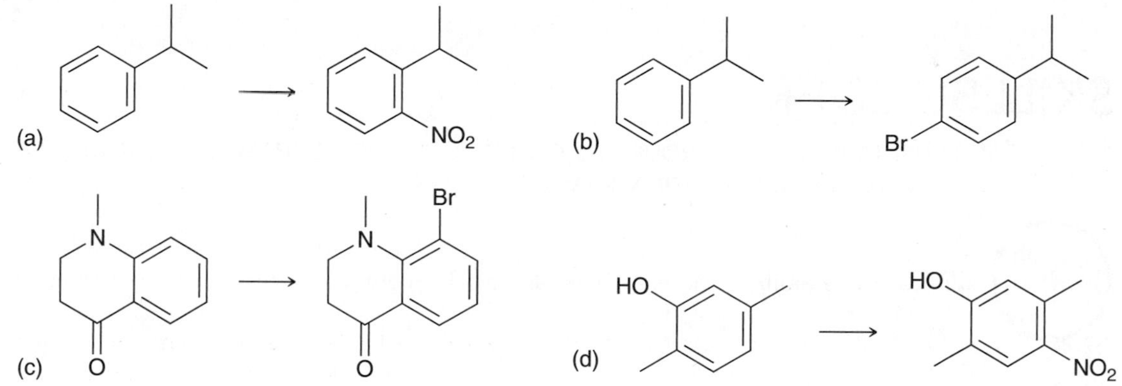

APPLY the skill **19.26** Predict the major product of the following reaction.

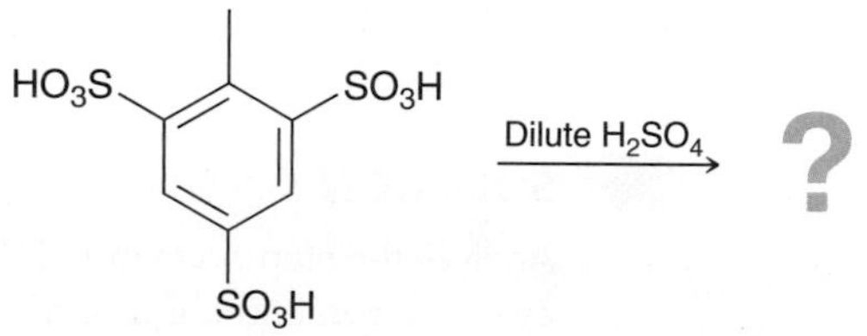

19.27 The following transformations cannot be accomplished, even with the help of blocking groups. In each case, explain why a blocking group will not help.

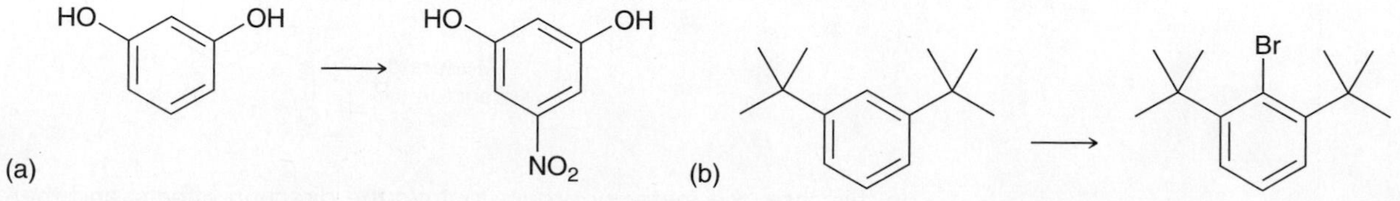

need more **PRACTICE?** **Try Problems 19.58d, 19.68c**

19.12 Synthesis Strategies

Monosubstituted Benzene Rings

The simplest kind of synthesis problem is one that requires formation of a monosubstituted benzene ring. Directing effects are irrelevant in such a case. You simply need to know what reagents are necessary to install the desired group. Figure 19.14 is a list of the reagents that we have seen thus far. This list should be committed to memory before moving on to more sophisticated synthesis problems. In total, we have seen 10 different groups that can be installed on an aromatic ring. Take special notice of the four groups shown in blue. Installation of these groups requires two steps.

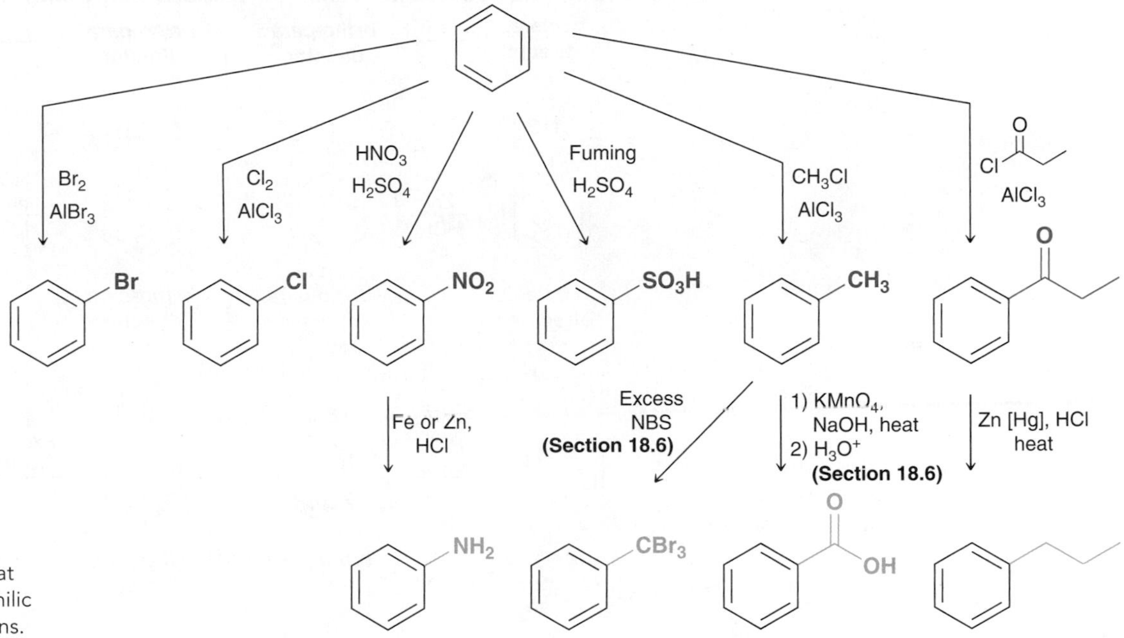

FIGURE 19.14 A list of functional groups that can be installed via electrophilic aromatic substitution reactions.

CONCEPTUAL CHECKPOINT

19.28 Identify the reagents necessary to convert benzene into each of the following compounds:

(a) Chlorobenzene (b) Nitrobenzene
(c) Bromobenzene (d) Ethylbenzene
(e) Propylbenzene (f) Isopropylbenzene
(g) Aniline (aminobenzene) (h) Benzoic acid
(i) Toluene

19.29 Identify the product obtained when benzene is treated with each of the following reagents:

(a) Fuming sulfuric acid (b) HNO_3/H_2SO_4
(c) Cl_2, $AlCl_3$ (d) Ethyl chloride, $AlCl_3$
(e) Br_2, Fe
(f) HNO_3/H_2SO_4 followed by Zn, HCl

Disubstituted Benzene Rings

Proposing a synthesis for a disubstituted benzene ring requires a careful analysis of directing effects to determine which group should be installed first. As an example, consider the following compound.

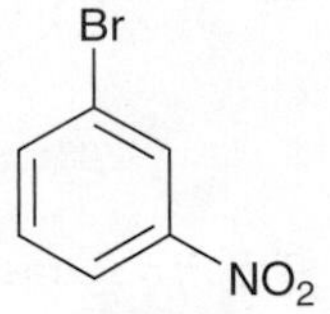

To make this compound from benzene requires two separate steps—bromination and nitration. Bromination followed by nitration will not produce the desired product, because a bromine substituent is *ortho-para* directing. In order to achieve the *meta* relationship between the two groups, nitration must be performed first. The nitro group is a *meta* director, which then directs the incoming bromine to the desired location.

The preceding example is fairly straightforward, because each group is installed with only one step. An extra consideration is necessary when installation of one of the groups requires two steps and involves a change in directing effects. These changes are summarized in Table 19.2.

TABLE 19.2 FUNCTIONAL GROUP CONVERSIONS THAT CHANGE DIRECTING EFFECTS

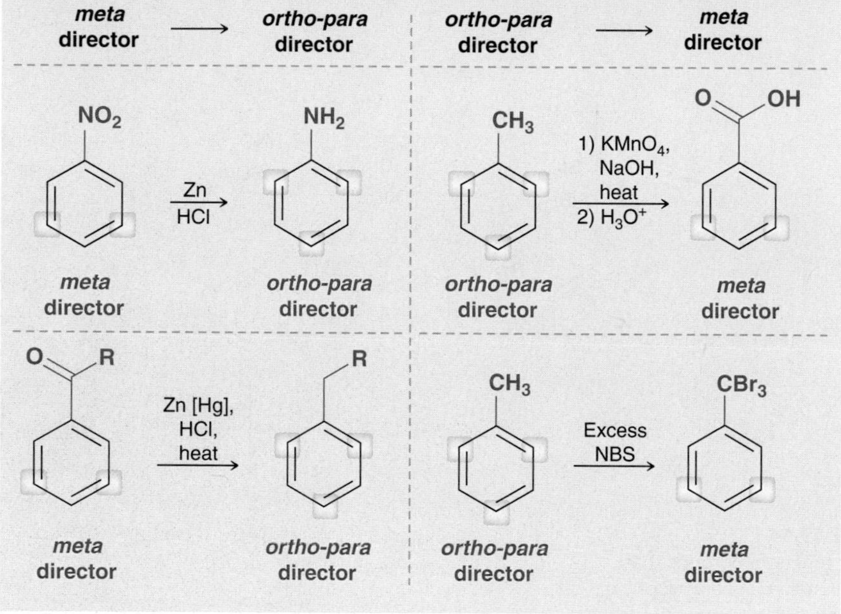

As an example, consider the installation of an amino group, which requires (1) nitration, followed by (2) reduction. The reduction converts a *meta*-directing nitro group into an *ortho-para*-directing amino group. This change in directing effects must be considered when planning a synthesis that requires installation of an amino group. To illustrate this point, consider the following example.

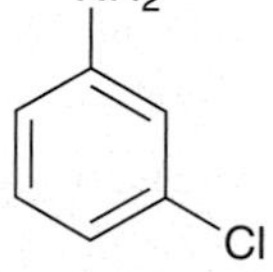

This compound has two groups that are *meta* to each other, and we must decide which group to install first. The problem is that both groups are *ortho-para* directing, so neither group will direct the other group into the correct location. This problem can be solved if we recognize that installation of the amino group involves a change in directing effects.

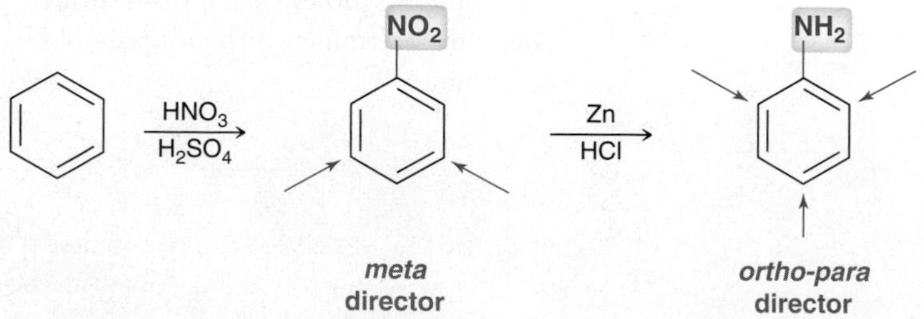

The nitro group is *meta* directing, while the amino group is *ortho-para* directing. The two steps above do not have to be consecutive, and we can exploit the *meta*-directing properties of the nitro group to install the chlorine in the correct position. Specifically, the correct regiochemical outcome is achieved with the following order of events: (1) nitration, (2) chlorination, and (3) reduction.

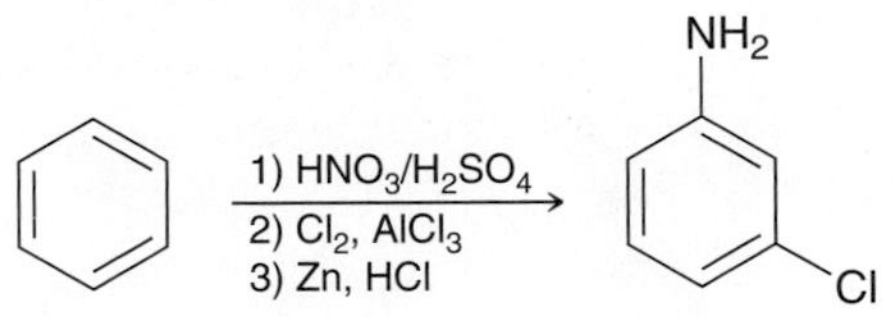

In addition to considering the order of events, two limitations must also be considered when planning a synthesis.

1. Nitration cannot be performed on a ring that contains an amino group.

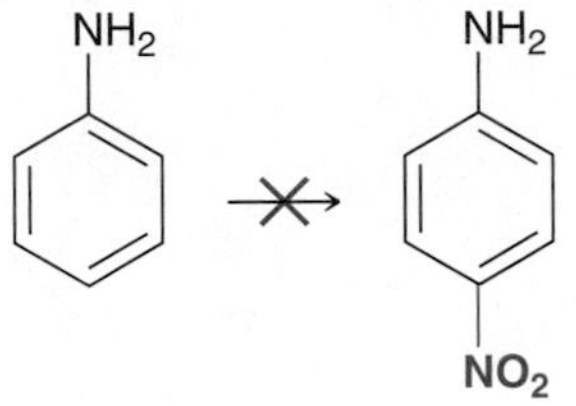

The reagents for nitration (a mixture of HNO_3 and H_2SO_4) can oxidize the amino group, often leading to a mixture of undesirable products. Attempts to perform this reaction often produce a tarry substance.

2. A Friedel-Crafts reaction (either alkylation or acylation) cannot be accomplished on rings that are either moderately or strongly deactivated. The ring must be either activated or weakly deactivated in order for a Friedel-Crafts reaction to occur.

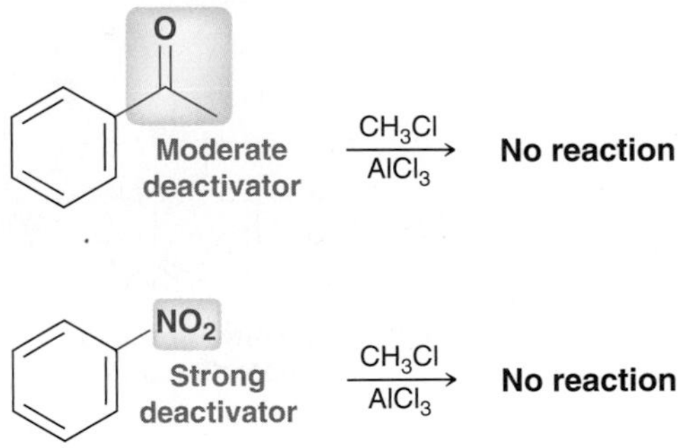

SKILLBUILDER

19.5 PROPOSING A SYNTHESIS FOR A DISUBSTITUTED BENZENE RING

LEARN the skill

Starting with benzene and using any other necessary reagents of your choice, design a synthesis of the following compound.

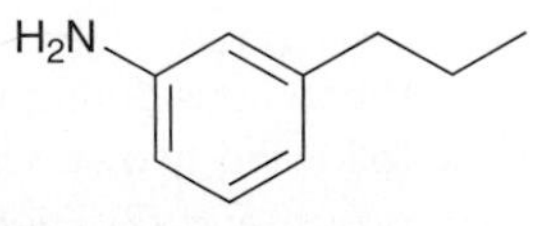

STEP 1
Identify the reagents necessary to install each substituent.

SOLUTION

Installation of the amino group requires a two-step process—nitration followed by reduction. Installation of the propyl group also requires a two-step process—acylation followed by reduction (in order to avoid carbocation rearrangements).

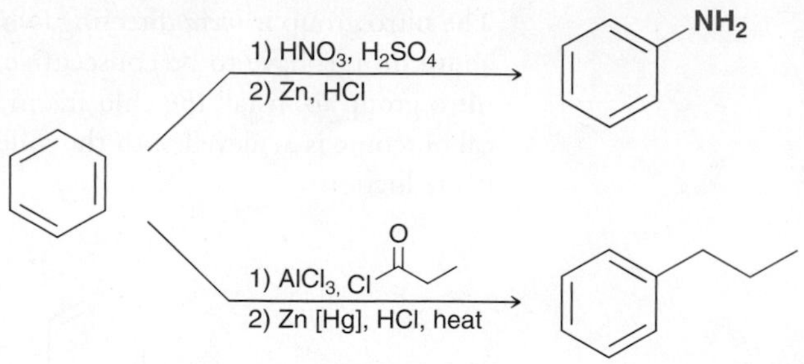

STEP 2
Determine the order of events that achieves the desired regiochemical outcome.

Now let's consider the order of events. These two groups must be installed in an order that places them *meta* to each other. The amino group is *ortho-para* directing, so it cannot be installed first. However, the propyl group is also *ortho-para* directing, so it too cannot be installed first. In this case, we are forced to exploit the *meta*-directing effects of either the nitro group or the acyl group.

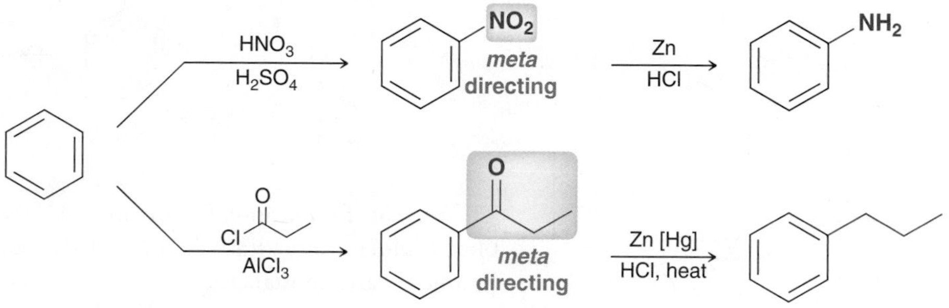

By taking advantage of the *meta*-directing effects of either the nitro group or the acyl group, we have two possible routes to consider:

Route 1

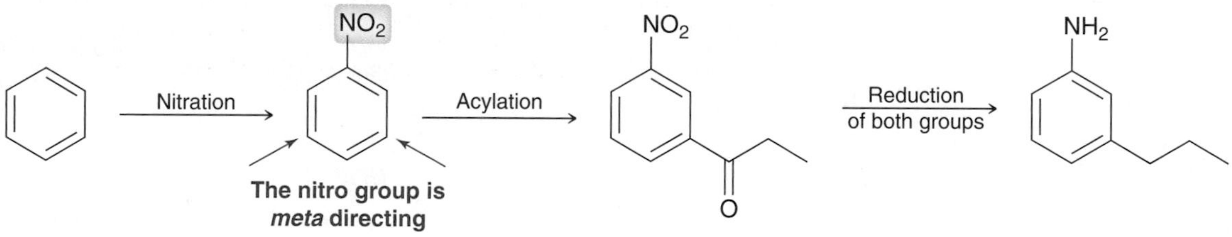

Route 2

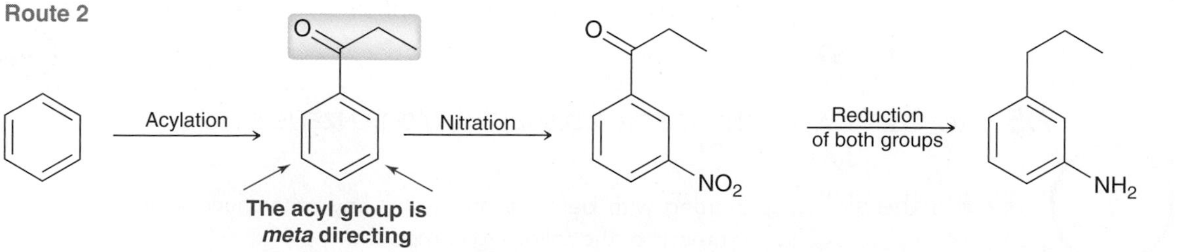

When considering the viability of each route, we must make sure not to violate either of the following two limitations: (1) Nitration cannot be performed on a ring possessing an amino group and (2) Friedel-Crafts reactions cannot be performed on a moderately or strongly deactivated ring.

Neither of the proposed routes violates the first limitation, but one of the routes does, indeed, violate the second limitation. Specifically, the first route involves a Friedel-Crafts acylation with a strongly deactivated ring (nitrobenzene). This will not work. Therefore, only the second route is viable. In the final step of this route, both groups are reduced under Clemmensen conditions.

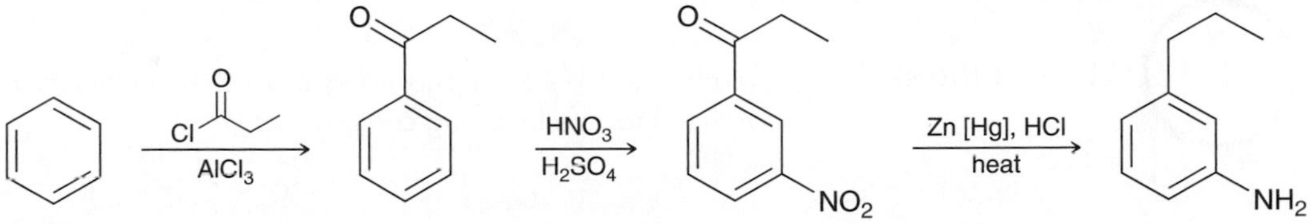

PRACTICE the skill **19.30** Starting with benzene and using any other necessary reagents of your choice, design a synthesis for each of the following compounds. Some of the problems have more than one plausible answer.

(a) 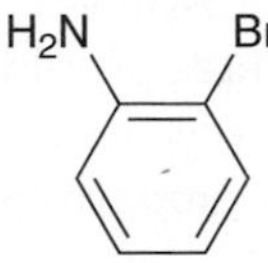

(b) 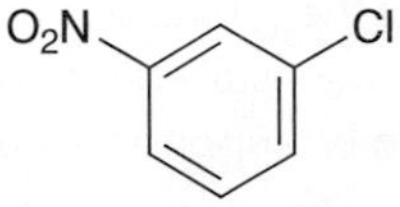

(c) 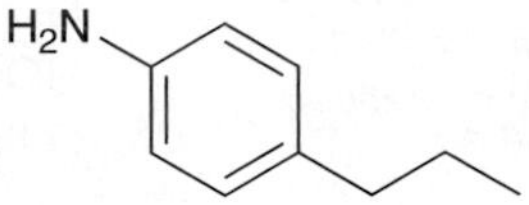

(d) 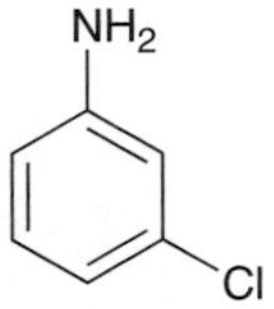

(e) 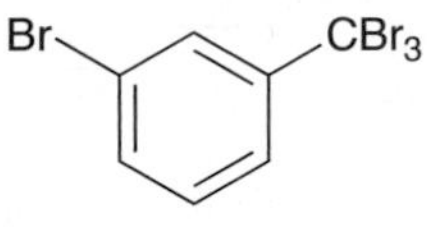

(f) 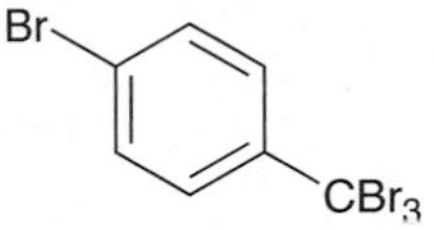

(g) 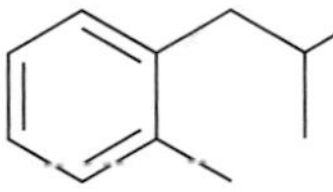

(h) 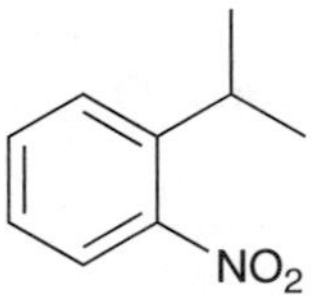

(i) 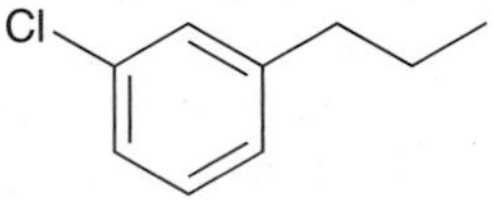

(j) 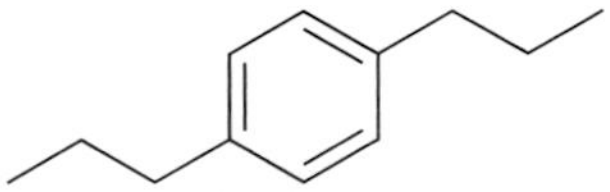

APPLY the skill

19.31 Using only reactions that we learned in this chapter, there are two different ways to prepare the following compound from benzene. Identify both ways, and then determine which way is likely to produce a better yield of the desired product. Explain your choice.

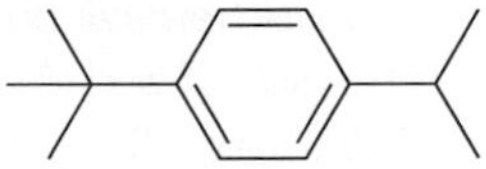

19.32 The following compounds cannot be made using only reactions that we learned in this chapter. For each compound, explain the issues that prevent its formation:

(a) 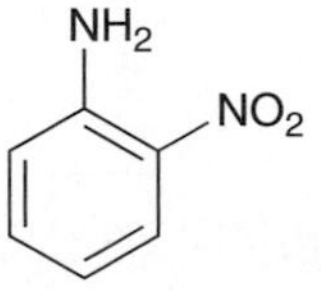

(b) 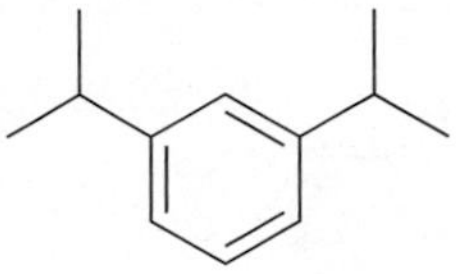

need more **PRACTICE?** **Try Problems 19.57, 19.58, 19.68, 19.75**

Polysubstituted Benzene Rings

When designing a synthesis for a polysubstituted benzene ring, it is often most efficient to utilize a retrosynthetic analysis, as discussed in Section 12.5. The following example illustrates the process.

SKILLBUILDER

19.6 PROPOSING A SYNTHESIS FOR A POLYSUBSTITUTED BENZENE RING

LEARN the skill

Starting with benzene and using any other necessary reagents of your choice, design a synthesis for the following compound:

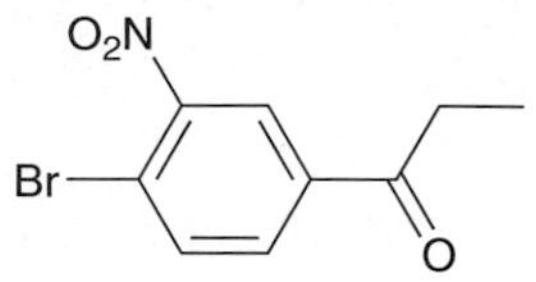

SOLUTION

STEP 1
Determine the last step of the synthesis.

Approaching this problem from a retrosynthetic point of view, we begin by determining the last step of the synthesis. There are three possibilities: (1) the Br group is installed last, (2) the NO_2 group is installed last, or 3) the acyl group is installed last.
Let's begin by supposing that the Br group is installed last.

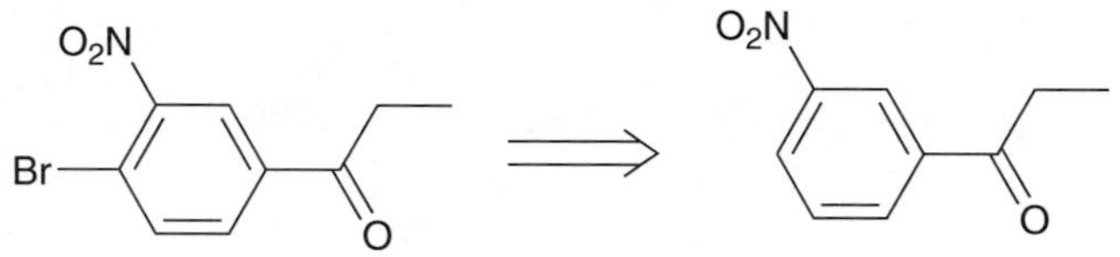

Recall that this arrow is a retrosynthetic arrow, and it means that the first compound might be made from the second compound. Our last step would therefore be a bromination reaction.

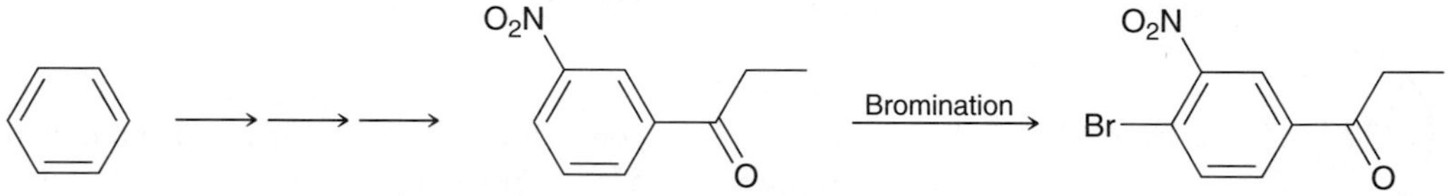

To consider the plausibility of this as our last step, we must first examine whether the desired regiochemical outcome will be achieved. In this case, we are trying to achieve the bromination of a disubstituted ring in which both groups (the nitro group and the acyl group) are *meta* directors. In such a case, the incoming Br group would be installed *meta* to both groups, which is not the desired location. Therefore, this step cannot be the last step of our synthesis.

Instead, let's consider installation of the acyl group as the last step. Once again, we must consider whether the desired regiochemical outcome would be achieved.

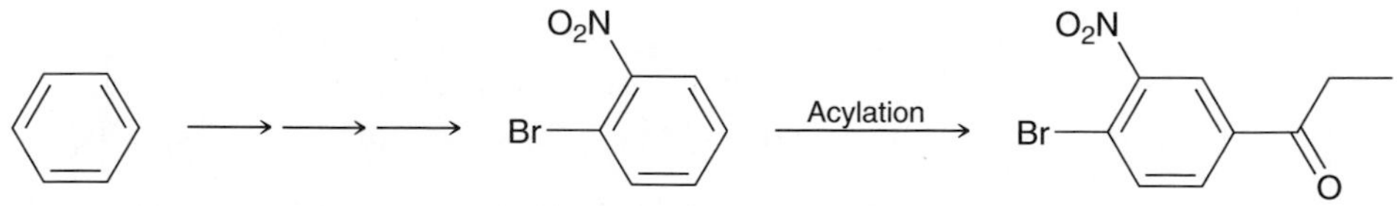

In this case, the desired regiochemical outcome would be achieved because both the Br group and the NO_2 group direct to the desired location. However, this transformation is a Friedel-Crafts acylation, so we must consider whether the proposed reaction violates any of the limitations for Friedel-Crafts acylation. In fact, there is a violation here because the ring is strongly deactivated by the presence of a nitro group, and the desired reaction simply cannot be achieved.

There is only one possibility left for the last step of our synthesis, which must be installation of the nitro group.

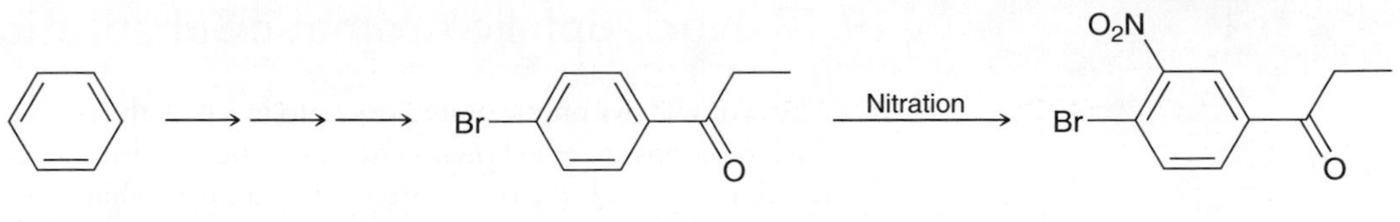

Both the Br group and the acyl group will direct to the desired location, and this reaction is plausible. This must be the last step.

STEP 2
Determine the penultimate (second-to-last) step of the synthesis.

Continuing to work backward, we must now consider the order in which the remaining two groups should be installed:

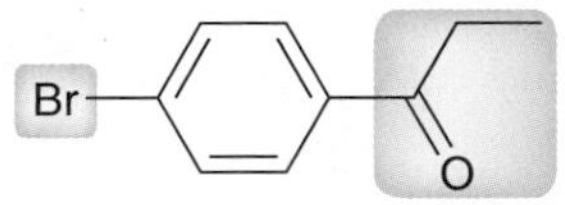

STEP 3
Consider any limitations for a proposed step.

STEP 4
Redraw the synthesis from start to finish.

Once again, we must choose a sequence of events that achieves the desired regiochemical outcome. The Br and acyl groups are *para* to each other, so we must consider which group is a *para* director. Indeed, the Br group is an *ortho-para* director (with a preference for *para*), while the acyl group is a *meta* director. Therefore, we conclude that bromination be carried out first followed by acylation. Whenever performing an acylation step, it is necessary to consider the limitations of acylation. This case requires the acylation of bromobenzene. The Br group is only *weakly* deactivating, which does not interfere with the acylation process (acylation is only unattainable with moderately deactivated or strongly deactivated rings). In summary, our proposed synthesis has the following sequence of events:

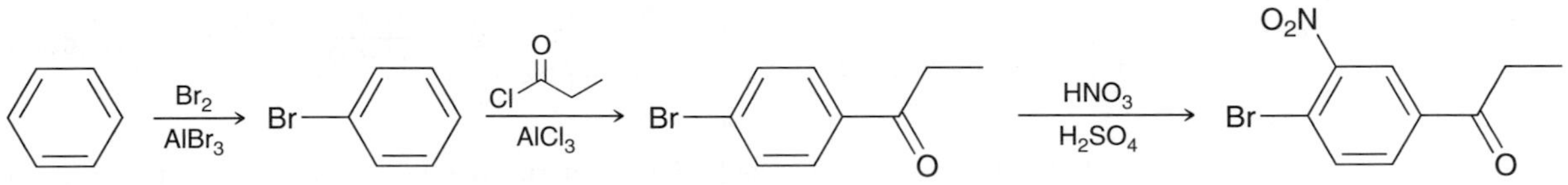

PRACTICE the skill

19.33 Starting with benzene and using any other necessary reagents of your choice, design a synthesis for each of the following compounds. In some cases, there may be more than one plausible answer.

(a) 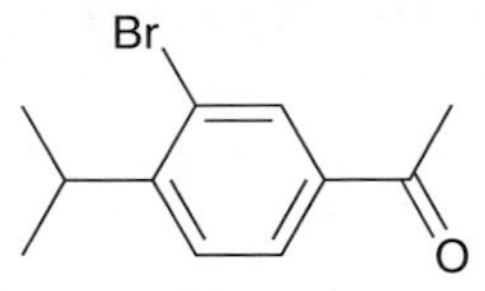

(b) 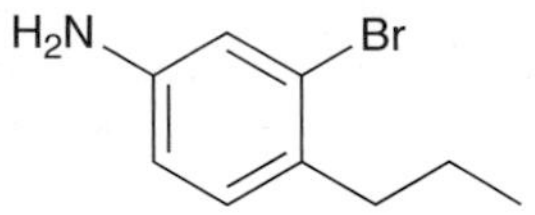

(c) 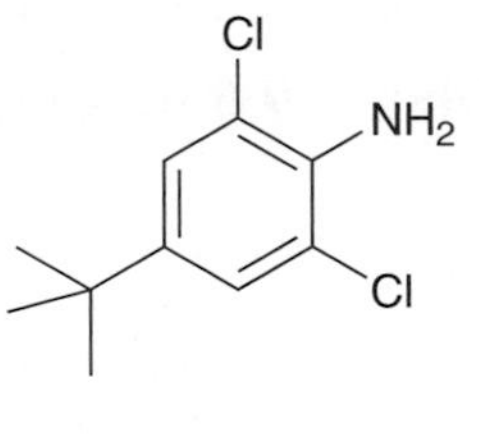

(d) 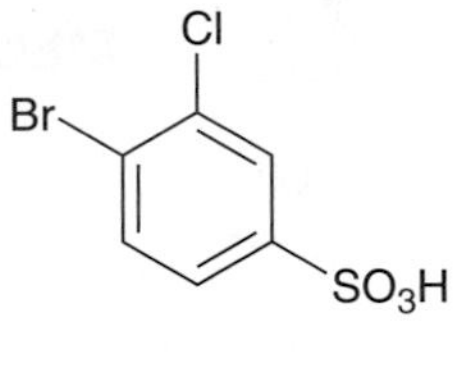

APPLY the skill

19.34 The following compound has a pentasubstituted benzene ring.

(a) Starting with benzene and using any other necessary reagents of your choice, design a synthesis for this compound.

(b) It is very difficult to install a sixth substituent. Explain.

(c) Is the ring activated or deactivated (relative to benzene)? Justify your answer.

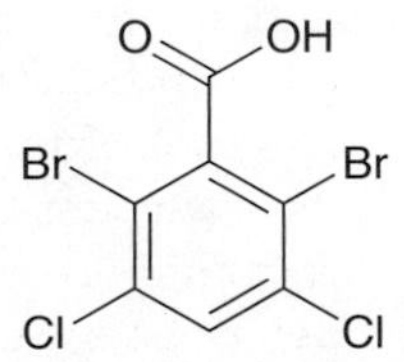

need more **PRACTICE?** **Try Problems 19.68, 19.73**

19.13 Nucleophilic Aromatic Substitution

Thus far, we have only explored reactions in which the aromatic ring attacks an electrophile (E^+). Such reactions are called *electrophilic* aromatic substitution reactions. In this section, we consider reactions in which the ring is attacked by a nucleophile. Such reactions are called **nucleophilic aromatic substitution** reactions. In the following example, an aromatic compound is treated with a strong nucleophile (hydroxide), which displaces a leaving group (bromide).

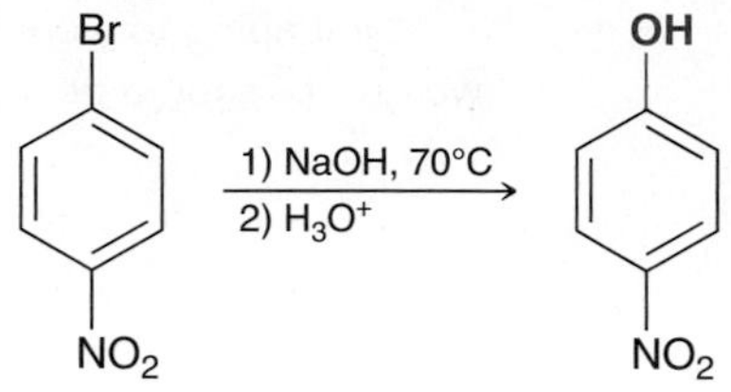
Br
1) NaOH, 70°C
2) H3O+
OH
NO2
NO2

In order for a reaction like this to occur, three criteria must be satisfied:

1. The ring must contain a powerful electron-withdrawing group (typically a nitro group).
2. The ring must contain a leaving group (usually a halide).
3. The leaving group must be either *ortho* or *para* to the electron-withdrawing group. If the leaving group is *meta* to the nitro group, the reaction is not observed.

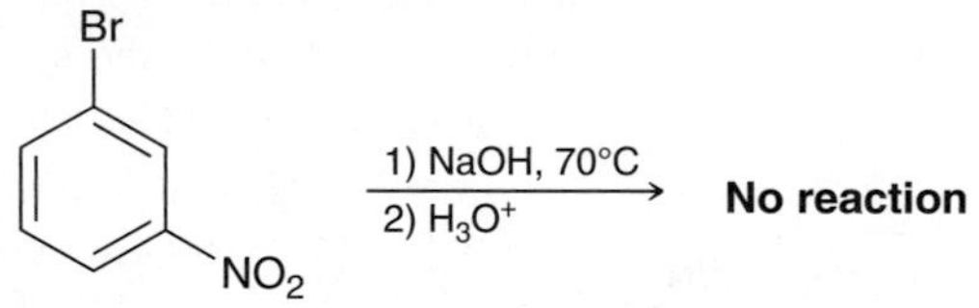
Br
NO2
1) NaOH, 70°C
2) H3O+
No reaction

In this example, the first two criteria are met, but the last criterion is not met.

Any mechanism that we propose for nucleophilic aromatic substitution must successfully explain the three criteria. Mechanism 19.8 accomplishes this task and is called the S_NAr mechanism.

MECHANISM 19.8 NUCLEOPHILIC AROMATIC SUBSTITUTION (S_NAr)

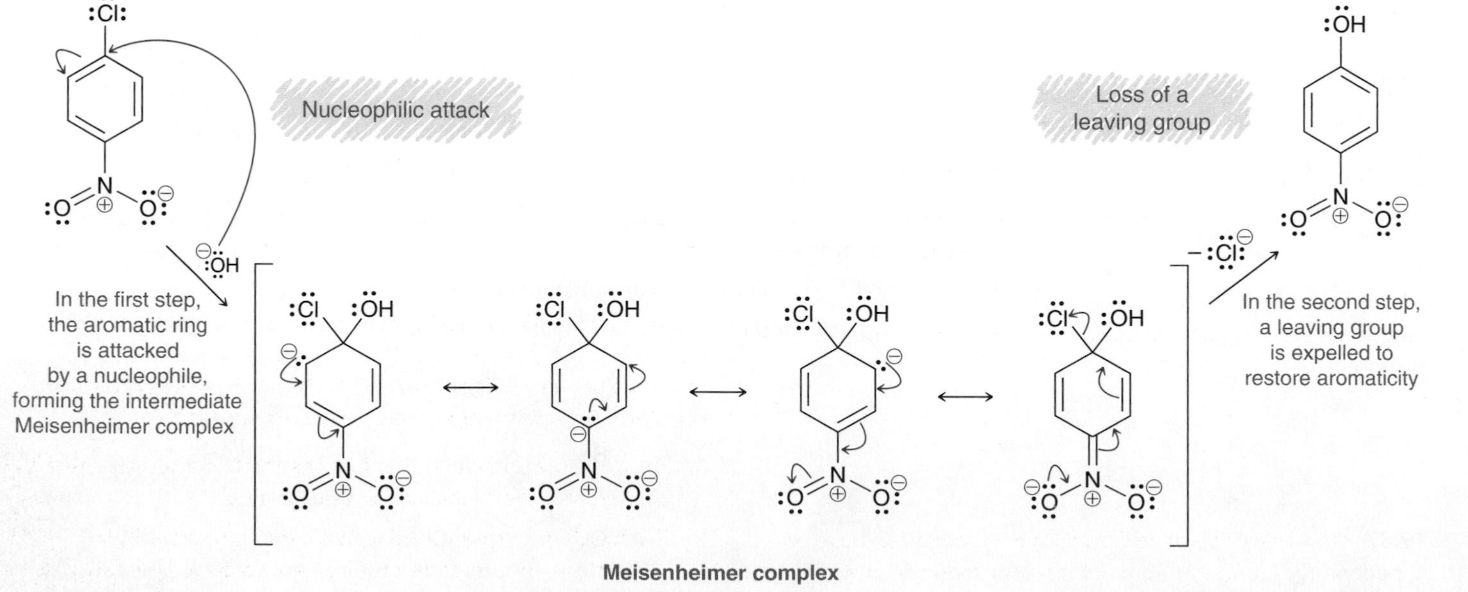
Nucleophilic attack
Loss of a leaving group
In the first step, the aromatic ring is attacked by a nucleophile, forming the intermediate Meisenheimer complex
In the second step, a leaving group is expelled to restore aromaticity
Meisenheimer complex

Much like the reactions we have seen thus far, this mechanism also involves two steps, but take special notice of the resonance-stabilized intermediate, called a **Meisenheimer complex**. This intermediate exhibits a negative charge that is resonance stabilized throughout the ring. This intermediate is very different from a sigma complex, which exhibits a positive charge that is resonance stabilized throughout the ring. The difference between these intermediates should make sense in that electrophilic aromatic substitution involves the ring attacking E^+, so the resulting intermediate will be positively charged; nucleophilic aromatic substitution involves the ring being attacked by a negatively charged nucleophile, so the resulting intermediate will be negatively charged. The second step of the S_NAr mechanism involves loss of a leaving group to restore aromaticity.

In order to understand the role of the nitro group in this reaction, consider the last resonance structure of the Meisenheimer complex in Mechanism 19.8. In that resonance structure, the negative charge is removed from the ring and resides on an oxygen atom. This resonance structure stabilizes the Meisenheimer complex, and we can think of the nitro group as a temporary reservoir for electron density. That is, the nucleophile attacks the ring, dumping its electron density into the ring, where it is temporarily stored on the nitro group. Then, the nitro group releases the electron density to expel a leaving group. With this in mind, we can understand the requirement for the nitro group as well as the requirement for the leaving group. In addition, the S_NAr mechanism also explains the requirement for the nitro group to be *ortho* or *para* to the leaving group. If the nitro group is *meta* to the leaving group, it cannot function as a reservoir. To convince yourself that this is the case, draw the structure of *meta*-chloronitrobenzene, and then attack that structure with hydroxide at the position containing the chlorine atom. Try to draw the resonance structures of the intermediate that is generated and you will see that the negative charge cannot be placed on the nitro group.

When hydroxide is used as the attacking nucleophile, the resulting product is a substituted phenol, which will be deprotonated by hydroxide to give a phenolate ion. Therefore, acid is required in a separate step to protonate the phenolate ion, and obtain a neutral product.

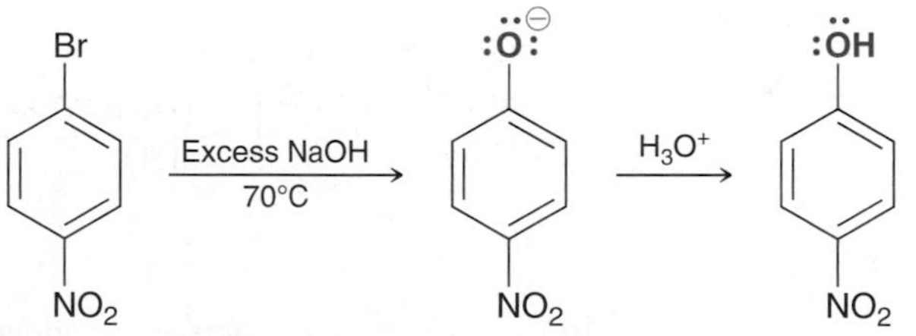

CONCEPTUAL CHECKPOINT

19.35 Predict the product of the following reaction.

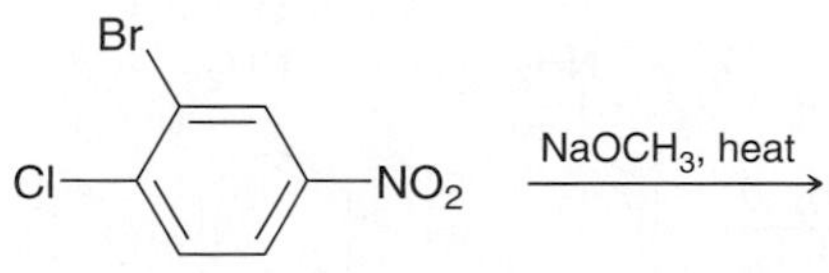

19.36 Starting with benzene and using any other necessary reagents of your choice, design a synthesis for the following compound.

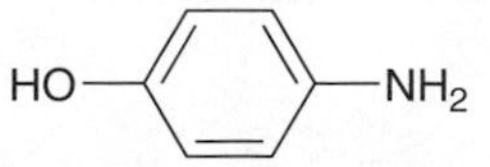

19.37 The presence of additional nitro groups can have an impact on the temperature at which a nucleophilic aromatic substitution will readily occur. Consider the following example.

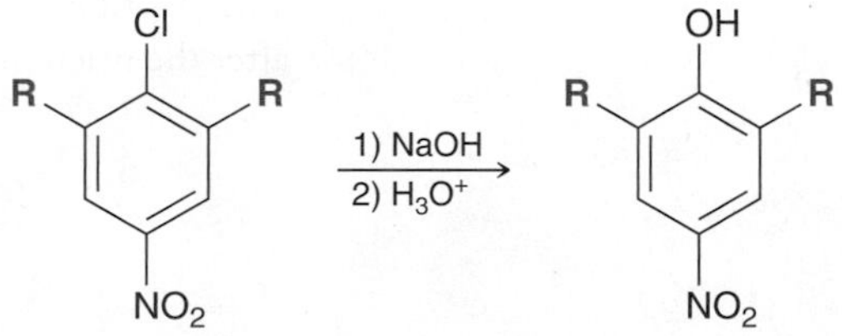

When both R groups are hydrogen atoms, the reaction readily occurs at 130°C. When one of the R groups is a nitro group, the reaction readily occurs at 100°C. When both R groups are nitro groups, the reaction readily occurs at 35°C.

(a) Provide an explanation that justifies the lower temperature requirement with additional nitro groups.

(b) If a fourth nitro group is placed on the ring, would you expect the temperature requirement to be further lowered? Explain your answer.

19.14 Elimination-Addition

In the previous section, we explained why a nitro group is required in order for a nucleophilic aromatic substitution reaction to proceed. In the absence of a powerful electron-withdrawing substituent, the reaction simply does not occur.

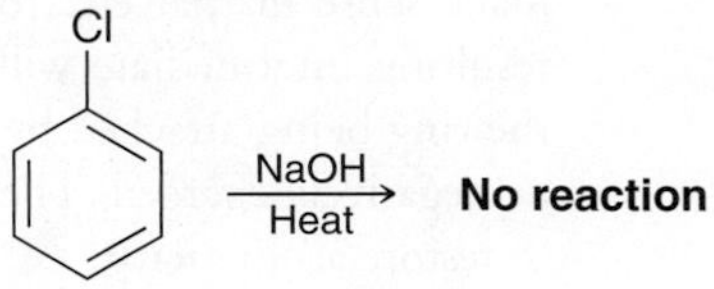

However, if the temperature and pressure are raised significantly, a reaction is in fact observed.

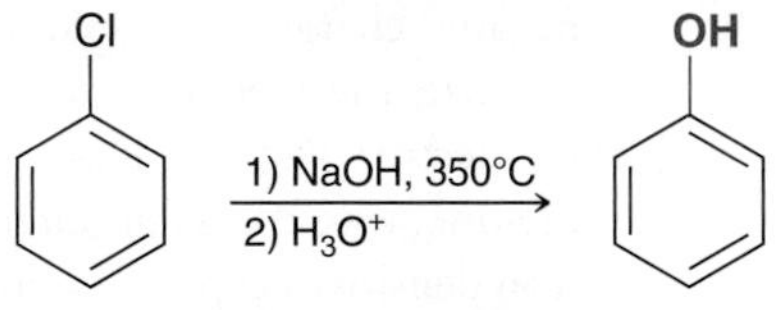

This reaction was first discovered in 1928 by scientists at the Dow Chemical Company. The reaction can also be performed at lower temperatures using the amide ion (H_2N^-) as a nucleophile.

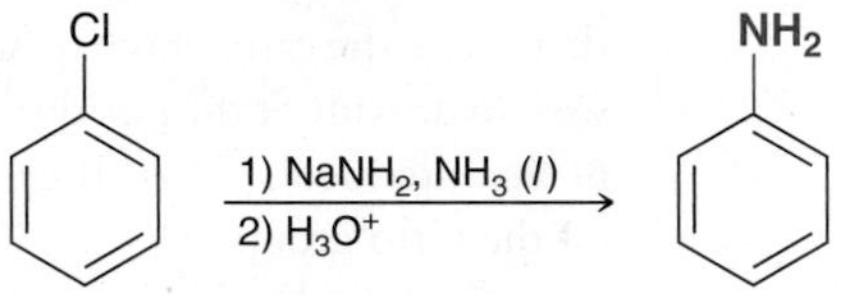

When other substituents are present on the ring, the regiochemical outcome is not what we might have expected.

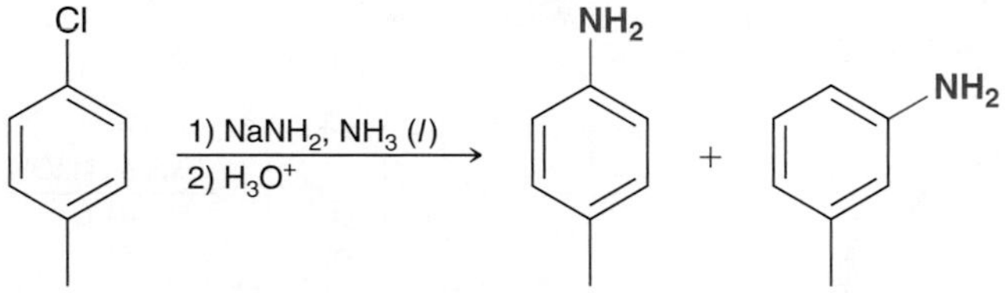

In this case, two products are obtained. This regiochemical outcome initially baffled chemists, as it cannot be explained with a simple S_NAr mechanism. Rather, a different mechanism is required to explain these results.

A clue to this puzzle comes from an isotopic labeling experiment. Chlorobenzene can be prepared such that the carbon bearing the chlorine atom is ^{14}C, a radioactive isotope of carbon. The position of the isotopic label (indicated with an asterisk) can then be tracked before and after the reaction.

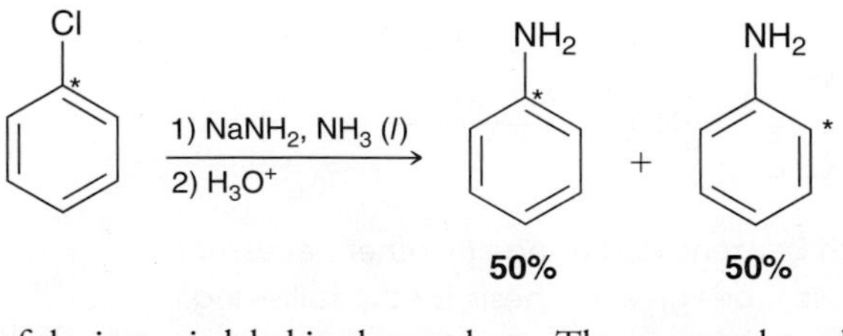

Notice the position of the isotopic label in the products. The proposed mechanism most consistent with these observations involves formation of a rather strange intermediate called **benzyne**.

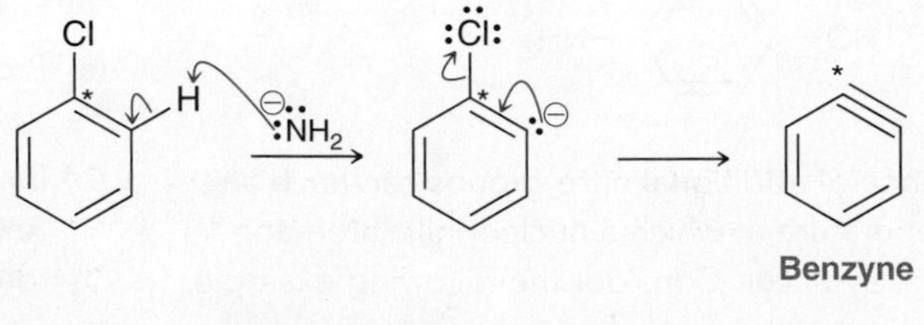

Elimination of H and Cl produces a very high energy intermediate called benzyne. This intermediate does not survive long because it is quickly attacked by the nucleophile, producing an addition reaction. Nucleophilic attack can take place at (a) the position of the isotopic label or (b) the other end of the triple bond.

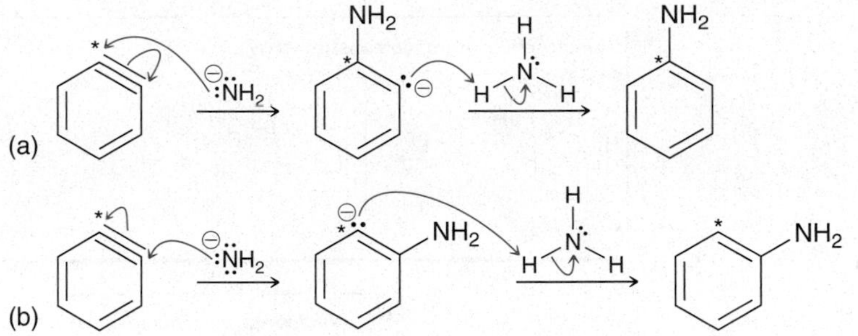
(a)
(b)
NH2
NH2
NH2
NH2

The attack can take place at either end of the triple bond with equal likelihood, explaining the observed results. This proposed mechanism is called **elimination-addition** (Mechanism 19.9).

MECHANISM 19.9 ELIMINATION-ADDITION

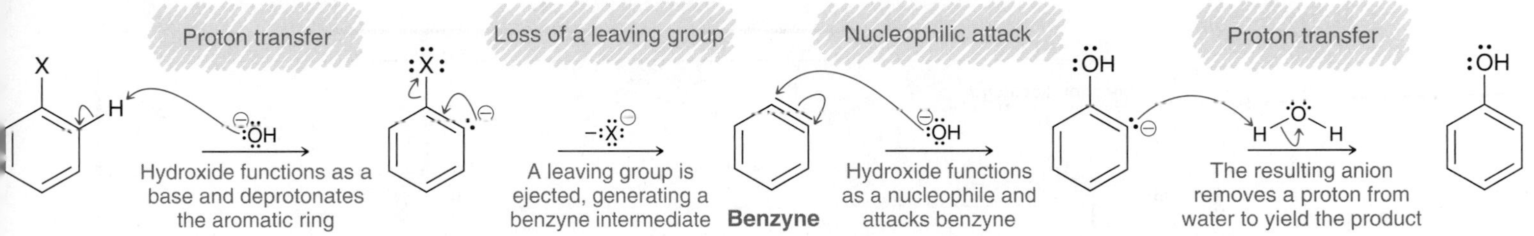
Proton transfer
Hydroxide functions as a base and deprotonates the aromatic ring
Loss of a leaving group
A leaving group is ejected, generating a benzyne intermediate
Benzyne
Nucleophilic attack
Hydroxide functions as a nucleophile and attacks benzyne
Proton transfer
The resulting anion removes a proton from water to yield the product

Evidence for this mechanism comes from a trapping experiment. When furan is added to the reaction mixture, a small amount of Diels-Alder cycloadduct is obtained.

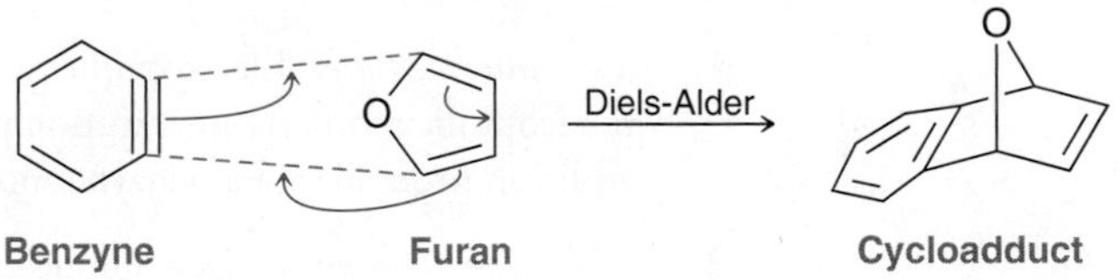
Diels-Alder
Benzyne
Furan
Cycloadduct

The presence of this cycloadduct can only be explained by invoking a benzyne intermediate, which is *trapped* by furan. The evidence requires that we explain how benzyne can exist, even for a brief moment. After all, a triple bond cannot be incorporated into a six-membered ring. This strange "triple bond" is best explained as resulting from the overlap of sp^2 orbitals rather than overlapping p orbitals.

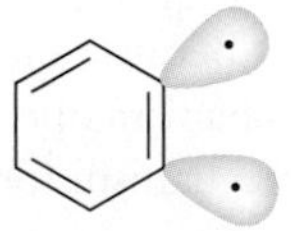

The overlap is very poor, and the intermediate more closely resembles a diradical than a triple bond. This explains why it is very unstable and so short-lived.

CONCEPTUAL CHECKPOINT

19.38 Draw both products that are obtained when 4-chloro-2-methyltoluene is treated with sodium amide followed by treatment with H_3O^+.

19.39 Starting with benzene and using any other necessary reagents of your choice, design a synthesis for anisole (methoxybenzene).

19.15 Identifying the Mechanism of an Aromatic Substitution Reaction

We have seen three different mechanisms for aromatic substitution reactions (Figure 19.15).

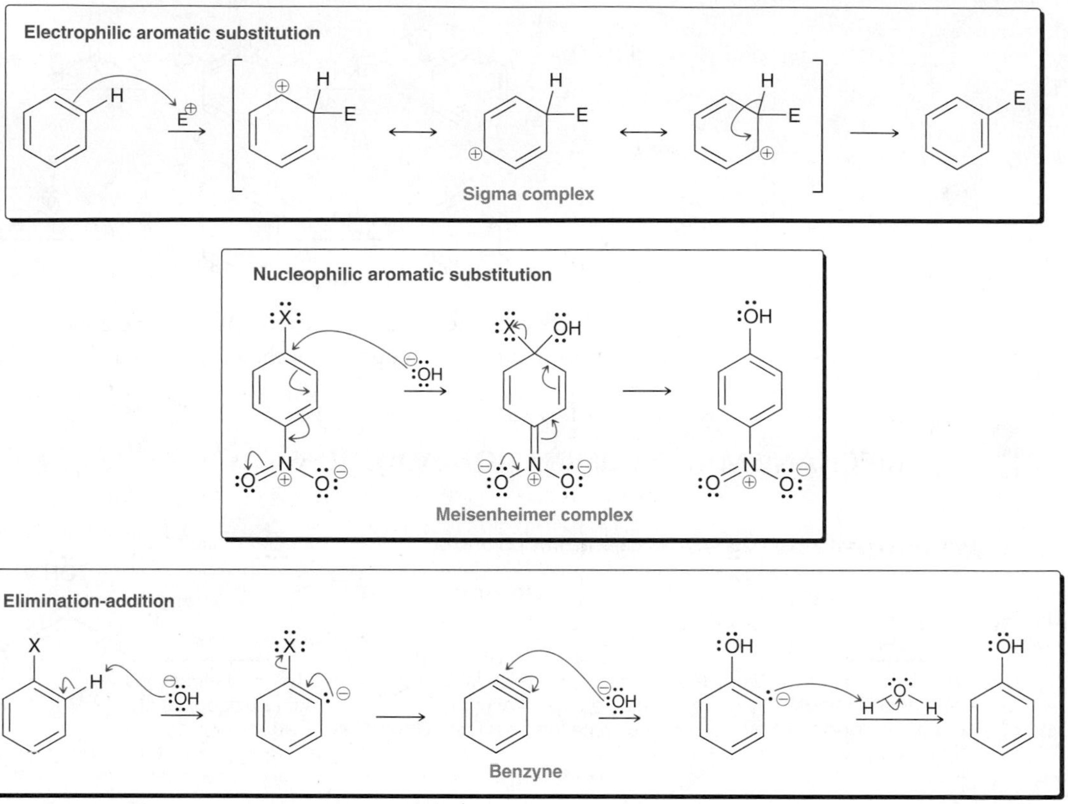

FIGURE 19.15 Three possible mechanisms for aromatic substitution.

All three mechanisms accomplish aromatic substitution, but there are a few key differences that warrant our attention:

1. *The intermediate:* Electrophilic aromatic substitution proceeds via a sigma complex, nucleophilic aromatic substitution proceeds via a Meisenheimer complex, and elimination-addition proceeds via a benzyne intermediate.
2. *The leaving group:* In electrophilic aromatic substitution, the incoming substituent replaces a proton. In the other two mechanisms, a negatively charged leaving group (such as a halide ion) is expelled.
3. *Substituent effects:* In electrophilic aromatic substitution, electron-withdrawing groups deactivate the ring toward attack, while in nucleophilic aromatic substitution, an electron-withdrawing group is required in order for the reaction to proceed.

Because of these fundamental differences, it is important to be able to determine which mechanism operates in any given situation. Figure 19.16 illustrates a decision tree for proposing a mechanism for an aromatic substitution.

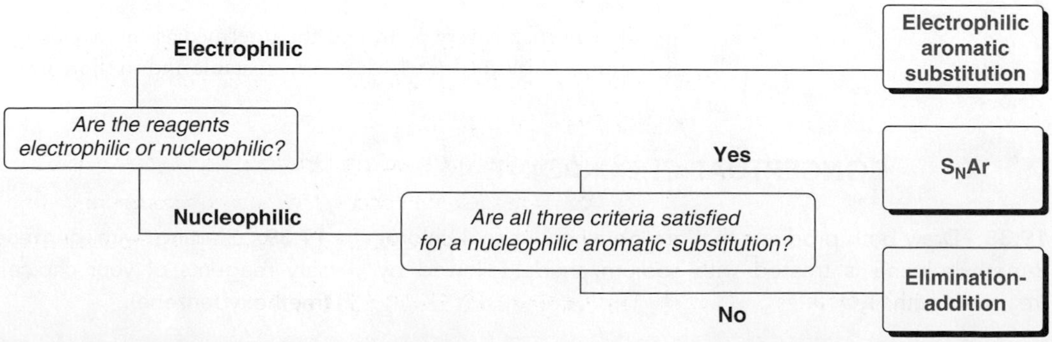

FIGURE 19.16 A decision tree for determining a mechanism for an aromatic substitution reaction.

SKILLBUILDER

19.7 DETERMINING THE MECHANISM OF AN AROMATIC SUBSTITUTION REACTION

LEARN the skill

Draw the most likely mechanism for the following transformation.

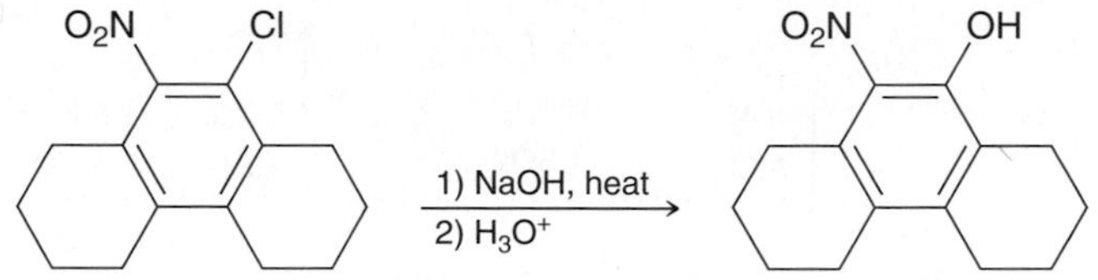

SOLUTION

STEP 1
Determine whether the reagents are electrophilic or nucleophilic.

First look at the reagents. NaOH is a common source of hydroxide ions (Na^+ is the counterion). Hydroxide is a nucleophile, not an electrophile, so we can rule out electrophilic aromatic substitution.

Next, look at the substrate to determine if all three criteria are present for a nucleophilic aromatic substitution: (1) there is a leaving group (chloride), (2) there is a nitro group, and (3) the nitro group is *ortho* to the leaving group. All three criteria are met, so the mechanism is likely to be S_NAr, which proceeds through a Meisenheimer complex.

STEP 2
If the reagent is nucleophilic, then determine if all three criteria are satisfied for a nucleophilic aromatic substitution.

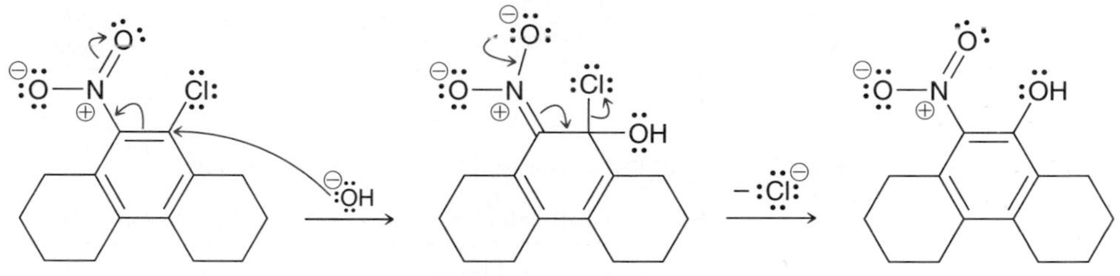

Meisenheimer complex

PRACTICE the skill

19.40 Draw the most likely mechanism for each of the following transformations.

(a)
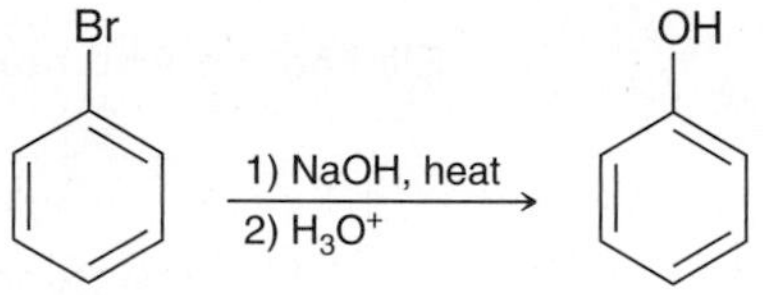

(b)
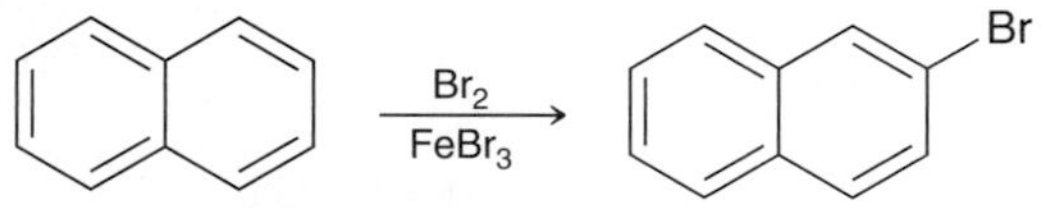

(c)
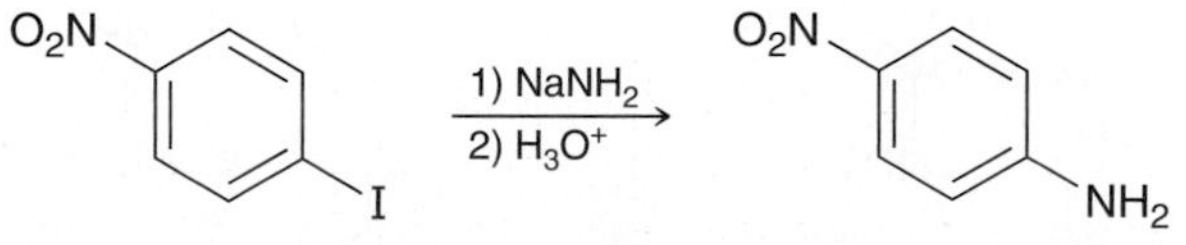

19.41 When 2-ethyl-5-chlorotoluene was treated with sodium hydroxide at high temperature, followed by treatment with H_3O^+, three constitutional isomers with molecular formula $C_9H_{12}O$ were obtained. Draw all three products.

APPLY the skill

19.42 When *ortho*-bromonitrobenzene is treated with NaOH at elevated temperature, only one product is formed.

(a) Draw the product.

(b) Identify the intermediate formed en route to the product.

(c) Would the reaction occur if the starting compound were *meta*-bromonitrobenzene?

(d) Would the reaction occur if the starting compound were *para*-bromonitrobenzene?

need more **PRACTICE?** **Try Problems 19.53–19.55, 19.60**

REVIEW OF REACTIONS

ELECTROPHILIC AROMATIC SUBSTITUTION

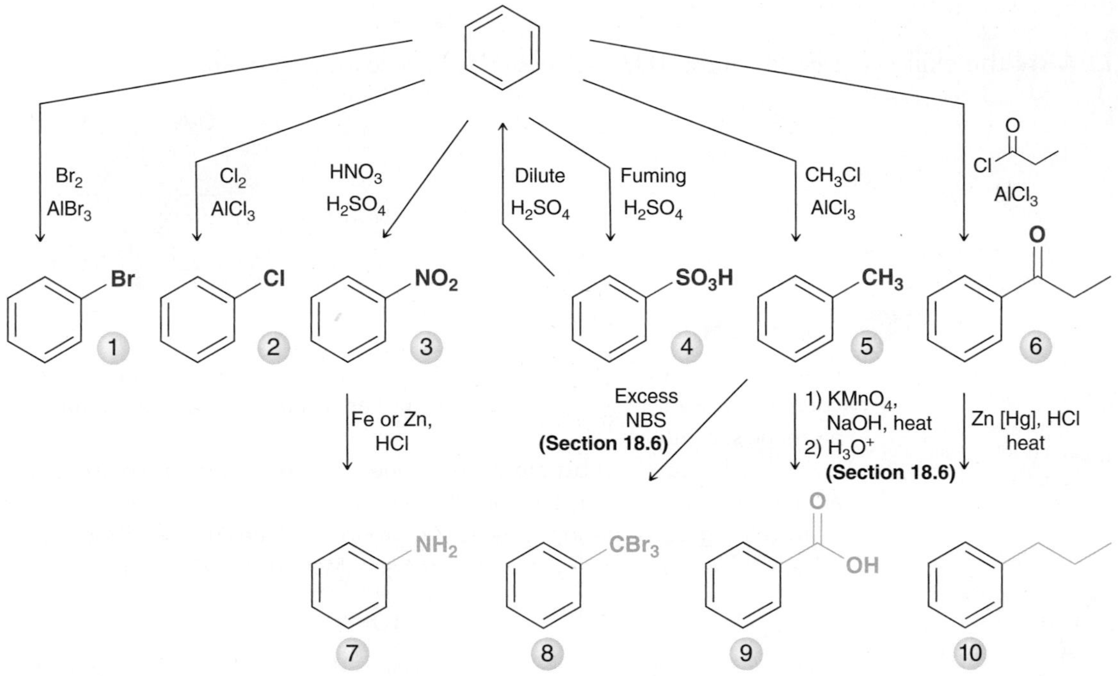

1. Bromination
2. Chlorination
3. Nitration
4. Sulfonation/desulfonation
5. Friedel-Crafts alkylation
6. Friedel-Crafts acylation
7. Reduction
8. Benzylic bromination
9. Oxidation
10. Clemmensen reduction

OTHER AROMATIC SUBSTITUTION REACTIONS

Nucleophilic Aromatic Substitution

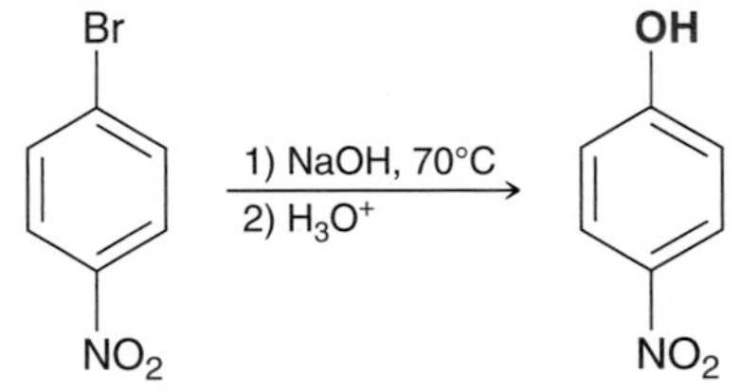

Elimination-Addition

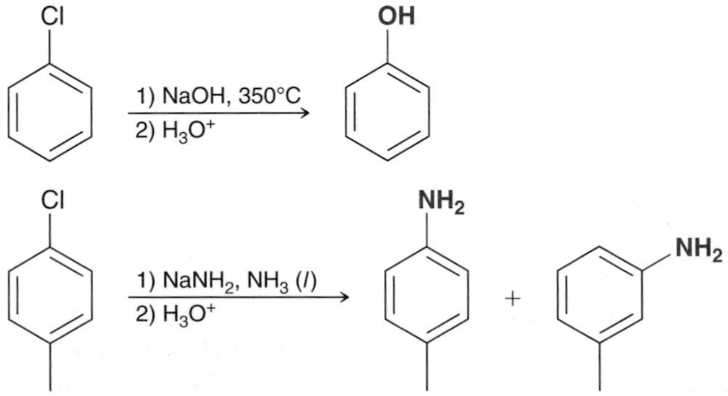

REVIEW OF CONCEPTS AND VOCABULARY

SECTION 19.1

- Alkenes undergo addition when treated with bromine, while benzene is inert under the same conditions.
- In the presence of iron, an **electrophilic aromatic substitution** reaction is observed between benzene and bromine.

SECTION 19.2

- Iron tribromide is a Lewis acid that interacts with Br_2 and generates Br^+, which is sufficiently electrophilic to be attacked by benzene.

- Electrophilic aromatic substitution involves two steps:
  - Formation of the **sigma complex**, or arenium ion. This step is endergonic.
  - Deprotonation, which restores aromaticity.
- Aluminum tribromide ($AlBr_3$) is another common Lewis acid that can serve as a suitable alternative to $FeBr_3$.
- Chlorination of benzene is accomplished with a suitable Lewis acid, such as aluminum trichloride.

SECTION 19.3

- Sulfur trioxide (SO_3) is a very powerful electrophile that is present in fuming sulfuric acid. Benzene reacts with SO_3 in a reversible process called **sulfonation**.

SECTION 19.4

- A mixture of sulfuric acid and nitric acid produces a small amount of **nitronium ion** (NO_2^+). Benzene reacts with the nitronium ion in a process called **nitration**.
- A nitro group can be reduced to an amino group, providing a two-step method for installing an amino group.

SECTION 19.5

- **Friedel-Crafts alkylation** enables the installation of an alkyl group on an aromatic ring.
- In the presence of a Lewis acid, an alkyl halide is converted into a carbocation, which can be attacked by benzene in an electrophilic aromatic substitution.
- A Friedel-Crafts alkylation is only efficient in cases where the carbocation cannot rearrange.
- When choosing an alkyl halide, the carbon atom connected to the halogen must be sp^3 hybridized.
- Polyalkylations are common and can generally be avoided by controlling the reaction conditions.

SECTION 19.6

- **Friedel-Crafts acylation** enables the installation of an **acyl group** on an aromatic ring.
- When treated with a Lewis acid, an acyl chloride will generate an **acylium ion**, which is resonance stabilized and not susceptible to carbocation rearrangements.
- When a Friedel-Crafts acylation is followed by a **Clemmensen reduction**, the net result is the installation of an alkyl group. This two-step process is a useful synthetic method for installing alkyl groups that cannot be installed efficiently with a direct alkylation process.
- Polyacylation is not observed, because introduction of an acyl group deactivates the ring toward further acylation.

SECTION 19.7

- An aromatic ring is **activated** by a methyl group, which is an ***ortho-para*** **director**.
- An aromatic ring is even more highly activated by a methoxy group, which is also an *ortho-para* director.
- All activators are *ortho-para* directors.

SECTION 19.8

- A nitro group **deactivates** an aromatic ring and is a *meta* director. Most deactivators are *meta* directors.

SECTION 19.9

- Halogens are an exception in that they are deactivators but are *ortho-para* directors.

SECTION 19.10

- **Strong activators** are characterized by the presence of a lone pair immediately adjacent to the aromatic ring.
- **Moderate activators** exhibit a lone pair that is already delocalized outside of the ring. Alkoxy groups are an exception and are moderate activators.
- Alkyl groups are **weak activators**.
- Halogens are **weak deactivators**.
- **Moderate deactivators** are groups that exhibit a π bond to an electronegative atom, where the π bond is conjugated with the aromatic ring.
- **Strong deactivators** are powerfully electron withdrawing, either by resonance or induction. There are three common groups that fall under this category.

SECTION 19.11

- When multiple substituents are present, the more powerful activating group dominates the directing effects.
- Steric effects often play an important role in determining product distribution.
- A **blocking group** can be used to control the regiochemical outcome of an electrophilic aromatic substitution.

SECTION 19.12

- Proposing a synthesis for a disubstituted benzene ring requires a careful analysis of directing effects to determine which group should be installed first.
- When designing a synthesis for a polysubstituted benzene ring, it is often most efficient to utilize a retrosynthetic analysis.

SECTION 19.13

- In a **nucleophilic aromatic substitution** reaction, the aromatic ring is attacked by a nucleophile. This reaction has three requirements:
  - The ring must contain a powerful electron-withdrawing group (typically a nitro group).
  - The ring must contain a leaving group.
  - The leaving group must be either *ortho* or *para* to the electron-withdrawing group.
- Nucleophilic aromatic substitution involves two steps:
  - Formation of a **Meisenheimer complex**
  - Loss of a leaving group to restore aromaticity

SECTION 19.14

- An **elimination-addition** reaction occurs via a **benzyne** intermediate. Evidence for this mechanism comes from isotopic labeling experiments as well as a trapping experiment.

SECTION 19.15

- The three mechanisms for aromatic substitution differ in (1) the intermediate, (2) the leaving group, and (3) substituent effects.

KEY TERMINOLOGY

activate 873
acyl group 871
acylium ion 871
arenium ion 860
benzyne 900
blocking group 888
Clemmensen reduction 872
deactivate 877
electrophilic aromatic substitution 859
elimination-addition 901
Friedel–Crafts acylation 871
Friedel–Crafts alkylation 867
meta-director 879
Meisenheimer complex 899
moderate activators 881
moderate deactivators 882
nitration 866
nitronium ion 866
nucleophilic aromatic substitution 898
ortho-para director 875
sigma complex 860
strong activators 881
strong deactivators 882
sulfonation 863
weak activators 882
weak deactivators 882

SKILLBUILDER REVIEW

19.1 IDENTIFYING THE EFFECTS OF A SUBSTITUENT

ACTIVATORS			DEACTIVATORS		
STRONG	**MODERATE**	**WEAK**	**WEAK**	**MODERATE**	**STRONG**
A lone pair immediately adjacent to the ring. **Examples:** $-\ddot{N}H_2$ $-\ddot{O}H$ **Exception:** $-\ddot{O}R$ (Moderate activator)	A lone pair that is already participating in resonance outside of the ring. **Example:** $-\ddot{O}-C(=O)-R$	Alkyl groups: —R	Halogens: —F —Cl —Br —I	A π bond to a heteroatom, where the π bond is conjugated to the ring. **Examples:** $-C(=O)R$ $-C{\equiv}N$	The following three groups: $-NO_2$ $-NR_3^{\oplus}$ $-CX_3$
← ***ORTHO-PARA*** **DIRECTORS** →				← ***META*** **DIRECTORS** →	

Try Problems 19.16–19.18, 19.44–19.46, 19.47a–c,f,h, 19.49a–d, 19.50a,b,d–g, 19.59a,b 19.64, 19.66

19.2 IDENTIFYING DIRECTING EFFECTS FOR DISUBSTITUTED AND POLYSUBSTITUTED BENZENE RINGS

STEP 1 Identify the nature of each group.

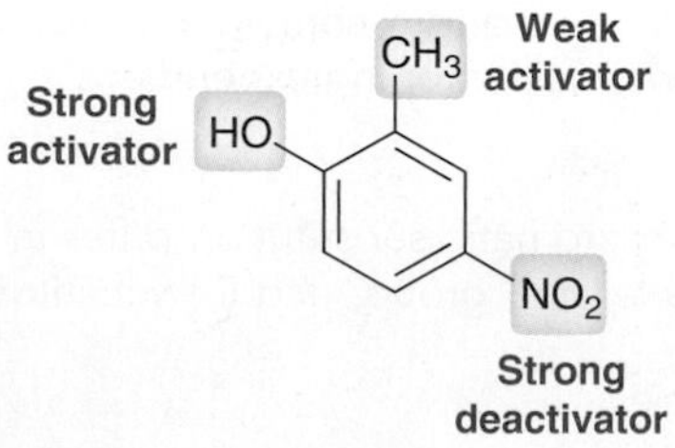

STEP 2 Select the most powerful activator, and then identify the positions that are *ortho* and *para* to that group.

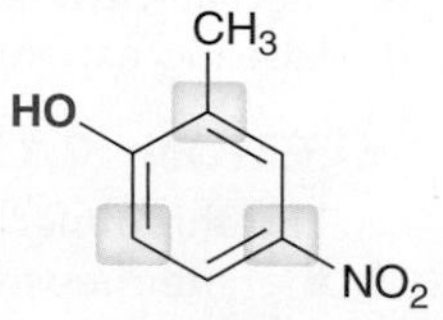

STEP 3 Identify the unoccupied positions.

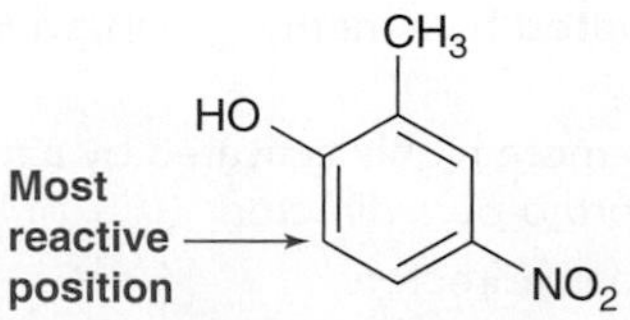

Try Problems 19.19–19.21, 19.47e, 19.50k, 19.56a,b, 19.59, 19.69

19.3 IDENTIFYING STERIC EFFECTS FOR DISUBSTITUTED AND POLYSUBSTITUTED AROMATIC BENZENE RINGS

STEP 1 Identify the nature of each group.

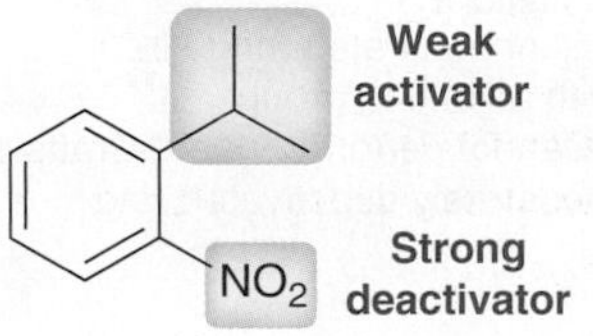

STEP 2 Select the most powerful activator, and then identify the positions that are *ortho* and *para* to that group.

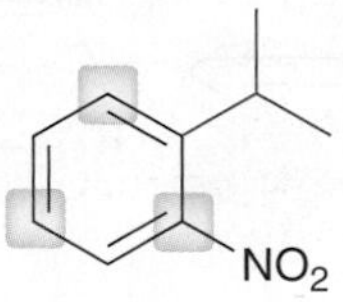

STEP 3 Identify the unoccupied positions that are the least sterically hindered.

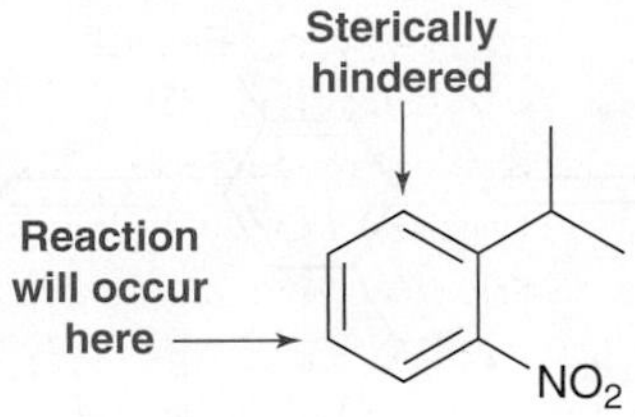

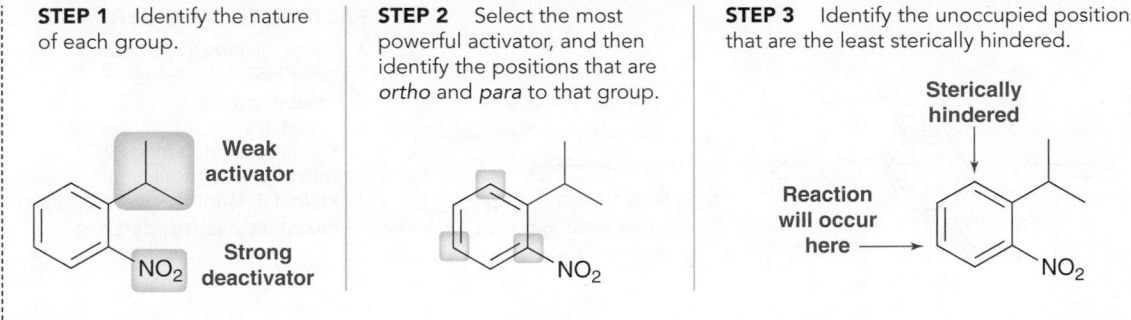

Try Problems 19.22–19.24, 19.59, 19.63, 19.69

19.4 USING BLOCKING GROUPS TO CONTROL THE REGIOCHEMICAL OUTCOME OF AN ELECTROPHILIC AROMATIC SUBSTITUTION REACTION

STEP 1 Determine whether or not a blocking group is necessary.

STEP 2 Install the blocking group.

STEP 3 Perform the desired reaction.

STEP 4 Remove the blocking group.

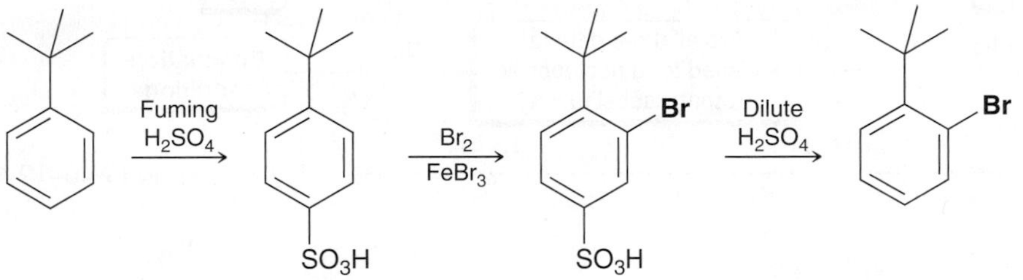

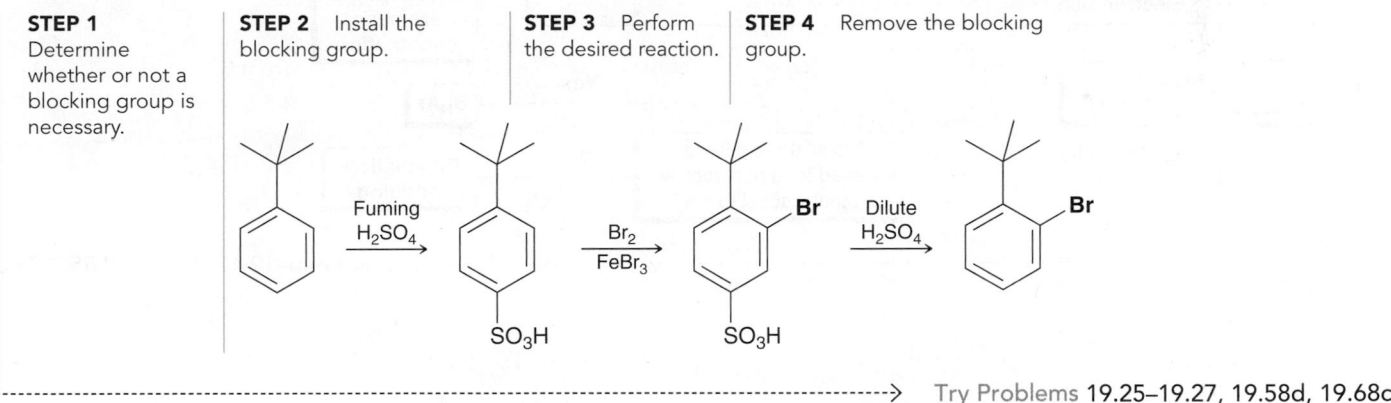

Try Problems 19.25–19.27, 19.58d, 19.68c

19.5 PROPOSING A SYNTHESIS FOR A DISUBSTITUTED BENZENE RING

EXAMPLE

Starting with benzene, propose a synthesis for the following compound:

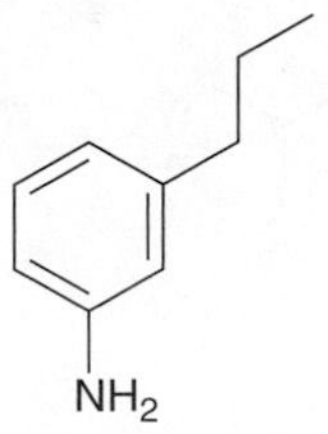

STEP 1 Identify the reagents necessary to install each group.

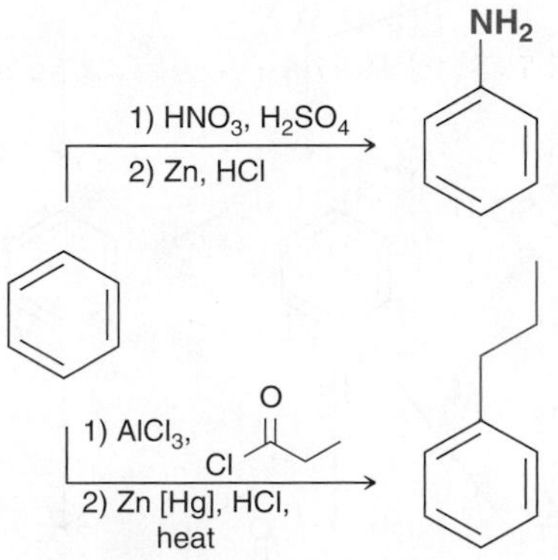

STEP 2 Determine the order of events that achieves the desired regiochemical outcome.

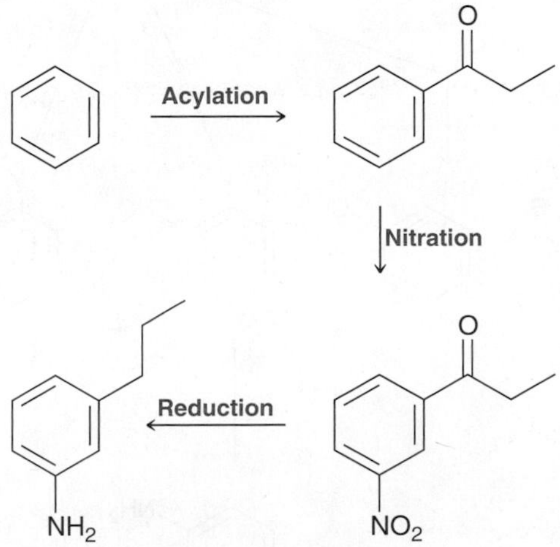

LIMITATIONS

Do not perform a nitration on a ring that possesses an amino group.

Do not perform a Friedel-Crafts reaction on a moderately or strongly deactivated ring.

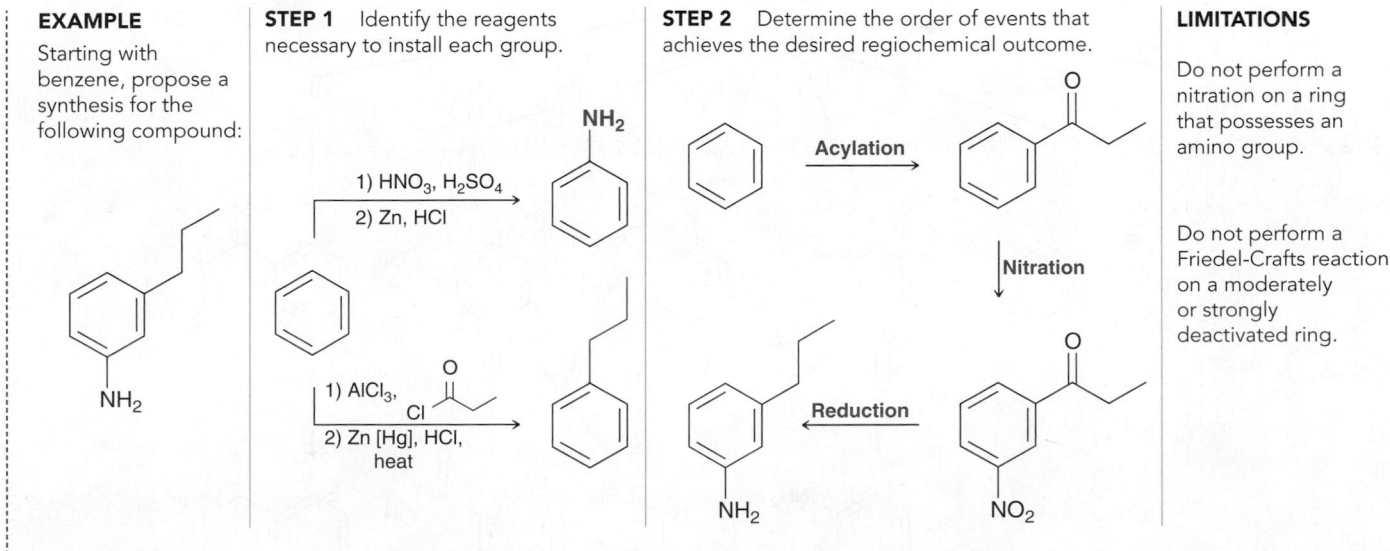

Try Problems 19.30 – 19.32, 19.57, 19.58, 19.68, 19.75

19.6 PROPOSING A SYNTHESIS FOR A POLYSUBSTITUTED BENZENE RING

RETROSYNTHETIC ANALYSIS

Example:

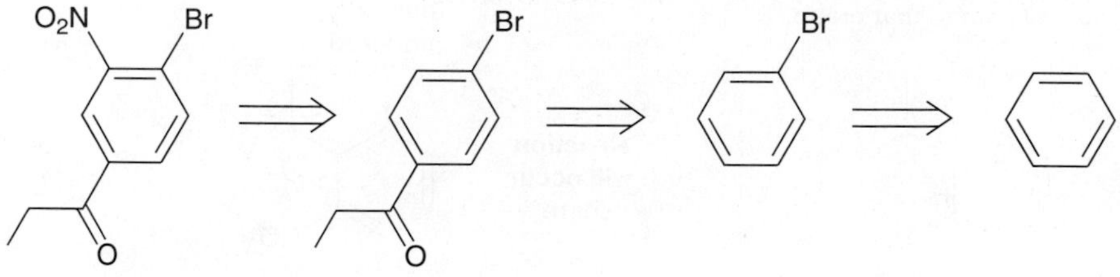

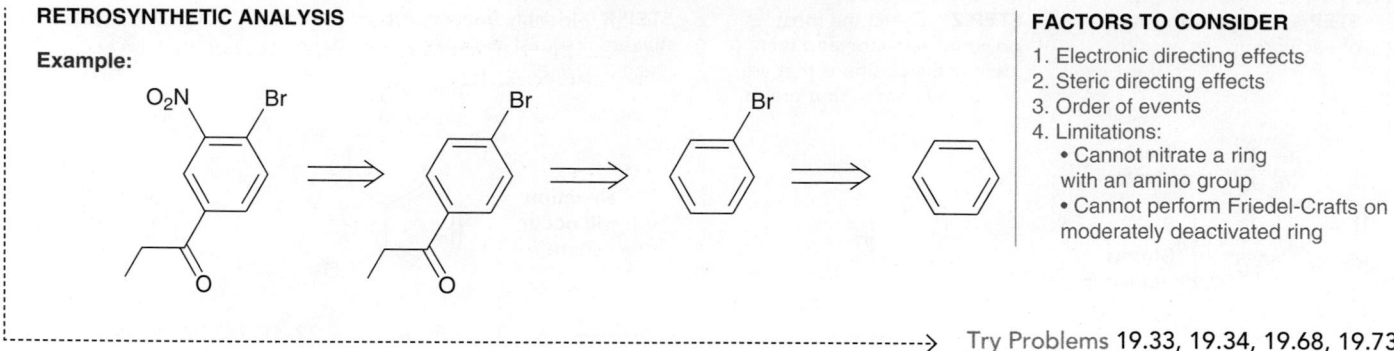

FACTORS TO CONSIDER

1. Electronic directing effects
2. Steric directing effects
3. Order of events
4. Limitations:
   - Cannot nitrate a ring with an amino group
   - Cannot perform Friedel-Crafts on moderately deactivated ring

Try Problems **19.33, 19.34, 19.68, 19.73**

19.7 DETERMINING THE MECHANISM OF AN AROMATIC SUBSTITUTION REACTION

DECISION TREE

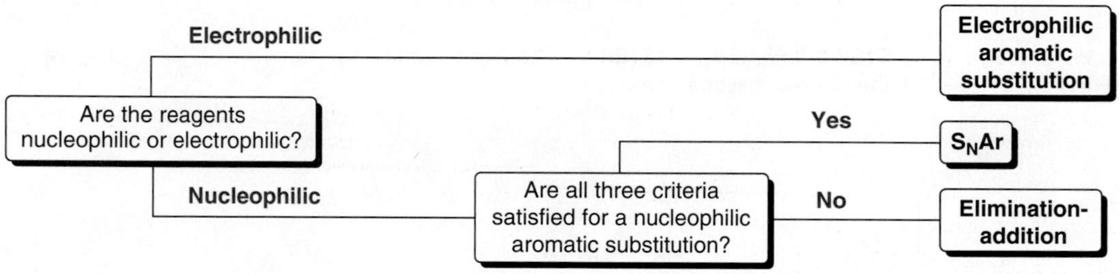

Try Problems **19.40–19.42, 19.53–19.55, 19.60**

PRACTICE PROBLEMS

19.43 Identify the reagents necessary to accomplish each of the following transformations:

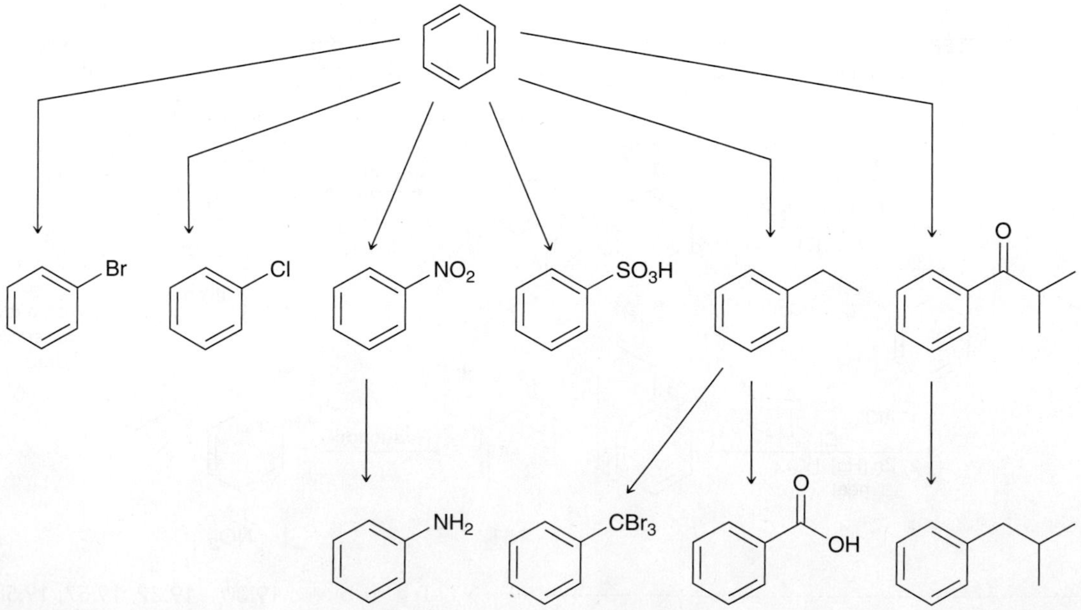

19.44 Rank the following compounds in order of increasing reactivity toward electrophilic aromatic substitution:

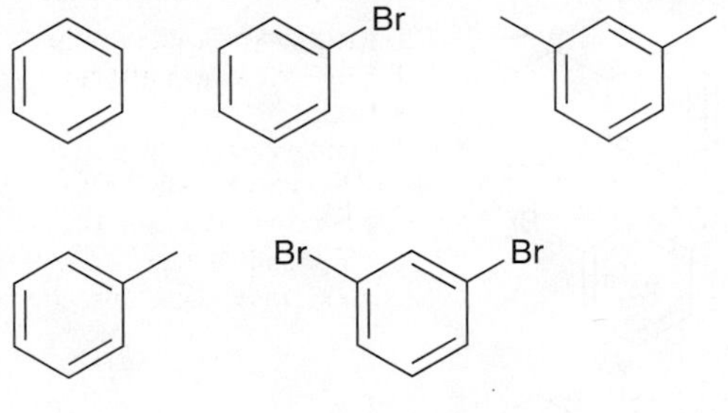

19.45 Identify which of the following compounds is most activated toward electrophilic aromatic substitution. Which compound is least activated?

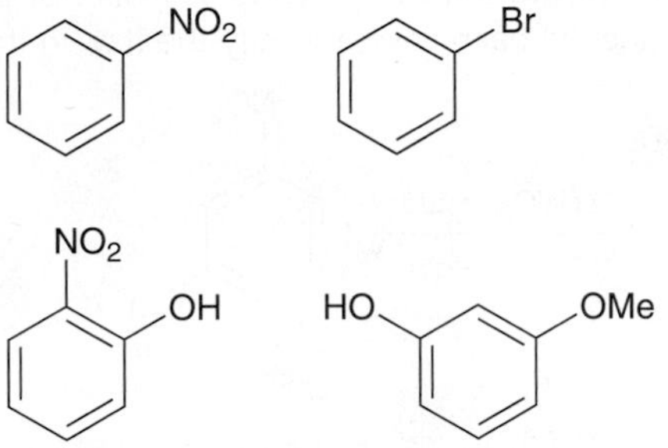

19.46 Predict the product(s) obtained when each of the following compounds is treated with a mixture of nitric acid and sulfuric acid:

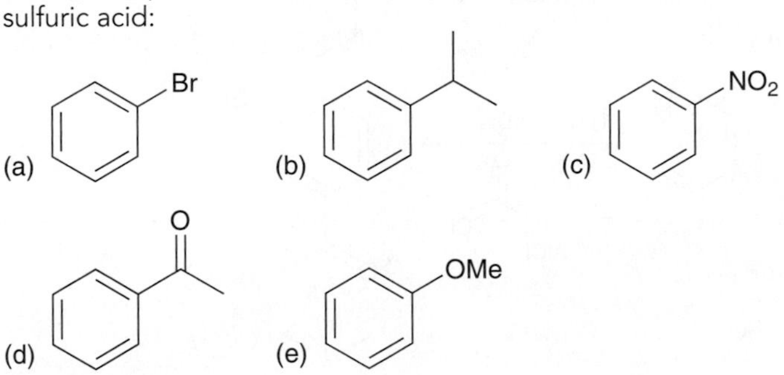

19.47 Predict the major product obtained when each of the following compounds is treated with fuming sulfuric acid:

(a) Chlorobenzene
(b) Phenol
(c) Benzaldehyde
(d) *ortho*-Nitrophenol
(e) *para*-Bromotoluene
(f) Benzoic acid
(g) *para*-Ethyltoluene
(h) Benzene

19.48 For each of the following groups, identify whether it is an activator or a deactivator, and determine its directing effects:

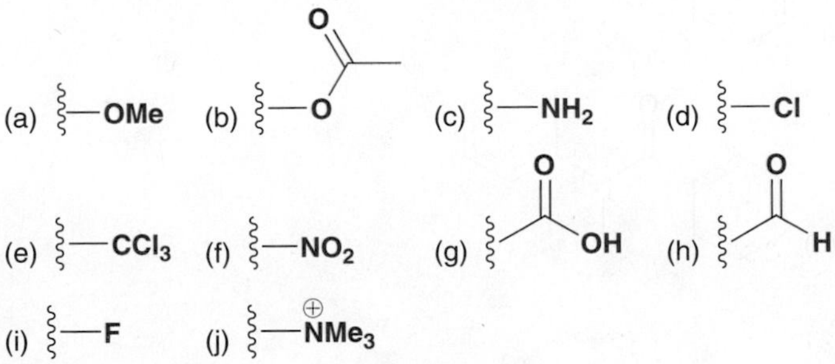

19.49 Predict the product(s) obtained when each of the following compounds is treated with chloromethane and aluminum trichloride. Some of the compounds might be unreactive. For those that are reactive, assume that conditions are controlled to favor monoalkylation.

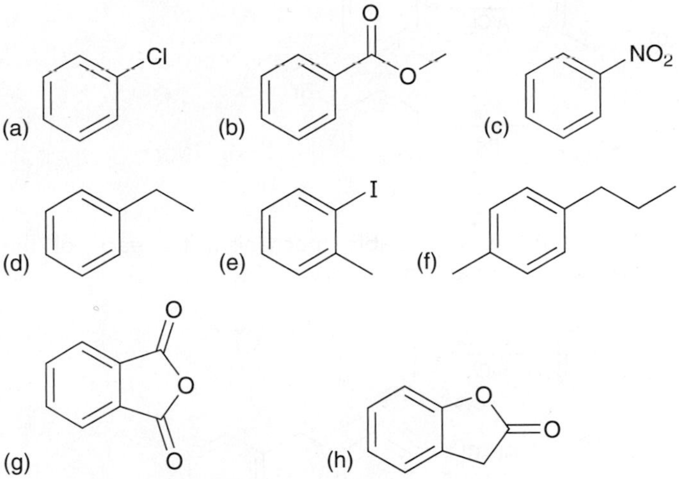

19.50 Predict the major product obtained when each of the following compounds is treated with bromine in the presence of iron tribromide.

(a) Bromobenzene
(b) Nitrobenzene
(c) *ortho*-Xylene
(d) *tert*-Butylbenzene
(e) Benzenesulfonic acid
(f) Benzoic acid
(g) Benzaldehyde
(h) *ortho*-Dibromobenzene
(i) *meta*-Nitrotoluene
(j) *meta*-Dibromobenzene
(k) *para*-Dibromobenzene

19.51 When the following compound is treated with a mixture of nitric and sulfuric acid at 50°C, nitration occurs to afford a compound with two nitro groups. Draw the structure of this product:

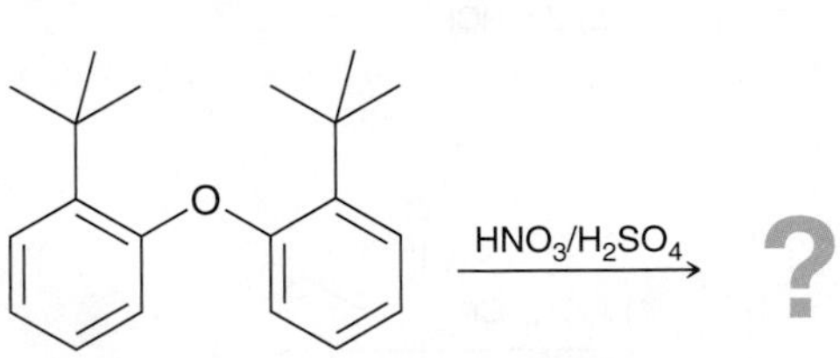

19.52 When benzene is treated with 2-methylpropene and sulfuric acid, the product obtained is *tert*-butylbenzene. Propose a mechanism for this transformation.

19.53 Draw a mechanism for each of the following transformations:

(a)

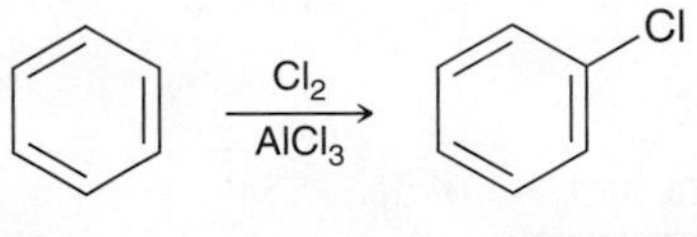

(b)

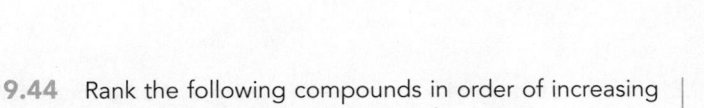

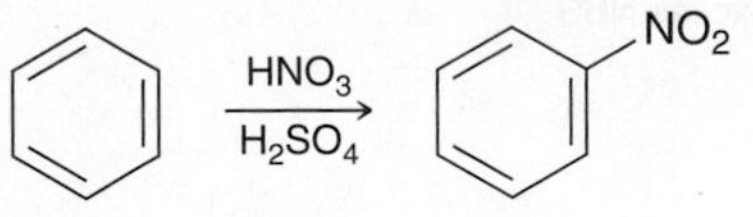

(Continued on the next page)

(c) 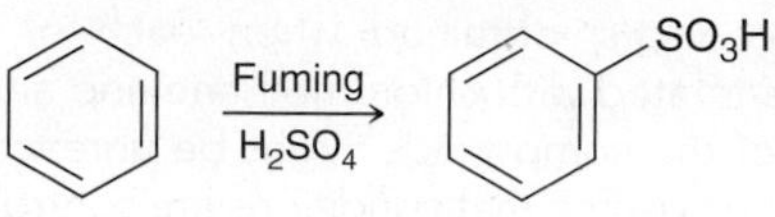

(d) 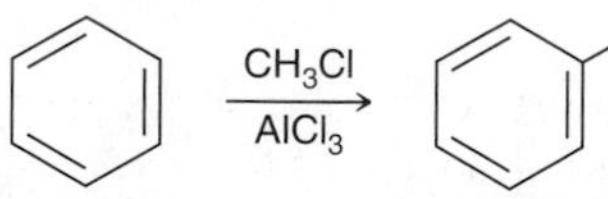

(e) 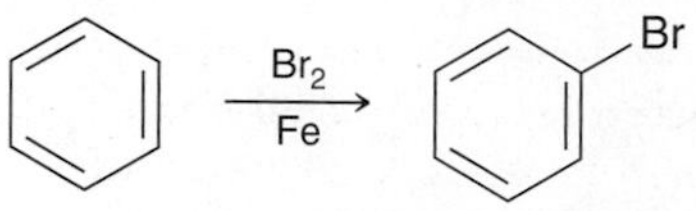

19.54 Propose a plausible mechanism for each of the following transformations:

(a) 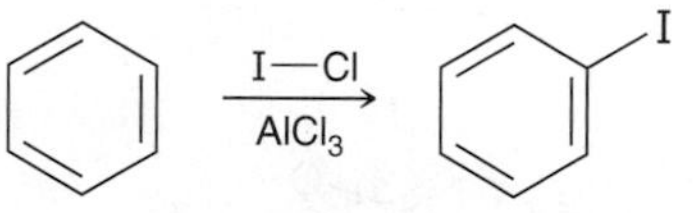

(b) 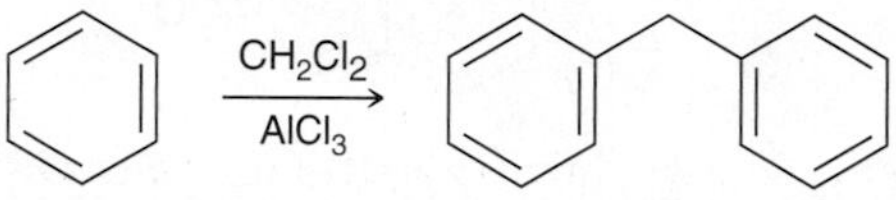

19.55 Propose a plausible mechanism for the following transformation:

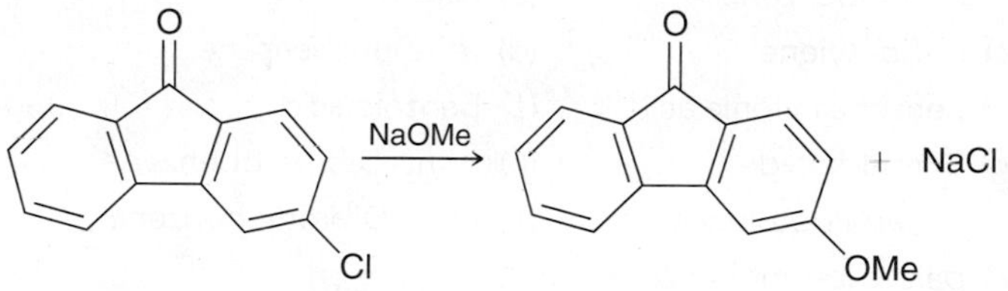

19.56 Predict the product(s) of the following reactions:

(a) 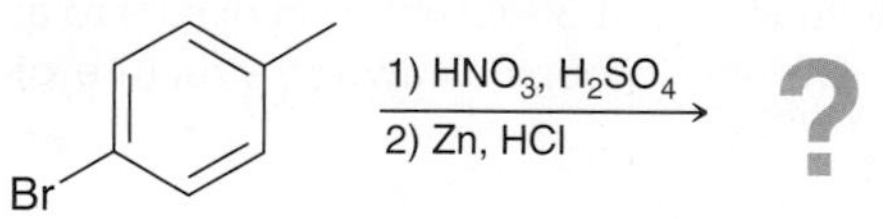

(b) 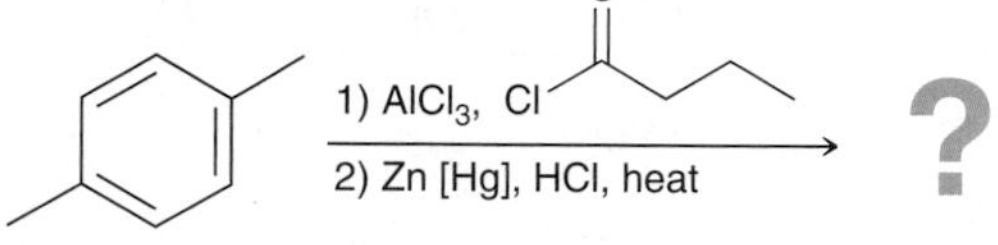

(c) 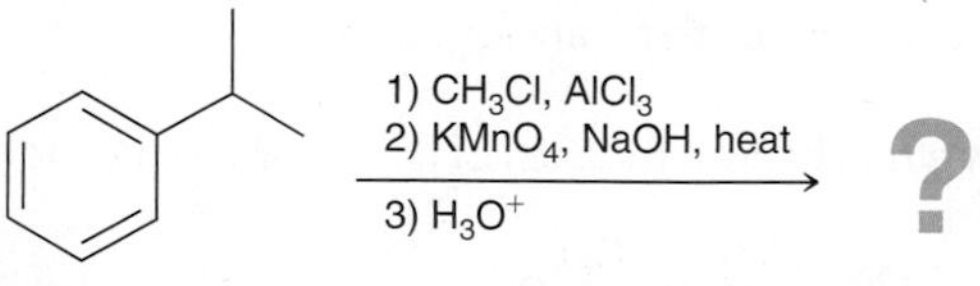

(d) 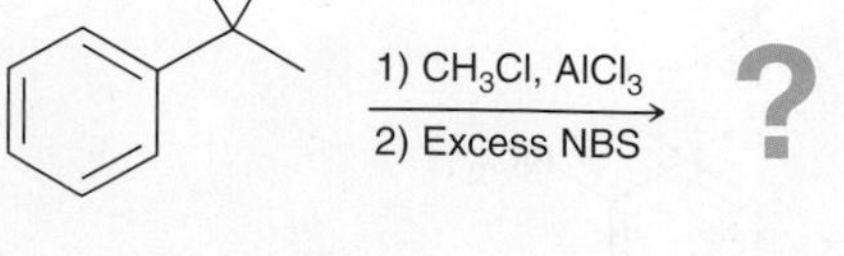

19.57 Starting with benzene and using any other necessary reagents of your choice, design a synthesis for each of the following compounds:

(a) 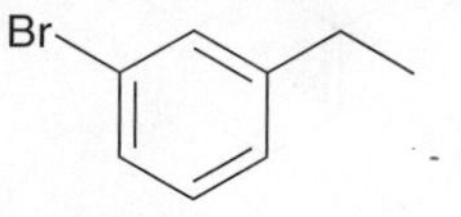

(b) 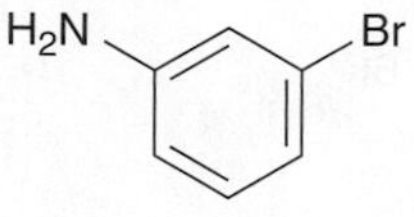

(c) 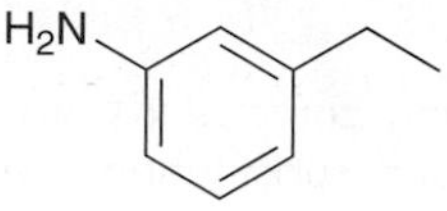

19.58 Each of the following syntheses will not produce the desired product. In each case, identify the flaw in the synthesis.

(a) 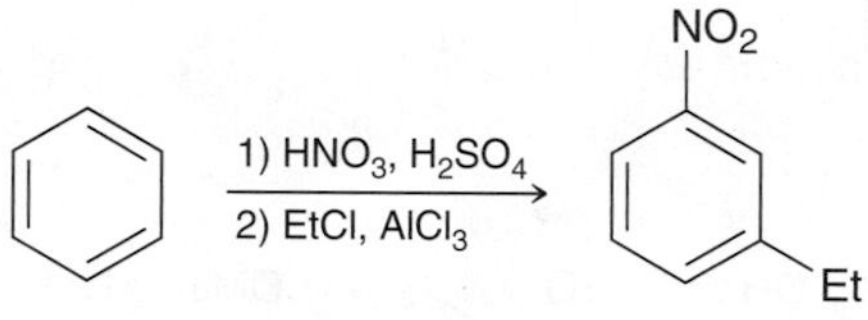

(b) 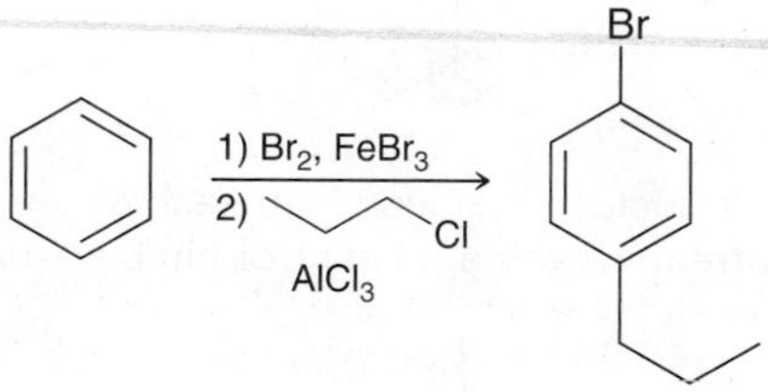

(c) 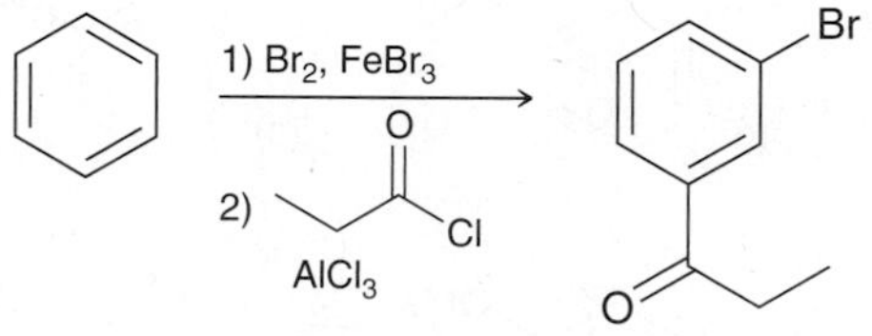

(d) 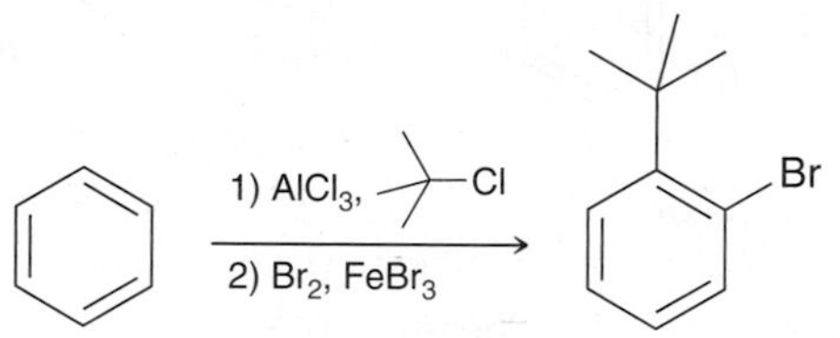

19.59 In each case, identify the most likely position at which monobromination would occur.

(a) 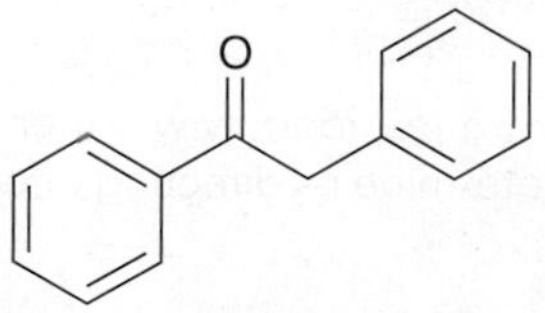

(b) 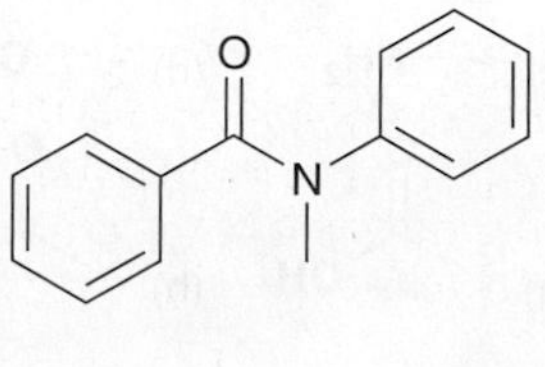

(c) 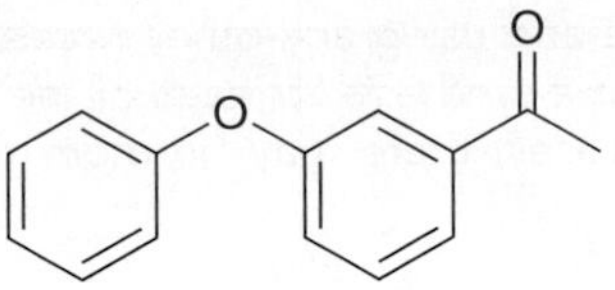

(d) 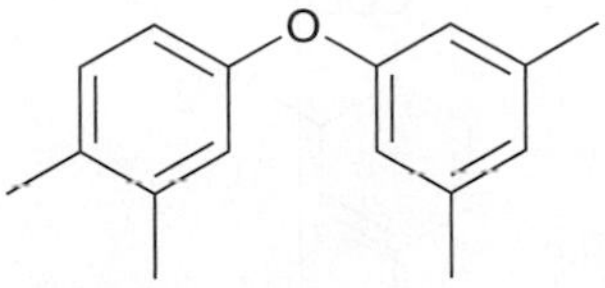

19.60 When *para*-bromotoluene is treated with sodium amide, two products are obtained. Draw both products, and propose a plausible mechanism for their formation.

19.61 Picric acid is a military explosive formed via the nitration of phenol under conditions that install three nitro groups. Draw the structure and provide an IUPAC name for picric acid.

19.62 Propose a plausible mechanism for the following transformation:

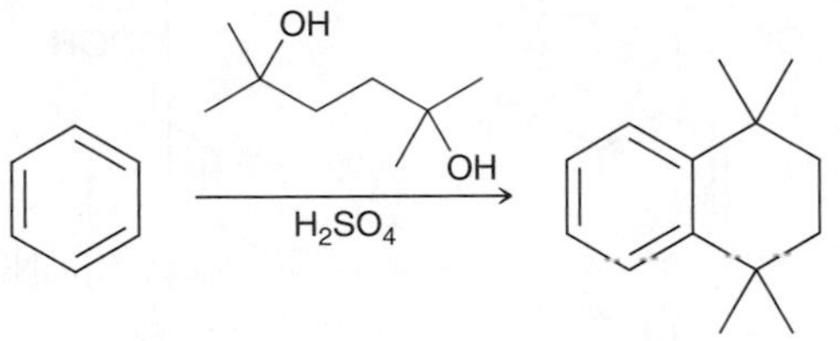

19.63 Benzene was treated with isopropyl chloride in the presence of aluminum trichloride under conditions that favor dialkylation. Draw the major product expected from this reaction.

19.64 Consider the structure of nitrosobenzene:

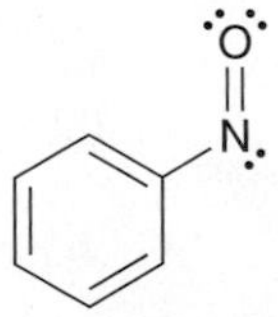

(a) Draw the resonance structures of the sigma complex formed when nitrosobenzene reacts with an electrophile (E^+) at the *ortho* position.

(b) Draw the resonance structures of the sigma complex formed when nitrosobenzene reacts with an electrophile (E^+) at the *meta* position.

(c) Draw the resonance structures of the sigma complex formed when nitrosobenzene reacts with an electrophile (E^+) at the *para* position.

(d) Compare the stability of the intermediates from parts a–c of this problem, and then predict whether the nitroso group is *ortho-para* directing or *meta* directing.

(e) The nitroso group has a lone pair adjacent to the ring (suggesting that it could be an activator), yet it also has a π bond to a heteroatom in conjugation with the ring (suggesting that it could be a deactivator). Experiments reveal that this group is a deactivator. Based on this information, identify which of the following groups is expected to have properties most similar to the nitroso group, and explain your choice:

19.65 Draw all resonance structures of the sigma complex formed when toluene undergoes chlorination at the *para* position.

19.66 For each of the following groups of compounds, identify which compound will react most rapidly with ethyl chloride in the presence of aluminum trichloride. Explain your choice in each case, and then predict the expected products of that reaction.

(a) 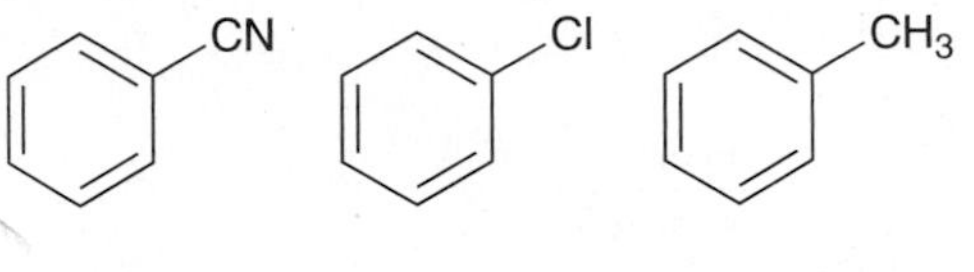

(b) 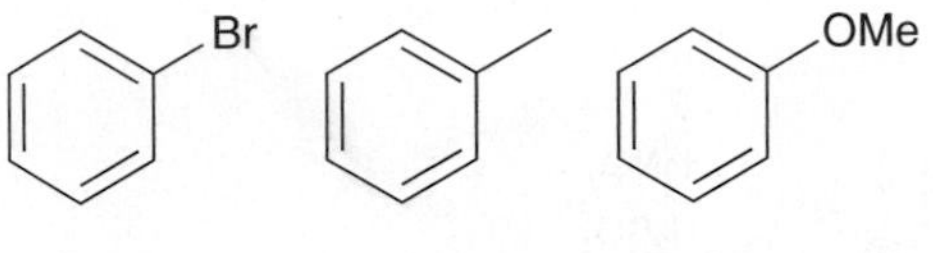

19.67 Compound **A** and compound **B** are both aromatic esters with molecular formula $C_8H_8O_2$. When treated with bromine in the presence of iron tribromide, compound **A** is converted into only one product, while compound **B** is converted into two different monobromination products. Identify the structure of compound **A** and compound **B**.

19.68 Propose a plausible synthesis for each of the following transformations:

(a) 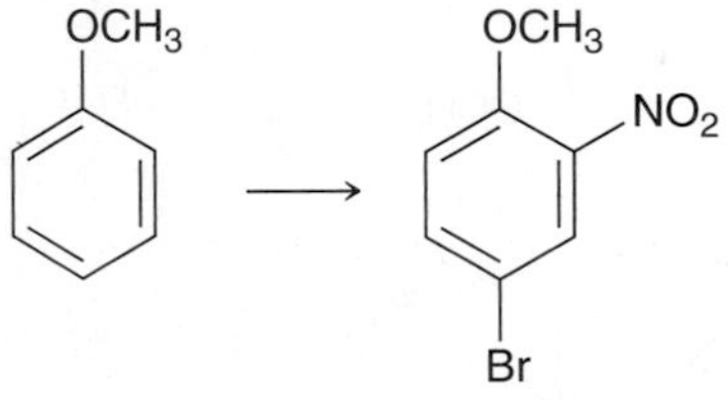

(b) 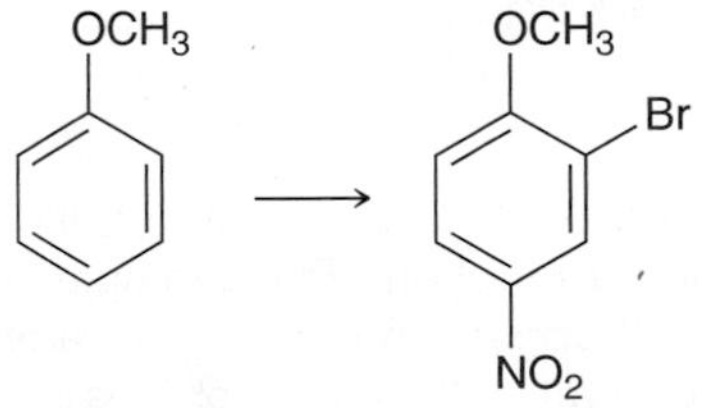

(c) 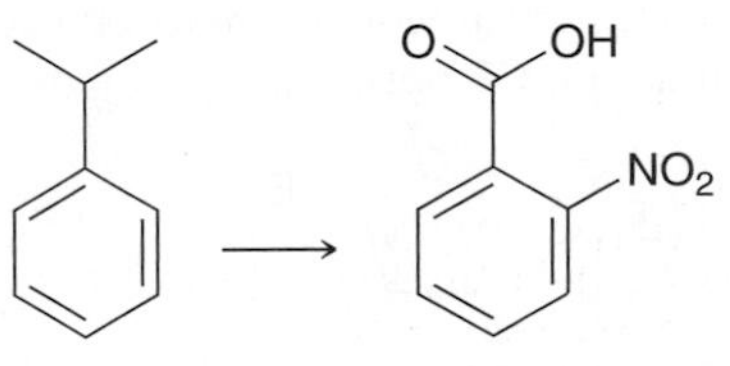

(d) 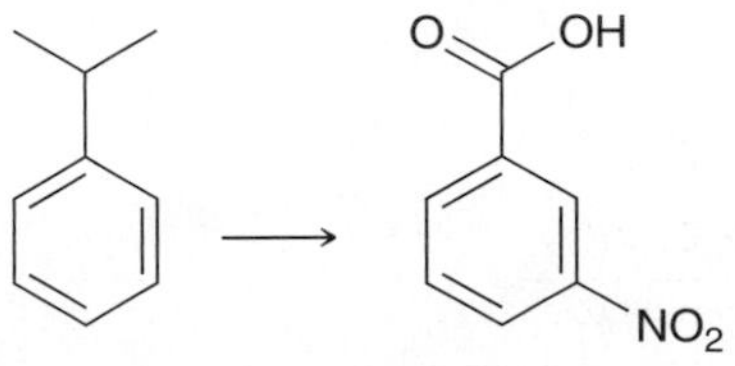

19.69 Predict the product(s) for each of the following reactions:

(a) 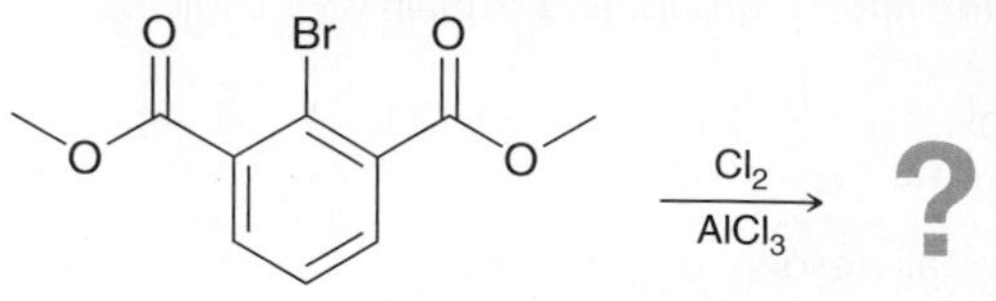

(b) 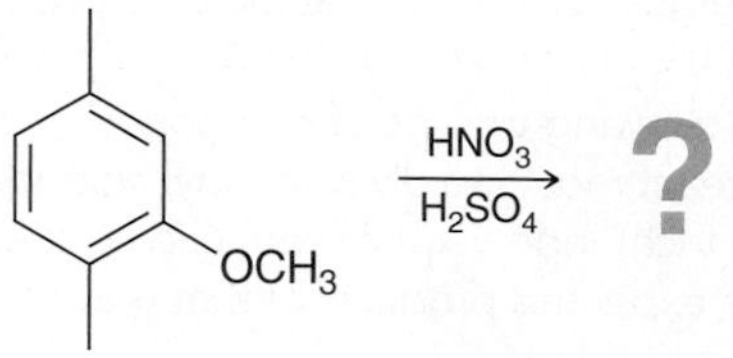

(c) 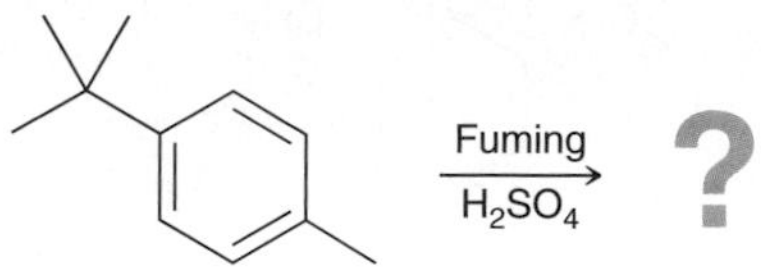

(d) 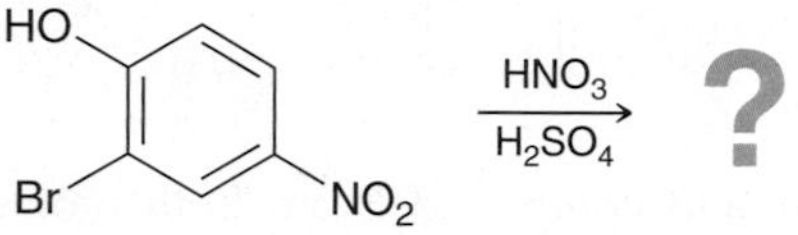

19.70 Each of the following compounds can be made with a Friedel-Crafts acylation. Identify the acyl chloride and the aromatic compound you would use to produce each compound.

(a) 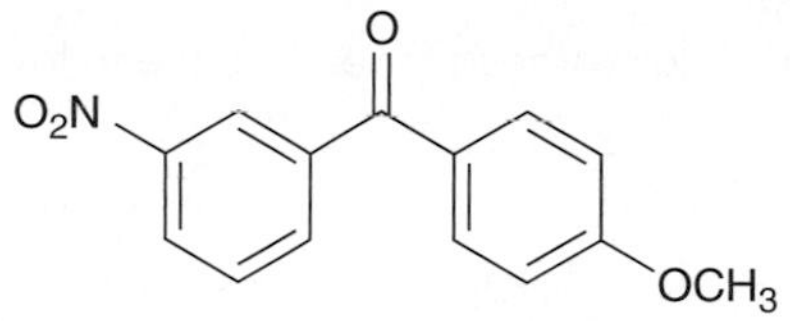

(b) 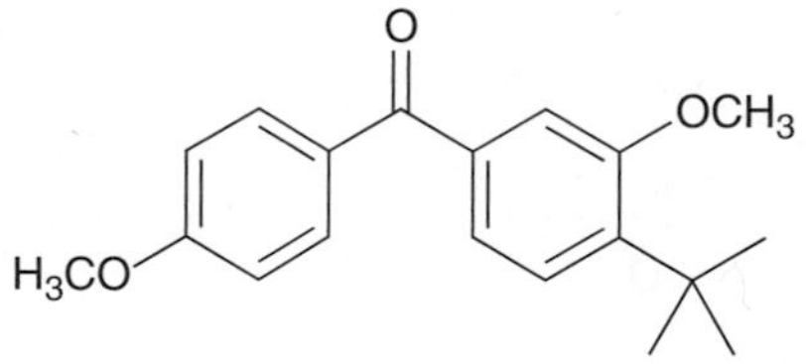

19.71 Aromatic heterocycles are also capable of undergoing electrophilic aromatic substitution. For example, when furan is treated with an electrophile, an electrophilic aromatic substitution reaction occurs in which the electrophile is installed exclusively at the C2 position. Explain why this reaction occurs at the C2 position, rather than the C3 position.

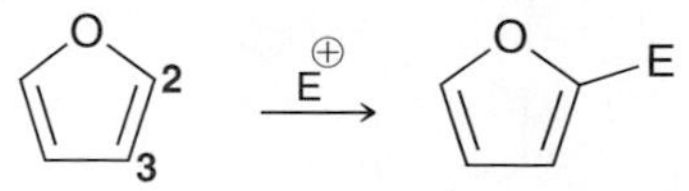

19.72 Starting with benzene and using any other necessary reagents of your choice, design a synthesis for each of the following compounds. In some cases, there may be more than one plausible answer.

(a) 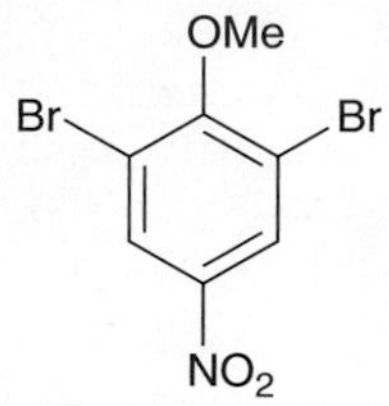

(b) 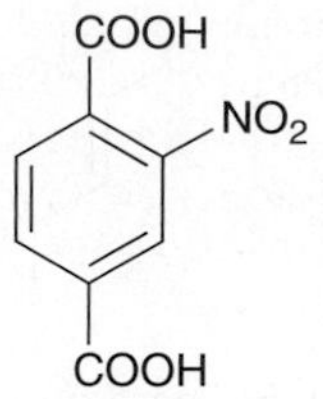

(c) 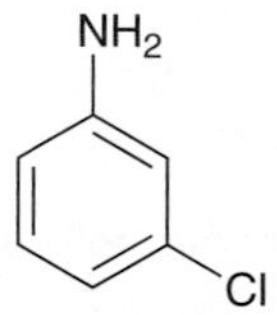

(d) 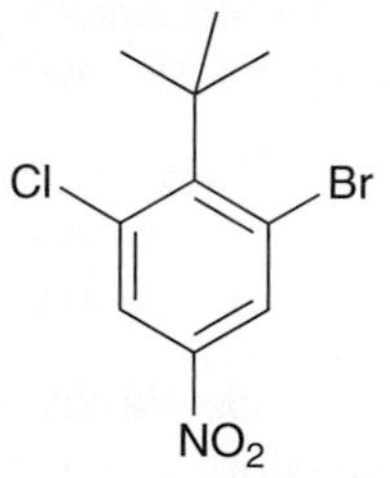

(e) 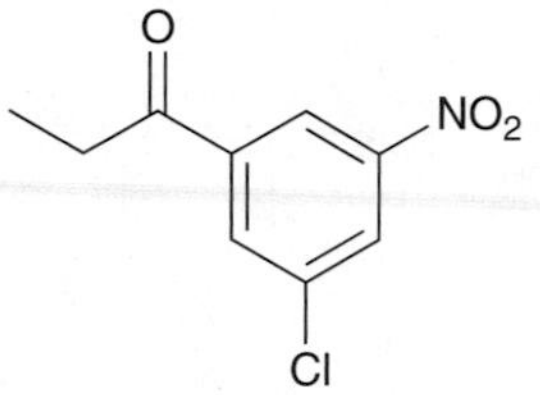

(f) 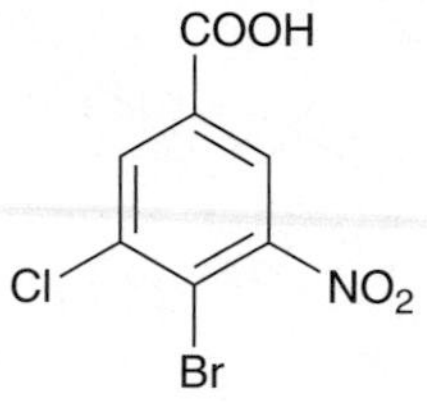

(g) 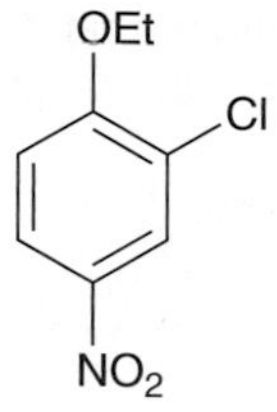

(h) 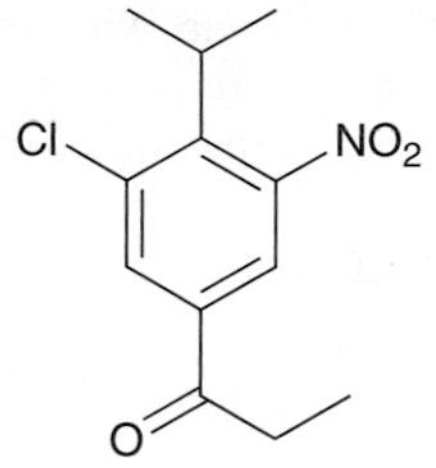

(i) 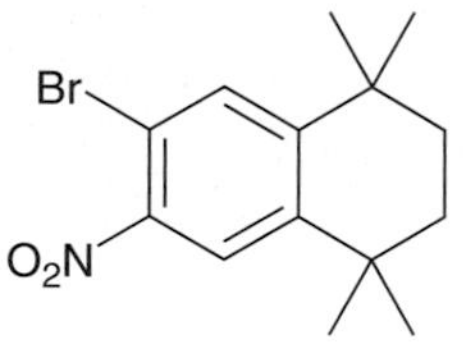

19.73 When benzene is treated with methyl chloride and aluminum trichloride under conditions that favor trialkylation, one major product is obtained. Draw this product, and provide an IUPAC name.

INTEGRATED PROBLEMS

19.74 Compound **A** has molecular formula C_8H_8O. An IR spectrum of compound **A** exhibits a signal at 1680 cm^{-1}. The ^1H NMR spectrum of compound **A** exhibits a group of signals between 7.5 and 8 ppm (with a combined integration of 5) and one upfield signal with an integration of 3. Compound **A** is converted into compound **B** under the following conditions:

$$\textbf{Compound A} \xrightarrow{\text{Zn [Hg], HCl, heat}} \textbf{Compound B}$$

When compound **B** is treated with Br_2 and $AlBr_3$, two different monobromination products are obtained. Identify both of these products, and predict which one will be the major product. Explain your reasoning.

19.75 Starting with benzene and using any other necessary reagents of your choice, design a synthesis for each of the following compounds. Each compound has a Br and one other substituent that we did not learn how to install. In each case, you will need to choose one of the substituents that we learned in this chapter and then modify that substituent using reactions from previous chapters. Be careful to consider the order of events in each case.

(a) 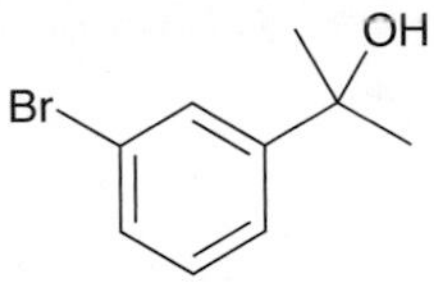

(b) 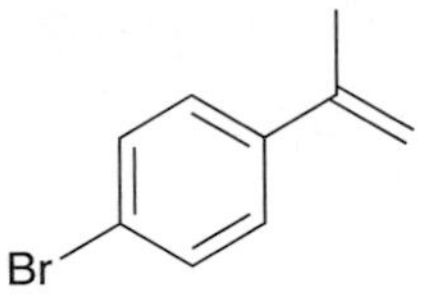

19.76 Benzene was treated with (*R*)-2-chlorobutane in the presence of aluminum trichloride, and the resulting product mixture was found to be optically inactive.

(a) What products are expected, assuming that conditions are chosen to favor monoalkylation?

(b) Explain why the product mixture is optically inactive.

19.77 The ^1H NMR spectrum of phenol exhibits three signals in the aromatic region of the spectrum. These signals appear at 6.7, 6.8, and 7.2 ppm. Use your understanding of shielding and deshielding effects (Chapter 16) to determine which signal corresponds with the *meta* protons. Explain your reasoning.

19.78 When toluene is treated with a mixture of excess sulfuric acid and nitric acid at high temperature, a compound is obtained that exhibits only two signals in its ^1H NMR spectrum. One signal appears upfield and has an integration of 3. The other signal appears downfield and has an integration of 2. Identify the structure of this compound, and assign an IUPAC name.

CHALLENGE PROBLEMS

19.79 Each of the following compounds is an aromatic compound bearing a substituent that we did not discuss in this chapter. Using the principles that we discussed in this chapter, predict the major product for each of the following reactions:

(a) 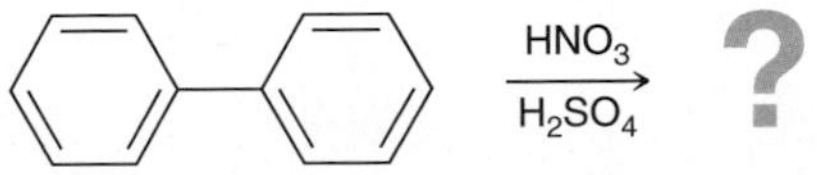

(b) 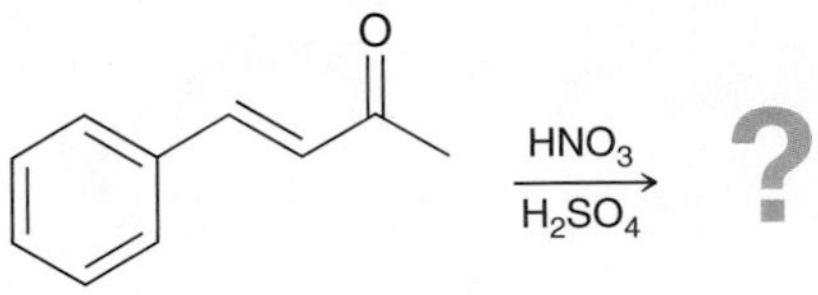

19.80 Predict the major product of the following reaction.

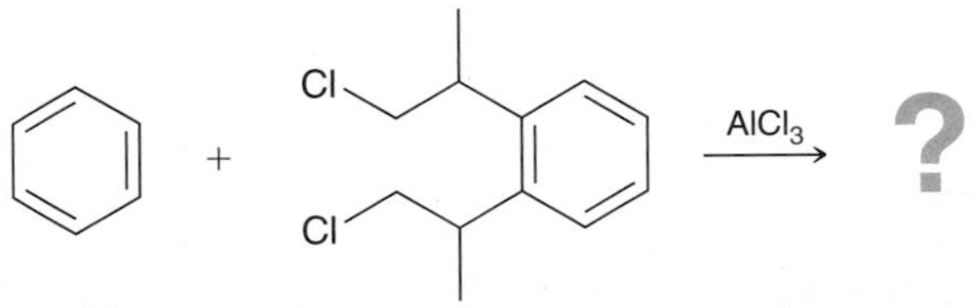

19.81 Bakelite is one of the first known synthetic polymers and was used to make radio and telephone casings in the early twentieth century. Bakelite is formed by treating phenol with formaldehyde under acidic conditions. Draw a plausible mechanism for the formation of Bakelite.

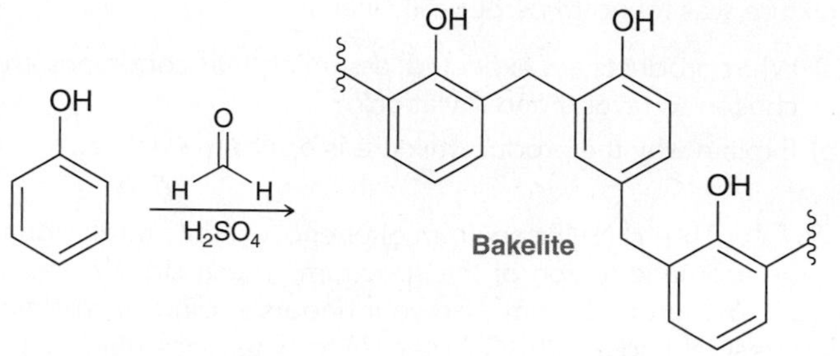

19.82 Propose a plausible mechanism for the following transformation:

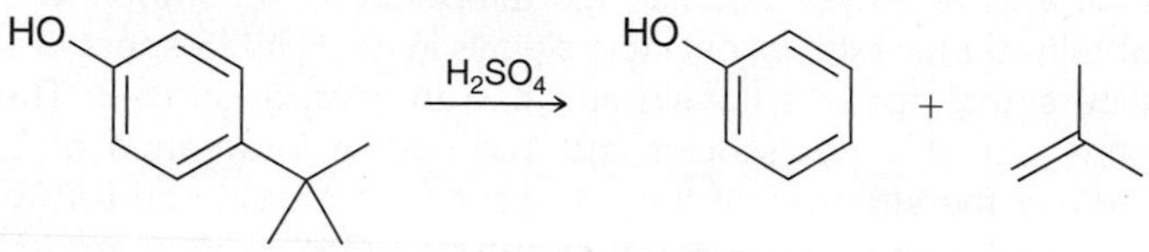

19.83 When *N*,*N*-dimethylaniline is treated with bromine, *ortho* and *para* products are observed. Yet, when *N*,*N*-dimethylaniline is treated with a mixture of nitric and sulfuric acid, only the *meta* product is observed. Explain these curious results.

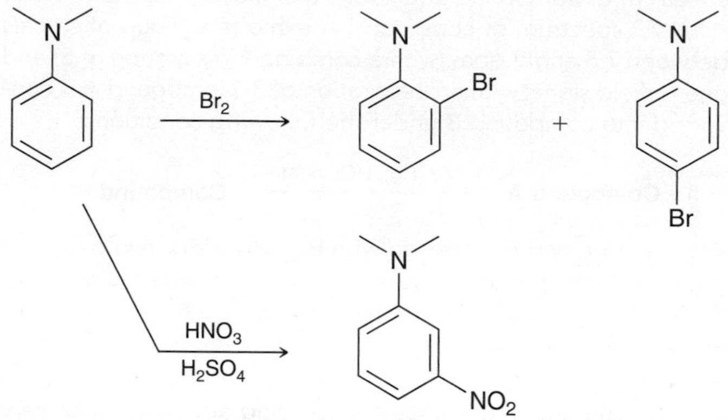

Aldehydes and Ketones

20

DID YOU EVER WONDER...

***why* beta-carotene, which makes carrots orange, is reportedly good for your eyes?**

This chapter will explore the reactivity of aldehydes and ketones. Specifically, we will see that a wide variety of nucleophiles will react with aldehydes and ketones. Many of these reactions are common in biological pathways, including the role that beta-carotene plays in promoting healthy vision. As we will see several times in this chapter, the reactions of aldehydes and ketones are also cleverly exploited in the design of drugs. The reactions and principles outlined in this chapter are central to the study of organic chemistry and will be used as guiding principles throughout the remaining chapters of this textbook.

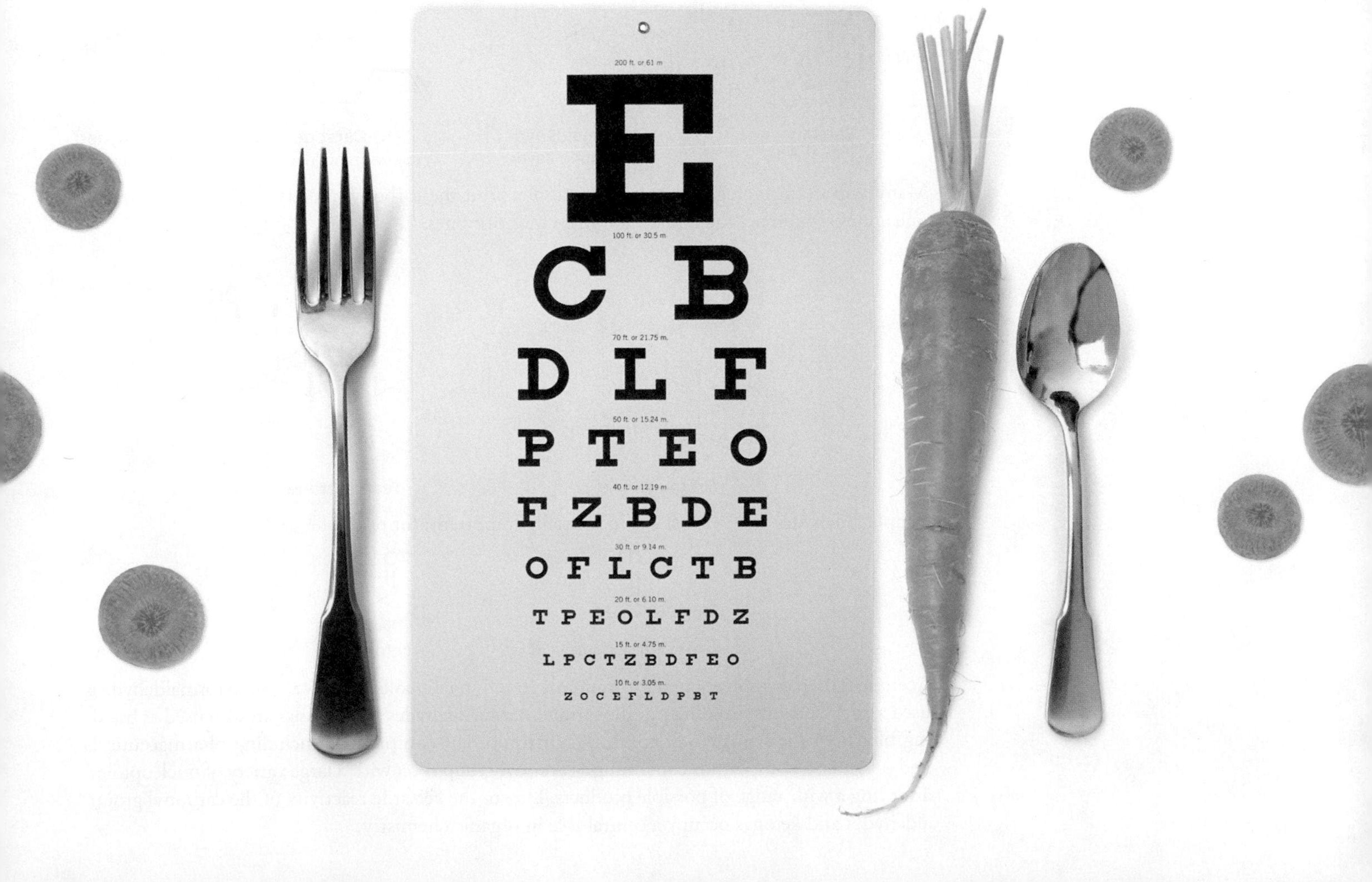

DO YOU REMEMBER?

Before you go on, be sure you understand the following topics. If necessary, review the suggested sections to prepare for this chapter:

- Grignard reagents (Section 13.6)
- Retrosynthetic analysis (Section 12.5)
- Oxidation of alcohols (Section 13.10)

WILEY PLUS Visit www.wileyplus.com to check your understanding and for valuable practice.

20.1 Introduction to Aldehydes and Ketones

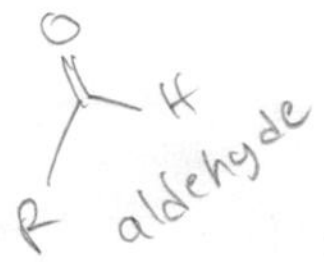

Aldehydes (RCHO) and ketones (R_2CO) are similar in structure in that both classes of compounds possess a C=O bond, called a **carbonyl group:**

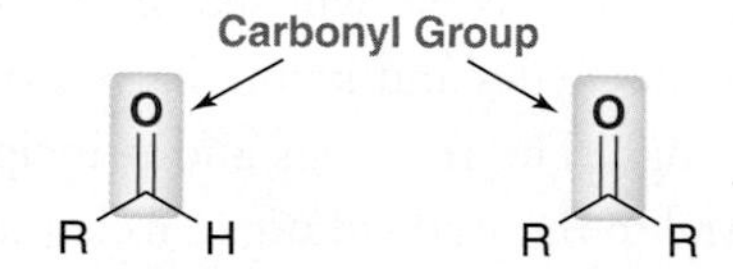

An aldehyde **A ketone**

The carbonyl group of an aldehyde is flanked by a hydrogen atom, while the carbonyl group of a ketone is flanked by two carbon atoms.

Aldehydes and ketones are responsible for many flavors and odors that you will readily recognize:

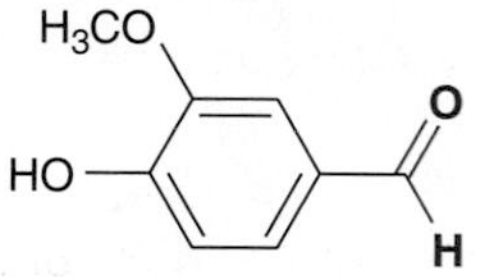

Vanillin
(Vanilla flavor)

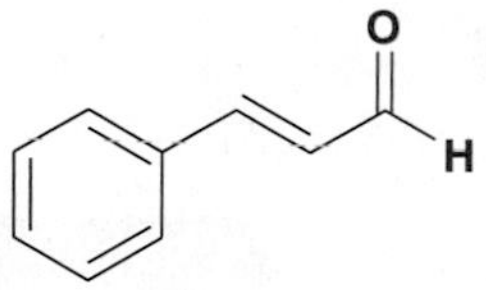

Cinnamaldehyde
(Cinnamon flavor)

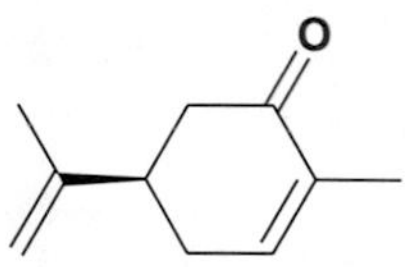

(*R*)-Carvone
(Spearmint flavor)

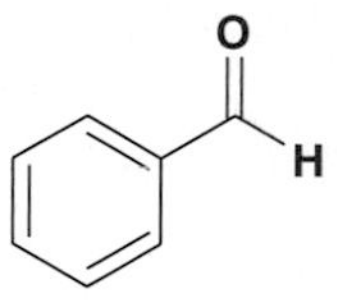

Benzaldehyde
(Almond flavor)

Many important biological compounds also exhibit the carbonyl moiety, including progesterone and testosterone, the female and male sex hormones.

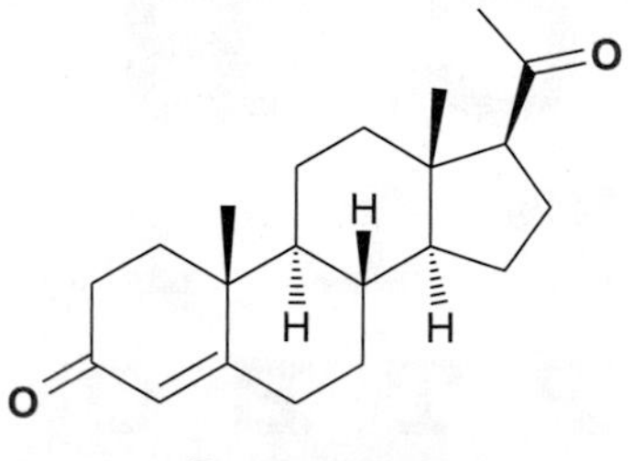

Progesterone

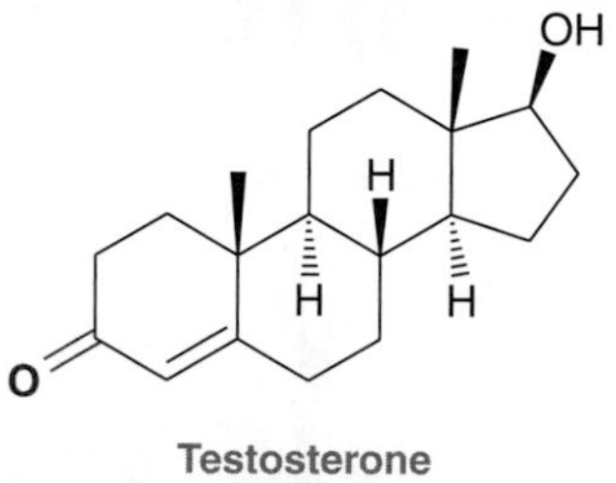

Testosterone

Simple aldehydes and ketones are industrially important; for example:

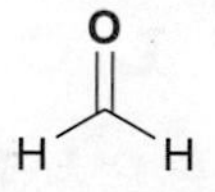

Formaldehyde

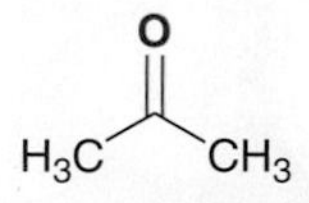

Acetone

Acetone is used as a solvent and is commonly found in nail polish remover, while formaldehyde is used as a preservative in some vaccine formulations. Aldehydes and ketones are also used as building blocks in the syntheses of commercially important compounds, including pharmaceuticals and polymers. Compounds containing a carbonyl group react with a large variety of nucleophiles, affording a wide range of possible products. Due to the versatile reactivity of the carbonyl group, aldehydes and ketones occupy a central role in organic chemistry.

20.2 Nomenclature

Nomenclature of Aldehydes

Recall that four discrete steps are required to name most classes of organic compounds (as we saw with alkanes, alkenes, alkynes, and alcohols):

1. Identify and name the parent.
2. Identify and name the substituents.
3. Assign a locant to each substituent.
4. Assemble the substituents alphabetically.

Aldehydes are also named using the same four-step procedure. When applying this procedure for naming aldehydes, the following guidelines should be followed:

When naming the parent, the suffix "-al" indicates the presence of an aldehyde group:

Butane **Butanal**

When choosing the parent of an aldehyde, identify the longest chain *that includes the carbon atom of the aldehydic group*:

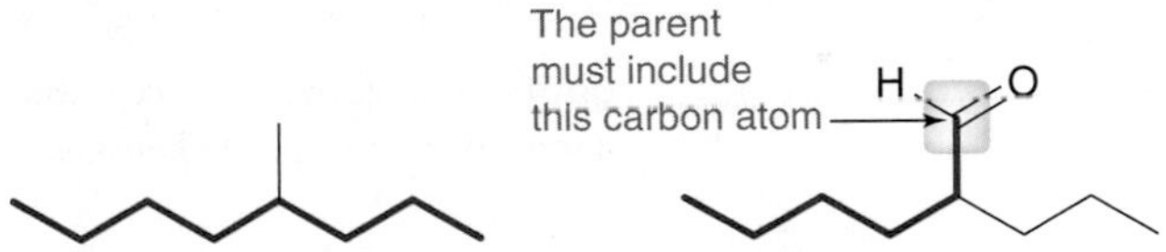

Parent = Octane **Parent = Hexanal**

When numbering the parent chain of an aldehyde, the aldehydic carbon is assigned number 1, despite the presence of alkyl substituents, π bonds, or hydroxyl groups:

Correct **Incorrect**

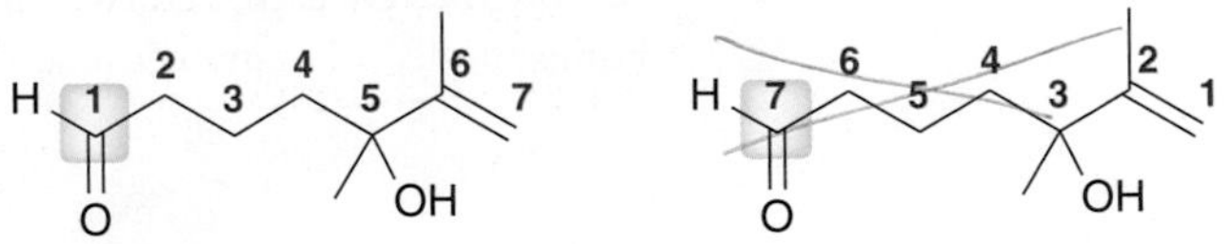

It is not necessary to include the locant in the name, because it is understood that the aldehydic carbon is the number 1 position.

As with all compounds, when a chirality center is present, the configuration is indicated at the beginning of the name; for example:

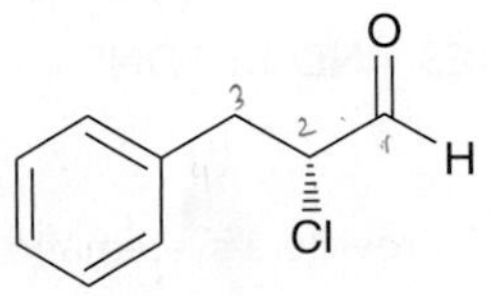

(*R*)-2-chloro-3-phenylpropanal

A cyclic compound containing an aldehyde group immediately adjacent to the ring is named as a carbaldehyde:

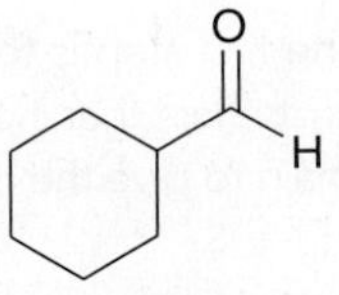

Cyclohexanecarbaldehyde

The International Union of Pure and Applied Chemistry (IUPAC) nomenclature also recognizes the common names of many simple aldehydes, including the three examples shown below:

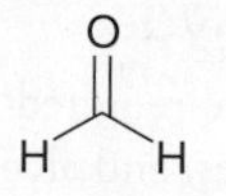
Formaldehyde

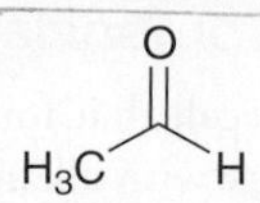
Acetaldehyde

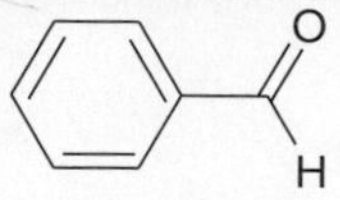
Benzaldehyde

Nomenclature of Ketones

Ketones, like aldehydes, are named using the same four-step procedure. When naming the parent, the suffix "-one" indicates the presence of a ketone group:

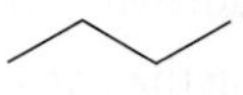
Butane

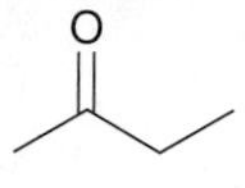
Butanone

The position of the ketone group is indicated using a locant. The IUPAC rules published in 1979 dictate that this locant be placed immediately before the parent, while the IUPAC recommendations released in 1993 and 2004 allow for the locant to be placed immediately before the suffix "-one":

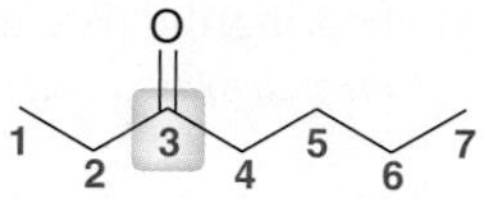
3-heptanone
or
heptan-3-*one*

Both names above are acceptable IUPAC names. IUPAC nomenclature recognizes the common names of many simple ketones, including the three examples shown below:

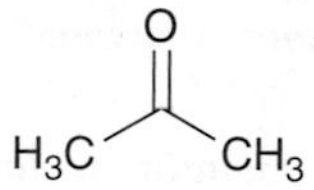
Acetone

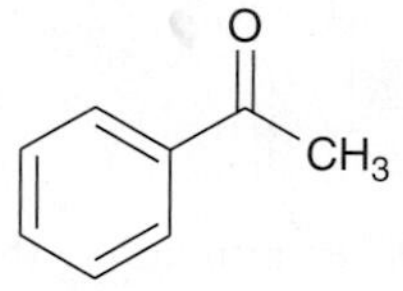
Acetophenone

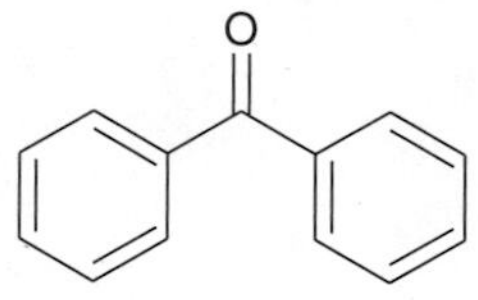
Benzophenone

Although rarely used, IUPAC rules also allow simple ketones to be named as *alkyl alkyl ketones*. For example, 3-hexanone can also be called ethyl propyl ketone:

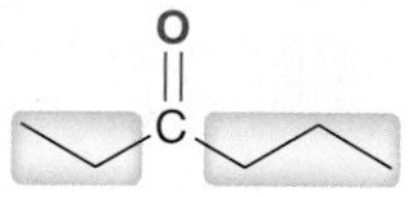
Ethyl propyl ketone

SKILLBUILDER

20.1 NAMING ALDEHYDES AND KETONES

LEARN the skill

Provide a systematic (IUPAC) name for the following compound:

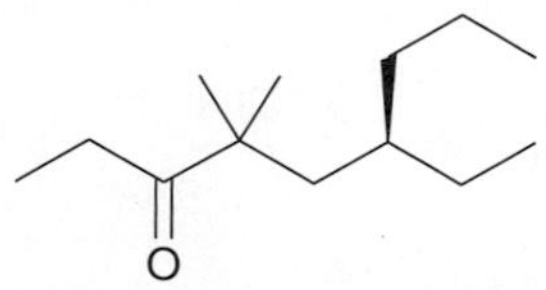

SOLUTION

STEP 1
Identify and name the parent.

The first step is to identify and name the parent. Choose the longest chain that includes the carbonyl group, and then number the chain to give the carbonyl group the lowest number possible:

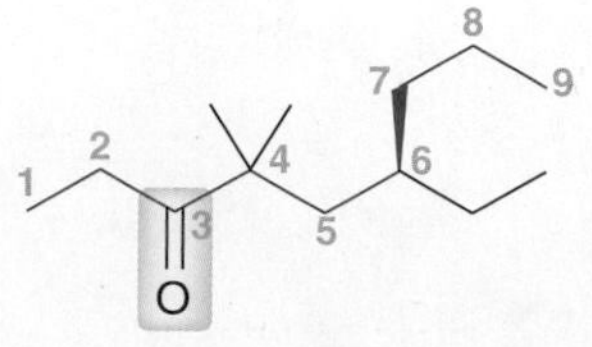
3-nonanone

STEP 2 Identify and name the substituents.
STEP 3 Assign a locant to each substituent.

Next, identify the substituents and assign locants:

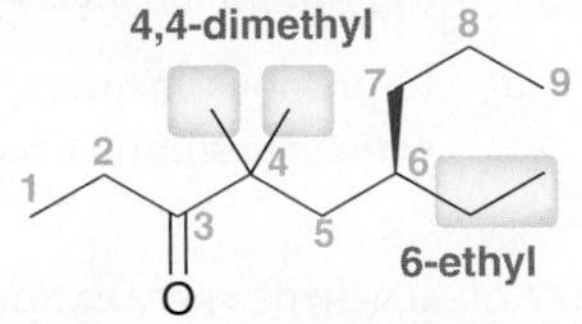

STEP 4 Assemble the substituents alphabetically.
STEP 5 Assign the configuration of any chirality centers.

Finally, assemble the substituents alphabetically: 6-ethyl-4,4-dimethyl-3-nonanone. Before concluding, we must always check to see if there are any chirality centers. This compound does exhibit one chirality center. Using the skills from Section 5.3, the *R* configuration is assigned to this chirality center:

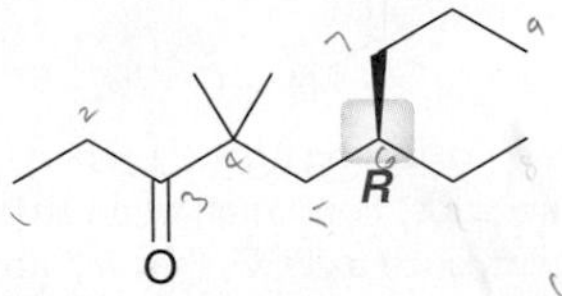

Therefore, the complete name is **(*R*)-6-ethyl-4,4-dimethyl-3-nonanone.**

PRACTICE the skill

20.1 Assign a systematic (IUPAC) name to each of the following compounds:

(a) 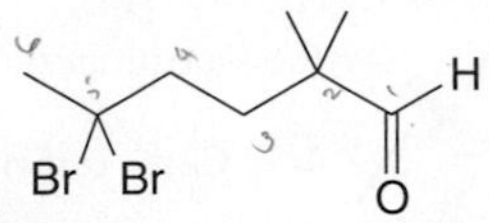 (b) 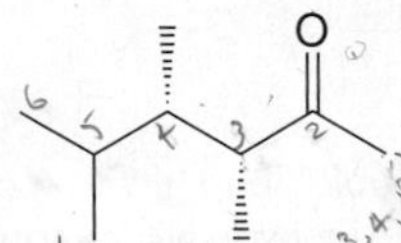 (c) 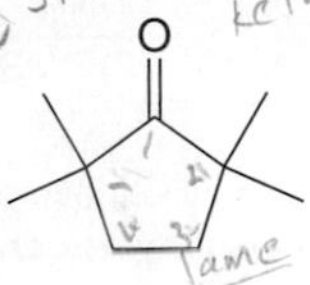

(d) 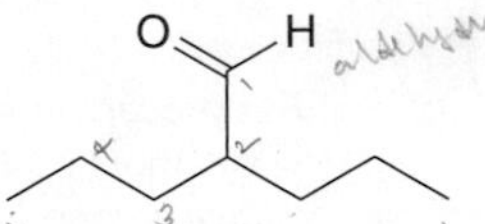 (e) 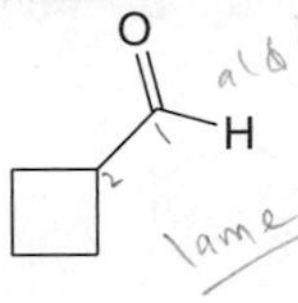

APPLY the skill

20.2 Draw the structure of each of the following compounds:

(a) (*S*)-3,3-dibromo-4-ethylcyclohexanone (b) 2,4-dimethyl-3-pentanone

(c) (*R*)-3-bromobutanal

20.3 Provide a systematic (IUPAC) name for the compound below. Be careful: This compound has two chirality centers (can you find them?).

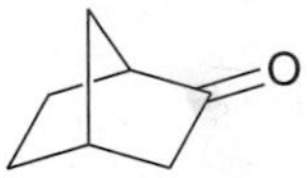

20.4 Compounds with two carbonyl moieties are named as alkane diones; for example:

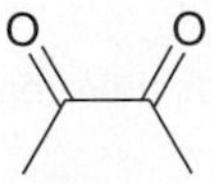

2,3-butanedione

The compound above is an artificial flavor added to microwave popcorn and movie-theater popcorn to simulate the butter flavor. Interestingly, this very same compound is also known to contribute to body odor. Name the following compounds:

(a) 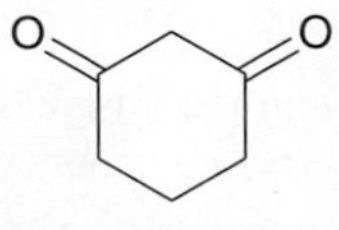 (b) 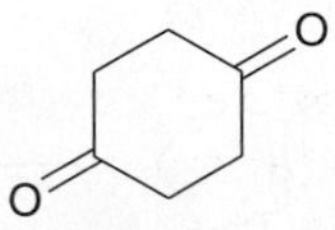 (c) 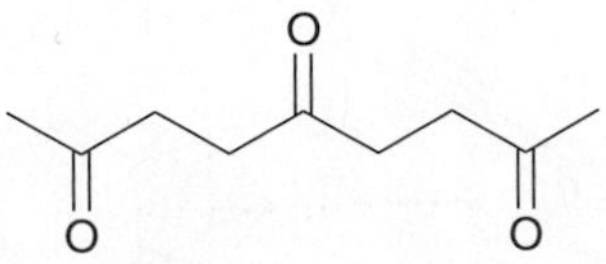

need more **PRACTICE?** **Try Problems 20.44–20.49**

20.3 Preparing Aldehydes and Ketones: A Review

In previous chapters, we have studied a variety of methods for preparing aldehydes and ketones, which are summarized in Tables 20.1 and 20.2, respectively.

TABLE 20.1 A SUMMARY OF ALDEHYDE PREPARATION METHODS COVERED IN PREVIOUS CHAPTERS

REACTION — SECTION

Oxidation of Primary Alcohols — 13.10

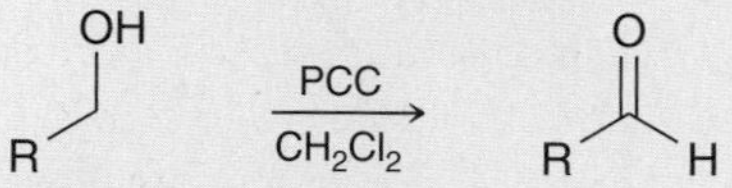

When treated with a strong oxidizing agent, primary alcohols are oxidized to carboxylic acids. Formation of an aldehyde requires an oxidizing agent, such as PCC, that will not further oxidize the resulting aldehyde.

Ozonolysis of Alkenes — 9.11

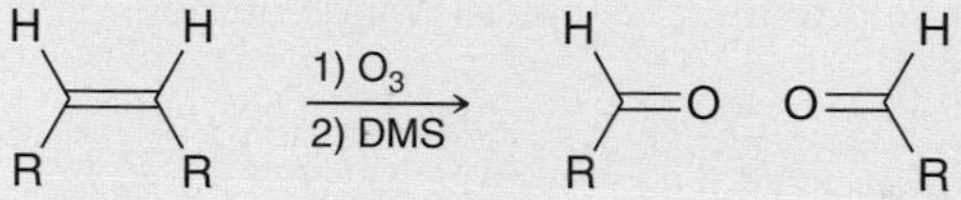

Ozonolysis will cleave a C=C double bond. If either carbon atom bears a hydrogen atom, an aldehyde will be formed.

Hydroboration-Oxidation of Terminal Alkynes — 10.7

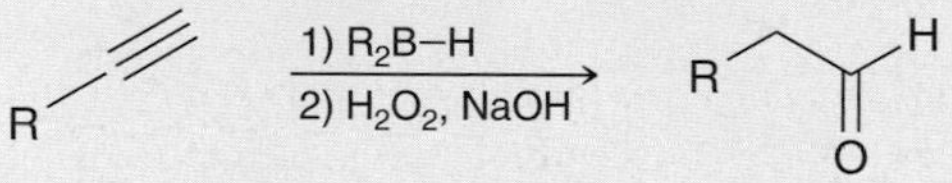

Hydroboration-oxidation results in an anti-Markovnikov addition of water across a π bond, followed by tautomerization of the resulting enol to form an aldehyde.

TABLE 20.2 A SUMMARY OF KETONE PREPARATION METHODS COVERED IN PREVIOUS CHAPTERS

REACTION — SECTION

Oxidation of Secondary Alcohols — 13.10

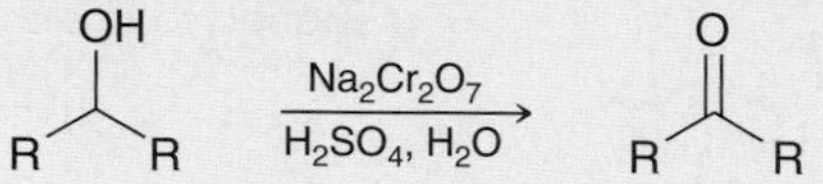

A variety of strong or mild oxidizing agents can be used to oxidize secondary alcohols. The resulting ketone does not undergo further oxidation.

Ozonolysis of Alkenes — 9.11

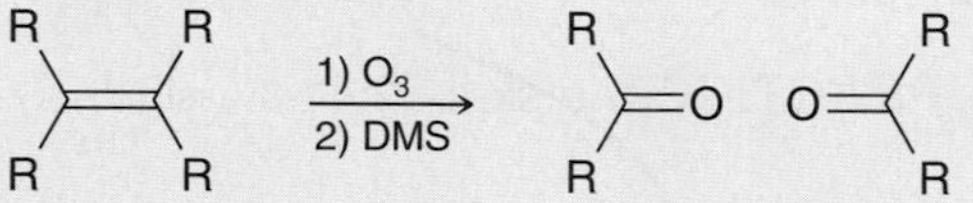

Tetrasubstituted alkenes are cleaved to form ketones.

Acid-Catalyzed Hydration of Terminal Alkynes — 10.7

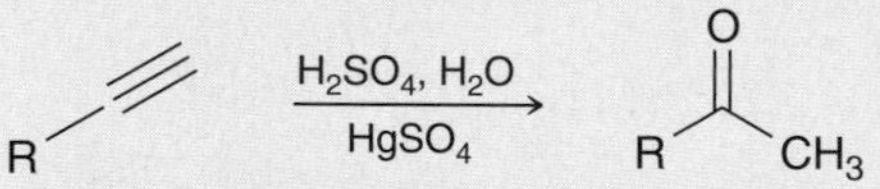

This procedure results in a Markovnikov addition of water across the π bond, followed by tautomerization to form a methyl ketone.

Friedel-Crafts Acylation — 19.6

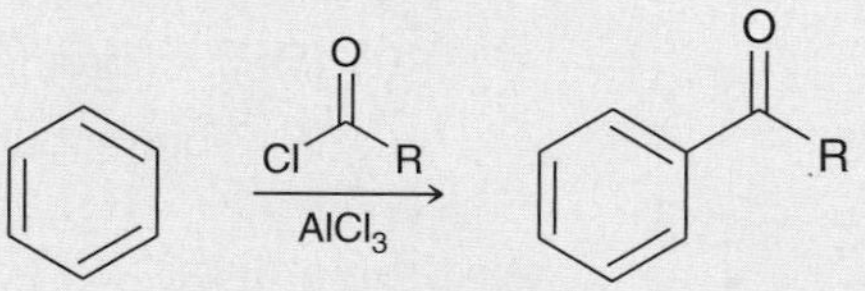

Aromatic rings that are not too strongly deactivated will react with an acid halide in the presence of a Lewis acid to produce an aryl ketone.

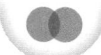

CONCEPTUAL CHECKPOINT

20.5 Identify the reagents necessary to achieve each of the following transformations:

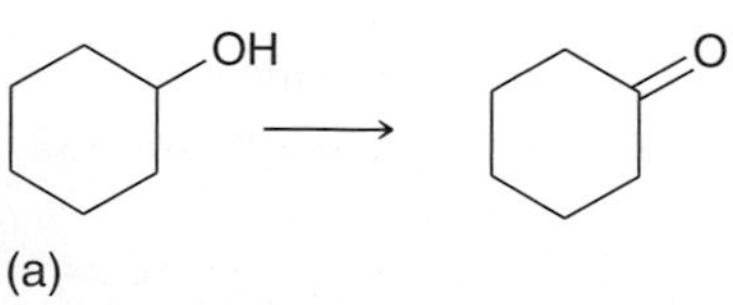
(a)

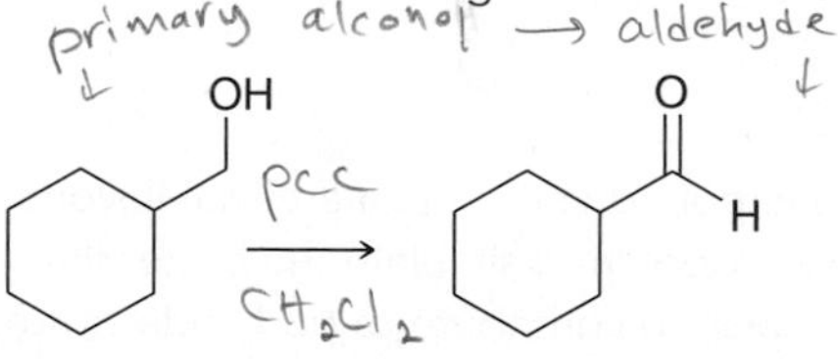
(b)

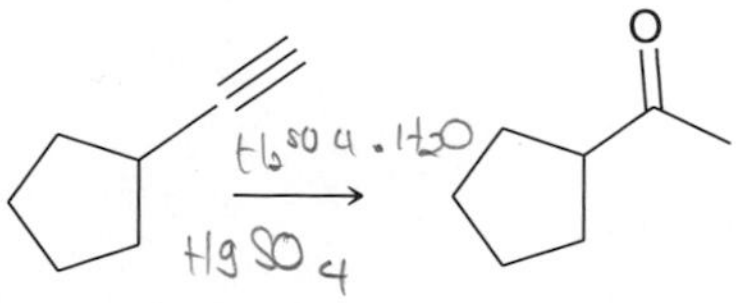
(c)

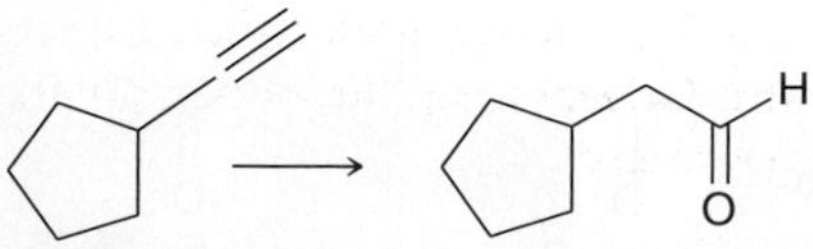
(d)

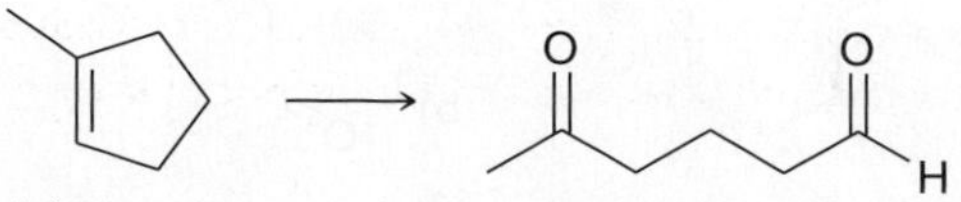
(e)

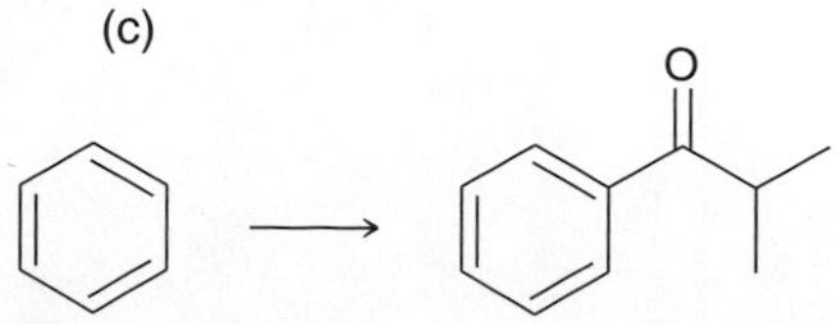
(f)

20.4 Introduction to Nucleophilic Addition Reactions

The electrophilicity of a carbonyl group derives from resonance effects as well as inductive effects:

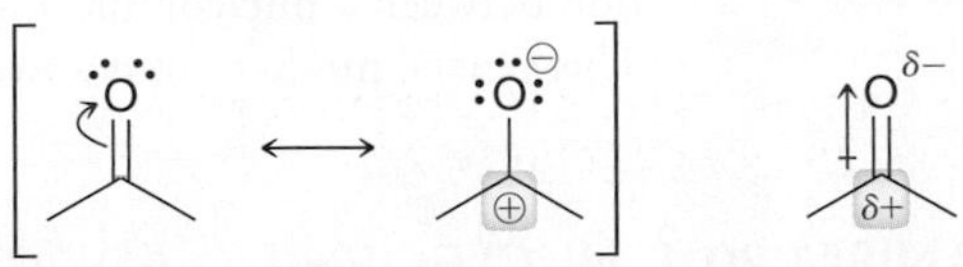

One of the resonance structures exhibits a positive charge on the carbon atom, indicating that the carbon atom is deficient in electron density (δ+). Inductive effects also render the carbon atom deficient in electron density. As a result, this carbon atom is particularly electrophilic and is susceptible to attack by a nucleophile. Molecular orbital calculations suggest that nucleophilic attack occurs at an angle of approximately 107° to the plane of the carbonyl group, and in the process, the hybridization state of the carbon atom changes (Figure 20.1).

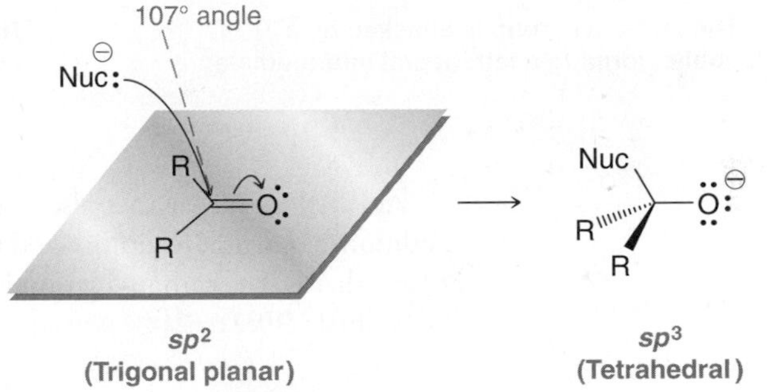

FIGURE 20.1 When a carbonyl group is attacked by a nucleophile, the carbon atom undergoes a change in hybridization and geometry.

The carbon atom is originally sp^2 hybridized with a trigonal planar geometry. After the attack, the carbon atom is sp^3 hybridized with a tetrahedral geometry. In recognition of this geometric change, the resulting alkoxide ion is often called a **tetrahedral intermediate.** This term appears many times throughout the remainder of this chapter.

In general, aldehydes are more reactive than ketones toward nucleophilic attack. This observation can be explained in terms of both steric and electronic effects:

1. *Steric effects.* A ketone has two alkyl groups (one on either side of the carbonyl) that contribute to steric hindrance in the transition state of a nucleophilic attack. In contrast, an aldehyde has only one alkyl group, so the transition state is less crowded and lower in energy.
2. *Electronic effects.* Recall that alkyl groups are electron donating. A ketone has two electron-donating alkyl groups that can stabilize the δ+ on the carbon atom of the carbonyl group. In contrast, aldehydes have only one electron-donating group:

A ketone has **two** electron-donating alkyl groups that stabilize the partial positive charge

An aldehyde has **only one** electron-donating alkyl group that stabilizes the partial positive charge

The δ+ charge of an aldehyde is less stabilized than a ketone. As a result, aldehydes are more electrophilic than ketones and therefore more reactive.

Aldehydes and ketones react with a wide variety of nucleophiles. As we will see in the coming sections of this chapter, some nucleophiles require basic conditions, while others require acidic conditions. For example, recall from Chapter 13 that Grignard reagents are very strong nucleophiles that will attack aldehydes and ketones to produce alcohols:

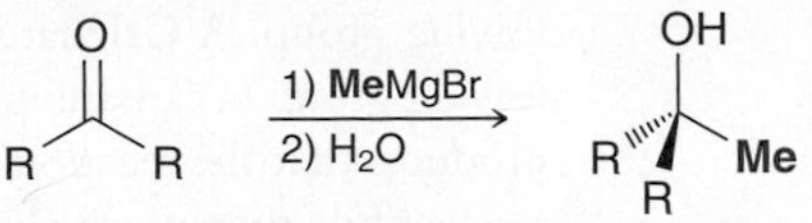

The Grignard reagent itself provides for strongly basic conditions, because Grignard reagents are both strong nucleophiles and strong bases. This reaction cannot be achieved under acidic conditions, because, as explained in Section 13.6, Grignard reagents are destroyed in the presence of an acid. The Grignard reaction above follows a general mechanism for the reaction between a nucleophile and a carbonyl group under basic conditions (Mechanism 20.1). This general mechanism has two steps: (1) nucleophilic attack followed by (2) proton transfer.

MECHANISM 20.1 NUCLEOPHILIC ADDITION UNDER BASIC CONDITIONS

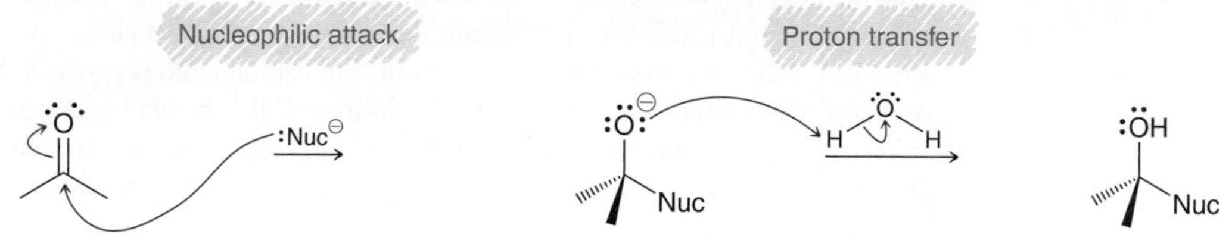

The carbonyl group is attacked by a nucleophile, forming a tetrahedral intermediate

The tetrahedral intermediate is protonated upon treatment with a mild proton source

Aldehydes and ketones also react with a wide variety of other nucleophiles under acidic conditions. In acidic conditions, the same two mechanistic steps are observed, but in reverse order—that is, the carbonyl group is first protonated and then undergoes a nucleophilic attack (Mechanism 20.2).

MECHANISM 20.2 NUCLEOPHILIC ADDITION UNDER ACIDIC CONDITIONS

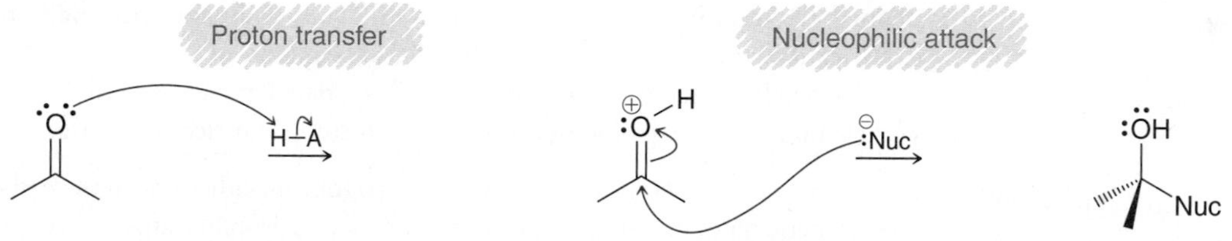

The carbonyl group is first protonated, rendering it even more electrophilic

The protonated carbonyl group is then attacked by a nucleophile

In acidic conditions, the first step plays an important role. Specifically, protonating the carbonyl group generates a very powerful electrophile:

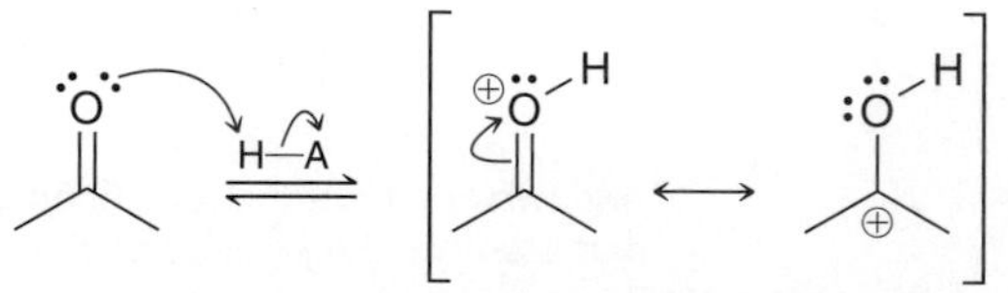

Very powerful electrophile

It is true that the carbonyl group is already a fairly strong electrophile; however, a protonated carbonyl group bears a full positive charge, rendering the carbon atom even more electrophilic. This is especially important when weak nucleophiles, such as H_2O or ROH, are employed, as we will see in the upcoming sections.

When a nucleophile attacks a carbonyl group under either acidic or basic conditions, the position of equilibrium is highly dependent on the ability of the nucleophile to function as a leaving group. A Grignard reagent is a very strong nucleophile, but it does not function as a leaving group (a carbanion is too unstable to leave). As a result, the equilibrium so greatly favors products that the reaction effectively occurs in only one direction. With a sufficient amount of nucleophile present, the ketone is not observed in the product mixture. In contrast, halides are good nucleophiles, but they are also good leaving groups. Therefore, when a halide functions as the nucleophile, the equilibrium actually favors the starting ketone:

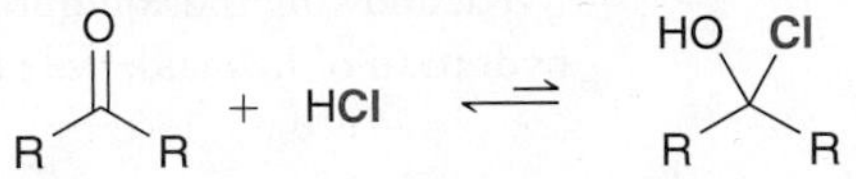

Once equilibrium has been achieved, the mixture consists primarily of the ketone, and only small quantities of the addition product.

In this chapter, we will explore a wide variety of nucleophiles, which will be classified according to the nature of the attacking atom. Specifically, we will see nucleophiles based on oxygen, sulfur, nitrogen, hydrogen, and carbon (Figure 20.2).

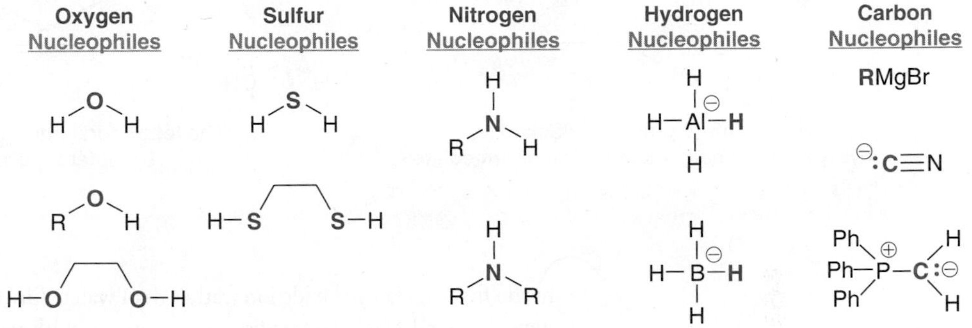

FIGURE 20.2 Various nucleophiles that can attack a carbonyl group.

The remainder of the chapter will be a methodical survey of the reactions that occur between the reagents in Figure 20.2 and ketones and aldehydes. We will begin our survey with oxygen nucleophiles.

CONCEPTUAL CHECKPOINT

20.6 Draw a mechanism for each of the following reactions:

(a)

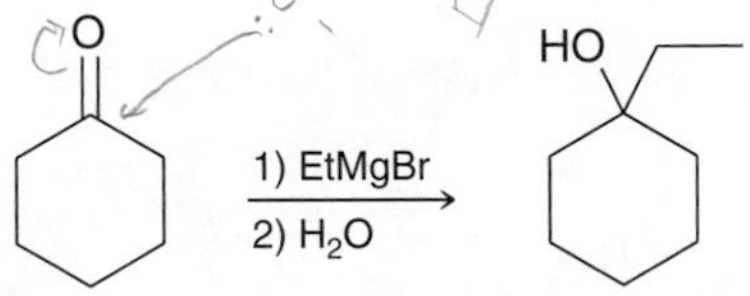

(b)

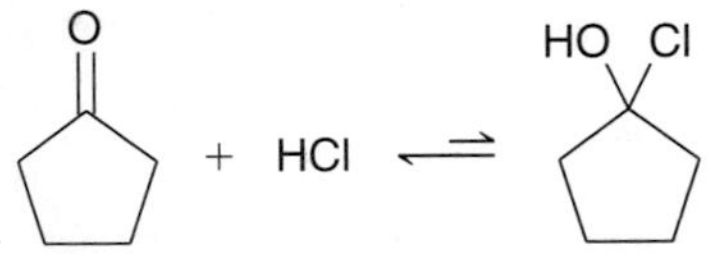

20.5 Oxygen Nucleophiles

Hydrate Formation

When an aldehyde or ketone is treated with water, the carbonyl group can be converted into a **hydrate:**

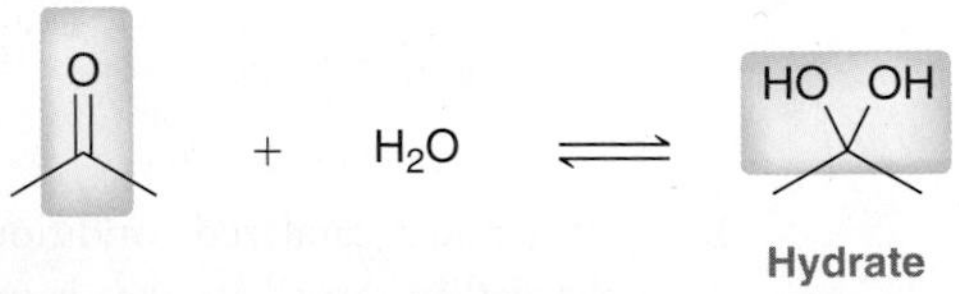

The position of equilibrium generally favors the carbonyl group rather than the hydrate, except in the case of very simple aldehydes, such as formaldehyde:

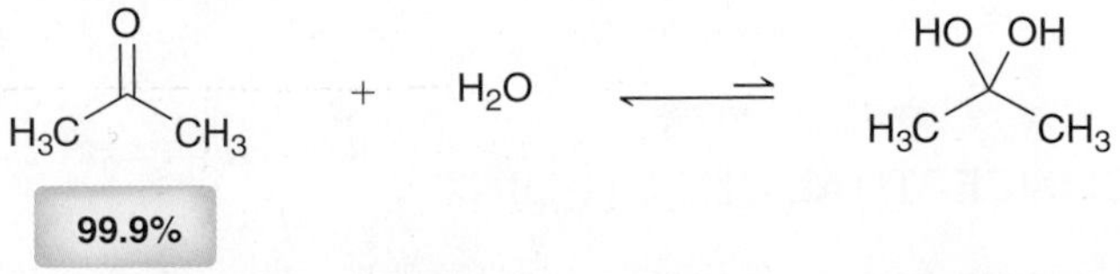

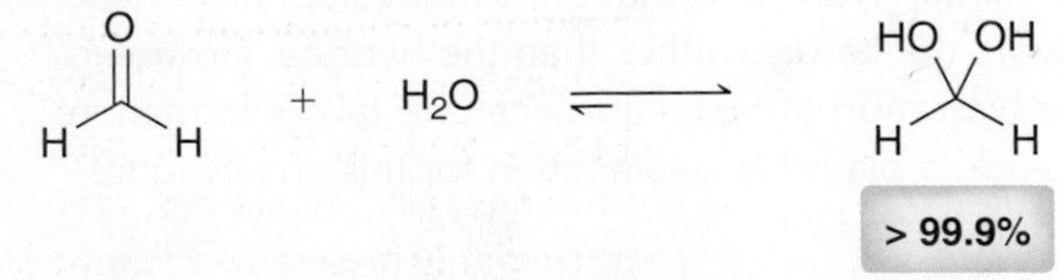

The rate of reaction is relatively slow under neutral conditions but is readily enhanced in the presence of either acid or base. That is, the reaction can be either acid catalyzed or base catalyzed, allowing the equilibrium to be achieved much more rapidly. Consider the base-catalyzed hydration of formaldehyde (Mechanism 20.3).

MECHANISM 20.3 BASE-CATALYZED HYDRATION

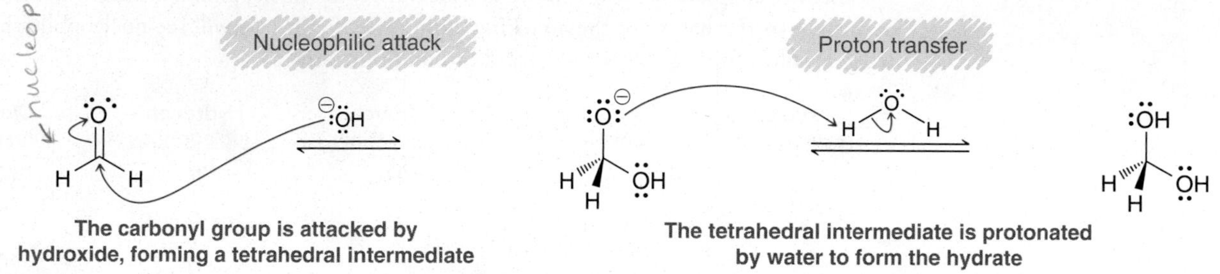
Nucleophilic attack
Proton transfer
The carbonyl group is attacked by hydroxide, forming a tetrahedral intermediate
The tetrahedral intermediate is protonated by water to form the hydrate

In the first step, a hydroxide ion (rather than water) functions as a nucleophile. Then, in the second step, the tetrahedral intermediate is protonated with water, regenerating a hydroxide ion. In this way, hydroxide serves as a catalyst for the addition of water across the carbonyl group.

Now consider the acid-catalyzed hydration of formaldehyde (Mechanism 20.4).

MECHANISM 20.4 ACID-CATALYZED HYDRATION

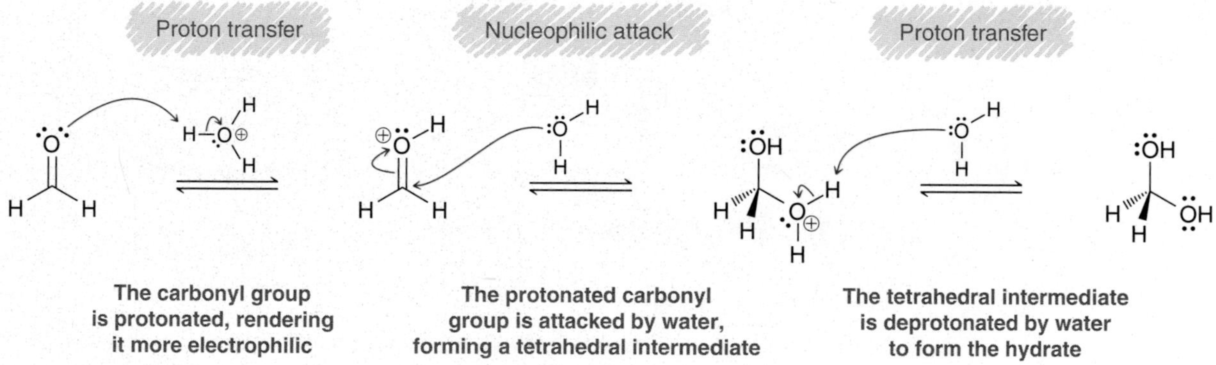
Proton transfer
Nucleophilic attack
Proton transfer
The carbonyl group is protonated, rendering it more electrophilic
The protonated carbonyl group is attacked by water, forming a tetrahedral intermediate
The tetrahedral intermediate is deprotonated by water to form the hydrate

Under acid-catalyzed conditions, the carbonyl group is first protonated, generating a positively charged intermediate that is extremely electrophilic (it bears a full positive charge). This intermediate is then attacked by water to form a tetrahedral intermediate, which is deprotonated to give the product.

CONCEPTUAL CHECKPOINT

20.7 For most ketones, hydrate formation is unfavorable, because the equilibrium favors the ketone rather than the hydrate. However, the equilibrium for hydration of hexafluoroacetone favors formation of the hydrate: Provide a plausible explanation for this observation.

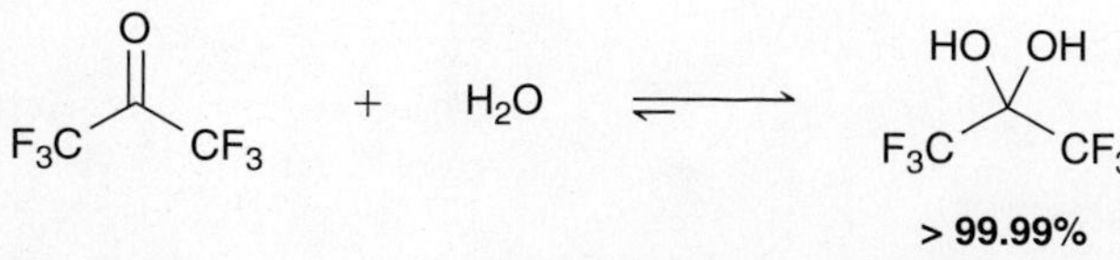
O
F3C
CF3
+
H2O
HO
OH
F3C
CF3
> 99.99%

Acetal Formation

The previous section discussed a reaction that can occur when water attacks an aldehyde or ketone. This section will explore a similar reaction, in which an alcohol attacks an aldehyde or ketone:

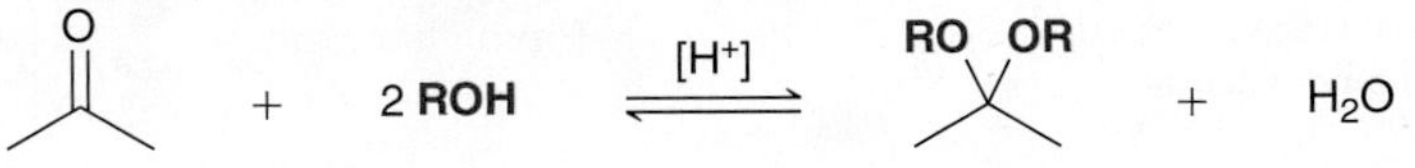

2 ROH
[H⁺]
RO OR
H₂O

Acetal

BY THE WAY
When the starting compound is a ketone, the product can also be called a "ketal." *Acetal* is a more general term, and it will be used exclusively for the remainder of this discussion.

In acidic conditions, an aldehyde or ketone will react with two molecules of alcohol to form an **acetal.** The brackets surrounding the H^+ indicate that the acid is a catalyst. Common acids used for this purpose include *para*-toluenesulfonic acid (TsOH) and sulfuric acid (H_2SO_4):

***p*-Toluenesulfonic acid** (TsOH)   **Sulfuric acid**

As mentioned earlier, the acid catalyst serves an important role in this reaction. Specifically, in the presence of an acid, the carbonyl group is protonated, rendering the carbon atom even more electrophilic. This is necessary because the nucleophile (an alcohol) is weak; it reacts with the carbonyl group more rapidly if the carbonyl group is first protonated. A mechanism for acetal formation is shown in Mechanism 20.5. This mechanism has many steps, and it is best to divide it conceptually into two parts: (1) The first three steps produce an intermediate called a **hemiacetal** and (2) the last four steps convert the hemiacetal into an acetal:

MECHANISM 20.5 ACETAL FORMATION

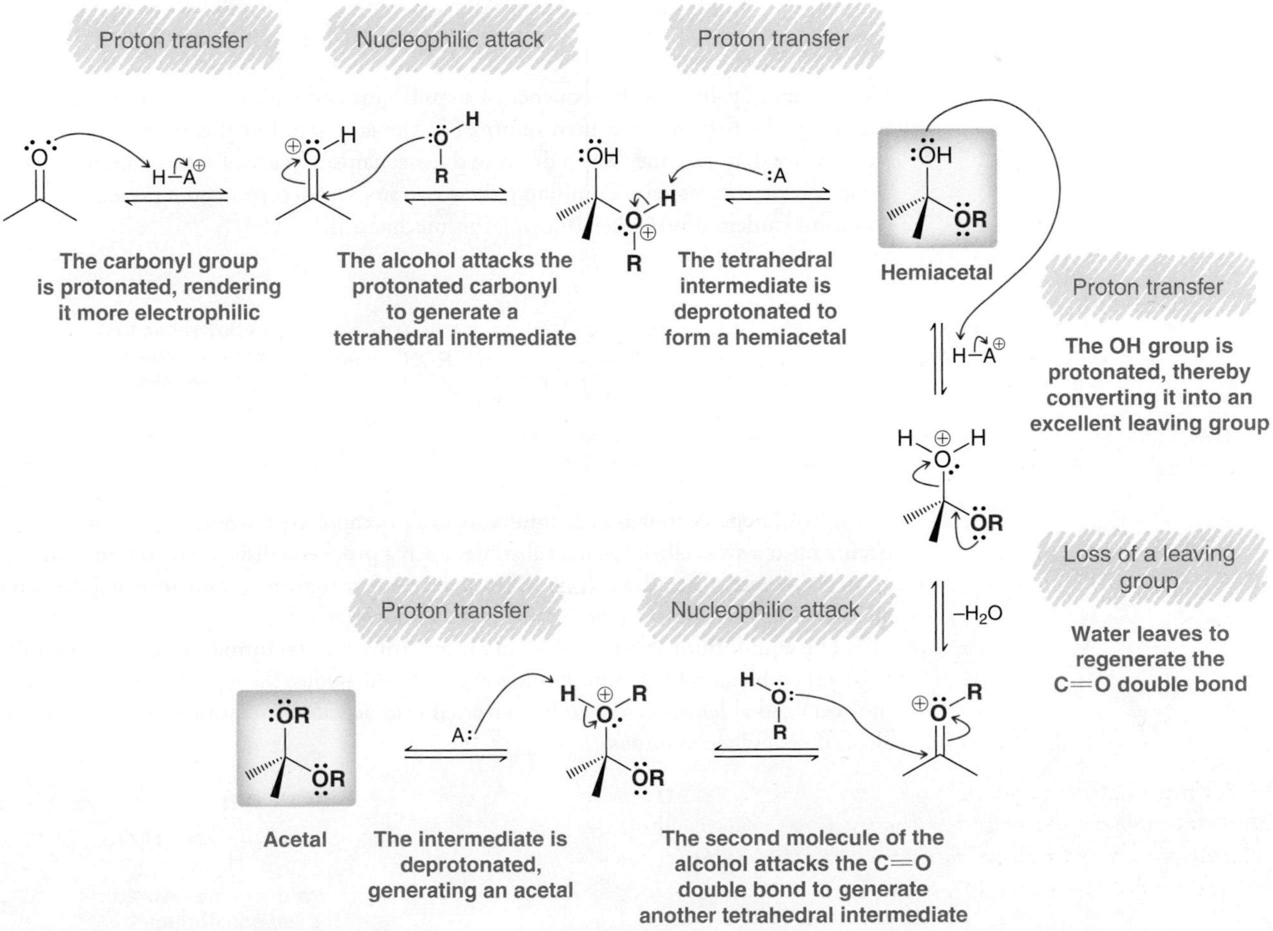

Proton transfer
The carbonyl group is protonated, rendering it more electrophilic
Nucleophilic attack
The alcohol attacks the protonated carbonyl to generate a tetrahedral intermediate
Proton transfer
The tetrahedral intermediate is deprotonated to form a hemiacetal
Hemiacetal
Proton transfer
The OH group is protonated, thereby converting it into an excellent leaving group
Loss of a leaving group
Water leaves to regenerate the C=O double bond
–H₂O
Nucleophilic attack
The second molecule of the alcohol attacks the C=O double bond to generate another tetrahedral intermediate
Proton transfer
The intermediate is deprotonated, generating an acetal
Acetal

Let's begin our analysis of this mechanism by focusing on the first part: formation of the hemiacetal, which involves the three steps in Figure 20.3.

FIGURE 20.3
The sequence of steps involved in formation of a hemiacetal.

Proton transfer → Nucleophilic attack → Proton transfer

Notice that the sequence of steps begins and ends with a proton transfer. This will be a recurring pattern in this chapter. Let's focus on the details of these three steps:

1. The carbonyl is protonated in the presence of an acid. The identity of the acid, HA^+, is most likely a protonated alcohol, which received its extra proton from the acid catalyst:

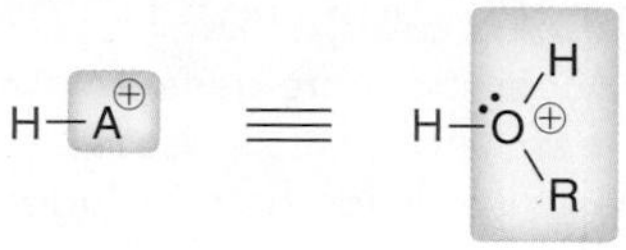

2. The protonated carbonyl is a very powerful electrophile and is attacked by a molecule of alcohol (ROH) to form a tetrahedral intermediate that bears a positive charge.

3. The tetrahedral intermediate is deprotonated by a weak base (A), which is likely to be a molecule of alcohol present in solution.

Notice that the acid is not consumed in this process. A proton is used in step 1 and then returned in step 3, confirming the catalytic nature of the proton in the reaction.

It is important to remember the specific order of these three steps, as we will soon encounter many other reactions that begin with the same three steps. These three steps are typical of reactions involving acid-catalyzed nucleophilic attack.

Now let's focus on the second part of the mechanism, conversion of the hemiacetal into an acetal, which is accomplished with the four steps in Figure 20.4.

FIGURE 20.4
The sequence of steps that convert a hemiacetal into an acetal.

Proton transfer → Loss of a leaving group → Nucleophilic attack → Proton transfer

Notice, once again, that the sequence of steps begins and ends with a proton transfer. A proton is used in the first step and then returned in the last step, but this time there are two middle steps rather than just one. When drawing the mechanism of acetal formation, make sure to draw these two steps separately. Combining these two steps is incorrect and represents one of the most common student errors when drawing this mechanism:

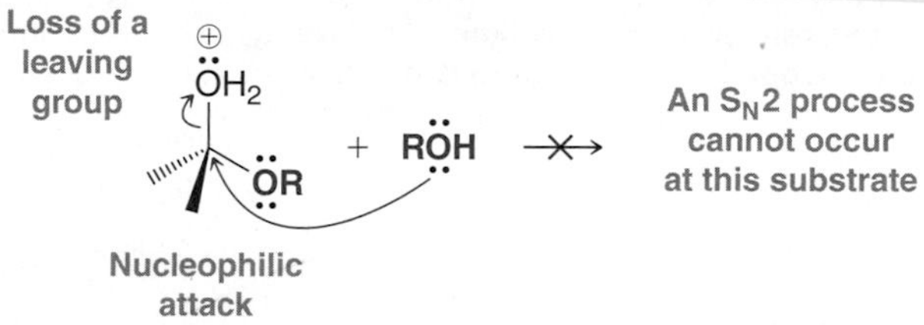

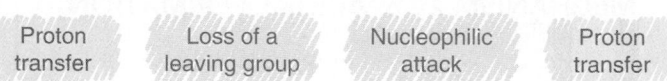

These two steps cannot occur simultaneously, because that would represent an S_N2 process occurring at a sterically hindered substrate. Such a process is disfavored and does not occur at an appreciable rate. Instead, the leaving group leaves first to form a resonance-stabilized intermediate, which is then attacked by the nucleophile in a separate step.

The equilibrium arrows in the full mechanism of acetal formation indicate that the process is governed by an equilibrium. For many simple aldehydes, the equilibrium favors formation of the acetal, so aldehydes are readily converted into acetals by treatment with two equivalents of alcohol in acidic conditions:

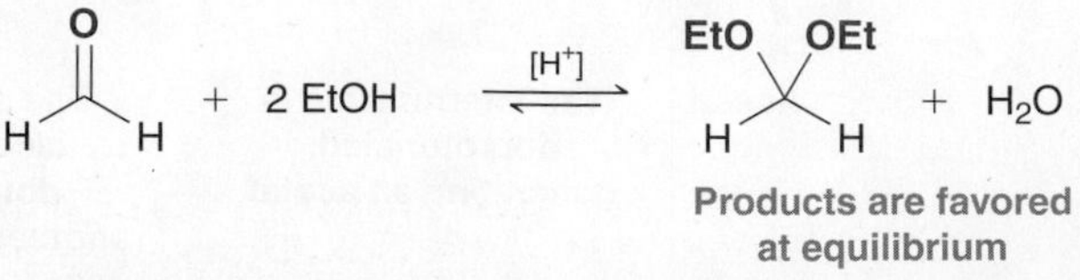

BY THE WAY
This technique exploits Le Châtelier's principle, which was covered in your general chemistry course. According to Le Châtelier's principle, if a system at equilibrium is upset by some disturbance, the system will change in a way that restores equilibrium.

However, for most ketones, the equilibrium favors reactants rather than products:

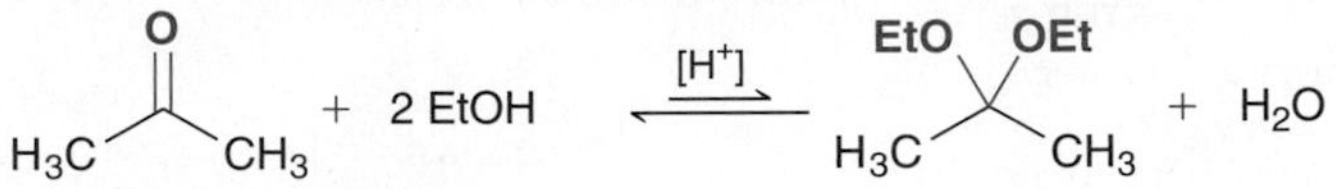

[H+]
+ 2 EtOH
+ H2O

Reactants are favored at equilibrium

In such cases, formation of the acetal can be accomplished by removing one of the products (water) via a special distillation technique. By removing water as it is formed, the reaction can be forced to completion.

Notice that acetal formation requires two equivalents of the alcohol. That is, two molecules of ROH are required for every molecule of ketone. Alternatively, a compound containing two OH groups can be used, forming a cyclic acetal.

This reaction proceeds via the regular seven-step mechanism for acetal formation: three steps for formation of the hemiacetal followed by four steps for formation of the cyclic acetal:

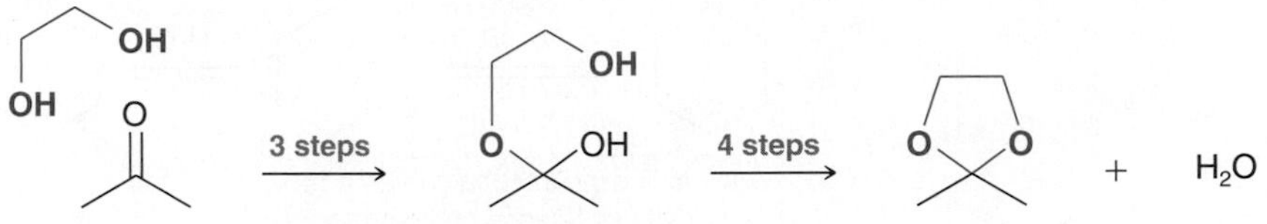

3 steps
4 steps
+ H2O
Hemiacetal
Cyclic acetal

The seven-step mechanism for acetal formation is very similar to other mechanisms that we will explore. It is therefore critical to master these seven steps. To help you draw the mechanism properly, remember to divide the entire mechanism into two parts, where each part begins and ends with proton transfers. Let's get some practice.

SKILLBUILDER

20.2 DRAWING THE MECHANISM OF ACETAL FORMATION

LEARN the skill

Draw a plausible mechanism for the following transformation:

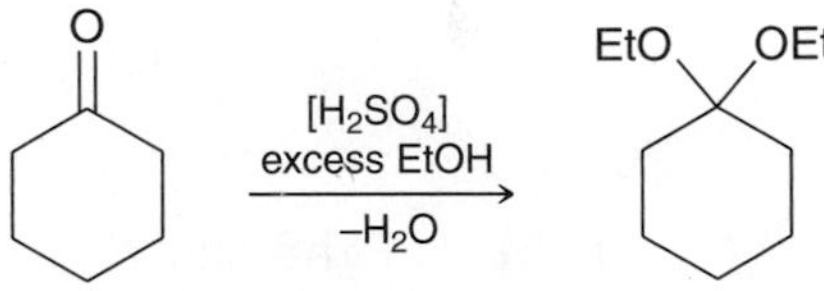

[H2SO4]
excess EtOH
–H2O

SOLUTION

The reaction above is an example of acid-catalyzed acetal formation, in which the product is favored by the removal of water. The mechanism can be divided into two parts: (1) formation of the hemiacetal and (2) formation of the acetal. Formation of the hemiacetal involves three mechanistic steps:

STEP 1
Draw the three steps necessary for hemiacetal formation.

Proton transfer → Nucleophilic attack → Proton transfer

When drawing these three steps, make sure to focus on proper arrow placement (as described in Chapter 6), and make sure to place all positive charges in their appropriate locations. Notice that every step requires two curved arrows.

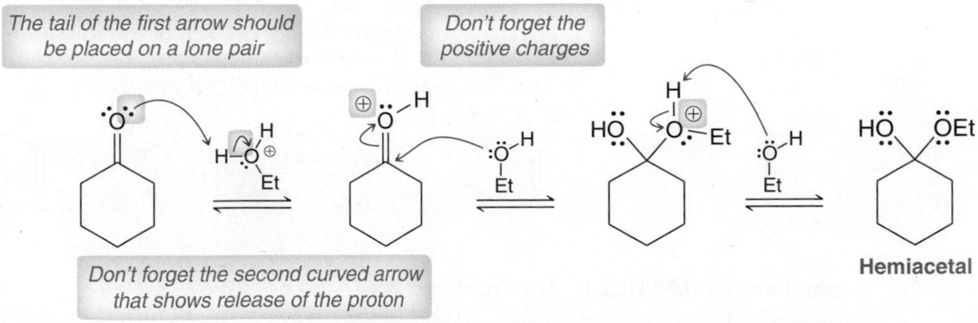

The tail of the first arrow should be placed on a lone pair
Don't forget the positive charges
Don't forget the second curved arrow that shows release of the proton
Hemiacetal

STEP 2
Draw the four steps necessary to convert the hemiacetal into an acetal.

Now let's focus on the last four steps of the mechanism, in which the hemiacetal is converted into an acetal:

Proton transfer → Loss of a leaving group → Nucleophilic attack → Proton transfer

Once again, this sequence of steps begins with a proton transfer and ends with a proton transfer. When drawing these four steps, make sure to draw the middle two steps separately, as discussed earlier. In addition, make sure to focus on proper arrow placement, and make sure to place all positive charges in their appropriate locations:

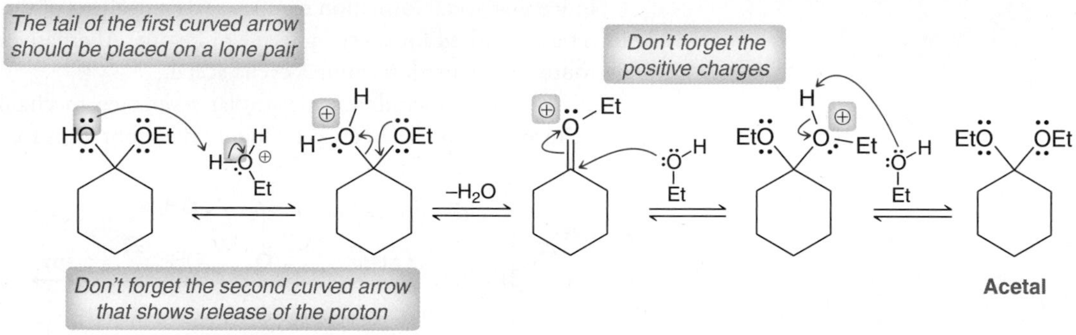

PRACTICE the skill

20.8 Draw a plausible mechanism for each of the following transformations:

(a)
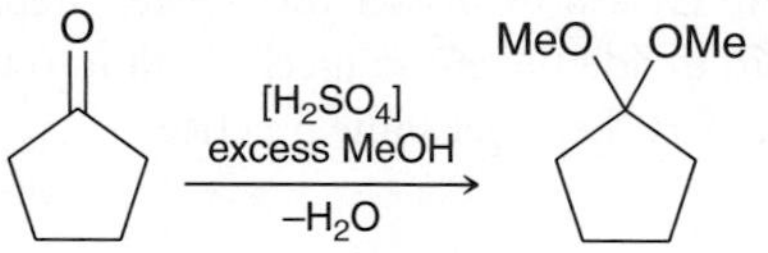

(b)
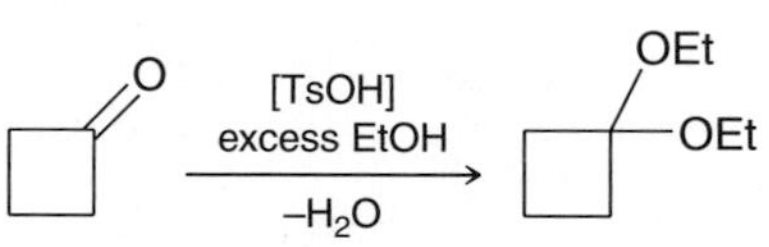

(c)
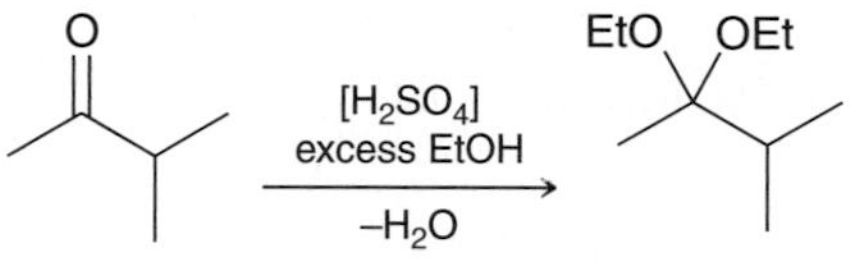

(d)
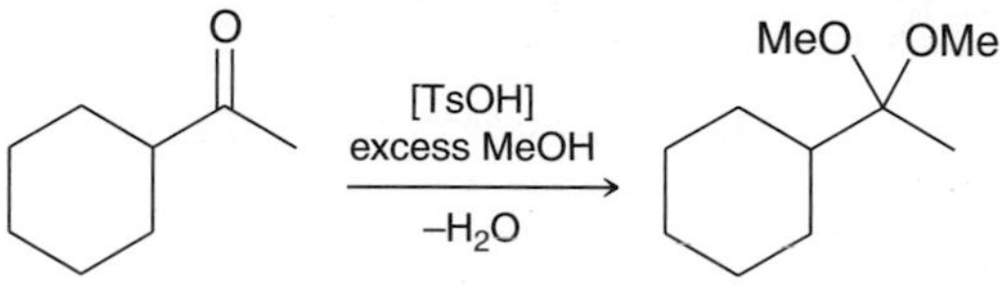

APPLY the skill

20.9 Draw a plausible mechanism for each of the following reactions:

(a)
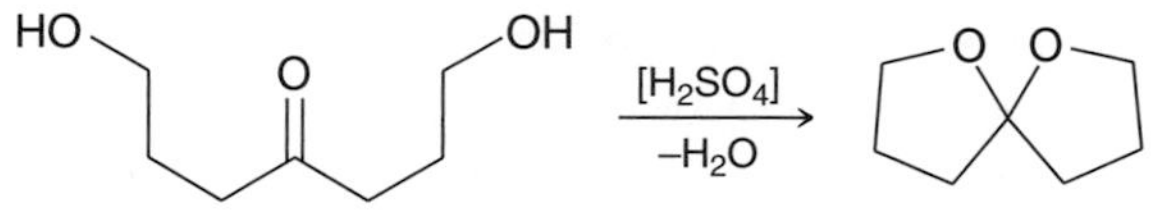

(b)
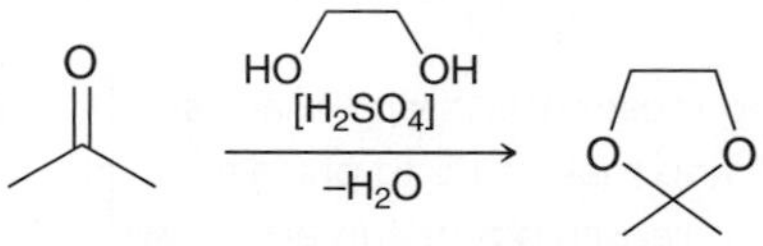

20.10 Predict the product of each of the following reactions:

(a)
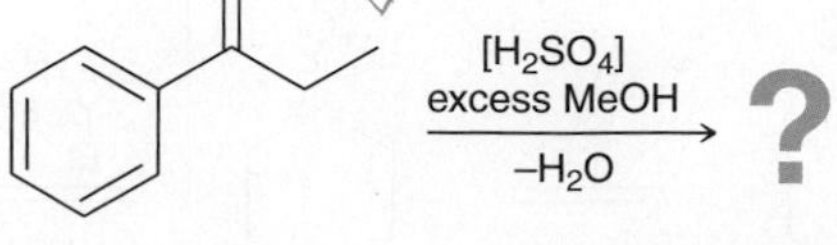

(b)
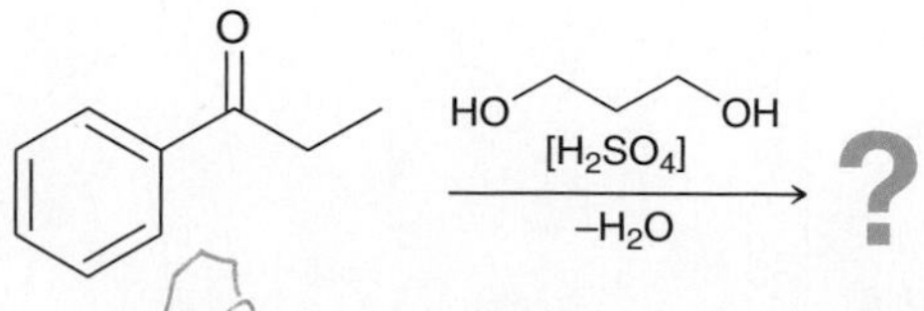

need more **PRACTICE?** **Try Problems 20.57, 20.62, 20.67**

Acetals as Protecting Groups

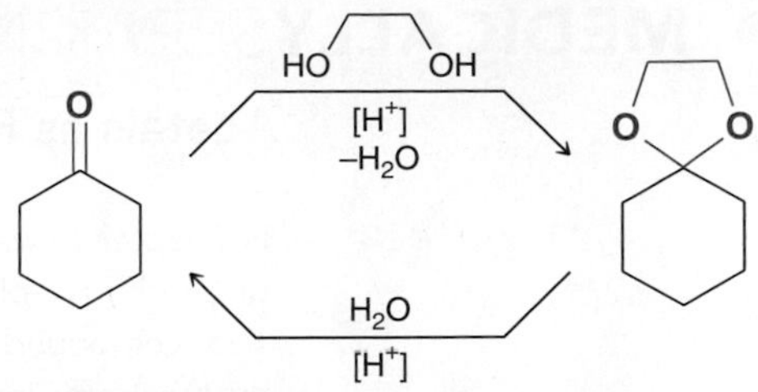

Acetal formation is a reversible process that can be controlled by carefully choosing reagents and conditions:

As mentioned in the previous section, acetal formation is favored by removal of water. To convert an acetal back into the corresponding aldehyde or ketone, it is simply treated with water in the presence of an acid catalyst. In this way, acetals can be used to protect ketones or aldehydes. For example, consider how the following transformation might be accomplished:

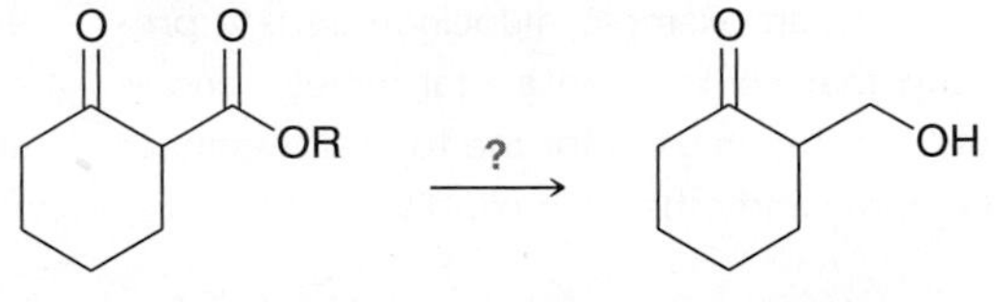

This transformation involves reduction of an ester to form an alcohol. Recall that lithium aluminum hydride (LAH) can be used to accomplish this type of reaction. However, under these conditions, the ketone moiety will also be reduced. The problem above requires reduction of the ester moiety without also reducing the ketone moiety. To accomplish this, a protecting group can be used. The first step is to convert the ketone into an acetal:

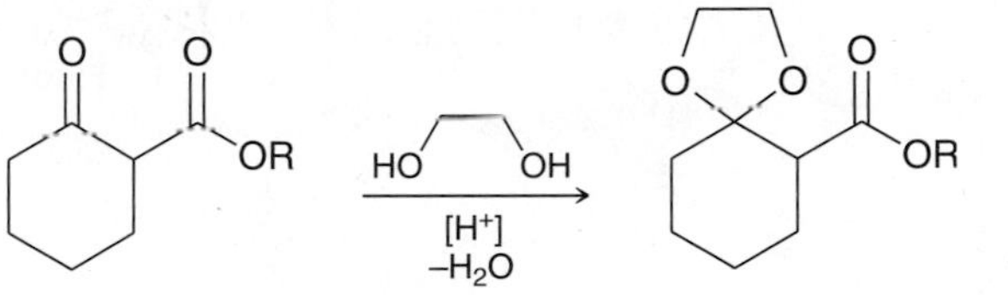

Notice that the ketone moiety is converted into an acetal, but the ester moiety is not. The resulting acetal group is stable under strongly basic conditions and will not react with LAH. This makes it possible to reduce only the ester, after which the acetal can be removed to regenerate the ketone: The three steps are summarized below:

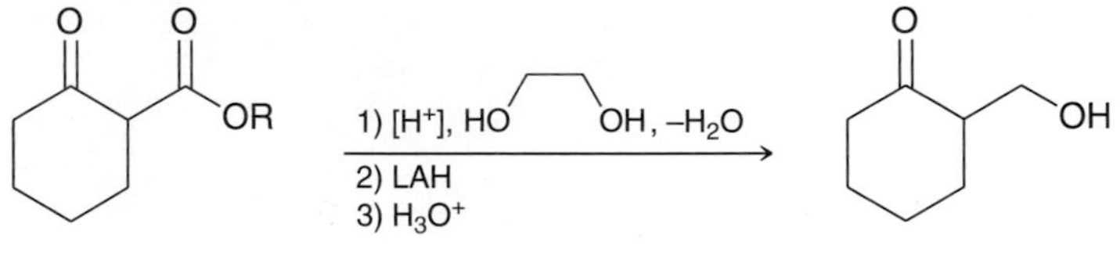

CONCEPTUAL CHECKPOINT

20.11 Propose an efficient synthesis for each of the following transformations:

(a) 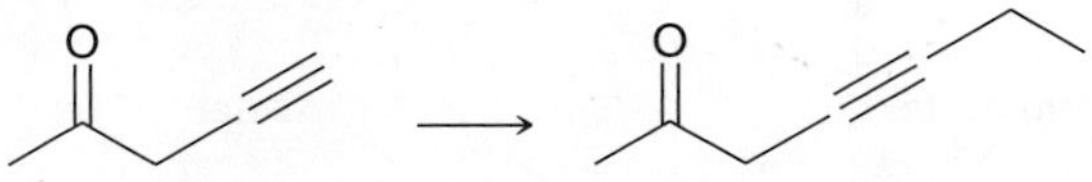

(c) 

(b) 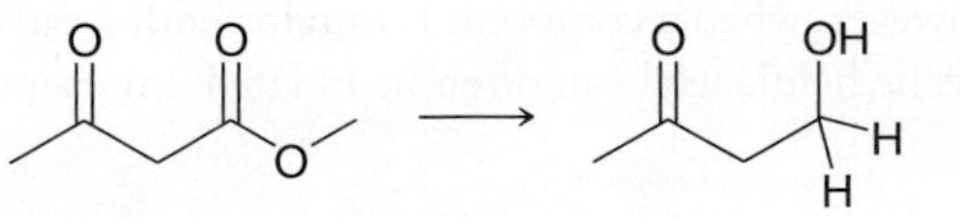

20.12 Predict the product(s) for each reaction below:

(a) 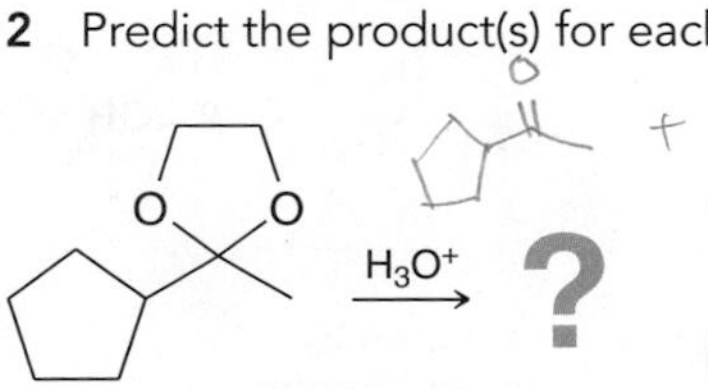

(b) 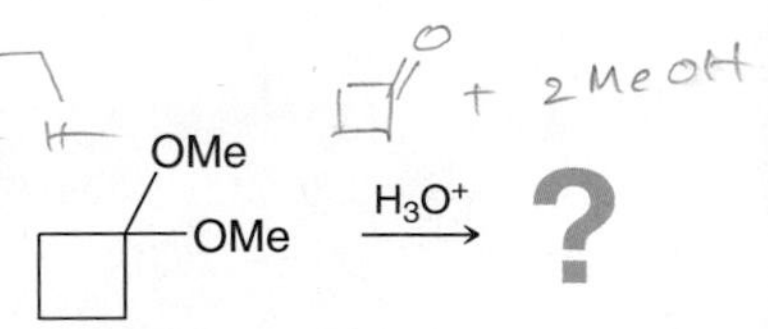

(c) 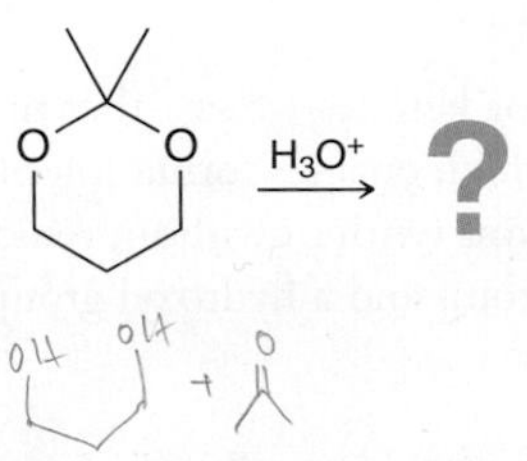

(d) 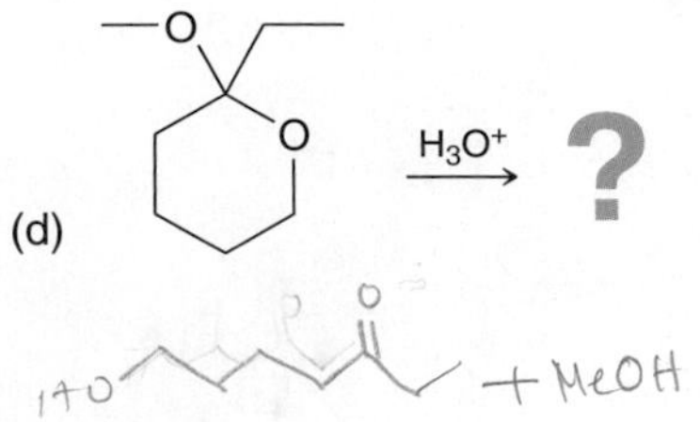

MEDICALLYSPEAKING)))

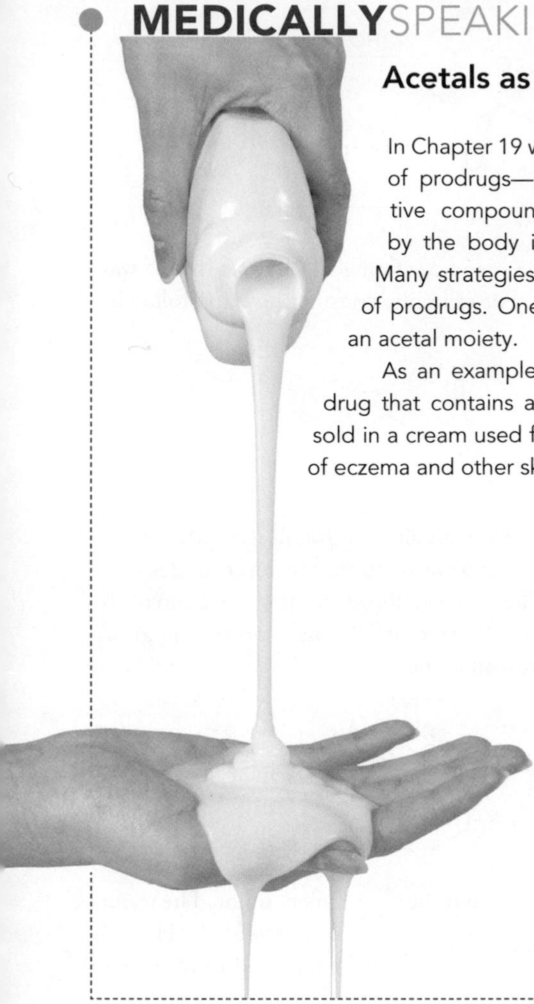

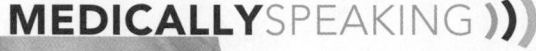

Acetals as Prodrugs

In Chapter 19 we explored the concept of prodrugs—pharmacologically inactive compounds that are converted by the body into active compounds. Many strategies are used in the design of prodrugs. One such strategy involves an acetal moiety.

As an example, fluocinonide is a prodrug that contains an acetal moiety, and is sold in a cream used for the topical treatment of eczema and other skin conditions.

Skin has several important functions, including preventing the absorption of foreign substances into the general circulation. This feature protects us from harmful substances, but it also prevents beneficial drugs from penetrating deep into the skin. This effect is most pronounced for drugs containing OH groups that can interact with binding sites on the skin's surface. To circumvent this problem, two OH groups can be temporarily converted into an acetal. The acetal prodrug is capable of penetrating the skin more deeply, because it lacks the OH groups that bind to the skin. Once the prodrug reaches its target, the acetal moiety is slowly hydrolyzed, thereby releasing the active drug:

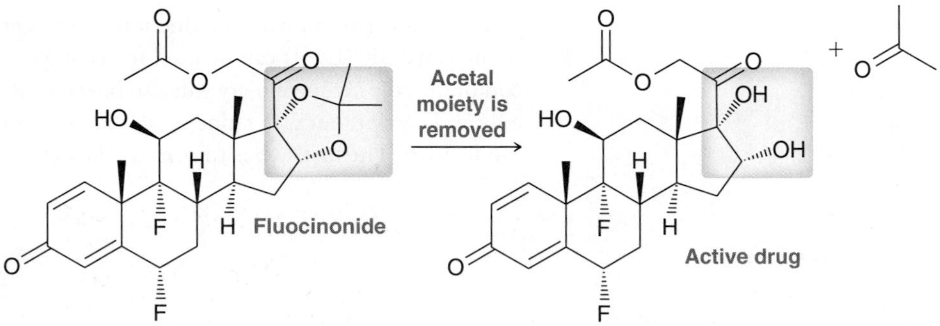

Treatment with fluocinonide is significantly more effective than direct treatment with the active drug, because the latter cannot reach all of the affected areas.

Stable Hemiacetals

In the previous section, we saw how to convert an aldehyde or ketone into an acetal. In most cases, it is very difficult to isolate the intermediate hemiacetal:

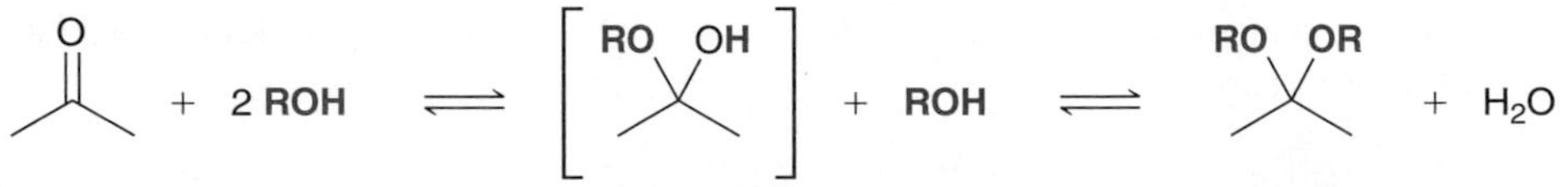

For ketones we saw that the equilibrium generally favors the reactants unless water is removed, which enables formation of the acetal. The hemiacetal is not favored under either set of conditions (with or without removal of water). However, when a compound contains both a carbonyl group and a hydroxyl group, the resulting cyclic hemiacetal can often be isolated; for example:

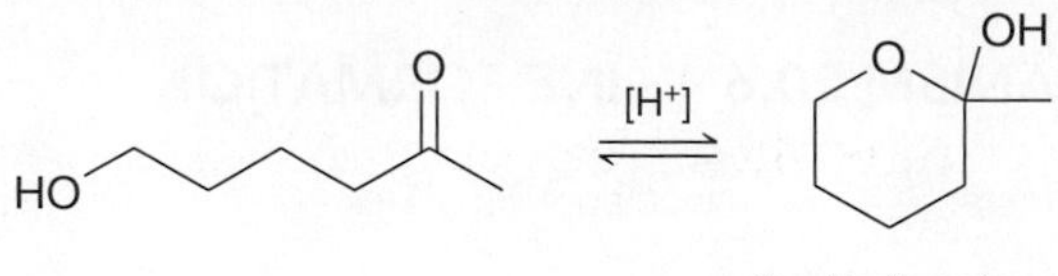

Cyclic hemiacetal

This will be important when we learn about carbohydrate chemistry in Chapter 24. Glucose, the major source of energy for the body, exists primarily as a cyclic hemiacetal:

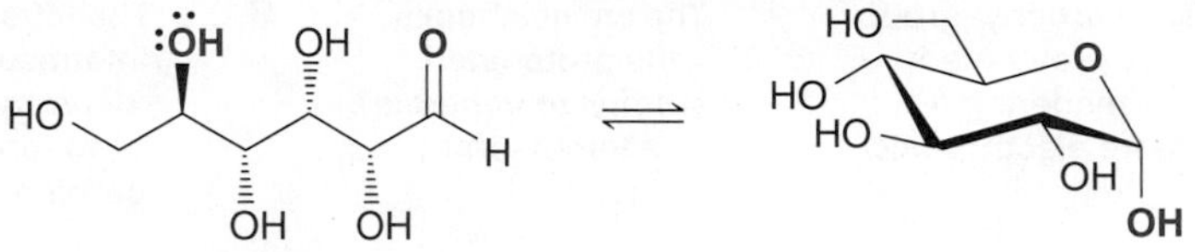

Glucose
(Open-chain)

Glucose
(Cyclic hemiacetal)

CONCEPTUAL CHECKPOINT

20.13 Draw a plausible mechanism for the following transformation:

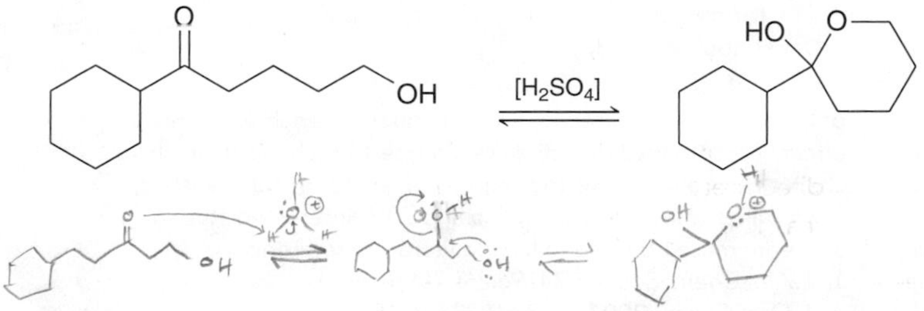

20.14 Compound A has molecular formula $C_8H_{14}O_2$. Upon treatment with catalytic acid, compound A is converted into the cyclic hemiacetal. Identify the structure of compound A.

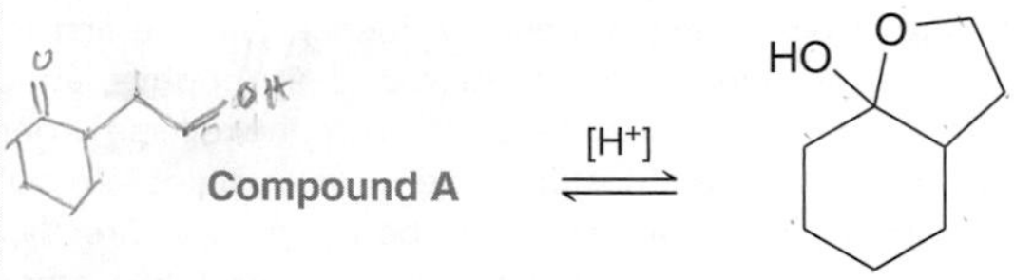

20.6 Nitrogen Nucleophiles

Primary Amines

In mildly acidic conditions, an aldehyde or ketone will react with a primary amine to form an **imine:**

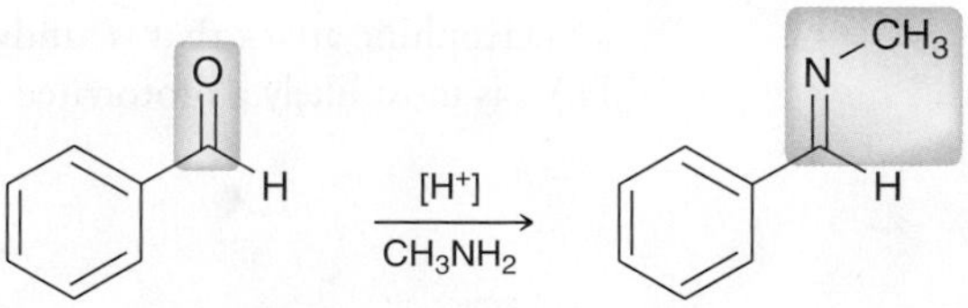

Imines are compounds that possess a C=N double bond and are common in biological pathways. Imines are also called Schiff bases, named after Hugo Schiff, a German chemist who first described their formation. A six-step mechanism for imine formation is shown in Mechanism 20.6. It is best to divide the mechanism conceptually into two parts (just as we did to conceptualize the mechanism of acetal formation): (1) The first three steps produce an intermediate called a **carbinolamine** and (2) the last three steps convert the carbinolamine into an imine:

MECHANISM 20.6 IMINE FORMATION

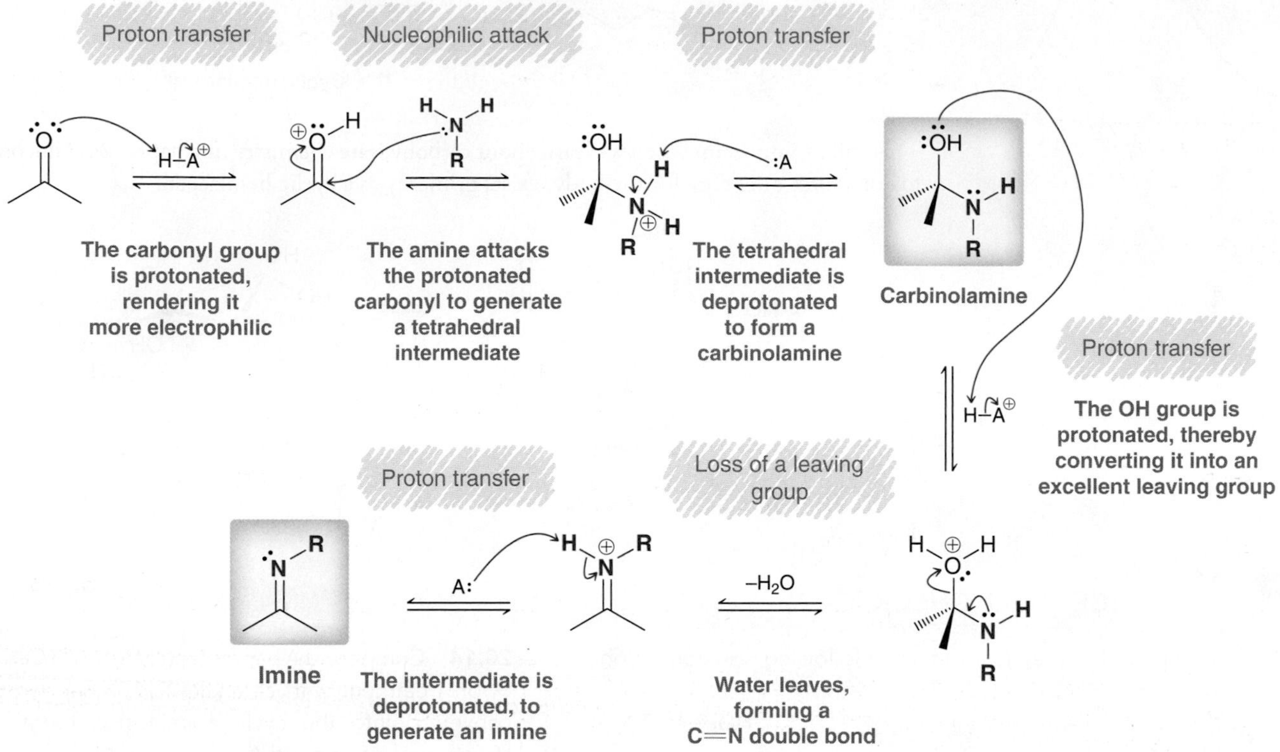
Proton transfer
Nucleophilic attack
Proton transfer
The carbonyl group is protonated, rendering it more electrophilic
The amine attacks the protonated carbonyl to generate a tetrahedral intermediate
The tetrahedral intermediate is deprotonated to form a carbinolamine
Carbinolamine
Proton transfer
The OH group is protonated, thereby converting it into an excellent leaving group
Loss of a leaving group
Water leaves, forming a C=N double bond
Proton transfer
The intermediate is deprotonated, to generate an imine
Imine
–H₂O

Note: There is experimental evidence that the first two steps of this mechanism (protonation and nucleophilic attack) more likely occur either simultaneously *or* in the reverse order of what is shown above. Most nitrogen nucleophiles are sufficiently nucleophilic to attack a carbonyl group directly, before protonation occurs. Nevertheless, the first two steps of the mechanism above have been drawn in the order shown (which only rarely occurs), because this sequence enables a more effective comparison of all acid-catalyzed mechanisms in this chapter and also unifies the rationale behind proton transfers, as we will discuss in Sections 20.6 and 20.7. Interested students can learn more from the following literature references:

1. *J. Am. Chem. Soc.*, **1974**, 96(26), 7998–09
2. *J. Org. Chem.*, **2007**, 72(22), 8202–8215

Proton transfer
Nucleophilic attack
Proton transfer

FIGURE 20.5
The sequence of steps involved in formation of a carbinolamine.

Let's begin our analysis of this mechanism by focusing on the first part: formation of the carbinolamine, which involves the three steps in Figure 20.5. Notice that these three steps are identical to the first three steps of acetal formation. Specifically, this sequence of steps involves a nucleophilic attack that is sandwiched between proton transfer steps. The identity of the acid, HA^+, is most likely a protonated amine, which received its extra proton from the acid source:

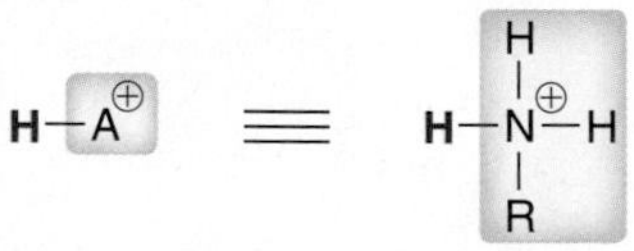

Once the carbinolamine has been formed, formation of the imine is accomplished with three steps (Figure 20.6). Notice again that this reaction sequence begins and ends with proton transfers.

Proton transfer
Loss of a leaving group
Proton transfer

FIGURE 20.6
The sequence of steps that convert a carbinolamine into an imine.

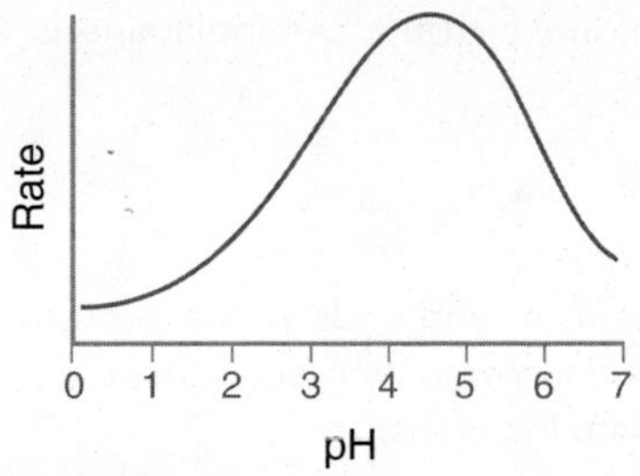
Rate
0 1 2 3 4 5 6 7
pH

FIGURE 20.7
The rate of imine formation as a function of pH.

The pH of the solution is an important consideration during imine formation, with the rate of reaction being greatest when the pH is around 4.5 (Figure 20.7). If the pH is too high (i.e., if no acid catalyst is used), the carbonyl group is not protonated (step 1 of the mechanism) and the carbinolamine is also not protonated (step 4 of the mechanism); so the reaction occurs more slowly. If the pH is too low (too much acid is used), most of the amine molecules will be protonated:

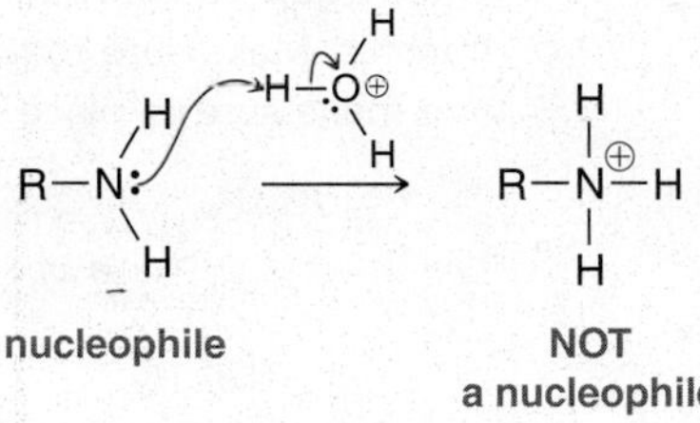
A nucleophile
NOT a nucleophile

Under these conditions, step 2 of the mechanism occurs too slowly. As a result, care must be taken to ensure optimal pH of the solution during imine formation.

SKILLBUILDER

20.3 DRAWING THE MECHANISM OF IMINE FORMATION

LEARN the skill

Draw a plausible mechanism for the following transformation:

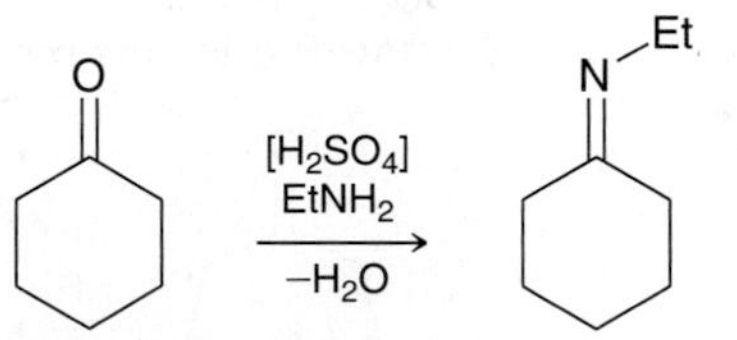
[H2SO4]
EtNH2
−H2O

SOLUTION

The reaction above is an example of imine formation. The mechanism can be divided into two parts: (1) formation of the carbinolamine and (2) formation of the imine.
Formation of the carbinolamine involves three mechanistic steps:

Proton transfer — Nucleophilic attack — Proton transfer

When drawing these three steps, make sure to place the head and tail of every curved arrow in its precise location, and make sure to place all positive charges in their appropriate locations. Notice that every step requires two curved arrows.

STEP 1
Draw the three steps necessary to form a carbinolamine.

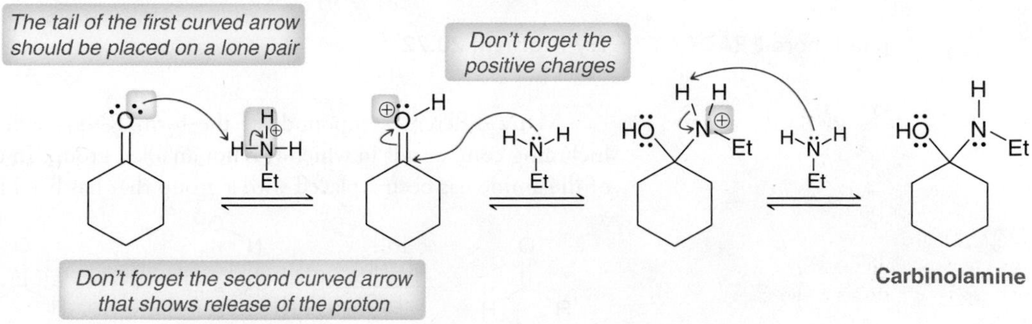
The tail of the first curved arrow should be placed on a lone pair
Don't forget the positive charges
Don't forget the second curved arrow that shows release of the proton
Carbinolamine

Now let's focus on the second part of the mechanism, in which the carbinolamine is converted into an imine. This requires three steps:

Proton transfer | Loss of a leaving group | Proton transfer

Once again, this sequence of steps begins with a proton transfer and ends with a proton transfer. Make sure to place the head and tail of every curved arrow in its precise location, and make sure to place all positive charges in their appropriate locations:

STEP 2
Draw the three steps necessary to convert the carbinolamine into an imine.

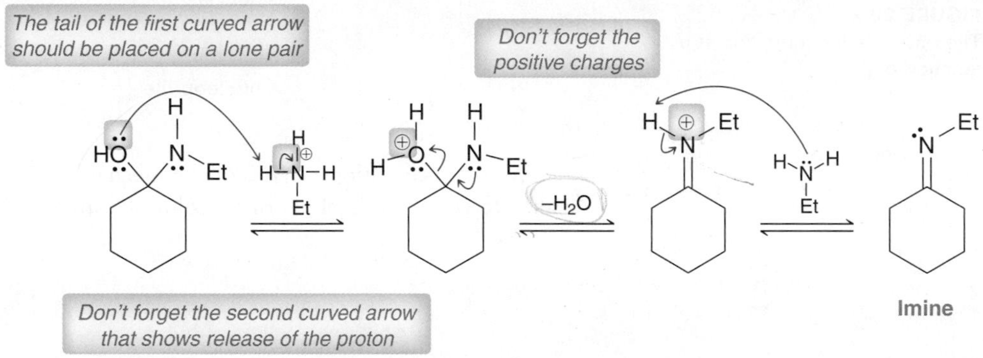

PRACTICE the skill **20.15** Draw a plausible mechanism for each of the following transformations:

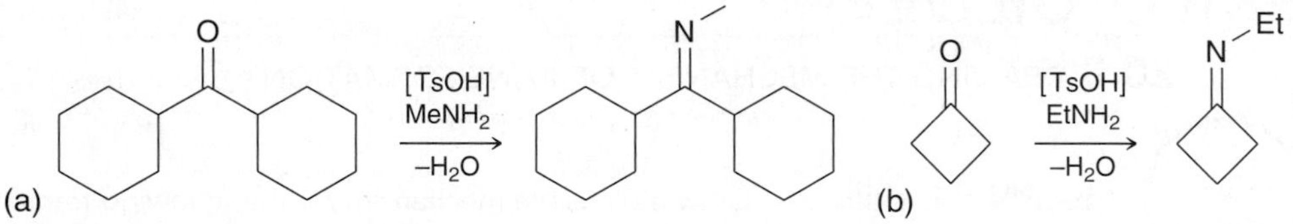

APPLY the skill

20.16 Predict the major product for each of the following reactions:

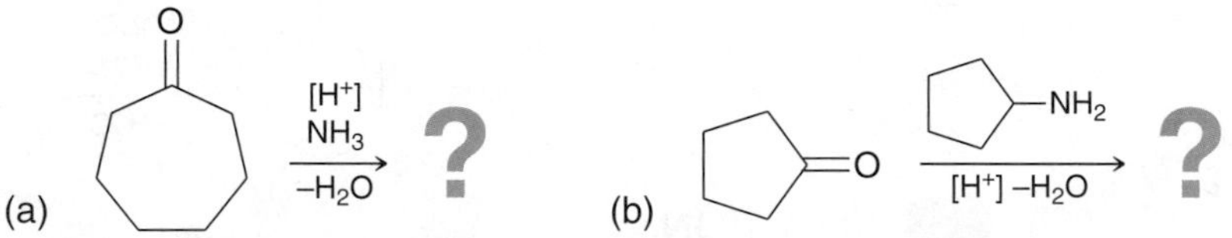

20.17 Predict the major product for each of the following intramolecular reactions:

20.18 Identify the reactants that you would use to make each of the following imines:

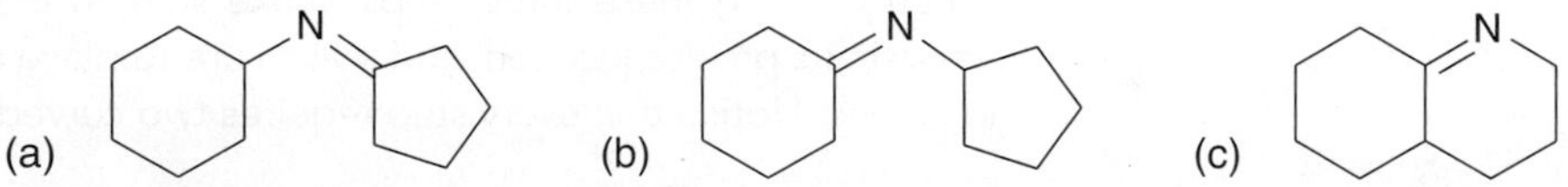

need more **PRACTICE?** **Try Problem 20.72**

Many different compounds of the form RNH_2 will react with aldehydes and ketones, including compounds in which R is not an alkyl group. In the following examples, the R group of the amine has been replaced with a group that has been highlighted in red:

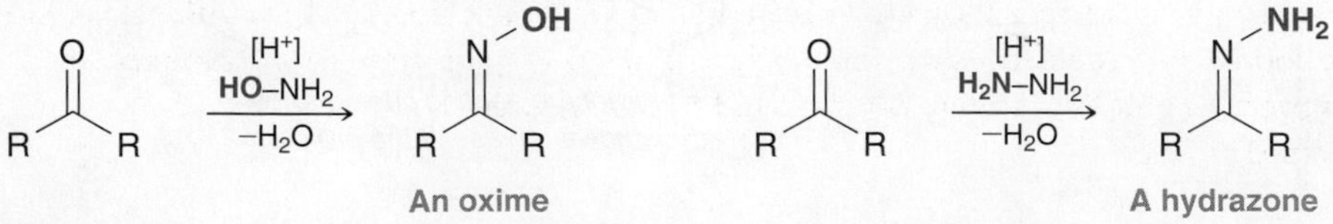

LOOKING AHEAD
Hydrazones are synthetically useful, as we will see in the discussion of the Wolff-Kishner reduction later in this chapter.

When hydroxylamine (NH_2OH) is used as a nucleophile, an **oxime** is formed. When hydrazine (NH_2NH_2) is used as a nucleophile, a **hydrazone** is formed. The mechanism for each of these reactions is directly analogous to the mechanism of imine formation.

PRACTICALLYSPEAKING

Beta-Carotene and Vision

Beta-carotene is a naturally occurring compound found in many orange-colored fruits and vegetables, including carrots, sweet potatoes, pumpkins, mangoes, cantaloupes, and apricots. As mentioned in the chapter opener, beta-carotene is known to be good for your eyes. To understand why, we must explore what happens to beta-carotene in your body. Imine formation plays an important role in the process.

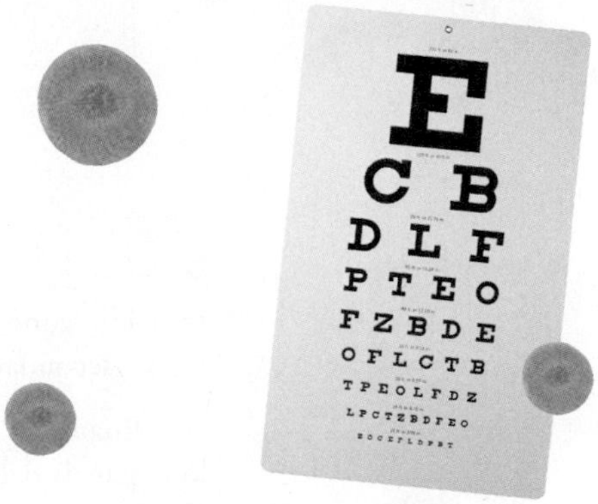

Beta-carotene is metabolized in the liver to produce vitamin A (also called retinol):

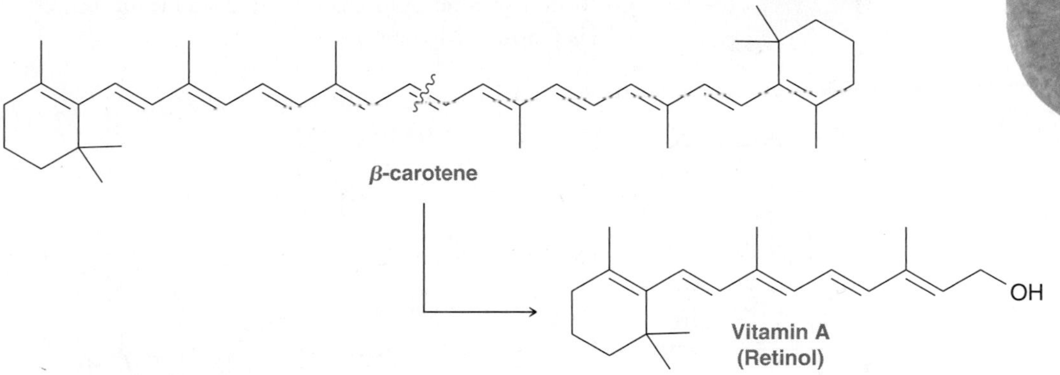

Vitamin A is then oxidized, and one of the double bonds undergoes isomerization to produce 11-*cis*-retinal:

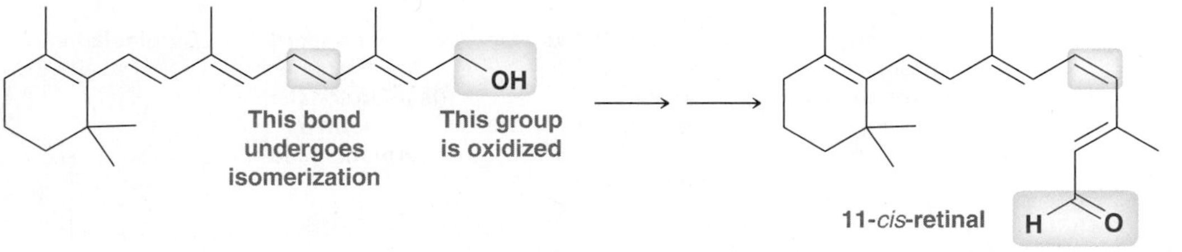

The resulting aldehyde then reacts with an amino group of a protein (called opsin) to produce rhodopsin, which possesses an imine moiety:

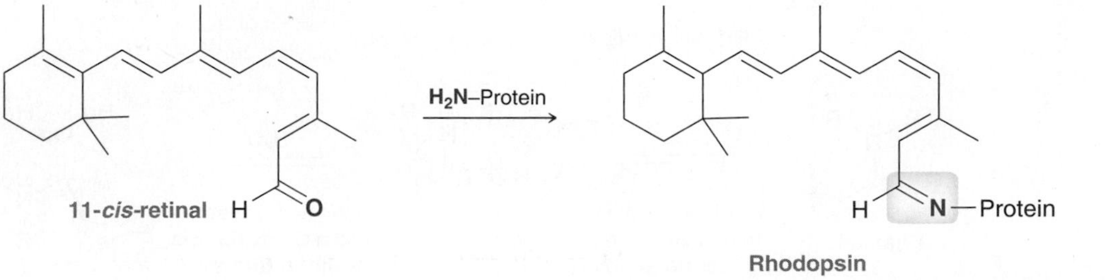

As described in Section 17.13, rhodopsin can absorb a photon of light, initiating a photoisomerization of the *cis* double bond to form a *trans* double bond. The resulting change in geometry triggers a signal that is ultimately detected by the brain and interpreted as vision.

A deficiency of vitamin A can lead to "night blindness," a condition that prevents the eyes from adjusting to dimly lit environments.

CONCEPTUAL CHECKPOINT

20.19 Predict the product of each of the following reactions:

(a) 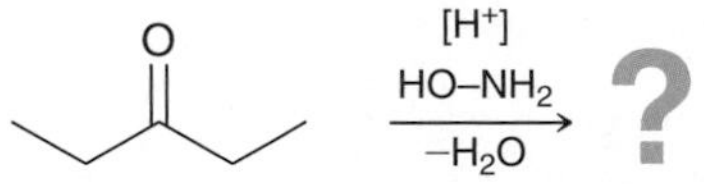

O
[H+]
HO–NH2
–H2O
?

(b) 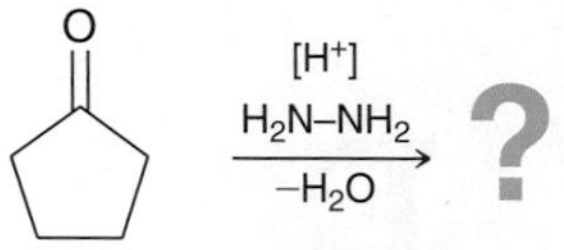

O
[H+]
H2N–NH2
–H2O
?

20.20 Identify the reactants that you would use to make each of the following compounds:

(a) 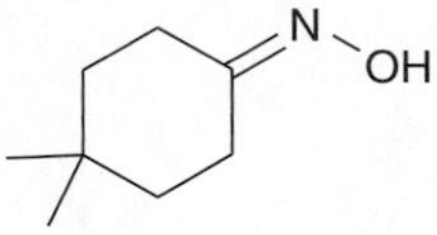

N
OH

(b) 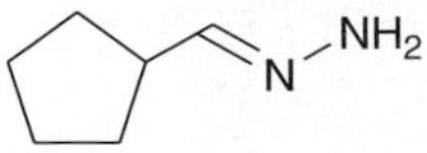

N
NH2

Secondary Amines

In acidic conditions, an aldehyde or ketone will react with a secondary amine to form an **enamine:**

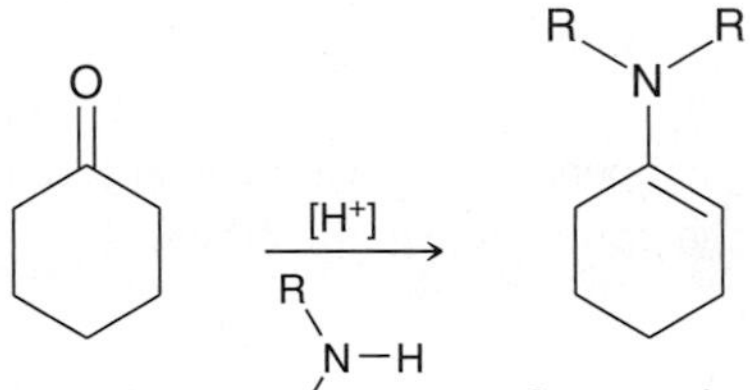

O
[H+]
R
N–H
R
R
N
R

An enamine

Enamines are compounds in which the nitrogen lone pair is delocalized by the presence of an adjacent C=C double bond. A mechanism for enamine formation is shown in Mechanism 20.7

MECHANISM 20.7 ENAMINE FORMATION

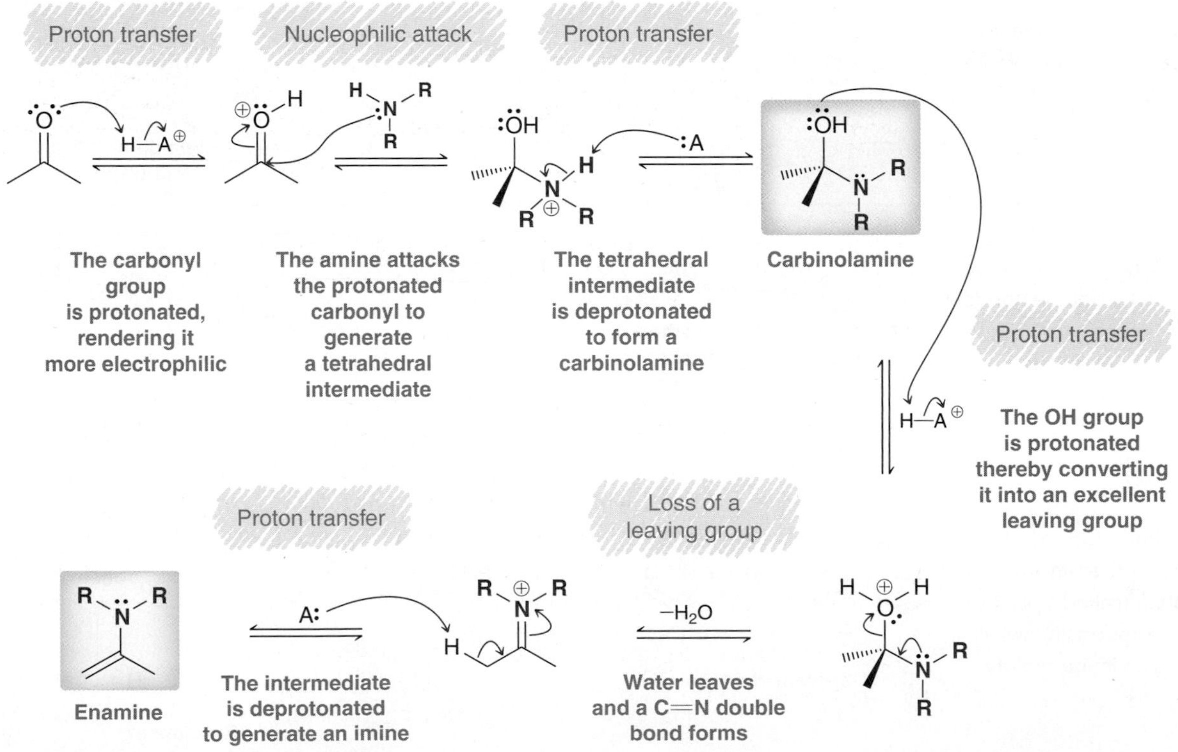

Proton transfer
Nucleophilic attack
Proton transfer
H–A⊕
:A
The carbonyl group is protonated, rendering it more electrophilic
The amine attacks the protonated carbonyl to generate a tetrahedral intermediate
The tetrahedral intermediate is deprotonated to form a carbinolamine
Carbinolamine
Proton transfer
H–A⊕
The OH group is protonated thereby converting it into an excellent leaving group
Loss of a leaving group
–H2O
Water leaves and a C=N double bond forms
Proton transfer
A:
The intermediate is deprotonated to generate an imine
Enamine

Note: There is experimental evidence that the first two steps of this mechanism (protonation and nucleophilic attack) more likely occur either simultaneously *or* in the reverse order of what is shown above. Most nitrogen nucleophiles are sufficiently nucleophilic to attack a carbonyl group directly, before protonation occurs. Nevertheless, the first two steps of the mechanism above have been drawn in the order shown (which only rarely occurs), because this sequence enables a more effective comparison of all acid-catalyzed mechanisms in this chapter and also unifies the rationale behind proton transfers, as we will discuss in Sections 20.6 and 20.7. Interested students can learn more from the following literature references:

1. *J. Am. Chem. Soc.*, **1974**, 96(26), 7998–09
2. *J. Org. Chem.*, **2007**, 72(22), 8202–8215

This mechanism of enamine formation is identical to the mechanism of imine formation except for the last step:

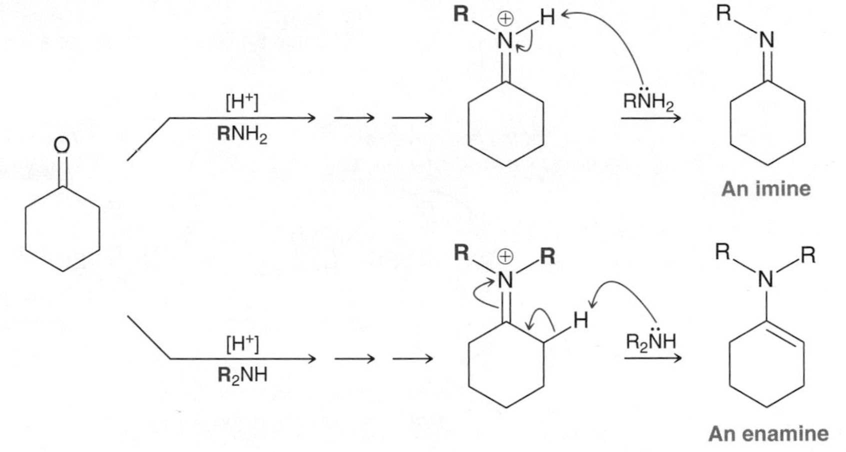
[H⁺]
RNH_2
$R\ddot{N}H_2$
[H⁺]
R_2NH
$R_2\ddot{N}H$
An imine
An enamine

The difference in the iminium ions explains the different outcomes for the two reactions. During imine formation, the nitrogen atom of the iminium ion possesses a proton that can be removed as the final step of the mechanism. In contrast, during enamine formation, the nitrogen atom of the iminium ion does not possess a proton. As a result, elimination from the adjacent carbon is necessary in order to yield a neutral species.

SKILLBUILDER

20.4 DRAWING THE MECHANISM OF ENAMINE FORMATION

LEARN the skill

Draw a plausible mechanism for the following reaction:

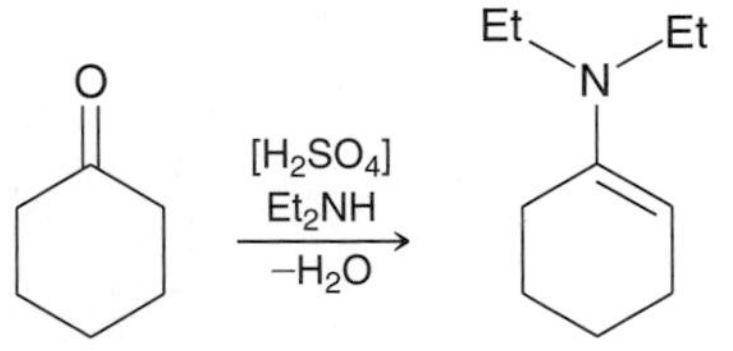
$[H_2SO_4]$
Et_2NH
$-H_2O$

SOLUTION

The reaction above is an example of enamine formation. The mechanism can be divided into two parts: (1) formation of the carbinolamine and (2) formation of the enamine.

Formation of the carbinolamine involves three mechanistic steps:

Proton transfer | Nucleophilic attack | Proton transfer

When drawing these three steps, make sure to place the head and tail of every curved arrow in its precise location, and make sure to place all positive charges in their appropriate locations. Notice that every step requires two curved arrows.

STEP 1
Draw the three steps necessary to form a carbinolamine.

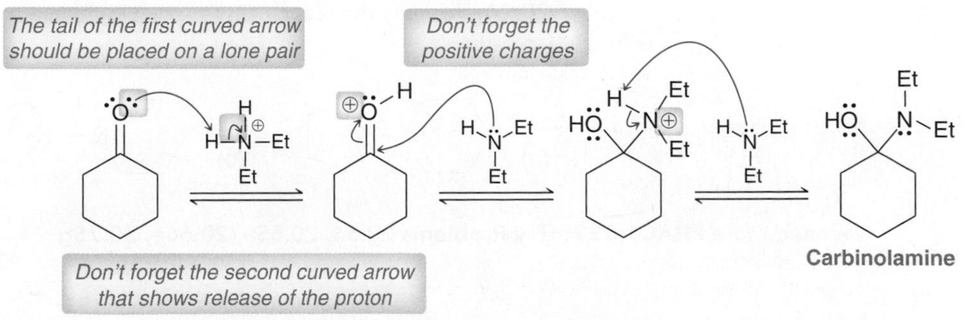
The tail of the first curved arrow should be placed on a lone pair
Don't forget the positive charges
Don't forget the second curved arrow that shows release of the proton
Carbinolamine

In the second part of the mechanism, the carbinolamine is converted into an enamine via a three-step process. Once again, this sequence begins and ends with proton transfers:

Proton transfer | Loss of a leaving group | Proton transfer

STEP 2
Draw the three steps necessary to convert the carbinolamine into an enamine.

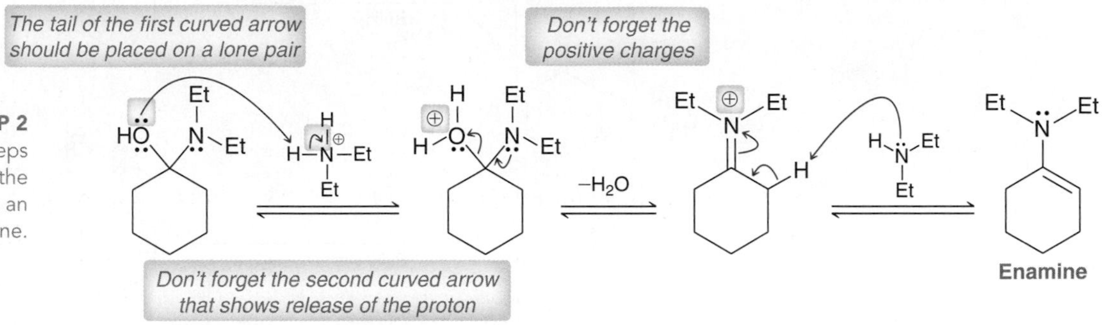

PRACTICE the skill

20.21 Draw a plausible mechanism for each of the following reactions:

(a) 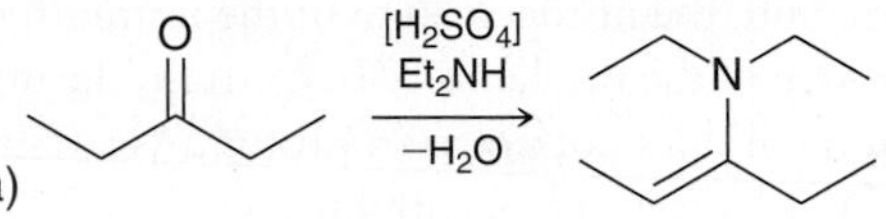

(b) 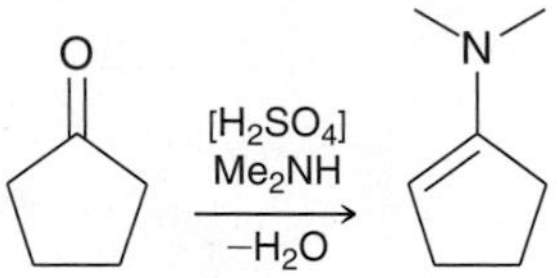

APPLY the skill

20.22 Predict the major product for each of the following reactions:

(a) 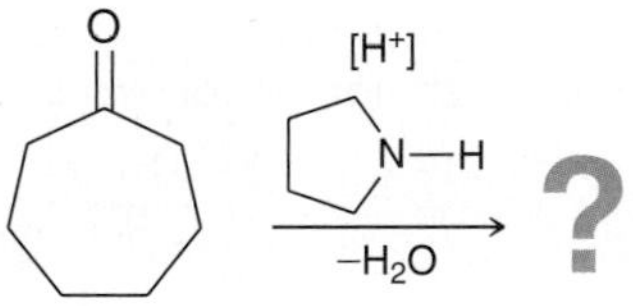

(b) 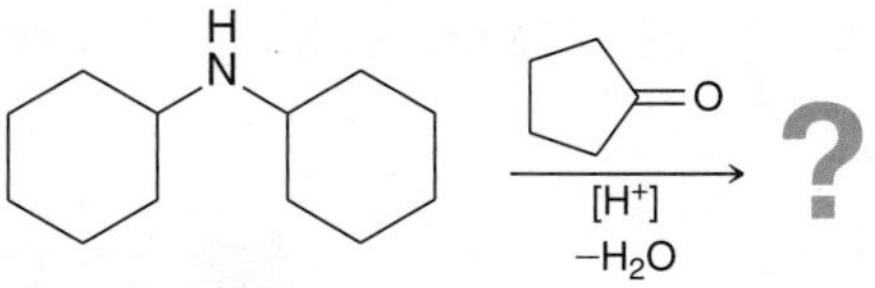

20.23 Predict the major product for each of the following intramolecular reactions:

(a) 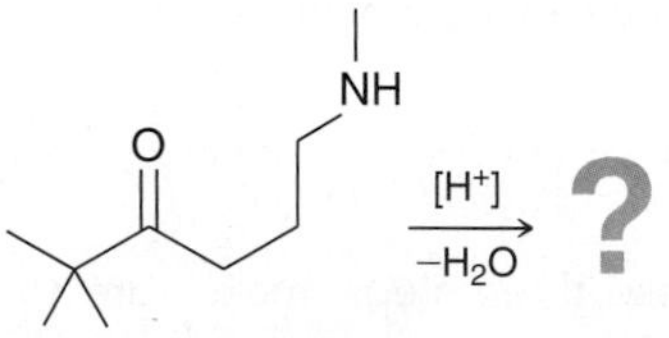

(b) 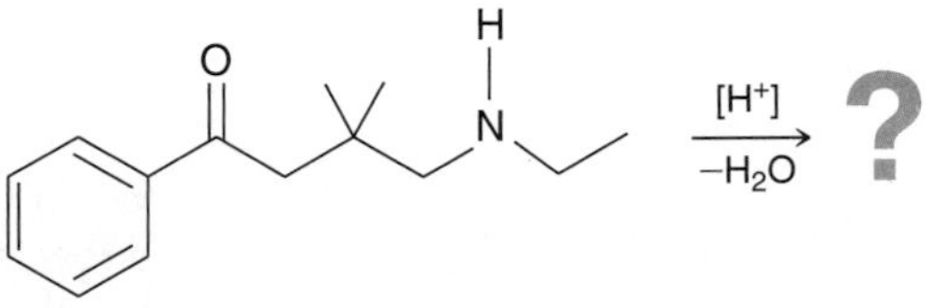

20.24 Identify the reactants that you would use to make each of the following enamines:

(a) 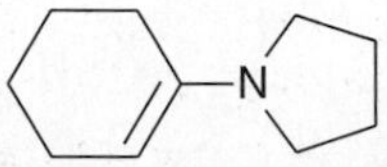 (b) 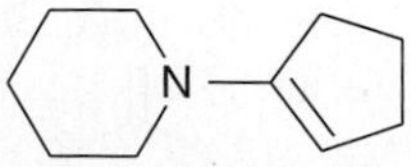 (c) 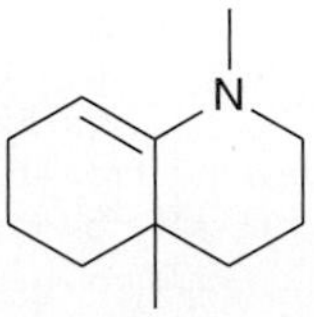

need more **PRACTICE?** **Try Problems 20.64, 20.65a, 20.66e, 20.75g**

Wolff-Kishner Reduction

At the end of the previous section, we noted that ketones can be converted into hydrazones. This transformation has practical utility, because hydrazones are readily reduced under strongly basic conditions:

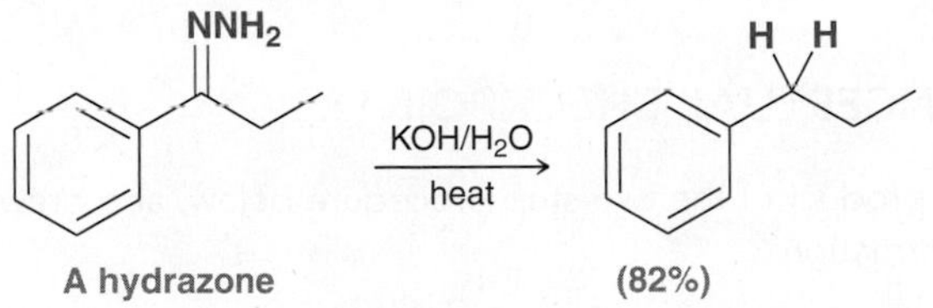
NNH2
KOH/H2O
heat
H H
A hydrazone
(82%)

This transformation is called the **Wolff-Kishner reduction,** named after the German chemist Ludwig Wolff (University of Jena) and the Russian chemist N. M. Kishner (University of Moscow). This provides a two-step procedure for reducing a ketone to an alkane:

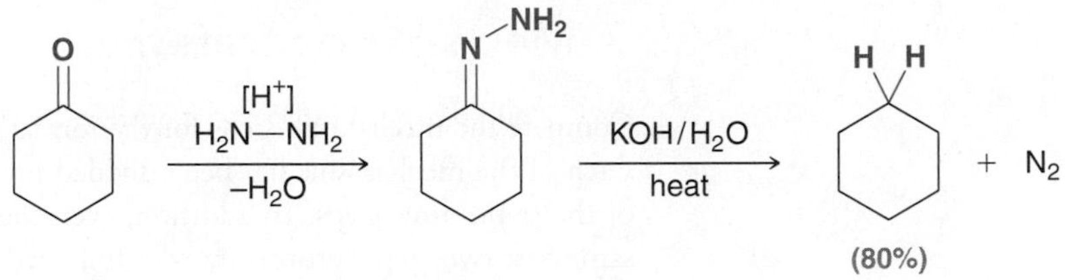
O
[H+]
H2N—NH2
−H2O
NH2
N
KOH/H2O
heat
H H
+ N2
(80%)

The second part of the Wolff-Kishner reduction is believed to proceed via Mechanism 20.8.

MECHANISM 20.8 THE WOLFF-KISHNER REDUCTION

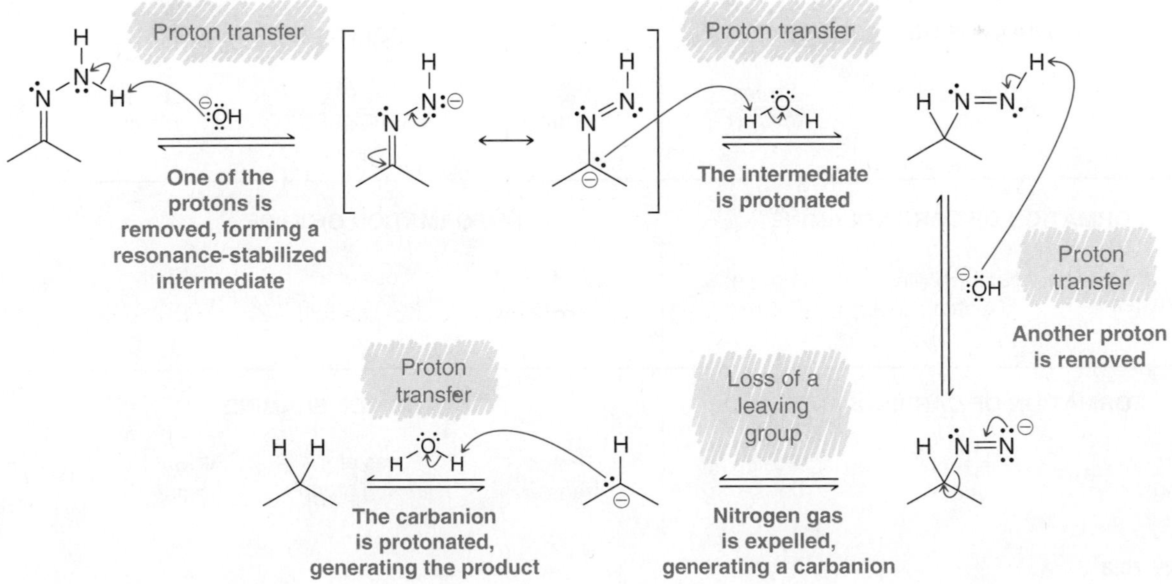
Proton transfer
One of the protons is removed, forming a resonance-stabilized intermediate
Proton transfer
The intermediate is protonated
Proton transfer
Another proton is removed
Loss of a leaving group
Nitrogen gas is expelled, generating a carbanion
Proton transfer
The carbanion is protonated, generating the product

Notice that four of the five steps in the mechanism are proton transfers, the exception being the loss of N_2 gas to generate a carbanion. This step warrants special attention, because formation of a carbanion in a solution of aqueous hydroxide is thermodynamically unfavorable (*significantly* uphill in energy). Why, then, does this step occur? It is true that the equilibrium for this step greatly disfavors formation of the carbanion, and therefore, only a very small number of molecules will initially lose N_2 to form the carbanion. However, the resulting N_2 gas then bubbles out of the reaction mixture, and the equilibrium is adjusted to form more nitrogen gas, which again leaves the reaction mixture. The evolution of nitrogen gas ultimately renders this step irreversible and forces the reaction to completion. As a result, the yields for this process are generally very good.

CONCEPTUAL CHECKPOINT

20.25 Predict the product of the two-step procedure below, and draw a mechanism for its formation:

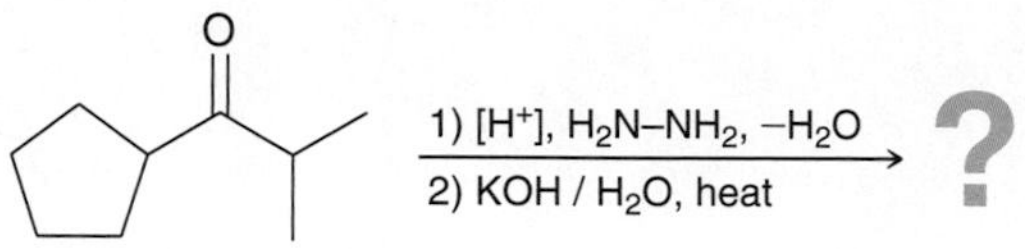

20.7 Mechanism Strategies

Compare the mechanistic steps for the formation of acetals, imines, and enamines (Figure 20.8). Each of the mechanisms has been divided into two parts, and in all cases, the first part consists of the same three steps. In addition, even the second part of each mechanism begins with the same first two steps (proton transfer followed by loss of a leaving group). In other words, these three mechanisms are identical until the fifth step, in which water is lost (loss of a leaving group), shown in blue in Figure 20.8. Rather than viewing these reactions as three separate, unrelated reactions, it is best to view them as nearly identical with different endings. Acetal formation has one additional nucleophile attack, giving a total of seven steps. In contrast, imine formation and enamine formation do not exhibit a nucleophilic attack during the second part of the mechanism, giving a total of only six steps.

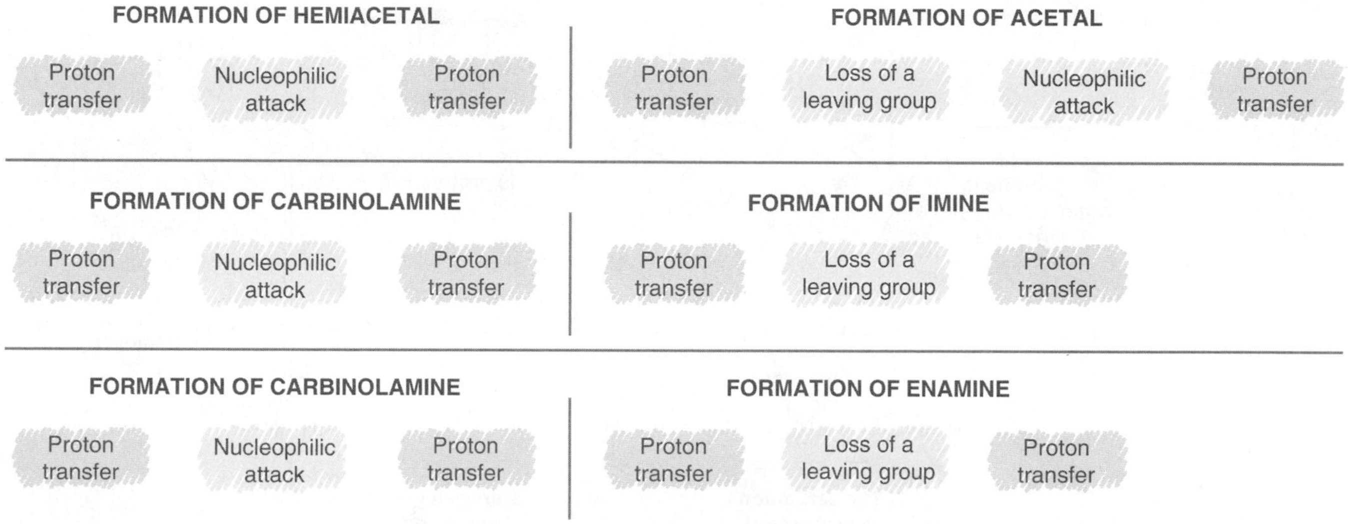

FIGURE 20.8
A comparison of the sequence of steps for acetal, imine, and enamine formation.

In each of these mechanisms, there are four proton transfer steps. In order to draw the mechanism correctly, it is critical to draw these proton transfers properly. To do so, it will be helpful to remember the following rules that dictate when and why proton transfers occur in acid-catalyzed conditions:

- The carbonyl group should be protonated before it is attacked. This generates a more powerful electrophile, and it avoids formation of a negative charge that would occur if the nucleophile attacked the carbonyl directly.

- Avoid formation of two positive charges on a single intermediate. This type of intermediate will generally be too high in energy to form.
- The leaving group should become neutral when it leaves. Do not expel hydroxide as a leaving group; rather, it should first be protonated so that it can leave as water.
- At the end of the mechanism, a proton transfer is used to form a neutral product.

The four rules above correspond with each of the four proton transfers, respectively. These four rules can be consolidated into one master rule that defines when and why proton transfer steps are utilized: *In acidic conditions, all reagents, intermediates, and leaving groups should either be neutral (no charge) or bear one positive charge.* All of the proton transfers in the mechanism occur in order to fulfill this requirement.

Acetals, imines, and enamines can be converted back into ketones by treatment with excess water under acid-catalyzed conditions:

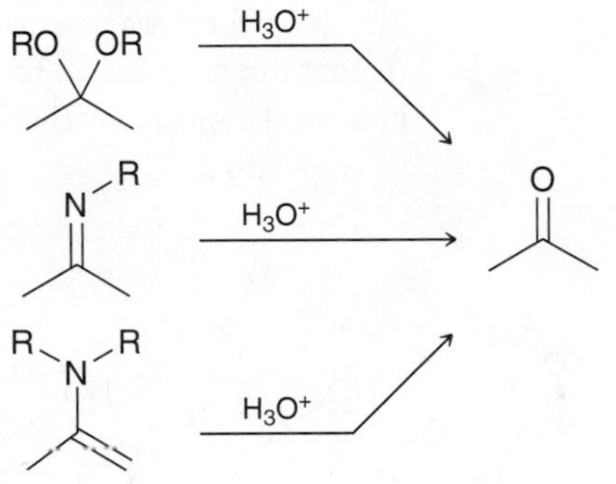

Each of the reactions above is called a **hydrolysis** reaction, because in each case bonds are cleaved by treatment with water. These three reactions are essentially the reverse of the reactions we have seen. The following procedure should be used in drawing the mechanisms for the above hydrolysis reactions:

1. Begin by drawing all of the intermediates without any curved arrows. For example, suppose that we want to draw the mechanism for hydrolysis of an acetal to form a ketone:

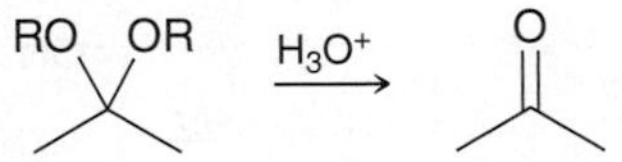

We did not learn the mechanism for this reaction; however, we did learn the mechanism for acetal formation. Think about the first intermediate in acetal formation (a protonated carbonyl), and then draw that intermediate as the last intermediate of the hydrolysis reaction:

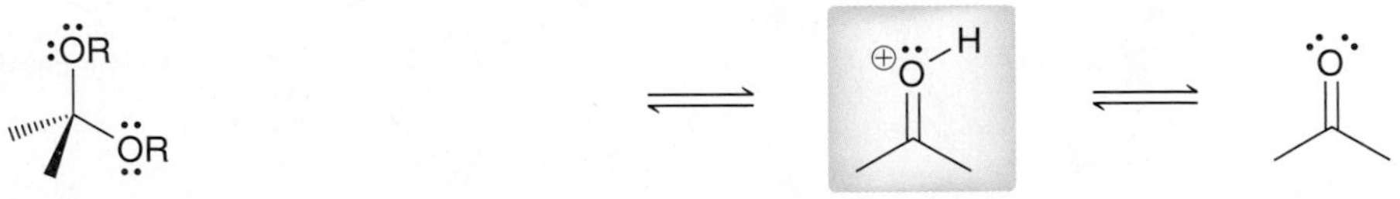

Continue working backward until you have drawn all of the intermediates.

2. Then, working forward, draw the curved arrows that are necessary to transform each intermediate into the next intermediate. At each stage, make sure you are following the master rule for proton transfers in acid-catalyzed conditions.

The skill of being able to draw the reverse of a known mechanism is incredibly important and will be used again for other reactions in the remaining chapters of this book. The following example illustrates this procedure.

SKILLBUILDER

20.5 DRAWING THE MECHANISM OF A HYDROLYSIS REACTION

LEARN the skill

Propose a mechanism for the following transformation:

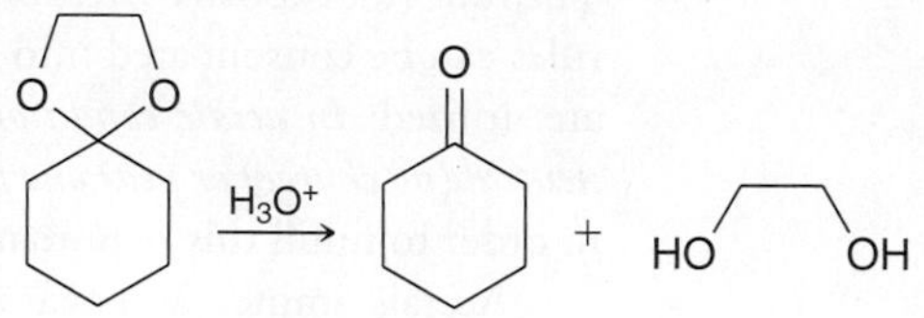

SOLUTION

This is a hydrolysis reaction in which a cyclic acetal is opened to form a ketone. We therefore expect the mechanism to be the reverse of acetal formation. Begin by considering all of the intermediates involved in acetal formation:

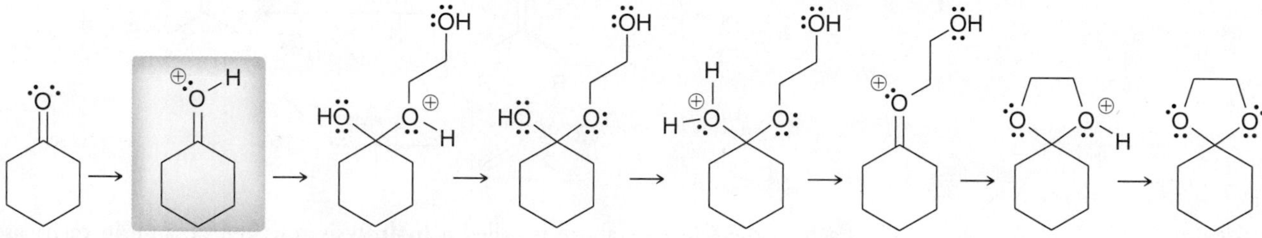

Ketone | First intermediate | Hemiacetal | Acetal

STEP 1
Draw all intermediates for acetal formation in reverse order.

We simply draw all of these intermediates in reverse order so that the first intermediate above (highlighted) becomes the last intermediate of the hydrolysis mechanism:

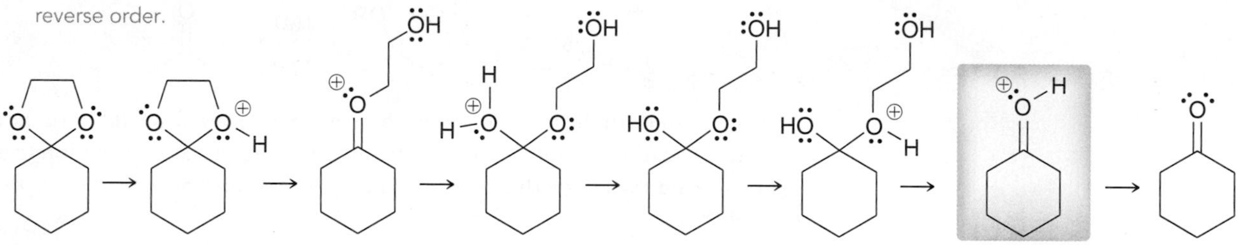

Acetal | Hemiacetal | Last intermediate | Ketone

Hydrolysis of the acetal must involve these intermediates, in the order shown above. If any of the intermediates has a negative charge, then you have made a mistake.

With the intermediates placed in the correct order, the final step is to draw the reagents and curved arrows that show how each intermediate is transformed into the next intermediate. Begin with the acetal, and work forward until reaching the ketone. This requires that your arrow-pushing skills are in good shape.

Make sure to use only the reagents that are provided, and obey the master rule for proton transfers. For example, this problem indicates that H_3O^+ is available. This means that H_3O^+ should be used for protonating, and H_2O should be used for deprotonating. Do not use hydroxide ions, as they are not present in sufficient quantity under acid-catalyzed conditions. Application of these rules gives the following answer:

LOOKING BACK
If you need to polish your arrow-pushing skills, go to Section 6.8.

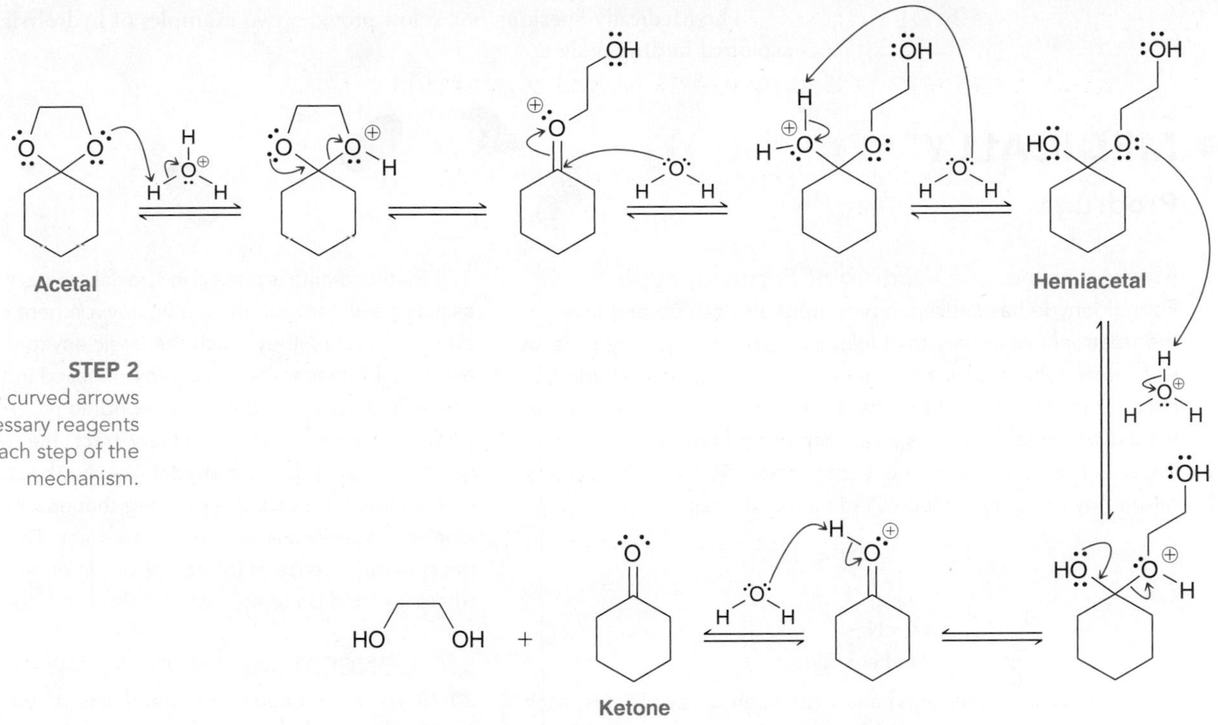

STEP 2
Draw the curved arrows and necessary reagents for each step of the mechanism.

Notice the use of equilibrium arrows, because the process is governed by an equilibrium, as noted in previous sections.

PRACTICE the skill **20.26** Propose a plausible mechanism for each of the following hydrolysis reactions:

(a) 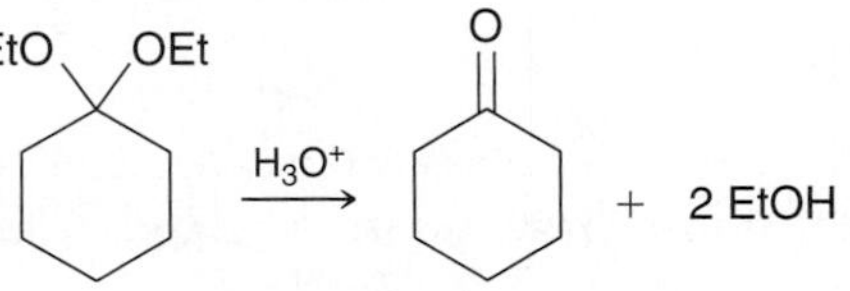

(b) 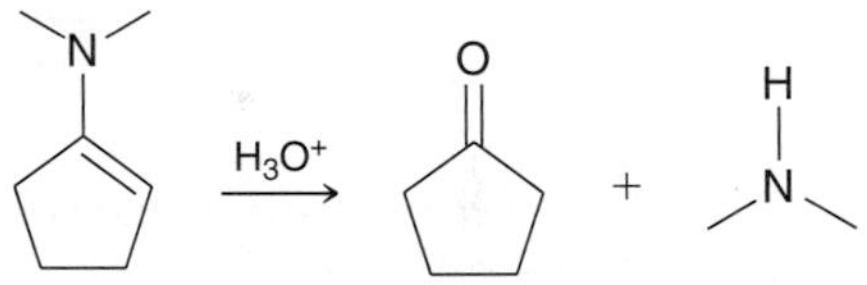

(c) 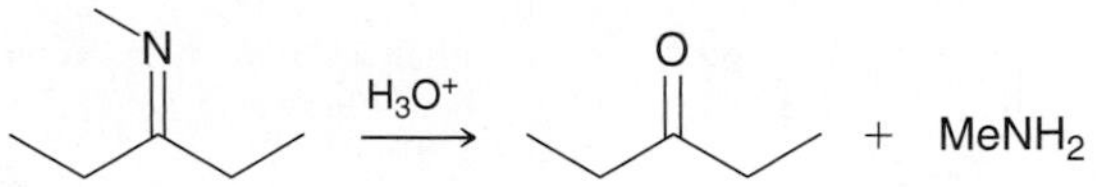

(d) 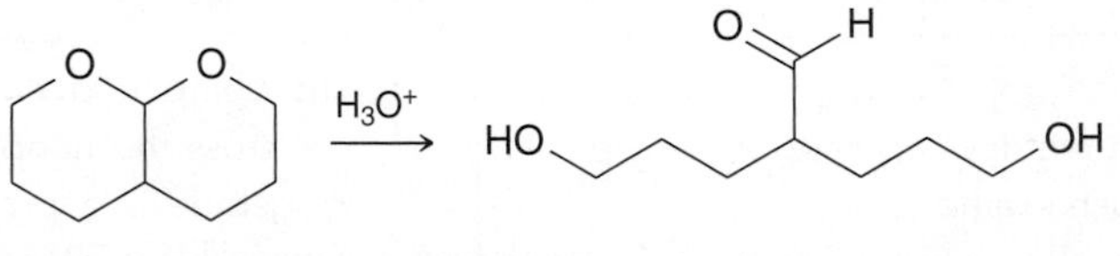

APPLY the skill **20.27** Propose a plausible mechanism for the reaction below:

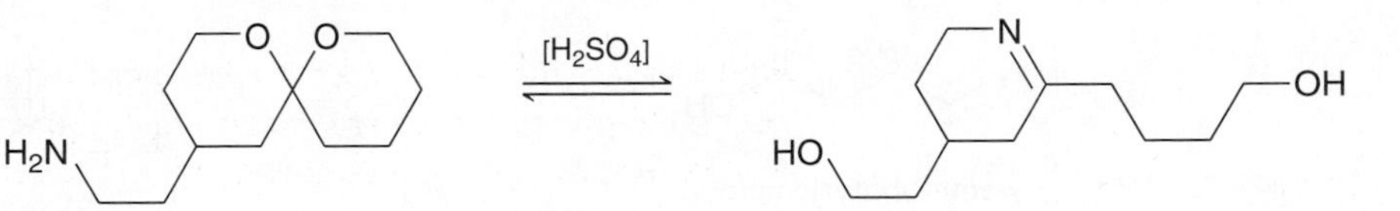

Hint: It is not necessary to open the acetal into a ketone (7 steps) followed by closing the imine (6 steps). A plausible mechanism can be drawn with fewer than 13 steps. Start working with the acetal (using the steps necessary to open an acetal), and look for an intermediate that can be attacked by the amino group. Make sure to avoid drawing an S_N2 process at a sterically hindered substrate!

need more **PRACTICE?** **Try Problem 20.65**

The Medically Speaking box below provides two examples of hydrolysis reactions that are exploited in drug design.

MEDICALLYSPEAKING

Prodrugs

Methenamine as a Prodrug of Formaldehyde

Formaldehyde has antiseptic properties and can be employed in the treatment of urinary tract infections due to its ability to react with nucleophiles present in urine. However, formaldehyde can be toxic when exposed to other regions of the body. Therefore, the use of formaldehyde as an antiseptic agent requires a method for selective delivery to the urinary tract. This can be accomplished by using a prodrug called methenamine:

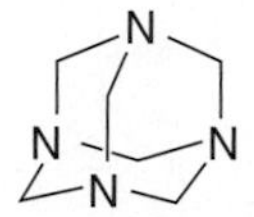

Methenamine

This compound is a nitrogen analogue of an acetal. That is, each carbon atom is connected to two nitrogen atoms, very much like an acetal in which a carbon atom is connected to two oxygen atoms. A carbon atom that is connected to two heteroatoms (O or N) can undergo acid-catalyzed hydrolysis:

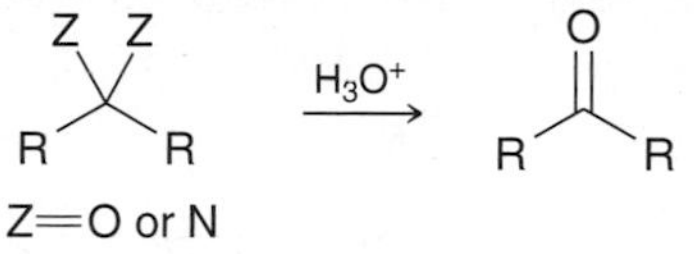

Each of the carbon atoms in methenamine can be hydrolyzed, releasing formaldehyde:

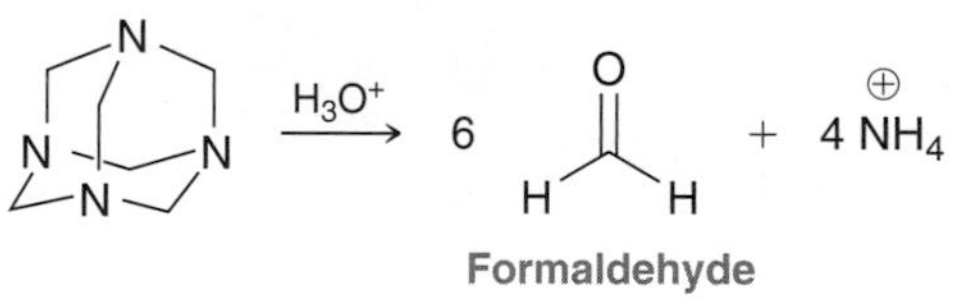

Formaldehyde

Methenamine is placed in special tablets that do not dissolve as they travel through the acidic environment of the stomach but do dissolve once they reach the basic environment of the intestinal tract. Methenamine is thereby released in the intestinal tract, where it is stable under basic conditions. Once it reaches the acidic environment of the urinary tract, methenamine is hydrolyzed, releasing formaldehyde, as shown above. In this way, methenamine is used as a prodrug that enables delivery of formaldehyde specifically to the urinary tract. This method prevents the systemic release of formaldehyde in other organs of the body where it would be toxic.

CONCEPTUAL CHECKPOINT

20.28 As shown above, methenamine is hydrolyzed in aqueous acid to produce formaldehyde and ammonia. Draw a mechanism showing formation of one molecule of formaldehyde (the remaining five molecules of formaldehyde are each released via a similar sequence of steps). The release of each molecule of formaldehyde is directly analogous to the hydrolysis of an acetal. To get you started, the first two steps are provided below:

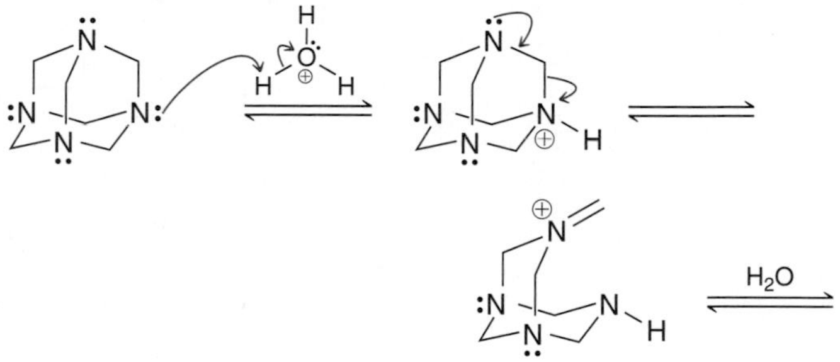

Imines as Prodrugs

The imine moiety is used in the development of many prodrugs. Here we will explore one such example.

The compound below, γ-aminobutyric acid, is an important neurotransmitter:

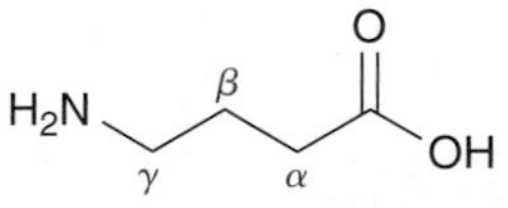

γ-aminobutyric acid

A deficiency of this compound can cause convulsions. Administering γ-aminobutyric acid directly to a patient is not an effective treatment, because the compound does not readily cross the blood-brain barrier. Why not? At physiological pH, the amino group is protonated and the carboxylic acid moiety is deprotonated:

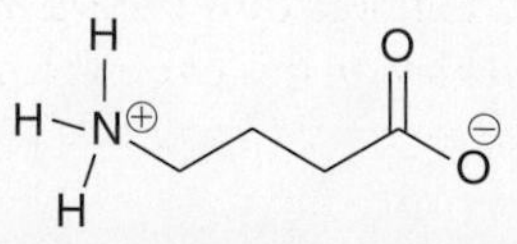

The compound exists primarily in this ionic form, which cannot cross the nonpolar environment of the blood-brain barrier. Progabide is a prodrug derivative used to treat patients who exhibit the symptoms of a deficiency of γ-aminobutyric acid:

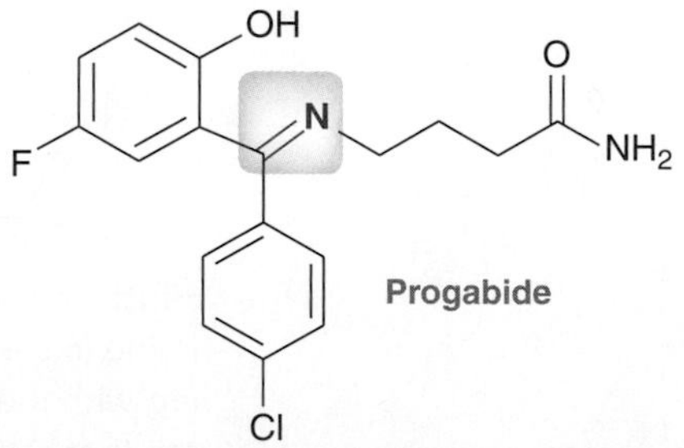

Progabide

The carboxylic acid has been converted to an amide, and the amino group has been converted into an imine (highlighted). At physiological pH, this compound exists primarily as a neutral compound (uncharged), and it can therefore cross the blood-brain barrier. Once in the brain, it is converted to γ-aminobutyric acid via hydrolysis of the imine and amide moieties:

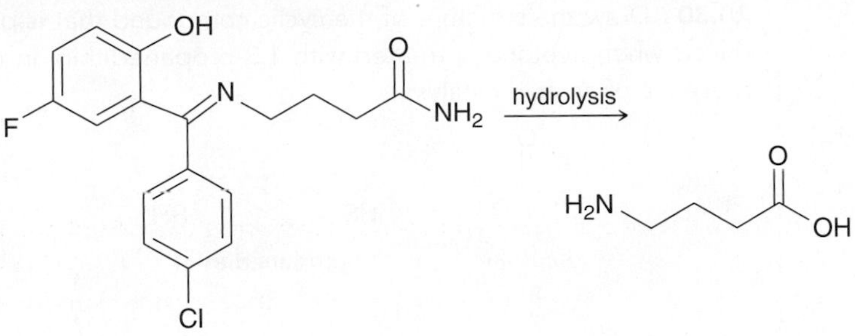

Progabide is just one example in which the imine moiety has been used in the development of a prodrug.

20.8 Sulfur Nucleophiles

In acidic conditions, an aldehyde or ketone will react with two equivalents of a thiol to form a **thioacetal:**

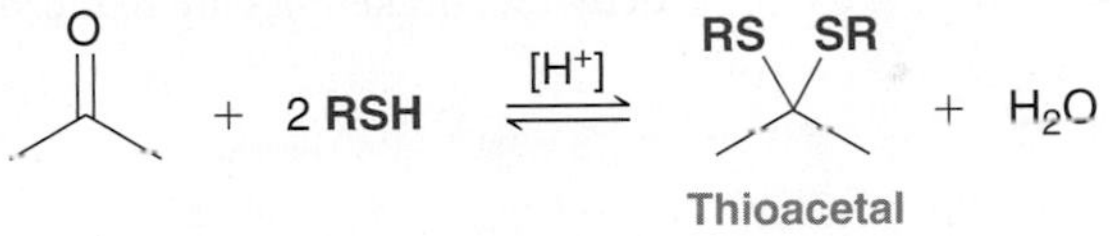

The mechanism of this transformation is directly analogous to acetal formation, with sulfur atoms taking the place of oxygen atoms. If a compound with two SH groups is used, a cyclic thioacetal is formed:

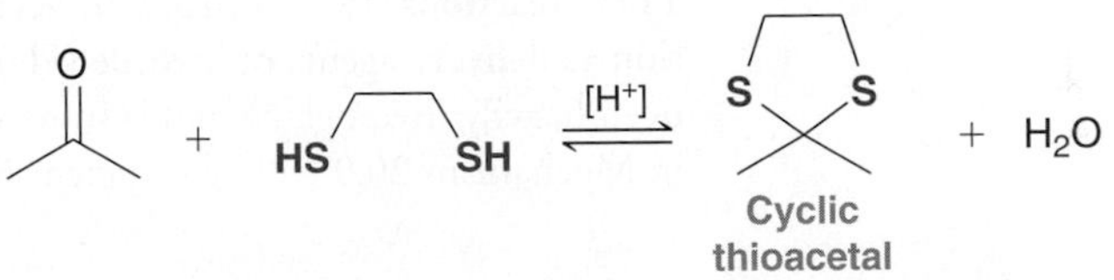

When treated with Raney nickel, thioacetals undergo **desulfurization,** yielding an alkane:

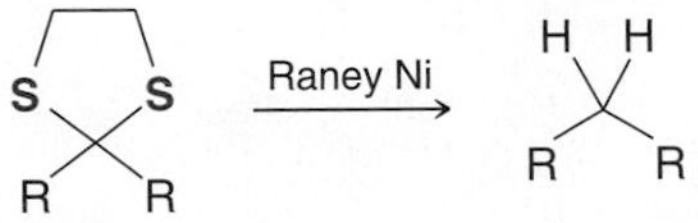

Raney Ni is a spongy form of nickel that has adsorbed hydrogen atoms. It is these hydrogen atoms that ultimately replace the sulfur atoms, although a discussion of the mechanism for desulfurization is beyond the scope of this text.

The reactions above provide us with another two-step method for the reduction of a ketone:

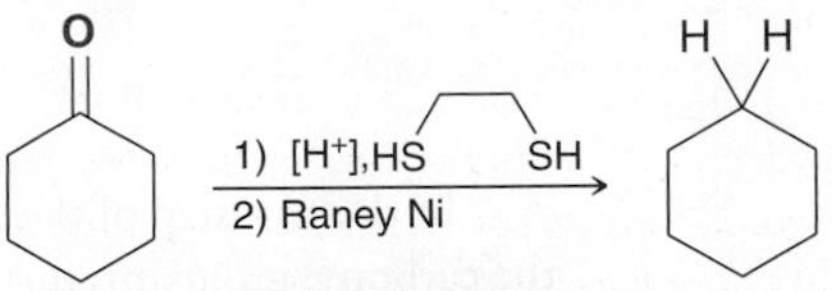

This method involves formation of the thioacetal followed by desulfurization with Raney nickel. It is the third method we have encountered for achieving this type of transformation. The other two methods are the Clemmensen reduction (Section 19.6) and the Wolff-Kishner reduction (Section 20.6).

CONCEPTUAL CHECKPOINT

20.29 Predict the major product for each reaction below:

(a)

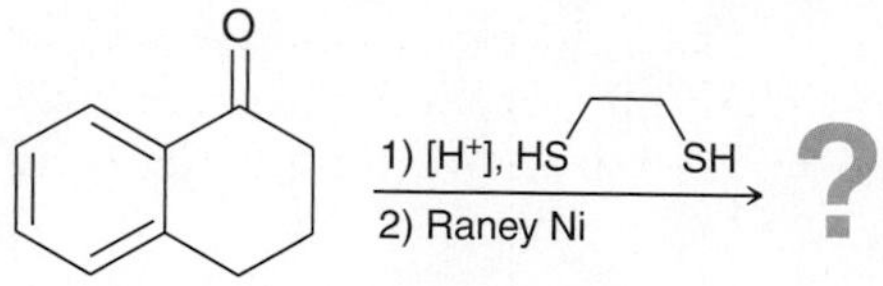

(b)

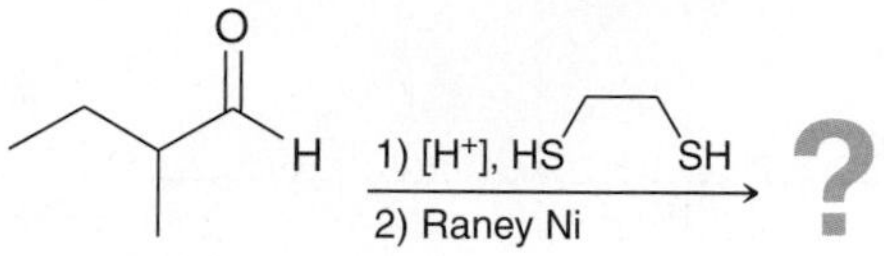

20.30 Draw the structure of the cyclic compound that is produced when acetone is treated with 1,3-propanedithiol in the presence of an acid catalyst.

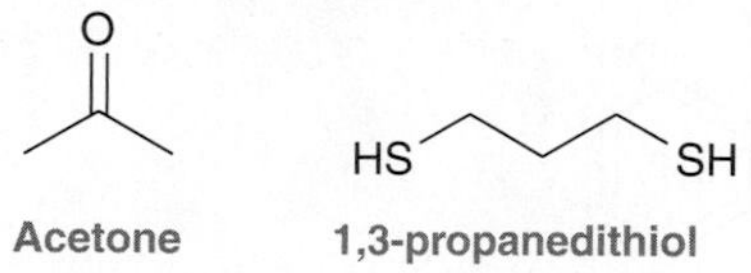

20.9 Hydrogen Nucleophiles

When treated with a hydride reducing agent, such as LAH or sodium borohydride ($NaBH_4$), aldehydes and ketones are reduced to alcohols:

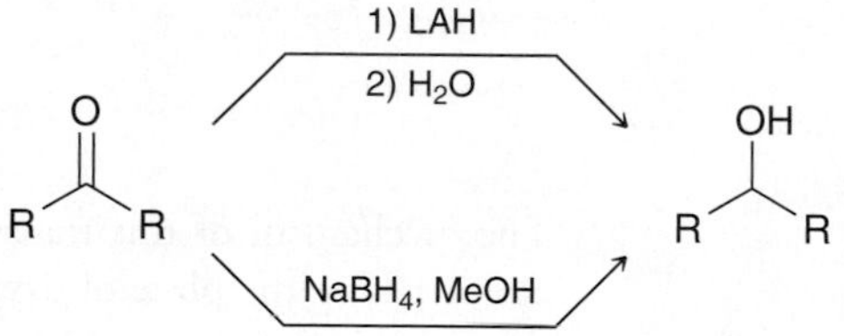

These reactions were discussed in Section 13.4, and we saw that LAH and $NaBH_4$ both function as delivery agents of hydride (H^-). The precise mechanism of action for these reagents has been heavily investigated and is somewhat complex. Nevertheless, the simplified version shown in Mechanism 20.9 will be sufficient for our purposes.

MECHANISM 20.9 THE REDUCTION OF KETONES OR ALDEHYDES WITH HYDRIDE AGENTS

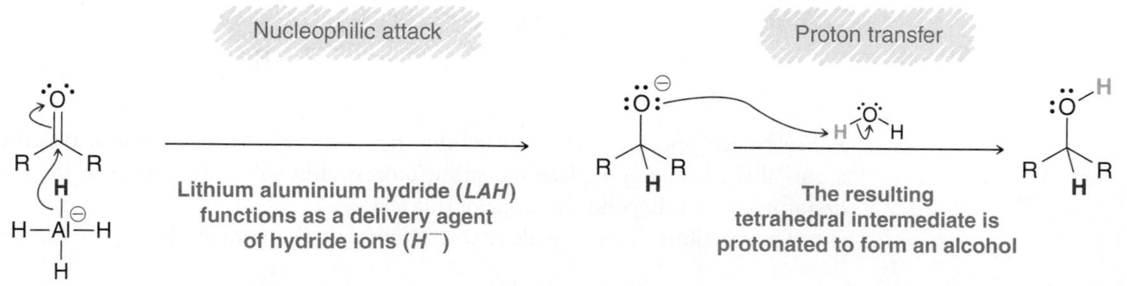

In the first step of the mechanism, the reducing agent delivers a hydride ion, which attacks the carbonyl group, producing a tetrahedral intermediate. This intermediate is then treated with a proton source to yield the product. This simplified mechanism does not take into account many important observations, such as the role of the lithium cation (Li^+). For example, when

LOOKING BACK
Hydride cannot function as a leaving group because it is too strongly basic. (See Section 7.8.)

12-crown-4 is added to the reaction mixture, the lithium ions are solvated (as described in Section 14.4), and reduction does not occur. Clearly, the lithium cation plays a pivotal role in the mechanism. However, a full treatment of the mechanism of hydride reducing agents is beyond the scope of this text, and the simplified version above will suffice.

The reduction of a carbonyl group with LAH or $NaBH_4$ is not a reversible process, because hydride does not function as a leaving group. Notice that the mechanism above employs one-way arrows (rather than equilibrium arrows) to signify that the reverse process is insignificant.

CONCEPTUAL CHECKPOINT

20.31 Predict the major product for each of the following reactions:

(a) 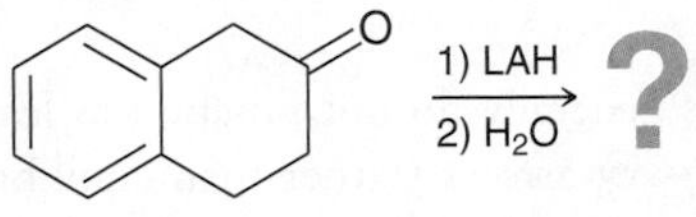

(b) 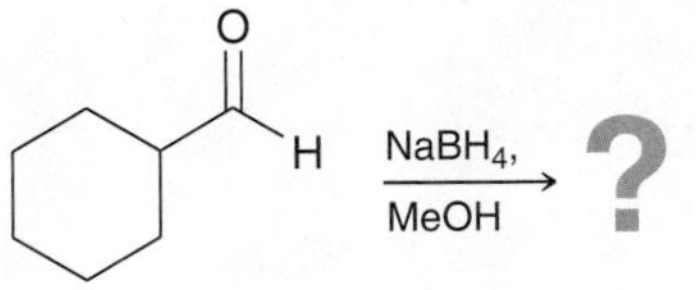

(c) 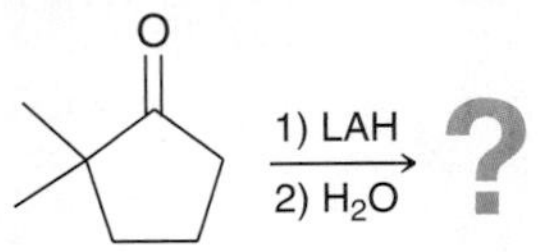

(d) 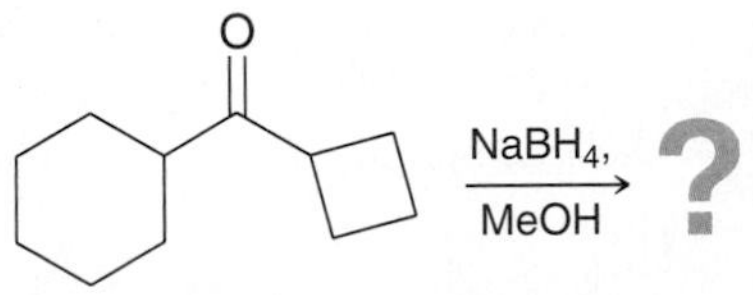

20.32 When 2 moles of benzaldehyde are treated with sodium hydroxide, a reaction occurs in which 1 mole of benzaldehyde is oxidized (giving benzoic acid) while the other mole of benzaldehyde is reduced (giving benzyl alcohol):

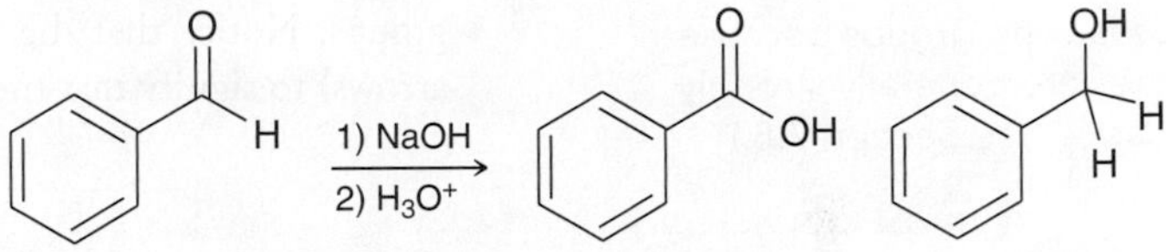

This reaction, called the Cannizzaro reaction, is believed to occur via the following mechanism: A hydroxide ion serves as a nucleophile to attack the carbonyl group of benzaldehyde, generating a tetrahedral intermediate. This tetrahedral intermediate then functions as a hydride reducing agent by delivering a hydride ion to another molecule of benzaldehyde. In this way, one molecule is reduced while the other is oxidized.

(a) Using the explanation above, draw the mechanism of the Cannizzaro reaction.

(b) What is the function of H_3O^+ in the second step?

(c) Water alone is not sufficient to accomplish the function of the second step. Explain.

20.10 Carbon Nucleophiles

Grignard Reagents

When treated with a Grignard reagent, aldehydes and ketones are converted into alcohols, accompanied by the formation of a new C–C bond:

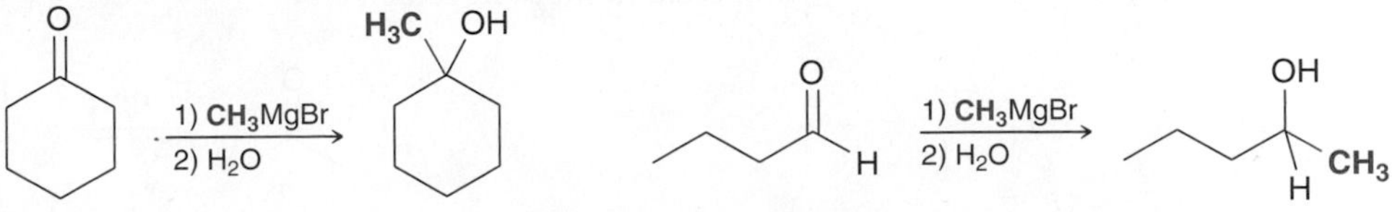

Grignard reactions were discussed in more detail in Section 13.6. The precise mechanism of action for these reagents has been heavily investigated and is fairly complex. The simplified version shown in Mechanism 20.10 will be sufficient for our purposes.

MECHANISM 20.10 THE REACTION BETWEEN A GRIGNARD REAGENT AND A KETONE OR ALDEHYDE

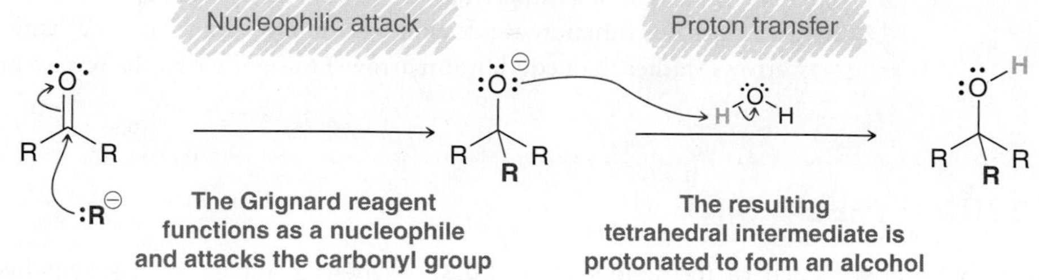

LOOKING BACK

Carbanions rarely function as leaving groups because they are generally strongly basic. (See Section 7.8.)

Grignard reactions are not reversible because carbanions generally do not function as leaving groups. Notice that the mechanism above employs one-way arrows (rather than equilibrium arrows) to signify that the reverse process is insignificant.

CONCEPTUAL CHECKPOINT

20.33 Predict the major product of each reaction below:

(a)

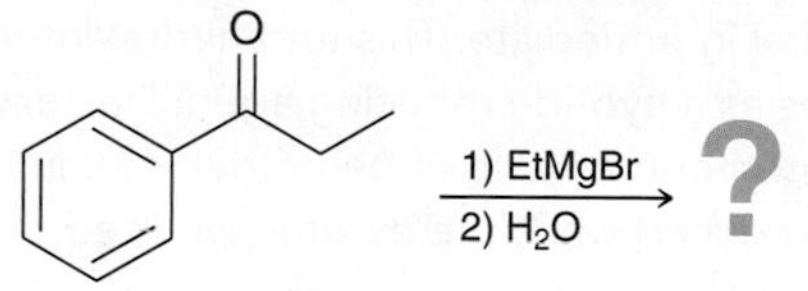

(b)

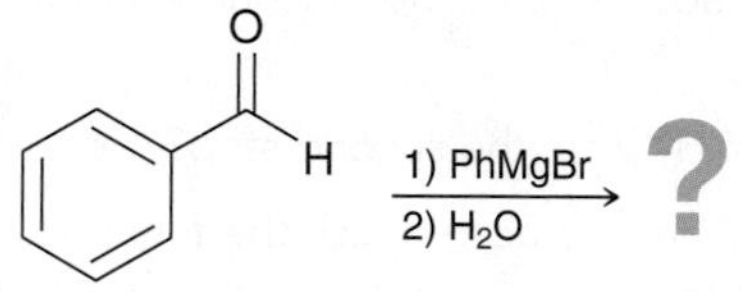

(c)

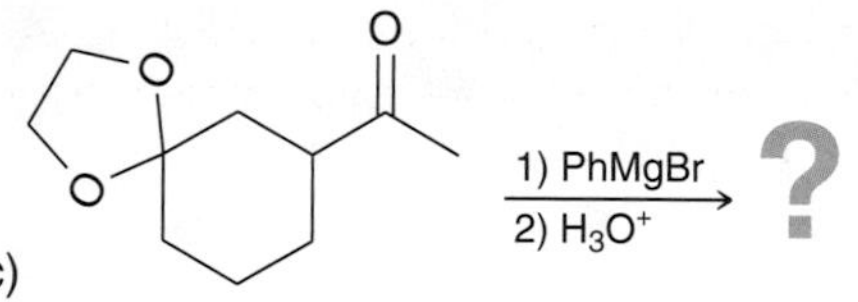

20.34 Identify the reagents necessary to accomplish each of the transformations below:

(a)

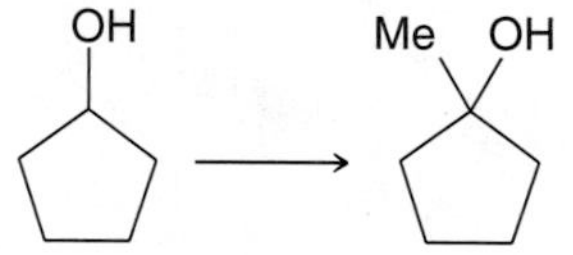

(b)

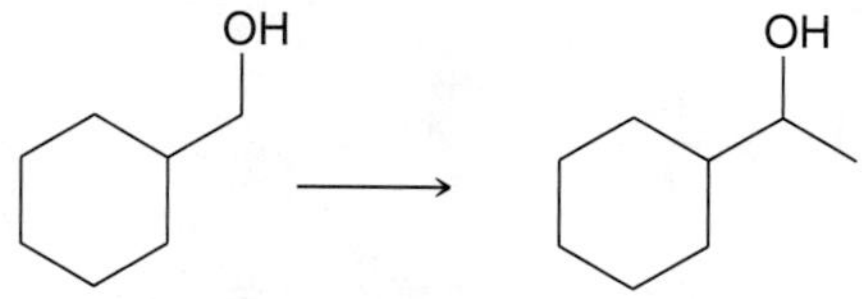

Cyanohydrin Formation

When treated with hydrogen cyanide (HCN), aldehydes and ketones are converted into **cyanohydrins,** which are characterized by the presence of a cyano group and a hydroxyl group connected to the same carbon atom:

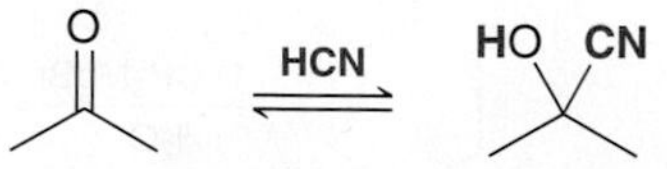

A cyanohydrin

This reaction was studied extensively by Arthur Lapworth (University of Manchester) and was found to occur more rapidly in mildly basic conditions. In the presence of a catalytic amount of base, a small amount of hydrogen cyanide is deprotonated to give cyanide ions, which catalyze the reaction (Mechanism 20.11).

MECHANISM 20.11 CYANOHYDRIN FORMATION

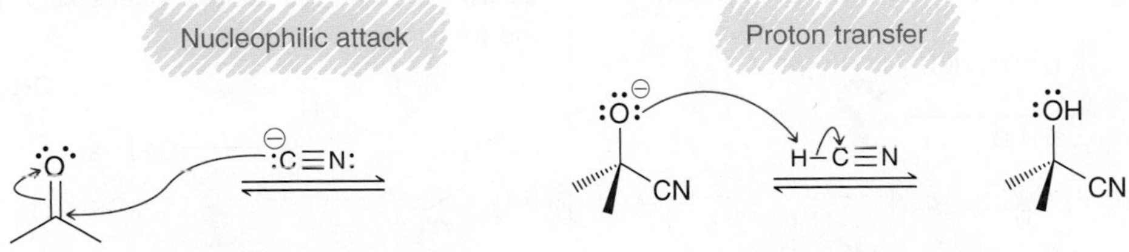

The cyanide ion functions as a nucleophile and attacks the carbonyl group, forming a tetrahedral intermediate

The tetrahedral intermediate is protonated, generating a cyanohydrin

In the first step, a cyanide ion attacks the carbonyl to produce a tetrahedral intermediate. This intermediate then abstracts a proton from HCN, regenerating a cyanide ion. In this way, cyanide functions as a catalyst for the addition of HCN to the carbonyl group.

Rather than using a catalytic amount of base to form cyanide ions, the reaction can simply be performed in a mixture of HCN and cyanide ions (from KCN). The process is reversible, and the yield of products is therefore determined by equilibrium concentrations. For most aldehydes and unhindered ketones, the equilibrium favors formation of the cyanohydrin:

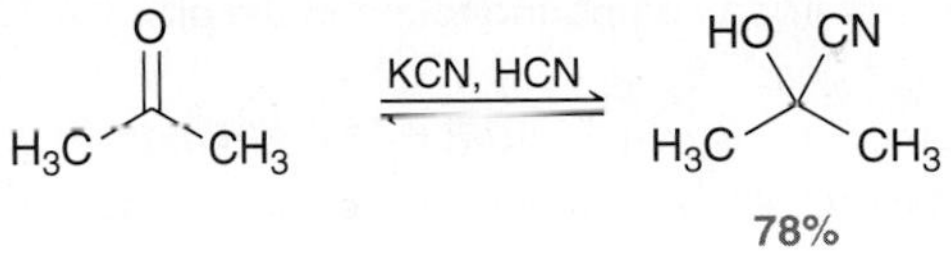

HCN is a liquid at room temperature and is extremely hazardous to handle because it is highly toxic and volatile (b.p. = 26°C). To avoid the dangers associated with handling HCN, cyanohydrins can also be prepared by treating a ketone or aldehyde with potassium cyanide and an alternate source of protons, such as HCl:

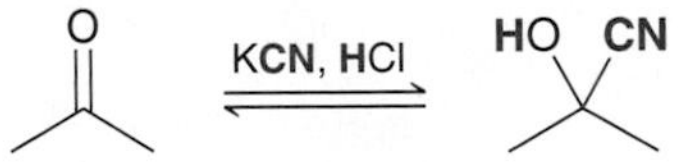

Cyanohydrins are useful in syntheses, because the cyano group can be further treated to yield a range of products. Two examples are shown below:

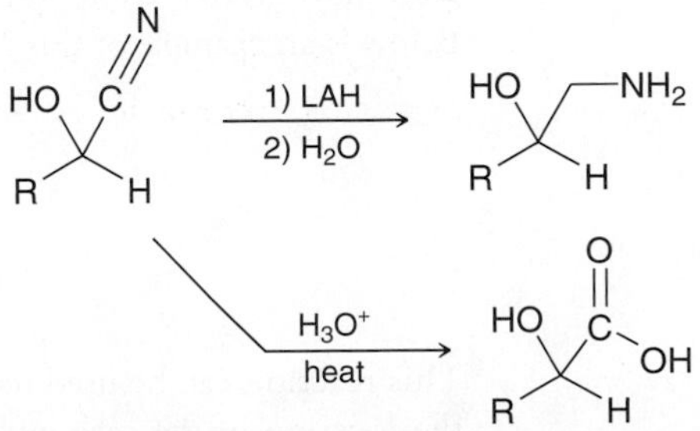

In the first example, the cyano group is reduced to an amino group. In the second example, the cyano group is hydrolyzed to give a carboxylic acid. Both of these reactions and their mechanisms will be explored in more detail in the next chapter.

CONCEPTUAL CHECKPOINT

20.35 Predict the major product for each reaction below:

(a)

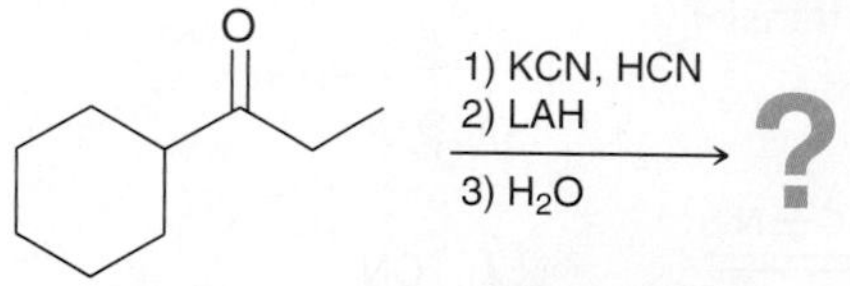

(b)

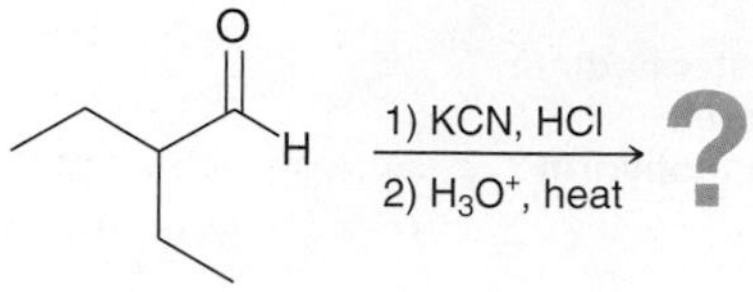

20.36 Identify the reagents necessary to accomplish each of the transformations below:

(a)

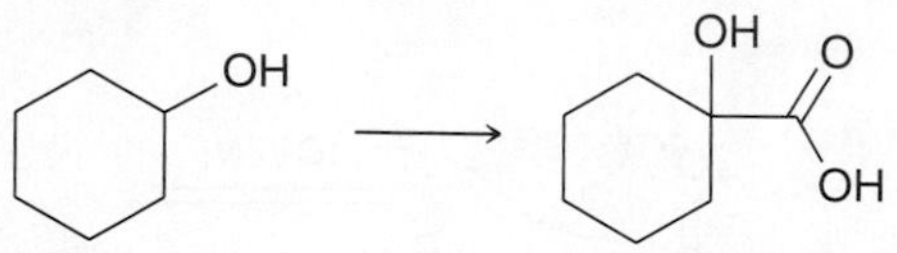

(b)

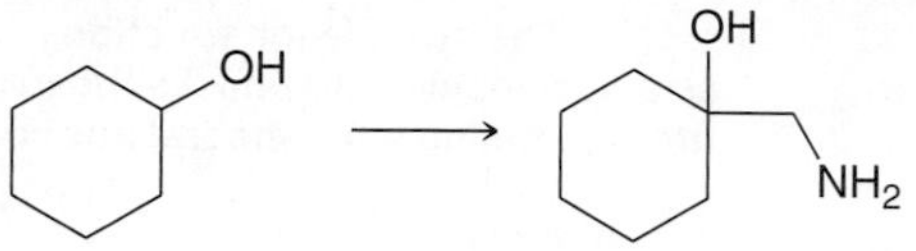

PRACTICALLYSPEAKING)))

Cyanohydrin Derivatives in Nature

Amygdalin is a naturally occurring compound found in the pits of apricots, wild cherries, and peaches.

If ingested, this compound is metabolized to produce mandelonitrile, a cyanohydrin, which is converted by enzymes into benzaldehyde and HCN gas, a toxic compound:

This last step (generation of HCN gas) is used as a defense mechanism by many species of millipedes. The millipedes manufacture and store mandelonitrile, and in a separate compartment, they store enzymes that are capable of catalyzing the conversion of mandelonitrile into benzaldehyde and HCN. To ward off predators, a millipede will mix the contents of the two compartments and secrete HCN gas.

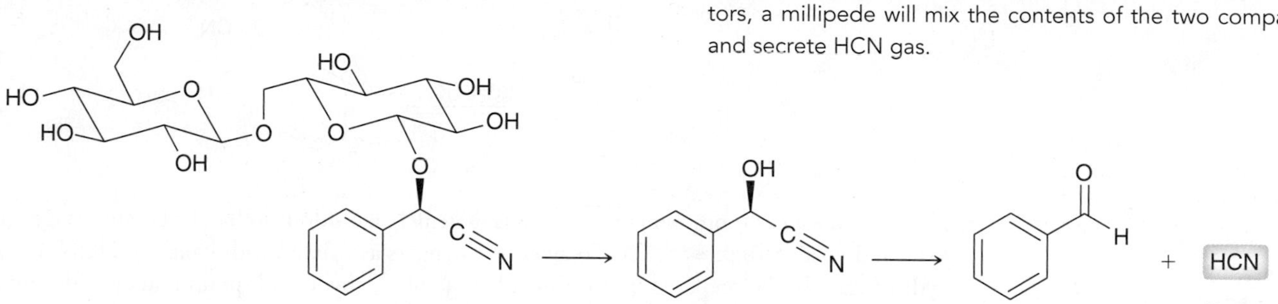

Amygdalin **Mandelonitrile** **Benzaldehyde** **Toxic**

Wittig Reaction

Georg Wittig, a German chemist, was awarded the 1979 Nobel Prize in Chemistry for his work with phosphorous compounds and his discovery of a reaction with enormous synthetic utility. Below is an example of this reaction, called the **Wittig reaction** (pronounced Vittig):

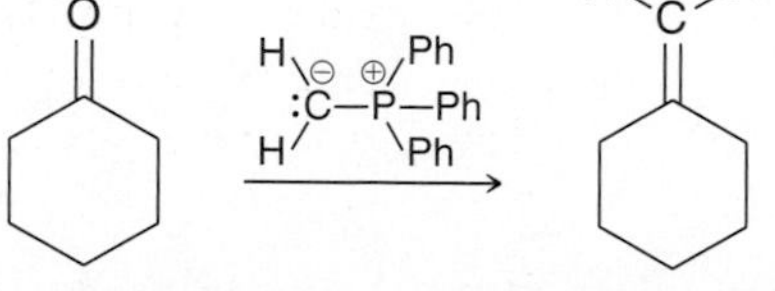

This reaction can be used to convert a ketone into an alkene by forming a new C–C bond at the location of the carbonyl moiety. The phosphorus-containing reagent that accomplishes this transformation is called a **phosphorane,** and it belongs to a larger class of compounds called **ylides.** An ylide is a compound with two oppositely charged atoms adjacent to each other. The phosphorane above exhibits a negative charge on the carbon atom and a positive charge on the phosphorous atom. This ylide does, in fact, have a resonance structure that is free of any charges:

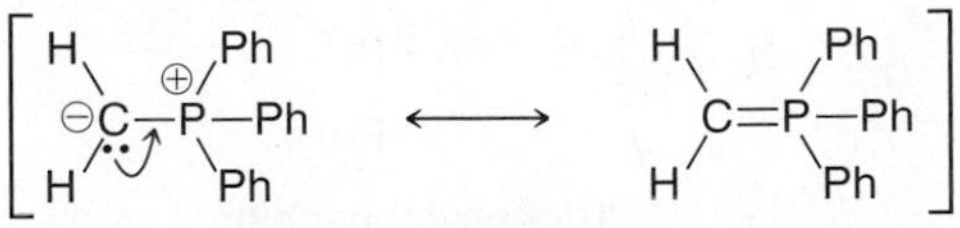

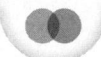

However, this resonance structure (with a C═P double bond) does not contribute much character to the overall resonance hybrid, because the *p* orbitals on C and P are vastly different in size and do not effectively overlap. A similar argument was used in describing S═O bonds in the previous chapter (Section 19.3). Despite this fact, the phosphorus ylide above, also called a **Wittig reagent,** is often drawn using either of the resonance structures shown above.

A mechanism for the Wittig reaction is shown in Mechanism 20.12.

MECHANISM 20.12 THE WITTIG REACTION

Nucleophilic attack	Nucleophilic attack	Rearrangement

A Betaine

An Oxaphosphetane

The Wittig reagent functions as a nucleophile and attacks the carbonyl group, forming a tetrahedral intermediate	A lone pair on the oxygen atom functions as a nucleophile and attacks the phosphorus atom in an intramolecular attack	The oxaphosphetane decomposes to produce an alkene and triphenylphosphine oxide

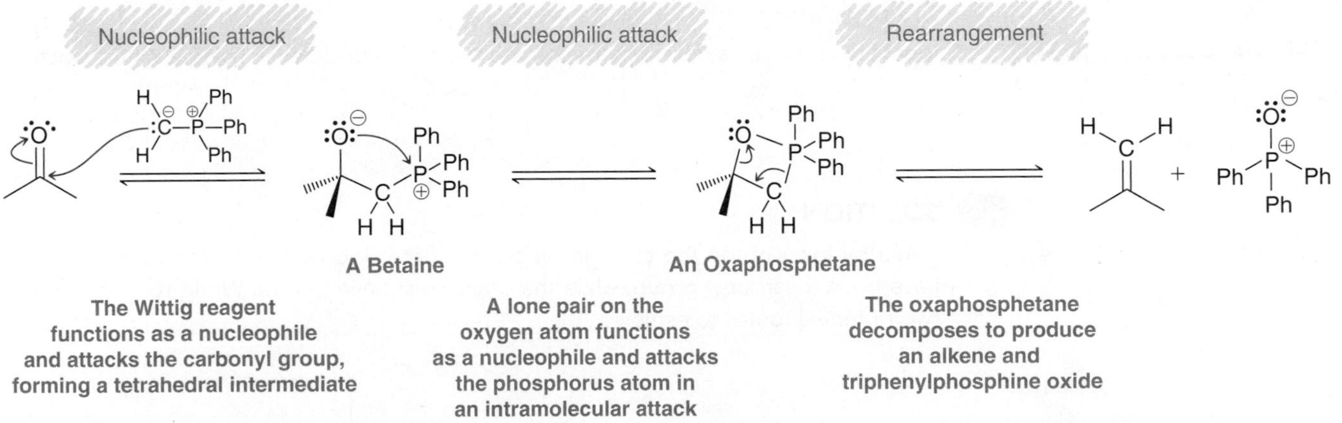

BY THE WAY

Experimental evidence suggests that the intermediate betaine is only formed in limited cases. In other cases, it appears that the Wittig reagent may react with the carbonyl compound in a [2+2] cycloaddition process, directly generating the oxaphosphetane. The mechanism for this reaction is still under investigation.

The Wittig reagent is a carbanion and can attack the carbonyl group in the first step of the mechanism, generating an intermediate called a **betaine** (pronounced "bay-tuh-een"). A betaine is a neutral compound with two oppositely charged atoms that are not adjacent to each other. The negatively charged oxygen atom then attacks the positively charged phosphorous atom in an intramolecular nucleophilic attack, generating an **oxaphosphetane.** This compound then rearranges to give the alkene product.

Wittig reagents are easily prepared by treating triphenylphosphine with an alkyl halide followed by a strong base:

$$\text{Ph—P:} \quad \xrightarrow[\text{2) BuLi}]{\text{1) CH}_3\text{I}} \quad \text{Ph—P—C:}$$

Triphenylphosphine **Wittig reagent**

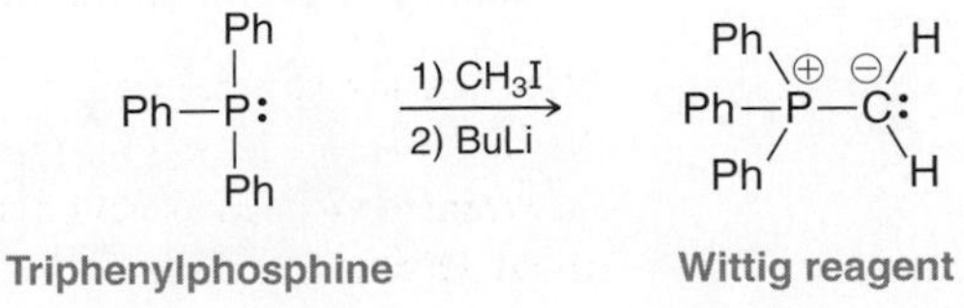

The mechanism of formation for Wittig reagents involves an S_N2 reaction followed by deprotonation with a strong base:

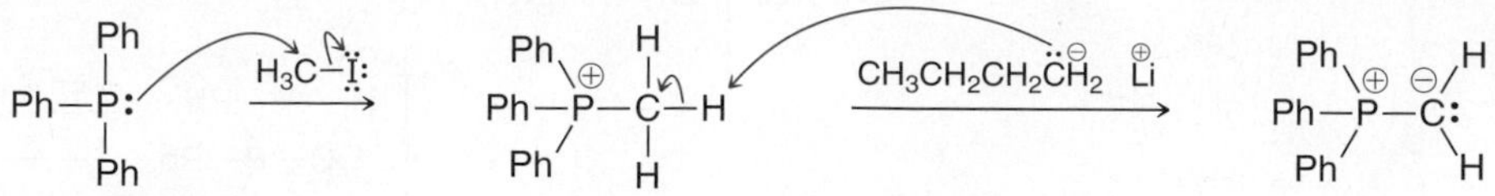

Since the first step is an S_N2 process, the regular restrictions of S_N2 processes apply. Specifically, primary alkyl halides will react more readily than secondary alkyl halides, and tertiary alkyl halides cannot be used. The Wittig reaction is useful for preparing mono-, di-, or trisubstituted alkenes. Tetrasubstituted alkenes are more difficult to prepare due to steric hindrance in the transition states.

The following exercise illustrates how to choose the reagents for a Wittig reaction.

SKILLBUILDER

20.6 PLANNING AN ALKENE SYNTHESIS WITH A WITTIG REACTION

LEARN the skill

Identify the reagents necessary to prepare the compound below using a Wittig reaction:

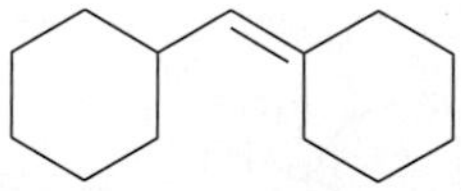

SOLUTION

Begin by focusing on the two carbon atoms of the double bond. One carbon atom must have been a carbonyl group, while the other must have been a Wittig reagent. This gives two potential routes to explore:

STEP 1
Using a retrosynthetic analysis, determine the two possible sets of reactants that could be used to form the C=C bond.

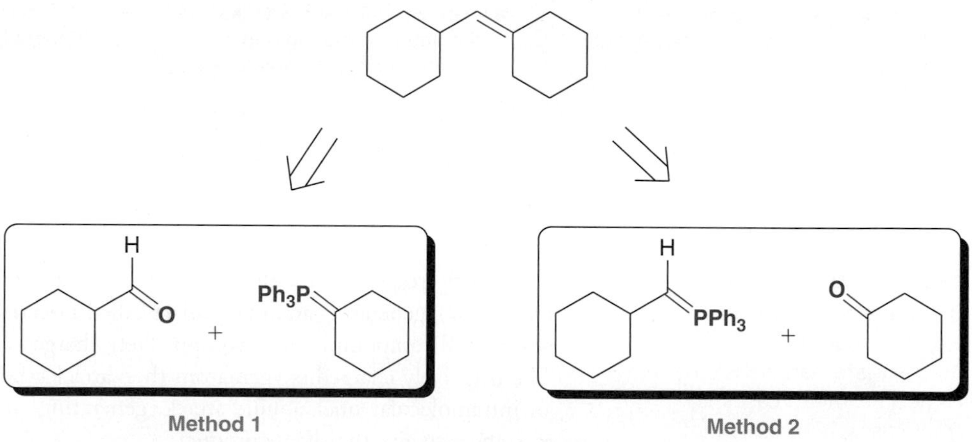

Let's compare these two methods by focusing on the Wittig reagent in each case. Recall that the Wittig reagent is prepared by an S_N2 process, and we therefore must consider steric factors during its preparation. Method 1 requires the use of a secondary alkyl halide:

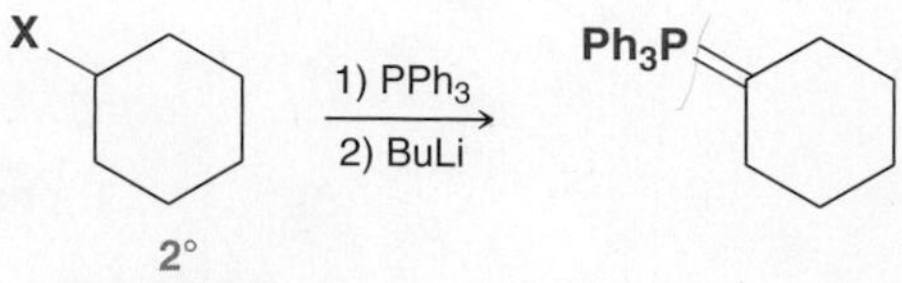

but method 2 requires the use of a primary alkyl halide:

STEP 2
Consider how you would make each possible Wittig reagent, and determine which method involves the less substituted alkyl halide.

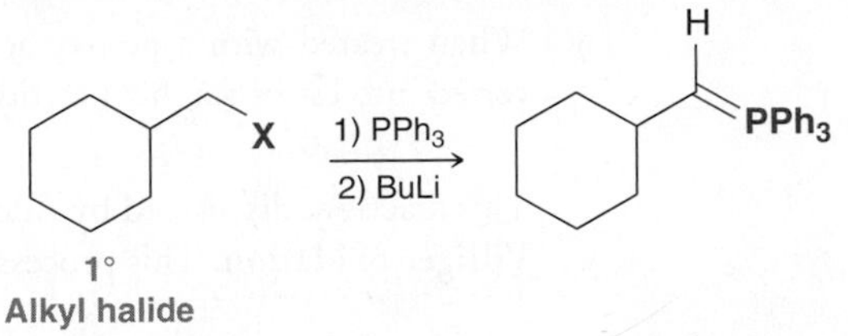

Method 2 is likely to be more efficient, because a primary alkyl halide will undergo S_N2 more rapidly than a secondary alkyl halide. Therefore, the following would be the preferred synthesis:

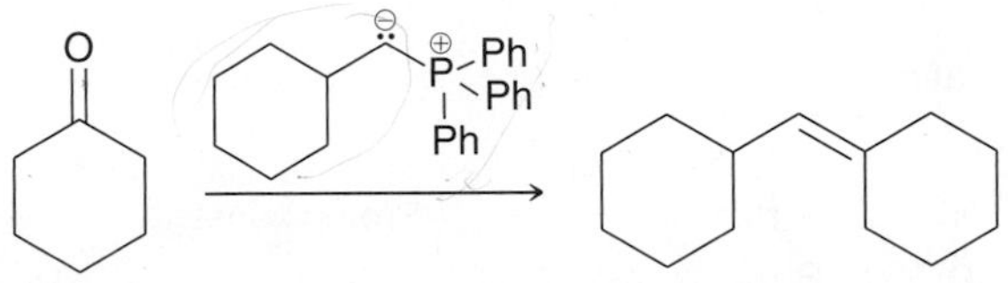

PRACTICE the skill **20.37** Identify the reagents necessary to prepare each of the following compounds using a Wittig reaction:

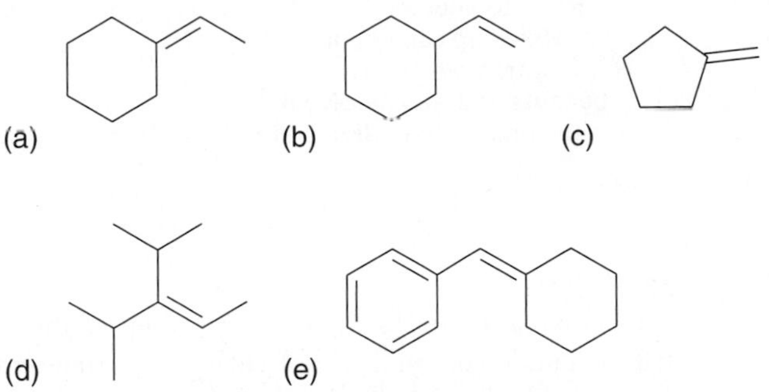

APPLY the skill **20.38** Consider the structure of beta-carotene, mentioned earlier in this chapter:

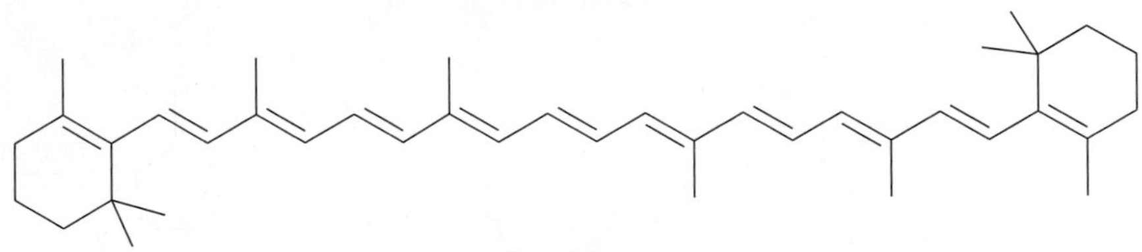

β-carotene

Design a synthesis of beta carotene using the compound below as your only source of carbon atoms:

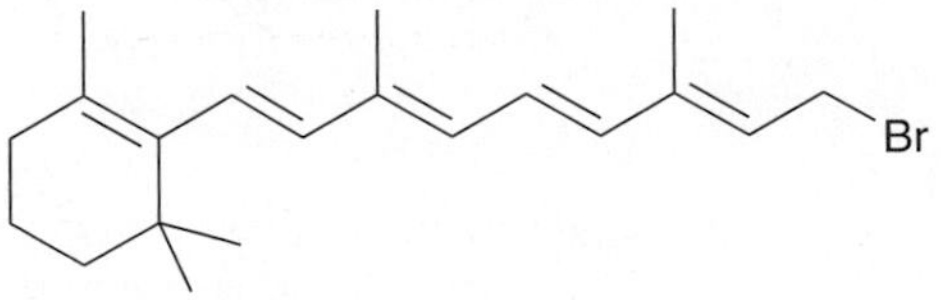

20.39 Identify the reagents necessary to accomplish each of the transformations below:

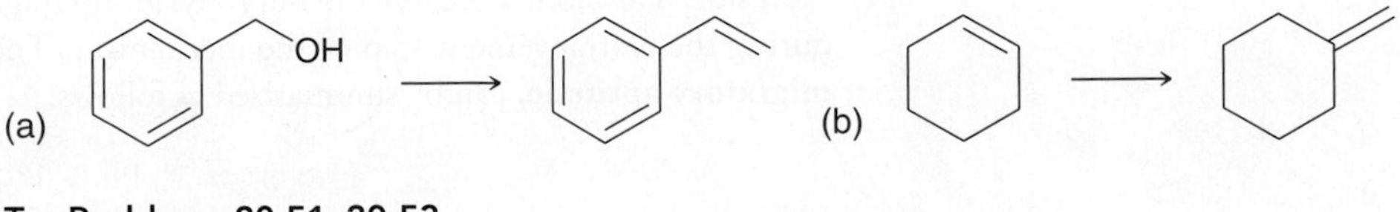

need more **PRACTICE?** **Try Problems 20.51–20.53**

20.11 Baeyer-Villiger Oxidation of Aldehydes and Ketones

When treated with a peroxy acid, ketones can be converted into esters via the insertion of an oxygen atom:

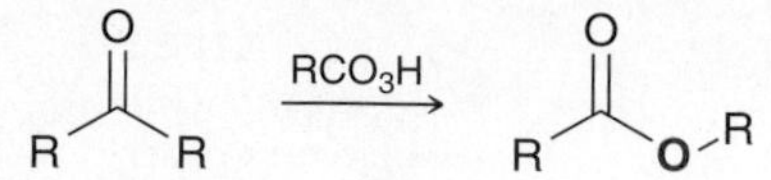

O
R
R
RCO_3H
O
R
O
R

This reaction, discovered by Adolf von Baeyer and Victor Villiger in 1899, is called the **Baeyer-Villiger oxidation.** This process is believed to proceed via Mechanism 20.13.

MECHANISM 20.13 THE BAEYER-VILLIGER OXIDATION

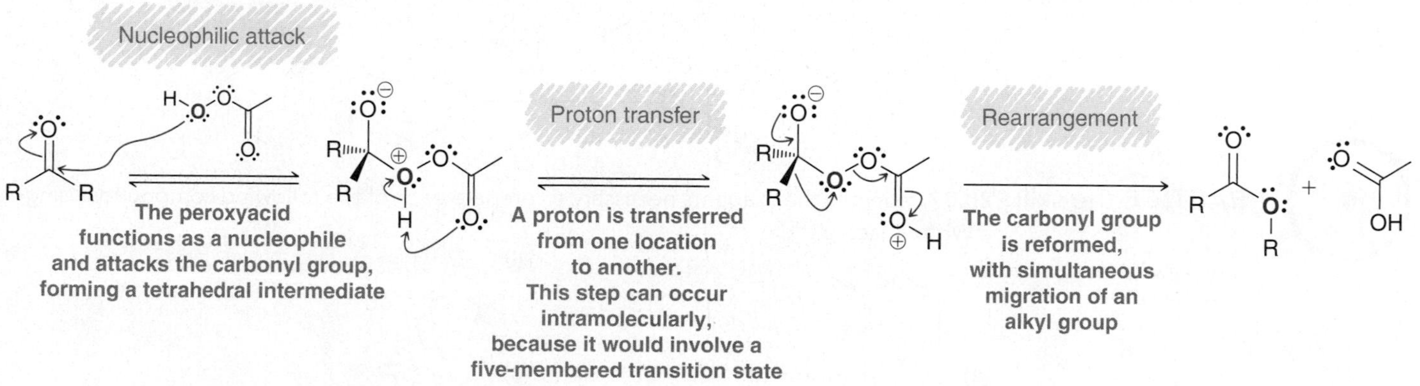

Nucleophilic attack
The peroxyacid functions as a nucleophile and attacks the carbonyl group, forming a tetrahedral intermediate
Proton transfer
A proton is transferred from one location to another. This step can occur intramolecularly, because it would involve a five-membered transition state
Rearrangement
The carbonyl group is reformed, with simultaneous migration of an alkyl group
OH

The peroxy acid attacks the carbonyl group of the ketone, giving a tetrahedral intermediate that then undergoes an intramolecular proton transfer (or two successive intermolecular proton transfers). Finally, the C=O double bond is re-formed by migration of an R group. This rearrangement produces the ester.

In much the same way, treatment of a cyclic ketone with a peroxy acid yields a cyclic ester, or **lactone.**

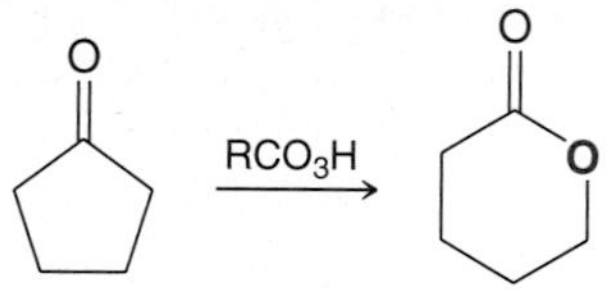

O
RCO_3H
O
O

A lactone

When an unsymmetrical ketone is treated with a peroxy acid, formation of the ester is regioselective; for example:

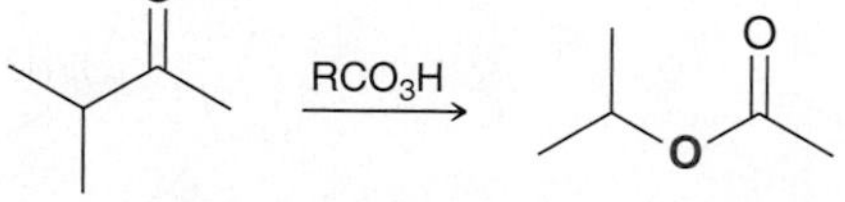

O
RCO_3H
O
O

In this case, the oxygen atom is inserted on the left side of the carbonyl group, rather than the right side. This occurs because the isopropyl group migrates more rapidly than the methyl group during the rearrangement step of the mechanism. The migration rates of different groups, or **migratory aptitude,** can be summarized as follows:

$$H > 3° > 2°, Ph > 1° > \text{methyl}$$

A hydrogen atom will migrate more rapidly than a tertiary alkyl group, which will migrate more rapidly than a secondary alkyl group or phenyl group. Below is one more example that illustrates this concept:

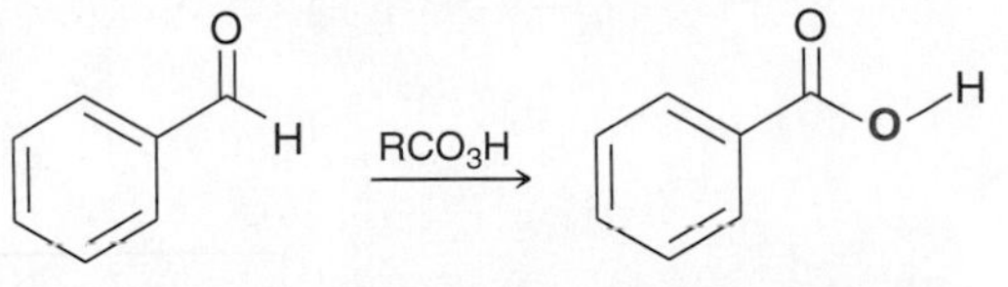

In this example, the oxygen atom is inserted on the right side of the carbonyl, because the hydrogen atom exhibits a greater migratory aptitude than the phenyl group.

CONCEPTUAL CHECKPOINT

20.40 Predict the major product of each reaction below:

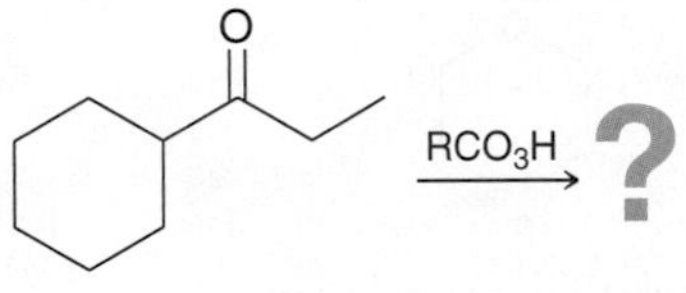

(a)

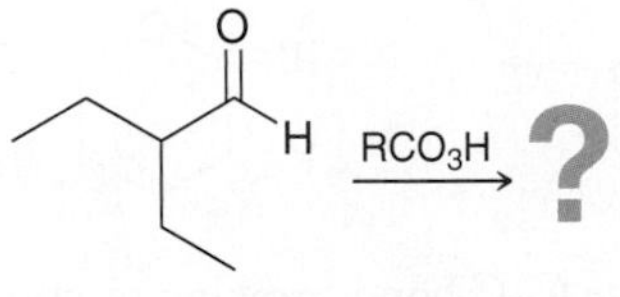

(b)

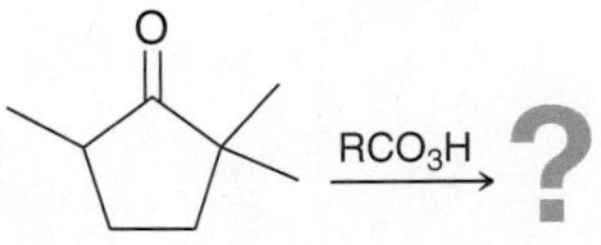

(c)

20.12 Synthesis Strategies

Recall from Chapter 12 that there are two main questions to ask when approaching a synthesis problem:

1. *Is there any change in the carbon skeleton?*
2. *Is there any change in the functional group?*

Let's focus on these issues separately, beginning with functional groups.

Functional Group Interconversion

In previous chapters, we learned how to interconvert many different functional groups (Figure 20.9). The reactions in this chapter expand the playing field by opening up the frontier of aldehydes and ketones. You should be able to fill in the reagents for each transformation in Figure 20.9. If you are having trouble, refer to Figure 13.13 for help. Then, you should be able to make a list of the various products than can be made from aldehydes and ketones and identify the required reagents in each case.

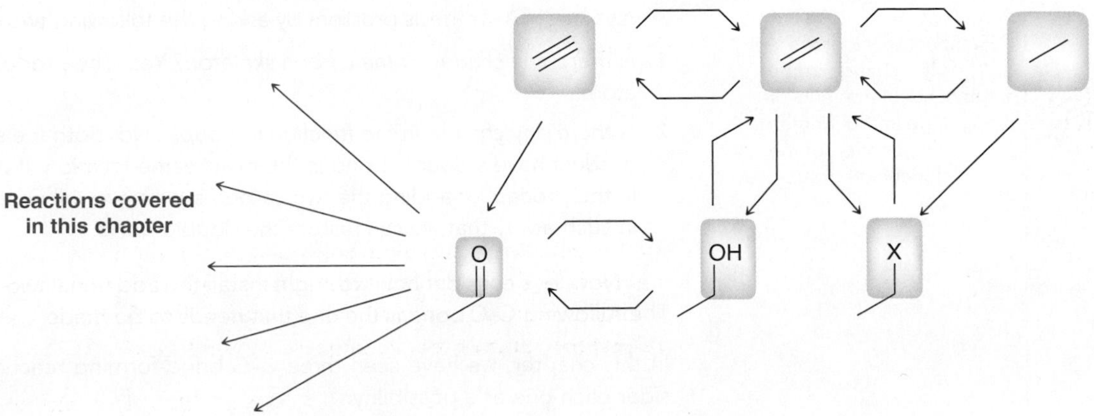

FIGURE 20.9
Functional groups that can be interconverted using reactions that we have learned thus far.

Reactions Involving a Change in Carbon Skeleton

In this chapter, we have seen three C–C bond-forming reactions: (1) a Grignard reaction, (2) cyanohydrin formation, and (3) a Wittig reaction:

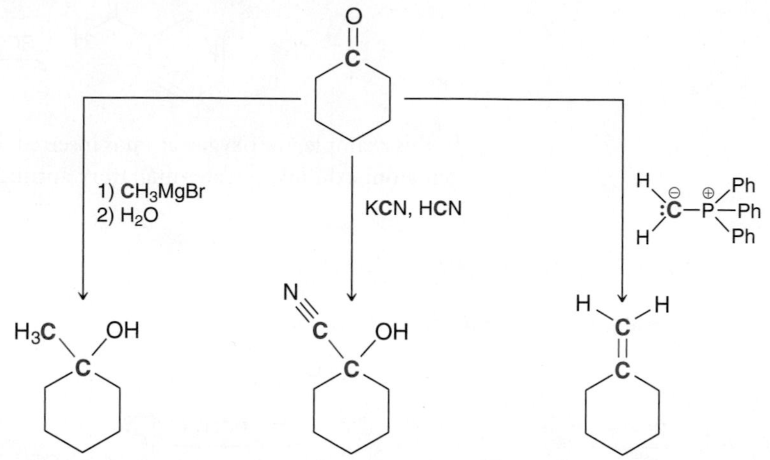

We have only seen one C–C bond-breaking reaction: the Baeyer-Villiger oxidation:

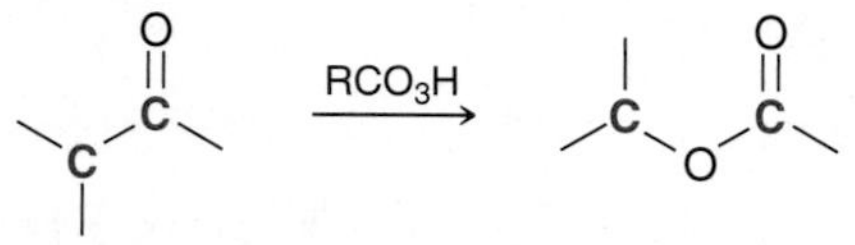

These four reactions should be added to your list of reactions that can change a carbon skeleton. Let's get some practice using these reactions.

SKILLBUILDER

20.7 PROPOSING A SYNTHESIS

LEARN the skill

Propose an efficient synthesis for the following transformation:

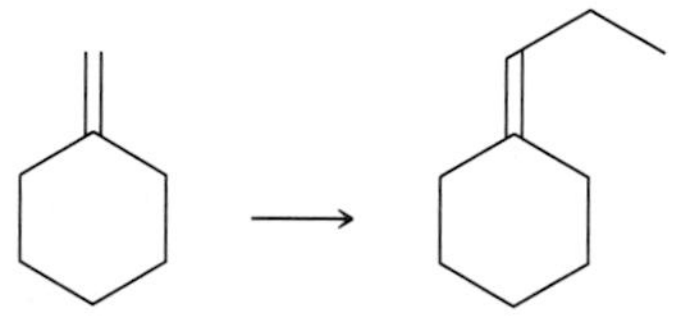

SOLUTION

STEP 1
Inspect whether there is a change in the carbon skeleton and/or a change in the identity or location of the functional groups.

Always begin a synthesis problem by asking the following two questions:

1. *Is there any change in the carbon skeleton?* Yes. The product has two additional carbon atoms.
2. *Is there any change in the functional groups?* No. Both the starting material and the product have a double bond in the exact same location. If we destroy the double bond in the process of adding the two carbon atoms, we will need to make sure that we do so in such a way that we can restore the double bond.

Now let's consider how we might install the additional two-carbon atoms. The following C–C bond is the one that needs to be made:

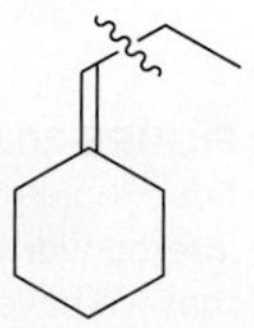

In this chapter, we have seen three C–C bond-forming reactions. Let's consider each one as a possibility.

STEP 2
When there is a change in the carbon skeleton, consider all of the C–C bond-forming reactions and all of the C–C bond-breaking reactions that you have learned so far.

We can immediately rule out cyanohydrin formation, as that process installs only one carbon atom, not two. So let's consider forming the C–C bond with either a Grignard reaction or a Wittig reaction.

A Grignard reagent won't attack a C═C double bond, so using a Grignard reaction would require first converting the C═C double bond into a functional group that can be attacked by a Grignard reagent, such as a carbonyl group:

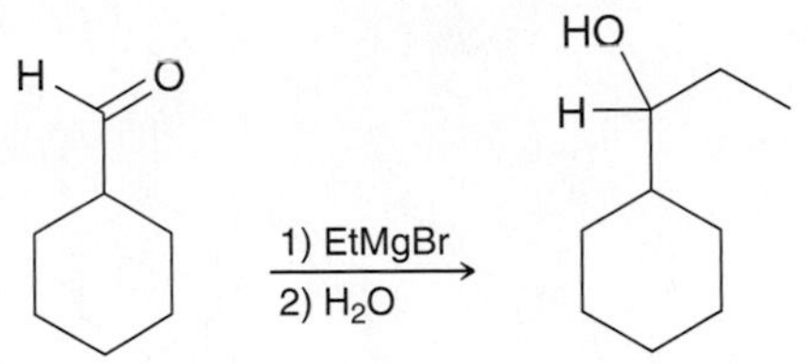

This reaction can indeed be used to form the crucial C–C bond. To use this method of C–C bond formation, we must first form the necessary aldehyde, then perform the Grignard reaction, and then finally restore the double bond in its proper location. This can be accomplished with the following reagents:

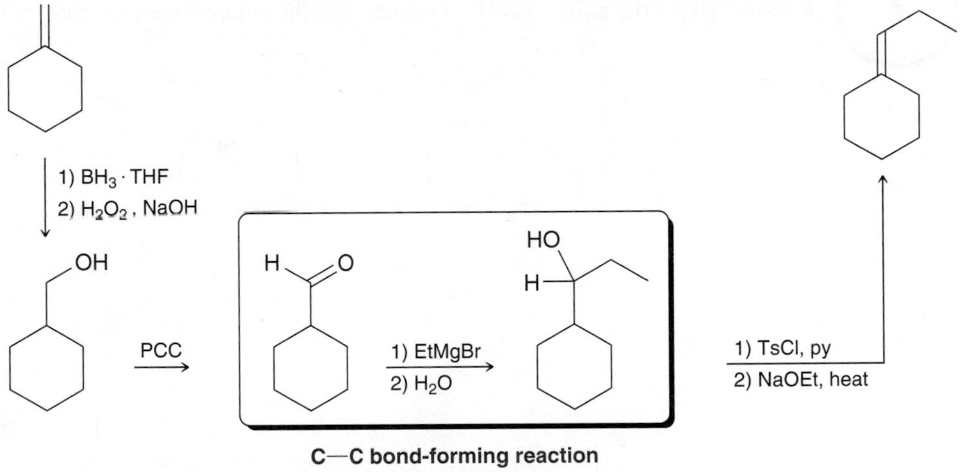

This provides us with a four-step procedure, and this answer is certainly reasonable.

Let's now explore the possibility of proposing a synthesis with a Wittig reaction. Recall that a Wittig reaction can be used to form a C═C bond, so we focus on formation of this bond:

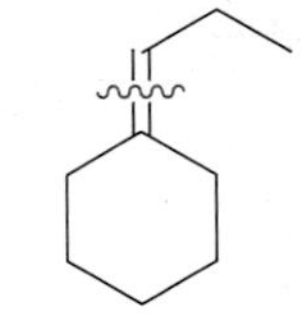

This bond can be formed if we start with a ketone and use the following Wittig reagent:

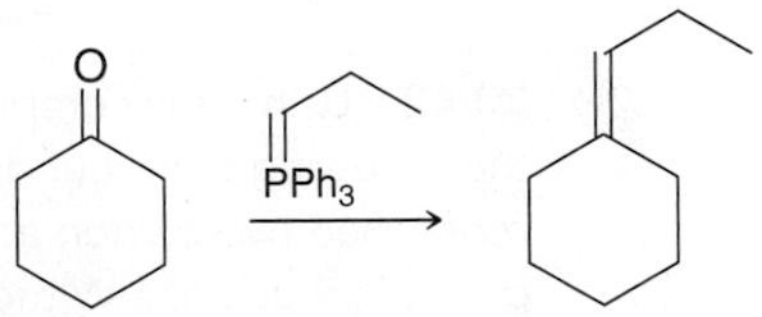

To use this reaction, we must first form the necessary ketone from the starting alkene:

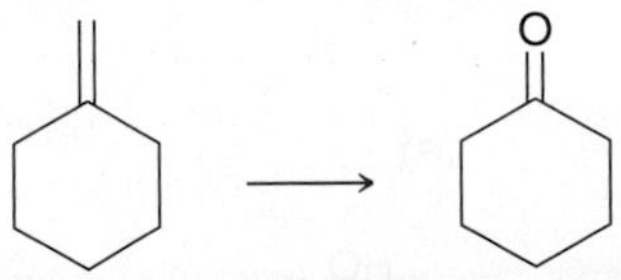

This can be accomplished with ozonolysis. This gives a two-step procedure for accomplishing the desired transformation: ozonolysis followed by a Wittig reaction. This approach is different than our first answer. In this approach, we are not attaching a two-carbon chain, but rather, we are first expelling a carbon atom and then attaching a three-carbon chain.

In summary, we have discovered two plausible methods. Both methods are correct answers to this problem, but the method employing the Wittig reaction is likely to be more efficient, because it requires fewer steps.

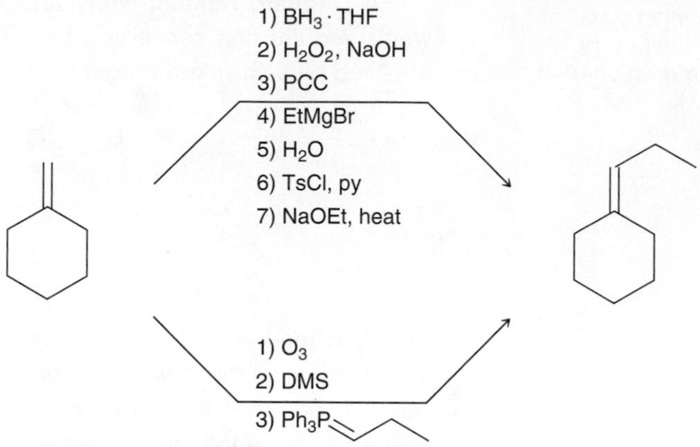

PRACTICE the skill **20.41** Propose an efficient synthesis for each of the following transformations:

(a) 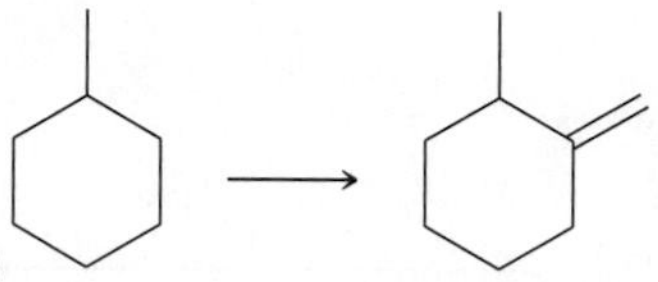

(b) 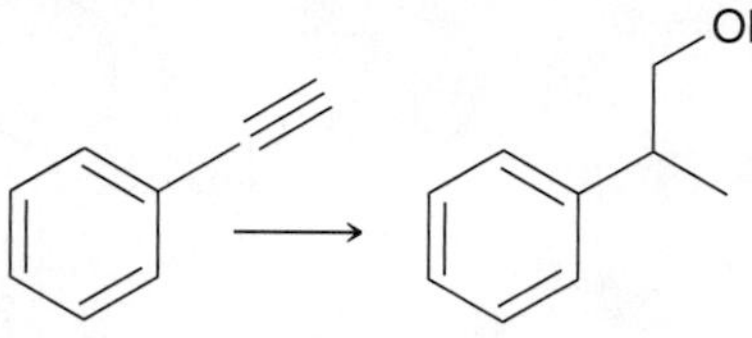

(c) 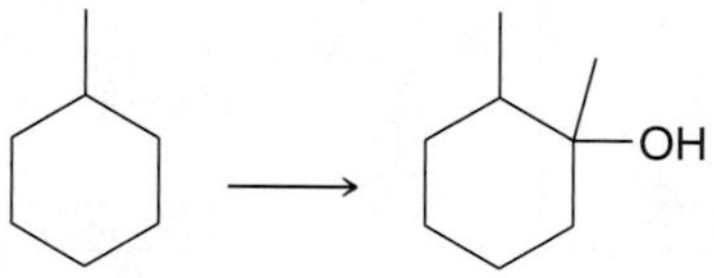

(d) 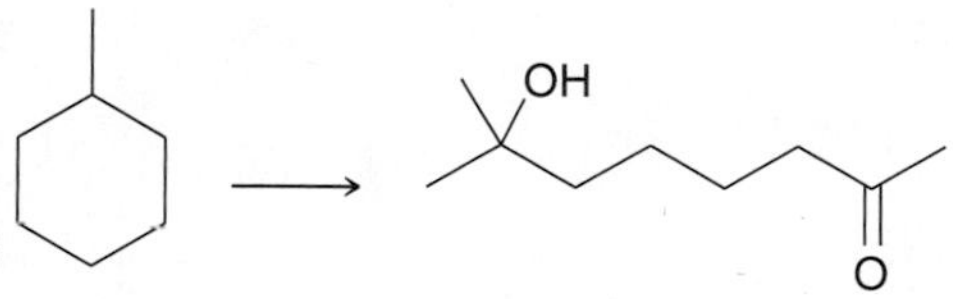

(e) 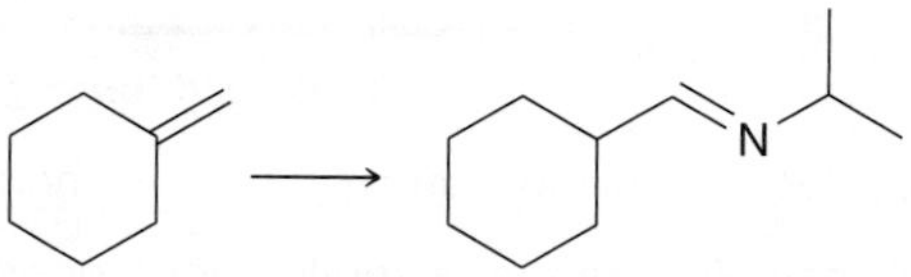

(f) 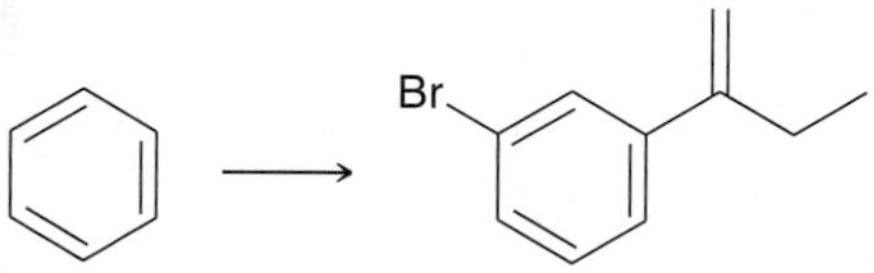

(g) 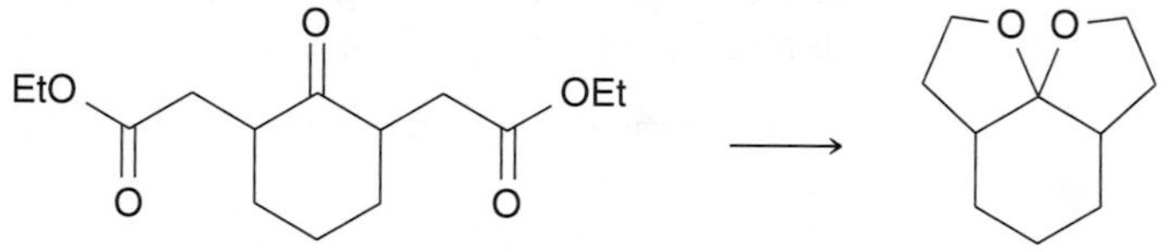

APPLY the skill

20.42 Using any compounds of your choosing, identify a method for preparing each of the following compounds. *Your only limitation is that the compounds you use can have no more than two carbon atoms.* For purposes of counting carbon atoms, you may ignore the phenyl groups of a Wittig reagent. That is, you are permitted to use Wittig reagents.

(a) 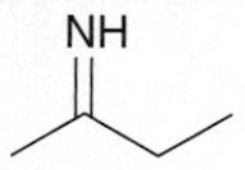

(b) 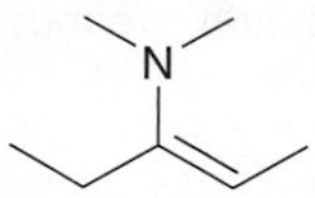

(c) 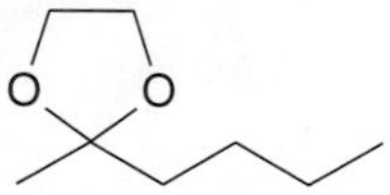

(d) 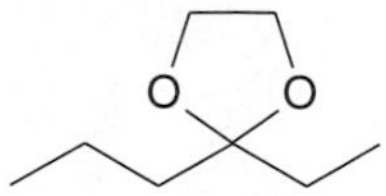

(e) 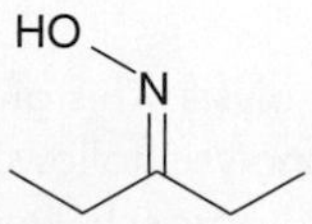

(f) 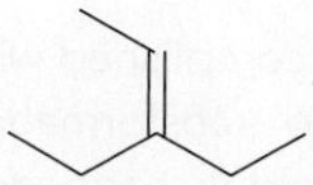

(g) 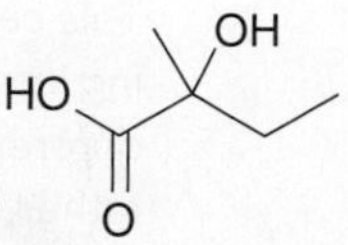

(h) 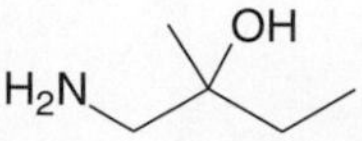

need more **PRACTICE?** **Try Problems 20.55, 20.58, 20.67–20.69, 20.71, 20.75**

20.13 Spectroscopic Analysis of Aldehydes and Ketones

Aldehydes and ketones exhibit several characteristic signals in their infrared (IR) and nuclear magnetic resonance (NMR) spectra. We will now summarize these characteristic signals.

IR Signals

The carbonyl group produces a strong signal in an IR spectrum, generally around 1715 or 1720 cm^{-1}. However, a conjugated carbonyl will produce a signal at a lower wavenumber as a result of electron delocalization via resonance effects:

LOOKING BACK
For an explanation of this effect, see Section 15.3.

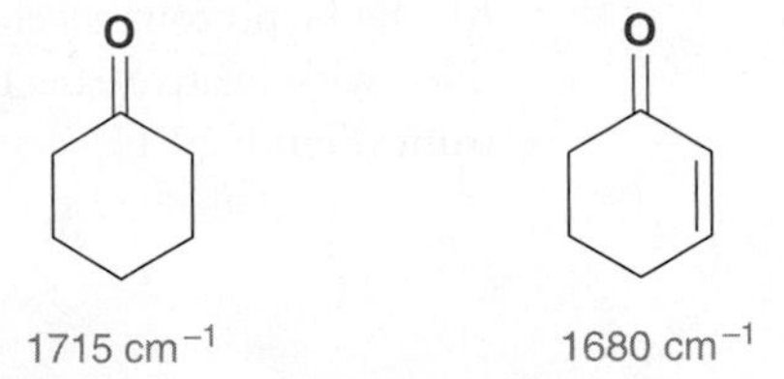

Ring strain has the opposite effect on a carbonyl group. That is, increasing ring strain tends to increase the wavenumber of absorption:

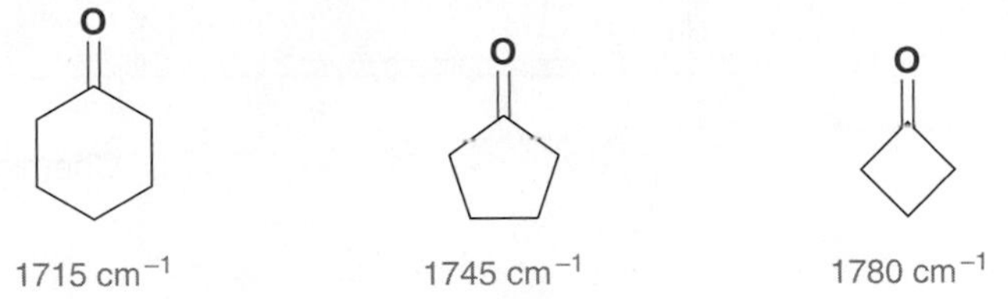

Aldehydes generally exhibit one or two signals (C–H stretching) between 2700 and 2850 cm^{-1} (Figure 20.10) in addition to the C=O stretch.

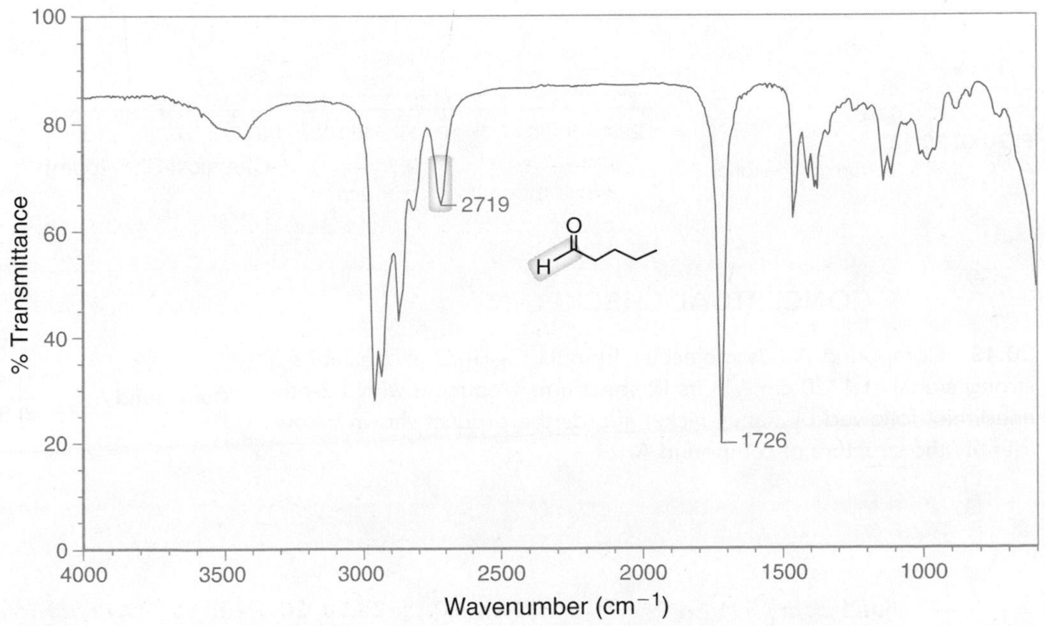

FIGURE 20.10
An IR spectrum of an aldehyde.

^{1}H NMR Signals

In a ^{1}H NMR spectrum, the carbonyl group itself does not produce a signal. However, it has a pronounced effect on the chemical shift of neighboring protons. We saw in Section 16.5 that a carbonyl group typically adds approximately +1 ppm to the chemical shift of its neighbors:

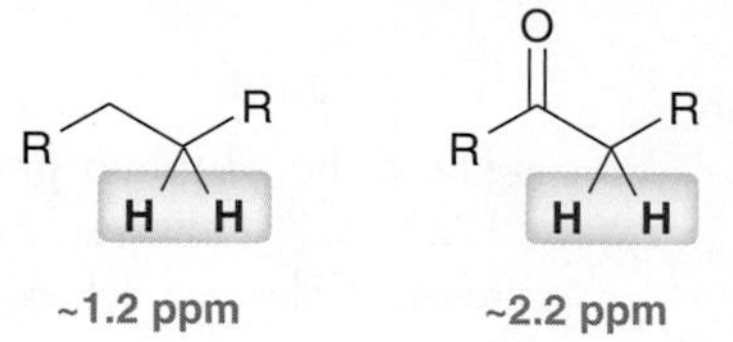

Aldehydic protons generally produce signals around 10 ppm. These signals can usually be identified with relative ease, because very few signals appear that far downfield in a ^{1}H NMR spectrum (Figure 20.11).

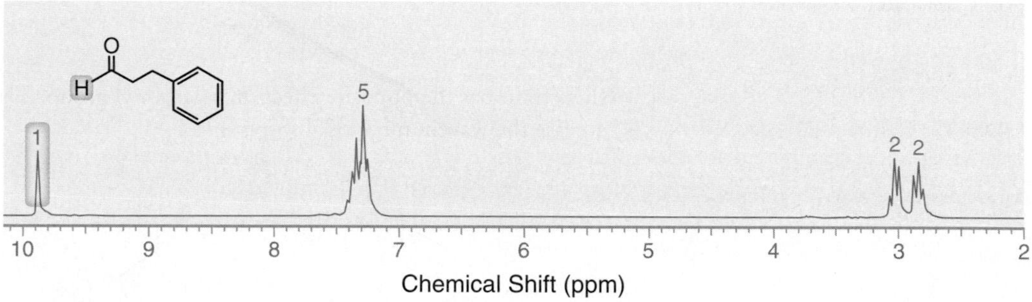

FIGURE 20.11
A ^{1}H NMR spectrum of an aldehyde.

^{13}C NMR Signals

In a ^{13}C NMR spectrum, a carbonyl group of a ketone or aldehyde will generally produce a weak signal near 200 ppm. This signal can often be identified with relative ease, because very few signals appear that far downfield in a ^{13}C NMR spectrum (Figure 20.12).

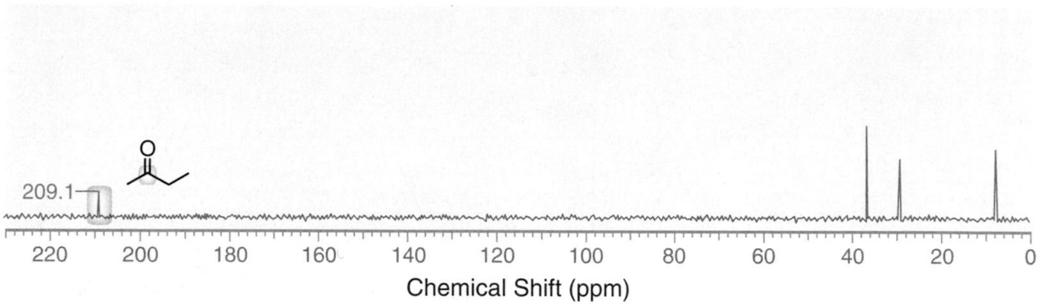

FIGURE 20.12
A ^{13}C NMR spectrum of a ketone.

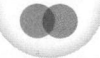

CONCEPTUAL CHECKPOINT

20.43 Compound A has molecular formula $C_{10}H_{10}O$ and exhibits a strong signal at 1720 cm^{-1} in its IR spectrum. Treatment with 1,2-ethanedithiol followed by Raney nickel affords the product shown below. Identify the structure of compound A.

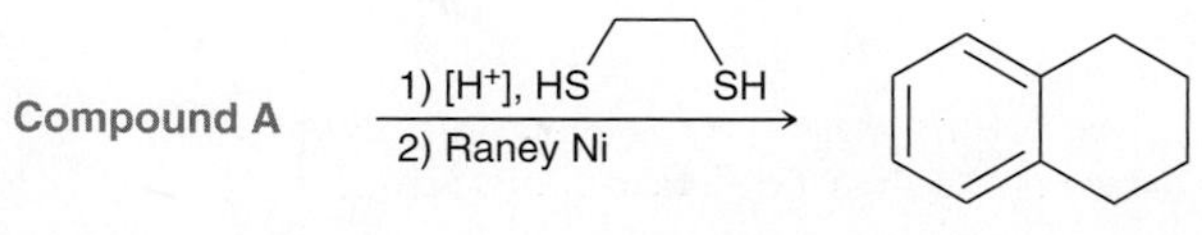

REVIEW OF REACTIONS

SYNTHETICALLY USEFUL REACTIONS

1. Hydrate Formation
2. Acetal Formation
3. Cyclic Acetal Formation
4. Cyclic Thioacetal Formation
5. Desulfurization
6. Imine Formation
7. Enamine Formation
8. Oxime Formation
9. Hydrazone Formation
10. Wolff-Kishner Reduction
11. Reduction of a Ketone
12. Grignard Reaction
13. Cyanohydrin Formation
14. Wittig Reaction
15. Baeyer-Villiger Oxidation

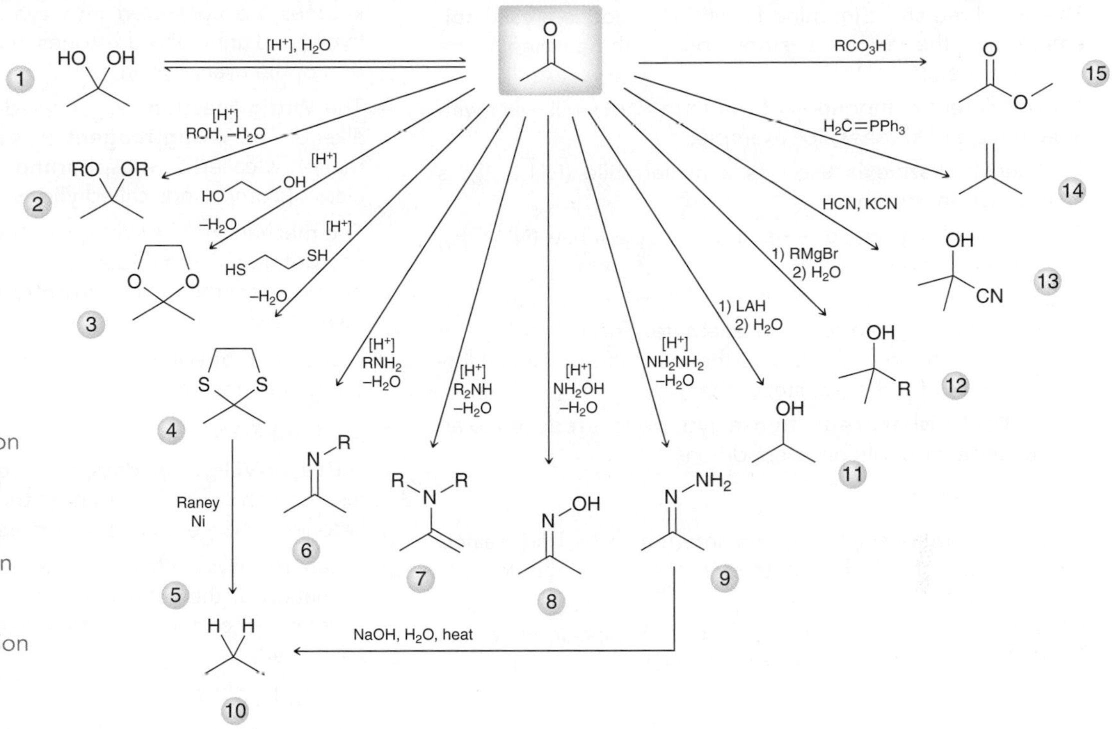

REVIEW OF CONCEPTS AND VOCABULARY

SECTION 20.1

- Both aldehydes and ketones contain a **carbonyl group**, and both are common in nature and industry and occupy a central role in organic chemistry.

SECTION 20.2

- The suffix "-al" indicates an aldehydic group, and the suffix "-one" is used for ketones.
- In naming aldehydes and ketones, locants should be assigned so as to give the carbonyl group the lowest number possible.

SECTION 20.3

- Aldehydes can be prepared via oxidation of primary alcohols, ozonolysis of alkenes, or hydroboration-oxidation of terminal alkynes.
- Ketones can be prepared via oxidation of secondary alcohols, ozonolysis of alkenes, acid-catalyzed hydration of terminal alkynes, or Friedel-Crafts acylation.

SECTION 20.4

- The electrophilicity of a carbonyl group derives from resonance effects as well as inductive effects.
- Aldehydes are more reactive than ketones as a result of steric effects and electronic effects.
- A general mechanism for nucleophilic addition under basic conditions involves two steps:
  1. Nucleophilic attack to generate a **tetrahedral intermediate**.
  2. Proton transfer
- The position of equilibrium is dependent on the ability of the nucleophile to function as a leaving group.

SECTION 20.5

- When an aldehyde or ketone is treated with water, the carbonyl group can be converted into a **hydrate**. The equilibrium generally favors the carbonyl group, except in the case of very simple aldehydes, or ketones with strong electron-withdrawing substituents.
- In acidic conditions, an aldehyde or ketone will react with two molecules of alcohol to form an **acetal**.
- In the presence of an acid, the carbonyl group is protonated to form a very powerful electrophile.
- The mechanism for acetal formation can be divided into two parts:
  1. The first three steps produce a **hemiacetal**.
  2. The last four steps convert the hemiacetal to an acetal.
- For many simple aldehydes, the equilibrium favors formation of the acetal; however, for most ketones, the equilibrium favors reactants rather than products.
- An aldehyde or ketone will react with one molecule of a diol to form a cyclic acetal.
- The reversibility of acetal formation enables acetals to function as protecting groups for ketones or aldehydes. Acetals are stable under strongly basic conditions.
- Hemiacetals are generally difficult to isolate unless they are cyclic.

SECTION 20.6

- In acidic conditions, an aldehyde or ketone will react with a primary amine to form an **imine**.
- The first three steps in imine formation produce a **carbinolamine**, and the last three steps convert the carbinolamine into an imine.
- Many different compounds of the form RNH_2 will react with aldehydes and ketones; for example:
  1. When hydrazine is used as a nucleophile (NH_2NH_2), a **hydrazone** is formed.
  2. When hydroxylamine is used as a nucleophile (NH_2OH), an **oxime** is formed.
- In acidic conditions, an aldehyde or ketone will react with a secondary amine to form an **enamine**. The mechanism of enamine formation is identical to the mechanism of imine formation except for the last step.
- In the **Wolff-Kishner reduction**, a hydrazone is reduced to an alkane under strongly basic conditions.

SECTION 20.7

- In acidic conditions, all reagents, intermediates, and leaving groups either should be neutral (no charge) or should bear one positive charge.
- **Hydrolysis** of acetals, imines, and enamines under acidic conditions produces ketones or aldehydes.

SECTION 20.8

- In acidic conditions, an aldehyde or ketone will react with two equivalents of a thiol to form a **thioacetal**. If a compound with two SH groups is used, a cyclic thioacetal is formed.
- When treated with Raney nickel, thioacetals undergo **desulfurization** to yield a methylene group.

SECTION 20.9

- When treated with a hydride reducing agent, such as lithium aluminum hydride (LAH) or sodium borohydride ($NaBH_4$), aldehydes and ketones are reduced to alcohols.
- The reduction of a carbonyl group with LAH or $NaBH_4$ is not a reversible process, because hydride does not function as a leaving group.

SECTION 20.10

- When treated with a Grignard agent, aldehydes and ketones are converted into alcohols, accompanied by the formation of a new C–C bond.
- Grignard reactions are not reversible, because carbanions do not function as leaving groups.
- When treated with hydrogen cyanide (HCN), aldehydes and ketones are converted into **cyanohydrins**. For most aldehydes and unhindered ketones, the equilibrium favors formation of the cyanohydrin.
- The **Wittig reaction** can be used to convert a ketone to an alkene. The **Wittig reagent** that accomplishes this transformation is called a **phosphorane**, which belongs to a larger class of compounds called **ylides**.
- The mechanism of a Wittig reaction involves initial formation of a **betaine**, which undergoes an intramolecular nucleophilic attack, generating an **oxaphosphetane**. Rearrangement gives the product.
- Preparation of Wittig reagents involves an S_N2 reaction, and the regular restrictions of S_N2 processes apply.

SECTION 20.11

- A **Baeyer-Villiger oxidation** converts a ketone to an ester by inserting an oxygen atom next to the carbonyl group. Cyclic ketones produce cyclic esters called **lactones**.
- When an unsymmetrical ketone is treated with a peroxy acid, formation of the ester is regioselective, and the product is determined by the **migratory aptitude** of each group next to the carbonyl.

SECTION 20.12

- This chapter explored three C–C bond-forming reactions: (1) a Grignard reaction, (2) cyanohydrin formation, and (3) a Wittig reaction.
- This chapter explored only one C–C bond-breaking reaction: the Baeyer-Villiger oxidation.

SECTION 20.13

- Carbonyl groups produce a strong IR signal around 1715 cm^{-1}. A conjugated carbonyl produces a signal at a lower wavenumber, while ring strain increases the wavenumber of absorption.
- Aldehydic C–H bonds exhibit one or two signals between 2700 and 2850 cm^{-1}.
- In a ^1H NMR spectrum, a carbonyl group adds approximately +1 ppm to the chemical shift of its neighbors, and an aldehydic proton produces a signal around 10 ppm.
- In a ^{13}C NMR spectrum, a carbonyl group produces a weak signal near 200 ppm.

KEY TERMINOLOGY

acetal 925
Baeyer-Villiger oxidation 954
betaine 951
carbinolamine 931
carbonyl group 916
cyanohydrins 948
desulfurization 945
enamine 936
hemiacetal 925
hydrate 923
hydrazone 935
hydrolysis 941
imine 931
lactone 954
migratory aptitude 954
oxaphosphetane 951
oxime 935
phosphorane 950
tetrahedral intermediate 921
thioacetal 945
Wittig reaction 950
Wittig reagent 951
Wolff-Kishner reduction 939
ylide 950

SKILLBUILDER REVIEW

20.1 NAMING ALDEHYDES AND KETONES

STEP 1 Choose the longest chain containing the carbonyl group, and number the chain starting from the end closest to the carbonyl group.

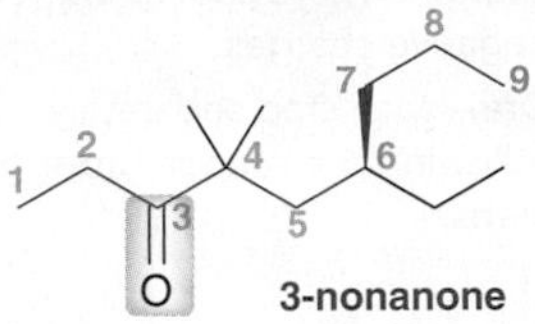

3-nonanone

STEP 2 AND 3 Identify the substituents, and assign locants.

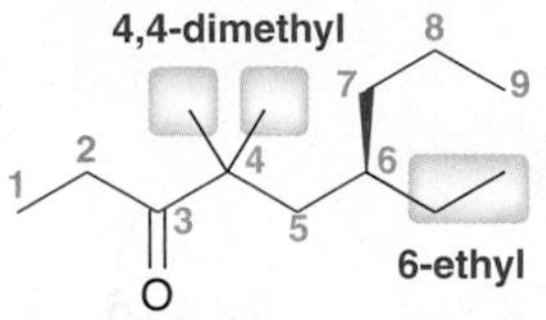

STEP 4 Assemble the substituents alphabetically.

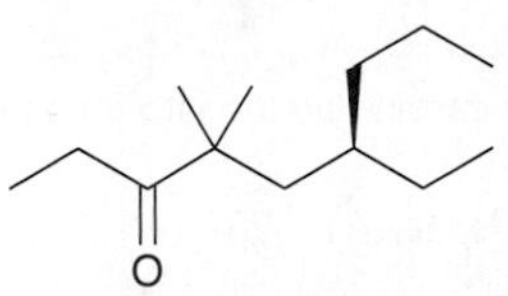

6-ethyl-4,4-dimethyl

STEP 5 Assign the configuration of any chirality center.

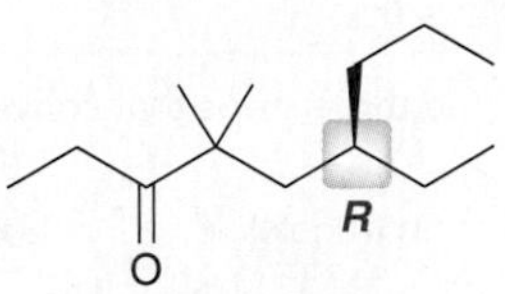

(*R*)-6-ethyl-4,4-dimethyl-3-nonanone

Try problems 20.1–20.4, 20.44–20.49

20.2 DRAWING THE MECHANISM OF ACETAL FORMATION

STEP 1 Draw the three steps necessary for hemiacetal formation.

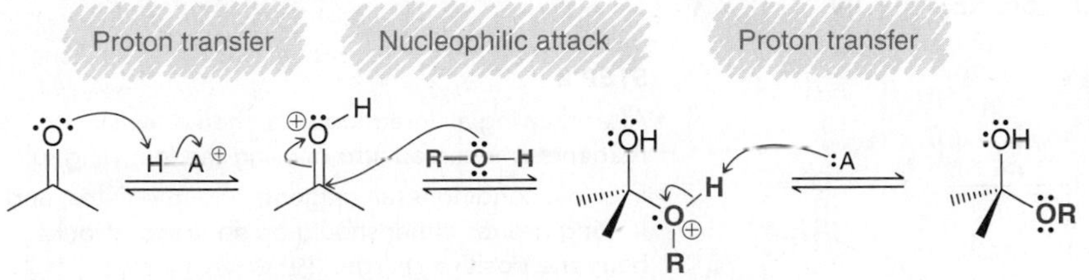

COMMENTS

- Every step has two curved arrows. Draw them precisely.
- Do not forget the charges.
- Draw each step separately.

STEP 2 Draw the four steps that convert the hemiacetal into an acetal.

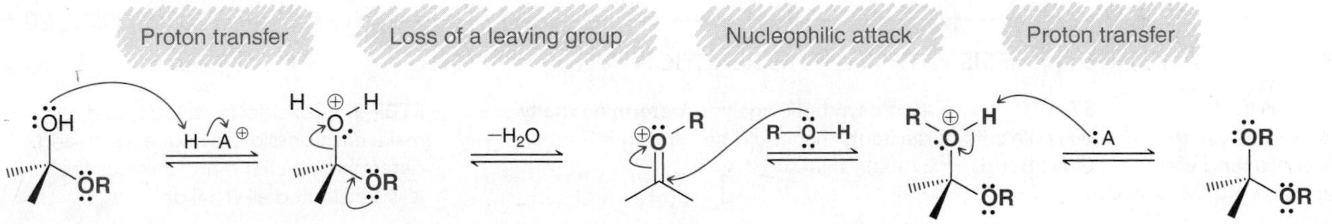

Try Problems 20.8–20.10, 20.57, 20.62, 20.67

20.3 DRAWING THE MECHANISM OF IMINE FORMATION

STEP 1 Draw the three steps necessary for carbinolamine formation.

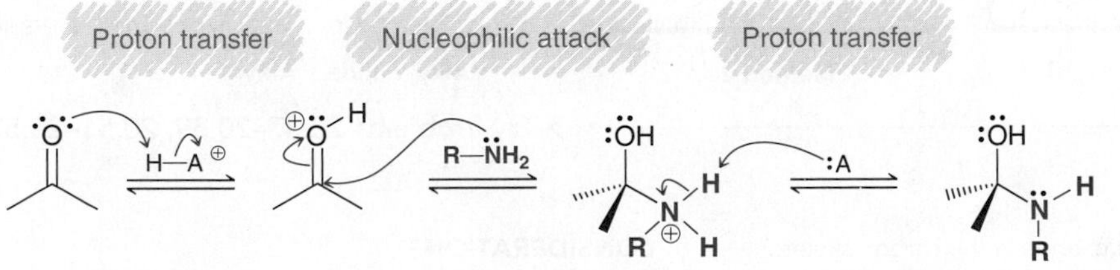

STEP 2 Draw the three steps that convert the carbinolamine into an imine.

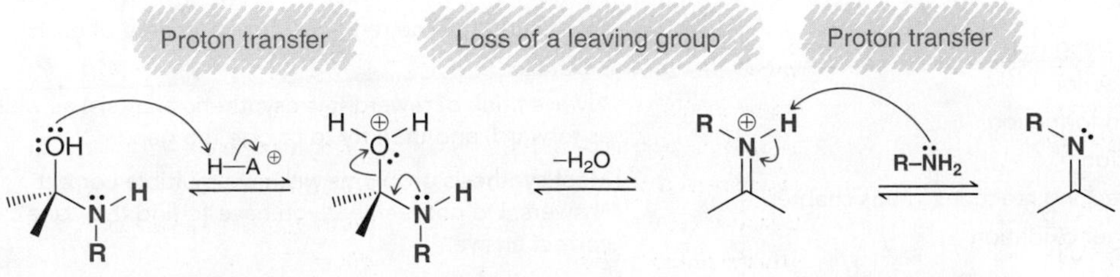

COMMENTS

- Each part of the mechanism begins with a proton transfer and ends with a proton transfer.
- Every step has two curved arrows. Make sure to draw them precisely.
- Do not forget the positive charges. There should be no negative charges.
- Draw each step separately, following the precise order of steps.

Try Problems 20.15–20.18, 20.72

20.4 DRAWING THE MECHANISM OF ENAMINE FORMATION

STEP 1 Draw the three steps necessary for carbinolamine formation.

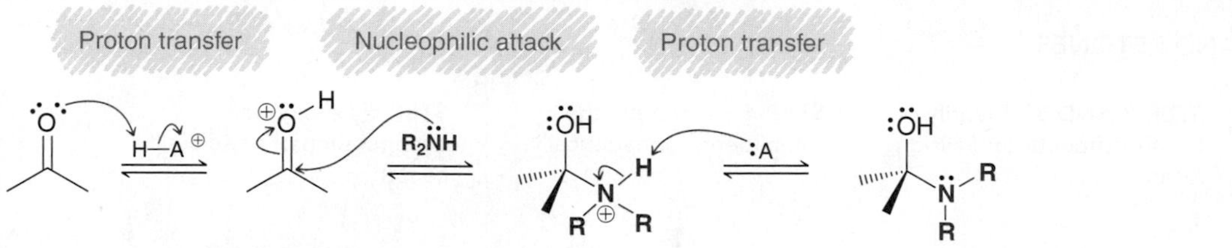

STEP 2 Draw the three steps that convert the carbinolamine into an enamine.

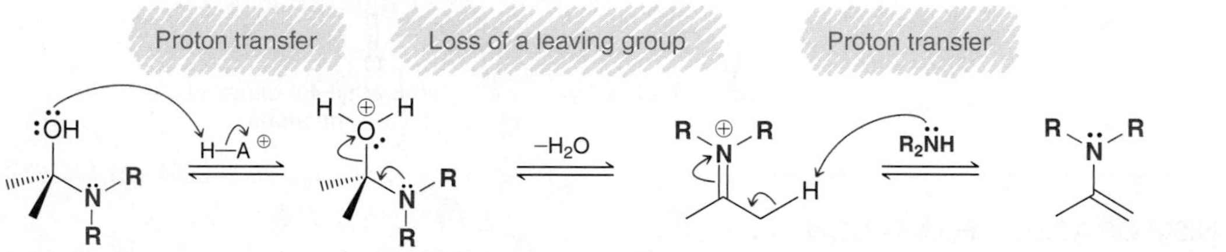

COMMENTS

- Each part of the mechanism begins with a proton transfer and ends with a proton transfer.
- Every step has two curved arrows. Make sure to draw them precisely.
- Do not forget the positive charges. There should be no negative charges.
- Draw each step separately following the precise order of steps.

Try Problems 20.21–20.24, 20.64, 20.65a, 20.66e, 20.75g

20.5 DRAWING THE MECHANISM OF A HYDROLYSIS REACTION

STEP 1 Working backwards, draw all intermediates.

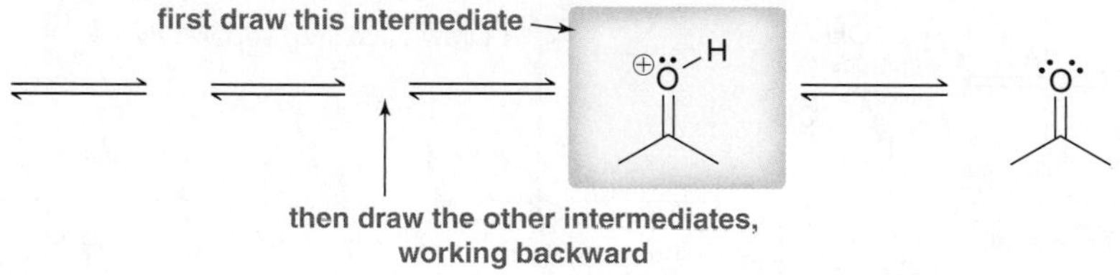

STEP 2

- After drawing all intermediates, then draw all reagents and curved arrows using the following rules:
- In acidic conditions, all reagents, intermediates, and leaving groups either should be neutral or should bear one positive charge.
- Use only those reagents that are already present.

Try Problems 20.26, 20.27, 20.65

20.6 PLANNING AN ALKENE SYNTHESIS WITH A WITTIG REACTION

EXAMPLE Identify the reactants you would use to prepare this compound via a Wittig reaction.

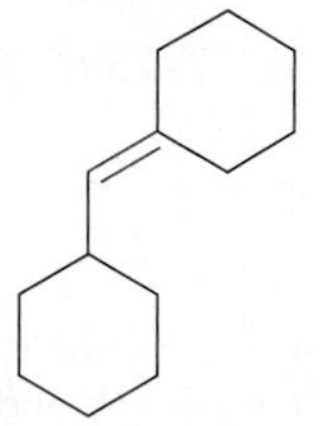

STEP 1 Using a retrosynthetic analysis, determine the two possible sets of reactants that could be used to form the C═C bond.

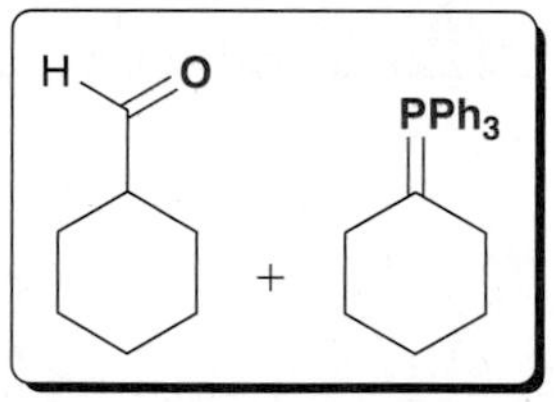

Method 1

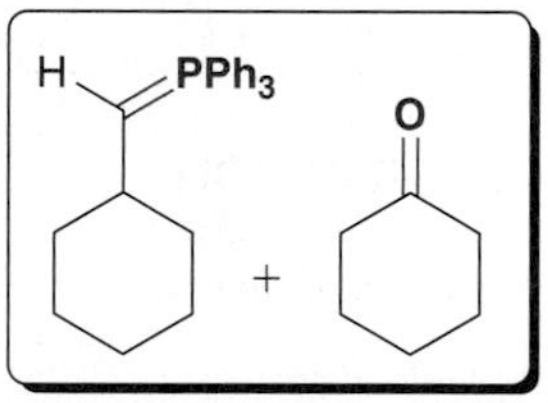

Method 2

STEP 2 Consider how you would make each possible Wittig reagent, and determine which method involves the less substituted alkyl halide.

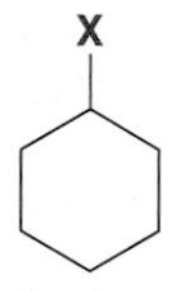

2°
Alkyl halide

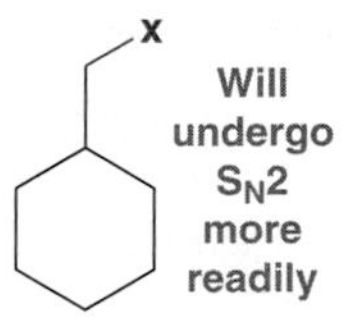

1°
Alkyl halide

Try Problems 20.37–20.39, 20.51–20.53

20.7 PROPOSING A SYNTHESIS

STEP 1 Begin by asking the following two questions:

1. Is there a change in the carbon skeleton?
2. Is there a change in the functional groups?

STEP 2 If there is a change in the carbon skeleton, consider all of the C–C bond-forming reactions and all of the C–C bond-breaking reactions that you have learned so far

C–C bond-forming reactions in this chapter

- Grignard reaction
- Cyanohydrin formation
- Wittig reaction

C–C bond-breaking reactions in this chapter

- Baeyer-Villiger oxidation

CONSIDERATIONS

Remember that the desired product should be the major product of your proposed synthesis.

Make sure that the regiochemical outcome of each step is correct.

Always think backward (retrosynthetic analysis) as well as forward, and then try to bridge the gap.

Most synthesis problems will have multiple correct answers. Do not feel that you have to find the "one" correct answer.

Try Problems 20.41, 20.42, 20.55, 20.58, 20.67–20.69, 20.71, 20.75

PRACTICE PROBLEMS

Note: Most of the Problems are available within *WileyPLUS*, an online teaching and learning solution.

20.44 Provide a systematic (IUPAC) name for each of the following compounds:

(a) 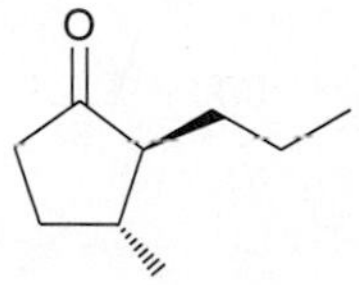

(b) 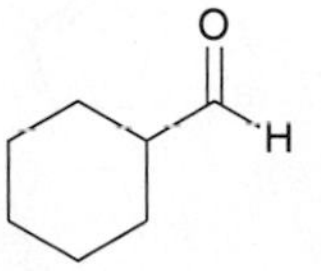

(c) 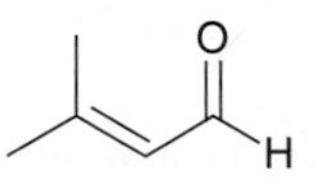

(d) 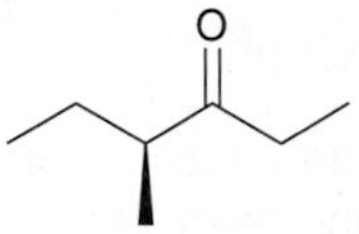

20.45 Draw the structure for each compound below:

(a) propanedial
(b) 4-phenylbutanal
(c) (*S*)-3-phenylbutanal
(d) 3,3,5,5-tetramethyl-4-heptanone
(e) (*R*)-3-hydroxypentanal
(f) *meta*-hydroxyacetophenone
(g) 2,4,6-trinitrobenzaldehyde
(h) tribromoacetaldehyde
(i) (3*R*,4*R*)-3,4-dihydroxy-2-pentanone

20.46 Draw all constitutionally isomeric aldehydes with molecular formula C_4H_8O, and provide a systematic (IUPAC) name for each isomer.

20.47 Draw all constitutionally isomeric aldehydes with molecular formula $C_5H_{10}O$, and provide a systematic (IUPAC) name for each isomer. Which of these isomers possesses a chirality center?

20.48 Draw all constitutionally isomeric ketones with molecular formula $C_6H_{12}O$, and provide a systematic (IUPAC) name for each isomer.

20.49 Explain why the IUPAC name of a compound will never end with the suffix "-1-one."

20.50 For each pair of the following compounds, identify which compound would be expected to react more rapidly with a nucleophile:

(a) 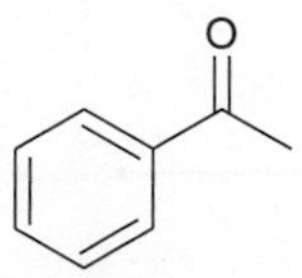 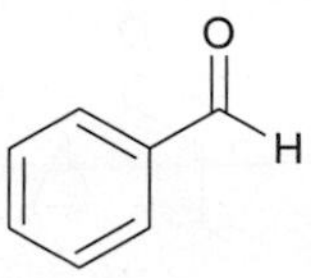

(b) 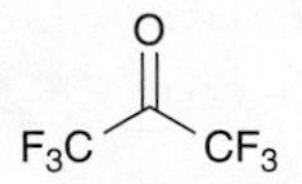 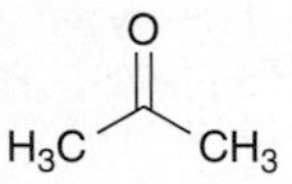

20.51 Draw the products of each Wittig reaction below. If two stereoisomers are possible, draw both stereoisomers:

(a) 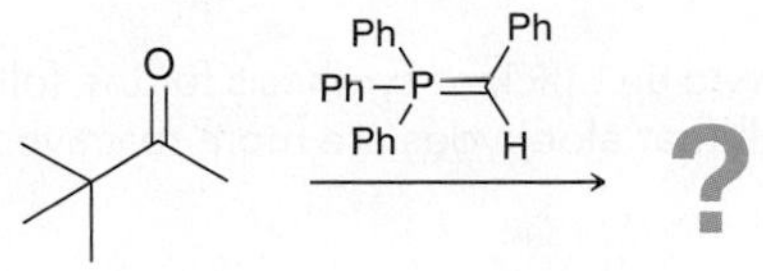

(b) 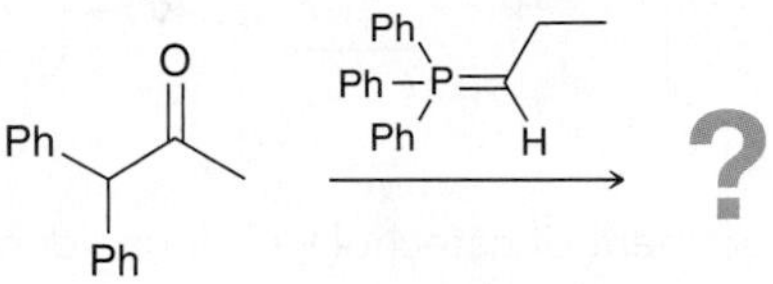

20.52 Draw the structure of the alkyl halide needed to prepare each of the following Wittig reagents, and then determine which Wittig reagent will be the more difficult to prepare. Explain your choice:

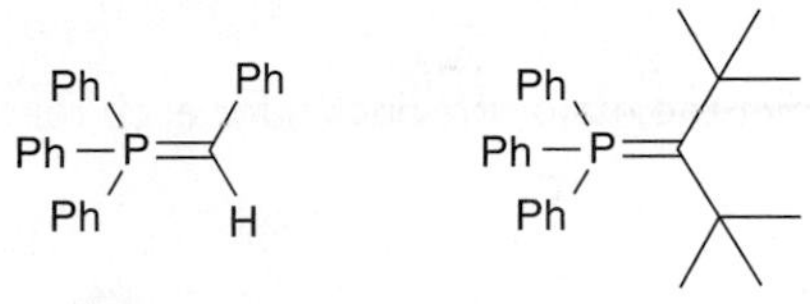

20.53 Show how a Wittig reaction can be used to prepare each of the following compounds. In each case, also show how the Wittig reagent would be prepared:

(a) 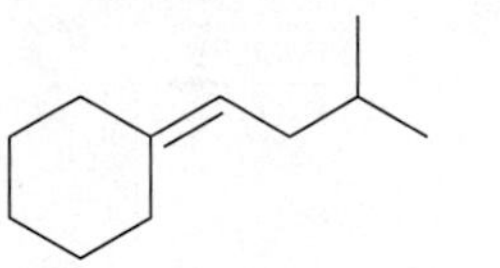

(b) 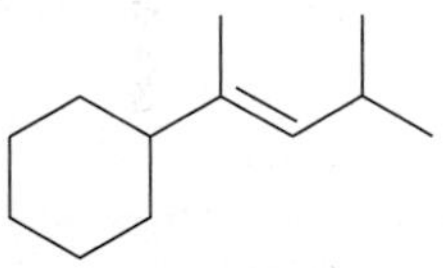

20.54 Choose a Grignard reagent and a ketone that can be used to produce each of the following compounds:

(a) 3-methyl-3-pentanol
(b) 1-ethylcyclohexanol
(c) triphenylmethanol
(d) 5-phenyl-5-nonanol

20.55 You are working in a laboratory, and you are given the task of converting cyclopentene into 1,5-pentanediol. Your first thought is simply to perform an ozonolysis followed by reduction with LAH, but your lab is not equipped for an ozonolysis reaction. Suggest an alternative method for converting cyclopentene into 1,5-pentanediol. For help, see Section 13.4 (reduction of esters to give alcohols).

20.56 Predict the major product(s) from the treatment of acetone with the following compounds:

(a) $[H^+]$, NH_3, $(-H_2O)$
(b) $[H^+]$, CH_3NH_2, $(-H_2O)$
(c) $[H^+]$, excess EtOH, $(-H_2O)$
(d) $[H^+]$, $(CH_3)_2NH$, $(-H_2O)$
(e) $[H^+]$, NH_2NH_2, $(-H_2O)$
(f) $[H^+]$, NH_2OH, $(-H_2O)$
(g) $NaBH_4$, MeOH
(h) MCPBA
(i) HCN, KCN
(j) EtMgBr followed by H_2O
(k) $(C_6H_5)_3P{=}CHCH_2CH_3$
(l) LAH followed by H_2O

20.57 Propose a plausible mechanism for the following transformation:

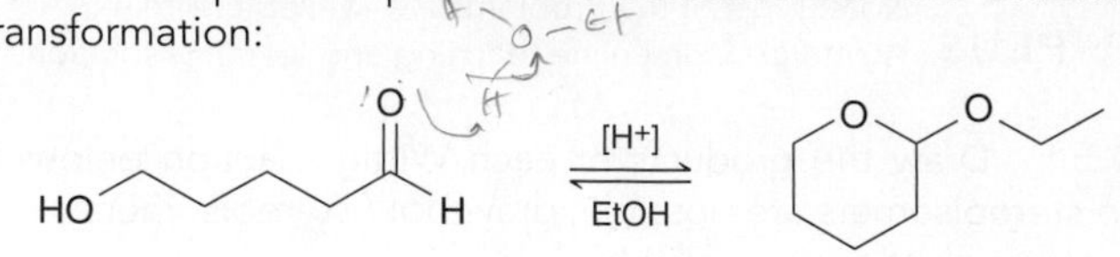

20.58 Devise an efficient synthesis for the following transformation (recall that aldehydes are more reactive than ketones):

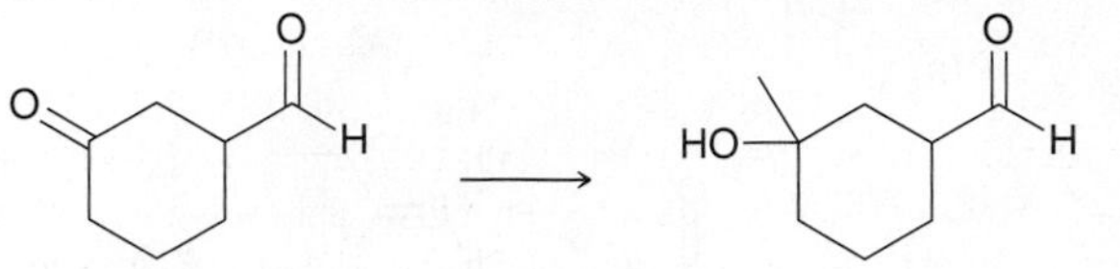

20.59 Treatment of catechol with formaldehyde in the presence of an acid catalyst produces a compound with molecular formula $C_7H_6O_2$. Draw the structure of this product.

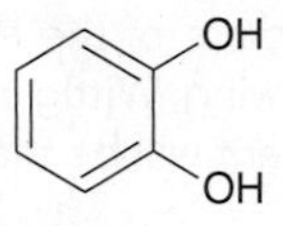

Catechol

20.60 Predict the major product(s) for each reaction below.

(a)

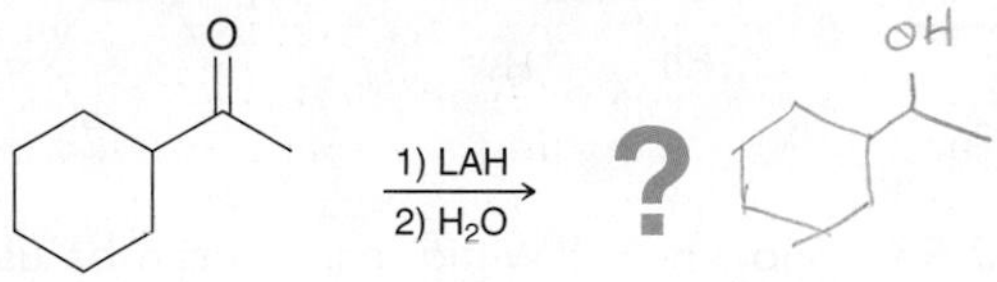

(b)

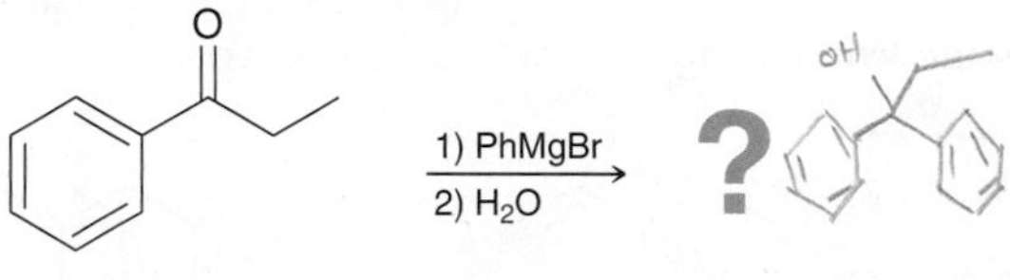

(c)

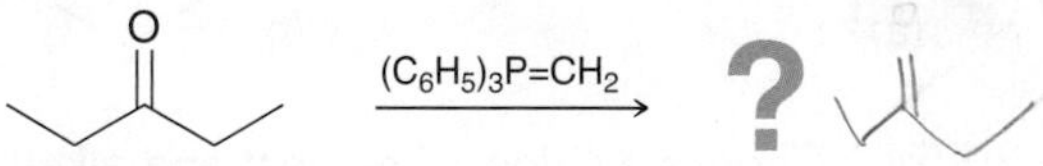

20.61 Starting with cyclopentanone and using any other reagents of your choosing, identify how you would prepare each of the following compounds:

(a) 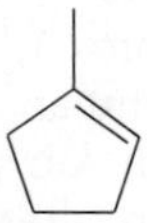 (b) 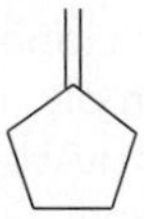 (c) 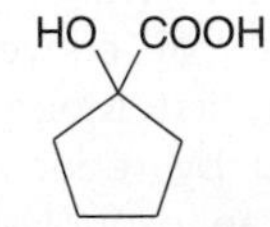 (d) 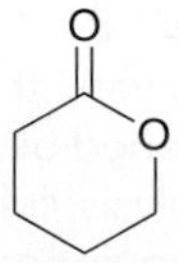

20.62 Glutaraldehyde is a germicidal agent that is sometimes used to sterilize medical equipment too sensitive to be heated in an autoclave. In mildly acidic conditions, glutaraldehyde exists in a cyclic form (below right). Draw a plausible mechanism for this transformation:

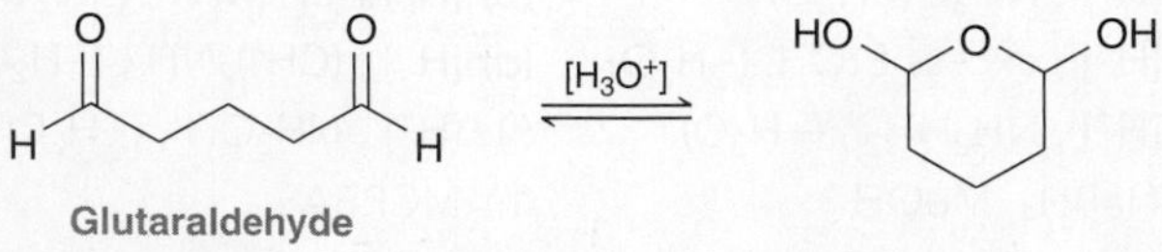

Glutaraldehyde

20.63 Predict the major product(s) obtained when each of the following compounds undergoes hydrolysis in the presence of H_3O^+:

(a) 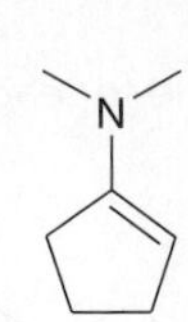 (b) 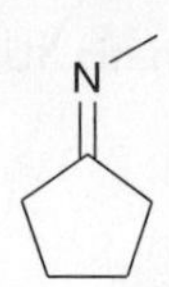 (c) 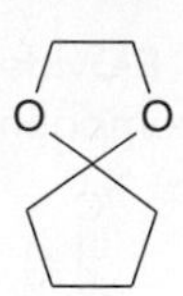

(d) 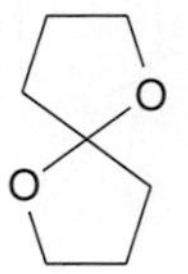 (e) 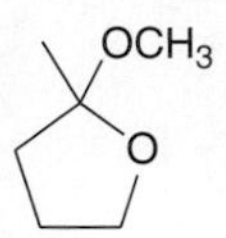

20.64 Identify all of the products formed when the compound below is treated with aqueous acid:

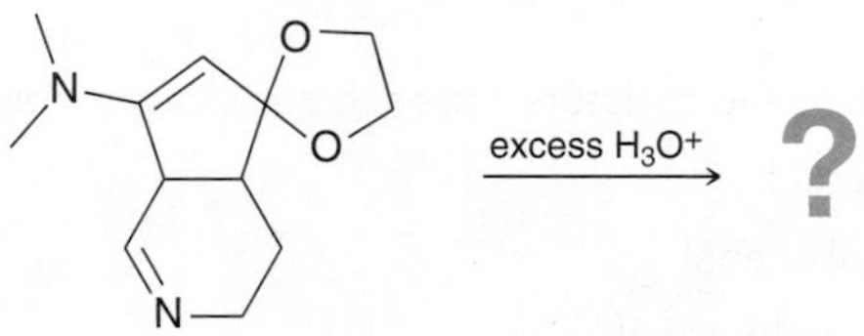

20.65 Draw a plausible mechanism for each of the following transformations:

(a)

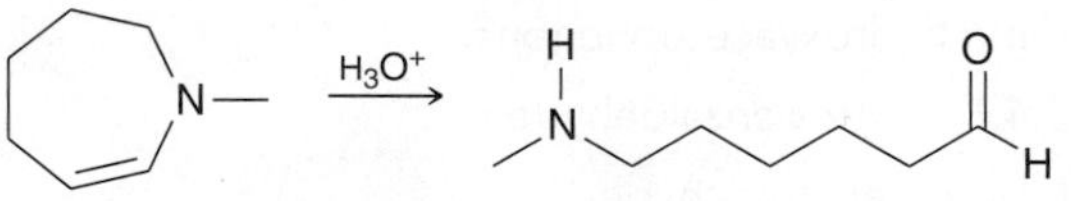

(b)

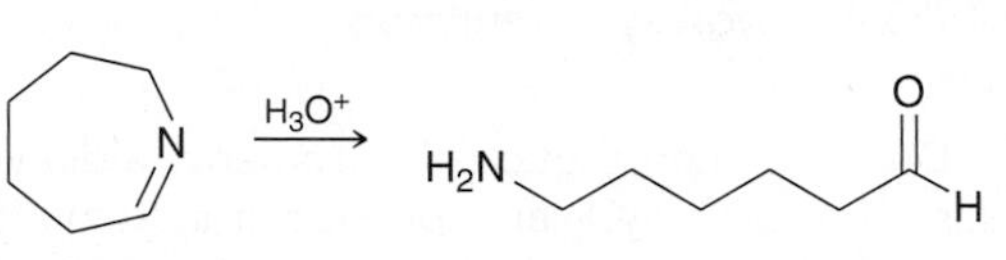

(c)

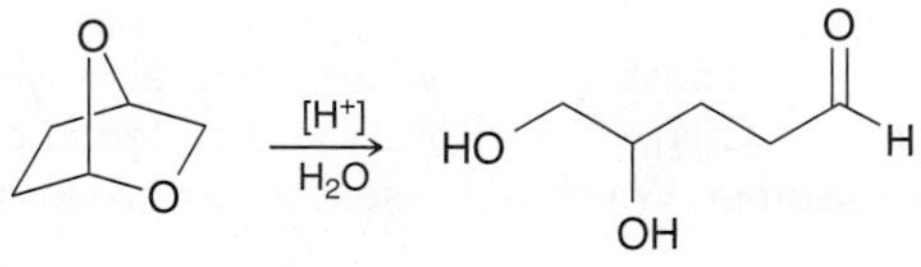

20.66 Predict the major product(s) for each of the following reactions:

(a)

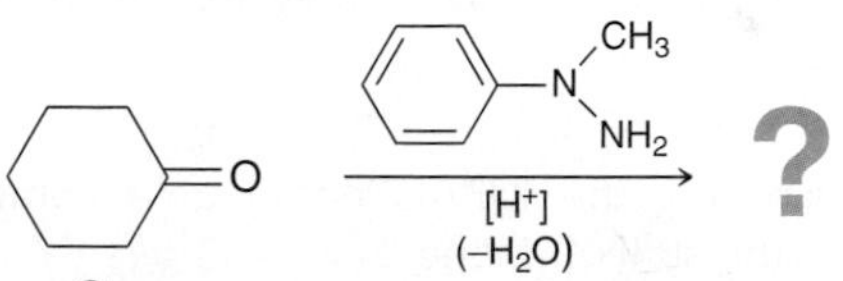

(b)

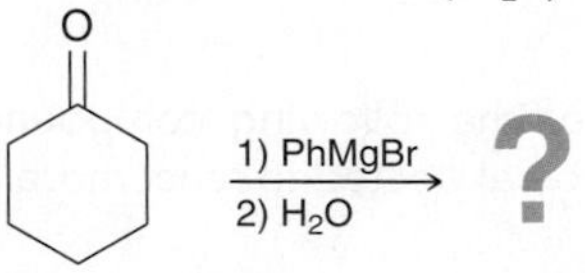

(c)

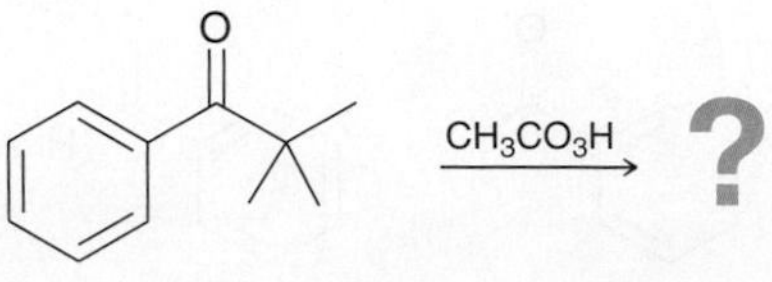

(d)

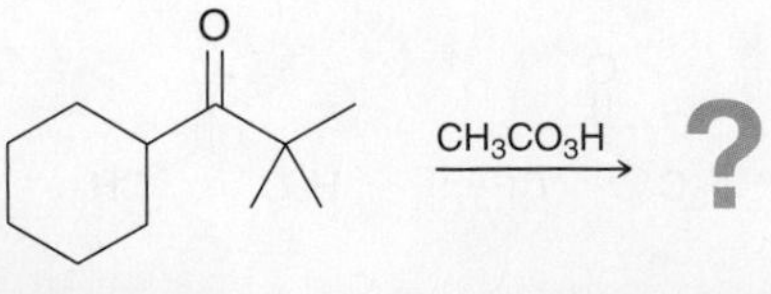

(e) 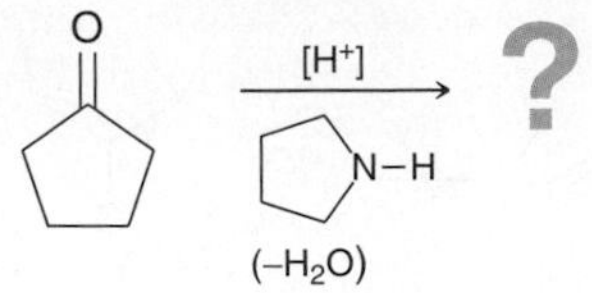

(f) 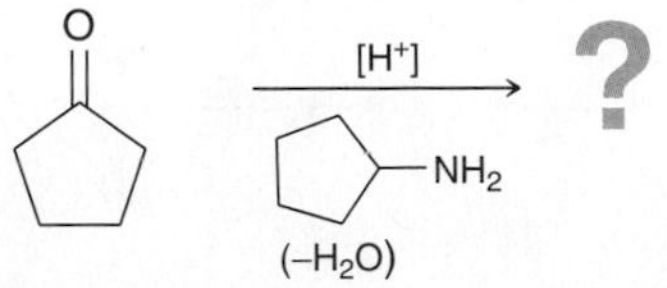

20.67 Identify the starting materials needed to make each of the following acetals:

(a) 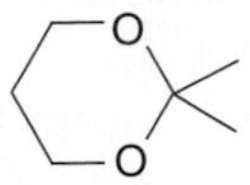 (b) 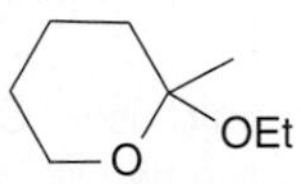 (c) 

20.68 Using ethanol as your only source of carbon atoms, design a synthesis for the following compound:

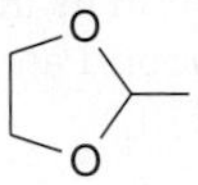

20.69 Propose an efficient synthesis for each of the following transformations:

(a) 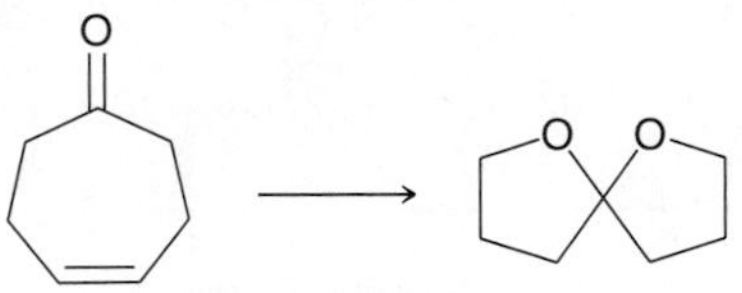

(b) 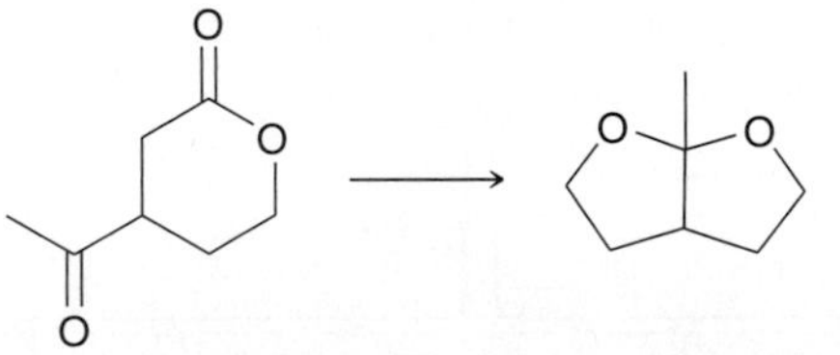

20.70 The compound below is believed to be a wasp pheromone. Draw the major product formed when this compound is hydrolyzed in aqueous acid:

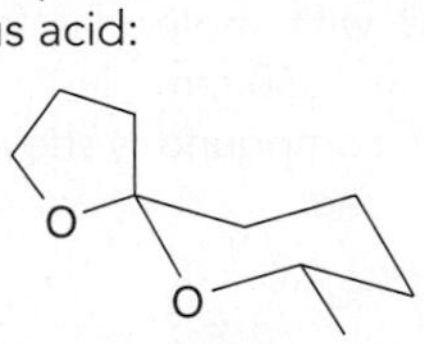

20.71 Propose an efficient synthesis for each of the following transformations:

(a) 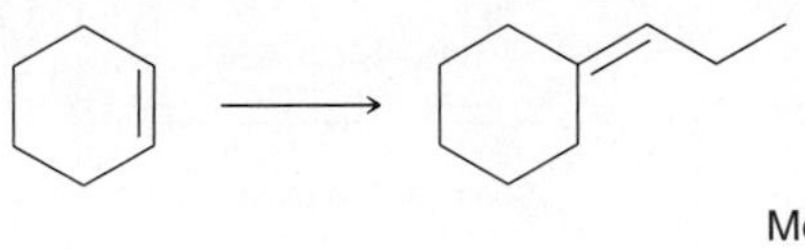

(b) 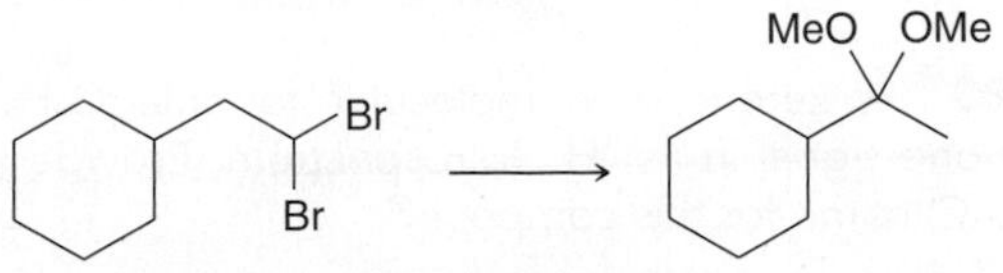

(c) 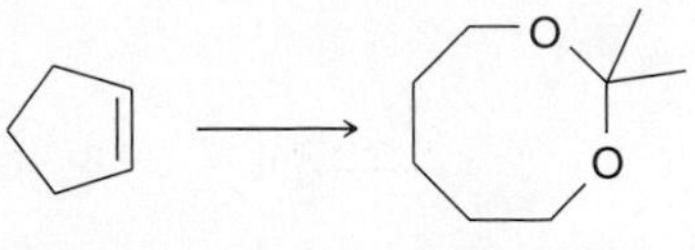

20.72 Draw a plausible mechanism for the following transformation:

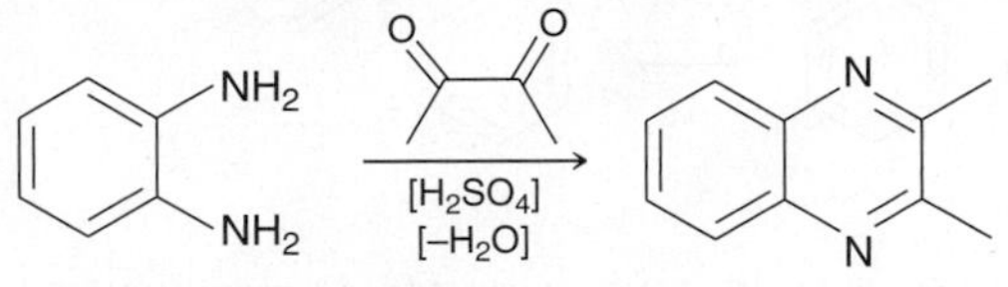

20.73 When cyclohexanone is treated with H_2O, an equilibrium is established between cyclohexanone and its hydrate. This equilibrium greatly favors the ketone, and only trace amounts of the hydrate can be detected. In contrast, when cyclopropanone is treated with H_2O, the resulting hydrate predominates at equilibrium. Suggest an explanation for this curious observation.

20.74 Consider the three constitutional isomers of dioxane ($C_4H_8O_2$):

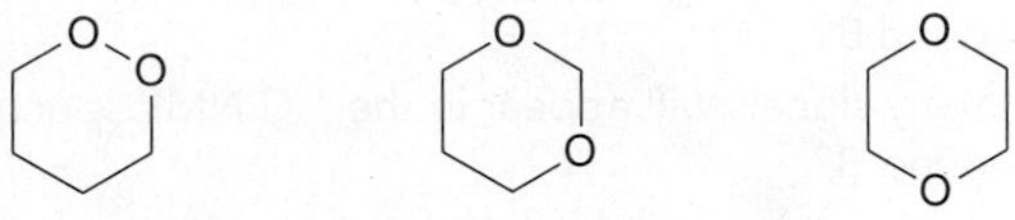

1,2-dioxane **1,3-dioxane** **1,4-dioxane**

One of these constitutional isomers is stable under basic conditions as well as mildly acidic conditions and is therefore used as a common solvent. Another isomer is only stable under basic conditions but undergoes hydrolysis under mildly acidic conditions. The remaining isomer is extremely unstable and potentially explosive. Identify each isomer, and explain the properties of each compound.

20.75 Propose an efficient synthesis for each of the following transformations:

(a) 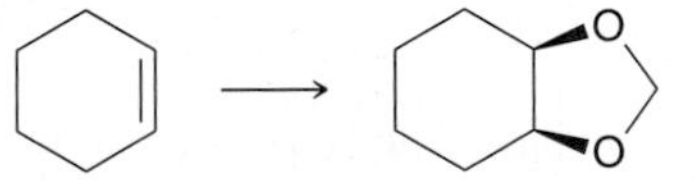

(b) 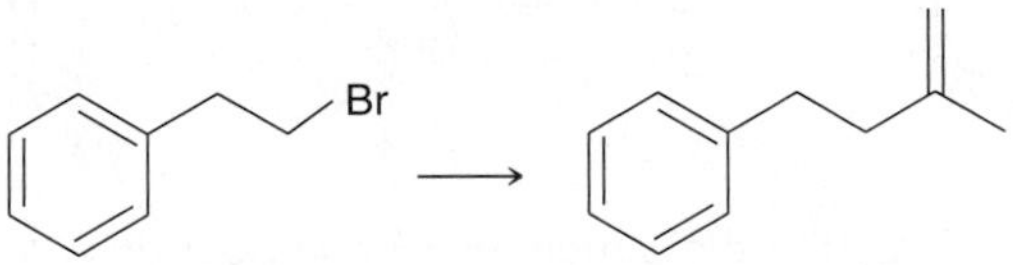

(c) 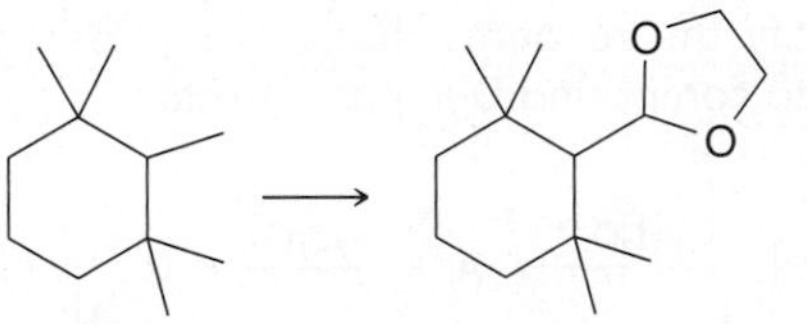

(d) 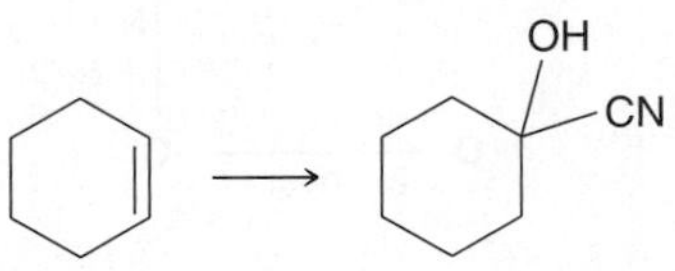

(e) 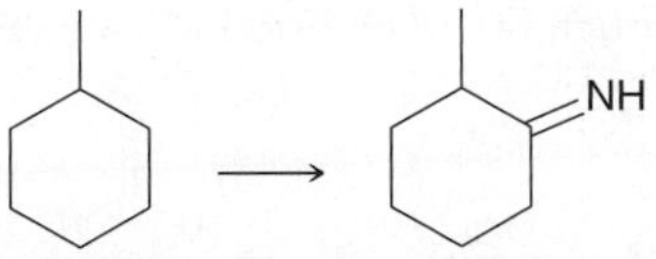

(f) 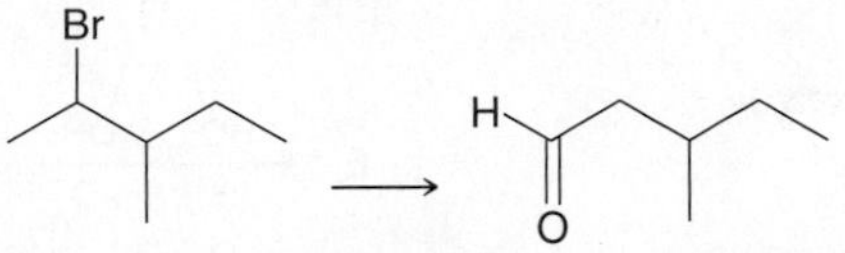

(g) 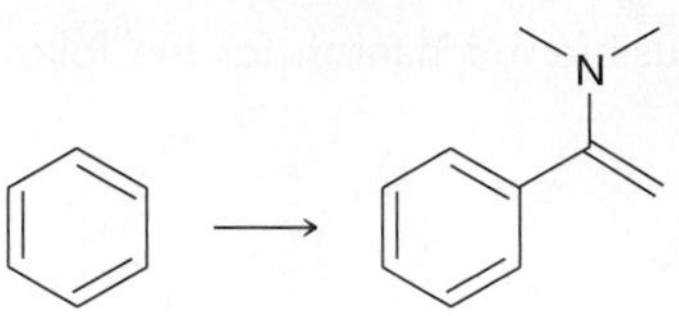

(h) 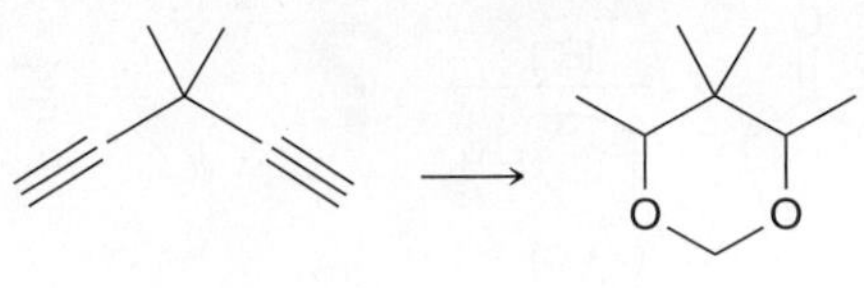

INTEGRATED PROBLEMS

20.76 Compound **A** has molecular formula $C_7H_{14}O$ and reacts with sodium borohydride in methanol to form an alcohol. The ^1H NMR spectrum of compound **A** exhibits only two signals: a doublet (I = 12) and a septet (I = 2). Treating compound **A** with 1,2-ethanedithiol ($HSCH_2CH_2SH$) followed by Raney nickel gives compound **B**.

(a) How many signals will appear in the ^1H NMR spectrum of compound **B**?

(b) How many signals will appear in the ^{13}C NMR spectrum of compound **B**?

(c) Describe how you could use IR spectroscopy to verify the conversion of compound **A** to compound **B**.

20.77 Using the information provided below, deduce the structures of compounds **A**, **B**, **C**, and **D**:

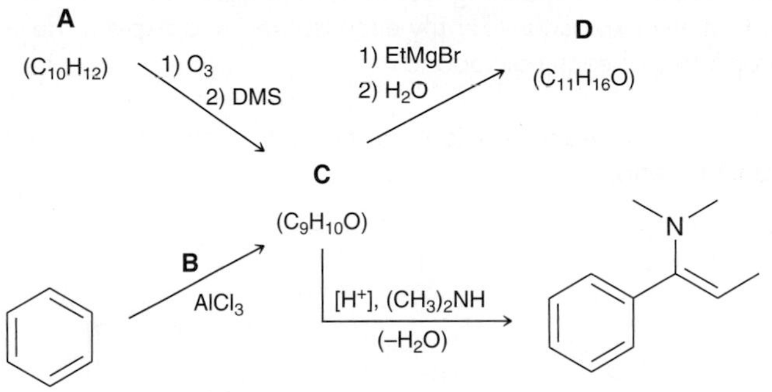

20.78 Identify the structures of compounds **A** to **D** below, and then identify the reagents that can be used to convert cyclohexene into compound **D** in just one step.

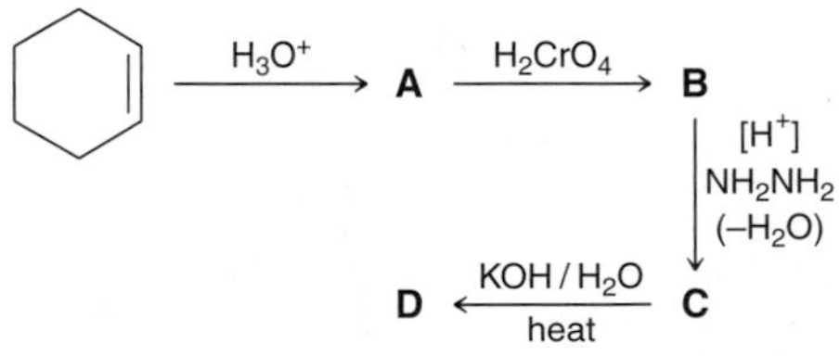

20.79 Identify the structures of compounds **A** to **E** below:

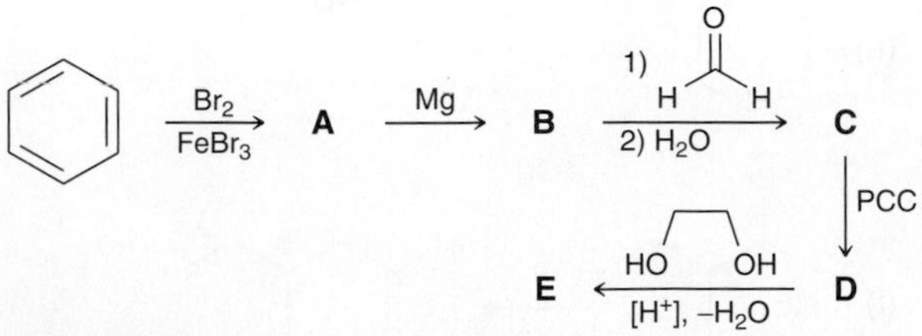

20.80 An aldehyde with molecular formula C_4H_6O exhibits an IR signal at 1715 cm^{-1}.

(a) Propose two possible structures that are consistent with this information.

(b) Describe how you could use ^{13}C NMR spectroscopy to determine which of the two possible structures is correct.

20.81 A compound with molecular formula $C_9H_{10}O$ exhibits a strong signal at 1687 cm^{-1} in its IR spectrum. The ^1H and ^{13}C NMR spectra for this compound are shown below. Identify the structure of this compound.

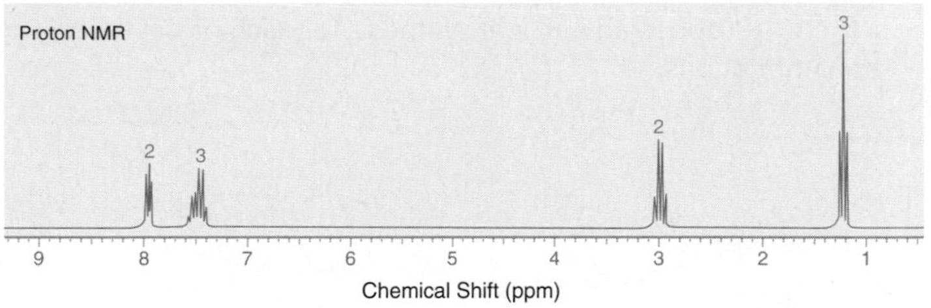

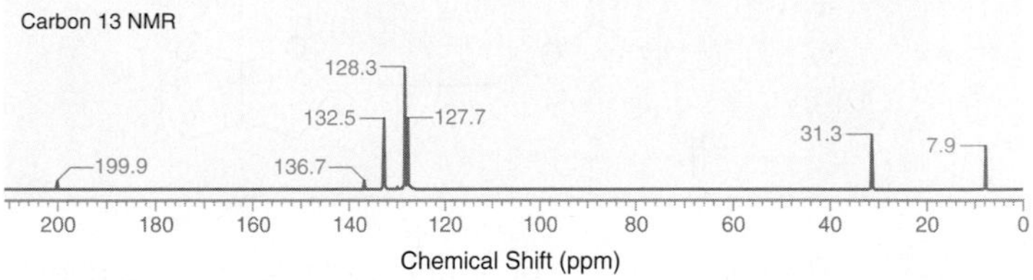

20.82 A compound with molecular formula $C_{13}H_{10}O$ produces a strong signal at 1660 cm^{-1} in its IR spectrum. The ^{13}C NMR spectrum for this compound is shown below. Identify the structure of this compound.

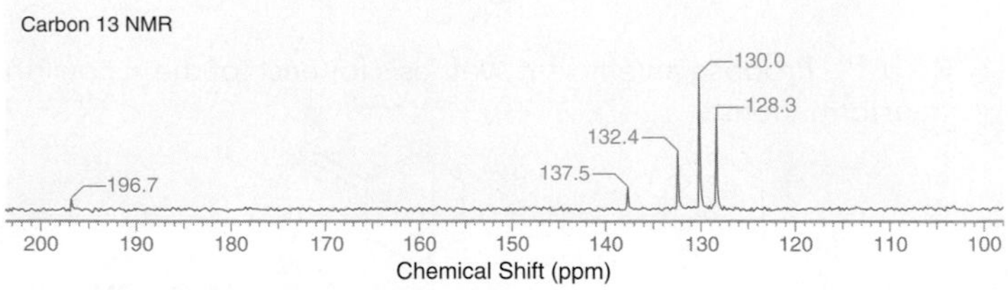

20.83 A ketone with molecular formula $C_9H_{18}O$ exhibits only one signal in its ^1H NMR spectrum. Provide a systematic (IUPAC) name for this compound.

CHALLENGE PROBLEMS

20.84 Draw a plausible mechanism for each of the following transformations:

(a)

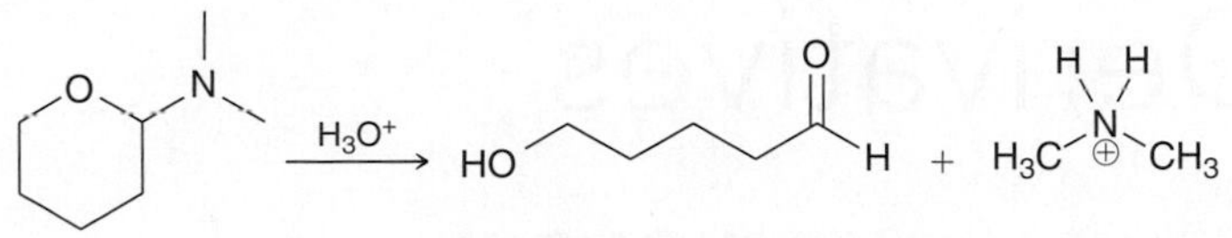

(b)

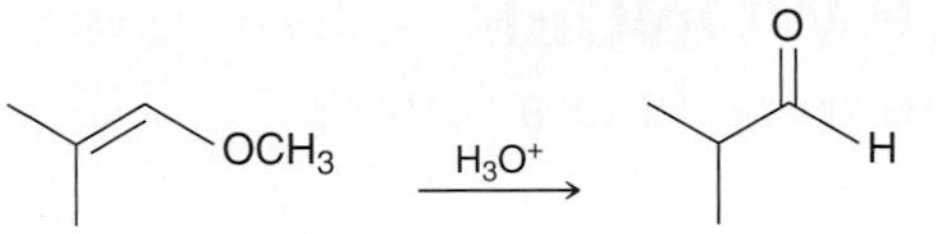

(c)

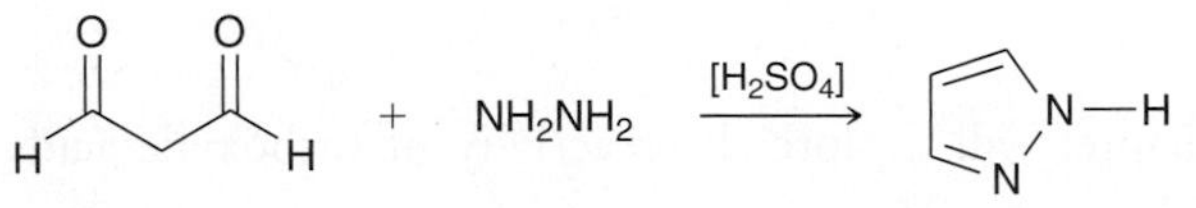

(d)

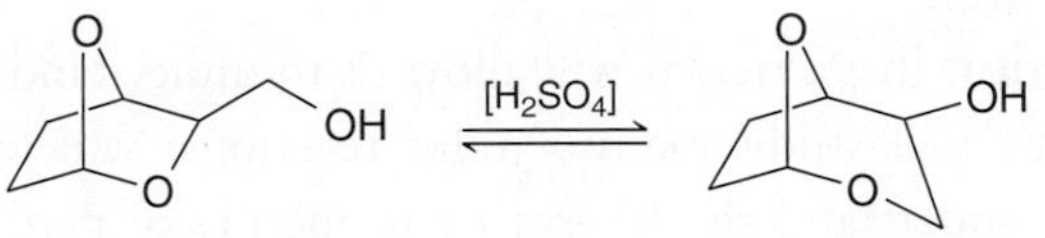

(e)

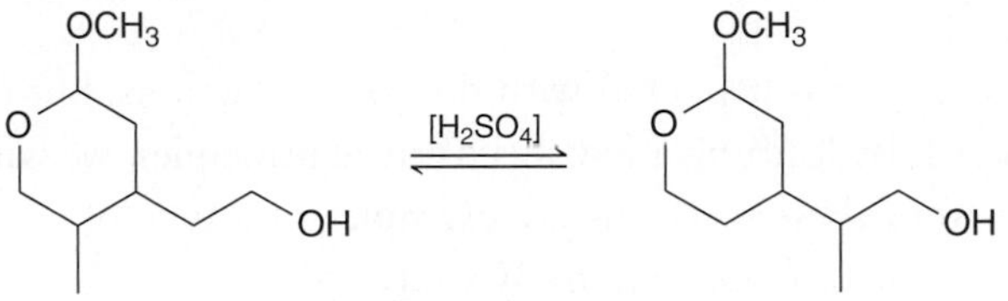

(f)

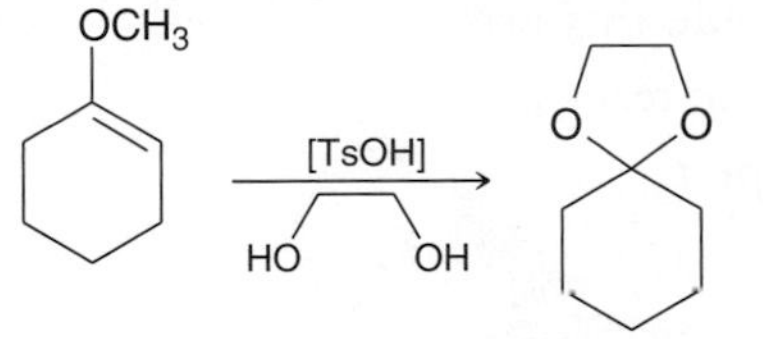

20.85 Under acid-catalyzed conditions, formaldehyde polymerizes to produce a number of compounds, including paraformaldehyde. Draw a plausible mechanism for this transformation:

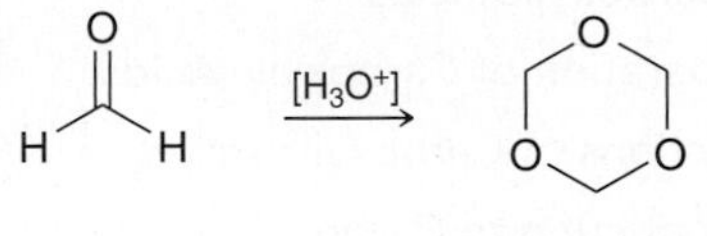

Paraformaldehyde

21

Carboxylic Acids and Their Derivatives

DID YOU EVER WONDER . . .
how aspirin is able to reduce a fever?

This chapter will explore the reactivity of carboxylic acids and their derivatives.

One reaction in particular will allow us to understand how aspirin works. We will study many similar reactions, which will also enable us to understand the life-saving properties of penicillin antibiotics.

This chapter is peppered with dozens of reactions, but don't be discouraged. By learning a few mechanistic principles, we will see that nearly all of these reactions are examples of a type of reaction called *nucleophilic acyl substitution*. By studying the mechanistic principles that guide this process, we will be able to unify the reactions presented in this chapter and greatly reduce the need for memorization.

DO YOU REMEMBER?

Before you go on, be sure you understand the following topics.
If necessary, review the suggested sections to prepare for this chapter.

- Mechanisms and Arrow Pushing (section 6.8)
- Loss of a Leaving Group (section 6.8)
- Drawing Curved Arrows (section 6.10)
- Nucleophilic Attack (section 6.8)
- Proton Transfers (section 6.8)

WILEY PLUS Visit www.wileyplus.com to check your understanding and for valuable practice.

21.1 Introduction to Carboxylic Acids

Carboxylic acids, which were introduced in Section 3.4, are compounds with a —COOH moiety. These compounds are abundant in nature, where they are responsible for some familiar odors.

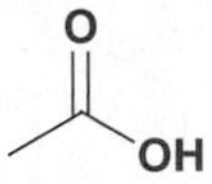

Acetic acid
(Responsible for the pungent smell of vinegar)

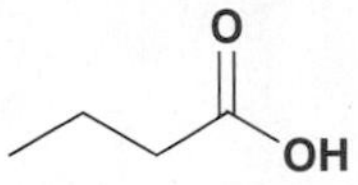

Butanoic acid
(Responsible for the rancid odor of sour butter)

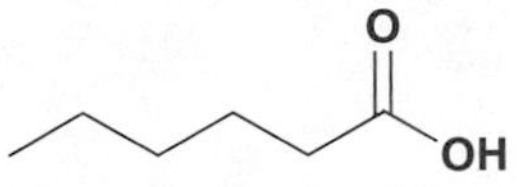

Hexanoic acid
(Responsible for the odor of dirty socks)

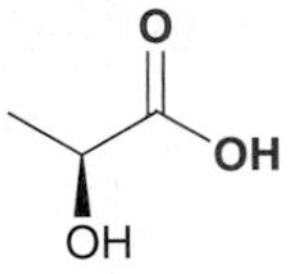

Lactic acid
(Responsible for the taste of sour milk)

Carboxylic acids are also found in a wide range of pharmaceuticals that are used to treat a variety of conditions.

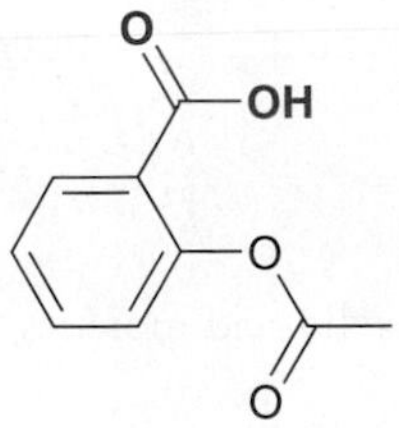

Acetylsalicylic acid
(Aspirin, a widely used analgesic)

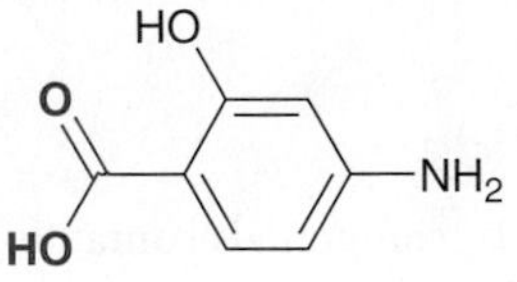

4-Aminosalicylic acid
(Used in the treatment of tuberculosis)

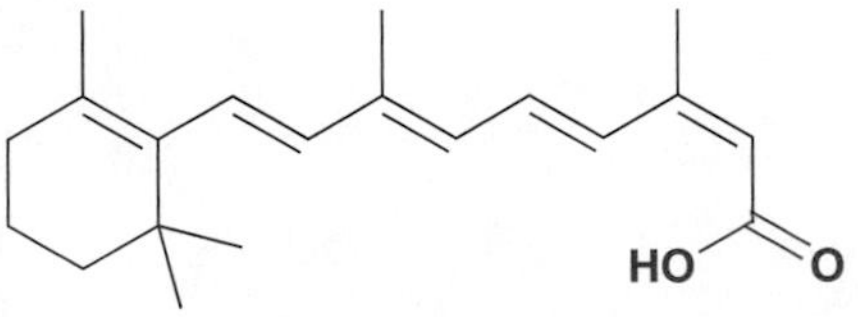

Isotretinoin
(Used in the treatment of acne)

Each year the United States produces over 2.5 million tons of acetic acid from methanol and carbon monoxide. The primary use of acetic acid is in the synthesis of vinyl acetate, which is used in paints and adhesives.

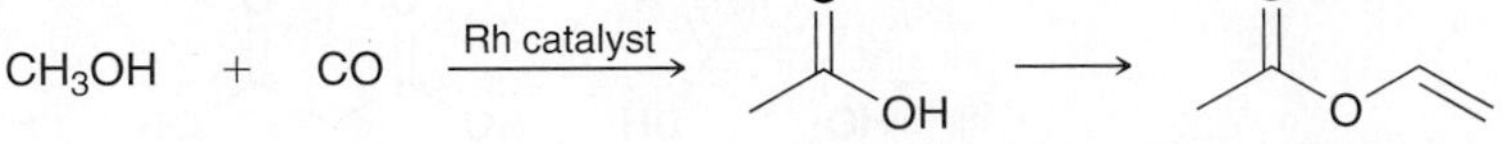

Acetic acid **Vinyl acetate**

Vinyl acetate is a derivative of acetic acid and is therefore said to be a **carboxylic acid derivative**. Carboxylic acids and their derivatives occupy a central role in organic chemistry, as we will see throughout this chapter.

21.2 Nomenclature of Carboxylic Acids

Monocarboxylic Acids

Monocarboxylic acids, compounds containing one carboxylic acid moiety, are named with the suffix "oic acid".

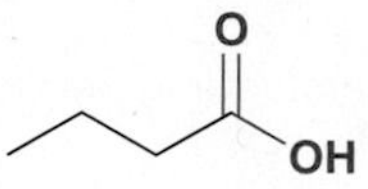

Butanoic acid

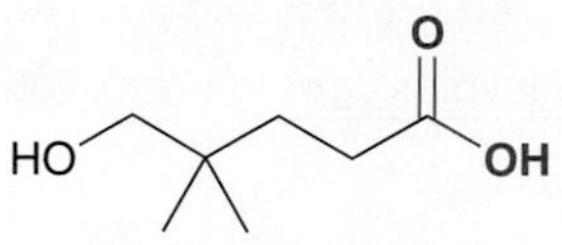

5-Hydroxy-4,4-dimethyl pentanoic acid

The parent is the longest chain that includes the carbon atom of the carboxylic acid moiety. That carbon atom is always assigned number 1 when numbering the parent.

When a carboxylic acid moiety is connected to a ring, the compound is named as an alkane carboxylic acid, for example;

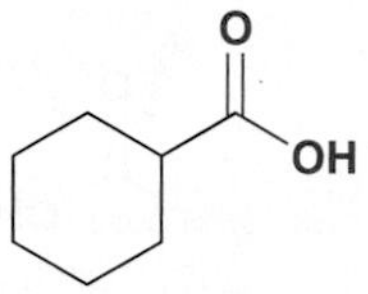

Cyclohexane carboxylic acid

Many simple carboxylic acids have common names accepted by IUPAC. The examples shown here should be committed to memory, as they will appear frequently throughout the chapter.

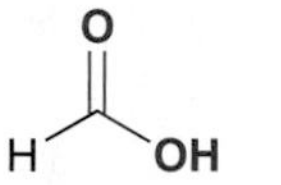

Formic acid

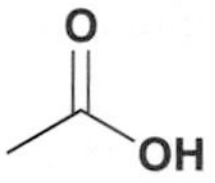

Acetic acid

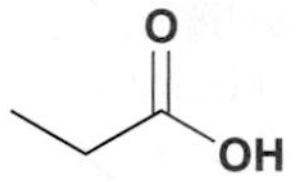

Propionic acid

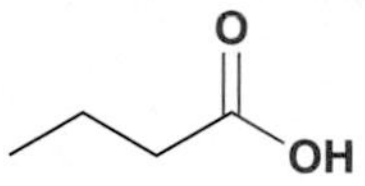

Butyric acid

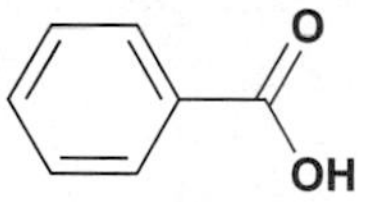

Benzoic acid

Diacids

Diacids, compounds containing two carboxylic acid moieties, are named with the suffix "dioic acid," for example;

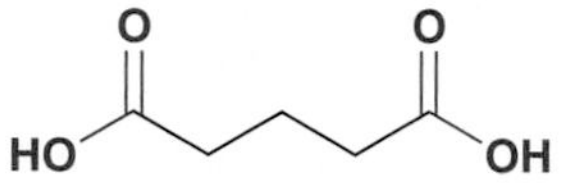

Pentanedioic acid

Many diacids have common names accepted by IUPAC.

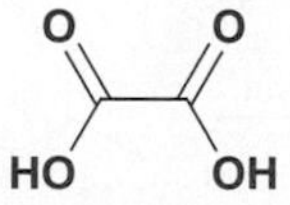

Oxalic acid

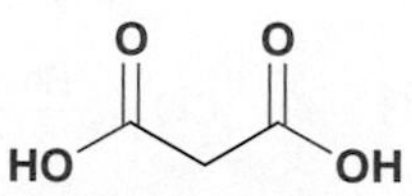

Malonic acid

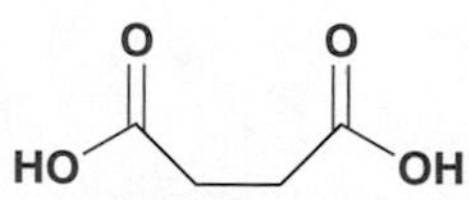

Succinic acid

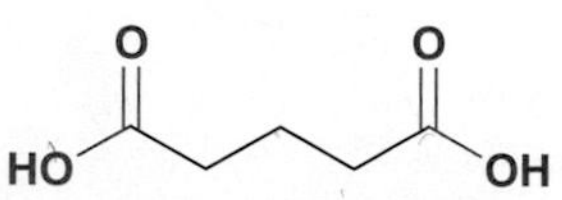

Glutaric acid

These compounds differ from each other only in the number of methylene (CH_2) groups separating the carboxylic acid moieties. These names are used very often in the study of biochemical reactions and should therefore be committed to memory.

CONCEPTUAL CHECKPOINT

21.1 Provide both an IUPAC name and a common name for each of the following compounds:

(a) $HO_2C(CH_2)_3CO_2H$ (b) $CH_3(CH_2)_2CO_2H$

(c) $C_6H_5CO_2H$ (d) $HO_2C(CH_2)_2CO_2H$

(e) CH_3COOH (f) HCO_2H

21.2 Draw the structure of each of the following compounds:

(a) Cyclobutanecarboxylic acid (b) 3,3-Dichlorobutyric acid

(c) 3,3-Dimethylglutaric acid

21.3 Provide an IUPAC name for each of the following compounds:

(a)
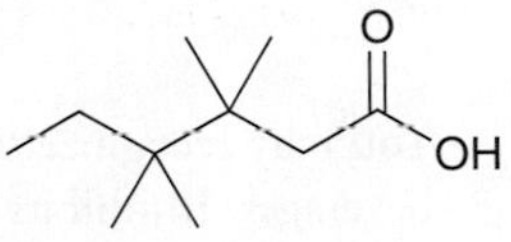

(b)
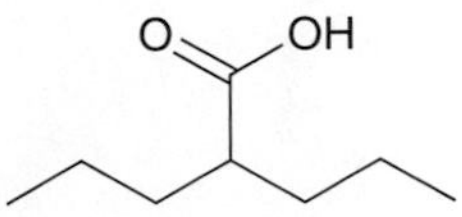

(c)
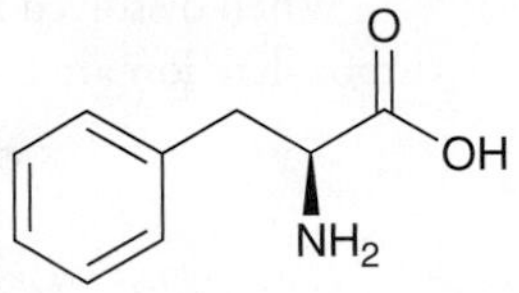

21.3 Structure and Properties of Carboxylic Acids

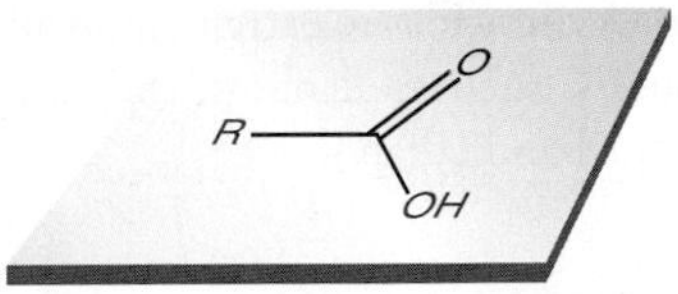

FIGURE 21.1
The carbonyl group as well as the two atoms attached to the carbonyl carbon all reside in a plane.

Structure

The carbon atom of a carboxylic acid moiety is sp^2 hybridized and therefore exhibits trigonal planar geometry with bond angles that are nearly 120° (Figure 21.1). Carboxylic acids can form two hydrogen-bonding interactions, allowing molecules to associate with each other in pairs.

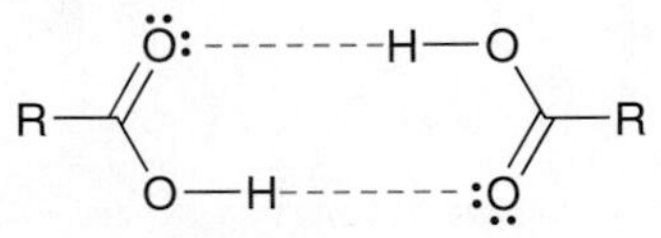

These hydrogen-bonding interactions explain the relatively high boiling points of carboxylic acids. For example, compare the boiling points of acetic acid and ethanol. Acetic acid has a higher boiling point as a result of stronger intermolecular forces.

Acetic acid b.p. = 118°C **Ethanol** b.p. = 78°C

Acidity of Carboxylic Acids

As their name implies, carboxylic acids exhibit mildly acidic protons. Treatment of a carboxylic acid with a strong base, such as sodium hydroxide, yields a carboxylate salt.

A carboxylate salt

LOOKING BACK
Carboxylate ions play an important role in how drugs are distributed throughout the body, as we saw in the Medically Speaking box at the end of Section 3.3 entitled "Drug Distribution and pK_a".

Carboxylate salts are ionic and are therefore more soluble in water than their corresponding carboxylic acids. Carboxylate ions are named by replacing the suffix "ic acid" with "ate", for example:

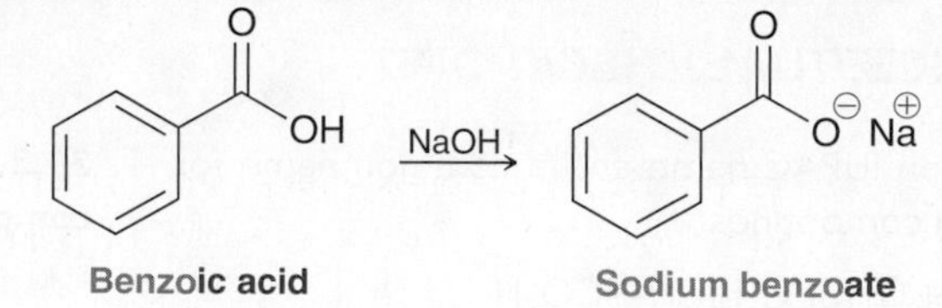

You may recognize the name sodium benzoate, as it is commonly found in food products and beverages. It inhibits the growth of fungi and serves as a food preservative.

When dissolved in water, an equilibrium is established in which the carboxylic acid and the carboxylate ion are both present.

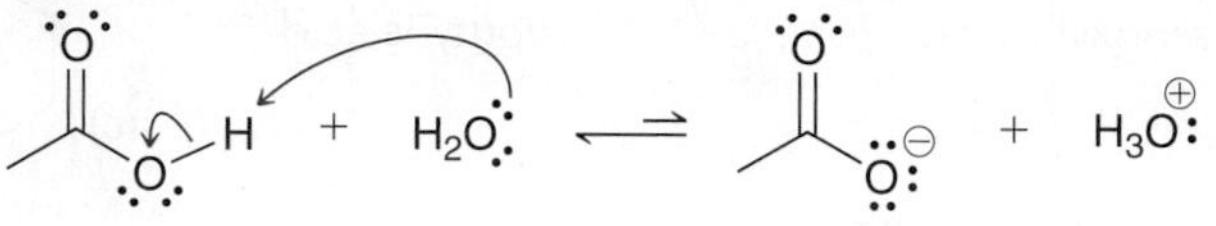

In most cases, the equilibrium significantly favors the carboxylic acid with a K_a usually around 10^{-4} or 10^{-5}. In other words, the pK_a of most carboxylic acids is between 4 and 5.

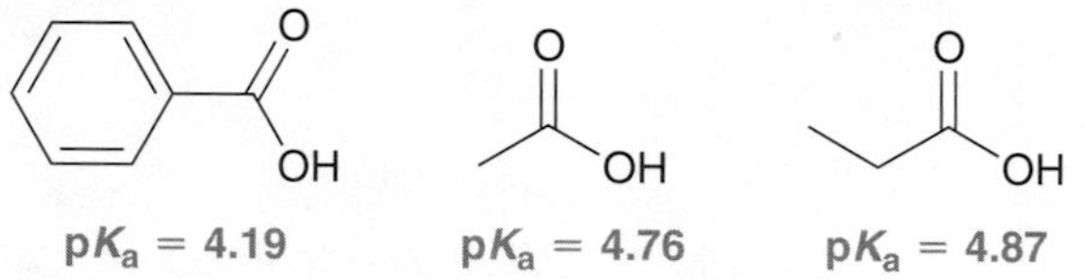

When compared to inorganic acids, such as HCl or H_2SO_4, carboxylic acids are extremely weak acids. But when compared to most classes of organic compounds, such as alcohols, they are relatively acidic. For example, compare the pK_a values of acetic acid and ethanol.

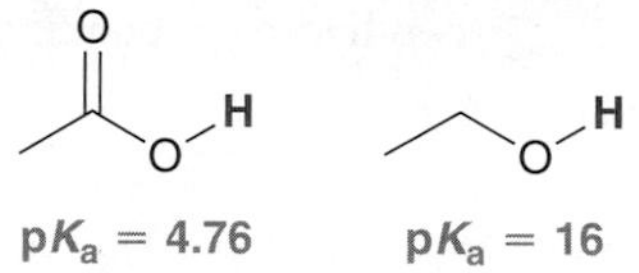

Acetic acid is 11 orders of magnitude more acidic than ethanol (over a hundred billion times more acidic). As explained in Section 3.4, the acidity of carboxylic acids is primarily due to the stability of the conjugate base, which is resonance stabilized.

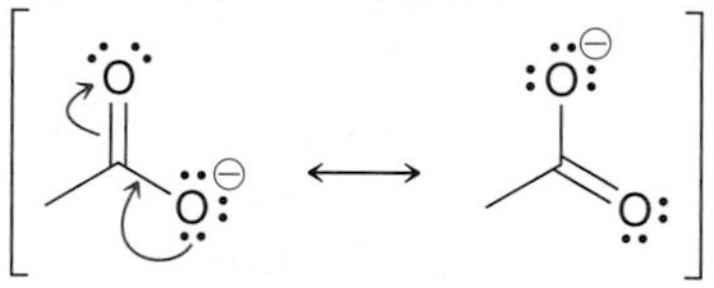

In the conjugate base of acetic acid, the negative charge is delocalized over two oxygen atoms, and it is therefore more stable than the conjugate base of ethanol. The delocalized nature of the charge can be seen in an electrostatic potential map of the acetate ion (Figure 21.2).

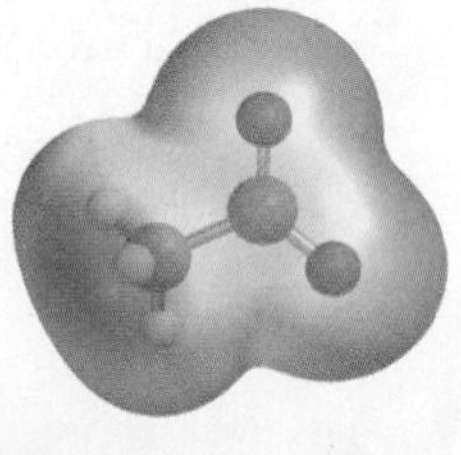

FIGURE 21.2
An electrostatic potential map of the acetate ion showing how the electron density is distributed over both oxygen atoms.

CONCEPTUAL CHECKPOINT

21.4 The following two compounds are constitutional isomers. Identify which of these is expected to be more acidic, and explain your choice.

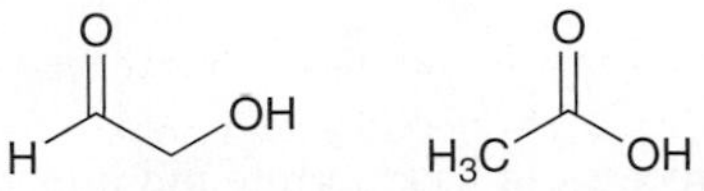

21.5 Consider the structure of *para*-hydroxyacetophenone, which has a pK_a value in the same range as a carboxylic acid, despite the fact that it lacks a COOH moiety. Offer an explanation for the acidity of *para*-hydroxyacetophenone.

***para*-Hydroxyacetophenone**
pK_a = 4.2

21.6 Based on your answer to the previous question, would you expect *meta*-hydroxyacetophenone to be more or less acidic than *para*-hydroxyacetophenone? Explain your answer.

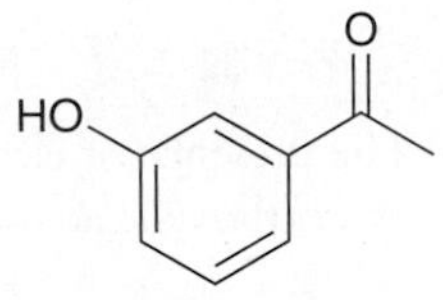

***meta*-Hydroxyacetophenone**

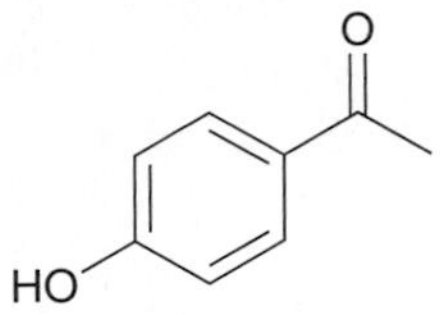

***para*-Hydroxyacetophenone**

21.7 When formic acid is treated with potassium hydroxide (KOH), an acid-base reaction occurs, forming a carboxylate ion. Draw the mechanism of this reaction, and identify the name of the carboxylate salt.

Carboxylic Acids at Physiological pH

Our blood is buffered to a pH of approximately 7.3, a value referred to as **physiological pH**. When dealing with buffered solutions, you may recall the **Henderson-Hasselbalch equation** from your general chemistry course.

$$\mathrm{pH} = \mathrm{p}K_\mathrm{a} + \log\frac{[\text{conjugate base}]}{[\text{acid}]}$$

This equation is often employed to calculate the pH of buffered solutions, although for our purposes we will rearrange the equation in the following way:

$$\frac{[\text{conjugate base}]}{[\text{acid}]} = 10^{(\mathrm{pH} - \mathrm{p}K_\mathrm{a})}$$

This rearranged form of the Henderson-Hasselbalch equation provides a method for determining the extent to which an acid will dissociate to form its conjugate base in a buffered solution. When the pK_a value of an acid is equivalent to the pH of a buffered solution into which it is dissolved, then

$$\frac{[\text{conjugate base}]}{[\text{acid}]} = 10^{(\mathrm{pH} - \mathrm{p}K_\mathrm{a})} = 10^{(0)} = 1$$

The ratio of the concentrations of conjugate base and acid will be 1. In other words, a carboxylic acid and its conjugate base will be present in approximately equal amounts when dissolved in a solution that is buffered such that pH = pK_a of the acid.

Now let's apply this equation to carboxylic acids at physiological pH (7.3), so that we can determine which form predominates (the carboxylic acid or the carboxylate ion). Recall that carboxylic acids generally have a pK_a value between 4 and 5. Therefore, at physiological pH:

$$\frac{[\text{conjugate base}]}{[\text{acid}]} = 10^{(\mathrm{pH} - \mathrm{p}K_\mathrm{a})} = 10^{(7.3 - \mathrm{p}K_\mathrm{a})} \approx 10^3$$

The ratio of the concentrations of the carboxylate ion and the carboxylic acid will be approximately 1000 : 1. That is, carboxylic acids will exist primarily as carboxylate salts at physiological pH. For example, pyruvic acid exists primarily as pyruvate ion at physiological pH.

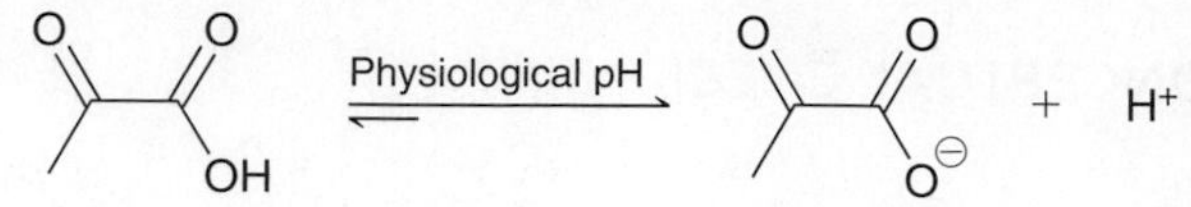

Pyruvic acid Pyruvate

Carboxylate ions play vital roles in many biological processes, as we will see in Chapter 25.

Substituent Effects on Acidity

The presence of electron-withdrawing substituents can have a profound impact on the acidity of a carboxylic acid.

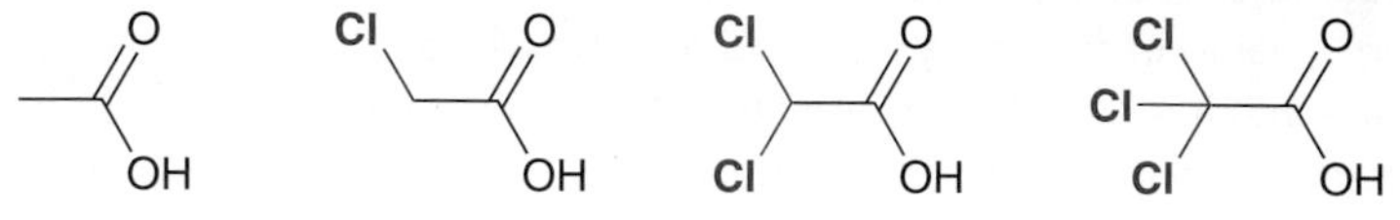

$pK_a = 4.8$ $pK_a = 2.9$ $pK_a = 1.3$ $pK_a = 0.9$

Notice that the pK_a decreases with each additional chlorine substituent. This trend is explained in terms of the inductive effects of the chlorine atoms, which can stabilize the conjugate base (as explained in Section 3.4). The effect of an electron-withdrawing group depends on its proximity to the carboxylic acid moiety.

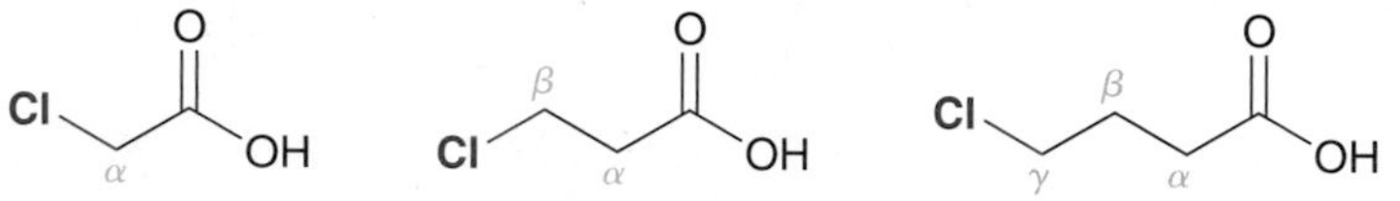

$pK_a = 2.9$ $pK_a = 4.1$ $pK_a = 4.5$

The effect is most pronounced when the electron-withdrawing group is located at the α position. As the distance between the chlorine atom and the carboxylic acid moiety increases, the effect of the chlorine atom becomes less pronounced.

The effects of electron-withdrawing substituents are also observed for substituted benzoic acids (Figure 21.3). In Sections 19.7–19.10, we discussed the electronic effects of each of the substitutents in Figure 21.3, and we saw that a nitro group is a powerful electron-withdrawing group. Consequently, the presence of the nitro group on the ring will stabilize the conjugate base, giving a low pK_a value (relative to benzoic acid). In contrast, a hydroxy group is a powerful electron-donating group (Section 19.10), and therefore, the presence of the hydroxy group will destabilize the conjugate base, giving a high pK_a value (relative to benzoic acid).

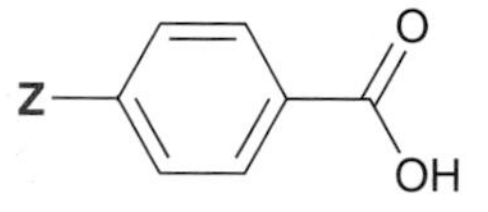

Z	—NO_2	—CHO	—Cl	—H	—CH_3	—OH
pK_a	3.4	3.8	4.0	4.2	4.3	4.5

FIGURE 21.3
The pK_a values of various *para*-substituted benzoic acids.

CONCEPTUAL CHECKPOINT

21.8 Acetic acid was dissolved in a solution buffered to a pH of 5.76. Determine the ratio of the concentrations of acetate ion and acetic acid in this solution. Which species predominates under these conditions?

21.9 Rank each set of compounds in order of increasing acidity:

(a) 2,4-Dichlorobutyric acid
2,3-Dichlorobutyric acid
3,4-Dimethylbutyric acid

(b) 3-Bromopropionic acid
2,2-Dibromopropionic acid
3,3-Dibromopropionic acid

21.4 Preparation of Carboxylic Acids

In previous chapters, we studied a variety of methods for preparing carboxylic acids (Table 21.1). In addition to the methods we have already seen, there are many other ways of preparing carboxylic acids. We will examine two of them.

TABLE 21.1 A REVIEW OF METHODS FOR PREPARING CARBOXYLIC ACIDS

REACTION	SECTION NUMBER	COMMENTS
Oxidative Cleavage of Alkynes 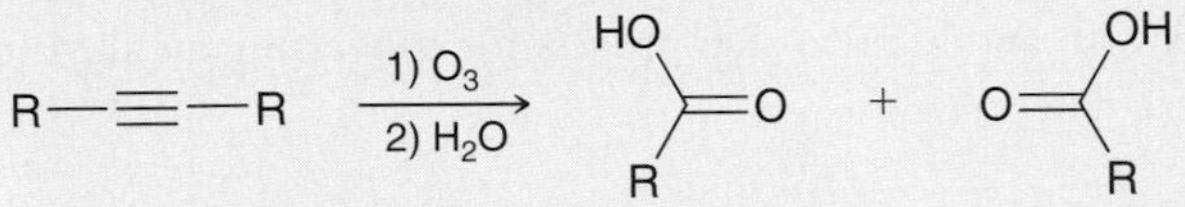	10.9	Oxidative cleavage will break a C≡C triple bond forming two carboxylic acids.
Oxidation of Primary Alcohols 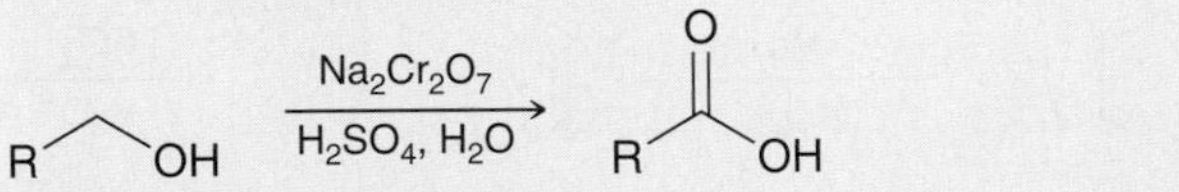	13.10	A variety of strong oxidizing agents can be used to oxidize primary alcohols and produce carboxylic acids.
Oxidation of Alkylbenzenes 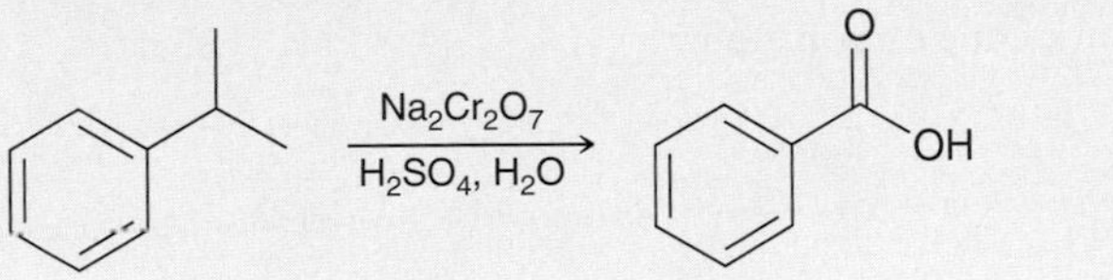	18.6	Any alkyl group on an aromatic ring will be completely oxidized to give benzoic acid, provided that the benzylic position has at least one hydrogen atom.

Hydrolysis of Nitriles

When treated with aqueous acid, a *nitrile* (a compound with a cyano group) can be converted into a carboxylic acid.

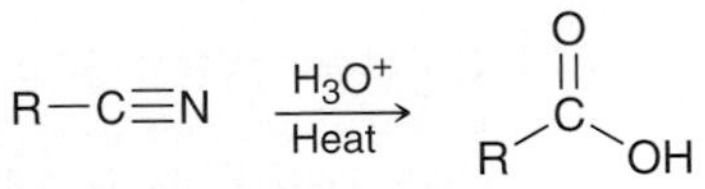

This process is called **hydrolysis**, and the mechanism for nitrile hydrolysis will be discussed later in this chapter. This reaction provides us with a two-step process for converting an alkyl halide to a carboxylic acid.

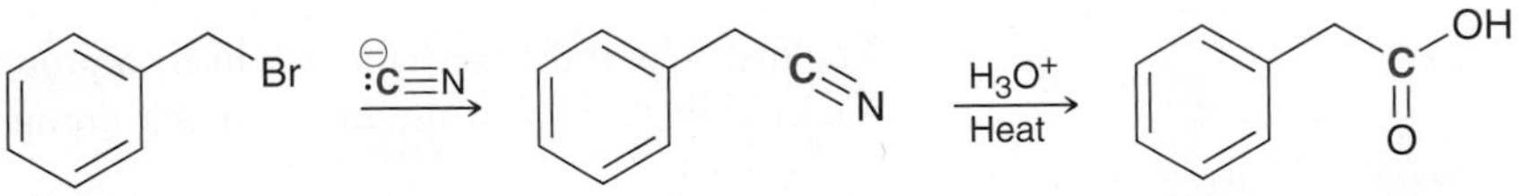

The first step is an S_N2 reaction in which cyanide acts as a nucleophile. The resulting nitrile is then hydrolyzed to yield a carboxylic acid that has one more carbon atom (shown in red) than the original alkyl halide. Since the first step is an S_N2 process, the reaction cannot occur with tertiary alkyl halides.

Carboxylation of Grignard Reagents

Carboxylic acids can also be prepared by treating a Grignard reagent with carbon dioxide:

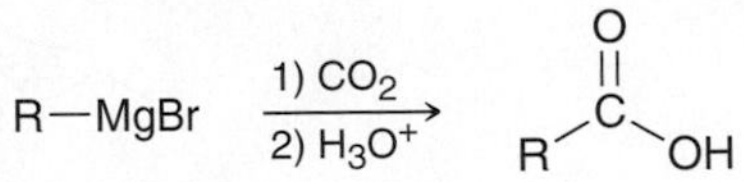

A mechanism for this process is shown below:

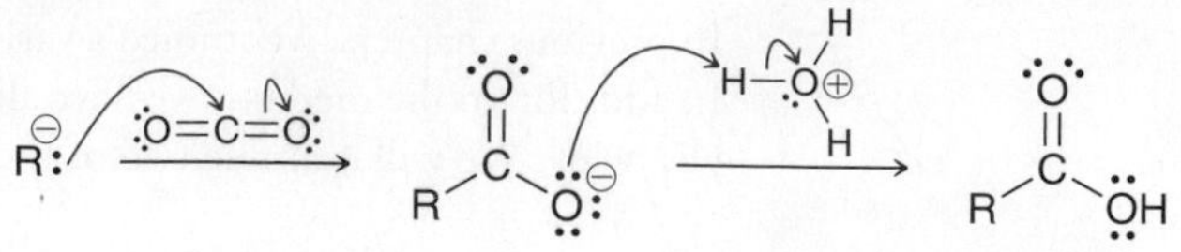

In the first step, the Grignard reagent attacks the electrophilic center of carbon dioxide, generating a carboxylate ion. Treating the carboxylate ion with a proton source affords the carboxylic acid. These two steps occur separately, as the proton source is not compatible with the Grignard reagent and can only be introduced after the Grignard reaction is complete. This reaction provides us with another two-step process for converting an alkyl (or vinyl or aryl) halide to a carboxylic acid.

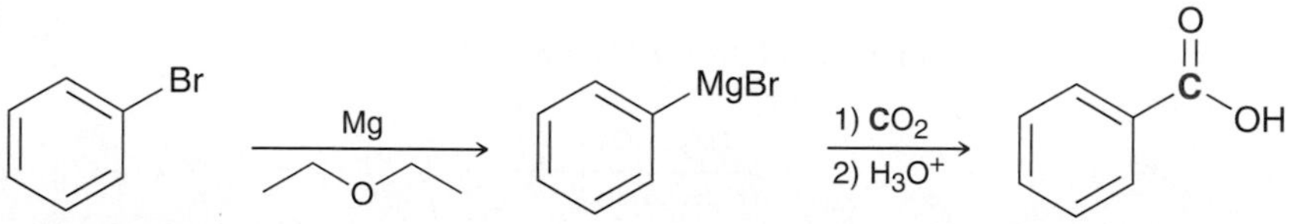

We have now seen two new methods for preparing carboxylic acids, both of which involve the introduction of one carbon atom.

CONCEPTUAL CHECKPOINT

21.10 Identify the reagents you would use to perform the following transformations:

(a) Ethanol → Acetic acid
(b) Toluene → Benzoic acid
(c) Benzene → Benzoic acid
(d) 1-Bromobutane → Pentanoic acid
(e) Ethylbenzene → Benzoic acid
(f) Bromocyclohexane → Cyclohexanecarboxylic acid

21.5 Reactions of Carboxylic Acids

Carboxylic acids are reduced to alcohols upon treatment with lithium aluminum hydride.

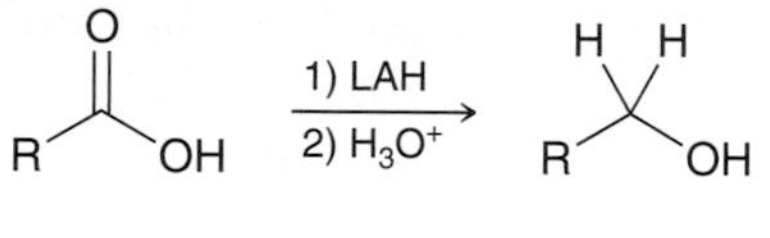

The first step of the mechanism is likely a proton transfer, because LAH is not only a powerful nucleophile, but it can also function as a strong base, forming a carboxylate ion.

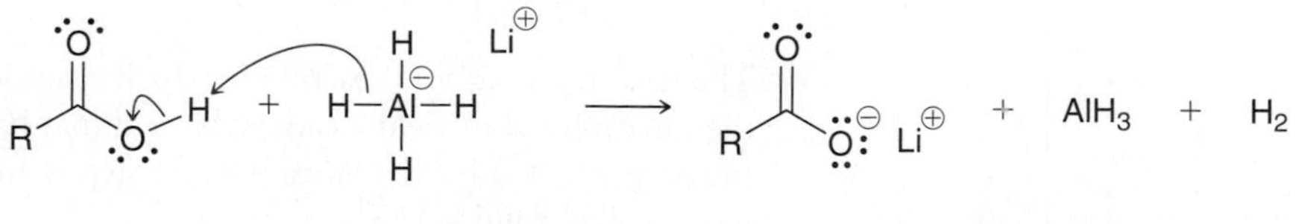

There are several possibilities for the rest of the mechanism. One possibility involves a reaction of the carboxylate ion with AlH_3 followed by elimination to form an aldehyde:

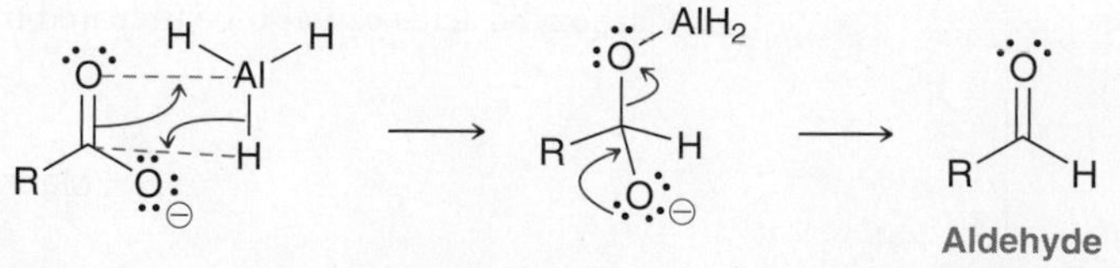

Under these conditions, the aldehyde cannot be isolated. Instead, it is further attacked by LAH to form an alkoxide, which is then protonated when H_3O^+ is added to the reaction flask.

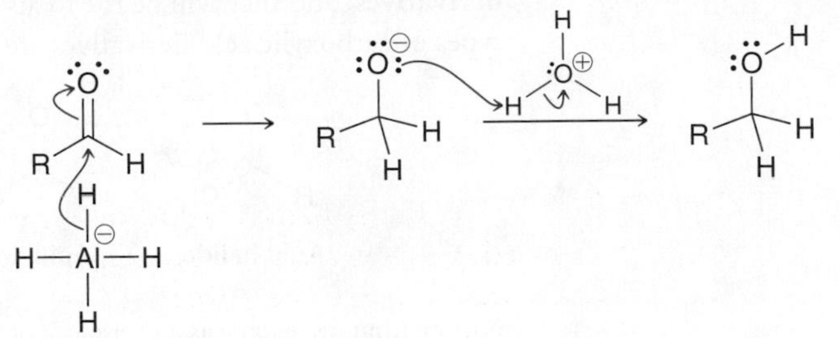

An alternative method for reducing carboxylic acids involves the use of borane (BH_3).

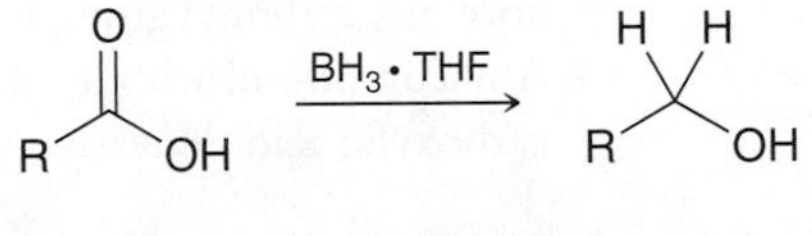

Reduction with borane is often preferred over reduction with LAH, because borane reacts selectively with a carboxylic acid moiety in the presence of another carbonyl group. As an example, if the following reaction were performed with LAH instead of borane, both carbonyl groups would be reduced.

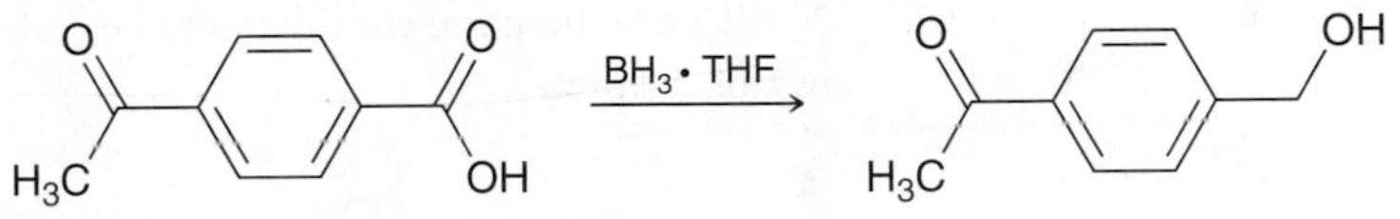

(80%)

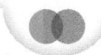

CONCEPTUAL CHECKPOINT

21.11 Identify the reagents you would use to achieve each of the following transformations:

(a) 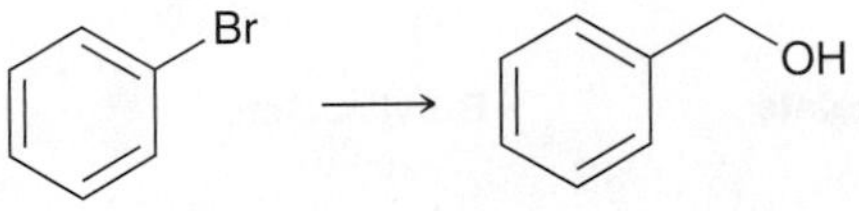

(b) 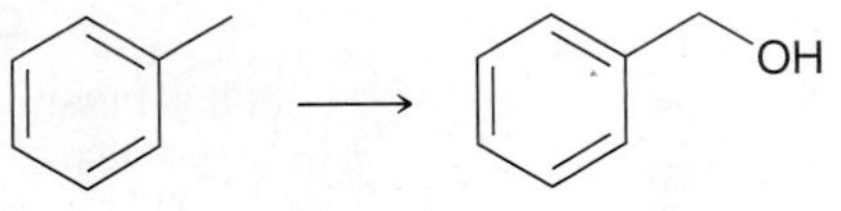

21.6 Introduction to Carboxylic Acid Derivatives

Classes of Carboxylic Acid Derivatives

In the previous section, we learned about the reaction between a carboxylic acid and LAH. This reaction is a reduction, because the carbon atom of the carboxylic acid moiety is reduced in the process:

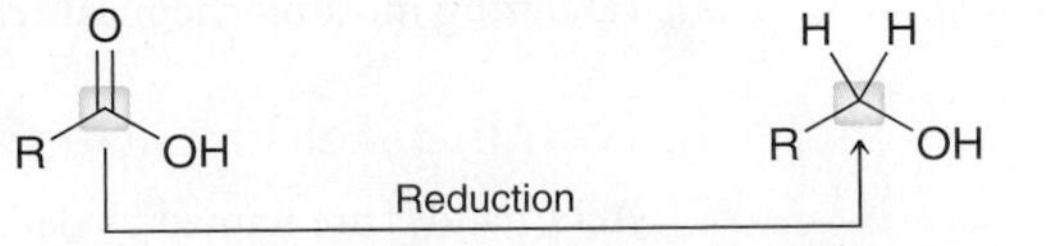

Carboxylic acids also undergo many other reactions that do not involve a change in oxidation state.

LOOKING BACK
For a review of oxidation states, see Section 13.4.

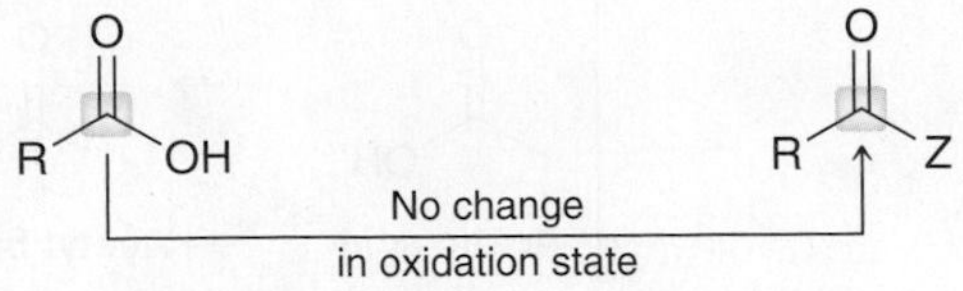

Replacement of the OH group with a different group (Z) does not involve a change in oxidation state if Z is a heteroatom (Cl, O, N, etc.). Compounds of this type are called **carboxylic acid derivatives**, and they will be the focus of the remainder of this chapter. The four most common types of carboxylic acid derivatives are shown below.

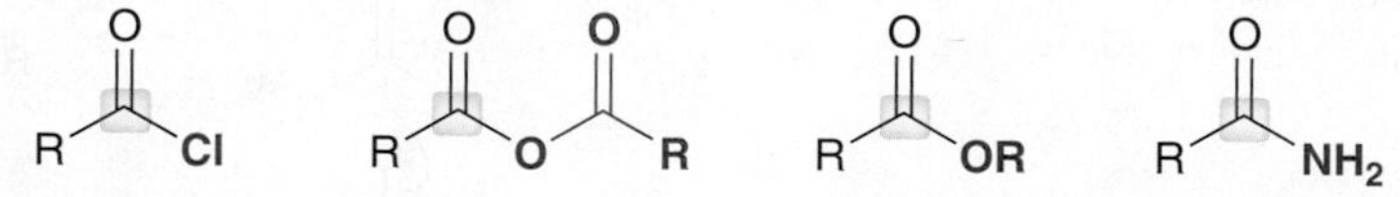

Acid halide **Acid anhydride** **Ester** **Amide**

Notice that in each case there is a carbon atom (highlighted in green) with three bonds to heteroatoms. As a result, each of these carbon atoms has the same oxidation state as the carbon atom of a carboxylic acid. Although all of these derivatives exhibit a carbonyl group, the presence of a carbonyl group is not a necessary requirement to qualify as a carboxylic acid derivative. Any compound with a carbon atom that has three bonds to heteroatoms will be classified as a carboxylic acid derivative. For example, consider the structure of nitriles.

$$R-C\equiv N$$

A nitrile

Nitriles exhibit a carbon atom with three bonds to a heteroatom (nitrogen). As a result, the conversion of a nitrile into a carboxylic acid (or vice versa) is neither a reduction nor an oxidation. Nitriles are therefore considered to be carboxylic acid derivatives, and they will also be discussed in this chapter.

Carboxylic Acid Derivatives in Nature

As we will soon see, acid halides and acid anhydrides are highly reactive and are therefore not very common in nature. In contrast, esters are more stable and are abundant in nature. Naturally occurring esters, such as the following three examples, often have pleasant odors and contribute to the aromas of fruits and flowers.

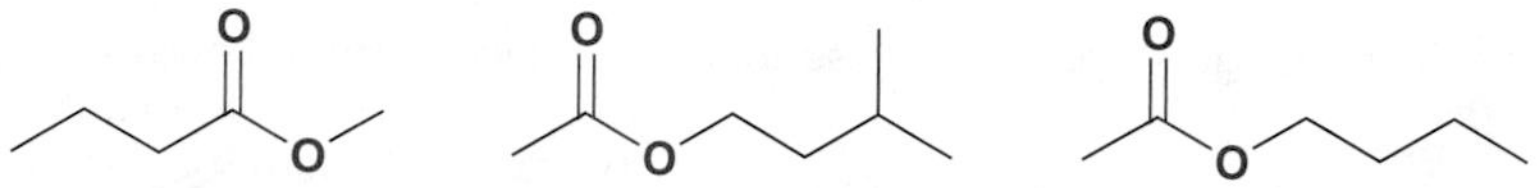

Methyl butanoate (pineapple) **Isopentyl acetate** (banana) **Butyl acetate** (pear)

Amides are abundant in living organisms. For example, proteins are comprised of repeating amide linkages.

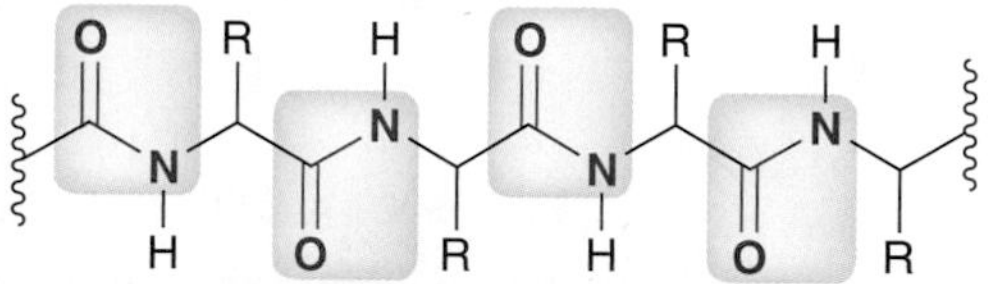

The structure of proteins

Chapter 26 will focus on the structure of proteins as well as the central role that proteins play in catalyzing most biochemical reactions.

Naming Acid Halides

Acid halides are named as derivatives of carboxylic acids by replacing the suffix "ic acid" with "yl halide":

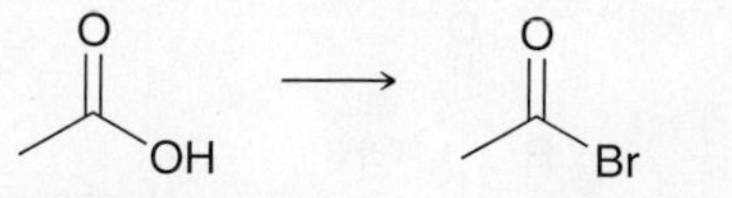

Acetic acid **Acetyl bromide**

Benzoic acid **Benzoyl chloride**

MEDICALLYSPEAKING

Sedatives

Sedatives are compounds that reduce anxiety and induce sleep. Our bodies utilize many natural sedatives, including melatonin.

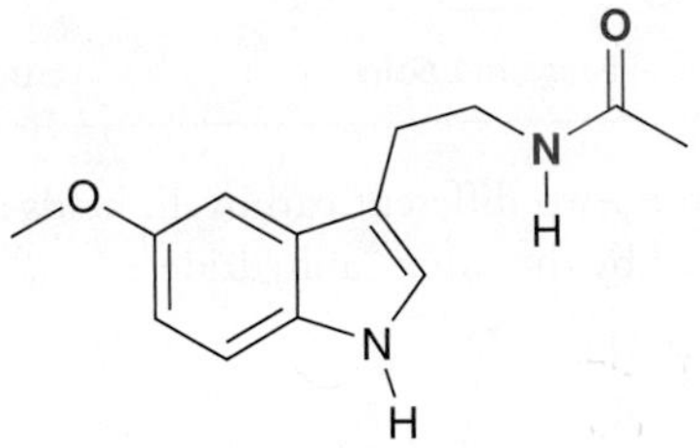

Melatonin

There is much evidence suggesting that melatonin plays an important role in regulating the body's natural sleep-wake cycle. For example, it has been observed that levels of melatonin for most people increase at night and then decrease in the morning. For this reason, many people take melatonin supplements to treat insomnia.

Notice the amide moiety in the structure of melatonin (shown in red). This moiety is a common feature in many drugs that are marketed as sedatives, for example,

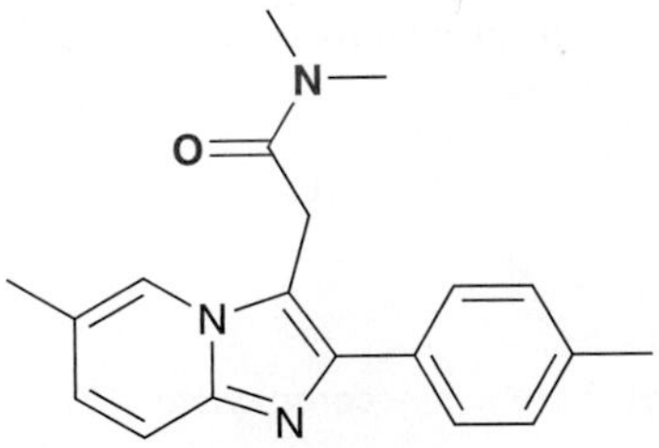

Zolpidem (Ambien™)

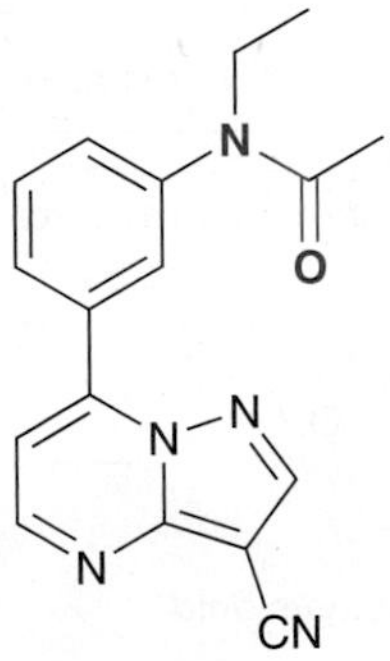

Zaleplon (Sonata™)

These drugs, which are similar in structure to melatonin, are used to treat insomnia. Other sedatives are used primarily in the treatment of excessive anxiety:

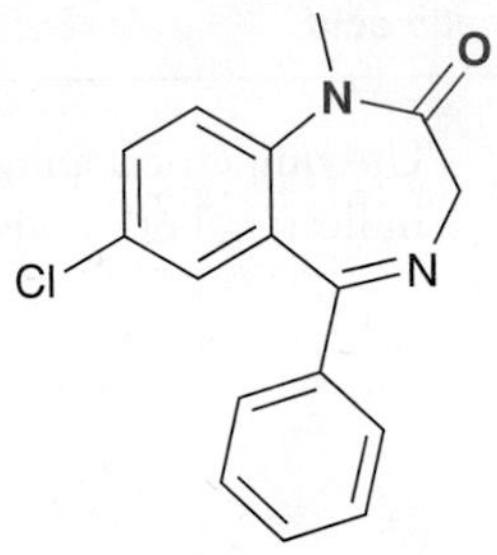

Diazepam (Valium™)

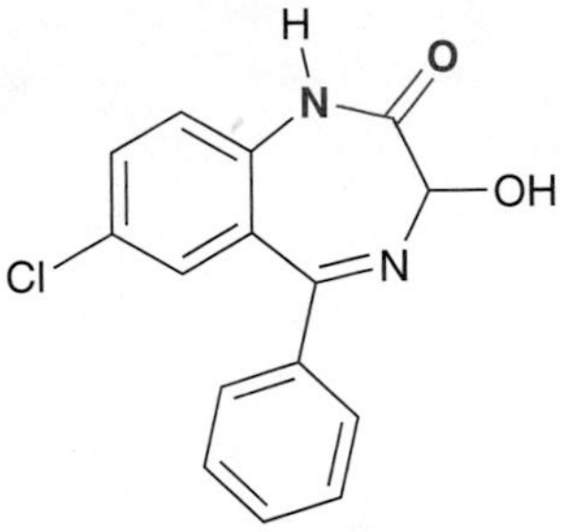

Oxazepam (Serax™)

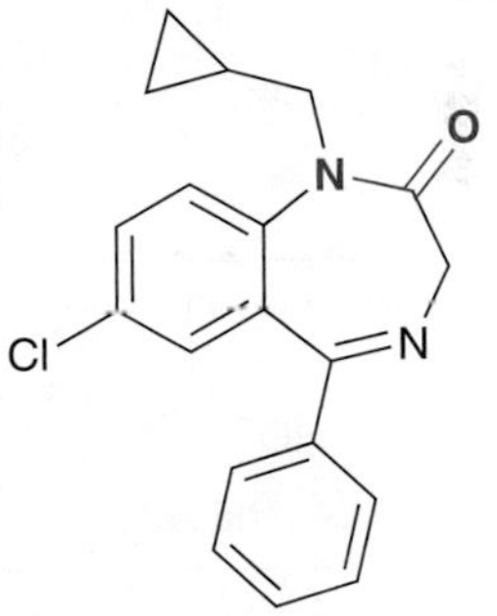

Prazepam (Verstran™)

Drugs used in the treatment of anxiety are called anxiolytic agents. The three examples shown are all similar in structure and belong to a class of compounds called benzodiazepines. Extensive research has been undertaken to elucidate the relationship between the structure and activity of benzodiazepines. It was found that the amide moiety is not absolutely necessary, but its presence does increase the potency of these agents.

When an acid halide moiety is connected to a ring, the suffix "carboxylic acid" is replaced with "carbonyl halide", for example,

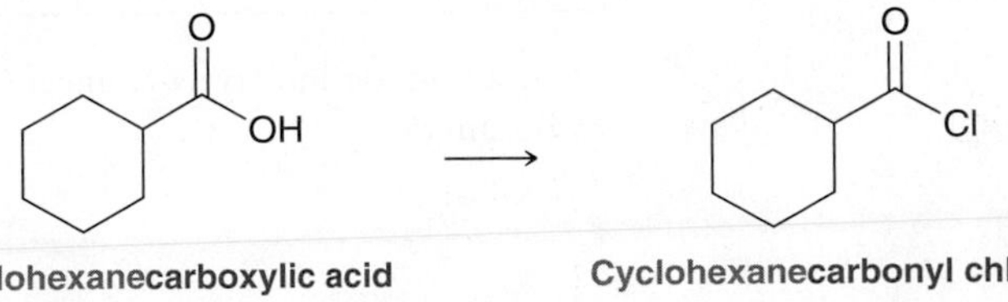

Cyclohexanecarboxylic acid → **Cyclohexanecarbonyl chloride**

Naming Acid Anhydrides

Acid anhydrides are named as derivatives of carboxylic acids by replacing the suffix "acid" with "anhydride."

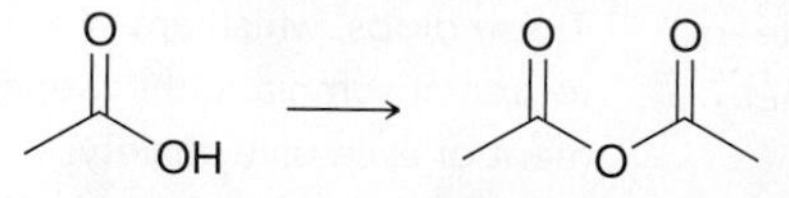

Acetic acid → Acetic **anhydride**

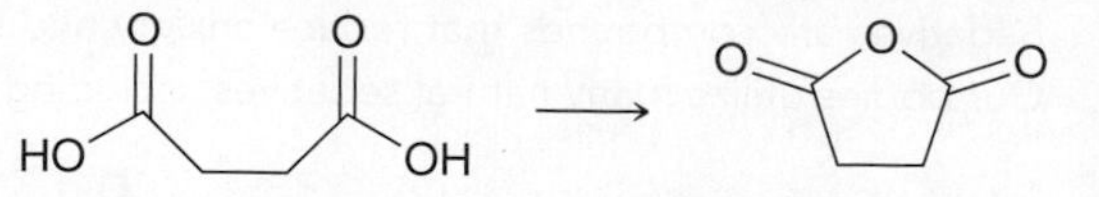

Succinic acid → Succinic **anhydride**

Unsymmetrical anhydrides are prepared from two different carboxylic acids and are named by indicating both acids alphabetically followed by the suffix "anhydride":

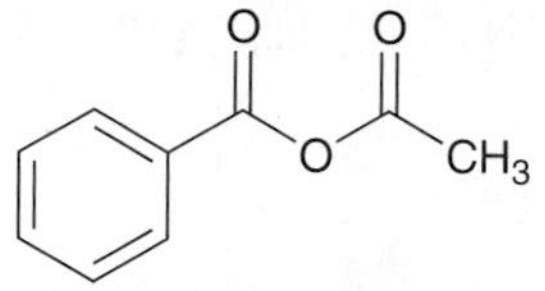

Acetic benzoic **anhydride**

Naming Esters

Esters are named by first indicating the alkyl group attached to the oxygen atom followed by the carboxylic acid, for which the suffix "ic acid" is replaced with "ate."

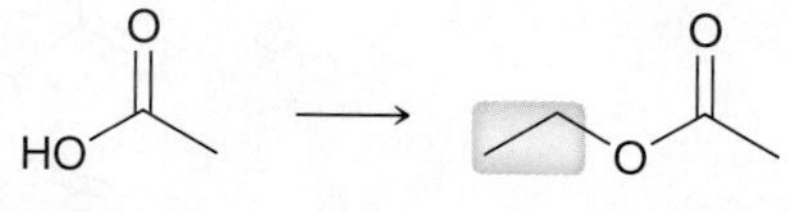

Acetic acid → Ethyl acet**ate**

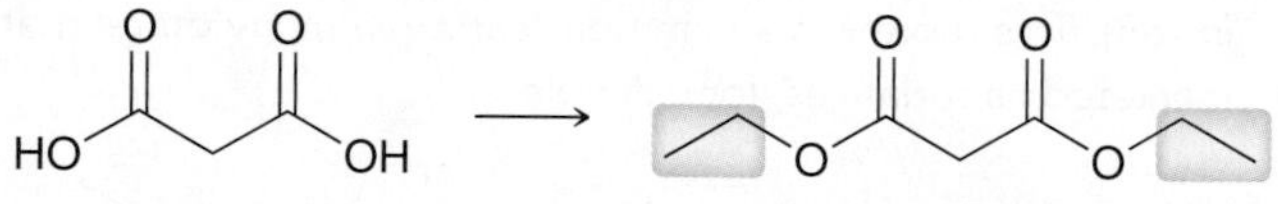

Malonic acid → Diethyl malon**ate**

The same methodology is applied when the ester moiety is connected to a ring, for example.

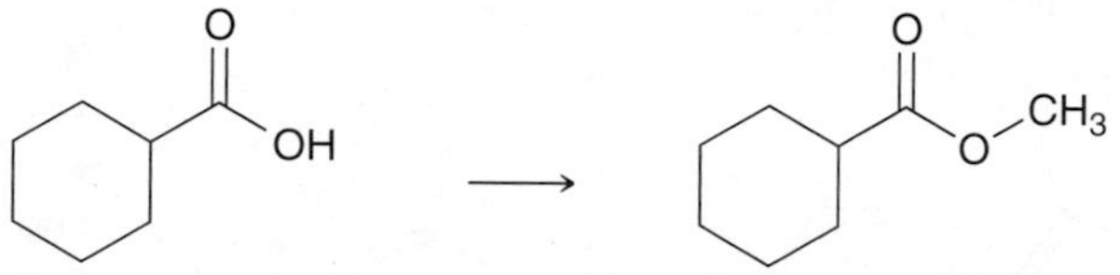

Cyclohexane**carboxylic acid** → Methyl cyclohexane**carboxylate**

Naming Amides

Amides are named as derivatives of carboxylic acids by replacing the suffix "ic acid" or "oic acid" with "amide."

Acetic acid → Acet**amide**

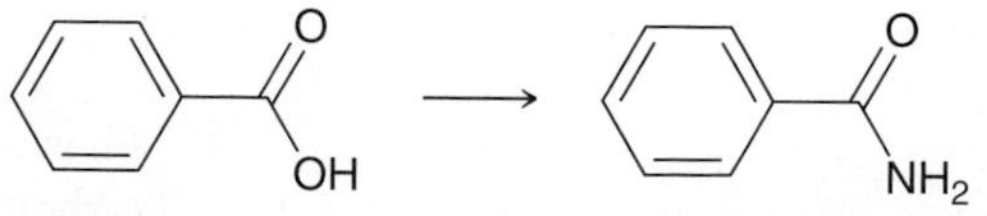

Benzoic acid → Benz**amide**

When an amide moiety is connected to a ring, the suffix "carboxylic acid" is replaced with "carboxamide."

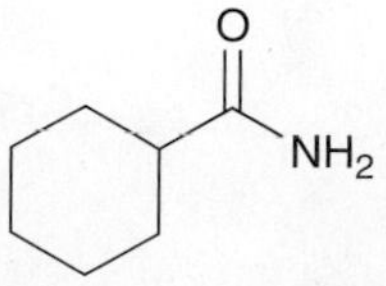

Cyclohexane**carboxamide**

If the nitrogen atom bears alkyl groups, these groups are placed at the beginning of the name, and the letter "*N*" is used as a locant to indicate that they are attached to the nitrogen.

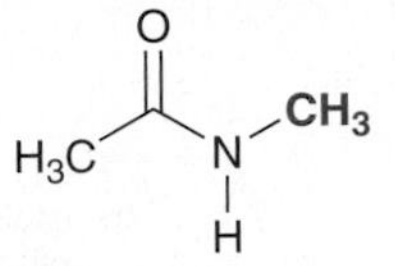

***N*-Methylacetamide**

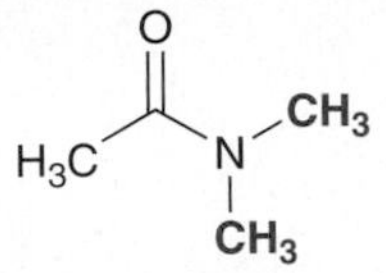

***N,N*-Dimethylacetamide**

Naming Nitriles

Nitriles are named as derivatives of carboxylic acids by replacing the suffix "ic acid" or "oic acid" with "onitrile."

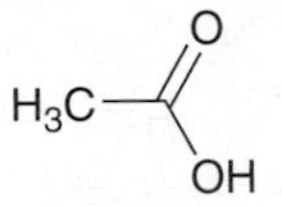

Acetic acid

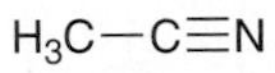

Acetonitrile

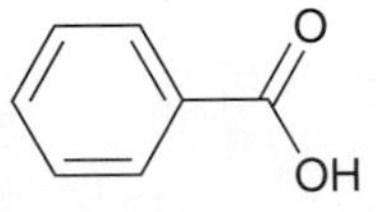

Benzoic acid

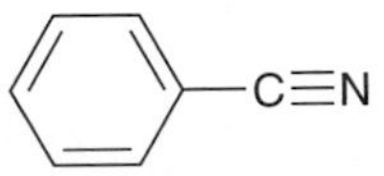

Benzonitrile

CONCEPTUAL CHECKPOINT

21.12 Provide a name for each of the following compounds:

(a) 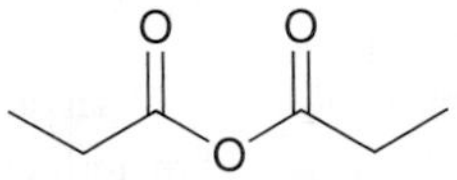

(b) 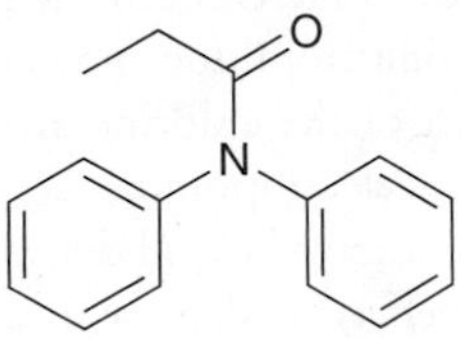

(c) 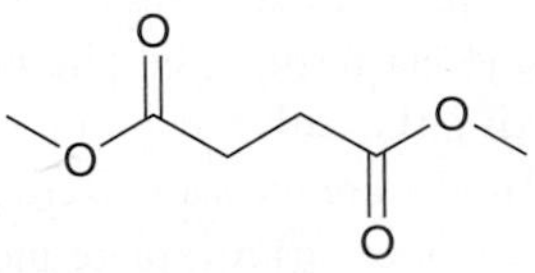

(d) 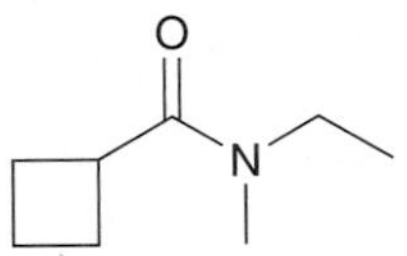

(e) 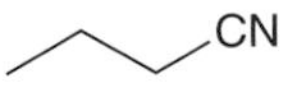

(f) 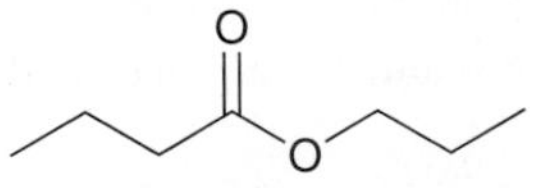

(g) 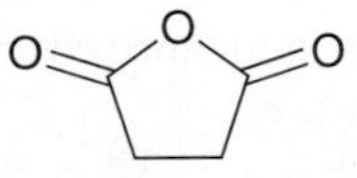

(h) 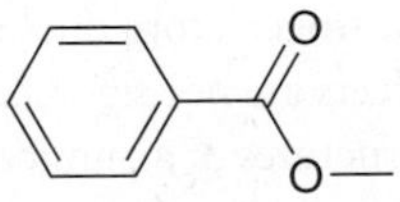

(i) 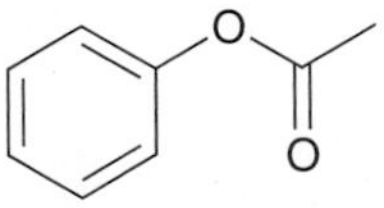

21.13 Draw a structure for each of the following compounds:

(a) Dimethyl oxalate (b) Phenyl cyclopentanecarboxylate (c) *N*-Methylpropionamide (d) Propionyl chloride

21.7 Reactivity of Carboxylic Acid Derivatives

Electrophilicity of Carboxylic Acid Derivatives

In the previous chapter, we saw that the carbon atom of a carbonyl group is electrophilic as a result of both inductive and resonance effects. The same is true of carboxylic acid derivatives, although there is a wide range of reactivity among the carboxylic acid derivatives, illustrated in Figure 21.4. Acid halides are the most reactive. To rationalize this, we must consider both inductive effects and resonance effects. Let's begin with induction. Chlorine is an electronegative atom and therefore withdraws electron density from the carbonyl group via induction.

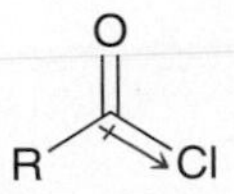

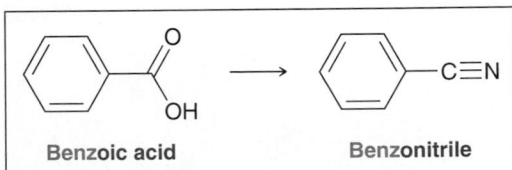

This effect renders the carbonyl group even more electrophilic when compared with the carbonyl group of a ketone.

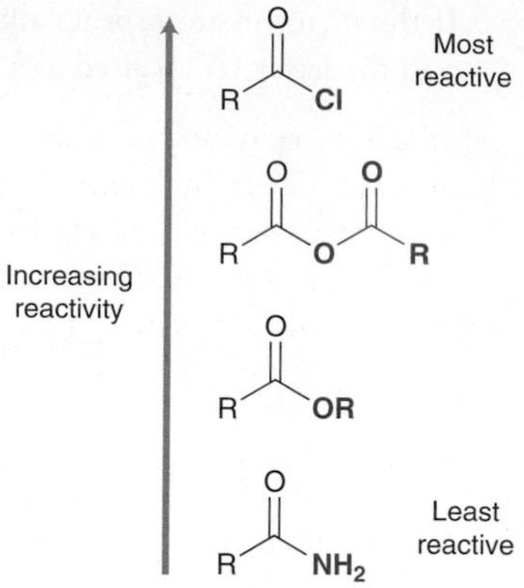

FIGURE 21.4
The relative order of reactivity of carboxylic acid derivatives.

Now let's consider resonance effects. An acid halide has three resonance structures.

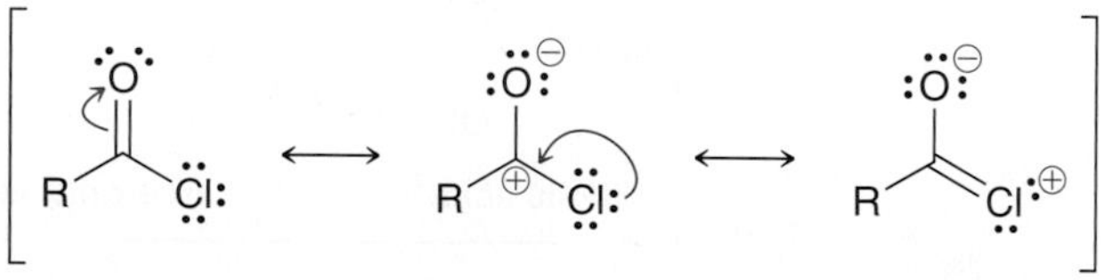

BY THE WAY
Note the similarity between this analysis and the analysis of the effect of a chlorine atom on the electronics of an aromatic ring (see Section 19.9).

The third resonance structure does not contribute much character to the overall resonance hybrid because the *p*-orbital overlap required for a C=Cl bond is not effective. This argument is similar to the argument provided in reference to the ineffective overlap of S=O bonds in Section 19.3. As a result, the chlorine atom does not donate much electron density to the carbonyl via resonance. The net effect of the chlorine atom is to withdraw electron density, rendering the carbonyl group extremely electrophilic.

Amides are the least reactive of the carboxylic acid derivatives. To rationalize this observation, we must once again explore inductive effects and resonance effects. Let's begin with induction. Nitrogen is less electronegative than chlorine or oxygen and is not an effective electron-withdrawing group. The nitrogen atom does not withdraw much electron density from the carbonyl group, and inductive effects are not significant. However, resonance effects are substantial. Consider the three resonance structures of an amide.

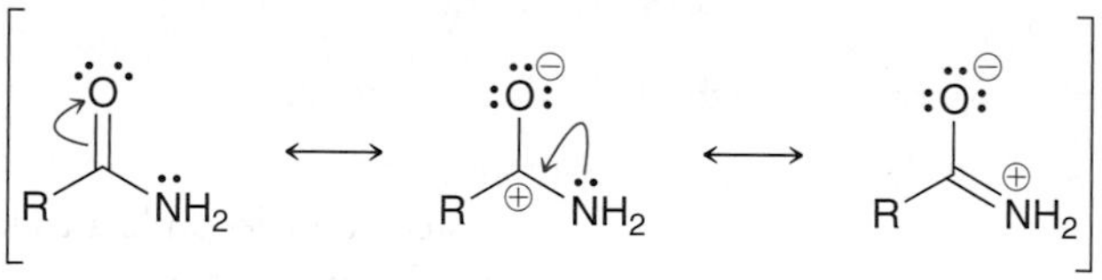

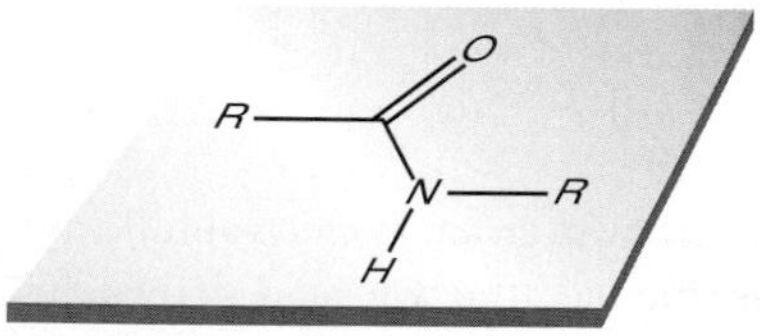

FIGURE 21.5
An illustration of the planar geometry of amides.

Unlike an acid halide, the third resonance structure of an amide contributes significant character to the overall resonance hybrid. The *p* orbital on the carbon atom effectively overlaps with a *p* orbital on the nitrogen atom, and the nitrogen atom can easily accommodate the positive charge. The nitrogen atom is sp^2 hybridized, and the geometry of the nitrogen atom is trigonal planar. As a result, the entire amide moiety lies in a plane (Figure 21.5). The C—N bond of an amide has significant double-bond character, which can be verified by observing the relatively high barrier to rotation for the C—N bond.

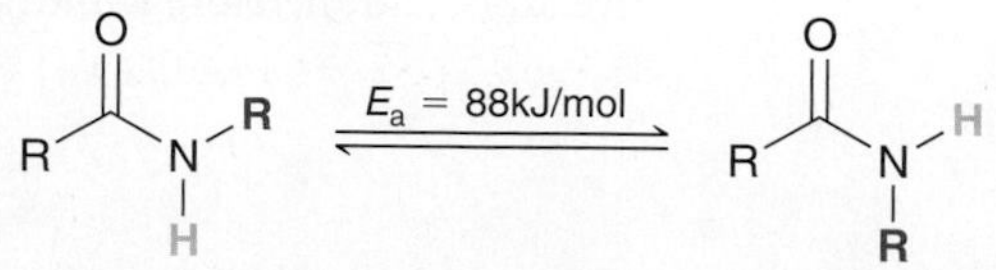

The restricted rotation of the C—N bond and the planar geometry of amide moieties will be important when we discuss protein structure in Chapter 27.

Nucleophilic Acyl Substitution

The reactivity of carboxylic acid derivatives is similar to the reactivity of aldehydes and ketones in a number of ways. In both cases, the carbonyl group is electrophilic and subject to attack by a nucleophile. In both cases, the same rules and principles govern the proton transfers that accompany the reactions, as we will soon see. Nevertheless, there is one critical difference between carboxylic acid derivatives and aldehydes/ketones. Specifically, carboxylic acid derivatives possess a heteroatom that can function as a leaving group, while aldehydes and ketones do not.

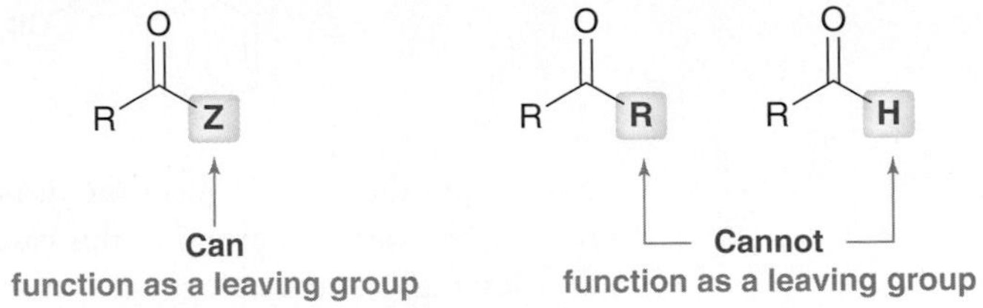
Can
function as a leaving group
Cannot
function as a leaving group

When a nucleophile attacks a carboxylic acid derivative, a reaction can occur in which the nucleophile replaces the leaving group:

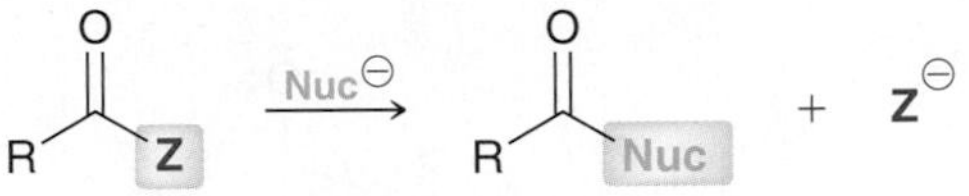

This type of reaction is called a **nucleophilic acyl substitution**, and the rest of this chapter will be dominated by various examples of this type of reaction. The general mechanism has two core steps (Mechanism 21.1).

MECHANISM 21.1 NUCLEOPHILIC ACYL SUBSTITUTION

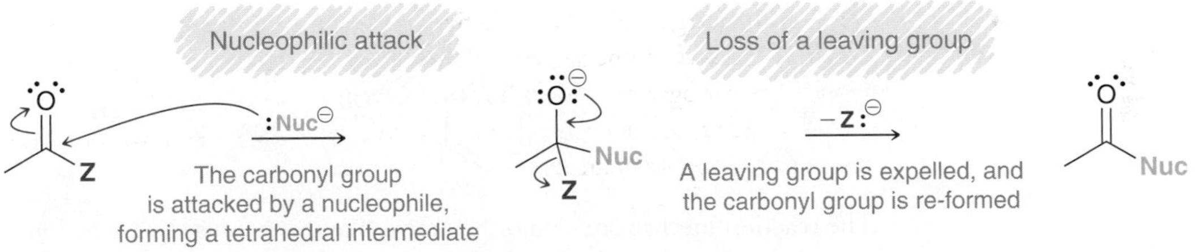
Nucleophilic attack
Loss of a leaving group
The carbonyl group is attacked by a nucleophile, forming a tetrahedral intermediate
A leaving group is expelled, and the carbonyl group is re-formed

In the first step, a nucleophile attacks the carbonyl, forming a tetrahedral intermediate. In the second step, the carbonyl group is re-formed via loss of a leaving group. Re-formation of the C=O double bond is a powerful driving force, and even poor leaving groups (such as RO^-) can be expelled under certain conditions. Hydride ions (H^-) and carbanions (C^-) cannot function as leaving groups under any conditions, so this type of reaction is not observed for ketones or aldehydes. There are only a few rare exceptions when H^- or C^- do function as leaving groups, and we will specifically explain why those cases are exceptions when we discuss them in Chapter 22. For our purposes, the following rule will guide our discussion for the remainder of this chapter: *When a nucleophile attacks a carbonyl group to form a tetrahedral intermediate, the carbonyl group will always be re-formed, if possible, but H^- and C^- are never expelled as leaving groups.* There will be no exceptions to this rule in this chapter.

Let's explore a specific example of a nucleophilic acyl substitution, so that we can see how the rule applies. Consider the following transformation:

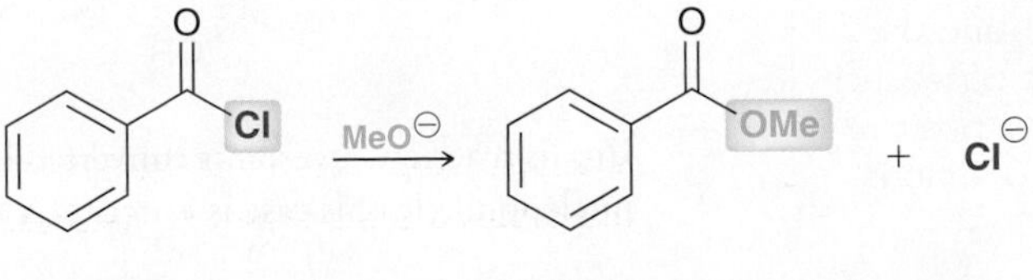

In this reaction, an acid chloride is converted into an ester. The mechanism of this transformation has two steps. In the first step, methoxide functions as a nucleophile and attacks the carbonyl group, forming a tetrahedral intermediate:

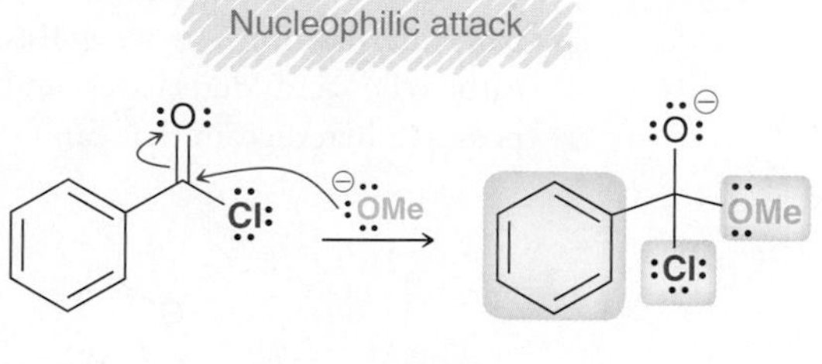

Now apply the rule: Re-form the carbonyl if possible, but never expel H^- or C^-. In order to re-form the carbonyl group in this case, one of the three highlighted groups must be expelled as a leaving group bearing a negative charge. The aromatic ring cannot leave, because that would involve expelling C^-. The remaining two choices (chloride or methoxide) are both viable options. Chloride is more stable than methoxide and is therefore a better leaving group, so the carbonyl will likely re-form to expel the chloride ion.

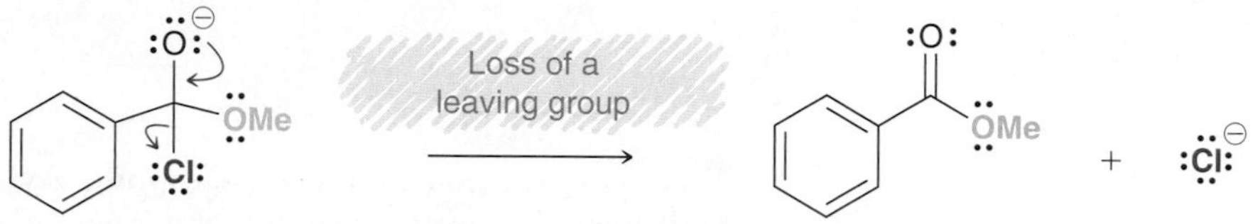

In summary, the mechanism for a nucleophilic acyl substitution reaction involves two core steps—nucleophilic attack and loss of a leaving group. Notice that these are the same two steps involved in an S_N2 process. However, there is one important difference. In an S_N2 process, the two steps occur in a concerted fashion (simultaneously), but in a nucleophilic acyl substitution reaction, the two steps must occur separately. It is a common mistake to draw these two steps as occurring together.

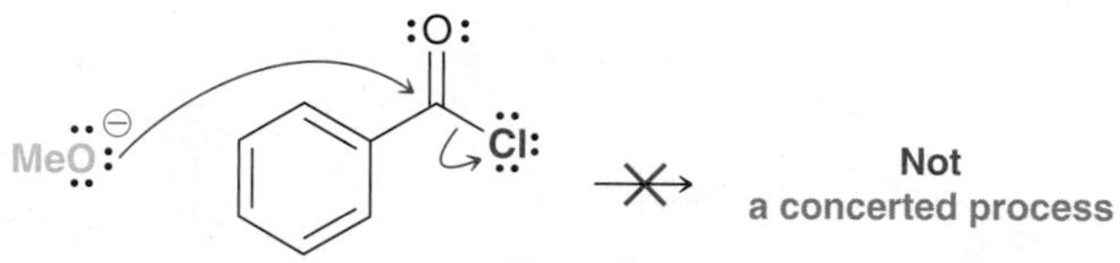

The reaction mechanism cannot be drawn like this because S_N2 reactions do not occur readily at sp^2-hybridized centers. When drawing a nucleophilic acyl substitution, make sure to draw the first step, which forms the tetrahedral intermediate, followed by the second step, which shows how the carbonyl group is re-formed.

The vast majority of the reactions in this chapter are nucleophilic acyl substitution reactions. All of these reactions will exhibit the two core steps of nucleophilic attack and loss of a leaving group to re-form the carbonyl group. But many of the reaction mechanisms will also exhibit proton transfers. In order to draw each mechanism properly, it is necessary to know why the proton transfers occur. The following rule will guide us in deciding whether or not to employ proton transfers in a particular mechanism: *In acidic conditions, avoid formation of negative charge. In alkaline conditions (strongly basic conditions such as hydroxide or methoxide), avoid formation of positive charge.*

This rule dictates that all participants in a reaction (reactants, intermediates, and leaving groups) should be consistent with the conditions employed. As an example, consider the following reaction.

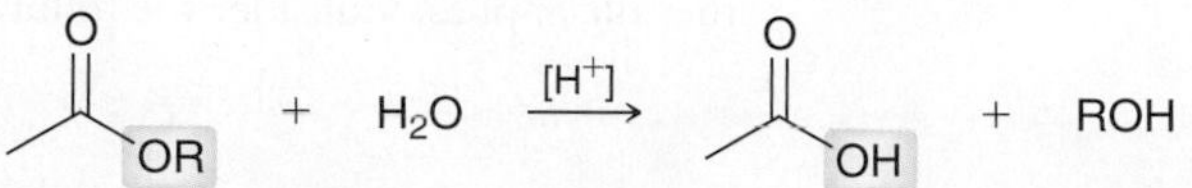

In this reaction, an ester is converted into a carboxylic acid under acid-catalyzed conditions. The nucleophile in this case is water (H_2O); however, the first step of the mechanism cannot simply

WATCH OUT

The "exception" to this rule is that the conjugate base of the acid (such as Cl^- or HSO_4^-) will have a negative charge in acidic conditions, and the counterion of the base (such as Na^+) will have a positive charge in basic conditions.

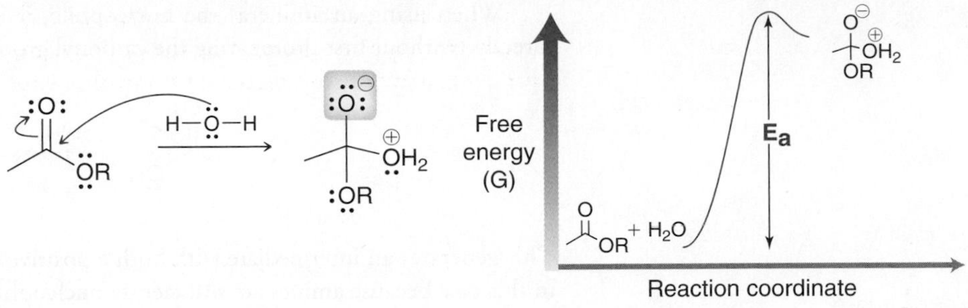

FIGURE 21.6
An energy diagram showing the large energy of activation associated with water directly attacking an ester.

be a nucleophilic attack (Figure 21.6). What is wrong with this step? The tetrahedral intermediate exhibits a negative charge, which is not consistent with acidic conditions. The energy diagram in Figure 21.7 illustrates a large energy of activation (E_a) for this step. To avoid formation of a negative charge, the acid catalyst is first employed to protonate the carbonyl group (just as we saw in the previous chapter with ketones and aldehydes).

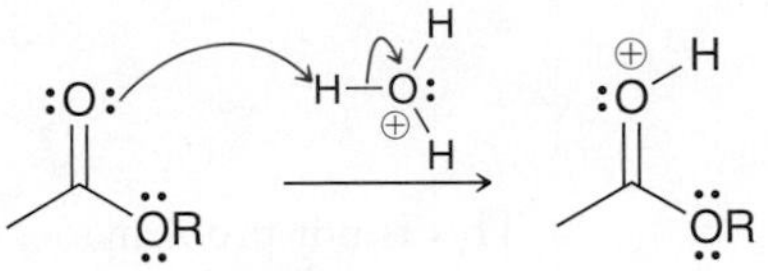

A protonated carbonyl group is significantly more electrophilic, and now when water attacks, no negative charge is formed. This step is now consistent with acidic conditions, because all participants are either neutral or positively charged. As seen in the energy diagram in Figure 21.7, the energy of activation (E_a) is now much smaller, because the reactants are already high in energy and no negative charge is being formed.

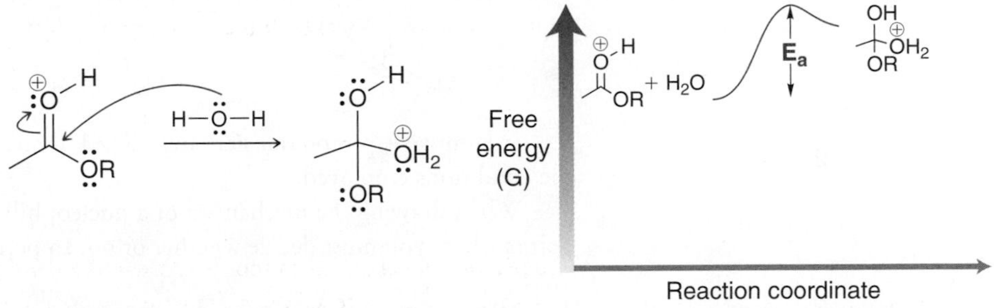

FIGURE 21.7
An energy diagram showing the small energy of activation associated with water attacking a protonated ester.

The rule about avoiding negative charges is only applied in acidic conditions. The story is different under basic conditions. For example, consider what happens when an ester is treated with hydroxide ion. In that case, the carbonyl is not protonated before the nucleophilic attack. Protonating the carbonyl group would involve formation of a positive charge, which is not consistent with basic conditions. Under basic conditions, hydroxide attacks the carbonyl directly to give a tetrahedral intermediate. The energy diagram in Figure 21.8 shows that the energy of activation (E_a) is not very large, because a negative charge is present in both the reactants and the intermediate. In other words, a negative charge is not being formed but is merely being transferred from one location to another.

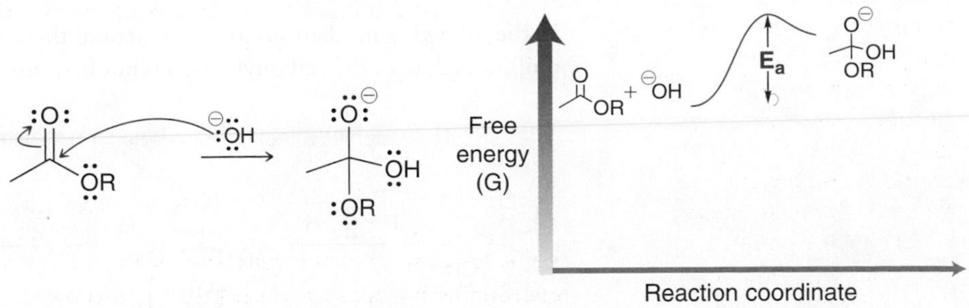

FIGURE 21.8
An energy diagram showing the small energy of activation associated with hydroxide attacking an ester.

When using an amine as the nucleophile, it is acceptable to attack the carbonyl group directly (without first protonating the carbonyl group).

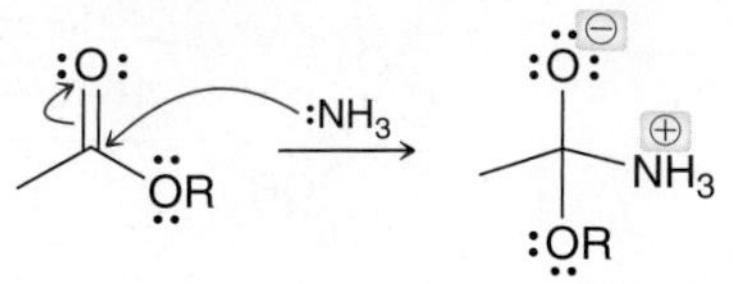

This generates an intermediate with both a positive charge and a negative charge, which is OK in this case because amines are sufficiently nucleophilic to attack a carbonyl group directly. Just avoid drawing an intermediate with two of the same kind of charge (two positive charges or two negative charges). This is true under all conditions: In acidic conditions, avoid an intermediate with two positive charges; in basic conditions, avoid an intermediate with two negative charges.

To illustrate the rule further, consider what happens when a tetrahedral intermediate re-forms. In basic conditions, it is acceptable to eject a methoxide ion.

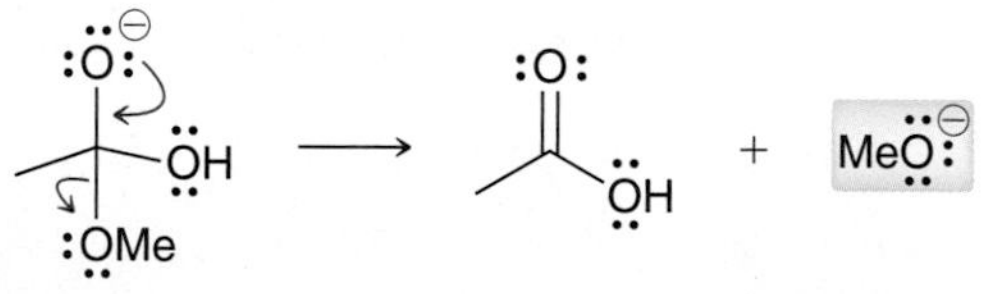

This is not problematic, because the intermediate already exhibits a negative charge. In other words, a negative charge is not being formed but is merely being transferred from one location to another. However, in acidic conditions, negative charges must be avoided. In acidic conditions, all participants (including the leaving group) should be either neutral or positive, so the methoxy group must first be protonated in order to function as a leaving group.

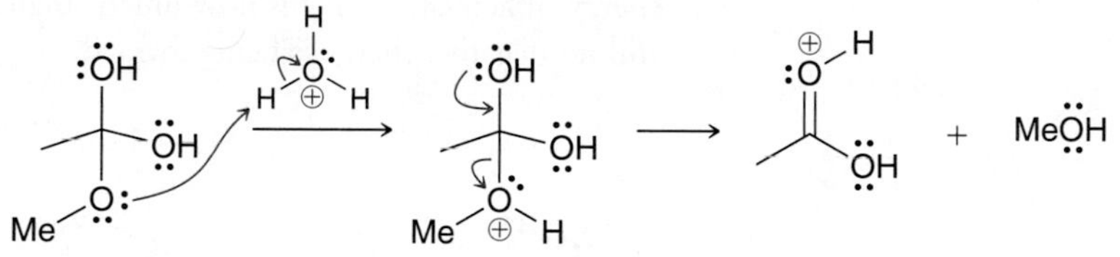

In summary, proton transfers are utilized in mechanisms in order to remain consistent with the conditions employed.

When drawing the mechanism of a nucleophilic acyl substitution reaction, there are three points where you must decide whether or not to perform a proton transfer.

Proton transfer ---- Nucleophilic attack ---- **Proton transfers** ---- Loss of a leaving group ---- **Proton transfer**

A proton transfer can occur (1) before the nucleophilic attack, (2) before loss of the leaving group, or (3) at the end of the mechanism. Some mechanisms will have proton transfers at all three points, while other mechanisms might only exhibit one proton transfer step at the end of the mechanism. As an example, consider the following reaction, which involves the conversion of an acid chloride into a carboxylic acid.

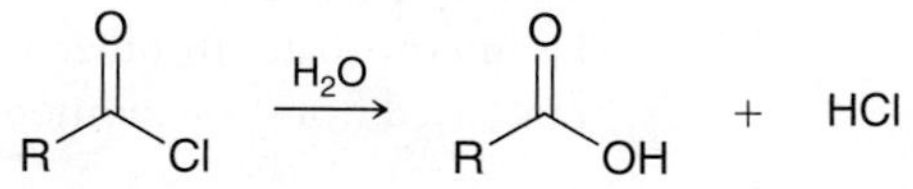

In the following mechanism for this reaction, there is no proton transfer step before the nucleophilic attack (i.e., the carbonyl group is not first protonated), because the reagents are not acidic.

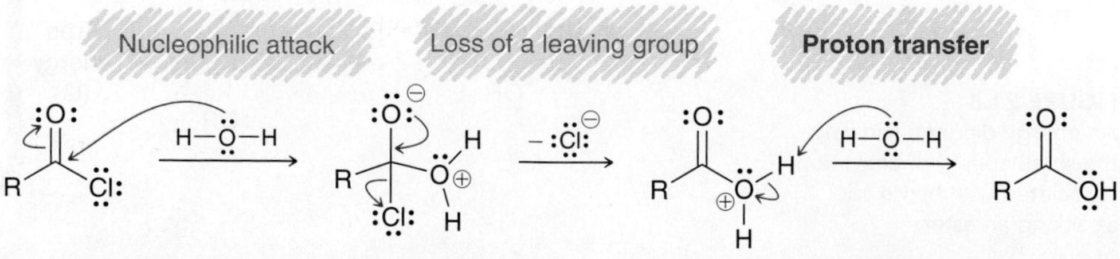

Similarly, there is no proton transfer step before loss of the leaving group, because the leaving group does not need to be protonated before it can leave. However, there is a proton transfer step at the end of the mechanism, in order to remove a proton and form the final product. The mechanism is comprised of the two core steps followed by a proton transfer. This pattern is typical for reactions of acid halides. We will see nearly a dozen reactions of acid halides, and all of their mechanisms will follow this pattern.

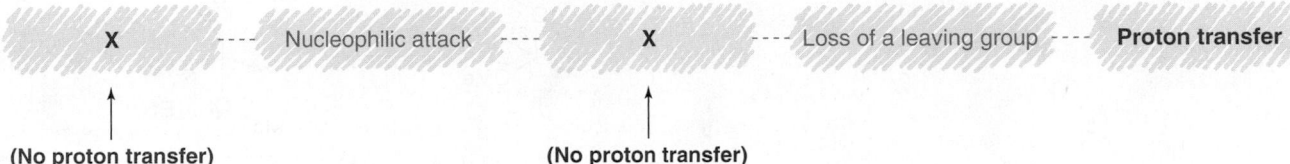

X ---- Nucleophilic attack ---- **X** ---- Loss of a leaving group ---- **Proton transfer**

(First X: **(No proton transfer)**; second X: **(No proton transfer)**)

In contrast, reactions performed under acid-catalyzed conditions commonly have proton transfer steps at all three possible points. The following SkillBuilder illustrates such a case.

SKILLBUILDER

21.1 DRAWING THE MECHANISM OF A NUCLEOPHILIC ACYL SUBSTITUTION REACTION

LEARN the skill

Propose a plausible mechanism for the following transformation:

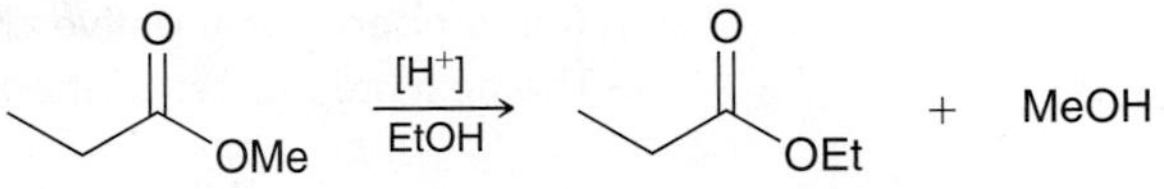

SOLUTION

In this reaction, a methoxy group is replaced by an ethoxy group:

As with all nucleophilic acyl substitution reactions, we expect the mechanism to have the following two core steps:

Nucleophilic attack ---- Loss of a leaving group

When drawing the mechanism, be sure to draw these steps separately. But we must also determine if any proton transfers are required. There are three points at which proton transfers might occur.

Proton transfer ---- Nucleophilic attack ---- **Proton transfers** ---- Loss of a leaving group ---- **Proton transfer**

Acidic conditions are employed in this case, so we must avoid formation of negative charges. This requirement dictates that we must perform proton transfer steps at all three possible points. The mechanism begins with a proton transfer in order to protonate the carbonyl group, rendering it a better electrophile:

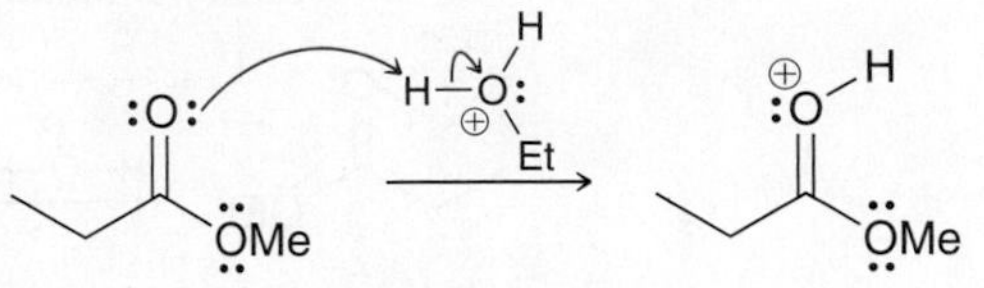

Notice the proton source that we use to protonate the carbonyl group. We cannot use EtOH as the proton source, because transfer of a proton from ethanol would involve creation of an ethoxide ion, which should be avoided in acidic conditions.

The next step is a nucleophilic attack, in which ethanol functions as a nucleophile and attacks the protonated carbonyl group, forming a tetrahedral intermediate that does not bear a negative charge.

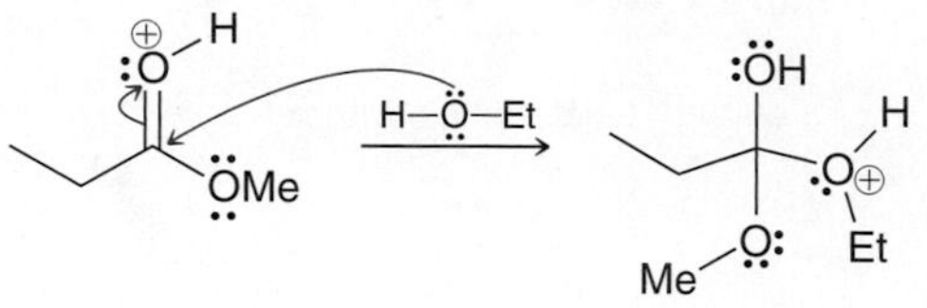

The tetrahedral intermediate formed in this step cannot immediately expel methoxide to re-form the carbonyl, because that would create a negative charge, which should be avoided in acidic conditions. So, we must first protonate the methoxy group. However, protonating the methoxy group would involve the formation of two positive charges, which should also be avoided. As a result, two separate proton transfers are required.

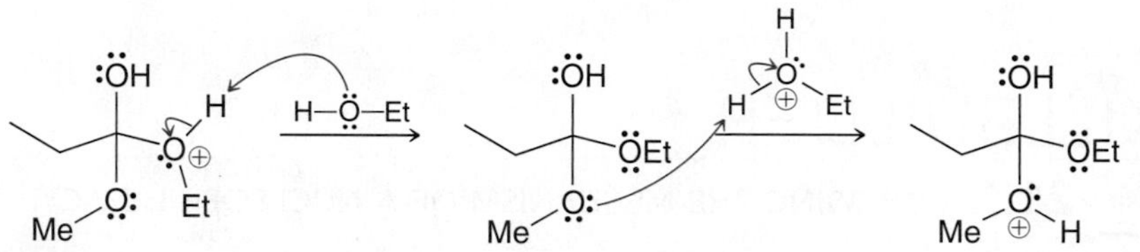

First, a proton is removed to form a new tetrahedral intermediate without a charge, followed by protonation of the methoxy group. Be careful not to use ethoxide as a base in the first step (remember—no negative charges in acidic conditions).

The next step is loss of the leaving group to re-form the carbonyl:

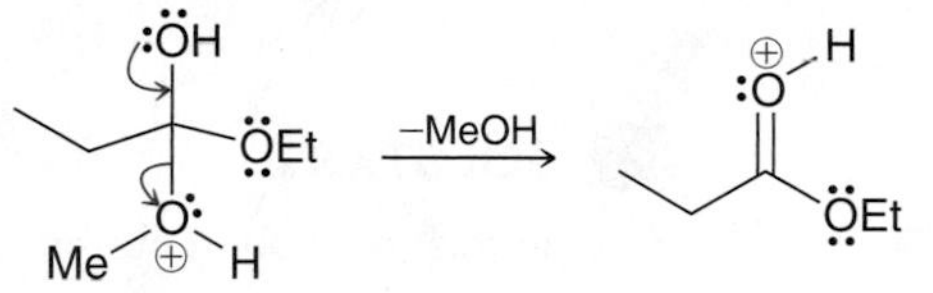

Finally, a proton transfer is used to remove the positive charge and form the product.

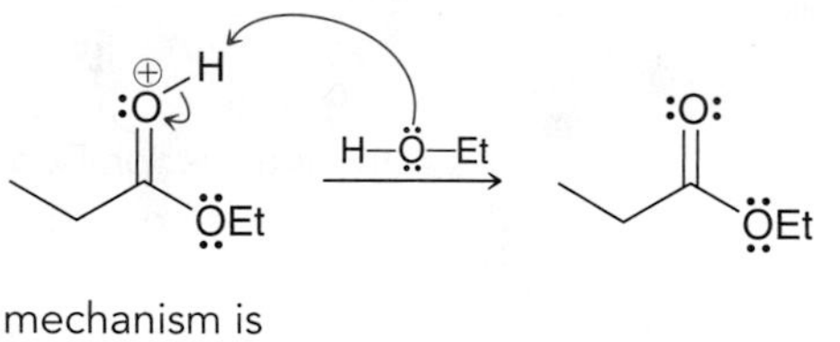

In summary, the complete mechanism is

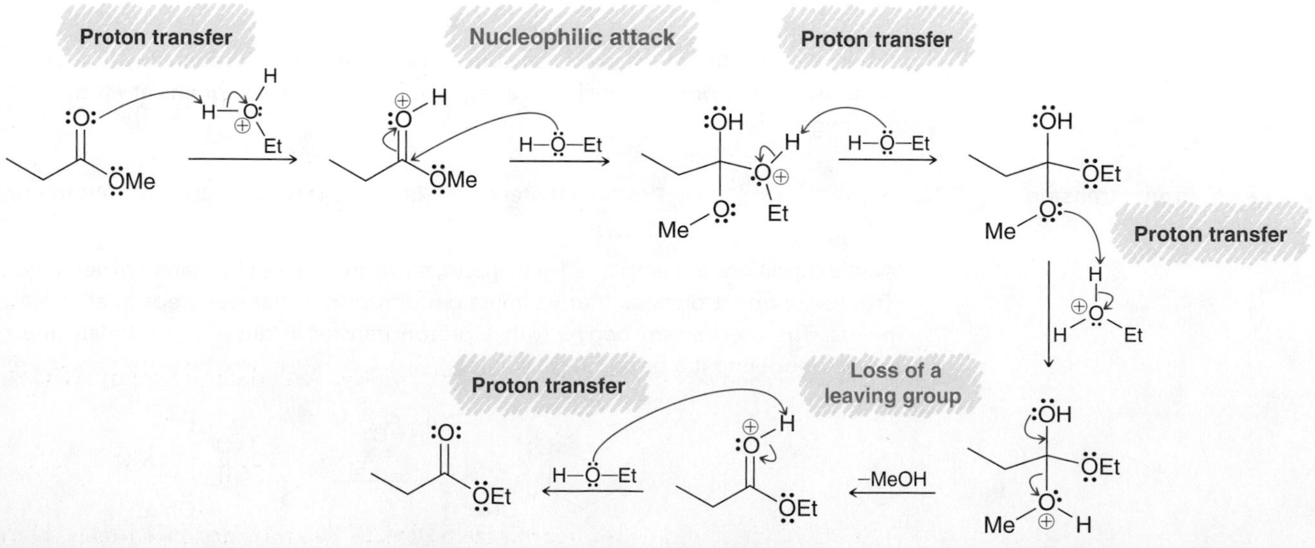

PRACTICE the skill **21.14** Propose a plausible mechanism for each of the following transformations. These reactions will all appear later in this chapter, so practicing their mechanisms now will serve as preparation for the rest of this chapter:

(a)

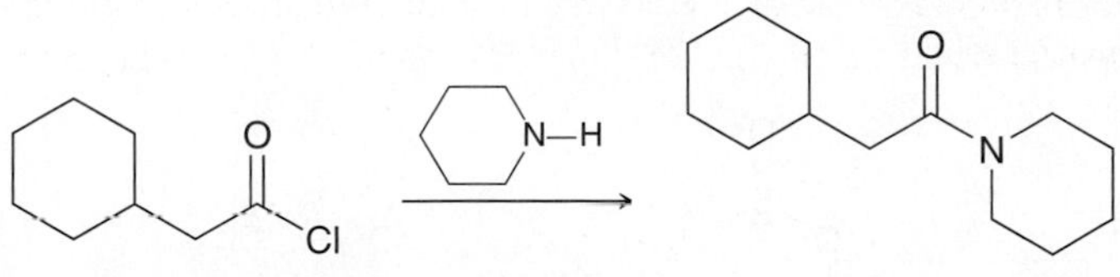

(b)

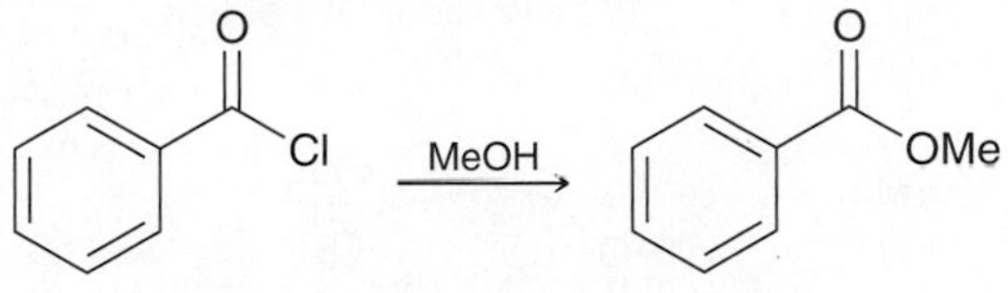

(c)

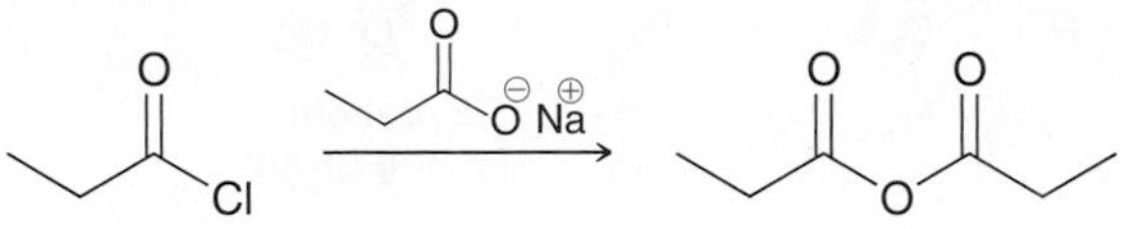

(d)

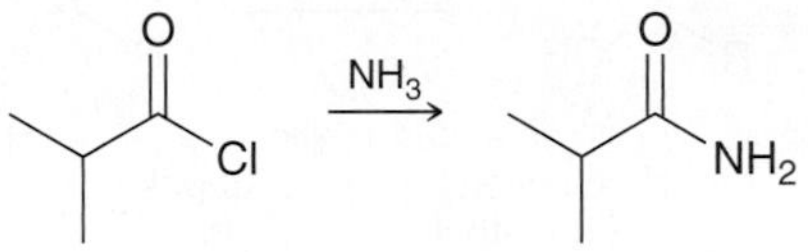

(e)

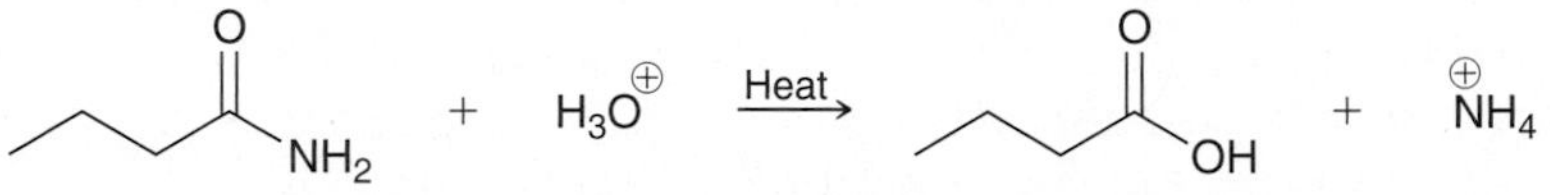

(f)

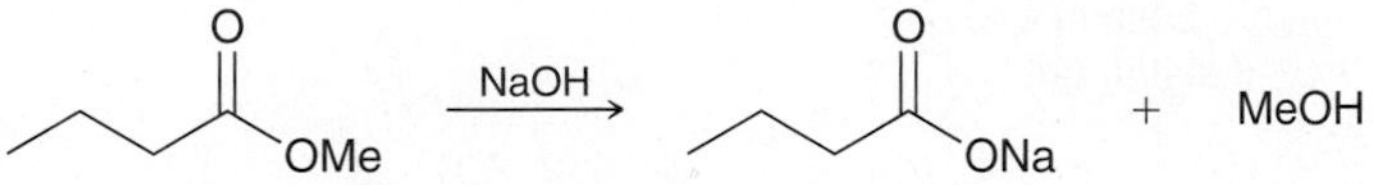

(g)

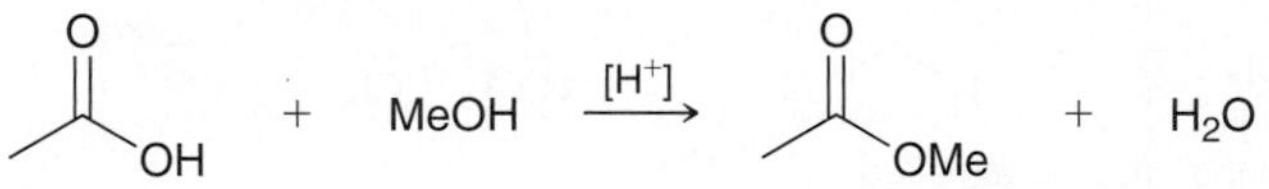

APPLY the skill **21.15** Propose a plausible mechanism for the following transformation:

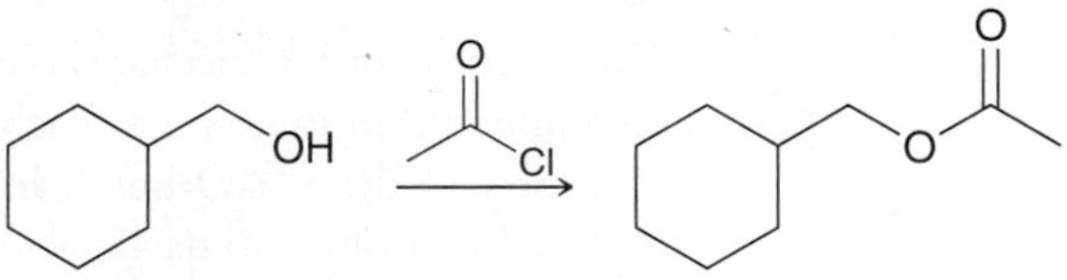

21.16 Propose a plausible mechanism for the following intramolecular transformation:

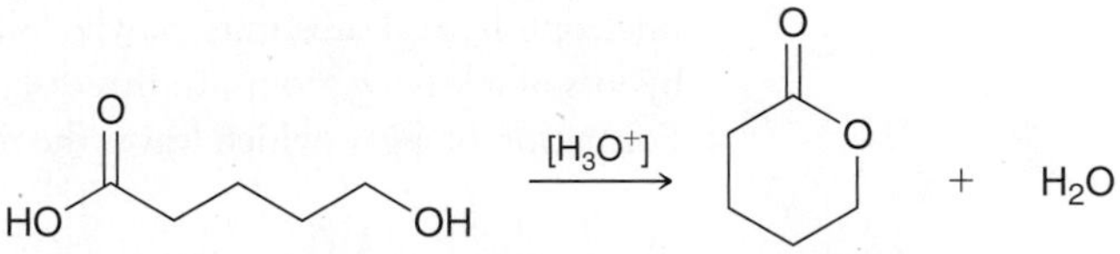

21.17 Propose a plausible mechanism for the following transformation:

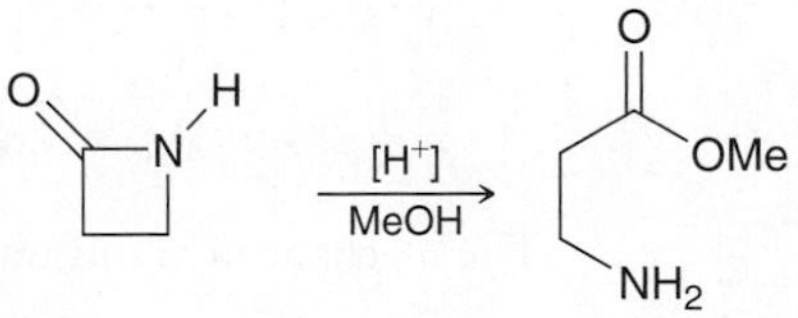

need more **PRACTICE?** **Try Problems 21.61, 21.72**

21.8 Preparation and Reactions of Acid Chlorides

Preparation of Acid Chlorides

Acid chlorides can be formed by treating carboxylic acids with thionyl chloride ($SOCl_2$):

$$R-COOH \xrightarrow{SOCl_2} R-COCl + SO_2 + HCl$$

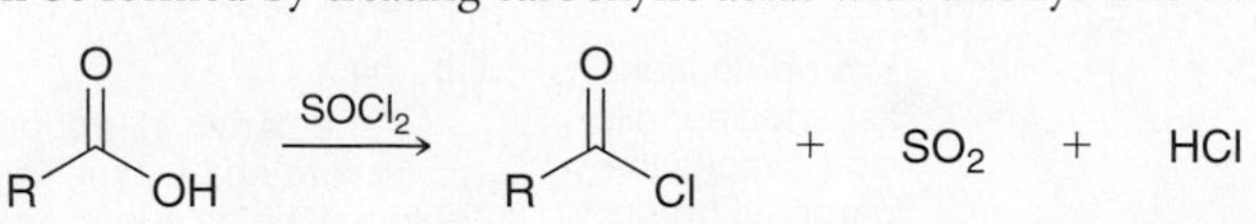

The mechanism for this transformation can be divided into two parts (Mechanism 21.2).

MECHANISM 21.2 PREPARATION OF ACID CHLORIDES VIA THIONYL CHLORIDE

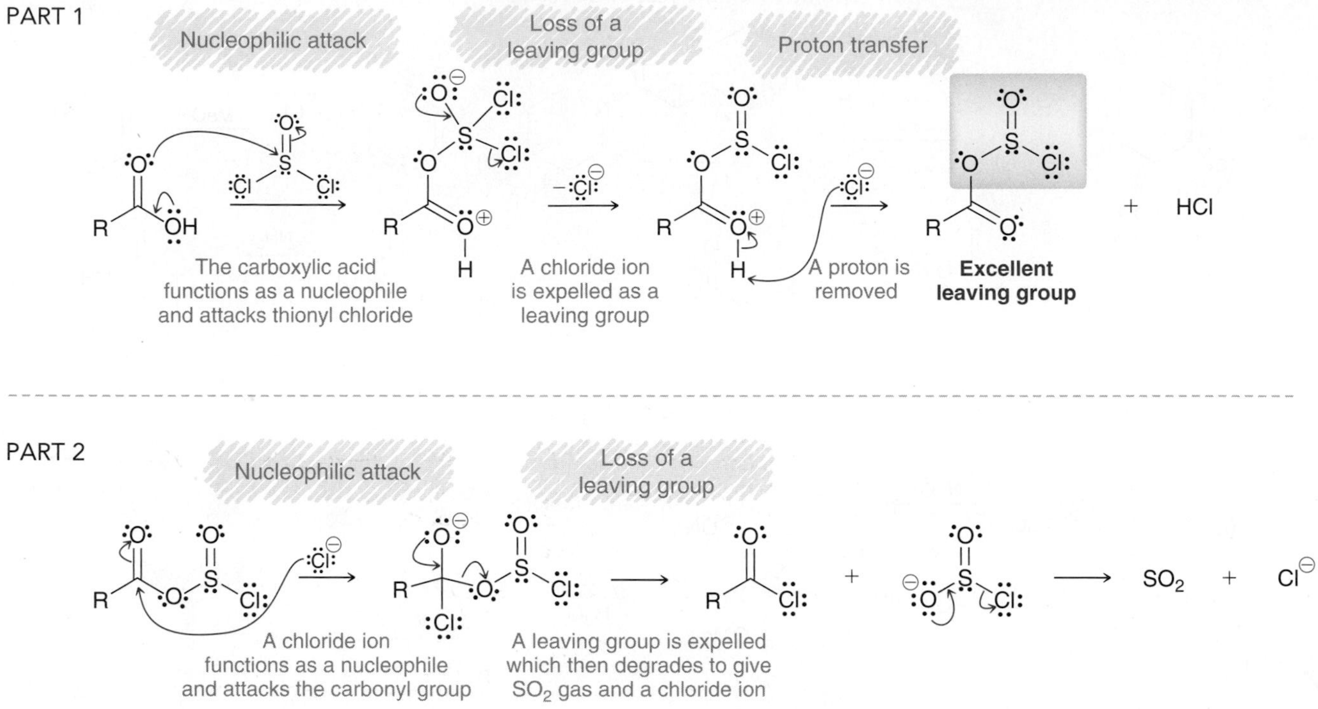
PART 1
Nucleophilic attack
Loss of a leaving group
Proton transfer
The carboxylic acid functions as a nucleophile and attacks thionyl chloride
A chloride ion is expelled as a leaving group
A proton is removed
Excellent leaving group
+ HCl
PART 2
Nucleophilic attack
Loss of a leaving group
A chloride ion functions as a nucleophile and attacks the carbonyl group
A leaving group is expelled which then degrades to give SO2 gas and a chloride ion
SO2 + Cl

The first part of the mechanism converts the OH group into a better leaving group, which is accomplished in three steps. Each of these three steps should seem familiar if we focus on the chemistry of the S=O bond. In the three steps, the S=O bond behaves very much like a C=O bond of a carboxylic acid derivative (as described in the previous section). First, the S=O bond is attacked by a nucleophile, then it is re-formed to expel a leaving group, and finally a proton transfer is used to remove the charge. Part 2 of the mechanism is a typical nucleophilic acyl substitution, which is accomplished in two steps: nucleophilic attack followed by loss of a leaving group. In this case, the leaving group further degrades to form gaseous SO_2. Formation of a gas (which leaves the reaction mixture) forces the reaction to completion.

Hydrolysis of Acid Chlorides

When treated with water, acid chlorides are hydrolyzed to give carboxylic acids.

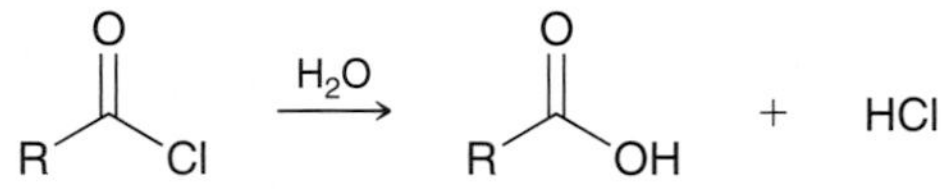
H2O
+ HCl

The mechanism of this transformation requires three steps (Mechanism 21.3).

MECHANISM 21.3 HYDROLYSIS OF AN ACID CHLORIDE

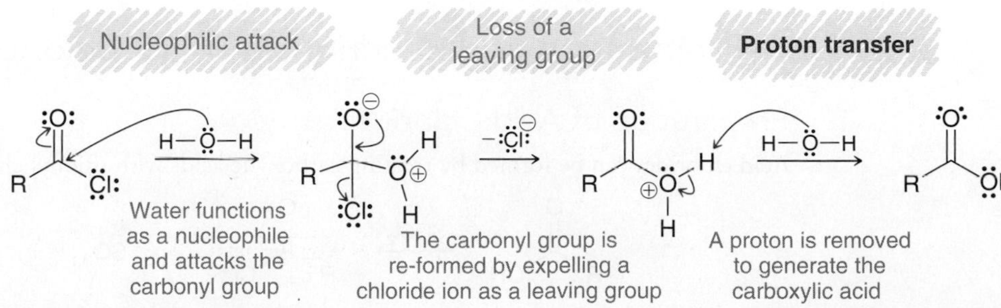
Nucleophilic attack
Loss of a leaving group
Proton transfer
Water functions as a nucleophile and attacks the carbonyl group
The carbonyl group is re-formed by expelling a chloride ion as a leaving group
A proton is removed to generate the carboxylic acid

These are the same three steps used in part 1 of the previous mechanism. This reaction produces HCl as a by-product. The HCl can often produce undesired reactions with other functional groups that might be present in the compound, so pyridine is used to remove the HCl as it is produced.

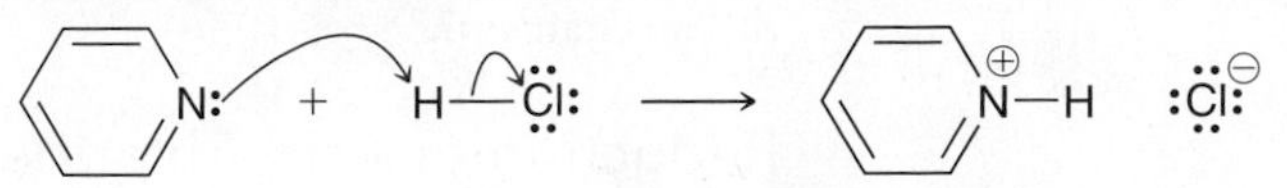

Pyridine is a base that reacts with HCl to form pyridinium chloride. This process effectively traps the HCl so that it is unavailable for any other side reactions.

Alcoholysis of Acid Chlorides

When treated with an alcohol, acid chlorides are converted into esters.

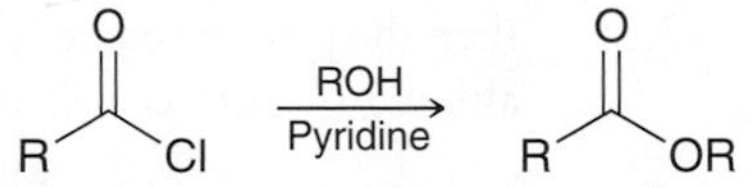

The mechanism of this transformation is directly analogous to hydrolysis of an acid chloride (three mechanistic steps), and pyridine is used as a base to neutralize the HCl as it is produced. This reaction is viewed from the perspective of the acid chloride, but the same reaction can be written from the perspective of the alcohol.

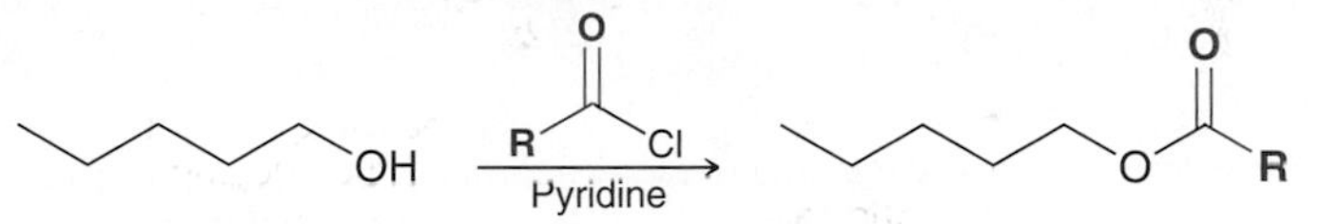

When shown like this, the OH group is said to undergo acylation, because an acyl group has been transferred to the OH group to produce an ester. This process is sensitive to steric effects, which can be exploited to selectively acylate a primary alcohol in the presence of a secondary (more hindered) alcohol.

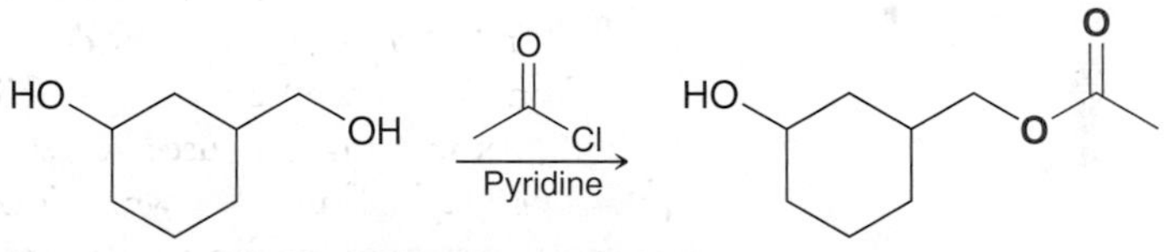

Aminolysis of Acid Chlorides

When treated with ammonia, acid chlorides are converted into amides.

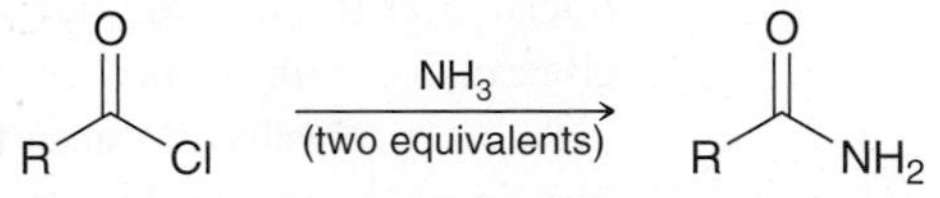

Pyridine is not used in this reaction, because ammonia itself is a sufficiently strong base to neutralize the HCl as it is produced. For this reaction, two equivalents of ammonia are necessary: one for the nucleophilic attack and the other to neutralize the HCl. This reaction also occurs with primary and secondary amines to produce *N*-substituted amides.

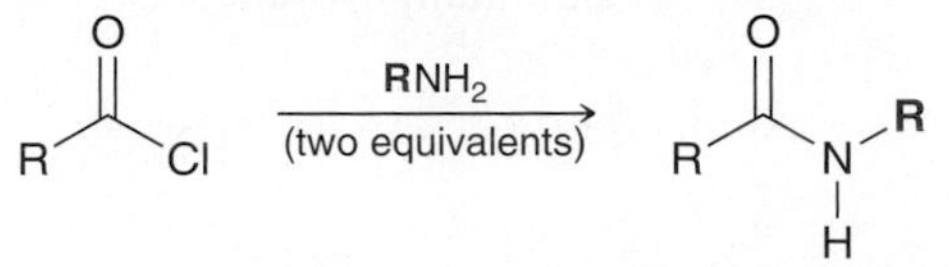

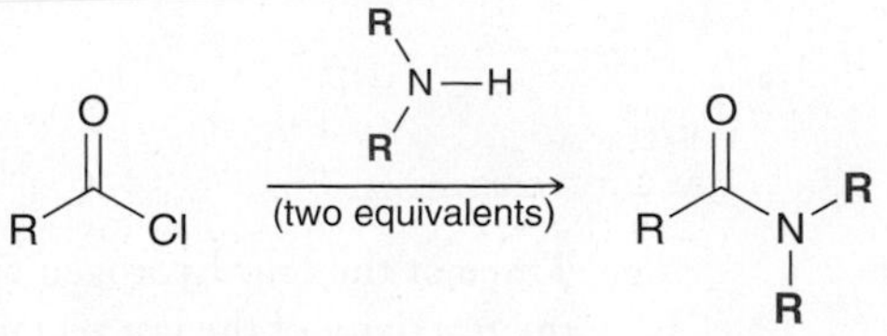

The mechanism for each of these reactions is directly analogous to the hydrolysis of an acid chloride. There are three steps: (1) nucleophilic attack, (2) loss of a leaving group to re-form the carbonyl, and (3) proton transfer to remove the positive charge. Can you draw the mechanism?

Reduction of Acid Chlorides

When treated with lithium aluminum hydride, acid chlorides are reduced to give alcohols:

$$
\underset{R}{\overset{O}{\|}}{C}\!-\!Cl \quad \xrightarrow[\text{2) } H_2O]{\text{1) Excess LAH}} \quad \underset{R}{\overset{OH}{|}}{CH}
$$

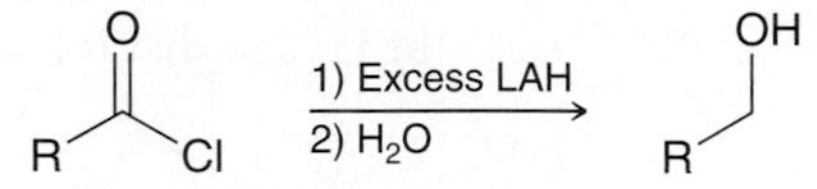

Notice that two separate steps are required. First the acid chloride is treated with LAH, and then the proton source is added to the reaction flask. Water (H_2O) can serve as a proton source, although H_3O^+ can also be used as a proton source (Mechanism 21.4).

MECHANISM 21.4 REDUCTION OF AN ACID CHLORIDE WITH LAH

Nucleophilic attack	Loss of a leaving group	Nucleophilic attack **(second time)**	Proton transfer
LAH delivers a hydride ion, which attacks the carbonyl group	The carbonyl group is re-formed by expelling a chloride ion as a leaving group	The carbonyl group is attacked again by hydride, generating an alkoxide	After the reaction is complete, the alkoxide ion is protonated with an acid

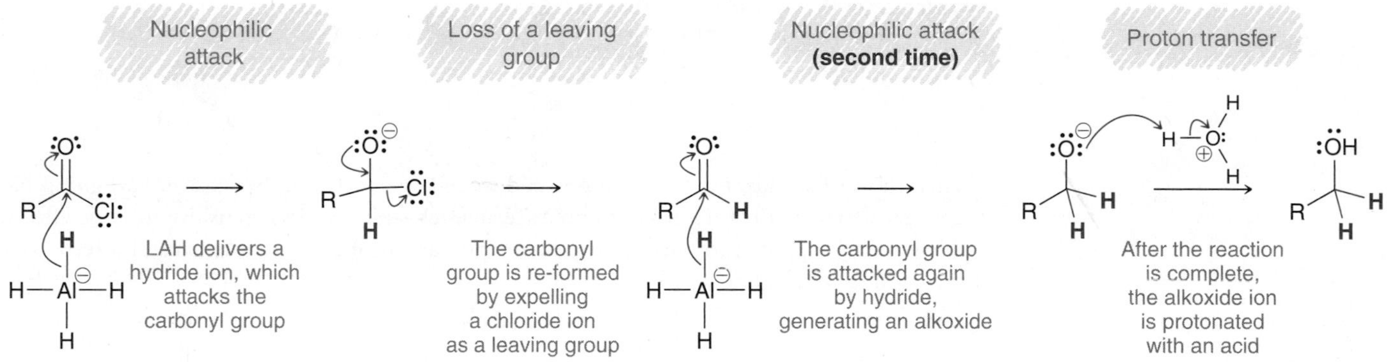

The first two steps of the mechanism are exactly what we might have expected: (1) nucleophilic attack followed by (2) loss of a leaving group to re-form the carbonyl. These two steps produce an aldehyde, which can be attacked again to give an alkoxide ion. Remember that the carbonyl should always be re-formed, if possible, without expelling H^- and C^-. The alkoxide ion produced after the second attack does not have any groups that can leave, and therefore, nothing else can occur until a proton source is provided to protonate the alkoxide ion. This reaction has little practical value, because acid chlorides are generally prepared from carboxylic acids, which can simply be treated directly with LAH to produce the alcohol.

The reaction between an acid chloride and LAH cannot be used to produce an aldehyde. Using one equivalent of LAH simply leads to a mess of products. Producing the aldehyde requires the use of a more selective hydride-reducing agent that will react with acid chlorides more rapidly than aldehydes. There are many such reagents, including lithium tri(*t*-butoxy) aluminum hydride.

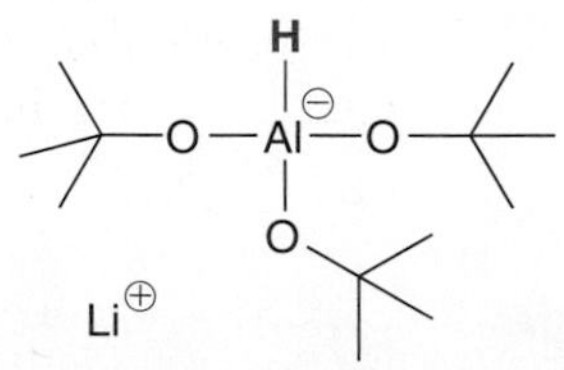

Lithium tri(*t*-butoxy) aluminium hydride

Three of the four hydrogen atoms have been replaced with *tert*-butoxy groups, which modify the reactivity of the last remaining hydride group. This reducing agent will react with the acid chloride rapidly but will react with the aldehyde more slowly, allowing the aldehyde to be isolated. These conditions can be used to convert an acid chloride into an aldehyde.

$$\underset{R}{\overset{O}{\|}}\underset{Cl}{\overset{}{}} \quad \underset{\text{2) H}_2\text{O}}{\overset{\text{1) LiAl(OR)}_3\text{H}}{\longrightarrow}} \quad \underset{R}{\overset{O}{\|}}\underset{H}{\overset{}{}}$$

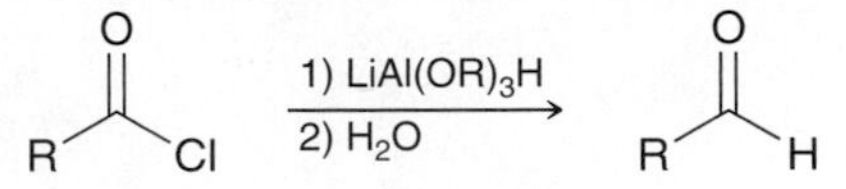

Reactions between Acid Chlorides and Organometallic Reagents

When treated with a Grignard reagent, acid chlorides are converted into alcohols, with the introduction of two alkyl groups.

$$\underset{Cl}{\overset{O}{\|}} \quad \underset{\text{2) H}_2\text{O}}{\overset{\text{1) Excess RMgBr}}{\longrightarrow}} \quad \underset{R}{\overset{OH}{\underset{R}{|}}}R$$

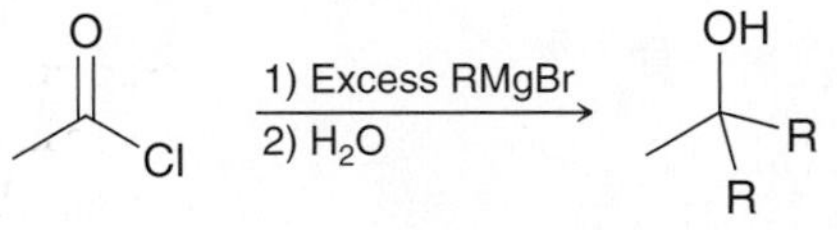

Just as with LAH, two separate steps are required. First the acid chloride is treated with the Grignard reagent, and then the proton source is added to the reaction flask. Water (H_2O) can serve as a proton source, although H_3O^+ can also be used (Mechanism 21.5).

MECHANISM 21.5 THE REACTION BETWEEN AN ACID CHLORIDE AND A GRIGNARD REAGENT

Nucleophilic attack	Loss of a leaving group	Nucleophilic attack **(second time)**	Proton transfer
A Grignard reagent functions as a nucleophile, and attacks the carbonyl group	The carbonyl group is re-formed by expelling a chloride ion as a leaving group	The carbonyl group is attacked again by a Grignard reagent, generating an alkoxide	After the reaction is complete, the alkoxide ion is protonated with an acid

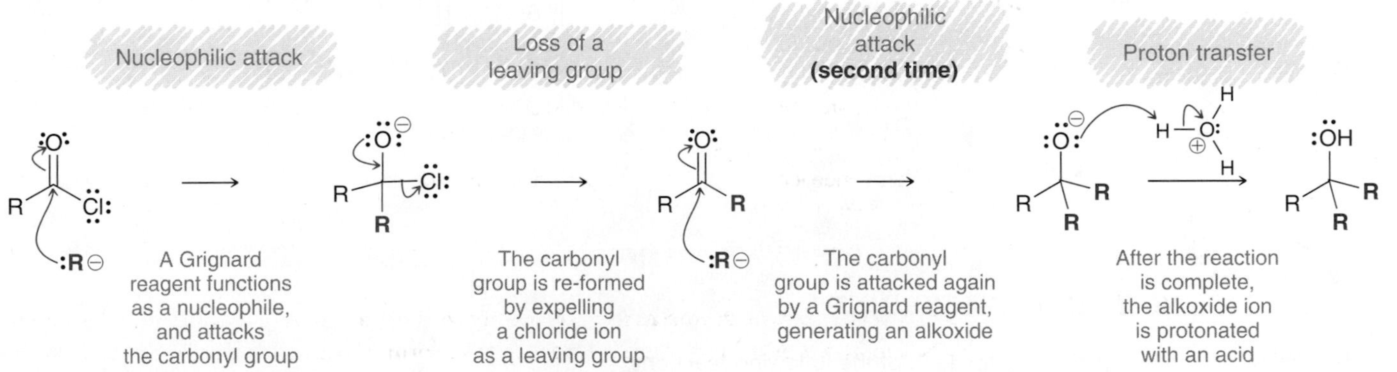

The first two steps of the mechanism are exactly what we would expect: (1) nucleophilic attack followed by (2) loss of a leaving group to re-form the carbonyl. These two steps produce a ketone, which can be attacked again by another Grignard reagent to give an alkoxide ion. Remember that the carbonyl should always be re-formed, if possible, but H^- and C^- should never be expelled. The alkoxide ion produced after the second attack does not have any groups that can leave, so nothing else can occur until a proton source is provided to protonate the alkoxide ion.

The reaction between an acid chloride and a Grignard reagent cannot be used to produce a ketone. Using one equivalent of the Grignard reagent simply leads to a mess of products. Producing the ketone requires the use of a more selective carbon nucleophile that will react with acid chlorides more rapidly than ketones. There are many such reagents. The most commonly used reagent for this purpose is a **lithium dialkyl cuprate**, also called a **Gilman reagent**.

$$\underset{R}{\overset{R}{\underset{|}{\overset{|}{\text{Cu}-\text{Li}}}}}$$

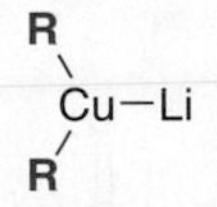

Lithium dialkyl cuprate

The alkyl groups in this reagent are attached to copper rather than magnesium, and their carbanionic character is less pronounced (a C—Cu bond is less polarized than a C—Mg bond). This reagent can be used to convert acid chlorides into ketones with excellent yields. The resulting ketone is not further attacked under these conditions.

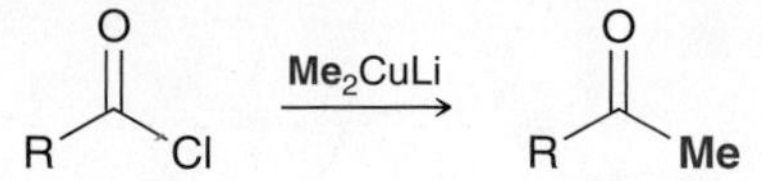

Summary of Reactions of Acid Chlorides

Figure 21.9 summarizes the reactions of acid chlorides discussed in this section.

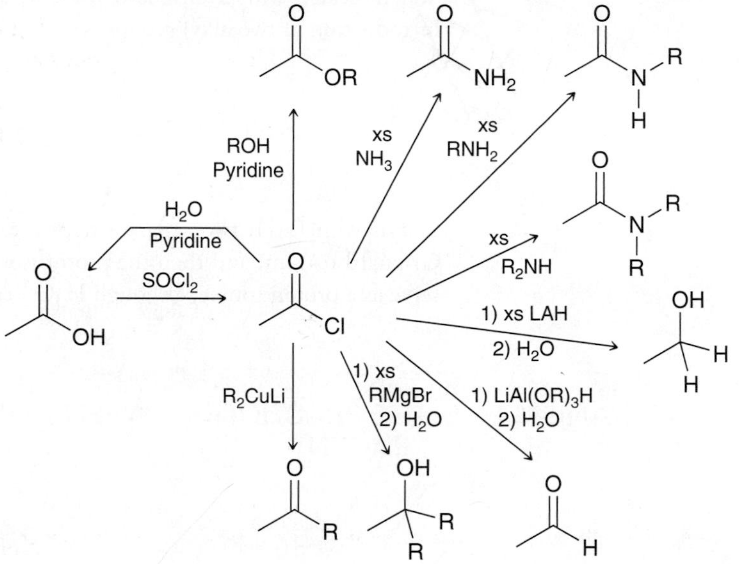

FIGURE 21.9
Reactions of acid chlorides.

CONCEPTUAL CHECKPOINT

21.18 Predict the major product(s) for each of the following reactions:

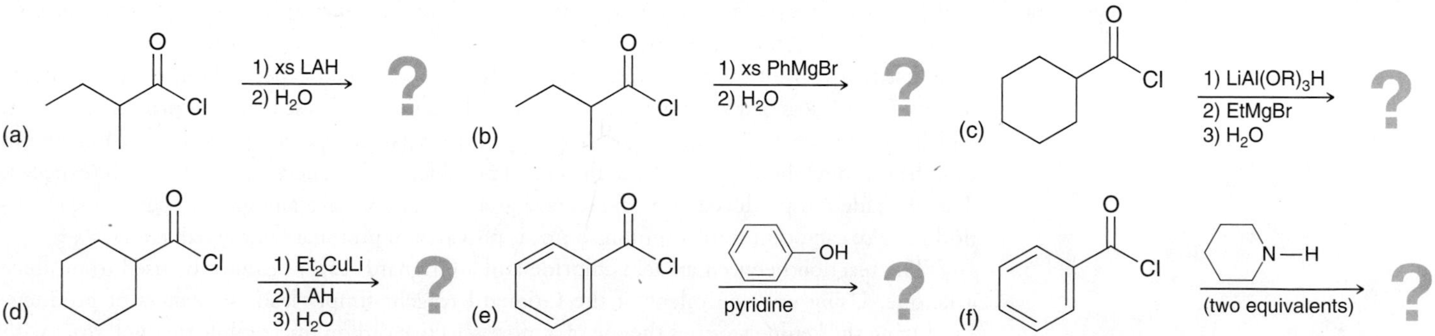

21.19 Identify the reagents necessary for the following transformation:

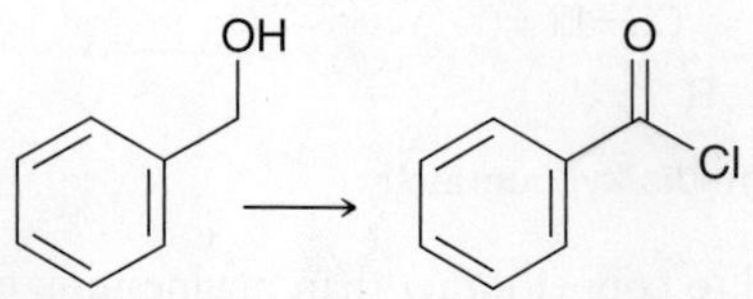

21.20 Propose a mechanism for the following transformation:

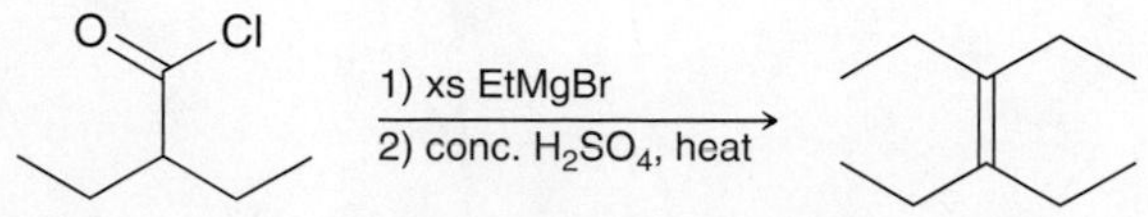

21.9 Preparation and Reactions of Acid Anhydrides

Preparation of Acid Anhydrides

Carboxylic acids can be converted into acid anhydrides with excessive heating.

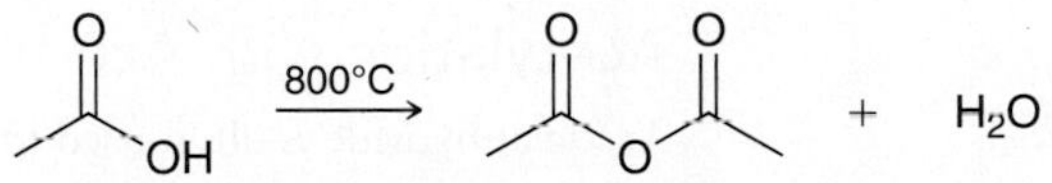

This method is only practical for acetic acid, as most other acids cannot survive the excessive heat. An alternative method for preparing acid anhydrides involves treating an acid chloride with a carboxylate ion, which functions as a nucleophile.

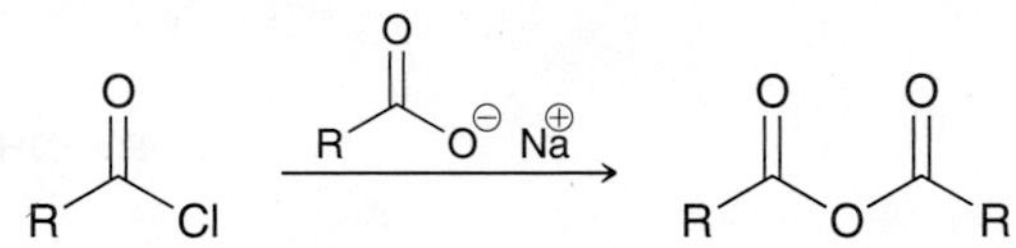

As we might expect, the mechanism of this transformation involves only two steps:

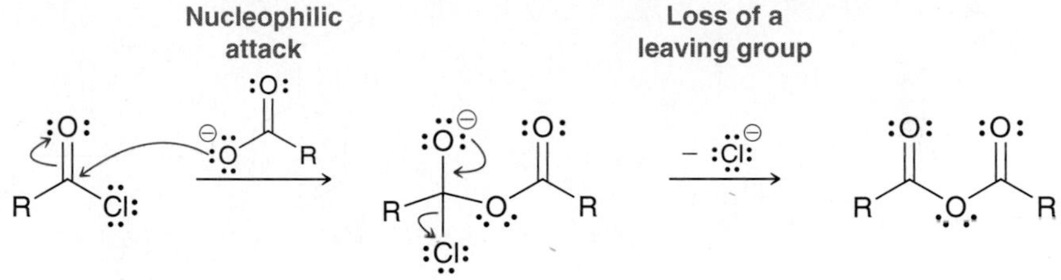

This method can be used to prepare symmetrical or unsymmetrical anhydrides.

Reactions of Acid Anhydrides

The reactions of anhydrides are directly analogous to the reactions of acid chlorides. The only difference is in the identity of the leaving group.

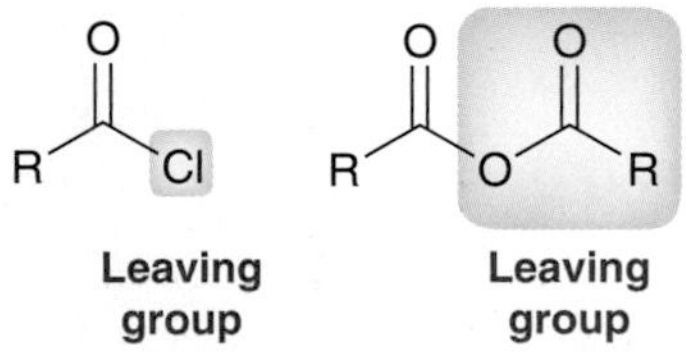

With an acid chloride, the leaving group is a chloride ion, and the by-product of the reaction is therefore HCl. With an acid anhydride, the leaving group is a carboxylate ion, and the by-product is therefore a carboxylic acid. As a result, it is not necessary to use pyridine in reactions with acid anhydrides, because HCl is not produced. Figure 21.10 summarizes the reactions of

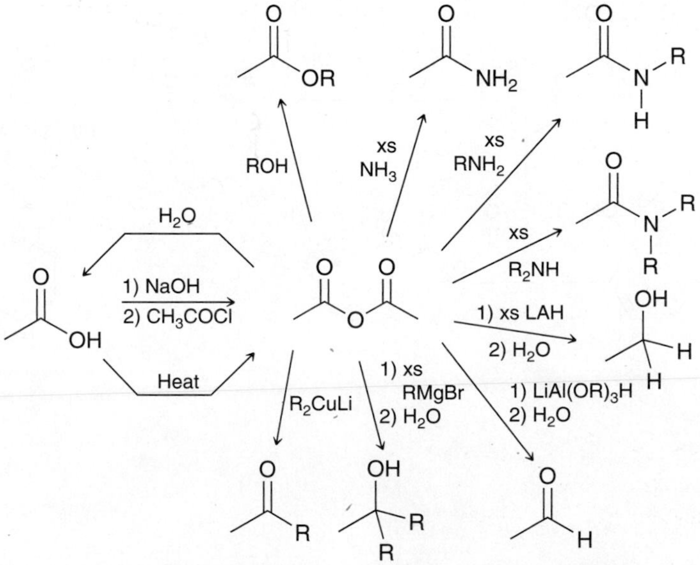

FIGURE 21.10
Reactions of acid anhydrides.

anhydrides. Each of these reactions produces acetic acid as a by-product. From a synthetic point of view, the use of anhydrides (rather than acid chlorides) involves the loss of half of the starting material, which is inefficient. For this reason, acid chlorides are more efficient as starting materials than acid anhydrides.

Acetylation with Acetic Anhydride

Acetic anhydride is often used to acetylate an alcohol or an amine.

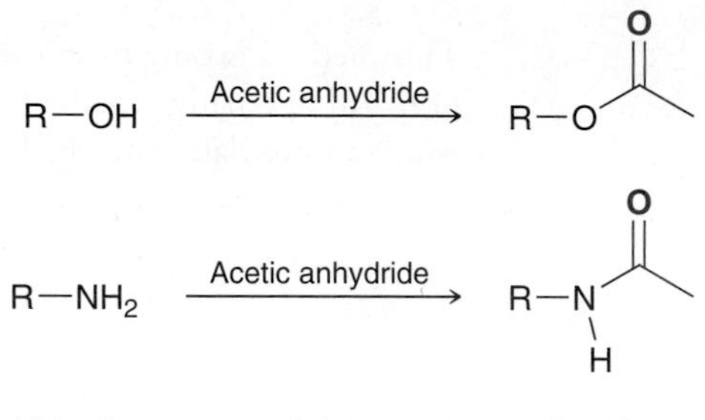

These reactions are utilized in the commercial preparation of aspirin and Tylenol™.

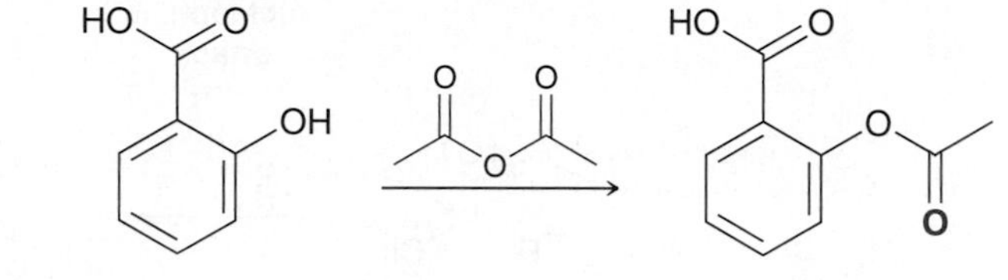

Aspirin

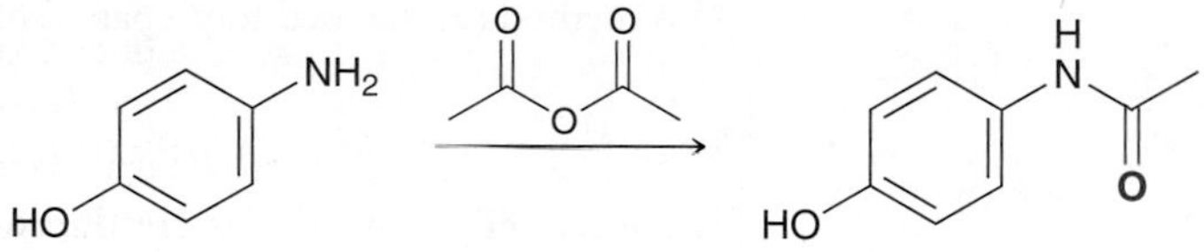

Tylenol™

CONCEPTUAL CHECKPOINT

21.21 Predict the major product(s) for each of the following reactions:

(a) 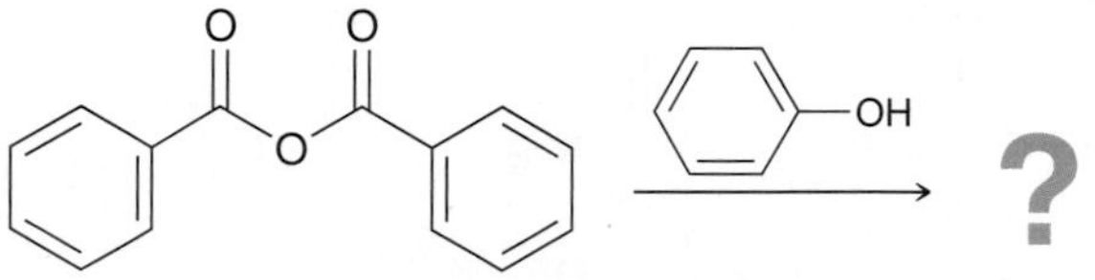

(b) 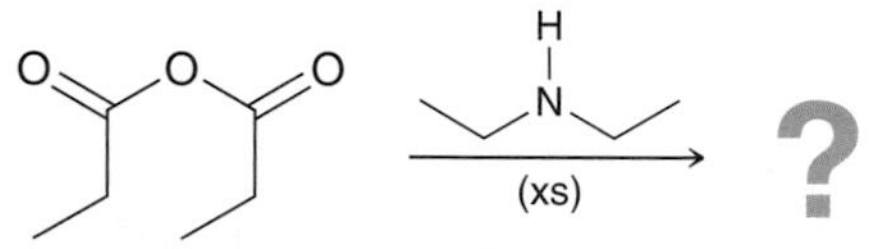

(c) 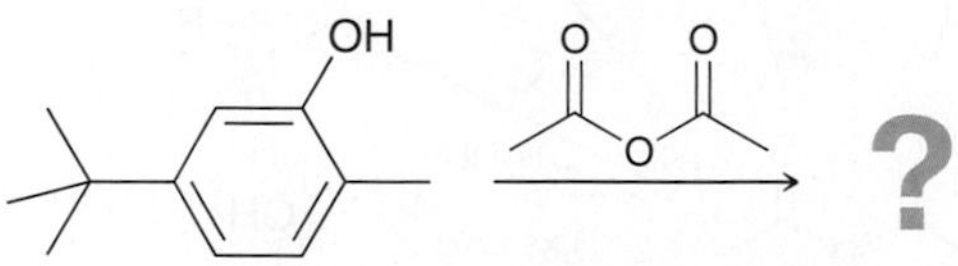

(d) 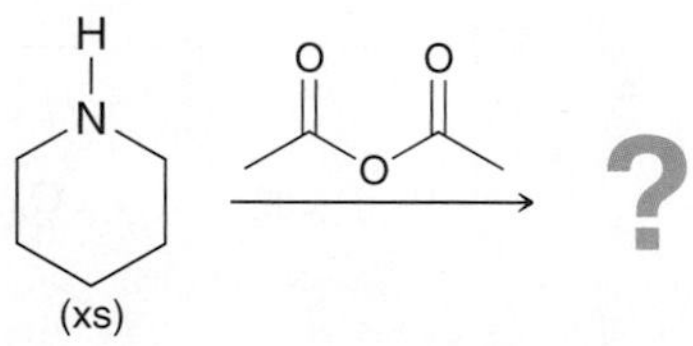

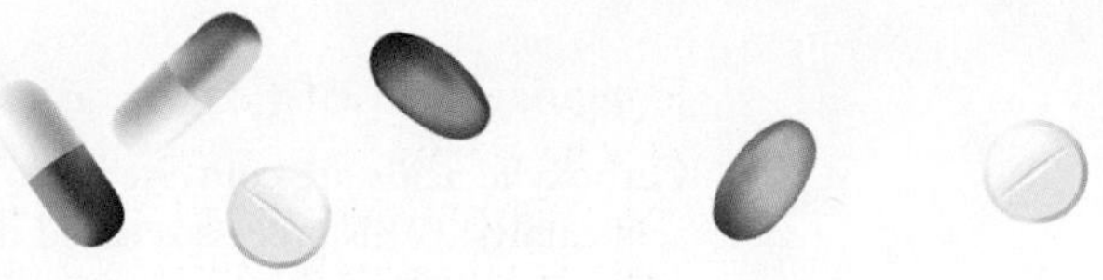

MEDICALLYSPEAKING **)))**

How Does Aspirin Work?

Aspirin is prepared from salicylic acid, a compound found in the bark of the willow tree that has been used for its medicinal properties for thousands of years. Aspirin's mechanism of action remained unknown until the early 1970s, when John Vane, Bengt Samuelsson, and Sune Bergström elucidated its role in blocking the synthesis of prostaglandins, for which they were awarded the 1982 Nobel Prize in Physiology or Medicine. Prostaglandins, which are compounds containing five-membered rings, will be discussed in greater detail in Chapter 26. Prostaglandins have many important biological functions, including stimulating inflammation and inducing fever. They are produced in the body from arachidonic acid via a process that is catalyzed by an enzyme called cyclooxygenase:

Arachidonic acid

Cyclooxygenase

PGG₂

Prostaglandins

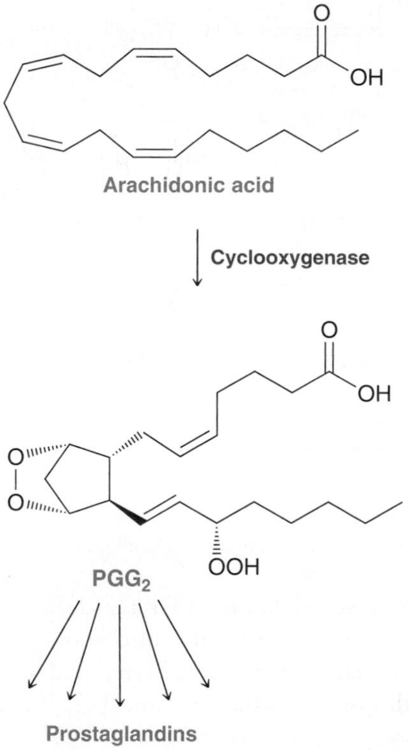

An OH group of cyclooxygenase reacts with aspirin, resulting in the transfer of an acetyl group from aspirin to cyclooxygenase:

Cyclo-oxygenase

Active enzyme

Cyclo-oxygenase

Acylated enzyme inactive

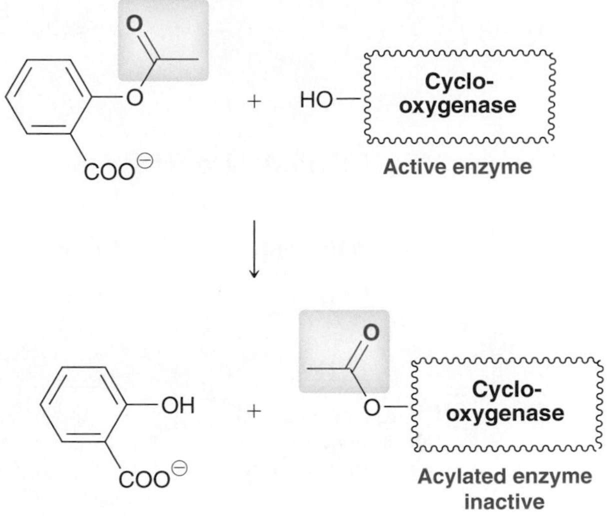

In this way, aspirin functions as an acetylating agent, much the way acetic anhydride functions as an acetylating agent in the preparation of aspirin. The same acetyl group that came from acetic anhydride (in the synthesis of aspirin) is ultimately transferred to cyclooxygenase. This process deactivates cyclooxygenase, thereby interfering with the synthesis of prostaglandins. With a decreased concentration of prostaglandins, the onset of inflammation is slowed and fevers are reduced.

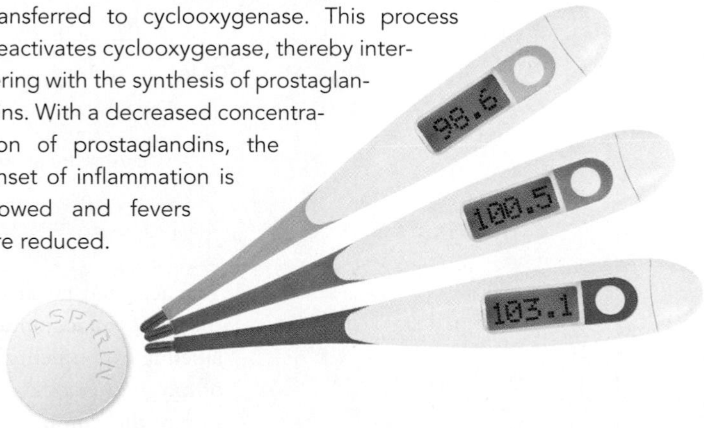

21.10 Preparation of Esters

Preparation of Esters via S_N2 Reactions

When treated with a strong base followed by an alkyl halide, carboxylic acids are converted into esters:

$$\underset{R}{\overset{O}{\parallel}}C-OH \quad \xrightarrow[\text{2) } CH_3I]{\text{1) NaOH}} \quad \underset{R}{\overset{O}{\parallel}}C-O-CH_3$$

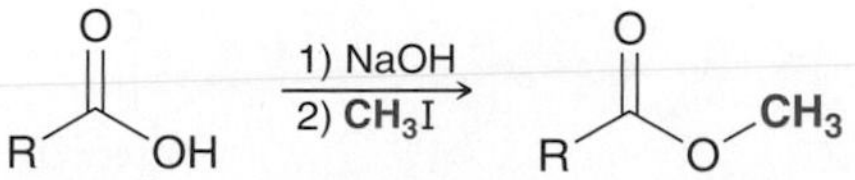

The carboxylic acid is first deprotonated to yield a carboxylate ion, which then functions as a nucleophile and attacks the alkyl halide in an S_N2 process. The expected limitations of S_N2 processes therefore apply. Specifically, tertiary alkyl halides cannot be used.

Preparation of Esters via Fischer Esterification

Carboxylic acids are converted into esters when treated with an alcohol in the presence of an acid catalyst. This process is called the **Fischer esterification** (Mechanism 21.6).

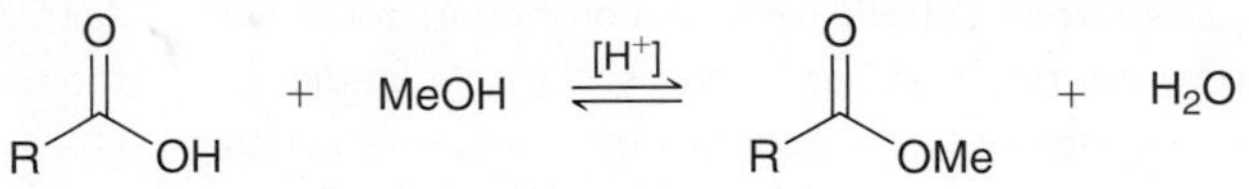

O
R
OH
+ MeOH
[H+]
O
R
OMe
+ H2O

MECHANISM 21.6 THE FISCHER ESTERIFICATION PROCESS

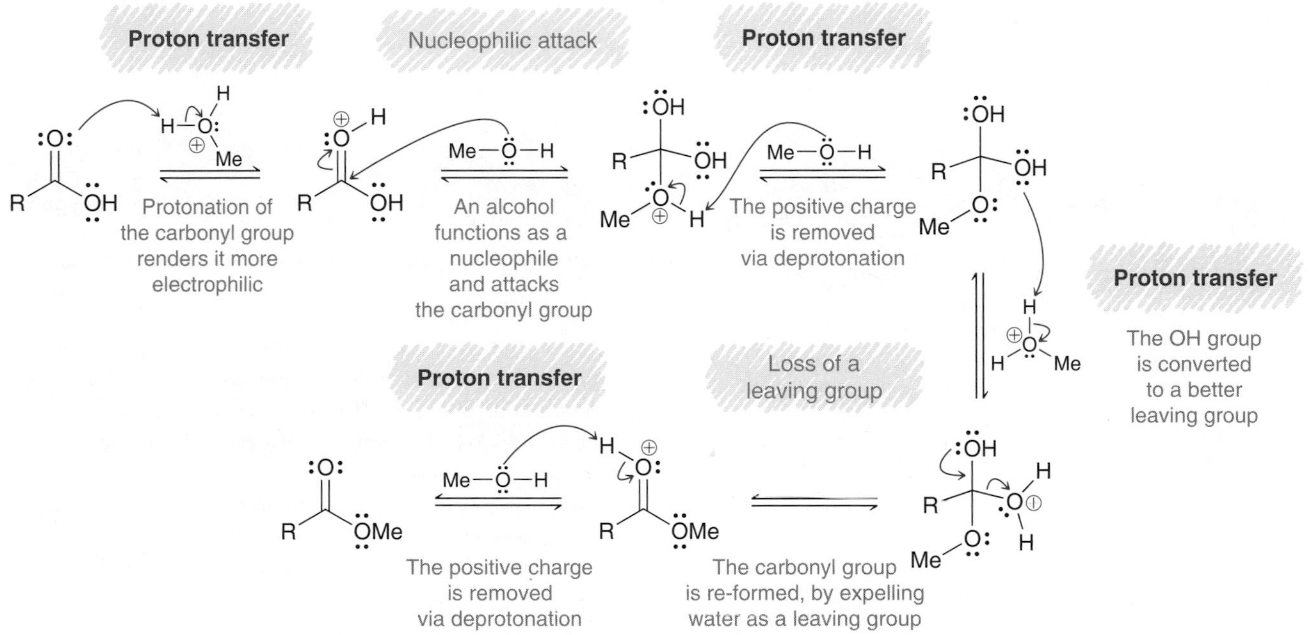

Proton transfer
Nucleophilic attack
Proton transfer
Protonation of the carbonyl group renders it more electrophilic
An alcohol functions as a nucleophile and attacks the carbonyl group
The positive charge is removed via deprotonation
Proton transfer
The OH group is converted to a better leaving group
Proton transfer
Loss of a leaving group
The positive charge is removed via deprotonation
The carbonyl group is re-formed, by expelling water as a leaving group
Me—O—H

The accepted mechanism is exactly what we would expect for a nucleophilic acyl substitution that takes place under acidic conditions. Evidence for this mechanism comes from isotopic labeling experiments in which the oxygen atom of the alcohol is replaced with a heavier isotope of oxygen (^{18}O), and the location of this isotope is tracked throughout the reaction. The location of the isotope in the product (shown in red) supports Mechanism 21.6.

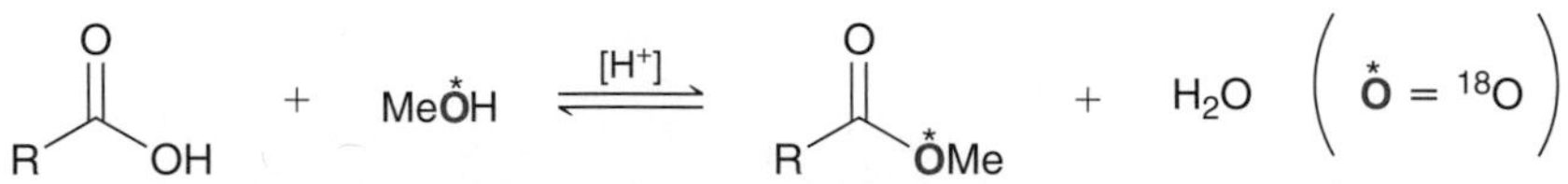

+ MeOH
[H+]
+ H2O
O* = 18O

LOOKING BACK
Le Châtelier's principle states that a system at equilibrium will adjust in order to minimize any stress placed on the system.

The Fischer esterification process is reversible and can be controlled by exploiting Le Châtelier's principle. That is, formation of the ester can be favored either by using an excess of the alcohol (i.e., using the alcohol as the solvent) or by removing water from the reaction mixture as it is formed.

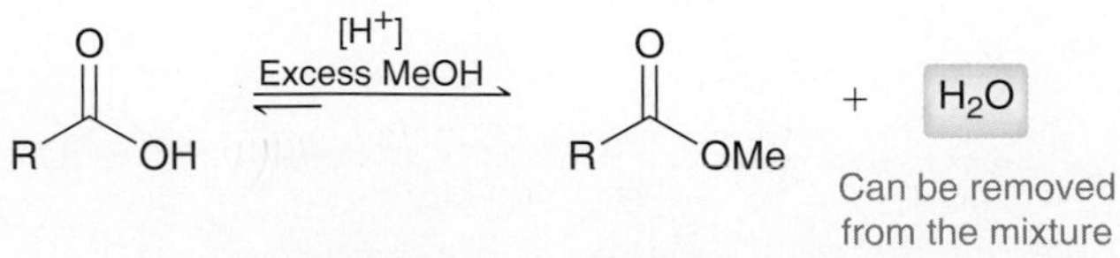

[H+]
Excess MeOH
+ H2O
Can be removed from the mixture

The reverse process, which is conversion of the ester into a carboxylic acid, can be achieved by using an excess of water, as we will see in Section 21.11.

Preparation of Esters via Acid Chlorides

Esters can also be prepared by treating an acid chloride with an alcohol. We already explored this reaction in Section 21.8.

$$\text{R—C(=O)—Cl} \xrightarrow[\text{Pyridine}]{\text{ROH}} \text{R—C(=O)—OR}$$

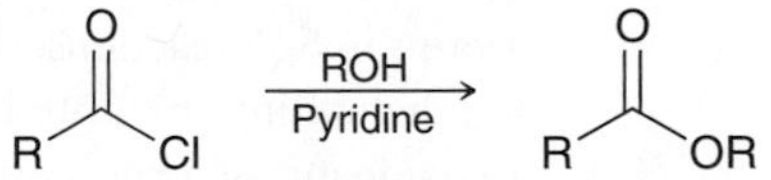

CONCEPTUAL CHECKPOINT

21.22 In this section, we have seen three ways to achieve the following transformation. Identify the reagents necessary for all three methods.

$$\text{PhC(=O)OH} \xrightarrow[?]{\substack{? \\ ?}} \text{PhC(=O)OEt}$$

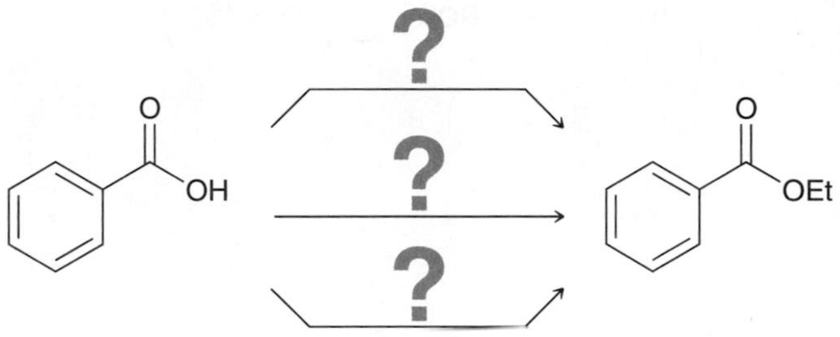

21.23 Identify reagents that can be used to accomplish each of the following transformations.

(a) benzyl alcohol → PhC(=O)OEt

(b) styrene → PhC(=O)OEt

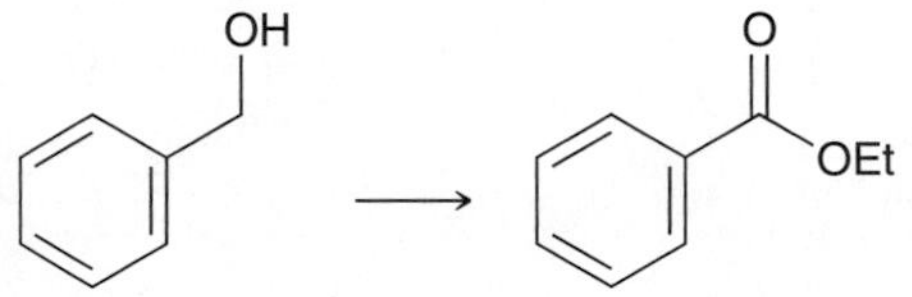

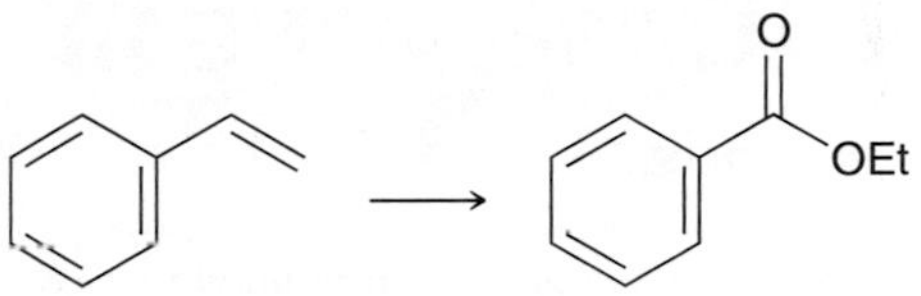

21.11 Reactions of Esters

Saponification

Esters can be converted into carboxylic acids by treatment with sodium hydroxide followed by an acid. This process is called **saponification** (Mechanism 21.7):

$$\text{R—C(=O)—OR} \xrightarrow[\text{2) H}_3\text{O}^+]{\text{1) NaOH}} \text{R—C(=O)—OH} \; + \; \text{ROH}$$

MECHANISM 21.7 SAPONIFICATION OF ESTERS

Nucleophilic attack	Loss of a leaving group	Proton transfer

Hydroxide functions as a nucleophile and attacks the carbonyl group

The carbonyl group is re-formed by expelling an alkoxide ion as a leaving group

The carboxylic acid is deprotonated by the alkoxide ion, generating a carboxylate ion

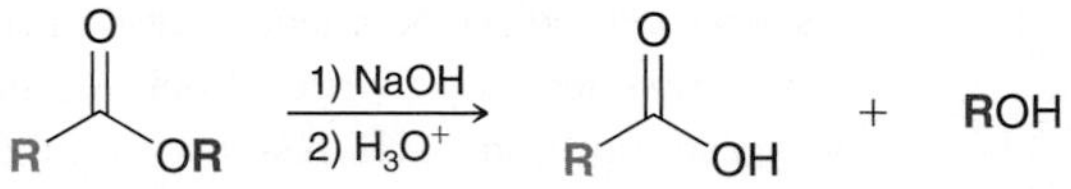

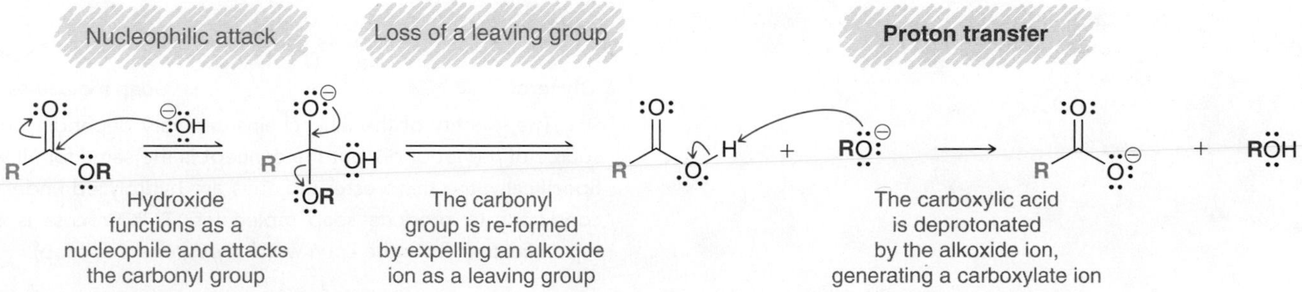

The first two steps of this mechanism are exactly what we would expect of a nucleophilic acyl substitution reaction occurring under basic conditions: (1) nucleophilic attack followed by (2) loss of a leaving group. In basic conditions, an alkoxide ion can function as a leaving group and is not protonated prior to its departure. Although alkoxide ions are not suitable leaving groups in S_N2 reactions, they can function as leaving groups in these circumstancess because the tetrahedral intermediate is sufficiently high in energy. The tetrahedral intermediate itself is an alkoxide ion, so expulsion of an alkoxide ion is not uphill in energy.

Under such strongly basic conditions, the carboxylic acid does not survive; it is deprotonated to produce a carboxylate salt. In fact, the formation of a stabilized carboxylate ion is a driving force that pushes the equilibrium to favor formation of products. After the reaction is complete, an acid is required to protonate the carboxylate ion to give the carboxylic acid.

Evidence for this mechanism comes from isotopic labeling experiments in which the oxygen atom of the alcohol is replaced with an oxygen isotope (^{18}O) that is tracked throughout the reaction. The location of the isotope in the alcohol by-product supports Mechanism 21.7.

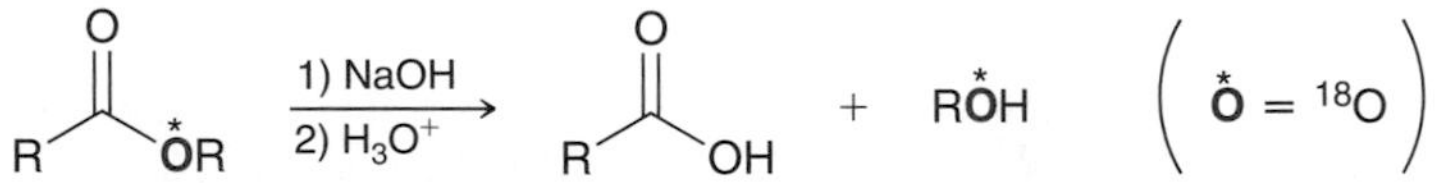

PRACTICALLYSPEAKING)))

How Soap Is Made

Recall from Chapter 1 that soaps are compounds that contain a polar group on one end of the molecule and a nonpolar group on the other end:

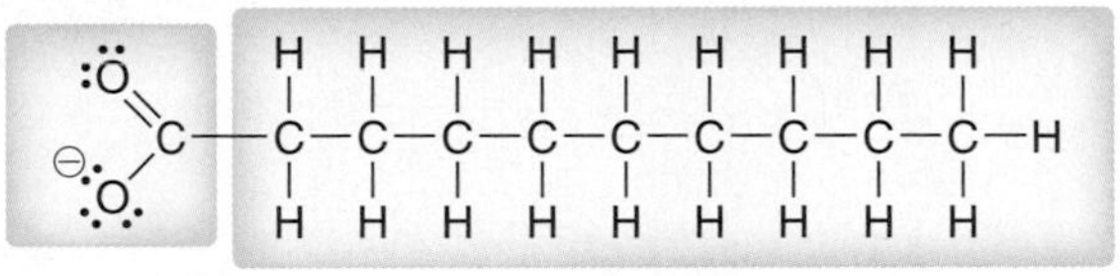

Polar group (hydrophilic) | Non-polar group (hydrophobic)

The hydrophobic tails of soap molecules surround oil molecules, forming a micelle, as described in Section 1.13.

Most soaps are produced from fats and oils, which contain three ester moieties. Upon treatment with a strong base, such as sodium hydroxide, the ester moieties are hydrolyzed, giving glycerol and three soap molecules:

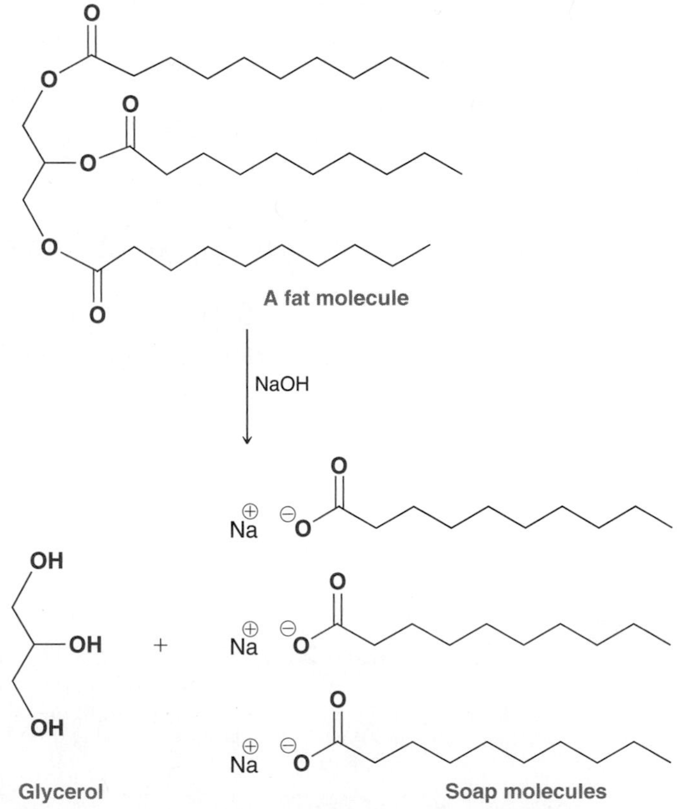

The identity of the alkyl chains can vary depending on the source of the fat or oil, but the concept is the same for all soaps. Specifically, the three ester moieties are hydrolyzed under basic conditions to produce soap molecules. This process is called *saponification*, from the Latin word *sapo* (meaning soap).

Acid-Catalyzed Hydrolysis of Esters

Esters can also be hydrolyzed under acidic conditions.

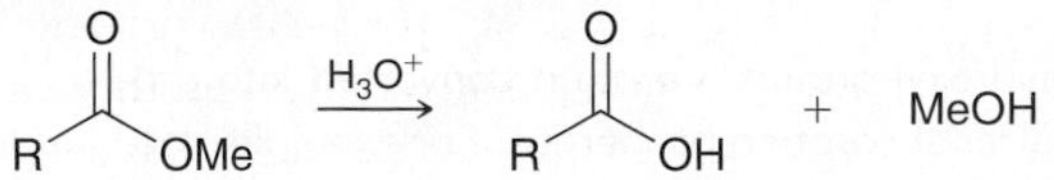

O
R
OMe
H_3O^+
O
R
OH
+
MeOH

This process (Mechanism 21.8) is the reverse of a Fischer esterification. The mechanism is exactly what we would expect for a nucleophilic acyl substitution that takes place under acidic conditions.

MECHANISM 21.8 ACID-CATALYZED HYDROLYSIS OF ESTERS

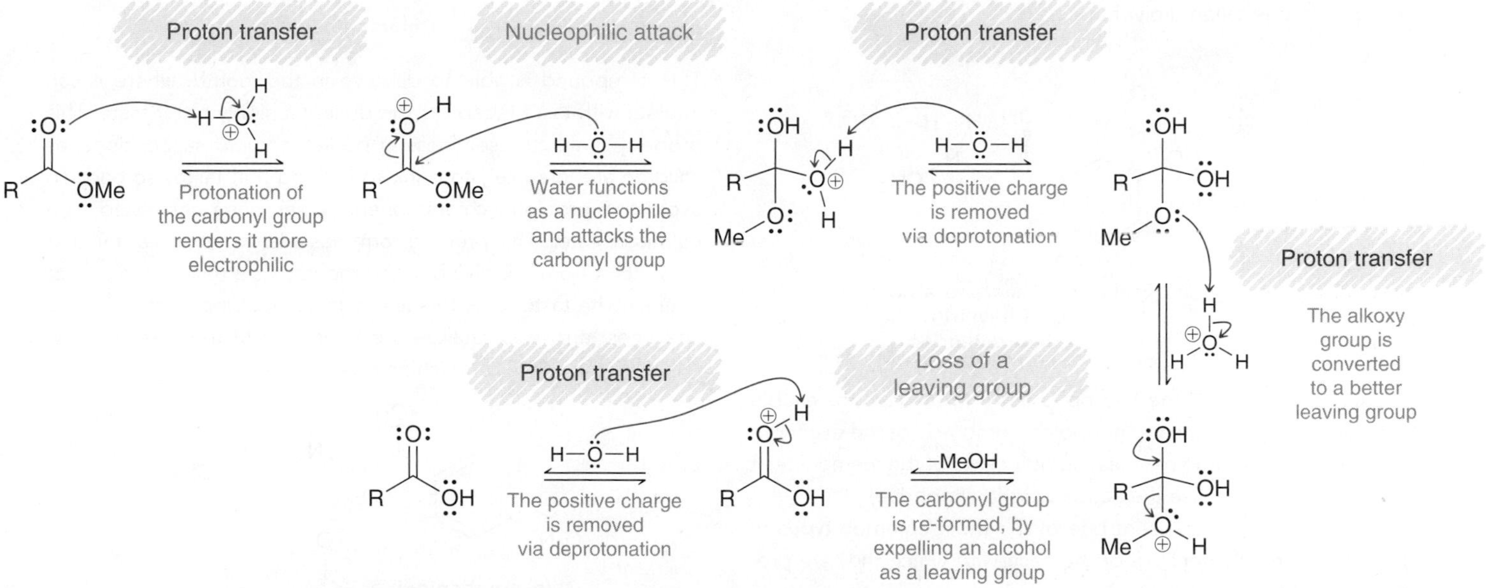

Proton transfer
Protonation of the carbonyl group renders it more electrophilic
Nucleophilic attack
Water functions as a nucleophile and attacks the carbonyl group
Proton transfer
The positive charge is removed via deprotonation
Proton transfer
The alkoxy group is converted to a better leaving group
Loss of a leaving group
−MeOH
The carbonyl group is re-formed, by expelling an alcohol as a leaving group
Proton transfer
The positive charge is removed via deprotonation

Aminolysis of Esters

Esters react slowly with amines to yield amides.

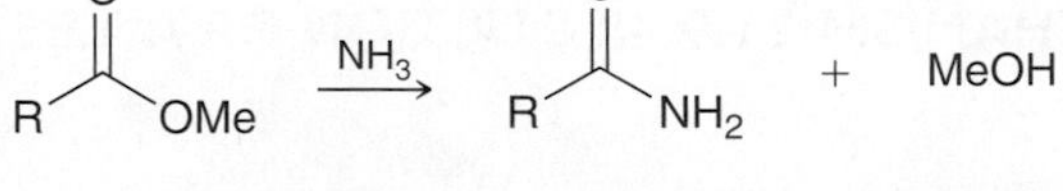

O
R
OMe
NH_3
O
R
NH_2
+
MeOH

This process has little practical utility, because preparation of amides is achieved more efficiently from the reaction between acid chlorides and ammonia or primary or secondary amines.

Reduction of Esters with Hydride-Reducing Agents

When treated with lithium aluminum hydride, esters are reduced to yield alcohols.

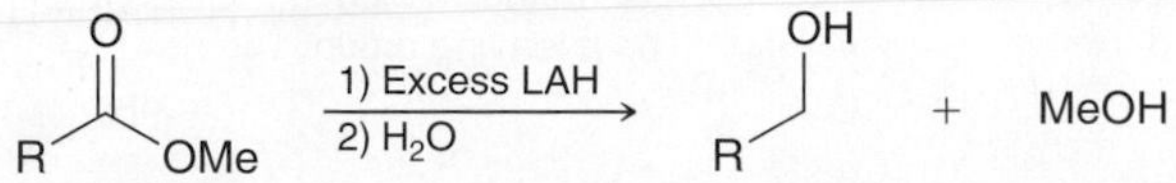

O
R
OMe
1) Excess LAH
2) H_2O
OH
R
+
MeOH

The mechanism for this process is somewhat complex, but the simplified version shown in Mechanism 21.9 will be sufficient for our purposes.

MEDICALLYSPEAKING

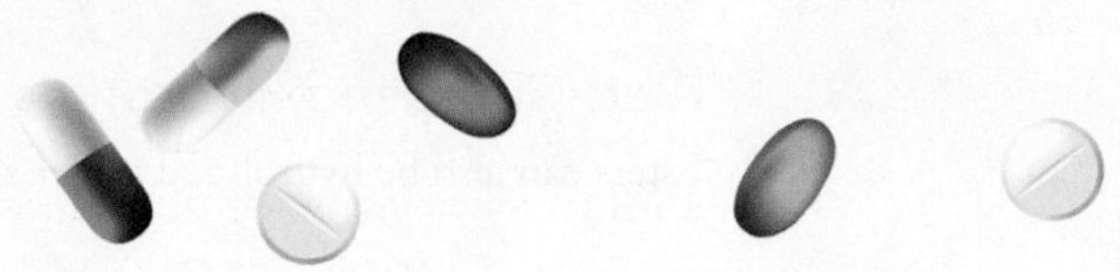

Esters As Prodrugs

Drugs containing hydroxyl groups are often converted into prodrugs to achieve better absorption properties. For example, consider the structure of epinephrine, which is used in the treatment of glaucoma:

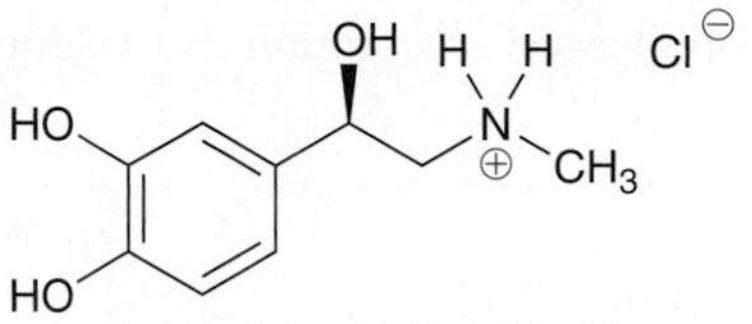

Epinephrine hydrochloride

A prodrug form of epinephrine has been developed in which the aromatic hydroxyl groups are acylated to form ester moieties. This prodrug is called dipivefrin:

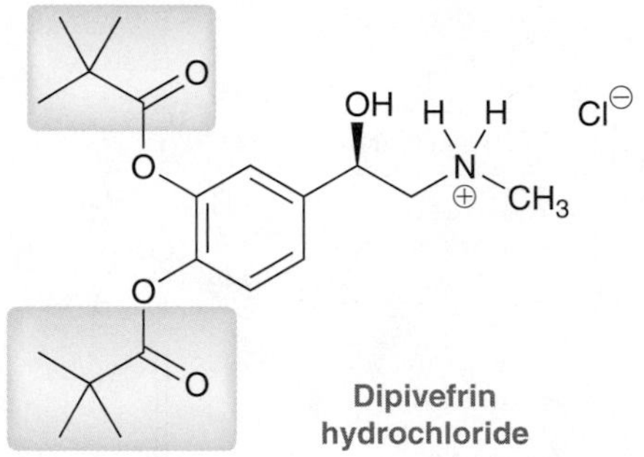

Dipivefrin hydrochloride

In this prodrug form, the hydrophobic *tert*-butyl groups enable the compound to cross the nonpolar membrane of the eye more readily. Once the drug reaches the other side of the membrane, it is hydrolyzed to release the active drug (epinephrine).

In fact, esters represent one of the most common types of prodrug, primarily because of the ease with which they are prepared and the ease with which they are hydrolyzed in the body. In addition, the precise ester group that is used in a particular situation can be chosen to control the rate and extent of hydrolysis. In the previous example, the *tert*-butyl groups were specifically chosen because of their steric bulk, which decreases the rate of hydrolysis, achieving the optimal delayed release of the drug.

Ester prodrugs are used for a variety of reasons, not just for increasing the rate of absorption for a drug. For example, consider the structure of the antibiotic agent chloramphenicol:

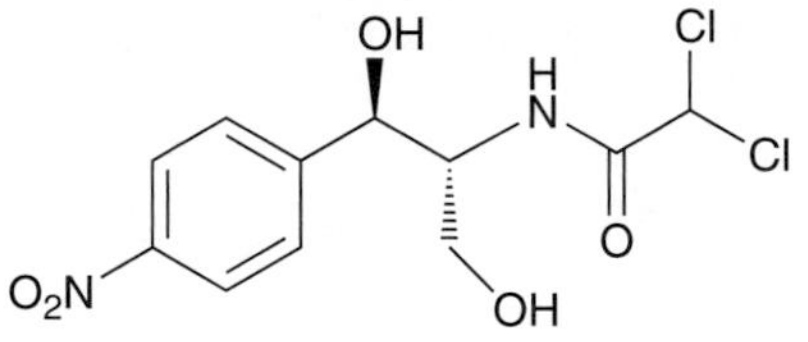

Chloramphenicol

This compound is able to dissolve in the mouth, where it can interact with taste receptors, producing a highly bitter taste. This property limits its usefulness in pediatric liquid suspensions, as children are less likely to drink something that tastes so bad. To avoid this problem, chloramphenicol has been converted to a palmitate ester. This prodrug form has a large, nonpolar tail that prevents it from dissolving in the mouth. As a result, it does not interact with taste receptors and is tasteless. Once it reaches the intestines, enzymes catalyze the hydrolysis of the ester moiety, releasing the active drug (chloramphenicol).

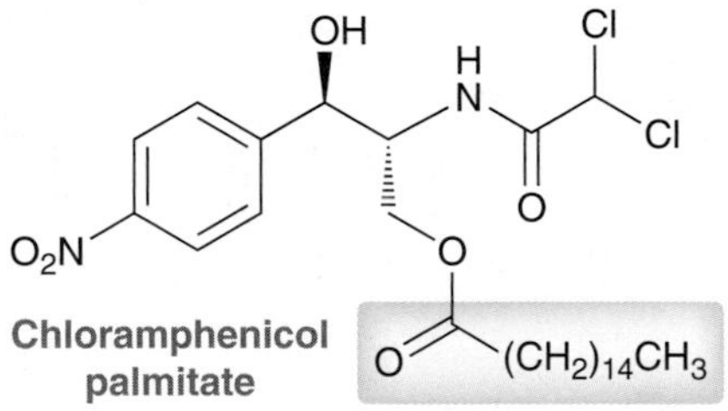

Chloramphenicol palmitate

MECHANISM 21.9 REDUCTION OF AN ESTER WITH LAH

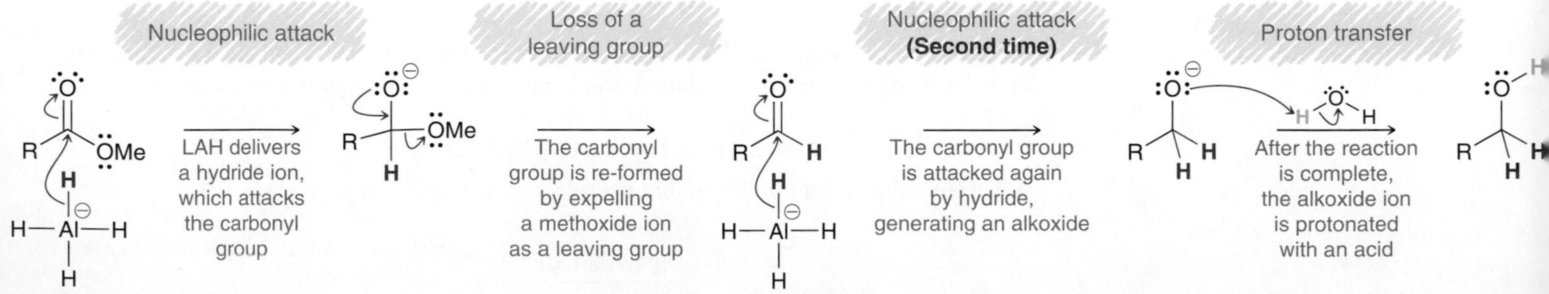
Nucleophilic attack
LAH delivers a hydride ion, which attacks the carbonyl group
Loss of a leaving group
The carbonyl group is re-formed by expelling a methoxide ion as a leaving group
Nucleophilic attack (Second time)
The carbonyl group is attacked again by hydride, generating an alkoxide
Proton transfer
After the reaction is complete, the alkoxide ion is protonated with an acid

This mechanism is directly analogous to the mechanism for reduction of an acid chloride with LAH. The first equivalent of LAH reduces the ester to an aldehyde, and the second equivalent of LAH reduces the aldehyde to an alcohol. Treating an ester with only one equivalent of LAH is not an efficient method for preparing an aldehyde, because aldehydes are more reactive than esters and will react with LAH immediately after being formed. If the desired product is an aldehyde, then DIBAH is used as a reducing agent instead of LAH. The reaction is performed at low temperature to prevent further reduction of the aldehyde.

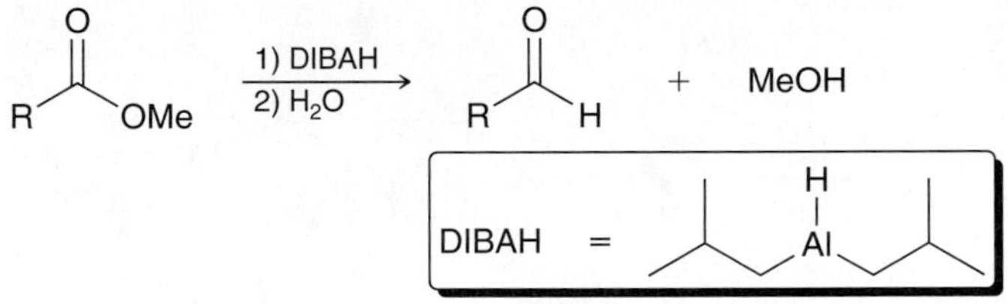

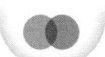

Reactions between Esters and Grignard Reagents

When treated with a Grignard reagent, esters are reduced to yield alcohols with the introduction of two alkyl groups.

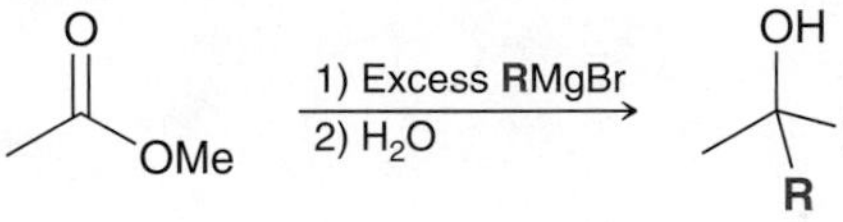

This process (Mechanism 21.10) is directly analogous to the reaction between a Grignard reagent and an acid chloride.

MECHANISM 21.10 THE REACTION BETWEEN AN ESTER AND A GRIGNARD REAGENT

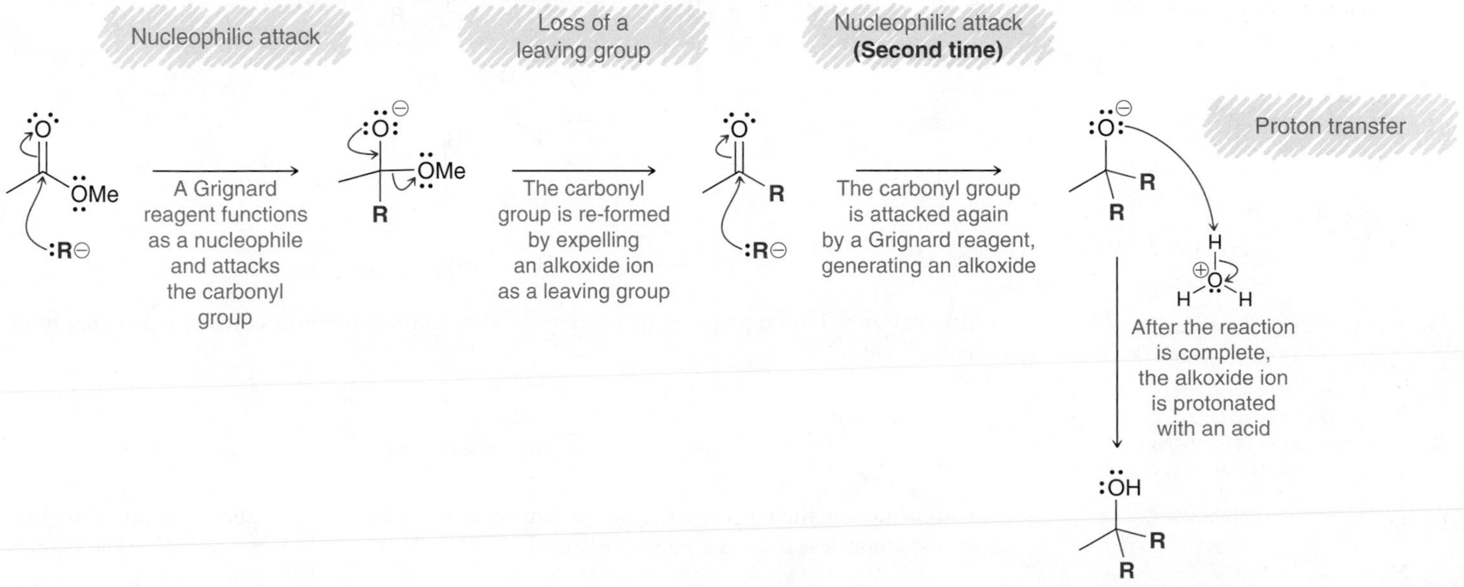

CONCEPTUAL CHECKPOINT

21.24 Predict the major product(s) for each of the following reactions:

(a) 1) xs LAH 2) H₂O → ?

(b) 1) xs EtMgBr 2) H₂O → ?

(c) 1) xs LAH 2) H₂O → ?

(d) H₃O⁺ → ?

(e) 1) NaOH 2) EtI → ?

(f) 1) xs EtMgBr 2) H₂O → ?

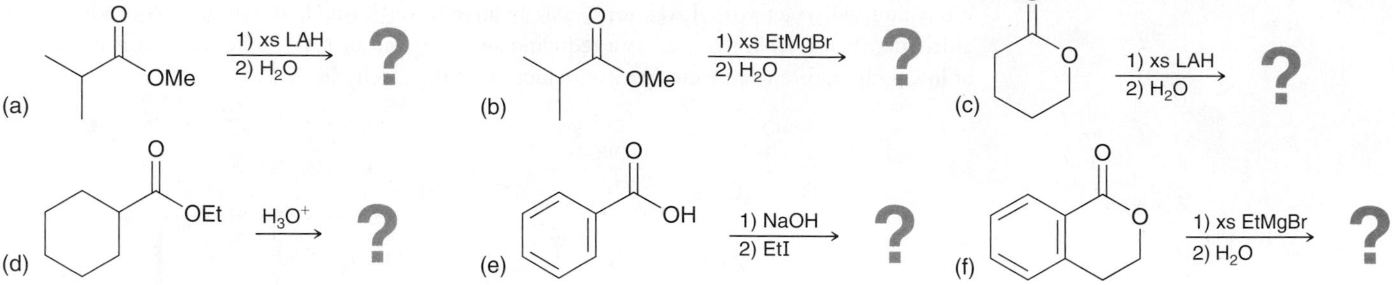

21.25 Propose a mechanism for the following transformation:

H₃O⁺ →

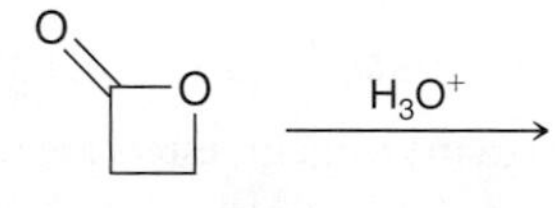

21.12 Preparation and Reactions of Amides

Preparation of Amides

Amides can be prepared from any of the carboxylic acid derivatives discussed earlier in this chapter.

$$R \overset{O}{-} Cl \quad \text{Most reactive}$$

$$R \overset{O}{-} O \overset{O}{-} R$$

$$R \overset{O}{-} OR$$

$$R \overset{O}{-} NH_2 \quad \text{Least reactive}$$

LOOKING AHEAD

Amides can also be prepared efficiently from carboxylic acids using a reagent called DCC. This reagent and its action are discussed in Section 25.6.

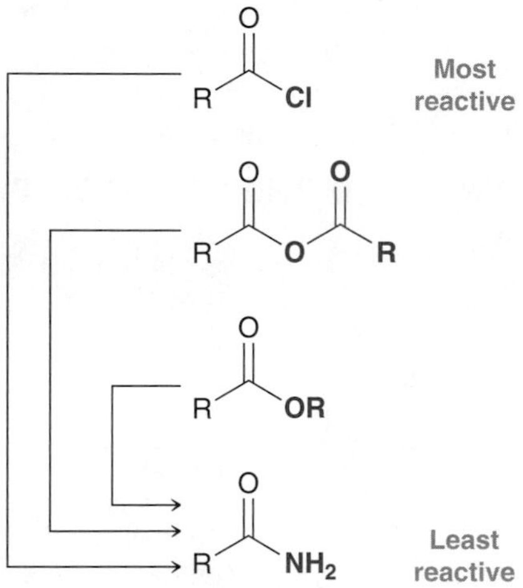

Although they can be prepared in a variety of ways, amides are most efficiently prepared from acid chlorides.

$$R \overset{O}{-} Cl \quad \xrightarrow[\text{(two equivalents)}]{NH_3} \quad R \overset{O}{-} NH_2$$

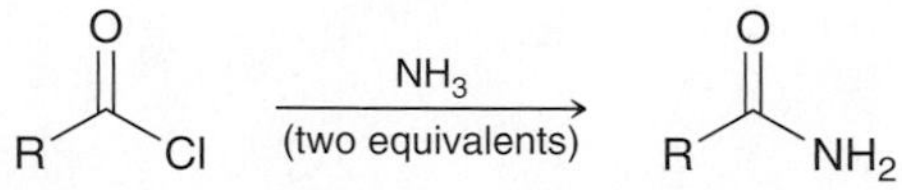

Acid halides are the most reactive of the carboxylic acid derivatives, so the yields are best when an acid chloride is used as a starting material.

PRACTICALLYSPEAKING

Polyamides and Polyesters

Consider what happens when a diacid chloride and a diamine react together:

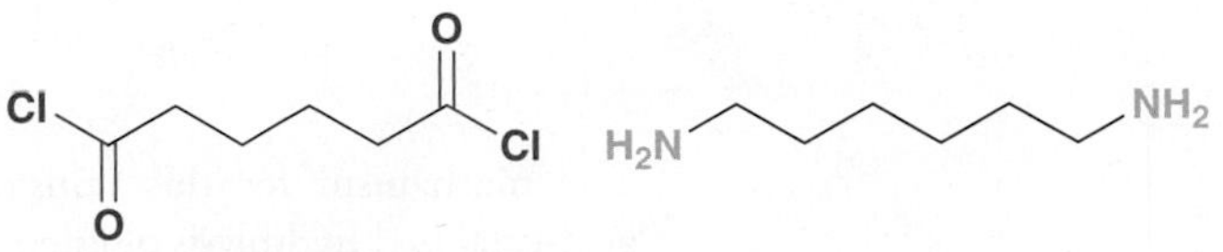

A diacid halide A diamine

Each molecule has two reactive ends, allowing formation of a polymer:

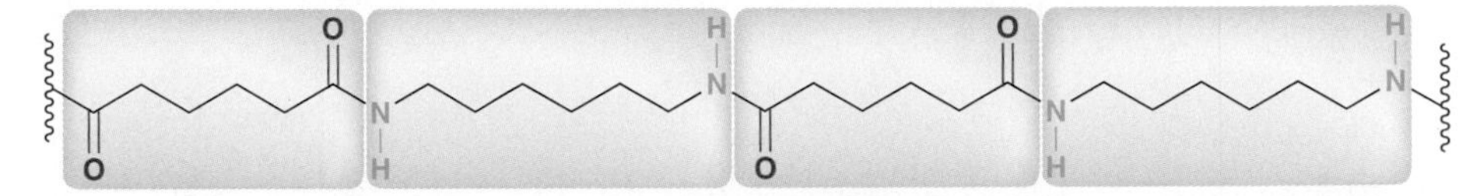

Nylon 6,6

The polymer exhibits multiple amide linkages and is therefore called a polyamide. This specific example is called Nylon 6,6 because it is created from two different compounds that both contain six carbon atoms. Nylon 6,6 was first used by the military to manufacture parachutes, but it quickly became popular as a replacement for silk in the manufacture of clothing.

Polyesters can be made in a similar way. Consider what happens when a diacid reacts with a diol:

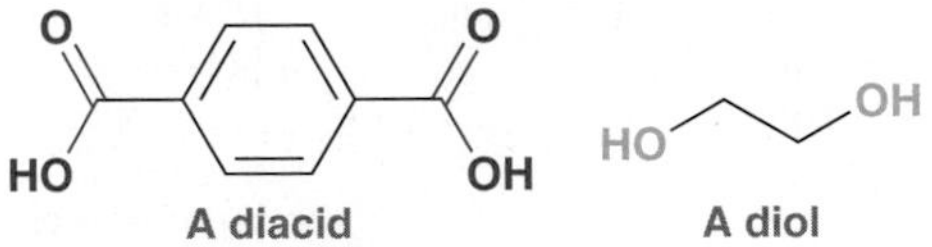

A diacid A diol

Each molecule has two reactive ends, allowing formation of a polymer:

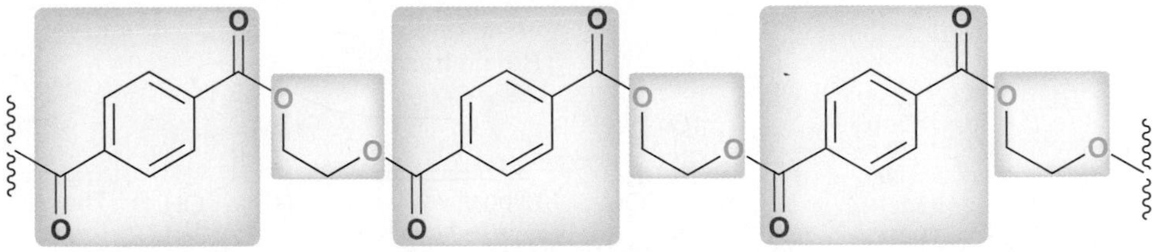

Polyethylene terephthalate (PET)

This polymer, polyethylene terephthalate (PET), exhibits multiple ester linkages and is therefore called a polyester. PET is sold under many trade names, including Dacron™ and Mylar™. It is primarily used in the manufacture of clothing. There are many different kinds of polyamides and polyesters, serving a variety of purposes. Kevlar, for example, is stronger than steel and is used in bulletproof vests:

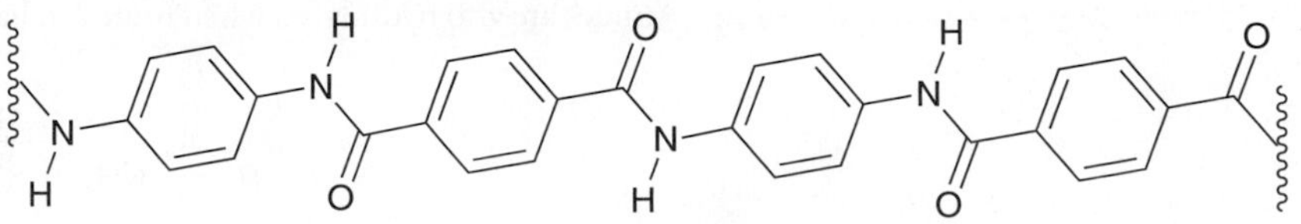

Kevlar™

Acid-Catalyzed Hydrolysis of Amides

Amides can be hydrolyzed to give carboxylic acids in the presence of aqueous acid, but the process is slow and requires heating to occur at an appreciable rate.

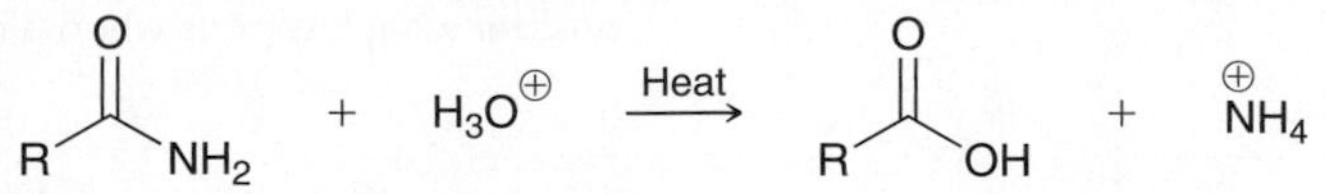

Heat

The mechanism for this transformation (Mechanism 21.11) is directly analogous to the acid-catalyzed hydrolysis of esters (Section 21.11).

In this reaction, notice that an ammonium ion (NH_4^+) is formed as a by-product. Since the ammonium ion ($pK_a = 9.2$) is a much weaker acid than H_3O^+ ($pK_a = -1.7$), the equilibrium greatly favors formation of products, rendering the process effectively irreversible.

MECHANISM 21.11 ACID-CATALYZED HYDROLYSIS OF AN AMIDE

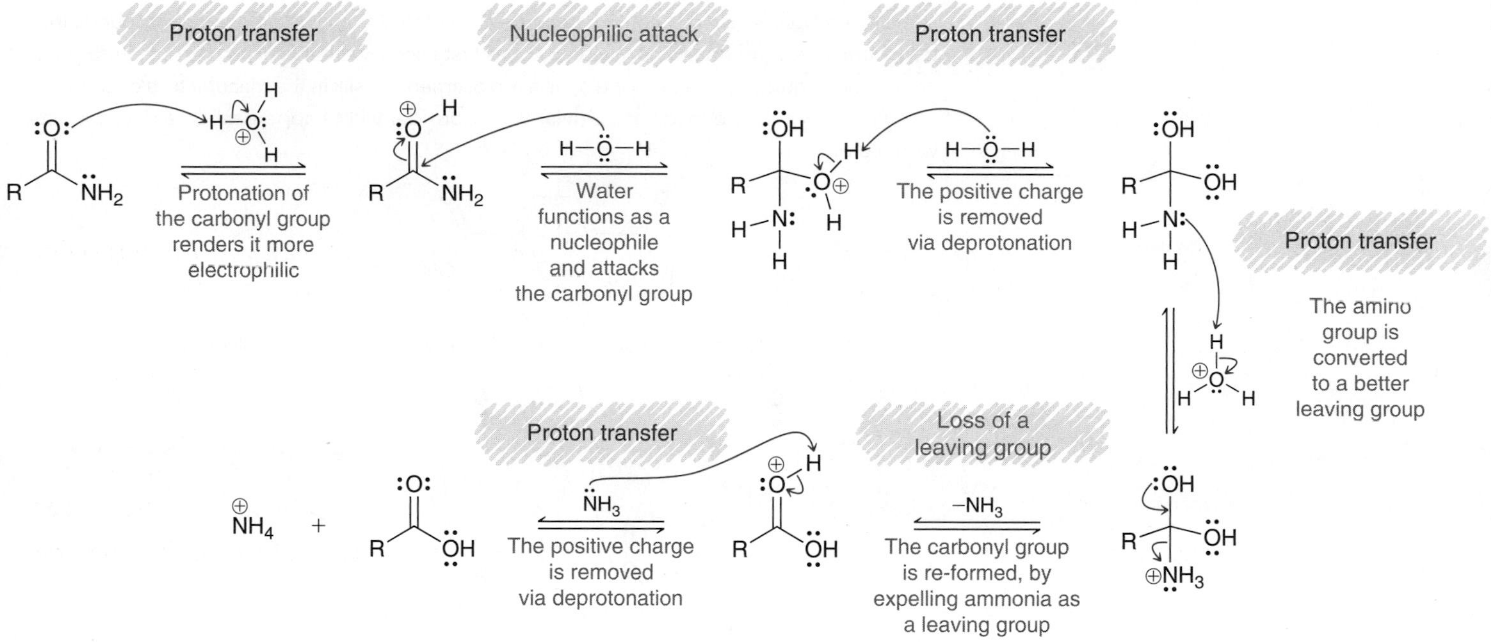

Proton transfer
Protonation of the carbonyl group renders it more electrophilic
Nucleophilic attack
Water functions as a nucleophile and attacks the carbonyl group
Proton transfer
The positive charge is removed via deprotonation
Proton transfer
The amino group is converted to a better leaving group
Loss of a leaving group
The carbonyl group is re-formed, by expelling ammonia as a leaving group
Proton transfer
The positive charge is removed via deprotonation

Base-Catalyzed Hydrolysis of Amides

Amides are also hydrolyzed when heated in basic aqueous solutions, although the process is very slow.

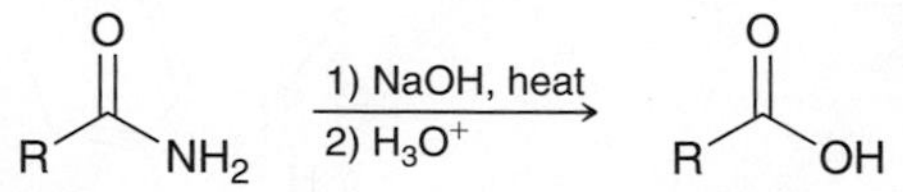

1) NaOH, heat
2) H3O+

The mechanism for this process (Mechanism 21.12) is directly analogous to the saponification of esters (Section 21.11).

MECHANISM 21.12 BASE-CATALYZED HYDRATION OF AMIDES

Nucleophilic attack	Loss of a leaving group	Proton transfer
Hydroxide functions as a nucleophile and attacks the carbonyl group	The carbonyl group is re-formed by expelling an amide ion as a leaving group	The carboxylic acid is deprotonated by the amide ion, generating a carboxylate ion

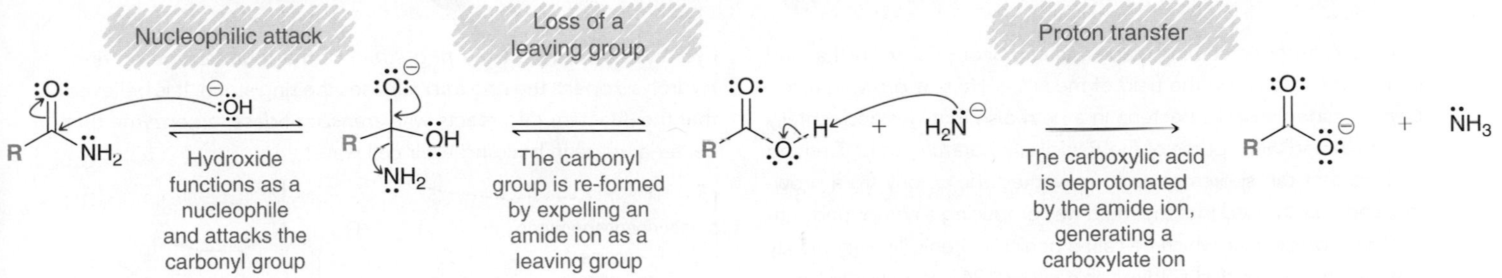

In the last step, formation of the carboxylate ion drives the reaction to completion and renders the process irreversible.

Reduction of Amides

When treated with excess LAH, amides are converted into amines.

$$
R\text{-}C(=O)\text{-}NH_2 \xrightarrow[\text{2) } H_2O]{\text{1) Excess LAH}} R\text{-}CH_2\text{-}NH_2
$$

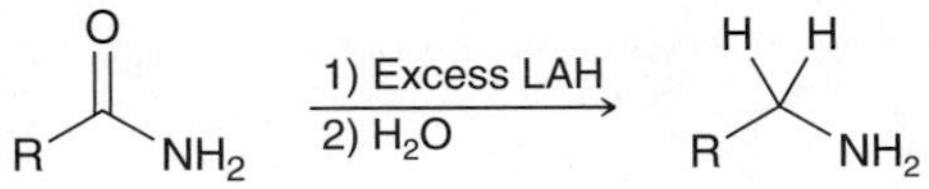

This is the first reaction we have seen that is somewhat different than the other reactions in this chapter. In this case, the carbonyl group is completely removed.

CONCEPTUAL CHECKPOINT

21.26 Predict the major product(s) for each of the following reactions:

(a)

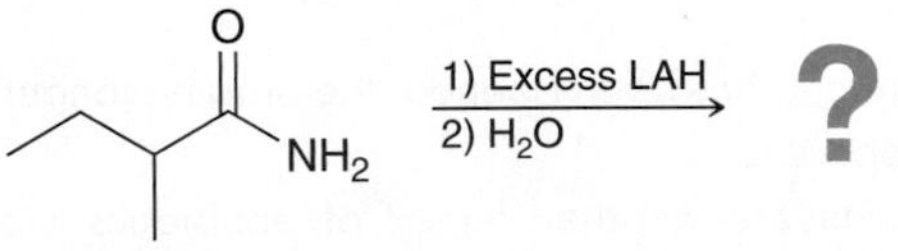

$$
\xrightarrow[\text{2) } H_2O]{\text{1) Excess LAH}} \; ?
$$

(b)

$$
\xrightarrow{\text{Excess } NH_3} \; ?
$$

(c)

$$
\xrightarrow[\text{Heat}]{H_3O^+} \; ?
$$

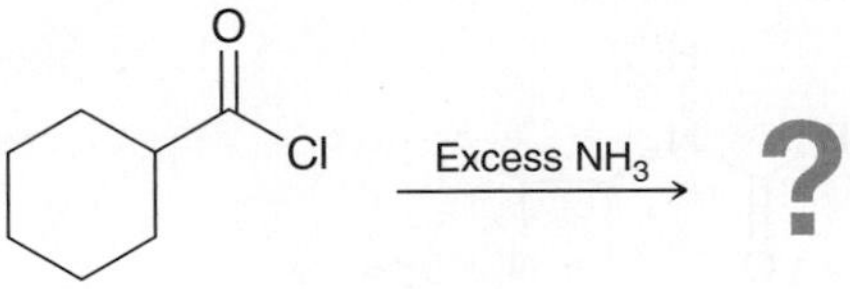

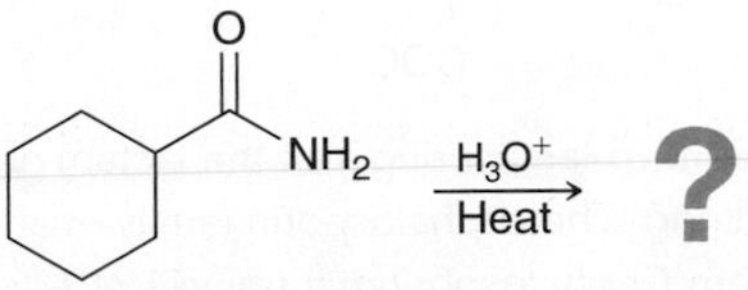

21.27 Identify the reagents necessary for the following transformation:

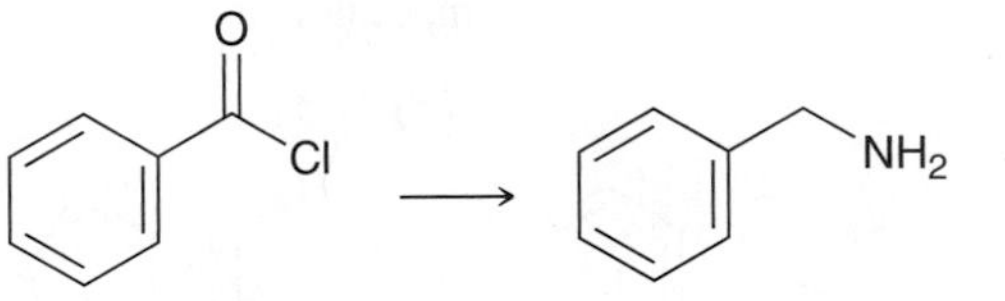

21.28 Propose a mechanism for each of the following transformations:

(a)

$$
\xrightarrow[\text{Heat}]{H_3O^+}
$$

(b)

$$
\xrightarrow[\text{2) } H_3O^+]{\text{1) NaOH, heat}}
$$

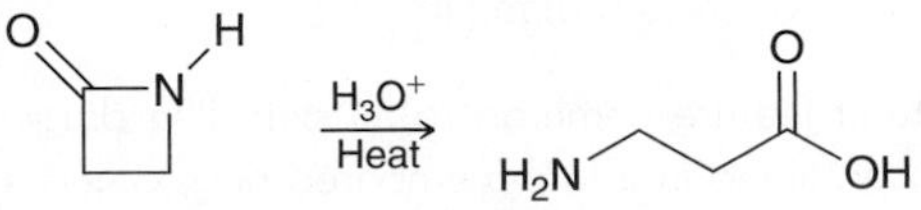

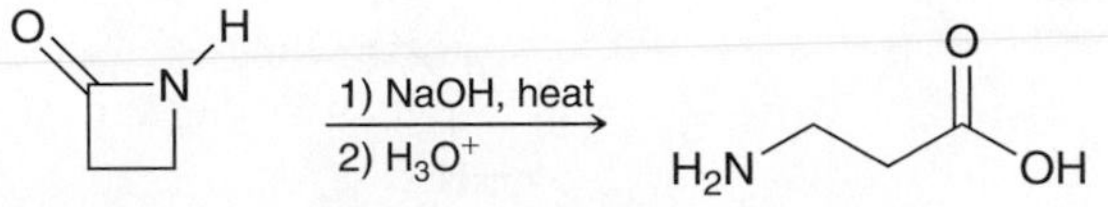

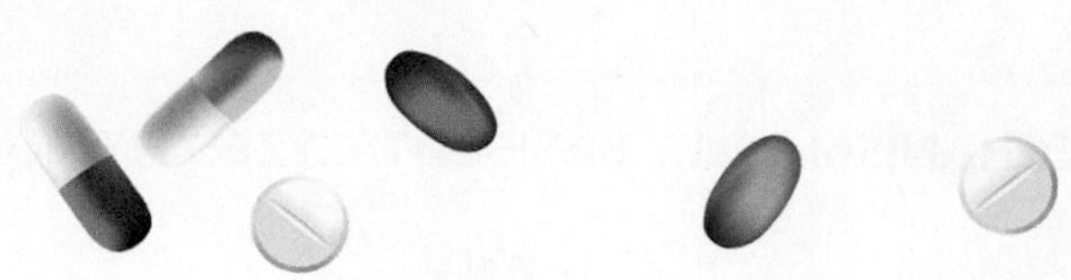

MEDICALLYSPEAKING
Beta-Lactam Antibiotics

In 1928, Alexander Fleming made a serendipitous discovery that had a profound impact on the field of medicine. He was growing a colony of *Staphylococcus* bacteria in a petri dish that was accidentally contaminated with spores of the *Penicillium notatum* mold. Fleming noticed that the spores of mold prevented the colony from growing, and he surmised that the mold was producing a compound with antibiotic properties, which he called penicillin. Penicillin was initially used to treat wounded soldiers as early as 1943 and shortly thereafter was used on the general population. It has been credited with saving millions of lives, and for his discovery, Fleming was a corecipient of the 1945 Nobel Prize in Physiology or Medicine.

Penicillin was initially believed to be one compound, as Fleming had suggested. However, in 1944, it became apparent that the *P. notatum* mold manufactures many structurally similar compounds, all of which exhibit antibacterial properties. This group of compounds is now said to belong to the penicillin family of drugs and can be represented with the following structural formula, where the R group can vary:

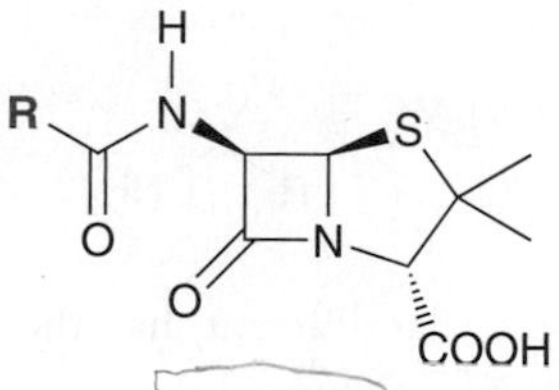

Penicillin

Over a dozen penicillin drugs are currently in clinical use, two of which are shown here:

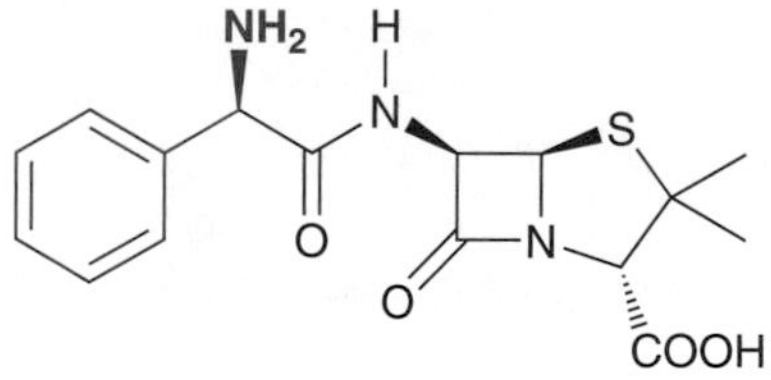

Ampicillin

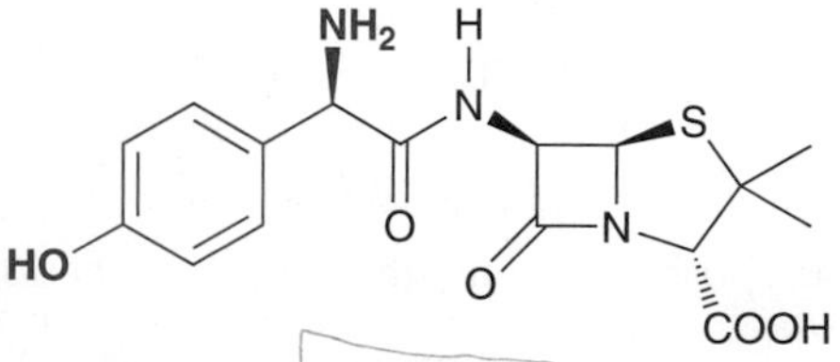

Amoxicillin

The key structural feature common to all penicillin drugs is an amide moiety contained in a four-membered ring, called a beta (β) lactam ring.

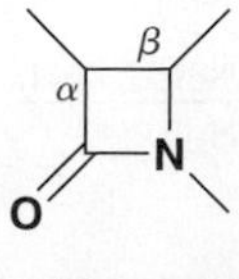

Amides are generally stable—they resist hydrolysis under most conditions. However β-lactams are particularly susceptible to hydrolysis because of the ring strain of the four-membered ring. Hydrolysis opens the ring and releases the ring strain. It is believed that the β-lactam ring reacts with *transpeptidase*, an enzyme that bacteria utilize in building their cell walls:

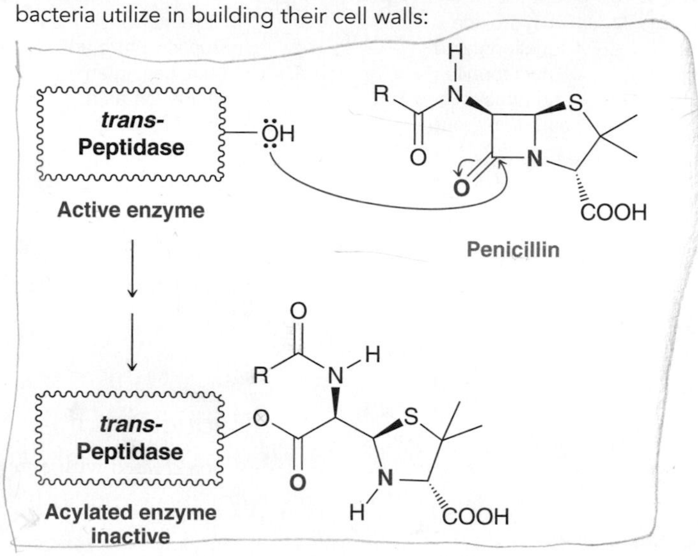

This reaction effectively acylates an OH group at the active site of the enzyme, and the acylated enzyme is inactive. By inactivating this enzyme, penicillin drugs are able to prevent bacteria from producing functional cell walls. Under these conditions, the bacteria are not able to reproduce, which allows the body's natural immune system to take control.

Some bacteria are resistant to penicillin drugs. These bacteria produce enzymes, called β-lactamases, that are capable of prematurely hydrolyzing the β-lactam ring before it has a chance to react with transpeptidase:

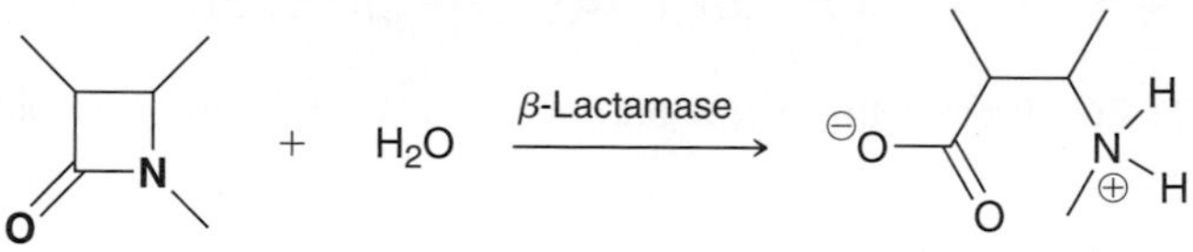

Once the β-lactam ring has been opened, the drug no longer has any antibiotic properties.

Cephalosporins are another family of antibiotics closely related in structure to penicillins:

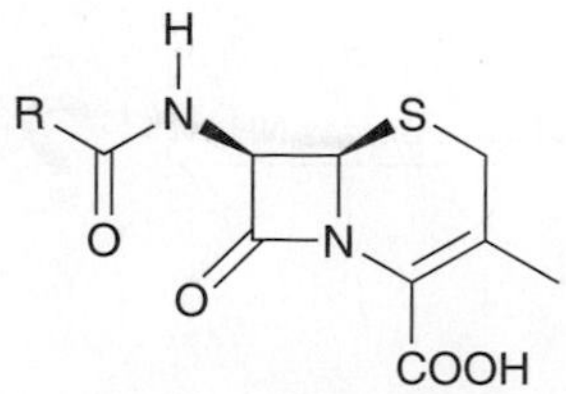

Cephalosporins also exhibit a β-lactam ring, but the lactam ring is fused to a six-membered ring. The cephalosporin family of antibiotics was first isolated from *Cephalosporium* fungi in 1945.

Extensive research is currently underway to identify other novel β-lactam antibiotics with superior properties. Over the decades to come, the list of known antibiotics is sure to grow.

21.13 Preparation and Reactions of Nitriles

Preparation of Nitriles via S$_N$2 Reactions

Nitriles can be prepared by treating an alkyl halide with a cyanide ion.

$$R\text{--}CH_2\text{--}Br \xrightarrow{\text{NaCN}} R\text{--}CH_2\text{--}C\equiv N + \text{NaBr}$$

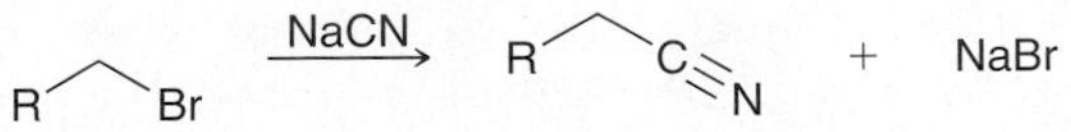

This process proceeds via an S$_N$2 mechanism, so tertiary alkyl halides cannot be used.

Preparation of Nitriles from Amides

Nitriles can also be prepared via the dehydration of an amide. Many reagents can be used to accomplish the transformation. One such reagent is thionyl chloride (SOCl$_2$).

$$R\text{--}C(=O)\text{--}NH_2 \xrightarrow{\text{SOCl}_2} R\text{--}C\equiv N + SO_2 + 2HCl$$

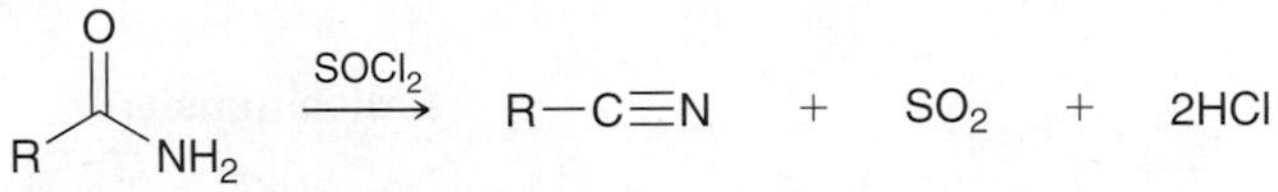

This process (Mechanism 21.13) is useful for preparing tertiary nitriles, which cannot be prepared via an S$_N$2 process.

MECHANISM 21.13 DEHYDRATION OF AMIDES

Nucleophilic attack	Loss of a leaving group	Proton transfer	Proton transfer
The amide functions as a nucleophile, and attacks thionyl chloride	Chloride is expelled as a leaving group	The positive charge on the nitrogen atom is removed via deprotonation	Elimination of a proton and a leaving group affords the product

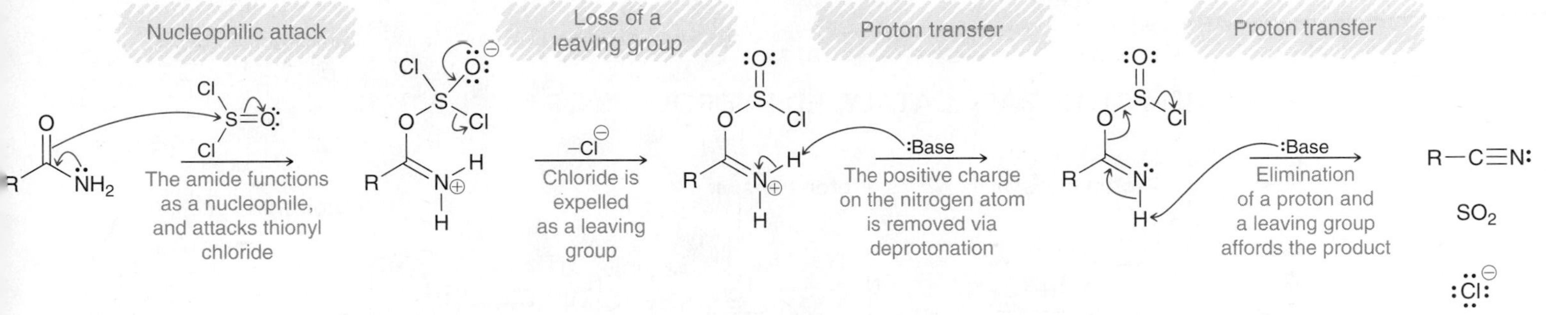

Hydrolysis of Nitriles

In aqueous acidic conditions, nitriles are hydrolyzed to afford amides, which are then further hydrolyzed to yield carboxylic acids.

$$R\text{--}C\equiv N \xrightarrow[\text{heat}]{H_3O^+} R\text{--}C(=O)\text{--}NH_2 \xrightarrow[\text{heat}]{H_3O^+} R\text{--}C(=O)\text{--}OH + \overset{\oplus}{N}H_4$$

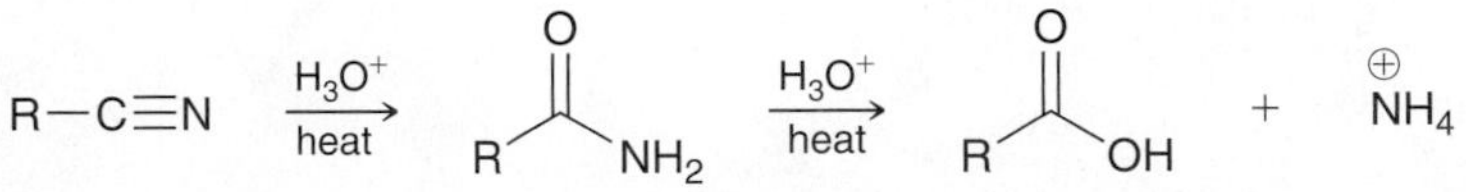

Formation of the amide occurs via Mechanism 21.14, and conversion of the amide into the carboxylic acid was discussed earlier (Mechanism 21.11).

Notice that, except for one nucleophilic attack, the mechanism consists entirely of proton transfers. Since the reaction is performed under acidic conditions, the proton transfer steps should follow the rule that says: *In acidic conditions, all reagents, intermediates, and leaving groups should either be neutral or bear only one positive charge.* Review Mechanism 21.14 and convince yourself that the function of each step is to obey this rule.

Alternatively, nitriles can also be hydrolyzed in aqueous base.

$$R\text{--}C\equiv N \xrightarrow[\text{2) }H_3O^+]{\text{1) NaOH, }H_2O} R\text{--}C(=O)\text{--}OH$$

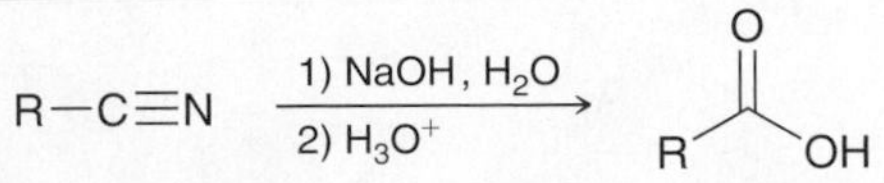

Once again, the nitrile is first converted to an amide (Mechanism 21.15), which is then converted to a carboxylic acid (see Mechanism 21.12).

MECHANISM 21.14 ACID-CATALYZED HYDROLYSIS OF NITRILES

Proton transfer

Protonation of
the nitrile group
renders it more
electrophilic

Nucleophilic attack

Water
functions as a
nucleophile
and attacks
the protonated nitrile

Proton transfer

The postiive charge
is removed
via deprotonation

Proton transfer

The nitrogen
atom is
protonated,
forming a
resonance –
stabilized
intermediate

Proton transfer

The positive charge
is removed
via deprotonation

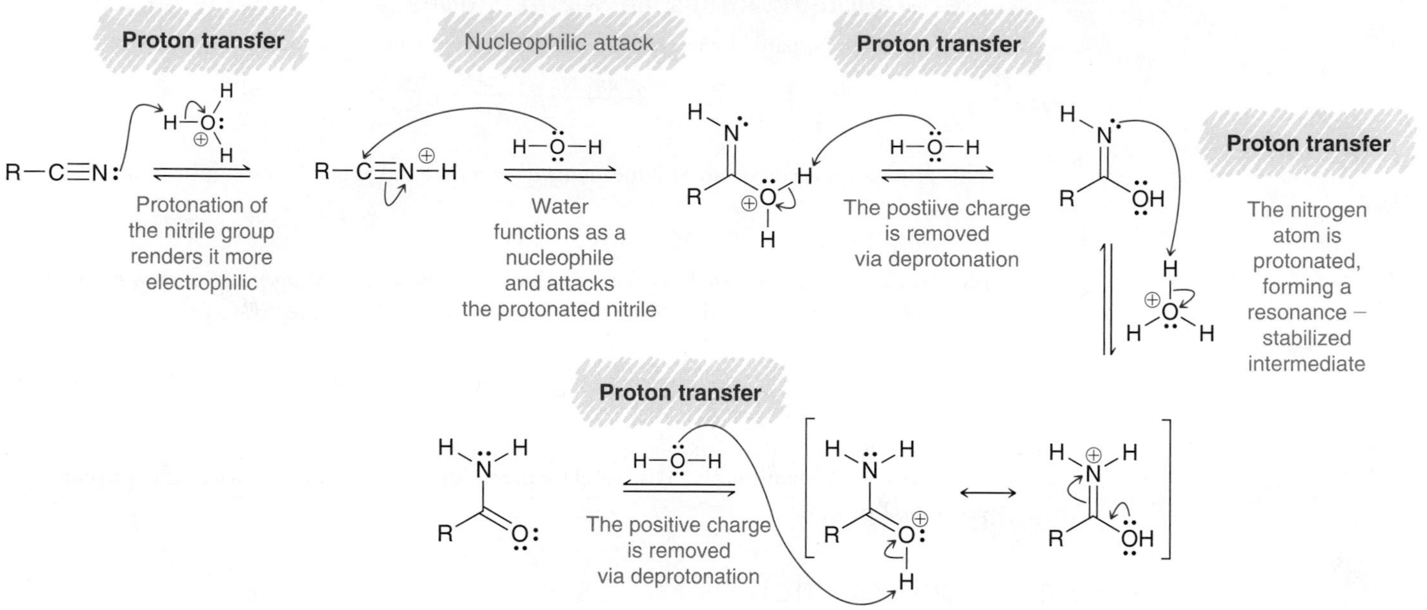

MECHANISM 21.15 BASE-CATALYZED HYDROLYSIS OF NITRILES

Nucleophilic attack

Hydroxide
functions as a
nucleophile
and attacks
the cyano group

Proton transfer

The negative charge
on the nitrogen atom
is removed via protonation

Proton transfer

Hydroxide
functions as a base
and removes a proton,
forming a resonance-
stabilized intermediate

Proton transfer

Protonation
affords the amide

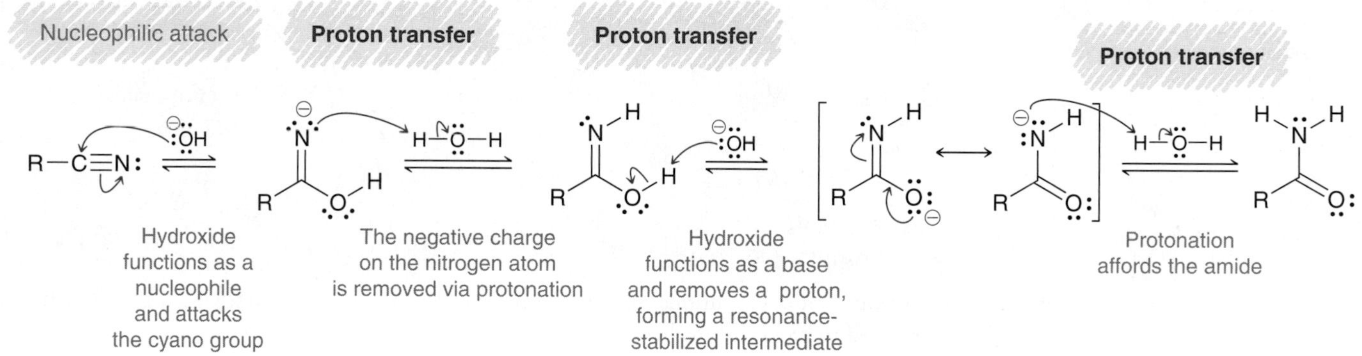

Reactions between Nitriles and Grignard Reagents

A ketone is obtained when a nitrile is treated with a Grignard reagent, followed by aqueous acid.

$$R-C{\equiv}N \xrightarrow[\text{2) H}_3\text{O}^+]{\text{1) RMgBr}}$$

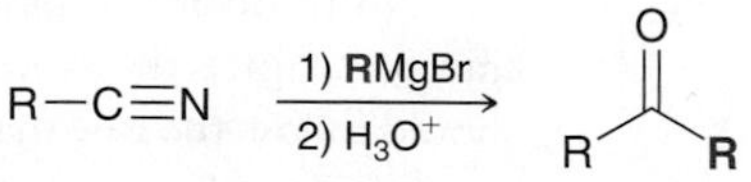

The Grignard reagent attacks the nitrile, much like it attacks a carbonyl group.

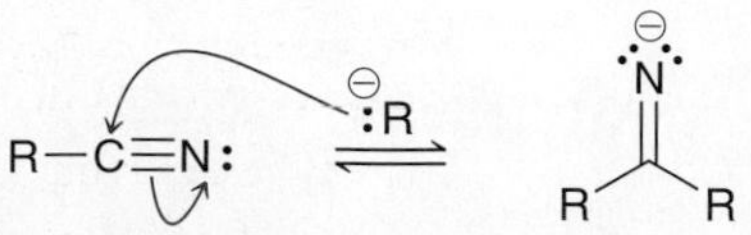

The resulting anion is then treated with aqueous acid to give an imine, which is then hydrolyzed
to a ketone under acidic conditions (see Section 20.6).

Reduction of Nitriles

At the beginning of this chapter, we saw that carboxylic acids can be reduced to alcohols by treatment with LAH. Similarly, nitriles are converted to amines when treated with LAH.

$$R-C\equiv N \xrightarrow[\text{2) H}_2\text{O}]{\text{1) xs LAH}} \underset{R}{\overset{H\ H}{\diagup}} NH_2$$

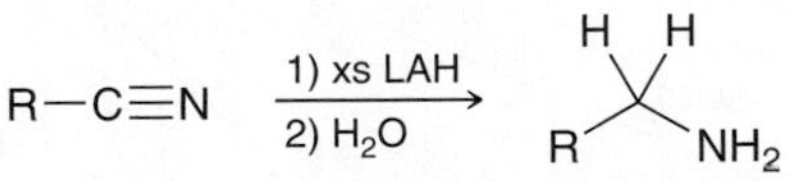

CONCEPTUAL CHECKPOINT

21.29 Predict the major product(s) for each of the following reactions:

(a)

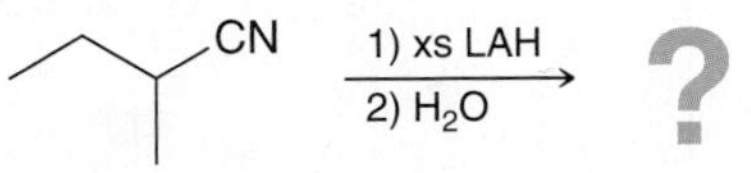

$$\xrightarrow[\text{2) H}_2\text{O}]{\text{1) xs LAH}} \quad ?$$

(b)

$$\xrightarrow[\text{3) H}_3\text{O}^+]{\substack{\text{1) NaCN} \\ \text{2) MeMgBr}}} \quad ?$$

(c)

$$\xrightarrow[\substack{\text{3) LAH} \\ \text{4) H}_2\text{O}}]{\substack{\text{1) EtMgBr} \\ \text{2) H}_2\text{O}}} \quad ?$$

(d)

$$\xrightarrow[\text{heat}]{\text{H}_3\text{O}^+} \quad ?$$

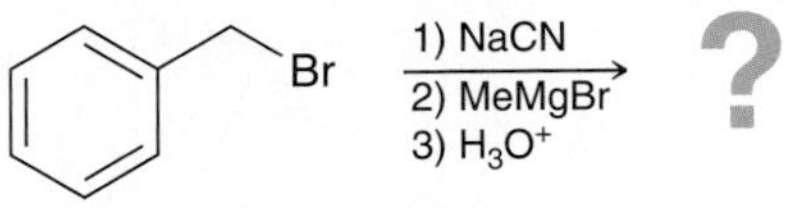

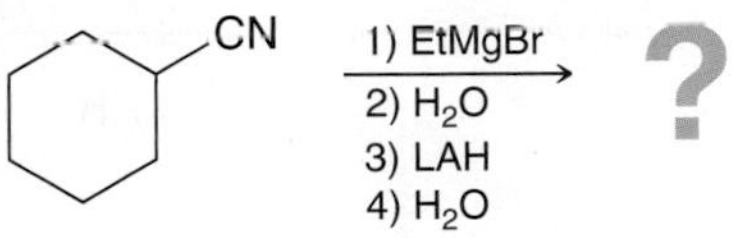

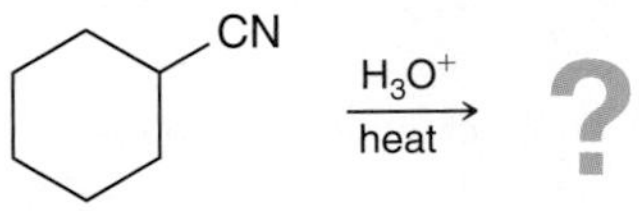

21.30 Identify the reagents necessary for each of the following transformations:

(a)

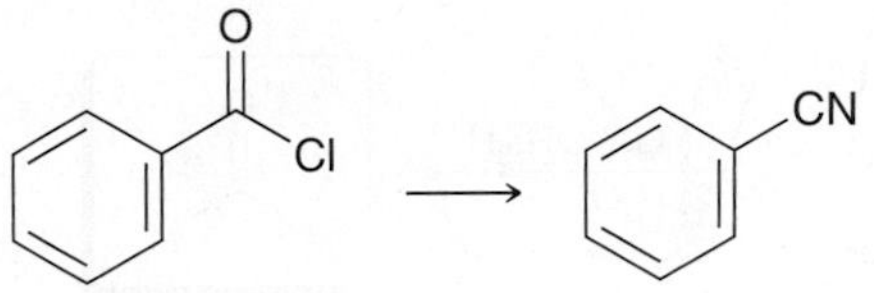

(b)

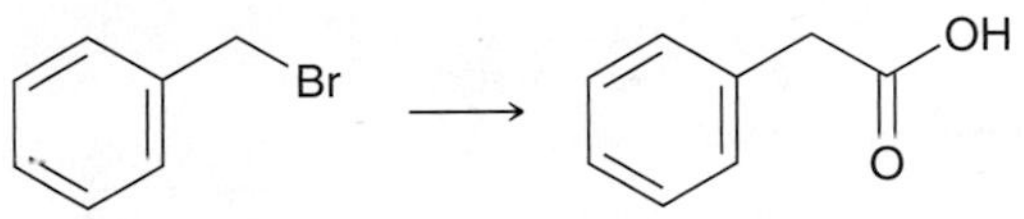

21.31. Propose a mechanism for the following transformation:

$$\xrightarrow[\text{Heat}]{\text{H}_3\text{O}^+}$$

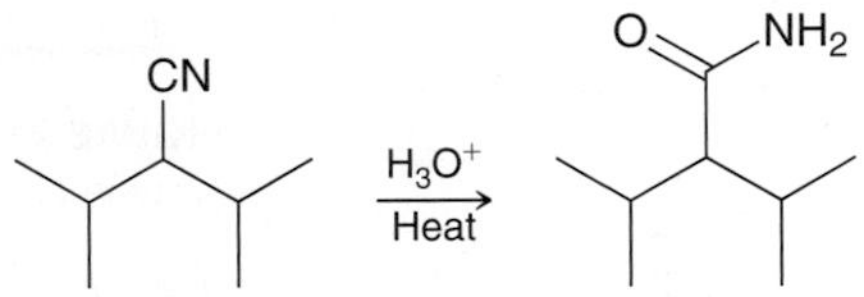

21.14 Synthesis Strategies

Recall from Chapter 12 that there are two primary considerations when approaching a synthesis problem: (1) a change in the carbon skeleton and (2) a change in the functional groups. Let's focus on each of these issues separately, beginning with functional groups.

Functional Group Interconversions

In this chapter, we have seen many different reactions that change the identity of a functional group without changing the location. Figure 21.11 summarizes how functional groups can be interconverted.

The figure shows many reactions, but there are a few key aspects of the diagram that are worth special attention:

- The figure is organized according to oxidation state. Carboxylic acids and their derivatives all have the same oxidation state and are shown at the top of the figure (for ease of presentation, carboxylic acids are shown above the derivatives but do have the same oxidation state). Aldehydes exhibit a lower oxidation state and are therefore shown below the derivatives. Alcohols and amines are at the bottom of the figure, since they have the lowest oxidation state.

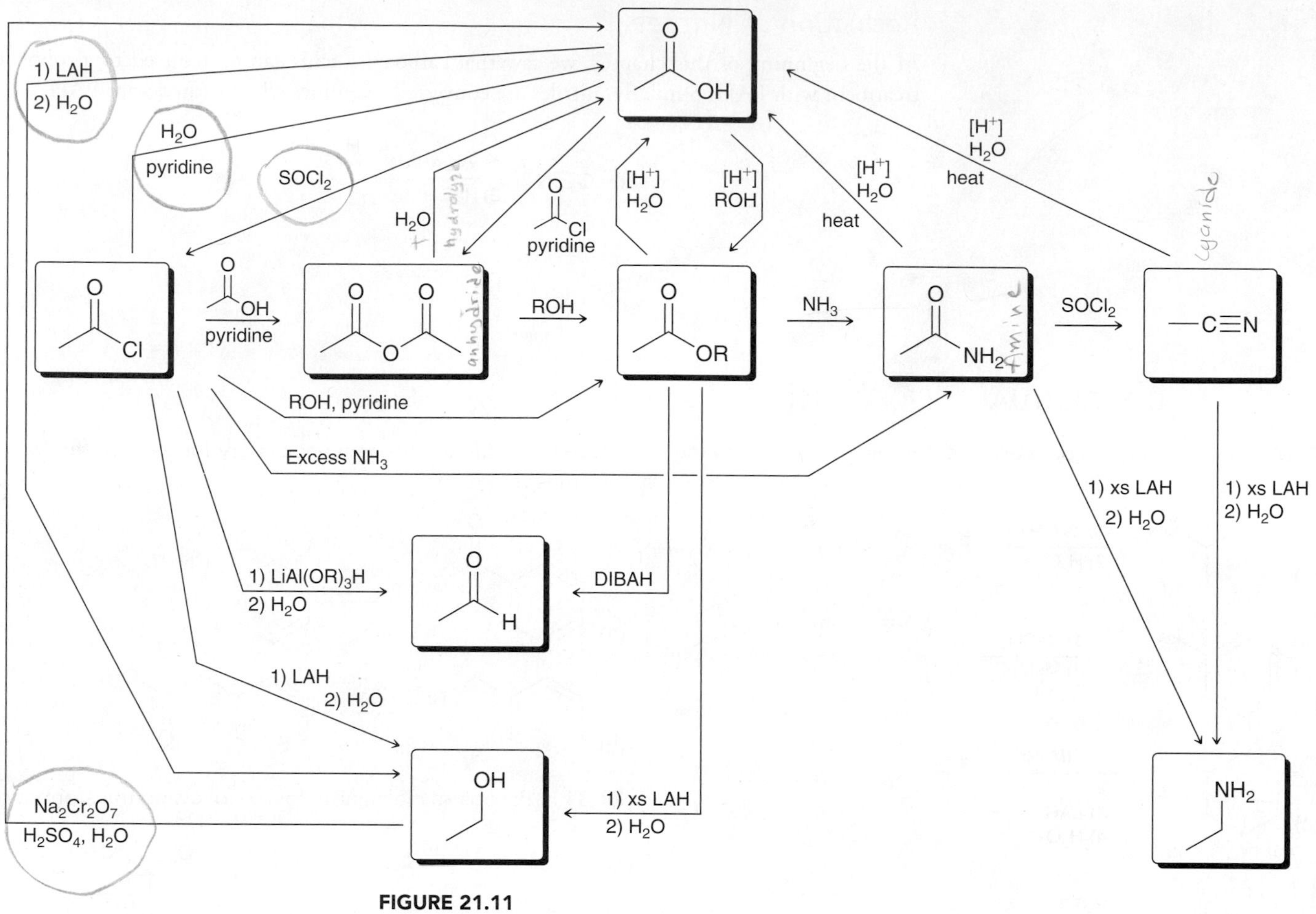

FIGURE 21.11
A summary of reactions from this chapter that enable the interconversion of functional groups.

- Among the carboxylic acid derivatives, the reactions are shown from left to right, from most reactive to least reactive. That is, acid chlorides (leftmost) are the most reactive and are most readily converted into the other derivatives:

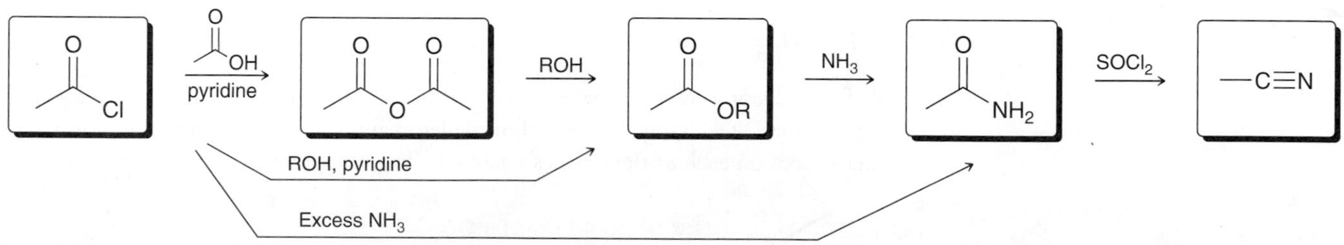

Going from right to left—for example, converting an ester into an acid chloride—requires two steps: first hydrolysis to form a carboxylic acid, then conversion of the acid into an acid chloride.

- We did not see a method for directly converting a carboxylic acid into either an amide or a nitrile. All other carboxylic acid derivatives (acid chlorides, acid anhydrides, and esters) can be made directly from a carboxylic acid.
- We have only seen two ways to make amines in this chapter (Chapter 24 will cover many other ways for making amines).

SKILLBUILDER

21.2 INTERCONVERTING FUNCTIONAL GROUPS

LEARN the skill

Propose an efficient synthesis for the following transformation:

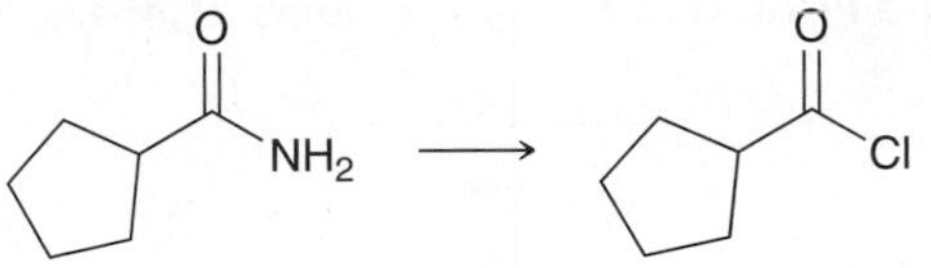

SOLUTION

This problem requires the transformation of an amide into an acid chloride. Amides are less reactive than acid chlorides, so this transformation requires a two-step process. The first step is hydrolysis of the amide to form a carboxylic acid:

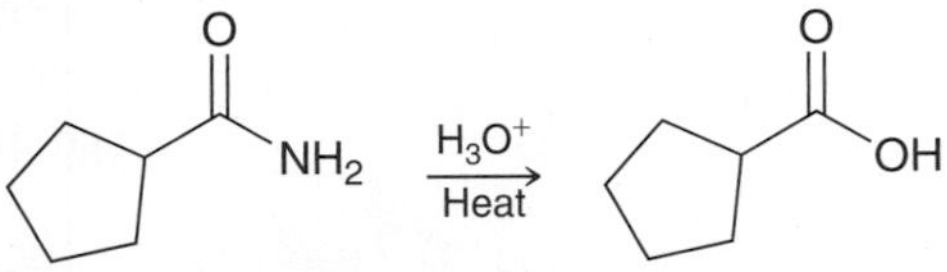

The second step is to convert the carboxylic acid into the desired acid chloride:

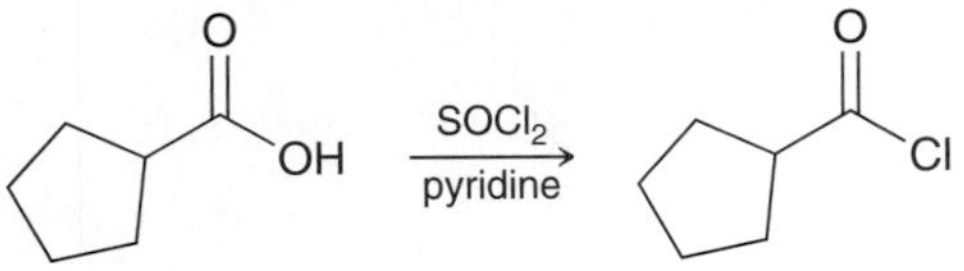

In summary, conversion of an amide into an acid chloride requires the following two steps:

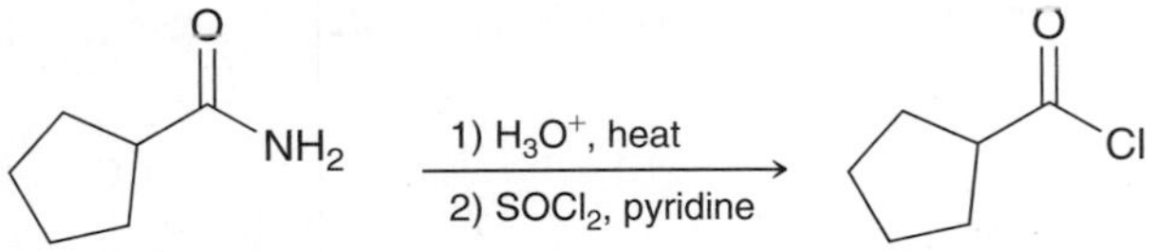

PRACTICE the skill **21.32** Identify reagents that will accomplish each of the following transformations:

(a)

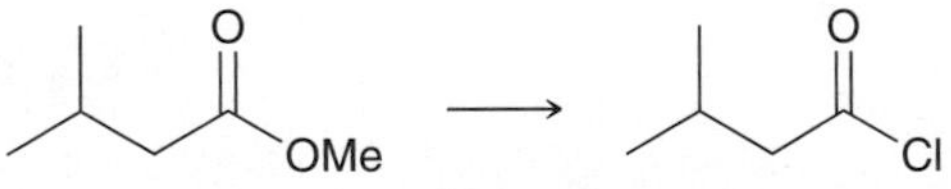

(b)

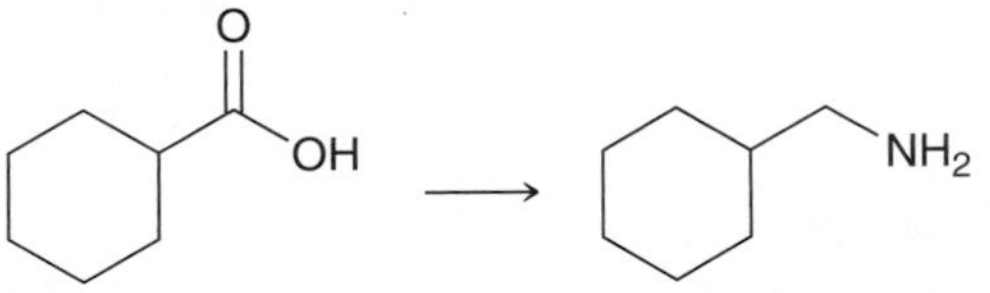

(c)

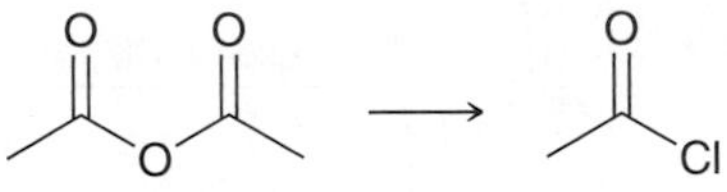

(d)

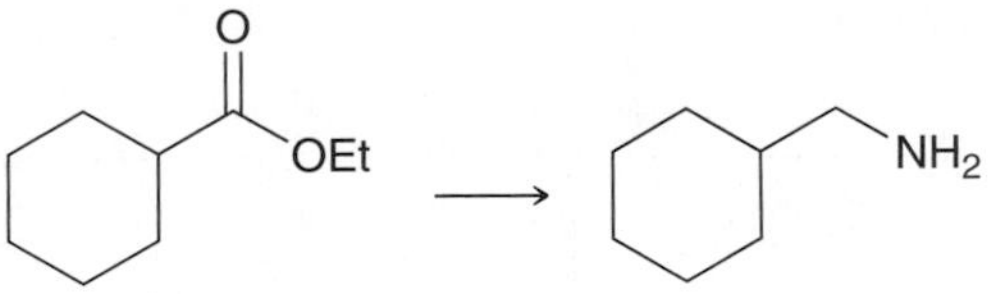

(e)

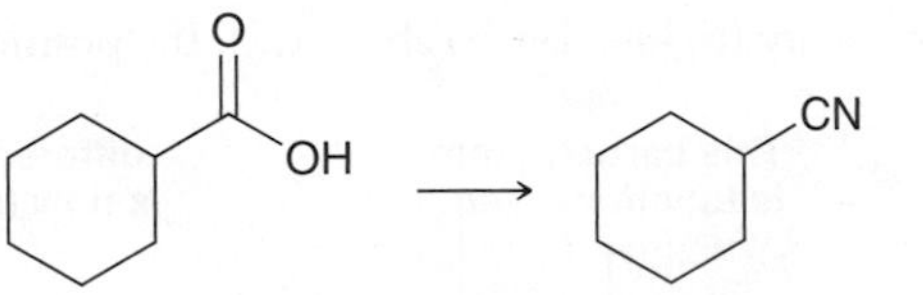

(f)

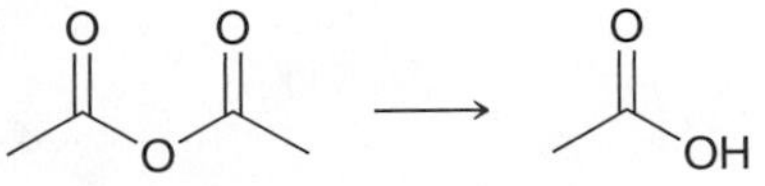

(g)

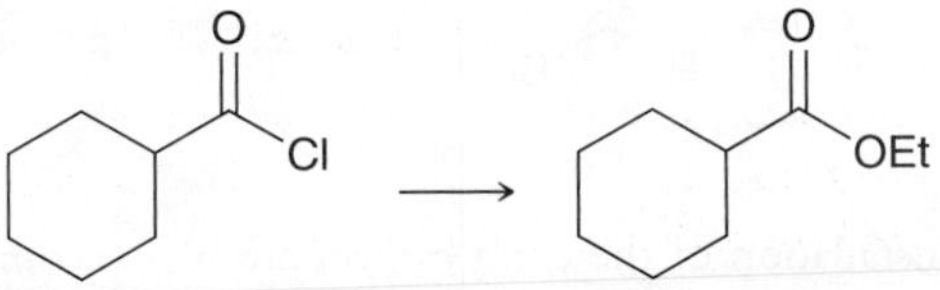

(h)

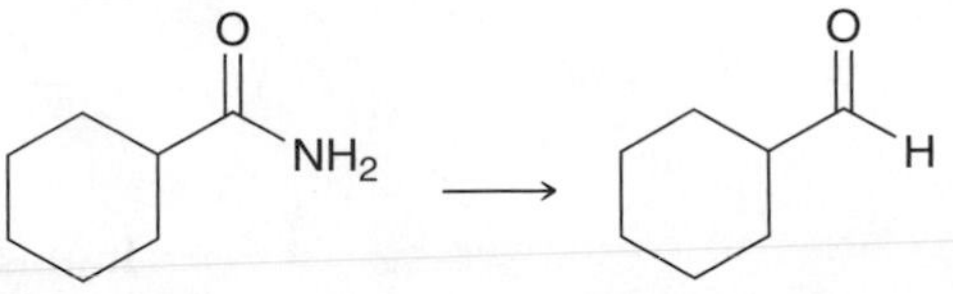

(i)

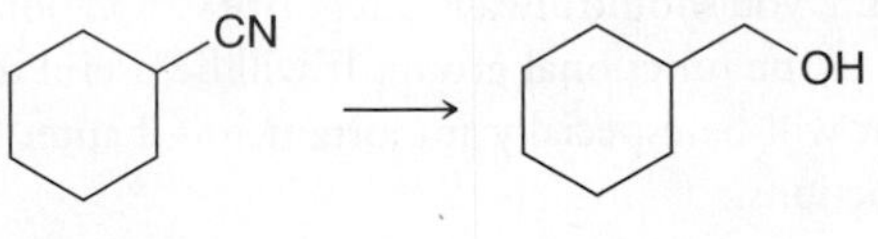

(j)

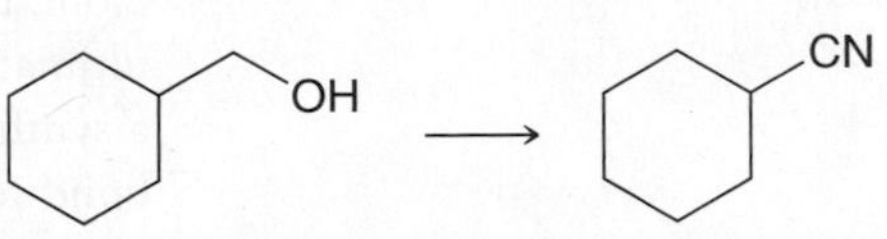

APPLY the skill

21.33 Using only reactions covered in this chapter (Figure 21.11), identify the minimum number of steps required to convert a primary alcohol into an amine (in Chapter 24, we will learn more efficient methods for achieving this type of transformation).

21.34 Propose an efficient method for converting 1-hexene into hexanoyl chloride. You will need to use reactions from previous chapters at the beginning of your synthesis.

need more **PRACTICE?** **Try Problems 21.45a,b, 21.46a, 21.52, 21.53a,c,e, 21.57**

C—C Bond-Forming Reactions

This chapter covered five new C—C bond-forming reactions, which can be placed into two categories (Table 21.2). In the first category (left), the location of the functional group remains unchanged. In each of the three reactions in this category, the product exhibits a functional group in the same location as the reactant.

TABLE 21.2 TWO KINDS OF C—C BOND-FORMING REACTIONS COVERED IN THIS CHAPTER

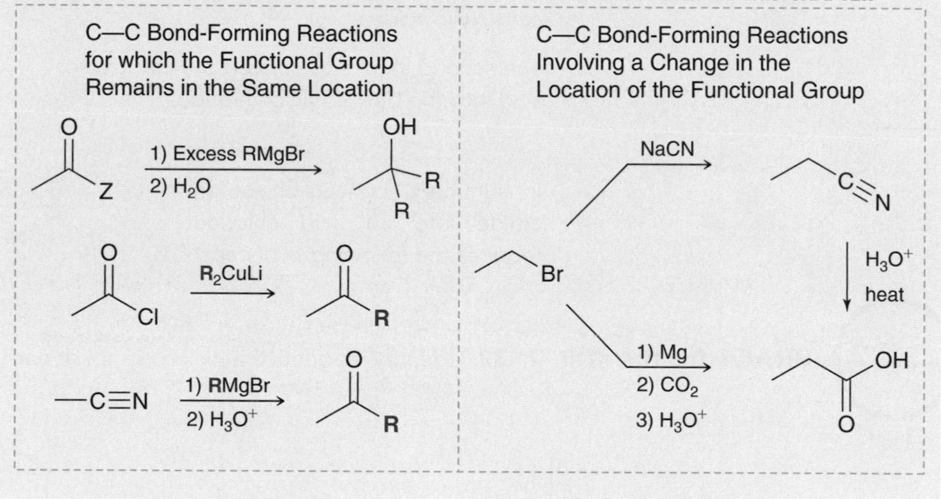

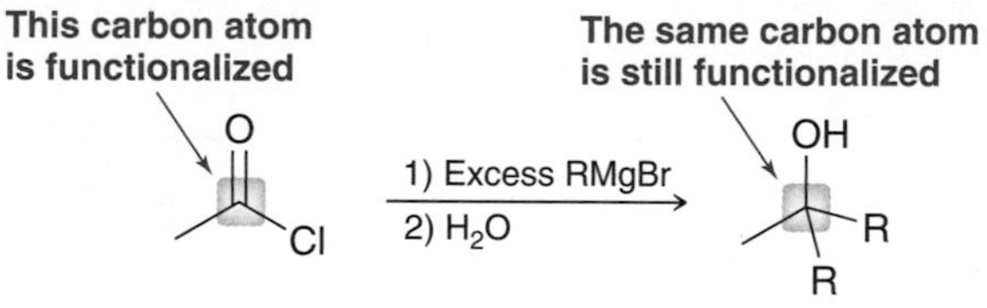

In the second category (right side of Table 21.2), the position of the functional group changes.

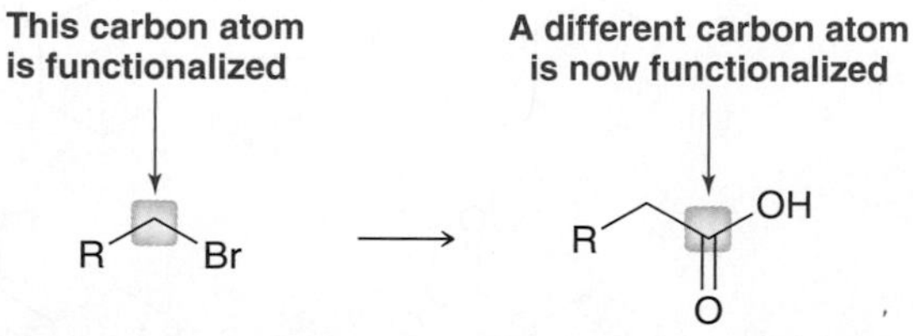

In this case, installation of the extra carbon atom is accompanied by a change in the location of the functional group. Specifically, the functional group moves to the newly installed carbon atom. In your mind, you should always categorize C—C bond-forming reactions in terms of the ultimate location of the functional group. It will be useful to keep this in mind when planning a synthesis, and it will be especially important in Chapter 22, which focuses on many C—C bond-forming reactions.

In Section 14.12, we also saw a method for installing an alkyl chain while simultaneously moving the position of a functional group by two carbon atoms.

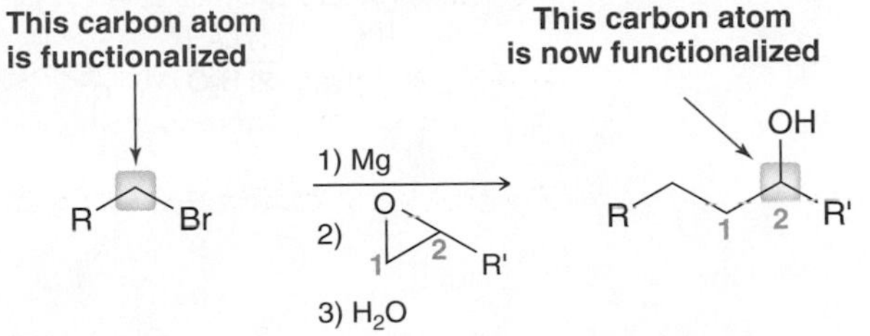

This is just another example of a C—C bond-forming reaction in which the ultimate location of the functional group is an important factor.

When planning a synthesis in which a C—C bond is formed, it is very important to consider the location of the functional group, as that will dictate which C—C bond-forming reaction to choose. Once the proper carbon atom has been functionalized, it is very easy to change the identity of the functional group at that carbon atom (as we saw in the previous section). However, it is slightly more complicated to change the location of a functional group if it is installed in the wrong location. The following example illustrates this idea.

SKILLBUILDER

21.3 CHOOSING THE MOST EFFICIENT C—C BOND-FORMING REACTION

LEARN the skill

Propose an efficient synthesis for the following transformation:

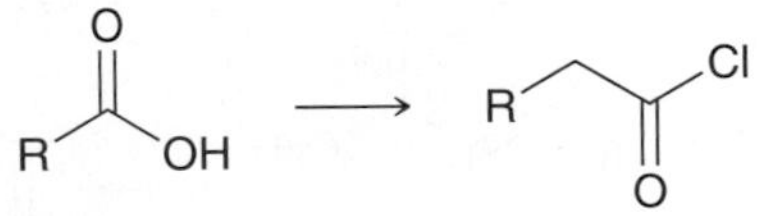

SOLUTION

Always begin a synthesis problem by asking the following two questions:

1. ***Is there any change in the carbon skeleton?*** Yes. The product has one additional carbon atom.
2. ***Is there any change in the identity or position of the functional group?*** Yes. The starting material is a carboxylic acid, and the product is an acid chloride. The position of the functional group has changed.

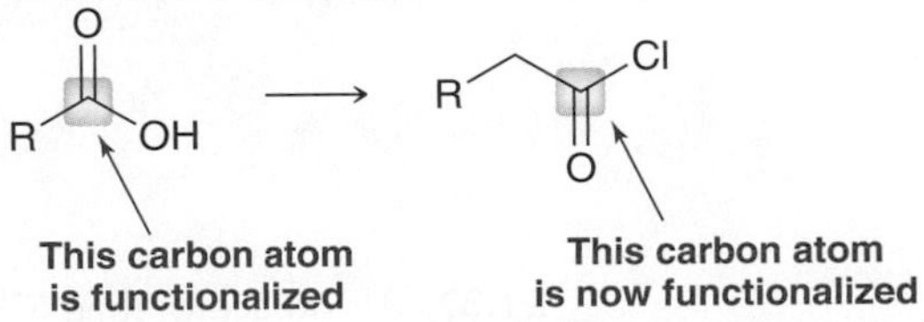

If we focus on the first question without considering the answer to the second question, we might end up going down a long, inefficient pathway. For example, imagine that we first install a methyl group without considering the desired location of the functional group.

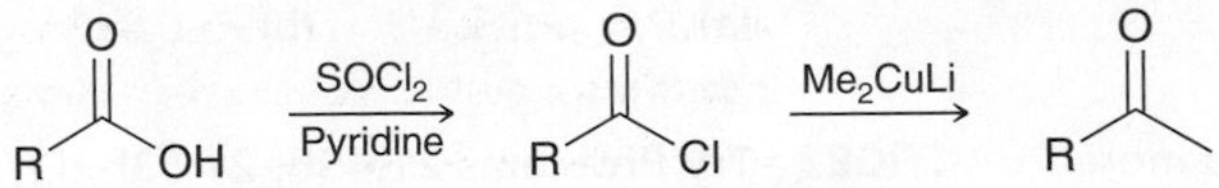

Now the carbon skeleton is correct, but the position of the functional group is wrong. Completing this synthesis involves moving the functional group with a multistep process.

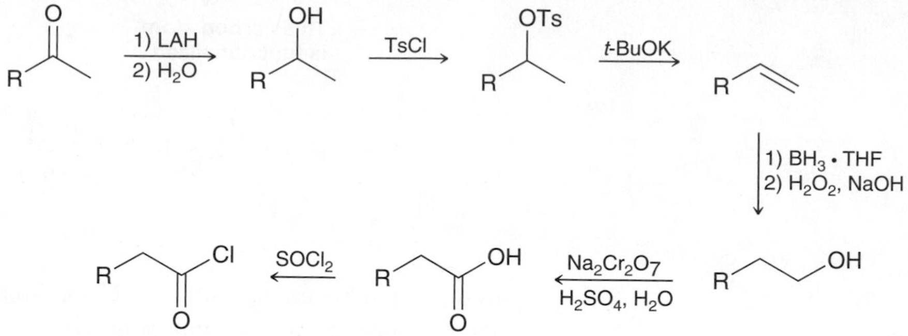

This synthesis is unnecessarily inefficient. It is more efficient to consider both questions at the same time (the change in carbon skeleton and the location of the functional group), because it is possible to install the carbon atom in such a way that the functional group is placed in the desired location.

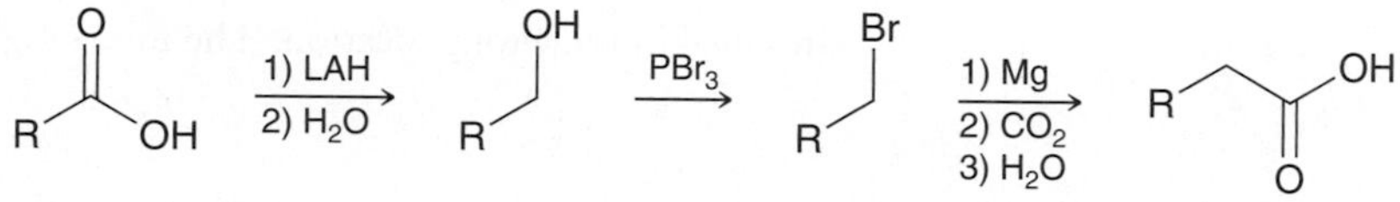

This approach forms the necessary C—C bond while simultaneously installing a functional group at the desired location. With this approach, it is possible to obtain the desired product using just one additional step. This provides for a shorter, more efficient synthesis without regiochemical complications.

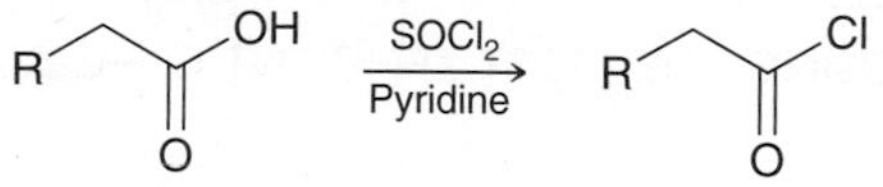

PRACTICE the skill **21.35** Propose an efficient synthesis for each of the following transformations:

(a) 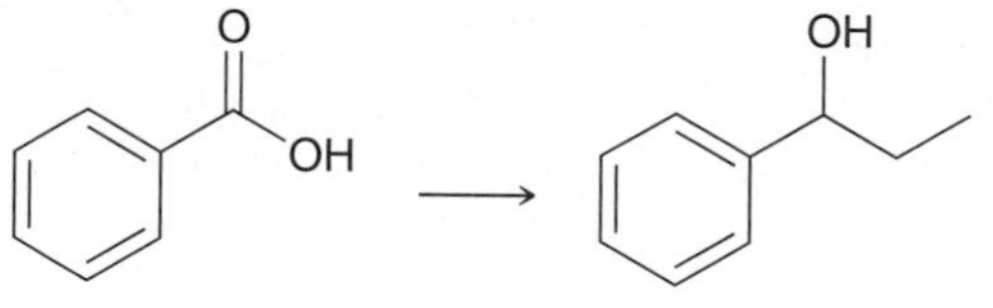

(b) 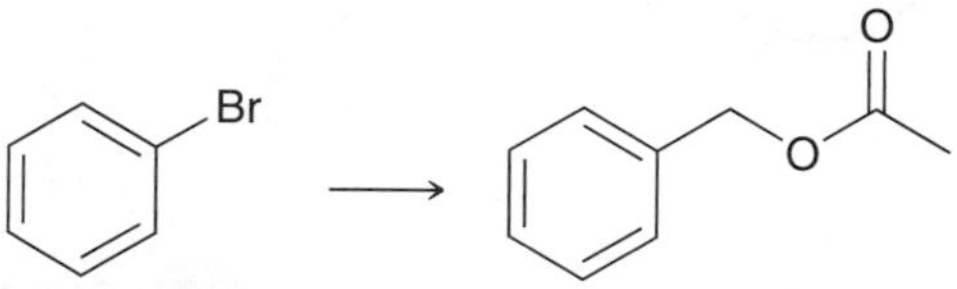

(c) 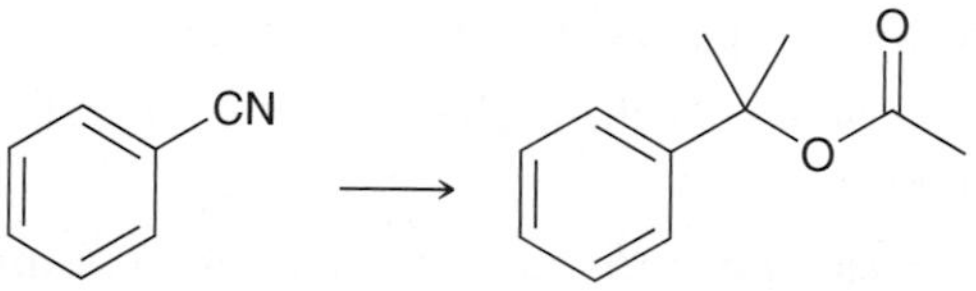

APPLY the skill **21.36** Using reactions from this chapter and the previous chapter (ketones and aldehydes), propose an efficient synthesis for the following transformation:

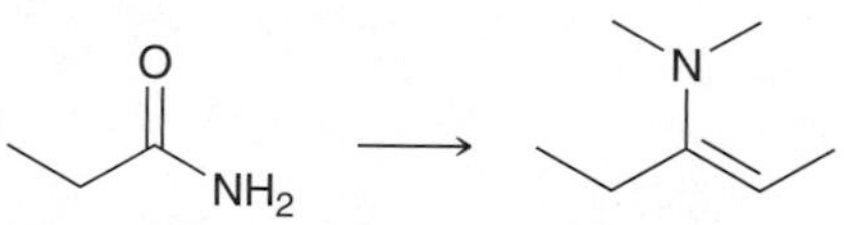

21.37 Using acetonitrile (CH_3CN) and CO_2 as your only sources of carbons, identify how you could prepare each of the following compounds:

(a) 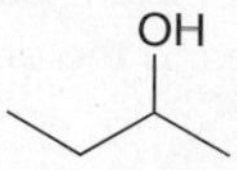

(b) 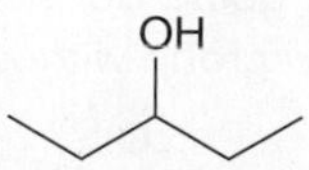

(c) 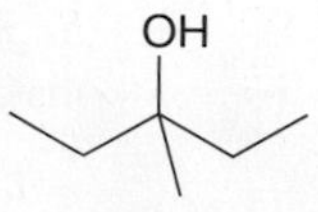

(d) 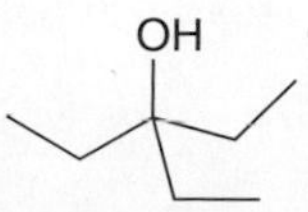

need more **PRACTICE?** **Try Problems 21.45b, 21.53b,d,f, 21.54, 21.55, 21.58, 21.74**

21.15 Spectroscopy of Carboxylic Acids and Their Derivatives

IR Spectroscopy

Recall that a carbonyl group produces a very strong signal between 1650 and 1850 cm^{-1} in an IR spectrum. The precise location of the signal depends on the nature of the carbonyl group. Table 21.3 gives carbonyl stretching frequencies for each of the carboxylic acid derivatives.

TABLE **21.3** IMPORTANT SIGNALS IN IR SPECTROSCOPY

Type of carbonyl group	$\underset{R}{\overset{O}{\|}}Cl$	$R\overset{O}{-}O\overset{O}{-}R$	$\underset{R}{\overset{O}{\|}}OH$	$\underset{R}{\overset{O}{\|}}OR$	$\underset{R}{\overset{O}{\|}}R$	$\underset{R}{\overset{O}{\|}}NH_2$
Wavenumber of absorption (cm^{-1})	~1800	1760, 1820 (two signals)	~1760	~1740	~1720	~1660

These numbers can be used to determine the type of carbonyl group in an unknown compound. When performing this type of analysis, recall that conjugated carbonyl groups will produce signals at lower frequencies (Section 15.33).

A ketone A conjugated ketone

1720 cm^{-1} 1680 cm^{-1}

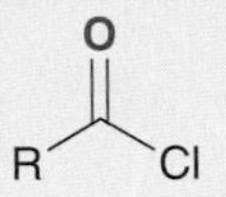

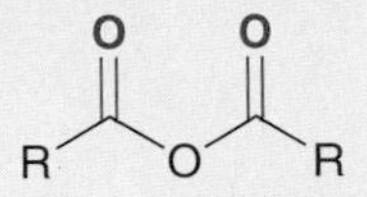

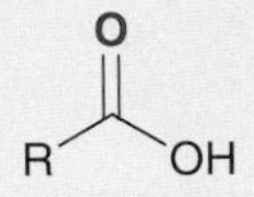

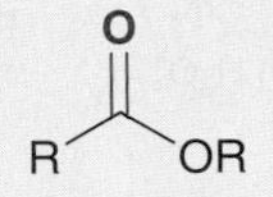

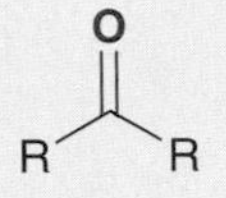

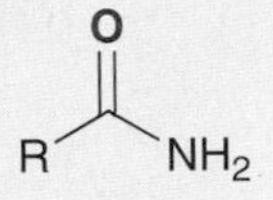

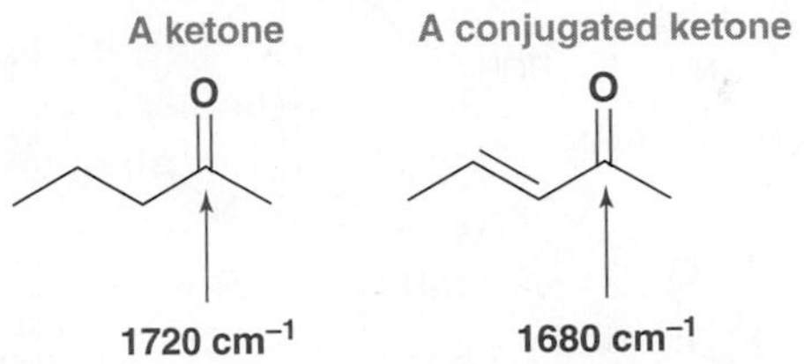

This shift to lower frequencies is observed for all conjugated carbonyl groups, including those present in carboxylic acid derivatives. For example, a conjugated ester will produce a signal below 1740 cm^{-1}, and this must be taken into account when analyzing a signal.

In addition, carboxylic acids show very broad O—H signals that span the distance between 2500 and 3300 cm^{-1} (Section 15.5). The C—N triple bond of a nitrile appears in the triple-bond region of the spectrum, at approximately 2200 cm^{-1} (Section 15.3).

^{13}C NMR Spectroscopy

The carbonyl group of a carboxylic acid derivative generally appears in the region between 160 and 185 ppm, and it is very difficult to use the precise location of a signal to determine the type of carbonyl group present in an unknown compound. The carbon atom of a nitrile typically produces a signal between 115 and 130 ppm in a ^{13}C NMR spectrum.

1H NMR Spectroscopy

As discussed in Section 16.5, the proton of a carboxylic acid typically produces a signal at approximately 12 ppm in a ^{1}H NMR spectrum.

CONCEPTUAL CHECKPOINT

21.38 Compound **A** has molecular formula $C_9H_8O_2$ and exhibits a strong signal at 1740 cm^{-1} in its IR spectrum. Treatment with two equivalents of LAH followed by water gives the following diol. Identify the structure of compound **A**.

Compound **A** $\xrightarrow[\text{2) H}_2\text{O}]{\text{1) LAH}}$

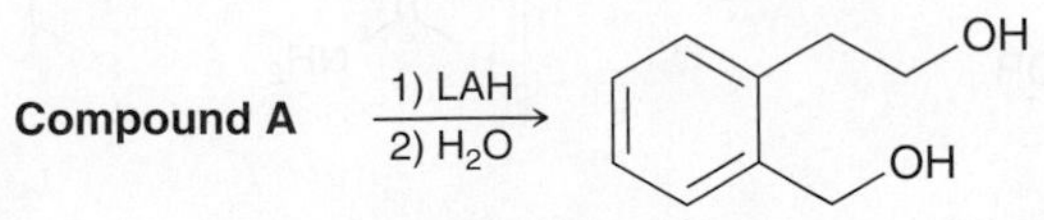

REVIEW OF REACTIONS

PREPARATION OF CARBOXYLIC ACIDS

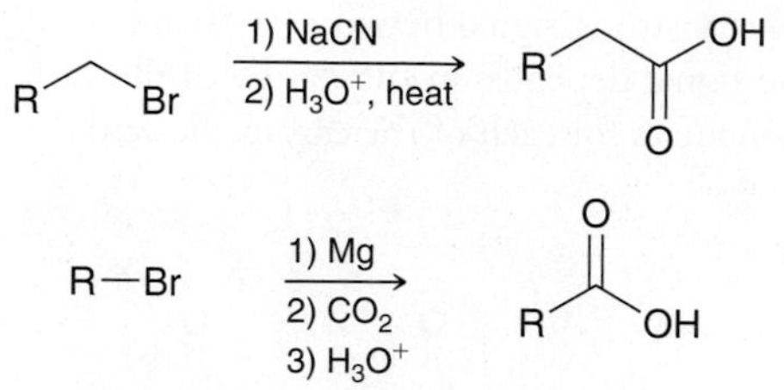

REACTIONS OF CARBOXYLIC ACIDS

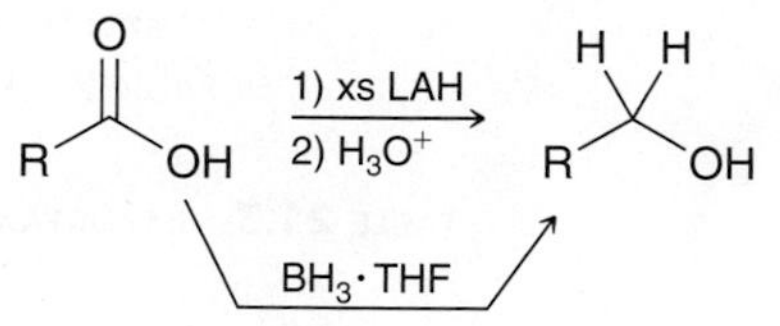

PREPARATION AND REACTIONS OF ACID CHLORIDES

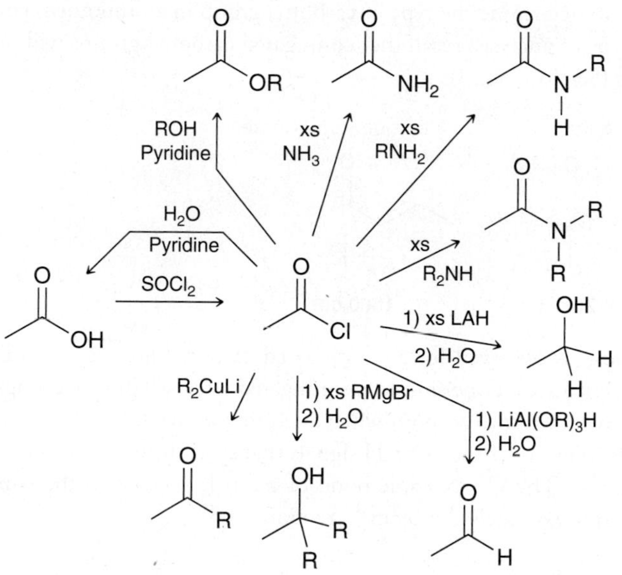

PREPARATION AND REACTIONS OF ACID ANHYDRIDES

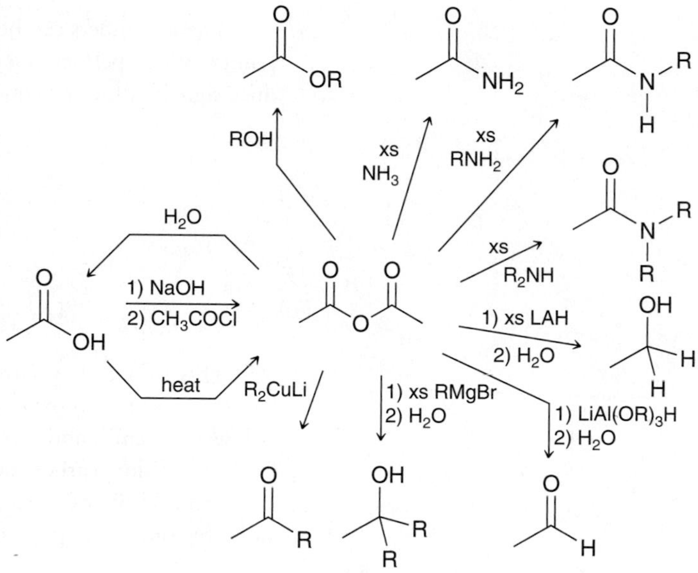

PREPARATION OF ESTERS

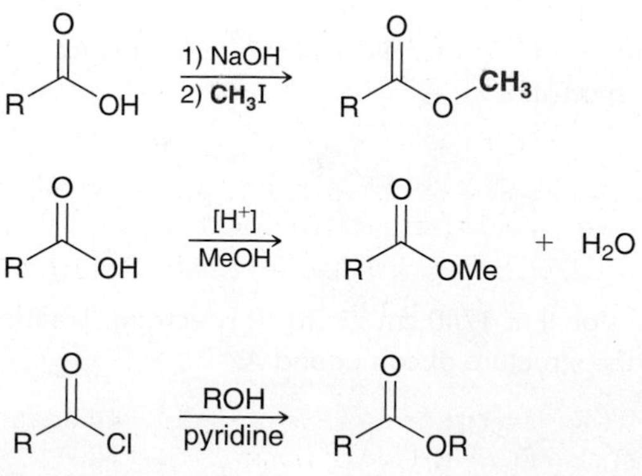

REACTIONS OF ESTERS

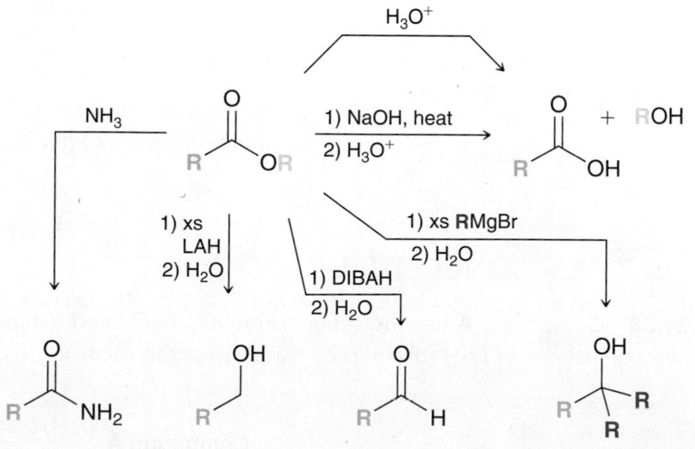

PREPARATION OF AMIDES

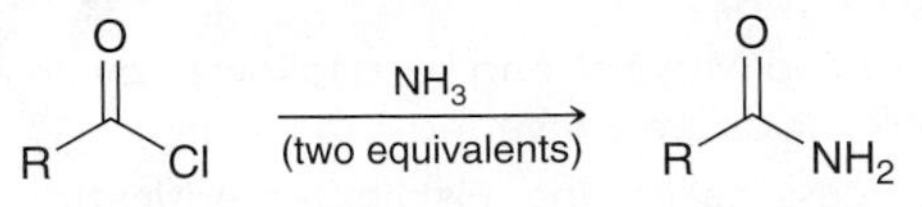

REACTIONS OF AMIDES

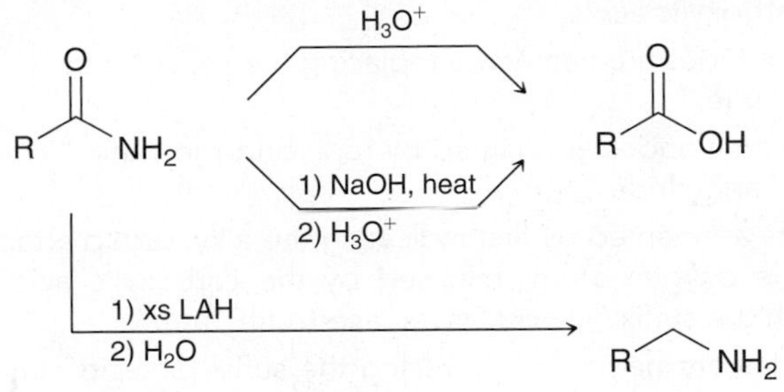

PREPARATION OF NITRILES

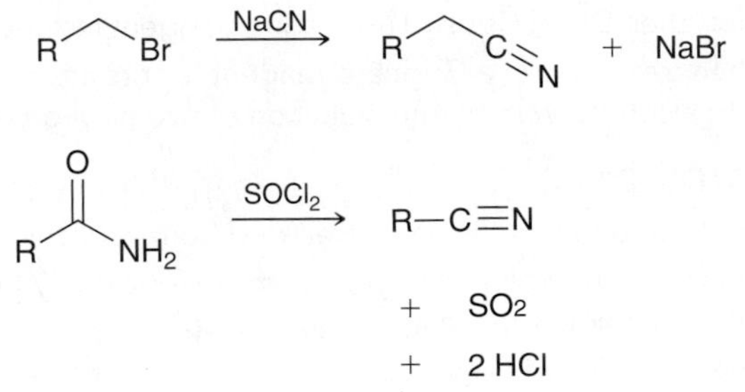

REACTIONS OF NITRILES

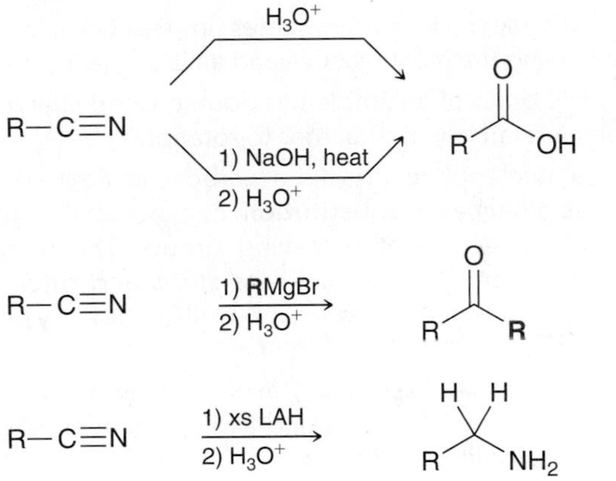

REVIEW OF CONCEPTS AND VOCABULARY

SECTION 21.1

- Carboxylic acids are abundant in nature, and they are widely used in the pharmaceutical and other industries.
- For industrial purposes, acetic acid is converted into vinyl acetate, which is a **carboxylic acid derivative**.

SECTION 21.2

- Compounds containing a carboxylic acid moiety are named with the suffix "oic acid."
- Compounds containing two carboxylic acid moieties are named with the suffix "dioic acid."
- Many simple carboxylic acids and diacids have common names accepted by IUPAC.

SECTION 21.3

- Carboxylic acids can form two hydrogen-bonding interactions.
- Treatment of a carboxylic acid with a strong base, such as sodium hydroxide, yields a carboxylate salt.
- Carboxylate ions are named by replacing the suffix "ic acid" with "ate."
- The pKa of most carboxylic acids is between 4 and 5.
- The acidity of carboxylic acids is due to the stability of the conjugate base, which is resonance stabilized.
- Using the **Henderson-Hasselbalch equation**, it can be shown that carboxylic acids exist primarily as carboxylate salts at **physiological pH**.
- Electron-withdrawing substituents can increase the acidity of a carboxylic acid; the strength of this effect depends on the distance between the electron-withdrawing substituent and the carboxylic acid moiety.

SECTION 21.4

- When treated with aqueous acid, a nitrile will undergo **hydrolysis,** yielding a carboxylic acid.
- Carboxylic acids can also be prepared by treating a Grignard reagent with carbon dioxide.

SECTION 21.5

- Carboxylic acids are reduced to alcohols upon treatment with lithium aluminum hydride or borane.

SECTION 21.6

- Carboxylic acid derivatives exhibit the same oxidation state as carboxylic acids.
- Acid halides are named by replacing the suffix "ic acid" with "yl halide."
- Acid anhydrides are named by replacing the suffix "ic acid" with "anhydride."
- Esters are named by first indicating the alkyl group attached to the oxygen atom, followed by the carboxylic acid, for which the suffix "ic acid" is replaced with "ate."
- Amides are named by replacing the suffix "ic acid" or "oic acid" with "amide."
- Nitriles are named by replacing the suffix "ic acid" with "nitrile."

SECTION 21.7

- Carboxylic acid derivatives differ in reactivity, with acid halides being the most reactive and amides the least reactive.
- The C—N bond of an amide has double-bond character and exhibits a relatively high barrier to rotation.
- When a nucleophile attacks a carboxylic acid derivative, a **nucleophilic acyl substitution** can occur in which the nucleophile replaces the leaving group. The mechanism of this reaction involves two core steps and often utilizes several proton transfer steps as well (especially in acidic conditions).
- When drawing a mechanism, avoid formation of negative charges in acidic conditions and avoid formation of positive charges in alkaline conditions.
- When a nucleophile attacks a carbonyl group to form a tetrahedral intermediate, always re-form the carbonyl if possible but never expel H^- or C^-.

SECTION 21.8

- Acid chlorides can be formed by treating carboxylic acids with thionyl chloride.
- When treated with water, acid chlorides are hydrolyzed to give carboxylic acids.
- When treated with an alcohol, acid chlorides are converted into esters.
- When treated with ammonia, acid chlorides are converted into amides. Two equivalents of ammonia are required: one to serve as a nucleophile and the other to serve as a base.
- When treated with excess LAH, acid chlorides are reduced to give alcohols because two equivalents of hydride attack. Selective hydride-reducing agents, such as lithium tri(*t*-butoxy) aluminum hydride, can be used to prepare the aldehyde.
- When treated with a Grignard reagent, acid chlorides are converted into alcohols with the introduction of two alkyl groups. Two equivalents of Grignard reagent attack. Preparing a ketone requires the use of a more selective organometallic reagent, such as a **lithium dialkyl cuprate**, also called a **Gilman reagent**.

SECTION 21.9

- Acetic acid can be converted into acetic anhydride with excessive heating.
- Acid anhydrides can be prepared by treating an acid chloride with a carboxylate ion.
- The reactions of anhydrides are the same as the reactions of acid chlorides except for the identity of the leaving group.

SECTION 21.10

- When treated with a strong base followed by an alkyl halide, carboxylic acids are converted into esters.
- In a process called the **Fischer esterification**, carboxylic acids are converted into esters when treated with an alcohol in the presence of an acid catalyst. This process is reversible.
- Esters can also be prepared by treating an acid chloride with an alcohol in the presence of pyridine.

SECTION 21.11

- Esters can be hydrolyzed to yield carboxylic acids by treatment with either aqueous base or aqueous acid. Hydrolysis under basic conditions is also called **saponification**.
- When treated with lithium aluminum hydride, esters are reduced to yield alcohols. If the desired product is an aldehyde, then DIBAH is used as a reducing agent instead of LAH.
- When treated with a Grignard reagent, esters are reduced to yield alcohols, with the introduction of two alkyl groups.

SECTION 21.12

- Amides are most efficiently prepared from acid chlorides.
- Amides are hydrolyzed to yield carboxylic acids by treatment with either aqueous base or aqueous acid.
- When treated with excess LAH, amides are converted into amines.

SECTION 21.13

- Nitriles can be prepared by treating an alkyl halide with a cyanide ion or via the dehydration of an amide.
- Nitriles can be hydrolyzed to yield carboxylic acids by treatment with either aqueous base or aqueous acid.
- A ketone is obtained when a nitrile is treated with a Grignard reagent, followed by aqueous acid.
- Nitriles are converted to amines when treated with LAH.

SECTION 21.14

- Carboxylic acids, their derivatives, aldehydes, alcohols, and amines can be readily interconverted using reactions covered in this chapter.
- When forming a C—C bond, always consider where you want the functional group to be located, as that will dictate which C—C bond-forming reaction to choose.

SECTION 21.15

- In IR spectroscopy, the precise location of a carbonyl stretching signal, which appears between 1650 and 1850 cm^{-1}, can be used to determine the type of carbonyl group in an unknown compound.
- Conjugated carbonyl groups produce signals at lower frequencies.
- In a carbon ^{13}C NMR spectrum, the carbonyl group of a carboxylic acid derivative will generally appear in the region between 160 and 185 ppm, and the carbon atom of a nitrile produces a signal between 115 and 130 ppm.
- In a ^{1}H NMR spectrum, the proton of a carboxylic acid produces a signal at approximately 12 ppm.

KEY TERMINOLOGY

carboxylic acid derivative 971
Fischer esterification 1000
Gilman reagent 995
Henderson-Hasselbalch equation 975
hydrolysis 977
lithium dialkyl cuprate 995
nucleophilic acyl substitution 985
physiological pH 975
saponification 1001

SKILLBUILDER REVIEW

21.1 DRAWING THE MECHANISM OF A NUCLEOPHILIC ACYL SUBSTITUTION REACTION

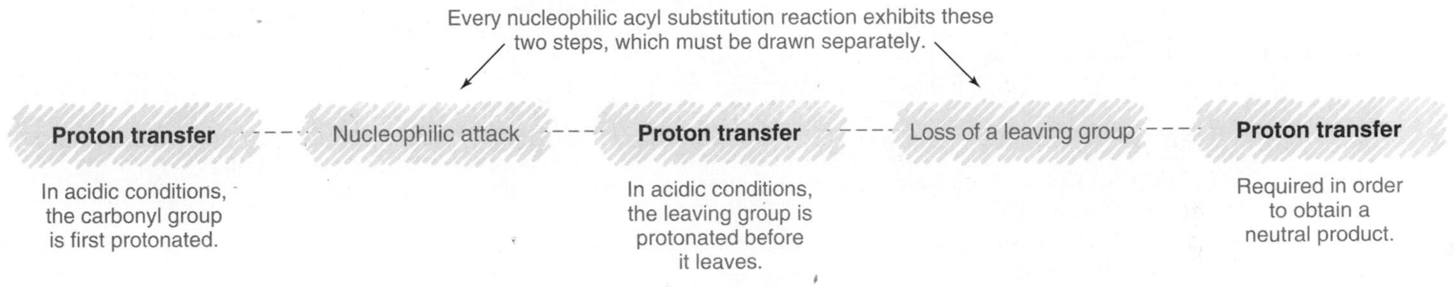

Try Problems **21.14–21.17, 21.61, 21.72**

21.2 INTERCONVERTING FUNCTIONAL GROUPS

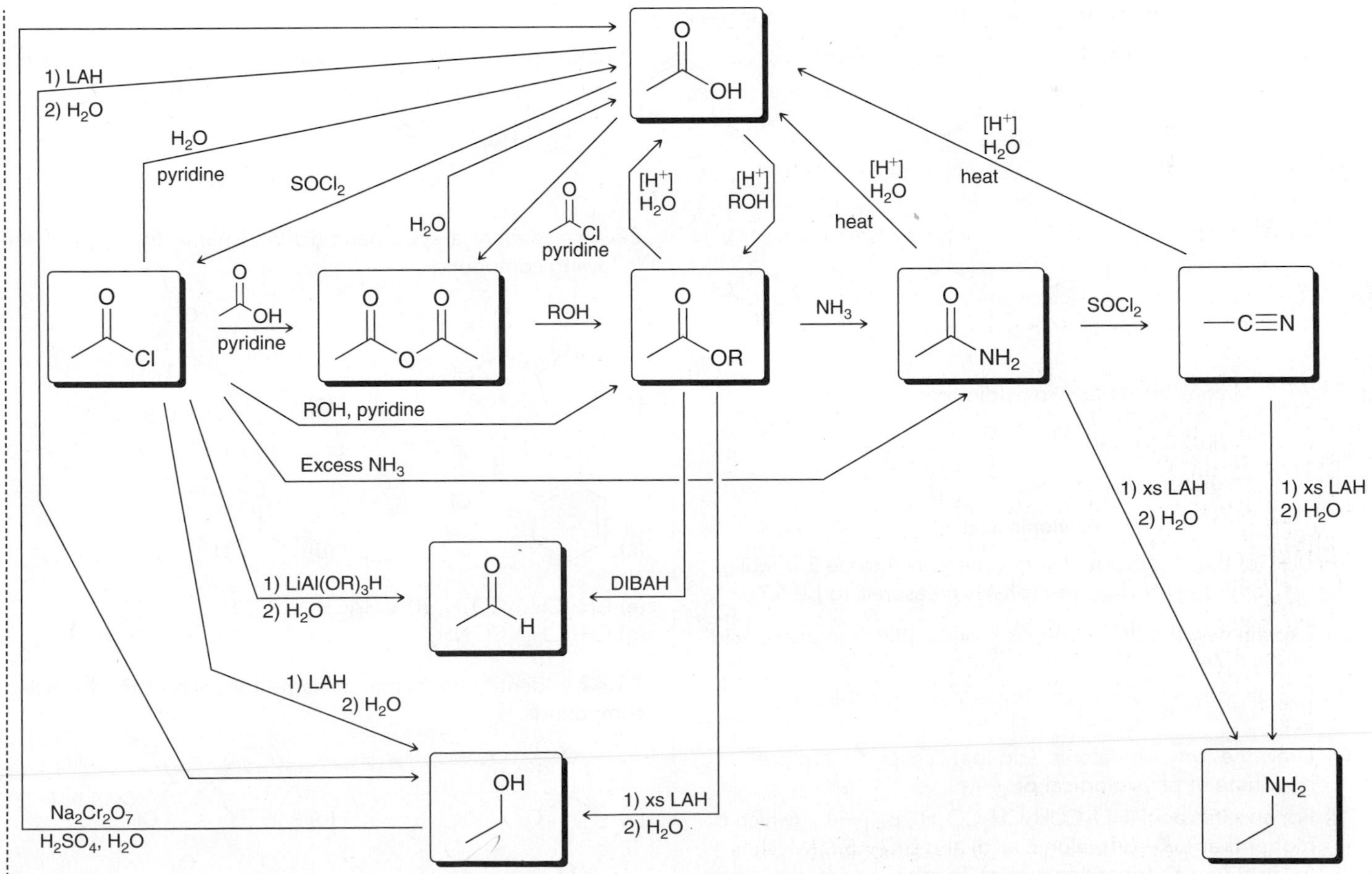

Try Problems **21.32–21.34, 21.45a,b, 21.46a, 21.52, 21.53a,c,e, 21.57**

21.3 CHOOSING THE MOST EFFICIENT C—C BOND-FORMING REACTION

C—C Bond-Forming Reactions for Which the Functional Group Remains in the Same Location

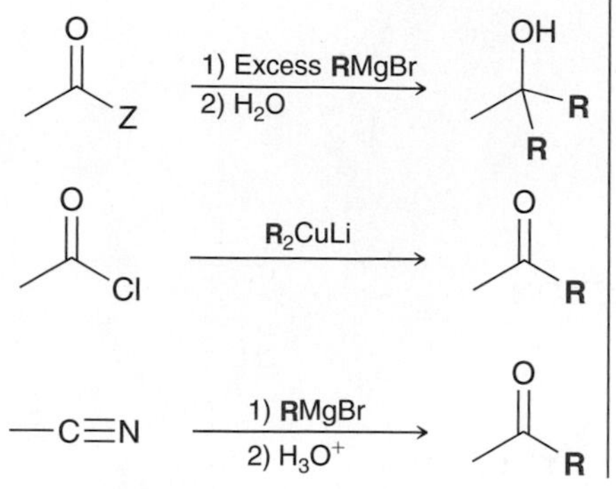

C—C Bond-Forming Reactions Involving a Change in the Location of the Functional Group

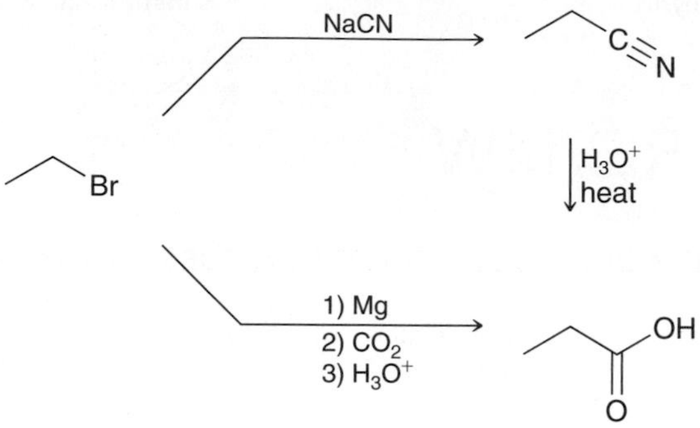

Try Problems 21.35–21.37, 21.45b, 21.53b,d,f, 21.54, 21.55, 21.58, 21.74

PRACTICE PROBLEMS

Note: Most of the Problems are available within *WileyPLUS*, an online teaching and learning solution.

21.39 Rank each set of compounds in order of increasing acidity:

(a)

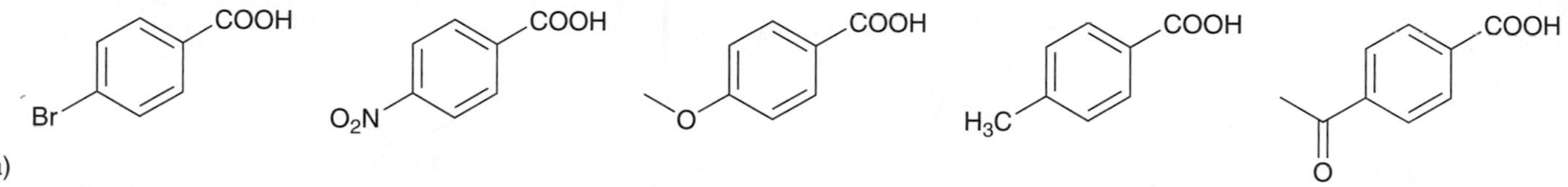

(b)

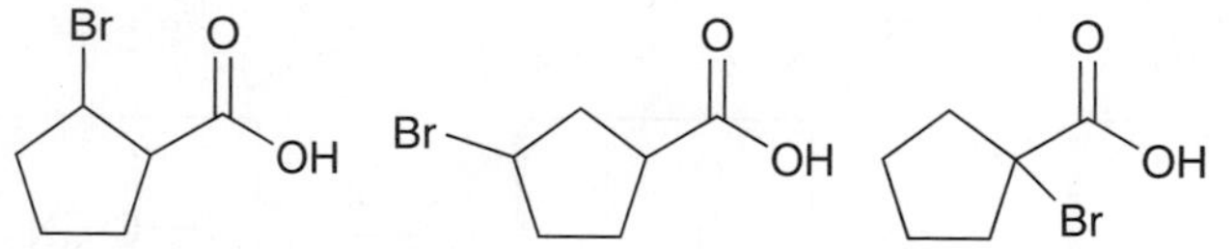

21.40 Malonic acid has two acidic protons:

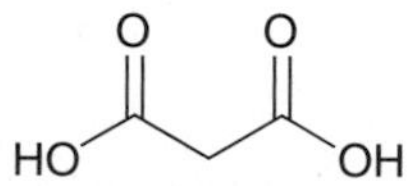

Malonic acid

The pK_a of the first proton (pK_1) is measured to be 2.8, while the pK_a of the second proton (pK_2) is measured to be 5.7.

(a) Explain why the first proton is more acidic than acetic acid ($pK_a = 4.76$).

(b) Explain why the second proton is less acidic than acetic acid.

(c) Draw the form of malonic acid that is expected to predominate at physiological pH.

(d) For succinic acid ($HO_2CCH_2CH_2CO_2H$), $pK_1 = 4.2$ (which is higher than pK_1 for malonic acid) and $pK_2 = 5.6$ (which is lower than pK_2 for malonic acid). In other words, the difference between pK_1 and pK_2 is not as large for succinic acid as it is for malonic acid. Explain this observation.

21.41 Identify a systematic (IUPAC) name for each of the following compounds.

(a) 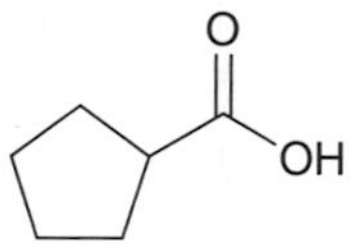
(b) 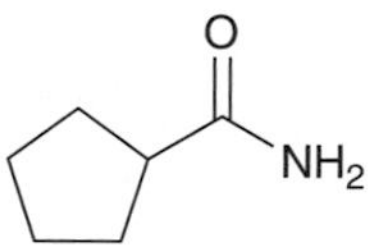
(c) 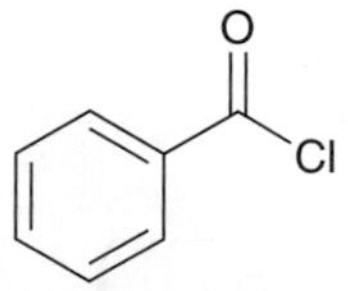
(d) 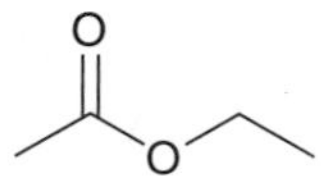

(e) $CH_3(CH_2)_4CO_2H$ (f) $CH_3(CH_2)_3COCl$
(g) $CH_3(CH_2)_4CONH_2$

21.42 Identify the common name for each of the following compounds:

(a) 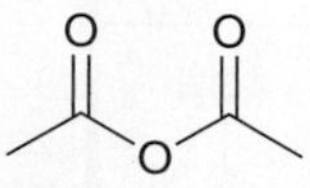
(b) 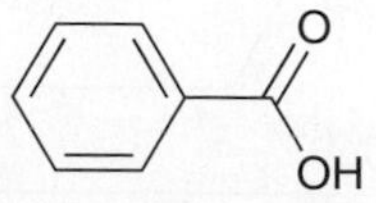
(c) 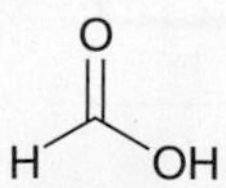
(d) 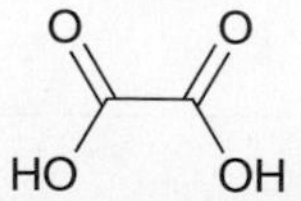

21.43 Draw the structures of eight different carboxylic acids with molecular formula $C_6H_{12}O_2$. Then, provide a systematic name for each compound, and identify which three isomers exhibit chirality centers.

21.44 Draw and name all constitutionally isomeric acid chlorides with molecular formula C_4H_7ClO. Then provide a systematic name for each isomer.

21.45 Identify the reagents you would use to convert pentanoic acid into each of the following compounds:

(a) 1-Pentanol (b) 1-Pentene (c) Hexanoic acid

21.46 Identify the reagents you would use to convert each of the following compounds into pentanoic acid:

(a) 1-Pentene (b) 1-Bromobutane

21.47 Careful measurements reveal that *para*-methoxybenzoic acid is less acidic than benzoic acid, while *meta*-methoxybenzoic acid is more acidic than benzoic acid. Explain these observations.

21.48 Predict the major product(s) formed when hexanoyl chloride is treated with each of the following reagents:

(a) $CH_3CH_2NH_2$ (excess)
(b) LAH (excess), followed by H_2O
(c) CH_3CH_2OH, pyridine
(d) H_2O, pyridine
(e) $C_6H_5CO_2Na$
(f) NH_3 (excess)
(g) Et_2CuLi
(h) EtMgBr (excess), followed by H_2O

21.49 Predict the major product(s) formed when cyclopentanecarboxylic acid is treated with each of the following reagents:

(a) $SOCl_2$
(b) LAH (excess), followed by H_2O
(c) NaOH
(d) $[H^+]$, EtOH

21.50 Predict the major product(s) for each of the following reactions:

(a) 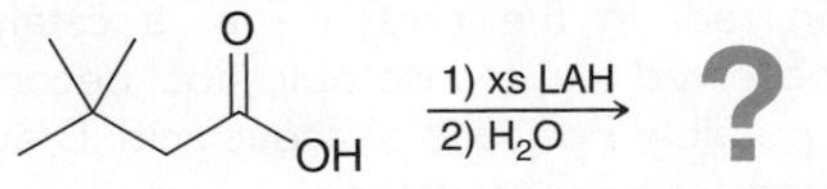

(b) 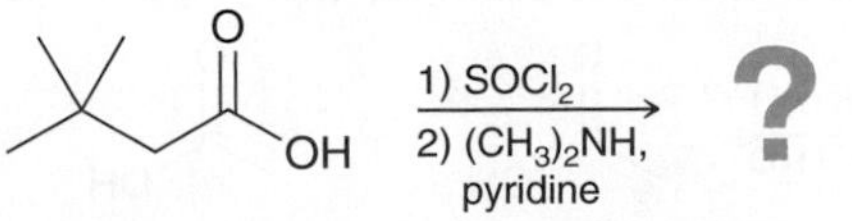

(c) 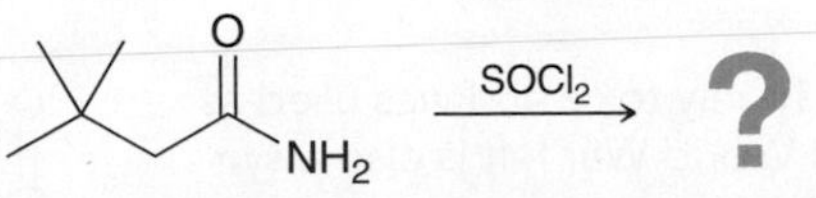

(d) 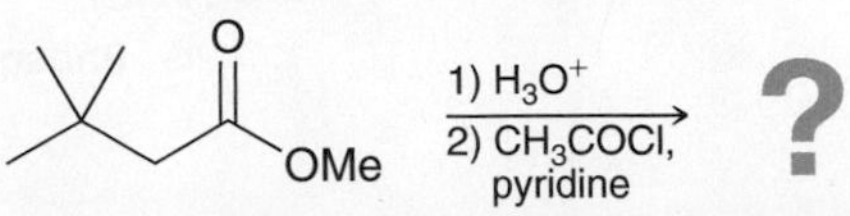

(e) 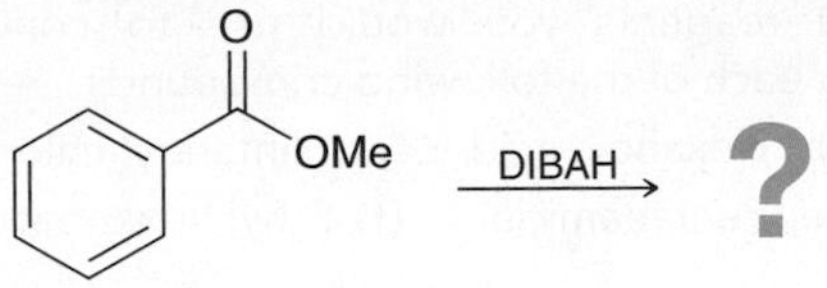

(f) 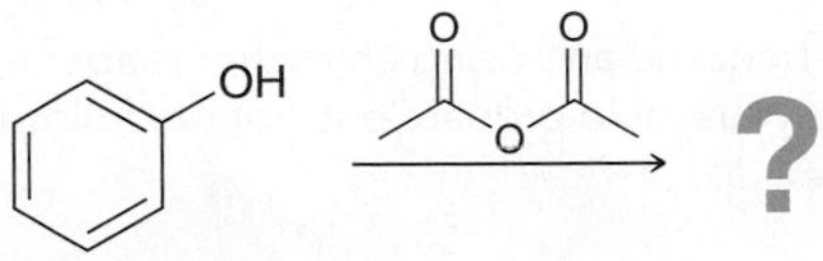

(g) 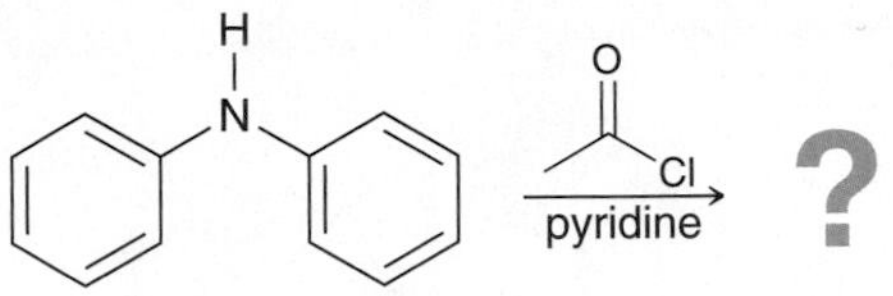

(h) 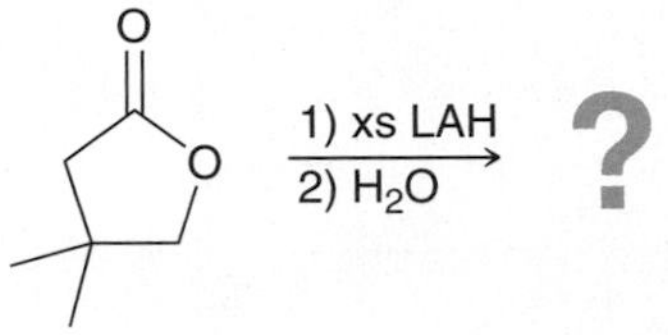

(i) 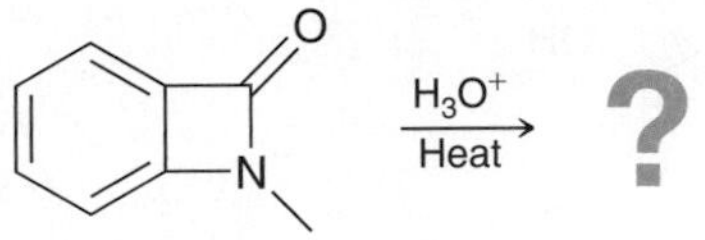

(j) 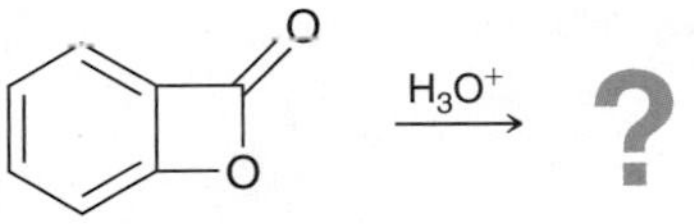

21.51 Identify the carboxylic acid and the alcohol that are necessary in order to make each of the following compounds via a Fischer esterification:

(a) 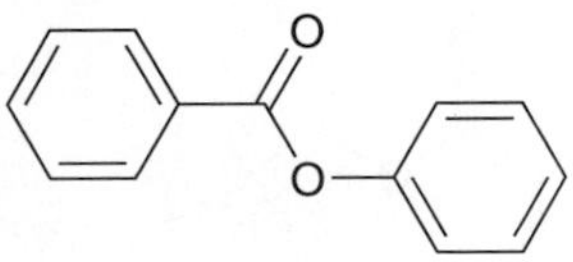

(b) 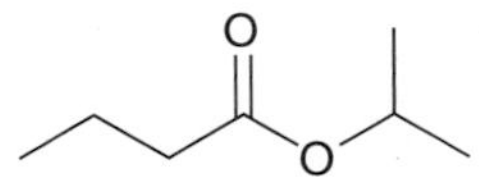

(c) $CH_3CH_2CO_2C(CH_3)_3$

21.52 Determine the structures of compounds **A** through **F**:

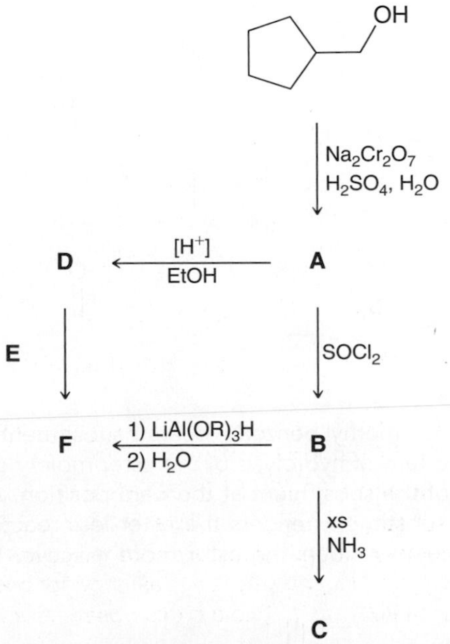

21.53 Identify the reagents you would use to convert 1-bromopentane into each of the following compounds:
(a) Pentanoic acid (b) Hexanoic acid (c) Pentanoyl chloride
(d) Hexanamide (e) Pentanamide (f) Ethyl hexanoate

21.54 Starting with benzene and using any other reagents of your choice, show how you would prepare each of the following compounds:

(a) 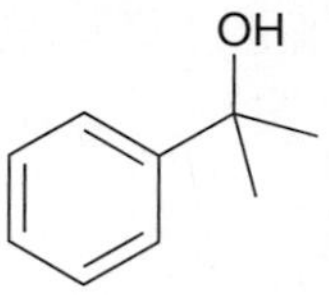

(b) 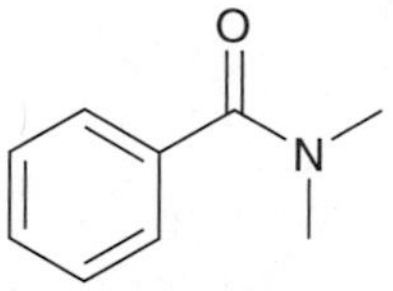

(c) 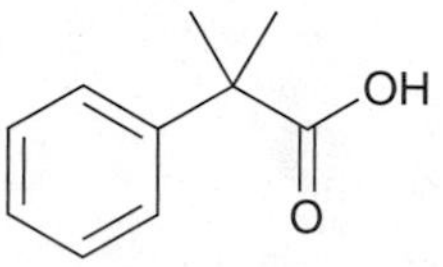

(d) 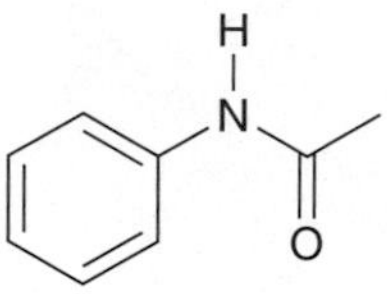

21.55 Propose an efficient synthesis for each of the following transformations:

(a) 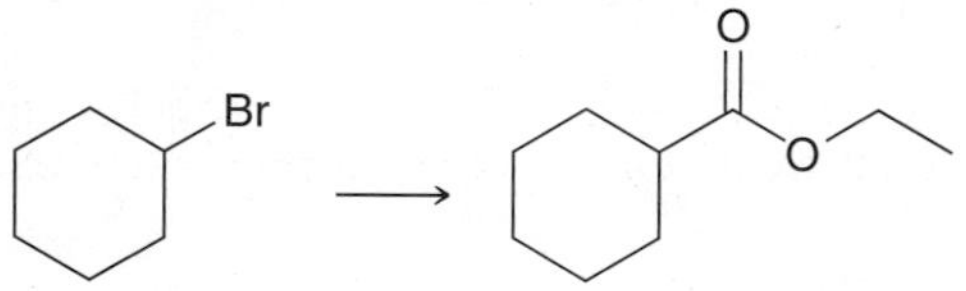

(b) 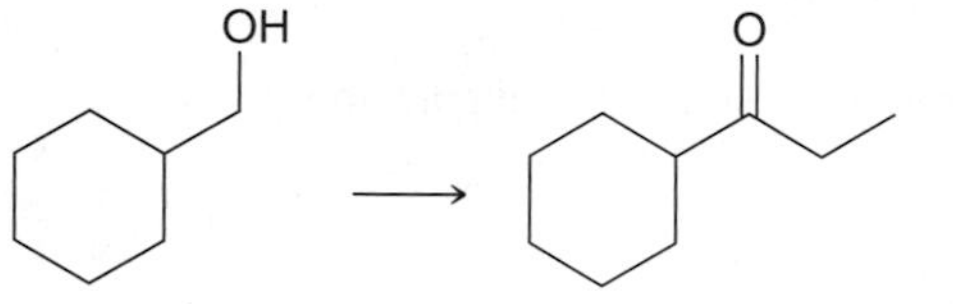

(c) 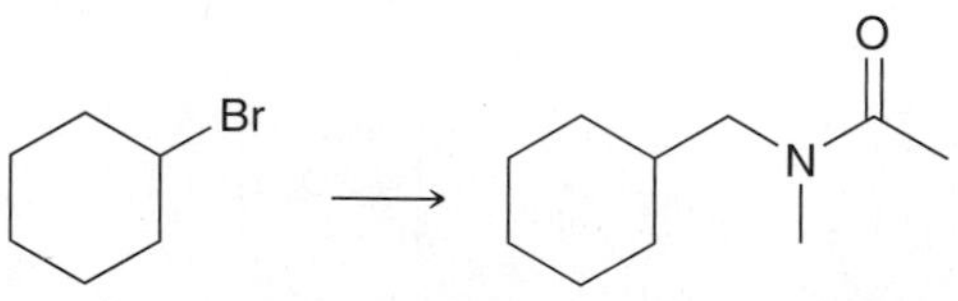

(d) 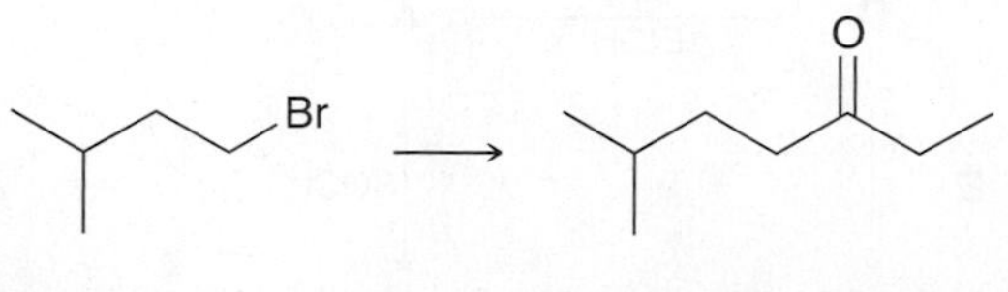

21.56 When methyl benzoate bears a substituent at the *para* position, the rate of hydrolysis of the ester moiety depends on the nature of the substituent at the *para* position. Apparently, a methoxy substituent renders the ester less reactive, while a nitro substituent renders the ester more reactive. Explain this observation.

21.57 Identify the reagents necessary to accomplish each of the following transformations:

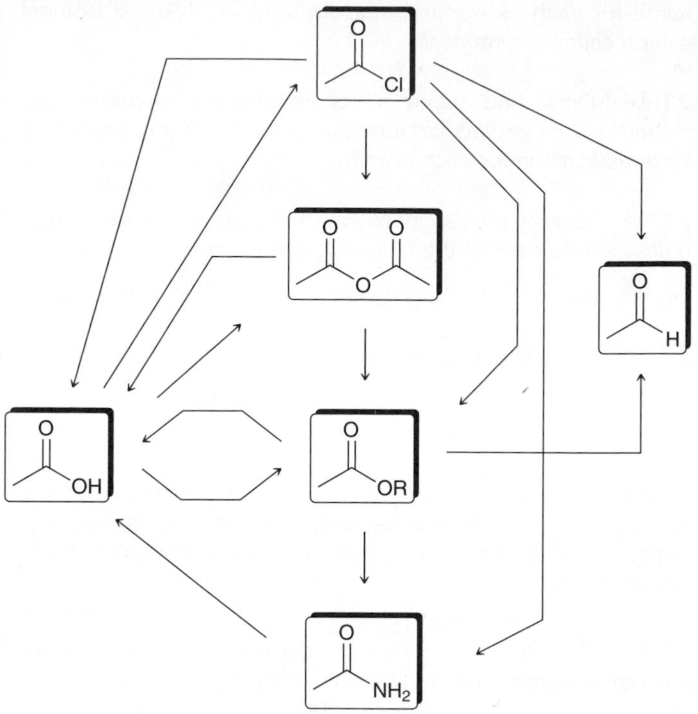

21.58 DEET is the active ingredient in many insect repellants, such as OFF™. Starting with *meta*-bromotoluene and using any other reagents of your choice, devise an efficient synthesis for DEET.

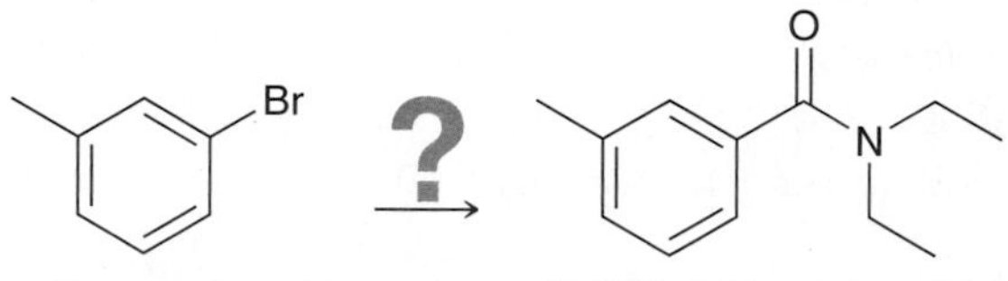

***m*-Bromotoluene** ***N,N*-Diethyl-*m*-toluamide (DEET)**

21.59 Predict the products that are formed when diphenyl carbonate is treated with excess methyl magnesium bromide.

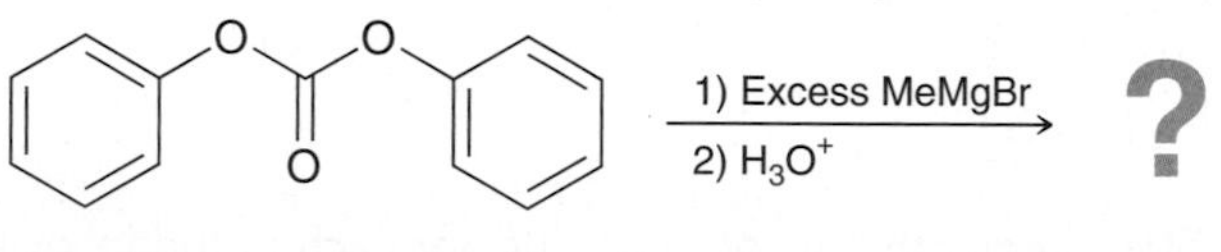

21.60 When acetic acid is treated with isotopically labeled water (^{18}O, shown in red) in the presence of a catalytic amount of acid, it is observed that the isotopic label becomes incorporated at both possible positions of acetic acid. Draw a mechanism that accounts for this observation.

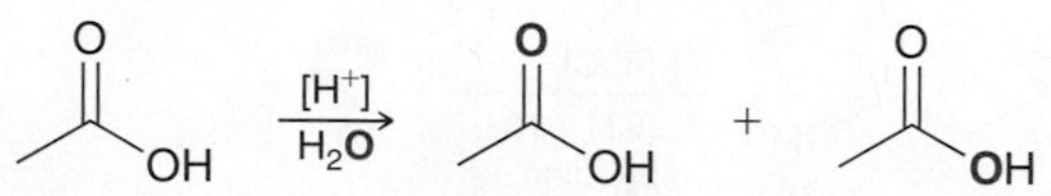

21.61 Phosgene is highly toxic and was used as a chemical weapon in World War I. It is also a synthetic precursor used in the production of many plastics.

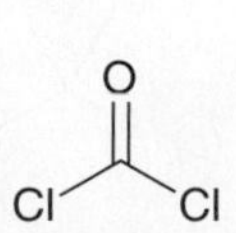

Phosgene

(a) When vapors of phosgene are inhaled, the compound rapidly reacts with any nucleophilic sites present (OH groups, NH_2 groups, etc.), producing HCl gas. Draw a mechanism for this process.
(b) When phosgene is treated with ethylene glycol ($HOCH_2CH_2OH$), a compound with molecular formula $C_3H_4O_3$ is obtained. Draw the structure of this product.
(c) Predict the product that is expected when phosgene is treated with excess phenylmagnesium bromide, followed by water.

21.62 Fluphenazine is an antipsychotic drug that is administered as an ester prodrug via intramuscular injection:

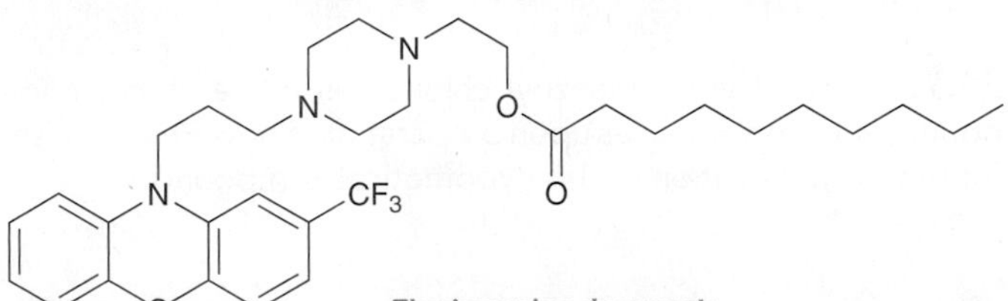

Fluphenazine decanoate

The hydrophobic tail of the ester is deliberately designed to enable a slow release of the prodrug into the bloodstream, where the prodrug is rapidly hydrolyzed to produce the active drug.
(a) Draw the structure of the active drug
(b) Draw the structure of and assign a systematic name for the carboxylic acid that is produced as a by-product of the hydrolysis step.

21.63 Benzyl acetate is a pleasant-smelling ester found in the essential oil of jasmine flowers and is used in many perfume formulations. Starting with benzene and using any other reagents of your choice, design an efficient synthesis for benzyl acetate.

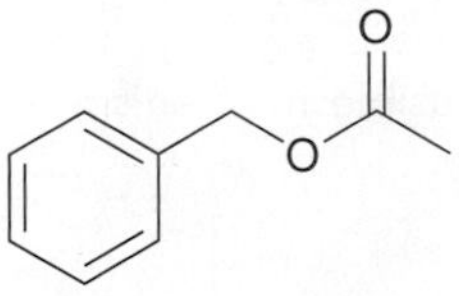

Benzyl acetate

21.64 Aspartame (below) is an artificial sweetener used in diet soft drinks and is marketed under many trade names, including Equal™ and Nutrasweet™. In the body, aspartame is hydrolyzed to produce methanol, aspartic acid, and phenylalanine. The production of phenylalanine poses a health risk to infants born with a rare condition called phenylketonuria, which prevents phenylalanine from being digested properly. Draw the structures of aspartic acid and phenylalanine.

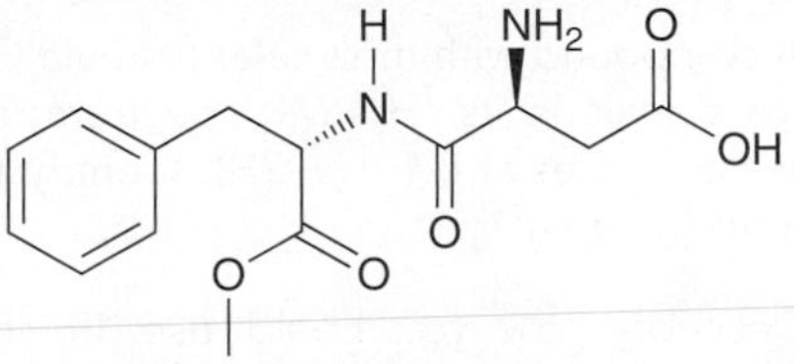

Aspartame

21.65 Draw a plausible mechanism for each of the following transformations:

(a) 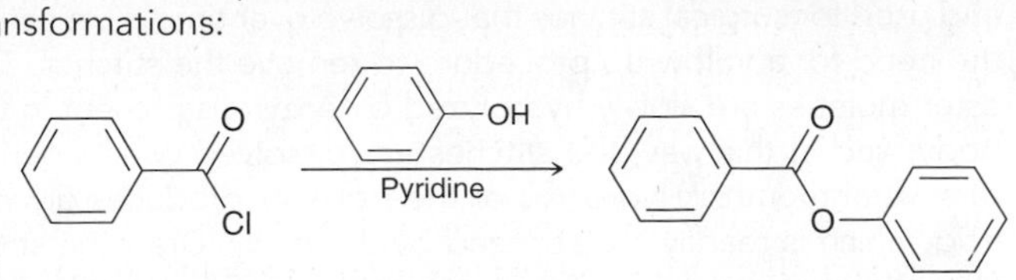

(b) 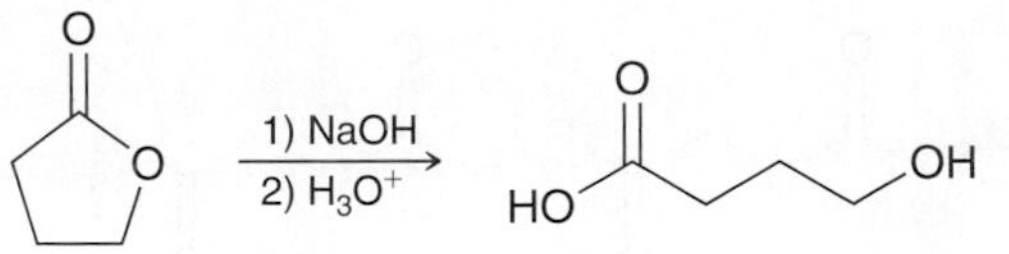

(c) 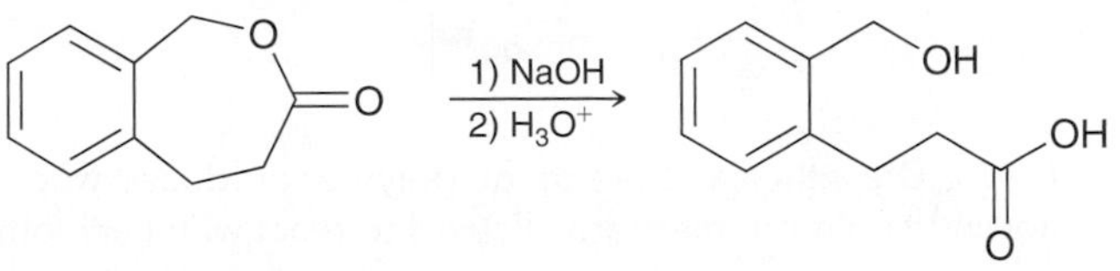

(d) 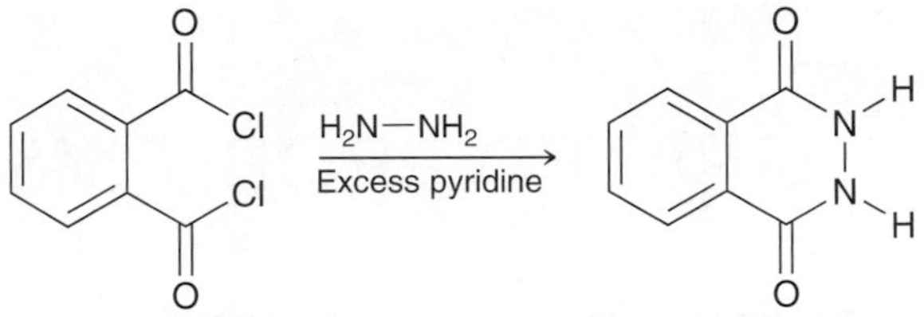

(e) 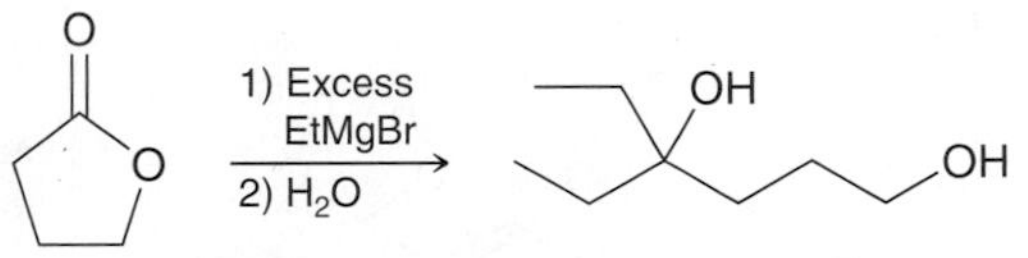

21.66 Ethyl trichloroacetate is significantly more reactive toward hydrolysis than ethyl acetate. Explain this observation.

21.67 Draw the structure of the diol that is produced when the following carbonate is heated under aqueous acidic conditions.

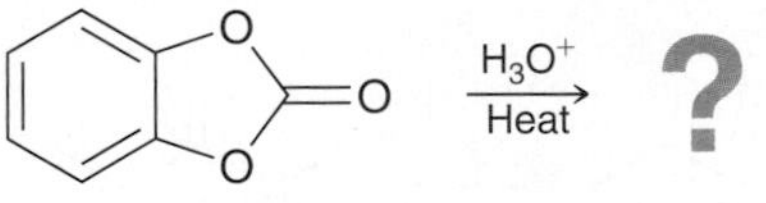

21.68 Pivampicillin is a penicillin prodrug:

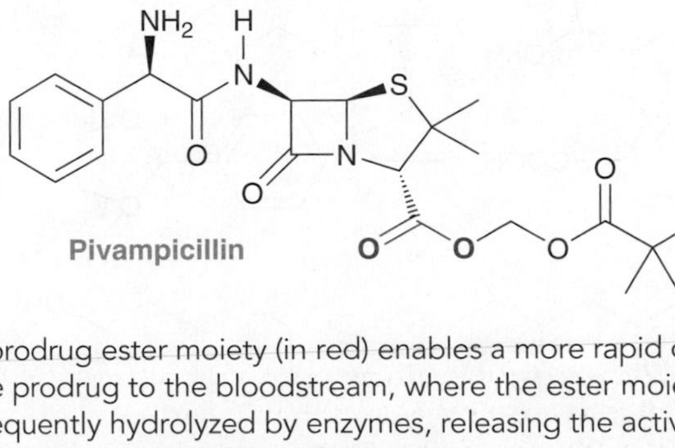

Pivampicillin

The prodrug ester moiety (in red) enables a more rapid delivery of the prodrug to the bloodstream, where the ester moiety is subsequently hydrolyzed by enzymes, releasing the active drug.
(a) Draw the structure of the active drug.
(b) What is the name of the active drug (see the Medically Speaking box at the end of Section 21.12)?

21.69 Dexon™ (below) is a polyester that is spun into fibers and used for surgical stitches that dissolve over time, eliminating the need for a follow-up procedure to remove the stitches. The ester moieties are slowly hydrolyzed by enzymes present in the body, and in this way, the stitches are dissolved over a period of several months. Hydrolysis of the polymer produces glycolic acid, which is readily metabolized by the body. Draw the structure of glycolic acid. Identify a systematic name for glycolic acid.

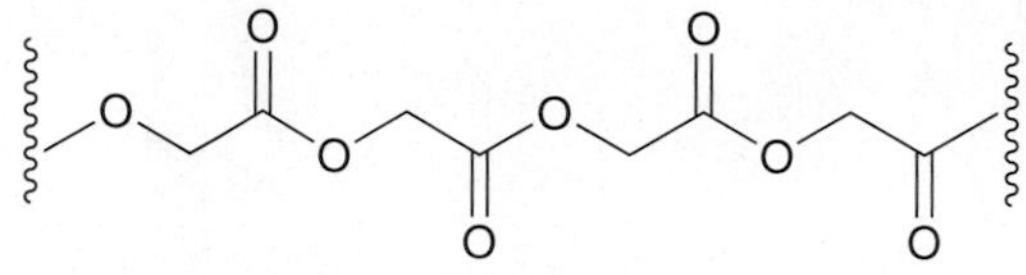

Dexon™

21.70 Draw the structure of the polymer produced when the following two monomers are allowed to react with each other:

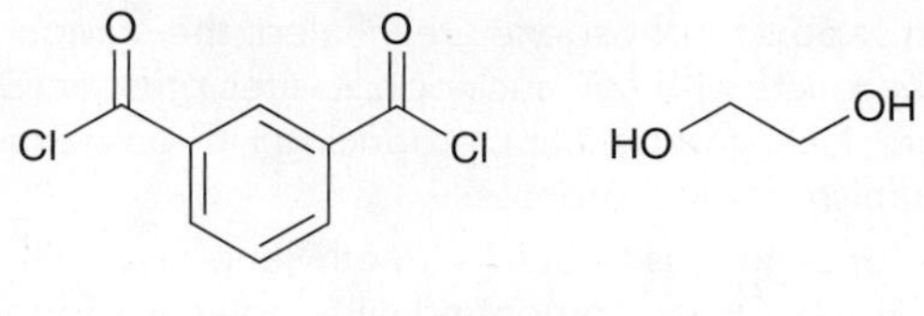

21.71 Identify what monomers you would use to produce the following polymer:

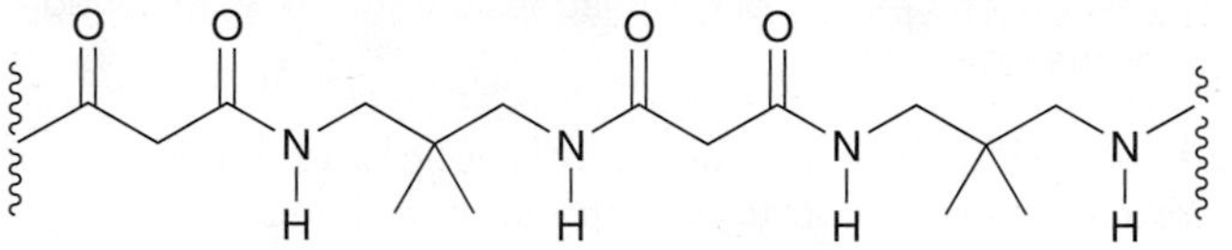

21.72 *meta*-Hydroxybenzoyl chloride is not a stable compound, and it polymerizes upon preparation. Show a mechanism for the polymerization of this hypothetical compound.

INTEGRATED PROBLEMS

21.73 Propose an efficient synthesis for each of the following transformations:

(a) 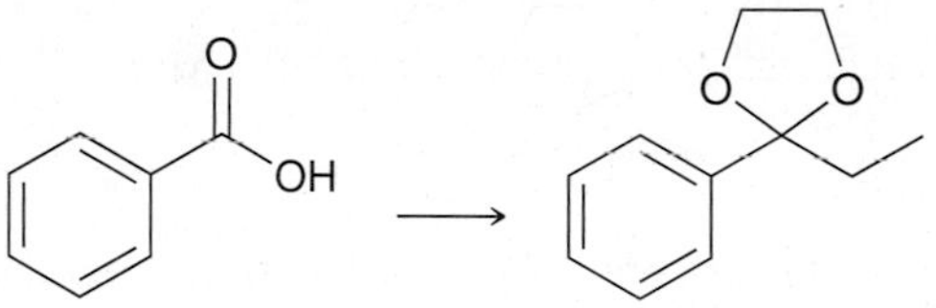

(b) 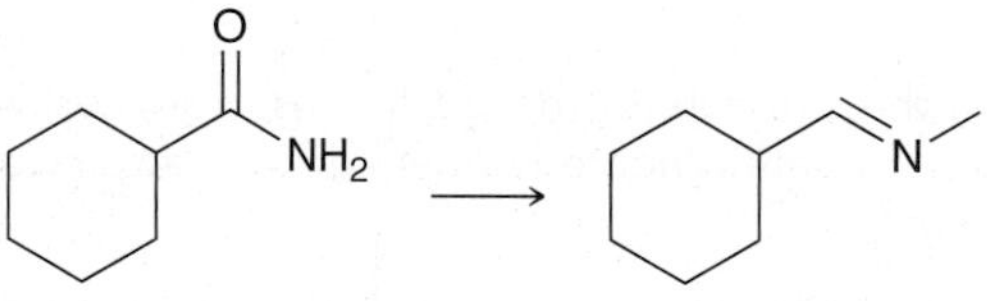

(c) 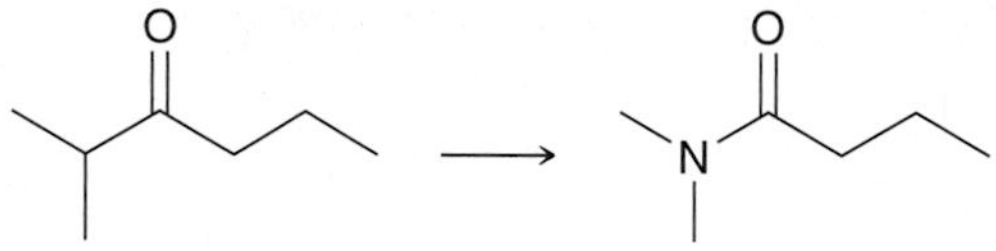

(d) 

(e) 

(f) 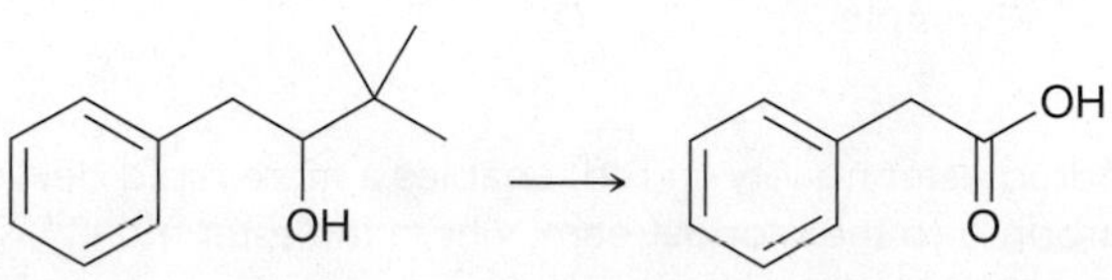

(g) 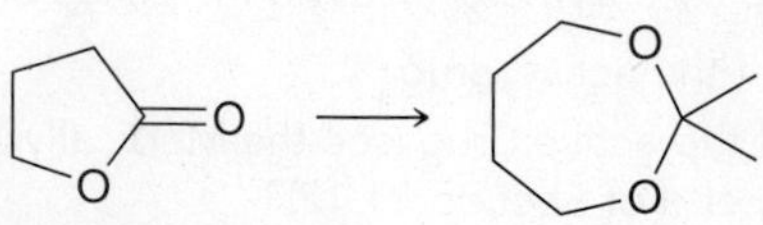

21.74 Starting with benzene and using any other reagents of your choice, devise a synthesis for acetaminophen:

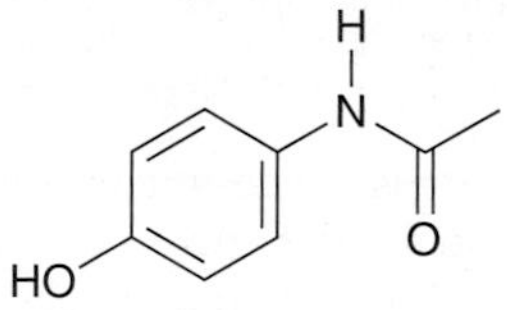

Acetaminophen (Tylenol™)

21.75 Draw a plausible mechanism for the following transformation

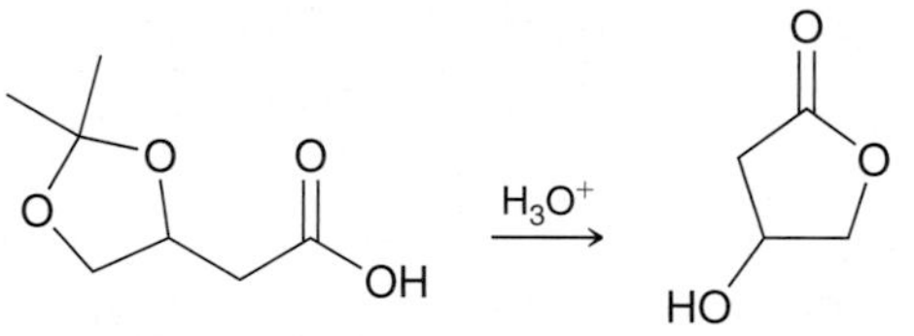

21.76 A carboxylic acid with molecular formula $C_5H_{10}O_2$ is treated with thionyl chloride to give compound **A**. Compound **A** has only one signal in its ^1H NMR spectrum. Draw the structure of the product that is formed when compound **A** is treated with excess ammonia.

21.77 A compound with molecular formula $C_{10}H_{10}O_4$ exhibits only two signals in its ^1H NMR spectrum: a singlet at 4.0 (I = 3H) and a singlet at 8.1 (I = 2H). Identify the structure of this compound.

21.78 Describe how you could use IR spectroscopy to distinguish between ethyl acetate and butyric acid.

21.79 Describe how you could use NMR spectroscopy to distinguish between benzoyl chloride and *para*-chlorobenzaldehyde.

21.80 A compound with molecular formula $C_8H_8O_3$ exhibits the following IR, ^1H NMR and ^{13}C NMR spectra. Deduce the structure of this compound.

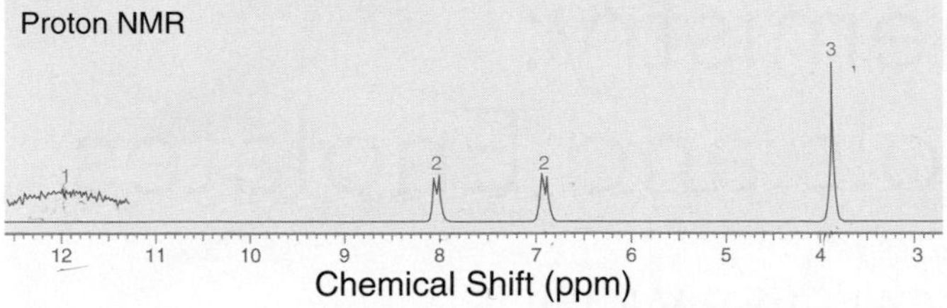

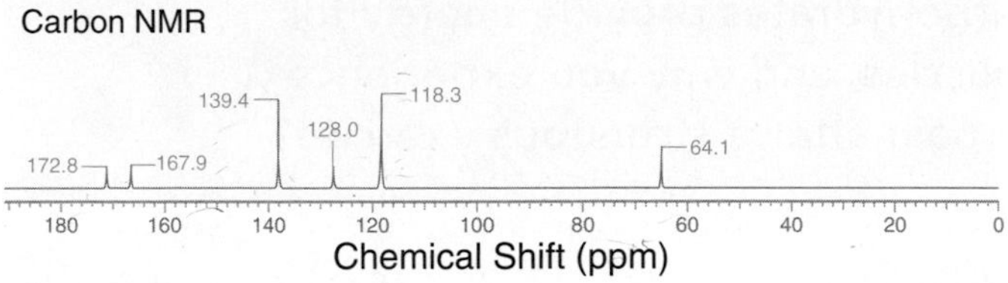

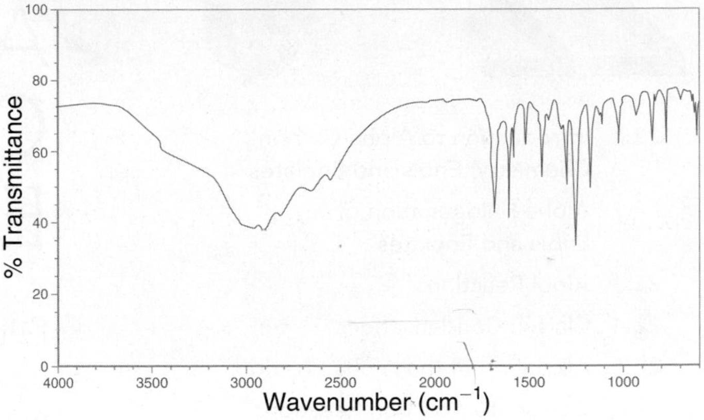

CHALLENGE PROBLEMS

21.81 Propose a mechanism for the following transformation, and explain how you could use an isotopic labeling experiment to verify your proposed mechanism:

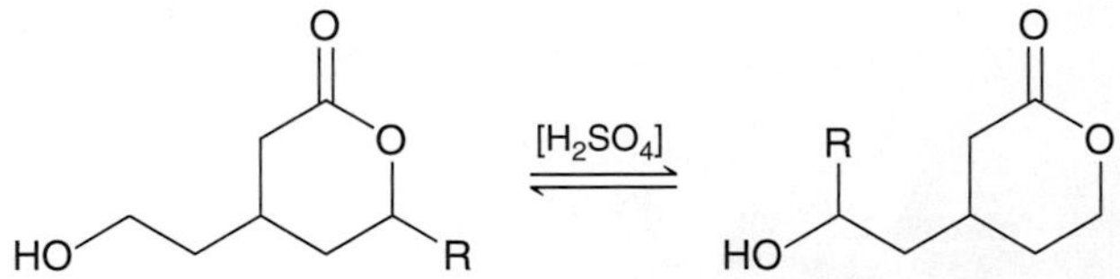

21.82 *N*-Acetylazoles undergo hydrolysis more readily than regular amides. Suggest a reason for the enhanced reactivity of *N*-acetylazoles toward nucleophilic acyl substitution:

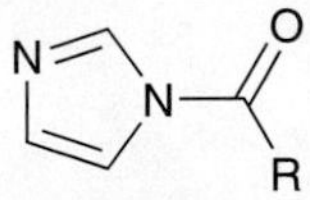

***N*-Acetylazole**

21.83 Dimethylformamide (DMF) is a common solvent:

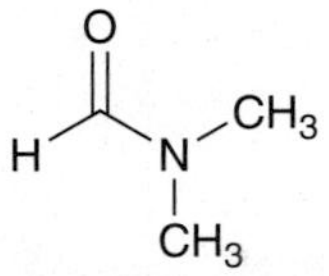

Dimethylformamide

(a) The ^1H NMR spectrum of DMF exhibits three signals. Upon treatment with excess LAH followed by water, DMF is converted into a new compound that exhibits only one signal in its ^1H NMR spectrum. Explain.

(b) Based on your answer to part a, how many signals do you expect in the ^{13}C NMR spectrum of DMF?

22

Alpha Carbon Chemistry: Enols and Enolates

DID YOU EVER WONDER...

how carbohydrates provide energy for your muscles, and why you experience muscle pain after a strenuous exercise?

In this chapter, we will explore a variety of C—C bond-forming reactions that are among the most versatile reactions available to synthetic organic chemists. Many of these reactions also occur in biochemical processes, including the metabolic pathways that produce energy for muscle contractions. The reactions in this chapter will greatly expand your ability to design syntheses for a wide variety of compounds.

DO YOU REMEMBER?

Before you go on, be sure you understand the following topics.
If necessary, review the suggested sections to prepare for this chapter:

- Brønsted-Lowry Acidity (section 3.4)
- Energy Diagrams: Thermodynamics vs. Kinetics (section 6.6)
- Nucleophilic Acyl Substitution (section 21.7)
- Position of Equilibrium and Choice of Reagents (section 3.5)
- Retrosynthetic Analysis (section 12.5)

WILEY PLUS Visit www.wileyplus.com to check your understanding and for valuable practice.

22.1 Introduction to Alpha Carbon Chemistry: Enols and Enolates

The Alpha Carbon

For compounds containing a carbonyl group, Greek letters are used to describe the proximity of each carbon atom to the carbonyl group.

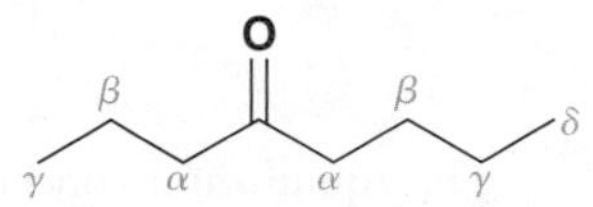

The carbonyl group itself does not receive a Greek letter. In this example there are two carbon atoms designated as alpha (α) positions. Hydrogen atoms are designated with the Greek letter of the carbon to which they are attached; for example, the hydrogen atoms (protons) connected to the α carbon atoms are called α protons. This chapter will explore reactions occurring at the α position.

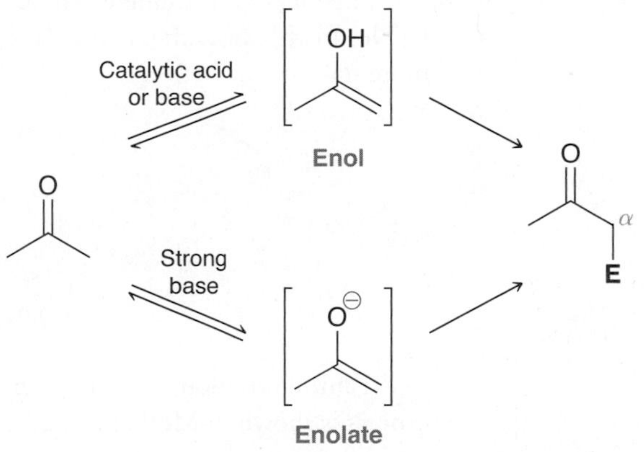

These reactions can occur via either an enol or an enolate intermediate. The vast majority of the reactions in this chapter will proceed via an enolate intermediate, but we will also explore some reactions that proceed via an enol intermediate.

Enols

In the presence of catalytic acid or base, a ketone will exist in equilibrium with an enol.

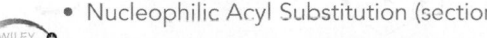

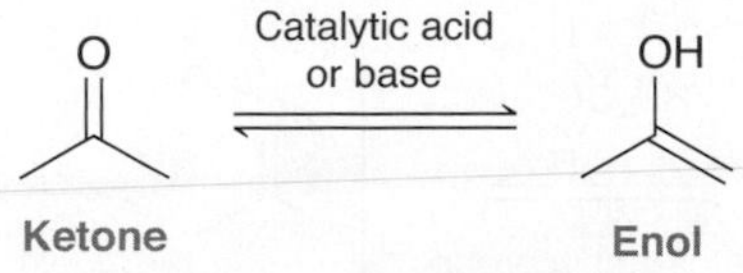

Recall that the ketone and enol shown are tautomers—rapidly interconverting constitutional isomers that differ from each other in the placement of a proton and the position of a double bond. Do not confuse tautomers with resonance structures. The two structures above are not resonance structures because they differ in the arrangement of their atoms. These structures represent two different compounds, both of which are present at equilibrium. In general, the position of equilibrium will significantly favor the ketone, as seen in the following example.

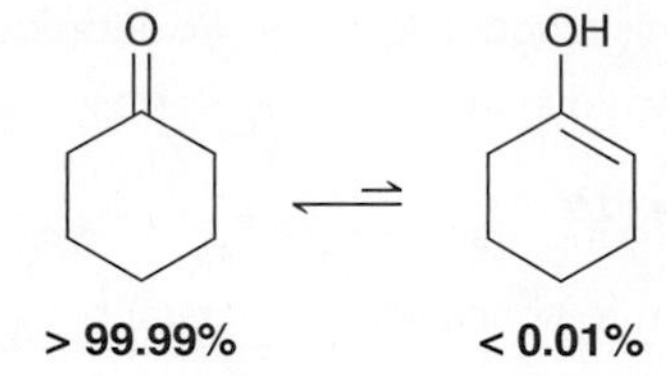

Cyclohexanone exists in equilibrium with its tautomeric enol form, which exhibits a very minor presence. This is the case for most ketones.

In some cases, the enol tautomer is stabilized and exhibits a more substantial presence at equilibrium. Consider, for example, the enol form of a beta-diketone, such as 2,4-pentanedione.

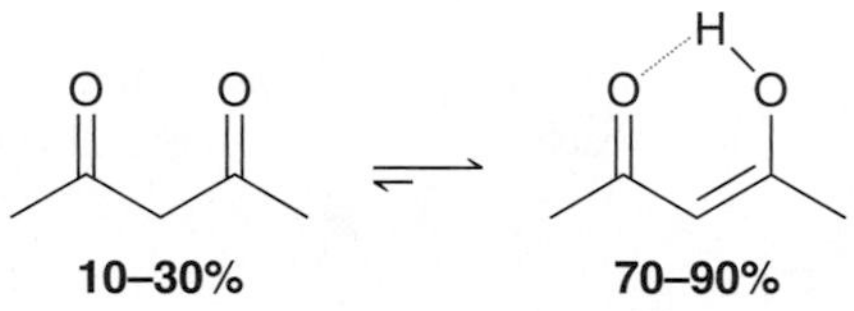

The equilibrium concentrations of the diketone and the enol depend on the solvent that is used, but the enol generally dominates. Two factors contribute to the remarkable stability of the enol in this case: (1) The enol has a conjugated π system, which is a stabilizing factor (see Section 17.2), and (2) the enol can form an intramolecular H-bonding interactions between the hydroxyl proton and the nearby carbonyl group (shown with a gray dotted line above). Both of these factors serve to stabilize the enol.

Phenol is an extreme example in that the concentration of the ketone is practically negligible. In this case, the ketone lacks aromaticity, while the enol is aromatic and significantly more stable.

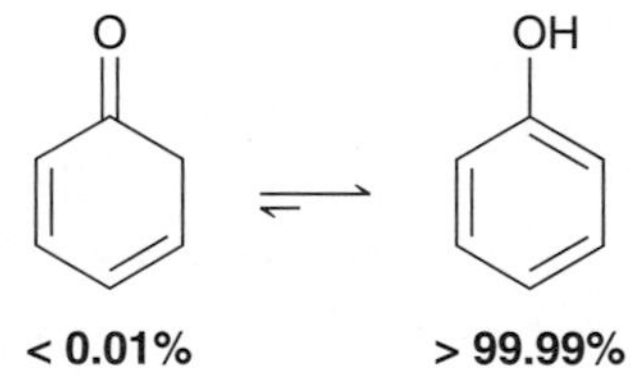

Tautomerization is catalyzed by trace amounts of either acid or base. The acid-catalyzed process is shown in Mechanism 22.1.

MECHANISM 22.1 ACID-CATALYZED TAUTOMERIZATION

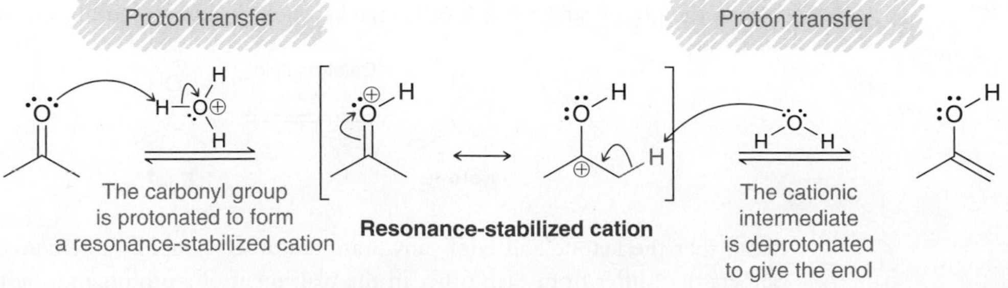

LOOKING BACK
This requirement was discussed in Section 21.7.

In the first step, the carbonyl group is protonated to form a resonance-stabilized cation, which is then deprotonated at the α position to give the enol. Notice that all reagents and intermediates are either neutral or positively charged. That is, the mechanism does not exhibit any negative charges, which is consistent with acidic conditions.

The base-catalyzed process for tautomerization is shown in Mechanism 22.2.

MECHANISM 22.2 BASE-CATALYZED TAUTOMERIZATION

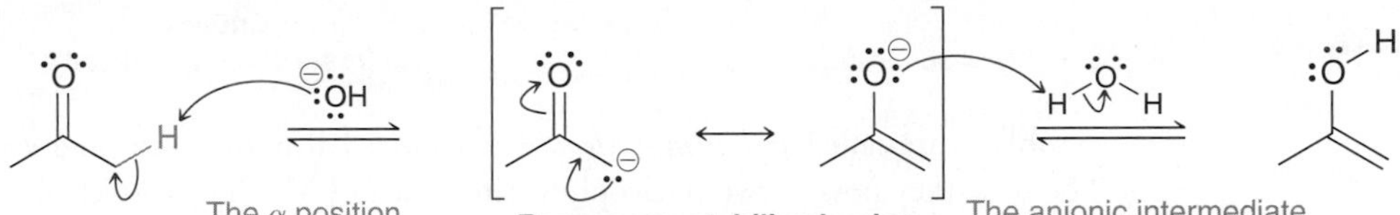
The α position is deprotonated to form a resonance-stabilized anion
Resonance-stabilized anion
The anionic intermediate is protonated to give the enol

In the first step, the α position is deprotonated to form a resonance–stabilized anion, which is then protonated to give the enol. Notice that all reagents and intermediates are either neutral or negatively charged. That is, the mechanism does not exhibit any positive charges, in order to be consistent with basic conditions.

The mechanisms for acid-catalyzed and base-catalyzed tautomerization involve the same two steps (protonation at the carbonyl group and deprotonation of the α position). The difference between these mechanisms is the order of events. In acid conditions, the first step is protonation of the carbonyl group, giving a positively charged intermediate. In basic conditions, the first step is deprotonation of the α position, giving a negatively charged intermediate.

It is difficult to prevent tautomerization even if care is taken to remove all acids and bases from the solution. Tautomerization can still be catalyzed by the trace amounts of acid or base that are adsorbed to the surface of the glassware (even after washing the glassware scrupulously). Unless extremely rare conditions are employed, you should always assume that tautomerization will occur if possible, and an equilibrium will quickly be established favoring the more stable tautomer. The enol tautomer is generally present only in small amounts, but it is very reactive. Specifically, the α position is very nucleophilic due to resonance.

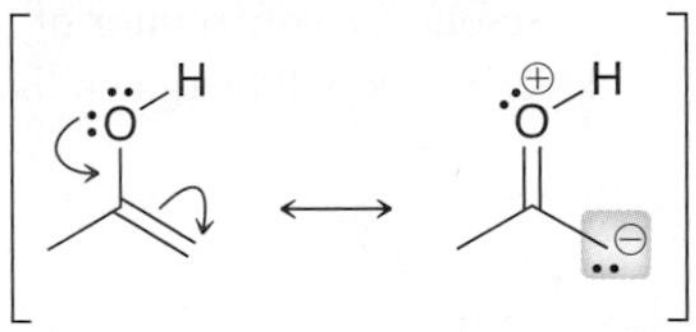

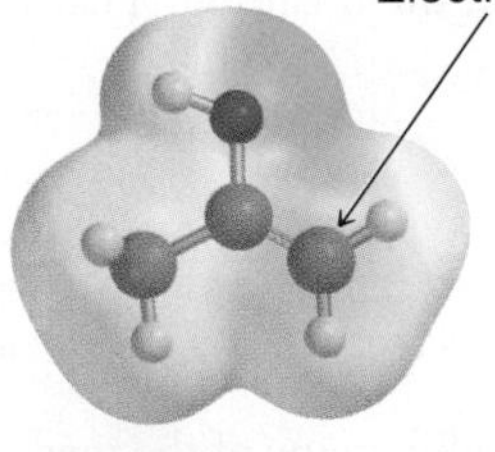
Electron rich

FIGURE 22.1
An electrostatic potential map of the enol of acetone, showing the nucleophilic character of the α position.

In the second resonance structure, the α position exhibits a lone pair, rendering that position nucleophilic. The effect of an OH group in activating the α position is similar to its effect in activating an aromatic ring (as seen in Section 19.10). The electron-donating effect of the OH group can be visualized with an electrostatic potential map of a simple enol (Figure 22.1). Notice that the α position is somewhat red, representing an electron-rich site. In Section 22.2, we will see two reactions in which the α position of an enol functions as a nucleophile.

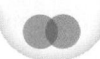

CONCEPTUAL CHECKPOINT

22.1 Draw a mechanism for the acid-catalyzed conversion of cyclohexanone into its tautomeric enol.

22.2 Draw a mechanism for the reverse process of the previous problem. In other words, draw the acid-catalyzed conversion of 1-cyclohexenol to cyclohexanone.

22.3 Draw the two possible enols that can be formed from 3-methyl-2-butanone, and show a mechanism of formation of each under base-catalyzed conditions.

Enolates

When treated with a strong base, the α position of a ketone is deprotonated to give a resonance-stabilized intermediate called an **enolate**.

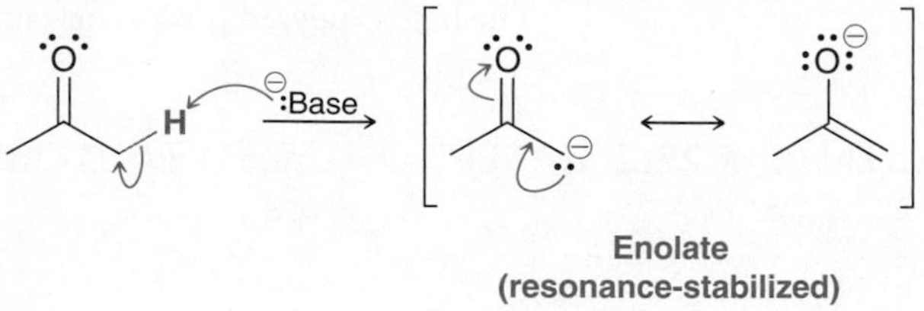

Enolates are called *ambident nucleophiles* (*ambi* is Latin for "both" and *dent* is Latin for "teeth"), because they possess two nucleophilic sites, each of which can attack an electrophile. When the oxygen atom attacks an electrophile, it is called O-attack; and when the α carbon attacks an electrophile, it is called C-attack.

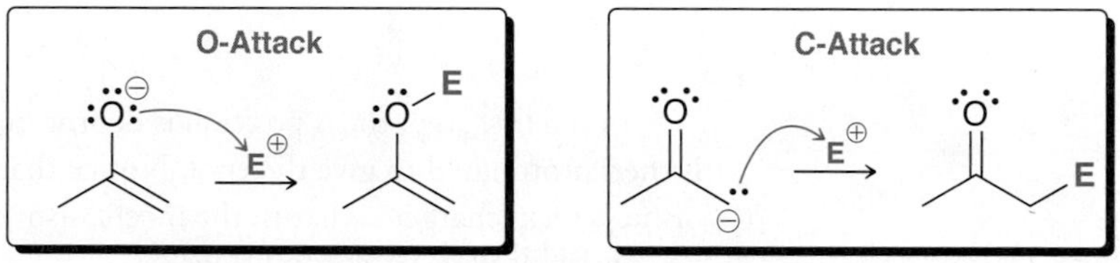

Although the oxygen atom of an enolate bears the majority of the negative charge, C-attack is nevertheless more common than O-attack. All of the reactions presented in this chapter will be examples of C-attack. When drawing the mechanism of an enolate undergoing C-attack, it is technically more appropriate to draw the resonance structure of the enolate in which the negative charge appears on the oxygen atom, because that drawing represents the more significant resonance contributor. Therefore, C-attack should be drawn like this:

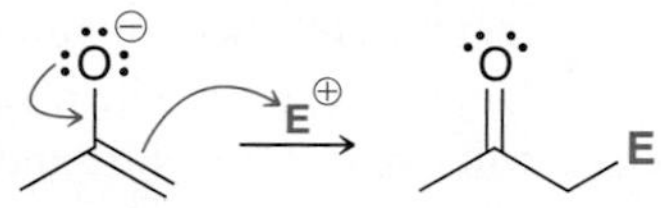

Nevertheless, for simplicity, when drawing mechanisms, we will often draw the less significant resonance contributor of the enolate, in which the negative charge is on the α carbon. That is, C-attack will be drawn like this:

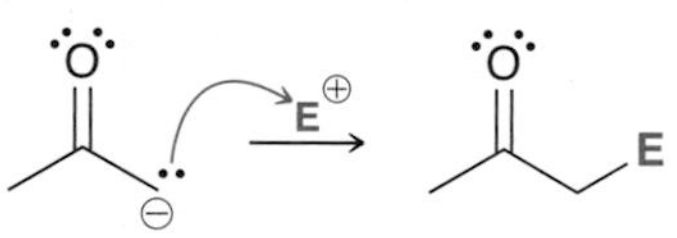

When drawn this way, fewer curved arrows are required, which will simplify many of the mechanisms in this chapter.

Enolates are more useful than enols because (1) enolates possess a full negative charge and are therefore more reactive than enols and (2) enolates can be isolated and stored for short periods of time, unlike enols, which cannot be isolated or stored. For these two reasons, the vast majority of reactions in this chapter proceed via enolate intermediates.

As we progress through the chapter, it is important to keep in mind that only the α protons of an aldehyde or ketone are acidic.

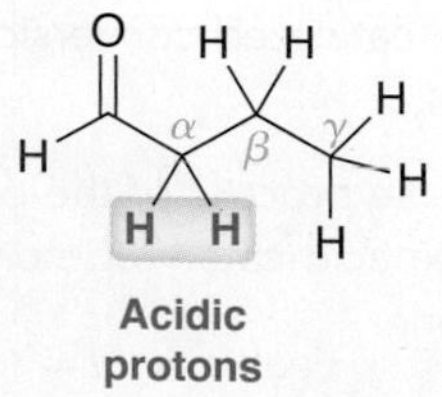

LOOKING BACK
For a review of the factors that affect the acidity of a proton, see Section 3.4.

In the example shown, the beta (β) and gamma (γ) protons are not acidic, and neither is the aldehydic proton. Deprotonation at any of those positions does not lead to a resonance-stabilized anion.

SKILLBUILDER

22.1 DRAWING ENOLATES

LEARN the skill

When the following ketone is treated with a strong base, an enolate ion is formed. Draw both resonance structures of the enolate.

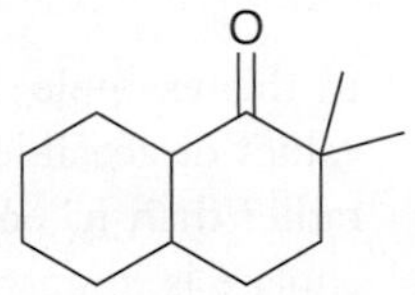

SOLUTION

STEP 1
Identify all α positions.

Begin by identifying all α positions. In this case, there are two α positions, but only one of them bears a proton. The α position on the right side of the carbonyl group has no protons, and therefore, an enolate cannot form on that side of the carbonyl group. The other α carbon (on the left side) does have a proton, and an enolate can be formed at that location.

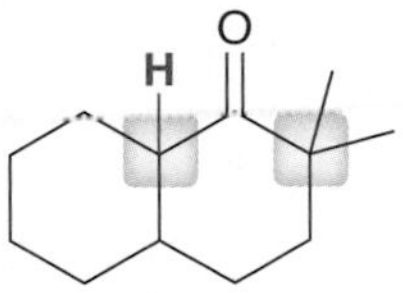

Using two curved arrows, remove the proton, and then draw the enolate with a lone pair and a negative charge at the α position (in place of the proton).

STEP 2
Remove the proton at the α position and draw the resulting anion.

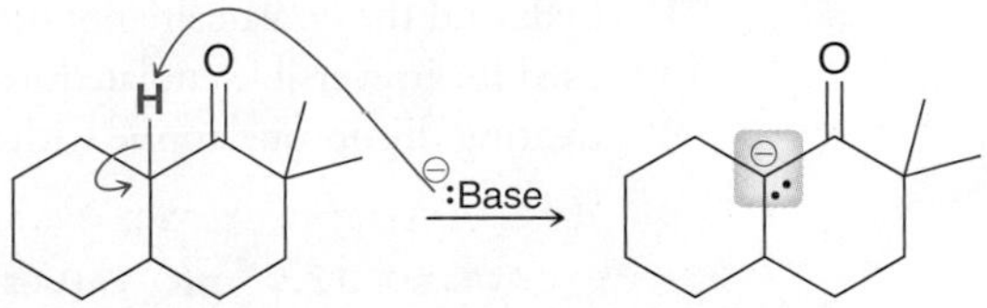

Finally, draw the other resonance structure using the skills developed in Section 2.10.

STEP 3
Draw the resonance structure, showing the charge on the oxygen atom.

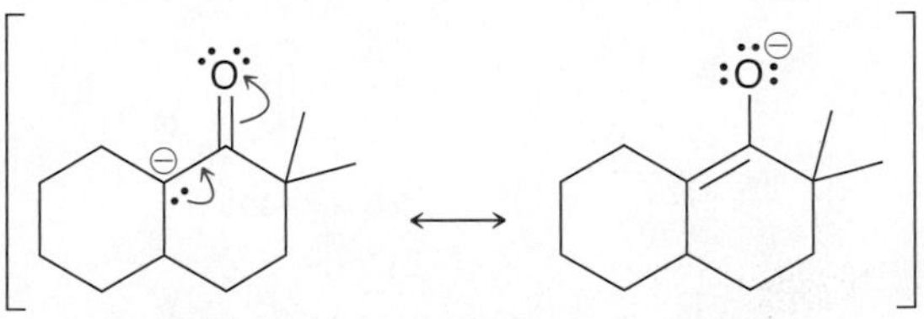

PRACTICE the skill

22.4 Draw both resonance structures of the enolate formed when each of the following ketones is treated with a strong base:

(a) 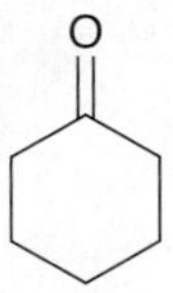
(b) 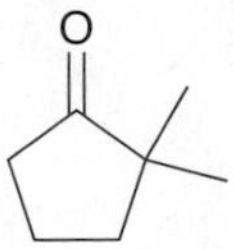
(c) 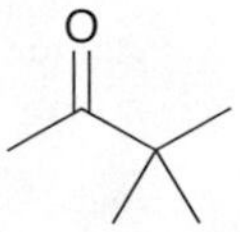
(d) 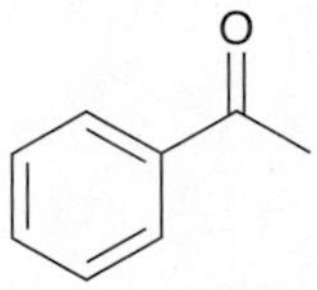
(e) 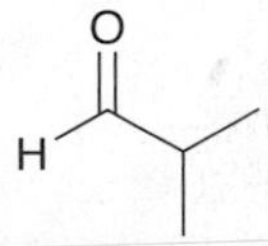

APPLY the skill

22.5 When 2-methylcyclohexanone is treated with a strong base, two different enolates are formed. Draw both of them.

need more **PRACTICE?** **Try Problem 22.59**

Choosing a Base for Enolate Formation

In this chapter, we will see many reactions involving enolates, and the choice of base will always be important. Some cases require a relatively mild base to form the enolate, while other cases demand a very strong base. In order to choose an appropriate base, we must take a careful look at pK_a values. Aldehydes and ketones typically have pK_a values in the range of 16–20, as seen in Table 22.1. The range of pK_a values is similar to the range exhibited by alcohols (ethanol has a pK_a of 16 and *tert*-butanol has a pK_a of 18). As a result, when an alkoxide ion is used as the base, an equilibrium is established in which the alkoxide ion and the enolate ion are both present. The relative amount of alkoxide and enolate ions is determined by the relative pK_a values, although there will usually be less of the enolate present at equilibrium. Consider the following example.

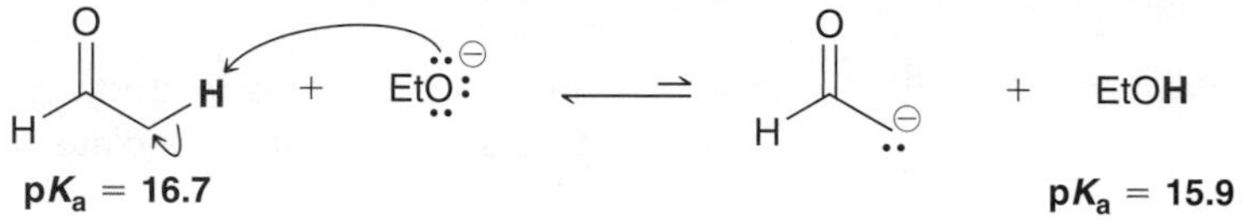

In this example, ethoxide is used as a base to deprotonate acetaldehyde. Notice that the pK_a values of acetaldehyde and ethanol are similar. The equilibrium slightly favors acetaldehyde, rather than its enolate, although both are present. The presence of both the aldehyde and its enolate is important because the enolate is a nucleophile and the aldehyde is an electrophile, and these two species will react with each other when they are both present, as we will see in Section 22.3.

In contrast, many other bases, such as sodium hydride, can irreversibly and completely convert the aldehyde into an enolate.

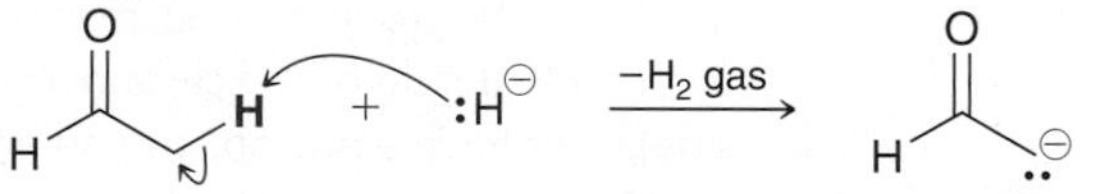

When sodium hydride is used as the base, hydrogen gas is formed. It bubbles out of solution, as all of the aldehyde molecules are converted into enolate ions. Under these conditions, the aldehyde and the enolate are not both present. Only the enolate is present. Another base commonly used for irreversible enolate formation is lithium diisopropylamide (LDA), which is prepared by treating diisopropylamine with butyllithium.

TABLE 22.1 pK_a VALUES OF SOME COMMON KETONES AND ALDEHYDES

Compound	pK_a	Enolate
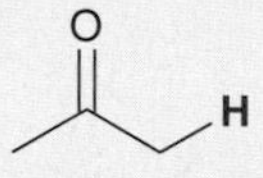 Acetone	19.2	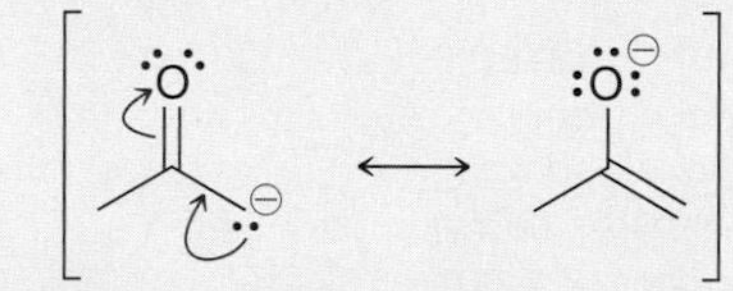
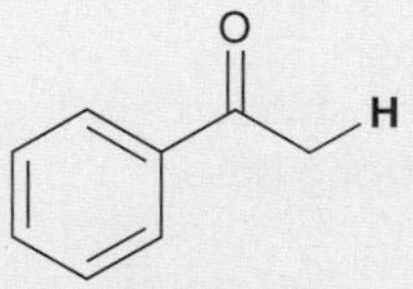 Acetophenone	18.3	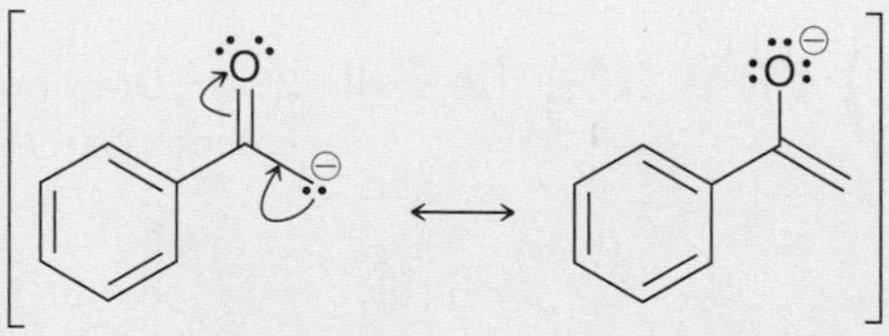
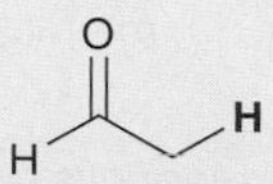 Acetaldehyde	16.7	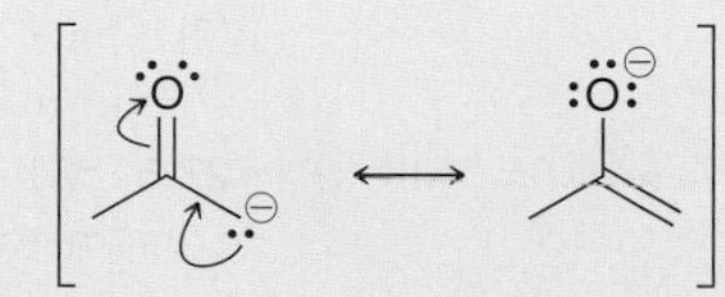

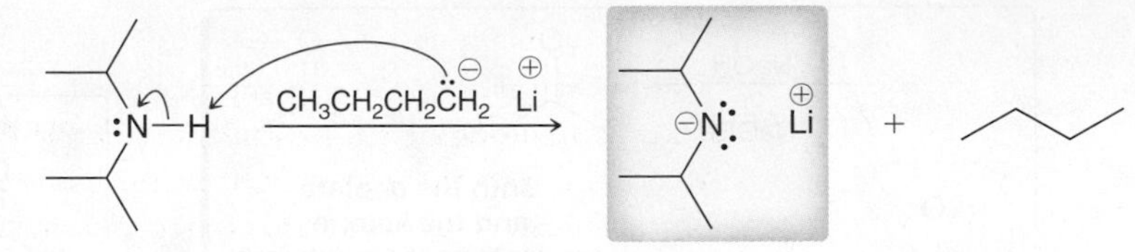

Diisopropylamine **Lithium diisopropylamide**

The pK_a of diisopropylamine is approximately 36, and therefore, LDA can be used to accomplish irreversible enolate formation.

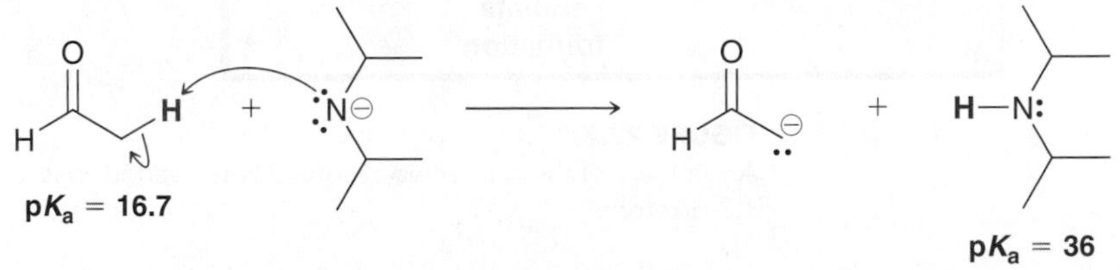

When LDA is used as a base, the amount of aldehyde present at equilibrium is negligible.

Now consider a compound with two carbonyl groups that are beta to each other:

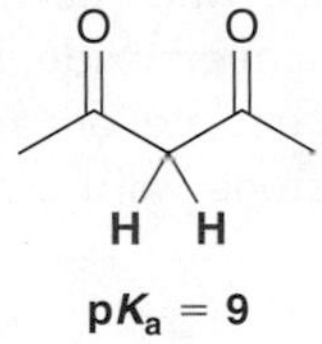

In this case, the central CH_2 group is flanked by two carbonyl groups. The protons of the CH_2 group (shown in red) are therefore highly acidic. As opposed to most ketones (with a pK_a somewhere in the range of 16–20), the pK_a of this compound is approximately 9. The acidity of these protons can be attributed to the highly stabilized anion formed upon deprotonation.

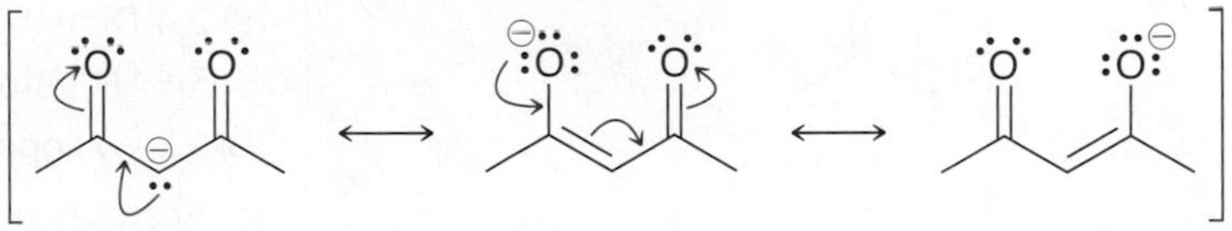

The anion is a doubly stabilized enolate ion with the negative charge being spread over two oxygen atoms and one carbon atom. Because of the relatively high acidity of beta diketones, LDA is not required to irreversibly deprotonate these compounds. Rather, treatment with hydroxide or an alkoxide ion is sufficient to ensure nearly complete enolate formation.

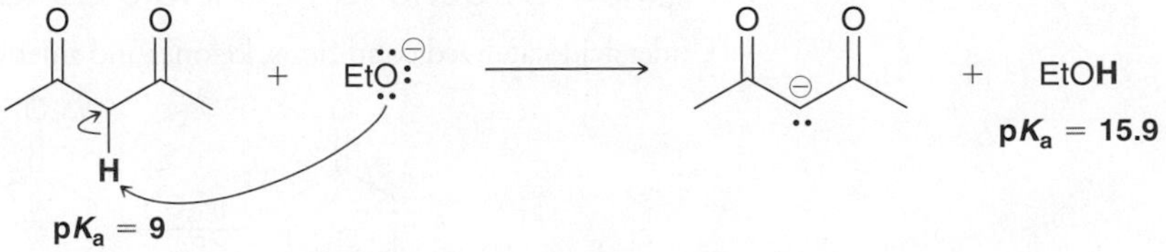

LOOKING BACK
To review how pK_a values can be used to determine equilibrium concentrations for an acid-base reaction, see Section 3.3.

The difference in pK_a values is about 7 (which represents seven orders of magnitude). In other words, when ethoxide is used as the base, 99.99999% of the diketone molecules are deprotonated to form enolates.

Figure 22.2 summarizes the relevant factors for choosing a base to form an enolate ion. The information summarized in this figure will be used several times in the upcoming sections of this chapter.

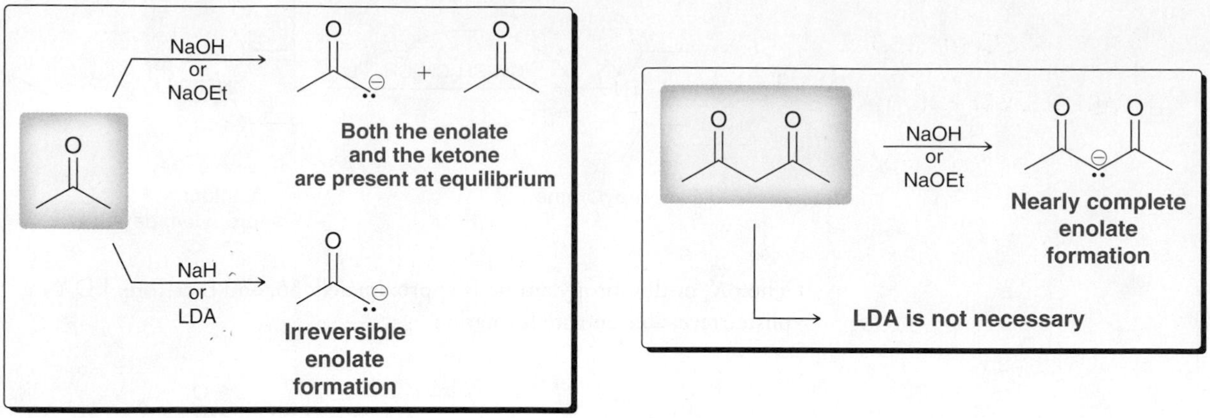

FIGURE 22.2
A summary of the outcomes observed when various bases are used to deprotonate ketones or 1,3-diketones.

CONCEPTUAL CHECKPOINT

22.6 Draw the enolate ion that is formed when each of the following compounds is treated with sodium ethoxide. In each case, draw all resonance structures of the enolate ion, and predict whether a substantial amount of ketone will be present together with the enolate at equilibrium.

(a) 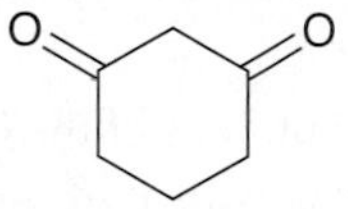

(b) 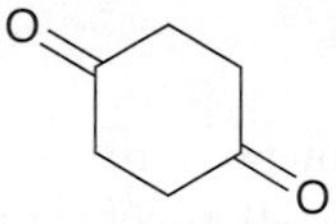

(c) 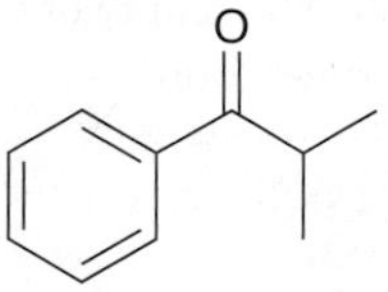

(d) 

22.7 When the following compound is treated with sodium ethoxide, nearly all of it is converted into an enolate. Draw the resonance structures of the enolate that is formed, and explain why enolate formation is nearly complete despite the use of ethoxide rather than LDA.

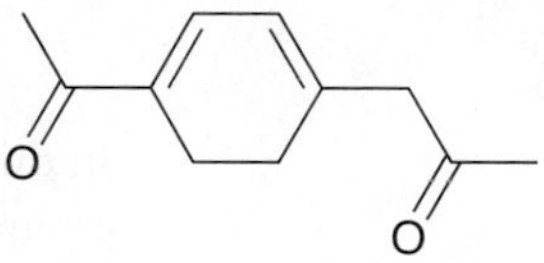

22.8 For each pair of compounds, identify which compound is more acidic and explain your choice.

(a) 2,4-Dimethyl-3,5-heptanedione or 4,4-Dimethyl-3,5-heptanedione

(b) 1,2-Cyclopentanedione or 1,3-cyclopentanedione

(c) Acetophenone or benzaldehyde

22.2 Alpha Halogenation of Enols and Enolates

Alpha Halogenation in Acidic Conditions

Under acid-catalyzed conditions, ketones and aldehydes will undergo halogenation at the α position.

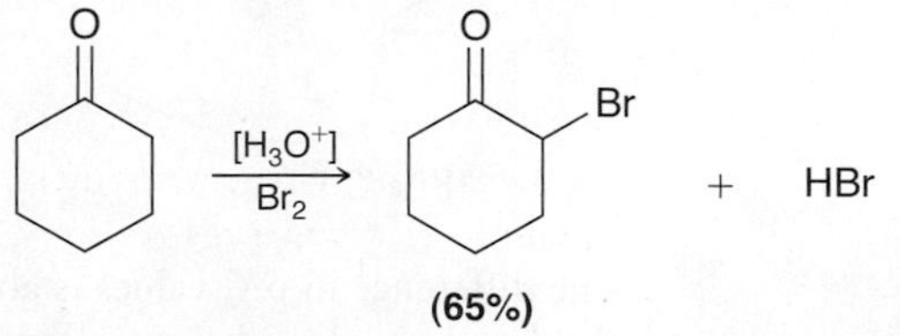

The reaction is observed for chlorine, bromine, and iodine, but not for fluorine. A variety of solvents can be used, including acetic acid, water, chloroform, and diethyl ether. The rate of halogenation is found to be independent of the concentration or identity of the halogen, indicating that the halogen does not participate in the rate-determining step. This information is consistent with Mechanism 22.3.

MECHANISM 22.3 ACID-CATALYZED HALOGENATION OF KETONES

PART 1: ENOL FORMATION

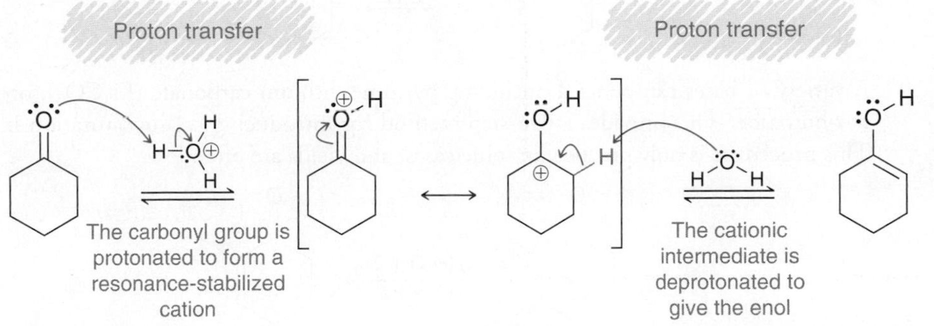
Proton transfer
Proton transfer
The carbonyl group is protonated to form a resonance-stabilized cation
The cationic intermediate is deprotonated to give the enol

PART 2: HALOGENATION

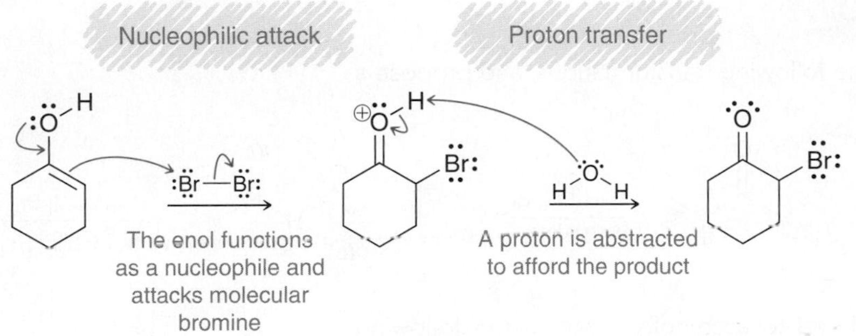
Nucleophilic attack
Proton transfer
The enol functions as a nucleophile and attacks molecular bromine
A proton is abstracted to afford the product

The mechanism has two parts. First the ketone undergoes tautomerization to produce an enol. Then, in the second part of the mechanism, the enol serves as a nucleophile and installs a halogen atom at the α position. The second part of the mechanism (halogenation) occurs more rapidly than the first part (enol formation), and therefore, enol formation represents the rate-determining step for the process. The halogen is not involved in enol formation, and therefore, the concentration of the halogen has no measurable impact on the rate of the overall process.

The by-product of α bromination is HBr, which is an acid and is capable of catalyzing the first part of the mechanism (enol formation). As a result, the reaction is said to be **autocatalytic**, that is, the reagent necessary to catalyze the reaction is produced by the reaction itself.

When an unsymmetrical ketone is used, bromination occurs primarily at the more substituted side of the ketone.

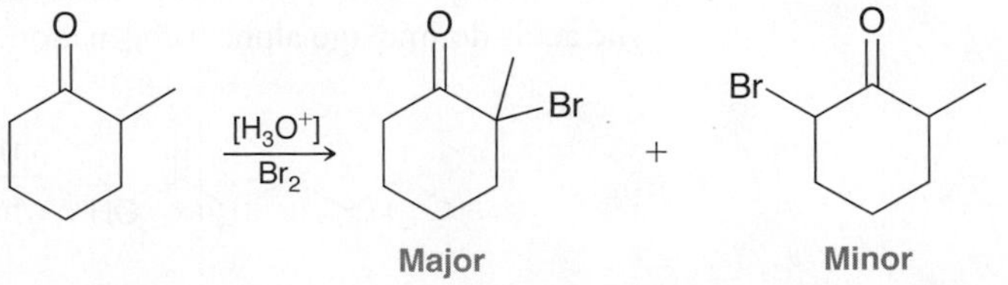
[H_3O^+]
Br_2
Major
Minor

In this example, a mixture of products is generally unavoidable. However, bromination occurs primarily at the side bearing the alkyl group, which can be attributed to the fact that the reaction proceeds more rapidly through the more substituted enol.

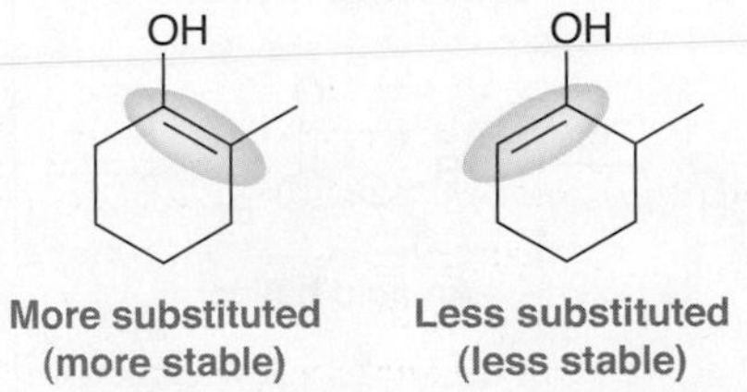
More substituted
(more stable)
Less substituted
(less stable)

The halogenated product can undergo elimination when treated with a base.

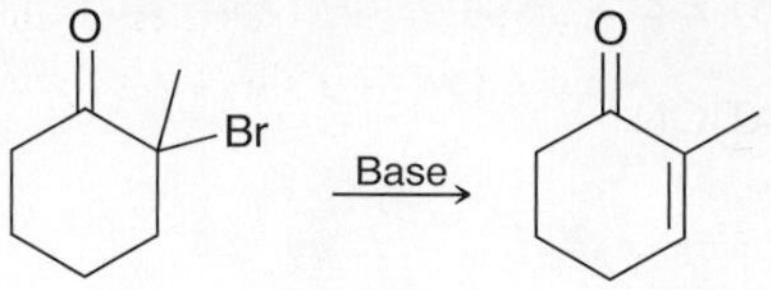

A variety of bases can be used, including pyridine, lithium carbonate (Li_2CO_3), or potassium *tert*-butoxide. This provides a two-step method for introducing α,β-unsaturation in a ketone. This procedure is only practical in some cases, and yields are often low.

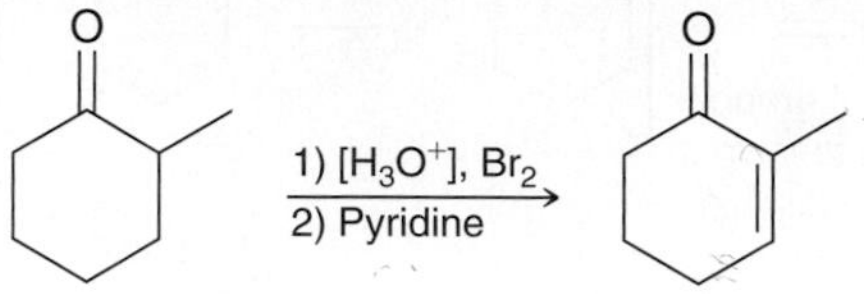

CONCEPTUAL CHECKPOINT

22.9 Predict the major product for each of the following transformations, and propose a mechanism for its formation:

(a)

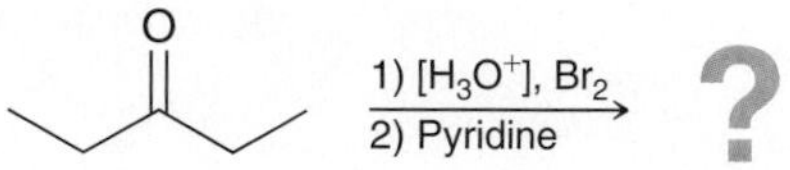

(b)

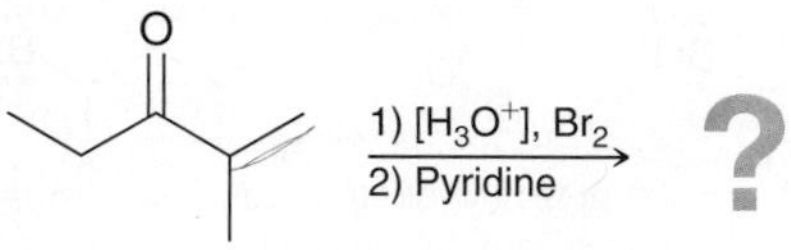

(c)

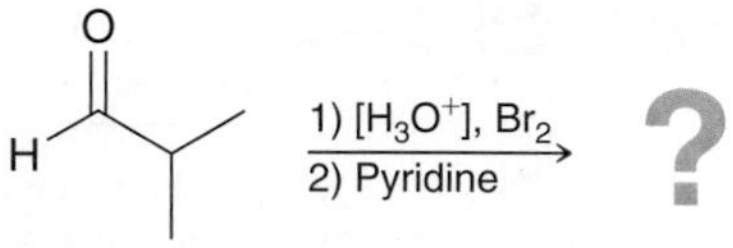

22.10 Identify the reagents that you would use to accomplish each of the following transformations:

(a)

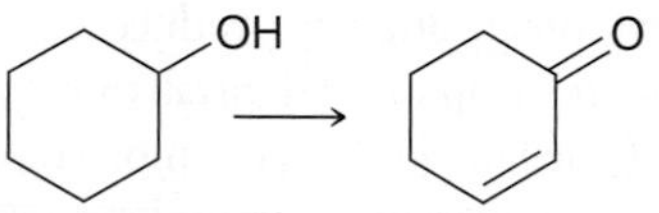

(b)

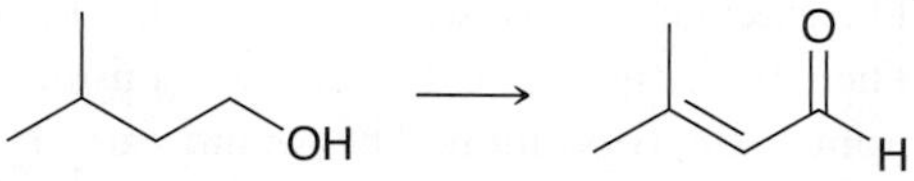

Alpha Bromination of Carboxylic Acids: The Hell-Volhard-Zelinski Reaction

Alpha halogenation, as described in the previous section, occurs readily with ketones and aldehydes, but not with carboxylic acids, esters, or amides. This is likely due to the fact that these functional groups are not readily converted to their corresponding enols. Nevertheless, carboxylic acids do undergo alpha halogenation when treated with bromine in the presence of PBr_3.

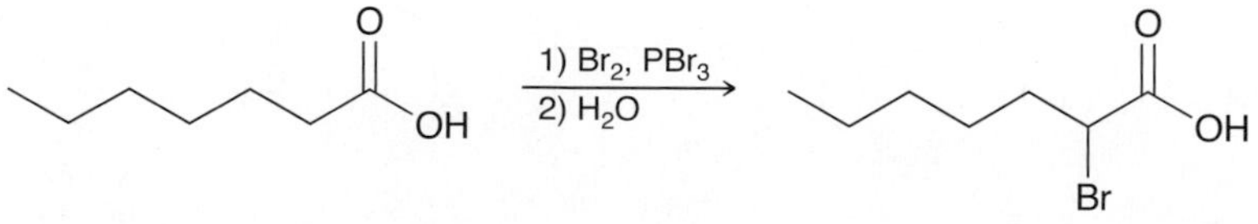

(90%)

This process, called the **Hell-Volhard-Zelinski reaction**, is believed to occur via the following sequence of events:

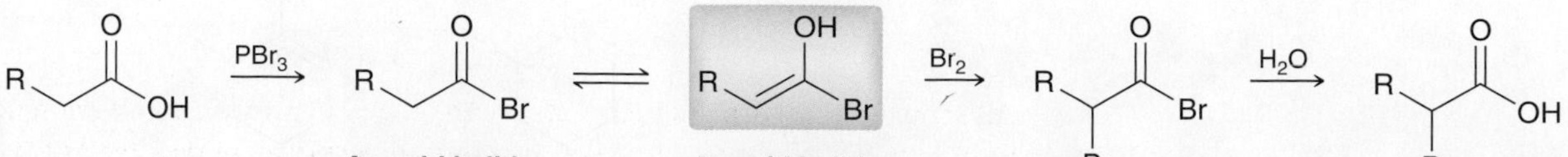

An acid halide **An acid halide enol**

The carboxylic acid first reacts with PBr_3 to form an acid halide, which exists in equilibrium with an enol. This enol then functions as a nucleophile and undergoes halogenation at the α position. Finally, hydrolysis regenerates the carboxylic acid.

CONCEPTUAL CHECKPOINT

22.11 Predict the major product for each of the following transformations:

(a) 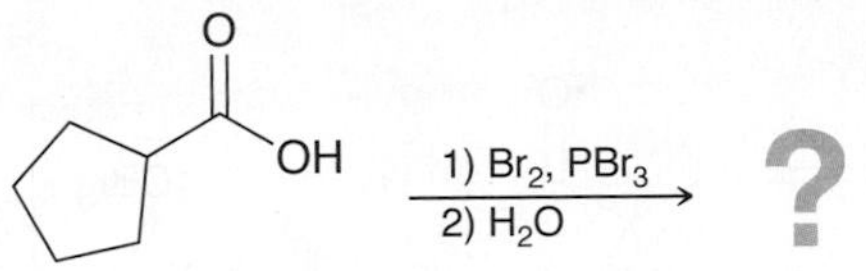

(b) 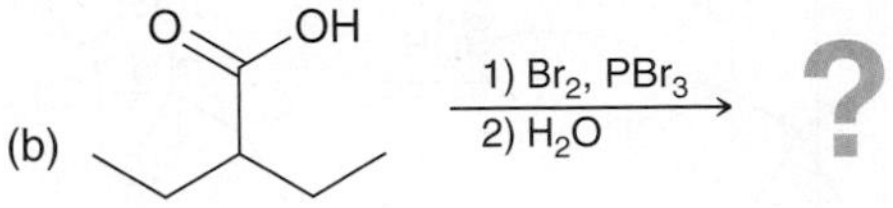

22.12 Identify the reagents that you would use to accomplish each of the following transformations (you will also need to use reactions from previous chapters).

(a) 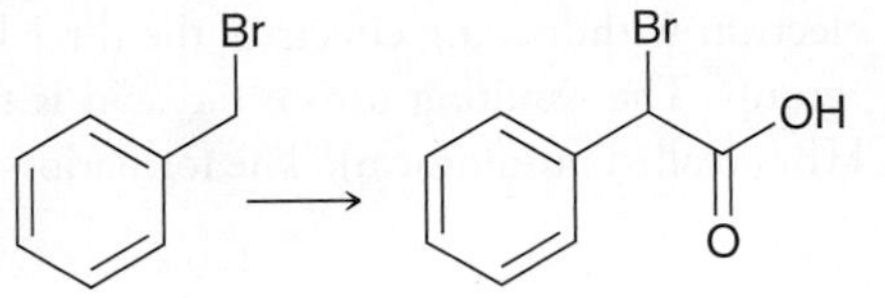

(b) 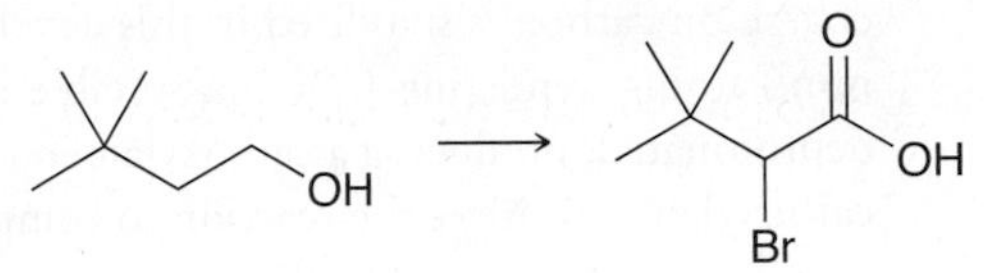

(c) 

Alpha Halogenation in Basic Conditions: The Haloform Reaction

We have seen that ketones will undergo alpha halogenation in acid-catalyzed conditions. A similar result can also be achieved in basic conditions:

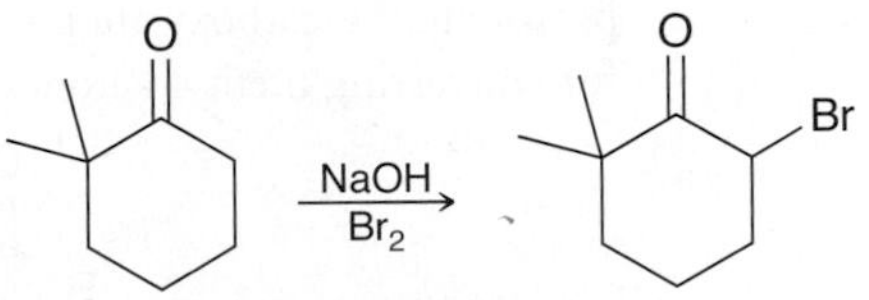

The base abstracts a proton to form the enolate, which then functions as a nucleophile and undergoes alpha halogenation.

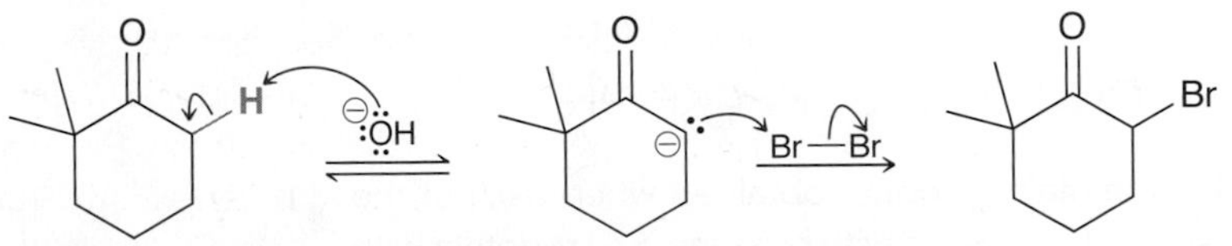

With hydroxide as the base, the concentration of enolate is always low but is continuously maintained by the equilibrium as the reaction proceeds.

When more than one α proton is present, it is difficult to achieve monobromination in basic conditions, because the brominated product is more reactive and rapidly undergoes further bromination.

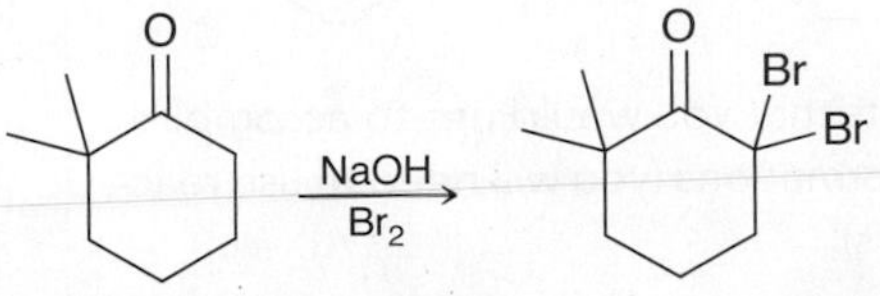

After the first halogenation reaction occurs, the presence of the halogen renders the α position more acidic, and the second halogenation step occurs more rapidly. As a result, it is often difficult to isolate the monobrominated product.

When a methyl ketone is treated with excess base and excess halogen, a reaction occurs in which a carboxylic acid is produced after acidic workup.

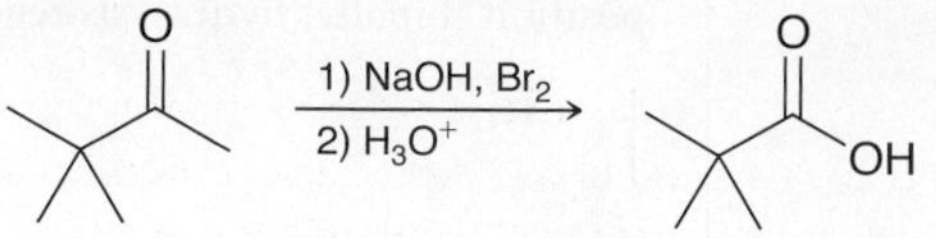

The mechanism is believed to involve several steps. First, the α protons are removed and replaced with bromine atoms, one at a time. Then, the tribromomethyl group can function as a leaving group, resulting in a nucleophilic acyl substitution reaction.

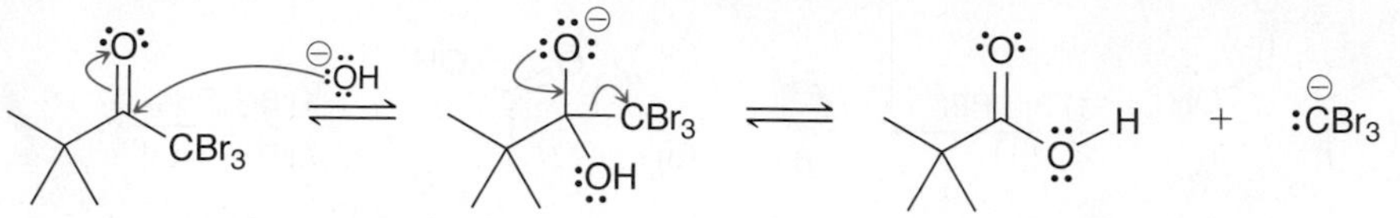

Notice that the leaving group is a carbanion, which violates the rule that we saw in Section 21.7 (re-form the carbonyl if possible but never expel C^-). The discussion in Chapter 21 indicated that there would be some exceptions to the rule, and here is the first exception. The negative charge on carbon is stabilized in this case by the electron-withdrawing effects of the three bromine atoms, rendering CBr_3^- a suitable leaving group. The resulting carboxylic acid is then deprotonated, producing a carboxylate ion and $CHBr_3$ (called bromoform). The formation of a carboxylate ion drives the reaction to completion.

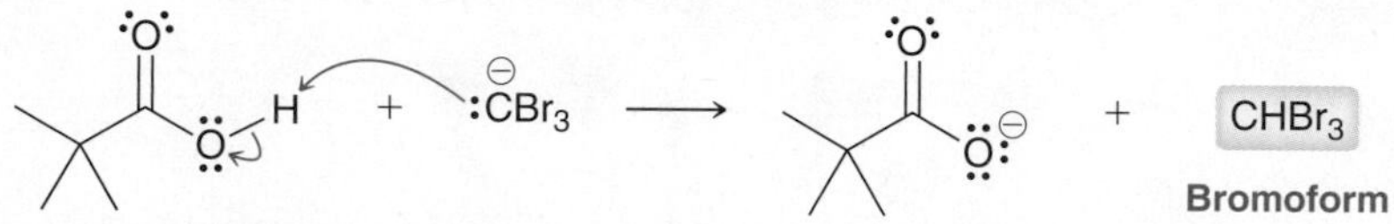

The same process occurs with chlorine and iodine, and the by-products are chloroform and iodoform, respectively. This reaction is named after the by-product that is formed and is called the **haloform reaction**. The reaction must be followed by treatment with a proton source to protonate the carboxylate ion and form the carboxylic acid. This process is synthetically useful for converting methyl ketones into carboxylic acids.

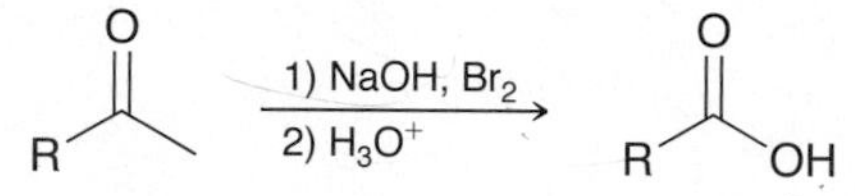

The haloform reaction is most efficient when the other side of the ketone has no α protons.

CONCEPTUAL CHECKPOINT

22.13 Predict the major product obtained when each of the following compounds is treated with bromine (Br_2) together with sodium hydroxide (NaOH) followed by aqueous acid (H_3O^+).

(a) 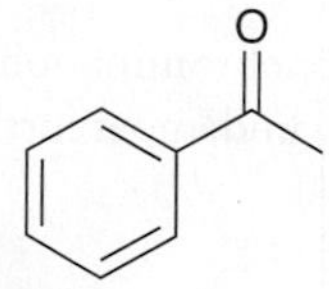 (b) 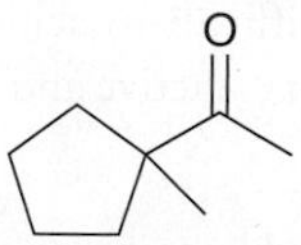 (c) 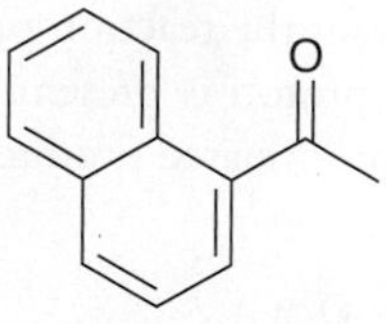

22.14 Identify the reagents that you would use to accomplish each of the following transformations (you will need to use reactions from previous chapters).

(a) 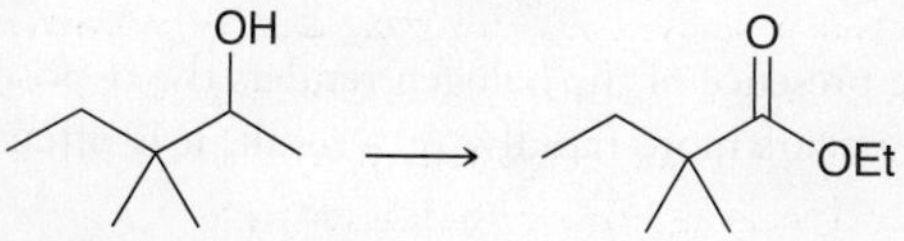

(b) 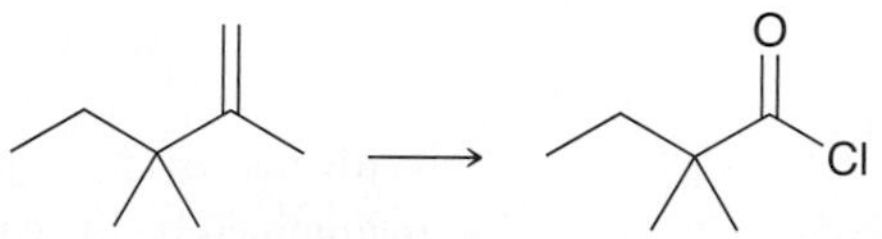

(c) 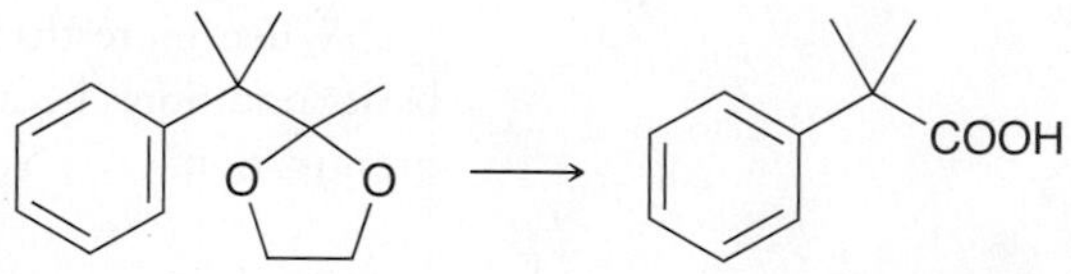

(d) 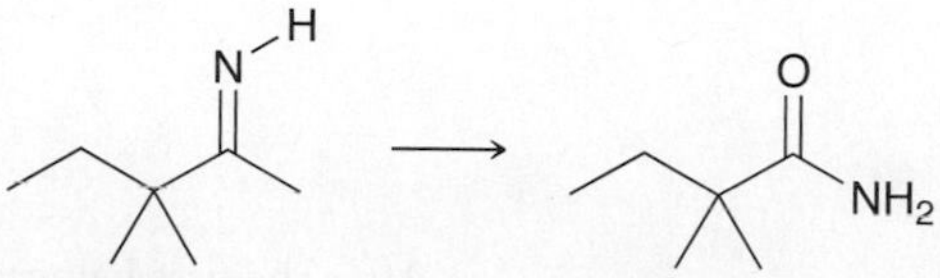

22.3 Aldol Reactions

Aldol Additions

Recall that when an aldehyde is treated with sodium hydroxide both the aldehyde and the enolate will be present at equilibrium. Under such conditions, a reaction can occur between these two species. For example, treatment of acetaldehyde with sodium hydroxide gives 3-hydroxybutanal.

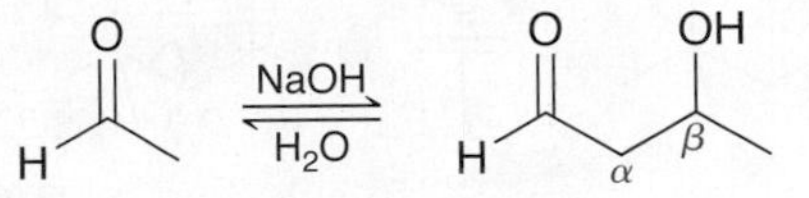

O
NaOH
H2O
H
O
OH
H
α
β

The product exhibits both an aldehydic group and a hydroxyl group, and it is therefore called an aldol (*ald* for "aldehyde" and *ol* for "alcohol"). In recognition of the type of product formed, the reaction is called an **aldol addition reaction**. Notice that the hydroxyl group is located specifically at the β position relative to the carbonyl group. The product of an aldol addition reaction is always a β-hydroxy aldehyde or ketone (Mechanism 22.4).

MECHANISM 22.4 ALDOL ADDITION

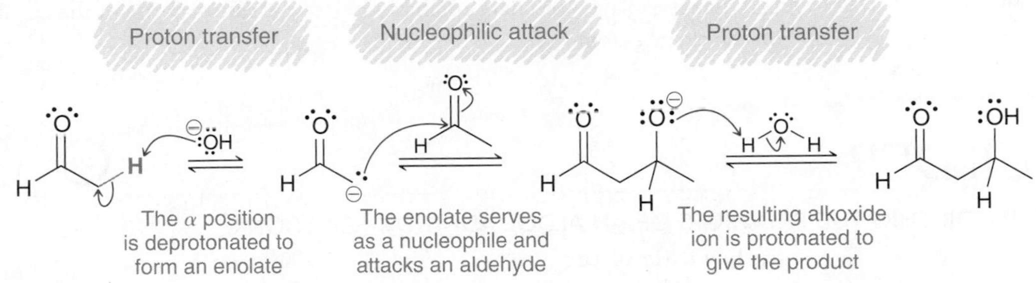

Proton transfer
Nucleophilic attack
Proton transfer
The α position is deprotonated to form an enolate
The enolate serves as a nucleophile and attacks an aldehyde
The resulting alkoxide ion is protonated to give the product

The mechanism for an aldol addition has three steps. In the first step, the aldehyde is deprotonated to form an enolate. Since hydroxide is used as the base, both the enolate and the aldehyde are present at equilibrium, and the enolate attacks the aldehyde. The resulting alkoxide ion is then protonated to yield the product. Notice the use of equilibrium arrows for every step of the mechanism. For most simple aldehydes, the position of equilibrium favors the aldol product.

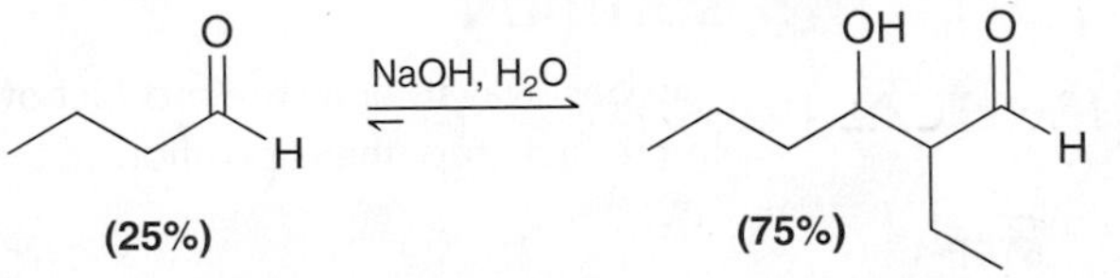

NaOH, H2O
(25%)
(75%)

However, for most ketones, the aldol product is not favored, and poor yields are common.

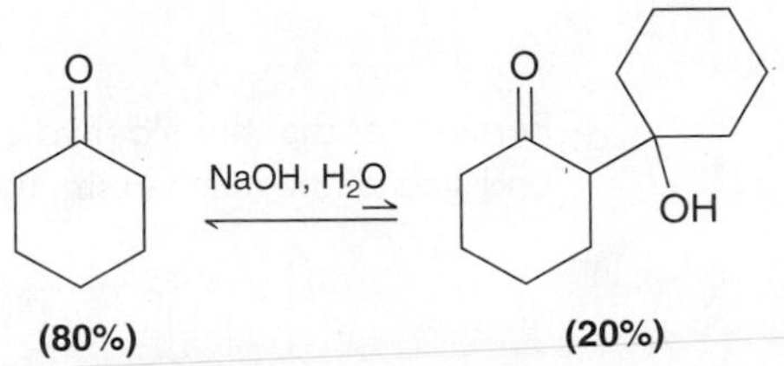

NaOH, H2O
OH
(80%)
(20%)

In this reaction, the reverse process is favored, that is, the β-hydroxy ketone is converted back into cyclohexanone more readily than the forward reaction. This reverse process, which is called a **retro-aldol reaction** (Mechanism 22.5), can be exploited in many situations (an example can be seen in the upcoming Practically Speaking box on muscle power).

MECHANISM 22.5 RETRO-ALDOL REACTION

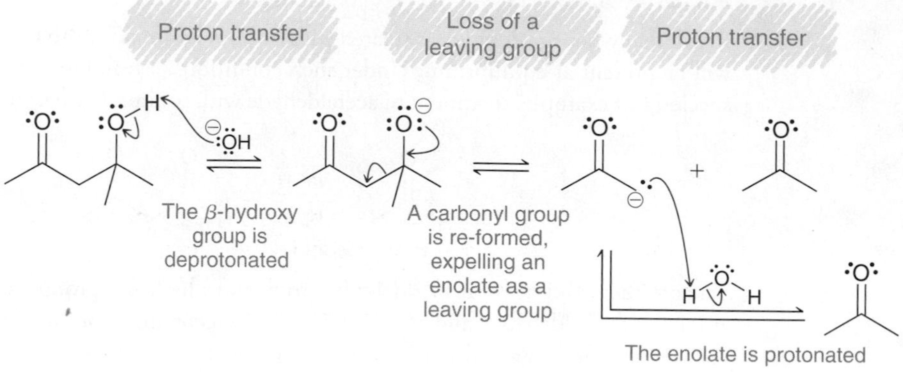
Proton transfer
Loss of a leaving group
Proton transfer
The β-hydroxy group is deprotonated
A carbonyl group is re-formed, expelling an enolate as a leaving group
The enolate is protonated

Once again, the mechanism has three steps. In fact, these three steps are simply the reverse of the three steps for an aldol addition. Notice that the second step in this mechanism involves re-formation of a carbonyl group to expel an enolate ion as a leaving group. This reaction therefore represents another exception to the rule about never expelling C^-. This exception is justified because an enolate is resonance stabilized, with the vast majority of the negative charge residing on the oxygen atom.

SKILLBUILDER

22.2 PREDICTING THE PRODUCTS OF AN ALDOL ADDITION REACTION

LEARN the skill

Predict the product of the aldol addition reaction that occurs when the following aldehyde is treated with aqueous sodium hydroxide.

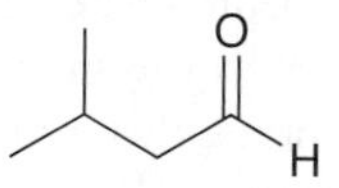

SOLUTION

The best way to draw the product of an aldol addition reaction is to consider the mechanism. In the first step, the α position of the aldehyde is deprotonated to form an enolate.

STEP 1
Consider all three steps of the mechanism.

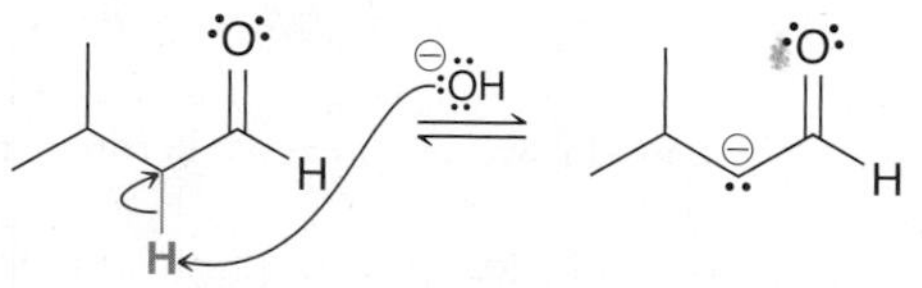

Remember that the aldehydic proton is not acidic. Only the α protons are acidic. In the second step of the mechanism, the enolate attacks an aldehyde, forming an alkoxide ion.

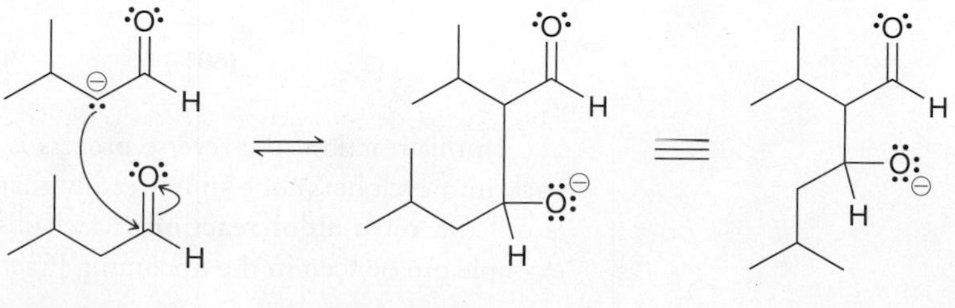

Finally, the alkoxide intermediate is protonated to give the product.

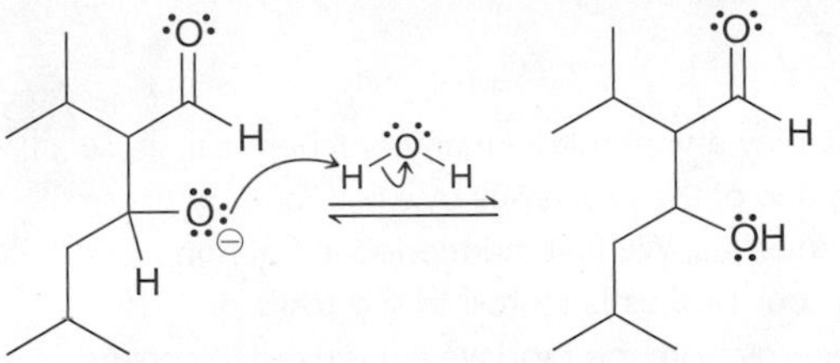

In total, there are three mechanistic steps: (1) deprotonation, (2) nucleophilic attack, and (3) protonation. Notice that the product is a β-hydroxy aldehyde. When drawing the product of an aldol addition reaction, always make sure that the hydroxyl group is located at the β position relative to the carbonyl group.

STEP 2
Double-check your answer to make sure that the product has an OH group at the β position.

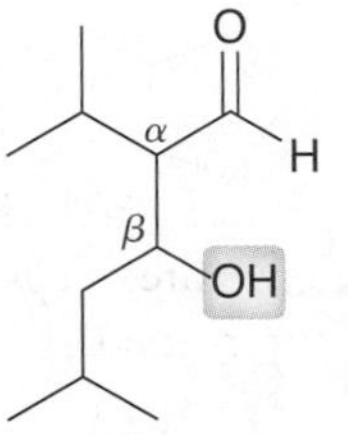

PRACTICE the skill **22.15** Predict the major product obtained when each of the following aldehydes is treated with aqueous sodium hydroxide:

(a) 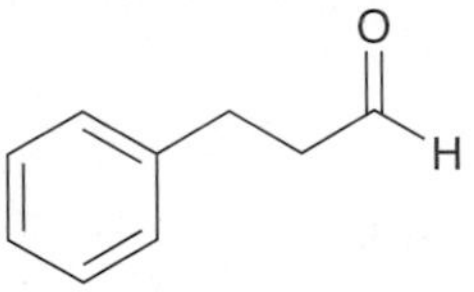 (b) 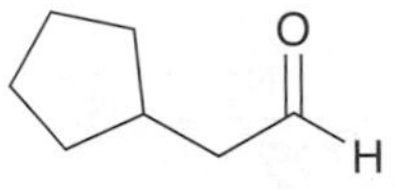 (c) 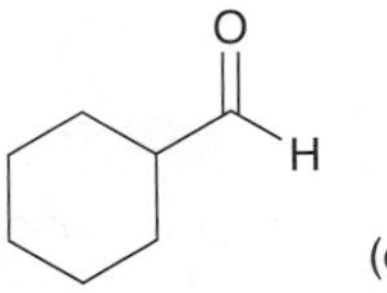 (d) 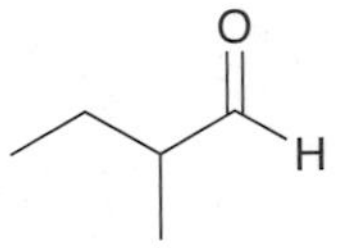

22.16 When each of the following ketones is treated with aqueous sodium hydroxide, the aldol product is obtained in poor yields. In these cases, special distillation techniques are used to increase the yield of aldol product. In each case, predict the aldol addition product that is obtained, and propose a mechanism for its formation:

(a) 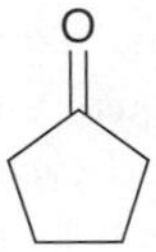 (b) 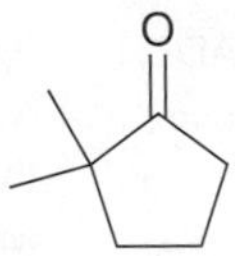 (c) 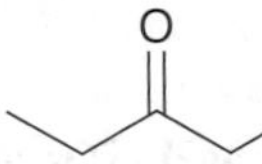 (d) 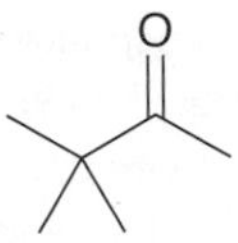

APPLY the skill **22.17** When treated with aqueous sodium hydroxide, 2,2-dimethylbutanal does not undergo an aldol addition reaction. Explain this observation.

22.18 An alcohol of molecular formula $C_4H_{10}O$ was treated with PCC to produce an aldehyde that exhibits exactly three signals in its ^1H NMR spectrum. Predict the aldol addition product that is obtained when this aldehyde is treated with aqueous sodium hydroxide.

22.19 Using acetaldehyde as your only source of carbon, show how you would prepare 1,3-butanediol.

need more **PRACTICE?** **Try Problem 22.72**

Aldol Condensations

When heated in acidic or basic conditions, the product of an aldol addition reaction will undergo elimination to produce unsaturation between the α and β positions:

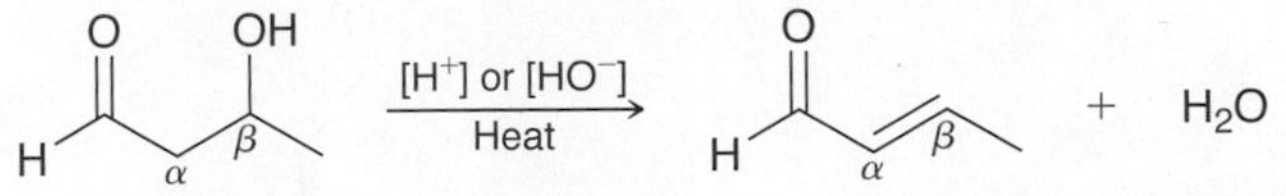

PRACTICALLYSPEAKING)))

Muscle Power

Retro-aldol reactions play a vital role in many biochemical processes, including one of the processes by which energy is generated for our muscles. We first mentioned in Section 13.11 that energy in our bodies is stored in the form of ATP molecules. That is, energy from the food we eat is used to convert ADP into ATP, which is stored. When energy is needed, ATP is broken down to ADP, and the energy that is released can be used for various life processes, such as muscle contraction.

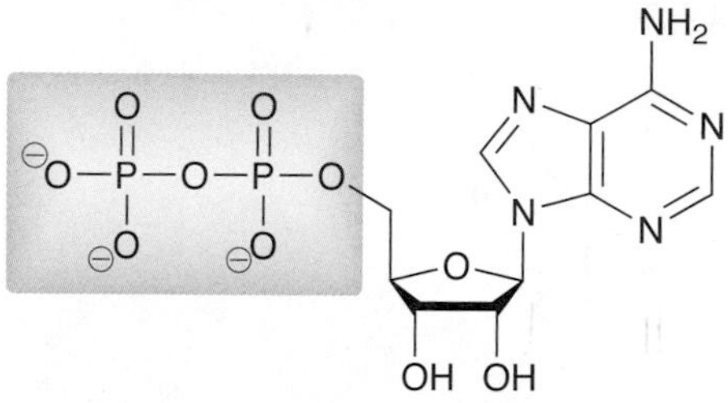

Adenosine diphosphate (ADP)

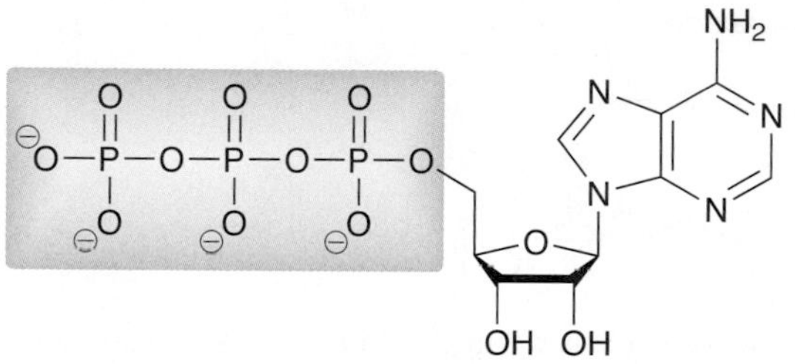

Adenosine triphosphate (ATP)

Notice that the structural difference between ATP and ADP is in the number of phosphate groups present (two in the case of ADP, three in the case of ATP). The ATP molecules in our muscles are used for any activity that requires a short burst of energy, such as a tennis serve, jumping, or throwing a ball. For activities that last longer than a second, such as sprinting, ATP molecules must be synthesized on the spot. This is initially achieved by a process called glycolysis, in which a molecule of glucose (obtained from metabolism of the carbohydrates we eat) is converted into two molecules of pyruvic acid.

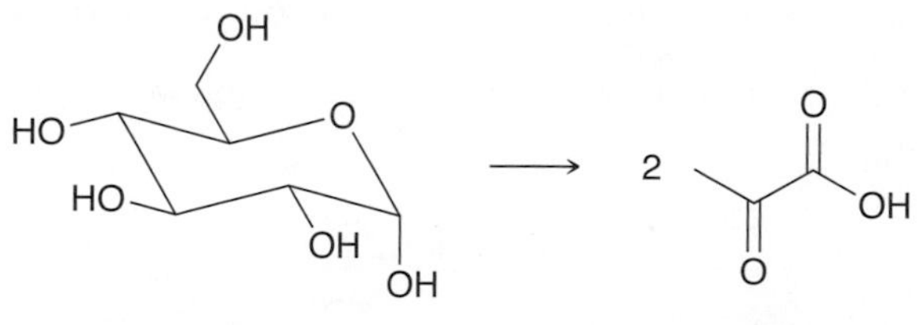

Glucose **Pyruvic acid**

Glycolysis involves many steps and is accompanied by the conversion of two molecules of ADP into ATP. This metabolic process therefore generates the necessary ATP for muscle contraction. One of the steps involved in glycolysis is a retro-aldol reaction, which is achieved with the help of an enzyme called aldolase.

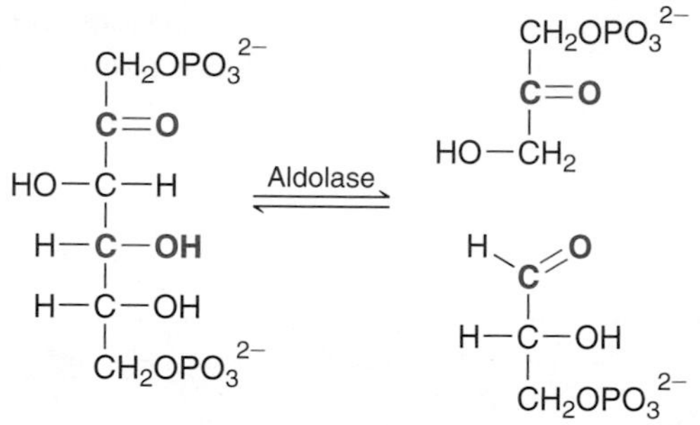

The end product of glycolysis is pyruvic acid, which is used as a starting material in a variety of biochemical processes. Some of the pyruvic acid produced by glycolysis is reduced to form lactic acid.

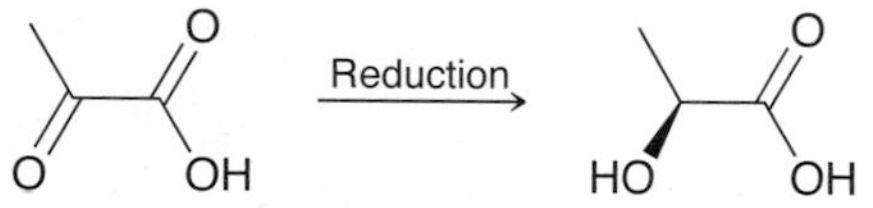

Pyruvic acid **Lactic acid**

The buildup of lactic acid in muscles is responsible for the pain we feel after a strenuous workout as well as muscle cramps.

Glycolysis provides ATP for activities lasting up to 1.5 min. Activities that exceed this time frame, such as long-distance running, require a different process for ATP generation, called the citric acid cycle. Unlike glycolysis, which can be achieved without oxygen (it is an anaerobic process), the citric acid cycle requires oxygen (it is an aerobic process). This explains why we breathe more rapidly during and after strenuous activity. The term "aerobic workout" is commonly used to refer to a workout that utilizes ATP that was generated by the citric acid cycle (an aerobic process) rather than glycolysis (an anaerobic process). A casual athlete can sense a shift from anaerobic ATP synthesis to aerobic ATP synthesis after about 1.5 min. Seasoned athletes will detect the shift occurring after about 2.5 min. If you watch the summer Olympics, you are likely aware of the distinction between sprinting and long-distance running. Athletes can run faster in a sprinting race, which relies mostly on glycolysis for energy production. Long-distance running requires the citric acid cycle for energy production.

Practically, this transformation is most readily achieved when an aldol addition is performed at elevated temperature. Under these basic conditions, the aldol addition reaction occurs, followed by dehydration to give an α,β-unsaturated product.

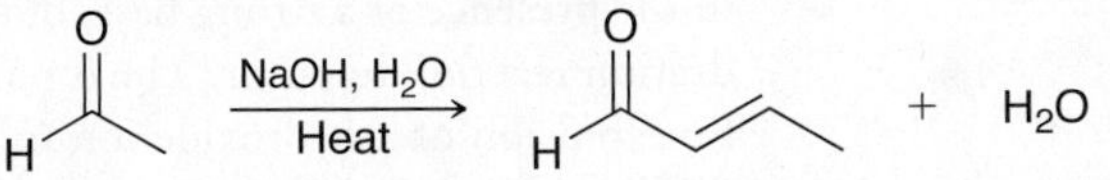
NaOH, H_2O
Heat
+ H_2O

This two-step process (aldol addition plus dehydration) is called an **aldol condensation**. The term *condensation* is used to refer to any reaction in which two molecules undergo addition accompanied by the loss of a small molecule such as water, carbon dioxide, or nitrogen gas. In the case of aldol condensations, water is the small molecule that is lost. Notice that the product of an aldol addition is a β-hydroxy aldehyde or ketone, while the product of an aldol condensation is an α,β-unsaturated aldehyde or ketone.

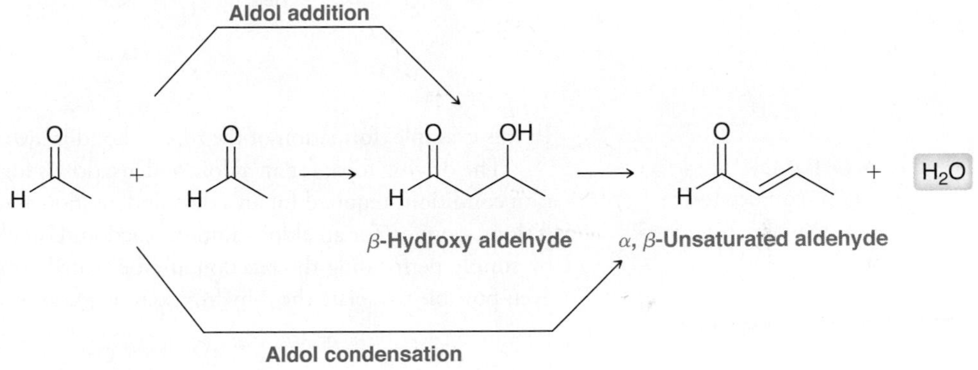
Aldol addition
β-Hydroxy aldehyde
α, β-Unsaturated aldehyde
H_2O
Aldol condensation

MECHANISM 22.6 ALDOL CONDENSATION

PART 1: ALDOL ADDITION

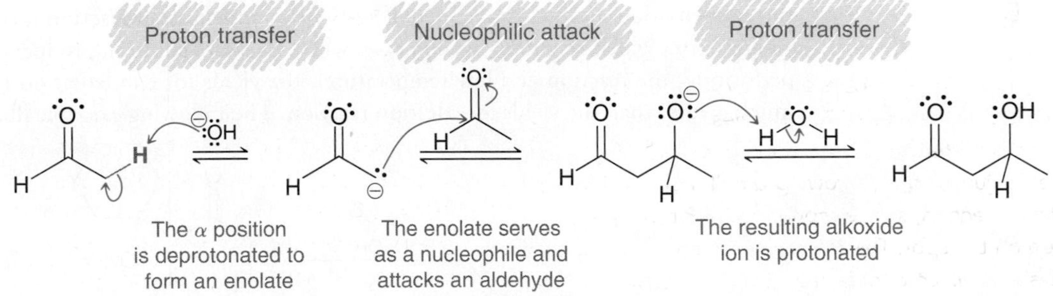
Proton transfer
Nucleophilic attack
Proton transfer
The α position is deprotonated to form an enolate
The enolate serves as a nucleophile and attacks an aldehyde
The resulting alkoxide ion is protonated

PART 2: ELIMINATION OF H_2O

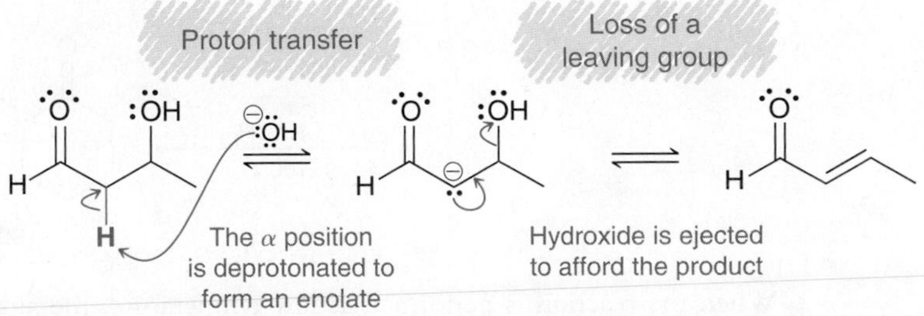
Proton transfer
Loss of a leaving group
The α position is deprotonated to form an enolate
Hydroxide is ejected to afford the product

The mechanism for an aldol condensation has two parts (Mechanism 22.6). The first part is just an aldol addition reaction, which has three mechanistic steps. The second part has two steps that accomplish the elimination of water. Normally, alcohols do not undergo dehydration in the presence of a strong base, but here, the presence of the carbonyl group enables the dehydration reaction to occur. The α position is first deprotonated to form an enolate ion, followed by expulsion of a hydroxide ion to produce α,β unsaturation. This two-step process, which is different from the elimination reactions we saw in Chapter 8, is called an **E1cb mechanism**. In an E1cb mechanism, the leaving group only leaves after deprotonation occurs.

In cases where two stereoisomeric π bonds can be formed, the product with the least steric hindrance is generally the major product.

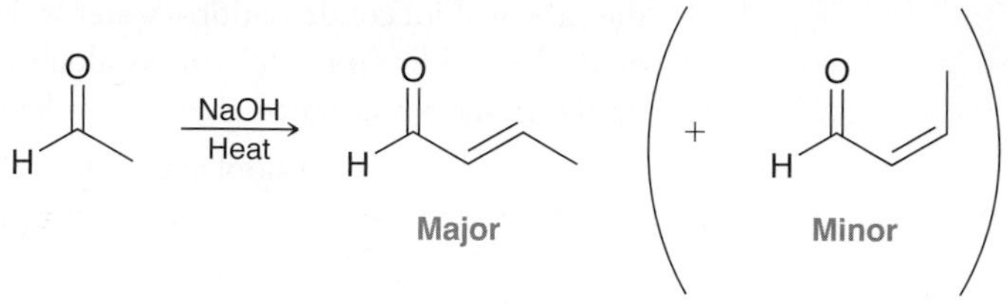

In this example, formation of the *trans* π bond is favored over formation of the *cis* π bond.

LOOKING BACK

The stability of conjugated π systems was discussed in Section 17.2.

The driving force for an aldol condensation is formation of a conjugated system. The reaction conditions required for an aldol condensation are only slightly more vigorous than the conditions required for an aldol addition reaction. Usually, an aldol condensation can be achieved by simply performing the reaction at an elevated temperature. In fact, in some cases, it is not even possible to isolate the β-hydroxyketone. As an example, consider the following case:

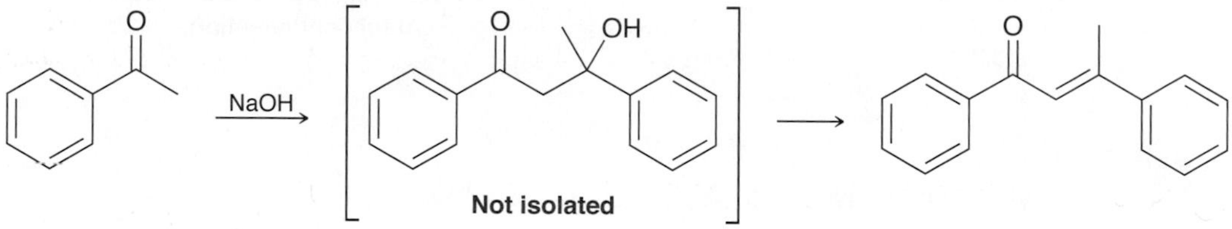

In this case, the aldol addition product cannot be isolated. Even at moderate temperatures, only the condensation product is obtained, because the condensation reaction involves formation of a highly conjugated π system. Even in cases where the aldol addition product can be isolated (by performing the reaction at a low temperature), the yields for condensation reactions are often much greater than the yields for addition reaction. The following example illustrates this point:

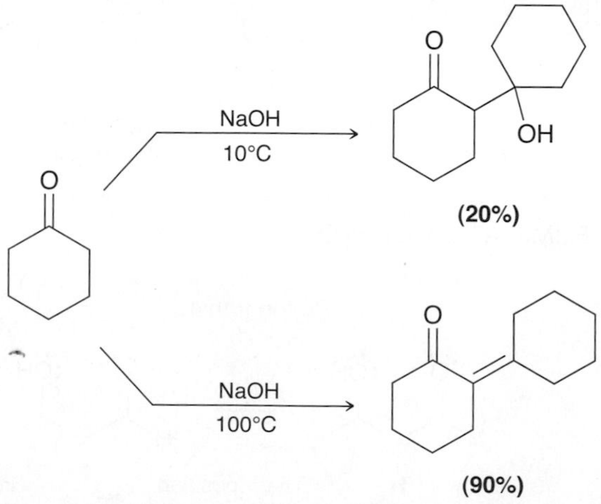

When the reaction is performed at low temperature, the aldol addition product is obtained, but the yield is very poor. As explained earlier, the starting material is a ketone, and the equilibrium does not favor formation of the aldol addition product. However, when the reaction is performed at an elevated temperature, the aldol condensation product is obtained in very good yield, because the equilibrium is driven by formation of a conjugated π system.

SKILLBUILDER

22.3 DRAWING THE PRODUCT OF AN ALDOL CONDENSATION

LEARN the skill

Predict the product of the aldol condensation that occurs when the following ketone is heated in the presence of aqueous sodium hydroxide.

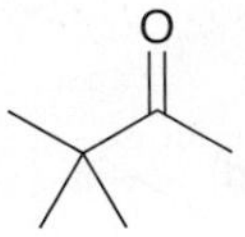

SOLUTION

One way to draw the product is to draw the entire mechanism, but for aldol condensations, there is a faster method for drawing the product. Begin by identifying the α protons:

STEP 1
Identify the α positions.

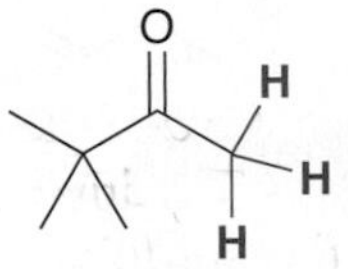

In order to achieve an aldol condensation, one of the α carbon atoms must bear at least two protons. In this case, one of the α positions possesses three protons, and the other α position possesses none.

Next draw two molecules of the ketone, oriented such that two α protons of one molecule are directly facing the carbonyl group of the other molecule:

STEP 2
Redraw two molecules of the ketone.

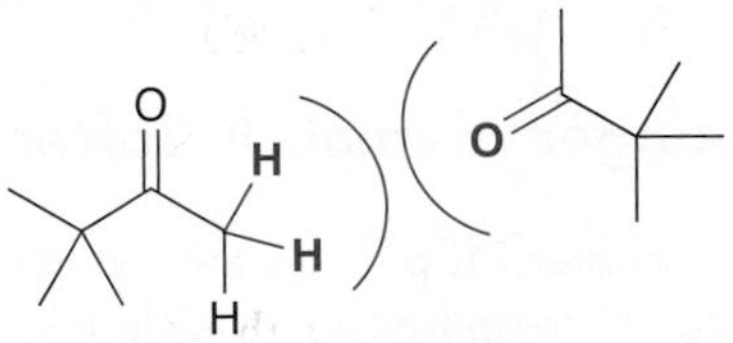

When drawn this way, it is easier to predict the product without having to draw the entire mechanism. Simply remove the two α protons and the oxygen and replace them with a double bond:

STEP 3
Remove H_2O and replace with a C=C bond.

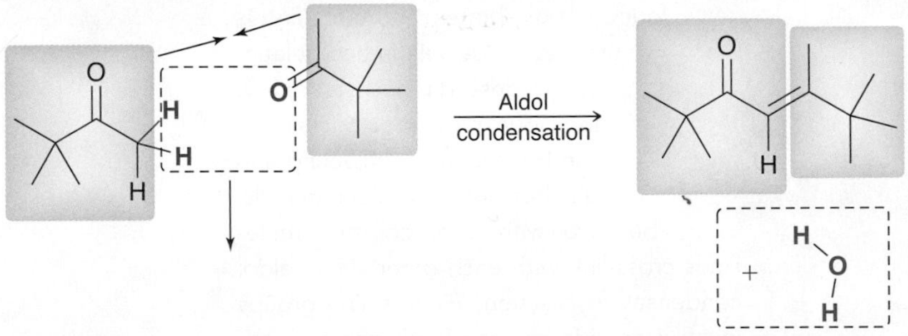

The product is an α,β-unsaturated ketone, and water is liberated as a by-product. In this case, two stereoisomers are possible, so we draw the product that exhibits the least steric hindrance:

STEP 4
Draw the isomer with the least steric hindrance.

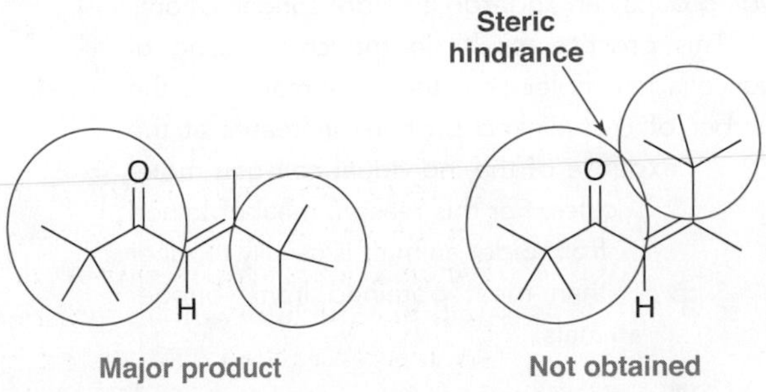

PRACTICE the skill **22.20** Draw the condensation product obtained when each of the following compounds is heated in the presence of aqueous sodium hydroxide.

(a) 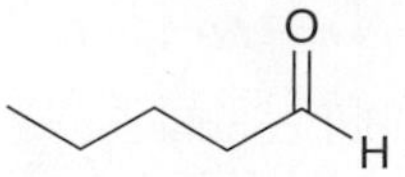
(b) 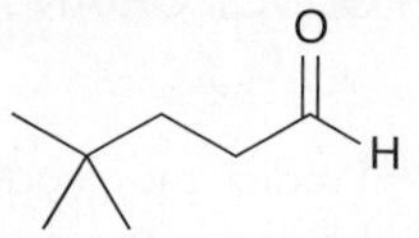
(c) 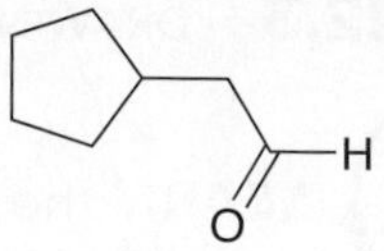
(d) 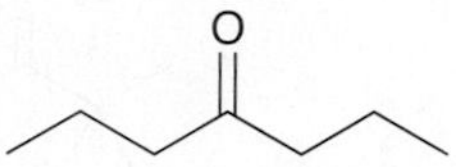
(e) 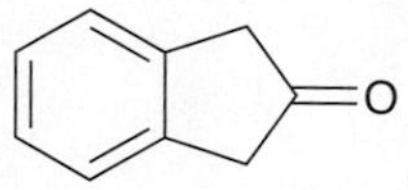
(f) 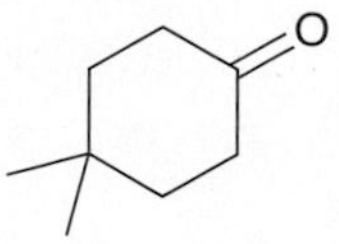

APPLY the skill **22.21** Identify the starting aldehyde or ketone needed to make each of the following compounds via an aldol condensation.

(a) 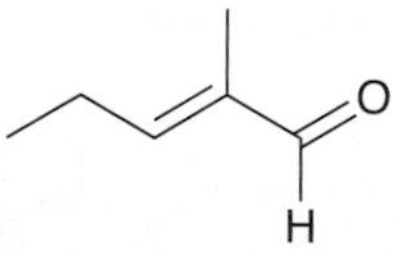
(b) 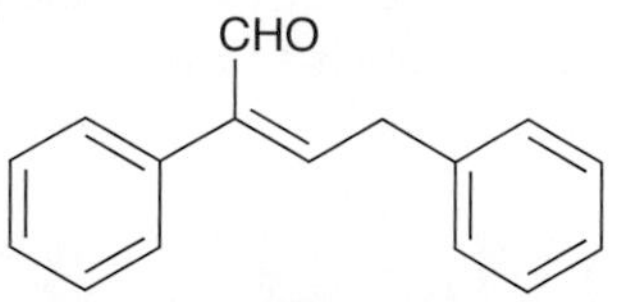
(c) 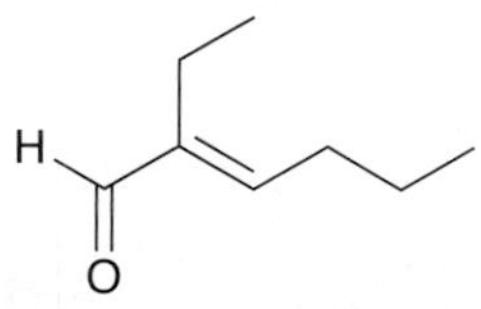

22.22 When 2-butanone is heated in the presence of aqueous sodium hydroxide, two constitutionally isomeric condensation products are obtained. Draw both products.

need more **PRACTICE?** **Try Problems 22.71, 22.84c**

PRACTICALLYSPEAKING

Why Meat from Younger Animals Is Softer

As mentioned in Chapter 21, proteins are important biological compounds comprised of multiple amide moieties. Proteins will be discussed in greater detail in Chapter 25. One of the most abundant proteins found in mammals is called collagen. It is found in skin, bones, and teeth and is used to make gelatin and gelatin-based desserts, such as Jell-O. Individual molecules of collagen can be isolated from young animals, but not from older animals, because with age, collagen molecules cross-link with each other via an aldol condensation reaction. First, amino groups located on side chains of collagen are converted to aldehyde groups via a process called oxidative deamination. Then, the resulting aldehyde groups can undergo an aldol condensation:

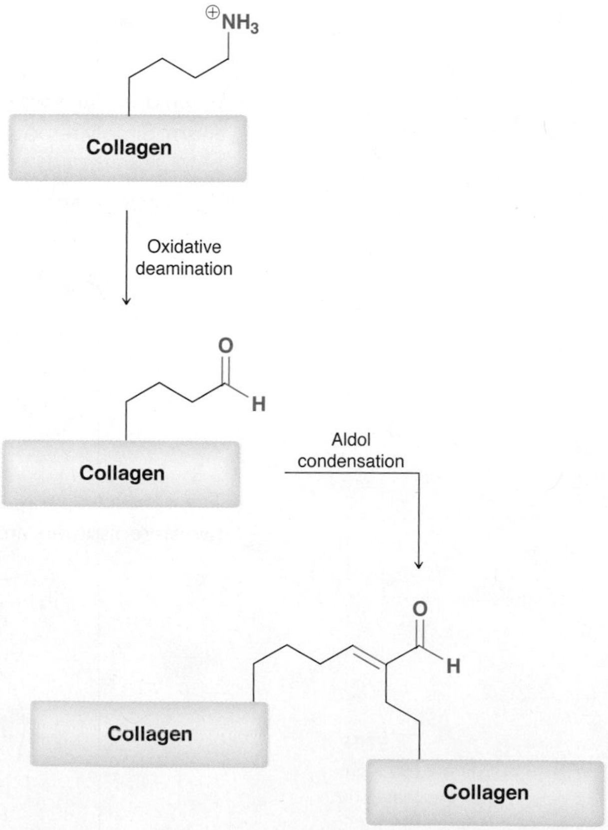

This process results in the cross-linking of two collagen molecules. As an animal ages, the number of cross-linked proteins increases at the expense of the individual collagen molecules. For this reason, meat obtained from older animals is usually tougher than meat obtained from younger animals.

Crossed Aldol Reactions

Until now, we have focused on symmetrical aldol reactions, that is, aldol reactions that occur between two identical partners. In this section, we explore **crossed aldol**, or **mixed aldol, reactions**, which are aldol reactions that can occur between different partners. As an example, consider what happens when a mixture of acetaldehyde and propionaldehyde is treated with a base. Under these circumstances, four possible aldol products can be formed (Figure 22.3). The

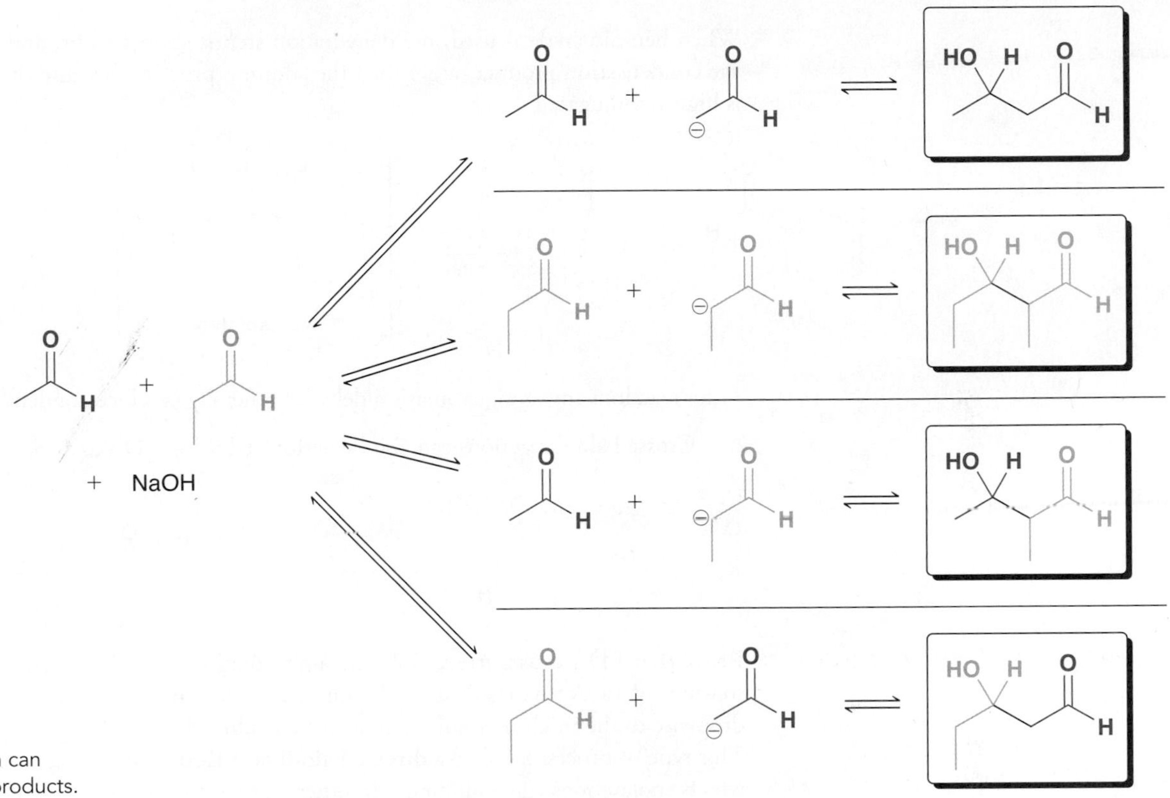

FIGURE 22.3
A crossed aldol reaction can produce four different products.

first two products are formed from symmetrical aldol reactions, while the latter two products are formed from crossed aldol reactions. Reactions that form mixtures of products are of little use, and therefore, crossed aldol reactions are only efficient if they can be performed in a way that minimizes the number of possible products. This is best accomplished in either of the following ways:

1. If one of the aldehydes lacks α protons and possesses an unhindered carbonyl group, then a crossed aldol can be performed. As an example, consider what happens when a mixture of formaldehyde and propionaldehyde is treated with a base.

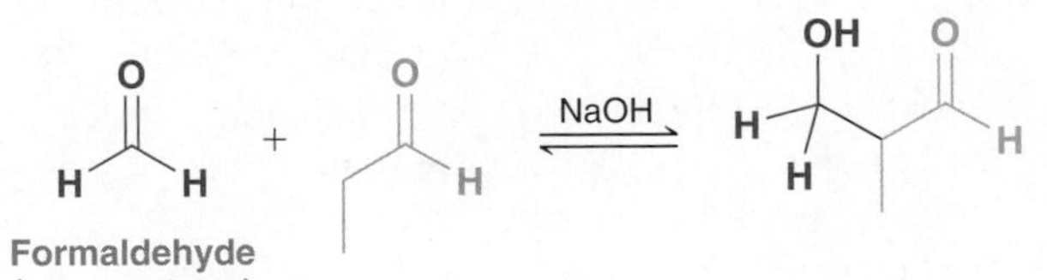

In this case, only one major aldol product is produced. Why? Formaldehyde has no α protons and therefore cannot form an enolate. As a result, only the enolate formed from propionaldehyde is present in solution. Under these conditions, there are only two possible products. The enolate can attack a molecule of propionaldehyde to produce a symmetrical aldol reaction, or the enolate can attack a molecule of formaldehyde to produce a crossed aldol reaction. The latter occurs more rapidly because the carbonyl group of formaldehyde is less hindered than the carbonyl group of propionaldehyde. As a result, one product predominates.

Crossed aldol reactions can also be performed with benzaldehyde.

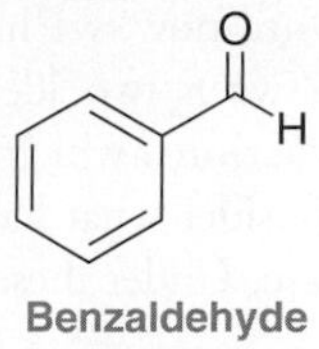

When benzaldehyde is used, the dehydration step is spontaneous, and the equilibrium favors the condensation product rather than the addition product, because the condensation product is highly conjugated.

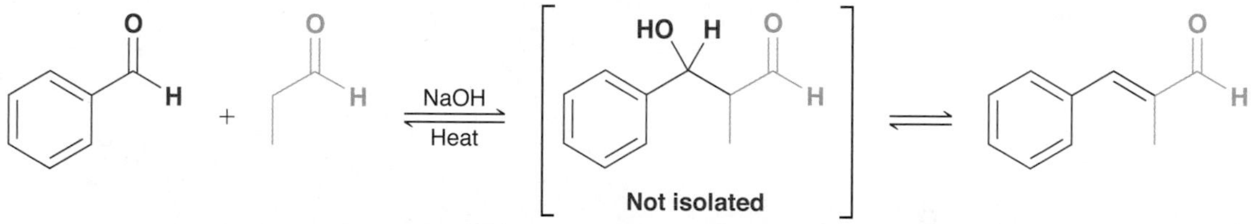

Aldol reactions involving aromatic aldehydes generally produce condensation reactions.

2. Crossed aldol reactions can also be performed using LDA as a base.

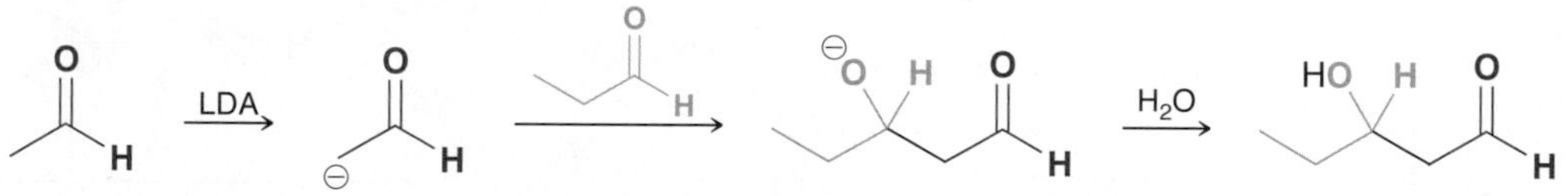

Recall that LDA causes irreversible enolate formation. If acetaldehyde is added dropwise to a solution of LDA, the result is a solution of enolate ions. Propionaldehyde can then be added dropwise to the mixture, resulting in a crossed aldol addition that produces one major product. This type of process is called a **directed aldol addition**, and its success is limited by the rate at which enolate ions can equilibrate. In other words, it is possible for an enolate ion to function as a base (rather than a nucleophile) and deprotonate a molecule of propionaldehyde. If this process occurs too rapidly, then a mixture of products will result.

SKILLBUILDER

22.4 IDENTIFYING THE REAGENTS NECESSARY FOR A CROSSED ALDOL REACTION

LEARN the skill

Identify the reagents necessary to produce the following compound via an aldol reaction.

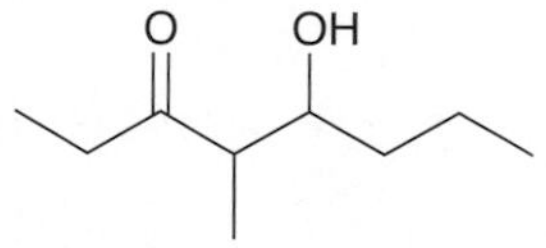

SOLUTION

Begin by identifying the α and β positions.

STEP 1
Identify the α and β positions.

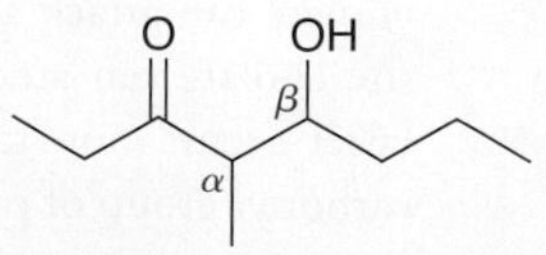

STEP 2
Use a retrosynthetic analysis that focuses on the α,β bond.

Now apply a retrosynthetic analysis. The bond between the α and β positions is the bond formed during an aldol reaction. To draw the necessary starting materials, break apart the bond between the α and β positions, drawing a carbonyl group in place of the OH group.

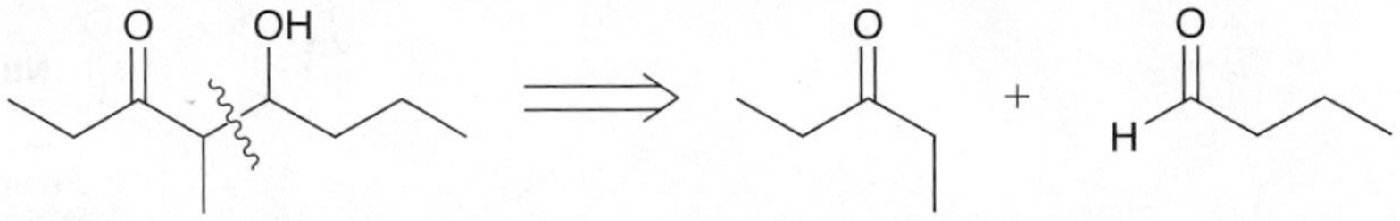

STEP 3
Identify an appropriate base.

These are the two starting carbonyl compounds. They are not identical, so a crossed aldol reaction is required.

The final step is to determine which base should be used. Both starting compounds have α protons, so hydroxide cannot be used because it would result in a mixture of four products. In this case, a crossed aldol can only be accomplished if LDA is used to achieve a directed aldol reaction.

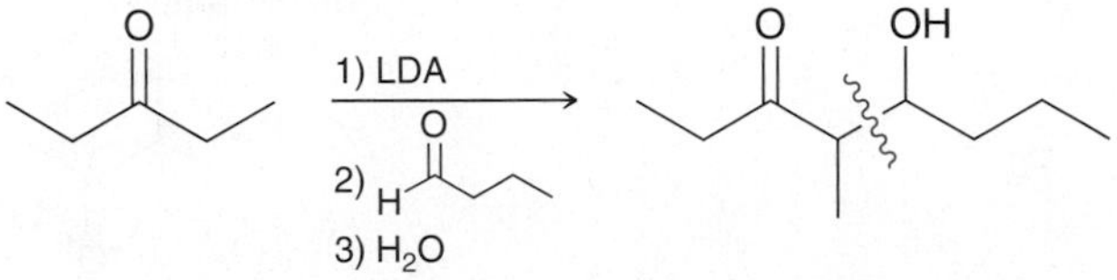

In the first step, the symmetrical ketone is irreversibly and completely deprotonated by LDA to produce a solution of enolate ions. Then, the aldehyde is added dropwise to the solution to achieve a directed aldol addition.

PRACTICE the skill **22.23** Identify the reagents necessary to produce each of the following compounds via an aldol reaction.

(a) 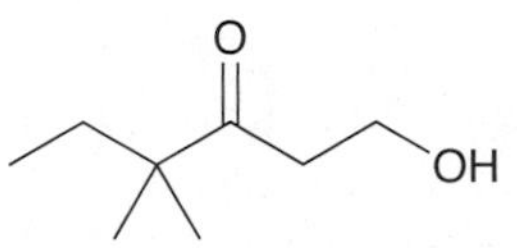 (b) 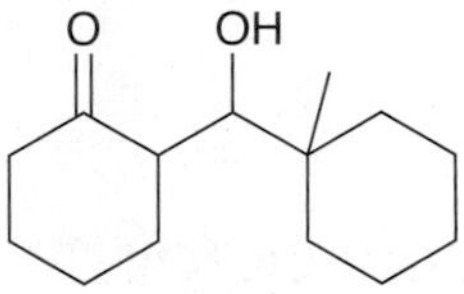 (c) 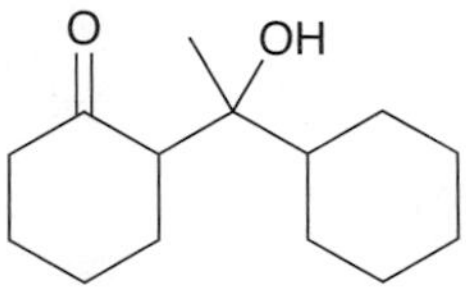

(d) 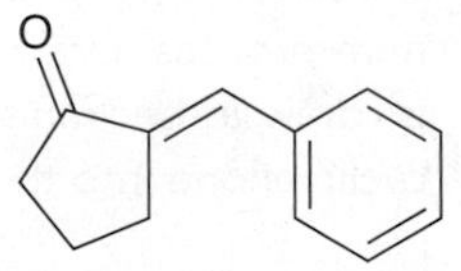 (e) 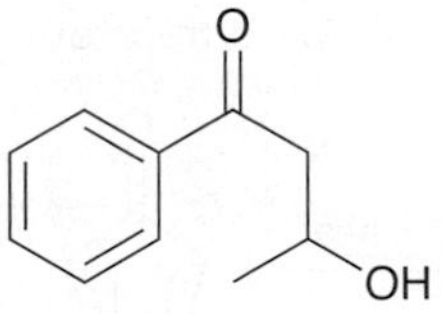

APPLY the skill **22.24** Using formaldehyde and acetaldehyde as your only sources of carbon atoms, show how you could make each of the following compounds. You may find it helpful to review acetal formation (Section 20.5).

(a) 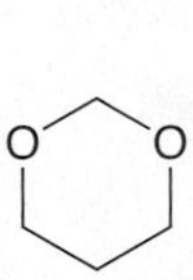 (b) 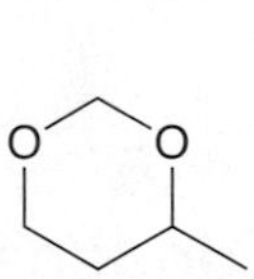 (c) 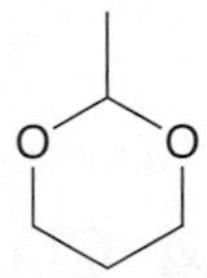 (d) 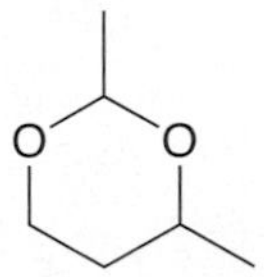

need more **PRACTICE?** **Try Problems 22.67, 22.68**

Intramolecular Aldol Reactions

Compounds that possess two carbonyl groups can undergo intramolecular aldol reactions. Consider the reaction that occurs when 2,5-hexanedione is heated in the presence of aqueous sodium hydroxide.

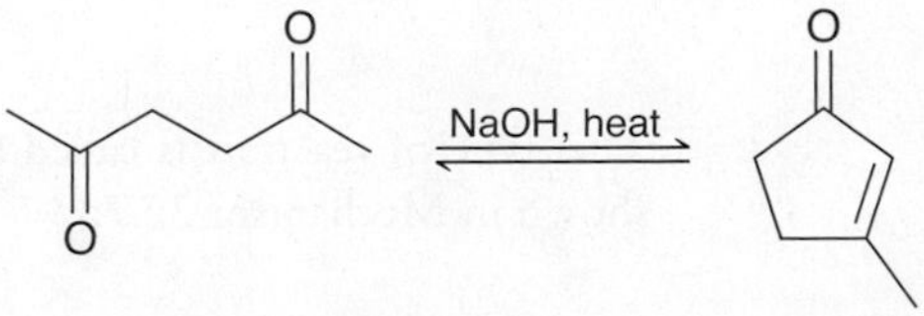

In this case, a cyclic product is formed. The mechanism for this process is nearly identical to the mechanism for any other aldol condensation but with one notable difference—the enolate and the carbonyl group are both present in the same molecule, resulting in an intramolecular attack.

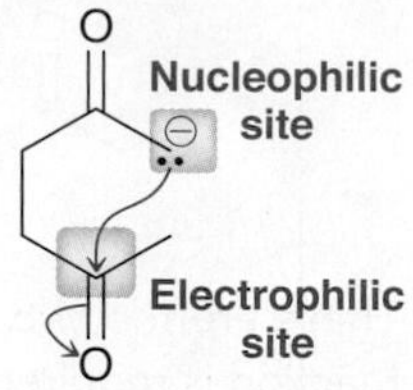

Intramolecular aldol reactions show a preference for formation of five- and six-membered rings. Smaller rings are possible but are generally not observed.

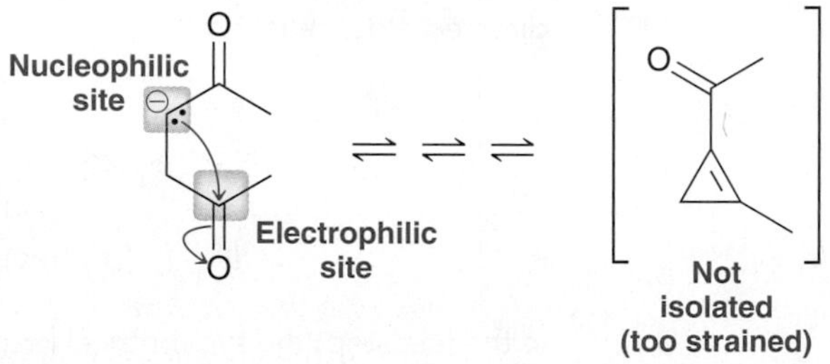

It is conceivable that the three-membered ring might be formed initially, but recall that the products of aldol reactions are determined by equilibrium concentrations. Once the equilibrium has been achieved, the strained, three-membered ring is not present in substantial amounts because the equilibrium favors formation of a nearly strain-free product.

CONCEPTUAL CHECKPOINT

22.25 Draw a mechanism for the following transformation:

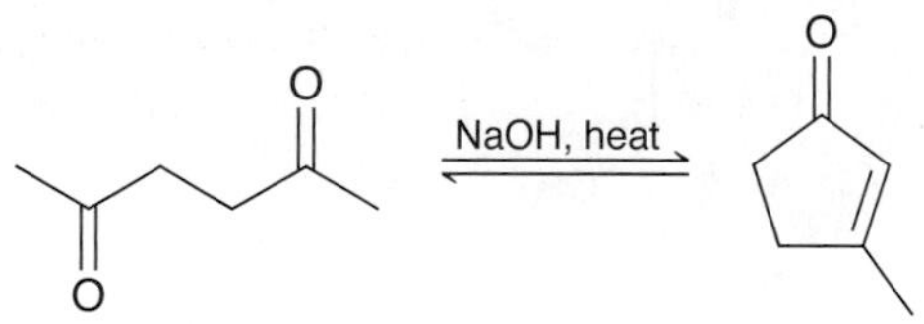

22.26 The reaction in the previous problem is an equilibrium process. Draw a mechanism of the reverse process. That is, draw a mechanism showing conversion of the conjugated, cyclic enone into the acyclic dione.

22.27 When 2,6-heptanedione is heated in the presence of aqueous sodium hydroxide, a condensation product with a six-membered ring is obtained. Draw the product and show a mechanism for its formation.

22.4 Claisen Condensations

The Claisen Condensation

Like aldehydes and ketones, esters also exhibit reversible condensation reactions.

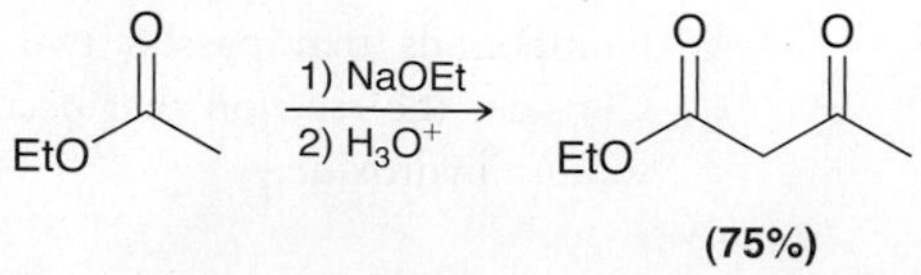

This type of reaction is called a **Claisen condensation**, and a mechanism for this process is shown in Mechanism 22.7.

MECHANISM 22.7 CLAISEN CONDENSATION

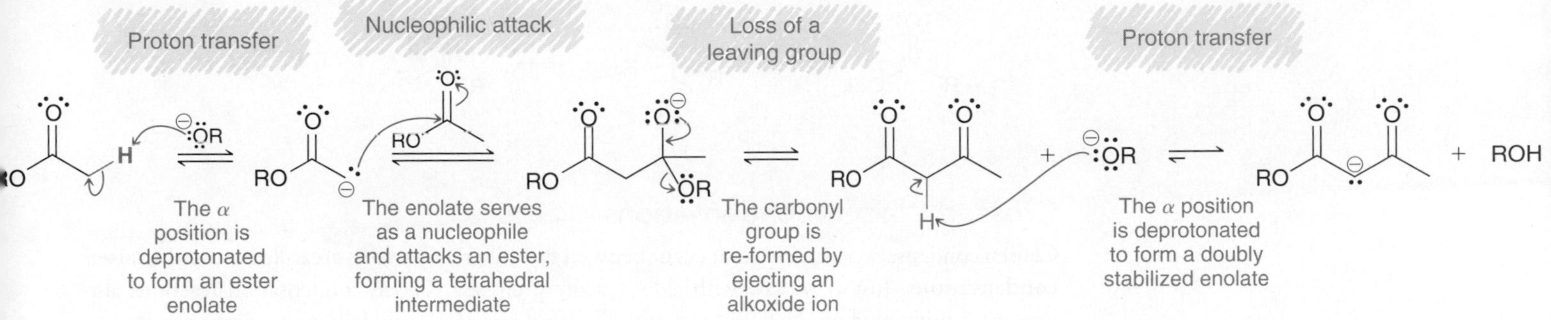
Proton transfer
Nucleophilic attack
Loss of a leaving group
Proton transfer
The α position is deprotonated to form an ester enolate
The enolate serves as a nucleophile and attacks an ester, forming a tetrahedral intermediate
The carbonyl group is re-formed by ejecting an alkoxide ion
The α position is deprotonated to form a doubly stabilized enolate
+ ROH

The first two steps of this mechanism are much like an aldol addition. The ester is first deprotonated to form an enolate, which then functions as a nucleophile and attacks another molecule of the ester. The difference between an aldol reaction and a Claisen condensation is the fate of the tetrahedral intermediate. In a Claisen condensation, the tetrahedral intermediate can expel a leaving group to re-form a C=O bond. The Claisen condensation is simply a nucleophilic acyl substitution reaction in which the nucleophile is an ester enolate and the electrophile is an ester. The product of this reaction is a β-keto ester.

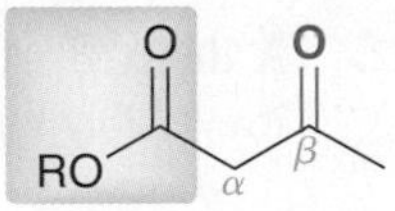

Notice that the last step in the mechanism is deprotonation of the β-keto ester to give a doubly stabilized enolate. This deprotonation step cannot be avoided, because the reaction occurs under basic conditions. Each molecule of base (alkoxide ion) is converted into the doubly stabilized enolate, which is a favorable transformation (downhill in energy). In fact, the deprotonation step at the end of the mechanism provides a driving force that causes the equilibrium to favor condensation. As a result, the base is not a catalyst but is actually consumed as the reaction proceeds. After the reaction is complete, it is necessary to use a mild acid in order to protonate the doubly stabilized enolate.

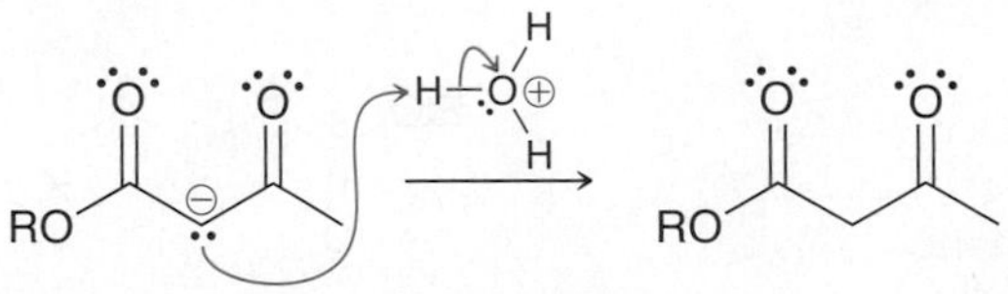

When performing a Claisen condensation, the starting ester must have two α protons. If it only has one α proton, then the driving force for condensation is absent (a doubly stabilized enolate cannot be formed).

Hydroxide cannot be used as the base for a Claisen condensation because it can cause hydrolysis of the starting ester.

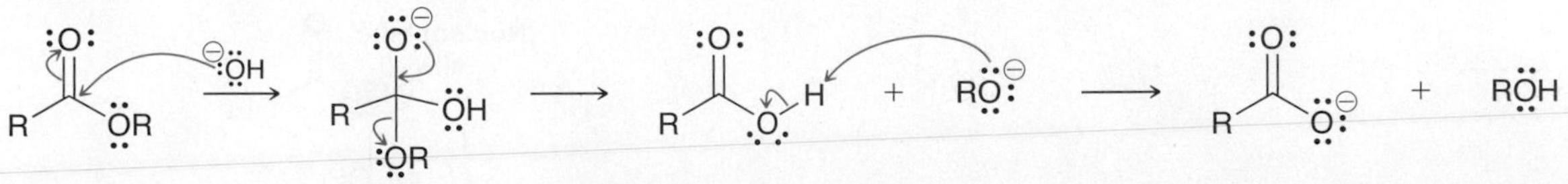

If hydroxide is used as a base, the ester is hydrolyzed to form a carboxylic acid, which is then irreversibly deprotonated to form a carboxylate salt, thereby preventing the Claisen condensation from occurring. Instead, Claisen condensations are achieved by using an alkoxide ion as the base.

Specifically, the alkoxide used must be the same OR group that is present in the starting ester in order to avoid *transesterification*. For example, if an ethyl ester is treated with methoxide, transesterification can convert the ethyl ester into a methyl ester.

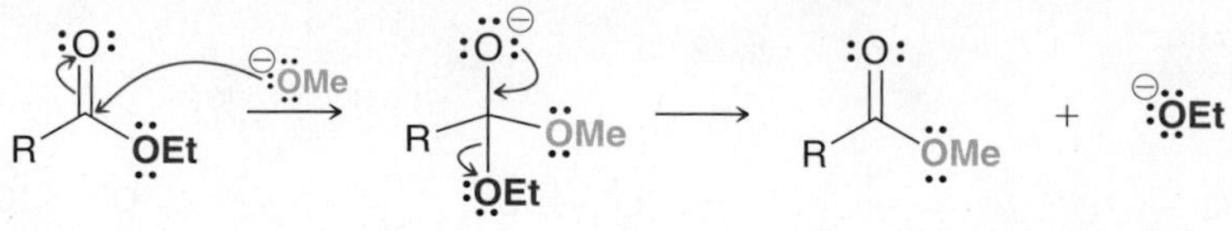

Crossed Claisen Condensations

Claisen condensation reactions that occur between two different partners are called **crossed Claisen condensations**. Just as we saw with aldol reactions, crossed Claisen condensation reactions also produce a mixture of products and are only efficient if one of the two following criteria are met:

1. If one ester has no α protons and cannot form an enolate, for example;

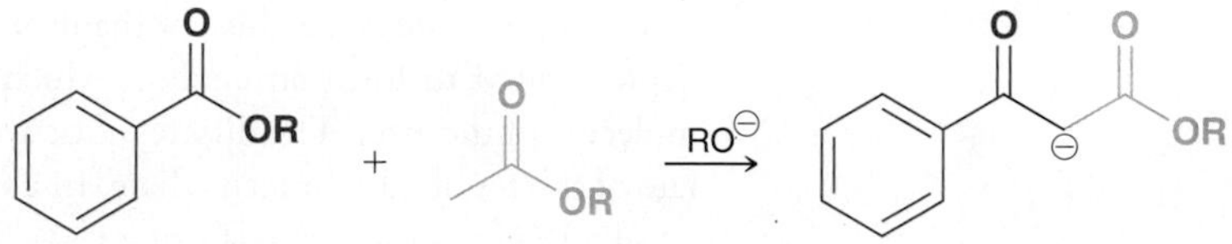

In this reaction, the aryl ester lacks α protons and cannot form an enolate. Only the other ester is capable of forming an enolate, which reduces the number of possible products.

2. A directed Claisen condensation can be performed in which LDA is used as a base to irreversibly form an ester enolate, which is then treated with a different ester.

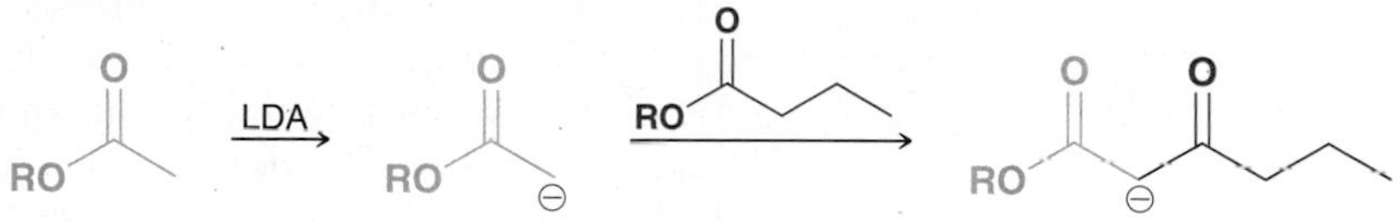

Intramolecular Claisen Condensations: The Dieckmann Cyclization

Just as we saw with aldol reactions, Claisen condensations can also occur in an intramolecular fashion, for example,

BY THE WAY
When drawing the product of a Dieckmann cyclization, it is wise to use a numbering system to keep track of all carbon atoms, as shown here.

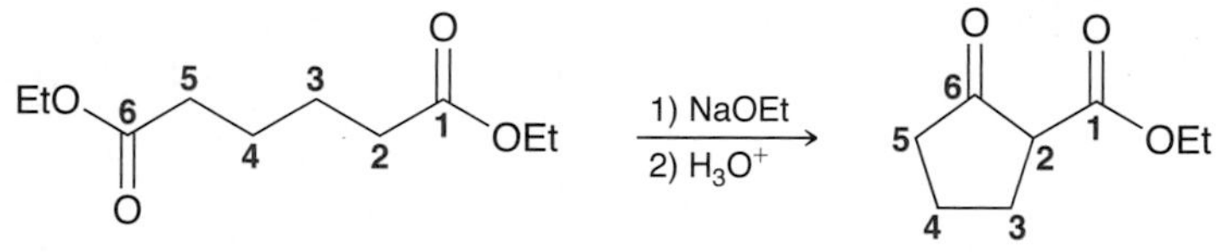

(80%)

This process is called a **Dieckmann cyclization**, and the product is a cyclic, β-keto ester. The one notable difference between this process and any other Claisen condensation is that the ester enolate and the ester moiety are both present in the same molecule, resulting in an intramolecular attack.

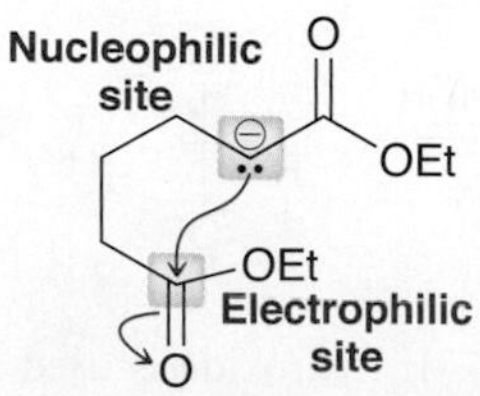

Intramolecular Claisen condensations show a preference for formation of five- and six-membered rings, just as we saw with intramolecular aldol reactions.

CONCEPTUAL CHECKPOINT

22.28 Identify the base you would use for each of the following transformations.

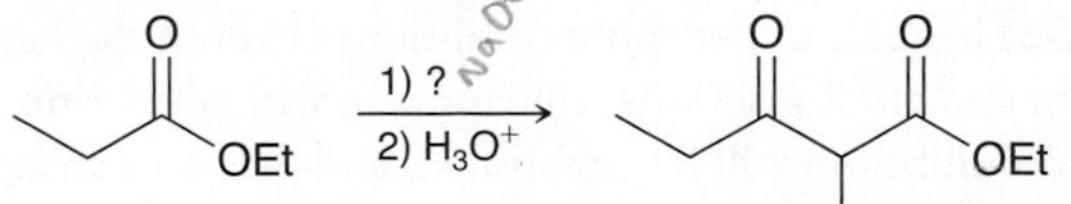

(a)

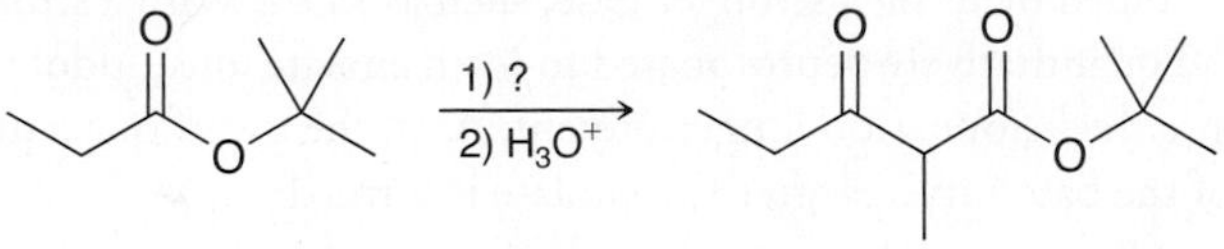

(b)

22.29 Predict the major product obtained when each of the following compounds undergoes a Claisen condensation.

(a) 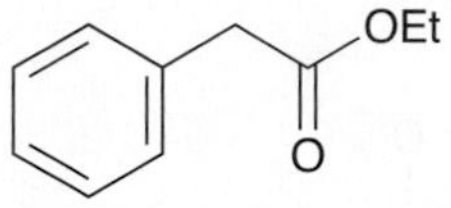

(b) 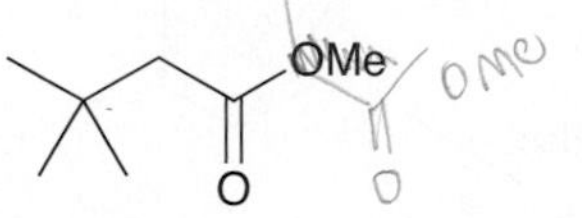

(c) 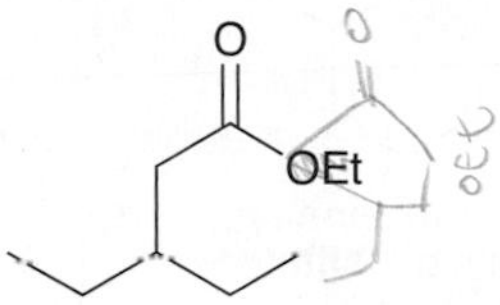

22.30 Identify the reagents that you would use to produce each of the following compounds using a Claisen condensation.

(a) 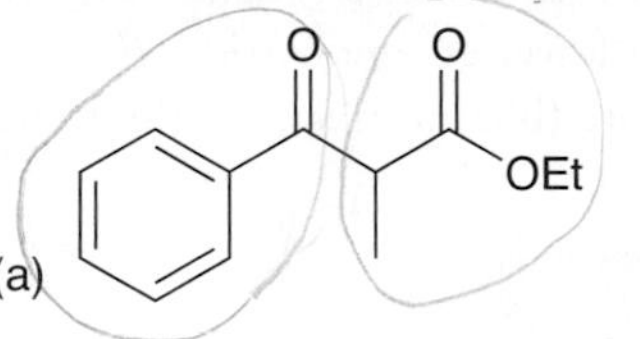

(b) 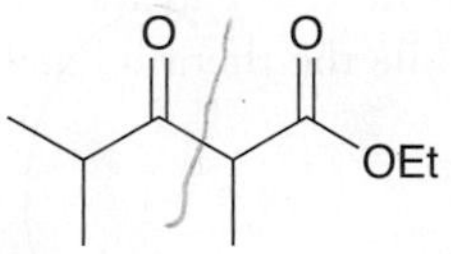

(c) 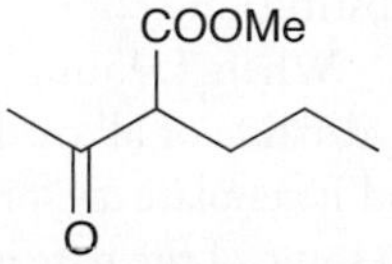

(d) 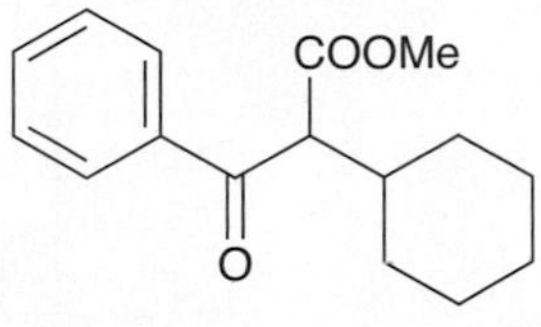

(e) 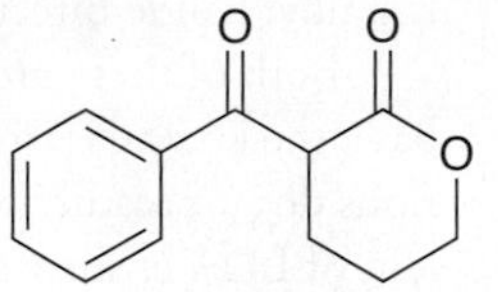

22.31 Predict the product of the Dieckmann cyclization that occurs when each of the following compounds is treated with sodium ethoxide.

(a) 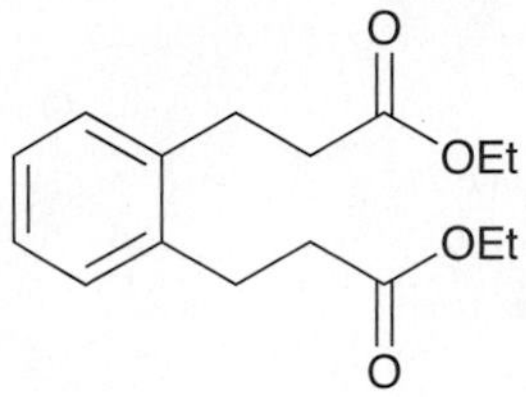

(b) 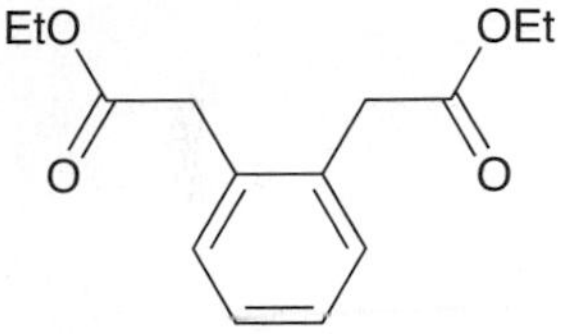

(c) 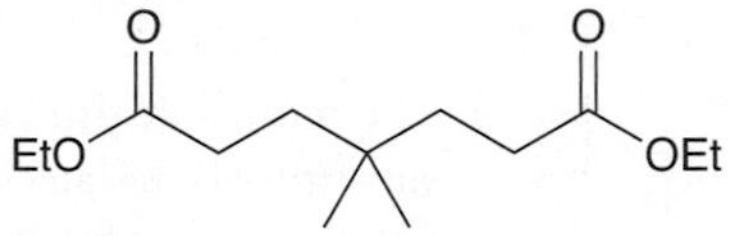

22.32 When the following compound is treated with sodium ethoxide, two condensation products are obtained, both of which are produced via Dieckmann cyclizations. Draw both products.

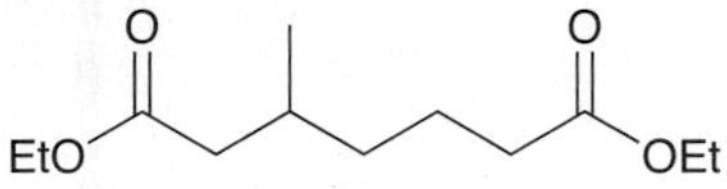

22.5 Alkylation of the Alpha Position

Alkylation via Enolate Ions

The α position of a ketone can be alkylated via a two-step process: (1) formation of an enolate followed by (2) treating the enolate with an alkyl halide.

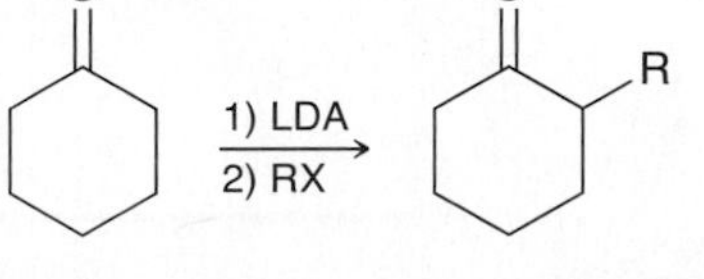

In this process, the enolate ion functions as a nucleophile and attacks the alkyl halide in an S_N2 reaction.

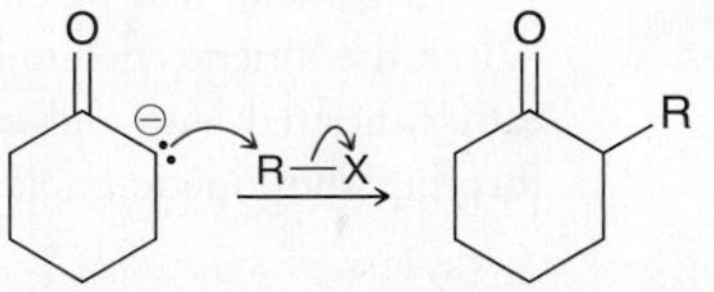

The usual restrictions for S_N2 reactions apply, so that the alkyl halide should be either a methyl halide or a primary halide. With a secondary or tertiary alkyl halide, the enolate functions as a base instead of a nucleophile, and the alkyl halide undergoes elimination rather than substitution.

When forming the enolate for an alkylation process, the choice of base is important. Hydroxide or alkoxide ions cannot be used because under those conditions (1) both the ketone and its enolate are present at equilibrium and aldol reactions will compete with alkylation and (2) some of the base is not consumed (an equilibrium will be established) and the base can attack the alkyl halide directly, providing for competing S_N2 and E2 reactions.

Both of these problems are avoided by using a stronger base, such as LDA. With a stronger base, the ketone is irreversibly and quantitatively deprotonated to form enolate ions. Aldol reactions do not readily occur, because the ketone is no longer present. Also, the use of one equivalent of LDA ensures that none of the base survives after the enolate is formed.

With an unsymmetrical ketone, two possible enolates can be formed.

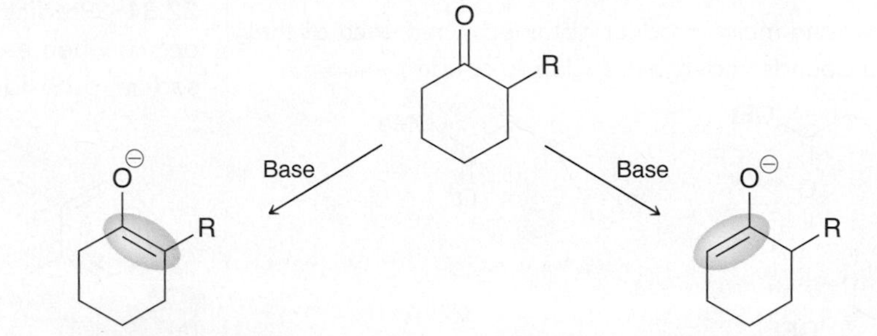

The more-substituted enolate is more stable and is called the *thermodynamic enolate*. The less-substituted enolate is less stable, but it is formed more rapidly and is therefore called the *kinetic enolate*. Figure 22.4 compares the pathways that lead to each of these enolates. Notice that formation of the kinetic enolate (red pathway) involves a lower energy barrier (E_a) and therefore occurs more rapidly, while the thermodynamic enolate (blue pathway) is the more stable enolate.

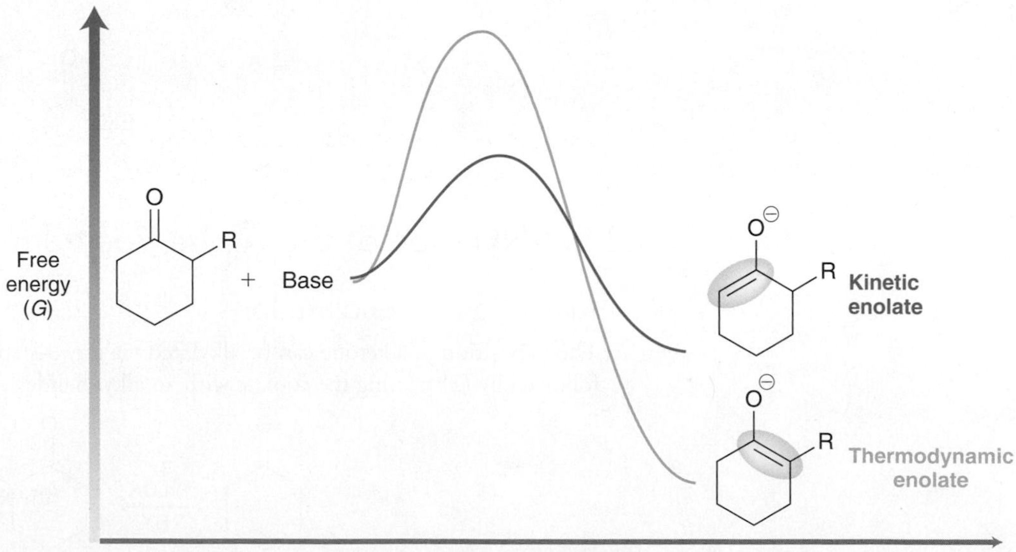

FIGURE 22.4 An energy diagram showing two possible pathways for the deprotonation of an unsymmetrical ketone. One pathway leads to the kinetic enolate, and the other pathway leads to the thermodynamic enolate. The differences in energy between these two pathways has been exaggerated for clarity of presentation.

In general, it is possible to choose conditions that will favor formation of either enolate. When the kinetic enolate is desired, LDA is used at low temperature (−78°C). LDA is a sterically hindered base and can more readily deprotonate the less hindered α position, thereby forming the kinetic enolate. Low temperature is necessary to favor formation of the kinetic enolate and to prevent equilibration of the enolates via proton transfers. When the thermodynamic enolate is desired, a nonsterically hindered base (such as NaH) is used, and the reaction is performed at room temperature. Under these conditions, both enolates are formed and an equilibration process occurs that favors the more stable, thermodynamic enolate. In this way, alkylation can be achieved at either α position by carefully choosing the reagents and conditions.

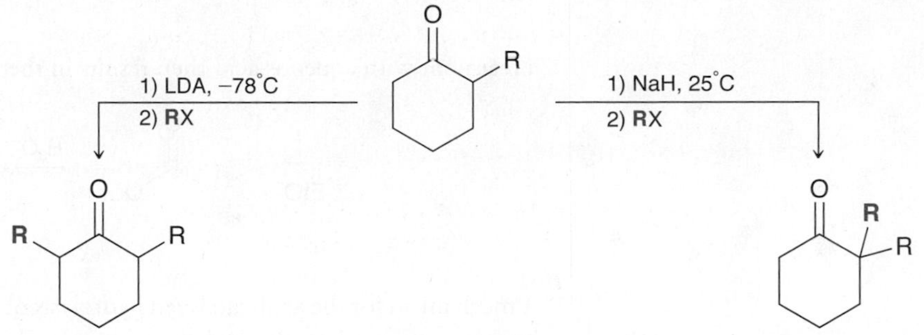

CONCEPTUAL CHECKPOINT

22.33 For each of the following reactions, predict the major product and propose a mechanism for its formation.

(a)

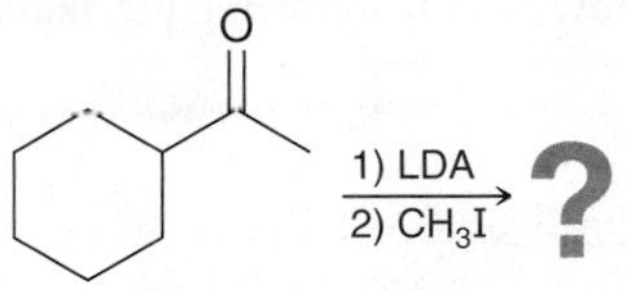

(b)

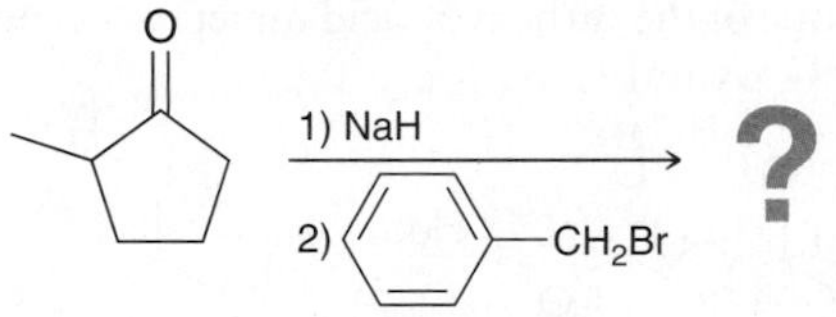

(c)

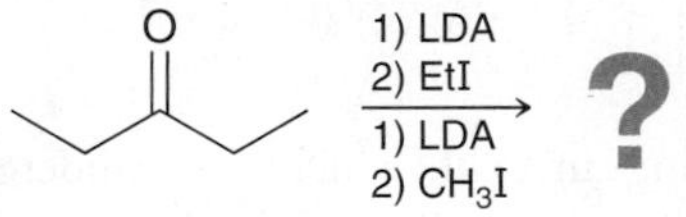

22.34 Identify the reagents you would use to achieve the following transformation:

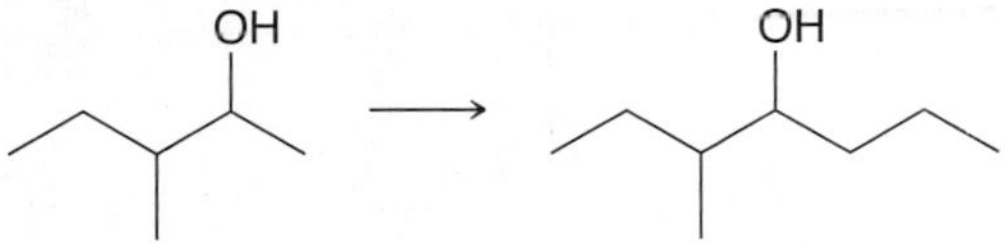

The Malonic Ester Synthesis

The **malonic ester synthesis** is a technique that enables the transformation of a halide into a carboxylic acid with the introduction of two new carbon atoms.

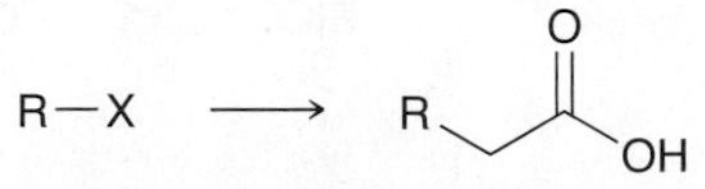

One of the key reagents used to achieve this transformation is diethyl malonate.

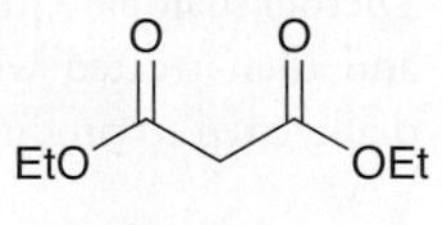

Diethyl malonate

This malonic ester is a fairly acidic compound. The first step of the malonic ester synthesis is the deprotonation of diethyl malonate, forming a doubly stabilized enolate.

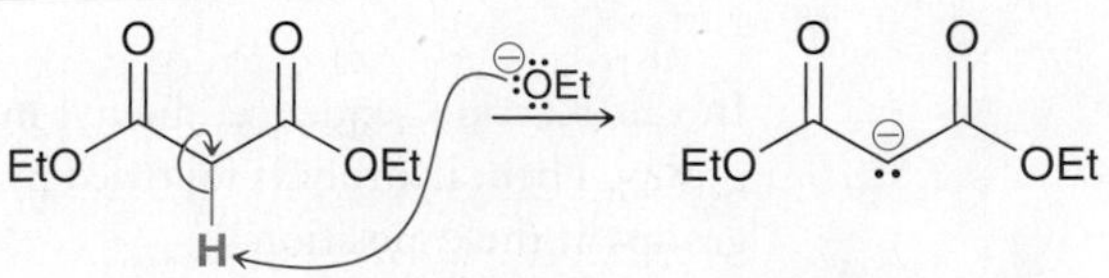

This enolate is then treated with an alkyl halide (RX), resulting in an alkylation step.

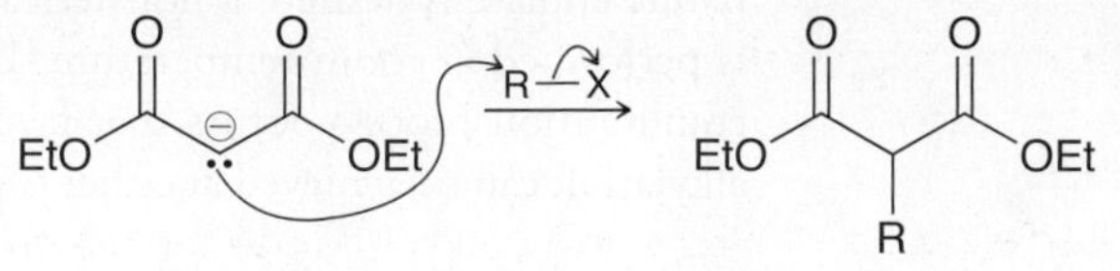

Treatment with aqueous acid then results in the hydrolysis of both ester moieties:

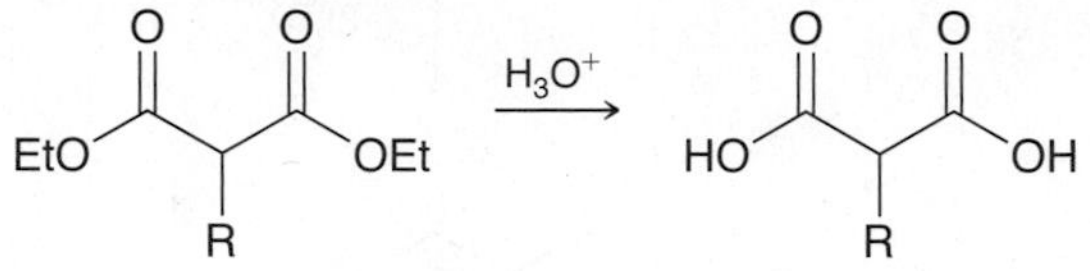

A mechanism for the acid-catalyzed hydrolysis of esters was discussed in Section 21.11. If hydrolysis is performed at elevated temperatures, the resulting 1,3-dicarboxylic acid will undergo **decarboxylation** to produce a monosubstituted acetic acid and carbon dioxide.

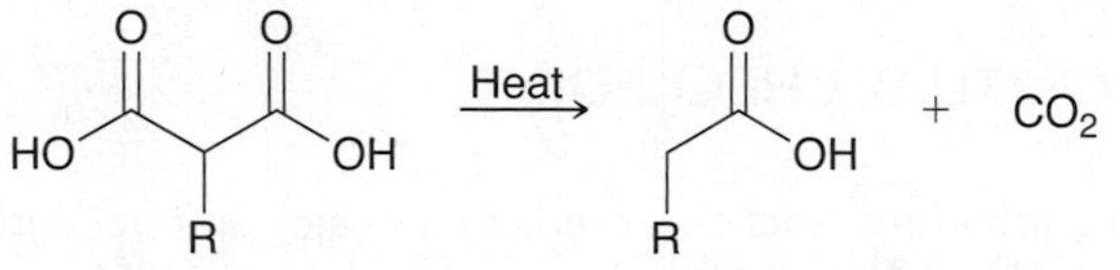

In this step, one of the carboxylic acid moieties is expelled as shown in the following mechanism.

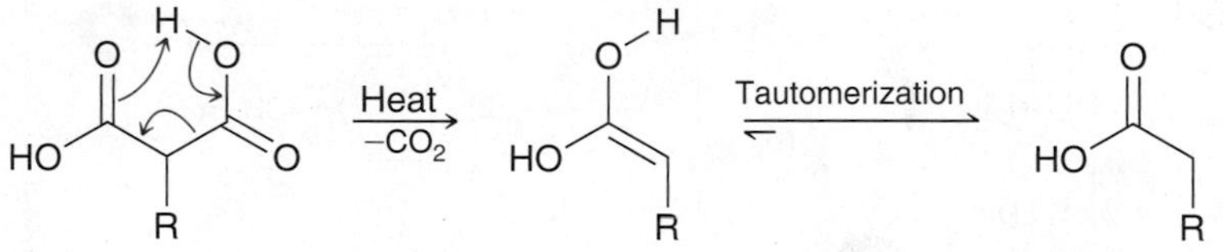

In the first step, a pericyclic reaction takes place, forming an enol, which then undergoes tautomerization. The process yields a monocarboxylic acid, which does not undergo further decarboxylation.

The following reaction is an example of a malonic ester synthesis.

LOOKING BACK
For a review of tautomerization, see Section 10.7.

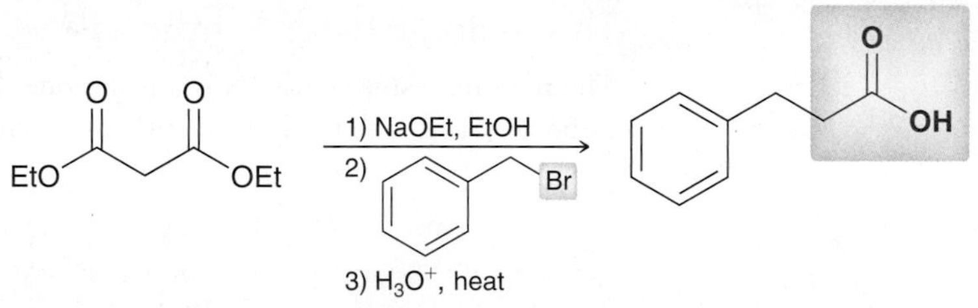

Diethyl malonate, the starting material, is first deprotonated, then treated with an alkyl halide, and then treated with aqueous acid at elevated temperature. Diethyl malonate can also be dialkylated to produce carboxylic acids.

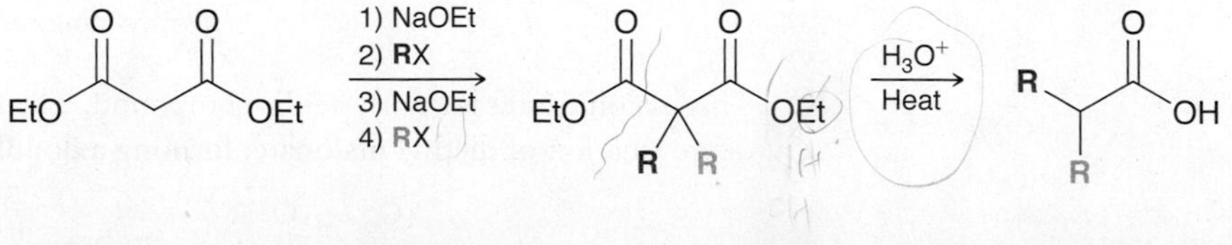

In this reaction sequence, diethyl malonate is alkylated twice, each time with a different alkyl group. Then, hydrolysis followed by decarboxylation produces a carboxylic acid with two alkyl groups at the α position.

SKILLBUILDER

22.5 USING THE MALONIC ESTER SYNTHESIS

LEARN the skill

Show how you would use the malonic ester synthesis to prepare the following compound:

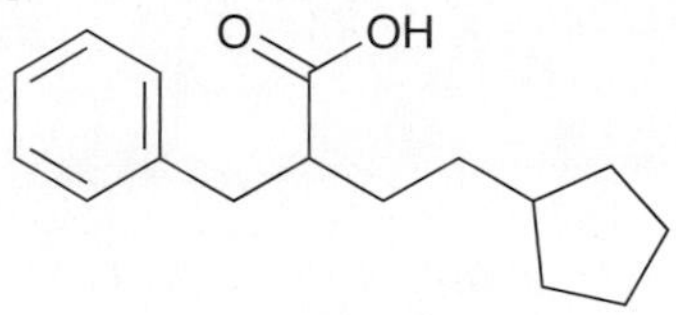

SOLUTION

Begin by finding the carboxylic acid moiety and the α position.

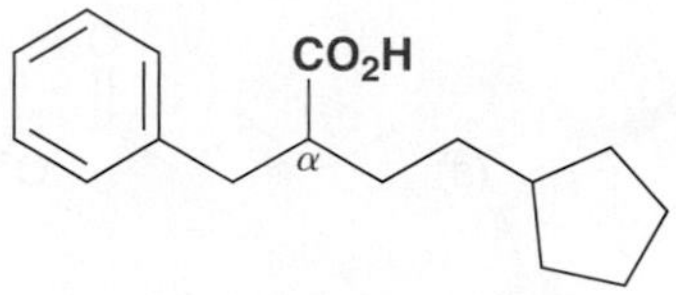

Identify the alkyl groups connected to the α position.

STEP 1 Identify the alkyl groups connected to the α position.

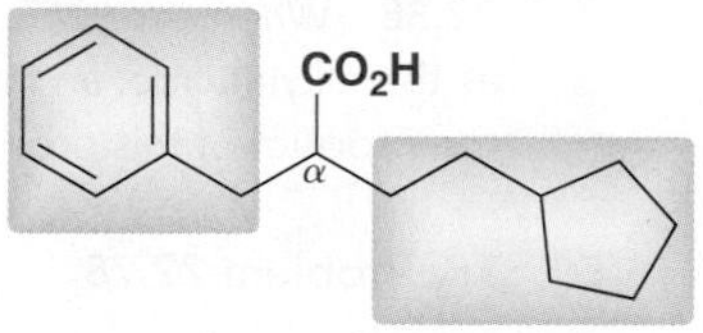

In this case, there are two. Next, identify the alkyl halides that are necessary in order to install those two alkyl groups. Just redraw the alkyl groups connected to some halide (iodide and bromide are more commonly used as leaving groups, although chloride can be used). Below are alkyl bromides that can be used:

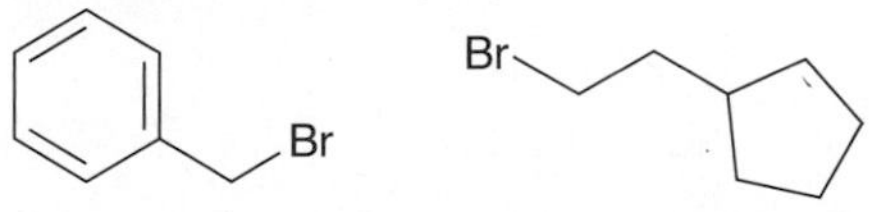

STEP 2 Identify the necessary alkyl halides and make sure they can serve as substrates in an S_N2 process.

Analyze the structures of these alkyl halides and make sure that both will readily undergo an S_N2 process. If either substrate is tertiary, then a malonic ester synthesis will fail, because the alkylation step will not occur. In this case, both alkyl halides are primary, so the malonic ester synthesis can be used.

In order to perform the malonic ester synthesis, we will need diethyl malonate and the two alkyl halides shown above. It does not matter which alkyl group is installed first. The synthesis can be summarized as follows.

STEP 3 Identify all reagents, beginning with diethyl malonate.

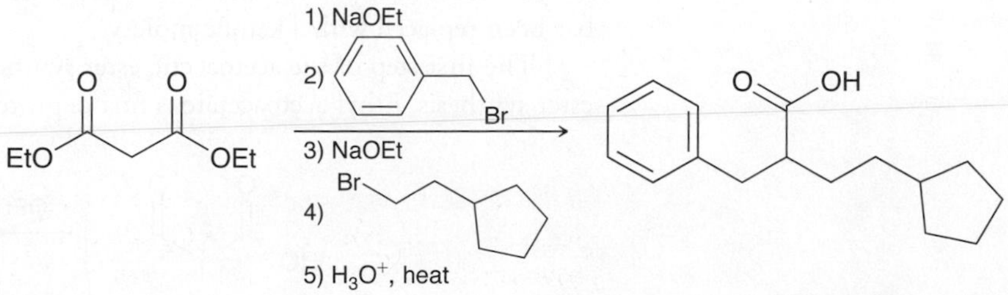

PRACTICE the skill **22.35** Propose an efficient synthesis for each of the following compounds using the malonic ester synthesis.

(a) 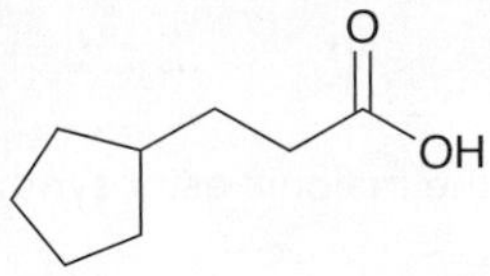

(b) 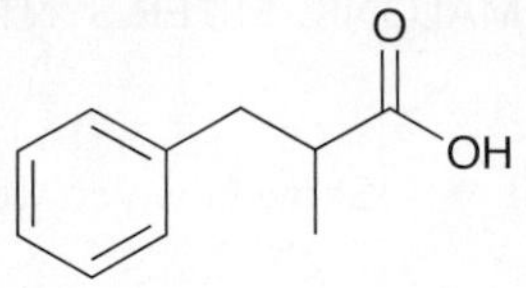

(c) 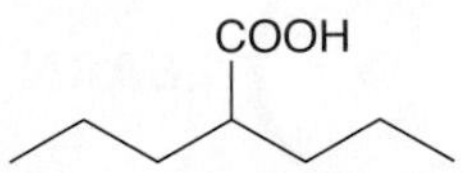

(d) 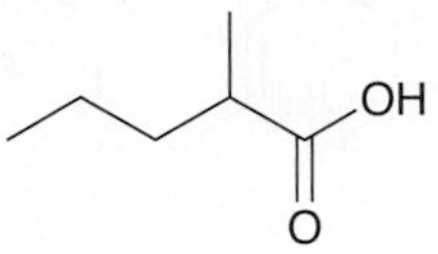

(e) 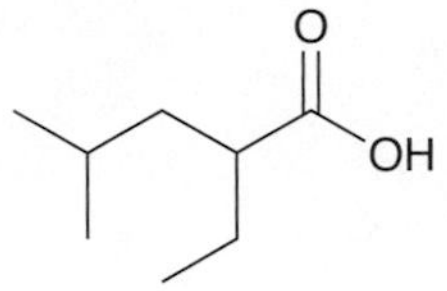

APPLY the skill **22.36** Starting with diethyl malonate and using any other reagents of your choice, propose an efficient synthesis for each of the following compounds:

(a) 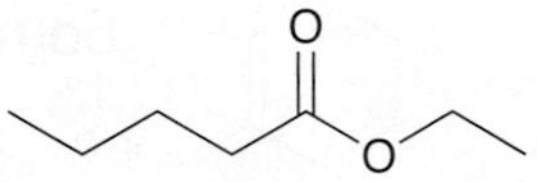

(b) 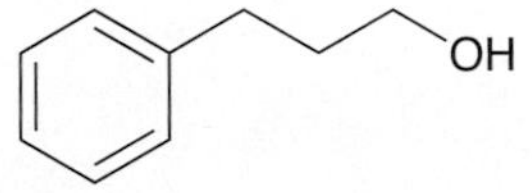

(c) 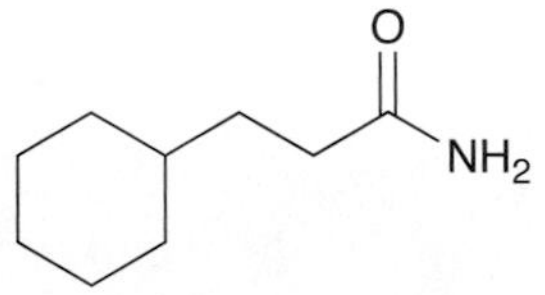

22.37 The malonic ester synthesis cannot be used to make 2,2-dimethylhexanoic acid. Explain why not.

22.38 When a malonic ester synthesis is performed using excess base and 1,4-dibromobutane as the alkyl halide, an intramolecular reaction occurs, and the product contains a ring. Draw the product of this process.

need more **PRACTICE?** **Try Problem 22.78**

The Acetoacetic Ester Synthesis

The **acetoacetic ester synthesis** is a useful technique that converts an alkyl halide into a methyl ketone with the introduction of three new carbon atoms.

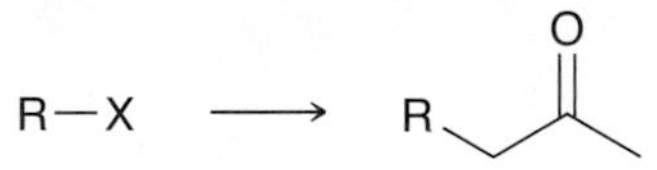

This process is very similar to the malonic ester synthesis except that the key reagent is ethyl acetoacetate rather than diethyl malonate.

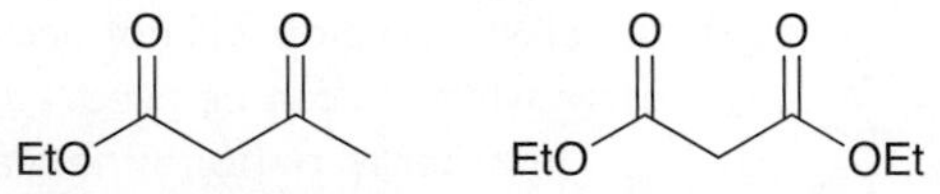

Ethyl acetoacetate **Diethyl malonate**

Ethyl acetoacetate is very similar in structure to diethyl malonate, but one of the ester moieties has been replaced with a ketone moiety.

The first step of the acetoacetic ester synthesis is analogous to the first step of the malonic ester synthesis. Ethyl acetoacetate is first deprotonated, forming a doubly stabilized enolate.

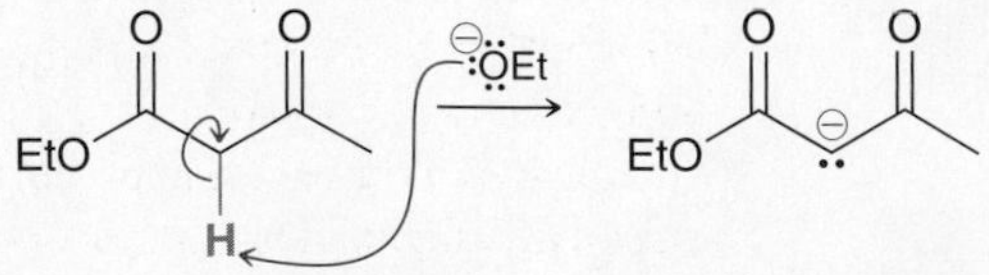

This enolate is then treated with an alkyl halide (RX), resulting in an alkylation step.

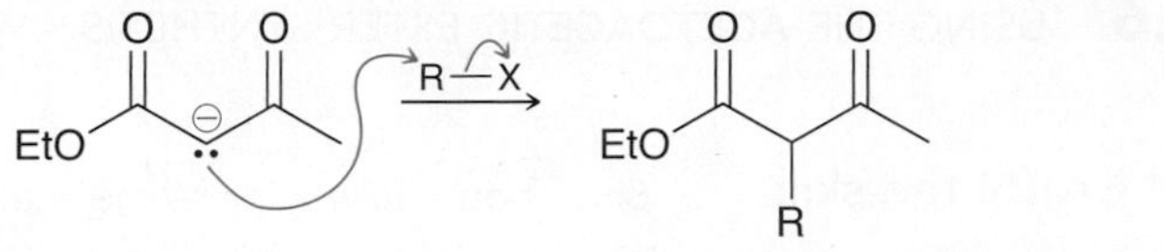

Treatment with aqueous acid then results in the hydrolysis of the ester moiety.

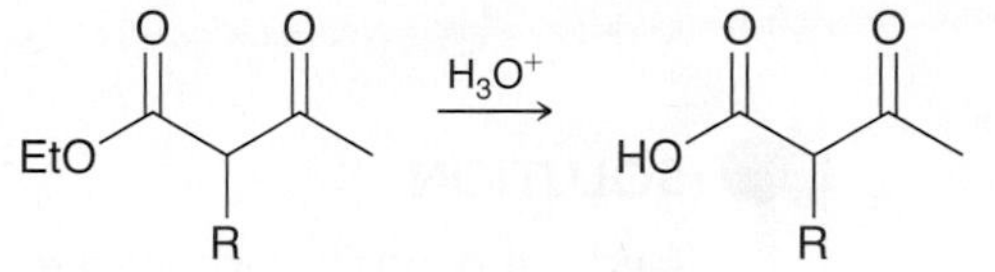

If hydrolysis is performed at an elevated temperature, the resulting β-keto acid will undergo decarboxylation to produce a ketone and carbon dioxide.

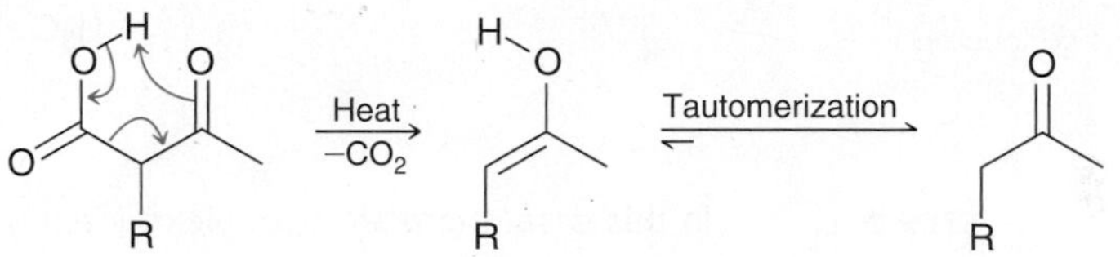

This process is very similar to the process we discussed for 1,3-dicarboxylic acids. In the first step, a pericyclic reaction forms an enol, which then undergoes tautomerization to form a ketone. Decarboxylation can occur because the carboxylic acid exhibits a carbonyl group in the β position, which enables the pericyclic reaction shown above. Below is an example of an acetoacetic ester synthesis.

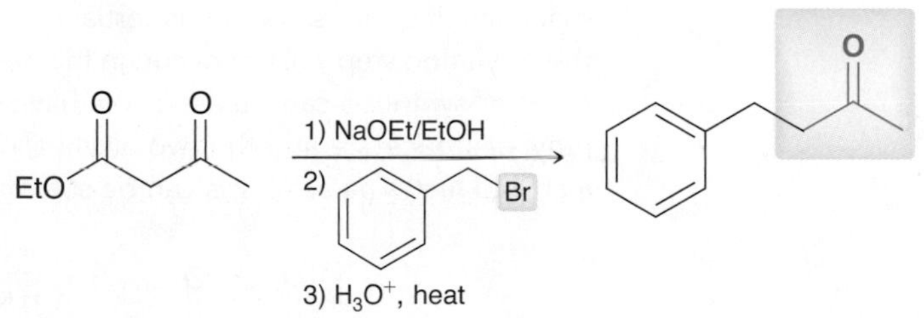

Notice the similarity to the malonic ester synthesis. Ethyl acetoacetate is the starting material. It is first deprotonated, then treated with an alkyl halide, and then treated with aqueous acid at elevated temperature. Ethyl acetoacetate can also be dialkylated to produce ketones as follows:

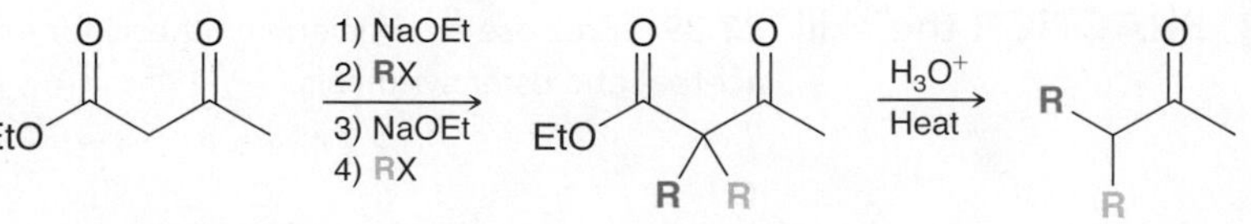

In this reaction sequence, ethyl acetoacetate is alkylated twice, each time with a different alkyl group. Then, hydrolysis followed by decarboxylation produces a derivative of acetone in which two alkyl groups are positioned at the α position. Why is it necessary to use an acetoacetic ester synthesis instead of simply treating the enolate of acetone with an alkyl halide? Wouldn't that be a more direct method for preparing substituted acetones? There are two answers to this question: (1) Direct alkylation of enolates is often difficult to achieve in good yields and (2) enolates are not only strong nucleophiles, but they are also strong bases, and as a result, enolates can react with alkyl halides to produce elimination products.

SKILLBUILDER

22.6 USING THE ACETOACETIC ESTER SYNTHESIS

LEARN the skill

Show how you would use the acetoacetic ester synthesis to prepare the following compound:

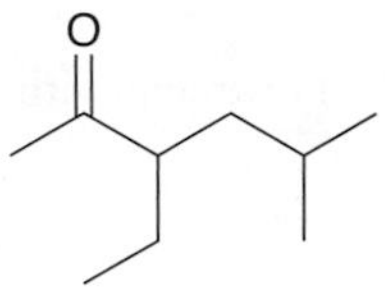

SOLUTION

Find the α position connected to the methyl ketone moiety and identify the alkyl groups connected to the α position:

STEP 1
Identify the alkyl groups connected to the α position.

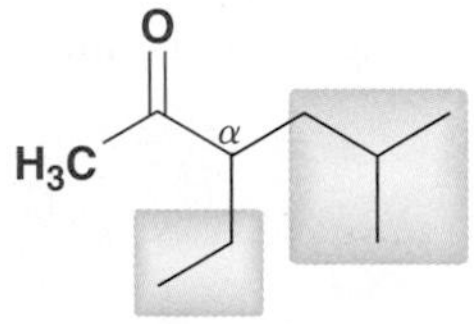

STEP 2
Identify the necessary alkyl halides and make sure they can serve as substrates in an S_N2 process.

In this case, there are two. Next, identify the alkyl halides needed to install those two alkyl groups. Just redraw those alkyl groups connected to some halide. Below are two alkyl iodides that can be used.

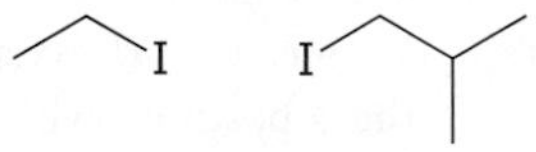

Analyze the structures of these alkyl halides and make sure that both will readily undergo S_N2 reactions. If either substrate is tertiary, then an acetoacetic ester synthesis will fail, because the alkylation step will not occur. In this case, both alkyl halides are primary, so the acetoacetic ester synthesis can be used. We have seen that the acetoacetic ester synthesis requires ethyl acetoacetate and the two alkyl halides shown. It does not matter which alkyl group is installed first. The synthesis can be summarized as follows:

STEP 3
Identify all reagents, beginning with ethyl acetoacetate.

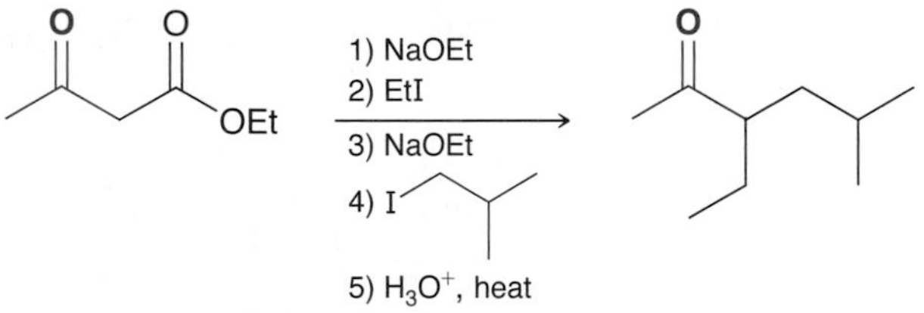

PRACTICE the skill

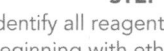

22.39 Propose an efficient synthesis for each of the following compounds using the acetoacetic ester synthesis.

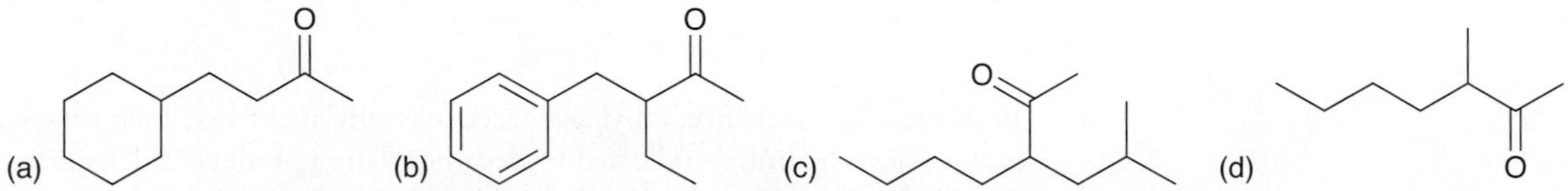

APPLY the skill

22.40 In Problem 22.38, we saw an intramolecular example of a malonic ester synthesis using excess base and 1,4-dibromobutane. If this dibromide is used in an acetoacetic ester synthesis, an intramolecular process can also occur. Predict the product of that reaction.

22.41 Starting with ethyl acetoacetate and using any other reagents of your choice, propose an efficient synthesis for each of the following compounds.

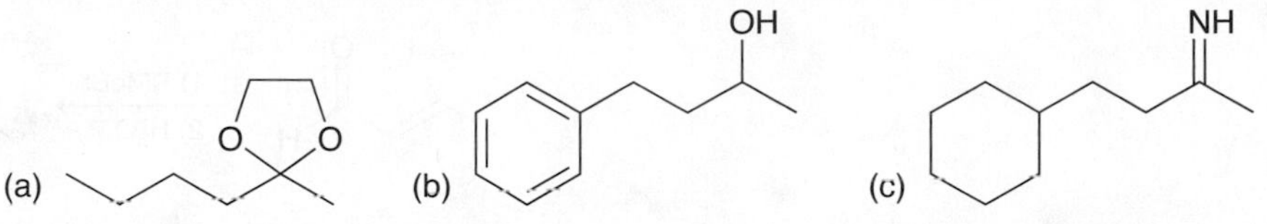

22.42 The acetoacetic ester synthesis cannot be used to make 3,3-dimethyl-2-hexanone. Explain why not.

22.43 The product of a Dieckmann cyclization can undergo alkylation, hydrolysis, and decarboxylation. This sequence represents an efficient method for preparing 2-substituted cyclopentanones and cyclohexanones (below). Using this information, propose an efficient synthesis of 2-propylcyclohexanone using 1,7-heptanediol and propyl iodide.

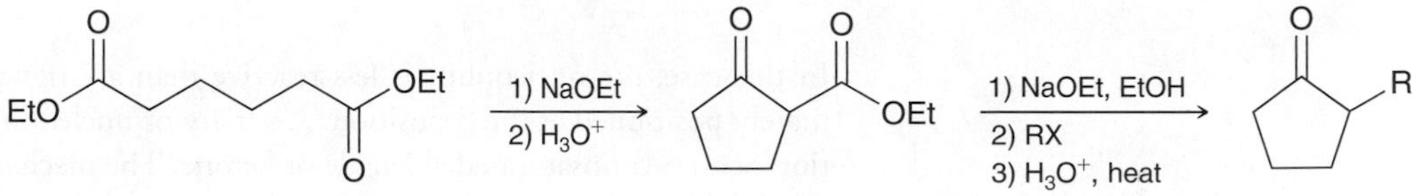

need more **PRACTICE?** **Try Problem 22.79**

22.6 Conjugate Addition Reactions

Michael Reactions

Recall that the product of an aldol condensation is an α,β-unsaturated aldehyde or ketone, for example,

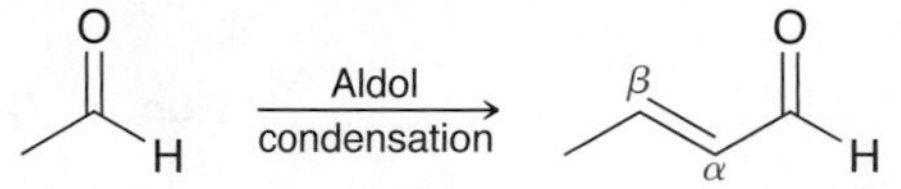

Aldehydes and ketones that possess α,β unsaturation, like the compound above, exhibit unique reactivity at the β position. In this section, we will explore reactions that can take place at the β position. To understand why the β position is reactive, we must draw the resonance structures of the compound.

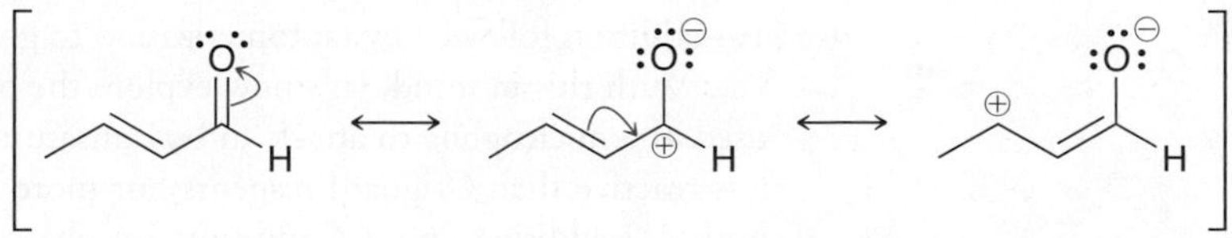

Notice that two of the resonance contributors exhibit positive charges. In other words, the compound will have two electrophilic positions, the carbon of the carbonyl group as well as the β position.

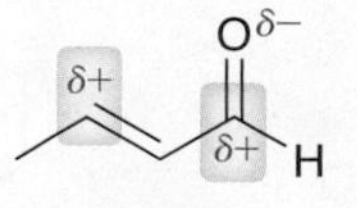

Two electrophilic positions

Both of these positions are therefore subject to attack by a nucleophile, and the nature of the nucleophile determines which position is attacked. For example, Grignard reagents tend to attack the carbonyl group rather than the β position.

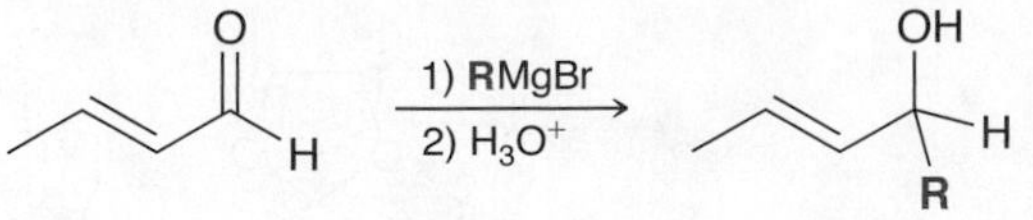

In this reaction, the nucleophile attacks the carbonyl group, giving a tetrahedral intermediate, which is then treated with a proton source in a separate step. In this process, two groups (R and H) have added across the C=O π bond in a 1,2-addition (for a review of the difference between 1,2-additions and 1,4-additions, see Section 17.4).

A different outcome is observed when a lithium dialkyl cuprate (R_2CuLi) is employed.

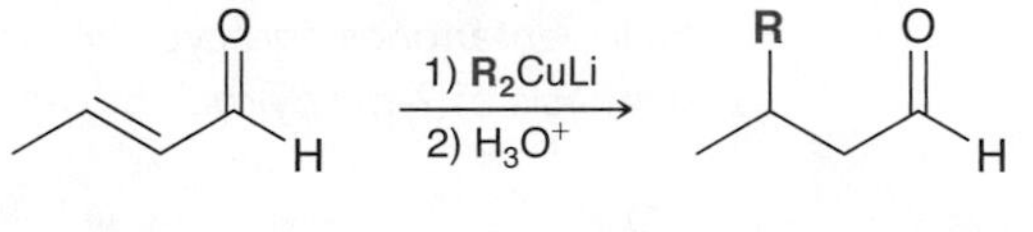

In this case, the nucleophile is less reactive than a Grignard reagent, and the R group is ultimately positioned at the β position. A variety of nucleophiles can be used to attack the β position of an α,β-unsaturated aldehyde or ketone. The mechanism involves attack at the β position followed by protonation to give an enol.

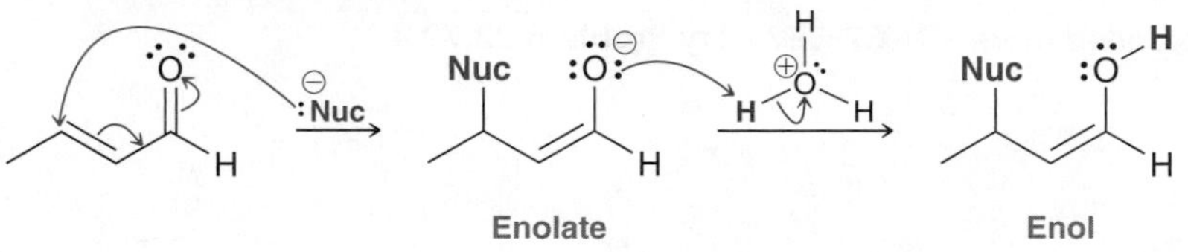

This type of reaction is called a **conjugate addition,** or a **1,4-addition**, because the nucleophile and the proton have added across the ends of a conjugated π system. Conjugate addition reactions were first discussed in Section 17.4 when we explored the reactivity of conjugated dienes. A major difference between conjugate addition across a diene and conjugate addition across an enone is that the latter produces an enol as the product, and the enol rapidly tautomerizes to form a carbonyl group.

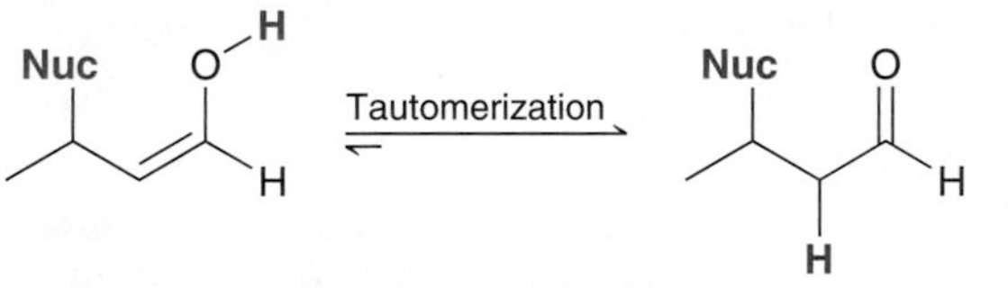

After the reaction is complete, it might appear that the two groups (shown in red) have added across the α and β positions in a 1,2-addition. However, the actual mechanism likely involves a 1,4-addition followed by tautomerization to give the ultimate product shown.

With this in mind, let's now explore the outcome of a reaction in which an enolate ion is used as a nucleophile to attack an α,β-unsaturated aldehyde or ketone. In general, enolates are less reactive than Grignard reagents but more reactive than lithium dialkyl cuprates. As such, both 1,2-addition and 1,4-addition are observed, and a mixture of products is obtained. In contrast, doubly stabilized enolates are sufficiently stabilized to produce 1,4 conjugate addition exclusively.

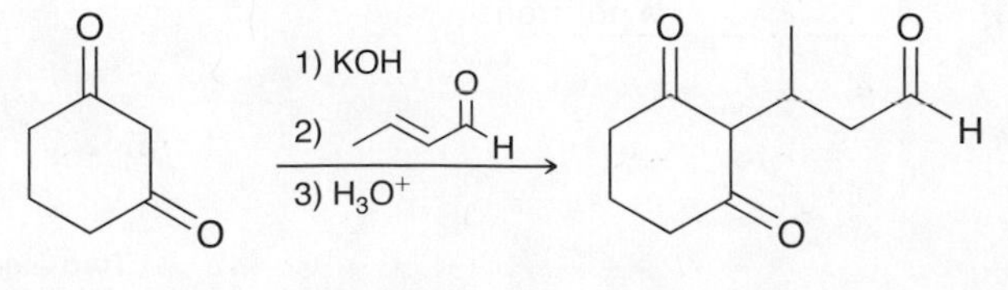

TABLE 22.2 **A LIST OF COMMON MICHAEL DONORS AND MICHAEL ACCEPTORS**

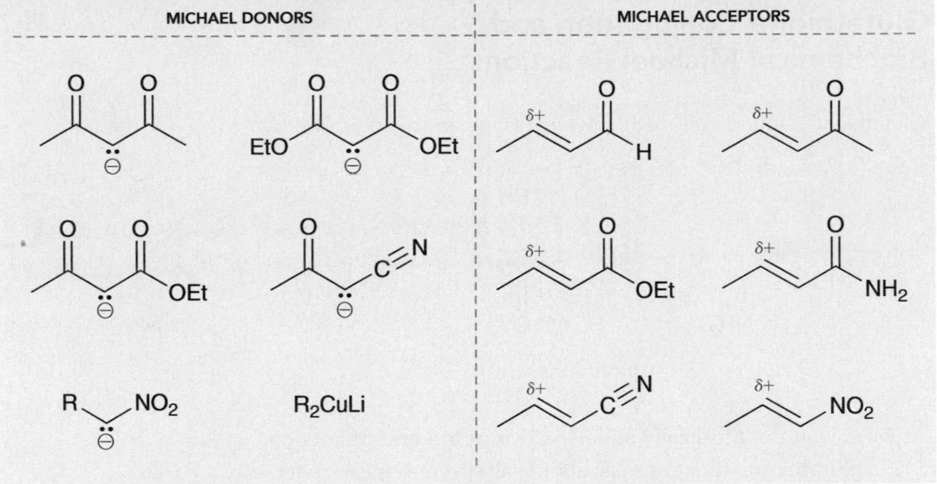

In this case, the starting diketone is deprotonated to form a doubly stabilized enolate ion, which then serves as a nucleophile in a 1,4-conjugate addition. This process is called a **Michael reaction**. The doubly stabilized enolate is called a **Michael donor**, while the α,β-unsaturated aldehyde is called a **Michael acceptor**.

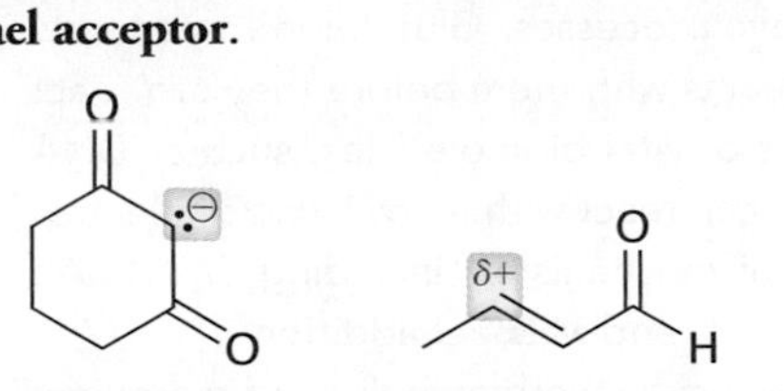

Michael donor Michael acceptor

A variety of Michael donors and acceptors are observed to react with each other to produce a Michael reaction. Table 22.2 shows some common examples of Michael donors and acceptors. Any one of the Michael donors in this table will react with any one of the Michael acceptors to give a 1,4-conjugate addition reaction.

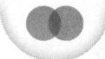

CONCEPTUAL CHECKPOINT

22.44 Identify the major product formed when each of the following compounds is treated with Et_2CuLi followed by mild acid.

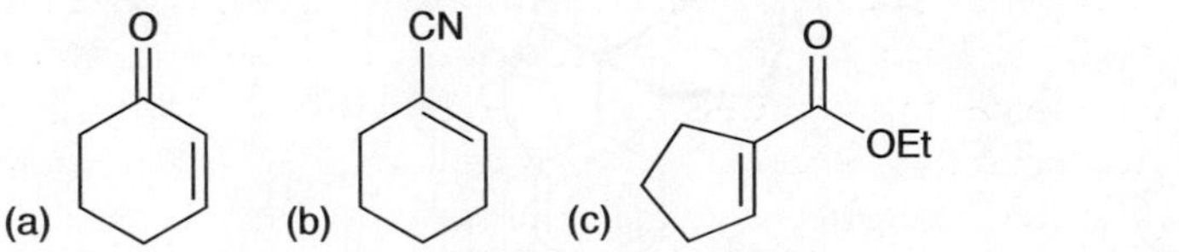

22.45 Predict the major product of the three following steps and show a mechanism for its formation.

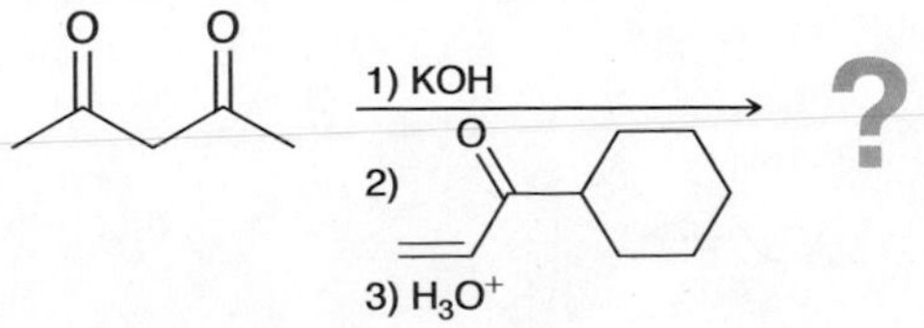

22.46 In the previous section, we learned how to use malonic ester as a starting material in the preparation of substituted carboxylic acids (the malonic ester synthesis). That method employed a step in which the enolate of malonic ester attacked an alkyl halide to give an alkylation product. In this section, we saw that the enolate of malonic ester can attack many electrophilic reagents other than simple alkyl halides. Specifically, the enolate of malonic ester can attack any of the Michael acceptors in Table 22.2. Using malonic ester as your starting material and any other reagents of your choice, show how you would prepare each of the following compounds.

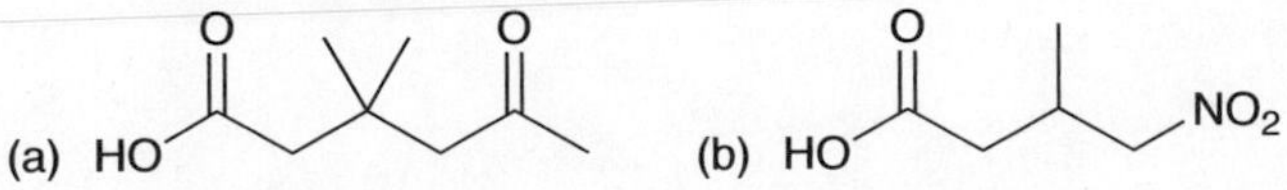

MEDICALLYSPEAKING

Glutathione Conjugation and Biochemical Michael Reactions

Glutathione is found in all mammals and plays many biological roles.

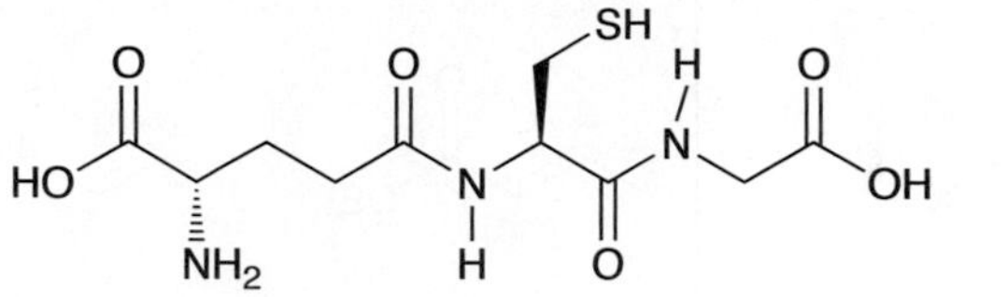

Glutathione

As we saw in the Medically Speaking box at the end of Section 11.9, glutathione functions as a radical scavenger by reacting with free radicals that are generated during metabolic processes. This compound has many other important biological functions. For example, the sulfur atom in glutathione is highly nucleophilic, which allows this compound to function as a scavenger for harmful *electrophiles* that are ingested or are produced by metabolic processes. Glutathione intercepts these electrophiles and reacts with them before they can react with the nucleophilic sites of vital biomolecules, such as DNA and proteins. Glutathione can react with many kinds of electrophiles through a variety of mechanisms, including S_N2, S_NAr, nucleophilic acyl substitution, and Michael addition.

A Michael addition occurs in the metabolism of morphine. One of the metabolites of morphine is generated via oxidation of the allylic alcohol moiety to produce an α,β-unsaturated ketone:

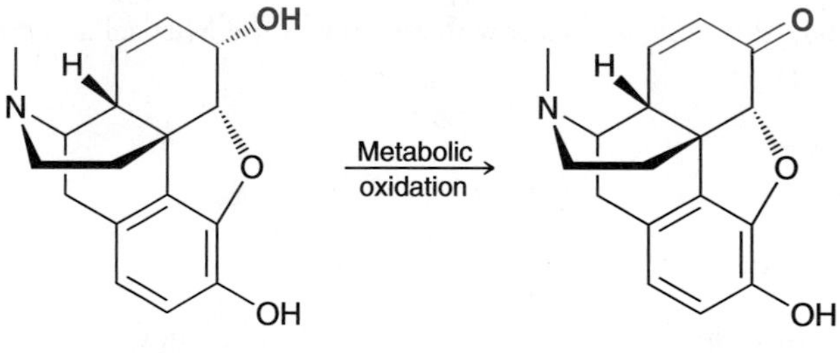

Morphine **A metabolite of morphine**

The β position of this metabolite is now highly electrophilic and subject to attack by a variety of nucleophiles. The sulfur atom of glutathione attacks at the β position, producing a Michael addition. This process is called glutathione conjugation, and the product is called a glutathione conjugate, which can be further metabolized to produce a mercapturic acid conjugate, a compound that is excreted in the urine.

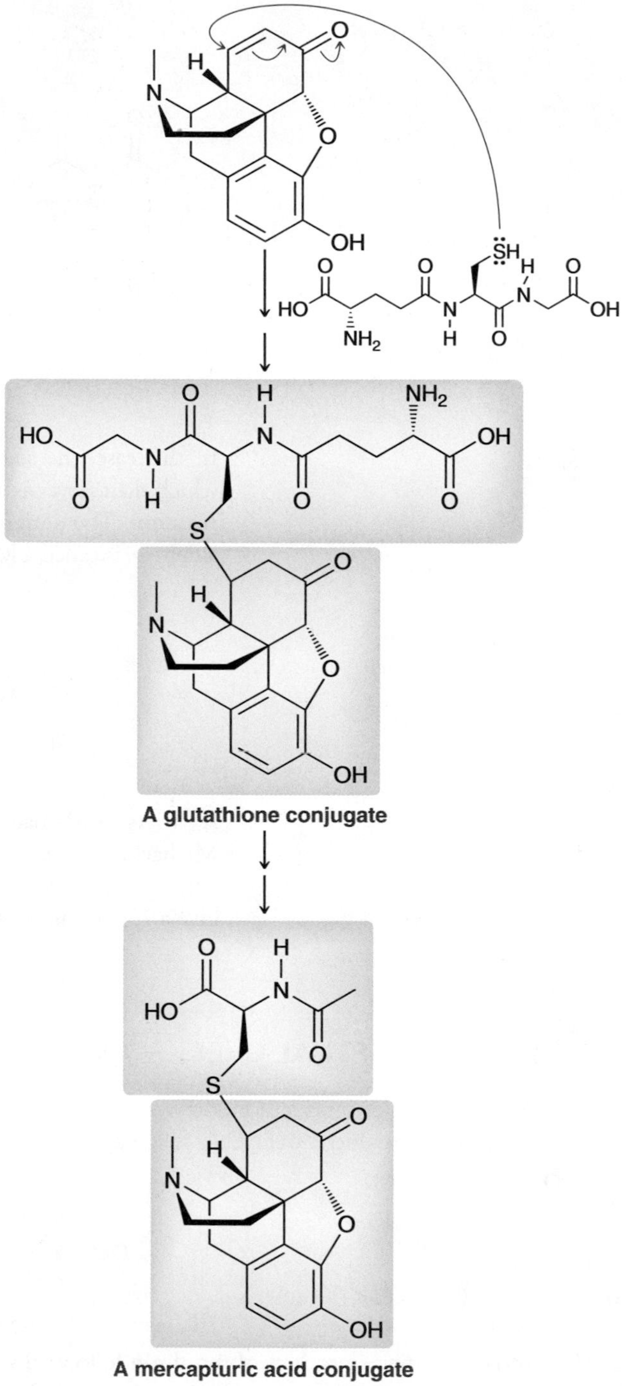

Glutathione conjugation also occurs in the metabolism of acetaminophen (Tylenol™). Metabolic oxidation produces an *N*-acetylimidoquinone intermediate, which is highly reactive.

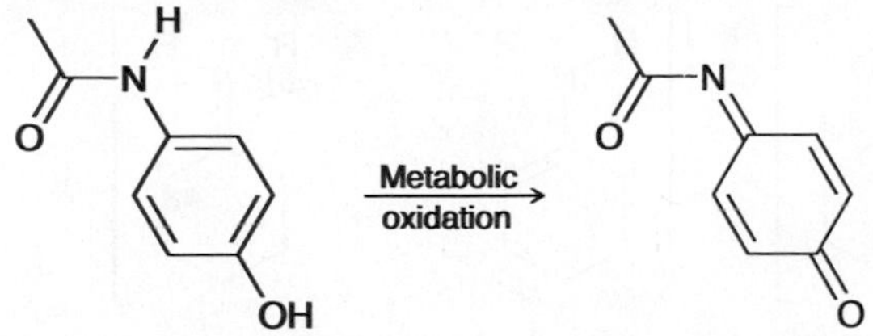

Acetaminophen **A metabolite of acetaminophen (*N*-acetylimidoquinone)**

This metabolite undergoes glutathione conjugation followed by further metabolism that generates a water-soluble mercapturic acid derivative which is excreted in the urine.

In our first discussion of glutathione and its role as a radical scavenger, we mentioned that an overdose of acetaminophen causes a temporary depletion of glutathione in the liver, during which time harmful radicals and electrophiles are free to run amok and cause permanent damage. Our current discussion of glutathione now explains why glutathione levels drop in the presence of too much acetaminophen. To review how an acetaminophen overdose is treated, see the Medically Speaking box at the end of Section 11.9.

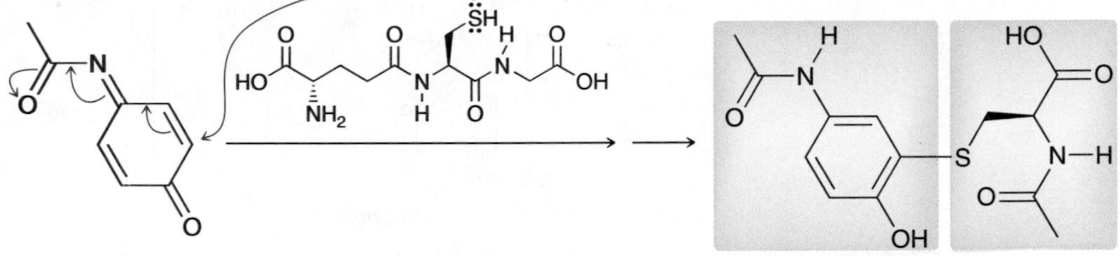

A mercapturic acid conjugate of acetaminophen

The Stork Enamine Synthesis

We saw in the previous section that doubly stabilized enolates can serve as Michael donors. It is important to point out that regular enolates do not serve as Michael donors.

A Michael donor **Not a Michael donor**

Therefore, the following synthesis will not be efficient.

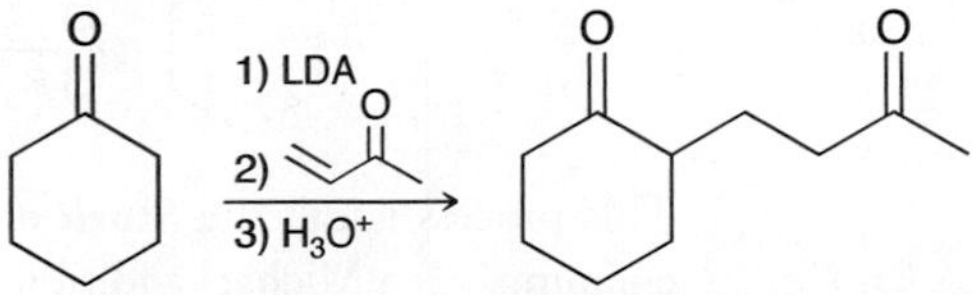

Not obtained in good yields

The synthetic route shown relies on an enolate serving as a Michael donor and attacking the β position of an α,β-unsaturated ketone. The process will not work because enolates are not stable enough to function as Michael donors. This transformation requires a stabilized enolate, or some species that will behave like a stabilized enolate. Gilbert Stork (Columbia University) developed a method for such a transformation in which the ketone is converted into an enamine by treatment with a secondary amine.

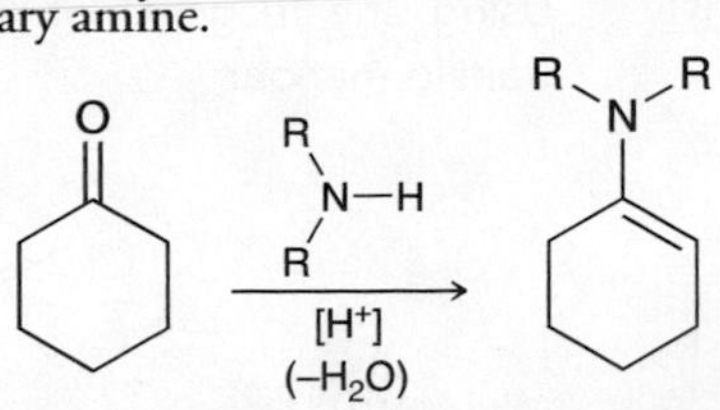

The mechanism for conversion of a ketone into an enamine was discussed in Section 20.6. To see the similarity between an enolate and an enamine, compare the resonance structures of an enolate with the resonance structures of an enamine.

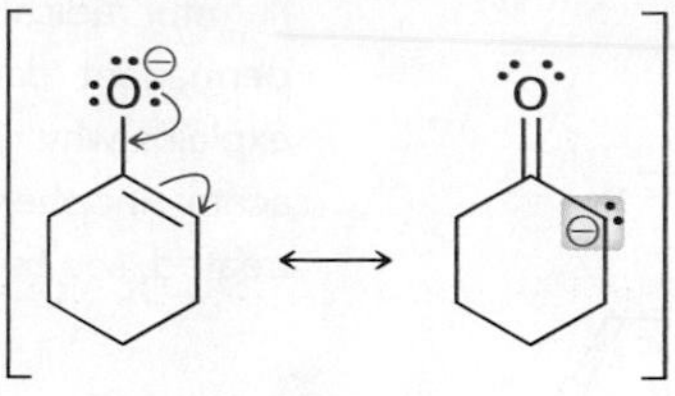

An enolate ion

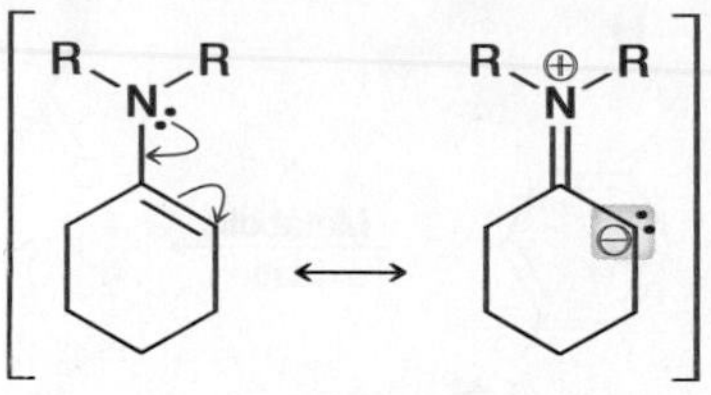

An enamine

Much like enolates, enamines are also nucleophilic at the α position. However, enamines do not possess a net negative charge, as enolates do, and therefore, enamines are less reactive than enolates. As such, enamines are effective Michael donors and will participate in a Michael reaction with a suitable Michael acceptor.

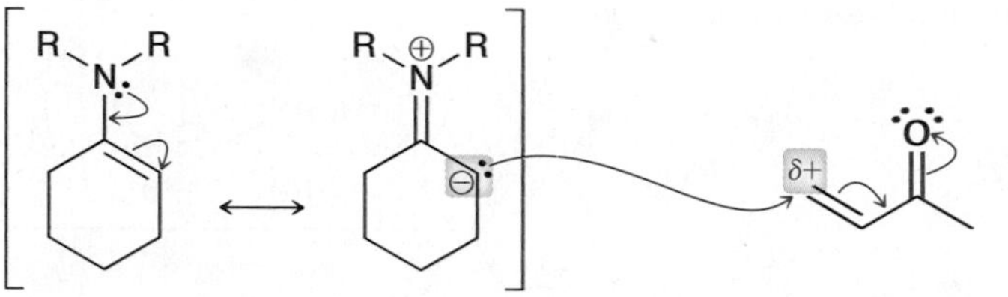

Michael donor Michael acceptor

This Michael reaction generates an intermediate that is both an iminium ion and an enolate ion, as shown below. When treated with aqueous acid, both groups are converted into carbonyl groups.

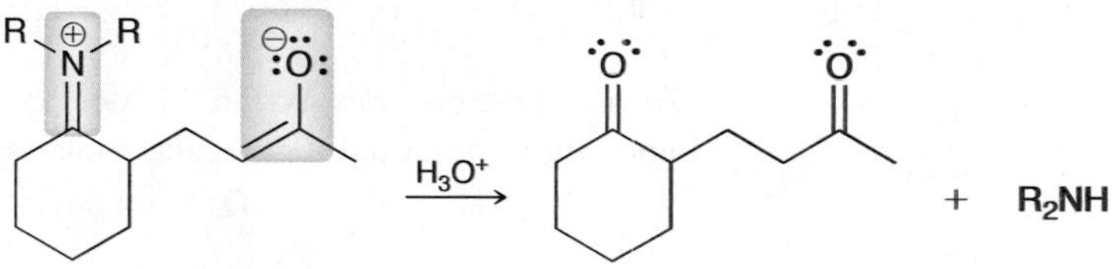

The iminium ion undergoes hydrolysis to form a carbonyl group, and the enolate ion is protonated to form an enol, which tautomerizes to form a carbonyl group.

The net result is a process for achieving the following type of transformation:

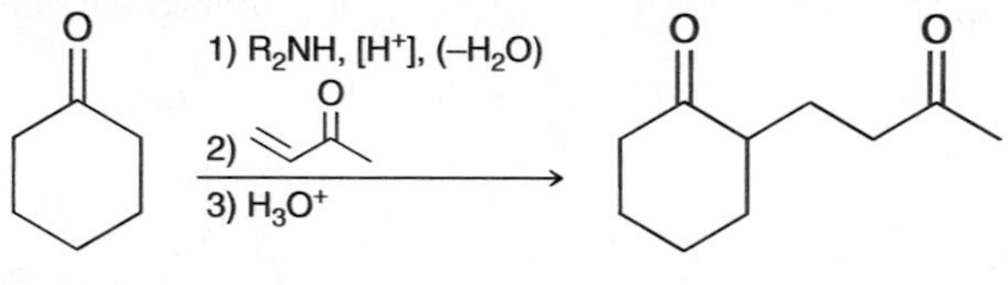

This process is called a **Stork enamine synthesis,** and it has three steps: (1) formation of an enamine, (2) a Michael addition, and (3) hydrolysis.

SKILLBUILDER

22.7 DETERMINING WHEN TO USE A STORK ENAMINE SYNTHESIS

LEARN the skill

Using any reagents of your choosing, show how you might accomplish the following transformation:

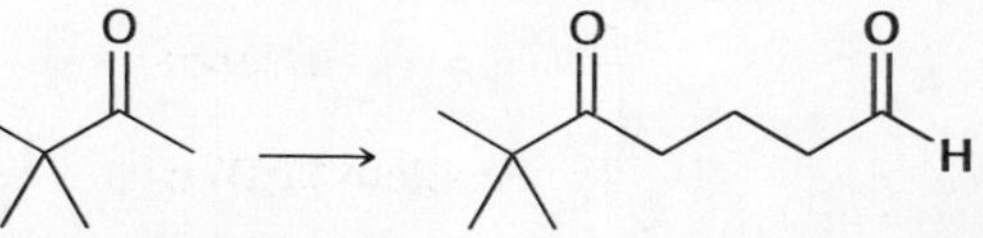

SOLUTION

This transformation requires that we install the following group at the α position of the starting material.

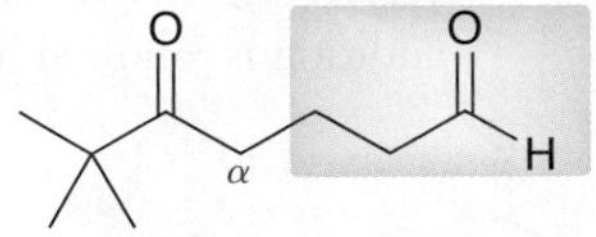

Using a retrosynthetic analysis, we can determine that this transformation is possible via a Michael addition:

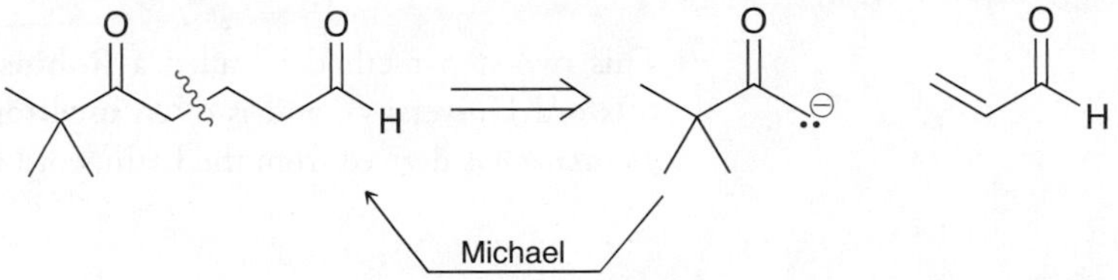

However, a Michael reaction won't work because the enolate is not sufficiently stabilized to function as a Michael donor. This case is therefore a perfect example of a situation in which a Stork enamine synthesis might be used to achieve the desired Michael addition.

The starting ketone is first converted into an enamine and then used as a Michael donor followed by hydrolysis.

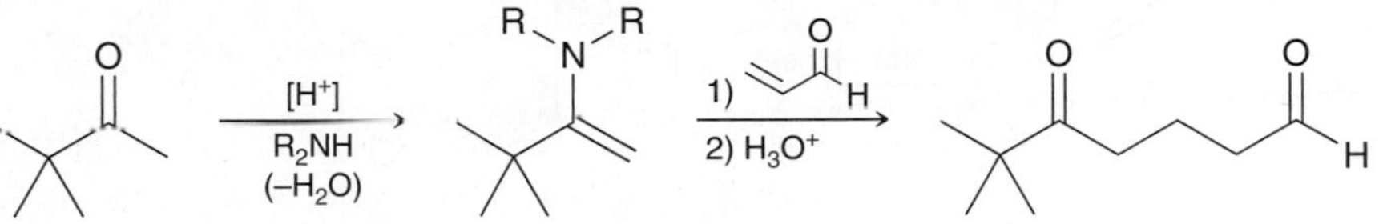

PRACTICE the skill **22.47** Using a Stork enamine synthesis, show how you might accomplish each of the following transformations.

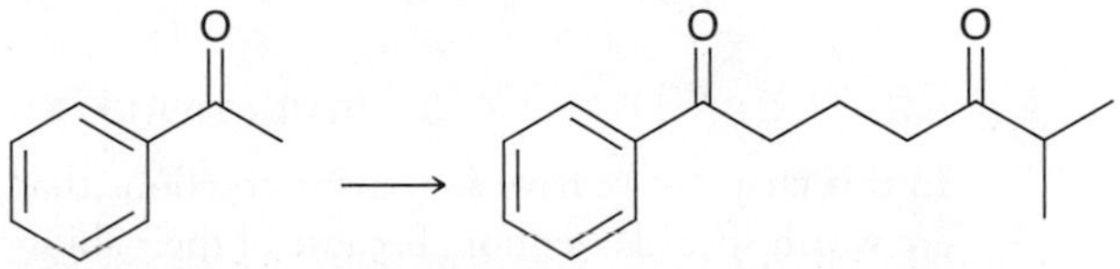

(a)

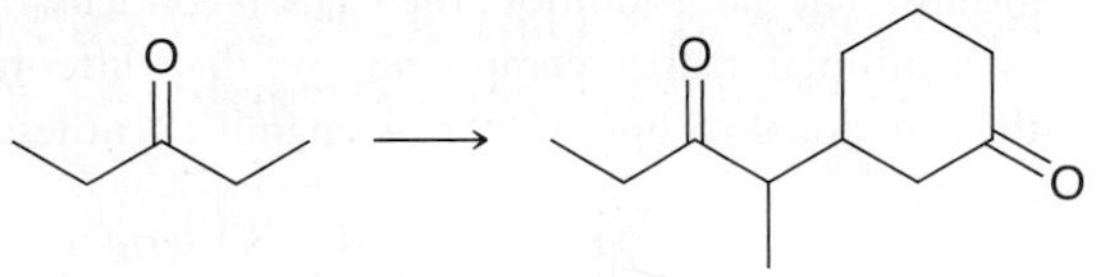

(b)

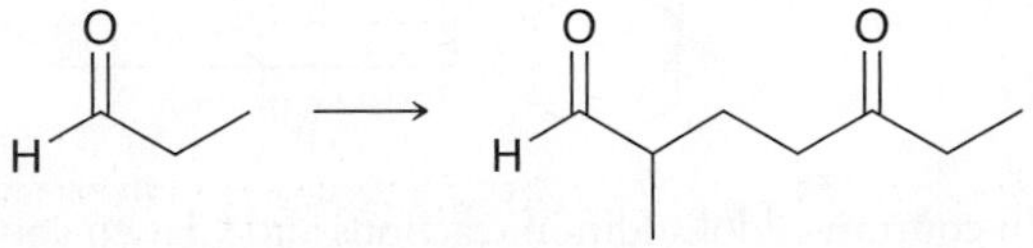

(c)

APPLY the skill **22.48** Using acetophenone as your only source of carbon atoms, propose a synthesis for the following compound.

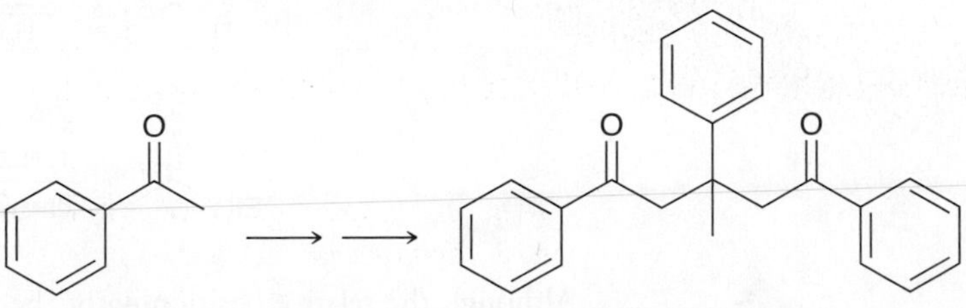

Acetophenone

need more **PRACTICE?** **Try Problem 22.87d**

The Robinson Annulation Reaction

In this chapter, we have seen many reactions. When performed in combination, they can be very versatile. One such example is a two-step method for forming a ring, in which a Michael addition is followed by an intramolecular aldol condensation.

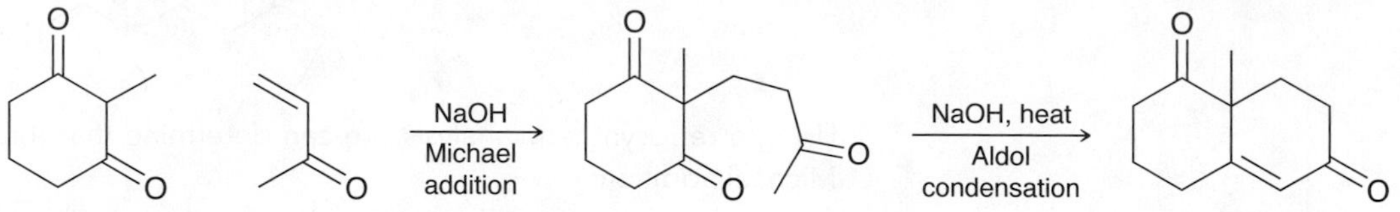

This two-step method is called a **Robinson annulation**, named after Sir Robert Robinson (Oxford University), and is often used for the synthesis of polycyclic compounds. The term *annulation* is derived from the Latin word for ring (*annulus*).

CONCEPTUAL CHECKPOINT

22.49 Draw a complete mechanism for the following transformation.

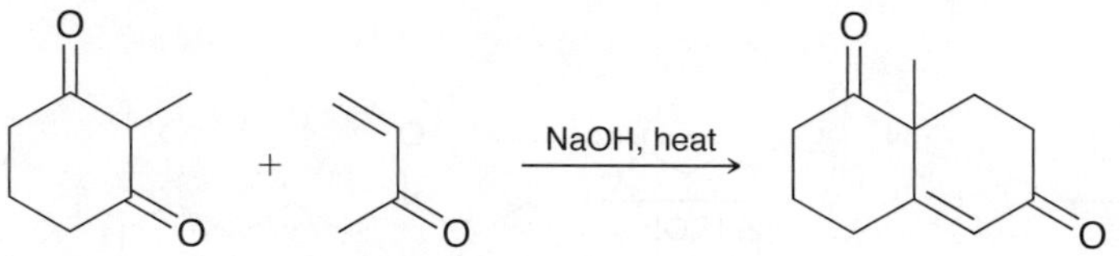

22.50 Identify the reagents you would use to prepare the following compound via a Robinson annulation.

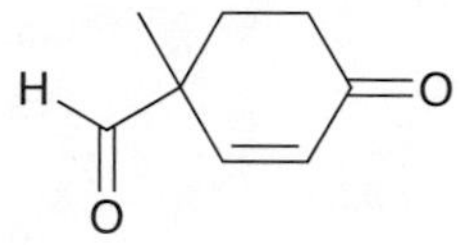

22.7 Synthesis Strategies

Reactions That Yield Difunctionalized Compounds

In this chapter, we have seen many reactions that form C—C bonds. Three of those reactions are worth special attention, because of their ability to produce compounds with two functional groups. The aldol addition, the Claisen condensation, and the Stork enamine synthesis all produce difunctionalized compounds, yet they differ from each other in the ultimate positioning of the functional groups. The Stork enamine synthesis produces 1,5-difunctionalized compounds.

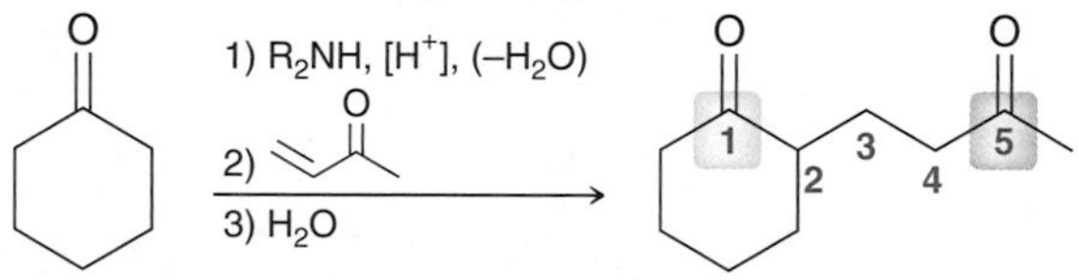

In contrast, aldol addition reactions and Claisen condensation reactions both produce 1,3-difunctionalized compounds.

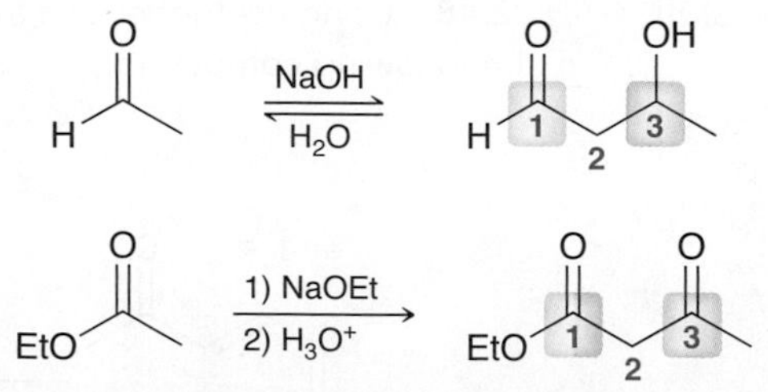

Although the relative positioning of the functional groups is similar for aldol additions and Claisen condensations, the oxidation states are different. An aldol addition produces a carbonyl group and a hydroxyl group, while a Claisen condensation produces an ester and a carbonyl group.

These considerations can be very helpful when designing a synthesis. If the target compound has two functional groups, then you should look at their relative positions. If the target compound is 1,5-difunctionalized, then you should think of using a Stork enamine synthesis. If it is 1,3-difunctionalized, then you should think of using an aldol addition or a Claisen condensation. The choice (aldol vs. Claisen) will be influenced by the oxidation states of the functional groups. The following exercise illustrates this type of analysis.

SKILLBUILDER

22.8 DETERMINING WHICH ADDITION OR CONDENSATION REACTION TO USE

LEARN the skill

Using 1-butanol as your only source of carbon, propose a synthesis for the following compound:

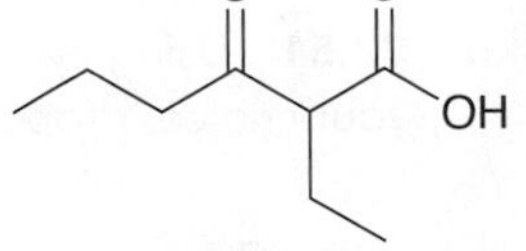

SOLUTION

Recall from Chapter 12 that there are always two things to analyze when approaching a synthesis problem: (1) any changes in the C—C framework and (2) the location and identity of the functional groups. Let's begin with the C—C framework. The target compound has a total of eight carbon atoms, and the starting material has only four carbon atoms. Therefore, our synthesis will require two molecules of butanol to construct the C—C framework of the target compound. In other words, we will need to use some kind of addition or condensation reaction. We have seen many such reactions. To determine which one will be most efficient in this case, we must analyze the location and identity of the functional groups.

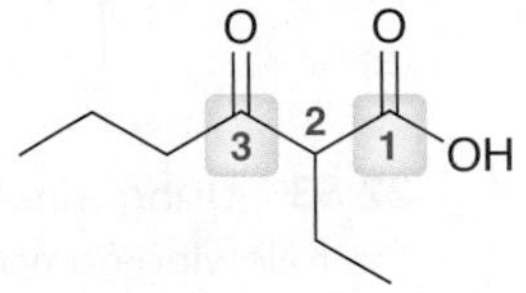

This compound is 1,3-difunctionalized, so we should focus our attention on either an aldol or a Claisen. To decide which reaction to use, look at the oxidation states of the two functional groups (a carbonyl group and a carboxylic acid moiety). In this case, a Claisen condensation seems like the likely candidate. The product of a Claisen condensation is always a β-keto ester, so if we use a Claisen condensation to form the carbon skeleton, then the last step of our synthesis will be hydrolysis of the ester to give a carboxylic acid.

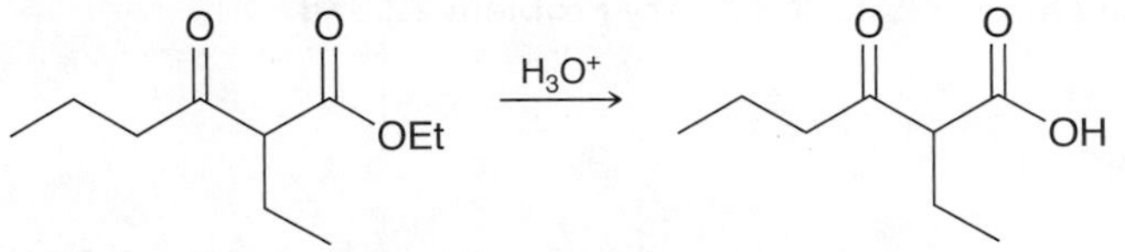

The penultimate step would be a Claisen condensation between two esters.

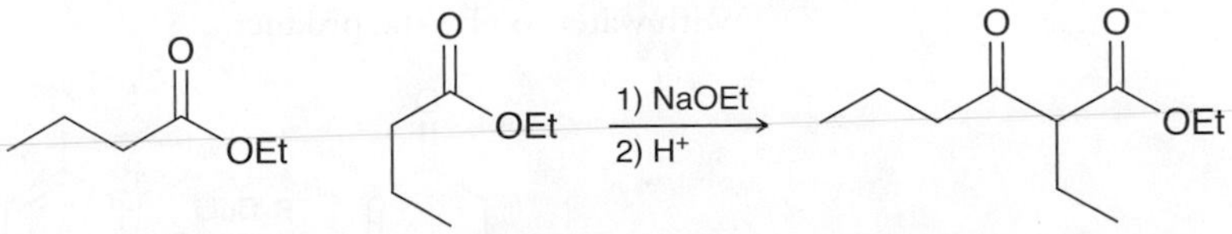

In this case, the two starting esters are identical, which simplifies the problem (it is not necessary to perform a mixed Claisen). This transformation is accomplished by simply treating the ester (ethyl butanoate) with sodium ethoxide.

To complete the synthesis, we need a method for making the necessary ester from 1-butanol. This can be accomplished by using reactions from previous chapters.

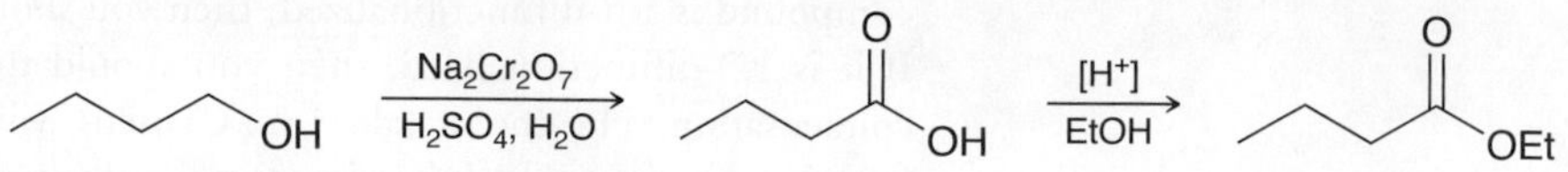

Our proposed synthesis is summarized as follows.

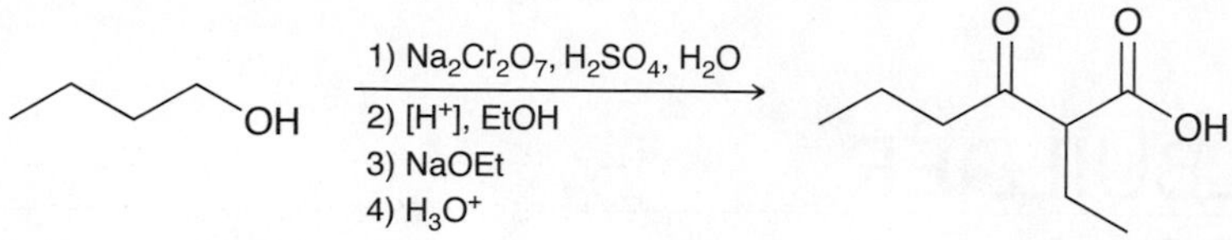

The last step has two functions: (1) to protonate the doubly stabilized enolate that is produced by the Claisen condensation and (2) to hydrolyze the resulting ester to give a carboxylic acid. Hydrolysis might require gentle heating, but excessive heating should be avoided as it would likely cause decarboxylation.

PRACTICE the skill **22.51** Using cyclopentanone as your starting material and using any other reagents of your choice, propose an efficient synthesis for each of the following compounds.

(a)
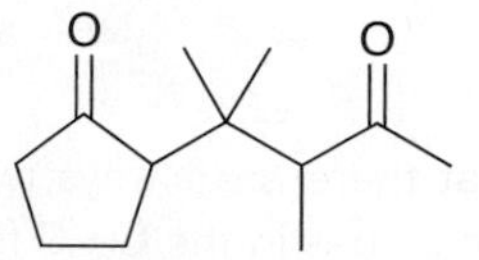

(b)
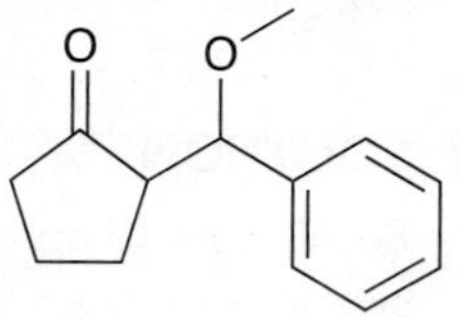

(c)
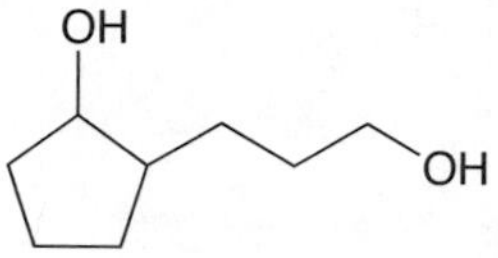

22.52 Using 1-propanol as your only source of carbon, propose an efficient synthesis for each of the following compounds.

(a)
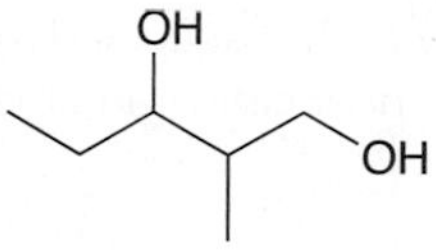

(b)
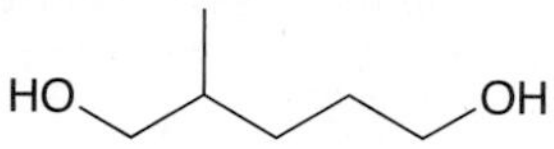

(c)
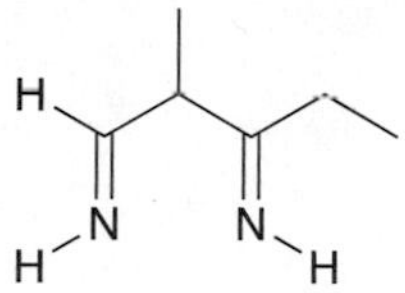

APPLY the skill **22.53** Using ethanol as your only source of carbon atoms, propose a synthesis for each of the following compounds.

(a)
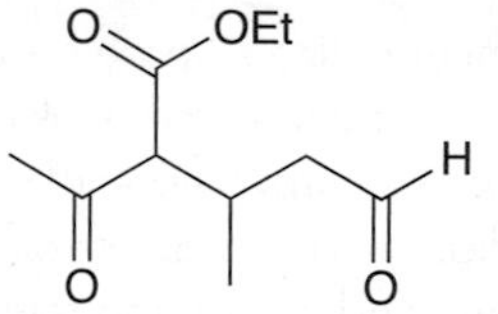

(b)
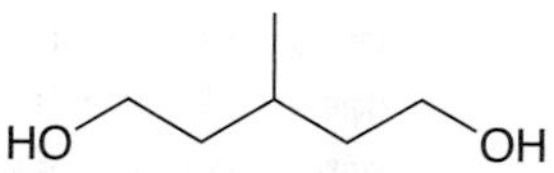

need more **PRACTICE?** **Try Problem 22.87d**

Alkylation of the Alpha and Beta Positions

Recall that the initial product of a Michael addition is an enolate ion, which is then quenched with water to give the product.

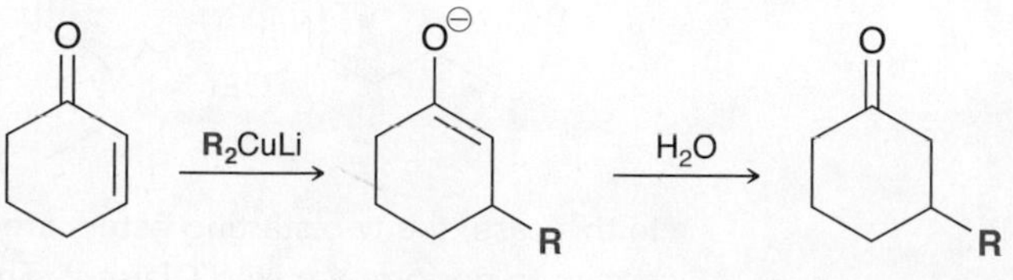

Enolate

Instead of quenching with water, the enolate generated from a Michael addition can be treated with an alkyl halide, which results in alkylation of the α position.

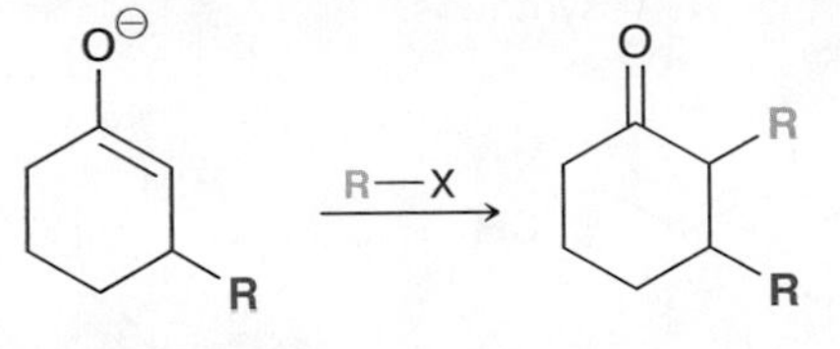

Enolate

This provides a method for alkylating both the α and β positions in one reaction flask. When using this method, the two alkyl groups need not be same. Below is an example of such a reaction.

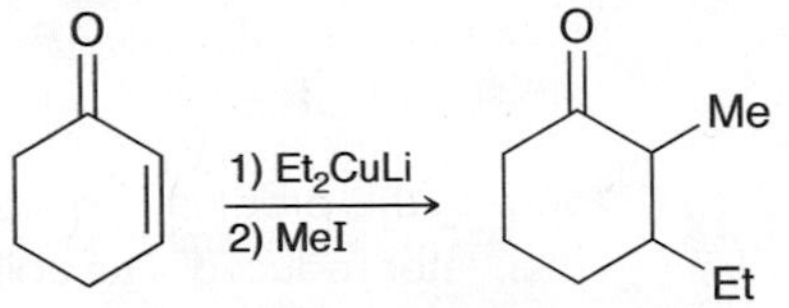

First, a carbon nucleophile is used to install an alkyl group at the β position, and then a carbon electrophile is used to install an alkyl group at the α position. The following exercise demonstrates the use of this technique.

SKILLBUILDER

22.9 ALKYLATING THE ALPHA AND BETA POSITIONS

LEARN the skill

Propose an efficient synthesis for the following transformation:

SOLUTION

Recall that there are always two things to analyze when approaching a synthesis problem: (1) any changes in the C—C framework and (2) the location and identity of functional groups. In this case, let's begin with functional groups, because very little change seems to have taken place. The hydroxyl group remains in the same position, but the double bond is destroyed. Now let's focus on the C—C framework. It appears that two methyl groups have added across the double bond.

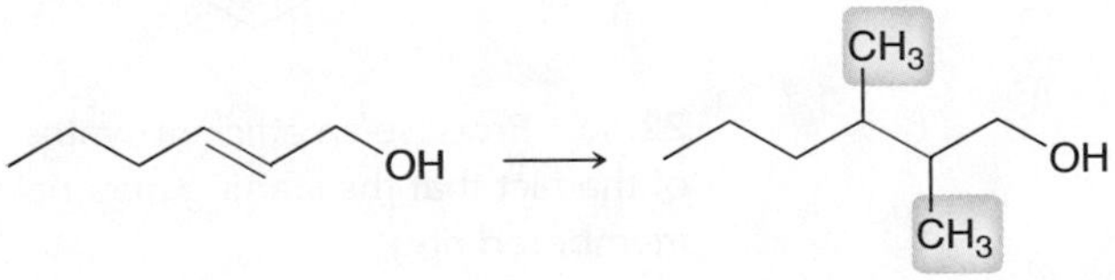

We did not see a way to add two alkyl groups across a double bond, but we did see a way to install two alkyl groups when the double bond is conjugated with a carbonyl group.

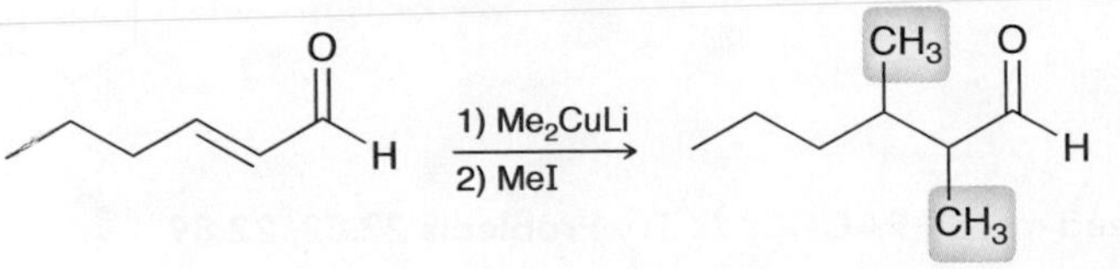

To use this method for installing the alkyl groups, we must first oxidize the alcohol to give a carbonyl group, which would then have to be reduced back to an alcohol at the end of the synthesis.

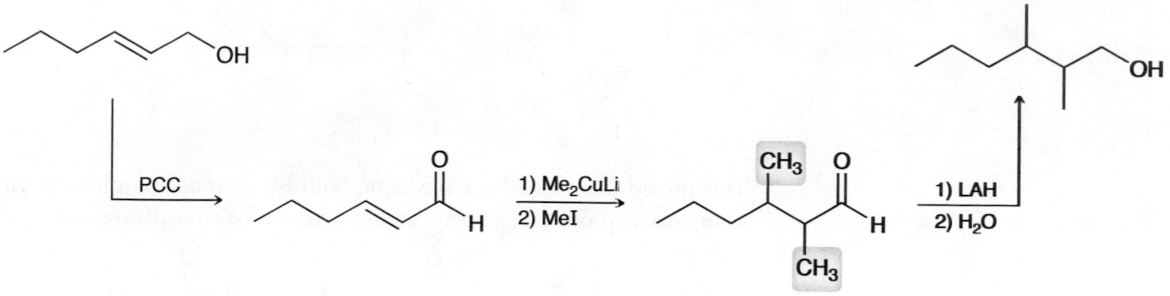

Solving this problem required us to recognize that the hydroxyl group could be oxidized and then later reduced. The ability to interchange functional groups is an important skill, as this problem demonstrates.

PRACTICE the skill **22.54** Propose an efficient synthesis for each of the following transformations:

(a)

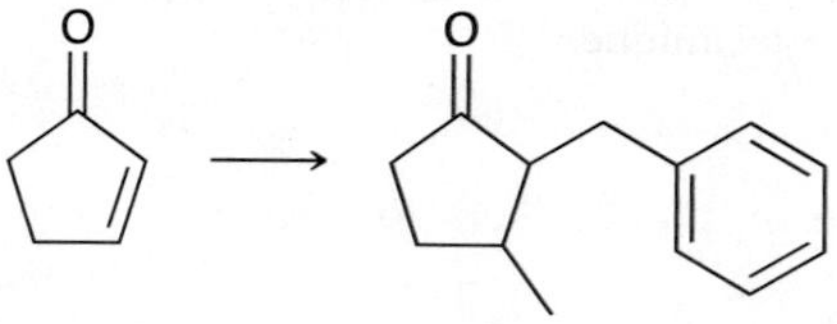

(b)

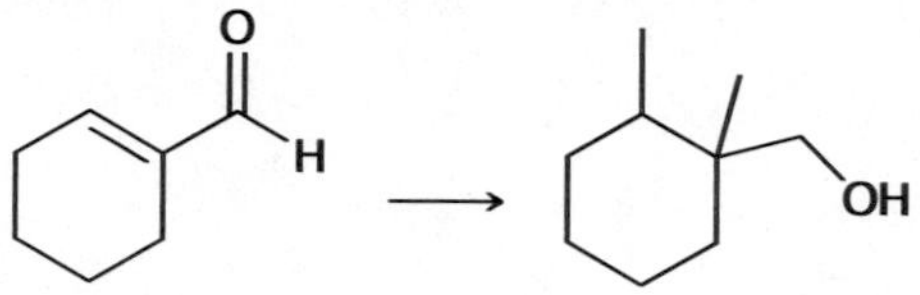

(c)

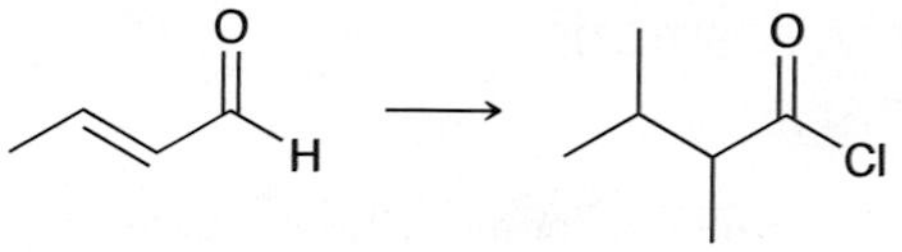

(d)

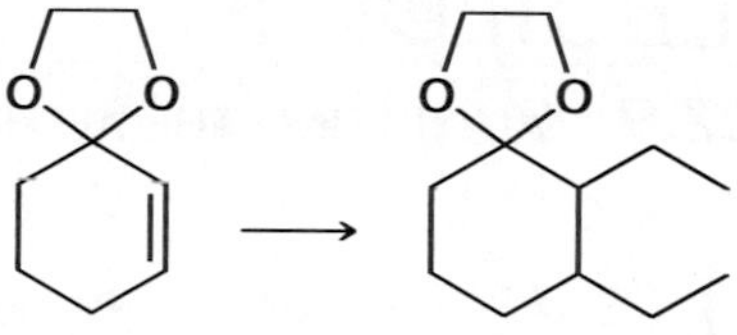

(e)

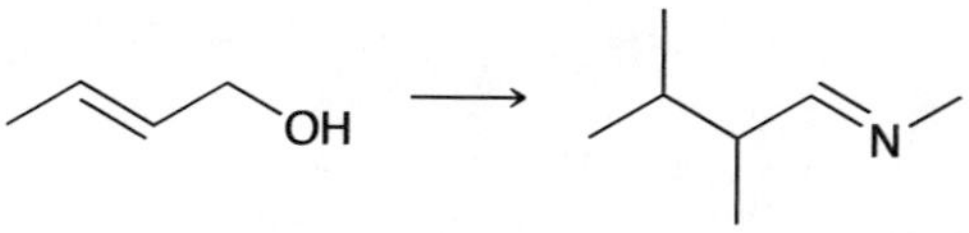

(f)

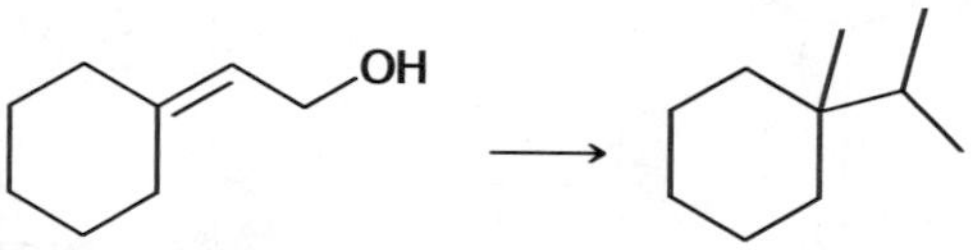

APPLY the skill **22.55** Propose an efficient synthesis for the following transformation.

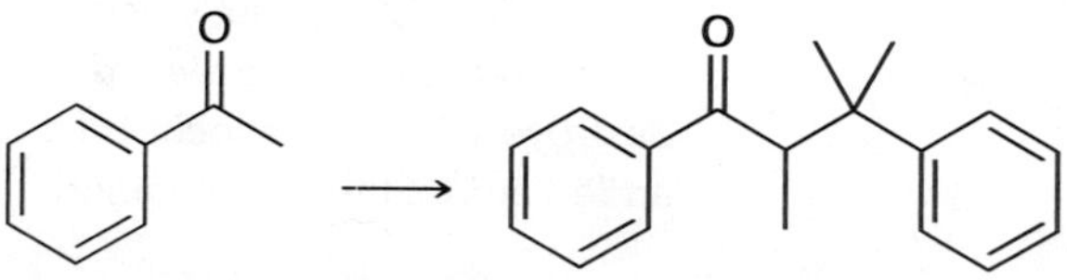

22.56 Propose an efficient synthesis for the following transformation. Take special notice of the fact that the starting material has a six-membered ring while the product has a five-membered ring.

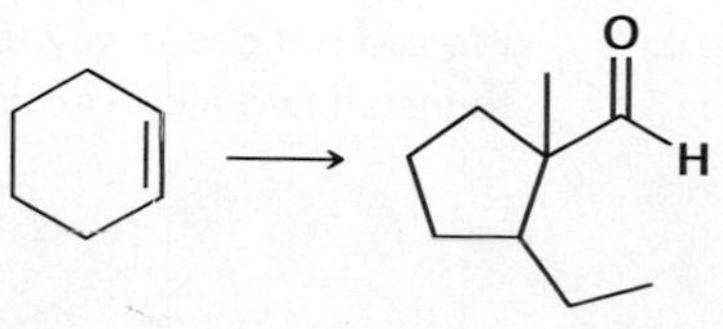

need more **PRACTICE?** **Try Problems 22.82, 22.89**

REVIEW OF REACTIONS

ALPHA HALOGENATION

Of Ketones

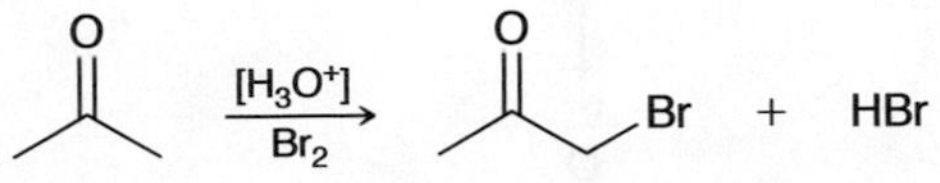

Haloform Reaction

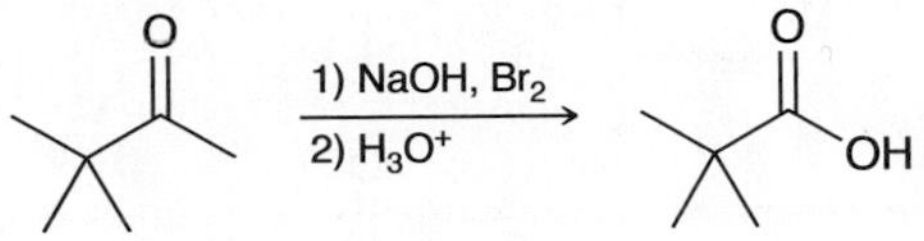

Of Carboxylic Acids (Hell-Volhard-Zelinski Reaction)

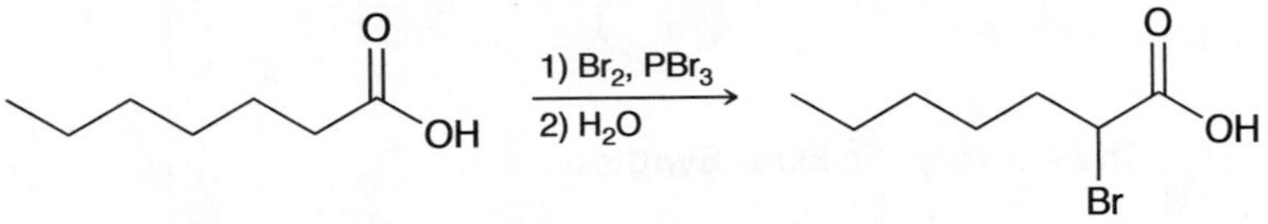

ALDOL REACTIONS

Aldol Addition and Condensation

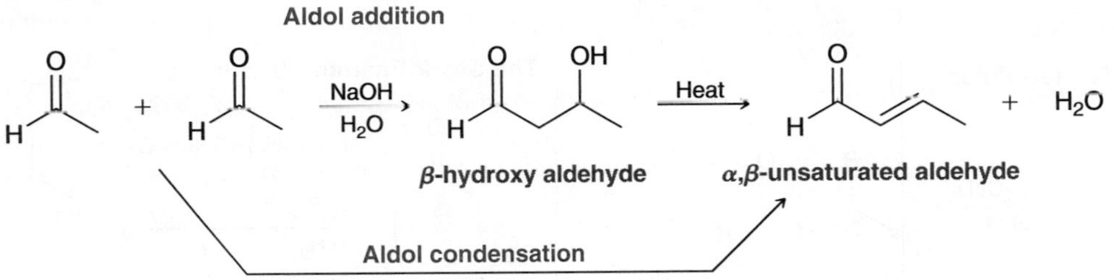

Crossed Aldol Condensation

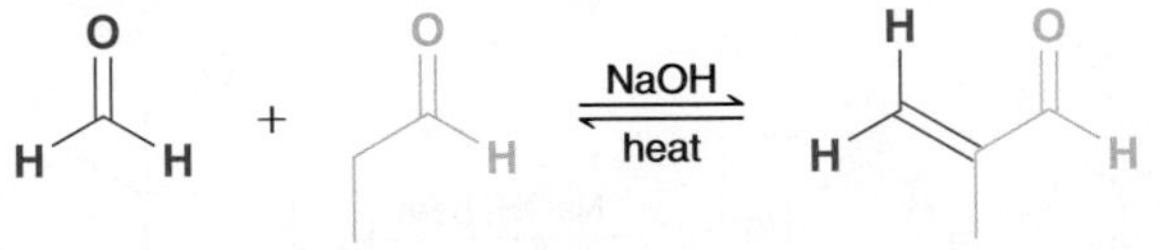

Intramolecular Aldol Condensation

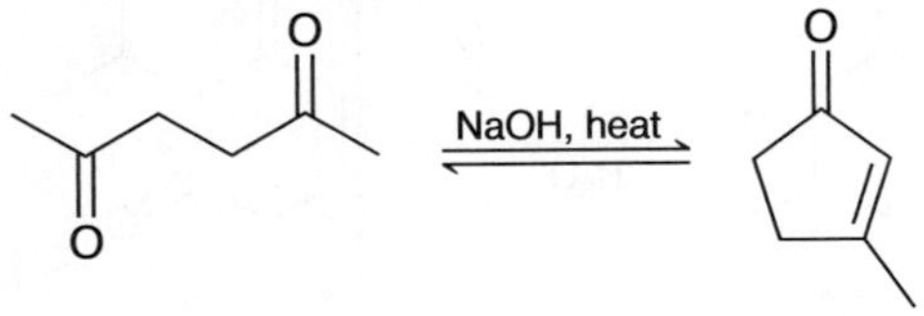

CLAISEN CONDENSATION

Claisen Condensation

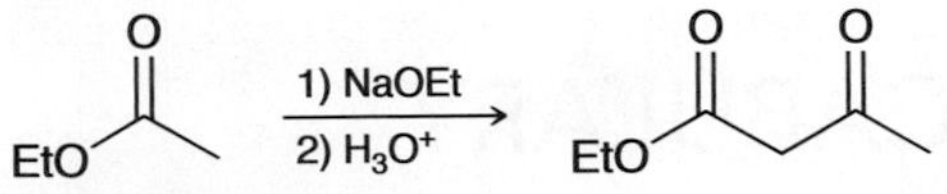

Intramolecular Claisen Condensation (Dieckmann Cyclization)

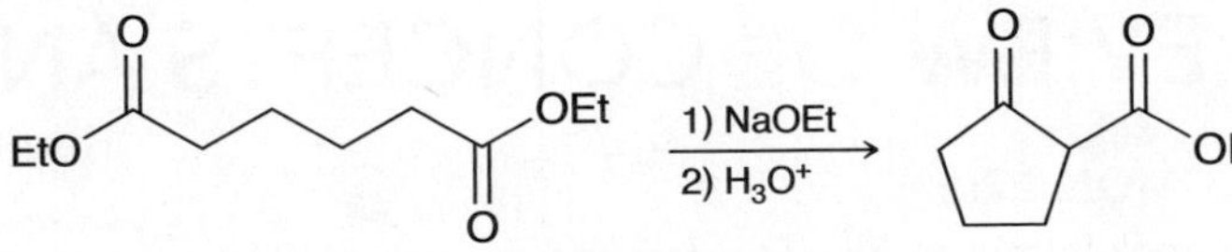

Crossed Claisen Condensations

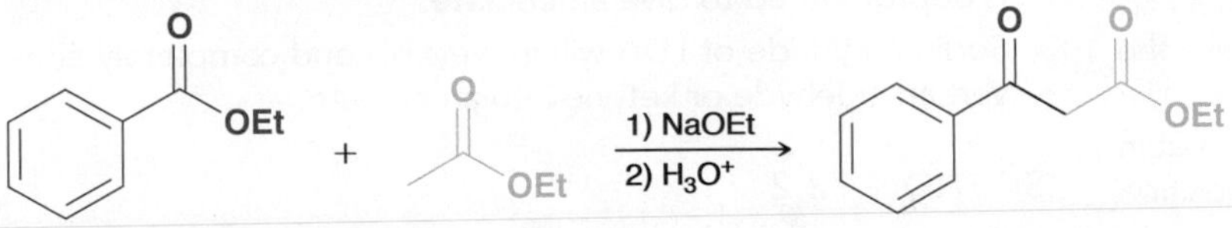

ALKYLATION

Via Enolate Ions

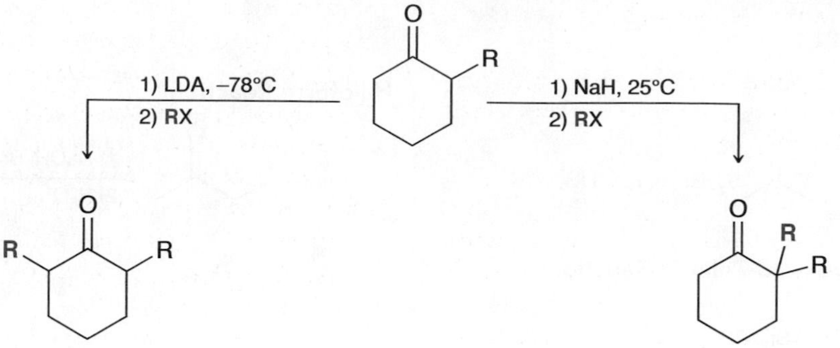

The Malonic Ester Synthesis

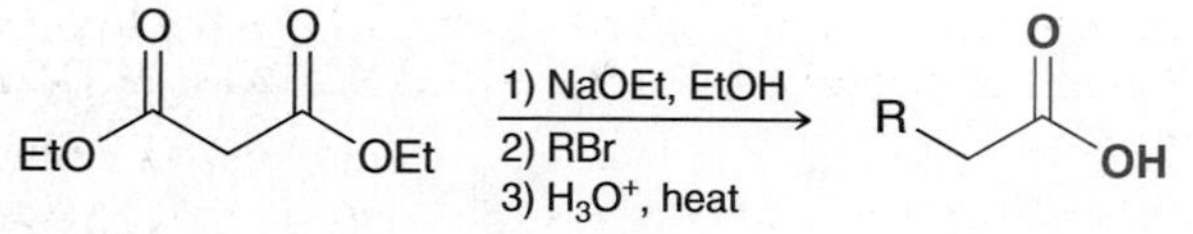

The Acetoacetic Ester Synthesis

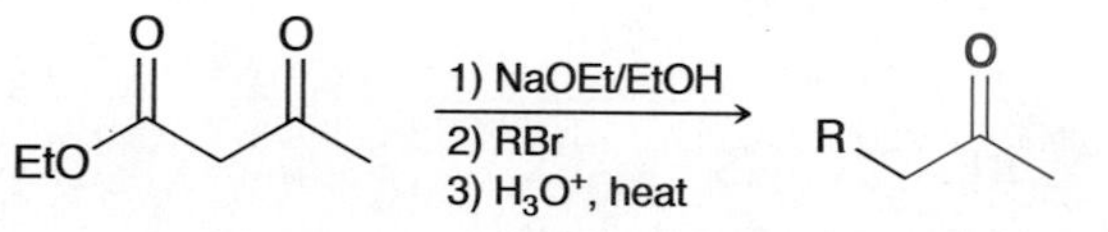

MICHAEL ADDITIONS

Stabilized Carbon Nucleophiles

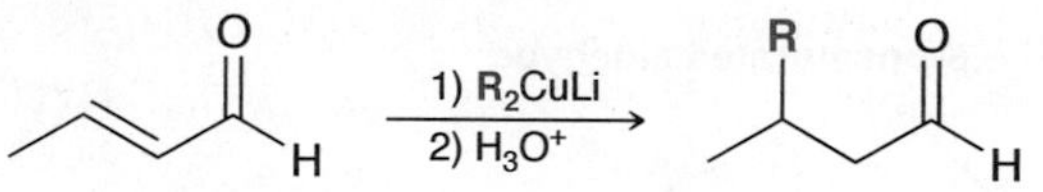

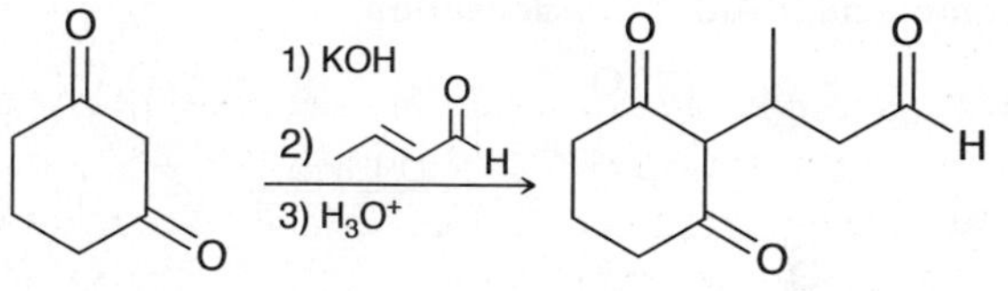

The Stork Enamine Synthesis

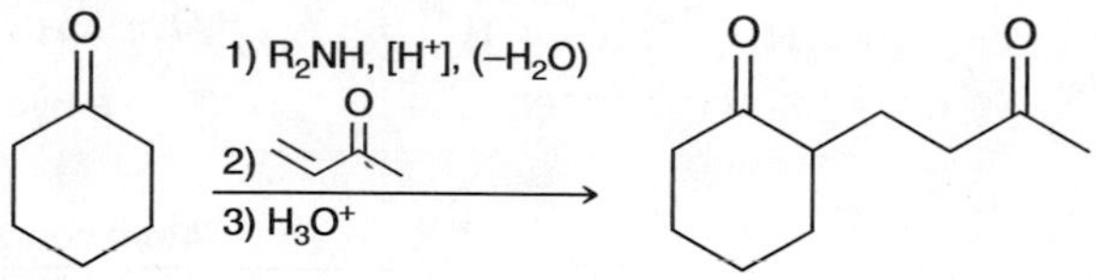

The Robinson Annulation

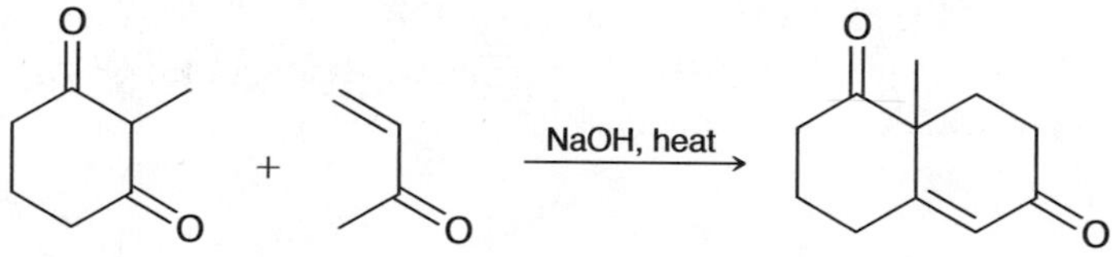

REVIEW OF CONCEPTS AND VOCABULARY

SECTION 22.1

- Greek letters are used to describe the proximity of each carbon atom to the carbonyl group. Alpha (α) protons are the protons connected to the α carbon.
- In the presence of catalytic acid or base, a ketone will exist in equilibrium with an enol. In general, the equilibrium position will significantly favor the ketone.
- The α position of an enol functions as a nucleophile.
- When treated with a strong base, the α position of a ketone is deprotonated to give an **enolate**.
- Sodium hydride or LDA will irreversibly and completely convert an aldehyde or ketone into an enolate.

SECTION 22.2

- Ketones and aldehydes will undergo alpha halogenation in acidic or basic conditions.

- The acid-catalyzed process produces HBr and is therefore **autocatalytic**.
- In the **Hell-Volhard-Zelinski reaction**, a carboxylic acid undergoes alpha halogenation when treated with bromine in the presence of PBr_3.
- In the **haloform reaction**, a methyl ketone is converted into a carboxylic acid upon treatment with excess base and excess halogen followed by acid workup.

SECTION 22.3

- When an aldehyde is treated with sodium hydroxide, an **aldol addition reaction** occurs, and the product is a β-hydroxy aldehyde or ketone.
- For most simple aldehydes, the position of equilibrium favors the aldol product.
- For most ketones, the reverse process, called a **retro-aldol reaction,** is favored.
- When an aldehyde is heated in aqueous sodium hydroxide, an **aldol condensation reaction** occurs, and the product is an α,β-unsaturated aldehyde or ketone. Elimination of water occurs via an **E1cb mechanism**.
- **Crossed aldol,** or **mixed aldol, reactions** are aldol reactions that occur between different partners and are only efficient if one partner lacks α protons or if a **directed aldol addition** is performed.
- Intramolecular aldol reactions show a preference for formation of five- and six-membered rings.

SECTION 22.4

- When an ester is treated with an alkoxide base, a **Claisen condensation reaction** occurs, and the product is a β-keto ester.
- A Claisen condensation between two different partners is called a **crossed Claisen condensation**.
- An intramolecular Claisen condensation, called a **Dieckmann cyclization**, produces a cyclic, β-keto ester.

SECTION 22.5

- The α position of a ketone can be alkylated by forming an enolate and treating it with an alkyl halide.
- For unsymmetrical ketones, reactions with LDA at low temperature favor formation of the kinetic enolate, while reactions with NaH at room temperature favor the thermodynamic enolate.
- When LDA is used with an unsymmetrical ketone, alkylation occurs at the less hindered position.
- The **malonic ester synthesis** enables the conversion of an alkyl halide into a carboxylic acid with the introduction of two new carbon atoms. The **acetoacetic ester synthesis** enables the conversion of an alkyl halide into a methyl ketone with the introduction of three new carbon atoms.
- **Decarboxylation** occurs upon heating a carboxylic acid with a β-carbonyl group.

SECTION 22.6

- Aldehydes and ketones that possess α,β unsaturation are susceptible to nucleophilic attack at the β position. This reaction is called a **conjugate addition, 1,4-addition,** or **Michael reaction.**
- The nucleophile is called a **Michael donor**, and the electrophile is called a **Michael acceptor**.
- Regular enolates do not serve as Michael donors, but the desired Michael reaction can be achieved with a **Stork enamine synthesis**.
- A **Robinson annulation** is a Michael addition followed by an intramolecular aldol and can be used to make cyclic compounds.

SECTION 22.7

- The Stork enamine synthesis produces 1,5-difunctionalized compounds.
- Aldol addition reactions and Claisen condensation reactions both produce 1,3-difunctionalized compounds.
- The initial product of a Michael addition is an enolate ion, which can be treated with an alkyl halide, thereby alkylating both the α and β positions.

KEY TERMINOLOGY

1,4-addition 1066
acetoacetic ester synthesis 1062
aldol addition reaction 1043
aldol condensation reaction 1047
autocatalytic 1039
Claisen condensation reaction 1054
conjugate addition 1066
crossed aldol reactions 1051
crossed Claisen condensation 1056
decarboxylation 1060
Dieckmann cyclization 1056
directed aldol addition 1052
E1cb mechanism 1048
enolate 1034
haloform reaction 1042
Hell-Volhard-Zelinski reaction 1040
malonic ester synthesis 1059
Michael acceptor 1067
Michael donor 1067
Michael reaction 1067
mixed aldol reactions 1051
retro-aldol reaction 1043
Robinson annulation 1072
Stork enamine synthesis 1070

SKILLBUILDER REVIEW

22.1 DRAWING ENOLATES

STEP 1 Identify all α protons.

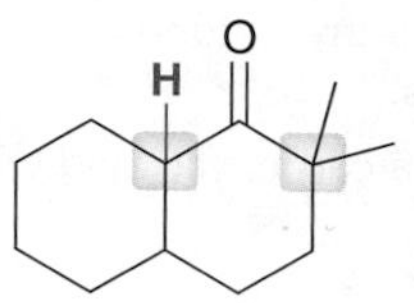

STEP 2 Using two curved arrows, remove the proton, and then draw the enolate with a lone pair and a negative charge at the α position.

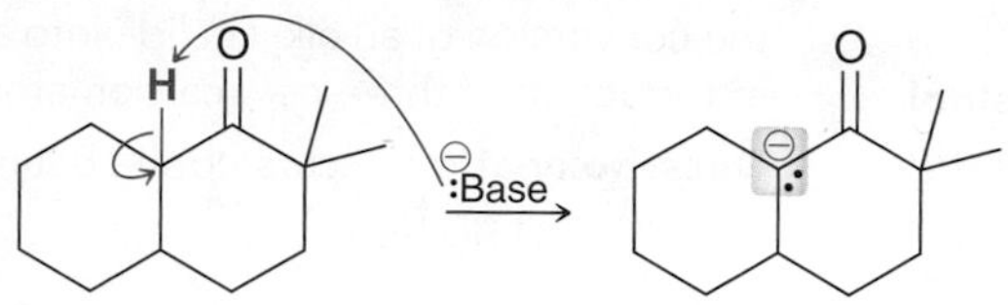

STEP 3 Draw the other resonance structure of the enolate.

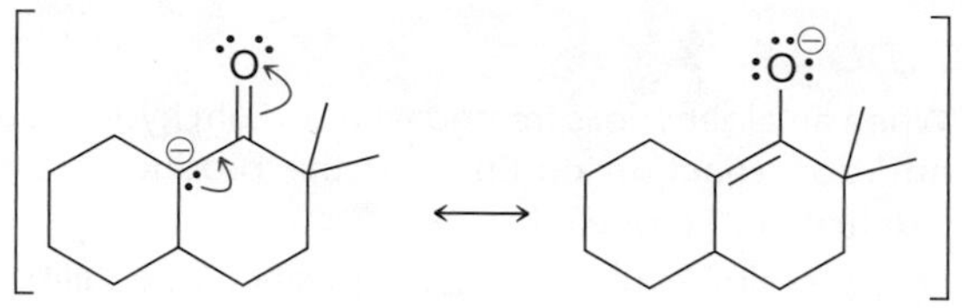

Try Problems 22.4, 22.5, 22.59

22.2 PREDICTING THE PRODUCTS OF AN ALDOL ADDITION REACTION

STEP 1 Draw all three steps of the mechanism as a guide for predicting the product.

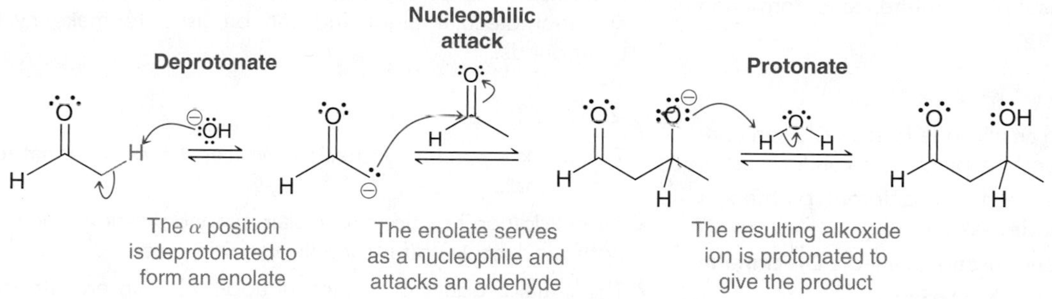

STEP 2 Double-check your answer to ensure that the product has a hydroxyl group at the β position.

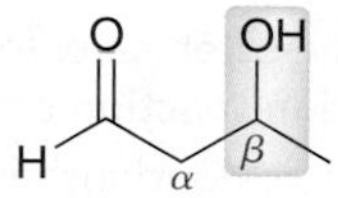

Try Problems 22.15–22.19, 22.72

22.3 DRAWING THE PRODUCT OF AN ALDOL CONDENSATION

STEP 1 Identify the α protons.

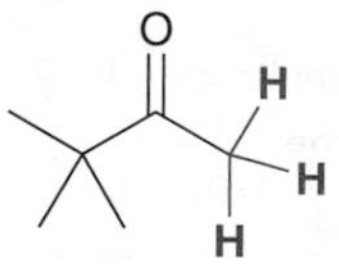

STEP 2 Draw two molecules of the ketone, oriented such that the two α protons of one molecule are directly facing the carbonyl group of the other molecule.

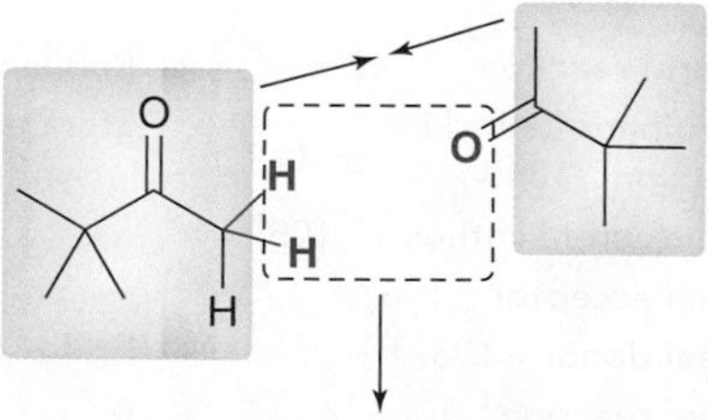

STEP 3 Remove the two α protons and the oxygen atom, forming a double bond in place of those groups.

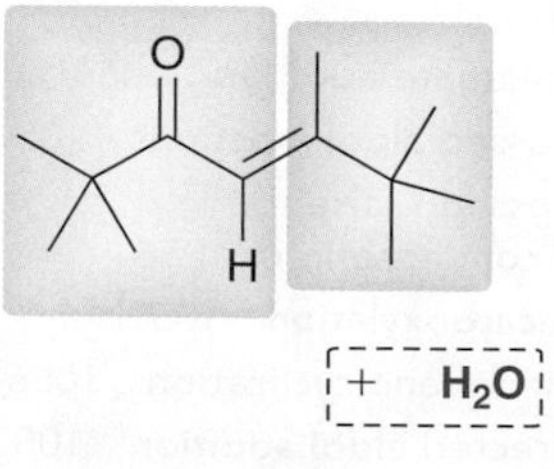

STEP 4 Make sure to draw the product with the least steric hindrance.

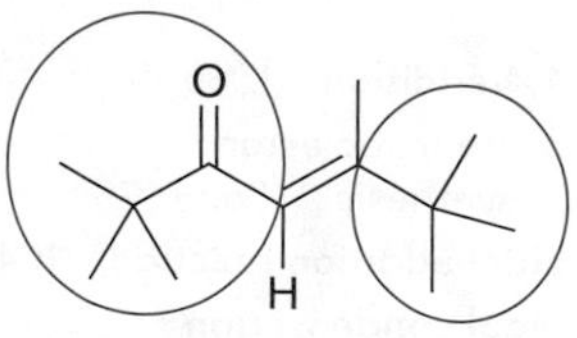

Try Problems 22.20–22.22, 22.71, 22.84c

22.4 IDENTIFYING THE REAGENTS NECESSARY FOR A CROSSED ALDOL REACTION

STEP 1 Identify the α and β positions.

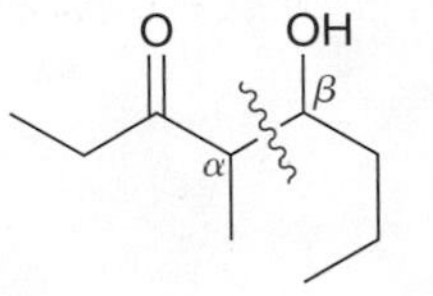

STEP 2 Using a retrosynthetic analysis, break apart the bond between the α and β positions, placing a carbonyl group in place of the hydroxyl group.

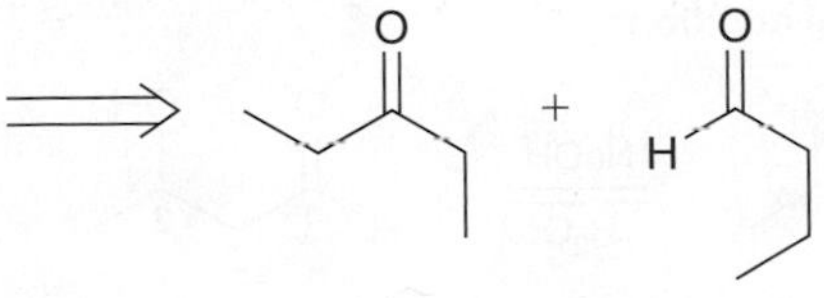

STEP 3 Determine which base should be used. A crossed aldol will require the use of LDA.

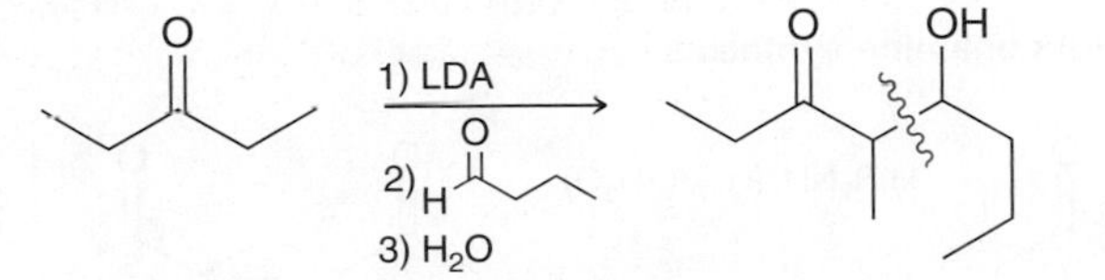

Try Problems 22.23, 22.24, 22.67, 22.68

22.5 USING THE MALONIC ESTER SYNTHESIS

STEP 1 Identify the alkyl groups that are connected to the α position of the carboxylic acid.

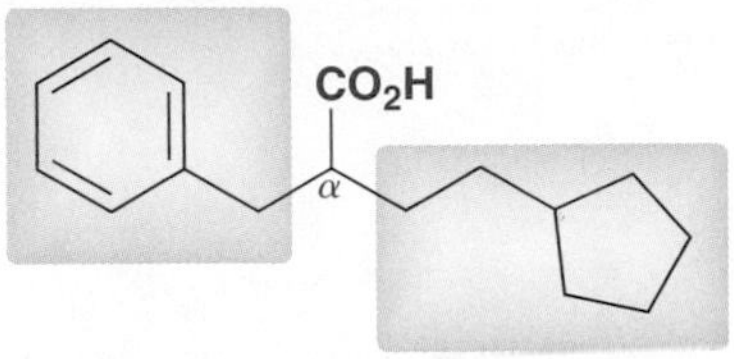

STEP 2 Identify the alkyl halides necessary, and ensure that both will readily undergo an S_N2 process.

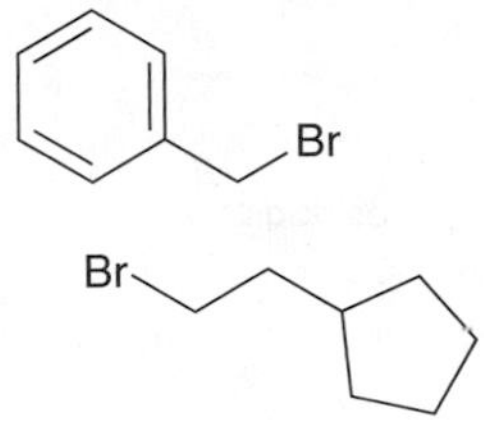

STEP 3 Identify the reagents. Begin with diethyl malonate as the starting material. Perform each alkylation, and then heat with aqueous acid.

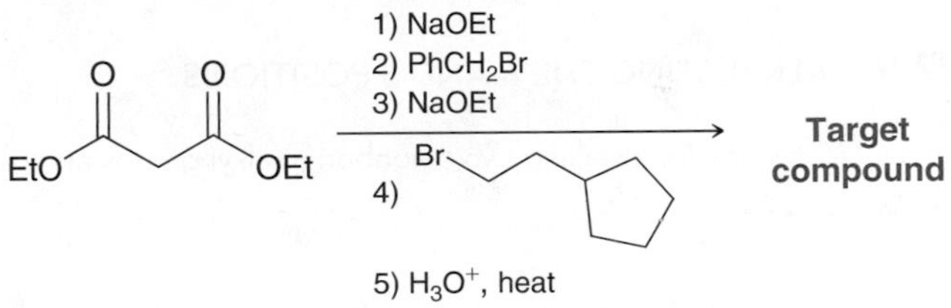

Try Problems 22.35–22.38, 22.78

22.6 USING THE ACETOACETIC ESTER SYNTHESIS

STEP 1 Identify the alkyl groups that are connected to the α position of the methyl ketone.

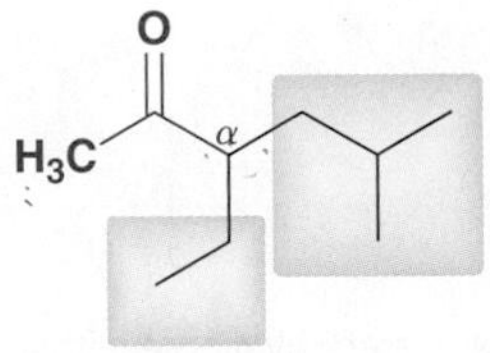

STEP 2 Identify the alkyl halides necessary, and ensure that both will readily undergo an S_N2 process.

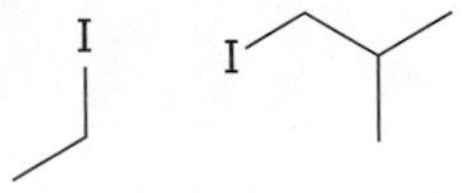

STEP 3 Identify the reagents. Begin with ethyl acetoacetate as the starting material. Perform each alkylation, and then heat with aqueous acid.

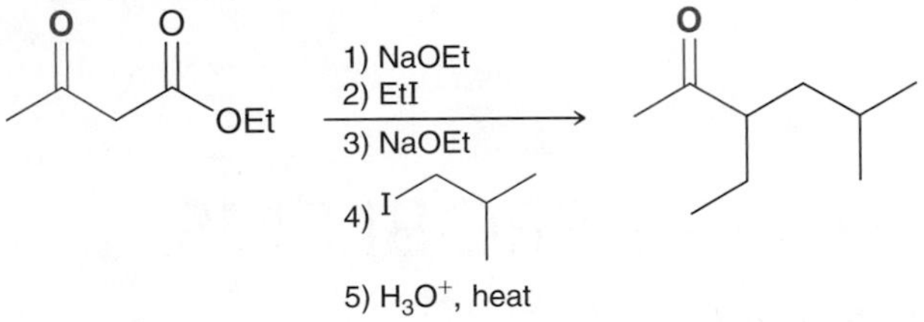

Try Problems 22.39–22.43, 22.79

22.7 DETERMINING WHEN TO USE A STORK ENAMINE SYNTHESIS

STEP 1 Using a retrosynthetic analysis, identify whether it is possible to prepare the target compound with a Michael addition.

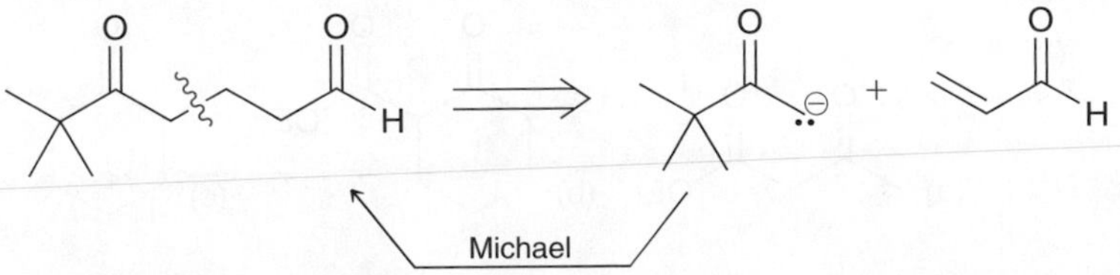

STEP 2 If the Michael donor must be an enolate, then a Stork enamine synthesis is required.

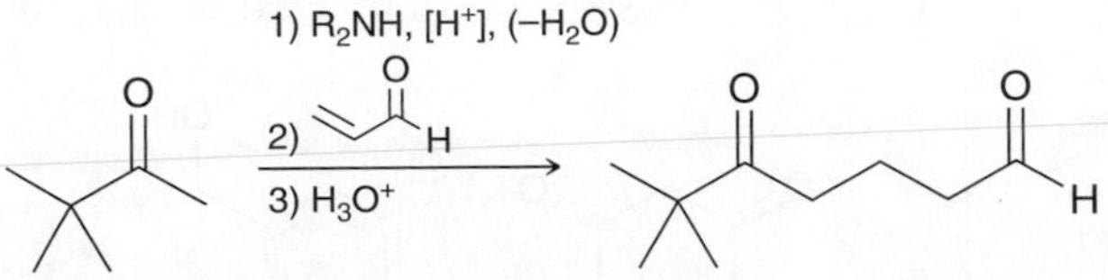

Try Problems 22.47, 22.48, 22.87d

22.8 DETERMINING WHICH ADDITION OR CONDENSATION REACTION TO USE

For compounds with two functional groups, the relative positioning of the groups, as well as their oxidation states, will dictate which addition or condensation reaction to use.

1,5-DIFUNCTIONALIZED COMPOUNDS

Stork enamine synthesis

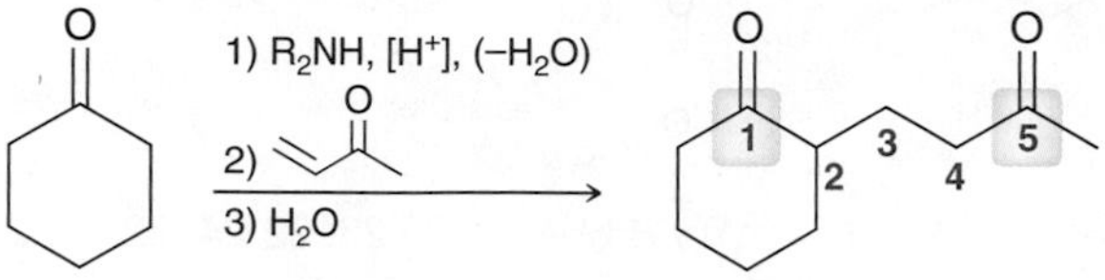

1,3-DIFUNCTIONALIZED COMPOUNDS

Aldol addition

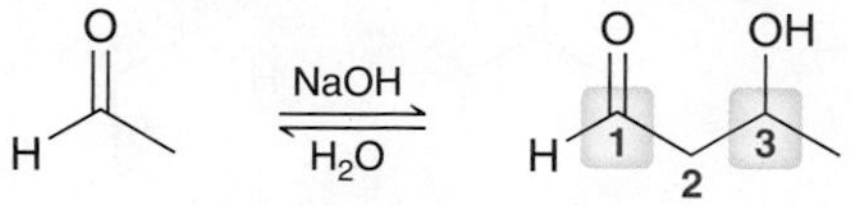

Claisen condensation

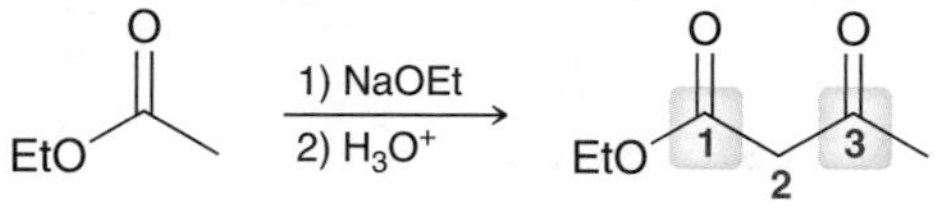

Try Problems 22.51–22.53, 22.87d

22.9 ALKYLATING THE α AND β POSITIONS

A strategy for installing two neighboring alkyl groups at the α and β positions.

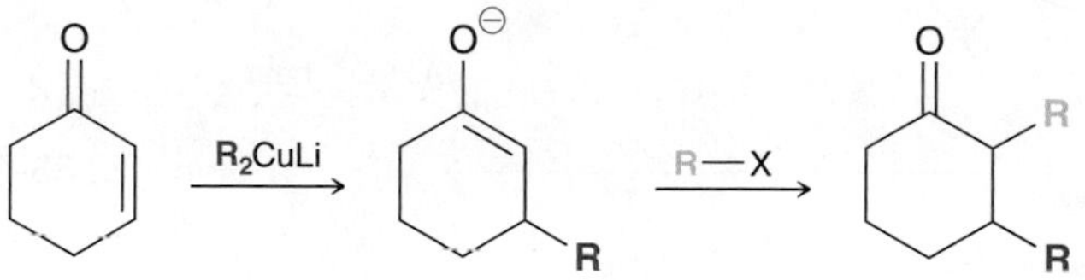

Enolate

Try Problems 22.54–22.56, 22.82, 22.89

PRACTICE PROBLEMS

Note: Most of the Problems are available within *WileyPLUS*, an online teaching and learning solution.

22.57 Identify which of the following compounds are expected to have $pK_a < 20$. For each compound with $pK_a < 20$, identify the most acidic proton in the compound.

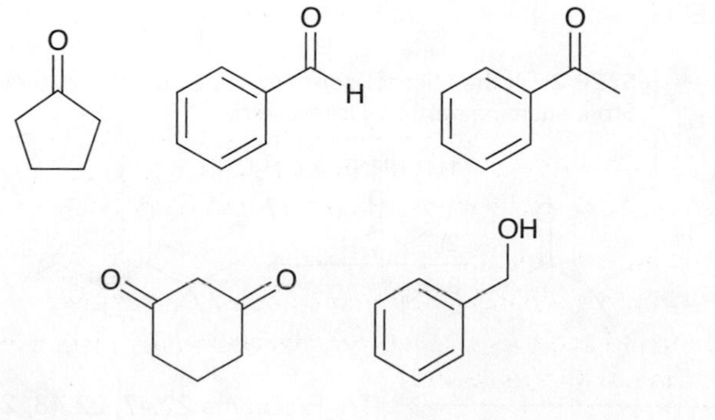

22.58 One of the compounds from the previous problem has $pK_a < 10$. Identify that compound, and explain why it is so much more acidic than all of the other compounds.

22.59 Draw resonance structures for the conjugate base that is produced when each of the following compounds is treated with sodium ethoxide.

(a) 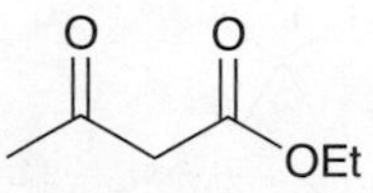

(b) 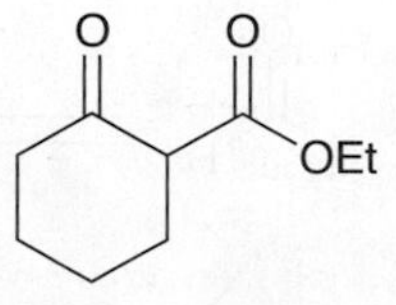

(c) 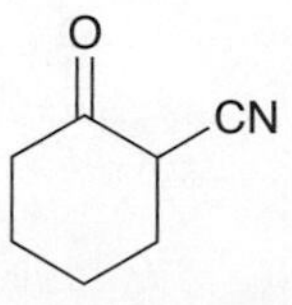

22.60 Rank the following compounds in terms of increasing acidity.

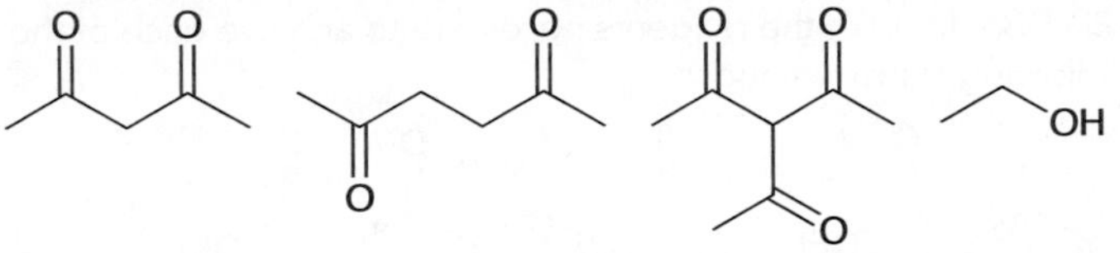

22.61 Draw the enol of each of the following compounds, and identify whether the enol exhibits a significant presence at equilibrium. Explain.

(a) 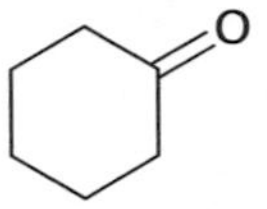
(b) 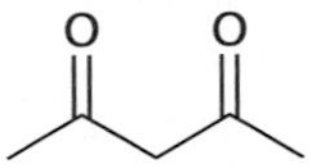
(c) 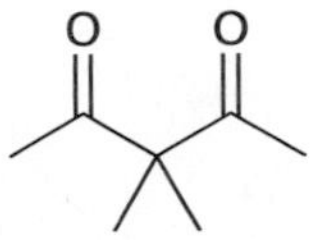
(d) 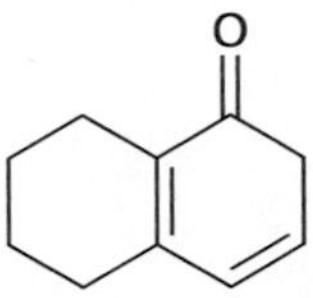

22.62 Ethyl acetoacetate has three enol isomers. Draw all three.

22.63 Draw the enolate that is formed when each of the following compounds is treated with LDA.

(a) 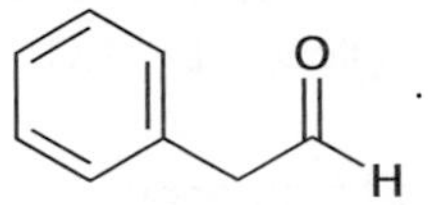
(b) 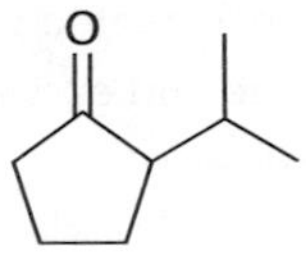
(c) 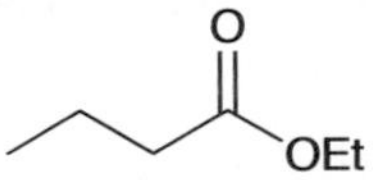
(d) 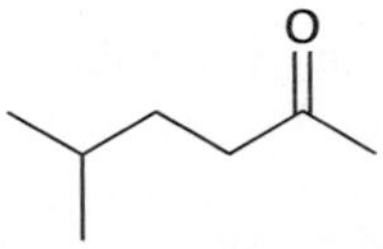

22.64 When 2-hepten-4-one is treated with LDA, a proton is removed from one of the gamma (γ) positions. Identify which γ position is deprotonated, and explain why the γ proton is the most acidic proton in the compound.

22.65 When optically active (*S*)-2-methylcyclopentanone is treated with aqueous base, the compound loses its optical activity. Explain this observation, and draw a mechanism that shows how racemization occurs.

22.66 The racemization process described in the previous problem also occurs in acidic conditions. Draw a mechanism for the racemization process in aqueous acid.

22.67 Draw all four β-hydroxyaldehydes that are formed when a mixture of acetaldehyde and pentanal is treated with aqueous sodium hydroxide.

22.68 Identify all of the different β-hydroxyaldehydes that are formed when a mixture of benzaldehyde and hexanal is treated with aqueous sodium hydroxide.

22.69 Propose a mechanism for the following isomerization, and explain the driving force behind this reaction. In other words, explain why the equilibrium favors the product.

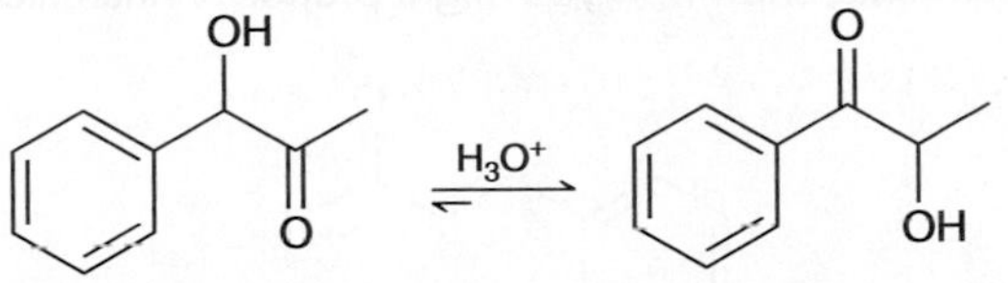

22.70 The isomerization in the previous problem can also occur in basic conditions. Draw a mechanism for the transformation in the presence of catalytic hydroxide.

22.71 Draw the product obtained when each of the following compounds is heated in the presence of a base to give an aldol condensation.

(a) 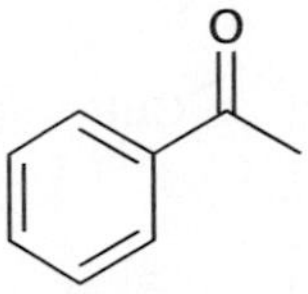
(b) 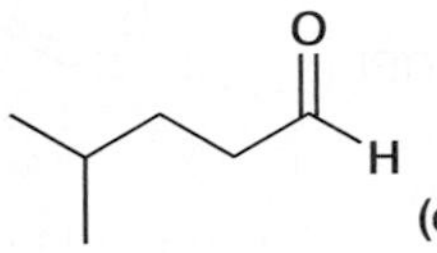
(c) 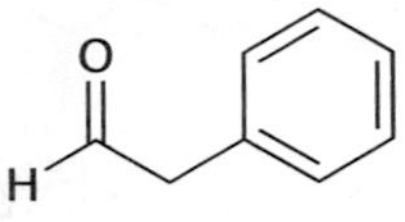

22.72 Trimethylacetaldehyde does not undergo an aldol reaction when treated with base. Explain why not.

22.73 Identify the reagents necessary to make each of the following compounds with an aldol condensation.

(a) 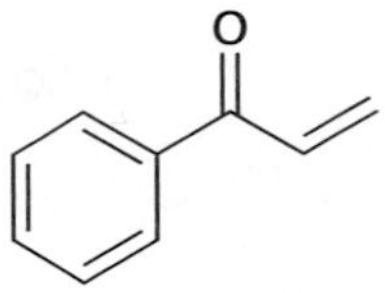
(b) 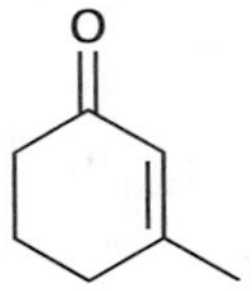
(c) 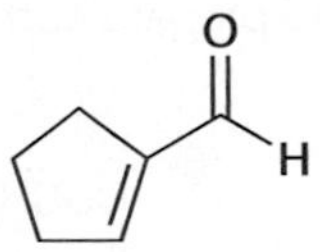
(d) 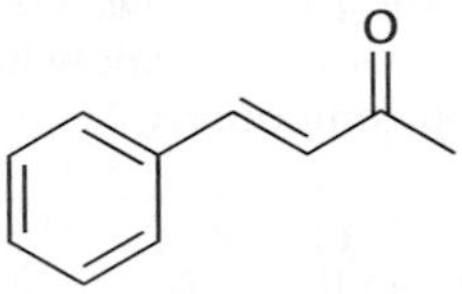

22.74 When acetaldehyde is treated with aqueous acid, an aldol reaction can occur. In other words, aldol reactions can also occur in acidic conditions, although the intermediate is different than the intermediate involved in the base-catalyzed reaction. Draw a mechanism for the acid-catalyzed process.

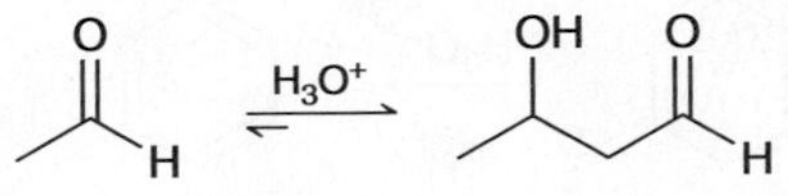

22.75 Diethyl malonate (the starting material for the malonic ester synthesis) reacts with bromine in acid-catalyzed conditions to form a product with molecular formula $C_7H_{11}BrO_4$.

(a) Draw the structure of the product.
(b) Draw a mechanism of formation for the product.
(c) Would you expect this product to be more or less acidic than diethyl malonate?

22.76 Cinnamaldehyde is one of the primary constituents of cinnamon oil and contributes significantly to the odor of cinnamon. Starting with benzaldehyde and using any other necessary reagents, show how you might prepare cinnamaldehyde.

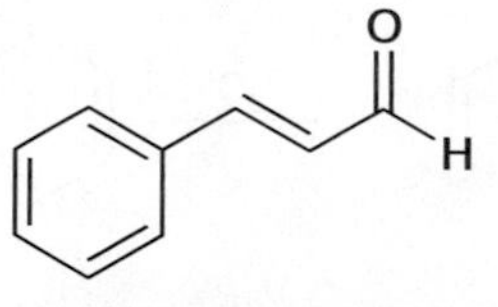

Cinnamaldehyde

22.77 Draw the condensation product that is expected when each of the following esters is treated with sodium ethoxide followed by acid workup.

(a) 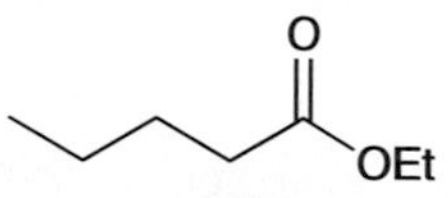

(b) 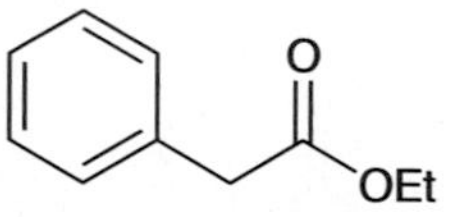

(c) 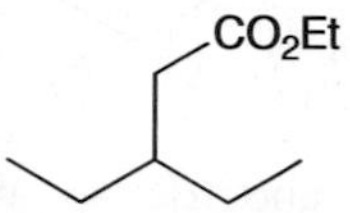

22.78 Starting with diethyl malonate, and using any other reagents of your choice, show how you would prepare each of the following compounds.

(a) 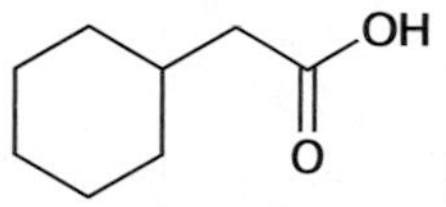

(b) 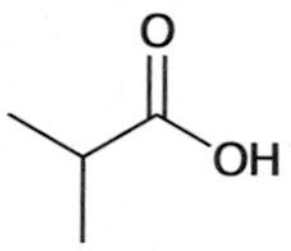

(c) 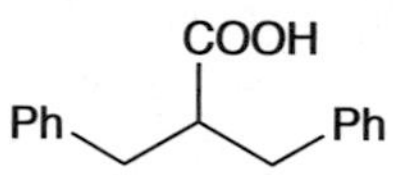

22.79 Starting with ethyl acetoacetate, and using any other reagents of your choice, show how you would prepare each of the following compounds.

(a) 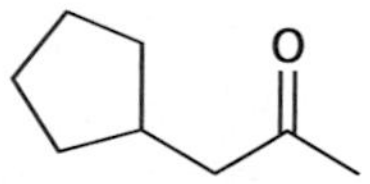

(b) 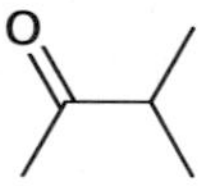

(c) 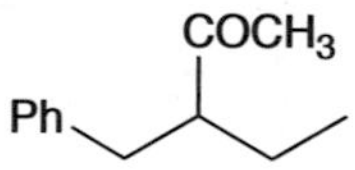

22.80 Propose a mechanism for the following transformation.

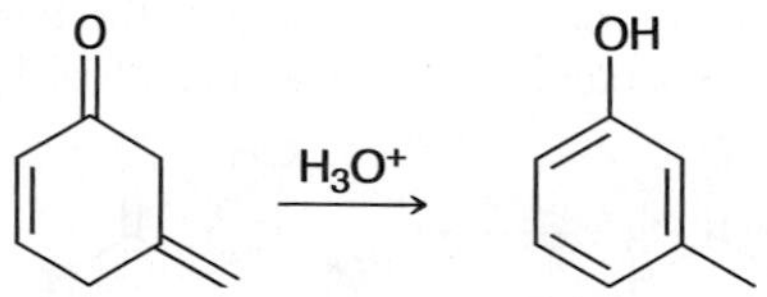

22.81 Draw the condensation product obtained when the following compound is heated in the presence of aqueous sodium hydroxide.

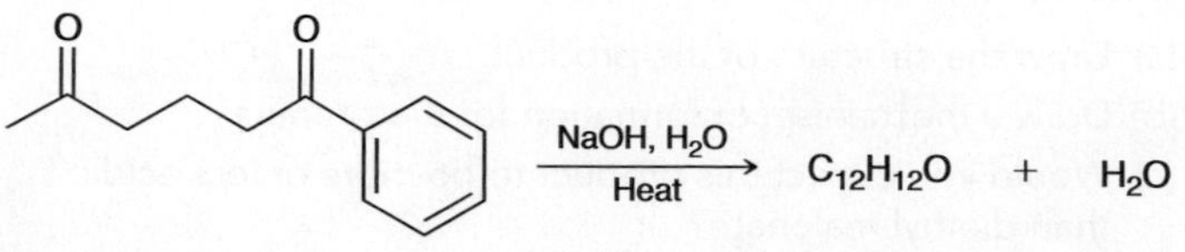

22.82 Identify the reagents you would use to convert 3-pentanone into 3-hexanone.

22.83 Identify the reagents necessary to achieve each of the following transformations.

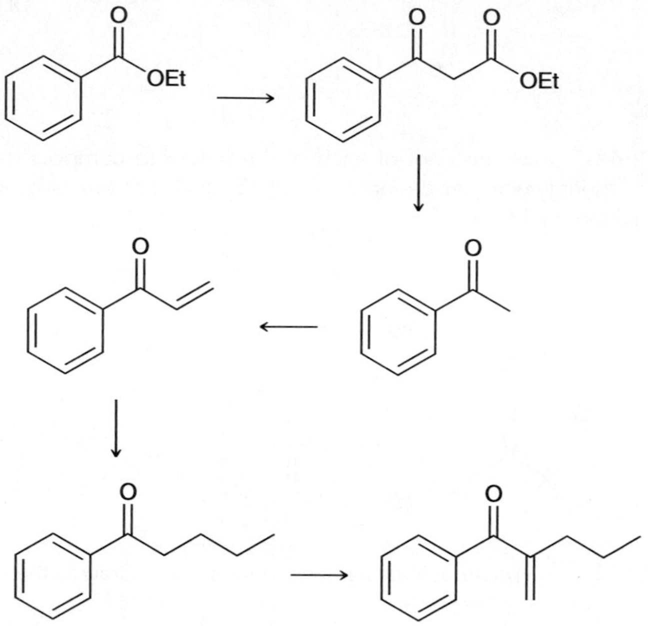

22.84 Draw the structure of the product that is obtained when acetophenone is treated with each of following reagents:

(a) Sodium hydroxide and excess iodine followed by H_3O^+

(b) Bromine in acetic acid

(c) Aqueous sodium hydroxide at elevated temperature

22.85 Draw a reasonable mechanism for the following transformation.

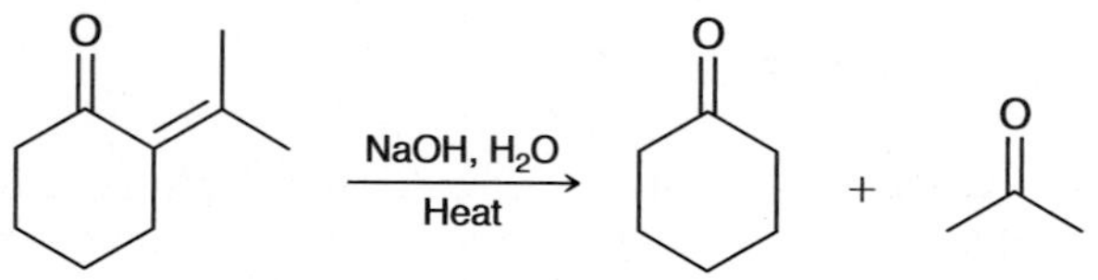

22.86 Predict the major product for each of the following transformations.

(a) 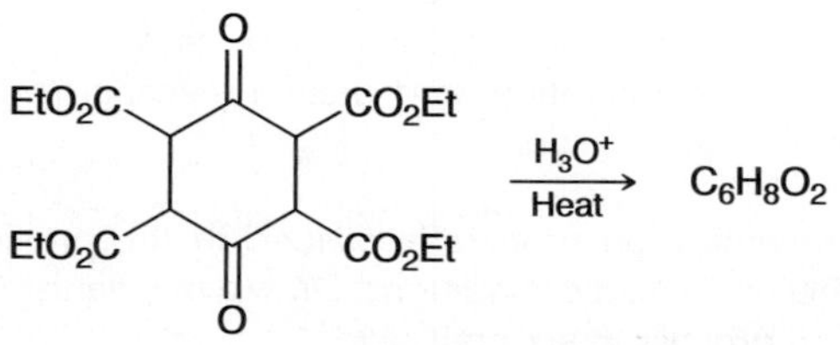

(b) 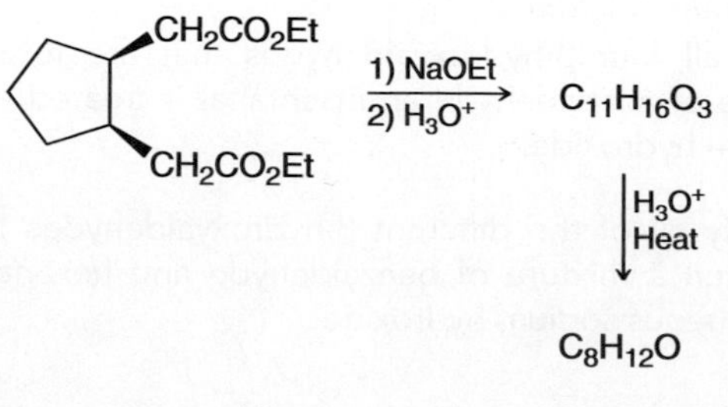

(c)

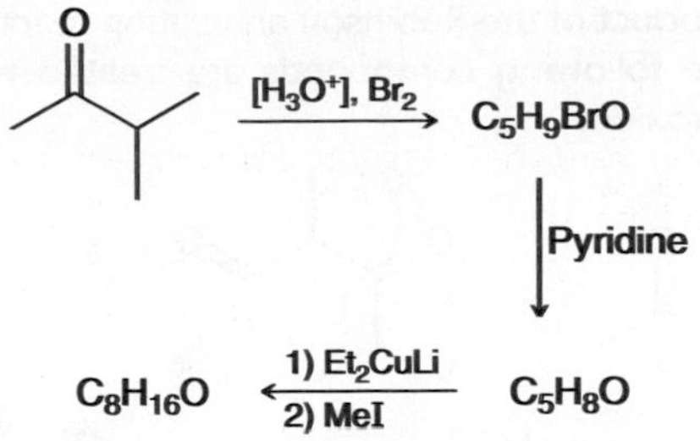

(d)

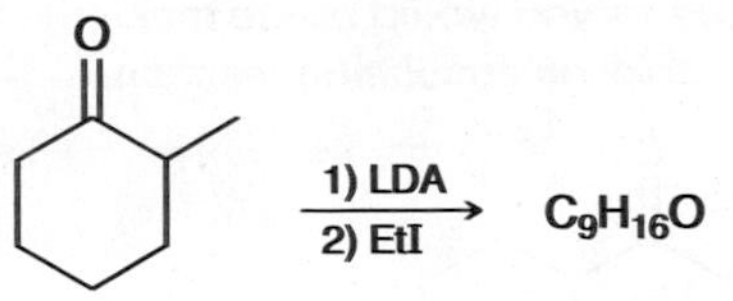

22.87 Propose an efficient synthesis for each of the following transformations.

(a)

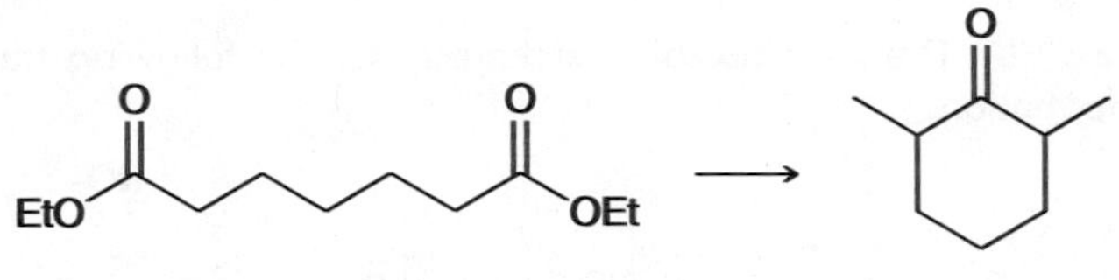

(b)

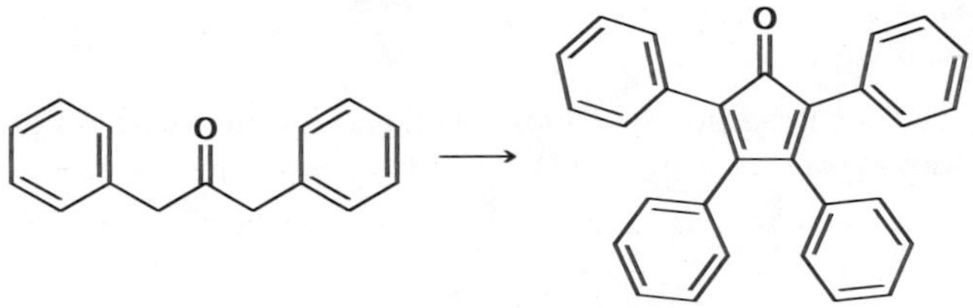

(c)

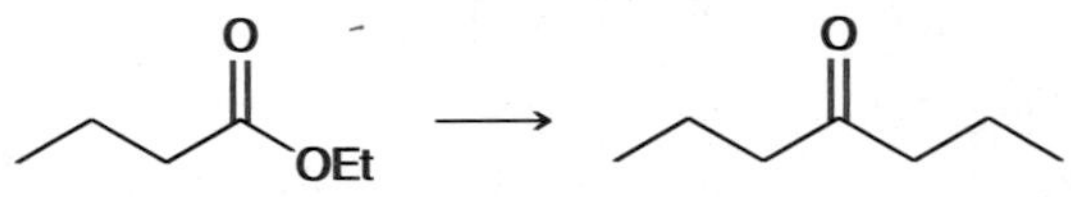

(d)

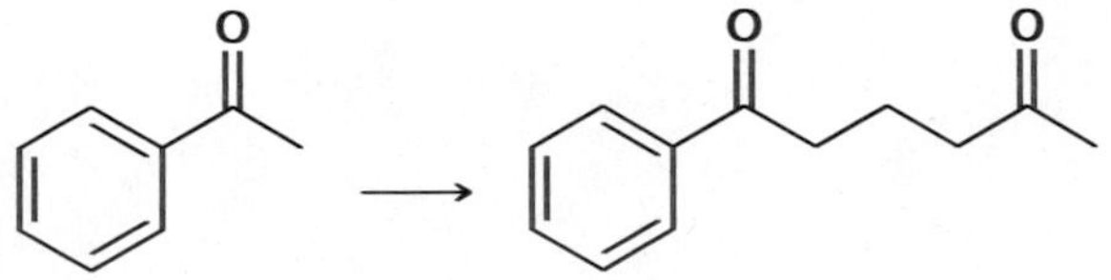

22.88 The product of an aldol condensation is an α,β-unsaturated ketone which is capable of undergoing hydrogenation to yield a saturated ketone. Using this technique, identify the reagents that you would need in order to prepare rheosmin via a crossed aldol reaction. Rheosmin is isolated from raspberries and is often used in perfume formulations for its pleasant odor. **HINT:** The presence of a phenolic proton will be problematic during an aldol reaction. (Can you explain why?) Consider using a protecting group (Section 13.7).

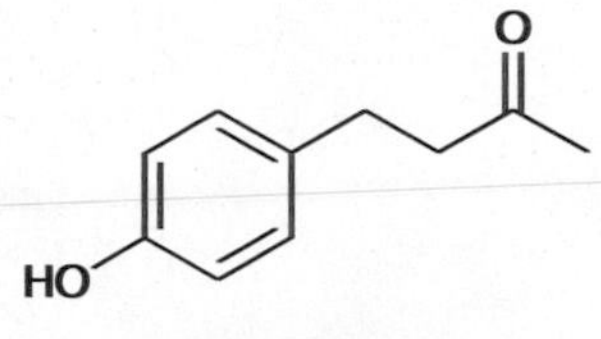

Rheosmin

22.89 Identify the reagents you would use to convert cyclohexanone into each of the following compounds.

(a)

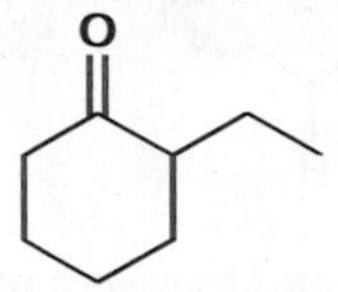

(b)

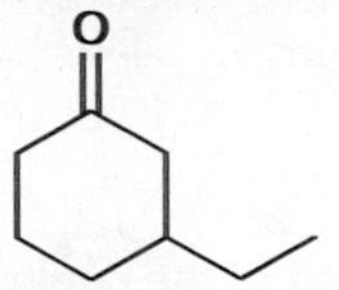

(c)

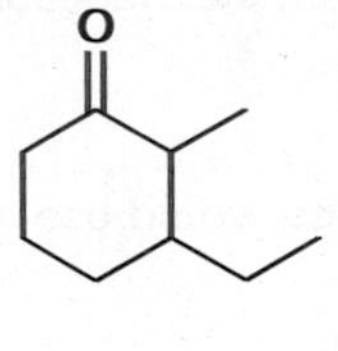

(d)

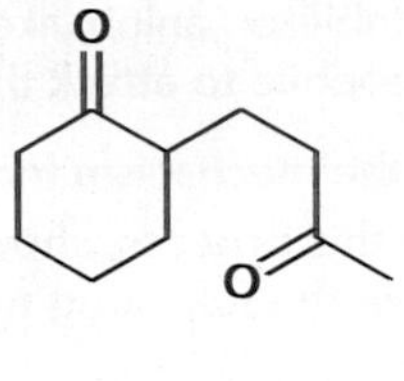

(e)

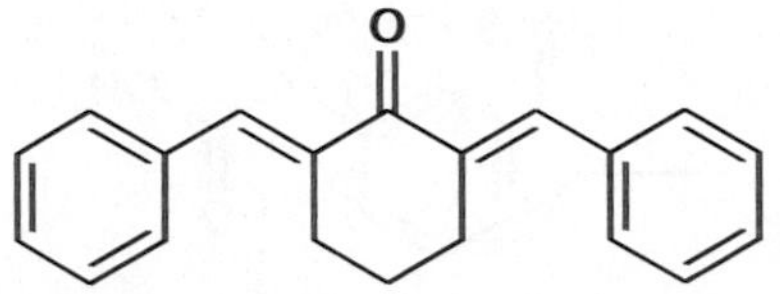

(f)

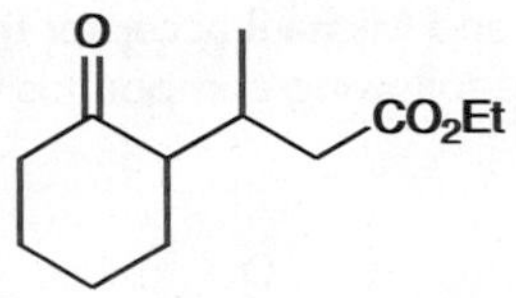

(g)

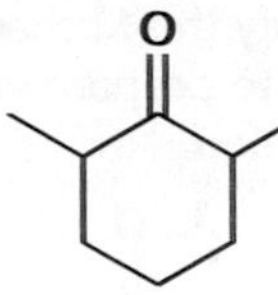

22.90 The enolate of a ketone can be treated with an ester to give a diketone. Draw a mechanism for this Claisen-like reaction, and explain why an acid source is required after the reaction is complete.

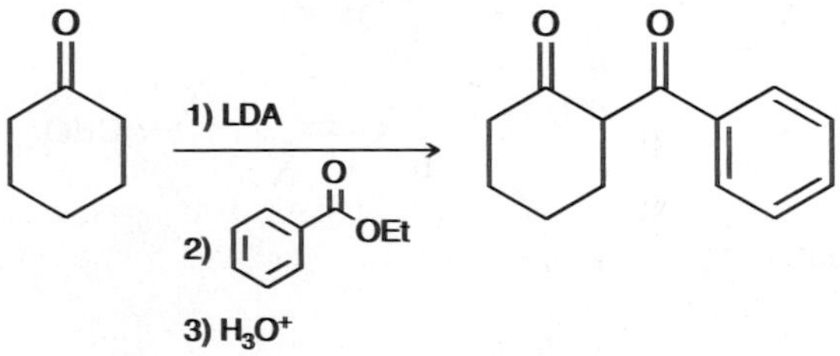

22.91 Beta-keto esters can be prepared by treating the enolate of a ketone with diethyl carbonate. Draw a plausible mechanism for this reaction.

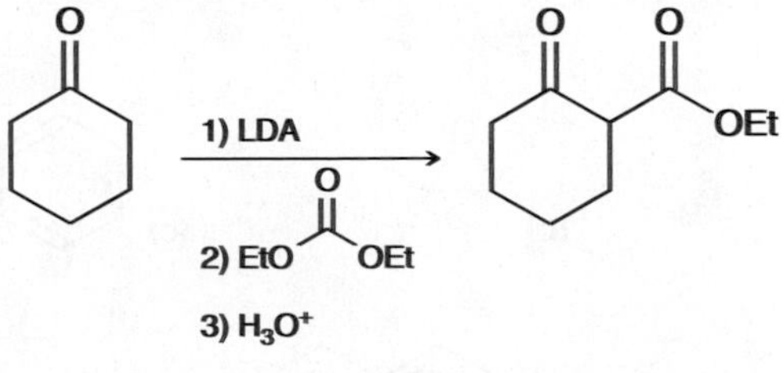

22.92 The enolate of an ester can be treated with a ketone to give a β-hydroxy ester. Draw a mechanism for this aldol-like reaction.

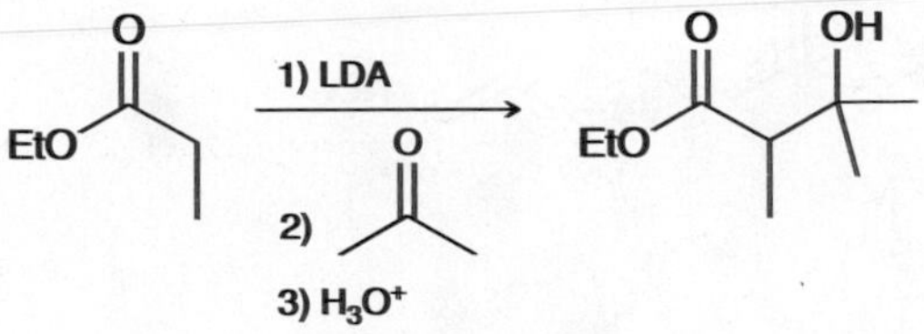

22.93 Nitriles undergo alkylation at the α position much like ketones undergo alkylation at the α position.

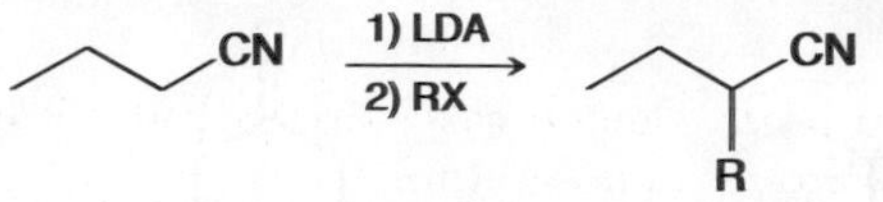

The α position of the nitrile is first deprotonated to give a resonance-stabilized anion (like an enolate), which then functions as a nucleophile to attack the alkyl halide.

(a) Draw the mechanism for this process.

(b) Using this process, show the reagents you would use to achieve the following transformation:

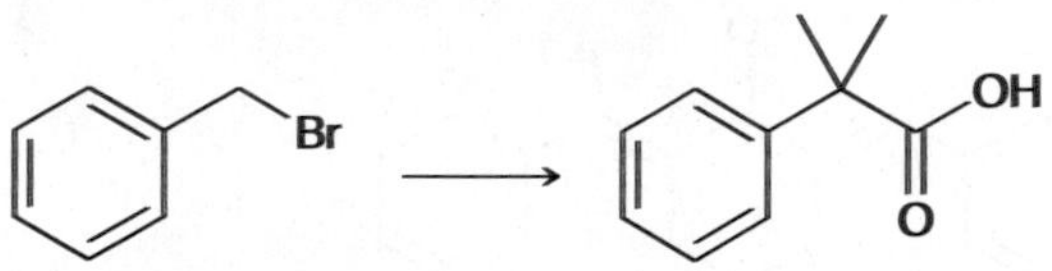

22.94 Identify the Michael donor and Michael acceptor that could be used to prepare each of the following compounds via a Michael addition.

(a) 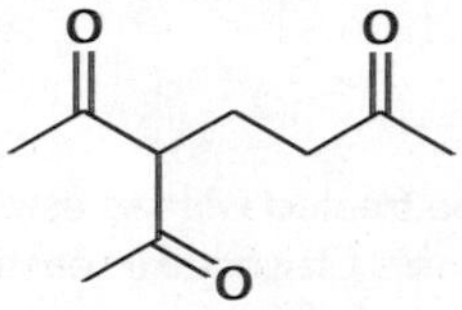

(b) 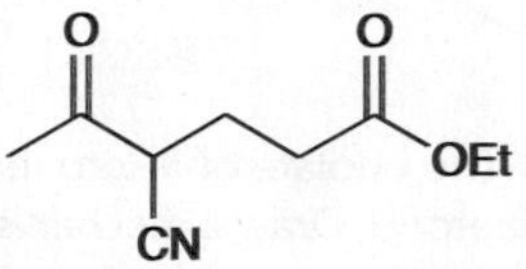

(c) 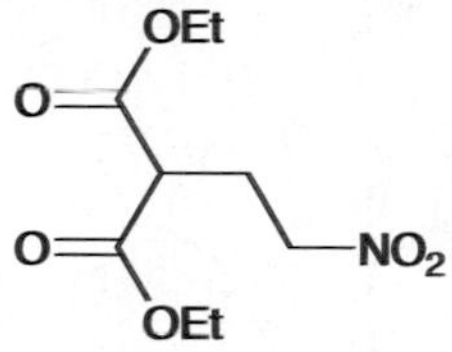

(d) 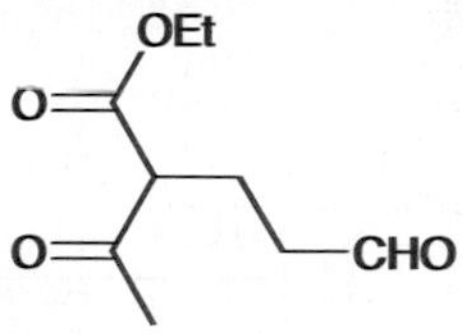

(e) 

22.95 The conjugate base of diethyl malonate can serve as a nucleophile to attack a wide range of electrophiles. Identify the product that is formed when the conjugate base of diethyl malonate reacts with each of the following electrophiles followed by acid workup.

(a) 

(b) 

(c) 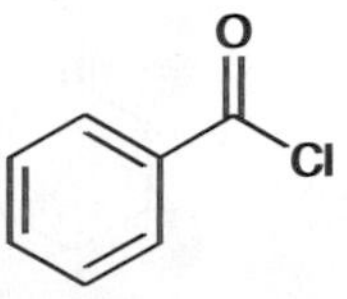

(d) 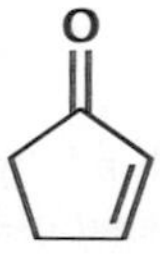

(e) 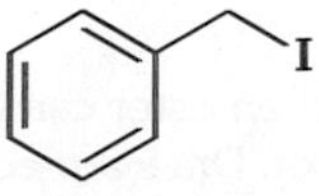

(f) 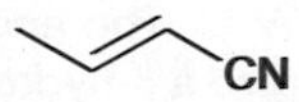

(g) 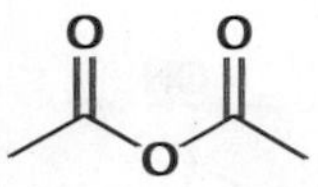

(h) 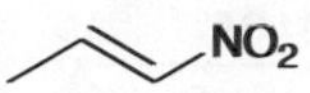

22.96 Draw the product of the Robinson annulation reaction that occurs when the following compounds are treated with aqueous sodium hydroxide.

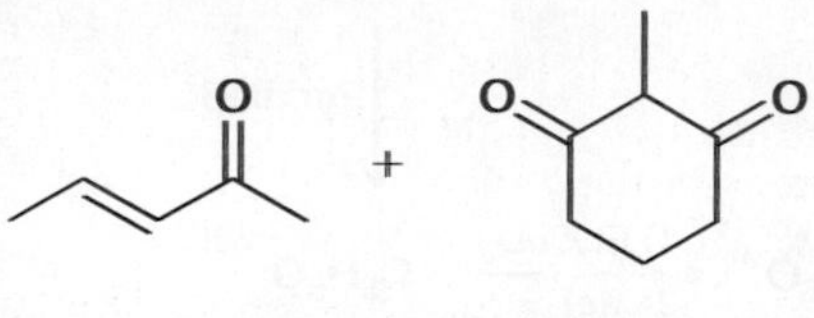

22.97 Identify what reagents you would use to make the following compound with a Robinson annulation reaction.

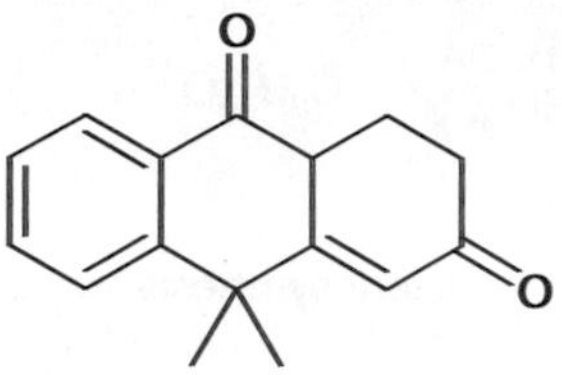

22.98 Draw a plausible mechanism for the following transformation.

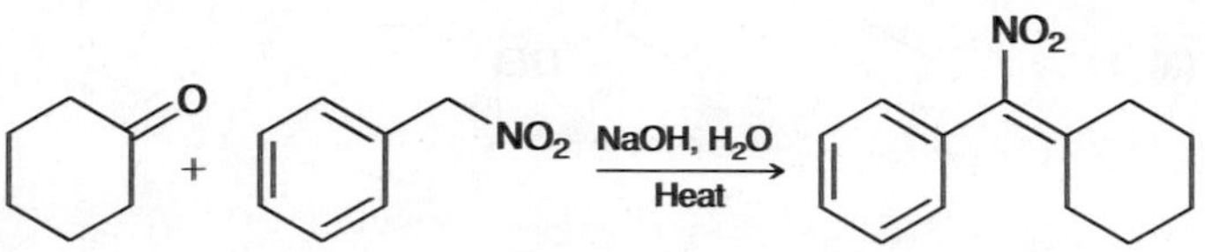

22.99 Propose an efficient synthesis for the following transformation.

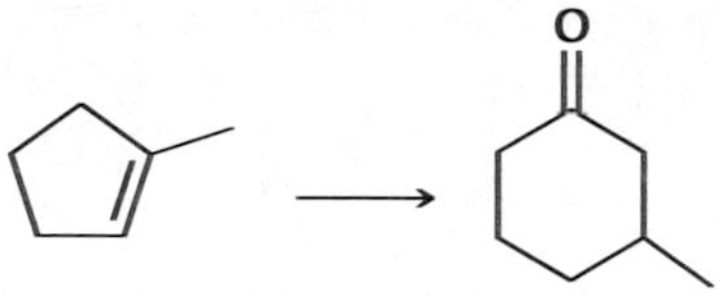

INTEGRATED PROBLEMS

22.100 For a pair of keto-enol tautomers, explain how IR spectroscopy might be used to identify whether the equilibrium favors the ketone or the enol.

22.101 Acrolein is an α,β-unsaturated aldehyde that is used in the production of a variety of polymers. Acrolein can be prepared by treating glycerol with an acid catalyst. Propose a plausible mechanism for this transformation.

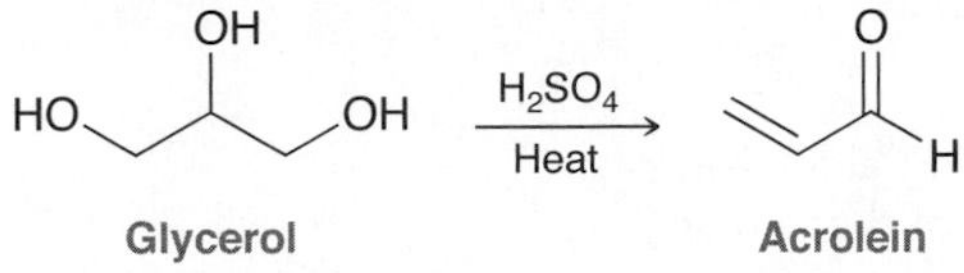

22.102 Draw the structure of the product with molecular formula $C_{10}H_{10}O$ that is obtained when the compound below is heated with aqueous acid.

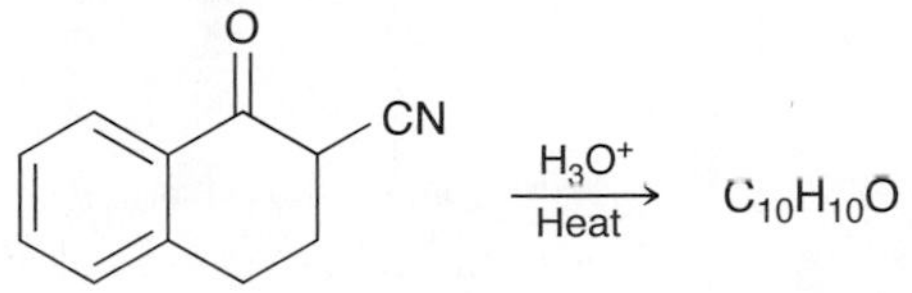

22.103 Lactones can be prepared from diethyl malonate and epoxides. Diethyl malonate is treated with a base, followed by an epoxide, followed by heating in aqueous acid:

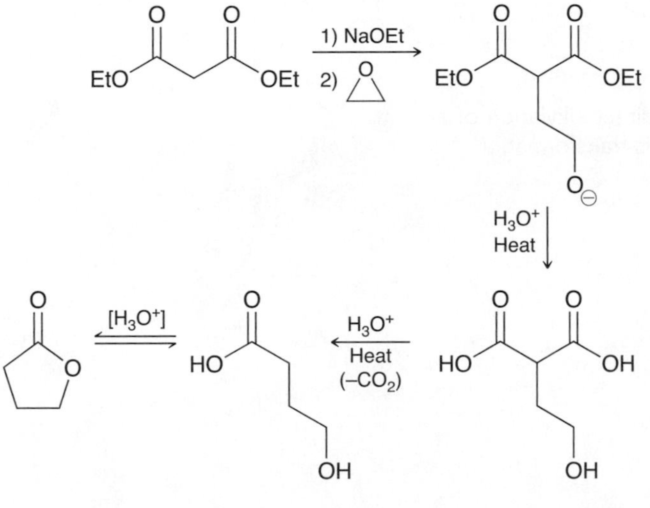

Using this process, identify what reagents you would need to prepare the following compound:

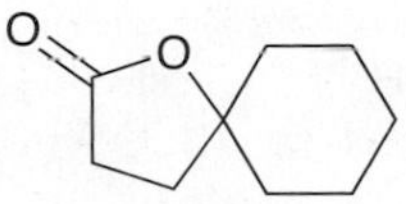

22.104 Predict the major product of the following transformation.

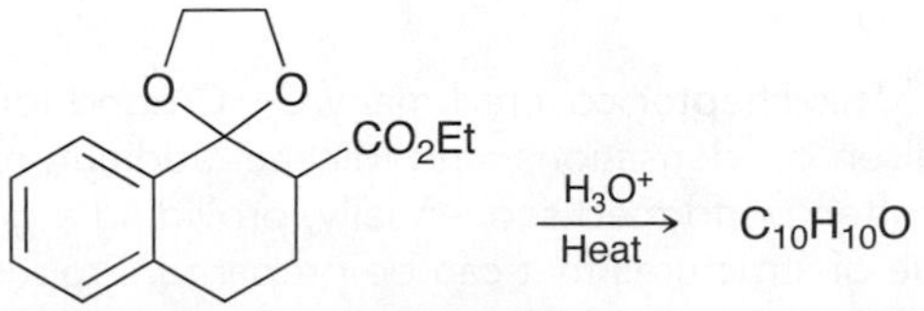

22.105 Consider the structures of the constitutional isomers, Compound **A** and Compound **B** (below). When treated with aqueous acid, Compound **A** undergoes isomerization to give a *cis* stereoisomer. In contrast, Compound **B** does not undergo isomerization when treated with the same conditions. That is, Compound **B** remains in the *trans* configuration. Explain the difference in reactivity between Compound **A** and Compound **B**.

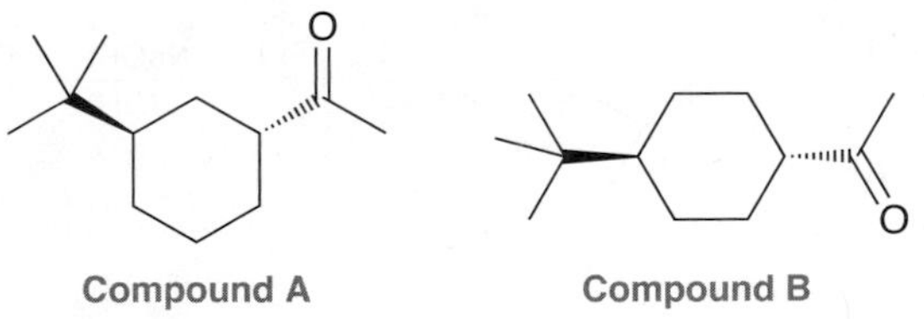

CHALLENGE PROBLEMS

22.106 Propose a plausible mechanism for the following transformation.

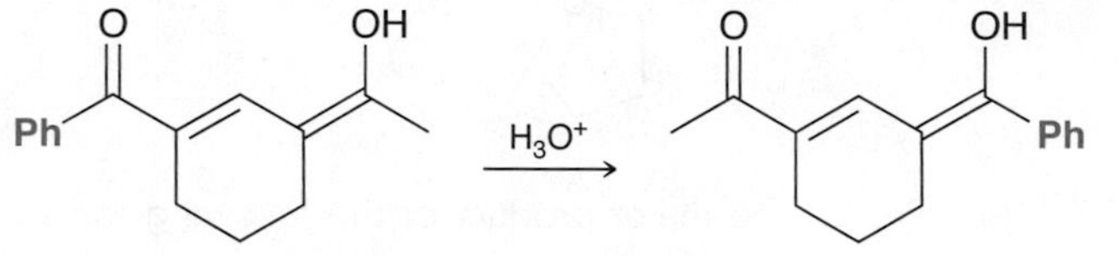

22.107 Propose a plausible mechanism for the following transformation.

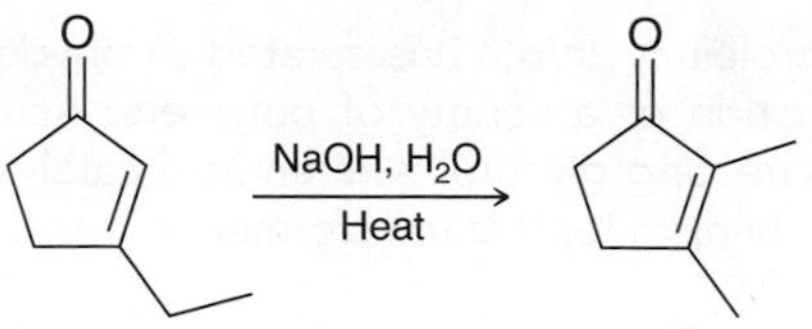

22.108 This chapter covered many C—C bond-forming reactions, including aldol reactions, Claisen condensations, and Michael addition reactions. Two or more of these reactions are often performed sequentially, providing a great deal of versatility and complexity in the type of structures that can be prepared. Propose a plausible mechanism for each of the following transformations.

(a)

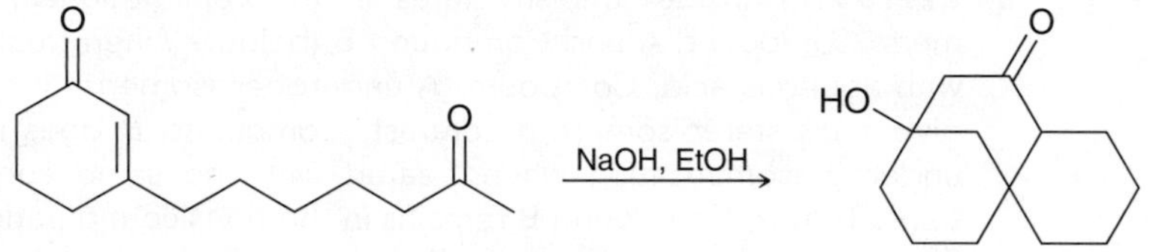

(b)

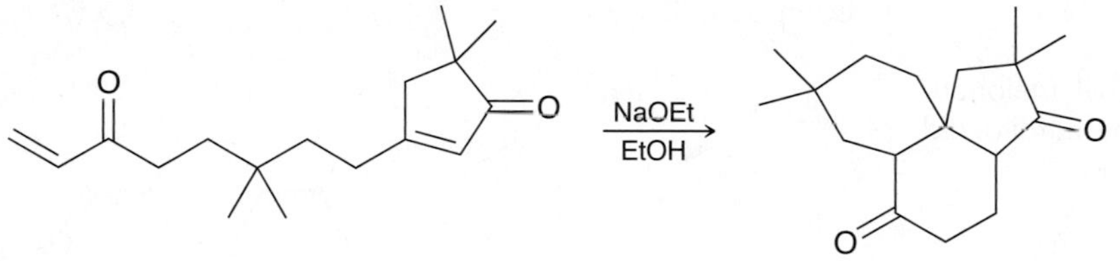

22.109 The following transformation cannot be accomplished by direct alkylation of an enolate. Explain why not, and then devise an alternate synthesis for this transformation.

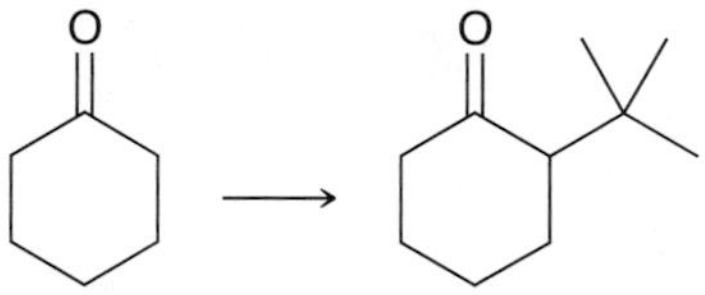

23

Amines

DID YOU EVER WONDER...

how drugs like Tagamet, Zantac, and Pepcid are able to control stomach acid production and alleviate the symptoms of acid reflux disease?

Tagamet, Zantac, and Pepcid are all compounds that contain several nitrogen atoms, which are important for the function of these drugs. In this chapter, we will explore the properties, reactions, and biological activity of many different nitrogen-containing compounds. The end of the chapter will revisit the structures and activity of the three drugs listed above, with a special emphasis on how medicinal chemists designed the first of these blockbuster drugs.

DO YOU REMEMBER?

Before you go on, be sure you understand the following topics.
If necessary, review the suggested sections to prepare for this chapter:

- Delocalized and Localized Lone Pairs (section 2.12)
- Aromatic Heterocycles (section 18.5)
- Brønsted-Lowry Acidity: A Quantitative Perspective (section 3.3)
- Activating Groups and Deactivating Groups (sections 19.7 and 19.8)

WILEY PLUS Visit www.wileyplus.com to check your understanding and for valuable practice.

23.1 Introduction to Amines

Classification of Amines

Amines are derivatives of ammonia in which one or more of the protons has been replaced with alkyl or aryl groups.

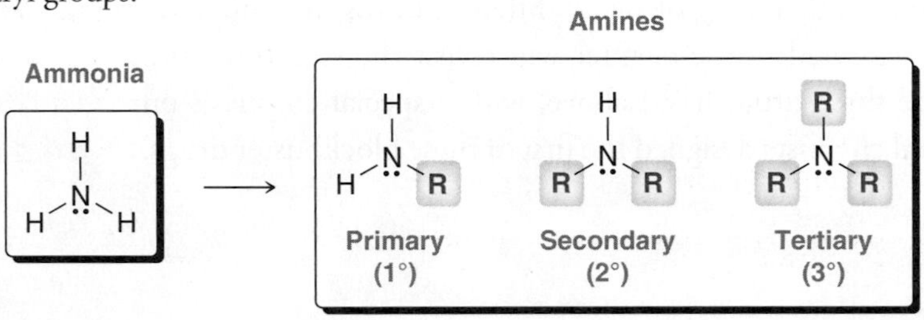

Amines are classified as primary, secondary, or tertiary, depending on the number of groups attached to the nitrogen atom. Note that these terms have a different meaning than when they were used in naming alcohols. A tertiary alcohol has three groups attached to the α carbon, while a tertiary amine has three groups attached to the nitrogen atom.

Amines are abundant in nature. Naturally occurring amines isolated from plants are called **alkaloids**. Below are examples of several alkaloids that have garnered public awareness as a result of their physiological activity:

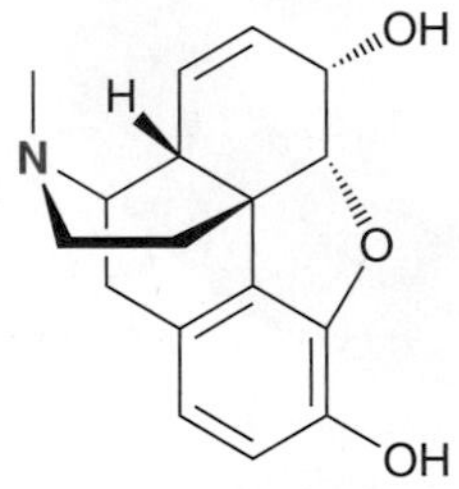

Morphine
(A potent analgesic isolated from the unripe seeds of the poppy plant *Papaver somniferum*)

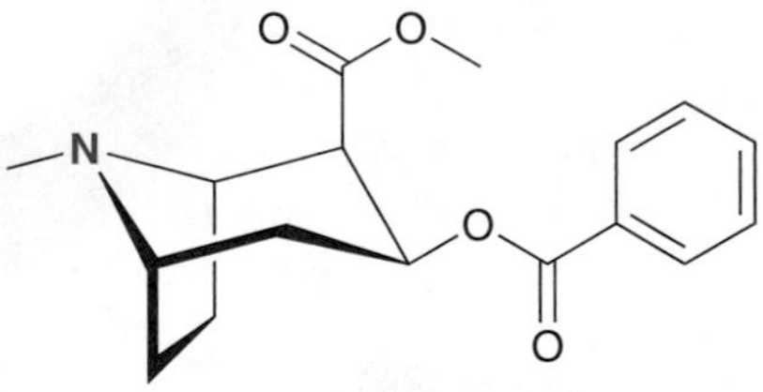

Cocaine
(A potent stimulant isolated from the leaves of the coca plant)

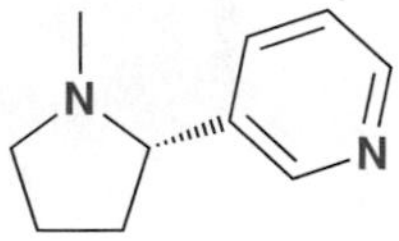

Nicotine
(An addictive and toxic compound found in tobacco)

Many amines also play vital roles in neurochemistry (chemistry taking place in the brain). Below are a few examples:

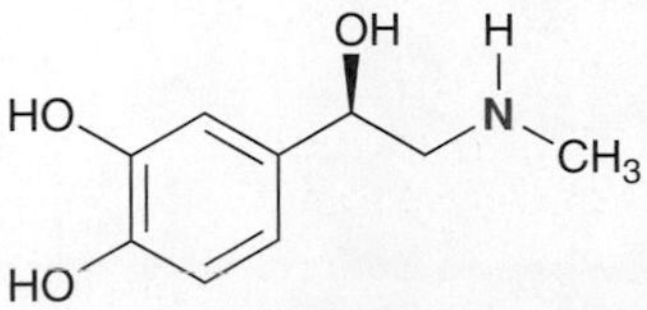

Adrenaline
(A "fight-or-flight" hormone, first discussed in Chapter 7)

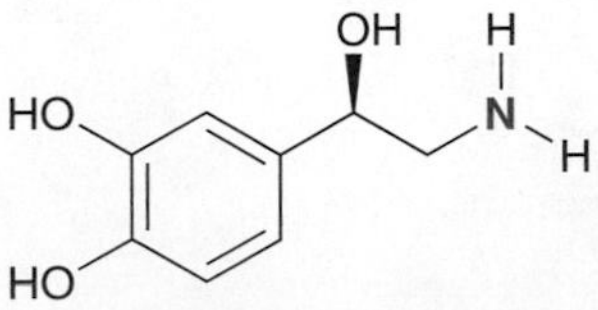

Noradrenaline
(Regulates heart rate and dilates air passages)

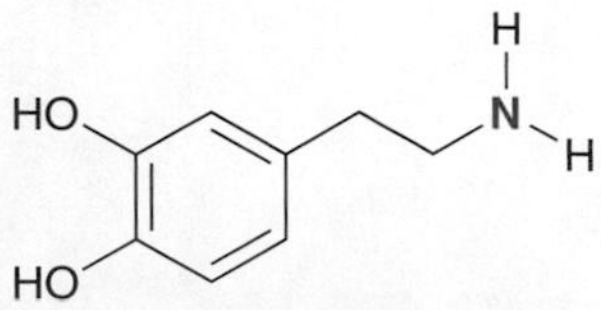

Dopamine
(Regulates motor skills and emotions)

Many pharmaceuticals also contain amine moieties, as we will see throughout this chapter.

MEDICALLYSPEAKING)))

Drug Metabolism Studies

Recall that many drugs are converted in our bodies to water-soluble compounds that can be excreted in the urine (see the Medically Speaking box in Section 13.9). This process, called drug metabolism, often involves the conversion of a drug into many different compounds called metabolites. Whenever a new drug is developed, medicinal chemists conduct drug metabolism studies in order to determine whether the observed activity is a property of the drug itself or whether the activity is derived from one of the drug's metabolites. Each of the metabolites is isolated, if possible, and tested for drug activity. In some cases, it is found that the drug and one of its metabolites both exhibit activity. In other cases, the drug itself has no activity, but one of its metabolites is found to be responsible for the observed activity. Drug metabolism studies have, in some cases, led to the development of safer drugs. One such example will now be described.

Terfenadine, marketed under the trade name Seldane was one of the first nonsedating antihistamines. After being widely used, it was discovered to cause cardiac arrhythmia (abnormal heart rhythm) among patients who were concurrently using antifungal agents. It is believed that antifungal agents inhibit the enzyme responsible for metabolizing terfenadine so that the drug remains in the body for a longer time. After several doses of terfenadine, the drug accumulates, and the high concentrations ultimately lead to the observed arrhythmia.

Medicinal chemists were able to sidestep this problem by carefully analyzing and testing the metabolites of terfenandine. Specifically, one of the metabolites was isolated and was found to possess a carboxylic acid moiety. Further studies showed that this metabolite, named fexofenadine, is the active compound, and terfenadine simply functions as a prodrug. But unlike the prodrug, this metabolite is further metabolized in the presence of antifungal agents. As a result, terfenadine was removed from the market and replaced with fexofenadine, which is now marketed in the United States under the trade name Allegra. This example illustrates how drug metabolism studies can lead to the development of safer drugs.

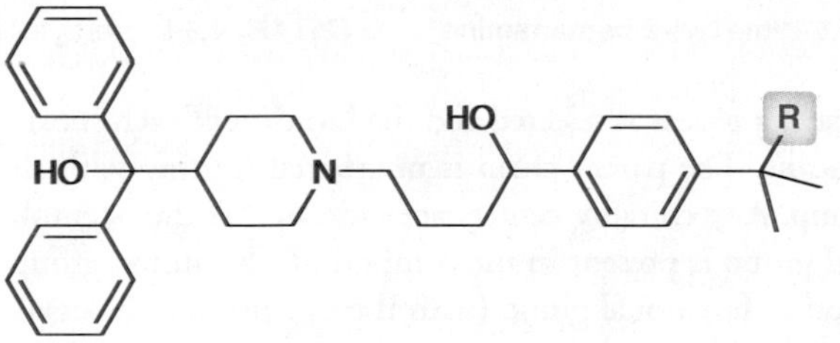

$R = CH_3$ **Seldane™** (terfenadine)

$R = COOH$ **Allegra™** (fexofenadine)

Reactivity of Amines

The nitrogen atom of an amine possesses a lone pair that represents a region of high electron density. This can be seen in an electrostatic potential map of trimethylamine (Figure 23.1). The presence of this lone pair is responsible for most of the reactions exhibited by amines. Specifically, the lone pair can function as a base or as a nucleophile:

Functioning as a base

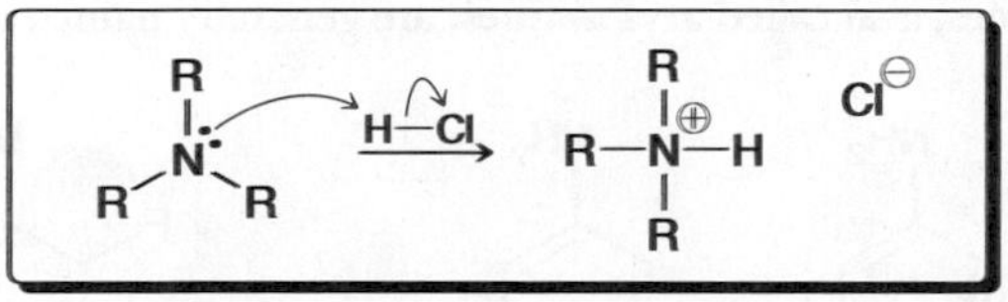

Functioning as a nucleophile

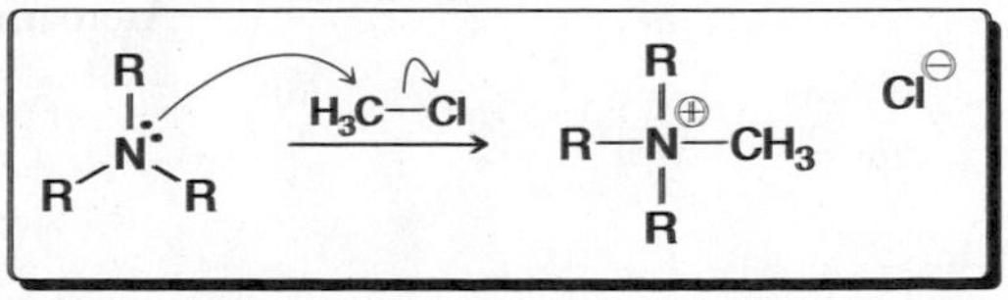

Section 23.3 explores the basicity of amines, while Sections 23.8–23.11 explore reactions in which amines function as nucleophiles.

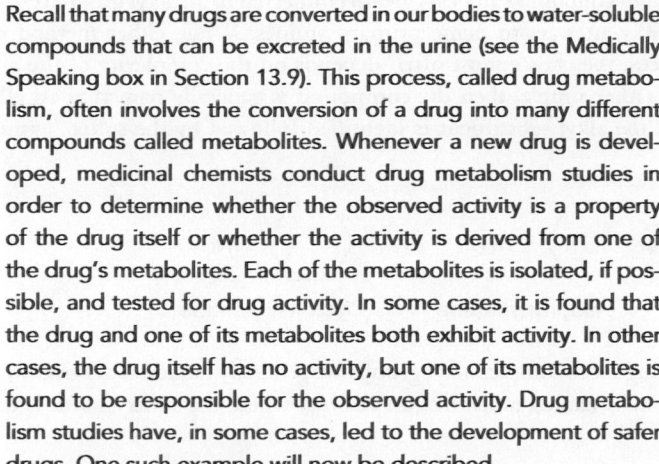

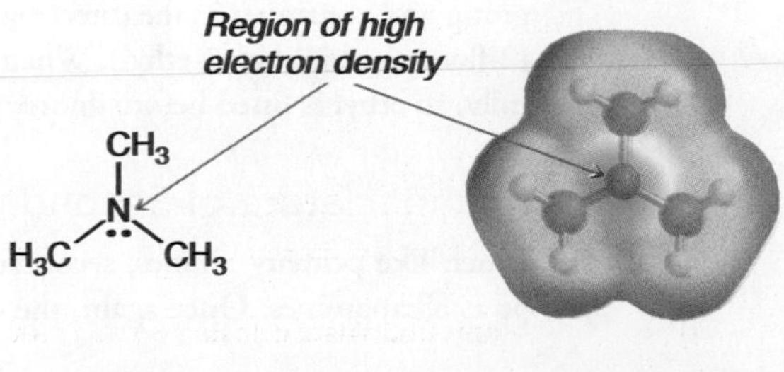

FIGURE 23.1 An electrostatic potential map of trimethylamine. The red area indicates a region of high electron density.

23.2 Nomenclature of Amines

Nomenclature of Primary Amines

A primary amine is a compound containing an NH_2 group connected to an alkyl group. IUPAC nomenclature allows two different ways to name primary amines. While either method can be used in most circumstances, the choice most often depends on the complexity of the alkyl group. If the alkyl group is rather simple, then the compound is generally named as an **alkyl amine**. With this approach, the alkyl substituent is identified followed by the suffix "amine." Below are several examples.

Ethylamine

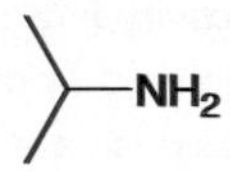
Isopropylamine

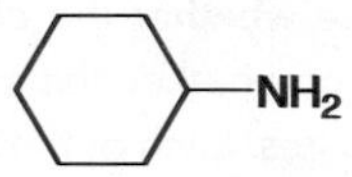
Cyclohexylamine

Primary amines containing more complex alkyl groups are generally named as **alkanamines**. With this approach, the amine is named much like an alcohol, where the suffix *-amine* is used in place of *-ol*.

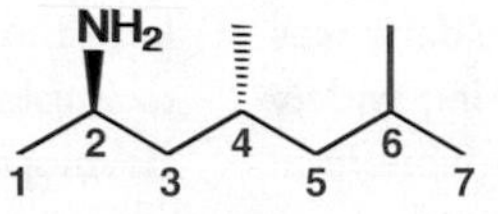
(2*R*,4*R*)-4,6-Dimethyl-2-heptanamine

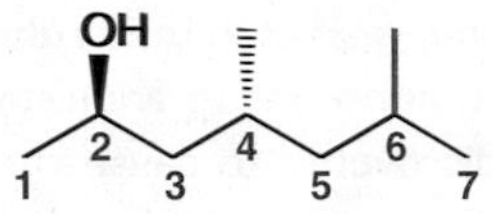
(2*R*,4*R*)-4,6-Dimethyl-2-heptanol

To use this approach, a parent is selected and the location of each substituent is identified with the appropriate locant. The parent chain is numbered starting with the side that is closer to the functional group. Any chirality centers are indicated at the beginning of the name. When another functional group is present in the compound, the amino group is generally listed as a substituent. The other functional group (with the exception of halogens) receives priority and its name becomes the suffix.

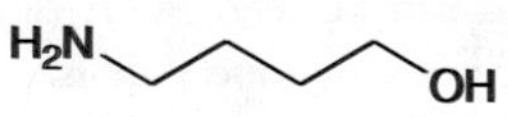
4-Aminobutanol

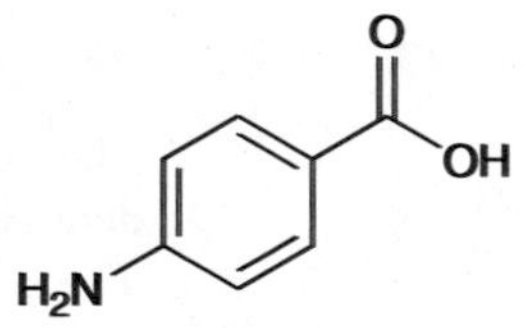
para-Aminobenzoic acid

Aromatic amines, also called **aryl amines**, are generally named as derivatives of aniline.

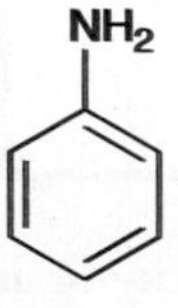
Aniline

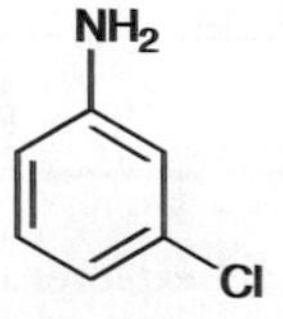
meta-Chloroaniline

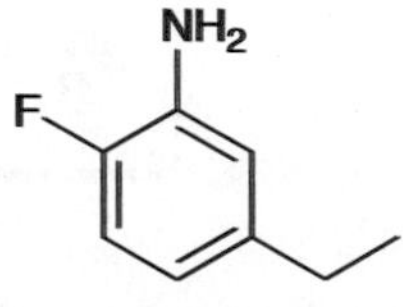
5-Ethyl-2-fluoroaniline

Notice in the last example that the numbering begins at the carbon attached to the amino group and continues in the direction that gives the lower number to the first point of difference (2-fluoro rather than 3-ethyl). When the name is assembled, all substituents are listed alphabetically, so ethyl is listed before fluoro.

Nomenclature of Secondary and Tertiary Amines

Much like primary amines, secondary and tertiary amines can also be named as alkyl amines or as alkanamines. Once again, the complexity of the alkyl groups typically determines which system is chosen. If all alkyl groups are rather simple in structure, then the groups are listed in alphabetical order. The prefixes "di" and "tri" are used if the same alkyl group appears more than once.

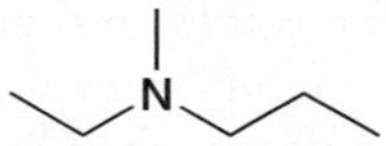

Ethylmethylpropylamine

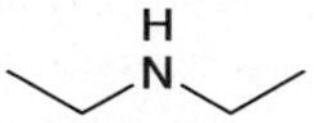

Diethylamine

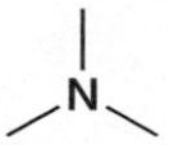

Trimethylamine

If one of the alkyl groups is complex, then the compound is typically named as an alkanamine, with the most complex alkyl group treated as the parent and the simpler alkyl groups treated as substituents.

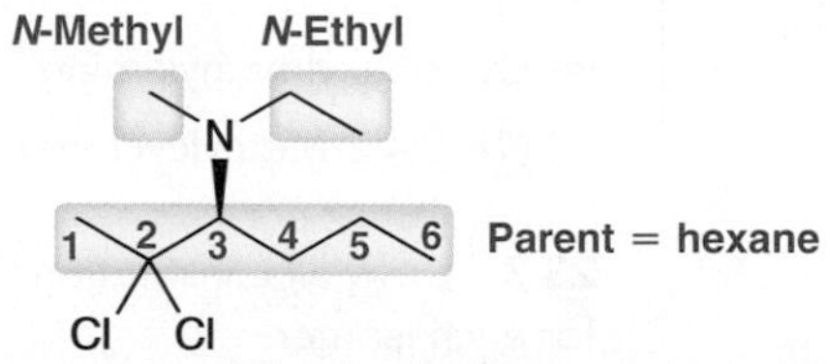

(*S*)-2,2-Dichloro-*N*-ethyl-*N*-methyl-3-hexanamine

In this example, the parent is a hexane chain and the suffix is *-amine*. The methyl and ethyl groups are listed as substituents using the locant "*N*" to identify that they are connected to the nitrogen atom.

SKILLBUILDER

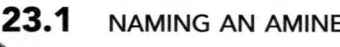

23.1 NAMING AN AMINE

LEARN the skill

Assign a name for the following compound.

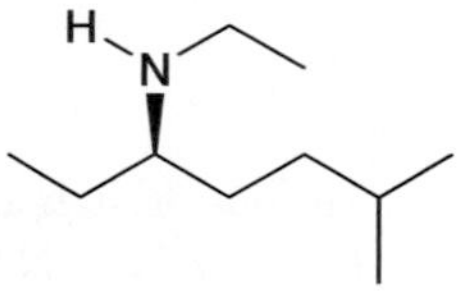

SOLUTION

STEP 1
Identify all of the alkyl groups connected to the nitrogen atom.

First identify all of the alkyl groups connected to the nitrogen atom: There are two alkyl groups, so this compound is a secondary amine.

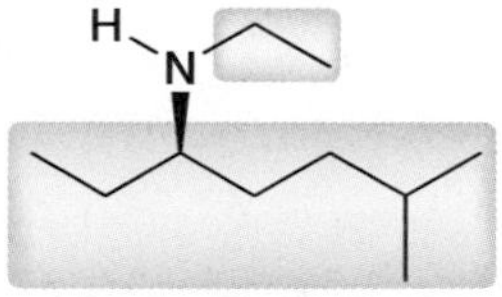

STEP 2
Determine what method to use.

STEP 3
If naming as an alkanamine, choose the more complex group to be the parent.

Next, determine whether to name the compound as an alkyl amine or as an alkanamine. If the alkyl groups are simple, then the compound can be named as an alkyl amine; otherwise, it should be named as an alkanamine. In this case, one of the alkyl groups is complex, so the compound will be named as an alkanamine. The complex alkyl group is chosen as the parent, and the other alkyl group is listed as a substituent, using the letter *N* as a locant. Notice that the parent is numbered starting from the side closer to the amine moiety.

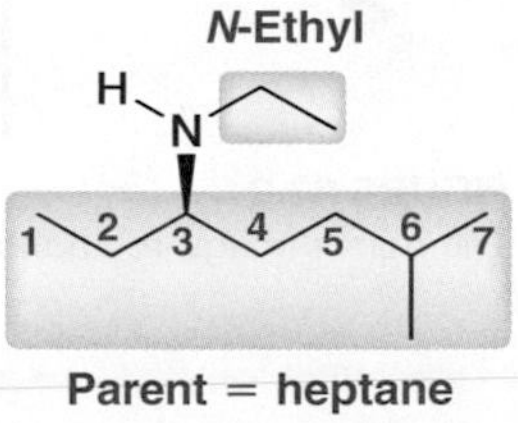

STEP 4
Assign locants, assemble the name, and assign configuration.

Finally, assign locants to each substituent, assemble the name, and assign the configuration of any chirality centers:

(*R*)-*N*-Ethyl-6-methyl-3-heptanamine

PRACTICE the skill **23.1** Assign a name for each of the following compounds:

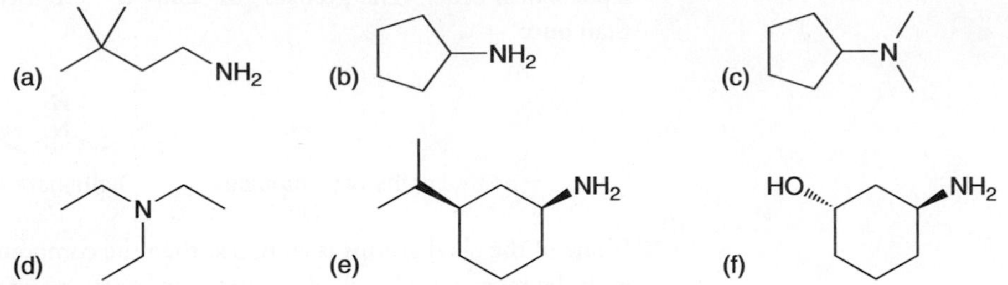

APPLY the skill **23.2** Draw the structure of each of the following compounds:

(a) Cyclohexylmethylamine (b) Tricyclobutylamine (c) 2,4-Diethylaniline
(d) (1*R*,2*S*)-2-Methylcyclohexanamine (e) *ortho*-Aminobenzaldehyde

23.3 Draw all constitutional isomers with molecular formula C_3H_9N, and provide a name for each isomer.

need more **PRACTICE?** **Try Problems 23.42, 23.46, 23.47**

23.3 Properties of Amines

Geometry

The nitrogen atom of an amine is typically sp^3 hybridized, with the lone pair occupying an sp^3-hybridized orbital. Consider trimethylamine as an example:

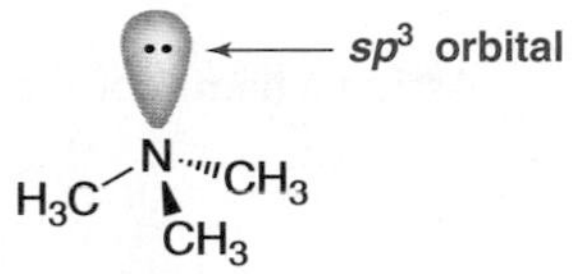

All four orbitals are arranged in a shape that approximates a tetrahedron, with bond angles of 108°. The C—N bond lengths are 147 pm, which is shorter than the average C—C bond of an alkane (153 pm) and longer than the average C—O bond of an alcohol (143 pm).

Amines containing three different alkyl groups are chiral compounds (Figure 23.2). The nitrogen atom has a total of four different groups (three alkyl groups and a lone pair) and is therefore a chirality center. Compounds of this type are generally not optically active at room temperature, because pyramidal inversion occurs quite rapidly, producing a racemic mixture of enantiomers (Figure 23.3). During pyramidal inversion, the compound passes through a transition state in which the nitrogen atom has sp^2 hybridization. This transition state is only about 25 kJ/mol (6 kcal/mol) higher in energy than the sp^3-hybridized geometry. This energy barrier is relatively small and is easily surmounted at room temperature. For this reason, most amines bearing three different alkyl groups cannot be resolved at room temperature.

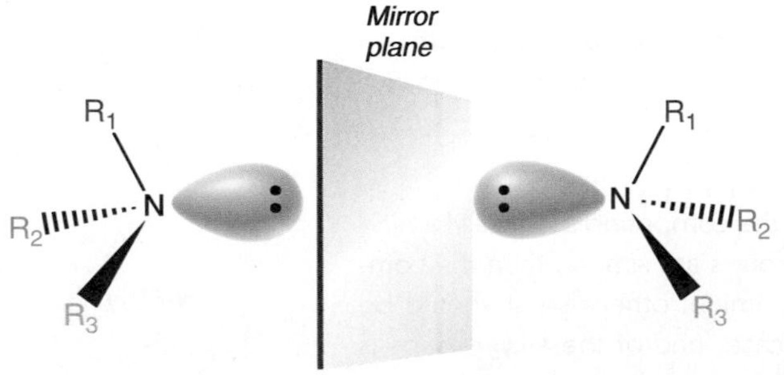

FIGURE 23.2
The mirror image-relationship of enantiomeric amines.

Colligative Properties

Amines exhibit solubility trends that are similar to the trends exhibited by alcohols. Specifically, amines with fewer than five carbon atoms per amino moiety will typically be water soluble, while amines with more than five carbon atoms per amino moiety will be only sparingly soluble. For example, ethylamine is soluble in water, while octylamine is not.

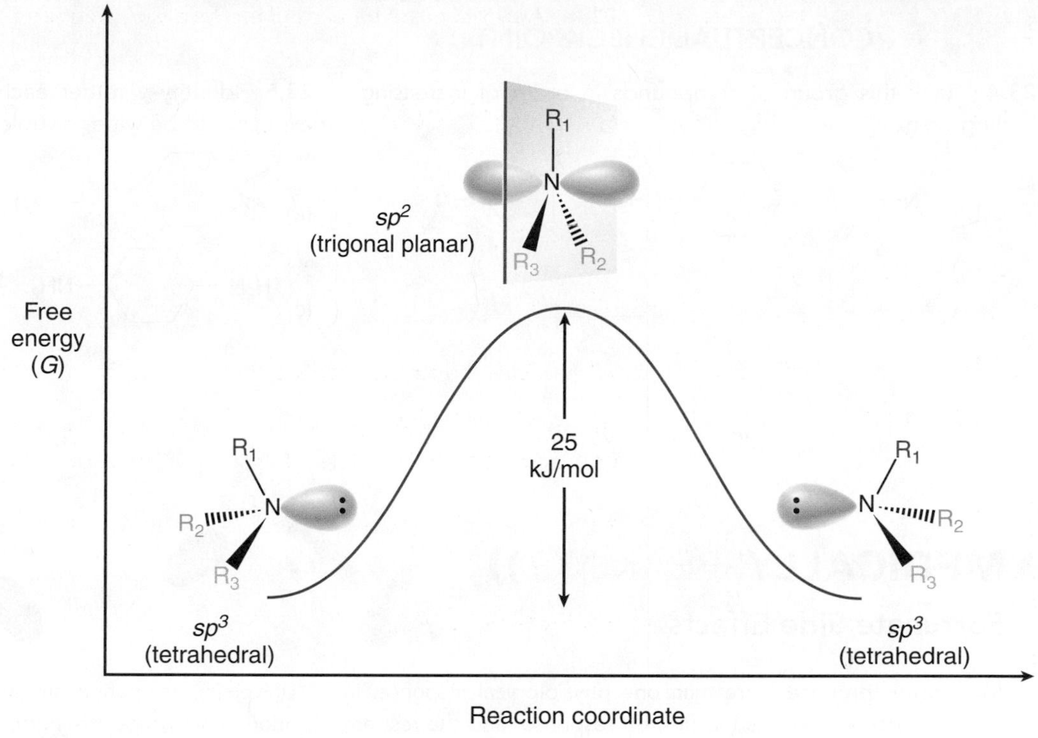

FIGURE 23.3
Pyramidal inversion of an amine enables the enantiomers to rapidly interconvert at room temperature.

Primary and secondary amines can form intermolecular H bonds and typically have higher boiling points than analogous alkanes but lower boiling points than analogous alcohols.

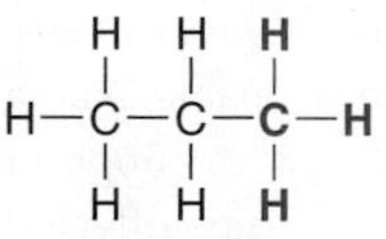

Propane
bp = −42°C

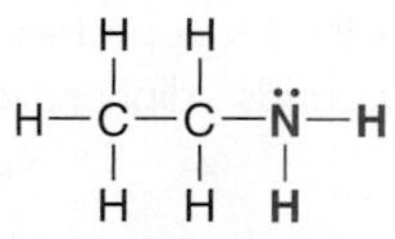

Ethylamine
bp = 17°C

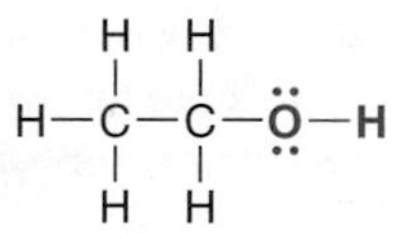

Ethanol
bp = 78°C

The boiling point of amines increases as a function of their capacity to form hydrogen bonds. As a result, primary amines typically have higher boiling points, while tertiary amines have lower boiling points. This trend can be observed by comparing the physical properties of the following three constitutional isomers.

Propylamine
bp = 50°C

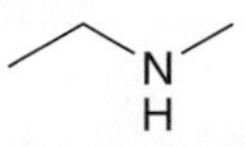

Ethylmethylamine
bp = 34°C

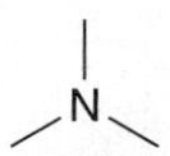

Trimethylamine
bp = 3°C

Other Distinctive Features of Amines

Amines with low molecular weights, such as trimethylamine, typically have a fishlike odor. In fact, the odor of fish is caused by amines that are produced when enzymes break down certain fish proteins. Putrescine and cadaverine are examples of compounds present in rotting fish. They are also present in urine, contributing to its characteristic odor.

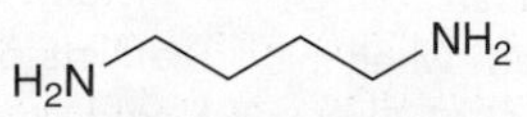

Putrescine
(1,4-butanediamine)

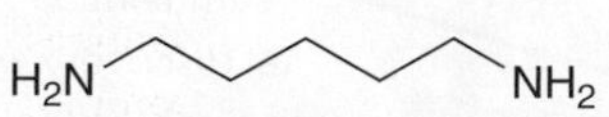

Cadaverine
(1,5-pentanediamine)

CONCEPTUAL CHECKPOINT

23.4 Rank this group of compounds in order of increasing boiling point.

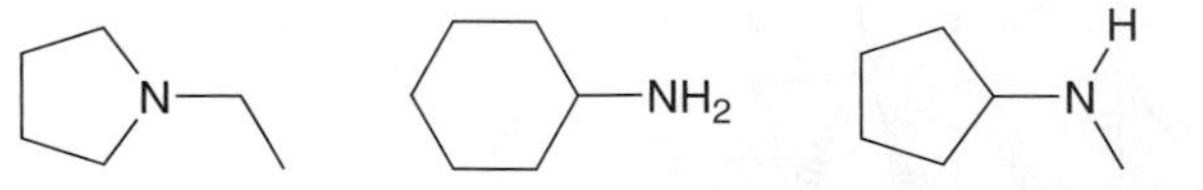

23.5 Identify whether each of the following compounds is expected to be water soluble:

(a) 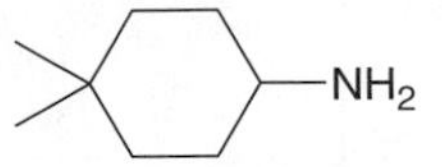

(b) 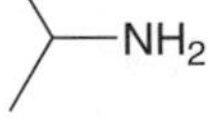

(c) 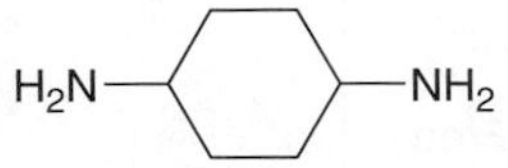

MEDICALLYSPEAKING

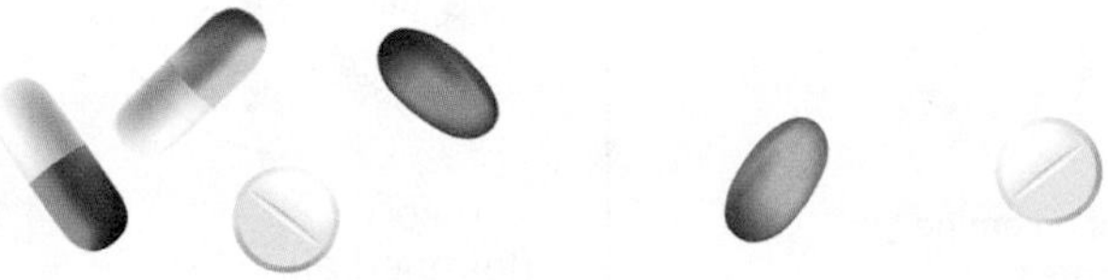

Fortunate Side Effects

Most drugs produce more than one physiological response. In general, one response is the desired response, and the rest are considered to be undesirable side effects. However, in some cases, the side effects have turned out to be fortunate discoveries. These accidental discoveries have led to the development of novel treatments for a wide range of ailments. The following are some of the better-known historical examples of this phenomenon.

Dimenhydrinate, which was originally developed as an antihistamine, is a mixture of two compounds: diphenhydramine (Benadryl) and 8-chlorotheophylline:

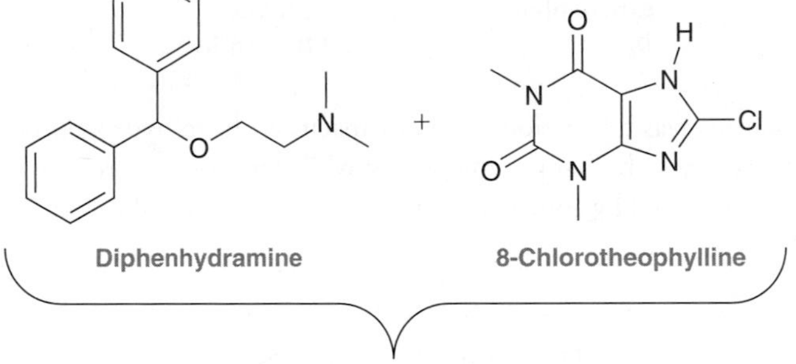

Diphenhydramine + **8-Chlorotheophylline**

Dimenhydrinate

The former is a first-generation (sedating) antihistamine, while the latter is a chlorinated derivative of caffeine (a stimulant). It was believed that the drowsiness of the former would be reduced by the effects of the latter. Unfortunately, the sedating effects of the antihistamine were too strong and overpowered the effects of the stimulant. However, an accidental discovery was made when this mixture was tested in 1947. During clinical studies at the Johns Hopkins University, one patient suffering from car sickness reported alleviation of his symptoms. Further studies were then conducted, and it was discovered that the mixture of compounds was effective in treating other forms of motion sickness, including air sickness and sea sickness. Marketed under the trade name Dramamine, this formulation quickly became one of the most widely used drugs for the treatment of motion sickness. It has since been replaced with a newer drug, called meclizine, which is sold under the trade name Dramamine II.

Bupropion and sildenafil are two other drugs that have useful side effects:

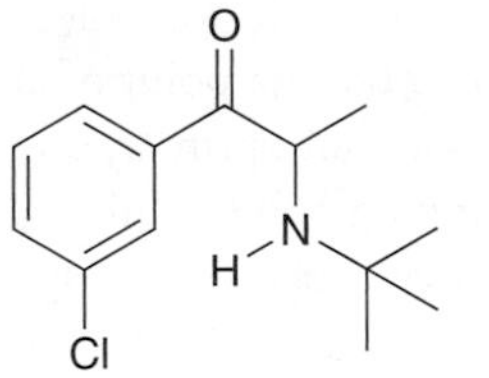

Zyban™ (bupropion)

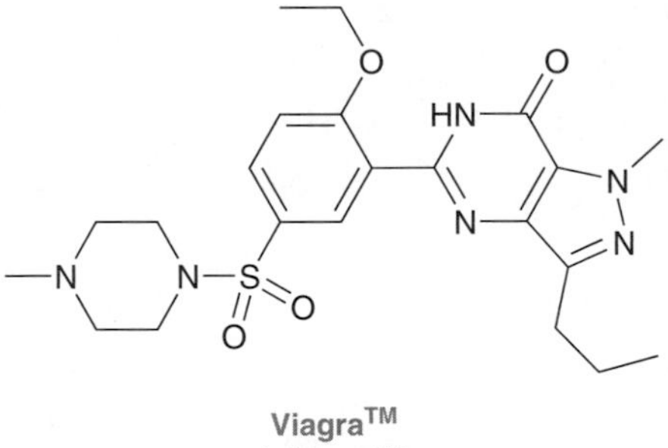

Viagra™ (sildenafil)

Bupropion was developed as an antidepressant but was found to help patients quit smoking during clinical trials. It is now marketed as a smoking cessation aid under the trade name Zyban. Sildenafil was designed to treat angina (chest pain that occurs when the heart muscle receives insufficient oxygen). In clinical studies, many of the volunteers experienced increased erectile function. Since sildenafil was found to be ineffective in treating angina, its manufacturer (Pfizer) decided to market the drug instead as a treatment for erectile dysfunction, a condition that afflicts one in nine adult males. Sildenafil is now marketed and sold under the trade name Viagra, with annual sales in excess of $1 billion.

Basicity of Amines

One of the most important properties of amines is their basicity. Amines are generally stronger bases than alcohols or ethers, and they can be effectively protonated even by weak acids.

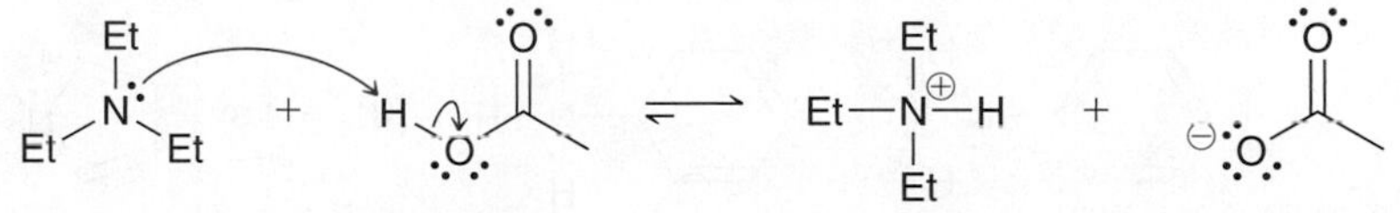

pK_a = 4.76 pK_a = 10.76

In this example, triethylamine is protonated using acetic acid. Compare the pK_a values of acetic acid (4.76) and the ammonium ion (10.76). Recall that the equilibrium will favor the weaker acid. In this case, the ammonium ion is six orders of magnitude weaker than acetic acid, and therefore, the amine will exist almost completely in protonated form (one in every million molecules will be in the neutral form). This example illustrates how the basicity of an amine can be quantified by measuring the pK_a of the corresponding ammonium ion. *A high pK_a indicates that the amine is strongly basic, while a low pK_a indicates that the amine is only weakly basic.* Table 23.1 shows pK_a values for the ammonium ions of many amines.

TABLE 23.1 THE pK_a VALUES OF THE AMMONIUM IONS OF SEVERAL AMINES

AMINE	pK_a OF AMMONIUM ION	AMINE	pK_a OF AMMONIUM ION
Alkyl amines		*Alkyl amines*	
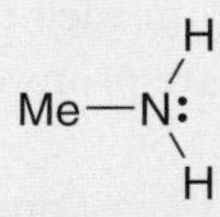 **Methylamine**	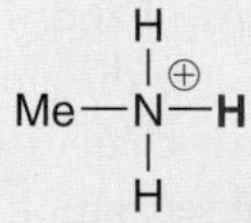 pK_a = 10.6	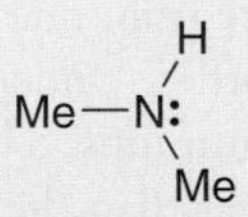 **Dimethylamine**	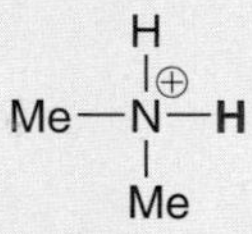 pK_a = 10.7
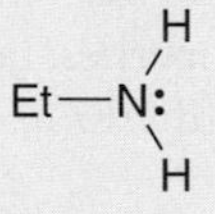 **Ethylamine**	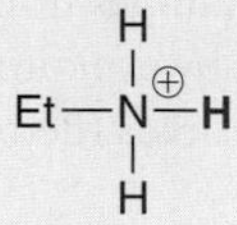 pK_a = 10.6	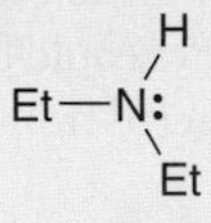 **Diethylamine**	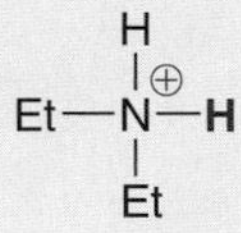 pK_a = 11.0
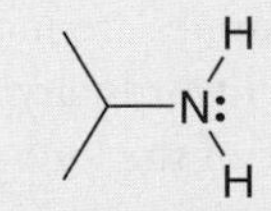 **Isopropylamine**	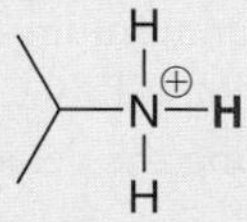 pK_a = 10.6	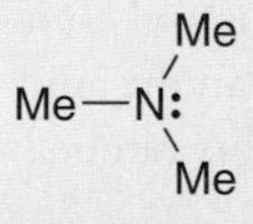 **Trimethylamine**	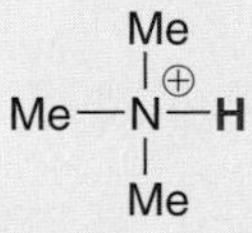 pK_a = 9.8
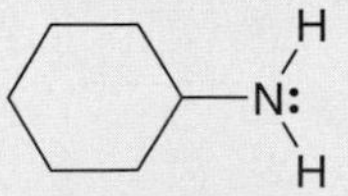 **Cyclohexylamine**	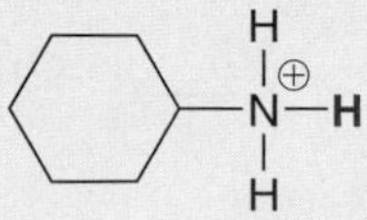 pK_a = 10.7	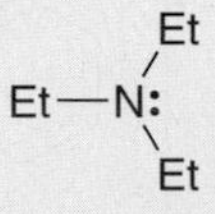 **Triethylamine**	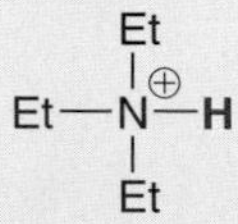 pK_a = 10.8

Continued

TABLE 23.1 ***(CONTINUED)***

AMINE		pKa OF AMMONIUM ION	AMINE		pKa OF AMMONIUM ION
Aryl amines			*Heterocyclic amines*		
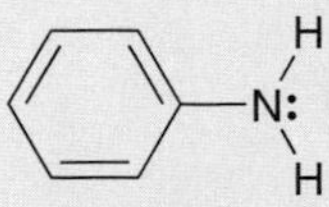 **Aniline**	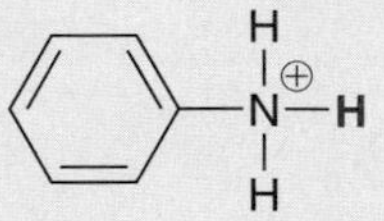	$\mathrm{p}K_a$ = 4.6	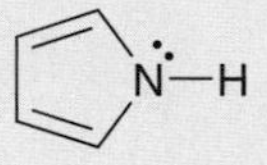 **Pyrrole**	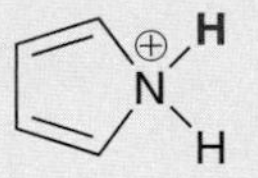	$\mathrm{p}K_a$ = 0.4
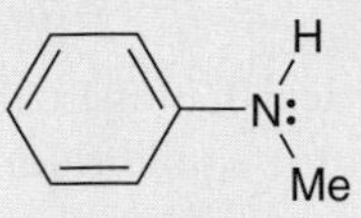 ***N*-Methylaniline**	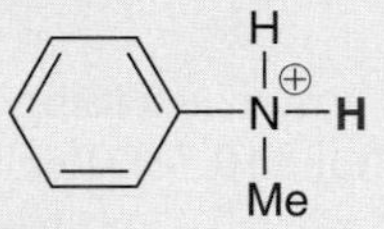	$\mathrm{p}K_a$ = 4.8	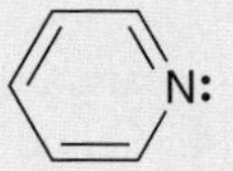 **Pyridine**	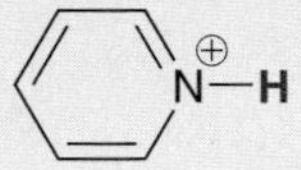	$\mathrm{p}K_a$ = 5.3
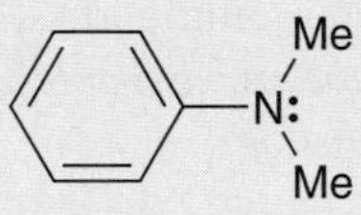 ***N,N*-Dimethylaniline**	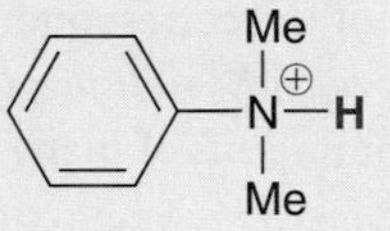	$\mathrm{p}K_a$ = 5.1			

The ease with which amines are protonated can be used to remove them from mixtures of organic compounds. This process is called **solvent extraction**. A mixture of organic compounds is first dissolved in an organic solvent. A solution of aqueous acid is then added to the mixture, and the contents are shaken vigorously. Upon standing, the organic and aqueous layers separate from onc another, much the way oil and water separate from each other in salad dressing bottles. Under these conditions, most of the amine molecules are protonated, forming ammonium ions. The ammonium ions bear charges and are therefore more soluble in the aqueous layer than the organic layer. Manual separation of the aqueous layer provides the ammonium ions which are then treated with base to regenerate the neutral amine. In this way, amines can be readily separated from a mixture of organic compounds. This process was used extensively in the nineteenth century to isolate amines from plant extracts. It was recognized early on that amines could be isolated from plants by exploiting their basic properties. For this reason, amines extracted from plants were called alkaloids (derived from the word *alkaline*, which means basic).

Delocalization Effects

The ammonium ions of most alkyl amines are characterized by a $\mathrm{p}K_a$ value between 10 and 11, but aryl amines are very different. Ammonium ions of aryl amines are more acidic (lower $\mathrm{p}K_a$) than the ammonium ion of alkyl amines (Table 23.2). In other words, aryl amines are less basic. This can be rationalized by considering the delocalized nature of the lone pair of an aryl amine:

LOOKING BACK
For a review of delocalized lone pairs, see Section 2.12.

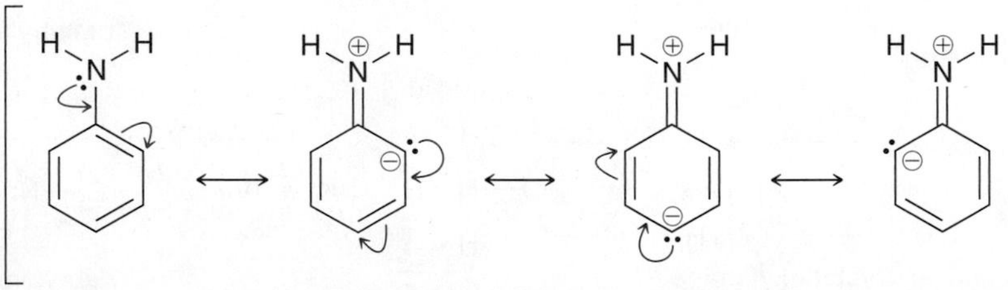

The lone pair occupies a *p* orbital and is delocalized by the aromatic system. This resonance stabilization is lost if the lone pair is protonated, and as a result, the nitrogen atom of an aryl amine is less basic than the nitrogen atom of an alkyl amine. If the aromatic ring bears a substituent, the basicity of the amino group will depend on the identity of the substituent (Table 23.2). Electron-donating groups, such as methoxy, slightly increase the basicity of aryl amines,

TABLE 23.2 **THE pKa VALUES OF THE AMMONIUM IONS OF SEVERAL PARA-SUBSTITUTED ANILINES**

para-substituted aniline	Ammonium ion	
H₂N—⟨ring⟩—N(H)(H):	H₂N—⟨ring⟩—N⁺(H)(H)H	$pK_a = 6.2$
MeO—⟨ring⟩—N(H)(H):	MeO—⟨ring⟩—N⁺(H)(H)H	$pK_a = 5.3$
Me—⟨ring⟩—N(H)(H):	Me—⟨ring⟩—N⁺(H)(H)H	$pK_a = 5.1$
⟨ring⟩—N(H)(H):	⟨ring⟩—N⁺(H)(H)H	$pK_a = 4.6$
Cl—⟨ring⟩—N(H)(H):	Cl—⟨ring⟩—N⁺(H)(H)H	$pK_a = 4.0$
NC—⟨ring⟩—N(H)(H):	NC—⟨ring⟩—N⁺(H)(H)H	$pK_a = 1.7$
O₂N—⟨ring⟩—N(H)(H):	O₂N—⟨ring⟩—N⁺(H)(H)H	$pK_a = 1.0$

Increasing basicity (left axis) — *Increasing acidity* (right axis)

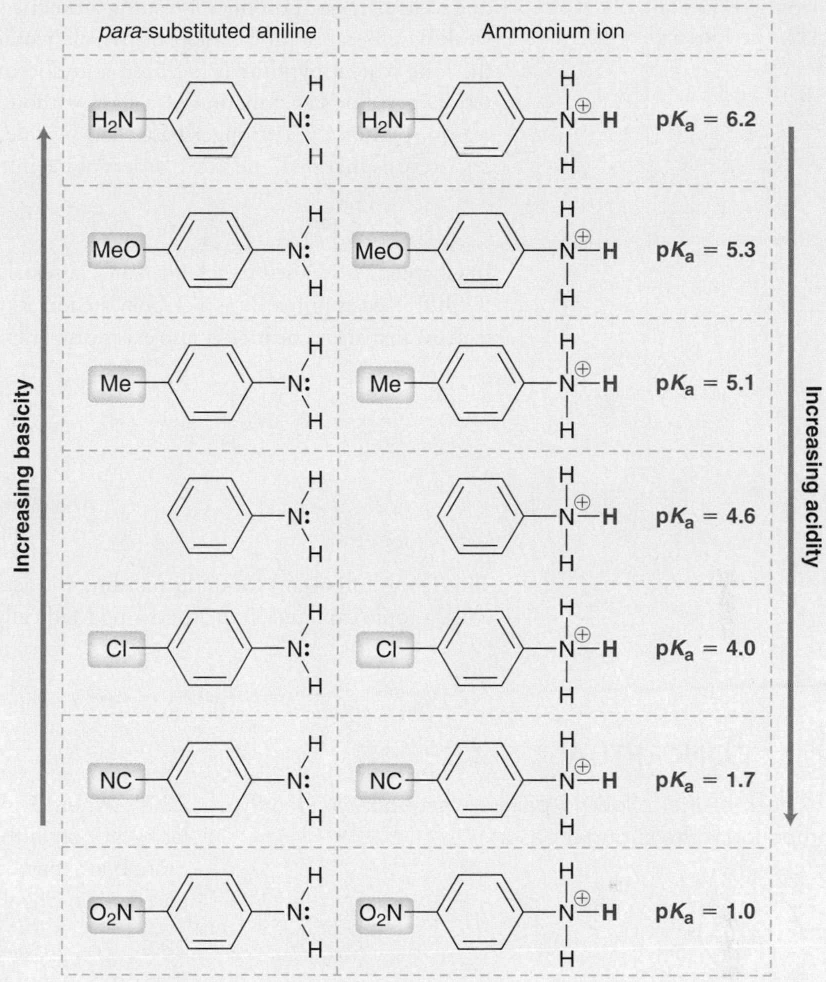

while electron-withdrawing groups, such as nitro, can significantly decrease the basicity of aryl amines. This profound effect is attributed to the fact that the lone pair in *para*-nitroaniline is extensively delocalized:

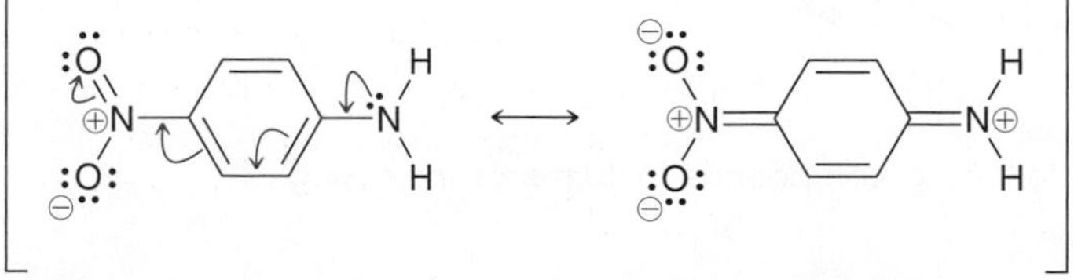

Amides also represent an extreme case of lone-pair delocalization. The lone pair on the nitrogen atom of an amide is highly delocalized by resonance. In fact, the nitrogen atom exhibits very little electron density, as can be seen in an electrostatic potential map (Figure 23.4). For this reason, amides do not function as bases and are very poor nucleophiles.

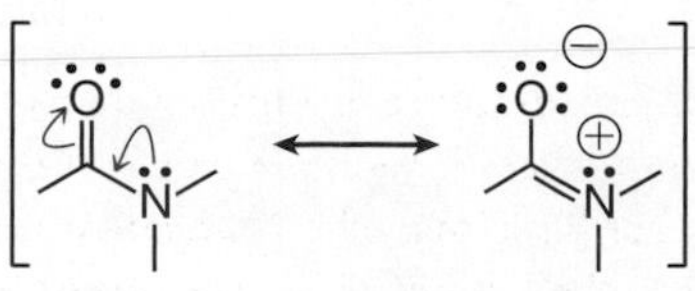

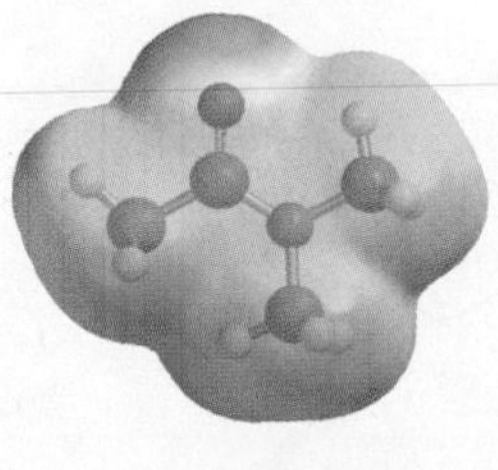

FIGURE 23.4

An electrostatic potential map of an amide. The nitrogen atom is not a region of high electron density.

LOOKING BACK
To visualize the orbitals occupied by the lone pairs in pyrrole and pyridine, see Figures 18.14 and 18.15.

Another example of electron delocalization can be seen in a comparison of the electrostatic potential maps of pyrrole and pyridine in Figure 18.16. The lone pair of pyrrole is delocalized due to its participation in establishing aromatic stabilization. As a result, pyrrole is an extremely weak base because protonation of pyrrole would result in the loss of aromaticity. In contrast, the lone pair in pyridine is localized and does not participate in establishing aromaticity. As a result, pyridine can function as a base without destroying its aromatic stabilization. For this reason, pyridine is a stronger base than pyrrole. In fact, inspection of the pK_a values in Table 23.1 reveals that pyridine is five orders of magnitude (~100,000 times) more basic than pyrrole.

Amines at Physiological pH

In Chapter 21, we used the Henderson-Hasselbalch equation to show that a carboxylic acid moiety (COOH) exists primarily as a carboxylate ion at physiological pH. A similar argument can be used to show that an amine moiety also exists primarily as a charged ammonium ion at physiological pH:

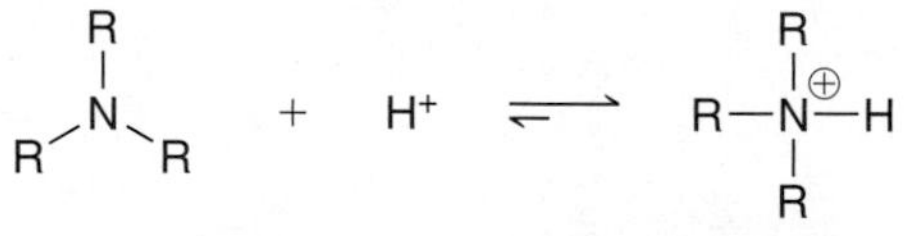

The ammonium ion predominates at physiological pH

This fact is important for understanding the activity of pharmaceuticals containing amine moieties, a topic explored in the following Medically Speaking box.

CONCEPTUAL CHECKPOINT

23.6 For each of the following pairs of compounds, identify which compound is the stronger base:

(a) 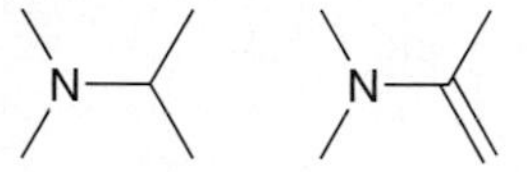

(b) 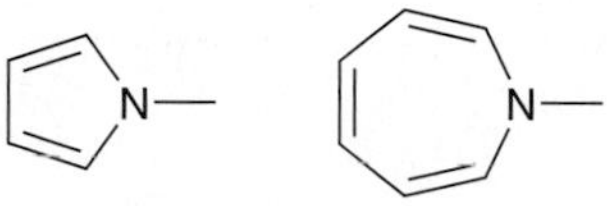

(c) 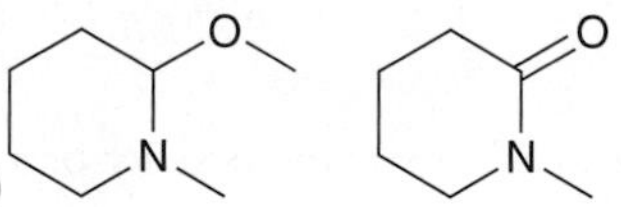

(d) 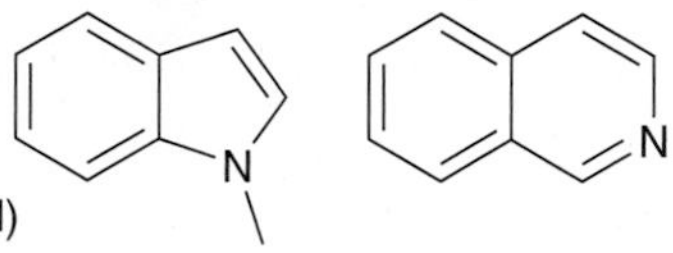

23.7 Rank the following compounds in terms of increasing basicity:

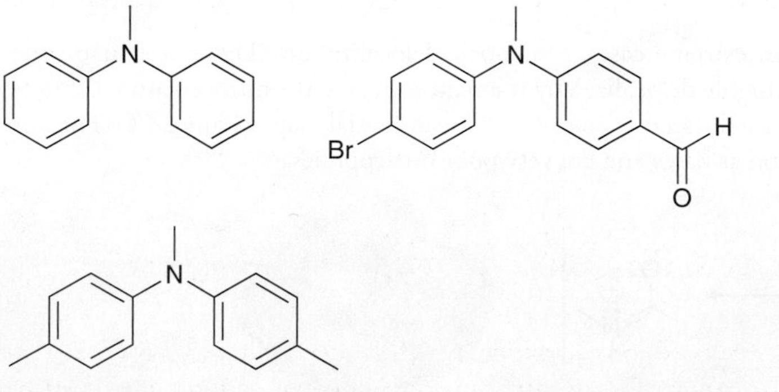

23.8 When (*E*)-4-amino-3-buten-2-one is treated with molecular hydrogen in the presence of platinum, the resulting amine is more basic than the reactant. Draw the reactant and the product, and explain why the product is a stronger base than the reactant.

23.9 For each of the following compounds, draw the form that predominates at physiological pH:

(a) 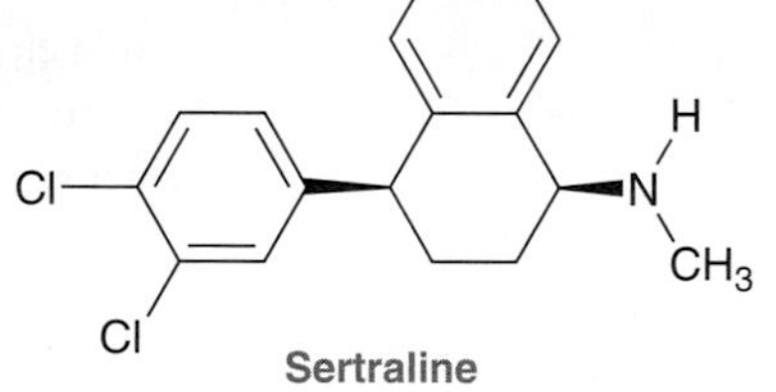

Sertraline (Zoloft)
An antidepressant

(b) 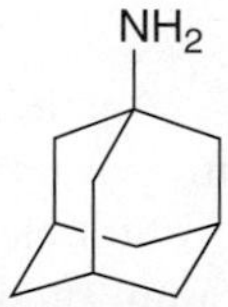

Amantadine
Used in the treatment of Parkinson's disease

(c) 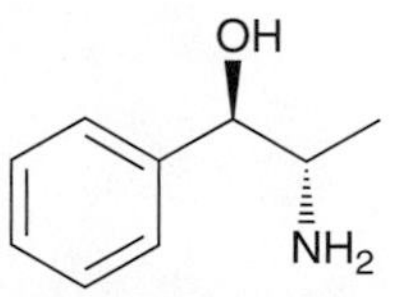

Phenylpropanolamine
A nasal decongestant

MEDICALLYSPEAKING))

Amine Ionization and Drug Distribution

In Chapter 18 we saw that allergic reactions are triggered by the binding of histamine to H_1 receptors. Compounds that compete with histamine to bind with H_1 receptors without triggering the allergic reaction are called H_1 antagonists and belong to the class of compounds called antihistamines. For a compound to exhibit H_1 antagonism it must possess two aromatic rings in close proximity as well as a tertiary amine moiety.

The amine moiety of an antihistamine exists as an equilibrium between neutral and ionic forms. The equilibrium concentrations of the two forms depend on the pK_a of the compound and the pH of the environment. Most antihistamines have an ammonium group with a pK_a between 8 and 9. As a result, both the neutral and ionic forms are present in substantial amounts at physiological pH, although the ionic form slightly predominates. This is important because it is the ionic form that binds to the receptor. The positively charged ammonium ion binds with a negative charge in the receptor. The neutral form of the drug cannot bind with the receptor.

Although the ionic form is required to bind with the receptor, it is incapable of crossing the nonpolar environment of cell membranes. So how does the drug reach its target? The answer stems from the equilibrium between ionic and neutral forms of the drug. The neutral form crosses the membrane, and once on the other side, it reestablishes equilibrium, producing a large concentration of the ionic form, which then binds to the receptor. On the other side of the membrane, the ionic forms that did not cross the membrane also reestablish an equilibrium, generating more of the neutral form that is capable of crossing the membrane. In this way, the drug can cross the membrane in the neutral form and interact with the receptor in the ionic form. The rate at which this process takes place depends on the equilibrium concentrations. If the equilibrium concentration of the neutral form is too small, then the drug will not cross the cell membrane at an appreciable rate. Similarly, if the equilibrium concentration of the ionic form is too small, then the drug will not bind at an appreciable rate with the receptor. This delicate balance between drug distribution and receptor binding is therefore highly dependent on the equilibrium concentrations at physiological pH, which in turn is dependent on the pK_a of the drug. Later in this chapter we will see an example of how pK_a values must be taken into account when new drugs are designed.

23.4 Preparation of Amines: A Review

In previous chapters, we have seen reactions that can be used to make amines from alkyl halides, carboxylic acids, or benzene. This section will briefly summarize each of these three methods.

Preparing an Amine from an Alkyl Halide

Amines can be prepared from alkyl halides in a two-step process in which the alkyl halide is converted to a nitrile, which then undergoes reduction.

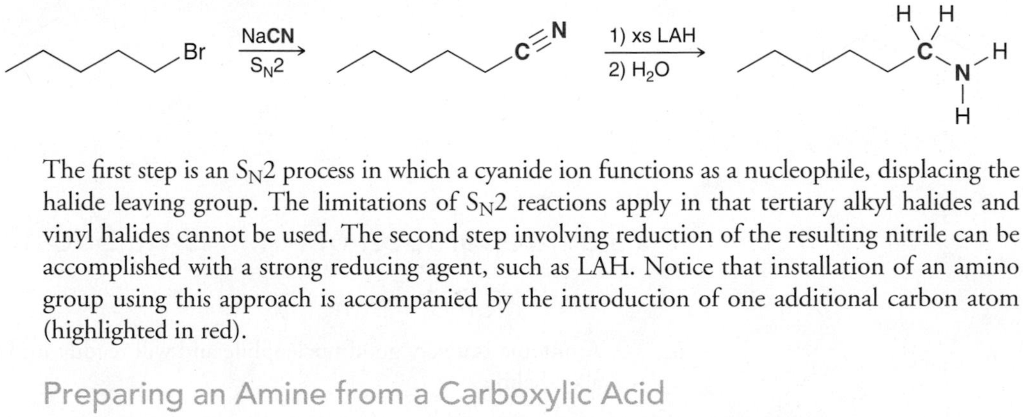

The first step is an S_N2 process in which a cyanide ion functions as a nucleophile, displacing the halide leaving group. The limitations of S_N2 reactions apply in that tertiary alkyl halides and vinyl halides cannot be used. The second step involving reduction of the resulting nitrile can be accomplished with a strong reducing agent, such as LAH. Notice that installation of an amino group using this approach is accompanied by the introduction of one additional carbon atom (highlighted in red).

Preparing an Amine from a Carboxylic Acid

Amines can be prepared from carboxylic acids using the following approach:

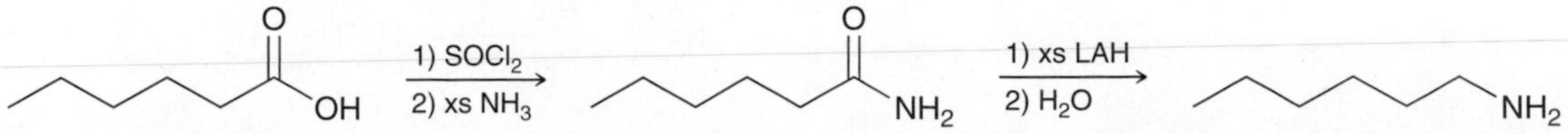

The carboxylic acid is first converted into an amide, which is then reduced to give an amine (Section 21.14). Notice that, using this approach, installation of an amino group does not involve introduction of any additional carbon atoms, that is, the carbon skeleton is not changed.

Preparing Aniline and Its Derivatives from Benzene

Aryl amines, such as aniline, can be prepared from benzene using the following approach:

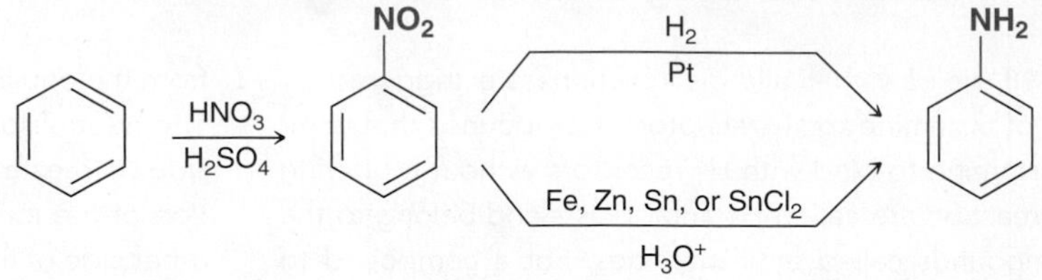

The first step involves nitration of the aromatic ring, and the second step involves reduction of the nitro group. Several reagents can be used to accomplish this reduction, including hydrogenation in the presence of a catalyst or reduction with iron, zinc, tin, or tin(II) chloride ($SnCl_2$) in the presence of aqueous acid. The latter method is a more mild approach that is used when other functional groups are present that would otherwise be susceptible to hydrogenation.

For example, a nitro group can be selectively reduced in the presence of a carbonyl group:

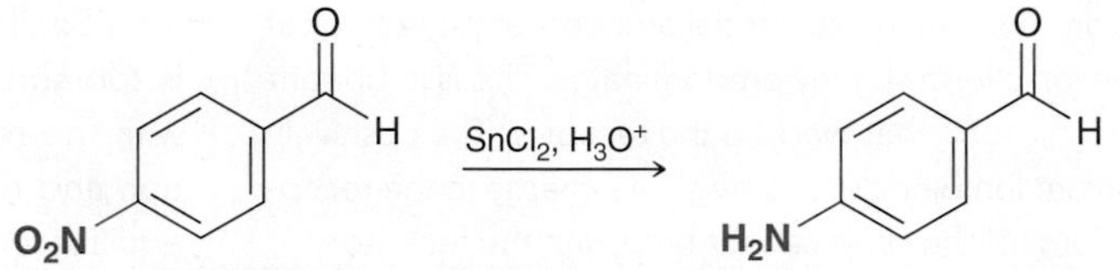

When reducing a nitro group in acidic conditions, the reaction must be followed up with a base, such as sodium hydroxide, because the resulting amino group will be protonated under acidic conditions (as we saw in Section 23.3).

CONCEPTUAL CHECKPOINT

23.10 Draw the structure of an alkyl halide or carboxylic acid that might serve as a precursor in the preparation of each of the following amines:

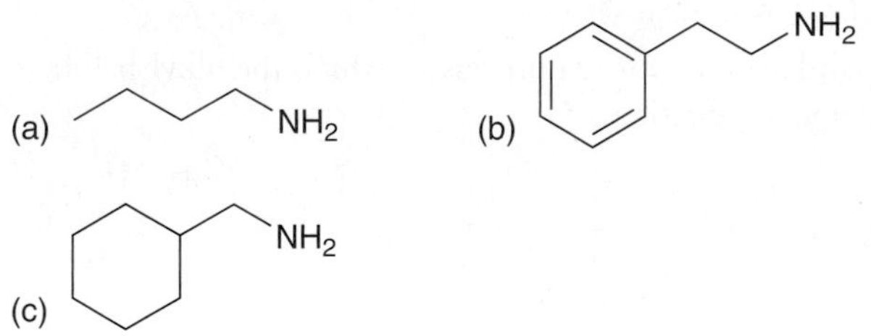

23.11 The following compound cannot be prepared from an alkyl halide or a carboxylic acid using the methods described in this section. Explain why each synthesis cannot be performed.

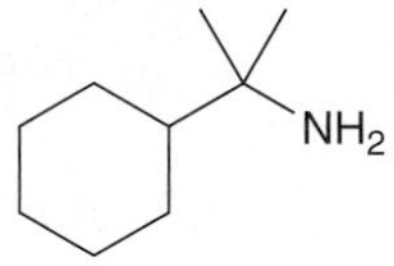

23.5 Preparation of Amines via Substitution Reactions

Alkylation of Ammonia

Ammonia is a very good nucleophile and will readily undergo alkylation when treated with an alkyl halide.

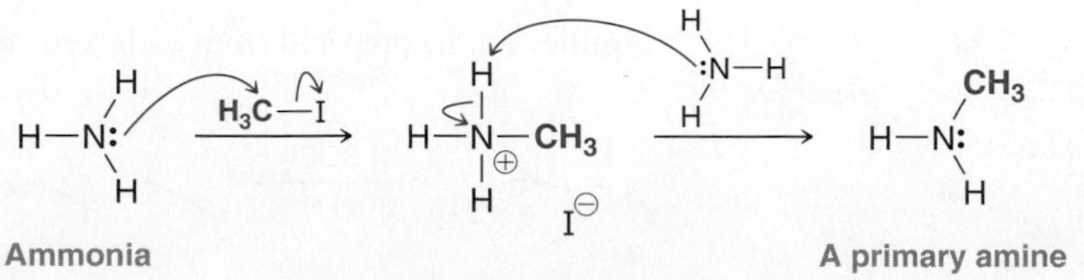

This reaction proceeds via an S_N2 process followed by deprotonation to give a primary amine. As the primary amine is formed, it can undergo further alkylation to produce a secondary amine, which undergoes further alkylation to produce a tertiary amine. Finally, the tertiary amine undergoes alkylation one more time to produce a **quaternary ammonium salt.**

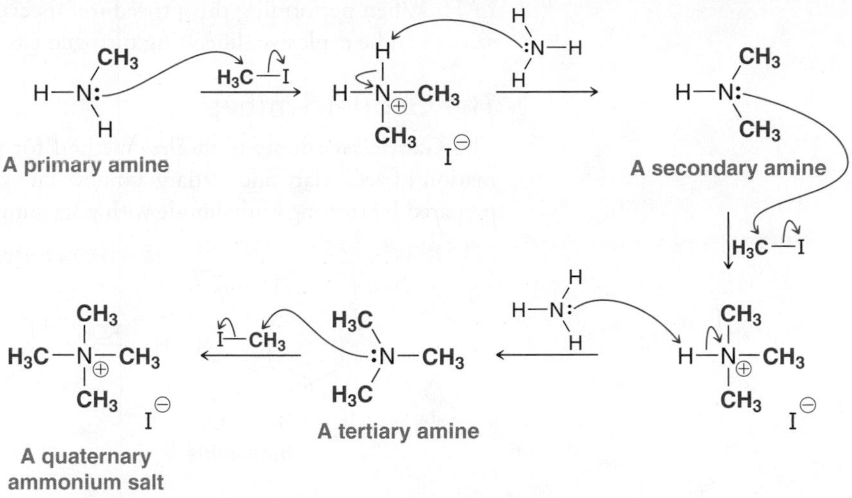
A primary amine
A secondary amine
A tertiary amine
A quaternary ammonium salt

If the quaternary ammonium salt is the desired product, then an excess of the alkyl halide is used, and ammonia is said to undergo **exhaustive alkylation**. However, monoalkylation is difficult to achieve because each successive alkylation renders the nitrogen atom more nucleophilic. If the primary amine is the desired product, then the process is generally not efficient because when 1 mol of ammonia is treated with 1 mol of the alkyl halide, a mixture of products is obtained. For this reason, the alkylation of ammonia is only useful when the starting alkyl halide is inexpensive and the desired product can be easily separated or when exhaustive alkylation is performed to yield the quaternary ammonium salt.

The Azide Synthesis

The **azide synthesis** is a better method for preparing primary amines than alkylation of ammonia because it avoids the formation of secondary and tertiary amines. This method involves treating an alkyl halide with sodium azide followed by reduction (Mechanism 23.1).

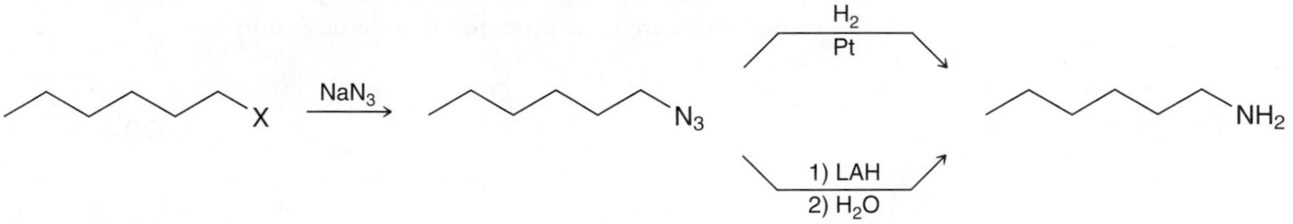
NaN3
H2
Pt
1) LAH
2) H2O

MECHANISM 23.1 THE AZIDE SYNTHESIS

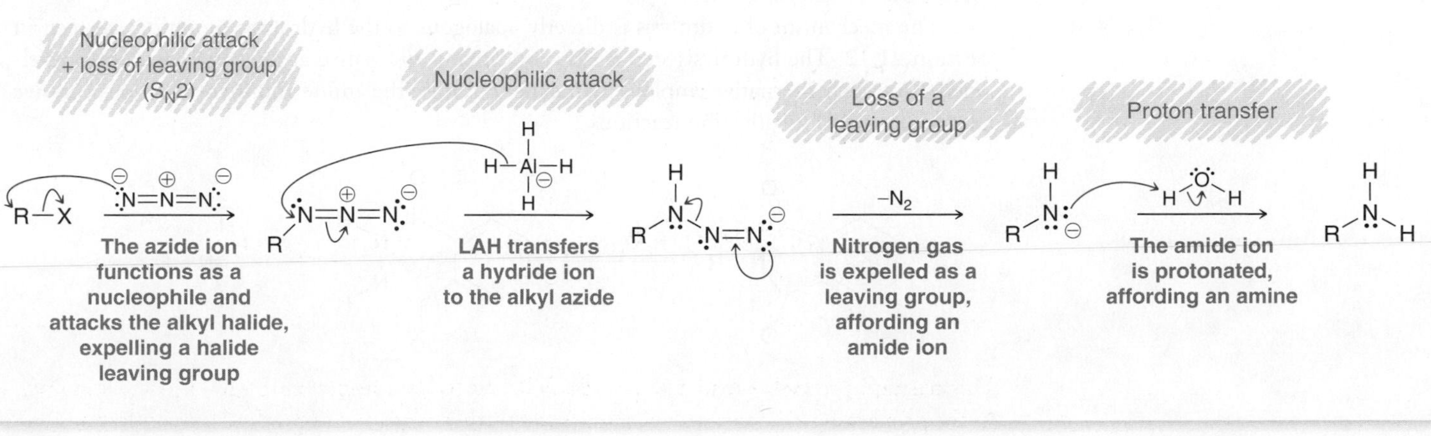
Nucleophilic attack + loss of leaving group (SN2)
Nucleophilic attack
Loss of a leaving group
Proton transfer
The azide ion functions as a nucleophile and attacks the alkyl halide, expelling a halide leaving group
LAH transfers a hydride ion to the alkyl azide
−N2
Nitrogen gas is expelled as a leaving group, affording an amide ion
The amide ion is protonated, affording an amine

In the first step of this mechanism, the azide ion functions as a nucleophile and attacks the alkyl halide in an S_N2 process, producing an alkyl azide. The alkyl azide is then reduced using LAH. When performing this procedure, special precautions must be observed, because alkyl azides can be explosive, liberating nitrogen gas.

The Gabriel Synthesis

The **Gabriel synthesis** is another method for preparing primary amines while avoiding formation of secondary and tertiary amines. The key reagent is potassium phthalimide, which is prepared by treating phthalimide with potassium hydroxide.

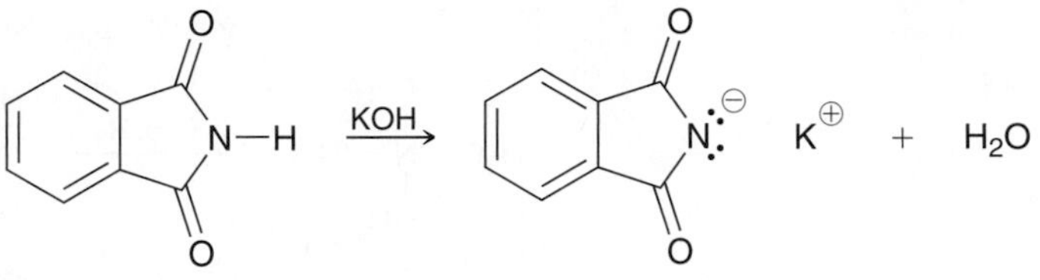

Hydroxide functions as a base and deprotonates phthalimide. The proton is relatively acidic ($pK_a = 8.3$), because it is flanked by two C═O groups (similar to a β-keto ester). Potassium phthalimide can function as a nucleophile and is readily alkylated to form a C—N bond (directly analogous to the acetoactetic ester synthesis, Section 22.5):

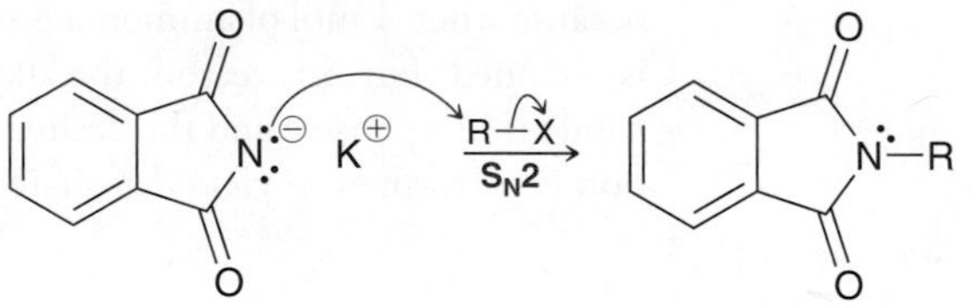

This reaction proceeds via an S_N2 process, so it works best with primary alkyl halides. It can be performed with secondary alkyl halides in many cases, but tertiary alkyl halides cannot be used. Acid-catalyzed or base-catalyzed hydrolysis is then performed to release the amine. Acidic conditions are more common than basic conditions.

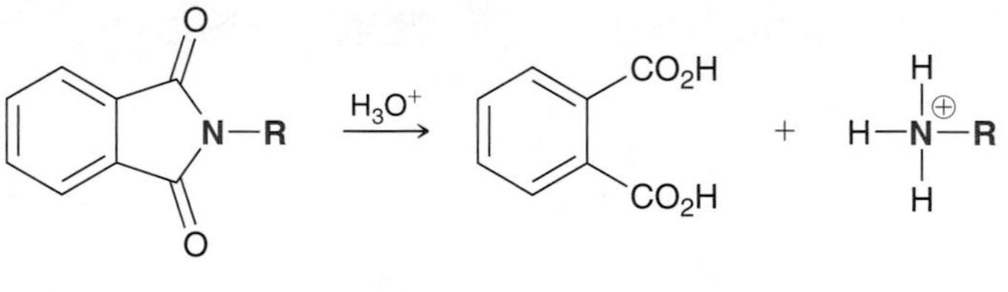

Under acidic conditions, an ammonium ion is generated, which must be treated with a base to release the uncharged amine.

The mechanism of hydrolysis is directly analogous to the hydrolysis of amides, as seen in Section 21.12. The hydrolysis step is slow, and many alternative approaches have been developed. One such alternative employs hydrazine to release the amine and involves two successive nucleophilic acyl substitution reactions.

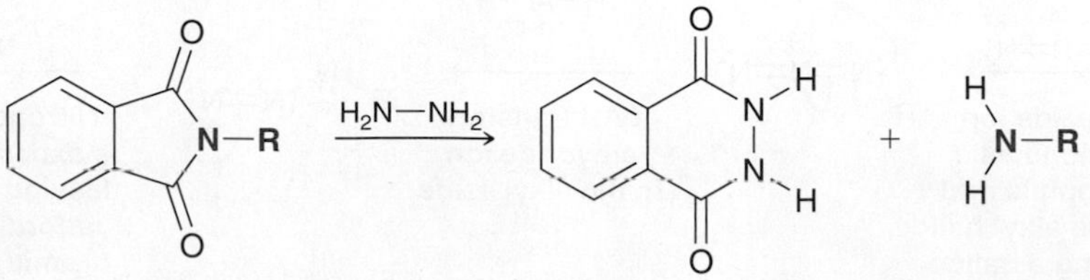

SKILLBUILDER

23.2 PREPARING A PRIMARY AMINE VIA THE GABRIEL REACTION

LEARN the skill

Amphetamine is a stimulant (stronger than caffeine) that increases alertness and decreases appetite. It was used extensively during World War II to prevent combat fatigue.

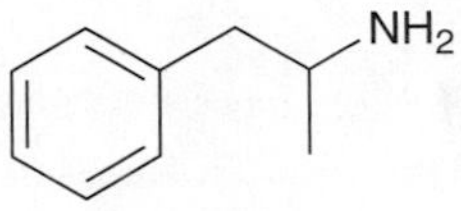

Amphetamine

Propose a method for preparing amphetamine that utilizes a Gabriel synthesis.

SOLUTION

First identify an alkyl halide that can serve as a precursor in the preparation of amphetamine. To draw this precursor, simply replace the amino group with a halide:

STEP 1 Identify the appropriate alkyl halide to use.

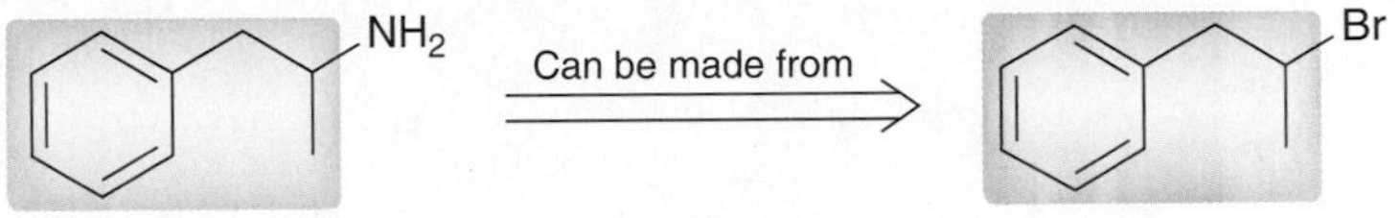

In the Gabriel synthesis, the nitrogen atom comes from phthalimide, which serves as the starting material. The process requires three steps: (1) deprotonation to form potassium phthalimide, (2) nucleophilic attack with an alkyl halide, and (3) hydrolysis. Draw the reagents for all three steps of the Gabriel synthesis:

STEP 2 Draw all three steps, starting with phthalimide.

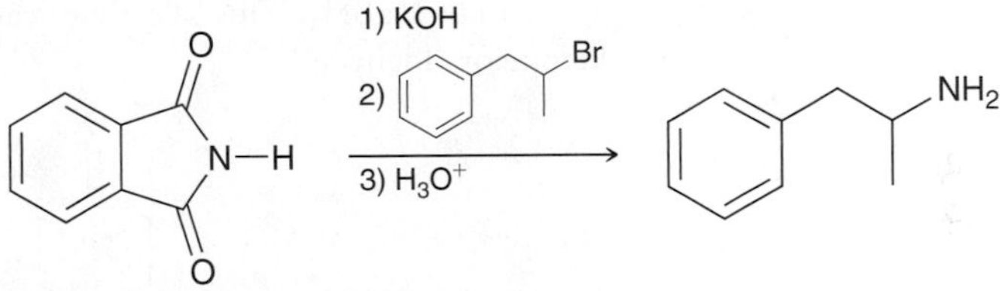

Note: Planning a Gabriel synthesis requires that you identify the alkyl halide to use in the second step of the process. The first and third steps remain the same regardless of the identity of the desired product.

PRACTICE the skill

23.12 Using a Gabriel synthesis, show how you would make each of the following compounds:

(a) 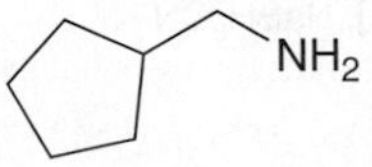

(b) 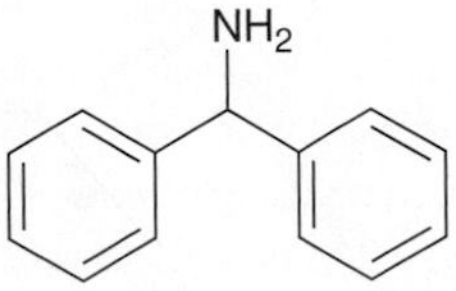

(c) 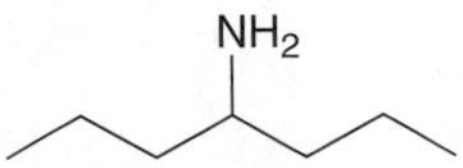

(d) 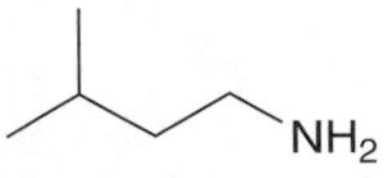

APPLY the skill

23.13 Using a Gabriel synthesis, propose an efficient synthesis for each of the following transformations. In each case, you will need to use reactions from previous chapters to convert the starting material into a suitable alkyl halide that can be used in a Gabriel synthesis.

(a) 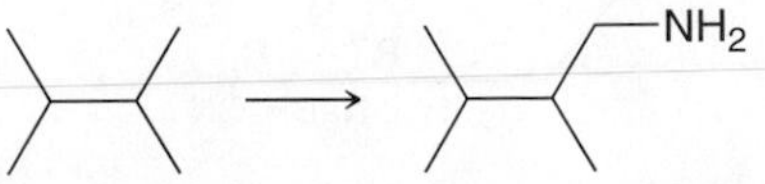

(b) 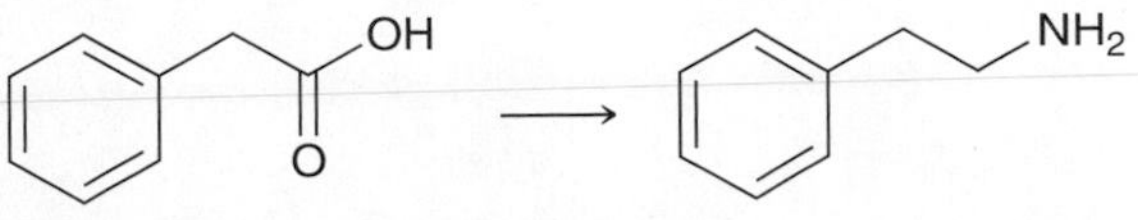

need more **PRACTICE?** **Try Problems 23.51a, 23.60, 23.61, 23.62b**

23.6 Preparation of Amines via Reductive Amination

Recall that ketones or aldehydes can be converted into imines when treated with ammonia in acidic conditions (Section 20.6). If the reaction is carried out in the presence of a reducing agent, such as H_2, then the imine is reduced as soon as it is formed, giving an amine.

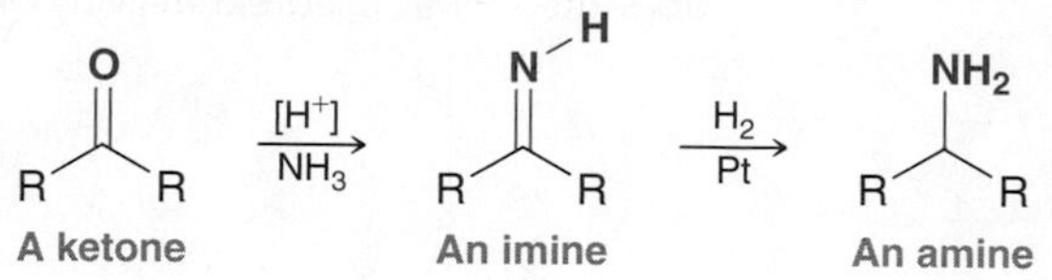

This process, called **reductive amination**, can be accomplished in one reaction flask by hydrogenating a solution containing the ketone, ammonia, and a proton source. With this approach, the imine is reduced under the conditions of its formation. Many other reducing agents are also frequently employed instead of hydrogenation. The most common reducing agent for this purpose is **sodium cyanoborohydride** ($NaBH_3CN$), which is similar in structure to sodium borohydride ($NaBH_4$), but a cyano group has replaced one of the hydrogen atoms.

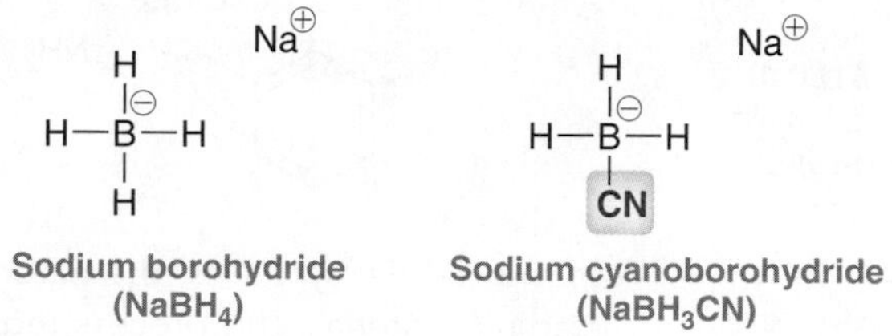

Recall that a cyano group is electron withdrawing (Section 19.10), and its presence stabilizes the negative charge on the boron atom. As a result, $NaBH_3CN$ is a more selective hydride-reducing agent than $NaBH_4$. This selectivity can be exploited to achieve the reductive amination of a ketone or aldehyde:

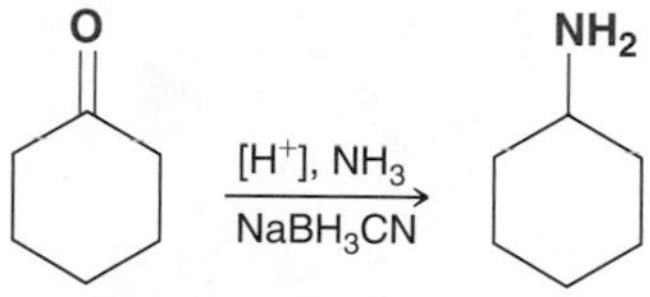

This process cannot be achieved with $NaBH_4$, which would simply reduce the ketone before it is converted into an imine. In contrast, $NaBH_3CN$ does not react with ketones but will reduce an iminium ion (a protonated imine). This process enables the conversion of a ketone or aldehyde into an amine in one reaction flask. The identity of the nucleophile (NH_3) can be changed, enabling the preparation of primary, secondary or tertiary amines:

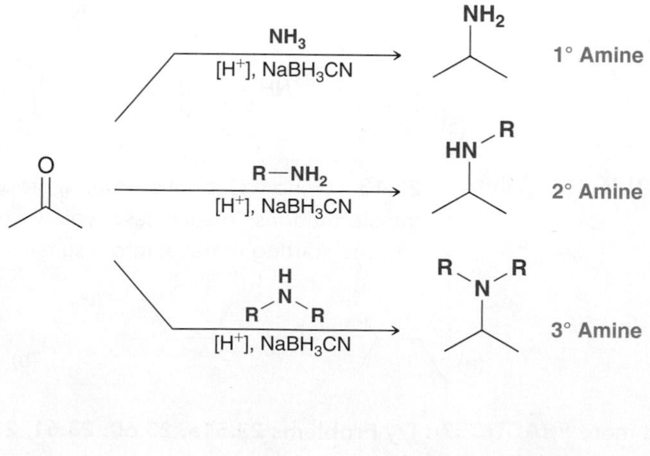

SKILLBUILDER

23.3 PREPARING AN AMINE VIA A REDUCTIVE AMINATION

LEARN the skill

Fluoxetine, sold under the trade name Prozac, is used in the treatment of depression and obsessive compulsive disorder. Show two different ways of preparing fluoxetine via reductive amination.

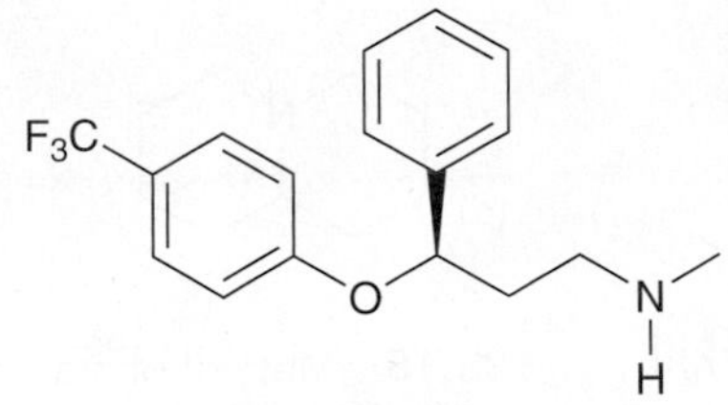

Fluoxetine

SOLUTION

STEP 1 Identify all C—N bonds.

STEP 2 Apply a retrosynthetic analysis of each C—N bond.

Begin by identifying all C—N bonds. Then, using a retrosynthetic analysis, identify the starting amine and the carbonyl compound that could be used to make each of the C—N bonds:

Method 1

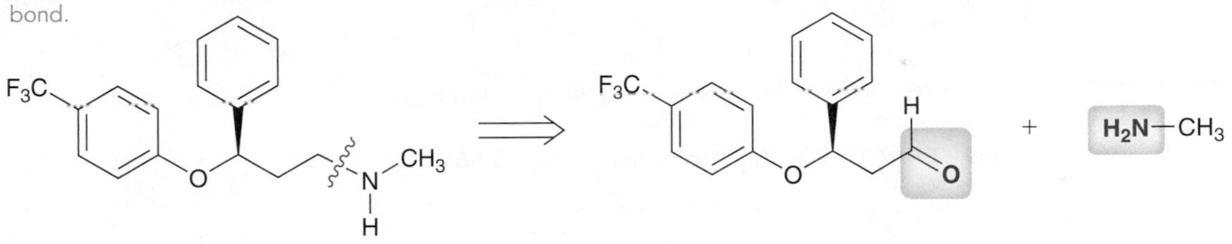

Method 2

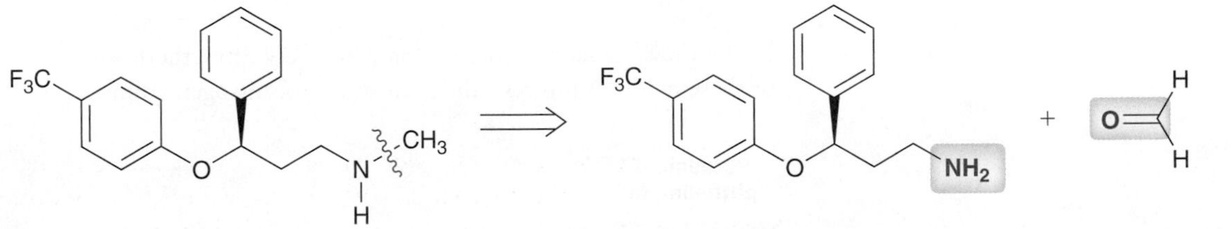

STEP 3 Draw all reagents necessary for each possible route.

Finally, show all of the reagents necessary for both of the synthetic routes shown. In each case, the desired compound can be produced in just one step, using $NaBH_3CN$:

Method 1

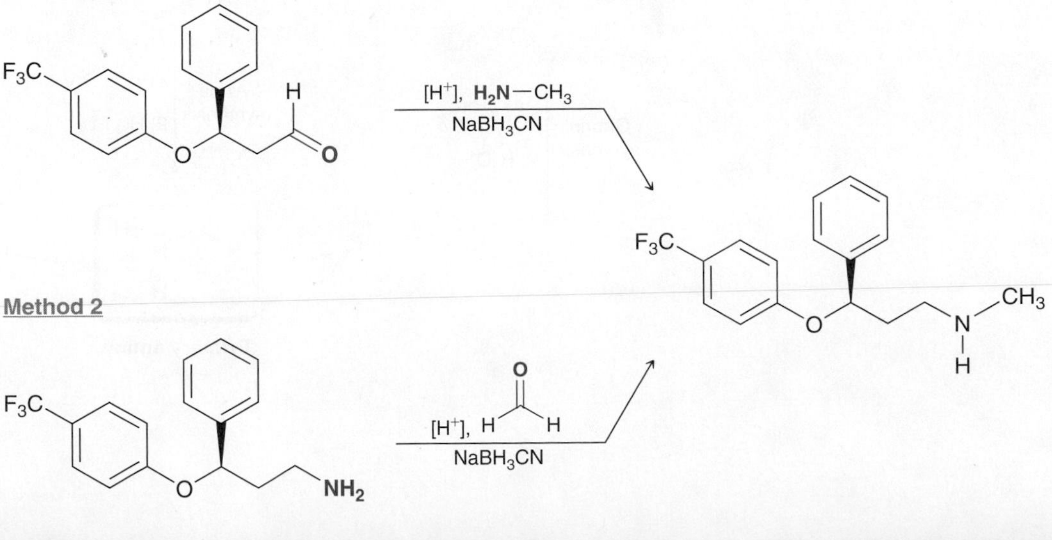

Method 2

PRACTICE the skill **23.14** Show two different ways of preparing each of the following compounds via a reductive amination:

(a) 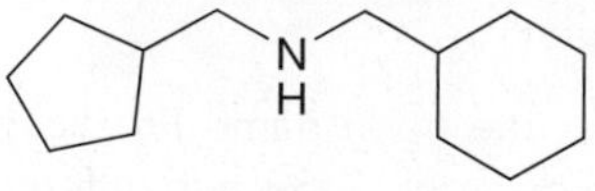

(b) 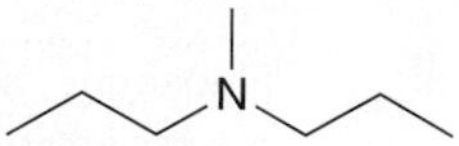

(c) 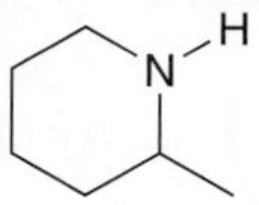

(d) 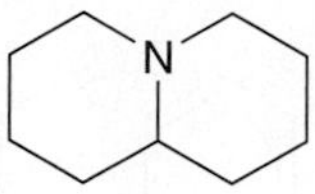

(e) 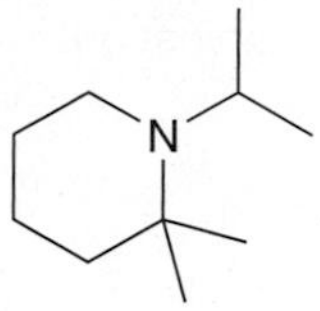

APPLY the skill **23.15** Methamphetamine is used in some formulations for the treatment of attention deficit disorder and can be prepared by reductive amination using phenylacetone and methylamine. Draw the structure of methamphetamine.

23.16 Tri-*tert*-butylamine cannot be prepared via a reductive amination. Explain.

23.17 Propose an efficient synthesis for the following transformation:

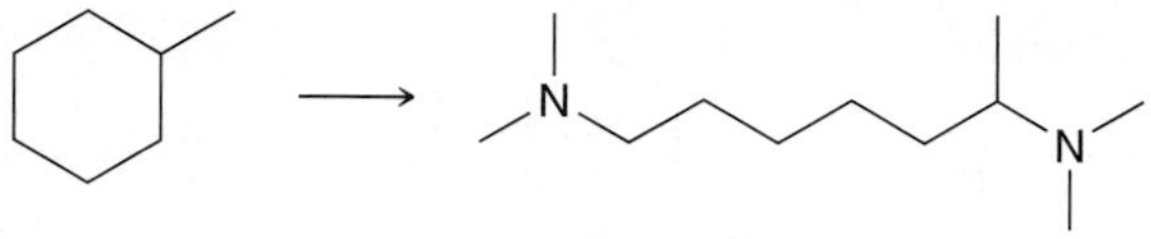

need more **PRACTICE?** **Try Problems 23.65, 23.66**

23.7 Synthesis Strategies

In the previous sections, we explored a variety of methods for preparing primary amines. Each of these methods utilizes a different source of nitrogen atoms.

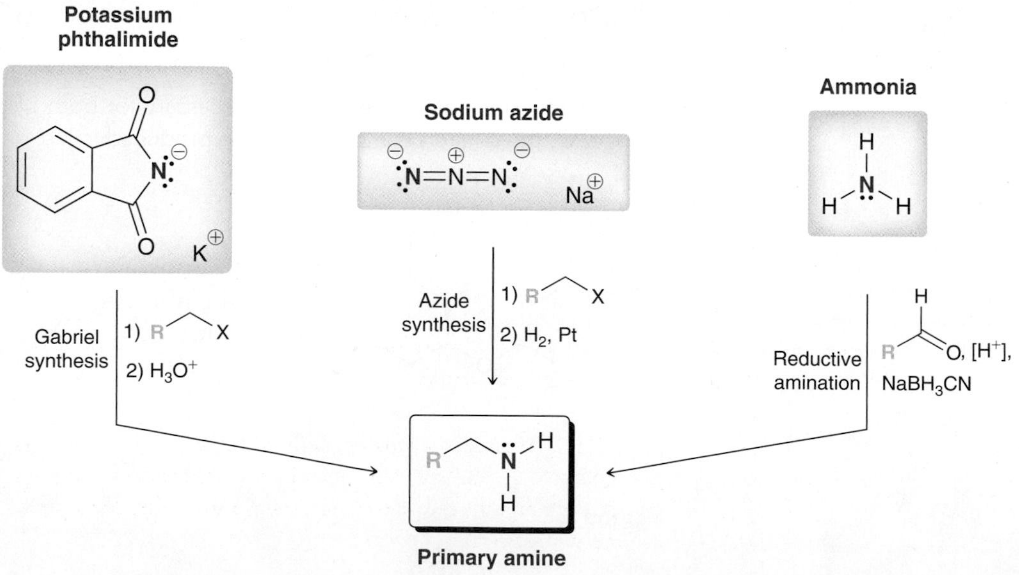

The Gabriel synthesis uses potassium phthalimide as the source of nitrogen, the azide synthesis uses sodium azide as the source of nitrogen, and reductive amination uses ammonia as the source of nitrogen. When preparing a primary amine, the available starting materials will dictate which of the three methods will be chosen.

Secondary amines are readily prepared from primary amines via reductive amination. Similarly, tertiary amines are readily prepared from secondary amines via reductive amination.

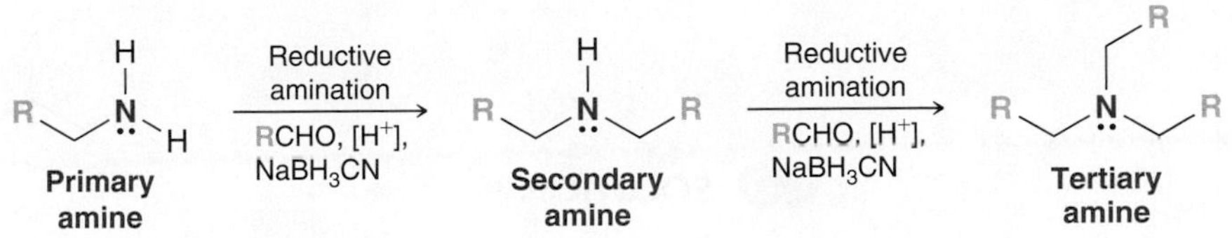

Direct alkylation of the primary amine is not efficient because polyalkylation is generally unavoidable, as seen in Section 23.5. Therefore, it is more efficient to use reductive amination as an indirect method for alkylating the nitrogen atom of an amine.

Tertiary amines can be converted into quaternary ammonium salts via alkylation. This process is efficient because polyalkylation is not possible with tertiary amines.

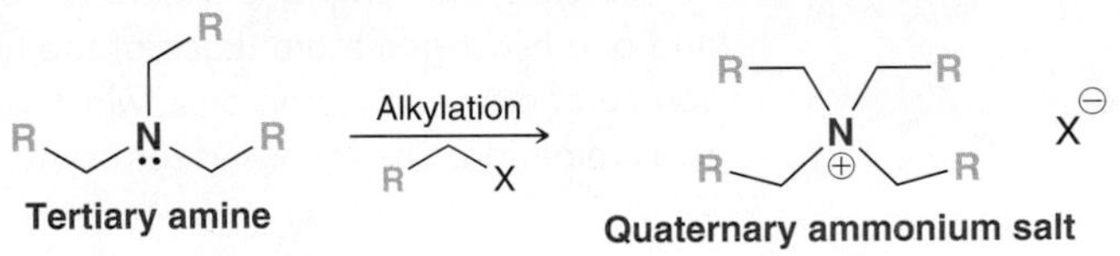

Using just a few reactions it is possible to prepare a wide variety of amines from simple starting materials.

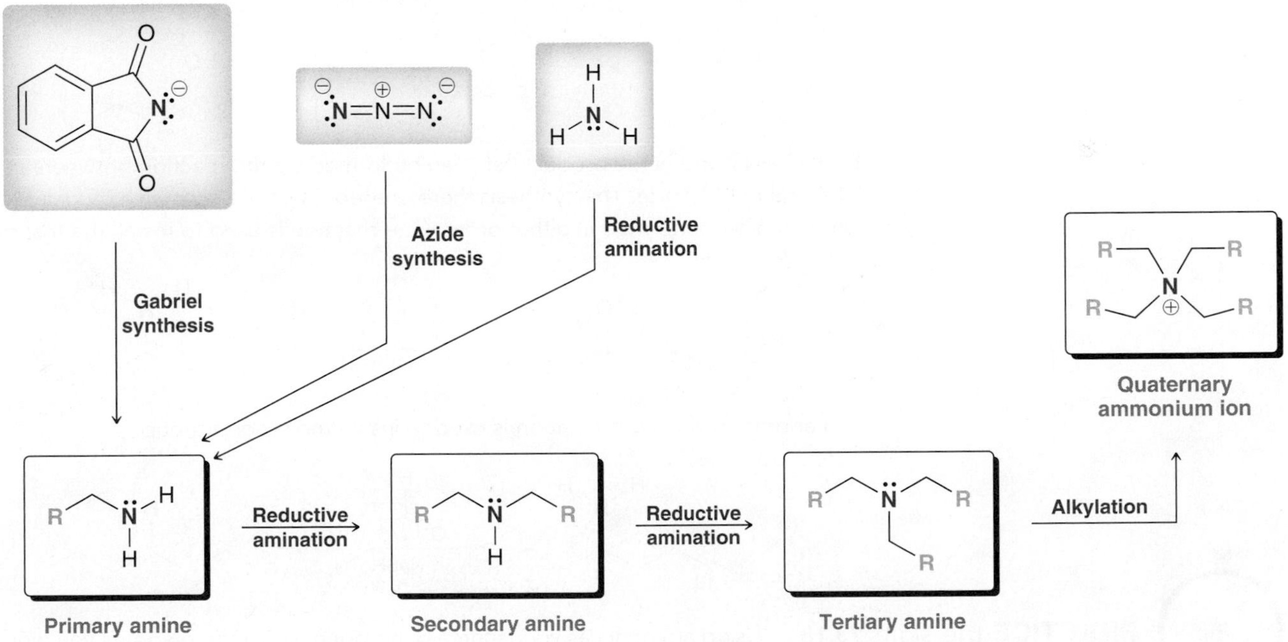

Using potassium phthalimide, sodium azide, or ammonia together with suitable alkyl halides, ketones, or aldehydes, it is possible to generate a large variety of primary, secondary, and tertiary amines as well as quaternary ammonium salts. Let's get some practice using these methods to prepare amines.

SKILLBUILDER

23.4 PROPOSING A SYNTHESIS FOR AN AMINE

LEARN the skill Using ammonia as your source of nitrogen, show what reagents you would use to prepare the following amine:

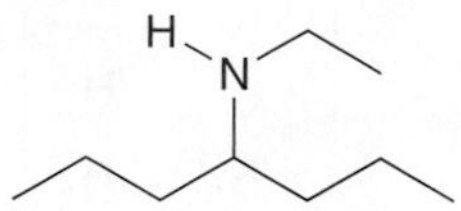

SOLUTION

Begin by identifying the alkyl groups connected to the nitrogen atom, and determine whether the amine is primary, secondary, or tertiary.

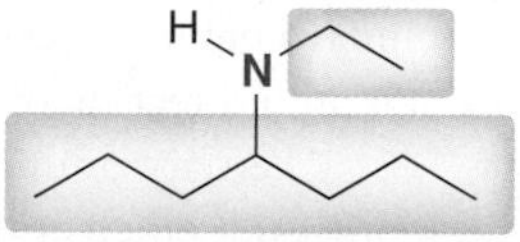

In this case, the amine is secondary because the nitrogen atom bears two alkyl groups and one hydrogen atom. Each of the highlighted groups must be installed separately. The source of nitrogen is ammonia, which dictates that both groups must be installed via reductive amination. The following retrosynthetic analysis reveals the necessary starting materials.

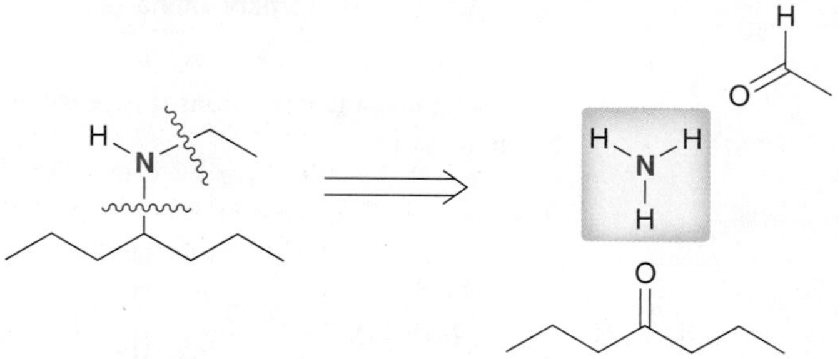

Each C—N bond in the product (left) can be formed via the reaction between ammonia and a ketone or aldehyde. This synthesis therefore requires two successive reductive aminations, which can be performed in either order. One reaction is used to install the first group,

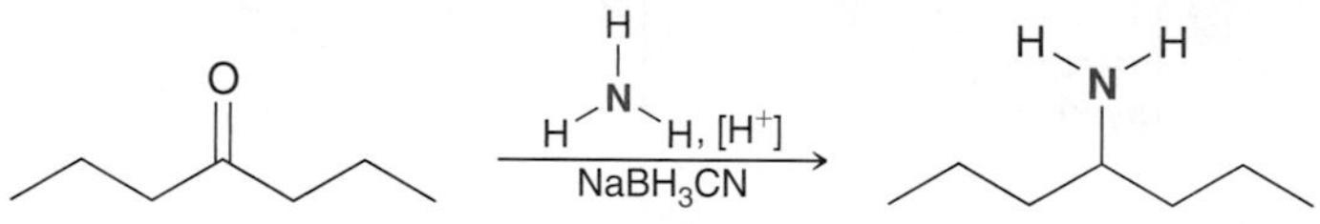

and another reductive amination is used to install the second group.

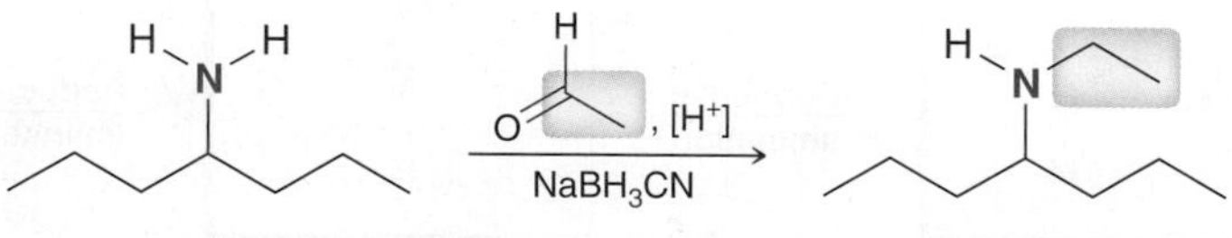

PRACTICE the skill **23.18** Using ammonia as your source of nitrogen, show the reagents you would use to prepare each of the following amines:

(a) 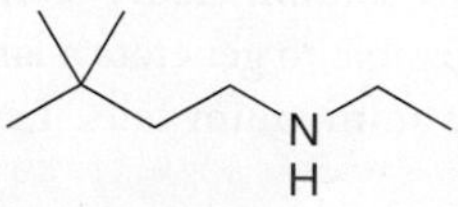 (b) 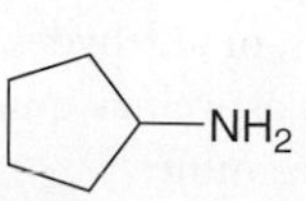 (c) 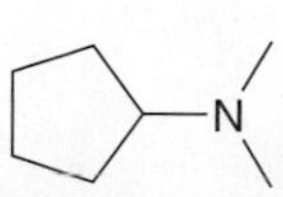

(d) 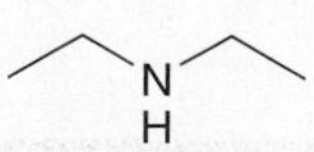 (e) 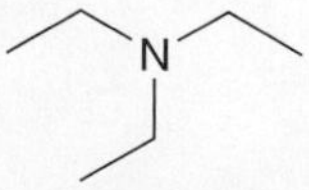 (f) 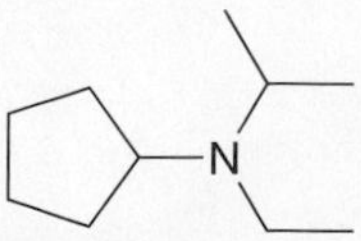

23.19 Starting with potassium phthalimide as your source of nitrogen and using any other reagents of your choice, show how you would prepare each of the compounds in Problem 23.18.

23.20 Starting with sodium azide as your source of nitrogen and using any other reagents of your choice, show how you would prepare each of the compounds in Problem 23.18.

APPLY the skill

23.21 Show the reagents you would use to achieve the following transformation:

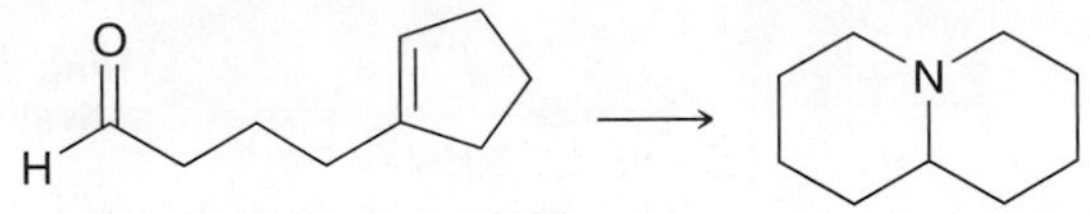

need more **PRACTICE?** **Try Problems 23.50, 23.51, 23.86**

23.8 Acylation of Amines

As discussed in Chapter 21, amines will react with acyl halides to yield amides.

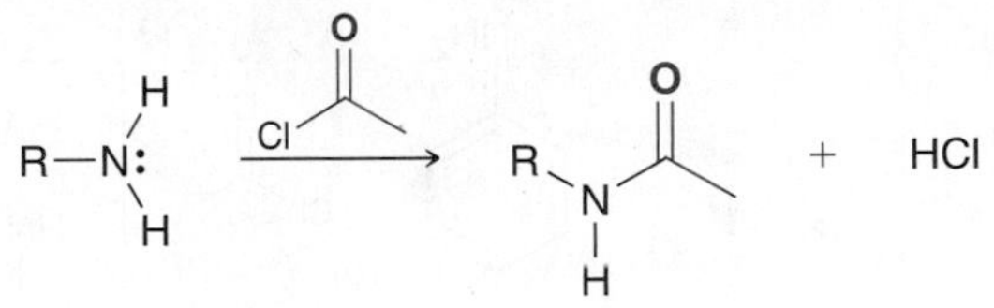

The reaction takes place via a nucleophilic acyl substitution process, and HCl is a by-product of the reaction. As HCl is produced, the starting amine is protonated to form an ammonium ion, which will not attack the acyl halide. Therefore, two equivalents of the amine are required. Polyacylation does not occur, because the resulting amide is not nucleophilic.

Acylation of an amino group is an extremely useful technique when performing electrophilic aromatic substitution reactions, because it enables transformations that are otherwise difficult to achieve. For example, consider what happens when aniline is treated with bromine.

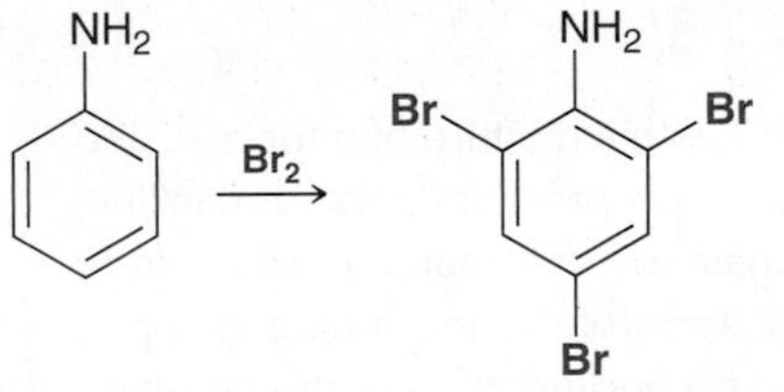

The amino group strongly activates the ring, and tribromination occurs readily. Monobromination of aniline is extremely difficult to achieve. Even when one equivalent of bromine is used, a mixture of products is obtained. This problem can be circumvented by first acylating the amino group. The resulting amide is less activating than an amino group, and monobromination becomes possible. After the bromination step, the amide group can be removed by hydrolysis. This three-step route provides a method for the monobromination of aniline, which would otherwise be difficult to achieve.

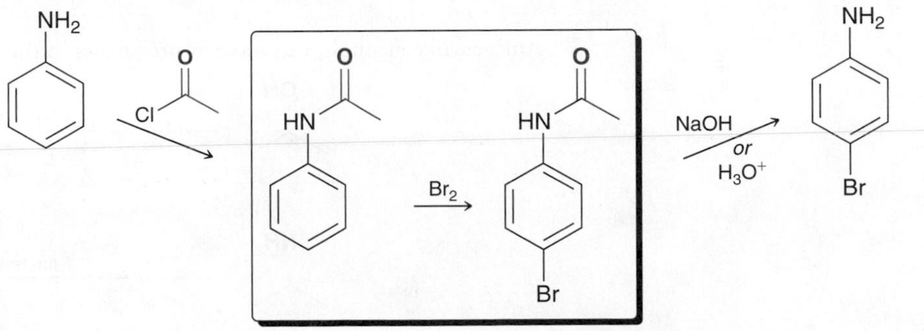

As another example, consider what happens if we try to perform a Friedel-Crafts reaction (alkylation or acylation) using aniline as a starting material.

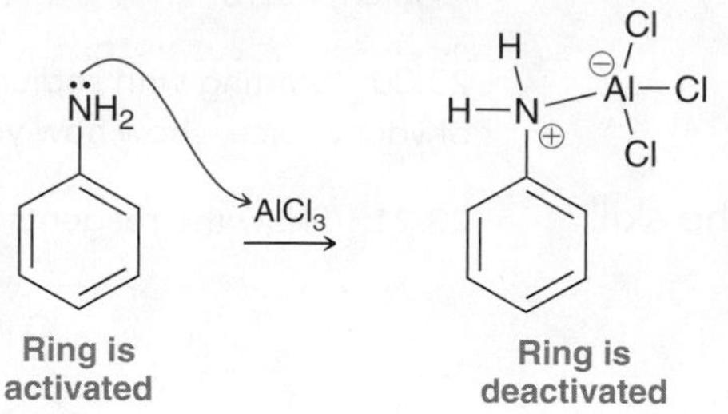

The amino group attacks the Lewis acid to form an acid-base complex in which the aromatic ring is now strongly deactivated. Recall that strongly deactivated rings are unreactive toward Friedel-Crafts reactions. As a result, it is not possible to achieve a Friedel-Crafts reaction with aniline as a starting material. Once again, this problem can be circumvented by first acylating the amino group. The nitrogen atom of the resulting amide is not nucleophilic. Under these conditions, a Friedel-Crafts reaction can be performed, and the amide group can then be removed via hydrolysis. This three-step route provides a method for alkylating aniline, something that cannot be accomplished directly.

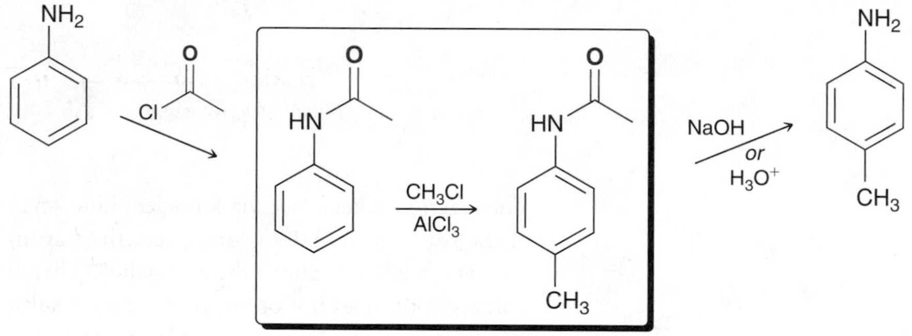

CONCEPTUAL CHECKPOINT

23.22 When aniline is treated with a mixture of nitric acid and sulfuric acid, the expected nitration product (*para*-nitroaniline) is obtained in poor yield. Instead, the major product from nitration is *meta*-nitroaniline. Apparently, the amino group is protonated under these acidic conditions, and the resulting ammonium group is a *meta*-director, rather than an *ortho-para* director. Propose a plausible method for converting aniline into *para*-nitroaniline.

23.23 Starting with nitrobenzene and using any other reagents of your choice, outline a synthesis of *para*-chloroaniline.

23.24 Using acetic acid as your only source of carbon atoms, show how you could make *N*-ethyl acetamide.

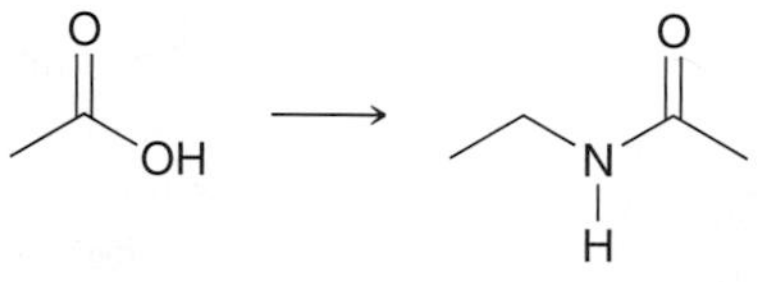

23.9 Hofmann Elimination

Amines, like alcohols, can serve as precursors in the preparation of alkenes.

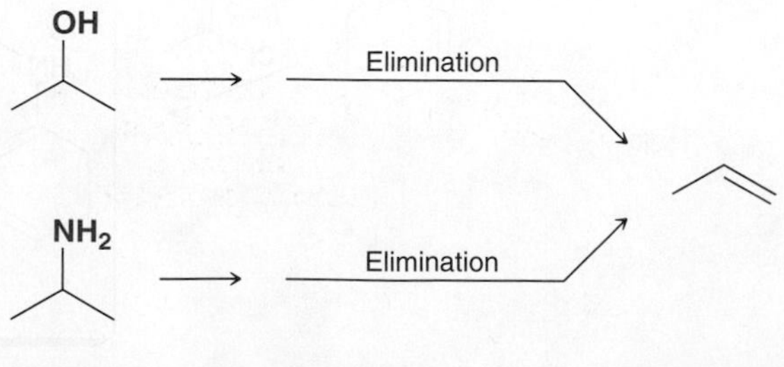

Recall that alcohols can only undergo an E2 process if the OH group is first converted into a better leaving group (such as a tosylate). Similarly, amines can also undergo an E2 reaction if the amino group is first converted into a better leaving group. This can be accomplished by treating the amine with excess methyl iodide.

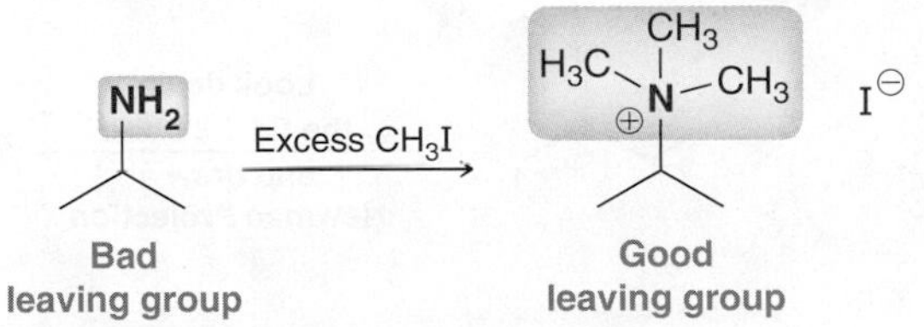

As we saw earlier in this chapter, amines will undergo exhaustive alkylation in the presence of excess methyl iodide, producing a quaternary ammonium salt. This process transforms the amino group into an excellent leaving group. Treating the quaternary ammonium salt with a strong base causes an E2 reaction that yields an alkene. The reagent most commonly used is aqueous silver oxide (Ag_2O):

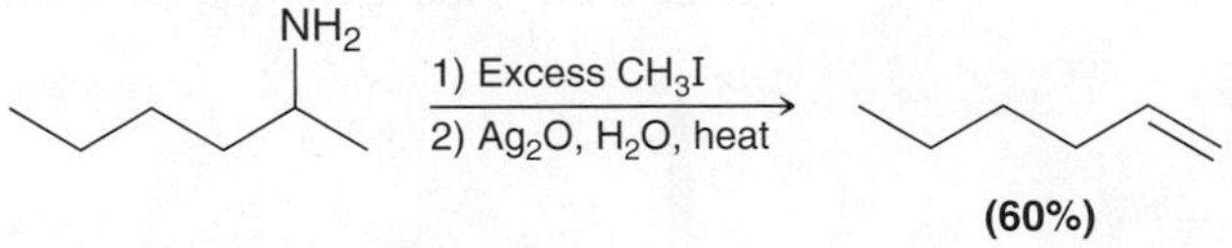

This process is called a **Hofmann elimination**. The function of silver oxide is to convert one ammonium salt into another ammonium salt by exchanging the iodide ion for a hydroxide ion.

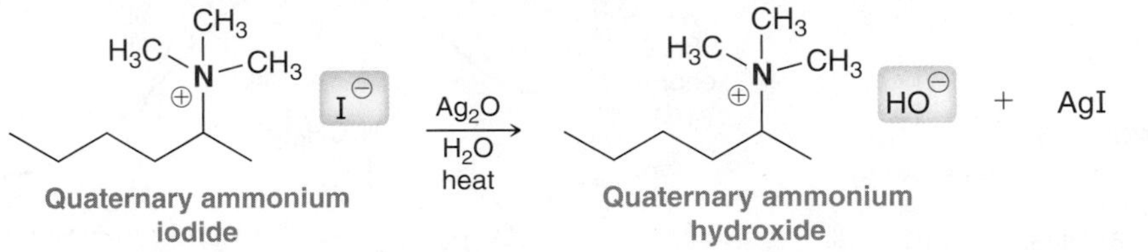

The newly formed hydroxide ion then serves as the base that triggers the E2 process:

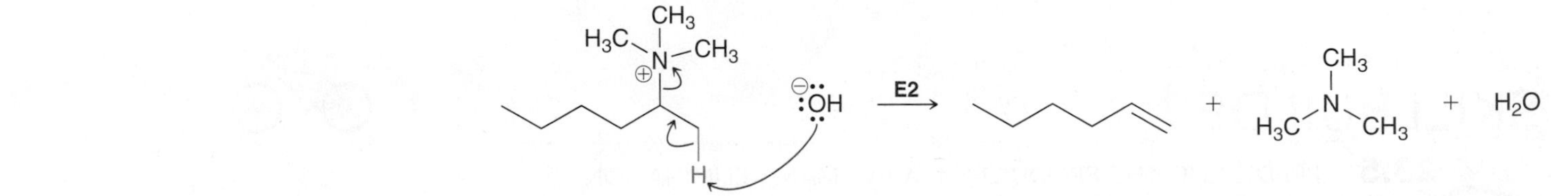

Take special notice of the regiochemical outcome; it is not what we might have expected. Recall that E2 reactions generally proceed to form the more substituted alkene. But in this case, the major product is the less substituted alkene, and only trace amounts of the more substituted product are formed. This observation can be rationalized with a steric argument, because the leaving group is very bulky. To understand the effect of the leaving group in controlling the regiochemical outcome, recall that E2 reactions take place from a conformation in which the leaving group is anticoplanar with the proton being removed. Consider the Newman projection showing the anticoplanar conformation that leads to the more substituted alkene.

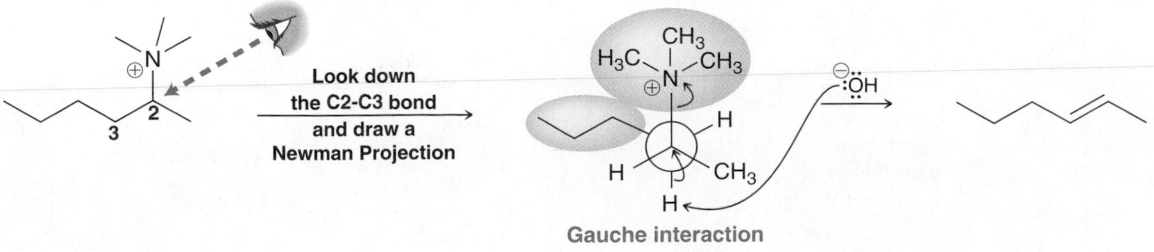

This conformation exhibits a gauche interaction, which raises the energy of the transition state. This *gauche* interaction is not present in the anticoplanar conformation, leading to the less substituted alkene.

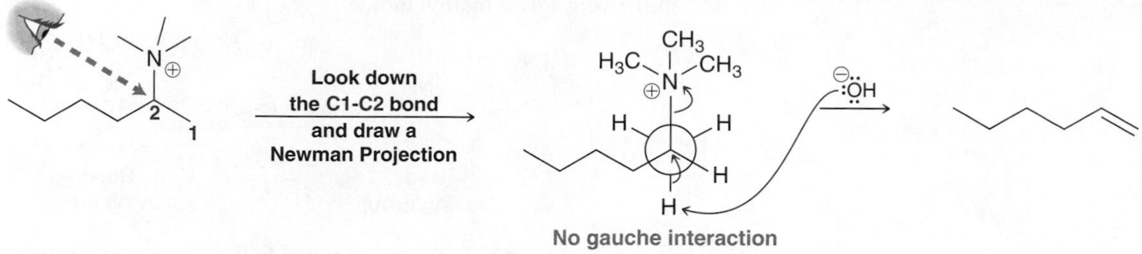

As a result, the transition state for formation of the less substituted alkene is lower in energy than the transition state for formation of the more substituted alkene (Figure 23.5). Formation of the less substituted alkene (shown in red) occurs more rapidly because it has a lower energy of activation, despite the fact that it is not the more stable product. This process is therefore said to be under kinetic control.

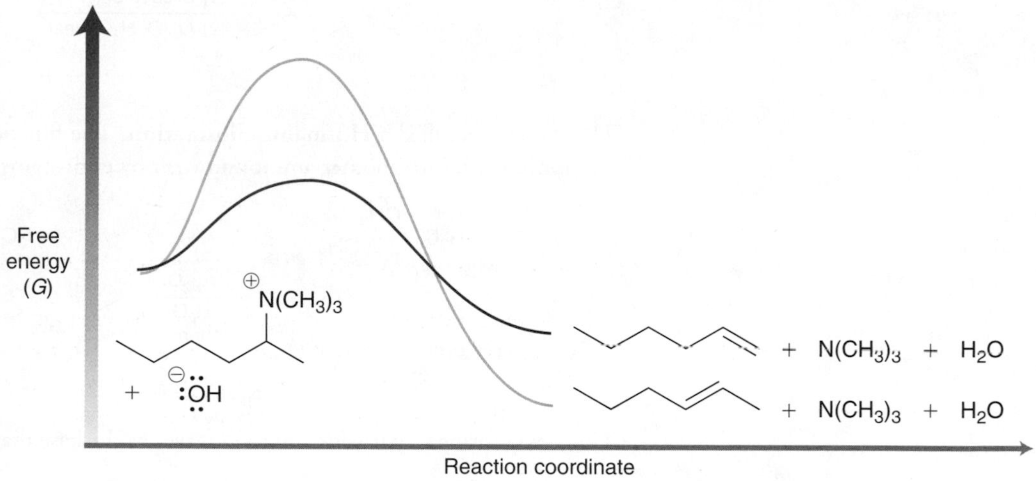

FIGURE 23.5 An energy diagram showing the two possible pathways for a Hofmann elimination. The less substituted alkene is formed because the activation energy is lower for that pathway.

SKILLBUILDER

23.5 PREDICTING THE PRODUCT OF A HOFMANN ELIMINATION

LEARN the skill

Draw the major product obtained when 3-methyl-3-hexanamine is treated with excess methyl iodide followed by aqueous silver oxide and heat.

SOLUTION

Begin by drawing the starting amine and identifying all α and β positions. Then, consider forming a C=C double bond between each possible pair of α and β positions:

STEP 1 Identify all α and β positions.

STEP 2 Consider the regiochemical possibilities.

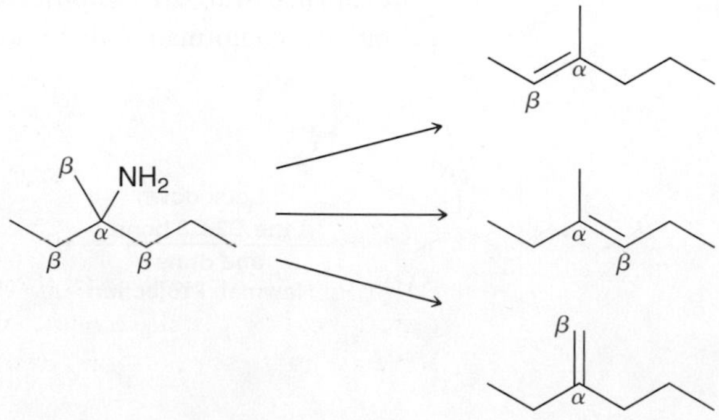

Compare the products and identify the alkene that is least substituted:

STEP 3
Identify the least substituted alkene.

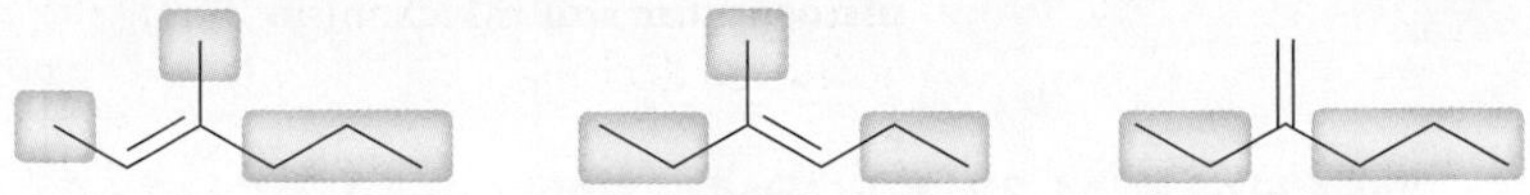

Trisubstituted **Tri**substituted **Di**substituted

The disubstituted product is expected to be the major product:

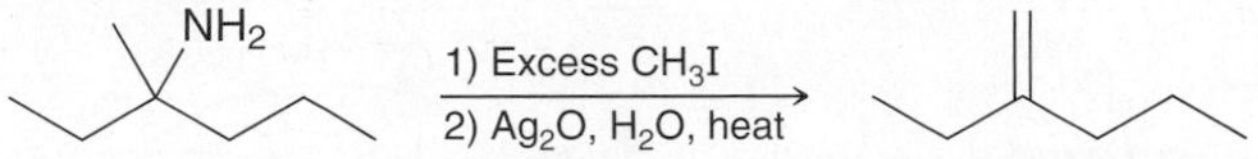

PRACTICE the skill **23.25** Draw the major product that is expected when each of the following compounds is treated with excess methyl iodide followed by aqueous silver oxide and heat:

(a) Cyclohexylamine (b) (*R*)-3-Methyl-2-butanamine (c) *N*,*N*-Dimethyl-1-phenylpropan-2-amine

APPLY the skill **23.26** Propose a synthesis for the following transformation (be sure to count the carbon atoms):

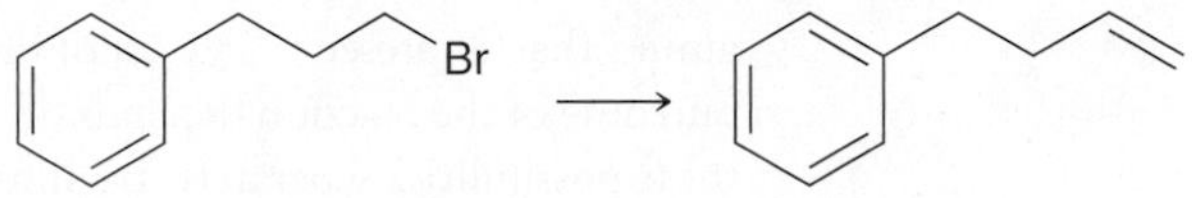

23.27 Compound **A** is an amine that does not possess a chirality center. Compound **A** was treated with excess methyl iodide and then heated in the presence of aqueous silver oxide to produce an alkene. The alkene was further subjected to ozonolysis to produce butanal and pentanal. Draw the structure of compound **A**.

23.28 Phencyclidine (PCP) was originally developed as an anesthetic for animals, but it has since become an illegal street drug because it is a powerful hallucinogen. Treatment of PCP with excess methyl iodide followed by aqueous silver oxide gives the following three principal products. Draw the structure of PCP.

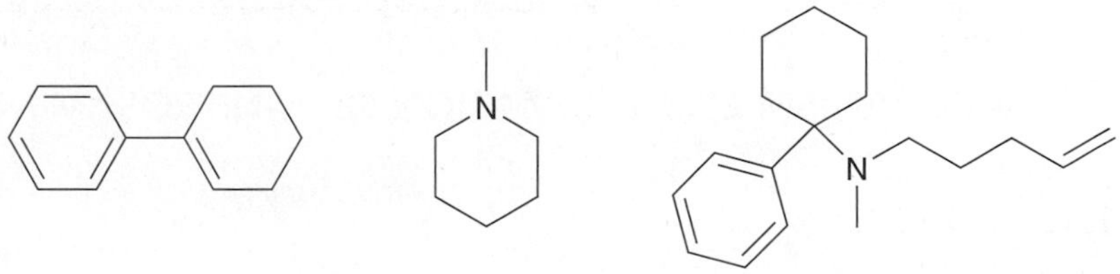

need more **PRACTICE?** **Try Problems 23.57, 23.69, 23.78**

23.10 Reactions of Amines with Nitrous Acid

This section will explore the reactions that occur when amines are treated with **nitrous acid**. Compare the structures of nitric acid (HNO_3) and nitrous acid (HNO_2):

Nitric acid (HNO_3) **Nitrous acid (HNO_2)**

Nitrous acid is unstable, and therefore, treating an amine with nitrous acid requires that the nitrous acid be prepared in the presence of the amine, called an *in situ* preparation. This is achieved by treating sodium nitrite ($NaNO_2$) with a strong acid, such as HCl or H_2SO_4.

Under these conditions nitrous acid is further protonated, followed by loss of water, to give a **nitrosonium ion** (Mechanism 23.2).

MECHANISM 23.2 FORMATION OF NITROUS ACID AND NITROSONIUM IONS

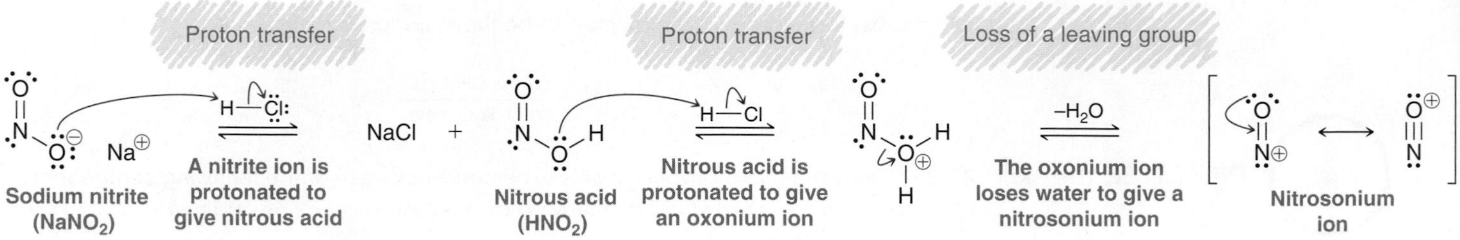

The nitrosonium ion is an extremely strong electrophile and is subject to attack by any amine that is present in solution. An equilibrium is established in which the concentration of nitrosonium ion is rather small. However, as the nitrosonium ion is formed, it can react with the amine that is present. The equilibrium then adjusts to produce more nitrosonium ion. The outcome of the reaction depends on whether the amine is primary or secondary. We will explore these possibilities separately, beginning with secondary amines.

Secondary Amines and Nitrous Acid

When a secondary amine is treated with sodium nitrite and HCl, the reaction produces an **N-nitrosamine** (Mechanism 23.3):

$$R_2N-H \xrightarrow[\text{HCl}]{\text{NaNO}_2} R_2N-N=O$$

An *N*-nitrosamine

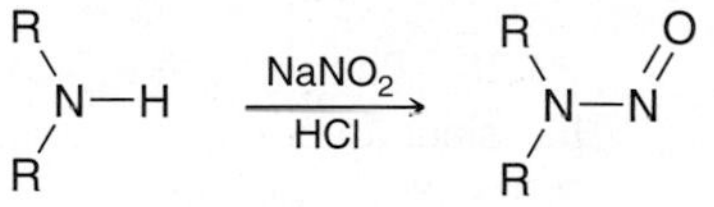

MECHANISM 23.3 FORMATION OF *N*-NITROSAMINES

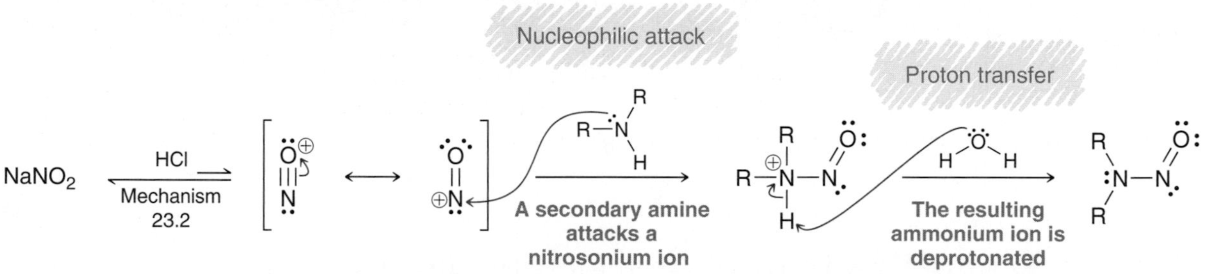

The amine functions as a nucleophile and attacks a nitrosonium ion that was generated *in situ* from sodium nitrite and HCl. The resulting ammonium ion is then deprotonated to give the product. Nitrosamines are known to be potent carcinogens. Several examples are shown below:

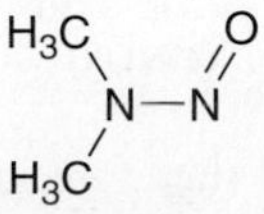

N-Nitrosodimethylamine
(Found in many food products, including cured meat, fish, and beer)

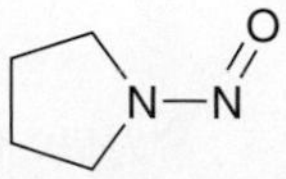

N-Nitrosopyrrolidine
(Found in fried bacon)

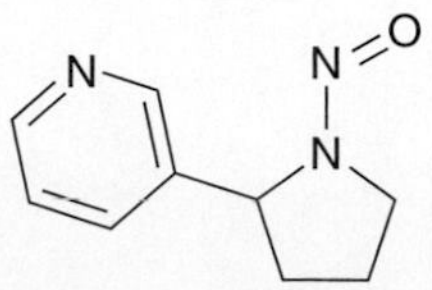

N-Nitrosonornicotine
(Found in tobacco smoke)

Primary Amines and Nitrous Acid

When a primary amine is treated with sodium nitrite and HCl, the reaction produces a **diazonium salt** (*azo* indicates a nitrogen atom, *diazo* indicates two nitrogen atoms, and *diazonium* indicates two nitrogen atoms with a positive charge).

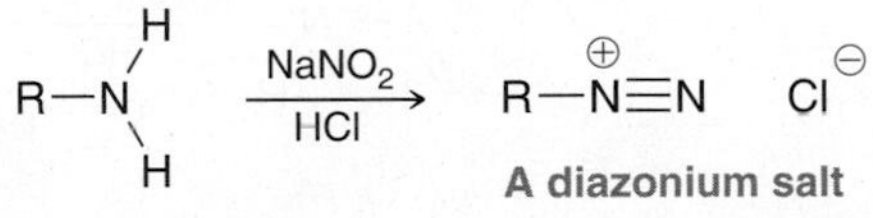

This process is called **diazotization** and is believed to proceed via Mechanism 23.4. The amine functions as a nucleophile and attacks a nitrosonium ion which was generated *in situ* from sodium nitrite and HCl. Several proton transfers follow, and the last step involves loss of water to generate the diazonium salt.

MECHANISM 23.4 DIAZOTIZATION

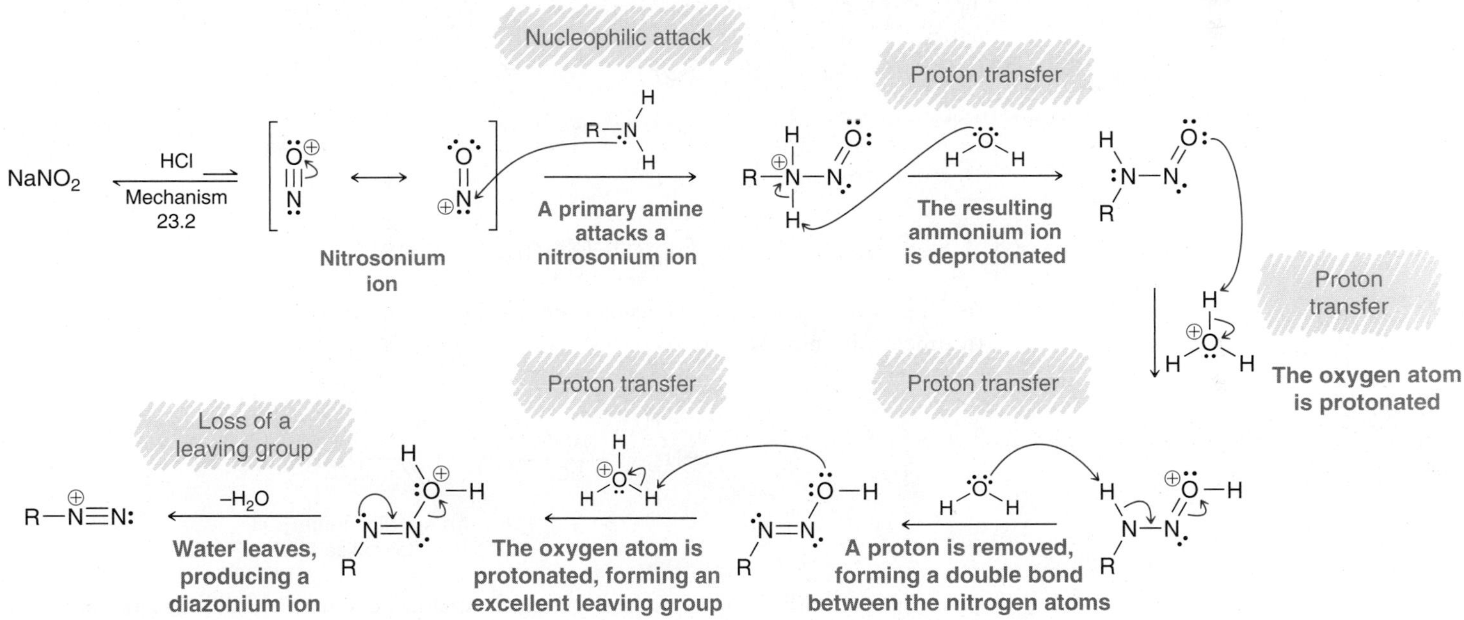

When the R group of the primary amine is an alkyl group as opposed to an aryl group, then the resulting diazonium salt is highly unstable and is too reactive to be isolated. It can spontaneously liberate nitrogen gas to form a carbocation, which then reacts in a variety of ways.

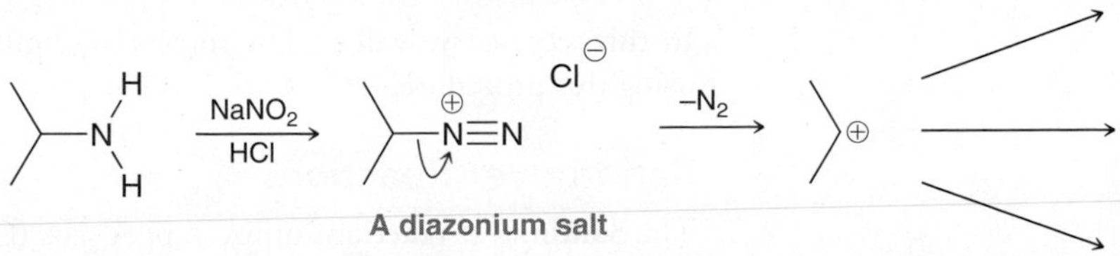

For example, the carbocation can be captured by water to form an alcohol or it can lose a proton to form an alkene. The reaction generates a mixture of products and is therefore not useful. In addition, the process is also dangerous, because the expulsion of nitrogen gas can be an explosive process.

If, however, the primary amine is an aryl amine, then the resulting aryldiazonium salt is stable enough to be isolated. It does not liberate nitrogen gas, because that would involve formation of an aryl cation, which is too high in energy to form.

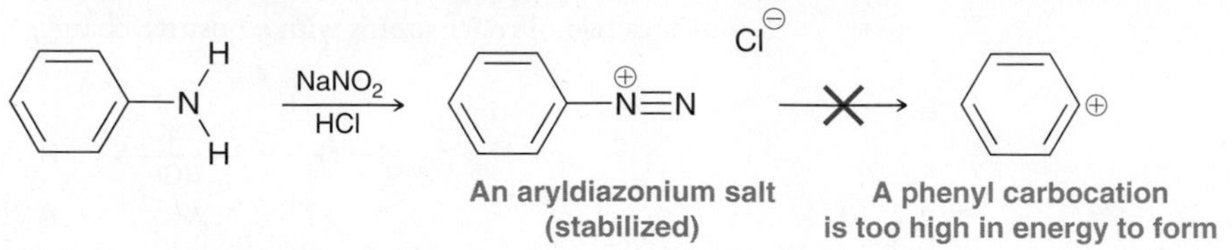

Aryldiazonium salts are extremely useful, because the diazonium group can be readily replaced with a number of other groups that are otherwise difficult to install on an aromatic ring. Reactions of aryldiazonium salts will be discussed in the next section.

CONCEPTUAL CHECKPOINT

23.29 Predict the major product obtained when each of the following amines is treated with a mixture of $NaNO_2$ and HCl:

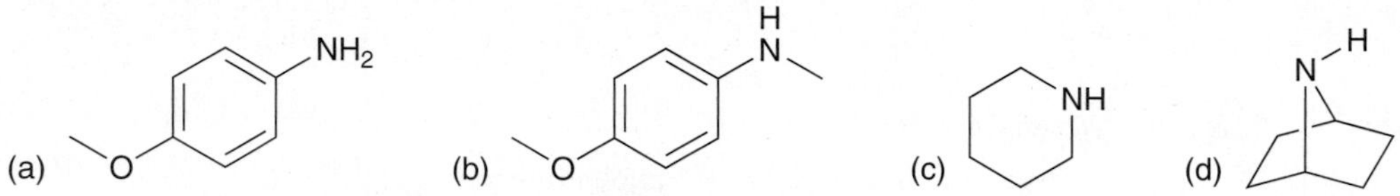

23.11 Reactions of Aryldiazonium Ions

In the previous section, we saw that aryl amines can be converted into aryldiazonium salts upon treatment with nitrous acid.

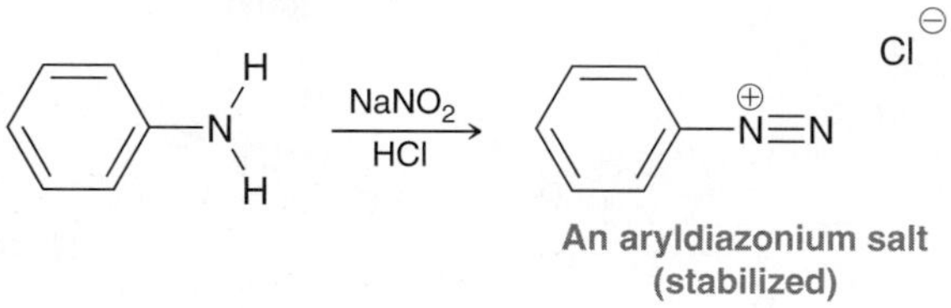

As mentioned earlier, this reaction is extremely useful, because many different reagents will replace the diazo group, allowing for a simple procedure for installing a wide variety of groups on an aromatic ring:

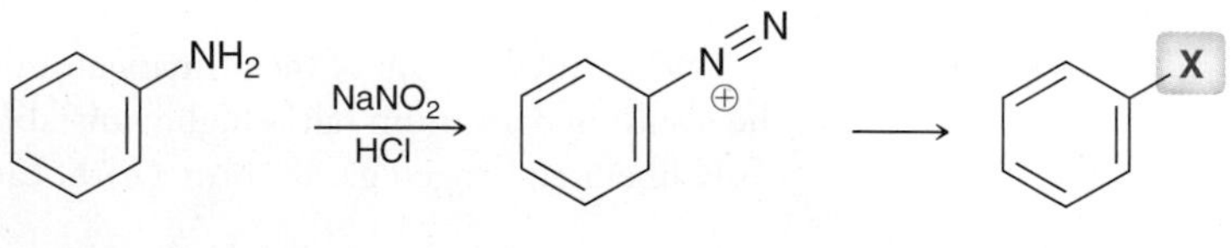

In this section, we will explore some of the groups that can be installed on an aromatic ring using this procedure.

Sandmeyer Reactions

The **Sandmeyer reactions** utilize copper salts (CuX) and enable the installation of a halogen or a cyano group on an aromatic ring:

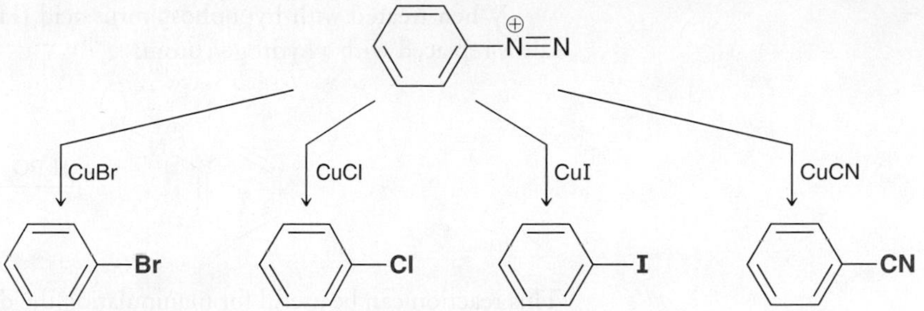

Notice the installation of a cyano group. Recall that a cyano group can be hydrolyzed in aqueous acid or base, which provides a method for installing a carboxylic acid moiety on an aromatic ring.

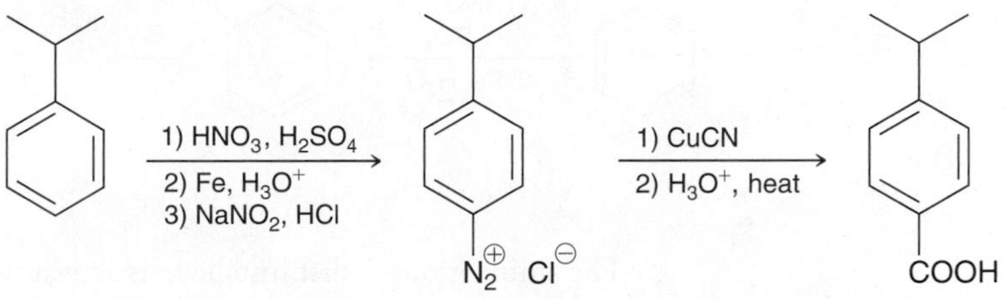

Fluorination

When treated with fluoroboric acid (HBF_4), an aryl diazonium salt is converted into a fluorobenzene. This reaction, called the **Schiemann reaction**, is useful for installing fluorine on an aromatic ring, which is not easy to accomplish with other methods.

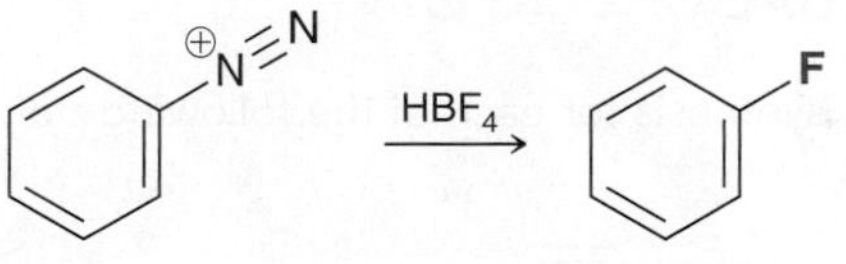

Other Substitution Reactions of Aryl Diazonium Salts

When an aryldiazonium salt is heated in the presence of water, the diazo group is replaced with a hydroxyl group.

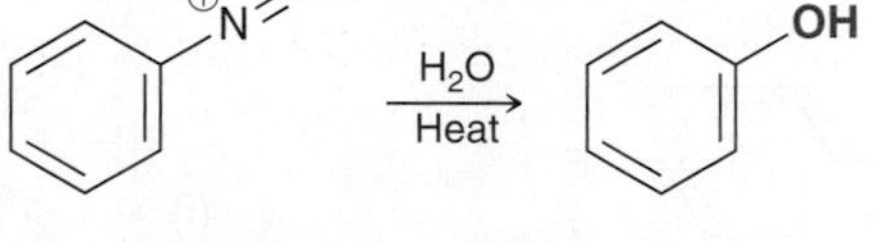

This procedure is very useful, because there are not many other ways to install an OH group on an aromatic ring. An example of this process is shown below.

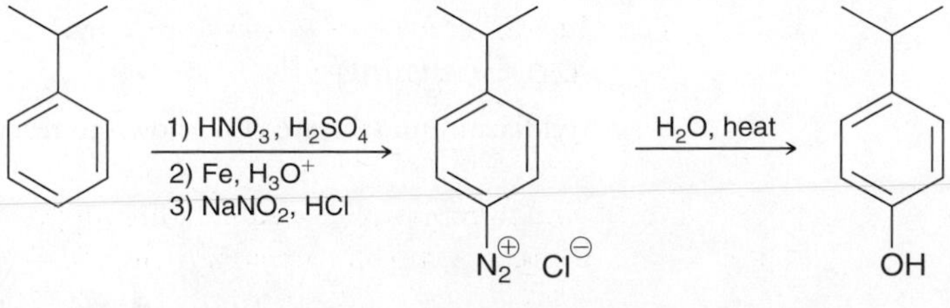

When treated with hypophosphorus acid (H_3PO_2), the diazo group of an aryldiazonium salt is replaced with a hydrogen atom:

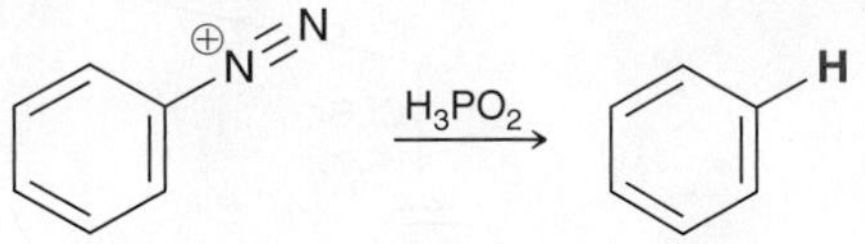

This reaction can be useful for manipulating the directing effects of a substituted aromatic ring. For example, consider the following synthesis of 1,3,5-tribromobenzene.

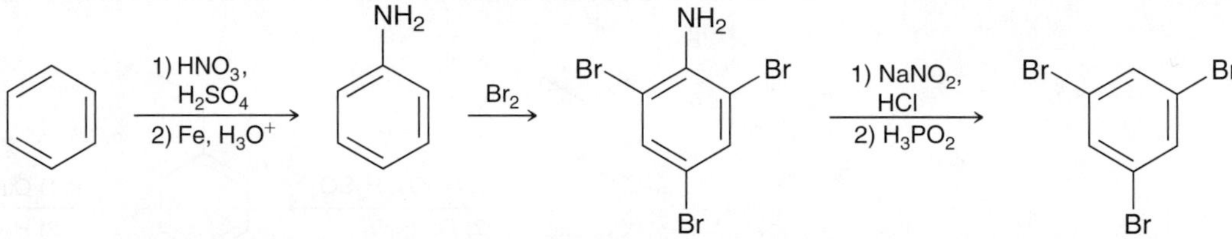

The amino group is first installed, its activating and directing effects are exploited, and then it is completely removed. The product of this sequence cannot be easily prepared from benzene via successive halogenation reactions because halogens are *ortho-para* directors.

The reactions of diazonium salts are believed to occur via radical intermediates, and their mechanisms are beyond the scope of our discussion.

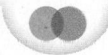

CONCEPTUAL CHECKPOINT

23.30 Propose an efficient synthesis for each of the following transformations:

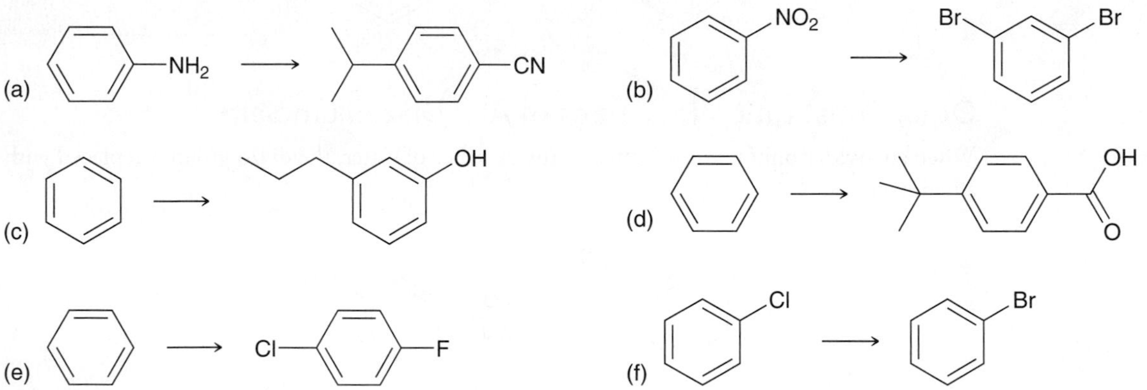

Azo Coupling

Aryldiazonium salts are also known to react with activated aromatic rings.

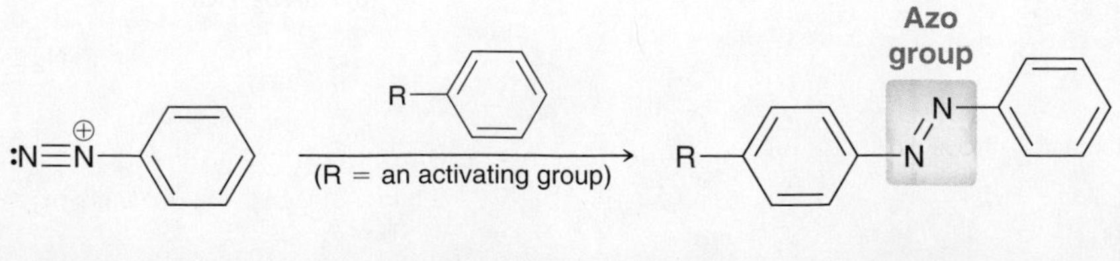

This process, called **azo coupling**, is believed to occur via an electrophilic aromatic substitution reaction.

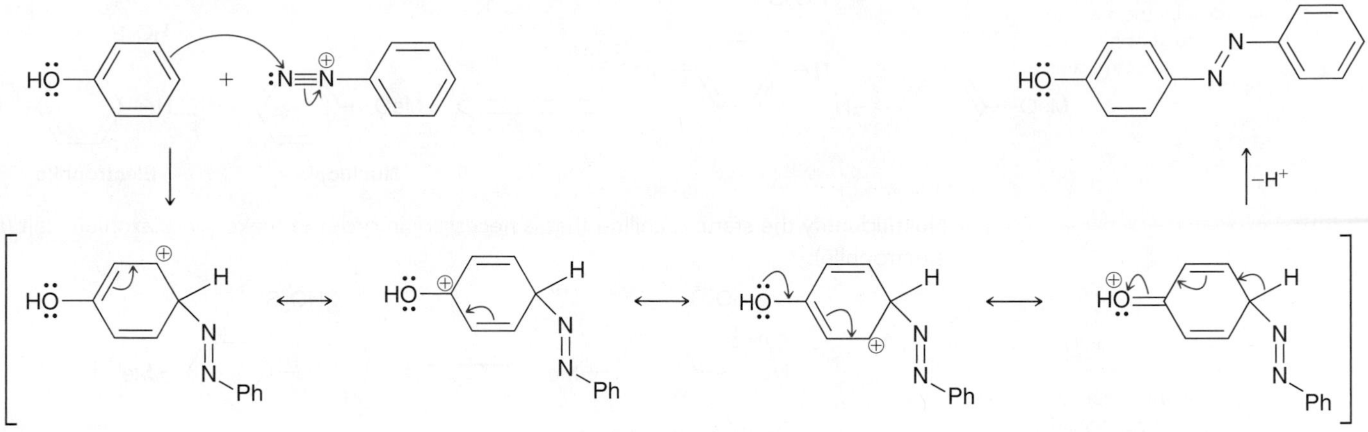

Sigma complex (resonance stabilized)

LOOKING BACK
For more on the source of color, see Section 17.12.

The reaction works best when the nucleophile is an activated aromatic ring. The resulting azo compound has extended conjugation and therefore exhibits color. By structurally modifying the starting materials (by placing substituents on the aromatic rings prior to azo coupling), a variety of products can be made, each of which will exhibit a unique color. These compounds, called **azo dyes**, were discussed in more detail in the Practically Speaking box in Section 19.3.

SKILLBUILDER

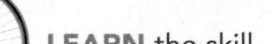

23.6 DETERMINING THE REACTANTS FOR PREPARING AN AZO DYE

LEARN the skill

Identify the reactants you would use to prepare the following azo dye via an azo coupling reaction.

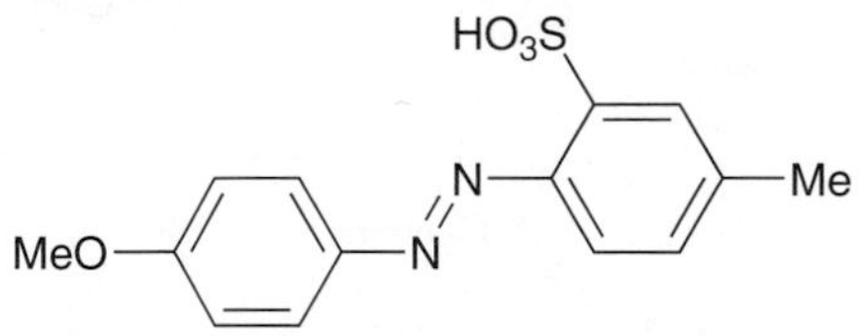

SOLUTION

Begin by identifying all substituents as activators or deactivators (for help, see Table 19.1)

STEP 1
Determine which ring is more strongly activated.

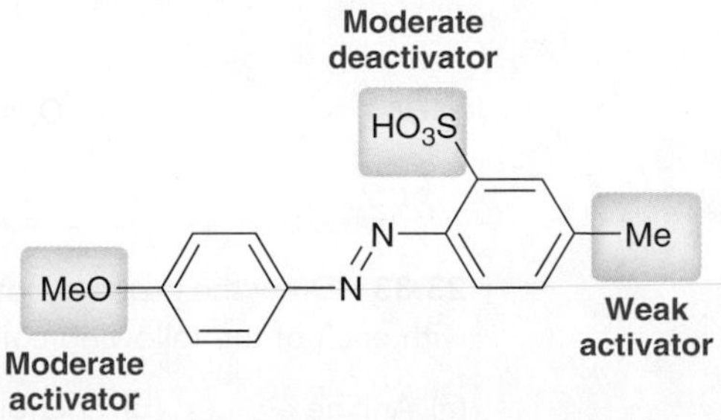

Based on this information, identify which ring is more strongly activated. In this case, it is the ring that bears the methoxy substituent. Using a retrosynthetic analysis, identify the starting nucleophile and electrophile needed to make this azo dye. The activated ring must be the nucleophile, and the other ring must function as the diazonium salt:

STEP 2
Identify the starting nucleophile and electrophile.

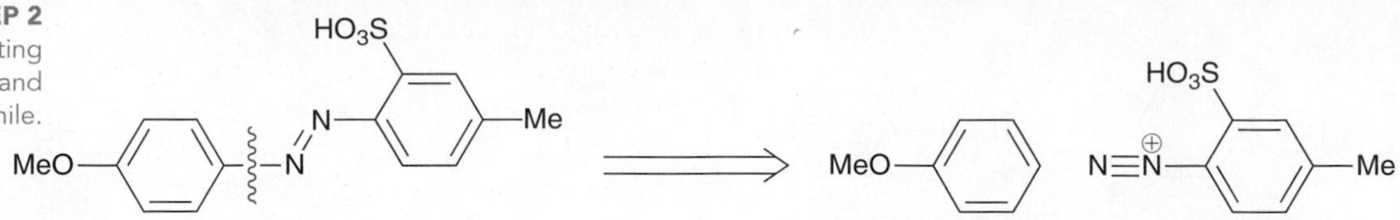

Nucleophile **Electrophile**

Next, identify the starting aniline that is necessary in order to make the diazonium salt (the electrophile):

STEP 3
Identify the substituted aniline that is necessary to make the desired electrophile.

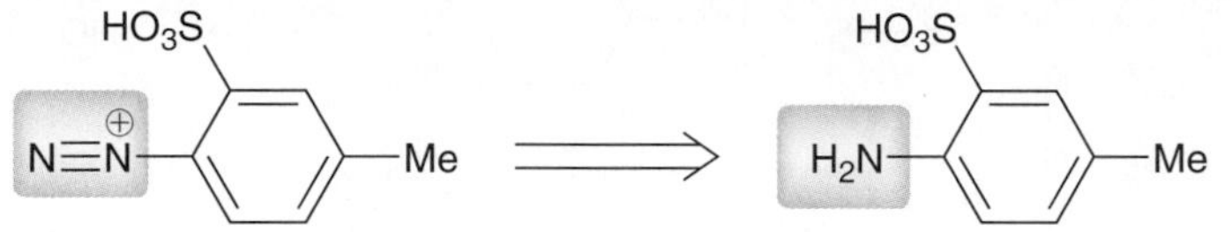

Electrophile

Finally, show all reagents. Draw the substituted aniline as the reactant, and then show the two steps necessary to convert this compound into the desired product. The first step utilizes $NaNO_2$ and HCl to produce the diazonium salt, and the second step utilizes an azo coupling reaction:

STEP 4
Show all reagents.

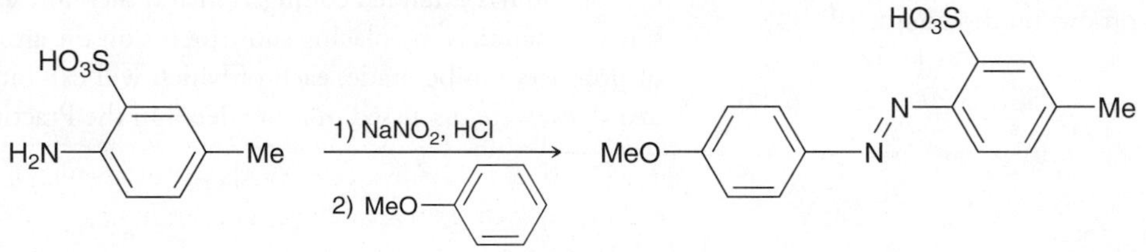

PRACTICE the skill

23.31 Identify the reactants you would use to prepare each of the following azo dyes via an azo coupling reaction:

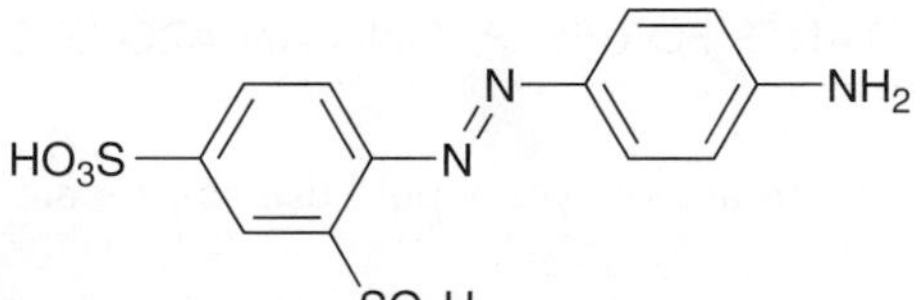

(a)

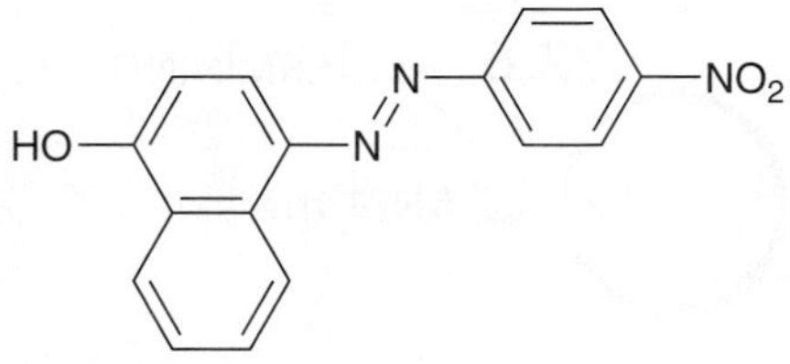

(b)

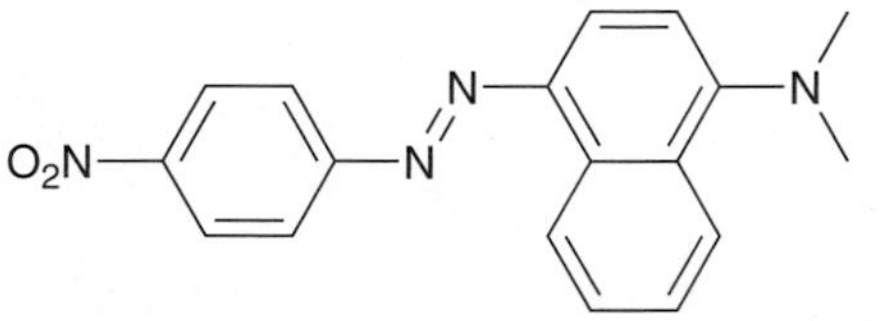

(c)

APPLY the skill

23.32 Starting with benzene and isopropyl chloride, show how you would prepare the following compound:

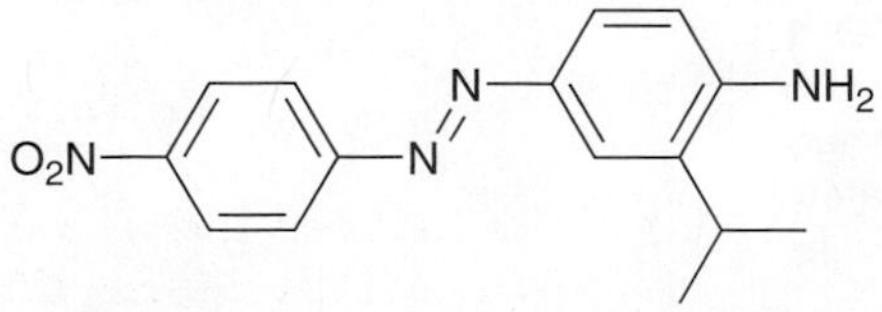

23.33 Draw the product obtained when the diazonium salt formed from aniline is treated with each of the following compounds:

(a) Aniline (b) Phenol (c) Anisole (methoxybenzene)

need more **PRACTICE?** **Try Problems 23.68, 23.73d**

23.12 Nitrogen Heterocycles

A **heterocycle** is a ring that contains atoms of more than one element. Common organic heterocycles are comprised of carbon and either nitrogen, oxygen, or sulfur. Consider, for example, the structures of Viagra and Nexium. Both of these compounds contain nitrogen heterocycles, highlighted in red. Many different kinds of heterocycles are commonly found in the structures of biological molecules and pharmaceuticals. We will discuss just a few simple heterocycles in this section.

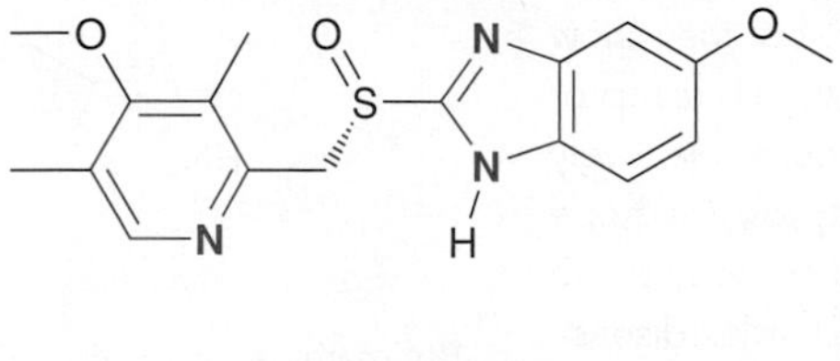

Esomeprazole (Nexium)
A proton pump inhibitor used in the treatment of ulcers and acid reflux disease

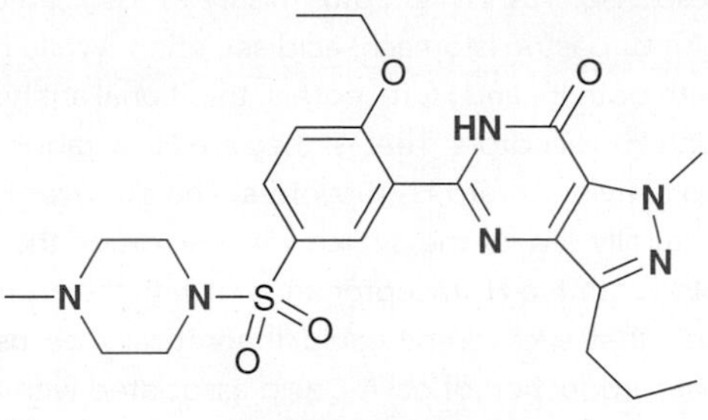

Sildenafil (Viagra)
Used in the treatment of erectile dysfunction and pulmonary arterial hypertension

Pyrrole and Imidazole

Pyrrole, a five-membered aromatic ring containing one nitrogen atom, is numbered starting with the nitrogen atom. As seen in Chapter 18, the lone pair of this nitrogen atom participates in aromaticity (Figure 18.15) and is therefore significantly less basic and less nucleophilic than a typical amine. Pyrrole undergoes reactions that are expected for an aromatic system, such as electrophilic aromatic substitution. In fact, pyrrole is even more reactive than benzene, and low temperatures are often required to control the reaction.

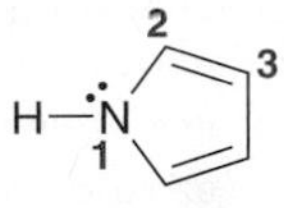

Pyrrole

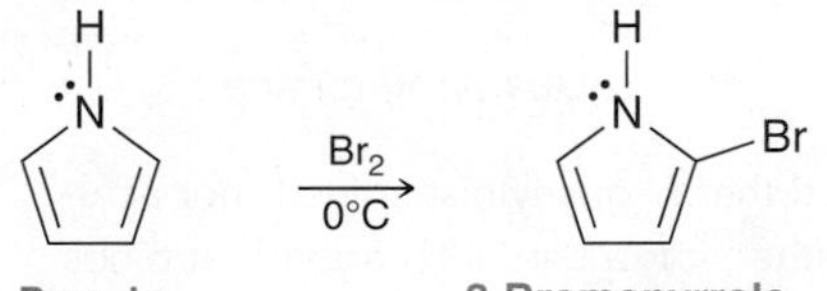

Pyrrole **2-Bromopyrrole (90%)**

Electrophilic aromatic substitution occurs primarily at C2, because the intermediate formed during attack at C2 is stabilized by resonance.

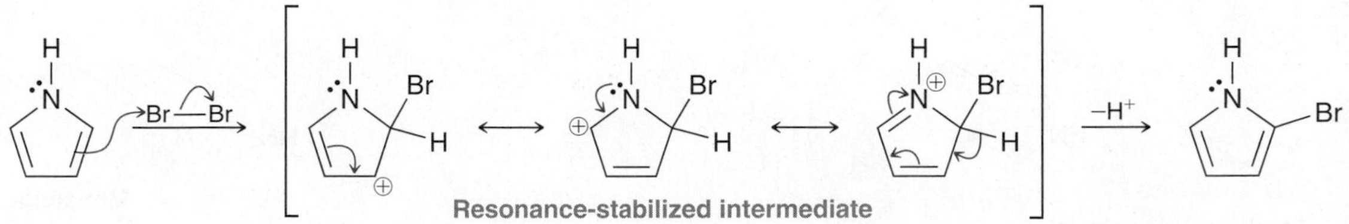

When attack takes place at C2, the intermediate has three resonance structures. In contrast, when attack takes place at C3, the intermediate has only two resonance structures.

An **imidazole** ring is like pyrrole but has one extra nitrogen atom at the 3 position. Histamine is an example of an important biological compound that contains an imidazole ring.

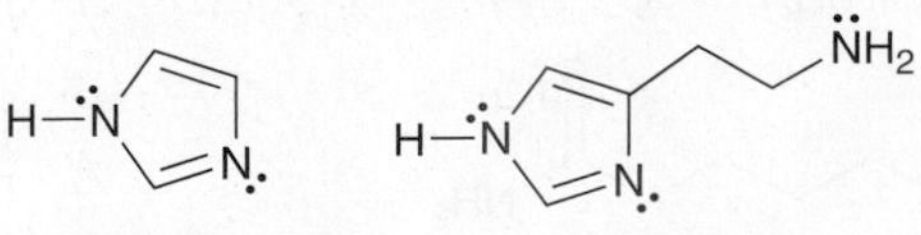

Imidazole **Histamine**

MEDICALLYSPEAKING

H_2-Receptor Antagonists and the Development of Cimetidine

Recall that histamine binds to the H_1 receptor, triggering allergic reactions (as seen in the Medically Speaking box in Section 18.5). It is now understood that histamine binds to other types of receptors as well, each of which elicits a different physiological response. The H_2 receptor has been associated with the stimulation of gastric (stomach) acid secretion. While histamine can bind with both H_1 and H_2 receptors, traditional antihistamines only bind with H_1 receptors. That is, they are H_1 antagonists, but they show no activity toward H_2 receptors. The discovery of the H_2 receptor naturally led to the search for new drugs that would selectively bind with the H_2 receptor and not with the H_1 receptor. The hope was that such drugs could theoretically be used to control the overproduction of gastric acid associated with acid reflux disease and ulcers. Years of research ultimately led to the development of cimetidine (Tagamet), the first selective H_2 antagonist.

The story of cimetidine begins with consideration of the features that might be required for a compound to exhibit selective H_2 antagonism. Medicinal chemists reasoned that the target compound must be sufficiently similar in structure to histamine to bind to the H_2 receptor, but it must be sufficiently different in order to be an antagonist, rather than an agonist. It was assumed that the imidazole ring was necessary for binding, so several hundred imidazole derivatives were prepared and tested. The first compound that showed any H_2 antagonistic activity was *N*-guanylhistamine.

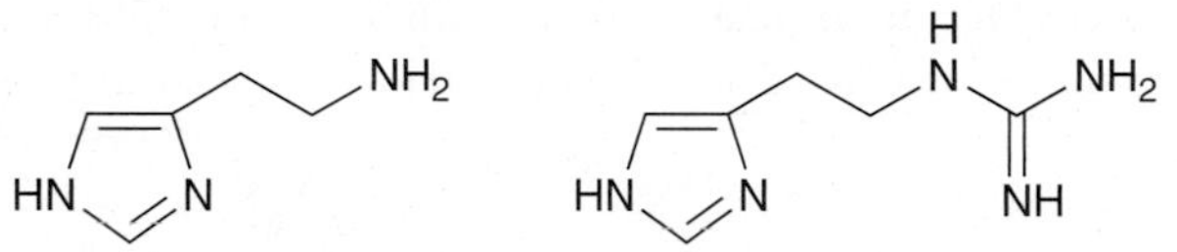

Histamine **N-Guanylhistamine**

Further studies revealed that *N*-guanylhistamine is not actually an H_2 antagonist, but rather, it is a partial H_2 agonist. It binds with the H_2 receptor and elicits the stimulation of gastric acid secretion, but to a smaller extent than histamine. This, of course, defeats the purpose. So the chemists continued their search, using *N*-guanylhistamine as a starting point. This compound was known to be protonated at physiological pH.

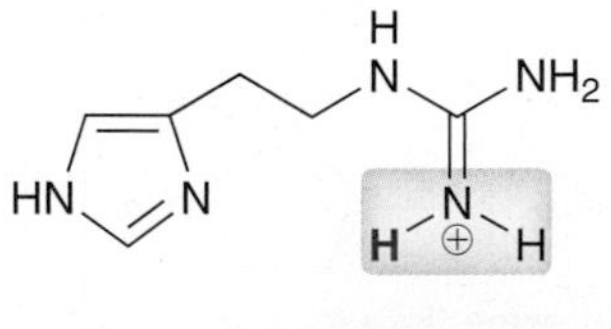

Medicinal chemists then tried to modify the guanyl group so that it would be neutral at physiological pH. After many different possibilities, a thiourea analogue was created that has one additional CH_2 group in the side chain, and the basic nitrogen atom of the guanyl group has been replaced with a sulfur atom that is not protonated at physiological pH.

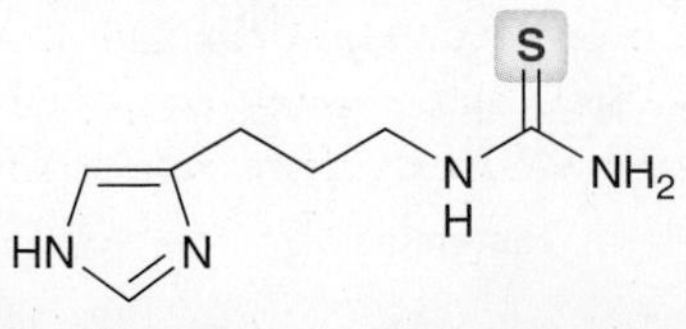

This compound did, in fact, exhibit weak antagonistic activity. The search then focused on creating analogues that might exhibit greater potency. An additional CH_2 was inserted in the side chain, and the terminal nitrogen atom was methylated. The resulting compound, called burimamide, was shown to be effective in inhibiting gastric acid secretion in rats.

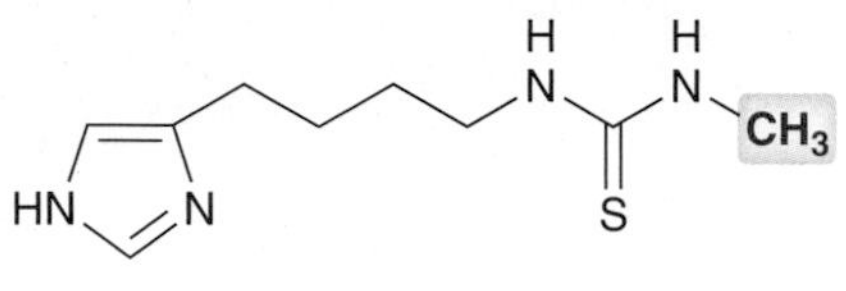

Burimamide

Burimamide was the first H_2 antagonist tested in humans. Unfortunately, it did not exhibit sufficient activity when taken orally, so the search continued, this time using the structure of burimamide as a starting point.

One approach was to compare the basicity of histamine and burimamide. In other words, we compare the pK_a values of the protonated imidazole ring in histamine and burimamide:

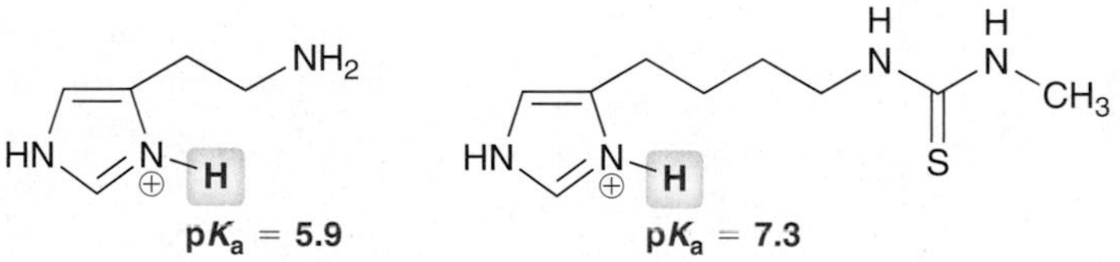

Histamine **Burimamide**

The protonated imidazole ring of histamine has a pK_a of 5.9, while the protonated imidazole ring of burimamide has a pK_a of 7.3. Therefore, at physiological pH, the equilibrium concentrations of neutral and ionic forms are different for histamine and burimamide. Research efforts were then directed toward developing an analogue of burimamide that more closely resembles the basicity of histamine. It was reasoned that the presence of an electron-withdrawing group in the side chain would lower the pK_a while an electron-donating group at the C4 position would increase the basicity of the nitrogen atom. This rationale led to the development of metiamide.

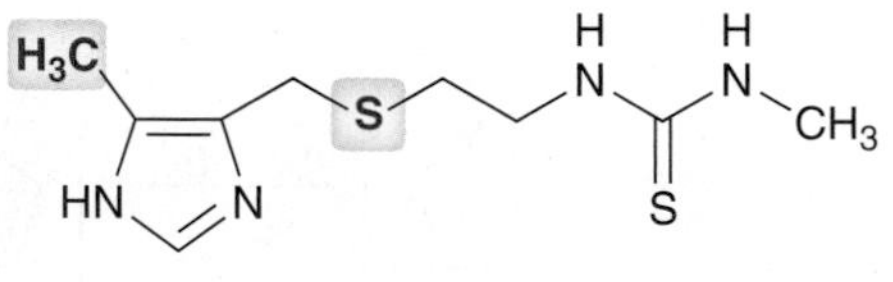

Metiamide

The structure of metiamide differs from that of burimamide in the presence of a methyl group at C4, and the replacement of one of the CH_2 groups in the side chain with an electron-withdrawing sulfur atom. This type of replacement is called an isosteric replacement, because the sulfur atom occupies roughly the same amount of space as a CH_2 group, but it has very different electronic properties. The net result is that the protonated form of metiamide has a pK_a that is nearly identical with the protonated form of histamine. Clinical trials showed that metiamide exhibited H_2 antagonistic activity that was nine times greater than burimamide, and it was effective in relieving ulcer symptoms; however, a few patients developed granulocytopenia (lowering of the white blood cell count). This side effect was not acceptable, because it would temporarily impair the ability of the immune system to function properly. It was believed that the cases of granulocytopenia were caused by the presence of the thiourea group, so alternative groups were considered. Ultimately, the thiourea group was replaced with a cyanoguanidine group.

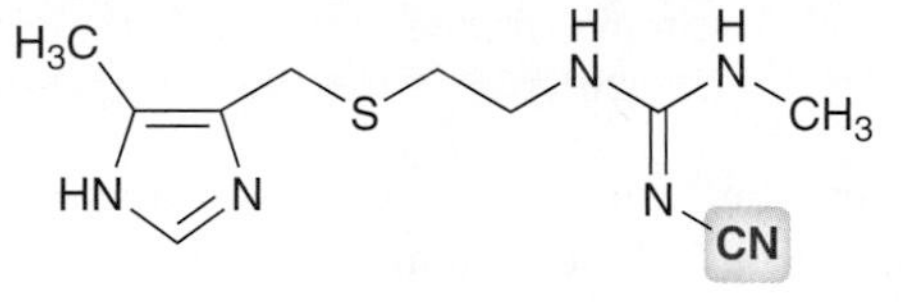

Tagamet™
(cimetidine)

Recall from earlier in our discussion that the presence of the guanyl group precluded antagonistic activity because the guanyl group is charged at physiological pH. However, in this derivative, the cyano group withdraws electron density from the neighboring nitrogen atom and delocalizes its lone pair. The resulting group is less basic and is not protonated at physiological pH. This compound, called cimetidine, was found to exhibit potent H_2 antagonistic properties and did not cause granulocytopenia in patients.

Cimetidine became available to the public in England in 1976 and in the United States in 1979, sold under the trade name Tagamet. Annual sales of this drug skyrocketed, ultimately reaching more than $1 billion in annual sales. Cimetidine earned its place in history as the first blockbuster drug. Since the development of cimetidine, other H_2 antagonists were developed and approved for use:

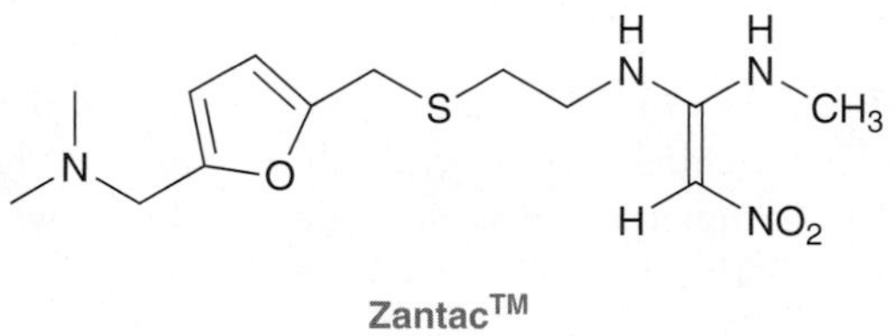

Zantac™
(ranitidine)

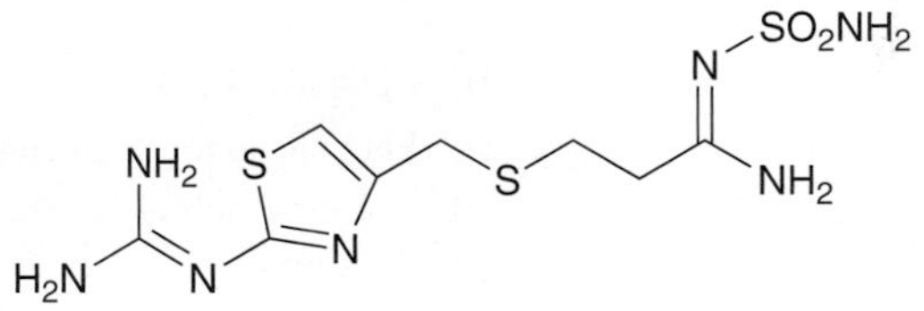

Pepcid™
(famotidine)

Ranitide (Zantac) soon overcame cimetidine (Tagamet) as the best-selling drug worldwide. Upon inspection of their structures, it is clear that the imidazole ring is not absolutely essential for H_2 antagonistic activity, as analogous heterocyclic aromatic rings can replace the imidazole ring. However, all of these analogues appear to have some derivative of the guanyl group on the side chain.

Pyridine and Pyrimidine

Pyridine is a six-membered aromatic ring containing one nitrogen atom and is numbered starting with the nitrogen atom:

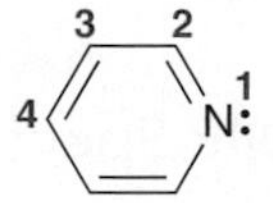

Pyridine

The lone pair of the nitrogen atom is localized and occupies an sp^2-hybridized orbital, and as a result, pyridine is a stronger base than pyrrole. Nevertheless, pyridine is still a weaker base than alkyl amines, because the lone pair is housed in an sp^2-hybridized orbital, rather than an sp^3-hybridized orbital (as seen in Figure 18.14). By occupying an sp^2-hybridized orbital, the electrons of the lone pair have more *s* character and are therefore closer to the positively charged nucleus, rendering them less basic.

Since pyridine is an aromatic compound, we would expect it to exhibit reactions that are characteristic of aromatic systems. In fact, pyridine undergoes electrophilic aromatic substitution reactions; however, the yields are generally quite low, because the inductive effect of the nitrogen atom renders the ring electron poor. High temperatures are required:

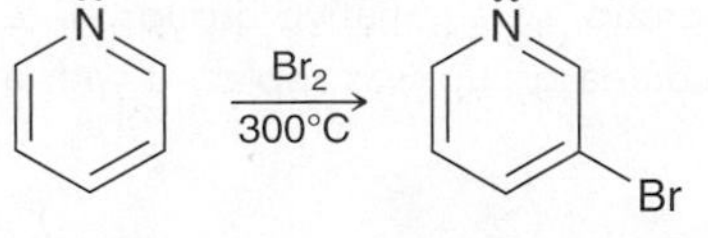

Pyrimidine is similar in structure to pyridine but contains one extra nitrogen atom at the 3-position. Pyrimidine rings are very common in biological molecules. Pyrimidine is less basic than pyridine due to the inductive effect of the second nitrogen atom:

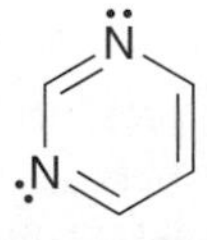

CONCEPTUAL CHECKPOINT

23.34 Pyridine undergoes electrophilic aromatic substitution at the C3 position. Justify this regiochemical outcome by drawing resonance structures of the intermediate produced from attack at C2, at C3, and at C4.

23.35 Predict the product obtained when pyrrole is treated with a mixture of nitric acid and sulfuric acid at 0°C.

23.13 Spectroscopy of Amines

IR Spectroscopy

In their IR spectra, primary and secondary amines exhibit signals between 3350 and 3500 cm^{-1}. These signals correspond with N—H stretching and are typically less intense than O—H signals. Primary amines give two peaks (symmetric and asymmetric stretching), while secondary amines give only one peak (Figure 15.19). This phenomenon was first explained in Section 15.5.

Tertiary amines lack an N—H bond and do not exhibit a signal in the region between 3350 and 3500 cm^{-1}. Tertiary amines can be detected with IR spectroscopy by treatment with HCl. The resulting N—H bond exhibits a characteristic signal between 2200 and 3000 cm^{-1}:

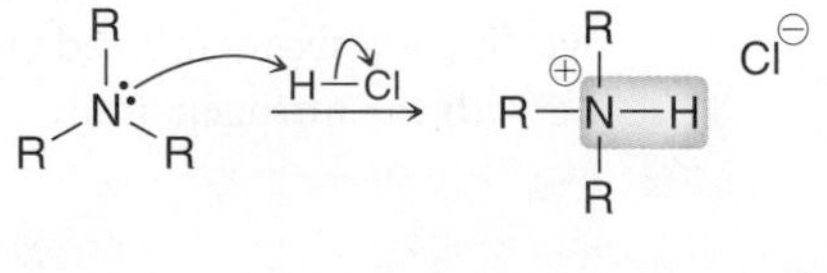

NMR Spectroscopy

In the ^1H NMR spectrum of an amine, any proton attached directly to the nitrogen atom (for primary and secondary amines) will typically appear as a broad signal between 0.5 and 5.0 ppm. The precise location is sensitive to many factors, including the solvent, concentration, and temperature. Splitting is generally not observed for these protons, because they are labile, and are exchanged at a rate that is faster than the timescale of the NMR spectrometer (as described in Section 16.7). The broad signal for these protons can generally be removed from the spectrum by dissolving the amine in D_2O, which results in a proton exchange that replaces these protons for deuterons.

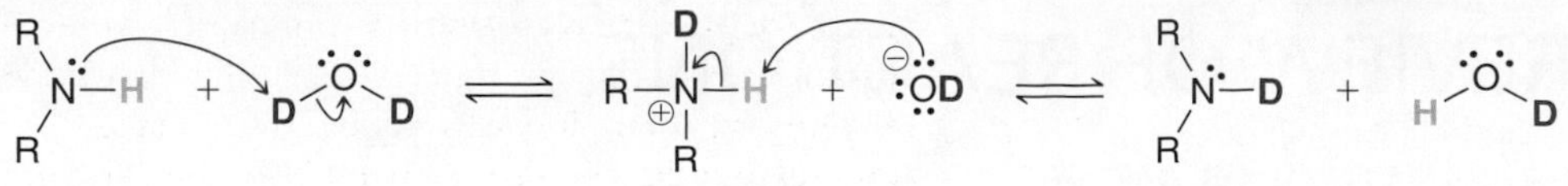

Protons connected to the α position typically appear between 2 and 3 ppm because of the deshielding effect of the nitrogen atom. As an example, consider the chemical shifts for the protons in propylamine:

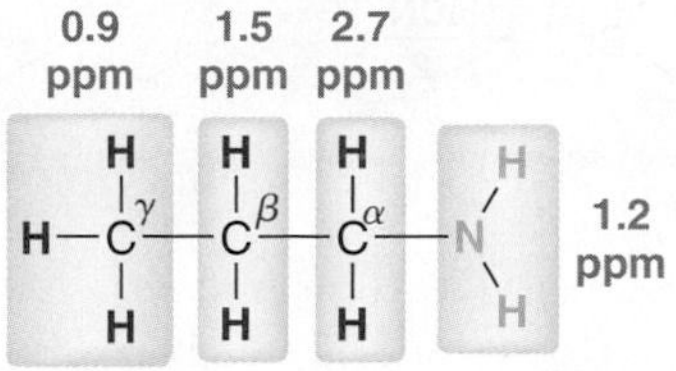

Notice that the deshielding effect of the nitrogen atom is greatest for the α protons (2.7 ppm) and tapers off with distance. The β protons are affected to a lesser extent (1.5 ppm), and the γ protons are not measurably affected (0.9 ppm).

In the ^{13}C NMR spectrum of an amine, the α carbon atoms typically appear between 30 and 50 ppm. That is, they are shifted about 20 ppm downfield due to the deshielding effect of the nitrogen atom.

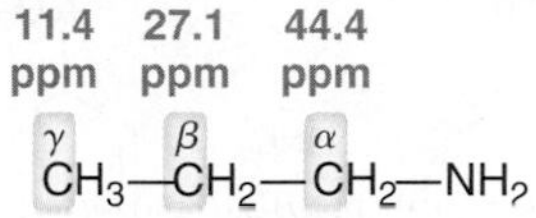

Once again, notice that the deshielding effect of the nitrogen atom is greatest for the α carbon (44.4 ppm) and tapers off with distance. The β carbon is affected to a lesser extent (27.1 ppm), and the γ carbon is not measurably affected (11.4 ppm).

Mass Spectrometry

The mass spectrum of an amine is characterized by the presence of a parent ion with an odd molecular weight. This follows the nitrogen rule (see Section 15.9), which states that a compound with an odd number of nitrogen atoms will produce a parent ion with an odd molecular weight. In addition, amines generally exhibit a characteristic fragmentation pattern. They undergo α cleavage to generate a radical and a resonance-stabilized cation (see Section 15.12).

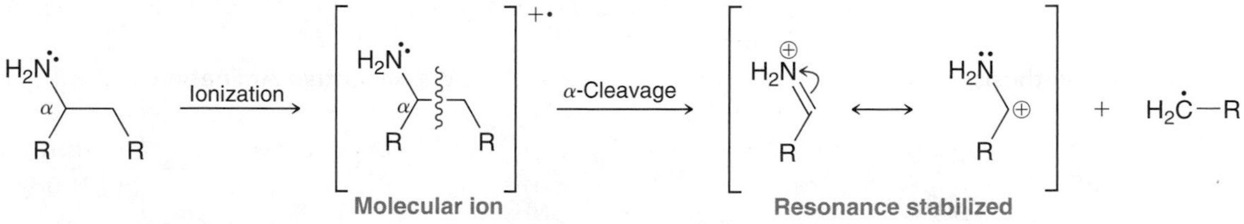

CONCEPTUAL CHECKPOINT

23.36 How would you use IR spectroscopy to distinguish between the following pairs of compounds?

(a) 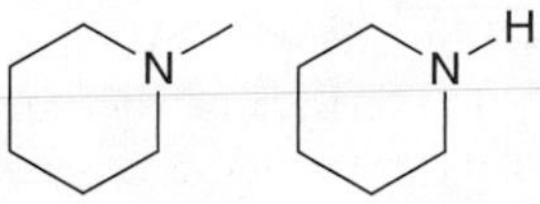 (b) 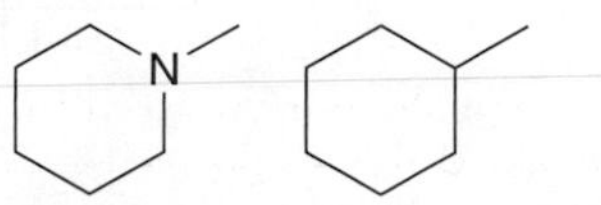

23.37 How would you use NMR spectroscopy to distinguish between the following pairs of compounds?

(a) 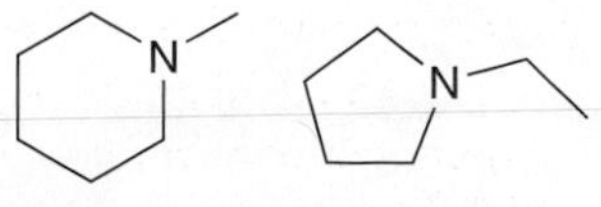

(b) 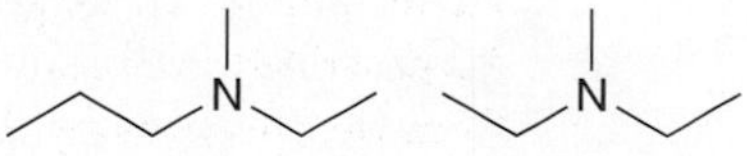

REVIEW OF REACTIONS

PREPARATION OF AMINES

From Alkyl Halides

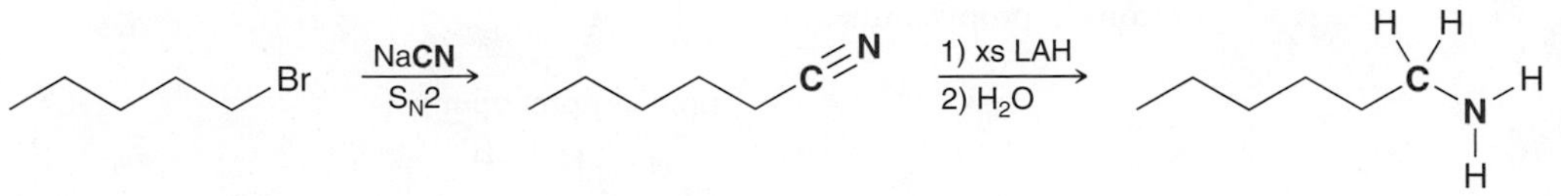

From Carboxylic Acids

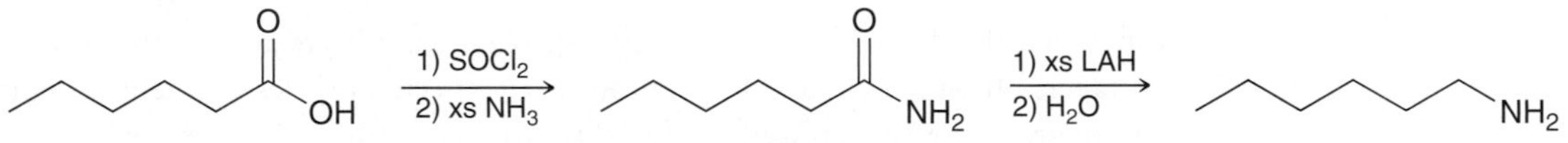

From Benzene

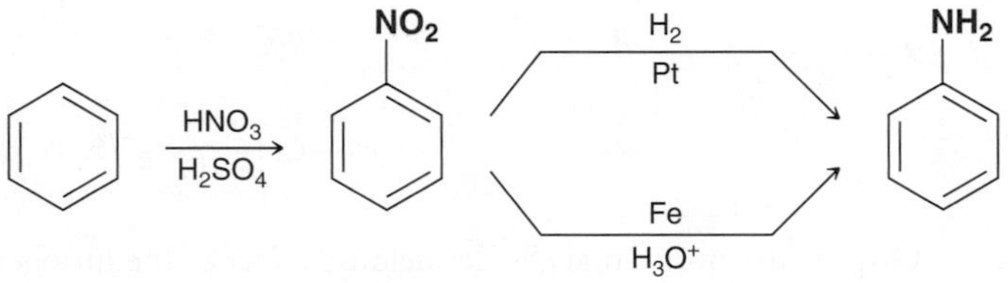

The Azide Synthesis

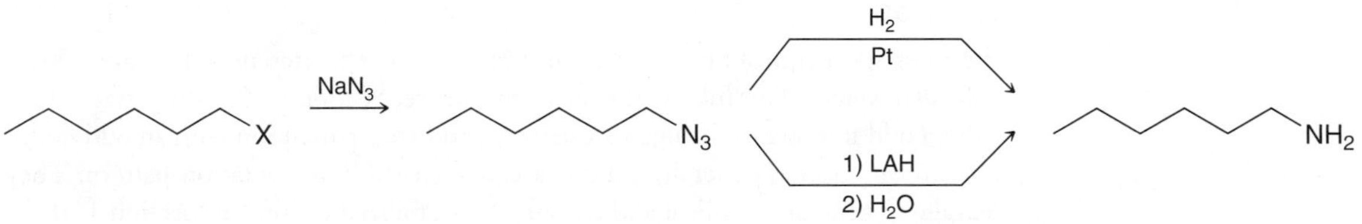

The Gabriel Synthesis

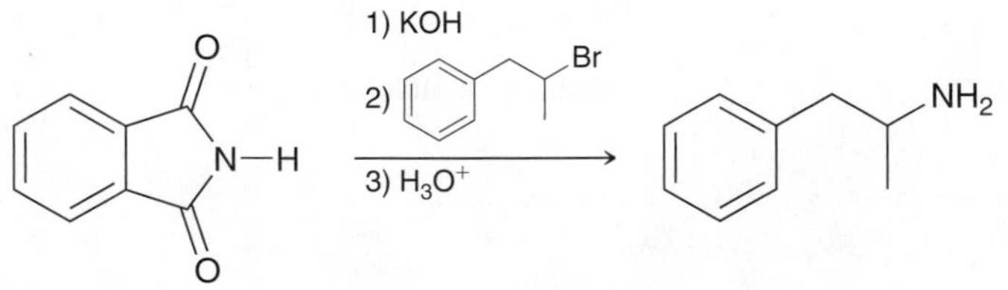

Via Reductive Amination

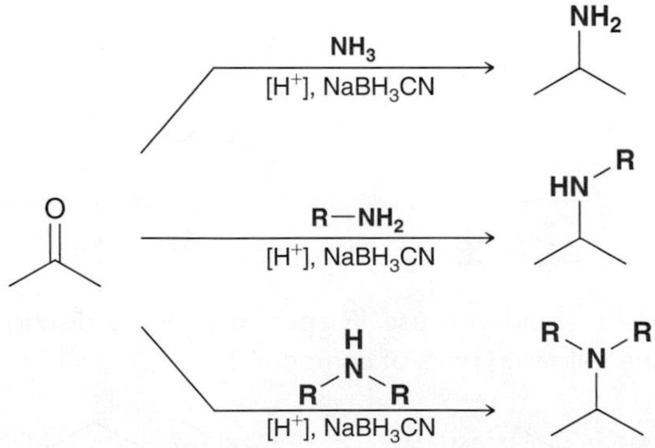

REACTIONS OF AMINES

Acylation

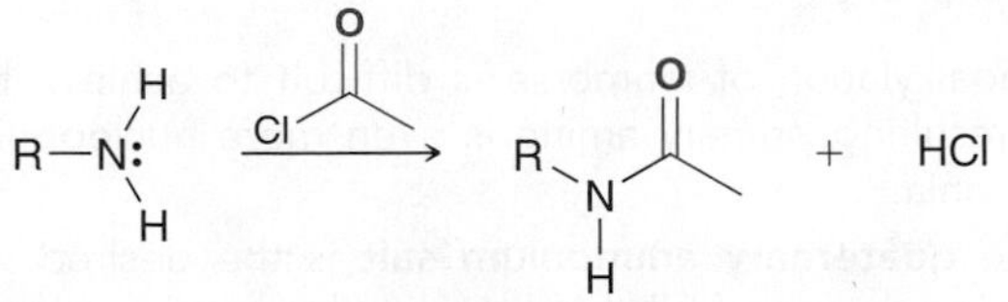

Hofmann Elimination

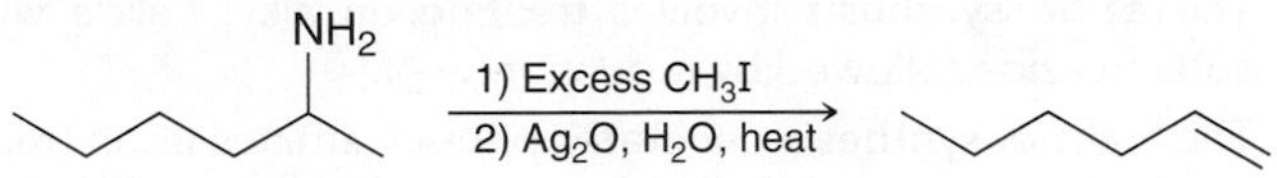

Reactions with Nitrous Acid

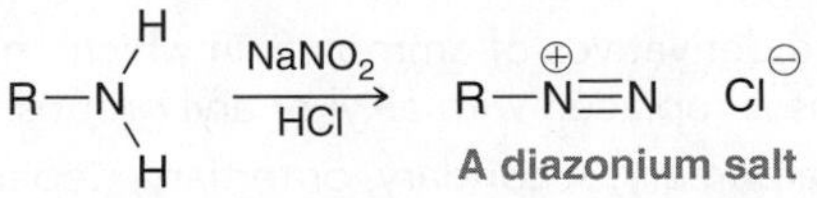

A diazonium salt

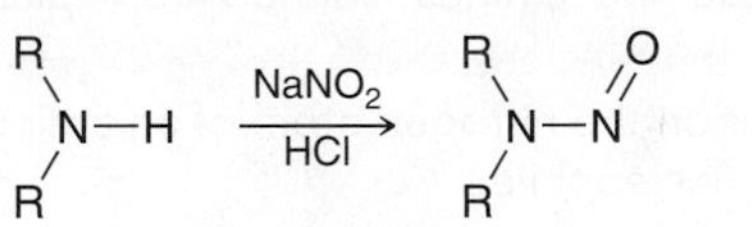

An *N*-nitrosamine

REACTIONS OF ARYLDIAZONIUM SALTS

Sandmeyer Reactions

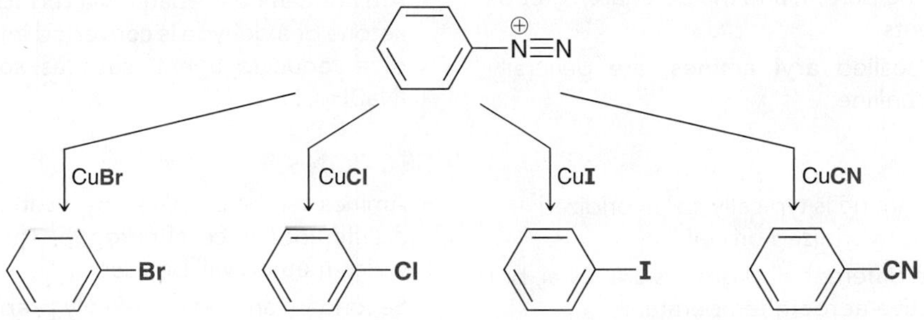

Fluorination (Schiemann Reaction)

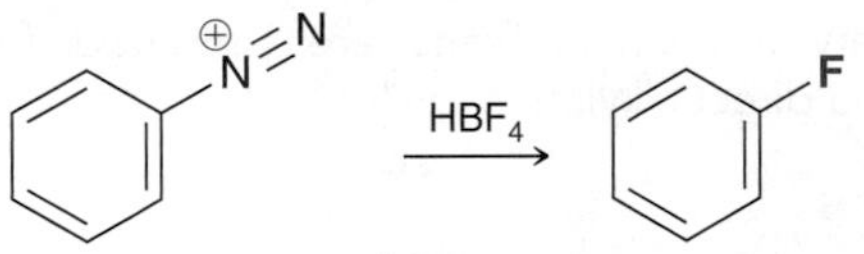

Other Reactions of Aryldiazonium Salts

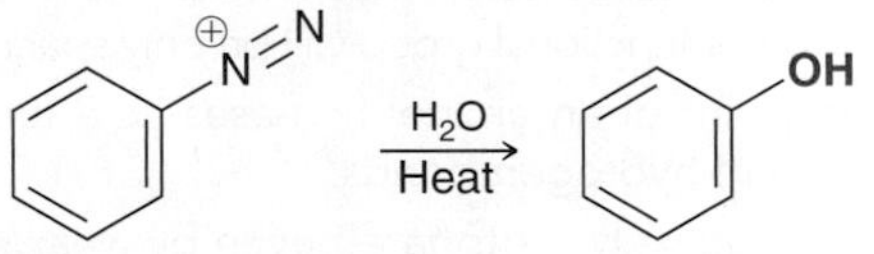

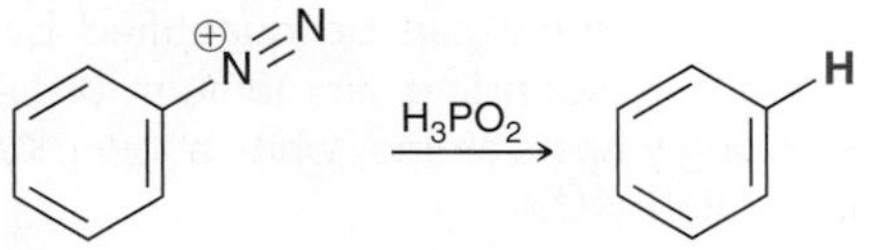

Azo Coupling

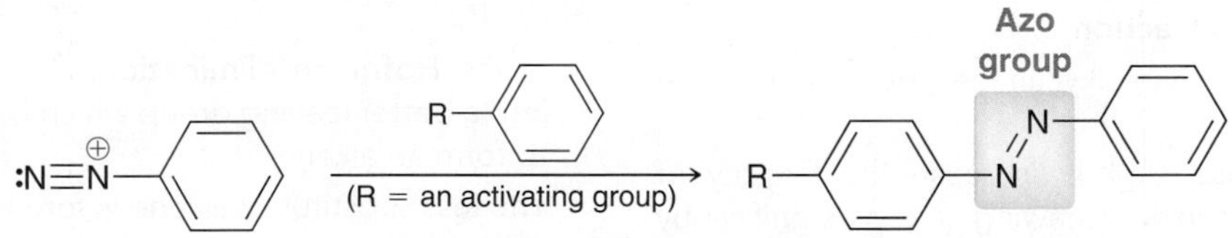

REACTIONS OF NITROGEN HETEROCYCLES

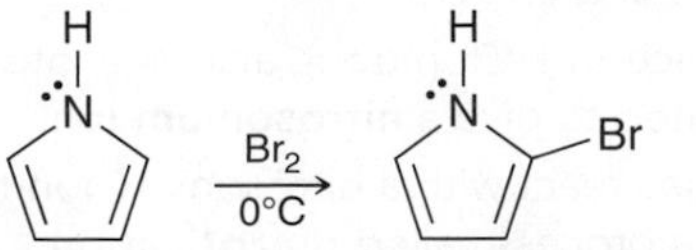

Pyrrole **2-Bromopyrrole**

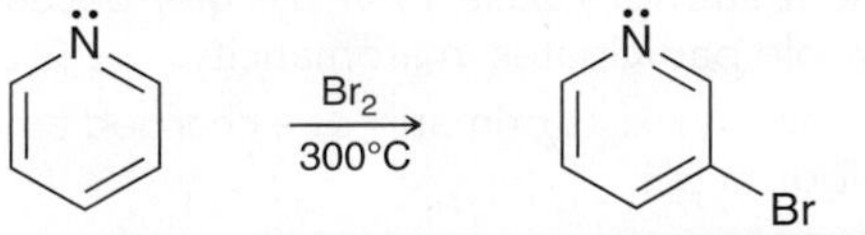

Pyridine **3-Bromopyridine**

REVIEW OF CONCEPTS AND VOCABULARY

SECTION 23.1

- **Amines** are derivatives of ammonia, in which one or more of the protons is replaced with alkyl or aryl groups.
- Amines are primary, secondary, or tertiary, depending on the number of groups attached to the nitrogen atom.
- Naturally occurring amines isolated from plants are called **alkaloids**.
- The lone pair on the nitrogen atom of an amine can function as a base or nucleophile.

SECTION 23.2

- Amines can be named as **alkyl amines** or as **alkanamines**, depending on the complexity of the alkyl groups.
- When naming an amine as an alkanamine, the most complex alkyl group is chosen as the parent, and the other alkyl groups are listed as N substituents.
- Aromatic amines, also called **aryl amines**, are generally named as derivatives of aniline.

SECTION 23.3

- The nitrogen atom of an amine is typically sp^3 hybridized; the lone pair occupies an sp^3-hybridized orbital.
- Amines containing three different alkyl groups are chiral, but they are not optically active at room temperature.
- Amines with fewer than five carbon atoms per functional group will typically be water soluble, while amines with more than five carbon atoms per functional group will be only sparingly soluble.
- The boiling point of an amine increases as a function of its capacity to form hydrogen bonds.
- Amines are effectively protonated even by weak acids.
- The basicity of an amine can be quantified by measuring the pK_a of the corresponding ammonium ion. A large pK_a indicates a strongly basic amine, while a low pK_a indicates a weakly basic amine.
- The ease with which amines are protonated can be used to remove them from mixtures of organic compounds in a process called **solvent extraction**.
- Aryl amines are less basic than alkyl amines, because the lone pair is delocalized.
- Electron-donating groups slightly increase the basicity of aryl amines, while electron-withdrawing groups significantly decrease the basicity of aryl amines.
- Amides do not function as bases, and they are very poor nucleophiles.
- Pyridine is a stronger base than pyrrole, because the lone pair in pyrrole participates in aromaticity.
- An amine moiety exists primarily as a charged ammonium ion at physiological pH.

SECTION 23.4

- Amines can be prepared from alkyl halides or from carboxylic acids.
- Aryl amines, such as aniline, can be prepared from benzene.

SECTION 23.5

- Monoalkylation of ammonia is difficult to achieve because the resulting primary amine is even more nucleophilic than ammonia.
- If the **quaternary ammonium salt** is the desired product, then an excess of alkyl halide can be used, and ammonia undergoes **exhaustive alkylation**.
- The **azide synthesis** involves treating an alkyl halide with sodium azide followed by reduction.
- The **Gabriel synthesis** generates primary amines upon treatment of potassium phthalimide with an alkyl halide, followed by hydrolysis or reaction with N_2H_4.

SECTION 23.6

- Amines can be prepared via **reductive amination**, in which a ketone or aldehyde is converted into an imine in the presence of a reducing agent, such as **sodium cyanoborohydride** ($NaBH_3CN$).

SECTION 23.7

- Amines can be prepared by a variety of methods, each using a different source of nitrogen. The starting materials dictate which method will be used.
- Secondary amines are readily prepared from primary amines via reductive amination. Tertiary amines are readily prepared from secondary amines via reductive amination.
- Quaternary ammonium salts are prepared from tertiary amines via direct alkylation.

SECTION 23.8

- Amines react with acyl halides to produce amides.
- Acylation of an amino group is a useful technique when performing electrophilic aromatic substitution reactions, because it enables transformations that are otherwise difficult to achieve.

SECTION 23.9

- In the **Hofmann elimination,** an amino group is converted into a better leaving group which is expelled in an E2 process to form an alkene.
- The less substituted alkene is formed for steric reasons.

SECTION 23.10

- **Nitrous acid** is unstable and must be prepared *in situ* from sodium nitrite and an acid.
- In the presence of HCl, nitrous acid is protonated, followed by loss of water, to give a **nitrosonium ion**.
- Primary amines react with a nitrosonium ion to yield a **diazonium salt** in a process called **diazotization**.
- Alkyldiazonium salts are unstable, but aryldiazonium salts can be isolated.
- Secondary amines react with a nitrosonium ion to produce an ***N*-nitrosamine**.

SECTION 23.11

- Aryldiazonium salts are very useful, because many different reagents will replace the diazo group.
- **Sandmeyer reactions** utilize copper salts (CuX), enabling the installation of a halogen or a cyano group.
- In the **Schiemann reaction**, an aryl diazonium salt is converted into a fluorobenzene by treatment with fluoroboric acid (HBF_4).
- An aryldiazonium salt can be treated with water to install a hydroxyl group and with hypophosphorus acid (H_3PO_2) to replace the diazo group with a hydrogen atom.
- Aryldiazonium salts react with activated aromatic rings in a process called **azo coupling** to produce colored compounds called **azo dyes**.

SECTION 23.12

- A **heterocycle** is a ring that contains atoms of more than one element.
- Pyrrole undergoes electrophilic aromatic substitution reactions, which occur primarily at C2.
- An **imidazole** ring is like pyrrole but has one extra nitrogen atom at the 3 position.
- Pyridine will undergo electrophilic aromatic substitution; however, the yields are quite low.
- **Pyrimidine** is similar in structure to pyridine but contains one extra nitrogen atom at the 3 position.

SECTION 23.13

- In their IR spectra, primary and secondary amines exhibit signals between 3350 and 3500 cm^{-1}. Primary amines give two peaks (symmetric and asymmetric stretching), while secondary amines give only one peak.
- When treated with HCl, tertiary amines can be readily detected with IR spectroscopy.
- In the ^1H NMR spectrum of an amine, any proton attached directly to the nitrogen atom will typically appear as a broad signal between 0.5 and 5.0 ppm.
- The protons connected to the α position typically appear between 2 and 3 ppm.
- In the ^{13}C NMR spectrum of an amine, the α carbon atoms typically appear between 30 and 50 ppm.

KEY TERMINOLOGY

alkaloids 1090
alkanamines 1092
alkyl amines 1092
amines 1090
aryl amines 1092
azide synthesis 1103
azo coupling 1121
azo dyes 1121
diazonium salts 1117
diazotization 1117
exhaustive alkylation 1103
Gabriel synthesis 1104
heterocycle 1123
Hofmann elimination 1113
imidazole 1123
***N*-nitrosamine** 1116
nitrosonium ion 1116
nitrous acid 1115
pyrimidine 1126
quaternary ammonium salt 1103
reductive amination 1106
Sandmeyer reactions 1118
Schiemann reaction 1119
sodium cyanoborohydride 1106
solvent extraction 1098

SKILLBUILDER REVIEW

23.1 NAMING AN AMINE

STEP 1 Identify all of the alkyl groups connected to the nitrogen atom.

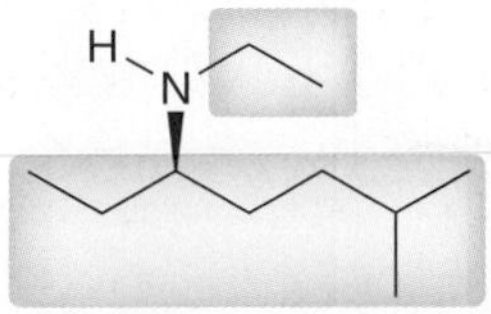

STEP 2 Determine whether to name the compound as an alkyl amine or as an alkanamine:

Alkyl amine = if all alkyl groups are simple

Alkanamine = if one of the alkyl groups is complex

STEP 3 If naming as an alkanamine, choose the complex alkyl group as the parent, and list the other alkyl groups as substituents (using the letter "N" as a locant).

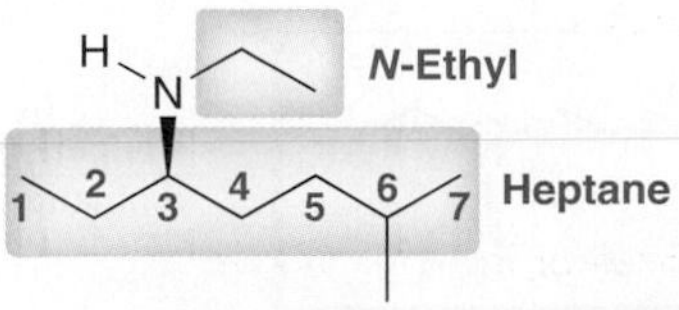

STEP 4 Assign locants to each substituent, assemble the name, and assign the configuration of any chirality centers.

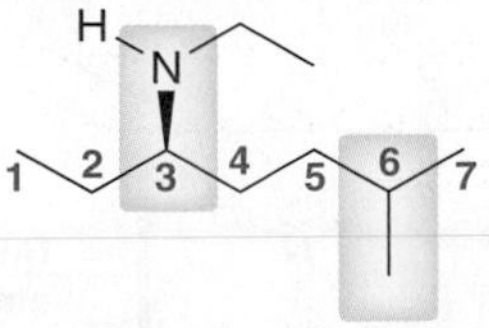

(*R*)-*N*-Ethyl-6-methyl-3-heptanamine

Try Problems 23.1–23.3, 23.42, 23.46, 23.47

23.2 PREPARING A PRIMARY AMINE VIA THE GABRIEL REACTION

STEP 1 Identify an alkyl halide that can serve as a precursor for preparing the desired amine. Simply replace the amino group with a halide.

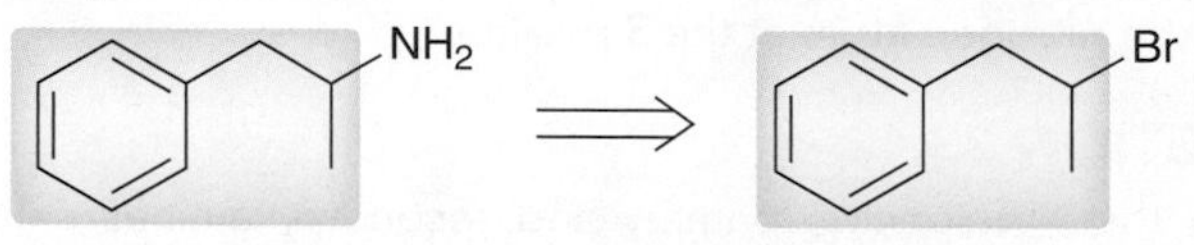

STEP 2 Draw the reagents for all three steps, starting with phthalimide as the starting material.

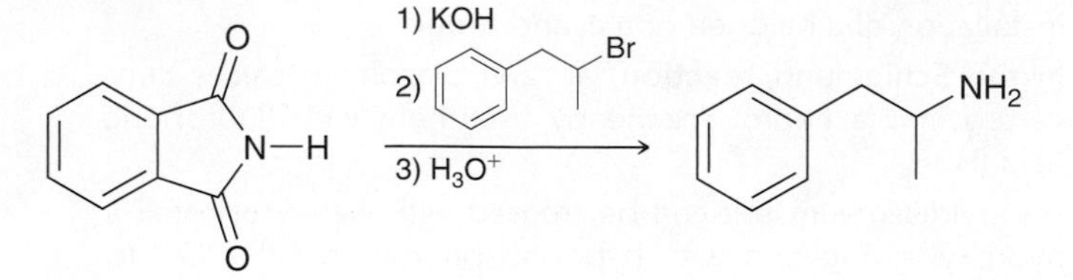

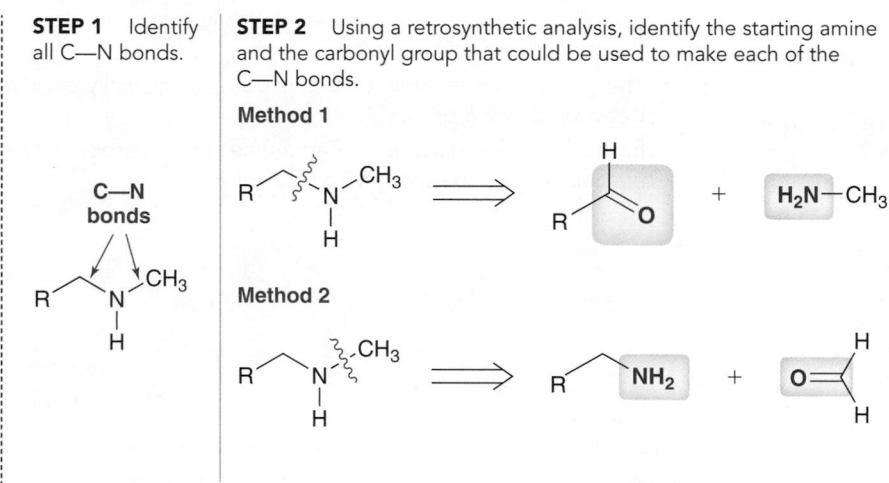

Try Problems 23.12, 23.13, 23.51a, 23.60, 23.61, 23.62b

23.3 PREPARING AN AMINE VIA A REDUCTIVE AMINATION

STEP 1 Identify all C—N bonds.

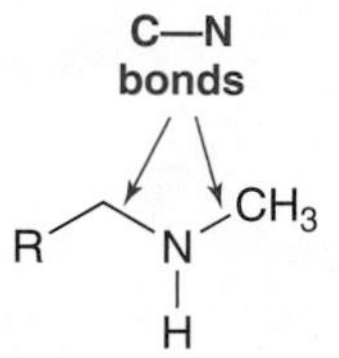

STEP 2 Using a retrosynthetic analysis, identify the starting amine and the carbonyl group that could be used to make each of the C—N bonds.

Method 1

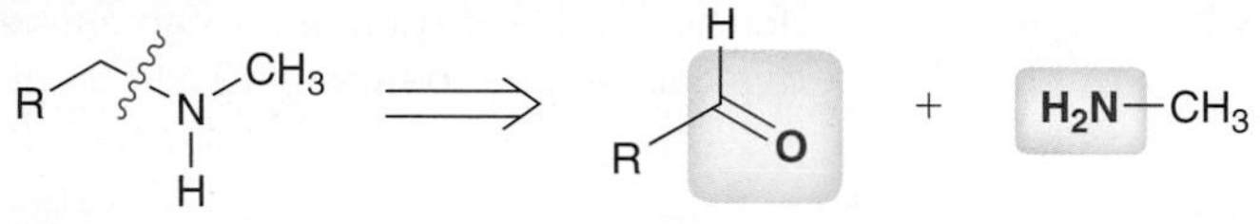

Method 2

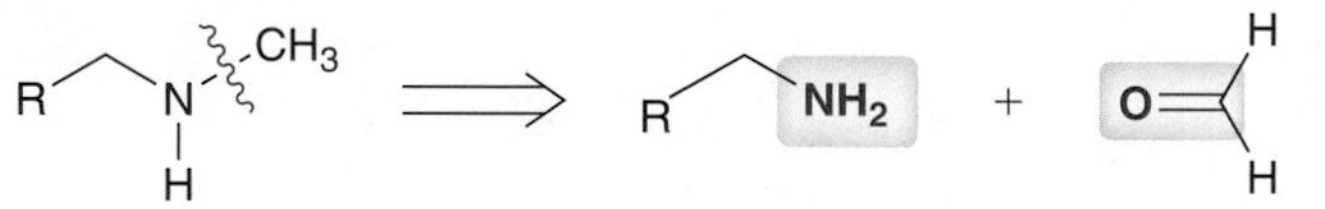

STEP 3 Show all reagents necessary for each synthetic route.

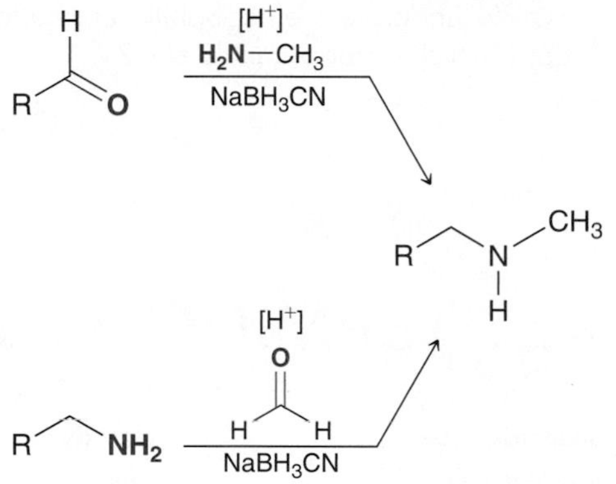

Try Problems 23.14–23.17, 23.65, 23.66

23.4 SYNTHESIS STRATEGIES

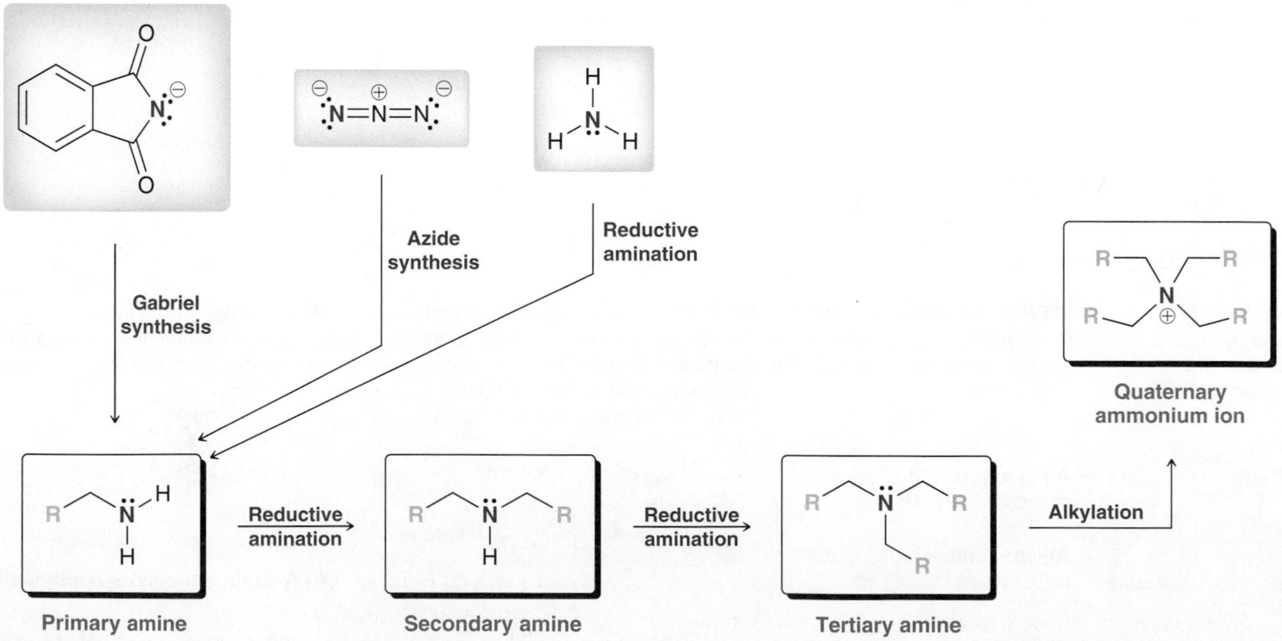

Try Problems 23.18–23.21, 23.50, 23.51, 23.86

23.5 PREDICTING THE PRODUCT OF A HOFMANN ELIMINATION

STEP 1 Identify all α and β positions.

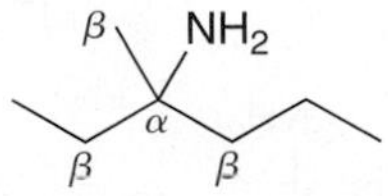

STEP 2 Consider forming a C=C bond between each pair of α and β positions.

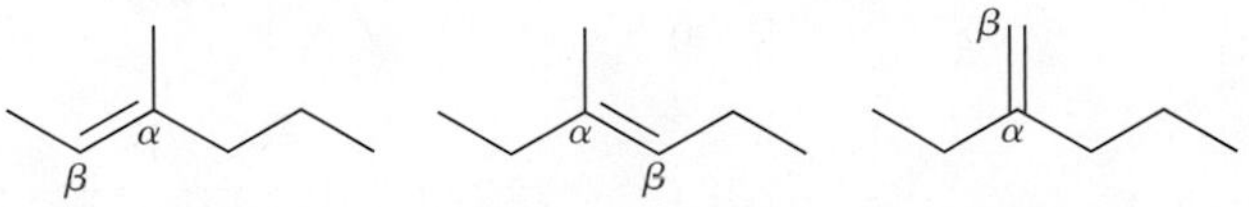

STEP 3 Compare all possible alkenes and identify which is the least substituted.

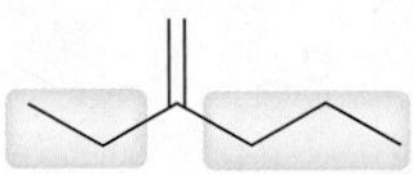

Disubstituted

Try Problems 23.25–23.28, 23.57, 23.69, 23.78

23.6 DETERMINING THE REACTANTS FOR PREPARING AN AZO DYE

STEP 1 Analyze all substituents and determine which ring is more activated.

Moderately activated

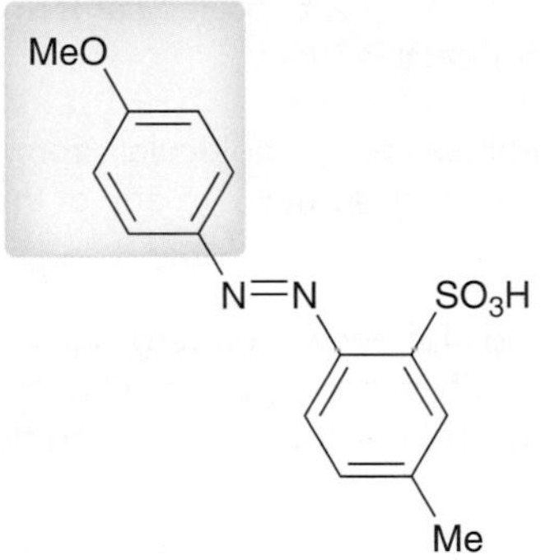

STEP 2 Using a retrosynthetic analysis, identify the starting nucleophile and electrophile.

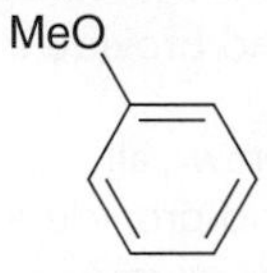

Nucleophile

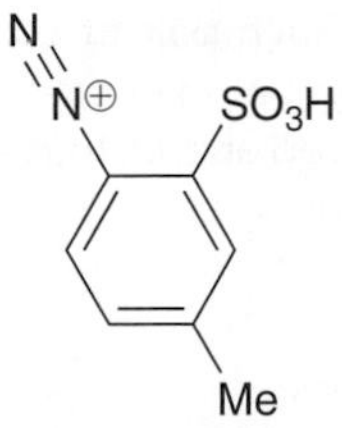

Electrophile

STEP 3 Identify the starting aniline that is necessary in order to make the diazonium salt.

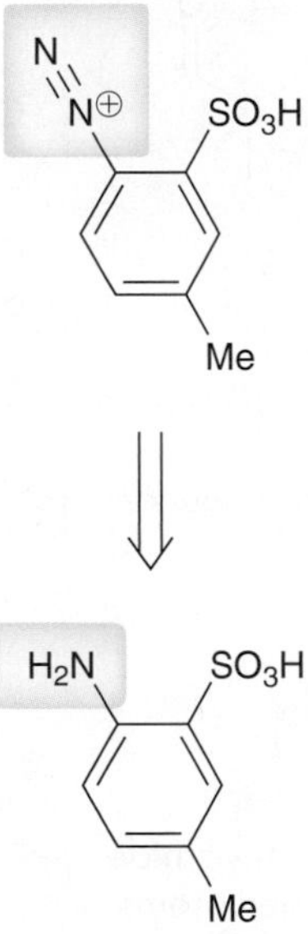

STEP 4 Show all reagents.

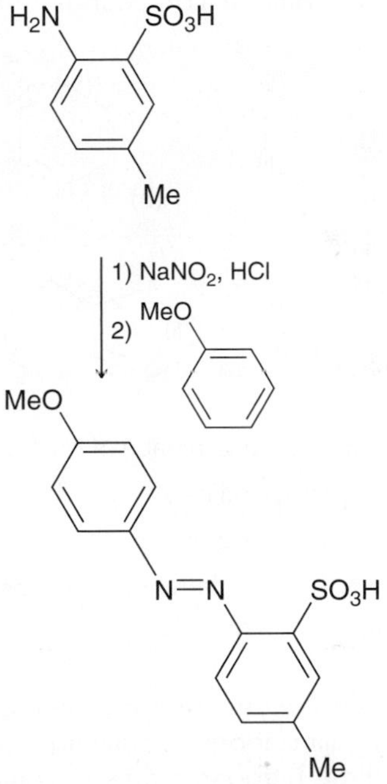

Try Problems 23.31–23.33, 23.68, 23.73d

PRACTICE PROBLEMS

Note: Most of the Problems are available within *WileyPLUS*, an online teaching and learning solution.

23.38 Spermine is a naturally occurring compound that contributes to the characteristic odor of semen. Classify each nitrogen atom in spermine as primary, secondary, or tertiary.

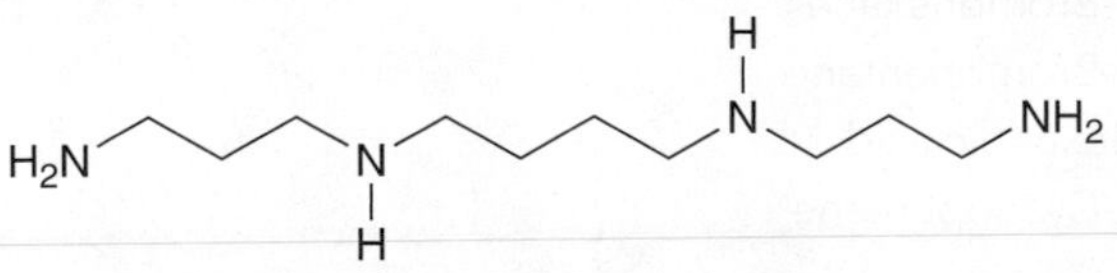

Spermine

23.39 Clomipramine is marketed under the trade name Anafranil and is used in the treatment of obsessive compulsive disorder.

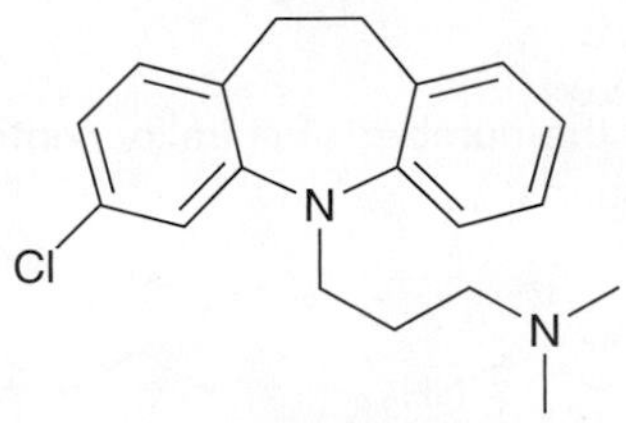

Clomipramine

(a) Identify which nitrogen atom in clomipramine is more basic, and justify your choice.

(b) Draw the form of clomipramine that is expected to predominate at physiological pH.

23.40 Cinchocaine is a long-acting local anesthetic used in spinal anesthesia. Identify the most basic nitrogen atom in cinchocaine.

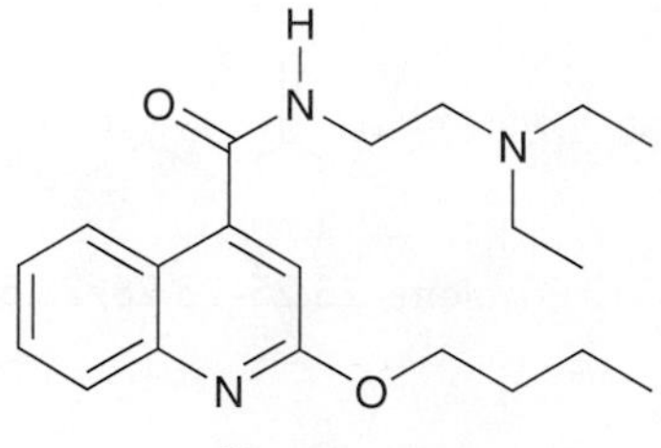

Cinchocaine

23.41 For each pair of compounds, identify the stronger base.

(a) 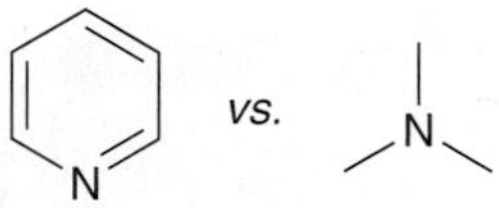

(b) 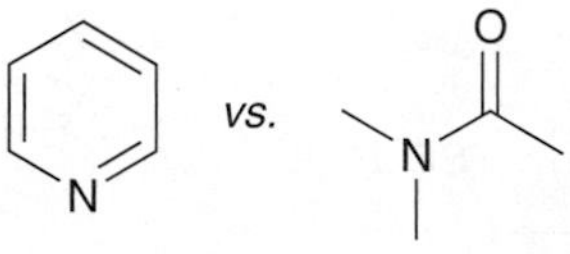

(c) 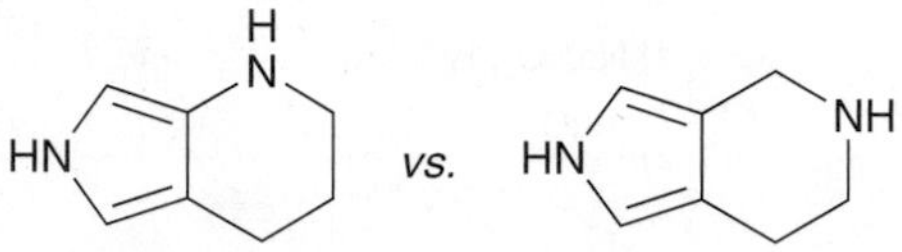

23.42 Draw the structure of each of the following compounds:

(a) *N*-Ethyl-*N*-isopropylaniline

(b) *N,N*-Dimethylcyclopropylamine

(c) (2*R*,3*S*)-3-(*N,N*-Dimethylamino)-2-pentanamine

(d) Benzylamine

23.43 Consider the structure of lysergic acid diethylamide (LSD), a potent hallucinogen containing three nitrogen atoms. One of these three nitrogen atoms is significantly more basic than the other two. Identify the most basic nitrogen atom in LSD, and explain your choice.

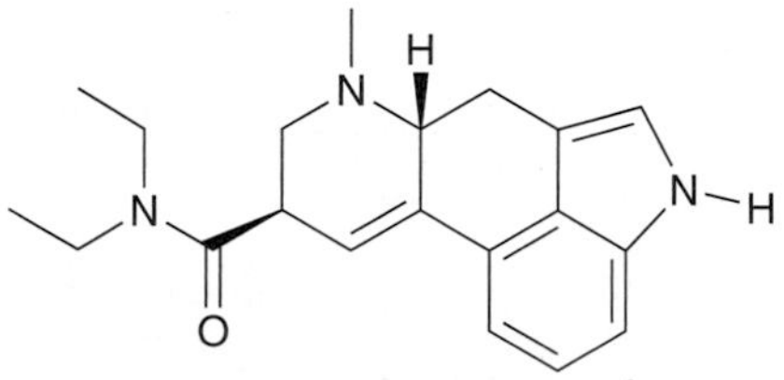

LSD

23.44 Identify the number of chirality centers in each of the following compounds:

(a) 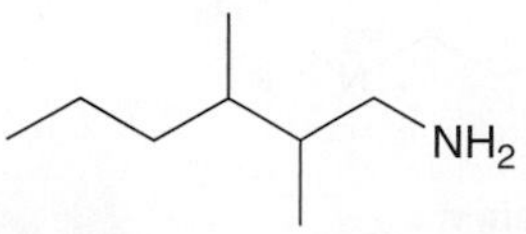

(b) 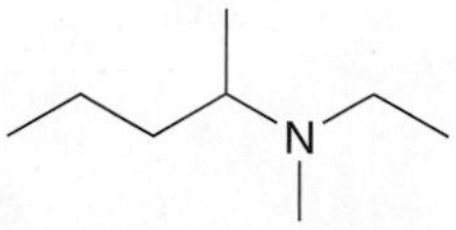

(c) 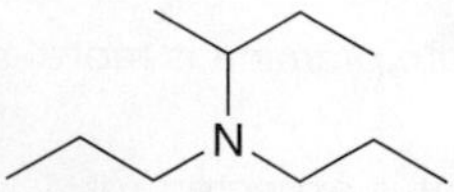

23.45 Assign a name for each of the following compounds:

(a) 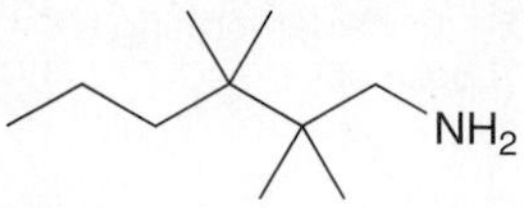

(b) 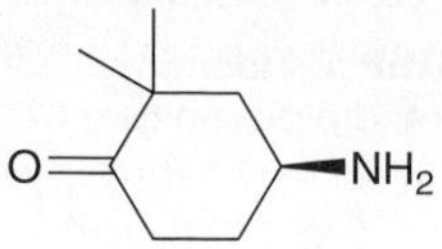

(c) 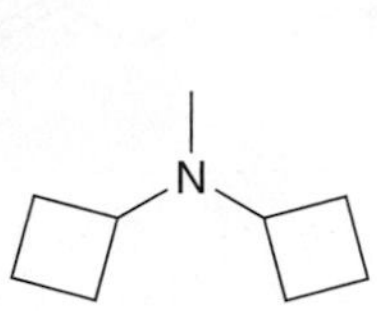

(d) 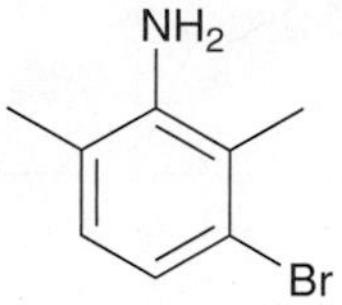

(e) 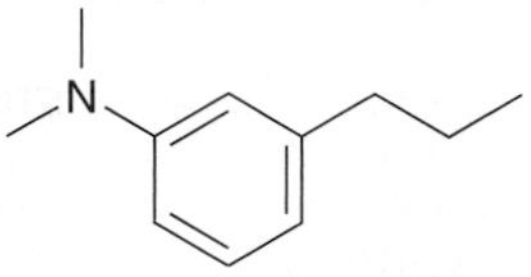

(f) 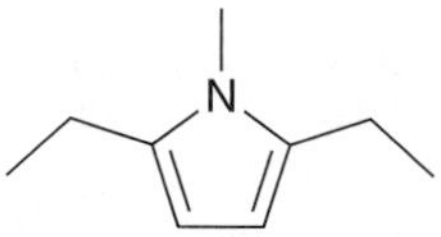

23.46 Draw all constitutional isomers with molecular formula $C_4H_{11}N$, and provide a name for each isomer.

23.47 Draw all tertiary amines with molecular formula $C_5H_{13}N$, and provide a name for each isomer. Are any of these compounds chiral?

23.48 Each pair of compounds below will undergo an acid-base reaction. In each case, identify the acid, identify the base, draw curved arrows that show the transfer of a proton, and draw the products.

(a) 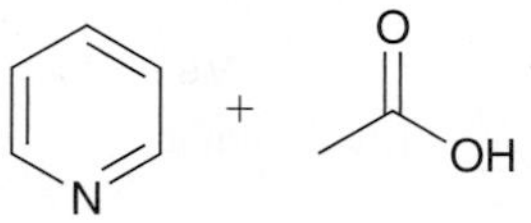

(b) 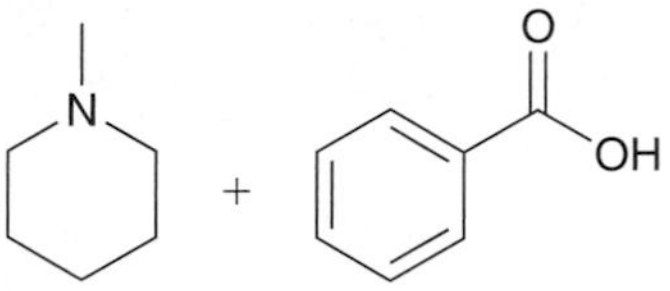

23.49 Draw the structure of the major product obtained when aniline is treated with each of the following reagents:

(a) Excess Br_2

(b) $PhCH_2COCl$, py

(c) Excess methyl iodide

(d) $NaNO_2$ and HCl followed by H_3PO_2

(e) $NaNO_2$ and HCl followed by CuCN

23.50 Identify how you would make each of the following compounds from 1-hexanol:

(a) Hexylamine (b) Heptylamine (c) Pentylamine

23.51 Identify how you would make hexylamine from each of the following compounds:

(a) 1-Bromohexane

(b) 1-Bromopentane

(c) Hexanoic acid

(d) 1-Cyanopentane

23.52 Tertiary amines with three different alkyl groups are chiral but cannot be resolved because pyramidal inversion causes racemization at room temperature. Nevertheless, chiral aziridines can be resolved and stored at room temperature.

Aziridine is a three-membered heterocycle containing a nitrogen atom. The following is an example of a chiral aziridine. In this compound, the nitrogen atom is a chirality center. Suggest a reason why chiral aziridines do not undergo racemization at room temperature.

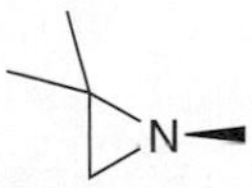

23.53 Lidocaine is one of the most widely used local anesthetics. Draw the form of lidocaine that is expected to predominate at physiological pH.

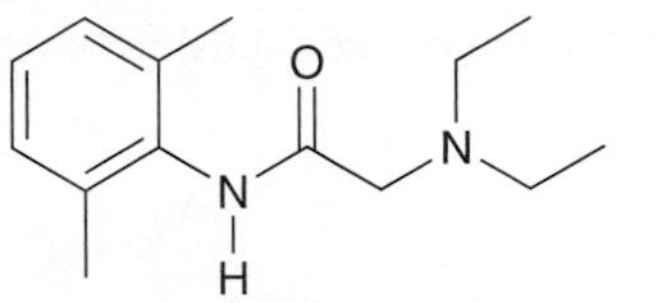

Lidocaine

23.54 Propose a mechanism for the following transformation:

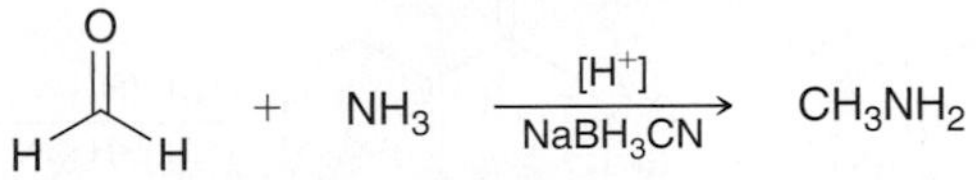

23.55 When aniline is treated with fuming sulfuric acid, an electrophilic aromatic substitution reaction takes place at the *meta* position instead of the *para* position, despite the fact that the amino group is an *ortho-para* director. Explain this curious result.

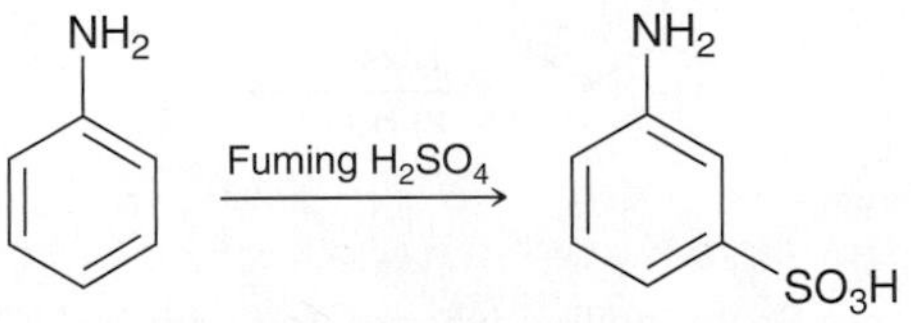

23.56 *para*-Nitroaniline is an order of magnitude less basic than *meta*-nitroaniline.

(a) Explain the observed difference in basicity.

(b) Would you expect the basicity of *ortho*-nitroaniline to be closer in value to *meta*-nitroaniline or to *para*-nitroaniline?

23.57 Methadone is a powerful analgesic that is used to suppress withdrawal symptoms in the rehabilitation of heroin addicts. Identify the major product that is obtained when methadone is subjected to a Hofmann elimination.

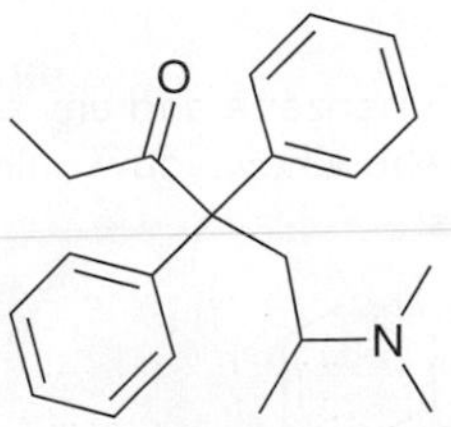

Methadone

23.58 In general, nitrogen atoms are more basic than oxygen atoms. However, when an amide is treated with a strong acid, such as sulfuric acid, it is the oxygen atom of the amide that is protonated, rather than the nitrogen atom. Explain this observation.

23.59 Propose a synthesis for each of the following transformations:

(a)

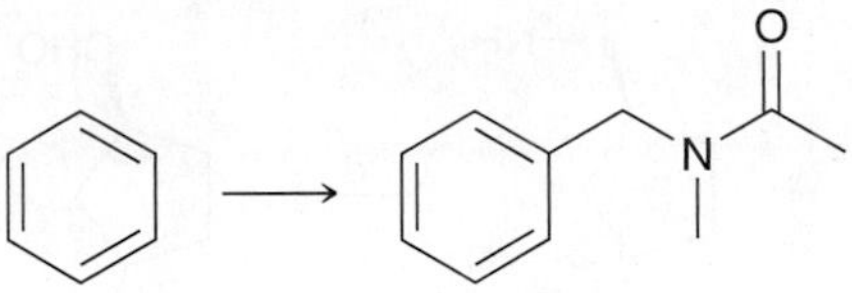

(b)

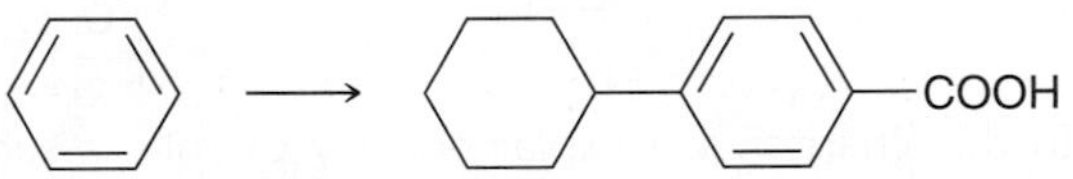

23.60 Draw a mechanism for the last step of the Gabriel synthesis, performed under basic conditions.

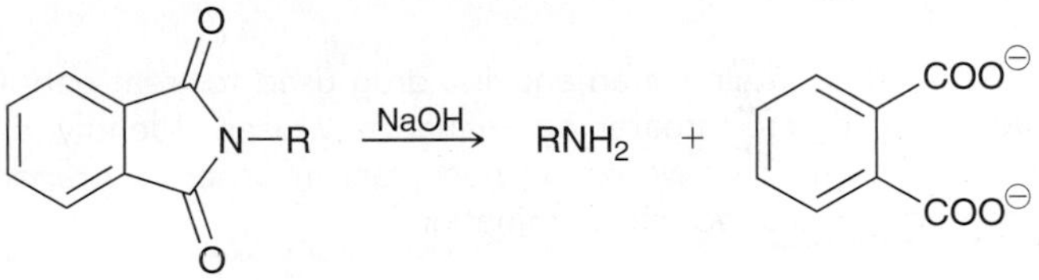

23.61 One variation of the Gabriel synthesis employs hydrazine to free the amine in the final step of the synthesis. Draw the by-product obtained in this process.

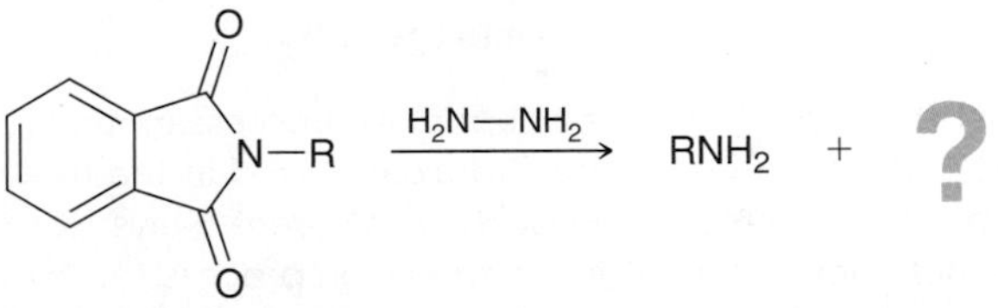

23.62 Predict the major product for each of the following reactions,

(a)

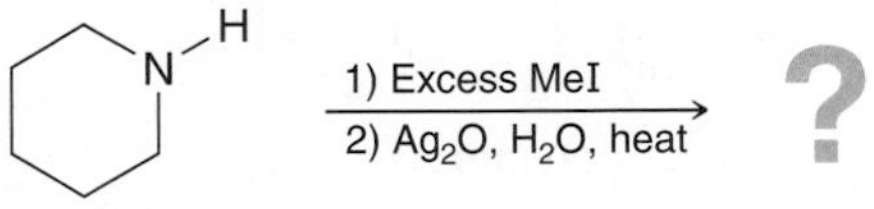

(b)

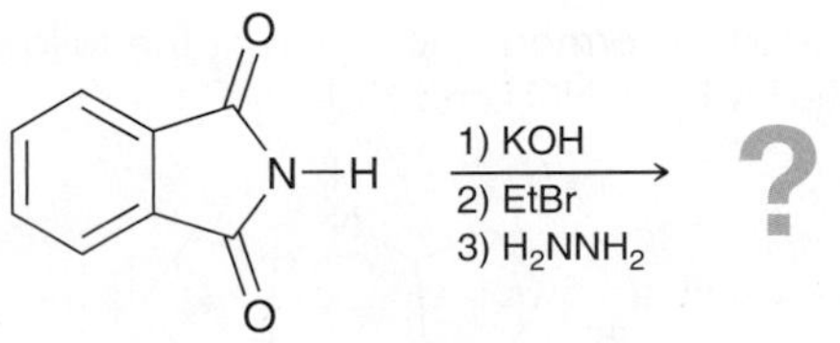

(c)

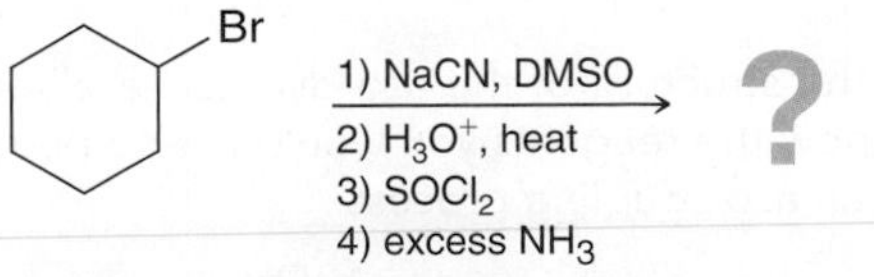

(d)

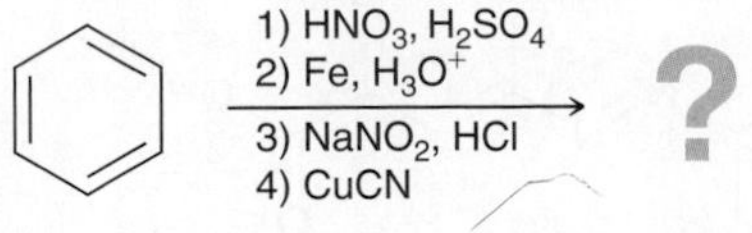

23.63 Fill in the missing reagents:

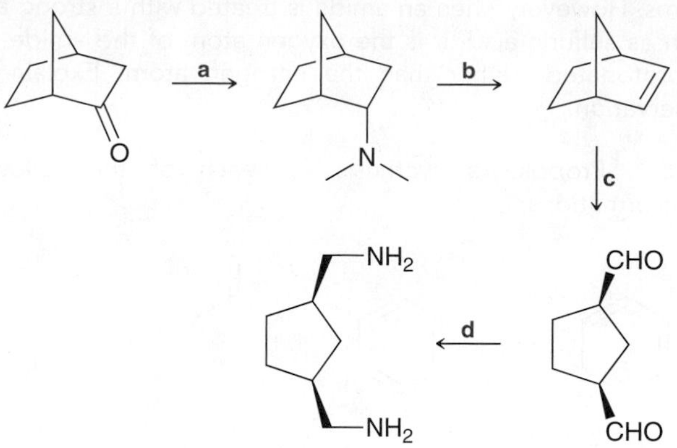

23.64 In this chapter, we explained why pyrrole is such a weak base, but we did not discuss the acidity of pyrrole. In fact, pyrrole is 20 orders of magnitude more acidic than most simple amines. Draw the conjugate base of pyrrole and explain its relatively high acidity.

23.65 Rimantadine is an antiviral drug used to treat people infected with life-threatening influenza viruses. Identify the starting ketone that would be necessary in order to prepare rimantadine via a reductive amination.

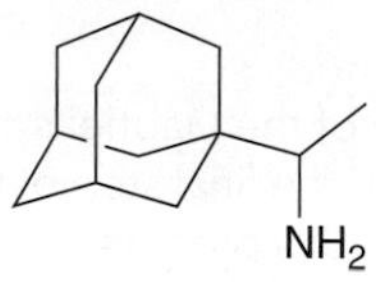

Rimantadine

23.66 Benzphetamine is an appetite suppressant that is marketed under the trade name Didrex and used in the treatment of obesity. Identify at least two different ways to make benzphetamine via a reductive amination process.

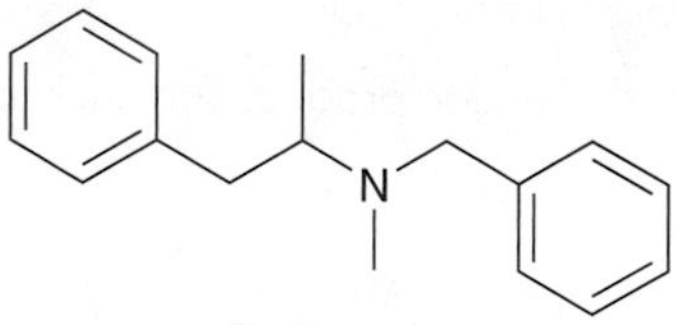

Benzphetamine

23.67 Draw the product formed when each of the following compounds is treated with $NaNO_2$ and HCl:

(a)
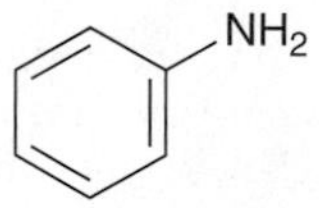

(b)
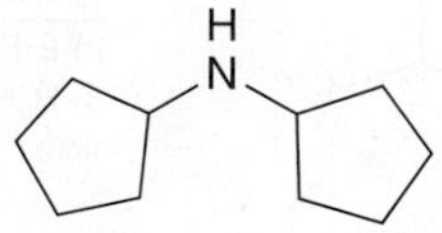

23.68 Consider the structure of the azo dye called alizarine yellow R (below). Show the reagents you would use to prepare this compound via an azo coupling process.

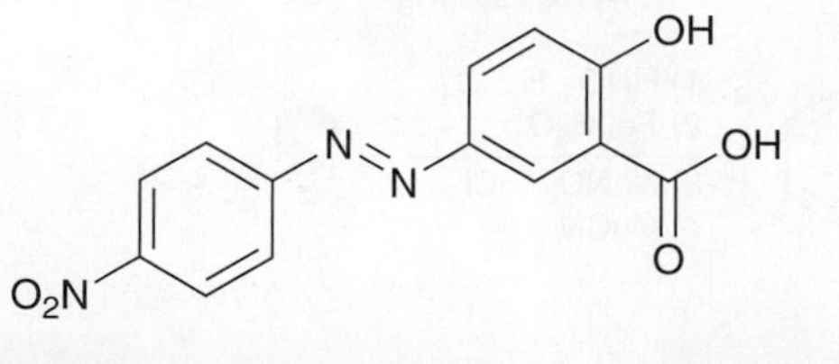

23.69 Draw the major product(s) that are expected when each of the following amines is treated with excess methyl iodide and then heated in the presence of aqueous silver oxide.

(a)
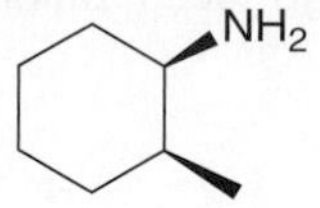

(b)
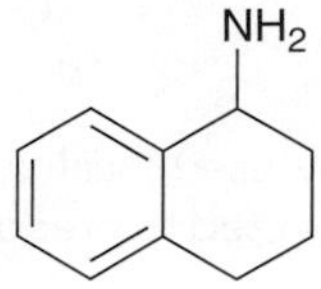

23.70 Predict the major product(s) for each of the following reactions:

(a)
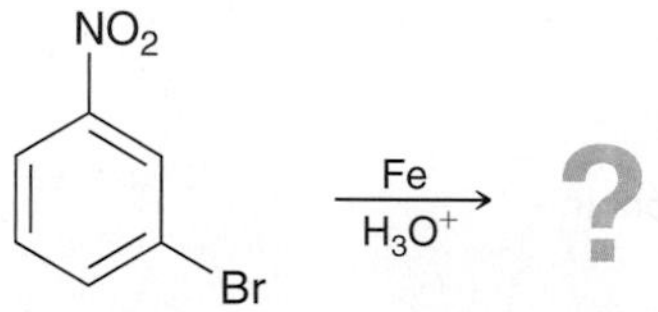

(b)
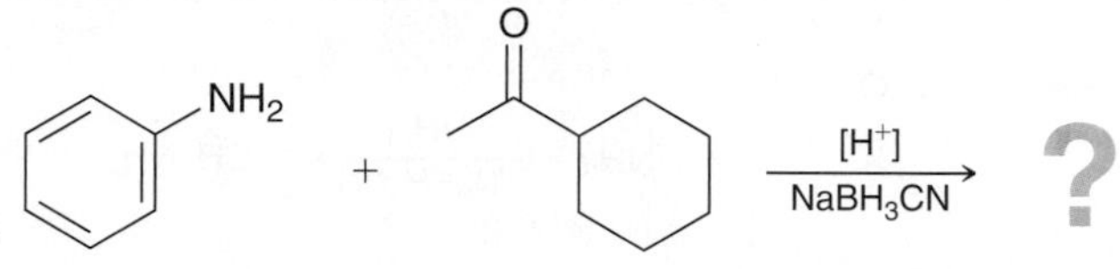

(c)
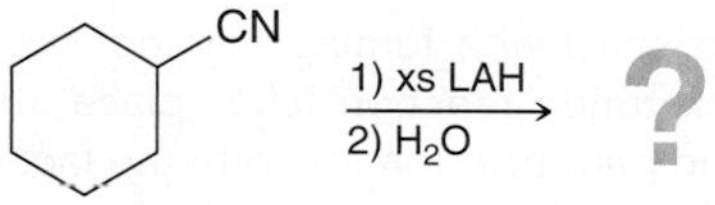

(d)
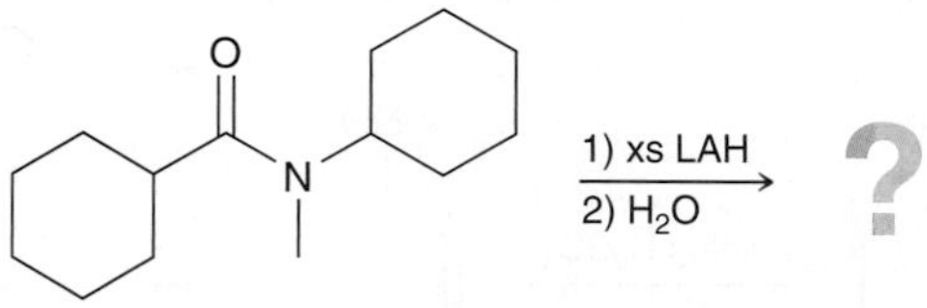

23.71 *meta*-Bromoaniline was treated with $NaNO_2$ and HCl to yield a diazonium salt. Draw the product obtained when that diazonium salt is treated with each of the following reagents:

(a) H_2O (b) HBF_4 (c) CuCN (d) H_3PO_2 (e) CuBr

23.72 Draw the expected product of the following reductive amination:

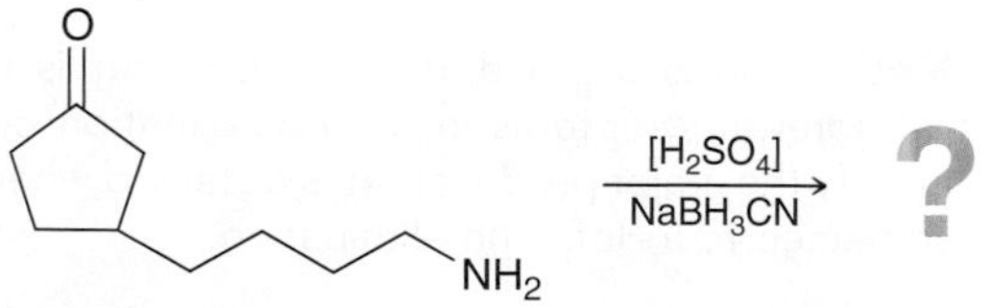

23.73 Starting with benzene and any reagents with three or fewer carbon atoms, show how you would prepare each of the following compounds:

(a)
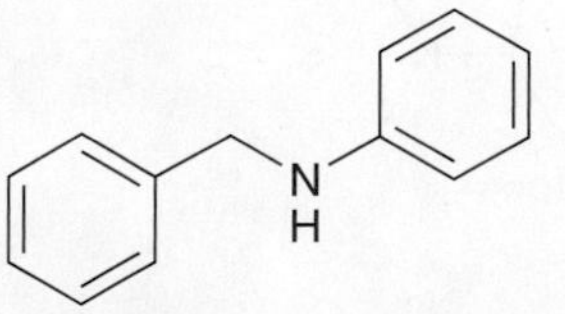

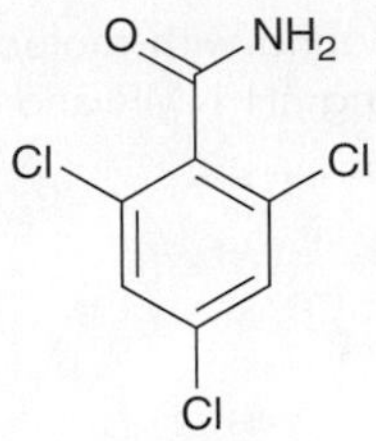

(b)

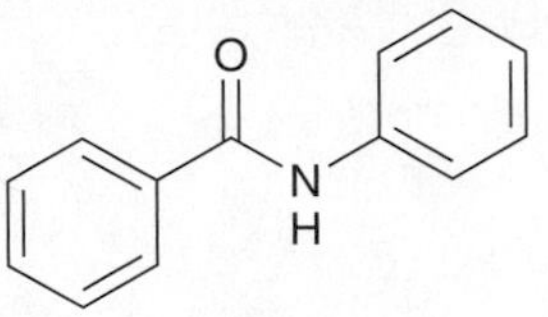

(c)

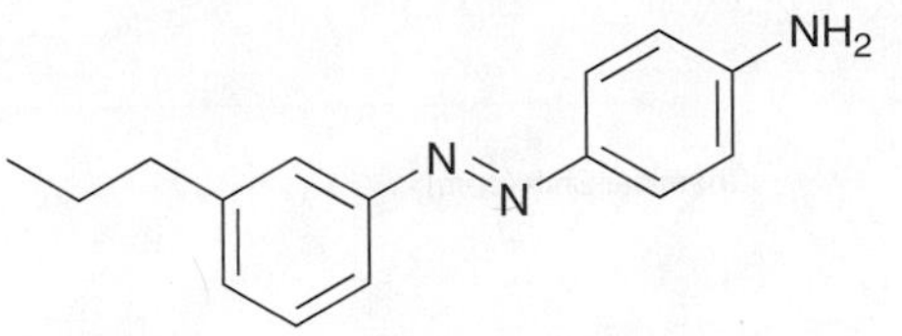

(d)

23.74 Draw the structures of all isomeric amines with molecular formula $C_6H_{15}N$ that are not expected to produce any signal above 3000 cm^{-1} in their IR spectra.

23.75 When the following compound is treated with excess methyl iodide, a quaternary ammonium salt is obtained that bears only one positive charge. Draw the structure of the quaternary ammonium salt.

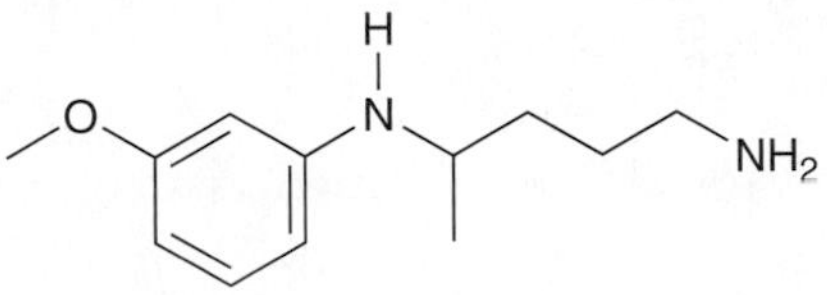

23.76 Draw a mechanism for the following transformation:

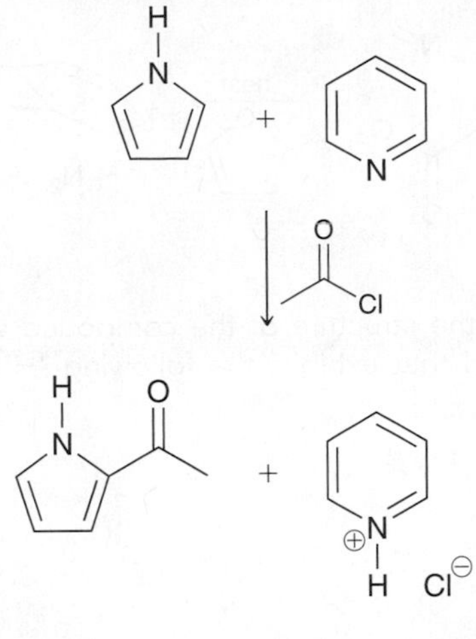

INTEGRATED PROBLEMS

23.77 A compound with molecular formula $C_5H_{13}N$ exhibits three signals in its proton NMR spectrum and no signals above 3000 cm^{-1} in its IR spectrum. Draw two possible structures for this compound.

23.78 Coniine has molecular formula $C_8H_{17}N$ and was present in the hemlock extract used to execute the Greek philosopher Socrates. Subjecting coniine to a Hofmann elimination produces (*S*)-*N*,*N*-dimethyloct-7-en-4-amine. Coniine exhibits one peak above 3000 cm^{-1} in its IR spectrum. Draw the structure of coniine.

23.79 Piperazine is an antihelminthic agent (a drug used in the treatment of intestinal worms) that has molecular formula $C_4H_{10}N_2$. The proton NMR spectrum of piperazine exhibits two signals. When dissolved in D_2O, one of these signals vanishes over time. Propose a structure for piperazine.

23.80 Primary or secondary amines will attack epoxides in a ring-opening process:

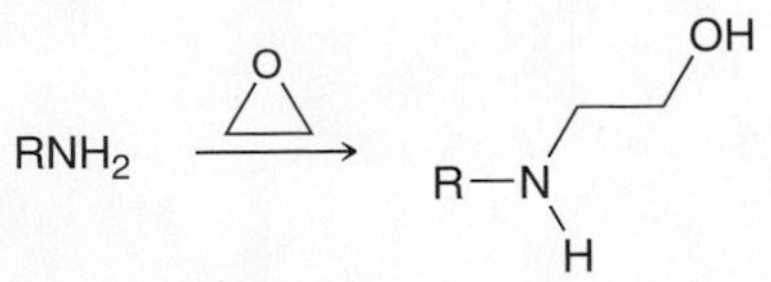

For substituted epoxides, nucleophilic attack generally takes place at the less sterically hindered side of the epoxide. Using this type of reaction, show how you might prepare the following compound from benzene, ammonia, and any other reagents of your choice.

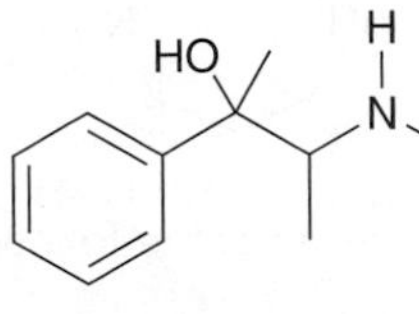

23.81 Phenacetin was widely used as an analgesic before it was removed from the market in 1983 on suspicion of being a carcinogen. It was widely replaced with acetaminophen (Tylenol), which is very similar in structure but is not carcinogenic. Starting with benzene and using any other reagents of your choice, outline a synthesis of phenacetin.

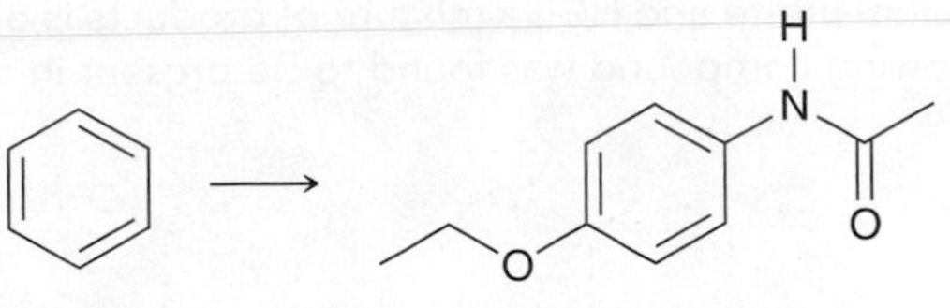

Phenacetin

23.82 Propose a mechanism for the following process:

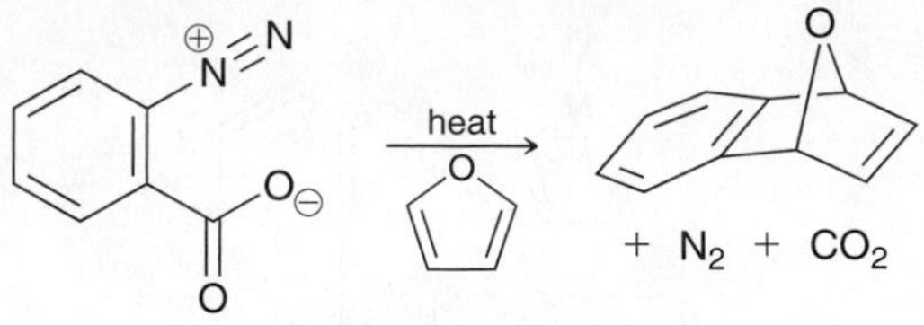

23.83 Draw the structure of the compound with molecular formula $C_6H_{15}N$ that exhibits the following ^1H NMR and ^{13}C NMR spectra:

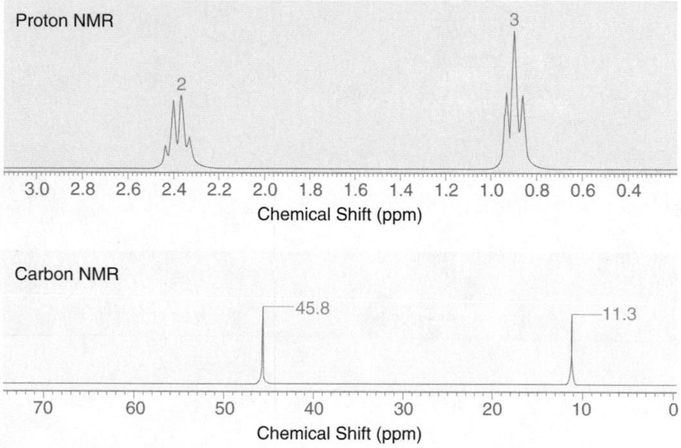

23.84 Draw the structure of the compound with molecular formula $C_8H_{11}N$ that exhibits the following ^1H NMR and ^{13}C NMR spectra:

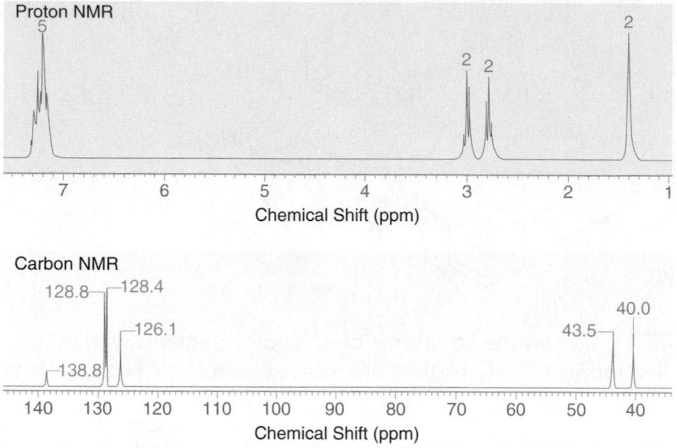

CHALLENGE PROBLEMS

23.85 Using benzene as your only source of carbon atoms and ammonia as your only source of nitrogen atoms, propose a synthesis for the following compound:

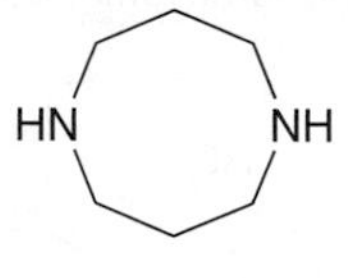

23.86 Propose a synthesis for the following transformation:

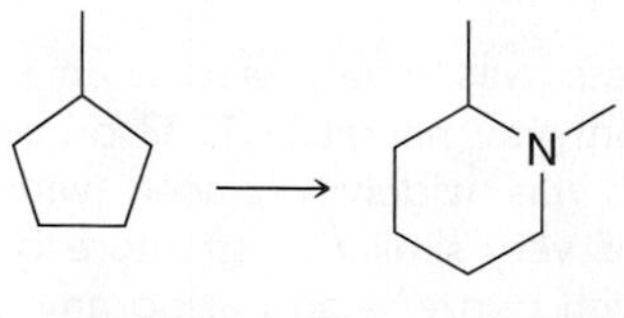

23.87 When 3-methyl-3-phenyl-1-butanamine is treated with sodium nitrite and HCl, a mixture of products is obtained. The following compound was found to be present in the reaction mixture. Account for its formation with a complete mechanism (make sure to show the mechanism of formation for a nitrosonium ion).

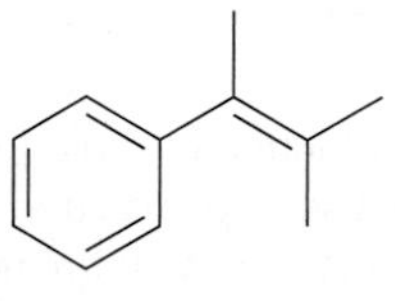

23.88 Guanidine is a neutral compound but is an extremely powerful base. In fact, it is almost as strong a base as a hydroxide ion. Identify which nitrogen atom in guanidine is so basic, and explain why guanidine is a much stronger base than most other amines.

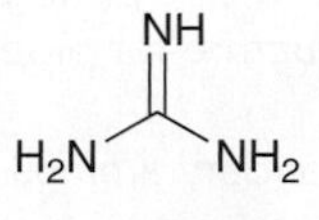

Guanidine

24

Carbohydrates

DID YOU EVER WONDER . . .
what Neosporin is and how it works?

Neosporin is an antibiotic cream that contains three active ingredients. One of them, called neomycin, is a carbohydrate derivative. In this chapter, we will investigate carbohydrates and the various roles they play in nature. After an introduction to the structure and reactivity of carbohydrates, we will return to analyze the structure of neomycin and related antibiotics in more detail.

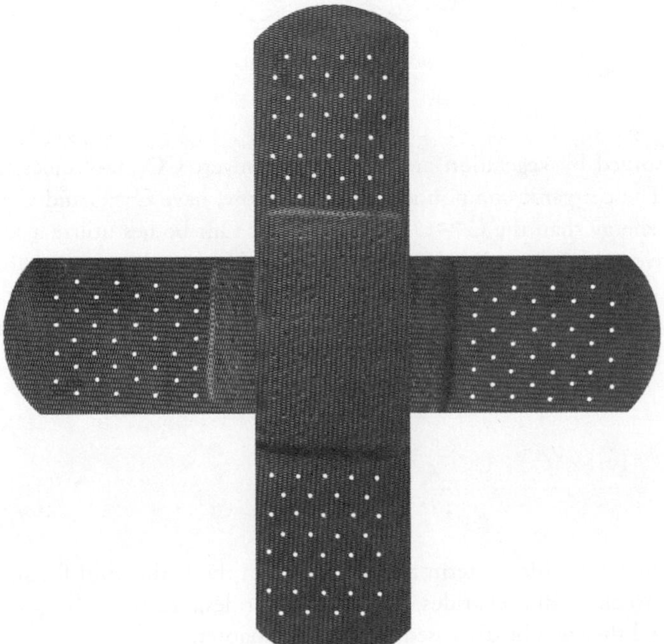

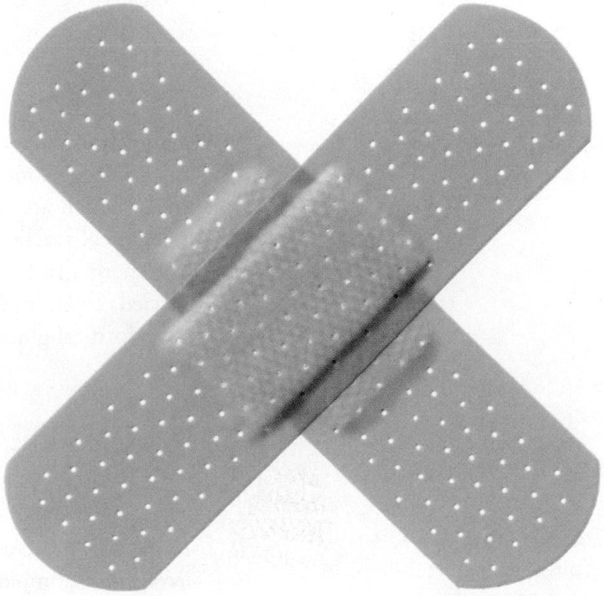

DO YOU REMEMBER?

Before you go on, be sure you understand the following topics.
If necessary, review the suggested sections to prepare for this chapter:

- Drawing Chair Conformations (Section 4.11)
- Fischer Projections (Section 5.7)
- Reduction of Aldehydes and Ketones (Section 20.9)
- Haworth Projections (Section 4.14)
- Hemiacetals and Acetals (Section 20.5)

24.1 Introduction to Carbohydrates

Carbohydrates, commonly referred to as sugars, are abundant in nature. They represent a significant portion of the foods that we eat, providing us with the energy necessary to drive most of the biochemical processes in our bodies. Carbohydrates are the building blocks used to provide structural rigidity for living organisms, including the wood found in trees and the shells of lobsters. Even our DNA is assembled from derivatives of carbohydrates, as we will see at the end of this chapter.

The earliest carbohydrates to be isolated and purified were originally considered to be hydrates of carbon. For example, glucose was known to have molecular formula $C_6H_{12}O_6$, which could be rearranged as $C_6(H_2O)_6$ to indicate six carbon atoms and six water molecules. As the structure of glucose was elucidated, this view was discarded, but the term *carbohydrate* persisted. Carbohydrates are now understood to be polyhydroxy aldehydes or ketones. For example, consider the structure of naturally occurring glucose. Trees and plants convert carbon dioxide and water into glucose during photosynthesis.

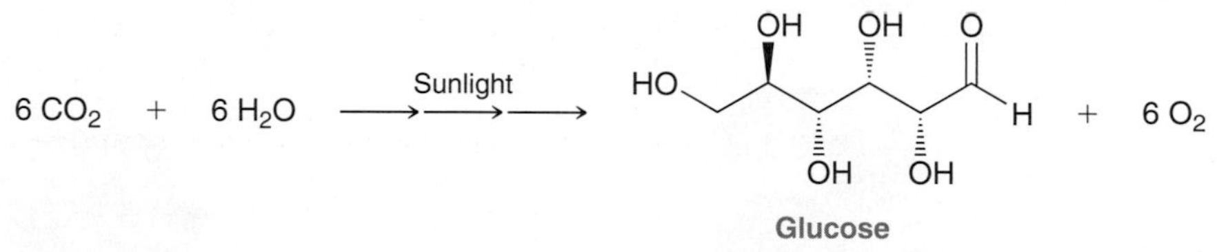

Energy from the sun is absorbed by vegetation and is used to convert CO_2 molecules into larger organic compounds. These organic compounds, such as glucose, have C—C and C—H bonds, which are higher in energy than the C=O bonds in CO_2. Our bodies utilize a series of chemical reactions to convert these compounds back into CO_2, thereby releasing the stored solar energy. In the process, we return carbon dioxide and water back to the environment to be recycled. In essence, our bodies are powered, in large part, by solar energy that has been stored in the form of glucose molecules.

24.2 Classification of Monosaccharides

Aldoses vs. Ketoses

LOOKING BACK
For a review of Fischer projections and the skills necessary to interpret these drawings refer to Section 5.7.

Simple sugars are called **monosaccharides**, a term that derives from the Latin word for sugar, *saccharum*. Complex sugars, such as disaccharides and polysaccharides, are made by joining monosaccharides together, and they will be discussed later in this chapter.

Monosaccharides generally contain multiple chirality centers, and Fischer projections are used to indicate the configuration at each chirality center. Glucose and fructose are two examples of simple sugars.

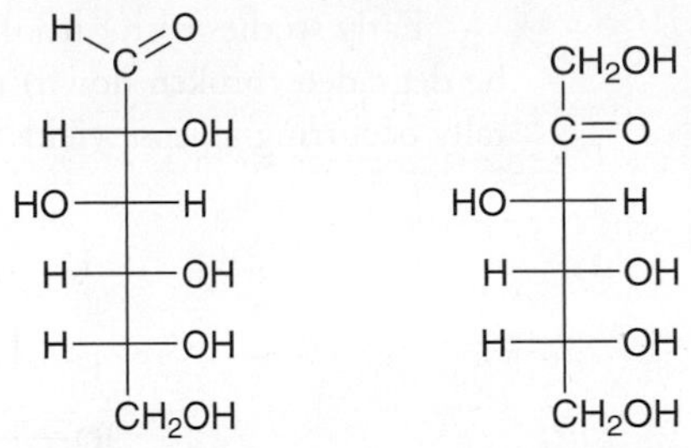

Glucose Fructose

The suffix *-ose* is used to signify a carbohydrate. Hundreds of different monosaccharides are known, each of which can generally be classified as either an *aldose* or a *ketose*. **Aldoses** contain an aldehyde moiety, while **ketoses** contain a ketone moiety. According to this classification scheme, glucose is an aldose and fructose is a ketose.

Aldoses and ketoses can be further classified based on the number of carbon atoms they contain. This is accomplished by inserting a term (tri-, tetra-, pent-, hex-, or hept-) immediately before the suffix *-ose*.

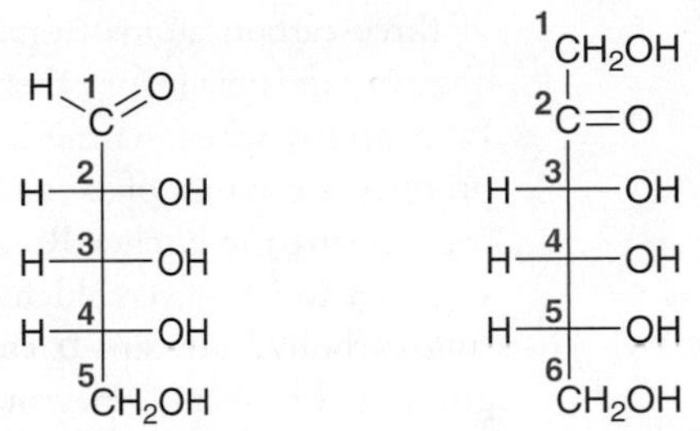

An aldopentose A ketohexose

The first compound is an aldose with five carbon atoms, and it is therefore called an *aldopentose*. The second compound is a ketose with six carbon atoms, and it is therefore called a *ketohexose*. In this way, carbohydrates are classified using three descriptors:

1. *aldo* or *keto*, indicating whether the compound is an aldehyde or a ketone
2. *tri-*, *tetr-*, *pent-*, *hex-*, or *hept-*, indicating the number of carbon atoms
3. *-ose*, indicating a carbohydrate

D and L Sugars

Glyceraldehyde is one of the smallest compounds considered to be a carbohydrate. It has only one chirality center and therefore can exist as a pair of enantiomers.

(+)-Glyceraldehyde (−)-Glyceraldehyde

As discussed in Section 5.4, enantiomers rotate plane-polarized light in opposite directions. One enantiomer rotates plane-polarized light in a clockwise fashion (dextrorotatory) and is designated as (+); the other enantiomer rotates plane-polarized light in a counterclockwise direction (levorotatory) and is designated as (−). Only (+) or dextrorotatory glyceraldehyde is abundant in nature, so glyceraldehyde obtained from natural sources is generally referred to as D-glyceraldehyde. Levorotatory or L-glyceraldehyde can be made in the laboratory, but it is generally not observed in nature.

Early studies with carbohydrates revealed that most naturally occurring carbohydrates can be degraded (broken down) to produce D-glyceraldehyde. For example, degradation of naturally occurring glucose yields D-glyceraldehyde (Figure 24.1). During the degradation process,

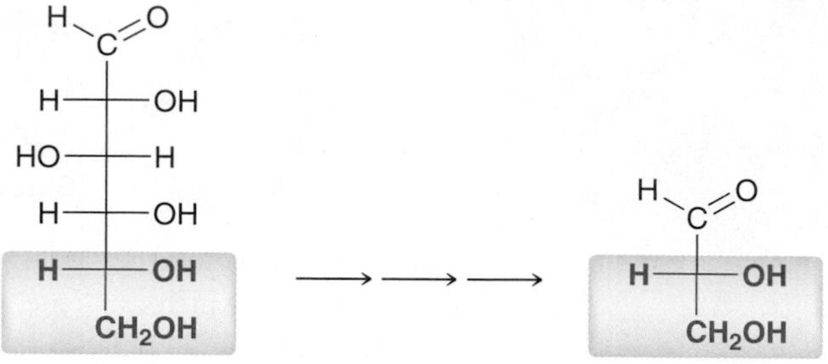

Glucose (naturally occuring) D-Glyceraldehyde

FIGURE 24.1
Degradation of most naturally occurring sugars produces D-glyceraldehyde.

the carbon atoms are removed one at a time (from the top of the Fischer projection). The loss of three carbon atoms from naturally occurring glucose yields D-glyceraldehyde. The same observation is made for other naturally occurring carbohydrates.

In contrast, when synthetic sugars (those prepared in the laboratory) are degraded, they produce a mixture of D- and L-glyceraldehyde. In response to these observations, chemists began using the Fischer-Rosanoff convention in which the letter D designates any sugar that degrades to (+)-glyceraldehyde. In accord with this convention, almost all naturally occurring carbohydrates are **D sugars**—that is, the chirality center farthest from the carbonyl group will have an OH group pointing to the right in the Fischer projection, as in the following examples.

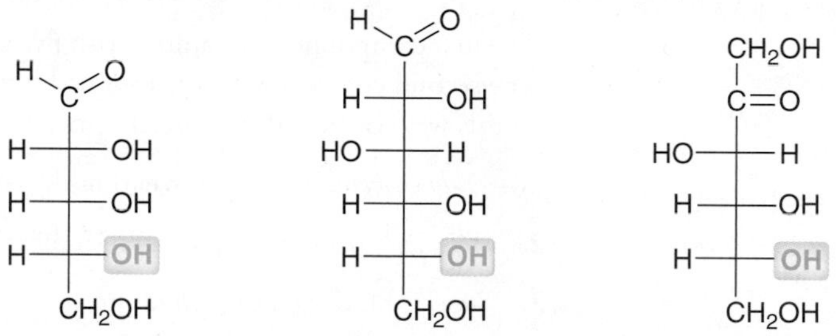

D-Ribose D-Glucose D-Fructose

While D-glyceraldehyde is dextrorotatory (by definition), other D sugars are not necessarily dextrorotatory. For example, D-erythrose and D-threose are actually levorotatory.

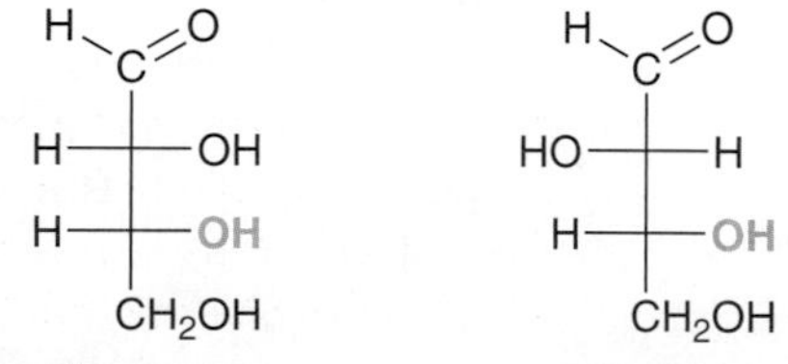

D-Erythrose $[\alpha]_D^{27} = -32.7°$ D-Threose $[\alpha]_D^{20} = -12.2°$

In this context, the D no longer refers to the direction in which plane-polarized light is rotated. Rather, it means that the lowest chirality center (the one farthest from the carbonyl group) has the *R* configuration just as it does in (+)-glyceraldehyde. Similarly, **L sugars** are not necessarily levorotatory, but rather an L sugar is simply the enantiomer of the corresponding D sugar.

CONCEPTUAL CHECKPOINT

24.1 Classify each of the following carbohydrates as an aldose or ketose, and then insert the appropriate term to indicate the number of carbon atoms present (e.g., an aldo*pent*ose):

(a) 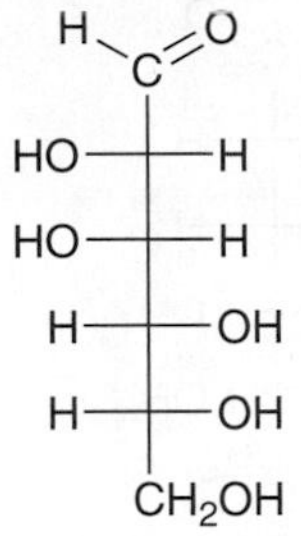

(b) 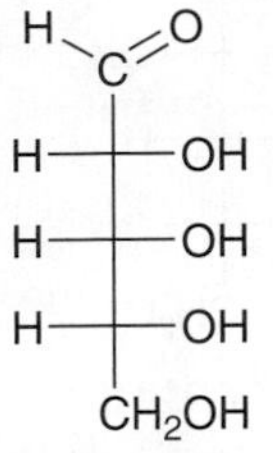

(c) 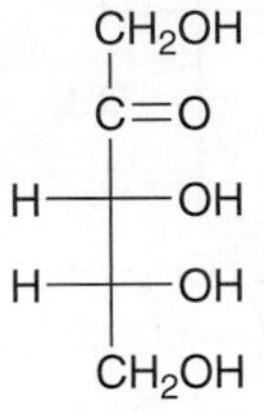

(d) 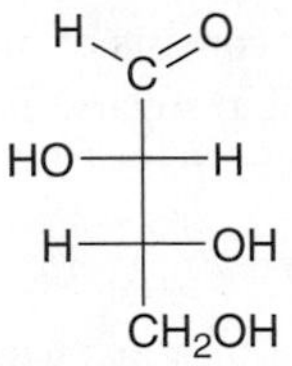

(e) 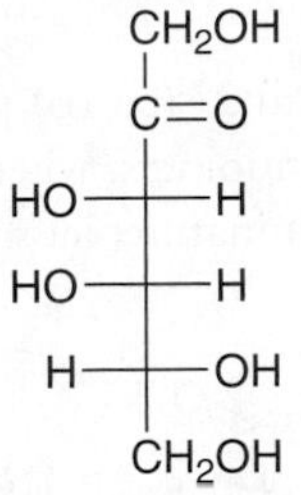

24.2 Would you expect an aldohexose and a ketohexose to be constitutionally isomeric? Explain why or why not.

24.3 Determine whether each of the following carbohydrates is a D sugar or an L sugar, and assign a configuration for each chirality center. After assigning the configuration for all of the chirality centers, do you notice any trend that would enable you to assign the configuration of a chirality center in a carbohydrate more quickly?

(a) 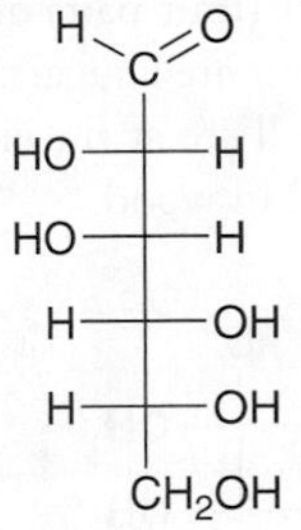

(b) 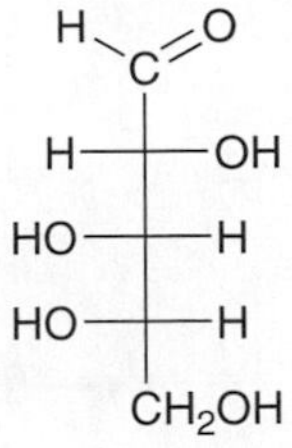

(c) 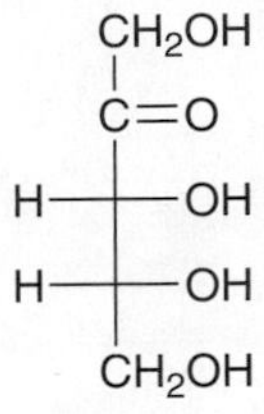

(d) 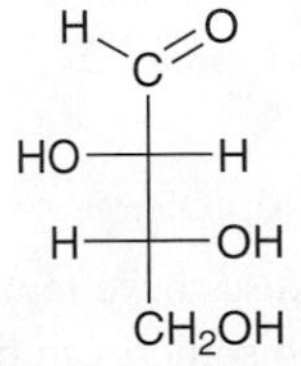

(e) 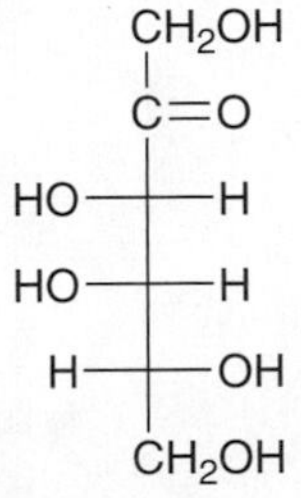

24.4 D-Allose is an aldohexose in which all four chirality centers have the *R* configuration. Draw a Fischer projection of each of the following compounds:

(a) D-Allose (b) L-Allose.

24.5 There are only two stereoisomeric ketotetroses.

(a) Draw both of them.

(b) Identify their stereoisomeric relationship.

(c) Identify which is a D sugar and which is an L sugar.

24.6 There are four stereoisomeric aldotetroses.

(a) Draw all four, and arrange them in pairs of enantiomers.

(b) Identify which stereoisomers are D sugars and which are L sugars.

24.3 Configuration of Aldoses

Aldotetroses

Aldotetroses have two chirality centers, and there are only four possible aldotetroses (two pairs of enantiomers). Two of the possible aldotetroses are D sugars, while the other two are L sugars. The D sugars are called D-erythrose and D-threose (Figure 24.2). The L sugars, called L-erythrose and L-threose, are the enantiomers of the D aldotetroses.

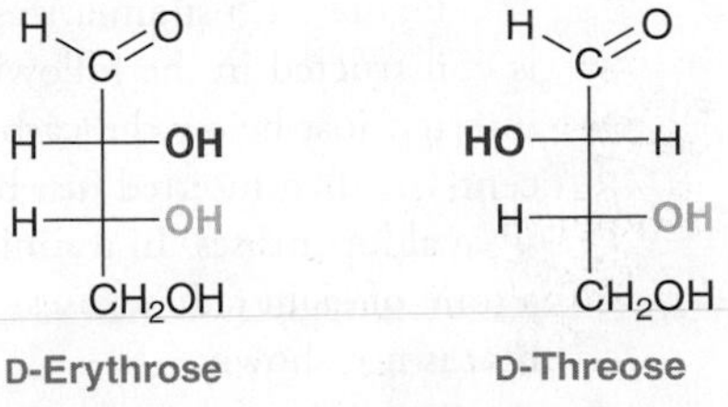

D-Erythrose D-Threose

FIGURE 24.2 The structures of the D aldotetroses.

Aldopentoses

LOOKING BACK
For a review of the 2^n rule, see Section 5.5.

Aldopentoses have three chirality centers, which means that there are eight (2^3) possible aldopentoses (four pairs of enantiomers). Four of the possible aldopentoses are D sugars, while the other four are L sugars. Figure 24.3 shows the D sugars. D-Ribose is a key building block of RNA, as we will see at the end of this chapter. D-Arabinose is produced by most plants, and D-xylose is found in wood.

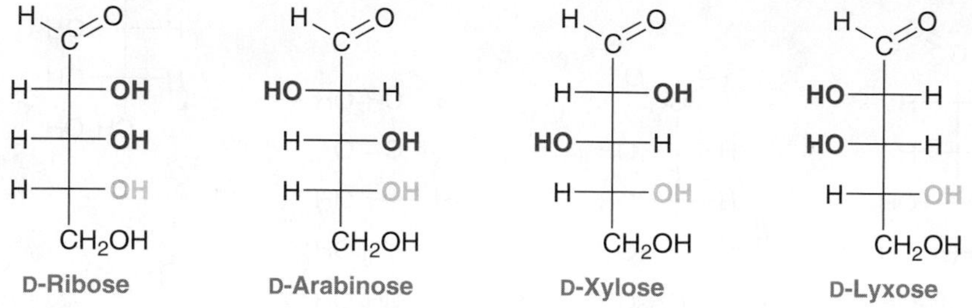

FIGURE 24.3
The structures of the D-aldopentoses.

Aldohexoses

Aldohexoses have four chirality centers, giving rise to 2^4 (or 16) possible stereoisomers. These 16 stereoisomers can be grouped into 8 pairs of enantiomers, where each pair consists of one D sugar and one L sugar. Since D sugars are observed in nature, let's focus on the D sugars. There are eight of them (Figure 24.4).

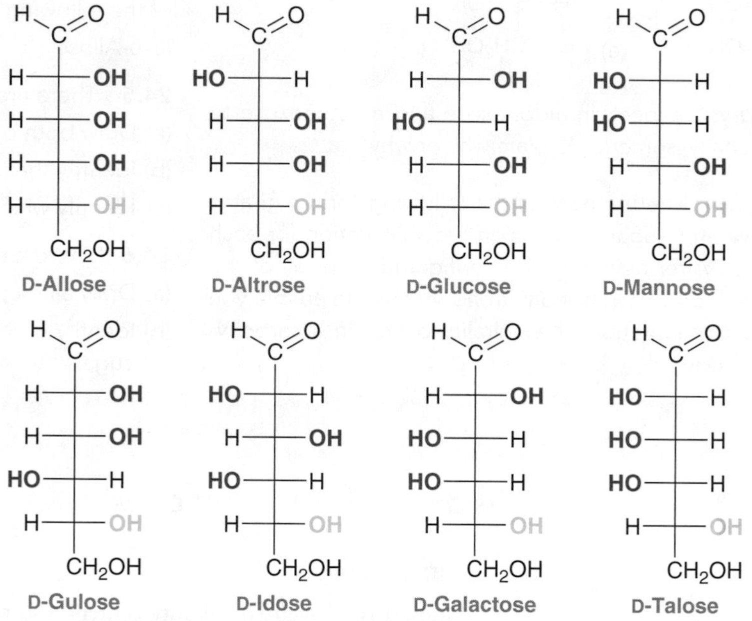

FIGURE 24.4
The structures of the D-aldohexoses.

Of the eight D aldohexoses, D-glucose is the most common and most important, and the rest of this chapter will focus primarily on D-glucose. Its structure should be at your fingertips.

Figure 24.5 summarizes the family of D aldoses discussed in this section. The family tree is constructed in the following way: Starting with D-glyceraldehyde, a new chirality center is inserted just below the carbonyl group, generating two possible aldotetroses. A new chirality center is then inserted just below the carbonyl group of each aldotetrose, generating four possible aldopentoses. In a similar way, each of the aldopentoses leads to two aldohexoses, giving a total of eight aldohexoses. Each compound in Figure 24.5 has a corresponding L enantiomer that is not shown.

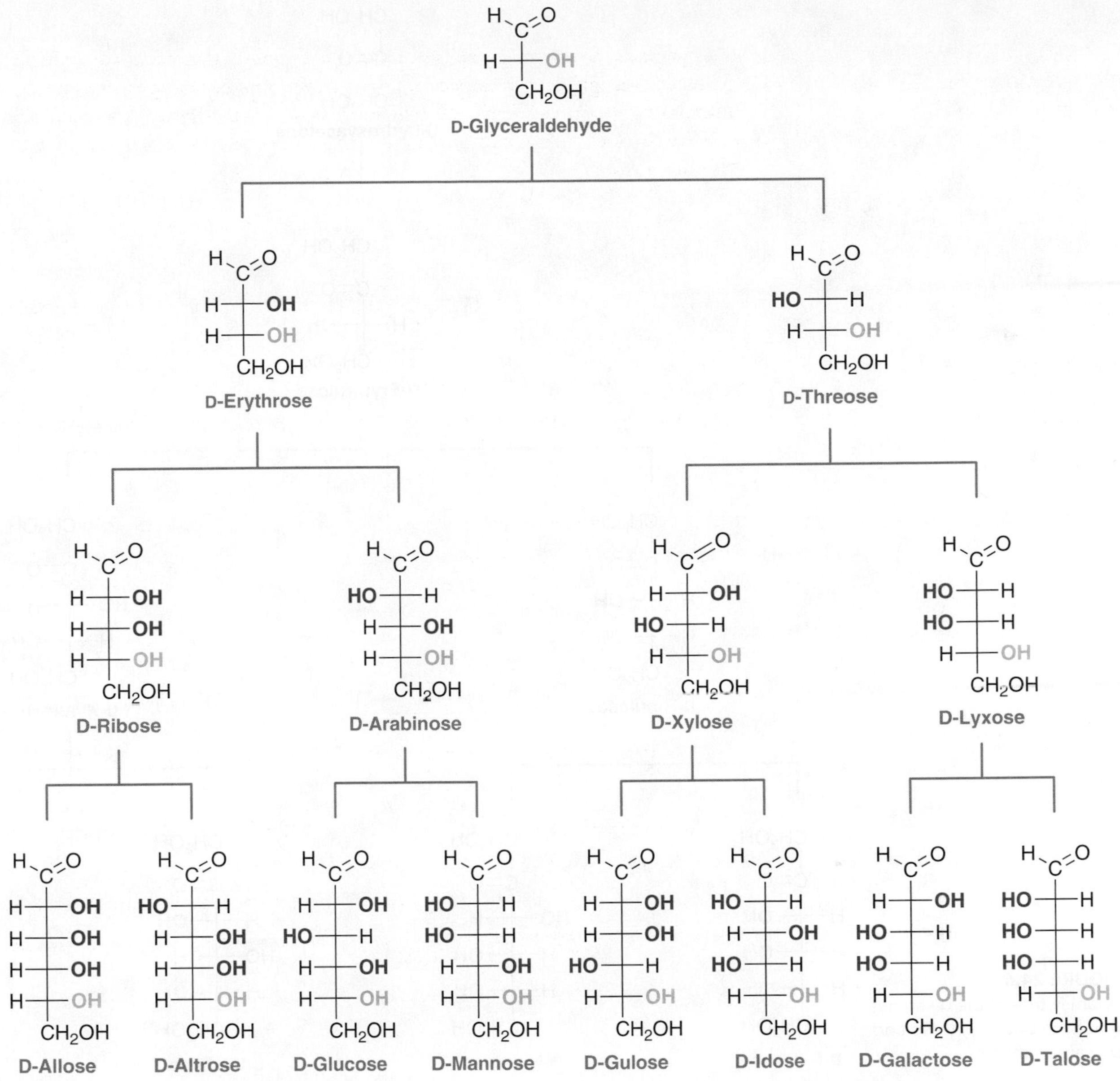

FIGURE 24.5
A family tree of the D-aldoses that have between three and six carbon atoms.

24.4 Configuration of Ketoses

Figure 24.6 summarizes the family of D ketoses that have between three and six carbon atoms. This family tree is constructed in much the same way as the family tree of aldoses, beginning with dihydroxyacetone as the parent. However, because the carbonyl group is at C2 rather than C1, ketoses have one less chirality center than aldoses of the same molecular formula. As a result, there are only four D ketohexoses, rather than eight. The most common naturally occurring ketose is D-fructose. Each compound in Figure 24.6 has a corresponding L enantiomer that is not shown.

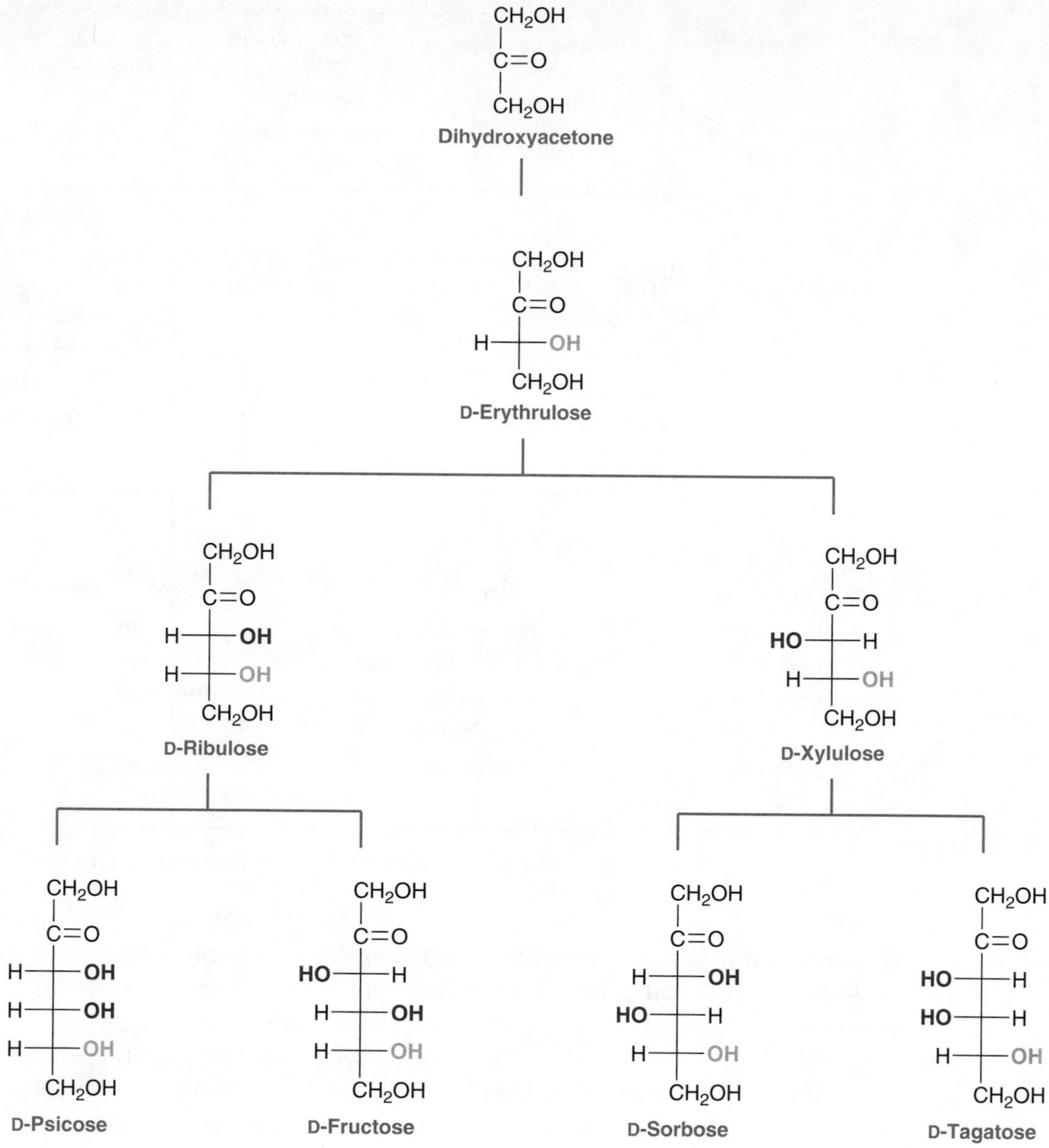

FIGURE 24.6 A family tree of the D ketoses that have between three and six carbon atoms.

CONCEPTUAL CHECKPOINT

24.7 Draw and name the enantiomer of D-fructose.

24.8 Which of the following terms best describes the relationship between D-fructose and D-glucose? Explain your choice.

(a) Enantiomers
(b) Diastereomers
(c) Constitutional isomers

24.5 Cyclic Structures of Monosaccharides

Cyclization of Hydroxyaldehydes

Recall from Section 20.5 (Mechanism 20.5) that an aldehyde can react with an alcohol in the presence of an acid catalyst to produce a hemiacetal.

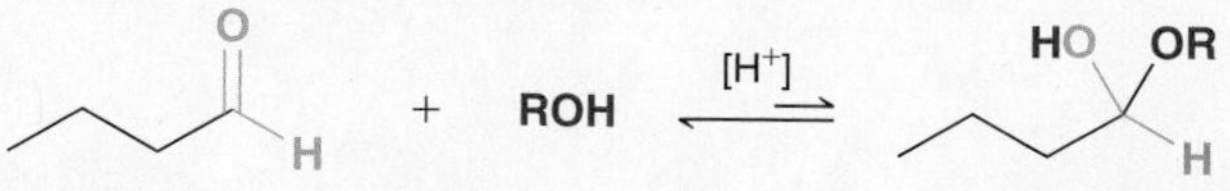

We saw that the equilibrium does not favor formation of the hemiacetal. However, when the aldehyde moiety and the hydroxyl group are contained in the same molecule, an intramolecular process can occur to form a cyclic hemiacetal with a more favorable equilibrium constant.

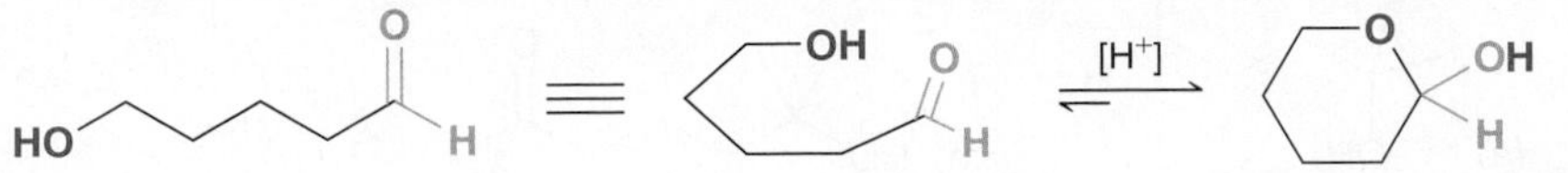

Cyclic hemiacetal

Six-membered rings are relatively strain free, and the equilibrium favors formation of the cyclic hemiacetal. This type of reaction is characteristic of bifunctional compounds containing both a hydroxyl group and a carbonyl group (an aldehyde or ketone moiety). When drawing the hemiacetal of a hydroxyaldehyde or hydroxyketone, it is crucial to keep track of all carbon atoms. It is a common mistake to draw the hemiacetal with either too many or too few carbon atoms. The following exercises are designed to help you avoid those mistakes.

SKILLBUILDER

24.1 DRAWING THE CYCLIC HEMIACETAL OF A HYDROXYALDEHYDE

LEARN the skill

Draw the cyclic hemiacetal that is formed when the following hydroxyaldehyde undergoes cyclization under acidic conditions.

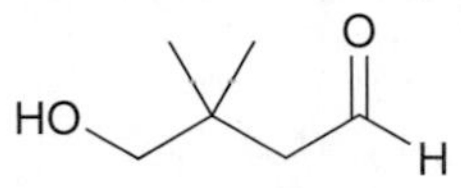

SOLUTION

This compound contains both a hydroxyl group and an aldehyde group, so it is expected to form a hemiacetal. We must first determine the size of the ring that is formed. In order to account for all carbon atoms and to avoid drawing an incorrect ring, it is best to assign numbers to the carbons so we can keep track of them. Begin with the carbon of the aldehyde group and continue until reaching the carbon connected to the OH group.

STEP 1
Using a numbering system, determine the size of the ring that is formed.

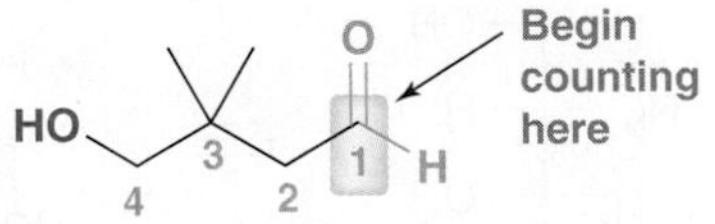

For now, we will not focus on the methyl groups as we will account for them after drawing the proper size ring. In this case, there are four carbon atoms that will be incorporated into the ring. The ring is formed when the oxygen atom of the OH group attacks the carbonyl, and as a result, the oxygen atom of the OH group must also be incorporated into the ring—that is, the ring will be comprised of four carbon atoms and one oxygen atom, generating a five-membered ring:

STEP 2
Draw the ring, making sure to incorporate an oxygen atom in the ring.

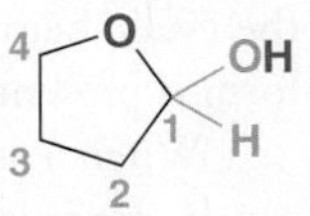

Notice that the first carbon atom (the C1 position) is connected to the oxygen of the ring as well as to the newly formed OH group. In other words, the hemiacetal group is located at C1. This numbering system not only helps with drawing the appropriate ring, but it also facilitates the proper placement of the substituents. In this case, there are two methyl groups, both attached to the C3 position.

STEP 3
Place the substituents.

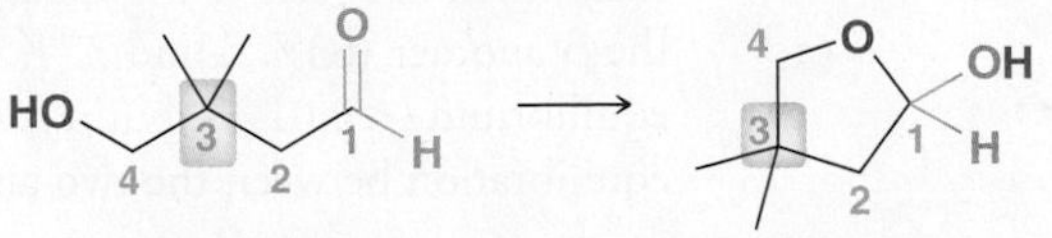

PRACTICE the skill **24.9** Draw the cyclic hemiacetal that is formed when each of the following bifunctional compounds is treated with aqueous acid:

(a) 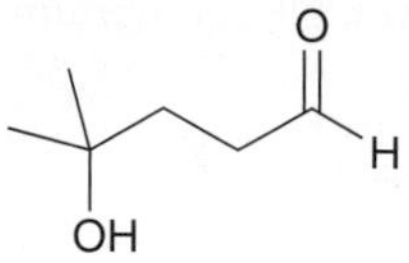

(b) 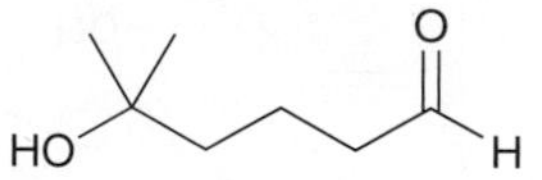

(c) 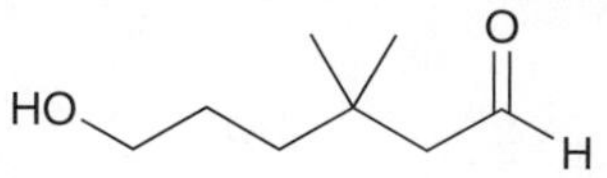

(d) 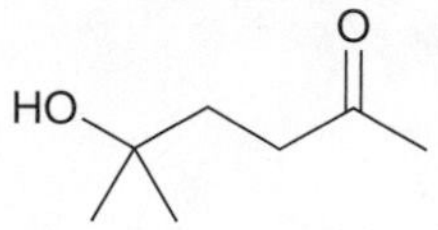

APPLY the skill **24.10** Identify the hydroxyaldehyde that will cyclize under acidic conditions to give the following hemiacetal:

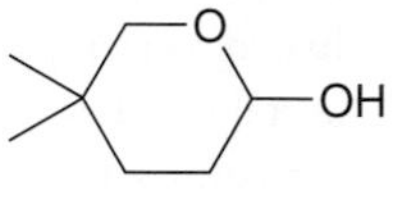

24.11 The following compound has one aldehyde group and two OH groups: Under acidic conditions, either one of the OH groups can function as a nucleophile and attack the carbonyl group, giving rise to two possible ring sizes.

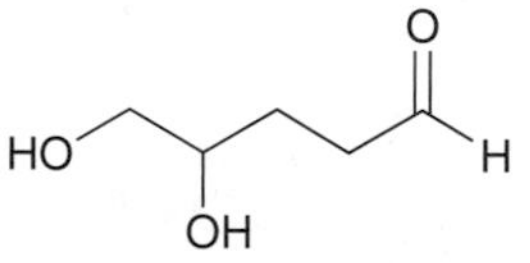

(a) Ignoring stereochemistry (for now), draw both possible rings.

(b) In Section 4.9, we discussed the ring strain associated with different size rings. Based on those concepts, predict which cyclic hemiacetal will be favored in this case.

need more **PRACTICE?** **Try Problems 24.46, 24.47**

Pyranose Forms of Monosaccharides

LOOKING BACK
For an introduction to Haworth projections, see Section 2.6.

We have seen that a compound containing both an OH group and an aldehyde moiety will undergo an intramolecular process to form a cyclic hemiacetal, with a favorable equilibrium constant. A similar equilibrium between open and closed forms is observed for carbohydrates. For example, D-glucose can exist in both open and closed forms. The following Haworth projections indicate the configuration at each chirality center:

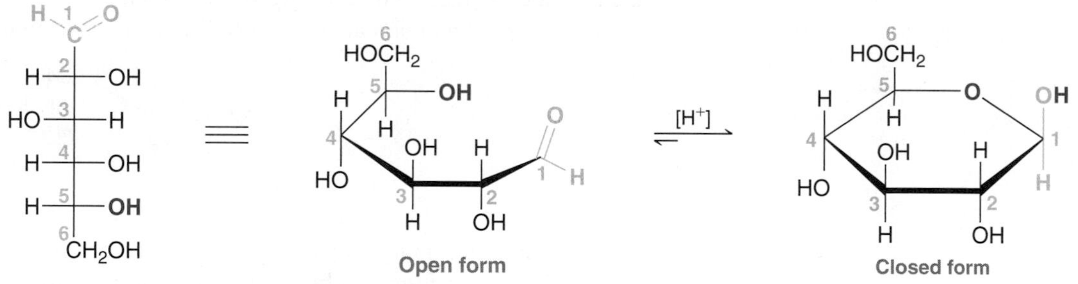

In the cyclic form, the ring is called a **pyranose ring**, named after pyran, a simple compound that possesses a six-membered ring with an oxygen atom incorporated in the ring. This equilibrium greatly favors formation of the cyclic hemiacetal. That is, D-glucose exists almost exclusively in the cyclic hemiacetal form called D-glucopyranose, and only a trace amount of the open-chain form is present at equilibrium.

When D-glucose undergoes cyclization, two different stereoisomers of D-glucopyranose can be formed (Figure 24.7). When glucose is closed into a cyclic hemiacetal, the aldehydic carbon (C1) becomes a chirality center, and both configurations for that new chirality center are possible. The C1 position is called the **anomeric carbon**, and the stereoisomers are called **anomers**. The two anomers are designated as α-D-glucopyranose and β-D-glucopyranose. In the **α anomer**, the hydroxyl group at the anomeric position is *trans* to the CH_2OH group, while in the **β anomer**, the hydroxyl group is *cis* to the CH_2OH group (Figure 24.8). An equilibrium is established between α-D-glucopyranose and β-D-glucopyranose with the equilibrium favoring the β anomer (63% β and 37% α). The open-chain form exhibits only a minimal presence at equilibrium (<0.01%), but it is important nonetheless, because it serves as the intermediate for equilibration between the two anomers.

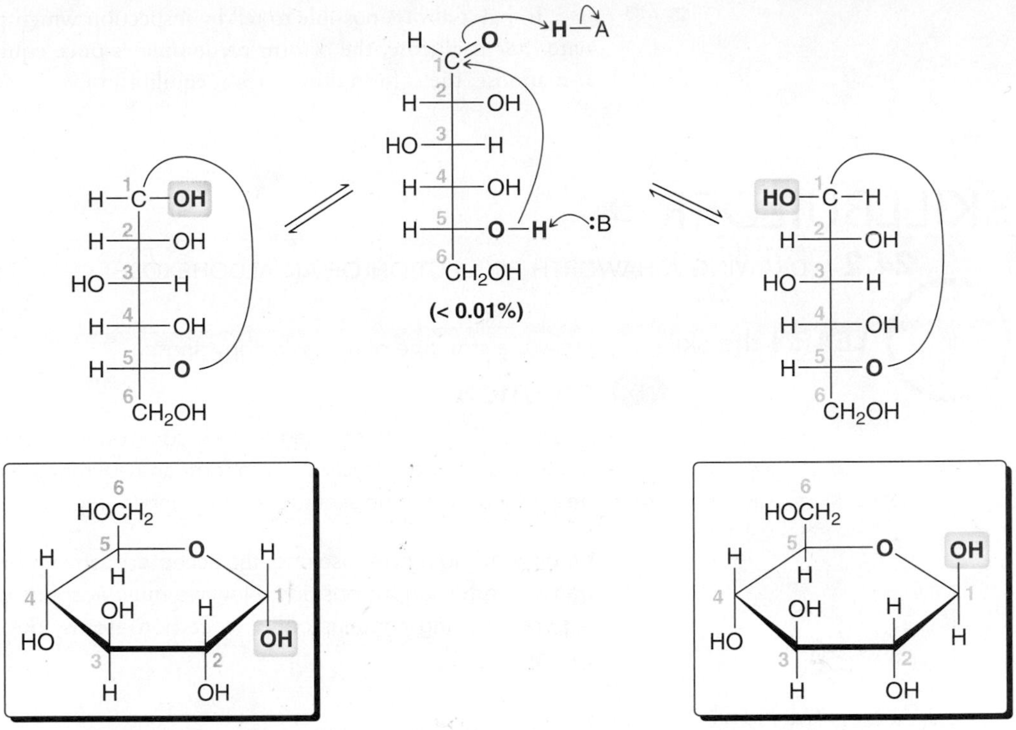

α-D-Glucopyranose $\qquad$ β-D-Glucopyranose

FIGURE 24.7 Formation of the two possible cyclic hemiacetals of D-glucose.

As we might expect, α-D-glucopyranose and β-D-glucopyranose have different physical properties because their relationship is diastereomeric. For example, compare the specific rotation for these two compounds:

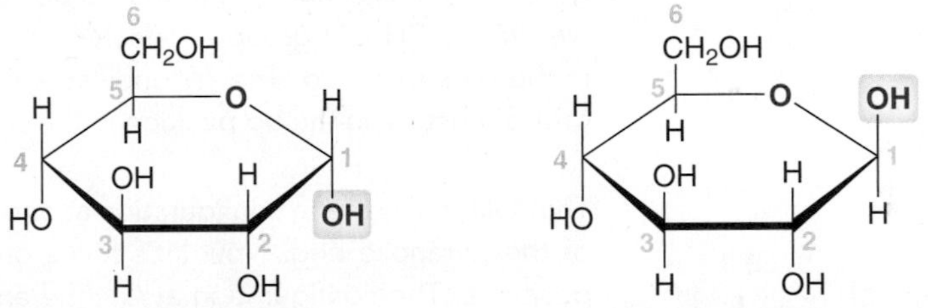

α-D-Glucopyranose $[\alpha]_D = +112.2°$ $\qquad$ **β-D-Glucopyranose** $[\alpha]_D = +18.7°$

When the pure α anomer is dissolved in water, it begins to equilibrate with the β anomer, resulting in a mixture that ultimately achieves the expected equilibrium concentrations. At first, the pure anomer exhibits a specific rotation of +112°, but as it equilibrates with the β anomer, the specific rotation changes until the equilibrium is established, and the specific rotation is measured to be +52.6°. A similar result is achieved when the pure β anomer is dissolved in water. At first, the pure anomer has a specific rotation of +18.7°, but as the mixture equilibrates, the specific rotation changes and ultimately is measured to be +52.6°. This phenomenon is called **mutarotation**, a term commonly used to describe the fact that α and β anomers can equilibrate via the open-chain form. Mutarotation occurs more rapidly in the presence of catalytic acid or base.

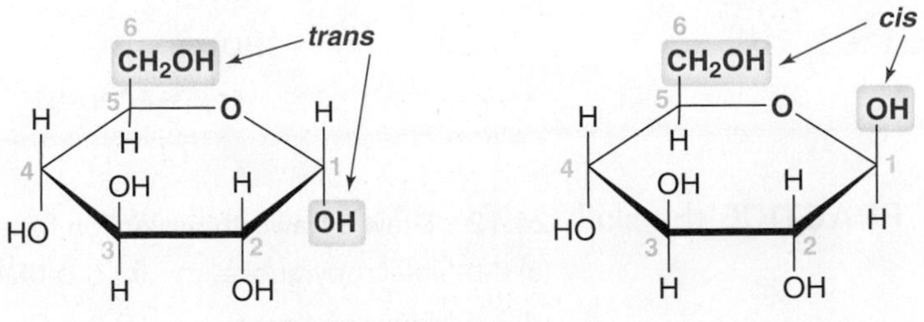

α $\qquad$ β

FIGURE 24.8 The α and β anomers are distinguished based on the relative position of the OH group at the anomeric position and the CH_2OH group at C5.

It is not always possible to tell by inspection which pyranose form (α or β) will predominate. For D-glucose, the β form predominates once equilibrium has been achieved, while for D-mannose, the α form dominates at equilibrium.

SKILLBUILDER

24.2 DRAWING A HAWORTH PROJECTION OF AN ALDOHEXOSE

LEARN the skill Draw the structure of α-D-galactopyranose.

SOLUTION

STEP 1 Draw the skeleton of a Haworth projection.

Analyze all parts of the name. This compound is D-galactose that has undergone cyclization to form an α-pyranose ring. Let's begin by drawing the skeleton of a pyranose ring:

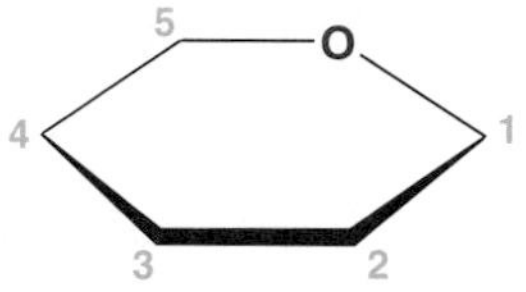

When drawing a pyranose ring, the accepted convention is to place the oxygen in the back right position. Now we must account for all carbon atoms. A pyranose ring contains only five carbon atoms, but D-galactose has six carbon atoms:

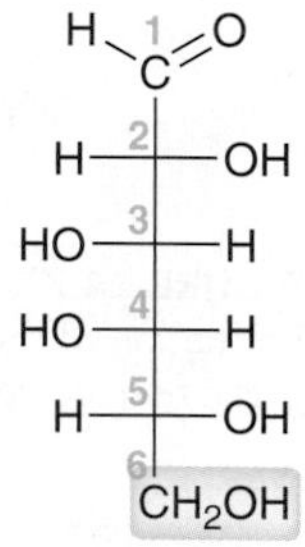

D-Galactose

STEP 2 Draw the CH_2OH group.

The sixth carbon atom (C6) in D-galactose is a CH_2OH group, and it is connected to the C5 position. Therefore, we draw a CH_2OH group connected to the C5 position in the pyranose ring. The "D" indicates that the CH_2OH group must be in the up position in the pyranose ring:

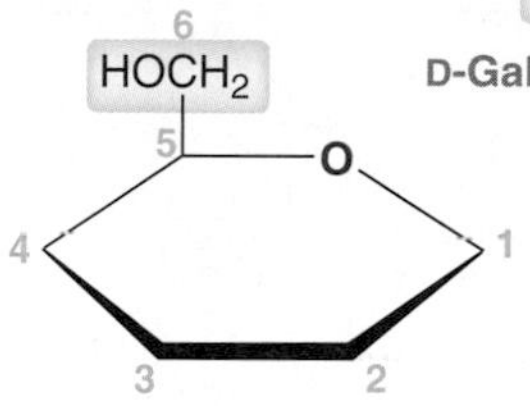

STEP 3 Draw the OH group at the anomeric position.

That takes care of the configuration at one of the positions of the pyranose ring. Now let's focus on the remaining positions. The configuration at C1 (the anomeric carbon) is conveyed in the name of the pyranose ring form. The term *alpha* means that the OH group at the C1 position must be down (*trans* to the CH_2OH group):

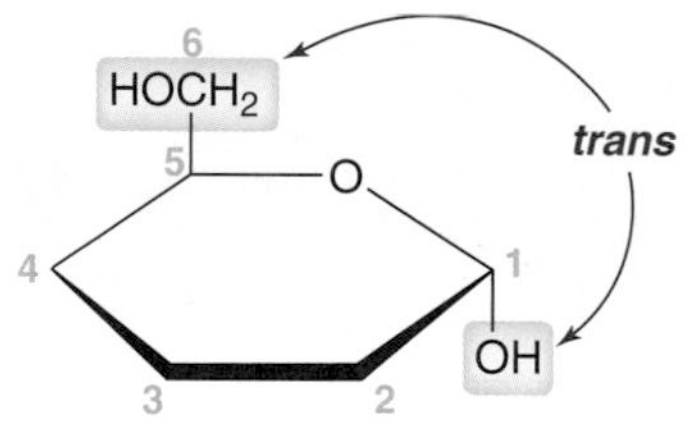

STEP 4 Draw the remaining groups.

The remaining chirality centers (C2, C3, and C4) are drawn using the following rule: Any OH groups on the right side of the Fischer projection of the open-chain form will be pointing down in the Haworth projection of the cyclic form, while any OH groups on the left side of the Fischer projection will be pointing up in the Haworth projection:

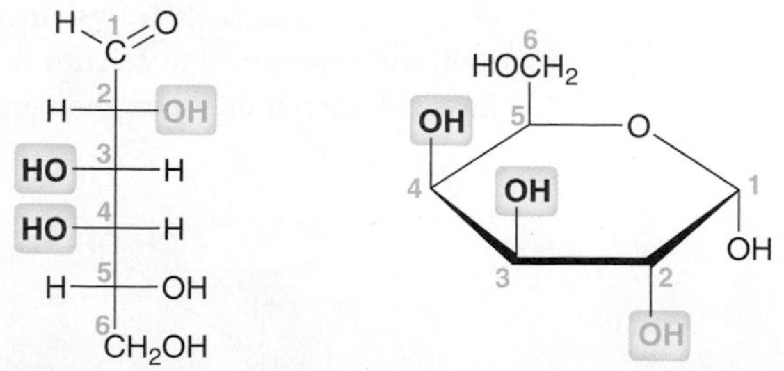

PRACTICE the skill **24.12** Draw a Haworth projection for each of the following compounds:

(a) β-D-Galactopyranose (b) α-D-Mannopyranose (c) α-D-Allopyranose
(d) β-D-Mannopyranose (e) β-D-Glucopyranose (f) α-D-Glucopyranose

24.13 Provide a complete name for the following compound:

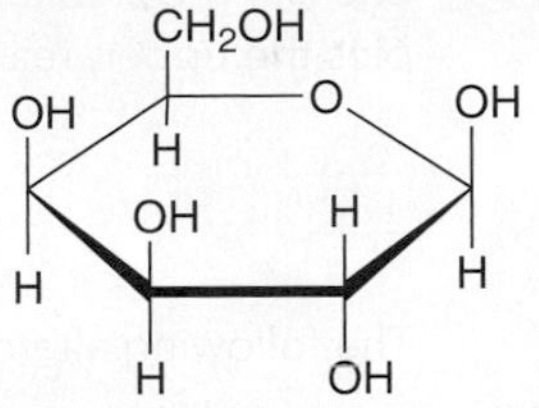

APPLY the skill

24.14 Mutarotation causes the conversion of β-D-mannopyranose to α-D-mannopyranose. Using Haworth projections, draw the equilibrium between the two pyranose forms and the open-chain form of D-mannose.

24.15 When D-talose is dissolved in water, an equilibrium is established in which two pyranose forms are present. Draw both pyranose forms and name them.

need more **PRACTICE?** **Try Problems 24.48a, 24.53b,c,d**

Three-Dimensional Representations of Pyranose Rings

We have seen that Haworth projections are often used for drawing the cyclic forms of D-glucose. While Haworth projections are extremely useful for showing configurations, they are not effective for communicating the conformation of a compound. A cyclic form of D-glucose will spend the majority of its time in a chair conformation.

LOOKING BACK
For a review of chair conformations and axial/equatorial positions, see Section 4.11.

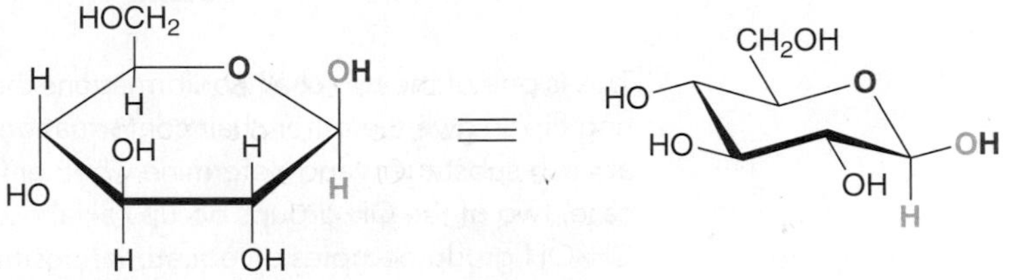

When drawing either a Haworth projection or a chair conformation, the convention is to place the oxygen atom in the upper, rear-right position, as highlighted in red. Notice that in a cyclic form of D-glucose all substituents can occupy equatorial positions, which renders the compound particularly stable and explains why D-glucose is the most common naturally occurring monosaccharide.

SKILLBUILDER

24.3 DRAWING THE MORE STABLE CHAIR CONFORMATION OF A PYRANOSE RING

LEARN the skill Draw the more stable chair conformation of α-D-galactopyranose.

SOLUTION

The first step is to draw the Haworth projection of the compound, which was done in the previous SkillBuilder:

STEP 1
Draw the Haworth projection.

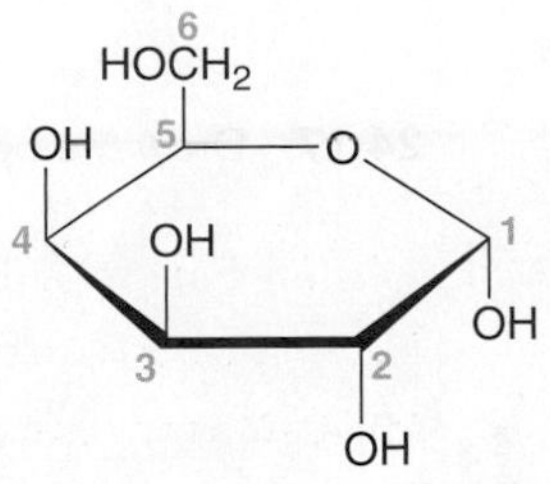

This compound can adopt two different chair conformations, and we must determine which one is more stable. We begin by drawing one of the chair conformations. The oxygen occupies the upper, rear-right position:

STEP 2
Draw the skeleton of a chair conformation with an oxygen atom in the upper, rear-right position.

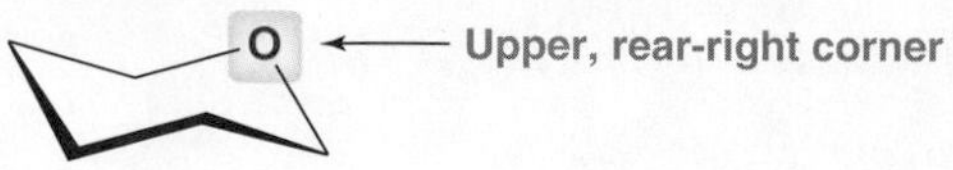

The following alternatives are not conventional for the reasons indicated:

Oxygen is in upper right corner, but is NOT in the rear

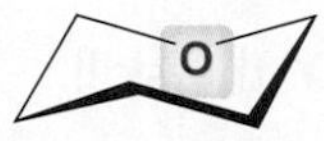

Oxygen is in rear right but is NOT in upper corner

Drawing the appropriate skeleton is critical for drawing the chair conformation properly. The next step is to draw all substituents on the chair using the skills we learned in SkillBuilders 4.12 and 4.13. Each substituent is labeled as UP or DOWN and then placed in the appropriate position on the chair conformation:

STEP 3
Label each substituent as UP or DOWN, and draw each substituent on the chair.

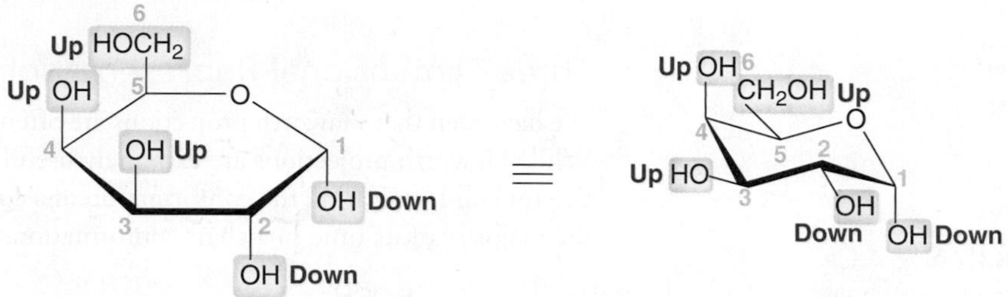

This is one of the two chair conformations that the compound can adopt. Before drawing the ring flip to give the other chair conformation, let's first analyze this one. Specifically, we look at each substituent and determine whether it occupies an axial or equatorial position. In this case, two of the OH groups occupy axial positions and two occupy equatorial positions; the CH_2OH group occupies an equatorial position.

STEP 4
Verify that this chair conformation is the more stable chair conformation.

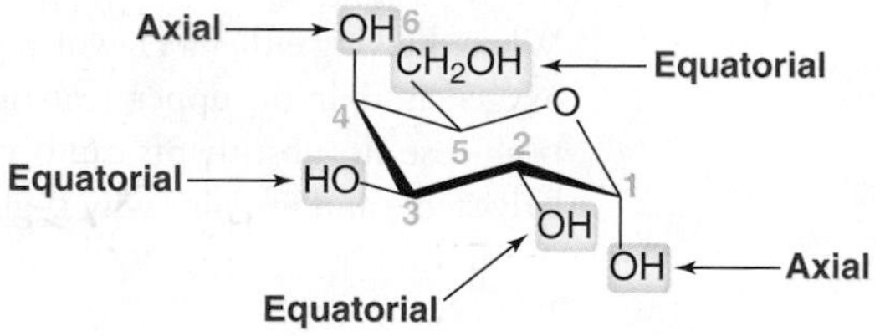

Recall that a ring flip converts all axial positions into equatorial positions and all equatorial positions into axial positions (see Section 4.10). Also recall that the more stable chair conformation will be the one in which the largest groups occupy equatorial positions. In this case, the CH_2OH is the largest group, so the most stable chair conformation will be the one in which this group occupies an equatorial position, as shown. The other chair conformation will be less stable, so it is not necessary to draw the ring flip. This is generally the case for D-aldohexoses (for an exception, see Problem 24.85).

PRACTICE the skill **24.16** Draw the more stable chair conformation for each of the following compounds:

(a) β-D-Galactopyranose (b) α-D-Glucopyranose (c) β-D-Glucopyranose

APPLY the skill **24.17** Draw the open-chain form of the following cyclic monosaccharide:

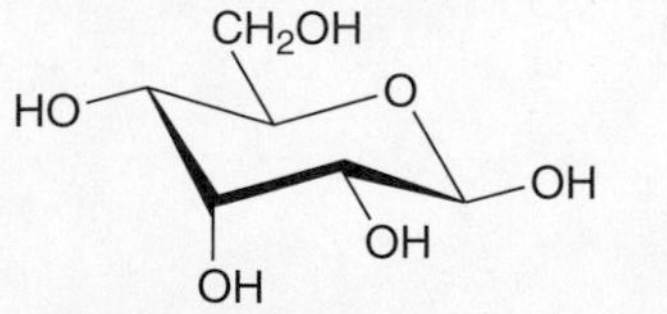

24.18 There are two chair conformations for β-D-glucopyranose. Draw the less stable chair conformation.

need more **PRACTICE?** **Try Problem 24.60**

Furanose Forms of Monosaccharides

As seen in Skillbuilder 24.1, hydroxyaldehydes can form five-membered cyclic hemiacetals.

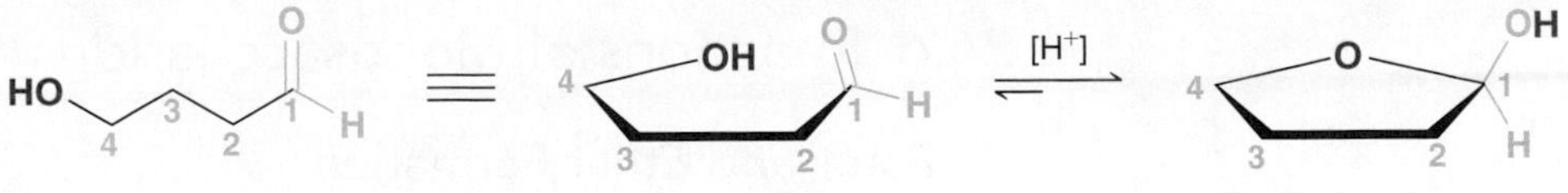

In much the same way, many carbohydrates can also form five-membered rings, called **furanose** rings, named after furan, a simple compound that possesses a five-membered ring with an oxygen atom incorporated in the ring. As an example, D-fructose can form a furanose ring that results from the reaction between the carbonyl group and the hydroxyl group connected to C5.

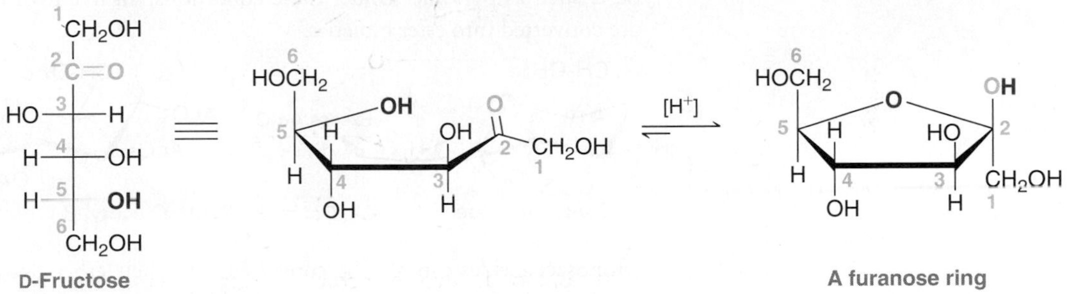

D-Fructose can also cyclize to give pyranose forms that result from the reaction involving the hydroxyl group at C6.

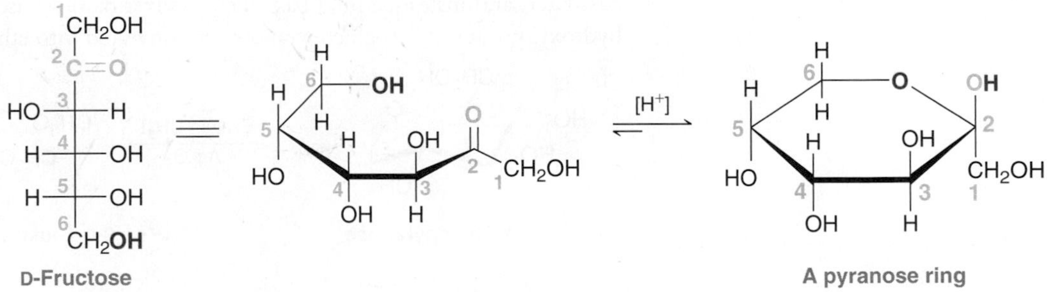

Therefore, D-fructose exists as an equilibrium between an open-chain form, two pyranose forms (α and β), and two furanose forms (α and β). When D-fructose is dissolved in water, the following equilibrium concentrations are observed: 70% β-pyranose, 2% α-pyranose, 23% β-furanose, 5% α-furanose, and 0.7% open chain. Despite the fact that the β-pyranose ring predominates at equilibrium as observed in the laboratory, it is the β-furanose form of D-fructose that participates in most biochemical pathways.

CONCEPTUAL CHECKPOINT

24.19 Consider the structures of the following two D-aldotetroses:

Each of these compounds exists as a furanose ring, which is formed when the OH at C4 attacks the aldehyde moiety. Draw each of the following furanose rings:

(a) α-D-Erythrofuranose (b) β-D-Erythrofuranose
(c) α-D-Threofuranose (d) β-D-Threofuranose

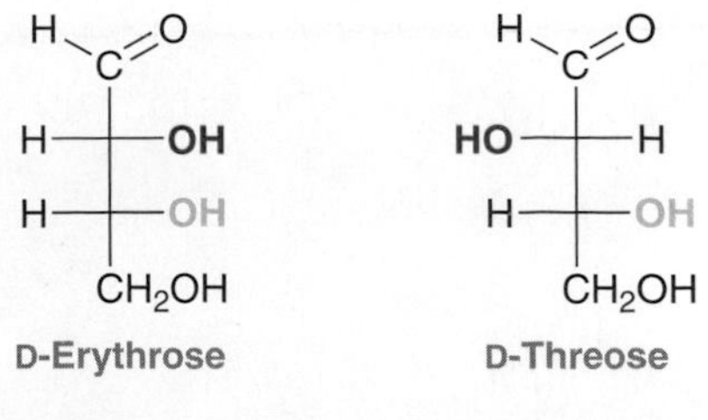

24.20 Draw a mechanism for the acid-catalyzed cyclization of L-threose to give β-L-threofuranose. (*Hint*: You may want to first review the mechanism for acid-catalyzed hemiacetal formation, Mechanism 20.5.)

24.21 Draw a mechanism for the acid-catalyzed cyclization of D-fructose to give β-D-fructofuranose.

24.22 Draw the open-chain form of the carbohydrate that can undergo acid-catalyzed cyclization to produce α-D-fructopyranose.

24.6 Reactions of Monosaccharides

Ester and Ether Formation

Monosaccharides are highly soluble in water (due to the presence of many hydroxyl groups) and are generally insoluble in most organic solvents. This property makes them difficult to purify by conventional methods. However, the ester derivatives of monosaccharides are soluble in most organic solvents and are easily purified. Monosaccharides are converted into their ester derivatives when treated with an excess of acid chloride or acid anhydride in the presence of a base, such as pyridine. Under these conditions, all five hydroxyl groups of β-D-glucopyranose are converted into ester moieties.

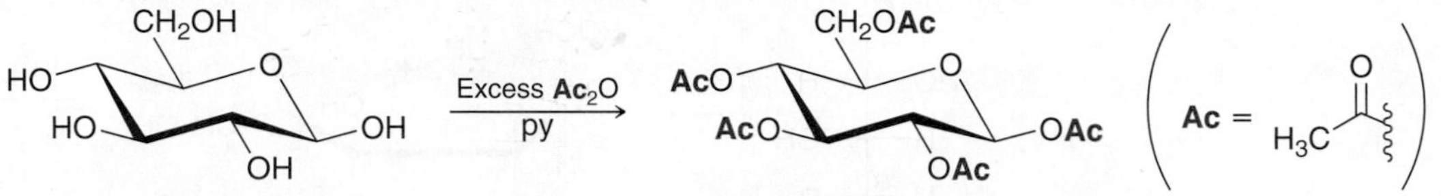

β-D-Glucopyranose **Penta-O-acetyl-β-D-glucopyranose**

Monosaccharides can also be converted into their ether derivatives via the Williamson ether synthesis. As discussed in Section 14.5, this process generally involves treating an alcohol with a strong base to form an alkoxide ion followed by treatment with an alkyl halide (an S_N2 process). When dealing with monosaccharides, a strong base cannot be used, for reasons that we will discuss later, and instead, a mild base such as silver oxide is used. Under these conditions, all five hydroxyl groups of β-D-glucopyranose are converted into ether moieties.

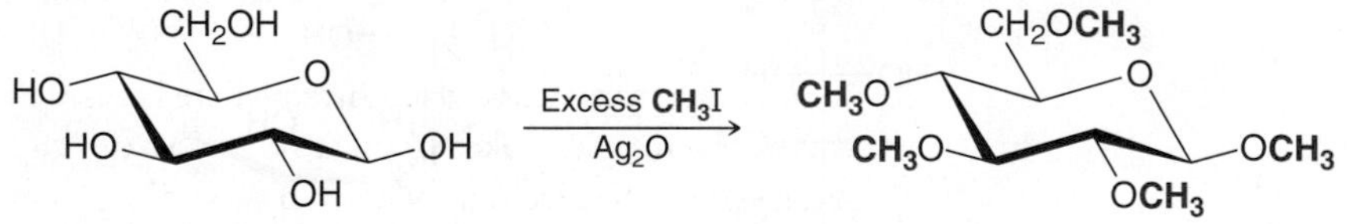

β-D-Glucopyranose **β-D-Glucopyranose pentamethyl ether**

CONCEPTUAL CHECKPOINT

24.23 Draw the product obtained when each of the following compounds is treated with acetic anhydride in the presence of pyridine:

(a) α-D-Galactopyranose (b) α-D-Glucopyranose

(c) β-D-Galactopyranose

24.24 Draw the product obtained when each of the compounds from the previous problem is treated with methyl iodide in the presence of silver oxide (Ag_2O).

Glycoside Formation

Recall from Section 20.5 that a hemiacetal will react with an alcohol in the presence of an acid catalyst to produce an acetal.

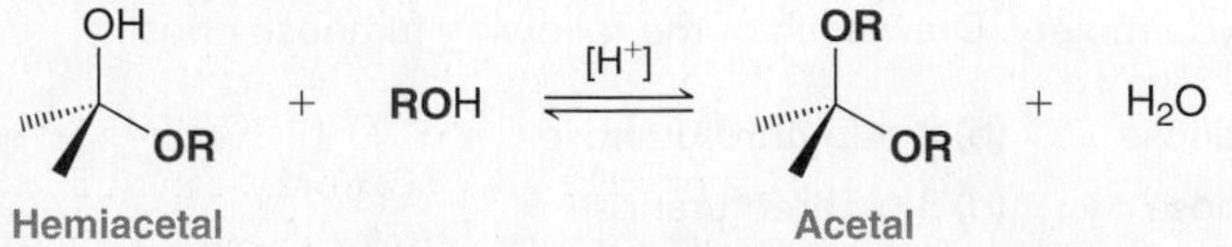

Hemiacetal **Acetal**

As mentioned earlier in this chapter, monosaccharides exist primarily as cyclic hemiacetals; they are therefore converted into acetals when treated with an alcohol under acid-catalyzed conditions. The acetal products obtained are called **glycosides.**

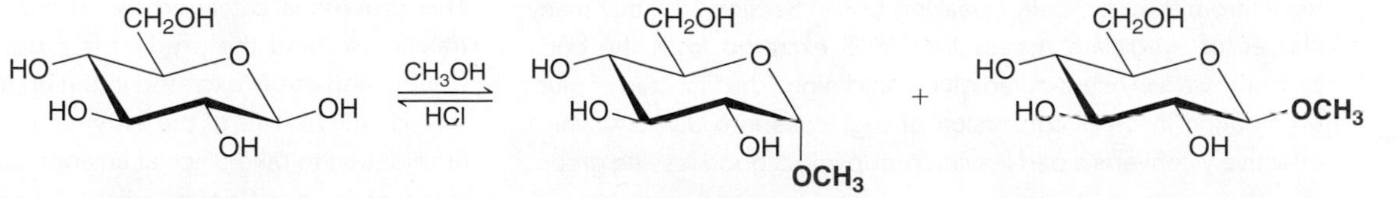

β-D-Glucopyranose (a cyclic hemiacetal) | **Methyl α-D-glucopyranoside (66%)** | **Methyl β-D-glucopyranoside (33%)**

Glycosides are named by placing the alkyl group as a prefix and the term *-oside* as a suffix. During glycoside formation, only the anomeric hydroxyl group is replaced. This is best explained by exploring the mechanism of glycoside formation, which is directly analogous to the mechanism of acetal formation described in Section 20.5. The anomeric hydroxyl group is protonated, followed by loss of water to generate a carbocation.

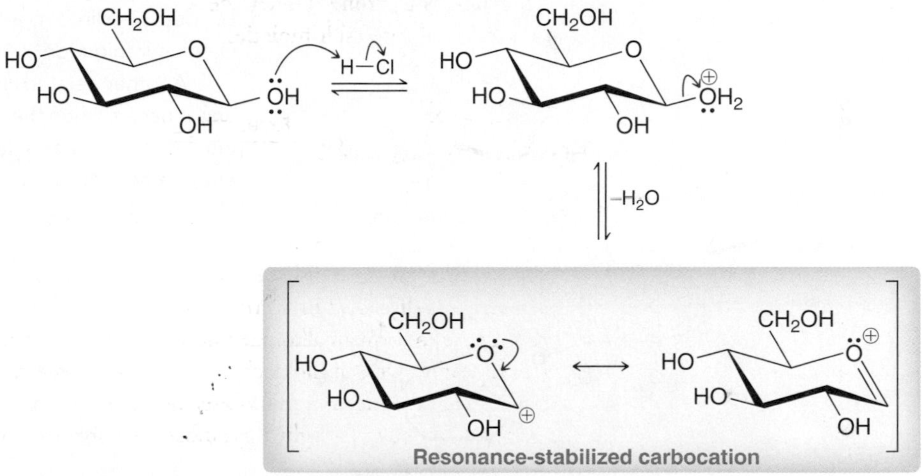

Notice that the carbocation intermediate is resonance stabilized. This stabilization only occurs when the reaction takes place at the anomeric position. The carbocation intermediate is then attacked by the alcohol, followed by loss of a proton to generate the product.

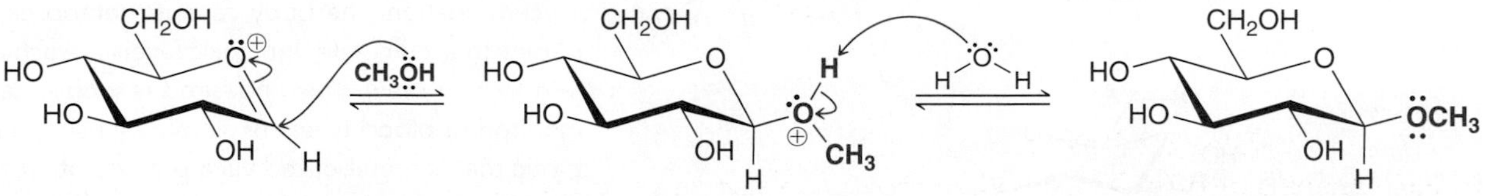

This mechanism shows formation of the β anomer. The α anomer is also observed, because the alcohol can attack the carbocation from the other face of the planar carbocation.

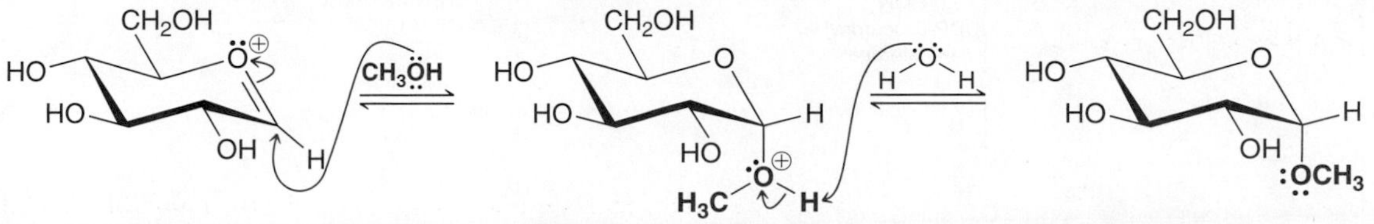

The product distribution is independent of the identity of the starting anomer—that is, α-D-glucopyranose and β-D-glucopyranose each yield the same ratio of anomeric glycosides when treated with an alcohol in acid-catalyzed conditions. Once isolated, the glycoside products are stable under neutral or basic conditions. However, upon treatment with aqueous acid, they are readily converted back to a hemiacetal.

MEDICALLY SPEAKING

Diseases Associated with Insufficient Glycoside Formation

Recall from the Practically Speaking box in Section 13.9 that many classes of drugs are metabolized and excreted from the body primarily via a process called glucuronidation. The first step of glucuronidation involves conversion of D-glucose into UDPGA, which effectively converts a bad leaving group into a good leaving group.

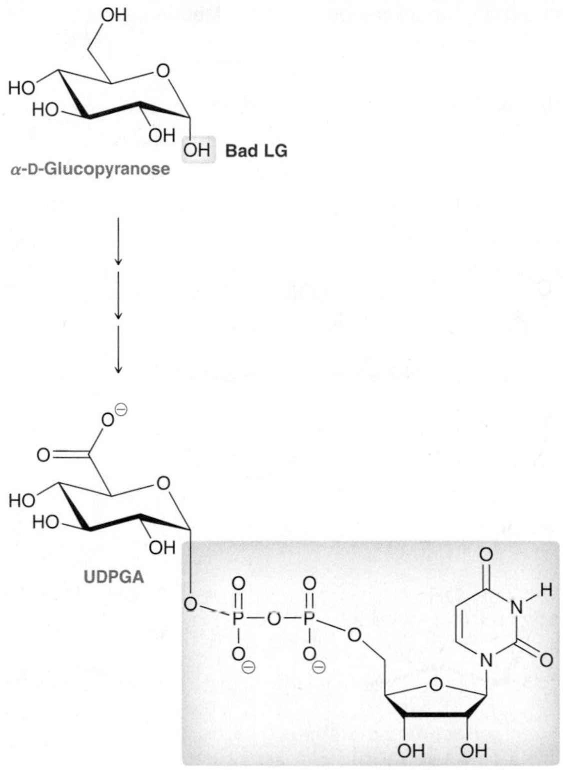

Good LG (called UDP)

UDPGA is a compound with a very good leaving group, called UDP, which is expelled when a drug such as an alcohol attacks UDPGA in an S_N2 process.

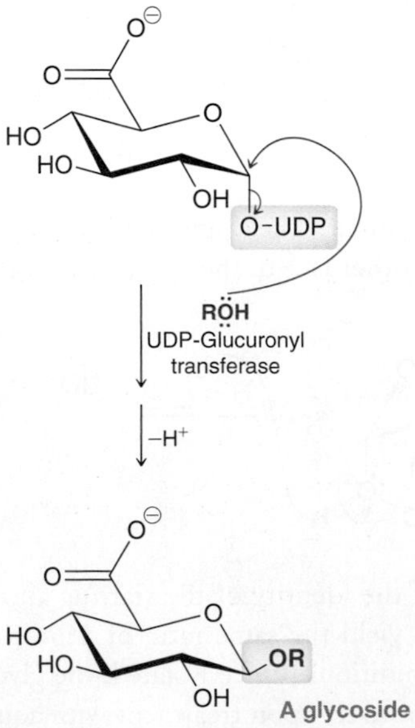

This process is catalyzed by an enzyme called UDP-glucuronyl transferase, and the product is a glycoside that is highly water soluble and easily excreted in the urine.

Healthy levels of the enzyme are required in order for glucuronidation to take place at an appreciable rate. There are several diseases, such as Crigler-Najjar syndrome and Gilbert's syndrome, that are associated with a deficiency of this important enzyme, which results in an inability to perform glucuronidation effectively. Similarly, newborn babies (especially premature babies) often have undeveloped systems that are incapable of producing sufficient quantities of the enzyme immediately after birth. It can take several days for the system to mature before glucuronidation can occur at an appreciable rate. This is one of the causes of neonatal jaundice, a condition that causes the yellowing of skin due to the buildup of bilirubin, a biochemical that is normally metabolized via glucuronidation.

Another related syndrome is called gray baby syndrome, and it occurs when the antibiotic chloramphenicol is administered intravenously to an infant or child with inadequate levels of the enzyme necessary for glucuronidation.

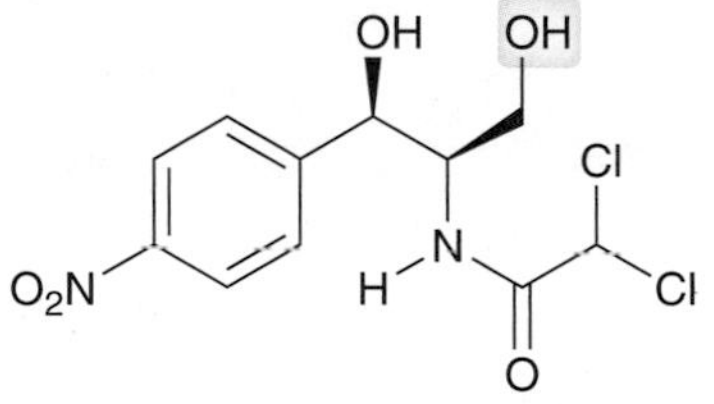

Chloramphenicol

This drug is primarily metabolized via glucuronidation when the highlighted OH group functions as a nucleophile and attacks UDPGA in an S_N2 process. Without the enzyme necessary for glucuronidation, the body cannot metabolize this drug, which begins to accumulate until toxic levels are achieved (causing the skin to turn gray). This undesired response can be avoided by monitoring blood levels or simply by using an alternative antibiotic that is metabolized via a process other than glucuronidation. Practicing physicians must be aware of these inherent dangers when prescribing antibiotics to newborn babies or to adults with syndromes associated with a deficiency of UDP-glucuronyl transferase.

CONCEPTUAL CHECKPOINT

24.25 When α-D-galactopyranose is treated with ethanol in the presence of an acid catalyst, such as HCl, two products are formed. Draw both products, and account for their formation with a mechanism.

24.26 Methyl α-D-glucopyranoside is a stable compound that does not undergo mutarotation under neutral or basic conditions. However, when subjected to acidic conditions, an equilibrium is established consisting of both methyl α-D-glucopyranoside and methyl β-D-glucopyranoside. Draw a mechanism that accounts for this observation.

Epimerization

When D-glucose is exposed to strongly basic conditions, it is converted into a mixture containing both D-glucose and D-mannose via the following process:

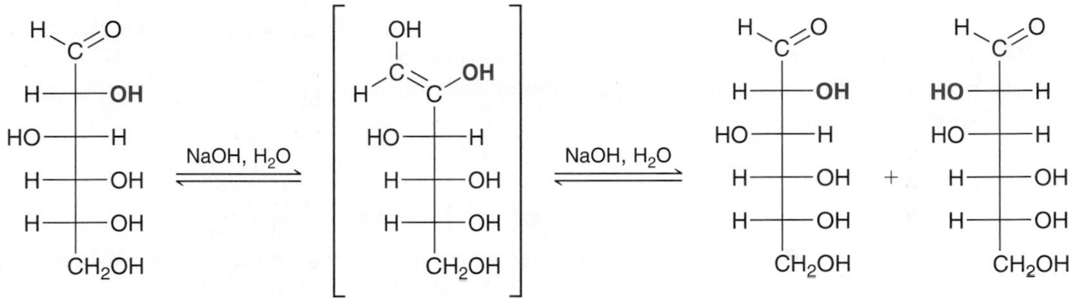

D-Glucose **An enediol** **D-Glucose** **D-Mannose**

D-Glucose first undergoes base-catalyzed tautomerization to form an enediol. This intermediate can undergo tautomerization once again to revert back to the aldose, but in the process, the configuration at C2 is lost, leading to a mixture of D-glucose and D-mannose. D-Glucose and D-mannose are said to be **epimers** because they are stereoisomers that differ from each other in the configuration of only one chirality center. When either pure D-glucose or pure D-mannose is treated with a strong base, epimerization occurs, giving a mixture containing both D-glucose and D-mannose. For this reason, chemists generally avoid exposing carbohydrates to strongly basic conditions.

CONCEPTUAL CHECKPOINT

24.27 Draw and name the structure of the aldohexose that is epimeric with D-glucose at each of the following positions:

(a) C2 (b) C3 (c) C4

Reduction of Monosaccharides

The carbonyl group of an aldose or ketose can be reduced upon treatment with sodium borohydride to yield a product called an **alditol**. Consider, for example, the reduction of D-glucose.

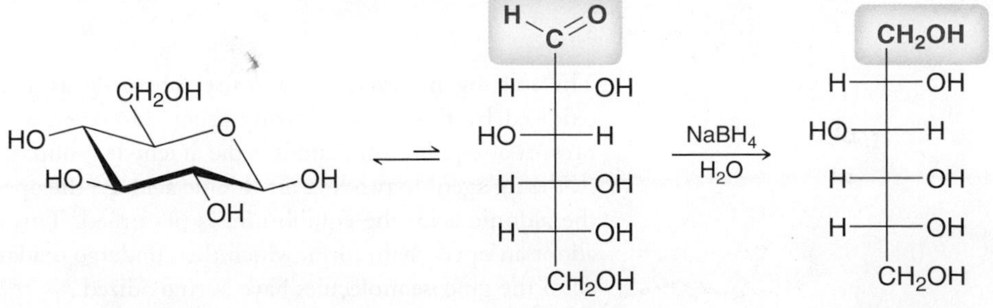

β-D-Glucopyranose **D-Glucose (open form)** **D-Glucitol (an alditol)**

The starting monosaccharide exists primarily as a hemiacetal, which does not react with sodium borohydride because it lacks a carbonyl group. However, a small amount of the open-chain form is present at equilibrium, and it is that form that reacts with sodium borohydride to produce an alditol. As the open-chain form of glucose is converted into the alditol, the equilibrium is perturbed. According to Le Châtelier's principle, this causes more of the glucose molecules to adopt an open-chain form, which then also undergo reduction. This process continues until nearly all of the glucose molecules have been reduced to D-glucitol. D-Glucitol is found in many fruits and berries. It is commonly referred to as D-sorbitol, or just sorbitol, and is often used as a sugar substitute in processed foods.

CONCEPTUAL CHECKPOINT

24.28 The same product is obtained when either D-altrose or D-talose is treated with sodium borohydride in the presence of water. Explain this observation.

24.29 The same product is obtained when either D-allose or L-allose is treated with sodium borohydride in the presence of water. Explain this observation.

24.30 Of the eight D-aldohexoses, only two of them form optically inactive alditols when treated with sodium borohydride in the presence of water. Identify these two aldohexoses and explain why their alditols are optically inactive.

Oxidation of Monosaccharides

When treated with a suitable oxidizing agent, the aldehyde moiety of an aldose can be oxidized to yield a compound called an **aldonic acid**.

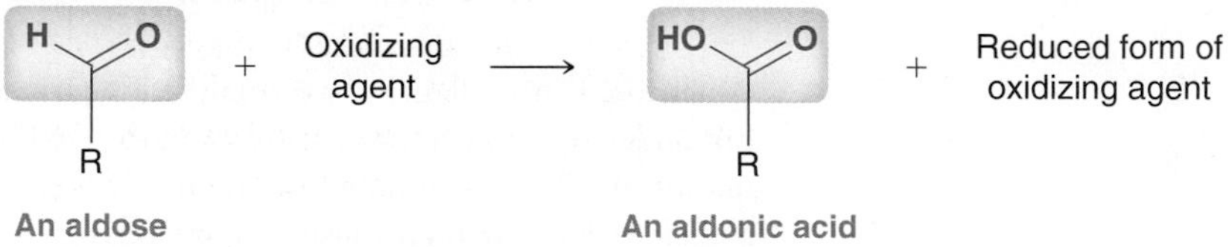

This reaction is observed for a wide variety of oxidizing agents. When a high yield is desired, a useful oxidizing agent is an aqueous solution of bromine buffered to a pH of 6. For example:

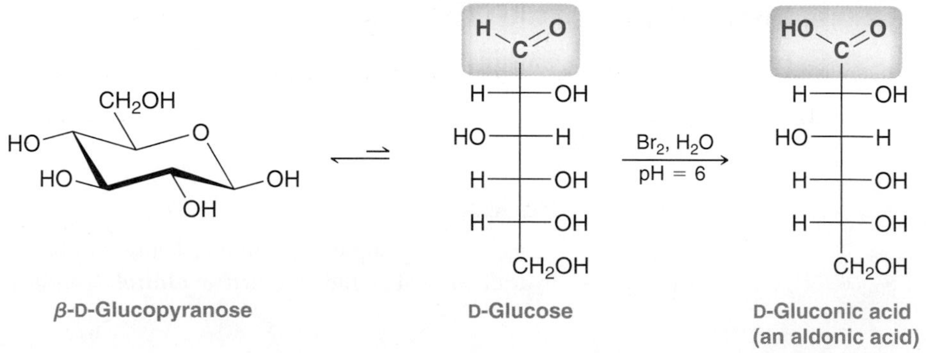

The starting monosaccharide exists primarily as a hemiacetal, and the OH groups are not oxidized by this mild oxidizing agent. However, a small amount of the open-chain form is present at equilibrium, and it is the aldehyde group of the open-chain form that reacts with the oxidizing agent to produce an aldonic acid. As the open-chain form of glucose is converted into the aldonic acid, the equilibrium is perturbed. This causes more of the glucose molecules to adopt an open-chain form, which then undergo oxidation, a process that continues until nearly all of the glucose molecules have been oxidized.

The oxidizing agent above will oxidize aldoses but does not react with ketoses. A ketose lacks the aldehydic hydrogen that is characteristic of an aldose.

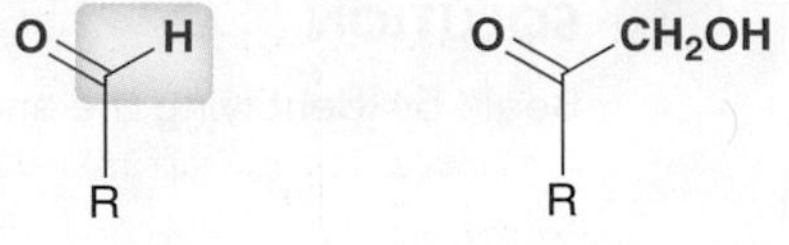

An aldose A ketose

It is true that ketoses are slowly converted into aldoses via epimerization and we might therefore expect that a small concentration of the ketose would be converted into an aldose, which would then undergo oxidation, providing a pathway for the ketose to be oxidized as well. However, the mildly acidic conditions employed (a buffered solution of $pH = 6$) prevent epimerization from occurring. Under these conditions, a ketose does not epimerize to give an aldose, and ketoses are therefore not oxidized by a buffered solution of aqueous bromine. This, in fact, provides a method for distinguishing between aldoses and ketoses. In contrast, there are other oxidizing agents that employ conditions under which epimerization can occur. These alternative reagents can oxidize ketoses as well as aldoses. Examples of such oxidizing agents include the Tollens' reagent (Ag^+ in aqueous ammonia), Fehlings' reagent (Cu^{2+} in aqueous sodium tartrate), and Benedict's reagent (Cu^{2+} in aqueous sodium citrate). All three can be used as chemical tests for the presence of an aldose or ketose.

The Tollens' reagent indicates the presence of an aldose or ketose when the surface of the reaction flask is coated with metallic silver (looking like a mirror), while the other two tests indicate the presence of an aldose or ketose with the formation of a reddish precipitate (Cu_2O). These three reactions are used only as tests to obtain structural information about unknown carbohydrates but are not efficient as preparative methods when sufficient quantities of the aldonic acid are desired. A carbohydrate that tests positively for any of these three tests is said to be a **reducing sugar** because the carbohydrate can reduce the oxidizing agent.

These chemical tests are useful for distinguishing between aldoses or ketoses and their glycosides. While aldoses and ketoses are reducing sugars, their corresponding glycosides are not. Glycosides do not exhibit ring-opening reactions under neutral or basic conditions and therefore do not produce even small quantities of an open-chain form containing an aldehyde group. As such they are not oxidized by any of the oxidizing agents discussed.

In the presence of a stronger oxidizing agent, such as HNO_3, the primary hydroxyl group is also oxidized, giving a dicarboxylic acid, called an **aldaric acid**. This reaction involves the oxidation of both of the highlighted functional groups.

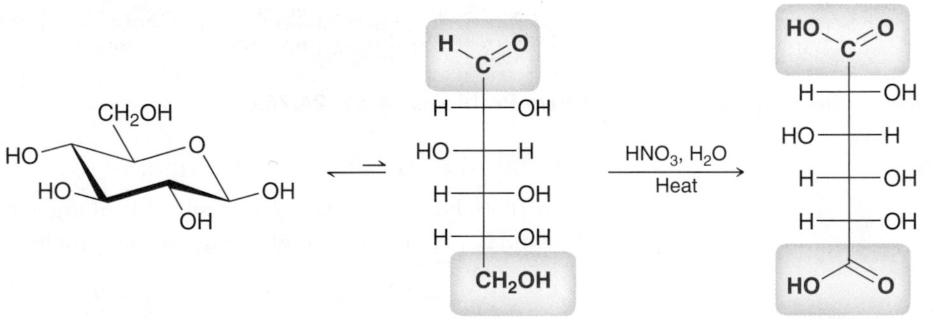

β-D-Glucopyranose D-Glucose D-Glucaric acid (an aldaric acid)

SKILLBUILDER

24.4 IDENTIFYING A REDUCING SUGAR

LEARN the skill

Identify whether the following compound is a reducing sugar:

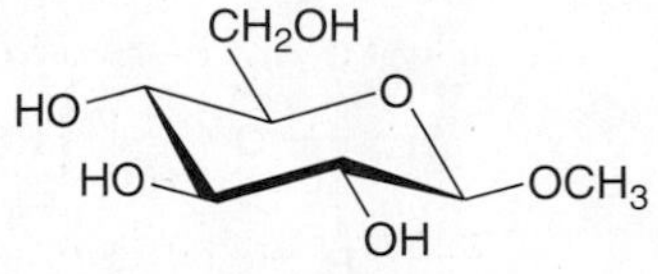

SOLUTION

Begin by identifying the anomeric position:

STEP 1
Identify the anomeric position.

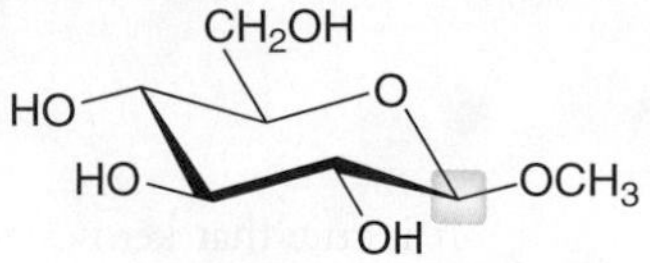

Now determine whether this compound is a hemiacetal or an acetal. Specifically, look to see if the group attached to the anomeric position is a hydroxy group (OH) or an alkoxy group (OR). If it is a hydroxy group, then the compound is a hemiacetal and will be a reducing sugar. If it is an alkoxy group, then the compound is an acetal and will not be a reducing sugar. In this case, the group attached to the anomeric position is a methoxy group. Therefore, this compound is an acetal and is not a reducing sugar.

STEP 2
Determine if the group at the anomeric position is a hydroxy group or alkoxy group.

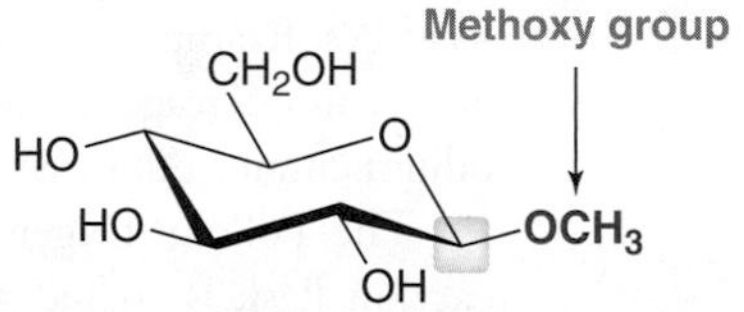

PRACTICE the skill **24.31** Determine whether each of the following compounds is a reducing sugar:

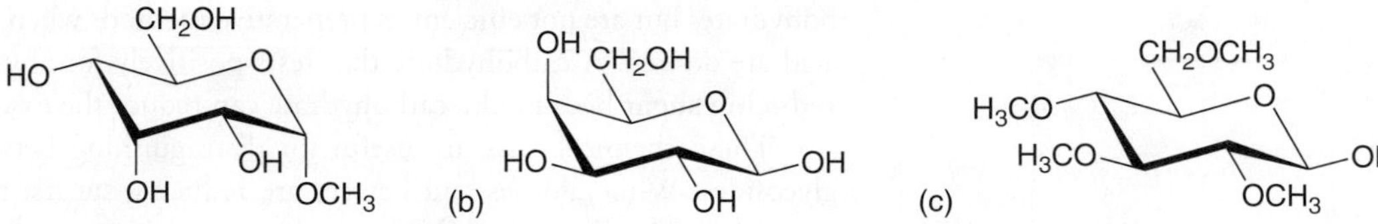

APPLY the skill **24.32** Draw and name the product obtained when each of the following compounds is treated with aqueous bromine (at pH = 6):

(a) α-D-Galactopyranose (b) β-D-Galactopyranose

(c) α-D-Glucopyranose (d) β-D-glucopyranose

24.33 Do you expect β-D-Glucopyranose pentamethyl ether to be a reducing sugar? Explain your reasoning.

need more **PRACTICE?** **Try Problems 24.69, 24.76a**

Chain Lengthening: The Kiliani-Fischer Synthesis

In 1886, Heinrich Kiliani (University of Freiburg, Germany) observed that an aldose will react with HCN to form a pair of stereoisomeric cyanohydrins.

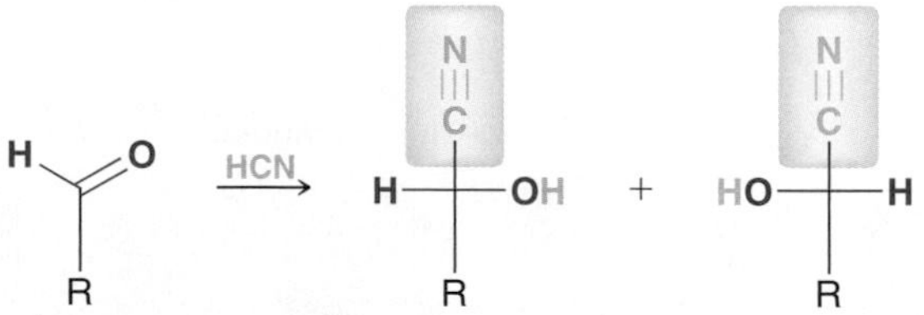

This process occurs via a nucleophilic acyl addition. A mechanism for cyanohydrin formation was discussed in Section 20.10. Building on Kiliani's observation, Emil Fischer (the same Fischer who gave us Fischer projections) then devised a multistep method for converting a cyano group to an aldehyde.

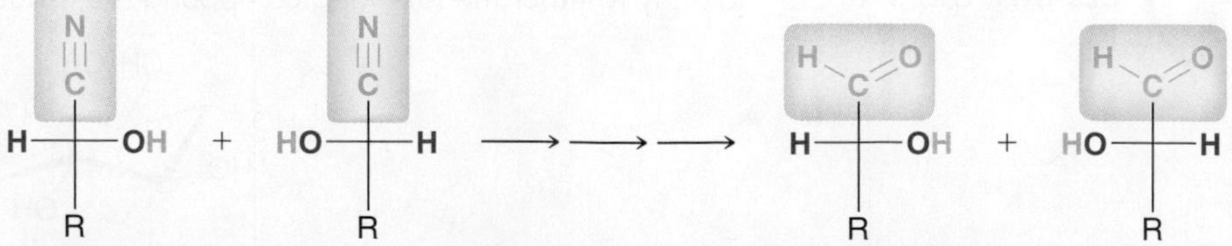

Overall, the net result is the ability to lengthen the chain of a carbohydrate by one carbon atom. For example, an aldopentose can be converted into an aldohexose using this process, called the **Kiliani-Fischer synthesis.** A more modern version of the Kiliani-Fischer synthesis enables the conversion of a cyano group into an aldehyde in just one step (rather than several steps). This is accomplished via hydrogenation of the cyano moiety using a poisoned catalyst in an aqueous solution. Under these conditions, the cyano group is reduced to give an imine, which then undergoes hydrolysis to give an aldehyde.

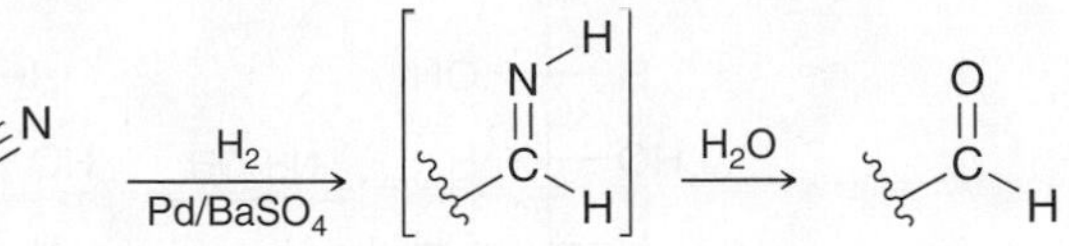

As an example of this process, consider the outcome achieved when the modern version of the Kiliani-Fischer synthesis is performed with D-arabinose as the starting material:

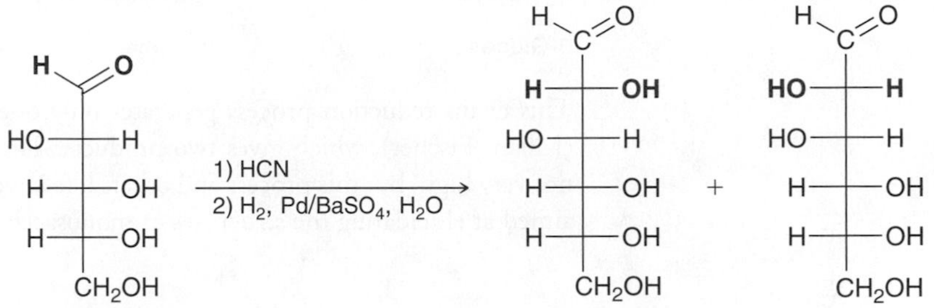

D-Arabinose D-Glucose D-Mannose

D-Arabinose has five carbon atoms (an aldopentose), and the products each have six carbon atoms (aldohexoses). In the process, C1 of the starting material is transformed into C2 in the products. Notice that two products are formed because the first step of the process is not stereoselective; that is, the new chirality center (C2) can have either configuration, giving D-glucose and D-mannose, which are C2 epimers.

CONCEPTUAL CHECKPOINT

24.34 Draw and name the pair of epimers formed when the following aldopentoses undergo a Kiliani-Fischer chain-lengthening process:

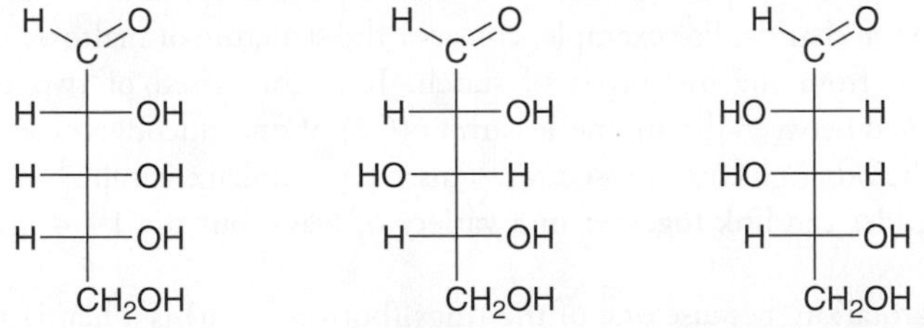

(a) D-Ribose (b) D-Xylose (c) D-Lyxose

24.35 Identify the reagents you would use to convert D-erythrose (Problem 24.22) into D-ribose. What other product is also formed in this process?

Chain Shortening: The Wohl Degradation

The **Wohl degradation** is the reverse of the Kiliani-Fischer synthesis and involves the removal of a carbon atom from an aldose. The aldehyde group is first converted to a cyanohydrin, followed by loss of HCN in the presence of a base.

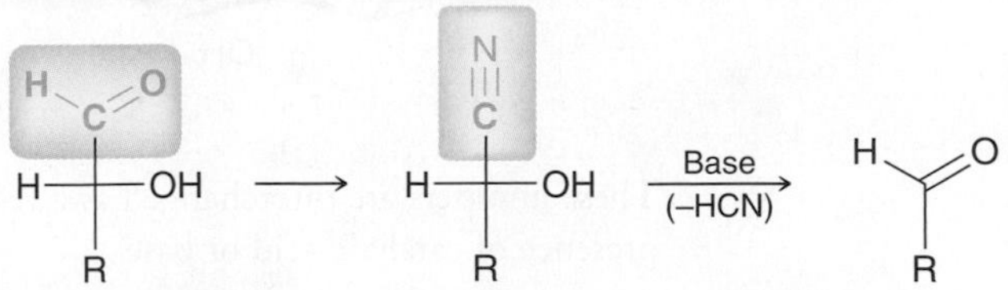

Overall, the net result is the ability to shorten a carbohydrate chain by one carbon atom. Conversion of the aldehyde into a cyano group is accomplished via oxime formation (as seen in Section 20.6) followed by dehydration. The resulting cyanohydrin then loses HCN when treated with a strong base to afford the new carbohydrate that has one less carbon atom than the starting carbohydrate:

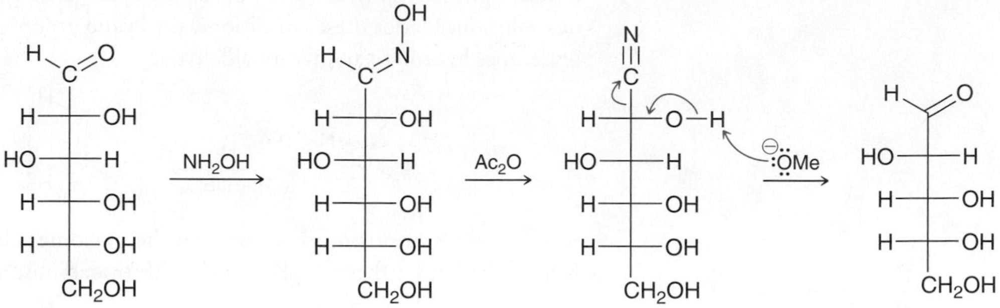

D-Glucose **An oxime** **A cyanohydrin** **D-Arabinose**

This chain reduction process generates only one product, in contrast with chain lengthening (Kiliani-Fischer), which gives two products. The yields of the Wohl degradation are generally not very high, but this process and others like it were incredibly useful during early investigations aimed at elucidating the structures of monosaccharides.

CONCEPTUAL CHECKPOINT

24.36 Draw and name the two aldohexoses that can be converted into D-ribose (Problem 24.37a) using a Wohl degradation.

24.37 Identify the reagents you would use to convert D-ribose into D-erythrose (Problem 24.22).

24.38 When D-glucose undergoes a Wohl degradation followed by a Kiliani-Fischer chain-lengthening process, a mixture of two epimeric products are obtained. Identify both epimers.

24.7 Disaccharides

Maltose

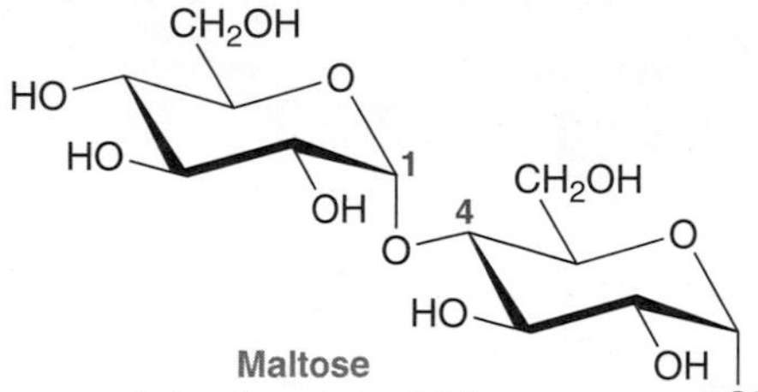

Maltose
(a 1→4 α-glycoside)

Disaccharides are carbohydrates comprised of two monosaccharide units joined via a glycosidic linkage between the anomeric carbon of one monosaccharide and a hydroxyl group of the other monosaccharide. For example, consider the structure of maltose.

Maltose is obtained from the hydrolysis of starch. It is comprised of two α-D-glucopyranose units joined between the anomeric carbon (C1) of one glucopyranose unit and the OH at C4 of the other glucopyranose unit. This type of linkage is called a 1→4 link. Two monosaccharides can link together in a variety of ways, but the 1→4 link is especially common.

Maltose undergoes mutarotation, because one of the rings (bottom right) is a hemiacetal. As a result, the anomeric position of that ring is capable of opening and closing, allowing for equilibration between the two possible anomers.

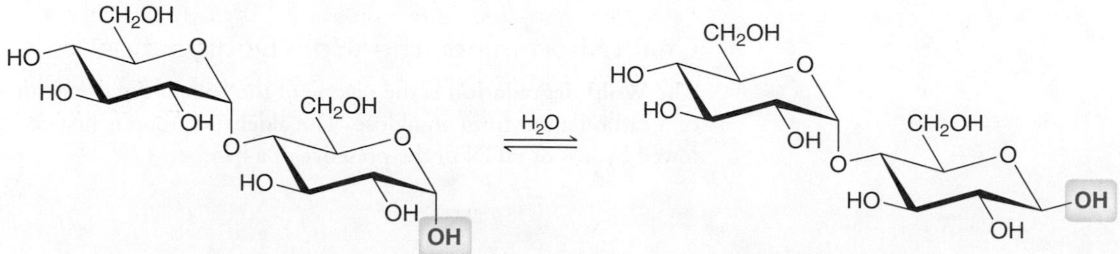

These anomers are interchanged as a result of mutarotation, which occurs more rapidly in the presence of catalytic acid or base.

Maltose is also a reducing sugar because one of the rings is a hemiacetal and exists in equilibrium with the open-chain form.

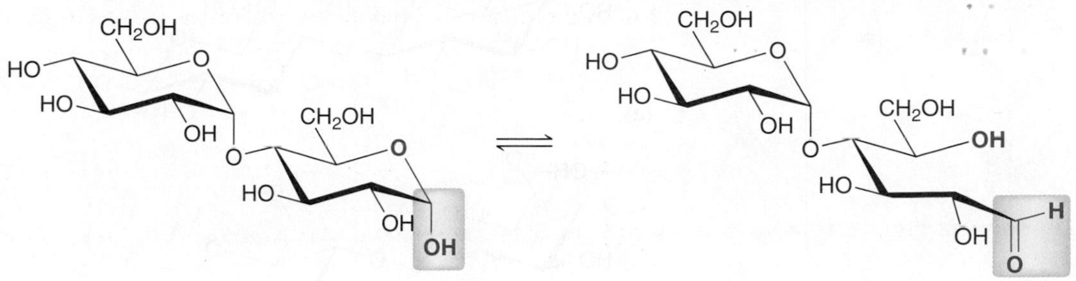

In the presence of an oxidizing agent, the aldehyde moiety of the open-chain form is oxidized to give an aldonic acid.

SKILLBUILDER

24.5 DETERMINING WHETHER A DISACCHARIDE IS A REDUCING SUGAR

LEARN the skill

Determine whether lactose is a reducing sugar:

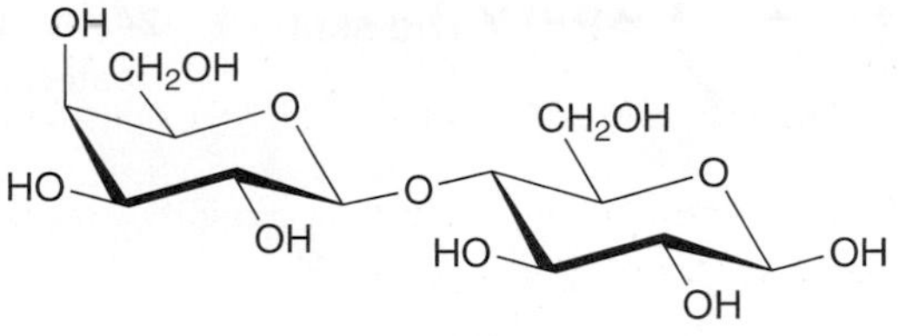

Lactose

SOLUTION

Begin by identifying the anomeric carbon on each ring. Remember the anomeric position of a ring is the position that is connected to two oxygen atoms.

STEP 1
Identify the anomeric position on each ring.

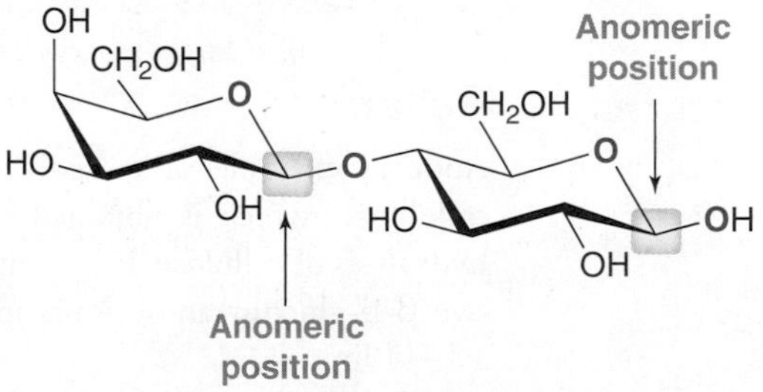

STEP 2
Determine if the group at each anomeric position is a hydroxy group or alkoxy group.

Now determine whether either of these positions is a hemiacetal. Specifically, identify whether the group attached to each anomeric position is a hydroxy group (OH) or an alkoxy group (OR). If even one of the anomeric positions has an OH group, then the compound has a hemiacetal moiety and will be a reducing sugar. If neither of the anomeric positions has an OH group, then the compound lacks a hemiacetal moiety and will not be a reducing sugar. In this case, one of the anomeric positions has an OH group. Therefore, this compound is a reducing sugar.

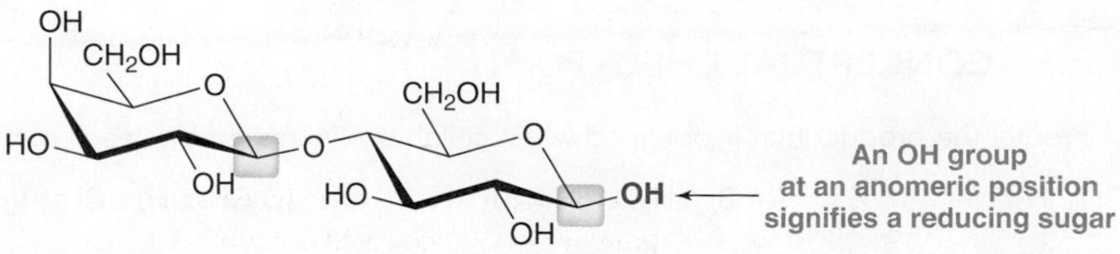

PRACTICE the skill **24.39** Determine whether each of the following disaccharides is a reducing sugar.

(a)

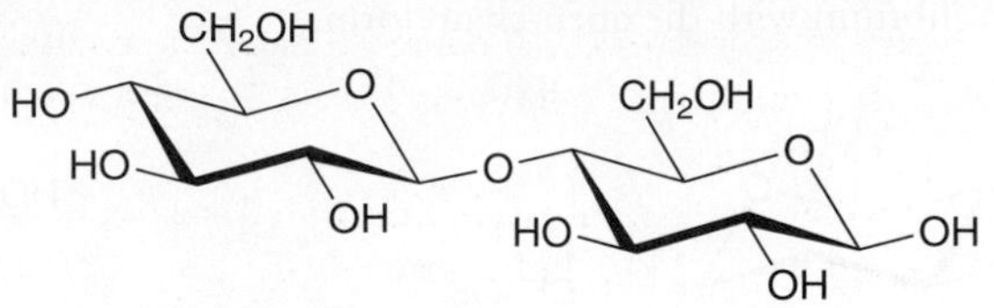

(b)

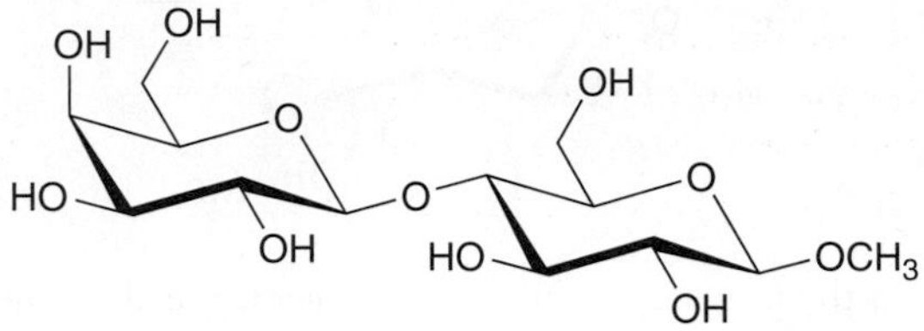

(c)

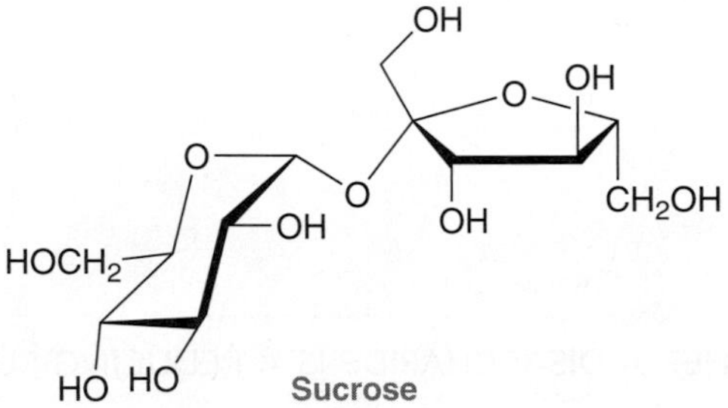

Sucrose

APPLY the skill **24.40** Draw the structure of the product obtained when the following disaccharide is treated with $NaBH_4$ in methanol.

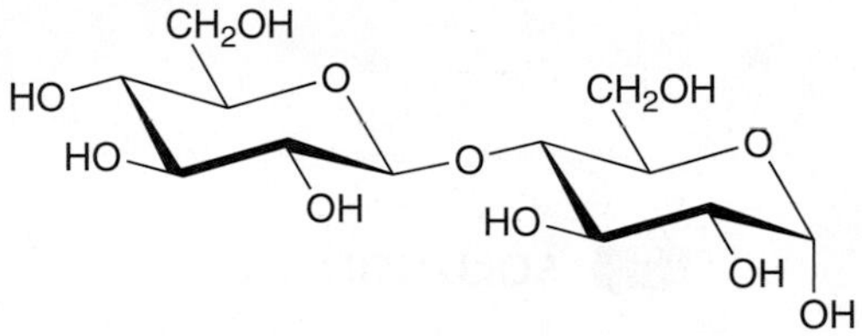

need more **PRACTICE?** **Try Problems 24.73**

Cellobiose

Another example of a disaccharide is cellobiose, which is obtained from the hydrolysis of cellulose. It is comprised of two β-D-glucopyranose units joined by a 1→4 link.

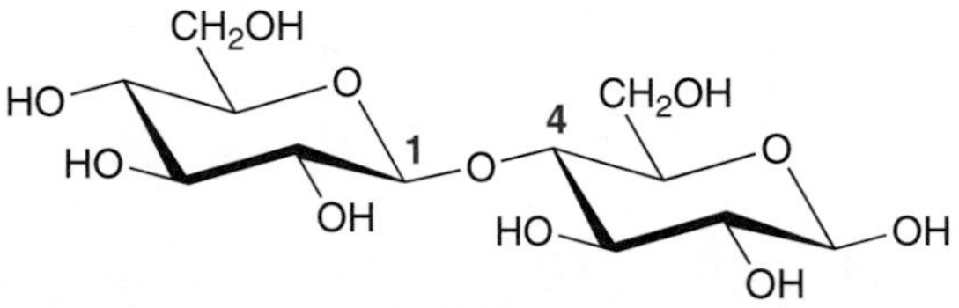

Cellobiose
(a 1→ 4 β-glycoside)

This compound is very similar in structure to maltose, only it is comprised of β-D-glucopyranose units rather than α-D-glucopyranose units. Much like maltose, cellobiose also exhibits mutarotation and is a reducing sugar, because the ring on the right is a hemiacetal and is therefore capable of opening and closing.

CONCEPTUAL CHECKPOINT

24.41 Predict the product that is obtained when cellobiose is treated with each of the following reagents:

(a) $NaBH_4$, H_2O (b) Br_2, H_2O (pH=6) (c) CH_3OH, HCl (d) Ac_2O, pyridine

Lactose

Lactose, commonly called milk sugar, is a naturally occurring disaccharide found in milk. Unlike maltose or cellobiose, lactose is comprised of two different monosaccharides—galactose and glucose.

The two monosaccharide units are joined by a 1→4 link, between C1 of galactose and C4 of glucose. Like the other disaccharides we have seen so far, lactose also exhibits mutarotation and is a reducing sugar, because the ring on the bottom right is a hemiacetal and is therefore capable of opening and closing.

Galactose (β-pyranose form)

Glucose (β-pyranose form)

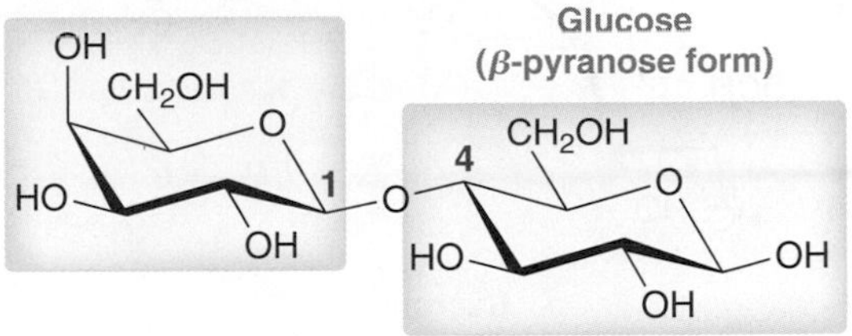

Lactose (a 1→ 4 β-glycoside)

MEDICALLYSPEAKING

Lactose Intolerance

In our bodies, an enzyme called lactase (β-galactosidase) catalyzes the hydrolysis of lactose to form glucose and galactose.

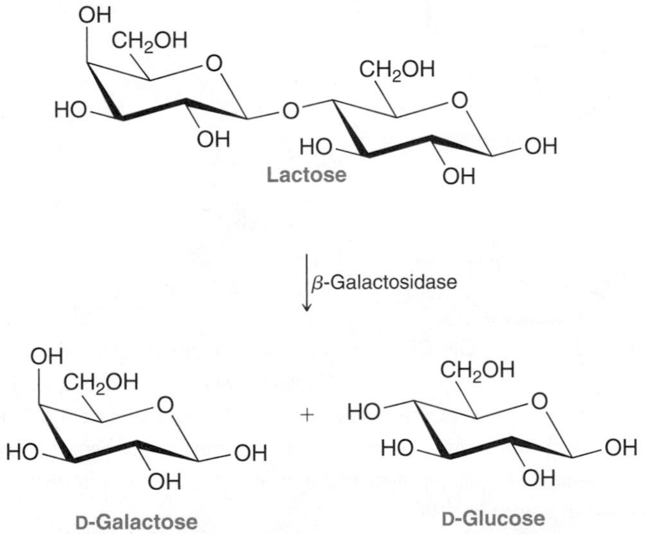

The galactose produced in this process is subsequently converted by the liver into additional glucose, which is then further metabolized to produce energy. Many people do not produce a sufficient amount of lactase and are incapable of hydrolyzing large quantities of lactose. Instead, lactose accumulates and is ultimately broken down into CO_2 and H_2 by bacteria present in the intestines. Bacterial degradation of lactose produces several by-products, including lactic acid.

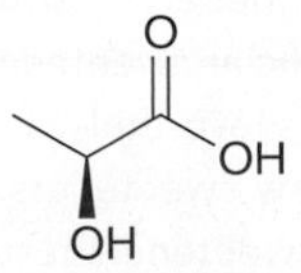

Lactic acid

The buildup of lactic acid and other acidic by-products causes cramps, nausea, and diarrhea. This condition is called lactose intolerance, and it is estimated that between 30 and 50 million Americans are lactose intolerant. Different ethnic and racial groups are affected to different extents, with the highest incidence of lactose intolerance occurring among Asians and the lowest incidence occurring among Europeans.

Lactose intolerance develops with time. Lactase production begins to decline for most children at about age two, although many people do not experience symptoms of lactose intolerance until later in life. It is estimated that 75% of the adult population worldwide will develop lactose intolerance. This condition also affects other species, including dogs and cats, both of which are particularly susceptible to developing lactose intolerance.

Lactose intolerance is easily treated through a controlled diet that minimizes the intake of food products containing lactose. Several dairy products are produced via processes that remove the lactose, and these products are marketed as being "lactose free." In addition, the enzyme lactase is available in tablet form without a prescription and can be taken prior to eating any products containing lactose.

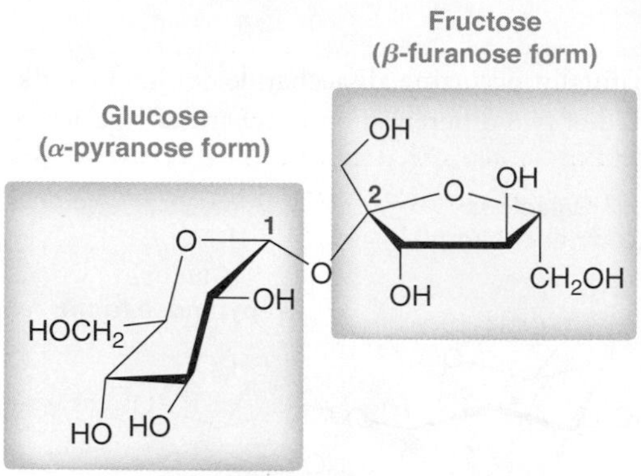

Sucrose
(a 1→2 glycoside)

Sucrose

Sucrose, commonly referred to as table sugar, is a disaccharide comprised of glucose and fructose linked at C1 of glucose and C2 of fructose. Hydrolysis of sucrose yields glucose and fructose. Honeybees have enzymes that catalyze this hydrolysis, allowing them to convert sucrose into honey, which is primarily a mixture of sucrose, glucose, and fructose. Honey is sweeter than table sugar, because fructose is sweeter than sucrose.

Unlike the other disaccharides we have seen thus far, sucrose is not a reducing sugar and does not undergo mutarotation. This can be explained by noting that sucrose is comprised of two units that are linked to each other via their anomeric carbons. As such, neither unit has a hemiacetal moiety and neither unit is capable of adopting an open-chain form.

PRACTICALLYSPEAKING)))

Artificial Sweeteners

A variety of health problems have been associated with excessive consumption of sucrose, including diabetes and tooth decay. These issues, together with the desire of many people to reduce their caloric intake, have fueled the development of many artificial sweeteners, such as the following compounds.

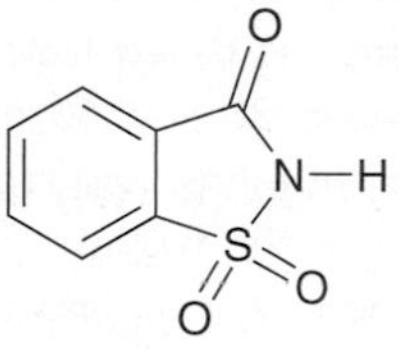

Saccharin

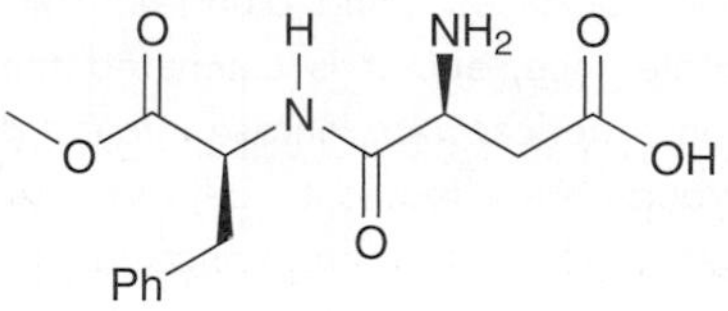

Aspartame

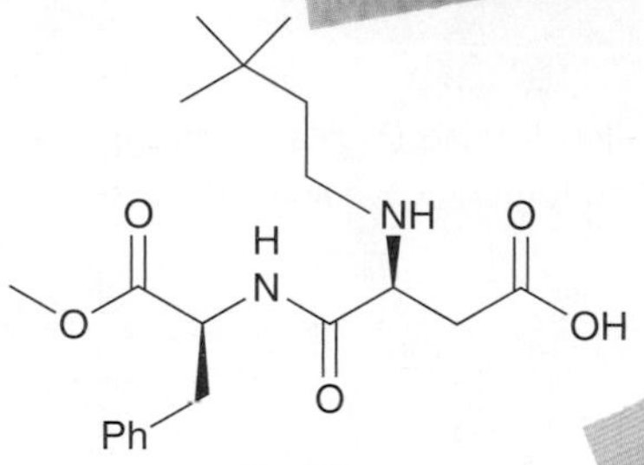

Neotame

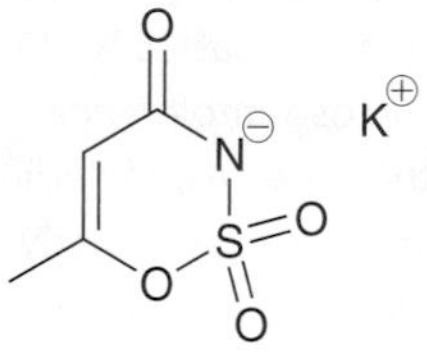

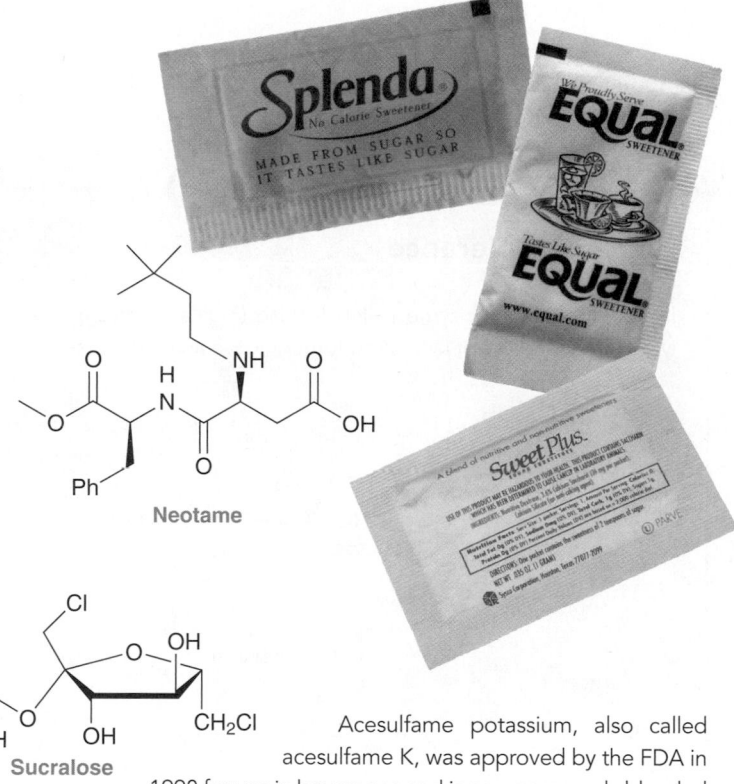

Acesulfame K

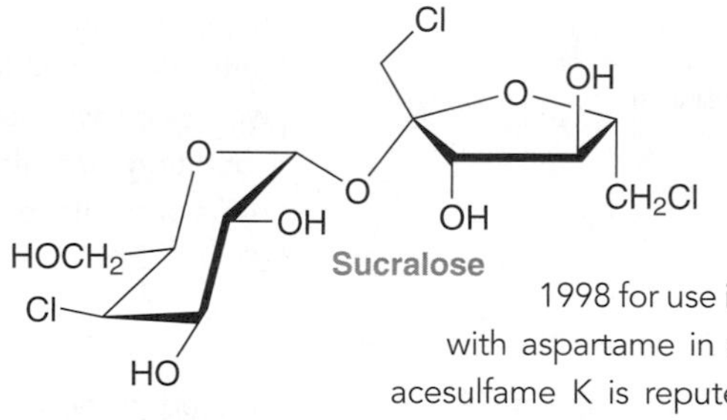

Sucralose

Saccharin is the oldest known artificial sweetener. It was discovered by accident in 1879 and was originally used by diabetics as an alternative to sucrose. In the 1970s some studies suggested that it might be carcinogenic. Extensive research since then has indicated that it is safe for consumption. Nevertheless, due to its metallic aftertaste, it has been largely replaced by many of the other artificial sweeteners on the market.

Aspartame (sold under the trade name NutraSweet) was approved by the FDA in 1981 and has been widely used in soft drinks ever since. It is about 200 times sweeter than sucrose. People suffering from a condition called *phenylketonuria* are incapable of fully metabolizing aspartame and must avoid consuming this compound, as discussed in Problem 21.61. A derivative of aspartame called Neotame was approved by the FDA in 2001 and is about 10,000 times sweeter than sucrose. Neotame must also be avoided by people with phenylketonuria.

Acesulfame potassium, also called acesulfame K, was approved by the FDA in 1998 for use in beverages and is now commonly blended with aspartame in soft drinks. The mixture of aspartame and acesulfame K is reputed to have less of a bitter aftertaste than either compound alone.

Among the common artificial sweeteners, sucralose (600 times sweeter than sucrose) is the only compound that resembles the structure of a carbohydyrate. It is prepared from sucrose, whereby three of the hydroxyl groups are replaced with chlorine atoms. The presence of the chlorine atoms prevents the body from effectively metabolizing the compound and releasing its energy content. Sucralose is especially popular for use in baked goods, because it does not decompose upon heating, as is the case for many of the other artificial sweeteners.

Many of the artificial sweeteners on the market were discovered by accident. Nevertheless, extensive ongoing research is aimed at finding new sweeteners with even better properties (none of the artificial sweeteners taste exactly like sucrose, to the dismay of many consumers). Over the decades to come, we are likely to see many other sugar substitutes enter the market.

24.8 Polysaccharides

Cellulose

Polysaccharides are polymers made up of repeating monosaccharide units linked together by glycoside bonds. For example, cellulose is comprised of several thousand D-glucose units linked by 1→4 β-glycoside bonds.

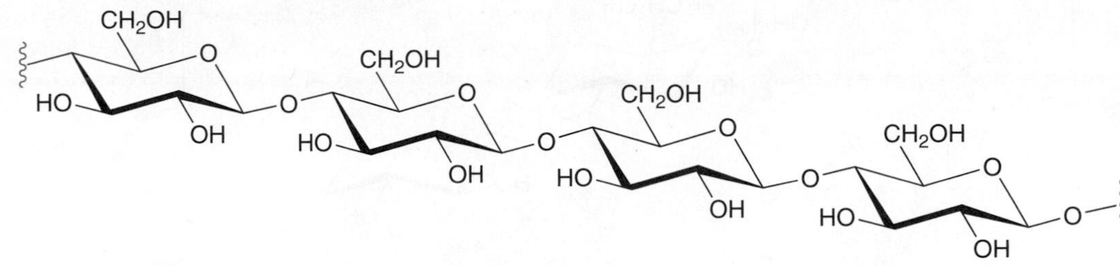

Cellulose
(a 1 → 4 *O*-(β-glucopyranoside) polymer

The average cellulose strand is comprised of approximately 7,000 glucose units but can be as large as 12,000 units. The strands of the polymer interact with each other via hydrogen bonding, providing structural rigidity for trees and plants. Wood is comprised of 30–40% cellulose, and cotton is comprised of 90% cellulose.

Starch

Starch is the major component of many of the foods that we eat, including potatoes, corn, and cereal grain. Starch can be separated into two components: amylose, which is insoluble in cold water, and amylopectin, which is soluble in cold water. Amylose is comprised of glucose units linked by 1→4 α-glycoside bonds.

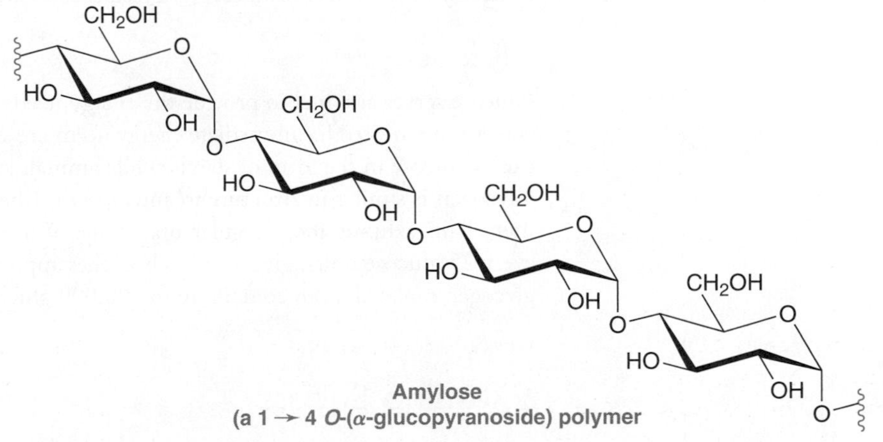

Amylose
(a 1 → 4 *O*-(α-glucopyranoside) polymer

A strand of amylose is linear, like cellulose, but is comprised of repeating α-D-glucopyranose units rather than β-D-glucopyranose units. Amylose accounts for approximately 20% of starch. The remaining 80% of starch is amylopectin, which is similar to amylose but also contains 1→6 α-glycoside branches approximately every 25 glucose units:

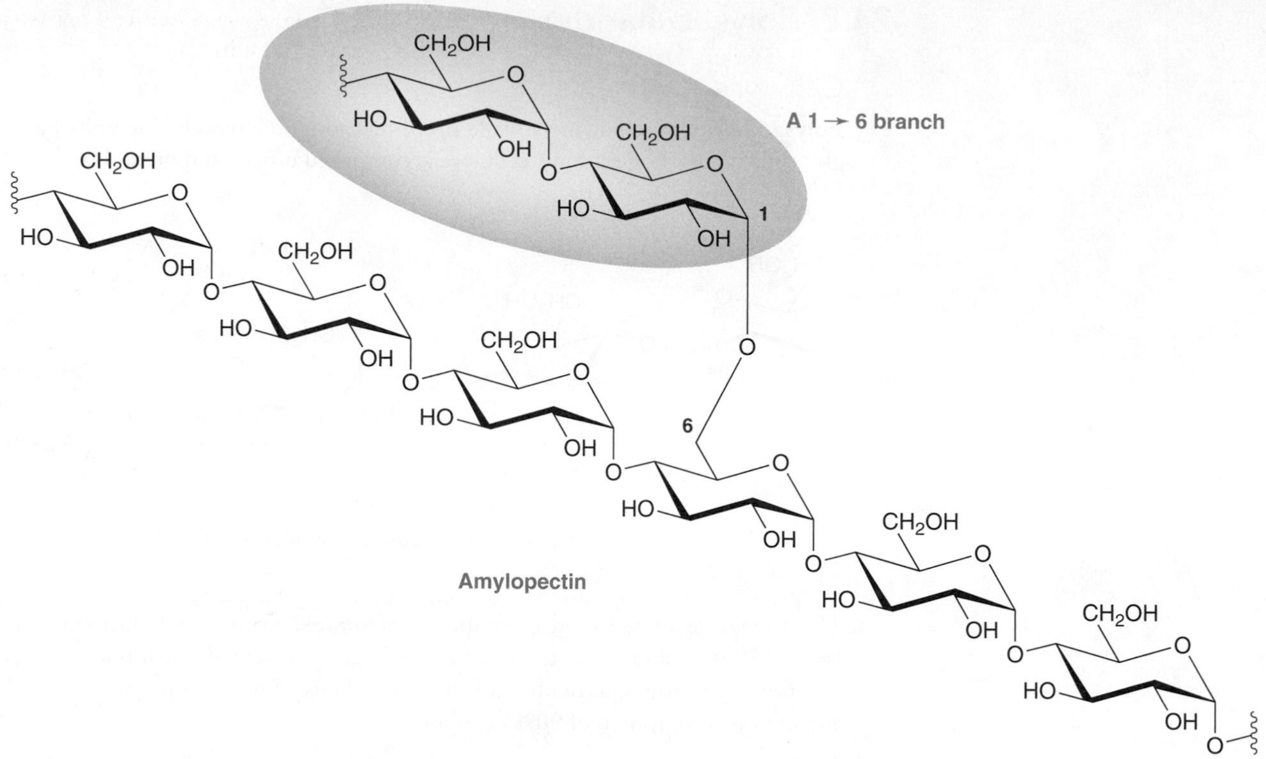

Amylopectin

Starch is digested by glycosidase enzymes that catalyze its hydrolysis, releasing individual molecules of glucose. These enzymes are highly selective in their activity and will not catalyze the hydrolysis of cellulose. There are organisms that have enzymes that can catalyze the hydrolysis of cellulose, but human beings lack those enzymes, which is why we can eat corn and potatoes but we cannot eat grass. Cows are also incapable of synthesizing the enzymes necessary to digest cellulose; however, the stomachs of cows contain microorganisms that do produce enzymes necessary to hydrolyze cellulose. As a result, cows are able to digest grass.

Glycogen

Glucose serves as a fuel to provide the energy needs for both plants and animals. Glucose molecules not required for immediate energy needs are stored in the form of a polymer. Plants store excess glucose in the form of starch, while animals store excess glucose in the form of glycogen. Glycogen is similar in structure to amylopectin (the major component of starch), but its polymer chain exhibits more regular branching. While amylopectin has branches approximately every 25 glucose units, glycogen has branches approximately every 10 glucose units. Individual glycogen molecules can contain up to 100,000 glucose units.

24.9 Amino Sugars

Amino sugars are carbohydrate derivatives in which an OH group has been replaced with an amino group. Amino sugars are quite common in nature and serve as important building blocks for biological polymers. One such example is β-D-glucosamine, which is biosynthesized from D-glucose:

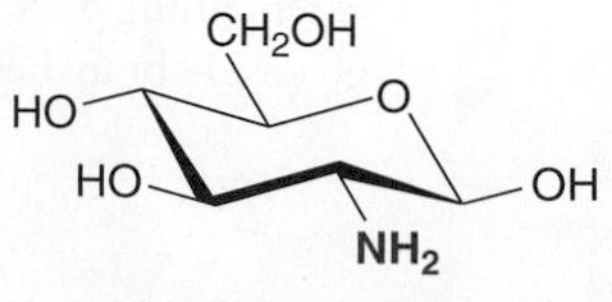

β-D-Glucosamine
(an amino sugar)

The *N*-acetyl derivative of this amino sugar serves as the repeating monosaccharide unit in the important biopolymer called chitin (pronounced "kite-in"):

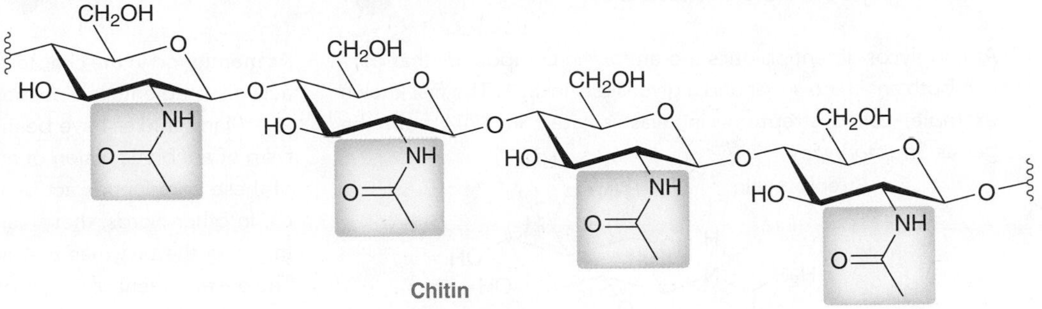

Chitin

Chitin is similar in structure to cellulose, but the amide moieties allow for more significant hydrogen bonding between neighboring strands, which renders the polymer even stronger than cellulose (wood). Chitin is the material used in the exoskeletons of arthropods and insects, and over a trillion pounds of this polymer are produced by living organisms each year.

24.10 *N*-Glycosides

When treated with an amine in the presence of an acid catalyst, monosaccharides are converted into their corresponding ***N*-glycosides**:

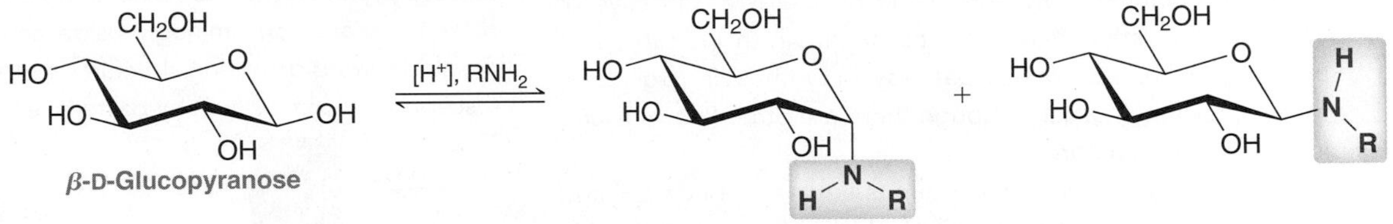

β-D-Glucopyranose   An α-*N*-glycoside   A β-*N*-glycoside

The mechanism for this process is analogous to that of glycoside formation. Two carbohydrates, in particular, D-ribose and 2-deoxy-D-ribose, form especially important *N*-glycosides.

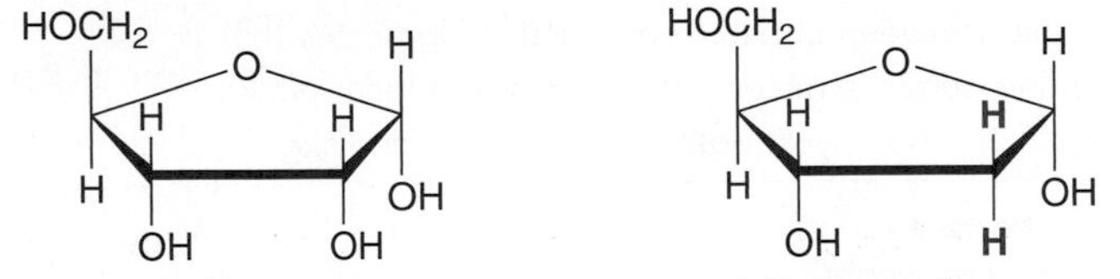

D-ribose (α-furanose form)   2-Deoxy-D-ribose (α-furanose form)

These two carbohydrates serve as the building blocks for RNA and DNA, respectively. These compounds are biologically coupled with certain nitrogen heterocycles (called bases) to yield special *N*-glycosides, called **nucleosides**. In each case, the β anomers are formed exclusively.

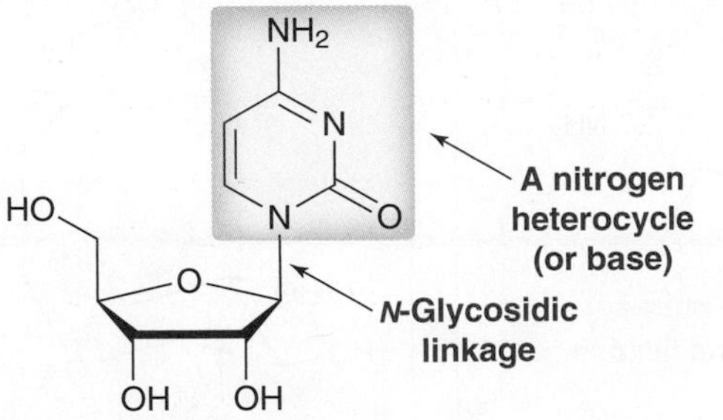

A ribonucleoside

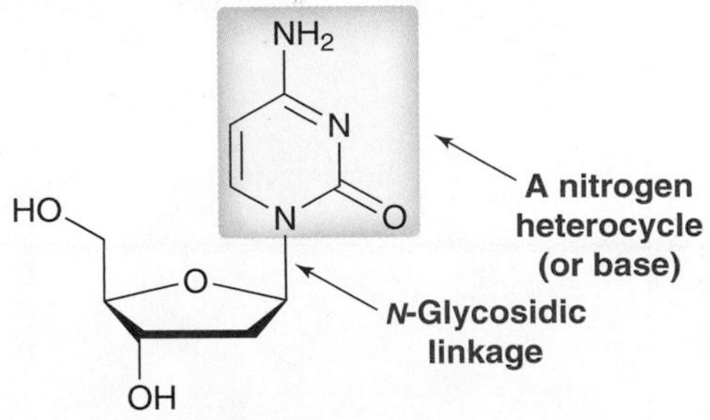

A deoxyribonucleoside

MEDICALLYSPEAKING)))

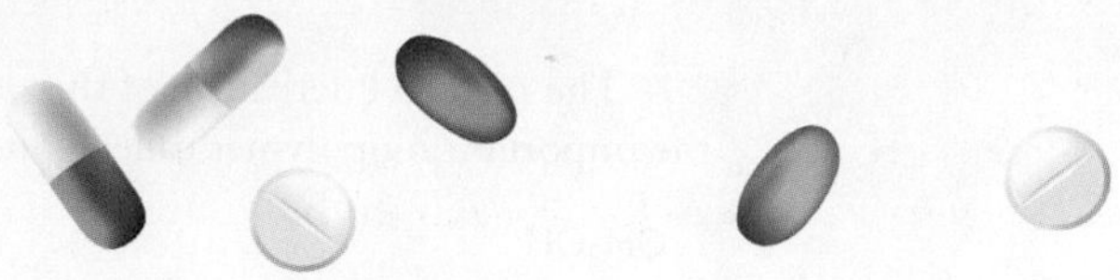

Aminoglycoside Antibiotics

Aminoglycoside antiobiotics are antibiotic compounds that contain both an amino sugar and a glycosidic linkage. The first known example, called streptomycin, was isolated in 1944 from the genus *Streptomyces*:

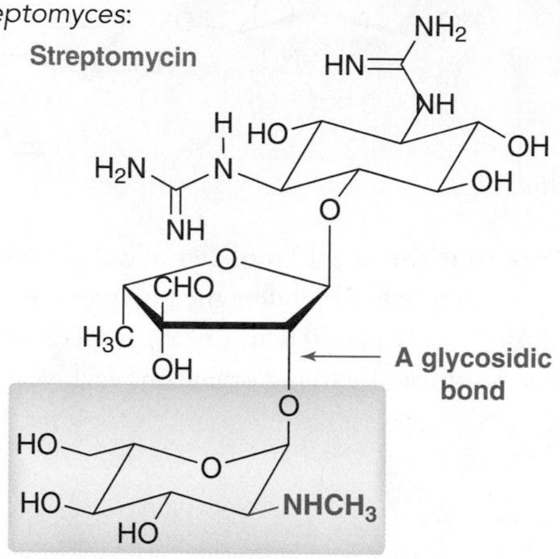

An amino hexose

Notice that the structure of streptomycin contains an amino sugar (highlighted). Interestingly, this amino hexose is a derivative of L-glucosamine, rather than D-glucosamine, which indicates that *Streptomyces* has developed a pathway for synthesizing L-glucose; this is somewhat unusual, although there are other known examples of L sugars in nature.

Many other antibiotic compounds have also been isolated from the genus *Streptomyces*, all of which are closely related in structure to streptomycin. Six of them are currently in use in the United States: kanamycin, neomycin, paromomycin, gentamicin, tobramycin, and netilmicin. All of these compounds exhibit at least one aminohexose in their structure, and extensive research on the relationship between structure and activity indicates that at least one amino sugar is necessary for antibiotic activity. As shown, kanamycin and neomycin both contain two amino sugars.

As mentioned in the chapter opener, neomycin is one of the three active ingredients in Neosporin.

Many studies have been undertaken to elucidate the mechanism of antibiotic action of aminoglycosides. It is believed that all of these compounds act by inhibiting protein synthesis in bacteria. In other words, these compounds prevent bacteria from synthesizing the enzymes that are necessary for proper functioning. The development of strains of antibiotic-resistant bacteria is a very serious problem in the field of medicine. The aminoglycosides are no exception to this problem. As clinical use of aminoglycoside antibiotics became more popular, some strains of bacteria have become resistant to them. Specifically, many strains of bacteria have developed the ability to produce enzymes that catalyze the modification of hydroxyl and amino groups of these antibiotics. For example, bacteria have emerged that are capable of transforming six different functional groups in kanamycin B. Some of these functional groups undergo enzyme-catalyzed acetylation, while others are phosphorylated. If any one of these groups is acetylated or phosphorylated, the modified antibiotic is no longer capable of binding to the bacterial RNA. Much of the current research on aminoglycoside antibiotics is directed toward designing structures that are less susceptible to inactivation by bacterial enzymes.

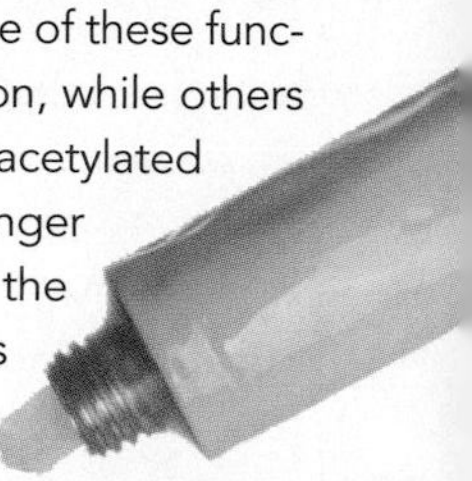

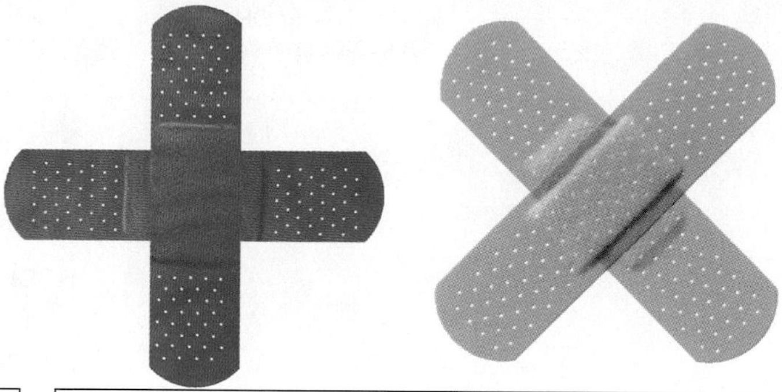

An amino hexose

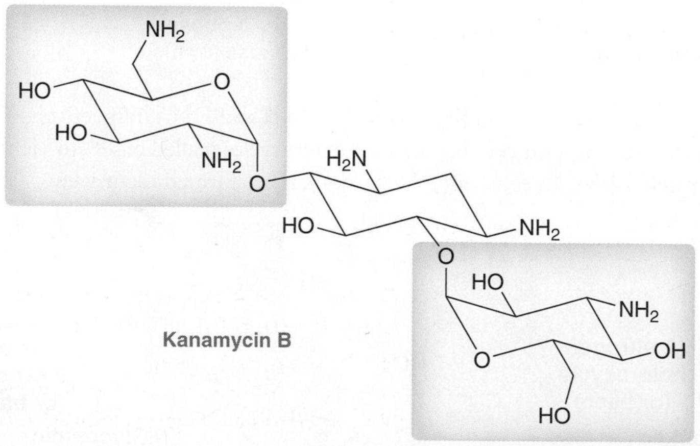

An amino hexose

An amino hexose

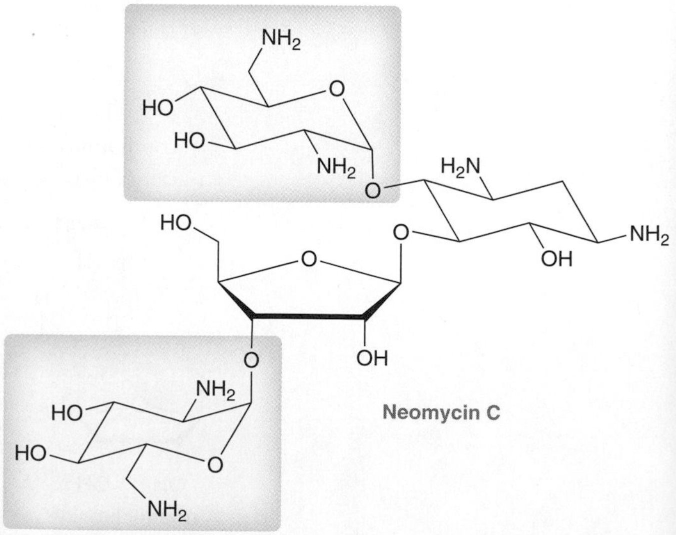

An amino hexose

DNA

Four kinds of heterocyclic amines are found as bases in DNA: cytosine (C), thymine (T), adenine (A), and guanine (G):

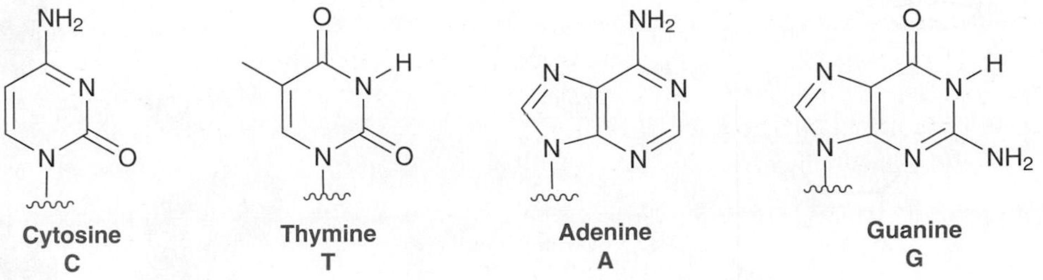

Each of these four bases can couple to 2-deoxyribose, giving rise to four possible deoxyribonucleosides (Figure 24.9).

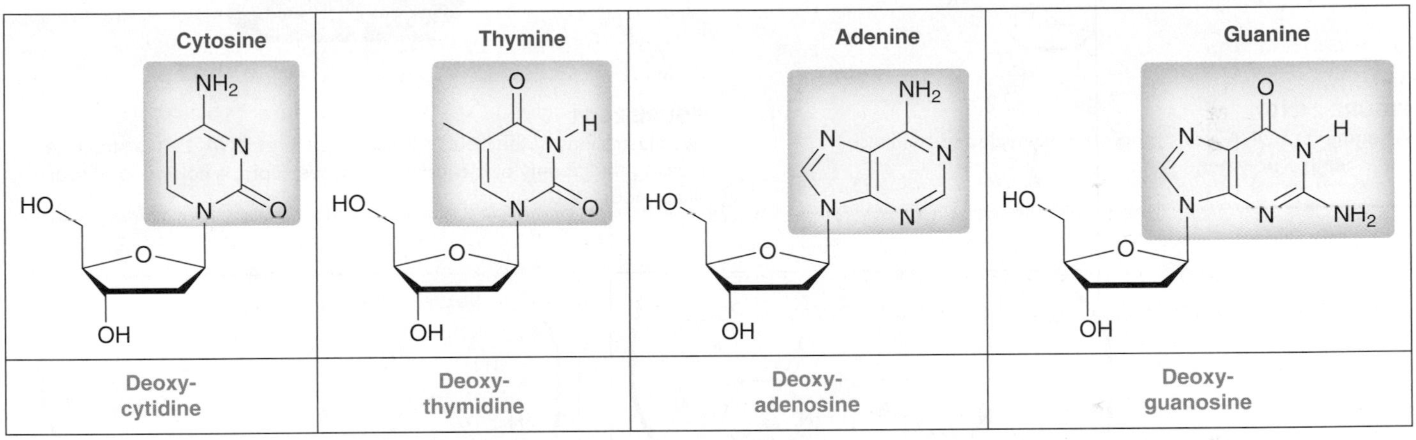

FIGURE 24.9
Structures of the four naturally occurring deoxyribonucleosides present in DNA.

Each of these four nucleosides can then be coupled to a phosphate group, giving compounds called **nucleotides** (rather than nucleosides). One of the four possible deoxyribonucleotides is shown at the right:

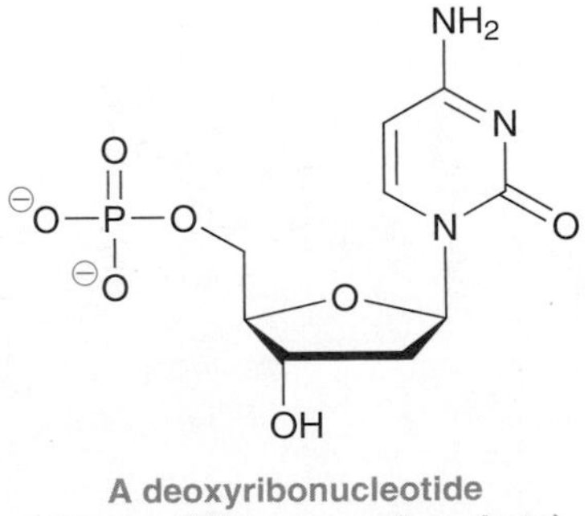

A deoxyribonucleotide (deoxycytidine monophosphate)

The difference between a nucleoside and a nucleotide is the presence of a phosphate group. Deoxyribonucleotides are comprised of three parts: deoxyribose, a nitrogen-containing base, and a phosphate group. When linked together, they serve as the building blocks of DNA (Figure 24.10).

The segment of DNA in Figure 24.10 shows nucleotides linked together in a polymer, or **polynucleotide**. Notice that each sugar is connected to two phosphate groups and serves as the backbone for DNA. The familiar double-helix form of DNA is formed from two polynucleotide strands twisted into a shape that resembles a spiral ladder (Figure 24.11). The rungs of the ladder are hydrogen bonds between the bases, which interact with each other in pairs. As shown in Figure 24.12, cytosine (C) forms hydrogen bonds with guanine (G), while adenine (A) forms hydrogen bonds with thymine (T). The two strands of the spiral ladder are therefore complementary strands. They can be separated from each other much the way a zipper can be opened. DNA encodes all of our genetic information and serves as a template for assembling RNA, described in the following section.

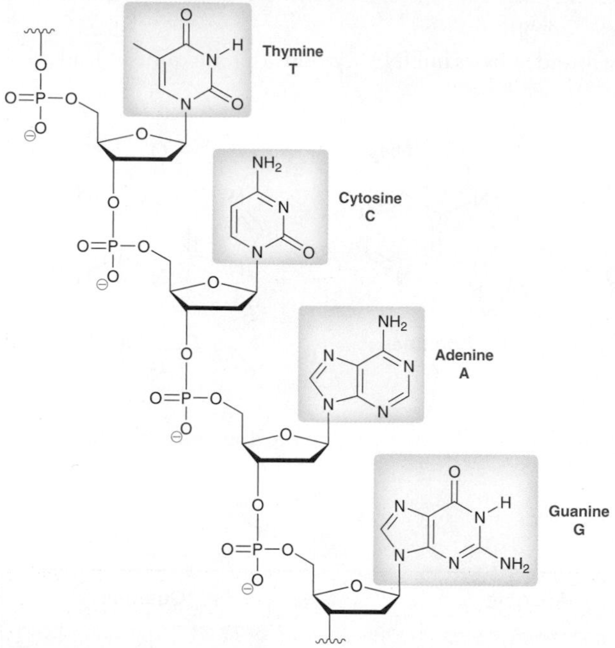

FIGURE 24.10
A segment of DNA, consisting of deoxynucleotides linked together in a polymer.

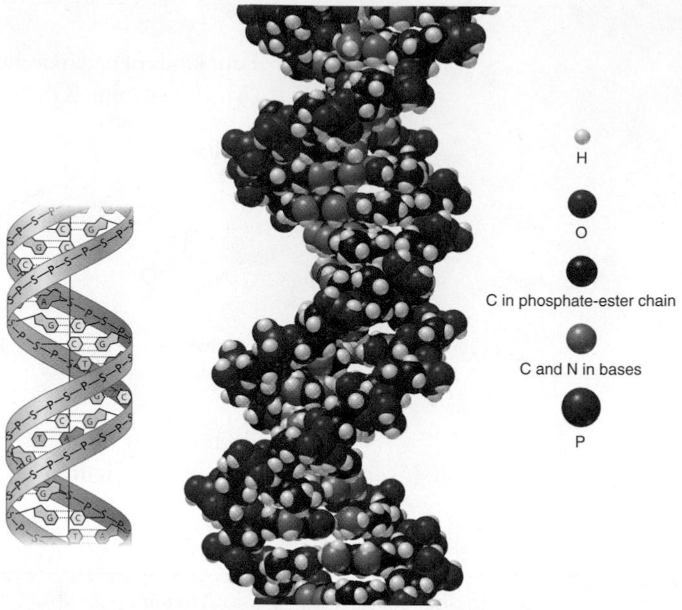

FIGURE 24.11
Two illustrations of the double-helix structure of DNA. Left, a structure showing the identify of the individual bases. Right, a color-coded space-filling model of DNA.

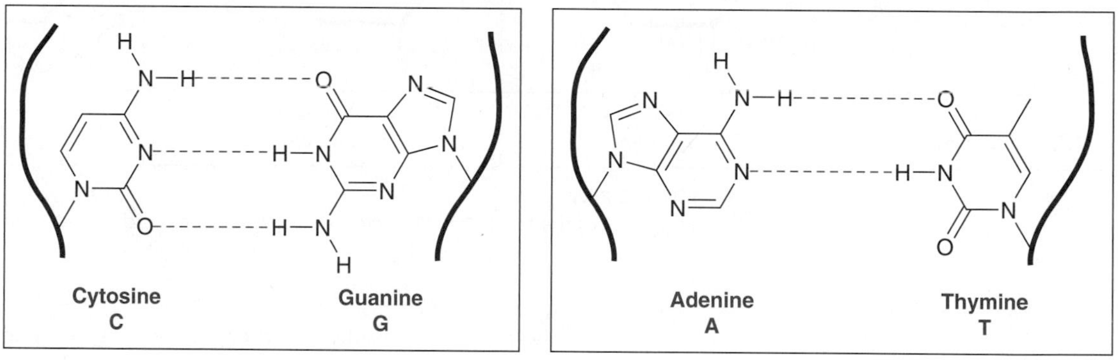

FIGURE 24.12
The hydrogen-bonding interactions that occur between complementary base pairs in DNA.

RNA

A strand of RNA is very similar in structure to a single strand of DNA. Both are comprised of repeating nucleotide units where each unit consists of a sugar, a phosphate, and a base. There are two important differences between DNA and RNA: (1) the sugar in DNA is D-deoxyribose while the sugar in RNA is D-ribose and (2) in place of the thymine found in DNA, RNA contains a base called uracil (U). The difference between thymine and uracil is that thymine contains a methyl group that is lacking in uracil. Like thymine, uracil forms hydrogen bonds with adenine (A).

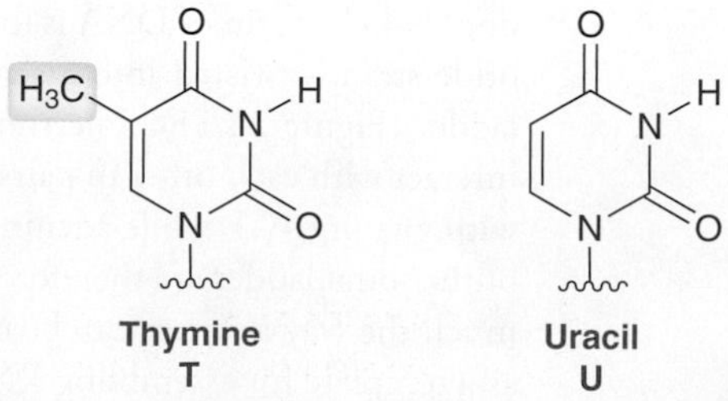

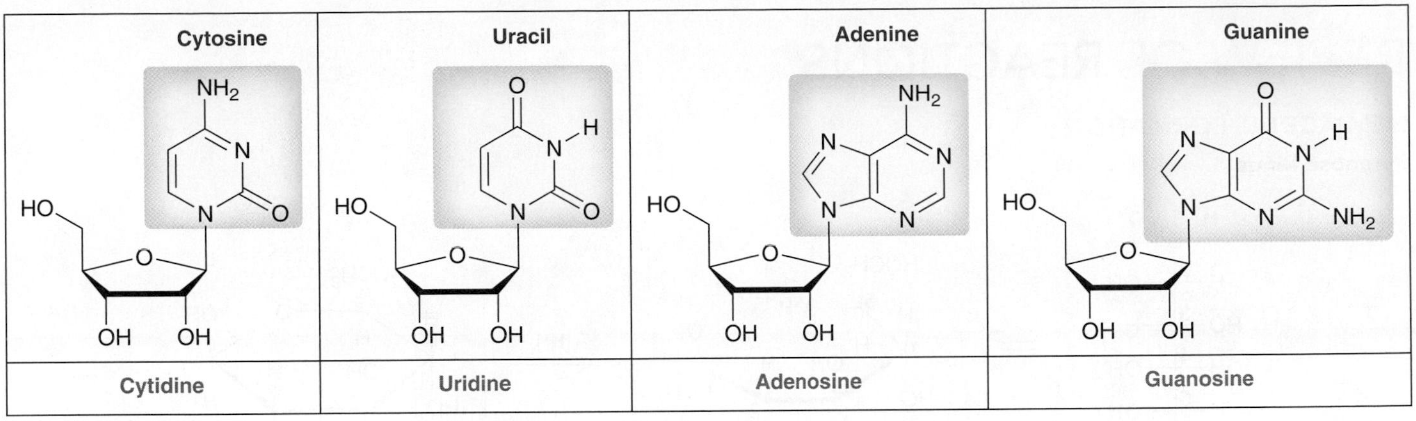

FIGURE 24.13
Structures of the four naturally occurring ribonucleosides present in RNA.

The four bases from which RNA is comprised are C, G, A, and U, giving rise to four possible nucleosides (Figure 24.13). Each of these nucleosides is coupled with a phosphate group, giving rise to four possible nucleotides which are the building blocks of RNA. A segment of RNA is shown in Figure 24.14. RNA directs the assembly of the proteins and enzymes that are used to catalyze the chemical reactions occurring in cells. Proteins and enzymes are the subjects of Chapter 26.

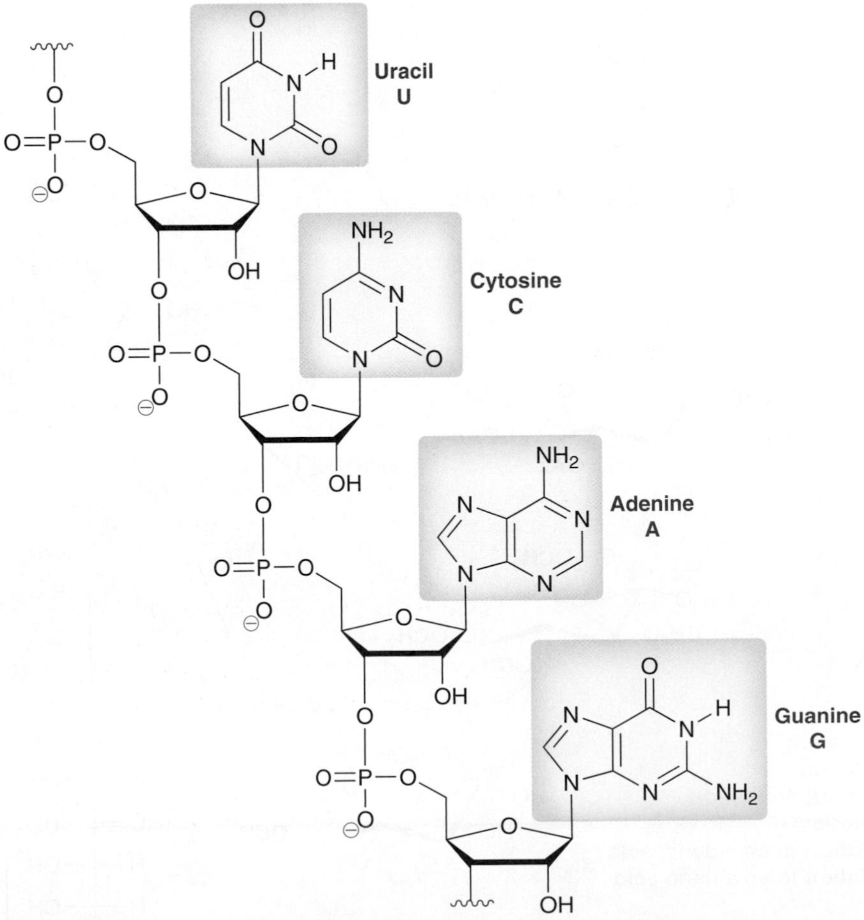

FIGURE 24.14
A segment of RNA, consisting of nucleotides linked together in a polymer.

REVIEW OF REACTIONS

HEMIACETAL FORMATION

Pyranose Rings

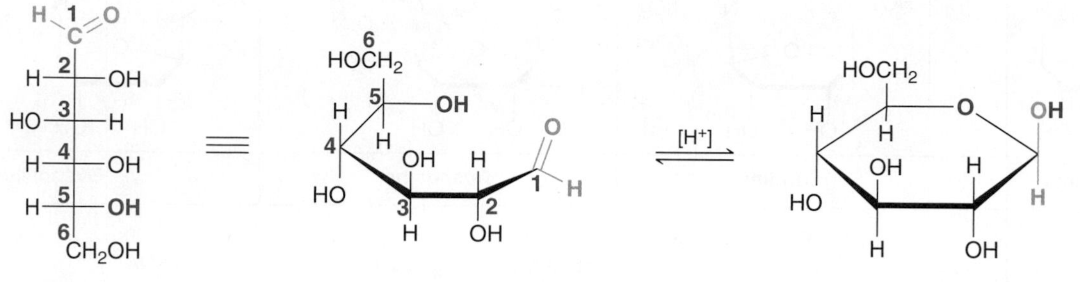

D-Glucose

A pyranose ring

FURANOSE RINGS

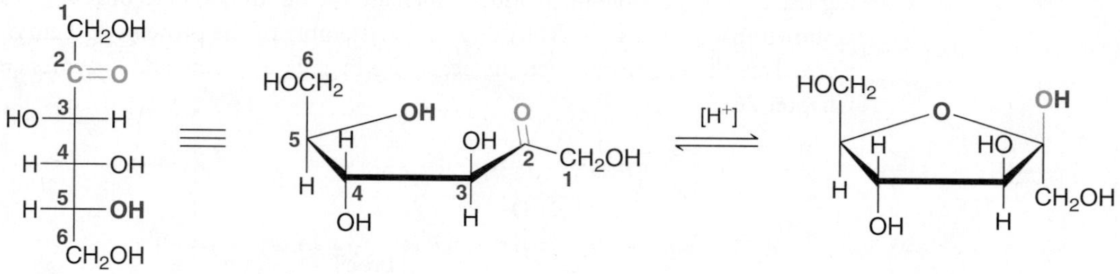

D-Fructose

A furanose ring

REACTIONS OF MONOSACCHARIDES

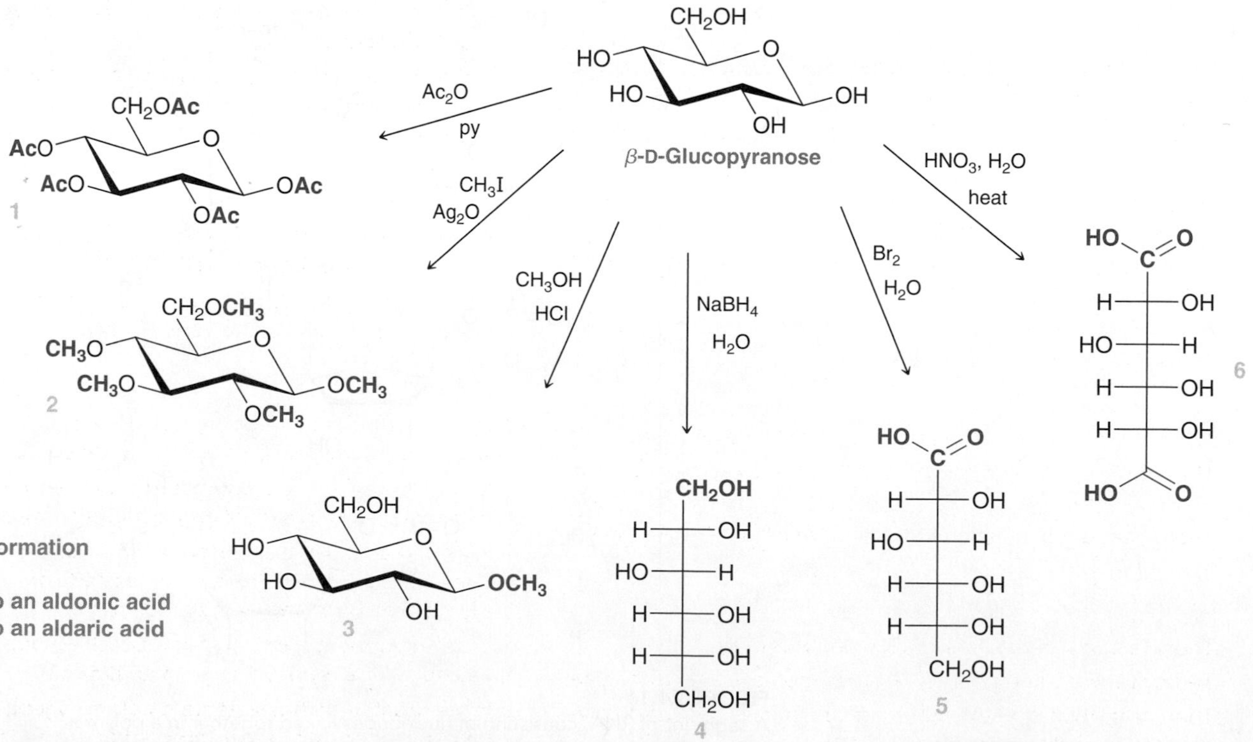

1. **Acetylation**
2. **Alkylation**
3. **Glycoside formation**
4. **Reduction**
5. **Oxidation to an aldonic acid**
6. **Oxidation to an aldaric acid**

CHAIN LENGTHENING AND CHAIN SHORTENING

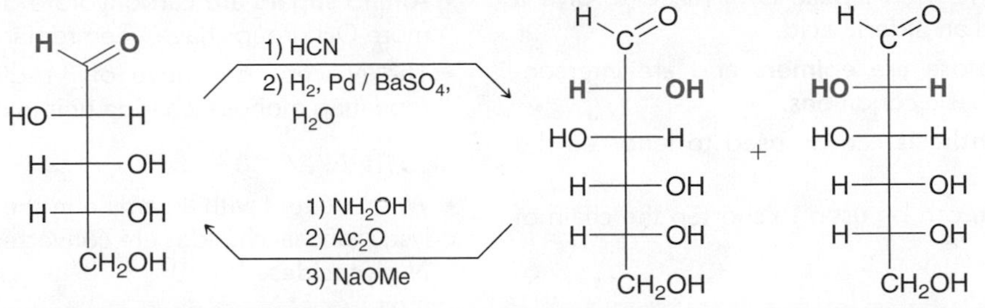

D-Arabinose — D-Glucose + D-Mannose

KEY TERMINOLOGY

α anomer 1148
aldaric acid 1159
alditol 1157
aldonic acid 1158
aldose 1141
amino sugars 1168
anomer 1148
anomeric carbon 1148
β anomer 1148
carbohydrates 1140
disaccharides 1162
D sugar 1142
epimer 1157
furanose 1153
glycosides 1155
ketose 1141
Kiliani-Fischer synthesis 1161
L sugar 1142
monosaccharides 1140
mutarotation 1149
N-glycosides 1169
nucleosides 1169
nucleotides 1171
polynucleotides 1171
polysaccharides 1167
pyranose ring 1148
reducing sugar 1159
Wohl degradation 1161

REVIEW OF CONCEPTS AND VOCABULARY

SECTION 24.1

- **Carbohydrates** are polyhydroxy aldehydes or ketones.
- In nature, carbohydrates are used for energy storage and structural rigidity.

SECTION 24.2

- Simple sugars are called **monosaccharides** and are generally classified as **aldoses** and **ketoses.**
- Aldoses and ketoses are further classified according to the number of carbon atoms they contain.
- (+)-glyceraldehyde, called D-glyceraldehyde, is abundant in nature, but its enantiomer is not.
- For all **D sugars,** the chirality center farthest from the carbonyl group has the *R* configuration.
- An **L sugar** is the enantiomer of the corresponding D sugar and is not necessarily levorotatory.

SECTION 24.3

- There are two D aldotetroses called D-erythrose and D-threose.
- There are four D aldopentoses called D-ribose, D-arabinose, D-xylose, and D-lyxose.
- There are eight D aldohexoses, of which D-glucose is the most abundant.

SECTION 24.4

- There is one D ketotetrose called D-erythrulose.
- There are two D ketopentoses called D-ribulose and D-xylulose.
- There are four D ketohexoses called D-psicose, D-fructose, D-sorbose, and D-tagatose.

SECTION 24.5

- Aldohexoses can form cyclic hemiacetals that exhibit a **pyranose** ring.
- Cyclization produces two stereoisomeric hemiacetals, called **anomers.** The newly created chirality center is called the **anomeric carbon.**
- In the α **anomer,** the hydroxyl group at the anomeric position is *trans* to the CH_2OH group, while in the β **anomer,** the hydroxyl group is *cis* to the CH_2OH group.
- The open-chain form exhibits only a minimal presence at equilibrium.
- Anomers equilibrate by a process called **mutarotation,** which is catalyzed by either acid or base.
- Some carbohydrates, such as D-fructose, can also form five-membered rings, called **furanose** rings.

SECTION 24.6

- Monosaccharides are converted into their ester derivatives when treated with excess acid chloride or anhydride.
- Monosaccharides are converted into their ether derivatives when treated with excess alkyl halide and silver oxide.
- When treated with an alcohol under acid-catalyzed conditions, monosaccharides are converted into acetals, called **glycosides.** Both anomers are formed.
- Upon treatment with sodium borohydride, an aldose or ketose can be reduced to yield an **alditol.**
- When treated with a suitable oxidizing agent, an aldose can be oxidized to yield an **aldonic acid.**

- Aldoses and ketoses are **reducing sugars.**
- When treated with HNO_3, an aldose is oxidized to give a dicarboxylic acid called an **aldaric acid.**
- D-Glucose and D-mannose are **epimers** and are interconverted under strongly basic conditions.
- The **Kiliani-Fischer synthesis** can be used to lengthen the chain of an aldose.
- The **Wohl degradation** can be used to shorten the chain of an aldose.

SECTION 24.7

- **Disaccharides** are comprised of two monosaccharide units joined together via a glycosidic linkage.
- Examples of disaccharides are maltose, cellobiose, lactose, and sucrose.

SECTION 24.8

- **Polysaccharides** are polymers consisting of repeating monosaccharide units linked by glycoside bonds.
- Examples of polysaccharides are starch, cellulose, and glycogen.

SECTION 24.9

- **Amino sugars** are carbohydrate derivatives in which one or more OH groups have been replaced with amino groups.
- The *N*-acetyl derivative of β-D-glucosamine serves as the repeating monosaccharide unit in chitin.

SECTION 24.10

- When treated with an amine in the presence of an acid catalyst, monosaccharides are converted into their corresponding ***N*-glycosides.**
- D-Ribose and D-2-deoxyribose are carbohydrates that form especially important *N*-glycosides called **nucleosides.**
- There are four naturally occurring nucleosides of 2-deoxyribose, each of which can be coupled to a phosphate group to form a **nucleotide.**
- DNA is comprised of nucleotides linked together in a polymer, or **polynucleotide**.
- RNA differs from DNA in that it contains D-ribose rather than deoxyribose, it contains uracil in place of thymine, and it is single stranded rather than double stranded like DNA.

SKILLBUILDER REVIEW

24.1 DRAWING THE CYCLIC HEMIACETAL OF A HYDROXYALDEHYDE

STEP 1 Assign numbers to the carbon atoms, beginning with the carbon of the aldehyde group, and continue until reaching the carbon connected to the OH group.

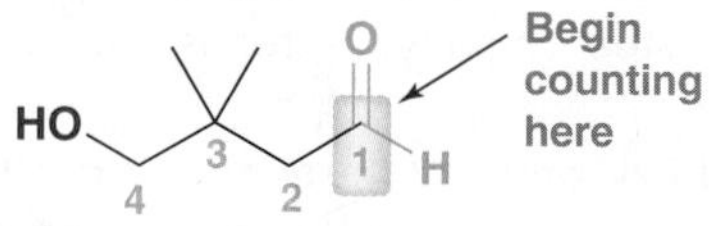

STEP 2 Using the numbering system from step 1, draw a ring that incorporates an oxygen atom in the ring.

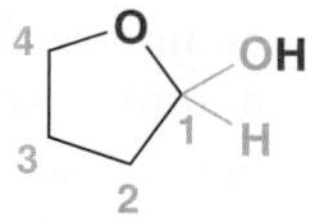

STEP 3 Using the numbering system from step 1, place the substituents in the correct locations.

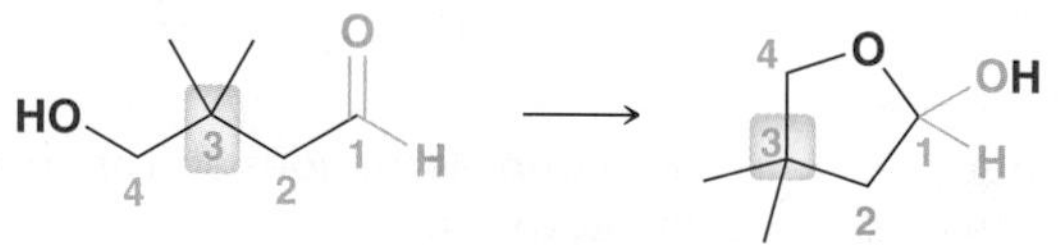

Try Problems **24.9–24.11, 24.46, 24.47**

24.2 DRAWING A HAWORTH PROJECTION OF AN ALDOHEXOSE

STEP 1 Draw the skeleton of a Haworth projection, with the ring oxygen in the back right corner. Number each position.

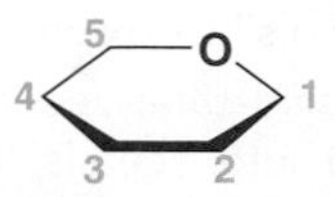

STEP 2 For D sugars, place the CH_2OH group pointing up at C5.

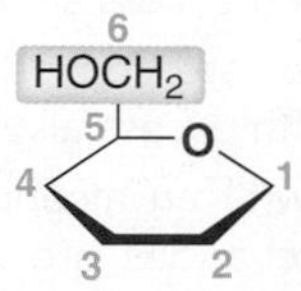

STEP 3 Place the OH group at the anomeric position. For the α anomer, the OH group should be *trans* to the CH_2OH group:

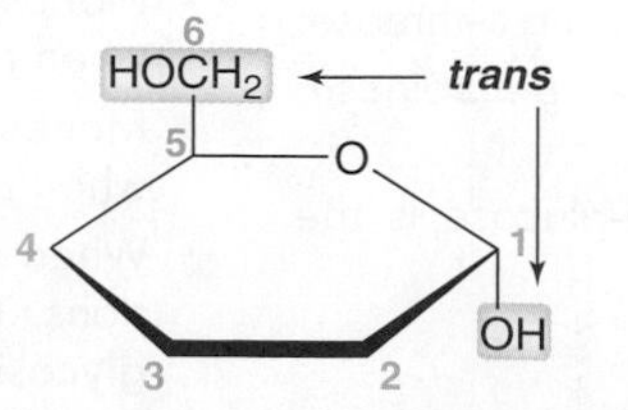

STEP 4 Place the remaining OH groups at C2, C3, and C4. Groups on the left side of the Fischer projection should point up. Groups on the right side of the Fischer projection should point down:

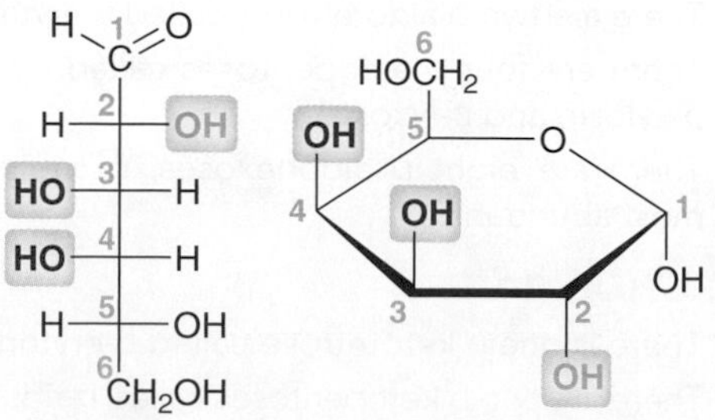

Try Problems **24.12–24.15, 24.48a, 24.53b,c,d**

24.3 DRAWING THE MORE STABLE CHAIR CONFORMATION OF A PYRANOSE RING

STEP 1 Draw the Haworth projection.

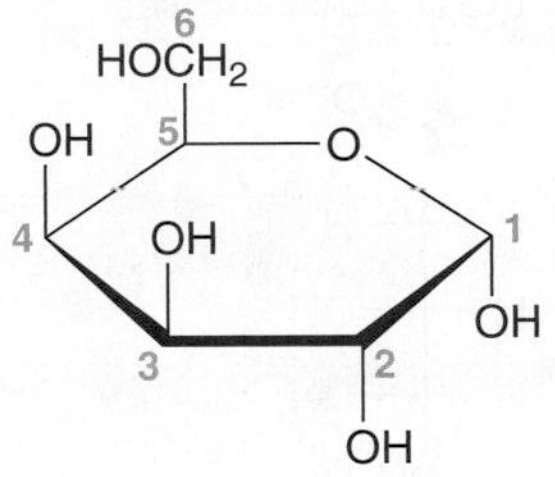

STEP 2 Draw the skeleton of a chair conformation, with the oxygen atom in the upper, rear-right corner.

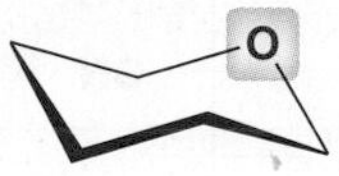

STEP 3 Label each substituent as either "up" or "down" and then place each substituent on the chair accordingly.

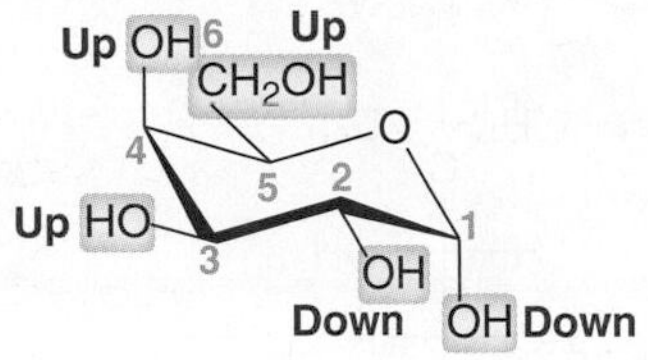

STEP 4 Verify that this chair conformation is more stable. Look for the largest group and make sure it is equatorial.

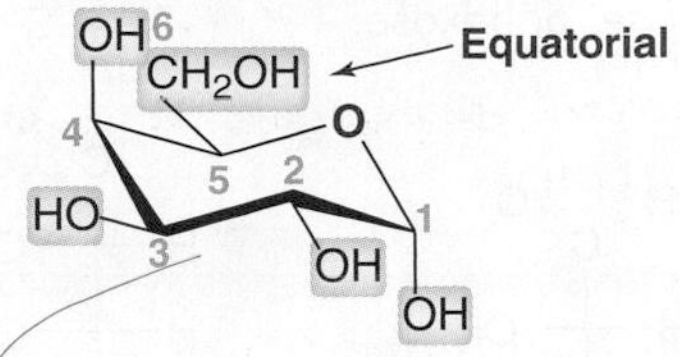

Try Problems 24.16–24.18, 24.60

24.4 IDENTIFYING A REDUCING SUGAR

STEP 1 Identify the anomeric positions.

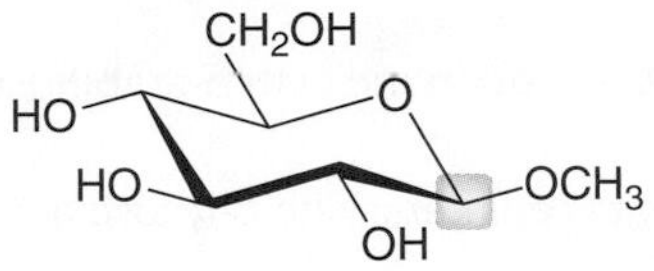

STEP 2 Determine if the group attached to the anomeric position is a hydroxy group or alkoxy group:

- If hydroxy group, then the compound is a reducing sugar. If alkoxy group, then the compound is not a reducing sugar.

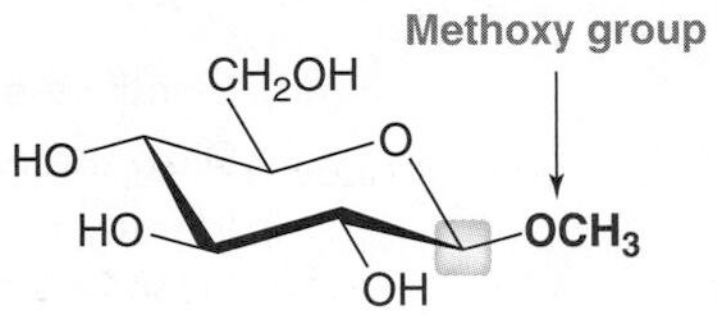

Try Problems 24.31–24.33, 24.69, 24.76a

24.5 DETERMINING WHETHER A DISACCHARIDE IS A REDUCING SUGAR

STEP 1 Identify the anomeric positions.

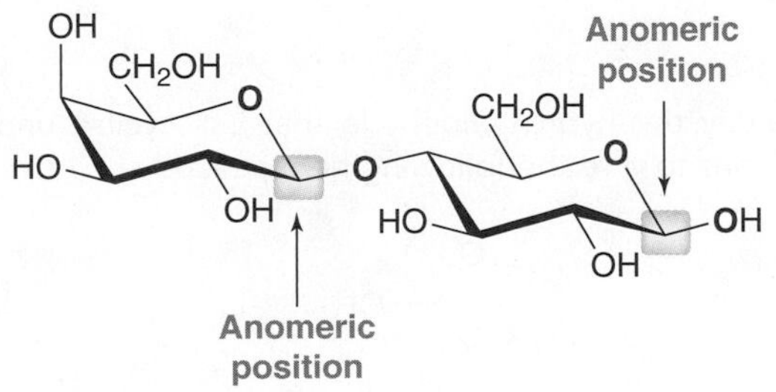

STEP 2 Determine if the groups attached to the anomeric positions are hydroxy groups or alkoxy groups:

- If at least one is a hydroxy group, then the compound is a reducing sugar.
- If neither are hydroxy groups, then the compound is not a reducing sugar.

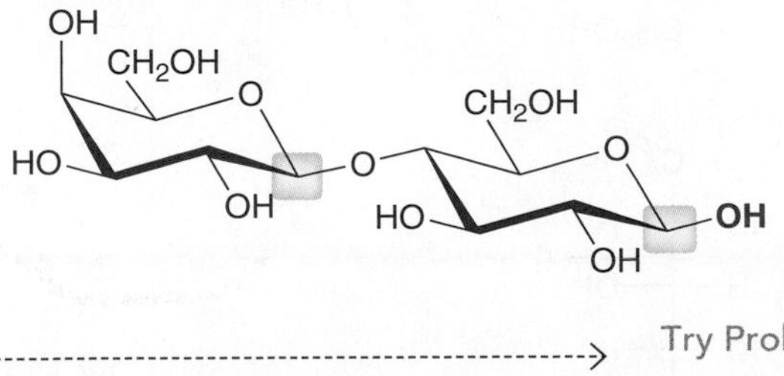

Try Problems 24.39, 24.40, 24.73

PRACTICE PROBLEMS

Note: Most of the Problems are available within *WileyPLUS*, an online teaching and learning solution.

24.42 Classify each of the following monosaccharides as either D or L, as either an aldo or a keto sugar, and as a tetrose, pentose, or hexose:

(a) 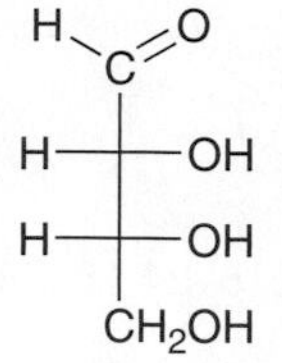

(b) 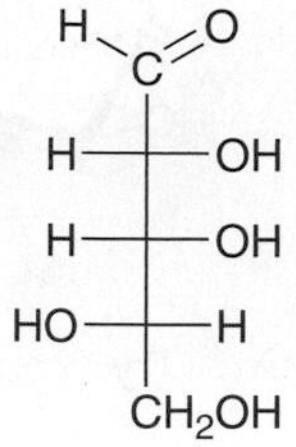

(c) 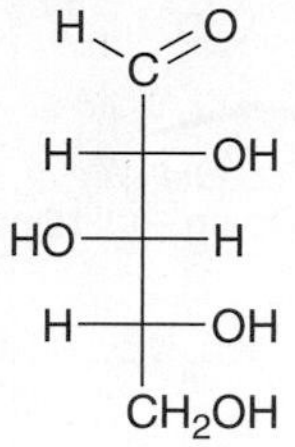

(d) 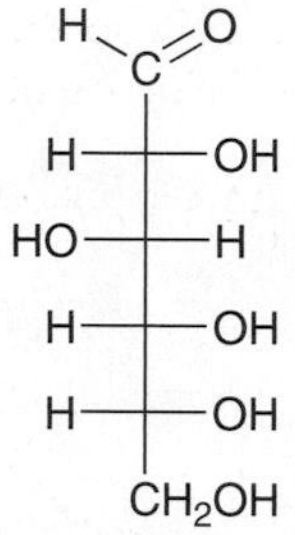

(e) 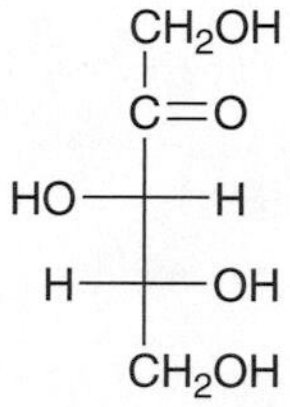

24.43 Identify each of the following structures as either D- or L-glyceraldehyde:

(a) 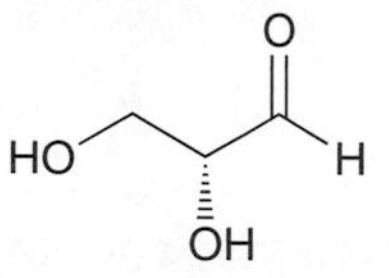

(b) 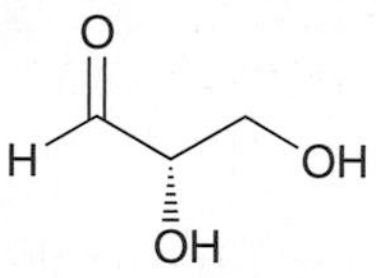

(c) 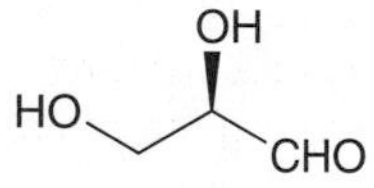

(d) 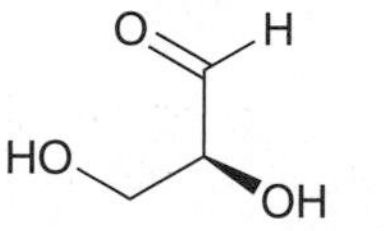

24.44 Name each of the following aldohexoses:

(a) 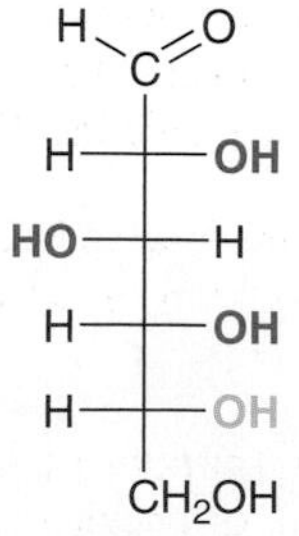

(b) 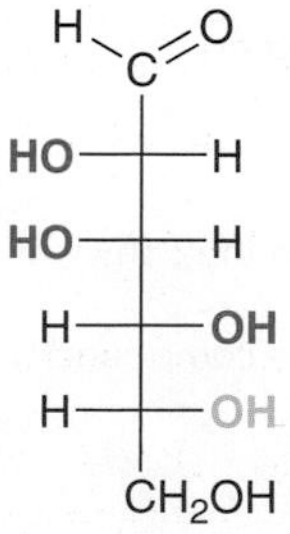

(c) 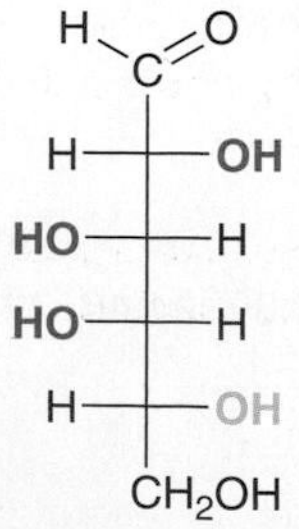

(d) 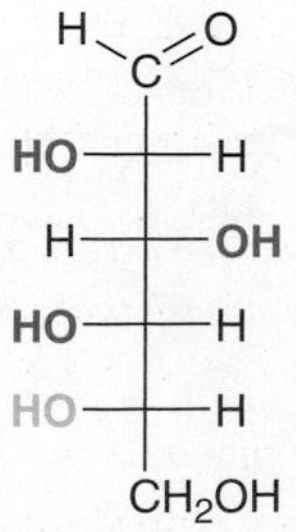

24.45 Consider the structures of the D aldopentoses:

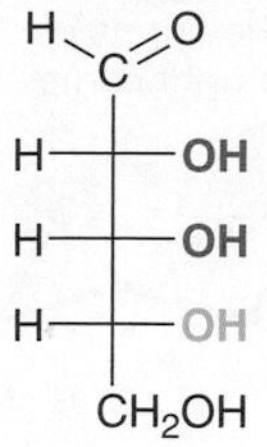

D-Ribose

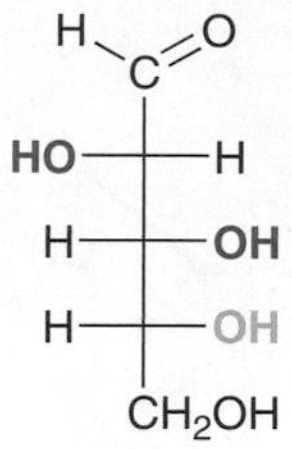

D-Arabinose

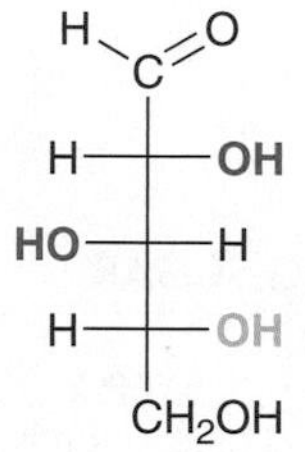

D-Xylose

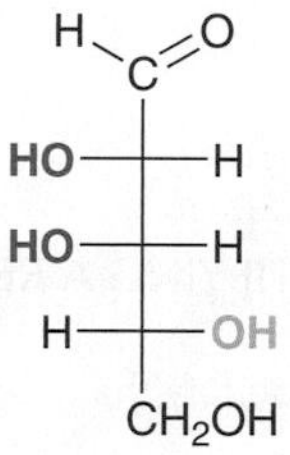

D-Lyxose

(a) Identify the aldopentose that is epimeric with D-arabinose at C2.

(b) Identify the aldopentose that is epimeric with D-lyxose at C3.

(c) Draw the enantiomer of D-ribose.

(d) Identify the relationship between the enantiomer of D-arabinose and the C2 epimer of L-ribose.

(e) Identify the relationship between D-ribose and D-lyxose

24.46 Draw the cyclic hemiacetal that is formed when each of the following bifunctional compounds is treated with aqueous acid.

(a) 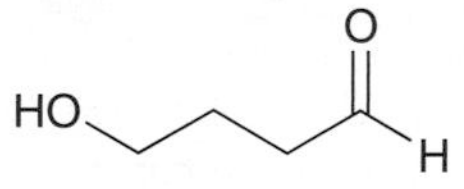

(b) 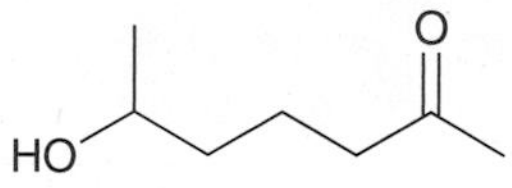

(c) 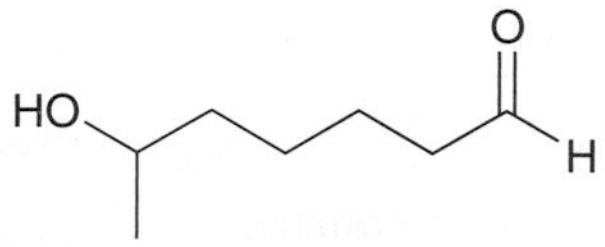

24.47 Identify the hydroxyaldehyde that will cyclize under acidic conditions to give the following hemiacetal:

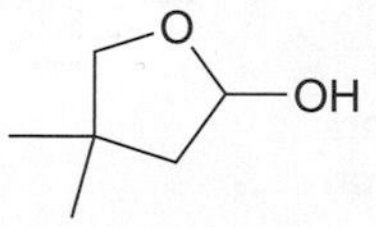

24.48 D-Ribose can adopt two pyranose forms and two furanose forms.

(a) Draw both pyranose forms of D-ribose, and identify each as α or β.

(b) Draw both furanose forms of D-ribose, and identify each as α or β.

24.49 For each of the following pairs of compounds, determine whether they are enantiomers, epimers, diastereomers that are not epimers, or identical compounds:

(a) 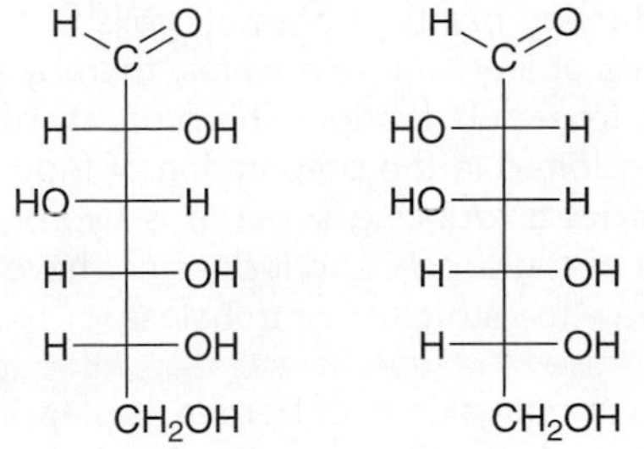

(b) 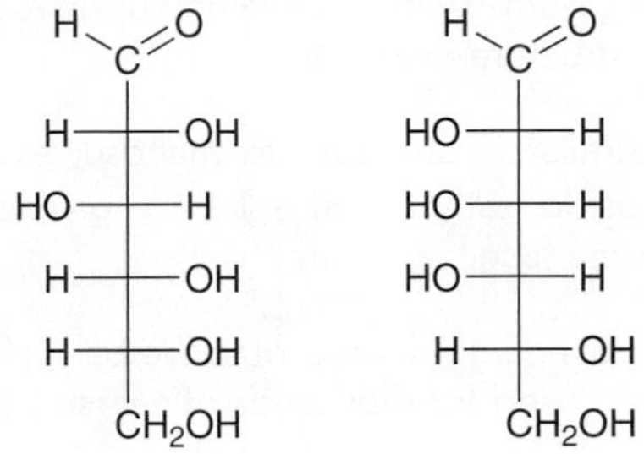

(c) 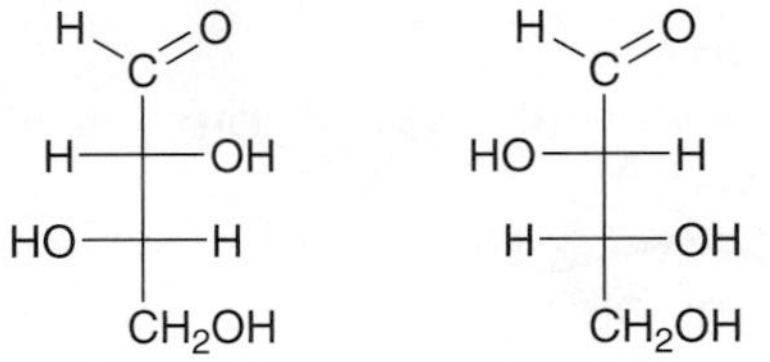

(d) 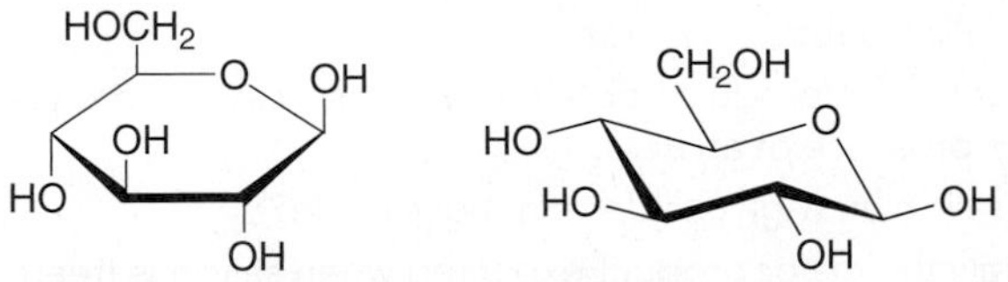

24.50 Draw the open-chain form of the compound formed when methyl β-D-glucopyranoside is treated with aqueous acid.

24.51 Assign the configuration of each chirality center in the following compounds:

(a) 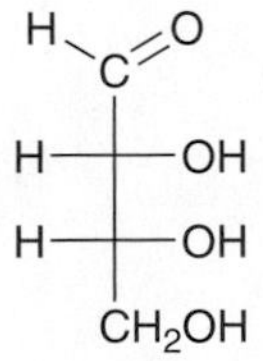

(b) 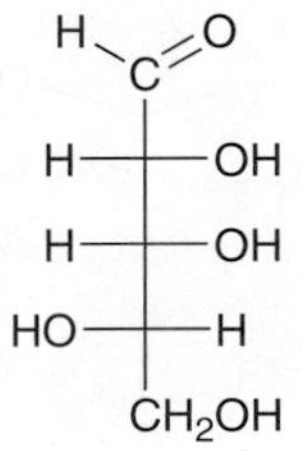

(c) 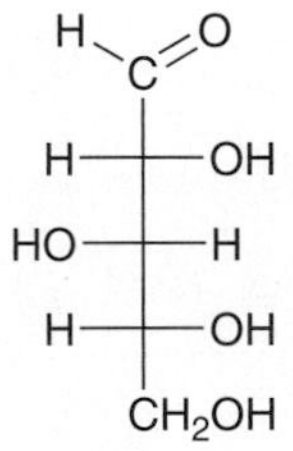

(d) 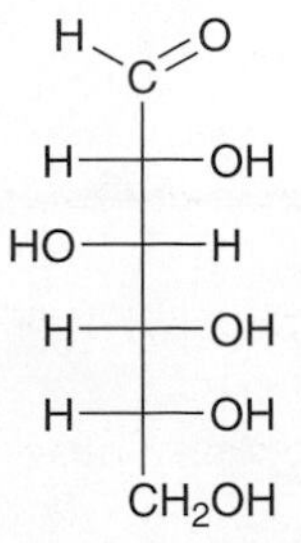

(e) 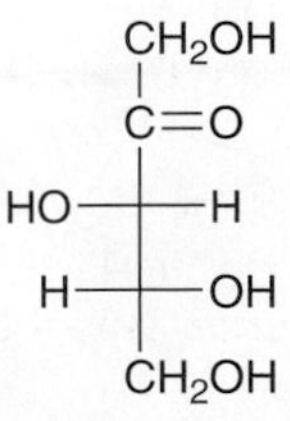

24.52 Draw a Fischer projection for each of the following compounds:

(a) D-Glucose

(b) D-Galactose

(c) D-Mannose

(d) D-Allose

24.53 Draw a Haworth projection for each of the following compounds:

(a) β-D-Fructofuranose

(b) β-D-Galactopyranose

(c) β-D-Glucopyranose

(d) β-D-Mannopyranose

24.54 Draw a Haworth projection showing the α-pyranose form of the D-aldohexose that is epimeric with D-glucose at C3.

24.55 Provide a complete name for each of the following compounds:

(a) 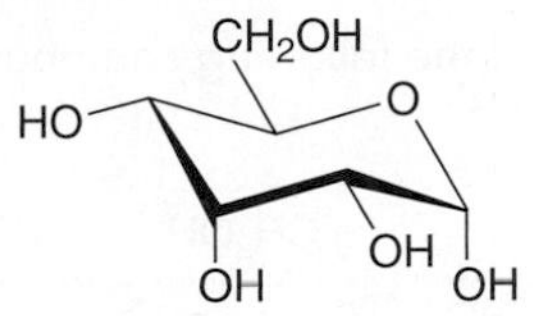

(b) 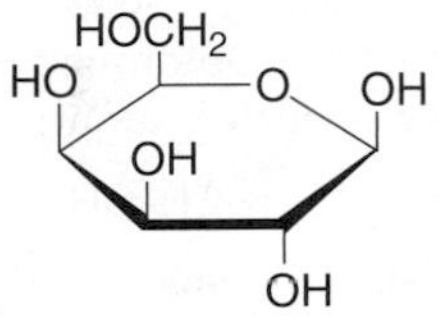

(c) 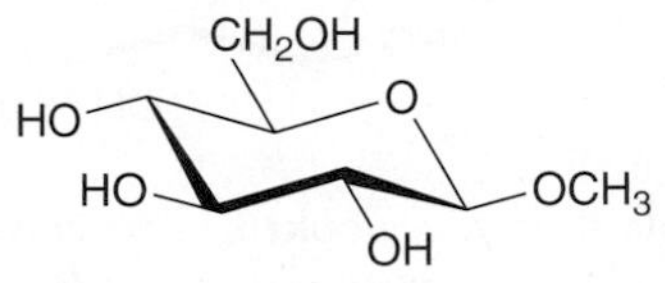

24.56 Draw the open-chain form of each of the compounds in the previous problem.

24.57 Draw the products that are expected when β-D-allopyranose is treated with each of the following reagents:

(a) Excess CH_3I, Ag_2O (b) Excess acetic anhydride, pyridine

(c) CH_3OH, HCl

24.58 When D-galactose is heated in the presence of nitric acid, an optically inactive compound is obtained. Draw the structure of the product and explain why it is optically inactive.

24.59 In addition to D-galactose, one other D-aldohexose also forms an optically inactive aldaric acid when treated with nitric acid. Draw the structure of this aldohexose.

24.60 Draw the more stable chair conformation of α-D-altropyranose and label all substituents as axial or equatorial.

24.61 Draw the products that are expected when α-D-galactopyranose is treated with excess methyl iodide in the presence of silver oxide, followed by aqueous acid.

24.62 For each of the following pairs of compounds, determine whether they are enantiomers, epimers, diastereomers that are not epimers, or identical compounds:

(a) D-Glucose and D-gulose

(b) 2-Deoxy-D-ribose and 2-deoxy-D-arabinose

24.63 Draw all possible 2-ketohexoses that are D sugars.

24.64 Identify the two aldohexoses that will undergo a Wohl degradation to yield D-ribose. Draw a Fischer projection of the open-chain form for each of these two aldohexoses.

24.65 Identify the two aldohexoses that are obtained when D-arabinose undergoes a Kiliani-Fischer synthesis.

24.66 Identify the two products obtained when D-glyceraldehyde is treated with HCN, and determine the relationship between these two products.

24.67 When treated with sodium borohydride, D-glucose is converted into an alditol.

(a) Draw the structure of the alditol.

(b) Which L-aldohexose gives the same alditol when treated with sodium borohydride?

24.68 Which of the D-aldohexoses are converted into optically inactive alditols upon treatment with sodium borohydride?

24.69 Determine whether each of the following compounds is a reducing sugar:

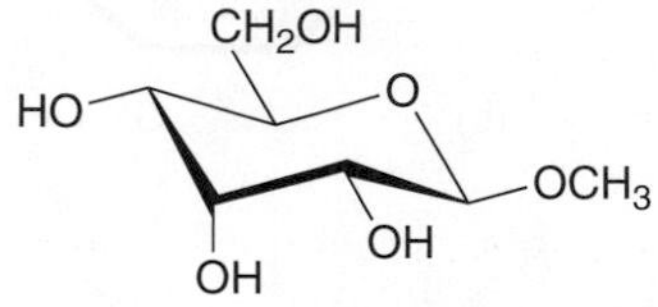

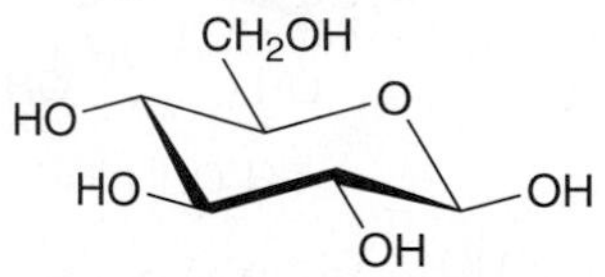

24.70 Identify the reagents that you would use to convert β-D-glucopyranose into each of the following compounds:

(a)
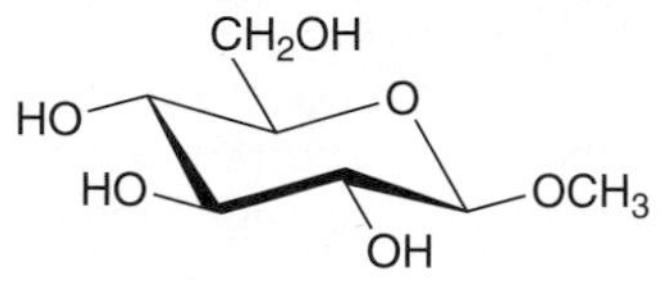

(b)
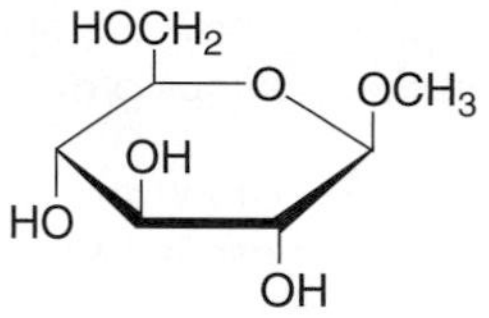

(c)
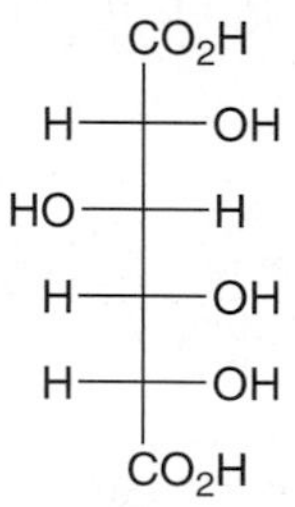

(d)
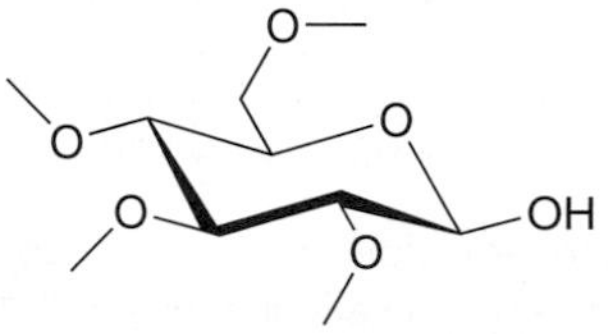

24.71 Identify the product(s) that would be formed when each of the following compounds is treated with aqueous acid:

(a) Methyl α-D-glucopyranoside (b) Ethyl β-D-galactopyranoside

24.72 Consider the structures of the four D-aldopentoses (Figure 24.3).

(a) Which D-aldopentose produces the same aldaric acid as D-lyxose?

(b) Which D-aldopentoses yield optically inactive alditols when treated with sodium borohydride?

(c) Which D-aldopentose yields the same alditol as L-xylose?

(d) Which D-aldopentose can close into a β-pyranose form in which all substituents are equatorial?

24.73 Trehalose is a naturally occurring disaccharide found in bacteria, insects, and many plants. It protects cells from dry conditions because of its ability to retain water, thereby preventing cellular damage from dehydration. This property of trehalose has also been exploited in the preparation of food and cosmetics. Trehalose is not a reducing sugar, it is hydrolyzed to yield two equivalents of D-glucose, and it does not have any β-glycoside linkages. Draw the structure of trehalose.

24.74 Xylitol is found in many kinds of berries. It is approximately as sweet as sucrose but with fewer calories. It is often used in sugarless chewing gum. Xylitol is obtained upon reduction of D-xylose. Draw a structure of xylitol.

24.75 Isomaltose is similar in structure to maltose, except that it is a 1→6 α-glycoside, rather than a 1→4 α-glycoside. Draw the structure of isomaltose.

24.76 Salicin is a natural analgesic present in the bark of willow trees, and it has been used for thousands of years to treat pain and reduce fevers.

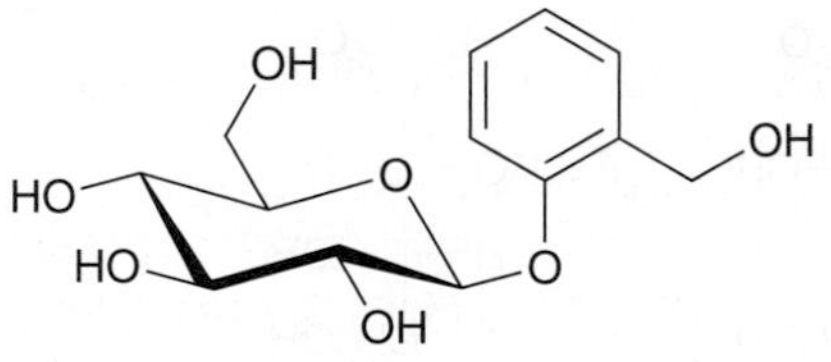

Salicin

(a) Is salicin a reducing sugar?

(b) Identify the products obtained when salicin is hydrolyzed in the presence of an acid.

(c) Is salicin an α-glycoside or a β-glycoside?

(d) Draw the major product expected when salicin is treated with excess acetic anhydride in the presence of pyridine.

(e) Would you expect salicin to exhibit mutarotation when dissolved in neutral water?

24.77 Draw a mechanism for the following transformation:

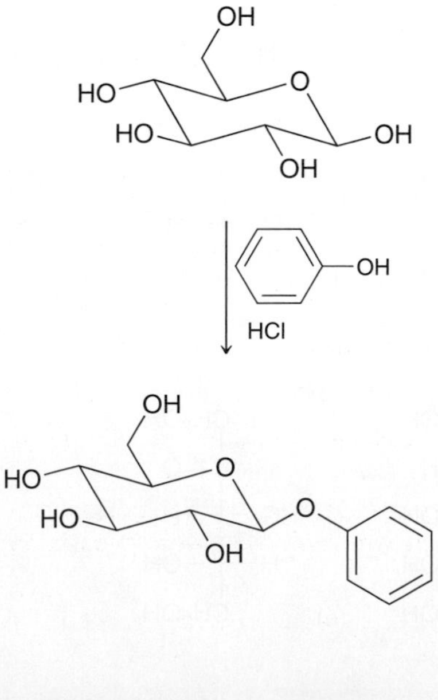

24.78 Draw the α-*N*-glycoside and the β-*N*-glycoside formed when D-glucose is treated with aniline ($C_6H_5NH_2$).

24.79 Draw the nucleoside formed from each of the following pairs of compounds and name the nucleoside:

(a) 2-Deoxy-D-ribose and adenine
(b) D-Ribose and guanine

24.80 Compound **A** is a D-aldopentose that is converted into an optically active alditol upon treatment with sodium borohydride. Draw two possible structures for compound **A**.

INTEGRATED PROBLEMS

24.81 When D-glucose is treated with an aqueous bromine solution (buffered to a pH of 6), an aldonic acid is formed called D-gluconic acid. Treatment of D-gluconic acid with an acid catalyst produces a lactone (cyclic ester) with a six-membered ring.

(a) Draw the structure of D-gluconic acid.
(b) Draw the structure of the lactone formed from D-gluconic acid, showing the configuration at each chirality center.
(c) Would you expect this lactone to be optically active?
(d) Explain how you would distinguish between the D-gluconic acid and the lactone using IR spectroscopy.

24.82 When each of the D-aldohexoses assumes an α-pyranose form, the CH_2OH group occupies an equatorial position in the more stable chair conformation. The one exception is D-idose, for which the CH_2OH group occupies an axial position in the more stable chair conformation. Explain this observation, and then draw the more stable chair conformation of L-idose in its α-pyranose form.

24.83 Draw the product that is expected when the β-pyranose form of compound **A** is treated with excess ethyl iodide in the presence of silver oxide. The following information can be used to determine the identity of compound **A**:

1. The molecular formula of compound **A** is $C_6H_{12}O_6$.
2. Compound **A** is a reducing sugar.
3. When compound **A** is subjected to a Wohl degradation two times sequentially, D-erythrose is obtained.
4. Compound **A** is epimeric with D-glucose at C3.
5. The configuration at C2 is *R*.

24.84 Explain why glucose is the most common monosaccharide observed in nature.

CHALLENGE PROBLEMS

24.85 When D-glucose is treated with aqueous sodium hydroxide, a complex mixture of carbohydrates is formed, including D-mannose and D-fructose. Over time, almost all aldohexoses will be present in the mixture. Even L-glucose can be detected, albeit in very small concentrations. Using the fewest number of mechanistic steps possible, draw a reasonable mechanism showing the formation of L-glucose from D-glucose:

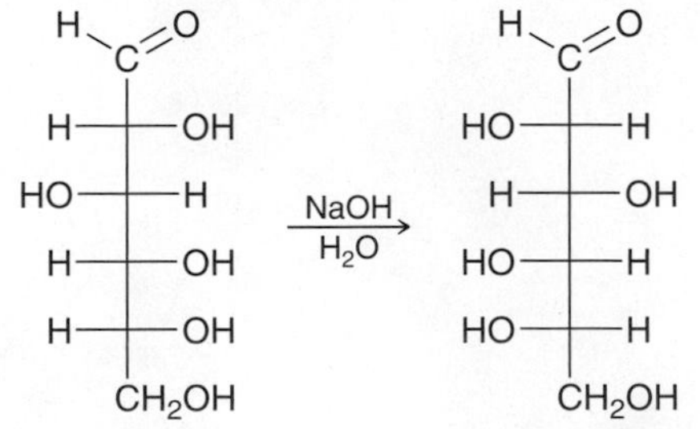

D-Glucose L-Glucose

24.86 Compound **X** is a D-aldohexose that can adopt a β-pyranose form with only one axial substituent. Compound **X** undergoes a Wohl degradation to produce an aldopentose, which is converted into an optically active alditol when treated with sodium borohydride. From the information presented, there are only two possible structures for compound **X**. Identify the two possibilities, and then propose a chemical test that would allow you to distinguish between the two possibilities and thereby determine the structure of compound **X**.

24.87 Compound **A** is a D-aldopentose. When treated with sodium borohydride, compound **A** is converted into an alditol that exhibits three signals in its ^{13}C NMR spectrum. Compound **A** undergoes a Kiliani-Fischer synthesis to produce two aldohexoses, compounds **B** and **C**. Upon treatment with nitric acid, compound **B** yields compound **D**, while compound **C** yields compound **E**. Both **D** and **E** are optically active aldaric acids.

(a) Draw the structure of compound **A**.
(b) Draw the structures of compounds **D** and **E**, and describe how you might be able to distinguish between these two compounds using ^{13}C NMR spectroscopy.

25

Amino Acids, Peptides, and Proteins

DID YOU EVER WONDER...

how crime-scene investigators are able to find invisible fingerprints and render them visible?

A fingerprint can often be the most important piece of evidence left behind at the scene of a crime. Police investigators use a variety of methods to visualize fingerprints. One such method involves the use of a chemical agent called ninhydrin, which reacts with the amino acids present in the fingerprint to produce colored compounds that can be seen. But what are amino acids, why are they present in our fingerprints, and what function do amino acids serve?

In this chapter, we will explore the structure and properties of amino acids, and we will see how they function as the building blocks that nature employs to assemble important biological compounds called peptides and proteins. These compounds serve a wide array of functions, as we will see later in this chapter. This chapter focuses on the structure, properties, function, and synthesis of amino acids, peptides, and proteins.

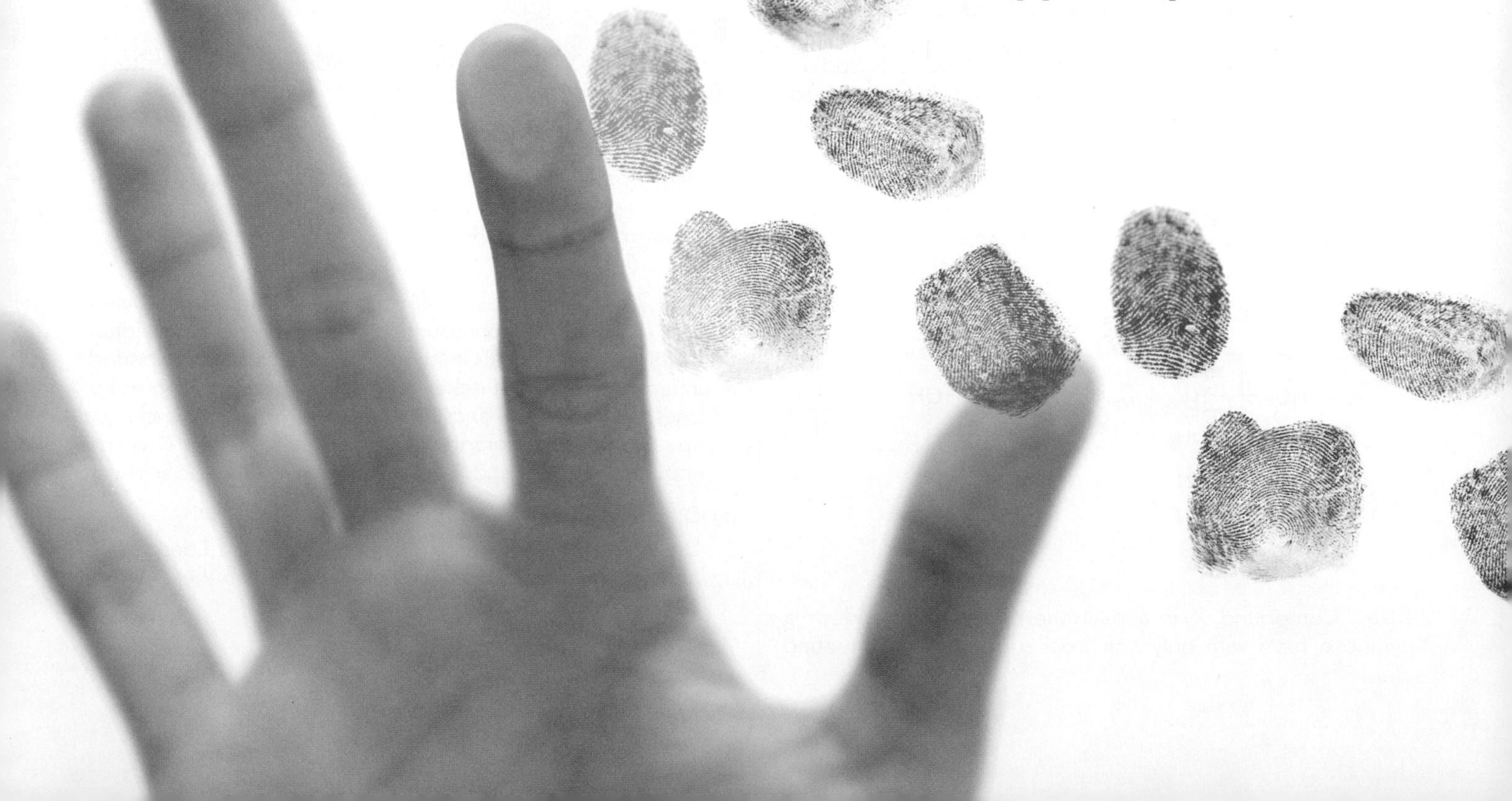

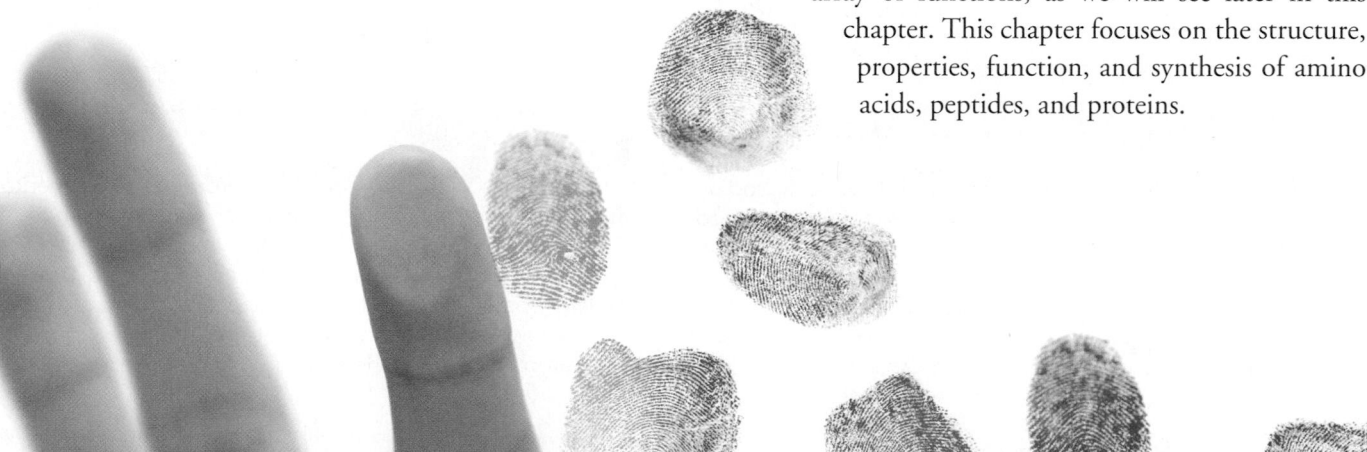

DO YOU REMEMBER?

Before you go on, be sure you understand the following topics.
If necessary, review the suggested sections to prepare for this chapter:

- Designating Configuration Using the Cahn-Ingold-Prelog System (section 5.3)
- Structure and Properties of Carboxylic Acids (section 21.3)
- Nucleophilic Acyl Substitution (section 21.7)
- Properties of Amines (section 23.3)

25.1 Introduction to Amino Acids, Peptides, and Proteins

Throughout this book, particularly in the Medically Speaking boxes, we have explored the relationship between the structure of a compound and its medicinal activity. The relationship between structure and activity is perhaps most striking for biological molecules called proteins. **Proteins** are polymers that are assembled from amino acid monomers that have been linked together, much like jigsaw puzzle pieces (Figure 25.1). Each **amino acid** contains an amino

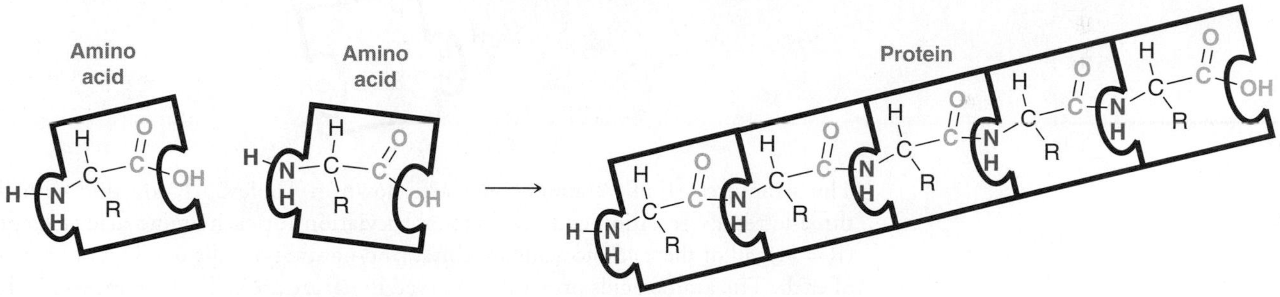

FIGURE 25.1
An illustration showing how amino acids serve as building blocks for proteins.

group and a carboxylic acid group. It is the presence of these two functional groups that enables amino acids to link together. An amino acid can have any number of carbon atoms separating the two functional groups, but of particular interest are the **alpha (α) amino acids** in which the two functional groups are separated by exactly one carbon atom.

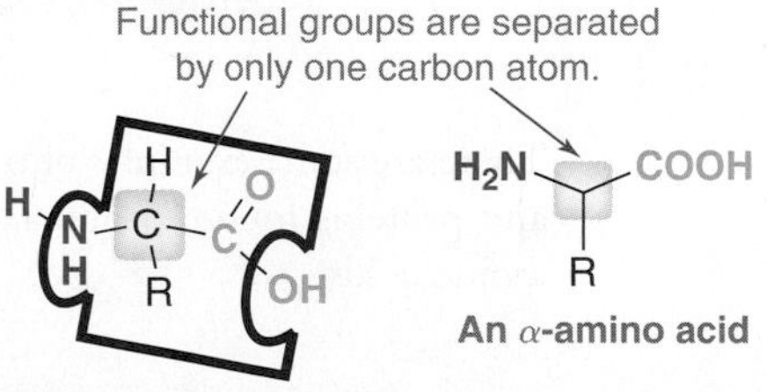

Amino acids of this type are called α-amino acids because the amino group is connected to the carbon atom that is alpha (α) to the carboxylic acid moiety. Notice that this α carbon is a chirality center, provided that the R group is not simply a hydrogen atom. The configuration of the α position will be discussed in the coming sections.

Amino acids are coupled together by amide linkages, also called **peptide bonds:**

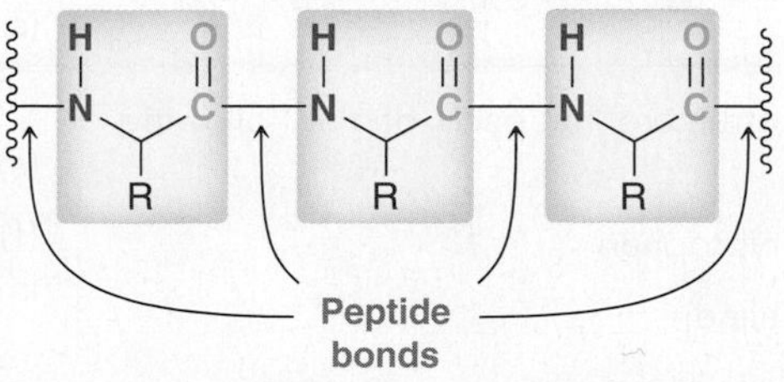

Relatively short amino acid chains are called **peptides**. A dipeptide is formed when two amino acids are coupled together, a tripeptide from three amino acids, a tetrapeptide from four, and so on. Chains comprised of fewer than 40 or 50 amino acids are often called polypeptides, while still larger chains are called proteins. Proteins serve a wide array of important biological functions, as we will discuss in the final section of this chapter. Certain proteins, called enzymes, serve as catalysts for most of the reactions that occur in living cells, and it is estimated that more than 50,000 different enzymes are needed for our bodies to function properly.

In order to understand the structure and function of proteins, we must first explore the structure and properties of the most basic building blocks, amino acids.

25.2 Structure and Properties of Amino Acids

Naturally Occurring Amino Acids

Hundreds of different amino acids are observed in nature, but only 20 amino acids are abundantly found in proteins. These twenty α-amino acids differ from each other only in the identity of the side chain (the R group, highlighted).

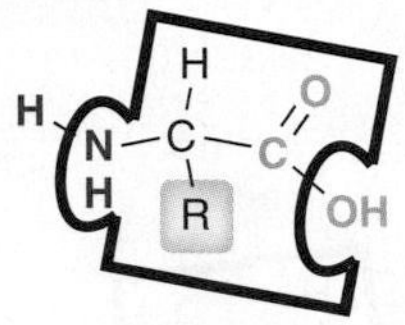

The structures of all 20 amino acids are shown in Table 25.1, together with the accepted three-letter abbreviation and one-letter abbreviation for each amino acid. Except for glycine (R = H), all of these amino acids are chiral, and nature typically employs only one enantiomer of each. The amino acids primarily observed in nature are called **L amino acids**, because their Fischer projections resemble the Fischer projections of L sugars.

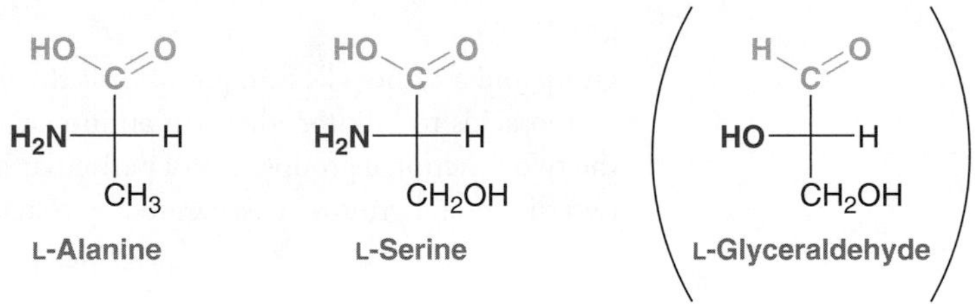

There are some examples of D amino acids found in nature, but for the most part, the peptides and proteins found in humans and other mammals are constructed almost exclusively from L amino acids.

CONCEPTUAL CHECKPOINT

25.1 Although most naturally occurring proteins are made up only of L amino acids, proteins isolated from bacteria will sometimes contain D amino acids. Draw Fischer projections for D-alanine and D-valine. In each case, assign the configuration (*R* or *S*) of the chirality center.

25.2 Draw a bond-line structure for each of the following amino acids.

(a) L-Leucine (b) L-Tryptophan

(c) L-Methionine (d) L-Valine

25.3 Of the 20 naturally occurring amino acids shown in Table 25.1, identify any amino acids that exhibit the following:

(a) A cyclic structure (b) An aromatic side chain

(c) A side chain with a basic group

(d) A sulfur atom

(e) A side chain with an acidic group

(f) A side chain containing a proton that will likely participate in hydrogen bonding

TABLE 25.1 THE STRUCTURES OF THE TWENTY NATURALLY OCCURRING AMINO ACIDS THAT ARE FOUND IN PROTEINS

Name	Structure	Abbreviation
Amino acids with nonpolar side chains		
Glycine	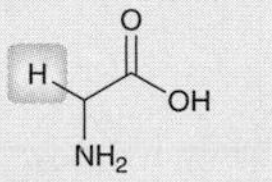	Gly G
Alanine	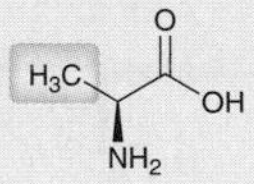	Ala A
Valine	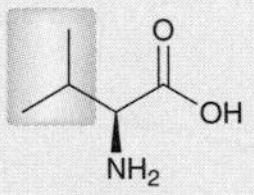	Val V
Leucine	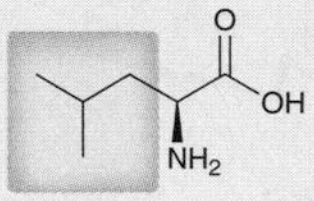	Leu L
Isoleucine	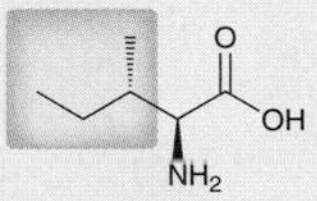	Ile I
Methionine	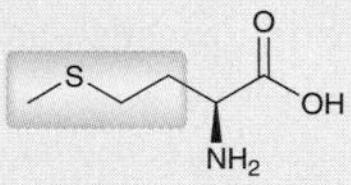	Met M
Proline	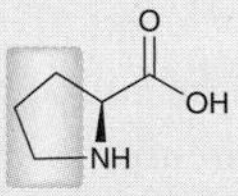	Pro P
Phenylalanine	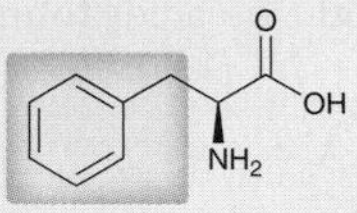	Phe F
Tryptophan	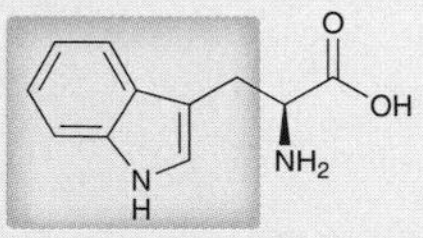	Trp W
Amino acids with polar side chains		
Asparagine	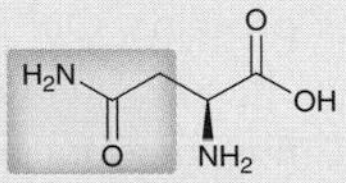	Asn N
Glutamine	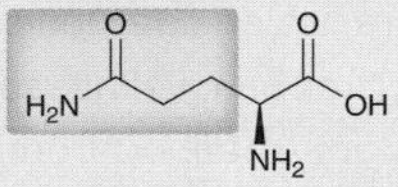	Gln Q
Serine	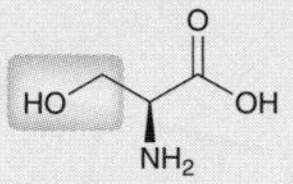	Ser S
Threonine	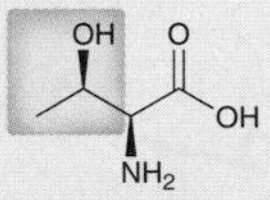	Thr T
Tyrosine	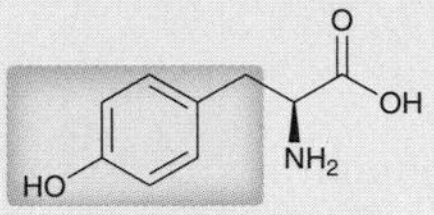	Tyr Y
Cysteine	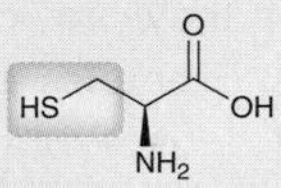	Cys C
Amino acids with acidic side chains		
Aspartic acid	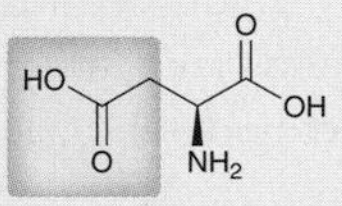	Asp D
Glutamic acid	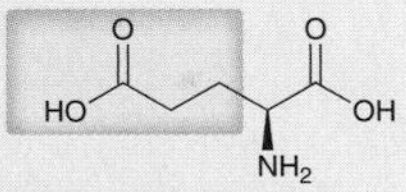	Glu E
Amino acids with basic side chains		
Arginine	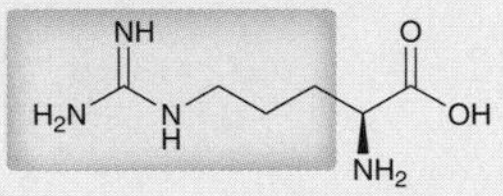	Arg R
Histidine	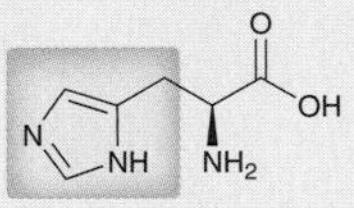	His H
Lysine	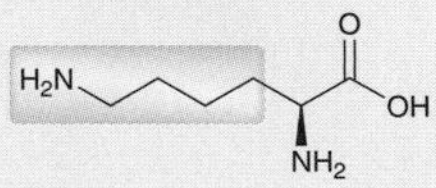	Lys K

Other Naturally Occurring Amino Acids

In addition to the 20 amino acids found in proteins (Table 25.1), there are other amino acids used by organisms for a variety of functions. For example, consider the structures of γ-aminobutyric acid (GABA) and thyroxine.

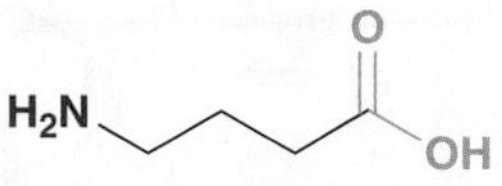

γ-Aminobutyric acid (GABA)

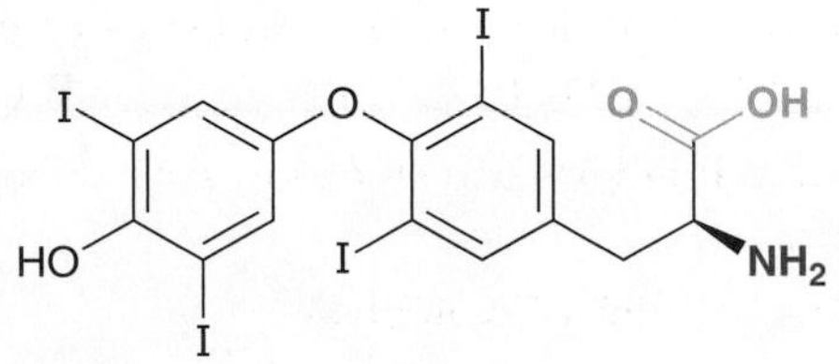

Thyroxine

PRACTICALLYSPEAKING

Nutrition and Sources of Amino Acids

In Section 25.1, we saw that 20 different L amino acids serve as the building blocks of proteins. Our bodies can synthesize 10 of these amino acids in sufficient quantity, but the other 10, called *essential amino acids*, must be obtained from our diet. The essential amino acids are isoleucine, leucine, methionine, phenylalanine, threonine, tryptophan, valine, arginine, histidine, and lysine. These amino acids are obtained from digesting foods containing proteins.

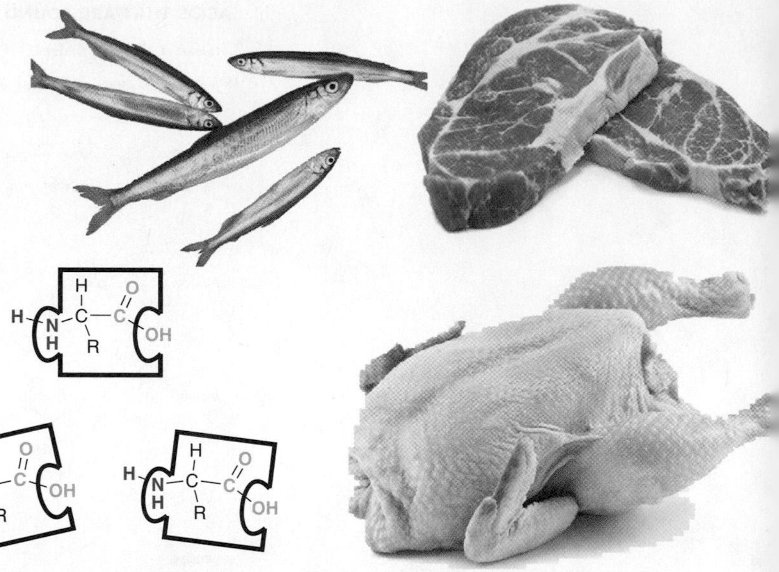

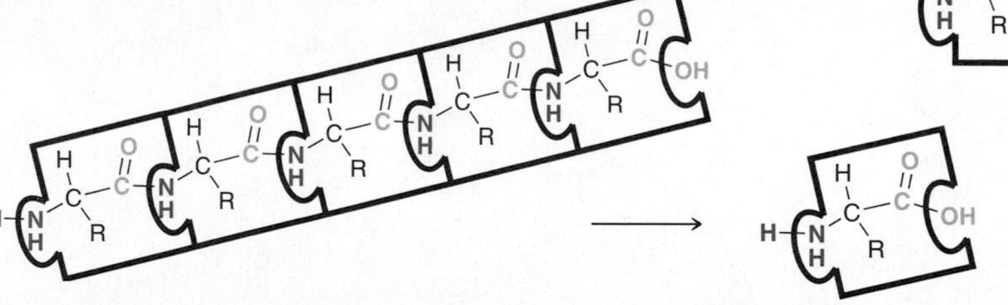

Proteins that contain all 10 essential amino acids are called *complete proteins*, as they provide us with all of the building blocks that we need. Examples of complete proteins include meat, fish, milk, and eggs. Proteins that are deficient in one or more of the essential amino acids are called *incomplete proteins*. Examples include rice (deficient in lysine and threonine), corn (deficient in lysine and tryptophan), and beans and peas (deficient in methionine).

Inadequate intake of the essential amino acids can lead to a host of diseases, which can be avoided with proper nutrition. Meat eaters receive all of the necessary amino acids from a piece of meat, while vegetarians must eat a variety of plant foods that complement each other. Alternatively, vegetarians can supplement their diet with a source of complete proteins, such as milk or eggs.

Both of these compounds are amino acids, but they are not found in proteins. GABA is found in the brain and acts as a neurotransmitter, while thyroxine is found in the thyroid gland and acts as a hormone. There are many other examples as well, but this chapter will focus almost exclusively on the 20 amino acids found in proteins.

Acid-Base Properties

When an amino acid is dissolved in a solution at a pH of 1, both functional groups exist primarily in their protonated forms.

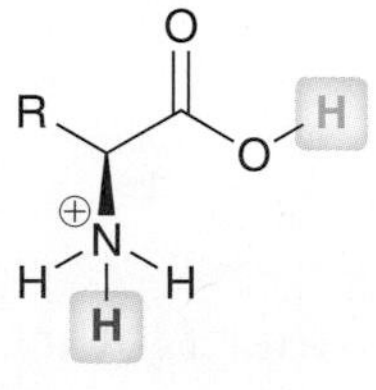

Each of the highlighted protons has its own unique pK_a value, often called pK_{a1} and pK_{a2}.

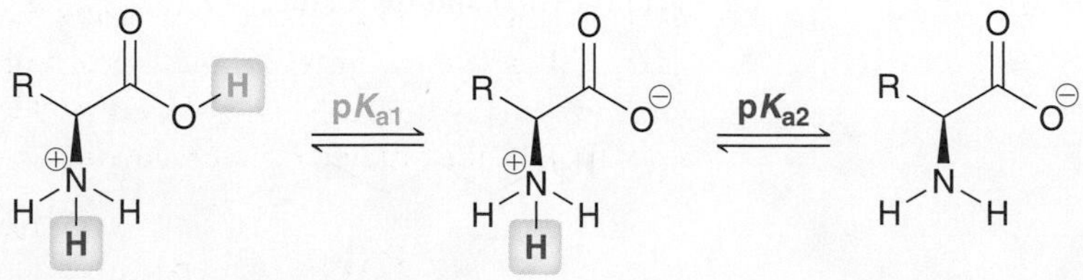

Notice that the carboxylic acid moiety is deprotonated first. That is, the first pK_a value refers to the acidity of the carboxylic acid moiety, while the second pK_a value refers to the acidity of the ammonium group. Some of the amino acids have side chains containing either basic or acidic groups. These amino acids will have a third pK_a value associated with the side chain, as can be seen in Table 25.2.

TABLE 25.2 THE PK_A VALUES FOR TWENTY NATURALLY OCCURRING AMINO ACIDS

AMINO ACID	α-COOH	α-NH_3^+	SIDE CHAIN
Alanine	2.34	9.69	—
Arginine	2.17	9.04	12.48
Asparagine	2.02	8.80	—
Aspartic acid	1.88	9.60	3.65
Cysteine	1.96	10.28	8.18
Glutamic acid	2.19	9.67	4.25
Glutamine	2.17	9.13	—
Glycine	2.34	9.60	—
Histidine	1.82	9.17	6.00
Isoleucine	2.36	9.60	—
Leucine	2.36	9.60	—
Lysine	2.18	8.95	10.53
Methionine	2.28	9.21	—
Phenylalanine	1.83	9.13	—
Proline	1.99	10.60	—
Serine	2.21	9.15	—
Threonine	2.09	9.10	—
Tryptophan	2.83	9.39	—
Tyrosine	2.20	9.11	10.07
Valine	2.32	9.62	—

Notice that the pK_a values for the carboxylic acid moieties are in the range of 2–3 for nearly all of the amino acids shown. For example, alanine has a pK_{a1} of 2.34. This value indicates that the uncharged form (COOH) and the anionic form (COO^-) will be present in equal amounts at a pH of 2.34. At any pH below 2.34 (highly acidic conditions), the uncharged form will predominate. At any pH above 2.34, the carboxylate anion will predominate. Indeed, this anionic form predominates at physiological pH.

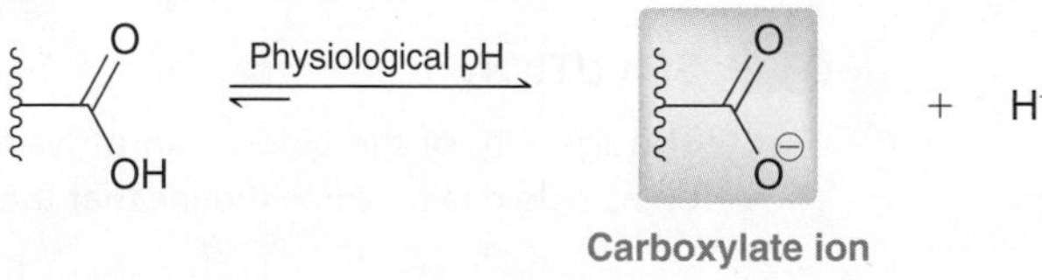

LOOKING BACK
Recall that the pH of blood is approximately 7.4, which is called the physiological pH (see Section 21.3).

Now let's focus on the pK_a values for the ammonium groups of amino acids. Table 25.2 indicates that these values are in the range of 9–10 for nearly all of the amino acids. For example, alanine has a pK_{a2} of 9.69. This value indicates that the uncharged and cationic forms of the ammonium group will be present in equal amounts at a pH of 9.69. At any pH above 9.69, the uncharged form will predominate. At any pH below 9.69, the cationic ammonium group will predominate. Indeed, the cationic form predominates at physiological pH.

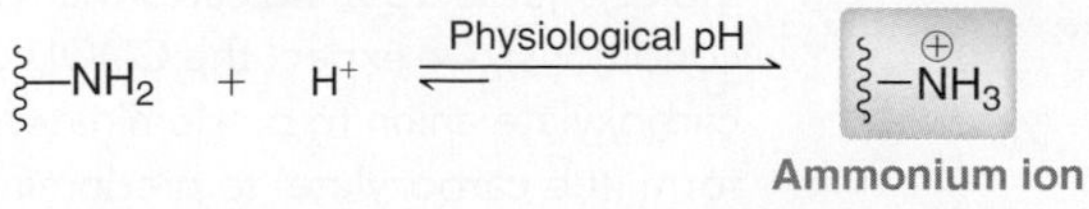

In summary, consider the structure of an amino acid at physiological pH. The amino group is protonated, while the carboxylic acid moiety is deprotonated.

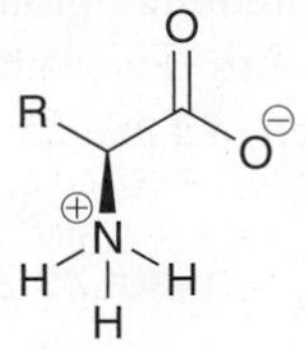

In this form, the amino acid is said to be a **zwitterion**, which is a net neutral compound that exhibits charge separation. An amino acid can be thought of as an internal salt, and as such, it will exhibit many of the physical properties of salts. For example, amino acids are highly soluble in water and have very high melting points. Amino acids are also said to be **amphoteric**, because they will react with either acids or bases. When treated with a base, an amino acid will function as an acid by giving up a proton (the ammonium group is deprotonated).

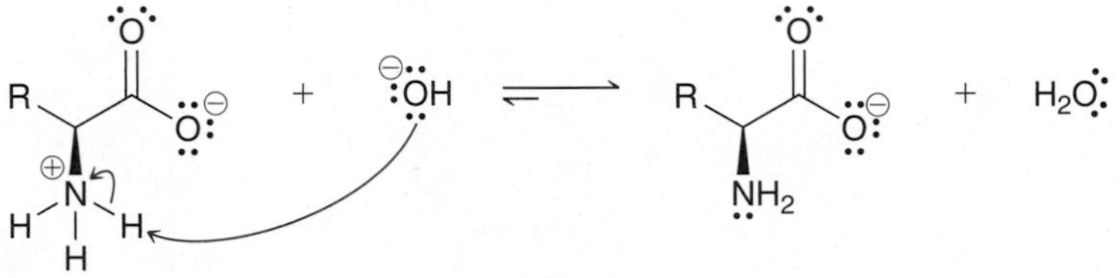

However, when treated with an acid, an amino acid will serve as a base by abstracting a proton (the carboxylate group is protonated).

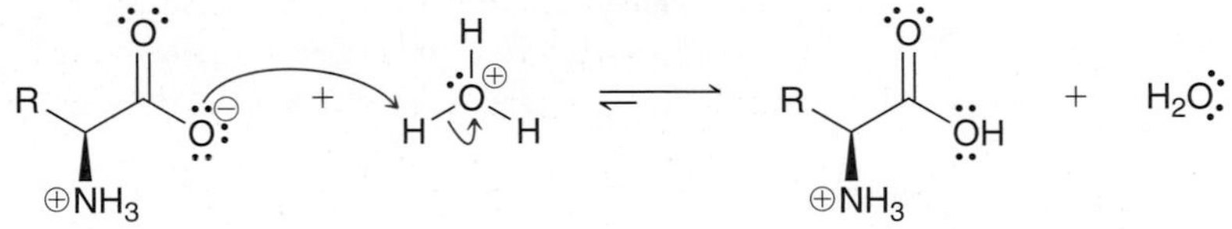

The two reactions show that, in its zwitterionic form, an amino acid can function either as an acid or as a base.

SKILLBUILDER

25.1 DETERMINING THE PREDOMINANT FORM OF AN AMINO ACID AT A SPECIFIC pH

LEARN the skill

Draw the form of lysine that predominates at a pH of 9.5.

SOLUTION

The identity of the side chain of lysine can be found in Table 25.1. This amino acid has a basic side chain, which means that the compound has three locations that we must consider.

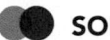

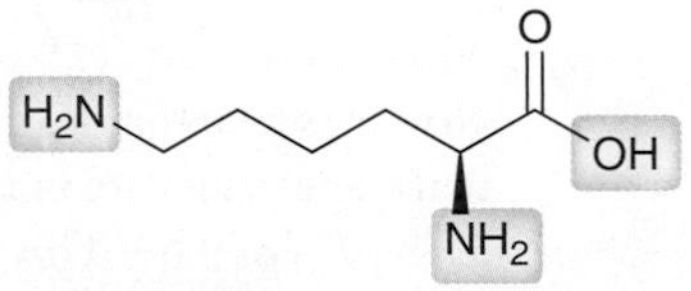

STEP 1
Identify the pK_a of the carboxylic acid moiety and determine which form predominates.

We must consider whether or not each amino group is protonated and whether or not the carboxylic acid moiety is deprotonated at the stated pH. Let's begin with the carboxylic acid moiety. Table 25.2 indicates that pK_{a1} for lysine is 2.18. At a pH below 2.18 (highly acidic conditions), we expect the COOH group to have its proton. But at higher pH, we expect the carboxylate anion to predominate. Accordingly, at a pH of 9.5, we expect the deprotonated form (the carboxylate) to predominate.

STEP 2
Identify the pK_a of the α-amino group and determine which form predominates.

Next consider the α-amino group. Table 25.2 indicates that pK_{a2} for lysine is 8.95. At a pH below 8.95 (lower pH = more acidic), we expect the amino group to be protonated, but at a higher pH, we expect the amino group to predominate in its uncharged form. Accordingly, at a pH of 9.5, we expect the uncharged form of the α amino group to predominate.

STEP 3
Identify the pK_a of the side chain, if necessary, and determine which form predominates.

Finally, consider the side chain, which has a pK_a of 10.53. At a pH below 10.53 (lower pH = more acidic), we expect the amino group to be protonated, but at a higher pH, we expect the amino group to predominate in its uncharged form. Accordingly, at a pH of 9.5, we expect the protonated form of the side-chain amino group to predominate. Therefore, the following form of lysine is expected to predominate at a pH of 9.5.

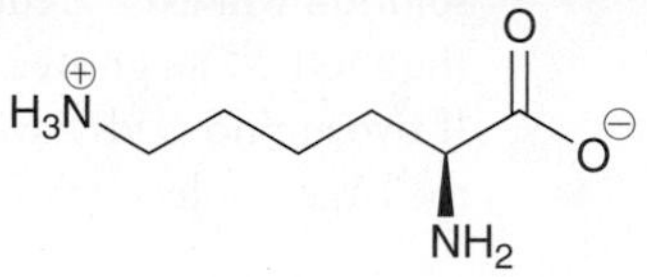

PRACTICE the skill **25.4** Draw the form of the amino acid that is expected to predominate at the stated pH.

(a) Alanine at a pH of 10
(b) Proline at a pH of 10
(c) Tyrosine at a pH of 9
(d) Asparagine at physiological pH
(e) Histidine at physiological pH
(f) Glutamic acid at a pH of 3

APPLY the skill **25.5** At a pH of 11, arginine is a more effective proton donor than asparagine. Explain.

25.6 The OH group on the side chain of serine is not deprotonated at a pH of 12. However, the OH group on the side chain of tyrosine is deprotonated at a pH of 12. This can be verified by inspecting the pK_a values in Table 25.2. Suggest an explanation for the difference in acid-base properties between these two OH groups.

need more **PRACTICE?** **Try Problems 25.40, 25.47, 25.48**

Isoelectric Point

For each amino acid, there is a specific pH at which the concentration of the zwitterionic form reaches its maximum value. This pH is called the **isoelectric point (pI)**, and each amino acid has its own unique pI. For amino acids that lack an acidic or basic side chain, the pI is simply the average of the two pK_a values. The following example shows the calculation for the pI of alanine.

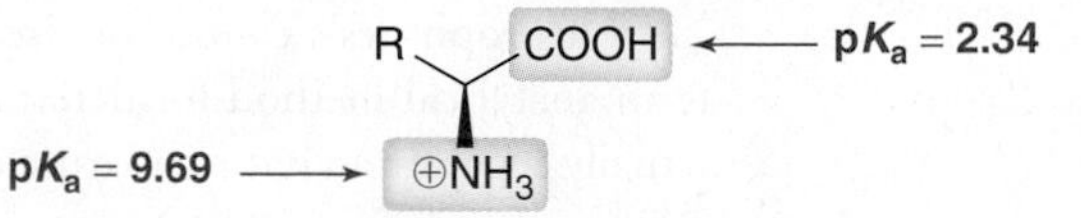

$$\text{pI} = \frac{2.34 + 9.69}{2} = 6.02$$

For amino acids with acidic or basic side chains, the pI is the average of the two pK_a values that correspond with the similar groups. For example, the pI of lysine is determined by the two amino groups, while the pI of glutamic acid is determined by the two carboxylic acid moieties.

Lysine

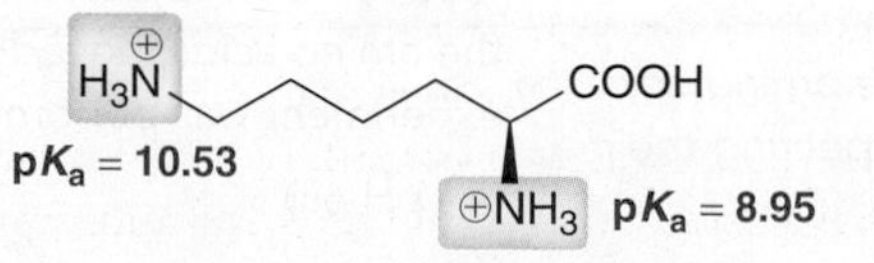

$$\text{pI} = \frac{10.53 + 8.95}{2} = 9.74$$

Glutamic acid

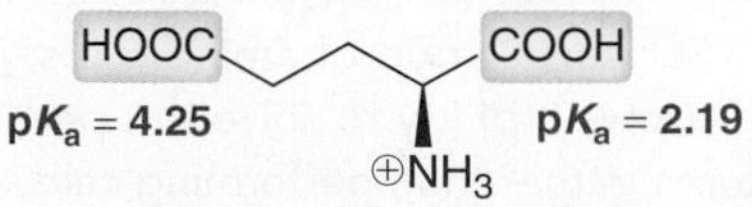

$$\text{pI} = \frac{4.25 + 2.19}{2} = 3.22$$

Separation of Amino Acids via Electrophoresis

Amino acids can be separated from each other by a variety of techniques. One method, called **electrophoresis**, relies on a difference in pI values and can be used to determine the number of different amino acids present in a mixture. In practice, a few drops of the mixture are applied to filter paper or gel that is placed in a buffered solution between two electrodes. When an electric field is applied, the amino acids separate based on their different pI values. If the pI of an amino acid is greater than the pH of the solution, the amino acid will exist predominantly in a form that bears a positive charge and will migrate toward the cathode. The greater the difference between pI and pH, the faster it will migrate. An amino acid with a pI that is lower than the pH of the solution will exist predominantly in a form that bears a negative charge and will migrate toward the anode. The greater the difference between pI and pH, the faster it will migrate (Figure 25.2). If two amino acids have very similar pI values (such as glycine and leucine), the amino acid with the larger molecular weight will move more slowly, because the charge has to carry a greater mass.

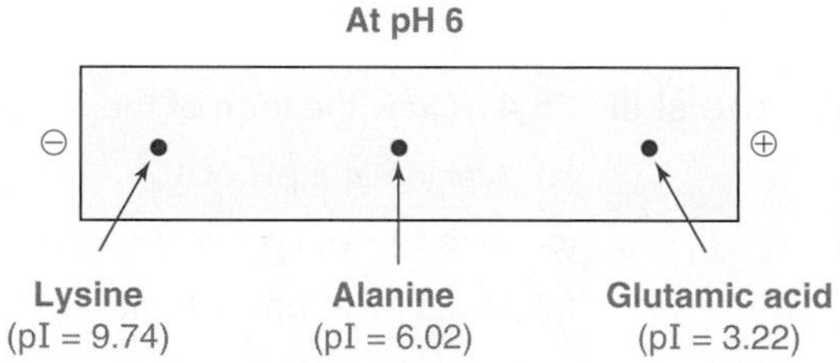

FIGURE 25.2
Separation of amino acids via electrophoresis.

Amino acids are colorless, so a detection technique is necessary in order to visualize the location of the various spots. The most common method involves treating the filter paper or gel with a solution containing ninhydrin followed by heating in an oven. Ninhydrin reacts with amino acids to produce a purple product.

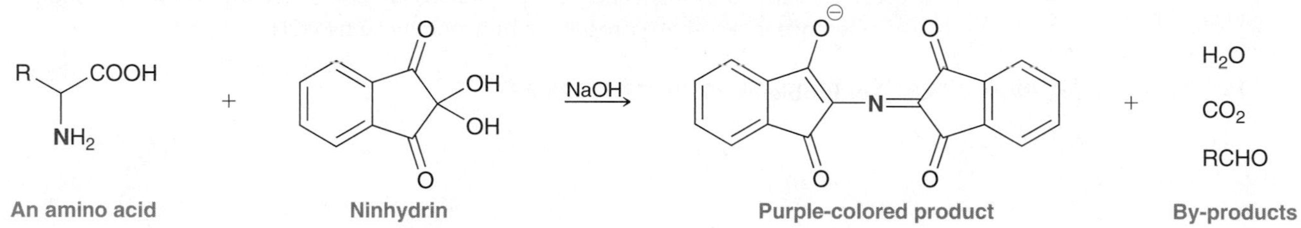

The nitrogen atom of the amino acid is ultimately incorporated into the purple product, and the rest of the amino acid is degraded into a few by-products (water, carbon dioxide, and an aldehyde). The purple compound is obtained regardless of the identity of the amino acid, provided that the amino acid is primary (i.e., not proline). The number of purple spots indicates the number of different kinds of amino acids present.

Electrophoresis cannot be used to separate large quantities of amino acids. It is used just as an analytical method for determining the number of amino acids in a mixture. In order to actually separate an entire mixture of amino acids, other laboratory techniques are used, such as column chromatography.

CONCEPTUAL CHECKPOINT

25.7 Using the data in Table 25.2, calculate the pI of the following amino acids.

(a) Aspartic acid (b) Leucine

(c) Lysine (d) Proline

25.8 For each group of amino acids, identify the amino acid with the lowest pI (try to solve this problem by inspecting their structures, rather than performing calculations).

(a) Alanine, aspartic acid, or lysine

(b) Methionine, glutamic acid, or histidine

25.9 Identify which 2 of the 20 naturally occurring amino acids are expected to have the same pI.

25.10 A mixture containing phenylalanine, tryptophan, and leucine was subjected to electrophoresis. Determine which of the amino acids moved the farthest distance assuming that the experiment was performed at the pH indicated:

(a) pH 6.0 (b) pH 5.0

25.11 Draw the aldehyde that is obtained as a byproduct when L-leucine is treated with ninhydrin.

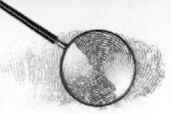

PRACTICALLYSPEAKING)))

Forensic Chemistry and Fingerprint Detection

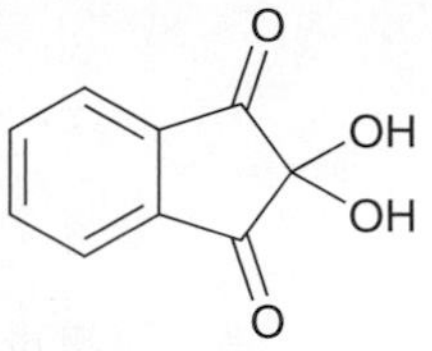

Ninhydrin

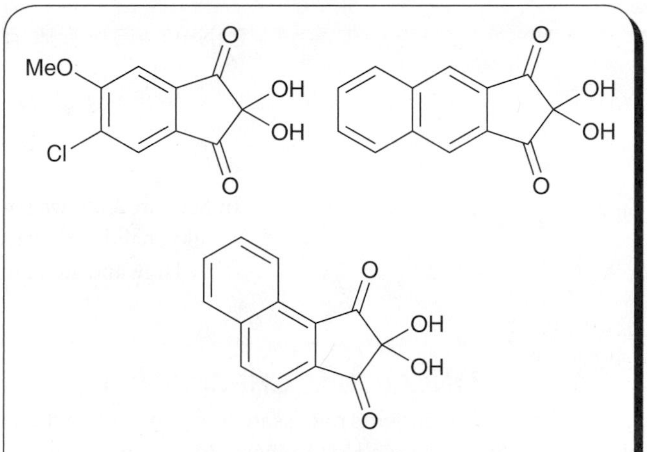

Ninhydrin analogues

There are many popular TV shows that portray the work of crime-scene investigators and the methods they employ to analyze evidence. The use of chemicals to visualize latent (invisible) fingerprints is a particularly common practice, and many compounds can be used for this purpose, including ninhydrin.

Latent fingerprints are created by the residue of sweat found on the surface of the skin. Sweat is comprised primarily of water (99%), but it also contains a variety of organic compounds, including amino acids. These amino acids are present only in very small concentrations, but they are relatively stable over long periods of time. When treated with ninhydrin, a reaction occurs between the amino acids and the ninhydrin, forming a fluorescent purple image.

To develop a fingerprint, a solution of ninhydrin is applied as a spray, and then mild heating is used to accelerate the reaction. The process has several undesirable features, including background staining (which reduces the contrast of the image) as well as light-induced fading of the image. For best results, the ninhydrin solution should be applied in the dark and at room temperature. Under such conditions, the development process can take up to two weeks, which is not feasible for most cases in which a rapid analysis is required.

Over the past 50 years, many attempts have been made to design ninhydrin analogues with better properties. The following are a few examples.

Some of these reagents do have slightly improved properties, but the improvements are not significant enough to justify the added expense associated with the replacement of ninhydrin as a standard method used by forensic chemists to develop latent fingerprints.

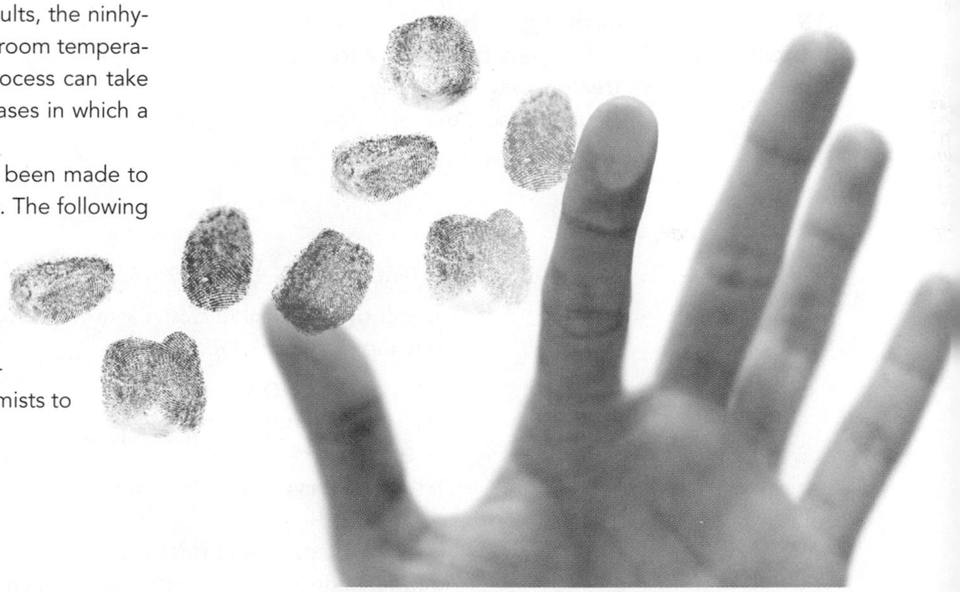

25.3 Amino Acid Synthesis

Over the last century, many methods have been used to prepare amino acids in the laboratory. In this section, we will explore a few of those methods.

Amino Acid Synthesis via α-Haloacids

One of the oldest methods for preparing racemic mixtures of α-amino acids involves the use of the Hell-Volhard-Zelinski reaction (Section 22.2) to functionalize the α position of a carboxylic acid.

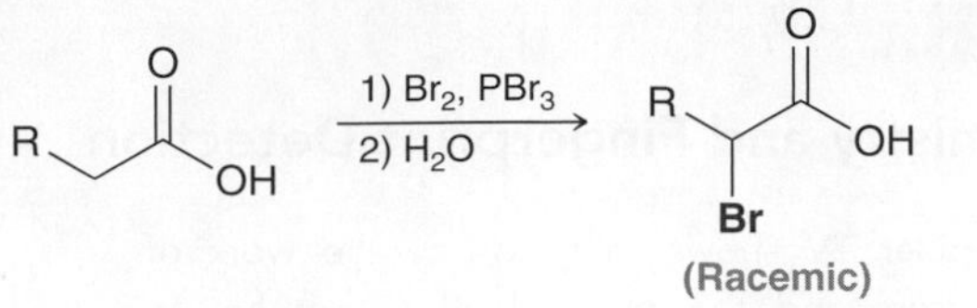

The halogen is then replaced with an amino group in an S_N2 reaction.

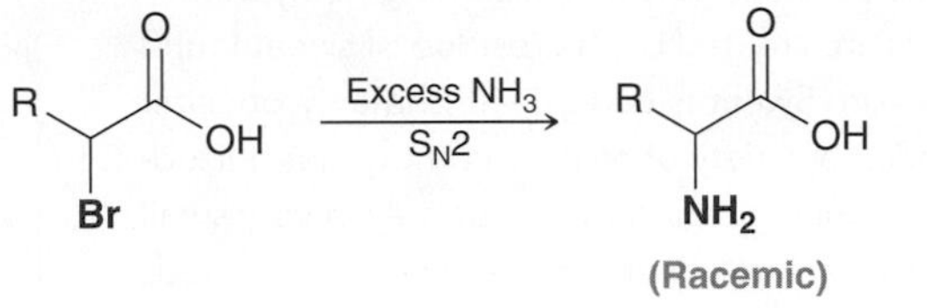

In Section 23.5, we saw that polyalkylation is often unavoidable when ammonia is treated with an alkyl halide. However, in this case, polyalkylation is not a problem because the alkyl halide is fairly large and steric hindrance prevents subsequent alkylations.

CONCEPTUAL CHECKPOINT

25.12 Identify the reagents necessary to make each of the following amino acids using a Hell-Volhard-Zelinski reaction.

(a) Leucine (b) Alanine (c) Valine

25.13 Each of the following carboxylic acids was treated with bromine and PBr_3 followed by water, and the resulting α-haloacid was then treated with excess ammonia. In each case, draw and name the amino acid that is produced.

(a) 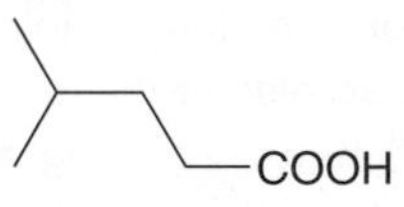

(b) 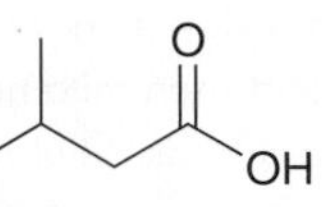

(c) 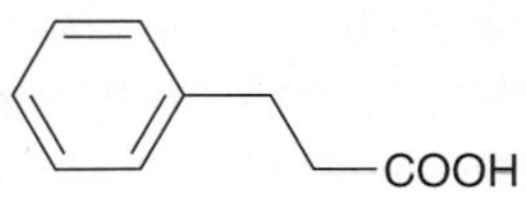

(d) Acetic acid

Amino Acid Synthesis via the Amidomalonate Synthesis

Recall that the malonic ester synthesis (Section 22.5) can be used to prepare carboxylic acids, starting with diethyl malonate.

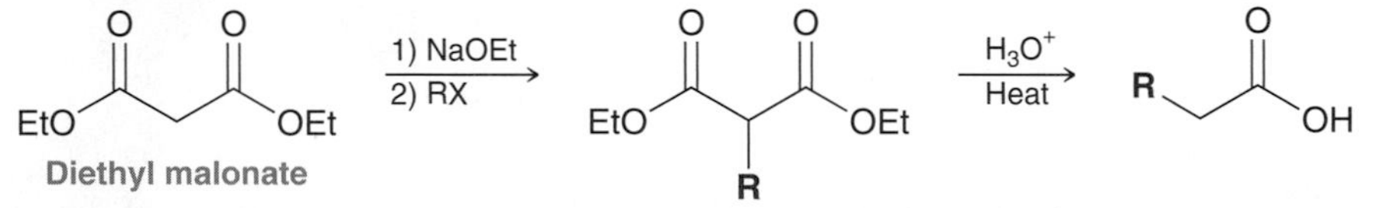

A clever adaptation of this process, called the **amidomalonate synthesis**, employs diethyl acetamidomalonate as the starting material, enabling the preparation of racemic α-amino acids.

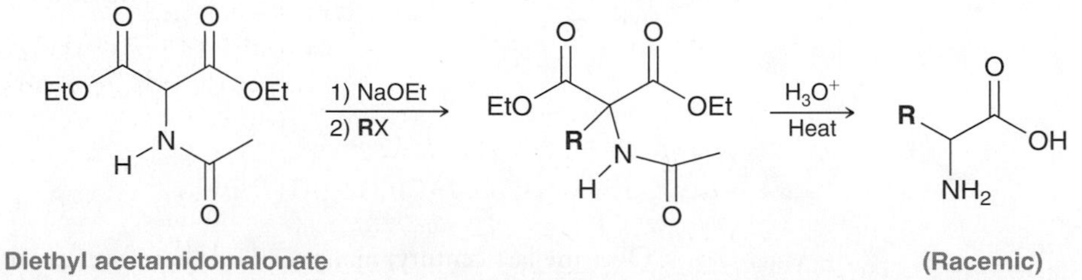

The process involves the same three steps used in the malonic ester synthesis: (1) deprotonation, (2) alkylation, and (3) hydrolysis and decarboxylation. During the hydrolysis step, the amide is hydrolyzed as well. As an example, consider the synthesis of phenylalanine via the amidomalonate synthesis.

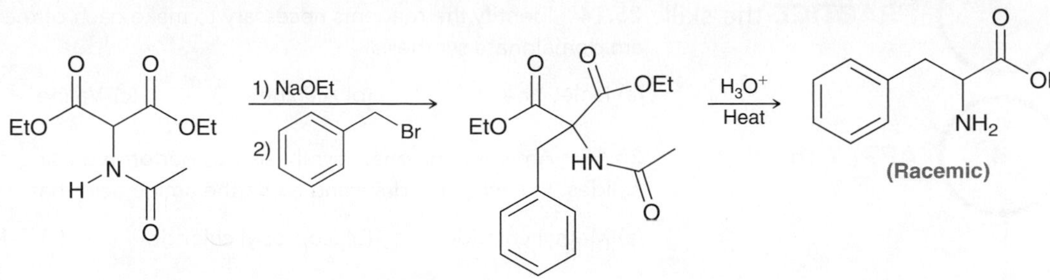

The identity of the amino acid obtained is determined by the choice of the alkyl halide in the second step.

SKILLBUILDER

25.2 USING THE AMIDOMALONATE SYNTHESIS

LEARN the skill

Show how you would use the amidomalonate synthesis to prepare a racemic mixture of tryptophan.

SOLUTION

Begin by identifying the side chain attached to the α position of tryptophan (see Table 25.1).

STEP 1
Identify the side chain connected to the α position.

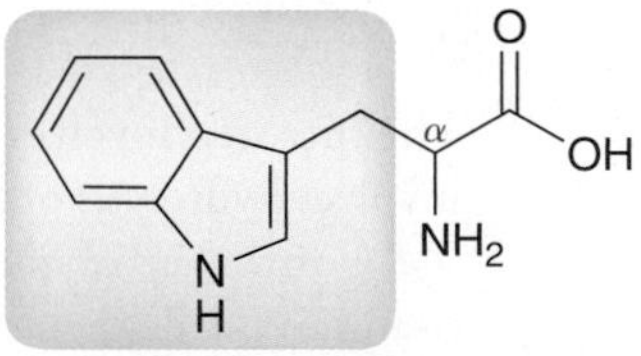

Next, identify the alkyl halide needed to install this group. Just redraw the alkyl group connected to a leaving group. Iodide and bromide are commonly used as leaving groups.

STEP 2
Identify the necessary alkyl halide and ensure that it will readily undergo S_N2.

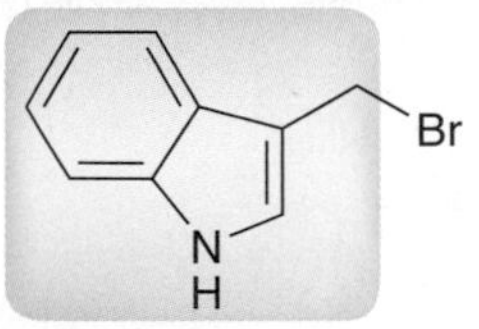

Analyze the structure of this alkyl halide and make sure that it will readily participate in an S_N2 reaction. If the substrate is tertiary, then an amidomalonate synthesis will fail, because the alkylation step will not occur. In this case, the alkyl halide is primary and would be expected to undergo an S_N2 process quite rapidly.

STEP 3
Identify the reagents beginning with acetamidomalonate.

In order to perform the amidomalonate synthesis, we must first deprotonate diethyl acetamidomalonate with a strong base and then treat the resulting anion with the alkyl halide. The resulting product is then heated in the presence of aqueous acid to achieve hydrolysis and decarboxylation.

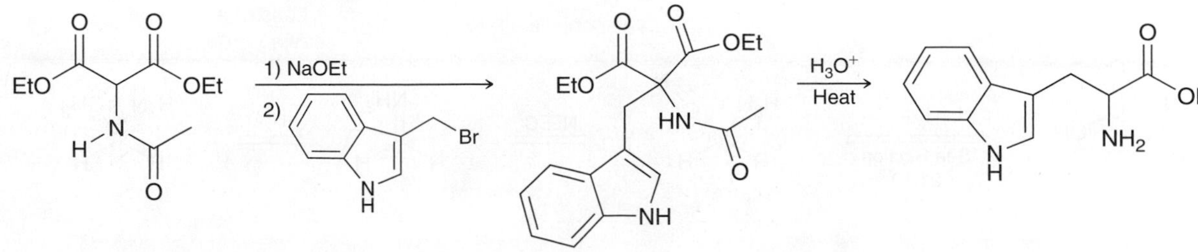

PRACTICE the skill **25.14** Identify the reagents necessary to make each of the following amino acids via the amidomalonate synthesis.

(a) Isoleucine (b) Alanine (c) Valine

APPLY the skill **25.15** An amidomalonate synthesis was performed using each of the following alkyl halides. In each case, draw and name the amino acid that is produced.

(a) Methyl chloride (b) Isopropyl chloride (c) 2-Methyl-1-chloropropane

25.16 Both leucine and isoleucine can be prepared via the amidomalonate synthesis, although one of these amino acids can be produced in higher yields. Identify the higher yield process and explain your choice.

need more **PRACTICE?** **Try Problems 25.57, 25.58, 25.59c, 25.60b, 25.84**

Amino Acid Synthesis via the Strecker Synthesis

Racemic α-amino acids can also be prepared from aldehydes via a two-step process called the **Strecker synthesis**.

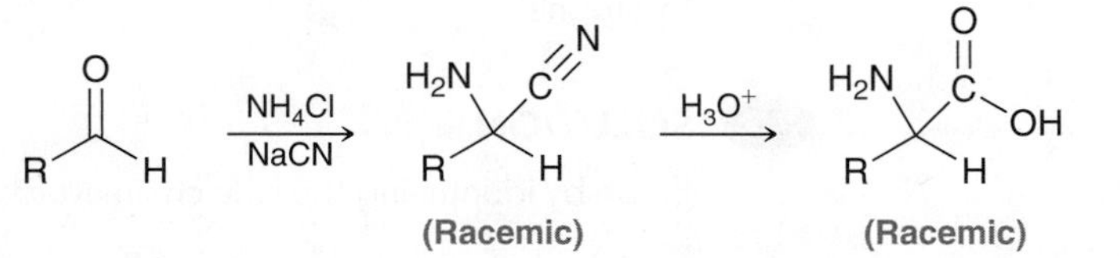

The first step involves conversion of the aldehyde into an α-amino nitrile, and the second step involves hydrolysis of the cyano group to yield a carboxylic acid moiety.

Formation of the α-amino nitrile likely proceeds via Mechanism 25.1. The resulting α-amino nitrile then undergoes hydrolysis via a mechanism found in Section 21.13.

MECHANISM 25.1 FORMATION OF AN α-AMINO NITRILE

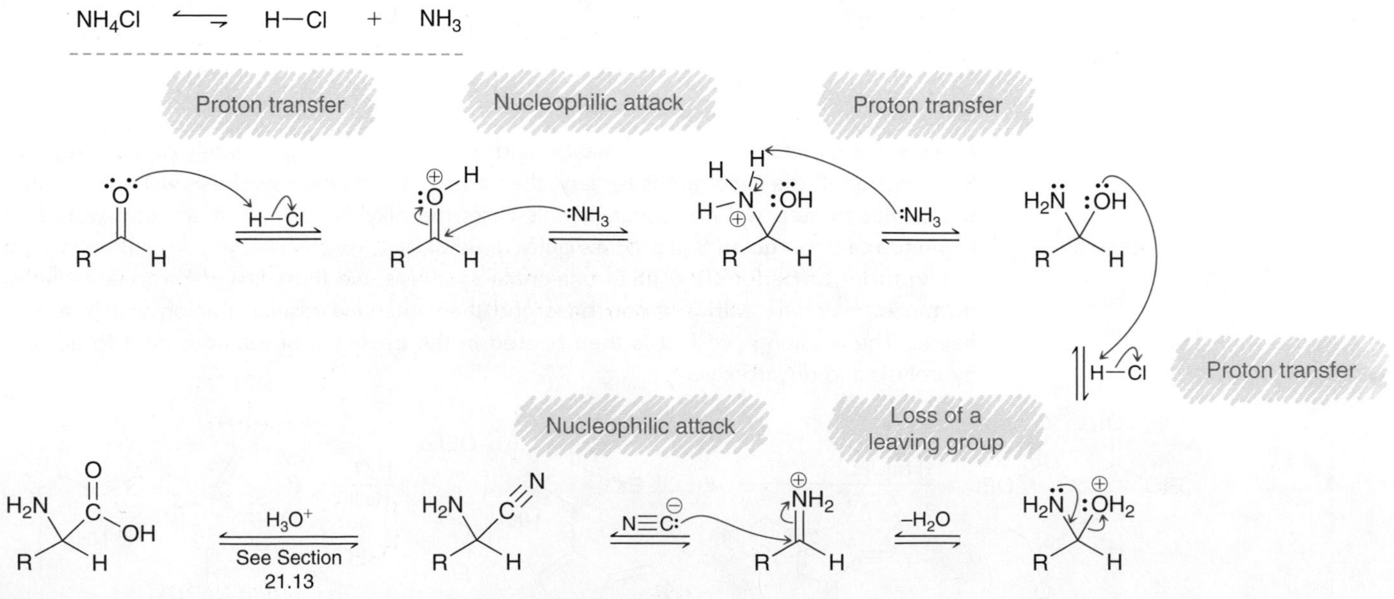

As an example, consider the synthesis of alanine via the Strecker synthesis:

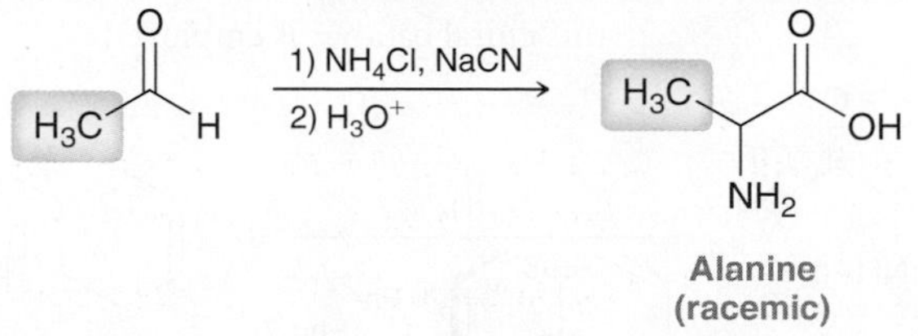

The identity of the amino acid obtained is determined by the choice of the starting aldehyde.

CONCEPTUAL CHECKPOINT

25.17 Identify the reagents necessary to make each of the following amino acids using a Strecker synthesis.

(a) Methionine (b) Histidine

(c) Phenylalanine (d) Leucine

25.18 Each of the following aldehydes was converted into an α-amino nitrile followed by hydrolysis to yield an amino acid. In each case, draw and name the amino acid that was produced.

(a) Acetaldehyde (b) 3-Methylbutanal

(c) 2-Methylpropanal

Enantioselective Synthesis of L Amino Acids

Racemic mixtures of amino acids are produced from each of the three methods described thus far. Obtaining optically active amino acids requires either resolution of the racemic mixture or enantioselective synthesis. Resolution is less efficient and more costly because half of the starting material is wasted in the process. Resolution is not necessary if an enantioselective synthesis is performed. In Section 9.7 we saw how Knowles developed a procedure for achieving asymmetric hydrogenation and then used that procedure for the enantioselective synthesis of L-dopa.

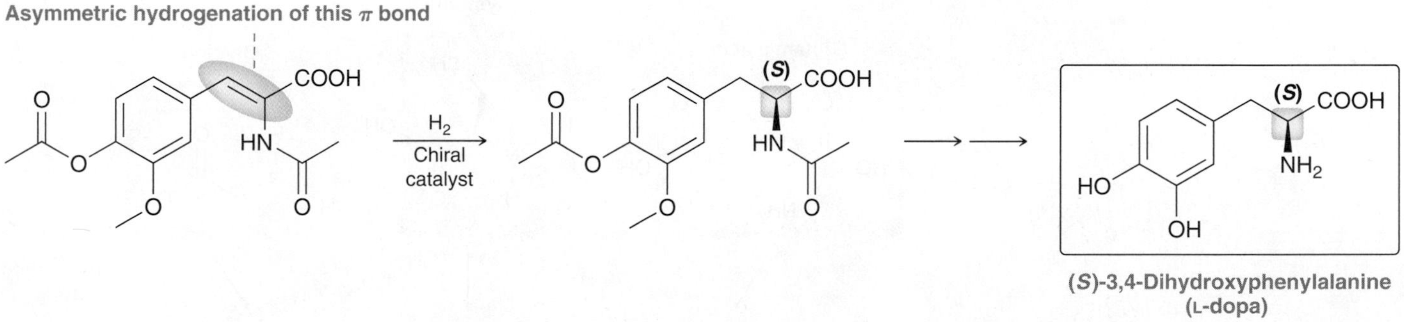

This technique employs a chiral catalyst and can be used to prepare amino acids with very high enantiomeric excess (% *ee*). Many of the chiral catalysts that have been developed contain a ruthenium atom (Ru) complexed to a chiral ligand, such as BINAP.

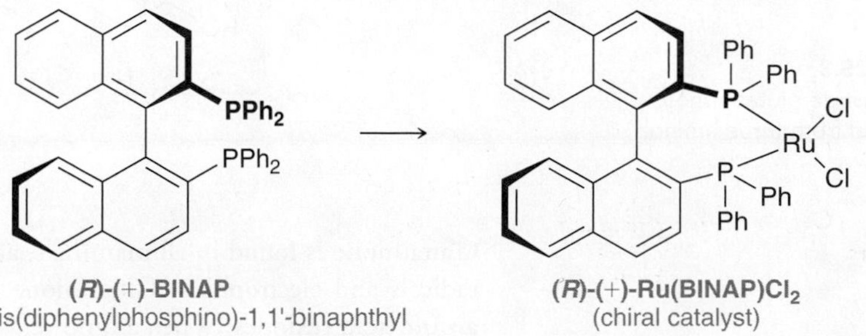

Only small amounts of a chiral catalyst are required to prepare amino acids with very high enantioselectivity. For example, preparation of D-phenylalanine proceeds with 99% *ee* when this chiral catalyst is employed.

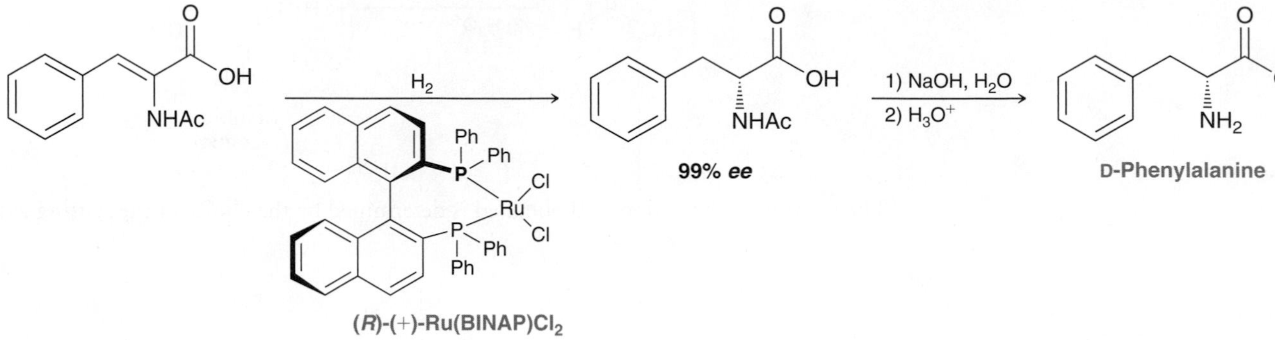

CONCEPTUAL CHECKPOINT

25.19 Identify the starting alkene necessary to make each of the following amino acids using an asymmetric catalytic hydrogenation.

(a) L-alanine (b) L-valine
(c) L-leucine (d) L-tyrosine

25.20 Explain why it is inappropriate to use a chiral catalyst in the preparation of glycine.

25.4 Structure of Peptides

The Assembly of Peptides from Amino Acids

As mentioned earlier in this chapter, amino acids are joined together to form peptides. For example, glutathione is a tripeptide assembled from glutamic acid, cysteine, and glycine (Figure 25.3).

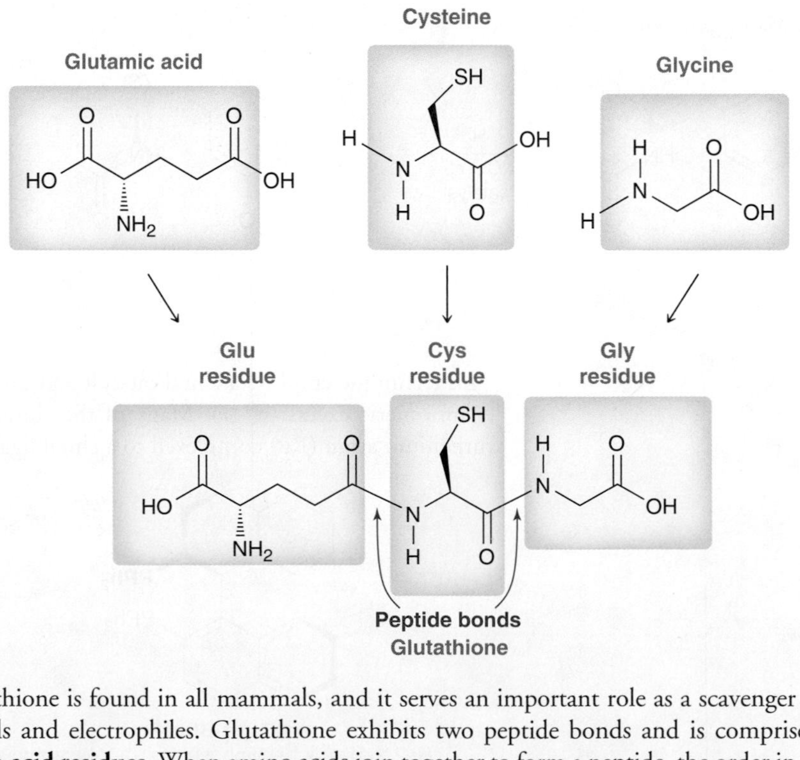

FIGURE 25.3 The tripeptide glutathione is assembled from three amino acids.

Glutathione is found in all mammals, and it serves an important role as a scavenger of harmful radicals and electrophiles. Glutathione exhibits two peptide bonds and is comprised of three **amino acid residues**. When amino acids join together to form a peptide, the order in which they are connected is important. For example, consider a simple dipeptide made by joining alanine and glycine. The peptide bond can be formed between the COOH group of alanine and the NH_2 group of glycine, or from the COOH group of glycine and the NH_2 group of alanine.

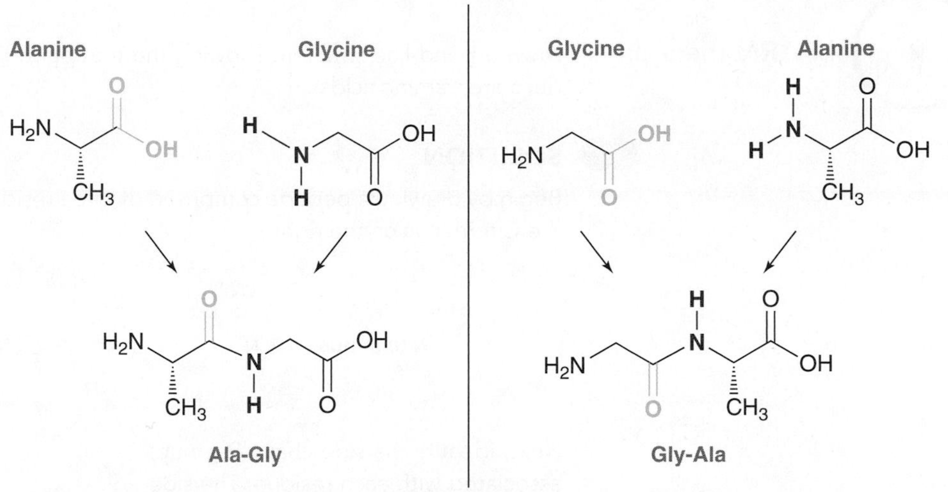

These two dipeptides are not the same compound. They are, in fact, constitutional isomers.

Peptide chains always have an amino group on one end, called the **N terminus,** and a COOH group on the other end, called the **C terminus** (Figure 25.4). By convention, peptides are always drawn with the N terminus on the left side.

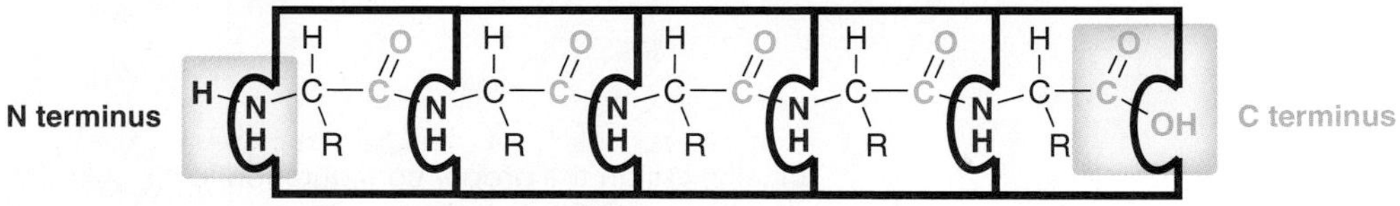

FIGURE 25.4
The N terminus and C terminus of a peptide chain.

The sequence of amino acid residues in a peptide can be abbreviated with one- or three-letter abbreviations, starting with the N terminus. For example, a dipeptide of glycine and alanine can be written as follows:

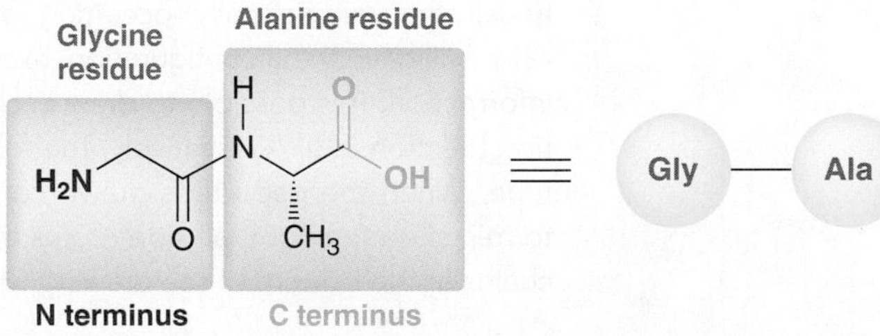

Simple peptide chains will have one N terminus and one C terminus. For example, consider the following decapeptide, for which the alanine residue is the N terminus and the leucine residue is the C terminus:

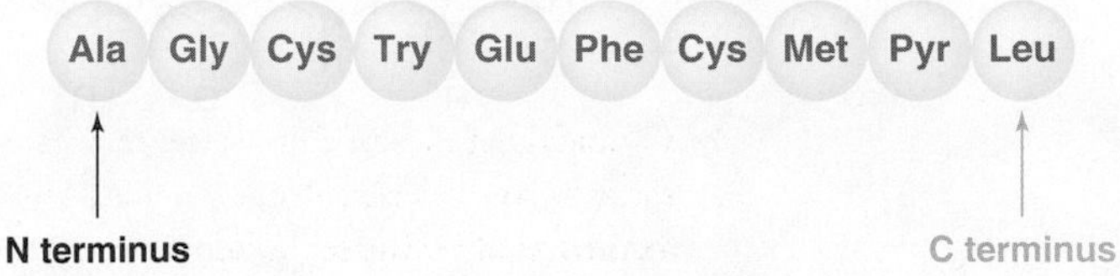

SKILLBUILDER

25.3 DRAWING A PEPTIDE

LEARN the skill

Draw a bond-line structure showing the tripeptide Phe-Val-Trp (assume that all three residues are L amino acids).

SOLUTION

Begin by drawing a peptide comprised of three residues with the N terminus on the left and the C terminus on the right.

STEP 1
Draw a peptide with the correct number of residues.

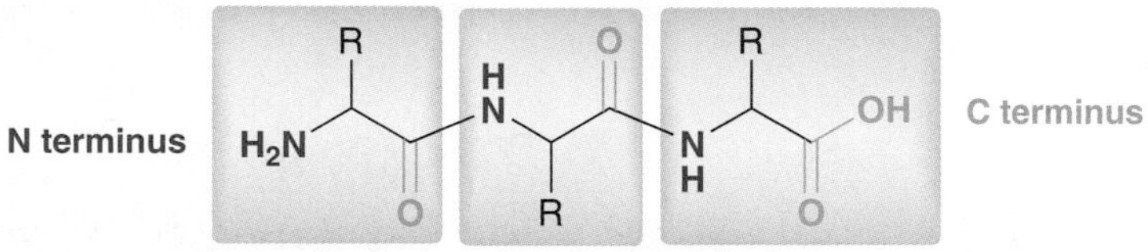

STEP 2
Identify the side chain associated with each residue.

Next, identify the side chain (R group) associated with each residue. The side chains associated with Phe, Trp and Val are highlighted.

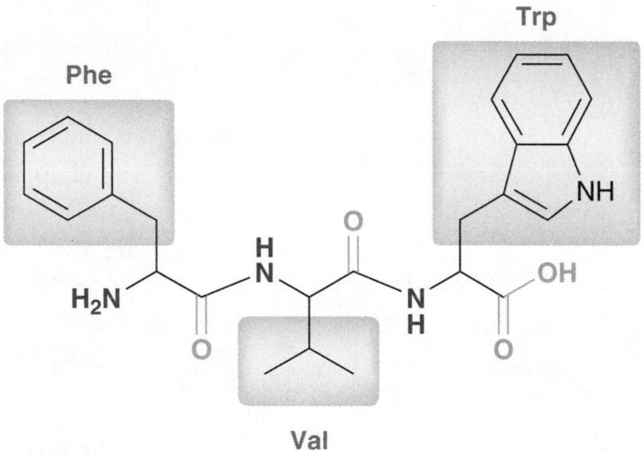

STEP 3
Assign the proper configuration for each α position.

Finally, assign the proper configuration for each side chain.

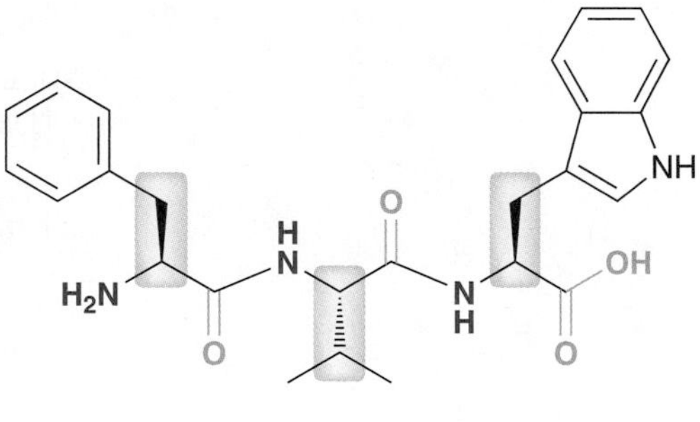

Recall that all naturally occurring amino acids will have the *S* configuration, except for cysteine, which has the *R* configuration. With this information, it is possible to draw each chirality center using the Cahn-Ingold-Prelog convention (Section 5.3). Alternatively, the following stereochemical shortcut can be applied to save time: When the peptide is drawn conventionally with the N terminus on the left and the C terminus on the right, all side chains at the top of the drawing will be on wedges, and all side chains at the bottom of the drawing will be on dashes.

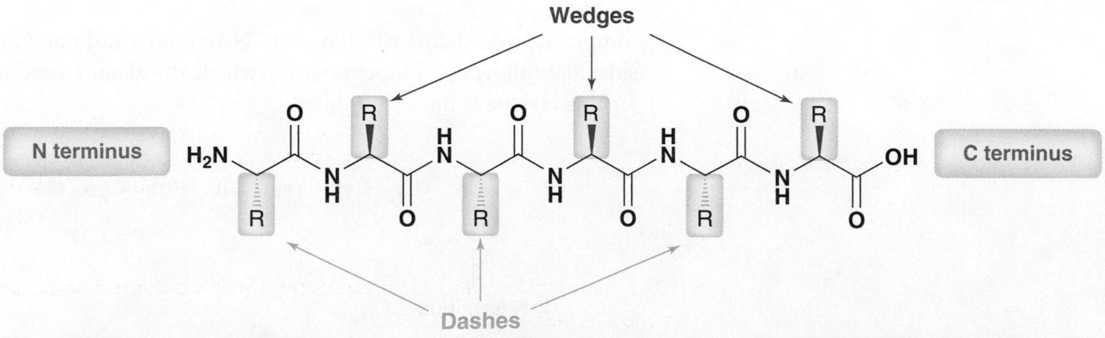

PRACTICE the skill **25.21** Draw the structure of each of the following peptides:

(a) Leu-Ala-Gly (b) Cys-Asp-Ala-Gly (c) Met-Lys-His-Tyr-Ser-Phe-Val

APPLY the skill **25.22** Using three- and one-letter abbreviations, show the sequence of amino acid residues in the following pentapeptide.

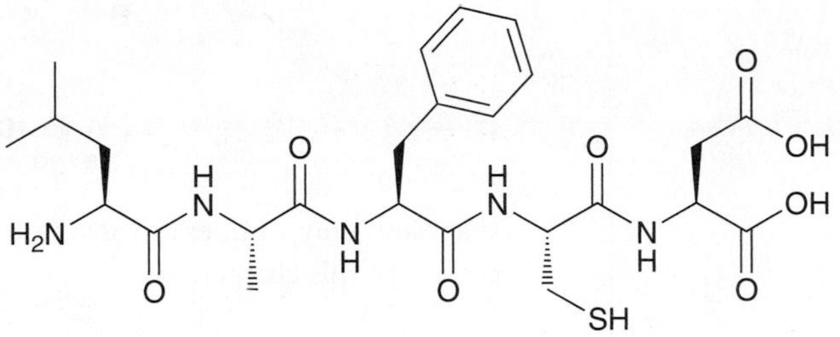

25.23 Determine which of the following peptides will have a higher molecular weight (*Hint*: It is not necessary to actually calculate the molecular weight of each peptide, but rather, just compare the side chains).

Cys-Tyr-Leu or Cys-Phe-Ile

25.24 Compare the following tripeptides and determine whether they are constitutional isomers or the same compound.

Ala-Gly-Leu and Leu-Gly-Ala

need more **PRACTICE?** **Try Problems 25.65, 25.66**

The Geometry of Peptide Bonds

In order to understand the three-dimensional geometry of peptides, we must first explore the geometry of peptide bonds. Recall that peptide bonds are amide linkages and that amides have a significant resonance structure that endows the C—N bond with some double-bond character.

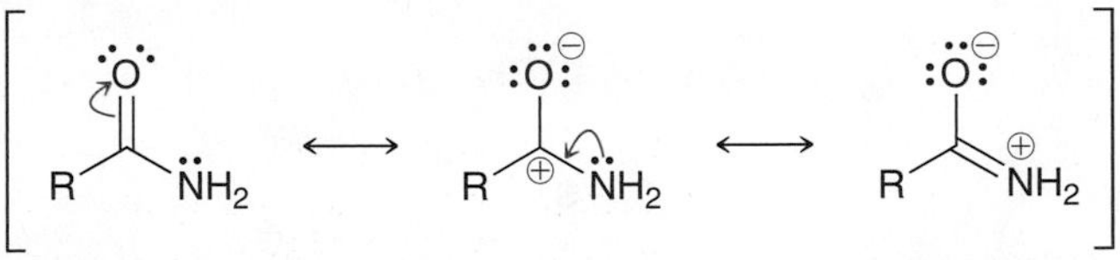

A significant contributor

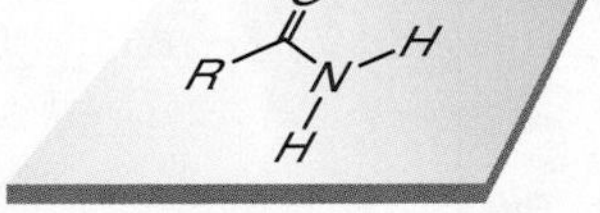

FIGURE 25.5
The planar structure of an amide.

As a result, the nitrogen atom is sp^2 hybridized and planar (Figure 25.5). Peptide bonds are simply amides, and therefore exhibit double-bond character just like regular amides. As such, a peptide bond experiences restricted rotation, giving rise to two possible conformations called ***s-trans*** and ***s-cis***.

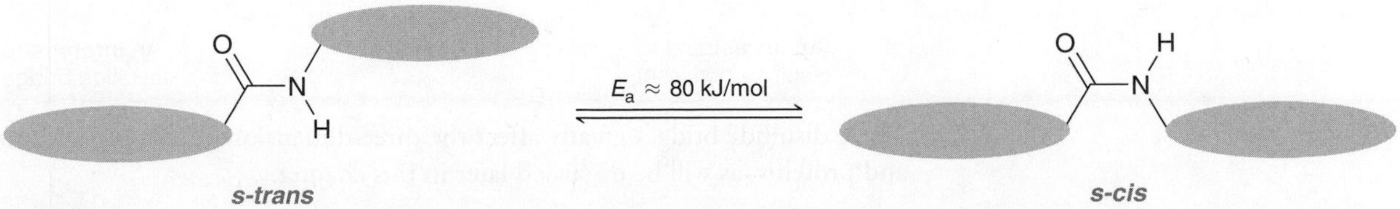

s-trans ***s-cis***

The *s-trans* conformation is usually more stable because it lacks steric hindrance that is present in the *s-cis* conformation.

In a polypeptide, each peptide bond will show a preference for adopting an *s-trans* conformation. Nevertheless, polypeptides are not entirely planar, because they still possess σ bonds that experience free rotation. Only the peptide bonds experience restricted rotation.

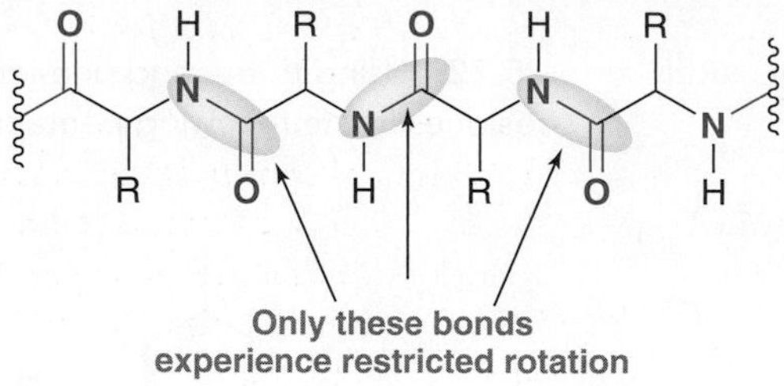

As a result, polypeptides can assume a variety of conformations, as will be discussed in upcoming sections of this chapter.

Disulfide Bridges

In Section 14.11 we saw that two thiols can be joined via an oxidation process to form a disulfide.

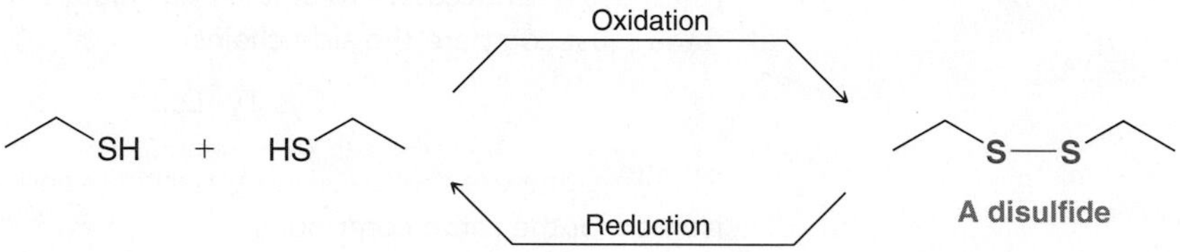

Of the 20 amino acids found in proteins, only cysteine contains a thiol moiety. As such, cysteine residues are uniquely capable of being joined to one another via **disulfide bridges**.

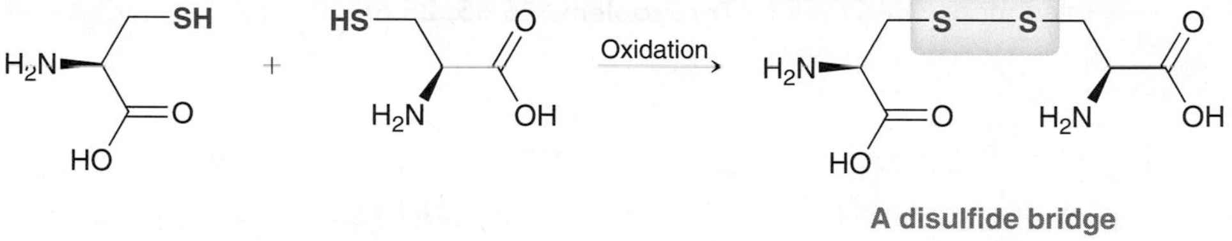

Disulfide bridges are commonly observed between cysteine residues in the same strand or between cysteine residues in different strands.

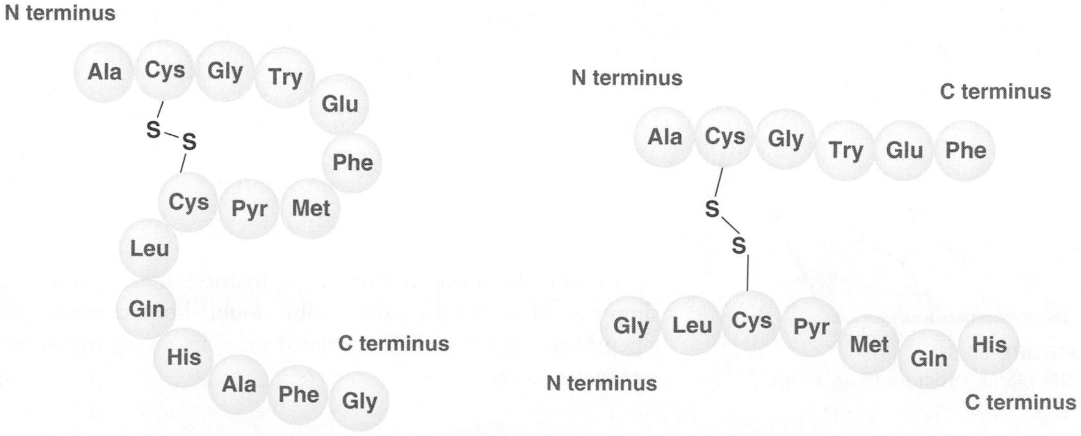

An **intra**strand disulfide bridge

An **inter**strand disulfide bridge

These disulfide bridges greatly affect the three-dimensional structure and properties of peptides and proteins, as will be discussed later in this chapter.

Some Interesting Peptides

Peptides play a variety of important biological roles. For example, consider the structures of the following two pentapeptides:

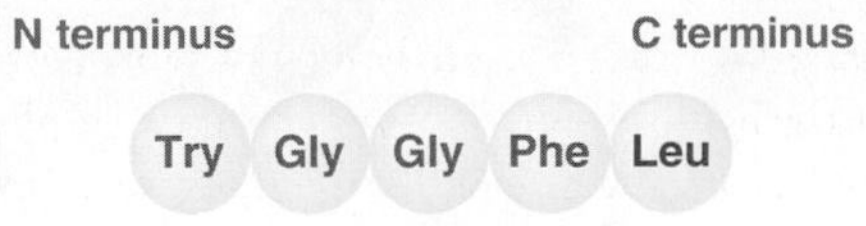

Leucine enkephalin

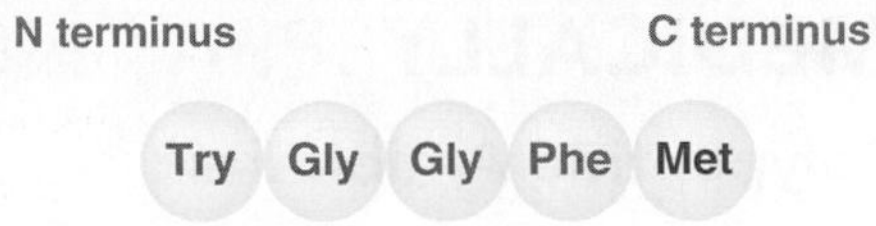

Methionine enkephalin

Leucine enkephalin and methionine enkephalin are both found in the brain, where they function in thc control of pain. It is believed that they interact with the same receptor as morphine. Notice that the sequences of these compounds are the same except for the last residue (Leu vs. Met). Bradykinin is another biologically important peptide. It functions as a hormone that dilates blood vessels and acts as an anti-inflammatory agent. Bradykinin is a nonapeptide (nine residues) with the following sequence:

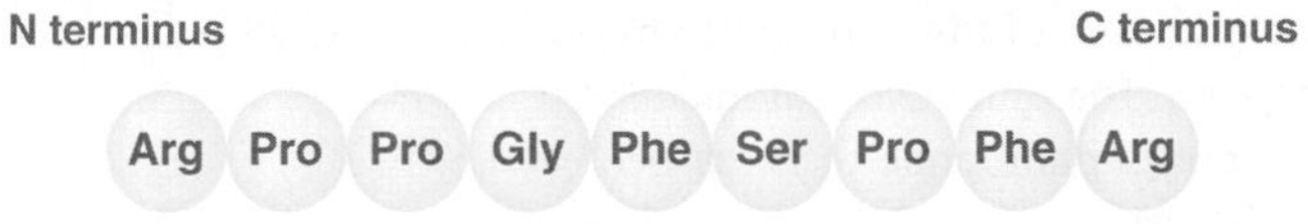

Bradykinin

Vasopressin and oxytocin are very similar in structure and represent two more biologically important peptides.

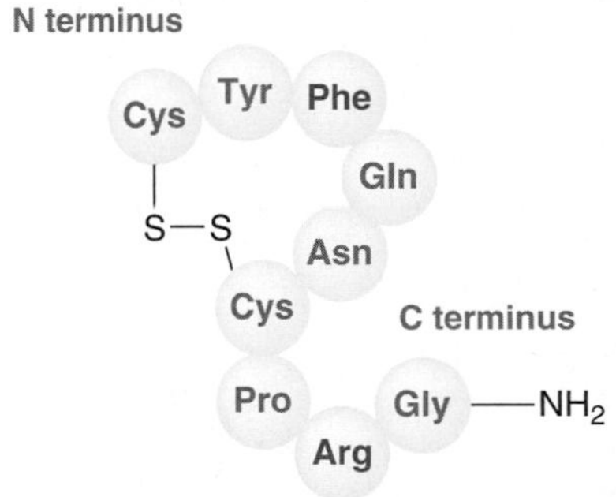

Vasopressin

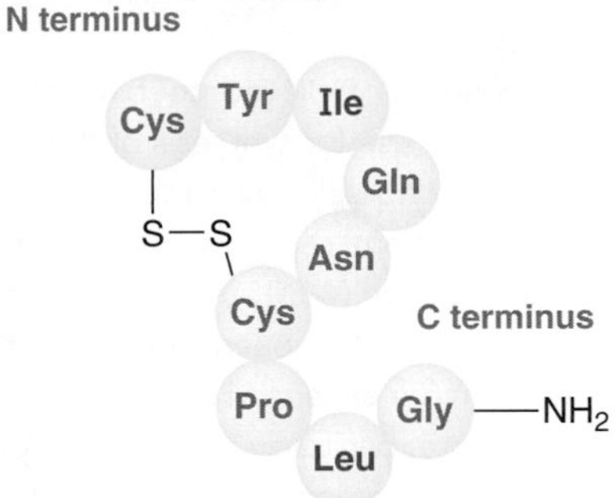

Oxytocin

Each of these peptides contains a disulfide bridge. Also notice that the C-terminal end in each case has been modified. The NH_2 group in each case indicates that the OH group of the C terminus has been replaced with an NH_2 group. In other words, the COOH group has been converted into a $CONH_2$ group (an amide moiety). While vasopressin and oxytocin are similar in structure (differing only in the identity of two residues, highlighted in red), they nevertheless have very different functions. Vasopressin is released from the pituitary gland and causes readsorption of water by the kidneys and controls blood pressure. Oxytocin induces labor and stimulates milk production in nursing mothers. These examples illustrate how a small difference in structure can produce a large difference in function. Throughout the rest of the chapter, we will see several other examples of the relationship between structure and function.

CONCEPTUAL CHECKPOINT

25.25 Draw the *s-trans* conformation of the dipeptide Phe-Leu and identify both the N terminus and the C terminus.

25.26 Draw the *s-cis* conformation of the dipeptide Phe-Phe and identify the source of steric hindrance.

25.27 Using a bond-line structure, show the tetrapeptide obtained when two molecules of Cys-Phe are joined by a disulfide bridge.

25.28 Aspartame is an artificial sweetener sold under the trade name NutraSweet™. Aspartame is the methyl ester of a dipeptide formed from L-aspartic acid and L-phenylalanine and can be summarized as **Asp-Phe**-OCH_3. Surprisingly, the analogous ethyl ester (**Asp-Phe**-OCH_2CH_3) is not sweet. If either L residue is replaced with its enantiomeric D residue, the resulting compound is bitter instead of sweet.

(a) Draw a bond-line structure of aspartame.

(b) Draw a bond-line structure for each of the three bitter stereoisomers of aspartame.

MEDICALLYSPEAKING

Polypeptide Antibiotics

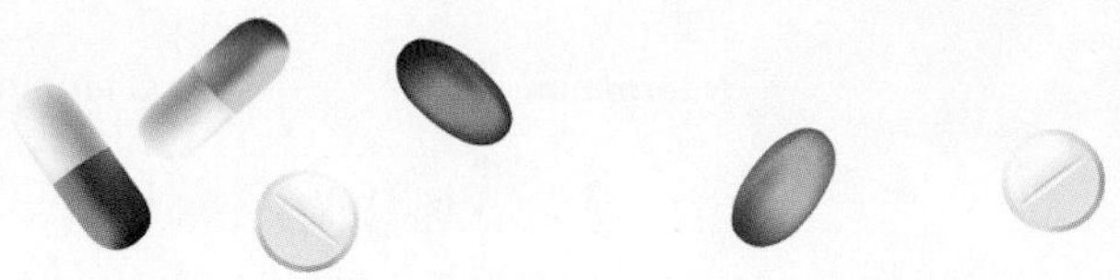

Some naturally occurring polypeptides are among the most powerful bactericidal antibiotics. Most have cyclic structures, and they often contain D amino acids not typically found in plants and animals. In addition, many polypeptide antibiotics contain regions comprised of carbohydrates or heterocycles instead of amino acids.

In the previous chapter, we explored the structure of neomycin, one of the principal components of Neosporin. The other two ingredients in Neosporin are polypeptide antibiotics, called bacitracin and polymyxin.

Bacitracin is a complex mixture of polypeptides that was first isolated in 1945 from a strain of *Bacillus subtilis*. The commercial product called bacitracin is a mixture comprised primarily of bacitracin A.

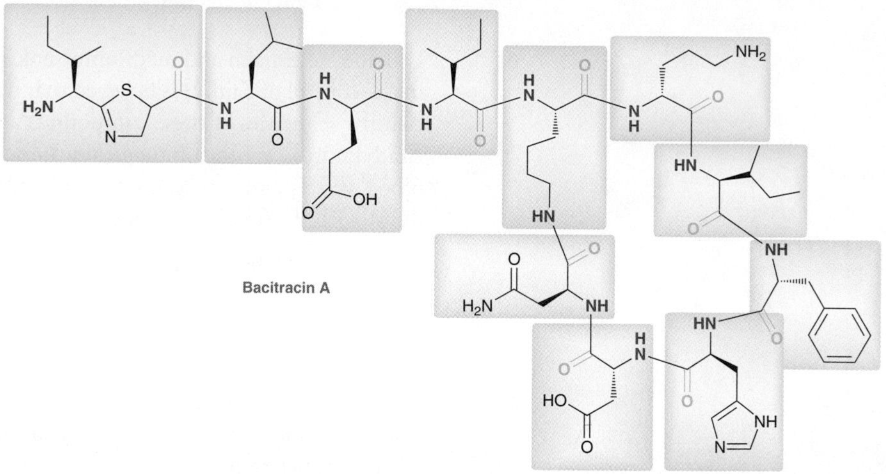

Bacitracin A

This compound is believed to bind to an enzyme that bacteria use for synthesis of their cell walls, thereby inhibiting cell wall synthesis. Under these conditions, bacteria cannot sustain the high internal osmotic pressure (4–20 atm), which causes the bacteria to burst. It has been found that the effects of bacitracin are enhanced by the presence of zinc, and therefore, bacitracin zinc is used commonly in topical preparations used to treat local infections. It is not absorbed by the gastrointestinal tract, so oral administration is generally ineffective.

Polymyxin B was first isolated from the bacterium *Bacillus polymyxa* in 1947.

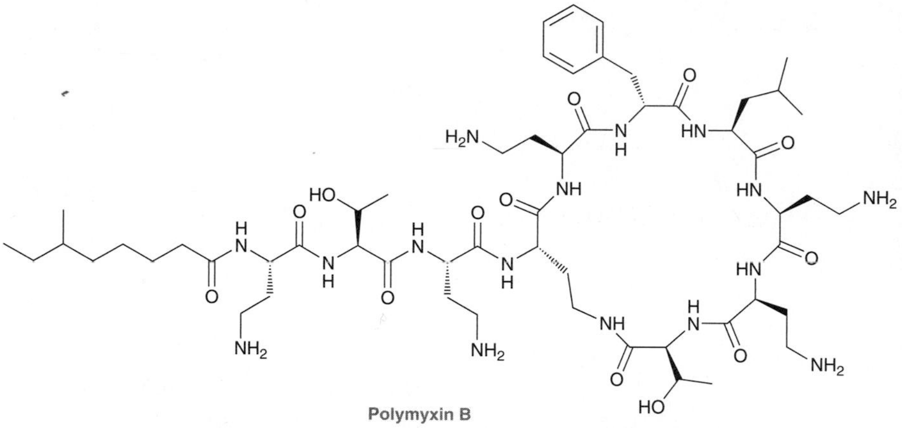

Polymyxin B

Polymyxin B is also commonly used in topical preparations to treat local infections. Like bacitracin A, it is also not absorbed by the gastrointestinal tract, so oral administration is ineffective except in treatment of intestinal infections. Bacitracin A and polymyxin B target different bacteria and are therefore commonly mixed together in topical preparations, such as Neosporin.

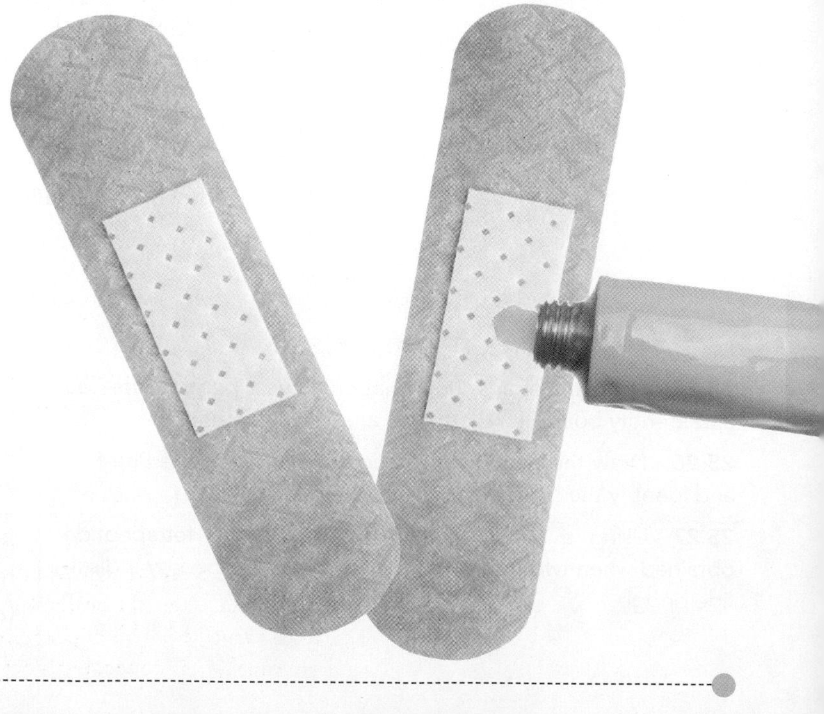

CONCEPTUAL CHECKPOINT

25.29 Bacitracin A is produced by bacteria and therefore contains some residues that are not from the list of 20 naturally occurring amino acids (Figure 25.1).

(a) Identify which amino acid residues are found in Table 25.1 and name them.

(b) Identify which of these amino acids are D rather than L.

(c) Identify all residues that are not listed in Table 25.1.

25.5 Sequencing a Peptide

The sequence of amino acids in a peptide can be analyzed using a variety of techniques. One such method is called the Edman degradation.

The Edman Degradation

The **Edman degradation** involves removing one amino acid residue at a time and identifying each residue as it is removed. The process is repeated until the entire peptide has been sequenced. To perform an Edman degradation, a peptide is first treated with phenyl isothiocyanate followed by trifluoroacetic acid. These two reagents remove the amino acid residue at the N terminus and convert it into a phenylthiohydantoin (PTH) derivative.

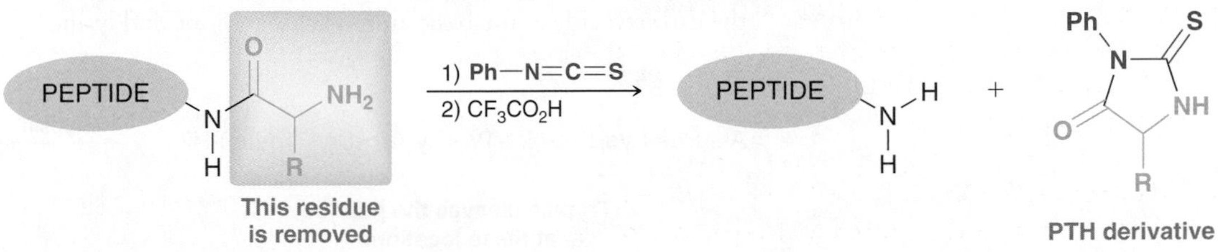

The PTH derivative can then be analyzed by a variety of different techniques to determine the identity of the amino acid residue that was removed. The transformation is believed to occur via Mechanism 25.2.

MECHANISM 25.2 THE EDMAN DEGRADATION

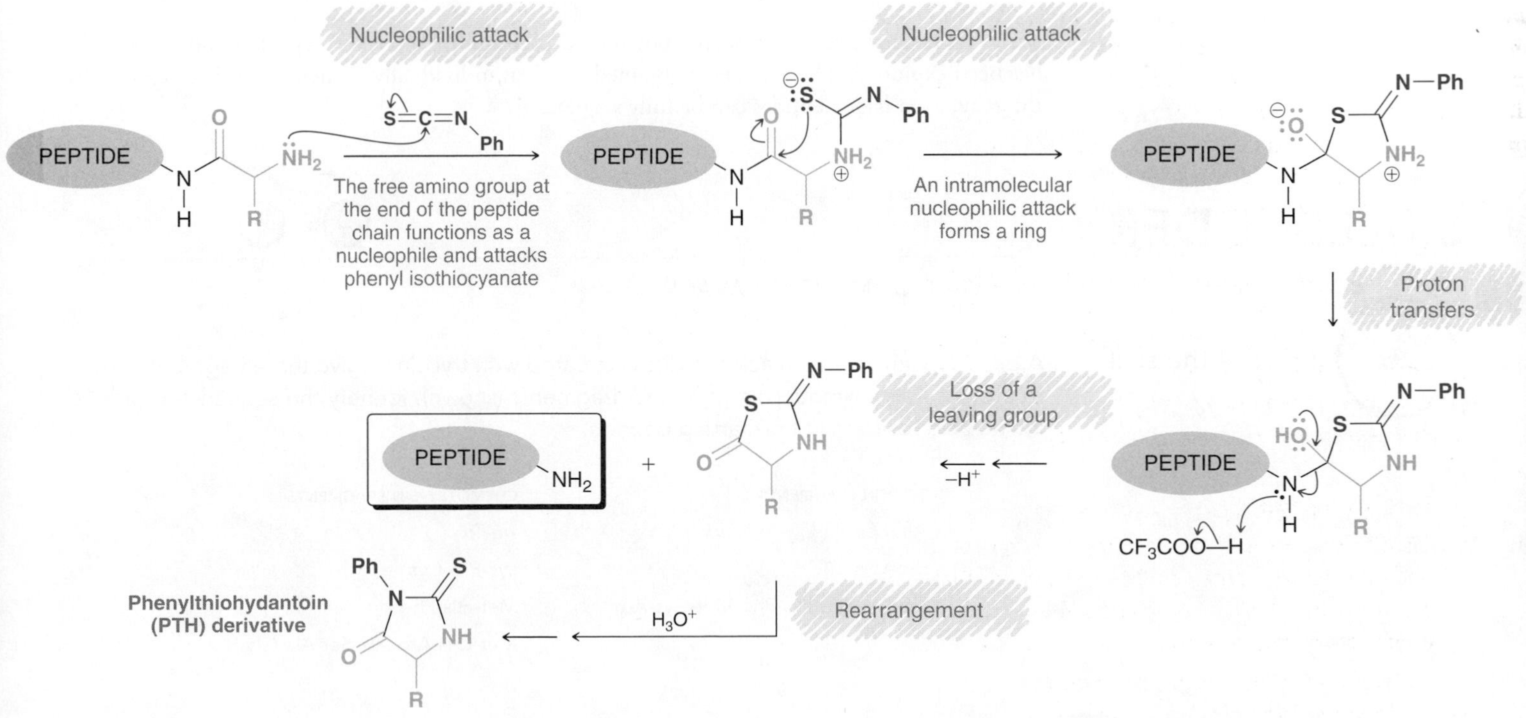

The process can be repeated on the chain-shortened peptide to remove another amino acid residue. In this way, the amino acid residues are removed one at a time until the entire sequence has been identified. This entire process has been fully automated, and automated peptide sequencers are capable of sequencing a peptide chain with as many as 50 amino acid residues.

CONCEPTUAL CHECKPOINT

25.30 Draw the structure of the initial PTH derivative formed when the tripeptide Ala-Phe-Val undergoes an Edman degradation.

Enzymatic Cleavage

For peptides containing more than 50 residues, the removal of each residue one at a time is not practical because unwanted side products accumulate and interfere with the results. Larger peptides are analyzed by first cleaving them into smaller fragments and then sequencing the fragments. A variety of enzymes, called **peptidases**, selectively hydrolyze specific peptide bonds. For example, trypsin is a digestive enzyme that catalyzes the hydrolysis of the peptide bond at the carboxyl side of the basic amino acids arginine and lysine.

Ala-Phe-**Lys** ⦚ Pro-Met-Tyr-Gly-**Arg** ⦚ Ser-Trp-Leu-His —Trypsin→ Ala-Phe-**Lys** + Pro-Met-Tyr-Gly-**Arg** + Ser-Trp-Leu-His

Trypsin cleaves the peptide at these locations

Chymotrypsin, another digestive enzyme, selectively hydrolyzes the carboxyl end of amino acids containing aromatic side chains, which includes phenylalanine, tyrosine, and tryptophan.

Ala-**Phe** ⦚ Lys-Pro-Met-**Tyr** ⦚ Gly-Arg-Ser-**Trp** ⦚ Leu-His —Chymotrypsin→ Ala-**Phe** + Lys-Pro-Met-**Tyr** + Gly-Arg-Ser-**Trp** + Leu-His

Chymotrypsin cleaves the peptide at these locations

Many other digestive enzymes are known, and their ability to cleave peptide bonds selectively has been exploited. The fragments obtained are then individually sequenced and compared. In this way, even large peptides can be fully sequenced.

SKILLBUILDER

25.4 PEPTIDE SEQUENCING VIA ENZYMATIC CLEAVAGE

LEARN the skill

A peptide with 16 amino acid residues is treated with trypsin to give three fragments, while treatment with chymotrypsin gives four fragments (shown). Identify the sequence of the 16 amino acid residues in the starting peptide.

TRYPSIN FRAGMENTS	CHYMOTRYPSIN FRAGMENTS
Ala-Ser-Ala-Gly-Phe-Lys	Ile-Trp
Pro-Cys	Lys-Pro-Cys
Ile-Trp-Met-His-Phe-Met-Cys-Arg	Met-His-Phe
	Met-Cys-Arg-Ala-Ser-Ala-Gly-Phe

SOLUTION

First consider the fragments generated upon cleavage with trypsin. Recall that trypsin will hydrolyze a peptide bond at the carboxyl side of arginine and lysine. We therefore expect that all fragments will have a C terminus that is either arginine or lysine, except for the fragment that contains the C terminus of the original peptide. This allows us to identify which fragment was last in the peptide sequence.

Pro—Cys
↑
This residue must have been the C terminus of the original peptide

There are only two other fragments, so there are only two possibilities for the sequence of the original peptide. The correct sequence can be determined by analyzing the chymotrypsin fragments. Specifically, we look for the fragment that ends with Pro-Cys, and we find that this fragment has a lysine residue immediately before the proline residue.

Lys—Pro—Cys

This information makes it possible to determine the order of the trypsin fragments.

Ile—Trp—Met—His—Phe—Met—Cys—Arg------Ala—Ser—Ala—Gly—Phe—Lys------Pro—Cys

↑ **N terminus** ↑ **C terminus**

PRACTICE the skill

25.31 A peptide with 22 amino acid residues is treated with trypsin to give four fragments, while treatment with chymotrypsin yields six fragments. Identify the sequence of the 22 amino acid residues in the starting peptide.

TRYPSIN FRAGMENTS		CHYMOTRYPSIN FRAGMENTS	
Trp-His-Phe-Met-Cys-Arg	Pro-Val-Ile-Leu-Arg	Lys-Pro-Val-Ile-Leu-Arg-Trp	His-Phe
Met-Phe-Val-Ala-Tyr-Lys	Gly-Pro-Phe-Ala-Val	Val-Ala-Tyr	Ala-Val
		Met-Cys-Arg-Gly-Pro-Phe	Met-Phe

APPLY the skill

25.32 The tetrapeptide Val-Lys-Ala-Phe is cleaved into two fragments upon treatment with trypsin. Identify the sequence of a tetrapeptide that will produce the same two fragments when treated with chymotrypsin.

25.33 Consider the structure of the following cyclic octapeptide. Would cleavage of this peptide with trypsin produce different fragments than cleavage with chymotrypsin? Explain.

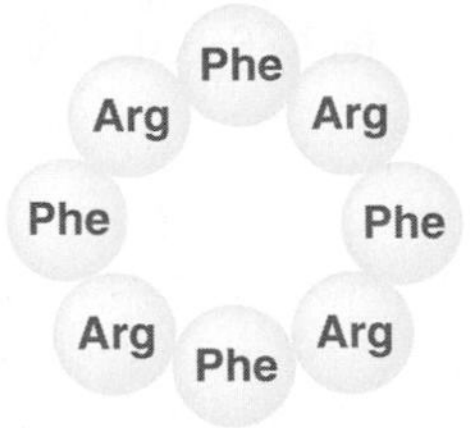

need more **PRACTICE?** **Try Problems 25.71, 25.74**

25.6 Peptide Synthesis

Peptide Bond Formation

Forming a peptide bond requires the coupling of a carboxylic acid group and an amino group.

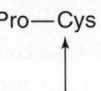

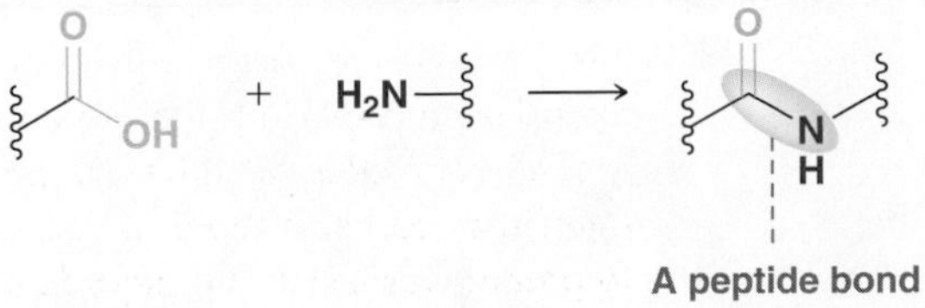

Several reagents are available to achieve this type of transformation. One common reagent is dicyclohexylcarbodiimide (DCC):

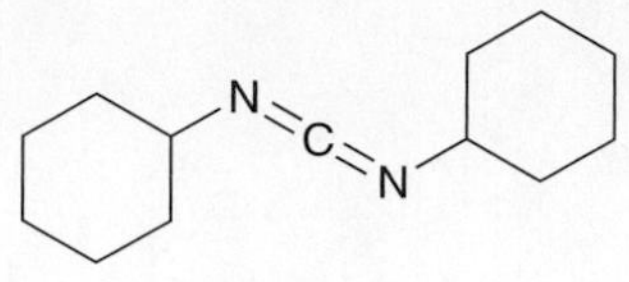

Dicyclohexylcarbodiimide (DCC)

Formation of a peptide bond via DCC is believed to proceed via Mechanism 25.3.

MECHANISM 25.3 PEPTIDE BOND FORMATION VIA DCC

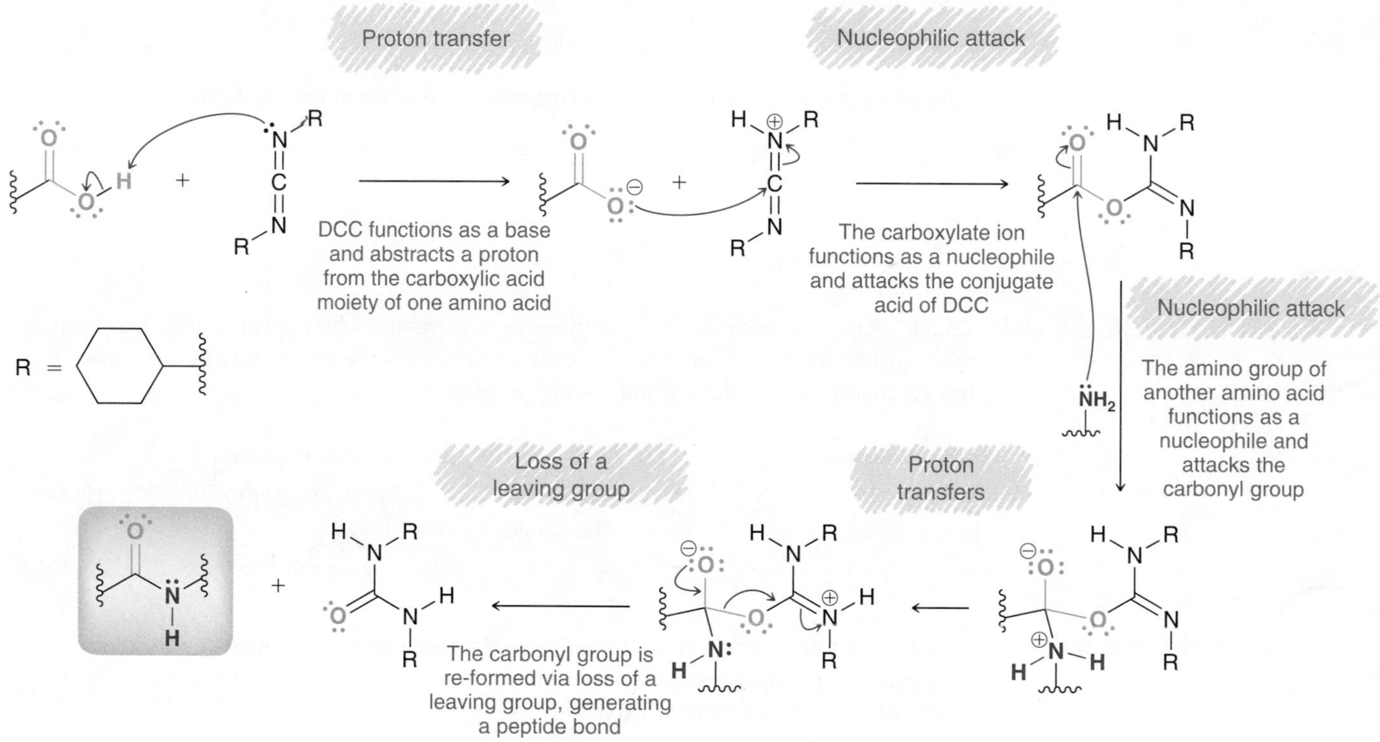
Proton transfer
DCC functions as a base and abstracts a proton from the carboxylic acid moiety of one amino acid
Nucleophilic attack
The carboxylate ion functions as a nucleophile and attacks the conjugate acid of DCC
R =
Nucleophilic attack
The amino group of another amino acid functions as a nucleophile and attacks the carbonyl group
Proton transfers
Loss of a leaving group
The carbonyl group is re-formed via loss of a leaving group, generating a peptide bond

A lack of regioselectivity becomes problematic when DCC is used to couple two different amino acids. Specifically, four different dipeptides are possible:

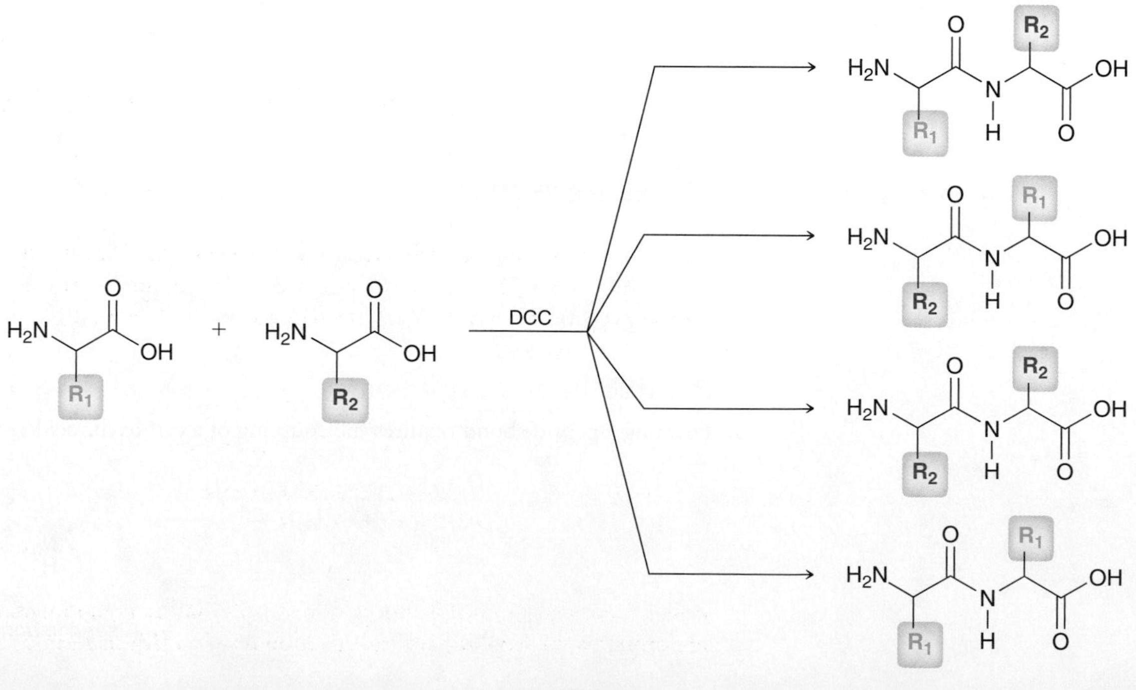
DCC

This problem can be circumvented by first protecting the NH_2 group of one amino acid and the COOH group of the other amino acid (Figure 25.6): The protected amino acids can be coupled regioselectively followed by removal of the protecting groups. The following sections describe how amino groups and carboxylic acid groups can be protected.

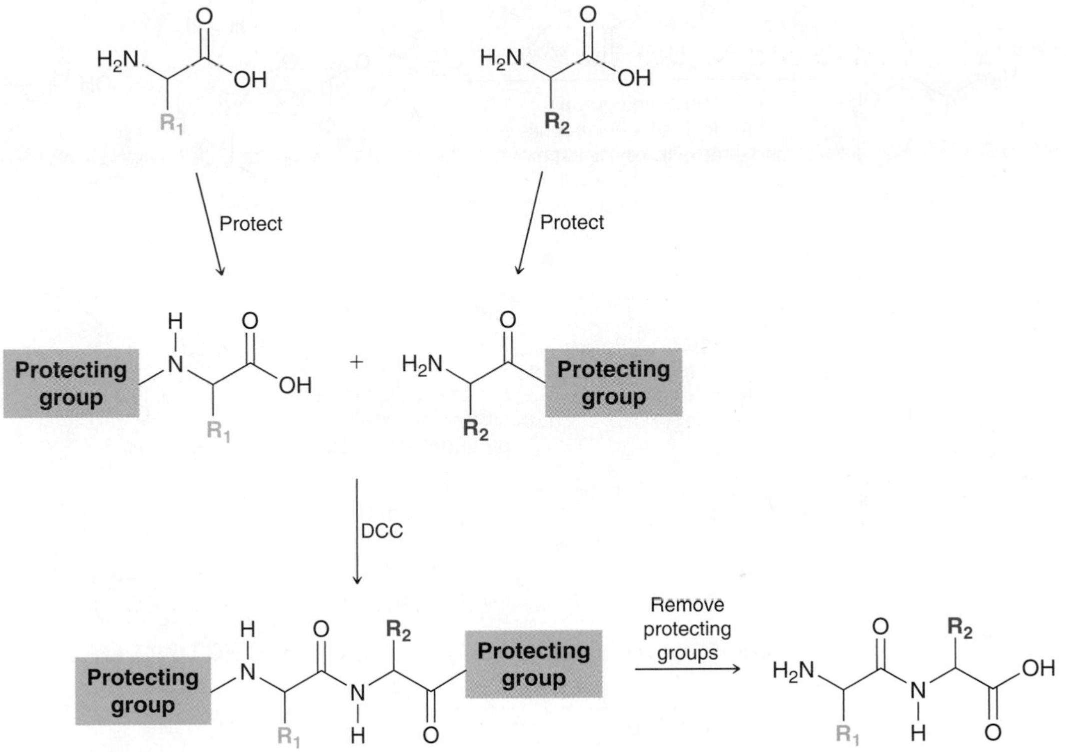

FIGURE 25.6
The overall strategy for synthesizing dipeptides involves protection, coupling, and then deprotection.

Protecting the Amino Group of an Amino Acid

An amino group can be protected by converting it into a carbamate.

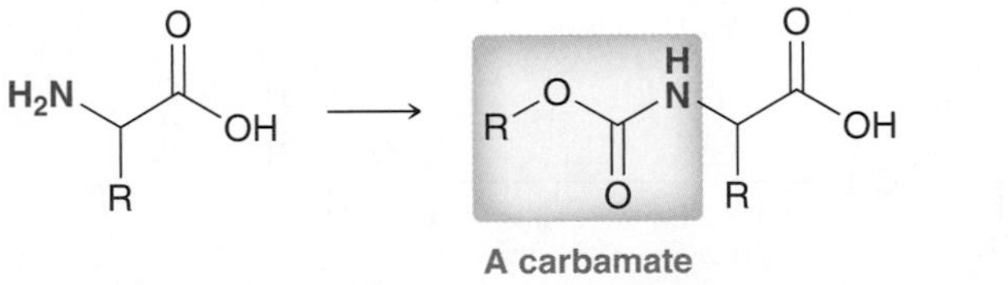

The nitrogen atom of a carbamate is less nucleophilic because its lone pair is delocalized by the neighboring carboxyl group. One of the most common carbamate protecting groups is the *tert*-butoxycarbonyl group (Boc), formed via treatment with di-*tert*-butyl dicarbonate.

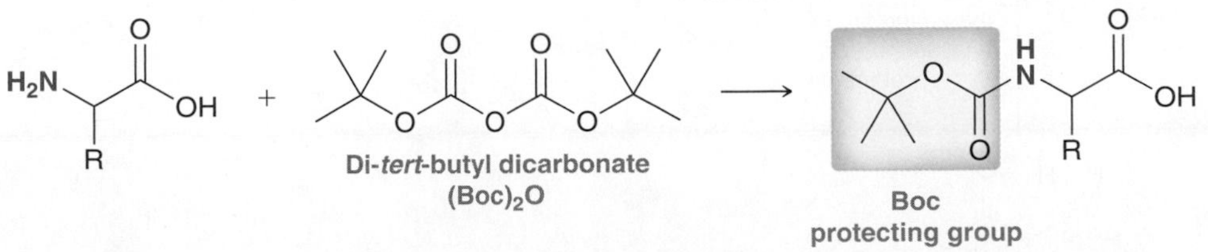

A likely mechanism for this process involves the amino group functioning as a nucleophile and performing a nucleophilic acyl substitution reaction (Mechanism 25.4).

MECHANISM 25.4 INSTALLATION OF A BOC PROTECTING GROUP

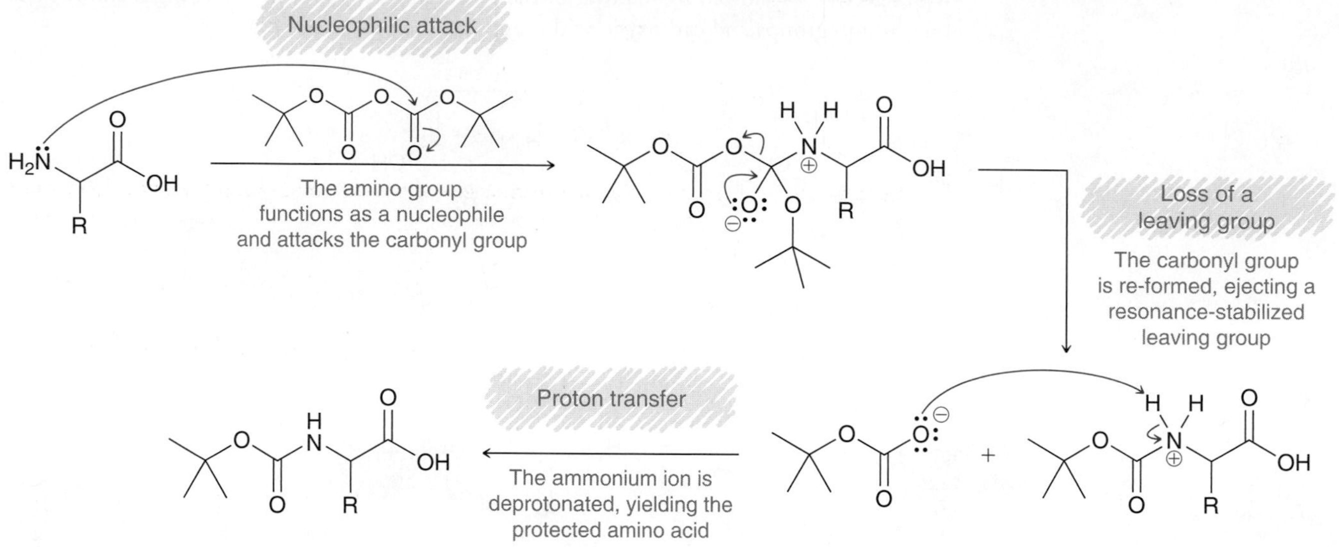
Nucleophilic attack
The amino group functions as a nucleophile and attacks the carbonyl group
Loss of a leaving group
The carbonyl group is re-formed, ejecting a resonance-stabilized leaving group
Proton transfer
The ammonium ion is deprotonated, yielding the protected amino acid

MECHANISM 25.5 REMOVAL OF A BOC PROTECTING GROUP

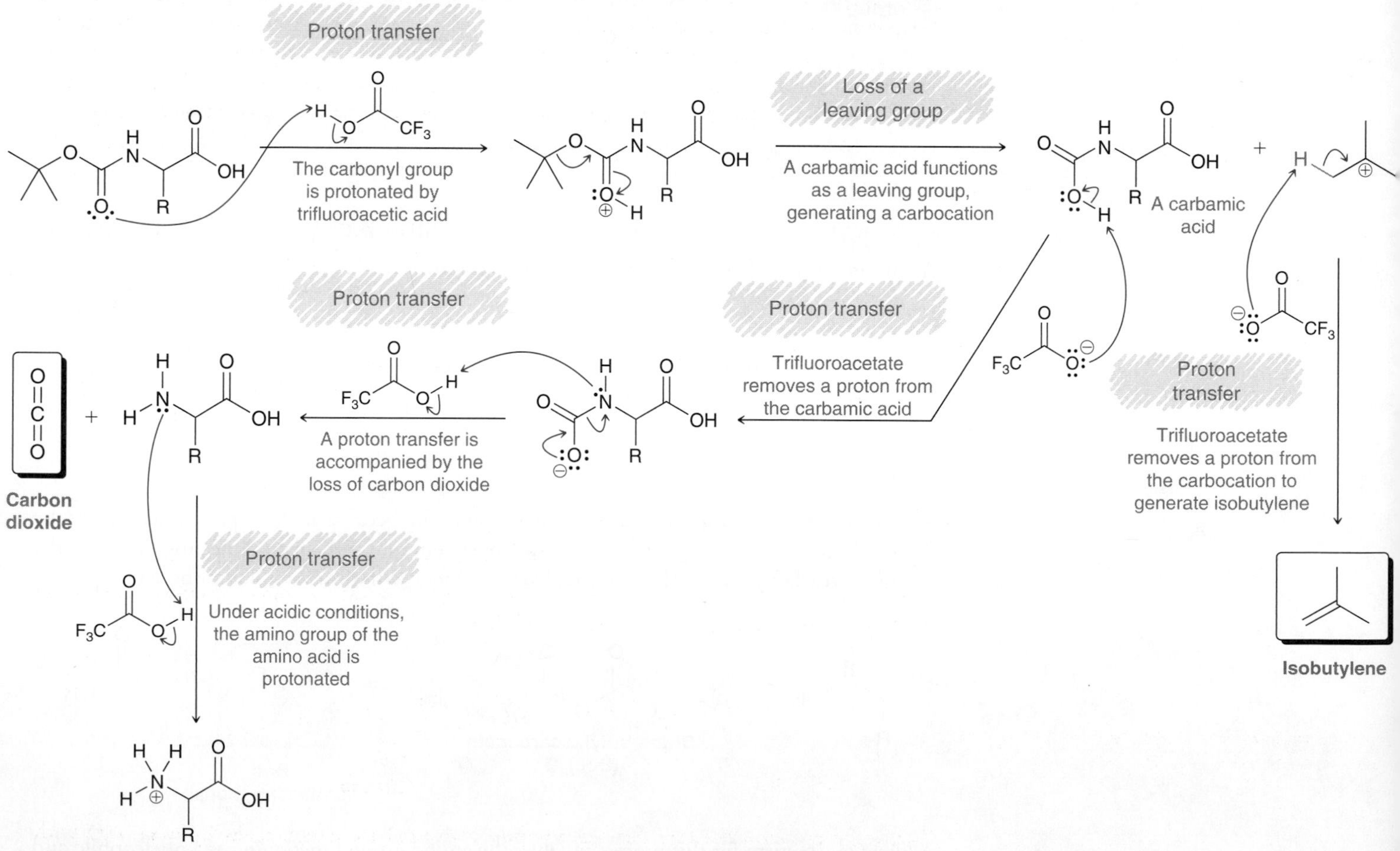
Proton transfer
The carbonyl group is protonated by trifluoroacetic acid
Loss of a leaving group
A carbamic acid functions as a leaving group, generating a carbocation
A carbamic acid
Proton transfer
Trifluoroacetate removes a proton from the carbocation to generate isobutylene
Isobutylene
Proton transfer
Trifluoroacetate removes a proton from the carbamic acid
Proton transfer
A proton transfer is accompanied by the loss of carbon dioxide
Carbon dioxide
Proton transfer
Under acidic conditions, the amino group of the amino acid is protonated

The Boc group is one of the most commonly used protecting groups because of the ease with which it can be removed. Several reagents can be used to remove a Boc group, including trifluoroacetic acid.

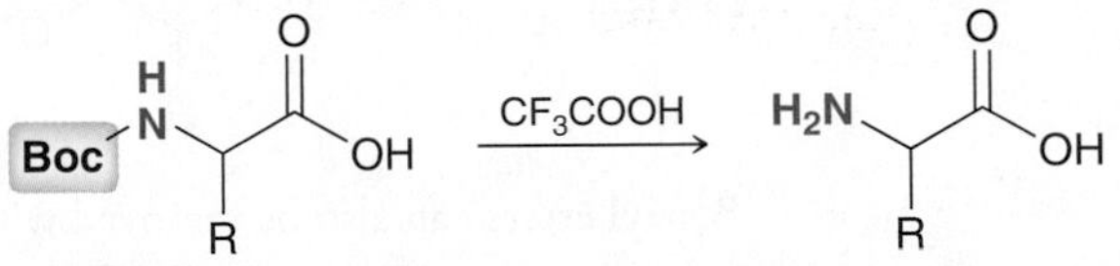

The mechanism is believed to involve protonation of the carbonyl group followed by formation of a carbocation and loss of carbon dioxide (Mechanism 25.5). This process generates isobutylene and carbon dioxide, both of which are gases.

The evolution of gases effectively drives the reaction to completion, and the yield for removal of a Boc group is often 100%.

An alternative protecting group is known as Fmoc; it is formed by treating an amino acid with 9-fluorenylmethyl chloroformate.

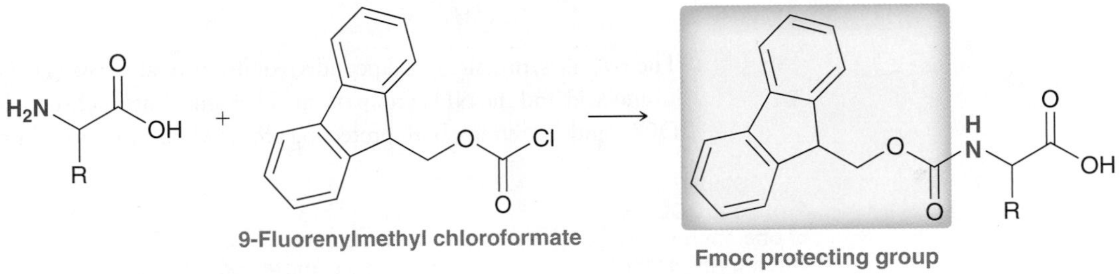

The mechanism for this process is similar to the mechanism for installing a Boc group; that is, the amino group functions as a nucleophile and performs a nucleophilic acyl substitution reaction (where chloride is the leaving group that is replaced). The Fmoc protecting group can be removed with a base, such as piperidine, which initiates the process by deprotonating the acidic benzylic position (Mechanism 25.6). Deprotonation at this position generates an anion that is similar to the anion of cyclopentadiene. The anion is aromatic and is therefore highly stabilized. Loss of carbon dioxide then drives the deprotection process to completion.

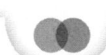

MECHANISM 25.6 REMOVAL OF A FMOC PROTECTING GROUP

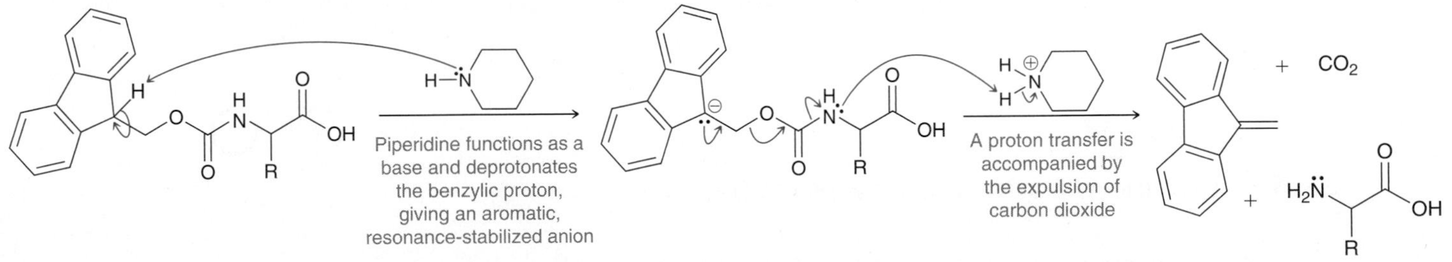

Protecting the Carboxylic Acid Moiety of an Amino Acid

A carboxylic acid moiety can be protected by converting it into an ester, which is achieved by treatment with an alcohol under acidic conditions.

LOOKING BACK
For a review of this process, see Section 21.10.

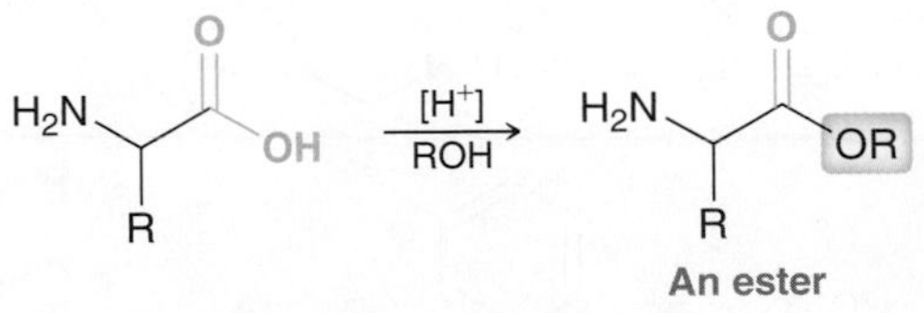

Carboxylic acid moieties are most commonly converted to methyl esters by treatment with MeOH or to benzyl esters by treatment with $PhCH_2OH$.

The ester protecting group can be removed with aqueous base (H_2O, NaOH).

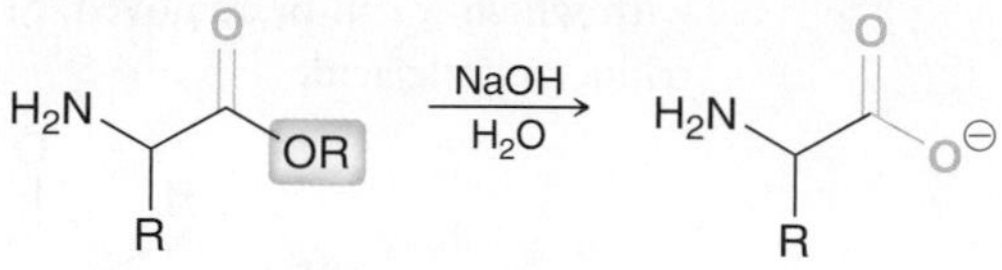

Benzyl esters can also be removed with hydrogenolysis or with HBr in acetic acid.

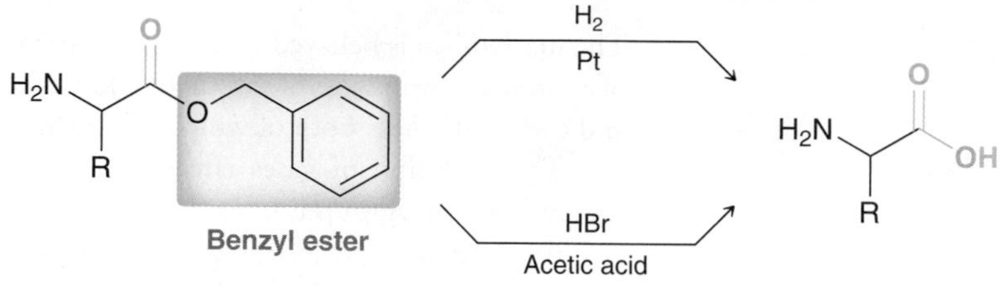

Preparing a Dipeptide

The overall synthesis of a dipeptide requires several steps: (1) protect the COOH group of one amino acid and the NH_2 group of the other amino acid, (2) couple the protected amino acids with DCC, and (3) remove both protecting groups. This process is illustrated in the synthesis of Ala-Gly.

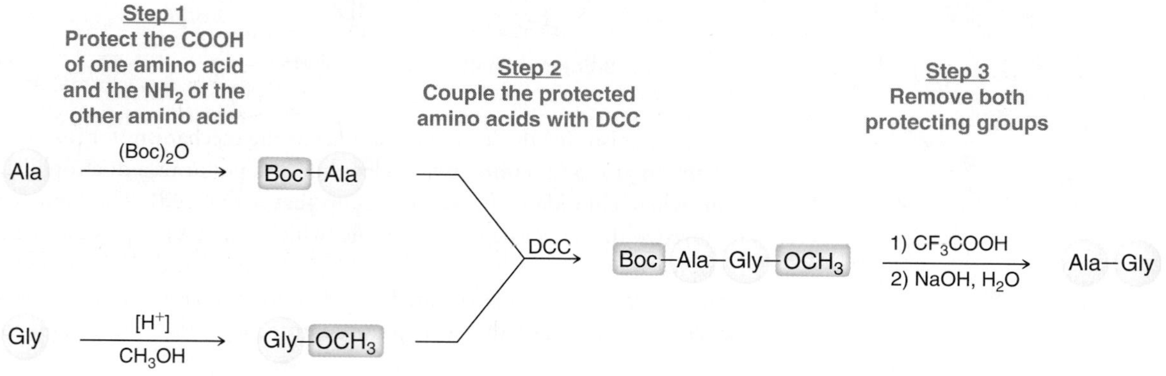

A similar method can be used to prepare tripeptides and tetrapeptides.

SKILLBUILDER

25.5 PLANNING THE SYNTHESIS OF A DIPEPTIDE

LEARN the skill

Propose a synthesis for the following dipeptide: Phe-Val

SOLUTION

Draw the two amino acids necessary and identify the functional groups that must be joined to form a peptide bond.

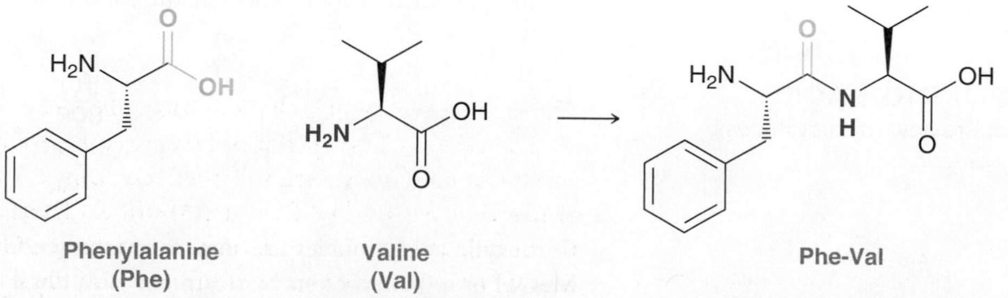

Identify the groups that must be protected (the groups that are not being coupled). The amino group is protected by installing a Boc protecting group.

STEP 1
Install the appropriate protecting groups.

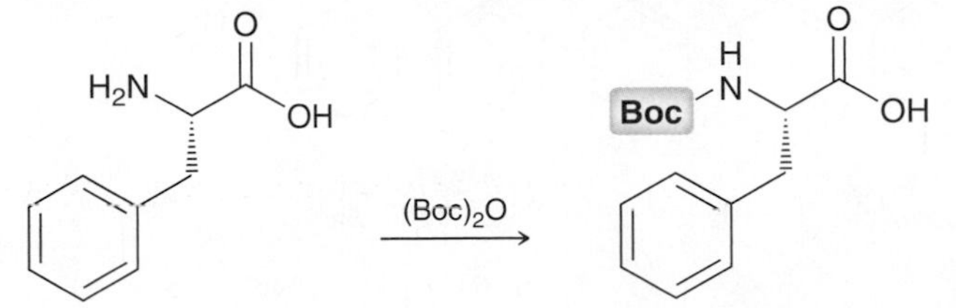

The carboxylic acid moiety is protected by converting it to an ester, such as a benzyl ester.

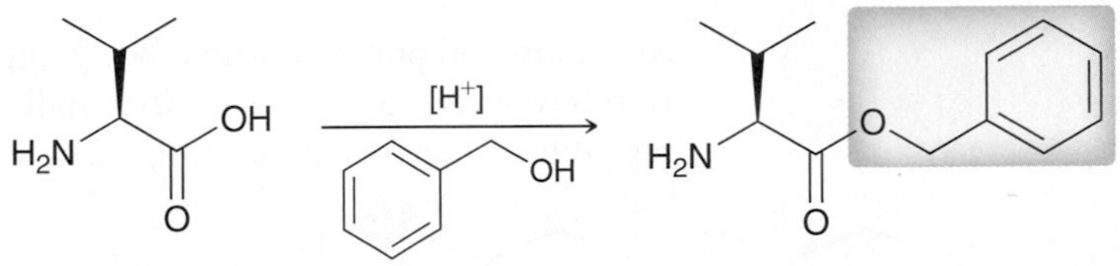

STEP 2
Couple the protected amino acids using DCC.

With the protecting groups in place, the next step is to form the desired peptide linkage using DCC.

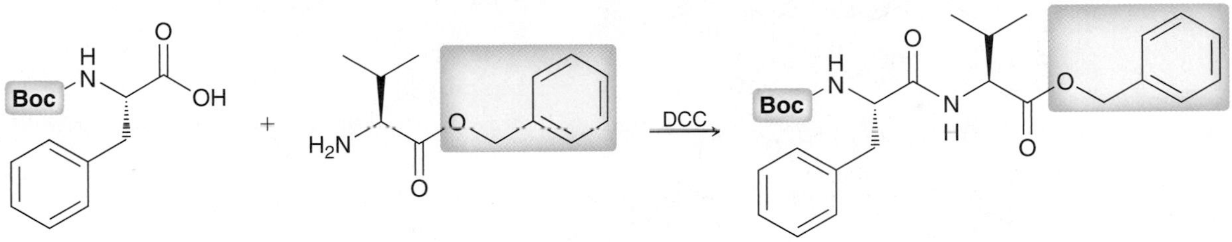

The final step is to remove the protecting groups to yield the desired product.

STEP 3
Remove the protecting groups.

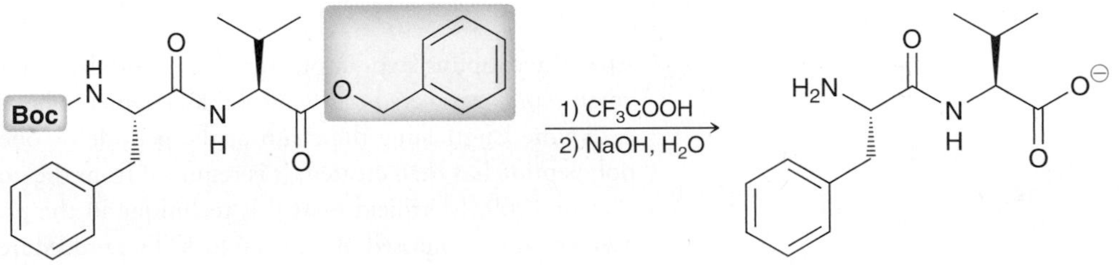

PRACTICE the skill **25.34** Draw all of the steps and reagents necessary to prepare each of the following dipeptides from their corresponding amino acids.

(a) Trp-Met (b) Ala-Ile (c) Leu-Val

APPLY the skill **25.35** Draw all of the steps and reagents necessary to prepare a tripeptide with the sequence Ile-Phe-Gly.

25.36 Draw all of the steps and reagents necessary to prepare a pentapeptide with the sequence Leu-Val-Phe-Ile-Ala.

need more **PRACTICE?** **Try Problems 25.77–25.79**

The Merrifield Synthesis

The method outlined in the previous section works well for very small peptides, but it is not feasible to use it for larger peptides, because each step requires isolation and purification due to the accumulation of unwanted side products. Larger peptides can be prepared with the **Merrifield synthesis**, developed by R. Bruce Merrifield (Rockefeller University). A protected amino acid is first tethered to beads of an insoluble polymer. One such polymer is a polystyrene derivative with CH_2Cl groups on some of the aromatic rings.

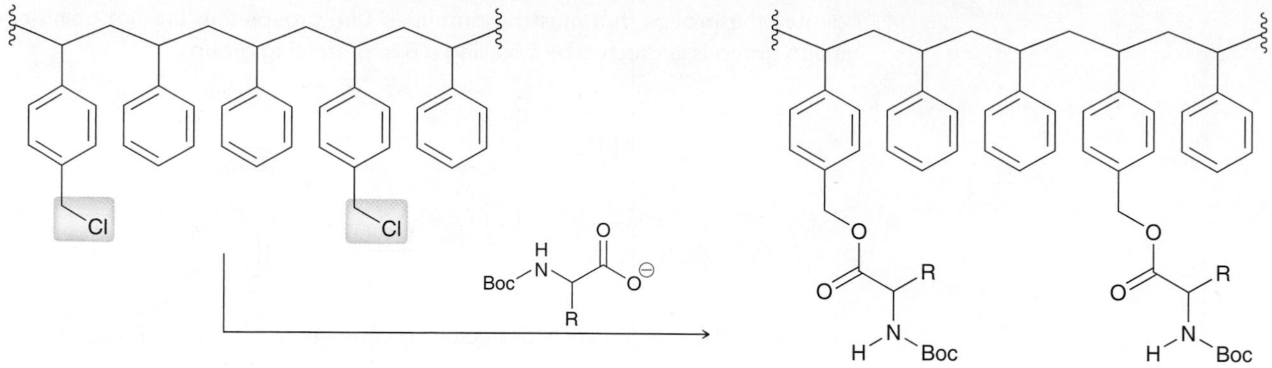

An amino acid protected with a Boc group is first attached to the polymer via an S_N2 reaction. In this way, the polymer serves as a handle that holds the amino acid in place. The Boc protecting group is then removed, and DCC is used to couple the next Boc-protected amino acid:

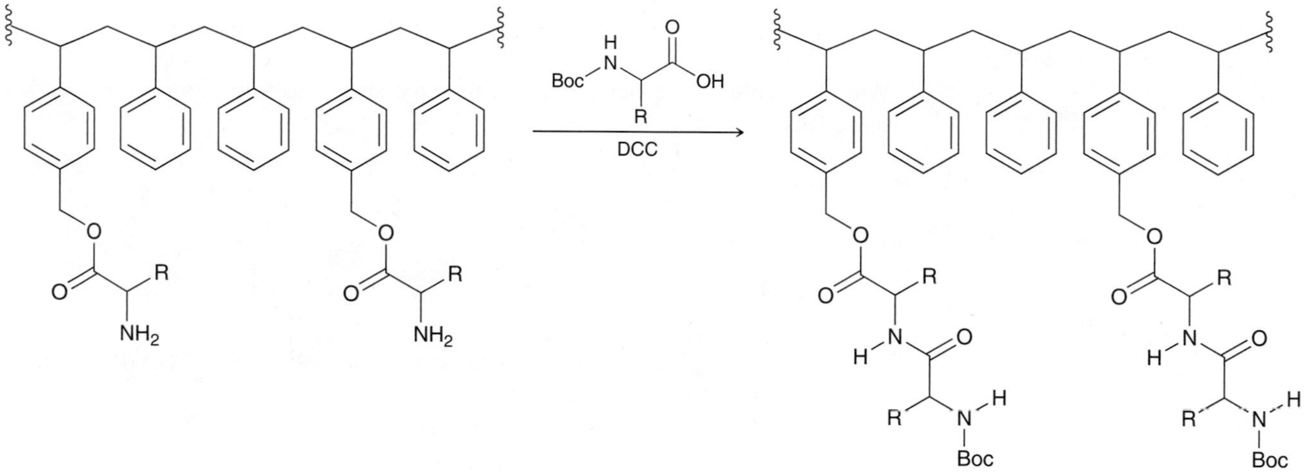

After the coupling step, impurities and by-products are simply washed away, while the peptide chain remains tethered to the insoluble polymer. The process can then be repeated many times, each time lengthening the chain of the peptide by one amino acid residue. When the desired polypeptide has been created, it is removed from the polymer using hydrofluoric acid (HF).

In 1969, Merrifield used this technique in the preparation of a protein called ribonuclease, which is comprised of 128 residues. The procedure required 369 separate reactions, which were performed in just six weeks. The overall yield of the process was 17%, implying that each individual step had a yield greater than 99%. Merrifield's ingenious method set the stage for an entirely new way of performing chemical reactions. For his groundbreaking work, he received the 1984 Nobel Prize in Chemistry. Merrifield's solid-phase method has now been completely automated and is accomplished by machines called peptide synthesizers.

SKILLBUILDER

25.6 PREPARING A PEPTIDE USING THE MERRIFIELD SYNTHESIS

LEARN the skill

Identify the steps you would use to prepare the following peptide via the Merrifield synthesis.

Ile-Gly-Leu-Ala-Phe

SOLUTION

Begin with the C terminus. In this case, the C terminus must be a phenylalanine residue, so the first step is to attach a Boc-protected phenylalanine to the polymer via an S_N2 process.

STEP 1
Attach the appropriate Boc-protected residue to the polymer.

The second step is to remove the protecting group, exposing the N terminus of the phenylalanine residue.

STEP 2
Remove the Boc protecting group.

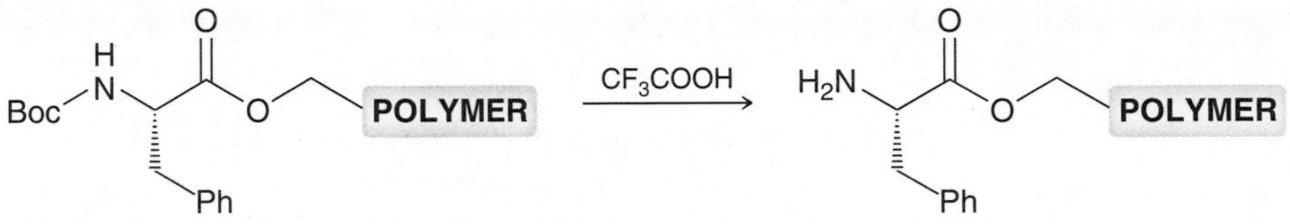

STEP 3
Using DCC, create a new peptide bond with a Boc-protected amino acid.

In the third step, the next residue (Ala) is installed as a Boc-protected residue using DCC to form the peptide bond.

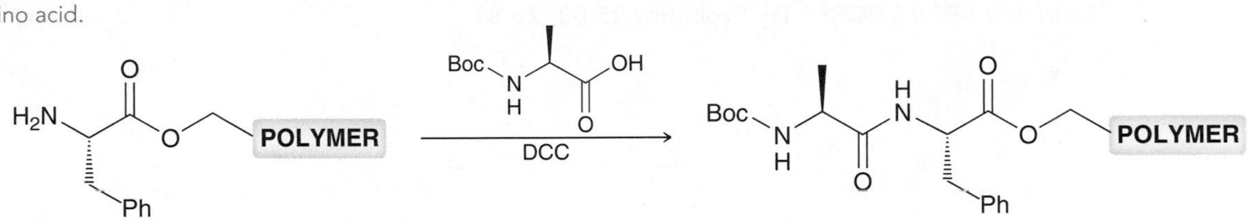

Steps 2 and 3 are then repeated for the installation of each additional residue until the desired peptide chain has been assembled.

STEP 4
Repeat steps 2 and 3 for each residue that is to be added to the growing peptide chain.

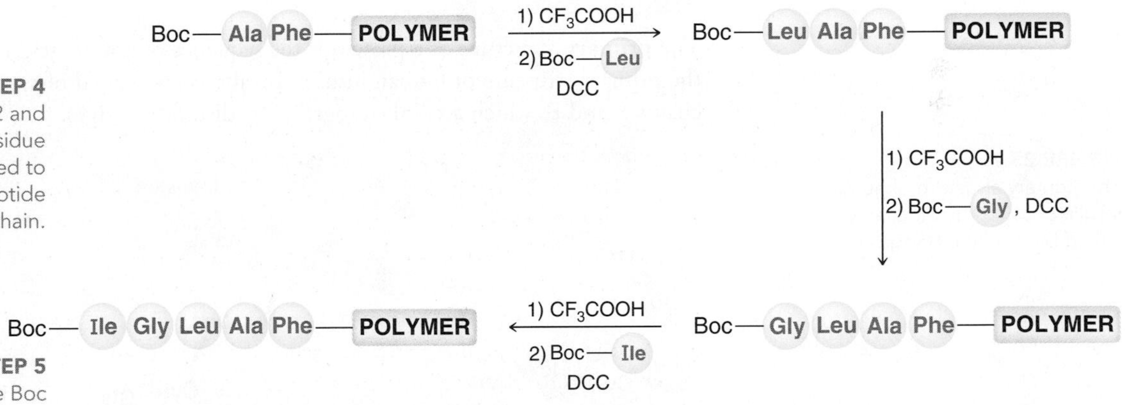

STEP 5
Remove the Boc protecting group and detach the peptide from the polymer.

Finally, the protecting group is removed, and the peptide is released from the polymer.

PRACTICE the skill **25.37** Identify all of the steps necessary to prepare each of the following peptides with a Merrifield synthesis.

(a) Phe-Leu-Val-Phe (b) Ala-Val-Leu-Ile

APPLY the skill **25.38** Identify the sequence of the tripeptide that would be formed from the following order of reagents. Clearly label the C terminus and N terminus of the tripeptide.

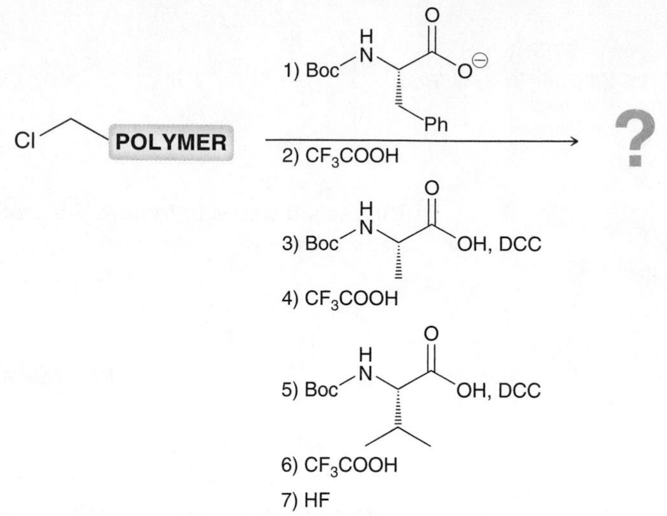

need more **PRACTICE?** **Try Problems 25.80, 25.81**

25.7 Protein Structure

Proteins are fairly large organic compounds, and as such, the structure of a protein is more complex than the structures of simple organic compounds. In fact, proteins are generally described in terms of four levels of structure: primary, secondary, tertiary, and quaternary structure.

Primary Structure

The **primary structure** of a protein is the sequence of amino acid residues. Figure 25.7 shows the primary structure of human insulin. Insulin is comprised of two polypeptide chains, called chains A and B, which are linked together by disulfide bridges.

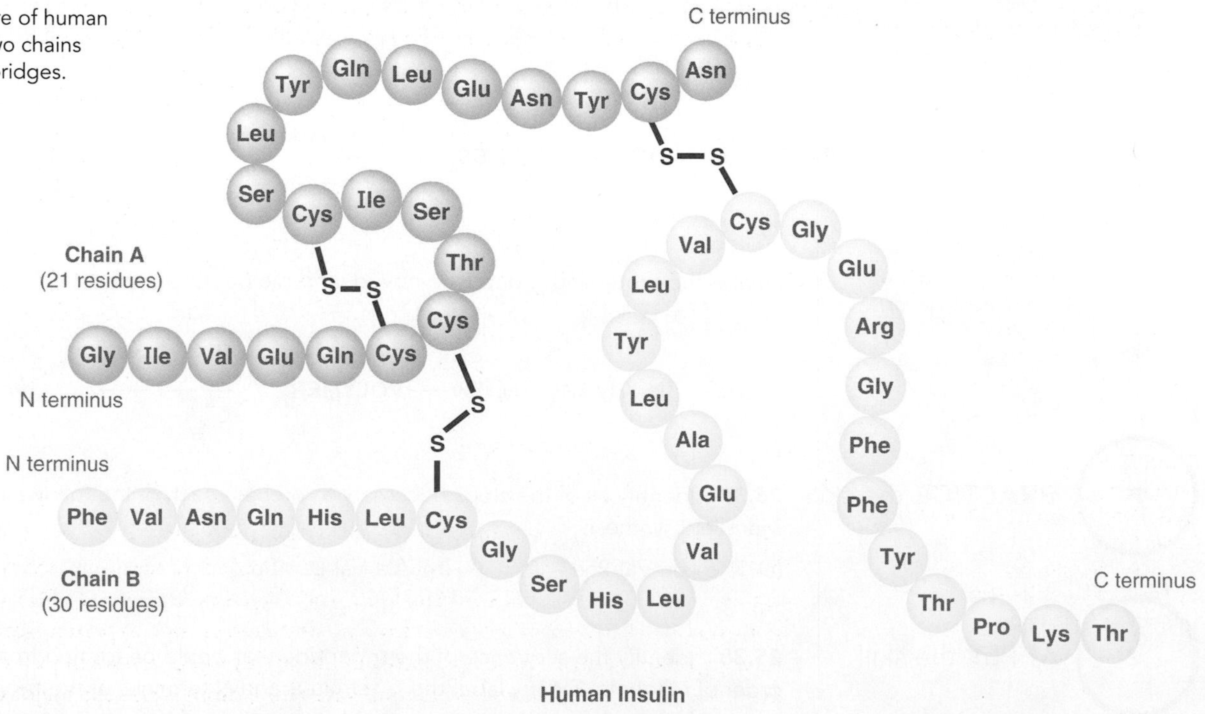

FIGURE 25.7
The primary structure of human insulin consists of two chains linked by disulfide bridges.

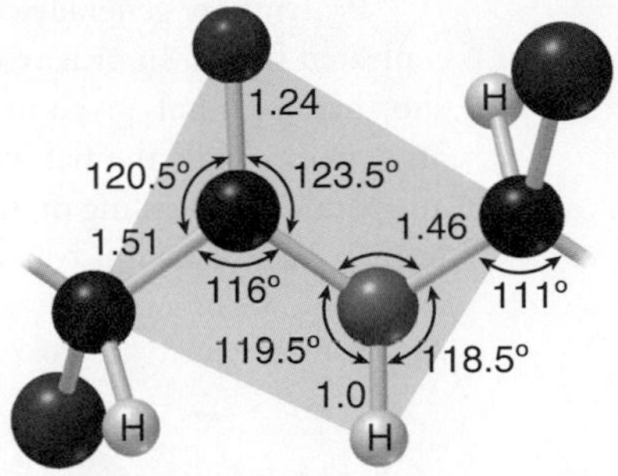

FIGURE 25.8
The planar geometry of a peptide bond.

Secondary Structure

The **secondary structure** of a protein refers to the three-dimensional conformations of localized regions of the protein. Recall that each peptide bond exhibits restricted rotation and planar geometry (Figure 25.8). This is the case for each amino acid residue in a protein (Figure 25.9).

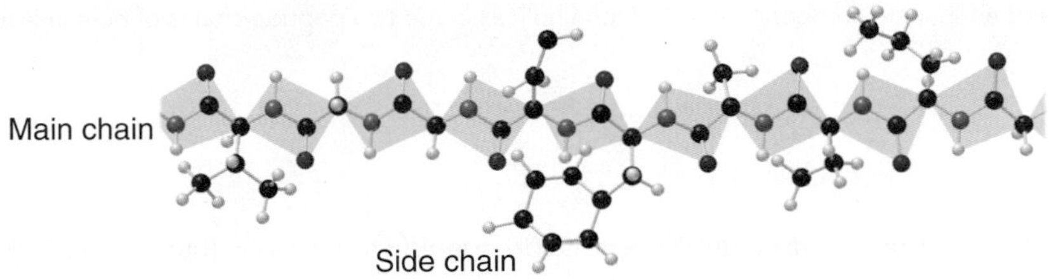

FIGURE 25.9
Proteins are comprised of repeating planar units, each of which can freely rotate with respect to the others.

Although the individual planes are free to rotate with respect to each other, their presence limits the conformations available to the protein. Depending on the sequence of amino acid residues, localized regions of peptides often adopt particular shapes. Two particularly stable arrangements are the **α helix** and **β pleated sheet**. An α helix forms when a portion of the protein twists into a clockwise spiral (Figure 25.10). Each turn has approximately four amino acid residues, and each C=O group experiences hydrogen bonding with an N—H group that is four residues farther along on the chain. In an α helix, the R groups (side chains) extend outward, away from the helix. A proline residue cannot be part of an α helix, because it lacks an N—H proton and does not participate in hydrogen bonding. Many proteins contain α helices. For example, α-keratin, the major component of hair, is comprised almost entirely of α helices. The composition of hair will be further discussed in Section 25.8.

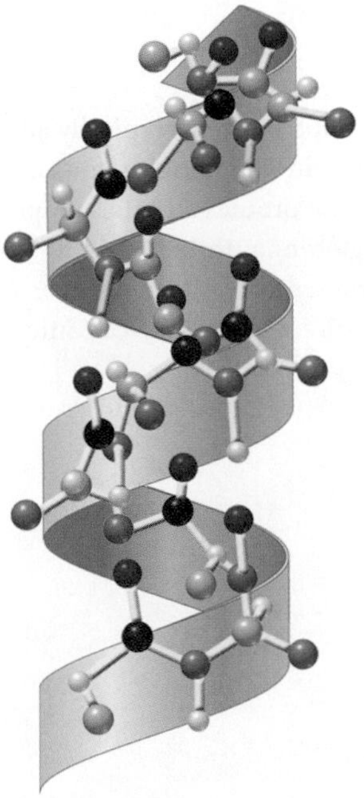

FIGURE 25.10
An α helix. (Illustration, Irving Geis. Image from Irving Geis Collection/Howard Hughes Medical Institute. Rights owned by ***HHMI***. Not to be reproduced without permission.)

Beta pleated sheets are formed when two or more protein chains line up side by side (Figure 25.11). In a β pleated sheet, hydrogen bonding occurs between the C=O group and N—H group of neighboring strands. The R groups (side chains) are positioned above and below the plane of the sheet, in an alternating pattern. Beta sheets are commonly formed from segments that bear residues with small R groups, such as Ala and Gly. With larger side chains, steric hindrance prevents the strands from getting close enough to participate in hydrogen bonding. Spider webs are comprised primarily of fibroin, a protein containing a pleated sheet arrangement that endows fibroin with structural strength. Fibroin is also the major component of silk.

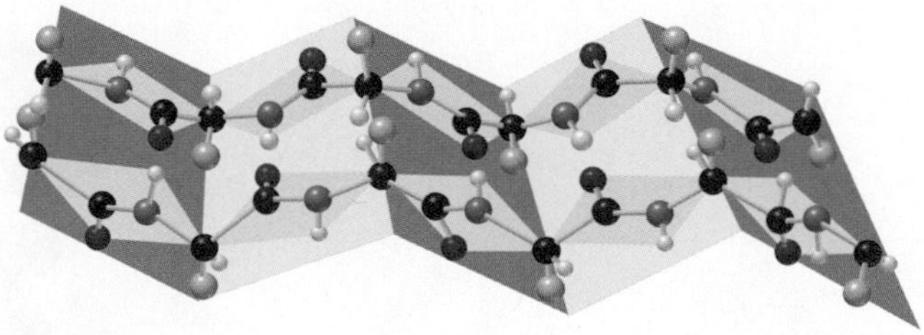

FIGURE 25.11
A β pleated sheet.

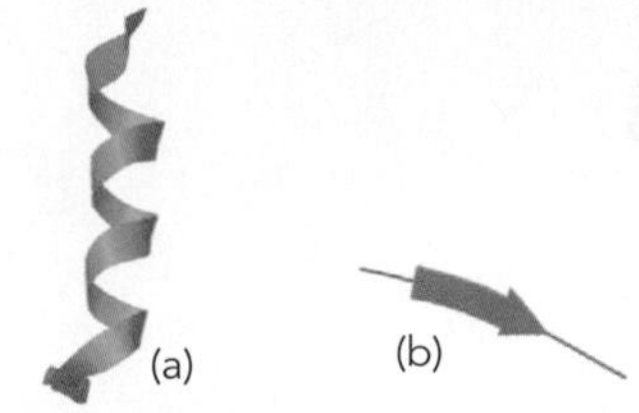

FIGURE 25.12
Symbols used when illustrating the secondary structure of proteins: (a) an α helix and (b) a β pleated sheet.

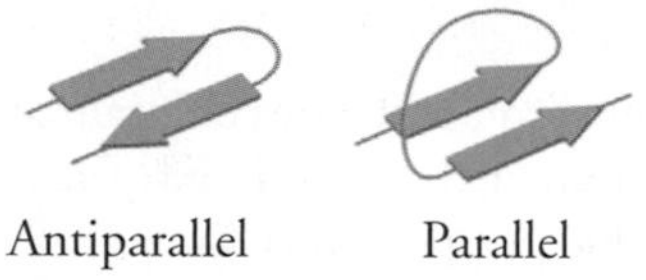

FIGURE 25.13
Parallel and antiparallel β sheets.

Proteins are generally comprised of many structural domains, including α helices as well as β pleated sheets. In order to draw a protein in a way that illustrates its secondary structure, the shorthand symbols given in Figure 25.12 are frequently used. The flat helical ribbon represents an α helix, while the flat wide arrow represents a β sheet. Beta sheets can be either parallel or antiparallel, depending on the direction of neighboring segments (Figure 25.13). Both chains of human insulin exhibit α helices (Figure 25.14).

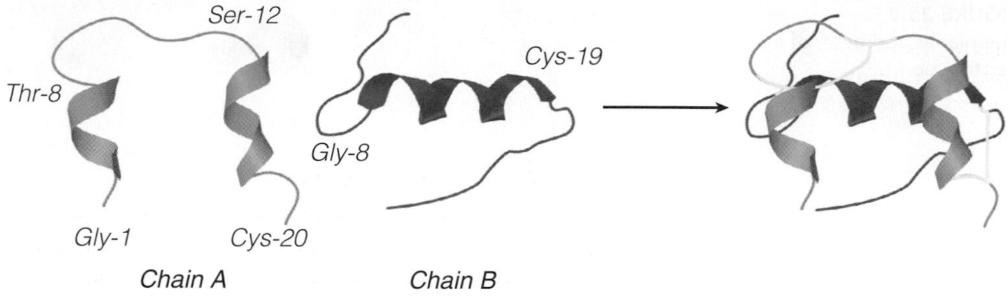

FIGURE 25.14
Alpha helices in the two peptide chains of human insulin.

CONCEPTUAL CHECKPOINT

25.39 Below is the primary structure for a peptide. Identify the regions that are most likely to form β pleated sheets.

Trp-His-Pro-Ala-Gly-Gly-Ala-Val-His-Cys-Asp-Ser-Arg-Arg-Ala-Gly-Ala-Phe

Tertiary Structure

The **tertiary structure** of a protein refers to its three-dimensional shape. Proteins typically adopt conformations that maximize stability. In aqueous solution, a protein will adopt a conformation in which the polar groups are pointing toward the exterior of the protein and the nonpolar groups are located in the interior. Consider the structure of myoglobin, a protein used to store molecular oxygen (Figure 25.15). The regions marked in red represent polar side groups, and they are positioned on the surface of the three-dimensional structure so that they can interact with water, thereby maximizing the stability of the protein.

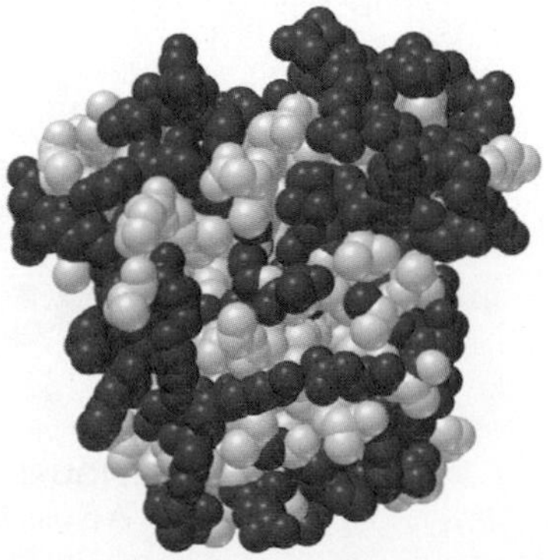

FIGURE 25.15
A space-filling model of myoglobin with hydrophilic groups highlighted in red.

The intramolecular forces holding a typical protein in its folded shape are relatively weak, and under conditions of mild heating, a protein can unfold, in a process called **denaturation** (Figure 25.16). A common example is the cooking of an egg, during which the protein albumin is denatured and changes from a clear solution to a white solid. The secondary and tertiary structures are lost, but the primary structure remains intact. That is, the sequence of amino acid

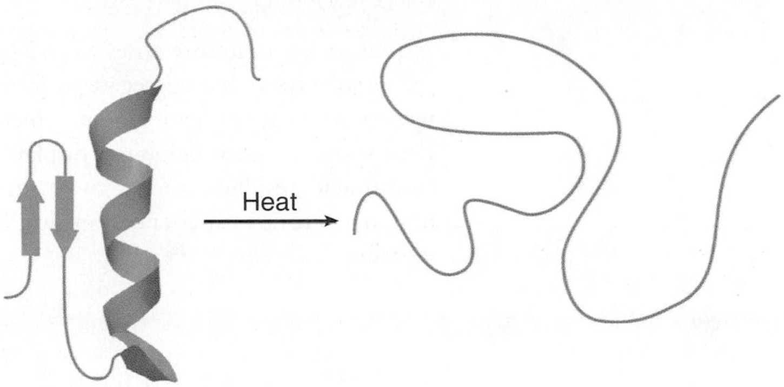

FIGURE 25.16
A protein loses its secondary and tertiary structures when it is denatured.

residues does not change. Loss of tertiary structure is typically accompanied by a loss of function, because the function of a protein is very much dependent on its three-dimensional shape. The denaturation process is irreversible in most cases. Factors other than heat can also cause denaturation, including a change in pH or a change in solvent.

MEDICALLYSPEAKING)))

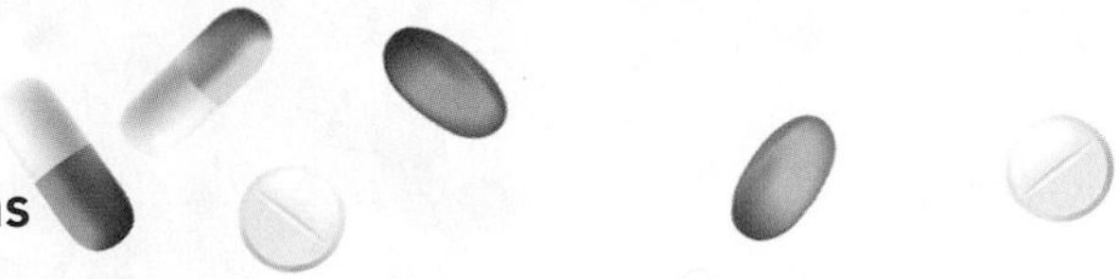

Diseases Caused by Misfolded Proteins

Until recently, it was widely believed that all contagious diseases were caused by living pathogens, such as viruses, bacteria, or fungi. For a small number of contagious diseases, scientists were baffled by their inability to identify the pathogen responsible. Examples include *Creutzfeldt-Jakob disease* in humans and *scrapie* in sheep. Both diseases are similar, causing loss of mental function and death. Autopsies of victims revealed plaques of amyloid protein surrounded by spongelike tissue in the brain. These diseases came to be known as spongiform encephalopathies. The prevailing belief was that some virus or other organism would soon be isolated and identified as the causitive pathogen.

In the 1980s Stanley Prusiner, a neurologist at the Univeristy of California at San Diego, systematically and scrupulously separated all of the components of scrapie-infected sheep brains. He found the infectious agent to be a particular protein. He suggested that scrapie and other related diseases are caused by *infectious proteins*, which he called *prions*. This conclusion was highly significant, because it represented the first example of a nonliving pathogen capable of causing deadly infections without employing DNA or RNA. For his groundbreaking work, Prusiner was awarded the 1997 Nobel Prize in Physiology or Medicine.

Prion diseases attracted significant public awareness in the 1990s when cows in England, Canada, and the United States became infected with a disease called bovine spongiform encephalopathy, which caused them to move around erratically and eventually die. This disease, commonly referred to as mad cow disease, was presumably caused by using the remains of scrapie-infected sheep to feed cows. Mad cow disease represents a significant risk to humans because eating infected meat could trigger a form of Creutzfeldt-Jakob disease in humans, which is a fatal disease with no cure.

Prions are believed to be normal proteins that have become misfolded. In other words, the primary structure of the protein remains unaltered, but the tertiary structure is different. The normal protein unfolds and then refolds into a new shape (a prion). When animals eat the remains of scrapie-infected sheep, the prion is not digested and is not detected by the immune system as a harmful substance. The infectious form of the prion then causes other normal proteins to misfold as well. As the amount of misfolded proteins increases, they begin to aggregate in the brain, causing the plaques and sponge like tissues associated with spongiform encephalopathies.

Over the last decade, progress in finding a cure for prion diseases has been slow but steady. Current research efforts are focused primarily on the design of drugs that can potentially inhibit misfolding. It is now understood that the body utilizes a variety of proteins called chemical *chaperones* to guide the folding of other proteins. By studying how these chaperones function, scientists believe that they will be able to synthesize chaperones that will prevent formation of the misfolded proteins characteristic of prion diseases. Prion diseases are an extreme example of the relationship between structure and function.

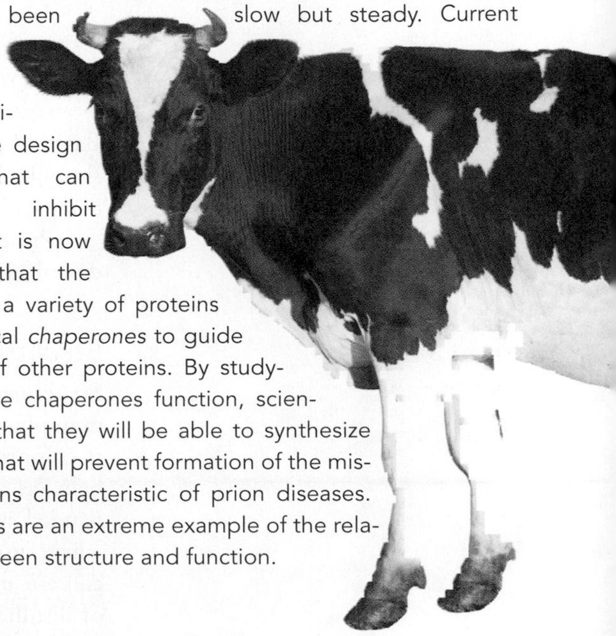

Quaternary Structure

A **quaternary structure** arises when a protein consists of two or more folded polypeptide chains, called **subunits**, that aggregate to form one protein complex. For example, consider the quaternary structure of hemoglobin, which is comprised of four different subunits (Figure 25.17). Each subunit in hemoglobin is displayed in a different color. The subunits have many nonpolar amino acid residues, and as such, there are many nonpolar groups on their surfaces. These nonpolar groups experience van der Waals interactions, which hold the quaternary structure together.

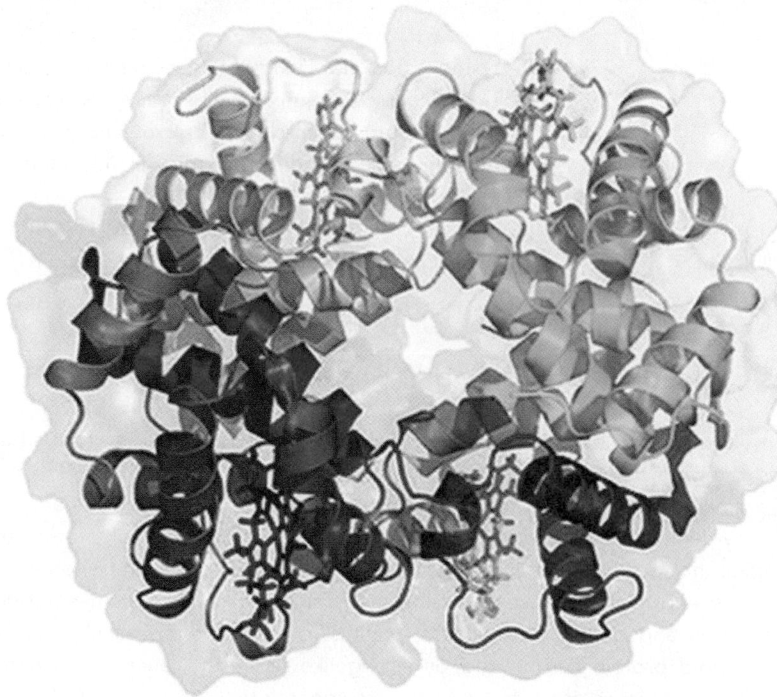

FIGURE 25.17
The quaternary structure of hemoglobin. For clarity, each subunit is presented in a different color.

25.8 Protein Function

Based on their shape, proteins are often classified as fibrous or globular. **Fibrous proteins** consist of linear chains that are bundled together. **Globular proteins** are chains that are coiled into compact shapes.

Proteins are also classified according to their function. A few different classes of proteins are described in the following sections.

Structural Proteins

Structural proteins are fibrous proteins that are used for their structural rigidity. For example, α-keratins are structural proteins found in hair, nails, skin, feathers, and wool. Hair is comprised of many polypeptide chains of α-keratin, each coiled into a right-handed α helix, as seen in Figure 25.18. Four of these right-handed helices are then wrapped into left-handed super-helices, similar to the way that rope is made from twisting individual threads. These super-helices are then clustered together to form individual hairs. In general, α-keratins contain many cysteine residues, which allows them to form disulfide bridges. The number of disulfide bridges determines the strength of the protein. For example, the α-keratins in fingernails have more disulfide bridges than the α-keratins in hair.

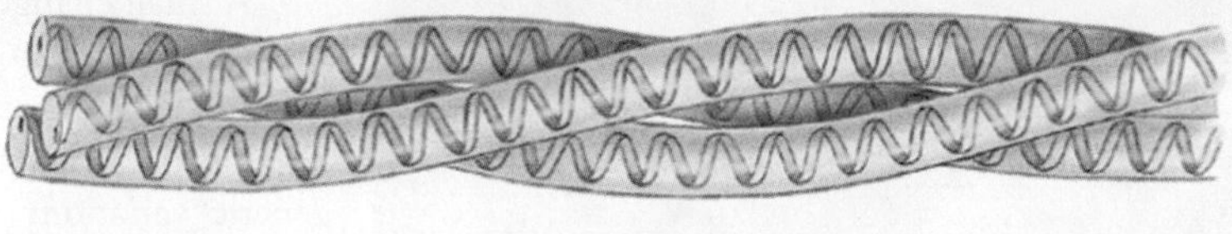

FIGURE 25.18
Polypeptide chains of α-keratin, coiled into a right-handed α helix.

Enzymes

Enzymes are among the most important biological molecules because they catalyze virtually all cellular processes. For example, glycosidase enzymes speed up the hydrolysis of polysaccharides by a factor 10^{17}. In other words, a reaction that would normally take millions of years is accomplished in milliseconds in the presence of an enzyme. Enzymes speed up reactions through their ability to bind to the transition state of a particular reaction, thereby lowering the energy of the transition state (Figure 25.19). By lowering the energy of the transition state,

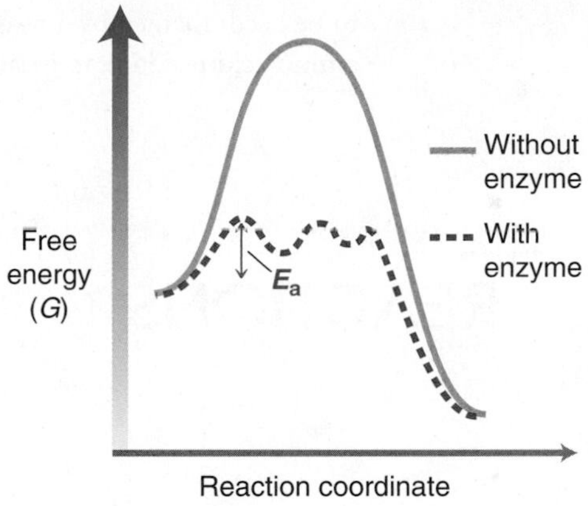

FIGURE 25.19
An energy diagram showing that an enzyme speeds up a reaction by lowering the energy of the transition state.

the energy of activation is lowered, and the reaction occurs more rapidly. Even the simplest bacteria require thousands of enzymes to carry out their life processes, and humans require over fifty thousand.

Transport Proteins

Transport proteins are used to transport molecules or ions from one location to another. Hemoglobin is a classic example of a transport protein, used to transport molecular oxygen from the lungs to all the tissues of the body. Hemoglobin consists of a protein unit bound to a nonprotein unit called a **prosthetic group**. The protein portion of hemoglobin was shown in Figure 25.17. The prosthetic group of hemoglobin binds oxygen.

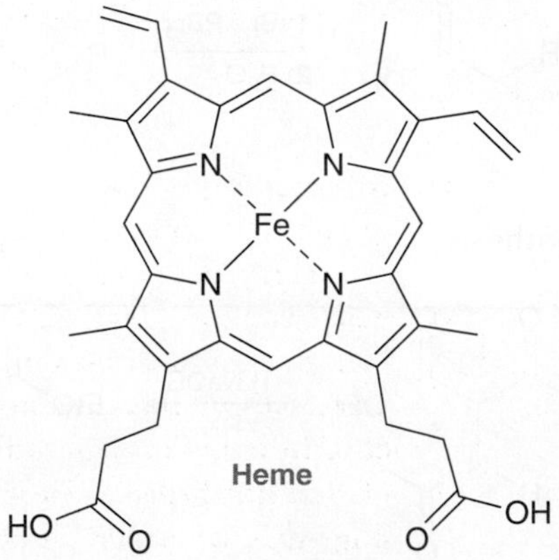

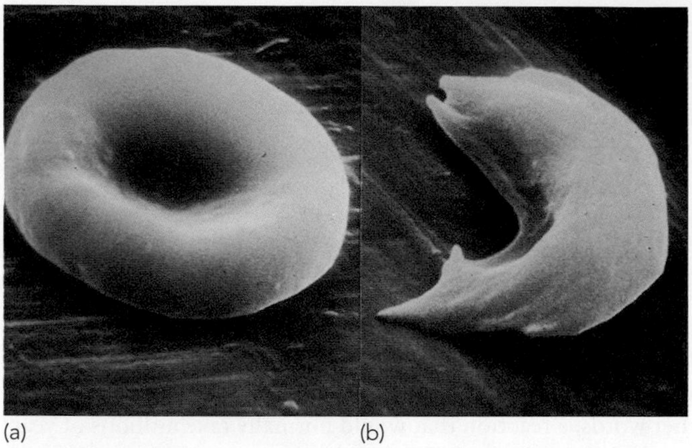

FIGURE 25.20
(a) Healthy red blood cells and (b) sickle-shaped red blood cells.

A small change in primary structure can result in a large change in function. Hemoglobin is a good example of this phenomenon. Hundreds of variations of hemoglobin have been discovered. Most humans have the common type, called HbA, but some people have genetic variations in which one amino acid residue is different. In many of these cases, the genetic variation has no effect on the structure or activity of hemoglobin. However, some variations can be deadly. In one such variation, called HbS, the sixth residue from the N terminus of the β chain (consisting of 146 residues) is valine instead of glutamic acid. This variation greatly affects the structure and activity of hemoglobin, causing red blood cells to be distorted, or sickle shaped (Figure 25.20). These distorted cells are highly susceptible to rupturing, and they interfere with the regular flow of blood. This condition is called sickle-cell anemia, and it can be fatal. A person with sickle-cell anemia has inherited two copies of the gene for abnormal hemoglobin, one from each parent. A person who carries only one copy of the sickle-cell gene is said to have the sickle-cell trait, which is a less severe form of the disease, but it can still cause serious problems under conditions of stress. Interestingly, the sickle-cell trait offers an advantage in that it seems to be accompanied by a resistance to malaria. This example illustrates how a change in a single amino acid residue can greatly affect the structure and function of a protein.

REVIEW OF REACTIONS

ANALYSIS OF AMINO ACIDS

Reaction with Ninhydrin

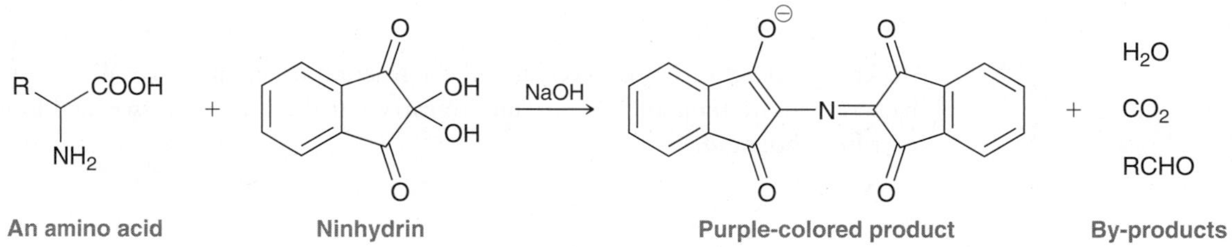

SYNTHESIS OF AMINO ACIDS

Via an α-Haloacid

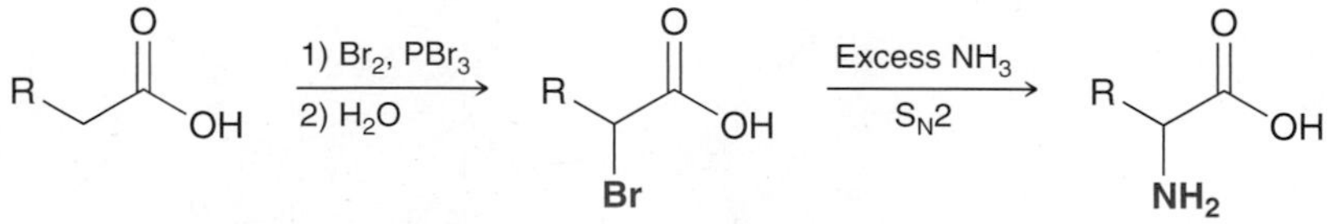

Via the amidomalonate Synthesis

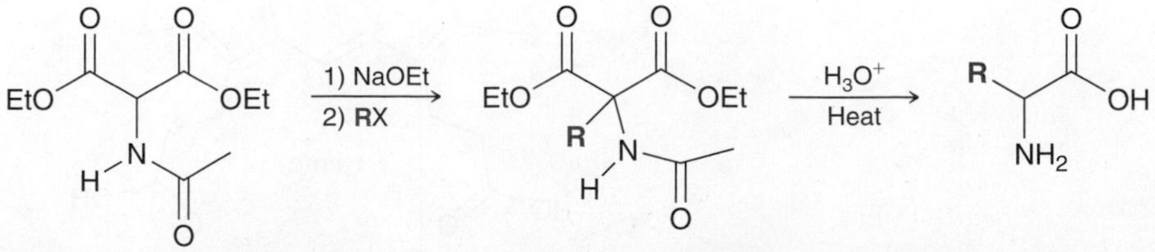

Via the Strecker Synthesis

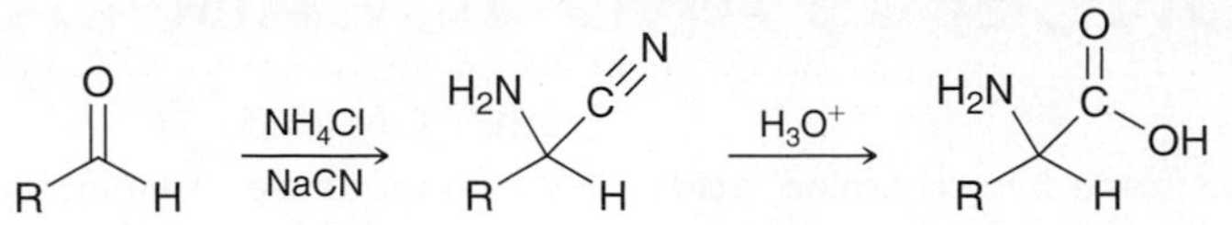

Enantioselective Synthesis

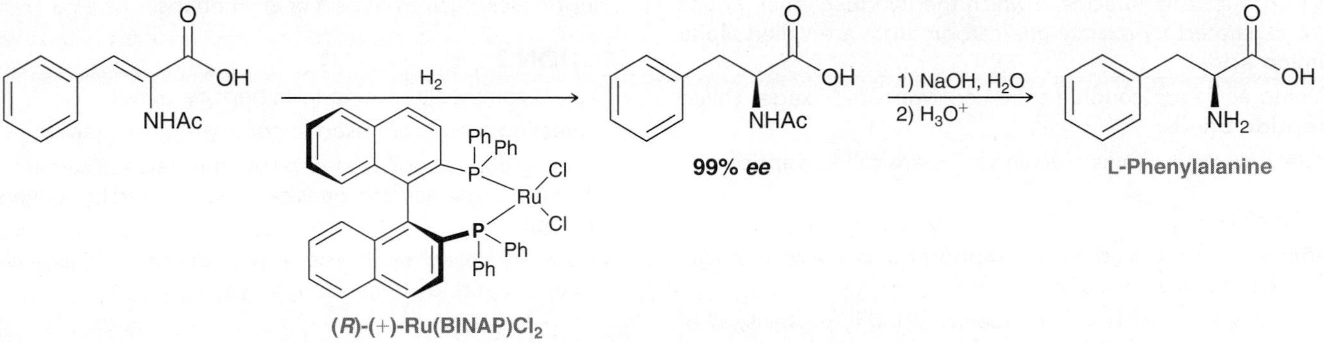

ANALYSIS OF AMINO ACIDS

Edman Degradation

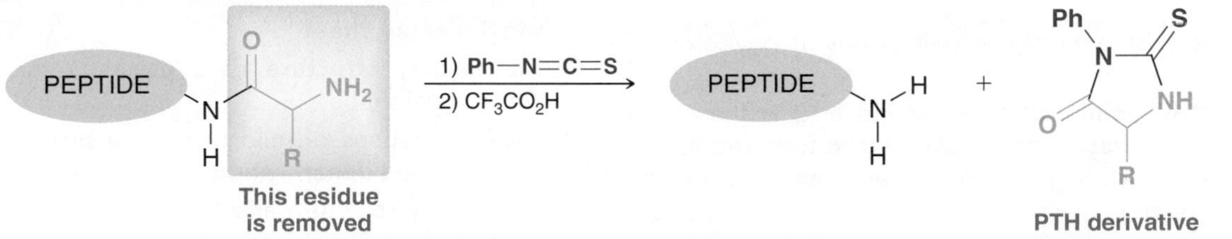

SYNTHESIS OF PEPTIDES

Peptide Bond Formation

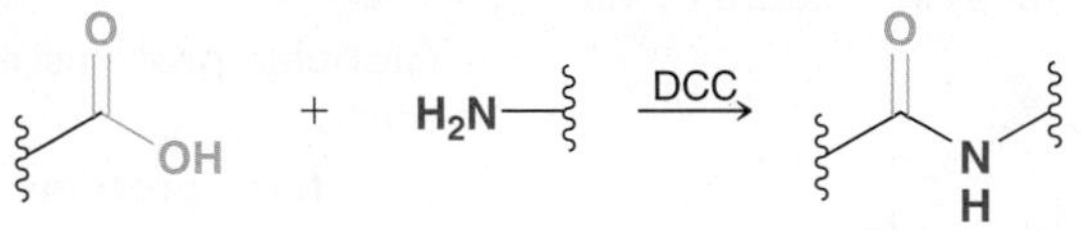

Protection and Deprotection of the N terminus

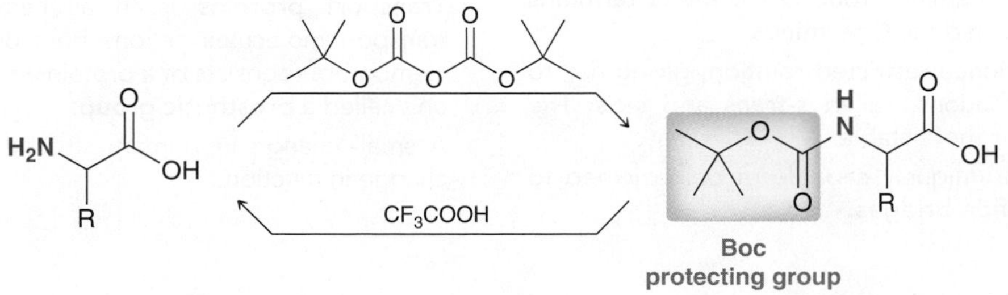

Protection and Deprotection of the C terminus

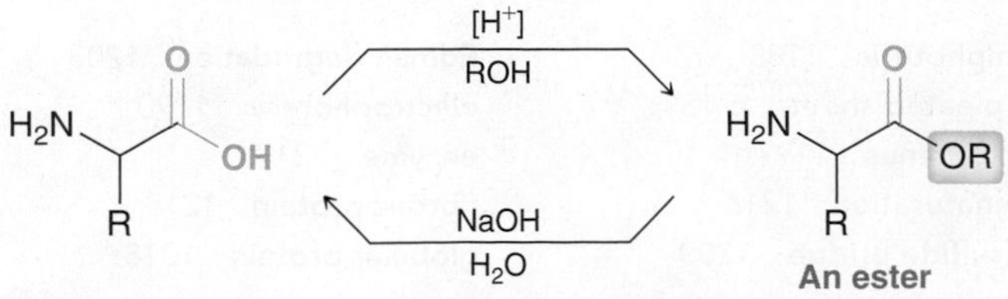

REVIEW OF CONCEPTS AND VOCABULARY

SECTION 25.1

- **Proteins** are large compounds formed from **amino acids** linked together.
- Each amino acid contains one amino group and one carboxylic acid group. Amino acids in which the two functional groups are separated by exactly one carbon atom are called **alpha amino acids**.
- Amino acids are coupled together by amide linkages called **peptide bonds.**
- Relatively short chains of amino acids are called **peptides**.

SECTION 25.2

- Amino acids are said to be **amphoteric**, because they can function either as acids or as bases.
- Only 20 amino acids are abundantly found in proteins, all of which are **L amino acids**, except for glycine which lacks a chirality center.
- Amino acids have two pK_a values, one for the carboxylic acid moiety and one for the ammonium group.
- Amino acids with basic or acidic side chains will exhibit a third pK_a value.
- Amino acids exist primarily as **zwitterions** at physiological pH.
- The **isoelectric point (pI)** of an amino acid is the pH at which the concentration of the zwitterionic form reaches its maximum value. During **electrophoresis**, amino acids are separated based on pI values.

SECTION 25.3

- Racemic mixtures of amino acids can be prepared in the laboratory via α-haloacids, via the **amidomalonate synthesis**, or via the **Strecker synthesis**. Optically active amino acids are obtained either via resolution of a racemic mixture or via enantioselective synthesis.

SECTION 25.4

- Peptides are comprised of **amino acid residues** joined by peptide bonds.
- Peptide chains have an amino group called the **N terminus** and a COOH group called the **C terminus**.
- Peptide bonds experience restricted rotation, giving rise to two possible conformations, called ***s-trans*** and ***s-cis***. The *s-trans* conformation is more stable.
- Cysteine residues are uniquely capable of being joined to one another via **disulfide bridges.**

SECTION 25.5

- The sequence of amino acids in a peptide can be analyzed using an **Edman degradation.**
- Large peptides are first cleaved into smaller fragments using **peptidases**, such as trypsin or chymotrypsin.

SECTION 25.6

- DCC is commonly used to form peptide bonds.
- Protecting groups are used to control regioselectivity.
- Amino groups can be protected by conversion to carbamates, while carboxylic acid moieties can be protected by conversion to esters.
- In the **Merrifield synthesis**, a peptide chain is assembled while tethered to an insoluble polymer.

SECTION 25.7

- The **primary structure** of a protein is the sequence of amino acid residues.
- The **secondary structure** of a protein refers to the three-dimensional conformations of localized regions of the protein. Two particularly stable arrangements are the **α helix** and **β pleated sheet**.
- The **tertiary structure** of a protein refers to its three-dimensional shape.
- Under conditions of mild heating, a protein can unfold, a process called **denaturation**.
- **Quaternary structure** arises when a protein consists of two or more folded polypeptide chains, called **subunits**, that aggregate to form one protein complex.

SECTION 25.8 PROTEIN FUNCTION

- **Fibrous proteins** consist of linear chains that are bundled together.
- **Globular proteins** are chains that are coiled into compact shapes.
- **Structural proteins**, such as α-keratin, are fibrous proteins that provide structural rigidity.
- **Enzymes** catalyze virtually all cellular processes.
- **Transport proteins**, such as hemoglobin, are used to transport molecules or ions from one location to another. Hemoglobin consists of a protein unit bound to a nonprotein unit called a **prosthetic group**.
- A small change in primary structure can result in a large change in function.

KEY TERMINOLOGY

peptide 1184
peptide bond 1183
primary structure 1214
prosthetic group 1219
proteins 1183
quaternary structure 1218
s-cis 1199
secondary structure 1215
s-trans 1199
Strecker synthesis 1194
structural protein 1218
subunit 1218
tertiary structure 1216
transport protein 1219
zwitterion 1188

SKILLBUILDER REVIEW

25.1 DETERMINING THE PREDOMINANT FORM OF AN AMINO ACID AT A SPECIFIC pH

STEP 1 Identify the pK_a associated with the carboxylic acid moiety, and determine whether this group will predominate as the uncharged form or as the carboxylate anion at the specified pH.

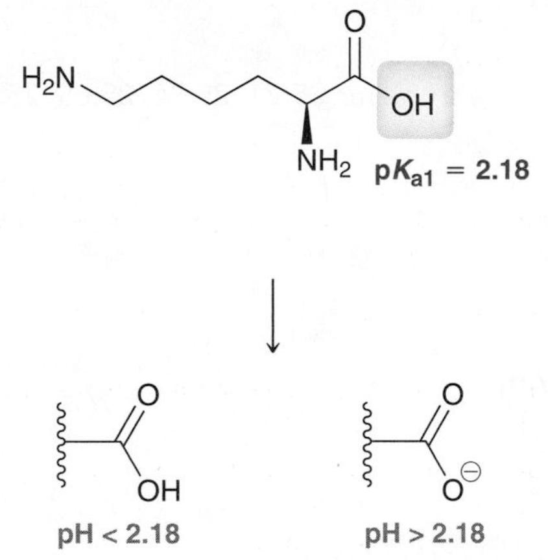

STEP 2 Identify the pK_a associated with the α amino group, and determine whether this group will predominate as the uncharged form or as the cationic form at the specified pH.

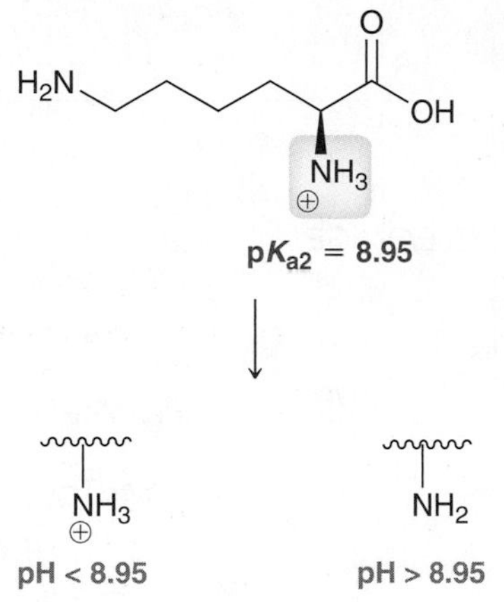

STEP 3 If the side chain has an ionizable group, identify the pK_a associated with the side chain, and determine the form of the side chain that will predominate at the specified pH.

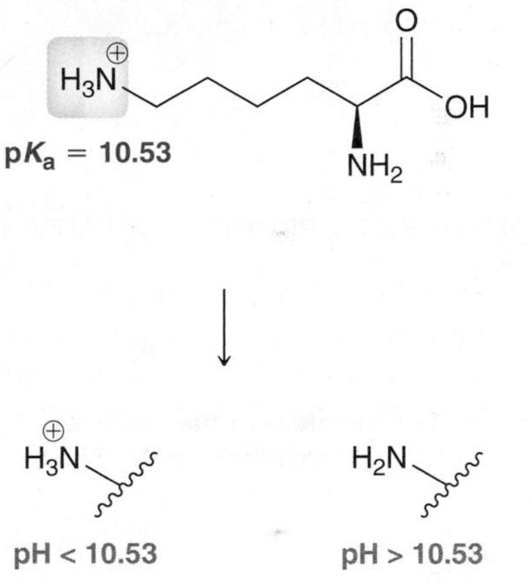

Try Problems 25.4–25.6, 25.40, 25.47, 25.48

25.2 USING THE AMIDOMALONATE SYNTHESIS

STEP 1 Identify the side chain that is connected to the α position.

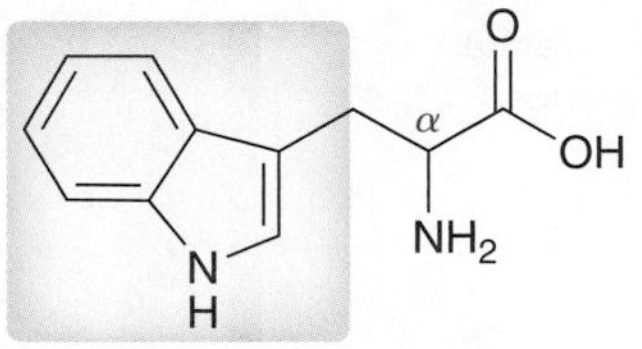

STEP 2 Identify the alkyl halide necessary, and ensure it will readily undergo an S_N2 process.

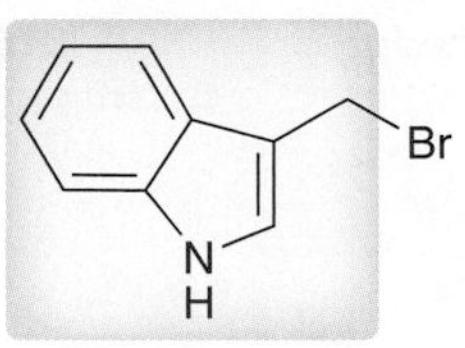

STEP 3 Identify the reagents. Begin with diethyl acetamidomalonate as the starting material. Perform the alkylation, and then heat with aqueous acid.

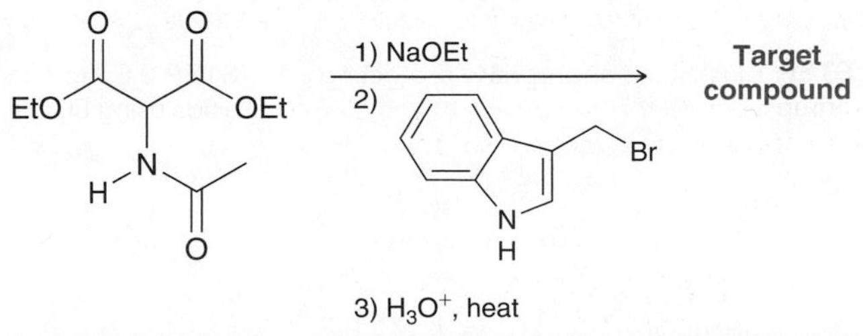

Try Problems 25.14–25.16, 25.57, 25.58, 25.59c, 25.60b, 25.84

25.3 DRAWING A PEPTIDE

STEP 1 Draw a peptide with the correct number of residues:

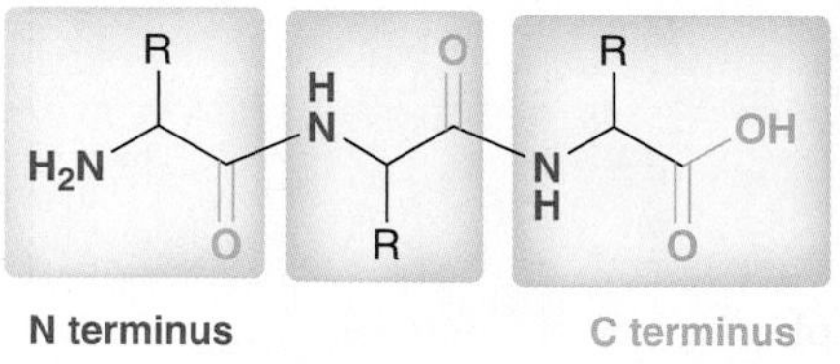

STEP 2 Identify the side chain associated with each residue:

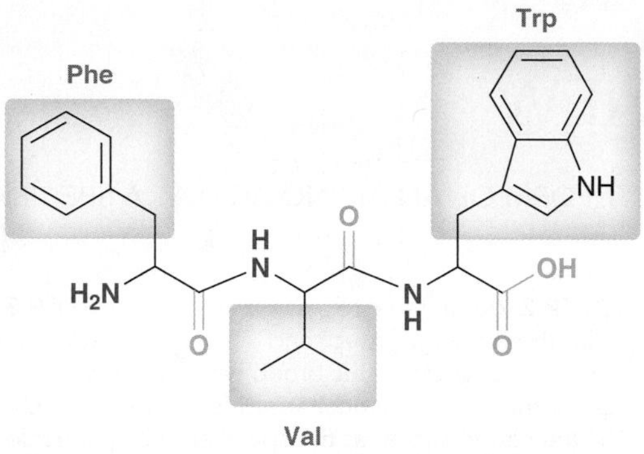

STEP 3 Assign the proper configuration for each α position.

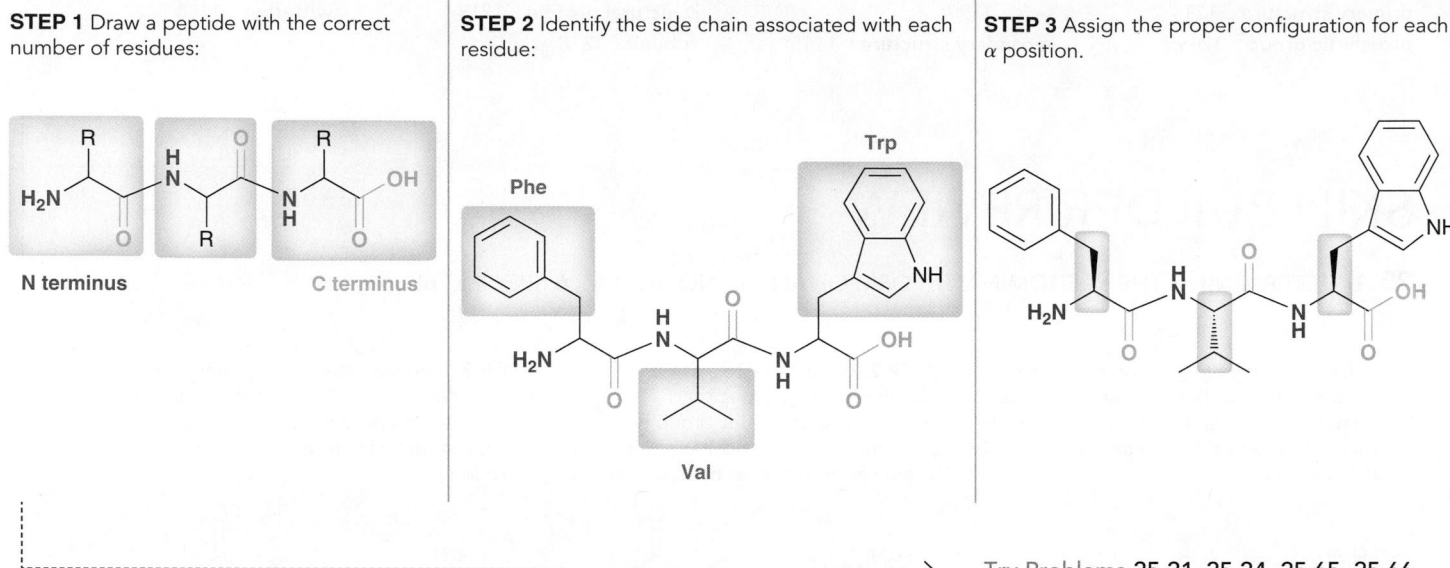

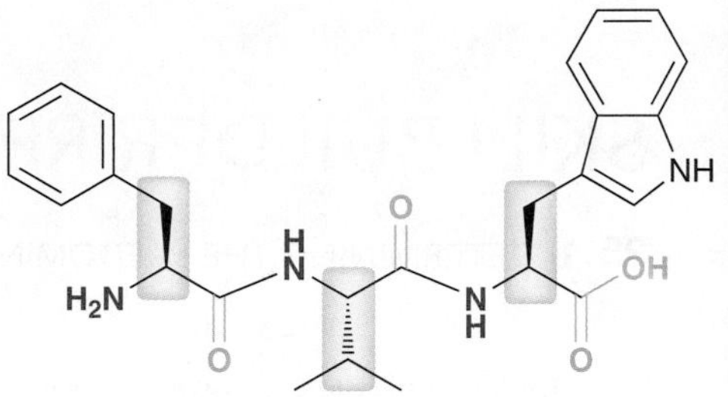

Try Problems **25.21–25.24, 25.65, 25.66**

25.4 SEQUENCING A PEPTIDE VIA ENZYMATIC CLEAVAGE

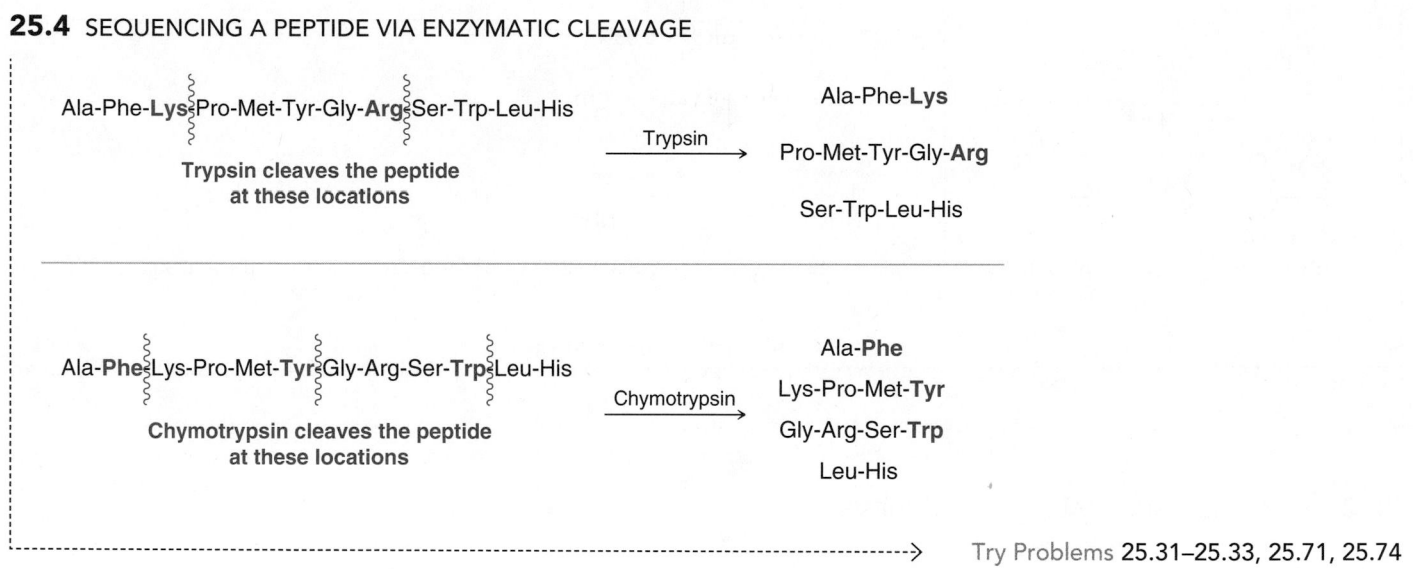

Ala-Phe-**Lys** ⦚ Pro-Met-Tyr-Gly-**Arg** ⦚ Ser-Trp-Leu-His

Trypsin cleaves the peptide at these locations

Trypsin →

Ala-Phe-**Lys**

Pro-Met-Tyr-Gly-**Arg**

Ser-Trp-Leu-His

Ala-**Phe** ⦚ Lys-Pro-Met-**Tyr** ⦚ Gly-Arg-Ser-**Trp** ⦚ Leu-His

Chymotrypsin cleaves the peptide at these locations

Chymotrypsin →

Ala-**Phe**

Lys-Pro-Met-**Tyr**

Gly-Arg-Ser-**Trp**

Leu-His

Try Problems **25.31–25.33, 25.71, 25.74**

25.5 PLANNING THE SYNTHESIS OF A DIPEPTIDE

STEP 1 Install the appropriate protecting groups.

STEP 2 Couple the protected amino acids using DCC.

STEP 3 Remove the protecting groups.

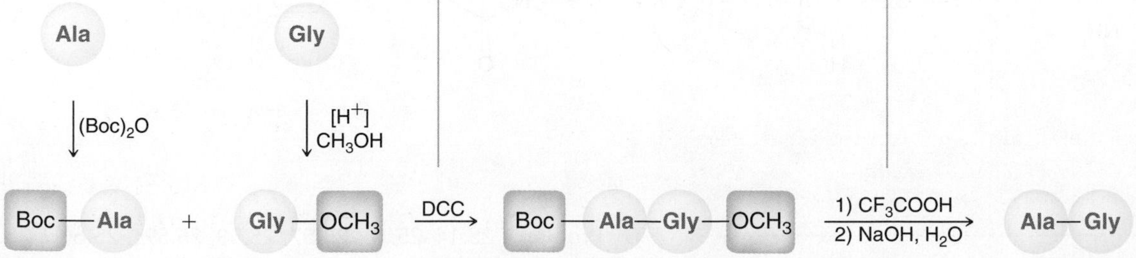

Try Problems **25.34–25.36, 25.77–25.79**

25.6 PREPARING A PEPTIDE USING THE MERRIFIELD SYNTHESIS

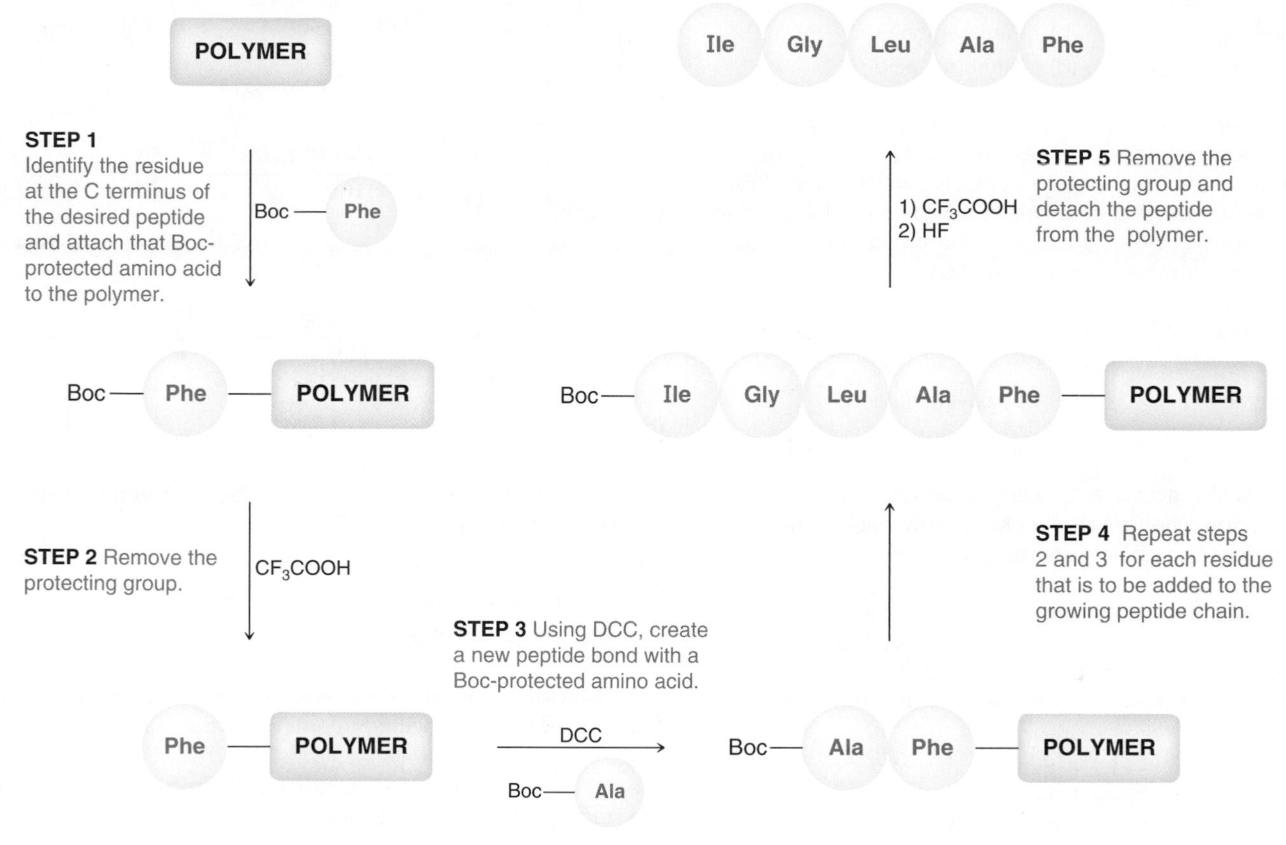
POLYMER
STEP 1
Identify the residue at the C terminus of the desired peptide and attach that Boc-protected amino acid to the polymer.
Boc — Phe
Boc — Phe — POLYMER
STEP 2 Remove the protecting group.
CF3COOH
Phe — POLYMER
STEP 3 Using DCC, create a new peptide bond with a Boc-protected amino acid.
DCC
Boc — Ala
Boc — Ala Phe — POLYMER
STEP 4 Repeat steps 2 and 3 for each residue that is to be added to the growing peptide chain.
Boc — Ile Gly Leu Ala Phe — POLYMER
1) CF3COOH
2) HF
STEP 5 Remove the protecting group and detach the peptide from the polymer.
Ile Gly Leu Ala Phe

Try Problems 25.37, 25.38, 25.80, 25.81

PRACTICE PROBLEMS:

Note: Most of the Problems are available within *WileyPLUS*, an online teaching and learning solution.

25.40 Draw a bond-line structure showing the zwitterionic form of each of the following amino acids:

(a) L-Valine (b) L-Tryptophan
(c) L-Glutamine (d) L-Proline

25.41 The 20 naturally occurring amino acids (Table 25.1) are all L amino acids, and they all have the *S* configuration, with the exception of glycine (which lacks a chirality center) and cysteine. Naturally occurring cysteine is an L amino acid, but it has the *R* configuration. Explain.

25.42 Draw a Fischer projection for each of the following amino acids:

(a) L-Threonine (b) L-Serine
(c) L-Phenylalanine (d) L-Asparagine

25.43 Seventeen of the 20 naturally occurring amino acids (Table 25.1) exhibit exactly one chirality center. Of the remaining three amino acids, glycine has no chirality center, and the other two amino acids each have two chirality centers.

(a) Identify the amino acids with two chirality centers.

(b) Assign the configuration of each chirality center in these two amino acids.

25.44 Draw all stereoisomers of L-isoleucine. In each stereoisomer, assign the configuration (*R* or *S*) of all chirality centers.

25.45 Arginine is the most basic of the 20 naturally occurring amino acids. At physiological pH, the side chain of arginine is protonated. Identify which nitrogen atom in the side chain is protonated. (*Hint*: Consider all three possibilities, and draw all resonance structures.)

25.46 Histidine possesses a basic side chain which is protonated at physiological pH. Identify which nitrogen atom in the side chain is protonated.

25.47 Draw the form of L-glutamic acid that predominates at each pH:

(a) 1.9 (b) 2.4 (c) 5.8 (d) 10.4

25.48 For each of the following amino acids, draw the form that is expected to predominate at physiological pH:

(a) L-Isoleucine (b) L-Tryptophan
(c) L-Glutamine (d) L-Glutamic acid

25.49 Using the data in Table 25.2, calculate the pI of the following amino acids:

(a) L-Alanine (b) L-Asparagine
(c) L-Histidine (d) L-Glutamic acid

25.50 Just as each amino acid has a unique pI value, proteins also have an overall observable pI. For example, lysozyme (present in tears and saliva) has a pI of 11.0 while pepsin (used in our stomachs to digest other proteins) has a pI of 1.0. What information does this give you about the types of amino acid residues that predominantly make up each of these proteins?

25.51 For each amino acid, draw the structure that predominates at the isoelectric point:

(a) L-Glutamine (b) L-Phenylalanine
(c) L-Proline (d) L-Threonine

25.52 Optically active amino acids undergo racemization at the α position when treated with strongly basic conditions. Provide a mechanism that supports this observation.

25.53 A mixture containing L-glycine, L-glutamine, and L-asparagine was subjected to electrophoresis. Identify which of the amino acids moved the farthest distance assuming that the experiment was performed at the pH indicated:

(a) 6.0 (b) 5.0

25.54 Draw the products that are expected when each of the following amino acids is treated with ninhydrin:

(a) L-Aspartic acid (b) L-Leucine
(c) L-Phenylalanine (d) L-Proline

25.55 A mixture of amino acids was treated with ninhydrin, and the following aldehydes were all observed in the product mixture:

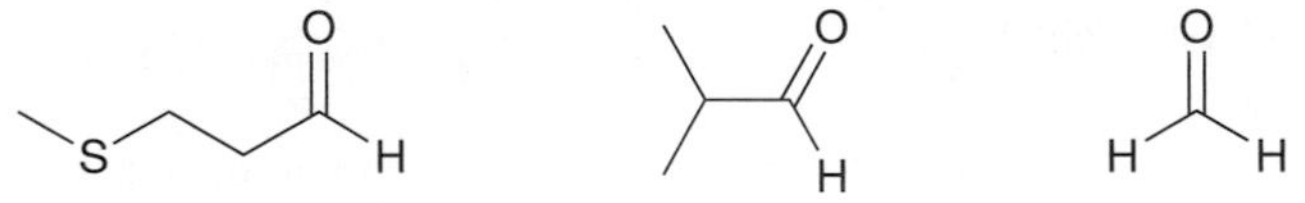

(a) Identify the structure and name all three amino acids in the starting mixture.
(b) In addition to the aldehydes, a purple-colored product is obtained. Draw the structure of the compound responsible for the purple color.
(c) Describe the origin of the purple color.

25.56 Show how you would use a Strecker synthesis to make valine.

25.57 Under similar conditions, alanine and valine were each prepared with an amidomalonate synthesis, and alanine was obtained in higher yields than valine. Explain the difference in yields.

25.58 The amidomalonate synthesis can be used to prepare amino acids from alkyl halides. When the amidomalonate synthesis is used to make glycine, no alkyl halide is required. Explain.

25.59 Predict the major product(s) for each of the following reactions:

(a)
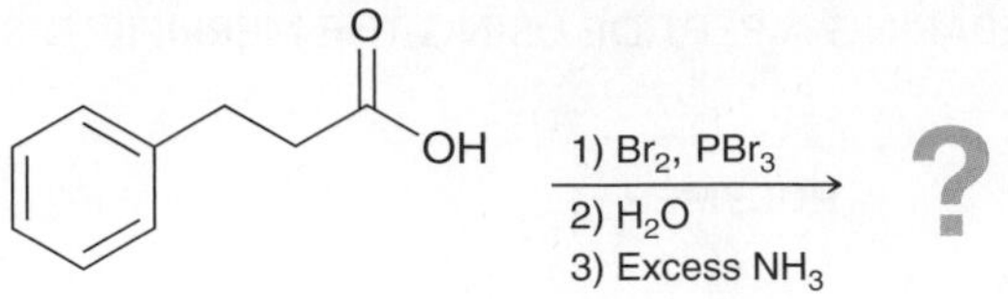

(b)
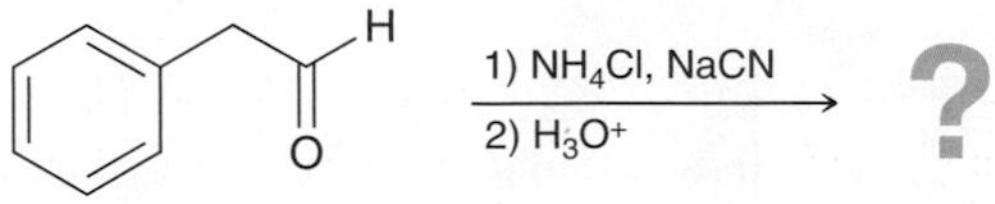

(c)
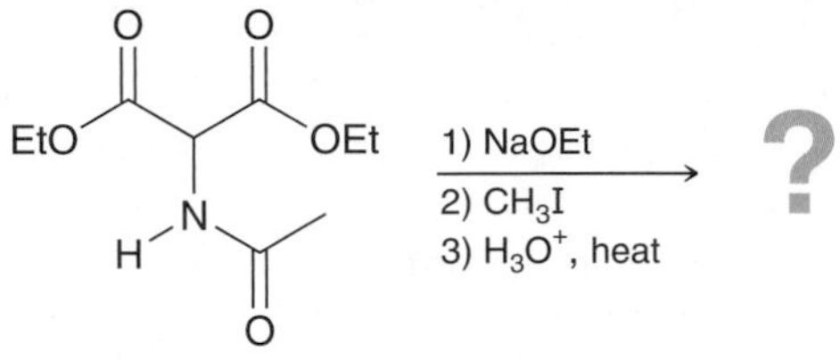

25.60 Show how racemic valine can be prepared by each of the following methods:

(a) The Hell-Volhard-Zelinski reaction
(b) The amidomalonate synthesis
(c) The Strecker synthesis

25.61 How many different pentapeptides can be constructed from the 20 naturally occurring amino acids in Table 25.1?

25.62 Using an asymmetric catalytic hydrogenation, identify the starting alkene that you would use to make L-histidine.

25.63 Using three-letter abbreviations, identify all possible acyclic tripeptides containing L-leucine, L-methionine, and L-valine.

25.64 Draw the predominant form of Asp-Lys-Phe at physiological pH.

25.65 Draw a bond-line structure of the peptide that corresponds with the following sequence of amino acid residues, and identify the N terminus and C terminus:

Trp-Val-Ser-Met-Gly-Glu

25.66 Methionine enkephalin is a pentapeptide that is produced by the body to control pain. From the sequence of its amino acid residues, draw a bond-line structure of methionine enkephalin.

N terminus **C terminus**

Tyr-Gly-Gly-Phe-Met

Methionine enkephalin

25.67 From its amino acid sequence, draw the form of aspartame that is expected to predominate at physiological pH:

Asp-Phe-OCH_3

25.68 It is believed that penicillin antibiotics are biosynthesized from amino acid precursors. Identify the two amino acids that are most likely utilized during the biosynthesis of penicillin antibiotics:

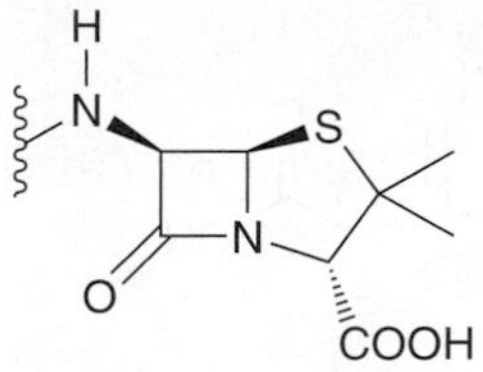

25.69 Green fluorescent protein (GFP), first isolated from bioluminescent jellyfish, is a protein containing 238 amino acid residues. The discovery of GFP has revolutionized the field of fluorescence microscopy, which enables biochemists to monitor the biosynthesis of proteins. The 2008 Nobel Prize in Chemistry was awarded to Martin Chalfie, Osamu Shimomura, and Roger Tsien for the discovery and development of GFP. The structural subunit of GFP responsible for fluorescence, called the fluorophore, results when three amino acid residues undergo cyclization. Identify the three amino acids that go into the biosynthesis of this fluorophore:

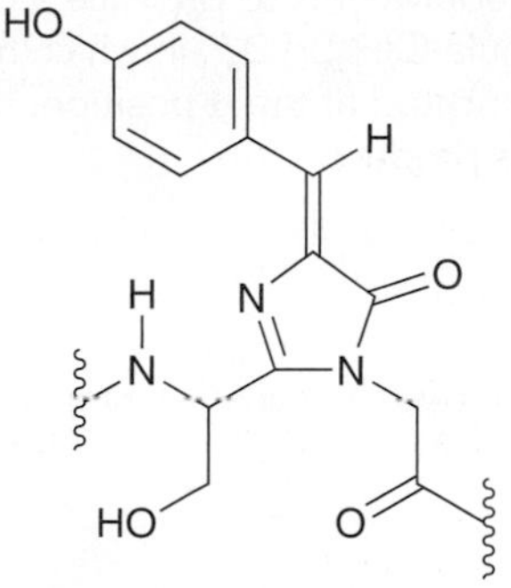

25.70 Propose two structures for a tripeptide that contains glycine, L-alanine, and L-phenylalanine but does not react with phenyl isothiocyanate.

25.71 Bradykinin has the following sequence:

(*N terminus*) ***Arg-Pro-Pro-Gly-Phe-Ser-Pro-Phe-Arg*** (*C terminus*)

Identify all fragments that will be produced when bradykinin, is treated with:

(a) Trypsin (b) Chymotrypsin

25.72 Identify the *N*-terminal residue of a peptide that yields the following PTH derivative upon Edman degradation:

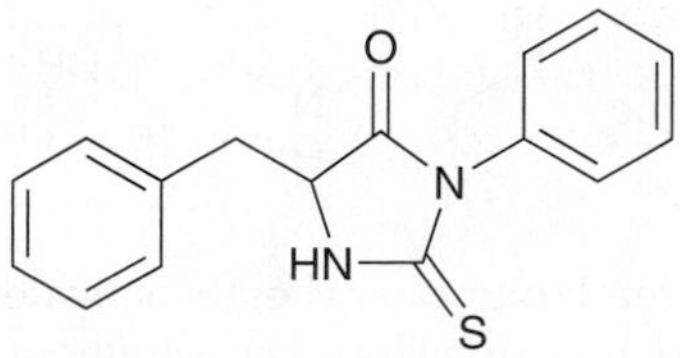

25.73 Treatment of a tripeptide with phenyl isothiocyanate yields compound **A** and a dipeptide. Treatment of the dipeptide with phenyl isothiocyanate yields compound **B** and glycine. Identify the structure of the starting tripeptide.

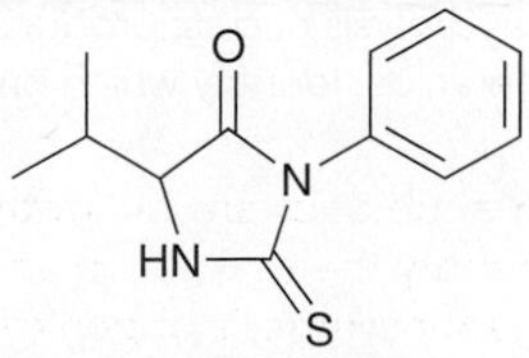

Compound A

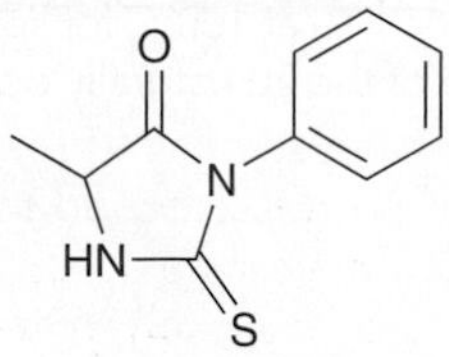

Compound B

25.74 Glucagon is a peptide hormone produced by the pancreas that, with insulin, regulates blood glucose levels. Glucagon is comprised of 29 amino acid residues. Treatment with trypsin yields four fragments, while treatment with chymotrypsin yields six fragments. Identify the sequence of amino acid residues for glucagon, and determine whether any disulfide bridges are present.

Trypsin fragments

His-Ser-Gln-Gly-Thr-Phe-Thr-Ser-Asp-Tyr-Ser-Lys
Ala-Gln-Asp-Phe-Val-Gln-Trp-Leu-Met-Asn-Thr
Tyr-Leu-Asp-Ser-Arg
Arg

Chymotrypsin fragments

His-Ser-Gln-Gly-Thr-Phe
Thr-Ser-Asp-Tyr
Leu-Met-Asn-Thr
Ser-Lys-Tyr
Leu-Asp-Ser-Arg-Arg-Ala-Gln-Asp-Phe
Val-Gln-Trp.

25.75 When the N terminus of a peptide is acetylated, the peptide derivative that is formed is unreactive toward phenyl isothiocyanate. Explain.

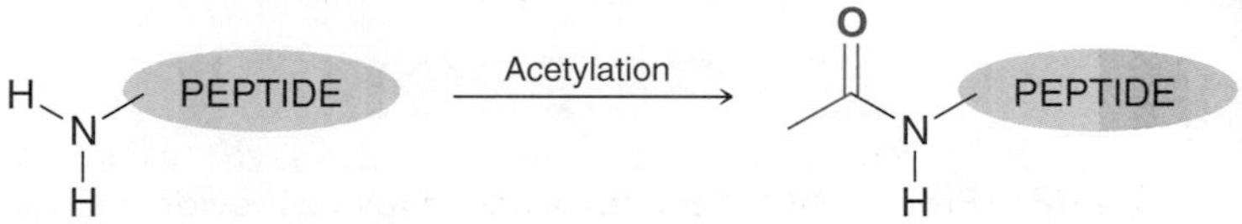

25.76 Predict the major product(s) of the reaction between L-valine and:

(a) MeOH, H^+
(b) Di-*tert*-butyl-dicarbonate
(c) NaOH, H_2O
(d) HCl

25.77 Show all steps necessary to make the dipeptide Phe-Ala from L-phenylalanine and L-alanine.

25.78 Draw all four possible dipeptides that are obtained when a mixture of L-phenylalanine and L-alanine is treated with DCC.

25.79 Identify the steps you would use to combine Val-Leu and Phe-Ile to form the tetrapeptide Phe-Ile-Val-Leu.

25.80 Identify all of the steps necessary to prepare the tripeptide Leu-Val-Ala with a Merrifield synthesis.

25.81 Draw the structure of the protected amino acid that must be anchored to the solid support in order to use a Merrifield synthesis to prepare leucine enkephalin.

(*N terminus*) ***Try-Gly-Gly-Phe-Leu*** (*C terminus*)

25.82 A proline residue will often appear at the end of an α helix but will rarely appear in the middle. Explain why proline generally cannot be incorporated into an α helix.

25.83 Draw a mechanism for the following reaction:

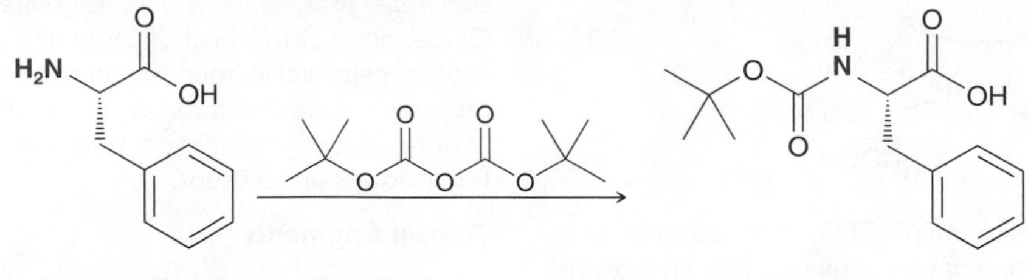

INTEGRATED PROBLEMS

25.84 Draw the alkyl halide that would be necessary to make the amino acid tyrosine using an amidomalonate synthesis. This alkyl halide is highly susceptible to polymerization. Draw the structure of the expected polymeric material.

25.85 When leucine is prepared with an amidomalonate synthesis, isobutylene (also called 2-methylpropene) is a gaseous byproduct. Draw a mechanism for the formation of this byproduct.

25.86 The side chain of tryptophan is not considered to be basic, despite the fact that it possesses a nitrogen atom with a lone pair. Explain.

25.87 Consider a process that attempts to prepare tyrosine using a Hell-Volhard-Zelinski reaction:

(a) Identify the necessary starting carboxylic acid.

(b) When treated with Br_2, the starting carboxylic acid can react with two equivalents to produce a compound with molecular formula $C_9H_8Br_2O_3$, in which neither of the bromine atoms is located at the α position. Identify the likely structure of this product.

CHALLENGE PROBLEMS

25.88 Proton NMR spectroscopy provides evidence for the restricted rotation of a peptide bond. For example, *N*,*N*-dimethylformamide exhibits three signals in its proton NMR spectrum at room temperature. Two of those signals are observed upfield, at 2.9 and 3.0 ppm. As the temperature is raised, the two signals begin to merge together. Over 180°C, these two signals combine into one signal. Explain how these results are evidence for the restricted rotation of peptide bonds.

25.89 We saw in Section 25.6 that DCC can be used to form a peptide bond. We explored the mechanism, and we saw that DCC activates the COOH moiety so that it readily undergoes nucleophilic acyl substitution. An alternative method for activating a COOH group involves converting it into an activated ester, such as a *para*-nitrophenyl ester:

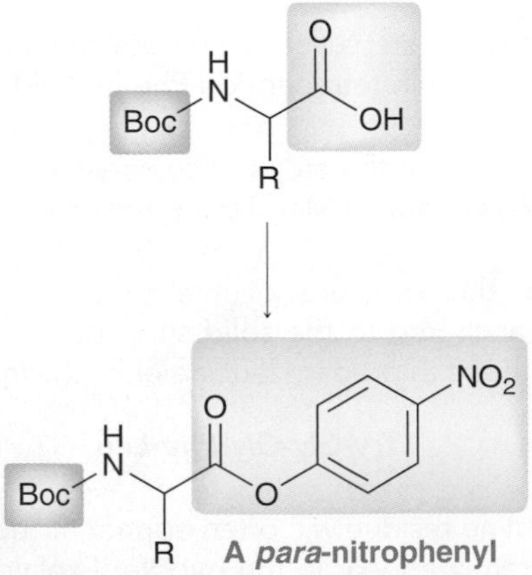

A *para*-nitrophenyl ester

The activated ester is readily attacked by a suitably protected amino acid to form a peptide bond:

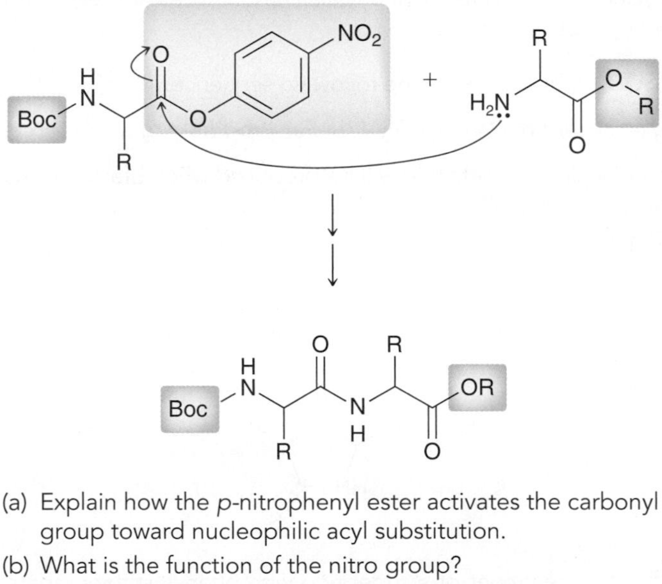

(a) Explain how the *p*-nitrophenyl ester activates the carbonyl group toward nucleophilic acyl substitution.

(b) What is the function of the nitro group?

(c) A *meta*-nitrophenyl ester is less activating than a *para*-nitrophenyl ester. Explain.

25.90 When the following compound is treated with concentrated HCl at 100°C for several hours, hydrolysis occurs, producing one of the 20 naturally occurring amino acids. Identify which one.

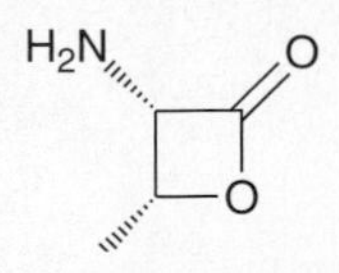

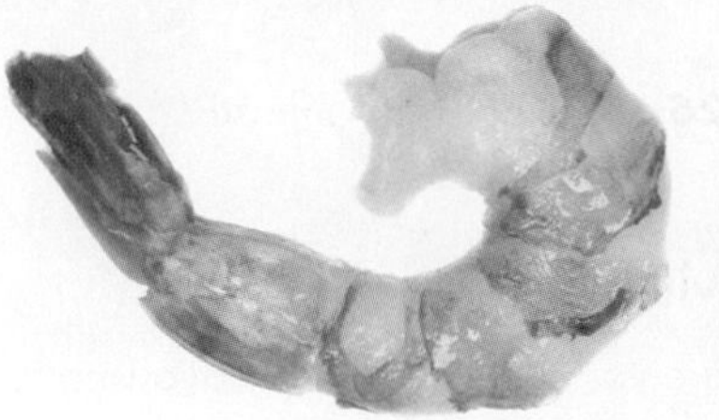

26

Lipids

DID YOU EVER WONDER...

why high cholesterol levels increase the risk of a heart attack?

Cholesterol is a type of steroid that belongs to a larger class of naturally occurring compounds called lipids. Cholesterol serves many important functions, as we will see in this chapter, and maintaining healthy cholesterol levels can truly be a matter of life and death. In this chapter, we will explore the properties and functions of steroids as well as the properties and functions of many other classes of lipids.

DO YOU REMEMBER?

Before you go on, be sure you understand the following topics.
If necessary, review the suggested sections to prepare for this chapter:

- Partially Hydrogenated Fats (section 9.7)
- Autooxidation and Antioxidants (section 11.9)
- Base-Catalyzed Hydrolysis of Esters: Saponification (section 21.11)

Visit www.wileyplus.com to check your understanding and for valuable practice.

26.1 Introduction to Lipids

Throughout the chapters of this book, we have regularly classified organic compounds based on their functional groups (e.g., alkenes, alkynes, alcohols). **Lipids** represent a somewhat unique category because they are not defined by the presence or absence of a particular functional group. Instead, they are defined by a physical property—solubility. Specifically, lipids are naturally occurring compounds that can be extracted from cells using nonpolar organic solvents. An extremely large number of biological compounds are considered to be lipids, so it is helpful to classify and categorize them. **Complex lipids** are lipids that readily undergo hydrolysis in aqueous acid or base to produce smaller fragments, while **simple lipids** do not readily undergo hydrolysis. The terms "simple" and "complex" can be somewhat misleading in this context, since many complex lipids have rather simple structures, while many simple lipids have more complex structures. Complex lipids are so named because they contain one or more ester moieties, which can be hydrolyzed to produce a carboxylic acid and an alcohol (see Section 21.11). In this chapter, we will explore three classes of complex lipids and three classes of simple lipids, as seen in Figure 26.1.

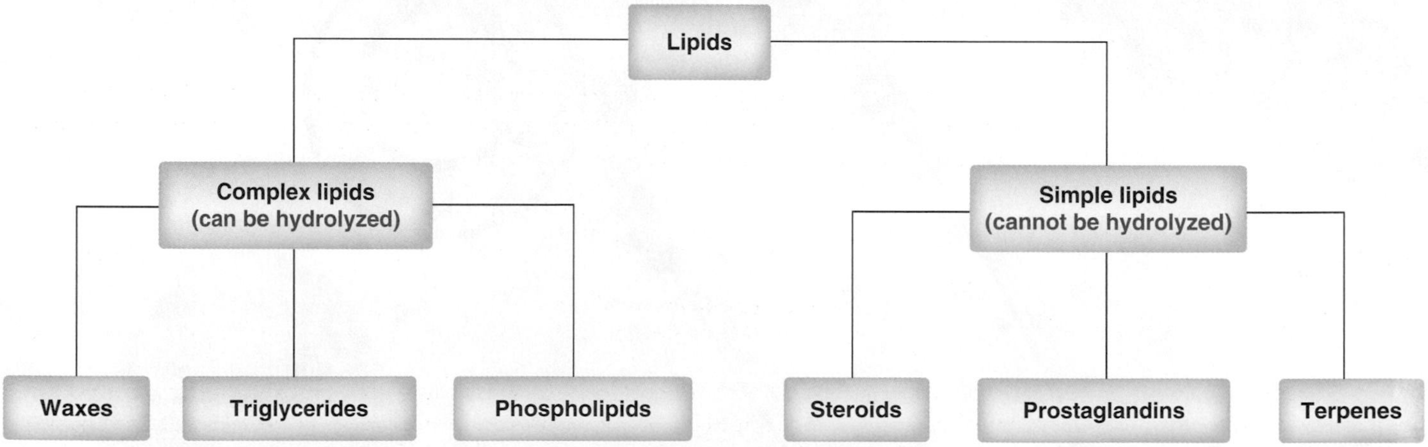

FIGURE 26.1
The six classes of lipids that will be discussed in this chapter.

The three major classes of complex lipids are waxes, triglycerides, and phospholipids. One example of each class is shown. Notice that all three classes contain ester moieties, rendering them easily hydrolyzed, and they also contain long, hydrocarbon chains, rendering them soluble in organic solvents.

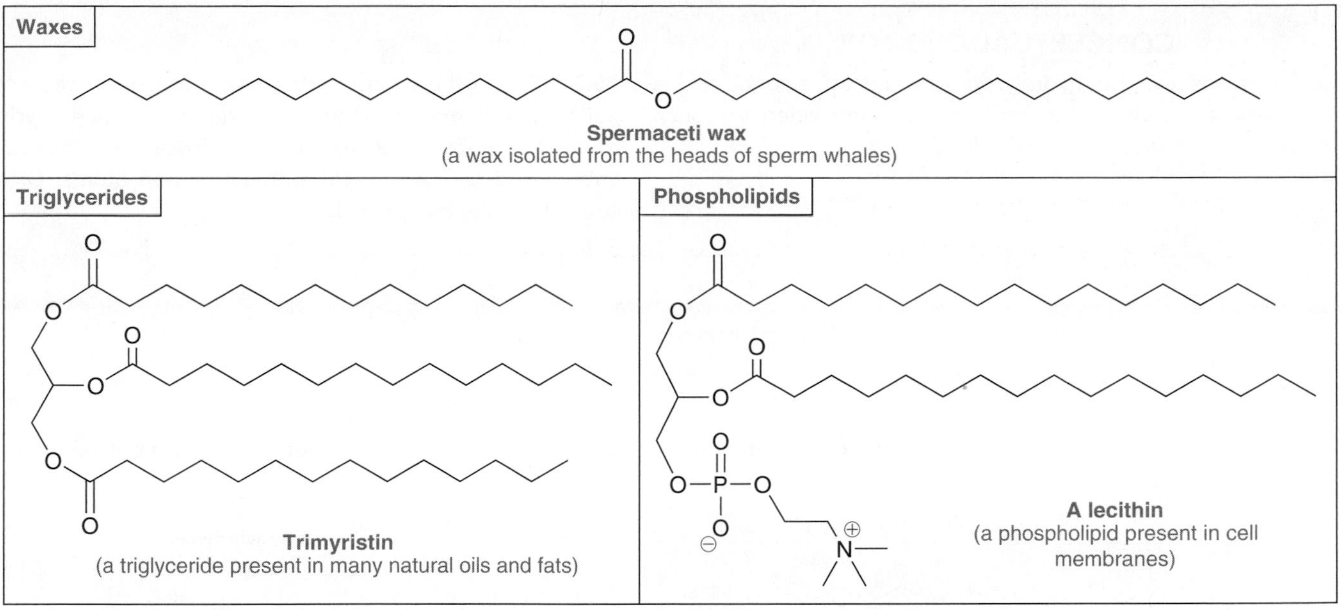

Just as complex lipids can be divided into several classes, simple lipids can also be divided into several classes. In this chapter, we will explore three such classes: steroids, prostaglandins, and terpenes. One example of each class is shown below.

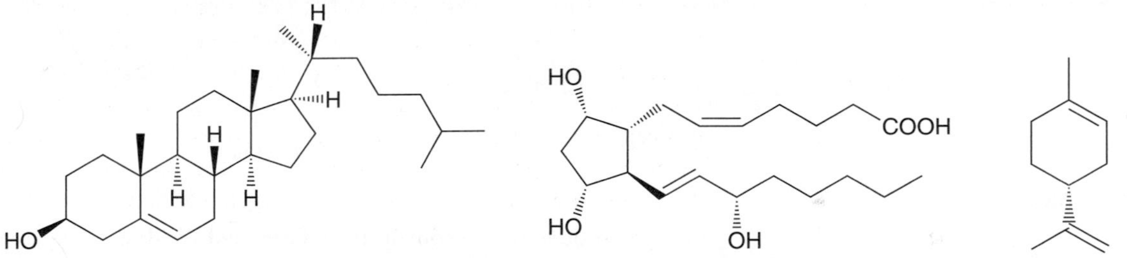

Cholesterol (a steroid) — **$PGF_{2\alpha}$** (a prostaglandin) — **Limonene** (a terpene)

26.2 Waxes

Waxes are high-molecular-weight esters that are constructed from carboxylic acids and alcohols. For example, triacontyl hexadecanoate, a major component of beeswax, is constructed from a carboxylic acid with 16 carbon atoms and an alcohol with 30 carbon atoms.

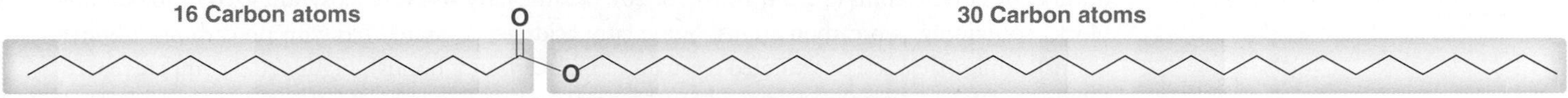

Triacontyl hexadecanoate
(a major component of beeswax)

LOOKING BACK
For a review of London dispersion forces, see Section 1.12.

Other waxes are similar in structure, differing only in the number of carbon atoms on either side of the ester moiety. These long hydrocarbon chains endow these compounds with high melting points as the result of extensive, intermolecular London dispersion forces between the hydrocarbon tails.

Waxes serve a wide variety of functions in living organisms. Sperm whales are believed to use spermaceti wax as an antenna for detecting sound waves (sonar), allowing the whale to map its environment. Many insects have protective coatings of wax on their exoskeletons. Birds utilize waxes on their feathers, rendering them water repellant. Similarly, the fur of some mammals, such as sheep, is coated with a mixture of waxes called lanolin. Waxes coat the surface of the leaves of many plants, thereby preventing evaporation and reducing water loss. For example, carnauba wax (a mixture of high-molecular-weight esters) is produced by the Brazilian palm tree and is commonly used in polish formulations for boats and automobiles.

CONCEPTUAL CHECKPOINT

26.1 Waxes can be hydrolyzed to yield an alcohol and a carboxylic acid. Draw the products obtained when triacontyl hexadecanoate undergoes hydrolysis.

26.2 One of the compounds present in lanolin was isolated, purified, and then treated with aqueous sodium hydroxide to yield an unbranched alcohol with 20 carbon atoms and an unbranched carboxylic acid with 22 carbon atoms. Draw the structure of the compound.

26.3 Triglycerides

The Structure and Function of Triglycerides

Triglycerides are triesters formed from glycerol and three long-chain carboxylic acids, or **fatty acids** (Figure 26.2).

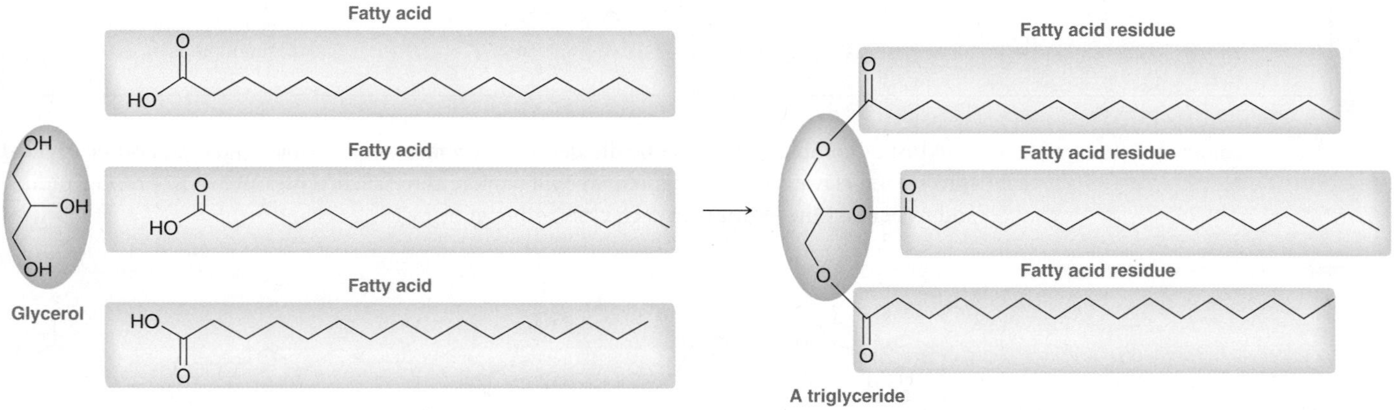

FIGURE 26.2
Triglycerides are triesters that are assembled from one equivalent of glycerol and three equivalents of fatty acids.

A triglyceride is said to contain three **fatty acid residues.** Triglycerides are used by mammals and plants for long-term energy storage. Triglycerides store approximately 9 kcal/g, which means that they are more than twice as efficient for energy storage than carbohydrates and proteins, which can only store approximately 4 kcal/g. The exact structure, physical properties, and energy content of a triglyceride depend on the identity of the three fatty acid residues. Let's take a closer look at the fatty acids that are commonly found in naturally occurring triglycerides.

Properties of Fatty Acids and Triglycerides

Fatty acids obtained from hydrolysis of naturally occurring triglycerides are long, unbranched carboxylic acids typically containing between 12 and 20 carbon atoms. They generally contain an even number of carbon atoms (12, 14, 16, 18, or 20), because fatty acids are biosynthesized from building blocks containing two carbon atoms. Some fatty acids are saturated (contain no carbon-carbon π bonds), while others are unsaturated (contain carbon-carbon π bonds). Figure 26.3 compares space-filling models of a saturated fatty acid (stearic acid) and an unsaturated fatty acid (oleic acid), both with 18 carbon atoms. In its lowest energy conformation, stearic acid is fairly linear (Figure 26.3a). In contrast, oleic acid exhibits a kink that prevents the compound from adopting a linear conformation (Figure 26.3b). The presence or absence of a kink has a profound impact on the melting point of fatty acids. Table 26.1 provides the melting points of several saturated and unsaturated fatty acids.

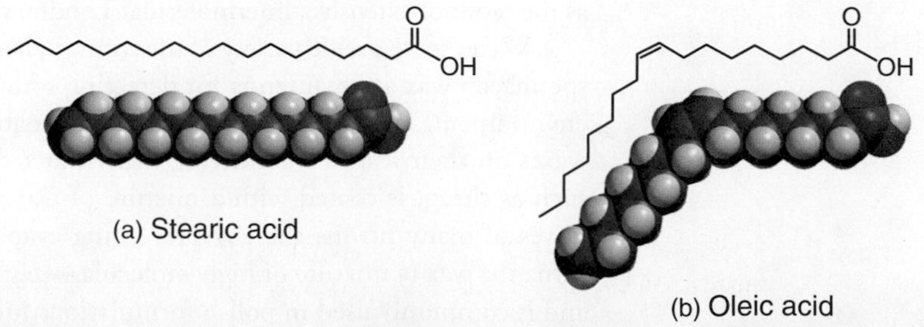

FIGURE 26.3
(a) A space-filling model for the lowest energy conformation of stearic acid. (b) A space-filling model for the lowest energy conformation of oleic acid.

TABLE 26.1 MELTING POINTS OF COMMON SATURATED AND UNSATURATED FATTY ACIDS

STRUCTURE AND NAME	NUMBER OF CARBON ATOMS	NUMBER OF CARBON-CARBON DOUBLE BONDS	MELTING POINT (°C)
SATURATED			
 Lauric acid	12	0	43
 Myristic acid	14	0	54
 Palmitic acid	16	0	63
 Stearic acid	18	0	69
 Arachidic acid	20	0	77
UNSATURATED			
 Palmitoleic acid	16	1	0
 Oleic acid	18	1	13
 Linoleic acid	18	2	−5
 Arachidonic acid	20	4	−50

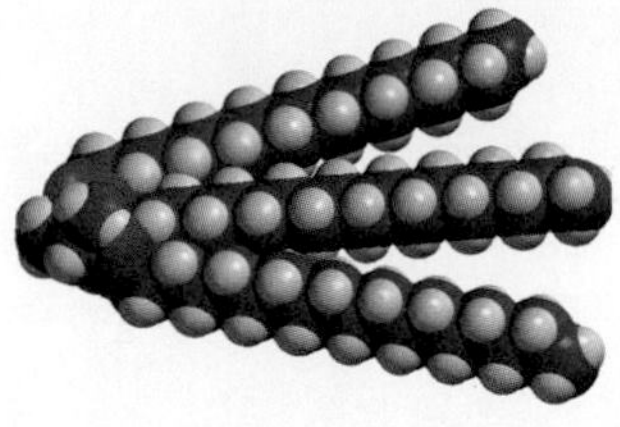

Tristearin
mp=72°C

Triolein
mp=−4°C

FIGURE 26.4
Space-filling models of tristearin and triolein.

Two important trends emerge from the data in Table 26.1:

1. For saturated fatty acids, the melting point increases with increasing molecular weight. For example, myristic acid exhibits a higher melting point than lauric acid, because myristic acid has more carbon atoms.

2. The presence of a *cis* double bond causes a decrease in the melting point. For example, oleic acid exhibits a lower melting point than stearic acid, because oleic acid has a *cis* double bond. Stearic acid molecules can pack more tightly in the solid phase, leading to stronger intermolecular London dispersion forces and, consequently, a higher melting point. In contrast, oleic acid molecules have a kink that prevents them from packing as efficiently, thereby decreasing the strength of the intermolecular London dispersion forces and lowering the melting point. Additional double bonds further lower the melting point, as can be seen when comparing the melting points of oleic acid and linoleic acid.

These two trends are also observed when comparing the melting points of triglycerides. That is, the melting point of a triglyceride depends on the number of carbon atoms in the fatty acid residues as well as the presence of any unsaturation. Triglycerides with unsaturated fatty acid residues have lower melting points than triglycerides with saturated fatty acid residues. For example, compare space-filling models of tristearin and triolein (Figure 26.4). Tristearin is a triglyceride formed from three molecules of stearic acid. Since all three fatty acids are saturated, the resulting triglyceride can pack efficiently in the solid state, allowing for strong intermolecular London dispersion forces. As a result, tristearin has a relatively high melting point (it is a solid at room temperature). In contrast, triolein is a triglyceride formed from three molecules of oleic acid. Since all three fatty acids are unsaturated, the resulting triglyceride cannot pack as efficiently in the solid state. As a result, triolein experiences weaker intermolecular London dispersion forces and therefore has a relatively low melting point (it is a liquid at room temperature). Triglycerides that are solids at room temperature are called **fats**, while those that are liquids at room temperature are called **oils**.

SKILLBUILDER

26.1 COMPARING MOLECULAR PROPERTIES OF TRIGLYCERIDES

LEARN the skill Which triglyceride would be expected to have the higher melting point?

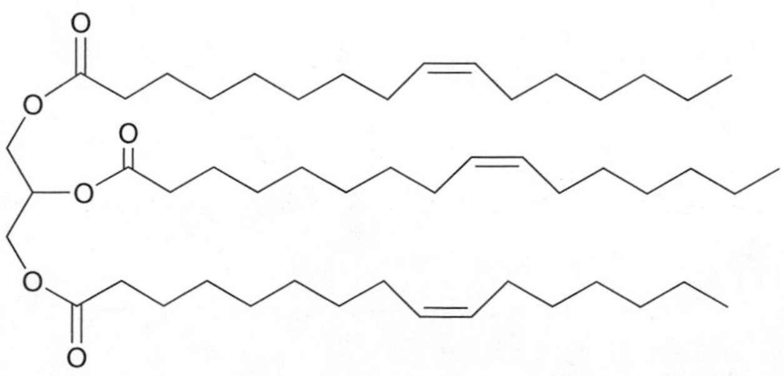

Tripalmitolein

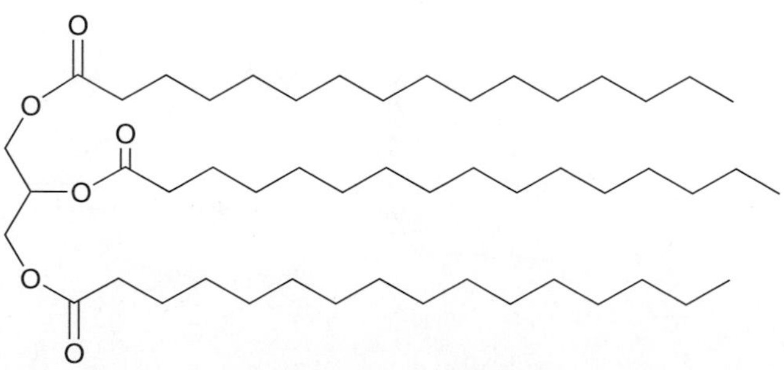

Tripalmitin

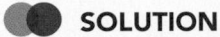

SOLUTION

First identify the fatty acid residues in each triglyceride:

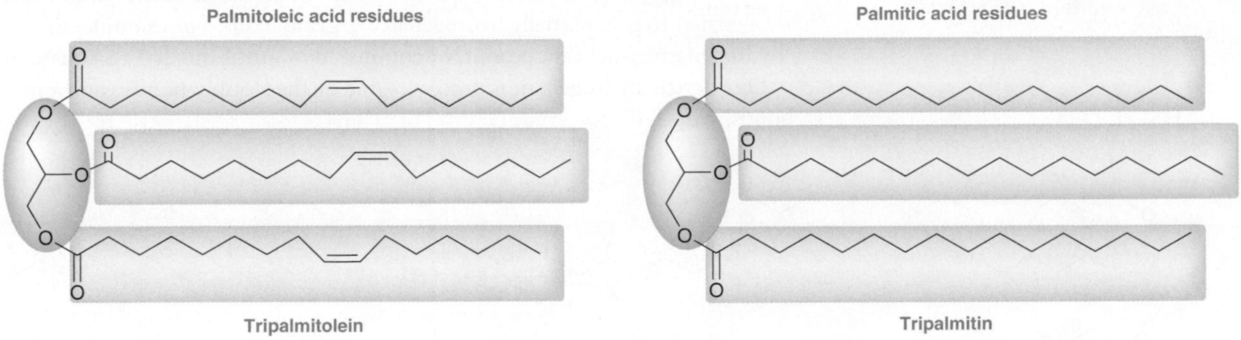

STEP 1
Identify the fatty acid residues.

As the names imply, tripalmitolein contains three palmitoleic acid residues, and tripalmitin contains three palmitic acid residues. We must compare these fatty acid residues, keeping in mind that two features contribute to a higher melting point.

STEP 2
Compare the length and saturation of the residues.

1. Length of fatty acid residues (longer chain = higher melting point)
2. Absence of C=C unsaturation (no unsaturation = higher melting point).

The first feature cannot be used to distinguish these triglycerides, because palmitoleic acid and palmitic acid each have 16 carbon atoms. However, the second feature does distinguish these triglycerides. Specifically, palmitoleic acid is an unsaturated carboxylic acid, while palmitic acid is a saturated carboxylic acid. We have seen that unsaturation causes inefficient packing, thereby lowering the melting point. Therefore, we expect palmitoleic acid will have the lower melting point and palmitic acid the higher melting point.

PRACTICE the skill

26.3 For each pair of triglycerides, identify the one that is expected to have the higher melting point. Consult Table 26.1 to determine which fatty acid residues are present in each triglyceride.

(a) Trilaurin and trimyristin (b) Triarachidin and trilinolein
(c) Triolein and trilinolein (d) Trimyristin and tristearin

APPLY the skill

26.4 Arrange the following three triglycerides in order of increasing melting point.

Tristearin, tripalmitin, and tripalmitolein

26.5 Tristearin has a melting point of 72 °C. Based on this information, would you expect triarachadin to be classified as a fat or as an oil?

26.6 Identify each of the following compounds as a fat or an oil. Explain your answers.

(a) A triglyceride containing one palmitic acid residue and two stearic acid residues
(b) A triglyceride containing one oleic acid residue and two linoleic acid residues

need more **PRACTICE?** **Try Problems 26.40a,d,e, 26.41a,d,e, 26.43**

26.4 Reactions of Triglycerides

Hydrogenation of Triglycerides

Triglycerides containing unsaturated fatty acid residues will undergo hydrogenation (see Section 9.7).

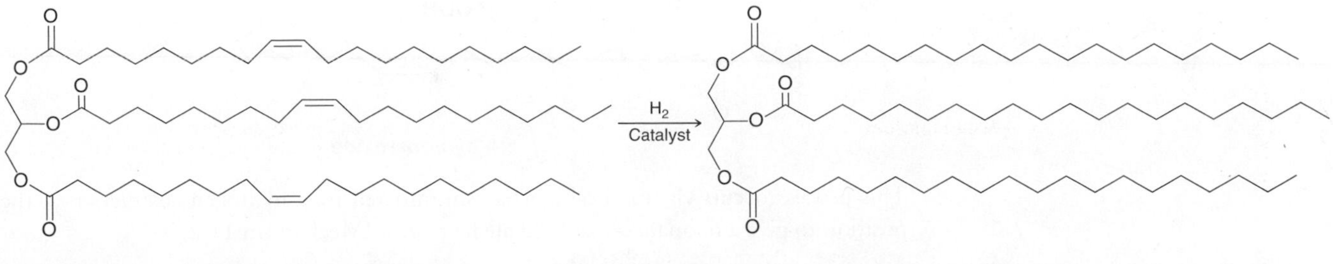

This type of transformation is generally achieved using high temperatures and a catalyst such as nickel (Ni). In the previous example, all of the carbon-carbon π bonds are hydrogenated, but it is also possible to control the conditions so that only some of the carbon-carbon π bonds are hydrogenated to give partially hydrogenated vegetable oils. For example, margarine is produced by hydrogenating soybean, peanut, or cottonseed oil until the desired consistency is achieved.

During the hydrogenation process, some of the double bonds can isomerize to give *trans* π bonds.

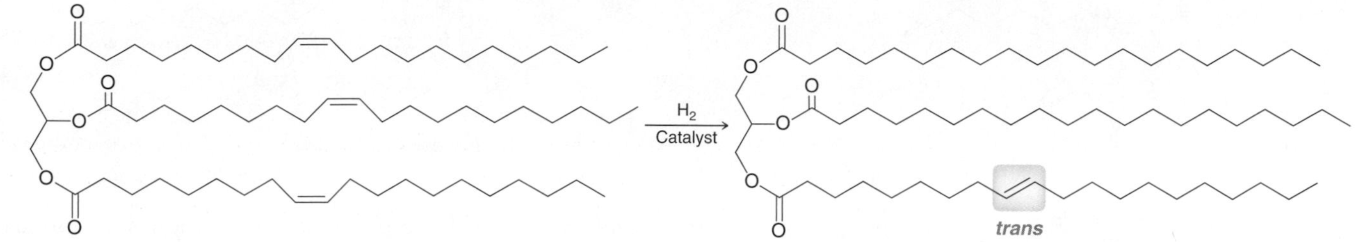

The hydrogenation process typically gives a mixture containing 10–15% *trans* unsaturated fatty acids. These "trans fats" have been implicated as culprits in raising cholesterol levels, thereby increasing the risk of heart attacks (as described later in this chapter). The FDA now requires food product labels to indicate the amount of trans fats present in food.

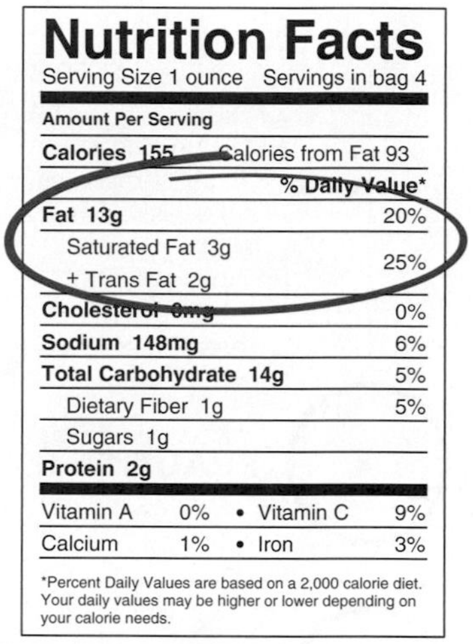

Nutrition Facts
Serving Size 1 ounce Servings in bag 4
Amount Per Serving
Calories 155 Calories from Fat 93
% Daily Value*
Fat 13g 20%
Saturated Fat 3g 25%
+ Trans Fat 2g
Cholesterol 0mg 0%
Sodium 148mg 6%
Total Carbohydrate 14g 5%
Dietary Fiber 1g 5%
Sugars 1g
Protein 2g
Vitamin A 0% • Vitamin C 9%
Calcium 1% • Iron 3%
*Percent Daily Values are based on a 2,000 calorie diet. Your daily values may be higher or lower depending on your calorie needs.

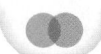

CONCEPTUAL CHECKPOINT

26.7 Triolein was treated with molecular hydrogen at high temperature in the presence of nickel. At completion, the reaction had consumed three equivalents of molecular hydrogen.

(a) Draw the structure of the product.

(b) Identify the name of the product.

(c) Determine whether the melting point of the product is higher or lower than triolein.

(d) When the product is treated with aqueous base, three equivalents of a fatty acid are produced. Identify this fatty acid.

26.8 Partial hydrogenation of triolein produces several different *trans* fats. Draw all possible *trans* fats that might be obtained in the process.

Autooxidation of Triglycerides

In the presence of molecular oxygen, triglycerides that contain unsaturated fatty acid residues are particularly susceptible to autooxidation at the allylic position to yield hydroperoxides (see Section 11.9)

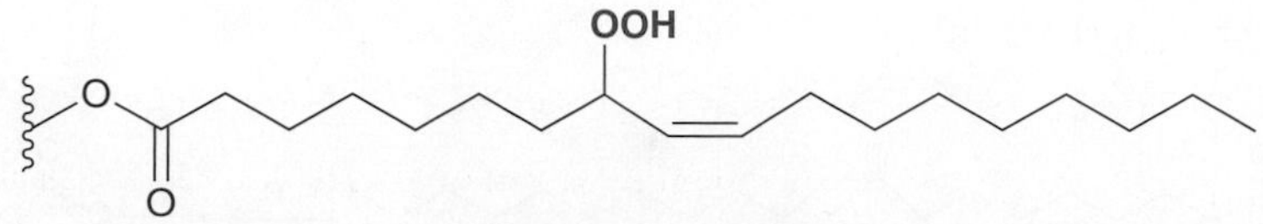

OOH

A hydroperoxide

This process occurs via a radical mechanism initiated by a hydrogen abstraction at the allylic position to give a resonance-stabilized allylic radical (Mechanism 11.2).

As with all radical mechanisms, the net reaction is the sum of the propagation steps:

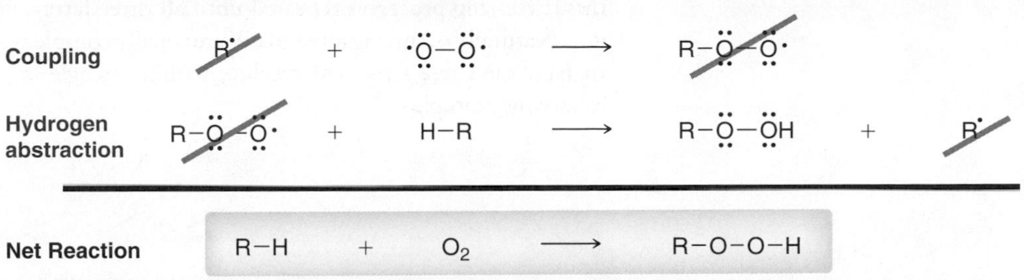
Coupling
Hydrogen abstraction
Net Reaction
R–H + O2 ⟶ R–O–O–H

The resulting hydroperoxide is responsible for the rancid smell that develops over time in foods containing unsaturated oils. In addition, hydroperoxides are also toxic. Therefore, food products containing unsaturated oils have a short shelf-life unless radical inhibitors are used to slow the formation of hydroperoxides. The role of radical inhibitors in food chemistry was discussed in Section 11.9.

Hydrolysis of Triglycerides

When treated with aqueous base, triglycerides undergo hydrolysis, yielding glycerol and three carboxylate ions.

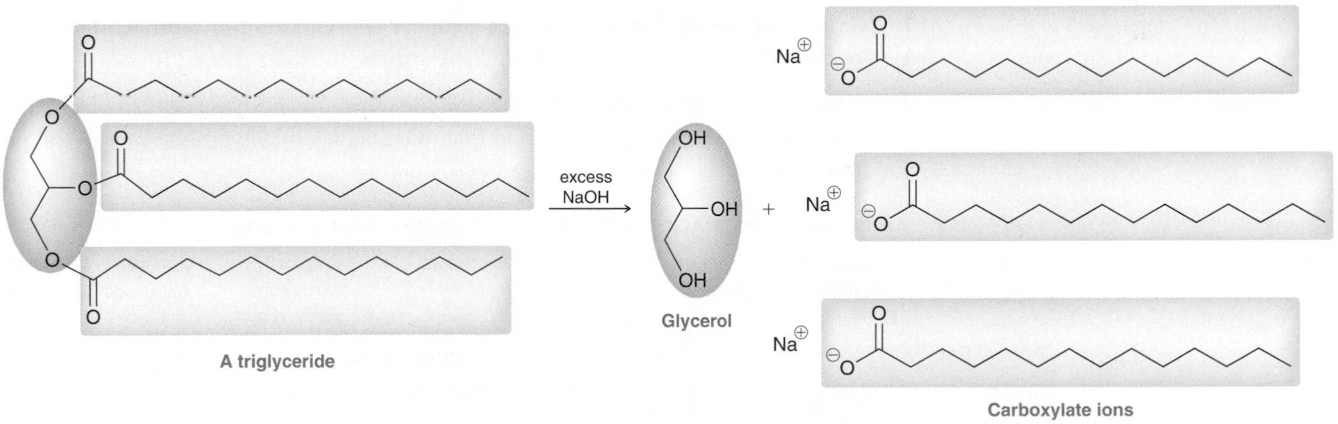
excess NaOH
A triglyceride
Glycerol
Carboxylate ions

Each of the three ester moieties of the triglyceride is hydrolyzed via a nucleophilic acyl substitution reaction (Mechanism 26.1).

MECHANISM 26.1 BASE-CATALYZED HYDROLYSIS OF AN ESTER MOIETY

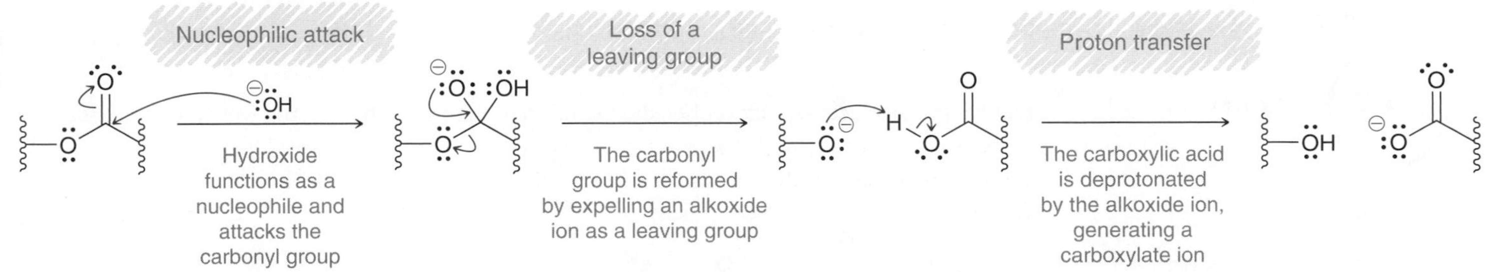
Nucleophilic attack
Hydroxide functions as a nucleophile and attacks the carbonyl group
Loss of a leaving group
The carbonyl group is reformed by expelling an alkoxide ion as a leaving group
Proton transfer
The carboxylic acid is deprotonated by the alkoxide ion, generating a carboxylate ion

In the first step, a hydroxide ion functions as a nucleophile and attacks the carbonyl group of the ester, generating a tetrahedral intermediate. This tetrahedral intermediate then re-forms the carbonyl group by expelling an alkoxide ion as a leaving group. Alkoxide ions are generally poor leaving groups, but in this case, the tetrahedral intermediate itself is an alkoxide ion, so expulsion of an alkoxide ion is not uphill in energy. In the final step, the alkoxide ion functions as a base and deprotonates the carboxylic acid, generating a carboxylate ion. For a triglyceride, this three-step process is repeated until all three fatty acid residues have been released.

Naturally occurring fats and oils are usually complex mixtures of many different triglycerides. In fact, the three fatty acid residues within a single triglyceride are often different, as in the following example.

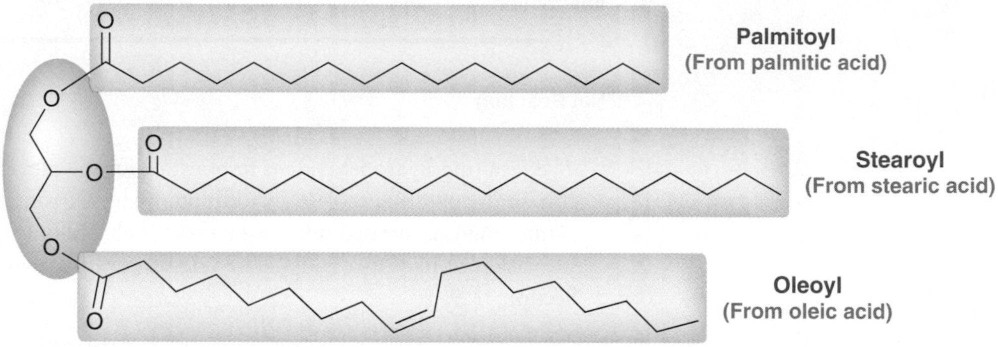

When heated with aqueous base, fats and oils can be completely hydrolyzed to provide a mixture of fatty acid carboxylates. Table 26.2 gives the approximate fatty acid composition of several common fats and oils. Two important observations emerge from the data in Table 26.2:

1. Triglycerides found in animals have a higher concentration of saturated fatty acids than triglycerides from vegetable origins.

2. The exact ratio of fatty acids differs from one source to another. For example, hydrolysis of corn oil produces more linoleic acid than oleic acid, while hydrolysis of olive oil produces more oleic acid.

TABLE 26.2 APPROXIMATE FATTY ACID COMPOSITIONS FOR SEVERAL FATS AND OILS

SOURCE	PERCENT SATURATED FATTY ACIDS	PERCENT OLEIC ACID	PERCENT LINOLEIC ACID	SOURCE	PERCENT SATURATED FATTY ACIDS	PERCENT OLEIC ACID	PERCENT LINOLEIC ACID
Animal Fat				**Vegetable Oil**			
Beef fat	55	40	3	Corn oil	14	34	48
Milk fat	37	33	3	Olive oil	11	82	5
Lard	41	50	6	Canola oil	9	54	30
Human fat	37	46	10	Peanut oil	12	60	20

SKILLBUILDER

26.2 IDENTIFYING THE PRODUCTS OF TRIGLYCERIDE HYDROLYSIS

LEARN the skill

Identify the products expected when the following triglyceride is hydrolyzed with aqueous base.

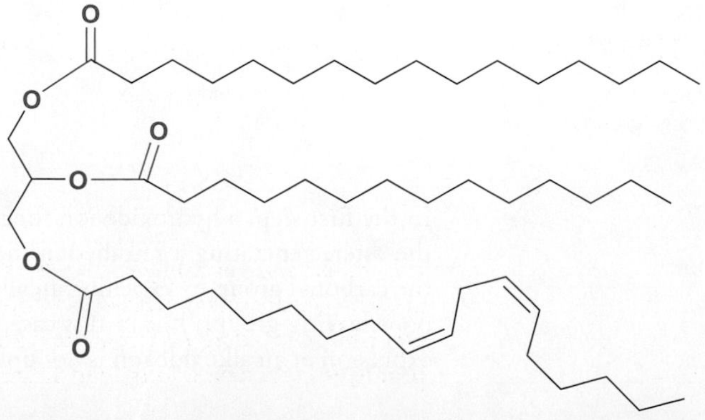

SOLUTION

STEP 1
Identify the ester moieties.

First identify the three ester moieties (highlighted in red). Each ester moiety is hydrolyzed to give an alcohol and a carboxylate ion, according to the following mechanism:

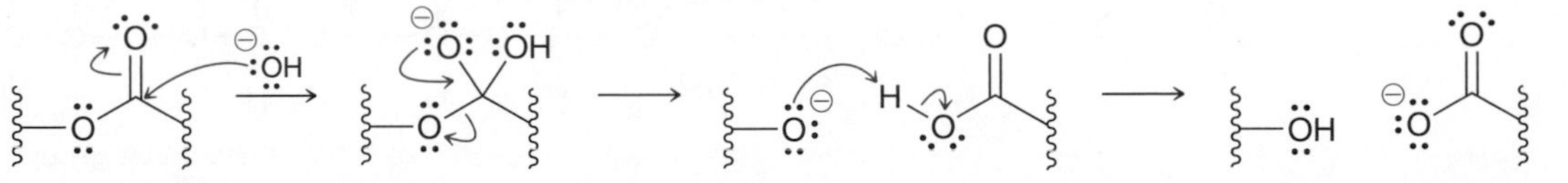

STEP 2
Draw glycerol and the three appropriate carboxylate ions.

Therefore, we expect the following products

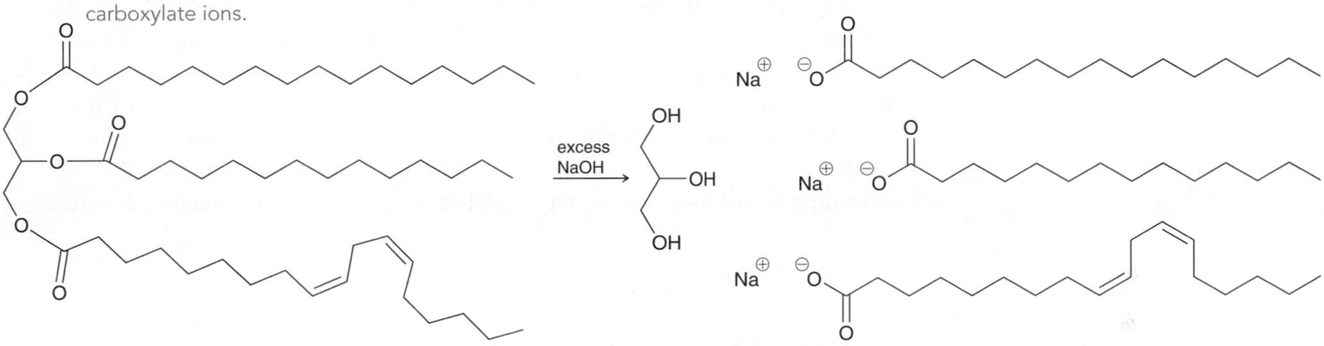

The products are glycerol and three carboxylate ions. Using Table 26.1, we can identify these carboxylate ions as the conjugate bases of palmitic acid, myristic acid, and linoleic acid.

PRACTICE the skill **26.9** Identify the products that are expected when the following triglyceride is hydrolyzed with aqueous sodium hydroxide.

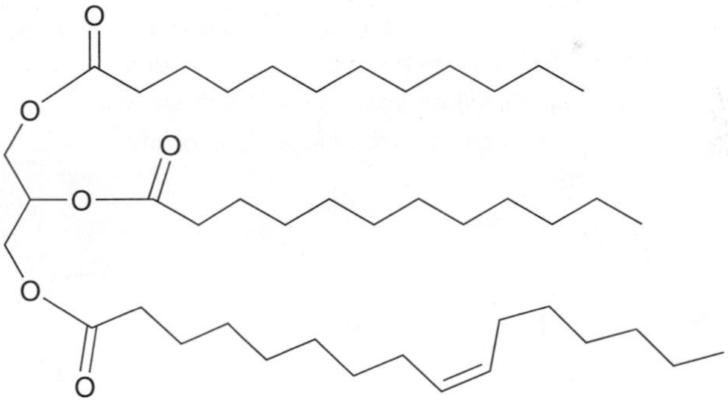

APPLY the skill

26.10 A triglyceride was treated with sodium hydroxide to yield glycerol and three equivalents of sodium laurate (the conjugate base of lauric acid). Draw the structure of the triglyceride.

26.11 An optically inactive triglyceride was hydrolyzed to yield one equivalent of palmitic acid and two equivalents of lauric acid. Draw the structure of the triglyceride.

need more **PRACTICE?** **Try Problems 26.30b, 26.40c, 26.41c**

Hydrolysis of triglycerides is commercially valuable, because soap is prepared via this process. Accordingly, the hydrolysis of triglycerides in the presence of aqueous base is also called saponification, derived from the Latin word for soap, *saponis*. Saponification of triglycerides produces carboxylate ions, which are the primary ingredients of soap. Recall from Chapter 1 that carboxylate ions function as soap because they have both a polar, hydrophilic group and a nonpolar, hydrophobic group.

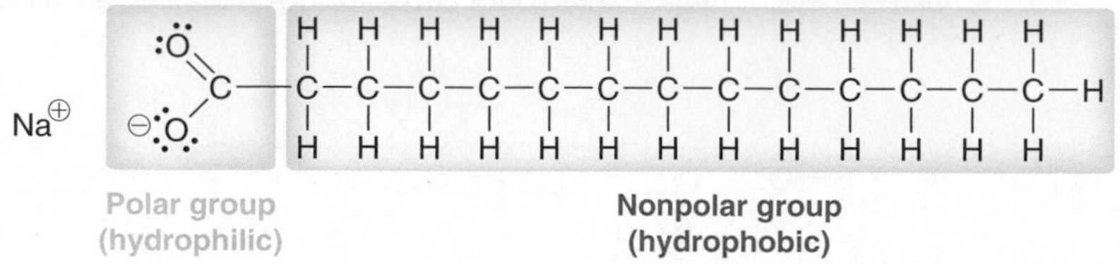

When dissolved in water, compounds of this type assemble around nonpolar substances to form spheres called *micelles* (Figure 1.54). The nonpolar substance is located at the center of the micelle, where it interacts with the nonpolar ends of the soap molecules via intermolecular London dispersion forces. The surface of the micelle is comprised of polar groups, which interact with the polar solvent. That is, the micelle acts as a unit that is solvated by the polar solvent. In this way, soap molecules can solvate nonpolar substances, such as grease, in polar solvents, such as water.

Most soaps are produced by boiling either animal fat or vegetable oil together with a strong alkaline solution, such as aqueous sodium hydroxide. The identity of the alkyl chains can vary, depending on the source of the fat or oil, but the concept is the same for all soaps.

PRACTICALLYSPEAKING)))

Soaps Versus Synthetic Detergents

Soap has been used for over two millennia, as people discovered long ago that soap could be made by heating animal fat together with wood ashes, which contain alkaline substances. Nevertheless, the usefulness of soap is diminished in the presence of water that contains high concentrations of calcium ions (Ca^{2+}) or magnesium ions (Mg^{2+}). When soap is used with such water, called hard water, a precipitate is formed as a result of the following ion exchange reaction.

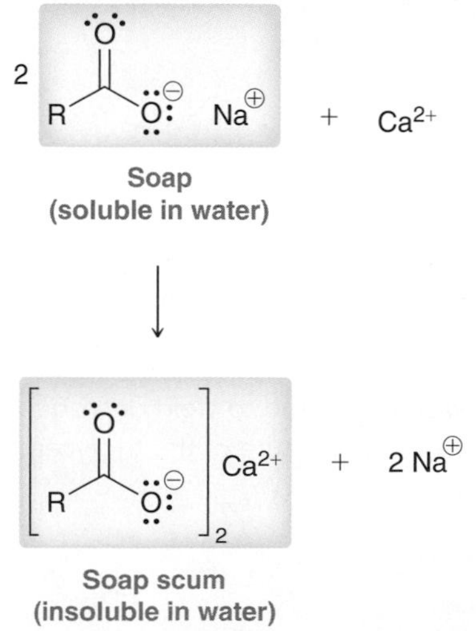

The generation of a precipitate, often called soap scum, limits the usefulness of soap. To circumvent this problem, chemists have developed synthetic detergents that do not form precipitates when used with hard water. Like soap, synthetic detergents also contain both hydrophobic and hydrophilic regions, but the identity of the hydrophilic region has been modified. Rather than using a carboxylate moiety, synthetic detergents use a different group. For example, consider the structure of sodium lauryl sulfate.

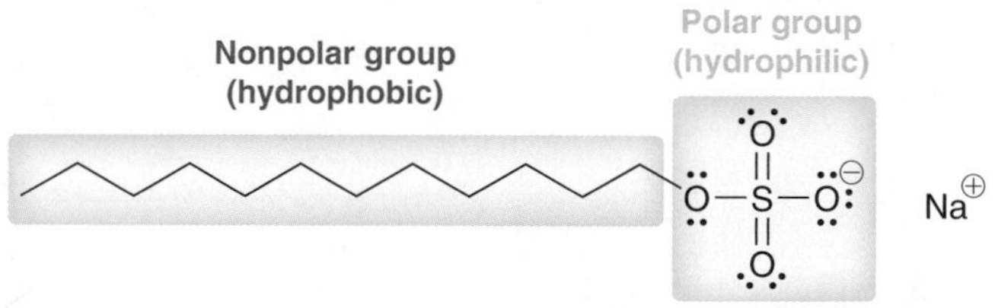

Sodium lauryl sulfate

Like soap molecules, this compound also has a hydrophobic group and a hydrophilic group. However, in this case, an ion exchange reaction does not generate a precipitate, because the calcium salt is water soluble, hence, no soap scum. Sodium lauryl sulfate is in fact a common ingredient found in many shampoo formulations.

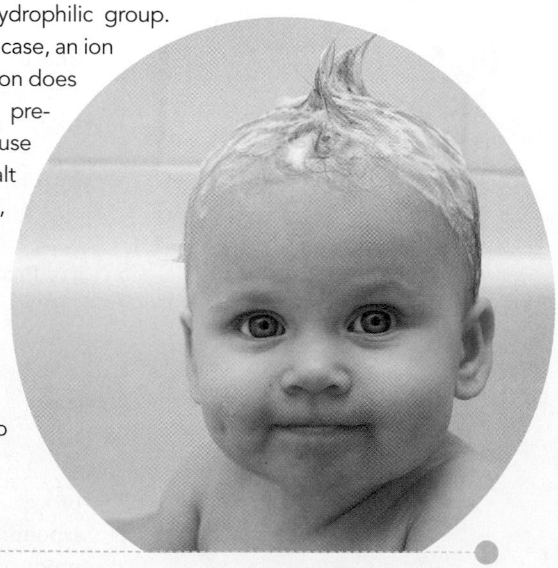

Transesterification of Triglycerides

We have seen that triglycerides are extremely efficient at storing energy. For this reason, our bodies use triglycerides as a source of fuel. It should therefore come as no surprise that diesel engines can be modified to use cooking oil as a fuel. In fact, coconut oil was used extensively as fuel for vehicles in World War I and World War II, when the supply of gasoline was scarce. This technique cannot be used in colder climates, because many oils will solidify at low temperatures. An alternative to vegetable oil is biodiesel, which is formed from the transesterification of vegetable oils to produce a mixture of fatty acid methyl esters.

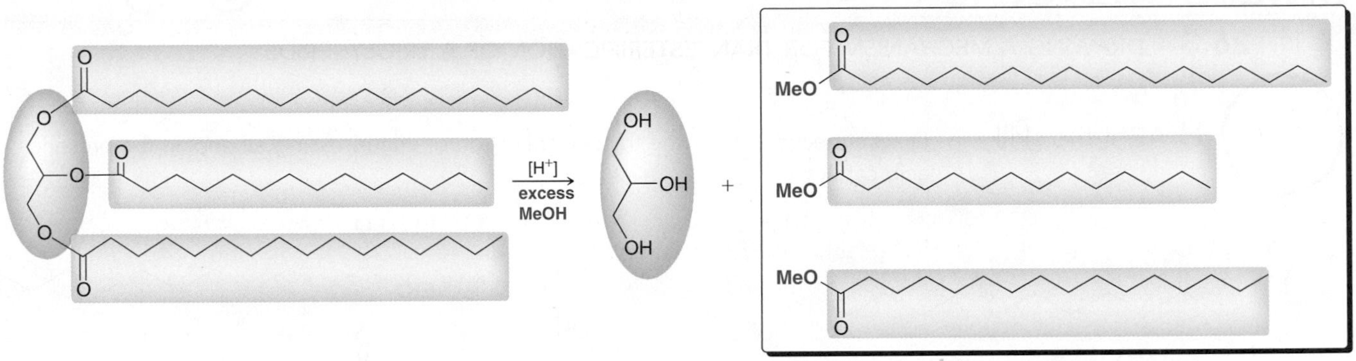
[H+]
excess
MeOH
OH
OH
OH
MeO
MeO
MeO

Biodiesel
(a mixture of fatty acid methyl esters)

This transformation can be achieved with either acid catalysis or base catalysis. Mechanism 26.2 shows the acid-catalyzed mechanism.

MECHANISM 26.2 ACID-CATALYZED TRANSESTERIFICATION

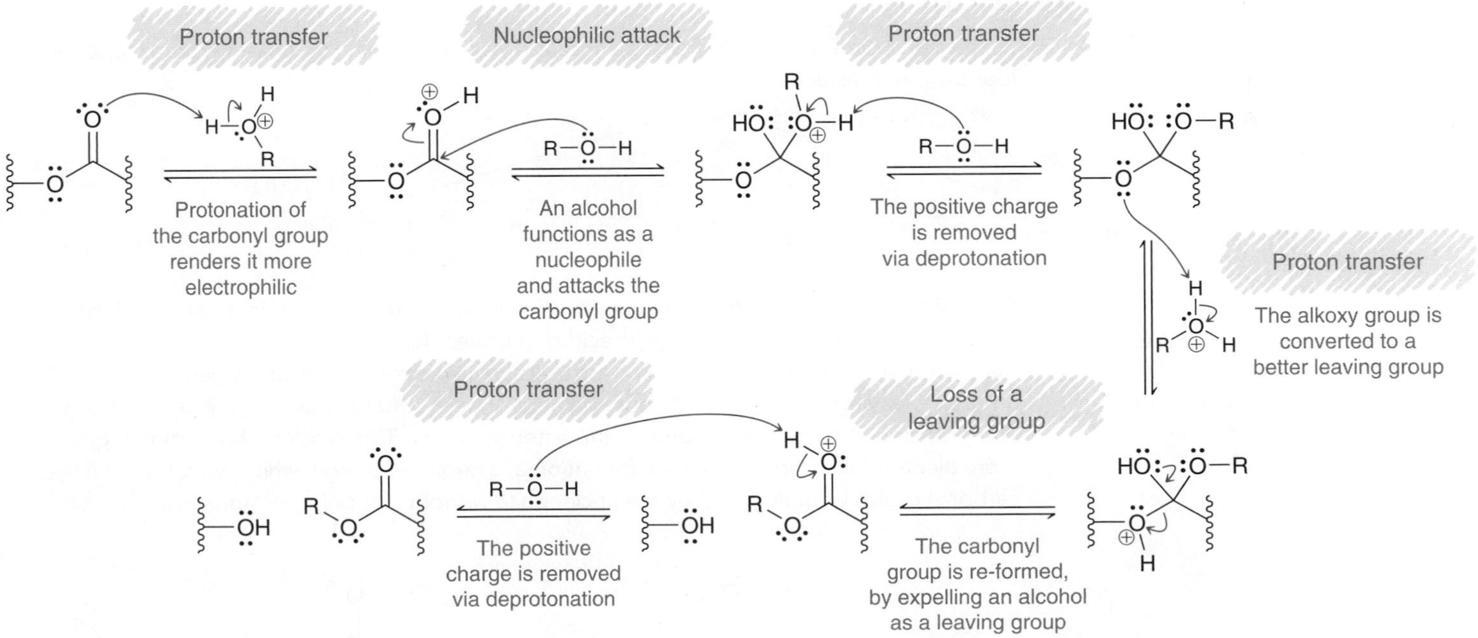
Proton transfer
Protonation of the carbonyl group renders it more electrophilic
Nucleophilic attack
R—O—H
An alcohol functions as a nucleophile and attacks the carbonyl group
Proton transfer
R—O—H
The positive charge is removed via deprotonation
Proton transfer
The alkoxy group is converted to a better leaving group
Loss of a leaving group
The carbonyl group is re-formed, by expelling an alcohol as a leaving group
Proton transfer
R—O—H
The positive charge is removed via deprotonation

LOOKING BACK
To review the function of these four proton transfers, see Section 21.7.

In the first step, the carbonyl group of the ester is protonated, which renders the carbonyl group more electrophilic. An alcohol, such as methanol, then functions as a nucleophile and attacks the protonated carbonyl group, giving a tetrahedral intermediate. After two proton transfer steps, the carbonyl group can re-form, followed by a final proton transfer step. This mechanism is identical to Mechanism 21.8 (Section 21.11). The mechanism consists of two core steps (nucleophilic attack and loss of a leaving group) and four proton transfers.

Transesterification makes it possible to convert triglycerides into biodiesel. Since biodiesel can be derived from plants (vegetable oil), it serves as a potential alternative to petroleum-based gasoline as a renewable source of energy. Unfortunately, the current cost associated with producing biodiesel outweighs the cost of producing an equivalent amount of gasoline, so it is unlikely that biodiesel will completely replace petroleum-based fuels in the near future. However, as the price of crude oil increases, biodiesel will become an attractive alternative.

SKILLBUILDER

26.3 DRAWING A MECHANISM FOR TRANSESTERIFICATION OF A TRIGLYCERIDE

LEARN the skill

Draw a mechanism for the transesterification of trilaurin using ethanol in the presence of an acid catalyst.

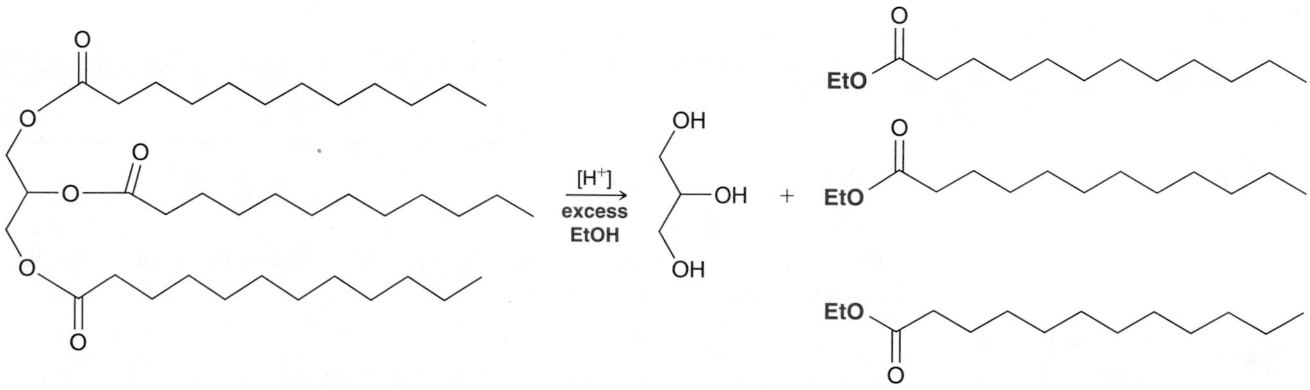

SOLUTION

Each ester moiety undergoes transesterification via a mechanism that has two core steps and four proton transfers.

Core steps

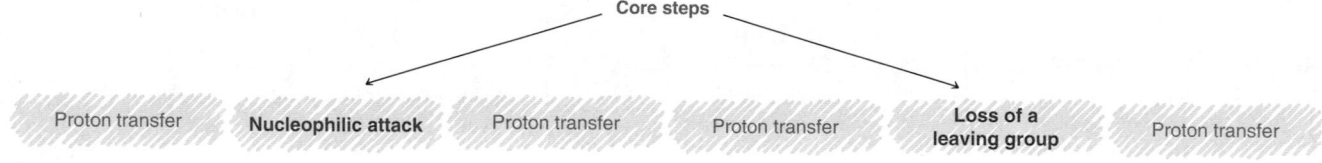

Proton transfer → **Nucleophilic attack** → Proton transfer → Proton transfer → **Loss of a leaving group** → Proton transfer

LOOKING BACK
For a review of how this rule is applied when drawing mechanisms, see Section 21.7.

Each of the four proton transfers serves a function, as we first saw in Chapter 21 when discussing the reactions of carboxylic acid derivatives. Recall that there is one guiding rule that determines when proton transfers are used: *In acidic conditions, all reagents, intermediates, and leaving groups either should be neutral or should bear only one positive charge.* The first step is to protonate one of the ester moieties. This renders the carbonyl group more electrophilic, and it avoids the formation of a negative charge, which would result if the carbonyl group were attacked by the nucleophile without first being protonated.

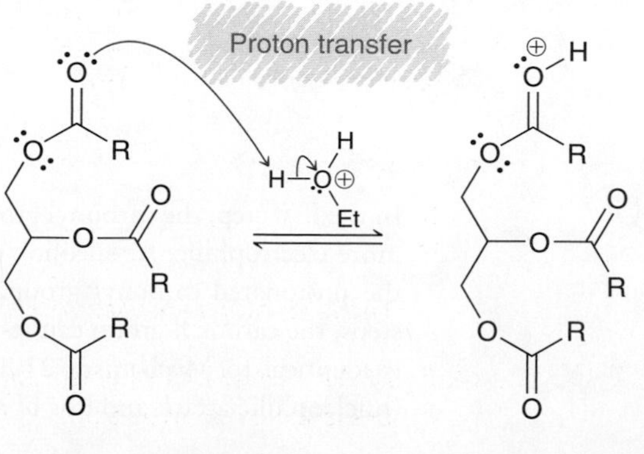

The second step is a nucleophilic attack. Ethanol functions as a nucleophile and attacks the protonated carbonyl group.

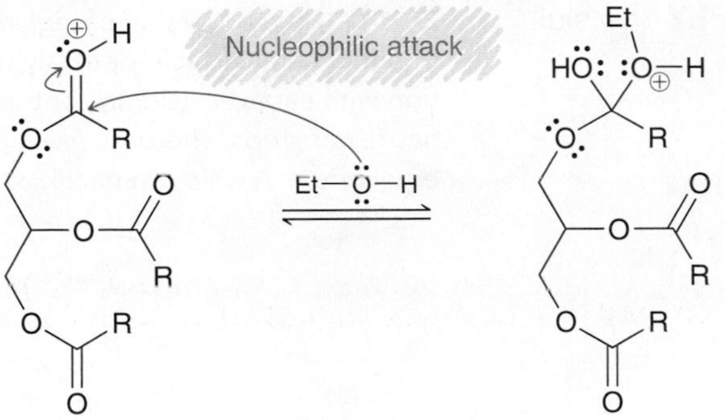

The next two steps are both proton transfers.

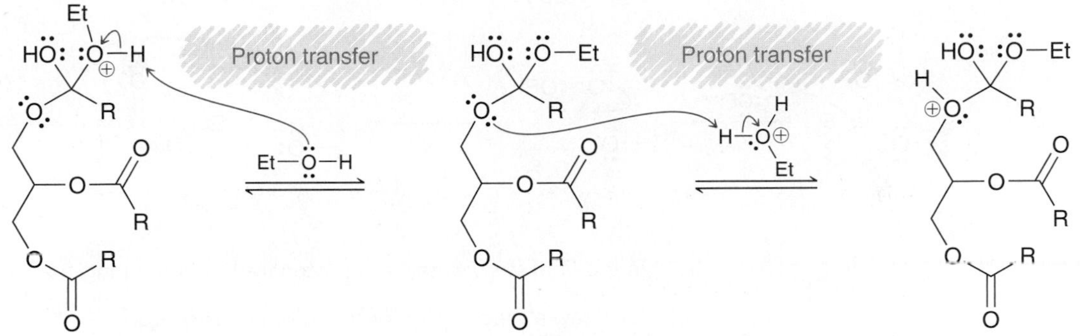

The first proton transfer removes the positive charge so as to avoid the formation of two positive charges that would result from the second proton transfer. The second proton transfer occurs to protonate the leaving group. This avoids the formation of a negative charge when the leaving group leaves in the next step.

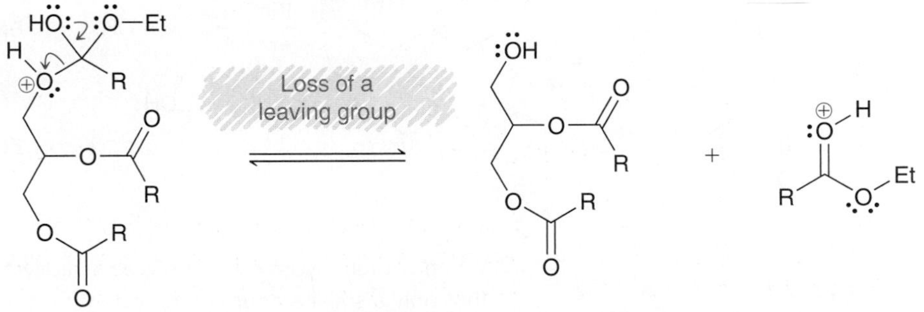

The final proton transfer removes the positive charge.

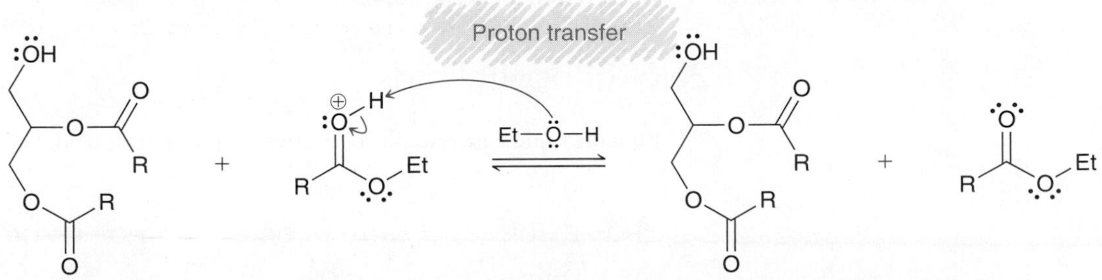

These six steps are then repeated for each of the other remaining fatty acid residues.

PRACTICE the skill **26.12** Draw a mechanism for the transesterification of tristearin using methanol in the presence of catalytic acid.

26.13 Draw the products obtained when triolein undergoes transesterification using isopropyl alcohol in the presence of catalytic sulfuric acid.

APPLY the skill

26.14 The conversion of triglycerides into biodiesel can be achieved in the presence of either catalytic acid or catalytic base. We have seen a mechanism for transesterification with catalytic acid. In contrast, the mechanism for base-catalyzed transesterification has fewer steps. The base, such as hydroxide, functions as a catalyst by establishing an equilibrium in which some alkoxide ions are present.

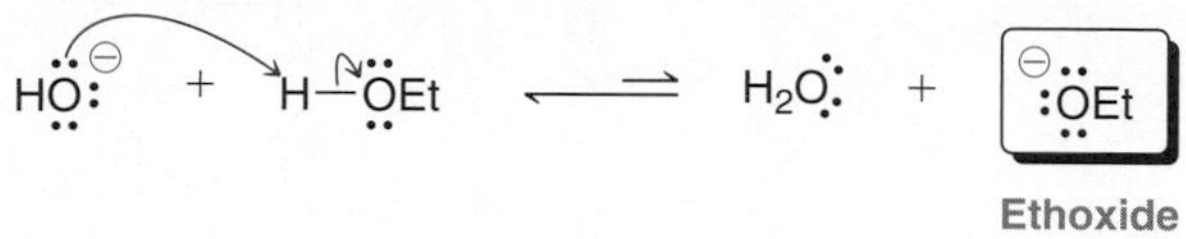

This equilibrium favors the hydroxide ions. Nevertheless, some ethoxide ions are present at equilibrium. These ethoxide ions are strong nucleophiles that can attack each ester moiety of the triglyceride according to the following mechanism.

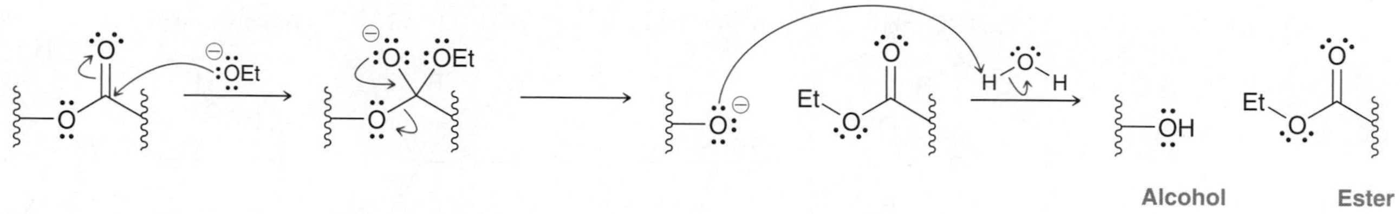

In the final step, water is deprotonated, regenerating the catalyst.

(a) Draw a mechanism for the following process.

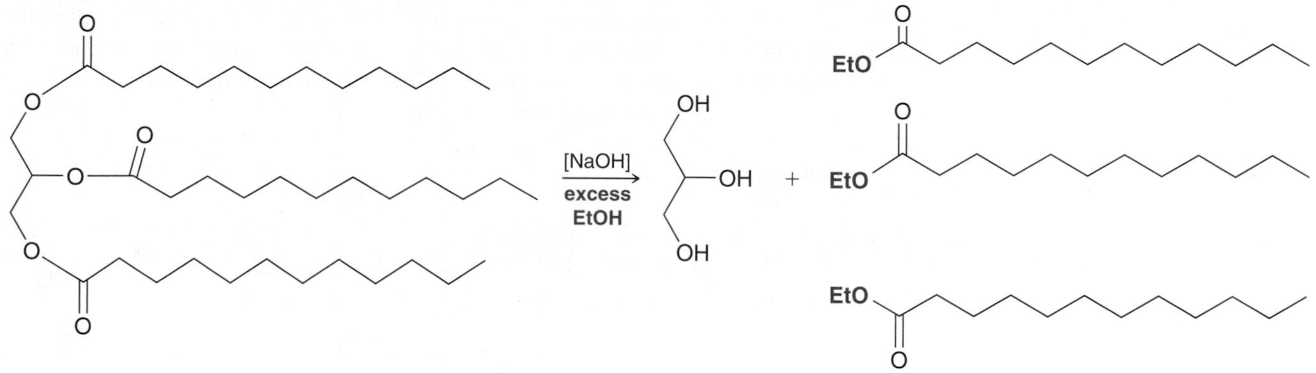

(b) When sodium hydroxide is used as a catalyst for transesterification, it is essential that only a small amount of the catalyst is present. Explain what would happen in the presence of too much sodium hydroxide.

need more **PRACTICE?** **Try Problems 26.44, 26.45**

26.5 Phospholipids

Phospholipids are esterlike derivatives of phosphoric acid.

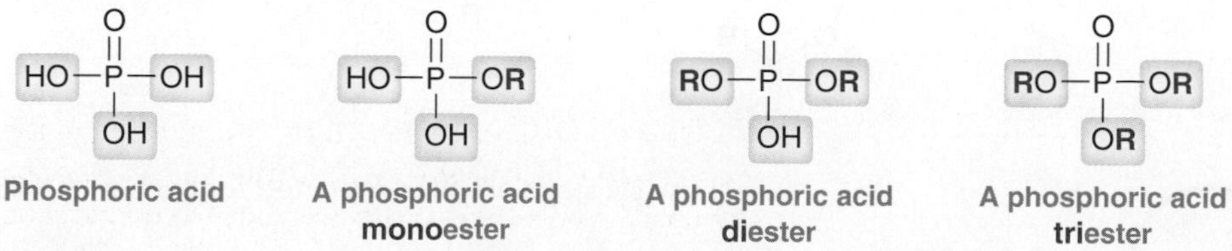

The most common phospholipids are phosphoglycerides.

Phosphoglycerides

Phosphoglycerides are very similar in structure to triglycerides, with the main difference being that in phosphoglycerides one of the three fatty acid residues is replaced by a phosphoester group. The simplest kind of phosphoglyceride is a phosphoric monoester called a **phosphatidic acid**.

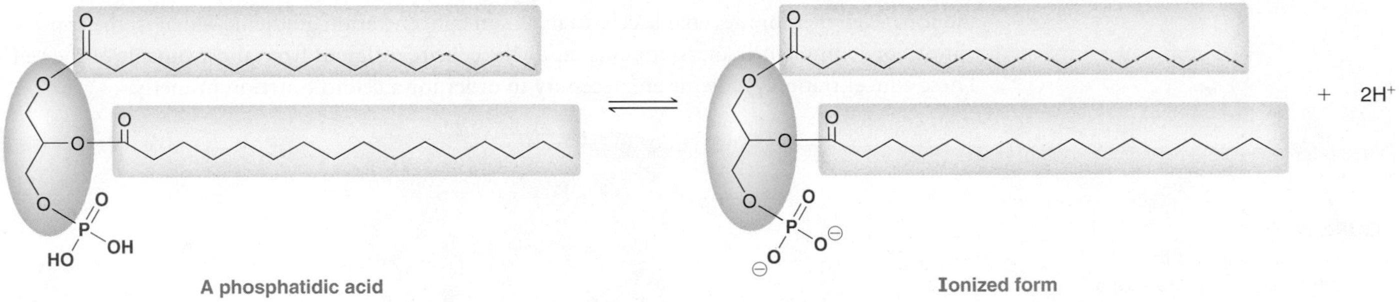

A phosphatidic acid **Ionized form**

At physiological pH, the ionized form of a phosphatidic acid predominates.

The most abundant phosphoglycerides are phosphoric acid diesters:

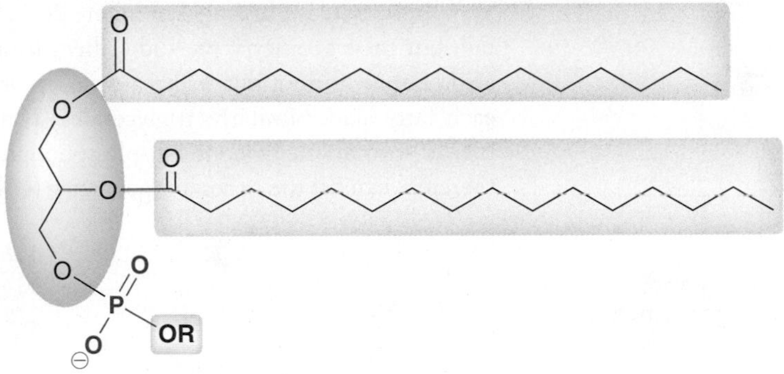

The identity of the alkoxy group can vary. Phosphoglycerides derived from ethanolamine and from choline are particularly abundant in the cells of plants and animals.

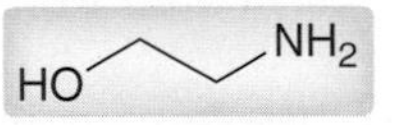

Ethanolamine

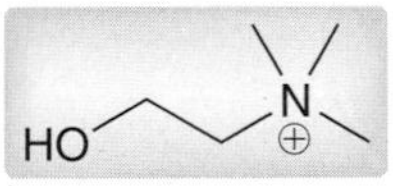

Choline

Phosphoglycerides that contain ethanolamine are called **cephalins**, while those that contain choline are called **lecithins.**

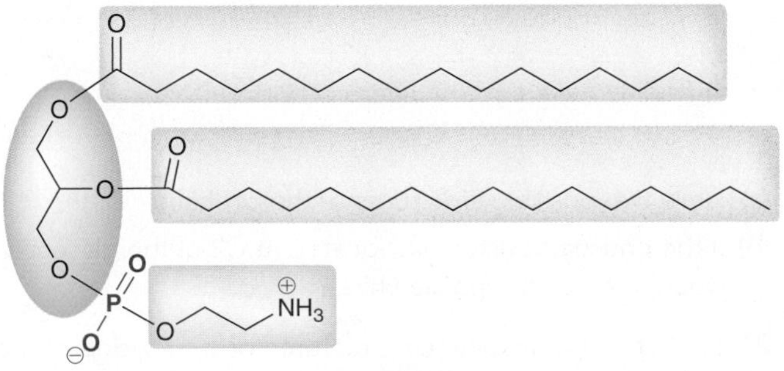

A cephalin

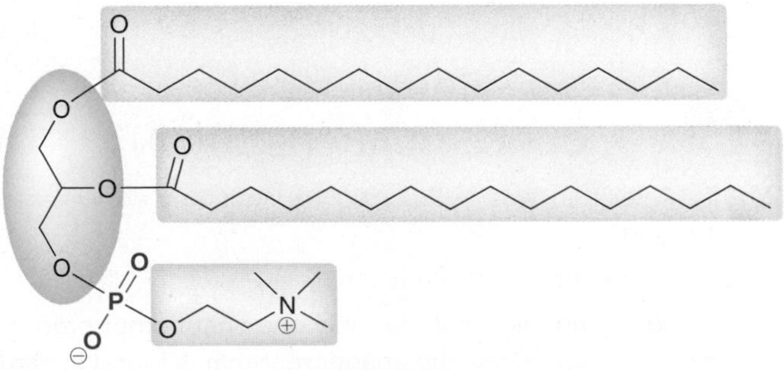

A lecithin

These compounds contain a chirality center (C2 of the glycerol unit) and generally exhibit the *R* configuration. Cephalins and lecithins have two nonpolar, hydrophobic tails and one polar head group. The polar head consists of the glycerol backbone as well as the phosphodiester, while the nonpolar tails are hydrocarbon chains. These features determine their function in cells, as will be described in the upcoming section.

Lipid Bilayers

In water, phosphoglycerides self-assemble into a **lipid bilayer** (Figure 26.5). In this way, the hydrophobic tails avoid contact with water and interact with each other via London dispersion forces. The surface of the bilayer is polar and therefore water soluble. Lipid bilayers constitute the main fabric of cell membranes, where they function as barriers that restrict the flow of water and ions. Cell membranes enable cells to maintain concentration gradients—that is, the concentrations of sodium and potassium ions inside the cell are different from those outside of the cell. These concentration gradients are necessary in order for a cell to function properly.

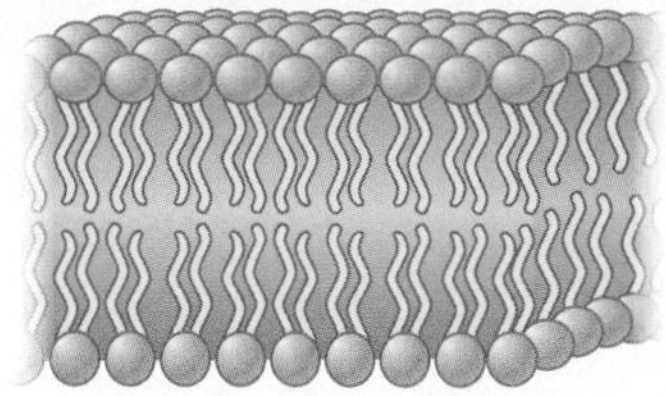

FIGURE 26.5
A graphic illustration of the three-dimensional structure of a lipid bilayer.

In order for compounds to self-assemble and form lipid bilayers, the compounds must have both a polar head and nonpolar tails. In addition, the three-dimensional shape of the compounds are also important, as illustrated in Figure 26.6. Fatty acids have a polar head group and a nonpolar tail, but their geometry precludes them from forming a bilayer. As seen in Figure 26.6a, fatty acids cannot form a bilayer because assembly of a bilayer would leave empty space in between each fatty acid. Similarly, triglycerides (Figure 26.6c) also lack the appropriate geometry for bilayer formation. In contrast, phospholipids (Figure 26.6b) have two hydrophobic tails and therefore exhibit the necessary geometry for bilayer formation.

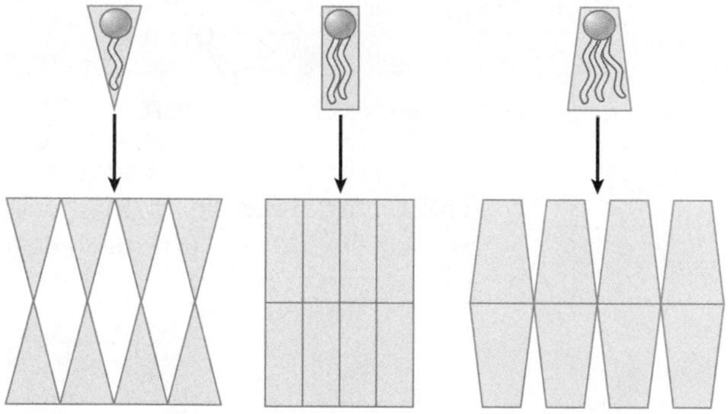

FIGURE 26.6
(a) The assembly of fatty acids in a lipid bilayer would leave empty space between the molecules, thereby destabilizing the potential bilayer.
(b) Phospholipids have just the right geometry to form lipid bilayers, with no empty space between the molecules.
(c) The assembly of triglycerides in a lipid bilayer would leave empty space between the molecules, thereby destabilizing the potential bilayer.

Many different phospholipids contribute to the fabric of the lipid bilayer, including cephalins and lecithins containing various fatty acid residues as well as a variety of other phospholipids.

CONCEPTUAL CHECKPOINT

26.15 A lecithin was hydrolyzed to yield two equivalents of myristic acid.

(a) Draw the structure of the lecithin.

(b) This compound is chiral, but only one enantiomer predominates in nature. Draw the enantiomer that is found in nature.

(c) The phosphodiester is generally located at C3 of the glycerol unit. If the phosphodiester was located at C2, would the compound be chiral?

26.16 A cephalin was hydrolyzed to yield one equivalent of palmitic acid and one equivalent of oleic acid.

(a) Draw two possible structures of the cephalin.

(b) If the phosphodiester was located at C2 of the glycerol unit, would the compound be chiral?

26.17 Draw the resonance structures of a fully deprotonated phosphatidic acid.

26.18 Octanol is more efficient than hexanol at crossing the cell membrane and entering a cell. Explain.

26.19 Would you expect glycerol to readily cross the membrane?

MEDICALLYSPEAKING)))

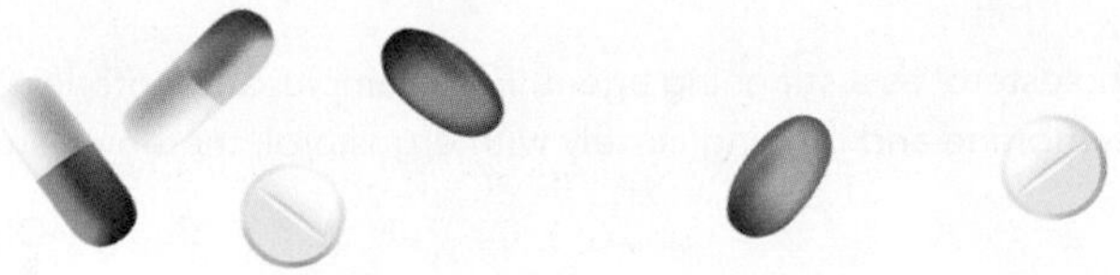

Selectivity of Antifungal Agents

Many different fungi are known to cause infections in human skin and nails. The conditions caused by these fungi are called *tinea*.

Type	Location
Tinea manuum	Hand
Tinea cruris	Groin
Tinea sycosis	Beard
Tinea capitis	Scalp
Tinea unguium	Nails

Treatment of these conditions can often be difficult because the cell membranes of fungi are nearly identical to the cell membranes of human cells, and many of the biochemical processes taking place at the cell membrane are also similar for both organisms. Therefore, any agents that interfere with the membrane integrity of fungi will generally also interfere with the membrane integrity of human cells, and agents that are toxic to fungi will also be toxic to humans. There are, however, small differences between the cell membranes of fungal and human cells. These differences have enabled medicinal chemists to develop agents that are toxic to fungal cells and less toxic to human cells. To understand how these agents work, we must focus on the nature of the hydrophobic region of the lipid bilayer.

The center of the lipid bilayer (the hydrophobic region) is highly fluid and resembles a liquid hydrocarbon.

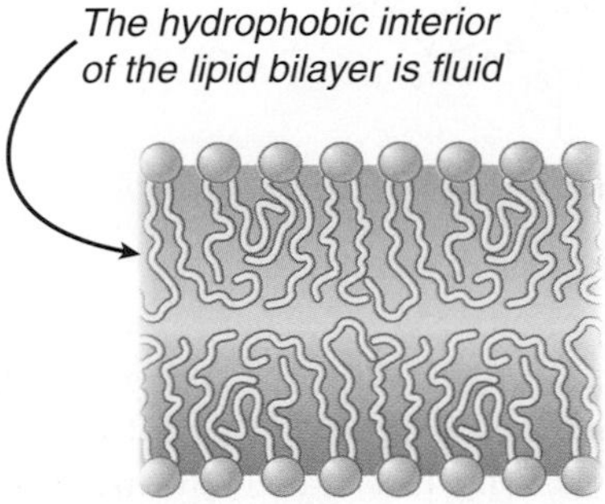

The hydrophobic tails are in rapid motion. This is an important feature, because it imparts flexibility to the lipid bilayer. However, if the bilayer is too fluid, then it becomes unstable. It would be incapable of holding its shape and carrying out its vital functions. In order to prevent the bilayer from becoming too fluid, it contains embedded stiffening agents. These agents are typically lipids with rigid geometry, and their effect is to limit the movement of the hydrocarbon tails.

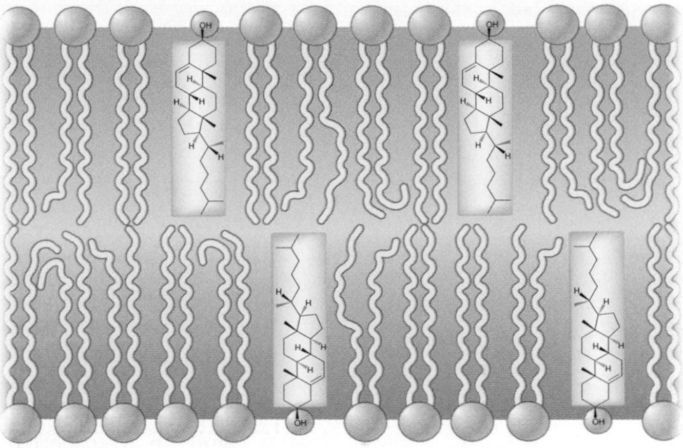

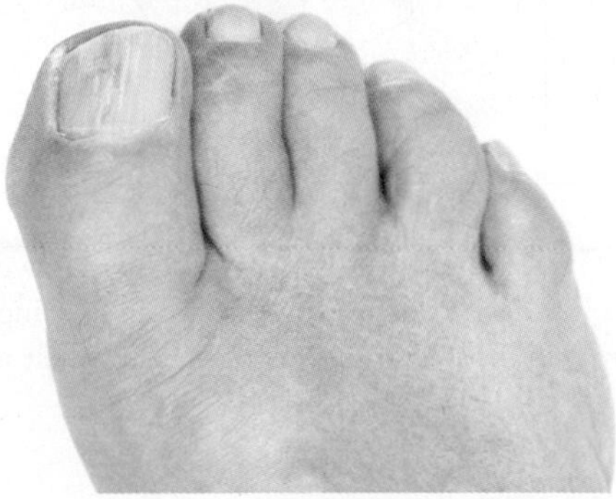

The hydrophobic region of the bilayer is still fluid, but motion is more limited, rendering the cell membrane more stiff. In human cells, the predominant stiffening agent is cholesterol, while in fungi, the predominant stiffening agent is ergosterol. Each of these compounds contains a small polar head and a large, rigid, nonpolar tail.

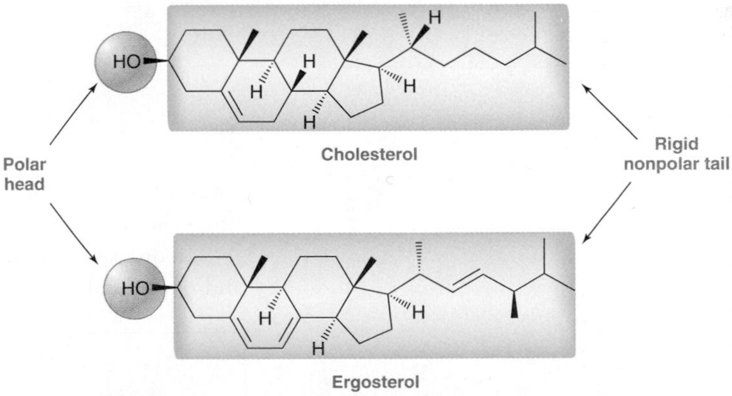

These stiffening agents are naturally occurring steroids, discussed in the upcoming section. Each of these compounds has an OH group that serves as a very small polar head group, while the rest of the compound is hydrophobic. Although they are very similar in structure, there are subtle differences. Specifically, ergosterol has two π bonds that are not present in cholesterol. These small differences provide a source of selectivity in treating fungal infections. Most antifungal drug research has focused on the development of agents that are capable of binding more effectively with ergosterol than cholesterol, thereby interfering with the function of ergosterol as a stiffening agent more so than they interfere with the function of cholesterol as a stiffening agent. For example, amphotericin B is a potent antifungal agent that is capable of penetrating the fungal cell membrane and binding closely with ergosterol, thereby disrupting the integrity of the fungal cell membranes.

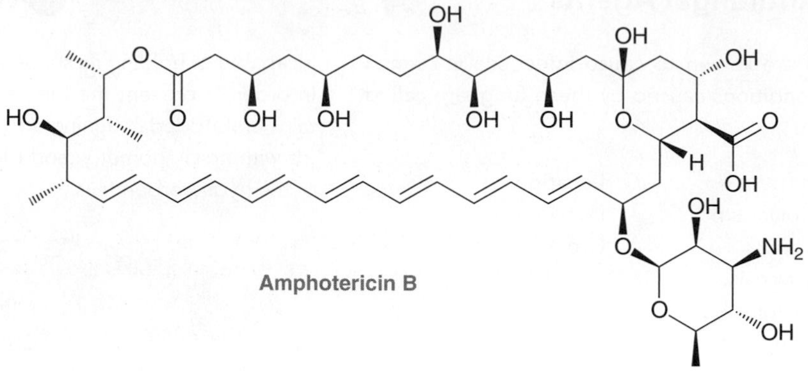

Amphotericin B

26.6 Steroids

Introduction to Steroids

Most steroids function as chemical messengers, or hormones, that are secreted by endocrine glands and transported through the bloodstream to their target organs. Steroids and their derivatives are also among the most widely used therapeutic agents. They are used in birth control and hormone replacement therapy and in the treatment of inflammatory conditions and cancer.

The structures of **steroids** are based on a tetracyclic ring system involving three six-membered rings and one five-membered ring.

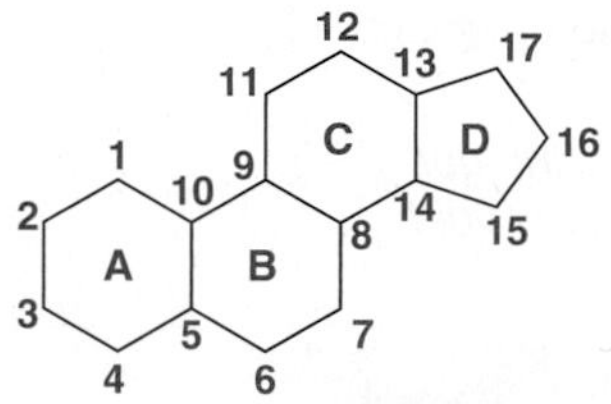

The tetracyclic skeleton of steroids

These rings are labeled using the letters A, B, C, and D, where the D ring is the five-membered ring. The carbon atoms in this system are numbered as shown.

In order to analyze the configuration of each ring fusion, recall the structures of *cis*-decalin and *trans*-decalin.

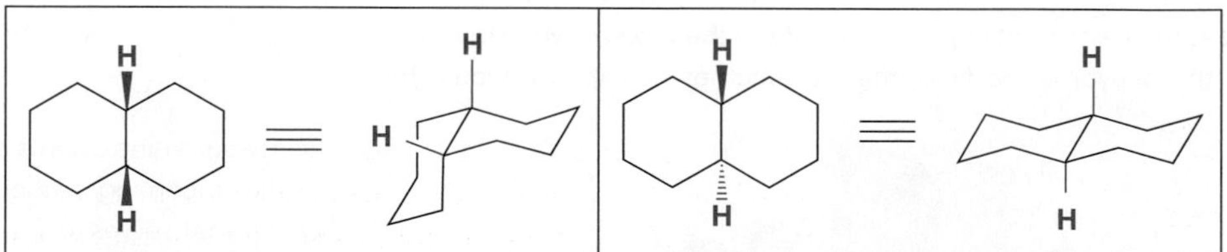

cis-Decalin

trans-Decalin

When six-membered rings are fused with a *cis* configuration, as in *cis*-decalin, the two rings are both free to exhibit ring flipping. In contrast, *trans*-decalin does not exhibit ring flipping and is a more rigid structure. The ring fusions are all *trans* in most steroids, giving steroids their rigid geometry.

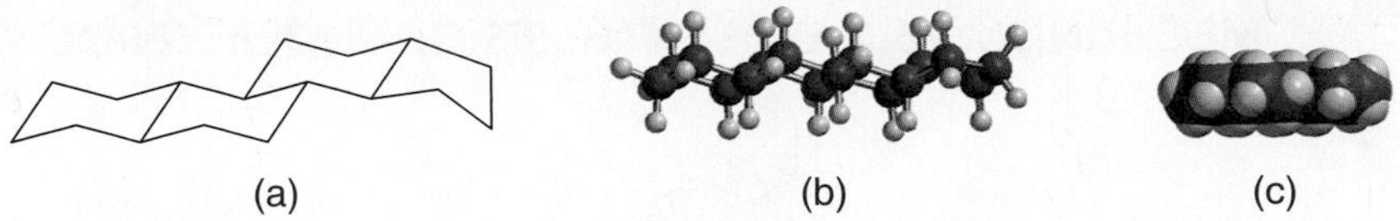

(a) (b) (c)

FIGURE 26.7
(a) The carbon skeleton of most steroids, (b) a ball-and-stick model of the carbon skeleton of a steroid, and (c) a space-filling model of the carbon skeleton of a steroid.

For some steroids, the A-B ring fusion is *cis*, although the B-C and C-D fusions are almost always *trans* in naturally occurring steroids. As a result, the carbon skeleton of most steroids provides for a fairly flat compound (Figure 26.7). The rigid geometry of cholesterol serves as an example.

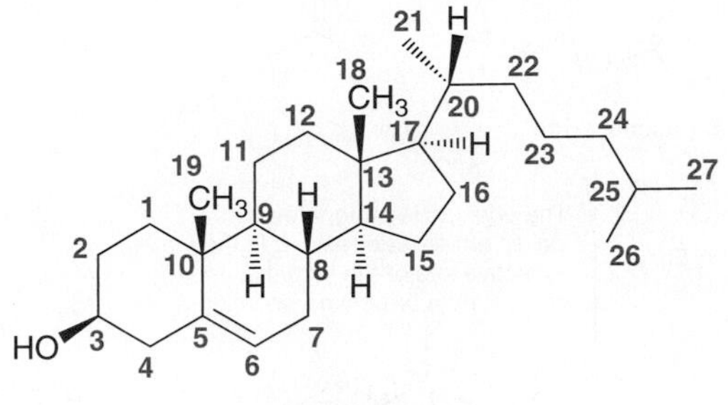

Cholesterol
(a steroid)

In the Medically Speaking box on antifungal drugs, we saw that the rigid structure of cholesterol enables the compound to function as a stiffening agent for lipid bilayers. Cholesterol contains two methyl groups (one attached to C10 and the other attached to C13) as well as a side chain attached to C17. This substitution pattern is common among many steroids. The two methyl groups occupy axial positions and are therefore perpendicular to the nearly planar skeleton, while the side chain occupies an equatorial position. This can be seen more clearly with three-dimensional representations of cholesterol (Figure 26.8). Cholesterol has eight chirality centers, giving rise to 2^8, or 256, possible stereoisomers. Nevertheless, only the one stereoisomer shown exists in nature.

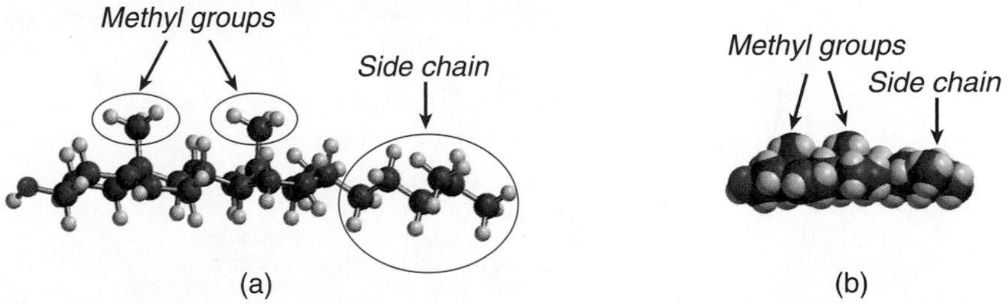

FIGURE 26.8
(a) A ball-and-stick representation of cholesterol which clearly shows that the two methyl groups occupy axial positions while the side chain occupies an equatorial position.
(b) A space-filling model of cholesterol which also shows the locations of the two methyl groups and the side chain.

Biosynthesis of Cholesterol

The biosynthesis of cholesterol involves many steps, some of which are shown in Mechanism 26.3. The starting material is squalene, which undergoes asymmetric epoxidation via enzymatic catalysis. The epoxide then opens to generate a carbocation, which undergoes a series of intramolecular cyclization reactions. All of these steps involve a π bond functioning as a nucleophile and attacking a carbocation in an intramolecular process. Each of these steps occurs within the cavity of an enzyme, where the configuration of each new chirality center is carefully controlled. Then, a series of rearrangements occurs that includes both methyl shifts and hydride shifts. Finally, deprotonation yields lanosterol, which is a precursor for cholesterol and all other steroids.

LOOKING BACK
For a discussion of the role of chiral catalysts in asymmetric epoxidation, see Section 14.9.

MECHANISM 26.3 BIOSYNTHESIS OF CHOLESTEROL

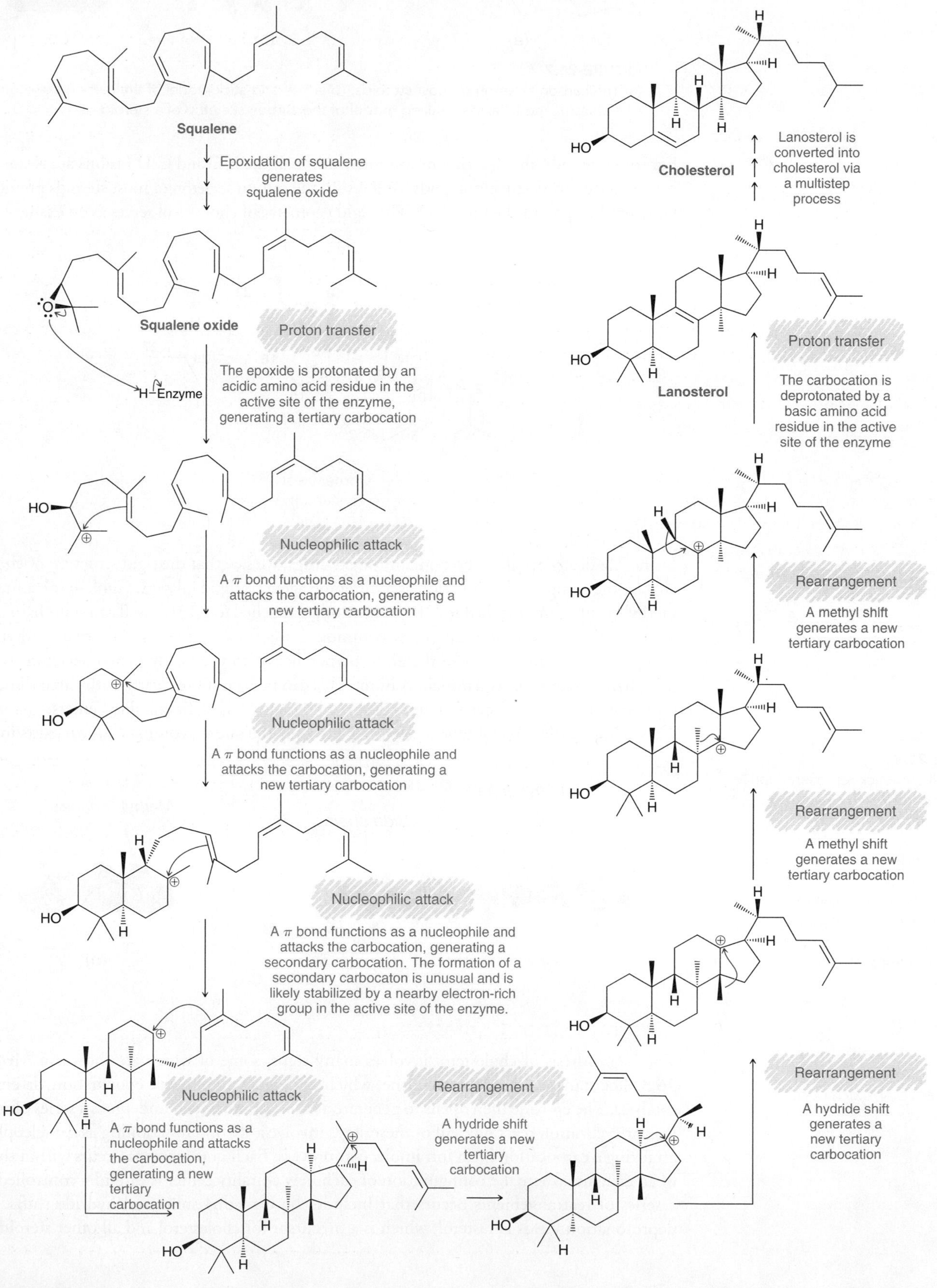
Squalene
Epoxidation of squalene generates squalene oxide
Squalene oxide
H–Enzyme
Proton transfer
The epoxide is protonated by an acidic amino acid residue in the active site of the enzyme, generating a tertiary carbocation
HO
Nucleophilic attack
A π bond functions as a nucleophile and attacks the carbocation, generating a new tertiary carbocation
Nucleophilic attack
A π bond functions as a nucleophile and attacks the carbocation, generating a new tertiary carbocation
Nucleophilic attack
A π bond functions as a nucleophile and attacks the carbocation, generating a secondary carbocation. The formation of a secondary carbocaton is unusual and is likely stabilized by a nearby electron-rich group in the active site of the enzyme.
Nucleophilic attack
A π bond functions as a nucleophile and attacks the carbocation, generating a new tertiary carbocation
Rearrangement
A hydride shift generates a new tertiary carbocation
Rearrangement
A hydride shift generates a new tertiary carbocation
Rearrangement
A methyl shift generates a new tertiary carbocation
Rearrangement
A methyl shift generates a new tertiary carbocation
Proton transfer
The carbocation is deprotonated by a basic amino acid residue in the active site of the enzyme
Lanosterol
Lanosterol is converted into cholesterol via a multistep process
Cholesterol

MEDICALLYSPEAKING)))

Cholesterol and Heart Disease

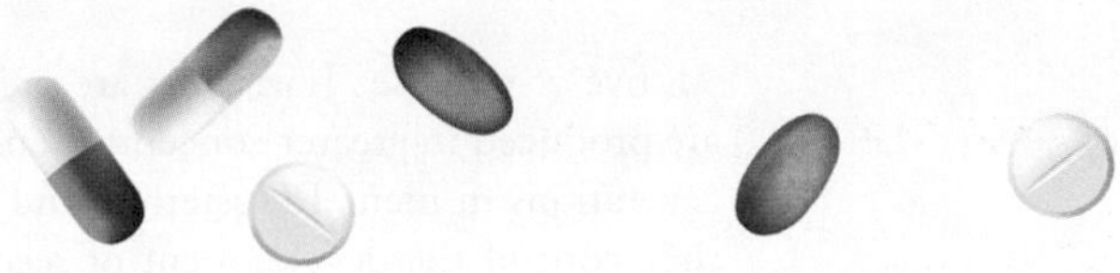

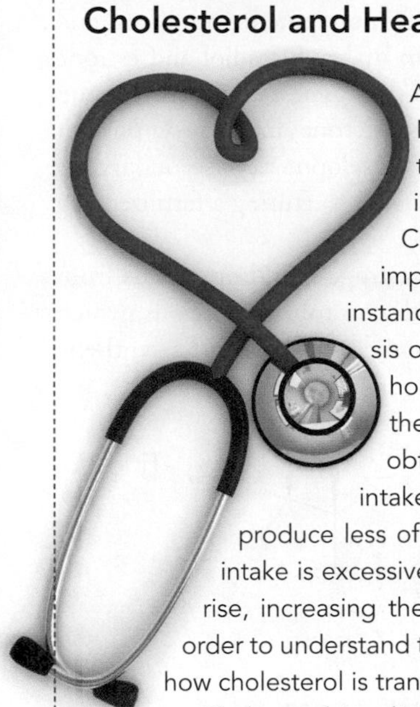

As described in the previous Medically Speaking box, cholesterol plays a vital role in maintaining the integrity of cell membranes. Cholesterol also has many other important biological functions. For instance, it is a precursor in the biosynthesis of most other steroids, including sex hormones. Our bodies can produce all the cholesterol we need, but we also obtain cholesterol in the diet. If dietary intake of cholesterol is high, our bodies produce less of it, in an attempt to compensate. If intake is excessively large, then cholesterol levels can rise, increasing the risk of heart attack and stroke. In order to understand the reason for this, we must consider how cholesterol is transported throughout the body.

Cholesterol is a lipid, which means that it is not water soluble. As a result, cholesterol cannot, by itself, be dissolved in the aqueous medium of the blood. Cholesterol is produced in the liver and then transported in the blood by large particles called lipoproteins. Lipoproteins consist of many proteins and lipids bound together via intermolecular interactions. Cholesterol and ester derivatives of cholesterol are contained in the center (the lipophilic region). Lipoproteins are soluble in the aqueous environment of the blood and are used by the body to transport lipids, such as cholesterol, throughout the body.

Lipoproteins are categorized into several different categories based on their density. High-density lipoproteins, or HDLs, have a higher proportion of proteins, while low-density lipoproteins, or LDLs, have a higher proportion of lipids. It is now understood that HDLs and LDLs have different functions. LDLs transport cholesterol from the liver to cells throughout the body, while HDLs transport cholesterol for the reverse journey, back to the liver, where they are used as precursors for the synthesis of other steroids. It is important that the concentration of HDLs is greater than the concentration of LDLs. If the LDL concentration is too high, then some of the LDLs will be unable to unload their cargo, and instead, they accumulate and form deposits in the arteries. These deposits restrict blood flow, which can lead to a heart attack (caused by an inadequate flow of blood to the heart) or a stroke (caused by an inadequate flow of blood to the brain).

A complete physical examination measures the relative concentrations of HDL cholesterol and LDL cholesterol. Low levels of HDL cholesterol indicate a greater risk for heart attack and stroke. In addition, recent studies also indicate that low levels of HDL cholesterol increase the risk of memory loss associated with Alzheimer's disease.

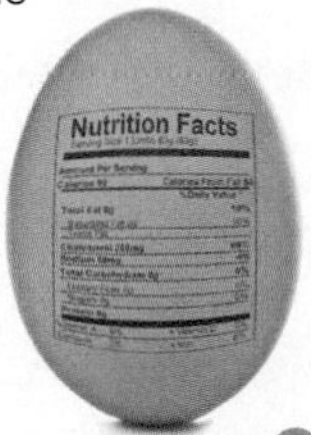

Sex Hormones

Human sex hormones are steroids that regulate tissue growth and reproductive processes. Male sex hormones are called **androgens**, and there are two types of female sex hormones called **estrogens** and **progestins**. Table 26.3 shows the most important male and female sex hormones.

TABLE 26.3 THE MOST IMPORTANT MALE AND FEMALE SEX HORMONES

Male sex hormones

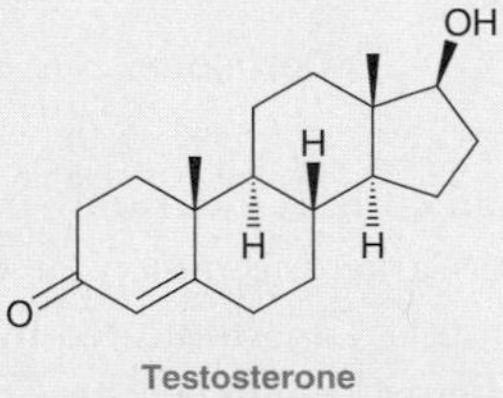

Testosterone

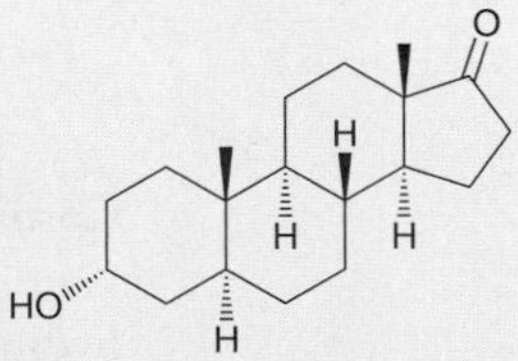

Androsterone

Androgens

Female sex hormones

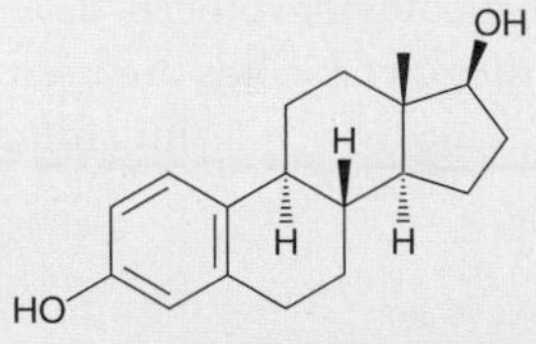

Estradiol

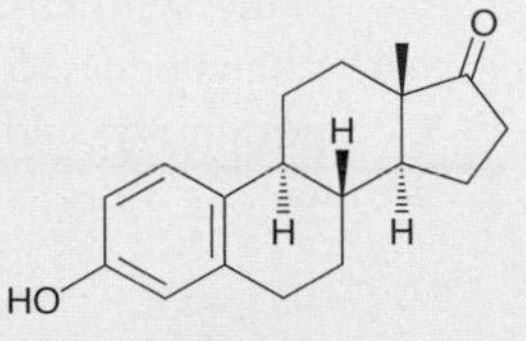

Estrone

Estrogens

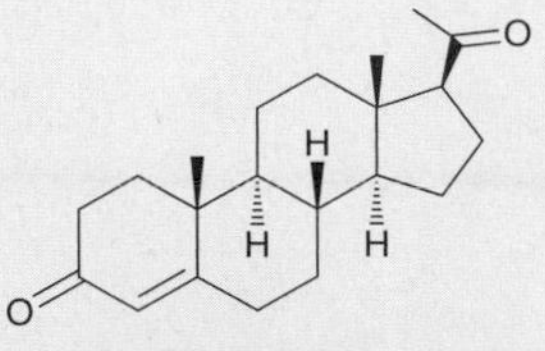

Progesterone

A progestin

All five of these sex hormones are present in both males and females. Estrogens and progestins are produced in greater concentrations in women, while androgens are produced in greater concentrations in men. Testosterone and androsterone are among the more potent androgens, and they control the development of secondary sex characteristics in males. Estradiol and estrone are estrogens, which are characterized by an aromatic A ring and the absence of a methyl group at C10. These hormones are produced in the ovaries from testosterone and play important roles in regulating a woman's menstrual cycle and controlling the development of secondary sex characteristics. Progesterone is a progestin that prepares the uterus for nurturing a fertilized egg during pregnancy.

During pregnancy, ovulation is inhibited by the release of estrogens and progestins from the placenta and ovaries. This process is mimicked by most birth control formulations, which generally contain a mixture of a synthetic estrogen (such as ethynyl estradiol) and a synthetic progestin (such as norethindrone).

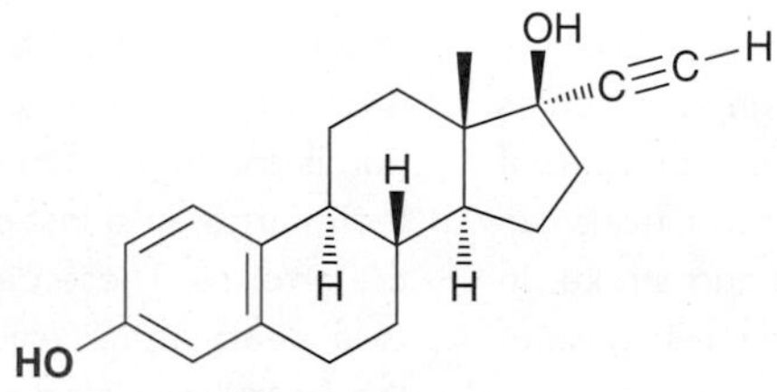

Ethynyl estradiol

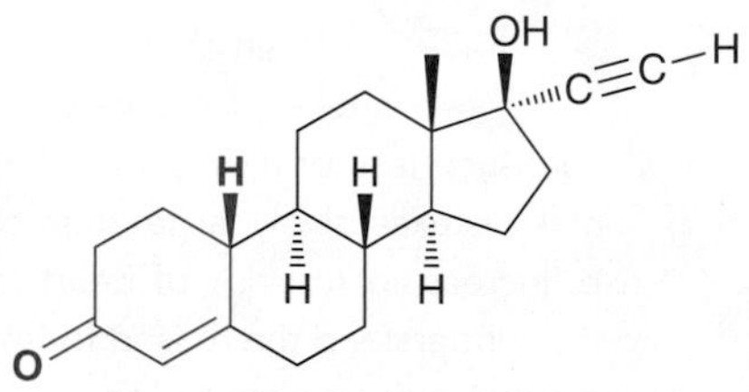

Norethindrone

This mixture of compounds inhibits ovulation in much the same way that the body naturally inhibits ovulation during pregnancy.

Adrenocortical Hormones

Adrenocortical hormones are thus named because they are secreted by the cortex (the outer layer) of the adrenal glands. Adrenocortical hormones are typically characterized by a carbonyl group or hydroxyl group at C11. Examples include cortisone and cortisol.

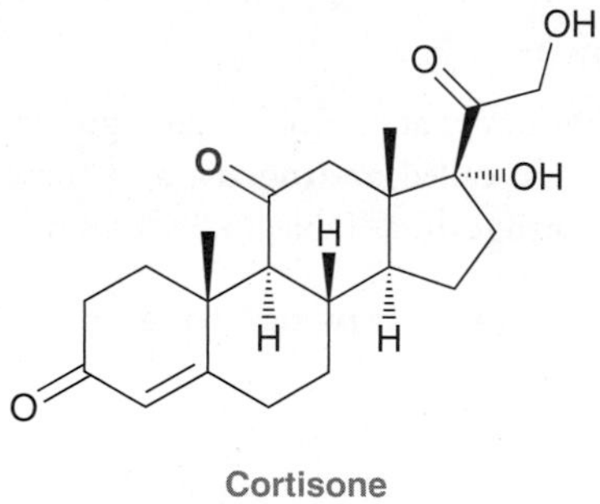

Cortisone

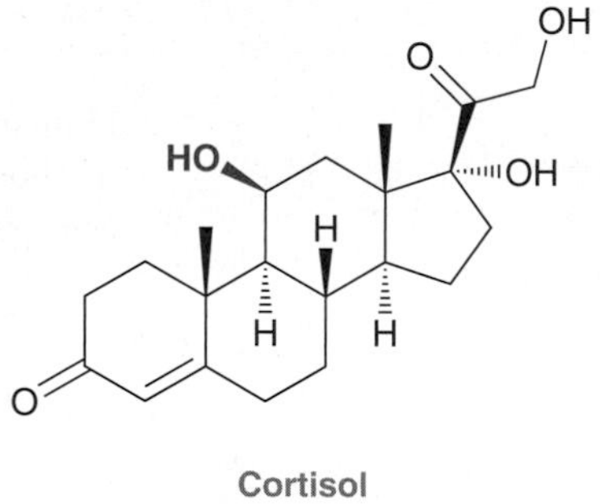

Cortisol

Cortisone and cortisol differ only in the identity of one functional group (highlighted in red). Cortisone has a ketone moiety at C11 (indicated by the "-one" suffix), while cortisol has a hydroxy group at C11 (indicated by the "-ol"suffix). Cortisol is more abundant in nature, but cortisone is better known because of its therapeutic use. Both agents are used to treat the effects of inflammatory diseases, including psoriasis (inflammation of the skin), arthritis (inflammation of the joints), and asthma (inflammation of the lungs).

Many synthetic corticoids have been developed that are even more potent than their natural analogues. Synthetic corticoids are used in the treatment of rashes caused by poison oak, poison ivy, and eczema as well as inflammatory diseases such as psoriasis, arthritis, and asthma.

MEDICALLYSPEAKING

Anabolic Steroids and Competitive Sports

Anabolic steroids are compounds that promote muscle growth by mimicking the tissue-building effect of testosterone. Many synthetic androgen analogues have been created that are even more potent than testosterone, including stanozolol, nandrolone, and methandrostenolone.

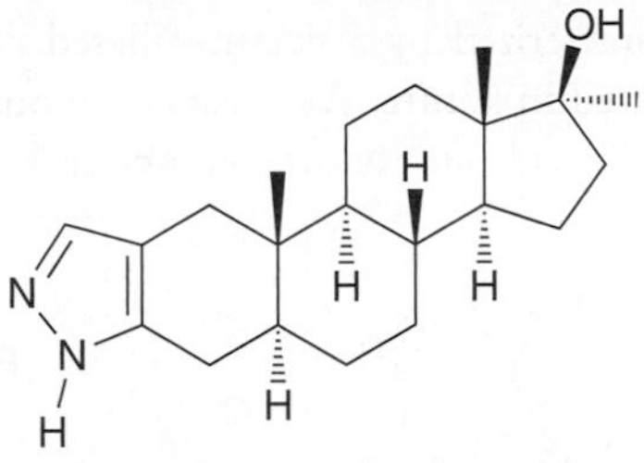

Stanozolol

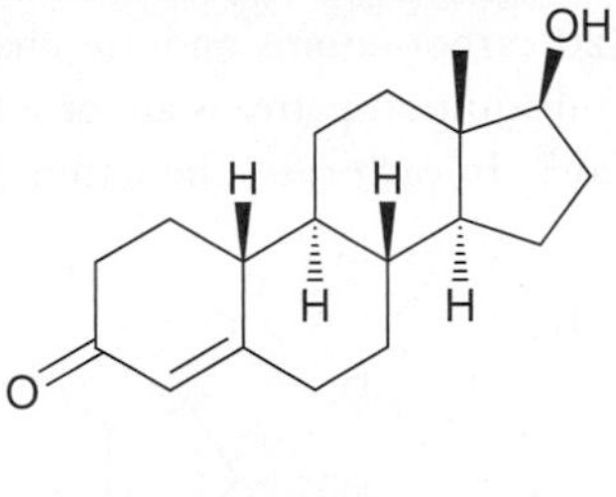

Nandrolone

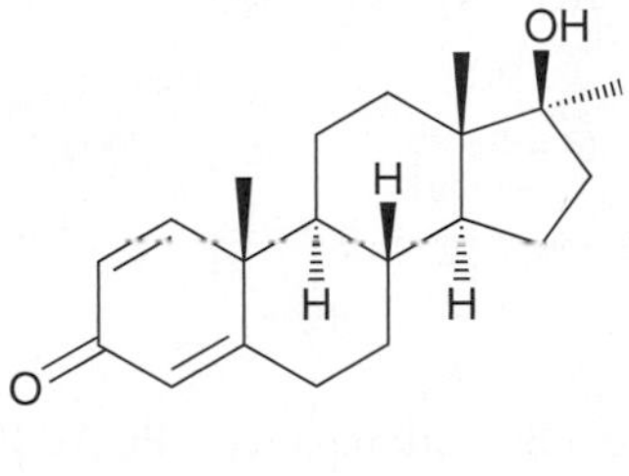

Methandrostenolone

Synthetic anabolic steroids were originally developed in the 1930s for use in treating illnesses and injuries that involved muscle deterioration. Unfortunately, the development of these drugs led to rampant abuse by athletes and bodybuilders, even as early as the 1940s. The problem received wide public attention during the 1988 Olympics when Ben Johnson (Canada) was disqualified as the gold medal winner of the 100-meter dash because his urine tested positive for traces of stanozolol.

Steroid abuse is still a major problem, particularly among bodybuilders and professional baseball and football players. Many studies have been conducted to determine whether there is any correlation between steroid use and increased athletic performance. Some studies have shown a small connection, while other studies show no connection whatsoever. It appears that the health risks associated with steroid use outweigh the uncertain benefits. The health risks include increased risk of heart disease, stroke, liver cancer, sterility, as well as adverse behavioral changes caused by increased aggressive tendencies (called "steroid rage").

CONCEPTUAL CHECKPOINT

26.20 The following compounds are steroids. One is an anabolic steroid called oxymetholone and the other, called norgestrel, is used in oral contraceptive formulations. Identify these compounds based on their structural features.

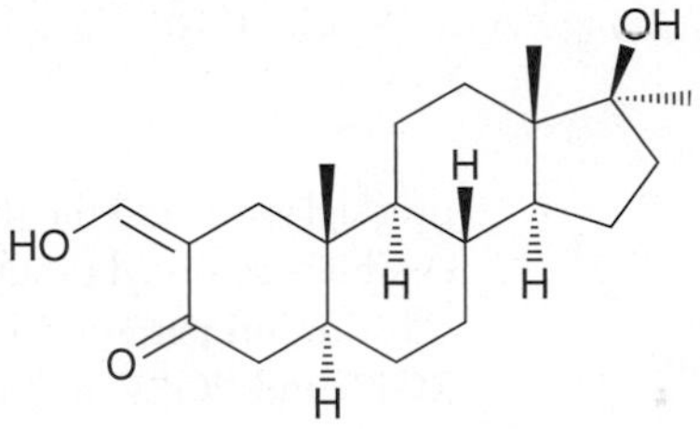

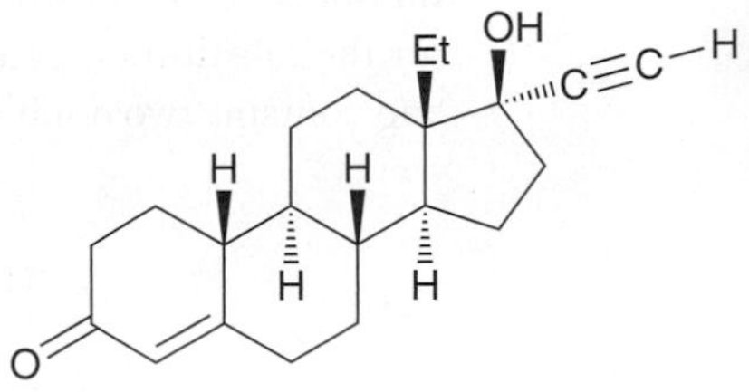

CONCEPTUAL CHECKPOINT

26.21 Draw the hypothetical ring-flip of *trans*-decalin, and explain why it does not occur. Use this analysis to explain why cholesterol has a fairly rigid three-dimensional geometry.

26.22 Draw chair conformations for each of the following compounds, and then identify whether each substituent is axial or equatorial:

(a) 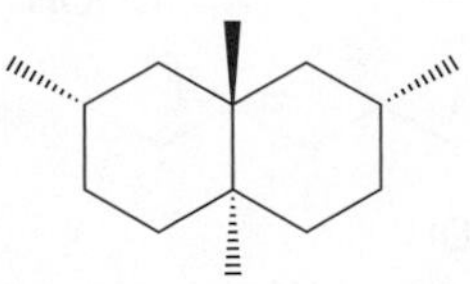

(b) 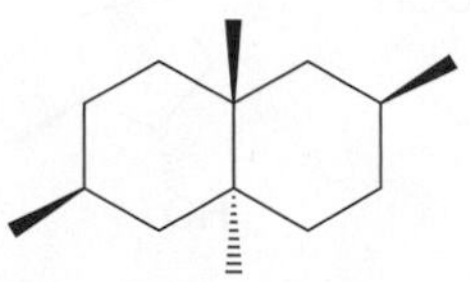

(c) 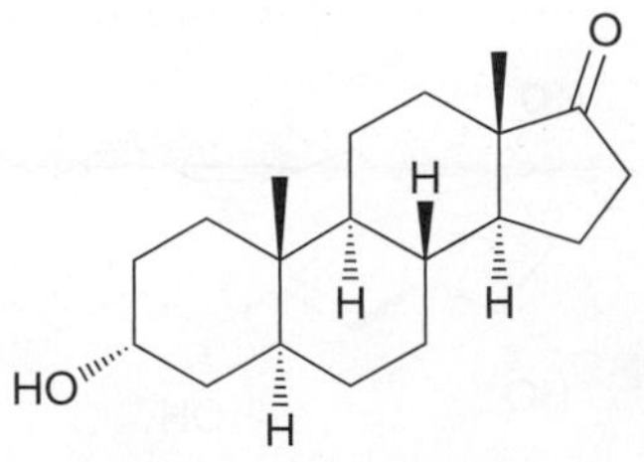

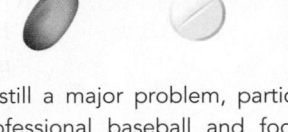

26.23 Prednisolone acetate is an anti-inflammatory agent in clinical use. It is similar in structure to cortisol, with the following two differences: (1) Prednisolone acetate exhibits a double bond between C1 and C2 of the A ring and (2) the primary hydroxyl group has been acetylated. Using this information, draw the structure of prednisolone acetate.

26.7 Prostaglandins

In the early 1930s, it was observed that human seminal fluid could induce muscle contractions in uterine tissue. It was believed that this phenomenon was caused by an acidic substance produced in the prostate gland. This unknown substance was called **prostaglandin** by Ulf von Euler (Karolina Institute in Sweden). In the 1950s, the acidic extract from sheep prostate glands was found to contain not one, but many structurally related prostaglandin substances. These prostaglandin compounds were separated, purified, and characterized and are now known to be present in all body tissues and fluid in very small concentrations. Despite our current understanding of the ubiquitous nature of these compounds, the original name is still used.

Prostaglandins contain 20 carbon atoms and are characterized by a five-membered ring with two side chains. Many substitution patterns are observed in nature, the most common of which are shown in Figure 26.9. In each case, the letters PG indicate that the compound is a

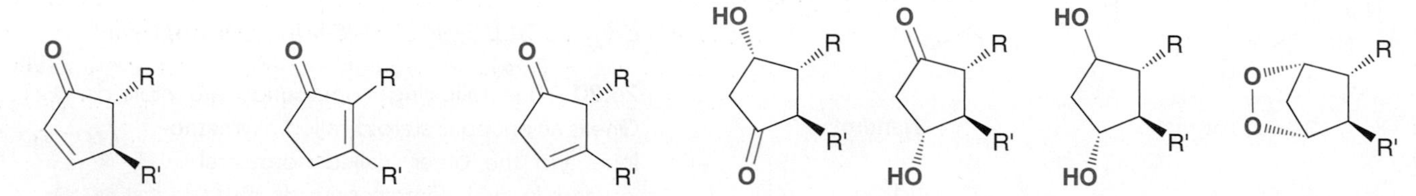

PGA	PGB	PGC	PGD	PGE	PGF	PGG
(an α,β-unsaturated ketone)	(an α,β-unsaturated ketone)	(a β,γ-unsaturated ketone)	(a β-hydroxy ketone)	(a β-hydroxy ketone)	(a 1,3-diol)	(an endoperoxide)

FIGURE 26.9 Common substitution patterns for prostaglandins.

prostaglandin and the third letter indicates the substitution pattern. PGAs, PGBs, and PGCs all exhibit a carbonyl group and a carbon-carbon π bond in the five-membered ring. These three substitution patterns differ from each other only in the location of the carbon-carbon π bond. PGDs and PGEs are β-hydroxyketones, PGFs are 1,3-diols, and PGGs are endoperoxides. The number of carbon-carbon π bonds in the side chains is indicated with a subscript after the letter for the substitution pattern. For example, PGE_2 (dinoprostone) has the E substitution pattern and contains two double bonds in the side chains.

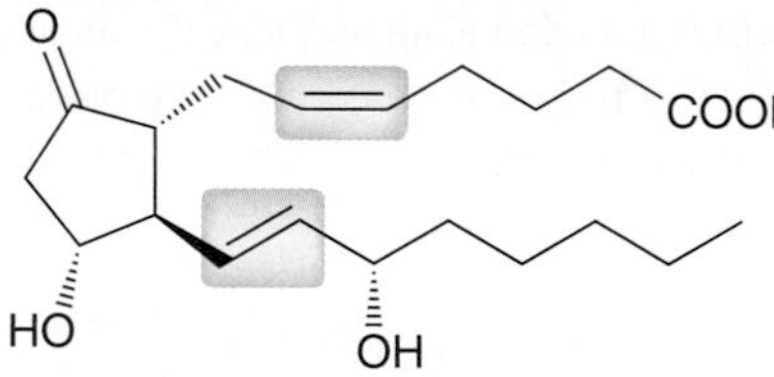

PGE_2

This particular prostaglandin regulates muscle contractions during labor and can be administered in larger doses to terminate pregnancies.

For the PGF substitution pattern, an additional descriptor is added to the name to indicate the configuration of the OH groups. A *cis* diol is designated as "α," while a *trans* diol is designated as "β".

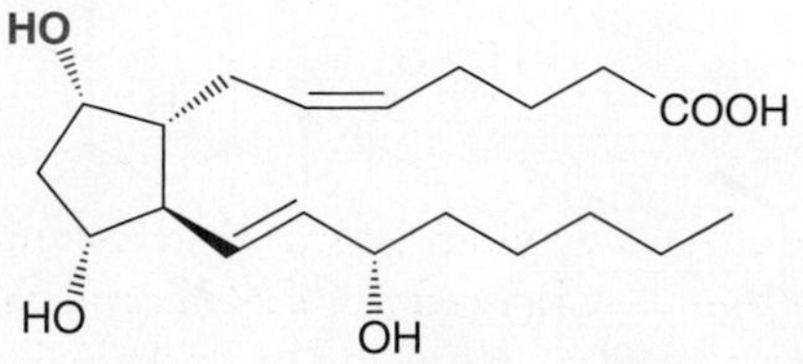

$PGF_{2\alpha}$

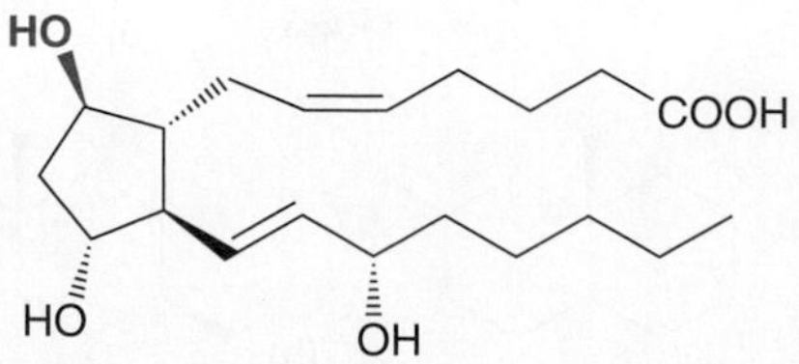

$PGF_{2\beta}$

CONCEPTUAL CHECKPOINT

26.24 Classify each prostaglandin according to the instructions provided in Section 26.7.

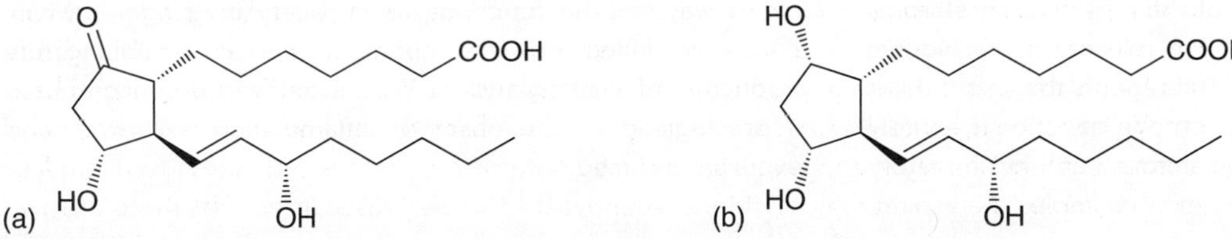

O
COOH
(a)
HO
OH
HO
COOH
(b)
HO
OH

Prostaglandins are biochemical regulators that are even more powerful than steroids. Unlike hormones, which are produced in one location and then transported to another location in the body, prostaglandins are called local mediators because they perform their function where they are synthesized. Prostaglandins exhibit a wide array of biological activity, including the regulation of blood pressure, blood clotting, gastric secretions, inflammation, kidney function, and reproductive systems. Prostaglandins are biosynthesized from arachidonic acid with the help of enzymes called cyclooxygenases. A few key steps of this process are outlined in Mechanism 26.4.

MECHANISM 26.4 BIOSYNTHESIS OF PROSTAGLANDINS FROM ARACHIDONIC ACID

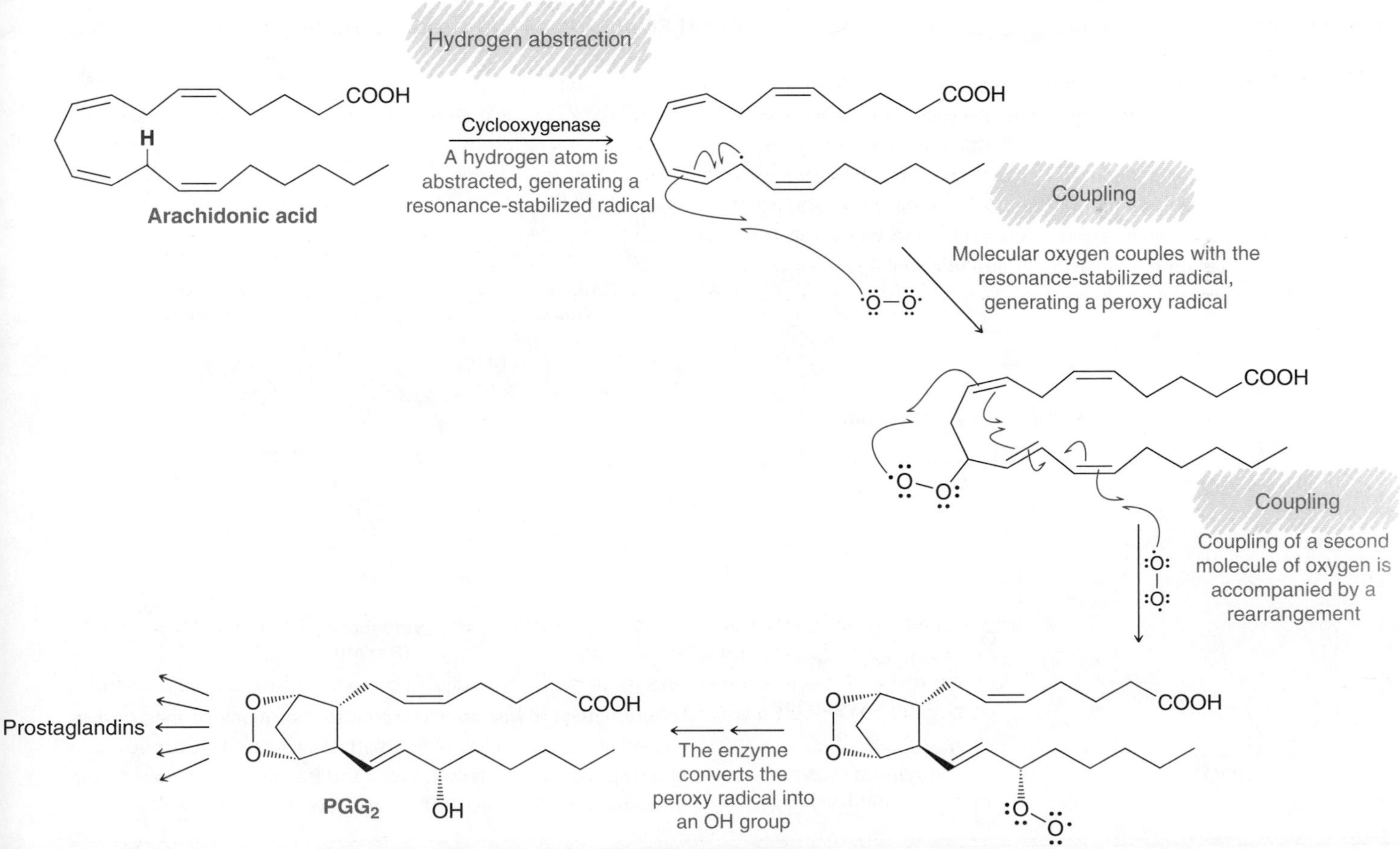

Hydrogen abstraction
COOH
H
Arachidonic acid
Cyclooxygenase
A hydrogen atom is abstracted, generating a resonance-stabilized radical
COOH
Coupling
Molecular oxygen couples with the resonance-stabilized radical, generating a peroxy radical
COOH
Coupling
Coupling of a second molecule of oxygen is accompanied by a rearrangement
COOH
The enzyme converts the peroxy radical into an OH group
COOH
Prostaglandins
PGG2
OH

The release of arachidonic acid is stimulated in response to trauma (tissue damage). It is believed that the anti-inflammatory effects of adrenocortical steroids derive from their ability to suppress the enzymes that cause the release of arachidonic acid, thereby preventing the biosynthesis of prostaglandins.

MEDICALLYSPEAKING

NSAIDs and COX-2 Inhibitors

As previously mentioned, the action of anti-inflammatory steroids stems from their ability to prevent the release of arachidonic acid. There is another class of therapeutic agents that also exhibit anti-inflammatory properties, but their mode of action is entirely different. These drugs are called non steroidal anti-inflammatory drugs, or NSAIDs, and the most common examples are aspirin, ibuprofen, and naproxen.

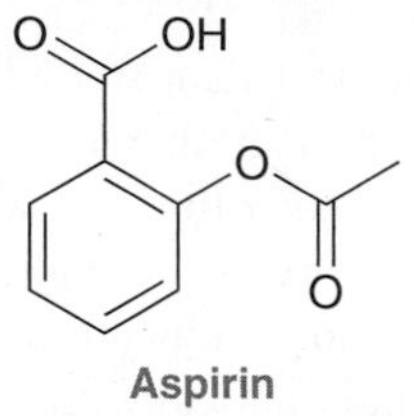

Aspirin

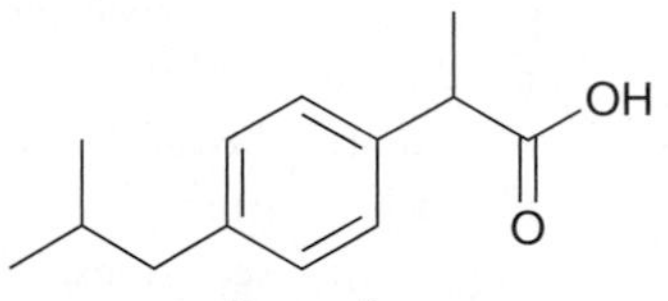

Ibuprofen

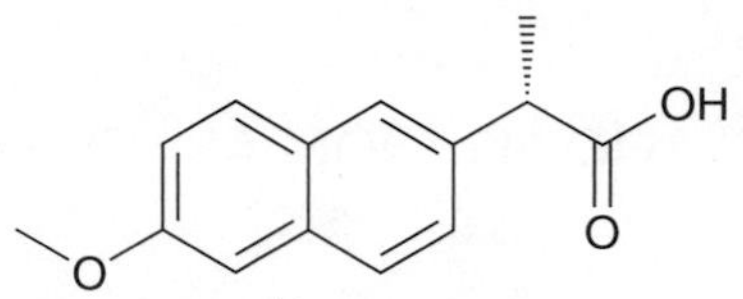

Naproxen

NSAIDs do not inhibit the enzymes that cause the release of arachidonic acid, but rather, they inhibit the cyclooxygenase enzymes that catalyze the conversion of arachidonic acid into prostaglandins. Ibuprofen and naproxen deactivate the cyclooxygenase enzymes by binding to them (thereby preventing the binding of arachidonic acid), while aspirin deactivates the cyclooxygenase enzymes by transferring an acetyl group to a serine residue within the active site of the enzymes.

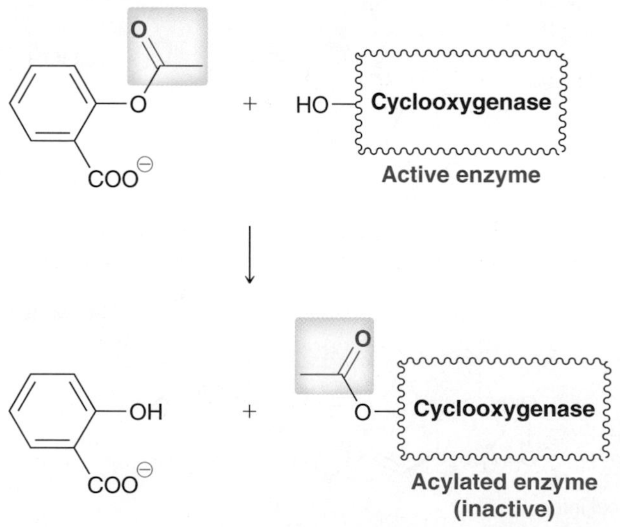

In this way, aspirin functions as an acetylating agent, which effectively deactivates the enzymes, thereby inhibiting the production of prostaglandins. With a decreased concentration of prostaglandins, the onset of inflammation is slowed, and fevers are reduced.

More recently, it has been discovered that there are two different kinds of cyclooxygenase enzymes, called COX-1 and COX-2. The primary function of COX-2 is to catalyze the synthesis of prostaglandins that cause inflammation and pain, while the primary function of COX-1 is to catalyze the synthesis of prostaglandins that help protect the stomach. NSAIDs inhibit the action of both COX-1 and COX-2 enzymes, and it has been discovered that the inhibition of COX-1 can result in gastric irritation. This understanding instigated an intensive search for therapeutic agents capable of selectively inhibiting COX-2 without also inhibiting COX-1. Extensive research efforts culminated in the release of several COX-2 inhibitors on the market in the late 1990s.

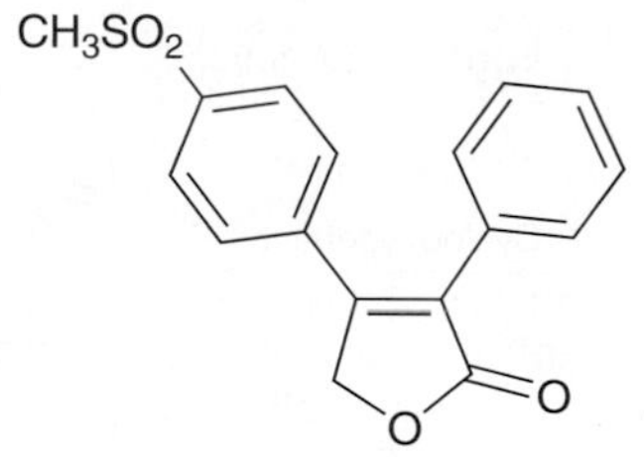

Rofecoxib (Vioxx)

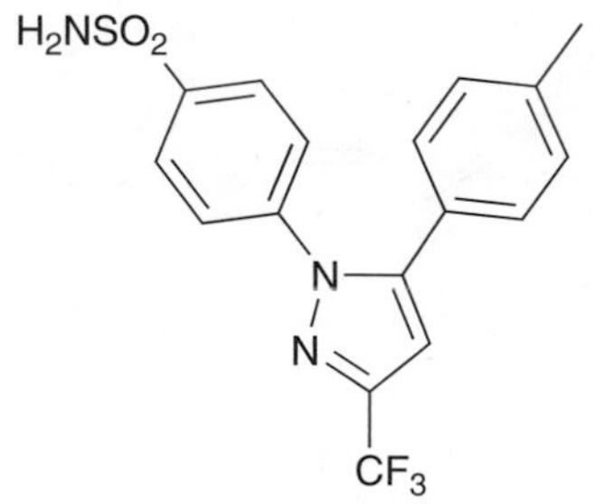

Celecoxib (Celebrex)

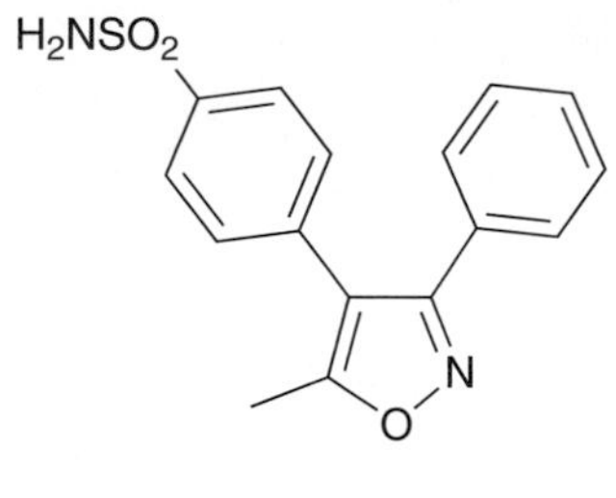

Valdecoxib (Bextra)

Unfortunately, it was later discovered that many of these drugs caused an increased risk of heart attacks and strokes, especially in elderly patients. As a result, Vioxx and Bextra were removed from the market in 2004 and 2005, respectively.

Prostaglandins belong to a larger class of compounds called **eicosanoids**, which include leukotrienes, prostaglandins, thromboxanes, and prostacyclins, all of which are biosynthesized from arachidonic acid.

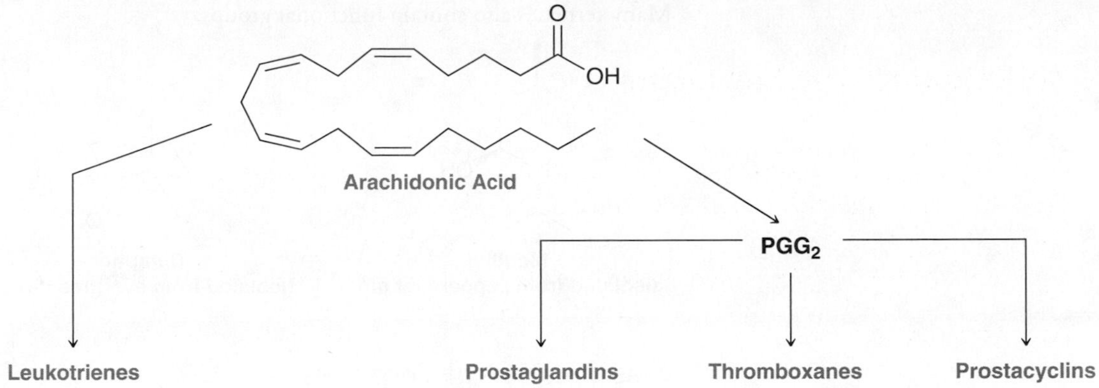

These four classes of compounds exhibit a wide array of biological activity. In some cases, their biological functions oppose each other. For example, thromboxanes are generally vasoconstrictors that trigger blood clotting, while prostacyclins are generally vasodilators that inhibit blood clotting. Our bodies are dependent on the proper balance between the effects of these compounds.

26.8 Terpenes

Terpenes are a diverse class of naturally occurring compounds that share one feature in common. According to the isoprene rule, all terpenes can be thought of as being assembled from **isoprene** units, each of which contains five carbon atoms.

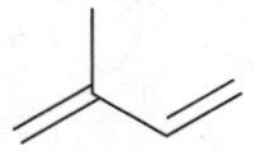

Isoprene

Consequently, the number of carbon atoms present in terpenes will be a multiple of five. The following examples have either 10 or 15 carbons.

Myrcene (isolated from bay and myrcia plants) **10 carbon atoms**

α-Pinene (isolated from pine trees) **10 carbon atoms**

β-Selinene (isolated from celery) **15 carbon atoms**

A wide variety of terpenes are isolated from the essential oils of plants. Terpenes generally have a strong fragrance and are often used as flavorants and odorants in a wide variety of applications, including food products and cosmetics.

During the biosynthesis of terpenes, isoprene units are generally connected head to tail.

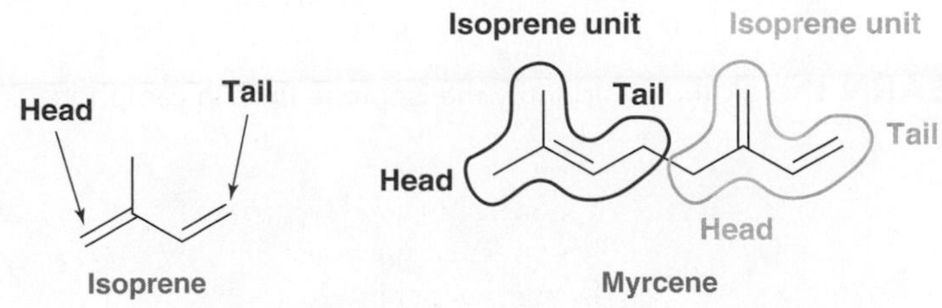

Many terpenes also contain functional groups:

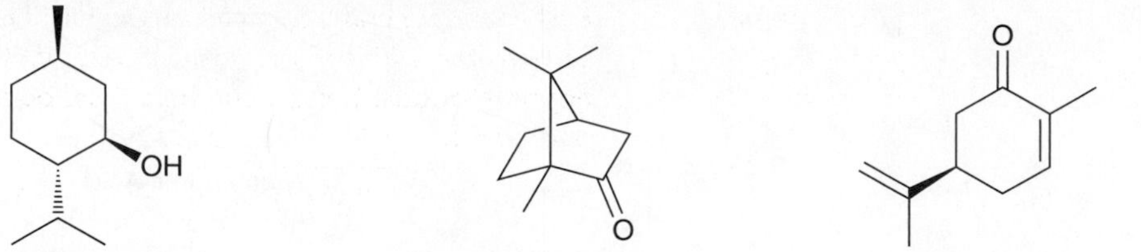

Menthol (isolated from peppermint oil) **Camphor** (isolated from evergreen trees) ***R*-Carvone** (flavor of spearmint)

Classification of Terpenes

TABLE 26.4 CLASSIFICATION OF TERPENES

CLASS	NO. OF CARBON ATOMS
Monoterpene	10
Sesquiterpene	15
Diterpene	20
Triterpene	30
Tetraterpene	40

Terpenes are classified based on units of 10 carbon atoms (two isoprene units). For example, a terpene with 10 carbon atoms is called a monoterpene, while a terpene with 20 carbon atoms is called a diterpene. As seen in Table 26.4, a compound with 15 carbon atoms is called a sesquiterpene. An example of a sesquiterpene is α-farnesene, found in the waxy coating on apple skins.

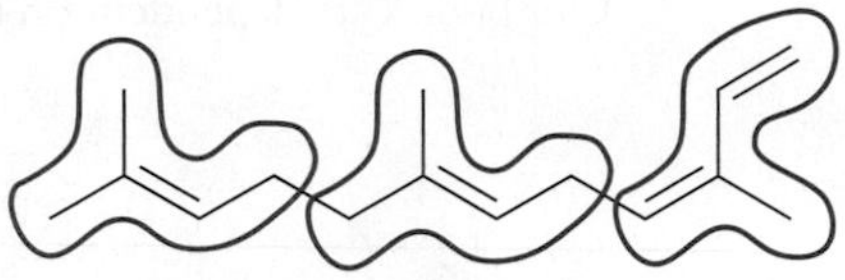

α-Farnesene

Beta-carotene and lycopene each contain 40 carbon atoms and are therefore classified as tetraterpenes.

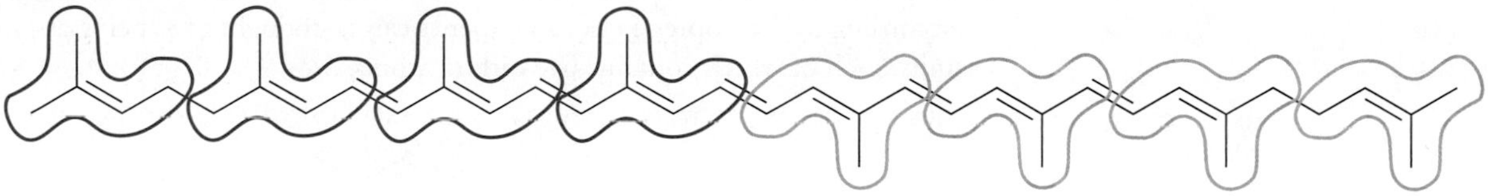

Lycopene

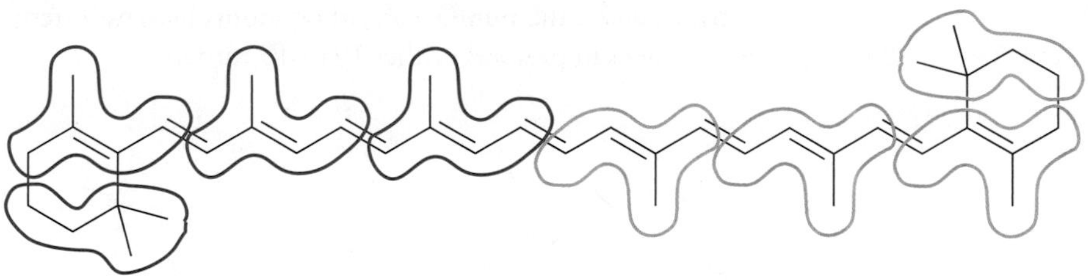

β-Carotene

Each of these compounds is assembled from two diterpenes.

SKILLBUILDER

26.4 IDENTIFYING ISOPRENE UNITS IN A TERPENE

LEARN the skill

Identify the isoprene units in camphor:

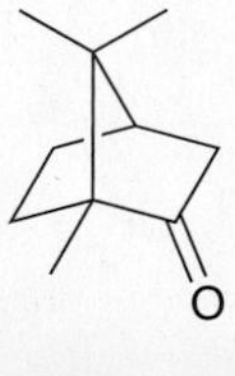

SOLUTION

STEP 1 Count the carbon atoms and identify the number of isoprene units.

First count the number of carbon atoms. This compound is a monoterpene, because it has 10 carbon atoms. Therefore, we are looking for two isoprene units. Each isoprene unit may or may not have double bonds, but there must be four carbon atoms in a straight chain with exactly one branch.

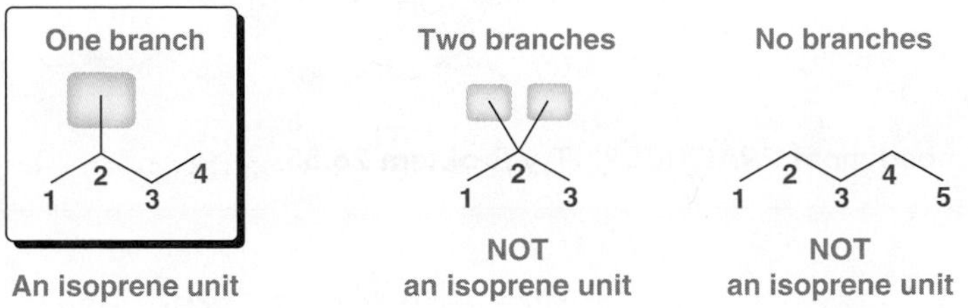

STEP 2 Look for methyl groups.

To identify the isoprene units, it is best to focus on any methyl groups as well as the carbon atoms to which they are attached. For example, the following carbon atoms must be grouped together:

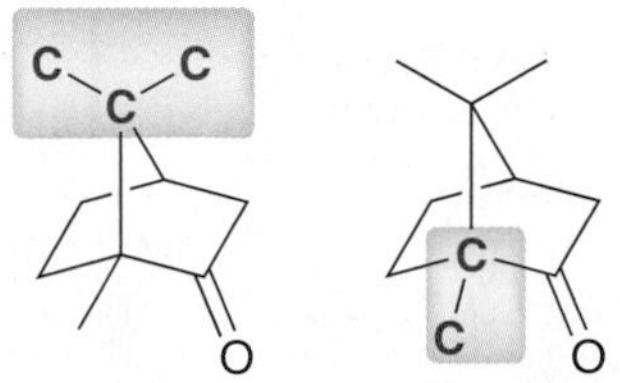

We now consider whether it is possible for the group of five carbon atoms to represent one isoprene unit. Are these five carbon atoms connected in a way that gives four carbon atoms in a straight chain with one branch?

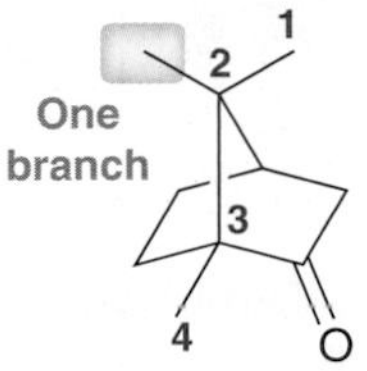

Indeed, these five carbon atoms could be an isoprene unit because they have the correct branching pattern. However, when we analyze the remaining five carbon atoms, they do not have the correct branching pattern.

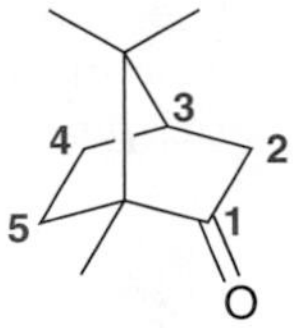

STEP 3 Use trial and error to find the isoprene units.

The remaining five carbon atoms cannot represent an isoprene unit because they do not exhibit a branch. Therefore, the original five carbon atoms that we first identified must in fact be part of two separate isoprene units.

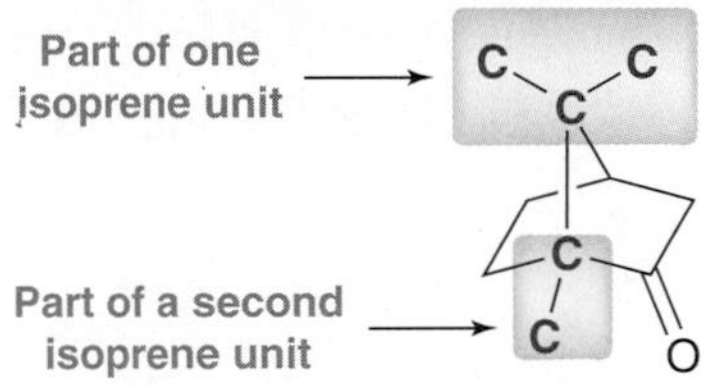

Finding each isoprene unit sometimes requires a little bit of trial and error. In this case, there are two acceptable solutions.

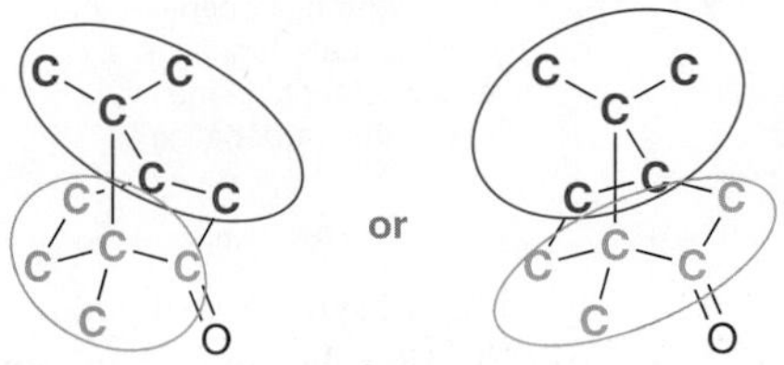

PRACTICE the skill **26.25** Circle the isoprene units in each of the following compounds.

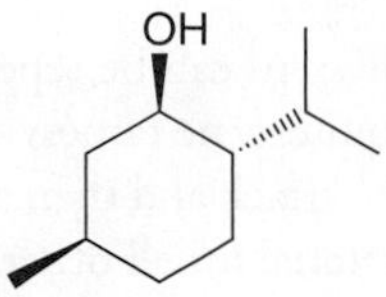

(a) **Menthol**

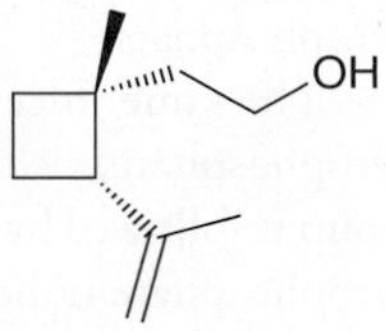

(b) **Grandisol**

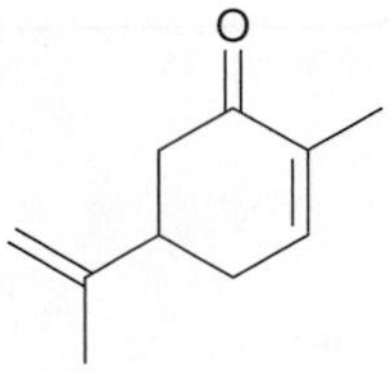

(c) **Carvone**

APPLY the skill

26.26 Determine whether each of the following compounds is a terpene.

(a) 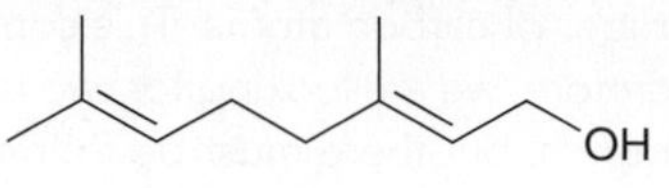
OH

(b) 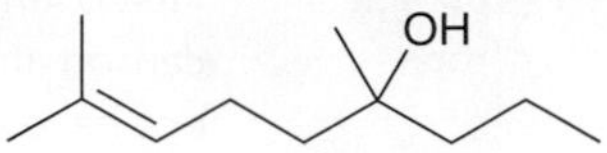
OH

(c) 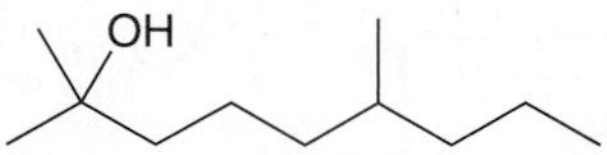
OH

(d) 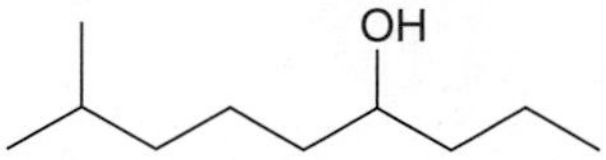
OH

need more **PRACTICE?** Try Problem 26.55

Biosynthesis of Terpenes

Although the isoprene rule considers terpenes to be constructed from isoprene units, the actual building blocks are dimethylallyl pyrophosphate and isopentenyl pyrophosphate.

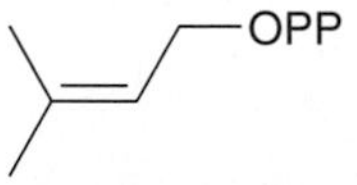
OPP

Dimethylallyl pyrophosphate

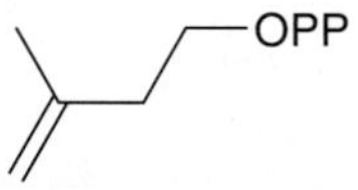
OPP

Isopentenyl pyrophosphate

All terpenes are biosynthesized from these two starting materials. In each of these compounds, the OPP group represents a biological leaving group, called pyrophosphate. OPP is a good leaving group because it is a weak base.

LOOKING BACK
For a review of the connection between leaving groups and basicity, see Section 7.8.

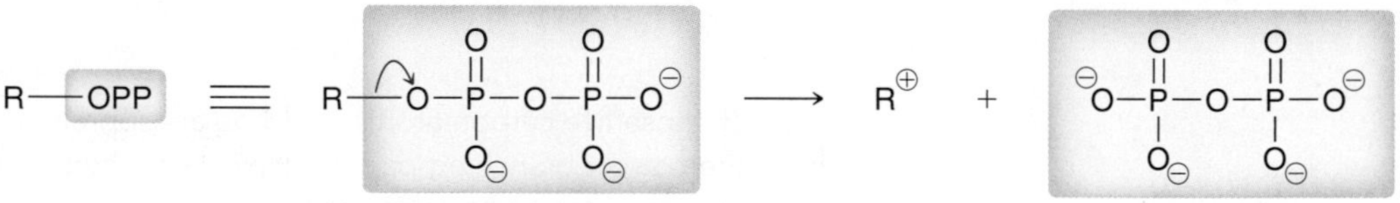
R—OPP
R
R⊕
+

Pyrophosphate (good leaving group)

Weak Base

The reaction between dimethylallyl pyrophosphate and isopentenyl pyrophosphate yields a monoterpene called geranyl pyrophosphate, which is the starting material for all other monoterpenes (Mechanism 26.5).

MECHANISM 26.5 BIOSYNTHESIS OF GERANYL PYROPHOSPHATE

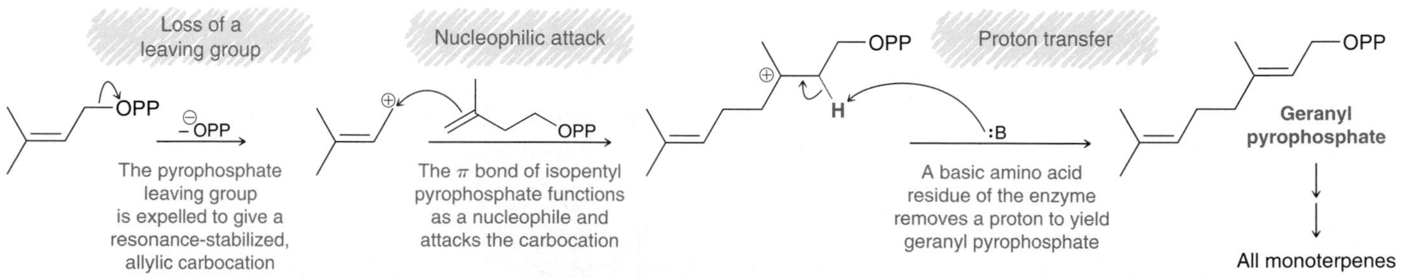
Loss of a leaving group
OPP
⊖OPP
The pyrophosphate leaving group is expelled to give a resonance-stabilized, allylic carbocation
Nucleophilic attack
OPP
The π bond of isopentyl pyrophosphate functions as a nucleophile and attacks the carbocation
OPP
H
Proton transfer
:B
A basic amino acid residue of the enzyme removes a proton to yield geranyl pyrophosphate
OPP
Geranyl pyrophosphate
All monoterpenes

The biosynthesis of geranyl pyrophosphate is achieved in just three steps. The first step involves loss of a leaving group (pyrophosphate) to generate a resonance-stabilized carbocation. The second step is a nucleophilic attack in which the carbocation is attacked by the π bond of isopentyl pyrophosphate to generate a new carbocation. Finally, a proton transfer yields geranyl pyrophosphate.

The same three mechanistic steps can be repeated to add another isoprene unit to geranyl pyrophosphate, yielding the sesquiterpene farnesyl pyrophosphate. Once again, loss of a leaving group is followed by nucleophilic attack and then a proton transfer (Mechanism 26.6). Farnesyl pyrophosphate is the starting material for all other sesquiterpenes and diterpenes.

MECHANISM 26.6 BIOSYNTHESIS OF FARNESYL PYROPHOSPHATE

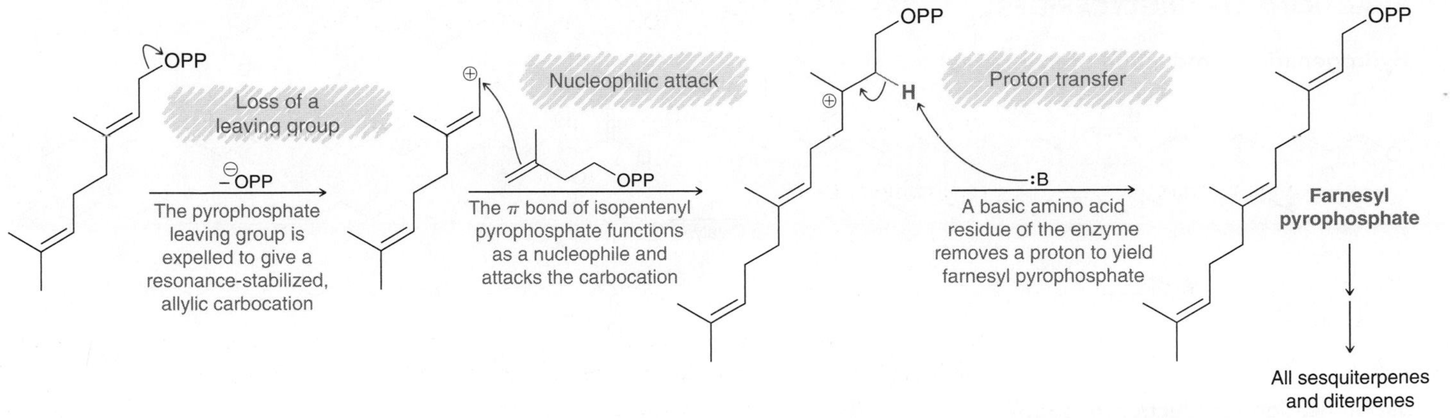
OPP
Loss of a leaving group
−OPP
The pyrophosphate leaving group is expelled to give a resonance-stabilized, allylic carbocation
Nucleophilic attack
OPP
The π bond of isopentenyl pyrophosphate functions as a nucleophile and attacks the carbocation
OPP
H
Proton transfer
:B
A basic amino acid residue of the enzyme removes a proton to yield farnesyl pyrophosphate
OPP
Farnesyl pyrophosphate
All sesquiterpenes and diterpenes

Squalene, the biological precursor for all steroids (as seen in Mechanism 26.3), is biosynthesized from the coupling of two molecules of farnesyl pyrophosphate.

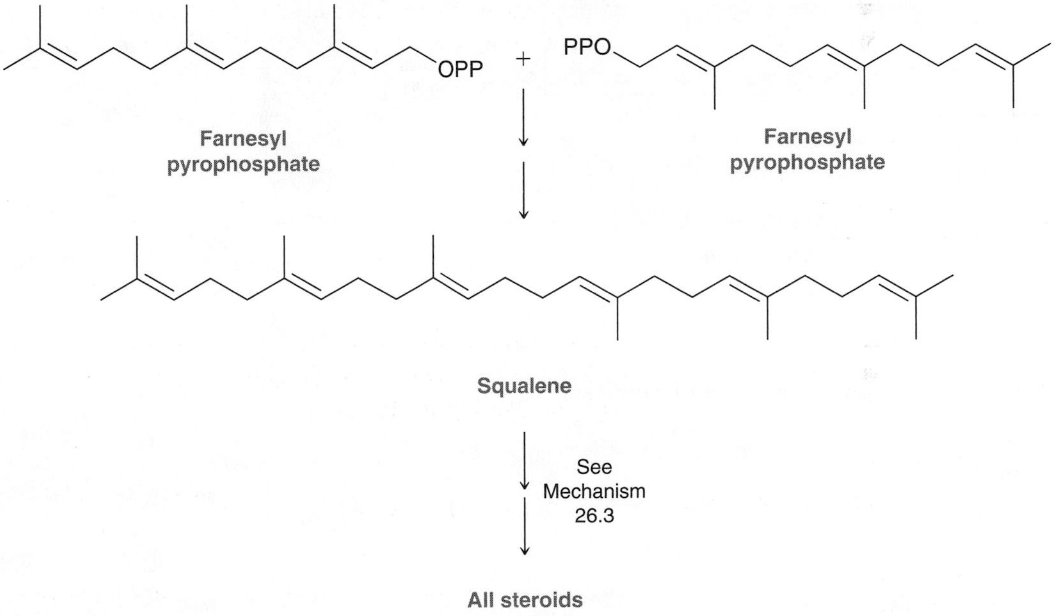
OPP
PPO
Farnesyl pyrophosphate
Farnesyl pyrophosphate
Squalene
See Mechanism 26.3
All steroids

CONCEPTUAL CHECKPOINT

26.27 Draw a mechanism for the following transformation.

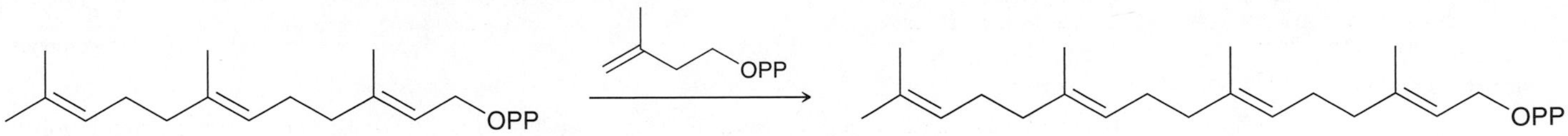
OPP
OPP
OPP

26.28 Draw a mechanism for the biosynthesis of α-farnesene starting with dimethylallyl pyrophosphate and isopentenyl pyrophosphate.

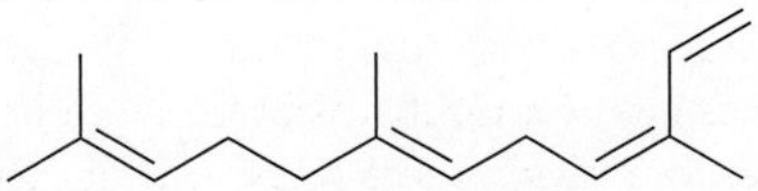

α-Farnesene

REVIEW OF REACTIONS SYNTHETICALLY USEFUL REACTIONS

REACTIONS OF TRIGLYCERIDES

Hydrogenation (production of margarine)

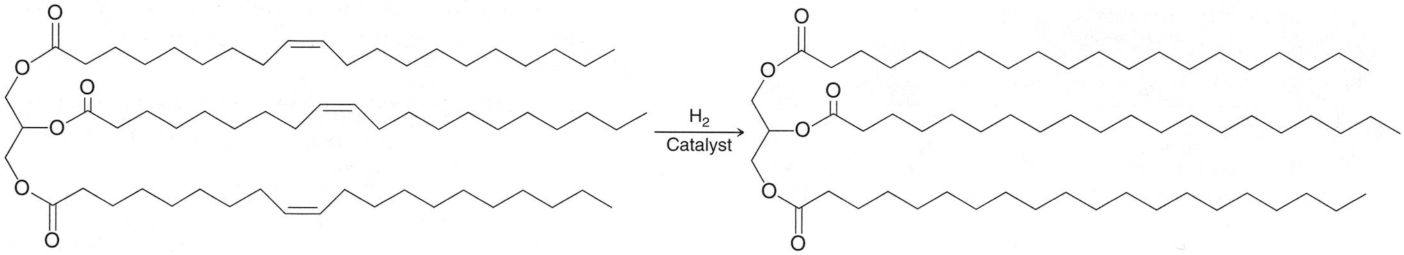

Saponification (production of soap)

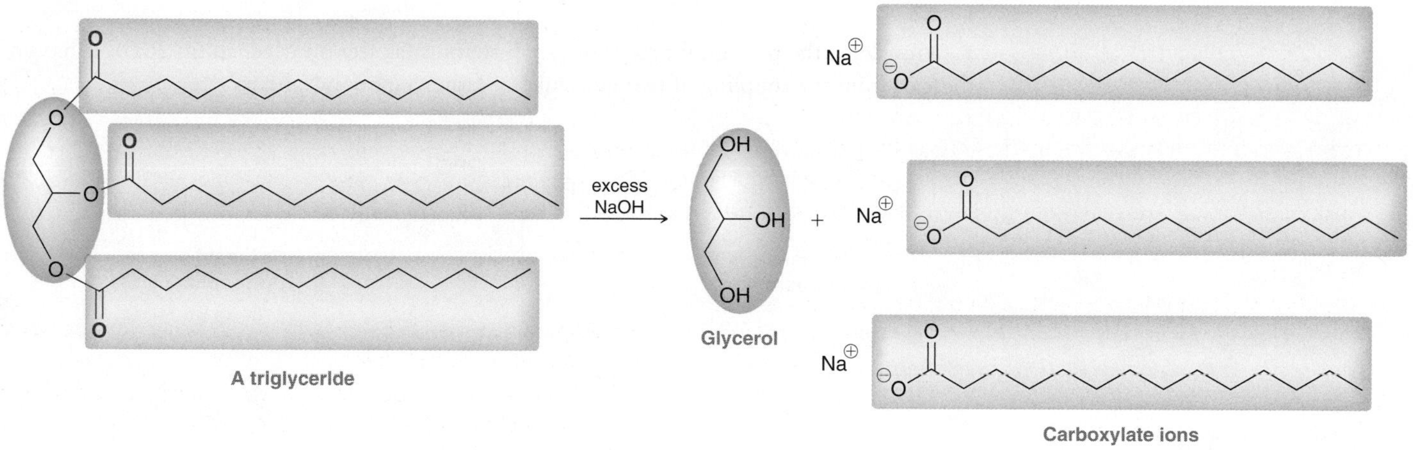

Transesterification (production of biodiesel)

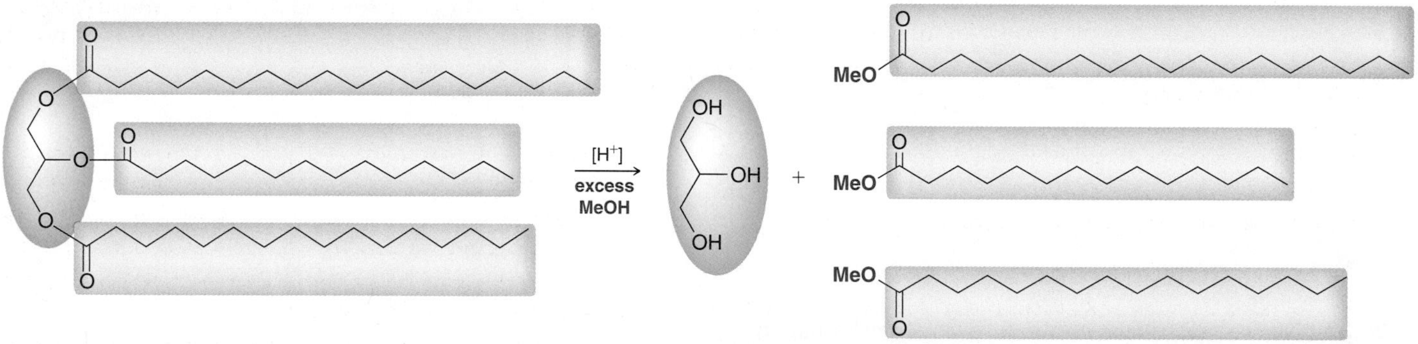

REVIEW OF CONCEPTS AND VOCABULARY

SECTION 26.1

- **Lipids** are naturally occurring compounds that are extracted from cells using nonpolar solvents.
- **Complex lipids** readily undergo hydrolysis, while **simple lipids** do not undergo hydrolysis.

SECTION 26.2

- **Waxes** are high-molecular-weight esters that are constructed from carboxylic acids and alcohols.

SECTION 26.3

- **Triglycerides** are the triesters formed from glycerol and three long-chain carboxylic acids, called **fatty acids**. The resulting triglyceride is said to contain three **fatty acid residues.**
- For saturated fatty acids, the melting point increases with increasing molecular weight. The presence of a *cis* double bond causes a decrease in the melting point.
- Triglycerides with unsaturated fatty acid residues have lower melting points than triglycerides with saturated fatty acid residues.
- Triglycerides that are solids at room temperature are called **fats**, while those that are liquids at room temperature are called **oils**.
- Triglycerides found in animals have a higher concentration of saturated fatty acids than triglycerides from vegetable origins.
- The exact ratio of fatty acids differs from one source to another.

SECTION 26.4

- Triglycerides containing unsaturated fatty acid residues will undergo hydrogenation. During the hydrogenation process, some of the double bonds can isomerize to give *trans* π bonds.
- In the presence of molecular oxygen, triglycerides are particularly susceptible to oxidation at the allylic position to produce hydroperoxides.
- When treated with aqueous base, triglycerides undergo hydrolysis, also called saponification.
- Transesterification of triglycerides can be achieved via either acid catalysis or base catalysis to produce biodiesel.

SECTION 26.5

- **Phospholipids** are esterlike derivatives of phosphoric acid.
- **Phosphoglycerides** are similar in structure to triglycerides except that one of the three fatty acid residues is replaced by a phosphoester group.
- The simplest kind of phosphoglyceride is a phosphoric monoester, called a **phosphatidic acid**.
- Phosphoglycerides that contain ethanolamine are called **cephalins,** while phosphoglycerides that contain choline are called **lecithins**.
- In water, phosphoglycerides will self-assemble to form a **lipid bilayer**.

SECTION 26.6

- The structures of **steroids** are based on a tetracyclic ring system, involving three six-membered rings and one five-membered ring.
- The ring fusions are all *trans* in most steroids, giving steroids their rigid geometry.
- All steroids, including cholesterol, are biosynthesized from squalene.
- Human sex hormones are steroids that regulate tissue growth and reproductive processes. Male sex hormones are called **androgens**, and the two types of female sex hormones are called **estrogens** and **progestins**.
- **Adrenocortical hormones** are employed by nature to treat the effects of inflammatory diseases.

SECTION 26.7

- **Prostaglandins** are biochemical regulators that are even more powerful than steroids.
- Prostaglandins are biosynthesized from arachidonic acid with the help of enzymes called cyclooxygenases.
- Prostaglandins contain 20 carbon atoms and are characterized by a five-membered ring with two side chains.
- Prostaglandins belong to a larger class of compounds called **eicosanoids**, which include leukotrienes, prostaglandins, thromboxanes, and prostacyclins, all of which exhibit a wide array of biological activity.

SECTION 26.8

- **Terpenes** are a class of naturally occurring compounds that can be thought of as being assembled from **isoprene** units.
- A terpene with 10 carbon atoms is called a monoterpene, while a terpene with 20 carbon atoms is called a diterpene.
- All terpenes are biosynthesized from dimethylallyl pyrophosphate and isopentenyl pyrophosphate.
- Geranyl pyrophosphate is the starting material for all monoterpenes, and farnesyl pyrophosphate is the starting material for all sesquiterpenes.

KEY TERMINOLOGY

SKILLBUILDER REVIEW

26.1 COMPARING MOLECULAR PROPERTIES OF TRIGLYCERIDES

STEP 1 Identify the fatty acid residues in each triglyceride.

Palmitoleic acid residues — Palmitic acid residues

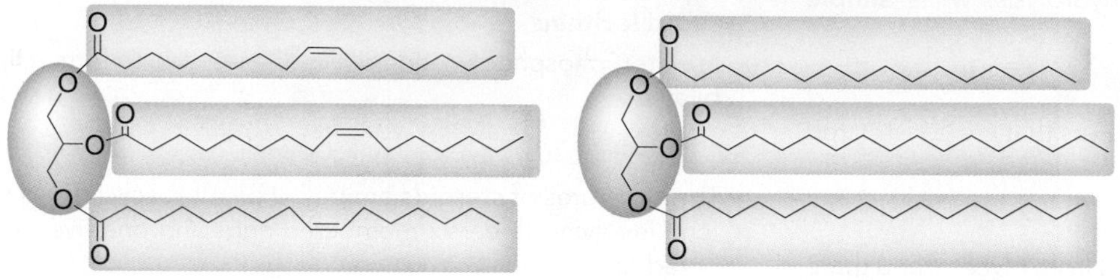

Tripalmitolein — Tripalmitin

STEP 2 Compare the residues, keeping in mind that the melting point is affected by:

1) Length of the chain (longer chain = higher melting point).

2) Saturated residues (no C—C π bonds) have higher melting points.

Try Problems **26.3–26.6, 26.40a,d,e, 26.41a,d,e, 26.43**

26.2 IDENTIFYING THE PRODUCTS OF TRIGLYCERIDE HYDROLYSIS

STEP 1 Identify the three ester moieties.

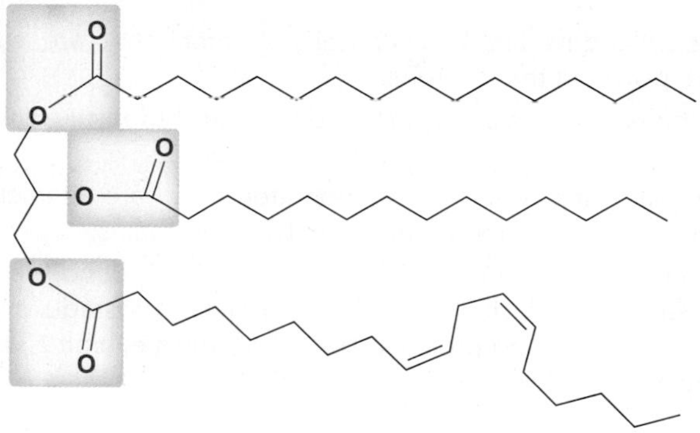

STEP 2 Each ester moiety is hydrolyzed, generating glycerol and three carboxylate ions.

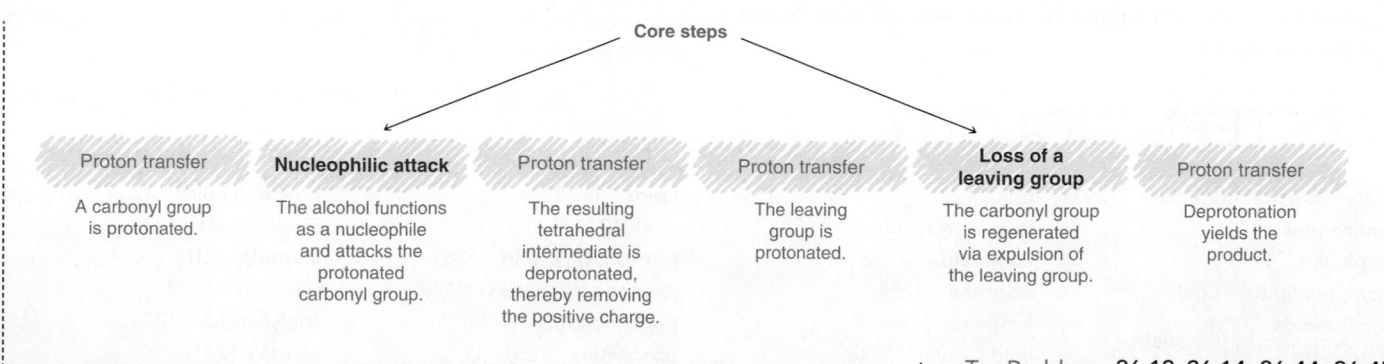

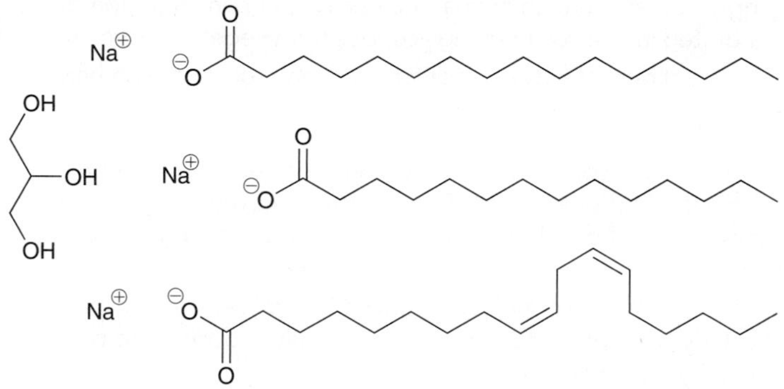

Try Problems **26.9–26.11, 26.30b, 26.40c, 26.41c**

26.3 DRAWING A MECHANISM FOR TRANSESTERIFICATION OF A TRIGLYCERIDE

Core steps

Proton transfer	**Nucleophilic attack**	Proton transfer	Proton transfer	**Loss of a leaving group**	Proton transfer
A carbonyl group is protonated.	The alcohol functions as a nucleophile and attacks the protonated carbonyl group.	The resulting tetrahedral intermediate is deprotonated, thereby removing the positive charge.	The leaving group is protonated.	The carbonyl group is regenerated via expulsion of the leaving group.	Deprotonation yields the product.

Try Problems **26.12–26.14, 26.44, 26.45**

26.4 IDENTIFYING ISOPRENE UNITS IN A TERPENE

STEP 1 Count the number of carbon atoms in order to identify the number of isoprene units.

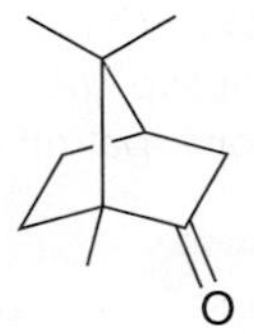

10 Carbon atoms = 2 Isoprene units

STEP 2 Look for any methyl groups and the carbon atoms to which they are attached.

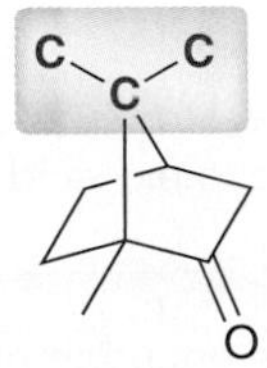

These three carbon atoms must be grouped together

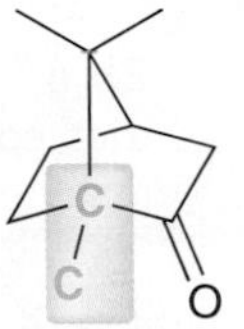

These two carbon atoms must be grouped together

STEP 3 Using trial and error, identify the isoprene units that have the correct branching structure.

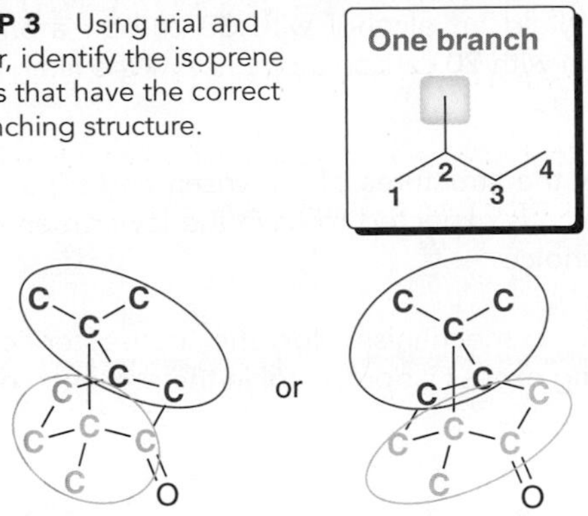

Try Problems 26.25, 26.26, 26.55

PRACTICE PROBLEMS

Note: Most of the Problems are available within *WileyPLUS*, an online teaching and learning solution.

26.29 Identify each of the following compounds as a wax, triglyceride, phospholipid, steroid, prostaglandin, or terpene.

(a) Stanozolol (b) Lycopene
(c) Tristearin (d) Distearoyl lecithin
(e) PGF_2 (f) Pentadecyl octadecanoate

26.30 Predict the product(s) formed when tripalmitolein is treated with each of the following reagents.

(a) Excess H_2, Ni

(b) Excess NaOH, H_2O

26.31 Draw two different cephalins that contain one lauric acid residue and one myristic acid residue. Are both of these compounds chiral?

26.32 Draw an optically active triglyceride that contains one palmitic acid residue and two myristic acid residues. Will this compound react with molecular hydrogen in the presence of a catalyst?

26.33 Which of the following compounds are lipids?

(a) L-Threonine (b) 1-Octanol
(c) Lycopene (d) Trimyristin
(e) Palmitic acid (f) D-Glucose
(g) Testosterone (h) D-Mannose

26.34 Draw the structure of *trans*-oleic acid.

26.35 Tristearin is less susceptible to becoming rancid than triolein. Explain.

26.36 Arrange the following compounds in order of increasing water solubility.

(a) A triglyceride constructed from one equivalent of glycerol and three equivalents of myristic acid

(b) A diglyceride constructed from one equivalent of glycerol and two equivalents of myristic acid

(c) A monoglyceride constructed from one equivalent of glycerol and one equivalent of myristic acid

26.37 Identify whether hexane or water would be more appropriate for extracting terpenes from plant tissues. Explain your choice.

26.38 Identify each of the following fatty acids as saturated or unsaturated.

(a) Palmitic acid (b) Myristic acid (c) Oleic acid
(d) Lauric acid (e) Linoleic acid (f) Arachidonic acid

26.39 Which of the fatty acids in the previous problem has four carbon-carbon double bonds?

26.40 Which of the following statements applies to triolein?

(a) It is a solid at room temperature.
(b) It is unreactive toward molecular hydrogen in the presence of Ni.
(c) It undergoes hydrolysis to produce unsaturated fatty acids.
(d) It is a complex lipid.
(e) It is a wax.
(f) It has a phosphate group.

26.41 Identify which of the following statements applies to tristearin.

(a) It is a solid at room temperature.
(b) It is unreactive toward molecular hydrogen in the presence of Ni.
(c) It undergoes hydrolysis to produce unsaturated fatty acids.
(d) It is a complex lipid.
(e) It is a wax.
(f) It has a phosphate group.

26.42 One of the compounds present in carnauba wax was isolated, purified, and then treated with aqueous sodium hydroxide to yield an alcohol with 30 carbon atoms and a carboxylate ion with 20 carbon atoms. Draw the likely structure of the compound.

26.43 Draw the structures of trimyristin and tripalmitin and determine which is expected to have the lower melting point. Explain your choice.

26.44 Draw a mechanism for the transesterification of trimyristin using excess isopropanol in the presence of an acid catalyst.

26.45 Draw a mechanism for the base-catalyzed transesterification of trimyristin using ethanol in the presence of sodium hydroxide.

26.46 Draw the structure of an optically inactive triglyceride that contains two oleic acid residues and one palmitic acid residue.

26.47 Draw the structure of an optically active triglyceride that contains two oleic acid residues and one palmitic acid residue.

26.48 Draw the enantiomer of cholesterol.

26.49 Circle the isoprene units in each of the following compounds

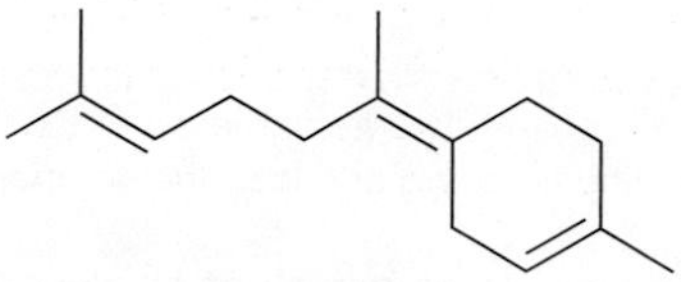

(a) **Bisabolene**

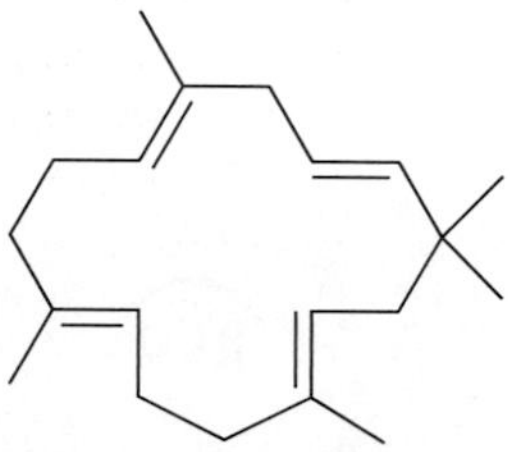

(b) **Flexibilene**

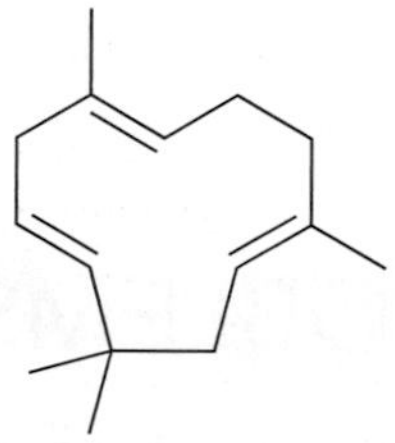

(c) **Humulene**

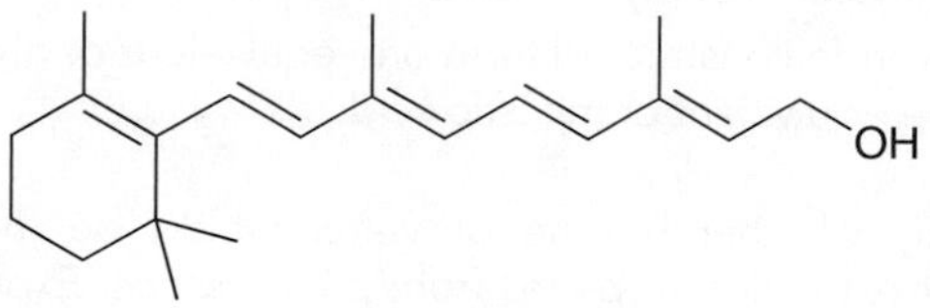

(d) **Vitamin A**

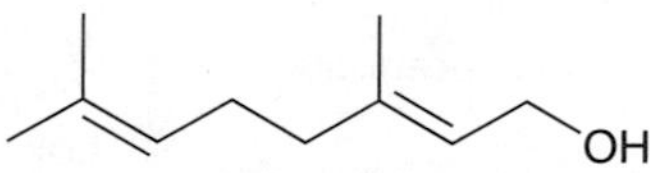

(e) **Geraniol**

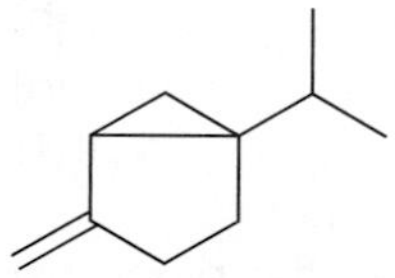

(f) **Sabinene**

26.50 Sphingomyelins are lipids with the following general structure:

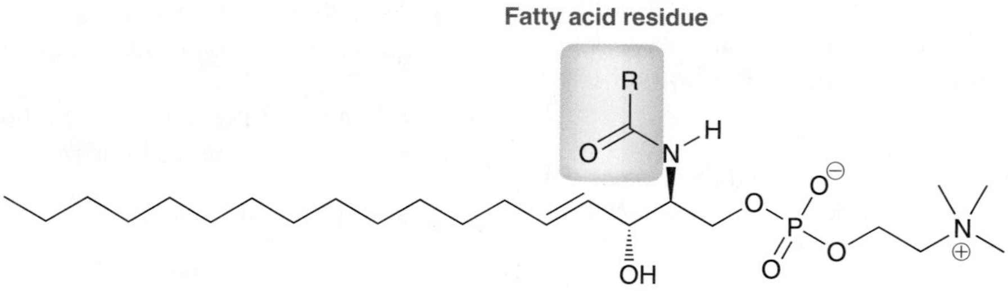

Sphingomyelins

(a) Identify the polar head and all hydrophobic tails in sphingomyelins.

(b) Do sphingomyelins have the appropriate structural features and three-dimensional geometry necessary to be major constituents of lipid bilayers?

INTEGRATED PROBLEMS

26.51 Treatment of cholesterol with MCPBA could potentially produce two diastereomeric epoxides.

(a) Draw both diastereomeric epoxides.

(b) Only one of these epoxides is formed. Predict which one, and explain why the other is not formed.

26.52 Identify the products expected when estradiol is treated with each of the following reagents.

(a) Excess Br_2

(b) PCC

(c) A strong base followed by excess ethyl iodide

(d) Excess acetyl chloride in the presence of pyridine

26.53 Olestra is a noncaloric oil substitute that is produced by esterifying sucrose with eight equivalents of fatty acids obtained from the hydrolysis of vegetable oils. The eight fatty acid residues give Olestra the consistency and flavor of cooking oil, but the steric bulk of the compound prevents digestive enzymes from hydrolyzing the ester moieties. As a result, Olestra passes through the digestive tract unaltered. Draw the structure of a molecule of Olestra that contains eight lauric acid residues. Is this compound chiral?

26.54 Identify the reagents you would use to convert oleic acid into each of the following compounds.

(a) Stearic acid

(b) Ethyl stearate

(c) 1-Octadecanol

(d) Nonanedioic acid

(e) 2-Bromostearic acid

26.55 Limonene is an optically active compound isolated from the peels of lemons and oranges.

(a) Is limonene a monoterpene or a diterpene?

(b) Treatment of limonene with excess HBr yields a compound with molecular formula $C_{10}H_{18}Br_2$. Identify the structure of this compound and determine whether it is chiral or achiral.

(c) Draw the products obtained when limonene is treated with O_3 followed by DMS.

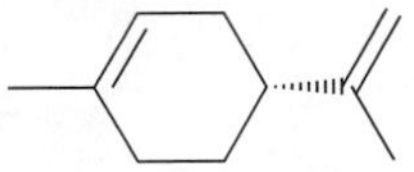

Limonene

CHALLENGE PROBLEMS

26.56 Starting with the following compound and using any other reagents of your choice, outline a synthesis for trimyristin.

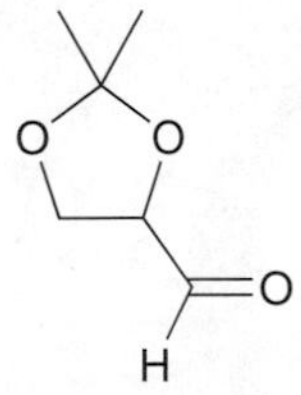

26.57 The following compound was isolated from nerve cells.

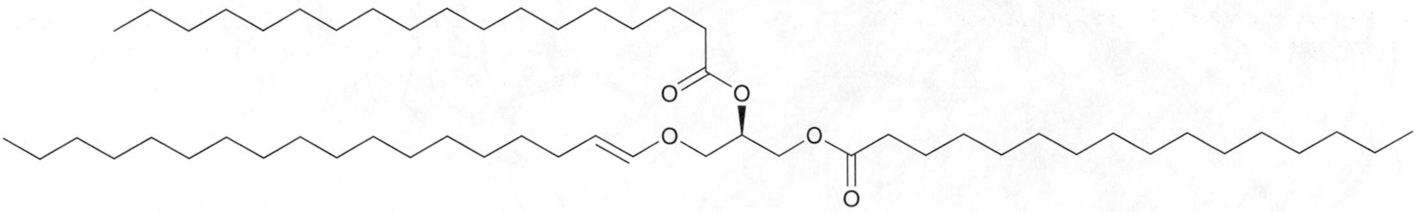

(a) Describe how this compound differs in structure from fats and oils.

(b) Three products are obtained when this compound is hydrolyzed with aqueous sodium hydroxide. Draw the structures of all three products.

(c) Four products are obtained when this compound is hydrolyzed with aqueous acid. Draw the structures of all four products.

27 Synthetic Polymers

DID YOU EVER WONDER... how bulletproof glass works?

Bulletproof glass, such as that used in armored vehicles, consists of several sheets of glass sandwiched between transparent polymer sheets. The resulting material is comprised of many layers, and the outer layers are capable of absorbing the impact of the bullet, thereby preventing it from penetrating the inner layers. The ability of the glass to resist bullets depends on many factors, including the type of polymer used, the thickness of the glass, the number of layers sandwiched together, the type of bullet, and the range from which the bullet is fired. For this reason, the glass is generally described as "bullet resistant" rather than "bulletproof."

This chapter will focus on the preparation, classification, and properties of a variety of synthetic polymers, including those found in bullet-resistant glass as well as a large variety of other applications.

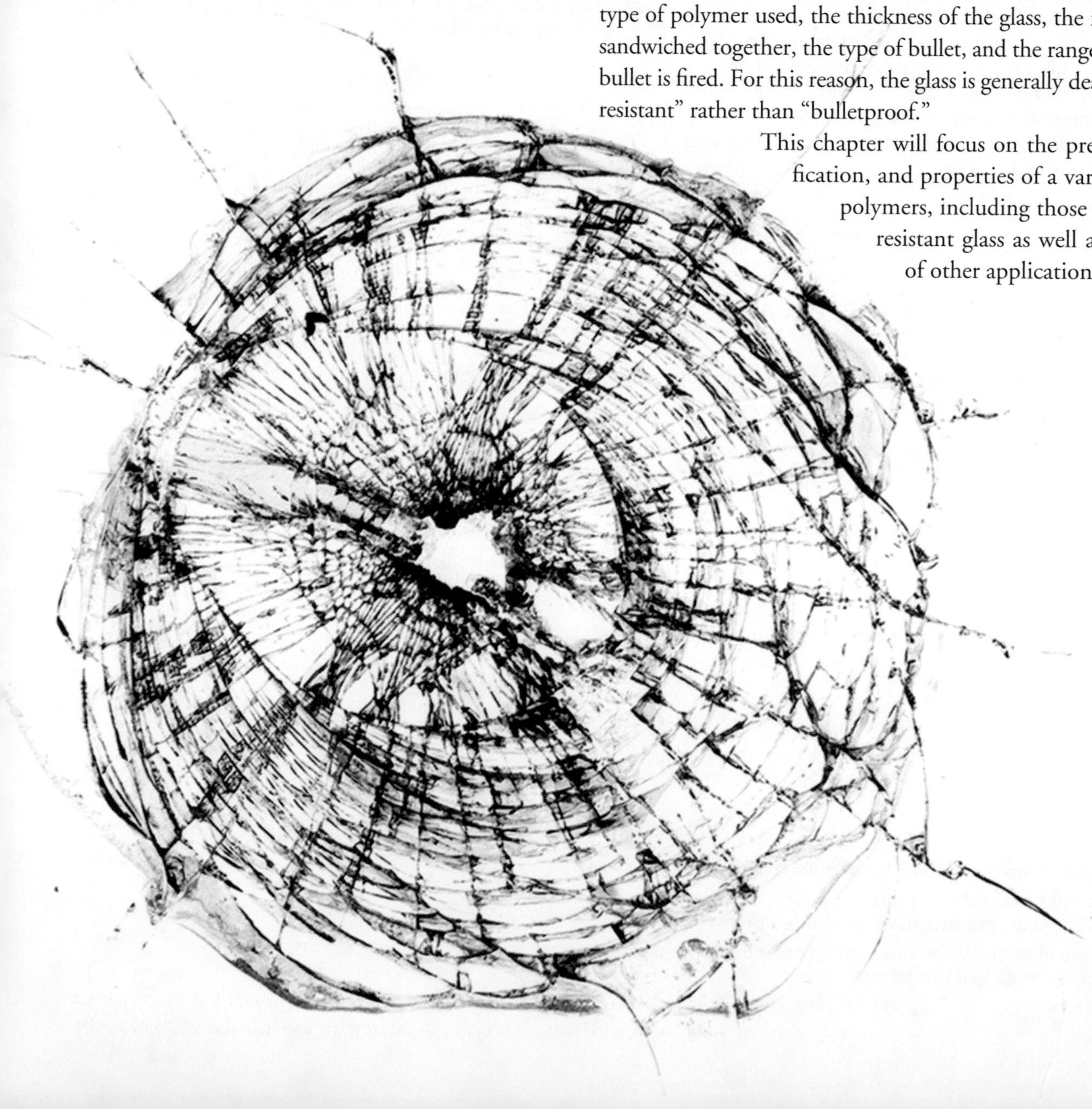

DO YOU REMEMBER?

Before you go on, be sure you understand the following topics.
If necessary, review the suggested sections to prepare for this chapter:

- Sources and Uses of Alkanes: An Introduction to Polymers (section 4.5)
- Cationic Polymerization and Polystyrene (section 9.3)
- Radical Polymerization (section 11.11)

WILEY PLUS Visit www.wileyplus.com to check your understanding and for valuable practice.

27.1 Introduction to Synthetic Polymers

Over the past century, our society has become heavily reliant on polymers. Coffee cups, soft-drink bottles, synthetic fabrics for clothing, CDs and DVDs, trash bags, artificial heart valves, and automobile parts are all manufactured from polymers. Polymers represent a multibillion-dollar industry, with more than 50 trillion pounds of synthetic polymers being manufactured each year in the United States alone.

Recall that polymers are comprised of repeating units that are constructed by joining monomers together.

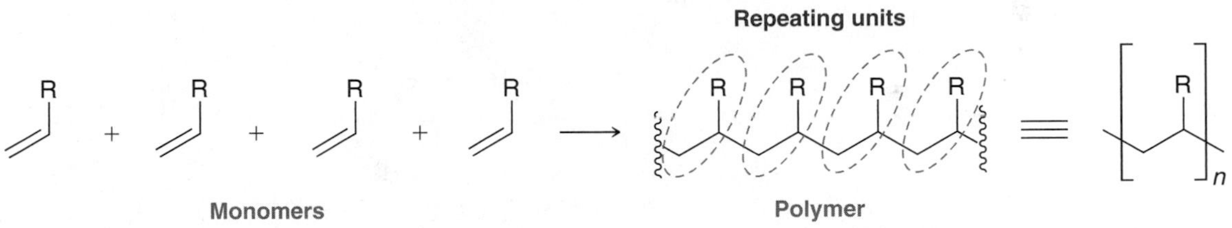

Polymers are drawn most efficiently by placing brackets around the repeating unit, where the subscript "*n*" indicates that the polymer is constructed from a large number of repeating units. Polymers have been discussed many times throughout this textbook, as indicated in Table 27.1.

TABLE 27.1 REFERENCES TO PRIOR DISCUSSIONS OF POLYMERS IN THIS TEXTBOOK

TOPIC	SECTION
Introduction to polymers	4.5
Cationic polymerization	9.3
Conducting organic polymers	10.1
Radical polymerization	11.11
Polymers from dienes	17.5
Polyesters and polyamides	21.12
Polysaccharides	24.8
N-Glycosides (DNA and RNA)	24.10
Proteins	25.1

Polymers can be divided into two major categories: synthetic polymers and biopolymers. The former are prepared by scientists in the laboratory or in factories, while the latter are produced by living organisms. Biopolymers include polysaccharides, DNA, RNA, and proteins, all of which were discussed in the previous chapters. This chapter will focus exclusively on synthetic polymers.

27.2 Nomenclature of Synthetic Polymers

IUPAC has established rules for assigning systematic names to polymers, but these names are rarely used by scientists. An alternative and more common system, also recognized by

IUPAC, involves naming the polymer based on the monomers from which it was derived, as shown in the following examples:

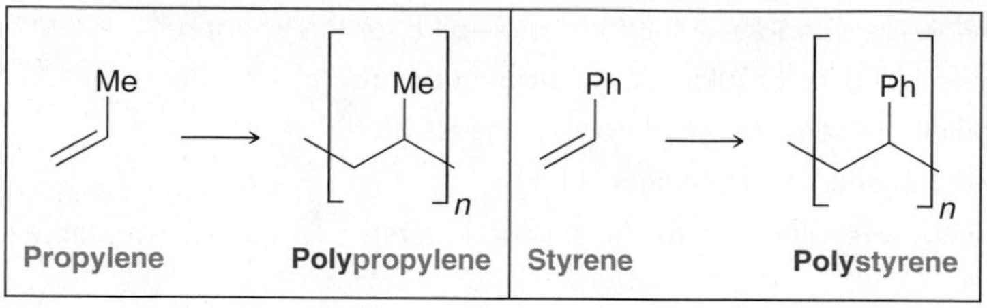

When writing the name of a polymer whose monomer contains two words, parentheses are used around the name of the monomer. Thus, polymerization of vinyl chloride yields poly(vinyl chloride).

Polymers are also often called by their trade name, such as Teflon and Kevlar. Table 27.2 provides a list of many common polymers and their uses.

TABLE 27.2 SEVERAL COMMON POLYMERS AND THEIR USES

NAME OF POLYMER	MONOMER STRUCTURE	POLYMER STRUCTURE	USES
Polyethylene	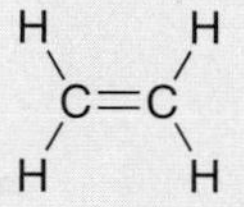	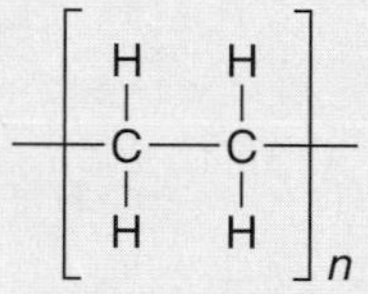	Bottles and trash bags
Polypropylene	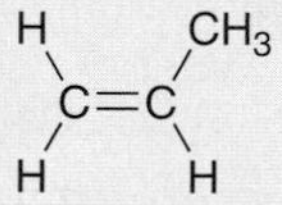	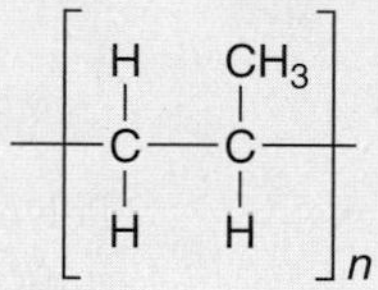	Carpet fibers, appliances, car tires
Polyisobutylene	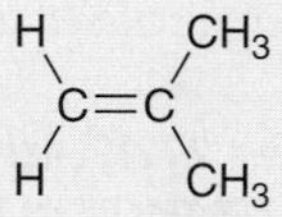	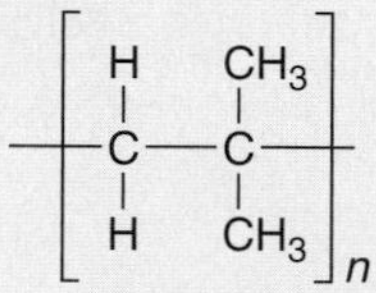	Caulks and sealants, bicycle inner tubes, basketballs
Polystyrene	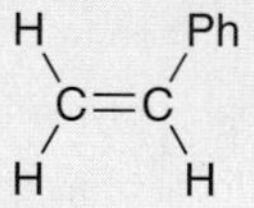	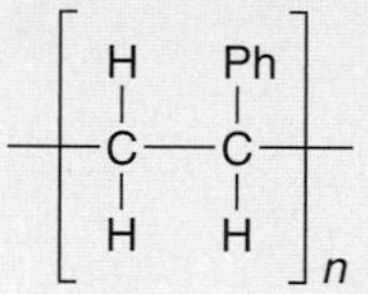	Foam insulation, televisions, radios
Poly(vinyl chloride)	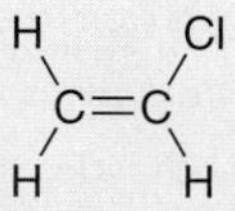	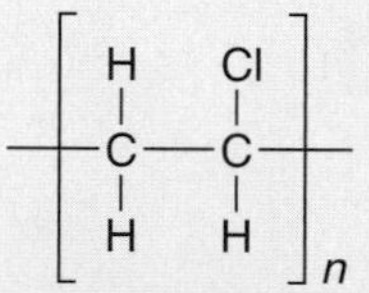	Water pipes, vinyl plastics
Poly(methyl-α-cyanoacrylate)	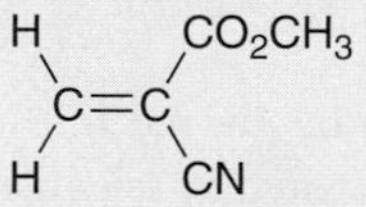	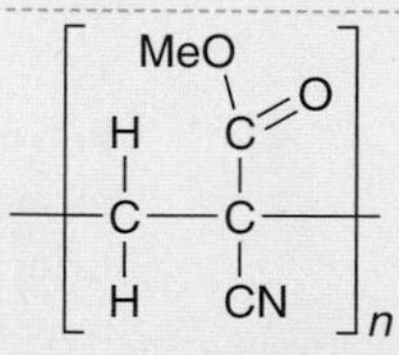	Superglue
Polytetrafluoroethylene	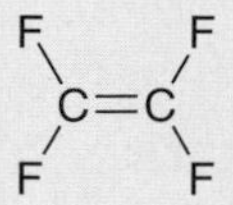	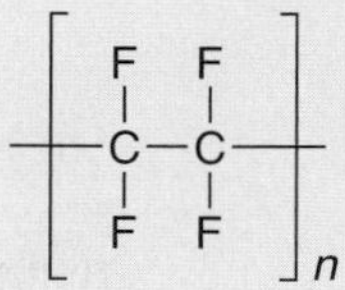	Nonstick coating for frying pans (Teflon®)

CONCEPTUAL CHECKPOINT

27.1 Draw and name the polymer that results when each of the following monomers undergoes polymerization.

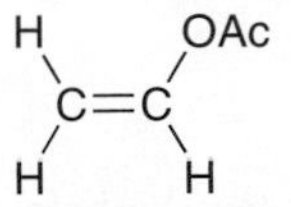

(a) **Vinyl acetate**

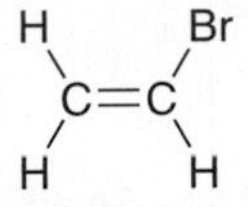

(b) **Vinyl bromide**

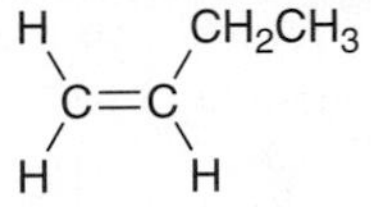

(c) **α-Butylene**

27.2 Sodium polyacrylate is a synthetic polymer used in diapers, because of its ability to absorb several hundred times its own mass of water. This extraordinary polymer is made from another polymer, poly(methyl acrylate), via hydrolysis of the ester moieties. Draw and name the monomer used to make poly(methyl acrylate).

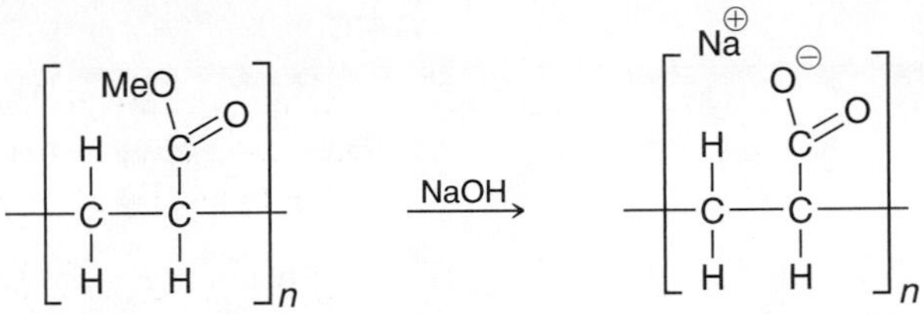

Poly(methyl acrylate) **Sodium polyacrylate**

27.3 Copolymers

The polymers discussed thus far are all called homopolymers. A **homopolymer** is a polymer constructed from a single type of monomer. In contrast, polymers constructed from two or more different types of monomers are called **copolymers**. A common example of a copolymer is Saran®, which is made from vinyl chloride and vinylidene chloride.

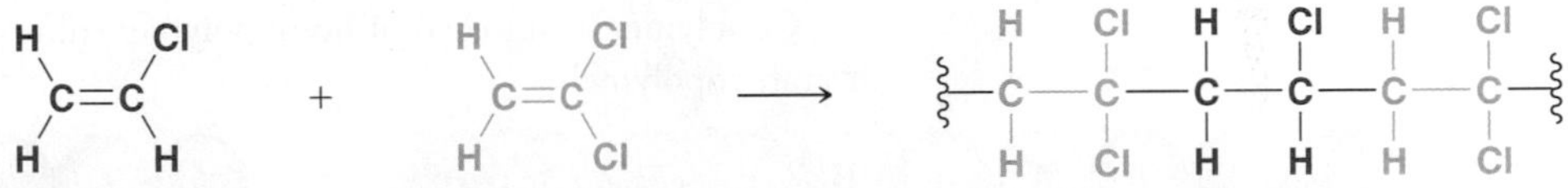

Vinyl chloride **Vinylidene chloride** **Saran®**

Copolymers often have different properties than either of their corresponding homopolymers. A list of some common copolymers and their applications is presented in Table 27.3. Copolymers are often classified based on the order in which the monomers are joined together. **Alternating copolymers** contain an alternating distribution of repeating units, while **random copolymers** contain a random distribution of repeating units.

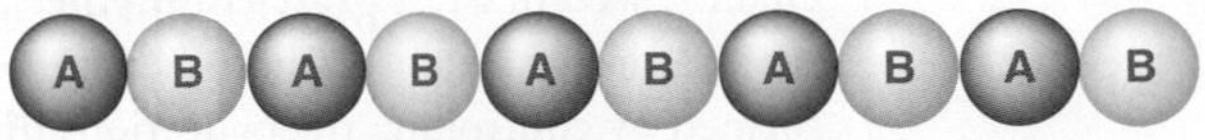

Alternating copolymer

Random copolymer

The precise distribution of any copolymer is dependent on the conditions of its formation. In practice, however, even alternating copolymers will exhibit regions in which the repeating units are distributed randomly.

TABLE 27.3 SOME COMMON COPOLYMERS AND THEIR USES

MONOMERS	COPOLYMER NAME	USES
 **Vinyl chloride** + **Vinylidene chloride**	Saran	Food packaging
 **Styrene** + **Acrylonitrile**	SAN	Dishwasher-safe kitchenware and battery cases
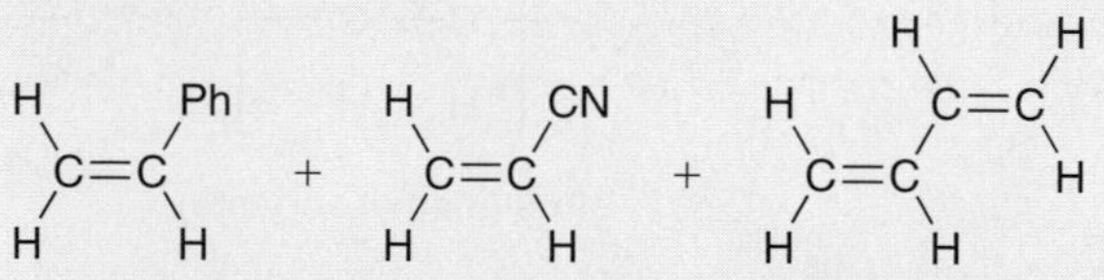 **Styrene** + **Acrylonitrile** + **1,3-Butadiene**	ABS	Crash helmets, luggage, and car bumpers
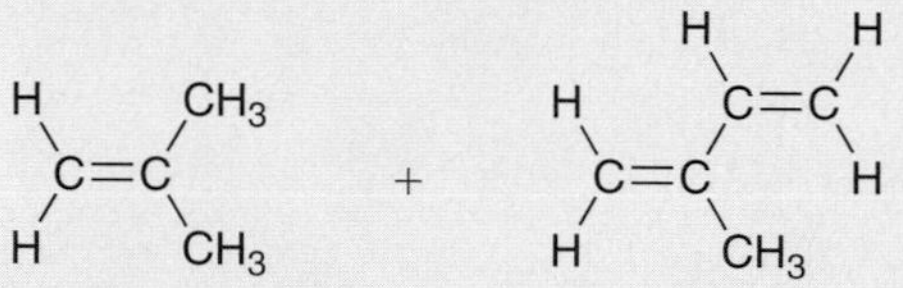 **Isobutylene** + **Isoprene**	Butyl rubber	Inner tubes, balls, sporting goods
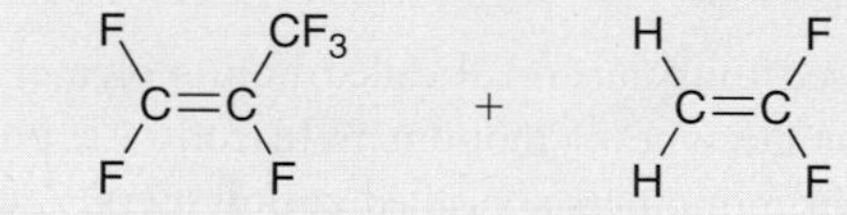 **Hexafluoropropylene** + **Vinylidene fluoride**	Viton	Gaskets, seals, and automotive fuel lines

Copolymers comprised of homopolymer subunits are classified as either block copolymers or graft copolymers.

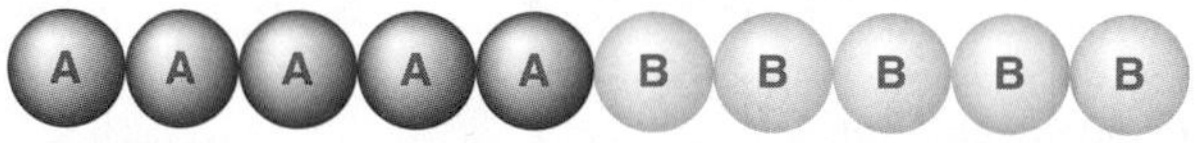

Block copolymer

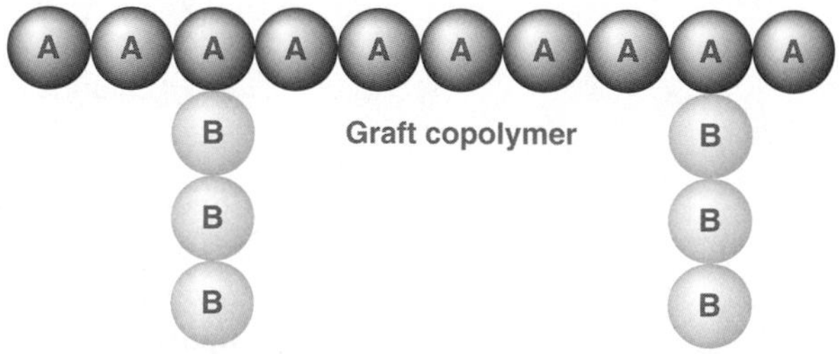

Graft copolymer

In a **block copolymer**, the different homopolymer subunits are connected together in one chain. In contrast, a **graft copolymer** contains sections of one homopolymer that have been grafted onto a chain of the other homopolymer. Block copolymers and graft copolymers are formed by controlling the conditions under which polymerization occurs.

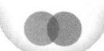

CONCEPTUAL CHECKPOINT

27.3 Draw a region of an alternating copolymer constructed from styrene and ethylene.

27.4 Draw a region of a block copolymer constructed from propylene and vinyl chloride.

27.5 Identify the monomers required to make the following alternating copolymer:

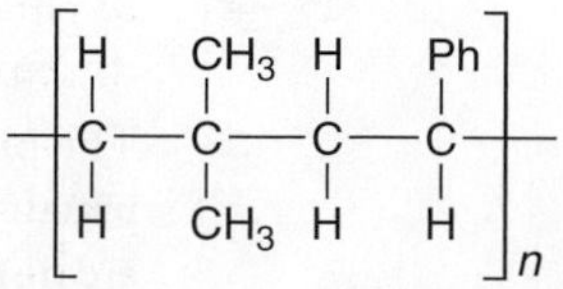

27.4 Polymer Classification by Reaction Type

Polymers can be classified in a variety of ways. Common classification schemes focus on either the type of reaction used to make the polymer, the mode of assembly, the structure, or the properties of the polymer. We will explore each of these classification schemes, beginning with the type of reaction used to make the polymer.

Addition Polymers

LOOKING BACK
Cationic addition was discussed in Chapter 9, free-radical addition was discussed in Chapter 11, and anionic addition was discussed in Chapter 22.

As seen in previous chapters, addition reactions involving π bonds can occur through a variety of mechanisms, including cationic addition, anionic addition, and free-radical addition. In much the same way, monomers can join together to form polymers via cationic addition, anionic addition, or free-radical addition. Polymers formed via any one of these processes are called **addition polymers**. A mechanism for radical polymerization was first discussed in Section 11.11 and is reviewed in Mechanism 27.1.

MECHANISM 27.1 RADICAL POLYMERIZATION

Initiation

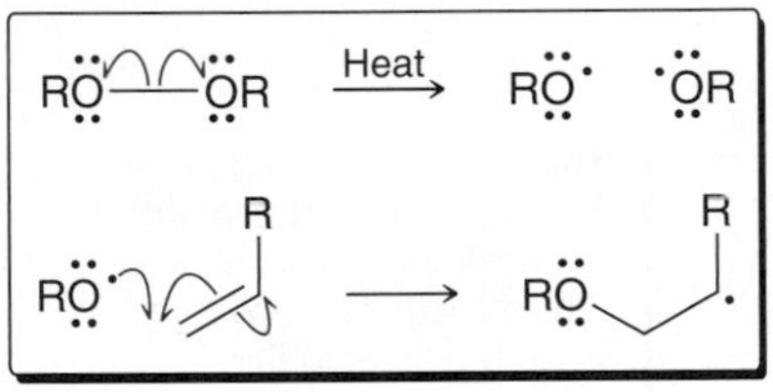

A radical initiator is formed, which reacts with a monomer to give a carbon radical

Propagation

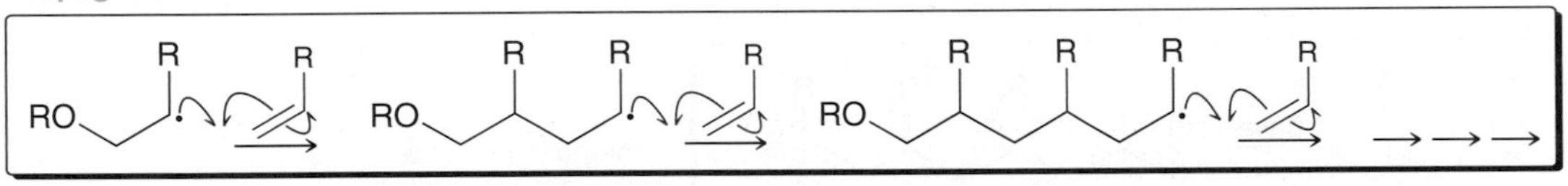

The radical addition process is repeated, building up the polymer chain

Termination

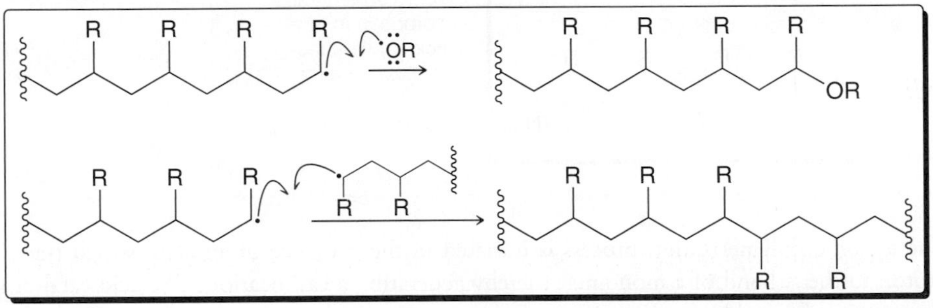

Each of these coupling reactions destroys radicals and is therefore an example of a termination step

In the initiation steps, a radical initiator is formed and then couples to one of the monomers, forming a carbon radical. This highly reactive intermediate then undergoes a propagation step in which it couples with another monomer. This process repeats itself, causing the polymer chain to grow. The process ends with a termination step in which two radicals couple together. Since the concentration of radicals is quite low at all times, the probability of two radicals coupling is rather small. As a result, thousands of monomers can be strung together in a single polymer chain before a termination step occurs. In this way, ethylene can be converted into polyethylene in the presence of a radical initiator. In fact, most derivatives of ethylene will also undergo radical polymerization under suitable conditions.

In contrast, cationic addition is only efficient with ethylene derivatives that contain an electron-donating group, such as the following examples.

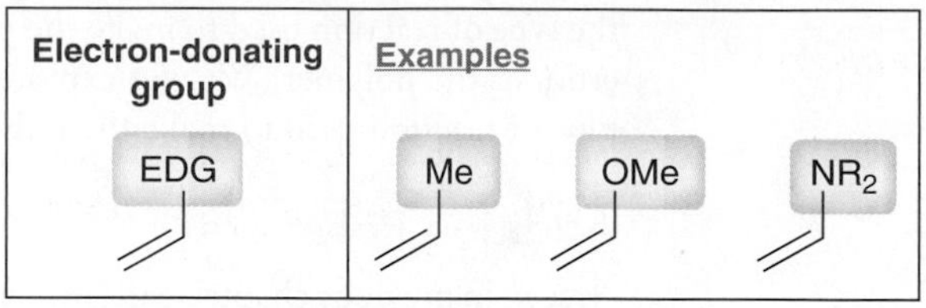

A mechanism for cationic polymerization was first discussed in the Practically Speaking box in Section 9.3 and is reviewed here in Mechanism 27.2.

MECHANISM 27.2 CATIONIC POLYMERIZATION

Initiation

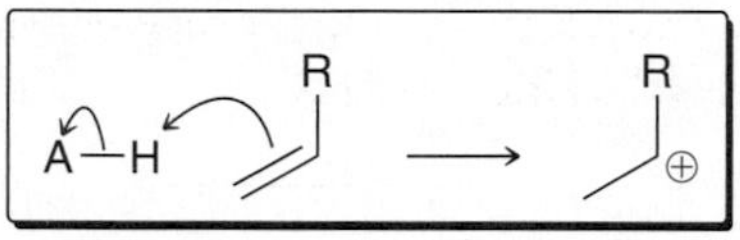

The π bond of a monomer is protonated, generating a carbocation intermediate

Propagation

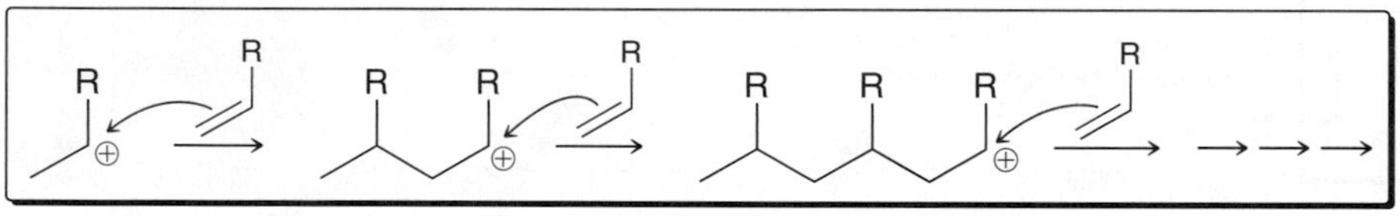

The π bond of a monomer acts as a nucleophile and attacks the carbocation. This process is repeated, building up the polymer chain

Termination

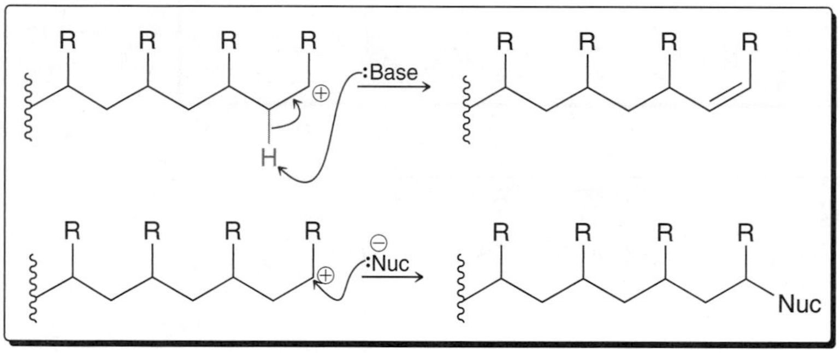

The polymerization process is terminated when either a base deprotonates the carbocation or a nucleophile attacks the carbocation

The cationic polymerization process is initiated in the presence of an acid, which transfers a proton to the π bond of a monomer, thereby generating a carbocation. The acid catalyst most often used is formed by treating BF_3 with water.

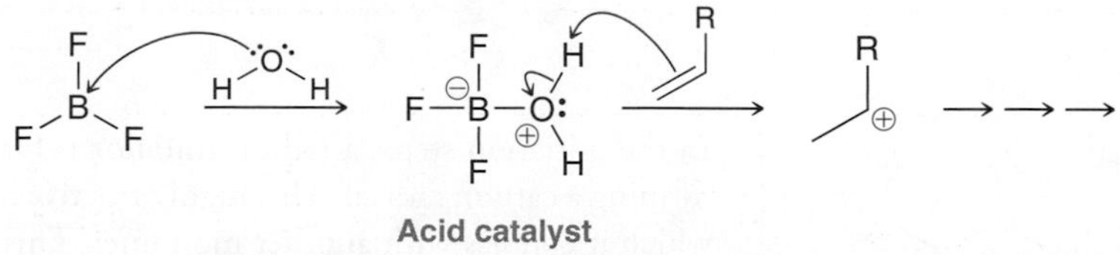

The carbocation generated during the initiation stage is then attacked by another monomer in a propagation step, and the process repeats itself, enabling the polymer chain to grow. The function of the electron-donating group is to stabilize the carbocation intermediate that is formed after the addition of each monomer to the growing polymer chain. For example, isobutylene readily undergoes cationic polymerization because a tertiary carbocation intermediate is formed during each propagation step.

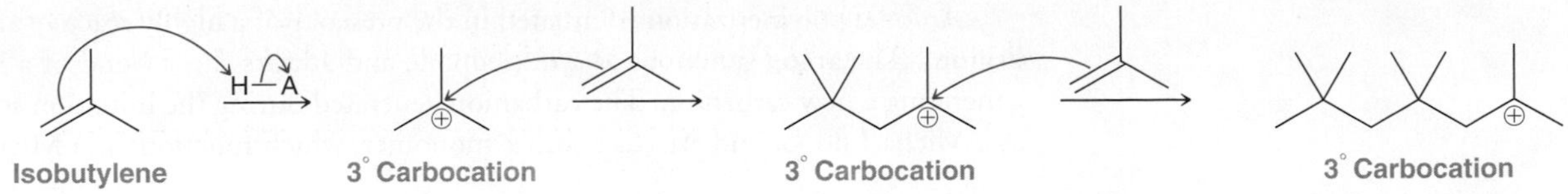

In contrast, ethylene does not readily undergo cationic polymerization, because the process would involve formation of a primary carbocation, which is not sufficiently stable to form at an appreciable rate.

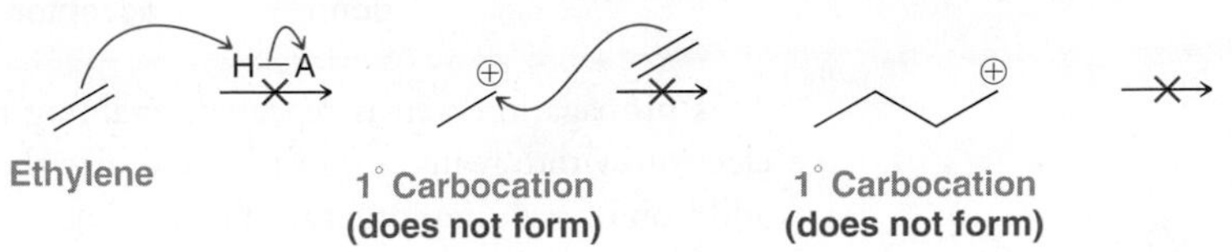

For monomers that undergo cationic polymerization, the process is terminated when the carbocation intermediate is deprotonated by a base or attacked by a nucleophile, as seen in Mechanism 27.2.

Unlike cationic addition, anionic addition is only efficient with ethylene derivatives that contain an electron-withdrawing group, such as the following examples.

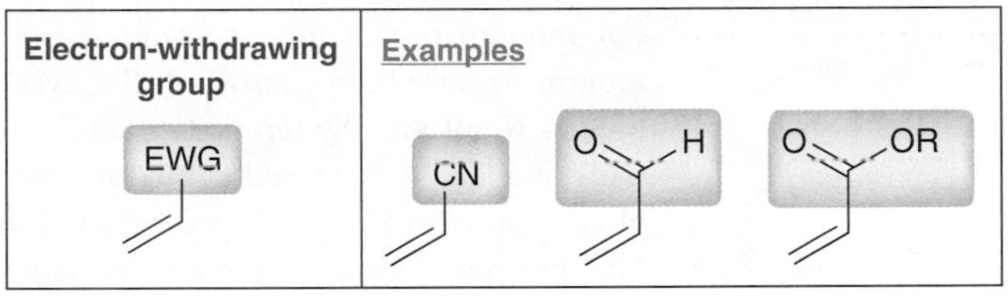

Electron-withdrawing groups were first discussed in Chapter 19, and a list can be found in Section 19.10. Anionic polymerization is believed to proceed via Mechanism 27.3.

MECHANISM 27.3 ANIONIC POLYMERIZATION

Initiation

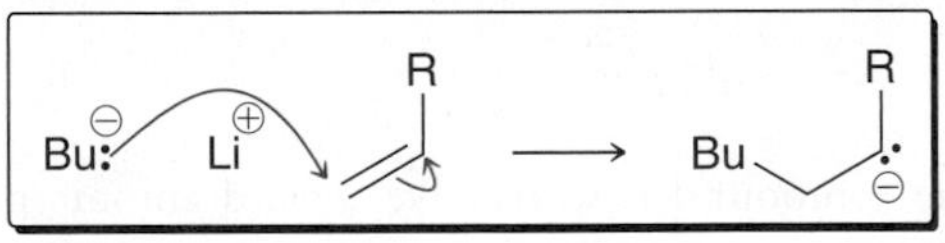

A carbon nucleophile initiates the process by attacking the π bond, generating a carbanion

Propagation

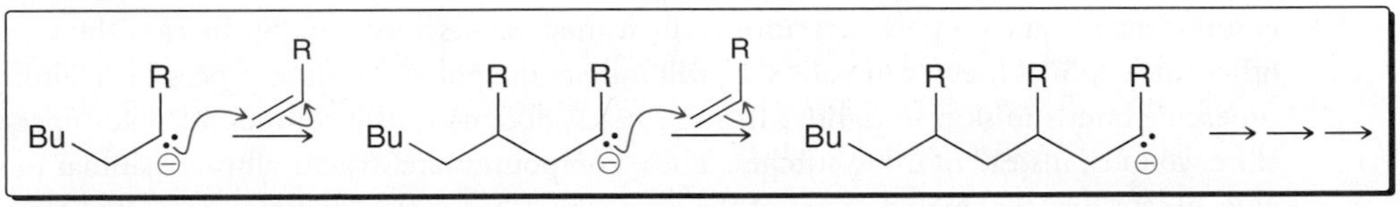

The carbanion acts as a nucleophile and attacks a monomer, creating a new carbanion. This process is repeated, building up the polymer chain

Termination

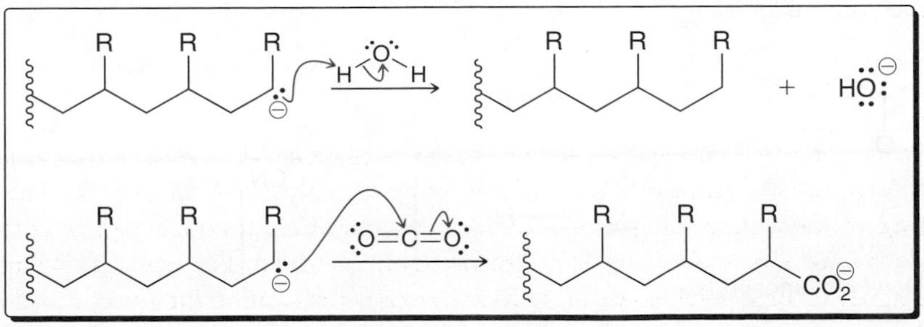

The polymerization process is terminated when the growing polymer is treated with a weak acid (such as H_2O) or with an electrophile (such as CO_2)

Anionic polymerization is initiated in the presence of a highly reactive anion, such as butyl lithium. The anion functions as a nucleophile and attacks the π bond of a monomer, thereby generating a new carbanion. The carbanion generated during the initiation stage then functions as a Michael donor and attacks another monomer, which functions as a Michael acceptor:

LOOKING BACK
For a review of Michael donors and Michael acceptors, see Section 22.6.

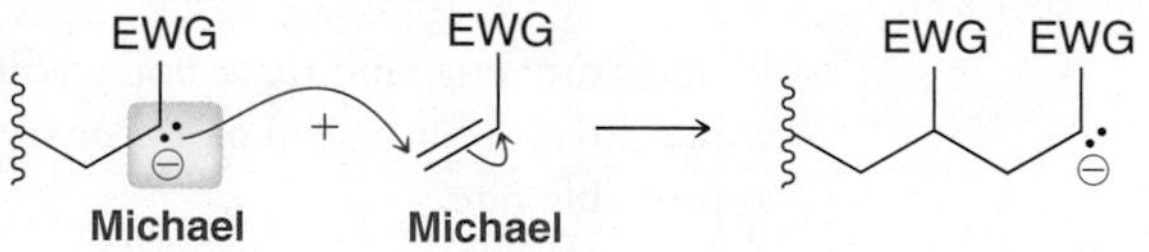

This propagation step is repeated, enabling the polymer chain to grow. The function of the electron-withdrawing group is to stabilize the carbanion intermediate that is formed after the addition of each monomer to the growing polymer chain.

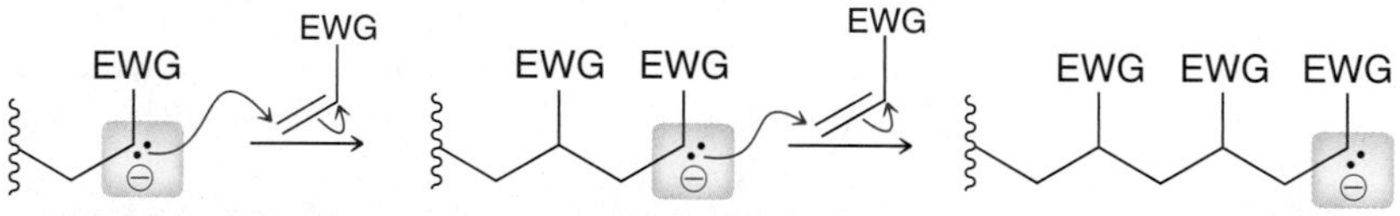

For monomers that undergo anionic polymerization, the process continues until all of the monomers have been consumed. But even after all of the monomers have been exhausted, the process is not actually terminated until a suitable acid (such as water) or electrophile (such as CO_2) is added to the reaction mixture. In the absence of a suitable acid or electrophile, the end of each polymer chain will possess a stabilized carbanion site, and the polymerization process can continue if more monomers are added to the reaction mixture. For this reason polymers generated through this process are often called **living polymers**.

Superglue is a common example of a monomer that readily polymerizes via an anionic addition process. Superglue is a pure solution of methyl-α-cyanoacrylate, which is a monomer containing two electron-withdrawing groups.

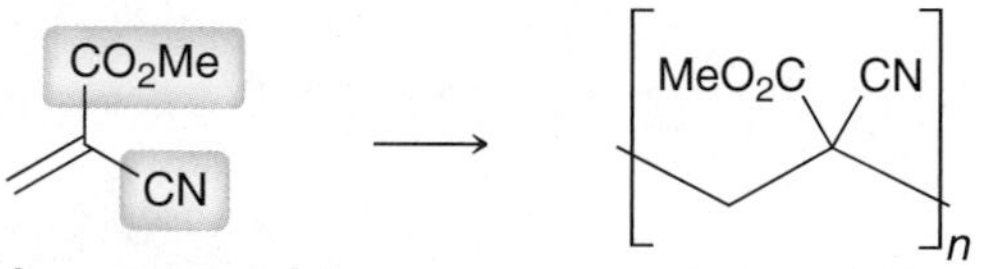

Methyl α-cyanoacrylate (superglue)

With two electron-withdrawing groups, the compound is so reactive toward anionic polymerization that even a weak nucleophile, such as water, will initiate the polymerization process. When superglue is applied to a metal surface, the moisture on the surface of the metal is sufficient to catalyze polymerization, which then occurs very rapidly. In fact, the water and other nucleophiles present in your skin will initiate the polymerization process, explaining why superglue bonds to skin so tightly. In some cases, doctors utilize compounds like superglue to close wounds, instead of using stitches. These compounds are structurally very similar to superglue, except that the methyl group of the ester is replaced with a slightly larger alkyl group. For example, Dermabond®, a cyanoacrylate ester with a 2-octyl group, was developed to replace stitches in certain situations.

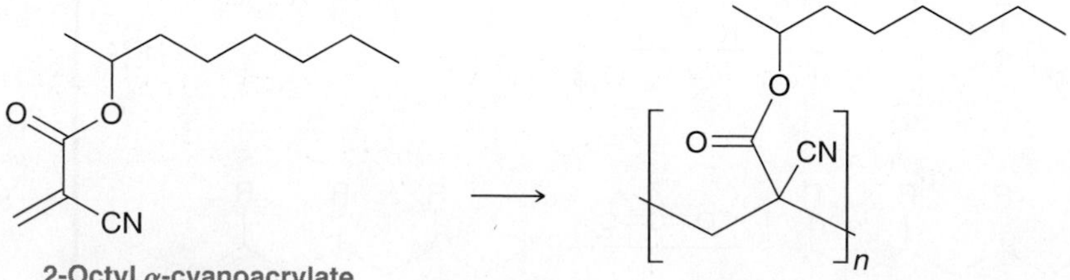

2-Octyl α-cyanoacrylate (Dermabond®)

SKILLBUILDER

27.1 DETERMINING THE MORE EFFICIENT POLYMERIZATION TECHNIQUE

LEARN the skill

Determine whether preparation of the following polymer would be best achieved via cationic addition or anionic addition.

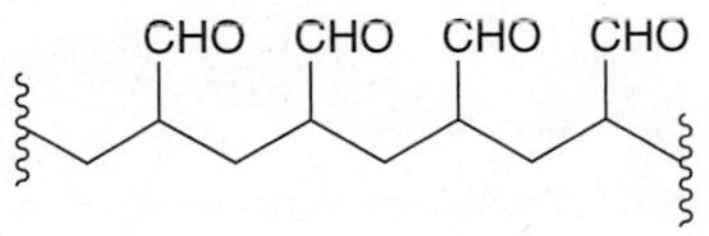

SOLUTION

First identify the repeating units in the polymer structure.

STEP 1 Identify the repeating units.

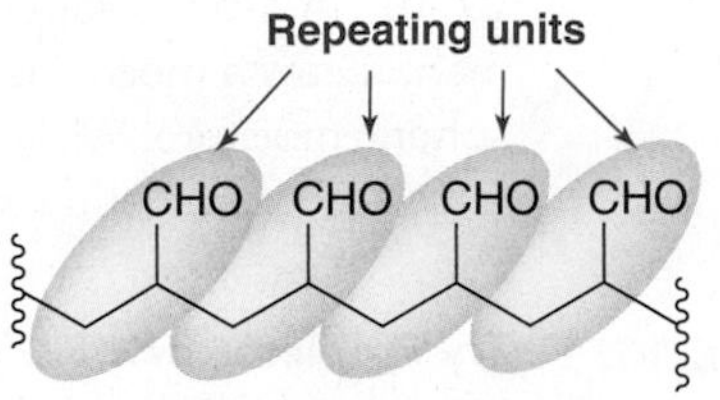

Next identify the monomer necessary to prepare a polymer with those repeating units.

STEP 2 Identify the required monomer.

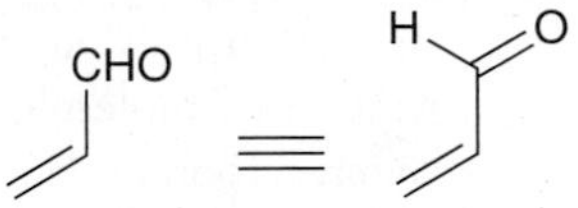

STEP 3 Determine the nature of the vinylic group.

Now determine whether the group attached to the vinyl position is an electron-withdrawing group or an electron-donating group, as first described in Section 19.10. In this case, the necessary monomer has a carbonyl group, which is an electron-withdrawing group because it can stabilize a negative charge via resonance. We therefore expect this monomer to polymerize most efficiently under basic conditions.

STEP 4 Use anionic conditions if the vinylic group is electron withdrawing and cationic conditions for an electron-donating group.

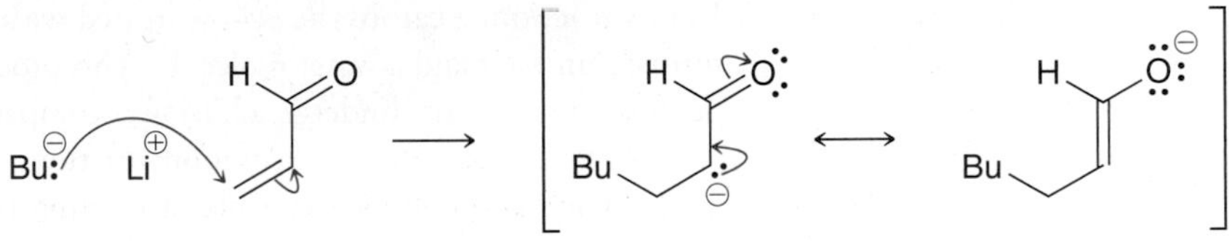

Resonance-stabilized

PRACTICE the skill **27.6** Determine whether preparation of each of the following polymers would best be achieved via cationic addition or anionic addition.

(a) 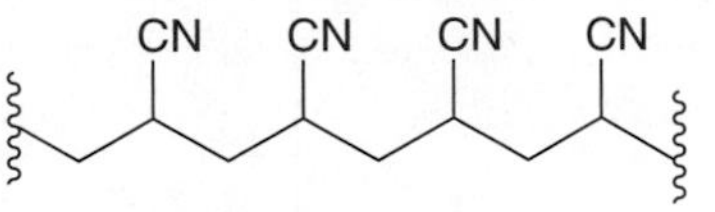

(b) 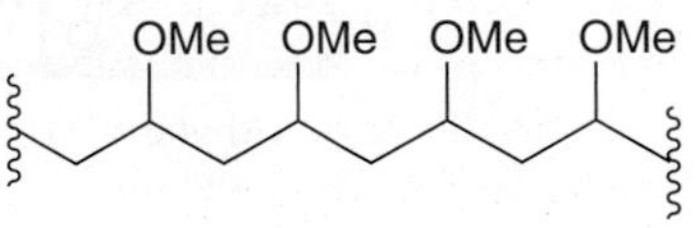

(c) 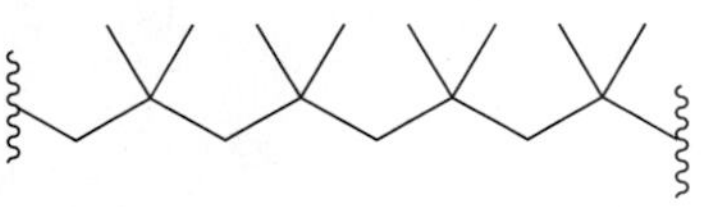

(d) 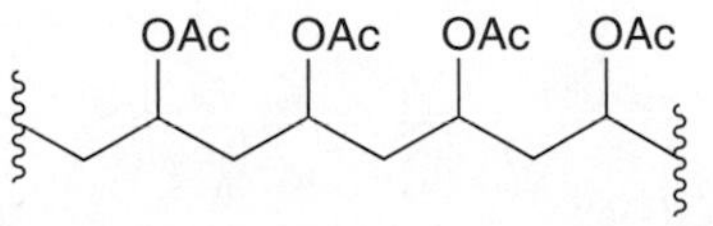

(e) 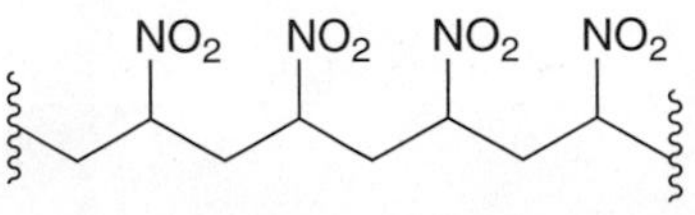

(f) 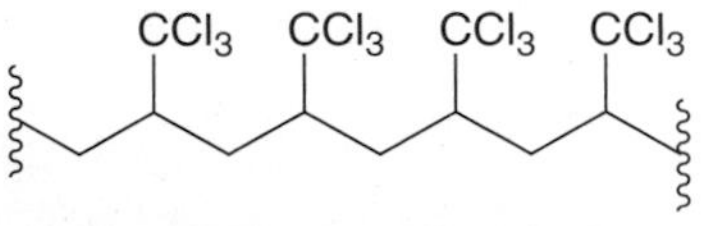

APPLY the skill

27.7 Arrange the following monomers in order of reactivity toward cationic polymerization.

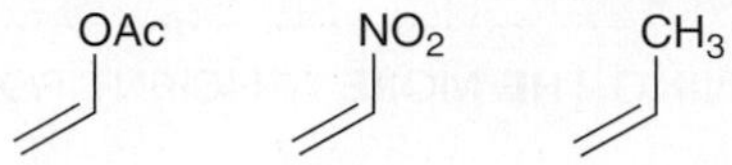

27.8 Arrange the following monomers in order of reactivity toward anionic polymerization.

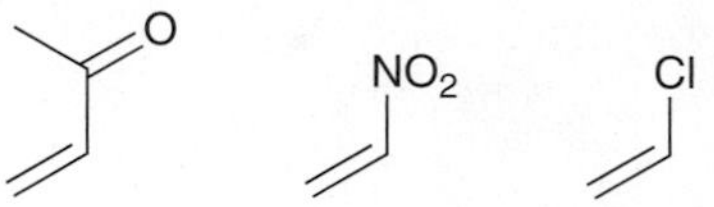

27.9 Many monomers that readily undergo cationic polymerization, such as isobutylene, will not readily undergo anionic polymerization. Styrene, however, can be effectively polymerized via cationic, anionic, or radical addition. Explain.

27.10 A tube of superglue will harden if it remains open too long in a humid environment. Draw a mechanism that shows how superglue polymerizes in the presence of atmospheric moisture. When drawing your mechanism, make sure to draw the initiation step, at least two propagation steps, and a termination step involving water as the acid.

need more **PRACTICE?** **Try Problems 27.27, 27.28, 27.34, 27.38–27.40**

Condensation Polymers

As seen in previous chapters, the term "condensation" is used to characterize any reaction in which two molecules undergo addition accompanied by the loss of a small molecule such as water, carbon dioxide, or nitrogen gas. Throughout this book, we have seen many condensation reactions. One such example is the Fischer esterification process, as seen in Section 21.10.

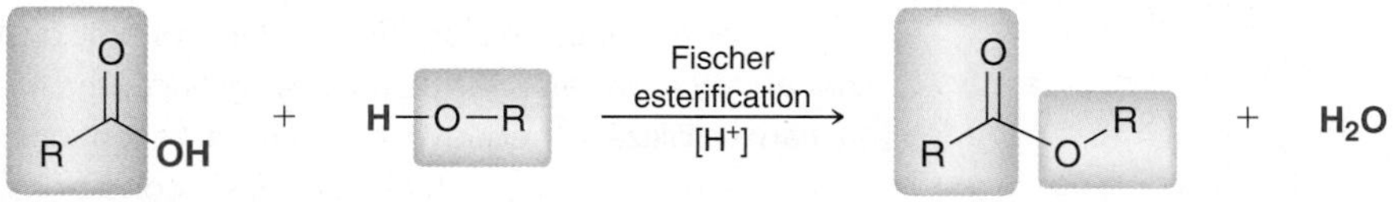

In this reaction, a carboxylic acid is treated with an alcohol in the presence of an acid catalyst, forming an ester and a water molecule. This process is considered to be a condensation reaction because the reactants undergo addition accompanied by the loss of a water molecule.

Now consider the condensation reactions that can occur when a diacid is treated with a diol. Each compound is capable of reacting twice, enabling formation of a polymer. As an example, consider the formation of poly(ethylene terephthalate), also known simply as PET, which is used to make soft-drink bottles.

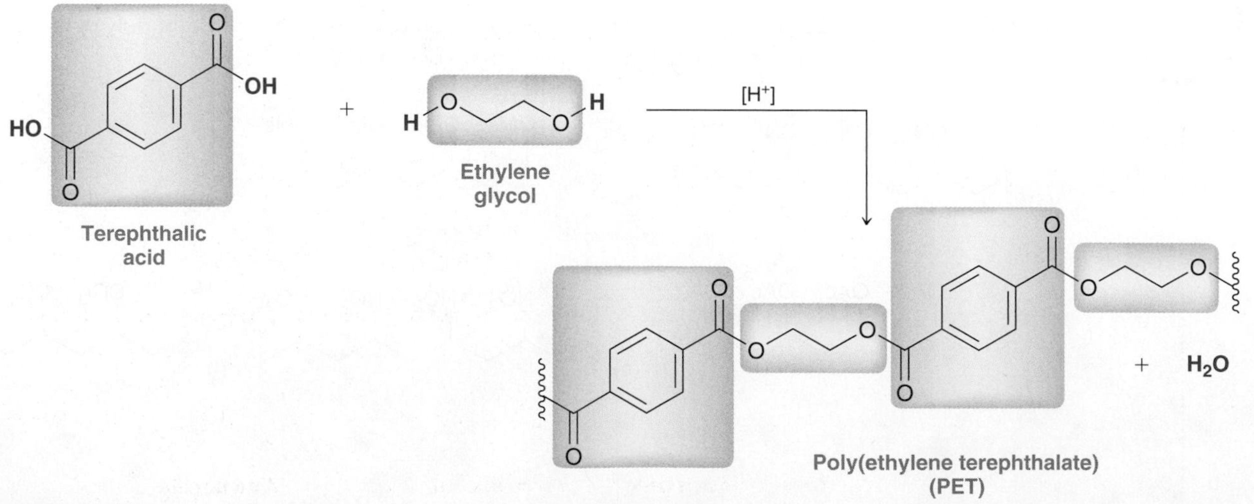

As shown, PET is prepared by successive Fischer esterification reactions. Since the polymer is generated via condensation reactions, it is called a **condensation polymer**. Because PET has repeating ester moieties, the polymer is also classified as a polyester. Like PET, nylon 6,6 is also prepared via condensation reactions and is also a condensation polymer. Nylon 6,6 is a polyamide, which is prepared from adipic acid and 1,6-hexanediamine.

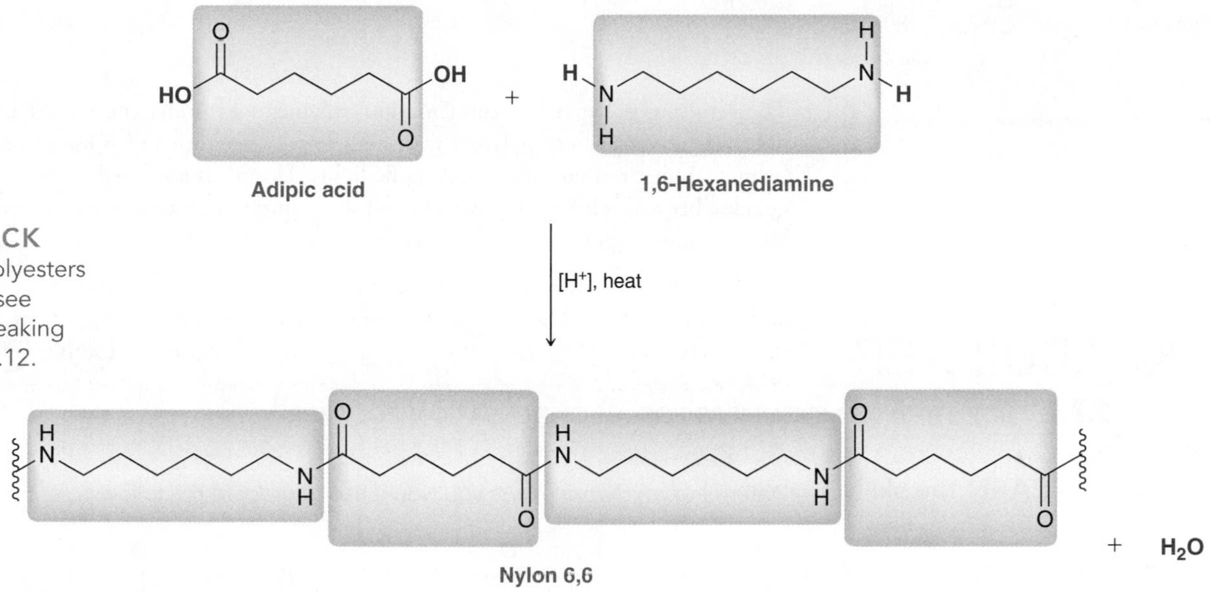

LOOKING BACK
For a review of polyesters and polyamides, see the Practically Speaking box in Section 21.12.

CONCEPTUAL CHECKPOINT

27.11 Draw the mechanism of formation of PET in acidic conditions. It might be helpful to first review the mechanism for the Fischer esterification process.

27.12 Draw the polymer that would be generated from the acid-catalyzed reaction between oxalic acid and resorcinol.

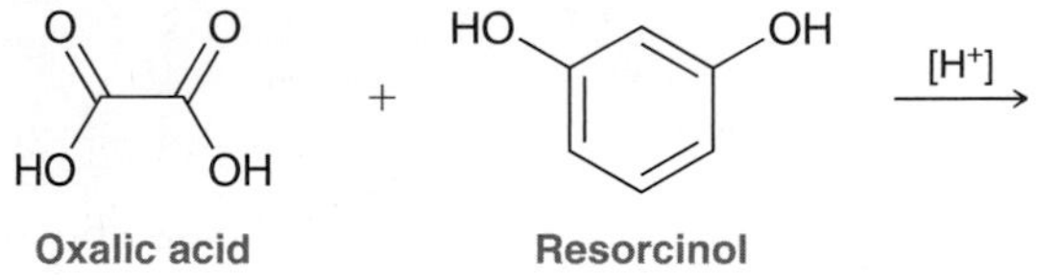

Polycarbonates are similar in structure to polyesters, but with repeating carbonate moieties instead of repeating ester moieties:

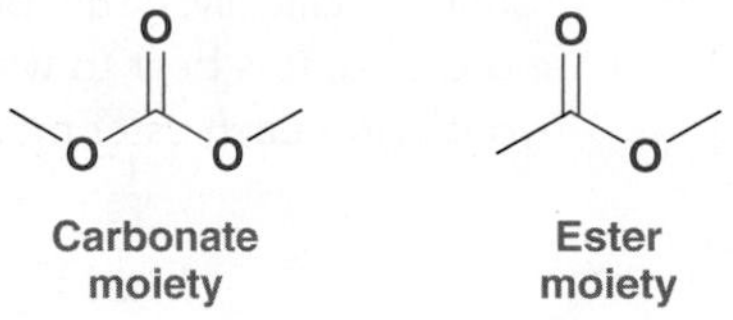

Carbonates can be formed from the reaction between phosgene and an alcohol.

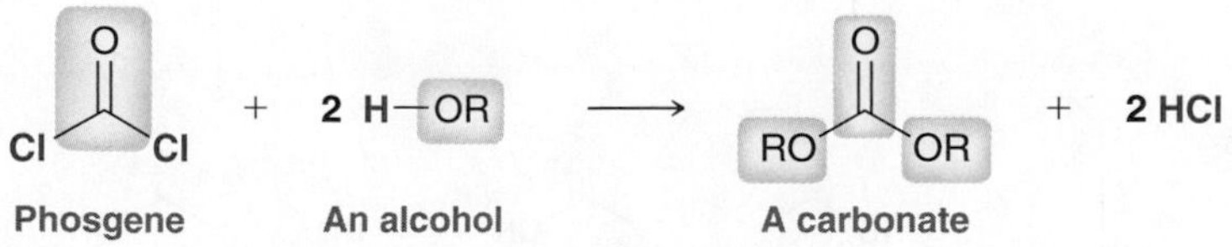

Polycarbonates are formed in a similar way, by treating phosgene with a diol. As an example, consider the reaction between phosgene and bisphenol A.

Phosgene

Bisphenol A → **Lexan** + **HCl**

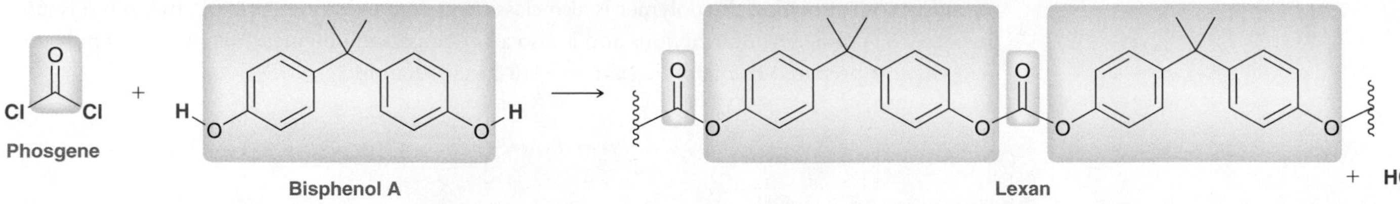

The resulting polymer is a condensation polymer sold under the trade name Lexan®. It is a lightweight, transparent polymer with high impact strength and is used to make bicycle safety helmets, bullet resistant glass, and traffic lights. Lexan® is also used in the production of safety goggles, because it is both lightweight and shatterproof. In recent years, polycarbonates have also become quite popular for the production of CDs and DVDs.

SKILLBUILDER

27.2 IDENTIFYING MONOMERS FOR A CONDENSATION POLYMER

LEARN the skill

Determine which monomers you would use to prepare the following condensation polymer:

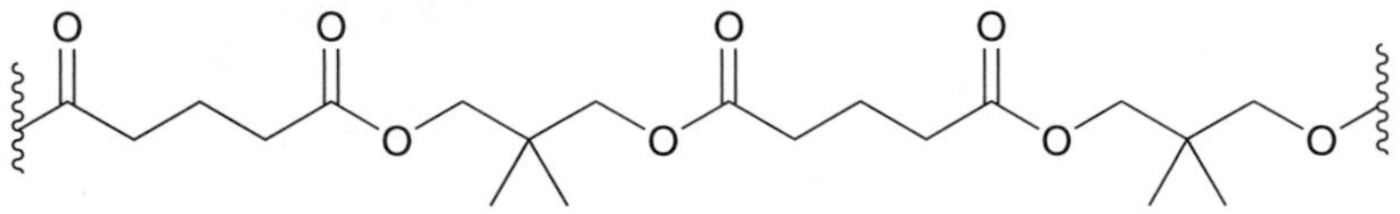

SOLUTION

First identify the type of functional group that repeats itself

STEP 1
Identify the repeating functional group.

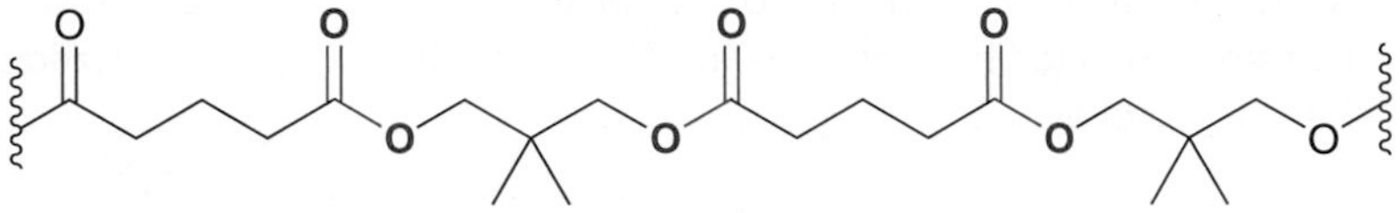

A polyester

In this case, the repeating functional group is an ester moiety, so this condensation polymer is a polyester. Recall that ester moieties can be prepared from the reaction between a carboxylic acid and an alcohol via a Fischer esterification process.

STEP 2
Identify the type of reaction that will produce the desired functional group.

$$R\text{—}COOH + H\text{—}O\text{—}R \xrightarrow[\text{[H}^+\text{]}]{\text{Fischer esterification}} R\text{—}CO\text{—}O\text{—}R + H_2O$$

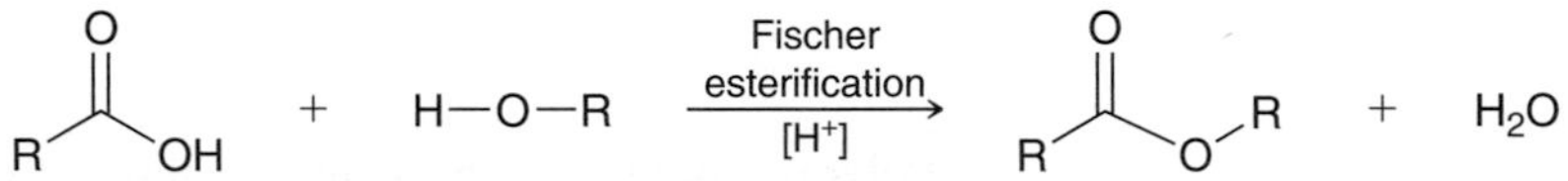

Preparing a polymer that exhibits repeating ester moieties requires reactants that are difunctional, specifically, a suitable diacid and diol must be selected. To determine the starting diol and diacid, it is best to work backward from the structure of the polyester, that is, imagine hydrolyzing each ester moiety (the reverse of a Fischer esterification).

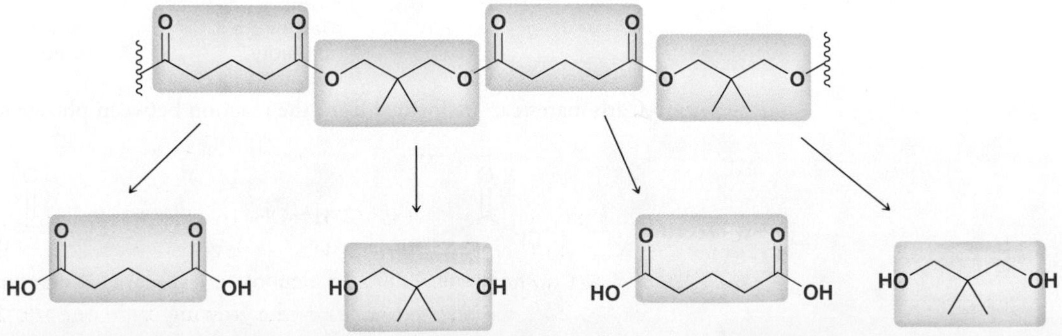

STEP 3
For a condensation polymer, identify the two difunctional monomers that are necessary.

Inspection of the individual monomers reveals that two different monomers are required. The following diacid and diol can be used to prepare the desired polymer.

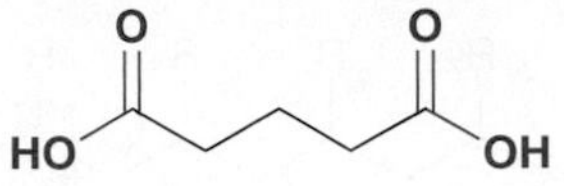

PRACTICE the skill **27.13** Kevlar is a condensation polymer used in the manufacture of bulletproof vests. Identify the monomers required for the preparation of Kevlar.

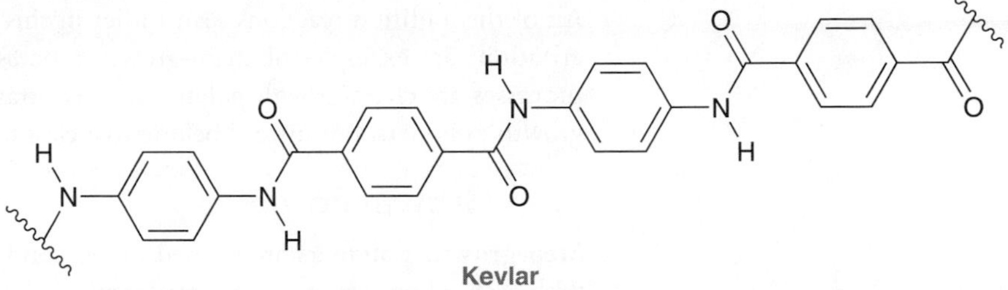

Kevlar

27.14 Identify the monomers required to make each of the following condensation polymers.

(a)

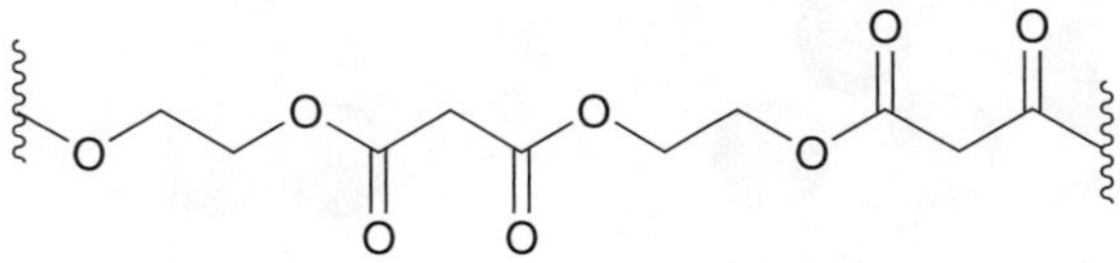

(b)

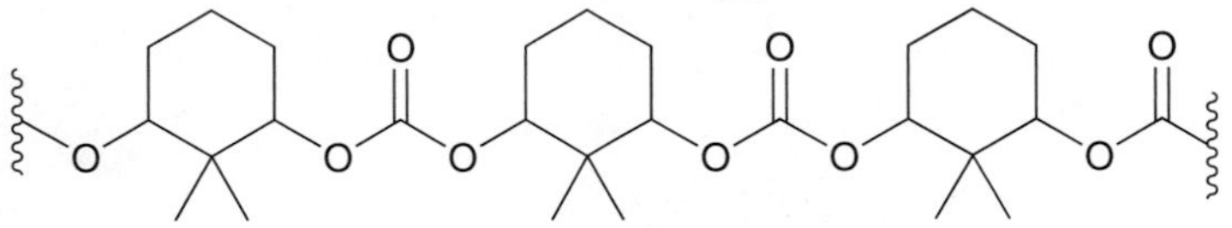

APPLY the skill

27.15 Identify the condensation polymer that would be produced when phosgene is treated with 1,4-butanediol.

27.16 Nylon 6 is a polyamide used in the manufacture of ropes. It can be prepared via hydrolysis of ε-caprolactam to form ε-aminocaproic acid followed by acid-catalyzed polymerization.

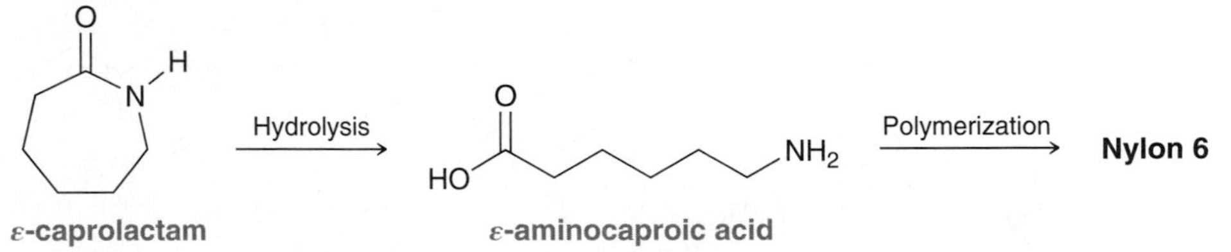

(a) Draw the structure of Nylon 6.

(b) Compare the structure of Nylon 6 with the structure of Nylon 6,6 (shown just before Conceptual Checkpoint 27.11). Which polymer exhibits a smaller repeating unit?

need more **PRACTICE?** **Try Problems 27.23, 27.30 a,b,d, 27.31, 27.32**

27.5 Polymer Classification by Mode of Assembly

Chain-Growth Polymers

Polymers are also often classified as either chain-growth or step-growth polymers. These terms are used to signify the way in which the polymers are assembled. **Chain-growth polymers** are formed under conditions in which the monomers do not react directly with each other, but rather, each monomer is added to the growing chain one at a time. The growing polymer chain generally has only one reactive site, called a *growth point*, and the monomers attach to the chain at the growth point.

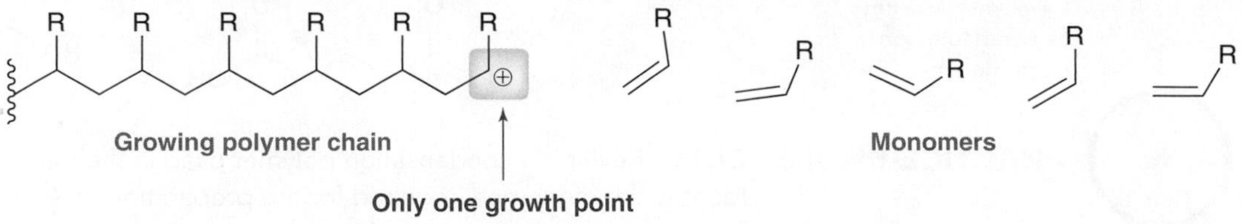

All of the addition reactions seen earlier in this chapter (radical, cationic, and anionic polymerization) are examples of chain-growth processes. Polymers created by any of these addition processes are chain-growth polymers. In contrast, most condensation polymers are not chain-growth polymers, but instead belong to a class called step-growth polymers.

Step-Growth Polymers

Step-growth polymers are formed under conditions in which the individual monomers react with each other to form **oligomers** (compounds constructed from just a few monomers), which are then joined together to form polymers.

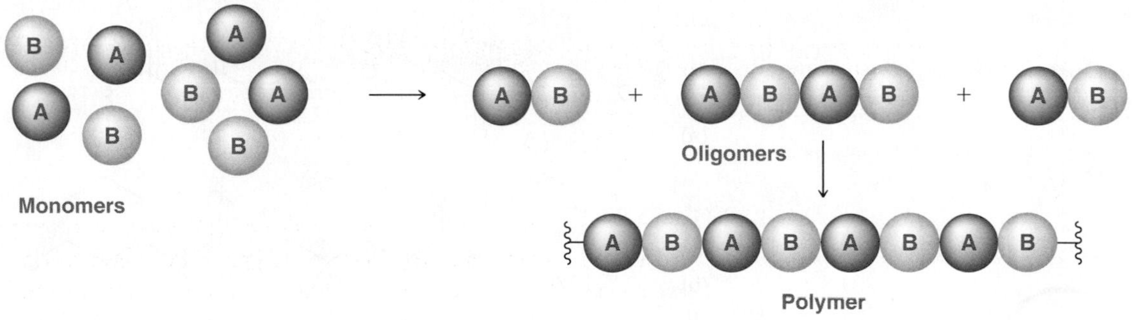

This mode of assembly occurs when difunctional monomers are used, just as we saw with condensation polymers in the previous section. In such a case, all monomers and oligomers have two growth points instead of just one. As an example, consider the formation of PET.

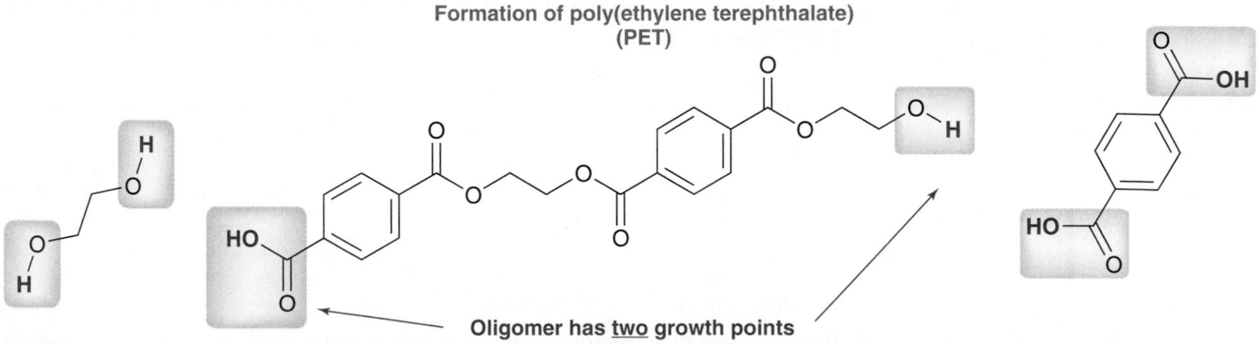

In general, step-growth polymers are usually condensation polymers, while chain-growth polymers are usually addition polymers, but there are exceptions, such as polyurethanes. **Polyurethanes** are made up of repeating urethane moieties, also sometimes called carbamate moieties.

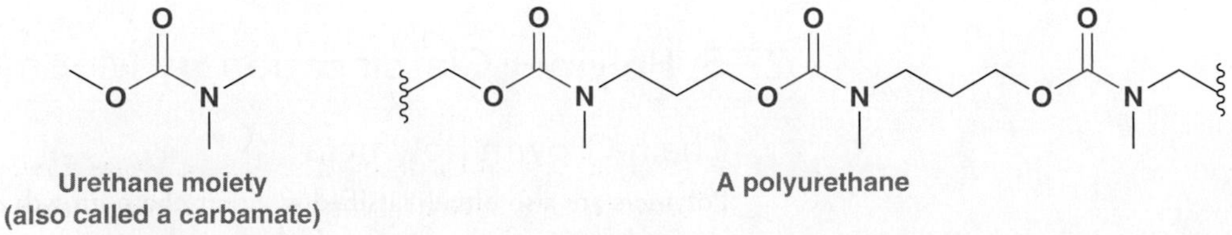

Urethanes can be prepared by treating an isocyanate with an alcohol.

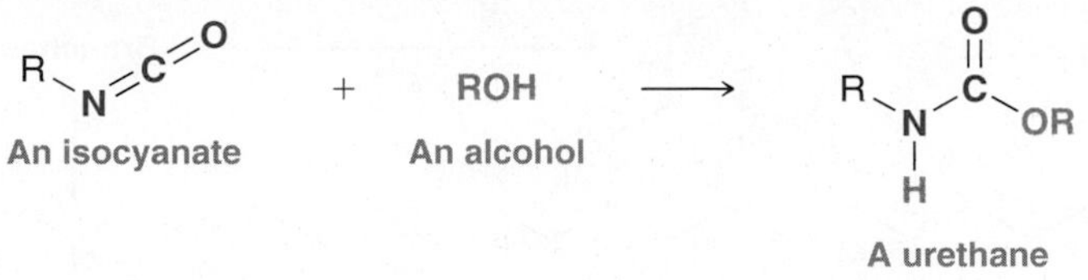

Polyurethanes can be prepared by treating a diisocyanate with a diol. In preparing polyurethanes, the diols most commonly used are small polymers that bear hydroxyl ends.

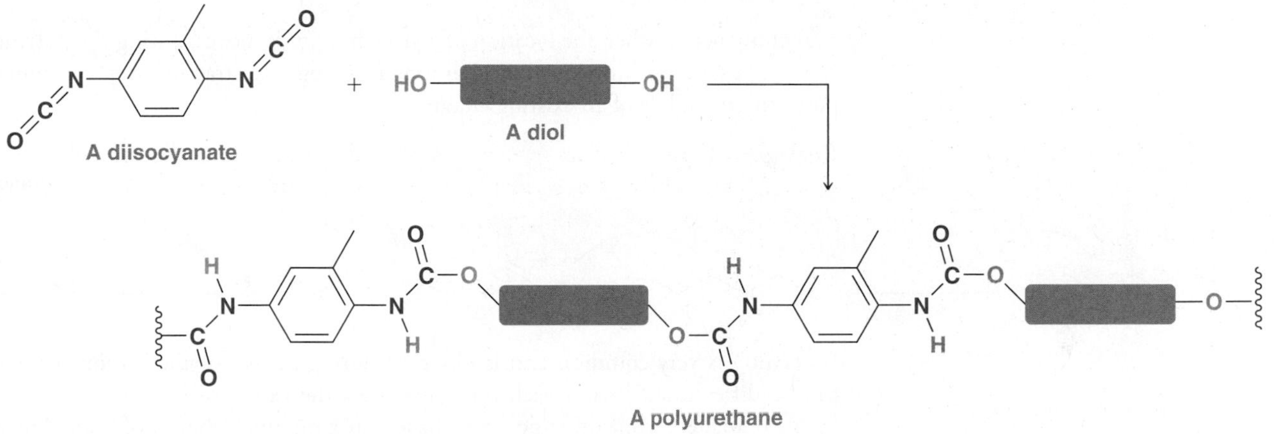

This process is not a condensation, because it does not involve the loss of a small molecule. Polyurethanes are technically addition polymers, but nevertheless, they are step-growth polymers because the growing polymer chain has two growth points rather than one. This example illustrates that step-growth polymers cannot always be classified as condensation polymers. Polyurethanes are used for insulation in the construction of homes and portable coolers.

CONCEPTUAL CHECKPOINT

27.17 Would these pairs of monomers form chain-growth or step-growth polymers?

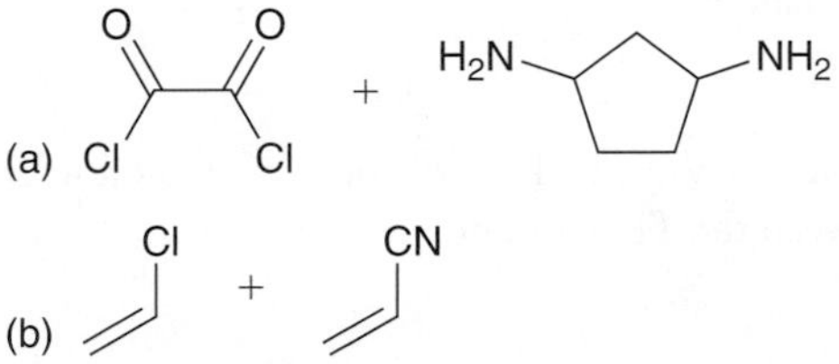

27.18 Should this polymer be classified as a chain-growth polymer or a step-growth polymer?

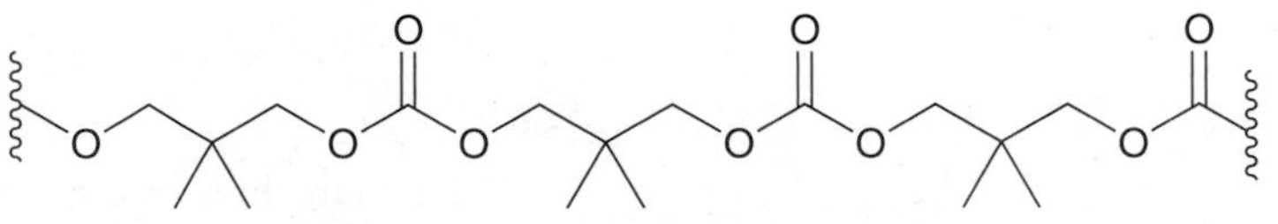

27.6 Polymer Classification by Structure

Branched Polymers

Polymers are also often classified by their structure. **Branched polymers** contain a large number of branches connected to the main chain of the polymer.

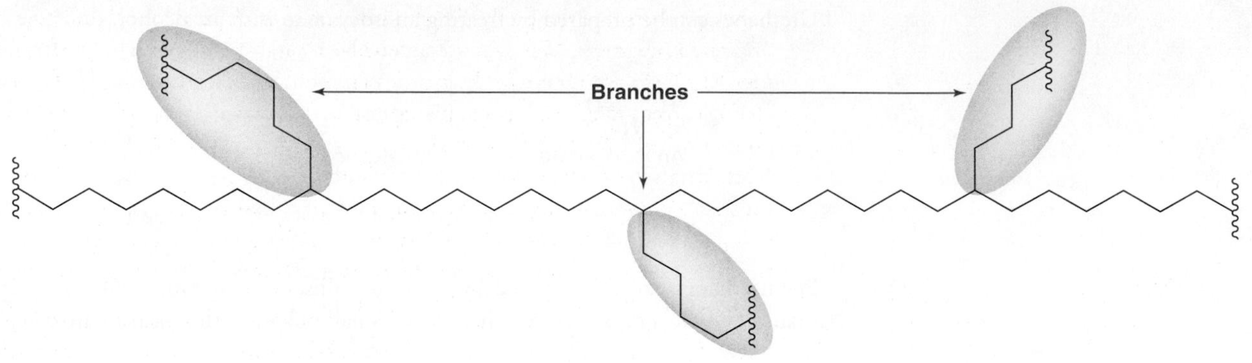

Branching occurs when the location of a growth point is moved during the polymerization process. For example, during radical addition, a hydrogen abstraction step can produce a growth point to the middle of an existing chain:

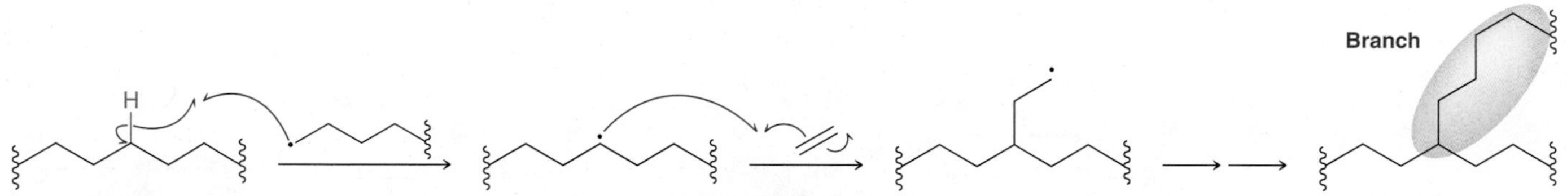

Branching is very common and is observed during many polymerization processes. Polymers can be differentiated from each other based on the extent of branching. Some polymers are highly branched, while other polymers have only a minimal amount of branching or no branching at all. The former are called branched polymers, while the latter are called **linear polymers**. The synthesis of linear polymers often requires the use of special catalysts, described in the following section.

Linear Polymers

The field of polymer chemistry was revolutionized in 1953 with the development of Ziegler-Natta catalysts. These catalysts are organometallic complexes that can be prepared by treating an alkyl aluminum compound, such as Et_3Al, with $TiCl_4$. Although the mechanism is not entirely understood, it is widely believed that the active form of the catalyst is an alkyltitanium intermediate with a vacant coordination site.

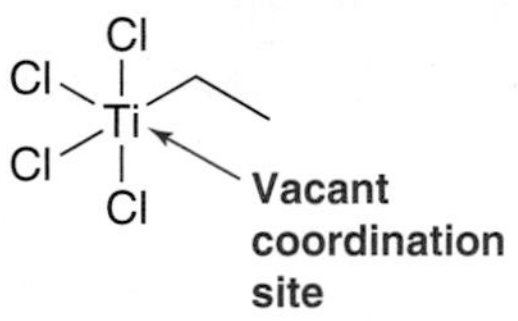

Titanium has six coordination sites, one of which is vacant. The π bond of a monomer coordinates with this site and is then inserted between the Ti—C bond.

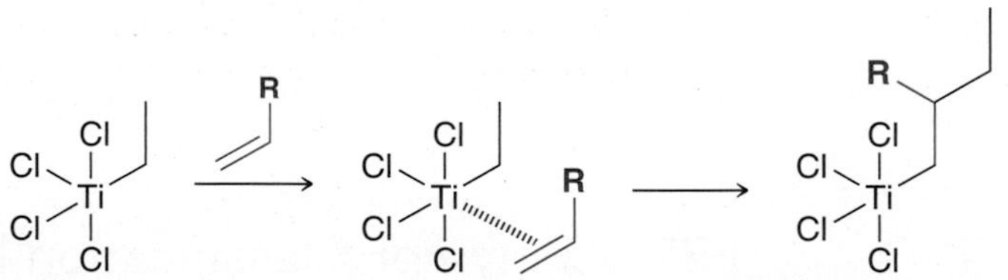

The insertion step creates a new vacant coordination site, allowing the process to be repeated indefinitely. The development of Ziegler-Natta catalysts enabled scientists to control the polymerization process in the following ways:

1. *Extent of branching*. Free radicals are not involved as intermediates when Ziegler-Natta catalysts are employed, and therefore very little branching occurs. Polymers produced with

these catalysts are generally linear polymers, which have very different properties than branched polymers. Without a Ziegler-Natta catalyst, ethylene polymerizes to form a polymer that exhibits about 20 branches per thousand carbon atoms. The resulting polymer chains cannot effectively pack, giving rise to a material with a relatively low density (~0.92 g/cm^3) called low-density polyethylene, or LDPE. However, in the presence of a Ziegler-Natta catalyst, ethylene polymerizes to form a polymer that exhibits about five branches per thousand carbon atoms. The resulting polymer chains can pack quite efficiently, giving rise to a material with a relatively high density, ~0.96 g/cm^3) called high-density polyethylene, or HDPE. LDPE is used to make trash bags, while HDPE has greater strength and is used to make plastic squeeze bottles and Tupperware.

2. *Stereochemical control.* When polymerization is performed using monosubstituted ethylenes as monomers, the resulting polymer has a large number of chirality centers. The relative configurations of these chirality centers are classified as isotactic, syndiotactic, or atactic.

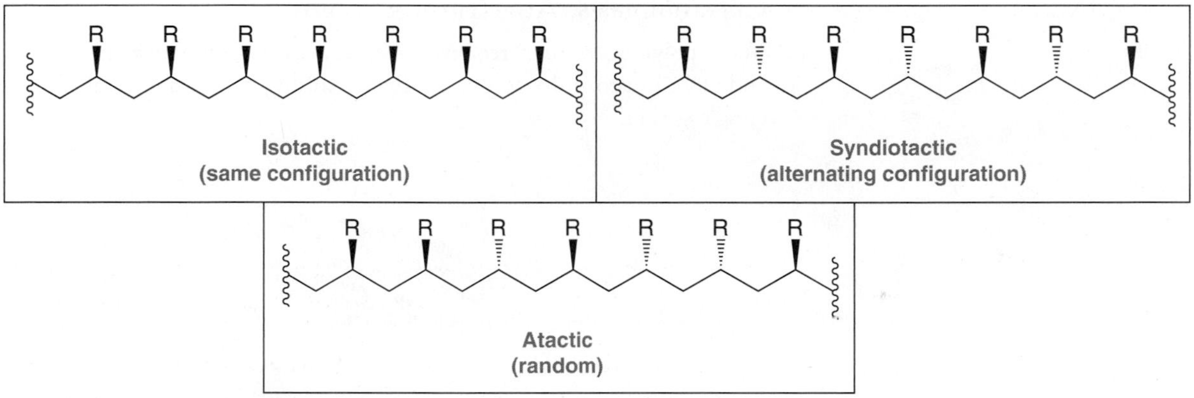

The term **isotactic** is used when all of the chirality centers have the same configuration, **syndiotactic** is used when the chirality centers have alternating configuration, and **atactic** is used when the chirality centers are not arranged in a pattern (they have random configurations). Isotactic, syndiotactic, and atactic polymers exhibit different properties, and all three can be made with the appropriate Ziegler-Natta catalyst.

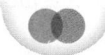

CONCEPTUAL CHECKPOINT

27.19 Polyisobutylene cannot be described as isotactic, syndiotactic, or atactic. Explain.

27.20 Polyethylene is used to make Ziploc bags and folding tables. Identify which of these applications is most likely to be made from HDPE and which is most likely to be made from LDPE.

Cross-Linked Polymers

In the Practically Speaking box at the end of Section 17.5, we discussed natural and synthetic rubbers. Recall from that discussion that the vulcanization process introduces disulfide bridges between neighboring chains within the polymer.

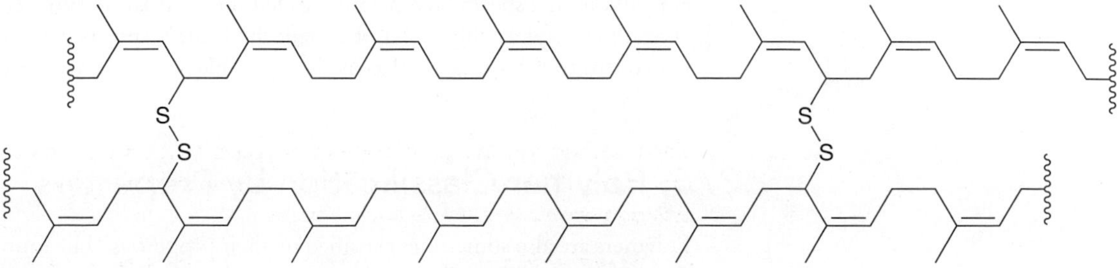

Vulcanized rubber

The resulting polymer is said to be a **crossed-linked polymer**, which markedly affects the properties of the polymer. Disulfide bridges are not the only way for chains to cross-link. For example, chains can also be cross-linked with branches.

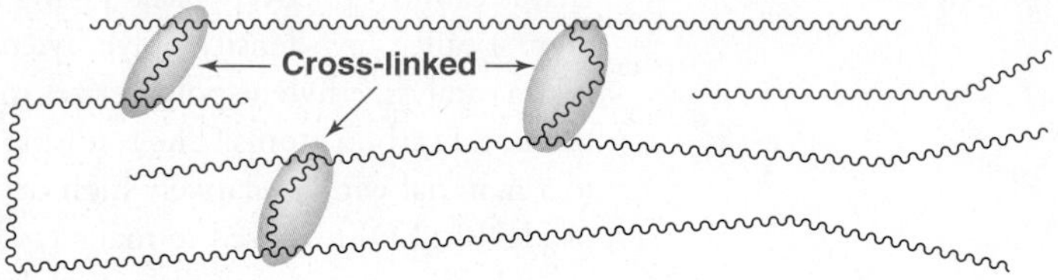

The conditions of polymer formation can be controlled so as to favor either a greater or lesser amount of cross-linking. This difference can have a profound impact on the properties of the resulting polymer, as we will see in the next section.

Crystalline vs. Amorphous Polymers

Many polymers contain regions, called **crystallites**, in which the chains are linearly extended and close in proximity to one another, resulting in van der Waals forces that hold the chains close together.

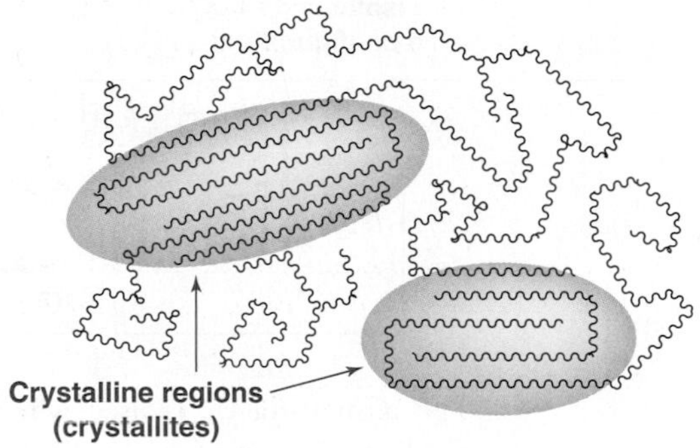

The regions that are not crystalline are called **amorphous** regions. Crystalline regions render a polymer hard and durable, while amorphous regions render a polymer flexible. The degree of crystallinity of a polymer, and therefore its physical properties, greatly depends on the steric requirements of the substituent(s) present in the repeating unit of the polymer. For example, compare the structures of polyethylene and polyisobutylene.

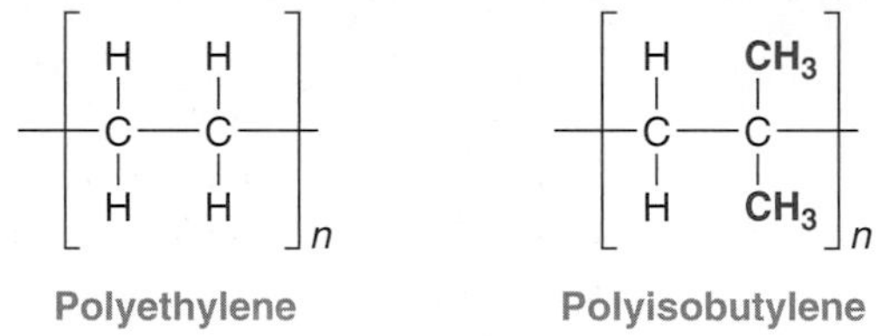

Linear polyethylene (HDPE) exhibits a high degree of crystallinity because there are no substituents to prevent the chains from closely packing. In contrast, polyisobutylene exhibits a low degree of crystallinity, because there are two methyl groups that provide steric bulk, preventing the chains from closely packing.

When highly crystalline polymers like HDPE are heated, the crystalline regions become amorphous at a specific temperature called the **melt transition temperature** (T_m). When noncrystalline polymers like polyisobutylene are heated, they become very soft. The temperature at which this transition takes place is called the **glass transition temperature** (T_g).

27.7 Polymer Classification by Properties

Polymers are also sometimes classified by their properties. The four most common categories are thermoplastics, elastomers, fibers, and thermosetting resins.

Thermoplastics

Thermoplastics are polymers that are hard at room temperature but soft when heated; that is, they exhibit a high T_g. Polymers in this category are very useful because they can be easily molded and are commonly used to make toys and storage containers. For example, PET is a thermoplastic used in the manufacture of soft-drink bottles. Other examples of thermoplastics include polystyrene, polyvinyl chloride (PVC), and low-density polyethylene.

Many thermoplastics become brittle at room temperature, which severely limits their utility. This is true of PVC. Pure PVC is highly susceptible to cracking at room temperature and is therefore useless for most applications. To avoid this problem, the polymer can be prepared in the presence of small molecules called **plasticizers**. These molecules become trapped between the polymer chains where they function as lubricants. Common plasticizers are dialkyl phthalates, such as di-2-ethylhexyl phthalate used in vinyl upholstery, raincoats, shower curtains, inflatable boats, and garden hoses. Some plasticizers evaporate slowly with time, and the polymer ultimately returns to a brittle state in which it can be easily cracked. Intravenous (IV) drip bags used in hospitals are typically made from PVC with plasticizers.

Elastomers

Elastomers are polymers that return to their original shape after being stretched. Elastomers are typically amorphous polymers that have a small degree of cross-linking. Natural rubber (described in Section 17.5) is one of the most common examples of an elastomer. A more contemporary example is Spandex, which is a polyurethane with a small degree of cross-linking. Spandex fibers are used to make bathing suits and athletic gear, among other applications.

Chewing gum is a complex mixture of polymers, many of which are elastomers. The polymer mixture, called the gum base, serves as a delivery vehicle for the flavors that are present in the mixture. Most gum manufacturers do not reveal the identity of the polymers present in their gum base, as this information is regarded as a trade secret. One thing is certain though: When you chew gum, you are chewing on a glob of synthetic polymers.

Fibers

Fibers are generated when certain polymers are heated, forced though small holes, and then cooled. The resulting fibers exhibit crystalline regions that are oriented along the axis of the fiber, which endow the fibers with significant tensile strength. Examples include Nylon, Dacron, and polyethylene, all of which have the appropriate degree of crystallinity to form fibers.

Thermosetting Resins

Thermosetting resins are highly cross-linked polymers that are generally very hard and insoluble. One such example is Bakelite, first produced in 1907 from the reaction between phenol and formaldehyde.

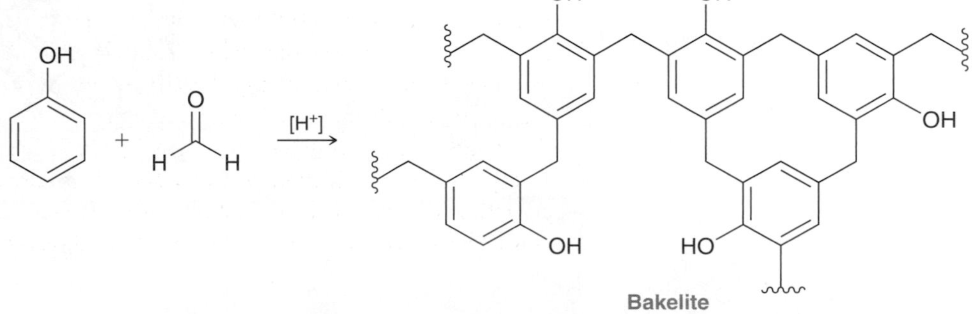

Phenol functions as a nucleophile, and formaldehyde (activated by protonation) functions as an electrophile in an electrophilic aromatic substitution reaction. The resulting alcohol then functions as an electrophile (again activated by protonation) and reacts with phenol in another electrophilic aromatic substitution reaction.

LOOKING BACK
For a justification of *ortho-para* directing effects, see Section 19.7.

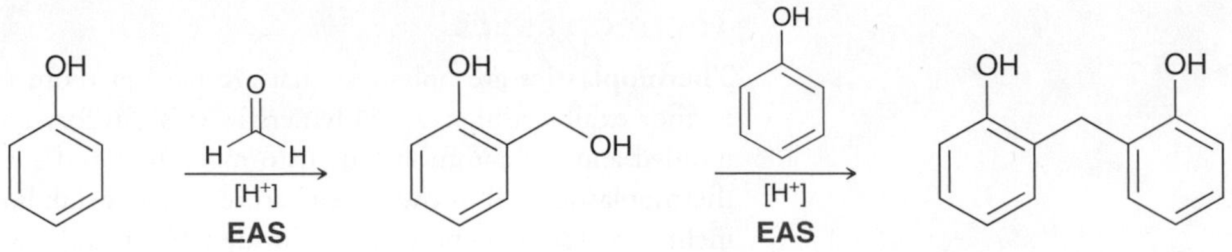

This process continues and can occur at the *ortho* and *para* positions of each ring,

Thermosetting resins can generally withstand high temperatures and are often used in high-temperature applications, such as missile nose cones.

PRACTICALLYSPEAKING

Safety Glass and Car Windshields

Automobile windshields are specifically designed so that they do not create free shards of glass when broken. This is accomplished in very much the same way that bullet-resistant glass is manufactured, as described in the chapter opener. Two sheets of glass are pressed together with a thin layer of polymer between them. The polymer most often used is poly(vinyl butyral), often called PVB.

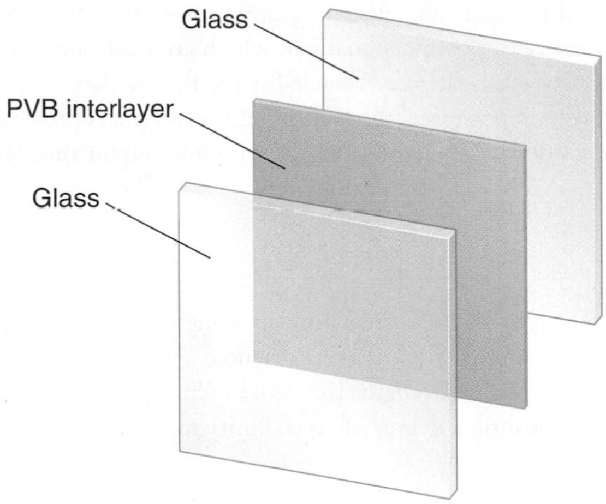

The PVB layer is generally prepared by treating polyvinyl alcohol with butyraldehyde.

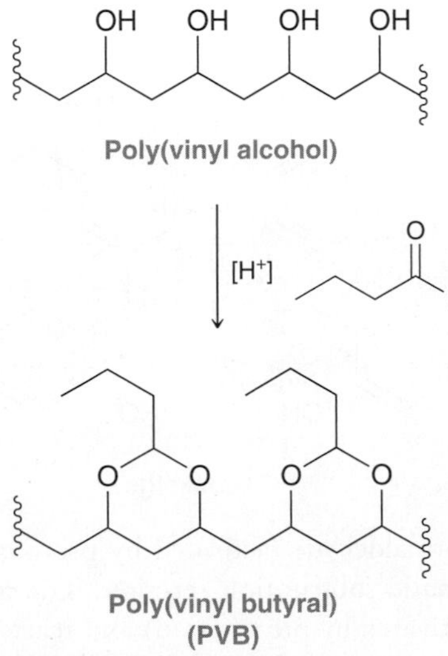

The carbonyl group of butyraldehyde reacts with two hydroxyl groups to form a cyclic acetal. The resulting polyacetal adheres to both sheets of glass. In the event of a collision, the glass breaks, but the individual pieces remain stuck to the polymer interlayer, thereby preventing the pieces of glass from causing injury. Bullet-resistant glass is very similar, but multiple sheets of glass are used, rather than just two. Alternatively, bullet-resistant glass can also be made using thicker polycarbonate interlayers such as Lexan.

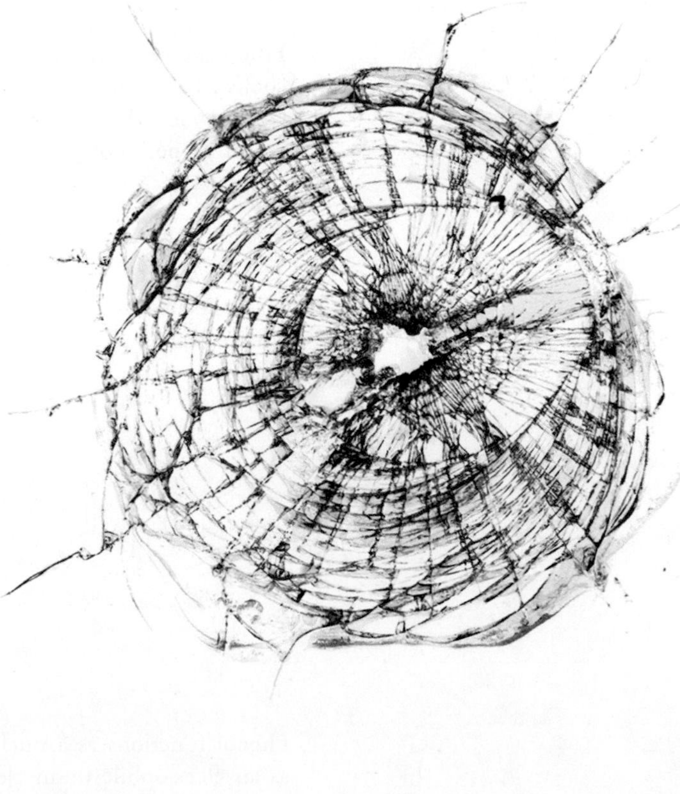

27.8 Polymer Recycling

Polymers have certainly increased our quality of life in countless ways, but their preparation and use have caused serious environmental concerns. Specifically, most synthetic polymers are not biodegradable, which means that they persist and accumulate in the environment. To address this growing problem, many polymers can be recycled. Some polymers are recycled with greater ease than others, depending on the identity of the polymer. For example, in the recycling of soft-drink bottles (made from PET) the bottles are chopped into small chips that are washed to remove the labels and adhesives, then treated with aqueous acid. A reaction occurs in which the polymer chips are broken down into the corresponding monomers. This process is essentially the reverse of the polymerization process that formed PET in the first place. The ester moieties undergo hydrolysis to produce terephthalic acid and ethylene glycol.

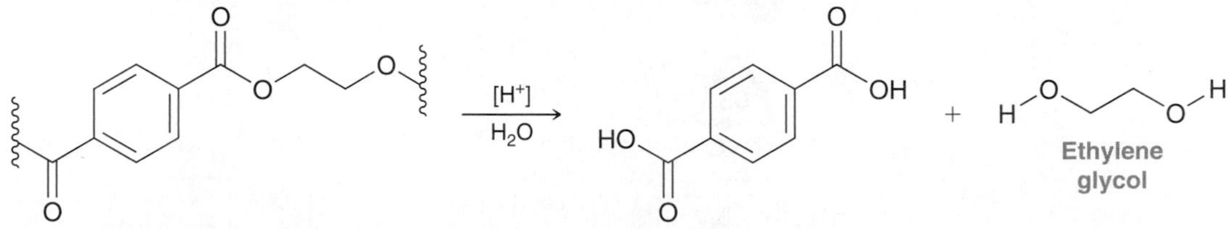

Poly(ethylene terephthalate) (PET)   Terephthalic acid

These monomers are purified and then used as feedstock for the production of new PET. PET is somewhat unique in its ease of recycling, because the polymer can be converted back into monomers. As of yet, there is no way to break down addition polymers, such as polyethylene or polystyrene, into their corresponding monomers. These polymers can be melted down and remolded into a new shape or they can be broken down into smaller fragments in a process that is similar to the cracking of petroleum.

There are many logistical problems that limit the effectiveness of polymer recycling, most significantly the collection and sorting of used polymer products. Different kinds of polymers must be recycled in different ways. When recycling PET, a small amount of a different polymer present in the batch will interfere with the recycling process. As such, polymer recycling requires that polymer products be sorted by hand. To facilitate the sorting process, most polymer products are labeled with recycling codes that indicate their composition. These codes (1–7) indicate the type of polymer used and are arranged in order of ease with which the polymer can be recycled (1 being the easiest and 7 being the most difficult). Table 27.4 indicates the seven recycling codes, the polymers that correspond with each code, and several uses for the recycled products. In many cases, the recycled polymer can be contaminated with adhesives and other materials that may have survived the washing stage. Therefore, recycled polymers cannot be used for food packaging.

CONCEPTUAL CHECKPOINT

27.21 Propose a mechanism for the acid-catalyzed hydrolysis of PET to regenerate the monomers terephthalic acid and ethylene glycol.

Another, more promising, method of reducing the accumulation of polymers in the environment is to develop polymers that are biodegradable and can be recycled by natural processes. **Biodegradable polymers** are polymers that can be broken down by enzymes produced by soil microorganisms. Much research is directed at the development of such polymers, and many have already been developed. Most biodegradable polymers exhibit ester or amide moieties, which can be hydrolyzed by enzymes. Examples include a class of compounds called polyhydroxyalkanoates (PHAs), which are polymers of β-hydroxy carboxylic acids.

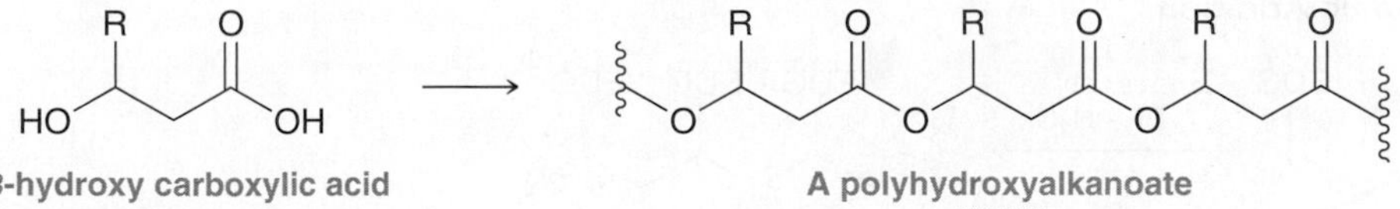
A β-hydroxy carboxylic acid   A polyhydroxyalkanoate (PHA)

TABLE 27.4 RECYCLING CODES AND USES OF RECYCLED PRODUCTS

RECYCLING CODE	POLYMER	STRUCTURE	RECYCLED PRODUCT
01 PET	Poly(ethylene terephthalate)	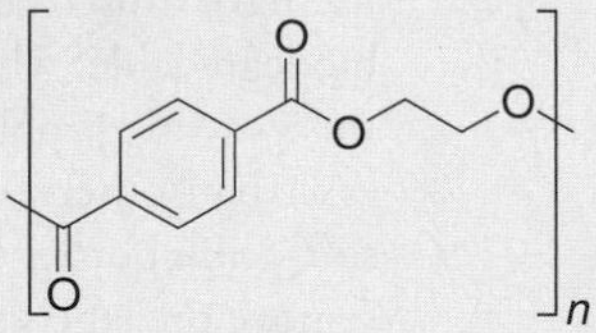	Clothing, carpet fibers, detergent bottles, audio and video tapes
02 PE-HD	High-density polyethylene	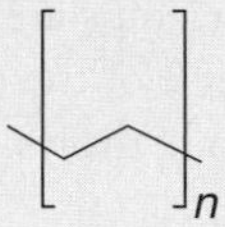	Tyvek insulation, clothing, picnic tables
03 PVC	Poly(vinyl chloride)	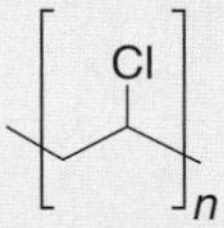	Floor mats, garden hoses, shower curtains, plumbing pipes
04 PE-LD	Low-density polyethylene	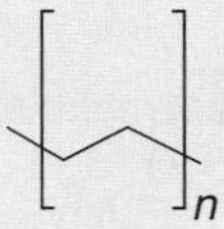	Trash bags
05 PP	Polypropylene	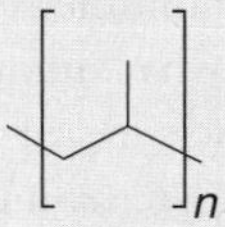	Rope, fishing nets, carpets
06 PS	Polystyrene	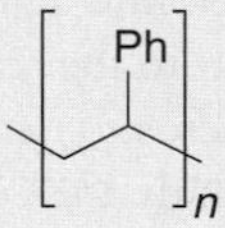	Styrofoam packaging materials, coat hangers, trash cans
07 O	All other polymers	Polycarbonates, polyurethanes, polyamides, etc.	Auto parts

There are many different PHAs, depending on the identity of the R group. The salient feature in all PHAs is the presence of the repeating ester moieties, which are readily hydrolyzed. These ester moieties serve as the "weak link" in the polymer. Much research has been devoted to creating polymers with weak links that render the polymers biodegradable. The decades to come are likely to see the emergence of new biodegradable polymers with properties that will rival the properties of traditional, nonbiodegradable polymers.

REVIEW OF REACTIONS

REACTIONS FOR FORMATION OF CHAIN-GROWTH POLYMERS

Radical Addition

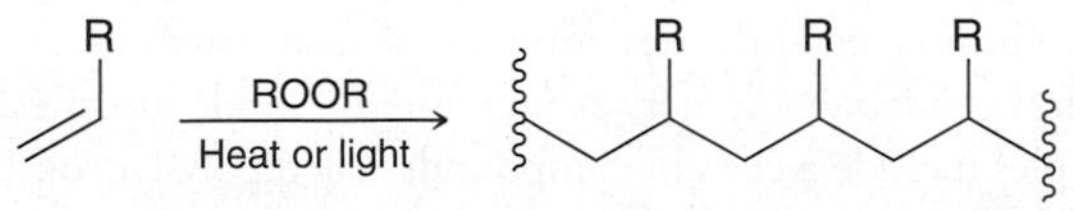

Anionic Addition

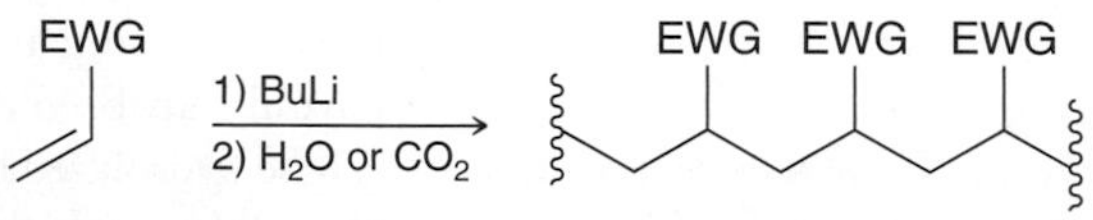

Cationic Addition

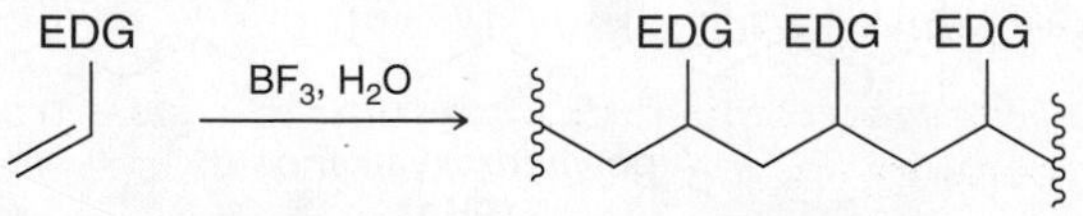

REACTIONS FOR FORMATION OF STEP-GROWTH POLYMERS

Fischer Esterification

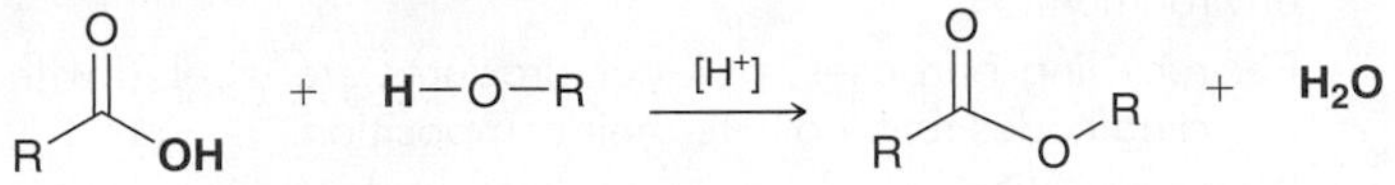

Carbonate Formation

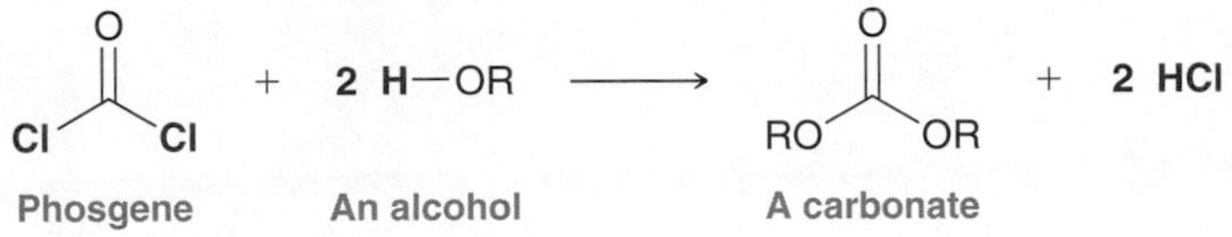

Amide Formation

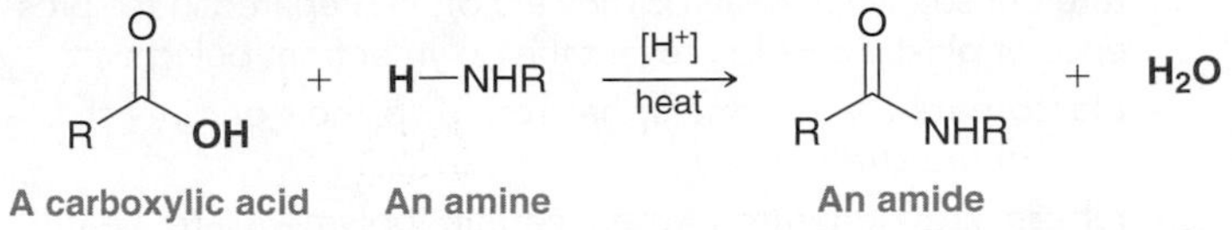

Urethane Formation

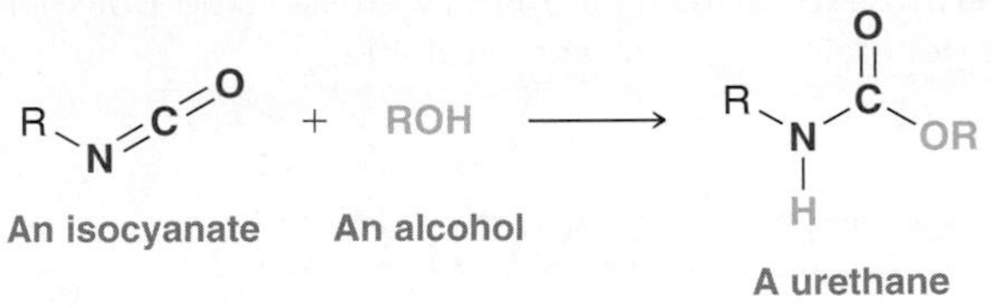

REVIEW OF CONCEPTS AND VOCABULARY

SECTION 27.1

- Polymers are comprised of repeating units that are constructed by joining monomers together.
- Synthetic polymers are prepared by scientists in the laboratory or in factories, while biopolymers are produced by living organisms.

SECTION 27.2

- The names of polymers are most commonly based on the monomers from which they are derived.

SECTION 27.3

- A **homopolymer** is a polymer made up of a single type of monomer. Polymers made from two or more different types of monomers are called **copolymers**.
- **Alternating copolymers** contain an alternating distribution of repeating units, while **random copolymers** contain a random distribution of repeating units.
- In a **block copolymer**, different homopolymer subunits are connected together in one chain. In a **graft copolymer**, sections of one homopolymer have been grafted onto a chain of another homopolymer.

SECTION 27.4

- Monomers can join together to form **addition polymers** by cationic, anionic, or free-radical addition.
- Most derivatives of ethylene will undergo radical polymerization under suitable conditions.
- Cationic addition is only efficient with derivatives of ethylene that contain an electron-donating group.
- Anionic addition is only efficient with derivatives of ethylene that contain an electron-withdrawing group, and polymers generated through this process are often called **living polymers**.
- Polymers generated via condensation reactions are called **condensation polymers**.
- **Polycarbonates** are similar in structure to polyesters, but with repeating carbonate moieties instead of repeating ester moieties.

SECTION 27.5

- **Chain-growth polymers** are formed under conditions in which each monomer is added to the growing chain one at a time. The monomers do not react directly with each other.
- **Step-growth polymers** are formed under conditions in which the individual monomers react with each other to form **oligomers**, which are then joined together to form polymers.
- In general, step-growth polymers are usually condensation polymers, while chain-growth polymers are usually addition polymers, but there are exceptions, such as **polyurethanes**.

SECTION 27.6

- **Branched polymers** contain a large number of branches connected to the main chain of the polymer, while **linear polymers** have only a minimal amount of branching.
- Polymers produced with Ziegler-Natta catalysts are generally linear polymers.
- When polymerization is performed using monosubstituted ethylenes as monomers, **isotactic**, **syndiotactic**, or **atactic** polymers can be formed, depending on the catalyst used.
- **Crossed-linked polymers** contain disulfide bridges or branches that connect neighboring chains.
- Most polymers contain some degree of cross-linking, and the conditions of polymer formation can be controlled so as to favor either more or less cross-linking.
- Many polymers contain regions called **crystallites** in which the chains are linearly extended and close in proximity to one another. The regions that are not crystalline are said to be **amorphous**.
- Crystalline regions become amorphous at the **melt transition temperature** (**T_m**).
- Noncrystalline polymers become soft at the **glass transition temperature** (**T_g**).

SECTION 27.7

- **Thermoplastics** are polymers that are hard at room temperature but soft when heated. They are often prepared in the presence of **plasticizers** to prevent the polymer from being brittle.
- **Elastomers** are polymers that return to their original shape after being stretched.
- **Fibers** are generated when certain polymers are heated, forced though small holes, and then cooled.
- **Thermosetting resins** are highly cross-linked polymers that are generally very hard and insoluble.

SECTION 27.8

- Most synthetic polymers persist and accumulate in the environment.
- For recycling purposes, polymer products are labeled with recycling codes that indicate their composition.
- **Biodegradable polymers** can be broken down by enzymes produced by microorganisms in the soil.

KEY TERMINOLOGY

addition polymers 1273
alternating copolymers 1271
amorphous 1256
atactic 1285
biodegradable polymers 1289
block copolymer 1272
branched polymers 1283
chain-growth polymers 1281
condensation polymers 1279
copolymers 1271
crossed-linked polymers 1286
crystallites 1286
elastomers 1287
fibers 1287
glass transition temperature (T_g) 1286
graft copolymer 1272
homopolymer 1271
isotactic 1285
linear polymers 1284
living polymers 1276
melt transition temperature (T_m) 1286
oligomers 1282
plasticizers 1287
polycarbonates 1279
polyurethanes 1282
random copolymers 1271
step-growth polymers 1282
syndiotactic 1285
thermoplastics 1287
thermosetting resins 1287

SKILLBUILDER REVIEW

27.1 DETERMINING WHICH POLYMERIZATION TECHNIQUE IS MORE EFFICIENT

STEP 1 Identify the repeating units.

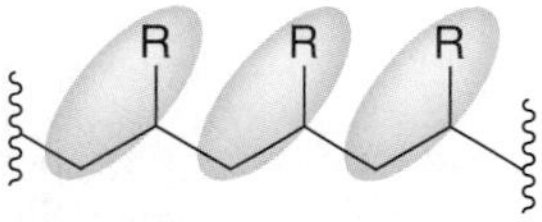

STEP 2 Identify the monomer necessary to make this polymer.

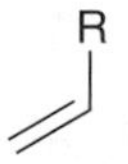

STEP 3 Determine whether the monomer bears an electron-withdrawing group or an electron-donating group:

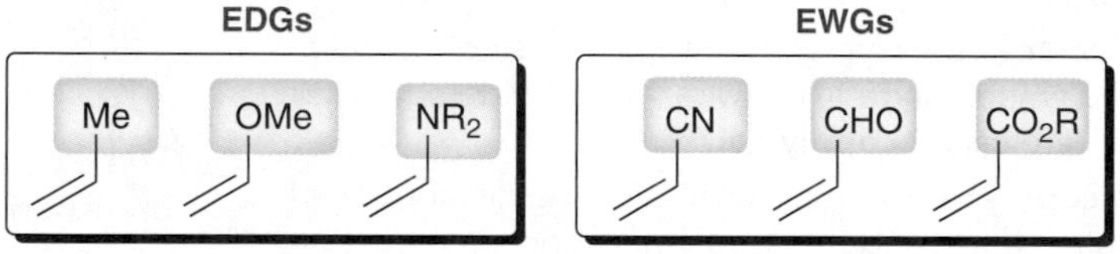

STEP 4 If EWG, then use anionic addition.

If EDG, then use cationic addition.

Try Problems 27.6–27.10, 27.27, 27.28, 27.34, 27.38–27.40

27.2 IDENTIFYING THE MONOMERS REQUIRED TO PRODUCE A DESIRED CONDENSATION POLYMER

STEP 1 Identify the functional group that repeats itself.

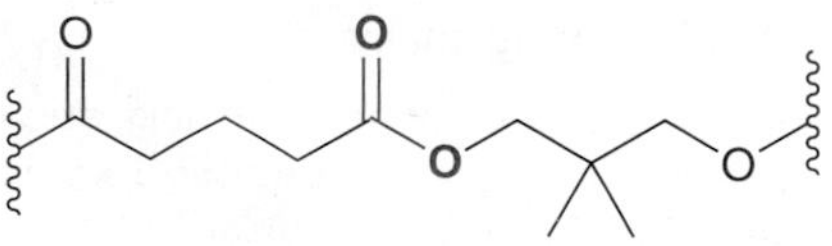

A polyester

STEP 2 Identify the type of reaction that will produce the desired functional group.

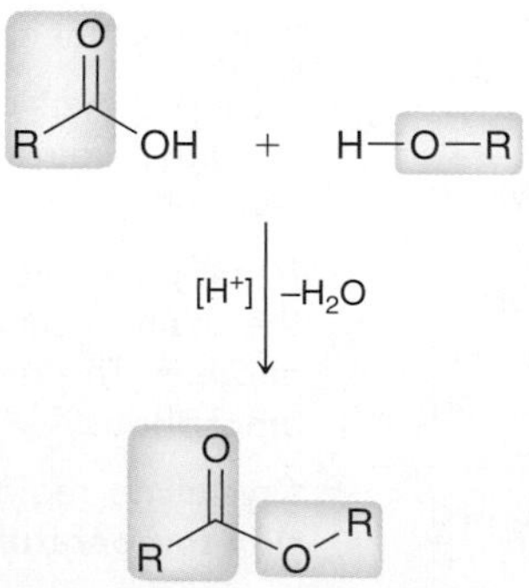

STEP 3 Identify the two difunctional monomers necessary to form the desired condensation polymer.

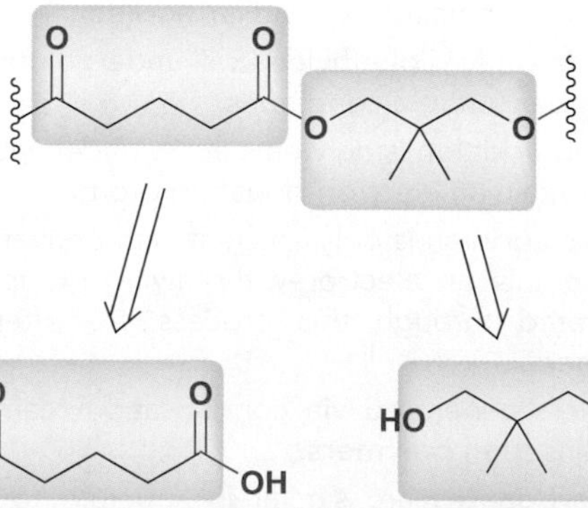

Try Problems 27.13–27.16, 27.23, 27.30a,b,d, 27.31, 27.32

PRACTICE PROBLEMS

PLUS *Note:* Most of the Problems are available within *WileyPLUS*, an online teaching and learning solution.

27.22 Draw and name the polymer that results when each of the following monomers undergoes polymerization.

(a) **Nitroethylene** (b) **Acrylonitrile** (c) **Vinylidene fluoride**

27.23 Kodel is a synthetic polyester with the following structure.

Kodel

(a) Identify what monomers you would use to make Kodel.

(b) Would you use acidic conditions or basic conditions for this polymerization process?

27.24 Draw the structures of the monomers required to make the following alternating copolymer.

$$\left[\begin{array}{cccc} H & Cl & H & Ph \\ | & | & | & | \\ C - & C - & C - & C \\ | & | & | & | \\ H & Cl & H & Ph \end{array}\right]_n$$

27.25 Draw a region of a block copolymer constructed from isobutylene and styrene.

27.26 Draw a region of an alternating copolymer constructed from vinyl chloride and ethylene.

27.27 Identify which of the following monomers would be most reactive toward cationic polymerization.

27.28 Identify which of the following monomers would be most reactive toward anionic polymerization:

27.29 Identify the repeating unit of the polymer formed from each of the following reactions, and then determine whether the polymer is a chain-growth or a step-growth polymer.

(a)

(b)

(c)

27.30 Quiana is a synthetic polymer that can be used to make fabric that mimics the texture of silk. It can be prepared from the following monomers:

(a) Draw the structure of Quiana.

(b) Is Quiana a polyester, a polyamide, a polycarbonate, or a polyurethane?

(c) Is Quiana a step-growth polymer or a chain-growth polymer?

(d) Is Quiana an addition polymer or a condensation polymer?

27.31 Draw the monomer(s) required to make each of the following condensation polymers:

(a) 

(b) 

27.32 Draw the condensation polymer produced when 1,4-cyclohexanediol is treated with phosgene.

27.33 Determine whether the following pairs of monomers would form chain-growth polymers or step-growth polymers.

(a) 

(b) 

27.34 Nitroethylene undergoes anionic polymerization so rapidly that it is difficult to isolate nitroethylene without it polymerizing. Explain.

27.35 Explain why vinyl shower curtains develop cracks over time.

27.36 Polyformaldehyde, sold under the trade name Delrin, is a strong polymer used in the manufacture of many guitar picks. It is prepared via the acid-catalyzed polymerization of formaldehyde.

(a) Draw the structure of polyformaldehyde.

(b) Would you describe polyformaldehyde as a polyether or a polyester?

(c) Is polyformaldehyde a step-growth polymer or a chain-growth polymer?

(d) Is polyformaldehyde an addition polymer or a condensation polymer?

27.37 The following monomer can be polymerized under either acidic or basic conditions. Explain.

27.38 Anionic polymerization of *para*-nitrostyrene occurs much more rapidly than anionic polymerization of styrene. Explain this difference in rate.

27.39 Cationic polymerization of *para*-methoxystyrene occurs much more rapidly than cationic polymerization of styrene. Explain this difference in rate.

27.40 Cationic polymerization of *para*-methoxystyrene occurs much more rapidly than cationic polymerization of *meta*-methoxystyrene. Explain this difference in rate.

27.41 Draw segments of the following polymers, indicating the stereochemistry with wedges and/or dashes.

(a) Syndiotactic poly(vinyl chloride)

(b) Isotactic polystyrene

27.42 Consider the structure of the following polymer.

(a) Draw the monomers you would use to prepare this polymer.

(b) Determine whether this polymer is a step-growth polymer or a chain-growth polymer.

(c) Determine whether this polymer is an addition polymer or a condensation polymer.

INTEGRATED PROBLEMS

27.43 Draw the polymer that is expected when the following monomers react under acidic conditions. You may find it helpful to review Section 20.6.

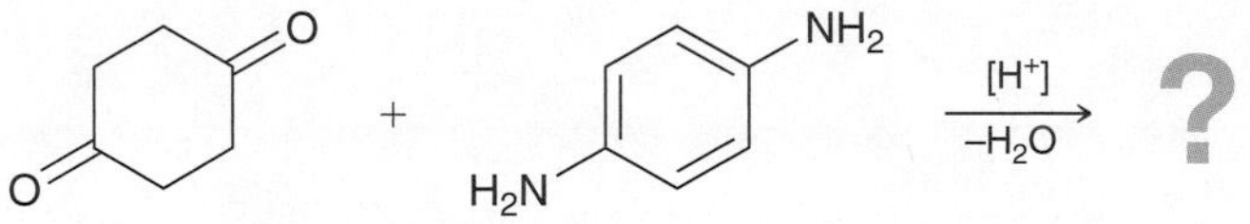

27.44 As described in the Practically Speaking box in Section 27.7, poly(vinyl alcohol) is used as a precursor to make PVB for use in automobile windshields. Poly(vinyl alcohol) can be prepared by polymerizing vinyl acetate and then removing the acetate groups via hydrolysis.

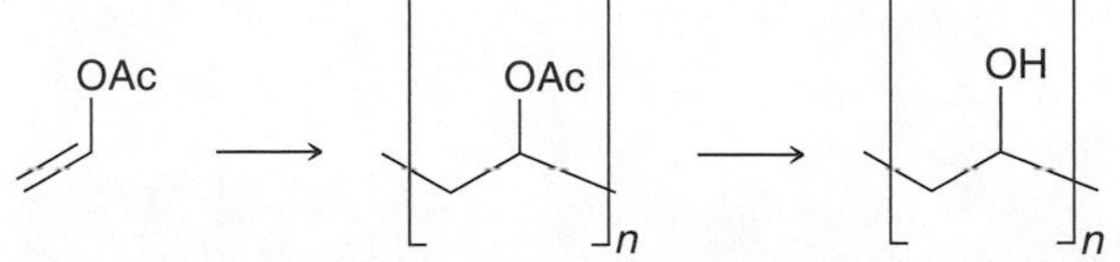

Vinyl acetate **Poly(vinyl acetate)** **Poly(vinyl alcohol)**

This two-step process is required because poly(vinyl alcohol) cannot be made from its corresponding monomer. Explain why vinyl alcohol will not directly undergo polymerization.

27.45 When a solution of aqueous sodium hydroxide is spilled on polyester clothing, a hole develops in the fabric. Describe how the polyester is destroyed.

CHALLENGE PROBLEMS

27.46 When 3-methyl-1-butene is treated with a catalytic amount of BF_3 and H_2O, a cationic polymerization process occurs, but the expected homopolymer is not formed. Instead, a random copolymer is obtained.

(a) Explain why there are two different repeating units.

(b) Draw a segment of the random copolymer that is formed, clearly showing the two different repeating units.

(c) Would you expect cationic polymerization of 3,3-dimethyl-1-butene to produce a random copolymer as well? Justify your answer.

27.47 When ethylene oxide is treated with a strong nucleophile, the epoxide ring is opened to form an alkoxide ion that can function as a nucleophile to attack another molecule of ethylene oxide. This process repeats itself, and a polymer is formed.

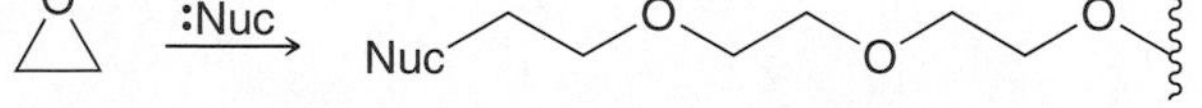

Ethylene oxide **Poly(ethylene oxide)**

The resulting polymer is called poly(ethylene oxide) or poly(ethylene glycol). It is sold under the trade name Carbowax and is as an adhesive and a thickening agent.

(a) Draw a mechanism showing the formation of a segment of poly(ethylene oxide).

(b) Ethylene oxide can also be polymerized when treated with an acid. Draw the mechanism of formation of a segment of poly(ethylene oxide) under acidic conditions.

(c) Identify the monomer you would use to prepare the following polymer:

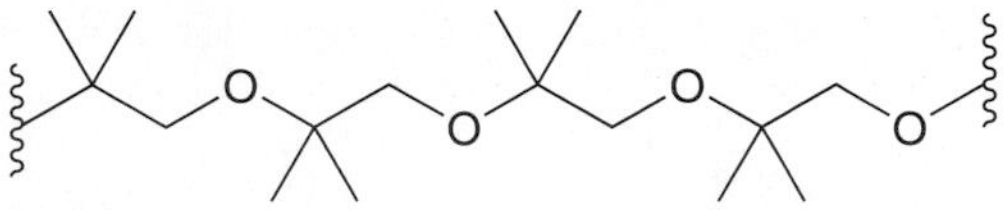

(d) Determine whether you would use basic conditions or acidic conditions to prepare the following polymer. Explain your choice.

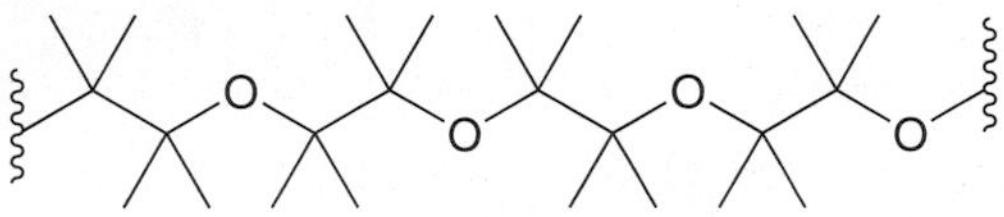

27.48 Using vinyl acetate as your only source of carbon atoms, design a synthesis for the following polymer:

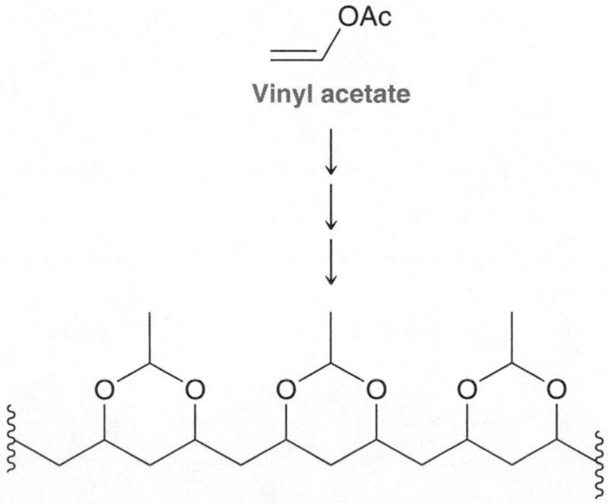

Glossary

[4+2]-cycloaddition (Sect. 17.7): A pericyclic reaction, also called a Diels-Alder reaction, that takes place between two different π systems, one of which is associated with four atoms while the other is associated with two atoms.

1,2-addition (Sect. 17.4): A reaction involving the addition of two groups to a conjugated π system in which one group is installed at the C1 position and the other group is installed at the C2 position.

1,2-adduct (Sect. 17.4): The product obtained from 1,2-addition across a conjugated π system.

1,2-elimination (Sect. 8.1): An elimination reaction in which a proton from the beta (β) position is removed together with the leaving group, forming a double bond.

1,3-diaxial interaction (Sect. 4.12): Steric interactions that occur in the chair conformation of a di- or polysubstituted cyclohexane when groups located at C1 and C3 both occupy axial positions.

1,4-addition (Sect. 17.4): A reaction involving the addition of two groups to a conjugated π system in which one group is installed at the C1 position and the other group is installed at the C4 position.

1,4-adduct (Sect. 17.4): The product obtained from 1,4-addition across a conjugated π system.

A

absorbance (Sect. 17.11): In UV-VIS spectroscopy, the value log (I_0/I) where I_0 is the intensity of the reference beam and I is the intensity of the sample beam.

absorption spectrum (Sect. 15.2): In IR spectroscopy as well as UV-VIS spectroscopy, a plot that measures the percent transmittance or absorption as a function of frequency.

acetal (Sect. 21.5): A functional group characterized by two alkoxy (OR) groups connected to the same carbon atom.Acetals can be used as protecting groups for aldehydes or ketones.

acetoacetic ester synthesis (Sect. 22.5): A three-step process that converts an alkyl halide into a methyl ketone with the introduction of three new carbon atoms.

acetylide ion (Sect. 10.3): The conjugate base of acetylene or any terminal alkyne.

acid-catalyzed hydration (Sect. 9.4): A reaction that achieves the addition of water across a double bond in the presence of an acid catalyst.

acidic cleavage (Sect. 14.6): A reaction in which bonds are broken in the presence of an acid. For example, in the presence of a strong acid, an ether is converted into two alkyl halides.

activation (Sect. 19.7): For a substituted aromatic ring, the effect of an electron-donating substituent that increases the rate of electrophilic aromatic substitution.

acyl group (Sect. 19.6): The term describing a carbonyl group (C═O bond) connected to an alkyl group or aryl group.

acylium ion (Sect. 19.6): The resonance-stabilized, cationic intermediate of a Friedel-Crafts acylation, formed by treating an acid halide with aluminum trichloride.

acyl peroxide (Sect. 11.3): A peroxide for which each oxygen atom is connected to an acyl group. Acyl peroxides are often used as radical initiators, because the O—O bond is especially weak.

addition polymers (Sect. 27.4): Polymers that are formed via cationic addition, anionic addition, or free-radical addition.

addition reactions (Sect. 9.1): Reactions that are characterized by the addition of two groups across a double bond. In the process, the pi (π) bond is broken.

addition to π bond (Sect. 11.2): One of the six kinds of arrow-pushing patterns used in drawing mechanisms for radical reactions. A radical adds to a π bond, destroying the π bond and generating a new radical.

adrenocortical hormones (Sect. 26.6): Hormones that are secreted by the cortex (the outer layer) of the adrenal glands. Adrenocortical hormones are typically characterized by a carbonyl group or hydroxyl group at C11 of the steroid skeleton.

alcohol (Sect. 13.1): A compound that possesses a hydroxyl group (OH).

aldaric acid (Sect. 24.6): A dicarboxylic acid that is produced when an aldose or ketose is treated with a strong oxidizing agent, such as HNO_3.

alditol (Sect. 24.6): The product obtained when the aldehyde moiety of an aldose is reduced.

aldol addition reaction (Sect. 22.3): A reaction that occurs when an aldehyde or ketone is attacked by an enolate ion. The product of an aldol addition reaction is always a β-hydroxy aldehyde or ketone.

aldol condensation (Sect. 22.3): An aldol addition followed by dehydration to give an α,β-unsaturated ketone or aldehyde.

aldonic acid (Sect. 24.6): The product obtained when the aldehyde moiety of an aldose is oxidized.

aldose (Sect. 24.2): A carbohydrate that contains an aldehyde moiety.

alkaloids (Sect. 23.1): Naturally occurring amines isolated from plants.

alkanamine (Sect. 23.2): A format for naming primary amines containing a complex alkyl group.

alkane (Sect. 4.1): A hydrocarbon that lacks π bonds.

alkene (Sect. 8.1): A compound that possesses a carbon-carbon double bond.

alkoxide (Sect. 13.2): The conjugate base of an alcohol.

alkoxy group (Sect. 14.2): An OR group.

alkoxymercuration-demercuration (Sect. 14.5): A two-step process that achieves Markovnikov addition of an alcohol (H and OR) across an alkene. The product of this process is an ether.

alkyl amines (Sect. 23.2): A format for naming amines containing simple alkyl groups.

alkylation (Sect. 10.10): A reaction that achieves the installation of an alkyl group. For example, an S_N2 reaction in which an alkyl group is connected to an attacking nucleophile.

alkyl group (Sect. 4.2): A substituent lacking π bonds and comprised of only carbon and hydrogen atoms.

alkyl halide (Sect. 7.2): An organic compound containing at least one halogen.

alkylthio group (Sect. 14.11): An SR group.

alkynide ion (Sect. 10.3): The conjugate base of a terminal alkyne.

allylic (Sect. 2.10): The positions that are adjacent to the vinylic positions of a carbon-carbon double bond.

allylic bromination (Sect. 11.7): A radical reaction that achieves installation of a bromine atom at an allylic position.

allylic carbocation (Sect. 6.11): A carbocation in which the positive charge is adjacent to a carbon-carbon double bond.

alpha (α) amino acid (Sect. 25.1): A compound containing a carboxylic acid moiety (COOH) as well as an amino group (NH_2), both of which are attached to the same carbon atom.

alpha (α) anomer (Sect. 24.5): The cyclic hemiacetal of an aldose in which the hydroxyl group at the anomeric position is *trans* to the CH_2OH group.

alpha (α) helix (Sect. 25.7): For proteins, a feature of secondary structure that forms when a portion of the protein twists into a spiral.

alpha (α) position (Sect. 7.2): The position immediately adjacent to a functional group.

alternating copolymers (Sect. 27.3): A copolymer that contains an alternating distribution of repeating units.

amidomalonate synthesis (Sect. 25.3): A synthetic method that employs diethyl acetamidomalonate as the starting material and enables the preparation of racemic α-amino acids.

amine (Sect. 23.1): Compounds containing a nitrogen atom that is connected to one, two, or three alkyl or aryl groups.

amino acid (Sect. 25.1): A compound containing a carboxylic acid moiety (COOH) as well as an amino group (NH_2).

amino acid residue (Sect. 25.4): The individual repeating units in a polypeptide chain or protein.

amino sugars (Sect. 24.9): Carbohydrate derivatives in which an OH group has been replaced with an amino group.

amorphous (Sect. 27.6): A region of a polymer in which nearby chains are not linearly extended and are not parallel to one another.

amphoteric (Sect. 25.2): Compounds that will react with either acids or bases. Amino acids are amphoteric.

androgens (Sect. 26.6): Male sex hormones.

angle strain (Sect. 4.9): The increase in energy associated with a bond angle that has deviated from the preferred angle of 109.5°.

annulenes (Sect. 18.5): Compounds consisting of a single ring containing a fully conjugated π system. Benzene is [6]annulene.

anomers (Sect. 24.5): Stereoisomeric cyclic hemiacetals of an aldose or ketose that differ from each other in their configuration at the anomeric carbon.

anomeric carbon (Sect. 24.5): The C1 position of the cyclic hemiacetal of an aldose or the C2 position of the cyclic hemiacetal of a ketose.

anti **addition** (Sect. 9.8): An addition reaction in which two groups are installed on opposite sides of a π bond.

anti **conformation** (Sect. 4.8): A conformation in which the dihedral angle between two groups is 180°.

antiaromatic (Sect. 18.4): Instability that arises when a planar ring of continuously overlapping *p* orbitals contains $4n$ π electrons.

antibonding MO (Sect. 1.8): A high-energy molecular orbital resulting from the destructive interference between atomic orbitals.

anti **coplanar** (Sect. 8.7): A conformation in which a hydrogen atom and a leaving group are separated by a dihedral angle of exactly 180°.

anti-Markovnikov addition (Sect. 9.3): An addition reaction in which a hydrogen atom is installed at the more substituted vinylic position and another group (such as a halogen) is installed at the less substituted vinylic position.

antioxidants (Sect. 11.9): Radical scavengers that prevent autooxidation by preventing radical chain reactions from beginning.

anti **periplanar** (Sect. 8.2): A conformation in which a hydrogen atom and a leaving group are separated by a dihedral angle of approximately 180°.

arenium ion (Sect. 19.2): The positively charged, resonance-stabilized, intermediate of an electrophilic aromatic substitution reaction. Also called a sigma complex.

aromatic compound (Sect. 18.1): A compound containing a planar ring of continuously overlapping *p* orbitals with $4n+2$ π electrons.

aryl amine (Sect. 23.2): An amine in which the nitrogen atom is connected directly to an aromatic ring.

asymmetric hydrogenation (Sect. 9.7): The addition of H_2 across only one face of a π bond.

asymmetric stretching (Sect. 15.5): In IR spectroscopy, when two bonds are stretching out of phase with each other.

atactic (Sect. 27.6): A polymer in which the repeating units contain chirality centers which are not arranged in a pattern (they have random configurations).

atomic mass unit (amu): (Sect. 15.13): A unit of measure equivalent to 1*g* divided by Avogadro's number.

atomic orbital (Sect. 1.6): A three-dimensional plot of ψ^2 of a wavefunction. It is a region of space that can accommodate electron density.

Aufbau principle (Sect. 1.6): A rule that determines the order in which orbitals are filled by electrons. Specifically, the lowest energy orbital is filled first.

autocatalytic (Sect. 22.1): A reaction for which the reagent necessary to catalyze the reaction is produced by the reaction itself.

autooxidation (Sect. 11.9): The slow oxidation of organic compounds that occurs in the presence of atmospheric oxygen.

auxochrome (Sect. 17.11) When applying Woodward-Fieser rules, the groups attached to the chromophore.

axial position (Sect. 4.11): For chair conformations of substituted cyclohexanes, a position that is parallel to a vertical axis passing through the center of the ring.

axis of symmetry (Sect. 5.6): An axis about which a compound possesses rotational symmetry.

azide synthesis (Sect. 23.5): A method for preparing primary amines that avoids the formation of secondary and tertiary amines.

azo coupling (Sect. 23.11): An electrophilic aromatic substitution reaction in which an aryldiazonium salt reacts with an activated aromatic ring.

azo dyes (Sect. 23.11): A class of colored compounds that are formed via azo coupling.

B

back-side attack (Sect. 7.4): In S_N2 reactions, the side opposite the leaving group, which is where the nucleophile attacks.

Baeyer-Villiger oxidation (Sect. 20.11): A reaction in which a ketone is treated with a peroxy acid and is converted into an ester via the insertion of an oxygen atom.

base peak (Sect. 15.8): In mass spectrometry, the tallest peak in the spectrum, which is assigned a relative value of 100%.

Beer's Law (Sect. 17.11): In UV-VIS spectroscopy, an equation describing the relationship between molar absorptivity (ε), absorbance (A), concentration (C) and path length (l):

$$\varepsilon = \frac{A}{(C \times l)}$$

bending (Sect. 15.2): In IR spectroscopy, a type of vibration that generally produces a signal in the fingerprint region of an IR spectrum.

bent (Sect. 1.10): A type of geometry resulting from an sp^3-hybridized atom that has two lone pairs. For example, the oxygen atom in H_2O.

benzylic position (Sect. 18.6): A carbon atom that is immediately adjacent to a benzene ring.

benzyne (Sect. 19.14): A high-energy intermediate formed during the elimination-addition reaction that occurs between chlorobenzene and either NaOH (at high temperature) or $NaNH_2$.

beta (β) anomer (Sect. 24.5): The cyclic hemiacetal of an aldose, in which the hydroxyl group at the anomeric position is *cis* to the CH_2OH group.

beta elimination (Sect. 8.1): An elimination reaction in which a proton from the beta (β) position is removed together with the leaving group, forming a double bond.

betaine (Sect. 20.10): A neutral compound with two oppositely charged atoms that are not adjacent to each other.

beta (β) pleated sheet (Sect. 25.7): For proteins, a feature of secondary structure that forms when two or more protein chains line up side-by-side.

beta (β) position (Sect. 7.2): The position immediately adjacent to an alpha (α) position.

bicyclic (Sect. 4.2): A structure containing two rings that are fused together.

bimolecular (Sect. 7.4): For mechanisms, a step that involves two chemical entities.

biodegradable polymers (Sect. 27.8): Polymers that can be broken down by enzymes produced by soil microorganisms.

Birch reduction (Sect. 18.7): A reaction in which benzene is reduced to give 1,4-cyclohexadiene.

block copolymer (Sect. 27.3): A copolymer in which the different homopolymer subunits are connected together in one chain.

blocking group (Sect. 19.11): A group that can be readily installed and uninstalled. Used for regiochemical control during synthesis.

boat conformation (Sect. 4.10): A conformation of cyclohexane in which all bond angles are fairly close to 109.5° and many hydrogen atoms are eclipsing each other.

bond cleavage (Sect. 12.3): The breaking of a bond, either homolytically or heterolytically.

bond dissociation energy (Sect. 6.1): The energy required to achieve homolytic bond cleavage (generating radicals).

bonding MO (Sect. 1.8): A low-energy molecular orbital resulting from the constructive interference between atomic orbitals.

bond-line structures (Sect. 2.2): The most common drawing style employed by organic chemists. All carbon atoms and most hydrogen atoms are implied but not explicitly drawn in a bond-line structure.

branched polymer (Sect. 27.6): A polymer that contains a large number of branches connected to the main chain of the polymer.

Bredt's rule (Sect. 8.4): A rule that states that it is not possible for a bridgehead carbon of a bicyclic system to possess a carbon-carbon double bond if it involves a *trans* π bond being incorporated in a small ring.

bridgeheads (Sect. 4.2): In a bicyclic system, the carbon atoms where the rings are fused together.

broadband decoupling (Sect. 16.11): In ^{13}C NMR spectroscopy, a technique in which all ^{13}C - ^{1}H splitting is suppressed with the use of two rf transmitters.

bromohydrin (Sect. 9.8): A compound containing a Br group and a hydroxyl group (OH) on adjacent carbon atoms.

bromonium ion (Sect. 9.8): A positively charged, bridged intermediate formed during the addition reaction that occurs when an alkene is treated with molecular bromine (Br_2).

Brønsted-Lowry acid (Sect. 3.1): A compound that can serve as a proton donor.

Brønsted-Lowry base (Sect. 3.1): A compound that can serve as a proton acceptor.

C

carbinolamine (Sect. 21.6): A compound containing a hydroxyl group (OH) and a nitrogen atom, both of which are conneceted to the same carbon atom.

carbocation (Sect. 6.7): A compound containing a positively charged carbon atom.

carbohydrates (Sect. 24.1): Polyhydroxy aldehydes or ketones with molecular formula $C_xH_{2x}O_x$.

carbonyl group (Sect. 13.4): A C═O bond.

carboxylic acid derivative (Sect. 21.6): A compound that is similar in structure to a carboxylic acid (RCOOH) but the OH group of the carboxylic acid has been replaced with a different group, Z, where Z is a heteroatom such as Cl, O, N, etc.

catalyst (Sect. 6.5): A compound that can speed up the rate of a reaction without itself being consumed by the reaction.

catalytic hydrogenation (Sect. 9.7): A reaction that involves the addition of molecular hydrogen (H_2) across a double bond in the presence of a metal catalyst.

cation (Sect. 3.8): A compound that bears a positive charge.

cellular respiration (Sect. 13.12): A process by which molecular oxygen is used to convert food into CO_2, water, and energy.

cephalins (Sect. 26.5): Phosphoglycerides that contain ethanolamine.

CFC (Sect. 11.8): A compound containing only carbon, chlorine, and fluorine.

chain branching (Sect. 11.11): During polymerization, the growth of a branch connected to the main chain.

chain-growth polymer (Sect. 27.5): A polymer that is formed under conditions in which the monomers do not react directly with each other, but rather, each monomer is added to the growing chain, one at a time.

chain reaction (Sect. 11.3): A reaction (generally involving radicals) in which one chemical entity can ultimately cause a chemical transformation for thousands of molecules.

chair conformation (Sect. 4.10): The lowest energy conformation for cyclohexane, in which all bond angles are fairly close to 109.5° and all hydrogen atoms are staggered.

chemically equivalent (Sect. 16.4): In NMR spectroscopy, protons (or carbon atoms) that occupy identical electronic environments and produce only one signal.

chemical shift (δ) (Sect. 16.5): In an NMR spectrum, the location of a signal, defined relative to the frequency of absorption of a reference compound, tetramethylsilane (TMS).

chiral (Sect. 5.2): An object that is not superimposable on its mirror image.

chirality center (Sect. 5.2): A tetrahedral carbon atom bearing four different groups.

chlorohydrin (Sect. 9.8): A compound containing a Cl group and a hydroxyl group (OH) on adjacent carbon atoms.

chromatogram (Sect. 15.14): In gas chromatography, a plot that identifies the retention time of each compound in the mixture.

chromophore (Sect. 17.11): In UV-VIS spectroscopy, the region of the molecule responsible for the absorption (the conjugated π system).

Claisen condensation reaction (Sect. 22.4): A nucleophilic acyl substitution reaction in which the nucleophile is an ester enolate and the electrophile is an ester.

Claisen rearrangement (Sect. 17.10): A [3,3] sigmatropic rearrangement that is observed for allylic vinylic ethers.

Clemmensen reduction (Sect. 19.6): A reaction in which a carbonyl group is completely reduced and replaced with two hydrogen atoms.

column chromatography (Sect. 5.9): A technique by which compounds are separated from each other based on a difference in the way they interact with the medium (the adsorbent) through which they are passed.

complex lipid (Sect. 26.1): A lipid that readily undergoes hydrolysis in aqueous acid or base to produce smaller fragments.

concerted (Sect. 6.9): A process during which all changes in bonding occur in a single step.

condensation polymer (Sect. 27.4): A polymer that is formed via a condensation reaction.

condensed structure (Sect. 2.1): A drawing style in which none of the bonds are drawn. Groups of atoms are clustered together when possible. For example, isopropanol has two CH_3 groups, both of which are connected to the central carbon atom, shown like this: $(CH_3)_2CHOH$.

conformation (Sect. 4.7): A three-dimensional shape that can be adopted by a compound as a result of rotation about single bonds.

conjugate acid (Sect. 3.1): In an acid-base reaction, the product that results when a base is protonated.

conjugate addition (Sect. 22.6): An addition reaction in which a nucleophile and a proton are added across the two ends of a conjugated π system.

conjugate base (Sect. 3.1): In an acid-base reaction, the product that results when an acid is deprotonated.

conjugated (Sect. 15.3): A compound in which two π bonds are separated from each other by exactly one single bond.

conjugated diene (Sect. 17.1): A compound in which two carbon-carbon π bonds are separated from each other by exactly one single bond.

conrotatory (Sect. 17.9): In electrocyclic reactions, a type of rotation in which the sets of lobes at the atoms forming a bond with each other must rotate in the same way.

conservation of orbital symmetry (Sect. 17.8): During a reaction, the requirement that the phases of the frontier MOs must be aligned.

constitutional isomers (Sect. 1.2): Compounds that have the same molecular formula but differ in the way the atoms are connected.

constructive interference (Sect. 1.7): When two waves interact with each other in a way that produces a wave with a larger amplitude.

continuous-wave (CW) spectrometer (Sect. 16.2): An NMR spectrometer that holds the magnetic field constant and slowly sweeps through a range of rf frequencies, monitoring which frequencies are absorbed.

Cope rearrangement (Sect. 17.10) A [3,3] sigmatropic rearrangement in which all six atoms of the cyclic transition state are carbon atoms.

coplanar (Sect. 8.7): Atoms that lie in the same plane.

copolymer (Sect. 27.3): A polymer that is constructed from more than one repeating unit.

coupling (of protons) (Sect. 16.7): A phenomenon observed most commonly for nonequivalent protons connected to adjacent carbon atoms in which the multiplicity of each signal is affected by the other.

coupling (of radicals) (Sect. 11.2): A radical process in which two radicals join together and form a bond.

coupling constant (Sect. 16.7): When signal splitting occurs in NMR spectroscopy, the distance between the individual peaks of a signal.

covalent bond (Sect. 1.5): A bond that results when two atoms share a pair of electrons.

crossed aldol reaction (Sect. 22.3): An aldol reaction that occurs between different partners.

crossed Claisen condensation (Sect. 22.4): A Claisen condensation reaction that occurs between different partners.

crossed-linked polymer (Sect. 27.6): A polymer in which neighboring chains are linked together, for example, by disulfide bonds.

crown ether (Sect. 14.4): Cyclic polyethers whose molecular models resemble crowns.

crystallite (Sect. 27.6): A region of a polymer in which the chains are linearly extended and close in proximity to one another, resulting in van der Waals forces that hold the chains close together.

C terminus (Sect. 25.4): For a peptide chain, the end that contains the COOH group.

cumulated diene (Sect. 17.1): A compound containing two adjacent π bonds.

curved arrows (Sect. 2.8): Tools that are used for drawing resonance structures and for showing the flow of electron density during each step of a reaction mechanism.

cyanohydrin (Sect. 20.10): A compound containing a cyano group and a hydroxyl group connected to the same carbon atom.

cycloaddition reactions (Sect. 17.6): A reaction in which two π systems are joined together in a way that forms a ring. In the process, two π bonds are converted into two σ bonds.

cycloalkane (Sect. 4.2): An alkane whose structure contains a ring.

D

dash (Sect. 2.6): In bond-line structures, a group going behind the page.

D sugar (Sect. 24.2): A carbohydrate for which the chirality center farthest from the carbonyl group will have an OH group pointing to the right in the Fischer projection.

deactivation (Sect. 19.8): For a substituted aromatic ring, the effect of an electron-withdrawing substituent that decreases the rate of electrophilic aromatic substitution.

debye (Sect. 1.11): A unit of measure for dipole moments, where 1 debye $= 10^{-18}$ esu · cm.

decarboxylation (Sect. 22.5): A reaction involving loss of CO_2, characteristic of compounds containing a carbonyl group that is beta to a COOH group.

degenerate (Sect. 4.7): Having the same energy.

degenerate orbitals (Sect. 1.6): Orbitals that have the same energy.

degree of substitution (Sect. 8.3): For alkenes, a classification method that refers to the number of alkyl groups connected to the double bond.

degree of unsaturation (Sect. 15.16): The absence of two hydrogen atoms associated with a ring or a π bond.

dehydration (Sect. 8.9): An elimination reaction involving the loss of H and OH.

dehydrohalogenation (Sect. 8.1): An elimination reaction involving the loss of H and a halogen (such as Cl, Br, or I).

delocalization (Sect. 2.7): The spreading of a charge or lone pair as described by resonance theory.

delocalized lone pair (Sect. 2.12): A lone pair that is participating in resonance.

denaturation (Sect. 25.7): A process during which a protein unfolds under conditions of mild heating.

DEPT ^{13}C NMR (Sect. 16.13): In ^{13}C NMR spectroscopy, a technique that utilizes two rf radiation emitters and provides information regarding the number of protons attached to each carbon atom in a compound.

deshielded (Sect. 16.1): In NMR spectroscopy, protons or carbon atoms whose surrounding electron density is poor.

desulfurization (Sect. 20.8): The conversion of a thioacetal into an alkane in the presence of Raney nickel.

dextrorotatory (Sect. 5.4): A compound that rotates plane-polarized light in a clockwise direction (+).

diagnostic region (Sect. 15.3): The region of an IR spectrum that contains signals that arise from double bonds, triple bonds, and X—H bonds.

diamagnetic anisotropy (Sect. 16. 5): An effect that causes different regions of space to be characterized by different magnetic field strengths.

diamagnetism (Sect. 16.1): The circulation of electron density in the presence of an external magnetic field, which produces a local (induced) magnetic field that opposes the external magnetic field.

diastereomers (Sect. 5.5): Stereoisomers that are not mirror images of one another.

diastereotopic (Sect. 16.4): Nonequivalent protons for which the replacement test produces diastereomers.

diazonium salt (Sect. 23.10): A compound that is formed upon treatment of a primary amine with $NaNO_2$ and HCl.

diazotization (Sect. 23.10): The process of forming a diazonium salt by treating a primary amine with $NaNO_2$ and HCl.

diborane (Sect. 9.6): B_2H_6. A dimeric structure formed when one borane molecule reacts with another.

Dieckmann cyclization (Sect. 22.4): An intramolecular Claisen condensation.

diene (Sect. 17.1): A compound containing two carbon-carbon π bonds.

dienophile (Sect. 17.7) A compound that reacts with a diene in a Diels-Alder reaction.

dihedral angle (Sect. 4.7): The angle by which two groups are separated in a Newman projection.

dihydroxylation (Sect. 9.9): A reaction characterized by the addition of two hydroxyl groups (OH) across an alkene.

diol (Sect. 13.5): A compound containing two hydroxyl groups (OH).

dipole-dipole interactions (Sect. 1.12): The resulting net attraction between two dipoles.

dipole moment (μ) (Sect. 1.11): The amount of partial charge (*d*) on either end of a dipole multiplied by the distance of separation (δ):

$$\mu = d \times \delta$$

directed aldol addition (Sect. 22.3): A technique for performing a crossed aldol addition that produces one major product.

disaccharide (Sect. 24.7): Carbohydrates comprised of two monosaccharide units joined via a glycosidic linkage between the anomeric carbon of one monosaccharide and a hydroxyl group of the other monosaccharide.

disrotatory (Sect. 17.9): In electrocyclic reactions, a type of rotation in which the sets of lobes at the atoms forming a bond with each other must rotate in opposite directions (one rotates clockwise while the other rotates counterclockwise).

dissolving metal reduction (Sect. 10.5): A reaction in which an alkyne is converted into a *trans* alkene.

disulfide (Sect. 14.11): A compound with the structure R—S—S—R.

disulfide bridge (Sect. 25.4): The moiety that is formed when two cysteine residues of a polypeptide or protein are joined together to form a disulfide.

divalent (Sect. 1.2): An element that forms two bonds, such as oxygen.

doublet (Sect. 16.7): In NMR spectroscopy, a signal that is comprised of two peaks.

downfield (Sect. 16.5): The left side of an NMR spectrum.

E

E (Sect. 8.4): For alkenes, a stereodescriptor that indicates that the two priority groups are on opposite sides of the π bond.

E1 (Sect. 8.9): A unimolecular elimination reaction.

E1cb mechanism (Sect. 22.3): An elimination reaction in which the leaving group only leaves after deprotonation occurs. This process occurs at the end of an Aldol condensation.

E2 (Sect. 8.7): A bimolecular elimination reaction.

eclipsed conformation (Sect. 4.7): A conformation in which groups are eclipsing each other in a Newman projection.

Edman degradation (Sect. 25.5): A method for analyzing the sequence of amino acids in a peptide by removing one amino acid residue at a time and identifying each residue as it is removed.

eicosanoids (Sect. 26.7): A class of lipids which includes leukotrienes, prostaglandins, thromboxanes, and prostacyclins.

elastomers (Sect. 27.7): Polymers that return to their original shape after being stretched.

electrocyclic reaction (Sect. 17.6): A pericyclic process in which a conjugated polyene undergoes cyclization. In the process, one π bond is converted into a σ bond, while the remaining π bonds all change their location. The newly formed σ bond joins the ends of the original π system, thereby creating a ring.

electromagnetic spectrum (Sect. 15.1): The range of all frequencies of electromagnetic radiation, which is arbitrarily divided into several regions, most commonly by wavelength.

electron density (Sect. 1.6): A term associated with the probability of finding an electron in a particular region of space.

electron impact ionization (EI) (Sect. 15.8): In mass spectrometry, an ionization technique that involves bombarding a compound with high-energy electrons.

electrophile (Sect. 6.7): A compound containing an electron-deficient atom that is capable of accepting a pair of electrons.

electrophilic aromatic substitution (Sect. 19.1) A substitution reaction in which an aromatic proton is replaced by an electrophile and the aromatic moiety is preserved.

electrophoresis (Sect. 25.2): A technique for separating amino acids from each other based on a difference in pI values.

electrospray ionization (ESI): (Sect. 15.15): In mass spectrometry, an ionization technique in which the compound is first dissolved in a solvent and then sprayed via a high-voltage needle into a vacuum chamber. The tiny droplets of solution become charged by the needle, and subsequent evaporation forms gas-phase molecular ions that typically carry one or more charges.

electrostatic potential maps (Sect. 1.5): A three-dimensional, rainbowlike image used to visualize partial charges in a compound.

elimination (of radicals) (Sect. 11.2): In radical reaction mechanisms, a step in which a bond forms between the alpha (α) and beta (β) positions. As a result, a single bond at the β position is cleaved, causing the compound to fragment into two pieces.

elimination-addition (Sect. 19.14): A reaction that occurs between chlorobenzene and either NaOH (at high temperature) or $NaNH_2$.

enamine (Sect. 21.6): A compound containing a nitrogen atom directly connected to a carbon-carbon π bond.

enantiomer (Sect. 5.2): A nonsuperimposable mirror image.

enantiomeric excess (Sect. 5.4): For a mixture containing two enantiomers, the difference between the percent concentration of the major enantiomer and the percent concentration of its mirror image.

enantiotopic (Sect. 16.4): Protons that are not interchangeable by rotational symmetry but are interchangeable by reflectional symmetry.

endergonic (Sect. 6.3): Any process with a positive ΔG.

endo (Sect. 17.7): In Diels-Alder reactions that produce bicyclic structures, the positions that are *syn* to the larger bridge.

endothermic (Sect. 6.1): Any process with a positive ΔH (the system receives energy from the surroundings).

energy of activation (Sect. 6.5): In an energy diagram, the height of the energy barrier (the hump) between the reactants and the products.

enol (Sect. 10.7): A compound containing a hydroxyl group (OH) connected directly to a carbon-carbon double bond.

enolate (Sect. 22.1): The resonance-stabilized conjugate base of a ketone, aldehyde, or ester.

enthalpy (Sect. 6.1): A measure of the exchange of energy between the system and its surroundings during any process.

entropy (Sect. 6.2): The measure of disorder associated with a system.

enzyme (Sect. 25.8): Important biological molecules that catalyze virtually all cellular processes.

epimer (Sect. 24.6): Stereoisomers that differ from each other in the configuration of only one chirality center.

epoxide (Sect. 9.9): A cyclic ether containing a three-membered ring system. Also called an oxirane.

equatorial position (Sect. 4.11): For chair conformations of substituted cyclohexanes, a position that is approximately along the equator of the ring.

equilibrium (Sect. 3.3): For a reaction, a state in which there is no longer an observable change in the concentrations of reactants and products.

estrogens (Sect. 26.6): Female sex hormones.

ether (Sect. 14.1): A compound with the structure R—O—R.

excited state (Sect. 17.3): A state that is achieved when a compound absorbs energy.

exergonic (Sect. 6.3): Any process with a negative ΔG.

exo (Sect. 17.7): In Diels-Alder reactions that produce bicyclic structures, the positions that are *anti* to the larger bridge.

exothermic (Sect. 6.1): Any process with a negative ΔH (the system gives energy to the surroundings).

F

fats (Sect. 26.3): Triglycerides that are solids at room temperature.

fatty acids (Sect. 26.3): Long-chain carboxylic acids.

fibers (Sect. 27.7): Strands of a polymer that are generated when the polymer is heated, forced through small holes, and then cooled.

fibrous proteins (Sect. 25.8): Proteins that consist of linear chains that are bundled together.

fingerprint region (Sect. 15.3): The region of an IR spectrum that contains signals resulting from the vibrational excitation of most single bonds (stretching and bending).

first order (Sect. 7.5): A rate equation where the rate is linearly dependent on the concentration of only one compound.

Fischer esterification (Sect. 21.10): A process in which a carboxylic acid is converted into an ester when treated with an alcohol in the presence of an acid catalyst.

Fischer projections (Sect. 5.7): A drawing style that is often used when dealing with compounds bearing multiple chirality centers, especially for carbohydrates.

fishhook arrow (Sect. 10.5): A curved arrow with only one barb, indicating the motion of just one electron.

flagpole interaction (Sect. 4.10): For cyclohexane, the steric interactions that occur between the flagpole hydrogen atoms in a boat conformation.

formal charge (Sect. 1.4): A charge associated with any atom that does not exhibit the appropriate number of valence electrons.

fragmentation (Sect. 15.8): In mass spectrometry, when the molecular ion breaks apart into fragments.

free induction decay (Sect. 16.2): In NMR spectroscopy, a complex signal which is a combination of all of the electrical impulses generated by each type of proton.

Freons (Sect. 11.8): CFCs that were heavily used for a wide variety of commercial applications, including as refrigerants, as propellants, in the production of foam insulation, as fire-fighting materials, and many other useful applications.

frequency (Sect. 15.1): For electromagnetic radiation, the number of wavelengths that pass a particular point in space per unit time.

Friedel-Crafts acylation (Sect. 19.6): An electrophilic aromatic substitution reaction that installs an acyl group on an aromatic ring.

Friedel-Crafts alkylation (Sect. 19.5): An electrophilic aromatic substitution reaction that installs an alkyl group on an aromatic ring.

frontier orbitals (Sect. 17.3): The highest occupied molecular orbital (HOMO) and lowest unoccupied molecular orbital (LUMO) that participate in a reaction.

frontier orbital theory (Sect. 17.3): The analysis of a reaction using MO theory, where only the frontier orbitals (HOMO and LUMO) are considered.

Frost circles (Sect. 18.4): A simple method for drawing the relative energy levels of the MO's for any conjugated ring system.

FT-NMR (Sect. 16.2): In nuclear magnetic resonance (NMR) spectroscopy, a technique in which the sample is irradiated with a short pulse that covers the entire range of relevant rf frequencies.

functional group (Sect. 2.3): A characteristic group of atoms/bonds that possess a predictable chemical behavior.

furanose (Sect. 24.5): A five-membered cyclic hemiacetal form of a carbohydrate.

G

Gabriel synthesis (Sect. 23.5): A method for preparing primary amines that avoids formation of secondary and tertiary amines.

***gauche* conformation** (Sect. 4.8): A conformation that exhibits a *gauche* interaction.

***gauche* interaction** (Sect. 4.8): The steric interaction that results when two groups in a Newman projection are separated by a dihedral angle of 60°.

GC-MS (Sect. 15.14): A gas chromatograph–mass spectrometer used for the analysis of a mixture that contains several compounds.

geminal (Sect. 10.4): Two groups connected to the same carbon atom. For example, a geminal dihalide is a compound with two halogens connected to the same carbon atom.

Gibbs free energy (*G*) (Sect. 6.3): The ultimate arbiter of the spontaneity of a reaction, where $\Delta G = \Delta H - T\Delta S$.

Gilman reagent (Sect. 21.8): A lithium dialkyl cuprate (R_2CuLi).

glass transition temperature (T_g) (Sect. 27.6): The temperature at which noncrystalline polymers become very soft.

globular protein (Sect. 25.8): Proteins that consist of chains that are coiled into compact shapes.

glycoside (Sect. 24.6): An acetal that is obtained by treating the cyclic hemiacetal form of a monosaccharide with an alcohol under acid-catalyzed conditions.

graft copolymer (Sect. 27.3): A polymer that contains sections of one homopolymer that have been grafted onto a chain of the other homopolymer.

Grignard reagent (Sect. 13.6): A carbanion with the structure RMgX.

H

haloalkane (Sect. 7.2): An organic compound containing at least one halogen.

haloform reaction (Sect. 22.2): A reaction in which a methyl ketone is converted into a carboxylic acid upon treatment with excess base and excess halogen.

halogen abstraction (Sect. 11.2): In radical reactions, a type of arrow-pushing pattern in which a halogen atom is abstracted by a radical, generating a new radical.

halogenation (Sect. 9.8): A reaction that involves the addition of X_2 (either Br_2 or Cl_2) across an alkene.

halohydrin formation (Sect. 9.8): A reaction which involves the addition of a halogen and a hydroxyl group (OH) across an alkene.

Hammond postulate (Sect. 6.6): In an exothermic process the transition state is closer in energy to the reactants than to the products, and therefore the structure of the transition state more closely resembles the reactants. In contrast, the transition state in an endothermic process is closer in energy to the products, and therefore the transition state more closely resembles the products.

Haworth projection (Sect. 4.14): For substituted cycloalkanes, a drawing style used to clearly identify which groups are above the ring and which groups are below the ring.

HCFCs (Sect. 11.8): Hydrochlorofluorocarbons. Compounds that are similar in structure to CFCs but also possess at least one C—H bond.

heat of combustion (Sect. 4.4): The heat given off during a reaction in which an alkane reacts with oxygen to produce CO_2 and water.

heat of reaction (Sect. 6.1): The heat given off during a reaction.

Hell-Volhard Zelinski reaction (Sect. 22.2): A reaction in which a carboxylic acid undergoes α-halogenation when treated with bromine in the presence of PBr_3.

hemiacetal (Sect. 20.5): A compound containing a hydroxyl group (OH) and an alkoxy group (OR) connected to the same carbon atom.

Henderson-Hasselbalch equation (Sect. 21.3): An equation that is often employed to calculate the pH of buffered solutions:

$$\mathrm{pH} = \mathrm{p}K_a + \log\frac{[\text{conjugated base}]}{[\text{acid}]}$$

heterocycle (Sect. 18.5): A cyclic compound containing at least one heteroatom (such as S, N, or O) in the ring.

heterogeneous catalyst (Sect. 9.7): A catalyst that does not dissolve in the reaction medium.

HFCs (Sect. 11.8): Compounds that contain only carbon, fluorine, and hydrogen (no chlorine).

high-resolution mass spectrometry (Sect. 15.13): A technique that involves the use of a detector that can measure the *m/z* values to four decimal places. This technique allows for the determination of the molecular formula of an unknown compound.

Hofmann elimination (Sect. 23.9): A reaction in which an amino group is treated with excess methyl iodide, thereby converting it into an excellent leaving group, followed by treatment with a strong base to give an E2 reaction that yields an alkene.

Hofmann product (Sect. 8.7): The less substituted product (alkene) of an elimination reaction.

HOMO (Sect. 1.8): The highest occupied molecular orbital.

homogeneous catalysts (Sect. 9.7): A catalyst that dissolves in the reaction medium.

homolytic bond cleavage (Sect. 6.1): Bond breaking that results in the formation of uncharged species called radicals.

homopolymer (Sect. 27.3): A polymer constructed from a single type of monomer.

homotopic (Sect. 16.4): Protons that are interchangeable by rotational symmetry.

Hückel's rule (Sect. 18.4): The requirement for an odd number of π electron pairs in order for a compound to be aromatic.

Hund's rule (Sect. 1.6): When considering electrons in atomic orbitals, a rule that states that one electron is placed in each degenerate orbital first, before electrons are paired up.

hydrate (Sect. 20.5): A compound containing two hydroxyl groups (OH) connected to the same carbon atom.

hydration (Sect. 9.4): A reaction in which a proton and a hydroxyl group (OH) are added across a π bond.

hydrazone (Sect. 20.6): A compound with the structure $R_2C{=}N{-}NH_2$.

hydride shift (Sect. 6.8): A type of carbocation rearrangement that involves the migration of a hydride ion (H^-).

hydroboration-oxidation (Sect. 9.6): A two-step process that achieves an anti-Markovnikov addition of a proton and a hydroxyl group (OH) across an alkene.

hydrocracking (Sect. 11.12): A process performed in the presence of hydrogen gas by which large alkanes in petroleum are converted into smaller alkanes that are more suitable for use as gasoline.

hydrogen abstraction (Sect. 11.2): In radical reactions, a type of arrow-pushing pattern in which a hydrogen atom is abstracted by a radical, generating a new radical.

hydrogen bonding (Sect. 1.12): A special type of dipole-dipole interaction that occurs between an electronegative atom and a hydrogen atom that is connected to another electronegative atom.

hydrogen deficiency index (HDI) (Sect. 15.16): A measure of the number of degrees of unsaturation in a compound.

hydrohalogenation (Sect. 9.3): A reaction that involves the addition of H and X (either Br or Cl) across an alkene.

hydrolysis (Sect. 20.7): A reaction in which bonds are cleaved by treatment with water.

hydroperoxide (Sect. 11.9): A compound with the structure R—O—O—H.

hydrophilic (Sect. 1.13): A polar group that has favorable interactions with water.

hydrophobic (Sect. 1.13): A nonpolar group that does not have favorable interactions with water.

hydroxyl group (Sect. 13.1): An OH group

hyperconjugation (Sect. 6.8): An effect that explains why alkyl groups stabilize a carbocation.

I

imidazole (Sect. 23.12): A compound containing a five-membered ring that is similar to pyrrole but has one extra nitrogen atom at the 3 position.

imine (Sect. 21.6): A compound containing a C=N bond.

induction (Sect. 1.5): The withdrawal of electron density that occurs when a bond is shared by two atoms of differing electronegativity.

initiation (Sect. 11.2): In radical reaction mechanisms, a step in which radicals are created.

integration (Sect. 16.6): In ^{1}H NMR spectroscopy, the area under a signal indicates the number of protons giving rise to the signal.

intermediate (Sect. 6.6): A structure corresponding to a local minimum (valley) in an energy diagram.

intermolecular forces (Sect. 1.12): The attractive forces between molecules.

internal alkyne (Sect. 10.2): A compound with the structure R—C≡C—R, where each R group is not a hydrogen atom.

inversion of configuration (Sect. 7.5): During a reaction, when the configuration of a chirality center is changed.

ionic bond (Sect. 1.5): A bond that results from the force of attraction between two oppositely charged ions.

ionic reaction (Sect. 6.7): A reaction that involves the participation of ions as reactants, intermediates, or products.

isoelectric point (pI) (Sect. 25.2): For an amino acid, the specific pH at which the concentration of the zwitterionic form reaches its maximum value.

isolated diene (Sect. 17.1): A compound containing two carbon-carbon π bonds that are separated by two or more single bonds.

isoprene (Sect. 26.8): 2-Methyl-1,3-butadiene.

isotactic (Sect. 27.6): A polymer in which the repeating units contain chirality centers which all have the same configuration.

IUPAC (Sect. 4.2): The International Union of Pure and Applied Chemistry.

J

***J* value** (Sect. 16.7): When signal splitting occurs in ^{1}H NMR spectroscopy, the distance between the individual peaks of a signal.

K

K_a (Sect. 3.3): A measure of the strength of an acid:

$$K_a = K_{eq}[H_2O] = \frac{[H_3O^+][A^-]}{[HA]}$$

K_{eq} (Sect. 3.3): A term that describes the position of equilibrium for a reaction:

$$K_{eq} = \frac{[H_3O^+][A^-]}{[HA][H_2O]}$$

keto-enol tautomerization (Sect. 10.7): The equilibrium that is established between an enol and a ketone in either acid-catalyzed or base-catalyzed conditions.

ketose (Sect. 24.2): A carbohydrate that contains a ketone moiety.

Kiliani-Fischer synthesis (Sect. 24.6): A process by which the chain of a carbohydrate is lengthened by one carbon atom.

kinetic control (Sect. 17.5): A reaction for which the product distribution is determined by the relative rates at which the products are formed.

kinetics (Sect. 6.5): A term that refers to the rate of a reaction.

L

L amino acid (Sect. 25.2): Amino acids with Fischer projections that resemble the Fischer projections of L sugars.

L sugar (Sect. 24.2): A carbohydrate for which the chirality center farthest from the carbonyl group will have an OH group pointing to the left in the Fischer projection.

labile (Sect. 16.7): Protons that are exchanged at a rapid rate.

lactone (Sect. 21.11): A cyclic ester.

lambda max (λ_{max}) (Sect. 17.11) In UV-VIS spectroscopy, the wavelength of maximum absorption.

leaving group (Sect. 7.1): A group capable of separating from a compound.

lecithins (Sect. 26.5): Phosphoglycerides that contain choline.

leveling effect (Sect. 3.6): An effect that prevents the use of bases stronger than hydroxide when the solvent is water.

levorotatory (Sect. 5.4): A compound that rotates plane-polarized light in a counterclockwise direction (−).

Lewis acid (Sect. 3.9): A compound capable of functioning as an electron pair acceptor.

Lewis base (Sect. 3.9): A compound capable of functioning as an electron pair donor.

Lewis structures (Sect. 1.3): A drawing style in which the electrons take center stage.

linear polymer (Sect. 27.6): A polymer that has only a minimal amount of branching or no branching at all.

lipid (Sect. 26.1): Naturally occurring compounds that can be extracted from cells using nonpolar organic solvents.

lipid bilayer (Sect. 26.5): The main fabric of cell membranes, assembled primarily from phosphoglycerides.

lithium dialkyl cuprate (Sect. 21.8): A nucleophilic compound with the general structure R_2CuLi.

living polymer (Sect. 27.4): A polymer that is formed via anionic polymerization.

localized lone pair (Sect. 2.12): A lone pair that is not participating in resonance.

locant (Sect. 4.2): In nomenclature, a number used to identify the location of a substituent.

London dispersion forces (Sect. 1.12): Attractive forces between transient dipole moments, observed in alkanes.

lone pair (Sect. 1.3): A pair of unshared, or nonbonding, electrons.

loss of a leaving group (Sect. 6.8): One of the four arrow-pushing patterns for ionic reactions.

LUMO (Sect. 1.8): The lowest unoccupied molecular orbital.

M

magnetic moment (Sect. 16.1): A magnetic field generated by a spinning proton.

malonic ester synthesis (Sect. 22.5): A synthetic technique that enables the transformation of a halide into a carboxylic acid with the introduction of two new carbon atoms.

Markovnikov addition (Sect. 9.3): In addition reactions, the observation that the hydrogen atom is generally placed at the vinylic position already bearing the larger number of hydrogen atoms.

mass spectrometer (Sect. 15.8): A device in which a compound is first vaporized and converted into ions, which are then separated and detected.

mass spectrometry (Sect. 15.8): The study of the interaction between matter and an energy source other than electromagnetic radiation. Mass spectrometry is used primarily to determine the molecular weight and molecular formula of a compound.

mass spectrum (Sect. 15.8): In mass spectrometry, a plot that shows the relative abundance of each cation that was detected.

mass-to-charge ratio(*m/z*) (Sect. 15.8): The determining factor by which ions are separated from each other in mass spectrometry.

Meisenheimer complex (Sect. 19.13): The resonance-stabilized intermediate of a nucleophilic aromatic substitution reaction.

melt transition temperature (T_m) (Sect. 27.6): The temperature at which the crystalline regions of a polymer become amorphous.

mercapto group (Sect. 14.11): An SH group.

mercurinium ion (Sect. 9.5): The intermediate formed during oxymercuration.

Merrifield synthesis (Sect. 25.6): A method for building a peptide from protected building blocks.

***meso* compound** (Sect. 5.6): A compound that possesses chirality centers and an internal plane of symmetry.

meta (Sect. 18.2): On an aromatic ring, the C3 position.

***meta* director** (Sect. 19.8): An electron-withdrawing group that directs the regiochemistry of an electrophilic aromatic substitution reaction such that the incoming electrophile is installed at the *meta* position.

methine group (Sect. 16.5): A CH group.

methylene group (Sect. 16.5): A CH_2 group.

methyl shift (Sect. 6.8): A type of carbocation rearrangement in which a methyl group migrates.

micelle (Sect. 1.13): A group of molecules arranged in a sphere such that the surface of the sphere is comprised of polar groups, rendering the micelle water soluble.

Michael acceptor (Sect. 22.6): The electrophile in a Michael reaction.

Michael donor (Sect. 22.6): The nucleophile in a Michael reaction.

Michael reaction (Sect. 22.6): A reaction in which a nucleophile attacks a conjugated π system, resulting in a 1,4-addition.

migratory aptitude (Sect. 20.11): In a Baeyer-Villiger oxidation, the migration rates of different groups, which determine the regiochemical outcome of the reaction.

miscible (Sect. 13.1): Two liquids that can be mixed with each other in any proportion.

mixed aldol reactions (Sect. 22.3): An aldol reaction that occurs between different partners.

moderate activator (Sect. 19.10) A group that moderately activates an aromatic ring toward an electrophilic aromatic substitution reaction.

moderate deactivators (Sect. 19.10): A group that moderately deactivates an aromatic ring toward an electrophilic aromatic substitution reaction.

molar absorptivity (Sect. 17.11): The amount of UV light absorbed at the λ_{max} of a compound, as described by Beer's law.

molecular dipole moment (Sect. 1.11): The vector sum of the individual dipole moments in a compound.

molecular ion (Sect. 15.8): In mass spectrometry, the ion that is generated when the compound is ionized.

molecular orbitals (Sect. 1.8): Orbitals associated with an entire molecule rather than an individual atom.

molecular orbital theory (Sect. 1.8): A description of bonding in terms of molecular orbitals, which are orbitals associated with an entire molecule rather than an individual atom.

monosaccharides (Sect. 24.2): Simple sugars that generally contain multiple chirality centers.

monovalent (Sect. 1.2): An element that is capable of forming one bond (such as hydrogen).

multiplet (Sect. 16.7): In ^{1}H NMR spectroscopy, a signal whose multiplicity is difficult to analyze.

multiplicity (Sect. 16.7): In ^{1}H NMR spectroscopy, the number of peaks in a signal.

mutarotation (Sect. 24.5): A term used to describe the fact that α and β anomers of carbohydrates can equilibrate via the open-chain form.

N

***n*+1 rule** (Sect. 16.7): In NMR spectroscopy, if n is the number of neighboring protons, then the multiplicity will be n+1.

N terminus (Sect. 25.4): For a peptide chain, the end that contains the amino group.

***N*-bromosuccinimide** (Sect. 11.7): A reagent used for allylic bromination to avoid a competing reaction in which bromine adds across the π bond.

Newman projection (Sect. 4.6): A drawing style that is designed to show the conformation of a molecule.

***N*-glycoside** (Sect. 24.10): The product obtained when a monosaccharide is treated with an amine in the presence of an acid catalyst.

nitration (Sect. 19.4): An electrophilic aromatic substitution reaction that involves the installation of a nitro group (NO_2) on an aromatic ring.

nitrogen rule (Sect. 15.9): In mass spectrometry, an odd molecular weight indicates an odd number of nitrogen atoms in the compound, while an even molecular weight indicates either an even number of nitrogen atoms or the absence of nitrogen.

nitronium ion (Sect. 19.4): The NO_2^+ ion, which is present in a mixture of nitric acid and sulfuric acid.

nitrosonium ion (Sect. 23.10): The NO^+ ion, which is formed when $NaNO_2$ is treated with HCl.

nitrous acid (Sect. 23.10): A compound with molecular formula HONO.

***N*-nitrosamine** (Sect. 23.10): A compound with the structure $R_2N—N{=}O$.

node (Sect. 1.6): In atomic and molecular orbitals, a location where the value of ψ is zero.

nomenclature (Sect. 4.2): A system for naming organic compounds.

nonaromatic (Sect. 18.5): A compound that lacks a ring with a continuous system of overlapping *p* orbitals or it lacks an odd number of π electron pairs.

norbornane (Sect. 4.15): The common name for bicyclo[2.2.1]heptane.

nuclear magnetic resonance (NMR) (Sect. 16.1): A form of spectroscopy that involves the study of the interaction between electromagnetic radiation and the nuclei of atoms.

nucleophile (Sect. 6.7): A compound containing an electron-rich atom that is capable of donating a pair of electrons.

nucleophilic acyl substitution (Sect. 21.7): A reaction in which a nucleophile attacks a carboxylic acid derivative.

nucleophilic aromatic substitution (Sect. 19.13): A substitution reaction in which an aromatic ring is attacked by a nucleophile, which replaces a leaving group.

nucleophilic attack (Sect. 6.8): One of the four arrow-pushing patterns for ionic reactions.

nucleosides (Sect. 24.10): The product formed when either D-ribose or 2-deoxy- D-ribose is coupled with certain nitrogen heterocycles (called bases).

nucleotides (Sect. 24.10): The product formed when a nucleoside is coupled to a phosphate group.

O

observed rotation (Sect. 5.4): The extent to which plane-polarized light is rotated by a solution of a chiral compound.

octet rule (Sect. 1.3): The observation that second-row elements (C, N, O, and F) will form the necessary number of bonds so as to achieve a full valence shell (eight electrons).

off-resonance decoupling (Sect. 16.11): In NMR spectroscopy, a technique in which only the one-bond couplings are observed. CH_3 groups appear as quartets, CH_2 groups appear as triplets, CH groups appear as doublets, and quaternary carbon atoms appear as singlets.

oils (Sect. 26.3): Triglycerides that are liquids at room temperature.

oligomers (Sect. 27.5): During the polymerization process, compounds constructed from just a few monomers.

optically active (Sect. 5.4): A compound that rotates plane-polarized light.

optically inactive (Sect. 5.4): A compound that does not rotate plane-polarized light.

optically pure (Sect. 5.4): A solution containing just one enantiomer, but not its mirror image.

organohalide (Sect. 7.2): An organic compound containing at least one halogen.

ortho (Sect. 18.2): On an aromatic ring, the C2 position.

***ortho-para* director** (Sect. 19.7): A group that directs the regiochemistry of an electrophilic aromatic substitution reaction such that the incoming electrophile is installed at the *ortho* or *para* positions.

oxaphosphetane (Sect. 20.10): An intermediate that is believed to be formed during Wittig reactions.

oxidation (Sect. 13.10): A reaction in which one compound undergoes an increase in oxidation state.

oxidation state (Sect. 13.4): A method of electron book-keeping in which all bonds are treated as if they were purely ionic.

oxime (Sect. 21.6): A compound with the structure $R_2C{=}N—OH$.

oxirane (Sect. 14.7): A cyclic ether containing a three-membered ring system. Also called an epoxide.

oxonium ion (Sect. 9.4): An intermediate with a positively charged oxygen atom.

oxymercuration-demercuration (Sect. 9.5): A two-step process for the Markovnikov addition of water across an alkene. With this process, carbocation rearrangements do not occur.

ozonolysis (Sect. 9.11): A reaction in which the C=C bond of an alkene is cleaved to form two C=O bonds.

P

para (Sect. 18.2): On an aromatic ring, the C4 position.

parent ion (Sect. 15.8): In mass spectrometry, the ion that is generated when the compound is ionized.

partially condensed structures (Sect. 2.1): A drawing style in which the CH bonds are not drawn explicitly, but all other bonds are drawn.

Pauli exclusion principle (Sect. 1.6): The rule that states that an atomic orbital or molecular orbital can accommodate a maximum of two electrons with opposite spin.

peptidases (Sect. 25.5): A variety of enzymes that selectively hydrolyze specific peptide bonds.

peptide (Sect. 25.1): A chain comprised of a small number of amino acid residues.

peptide bond (Sect. 25.1): The amide linkage by which two amino acids are coupled together to form peptides.

pericylic reactions (Sect. 17.6): Reactions that occur via a concerted process and do not involve either ionic or radical intermediates.

periplanar (Sect. 8.7): A conformation in which a hydrogen atom and a leaving group are approximately coplanar.

peroxides (Sect. 11.3): Compounds with the general structure R—O—O—R.

phenolate (Sect. 13.2): The conjugate base of phenol or a substituted phenol.

phenoxide (Sect. 13.2): The conjugate base of phenol or a substituted phenol.

phenyl group (Sect. 18.2): A C_6H_5 group.

phosphatidic acid (Sect. 26.5): A phosphoric monoester, which is the simplest kind of phosphoglyceride.

phosphoglycerides (Sect. 26.5): Compounds that are very similar in structure to triglycerides, with the main difference being that one of the three fatty acid residues is replaced by a phosphoester group.

phospholipids (Sect. 26.5): Esterlike derivatives of phosphoric acid.

phosphorane (Sect. 20.10): The phosphorus-containing reagent used to achieve a Wittig reaction.

photochemical reaction (Sect. 17.3): A reaction that is performed with photochemical excitation (usually UV light).

photon (Sect. 15.1): When electromagnetic radiation is viewed as a particle, an individual packet of energy.

physiological pH (Sect. 21.3): The pH of blood (approximately 7.3).

pi (π) bond (Sect. 1.9): A bond formed from adjacent, overlapping *p* orbitals.

plane of symmetry (Sect. 5.6): A plane that bisects a compound into two halves that are mirror images of each other.

plane-polarized light (Sect. 5.4): Light for which all photons have the same polarization, generally formed by passing light through a polarizing filter.

plasticizers (Sect. 27.7): Small molecules that are trapped between polymer chains where they function as lubricants, preventing the polymer from being brittle.

polar aprotic solvent (Sect. 7.8): A solvent that lacks hydrogen atoms connected directly to an electronegative atom.

polar covalent bond (Sect. 1.5): A bond in which the difference in electronegative values of the two atoms is between 0.5 and 1.7.

polarimeter (Sect. 5.4): A device that measures the rotation of plane-polarized light caused by optically active compounds.

polarizability (Sect. 6.7): The ability of an atom or molecule to distribute its electron density unevenly in response to external influences.

polarization (Sect. 5.4): Light for which all photons have the same orientation of their electric fields.

polar protic solvent (Sect. 7.8): A solvent that contains at least one hydrogen atom connected directly to an electronegative atom.

polar reaction (Sect. 6.7): A reaction that involves the participation of ions as reactants, intermediates, or products.

polycarbonates (Sect. 27.4): Polymers that are similar in structure to polyesters but with repeating carbonate moieties (—O—CO_2—) instead of repeating ester moieties (—CO_2—).

polycyclic aromatic hydrocarbons (PAHs) (Sect. 18.5): Compounds containing multiple aromatic rings fused together.

polyether (Sect. 14.4): A compound containing several ether moieties.

polynucleotide (Sect. 24.10): A polymer constructed from nucleotides linked together.

polysaccharides (Sect. 24.8): Polymers made up of repeating monosaccharide units linked together by glycoside bonds.

polyurethanes (Sect. 27.5): Polymers made up of repeating urethane moieties, also sometimes called carbamate moieties (—N—CO_2—).

primary alkyl halide (Sect. 7.2): An organohalide in which the alpha (α) position is connected to only one alkyl group.

primary carbocation (Sect. 6.8): A carbocation in which the electrophilic carbon atom is connected to only one alkyl group.

primary structure (Sect. 25.7): For proteins, the sequence of amino acid residues.

progestins (Sect. 26.6): Female sex hormones.

propagation (Sect. 11.2): For radical reactions, the steps whose sum gives the net chemical reaction.

prostaglandins (Sect. 26.7): Lipids that contain 20 carbon atoms and are characterized by a five-membered ring with two side chains.

prosthetic group (Sect. 25.8): A nonprotein unit attached to a protein, such as heme in hemoglobin.

protecting group (Sect. 13.7): A group that is used during synthesis to protect a functional group from the reaction conditions.

proteins (Sect. 25.1): Polypeptide chains comprised of more than 40 or 50 amino acids.

proton transfer (Sect. 6.8): One of the four arrow-pushing patterns for ionic reactions.

PVC (Sect. 11.11): A polymer called poly(vinyl chloride), formed from the polymerization of vinyl chloride ($H_2C{=}CHCl$).

pyranose (Sect. 24.5): A six-membered cyclic hemiacetal form of a carbohydrate.

pyrimidine (Sect. 23.11): A compound that is similar in structure to pyridine but contains one extra nitrogen atom at the 3 position.

Q

quantum mechanics (Sect. 1.6): A mathematical description of an electron that incorporates its wavelike properties.

quartet (Sect. 16.7): In NMR spectroscopy, a signal that is comprised of four peaks.

quaternary ammonium salt (Sect. 23.5): An ionic compound containing a positively charged nitrogen atom connected to four alkyl groups.

quaternary structure (Sect. 25.7): The structure that arises when a protein consists of two or more folded polypeptide chains that aggregate to form one protein complex.

quintet (Sect. 16.7): In NMR spectroscopy, a signal that is comprised of five peaks.

R

R (Sect. 5.3): A term used to designate the configuration of a chirality center, determined in the following way: Each of the four groups is assigned a priority, and the molecule is then rotated (if necessary) so that the #4 group is directed behind the page (on a dash). A clockwise sequence for 1-2-3 is designated as *R*.

racemic mixture (Sect. 5.4): A solution containing equal amounts of both enantiomers.

radical (Sect. 6.1): A chemical entity with an unpaired electron.

radical anion (Sect. 10.5): An intermediate that has both a negative charge and an unpaired electron.

radical inhibitor (Sect. 11.3): A compound that prevents a radical chain process from either getting started or continuing.

radical initiator (Sect. 11.3): A compound with a weak bond that undergoes homolytic bond cleavage with great ease, producing radicals that can initiate a radical chain process.

random copolymer (Sect. 27.3): A polymer, comprised of more than one kind of repeating unit, in which there is a random distribution of repeating units.

rate-determining step (Sect. 7.5): The slowest step in a multistep reaction which determines the rate of the reaction.

rate equation (Sect. 6.5): An equation that describes the relationship between the rate of a reaction and the concentration of reactants.

reaction mechanism (Sect. 3.2): A series of intermediates and curved arrows that show how the reaction occurs in terms of the motion of electrons.

rearrangement (Sect. 6.8): One of the four arrow-pushing patterns for ionic reactions.

reducing agent (Sect. 13.4): A compound that reduces another compound and in the process is itself oxidized. Sodium borohydride and lithium aluminum hydride are reducing agents.

reducing sugar (Sect. 24.6): A carbohydrate that is oxidized upon treatment with Tollens' reagent, Fehling's reagent, or Benedict's reagent.

reduction (Sect. 13.4): A reaction in which a compound undergoes a decrease in oxidation state.

reductive amination (Sect. 23.6): The conversion of a ketone or aldehyde into an imine under conditions in which the imine is reduced as soon as it is formed, giving an amine.

regiochemistry (Sect. 8.7): A term describing a consideration that must be taken into account for a reaction in which two or more constitutional isomers can be formed.

regioselective (Sect. 8.7): A reaction that can produce two or more constitutional isomers but nevertheless produces one as the major product.

replacement test (Sect. 16.4): A test for determining the relationship between two protons. The compound is drawn two times, each time replacing one of the protons with deuterium. If the two compounds are identical, the protons are homotopic. If the two compounds are enantiomers, the protons are enantiotopic. If the two compounds are diastereomers, the protons are diastereotopic.

resolution (Sect. 5.9): The separation of enantiomers from a mixture containing both enantiomers.

resolving agents (Sect. 5.9): A compound that can be used to achieve the resolution of enantiomers.

resonance (Sect. 2.7): A method that chemists use to deal with the inadequacy of bond-line drawings.

resonance stabilization (Sect. 2.7): The stabilization associated with the delocalization of electrons via resonance.

resonance structures (Sect. 2.7): A series of structures that are melded together (conceptually) to circumvent the inadequacies of bond-line drawings.

retention of configuration (Sect. 7.5): During a reaction, when the configuration of a chirality center remains unchanged.

retention time (Sect. 15.14): The amount of time required for a compound to exit from a gas chromatograph.

retro-aldol reaction (Sect. 22.3): The reverse of an aldol reaction. A β-hydroxyketone or aldehyde is converted into two ketones or aldehydes.

retro Diels-Alder (Sect. 17.7): The reverse of a Diels-Alder reaction, achieved at high temperature. A cyclohexene derivative is converted into a diene and a dienophile.

retrosynthetic analysis (Sect. 12.5): A systematic set of principles that enable the design of a synthetic route by working backward from the desired product.

ring flip (Sect. 4.12): A conformational change in which one chair conformation is converted into the other.

Robinson annulation (Sect. 22.6): The combination of a Michael addition followed by an aldol condensation to form a ring.

S

S (Sect. 5.3): A term used to designate the configuration of a chirality center, determined in the following way: Each of the four groups is assigned a priority, and the molecule is then rotated (if necessary) so that the #4 group is directed behind the page (on a dash). A counterclockwise sequence for 1-2-3 is designated as *S*.

Sandmeyer reactions (Sect. 23.11): Reactions that utilize copper salts (CuX) and enable the installation of a halogen or a cyano group on an aromatic ring.

saponification (Sect. 21.11): The base-catalyzed hydrolysis of an ester. This method is used to make soap.

saturated (Sect. 15.16): A compound that contains no π bonds.

saturated hydrocarbon (Sect. 4.1): A hydrocarbon that contains no π bonds.

Schiemann reaction (Sect. 23.11): The conversion of an aryl diazonium salt into fluorobenzene upon treatment with fluoroboric acid (HBF_4).

s-cis (Sect. 17.2): A conformation of a conjugated diene in which the disposition of the two π bonds with regard to the connecting single bond is *cis-like* (a dihedral angle of 0°).

secondary alkyl halide (Sect. 7.2): An organohalide in which the alpha (α) position is connected to exactly two alkyl groups.

secondary carbocation (Sect. 6.8): A carbocation in which the electrophilic carbon atom is connected to exactly two alkyl groups.

secondary structure (Sect. 25.7): The three-dimensional conformations of localized regions of a protein, including helices and β-pleated sheets.

second order (Sect. 7.4): A rate equation where the rate is linearly dependent on the concentration of two compounds.

Sharpless asymmetric epoxidation (Sect. 14.9): A reaction that converts an alkene into an epoxide via a stereospecific pathway.

shielded (Sect. 16.1): In NMR spectroscopy, protons or carbon atoms whose surrounding electron density is rich.

sigma (σ) bond (Sect. 1.7): A bond that is characterized by circular symmetry with respect to the bond axis.

sigma complex (Sect. 19.2): The positively charged intermediate of an electrophilic aromatic substitution reaction.

sigmatropic rearrangements (Sect. 17.6): A pericyclic reaction in which one σ bond is formed at the expense of another.

simple lipids (Sect. 26.1): A lipid that does not undergo hydrolysis in aqueous acid or base to produce smaller fragments.

singlet (Sect. 16.7): In NMR spectroscopy, a signal that is comprised of only one peak.

S_N1 (Sect. 7.5): A unimolecular nucleophilic substitution reaction.

S_N2 (Sect. 7.4): A bimolecular nucleophilic substitution reaction.

sodium cyanoborohydride (Sect. 23.6): A selective reducing agent ($NaBH_3CN$) that can be used for reductive amination.

soluble (Sect. 13.1): A term used to indicate that a certain volume of a compound will dissolve in a specified amount of a liquid at room temperature.

solvent extraction (Sect. 23.3): A process by which one or more compounds are removed from a mixture of organic compounds, based on a difference in solubility and/or acid-base properties.

solvolysis (Sect. 7.6): A substitution reaction in which the solvent functions as the nucleophile.

specific rotation (Sect. 5.4): For a chiral compound that is subjected to plane-polarized light, the observed rotation when a standard concentration (1 g/mL) and a standard path length (1 dm) are used.

spectroscopy (Sect. 15.1): The study of the interaction between matter and electromagnetic radiation.

***sp*-hybridized orbitals** (Sect. 1.9): Atomic orbitals that are achieved by mathematically averaging one *s* orbital with only one *p* orbital to form two hybridized atomic orbitals.

sp^2-hybridized orbitals (Sect. 1.9): Atomic orbitals that are achieved by mathematically averaging one *s* orbital with two *p* orbitals to form three hybridized atomic orbitals.

sp^3-hybridized orbitals (Sect. 1.9): Atomic orbitals that are achieved by mathematically averaging one *s* orbital with three *p* orbitals to form four hybridized atomic orbitals.

spin-spin splitting (Sect. 16.7): A phenomenon observed most commonly for nonequivalent protons connected to adjacent carbon atoms, in which the multiplicity of each signal is affected by the other.

spontaneous (Sect. 6.2): A reaction with a negative ΔG, which means that products are favored at equilibrium.

staggered conformation (Sect. 4.7): A conformation in which nearby groups in a Newman projection have a dihedral angle of 60°.

standard atomic weight (Sect. 15.13): The weighted averages for each element, which takes into account isotopic abundance.

step-growth polymers (Sect. 27.5): Polymers that are formed under conditions in which the individual monomers react with each other to form oligomers, which are then joined together to form polymers.

stereoisomers (Sect. 4.14): Compounds that have the same constitution but differ in the 3D arrangement of atoms.

stereoselective (Sect. 8.7): A reaction in which one substrate produces two stereoisomers in unequal amounts.

stereospecific (Sect. 7.4): A reaction in which the configuration of the product is dependent on the configuration of the starting material.

sterically hindered (Sect. 3.7): A compound or region of a compound that is very bulky.

steric number (Sect. 1.10): The total of (single bonds + lone pairs) for an atom in a compound.

steroids (Sect. 26.6): Lipids that are based on a tetracyclic ring system involving three six-membered rings and one five-membered ring. Cholesterol is an example.

Stork enamine synthesis (Sect. 22.6): A Michael reaction in which an enamine functions as a nucleophile.

s-trans (Sect. 17.2): A conformation of a conjugated diene in which the disposition of the two π bonds with regard to the connecting single bond is *trans-like* (a dihedral angle of 180°).

Strecker synthesis (Sect. 25.3): A synthetic technique for preparing racemic α-amino acids from aldehydes.

stretching (of bonds) (Sect. 15.2): In IR spectroscopy, a type of vibration that generally produces a signal in the diagnostic region of an IR spectrum.

strong activator (Sect. 19.10): A group that strongly activates an aromatic ring toward electrophilic aromatic substitution, thereby significantly enhancing the rate of the reaction.

strong deactivator (Sect. 19.10): A group that strongly deactivates an aromatic ring toward electrophilic aromatic substitution, thereby significantly decreasing the rate of the reaction.

structural protein (Sect. 25.8): Fibrous proteins that are used for their structural rigidity. Examples include α-keratins found in hair, nails, skin, feathers, and wool.

substituents (Sect. 4.2): In nomenclature, the groups connected to the parent chain.

substitution reaction (Sect. 7.1): A reaction in which one group is replaced by another group.

substrate (Sect. 7.1): The starting alkyl halide in a substitution or elimination reaction.

sulfide (Sect. 14.11): A compound that is similar in structure to an ether, but the oxygen atom has been replaced with a sulfur atom. Also called a thioether.

sulfonate ions (Sect. 7.8): Common leaving groups. Examples include tosylate, mesylate, and triflate ions.

sulfonation (Sect. 19.3): An electrophilic aromatic substitution reaction in which an SO_3H group is installed on an aromatic ring.

sulfone (Sect. 14.11): A compound that contains a sulfur atom that has double bonds with two oxygen atoms and is flanked on both sides by R groups.

sulfoxide (Sect. 14.11): A compound containing an S═O bond that is flanked on both sides by R groups.

superimposable (Sect. 5.2): Two objects that are identical.

symmetrical ether (Sect. 14.2): An ether (R—O—R) where both R groups are identical.

symmetric stretching (Sect. 15.5): In IR spectroscopy, when two bonds are stretching in phase with each other.

symmetry-allowed (Sect. 17.8): A reaction that obeys conservation of orbital symmetry.

symmetry-forbidden (Sect. 17.8): A reaction that disobeys conservation of orbital symmetry.

***syn* addition** (Sect. 9.6): An addition reaction in which two groups are added to the same face of a π bond.

***syn* coplanar** (Sect. 8.7): A conformation in which a hydrogen atom and a leaving group are separated by a dihedral angle of exactly 0°.

syndiotactic (Sect. 27.6): A polymer in which the repeating units contain chirality centers which have alternating configuration.

T

tautomers (Sect. 10.7): Constitutional isomers that rapidly interconvert via the migration of a proton.

terminal alkynes (Sect. 10.2): Compounds with the following structure: R—C≡C—H

termination (Sect. 11.2): In radical reactions, a step in which two radicals are joined to give a compound with no unshared electrons.

termolecular (Sect. 10.6): For mechanisms, a step that involves three chemical entities.

terpenes (Sect. 26.8): A diverse class of naturally occurring compounds that can be thought of as being assembled from isoprene units, each of which contains five carbon atoms.

tertiary alkyl halide (Sect. 7.2): An organohalide in which the alpha (α) position is connected to three alkyl groups.

tertiary carbocation (Sect. 6.8): A carbocation in which the electrophilic carbon atom is connected to three alkyl groups.

tertiary structure (Sect. 25.7): The three-dimensional shape of a protein.

tetrahedral (Sect. 1.10): The geometry of an atom with four bonds separated from each other by 109.5°.

tetrahedral intermediate (Sect. 21.4): An intermediate with tetrahedral geometry. This type of intermediate is formed when a nucleophile attacks a carbonyl group.

tetravalent (Sect. 1.2): An element, such as carbon, that forms four bonds.

thermodynamic control (Sect. 17.5): A reaction for which the ratio of products is determined solely by the distribution of energy among the products.

thermodynamics (Sect. 6.3): The study of how energy is distributed under the influence of entropy. For chemists, the thermodynamics of a reaction specifically refers to the study of the relative energy levels of reactants and products.

thermoplastics (Sect. 27.7): Polymers that are hard at room temperature but soft when heated.

thermosetting resins (Sect. 27.7): Highly cross-linked polymers that are generally very hard and insoluble.

thioacetal (Sect. 20.8): A compound that contains two SR groups, both of which are connected to the same carbon atom.

thiolate (Sect. 14.11): The conjugate base of a thiol.

thiols (Sect. 14.11): Compounds containing a mercapto group (SH).

three-center, two-electron bonds (Sect. 9.6): A bond in which two electrons are associated with three atoms, such as in diborane (B_2H_6).

torsional angle (Sect. 4.7): The angle between two groups in a Newman projection, also called the dihedral angle.

torsional strain (Sect. 4.7): The difference in energy between staggered and eclipsed conformations (for example, in ethane).

tosylate (Sect. 7.8): An excellent leaving group (OTs).

transition state (Sect. 6.6): A state through which a reaction passes. On an energy diagram, a transition state corresponds with a local maximum.

transport protein (Sect. 25.8): A protein used to transport molecules or ions from one location to another. Hemoglobin is a classic example of a transport protein, used to transport molecular oxygen from the lungs to all the tissues of the body.

triglyceride (Sect. 26.3): A triester formed from glycerol and three long-chain carboxylic acids.

trigonal planar (Sect. 1.10): A geometry adopted by an atom with a steric number of 3. All three groups lie in one plane and are separated by 120°.

trigonal pyramidal (Sect. 1.10): A geometry adopted by an atom that has one lone pair and a steric number of 4.

triplet (Sect. 16.7): In NMR spectroscopy, a signal that is comprised of three peaks.

trivalent (Sect. 1.2): An element, such as nitrogen, that forms three bonds.

twist boat conformation (Sect. 4.10): A conformation of cyclohexane that is lower in energy than a boat conformation but higher in energy than a chair conformation.

U

unimolecular (Sect. 7.5): For mechanisms, a step that involves only one chemical entity.

unsaturated (Sect. 15.16): A compound containing one or more π bonds.

unsymmetrical ether (Sect. 14.2): An ether (R—O—R) where the two R groups are not identical.

V

valence (Sect. 1.2): The number of bonds usually formed by an element. For example, oxygen is divalent because it generally forms two bonds.

valence bond theory (Sect. 1.7): A theory that treats a bond as the sharing of electrons that are associated with individual atoms, rather than being associated with the entire molecule.

vibrational excitation (Sect. 15.1): In IR spectroscopy, the energy of a photon is absorbed and temporarily stored as vibrational energy.

vicinal (Sect. 10.4): A term used to describe two identical groups attached to adjacent carbon atoms.

vinylic (Sect. 2.10): The carbon atoms of a carbon-carbon double bond.

vinylic carbocation (Sect. 10.6): A carbocation in which the positive charge resides on a vinylic carbon atom. This type of carbocation is very unstable and will not readily form in most cases.

VSEPR theory (Sect. 1.10): Valence Shell Electron Pair Repulsion Theory, which can be used to predict the geometry around an atom.

W

wavelength (Sect. 15.1): The distance between adjacent peaks of an oscillating magnetic or electric field.

wavenumber (Sect. 15.2): In IR spectroscopy, the location of each signal is reported in terms of this frequency-related unit.

waxes (Sect. 26.2): High-molecular-weight esters that are constructed from carboxylic acids and alcohols.

weak activators (Sect. 19.10): A group that weakly activates an aromatic ring toward electrophilic aromatic substitution, thereby enhancing the rate of the reaction.

weak deactivators (Sect. 19.10): A group that weakly deactivates an aromatic ring toward electrophilic aromatic substitution, thereby decreasing the rate of the reaction.

wedge (Sect. 2.6): In bond-line structures, a group in front of the page.

Williamson ether synthesis (Sect. 14.5): A method for preparing an ether from an alkoxide ion and an alkyl halide (via an S_N2 process).

Wittig reaction (Sect. 20.10): A reaction that converts an aldehyde or ketone into an alkene, with the introduction of one or more carbon atoms.

Wittig reagent (Sect. 20.10): A reagent used to perform a Wittig reaction.

Wohl degradation (Sect. 24.6): A process that involves the removal of a carbon atom from an aldose. The aldehyde group is first converted to a cyanohydrin, followed by loss of HCN in the presence of a base.

Wolff-Kishner reduction (Sect. 21.6): A method for converting a carbonyl group into a methylene group (CH_2) under basic conditions.

Woodward-Fieser rules (Sect. 17.11): Rules for predicting the wavelength of maximum absorption for a compound with extended conjugation.

Y

ylide (Sect. 20.10): A compound with two oppositely charged atoms adjacent to each other.

Z

Z (Sect. 8.4): For alkenes, a stereodescriptor that indicates that the two priority groups are on the same side of the π bond.

Zaitsev product (Sect. 8.7): The more substituted product (alkene) of an elimination reaction.

zwitterion (Sect. 25.2): A net neutral compound that exhibits charge separation. Amino acids exist as zwitterions at physiological pH.

Credits

All chapters SkillBuilders: *road bicycle front wheel*, jules2000/Shutterstock; *bicycle*, Igor Shikov/Shutterstock. Medically Speaking: *pills*, Norph/Shutterstock.

Chapter 1 Opener: *kite string*, Dave King/Dorling Kindersley/Getty Images, Inc.; *bolt of lightning striking Empire State Building at night*, Paul Katz/Photolibrary/Getty Images, Inc.; *brass skeleton key hanging from string*, Gary S Chapman/Photographer's Choice/Getty Images, Inc.; *lightning striking the Empire State Building during a storm*, Keystone/Hulton Archive/Getty Images, Inc.; *polaroid frame*, Cole Vineyard/iStockphoto. Practically Speaking, p. 36: *Tokay gecko*, Eric Isselée/iStockphoto. Figures 1.6, 1.13, 1.23–1.25, 1.27–1.29, 1.32, 1.33, 1.36–1.40, 1.42, 1.44, 1.47, 1.48, Table 1.1, and Table 1.2 are Reprinted with permission of John Wiley & Sons, Inc. from Solomons, G., *Organic Chemistry, 10e*, © 2011. Running head: *lightning strikes*, Justin Horrocks/iStockphoto; *brass key*, J. R. Bale/Stockphotopro, Inc.

Chapter 2 Opener: *poppy flower*, Neil Fletcher/Getty Images, Inc.; *hand drawing molecule*, Vladislav Susoy/Shutterstock; *capsule, tablet, gel tab*, Norph/Shutterstock. Medically Speaking, p. 65: *poppy flower*, Neil Fletcher/Getty Images, Inc. Running head: *capsule, tablet, gel tab*, Norph/Shutterstock.

Chapter 3 Opener: *spoons of assorted raising agents*, Z. Sandmann-StockFood Munich/StockFood America; *batch of baking powder biscuits*, Glenn Peterson/StockFood America; *lemon*, Shane White/iStockpho; *whole meal flour*, Masterfile. Medically Speaking, p. 98: *pink antacid*, ©Lawrence Manning/Corbis/Glow Images; *spoonful of medicine (pink antacid)*, ©Radius Images/Glow Images. Practically Speaking, p. 126: *spoons of assorted raising agents*, Z. Sandmann-StockFood Munich/StockFood America; *batch of baking powder biscuits*, Glenn Peterson/StockFood America. Running head: *lemon slice*, Susan Trigg/ iStockphoto.

Chapter 4 Opener: *hiv virus(3 photos)*, Sebastian Kaulitzki/Shutterstock; *AIDS Awareness Ribbon*, Gary Woodard/iStockphoto. Practically Speaking, p. 142: *moth*, Tanya Back/iStockphoto. Practically Speaking, p. 155: *Empty blue plastic bag*, iStockphoto; *plastic grocery bags*, iStockphoto. Medically Speaking, 163: *hiv virus(3 photos)*, Sebastian Kaulitzki/Shutterstock. Figure 4.19 Reprinted with permission of John Wiley & Sons, Inc. from Solomons, G., *Organic Chemistry, 10e*, © 2011. Running head: *hiv virus (3 photos)*, Sebastian Kaulitzki/Shutterstock.

Chapter 5 Opener: *blue med box*, Jim Jurica/iStockphoto; *pill capsule*, iStockphoto; *Vioxx structure*, Horst Puschmann/iStockphoto; *drug capsule*, Anthony Coulson/iStockphoto. Practically Speaking, p. 198: *spice mix*, Tatyana Nyshko/iStockphoto; *spices*, Adam Gryko/iStockphoto. Medically Speaking, p. 205: *blue box*, Jim Jurica/iStockphoto; *drug spills*, Anthony Coulson/iStockphoto. Figures 5.2, 5.5, 5.6, 5.10, and the art in the Medically Speaking box on p. 204 Reprinted with permission of John Wiley & Sons, Inc. from Solomons, G., *Organic Chemistry, 10e*, © 2011. Figure 5.9 Reprinted with permission of John Wiley & Sons, Inc. from Holum, J.R., *Organic Chemistry: A Brief Course*, p.316, © 1975. Running head: *capsule*, Anthony Coulson/iStockphoto.

Chapter 6 Opener: *dynamite bomb*, O.V.D./Shutterstock; *floating dollars*, Lukiyanova Natalia/frenta/Shutterstock. Practically Speaking, p. 241: *round bomb with fuse*, Andrzej Tokarski/iStockphoto. Practically Speaking, p. 242: *growing plant*, Loren Evans/iStockphoto. Medically Speaking, p. 248: *dynamite*, O.V.D./Shutterstock. Practically Speaking, p. 249: *wheat overfilled glass of beer*, Dinamir Predov/iStockphoto; *wheat berries with ears*, Kenan Savas/iStockphoto. Figure 6.27 Reprinted with permission of John Wiley & Sons, Inc. from Solomons, G., *Organic Chemistry, 10e*, © 2011. Running head: *dynamite*, O.V.D./Shutterstock.

Chapter 7 Opener and running head: *peritoneal macrophage phagocytosis of E. Coli*, Phototake. Figure 7.13 Reprinted with permission of John Wiley & Sons, Inc. from Solomons, G., *Organic Chemistry, 10e*, © 2011.

Chapter 8 Opener: *clock hands*, Luis Carlos Torres/iStockphoto; *green tomato*, Tarek El Sombati/iStockphoto; *tomato leaves*, Jennifer Stone/iStockphoto; paper bag, ©Media Bakery; *red tomato*, Yellowj/Shutterstock. Practically Speaking, p.335: *codling moth*, P. Hartmann/WILDLIFE/Peter Arnold, Inc.; *apple on a branch*, hans slegers/iStockphoto; *branch of apple tree blossoms*, BORTEL Pavel/Shutterstock; *codling moth larvae*, N A Callow/Photoshot. Figure 8.4 Reprinted with permission of John Wiley & Sons, Inc. from Solomons, G., *Organic Chemistry, 10e*, © 2011. Running head: *clock hands*, Luis Carlos Torres/iStockphoto; *green tomato*, Tarek El Sombati/iStockphoto.

Chapter 9 Opener: *Styrofoam peanuts*, CrackerClips/Shutterstock. Practically Speaking, p. 405: *Styrofoam peanuts*, CrackerClips/Shutterstock. Practically Speaking, p. 410: *fuel pump with last drop of gasoline*, Grafissimo/iStockphoto. Practically Speaking, p. 424: *Kentucky Fried Chicken bucket and McDonald's french fries*, ©AP/Wide World Photos. Running head: *Styrofoam peanuts*, CrackerClips/Shutterstock.

Chapter 10 Opener: *headache*, Sebastian Kaulitzki/Shutterstock; *neuron*, Imrich Farkas/Shutterstock. Margin, p. 455: *welding torch with flying sparks*, Fertnig/iStockphoto. Medically Speaking, p. 456: *headache*, Sebastian Kaulitzki/Shutterstock; *neuron*, Imrich Farkas/Shutterstock. Practically Speaking, p. 457: *digital camera*, Marek Mnich/iStockphoto; *LEDs in red, green and blue*, ©Media Bakery. Running head: *neuron*, Imrich Farkas/Shutterstock.

Chapter 11 Opener: *hand holding fire extinguisher*, Sabine Scheckel/Photodisc/Getty Images, Inc.; *burn hole in white paper*, Robyn Mackenzie/Shutterstock; *fire flame*, Valeev/Shutterstock; *molecules*, David Klein. Practically Speaking, p. 516: *hand holding fire extinguisher*, Sabine Scheckel/Photodisc/Getty Images, Inc.; *fire flame*, Valeev/Shutterstock. Margin, p. 521: *anti-aging nourishing cream*, Jakub Pavlinec/iStockphoto. Running head: *fire flame*, Valeev/Shutterstock.

Chapter 12 Opener: *pills*, Matej Pribelsky/iStockphoto, Josiah Lewis/iStockphoto, paul kline/iStockphoto, Miroslav Tolimir/Shutterstock, Photolink/Shutterstock. Running head: *pills*, Matej Pribelsky/iStockphoto, Josiah Lewis/iStockphoto, paul kline/iStockphoto, Miroslav Tolimir/Shutterstock, Photolink/Shutterstock.

Chapter 13 Opener: *glass, blue umbrella,* Michael Gray/iStockphoto; *yellow umbrella,* Doug Cannell/iStockphoto; *icebag*, Richard Cano/iStockphoto. Practically Speaking, p. 584: *overheated car*, © Media Bakery. Medically Speaking, p. 591: *legs*, Jens Handt/iStockphoto. Practically Speaking, p. 598: *breathalyzer*, Bridget McGill/iStockphoto. Practically Speaking, p. 602: *glass/umbrella*, Michael Gray/iStockphoto; *icebag*, Richard Cano/iStockphoto. Spectra in Problems 13.53-13.56 © Dr. Richard A. Tomasi. Running head: *glass, blue umbrella,* Michael Gray/iStockphoto.

Chapter 14 Opener: *cigarette butt*, Peter vd Rol/Shutterstock; *smoke trail*, stavklem/Shutterstock; *cigarette*, Oliver Hoffmann/Shutterstock. Practically Speaking, p. 646: *instruments/tray*, Ruslan Kerlmov/iStockphoto. Spectra in Problems 14.52-14.55 © Dr. Richard A. Tomasi. Running head: *cigarette butt*, Peter vd Rol/Shutterstock.

Chapter 15 Opener: *stages of kernel popping,* Bill Grove/iStockphoto; *microwave oven*, 2happy/Shutterstock; *popcorn in bag*, Joao Virissimo/Shutterstock; *unpopped kernel*, Charles B. Ming Onn/Shutterstock. Practically Speaking, p. 674: *microwave oven*, 2happy/Shutterstock; *popcorn in bag*, Joao Virissimo/Shutterstock; *unpopped kernel*, Charles B. Ming Onn/Shutterstock. Medically Speaking, p. 675: *house*, Ted Kinsman/Photo Researchers, Inc.; *torso*, SPL/Photo Researchers, Inc.. Margin, p. 676: *salt plate*, Martyn F. Chillmaid/Photo Researchers, Inc.. Practically Speaking, p. 684: *Intoxilizer 5000*, ©AP/Wide World Photos. p. 696, *people*, ©Media Bakery. Figures 15.1, 15.2, 15.31, and art on pages 678, 681, 686, and 692 Reprinted with permission of John Wiley & Sons, Inc. from Solomons, G., *Organic Chemistry 10e*, © 2011. Spectra in Problems 15.9, 15.10, and 15.12 © Dr. Richard A. Tomasi. Figure 15.21 Reprint with the permission of John Wiley & Sons, Inc. from Holum, J.R., *Organic Chemistry: A Brief Course*, © 1975. Spectra in Problems 15.24, 15.46, 15.53, 15.59, 15.60, and Figure 15.30 Reprinted with permission of the SDBS, National Institute of Advanced Industrial Science and Technology. Running head: *popcorn in bag*, Joao Virissimo/Shutterstock.

Chapter 16 Opener: *magnet*, Sideways Design/Shutterstock; *blue sky/clouds*, Andresr/Shutterstock; *strawberry*, ranplett/iStockphoto; *strawberry with stem*, Stephen Rees/iStockphoto; *frog*, Marcus Jones/iStockphoto; *hazelnuts*, fotosav/iStockphoto; *hot-air balloon*, Mike Sonnenberg/iStockphoto. Medically Speaking, p. 771: *MRI*, Chris Bjornberg/Photo Researchers. Figures 16.1, 16.2, 16.4, 16.5, 16.6, 16.9, 16.13, 16.14, and 16.21 Reprinted with permission of John Wiley & Sons, Inc. from Solomons, G., *Organic Chemistry 10e*, © 2011. Spectra on pages 731, 739, 744-748, 757, 759, 762, 763, 766 and in Problems 16.11-16.13, 16.17, 16.23, 16.56-16.58. 16.62-16.64 © Dr. Richard A. Tomasi. Running head: *magnet*, Sideways Design/Shutterstock.

Chapter 17 Opener: *glass and spill*, Jan Martin Will/Shutterstock; *letters spelling BLEACH*, Aqua/Shutterstock; *wineglass ring*, Thomas Northcut/Lifesize/Getty Images, Inc.. Practically Speaking, p. 800: *rubberband ball*, Sam Cornwell/Shutterstock. Practically Speaking, p. 828 *spill*, Jan Martin Will/Shutterstock; *letters spelling BLEACH*, Aqua/Shutterstock. Figure 17.7 Reprinted with permission of John Wiley & Sons, Inc. from Solomons, G., *Organic Chemistry 10e*, © 2011. Running head: *wineglass ring*, Thomas Northcut/Lifesize/Getty Images, Inc.

Chapter 18 Opener: *foil pill pack*, Loic Bernard/iStockphoto; *red pills*, Ana de Sousa/Shutterstock; *light brown bean*, Fedorov Oleksiy/Shutterstock; *medium brown bean*, David W. Leindecker/Shutterstock; *dark brown bean*, Al Mueller/Shutterstock; *white pills*, Roman Sigaev/Shutterstock. Practically Speaking, p.842: *coal*, Marek Kosmal/iStockphoto; *charcoal/flame*, Don Nichols/iStockphoto. Medically Speaking, p. 860: *foil pill pack*, Loic Bernard/iStockphoto; *red pills*, Ana de Sousa/Shutterstock; *light brown bean*, Fedorov Oleksiy/Shutterstock; *medium brown bean*, David W. Leindecker/Shutterstock; *dark brown bean*, Al Mueller/Shutterstock; *white pills*, Roman Sigaev/Shutterstock; p. 861, *tissue*, © Media Bakery; *pill pack/red pill*, Ana de Sousa/Shutterstock; *white pill*, Roman Sigaev/Shutterstock; *brown bean*, Fedorov Oleksiy/Shutterstock. Practically Speaking, p. 872: *gems*, Evgeny Terentev/iStockphoto; *green ball model*, Fabrizio Denna/iStockphoto. Figures 18.2-18.4 Reprinted with permission of John Wiley & Sons, Inc. from

Solomons, G., *Organic Chemistry 10e*, © 2011. Figure 18.19 and art in Problems 18.59 and 18.60 © Dr. Richard A. Tomasi. Art on p. 872 Reprinted with permission from Diederic, F., and Whette, R.L. Accounts of Chemical Research, **Vol 25**, pp. 119-126, 1992. Copyright 1992 by American Chemical Society. Running head: *pill pack/red pill*, Ana de Sousa/Shutterstock; *white pill*, Roman Sigaev/Shutterstock; *brown bean*, Fedorov Oleksiy/Shutterstock.

Chapter 19 Opener: *cereal*, Andy Waashnik; *blue spurt*, Auke Holwerda/iStockphoto; *blue liquid*, Pulse/Corbis/Jupiter Images Corp.; *green liquid*, Pulse/Corbis/Jupiter Images Corp.; *yellow liquid*, Pulse/Corbis/Jupiter Images Corp.; *red liquid*, Stuart Burford/iStockphoto; *bottles*, John A. Meents/iStockphoto. Practically Speaking, p. 888, *cereal*, Andy Waashnik; *blue liquid*, Pulse/Corbis/Jupiter Images Corp.; *green liquid*, Pulse/Corbis/Jupiter Images Corp.; *yellow liquid*, Pulse/Corbis/Jupiter Images Corp.; *red liquid*, Stuart Burford/iStockphoto; *bottles*, John A. Meents/iStockphoto. Running head: *blue spurt*, Auke Holwerda/iStockphoto.

Chapter 20 Opener: *eye chart*, © Media Bakery; *large whole carrot*, iStockphoto; *smaller carrot*, Ints Vikmanis/Shutterstock; *flatware*, Lane V. Erickson/Shutterstock; *carrot slice*, ZTS/Shutterstock. Medically Speaking, p. 954: *hands/lotion*, LoaDao/Shutterstock. Practically Speaking, p. 959: *eyechart*, © Media Bakery; *carrot slice*, ZTS/Shutterstock. Practically Speaking, p. 974: *worm*, © Media Bakery. Figures 20.10-20.12 and spectra in Problems 20.81 and 20.82 © Dr. Richard A. Tomasi. Running head: *eye chart*, © Media Bakery; *carrot*, iStockphoto.

Chapter 21 Opener: *thermometers*, IGLIRA/Veer; *aspirin*, Bruce Roiff/Shutterstock. Medically Speaking, p. 1024, *thermometers*, IGLIRA/Veer; *aspirin*, Bruce Roiff/Shutterstock. Practically Speaking, p. 1027, *soap*, Gabor Izso/iStockphoto. Spectra in Problem 21.80 © Dr. Richard A. Tomasi. Running head: *thermometers*, IGLIRA/Veer.

Chapter 22 Opener: *pineapple barbell*, Daniel Loiselle/iStockphoto; *pineapple and watermelon barbells*, Daniel Loiselle/iStockphoto. Practically Speaking, p. 1072, *pineapple and watermelon barbells*, Daniel Loiselle/iStockphoto. Practically Speaking, p. 1076, *jello*, Lauri Patterson/iStockphoto. Running head: *pineapple and watermelon barbells*, Daniel Loiselle/iStockphoto.

Chapter 23 Opener: *pair red peppers*, Doug Cannell/iStockphoto; *single red chili*, Katja Zgonc/iStockphoto; *pink tablets short stack*, Bryan Sikora/Shutterstock; *pink tablets tall stack*, Bryan Sikora/Shutterstock; *green pepper*, Reinhold Leitner/Shutterstock; *yellow pepper*, J.D. Rogers/Shutterstock. Medically Speaking, p. 1122, *boat*, Carlos Arranz/iStockphoto. Medically Speaking box, p. 1151, *single red chili*, Katja Zgonc/iStockphoto; *pink tablets short stack*, Bryan Sikora/Shutterstock; *pink tablets tall stack*, Bryan Sikora/Shutterstock; *green pepper*, Reinhold Leitner/Shutterstock; *yellow pepper*, J.D. Rogers/Shutterstock. Spectra in Problems 23.83 and 23.84 © Dr. Richard A. Tomasi. Running head: *single red chili*, Katja Zgonc/iStockphoto; *pink tablets short stack*, Bryan Sikora/Shutterstock.

Chapter 24 Opener: *single bandage*, Ales Veluscek/iStockphoto; *crossed tan bandages*, blackred/iStockphoto; *crossed red bandages*, Jonathan Brizendine/iStockphoto; *tube*, motorolka/Shutterstock. Medically Speaking, p. 1191, *ice cream*, Donald Erickson/iStockphoto. Practically Speaking, p. 1193: *Sweetplus*, Martin Bond/Photo Researchers; *Equal*, Rachel Epstein/PhotoEdit; *Splenda*, KRT Photograph by Ken Lambert/Seattle Times/NewsCom. Medically Speaking, p. 1196: *crossed tan bandages*, blackred/iStockphoto; *crossed red bandages*, Jonathan Brizendine/iStockphoto; *tube*, motorolka/Shutterstock. Figure 21.11 (left) Reprinted with permission of John Wiley & Sons, Inc. from Solomons, G., *Organic Chemistry 10e*, © 2011, (right) Reprinted with permission of The McGraw-Hill Companies from Neal, L., *Chemistry and Biochemistry: A COmprehensive Introduction*, © 1971.; Running head: *crossed tan bandages*, blackred/iStockphoto; *crossed red bandages*, Jonathan Brizendine/iStockphoto.

Chapter 25 Opener: *magnifying glass*, dem10/iStockphoto; *fingerprints*, blaneyphoto/iStockphoto; *hand*, TommL/iStockphoto. Practically Speaking, p. 1212: *chicken*, Ivan Kmit/iStockphoto; *steak*, Ivan Kmit/iStockphoto; *fish*, Valery Kirsanov/iStockphoto. Practically Speaking, p. 1217: *fingerprints*, blaneyphoto/iStockphoto; *hand*, TommL/iStockphoto. Medically Speaking, p. 1228: *bandage*, Ales Veluscek/iStockphoto; *tube*, motorolka/Shutterstock. Medically Speaking, p. 1243: *cow*, Supertrooper/Shutterstock. Figure 25.8 Reprinted with permission of John Wiley & Sons, Inc. from Voet, D. and Voet, J.G., *Biochemistry*, Second edition. Copyright 1995 Voet, D. and Voet, J.G. Figure 25.9 Reprinted with permission of John Wiley & Sons, Inc. from Solomons, G., *Organic Chemistry 10e*, © 2011. Figures 25.10 and 25.11: Illustration, Irving Geis. Image from Irving Geis Collection/Howard Hughes Medical Institute. Rights owned by HHMI. Not to be reproduced without permission. Figure 25.15 Reprinted with permission of John Wiley & Sons, Inc. from Pratt, C. and Cornely, K., *Essential Biochemistry, 1e*, © 2004. Figure 25.18 Reprinted with permission of John Wiley & Sons, Inc., from Black, Jacquelyn G., *Microbiology, 7e*, © 2008. Figure 25.20, Stanley Fledger/Visuals Unlimited. Running head: *magnifying glass*, dem10/iStockphoto.

Chapter 26 Opener: *bacon*, Craig Veltri/iStockphoto; *stethoscope*, Mustafa Deliormanili/iStockphoto; *egg*, Debbi Smirnoff/iStockphoto; *shrimp*, Alex Staroseitsev/Shutterstock. Practically Speaking, p. 1266: *baby*, Tim Kimberley/iStockphoto. Medically Speaking, p. 1273: *foot*, Ziga Lisjak/iStockphoto. Medically Speaking, p. 1276: *stethoscope*, Mustafa Deliormanili/iStockphoto; *egg*, Debbi Smirnoff/iStockphoto. Medically Speaking, p. 1279: *arm*, Damir Spanic/iStockphoto. Figures 26.5 and 26.6 Reprinted with permission of John Wiley & Sons, Inc. from Pratt, C. and Cornely, K., *Essential Biochemistry, 1e*, © 2004. Running head: *stethoscope*, Mustafa Deliormanili/iStockphoto.

Chapter 27 Opener: *glass*, Irochka/Dreamstime.com; *bullet*, Vladimir Kirlenko/Shutterstock. Practically Speaking, p.1314, *glass*, Irochka/Dreamstime. Running head: *bullet*, Vladimir Kirlenko/Shutterstock.

Index

Note: In this index, *f* and *t* following the page number indicate figures and tables, respectively.

A

C

D

F

G

M

N

O

U

V

W

Y

Z

PERIODIC TABLE OF THE ELEMENTS

Atomic number→ 6
Symbol → C
Name (IUPAC) → Carbon
Atomic mass → 12.011

IUPAC recommendations→
Chemical Abstracts Service group notation →

1 IA	2 IIA	3 IIIB	4 IVB	5 VB	6 VIB	7 VIIB	8 VIIIB	9 VIIIB	10 VIIIB	11 IB	12 IIB	13 IIIA	14 IVA	15 VA	16 VIA	17 VIIA	18 VIIIA
1 H Hydrogen 1.0079																	2 He Helium 4.0026
3 Li Lithium 6.941	4 Be Berylium 9.0122											5 B Boron 10.811	6 C Carbon 12.011	7 N Nitrogen 14.007	8 O Oxygen 15.999	9 F Fluorine 18.998	10 Ne Neon 20.180
11 Na Sodium 22.990	12 Mg Magnesium 24.305											13 Al Aluminum 26.982	14 Si Silicon 28.086	15 P Phosphorus 30.974	16 S Sulfur 32.065	17 Cl Chlorine 35.453	18 Ar Argon 39.948
19 K Potassium 39.098	20 Ca Calcium 40.078	21 Sc Scandium 44.956	22 Ti Titanium 47.867	23 V Vanadium 50.942	24 Cr Chromium 51.996	25 Mn Manganese 54.938	26 Fe Iron 55.845	27 Co Cobalt 58.933	28 Ni Nickel 58.693	29 Cu Copper 63.546	30 Zn Zinc 65.409	31 Ga Gallium 69.723	32 Ge Germanium 72.64	33 As Arsenic 74.922	34 Se Selenium 78.96	35 Br Bromine 79.904	36 Kr Krypton 83.798
37 Rb Rubidium 85.468	38 Sr Strontium 87.62	39 Y Yttrium 88.906	40 Zr Zirconium 91.224	41 Nb Niobium 92.906	42 Mo Molybdenum 95.94	43 Tc Technetium (98)	44 Ru Ruthenium 101.07	45 Rh Rhodium 102.91	46 Pd Palladium 106.42	47 Ag Silver 107.87	48 Cd Cadmium 112.41	49 In Indium 114.82	50 Sn Tin 118.71	51 Sb Antimony 121.76	52 Te Tellurium 127.60	53 I Iodine 126.90	54 Xe Xeno 131.29
55 Cs Caesium 132.91	56 Ba Barium 137.33	57 *La Lanthanum 138.91	72 Hf Hafnium 178.49	73 Ta Tantalum 180.95	74 W Tungsten 183.84	75 Re Rhenium 186.21	76 Os Osmium 190.23	77 Ir Iridium 192.22	78 Pt Platinum 195.08	79 Au Gold 196.97	80 Hg Mercury 200.59	81 Tl Thallium 204.38	82 Pb Lead 207.2	83 Bi Bismuth 208.98	84 Po Polonium (209)	85 At Astatine (210)	86 Rn Radon (222)
87 Fr Francium (223)	88 Ra Radium (226)	89 #Ac Actinium (227)	104 Rf Rutherfordium (261)	105 Db Dubnium (262)	106 Sg Seaborgium (266)	107 Bh Bohrium (264)	108 Hs Hassium (277)	109 Mt Meitnerium (268)	110 Ds Darmstadtium (281)	111 Rg Roentgenium (272)							

*Lanthanide Series	58 Ce Cerium 140.12	59 Pr Praseodymium 140.91	60 Nd Neodymium 144.24	61 Pm Promethium (145)	62 Sm Samarium 150.36	63 Eu Europium 151.96	64 Gd Gadolinium 157.25	65 Tb Terbium 158.93	66 Dy Dysprosium 162.50	67 Ho Holmium 164.93	68 Er Erbium 167.26	69 Tm Thulium 168.93	70 Yb Ytterbium 173.04	71 Lu Lutetium 174.97
# Actinide Series	90 Th Thorium 232.04	91 Pa Protactinium 231.04	92 U Uranium 238.03	93 Np Neptunium (237)	94 Pu Plutonium (244)	95 Am Americium (243)	96 Cm Curium (247)	97 Bk Berkelium (247)	98 Cf Californium (251)	99 Es Einsteinium (252)	100 Fm Fermium (257)	101 Md Mendelevium (258)	102 No Nobelium (259)	103 Lr Lawrencium (262)